| JPC | J. Phys. Chem. | PIA | Proc. Indian Acad. Sci. |
|-----|----------------|-----|-------------------------|
| JPR | J. Prakt. Chem. | PIA(A) | Proc. Indian Acad. Sci., Sect. A |
| JPS | J. Pharm. Sci. | PMH | Phys. Methods Heterocycl. Chem. |
| JSP | J. Mol. Spectrosc. | PNA | Proc. Natl. Acad. Sci. USA |
| JST | J. Mol. Struct. | PS | Phosphorus Sulfur |
| K | Kristallografiya | QR | Q. Rev., Chem. Soc. |
| KGS | Khim. Geterotsikl. Soedin. | RCR | Russ. Chem. Rev. (Engl. Transl.) |
| LA | Liebigs Ann. Chem. | RRC | Rev. Roum. Chim. |
| M | Monatsh. Chem. | RTC | Recl. Trav. Chim. Pays-Bas |
| MI | Miscellaneous [book or journal] | S | Synthesis |
| MIP | Miscellaneous Pat. | SA | Spectrochim. Acta |
| MS | Q. N. Porter and J. Baldas, 'Mass Spectrometry of Heterocyclic Compounds', Wiley, New York, 1971 | SA(A) | Spectrochim. Acta, Part A |
| | | SAP | S. Afr. Pat. |
| | | SC | Synth. Commun. |
| | | SH | W. L. F. Armarego, 'Stereochemistry of Heterocyclic Compounds', Wiley, New York, 1977, parts 1 and 2 |
| N | Naturwissenschaften | | |
| NEP | Neth. Pat. | | |
| NJC | Nouv. J. Chim. | | |
| NKK | Nippon Kagaku Kaishi | SST | Org. Compd. Sulphur, Selenium, Tellurium [R. Soc. Chem. series] |
| NMR | T. J. Batterham, 'NMR Spectra of Simple Heterocycles', Wiley, New York, 1973 | T | Tetrahedron |
| | | TH | Thesis |
| OMR | Org. Magn. Reson. | TL | Tetrahedron Lett. |
| OMS | Org. Mass Spectrom. | UKZ | Ukr. Khim. Zh. (Russ. Ed.) |
| OPP | Org. Prep. Proced. Int. | UP | Unpublished Results |
| OR | Org. React. | USP | U.S. Pat. |
| OS | Org. Synth. | YZ | Yakugaku Zasshi |
| OSC | Org. Synth., Coll. Vol. | ZC | Z. Chem. |
| P | Phytochemistry | ZN | Z. Naturforsch. |
| PAC | Pure Appl. Chem. | ZN(B) | Z. Naturforsch., Teil B |
| PC | Personal Communication | ZOB | Zh. Obshch. Khim. |
| PH | 'Photochemistry of Heterocyclic Compounds', ed. O. Buchardt, Wiley, New York, 1976 | ZOR | Zh. Org. Khim. |
| | | ZPC | Hoppe-Seyler's Z. Physiol. Chem. |

# COMPREHENSIVE
# HETEROCYCLIC CHEMISTRY
## IN 8 VOLUMES

# COMPREHENSIVE
# HETEROCYCLIC CHEMISTRY

*The Structure, Reactions, Synthesis
and Uses of
Heterocyclic Compounds*

## Volume 3

**Chairman of the Editorial Board**
## ALAN R. KATRITZKY, FRS
*University of Florida*

**Co-Chairman of the Editorial Board**
## CHARLES W. REES, FRS
*Imperial College of Science and Technology
University of London*

## Part 2B
Six-membered Rings with Oxygen, Sulfur or
Two or More Nitrogen Atoms

EDITORS
## A. JOHN BOULTON and
## ALEXANDER McKILLOP
*University of East Anglia*

## PERGAMON PRESS
**OXFORD · NEW YORK · TORONTO · SYDNEY · PARIS · FRANKFURT**

| | |
|---|---|
| U.K. | Pergamon Press Ltd., Headington Hill Hall, Oxford OX3 0BW, England |
| U.S.A. | Pergamon Press Inc., Maxwell House, Fairview Park, Elmsford, New York 10523, U.S.A. |
| CANADA | Pergamon Press Canada Ltd., Suite 104, 150 Consumers Road, Willowdale, Ontario M2J 1P9, Canada |
| AUSTRALIA | Pergamon Press (Aust.) Pty. Ltd., P.O. Box 544, Potts Point, N.S.W. 2011, Australia |
| FRANCE | Pergamon Press SARL, 24 rue des Ecoles, 75240 Paris, Cedex 05, France |
| FEDERAL REPUBLIC OF GERMANY | Pergamon Press GmbH, Hammerweg 6, D-6242 Kronberg-Taunus, Federal Republic of Germany |

First edition 1984

**Library of Congress Cataloging in Publication Data**

Main entry under title:

Comprehensive heterocyclic chemistry.

Includes indexes.
Contents: v. 1. Introduction, nomenclature, literature, biological aspects, industrial uses, less-common heteroatoms –
v. 2. Six-membered rings with one nitrogen atom – [etc.] –
v. 8. Indexes.
1. Heterocyclic compounds.   I. Katritzky, Alan R. (Alan Roy)
II. Rees, Charles W. (Charles Wayne)
QD400.C65   1984   547′.59   83-4264

**British Library Cataloguing in Publication Data**

Comprehensive heterocyclic chemistry
1. Heterocyclic compounds.
I. Katritzky, Alan R.   II. Rees, Charles W.
547′.59   QD400

ISBN 0-08-030703-5 (vol. 3)
ISBN 0-08-026200-7 (set)

Typeset by J. W. Arrowsmith Ltd., Winterstoke Road, Bristol
Printed in Great Britain by A. Wheaton & Co. Ltd., Exeter

# Contents

Foreword      vii

Contributors to Volume 3      ix

Contents of All Volumes      xi

2.12   Pyridazines and their Benzo Derivatives      1
M. TIŠLER and B. STANOVNIK, *E. Kardelj University, Ljubljana*

2.13   Pyrimidines and their Benzo Derivatives      57
D. J. BROWN, *Australian National University*

2.14   Pyrazines and their Benzo Derivatives      157
A. E. A. PORTER, *University of Stirling*

2.15   Pyridodiazines and their Benzo Derivatives      199
E. LUNT and C. G. NEWTON, *May & Baker Ltd., Dagenham*

2.16   Pteridines      263
W. PFLEIDERER, *Universität Konstanz*

2.17   Other Diazinodiazines      329
R. N. CASTLE, *University of South Florida* and S. D. PHILLIPS, *Olin Research Center, New Haven*

2.18   1,2,3-Triazines and their Benzo Derivatives      369
H. NEUNHOEFFER, *Technische Hochschule Darmstadt*

2.19   1,2,4-Triazines and their Benzo Derivatives      385
H. NEUNHOEFFER, *Technische Hochschule Darmstadt*

2.20   1,3,5-Triazines      457
J. M. E. QUIRKE, *University of Durham* and *Florida International University*

2.21   Tetrazines and Pentazines      531
H. NEUNHOEFFER, *Technische Hochschule Darmstadt*

2.22   Pyrans and Fused Pyrans: (i) Structure      573
P. J. BROGDEN, C. D. GABBUTT and J. D. HEPWORTH, *Preston Polytechnic*

2.23   Pyrans and Fused Pyrans: (ii) Reactivity      647
G. P. ELLIS, *University of Wales Institute of Science and Technology*

2.24   Pyrans and Fused Pyrans: (iii) Synthesis and Applications      737
J. D. HEPWORTH, *Preston Polytechnic*

2.25  Thiopyrans and Fused Thiopyrans                                                     885
      A. H. INGALL, *Fison's Pharmaceuticals Ltd., Loughborough*

2.26  Six-membered Rings with More than One Oxygen or Sulfur Atom                          943
      M. J. COOK, *University of East Anglia*

2.27  Oxazines, Thiazines and their Benzo Derivatives                                      995
      M. SAINSBURY, *University of Bath*

2.28  Polyoxa, Polythia and Polyaza Six-membered Ring Systems                             1039
      C. J. MOODY, *Imperial College of Science and Technology*

      References                                                                          1087

# Foreword

## Scope

Heterocyclic compounds are those which have a cyclic structure with two, or more, different kinds of atom in the ring. This work is devoted to organic heterocyclic compounds in which at least one of the ring atoms is carbon, the others being considered the heteroatoms; carbon is still by far the most common ring atom in heterocyclic compounds. As the number and variety of heteroatoms in the ring increase there is a steady transition to the expanding domain of inorganic heterocyclic systems. Since the ring can be of any size, from three-membered upwards, and since the heteroatoms can be drawn in almost any combination from a large number of the elements (though nitrogen, oxygen and sulfur are the most common), the number of possible heterocyclic systems is almost limitless. An enormous number of heterocyclic compounds is known and this number is increasing very rapidly. The literature of the subject is correspondingly vast and of the three major divisions of organic chemistry, aliphatic, carbocyclic and heterocyclic, the last is much the biggest. Over six million compounds are recorded in *Chemical Abstracts* and approximately half of these are heterocyclic.

## Significance

Heterocyclic compounds are very widely distributed in Nature and are essential to life; they play a vital role in the metabolism of all living cells. Thus, for example, the following are heterocyclic compounds: the pyrimidine and purine bases of the genetic material DNA; the essential amino acids proline, histidine and tryptophan; the vitamins and coenzyme precursors thiamine, riboflavine, pyridoxine, folic acid and biotin; the $B_{12}$ and E families of vitamin; the photosynthesizing pigment chlorophyll; the oxygen transporting pigment hemoglobin, and its breakdown products the bile pigments; the hormones kinetin, heteroauxin, serotonin and histamine; together with most of the sugars. There are a vast number of pharmacologically active heterocyclic compounds, many of which are in regular clinical use. Some of these are natural products, for example antibiotics such as penicillin and cephalosporin, alkaloids such as vinblastine, ellipticine, morphine and reserpine, and cardiac glycosides such as those of digitalis. However, the large majority are synthetic heterocyclics which have found widespread use, for example as anticancer agents, analeptics, analgesics, hypnotics and vasopressor modifiers, and as pesticides, insecticides, weedkillers and rodenticides.

There is also a large number of synthetic heterocyclic compounds with other important practical applications, as dyestuffs, copolymers, solvents, photographic sensitizers and developers, as antioxidants and vulcanization accelerators in the rubber industry, and many are valuable intermediates in synthesis.

The successful application of heterocyclic compounds in these and many other ways, and their appeal as materials in applied chemistry and in more fundamental and theoretical studies, stems from their very complexity; this ensures a virtually limitless series of structurally novel compounds with a wide range of physical, chemical and biological properties, spanning a broad spectrum of reactivity and stability. Another consequence of their varied chemical reactivity, including the possible destruction of the heterocyclic ring, is their increasing use in the synthesis of specifically functionalized non-heterocyclic structures.

## Aims of the Present Work

All of the above aspects of heterocyclic chemistry are mirrored in the contents of the present work. The scale, scope and complexity of the subject, already referred to, with its correspondingly complex system of nomenclature, can make it somewhat daunting initially. One of the main aims of the present work is to minimize this problem by presenting a comprehensive account of fundamental heterocyclic chemistry, with the emphasis on basic principles and, as far as possible, on unifying correlations in the properties, chemistry and synthesis of different heterocyclic systems and the analogous carbocyclic structures. The motivation for this effort was the outstanding biological, practical and theoretical importance of heterocyclic chemistry, and the absence of an appropriate major modern treatise.

At the introductory level there are several good textbooks on heterocyclic chemistry, though the subject is scantily treated in most general textbooks of organic chemistry. At the specialist, research level there are two established ongoing series, 'Advances in Heterocyclic Chemistry' edited by Katritzky and 'The Chemistry of Heterocyclic Compounds' edited by Weissberger and Taylor, devoted to a very detailed consideration of all aspects of heterocyclic compounds, which together comprise some 100 volumes. The present work is designed to fill the gap between these two levels, *i.e.* to give an up-to-date overview of the subject as a whole (particularly in the General Chapters) appropriate to the needs of teachers and students and others with a general interest in the subject and its applications, and to provide enough detailed information (particularly in the Monograph Chapters) to answer specific questions, to demonstrate exactly what is known or not known on a given topic, and to direct attention to more detailed reviews and to the original literature. Mainly because of the extensive practical uses of heterocyclic compounds, a large and valuable review literature on all aspects of the subject has grown up over the last few decades. References to all of these reviews are now immediately available: reviews dealing with a specific ring system are reported in the appropriate monograph chapters; reviews dealing with any aspect of heterocyclic chemistry which spans more than one ring system are collected together in a logical, readily accessible manner in Chapter 1.03.

The approach and treatment throughout this work is as ordered and uniform as possible, based on a carefully prearranged plan. This plan, which contains several novel features, is described in detail in the Introduction (Chapter 1.01).

ALAN R. KATRITZKY                                    CHARLES W. REES
*Florida*                                            *London*

# Contributors to Volume 3

Dr P. J. Brodgen
School of Sciences, Preston Polytechnic, Corporation Street, Preston PR1 2TQ, UK

Dr D. J. Brown
The John Curtin School of Medical Research, PO Box 334, Canberra City, ACT 2601, Australia

Professor R. N. Castle
Department of Chemistry, University of South Florida, Tampa, FL 33620, USA

Dr M. J. Cook
School of Chemical Sciences, University of East Anglia, Norwich NR4 7TJ, UK

Dr G. P. Ellis
Chemistry Department, UWIST, King Edward VII Avenue, Cardiff CF1 3NU, UK

Dr C. D. Gabbutt
School of Sciences, Preston Polytechnic, Corporation Street, Preston PR1 2TQ, UK

Dr J. D. Hepworth
School of Sciences, Preston Polytechnic, Corporation Street, Preston PR1 2TQ, UK

Dr A. H. Ingall
17 Sandringham Drive, Loughborough, Leicestershire, UK

Dr. E. Lunt
May & Baker Ltd., Dagenham, Essex RM10 7XS, UK

Dr C. J. Moody
Department of Chemistry, Imperial College of Science and Technology, South Kensington, London SW7 2AY, UK

Professor H. Neuhoeffer
Institut für Organische Chemie, Technische Hochschule Darmstadt, Fachgebiet Chemie de Technischen Gewerbe, Petersenstrasse 22, D-6110 Darmstadt, Federal Republic of Germany

Dr C. G. Newton
May & Baker Ltd., Dagenham, Essex RM10 7XS, UK

Professor W Pfleiderer
Fakultät für Chemie, Universität Konstanz, Postfach 5560, D-7750 Konstanz 1, Federal Republic of Germany

Dr S. E. Phillips
Olin Research Center, 275 Winchester Avenue, New Haven, CT 06511, USA

Dr A. E. A. Porter
Department of Chemistry, University of Stirling, Stirling FK9 4LA, UK

Dr J. M. E. Quirke
Department of Physical Sciences, College of Arts and Sciences, Florida International University, Tamiami Trial, Miami, FL 33199, USA

Dr M. Sainsbury
School of Chemistry, University of Bath, Claverton Down, Bath BA2 7AY, UK

Professor B. Stanovnik
Department of Chemistry and Chemical Technology, E. Kardelj University, 6 Murnikova, 61000 Ljubljana, Yugoslavia

Professor M. Tišler
Department of Chemistry and Chemical Technology, E. Kardelj University, 6 Murnikova, 61000 Ljubljana, Yugoslavia

# Contents of All Volumes

**Volume 1** (Part 1: Introduction, Nomenclature, Review Literature, Biological Aspects, Industrial Uses, Less-common Heteroatoms)

1.01  Introduction
1.02  Nomenclature of Heterocyclic Compounds
1.03  Review Literature of Heterocycles
1.04  Biosynthesis of Some Heterocyclic Natural Products
1.05  Toxicity of Heterocycles
1.06  Application as Pharmaceuticals
1.07  Use as Agrochemicals
1.08  Use as Veterinary Products
1.09  Metabolism of Heterocycles
1.10  Importance of Heterocycles in Biochemical Pathways
1.11  Heterocyclic Polymers
1.12  Heterocyclic Dyes and Pigments
1.13  Organic Conductors
1.14  Uses in Photographic and Reprographic Techniques
1.15  Heterocyclic Compounds as Additives
1.16  Use in the Synthesis of Non-heterocycles
1.17  Heterocyclic Rings containing Phosphorus
1.18  Heterocyclic Rings containing Arsenic, Antimony or Bismuth
1.19  Heterocyclic Rings containing Halogens
1.20  Heterocyclic Rings containing Silicon, Germanium, Tin or Lead
1.21  Heterocyclic Rings containing Boron
1.22  Heterocyclic Rings containing a Transition Metal


**Volume 2** (Part 2A: Six-membered Rings with One Nitrogen Atom)

2.01  Structure of Six-membered Rings
2.02  Reactivity of Six-membered Rings
2.03  Synthesis of Six-membered Rings
2.04  Pyridines and their Benzo Derivatives: (i) Structure
2.05  Pyridines and their Benzo Derivatives: (ii) Reactivity at Ring Atoms
2.06  Pyridines and their Benzo Derivatives: (iii) Reactivity of Substituents
2.07  Pyridines and their Benzo Derivatives: (iv) Reactivity of Non-aromatics
2.08  Pyridines and their Benzo Derivatives: (v) Synthesis
2.09  Pyridines and their Benzo Derivatives: (vi) Applications
2.10  The Quinolizinium Ion and Aza Analogs
2.11  Naphthyridines, Pyridoquinolines, Anthyridines and Similar Compounds


**Volume 3** (Part 2B: Six-membered Rings with Oxygen, Sulfur or Two or More Nitrogen Atoms)

2.12  Pyridazines and their Benzo Derivatives
2.13  Pyrimidines and their Benzo Derivatives

2.14   Pyrazines and their Benzo Derivatives
2.15   Pyridodiazines and their Benzo Derivatives
2.16   Pteridines
2.17   Other Diazinodiazines
2.18   1,2,3-Triazines and their Benzo Derivatives
2.19   1,2,4-Triazines and their Benzo Derivatives
2.20   1,3,5-Triazines
2.21   Tetrazines and Pentazines
2.22   Pyrans and Fused Pyrans: (i) Structure
2.23   Pyrans and Fused Pyrans: (ii) Reactivity
2.24   Pyrans and Fused Pyrans: (iii) Synthesis and Applications
2.25   Thiopyrans and Fused Thiopyrans
2.26   Six-membered Rings with More than One Oxygen or Sulfur Atom
2.27   Oxazines, Thiazines and their Benzo Derivatives
2.28   Polyoxa, Polythia and Polyaza Six-membered Ring Systems

**Volume 4** (Part 3: Five-membered Rings with One Oxygen, Sulfur or Nitrogen Atom)

3.01   Structure of Five-membered Rings with One Heteroatom
3.02   Reactivity of Five-membered Rings with One Heteroatom
3.03   Synthesis of Five-membered Rings with One Heteroatom
3.04   Pyrroles and their Benzo Derivatives: (i) Structure
3.05   Pyrroles and their Benzo Derivatives: (ii) Reactivity
3.06   Pyrroles and their Benzo Derivatives: (iii) Synthesis and Applications
3.07   Porphyrins, Corrins and Phthalocyanines
3.08   Pyrroles with Fused Six-membered Heterocyclic Rings: (i) *a*-Fused
3.09   Pyrroles with Fused Six-membered Heterocyclic Rings: (ii) *b*- and *c*-Fused
3.10   Furans and their Benzo Derivatives: (i) Structure
3.11   Furans and their Benzo Derivatives: (ii) Reactivity
3.12   Furans and their Benzo Derivatives: (iii) Synthesis and Applications
3.13   Thiophenes and their Benzo Derivatives: (i) Structure
3.14   Thiophenes and their Benzo Derivatives: (ii) Reactivity
3.15   Thiophenes and their Benzo Derivatives: (iii) Synthesis and Applications
3.16   Selenophenes, Tellurophenes and their Benzo Derivatives
3.17   Furans, Thiophenes and Selenophenes with Fused Six-membered Heterocyclic Rings
3.18   Two Fused Five-membered Rings each containing One Heteroatom

**Volume 5** (Part 4A: Five-membered Rings with Two or More Nitrogen Atoms)

4.01   Structure of Five-membered Rings with Two or More Heteroatoms
4.02   Reactivity of Five-membered Rings with Two or More Heteroatoms
4.03   Synthesis of Five-membered Rings with Two or More Heteroatoms
4.04   Pyrazoles and their Benzo Derivatives
4.05   Pyrazoles with Fused Six-membered Heterocyclic Rings
4.06   Imidazoles and their Benzo Derivatives: (i) Structure
4.07   Imidazoles and their Benzo Derivatives: (ii) Reactivity
4.08   Imidazoles and their Benzo Derivatives: (iii) Synthesis and Applications
4.09   Purines
4.10   Other Imidazoles with Fused Six-membered Rings
4.11   1,2,3-Triazoles and their Benzo Derivatives
4.12   1,2,4-Triazoles
4.13   Tetrazoles
4.14   Pentazoles
4.15   Triazoles and Tetrazoles with Fused Six-membered Rings

**Volume 6** (Part 4B: Five-membered Rings with Two or More Oxygen, Sulfur or Nitrogen Atoms)

4.16 Isoxazoles and their Benzo Derivatives
4.17 Isothiazoles and their Benzo Derivatives
4.18 Oxazoles and their Benzo Derivatives
4.19 Thiazoles and their Benzo Derivatives
4.20 Five-membered Selenium–Nitrogen Heterocycles
4.21 1,2,3- and 1,2,4-Oxadiazoles
4.22 1,2,5-Oxadiazoles and their Benzo Derivatives
4.23 1,3,4-Oxadiazoles
4.24 1,2,3-Thiadiazoles and their Benzo Derivatives
4.25 1,2,4-Thiadiazoles
4.26 1,2,5-Thiadiazoles and their Benzo Derivatives
4.27 1,3,4-Thiadiazoles
4.28 Oxatriazoles and Thiatriazoles
4.29 Five-membered Rings (One Oxygen or Sulfur and at least One Nitrogen Atom) Fused with Six-membered Rings (at least One Nitrogen Atom)
4.30 Dioxoles and Oxathioles
4.31 1,2-Dithioles
4.32 1,3-Dithioles
4.33 Five-membered Rings containing Three Oxygen or Sulfur Atoms
4.34 Dioxazoles, Oxathiazoles and Dithiazoles
4.35 Five-membered Rings containing One Selenium or Tellurium Atom and One Other Group VI Atom and their Benzo Derivatives
4.36 Two Fused Five-membered Heterocyclic Rings: (i) Classical Systems
4.37 Two Fused Five-membered Heterocyclic Rings: (ii) Non-classical Systems
4.38 Two Fused Five-membered Heterocyclic Rings: (iii) $1,6,6a\lambda^4$-Trithiapentalenes and Related Systems

**Volume 7** (Part 5: Small and Large Rings)

5.01 Structure of Small and Large Rings
5.02 Reactivity of Small and Large Rings
5.03 Synthesis of Small and Large Rings
5.04 Aziridines, Azirines and Fused-ring Derivatives
5.05 Oxiranes and Oxirenes
5.06 Thiiranes and Thiirenes
5.07 Fused-ring Oxiranes, Oxirenes, Thiiranes and Thiirenes
5.08 Three-membered Rings with Two Heteroatoms and Fused-ring Derivatives
5.09 Azetidines, Azetines and Azetes
5.10 Cephalosporins
5.11 Penicillins
5.12 Other Fused-ring Azetidines, Azetines and Azetes
5.13 Oxetanes and Oxetenes
5.14 Thietanes, Thietes and Fused-ring Derivatives
5.15 Four-membered Rings with Two or More Heteroatoms and Fused-ring Derivatives
5.16 Azepines
5.17 Oxepanes, Oxepins, Thiepanes and Thiepins
5.18 Seven-membered Rings with Two or More Heteroatoms
5.19 Eight-membered Rings
5.20 Larger Rings except Crown Ethers and Heterophanes
5.21 Crown Ethers and Cryptands
5.22 Heterophanes

**Volume 8** (Part 6: Indexes)

Subject Index
Author Index
Ring Index
Data Index

# 2.12

# Pyridazines and their Benzo Derivatives

## M. TIŠLER and B. STANOVNIK
*E. Kardelj University, Ljubljana*

| | | |
|---|---|---|
| 2.12.1 | INTRODUCTION | 2 |
| 2.12.2 | STRUCTURE | 2 |
| *2.12.2.1* | *Physical Properties* | 3 |
| *2.12.2.1.1* | *Ionization constants, $pK_a$ and tautomerism* | 4 |
| *2.12.2.1.2* | *Spectroscopic properties* | 6 |
| 2.12.3 | FULLY CONJUGATED RINGS: REACTIVITY AT RING ATOMS | 9 |
| *2.12.3.1* | *Thermal and Photochemical Reactions Involving No Other Species* | 9 |
| *2.12.3.1.1* | *Thermal rearrangements* | 9 |
| *2.12.3.1.2* | *Photochemical reactions* | 10 |
| *2.12.3.2* | *Electrophilic Attack at Nitrogen* | 14 |
| *2.12.3.2.1* | *Alkylation* | 14 |
| *2.12.3.2.2* | *O → N Rearrangement* | 15 |
| *2.12.3.2.3* | *Quaternization* | 17 |
| *2.12.3.2.4* | *N-Oxidation* | 18 |
| *2.12.3.2.5* | *N-Amination* | 20 |
| *2.12.3.3* | *Electrophilic Attack at Carbon* | 20 |
| *2.12.3.4* | *Electrophilic Hydrogen Exchange* | 22 |
| *2.12.3.5* | *Nucleophilic Attack at Carbon* | 22 |
| *2.12.3.5.1* | *Pyridazines* | 22 |
| *2.12.3.5.2* | *Cinnolines* | 25 |
| *2.12.3.5.3* | *Phthalazines* | 25 |
| *2.12.3.5.4* | *Nucleophilic substitution of other groups* | 25 |
| *2.12.3.5.5* | *Ring opening and ring contractions* | 29 |
| *2.12.3.6* | *Nucleophilic Attack at Hydrogen* | 29 |
| *2.12.3.7* | *Reaction with Radicals and Electron-deficient Species* | 30 |
| *2.12.3.8* | *Reactions with Cyclic Transition States* | 30 |
| 2.12.4 | FULLY CONJUGATED RINGS: REACTIVITY OF SUBSTITUENTS | 31 |
| *2.12.4.1* | *Fused Benzene Rings* | 31 |
| *2.12.4.2* | *C-Linked Substituents* | 31 |
| *2.12.4.2.1* | *Methyl groups* | 31 |
| *2.12.4.2.2* | *Hydroxymethyl groups* | 32 |
| *2.12.4.2.3* | *Carbonyl groups* | 32 |
| *2.12.4.2.4* | *Carboxyl groups and derivatives* | 32 |
| *2.12.4.2.5* | *Cyano groups* | 34 |
| *2.12.4.3* | *N-Linked Substituents* | 34 |
| *2.12.4.3.1* | *Reduction of nitro groups* | 34 |
| *2.12.4.3.2* | *Acylation* | 35 |
| *2.12.4.3.3* | *Condensation with carbonyl compounds* | 35 |
| *2.12.4.3.4* | *Diazotization* | 35 |
| *2.12.4.3.5* | *Hydrazinopyridazines* | 35 |
| *2.12.4.4* | *O-Linked Substituents* | 36 |
| *2.12.4.5* | *S-Linked Substituents* | 36 |
| *2.12.4.6* | *Metal- and Metalloid-linked Substituents* | 37 |
| *2.12.4.7* | *Substituents Attached to Ring Nitrogens* | 37 |
| 2.12.5 | SATURATED AND PARTIALLY SATURATED RINGS | 37 |
| *2.12.5.1* | *Dihydro-pyridazines and -pyridazinones* | 37 |
| *2.12.5.2* | *Tetrahydropyridazines* | 39 |
| *2.12.5.3* | *Hexahydropyridazines* | 40 |
| 2.12.6 | SYNTHESIS | 40 |
| *2.12.6.1* | *Introduction* | 40 |
| *2.12.6.2* | *Formation of One Bond* | 40 |
| *2.12.6.2.1* | *Between two heteroatoms* | 40 |

| 2.12.6.2.2 | *Adjacent to a heteroatom* | 41 |
| 2.12.6.2.3 | *γ to the heteroatom* | 42 |
| 2.12.6.3 | *Formation of Two Bonds* | 43 |
| 2.12.6.3.1 | *From [5 + 1] atom fragments* | 43 |
| 2.12.6.3.2 | *From [4 + 2] atom fragments* | 44 |
| 2.12.6.3.3 | *From [3 + 3] atom fragments* | 51 |
| 2.12.6.4 | *Formation of Three or Four Bonds* | 52 |
| 2.12.6.5 | *From Other Heterocycles* | 52 |
| 2.12.6.6 | *Best Methods for the Synthesis of the Parent Compounds and Some Important Derivatives* | 55 |
| 2.12.7 | APPLICATIONS | 56 |

## 2.12.1 INTRODUCTION

Pyridazine (1,2-diazine) (**1**) and its benzo analogs cinnoline (1,2-diazanaphthalene) or benzo[*c*]pyridazine (**2**) and phthalazine (benzo[*d*]pyridazine) (**3**) have been known since the nineteenth century. Although the basic synthetic principles and reactivity were investigated in the early years, interest in these compounds was not very intense, compared with pyrimidines and their bicyclic analogs, as they were not found in nature. However, during the last three decades intensive research has been stimulated because many derivatives have found application as a result of their biological activity.

(**1a**)          (**1b**)          (**2**)          (**3a**)          (**3b**)

The chemistry of pyridazine and of its benzo analogs is presented in detail in several books and monographs. In addition to the older books which cover pyridazines ⟨B-57MI21200, B-59MI21200⟩, cinnolines ⟨B-57MI121201, B-59MI21201, 53HC(5)3⟩ and phthalazines ⟨53HC(5)69, B-57MI21202, B-59MI21202⟩, the literature has been updated to mid-1971 for pyridazines ⟨73HC(28)1⟩, cinnolines ⟨73HC(27)1⟩ and phthalazines ⟨73HC(27)323⟩.

A review article on pyridazines ⟨68AHC(9)211⟩ was updated with literature to 1977 ⟨79AHC(24)363⟩ and there are also several monographs which deal with specific aspects of the chemistry of pyridazines or their activity. These reviews include the chemistry of pyridazines and pyridazinones ⟨58AG5, 65AG(E)292⟩, pyridazine *N*-oxides ⟨66MI21200⟩, pyridazine *N*-ylides ⟨76T2647⟩, formation of pyridazines from 1,2,4,5-tetrazines ⟨78HC(33)1073⟩, the rearrangement of pyridazines ⟨66RCR9⟩, the azuleno[1,8-*cd*]pyridazines ⟨78H(11)387⟩, conformational studies of hexahydropyridazine derivatives ⟨78ACR14⟩, π-complexes of transition metals with pyridazines ⟨78JHC1057⟩, antimicrobial activity of pyridazines ⟨77MI21200⟩, diazaquinones such as pyridazine-3,6-dione and phthalazine-1,4-dione ⟨78H(9)1771⟩ and thermal and photochemical decomposition with elimination of nitrogen from pyridazine and phthalazine derivatives ⟨80AG815⟩.

There are also reviews on the biological activity of certain phthalazines ⟨59MI21203⟩ and on Reissert compounds from phthalazines ⟨80H(14)1033⟩. The chemistry of benzo[*c*]cinnolines has also been covered ⟨79AHC(24)151⟩.

## 2.12.2 STRUCTURE

Pyridazine, one of the three possible isomeric diazines, is assumed to be a planar six-membered ring and is represented as a resonance hybrid of two structures (**1a**) and (**1b**) with a greater contribution from the canonical structure (**1a**).

This is supported by the results of electron diffraction, microwave spectroscopy data and X-ray analysis, which all indicate that the N—N bond has single bond character. Bond lengths and angles for pyridazine and its 3,6-dichloro derivative have been calculated from electron diffraction and microwave data ⟨77ACS(A)63⟩ (Figure 1). The X-ray analyses of several pyridazine derivatives, such as pyridazine-3(2*H*)-thione ⟨66AX249⟩, 6-oxo-1,6-dihydropyridazine-3-carboxamide ⟨63AX318⟩, maleic hydrazide ⟨76JCS(P2)1386⟩, 4,5-diamino-1-

**Figure 1**

methyl-3-(methylthio)pyridazinium iodide ⟨81H(16)9⟩ and others have been published. These data indicate that the pyridazine ring is planar and that protonation or quaternization does not affect the bond lengths within the ring.

N—N Bond lengths, bond angles and electron densities have been calculated by different methods, such as variable electronegativity-self-consistent field (VESCF), EHT, CNDO, LCAO, LCAO-FE, SCMO ⟨73HC(28)1⟩, PPP-CI based on regular hexagon, energy weighted maximum overlap (EWMO), CNDO/S-CI, MINDO, MINDO/2 ⟨79MI21200, 79CPB2105⟩, especially in combination with photoelectron and $^{13}$C NMR spectra. CNDO/2 and INDO calculations have been performed and the lone-pair orbital energies compared with those obtained by *ab initio* calculations. In pyridazine the symmetric lone-pair combination is lower in energy than the antisymmetric one ⟨78BCJ3443⟩.

## 2.12.2.1 Physical Properties

In general, pyridazine can be compared with pyridine. It is completely miscible with water and alcohols, as the lone electron pairs on nitrogen atoms are involved in formation of hydrogen bonds with hydroxylic solvents, benzene and ether. Pyridazine is insoluble in ligroin and cyclohexane. The solubility of pyridazine derivatives containing OH, SH and NH$_2$ groups decreases, while alkyl groups increase the solubility. Table 1 lists some physical properties of pyridazine.

**Table 1** Physical Properties of Pyridazine

| | |
|---|---|
| Melting point | −8 °C |
| Boiling point | 208 °C (760 mm), 207, 4 °C (762.5 mm) |
| | 87 °C (14 mm), 48 °C (1 mm) |
| Density: $d_4^{20} = 1.1054$, $d_4^{23.5} = 1.1035$, $d_4^{18} = 1.107$ | |
| Index of refraction: $n_D^{23.5} = 1.5231$ | |
| Surface tension: $5.015 \times 10^{10}$ N m$^{-2}$ at 0 °C | |
| Salts: | |
| Hydrochloride, yellow solid, m.p. 161–163 °C | |
| Monopicrate, yellow solid, dec. 170–175 °C | |

Dipole moments have been calculated and determined experimentally for many pyridazine derivatives containing halogen, alkyl and other groups at different positions. The calculated and experimental values are generally in good agreement (Table 2). The high boiling point

**Table 2** Dipole Moments

| | μ (D) | |
|---|---|---|
| *Compound* | *Experimental* | *Calculated* |
| Pyridazine | 3.95 | 4.00 |
| 3-Methylpyridazine | 3.86 | 3.96 |
| 4-Methylpyridazine | 4.34 | 4.29 |
| 3-Chloropyridazine | 4.42 | 4.24 |
| 3,6-Dichloropyridazine | 4.11 | 3.94 |
| 3-Styrylpyridazine | 5.82 | — |
| 3-Acetylpyridazine | 2.48 | 4.89[a] |
| | | 2.19[b] |
| 3-Ethoxycarbonylpyridazine | 3.33 | 4.34[a] |
| | | 2.30[b] |

[a] Calculated allowing for free rotation.
[b] Calculated *trans* configuration.

of pyridazine is due to intermolecular attractions, which are attributed to electrostatic forces arising from the high permanent dipole.

### 2.12.2.1.1 *Ionization constants, pK$_a$ and tautomerism*

When a second nitrogen atom is introduced into the pyridine ring the basicity is reduced (p$K_a$ 5.23 for pyridine and 2.33 for pyridazine). The effect of the additional substituents on p$K_a$ depends on the position of the substituents (Table 3). An extensive set of p$K_a$ values of pyridazine derivatives has been submitted to correlation analysis using the Hammett and the two Taft equations, which shows that the p$K_a$ values are most sensitive to the effect of a 2-substituent followed by the effects of 3- and 4-substituents. The interactions between nitrogen atom and 2-substituents represent over 70% of the inductive character. The composition of the effects of $+M$ 4-substituents is significantly enriched in the resonance interactions, whereas $-M$ 4-substituents interact with the nitrogen atom mainly by induction 〈77MI21201〉.

**Table 3**  p$K_a$ Values for Pyridazines (20 °C)

| Compound | p$K_a$ | Compound | p$K_a$ |
|---|---|---|---|
| Pyridazine | 2.33 | 3-Methylmercaptopyridazine | 2.26 |
| 4-Methylpyridazine | 2.92 | 4-Methylmercaptopyridazine | 3.26 |
| 3-Methoxypyridazine | 2.52 | 3-Aminopyridazine | 5.19 |
| 4-Methoxypyridazine | 3.70 | 4-Aminopyridazine | 6.69 |
| 3,6-Dimethylpyridazine | 1.61 | 3-Amino-6-methylpyridazine | 5.32 |

At ring nitrogen, unsubstituted pyridazinones are weak acids, but maleic hydrazide is rather a strong acid. Ionization constants of some hydroxy, amino and sulfur containing pyridazines are given in Tables 4 and 5. Pyridazinethiones are weaker bases than the corresponding pyridazinones. The same is also true for the *N*-methyl derivatives, but methylthio derivatives are slightly weaker bases than the corresponding methoxy derivatives.

**Table 4**  p$K_a$ Values for Hydroxypyridazines

| Compound | Proton gain | Proton loss |
|---|---|---|
| Pyridazin-3(2*H*)-one | −1.8 | 10.46 |
| Pyridazin-4(1*H*)-one | 1.07 | 8.68 |
| 3-Methoxypyridazine | 2.52 | — |
| 4-Methoxypyridazine | 3.70 | — |
| 2-Methylpyridazin-3(2*H*)-one | −2.1 | — |
| 1-Methylpyridazin-4(1*H*)-one | 1.02 | — |
|  | 1.1 | — |
| 6-Hydroxypyridazin-3(2*H*)-one | 5.5 | 13 |
|  | 5.65 | — |
|  | 5.67 | — |
| 3,6-Dimethoxypyridazine | 1.61 | — |

Pyridazines with a 'hydroxy' group at an α- or γ-position to a ring nitrogen atom, *i.e.* 3- and 4-hydroxypyridazines (**4**) and (**5**), exist predominantly in the oxo form. This conclusion is based on spectroscopic evidence from UV spectra of unsubstituted compounds and their *N*-methyl and *O*-methyl derivatives in alkaline, neutral and acidic solutions. In some instances, as for example for 6-oxo-1,6-dihydropyridazine-3-carboxamide, there is also evidence from X-ray analysis 〈54AX199, 63AX318〉. Maleic hydrazide and substituted maleic hydrazides exist in the monohydroxymonooxo form (**6**).

Similarly, 3- and 4-hydroxycinnolines and their derivatives exist predominantly in the oxo forms (**7**) and (**8**), as supported by both p$K_a$ values and UV spectral data. The oxo forms for 3-hydroxycinnolines have been further supported by IR data 〈76MI21200〉. Phthalazin-4-ones exist in the oxo form and phthalic hydrazide in the monohydroxymonooxo

**Table 5** p$K_a$ Values for Sulfur-containing Pyridazines

| Compound | Proton gain | Proton loss | |
|---|---|---|---|
| | | first | second |
| Pyridazine-3(2H)-thione | −2.68 | 8.30 | — |
| 2-Methylpyridazine-3(2H)-thione | −2.95 | — | — |
| 3-Methylmercaptopyridazine | 2.26 | — | — |
| Pyridazine-4(1H)-thione | −0.75 | 6.54 | — |
| 1-Methylpyridazine-4(1H)-thione | −0.83 | — | — |
| 4-Methylmercaptopyridazine | 3.26 | — | — |
| 6-Mercaptopyridazine-3(2H)-thione | −0.5 | 2.1 | 10.4 |
| 3,6-Bis(methylmercapto)pyridazine | −6.0 | — | — |
| 6-Hydroxypyridazine-3(2H)-thione | −1.7 | 3.6 | >12 |
| | −1.39 | 3.32 | — |
| 6-Methylthiopyridazine-3(2H)-one | — | 10.11 | — |
| 6-Methoxypyridazine-3(2H)-thione | −2.36 | 6.95 | — |
| | −2.3 | 8.5 | — |
| 3-Methoxy-6-methylthiopyridazine | 1.84 | — | — |
| 6-Aminopyridazine-3(2H)-thione | −0.14 | 9.05 | — |
| 6-Amino-3-methylthiopyridazine | 5.61 | — | — |
| 6-Methylaminopyridazine-3(2H)-thione | −0.04 | 9.46 | — |
| 3-Methylthio-6-methylaminopyridazine | 5.94 | — | — |
| 6-Piperidinopyridazine-3(2H)-thione | −0.06 | 9.31 | — |
| 3-Methylthio-6-piperidinopyridazine | 5.13 | — | — |

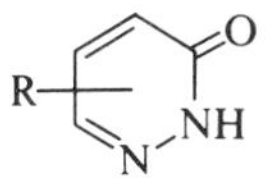

(4) 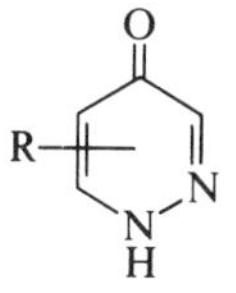    (5) 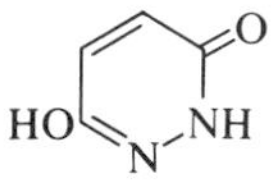    (6)

form (**9**); for details see ⟨63AHC(1)339⟩. Both 4- and 6-hydroxypyridazine 1-oxides exist predominantly in the *N*-hydroxypyridazinone forms (**10**) and (**11**), while 3- and 5-hydroxypyridazine 1-oxides exist in the hydroxy *N*-oxide forms (**12**) and (**13**).

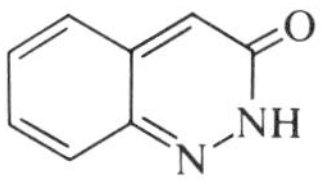

(7) 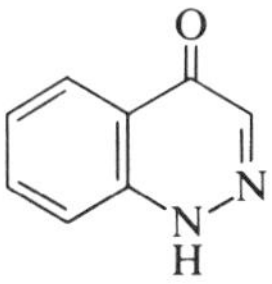    (8) 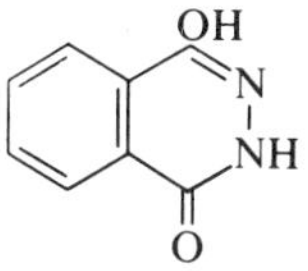    (9) 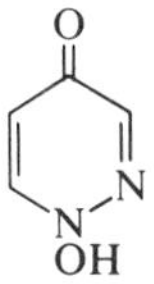    (10)

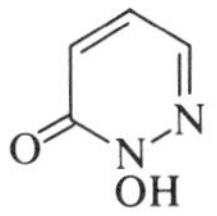

(11) 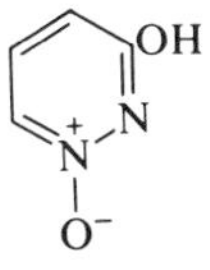    (12) 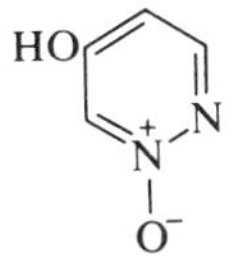    (13)

Pyridazine-3(2H)-thiones exist in the thione form (**14**), as is evident from an X-ray structure analysis of pyridazine-3(2H)-thione. 6-Mercaptopyridazine-3(2H)-thione is predominantly in the monothiolmonothione form (**15**) in aqueous solution and in the solid state, 6-hydroxypyridazine-3(2H)-thiones are in the hydroxythione form (**16**) and 6-aminopyridazine-3(2H)-thiones exist in the aminothione form (**17**); for further details see ⟨73HC(28)755⟩. Cinnoline-4(1H)-thiones and phthalazine-1(2H)-thione have been shown on the basis of UV data and ionization constants to exist in the thione forms.

Both 3- and 4-aminopyridazines exist in the amino form; 4-aminocinnoline was formerly claimed to be anomalous and to be best represented in the imino form (**18**), while 4-alkylamino-3-phenylcinnolines are in the amino form. There seems no doubt, however, that amino forms predominate in both the 3- and 4-aminocinnoline series ⟨71JCS(B)2344⟩.

For further details and examples of prototropic tautomerism in pyridazines and condensed pyridazines, see ⟨63AHC(1)339, B-76MI21200⟩.

| (14) | (15) | (16) | (17) | (18) |

### 2.12.2.1.2 Spectroscopic properties

The UV spectra of pyridazine, its anion and cation have been measured in different solvents and the results discussed. The blue shift on changing from nonpolar to more polar hydroxylic solvents is due mainly to the hydrogen bonding of a hydroxylic solvent to a nitrogen electron lone-pair. The association constant of hydrogen bonding is 17.5 kJ mol$^{-1}$ or 19 kJ mol$^{-1}$, obtained from UV and IR spectral data respectively. The relatively large solvent shift on changing from hydrocarbon to water is ascribed to hydrogen bonding (42 kJ mol$^{-1}$) between one molecule of water and both pyridazine nitrogen atoms. IR and Raman spectra have been reviewed ⟨63PMH(2)161⟩. The changes in the Raman spectra of pyridazine adsorbed on silica are interpreted as the result of hydrogen bond formation between the surface hydroxy groups and the nitrogen atoms of pyridazine ⟨78BCJ3063⟩.

The $^1$H NMR spectrum of pyridazine shows two symmetrical quartets of an $A_2X_2$ or $A_2B_2$ type dependent on the solvent and concentration. The $^{13}$C satellites have been used to obtain all coupling constants. Spectra of *C*-substituted pyridazines, methylthio- and methylsulfonyl-pyridazines, both as neutral molecules and as cations, N-1 and N-2 quaternized species, pyridazinones, hydroxypyridazinones, *N*-oxides and 1,2-dioxides have been reviewed ⟨B-73NMR88⟩ and are summarized in Tables 6, 7 and 8.

The $^1$H NMR spectra of cinnoline and its derivatives are complex. The unequivocal assignments are based on the complete iterative analysis of the spectra of a large number

**Table 6**   Spectral Data for Some Pyridazines (CDCl₃)

| Substituents | Chemical shifts[a] | | | | Coupling constants[a] | | | | | |
|---|---|---|---|---|---|---|---|---|---|---|
| | $\tau(3)$ | $\tau(4)$ | $\tau(5)$ | $\tau(6)$ | $J_{34}$ | $J_{35}$ | $J_{36}$ | $J_{45}$ | $J_{46}$ | $J_{56}$ |
| None | 0.76 | 2.46 | — | — | 4.9 | 2.0 | 3.0 | 8.4 | — | — |
| 3-Me | (7.26) | 2.62 | 2.60 | 0.94 | — | 2.2 | 3.0 | 8.6 | 1.8 | 4.7 |
| 4-Me | 0.92 | (7.60) | 2.67 | 0.96 | — | 2.2 | 3.0 | (1.0) | — | 5.0 |
| 3-Cl | — | 2.41 | 2.45 | 0.83 | — | — | — | 8.8 | 1.8 | 4.7 |
| 3-OMe | (5.92) | 2.98 | 2.59 | 1.12 | — | — | — | 9.0 | 1.7 | 4.5 |
| 3-Me,6-Cl | (7.29) | 2.65 | 2.57 | — | — | — | — | 8.8 | — | — |
| 3-Cl,4-Me | — | (7.54) | 2.58 | 1.01 | — | — | — | (1.0) | — | 4.9 |
| 4-Me,6-Cl | 1.01 | (7.59) | 2.60 | — | — | 2.2 | — | (1.0) | — | — |
| 3-Cl,6-Me | — | 2.57 | 2.64 | (7.30) | — | — | — | 8.8 | — | — |
| 3-Me,6-OMe | (7.38) | 2.74 | 3.10 | (5.88) | — | — | — | 8.8 | — | — |
| 3,6-Di-Cl | — | 2.43 | 2.43 | — | — | — | — | — | — | — |
| 3-OMe,6-Cl | (5.99) | 3.01 | 2.62 | — | — | — | — | 9.0 | — | — |
| 3-Cl,6-OMe | — | 2.62 | 3.04 | (5.89) | — | — | — | 9.1 | — | — |

[a] Chemical shifts in parentheses are for the methyl group of the substituent. Coupling constants in parentheses are between the substituent methyl group and the indicated ring proton.

**Table 7**   Spectral Data for Maleic Hydrazide and Derivatives (DMSO)

| Compound | $\tau(4)$ | $\tau(5)$ | $J_{45}$ | $\tau(NMe)$ | $\tau(OMe)$ |
|---|---|---|---|---|---|
| Maleic hydrazide | 3.04 | 3.04 | — | — | — |
| Pyridazin-3(2*H*)-one | | | | | |
| 6-Cl | 3.00 | 2.49 | 9.9 | — | — |
| 2-Me,6-Cl | 2.96 | 2.46 | 9.8 | 6.36 | — |
| 2-Me,6-OH | 3.13 | 2.96 | 9.4 | 6.51 | — |
| 2-Me,6-OMe | 3.09 | 2.84 | 9.0 | 6.47 | 6.22 |
| Pyridazine-3,6(1*H*,2*H*)-dione | 3.09 | 3.09 | — | — | — |
| 3,6-Dimethoxypyridazine | 2.84 | 2.84 | — | — | 6.01 |

**Table 8** Spectral Data for Pyridazine *N*-Oxides (CDCl$_3$)

(a) *Monooxides*

| Substituents | $\tau(3)$ | $\tau(4)$ | $\tau(5)$ | $\tau(6)$ | $\tau$(Me) | $J_{34}$ | $J_{35}$ | $J_{36}$ | $J_{45}$ | $J_{46}$ | $J_{56}$ | $\tau$(OMe) |
|---|---|---|---|---|---|---|---|---|---|---|---|---|
| None | 1.46 | 2.78 | 2.17 | 1.74 | — | 5.3 | 2.5 | 1.0 | 8.0 | 1.0 | 6.5 | — |
| 3-Cl | — | 2.77 | 2.25 | 1.82 | — | — | — | — | 8.3 | 1.0 | 6.5 | — |
| 3-Me | — | 3.02 | 2.42 | 1.90 | 7.48 | — | — | — | 8.2 | 0.5 | 6.1 | — |
| 4-Me | 1.65 | — | 2.43 | 1.83 | 7.64 | 0.2 | 2.8 | 0.5 | 0.5 | 0.2 | 6.2 | — |
| 5-Me | 1.62 | 3.01 | — | 1.92 | 7.63 | 5.6 | 0.2 | 0.5 | 0.7 | 0.7 | 0.7 | — |
| 6-Me | 1.63 | 2.88 | 2.27 | — | 7.49 | 5.6 | 2.5 | — | 8.2 | — | — | — |
| 3,6-Di-Cl | — | 2.78 | 2.10 | — | — | — | — | — | 8.4 | — | — | — |
| 3-Cl,4-Me | — | — | 2.41 | 1.90 | 7.61 | — | — | — | 0.7 | 0.2 | 6.2 | — |
| 3-Cl,5-Me | — | 2.95 | — | 2.01 | 7.64 | — | — | — | 0.7 | 0.7 | 0.7 | — |
| 3-Cl,6-Me | — | 2.83 | 2.27 | — | 7.51 | — | — | — | 8.3 | — | — | — |
| 4-Cl,6-Me | 1.61 | — | 2.32 | — | 7.49 | — | 3.0 | — | — | — | — | — |
| 3,4-Di-Cl,6-Me | — | — | 2.23 | — | 7.50 | — | — | — | — | — | — | — |
| 3-OMe | — | 3.31 | 2.47 | 2.05 | — | — | — | — | 8.6 | 0.7 | 5.8 | 5.98 |
| 3-OMe,6-Me | — | 3.33 | 2.48 | — | 7.55 | — | — | — | 8.5 | — | — | 5.98 |
| 4-OMe,6-Me | 1.92 | — | 2.80 | — | 7.48 | — | 3.7 | — | — | — | — | 6.07 |
| 3-OMe,6-Cl | — | 3.28 | 2.29 | — | — | — | — | — | 8.8 | — | — | 5.96 |
| 3,6-Di-OMe | — | 3.27 | 2.64 | — | — | — | — | — | 8.7 | — | — | 5.92, 6.02 |
| 3-OMe,4-Cl,6-Me | — | — | 2.46 | — | 7.56 | — | — | — | — | — | — | 5.92 |
| 3,4-Di-OMe,6-Me | — | — | 3.05 | — | 7.57 | — | — | — | — | — | — | 5.90, 6.02 |
| 4-NO$_2$ | 0.70 | — | 1.35 | 1.84 | — | — | 2.3 | 0.5 | — | — | 7.0 | — |
| 4-NO$_2$,6-Me | 0.80 | — | 1.55 | — | 7.41 | — | 3.3 | — | — | — | — | — |

(b) *Dioxides*

| Substituents | $\tau(3)$ | $\tau(4)$ | $\tau(5)$ | $\tau(6)$ | $J_{34}$ | $J_{35}$ | $J_{36}$ | $J_{45}$ | $J_{46}$ | $J_{56}$ |
|---|---|---|---|---|---|---|---|---|---|---|
| None | 1.85 | 2.92 | 2.92 | 1.85 | — | — | — | — | — | — |
| 3-Me | 7.44 | 2.90 | 2.95 | 1.86 | ~0.4 | — | ~0.6 | 8.1 | 2.4 | 6.1 |
| 4-Me | 1.93 | 7.65 | 3.02 | 1.89 | ~0.8 | 2.4 | 0.7 | 0.7 | — | 6.8 |
| 3,6-Di-Me | 7.47 | 3.04 | 3.04 | 7.47 | <0.2 | — | — | — | — | <0.2 |

of derivatives, assisted by deuteration studies; in general, H-3 and H-5 are easily identified. For a review see ⟨B-73NMR324⟩; the data are presented in Table 9. The spectra of phthalazine have been analyzed ⟨65AJC707⟩. On the basis of linear correlation of $^{13}$C and $^{1}$H chemical shifts of 4-substituted pyridazines with those of monosubstituted benzenes and monosubstituted pyridines, the corresponding chemical shifts of substituted pyridazines can be predicted ⟨79CPB1169⟩. Correlation of the substituent-induced $^{13}$C chemical shifts of 4-substituted pyridazines with total charge densities, calculated by the MINDO/2 method, showed a linear relationship between these two parameters ⟨79CPB2105⟩. The changes in the $^{13}$C chemical shifts due to *N*-methylation and *N*-protonation could be predicted by the corresponding $\Delta\delta$ values of pyridines. Both reactions have a similar effect, although for

**Table 9** Spectral Data for Cinnolines[a]

| Substituents | Solvent | $\tau(3)$ | $\tau(4)$ | $\tau(5)$ | $\tau(6)$ | $\tau(7)$ | $\tau(8)$ |
|---|---|---|---|---|---|---|---|
| None[a] | Me$_2$CO | 0.67 | 1.91 | 1.98 | 2.16 | 2.06 | 1.51 |
| | CCl$_4$ | 0.90 | 2.27 | 2.43 | 2.43 | 2.43 | 1.70 |
| | CDCl$_3$ | 0.72 | — | — | — | — | 1.47 |
| 4-Me | CDCl$_3$ | 1.04 | 7.47 | 2.32 | 2.32 | 2.32 | 1.70 |
| 4-CO$_2$H | DMSO | 0.15 | — | 1.50 | 2.07 | 2.07 | 1.37 |
| 3-NH$_2$ | DMSO | — | 3.00 | 2.47 | 2.47 | 2.47 | 1.83 |
| 4-NH$_2$ | DMSO | 1.27 | 2.63 | 2.35 | 2.35 | 1.83 | 1.83 |
| 5-NH$_2$ | DMSO | 0.87 | 1.75 | — | 3.18 | 2.48 | 2.48 |
| 8-NH$_2$ | CDCl$_3$ | 0.97 | 3.15 | 2.45 | 3.15 | 2.70 | 4.62 |
| 3-NO$_2$ | DMSO | — | 0.77 | 1.70 | 1.93 | 1.93 | 1.40 |
| 5-NO$_2$ | CDCl$_3$ | 0.60 | 1.20 | — | 1.17 | 2.10 | 1.38 |
| 8-NO$_2$ | CDCl$_3$ | 0.60 | 2.05 | 2.80 | 2.05 | 2.05 | — |

[a] Coupling constants: $J_{34}$ 5.7, $J_{48}$ 0.8, $J_{56}$ 7.8, $J_{57}$ 1.5, $J_{58}$ 0.8, $J_{67}$ 6.9, $J_{68}$ 1.3, $J_{78}$ 8.6.

*N*-protonation the absolute magnitude of the $\Delta\delta$ values is larger than for *N*-methylation $\langle$77OMR(9)53$\rangle$. Nuclear relaxation rates of $^{13}$C and $^{14}$N have been measured as a function of temperature for neat liquid pyridazine, and nuclear Overhauser enhancement has been used to separate the dipolar and spin rotational contributions to $^{13}$C relaxation. $^{13}$C Dipolar relaxation rates have been combined with quadrupole $^{14}$N relaxation rates to determine rotational correlation times for motion about each principal molecular axis $\langle$78MI21200$\rangle$. $^{13}$C NMR analysis has been used to determine the structure of phenyllithium–pyridazine adducts and of the corresponding dihydropyridazines obtained by hydrolysis of the adducts $\langle$78RTC116$\rangle$.

The mass spectrum of pyridazine is simple and high resolution measurements have shown that the ion at *m/e* 52 is composed of both $C_4H_4^+$ (73.5%) and $C_3H_2N^+$ (26.5%) ions; the ions at *m/e* 51 and *m/e* 50 are due to the fragments $C_4H_3^+$ and $C_4H_2^+$, respectively (Scheme 1).

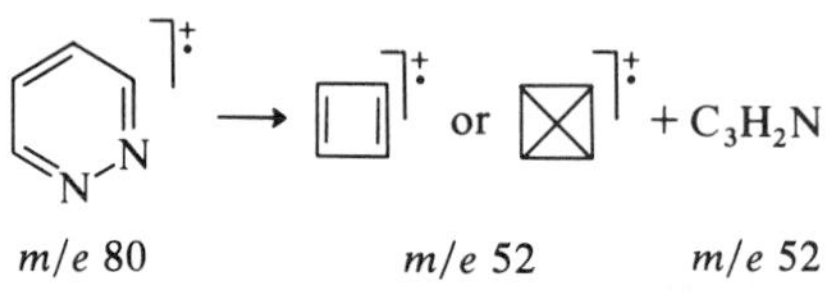

*m/e* 80                    *m/e* 52           *m/e* 52

**Scheme 1**

A simple fragmentation pattern is also characteristic for chloro-, methyl- and amino-pyridazines. Pyridazinone fragments by loss of carbon monoxide followed by loss of $N_2$ (Scheme 2).

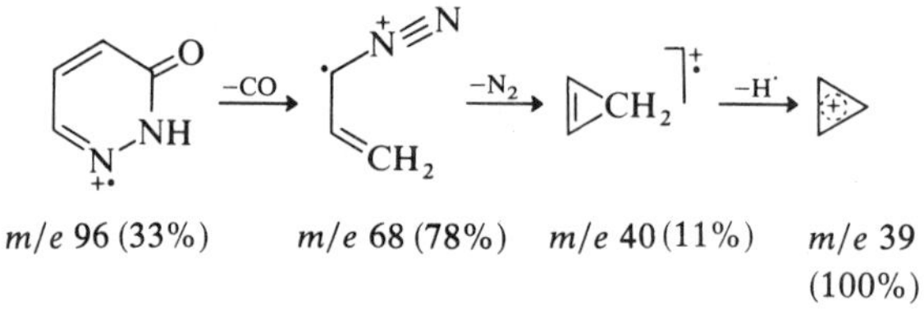

*m/e* 96 (33%)       *m/e* 68 (78%)    *m/e* 40 (11%)    *m/e* 39 (100%)

**Scheme 2**

The tendency for fragmentation by loss of $N_2$, when compared with pyridazine, is reduced in the case of phthalazine where losses of HCN become important (Scheme 3). This is supported by quantum mechanical calculations which show that the structure (**3a**) better represents the ground state of the molecule than structure (**3b**).

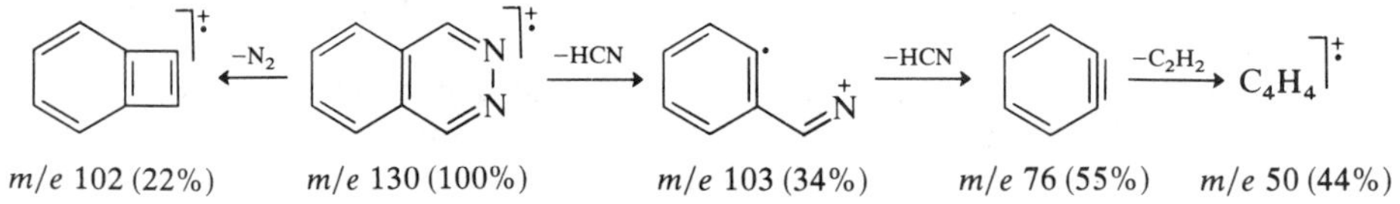

*m/e* 102 (22%)    *m/e* 130 (100%)    *m/e* 103 (34%)    *m/e* 76 (55%)    *m/e* 50 (44%)

**Scheme 3**

The electron impact mass spectrometric fragmentations of (*E*)-3- and (*E*)-4-styryl-pyridazines show that the intensity ratio of the $M^+$ and $(M-1)^+$ ions, the general degree of fragmentation and the elimination pathways of nitrogen are the most characteristic features distinguishing between the two isomeric compounds $\langle$81JHC255$\rangle$.

The ESR spectrum of the pyridazine radical anion, generated by the action of sodium or potassium, has been reported, and oxidation of 6-hydroxypyridazin-3(2*H*)-one with cerium(IV) sulfate in sulfuric acid results in an intense ESR spectrum $\langle$79TL2821$\rangle$. The self-diffusion coefficient and activation energy, the half-wave potential ($-2.16$ eV), magnetic susceptibility and room temperature fluorescence in solution ($\lambda_{max} = 23\,800$ cm$^{-1}$, life time $2.6 \times 10^{-9}$ s) are reported.

The photoelectron spectra of pyridazine have been interpreted on the basis of many-body Green's function calculations both for the outer and the inner valence region. The calculations confirm that ionization of the first *n*-electron occurs at lower energy than of the first $\pi$-electron $\langle$79MI21201$\rangle$. A large number of bands in the photoelectron spectrum of 3,6-diphenylpyridazine in stretched polymer sheets have been assigned to transitions predicted by semiempirical models, and large MCD B-terms are predicted for the two lowest $\pi \rightarrow \pi^*$ transitions ⟨79MI21200⟩.

## 2.12.3 FULLY CONJUGATED RINGS: REACTIVITY AT RING ATOMS

### 2.12.3.1 Thermal and Photochemical Reactions Involving No Other Species

#### 2.12.3.1.1 Thermal rearrangements

3-Azidopyridazine 2-oxides give on thermolysis either maleonitrile (**19**), $\beta$-cyanoacrylate (**20**) or 2-cyano-1-hydroxypyrazole (**21**), dependent on the substituent at the 6-position (Scheme 4) ⟨75CC703⟩.

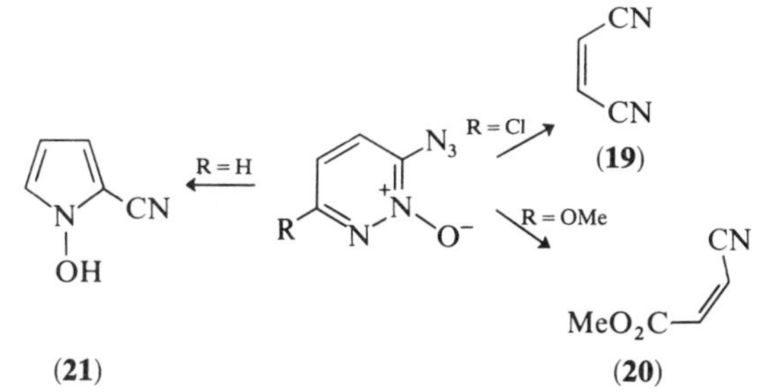

**Scheme 4**

Rearrangement of perfluoro(4,5-di-s-butylpyridazine) to a mixture of perfluoro(4,5-di-s-butylpyrimidine) and perfluoro(2,5-di-s-butylpyrazine) occurs at 300 °C. Cross-over experiments between various fluorinated pyridazine derivatives and doubly $^{15}$N-labelled derivatives rule out any rearrangement mechanism involving a cycloaddition reaction. The process is regarded as a free-radical promoted formation of valence isomers. Some fluorinated compounds act as promoters of this rearrangement. From perfluoro(4,5-diethylpyridazine), perfluoro(4,5-diethylpyrazine) is also formed (Scheme 5) ⟨81JCS(P1)1071, 79CC445⟩. The rearrangement of perfluoro(4,5-diisopropylpyridazine) is sensitized in the presence of perfluoro(4,5-di-s-butylpyridazine) or tetrafluoropyridazine ⟨79CC446⟩.

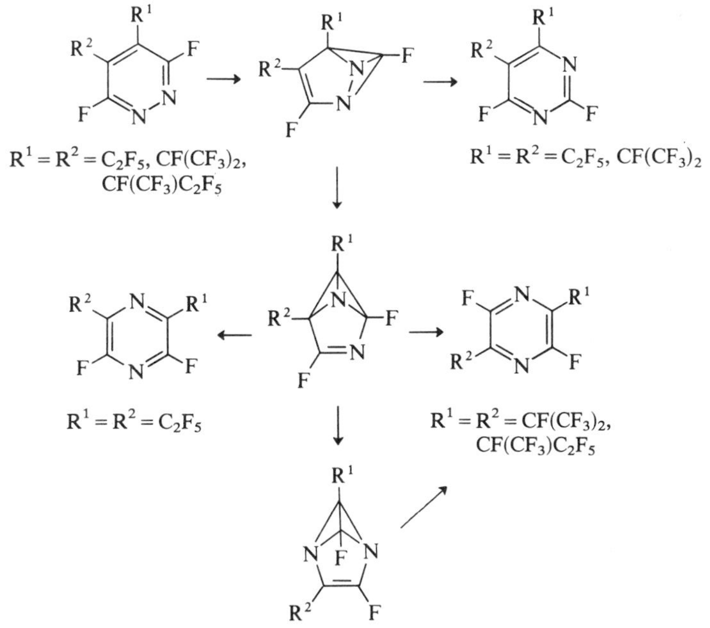

**Scheme 5**

Perfluoropyridazines and perfluoro(alkylpyridazines) isomerize thermally to perfluoropyrimidines as the main products. In some instances traces of perfluoropyrazines are also formed (Scheme 6) ⟨74JCS(P1)1513⟩.

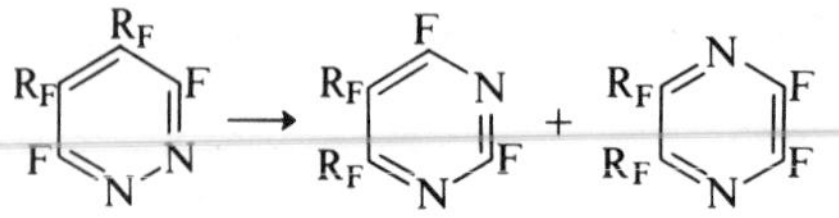

**Scheme 6**

On the other hand, perfluoro(tetraphenylpyridazines) and some perfluoro(alkyl-pyridazines) undergo nitrogen elimination, or both nitrogen elimination and rearrangement, to give perfluoroalkynes and perfluoropyrimidines (Scheme 7) ⟨74JCS(P1)1513⟩.

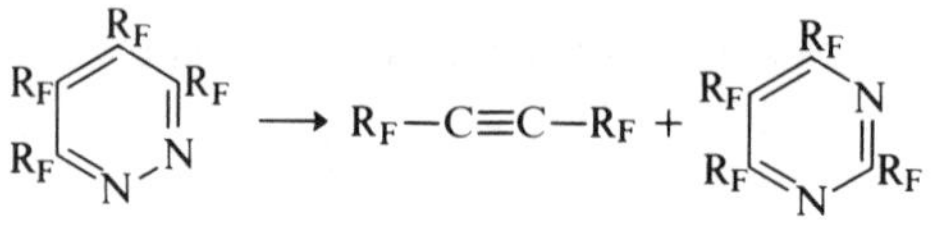

**Scheme 7**

4-Diazo-5,6-dioxo-1-phenyl-1,4,5,6-tetrahydropyridazine (**22**) rearranges at room temperature in a dark place in approximately ten days under strictly anhydrous conditions to an intermediate to which the structure of a 'pyrazololactone' (**23**) is assigned, on the basis of physical and chemical evidence compared with benzopropiolactone or 2-thioben-zopropiolactone. The structure is represented best by resonance structures with charge-separated forms (**23a–c**), stabilized by an aromatic ring attached to N-1 of the pyrazole ring. In the presence of atmospheric moisture the 'pyrazololactone' is converted to 5-hydroxy-1-phenylpyrazole-4-carboxylic acid (**24**; R = CO$_2$H), which slowly decarboxylates to 5-hydroxy-1-phenylpyrazole (**24**; R = H) (Scheme 8) ⟨79H(12)457⟩. For further examples see ⟨68AHC(9)211, 79AHC(24)363⟩.

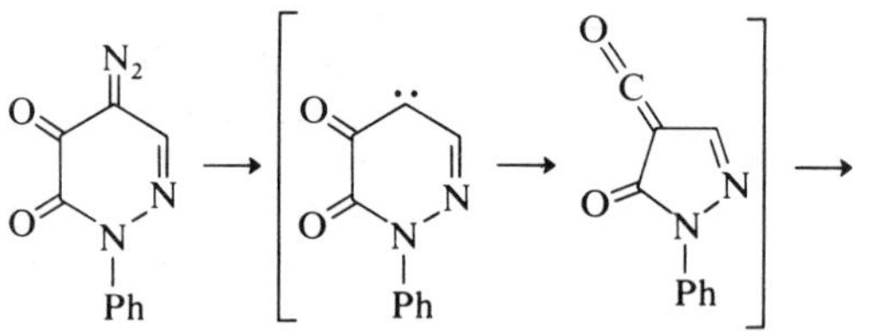

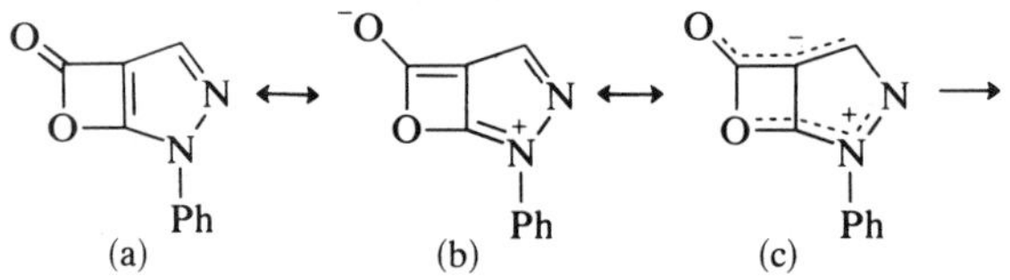

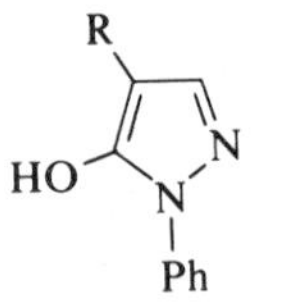

**Scheme 8**

### 2.12.3.1.2 Photochemical reactions

There are various photochemical transformations of pyridazines, their corresponding benzo analogs, *N*-oxides and *N*-imides. Gas-phase photolysis of pyridazine affords nitrogen and vinylacetylene as the main products. Perfluoropyridazine gives first perfluoropyrazine, which isomerizes slowly into perfluoropyrimidine.

Photoisomerization of perfluoro(4,5-diisopropylpyridazine) is postulated to proceed through 'Dewar diazabenzenes' (**25**) and (**26**) to perfluoro(2,5-diisopropylpyrazine) (**27**), which is in equilibrium with the isomeric perfluoro(2,6-diisopropylpyrazine) (**28**) after prolonged irradiation in the liquid phase (Scheme 9) ⟨75JCS(P1)1130⟩. Benzo-fused pyridazines do not isomerize readily under photochemical conditions. An exception is perfluorocinnoline which rearranges to perfluoroquinazoline.

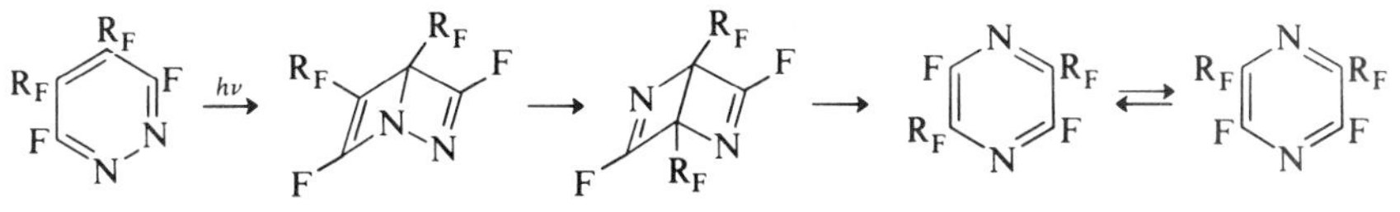

**Scheme 9**

Pyridazin-3(2*H*)-ones rearrange to 1-amino-3-pyrrolin-2-ones (**29**) and (**30**) upon irradiation in neutral methanol (Scheme 10), while photolysis of 5-amino-4-chloro-2-phenylpyridazin-3(2*H*)-one gives the intermediate (**31**) which cyclizes readily to the bis-pyridazinopyrazine derivative (**32**; Scheme 11).

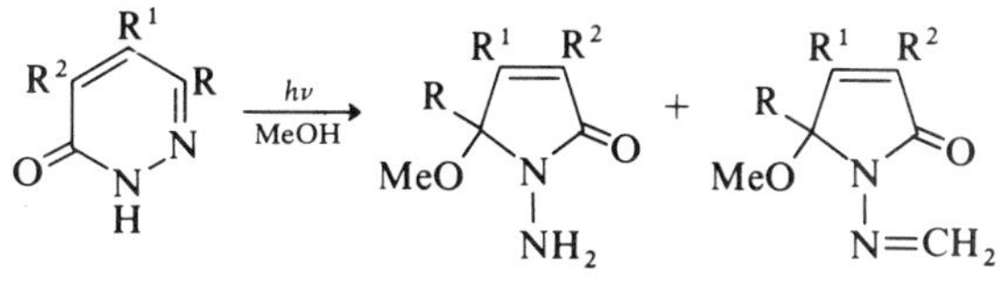

**Scheme 10**

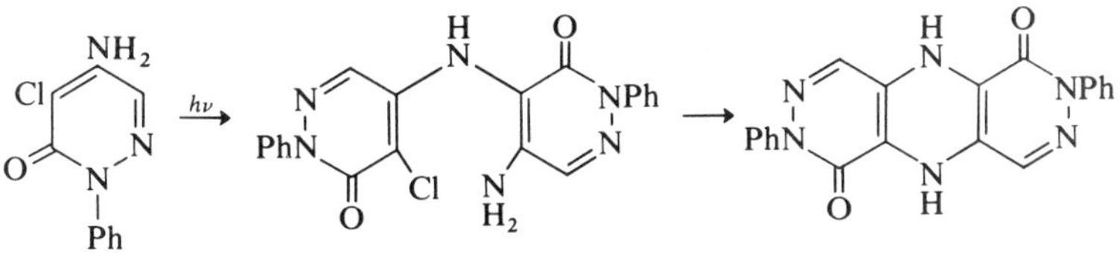

**Scheme 11**

5-Oxidopyridazinium betaines isomerize photochemically into pyrimidin-4(3*H*)-ones (**33**). Irradiation of 3-oxidopyridazinium betaine or 1-oxidophthalazinium betaine in water affords similarly the corresponding pyridazin-3(2*H*)-one (**35**) and phthalazin-1(2*H*)-one derivative (**37**). However, photolysis in acetonitrile affords stable diaziridines (**34**) and (**36**) which can be converted in the presence of water to the final products (**35**) and (**37**) (Scheme 12) ⟨79JCS(P1)1199⟩.

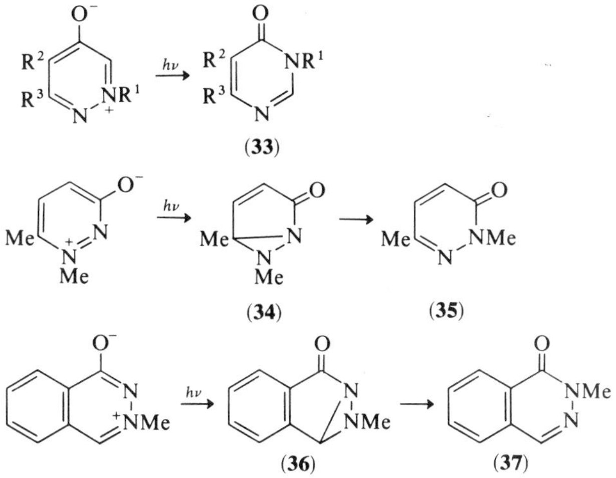

**Scheme 12**

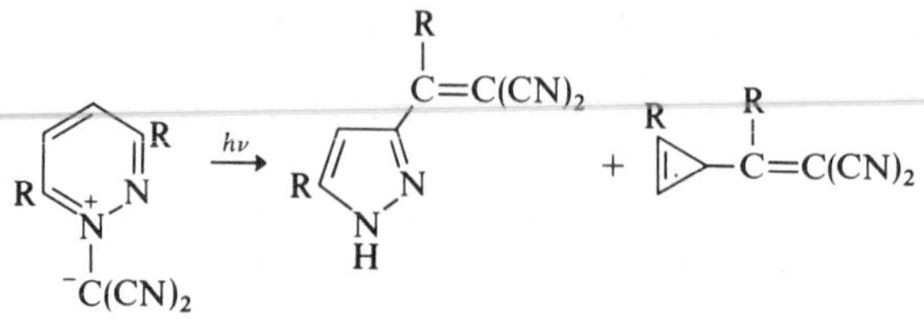

**(38)**          **(39)**

**Scheme 13**

Pyridazinium dicyanomethylide gives a mixture of the pyrazole (**38**) and the substituted cyclopropene (**39**; Scheme 13).

Photolysis of pyridazine *N*-oxide and alkylated pyridazine *N*-oxides results in deoxygenation. When this is carried out in the presence of aromatic or methylated aromatic solvents or cyclohexane, the corresponding phenols, hydroxymethyl derivatives or cyclohexanol are formed in addition to pyridazines. In the presence of cyclohexene, cyclohexene oxide and cyclohexanone are generated.

Transformation of pyridazine 1-oxides and their methyl derivatives into cyclopropyl ketones and/or substituted furans can also occur (Scheme 14).

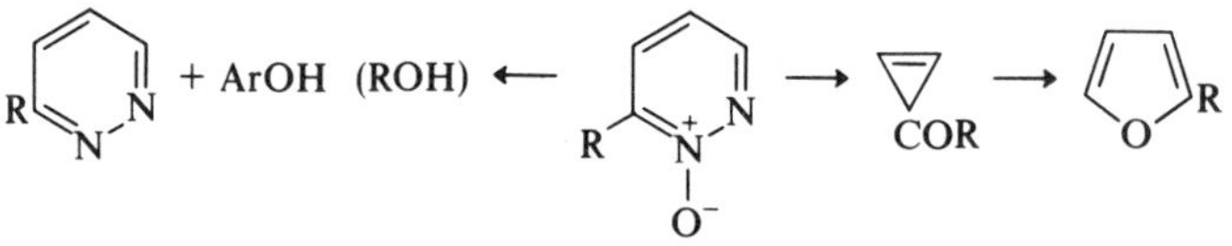

**Scheme 14**

3,6-Diphenylpyridazine *N*-oxide gives, in addition to the deoxygenated product, the corresponding diazoketone (**40**) as the only product of photoisomerization. The latter reacts further to form 2,5-diphenylfuran (**41**) or 5-benzoyl-3-phenylpyrazole (**42**) ⟨73JA7402⟩. Higher arylated pyridazine *N*-oxides react in an analogous manner, although irradiation of 3,4,5,6-tetraphenylpyridazine 1-oxide is more complex. Here, a mixture of tetraphenylpyridazine and (**43**) together with 2,3,4,5-tetraphenylfuran (**44**) and *cis*- and *trans*-dibenzoyldiphenylethylene (**45**) is produced (Scheme 15).

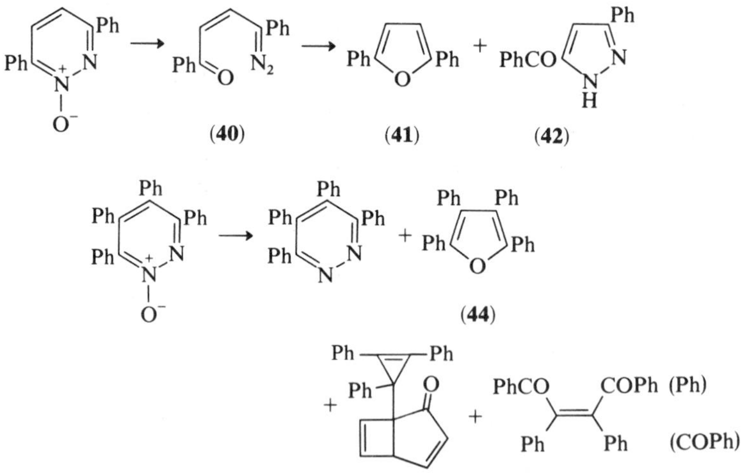

**(40)**          **(41)**          **(42)**

**(44)**

**(43)**          **(45)**

**Scheme 15**

2-Methylpyridazin-3(2*H*)-one 1-oxides unsubstituted at position 6 afford photochemically a mixture of the corresponding 6-hydroxypyridazin-3(2*H*)-one, lactone (**46**) and isomaleimide (**47**). 4-Methylcinnoline 1-oxide gives a mixture of the methyl substituted benzofuran (**48**), the indazole (**49**) and the indole (**50**), while 4-methylcinnoline 2-oxide gives a mixture of the deoxygenated starting compound, benzisoxazole (**51**) and *o*-aminoacetophenone (Scheme 16).

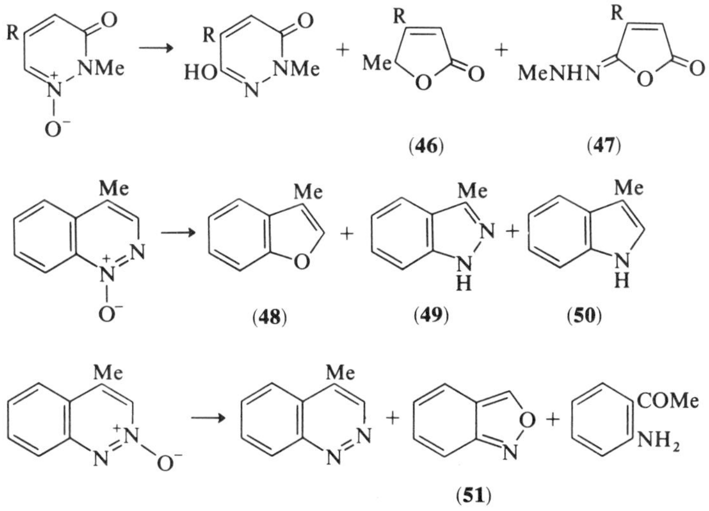

**(46)**  **(47)**

**(48)**  **(49)**  **(50)**

**(51)**

**Scheme 16**

Photolysis of 1,4-diphenylphthalazine 2-oxide in various solvents gives 1,3-diphenyl-isobenzofuran (**52**) as the primary product. In the presence of oxygen, deoxygenation to 1,4-diphenylphthalazine and oxidation of the initially formed 1,3-diphenylisobenzofuran (**52**) to 1,2-dibenzoylbenzene also take place (Scheme 17).

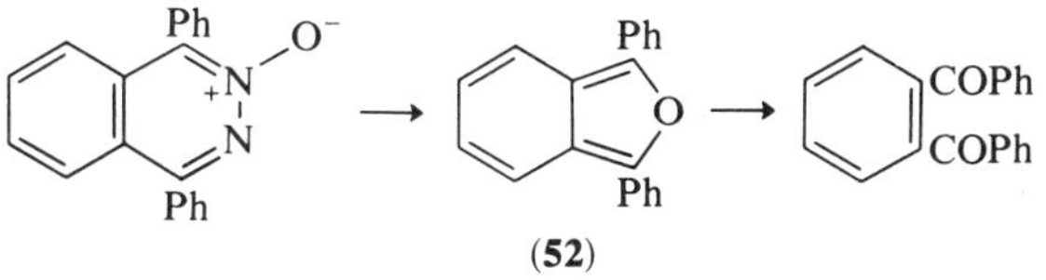

**(52)**

**Scheme 17**

Phototransformation of pyridazine 1,2-dioxides sharply contrasts with that of pyridazine 1-oxides. Pyridazine 1,2-dioxide derivatives give 3a,6a-dihydroisoxazolo[5,4-*d*]isoxazoles (**53**) through postulated bisiminoxyl radicals. 3,6-Diphenylpyridazine 1,2-dioxide gives, besides the corresponding bicyclic derivative (**53**), 3-phenylisoxazole (**54**) and 4,5-diphenyl-furoxan (**55**). The last two products can be explained by generation of the nitrile oxide from the intermediate (**53**) with subsequent dimerization to the furoxan (**55**; Scheme 18) ⟨79T1267⟩.

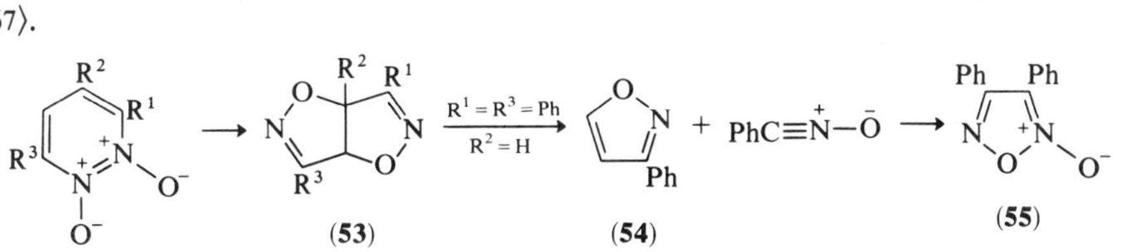

**(53)**  **(54)**  **(55)**

**Scheme 18**

Photolysis of pyridazine *N*-ethoxycarbonylimide results in the formation of the pyrrole derivative (**56**). The rearrangement is postulated to proceed *via* a diaziridine, followed by ring expansion to the corresponding 1,2,3-triazepine derivative and rearrangement to a triazabicycloheptadiene, from which finally a molecule of nitrogen is eliminated (Scheme 19) ⟨80CPB2676⟩.

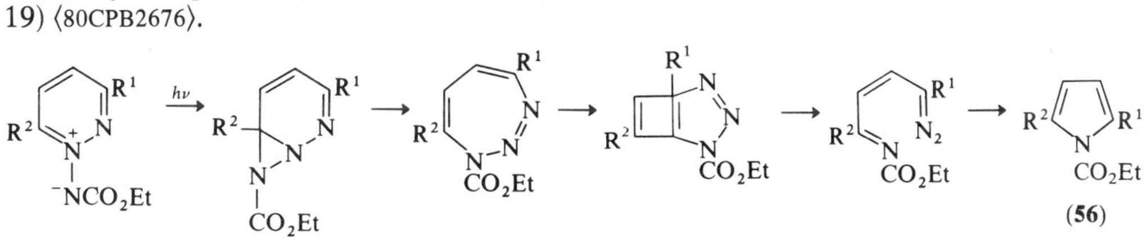

**(56)**

**Scheme 19**

In some instances the solvents may react with the substrate during the irradiation. For example, 3,6-dichloropyridazine, when irradiated in acidified methanol, gives a mixture of monomethylated (56), dimethylated (57) and hydroxymethylated (58) compounds. Further transformation of the hydroxymethyl compound (58) results in the formation of γ-lactones (59) and succinates (60; Scheme 20).

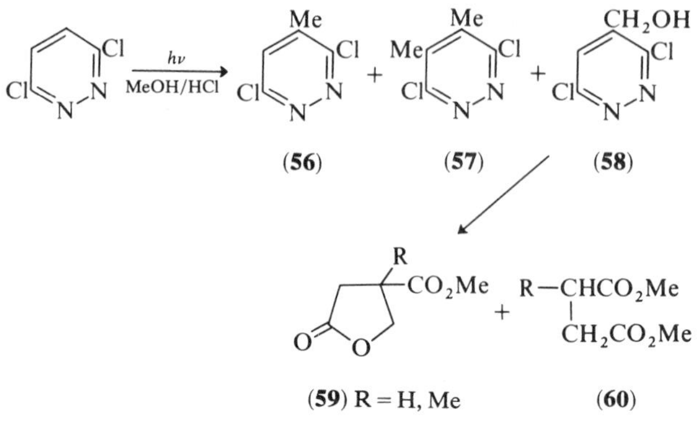

**Scheme 20**

### 2.12.3.2 Electrophilic Attack at Nitrogen

#### 2.12.3.2.1 Alkylation

Alkylation of pyridazinones results in either *N*- or *O*-substitution and is usually carried out with an alkyl halide or dialkyl sulfate in the presence of a base. Besides methyl iodide, α-halo acids and esters and alkylaminoalkyl halides are frequently employed to produce *N*-alkylated products. Diazomethane gives both *N*-methylated and *O*-methylated products; 3-methylpyridazin-6(1*H*)-one, for example, gives a mixture of 6-methoxy-3-methyl-pyridazine and 1,3-dimethylpyridazin-6(1*H*)-one. Highly substituted pyridazinones can be methylated to give betaines. For example, 3-substituted 5-aminopyridazin-6(1*H*)-ones are transformed into a mixture of 3-substituted 5-amino-1-methylpyridazin-6(1*H*)-one (61) and the 2-methyl betaine (62) on treatment with dimethyl sulfate. The ratio of the products (61) and (62) is dependent on the size of the substituent at position 3, and with bulky groups less betaine is formed. Treatment of 5-ethoxycarbonyl-3-methyl-6-phenylpyridazin-4(1*H*)-one with diazomethane gives the $N^1$-methyl compound (63), the 4-methoxy derivative (64) and the corresponding betaine (65; Scheme 21) ⟨79JOC3053⟩.

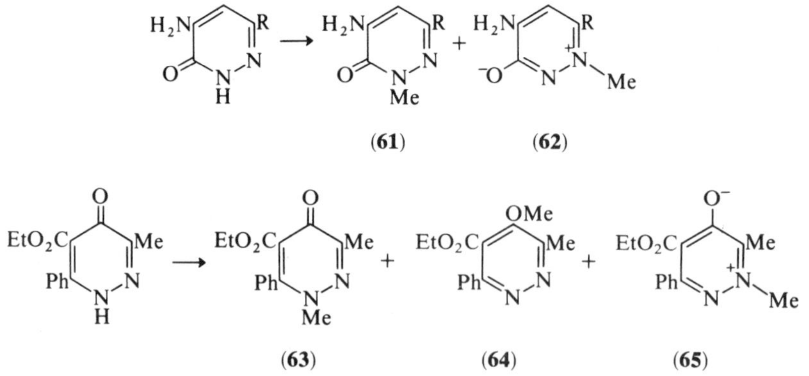

**Scheme 21**

If reagents with large alkyl groups are used, mixtures of *N*- and *O*-alkylated compounds are usually obtained. Thus, benzylation of pyridazin-3(2*H*)-one affords a mixture of *N*- and *O*-benzyl derivatives in the ratio of 2:1, while ethoxycarbonylmethylation with diazoacetic ester gives the *N*- and *O*-carboxymethyl derivatives in the ratio of 1:10. In some instances alkylation can be directed by the reaction conditions. For example, glucosidation of pyridazinones with tetraacetylbromoglucose gives exclusively *N*-glucosyl derivatives, while glucosidation of the silver salts of the pyridazinones gives only *O*-alkylated products. Pyridazinones with additional hydroxy groups give various mono- and di-alkylated products, dependent on the alkylating agent and reaction conditions, mainly temperature and pH. For example, methylation of maleic hydrazide with diazomethane gives first the mono-*O*-methyl derivative (**66**), which can be further methylated with the same alkylating agent to produce the *N,O*-dimethyl derivative (**67**). On the other hand, methylation with dimethyl sulfate in the presence of aqueous base gives first the *N*-methyl compound (**68**), and prolonged heating at 150 °C affords a mixture of the 1,2-dimethyl compound (**70**) and 3-methoxy-1-methyl derivative (**69**); the ratio of (**69**) to (**70**) is dependent on the reaction time. On prolonged heating the 1,2-dimethyl isomer (**70**) predominates, as a result of thermal *O → N* methyl group migration (Scheme 22).

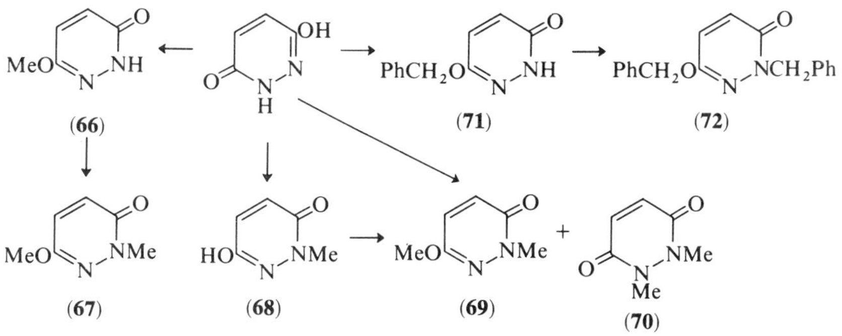

**Scheme 22**

The methylation of *N*-methyl derivatives of maleic hydrazide gives in general *O*-alkylated products. The opposite results are obtained with benzyl halides as alkylating agents. In this case the *O*-benzyl derivative (**71**) is formed, which is then further benzylated to the *N,O*-dibenzyl derivative (**72**). When ethyl chloroacetate is used, the direction of alkylation is dependent on pH. At pH above 8, *O*-alkylation occurs; at pH below 8, *N*-alkylation takes place exclusively; in neutral and acidic solutions only *N*-alkylated products are formed.

### 2.12.3.2.2 *O → N Rearrangement*

The second method for the preparation of *N*-alkylated pyridazine derivatives is *O → N* rearrangement of alkyl groups in alkoxy and dialkoxy derivatives. 3,6-Dialkoxypyridazines rearrange to the corresponding 3-alkoxy-1-alkylpyridazin-6(1*H*)-ones, usually in the presence of a catalyst such as mineral acid, *p*-toluenesulfonic acid, aluminum chloride and iron(III) chloride at high temperatures (up to 300 °C). If *O*-alkylated compounds decompose in the presence of acid, mercury(II) bromide in an aprotic solvent such as toluene is used. In this way, 4-glycosyloxy- and 3,6-diglycosyloxy-pyridazines are converted to the corresponding *N*-substituted derivatives. 3,6-Diallyloxypyridazine rearranges at 200 °C in the presence or absence of a catalyst in a Claisen type of rearrangement to 2-allyl-6-allyloxy-pyridazin-3(2*H*)-one as the main product and 1,2-diallylpyridazine-3,6(1*H*,2*H*)-dione as by-product.

*N*-Methylated pyridazinones can be obtained from 3,6-dialkoxypyridazines by treatment with alkyl halides or dialkyl sulfates. Methyl iodide and dimethyl sulfate are most frequently used. According to the proposed mechanism, an intermediate quaternary pyridazinium salt is formed, followed by elimination of a group R from the alkoxy group. At higher temperature, 1,2-dimethylpyridazine-3,6(1*H*,2*H*)-dione is formed with dimethyl sulfate.

The Michael-type addition of maleic hydrazide and other pyridazinones to activated alkenes, such as methyl acrylate, acrylonitrile, methyl vinyl ketone and other *α,β*-unsaturated carbonyl compounds, results in the formation of mono-*N*-substituted products.

Since pyridazinones are acidic compounds, they undergo *N*-hydroxy- and *N*-amino-methylation. They react also with aromatic aldehydes in the presence of acetic anhydride in the ratio of 2 : 1 to give the condensation product (**73**; Scheme 23).

*N*-Methylation of polysubstituted pyridazinones is frequently accompanied by some side reactions, mainly substitutions. For example, methylation of 4-nitro-5,6-diphenyl-pyridazin-3(2*H*)-one with methyl iodide in the presence of sodium methoxide affords

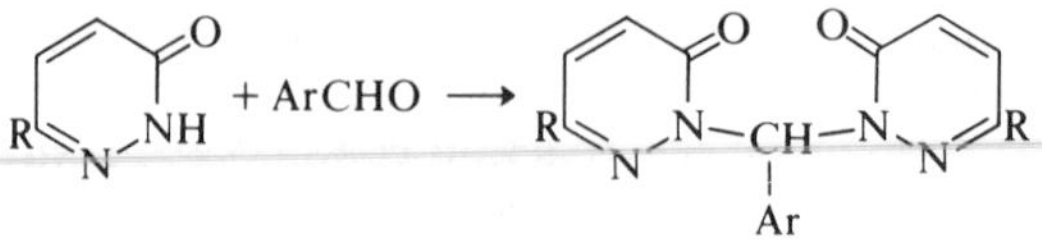

(73)

**Scheme 23**

2-methyl-4-nitro-5,6-diphenylpyridazin-3(2*H*)-one (**74**) and 4-methoxy-2-methyl-5,6-diphenylpyridazin-3(2*H*)-one (**75**). Methylation of the same starting compound with dimethyl sulfate in alkaline medium at 0–5 °C, however, provides only the 2-methyl-4-nitro derivative (**74**), while at 80–85 °C three products, *i.e.* 2-methyl-4-nitro- (**74**), 4-methoxy-2-methyl- (**75**) and 4-hydroxy-2-methyl-5,6-diphenylpyridazin-3(2*H*)-one (**76**) are formed. Methylation of 4-methoxy-5,6-diphenylpyridazin-3(2*H*)-one (**77**) with dimethyl sulfate at 5 °C in alkaline medium gives only the 4-methoxy-2-methyl derivative (**75**), while methylation of 4-hydroxy-5,6-diphenylpyridazin-3(2*H*)-one (**78**) produces four compounds, namely 4-methoxy-2-methyl- (**75**), 4-hydroxy-2-methyl- (**76**), 4-methoxy-1-methyl-5,6-diphenylpyridazinium 3-oxide (**79**) and 3,4-dimethoxy-5,6-diphenylpyridazine (**80**). 4-Amino-5,6-diphenylpyridazin-3(2*H*)-one gives with dimethyl sulfate in the presence of sodium hydroxide the corresponding 4-amino-2-methyl compound (**81**) and 4-amino-1-methyl-5,6-diphenylpyridazinium 3-oxide (**82**). 4-Phenoxy-, 4-methylthio-, 4-phenylthio, 4-pyrrolidinyl-, 4-piperidinyl- and 4-morpholino-5,6-diphenylpyridazin-3(2*H*)-ones are transformed with dimethyl sulfate in the presence of a base exclusively into the 2-methylated derivatives ⟨80MI21200⟩.

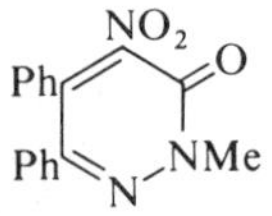

(74) 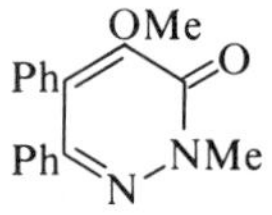          (75) 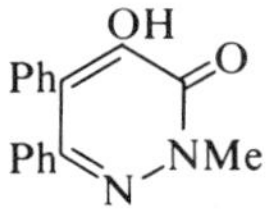          (76) 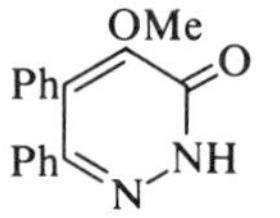          (77) 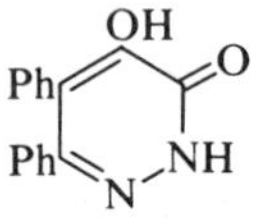          (78)

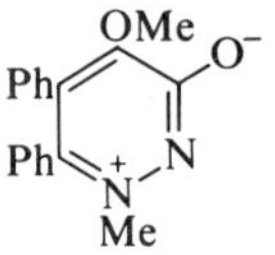

(79) 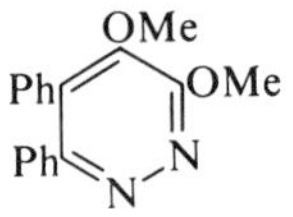          (80) 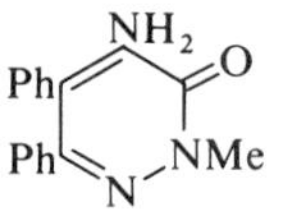          (81) 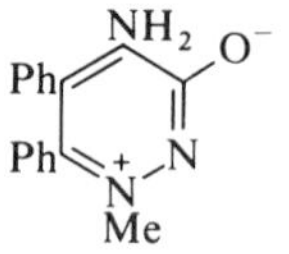          (82)

Acylation of pyridazinones and related compounds in the presence of weakly basic catalysts such as pyridine or sodium acetate produces *N*-acylated products, while *O*-acylated products are obtained under strongly basic conditions. However, the reaction between 6-chloropyridazin-3(2*H*)-one with chlorocarbonates and that of maleic hydrazide with unsaturated acid chlorides or chloromethylsulfonyl chloride gives preferentially *N*-substituted products.

Cinnolin-3(2*H*)-one (**7**) is methylated with diazomethane or methyl sulfate to give 2-methylcinnolin-3(2*H*)-one. In a similar manner, benzylation with benzyl chloride, cyanoethylation with acrylonitrile in the presence of benzyltrimethylammonium hydroxide and glucosidation with tetra-*O*-acetyl-α-D-glucopyranosyl bromide in the presence of a base affords the corresponding 2-substituted cinnolin-3(2*H*)-ones. However, glucosidation of the silver salt of cinnolin-3(2*H*)-one produces the corresponding *O*-substituted compound.

Alkylations of cinnolin-4(1*H*)-one (**8**) with methyl iodide, ethyl iodide, dimethyl and diethyl sulfates, isopropyl bromide, benzyl chloride, *etc.* take place predominantly at position 2 to give 2-alkyl-4-hydroxycinnolinium anhydro salts (**83**), together with small amounts of 1-methylcinnolin-4-one (**84**).

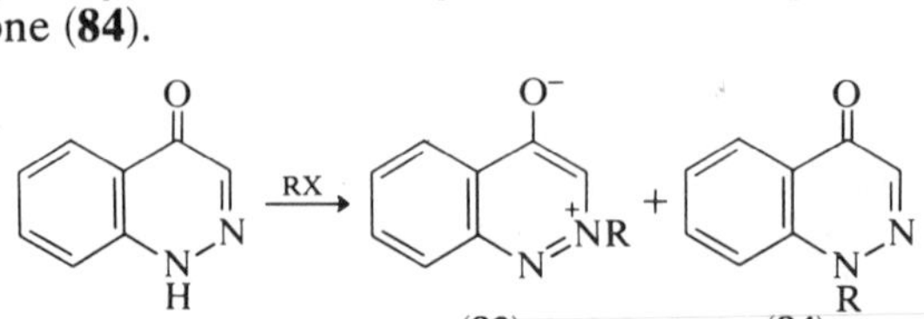

(8)          (83)          (84)

When large groups, such as phenyl, bromo, ethoxycarbonyl or nitro are attached at position 3, the principal products are 1-alkylcinnolin-4(1*H*)-ones. Cyanoethylation and acetylation of cinnolin-4(1*H*)-one takes place exclusively at N-1. Phthalazin-1(2*H*)-ones give 2-substituted derivatives on alkylation and acylation. Alkylation of 4-hydroxyphthalazin-1(2*H*)-one with an equimolar amount of primary halide in the presence of a base leads to 2-alkyl-4-hydroxyphthalazin-1(2*H*)-one and further alkylation results in the formation of 4-alkoxy-2-alkylphthalazinone. Methylation of 4-hydroxy-2-methyl-phthalazinone with dimethyl sulfate in aqueous alkali gives a mixture of 4-methoxy-2-methylphthalazin-1(2*H*)-one and 2,3-dimethylphthalazine-1,4(2*H*,3*H*)-dione, whereas methylation of 4-methoxyphthalazin-1(2*H*)-one under similar conditions affords only 4-methoxy-2-methylphthalazinone.

Michael-type addition, hydroxymethylation and Mannich reaction take place at nitrogen to give the corresponding 2-substituted 4-hydroxyphthalazin-1(2*H*)-ones.

### *2.12.3.2.3 Quaternization*

In general, monoquaternization of pyridazines proceeds easily as expected from the relatively high basicity compared with other six-membered heterocycles. The most common reagents used for the preparation of pyridazinium salts are alkyl halides, $\alpha,\omega$-dihalogeno-alkanes, $\alpha$- and $\beta$-halo esters, dimethyl sulfate and methyl *p*-toluenesulfonate. Since the pyridazine system possesses two nitrogen atoms whose free electron pairs are not part of the $\pi$-electron system, it can form diquaternary salts without loss of aromaticity. In practice, diquaternary salts can be prepared only by using oxonium salts as alkylating agents.

The position of monoquaternization is determined by the substituents. Both 3- and 4-methylpyridazines quaternize at N-1 and N-2. On the basis of quaternization of numerous 3,6-disubstituted pyridazines the substituents can be placed in the following order of activation of the quaternization process: Me > H > Cl > SMe > Ph > OMe. The reaction rates of quaternization of 3,6-dialkylpyridazines with methyl iodide, which provides a measure of the steric effects of alkyl groups, show that the product distribution is kinetically controlled, solvent-dependent, and that steric effects of substituents overrule electronic factors. A method has been devised which enables calculation of isomer ratios resulting from the *N*-methylation of pyridazines with different substituents at positions 3 and 6. Observed and calculated isomer ratios are presented in Table 10 ⟨72T1983⟩. Attempts to prepare diquaternary salts from pyridazines and methyl iodide, ethyl bromide or *cis*-1,4-dibromobut-2-ene have been unsuccessful, due to reduction of nucleophilicity of the second nitrogen atom upon quaternization of the first. Diquaternary salts can be prepared by using oxonium salts, the effectiveness of which is in the following order: $Me_3O^+BF_4^-$ > $Et_3O^+BF_4^-$ > $(EtO)_2CH^+BF_4^-$ ⟨72JOC2259⟩.

**Table 10** Experimental and Calculated $N_1/N_2$ Methylation Product Ratios for the Reaction of 3-X,6-Y-Pyridazines with Methyl Iodide in Acetonitrile

| X | Y | Ratio $N_1/N_2$ (%) Expt. | Ratio $N_1/N_2$ (%) Calc. |
|---|---|---|---|
| Me | H | 72/28 | 81/19 |
| CO$_2$Me | H | >92/<2 | 94/6 |
| Me | Cl | 21/79 | 11/89 |
| Me | Br | 23/77 | 12/88 |
| Me | NH$_2$ | 45/55 | 58/42 |
| Me | MeCONH | 11/89 | 10/90 |
| Cl | NH$_2$ | 69/31 | 92/8 |

Quaternization of various pyridazinethiones and pyridazinyl sulfides is dependent on the substituents attached to the pyridazine ring. For example, 3-methylthiopyridazine and 6-methyl-3-methylthiopyridazine react with methyl iodide to form the corresponding 1-methyl-3-methylthio and 1,6-dimethyl-3-methylthio derivatives (**85**). On the other hand, if a larger group, such as methoxy or phenyl, is attached at the 6-position, quaternization takes place at position 2 to give 6-substituted 2-methyl-3-methylthiopyridazines (**86**; Scheme 24).

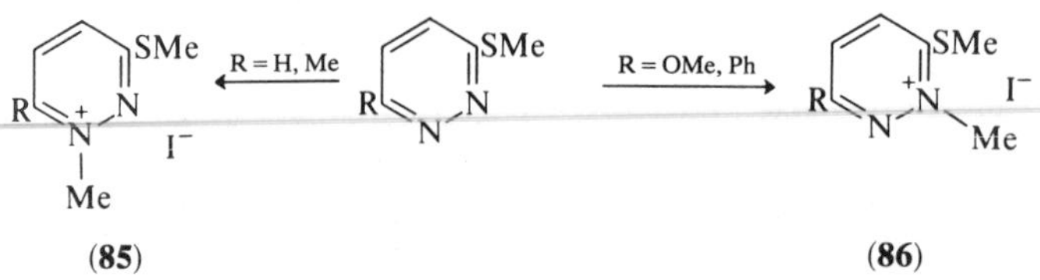

**Scheme 24**

2-Alkyl-substituted pyridazine-3(2H)-thiones undergo reaction with methyl iodide at the sulfur atom. Methylation of 4,5-diaminopyridazine-3(2H)-thione with excess methyl iodide produces 4,5-diamino-1-methylthiopyridazinium iodide ⟨81JOC2467⟩.

Cinnoline is protonated at N-2. Alkylation of alkyl- and aryl-cinnolines with alkyl halides and sulfates yields predominantly N-2 quaternary salts. When large groups are attached at position 3, the alkylation site is shifted towards N-1. For example, 4-methylcinnoline gives with methyl iodide the corresponding N-2 and N-1 methiodides in the ratio of 10:1, 4-ethyl-3-methylcinnoline in the ratio of 1:1 and 3-phenylcinnoline in the ratio of 1:4. 4-Alkoxy- and aryloxy-cinnolines are quaternized at N-2, but when large groups are attached at position 3, quaternization occurs predominantly at N-1. 4-Methylthiocinnoline yields a mixture of 2-methyl-4-methylthiocinnolinium iodide (59%) and the 1-methyl isomer (13%). Protonation of 3-aminocinnolines takes place predominantly at N-1. Methylation of 3-aminocinnoline gives N-1 and N-2 methiodides in equal amounts; with 4-aminocinnoline the ratio is 1:2.

Phthalazine forms monoquaternary salts with alkyl halides. 1-Substituted phthalazines are quaternized predominantly at position 3, and the site of quaternization of 1,4-disubstituted phthalazines is determined by the steric hindrance of both substituents.

### 2.12.3.2.4 N-Oxidation

Numerous oxidizing agents have been used for *N*-oxidation of pyridazine and its derivatives. The most widely used is peracetic acid, prepared *in situ*. Trifluoroperacetic acid, monoperoxyphthalic acid in ether solution, perbenzoic or *m*-chloroperbenzoic acid in chloroform solution, monoperoxymaleic acid in dichloromethane or other suitable organic solvents, hydrogen peroxide in the presence of sodium tungstate, peroxydichloromaleic acid, and hydrogen peroxide in sulfuric or polyphosphoric acid can also be employed for this purpose. The reaction course is dependent on the ring nitrogen electron density. If an electron-donating group is present at a position adjacent to, or conjugated with, the ring nitrogen, the electron density is increased and *N*-oxidation is easier. On the other hand, electron withdrawing groups at the corresponding positions decrease the electron density and *N*-oxidation is more difficult.

Pyridazine *N*-oxide is formed in 89% yield by oxidation of the parent system with peracetic acid. Two isomeric *N*-oxides are generally formed by *N*-oxidation of unsymmetrically substituted pyridazines. 3-Substituted pyridazines, such as 3-methyl- and 3-phenylpyridazine, give a mixture of the corresponding 1- and 2-oxides in ratios depending on the reaction conditions. 4-Methylpyridazine yields the corresponding 2-oxide and 1-oxide in a ratio of about 4:1. 3-Aminopyridazines and 3-amino-substituted pyridazines are converted mainly into 2-oxides with the corresponding 1-oxides as side products. Some 3-aminopyridazines can be oxidized with 85% hydrogen peroxide in polyphosphoric acid to 3-nitropyridazine 1-oxides. These results can be explained by the electronic effects of the substituents on the ring nitrogen atom.

On the other hand, 3-methoxy- and other 3-alkoxy-pyridazines, and 3-benzoyloxy-pyridazines, are converted into 1-oxides. This is apparently anomalous, but dipole moment measurements show that the alkoxy groups are in a *cis* orientation and therefore sterically hinder attack at the nitrogen at position 2. Oxidation of 3,6-dimethylpyridazine gives the mono-*N*-oxide. Pyridazine derivatives with methyl and phenyl groups at positions 3 and 6 are always *N*-oxidized at the nitrogen atom adjacent to the methyl group. 3,4-Dimethylpyridazine is converted by hydrogen peroxide in acetic acid into a mixture of 2-oxide (16%) and 1-oxide (36%). By contrast, 6-chloro-3,4-dimethylpyridazine gives predominantly its 2-oxide (83%) and only traces of 1-oxide (0.7%), whereas from 6-methoxy-3,4-dimethyl-pyridazine the corresponding 2-oxide was obtained as the only product. 2-Oxides were also isolated as the only products by the oxidation of 6-methoxy-3-methylpyridazine, 6-ethoxy-3-methylpyridazine and 3-methylpyridazin-6(1*H*)-one.

*N*-Oxidation of 3,6-dialkoxypyridazines (OMe, OEt, OPr$^n$, OBu$^n$) gives monoxides, while 3,6-di-*t*-butoxy- and 3,6-dibenzyloxy-pyridazine cannot be *N*-oxidized, but give 6-hydroxy-pyridazin-3(2*H*)-one and 6-benzyloxypyridazin-3(2*H*)-one as hydrolysis products. Methyl-thiopyridazines give both *S*-oxidation and *N*-oxidation products with various oxidizing agents in some instances.

*N*-Oxidation of 3,6-dichloropyridazine is more difficult because of its low basicity, and the hydrolysis product 6-chloropyridazin-3(2*H*)-one is also formed together with the corresponding *N*-oxide.

3-Alkoxy-6-chloropyridazine is oxidized at the nitrogen next to the chlorine. On the other hand, *N*-oxidation of compounds with an amino or acylamino group and either an alkoxy group or a chlorine atom at positions 3 and 6 always takes place at the nitrogen next to the amino group.

In some instances the migration of a methyl group from the 6-methoxy group to the oxygen of the *N*-oxide group occurs. For example, when 3,6-dimethoxypyridazine is oxidized with hydrogen peroxide in acetic acid to give 3,6-dimethoxypyridazine 1-oxide (**87**), 1,3-dimethoxypyridazin-6(1*H*)-one (**88**) and 6-methoxypyridazin-3(2*H*)-one (**89**) are also formed. Oxidation with monoperoxymaleic acid in dichloromethane affords only 3,6-dimethoxypyridazine 1-oxide (**87**) and 1,3-dimethoxypyridazin-6(1*H*)-one (**88**) in the ratio of 5:3. *N*-Oxidation of 3-alkynylpyridazines with *m*-chloroperbenzoic acid occurs exclusively at N-1 ⟨78H(9)1397⟩.

(**87**)  (**88**)  (**89**)

*N*-Oxidation of 3,3'-bipyridazines (**90**) is dependent on the substituents attached at positions adjacent to ring nitrogen in an analogous manner to simple pyridazines. For example, the 6,6'-dimethyl compound gives with hydrogen peroxide the 1,1'-dioxide and with perbenzoic acid a mixture of 1-oxide and traces of 1,1'-dioxide. From the 6,6'-dimethoxy compound the 2,2'-dioxide or a mixture of 2-oxide and 2,2'-dioxide are formed under similar conditions, while the 5,5'-dimethyl analog gives both 1,1'- and 2,2'-dioxides ⟨70CPB1340⟩.

(**90**)  (**91**)

[2.2](2,5)-Furano(3,6)pyridazinophane gives with *m*-chloroperbenzoic acid the corresponding chiral mono-*N*-oxide (**91**) ⟨80JOC4584⟩.

Direct oxidation of pyridazine and its derivatives with hydrogen peroxide (50–90%) in acetic acid gives, besides the isomeric mono 1- and 2-oxides, also the corresponding 1,2-dioxides in low yield. In this way, pyridazine 1,2-dioxide, 3-methyl-, 4-methyl- and 3,6-dimethyl-pyridazine 1,2-dioxides can be obtained.

*N*-Oxidation of cinnoline (**2**) with hydrogen peroxide in acetic acid results in the formation of four products: cinnoline 1-oxide (**92**; 26%), cinnoline 2-oxide (**93**; 50%), cinnoline 1,2-dioxide (**94**; 0.3%) and indazole (**95**; 3%) (Scheme 25). The yield of 1,2-dioxide can be increased, under forcing conditions, up to 13%. 4-Methylcinnoline gives the corresponding 1-oxide and 2-oxide in the ratio of 1:2 with a low yield of the 1,2-dioxide (4%), while 4-ethyl-3-methylcinnoline and 3,4-dimethylcinnoline form the corresponding 2-oxides and 1,2-dioxides in the ratio of 2:1. 4-Methoxycinnoline affords a mixture of 2-oxide (33%) and 1-oxide (18%). In some instances, thermal $O \rightarrow N$ rearrangement of alkyl groups of 4-alkoxy-3-phenylcinnolines produced 1-alkylcinnolin-4(1*H*)-ones.

**Scheme 25**

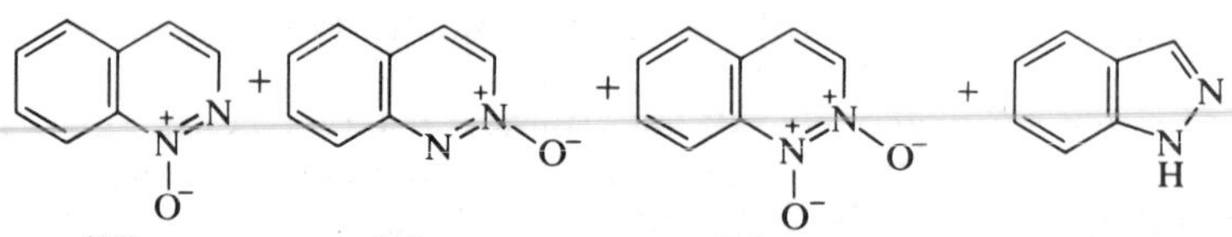

*N*-Oxidation of phthalazine with monoperoxyphthalic acid produces the corresponding 2-oxide in 73% yield. 1-Alkyl-substituted phthalazines give mixtures of 2-oxides and 3-oxides, the proportion of which is mainly dependent on the steric hindrance of the adjacent groups. 1-Alkoxy- and 1-phenyl-substituted phthalazines give 3-oxides exclusively. In 1,4-disubstituted derivatives the direction of *N*-oxidation depends on the size of groups at positions 1 and 4. For example, 1-methyl-4-phenylphthalazine gives only 2-oxide, 1-chloro- and 1-methoxy-4-methylphthalazine give 3-oxides and 1,4-dialkoxyphthalazines do not react.

### 2.12.3.2.5 N-Amination

Several *N*-aminopyridazines can be prepared by *N*-amination of pyridazines with hydroxylamine-*O*-sulfonic acid or *O*-mesitylenesulfonylhydroxylamine. *N*-Amination of 3-aminopyridazines takes place preferentially at the nitrogen atom adjacent to the amino group, giving the corresponding *N*-aminopyridazinium salts usually in high yields. The compounds are valuable synthons as precursors for bicyclic heterocyclic compounds with a bridgehead nitrogen atom ⟨75JHC107, 81T1787⟩. A number of *N*-aminopyridazinium halides can also be prepared by *N*-amination with potassium hydroxylamine-*O*-sulfonate.

3-Methyl- and 6-methoxy-3-methylpyridazine are aminated at N-2, while 3-methoxy-pyridazine reacts at N-1.

### 2.12.3.3 Electrophilic Attack at Carbon

The 3-, 4-, 5- and 6-positions in the pyridazine nucleus are electron deficient due to the negative mesomeric effect of the nitrogen atoms. Therefore, electrophilic substitution in pyridazines is difficult even in the presence of one or two electron-donating groups. The first reported example is nitration of 4-amino-3,6-dimethoxypyridazine to yield the corresponding 5-nitro derivative. Nitration of 3-methoxy-5-methylpyridazine gives the 6-nitro-, 4-nitro- and 4,6-dinitro derivatives. In the case of pyridazinones, however, several electrophilic reactions are known. Nitration of 4,5-dichloro-2-methylpyridazin-3(2*H*)-one occurs at position 6.

Direct chlorination of 3,6-dichloropyridazine with phosphorus pentachloride affords 3,4,5,6-tetrachloropyridazine. The halogen is usually introduced next to the activating oxo group. Thus, 1,3-disubstituted pyridazin-6(1*H*)-ones give the corresponding 5-chloro derivatives, frequently accompanied by 4,5-dichloro compounds as by-products on treatment with chlorine, phosphorus pentachloride or phosphoryl chloride–phosphorus pentachloride.

1,2-Disubstituted pyridazine-3,6(1*H*,2*H*)-diones add halogens to the 4,5-double bond, followed by dehydrohalogenation to give 4-halo derivatives. 1,2-Disubstituted 5-bromopyridazine-3,6(1*H*,2*H*)-diones react with bromine to give the corresponding 4,5-dibromo derivative. The Mannich reaction with 2-arylpyridazin-3(2*H*)-one occurs at position 4.

The behavior of the pyridazine nucleus towards electrophilic reagents changes after *N*-oxidation, due to the mesomeric effect of the *N*-oxide group, according to which the α- and γ-positions, with respect to the *N*-oxide group, become more reactive ⟨B-71MI21200⟩. Nitration of pyridazine 1-oxide and many 3- and 6-monosubstituted and 3,6-disubstituted derivatives with a mixture of nitric and sulfuric acids occurs at position 4 to give the corresponding 4-nitropyridazine 1-oxides. When position 4 is occupied, nitration either occurs at the α-, *i.e.* 6-position, or fails completely.

When nitration of pyridazine *N*-oxides is carried out with acyl nitrates (prepared *in situ* from acyl chlorides and silver nitrate) the reaction takes place at the β-position relative to the *N*-oxide group. Under these circumstances only mononitro derivatives are formed. For example, nitration of pyridazine 1-oxide with acetyl nitrate yields 3-nitropyridazine 1-oxide (17%) and 5-nitropyridazine 1-oxide (0.8%), whereas with benzoyl nitrate a better yield of 5-nitropyridazine 1-oxide is obtained.

Pyridazine 1-oxides substituted at position 3 or positions 3 and 6 afford the corresponding 5-nitro derivatives. A methyl group at position 6 (α with respect to the *N*-oxide group) is frequently converted into the cyano group, and a methoxy group at position 6 is demethylated by benzoyl chloride/silver nitrate. For example, 3-substituted 6-methylpyridazine 1-oxides give the 5-nitro derivatives (**96**) and the 6-cyano-5-nitro derivatives (**97**), whereas 3,6-dimethoxypyridazine 1-oxide is converted by benzoyl chloride/silver nitrate into 1-benzoyloxy-3-methoxypyridazin-6(1*H*)-one (**98**) and 1-hydroxy-3-methoxypyridazin-6(1*H*)-one (**99**; Scheme 26).

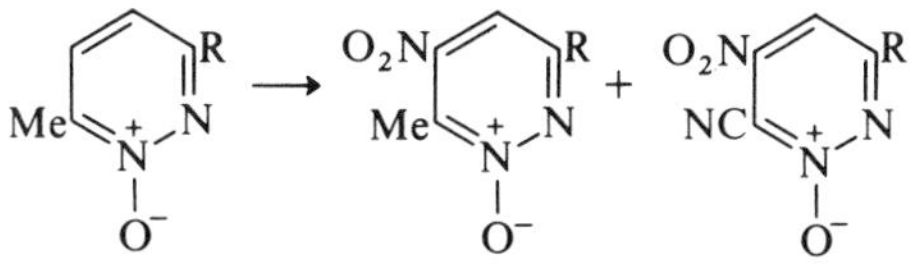

**Scheme 26**

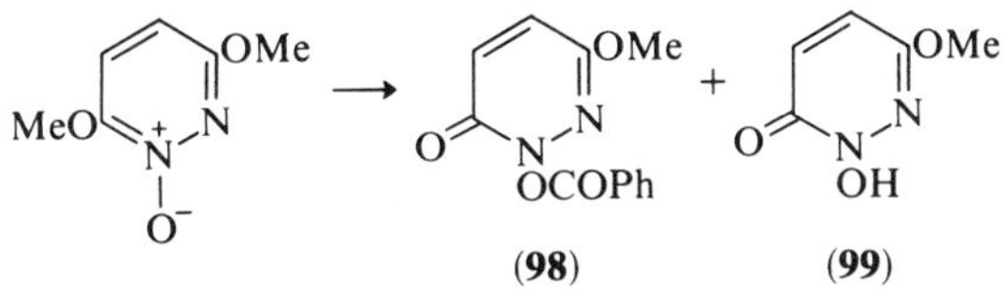

3,6-Dimethylpyridazine 1,2-dioxide gives the 4-nitro derivative in good yield with nitric acid, while with benzoyl nitrate the yield is considerably lower.

Mannich reaction with pyridazinone 1-oxides takes place at the α- or γ-positions relative to the *N*-oxide group, in contrast to the reaction in the pyridazinone series, where *N*-substituted products are formed. Pyridazin-3(2*H*)-one 1-oxide gives first the corresponding 6-substituted derivative; with excess of the reagents, 4,6-disubstituted products are obtained. When position 6 is blocked the corresponding 4-dialkylaminomethyl derivatives are obtained.

Bromination of pyridazin-3(2*H*)-one 1-oxide and 5-hydroxypyridazine 1-oxide affords 4,6-dibromo derivatives as the sole products. Pyridazin-3(2*H*)-one 1-oxides with position 6 blocked afford the 4-monobromo derivatives. Chlorination proceeds similarly to give dichloro products, while after nitration only 4-nitro derivatives are obtained.

MO calculations of the cinnoline ring system show that the relative order of reactivities for electrophilic substitution is 5=8>6=7>3≫4. This is confirmed experimentally, as nitration of cinnoline with a mixture of nitric and sulfuric acids affords 5-nitrocinnoline (33%) and 8-nitrocinnoline (28%). Similarly, 4-methylcinnoline gives a mixture of 4-methyl-8-nitrocinnoline (28%) and 4-methyl-5-nitrocinnoline (13%).

Nitration of cinnolin-4(1*H*)-one yields a mixture of the 3-nitro (0.9%), 5-nitro (0.38%), 6-nitro (58.4%), 7-nitro (0.36%) and 8-nitro (39.9%) derivatives. The 3-nitro isomer is postulated to result from nitration of the free base, while the other mononitro isomers are formed from the protonated molecule.

4-Aminocinnoline is nitrated in fuming nitric and concentrated sulfuric acids to give 4-amino-6-nitrocinnoline.

Cinnolin-4(1*H*)-one and its 6-chloro, 6-bromo, 6-nitro and 8-nitro derivatives react with sulfuryl chloride or bromine in acetic acid to give the corresponding 3-halo derivatives in about 20% yields. Iodination of 8-hydroxycinnolin-4(1*H*)-one with a mixture of potassium iodide and potassium iodate gives the 5,7-diiodo derivative; the 6,8-diiodo derivative is formed from 5-hydroxycinnolin-4(1*H*)-one.

When cinnoline 1-oxide is treated with nitric and sulfuric acids or potassium nitrate in sulfuric acid, 4-nitrocinnoline 1-oxide is obtained in yields ranging from 3–64% depending on the reaction conditions. With a mixture of fuming nitric and sulfuric acids, the corresponding 4-nitro-, 4,5-dinitro- and a small amount of the 5-nitro-cinnoline derivatives are obtained.

Both 4-nitrocinnoline 1-oxide and the 5-nitro isomer give 4,5-dinitrocinnoline 1-oxide when treated with fuming nitric and sulfuric acids. Cinnoline 1-oxide also reacts with benzoyl chloride/silver nitrate to give 3-nitrocinnoline 1-oxide in 71% yield.

Nitration of cinnoline 2-oxide takes a different course. With nitric and sulfuric acids or with potassium nitrate and sulfuric acid a mixture of 8-nitrocinnoline 2-oxide, 6-nitrocinnoline 2-oxide and 5-nitrocinnoline 2-oxide is obtained, while with benzoyl nitrate in chloroform only a low yield (1.5%) of the 5-nitro derivative is obtained.

Upon nitration of phthalazines a nitro group is introduced only into the carbocyclic ring. Side reactions occur frequently. For example, nitration of phthalazine or its 1-methyl analog with potassium nitrate in concentrated sulfuric acid gives the 5-nitro derivative (**100**) as the main product together with 5-nitrophthalazin-1(2H)-one (**101**) as a by-product (Scheme 27).

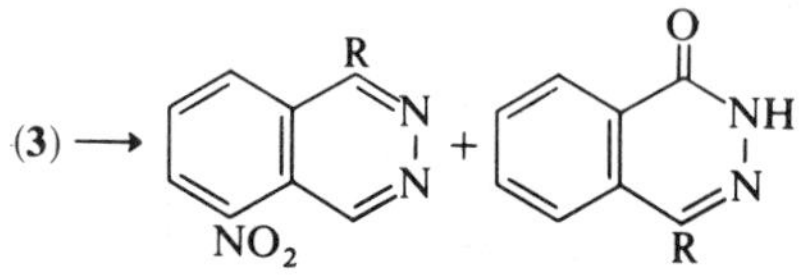

(**100**) R=H, Me    (**101**) R=H, Me

**Scheme 27**

### 2.12.3.4 Electrophilic Hydrogen Exchange

4-Aminopyridazine in the form of its conjugate acid exchanges hydrogen in acidic solution at position 5. In the low acidity region the conjugate acid exchanges hydrogen by the ylide mechanism at positions 3 and 6 at approximately equal rates. Exchange in deuteriosulfuric acid is accompanied by competitive hydrolysis to pyridazin-4(1H)-one. In strongly acidic media, only the proton at position 5 of pyridazin-4(1H)-one is exchanged, while in weakly acidic media, all protons undergo exchange with deuterium in the order 3>6>5. Pyridazin-3(2H)-one undergoes exchange first at position 5, followed by position 4. Proposed extrapolation of exchange rates data to pH=0 and 100 °C allows rate constants for acid-catalyzed hydrogen exchange to be compared ⟨73JCS(P2)1065⟩.

### 2.12.3.5 Nucleophilic Attack at Carbon

#### 2.12.3.5.1 Pyridazines

Addition of Grignard reagents and organolithium compounds to the pyridazine ring proceeds as a nucleophilic attack at one of the electron-deficient positions to give initially 1,4-addition products which rearrange to 1,2-dihydro products. 3-Methoxy-6-phenyl-pyridazine reacts with *t*-butylmagnesium chloride to give 4-*t*-butyl-3-methoxy-6-phenyl-4,5-dihydropyridazine (**102**) as the major product together with the isomeric pyridazinones (**103**) and (**104**) (Scheme 28). Similarly, *n*-butyllithium and *t*-butyllithium add to the 4,5-double bond of 3,6-dimethoxypyridazine to give 4-substituted 4,5-dihydropyridazine derivatives.

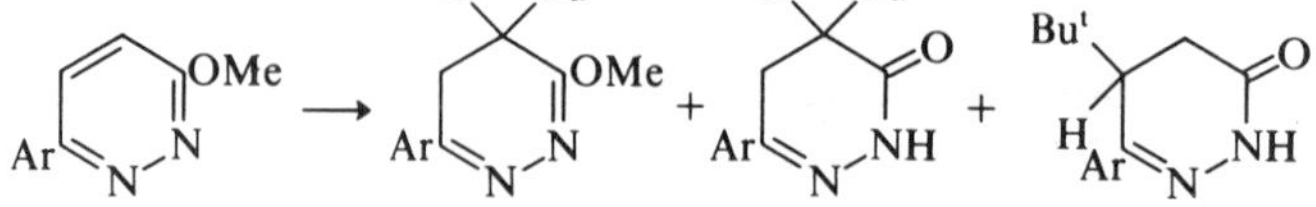

(**102**)          (**103**)          (**104**)

**Scheme 28**

Phenyllithium in ether adds to pyridazine and 6-substituted pyridazines at position 3. By using TMEDA, addition at position 4 is strongly promoted ⟨78RTC116⟩.

The addition of phenyllithium to 6-arylpyridazin-3(2H)-one takes place at position 6 to give 6-aryl-3-oxo-6-phenyl-1,2,3,4-tetrahydropyridazine and the reaction of 6-aryl-2,4-diphenylpyridazin-3(2H)-one with phenyllithium or phenylmagnesium bromide affords 6-aryl-2,3,4,6-pentaphenyl-1,2,3,4-tetrahydropyridazine ⟨80S457⟩.

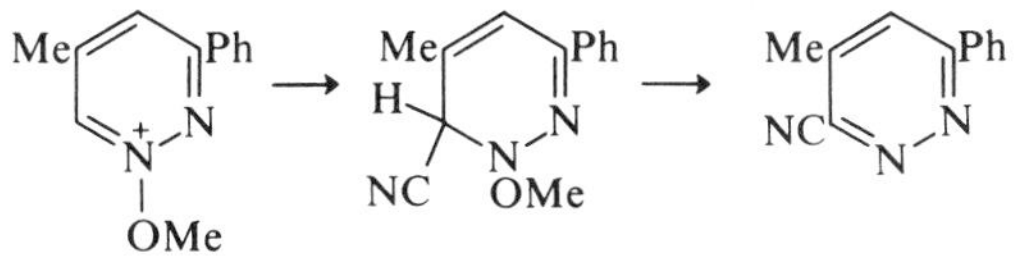

**Scheme 29**

1-Methoxypyridazinium salts are converted into cyanopyridazines on treatment with potassium cyanide, with the cyano group entering the $\alpha$-position with respect to the $N$-oxide group of the starting compound (Scheme 29).

3-Methylpyridazine gives the pyridazine Reissert compound (**105**) with trimethylsilyl cyanide and freshly distilled benzoyl chloride. On the other hand, when pyridazine or 3-methylpyridazine reacts with undistilled benzoyl chloride the bicyclic compounds (**106**) are formed and these react with dimethyl acetylenedicarboxylate in anhydrous DMF to give pyrrolopyridazine derivatives (**107**; Scheme 30) ⟨81JHC443⟩.

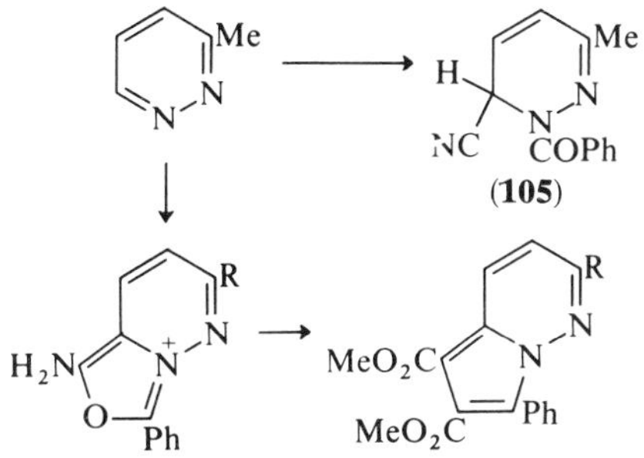

**Scheme 30**

Reaction of various pyridazine derivatives with nitromethane or nitroethane in DMSO affords the corresponding 5-methyl and 5-ethyl derivatives. The reaction proceeds as a nucleophilic attack of the nitroalkane at the position 5. In this way, 3,6-dichloro-4-cyanopyridazine, 4-carboxy- and 4-ethoxycarbonyl-pyridazin-3(2$H$)-ones and 4-carboxy- and 4-ethoxycarbonyl-pyridazin-6(1$H$)-ones can be alkylated at position 5 ⟨77CPB1856⟩.

Nucleophilic substitution in the pyridazine 1-oxide series takes place either according to pathway (a) or pathway (b) (Scheme 31). Pathway (a) operates when position 6 is unsubstituted.

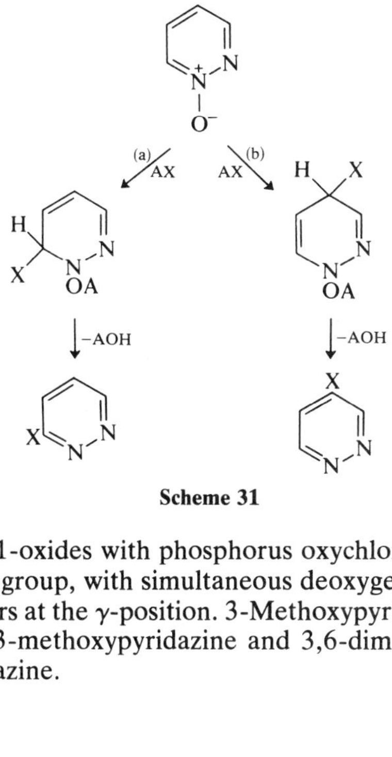

**Scheme 31**

Treatment of pyridazine 1-oxides with phosphorus oxychloride results in $\alpha$-chlorination with respect to the $N$-oxide group, with simultaneous deoxygenation. When the $\alpha$-position is blocked, substitution occurs at the $\gamma$-position. 3-Methoxypyridazine 1-oxide, for example, is converted into 6-chloro-3-methoxypyridazine and 3,6-dimethylpyridazine 1-oxide into 4-chloro-3,6-dimethylpyridazine.

Pyridazine 1-oxides unsubstituted at position 6 rearrange in the presence of acetic anhydride to pyridazin-6(1*H*)-ones. Pyridazin-3-ol 1-oxide and 3-aminopyridazine 1-oxide are transformed by boiling acetic anhydride into 6-hydroxypyridazin-3(2*H*)-one and 6-aminopyridazin-3(2*H*)-one, respectively. When the α-oxide group is substituted with a methyl, alkoxy or halogen atom, it may be attacked by the reagent. In this manner, 3-methoxy-6-methylpyridazine 1-oxide gives 6-acetoxymethyl-3-methoxypyridazine, but 3,6-dimethoxypyridazine 1-oxide is transformed into three products: 1-acetoxy-3-methoxy-pyridazin-6(1*H*)-one (**108**), 1,3-dimethoxypyridazin-6(1*H*)-one (**109**) and 1-hydroxy-3-methoxypyridazin-6(1*H*)-one (**110**; Scheme 32). 6-Chloropyridazine 1-oxides are converted into 1-hydroxypyridazin-6(1*H*)-ones.

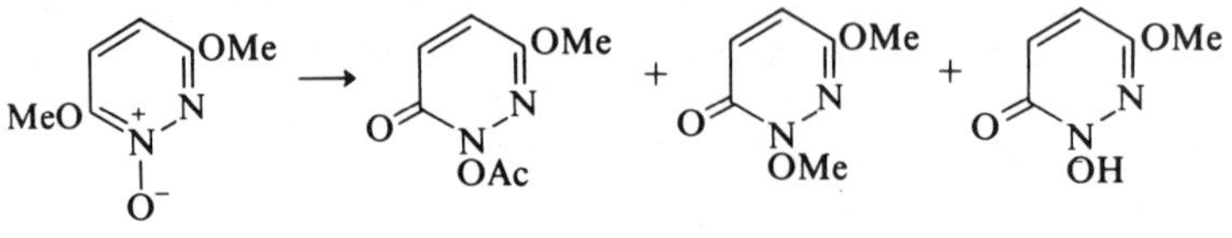

**(108)**      **(109)**      **(110)**

**Scheme 32**

6-Chloro-4-methyl- and 6-methoxy-4-methyl-pyridazine 1-oxides give the corresponding 4-acetoxymethyl derivatives, while the corresponding 6-substituted 5-methylpyridazine 1-oxides do not react at the methyl group.

A substituted acylamino group can be introduced by reaction of pyridazine 1-oxide with *N*-phenylbenzonitrilium hexachloroantimonate; 3-*N*-benzoylanilinopyridazine is formed ⟨75JOC41⟩.

A very useful procedure for introducing a cyano group into a pyridazine ring is the Reissert-type reaction of the *N*-oxide with cyanide ion in the presence of an acyl halide or dimethyl sulfate. The cyano group is introduced into the α-position with respect to the *N*-oxide function of the starting compound. The yields are, however, generally poor. In this way, 6-cyanopyridazines (**111**) can be obtained from the corresponding pyridazine 1-oxides (Scheme 33).

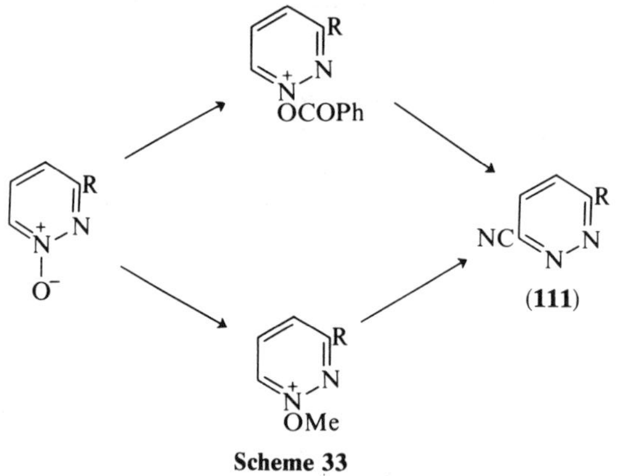

**(111)**

**Scheme 33**

3-Methoxypyridazine 1-oxide reacts with methyl β-aminocrotonate in the presence of benzoyl chloride to give α-(6-methoxy-3-pyridazinyl) β-aminocrotonate (**112**) which can be converted by mild hydrolysis into the corresponding acetate (**113**; Scheme 34) ⟨78JHC1425⟩.

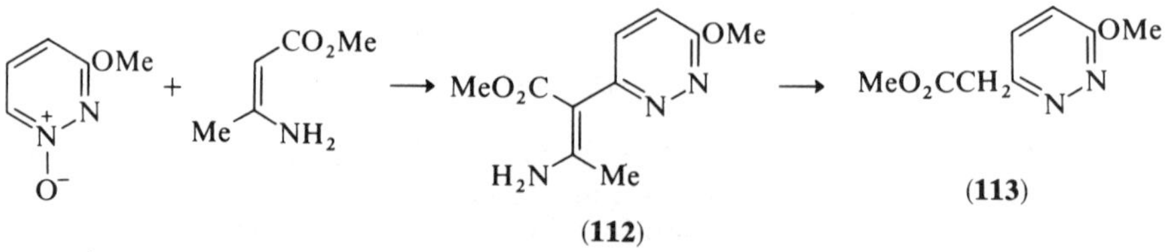

**(112)**      **(113)**

**Scheme 34**

### 2.12.3.5.2 Cinnolines

The reaction of cinnoline 2-oxide with phenylmagnesium bromide gives phenanthrene, *trans*- and *cis*-stilbene, 2,3-diphenyl-1,2-dihydrocinnoline and 2-styrylazobenzene in yields of 1–15%. Analogous results are also obtained from 4-methylcinnoline 2-oxide.

### 2.12.3.5.3 Phthalazines

Addition of phenylmagnesium bromide to phthalazin-1(2*H*)-one or derivatives like 4-phenylphthalazin-1(2*H*)-one, 4-phenylphthalazine-1(2*H*)-thione and 2-substituted phthalazin-1(2*H*)-ones results in the formation of 1,4-diphenylphthalazines, while addition of organolithium compounds to phthalazin-1(2*H*)-one gives the 4-substituted derivative. 1-Phenylphthalazine 3-oxide undergoes a Reissert-type reaction with benzoyl chloride in the presence of potassium cyanide to give 4-cyano-1-phenylphthalazine. Phthalazine and substituted phthalazines undergo 1,2-addition of phenylmagnesium bromide and other Grignard reagents across the C=N bond, followed by autoxidation to give 1-monosubstituted or 1,4-disubstituted phthalazines.

The Reissert compound obtained from phthalazine and potassium cyanide in the presence of benzoyl chloride can be converted with methyl iodide and sodium hydride in DMF into 2-benzoyl-1-cyano-1-methyl-1,2-dihydrophthalazine, which is transformed after alkaline hydrolysis into 1-methylphthalazine. Addition of Grignard reagents to 1-phenyl- or 1-alkyl-phthalazine 2- and 3-oxides gives 1,4-disubstituted phthalazines.

### 2.12.3.5.4 Nucleophilic substitution of other groups

Electronegative substituents on the pyridazine ring are activated towards nucleophilic substitution. The reactions are dependent on the position of the leaving group, the nucleophile, the influence of other groups attached to the pyridazine ring and on the reaction conditions. The most common leaving group is the chloride ion. Halopyridazinones react easily with a wide variety of nucleophiles, such as ammonia, amines, hydrazines, aqueous solutions of inorganic bases, alkoxides and phenoxides, hydrogen sulfide, and alkyl and aryl thiols. *N*-Unsubstituted pyridazinones react with more difficulty with strongly basic nucleophiles as the negative charge, due to ionization, deactivates the pyridazine system for nucleophilic attack. However, this problem can be overcome in some instances by using weakly basic catalysts. The reactivity of halogens in halopyridazin-6(1*H*)-ones is 4 > 5 > 3, so that the most reactive halogen in halopyridazin-6(1*H*)-one is at the *meta* position with regard to the oxo group. The reaction of 3,4-dihalo-, 4,5-dihalo- and 3,4,5-trihalo-pyridazin-6(1*H*)-ones with nucleophiles stops at monosubstitution to give 4-substituted products, while 3,5-dichloropyridazin-6(1*H*)-ones yield 5-substituted derivatives.

Halo-substituted 3-hydroxypyridazin-6(1*H*)-ones react in some instances by *cine* substitution. For example, 4-chloro-1-methyl-2-phenylpyridazin-6(1*H*)-one, when treated with piperidine, yields a mixture of the corresponding 4- and 5-piperidino isomers in nearly equal amounts.

Substitution of a halogen atom in monohalogenated pyridazines presents no particular problem, even in the presence of other substituents such as alkyl, aryl, alkoxy, nitro, methylthio, methylsulfonyl, carboxy, carboxamido, *etc.* In dihalopyridazines only one halogen atom can be replaced with ammonia or primary amines. In some instances, 3,6-disubstituted products are obtained in the presence of catalysts, such as copper bronze or copper salts, or at higher temperatures. Secondary amines, such as dialkylamines, piperidine, piperazine and pyrrole, give 3,6-disubstituted pyridazines more readily. Unsymmetrically substituted 3,6-dihalopyridazines, when treated with nucleophiles, give in general two isomeric monosubstitution products. For example, 3,6-dichloro-4-methylpyridazine is transformed with ammonia into a mixture of 6-amino-3-chloro-4-methyl- and 3-amino-6-chloro-4-methyl-pyridazine in the ratio of 10:1. In 3,4,5- and 3,4,6-trichloropyridazines the chlorine at position 4 is usually the most reactive for nucleophilic reaction and 4-substituted products are obtained. 3,4,5-Trichloropyridazine, when treated with ethanolic ammonia, is transformed into a mixture of the corresponding 4- and 5-substituted amino compounds. With an equimolar amount of methoxide the chlorine at the 5-position is displaced, while with two equivalents of methoxide ion a mixture of 4-chloro-3,5-dimethoxy- and 3-chloro-4,5-dimethoxy-pyridazine is obtained. The exceptions are the reactions in which the chlorine is replaced with a 'hydroxy' group. When 3,4,6-trichloropyridazine is heated under reflux in acetic acid, 4,6- and 5,6-dichloro-3(2H)-pyridazinones are formed and 3,4,5-trichloropyridazine yields 4,5-dichloropyridazin-3(2H)-one. A large number of various alkoxy and aryloxy derivatives with additional substituents such as alkyl, aryl, amino, substituted amino, substituted sulfonamido, alkoxy, alkylthio, methylsulfonyl, cyano, nitro, *etc.*, can be prepared by replacement of halogen with sodium alkoxides and phenoxides. However, some groups can be affected by this procedure; in most cases deacetylation of an acetylamino group and hydrolysis of a cyano group takes place. In 5-amino-3,4-dichloropyridazine and 4-amino-3,5-dichloropyridazine the chlorine at position 3 is replaced by an alkoxy group. When 3,6-dichloro-4-methylpyridazine is treated with one equivalent of sodium alkoxide, the ratio of 3-alkoxy to 6-alkoxy derivative is dependent on the size of the alkoxy group. With methoxide approximately equal amounts of both isomers are obtained, while with increasing size of the alkoxide, position 6 is favored for attack. The order of substitution in tetrachloro- and tetrafluoro-pyridazines with methoxide is 4 > 5 > 3,6, whereas with methanol the order is reversed. Tetrafluoropyridazine is transformed by methanol at 0 °C to 3,4,5-trifluoro-6-methoxypyridazine, and at 20 °C to 4,5-difluoro-3,6-dimethoxypyridazine. Further treatment of this 6-methoxypyridazine with one equivalent of sodium methoxide gives a mixture of 4,6-dimethoxy and 5,6-dimethoxy derivatives, while with two equivalents the 3,5,6-trimethoxy compound is obtained.

A thiol or thioxo group is introduced into the pyridazine ring by nucleophilic substitution of chlorine by a number of reagents. The most commonly used are sodium or potassium hydrogen sulfide, thiourea, hydrogen sulfide in pyridine or DMF and phosphorus pentasulfide in pyridine. The thiol groups can be introduced stepwise into polychloropyridazines by selecting the appropriate conditions. However, when two exchangeable halogen atoms are *ortho* to each other, besides dimercapto compounds, some polycyclic systems are also formed. For example, 4,5-dichloro-1-phenylpyridazin-6(1H)-one gives with ethanolic sodium hydrogen sulfide at room temperature 5-chloro-1-phenyl-4-thiopyridazin-6(1H)-one, which is further transformed at higher temperatures into the 4,5-dithio derivative. On the other hand, 4,5-dichloropyridazin-6(1H)-one and its 1-phenyl analog afford with sodium hydrogen sulfide in absolute ethanol dipyridazino[4,5-*b*:4',5'-*e*][1,4]dithiin-1,6(2H,7H)-dione (**114**; R = H) and its 2,7-diphenyl analog (**114**; R = Ph). From 1-benzyl-5-chloro-4-thiopyridazin-6(1H)-one two isomeric products, *i.e.* the 2,7-dibenzyl derivative (**114**; R = CH₂Ph) and its 2,8-dibenzyl-1,9-dione isomer (**115**), are obtained (Scheme 35).

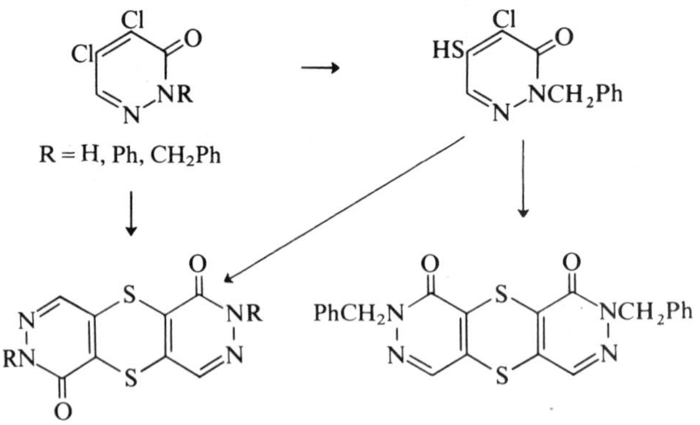

**Scheme 35**

When an acetylamino group is attached at an *ortho* position the replacement of chlorine is followed by cyclization. For example, 4-acetylamino-5-chloro-1-phenylpyridazin-6(1H)-one is converted with hydrogen sulfide in DMF to 2-methyl-6-phenylthiazolo[4,5-*d*]pyridazin-7(6H)-one (**116**).

When thiourea or thiosemicarbazide are used for substitution of a chlorine atom with a mercapto group, thiouronium salts are formed first, and they are hydrolysed by aqueous base. Other groups, such as alkoxy, alkyl, amino and cyano, if present, are usually not affected. However, in some instances thiuronium intermediates cyclize to bicyclic systems. For example, 4-bromo-1-methyl-2-phenylpyridazine-3,6(1*H*,2*H*)-dione cyclizes to 2-amino-6-methyl-5-phenylthiazolo[4,5-*d*]pyridazine-4,7(5*H*,6*H*)-dione (**117**; Scheme 36).

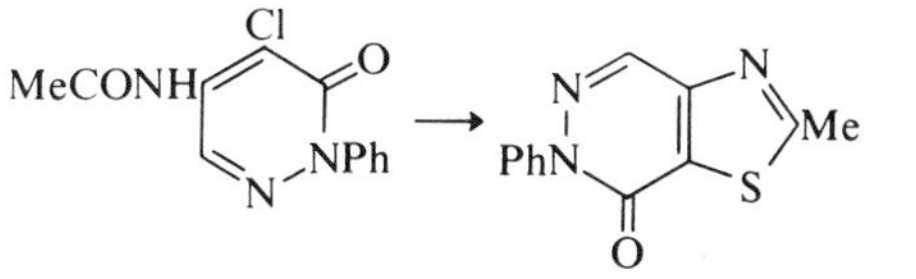

(**116**)

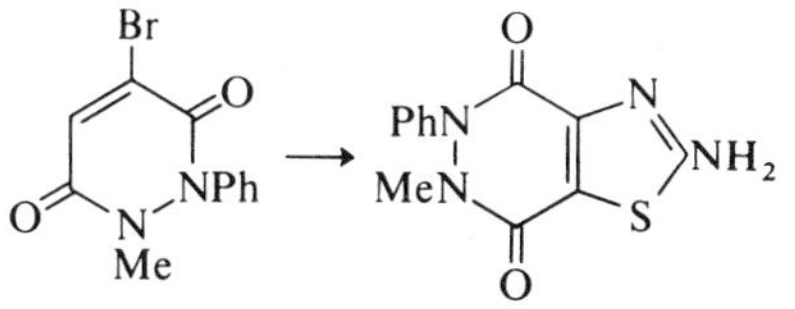

(**117**)

**Scheme 36**

Phosphorus pentasulfide in pyridine can be used also for simultaneous substitution of oxygen and chlorine in polysubstituted pyridazinones. For example, 4,5-dichloro- and 4,5-dibromo-pyridazin-6(1*H*)-one give 3,4,5-trithiopyridazine in this way.

Alkylthio- and arylthio-pyridazines can be prepared from the corresponding halo-substituted pyridazines by using appropriate alkyl and aryl thiolates.

A thiocyanato group can be introduced by treatment of the corresponding chloro or bromo compounds with ammonium, sodium or potassium thiocyanate. In polychloro compounds only one halogen is replaced.

When chloro compounds are treated with sodium azide in ethanol or aqueous acetone the corresponding azides or tetrazolo[1,5-*b*]pyridazines are obtained. For example, 3-azido- and 4-azido-pyridazine 1-oxides are obtained from the corresponding chloro compounds; 3,6-dichloropyridazine exchanges both chlorine atoms to give 6-azidotetrazolo[1,5-*b*]pyridazine. Valence isomerization occurs when another group is attached at position 4 of 3,6-dichloropyridazine. In this way, 3,6-dichloro-4-methylpyridazine is converted with sodium azide into a mixture of 6-azido-7-methyl- (**118**) and 6-azido-8-methyl-tetrazolo[1,5-*b*]pyridazine (**119**; Scheme 37).

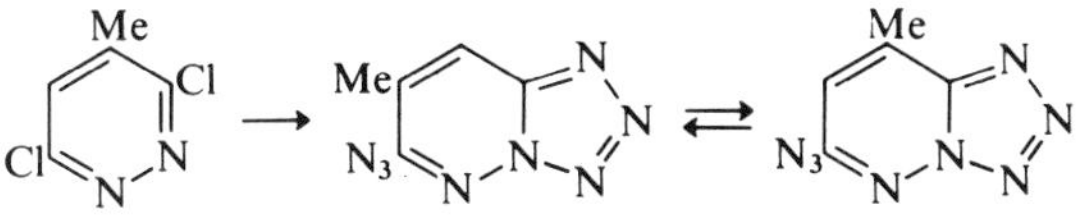

(**118**)      (**119**)

**Scheme 37**

The reactivity of halogens in pyridazine *N*-oxides towards nucleophilic substitution is in the order $5 > 3 > 6 > 4$. This is supported by kinetic studies of the reaction between the corresponding chloropyridazine 1-oxides and piperidine. In general, the chlorine atoms in pyridazine *N*-oxides undergo replacement with alkoxy, aryloxy, piperidino, hydrazino, azido, hydroxylamino, mercapto, alkylmercapto, methylsulfonyl and other groups.

3,6-Dichloropyridazine 1-oxide produces both isomers with alkoxides. However, the ratio is dependent on the size of the alkoxy group. In the reaction with sodium methoxide 80% of 6-chloro-3-methoxypyridazine 1-oxide and 7.5% of 3-chloro-6-methoxypyridazine 1-oxide are formed. Similar results are also obtained with sodium ethoxide, while sodium propoxide affords only 6-chloro-3-propoxypyridazine 1-oxide. Amines react similarly, while only chlorine at the 3-position can be substituted with an azido group to give 3-azido-6-chloropyridazine 1-oxide.

The second most important nucleophilic substitution in pyridazine *N*-oxides is the replacement of a nitro group. Nitro groups at the 3-, 4-, 5- and 6-position are easily substituted thermally with a chlorine or bromine atom, using acetyl chloride or hydrobromic acid respectively. Phosphorus oxychloride and benzoyl chloride are used less frequently for this purpose. Nitro groups in nitropyridazine *N*-oxides are easily replaced by alkoxide. The order of reactivity is position $5 > 4 > 3$. In some instances, besides the nitro group, other groups may be exchanged at the same time. For example, 3,4-dimethoxy-6-methylpyridazine is obtained from 3-chloro-6-methyl-4-nitropyridazine 1-oxide with sodium methoxide under reflux.

There are several examples of replacement of methoxy and ethoxy groups in substituted pyridazine N-oxides, with ammonia or hydrazine producing the corresponding amino and hydrazino compounds.

Amino groups in pyridazine N-oxides can be diazotized and the diazonium group further replaced by halogens, hydroxy group or hydrogen. So, 3-, 4-, 5- and 6-bromopyridazine 1-oxides can be prepared from the corresponding amino N-oxides.

When 6-amino-3-chloropyridazine 1-oxide is diazotized in 50% sulfuric acid, 6-hydroxy-3-pyridazinediazonium anhydro salt is formed. An azido group at either position in pyridazine N-oxides can readily be replaced with sodium alkoxides.

In some instances a carbon–carbon bond can be formed with C-nucleophiles. For example, 3-carboxamido-6-methylpyridazine is produced from 3-iodo-6-methylpyridazine by treatment with potassium cyanide in aqueous ethanol and 1,3-dimethyl-6-oxo-1,6-dihydropyridazine-4-carboxylic acid from 4-chloro-1,3-dimethylpyridazin-6-(1$H$)-one by reaction with a mixture of cuprous chloride and potassium cyanide. Chloro-substituted pyridazines react with Grignard reagents. For example, 3,4,6-trichloropyridazine reacts with $t$-butyl-magnesium chloride to give 4-$t$-butyl-3,5,6-trichloro-1,4-dihydropyridazine (**120**) and 4,5-di-$t$-butyl-3,6-dichloro-1,4-dihydropyridazine (**121**) and both are converted into 4-$t$-butyl-3,6-dichloropyridazine (**122**; Scheme 38).

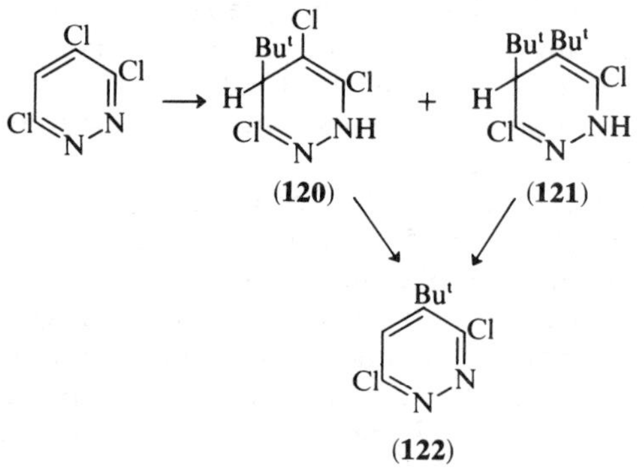

**Scheme 38**

Alkyl- and aryl-pyridazines can be prepared by cross-coupling reactions between chloropyridazines and Grignard reagents in the presence of nickel–phosphine complexes as catalysts. Dichloro[1,2-bis(diphenylphosphino)propane]nickel is used for alkylation and dichloro[1,2-bis(diphenylphosphino)ethane]nickel for arylation ⟨78CPB2550⟩. 3-Alkynyl-pyridazines and their N-oxides are prepared from 3-chloropyridazines and their N-oxides and alkynes using a Pd(PPh$_3$)Cl$_2$–Cu complex and triethylamine ⟨78H(9)1397⟩.

3,6-Dichloropyridazine gives 3-chloro-6(1$H$)-dicyanomethylenepyridazine (**123**) in high yield with sodiomalononitrile in THF.

3-Alkylidene-2,3-dihydropyridazines (**124**) are synthesized by coupling of 3-thiomethyl-pyridazinium salts with active methylene compounds in the presence of potassium carbonate in DMF (Scheme 39) ⟨79TL4837⟩.

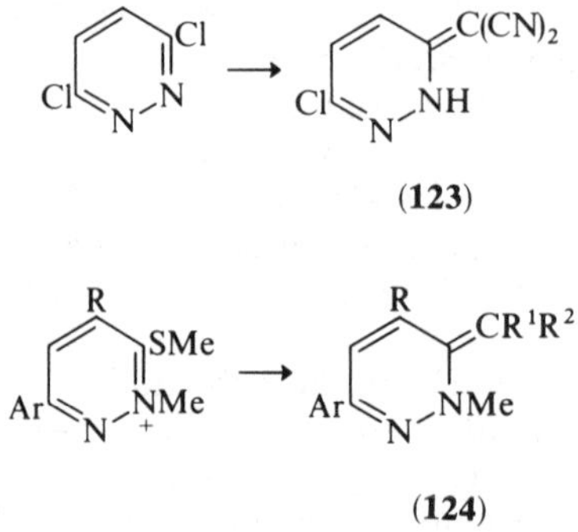

(**123**)

(**124**)

**Scheme 39**

Halogenated pyridazines are generally inert as arylating agents in Friedel–Crafts reactions. The only example is the reaction of 3,6-dichloropyridazine with resorcinol and hydroquinone to give 3-aryl-6-chloropyridazines.

### *2.12.3.5.5 Ring opening and ring contractions*

Reaction of pyridazine 1-oxide with phenylmagnesium bromide gives 1,4-diphenyl-butadiene as the main product and 1-phenylbut-1-en-3-yne and 3,6-diphenylpyridazine as by-products, while alkyl Grignard reagents lead to the corresponding 1,3-dienes exclusively ⟨79JCS(P1)2136⟩.

Pyridazinones may undergo ring contraction to pyrroles, pyrazoles and indoles, the process being induced either by an acid or base. The structure of the final product is strongly dependent on the reaction conditions. For example, 4,5-dichloro-1-phenylpyridazin-6(1*H*)-one rearranges thermally to 4-chloro-1-phenylpyrazole-5-carboxylic acid (**125**), while in aqueous base the corresponding 4-hydroxy acid (**126**) is formed (Scheme 40).

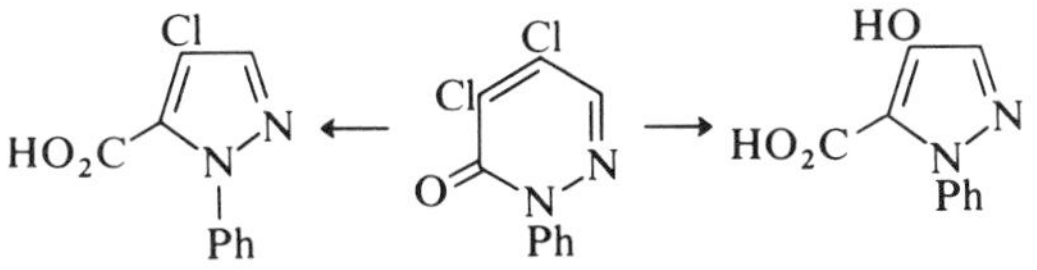

(**125**)  (**126**)

**Scheme 40**

4-Hydroxy-6-methyl-2-phenylpyridazin-3(2*H*)-one and 4-hydroxy-5-nitropyridazin-3(2*H*)-one rearrange in acidic medium to 3-methyl-1-phenylpyrazole-5-carboxylic acid and 4-nitropyrazole-5-carboxylic acid. 4-Hydroxypyridazin-3(2*H*)-ones with a hydroxy group or other group at positions 5 or 6, which is easily replaced in alkaline medium, are transformed into 5-(or 3-)pyrazolones with hot alkali. An interesting example is ring contraction of 5-chloro-4-(methylthio)-1-phenylpyridazin-6(1*H*)-one which gives, besides pyrazole derivative (**127**), 4-hydroxy-5-methylthio-1-phenylpyridazin-6(1*H*)-one (**128**; Scheme 41).

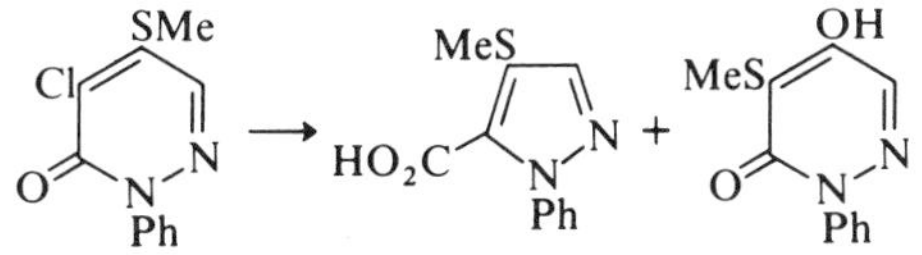

(**127**)  (**128**)

**Scheme 41**

4-Cyanopyrazole is obtained from treatment of 4-amino-3-halopyridazines with potassium amide in liquid ammonia, while 4-amino-3,6-dihalopyridazines are rearranged under the same conditions to 3-cyanomethyl-1,2,4-triazole.

1,3,4-Triphenyl-1,6-dihydropyridazine is transformed into 1,3,4-triphenylpyrazole in acidic solution, and desulfurization of 4,5,6-triphenylpyridazine-3(2*H*)-thione with Raney nickel affords 2,3,4-triphenylpyrrole.

5-Phenylpyridazin-4-oxime is transformed into 5-methyl-2-phenylpyrrole when treated with zinc in acetic acid.

In some instances, ring contraction is accompanied by cyclization to indole derivatives. For example, 1-aryl-6-oxo-1,4,5,6-tetrahydropyridazines with a carboxyl or methyl group at position 3 give indoles when treated with an ethanolic solution saturated with hydrogen chloride or in the presence of BF$_3$ etherate.

3,6-Dibenzoylpyridazine-4,5-dione 1,2-dioxide is transformed into 3,5-dibenzoyl-1,2,4-oxadiazole in the presence of an alcohol.

### 2.12.3.6 Nucleophilic Attack at Hydrogen

In pyridazine, base-catalyzed hydrogen–deuterium exchange takes place at positions 4 and 5 more easily than at positions 3 or 6. Deuteration of pyridazine 1-oxide in NaOD/D$_2$O

(1%) occurs stepwise at all four positions, first at positions 6 and 5 and then positions 4 and 3.

In pyridazin-3(2H)-one only the hydrogen at position 6 is replaced at 180 °C, while in 3-hydroxypyridazine 1-oxide the deuterium exchange takes place first at position 6 and then at position 4 at 140 °C. 5-Hydroxy- and 5-methoxy-pyridazine 1-oxide exchange only the hydrogen at position 6 at 150 °C.

### 2.12.3.7 Reactions with Radicals and Electron-deficient Species

It has already been mentioned that some radical reactions can occur as side reactions by irradiation of pyridazine derivatives, especially in hydroxylic solvents.

Homolytic acylation of ethyl pyridazine-4-carboxylate is a convenient general method for preparation of 4-acylpyridazines (Scheme 42) ⟨79M365⟩.

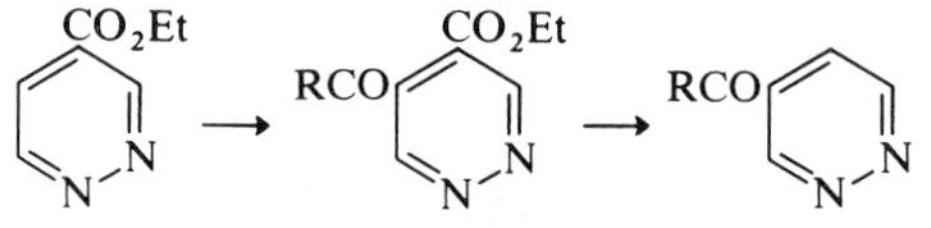

Scheme 42

α-N-Amido radicals prepared from 1-formylpyrrolidine and 1-acetylpyrrolidine react with pyridazine and 4-methylpyridazine to give the corresponding 5-substituted pyridazine derivatives.

Protonated pyridazine is attacked by nucleophilic acyl radicals at positions 4 and 5 to give 4,5-diacylpyridazines. When acyl radicals with a hydrogen atom at the α-position to the carbonyl group are used, the diacylpyridazines are mainly converted into cyclopenta[d]pyridazines by intramolecular aldol reactions (Scheme 43).

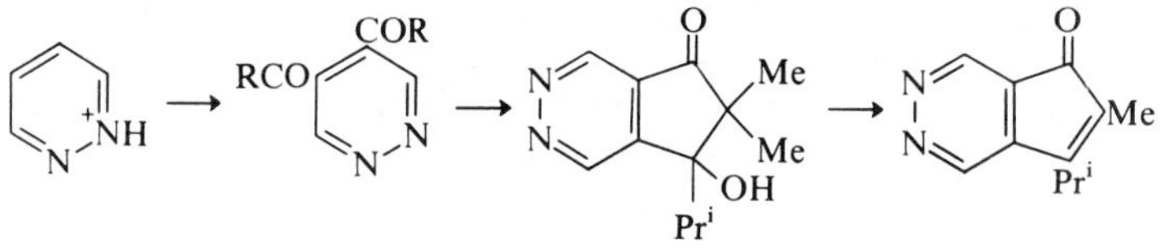

Scheme 43

Reactions of 3-chloro-6-methoxypyridazine with ketone enolates in liquid ammonia exhibit characteristics consistent with a radical chain mechanism for substitution ⟨81JOC294⟩.

Pyridazine and its 3-methyl and 4-methyl derivatives react with *sym*-trioxanyl radicals. The reaction takes place selectively at positions 4 and/or 5, and partially at the α-position to a ring nitrogen atom, leading thus to the formylmethylpyridazine derivatives ⟨80JHC1501⟩.

### 2.12.3.8 Reactions with Cyclic Transition States

Pyridazine N-oxides react with benzyne to give a mixture of 1-benzoxepin (129) and arylpyridazine (130), while N-acetylpyridaziniumimide forms a cycloadduct (131) which is further transformed into (132) and (133) (Scheme 44).

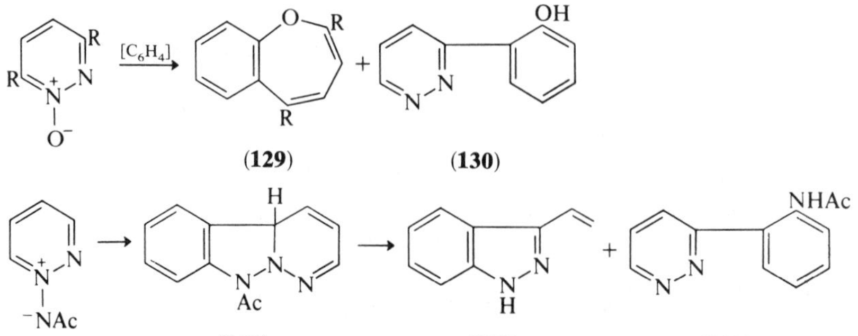

Scheme 44

Pyridazine carboxylates and dicarboxylates undergo cycloaddition reactions with unsaturated compounds with inverse electron demand to afford substituted pyridines and benzenes respectively (Scheme 45).

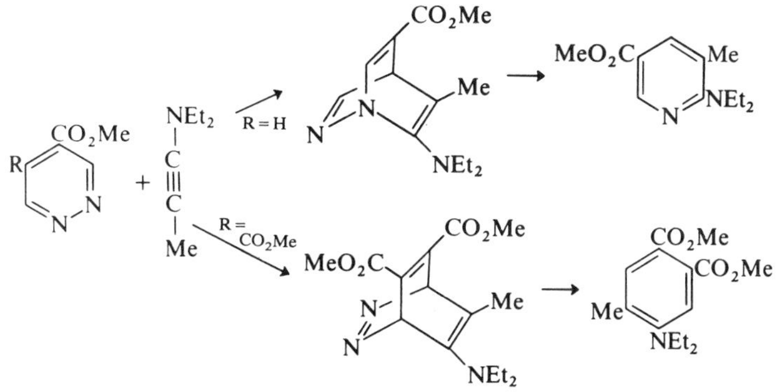

**Scheme 45**

The dipolar cycloaddition of 2-diazopropane to 1-methyl-3-phenylpyridazin-6(1*H*)-one takes place through an unstable adduct which thermally decomposes to a 1,2-diazepinone, a pyridazinone and diazanorcaradiene derivative (Scheme 46).

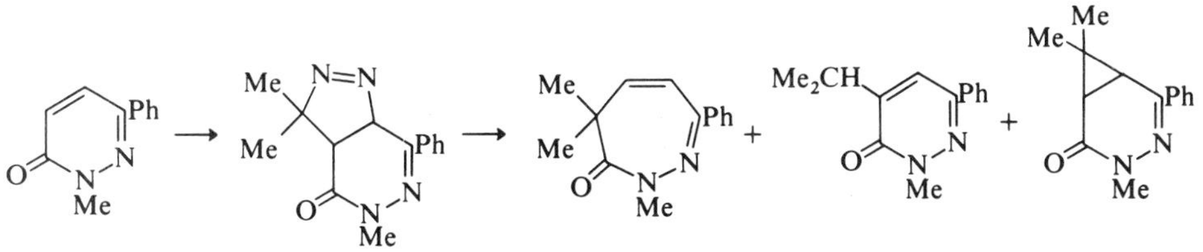

**Scheme 46**

An intramolecular [4+2] cycloaddition is represented by a thermal conversion of allylphenoxy- and allyloxyphenoxy-pyridazines into xanthenes ⟨80CPB198⟩.

Pyridazinium and phthalazinium dicyanomethylides give indolizines as primary adducts with 1,2,3-triphenylcyclopropene, either by [4+2] cycloaddition or by 1,3-dipolar addition of the ylide to triphenylcyclopropene ⟨81JCS(P1)73⟩.


## 2.12.4 FULLY CONJUGATED RINGS: REACTIVITY OF SUBSTITUENTS

### 2.12.4.1 Fused Benzene Rings

Substituents on benzene or benzenoid rings in fused pyridazines, *i.e.* in cinnolines and phthalazines, usually exhibit reactivity which is similar to that found in the correspondingly substituted fused aromatic compounds, such as naphthalene, and is therefore not discussed here.

Since the pyridazine ring is generally more stable to oxidation than a benzene ring, oxidation of alkyl and aryl substituted cinnolines and phthalazines can be used for the preparation of pyridazinedicarboxylic acids. For example, oxidation of 4-phenylcinnoline with potassium permanganate yields 5-phenylpyridazine-3,4-dicarboxylic acid, while alkyl substituted phthalazines give pyridazine-4,5-dicarboxylic acids under essentially the same reaction conditions.


### 2.12.4.2 *C*-Linked Substituents

#### 2.12.4.2.1 Methyl groups

3-Methylpyridazine can be oxidized with selenium dioxide to give 3-formylpyridazine, and methyl groups attached to any position in pyridazine *N*-oxides are transformed with pentyl nitrite in the presence of sodium amide in liquid ammonia into the corresponding aldehyde oxime. A large variety of alkyl-substituted pyridazines can be oxidized to pyridazinecarboxylic acids. Potassium permanganate is the most frequently used oxidizing agent, but potassium or sodium dichromate in sulfuric acid, chromic acid in acetic acid, and concentrated nitric acid are also used. Other substituents, such as halogen and methoxy groups, usually remain unchanged.

Because of the electron-attracting properties of the ring nitrogen atoms, methyl groups undergo aldol-like condensations. For example, 3- and 4-methylpyridazine react with chloral to give 3- or 4-(2-hydroxy-3,3,3-trichloropropyl)pyridazine, and 4-methylpyridazine reacts with anisaldehyde to yield 4-(*p*-methoxystyryl)pyridazine.

For the reactions of methylpyridazine 1-oxides with benzaldehyde in the presence of sodium methoxide, the order of reactivity of methyl groups at various positions is 5 > 4,6 > 3. 3-Methylpyridazine 1-oxide is converted by acetic anhydride into the 3-acetoxymethyl compound, which is easily hydrolyzed to 3-hydroxymethylpyridazine.

Side-chain lithiation with lithium diisopropylamide and subsequent alkylation or acylation is a practical method for the preparation of various alkyl-, alkenyl- and acyl-methylpyridazines ⟨78CPB2428, 78CPB3633, 79CPB916⟩ (Scheme 47).

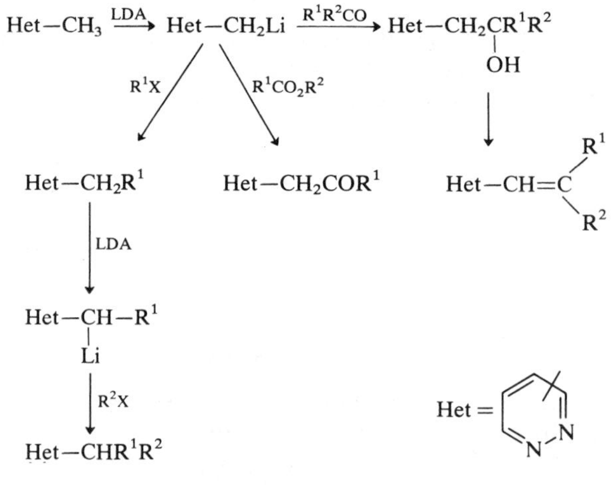
Het—CH3
LDA
Het—CH2Li
R1R2CO
Het—CH2CR1R2
OH
R1X
R1CO2R2
Het—CH2R1
Het—CH2COR1
Het—CH=C
R1
R2
LDA
Het—CH—R1
Li
R2X
Het—CHR1R2
Het =
N
N

**Scheme 47**

### *2.12.4.2.2 Hydroxymethyl groups*

Hydroxymethylpyridazines are easily oxidized with selenium dioxide to the corresponding aldehydes. Oxidation of the corresponding secondary alcohols with chromic acid in aqueous sulfuric acid gives ketones, while oxidation of a hydroxymethyl group with permanganate leads to the pyridazinecarboxylic acids. Hydroxymethyl groups are converted into chloromethyl groups with thionyl chloride or phosphorus oxychloride.

### *2.12.4.2.3 Carbonyl groups*

Pyridazine aldehydes and ketones with the carbonyl group at the ring or in a side chain react in the usual manner. They form hydrazones, semicarbazides, oximes, *etc*. Side-chain aldehydes can be easily oxidized to pyridazinecarboxylic acids with silver nitrate and side-chain ketones are oxidized to carboxylic acids by treatment with potassium permanganate or hydrogen peroxide.

Ketones can be reduced by the Wolff–Kishner method to the corresponding alkyl compounds, or by sodium in ethanol to the corresponding alcohols. An alkali-catalyzed deacylation of 3-acetyl-6-methoxypyridazine 1-oxide occurs quantitatively on treatment with dilute sodium hydroxide.

### *2.12.4.2.4 Carboxyl groups and derivatives*

Pyridazinecarboxylic acids can also be prepared by hydrolysis of esters, nitriles and amides in the presence of acids or alkali. Another interesting method is partial decarboxylation of pyridazinepolycarboxylic acids. For example, pyridazine-3,4,5,6-tetracarboxylic acid is decarboxylated in the presence of an acid to pyridazine-4,5-dicarboxylic acid. The carboxy group at position 3 is lost more easily than that at position 4.

Pyridazinecarboxylic acids are more acidic than benzoic acids, due to the electronegativity of the pyridazine ring, but oxopyridazinecarboxylic acids are weaker than the corresponding pyridazinecarboxylic acids (Table 11).

**Table 11**   Acidity of Some Pyridazinecarboxylic Acids

| Compound | $pK_a$ |
|----------|--------|
| Pyridazine-3-carboxylic acid | 3.0 |
| Pyridazine-4-carboxylic acid | 2.8 |
| 6-Oxo-1,6-dihydropyridazine-3-carboxylic acid | 3.2 |
| 6-Oxo-1,6-dihydropyridazine-4-carboxylic acid | 3.7 |

Practically all pyridazine-carboxylic and -polycarboxylic acids undergo decarboxylation when heated above 200 °C. As the corresponding products are usually isolated in high yields, decarboxylation is frequently used as the best synthetic route for many pyridazine and pyridazinone derivatives. For example, pyridazine-3-carboxylic acid eliminates carbon dioxide when heated at reduced pressure to give pyridazine in almost quantitative yield, but pyridazine is obtained in poor yield from pyridazine-4-carboxylic acid. Decarboxylation is usually carried out in acid solution, or by heating dry silver salts, while organic bases such as aniline, dimethylaniline and quinoline are used as catalysts for monodecarboxylation of pyridazine-4,5-dicarboxylic acids.

Esters of pyridazine- and oxopyridazine-carboxylic acids are prepared either by esterification with diazomethane, with an alcohol in the presence of an acid, or by reaction of acid chlorides with an alcohol. The choice of the reagent used for the preparation of pyridazinecarboxylic acid chlorides depends on the presence of other groups. For example, thionyl chloride converts only a carboxylic group to the acid chloride, while a hydroxy (or potential hydroxy) group in pyridazinones or hydroxypyridazinones is not replaced. On the other hand, if a hydroxy (or potential hydroxy) group is to be replaced simultaneously, phosphorus oxychloride and/or phosphorus pentachloride can be employed. In some instances, hydrolysis of cyanopyridazines with sulfuric acid in anhydrous alcohol gives esters in good yield. Claisen condensation of ethyl pyridazine-3-carboxylate with methyl acetate followed by hydrolysis gives 3-acetylpyridazine.

Pyridazinecarboxamides are prepared from the corresponding esters or acid chlorides with ammonia or amines or by partial hydrolysis of cyanopyridazines. Pyridazinecarboxamides with a variety of substituents are easily dehydrated to nitriles with phosphorus oxychloride and are converted into the corresponding acids by acid or alkaline hydrolysis. They undergo Hofmann degradation to give the corresponding amines, while in the case of two *ortho* carboxamide groups pyrimidopyridazines are formed.

Some 4,5-disubstituted pyridazines exhibit ring–chain isomerism involving heterospiro compounds. For example, 5-(*o*-aminophenylcarbamoyl)pyridazine-4-carboxylic acid exists in a zwitterionic form in the solid state, but in a solution of DMSO it is almost exclusively 3′,4′-dihydro-3′-oxospiro[pyridazine-5(2*H*),2′(1′*H*)-quinoxaline]-4-carboxylic acid (**134**). The equilibrium is strongly influenced by the nature of the solvent, the substituents on the pyridazine ring and the nucleophilicity of the group attached to the phenyl ring (Scheme 48) ⟨80JCS(P2)1339⟩.

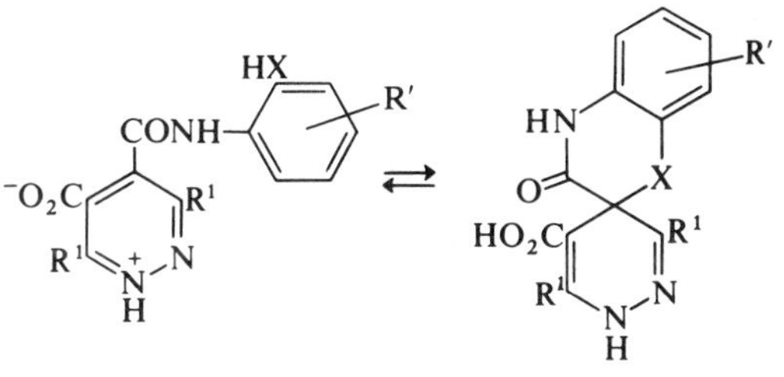

(**134**) X = NH, O

**Scheme 48**

Pyridazinecarbohydrazides are prepared in the normal way from an ester or acid chloride and hydrazine or a substituted hydrazine, generally in good yields. Pyridazines with two *ortho* alkoxycarbonyl groups give cyclic hydrazides with hydrazine, which are pyridazinopyridazines.

### 2.12.4.2.5 Cyano groups

Cyanopyridazines add ammonia, primary and secondary amines and hydroxylamine to give amidines or amidoximes. Substituted amides, thioamides and carboximidates can be also prepared. With hydrazine, 3-pyridazinylcarbohydrazide imide is formed and addition of methylmagnesium iodide with subsequent hydrolysis of the imine affords the corresponding pyridazinyl methyl ketone.

### 2.12.4.3 N-Linked Substituents

### 2.12.4.3.1 Reduction of nitro groups

Nitropyridazines are reduced catalytically either over platinum, Raney nickel or palladium–charcoal catalyst. When an *N*-oxide function is present, palladium–charcoal in neutral solution is used in order to obtain the corresponding amino *N*-oxide. On the other hand, when hydrogenation is carried out in aqueous or alcoholic hydrochloric acid and palladium–charcoal or Raney nickel are used for the reduction of the nitro group, deoxygenation of the *N*-oxide takes place simultaneously. Halonitropyridazines and their *N*-oxides are reduced, dehalogenated and deoxygenated to aminopyridazines or to aminopyridazine *N*-oxides under analogous conditions.

Hydroxyaminopyridazine 1-oxides are usually formed by catalytic hydrogenation of the corresponding nitro derivatives over palladium–charcoal in methanol, provided that the reaction is stopped after absorption of two moles of hydrogen. 3-Hydroxyaminopyridazine 1-oxide and 6-amino-4-hydroxyamino-3-methoxypyridazine 1-oxide are prepared in this way, while 5-hydroxyamino-3-methylpyridazine 2-oxide and 5-hydroxyamino-6-methoxy-3-methylpyridazine 2-oxide are obtained by chemical reduction of the corresponding nitro compounds with phenylhydrazine.

Sometimes bimolecular products are formed from reduction of nitropyridazine *N*-oxides. For example, 3-acetamido-6-methoxy-5-nitropyridazine 2-oxide, on reduction over palladium–charcoal, affords two different products, depending on the reaction conditions (Scheme 49).

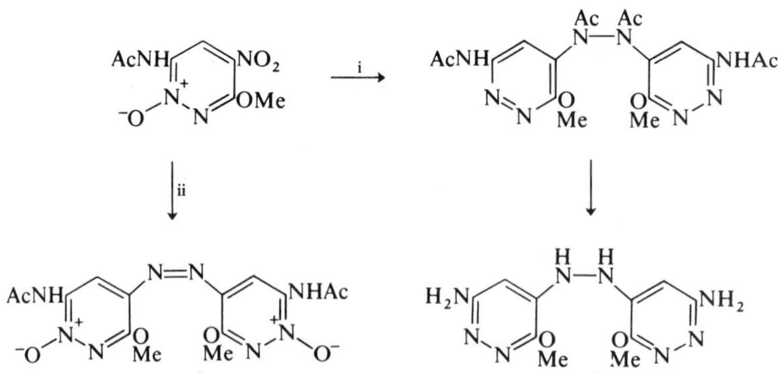

i, H$_2$ (3.5 moles), Ac$_2$O; ii, H$_2$ (2.5 moles), AcOH

**Scheme 49**

Reduction of nitroaminopyridazines yields the corresponding aminopyridazines. Reductive cleavage of hydrazinopyridazines to give amino compounds is of practical significance in cases when halogen atoms are resistant to ammonolysis. Many substituted 3,4-diamino-, 4,5-diamino- and 3,5-diamino-pyridazines can be prepared in this way.

### 2.12.4.3.2 Acylation

Aminopyridazines can be formylated, and the corresponding diamines are either mono- or di-formylated with formic acid. Acetylation is usually carried out with acetic anhydride; sulfonyl derivatives are obtained by treatment of aminopyridazines with methanesulfonyl chloride in the presence of trimethylamine or with arenesulfonyl chlorides in pyridine. Amino-substituted pyridazinones give *N*- and *O*-substituted products. For example, 4-aminopyridazin-3(2*H*)-ones give with *p*-toluenesulfonyl chloride a mixture of 4-tosyl and 3-*O*-tosyl derivatives, whereas tosylation of 4-amino-3,6-dimethoxypyridazine in pyridine gives several products, tosylated at the amino group and at the ring nitrogen atom.

The most commonly used methods for the preparation of pyridazinesulfonamides are the condensation of aminopyridazines with *p*-acylaminobenzenesulfonyl chloride and the reaction of halosubstituted pyridazines with sulfanilamide by fusion or in an appropriate solvent.

### 2.12.4.3.3 Condensation with carbonyl compounds

Amino-pyridazines and -pyridazinones react with monomethyl- or *N,N*-dimethyl-formamide and other aliphatic amides in the presence of phosphorus trichloride, thionyl chloride, phosgene or benzenesulfonyl chloride to give mono- or di-alkylaminomethyl-eneamino derivatives. The same compounds can be prepared conveniently with *N,N*-dimethylformamide dimethyl acetal in high yield (Scheme 50).

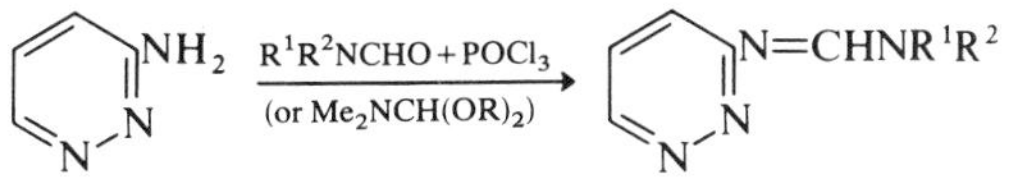

**Scheme 50**

### 2.12.4.3.4 Diazotization

3-Amino-6-chloro-4-methyl- and 3-amino-6-chloro-5-methyl-pyridazine and 3-amino-6-methylpyridazin-4(1*H*)-one are transformed with sodium nitrite in the presence of acid into the corresponding oxo compounds. If concentrated hydrochloric acid is used, in some instances the corresponding chloro derivatives are obtained as side products. On the other hand, 3-, 4-, 5- and 6-aminopyridazine 1-oxides and derivatives are transformed into stable diazonium salts, which can easily be converted into the corresponding halo derivatives. In this way 3-, 4-, 5- and 6-bromopyridazine 1-oxides, 5-chloropyridazine 1-oxide, 3,4,5-trichloropyridazine 1-oxide and 6-chloropyridazine 1-oxide can be obtained.

### 2.12.4.3.5 Hydrazinopyridazines

Hydrazinopyridazines are easily formylated with formic acid or ethyl formate and acetylated with acetic anhydride. *N*-Pyridazinylthiosemicarbazides are obtained from thiocyanates or alkyl- and aryl-isothiocyanates. Hydrazinopyridazines condense with aliphatic and aromatic aldehydes and ketones to give hydrazones.

Treatment of hydrazinopyridazines with nitrous acid gives the corresponding azido compounds. In this way, 3-azidopyridazine 1-oxide and 2-oxide, 4-azidopyridazine 1-oxide and 5-azidopyridazine 1-oxide, 4-azido-3,6-dimethoxypyridazine and others are prepared. The azido group at position 3 cyclizes with the adjacent nitrogen to form tetrazolo[1,5-*b*]pyridazine if the nitrogen at the 1-position of the original compound is not substituted. 3-Azidopyridazine 1-oxide and 2-oxide cyclize to tetrazolo[1,5-*b*]pyridazine after deoxygenation with phosphorus trichloride. On the other hand, the tetrazole ring is opened and transformed into an azido group on *N*-oxidation with hydrogen peroxide in polyphosphoric acid (Scheme 51).

The aza-transfer reaction between 3-hydrazinopyridazines and aromatic diazonium salts or heterocyclic diazo compounds affords the corresponding tetrazolo[1,5-*b*]pyridazines, while 3-hydrazinopyridazine 1-oxide gives 3-azidopyridazine 1-oxide ⟨76TL3193, 76T725⟩.

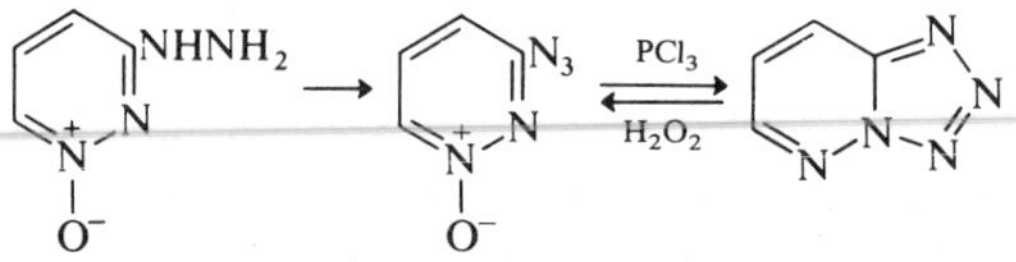

**Scheme 51**

On the other hand, elimination of the hydrazino group in 3-hydrazinopyridazines is possible with *p*-toluenesulfonamide under phase transfer reaction conditions ⟨78TL3059⟩.

### 2.12.4.4 *O*-Linked Substituents

Alkoxy substituents at various positions in the pyridazine, pyridazinone or pyridazine *N*-oxide ring are hydrolyzed in the presence of concentrated hydrochloric acid at about 130–150 °C or in dilute sodium or potassium hydroxide to give the corresponding hydroxy-pyridazines or pyridazinones. Acid hydrolysis is generally preferred, since under basic conditions other sensitive groups may be displaced or hydrolyzed. Hydrogenolysis of 3-benzoyloxypyridazine 1-oxide over palladium–charcoal selectively removes the benzyl group to give the corresponding 3-hydroxypyridazine 1-oxide. However, on prolonged reaction deoxygenation of the *N*-oxide also takes place.

When alkoxypyridazine 1-oxides are heated alone or in the presence of *p*-toluenesulfonic acid the methyl group migrates from the methoxy group to the *N*-oxide group. In this manner, 4-methoxypyridazine 1-oxide rearranges to 1-methoxypyridazin-4(1*H*)-one, 5-methoxypyridazine 1-oxide to 2-methylpyridazin-5(2*H*)-one 1-oxide and substituted 3,6-dimethoxypyridazine 1-oxides to 1,3-dimethoxypyridazin-6(1*H*)-ones.

### 2.12.4.5 *S*-Linked Substituents

Various alkylating agents are used for the preparation of pyridazinyl alkyl sulfides. Methyl and ethyl iodides, dimethyl and diethyl sulfate, α-halo acids and esters, β-halo acids and their derivatives, α-halo ketones, benzyl halides and substituted benzyl halides and other alkyl and heteroarylmethyl halides are most commonly used for this purpose. Another method is the addition of pyridazinethiones and pyridazinethiols to unsaturated compounds, such as 2,3(4*H*)-dihydropyran or 2,3(4*H*)-dihydrothiopyran, and to compounds with activated double bonds, such as acrylonitrile, acrylates and quinones.

Besides displacement reactions, oxidations, rearrangements and cleavage of the sulfide linkage, the most important reactions take place at the sulfur atom.

The *S* → *N* rearrangement of pyridazinethione glycosides proceeds smoothly under the influence of mercury(II) bromide. For example, 3-(tetraacetyl-1-β-D-glucosylmercapto)pyridazines rearrange to 2-(tetraacetyl-1-β-D-glucosyl)pyridazine-3(2*H*)-thiones.

Pyridazines with an appropriate side chain attached to the sulfur atom at position 3 can be transformed into bicyclic systems. For example, pyridazinyl β-ketoalkyl sulfides are cyclodehydrated in sulfuric acid to give thiazolopyridazinium salts, and 3-carboxymethyl-thiopyridazines are transformed by acetic anhydride in pyridine into 3-hydroxythiazolo[3,2-*b*]pyridazinium anhydro salts (Scheme 52).

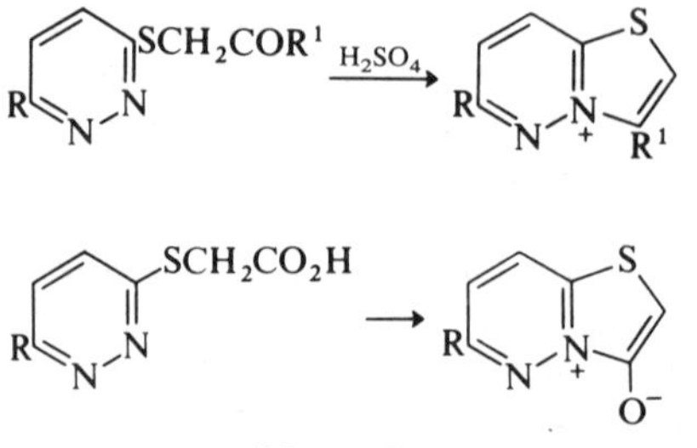

**Scheme 52**

Alkyl- and aryl-thiopyridazines are oxidized to sulfoxides, sulfones or sulfonic acids, depending on the reaction conditions. *N*-Oxidation can take place simultaneously.

A 3-, 4- and 5-methylthio group can be converted into the corresponding sulfoxide or sulfone by oxidation with hydrogen peroxide in aqueous or acetic acid solution, *m*-chloroperbenzoic acid, potassium permanganate in acidic solution or with chlorine at low temperatures. If alkylpyridazines with additional methoxy or phenoxy substituents such as a halogen group are oxidized with potassium permanganate in dilute sulfuric acid, hydrolysis can occur to give the corresponding pyridazinones.

Acylation of pyridazinethiones with acetyl chloride or benzoyl chloride gives the corresponding *S*-acylated products. 6-Mercaptopyridazine-3(2*H*)-thione gives either mono- or di-*S*-acylated products. A bispyridazinyl derivative is formed when phosgene or thiophosgene is used as acylating agent.

Pyridazinethiones are readily oxidized to the corresponding disulfides with iodine, aqueous iron(III) chloride, hydrogen peroxide in acetic acid, potassium permanganate in acetic acid, and upon long exposure to air.

### 2.12.4.6 Metal- and Metalloid-linked Substituents

Several 3-substituted 6-methylmercuriothiopyridazines and complexes of perfluoropyridazine with metal carbonyl anions have been prepared ⟨67MI21200⟩.

Pyridazines form complexes with iodine, iodine monochloride, bromine, nickel(II) ethyl xanthate, iron carbonyls, iron carbonyl and triphenylphosphine, boron trihalides, silver salts, mercury(I) salts, iridium and ruthenium salts, chromium carbonyl and transition metals, and pentammine complexes of osmium(II) and osmium(III) ⟨79ACS(A)125⟩. Pyridazine *N*-oxide and its methyl and phenyl substituted derivatives form copper complexes ⟨78TL1979⟩.

### 2.12.4.7 Substituents Attached to Ring Nitrogens

The reactivity of substituents is in most cases similar to that of *C*-linked substituents and is therefore not discussed here in detail. The most important reaction in this group is deoxygenation of the *N*-oxides. This can be achieved with complex metal hydrides, phosphite esters, phosphorus trihalide, sulfide ions, dissolving metals and metal ions of lower valency, *etc*. Catalytic hydrogenation is used extensively for this purpose. However, many side reactions can occur, especially various nucleophilic substitutions, dehalogenation and reduction of nitro groups. A selective deoxygenation takes place with a molybdenum(III) species. This method is advantageous in the presence of halogens and if partial hydrogenation of the ring system may take place on catalytic hydrogenation ⟨80S129⟩.

Catalytic hydrogenation of 3,6-diphenylpyridazine 1,2-dioxide over palladium–charcoal affords 3,6-diphenylpyridazine 1-oxide and 3,6-diphenylpyridazine ⟨79JOC3524⟩.

## 2.12.5 SATURATED AND PARTIALLY SATURATED RINGS

### 2.12.5.1 Dihydro-pyridazines and -pyridazinones

6-Aryl-4,5-dihydropyridazine-3(2*H*)-one undergoes ring opening when submitted to Wolff–Kishner reduction, while with lithium aluminum hydride the corresponding 2,3,4,5-tetrahydro product is obtained.

6-Aryl-2-phenyl-4,5-dihydropyridazin-3(2*H*)-ones react either with phenylmagnesium bromide or with phenyllithium to give 6-aryl-2,6-diphenyl-1,4,5,6-tetrahydropyridazin-3(2*H*)-ones (**135**) (products of 1,2-addition to the azomethine bond), while 2-methyl-6-phenyl-4,5-dihydropyridazine-3(2*H*)-one reacts with two equivalents of phenylmagnesium bromide at the carbonyl and azomethine group to produce 2-methyl-3,3,6,6-tetraphenyl-hexahydropyridazine (**136**) (Scheme 53) ⟨80JPR617⟩.

6-Aryl-4,5-dihydropyridazin-3(2*H*)-ones react with pyrrolylmagnesium bromide to give 6-aryl-3(1-pyrrolyl)pyridazines or, when 1:4 molar amounts of reagents are used, a mixture of 6-aryl-3(1-pyrrolyl)pyridazines and 3,4-di(1-pyrrolyl)-4,5-dihydropyridazines (Scheme 54 ⟨79RRC453⟩.

Addition of bromine to 5-*t*-butyl-3,6-dimethoxy-4,5-dihydropyridazine produces 5-bromo-4-*t*-butyl-3-methoxy-4,5-dihydropyridazin-6(1*H*)-one.

**(135)**                                    **(136)**

**Scheme 53**

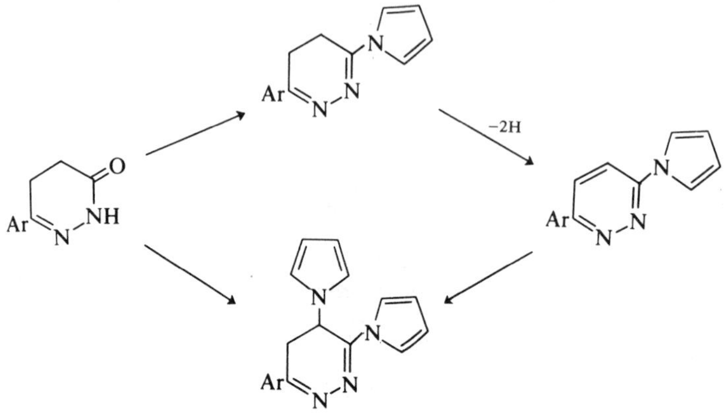

**Scheme 54**

Dehydrogenation of 4,5-dihydropyridazin-3(2$H$)-ones with bromine in acetic acid is accomplished in high yield only when the phenyl ring is substituted with alkyl groups or halogen atoms. If an alkoxy group is attached to the phenyl ring, bromination of the aromatic ring occurs. 4,5-Dihydropyridazin-3(2$H$)-ones are labile in acidic and alkaline solutions, and substituted butyric acids are formed. Ring inversion of some diazanorcaradienes occurs at elevated temperatures, presumably involving a monocyclic diazepine (**137**) as intermediate ⟨79BSB905⟩. 7,7-Diethoxycarbonyl-2,5-diphenyl-3,4-diazanorcaradiene, when hydrolyzed to the half-ester (**138**), undergoes decarboxylation and ring-opening upon heating to 130–140 °C in the absence of solvent to give 4-ethoxycarbonylmethyl-3,6-diphenyl-pyridazine (Scheme 55) ⟨77BCJ2153⟩.

**(137)**

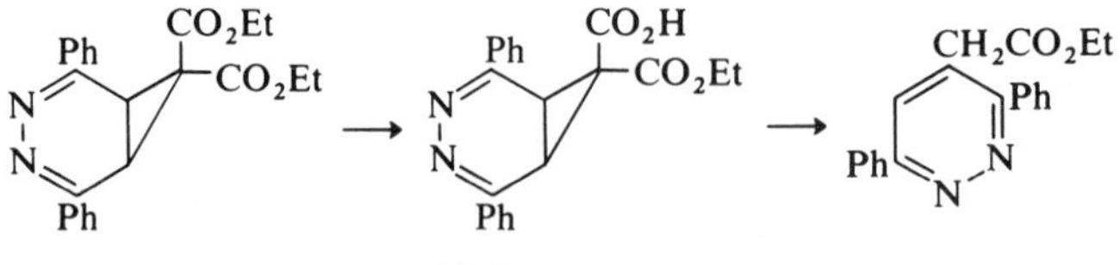

**(138)**

**Scheme 55**

Pyridazine-3,6-diones (diazaquinones) are prepared from cyclic hydrazides by oxidation with lead tetraacetate or other oxidizing agents, such as *t*-butyl hypochlorite, chlorine or nickel peroxide.

Since diazaquinones are among the most powerful dienophiles, they undergo [4+2] cycloaddition (Diels–Alder) reactions with a great variety of dienes to give various heterocyclic systems accessible with difficulty by other methods. Diazaquinone reacts with butadiene and substituted butadienes, carbocyclic and heterocyclic dienes, 1-vinylcycloalkenes, polyaromatic compounds and vinylaromatic compounds to afford bicyclic and polycyclic bridgehead diaza systems, including diazasteroids (Scheme 56).

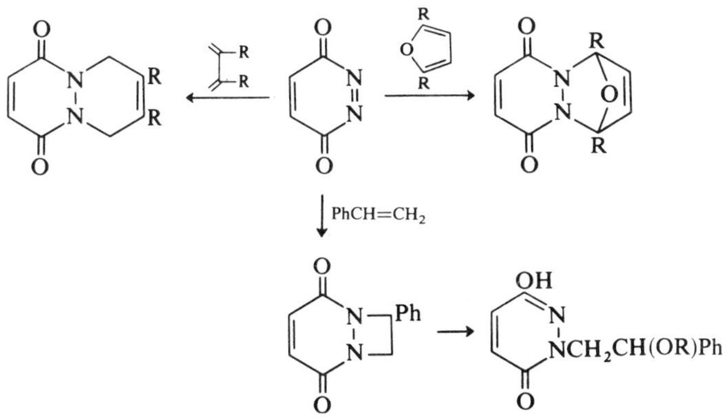

**Scheme 56**

[2+2] Cycloadditions of diazaquinones with unsaturated compounds yield diazacyclobutanes, from which N-substituted 3-hydroxypyridazin-6(1H)-ones are formed after addition of water, t-butanol or acetic acid (Scheme 56). The same types of compound are also obtained from enamines.

Benzo analogs, 1,4-phthalazinediones and derivatives, react in the same way, while the corresponding diazaorthoquinones, *i.e.* 3,4-diketocinnolines, are not known.

3,6-Dihydropyridazines are known as labile intermediates, and lose nitrogen at −78 °C with a half-life of 30 seconds or less. On the other hand, the corresponding 2-oxides are stable compounds which lose $N_2O$ only at 300 °C or above ⟨77JA8505⟩.

## 2.12.5.2 Tetrahydropyridazines

1,2-Diethoxycarbonyl-1,2,3,6-tetrahydropyridazines are reduced with lithium aluminum hydride to give 1,2-dimethylhexahydropyridazines. Similarly, 1,4,5,6-tetrahydropyridazines afford the corresponding hexahydropyridazines. Free radical addition of thioacetic acid to the double bond of 1,2-diethoxycarbonyl-1,2,3,6-tetrahydropyridazine produces 1,2-diethoxycarbonyl-4-S-thioacetoxyhexahydropyridazine. Selective hydrolysis affords 1,2-diethoxycarbonyl-4-thio- and then 4-thio-hexahydropyridazines ⟨80JHC1465⟩.

Thermal decomposition of unsubstituted 3,4,5,6-tetrahydropyridazine at 439 °C in the gas phase proceeds 55% *via* tetramethylene and 45% *via* a stereospecific alkene forming pathway. The thermal decomposition of labelled *cis*-3,4,5,6-tetrahydropyridazine-3,4-$d_2$ affords *cis*-ethylene-1,2-$d_2$, *trans*-ethylene-1,2-$d_2$, *cis*-cyclobutane-1,2-$d_2$ and *trans*-cyclobutane-1,2-$d_2$ (Scheme 57) ⟨79JA3663, 80JA3863⟩.

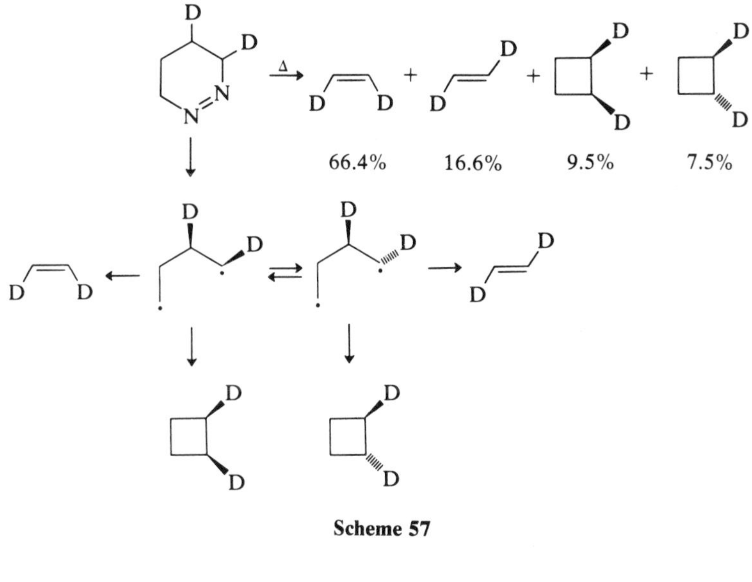

**Scheme 57**

Thermal decomposition of *cis-* and *trans-*3,6-dimethyl-3,4,5,6-tetrahydropyridazines affords propene, *cis-* and *trans-*1,2-dimethylcyclobutanes and 1-hexene. The stereochemistry of the products is consistent with the intermediacy of the 1,4-biradical 2,5-hexadienyl. The results indicate that thermal reactions of cyclic azo compounds and cyclobutanes of similar substitution proceed with similar stereospecificity when compared at similar temperatures ⟨79JA2069⟩.

### 2.12.5.3 Hexahydropyridazines

A considerable amount of attention has been devoted to conformational studies of 1,2-disubstituted hexahydropyridazines because of the equilibria involving eclipsing and non-eclipsing ring inversion, and eclipsing on one side and non-eclipsing nitrogen inversion on the other side (Scheme 58). The particular equilibrium depends on the type and number of substituents attached to the carbon atoms of the hexahydropyridazine system. Some inconsistencies arising from dipole moment measurements and $^1$H and $^{13}$C NMR measurements have been explained and corrected by other methods, mainly by variable temperature photoelectron measurements ⟨80JA7438⟩. (For reviews on conformational analysis, see references ⟨78ACR14, B-75MI21200⟩.)

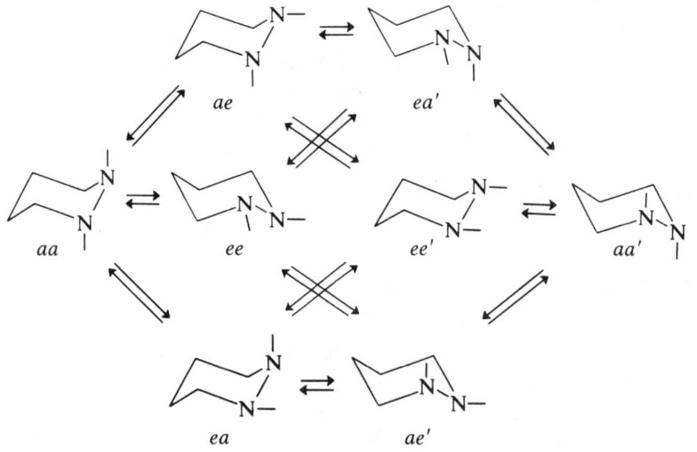

**Scheme 58**

## 2.12.6 SYNTHESIS

### 2.12.6.1 Introduction

There are some characteristic features of pyridazine and benzopyridazine syntheses. Almost all important and efficient synthetic approaches for pyridazines and phthalazines use hydrazine or substituted hydrazines as the source of the two ring nitrogens, while the carbon part of the skeleton may originate from starting compounds of different functionality. In the case of most cinnoline syntheses, both ring nitrogens originate from a diazotized amino group or a substituted hydrazino group. In general, syntheses from [4 + 2] fragments constitute the major part of syntheses of pyridazines and their benzo analogs.

Recently, many transformations of various heterocycles into pyridazines have been reported. From the synthetic point of view it appears that furan derivatives are the most valuable.

### 2.12.6.2 Formation of One Bond

#### 2.12.6.2.1 *Between two heteroatoms*

Syntheses of this type have been reported only recently ⟨78H(9)1367, 79JOC3524⟩. Unsaturated 1,4-dioximes are transformed by oxidative cyclization into pyridazine 1,2-dioxides

(Scheme 59). If phenyliodoso bistrifluoroacetate is used as the oxidizing agent the major product is an isoxazoloisoxazole, but pyridazine 1,2-dioxides are formed in minor amounts together with other products. With lead tetraacetate, in general, pyridazine 1,2-dioxides are the major products.

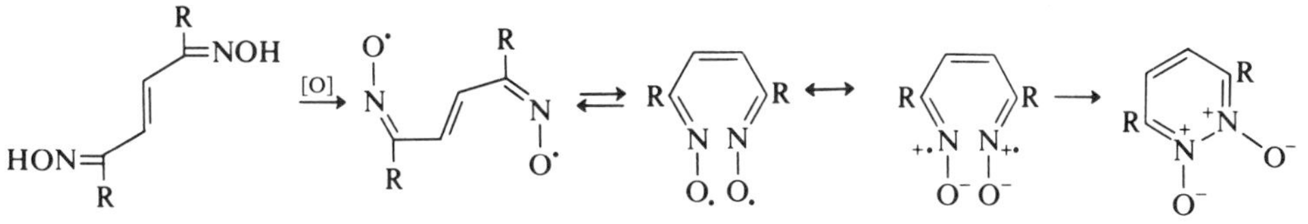

**Scheme 59**

Similarly, reduced pyridazines have been prepared by pertungstate oxidation of the corresponding diamines, as shown in Scheme 60 ⟨75JOC1395⟩.

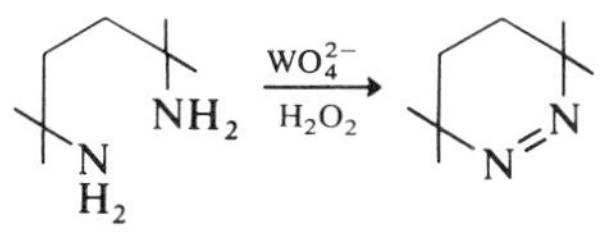

**Scheme 60**

### 2.12.6.2.2 *Adjacent to a heteroatom*

Unsaturated hydrazones, unsaturated diazonium salts or hydrazones of 2,3,5-triketones can be used as suitable precursors for the formation of pyridazines in this type of cyclization reaction. As shown in Scheme 61, pyridazines are obtainable in a single step by thermal cyclization of the tricyanohydrazone (**139**), prepared from cyanoacetone phenylhydrazone and tetracyanoethylene ⟨76CB1787⟩. Similarly, in an attempted Fischer indole synthesis the hydrazone of the cyano compound (**140**) was transformed into a pyridazine (Scheme 61)

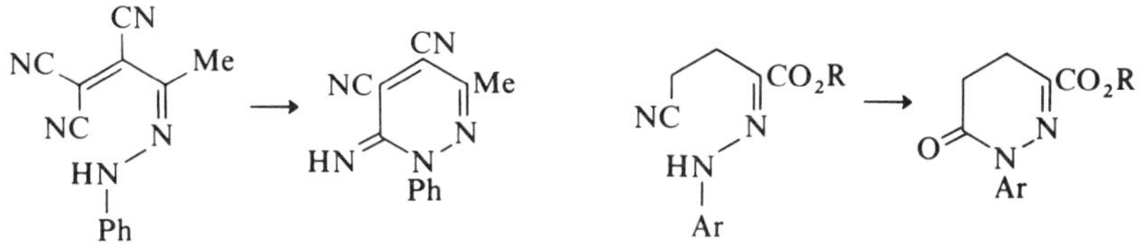

(**139**)                (**140**)

**Scheme 61**

⟨69LA(726)81⟩. γ,δ-Unsaturated β-diketones, when treated with diazonium salts in a Japp–Klingemann reaction, also give the corresponding hydrazones. These are transformed upon dissolution in hot ethanol and upon standing at room temperature to 1,4,5,6-tetrahydropyridazin-4-ones in good yield ⟨80S623⟩. However, an earlier report ⟨56JA2144⟩ of a similar reaction suggested that 1,2,3,4-tetrahydropyridazin-4-ones are formed (Scheme 62).

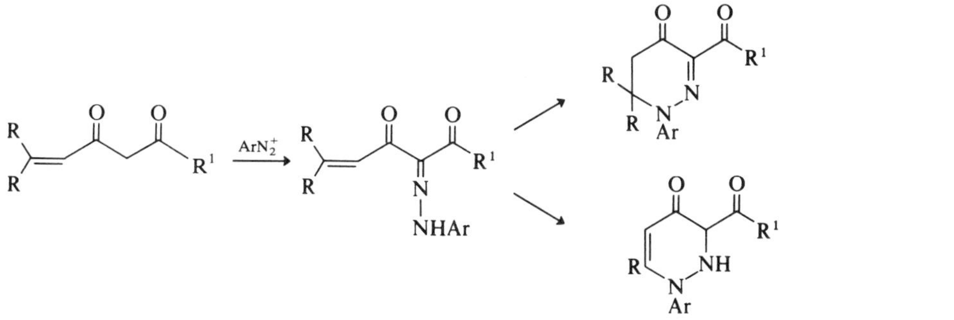

**Scheme 62**

On the other hand, unsaturated diazo compounds are thermally transformed by 1,1-cycloaddition into a bicyclic pyrazole (**141**). Although reversibility of this cycloaddition is

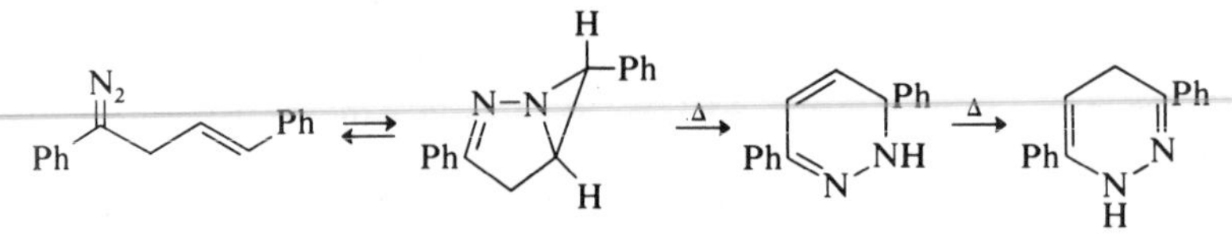

**Scheme 63**

observed on heating, the cycloadduct is also transformed into a 2,3-dihydropyridazine
derivative (**142**), which subsequently forms the corresponding 1,4-dihydro compound (**143**;
Scheme 63) ⟨80TL1009⟩. 2-Arylhydrazones of 2,3,5-triketones, which are obtained by ring
cleavage of the coupling product between diazonium salts and 3-furanones, are converted
thermally or in the presence of acetic acid or sodium hydrogen carbonate in good yields
into the corresponding pyridazin-4(1$H$)-ones (Scheme 64) ⟨79S790⟩.

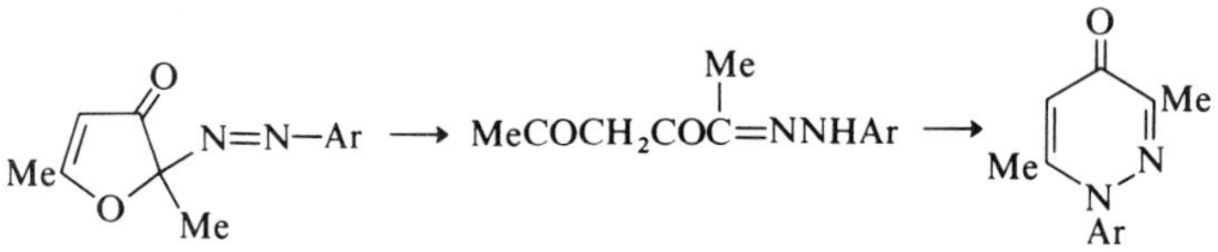

**Scheme 64**

### 2.12.6.2.3 γ to the heteroatom

There are some recent examples of this type of synthesis of pyridazines, but this approach
is more valuable for cinnolines. Alkyl and aryl ketazines can be transformed with lithium
diisopropylamide into their dianions, which rearrange to tetrahydropyridazines, pyrroles
or pyrazoles, depending on the nature of the ketazine. It is postulated that the reaction
course is mainly dependent on the electron density on the carbon termini bearing anionic
charges (Scheme 65) ⟨78JOC3370⟩.

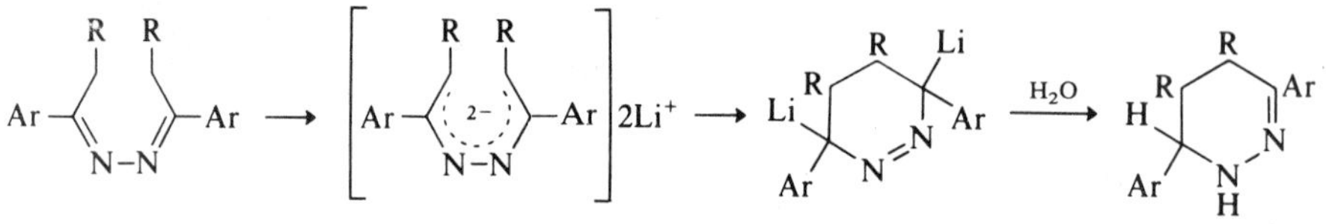

**Scheme 65**

Phosphacumulenylides have been employed for the synthesis of pyridazines. The coupling
products between 1,3-dicarbonyl compounds and benzenediazonium chloride react with
phosphacumuleneylides to give the corresponding phosphoranes (**144**) which subsequently
undergo an intramolecular Wittig reaction to form the corresponding pyridazines in good
yield (Scheme 66) ⟨80TL2939⟩.

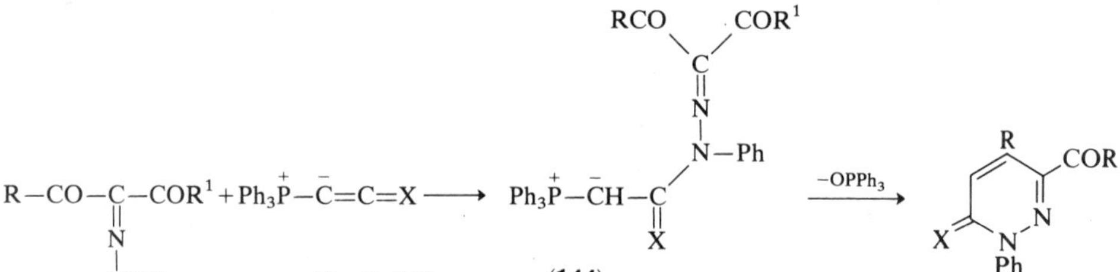

**Scheme 66**

3,4-Diphenylcinnoline can be prepared from benzil monophenylhydrazone in the
presence of about 80% sulfuric acid ⟨49MI21200⟩. Synthetically more important, however,
is the cyclization of mesoxalyl chloride phenylhydrazones under Friedel–Crafts conditions
⟨61JCS2828⟩. As outlined in Scheme 67, the starting mesoxalate phenylhydrazones are
obtained by coupling diazotized aromatic amines with diethyl malonate. After conversion

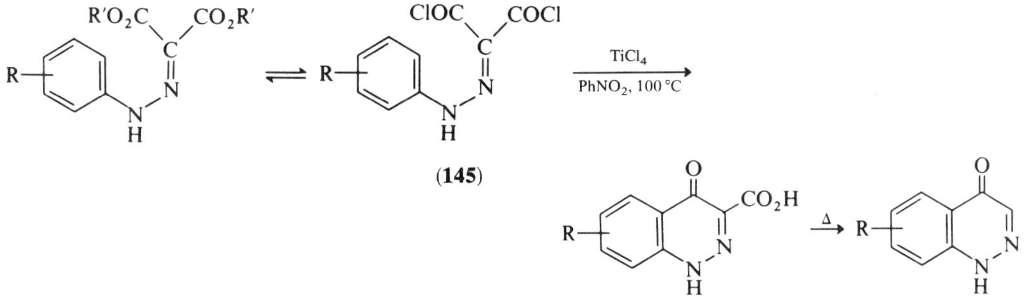

**Scheme 67**

into the diacids and dichlorides (145), the titanium tetrachloride-catalyzed cyclization gives the corresponding 4-oxo-1*H*-cinnoline-3-carboxylic acids (146), which can be decarboxylated into cinnolin-4(1*H*)-ones. Cinnolines have been prepared recently by a new synthetic approach (Scheme 68) ⟨81T3513⟩. Enamine esters and amides react with arenediazonium salts to give the corresponding iminium hydrazones (147) and these cyclize thermally to the corresponding cinnoline-3-carboxylic acid esters or amides.

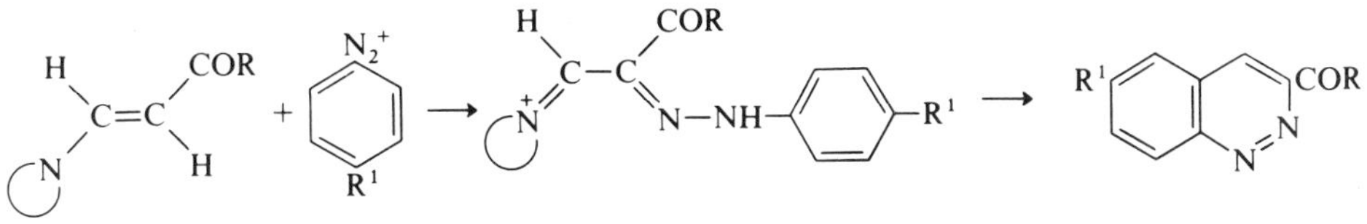

**Scheme 68**

### 2.12.6.3 Formation of Two Bonds

#### 2.12.6.3.1 From [5 + 1] atom fragments

This type of cyclization is important only for the formation of cinnolines. In all cases, the starting compounds have an *ortho* amino group, which upon diazotization undergoes ring closure with the other functionality, most frequently with a multiple bond.

One of the widely used cinnoline syntheses is the transformation of diazotized *o*-aminoarylethylenes into this bicyclic system (Widman–Stoermer synthesis) (Scheme 69).

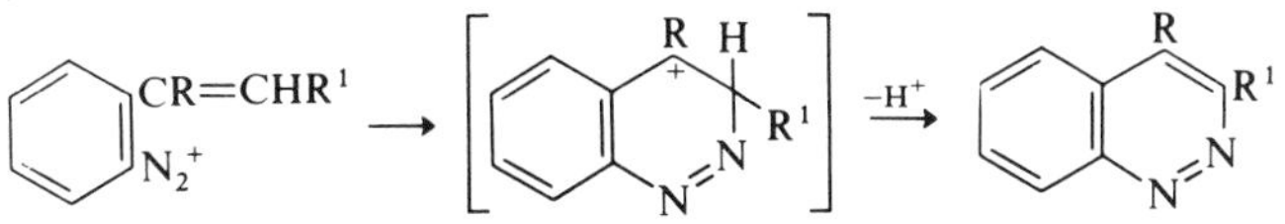

**Scheme 69**

This method is suitable only for the preparation of 4-substituted and/or 3,4-disubstituted derivatives, the substituents being only alkyl, aryl or heteroaryl groups. The presence of electron-withdrawing groups in the unsaturated side chain prevents the cyclization step. This is understandable if the influence of such groups on the stability of the intermediate carbonium ion is considered. Of more limited application is the analogous cyclization of diazotized *o*-aminophenylpropiolic acids, the reaction being referred to as the Richter synthesis (Scheme 70). A related synthesis (also referred to as the Neber–Bossel synthesis)

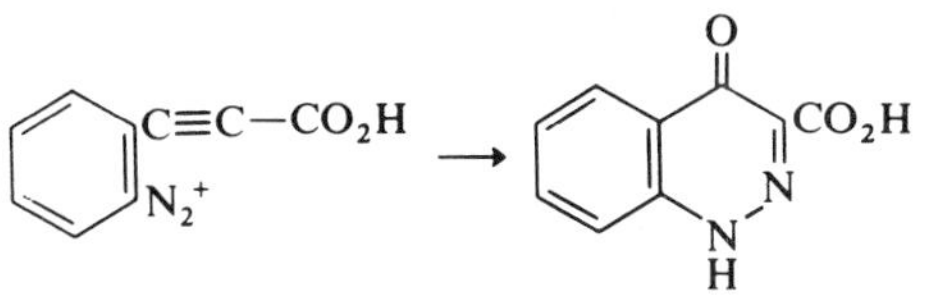

**Scheme 70**

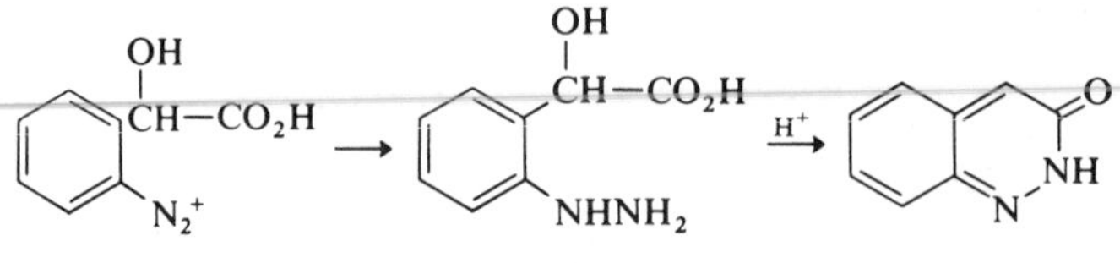

**(148)**

**Scheme 71**

involves the diazotization of sodium *o*-aminomandelate in concentrated hydrochloric acid and subsequent reduction of the diazonium group to a hydrazino group (**148**). Further cyclization gives cinnolin-3(2*H*)-one (Scheme 71). Only a few other derivatives have been prepared in this manner. If α-substituted *o*-aminomandelates are used as starting compounds, the first products of diazotization and reduction are 3-substituted 1-aminodioxindoles (**149**). These undergo hydrolytic ring opening in the presence of a base to give *o*-hydrazinomandelates (**150**) which cyclize to the corresponding cinnolines upon neutralization (Scheme 72).

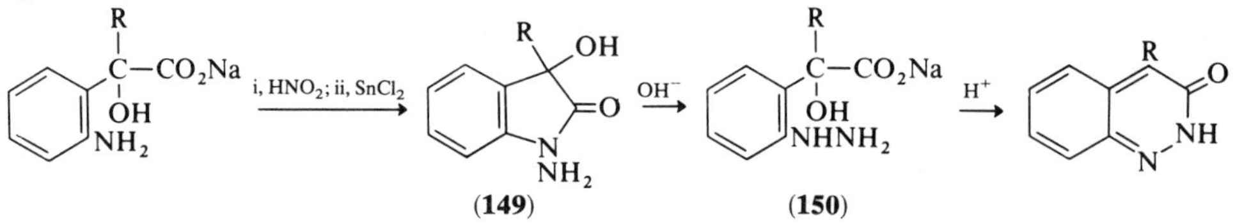

**(149)**                **(150)**

**Scheme 72**

Diazotized 2-aminoacetophenones are also cyclized to cinnolin-4(1*H*)-ones (Borsche synthesis). This is a general method for the preparation of compounds with alkyl and aryl groups or other substituents in the benzene ring of the bicyclic system. A variant of this method, limited by the availability of the corresponding *o*-aminocarbonyl compounds, consists of coupling of the diazotized *o*-aminobenzaldehyde or *o*-aminoacetophenone to nitromethane, the resulting hydrazone (**151**) then being cyclized in the presence of aluminum oxide, sodium hydroxide or an anion exchange resin. This reaction is known as the Baumgarten method ⟨58JA1977⟩ and is outlined in Scheme 73 together with the Borsche synthesis.

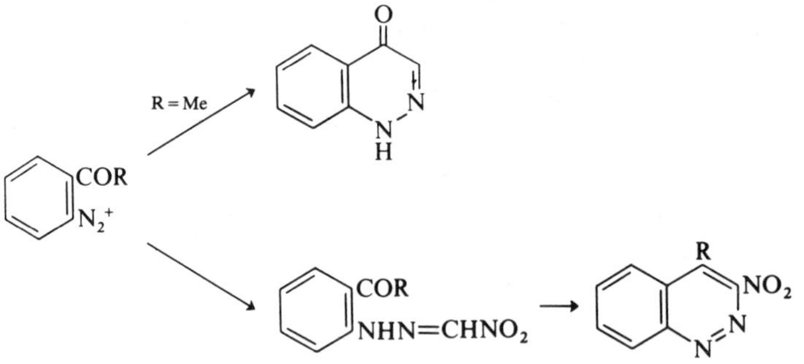

**(151)**

**Scheme 73**

### 2.12.6.3.2 From [4+2] atom fragments

This type of reaction includes the most important and versatile synthetic approaches to numerous pyridazines and phthalazines. They can be divided into two general groups, the first being condensation between a dicarbonyl compound and a hydrazine, and the second [4+2] cycloaddition reactions of the Diels–Alder type. There are only a few examples of pyridazine ring formation from a diamino compound and a dihalo-substituted organic molecule. In this way tetrahydrophthalazines are obtainable (Scheme 74) ⟨58CB1982, 62CB2012, 63JA2144⟩.

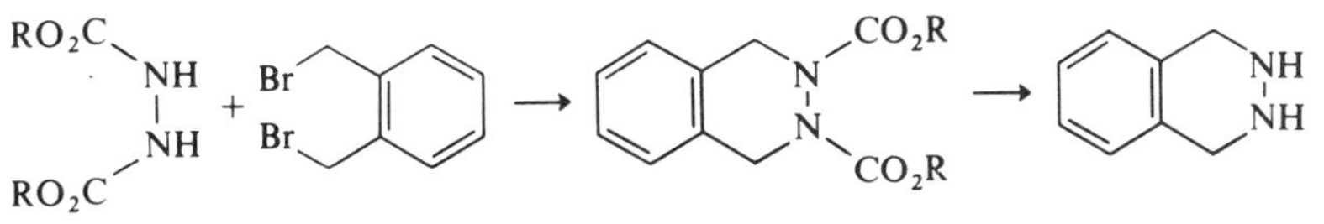

**Scheme 74**

A widely used synthetic approach for pyridazines or phthalazines is the reaction between hydrazine or a substituted hydrazine and 1,4-dicarbonyl compounds, such as 1,4-diketones, 1,4-keto aldehydes, 1,4-dialdehydes, 1,4-keto acids or derivatives, and 1,4-dicarboxylic acids and derivatives, in particular their anhydrides. In general, the reaction may proceed in a single step or *via* the corresponding intermediate hydrazones. The reaction products are usually the dihydro derivatives in the pyridazine series and these can easily be dehydrogenated to the corresponding pyridazines or pyridazinones. Most frequently bromine in glacial acetic is used as the oxidizing agent and successful dehydrogenation is dependent upon the substituents present in the ring. Oxidation with $PtO_2$ is also very efficient ⟨79JA766⟩. When unsaturated 1,4-dicarbonyl compounds are used, pyridazines are obtained directly and the *cis*-isomers react more readily. Some examples are given in Scheme 75.

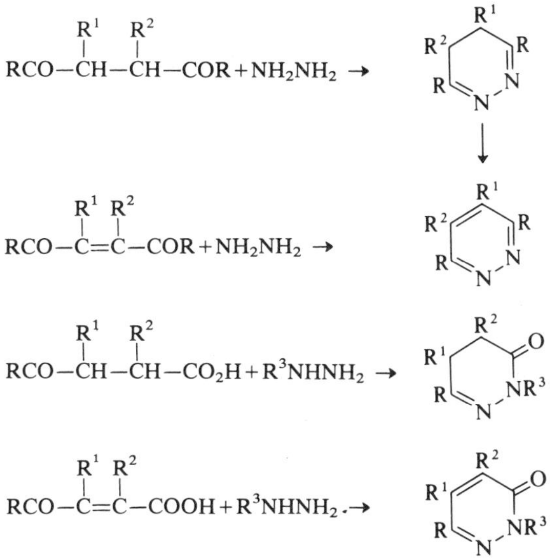

**Scheme 75**

It is customary to perform the condensation of 1,4-dicarbonyl compounds with hydrazines in the presence of mineral acid to avoid the formation of *N*-aminopyrroles. Contrary to early claims that 4,5-dihydropyridazines are formed ⟨07CB4598⟩, these compounds are now regarded as 1,4-dihydro derivatives ⟨81CB564⟩.

Recently, various monosaccharides have been employed as a source of 1,4-dicarbonyl compounds for pyridazine syntheses ⟨70CB1846, 74CPB1732, 75JHC957⟩. Moreover, diacylcyclopropanes have been also reported to give pyridazines; the bicyclic products are transformed by $P_4S_{10}$ into the corresponding 3-thioxopyridazines ⟨80JHC541⟩.

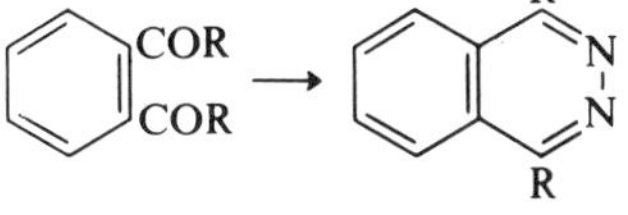

**Scheme 76**

In a similar manner, phthalazine or its alkyl- or aryl-substituted derivatives are obtainable from 1,2-diacylarenes (Scheme 76). Phthalaldehydic acid and its analogs are transformed by hydrazines into the corresponding phthalazin-1(2*H*)-ones. Phthalazin-1(2*H*)-one itself is prepared from naphthalene by oxidation, subsequent treatment with hydrazine and decarboxylation as shown in Scheme 77 ⟨55YZ1423, 64FRP1335759⟩. 4-Substituted phthalazin-1(2*H*)-ones are prepared in a similar way from 2-acylbenzoic acids. 3-Hydroxyphthalides,

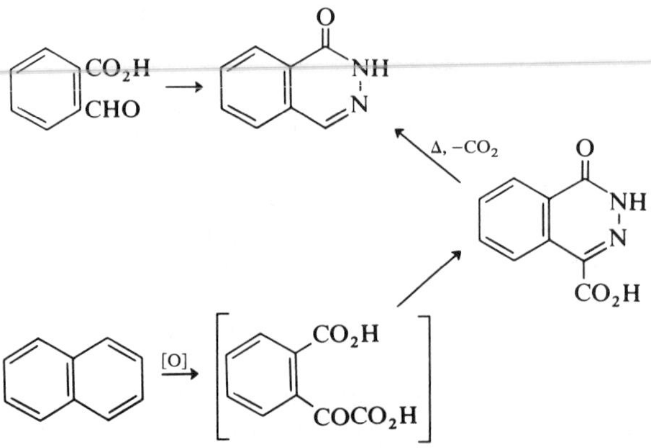

**Scheme 77**

3-alkylidenephthalides, phthalamidine and 3-hydroxyphthalamidine can also be converted into the corresponding phthalazin-1(2*H*)-ones (Scheme 78). 1,2-Dicarboxylic acids and their derivatives are also widely used for the preparation of pyridazines. A great variety

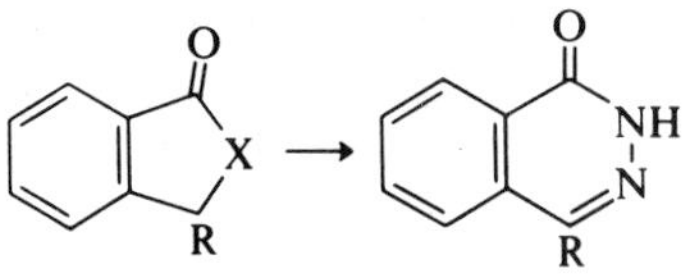

X = O, NH
R = H, OH, =CHR'

**Scheme 78**

of substituted pyridazines have been prepared from maleic anhydride and its substituted analogs by reactions with hydrazine or mono- or di-substituted hydrazines. Suitable reaction conditions can give either the corresponding pyridazinones directly or the intermediate 3-carboxyacryloylhydrazines which are cyclized thermally in a subsequent step. Hydrazines with a strong electron-donor group form pyridazines directly. If 2-carboxyacryloylhydrazines are dehydrated in acidic media, *N*-aminomaleimides (**152**) or pyridazinones (**153**) are formed. The former are isomerized under the influence of acid to pyridazinones and the formation of aminomaleimides can be suppressed if the condensation is performed in strongly acidic solution (Scheme 79).

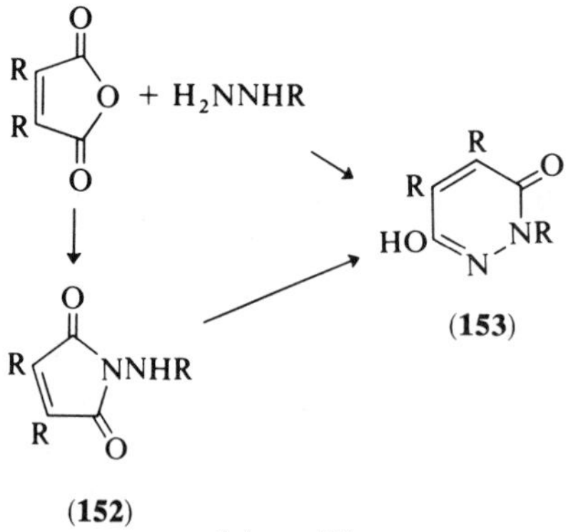

**Scheme 79**

With the saturated analogs, *i.e.* succinic anhydride and its derivatives, pyridazines are formed in only a few cases. The reaction has been applied to the preparation of perhydro-pyridazines and their 3,6-diones ⟨68MI21200, 70JOC1468⟩. For the synthesis of 4,5-dihalopyridazinones, β-formylacrylic acids, for example mucochloric acid, are useful synthons (Scheme 80).

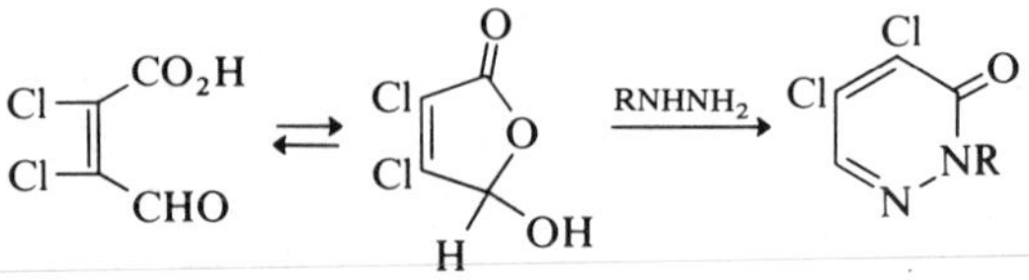

**Scheme 80**

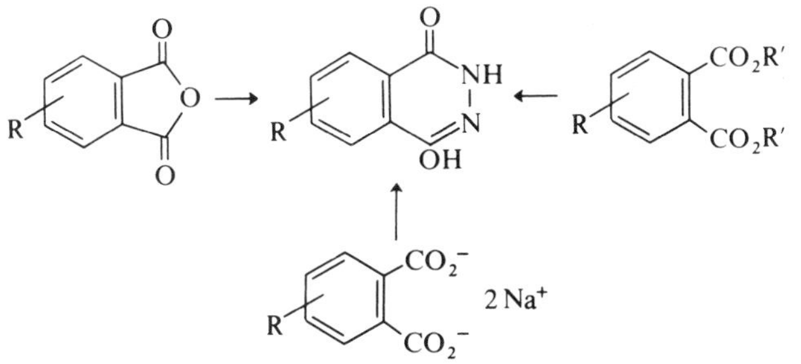

**Scheme 81**

Phthalic anhydride and diethyl phthalate are easily converted with hydrazine into 4-hydroxyphthalazin-1(2*H*)-one. Its substituted derivatives have been prepared using substituted hydrazines, substituted phthalic anhydrides, or diesters or disodium salts of substituted phthalic acids (Scheme 81). However, condensation of phenylhydrazine with phthalic anhydride gives only a small amount of the corresponding phthalazine, the main product being 2-anilinophthalimide. This can be rearranged in the presence of base into the phthalazine derivative. For the preparation of 2,3-disubstituted derivatives, 1,2-disubstituted hydrazines are reacted with the appropriate phthalic anhydrides or phthaloyl chlorides. Derivatives of 4-amino- or 4-hydrazino-phthalazin-1(2*H*)-one have been prepared either from the corresponding monothiophthalimide and 3-aminoisoindolin-3-one (**154**) or from ethyl 2-cyanobenzoate (**155**) and hydrazine hydrate (Scheme 82). Similarly,

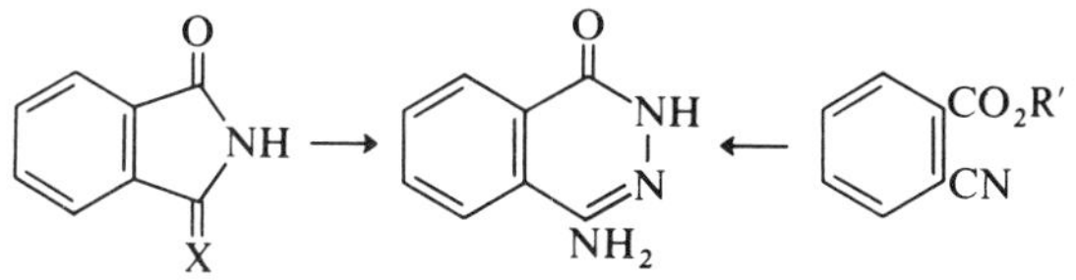

**(154)** X = S, NH　　　　　　　　　　　　　　　　**(155)**

phthalodinitrile is converted by methanolic hydrazine into 1,4-diaminophthalazine (**156**) ⟨68JHC111⟩, whereas with excess of hydrazine in dioxane–acetic acid 1,4-dihydrazinophthalazine (**157**) is obtained, the reaction taking place *via* the 1,4-diamino compound (**156**; Scheme 83) ⟨59JOC1205⟩.

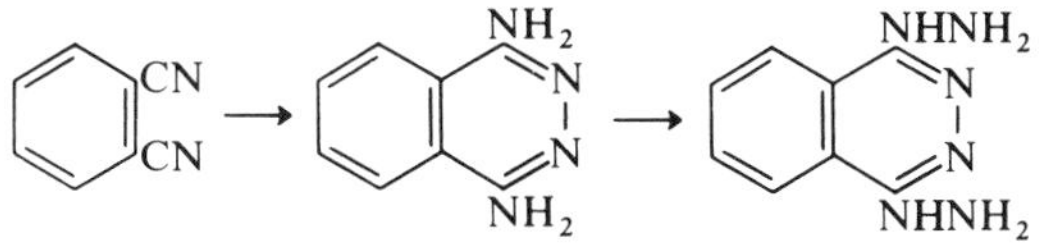

**(156)**　　　　　　**(157)**

**Scheme 83**

As masked dicarbonyl compounds, enaminoketones (**158**) or vinylamidinium salts (**159**; **160**) can be employed (Scheme 84) ⟨79S385⟩. Similarly, 3-phenacylpyridinium methiodides

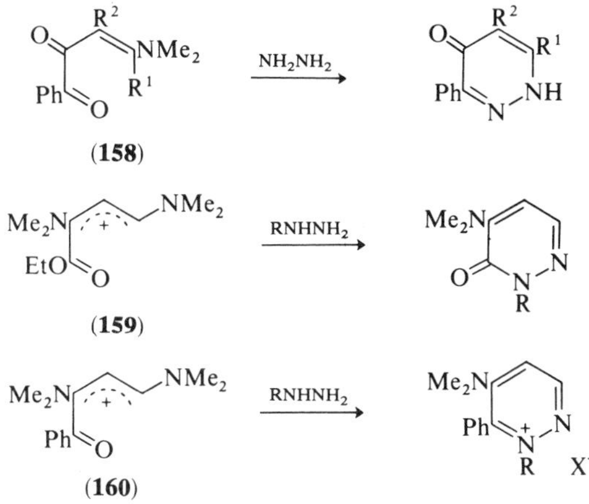

**(158)**

**(159)**

**(160)**

**Scheme 84**

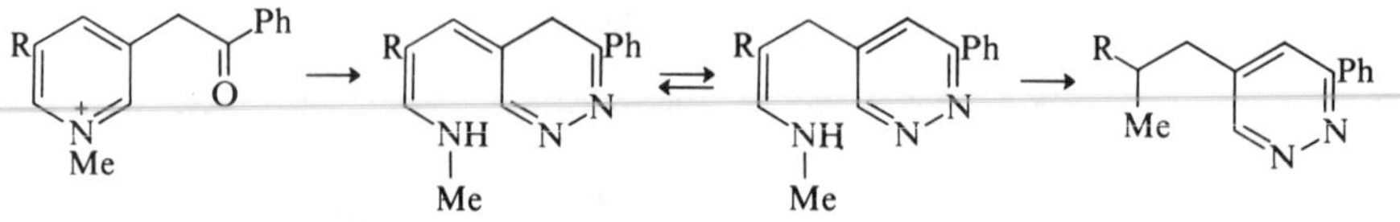

(161)

**Scheme 85**

(161) can be converted into pyridazines as shown in Scheme 85 ⟨80JCS(P1)72⟩. The reaction can be applied also to more complex systems, *i.e.* polycyclic pyridines, and is interpreted as proceeding by ring opening and recyclization, aromatization and Wolff-Kishner reduction.

For the synthesis of 4-nitropyridazines a new approach has been developed in which the hydrazide of nitrothiolacetic acid or the corresponding amidrazone (162) reacts with glyoxal or methylglyoxal (Scheme 86) ⟨77TL3619⟩.

(162)

**Scheme 86**

Numerous reduced cinnolines (163) are prepared from cyclohexanones substituted in the 2-position with a suitable functional group, such as acetonyl, glyoxyl, hydroxyacetonyl, carboxymethyl, ethoxycarbonylmethyl, malonyl, *etc.* These compounds react with hydrazines and aromatization of the pyridazine ring may proceed spontaneously during the reaction or the partially reduced pyridazine rings can be aromatized with common dehydrogenation agents. The various types of starting material and the reaction course are

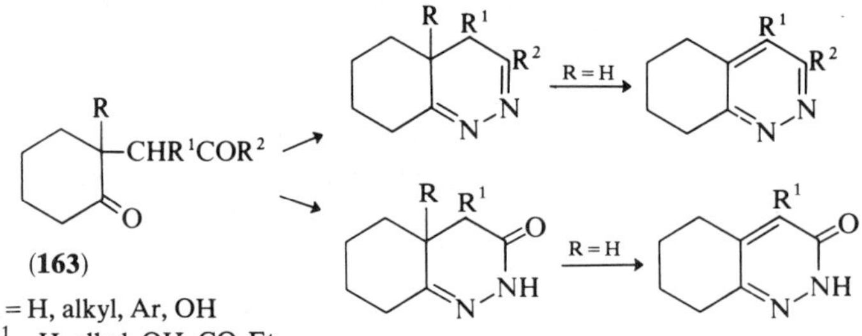

(163)

R = H, alkyl, Ar, OH
$R^1$ = H, alkyl, OH, $CO_2Et$
$R^2$ = H, alkyl, Ar, OH, OEt

**Scheme 87**

presented in Scheme 87. 1-Hydroxy-2-oxocyclohexylacetic acids (164) give first the corresponding γ-lactones (165) and these are subsequently transformed into reduced cinnolines (Scheme 88).

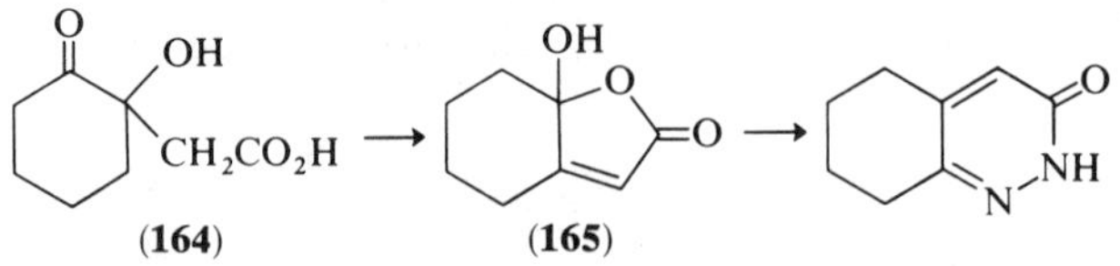

(164)                    (165)

**Scheme 88**

A large number of pyridazines are synthetically available from [4+2] cycloaddition reactions. In one general method, azo or diazo compounds are used as dienophiles, and a second approach is based on the reaction between 1,2,4,5-tetrazines and various unsaturated compounds. The most useful azo dienophile is a dialkyl azodicarboxylate which reacts with appropriate dienes to give reduced pyridazines and cinnolines (Scheme 89). With highly substituted dienes the normal cycloaddition reaction is prevented, and, if the ethylenic group in styrenes is substituted with aryl groups, indoles are formed preferentially. The cycloadduct with 2,3-pentadienal acetal is a tetrahydropyridazine derivative which has been used for the preparation of 2,5-diamino-2,5-dideoxyribose ⟨80LA1307⟩.

Various azo compounds have been introduced recently for these cycloadditions. Azoalkenes react with alkenes to give in some cases a mixture of regioisomeric products

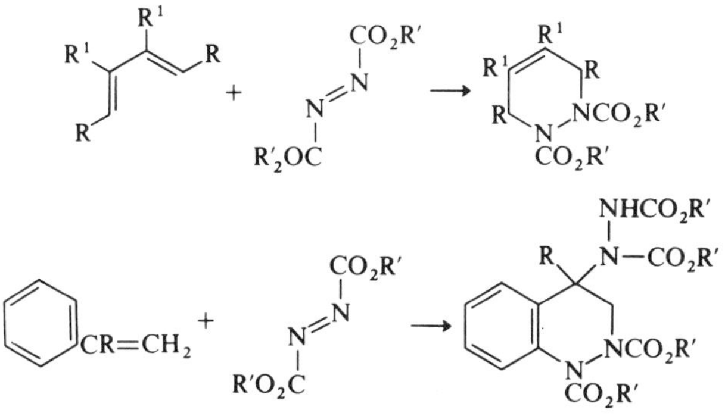

**Scheme 89**

(**166, 167**), as shown in Scheme 90 ⟨77TL117, 77ZN(B)72⟩. Diimine generated *in situ* has been used for the formation of pyridazines on a polymer support containing a diene group.

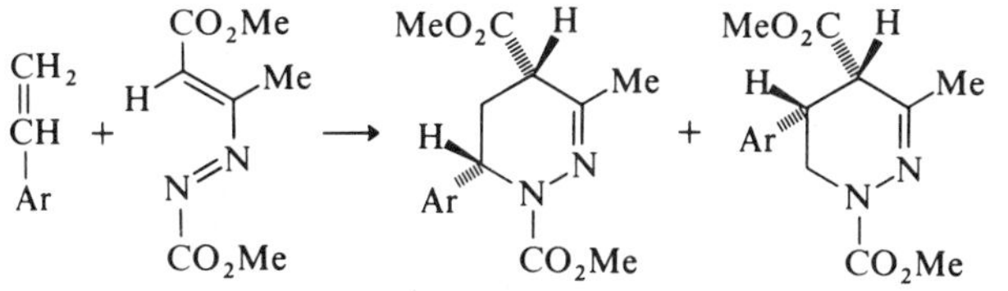

(**166**)      (**167**)

**Scheme 90**

3-Hydroxyhexahydropyridazine was freed from the polymer (**168**) by hydrolysis (Scheme 91) ⟨79TL1333⟩. Cyclic azo compounds may also be employed; for example with 1,2,4-triazole-3,5-dione, adducts (**169, 170**) are formed first and these are subsequently transfor-

(**168**)

**Scheme 91**

med by hydrolysis and oxidation into the corresponding reduced pyridazines (Scheme 92) ⟨76JA1875⟩.

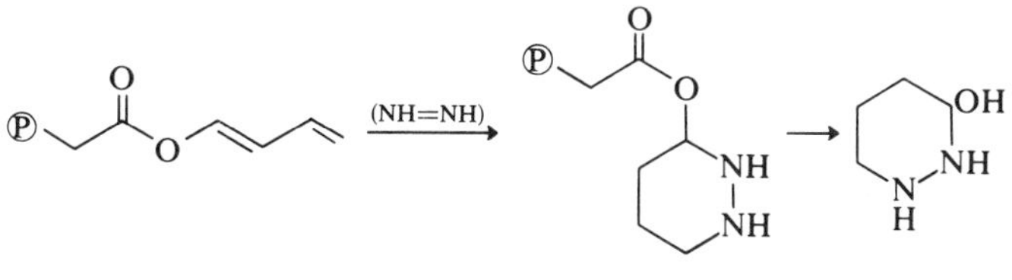

(**169**)      (**170**)

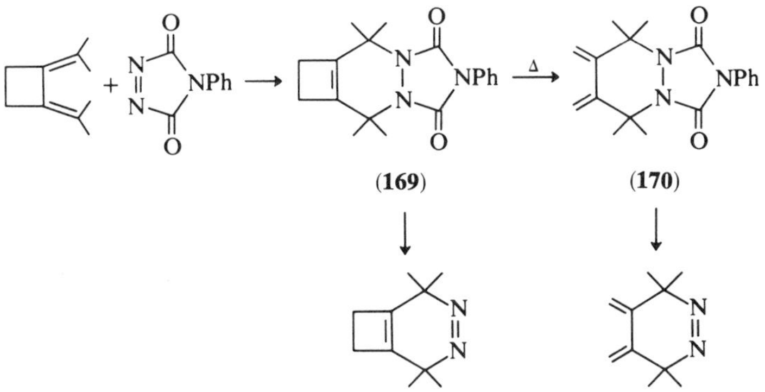

**Scheme 92**

Phthalazinedione has been used in a similar manner for the synthesis of (3*S*,5*S*)-5-hydroxyhexahydropyridazine-3-carboxylic acid (**171**), a naturally occurring amino acid which is a constituent of the monamycin antibiotics (Scheme 93) ⟨77H(7)119⟩.

Although the most general cycloaddition reaction of diazo compounds is that they react as 1,3-dipoles, recently some reactions have been reported in which they react as 1,2-dipoles,

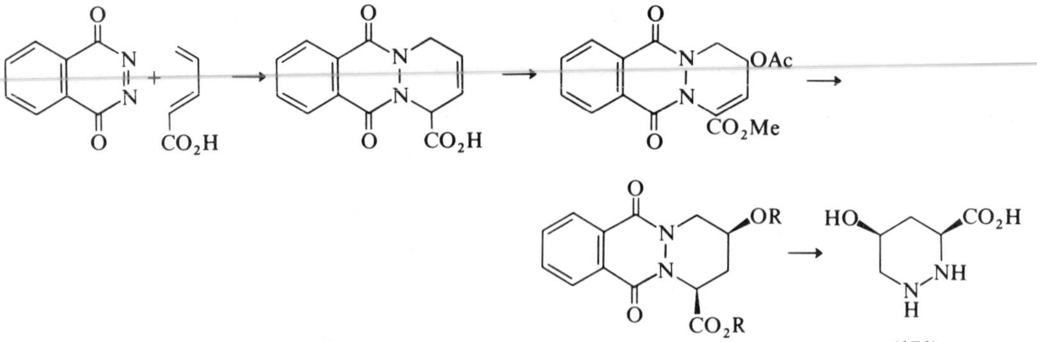

**Scheme 93**

and with dienes the corresponding pyridazines are formed as shown in Scheme 94 ⟨77H(6)681, 75JA5291⟩. Arenediazo cyanides react with dienes in a similar manner to form pyridazines (Scheme 95) ⟨79CC1019⟩.

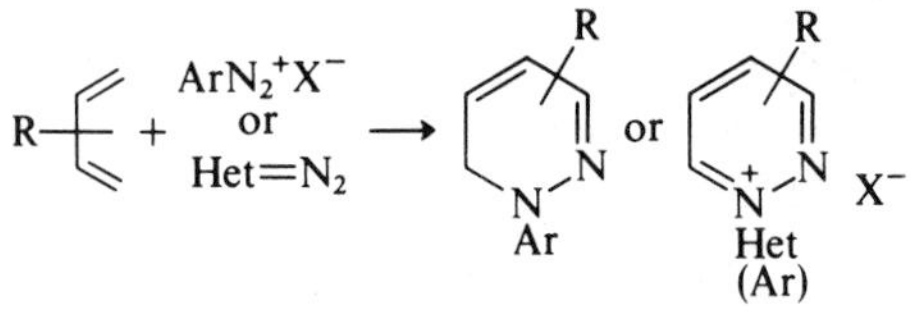

**Scheme 94**

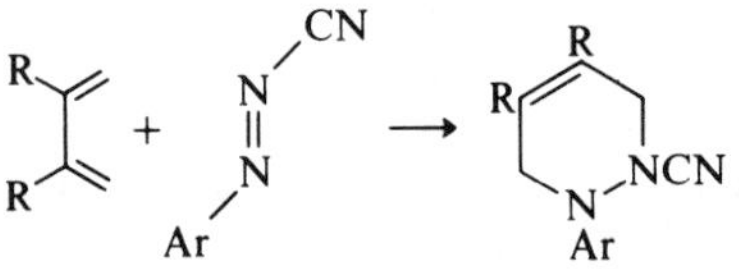

**Scheme 95**

In 1959 Carboni and Lindsay first reported the cycloaddition reaction between 1,2,4,5-tetrazines and alkynes or alkenes ⟨59JA4342⟩ and this reaction type has become a useful synthetic approach to pyridazines. In general, the reaction proceeds between 1,2,4,5-tetrazines with strongly electrophilic substituents at positions 3 and 6 (alkoxycarbonyl, carboxamido, trifluoromethyl, aryl, heteroaryl, *etc.*) and a variety of alkenes and alkynes, enol ethers, ketene acetals, enol esters, enamines ⟨78HC(33)1073⟩ or even with aldehydes and ketones ⟨79JOC629⟩. With alkenes 1,4-dihydropyridazines (**172**) are first formed, which in most cases are not isolated but are oxidized further to pyridazines (**173**). These are obtained directly from alkynes which are, however, less reactive in these cycloaddition reactions. In general, the overall reaction which is presented in Scheme 96 is strongly

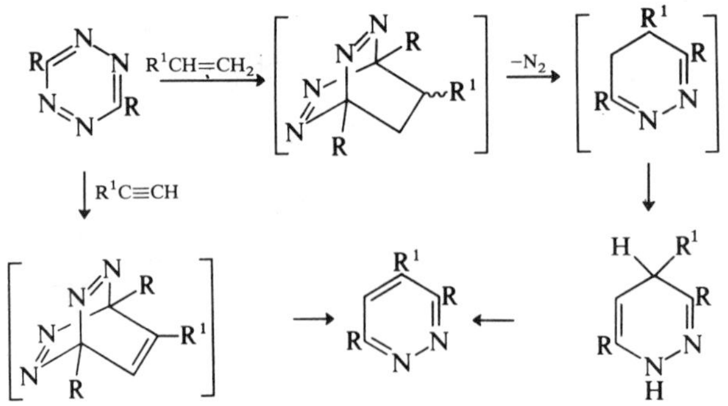

**Scheme 96**

dependent on the electronic effects of the substituents present. Recently, the reaction has been extended for the introduction of trifluoromethyl groups in the pyridazine ring ⟨78JCS(P1)378⟩, or with silylalkynes a trimethylsilyl group can be introduced ⟨81CB3154⟩. With fulvenes, cyclopenta[*d*]pyridazines (**174**) are formed ⟨78T2509⟩; with benzyne, diethyl phthalazine-1,4-dicarboxylate was obtained ⟨66TL4979⟩; and with dienamines, substituted bispyridazinylmethanes (**175**) are formed ⟨79LA675⟩. These transformations are shown in Scheme 97.

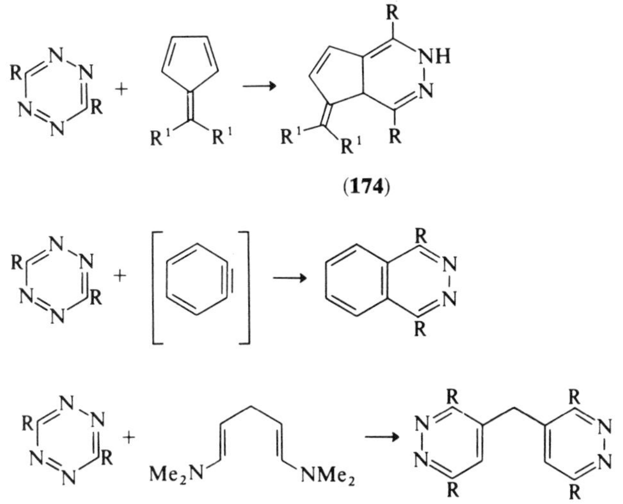

**Scheme 97**

### 2.12.6.3.3 From [3 + 3] fragments

This synthetic appproach has been used in a few cases for the preparation of pyridazines from diazo compounds and cyclopropenes. In general, cycloadducts (**176**) are formed first and these rearrange in the presence of acid or alkali to pyridazines (Scheme 98) ⟨69TL2659, 76H(5)401⟩. Tetrachlorocyclopropene reacts similarly and it was found that the stability of the bicyclic intermediates is mainly dependent on substitution ⟨78JCR(S)40, 78JCR(M)0582⟩.

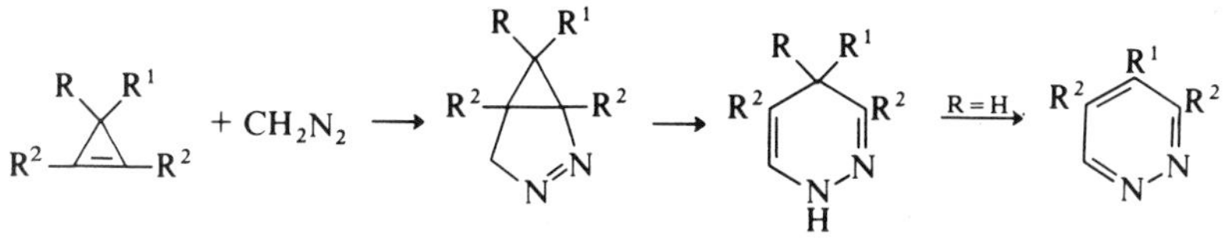

**Scheme 98**

With unsymmetrically substituted cyclopropenes, isomeric cycloadducts (**177**) and (**178**) and pyridazines are formed (Scheme 99) ⟨80LA590⟩.

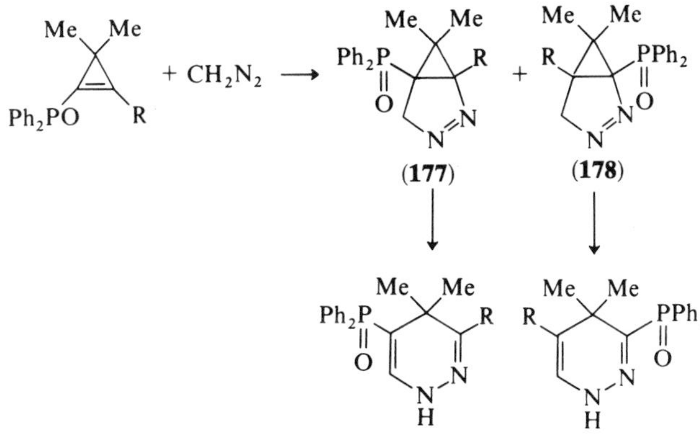

**Scheme 99**

### 2.12.6.4 Formation of Three or Four Bonds

Pyridazines are available also from 1,2-dicarbonyl compounds in a synthetic principle developed by Schmidt and Druey ⟨54HCA134, 54HCA1467⟩. There are several variants and the most simple is a one-pot condensation of three components: a 1,2-dicarbonyl compound, an ester with a reactive α-methylene group and a hydrazine, at the most monosubstituted. However, the preferred synthetic approach is first to prepare the monohydrazone of a 1,2-dicarbonyl compound or of an α-ketocarboxylic ester (**179**), and then to treat with an ester containing a reactive α-methylene group. In a recent extension of this procedure, malonic ester ⟨79MI21202⟩ or malononitrile ⟨79JPR71⟩ has been used. Another variant consists of the preparation of an acid hydrazide (**180**) which is subsequently condensed with the appropriate 1,2-dicarbonyl compound. In the presence of sodium ethoxide cyclization occurs and the pyridazinone is obtained directly, whereas in the absence of base the hydrazone (**181**) is formed. This can be isolated and cyclized in a separate step. All these reaction types are used for the syntheses of substituted pyridazin-3(2*H*)-ones. From arylglyoxal hydrazones and malononitrile, however, 3-amino-4-cyanopyridazines (**182**) are formed. These transformations are presented in Scheme 100. Detailed investigations of the reaction between benzoin and hydrazine have shown that, besides 3,4,5,6-tetraphenylpyridazine, a complex mixture of compounds is formed ⟨71JCS(C)2807⟩.

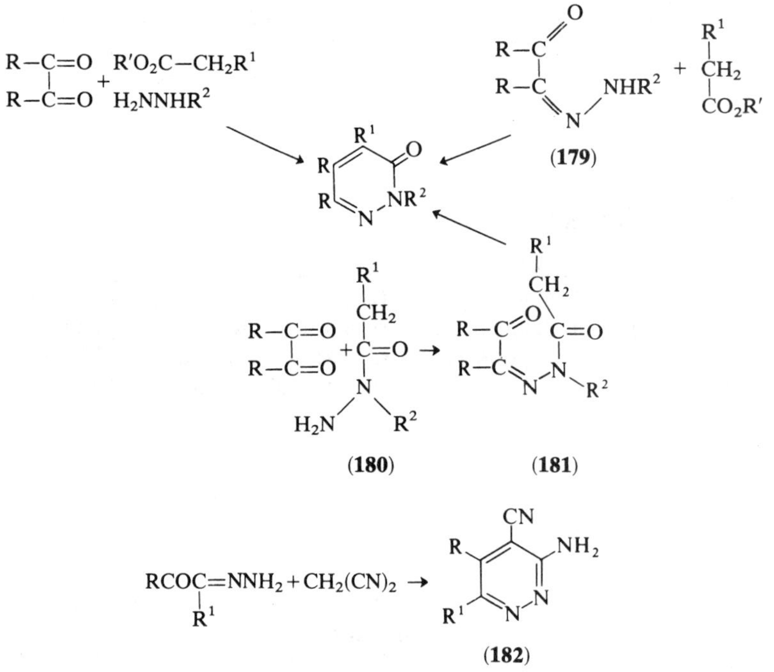

**Scheme 100**

### 2.12.6.5 From Other Heterocycles

There are several examples of the formation of pyridazines from other heterocycles, such as azirines, furans, pyrroles, isoxazoles, pyrazoles or pyrans and by ring contraction of 1,2-diazepines. Their formation is mentioned in Section 2.12.6.3.2.

In a titanium tetrachloride-induced cleavage of arylazirines at low temperature, arylpyridazines are formed ⟨77JA4330⟩. Furans can serve as useful precursors for unsaturated 1,4-dicarbonyl compounds. These are obtained from the dialkoxy-2,5-dihydrofurans (**183**) which are formed through addition of bromine and an alcohol, usually methanol, to the corresponding furans. Subsequent hydrolytic ring opening of the furan ring and reaction with hydrazine gives the corresponding pyridazines. Steric factors have a strong influence and polyalkylpyridazines are not formed according to this reaction sequence. An example of this type of synthesis is given in Scheme 101. There are also few examples of direct reaction between furans and hydrazine to give pyridazines. It is most likely that addition of hydrazine to the furan ring is followed by ring opening and recyclization ⟨77JHC75,

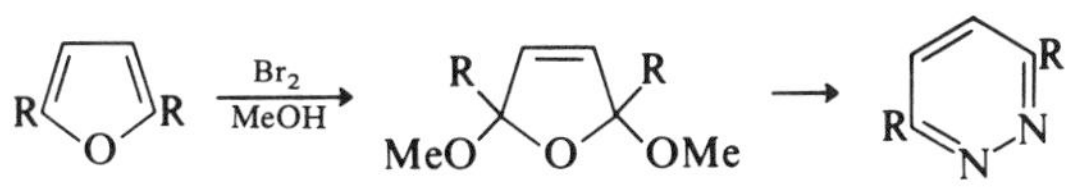

(183)

**Scheme 101**

69AK(30)261⟩. 2-Acetoxyfuran-3(2*H*)-ones react with hydrazine to give 3,6-disubstituted-4-ethoxycarbonylpyridazin-4(1*H*)-ones (**184**) as the main product, but with mono-substituted hydrazines in addition to these pyridazines anhydro-5-hydroxypyridazinium hydroxide (**185**) derivatives and some pyrazole derivatives are also formed (Scheme 102) ⟨79JOC3053⟩. The

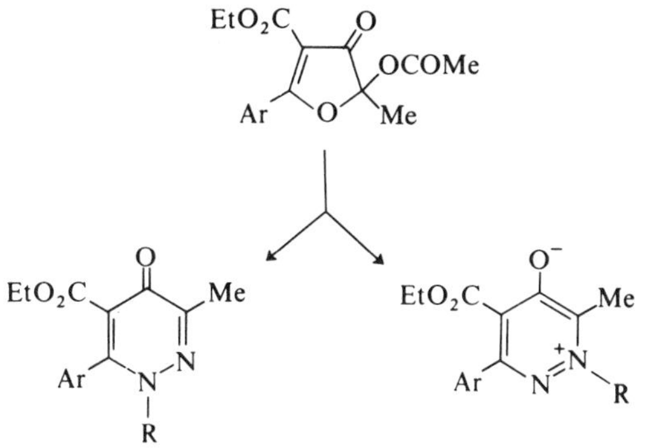

(184)                    (185)

**Scheme 102**

formation of these compounds is interpreted as occurring by a competition in the ring-opening and ring-closure steps and to depend on the relative nucleophilicities of the hydrazine nitrogens. 1,4,5,6-Tetrahydropyridazines are formed by ring enlargement from 1-aminopyrrolidines in the presence of silica gel and chloroform. It is postulated that the reaction involves oxidation on the silica gel surface to give the intermediate diazenes (**186**)

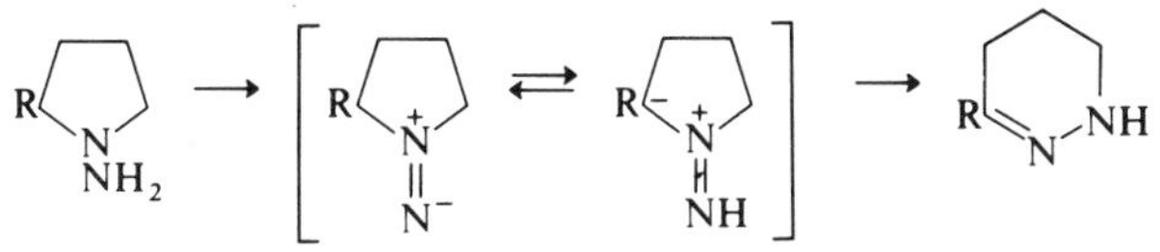

(186)

**Scheme 103**

which then rearrange (Scheme 103) ⟨79TL5025⟩. Similarly, a derivative of 2-aminoisoindolone is rearranged in a base-catalyzed reaction into a pyridazin-4(1*H*)-one derivative (Scheme 104) ⟨79TL2921⟩.

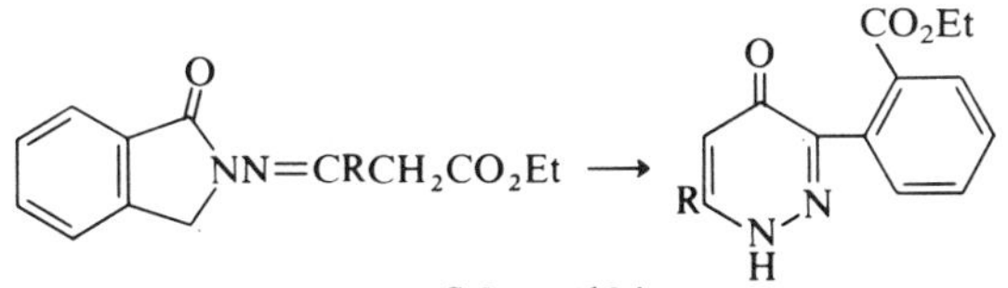

**Scheme 104**

In a similar manner to the formation of pyridazines from *N*-aminopyrroles, cinnolines or phthalazines are obtainable from the corresponding 1-aminooxindoles or 2-aminophthalimides. If the relatively inaccessible 1-aminooxindoles are treated with lead tetraacetate, mercuric acetate, *t*-butyl hypochlorite ⟨69JCS(C)772⟩ or other agents, cinnolones are formed as shown in Scheme 105. The reaction was postulated to proceed *via* an intermediate

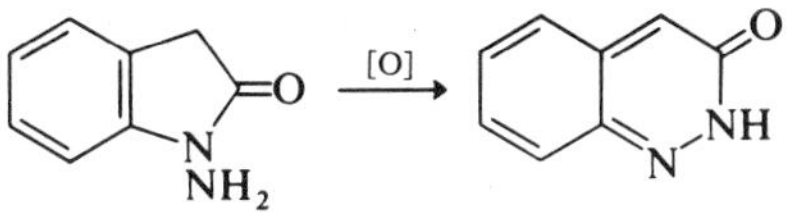

**Scheme 105**

nitrene or nitrenoid species ⟨60JA3977⟩, but later such an intermediate was questioned ⟨69JHC333⟩. 2-Aminophthalimide when heated with dilute alkali or acid is rearranged almost quantitatively into 4-hydroxyphthalazin-1(2*H*)-one (Scheme 106). Similarly, 2-anilino-phthalimide gives 2-phenyl-4-hydroxyphthalazin-1(2*H*)-one ⟨55JCS852⟩.

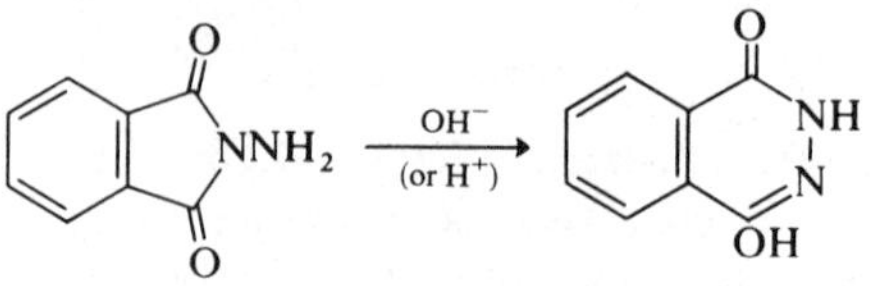

**Scheme 106**

Pyridazines are formed from pyrones or their thioxo analogs or from appropriate pyridones. Pyrones or pyridones react with diazonium salts to give the corresponding hydrazones (**187**) and (**188**) which are rearranged under the influence of acid or base into pyridazinones as shown in Scheme 107. On the other hand, kojic acid is transformed with hydrazine into a 1,4-dihydropyridazine and a pyrazole derivative. 4*H*-Pyran-4-thiones

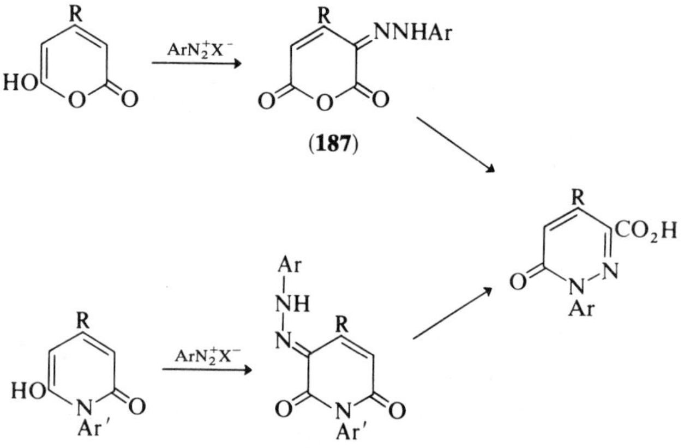

**Scheme 107**

with a β-hydroxy group are also transformed with hydrazine into a mixture of pyridazines (**189**), pyrazoles (**190**) and 1-aminopyridine-4(1*H*)-thiones (**191**). Depending upon ring substituents, the attack of hydrazine occurs either at position 2 or position 6. It appears

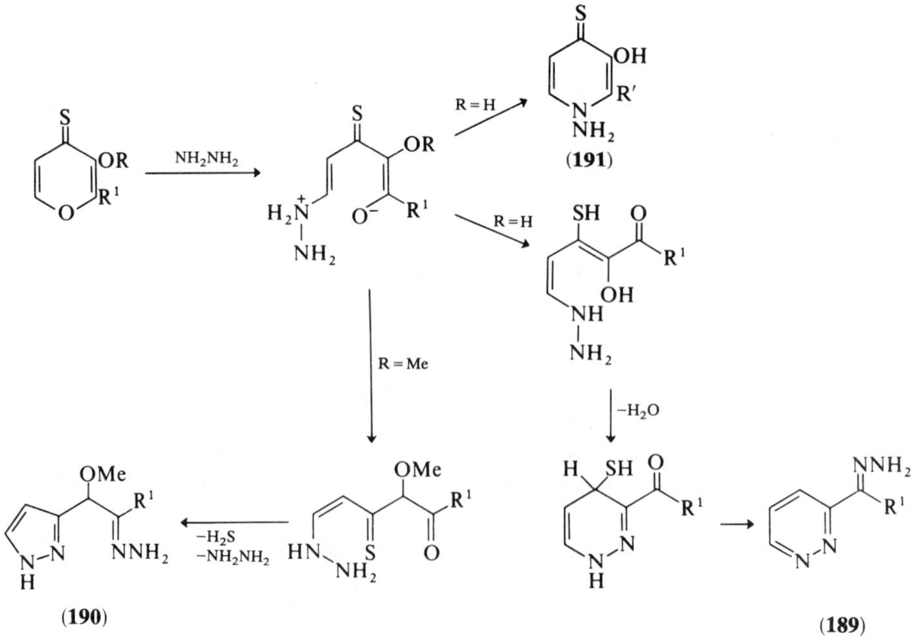

**Scheme 108**

that the presence of a $\beta$-hydroxy group is a prerequisite for pyridazine ring formation since the corresponding $\beta$-methoxy compounds were transformed exclusively into the corresponding pyrazoles (Scheme 108) ⟨78BCJ179⟩.

Some 1,2-diazepines are also transformed into pyridazines. Upon bromination they are transformed into diazanorcaradiene derivatives (**192**), which upon partial saponification and heating at 130–140 °C in the absence of solvent give the corresponding pyridazines. Alternatively, dehalogenation with zinc in boiling ethanol gives diarylpyridazines (**193**) as the main products, together with the dihydrodiazepine (**194**; Scheme 109) ⟨77BCJ2153⟩.

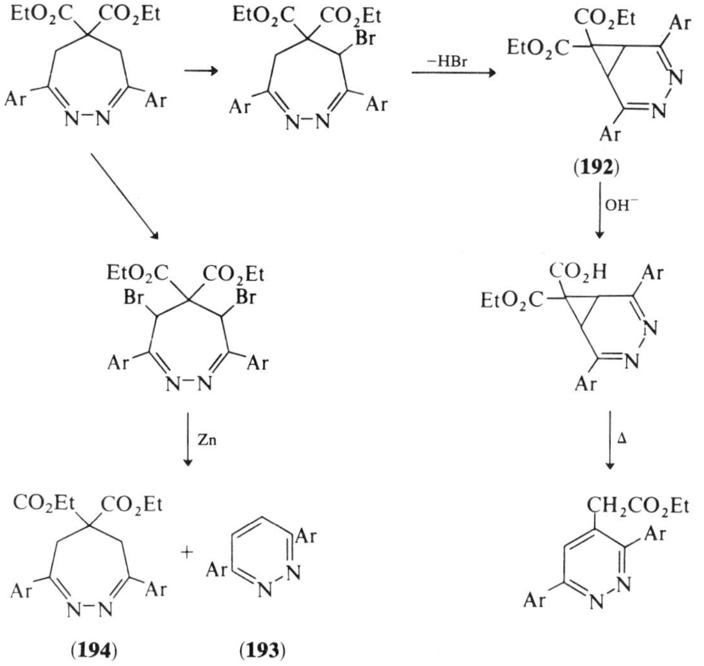

**Scheme 109**

### 2.12.6.6 Best Methods for the Synthesis of the Parent Compounds and Some Important Derivatives

The most useful syntheses of pyridazines and their alkyl and other derivatives begins with the reaction between maleic anhydride and hydrazine to give maleic hydrazide. This is further transformed into 3,6-dichloropyridazine which is amenable to nucleophilic substitution of one or both halogen atoms; alternatively, the halogen(s) can be replaced by hydrogen as shown in Scheme 110. In this manner a great number of pyridazine derivatives are prepared.

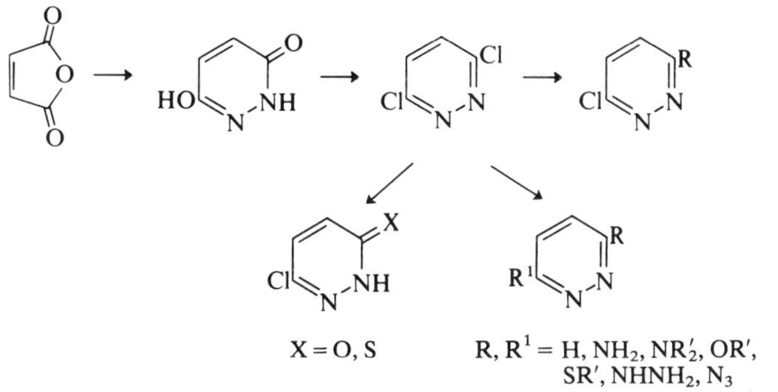

X = O, S          R, R$^1$ = H, NH$_2$, NR$_2'$, OR',
                            SR', NHNH$_2$, N$_3$

**Scheme 110**

Pyridazine itself is best prepared (in about 60–67% yield) from 2,5-diacetoxy- or 2,5-dimethoxy-2,5-dihydrofuran ⟨50ACS1233, 56JOC764⟩ or by hydrodehalogenation of 3-chloro-

or 3,6-dichloro-pyridazine in the presence of palladized charcoal ⟨51JA1873, 54YZ1195⟩. Pyridazines with a side chain forming a C—C bond with a ring carbon atom are usually prepared from methylpyridazines or other alkyl derivatives as shown in Scheme 111.

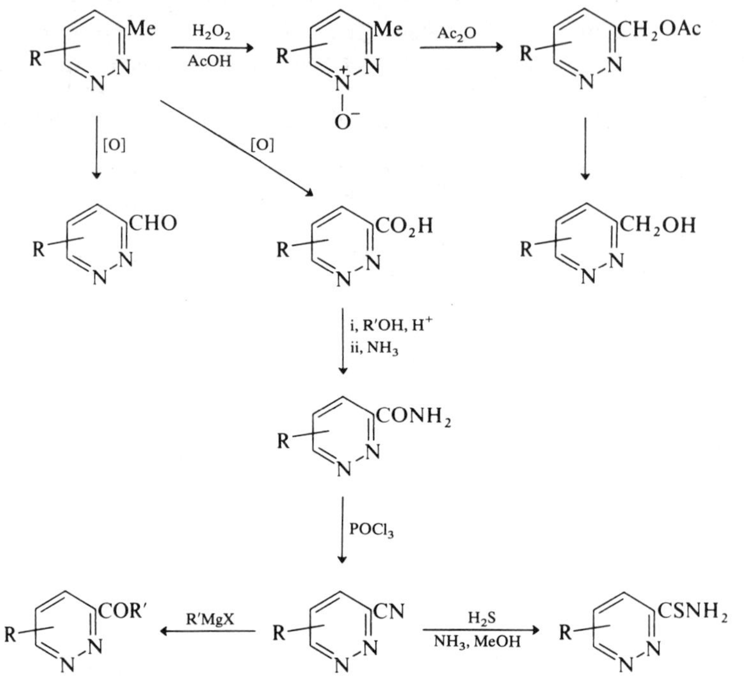

**Scheme 111**

Alkyl- or aryl-cinnolines substituted in the benzene ring are best prepared by the Widman–Stoermer synthesis or by the Borsche synthesis, the latter leading to 4-cinnolinones. From these, 4-chlorocinnolines are prepared in the usual way and further nucleophilic substitution gives the corresponding cinnolines with an alkoxy, aryloxy, amino, thio, alkylthio or other groups. The best method for the preparation of cinnoline is thermal decarboxylation of cinnoline-4-carboxylic acid; the parent compound is obtained in about 70% yield together with some 4,4'-bicinnoline ⟨51JCS1971⟩.

4-Hydroxyphthalazin-1(2H)-one is obtained in a smooth reaction between phthalic anhydride and hydrazine hydrate and this is again the starting compound for many 1-substituted and/or 1,4-disubstituted phthalazines. The transformations of 1,4-dichloro-phthalazine, which is prepared in the usual manner, follow a similar pattern as shown for pyridazines in Scheme 110. On the other hand, phthalonitrile is the preferential starting compound for amino- and hydrazino-phthalazines. The most satisfactory synthesis of phthalazine is the reaction between $\alpha,\alpha,\alpha',\alpha'$-tetrachloro-o-xylene and hydrazine sulfate in sulfuric acid ⟨67FRP1438827⟩, although catalytic dehalogenation of 1-chloro- or 1,4-dichloro-phthalazine or oxidation of 1-hydrazinophthalazine also provides the parent compound in moderate yield.

## 2.12.7 APPLICATIONS

Many pyridazines, cinnolines and phthalazines exhibit biological activity and some are used as drugs ⟨B-78MI21201⟩. Only a few pyridazine derivatives have been found in nature. From *Streptomyces jamaicensis* antibacterial monamycins have been isolated; these are 18-membered ring cyclohexadepsipeptides, containing as a structural unit hexahy-dropyridazine-3-carboxylic acid or its derivatives ⟨71JCS(C)526⟩. Pyridazines influence plant growth and maleic hydrazide has been used extensively as a sucker-inhibiting agent in tobacco fields in the USA. Many halopyridazines have found use as herbicides or fungicides, in particular those with thiophosphate or dithiophosphate groups.

In the cinnoline series only derivatives of 3- or 4-aminocinnoline have been found to exhibit biological activity. Some hydrazinophthalazines, in particular 1-hydrazino-(Hydralazin) and 1,4-dihydrazino-phthalazine (Dihydralazin), are excellent hypotensive and antihypertensive agents.

# 2.13

# Pyrimidines and their Benzo Derivatives

D. J. BROWN

*Australian National University*

| | | |
|---|---|---|
| 2.13.1 | STRUCTURE OF PYRIMIDINES | 58 |
| 2.13.1.1 | The Geometry of Pyrimidines | 58 |
| 2.13.1.2 | The Ionization of Pyrimidines | 59 |
| 2.13.1.2.1 | The $pK_a$ of pyrimidines | 60 |
| 2.13.1.2.2 | The acidic $pK_a$ values of pyrimidines | 60 |
| 2.13.1.2.3 | The basic $pK_a$ values of pyrimidinamines | 60 |
| 2.13.1.2.4 | The $pK_a$ values of aminopyrimidinones | 61 |
| 2.13.1.2.5 | The $pK_a$ values of quinazolines | 61 |
| 2.13.1.3 | The NMR Spectra of Pyrimidines | 61 |
| 2.13.1.3.1 | Proton NMR spectra | 62 |
| 2.13.1.3.2 | Non-proton NMR spectra | 63 |
| 2.13.1.4 | The IR and Raman Spectra of Pyrimidines | 64 |
| 2.13.1.5 | The UV Spectra of Pyrimidines | 65 |
| 2.13.1.6 | The Mass Spectra of Pyrimidines | 65 |
| 2.13.1.7 | The Stereochemistry of Pyrimidines and Hydropyrimidines | 66 |
| 2.13.1.8 | Tautomerism in Pyrimidines | 66 |
| 2.13.1.8.1 | The structure of pyrimidinones and quinazolinones | 66 |
| 2.13.1.8.2 | The structure of pyrimidinethiones and quinazolinethiones | 67 |
| 2.13.1.8.3 | The structure of pyrimidinamines and quinazolinamines | 67 |
| 2.13.1.8.4 | The structure of pyrimidines with two or more tautomeric groups | 67 |
| 2.13.2 | REACTIVITY OF PYRIMIDINES | 68 |
| 2.13.2.1 | Reactivity of Ring Atoms of Pyrimidines | 68 |
| 2.13.2.1.1 | Electrophilic attack | 68 |
| 2.13.2.1.2 | Nucleophilic attack | 71 |
| 2.13.2.1.3 | Free radical attack | 73 |
| 2.13.2.1.4 | Photochemical reactions | 73 |
| 2.13.2.1.5 | Oxidative and reductive reactions | 74 |
| 2.13.2.2 | Reactivity of Substituents on Pyrimidine | 75 |
| 2.13.2.2.1 | Reactivity of the fused benzene ring in quinazoline | 76 |
| 2.13.2.2.2 | Reactivity of alkyl groups | 76 |
| 2.13.2.2.3 | Reactivity of aryl and heteroaryl substituents | 78 |
| 2.13.2.2.4 | Reactivity of acyl groups | 79 |
| 2.13.2.2.5 | Reactivity of carboxylic acids and derivatives | 80 |
| 2.13.2.2.6 | Reactivity of amino, nitro and related groups | 84 |
| 2.13.2.2.7 | Reactivity of hydroxy, alkoxy and related groups | 89 |
| 2.13.2.2.8 | Reactivity of mercapto, alkylthio and related groups | 93 |
| 2.13.2.2.9 | Reactivity of halogeno substituents | 98 |
| 2.13.2.2.10 | Reactivity of metallo derivatives | 104 |
| 2.13.2.2.11 | Reactivity of N-oxides | 105 |
| 2.13.3 | SYNTHESIS OF PYRIMIDINES | 106 |
| 2.13.3.1 | Primary Synthesis of the Pyrimidine Ring | 106 |
| 2.13.3.1.1 | Primary syntheses involving formation of one bond | 106 |
| 2.13.3.1.2 | Primary syntheses involving formation of two bonds | 107 |
| 2.13.3.1.3 | Primary syntheses involving formation of three bonds | 116 |
| 2.13.3.1.4 | Primary syntheses involving formation of four or more bonds | 118 |
| 2.13.3.2 | Syntheses of the Pyrimidine Ring from Other Heterocycles | 119 |
| 2.13.3.2.1 | Pyrimidines from pyrroles | 119 |
| 2.13.3.2.2 | Pyrimidines from imidazoles | 119 |
| 2.13.3.2.3 | Pyrimidines from isoxazoles and oxazoles | 119 |
| 2.13.3.2.4 | Pyrimidines from pyridines, pyrazines and triazines | 120 |
| 2.13.3.2.5 | Pyrimidines from oxazines and thiazines | 121 |
| 2.13.3.2.6 | Pyrimidines from benzofurans and other O-heterocycles | 121 |

2.13.3.2.7  *Pyrimidines from purines and related fused systems*                                   121
2.13.3.2.8  *Pyrimidines from pteridines and related fused systems*                                122
  2.13.3.3  *Preferred Synthetic Routes to Pyrimidines*                                   123
    2.13.3.3.1  *Synthesis of pyrimidine*                                         123
    2.13.3.3.2  *Synthesis of quinazoline*                                        123
    2.13.3.3.3  *Synthesis of alkyl and aryl derivatives*                         124
    2.13.3.3.4  *Synthesis of acyl derivatives*                                   125
    2.13.3.3.5  *Synthesis of carboxy and related derivatives*                    126
    2.13.3.3.6  *Synthesis of amino, nitro and related derivatives*               129
    2.13.3.3.7  *Synthesis of hydroxy, alkoxy and related derivatives*            132
    2.13.3.3.8  *Synthesis of mercapto, alkylthio and related derivatives*        135
    2.13.3.3.9  *Synthesis of halogeno derivatives*                              139
    2.13.3.3.10 *Synthesis of N-oxides and derivatives*                           141
2.13.4  SOME IMPORTANT PYRIMIDINES                                                                 142
  2.13.4.1  *Naturally Occurring Pyrimidines*                                             142
    2.13.4.1.1  *Uracil [pyrimidine-2,4(1H,3H)-dione]*                            142
    2.13.4.1.2  *Thymine [5-methylpyrimidine-2,4(1H,3H)-dione]*                   143
    2.13.4.1.3  *Vicine and divicine [2,6-diamino-5-hydroxypyrimidin-4(3H)-one]*  143
    2.13.4.1.4  *Convicine and isouramil [6-amino-5-hydroxypyrimidine-2,4(1H,3H)-dione]*  144
    2.13.4.1.5  *Cytosine [4-aminopyrimidin-2(1H)-one]*                           144
    2.13.4.1.6  *5-Methylcytosine*                                               145
    2.13.4.1.7  *5-Hydroxymethylcytosine*                                        145
    2.13.4.1.8  *Orotic acid (2,6-dioxo-1,2,3,6-tetrahydropyrimidine-4-carboxylic acid)*  145
    2.13.4.1.9  *Willardiine*                                                    146
    2.13.4.1.10 *Lathyrine or tingitanin*                                        146
    2.13.4.1.11 *Some pyrimidine antibiotics*                                    147
    2.13.4.1.12 *Tetrodotoxin or tarichatoxin*                                   147
    2.13.4.1.13 *The quinazoline alkaloids*                                      148
    2.13.4.1.14 *Other pyrimidines from nature*                                  149
  2.13.4.2  *Pyrimidine Drugs*                                                            150
    2.13.4.2.1  *The barbiturates*                                               150
    2.13.4.2.2  *The pyrimidine sulfonamides*                                    150
    2.13.4.2.3  *The antimicrobial pyrimidine-2,4-diamines*                      151
    2.13.4.2.4  *Halogenopyrimidines as antitumour agents*                       152
    2.13.4.2.5  *Other pyrimidine drugs*                                         152
  2.13.4.3  *Some Pyrimidines of Veterinary or Agricultural Interest*                     154
    2.13.4.3.1  *Diaveridine*                                                    154
    2.13.4.3.2  *Dimpylate*                                                      154
    2.13.4.3.3  *'Enheptin-P'*                                                   154
    2.13.4.3.4  *Bromacil, isocil, etc.*                                         154
  2.13.4.4  *Trivial Names for Pyrimidines*                                               155

## 2.13.1 STRUCTURE OF PYRIMIDINES

The replacement of two CH units in benzene by nitrogen atoms to give pyrimidine (**1**) naturally results in reduced symmetry, so that all bonds are no longer of the same length and all bond angles no longer equal; however, pyrimidine does retain symmetry about the 2,5-axis so three differing pairs of equal bond lengths and three differing pairs of equal bond angles result. In addition, the nitrogen atoms constitute havens for $\pi$-electrons, which are equally distributed about the ring in benzene, so that the reactivities of the 2-, 4/6- and 5-carbon atoms, as well as of the substituents attached to them, vary considerably (see Sections 2.13.2.1 and 2). The remaining symmetry of pyrimidine, apart from the molecular plane, is destroyed by unsymmetrical substitution at positions 4 and/or 6; for example, 4-methylpyrimidine is totally unsymmetrical whereas 2- or 5-methylpyrimidine and 4,6-dimethylpyrimidine retain 2,5-symmetry, although the absolute reactivities at other positions will differ, of course, from those at the corresponding positions of pyrimidine itself on account of inductive, mesomeric and/or steric factors.

### 2.13.1.1 The Geometry of Pyrimidines

The pyrimidine ring is virtually flat. Its corrected bond lengths, as determined by a least-squares analysis of the crystal structure data for a unit cell of four molecules, are shown in formula (**2**) ⟨60AX80⟩, and the bond angles derived from these data show good agreement with those (**3**) derived by other means ⟨63JCS5893⟩; for comparison, each bond length in benzene is 1.40 Å and each bond angle is 120°. Crystallographic data on pyrimidines and quinazolines have been summarized, with references to 1970 ⟨72PMH(5)1⟩.

The electron-distribution diagram (**4**) shows the gain or loss of $\pi$-electrons at each atom of pyrimidine ⟨67MI21300⟩, as calculated by the VESCF method: this procedure ⟨59AJC554⟩ calculates effective nuclear charges in the field wherein the $\pi$-electrons move and the occupation number of the $2p-\pi$ atomic orbital on a conjugated atom is taken to be the $\pi$-electron density for that atom in the molecule, rather than that for the neutral isolated atom. It is evident that there is considerable depletion of electron density at the 2- and 4/6-positions, a slight depletion at the 5-position and a greatly enhanced density at the N-atoms. The depletions at positions $\alpha$ or $\gamma$ to the ring-nitrogens are naturally more marked in pyrimidine, where the ring-nitrogens act in unison, than in pyridazine or pyrazine, where they do not so act, or in pyridine which has only one electron-withdrawing centre. Indeed, on this depends the unique character of pyrimidine chemistry.

(1)  (2)  (3)  (4)

Such calculations have been made also for pyrimidines of biological interest ⟨B-60MI21302⟩. That for uracil (**5**) is interesting in that a figure of $-0.22$ is assigned to the 5-position, compared with almost zero in pyrimidine: this immediately explains the ease of electrophilic attack at the 5-position of uracil as well as the lack of nucleophilic activity at the same position.

The dipole moment of pyrimidine is calculated by various workers as 2.13, 2.19 or 2.25 D; experimental determinations have given 2.10 to 2.40 D ⟨61MI21301⟩. No such agreement is achieved in resonance energy values for pyrimidine which have been given as 8, 14, 20, 26, 33, 38 or 40 kcal mol$^{-1}$, compared with a value of 41 for benzene: such disagreement makes the figures all but valueless ⟨55MI21301; 62ACS916⟩; that for quinazoline is given as 30.4 (*cf.* naphthalene 30.5) ⟨69JA6321⟩.

Electron-density calculations for quinazoline (which has no symmetry) vary markedly with the method used. The diagram (**6**) has the same bases as that given for pyrimidine above: it will be observed that the 2- and 4-positions in quinazoline are comparable with the corresponding positions in pyrimidine and that the 'aromatic' carbon atoms (C-5–C-8) in quinazoline are roughly comparable with C-5 in pyrimidine ⟨67MI21300⟩. The dipole moment of quinazoline does not appear to have been measured, but that of 2-methylquinazoline is 2.2 D.

(5)  (6)  (7)  (8)

### 2.13.1.2 The Ionization of Pyrimidines

The ability of a pyrimidine to shed a proton and become an anion and/or to accept a proton and become a cation, is governed by its ionization constant, usually expressed as the $pK_a$ value. These may be measured conveniently by following the pH as deprotonation or protonation occurs on progressive addition of base or acid (potentiometric titration), by following the change in UV absorption at a suitable wavelength during the same process (spectrometric titration) or by other such means ⟨B-71MI21303⟩. Such $pK_a$ values are useful criteria of identity and homogeneity. More important, they are invaluable in structural studies: for example, pyrimidin-2-amine (**7**; R = H) and methyl iodide give the imine (**8**) in the absence of base but the isomeric amine (**7**; R = Me) in the presence of base; the former product, a fairly strong base of $pK_a$ 10.75, can be easily so distinguished from its isomer, a moderately weak base of $pK_a$ 3.82, comparable with that of the starting material ($pK_a$ 3.54) ⟨55JCS4035⟩. Knowledge of the $pK_a$ value is also an essential prerequisite for choosing an appropriate aqueous buffer in which to measure the UV spectrum of a pure ionic (or neutral) species (see Section 2.13.1.5). Finally, ionization constants may be used in simple cases to determine the ratio of tautomers in an aqueous solution of, for example, pyrimidin-2-ol $\rightleftharpoons$ pyrimidin-2(1$H$)-one (see Section 2.13.1.8). Extended discussions and tables of $pK_a$ values are available ⟨62HC(16)472, 70HC(16-S1)368, B-73MI21304⟩ and methods based on quantified analogy may be used reasonably successfully to estimate $pK_a$ values without measurement ⟨B-81MI21302⟩.

### 2.13.1.2.1 The pK_a of pyrimidines

Pyrimidine is a much weaker base ($pK_a$ 1.31 and *ca.* −6.3) than pyridine ($pK_a$ 5.2) because the second ring-nitrogen shares the available $\pi$-electrons with the first and the system therefore approximates to 3-nitropyridine (9) ($pK_a$ 0.8). The addition of a mildly electron-releasing methyl group partly redresses the deficiency so that 4-methylpyrimidine has $pK_a$ 2.0 and 4,6-dimethylpyrimidine has $pK_a$ 2.8; the same factor is seen in 4-methoxypyrimidine ($pK_a$ 2.5). The basic strength of pyrimidin-2(1$H$)-one (10; R = H) ($pK_a$ 2.24) is also greater than that of pyrimidine, because the nitrogen involved in cyclic amide formation no longer has the capacity of a ring-nitrogen to attract $\pi$-electrons; in uracil (5), both ring-nitrogen atoms are so involved and, being protonated, the oxygen atoms are the next available basic centres; the basic strength again falls, to $pK_a$ −3.4.

### 2.13.1.2.2 The acidic pK_a values of pyrimidines

The pyrimidinones, pyrimidinethiones, pyrimidin-5-ols, pyrimidinecarboxylic acids and pyrimidinesulfonic acids all ionize as acids, but because pyrimidine is such a weak base, the formation of zwitterionic species seldom occurs. Because of powerful electron withdrawal by the second ring-nitrogen atom, the cyclic amide, pyrimidin-2(1$H$)-one (10; R = H), is a stronger acid ($pK_a$ 9.17) than the corresponding pyridinone ($pK_a$ 11.7); as usual, addition of a *C*-methyl group is acid weakening, so that 4-methylpyrimidin-2(1$H$)-one (10; R = Me) has $pK_a$ 9.8. Uracil (5) ($pK_a$ 9.4 and *ca.* 12) is slightly less acidic than the pyrimidinones because both nitrogen atoms are involved as cyclic amides and are therefore less electron withdrawing: however, barbituric acid (11; R = H) and dialuric acid (11; R = OH) have potential hydroxyl groups and their acidity ($pK_a$ 2.8) is considerable. Thiones are rather stronger acids than the corresponding pyrimidinones by about 2 units. Both types are increased in acidity by electron-withdrawing substituents: compare uracil (5) ($pK_a$ 9.4) with 5-nitrouracil ($pK_a$ 5.6) or 5-bromouracil ($pK_a$ 8.0). As might be expected, pyrimidine-2-carboxylic acid ($pK_a$ 2.85) is stronger than benzoic acid ($pK_a$ 4.21) and pyrimidine-2-sulfonic acid (12; X = N) ($pK_a$ −1.7) is stronger than benzenesulfonic acid (12; X = CH) ($pK_a$ 2.5).

(9)          (10)                    (11)                    (12)

(13)                              (14)

### 2.13.1.2.3 The basic pK_a values of pyrimidinamines

The simple pyrimidinamines, *e.g.* (7; R = H), are bases of moderate strength, $pK_a$ 3–6. Thus the rise in basic strength from pyrimidine ($pK_a$ 1.31) to pyrimidin-2-amine ($pK_a$ 3.54) or to pyrimidin-4-amine ($pK_a$ 5.71) is far more than the one unit rise observed from aniline to the three phenylenediamines; on the other hand, pyrimidin-5-amine ($pK_a$ 2.8) is of the expected order. The reason for the high values in the 2- and 4-isomers is the increase of resonance in the cations, *e.g.* (**13**; R = H), compared with the neutral molecule, *e.g.* (**7**; R = H): this stabilizes the cation, thus increasing the basic strength. No such increase in resonance is possible in the cation of pyrimidin-5-amine. Addition of *C*-methyl groups has the expected base-strengthening effect: 4-methyl- and 4,6-dimethyl-pyrimidin-2-amine have $pK_a$ 4.15 and 4.85, respectively. Insertion of a methyl group on the amine does likewise: the methylamine (**7**; R = Me) has $pK_a$ 3.82. Other electron-releasing groups also increase the basic strength: pyrimidine-4,5-diamine ($pK_a$ 6.03), pyrimidine-2,4-diamine ($pK_a$ 7.3) and pyrimidine-2,4,5-triamine ($pK_a$ 7.63) all show this effect. In contrast, electron-withdrawing groups reduce basic strength: compare pyrimidin-2-amine ($pK_a$ 3.54) with its 5-bromo ($pK_a$ 1.95), 5-cyano ($pK_a$ 0.7) and 5-nitro ($pK_a$ 0.35) derivatives.

Alkylation of pyrimidin-2(or 4)-amine on a ring-nitrogen gives an imine, *e.g.* (**8**), of quite high basic strength ($pK_a$ 10.7) because its cation, *e.g.* (**13**; R = Me), has typical and effective resonance stabilization; indeed, methylation of pyrimidine-2,4-diamine gives a still stronger base ($pK_a > 13$) due to an even more resonance-stabilized cation (**14**).

#### 2.13.1.2.4 The pK_a values of aminopyrimidinones

In general, the $pK_a$ values of aminopyrimidinones indicate that the amino group has weakened the acid function and the oxo substituent has weakened the basic strength. However, there are apparent anomalies in seeking to quantify these changes ⟨62HC(16)470⟩. For example, 2-aminopyrimidin-4(3*H*)-one (isocytosine: **15**; R = NH₂) is only 0.8 units weaker as an acid than the pyrimidinone (**15**; R = H) whereas 4-aminopyrimidin-2(1*H*)-one (cytosine: **16**; R = NH₂) is no less than three units weaker than the corresponding pyrimidinone (**16**; R = H). The effect of addition of further groups is also unusual. For example, both isocytosine (**15**; R = NH₂) and cytosine (**16**; R = NH₂) are virtually unaffected in acidic or basic strength by additional amino groups, although in each case an additional nitro group does increase acidity and decrease basicity as expected. The aminopyrimidine-thiones are stronger acids and weaker bases than the corresponding aminopyrimidinones.

#### 2.13.1.2.5 The pK_a values of quinazolines

Quinazoline is a stronger base (equilibrium $pK_a$ 3.51) than pyrimidine ($pK_a$ 1.31) because its cation is stabilized as a covalent 3,4-hydrate (**17**). The insertion of a 4-methyl group interferes with hydration so that 4-methylquinazoline (**18**) forms a regular cation and has $pK_a$ 2.52 (*cf.* 4-methylpyrimidine, $pK_a$ 2.0); in contrast, 2-methylquinazoline does form a covalently hydrated cation and its $pK_a$ is 4.52 ⟨63PMH(1)1, 79AHC(24)1⟩. The quinazolinones are weaker acids than the corresponding pyrimidinones by about 1.5 units and the four quinazolinols have $pK_a$ values 7.3–8.6, *i.e.* appreciably weaker than pyrimidin-5-ol ($pK_a$ 6.8). Most quinazolinones, quinazolinamines and methoxyquinazolines are appreciably stronger bases than the corresponding pyrimidines, mainly on account of covalent hydration in their cations; where no hydration is possible, the values are comparable in both series ⟨67HC(24-1)19⟩.

(**15**)          (**16**)          (**17**)          (**18**)

### 2.13.1.3 The NMR Spectra of Pyrimidines

The bulk of available NMR data on pyrimidines are naturally proton spectra but of recent years [19]F, [13]C, [15]N and even [14]N spectra have gradually become more common.

### 2.13.1.3.1  Proton NMR spectra

The first proton NMR spectrum of pyrimidine (1) appeared in 1960 and the data were confirmed subsequently ⟨62JA336⟩. The relative deshielding of the four protons is H-2 > H-4 = H-6 > H-5 as might be expected [$\delta$(CDCl$_3$): 9.26, $J_{2,5}$ 1.5 Hz, H-2; 8.78, $J_{4,5(5,6)}$ 5.0, H-4/6; 7.46, H-5]; the spectrum shows little change from neat liquid to dilute carbon tetrachloride solution, suggesting that association effects have little influence. The three mono-*C*-methylpyrimidines do show long range coupling between the methyl protons and *ortho* or *para* ring protons. For example, 2-methylpyrimidine has $J_{Me,5}$ 0.6 Hz and 4-methylpyrimidine has $J_{Me,5}$ 0.4 Hz (both in CDCl$_3$) ⟨64AK(22)65⟩. The NMR spectra of most other pyrimidines, which are monosubstituted by a non-tautomeric group, are unexceptional. Thus, 2-substituted pyrimidines show a doublet and a triplet typical of such an A$_2$X system with H-4/6 broadened by coupling to the adjacent nitrogen; the chemical shifts and $J_{4,5}$ naturally vary with the nature of the 2-substituent: compare pyrimidine-2-carbonitrile (19; R = CN) [$\delta$(Me$_2$CO): 9.04, H-4/6; 7.86, $J_{4,5}$ 5.1 Hz, H-5] and 2-bromopyrimidine (19; R = Br) [$\delta$(Me$_2$CO): 8.72, H-4/6; 7.57, $J_{4,5}$ 4.8 Hz, H-5] ⟨64AK(22)65⟩. The 4-substituted pyrimidines have no symmetry and show typical ABX patterns with small *para*-couplings involving H-2 as exemplified in 4-methylthiopyrimidine (20) [$\delta$(CDCl$_3$): 9.02, $J_{2,5}$ 1.3 Hz, H-2; 8.43, $J_{5,6}$ 5.6, H-6; 7.25, H-5; 2.58, SMe] ⟨B-73NMR97⟩. The spectra of 5-substituted pyrimidines simply show two broad singlets with no *meta*-coupling evident across the ring-nitrogen atoms, except in 5-nitropyrimidine (21) in which a small such coupling is manifest [$\delta$(CDCl$_3$): 9.67, H-4/6; 9.64, $J_{2,4}$ 0.3 Hz, H-2] ⟨67JCS(C)573⟩. There is little change in the spectrum of 5-nitropyrimidine in D$_2$O, but when acidified the spectrum becomes three singlets [$\delta$(D$_2$O/DCl): 8.66, 8.31, 6.49], indicating a covalently hydrated cation (22; R = H): the position of hydration is proven by the similarity in spectrum of the 2-methyl-5-nitropyrimidine cation (22; R = Me), in which 1,2(2,3)-hydration is precluded ⟨67JCS(C)573⟩.

The NMR spectra of pyrimidines bearing tautomeric groups (NH$_2$, OH, SH) in the 2-, 4- or 6-position are discussed in connexion with the tautomerism of such pyrimidines (Section 2.13.1.8).

(19)          (20)          (21)          (22)

The NMR spectra of simple dihydropyrimidines are not easy to interpret ⟨62JOC4090⟩ but the spectrum of the 2,5-dihydropyrimidine (23) is particularly interesting in having $J_{2,5}$ 5.5 Hz (as shown by decoupling procedures), a large value for long range coupling ⟨65JCS6695⟩. Work on dihydropyrimidinones has been more popular and rewarding. For example, the spectrum of 5-cyano-1-methyl-3,4-dihydropyrimidin-2(1*H*)-one (24) [$\delta$(CDCl$_3$): 6.80, $J_{4,6}$ 1.0 Hz, H-6; 4.13, H-4; 3.12, NMe] clearly shows reduction of the 3,4-bond, as do also those of its 3-methyl isomer and 1,3-dimethyl homologue ⟨64JOC1740⟩. The preferred conformation of 5,6-dihydrouracils may be determined by means of NMR, the coupling between H-1 and H-6 being used as the diagnostic probe: at room temperature, the dihydrouracil (25), as well as its 1-methyl and 1,3-dimethyl derivatives, exist as two rapidly flipping half-chair conformers; however, on 5- or 6-substitution, the resulting dihydropyrimidine is 'frozen' into a single conformation ⟨66TL4189, 69T3807⟩.

In simple 1,4,5,6-tetrahydropyrimidines at room temperature, the distinction between axial and equatorial protons is lost because of rapid interconversion and spin multiplets of a simple type are seen: thus the hydrochloride (26) in D$_2$O/DCl shows a four-proton triplet at $\delta$ 3.42 (2 × NCH$_2$) and a two-proton quintet at $\delta$ 1.99 (5-CH$_2$); the pentamethyltetrahydropyrimidine base (27) also shows a single peak for 5-CH$_2$ at $\delta$ 1.74, indicating rapid equilibrium of conformers ⟨67AJC1643⟩. A careful study of ring inversion of 1,3-dimethylhexahydropyrimidine (28) and related compounds at various temperatures can give thermodynamic activation constants: at temperatures below *ca.* −30 °C, chair to chair isomerization becomes slow on the NMR time scale and the CH$_2$ signal becomes an AB-quartet, $J_{AB}$ *ca.* 10 Hz ⟨67JCS(B)560, 68T829⟩.

### 2.13.1.3.2 Non-proton NMR spectra

The early [19]F (natural abundance) data for a number of mono- to tetra-fluoropyrimidines ⟨62JOC2580, 67JCS(C)1822⟩ serve to illustrate such spectra. For example, 2,4,5,6-tetrafluoropyrimidine shows the following signals shifted in p.p.m. from the externa' trifluoroacetic acid signal: $-30.6$, $J_{2,5}$ 26 Hz, F-2; $-3.8$, $J_{4,5}$ 18, F-4/6; $+99$, F-5. Under similar conditions, 2,4,6-trifluoropyrimidine ($-35.2$, $J_{F-2,H-5}$ 1.1 Hz, F-2; $-24.0$, $J_{F-4,H-5}$ 1.8, F-4/6] and 4-anilino-2,5,6-trifluoropyrimidine ($-29.8$, $J_{2,5}$ 26.6 Hz, $J_{2,6}$ 2.8, F-2; $+10.0$, $J_{5,6}$ 16.9, F-6; $+101.2$, $J_{F-5,NH}$ 2.8, F-5) show two types of F–H coupling which complicate [19]F spectra. Such complication also occurs in reverse when proton spectra of fluoropyrimidines are measured, especially as *meta*-coupling across a ring-nitrogen is not precluded. For example, 2-fluoropyrimidine [$\delta$(CDCl$_3$): 8.75, $J_{4,5}$ 6 Hz, $J_{4,F}$ 1.7, H-4/6; 7.40, $J_{4,5}$ 6, $J_{5,F}$ 6, H-5] and 4-fluoro-2-methylpyrimidine [$\delta$(CDCl$_3$): 8.97, $J_{5,6}$ 6 Hz, $J_{6,F}$ 12, H-6; 6.96, $J_{5,F}$ 3, H-5; 2.73, Me] show *ortho*-, *meta*- and *para*-coupling between fluorine and protons, as well as the usual proton–proton couplings ⟨74JCS(P2)204⟩.

(23)          (24)          (25)          (26)

(27)          (28)          (29)          (30)

The [13]C NMR spectra of pyrimidines can be measured using the natural abundance of the isotope in regular samples. Early work is complicated by the use of benzene or carbon disulfide as standard from which chemical shifts are measured but all the shifts given below are converted into p.p.m. from TMS for convenience. The data on pyrimidine in per-deuteroacetone for chemical shifts ⟨68JA697⟩, short-range coupling constants ⟨70MI21301⟩, and long-range coupling constants ⟨73JOC1313⟩, may be summarized as follows: 159.7, $J_{C-2,H-2}$ 203 Hz, $J_{C-2,H-4}$ 10.3, $J_{C-2,H-5}$ 0, C-2; 157.7, $J_{C-4,H-2}$ 9.1, $J_{C-4,H-4}$ 183, $J_{C-4,H-5}$ 1.9, $J_{C-4,H-6}$ 5.3, C-4/6; 122.3, $J_{C-5,H-2}$ 1.9, $J_{C-5,H-4}$ 9.5, $J_{C-5,H-5}$ 169, C-5. The spectrum of quinazoline is also available but without coupling data ⟨75AG356⟩. Complete data for a series of 2-monosubstituted pyrimidines ⟨76OMR(8)357⟩ and for their 4- and 5-isomers ⟨79OMR(12)212⟩ are available and the variations of chemical shift and coupling constant follow reasonably predictable patterns; indeed, the chemical shifts may be correlated with $\pi$-electron densities but the [13]C–[1]H coupling constants do not correlate with the electronegativity of the substituents. Data for 21 mono- to tetra-methylpyrimidines and for mono- to tri-methylpyrimidinamines are also available ⟨74TL3123⟩.

Until relatively recent improvements in instrumentation, the acquisition of [15]N NMR natural abundance spectra was virtually impossible; even today, [15]N-enriched samples are highly desirable. Spectra are complicated because both [15]N–[1]H and [15]N–[13]C coupling occurs; thus, they are useful for studying tautomerism in pyrimidinones, *etc.* (see Section 2.13.1.8). The first compounds to be so studied were the [15]N-tagged cytosine derivative (29) ⟨65JA5575⟩, the fully labelled analogue and several labelled uracil derivatives, *e.g.* (30) ⟨65JA5439, 79JOC1627⟩. Some very satisfactory work on [15]N(1)-tagged samples of 2-methylthio- and 2-chloro-pyrimidine, pyrimidin-2-one and its 3-methyl derivative, pyrimidine-2-thione and its 3-methyl derivative, and their cations ⟨72TH21300⟩ has never been published in full, although some data are available in tabulated form ⟨B-73NMR97⟩. In such derivatives, the [15]N-signal appears 167–290 p.p.m. downfield from the ammonium ion, according to the substitution. Introduction of the label removes symmetry so that the usual doublet for H-4/6 becomes a well-resolved 16-peak multiplet and the usual triplet for H-5 is doubled by coupling with [15]N; as might be expected, protonation introduces large changes in [15]N–[1]H coupling constants and indeed in the whole pattern.

By using exceptionally wide probes, natural abundance $^{15}$N-spectra can be obtained providing the compound is sufficiently soluble and that it is stable enough for long accumulation runs. Thus, without the tedious necessity of tagging, pyrimidine (base) shows a single $^{15}$N-peak at −85 p.p.m. (relative to nitromethane), while as a monocation it appears at −135 p.p.m. and as a dication at −183 p.p.m. ⟨80HCA504⟩. Data and analytical discussion for a variety of mono-, di- and tri-aminopyrimidines and for some of their *N*-oxides, with and without methyl groups, are also available ⟨80HCA504⟩: the one-bond $^{15}$N−$^{1}$H coupling constants for most amino groups of pyrimidinamines fall within the range 87–90 Hz except for a 5-amino group (81 Hz) and the amino group (92 Hz) in 2-amino-1-methylpyrimidinium iodide. The spectrum of 2,5,6-triaminopyrimidin-4-one as dication in water (shifts in p.p.m. relative to nitromethane: −256, N-1; −243, 2-NH$_2$; −350, N-3; −303, 5-NH$_2$; −298, 6-NH$_2$ ⟨78HCA2108⟩) and, more recently, those of pyrimidine-2,4-diamine in DMSO (shifts in p.p.m. downfield from $^{15}$NH$_4$ ion: 196, $J_{N-1,H-6}$ 11 Hz, $J_{N-1,H-5}$ 1.8, N-1; 187, $J_{N-3,H-5}$ 3.7, N-3; 60, 2-NH$_2$; 62, 4-NH$_2$ ⟨81AJC1539⟩), quinazoline in DMSO (shifts in p.p.m. down from $^{15}$NH$_4$ ion: 263, $J_{N-1,H-2}$ 15.9 Hz, N-1; 274, $J_{N-3,H-2}$ 10.9, $J_{N-3,H-4}$ 14.5, N-3 ⟨81AJC1539⟩) and several other pyrimidines, all exemplify the natural abundance approach. The use of $^{14}$N NMR spectra is impeded by the very low sensitivity to NMR detection (*ca.* 10$^{-3}$ of protons) and the broad signals observed, but the technique may be used for 2-substituted pyrimidines by employing a $^{1}$H–$^{14}$N double resonance method with long accumulation times; correlations between $^{14}$N chemical shifts and $\pi$-electron densities at N are evident ⟨78BSB271⟩.

### 2.13.1.4  The IR and Raman Spectra of Pyrimidines

Since IR spectra are essentially due to vibrational transitions, many substituents with single bonds or isolated double bonds give rise to characteristic absorption bands within a limited frequency range; in contrast, the absorption due to conjugated multiple bonds is usually not characteristic and cannot be ascribed to any particular grouping. Thus IR spectra afford reference data for identification of pyrimidines, for the identification of certain attached groups and as an aid in studying qualitatively the tautomerism (if any) of pyrimidinones, pyrimidinethiones and pyrimidinamines in the solid state or in non-protic solvents (see Section 2.13.1.8).

Naturally, pyrimidin-2(and 4)-one give strong absorption bands for C=O and N—H bond-stretching vibrations; moreover, it is well established ⟨57JCS4874⟩ that the six-membered heteroaromatic 'hydroxy' compounds with oxygen $\alpha$ to a ring-nitrogen give N—H stretching bands in the range 3360–3420 cm$^{-1}$, while those with oxygen $\gamma$ to a ring-nitrogen have such bands in the range 3415–3445 cm$^{-1}$: application of this correlation to pyrimidin-4-one shows that the major N—H band arises from the *ortho*-quinonoid amide (**31**) rather than from its *para*-quinonoid isomer. Further, the appropriate IR (solid) and Raman (aqueous solution) vibrations of the pyrimidinones, as hydrochlorides and sodium salts, suggest that the extra proton is attached to the doubly bound nitrogen atom in the cations ⟨60JCS1226, 60JCS1232⟩.

Pyrimidin-2(and 4)-amine, *e.g.* (**32**; R = H), show two N—H stretching bands at 3400 and 3500 cm$^{-1}$ ⟨55JCS4035⟩, representing respectively the symmetric and antisymmetric vibrations; imines, *e.g.* (**8**), absorb at distinctly lower frequencies (*ca.* 3300 cm$^{-1}$). The H—N—H bond angle, calculated from these frequencies, suggests that the amino lone pair is more conjugated with the nucleus in pyrimidin-2(and 4)-amine than in pyrimidin-5-amine: the bond angle in the former (119°) is consistent with trigonal hybridization in the $\sigma$-bonds of the amino group, leaving the lone pair in a nitrogen 2$p$ orbital to conjugate with the $\pi$-electron system; in the 5-isomer, the bond angle (112°) is akin to that of aniline, suggesting tetrahedral hybridization in the $\sigma$-bonds of the amino group and the lone pair in an $sp$-hybrid orbital of which only the $p$ component can conjugate ⟨58JCS3619⟩. The methyl-aminopyrimidine (**32**; R = Me) also has two N—H stretching bands but quite close together (3443, 3466 cm$^{-1}$); it has been suggested that these represent not amino and imino tautomers, but the hydrogen-bonded form (**33**) and the non-hydrogen-bonded form (**32**; R = Me).

Most of the bands in the IR and Raman spectra of unsubstituted pyrimidine can be assigned to the vibrations of the nucleus ⟨57SA113⟩ and the spectra of its cation are rather similar ⟨60JCS1226, 60JCS1232⟩; salient points have been discussed ⟨62HC(16)477⟩. Similar treatment may be accorded to quinazoline ⟨70JSP(36)310⟩. It must be remarked in conclusion that little notable IR or Raman work in the pyrimidine area has appeared in the last twenty years, since the advent of NMR spectroscopy.

(31)          (32)          (33)

## 2.13.1.5 The UV Spectra of Pyrimidines

In practice, the UV spectrum of a pyrimidine will depend on the state of the nucleus (fully 'aromatic', partly reduced or fully reduced), on the position, number and nature of any chromophoric substituents attached to it, and on the ionic species present in the solution being measured. Strangely enough, the last factor was all but ignored until about 1950, so that many spectra recorded prior to that date are quite meaningless, even for characterization of the pyrimidine in question. Besides characterization, UV spectroscopy may be used for structural clues, for example detecting covalent hydration ⟨61JCS2689⟩, for quantitative analysis even in very dilute solution, for following the kinetics of reactions or rearrangements ⟨63JCS1276⟩, for studying tautomeric equilibria (see Section 2.13.1.8), for measuring ionization constants ⟨B-71MI21303⟩ and for many other purposes. A full theoretical introduction to the UV spectra of heterocycles ⟨63PMH(2)1⟩ and an invaluable collection of data to illustrate it ⟨71PMH(3)67⟩, as well as two specialized reviews of such pyrimidine spectra ⟨62HC(16)477, B-73MI21305⟩ makes it unnecessary to do more than point out a few fundamental facts below.

The UV absorption of pyrimidine occurs in two bands centred at 243 and 298 nm in cyclohexane. The second band is ascribed to the electronic transition from a nitrogen lone pair non-bonding orbital to an empty ring $\pi$-orbital, in short an $n \rightarrow \pi^*$ transition, on account of the hypsochromic shift observed on changing solvent from cyclohexane to water ⟨59JCS1240, 59JCS1247⟩. The lone pairs become engaged in hydrogen bonding in water so that the absorbed radiation must be of higher energy (*i.e.* of lower wavelength) to provide for breaking the hydrogen bonds as well as bringing about the electronic transition. As might be expected, this band is represented only by a residual inflexion in the spectrum of pyrimidine cation in aqueous buffer. The more intense band at 243 nm is ascribed to a transition from the occupied $\pi$-orbital of the ring having the highest energy to the empty $\pi$-orbital of lowest energy, a $\pi \rightarrow \pi^*$ transition akin to that which accounts for the 250 nm band of benzene; such bands are unaffected by solvent changes.

In general, electron-releasing substituents cause a bathochromic shift of the $n \rightarrow \pi^*$ band while electron-withdrawing substituents do the reverse. However, the $\pi \rightarrow \pi^*$ band undergoes a bathochromic shift by either type of substituent and its intensity is increased. These effects are roughly additive in di- or poly-substituted pyrimidines, except when the addition of one or more tautomeric groups affects the fundamental structure, *i.e.* the double bond arrangement, of the nucleus ⟨52JCS3716, 52JCS3722⟩.

The UV spectrum of quinazoline is less simple: bands at 220, 267 and 311 all represent $\pi \rightarrow \pi^*$ transitions and a characteristic inflexion of low intensity at 330 nm represents the $n \rightarrow \pi^*$ transition ⟨63PMH(2)1⟩. Quinazoline as cation (17) is 3,4-hydrated covalently, so that the spectrum ($\lambda_{max}$ 208, 260 nm) is quite abnormal ⟨61JCS2689⟩, as indeed are those of many of its derivatives ⟨62JCS561⟩. The spectra of quinazolines have been collected ⟨71PMH(3)67⟩.

## 2.13.1.6 The Mass Spectra of Pyrimidines

The mass spectra of pyrimidines and quinazolines are generally simple and of no great interest, factors which account for the paucity of data available ⟨B-71MS468⟩.

The dominant fragmentation mode of pyrimidine is loss of HCN twice, to give ionized acetylene, *m/e* 26, as base peak; whether C-2 or C-4 is involved in the initial loss of HCN appears to be unknown. Quinazoline behaves similarly to give ionized benzyne (**34**; R = H), but in this case the spectrum of 4-deuterio-quinazoline indicates that N-3 and C-4 are eliminated in the initial loss of HCN. The 5-, 6-, 7- and 8-methylquinazolines fragment similarly to give the homologous benzynes (**34**; R = Me); 2-methylquinazoline eliminates HCN followed by MeCN, and 4-methylquinazoline does the same but in reverse order ⟨67JCS(B)892⟩.

Pyrimidin-2-amine also fragments by sequential loss of two molecules of HCN but the product from the first loss appears to be the pyrazole radical cation (**35**) or possibly an imidazole, both of which are known to fragment according to the subsequent pattern for pyrimidin-2-amine ⟨65JA4569⟩. Substituted pyrimidin-2-amines behave similarly but pyrimidin-4-amine and its derivatives follow a greatly modified and quite complicated pattern ⟨66T3117⟩. Both uracil (**36**; R = H) and thymine (**36**; R = Me) undergo an initial retro-Diels–Alder reaction by losing HCNO to give the fragments (**37**; R = H or Me); the subsequent fragmentation is logical enough ⟨65JA4569⟩. Cytosine (**38**) fragments by at least three pathways, of which one involves an initial loss of CO, but barbituric acid proceeds by a logical loss of 2 × HNCO in the main: for summaries see ⟨B-71MS468⟩. The fragmentations of several hexahydropyrimidines are available ⟨67AJC1643, 70HC(16-S1)322⟩.

(**34**)        (**35**)        (**36**)        (**37**)        (**38**)

### 2.13.1.7   The Stereochemistry of Pyrimidines and Hydropyrimidines

There is an admirable summary of the stereochemistry of barbiturates and di- to hexa-hydropyrimidines ⟨B-77SH(1)177⟩. Further information on reduced pyrimidines is collected ⟨70HC(16-S1)322⟩ and some examples of the use of proton NMR spectra in elucidating the conformations of hydropyrimidines is given elsewhere (Section 2.13.1.3.1), based on the general principles of such work ⟨65QR426⟩.

### 2.13.1.8   Tautomerism in Pyrimidines

In view of the excellent detailed reviews on tautomerism of potentially hydroxy-, mer-capto- and amino-heterocycles, including pyrimidines ⟨63AHC(1)339; 76AHC(S1)71⟩, this survey is quite brief. It is now considered axiomatic that the reactions of a tautomeric system can give virtually no information about its composition; early laudable but misguided experiments along such lines have proven meaningless. Information can be obtained only by physical means which do not upset the delicate equilibria in tautomeric systems and, for pragmatic reasons, it is best to compare results with those for the models of extreme forms in which the mobile proton is 'frozen' in position as a methyl group. Although many physical techniques may be used, IR, Raman, UV and NMR spectroscopy, as well as ionization constant measurements, are the most useful in the pyrimidine series.

#### *2.13.1.8.1   The structure of pyrimidinones and quinazolinones*

Preliminary IR spectral studies were said to suggest that pyrimidinones existed as pyrimidinols ⟨50JCS3062⟩ but this conclusion was promptly reversed ⟨52JCS168⟩ on better experimental evidence; subsequent comparison with their *N*- and *O*-methyl derivatives showed that the pyrimidinones (**39a**; R = H) and (**40a**; R = H) along with their *N*-methyl derivatives (**39a**; R = Me), (**40a**; R = Me) and (**40b**; R = Me) all exhibited $\nu_{CO}$ in the range 1600–1700 cm$^{-1}$, whereas the methoxypyrimidines (**39b**; R = Me) and (**40c**; R = Me) showed no such absorptions ⟨53JCS331, 55JCS211⟩. Closer analysis of the spectra for pyrimidin-4-one (**40a**; R = H) showed that the *ortho*-quinonoid form (**40a**; R = H) is the predominant tautomer (see Section 2.13.1.4).

Similar comparisons were also made of the various UV spectra in aqueous buffer ⟨51JCS1004, 53JCS331, 55JCS211⟩ with exactly similar conclusions; the p$K_a$ values for pyrimidin-

4-one (**40a**; R = H) and its methylated derivatives (**40a–c**; R = Me) confirmed the conclusion on the dominance of form (**40a**; R = H) in aqueous solution ⟨58JCS674⟩. More recently, the comparisons were made yet again, this time by NMR spectra in $D_2O$ ⟨66JOC175⟩ and in DMSO ⟨65JHC447⟩: the same conclusions emerged.

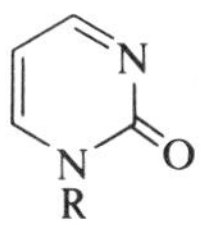

| (39a) | (39b) | (40a) | (40b) | (40c) | (41) |

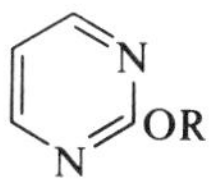

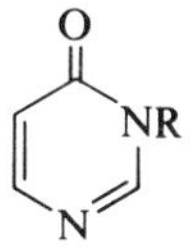

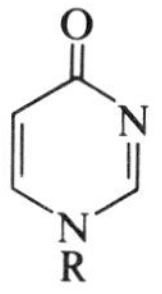

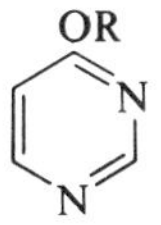

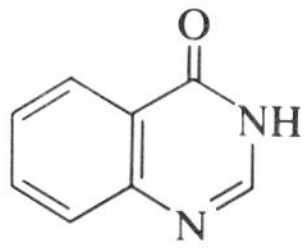

Quinazolin-2(and 4)-one were first shown to exist as oxo tautomers by IR and UV spectroscopic comparisons akin to those above ⟨52JA4834, 51JCS3318⟩; like its pyrimidinone analogue, the quinazolinone (**41**) prefers that configuration to the *para*-quinonoid tautomer ⟨57JCS4874⟩, a conclusion upheld by the NMR study of simple analogues ⟨69T783⟩.

### 2.13.1.8.2 *The structure of pyrimidinethiones and quinazolinethiones*

Pyrimidine-2-thione (**42**) was shown quite early, by comparison of its UV spectrum with those of *N*- and *S*-methylated derivatives, to exist as such ⟨51JCS1004, 52JCS3716, 52JCS3722⟩; at the same time, pyrimidine-4-thione was shown on similar criteria to exist in one or both of the thioxo forms (**43a**) and (**43b**) and this was confirmed by IR measurements ⟨60JCS1237⟩. Soon after, repetition of the UV approach coupled with careful analysis of ionization constants showed that the thione (**43a**) predominates in aqueous solution ⟨62JCS3129⟩ and this was confirmed by dipole moment measurements in benzene ⟨73CR(C)(276)1341⟩. The general situation is similar for quinazoline-4(3*H*)-thione (**44**) which exists as such ⟨62JCS3129⟩.

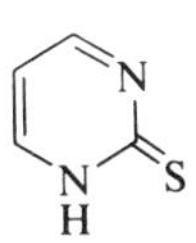

| (42) | (43a) | (43b) | (44) |

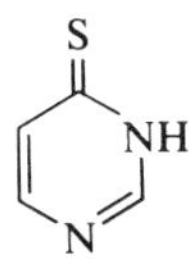

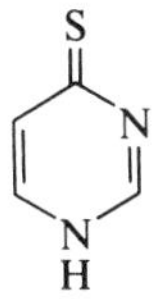

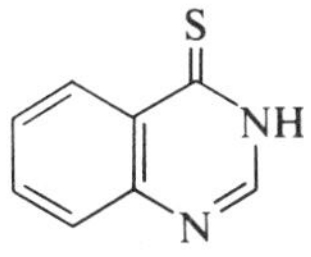

### 2.13.1.8.3 *The structure of pyrimidinamines and quinazolinamines*

Despite earlier suggestions to the contrary, by the fifties it had become evident that, in the UV spectra and p$K_a$, pyrimidin-2-amine (**45**; R = H) resembled closely the dimethyl-amino analogue (**45**; R = Me) but differed markedly from the 'frozen' imine (**46**) ⟨53JCS331, 55JCS4035⟩. The same was true of pyrimidin-4-amine (**47**; R = H) in comparison with the dimethylamino model (**47**; R = Me) and the one imine (**48**) then available; the comparison was formally completed years later when the missing imine (**49**) was synthesized ⟨71JCS(C)2507⟩. The UV spectrum of quinazolin-4-amine is consistent with that formulation ⟨51JCS3318⟩. Thus the simple pyrimidinamines exist as such to an overwhelming extent; on similar criteria, the substituted pyrimidin-4-amines (**50**; R = CN) and (**50**; R = CHO) exist as such ⟨67CB3664⟩; moreover, pyrimidine-2,4-diamine (**51**) exists as the diamino tautomer on UV ⟨65JCS755⟩ and NMR spectral evidence ⟨64AK(22)65⟩.

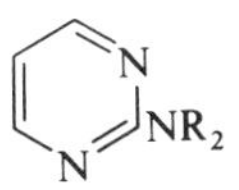

| (45) | (46) | (47) | (48) | (49) | (50) | (51) |

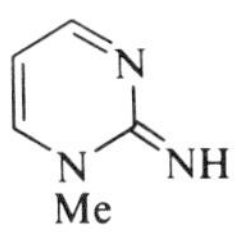

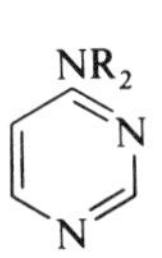

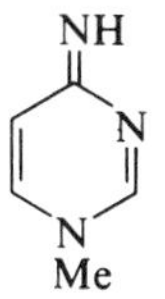

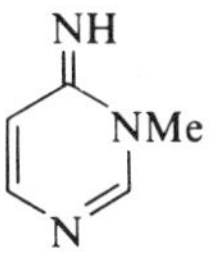

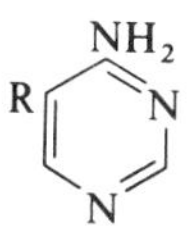

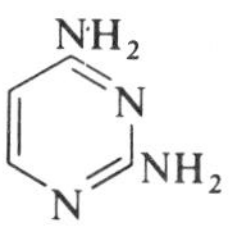

### 2.13.1.8.4 *The structure of pyrimidines with two or more tautomeric groups*

Uracil exists as the dioxo tautomer (**52**; R = H) in the solid state on the evidence of refined X-ray analyses in which the positions of hydrogen atoms are determined directly

⟨54AX313, 67AX1102⟩; many spectral data and comparisons indicate that the same tautomer predominates in solution: for example, UV ⟨61JCS504⟩, Raman ⟨67SA(A)2551⟩ and $^{15}$N NMR ⟨65JA5439⟩. Thymine (**52**; R = Me) ⟨61AX333, 69AX(B)1038⟩ and 5-nitrouracil (**52**; R = NO$_2$) ⟨59JCS3647⟩ are also dioxo in structure. The sites of protonation and deprotonation in uracils have been examined carefully and the results are summarized ⟨76AHC(S1)71⟩ as uracil cation (**53**) and dication (**54**); uracil anion (**55**); and, not surprisingly, 5-aminouracil cation (**56**).

The structure of barbituric acid was the subject of disagreement for many years, but since 1952 ⟨52BSB44⟩ the trioxo formulation (**57**; R = H) has been accepted generally, along with the fact that barbituric acid loses a proton, first from carbon (anion) and subsequently from nitrogen (dianion). Barbital (5,5-diethylbarbituric acid) adopts a similar trioxo form (**57**; R = Et) ⟨69AX(B)1978⟩.

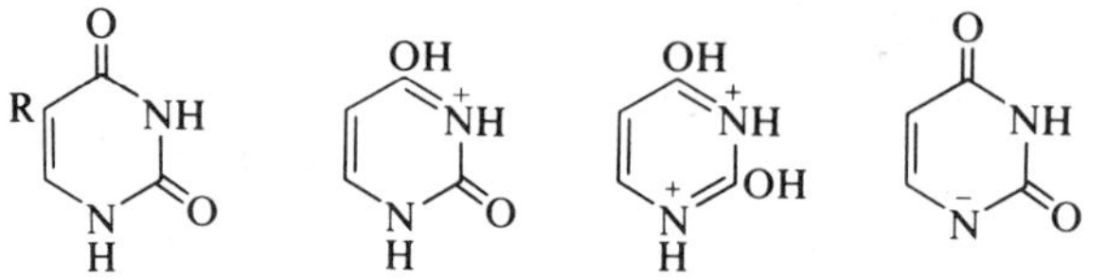

Although the UV spectra of 'pseudouracil', in comparison with some of its methylated derivatives, suggested that it might be a mixture of the dioxo form (**58**) and a hydroxyoxo form ⟨64AJC567⟩ and NMR data seemed to uphold this in general terms ⟨66JOC175⟩, it now seems clear on extended criteria that the betaine (**59**) is probably the main contributor in solution ⟨66JCS(B)565⟩.

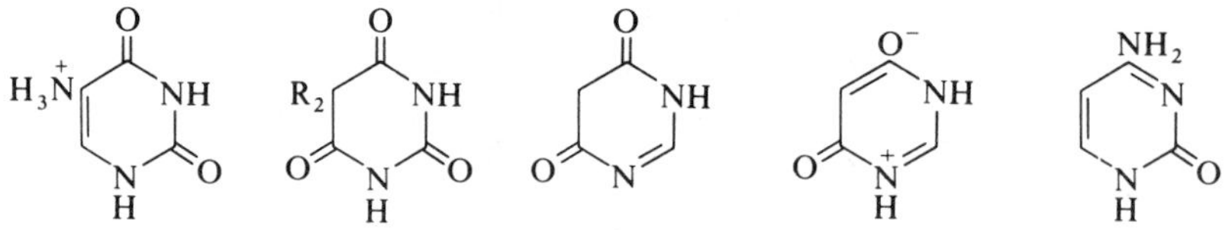

The structure of cytosine is certainly the aminooxo form (**60**) which has been confirmed by X-ray analyses ⟨73AX(B)1234⟩, UV and NMR data ⟨63JCS3046⟩, Raman spectra ⟨67SA(A)2551⟩ and even by $^{14}$N NMR spectra ⟨72JPC5087⟩.

The structural work on many other tautomeric pyrimidines is summarized in the reviews previously quoted ⟨63AHC(1)339, 76AHC(S1)71⟩.

## 2.13.2 REACTIVITY OF PYRIMIDINES

Much of the reactivity shown by the ring atoms and substituents of pyrimidine is akin to that of the corresponding parts of 1,3-dinitrobenzene and 3-nitropyridine. This arises from the quantitatively similar electron-withdrawing effects of doubly-bound ring nitrogen atoms and of nitro groups in reducing sharply the aromaticity of the cyclic system.

### 2.13.2.1 Reactivity of Ring Atoms of Pyrimidines

In pyrimidines, the above effect produces significant electron depletion at C-2, C-4 and C-6 but only relatively minor depletion at C-5. Accordingly, nucleophilic attack will take place at the former group of positions whereas electrophilic attack will be confined to C-5 or the ring nitrogen atoms. The addition of electron-withdrawing substituents will increase this effect; electron-releasing substituents will decrease it by making all positions more nearly equal.

#### 2.13.2.1.1 Electrophilic attack

(*a*) Pyrimidine and its simple derivatives are quaternized with alkyl halides to give 1-methylpyrimidinium iodide (**61**; R = Me, X = I) ⟨53JCS1646⟩ or its appropriate analogue, although the ease of quaternization varies according to the substituents present. Pyrimidine-

2(or 4)-amines, pyrimidin-2(or 4)-ones and pyrimidine-2(or 4)-thiones are all special cases. Thus pyrimidin-2-amine (**62**) forms the methiodide (**63**) which, on treatment with cold alkali, gives pyrimidin-2(1*H*)-imine (**64**), the anhydro base of a quaternary hydroxide corresponding to the iodide (**63**). Such strongly basic imines are unstable: the above undergoes Dimroth rearrangement (see Section 2.13.2.2.6) to *N*-methylpyrimidin-2-amine (**65**); others, *e.g.* 1-methylpyrimidin-4(1*H*)-imine, which cannot rearrange, usually hydrolyze readily to the corresponding *N*-methylpyrimidinones ⟨55JCS4035⟩. Simple pyrimidinones are rather weak bases and appear to resist quaternization; in contrast, their anionic salts undergo rapid *N*-alkylation with alkyl halides and the product, *e.g.* (**66**), may subsequently form a stable quaternary salt, *e.g.* (**67**) ⟨66JOC3969⟩. Simple pyrimidinethiones resist direct quaternization less strongly and their anions give *S*-alkylated, *e.g.* (**68**), rather than *N*-alkylated derivatives ⟨59JCS525⟩, although these may subsequently undergo quaternization as above. Quinazoline is quaternized at N-3.

(*b*) Like the introduction of a third nitro group into benzene or toluene, nitration of pyrimidine and its alkyl derivatives would be expected to be difficult: in fact, it cannot be done, probably because the cation of pyrimidine cannot survive the extreme conditions needed. However, the presence of even one electron-releasing substituent makes nitration at the 5-position just possible under quite vigorous conditions: for example, treatment of pyrimidin-2(1*H*)-one (**69**; R = H) in concentrated sulfuric acid at >90 °C with potassium nitrate gives the 5-nitropyrimidin-2(1*H*)-one (**69**; R = NO$_2$) ⟨69JHC593⟩. Two or more electron-releasing substituents make 5-nitration relatively easy, so that 1-methyl-5-nitropyrimidine-2,4-dione (**70**) and similar compounds may be formed by nitration at room temperature ⟨55JCS211⟩. Nitration of aminopyrimidines sometimes gives an intermediate *N*-nitro derivative: for example, 6-chloropyrimidine-2,4-diamine (**71**; R = H) affords 4-chloro-6-nitroaminopyrimidin-2-amine (**71**; R = NO$_2$) and only in the presence of an excess of concentrated sulfuric acid does this yield the 5-nitro isomer (**72**) ⟨66JMC573⟩. Quinazoline does nitrate, but only in the benzene ring (see Section 2.13.2.2.1).

(*c*) Like nitration, nitrosation of pyrimidines requires the presence of electron-releasing substituents, usually three in number and consisting of oxo, thioxo, or amino groups. However, pyrimidine-4,6-diamine (**73**; R = H) is nitrosated to give the blue 5-nitroso derivative (**73**; R = NO) proven in structure by reduction to the triamine (**73**; R = NH$_2$) ⟨56JCS4106⟩. A typical nitrosation is that yielding 6-(2'-hydroxyethyl)amino-5-nitrosopyrimidine-2,4(1*H*,3*H*)-dione (**74**; R = NO) by treatment of the substrate (**74**; R = H) with aqueous nitrous acid at room temperature ⟨60JCS4768⟩. Unlike nitration, nitrosation can occur at the 4/6-position, providing there are 3 electron-releasing groups present to render the remaining position susceptible, as in the case of 2-amino-5-hydroxy-6-nitrosopyri-midin-4(3H)-one (75) ⟨64JCS1001⟩. In employing nitrosation, it must be remembered that N-nitrosation of secondary amino substituents is possible as well as C-nitrosation of appropriately activated methyl substituents; an example of the latter is the formation of the nitroso oxime (76) ⟨47JCS943⟩.

(d) Pyrimidines with two or more electron-donating groups undergo diazo coupling at the 5-position or, if that is occupied, sometimes at the 4/6-position ⟨64JCS1001⟩; exceptionally, 4,6-dimethylpyrimidin-2(1H)-one (77; R = H) does couple with diazotized aniline (see Section 2.13.2.2.2) but not to give the 5-phenylazo derivative (77; R = N₂Ph). However, the hydroxypyrimidinone (78; R = H) does so in a regular way to give the azo derivative (78; R = N₂Ph) (or its phenylhydrazone tautomer) ⟨48BSF688⟩. Some useful comparisons of diazonium reagent, yields, and procedures have been made ⟨66JPS568⟩.

(77)              (78)

(e) Even unsubstituted pyrimidine undergoes 5-halogenation, albeit only under vigorous conditions: vapor-phase bromination of the free base occurs best at about 230 °C but a better procedure to obtain 5-bromopyrimidine (79) is to drop bromine into a solution of pyrimidine hydrochloride in nitrobenzene at 130 °C ⟨73JHC153⟩. 5-Chloro, 5-fluoro- and 5-iodo-pyrimidine are all known but have not been made by direct halogenation. The direct 5-chlorination and bromination of 2- and 4-methylpyrimidine (using the halogenosucc-inimide) is unsatisfactory because the methyl groups are more susceptible than the 5-position: thus the mono-, di- and tri-halogenomethylpyrimidines are the main products although a little 5-chloro-2- and 5-chloro-4-methylpyrimidine (80) can be isolated from the appropriate halogenation reaction mixtures ⟨74AJC2251⟩. Halogenation of pyrimidines bearing one or more electron-releasing groups is very easy: bromine or chlorine in acetic acid, or indeed practically any compatible solvent, gives good yields. For iodination, as well as for bromination or chlorination in the presence of an acid-labile group, the appropriate halogenosuccinimide is the reagent of choice ⟨66T2401⟩. Sulphuryl chloride is sometimes more convenient than elemental chlorine for 5-chlorinations. Although 5-fluoropyrimidines are usually made indirectly, 5-fluorouracil (81) may be made in good yield by passing fluorine gas into an aqueous suspension of uracil at room temperature or, in even better yield but less conveniently, by using perfluoromethanol as the fluorinating reagent ⟨72JOC329⟩, or, in 90% yield, by treatment of uracil with the graphite intercalate $C_{19}XeF_6$ ⟨80TL277⟩.

(79)        (80)        (81)        (82)        (83)        (84)

It has long been known ⟨62HC(16)172⟩ that barbituric acid (82; R¹ = R² = H) reacts with bromine to give initially the 5-bromo derivative (82; R¹ = Br, R² = H) which then reacts with more bromine to give the 5,5-dibromo derivative (82; R¹ = R² = Br); treatment of the latter with ammonia or a reducing agent causes reversion to the monobromo derivative. N-Alkylbarbituric acids behave similarly and the scenario for chlorination is similar. Direct fluorination of barbituric acid gives a stable 5,5-difluoro derivative (82; R¹ = R² = F) but the monofluoro derivative has not been prepared; nor have iodinated derivatives ⟨74JCS(P1)2095⟩. The bromination of uracil (83; R = H) in aqueous media is somewhat different: the first product is indeed the 5-bromo derivative (83; R = Br) which subsequently appears to add hypobromous acid to give the dibromodihydro derivative (84); this is quite stable but reverts to the monobromo derivative on prolonged boiling in mineral acid. Cytosine and related compounds behave similarly towards bromine and chlorine. Mechanisms have been discussed ⟨59JOC11⟩.

(*f*) The direct formation of dipyrimidin-5-yl sulfides occurs on treatment of appropriate 5-unsubstituted pyrimidine substrates with sulfur mono- or di-chloride. Thus, reaction of uracil (**83**; R = H) with sulfur monochloride in boiling formic acid gives diuracil-5-yl sulfide in good yield; sulfur dichloride gives a poor yield. Simple derivatives of uracil and barbituric acid undergo similar reactions but not cytosine, isocytosine, 2,4-bismethylthiopyrimidine or pyrimidine-4,6-dione (**59**). The mechanism is unknown ⟨72AJC2275⟩.

(*g*) The direct 5-sulfonation of pyrimidines which already bear at least one electron-releasing group is possible but has been used sparingly; chlorosulfonic acid is the reagent of choice and the resulting pyrimidinesulfonyl chloride is easily converted into the sulfonic acid. Vigorous conditions are required: pyrimidin-2-amine and chlorosulfonic acid refluxed together for 8 hours yield the sulfonic acid (**85**) on treatment of the reaction mixture with ice ⟨63JMC58⟩.

(*h*) Direct *C*-formylation at the 5-position of pyrimidines is possible by the Reimer–Tiemann or Vilsmeier reactions. The former is illustrated by the treatment of uracil with sodium hydroxide/chloroform to give 5-formyluracil (**86**) ⟨67B2168⟩ and numerous other pyrimidines with 2 or 3 electron-releasing groups behave similarly. Vilsmeier reagents, *e.g.* DMF/phosphoryl chloride, are less convenient because they usually (but not always) convert pyrimidinones into the corresponding chloropyrimidine as well as introducing the *C*-formyl group: for example, 2-amino-6-hydroxypyrimidin-4(3*H*)-one (**87**) gives 2-amino-4,6-dichloropyrimidine-5-carbaldehyde (**88**). Another complication is the occasional attack of an amino group instead of *C*-formylation as when 2,6-dimethoxypyrimidin-2-amine gives only its dimethylaminomethylene derivative (**89**), possibly due to insufficient activation at the 5-position ⟨65M1567⟩.

(**85**)      (**86**)      (**87**)      (**88**)

(**89**)      (**90**)      (**91**)      (**92**)

(*i*) Direct (*N*-substituted)-aminomethylation at the 5-position of pyrimidines with at least two electron-releasing groups can be done by the Mannich reaction; a methyl or an amino substituent may also (or preferentially) undergo the reaction, especially if the 5-position is insufficiently activated. For example, reaction of 6-amino-2-methylpyrimidin-4(3*H*)-one (**90**; R = H) with formaldehyde and piperidine gives initially the 5-piperidinomethyl derivative (**90**; R = CH$_2$N(CH$_2$)$_5$) and subsequently, with an excess of reagents, the product (**91**) ⟨56YZ234⟩. Isobarbituric acid (**92**; R = H) and related compounds, with an occupied 5-position and lacking reactive substituents, have a sufficiently reactive 6-position to permit Mannich attack at that point to yield, for example, 5-hydroxy-6-piperidinomethylpyrimidine-2,4(1*H*,3*H*)-dione (**92**; R = CH$_2$N(CH$_2$)$_5$) ⟨67JHC49⟩.

(*j*) The phenomenon of 5-hydroxymethylation is a standard case of electrophilic attack. Thus uracil (**83**; R = H) and paraformaldehyde in aqueous alkali furnish 5-hydroxymethylpyrimidine-2,4(1*H*,3*H*)-dione (**83**; R = CH$_2$OH) in good yield ⟨59JA2521⟩. Aromatic aldehydes react differently to yield 5-benzylidene derivatives of, for example, 1-methylbarbituric acid ⟨78CC764⟩.

(*k*) The *N*- and 5-oxidation of pyrimidines, which might be considered electrophilic attacks, are both treated in Section 2.13.2.1.5 along with nuclear reduction. It is surprising that pyrimidine *N*-oxides appear to be no more reactive towards electrophiles than the parents.

### 2.13.2.1.2 *Nucleophilic attack*

Despite the fact that the 2-, 4- and 6-positions of pyrimidine are predisposed to direct nucleophilic attack, relatively few simple examples of such reactions are recorded.

(a) Amination of 4-methylpyrimidine with sodium amide in decalin yields 4-methyl-pyrimidin-2-amine (**93**, R = H), 6-methylpyrimidine-2,4-diamine (**93**, R = NH₂) and other products ⟨39YZ97⟩. The complicated reactions of halogenopyrimidines with sodium amide appear to involve a direct amination step in some cases (see Section 2.13.2.2.9). Quinazoline reacts with sodium amide to give quinazolin-4-amine (**94**; R = H) and with hydrazine to give quinazolin-4-ylhydrazine (**94**; R = NH₂); the mechanism may involve initial 3,4-addition ⟨60YZ245⟩.

(93)    (94)    (95)    (96)

(97)    (98)    (99)    (100)

(b) Simple pyrimidines undergo some nucleophilic additions at the 3,4- or 4,5-bond and quinazolines do so quite readily at the 3,4-bond. Although covalent hydration has been detected in solutions of pyrimidin-2(1*H*)-one, which exists partly as the adduct (**95**) ⟨68CC289⟩, only in pyrimidines bearing a strongly electron-withdrawing group does the hydration reach completion. Thus the cation of 5-nitropyrimidine exists as the adduct (**96**) ⟨67JCS(C)573⟩ and, in acidic media, 5-methylsulfonyl- and 5-methylsulfinyl-pyrimidine do likewise ⟨68JCS(C)1452⟩. In contrast, even the unsubstituted quinazoline cation exists largely as hydrate (**97**). Thus the UV spectrum of quinazoline cation (in aqueous buffer) is displaced to shorter wavelengths (by *ca.* 45 nm) compared with that of the neutral species; this relationship does not pertain with 4-methylquinazoline in which the substituent provides effective steric hindrance to covalent hydration and the spectra of both species are therefore closely similar. The structure (**97**) has been confirmed quite adequately by NMR and other means, the effects of substitution have been explored, and oxidation of the hydrate in aqueous acid yields quinazolin-4(3*H*)-one (**98**) ⟨B-76MI21300⟩. On irradiation, uracil and related pyrimidines undergo, among other important reactions (see Section 2.13.2.1.4), a photohydration across the 5,6-bond. A typical product (from 1,3-dimethylpyrimidine-2,4(1*H*, 3*H*)-dione) is the 6-hydroxy-1,3-dimethyl-5,6-dihydropyrimidine-2,4(1*H*, 3*H*)-dione (**99**), carefully proven in structure, which reverts to its precursor on treatment with acid or alkali, and undergoes ring opening to yield *N,N'*-dimethylmalonamide (**100**) on excessive irradiation ⟨B-76MI21300⟩.

(c) Simple pyrimidines easily undergo addition by Grignard and a'kyllithium reagents, usually across the 3,4-bond if the site is available. Thus pyrimidine and phenylmagnesium bromide give the adduct (**101**; R = MgBr) which decomposes in water to 4-phenyl-3,4-dihydropyrimidine (**101**; R = H), and subsequent oxidation gives 4-phenylpyrimidine; the same result may be achieved by using phenyllithium *via* the intermediate (**101**; R = Li) ⟨58CB2832⟩. Having 4- and 6-positions which are occupied, 4,6-dimethoxypyrimidine undergoes 2,5-addition by *t*-butyllithium to give the intermediate (**102**; R = Li) which, with water, gives 2-*t*-butyl-4,6-dimethoxy-2,5-dihydropyrimidine (**102**; R = H); in contrast, the analogous 4-chloro-6-methoxypyrimidine is attacked by the same reagent to yield the 5-lithio derivative (**103**) without the introduction of a *t*-butyl group ⟨65JCS6695⟩. The important reactions of halogenopyrimidines with metal alkyls, in which the halogeno substituent is involved, are covered in Sections 2.13.2.2.9 and 2.13.2.2.10. Quinazoline behaves normally towards such reagents to give 3,4-adducts, *e.g.* (**104**; R = Li or MgBr) ⟨60YZ245⟩.

(101)    (102)    (103)    (104)

**(105)**        **(106)**        **(107)**        **(108)**

(*d*) Pyrimidines form 3,4-adducts with primary and sometimes secondary amines but these usually react further. Thus pyrimidine and hydrazine yield pyrazole (**106**) by ring fission of the initial adduct (**105**) at the 3,4-bond followed by alternative reclosure and final loss of a fragment representing the N-1, C-2 and N-3 portion of the original pyrimidine ⟨68RTC1065⟩; this mechanism is confirmed by the fact that similar treatment of the imine (**107**) also gives the pyrazole ⟨68JCS(C)1452⟩.

(*e*) Some other additions are represented best in the quinazolines. Sodium hydrogen sulfite, hydrogen cyanide and a variety of ketones all add across the 3,4-bond of quinazoline to yield the adducts (**108**; R = $SO_3Na$, CN, $CH_2Ac$, *etc.*) ⟨61JCS2689⟩.

### 2.13.2.1.3 Free radical attack

The concept and use of free radical attack on pyrimidines has been little developed. However, pyrimidine does react slowly with *p*-nitrobenzenediazonium chloride to yield some 2- and 4-*p*-nitrophenylpyrimidines ⟨51JCS2323⟩; in addition, 2,4-and 4,6-dimethyl-pyrimidine are converted by hydroxymethylene radicals (from ammonium peroxydisulfate/methanol) into 6- and 2-hydroxymethyl derivatives, respectively ⟨77H(6)525⟩. Certain bipyrimidine photoproducts appear to be formed from two similar or dissimilar pyrimidinyl radicals (see Section 2.13.2.1.4).

### 2.13.2.1.4 Photochemical reactions

There is a scattered body of data in the literature on 'ordinary' photochemical reactions in the pyrimidine and quinazoline series: in most cases the mechanisms are unclear. For example, UV irradiation of 4-aminopyrimidine-5-carbonitrile (**109**; R = H) in methanolic hydrogen chloride gives the 2,6-dimethyl derivative (**109**; R = Me) in good yield; the 5-aminomethyl analogue is made similarly ⟨68T5861⟩. Another random example is the irradiation of 4,6-diphenylpyrimidine 1-oxide in methanol to give 2-methoxy-4,6-diphenylpyrimidine, probably by addition of methanol to an intermediate oxaziridine (**110**) followed by dehydration ⟨76JCS(P1)1202⟩.

**(109)**        **(110)**        **(111)**        **(112)**

In contrast, the photochemistry of uracil, thymine and related 'bases' has a large and detailed literature because most of the adverse effects produced by UV irradiation of tissues seem to result from dimer formation involving adjacent thymine residues in DNA. Three types of reaction are recognizable: (i) photohydration of uracil but not thymine (see Section 2.13.2.1.2), (ii) the oxidation of both bases during irradiation and (iii) photodimer formation.

Irradiation of an aqueous solution of thymine (**111**; R = Me) in the presence of air produces (irreversibly) at least four products: (**111**; R = $CH_2OH$, CHO, $CO_2H$ or H), possibly *via* 'dithymine peroxide', a linear dimeric molecule (**112**). The significance of these reactions has been discussed ⟨65MI21300⟩.

The formation of photodimers of the 'cyclobutane' type from thymine (**111**; R = Me) occurs most effectively when a frozen aqueous solution of the substrate is irradiated. After the independent recognition of such dimers in two laboratories about 1960, it was some six years before the main constituent was identified beyond doubt ⟨66JCS(C)2239⟩ as the *cis–syn* entity (**113**), a U-shaped molecule in which the planes of the six-membered rings

are parallel and at 90° to that of the cyclobutane ring. Similar dimers are formed by
irradiation of frozen solutions or thin (solid) films of uracil (**111**; R = H), 1,3-dimethyluracil,
5-bromouracil (**111**; R = Br), cytosine (4-aminopyrimidin-2(1*H*)-one), orotic acid (2,6-
dioxo-1,2,3,6-tetrahydropyrimidine-4-carboxylic acid) and related pyrimidines; in most
cases, two or more isomeric dimers are formed. When irradiated in unfrozen aqueous
solution or suspension, the thymine dimer (**113**) reverts to its precursor, thymine. It seems
that the formation of crystalline ice forces the solute into solid aggregates in which the
molecules of thymine hydrate are arranged suitably for dimerization on irradiation; if other
frozen solvents are used, no dimerization takes place. In contrast, when the dimer is
irradiated in liquid water, the equilibrium is reversed because the monomeric molecules
distribute themselves randomly in the solution as they are formed and hence have little
tendency to dimerize again ⟨B-76MI21301⟩.

(**113**)              (**114**)              (**115**)

As mentioned above (Section 2.13.2.1.3), bipyrimidine photoproducts can arise, probably
by reaction between two radicals. Thus, irradiation of an aqueous solution of 5-bromouracil
(**111**; R = Br) in the absence of oxygen produces a variety of products including uracil,
barbituric acid, 5-carboxyuracil (**111**; R = CO₂H), several non-pyrimidine compounds and,
as a stable end-product, the biuracil (**114**; R = H). A similar product (**114**; R = Me) is
formed from 5-bromo-1,3-dimethyluracil (**115**). When two such related uracil derivatives
are irradiated together, a mixed bipyrimidine product is formed, *inter alia* ⟨B-76MI21302⟩.

### 2.13.2.1.5 Oxidative and reductive reactions

(*a*) The Elbs persulfate reaction has been applied successfully to pyrimidin-2(or 4)-ones
and pyrimidine-2,4-diamines. Thus 4,6-dimethylpyrimidin-2(1*H*)-one (**116**; R = H) reacts
with cold alkaline persulfate to give the sulfate ester (**116**; R = OSO₃H) which undergoes
hydrolysis in boiling hydrochloric acid to afford 5-hydroxy-4,6-dimethylpyrimidin-2(1*H*)-
one (**116**; R = OH). Other typical products from such oxidation of appropriate substrates
include 2,4-diaminopyrimidin-5-ol, 2-amino-5-hydroxypyrimidin-4(3*H*)-one and 5-
hydroxy-6-methylpyrimidine-2,4-(1*H*,3*H*)-dione ⟨56JCS2033⟩.

(*b*) Although it has not been used for simple pyrimidines, the oxidation of quinazoline
covalent hydrates may be used to introduce a 4-oxo (hydroxy) group. For example,
quinazoline, existing in acid solution as the hydrated cation (**97**), is converted by peroxide
or peroxyacetic acid into quinazolin-4(3*H*)-one (**98**) ⟨61JCS2689⟩; in alkaline solution, where
quinazoline is not hydrated, permanganate oxidation takes a different course by destroying
the benzene ring to yield pyrimidine-4,5-dicarboxylic acid (**117**).

(*c*) Elevation of oxidation state is possible in some pyrimidines. 'Dialuric acid' (5,6-
dihydroxypyrimidin-2,4(1*H*,3*H*)-dione) (**118**) may be oxidized to 'alloxan' (**119**); the

(**116**)              (**117**)              (**118**)              (**119**)

(**120**)                    (**121**)         (**122**)

oxidation probably proceeds through 'alloxantin' (120) and all stages are reversible with appropriate reducing agents ⟨43OS(23)3⟩.

(*d*) One route to pyrimidine *N*-oxides is the direct oxidation of the corresponding pyrimidines, usually with peracetic acid. Although pyrimidine and its simple alkyl derivatives can be so *N*-oxidized, yields tend to be rather low, *e.g.* 9% for pyrimidine *N*-oxide (121; R = H) but 57% for 4,6-dimethylpyrimidine *N*-oxide (121; R = Me) ⟨59JCS525⟩. It has been established that a 2-phenyl group interferes with *N*-oxidation by steric hindrance, that a 4-phenyl group has no such effect on 1-oxidation and that 4-substituted pyrimidines lacking an additional 6-substituent tend to decompose during oxidation ⟨67YZ1096⟩. Because they undergo *N*-oxidation satisfactorily, many alkoxypyrimidine *N*-oxides are known; monoperoxymaleic acid has been found uniquely suitable for such oxidation of chloropyrimidines to give, for example, 4-chloro-2,6-dimethylpyrimidine 1-oxide (122) ⟨68CPB1337⟩. The sometimes difficult task of distinguishing 1- from 3-oxides of unsymmetrical pyrimidines, even by NMR spectra, may be assisted by addition of a lanthanide shift reagent ⟨77H(8)257⟩. Few quinazoline *N*-oxides have been made by direct oxidation. However, reaction of 4-alkoxyquinazolines with peroxyphthalic acid in ether gives 1-oxides in reasonable yield ⟨59CPB(7)152⟩.

(*e*) The nuclear reduction of many pyrimidines may be done by hydrogenation over palladium or platinum catalysts, especially in acidic media. Such treatment usually reduces those double bonds free of an attached group (other than halogeno, which is removed anyway) but it usually stops at the tetrahydro stage leaving the 1,2/2,3-double bond untouched. Thus uracil is reduced to 5,6-dihydropyrimidine-2,4(1*H*,3*H*)-dione (123), 4,6-dichloropyrimidin-2-amine gives 1,4,5,6-tetrahydropyrimidin-2-amine (124; R = NH₂) and pyrimidine itself gives the tetrahydro derivative (124; R = H). Likewise, boiling with Raney nickel gives the dihydro derivative (123) from uracil and a 3,4,5,6-tetrahydro derivative (125) from pyrimidin-2(1*H*)-one; pyrimidin-4-amine and pyrimidin-4(3*H*)-one resist such treatment but yield to hydrogenation. Vigorous hydrogenation of pyrimidin-2-amine gives, rather exceptionally, the hexahydro derivative (126). Pyrimidin-5-amine and 5-hydroxypyrimidine each give only a dihydro derivative. Hydrogenation in (buffered) neutral or alkaline solution seldom reduces the ring and can be used to modify substituents without much fear of nuclear reduction ⟨62HC(16)430⟩.

(123)    (124)    (125)    (126)    (127)    (128)    (129)

Most classical reducing agents leave the pyrimidine nucleus unaffected, as does sodium borohydride; lithium aluminium hydride usually gives a di- or tetra-hydro derivative, according to substituent(s) present ⟨70HC(16-S1)322⟩.

Hydrogenation of quinazoline stops normally at the 3,4-dihydro derivative (127; R = H) ⟨61JCS2697⟩ but under forcing conditions the 1,2,3,4-tetrahydro derivative (128) may be formed ⟨57YZ507⟩. Even 4-substituted quinazolines give 3,4-dihydro derivatives (127) but the rates of reduction are much lower than those of 4-unsubstituted or 2-substituted quinazolines. In contrast with simple pyrimidines, many quinazolines may be reduced to their tetrahydro derivatives with sodium amalgam, sodium/alcohol, lithium aluminium hydride and sometimes metal/acid reagents. Electrolytic reduction has also been used with some success. For example, 3-phenylquinazolin-4(3*H*)-one gave either 4-hydroxy-3-phenyl-1,2,3,4-tetrahydroquinazoline (129; R = OH) or the dehydroxylated derivative (129; R = H) according to the type of electrode. Quinazolines reduced in the benzene ring are known but they cannot be made by reductive processes ⟨67HC(24-1)391⟩.

### 2.13.2.2 Reactivity of Substituents

Metathetical reactions of pyrimidines have been studied widely over the last hundred years. Indeed, probably 90% of all known pyrimidines have been made from a relatively few pyrimidine substrates available by primary synthesis.

### 2.13.2.2.1 Reactivity of the fused benzene ring in quinazolines

Electrophilic attack on quinazoline should occur in the benzene ring; theoretical considerations suggest the following order for positional reactivity: $8 > 6 > 5 > 7 \gg 4 > 2$ ⟨57JCS2521⟩. In the event, nitration of quinazoline yields only the 6-nitro derivative (**130**) because the substrate exists in the nitrating acid as the anhydrous dication ⟨67MI21300⟩. Nitration of 2- and 4-substituted quinazolines also gives 6-nitro derivatives, *e.g.* 1-methyl-6-nitroquinazolin-4(1*H*)-one (**131**) ⟨49JCS1354⟩. Chlorination of quinazolin-4(3*H*)-one (**132**; $R^1 = R^2 = H$) under vigorous conditions gives 6-chloro- (**132**; $R^1 = Cl$, $R^2 = H$), 8-chloro- (**132**; $R^1 = H$, $R^2 = Cl$) and 6,8-dichloro-quinazolin-4(3*H*)-one (**132**; $R^1 = R^2 = Cl$). Chlorosulfonation of the same substrate gives the 6-chlorosulfonyl derivative (**132**; $R^1 = SO_2Cl$, $R^2 = H$) ⟨60JA2731⟩. Oxidation of quinazoline by alkaline permanganate destroys the benzene ring to give pyrimidine-4,5-dicarboxylic acid (**117**).

(**130**)          (**131**)          (**132**)

### 2.13.2.2.2 Reactivity of alkyl groups

Methyl groups in the 2-, 4- or 6-position of pyrimidine are 'active' in the sense of that in 2,4-dinitrotoluene; in the 5-position, a methyl group resembles that of toluene itself.

(*a*) The conversion of appropriate methyl- into styryl-pyrimidines can be done by condensation with aldehydes in the presence of an acid, Lewis acid, dehydrating agent or strong organic base as catalyst. For example, 4-methylpyrimidine and benzaldehyde with acetic anhydride give 4-styrylpyrimidine (**133**) ⟨67JCS(C)1343⟩. Other commonly used catalysts are zinc chloride, piperidine, hydrochloric acid and concentrated sulfuric acid. Methyl groups in the 4/6-position appear to be marginally more reactive to styrylation than are 2-methyl groups; 5-methyl groups and those on the benzene ring of quinazolines are invariably unaffected. Like 2- and 4-methylpyrimidine, 2-methylquinazoline reacts with chloral to yield 2-(quinazolin-2'-yl)-1-trichloromethylethanol (**134**) which, on treatment with alkali, undergoes dehydration and hydrolysis to give 3-(quinazolin-2'-yl)acrylic acid (**135**) ⟨56CB2578⟩.

(**133**)          (**134**)          (**135**)

(*b*) The oxidation of alkyl- to carboxy-pyrimidines is a common reaction; it is sometimes advantageous to convert methyl- into styryl-pyrimidines prior to such oxidation but directly attached phenyl groups are resistant. For example, treatment of 4-methyl-2-phenyl-pyrimidine (**136**; $R = Me$) with permanganate gives 2-phenylpyrimidine-4-carboxylic acid (**136**; $R = CO_2H$), but 5-bromo-2-methylpyrimidine (**137**; $R = Me$) resists both permanganate and nitric acid oxidation and only after styrylation to (**137**; $R = CH=CHPh$) does it yield 5-bromopyrimidine-2-carboxylic acid (**137**; $R = CO_2H$) ⟨53JCS3129⟩. Controlled oxidation of *C*-methyl to *C*-formyl groups can be done with selenium dioxide in acetic acid but only 'active' methyl groups are affected: 5,6-dimethylpyrimidine-2,4(1*H*,3*H*)-dione (**138**; $R = Me$) yields the 6-formyl-5-methyl analogue (**138**; $R = CHO$) ⟨67JHC163⟩. Other types of oxidation are the conversion of 1,3,6-trimethylpyrimidine-2,4(1*H*,3*H*)-dione (**139**; $R = Me$) to 1,3-dimethyl-2,6-dioxo-1,2,3,6-tetrahydropyrimidine-4-carboxamide (**139**; $R = CONH_2$) by treatment with potassium ferricyanide and ammonia ⟨57BBA(23)295⟩ and of 4-methylpyrimidine to *N*-phenylpyrimidine-4-thiocarboxamide (**140**) by boiling with sulfur in aniline ⟨60CB2410⟩.

(*c*) Direct halogenation of *C*-alkyl groups attached to pyrimidine is possible; the reaction is unaffected by the position of the group and it appears to be catalyzed by light. The progressive bromination of 2- and 4-methylpyrimidine has been mentioned (Section

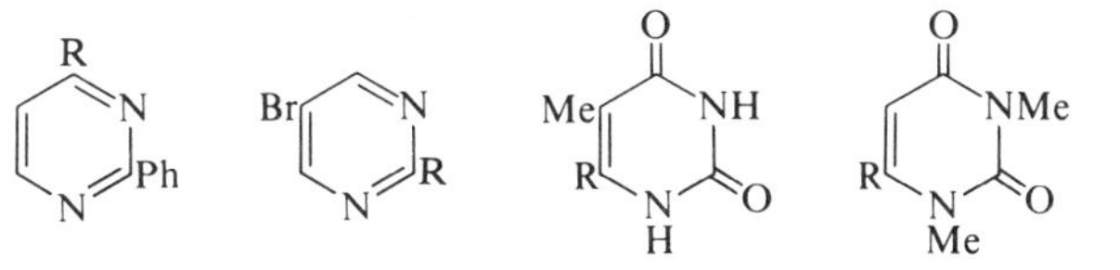

(136)　(137)　(138)　(139)

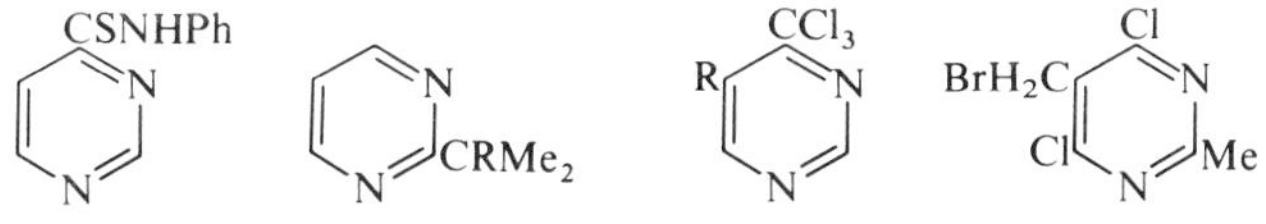

(140)　(141)　(142)　(143)

2.13.2.1.1) but 2-isopropylpyrimidine (**141**; R = H) undergoes only monobromination by NBS to give 2-(1′-bromo-1′-methylethyl)pyrimidine (**141**; R = Br) ⟨77AJC1785⟩. Exhaustive chlorination of 4-methylpyrimidine yields the trichloromethyl analogue (**142**; R = H) and similar treatment of 4,5-dimethylpyrimidine results in preferential trichlorination at the 4-position to give the analogue (**142**; R = Me) ⟨60CB2405⟩; however, bromination of 4,6-dichloro-2,5-dimethylpyrimidine results in attack at the 5- rather than the 2-methyl substituent to yield 5-bromomethyl-4,6-dichloro-2-methylpyrimidine (**143**) ⟨53CPB387⟩. Iodination of 4-methylpyrimidine by elemental iodine in pyridine probably gives the expected iodomethylpyrimidine initially: the actual product isolated is pyrimidin-4-ylpyridinium iodide ⟨66JPR(33)50⟩. Many fluoro- and polyfluoro-alkylpyrimidines are known but most have been made by high temperature isomerizations or near-pyrolyses in the presence of fluoride ion. When 2,4-dimethylquinazoline reacts with hypobromite, preferential bromination of the 4-substituent occurs to yield 2-methyl-4-tribromomethylquinazoline ⟨51JA5777⟩.

(*d*) The nitration of alkyl groups is not possible but they can be oxidized by nitric acid during the 5-nitration or attempted 5-nitration of alkylpyrimidines bearing electron-releasing groups. Thus too vigorous nitration of 6-methylpyrimidine-2,4(1*H*,3*H*)-dione (**144**) leads to a poor or no yield of the required 5-nitro derivative (**145**) ⟨54JCS3832⟩ although some 5-nitropyrimidine-2,4(1*H*,3*H*)-dione (**147**) may be formed by oxidation to the 6-carboxylic acid (**146**) and subsequent decarboxylation.

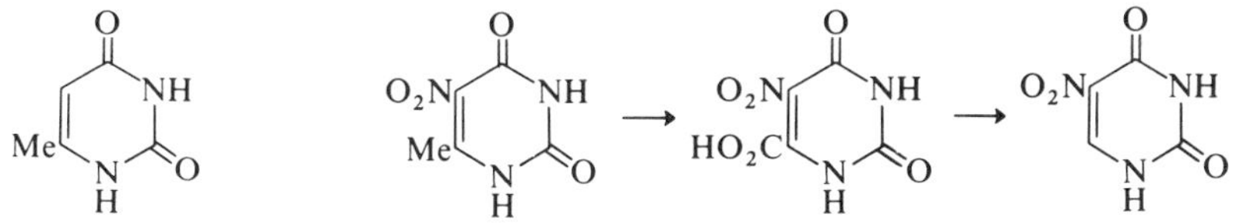

(144)　(145)　(146)　(147)

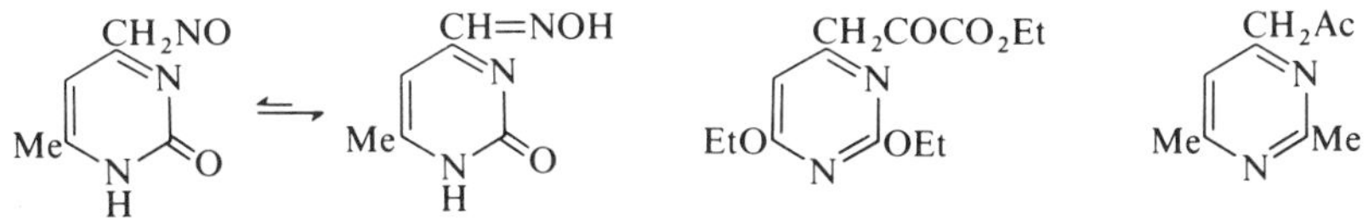

(148)　(149)　(150)　(151)

(*e*) Direct nitrosation of (activated) 2- or 4/6-methyl groups can occur, especially if there are insufficient electron-releasing substituents to permit 5-nitrosation. Thus 4,6-dimethylpyrimidin-2(1*H*)-one yields the nitroso derivative (**148**) which is tautomeric with the aldoxime (**149**) ⟨64JHC130⟩. The same type of product can be obtained by the Claisen condensation using nitrite esters: 4-methylpyrimidine, pentyl nitrite and potassium *t*-butoxide give the oxime of pyrimidine-4-carbaldehyde, which may also be made by nitrosation in acidic media ⟨70LA(737)39⟩. More conventional Claisen reactions with carboxylic esters are well known; for example, treatment of 2,4-diethoxy-6-methylpyrimidine with diethyl oxalate and potassium ethoxide in ether/pyridine gives the ethoxalylmethyl derivative (**150**) ⟨65JHC49⟩. In such reactions, 4/6-methyl groups are more reactive than 2-methyl: 2,4,6-trimethylpyrimidine, sodium amide and ethyl acetate give only 2,6-dimethylpyrimidin-4-ylacetone (**151**) ⟨55JA1559⟩.

(*f*) Those 2- or 4/6-methylpyrimidines with less than two other electron-releasing groups undergo Mannich reactions on a methyl group; otherwise, the reaction will involve the 5-position, providing it is free (for the latter see Section 2.13.2.1.1). Accordingly,

4-methylpyrimidine, dimethylamine, formalin and hydrochloric acid give *N,N*-dimethyl-2-(pyrimidin-4'-yl)ethylamine (**152**) ⟨54JA1879⟩. Exceptionally, 5-methylpyrimidine-2,4(1*H*,3*H*)-dione (thymine) reacts at its methyl group to yield, for example, 5-morpholinomethylpyrimidine-2,4(1*H*,3*H*)-dione (**153**) ⟨60JA991⟩. A 4-methyl group is more reactive than a 2-methyl group in quinazoline: 2,4-dimethylquinazoline reacts with dimethylamine and formalin only at the 4-position to yield the Mannich base (**154**); 2-methylquinazoline is unaffected by the same reagents ⟨51JA5777⟩.

(g) Hydroxymethylation of a methyl group on pyrimidines can be done using paraformaldehyde: for example, 4-methylpyrimidine gives (at 150 °C) 2-(pyrimidin-4'-yl)ethanol (**155**) ⟨61JPR(12)206⟩.

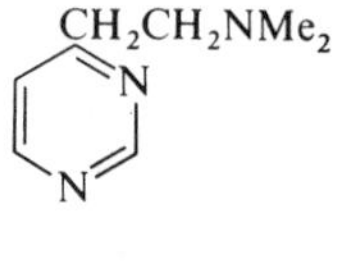

| (152) | (153) | (154) |

| (155) | (156) | (157) | (158) |

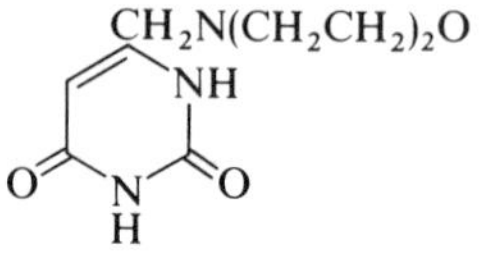

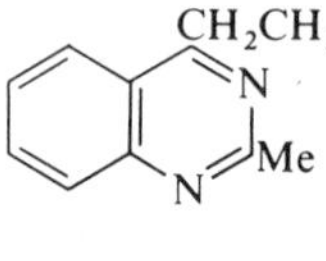

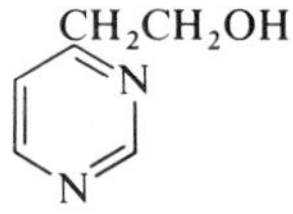

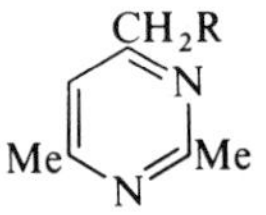

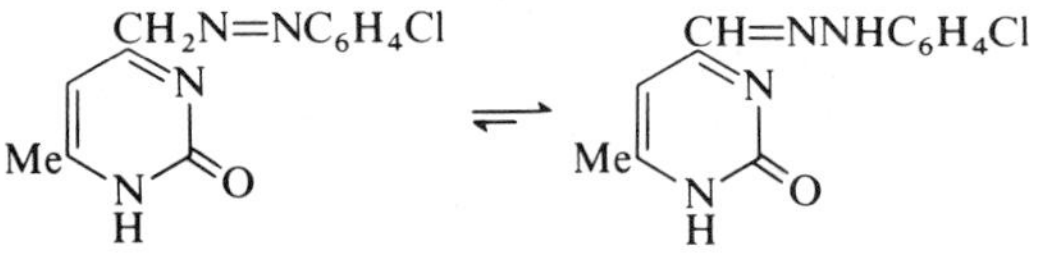

(h) Methyl groups can react with alkylmetals and the resulting complex can undergo further reactions. In this way, 2,4,6-trimethylpyrimidine (**156**; R = H) is converted by phenyllithium into its lithio derivative (**156**; R = Li) which reacts with methyl iodide to furnish 4-ethyl-2,6-dimethylpyrimidine (**156**; R = Me) ⟨52JCS3065⟩.

(i) Although *C*-alkylpyrimidinones do not undergo diazo coupling at their 5-position for lack of sufficient electron release, at least some do so at an alkyl group. Thus 4,6-dimethylpyrimidin-2(1*H*)-one (**77**; R = H) couples with diazotized *p*-chloroaniline to give the azo derivative (**157**), which tautomerizes to the corresponding hydrazone (**158**). Several related examples have been described ⟨77JCS(P1)1985⟩.

(j) Deuteration of *C*-methyl protons in simple methylpyrimidines and their amino and hydroxy derivatives has been studied under acidic and basic conditions. The exchange is acid/base catalyzed with, for example, a minimal rate at pH 4 for 1,4,6-trimethylpyrimidin-2(1*H*)-imine ⟨67JCS(B)171⟩.

### 2.13.2.2.3 Reactivity of aryl or heteroaryl substituents

There are few reports on the reactivity of aryl substituents attached to pyrimidine or quinazoline. However, nitration may be carried out with some success.

Although early reports on nitration of phenylpyrimidines describe only *m*-nitro derivatives, the situation is far less simple because the nature of the products depends on the conditions and reagents. Thus 4-phenylpyrimidine (**159**) in nitric/sulfuric acid gives a 2:3 mixture of 4-*o*- (**160**) and 4-*m*-nitrophenylpyrimidine (**161**); in nitric/trifluoroacetic acid, all three isomers (**160–162**) are formed in the ratio 9:6:5 ⟨67CJC1431⟩. Nitration of 4,6-dimethyl-5-phenylpyrimidin-2-amine by sulfuric acid/potassium nitrate gives mainly the *m*-nitro derivative (**163**), accompanied by some 4,6-dimethyl-5-*p*-nitrophenylpyrimidin-2(1*H*)-one (**164**; R = Me) ⟨71JCS(C)250⟩; under similar conditions, 5-phenylpyrimidin-2-amine gives only the *p*-nitro derivative (**165**; R = NH₂) and 2-methylsulfonyl-5-phenylpyrimidine gives a similar derivative (**165**; R = SO₂Me) ⟨71JCS(C)425⟩. Both 5-phenyl-

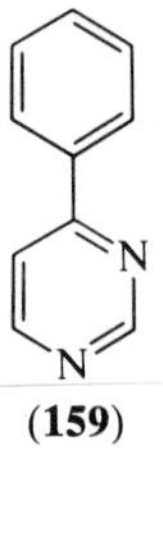

| (159) | (160) | (161) | (162) |

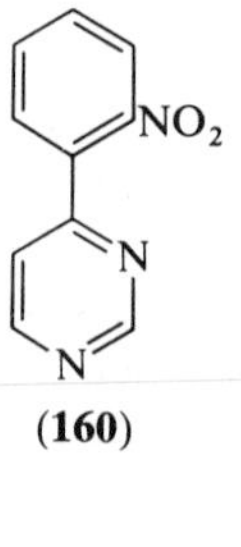

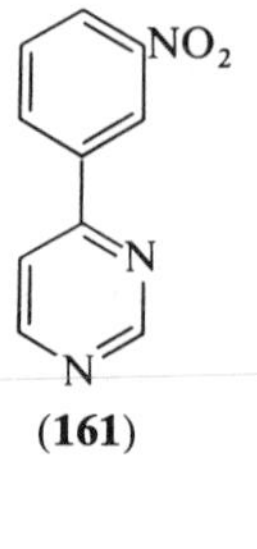

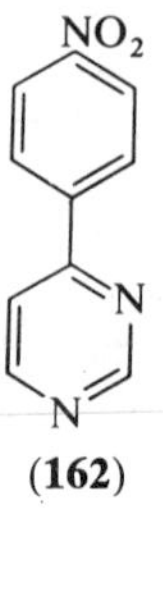

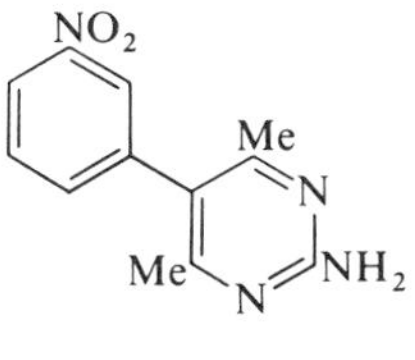

(163) 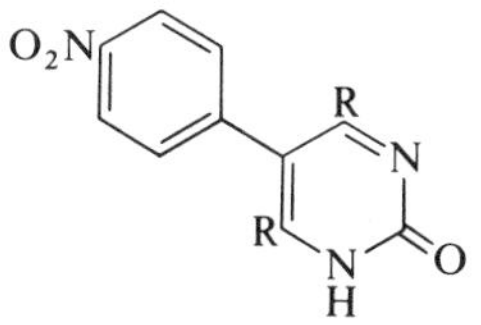  (164) 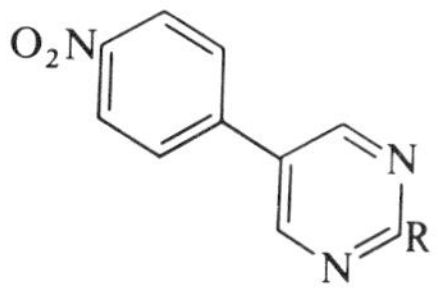  (165)

pyrimidin-2(1*H*)-one and 2-methoxy-5-phenylpyrimidine give their respective *p*-nitro derivatives (**164**; R = H) and (**165**; R = OMe) ⟨70JCS(C)214⟩.

### 2.13.2.2.4 *Reactivity of acyl groups*

(*a*) The oxidation of pyrimidinecarbaldehydes to the corresponding carboxylic acids may be done by alkaline permanganate or acidic dichromate, as exemplified in the conversion of 6-oxo-1,4-dihydropyrimidine-4-carbaldehyde (**166**) into the acid (**167**); the Cannizzaro disproportionation reaction on the same substrate gives the acid (**167**) and 6-hydroxymethyl-pyrimidin-4(3*H*)-one (**168**) in equal amounts ⟨57BSB292⟩. Alkaline hypochlorite oxidizes 6-chloro-4-phenylquinazoline-2-carbaldehyde (**169**; R = CHO) to the acid (**169**; R = CO$_2$H), albeit in poor yield ⟨64JOC332⟩.

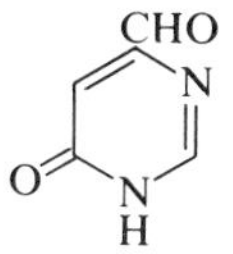

(166) 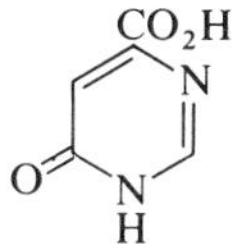  (167) 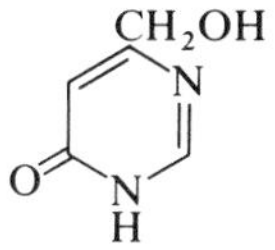  (168) 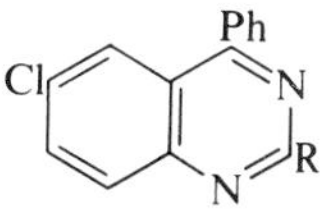  (169)

(*b*) The reduction of pyrimidinecarbaldehydes can be done by sodium borohydride: the primary alcohol (**168**) is better made this way from the aldehyde (**166**) than by the above Cannizzaro reaction, and the aldehyde (**170**; R = CHO) gives 2-methylthiopyrimidine-4,5-dimethanol (**170**; R = CH$_2$OH) in good yield ⟨68JCS(C)1203⟩. The same reagent may reduce ketones to the secondary alcohols or even further: the keto ester (**171**; R = COCO$_2$Et) thus gives a mixture of the products (**171**; R = CHOHCH$_2$OH) and (**171**; R = CH$_2$CH$_2$OH) ⟨65JOC2398⟩. Hydrogenation over palladium or treatment with lithium aluminum hydride also reduces aldehydes ⟨72LA(766)73⟩.

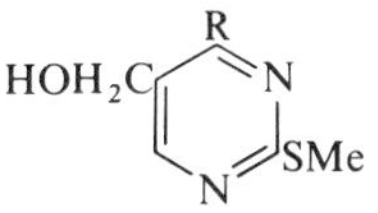

(170) 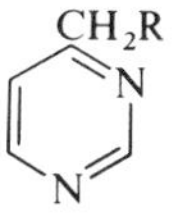  (171) 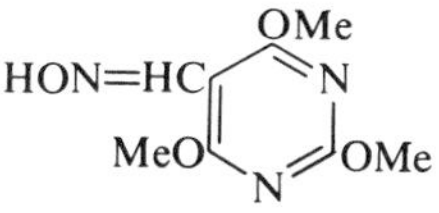  (172) 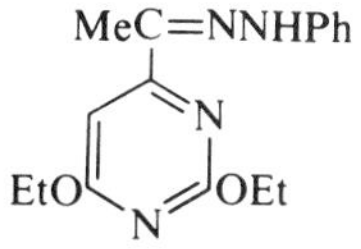  (173)

(*c*) The usual derivatives of pyrimidine aldehydes and ketones are formed normally. For example, the aldoxime (**172**) is formed from the corresponding aldehyde in the usual way ⟨67JOC2308⟩, the phenylhydrazone (**173**) from the appropriate ketone likewise ⟨56JA2136⟩, and the literature contains innumerable hydrazones, dinitrophenylhydrazones, semicarbazones, Schiff bases and related derivatives of both aldehydes and ketones in the pyrimidine and quinazoline series ⟨70HC(16-S1)315⟩.

(*d*) Pyrimidinecarbaldehydes condense with compounds with an activated methylene group to yield styryl- or other (substituted) vinyl-pyrimidines. Thus the aldehyde (**174**) and *p*-nitrophenylacetic acid give (after spontaneous decarboxylation) the *p*-nitrostyryl-pyrimidine (**175**; R = C$_6$H$_4$NO$_2$) and reaction of the same substrate with nitromethane gives the nitrovinylpyrimidine (**175**; R = NO$_2$) ⟨57JCS4845⟩.

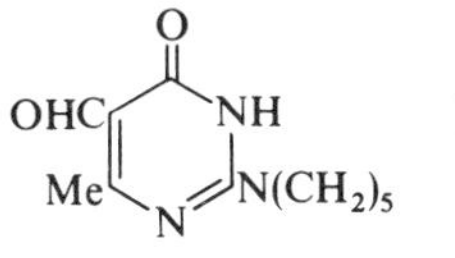

(174) 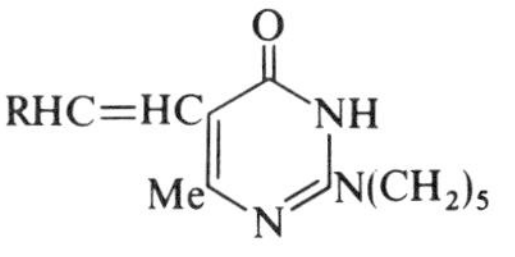  (175) 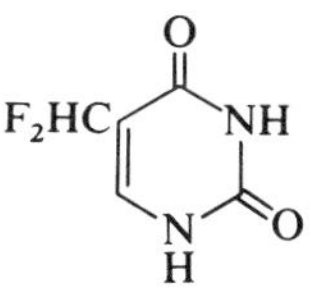  (176) 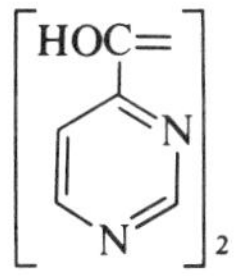  (177)

(*e*) Aldehydes may also undergo a specialized fluorodeoxygenation by heating with sulfur tetrafluoride, so that uracil-5-carbaldehyde gives 5-difluoromethylpyrimidine-2,4(1*H*,3*H*)-dione (**176**) ⟨66JMC876⟩. Pyrimidine-4-carbaldehyde reacts normally with potassium cyanide to yield a 'pyrimidoin', formulated in the enediol form (**177**) ⟨64CB3407⟩.

### 2.13.2.2.5 *Reactivity of carboxylic acids and derivatives*

(*a*) The most useful reaction of pyrimidinecarboxylic acids is decarboxylation. In this way, unwanted alkyl groups, which are often obtained conveniently during primary synthesis, may be oxidized to carboxy groups (see Section 2.13.2.2.2) and then removed. The oxidation of 4-methylpyrimidine to pyrimidine-4-carboxylic acid (**178**; R = H) and subsequent decarboxylation constituted a classical route to pyrimidine itself in the last century; however, a more practical route was oxidation of 4,6-dimethylpyrimidine to pyrimidine-4,6-dicarboxylic acid (**178**; R = $CO_2H$) followed by decarboxylation in diphenyl ether at 240 °C ⟨59JCS525⟩. Although the last decarboxylation can be done effectively on a considerable laboratory scale, yields are often better when such reactions are done in small quantities. Thus 5-bromopyrimidine-2,4(1*H*,3*H*)-dione (**179**; R = H) is best obtained from its carboxylic acid (**179**; R = $CO_2H$) by dry heating 0.5 g portions at 285 °C for 5 minutes ⟨63CCC2491⟩. Conditions for decarboxylation vary considerably according to the compound: for example, recrystallization from ethanol, dry heating at 240–280 °C with or without vacuum, in quinoline at *ca.* 250 °C with or without copper powder, in benzophenone at 280 °C, in acetic anhydride at 120 °C, in refluxing *N*-ethylaniline or morpholine, *etc.* 5-Carboxylic acids seem to be rather less easily decarboxylated than others. The dicarboxylic acid (**180**) loses one of its carboxyl groups on mere acidification of its dipotassium salt ⟨65JHC1⟩. Quinazolines with carboxy groups on the pyrimidine ring decarboxylate normally but those with carboxy groups on the benzene ring do not, or only with the greatest difficulty.

(**178**)          (**179**)          (**180**)

Although it is seldom used, esterification of pyrimidinecarboxylic acids proceeds normally. Conditions are illustrated by the conversion of pyrimidine-4-carboxylic acid (**181**; R = H) into its methyl ester (**181**; R = Me) by methanol/sulfuric acid (47%), methanol/hydrogen chloride (80%), or by diazomethane (*ca.* 100%) ⟨60MI21300⟩. The isomeric methyl pyrimidine-2-carboxylate is formed by treatment of the silver salt of the acid with methyl iodide. Higher esters, *e.g.* (**182**; R = Bu), are usually made by warming the acid (**182**; R = H) with the appropriate alcohol and sulfuric acid ⟨60JOC1950⟩.

(*b*) Acid chloride formation is usually performed by heating the carboxylic acid with thionyl chloride. Quite often the product is not characterized but immediately used, as in amide preparation. However, an example of a pure acid chloride is 2-methylthio-5-phenylthiopyrimidine-4-carbonyl chloride (**183**; R = Cl), prepared by refluxing the corresponding acid (**183**; R = OH) with thionyl chloride ⟨76CCC2771⟩. Phosphoryl chloride and/or phosphorus pentachloride may be used to convert acids into acid chlorides but, unlike thionyl chloride, they also convert any potential hydroxy groups present into chloro substituents. For example, orotic acid (**184**) in this way gives 2,6-dichloropyrimidine-4-carbonyl chloride (**185**) which is a useful intermediate for further reactions ⟨62JOC3507⟩. Quinazolinecarboxylic acids behave similarly with thionyl and phosphoryl chloride ⟨67HC(24-1)475⟩.

(**181**)          (**182**)          (**183**)          (**184**)

(*c*) Pyrimidinecarboxylic acids undergo a few other reactions. Reaction of the uracil carboxylic acid (**182**; R = H) with sulfur tetrafluoride at 25 °C gives the trifluoromethyluracil (**186**) ⟨66JMC876⟩; orotic acid (**184**) is converted directly into the amide (**187**) by salt formation with butylamine followed by dehydration ⟨64JPR(26)43⟩; and treatment of the acid (**188**; R = OH) with methyllithium gives the ketone (**188**; R = Me), quite adequately confirmed in structure ⟨56JA2136⟩.

(**185**)      (**186**)      (**187**)      (**188**)

The reactions of esters are practically confined to reduction, hydrolysis and aminolysis.

(*d*) The reduction of esters to the corresponding alcohol is done with lithium aluminium hydride and can be applied to ester groups at any position or with the ester group attached to a side-chain. The reaction, usually done in THF or ether, is illustrated by the conversion of ethyl 2-methyl-4-methylaminopyrimidine-5-carboxylate (**189**; R = H) and its 4-dimethyl-amino analogue (**189**; R = Me) into the corresponding alcohols (**190**; R = H or Me) ⟨66ZPC(344)16⟩, and by the reduction of the ester (**191**; R = CO$_2$Et) to the alcohol (**191**; R = CH$_2$OH) ⟨64LA(673)153⟩. With care, an ester can be reduced by lithium aluminium hydride to the aldehyde: ethyl pyrimidine-4-carboxylate (**181**; R = Et) gives at −70 °C pyrimidine-4-carbaldehyde (**192**), albeit in poor yield ⟨65JOC2398⟩.

(**189**)      (**190**)      (**191**)      (**192**)

(**193**)      (**194**)      (**195**)      (**196**)

(*e*) The hydrolysis of esters in the pyrimidine series is quite normal. Because esters are more easily made by primary synthesis than are carboxylic acids, the hydrolysis of esters is a common procedure. Such hydrolysis may be done in acidic or basic media and the main factors in choosing conditions are the stability of other groups in the molecule and the possibility of losing the resulting carboxyl group by decarboxylation. Thus ethyl pyrimidine-5-carboxylate (**193**; R = Et) in 2M sodium hydroxide at 50 °C gives the carboxylic acid (**193**; R = H) in excellent yield ⟨62JOC2264⟩. Likewise, ethyl 1-benzyloxyuracil-5-carboxylate (**194**; R = Et) in alkali gives the corresponding acid (**194**; R = H) but in hydrobromic acid/acetic acid it gives 1-hydroxyuracil-5-carboxylic acid (**195**) ⟨64M265⟩. Acid chlorides are hydrolyzed to their parent carboxylic acids under quite mild conditions.

′(*f*) The aminolysis of esters of pyrimidine occurs normally to yield amides. The reagent is commonly alcoholic ammonia or alcoholic amine, usually at room temperature for 20–24 hours, but occasionally under reflux; aqueous amine or even undiluted amine are used sometimes. The process is exemplified in the conversion of methyl pyrimidine-5-carboxylate (**193**; R = Me) or its 4-isomer by methanolic ammonia at 25 °C into the amide (**196**) or pyrimidine-4-carboxamide, respectively ⟨60MI21300⟩, and in the butylaminolysis of butyl uracil-6-carboxylate (butyl orotate) by ethanolic butylamine to give *N*-butyluracil-5-carboxamide (**187**) ⟨60JOC1950⟩. Hydrazides are made similarly from esters with ethanolic hydrazine hydrate.

The reactions of amides fall into hydrolysis, dehydration and 'degradation' (to amines); acid hydrazides and acid azides undergo additional reactions.

(*g*) The hydrolysis of amides is an unusual reaction in the pyrimidine series although many nitriles have been hydrolyzed to carboxylic acids, presumably *via* unisolated amides. However, 4-aminopyrimidine-5-carboxamide (**197**; R = NH$_2$), is converted into the corresponding carboxylic acid (**197**; R = OH) by hot 10% aqueous alkali ⟨53JCS331⟩ and 6-aminopyrimidine-4-carboxamide with warm sulfuric acid plus sodium nitrite undergoes two reactions to give the carboxylic acid (**198**) ⟨56CB12⟩.

(*h*) The dehydration of amides to yield nitriles has not been used widely in the pyrimidines. Simple examples include the conversion of pyrimidine-2-carboxamide (**199**) and its 4- and 5-isomers into the nitrile (**200**) and respective isomers in good yield by heating with phosphoryl chloride ⟨60MI21300⟩; the 5-isomer is also available by dry distillation of the amide with phosphorus pentoxide at *ca.* 250 °C ⟨62JOC2264⟩. Other successful dehydrating mixtures are phosphoryl chloride in xylene and the phosphoryl chloride/pyridine complex.

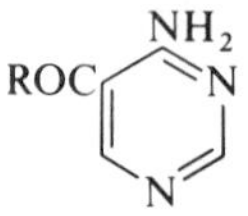

(**197**) 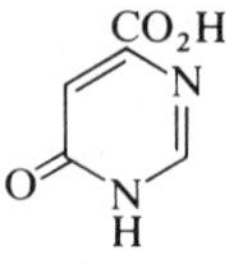      (**198**) 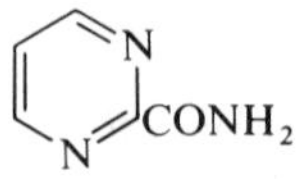      (**199**) 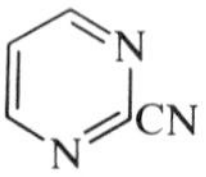      (**200**)

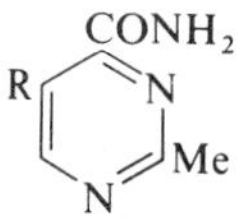

(**201**) 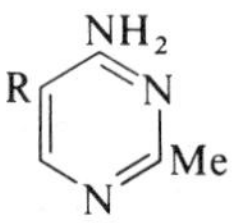      (**202**) 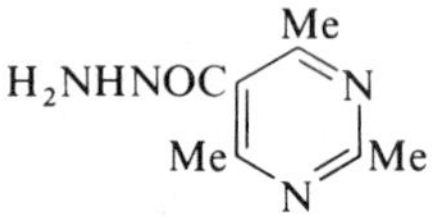      (**203**) 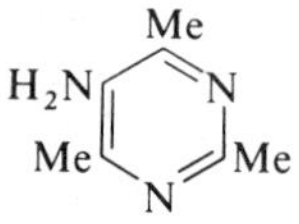      (**204**)

(*i*) Although they are frankly unattractive procedures, the Hofmann or Curtius 'degradation' of amides (or hydrazides) to amines has been used in all positions of pyrimidine. For example, 5-bromo-2-methylpyrimidine-4-carboxamide (**201**; R = Br) undergoes the Hofmann reaction with hypochlorous acid to give 5-bromo-2-methylpyrimidin-4-amine (**202**; R = Br) and the chloro analogue (**201**; R = Cl) likewise gives the amine (**202**; R = Cl) ⟨49CCC223⟩. A short Curtius reaction is represented in the treatment of 2,4,6-trimethyl-pyrimidine-5-carboxhydrazide (**203**) with nitrous acid to give the amine (**204**) ⟨58HCA1806⟩ and there are examples in the older literature of both reactions being used to degrade carbamoylalkyl- to aminoalkyl-pyrimidines ⟨62HC(16)316⟩.

(*j*) Special reactions of hydrazides and azides are illustrated by the conversion of the hydrazide (**205**) into the azide (**206**) by nitrous acid ⟨60JOC1950⟩ and thence into the urethane (**207**) by ethanol ⟨64FES(19)1050⟩; the conversion of the same azide (**206**) into the *N*-alkylamide (**208**) by ethylamine; the formation of the hydrazone (**209**) from acetaldehyde and the hydrazide (**205**); and the *N*-acylation of the hydrazide (**205**) to give, for example, the formylhydrazide (**210**) ⟨65FES(20)259⟩. It is evident that there is an isocyanate intermediate between (**206**) and (**207**): such compounds have been isolated sometimes, *e.g.* (**211**). Several of the above reactions are involved in some Curtius degradations.

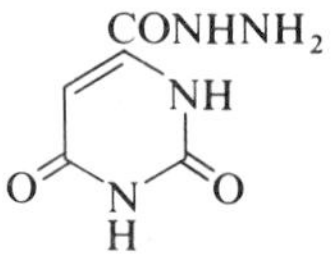

(**205**) 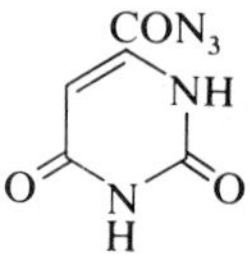      (**206**) 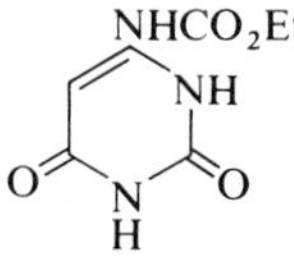      (**207**) 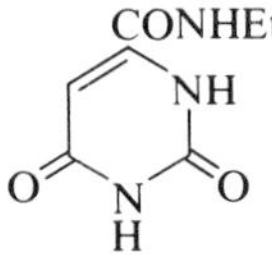      (**208**)

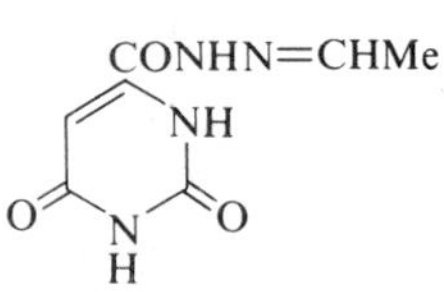

(**209**) 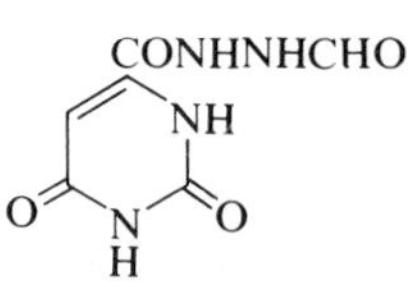      (**210**) 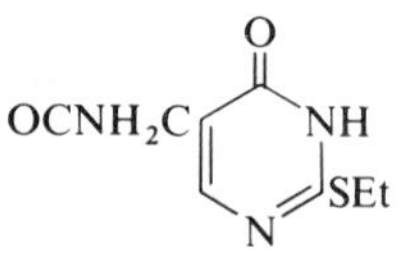      (**211**)

The reactions of pyrimidine nitriles are quite numerous.

(*k*) The reduction of nitriles to aminomethyl derivatives is usually done by hydrogenation under pressure over Raney nickel (or sometimes platinum or palladium) in the presence of an excess of alcoholic ammonia, which inhibits the formation of secondary amines at least in part. When groups sensitive to hydrogenation are present the reagent of choice is lithium aluminium hydride. For example, hydrogenation of 4-dimethylamino(or methoxy)-2-methylpyrimidine-5-carbonitrile (**212**; R = NMe$_2$ or OMe) gives the aminomethyl derivatives (**213**; R = NMe$_2$ or OMe) in excellent yield although the simpler nitrile (**212**; R = H) gives only the tetrahydro derivative of the expected product (**213**; R = H) ⟨66JCS(C)649⟩. However, 4-chloro-2-ethylthiopyrimidine-5-carbonitrile (**214**; R = CN), which has two groups inimical to hydrogenation, is reduced by LAH to the aminomethyl derivative (**214**; R = CH$_2$NH$_2$) ⟨54LA(588)45⟩.

(212)  (213)  (214)  (215)

(216)  (217)  (218)  (219)

(*l*)  The hydrolysis of nitriles usually occurs at the C—N bond to give amides or carboxylic acids according to conditions but, with appropriate activation, nucleophilic displacement of the cyano group may occur to give the pyrimidinone. Hydrolysis to the carboxylic acid is usually done in hot aqueous alkali: 4,6-dimethylpyrimidine-2-carbonitrile (215; R = CN) gives the corresponding acid (215; R = CO$_2$H) in 1M aqueous alkali at 60 °C ⟨56M526⟩; and 2-methylthiopyrimidine-4-carbonitrile (216; R = CN) gives the acid (216; R = CO$_2$H) under moderate conditions (boiling 2M sodium hydroxide for 2 h) although too vigorous conditions lead to 2-oxo-1,2-dihydropyrimidine-4-carboxylic acid ⟨64JHC201⟩. Hydrolysis to the amide requires more controlled conditions: the nitrile (215; R = CN) in warm dilute ammonia gives the amide (215; R = CONH$_2$) ⟨56M526⟩. Such control is better achieved by hydrolysis in aqueous alkali containing hydrogen peroxide (the Radziszewski reaction); in this way, 4-aminopyrimidine-5-carbonitrile (217; R = CN) at 50 °C gives the amide (217; R = CONH$_2$), but without the peroxide it gives mainly the acid (217; R = CO$_2$H) ⟨53JCS331⟩. Alternatively, hydrolysis of nitrile to amide can be done in strong sulfuric acid, usually at room temperature, thus making use of the powerful limitation on water in such media: 2-amino-4-methylpyrimidine-5-carbonitrile (218; R = CN) gives the amide (218; R = CONH$_2$) ⟨67JCS(C)1928⟩. The alternative hydrolysis (and alcoholysis) of pyrimidine nitriles is not well represented but 2-methylpyrimidine-4,5-dicarbonitrile (212; R = CN) reacts with hot water to give 2-methyl-4-oxo-3,4-dihydropyrimidine-5-carbonitrile (219) and with boiling methanol to give the 4-methoxy analogue (212; R = OMe) ⟨68JCS(C)1203⟩; quinazoline-2(and 4)-carbonitriles are displaced by alkoxide ion and other nucleophiles ⟨64CPB43⟩.

(*m*) The reaction of nitriles with hydrogen sulfide leads to thioamides. The conditions are typified in the conversion of 2-(4′,6′-dimethylpyrimidin-2′-yl)acetonitrile (220; R = CN) into the thioamide (220; R = CSNH$_2$) by heating in pyridine containing some triethylamine and presaturated with hydrogen sulfide ⟨71JMC244⟩. The pyridine may be replaced by ethanol and the triethylamine by ammonia: pyrimidine-2-carbonitrile and its 4- and 5-isomers all give, in excellent yield, the corresponding thioamides by warming in alcoholic ammonium hydrogen sulfide ⟨60MI21300⟩.

(*n*) As in other series, the aminolysis and alcoholysis of nitriles has been much used. Direct aminolysis can be done by warming with an aromatic amine in the presence of aluminium chloride: pyrimidine-2-carbonitrile thus gives the *N*-phenylcarboxamidine (221) ⟨60MI21300⟩. Another procedure, often more convenient, is to heat the nitrile (>200 °C) with ammonium benzenesulfonate to give the amidinium benzenesulfonate salt: in this way the same product (221) can be obtained, albeit in rather lower yield ⟨67JCS(C)1204⟩. However, the traditional Pinner method of making amidines may be used. This involves initial alcoholysis of the nitrile in alcoholic hydrogen chloride to give the 'imino ether' followed by treatment with ammonia or an amine to give the amidine. The process is illustrated by

(220)  (221)  (222)  (223)

(224)  (225)  (226)  (227)

the (cold) ethanolysis of pyrimidine-5-carbonitrile to the imino ether (**222**) (as hydrochloride) and subsequent treatment with cold ethanolic ammonia to yield the amidine (**223**) ⟨60MI21300⟩.

(*o*) Some reactions of nitriles include the conversion of pyrimidine-4-carbonitrile by methylmagnesium iodide into the ketone (**224**) ⟨60MI21300⟩; of the nitrile (**215**; R = CN) into the ketone (**215**; R = COPh) using phenylmagnesium bromide ⟨56M526⟩; of 4-aminopyrimidine-5-carbonitrile (**217**; R = CN) into the aldehyde (**217**; R = CHO) by partial reduction over palladium in acidic media to the imine (**225**) followed by gentle hydrolysis in aqueous ammonia ⟨67CB3664⟩; and of 2-methylpyrimidine-4,5-dicarbonitrile (**212**; R = CN) into the ester (**212**; R = CO$_2$Me) (rather than an imino ether) by using warm methanolic hydrogen chloride ⟨65JHC202⟩. The related 'nitrile oxide' (**226**), made by dehydrogenation of the appropriate oxime with NBS/sodium methoxide, reacts with aniline to give the oxime (**227**) ⟨67JOC2308⟩. It is interesting that quinazoline-2-carbonitrile reacts with Grignard reagents normally to yield the corresponding ketones ⟨64CPB43⟩ but quinazoline-4-carbonitrile undergoes displacement of the cyano group by ethylmagnesium bromide to yield 4-ethylquinazoline ⟨62CPB1043⟩.

### 2.13.2.2.6 *Reactivity of amino, nitro and related groups*

(*a*) The hydrolysis of pyrimidinamines can be done directly under acidic or alkaline conditions or indirectly by treatment with nitrous acid. Acidic hydrolysis usually requires quite vigorous conditions and 4/6-amino groups are most susceptible. For example, 2-phenylpyrimidin-4-amine (**228**) requires 33% hydrochloric acid at 150 °C to give the corresponding pyrimidinone (**229**) ⟨67RTC567⟩; the older literature contains examples of the acidic hydrolysis of 2-amino groups; and pyrimidine-4,5-diamine in boiling 6M hydrochloric acid for 4 hours gives 5-hydroxypyrimidin-4(3*H*)-one (**230**) ⟨63JCS5590⟩. Extranuclear amino groups are resistant to acid treatment. Alkaline hydrolysis of amino groups is more common, although the ease varies with other substituents present. Thus pyrimidin-2-amine (**231**; R = H) requires 10M alkali at 120 °C to give pyrimidin-2(1*H*)-one ⟨50MI21300⟩. *N*-Methylpyrimidin-2-amine requires 160 °C to give the same product ⟨63CB534⟩, but 5-nitropyrimidin-2-amine (**231**; R = NO$_2$) requires only dilute alkali or even aqueous ammonia at 90–100 °C to give 5-nitropyrimidin-2(1*H*)-one. A fascinating kinetic study of 2- and 4-'deamination', *i.e.* hydrolysis, in alkali ⟨69JCS(B)96⟩ indicates that the rates for 4-amino derivatives are higher than those for 2-amino derivatives and that both are increased by an (inert) electron-withdrawing substituent, *e.g.* bromo, but decreased by an electron-releasing group, *e.g.* methyl. 5-Amino and extranuclear amino groups appear to withstand alkaline treatment. The indirect hydrolysis of a primary amino group by initial treatment with nitrous acid probably proceeds *via* a diazonium salt. A good example is the treatment of 4,6-dimethylpyrimidin-2-amine (**215**; R = NH$_2$) in aqueous acetic acid at about 50 °C with sodium nitrite to yield 4,6-dimethylpyrimidin-2(1*H*)-one (**232**) ⟨66JCS(C)2031⟩; 2-benzylthio-6-chloropyrimidin-4-amine (**233**) is converted similarly into the corresponding pyrimidin-4(3*H*)-one (**234**), indicating the gentle nature of a method which leaves both a chloro and a benzylthio substituent intact ⟨62JOC1462⟩. The reaction is also applicable to 5-amino groups ⟨56YZ776⟩ and its application to an extranuclear amino group is illustrated by the conversion of 2,4-dimethylpyrimidin-5-ylmethylamine (**235**; R = NH$_2$) into the corresponding methanol (**235**; R = OH) ⟨65JMC750⟩. The reaction is contraindicated if the

(**228**)          (**229**)          (**230**)          (**231**)

(**232**)          (**233**)          (**234**)          (**235**)

5-position is both free and suitably activated for nitrosation, if there is a secondary amino group ripe for nitrosation, and for pyrimidine-4,5-diamines with which nitrous acid will form $v$-triazolopyrimidines.

(*b*) The transformation of amino to halogeno substituents is done also *via* diazotization. The reaction is most used for 2-amino groups and there is always a substantial proportion of the corresponding pyrimidin-2(1*H*)-one formed as a byproduct. Thus when 4-*t*-butyl-pyrimidin-2-amine (**236**; R = NH$_2$) is treated in concentrated hydrochloric acid at −10 °C with sodium nitrite, 4-*t*-butyl-2-chloropyrimidine (**236**; R = Cl) results along with a comparable yield of 4-*t*-butylpyrimidin-2(1*H*)-one ⟨67JCS(C)1928⟩. Likewise, when an excess of bromide ion is present, a bromopyrimidine results, as in the transformation of 4,6-dimethyl-pyrimidin-2-amine (**237**; R = NH$_2$) into 2-bromo-4,6-dimethylpyrimidine (**237**; R = Br) with 4,6-dimethylpyrimidin-2(1*H*)-one (**232**) as byproduct ⟨64JOC943⟩. 2-Amino- yields 2-fluoro-pyrimidine by diazotization in hydrofluoric acid ⟨75CCC1390⟩. Similar Sandmeyer-type reactions can be used to convert 5-amino-6-methylpyrimidine-2,4(1*H*,3*H*)-dione (**238**; R = NH$_2$) into its 5-chloro (cuprous chloride), 5-bromo (cuprous bromide) or 5-iodo (potassium iodide) analogues (**238**; R = Cl, Br or I) ⟨66MI21300⟩. The reaction can be applied to extranuclear amino groups ⟨61JOC598⟩ and occasionally to 4-amino groups ⟨67JCS(C)1204⟩.

(236)    (237)    (238)    (239)

(240)    (241)    (242)

(*c*) The formation of simple Schiff bases is easiest between pyrimidin-5-amines and aldehydes. Thus 5-aminopyrimidine-2,4(1*H*,3*H*)-dione (5-aminouracil) and cinnamalde-hyde in warm dilute hydrochloric acid give the cinnamylideneamino derivative (**239**) ⟨67JCS(C)1745⟩, while 5,6-diaminouracil reacts only at the 5-position on treatment with benzaldehyde in ethanol to give 4-amino-5-benzylideneaminouracil (**240**) ⟨77JCS(P1)1336⟩. DMF reacts as an aldehyde with pyrimidinamines: 2,6-dichloropyrimidin-4-amine gives the Schiff base (**241**; R = Cl) ⟨65M1567⟩; however, it is better to use dimethylformamide dimethyl acetal, which with pyrimidin-4-amine gives the product (**241**; R = H) under very gentle conditions. Pyrimidin-4-ylhydrazine reacts similarly to give the analogue (**242**) ⟨76S833⟩. Most Schiff bases undergo conversion to the parent amines under hydrolytic conditions appropriate to the particular case.

(*d*) Most primary and secondary pyrimidinamines undergo acylation normally but the ease of reaction varies with the position and type of amino group: thus, any hydrazino group > 5-amino > 2- or 4-amino. Under vigorous conditions, two acyl groups can be accommodated on a single primary amino group. Some typical examples of acylation are 4,6-diacetamidopyrimidine (**243**; R = Ac) and 4-acetamido-6-aminopyrimidine (**243**; R = H) (2 and 1 mol of acetic anhydride, respectively, at reflux) ⟨62JMC808⟩; 5-acetamido-2-amino-4,6-dimethylpyrimidine (acetic anhydride at 50 °C); and 5-formamido-4,6-bismethylaminopyrimidine (**244**) (90% formic acid at 100 °C) ⟨65JCS3770⟩. The ease of formylating a hydrazinopyrimidine is illustrated in the derivative (**245**), prepared by warming the parent amine with butyl formate ⟨63JOC923⟩. Treatment of an amino or hydrazino compound with an orthoester leads to the enol ether of an acyl derivative: for example, pyrimidin-4-ylhydrazine and triethyl orthoacetate at 75 °C give the ethoxyethylidene deriva-tive (**246**); under more vigorous conditions, this cyclizes to 3-methyl-*s*-triazolo[4,3-*c*]pyrimidine ⟨78AJC2505⟩. Benzoylation is usually done with benzoyl chloride in pyridine, and trichloro- or trifluoro-acetylation is commonly done with the halogenated acid or its anhydride. Deacylation often needs quite vigorous treatment, such as boiling hydrochloric acid, hot methanolic hydrogen chloride or boiling methanolic sodium methoxide; deacylation of trihalogenoacetyl derivatives is somewhat easier ⟨70HC(16-S1)245⟩.

Structures (243), (244), (245), (246) with labels NHR/AcHN, OHCHN/NHMe/MeHN, NHNHCHO/H₂N/Cl, NHN=CMeOEt

Structures (247), (248), (249) with labels NHSO₂C₆H₄NO₂-*p*, NHR/Me/Me, R/O₂N/NMe₂

The much-used sulfonylation by aromatic sulfonyl chlorides is illustrated by the conversion of quinazolin-4-amine into the sulfonamide (247) using *p*-nitrobenzenesulfonyl chloride at 60 °C ⟨56JCS3509⟩.

(*e*) The process of transamination can be used in the pyrimidine and quinazoline series; the reaction is applicable to most 4/6- and some 2-amino groups. For example, heating 2,6-dimethylpyrimidin-4-amine (248; R = H) hydrochloride with butylamine at 170 °C for 20 hours gives an excellent yield of *N*-butyl-2,6-dimethylpyrimidin-4-amine (248; R = Bu). Similarly, 4,6-dimethylpyrimidin-2-amine with benzylamine gives *N*-benzyl-4,6-dimethyl-pyrimidin-2-amine ⟨60JA3971⟩. The substrate need not be a primary amine and diamines can undergo partial transamination if the conditions are appropriate: thus *N,N,N',N'*-tetramethyl-5-nitropyrimidine-2,4-diamine (249; R = NMe₂) on warming with hydrazine hydrate for a few minutes gives 2-dimethylamino-5-nitropyrimidin-4-ylhydrazine (249; R = NHNH₂) ⟨64JHC175⟩.

(*f*) Treatment of a simple pyrimidin-2(or 4/6)-amine with an alkyl halide invariably leads first to alkylation at a ring-nitrogen atom with the formation of an *N*-alkylated pyrimidinimine, which may rearrange subsequently to an alkylaminopyrimidine ⟨70HC(16-S1)284⟩. For example, pyrimidin-2-amine (250) and methyl iodide give the imine (251) as hydriodide which, in alkaline solution, ring-opens to the intermediate (252); this subsequently undergoes reclosure involving the imino group to give the methylamino compound (253). The change (251) → (253) is known as a Dimroth rearrangement: such rearrangements are facilitated by additional electron-withdrawing groups but they are retarded or even prevented by additional electron-releasing groups; their mechanism is well investigated and they occur in many π-deficient heteroaramatic systems ⟨B-68MI21300⟩. In contrast, pyrimidin-4-amine undergoes *N*-alkylation at the 1-position to give the imine (254) in which Dimroth rearrangement is impossible: in alkaline media, the imine simply hydrolyzes to afford 1-methylpyrimidin-4(3*H*)-one (255) ⟨55JCS4035⟩. The rearrangement (251) → (253) proceeds not only in aqueous media but also in neat secondary amines which form analogous intermediates, *e.g.* (256); no reaction occurs in tertiary amines and primary amines cause irreparable ring fission ⟨68JCS(C)1452⟩. 2-Aminoquinazolin-4(3*H*)-one (257; R = H) gives 3-alkyl derivatives which undergo Dimroth rearrangement to 2-alkyl-aminoquinazolin-4(3*H*)-ones (257; R = alkyl) ⟨60JCS3540⟩.

Structures (250), (251), (252), (253) — reaction scheme with NH₂ → N-Me/NH → HOHC/HN-Me/NH → NHMe

Structures (254), (255), (256), (257) — reaction scheme with NH/Me → O/Me → Et₂NHC/HN-Me/NH → O/NH/NHR

Unlike simple alkyl halides, ethyl chloroformate appears to react with primary and secondary amino groups in any position to give directly the corresponding urethane, *e.g.* (258) ⟨64JMC364⟩. Such alkylations proceed in pyridine, aqueous alkali or even warm benzene ⟨62JOC982⟩.

(*g*) Although regular amino groups are not easily displaced, the (quaternary) trialkylammonio substituent is an excellent leaving-group. However, vigorous conditions must be avoided because the group is prone to split out an alkyl halide leaving a dimethylamino substituent, useless for displacement. For example, reaction of *N,N,N*-trimethyl-2-methyl-thiopyrimidin-4-ylammonium chloride (**259**) with potassium cyanide in acetamide at 90 °C gives 2-methylthiopyrimidine-4-carbonitrile (**260**) ⟨64JHC130⟩; likewise, 2,6-dimethoxy-pyrimidin-4-yl-*N,N,N*-trimethylammonium chloride reacts with aqueous sodium azide or potassium fluoride in ethylene glycol to give 4-azido- (**261**; R = N$_3$) or 4-fluoro-2,6-dimethoxypyrimidine (**261**; R = F), respectively ⟨61JOC3392⟩. Other nucleophilic displacement reactions also occur ⟨72JCS(P1)1269⟩. Extranuclear but not 5-ammonio groups may be made easily and used similarly ⟨70HC(16-S1)262⟩.

(**258**)    (**259**)    (**260**)    (**261**)

(*h*) Pyrimidinamines undergo a variety of other reactions of less general significance. Pyrimidin-5-amine condenses with nitrosobenzene to give 5-phenylazopyrimidine ⟨51JCS1565⟩. Most pyrimidine-4,5-diamines which bear an additional amino, alkoxy or alkylthio group are prone to rapid oxidation as free bases in air. Thus they very soon appear almost black, although the coloured material (of yet unknown composition) forms only a minute fraction of the whole. By working in strictly anaerobic conditions, 4,6-bismethyl-aminopyrimidin-5-amine (**262**) can be obtained as a colourless solid ⟨60JCS1978⟩; its salts and those of similar pyrimidines are relatively stable in air. Hydrazino groups may be removed from pyrimidines oxidatively; silver oxide in methanol is particularly effective. In this way 4,6-dihydrazino-5-nitropyrimidine (**263**; R = NHNH$_2$) gives 5-nitropyrimidine (**263**; R = H) ⟨67JCS(C)573⟩ and other examples are known. Hydrazino groups may be partly degraded by hydrogenation to an amino group: thus 2,6-dimethylpyrimidin-4-ylhydrazine gives 2,6-dimethylpyrimidin-4-amine by hydrogenation in ethanol over Raney nickel or by boiling with an excess of Raney nickel (catalyst) in ethanol; likewise, 1-(2′,6′-dimethoxy-pyrimidin-4′-yl)-1-methylhydrazine (**264**) (or a 2-acylated derivative) gives the corresponding methylamino derivative (**261**; R = NHMe) ⟨62YZ528⟩. Pyrimidinylhydrazines are converted into the corresponding azides by treatment with nitrous acid. Such azido derivatives are in equilibrium with the corresponding tetrazolopyrimidines, *e.g.* (**265**) ⟨65JOC826⟩.

(**262**)    (**263**)    (**264**)    (**265**)

(*i*) The reduction of 5-nitropyrimidines represents the only important reaction of such derivatives. The choice of method is determined by the other groups present. Hydrogenation over Raney nickel or palladium has the widest laboratory application and it is contraindicated only by the presence of a thioxo, halogeno or a few other substituents, sensitive to such treatment. The use of hydrogen pressures above 1 atmosphere is seldom required and even quite insoluble nitropyrimidines may be hydrogenated, albeit slowly, in ethanol because the product is usually more soluble: even 4-amino-5-nitropyrimidin-2(1*H*)-one (**266**), which is virtually insoluble in methanol, may be hydrogenated as a suspension therein, and this despite the fact that the resulting 4,5-diaminopyrimidin-2(1*H*)-one (**267**) is also poorly soluble in methanol ⟨57MI21300⟩. Although the choice of solvent is limited with Raney nickel, palladium catalysts may be used in a variety of solvents such as DMF, 2-methoxyethanol, acetic acid or hydrochloric acid; platinum catalysts may be used in dilute ammonia which is sometimes convenient if a substrate is soluble as its anion ⟨70HC(16-S1)99⟩.

(**266**)    (**267**)    (**268**)    (**269**)

When hydrogenation is inapplicable, or for large scale working, several convenient classical reducing agents are available. They are illustrated by the reduction of 6-methoxy-5-nitropyrimidine-4(3$H$)-thione (**268**) by dithionite ⟨61JOC4961⟩, 5-nitro-2-trifluoromethyl-pyrimidine-4,6-diamine by iron/acetic acid ⟨61JOC4504⟩, 6-methylamino-5-nitropyrimidine-2,4(1$H$,3$H$)-dione by zinc/formic acid to the formamido analogue (**269**) directly ⟨60JCS4776⟩, and 5-nitro-2-styrylpyrimidin-4-amine by stannous chloride in hydrochloric acid ⟨67AJC1041⟩; other agents include zinc dust/water, hydriodic acid/red phosphorus, ferrous hydroxide and sodium hydrogen sulfide ⟨70HC(16-S1)99⟩.

(*j*) The reduction of 5- and 4-nitrosopyrimidines is an often-used reaction. Hydrogenation is seldom used but is effective over Raney nickel for preparing the corresponding 5-aminopyrimidine from 6-amino-2-dimethylamino-5-nitrosopyrimidin-4(3$H$)-one (**270**), its 6-methylamino homologue and related nitrosopyrimidines ⟨62CB(95)1597⟩. The classical reducing agent sodium or ammonium hydrogen sulfide is still used occasionally, for example in reducing 2,6-diamino-5-nitrosopyrimidine-4(3$H$)-thione (**271**) or its *S*-methyl derivative ⟨65JPS1626⟩. In fact, sodium dithionite is nearly always used for such reductions, usually in mildly alkaline solution when this not contraindicated. Typical procedures are described for so reducing 6-amino-5-nitroso-2-thiouracil (**272**) ⟨60JOC148⟩, 2-amino-6-ethylamino-5-nitrosopyrimidin-4(3$H$)-one ⟨62RTC443⟩ and many other analogues ⟨70HC(16-S1)103⟩. The reduction of 2-amino-5-hydroxy-6-nitrosopyrimidin-4(3$H$)-one (**273**) to 'divicine' is more effective with dithionite than with hydrogenation over palladium ⟨64JCS1001⟩.

(270)            (271)            (272)            (273)

(*k*) The ability of a nitroso group to condense with an activated methylene group has been used in the Timmis synthesis of pteridines. For example, benzyl phenyl ketone (**274**) condenses with 5-nitrosopyrimidine-2,4,6-triamine (**275**) to give 6,7-diphenylpteridine-2,4-diamine (**276**); preparative and other aspects have been reviewed ⟨B-64MI21300⟩.

(*l*) 4-Alkylamino-5-nitrosopyrimidines undergo acylation in their isonitroso forms to yield acyloxyimino derivatives, such as the 5-benzoyloxyimino-6-methylimino-5,6-dihydropyrimidine-2,4(1$H$,3$H$)-dione (**277**) ⟨70CB900⟩, which are powerful acylating agents in their own right.

(274)       (275)            (276)            (277)

(278)            (279)            (280)

(*m*) Reduction of 4- and 5-arylazopyrimidines is the only reaction which has been studied extensively. Catalytic hydrogenation is commonly used but all manner of other reagents are also effective in cleaving the azo bond to produce pyrimidin-5-amines. For example, reaction of the *p*-chlorophenylazopyrimidine (**278**) with Raney nickel at 95 °C and 50 atmospheres of hydrogen gives 6-methylaminopyrimidine-2,4,5-triamine ⟨57JCS2146⟩, but the use of pressure in such hydrogenations appears to be more a convenience than a necessity. Other reagents ⟨70HC(16-S1)109⟩ include sodium hydrogen sulfide, tin/hydrochloric acid for the thioether (**279**), zinc dust/sulfuric acid, sodium dithionite, zinc/acetic acid, stannous chloride, electrolytic reduction, and Raney nickel (without hydrogen) to convert the phenylazopyrimidinethione (**280**) into 5-amino-6-methylpyrimidin-4(3$H$)-one in one operation ⟨66MI21301⟩.

### 2.13.2.2.7 Reactivity of hydroxy, alkoxy and related groups

(a) The most important reaction of pyrimidinones (2- and 4/6-'hydroxy'pyrimidines) and quinazolinones is replacement by halogeno substituents. In simple cases, this is achieved by boiling in phosphoryl chloride: for example, 2-ethylpyrimidin-4(3H)-one (281) gives 4-chloro-2-ethylpyrimidine (282) ⟨65RTC1101⟩ and unsaturated alkyl groups are no bar to success ⟨67JCS(C)1922⟩. The same procedure is used sometimes for pyrimidinediones and even pyrimidinetriones ⟨64JCS3204⟩ but it is necessary often to add a tertiary organic base as a 'catalyst' in order to obtain good yields: 4,6-dichloro-5-phenylpyrimidine (283) is made thus ⟨65CCC3730⟩ using added diethylaniline. A similar procedure is nearly always necessary in the presence of nitro, amino, cyano, trifluoromethyl or an ether grouping attached to the original pyrimidinone ⟨70HC(16-S1)110⟩. Although phosphorus pentachloride can be used also for such reactions, it is less convenient and can cause incidental unwanted chlorinations, especially if the 5-position is free or if there is a methyl group present. For example, treatment of 4-methylpyrimidin-2(1H)-one (284) with phosphorus pentachloride in phosphoryl chloride at 135 °C gives 2-chloro-4-trichloromethylpyrimidine (285) ⟨61JOC4419⟩. Phosphoryl bromide may be used to convert pyrimidinones into bromopyrimidines, e.g. 2-bromo-4,6-diphenylpyrimidine (286) ⟨74RTC227⟩.

(281)　　　(282)　　　(283)　　　(284)

(285)　　　(286)　　　(287)　　　(288)

5-Hydroxy groups are usually unaffected by phosphorus halides but extranuclear (alcoholic) hydroxy groups are so converted into chloro/bromo substituents: thus, 5-hydroxymethylpyrimidine-2,4(1H,3H)-dione (287; R = OH) gives 2,4-dichloro-5-chloromethylpyrimidine (288) in excellent yield ⟨66LA(692)119⟩. However, thionyl chloride is more convenient, especially if 'nuclear' hydroxy groups are not to be affected: in this way, the above substrate (287; R = OH) gives only the monochloro derivative (287: R = Cl) ⟨61CCC893⟩. Finally, the same transformation (287; R = OH → Cl) can be performed by stirring the substrate in concentrated hydrochloric acid at room temperature for 30 minutes ⟨B-78MI21300⟩. The bromo (287; R = Br) and iodo (287; R = I) analogues may be made similarly ⟨66JMC97⟩.

(b) The thiation of pyrimidinones (whether tautomeric with pyrimidinols or not) can be done by heating with good quality phosphorus pentasulfide in xylene, tetralin or (best) pyridine or one of its homologues; pyrimidin-4/6-ones react more readily than pyrimidin-2-ones. The process is illustrated in the conversion of pyrimidin-4(3H)-one (289; X = O) into the corresponding thione (289; X = S) ⟨65JCS2778⟩, of 5-bromopyrimidin-2(1H)-one (290; X = O) into the thione (290; X = S) in only 25% yield ⟨66AJC2321⟩, and of pyrimidine-2,4(1H,3H)-dione (291; X = O) into the monothione (291; X = S) ⟨62CPB647, 80AJC1147⟩. The peculiar efficacy of α-picoline as solvent is shown in the conversion of 2-thiouracil into pyrimidine-2,4(1H,3H)-dithione (292) within 15 minutes, as compared with 8 hours when xylene is used ⟨B-68MI21301⟩. Thiation in the presence of an amino group sometimes gives poor yields, but many otherwise successful examples are known. The formation of 3-methylpyrimidine-4(3H)-thione (293; X = S) from the corresponding methyl-pyrimidinone (293; X = O) and the thione (294; X = S) from the quinazolinone (294; X = O) are two examples of the thiation of non-tautomeric oxo substituents ⟨62JCS3129⟩. Extra-nuclear and 5-hydroxypyrimidines, as well as 5-, 6-, 7- and 8-hydroxyquinazolines, resist thiation.

(c) Pyrimidin- and quinazolin-2(and 4/6)-ones resist acylation but pyrimidin- and quinazolin-ols undergo O-acylation easily. Thus 5-hydroxypyrimidin-2(1H)-one (295; R = H) and hot benzoyl chloride give only the monobenzoyl derivative (295; R = PhCO) ⟨65JCS7116⟩

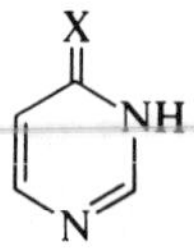

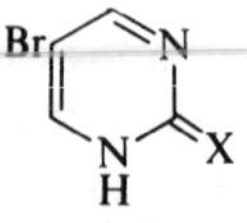

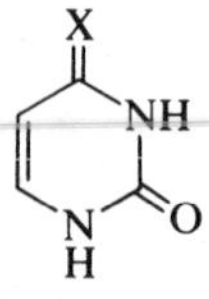

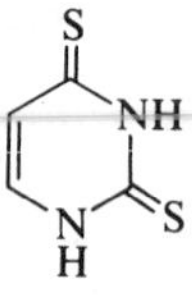

(289)                (290)                (291)                (292)

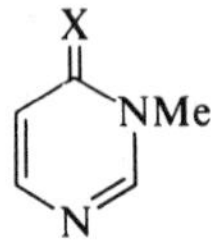

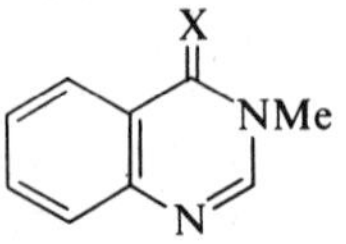

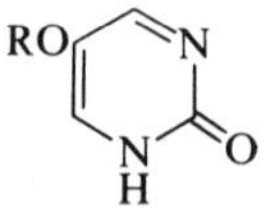

(293)                (294)                (295)

and 5-β-hydroxyethylpyrimidine-2,4(1H,3H)-dione gives the corresponding acetoxyethyl derivative on treatment with acetic anhydride in pyridine ⟨64JOC2670⟩.

(d) The alkylation of pyrimidinones usually gives N(1)- and/or N(3)-alkyl derivatives with very little O-alkyl derivative, even when diazomethane is used. For example, pyrimidin-2(1H)-one (296) and diazomethane give 52% of 1-methylpyrimidin-2(1H)-one (297) but only 17% of 2-methoxypyrimidine (298); pyrimidin-4-(3H)-one (289; X = O) behaves similarly to give mainly the N-methyl derivative (299) with a little 4-methoxypyrimidine. When dimethyl sulfate/alkali is used to methylate 2-thiouracil (300), the S-methyl derivative (301) is formed immediately, followed by two N-methyl derivatives (302) and (303) in almost equal quantities; no O-methyl derivative is detectable ⟨55JCS211⟩. Pyrimidine-2,4(1H,3H)-dione (304; R = H) with diazomethane, methyl iodide/sodium methoxide or dimethyl sulfate/alkali gives only its 1,3-dimethyl derivative (304; R = Me); only in solvents of low polarity, e.g. toluene, can monomethylation of the potassium salt of uracil with methyl iodide be achieved to give 1-methylpyrimidine-2,4(1H,3H)-dione (305) ⟨67IZV1811⟩. The presence and position(s) of other groups exert profound effects on such alkylations: reviews should be consulted for details ⟨70HC(16-S1)269⟩.

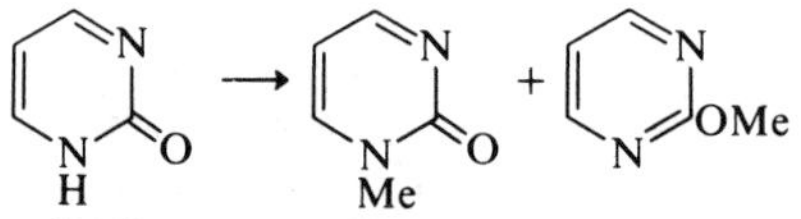

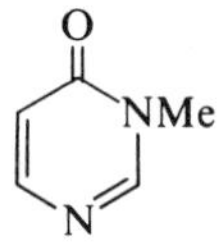

(296)           (297)        (298)           (299)

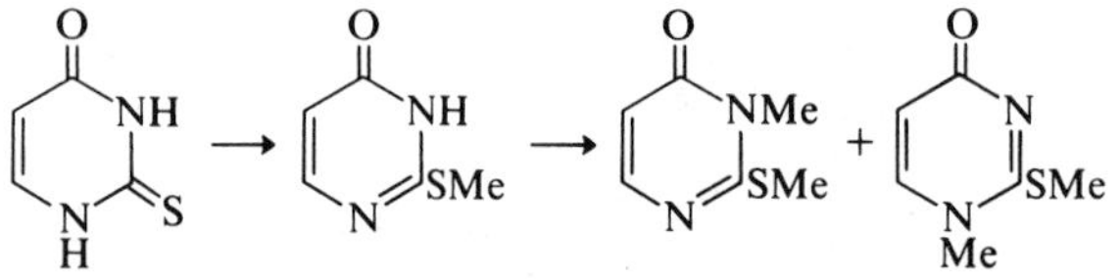

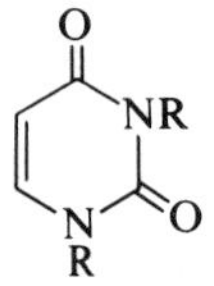

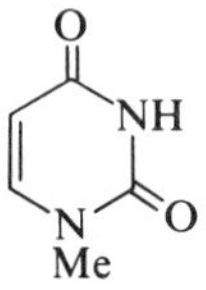

(300)           (301)           (302)        (303)           (304)           (305)

Being phenolic, pyrimidin-5-ols and quinazolin-5(6, 7 or 8)-ols should be O-alkylated readily but there are few examples apart from the complete methylation of 5-hydroxypyrimidine-2,4(1H,3H)-dione (isobarbituric acid) with either diazomethane or dimethyl sulfate/alkali, to give 5-methoxy-1,3-dimethylpyrimidine-2,4(1H,3H)-dione (306) ⟨47JA2138⟩ and the allylation of quinazolin-8-ol to 8-allyloxyquinazoline (307) by allyl bromide in methanolic sodium methoxide ⟨54JCS505⟩. The alkylation of extranuclear hydroxypyrimidines is also easy but the usual methods often induce N-alkylation as well, especially when there are other tautomeric substituents on the ring. This can be avoided by using alcoholic hydrogen chloride in boiling toluene (to entrain the water formed): thus 5-hydroxymethylpyrimidine-2,4(1H,3H)-dione (308; R = H) and benzyl alcohol containing hydrogen chloride give the corresponding 5-benzyloxymethyl derivative (308; R = CH₂Ph) in good yield ⟨66CCC1053⟩.

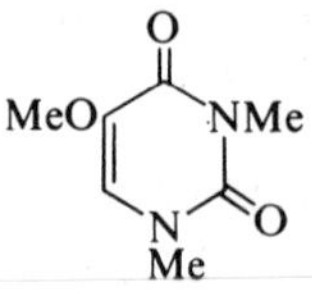

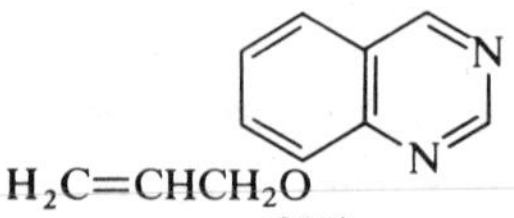

(306)                          H₂C=CHCH₂O        (307)

(308)            (309)            (310)

(*e*) All types of hydroxypyrimidine can undergo *O*-trimethylsilylation, either to increase volatility for GLC, *etc.*, or to protect the oxygen during *N*-alkylation or such processes. For example, quinazoline-2,4(1*H*,3*H*)-dione and hexamethyldisilazane give the 2,4-bistrimethylsilyloxyquinazoline (309) which undergoes methylation to give (after acidic hydrolysis) 1-methylquinazoline-2,4(1*H*,3*H*)-dione (310); nucleosides can be made similarly using an acetylated chloro sugar in place of methyl iodide: a final alkaline hydrolysis removes both the acetyl and silicon-containing protecting groups. The same processes are applicable to uracil and other simple pyrimidines ⟨71CCC246⟩.

(*f*) The direct removal of 4-hydroxy (oxo) substituents from pyrimidine is possible occasionally but is never satisfactory: for example, sodium amalgam reduction of uracil (304; R = H) gives <5% yield of pyrimidin-2(1*H*)-one ⟨64BJ(90)76⟩. However, extranuclear (alcoholic) hydroxy groups can be removed reasonably well by hydriodic acid/red phosphorus or better by hydrogenolysis. The latter process is exemplified by the conversion of 5-hydroxymethyl-2-methylpyrimidin-4(3*H*)-one (311; R = OH) into 2,5-dimethyl-pyrimidin-4(3*H*)-one (311; R = H), using a palladium catalyst ⟨64BSF936⟩.

(*g*) Such hydroxyalkylpyrimidines also undergo oxidation like any alcohol to the corresponding aldehyde or carboxylic acid. For example, the hydroxymethyluracil (308; R = H) is oxidized to the aldehyde (312) by manganese dioxide in water (or better, DMSO), by ceric sulfate or by potassium persulfate with silver ion as a catalyst ⟨66TL5253⟩.

(*h*) A peculiar aminolysis of pyrimidinones is possible with the introduction of secondary or tertiary amino groups in place of the 'hydroxy' group, using appropriate phosphoramides as reagents at high temperatures. For example, uracil and tris(dimethylamino)phosphine oxide at 235 °C give *N,N,N',N'*-tetramethylpyrimidine-2,4-diamine (313) ⟨69IZV655⟩. However, selectivity is possible: thus 1-methyluracil (305) and tris(phenylamino)phosphine oxide at 200 °C give only 4-anilino-1-methylpyrimidin-2(1*H*)-one (314). It is evident that both substrate and product must be exceptionally thermostable for this reaction to be practical ⟨70IZV1198⟩.

(311)          (312)          (313)          (314)

(315)          (316)          (317)          (318)

(*i*) Alkoxy- and aryloxy-pyrimidines undergo hydrolysis to the appropriate hydroxy compounds. The 2- and 4/6-alkoxypyrimidines do so very easily in acidic, or sometimes alkaline, media; indeed, it is often easier to hydrolyze a methoxypyrimidine than the corresponding chloropyrimidine, so that the indirect conversion of chloropyrimidine to pyrimidinone *via* an alkoxypyrimidine is used from time to time, especially in the presence of other hydrolysis-prone groups. This process is illustrated by the conversion of *N*-benzyl-2-chloropyrimidin-4-amine (315; R = Cl) into the ethoxy analogue (315; R = OEt) followed by brief acidic hydrolysis to the benzylaminopyrimidinone (316) ⟨64LA(673)153⟩. Partial cleavage of a dialkoxypyrimidine is possible by heating with a single equivalent of hydrogen chloride in acetonitrile: in this way, 4,6-dibenzyloxypyrimidine gives 6-benzyloxypyrimidin-4(3*H*)-one (317) in good yield ⟨67CCC1298⟩. The relative ease of hydrolysis of a methoxy and a methylthio group (neglecting the mutual minor electronic effects) is seen in the alkaline hydrolysis of 4-methoxy-6-methylthiopyrimidine to give 6-methylthiopyrimidin-4(3*H*)-one (**318**) ⟨64AJC567⟩.

The conversion of 5- and extranuclear-alkoxypyrimidines to the corresponding hydroxy compounds needs much more vigorous conditions. 5-Methoxypyrimidine gives pyrimidin-5-ol only by using potassium hydroxide in boiling ethylene glycol ⟨60JCS4590⟩ while 5-*β*-ethoxyethyl-6-methylpyrimidin-4(3*H*)-one (**319**; R = OEt) requires concentrated hydro-chloric acid at 150 °C to give the 5-*β*-chloroethyl analogue (**319**; R = Cl) and thence in water at 150 °C, the 5-*β*-hydroxyethyl analogue (**319**; R = OH) ⟨42JOC309⟩. As in other series, benzyloxy groups uniquely undergo hydrogenolysis to the corresponding hydroxy derivatives. This can be very useful in the presence of other hydrolysis-sensitive, but reduction-insensitive, groups: 4-benzyloxy-6-fluoropyrimidin-2-amine (**320**) gives 2-amino-6-fluoropyrimidin-4(3*H*)-one (**321**) without affecting the active fluoro substituent ⟨63JMC688⟩.

(**319**)        (**320**)        (**321**)        (**322**)

(*j*) Some 2- and 4/6-alkoxypyrimidines undergo transalkoxylation. The older literature contains examples using sodium alkoxide in the appropriate alcohol under very vigorous conditions ⟨62HC(16)245⟩ but more recently silver oxide has proven to be an excellent catalyst, at least for activated alkoxy groups. Thus, 2-methoxypyrimidine resists transalkoxylation but 2-methoxy-5-nitropyrimidine (**322**; R = H) is converted easily into its propoxy homologue (**322**; R = Et) by boiling in propanol containing silver oxide; likewise, 2,4-dimethoxy-5-nitro- gives 5-nitro-2,4-dipropoxy-pyrimidine, although 2,4-dimeth-oxy-6-methyl-5-nitro- gives only 4-methoxy-6-methyl-5-nitro-2-propoxy-pyrimidine ⟨70JCS(C)2661⟩.

(*k*) Most 2- or 4/6-alkoxypyrimidines are satisfactory substrates for aminolysis. The conditions needed to bring about the reaction are affected profoundly by any activation or deactivation afforded by additional electron-withdrawing or electron-releasing groups pres-ent and also by the size of the alkoxy group being replaced. Thus, butylaminolysis of 4-methoxypyrimidine is 10 times faster than that of 2-methoxypyrimidine and 800 times faster than that of 2-isopropoxypyrimidine (for comparison, 2-chloropyrimidine undergoes aminolysis *ca.* 40 000 times faster than 2-methoxypyrimidine); in practical terms, 4-methoxypyrimidine (**323**) and butylamine require 150 °C for 3 hours to produce a good yield of *N*-butylpyrimidin-4-amine (**324**) ⟨66AJC1487⟩. In contrast, 4-methoxy-5-nitropyrimidine requires only 3 hours at 60 °C in methanolic *t*-butylamine (a reagent with *ca.* 0.1% of the nucleophilic activity of *n*-butylamine) to produce a good yield of *N*-*t*-butyl-5-nitropyrimidine-4-amine ⟨67AJC1041⟩. Many other examples are known and such reactions are of great preparative value ⟨70HC(16-S1)189⟩.

(**323**)        (**324**)        (**325**)        (**326**)

(**327**)        (**328**)        (**329**)

(*l*) The rearrangement of 2- and 4/6-alkoxypyrimidines into *N*-alkylpyrimidinones can be done in two ways, by thermal means or by the Hilbert–Johnson reaction. Although thermal rearrangement is of limited preparative value, it has been studied in some detail. The rearrangement of 2-methoxypyrimidine (**325**) into 1-methylpyrimidin-2(1*H*)-one (**326**) is a first-order reaction at 160 °C strongly catalyzed by tertiary bases whose efficiency varies according to their basic strength ⟨65JCS4911⟩. Higher alkoxypyrimidines rearrange more slowly and the reaction is accelerated by electron-withdrawing groups and retarded by electron-releasing groups: 2-methoxy-5-nitropyrimidine rearranges 2000 times more rapidly than 2-methoxypyrimidine but 2-methoxy-4-methylpyrimidine does so 7 times less rapidly than 2-methoxypyrimidine ⟨68AJC243⟩. The mechanism is intermolecular but not free-radical. 4-Allenyloxy- and 4-allynyloxy-pyrimidines also undergo intramolecular *ortho* Claisen rearrangement: for example, 4-allyloxy-2-phenylpyrimidine (**327**) gives 5-allyl-2-phenylpyrimidin-4(3*H*)-one (**328**) as well as 3-allyl-2-phenylpyrimidin-4(3*H*)-one (**329**) ⟨66JOC406⟩.

The Hilbert–Johnson reaction is certainly not a rearrangement, although it often appears as such. It is illustrated by the reaction of 5-bromo-2,4-dimethoxypyrimidine (**330**) with methyl iodide to yield 5-bromo-4-methoxy-1-methylpyrimidin-2(1*H*)-one (**331**) of proven structure ⟨62JCS1540⟩. However, the mechanism is quickly revealed by the conversion of 2,4-diethoxypyrimidine (**332**) by methyl iodide into 2,4-diethoxy-1-methylpyrimidinium iodide (**333**) of proven structure and its subsequent loss of ethyl iodide to give 4-ethoxy-1-methylpyrimidin-2(1*H*)-one (**334**) ⟨68JA1678⟩. The reaction is extensively used in the preparation of nucleosides by initial quaternization with a halogeno sugar ⟨67AHC(8)115⟩.

Both thermal and Hilbert–Johnson reactions occur quite readily in 2- and 4/6-alkoxyquinazolines. For example, distillation of 2,4-bis-β-diethylaminoethoxyquinazoline *in vacuo* yields 1,3-bis-β-diethylaminoethylquinazolin-2,4(1*H*,3*H*)-dione ⟨60JCS3546⟩. A complicated example is furnished by heating a tetramethoxybipyrimidine with methyl iodide to give four separable products progressively ⟨81AJC1157⟩.

(330)    (331)    (332)    (333)    (334)

### 2.13.2.2.8 *Reactivity of mercapto, alkylthio and related groups*

Sulfur-containing derivatives are of immense importance in the pyrimidine series, mainly because their diverse reactions make them convenient intermediates.

(*a*) Mercapto groups in any position may be removed in favour of hydrogen by reductive or oxidative desulfurization. The reductive process is carried out by boiling with 4–10 parts of Raney nickel catalyst in water, aqueous base (usually ammonia) or an alcohol; sulfur is eliminated as nickel sulfide. For example, 5,6-diaminopyrimidine-2,4(1*H*,3*H*)-dithione (**335**) gives pyrimidine-4,5-diamine (**336**) in excellent yield ⟨62JOC986⟩. The oxidative process is much less reliable but it may be essential when a reduction-sensitive group is present or when the desulfurization has to be done on a large scale. It is customary to use alkaline hydrogen peroxide, which probably gives an intermediate sulfinic acid; this either loses sulfur dioxide to give the required product or it may be hydrolyzed to yield the corresponding pyrimidinone. Sometimes it does both, giving a mixture of products ⟨62HC(16)277⟩. A successful example is the oxidation of a 5-substituted 2-thiouracil (**337**) with lead tetracetate and hydrogen peroxide to afford the desulfurized product (**338**) in good yield ⟨66AP362⟩. The intermediate sulfinic acid can sometimes be isolated, as was 4,6-diamino-(**339**; R = H) and 4,5,6-triaminopyrimidine-2-sulfinic acid (**339**; R = NH₂) prior to acidic hydrolysis to yield the diamine (**340**; R = H) and triamine (**340**; R = NH₂) respectively ⟨56JCS4106⟩.

(335)    (336)    (337)    (338)

(339)    (340)    (341)    (342)

(*b*) All mercaptopyrimidines may be *S*-alkylated easily, usually by dissolution in aqueous or alcoholic base and shaking with an alkyl halide or dimethyl sulfate at room temperature. In this way, 4,5-diaminopyrimidine-2-thione (**341**) gives 2-methylthiopyrimidine-4,5-diamine (**342**) in a few minutes ⟨54JCS2060⟩.

Thiouracil and its *C*-alkyl derivatives are exceptionally sluggish towards *S*-methylation. For example, under the usual conditions, thiouracil (**343**) gives 2-methylthiopyrimidin-4(3*H*)-one (**344**) in only 5% yield, but by using methyl iodide (2 mol) in anhydrous DMF as solvent the yield is satisfactory; other *S*-alkylations of thiouracil may be done similarly ⟨73CB3039⟩. *S*-Alkylations can be done with substituted alkyl halides; for example, reaction of pyrimidine-2(1*H*)-thione with chloroacetamide in aqueous sodium hydrogen carbonate at 90 °C gives 2-(pyrimidin-2'-ylthio)acetamide (**345**) ⟨79AJC2713⟩. When an aprotic solvent is desirable, *S*-methylation may be done conveniently by boiling the thione with dimethyl-formamide dimethyl acetal in toluene; for example, pyrimidine-2,4(1*H*,3*H*)-dithione gives 2,4-bismethylthiopyrimidine (**346**) ⟨81AJC1729⟩.

(*c*) The direct hydrolysis of pyrimidinethiones is seldom possible although there are several examples in the older literature ⟨62HC(16)233⟩. In practice, pyrimidinethiones are converted into pyrimidinones by (i) *S*-alkylation and subsequent hydrolysis, (ii) oxidation to a sulfinic or sulfonic acid with subsequent hydrolysis, or (iii) *S*-alkylation followed by oxidation to a sulfoxide or sulfone and final hydrolyis. Examples of individual reactions are given in appropriate parts of this section.

(**343**)         (**344**)         (**345**)         (**346**)

(**347**)         (**348**)         (**349**)         (**350**)

(*d*) The aminolysis of pyrimidinethiones occurs with some reluctance and it is often preferable to convert the thione into an alkylthio or alkylsulfonyl derivative prior to aminolysis. However, the direct reaction is used, especially with pyrimidine-2,4-dithiones which undergo preferential aminolysis of the 4-mercapto group. Thus, the parent dithione (**347**; R = H) reacts with aqueous ammonia at 100 °C to yield 4-aminopyrimidine-2(1*H*)-thione (**348**) ⟨B-68MI21301⟩; with the 5-nitrodithione (**347**; R = NO₂) as substrate, both mercapto groups are so activated that the product is 5-nitropyrimidine-2,4-diamine (**349**) under comparable conditions but with the 5-aminodithione (**347**; R = NH₂), the reverse pertains and no reaction occurs ⟨54MI21300⟩. However, the amine substrate (**347**; R = NH₂) does yield at 130 °C to the better nucleophile methylamine, to afford 5-amino-4-methyl-aminopyrimidine-2(1*H*)-thione (**350**) ⟨66JCS(C)1065⟩. Likewise, quinazoline-4-thiones react readily with alkylamines, but sluggishly with aniline or dialkylamines, to give the *N*-substituted quinazolin-4-amines ⟨46JOC349⟩. Hydrazine is a particularly good nucleophile in this context ⟨80AJC1147⟩ and will attack even unactivated pyrimidine-2-thiones ⟨63ZOB2673⟩.

(*e*) The oxidation of mercaptopyrimidines has been mentioned above in respect of desulfurization. In fact, the pyrimidine-2(and 4/6)-thiones are reasonably stable in air but pyrimidine-5-thiols and extranuclear thiols are oxidized in air to disulfides. Thus, the thiones (**351**; R = Ac or CO₂Et) are oxidized by bromine to the disulfides (**352**; R = Ac or CO₂Et) ⟨63CB526⟩; similar oxidations may be effected with air (in alkali), iodine (in alkali) ⟨72JCS(P1)522⟩, nitrous acid or peroxides. The more profound oxidation of pyrimidinethiones to pyrimidinesulfinic or pyrimidinesulfonic acids (*via* disulfides) is possible. For example, the diamino-2-thiouracil (**353**) and a limited amount of alkaline peroxide give 4,5-diamino-6-oxo-1,6-dihydropyrimidine-2-sulfinic acid (**354**) ⟨60JOC148⟩ and 6-methyl-2-thiouracil with alkaline permanganate gives 4-methyl-6-oxo-1,6-dihydropyrimidine-2-sulfonic acid ⟨62ZOB(32)1709⟩. The oxidation of pyrimidine-2(1*H*)-thione by permanganate gives potassium pyrimidine-2-sulfonate ⟨72JCS(P1)522⟩ but oxidation by chlorine at a low temperature yields pyrimidine-2-sulfonyl chloride and several derivatives may be made similarly ⟨50JA4890⟩. Similar oxidations in the presence of an excess of fluoride ion give pyrimidinesulfonyl fluorides ⟨72JCS(P1)522⟩ and all such reactions are applicable to simple quinazolines ⟨72AJC2641⟩.

(*f*) The conversion of pyrimidinethiones into the corresponding pyrimidinyl thiocyanates may be done by treatment with cyanogen bromide: for example, pyrimidin-2-yl thiocyanate (**355**) and 6-methyl-2-methylthiopyrimidin-4-yl thiocyanate are made by allowing the appropriate thiones (as sodium salts) to react with cyanogen bromide in aqueous solution at 10 °C ⟨63YZ1086⟩.

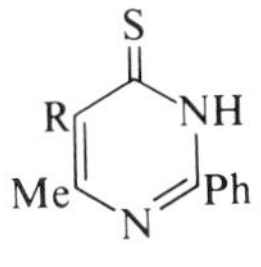

(**351**) 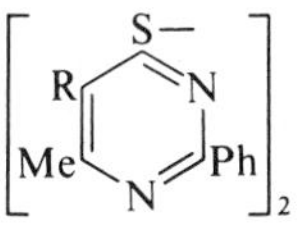    (**352**) 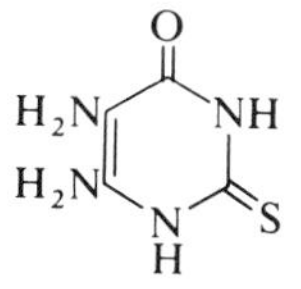    (**353**) 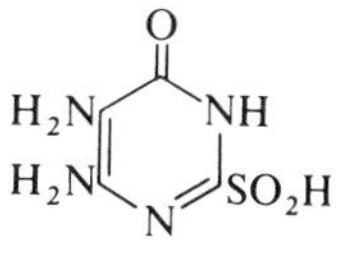    (**354**)

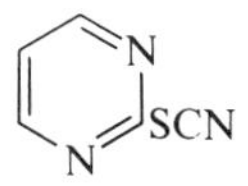

(**355**) 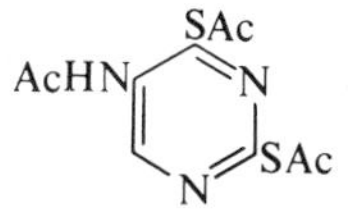    (**356**) 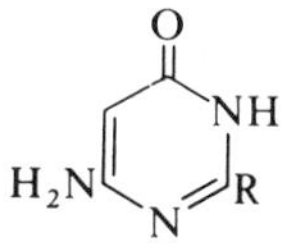    (**357**) 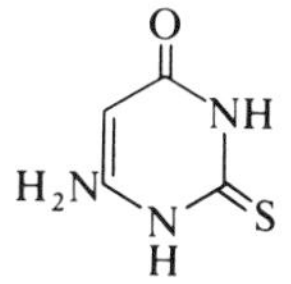    (**358**)

(*g*) Somewhat similar *S*-acylation of most pyrimidinethiols and thiones can be done under Schotten–Baumann conditions or by warming with an acid anhydride. For example, the aminodithione (**347**; R = NH$_2$) yields the *N,S,S*-triacetyl derivative (**356**) ⟨58CPB346⟩.

(*h*) Alkyl- and aryl-thiopyrimidines undergo direct desulfurization by Raney nickel, usually by boiling in an alcoholic suspension; yields are generally lower than in desulfurization of the corresponding mercapto compounds. For example, 6-aminopyrimidin-4(3*H*)-one (**357**; R = H) is obtained from its 2-methylthio derivative (**357**; R = SMe) in 70–80% yield, but from the corresponding mercapto compound (**358**) in >90% yield ⟨64CI(L)418⟩.

(*i*) The dealkylation of alkylthiopyrimidines may be done in three ways: (i) acidic hydrolysis, as in the conversion of 1-benzyl-2-benzylthiopyrimidin-4-(1*H*)-one (**359**) by hot hydrochloric acid into 1-benzyl-2-thiouracil (**360**) ⟨65JHC447⟩, (ii) by means of a reduction, as in the transformation of 6-benzylthiopyrimidine-2,4,5-triamine by sodium/liquid ammonia into 2,5,6-triaminopyrimidine-4(3*H*)-thione (**361**) ⟨64JOC3370⟩, and (iii) cleavage by a Lewis acid in an aprotic solvent, as in the debenzylation of 2-benzylthio-5-iodopyrimidin-4(3*H*)-one by aluminium bromide in toluene to give 5-iodo-2-thiouracil (**362**) ⟨54JA3666⟩. In addition, there are several examples of treating an alkylthiopyrimidine with phosphorus pentasulfide in xylene to give (albeit in poor yield) the corresponding thione: in this way, 3-methyl-2-methylthiopyrimidine-4(3*H*)-one (**363**) gives 3-methylpyrimidine-2,4(1*H*,3*H*)-dithione (**364**) but the mechanism is obscure ⟨59MI21300⟩.

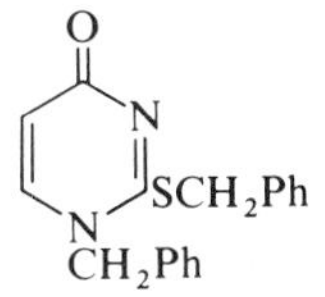

(**359**) 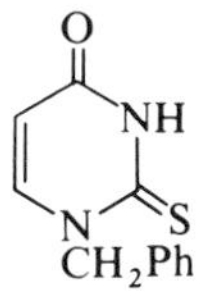    (**360**) 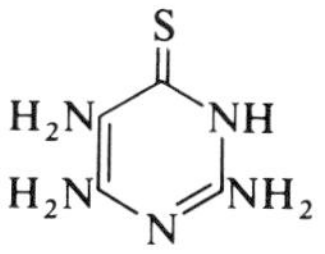    (**361**) 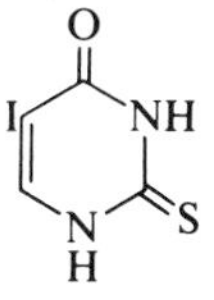    (**362**) 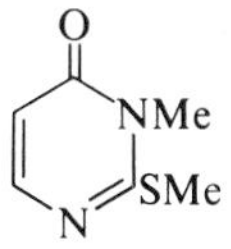    (**363**)

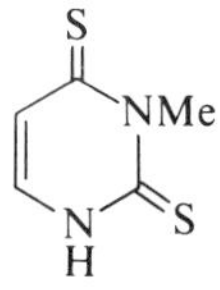

(**364**) 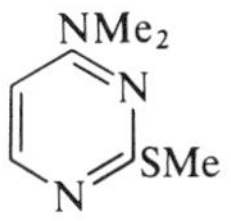    (**365**) 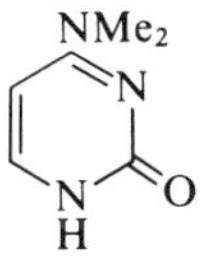    (**366**) 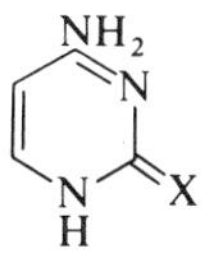    (**367**) 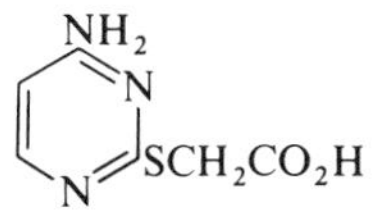    (**368**)

(*j*) The hydrolysis of 2- and 4/6-alkylthiopyrimidines to the corresponding pyrimidinones is best done in hot acidic media (within a fume hood). For example, *N,N*-dimethyl-2-methylthiopyrimidin-4-amine (**365**) in boiling 6M hydrochloric acid gives

4-dimethylaminopyrimidin-2(1*H*)-one (**366**) ⟨62AJC851⟩. This process forms part of an 'integrated procedure' for converting a pyrimidinethione into a pyrimidinone *via* an (unisolated) carboxymethylthio derivative: this important process is illustrated by the sequence thiocytosine (**367**; X = S) to 2-(4'-aminopyrimidin-2'-ylthio)acetic acid (**368**) to cytosine (**367**; X = O) ⟨B-68MI21301⟩. Alkylthiopyrimidines also undergo alcoholysis in alcoholic sodium alkoxide; for example, 5-fluoro-1-methyl-4-methylthiopyrimidin-2(1*H*)-one and methanolic sodium methoxide at 20 °C for 24 hours gives the corresponding 4-methoxy analogue in 60% yield ⟨82ACS(B)15⟩.

(*k*) The aminolysis of 2- and 4/6-alkylthiopyrimidines has a large literature. Kinetic studies show that 4/6-methylthiopyrimidines undergo butylaminolysis more rapidly than the corresponding 2-methylthiopyrimidines under the same conditions, and that ethylthiopyrimidines react more slowly than their methylthio analogues under comparable conditions ⟨66AJC1487⟩. In addition, the usual substituent effects are evident in that electron-release by additional *C*-methyl groups decreases the rate of aminolysis and electron-withdrawal by a 5-bromo or 5-nitro substituent increases the rate markedly. Such differences can be quite spectacular on a preparative basis: 4-methyl-2-methylthiopyrimidine (**369**; R = SMe) requires butylamine at 180 °C for 65 hours to give a 40% yield of *N*-butyl-4-methylpyrimidin-2-amine (**369**; R = NHBu) whereas 2-methylthio-5-nitropyrimidine (**370**; R = SMe) requires butylamine only at 20 °C for 1 minute to give a 95% yield of the corresponding butylamino product (**370**; R = NHBu) ⟨66AJC2321⟩. It is therefore wise to consult summaries of the old ⟨62HC(16)288⟩ and the more recent literature ⟨70HC(16-S1)217⟩ prior to underaking such aminolyses.

(**369**)　　　(**370**)　　　(**371**)　　　(**372**)　　　(**373**)　　　(**374**)

(*l*) The oxidation of alkylthiopyrimidines gives the corresponding sulfoxides or sulfones, according to the proportion of oxidizing agent, which is usually potassium permanganate, sodium periodate, chlorine or a peroxy acid. The process is usually carried out because the sulfoxides and sulfones provide much better leaving-groups in nucleophilic displacements than do alkylthiopyrimidines. Oxidation to sulfoxide is illustrated by the periodate or *m*-chloroperbenzoic acid oxidation of 2-, 4- or 5-methylthiopyrimidine to the corresponding methylsulfinylpyrimidine (**371**) ⟨67JCS(C)568⟩ and by the conversion of 5-methylthio- into 5-methylsulfinyl-pyrimidine-2,4(1*H*,3*H*)-dione using peracetic acid as oxidant ⟨67M1577⟩. Sulfones are easier to make, simply because a slight excess of oxidizing agent may be used. 2-Methylsulfonylpyrimidine (**372**; R = Me) and its phenyl analogue (**372**; R = Ph) can be made from the thioethers using chlorine or from the sulfoxides with *m*-chloroperbenzoic acid ⟨67JCS(C)568⟩; permanganate is convenient for the conversion of 5-ethylthio- or 5-pentylthiopyrimidine-2,4(1*H*,3*H*)-dione (**373**; R = Et or C₅H₁₁) into the corresponding sulfones (**374**) ⟨65JCS3987⟩.

(*m*) Sulfides and disulfides undergo oxidation and reduction but few other reactions. Illustration is afforded by the oxidation of di-(2-amino-6-methylpyrimidin-4-yl) sulfide (**375**; X = S) by acidic permanganate to the corresponding sulfone (**375**; X = SO₂) ⟨50BSF616⟩; the performic acid oxidation of di-(2,6-dimethoxypyrimidin-4-yl) disulfide to 2,6-dimethoxypyrimidine-4-sulfonic acid (**376**) ⟨54JA2899⟩; and the reduction of the disulfide (**377**) by sodium borohydride to 6-anilino-5-mercaptopyrimidin-4(3*H*)-one (**378**) ⟨66JPR(32)26⟩. Other effective reducing agents include LAH, sodium hydrogen sulfite and zinc/acetic acid ⟨62HC(16)291⟩.

(**375**)　　　　(**376**)　　　　(**377**)　　　　(**378**)

(*n*) Pyrimidinesulfinic acids are so rare that few reactions are recorded but pyrimidinesulfonic acids are known to undergo a variety of reactions. The typical sulfinic acid (**379**; R = SO₂H) reacts with ethanolic hydrogen chloride to give 5,6-diaminopyrimidin-4(3*H*)-one

(**379**; R = H) but with aqueous hydrochloric acid to give 5,6-diaminopyrimidine-2,4(1*H*,3*H*)-dione (**380**) ⟨60JOC148⟩. Most of the reactions of sulfonic acids are illustrated in the following simple examples. Pyrimidine-2-sulfonic acid (**381**) is reasonably stable (as anion) at pH 7 but it is rapidly hydrolyzed in acid (3M hydrochloric acid) or alkali (1M sodium hydroxide) to give pyrimidin-2(1*H*)-one ($t_{1/2}$ values at 25 °C 45 minutes and <1 minute, respectively); addition of *C*-methyl groups slows the hydrolysis rates but the isomeric pyrimidine-4-sulfonic acid is hydrolyzed even more rapidly ⟨71JCS(B)2214⟩. 2,6-Dimethyl-pyrimidine-4-sulfonic acid (**382**; R = SO₃H) undergoes displacement of the sulfo group by hydrazine to give 2,6-dimethylpyrimidin-4-ylhydrazine (**382**; R = NHNH₂) ⟨72JCS(P1)522⟩ and by cyanide ion to give 2,6-dimethylpyrimidine-4-carbonitrile (**382**; R = CN) ⟨55CPB173⟩. 2,6-Dimethoxypyrimidine-4-sulfonic acid (**383**; R = SO₃H) reacts with phosphorus pentachloride to give 4-chloro-2,6-dimethoxypyrimidine (**383**; R = Cl) or the 4,5-dichloro analogue, according to conditions ⟨54JA6052⟩, and with sodium sulfanilamide to give the product (**383**; R = HNO₂SC₆H₄NH₂-*p*) ⟨61M1212⟩. The somewhat analogous reactions of pyrimidinesulfonyl halides are typified by the fluorides. Thus, 4,6-dimethyl-pyrimidine-2-sulfonyl fluoride (**384**; R = F) reacts with liquid ammonia to give the sulfonamide (**384**; R = NH₂), with hydrazine in methanol at −10 °C to give the hydrazide (**384**; R = NHNH₂), with diethylamine at 150 °C to give *N,N*-diethyl-4,6-dimethyl-pyrimidin-2-amine (**385**; R = NEt₂), with boiling water to give 4,6-dimethylpyrimidin-2(1*H*)-one, with sodium azide to give the tetrazole (**265**) corresponding to the azide (**385**; R = N₃) and with methanolic methoxide to give 2-methoxy-4,6-dimethylpyrimidine (**385**; R = OMe) ⟨72JCS(P1)522⟩. In contrast to the last example, the sulfonyl chloride (**386**; R = Cl) reacts with methanol to give methyl 4-amino-2,6-dioxo-1,2,3,6-tetrahy-dropyrimidine-5-sulfonate (**386**; R = OMe) ⟨63JOC1994⟩. The sulfonyl chloride (**387**) may be reduced to 5-mercaptopyrimidine-2,4(1*H*,3*H*)-dione (**388**) by an excess of zinc in cool sulfuric acid, but with less zinc and warm acid the disulfide (**389**) is formed ⟨56JA401⟩. The pyrimidin-5-ylsulfates, *e.g.* (**390**), are intermediates in the Elbs persulfate oxidation of pyrimidines (see Section 2.13.2.1.5).

(**379**)   (**380**)   (**381**)   (**382**)

(**383**)   (**384**)   (**385**)   (**386**)

(**387**)   (**388**)   (**389**)   (**390**)

(*o*) The 2- and 4/6-alkylsulfinyl- and corresponding alkylsulfonyl-pyrimidines are all excellent substrates for nucleophilic displacement reactions, as illustrated by the following examples. 2-Methylsulfonylpyrimidine (**391**) with aqueous alkali at 25 °C gives pyrimidin-2(1*H*)-one, with ethanolic sodium ethoxide gives 2-ethoxypyrimidine, with hydriodic acid gives 2-methylthiopyrimidine, with potassium cyanide in DMF at 100 °C gives pyrimidine-2-carbonitrile, with sodium azide in DMF gives pyrimidin-2-yl azide, with hydrazine gives pyrimidin-2-ylhydrazine and with cyclohexylamine gives *N*-cyclohexylpyrimidin-2-amine. 2-Methylsulfinylpyrimidine (**392**) undergoes similar reactions and, in addition, may be oxidized by *m*-chloroperbenzoic acid in chloroform to the sulfone (**391**). A kinetic study of the pentylaminolysis of several simple sulfoxides, sulfones and (for comparison) 2-chloropyrimidine, indicates that there is little to choose between corresponding 2-methylsulfinyl-, 2-methylsulfonyl-, 2-phenylsulfonyl- or 2-chloro-pyrimidines in that the rates differ by <5-fold; 4-methylsulfinyl- and 4-methylsulfonyl-pyrimidine react faster than their respective 2-isomers ⟨67JCS(C)568⟩. 5-Methylsulfinyl- and 5-methylsulfonyl-pyrimidine (**393**) do not undergo normal pentylaminolysis: instead, the sequence (**393**) → (**397**) occurs. The product (**397**) and the corresponding sulfoxide are of proven structure ⟨68JCS(C)1452⟩.

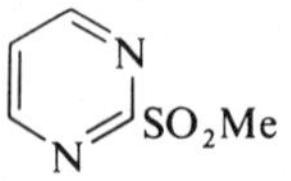

(**391**) 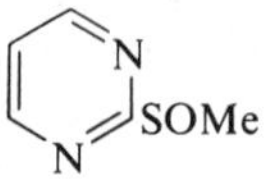          (**392**)

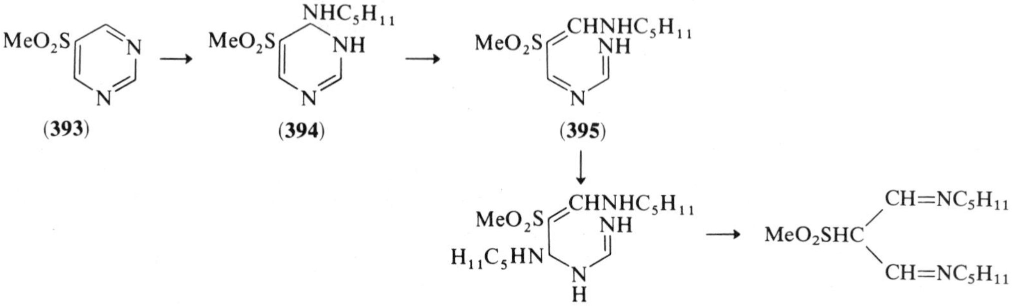

(**393**)          (**394**)          (**395**)

(**396**)          (**397**)

### 2.13.2.2.9 Reactivity of halogeno substituents

(*a*) Chloropyrimidines are the most versatile intermediates in the series. The removal of halogeno substituents from any position can be done directly by hydrogenolysis, providing other groups will withstand the reductive conditions. The usual method is hydrogenation over palladium in the presence of a base (magnesium oxide, calcium carbonate, sodium hydroxide, *etc.*), without which there is a distinct danger of at least partial reduction of the pyrimidine ring. Typical examples are the preparation of 5-acetylpyrimidine (**398**; R = H) from its 2,4-dichloro derivative (**398**; R = Cl), of ethyl 2-(pyrimidin-5′-yl)acetate (**399**; R = H) from its 4-chloro derivative (**399**; R = Cl) ⟨66AP362⟩, of 2,4-diamino-1-methyl-pyrimidinium chloride (**400**; R = H) from its 5-bromo derivative (**400**; R = Br) ⟨65JCS755⟩, and of 4-methylpyrimidine (**401**; R = H) from the trichloromethyl analogue (**401**; R = Cl) ⟨60CB2405⟩. For NMR and MS studies, specific deuterations may be done by shaking a chloropyrimidine with deuterium over a palladium catalyst; 2-deuteropyrimidine-4,5-diamine is so made from the 2-chlorodiamine ⟨65JCS623⟩.

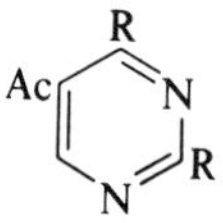

(**398**) 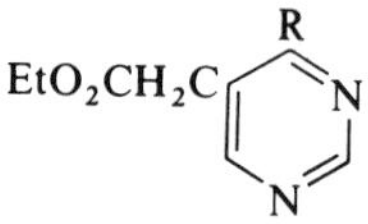          (**399**) 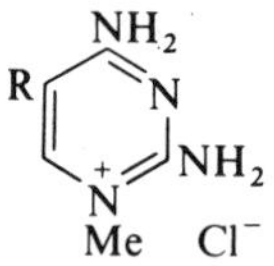          (**400**) 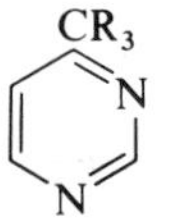          (**401**)

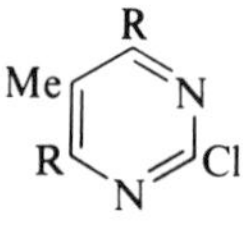

(**402**) 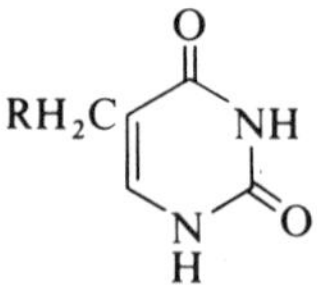          (**403**) 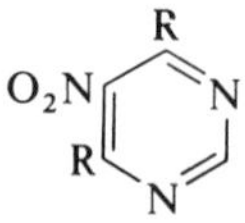          (**404**) 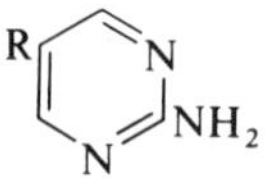          (**405**)

Zinc dust in water or mild alkali appears to be rather selective in removing 4/6- and extranuclear-halogeno substituents: for example, 2,4,6-trichloro-5-methylpyrimidine (**402**; R = Cl) with zinc dust in aqueous ammonia gives 2-chloro-5-methylpyrimidine (**402**; R = H) in good yield ⟨68AJC243⟩ and zinc in deuterium chloride/deuterium oxide converts 5-chloromethylpyrimidine-2,4(1*H*,3*H*)-dione (**403**; R = Cl) into α-deutero-thymine (**403**; R = D) ⟨63CCC2491⟩. Hot hydriodic acid removes 2-chloro, bromo or even iodo substituents very conveniently but not 4-halogeno substituents; there is one example of a 5-chloro substituent being removed in this way ⟨73AJC443⟩.

When reductive conditions are contraindicated, the indirect removal of chlorine by oxidation of a derived hydrazino derivative with silver oxide, may be used; this is illustrated in the conversion of 4,6-dichloro-5-nitropyrimidine (**404**; R = Cl) into 5-nitropyrimidine (**404**; R = H) *via* the dihydrazino intermediate (**404**; R = NHNH$_2$) ⟨67JCS(C)573⟩. When submitted to a Busch reaction (methanolic alkali/hydrazine hydrate/palladium), 5-bromopyrimidin-2-amine (**405**; R = Br) gives, not the expected 5,5'-bipyrimidine-2,2'-diamine, but pyrimidin-2-amine (**405**; R = H) by an unknown mechanism ⟨67JCS(C)1204⟩. Dehalogenation reactions are equally applicable to chloroquinazolines ⟨67HC(24-1)227⟩.

(*b*) The aminolysis of halogenopyrimidines has a vast literature, of which only a small part is quantitative. The first classical kinetic studies on the aminolysis of a few simple 2- and 4-chloropyrimidines ⟨54JCS1190⟩ were followed by more pragmatic studies under preparative conditions, later summarized ⟨B-69MI21300⟩. The following facts emerge. (i) The rate constant for aminolysis of a given chloropyrimidine by an *n*-alkylamine is almost unaffected by lengthening the alkyl chain or by branching at or beyond the γ-carbon atom. (ii) A β-branch has a small and an α-branch a large rate retarding effect: one α-branch reduces the rate to *ca.* 5%, and two α-branches reduce it to *ca.* 0.1% of that for the *n*-alkylamine. (iii) Di-*n*-alkylamines react at about the same rate as an α-branched alkylamine. (iv) A 4-chloropyrimidine is usually marginally more reactive than the corresponding 2-chloropyrimidine ⟨65AJC1811⟩. The effect of additional substituents on the reactivity of a 2-chloropyrimidine towards aminolysis is illustrated in the following relative rates: 2-chloro-4,6-dimethylpyrimidine, 1; 2-chloropyrimidine, 10; 5-bromo-2-chloropyrimidine, 200; 2-chloro-5-nitropyrimidine, 3 000 000. The almost universal practice of 'overkill', in guessing conditions for an aminolysis, can be avoided by doing a single small-scale experiment culminating in one chloride-ion titration; from the result, optimal conditions of time and temperature may be determined with fair precision from a nomograph ⟨64AJC794⟩.

The ease of displacing a chloro, bromo or iodo substituent in comparable pyrimidines by aminolysis is disappointingly similar. Using 2-halogeno-4,6-dimethyl- and 4-halogeno-2,6-dimethyl-pyrimidines as substrates in a kinetic study of isopentylaminolysis, the order of activity is Br > I > Cl but the maximal difference in each series is only three-fold ⟨71JCS(C)1889⟩.

Preparative procedures are exemplified in the conversion of 4-chloro-2,6-dimethylpyrimidine (**406**; R = Cl) into 2,6-dimethylpyrimidin-4-amine (**406**; R = NH$_2$), its *N*-methyl (**406**; R = NHMe) and its *N,N*-dimethyl derivative (**406**; R = NMe$_2$), all in *ca.* 80% yield ⟨66JCS(C)226⟩. The aminolysis of dichloropyrimidines will replace one or both chloro substituents according to conditions. Reaction of 4,6-dichloropyrimidine (**407**; R = Cl) and ethanolic methylamine at 90–100 °C gives only 6-chloro-*N*-methylpyrimidin-4-amine (**407**; R = NHMe) ⟨55MI21300⟩; the second chlorine is deactivated by the methylamino group and consequently needs more vigorous conditions for replacement, *e.g.* aqueous dimethylamine at 120–130 °C, to yield *N,N,N'*-trimethylpyrimidine-4,6-diamine (**408**) ⟨63CB2977⟩. Being unsymmetrical, a 2,4-dichloropyrimidine initially undergoes aminolysis to give a mixture of two monoamino compounds (with the 4-amino isomer predominating), both of which give the same diamine on more vigorous treatment with the same amine. The use of aqueous–alcoholic amine appears to enhance the preference for 4-aminolysis at the first stage ⟨64LA(673)153⟩, thereby producing, for example, 2-chloropyrimidin-4-amine or the analogue (**409**) in good yield from appropriate reactants. The aminolysis of a 2,4,6-trihalogenopyrimidine is typified by trifluoropyrimidine (**410**) and alcoholic ammonia: at 25 °C they give a mixture of the difluoropyrimidinamines (**411**) and (**412**) which, in turn, at *ca.* 100 °C give a single product, 6-fluoropyrimidine-2,4-diamine (**413**; R = F) and finally at 160 °C pyrimidine-2,4,6-triamine (**413**; R = NH$_2$) ⟨63JMC688⟩. The effects of additional groups on aminolyses of innumerable halogenopyrimidines have been reviewed in detail ⟨70HC(16-S1)129⟩.

(**406**)  (**407**)  (**408**)  (**409**)

(**410**) → (**411**) + (**412**) → (**413**)

Although halogeno substituents in the benzene ring of quinazoline appear to be completely resistant to aminolysis, 5-halogenopyrimidines are by no means so. Thus 5-bromo-4-*t*-butylpyrimidine (**414**; R = Br) and ammonia yield 4-*t*-butylpyrimidin-5-amine (**414**; R = NH$_2$), but only on heating together at 135 °C for 50 hours ⟨66RTC1101⟩; aniline or its derivatives react with 5-bromopyrimidine-2,4(1*H*,3*H*)-dione (**415**; R = Br) at 90 °C in ethylene glycol to afford 5-anilinouracil (**415**; R = NHPh) or appropriate derivatives in good yield ⟨66JMC108⟩. Amination of such halogenopyrimidines by sodium amide in liquid ammonia gives apparently anomalous results; for example, the bromopyrimidine (**414**; R = Br) gives 6-*t*-butylpyrimidin-4-amine (**416**). The mechanism of such reactions is quite complicated but virtually proven ⟨78JHC1121⟩.

Extranuclear halogeno substituents are easily aminolyzed. For example, 5-chloromethyl-pyrimidine-2,4(1*H*,3*H*)-dione (**417**; R = Cl) and cold dimethylamine give 5-dimethyl-aminomethyluracil (**417**; R = NMe$_2$) ⟨61CCC893⟩, but 5-*β*-bromoethyluracil (**417**; R = CH$_2$Br) is best allowed to react with potassium phthalimide initially, followed by acidic hydrolysis to afford 5-*β*-aminoethyluracil (**417**; R = CH$_2$NH$_2$) if complications are to be avoided ⟨64ZOB2577⟩. The chloromethyl compound (**417**; R = Cl) reacts with hydrazine to give, not 5-hydrazinomethyl- (**417**; R = NHNH$_2$), but 5-hydrazonomethyl-uracil (**418**), a reaction which involves oxidation by a mechanism unknown; hydroxylamine reacts similarly to give 5-hydroxyiminomethyluracil (**419**), the oxime of uracil-5-carbaldehyde ⟨66JOC4239⟩.

(**414**)   (**415**)   (**416**)   (**417**)

(**418**)   (**419**)   (**420**)   (**421**)

Active chloropyrimidines react with trimethylamine in benzene to give the trimethylammonio derivative as chloride. Thus, 2-chloropyrimidine or its 5-nitro derivative give *N*,*N*,*N*-trimethylpyrimidin-2-ylammonium chloride (**420**; R = H) or its 5-nitro derivative (**420**; R = NO$_2$) ⟨71JCS(B)1675⟩; 2-bromopyrimidine gives the bromide corresponding to (**420**; R = H) ⟨73JHC47⟩; and 2-chloromethylpyrimidine gives trimethyl(pyrimidin-2-ylmethyl)ammonium chloride (**421**) ⟨74AJC2251⟩.

(*c*) Most halogenopyrimidines undergo alcoholysis by treatment with alcoholic alkoxide, although very deactivated substrates may require severe conditions. Thus, brief warming of 2- or 4-chloropyrimidine with methanolic sodium methoxide affords 2- or 4-methoxypyrimidine, respectively ⟨53JCS331⟩; 2-phenoxypyrimidine may be made similarly but it is better to reflux 2-chloropyrimidine with phenol and potassium carbonate in toluene ⟨51YZ1420⟩ and 2-phenoxyquinazoline (**422**) can be made rather similarly ⟨56JCS4191⟩. Considerable selectivity may be achieved from di- and tri-chloropyrimidines by controlling conditions carefully: 4,6-dichloro- (**423**; R = Cl) gives 4-chloro-6-methoxy- (**424**) (using 1 mol of methoxide) or 4,6-dimethoxy-pyrimidine (**423**; R = OMe) (using 5 mol of methoxide) ⟨61JOC2764⟩. 2,4-Dichloro-6-methylpyrimidine gives 2-chloro-4-methyl-6-propoxypyrimidine (**425**) free of the 2-propoxy isomer (using 1 mol sodium propoxide in propanol at 40 °C) or 4-methyl-2,6-dipropoxypyrimidine (using 2 mol of propoxide at 80 °C) ⟨62AP649⟩. The conversion of 6-chloro- (**426**; R = Cl) into 6-benzyloxy-pyrimidine-2,4-diamine (**426**; R = OCH$_2$Ph) requires heating with sodium benzyl alcoholate at 160 °C for 3 hours to displace the deactivated chloro substituent of the substrate ⟨61CB12⟩. Few 5-halogeno substituents are recorded as undergoing alcoholysis but 5-chloro- (**427**; R = Cl) gives 5-ethoxy-6-hydroxy-2-hydroxymethylpyrimidin-4(3*H*)-one (**427**; R = OEt) on heating with ethanolic sodium ethoxide for 12 hours ⟨61JOC1874⟩. In contrast, extranuclear halogeno substituents undergo alcoholysis very readily. For example, reaction of 2,4-dichloro-5-chloromethylpyrimidine (**428**; R = Cl) with 1 mol of sodium ethoxide gives 2,4-dichloro-5-ethoxymethyl- (**428**; R = OEt), with 2 mol of ethoxide gives 2-chloro-4-ethoxy-

5-ethoxymethyl- (**429**; R = Cl) and with >3 mol of ethoxide gives 2,4-diethoxy-5-ethoxy-methyl-pyrimidine (**429**; R = OEt), all in good yield ⟨66LA(692)119⟩.

(**422**)   (**423**)   (**424**)   (**425**)

(**426**)   (**427**)   (**428**)   (**429**)

(*d*) The hydrolysis of halogenopyrimidines may usually be done directly in aqueous acid or alkali, although some workers still prefer to convert the halogeno- into an alkoxy-pyrimidine first and then perform hydrolysis. Examples of direct hydrolysis are seen in the conversion of 6-fluorouracil (**430**; R = F) by 1 M hydrochloric acid at 100 °C into barbituric acid (**430**; R = OH) ⟨64JMC207⟩; the 2-chloropyrimidine (**431**) into the pyrimidinone (**432**) by 1 M sodium hydroxide at 100 °C ⟨61JCS5131⟩; 5-bromo-1,3-dimethylpyrimidine-2,4(1*H*,3*H*)-dione (**433**; R = Br) into the 5-hydroxy analogue (**433**; R = OH) by boiling in aqueous sodium hydrogen carbonate, an abnormally facile hydrolysis among 5-halogenated pyrimidines, which usually require quite drastic conditions ⟨59JA3786⟩; and 4-bromomethyl-pyrimidine (**434**; R = Br) into pyrimidin-4-ylmethanol (**434**; R = OH) by 1 M alkali at 40 °C ⟨74AJC2251⟩.

(**430**)   (**431**)   (**432**)   (**433**)

(**434**)   (**435**)   (**436**)   (**437**)

(*e*) The thioalcoholysis of a halogeno substituent is seldom required in the 2- or 4/6-position because *S*-alkylation of the corresponding thione is usually preferable. However, it can be done and, in the case of arylthiopyrimidines, it is the route of choice because *S*-arylation is normally impossible. For example, 2-chloro- yields 2-phenylthio-pyrimidine (**435**) by warming in ethanolic sodium thiophenate ⟨67JCS(C)568⟩; and 2,6-dichloropyrimidin-4-amine (**436**; R = Cl) yields (with 1 mol of sodium ethanethiolate at 50 °C) 6-chloro-2-ethylthiopyrimidin-4-amine (**436**; R = SEt) or (with an excess of reagent at 110 °C) 2,6-bisethylthiopyrimidin-4-amine (**437**) ⟨61M183⟩. 5-Halogenopyrimidines undergo the reaction but less readily. For example, 5-bromo- is converted into 5-methylthio-pyrimidine by boiling for 12 hours in ethanolic sodium methanethiolate ⟨67JCS(C)568⟩, although other 5-halogenopyrimidines appear to need temperatures up to 150 °C. 7-Fluoroquinazoline reacts with sodium ethanethiolate in boiling ethanol to give 7-ethylthioquinazoline ⟨67JCS(B)449⟩. Extranuclear halogenopyrimidines undergo thioalcoholysis quite easily, as in the conversion of 2-chloromethyl-4,6-dimethylpyrimidine into 4,6-dimethyl-2-methyl-thiomethylpyrimidine by aqueous sodium methanethiolate at 70 °C ⟨80CPB3362⟩.

(*f*) The thiolysis of halogenopyrimidines is possible in all but the 5-position. Examples are furnished by the conversion of 6-chloropyrimidin-4-amine (**438**) by sodium hydrogen sulfide into 6-aminopyrimidine-4(3*H*)-thione (**439**; R = H) or by 6-chloro-4-dimethyl-amino-1-methylpyrimidinium iodide into 6-dimethylamino-3-methylpyrimidine-4(3*H*)-thione (**439**; R = Me) ⟨65AJC559⟩. Selenolysis can be done similarly, *e.g.* 2,4-dichloro-5-methylpyrimidine into 5-methylpyrimidine-2,4(1*H*,3*H*)-diselone (**440**) ⟨63JMC36⟩. It must be remembered that sodium hydrogen sulfide will reduce a nitro group and may sometimes replace an alkylthio group by a mercapto group. For example, 2-chloro-5-nitropyrimidin-4-amine (**441**) gives 4,5-diaminopyrimidine-2(1*H*)-thione (**442**) ⟨52MI21300⟩ and 4-chloro-2-methylthiopyrimidine (**443**) gives (with sodium hydrogen sulfide in ethylene glycol at 150 °C) pyrimidine-2,4(1*H*,3*H*)-dithione ⟨61JOC792⟩. Under these circumstances an indirect thiolysis may be done, as exemplified in the conversion of 2-chloro-5-nitropyrimidine, by boiling alcoholic thiourea in 1 minute, into the thiouronium chloride (**444**) and thence, by dissolution in warm alkali and acidification, into 5-nitropyrimidine-2(1*H*)-thione (**445**) ⟨66AJC2321⟩. The replacement of an extranuclear halogeno substituent by a mercapto group is not straightforward; use of thiourea or sodium hydrogen sulfide invariably leads to a sulfide or disulfide, but thioacetic acid in pyridine does give some of the mercaptan. For example, 5-bromomethyl-2-methylpyrimidin-4-amine gives a mixture of 4-amino-2-methyl-pyrimidin-5-ylmethanethiol and the corresponding disulfide ⟨58JOC1738⟩; 5-chloromethyl-is converted somewhat similarly into 5-mercaptomethyl-uracil ⟨66JMC97⟩.

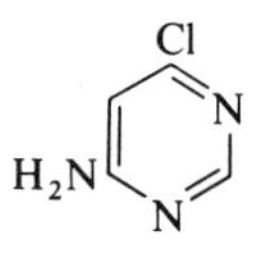

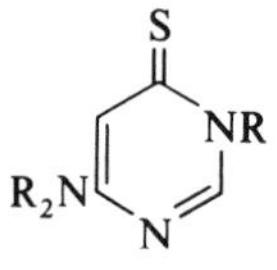

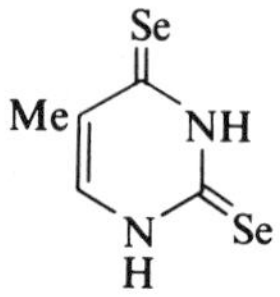

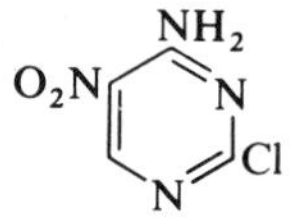

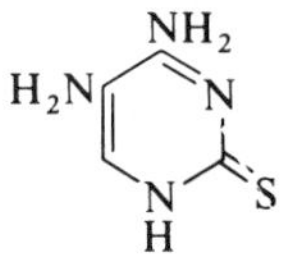

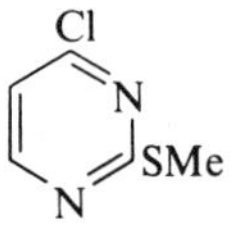

(*g*) Halogeno substituents may be replaced by other sulfur-containing groups. Thus, an alternative to the oxidation of thiones (Section 2.13.2.2.8) to yield sulfonic acids is to treat a chloropyrimidine with potassium sulfite. In this way, 4-chloro-2,6-dimethylpyrimidine (**446**) and freshly prepared aqueous potassium sulfite at 90 °C for 20 minutes gives potassium 2,6-dimethylpyrimidine-4-sulfonate (**447**); the isomeric potassium 4,6-dimethylpyrimidine-2-sulfonate and related compounds can be made similarly ⟨71JCS(B)2214⟩. Likewise, the oxidative route to some sulfones can be avoided by treating a chloropyrimidine with potassium benzenesulfinate or a related compound; thus, 2-chloropyrimidine and sodium benzenesulfinate, heated in DMSO at 100 °C give 2-phenylsulfonylpyrimidine (**448**) ⟨67JCS(C)568⟩. The replacement of an active chloro by a thiocyanato substituent is illustrated in the conversion by ammonium thiocyanate of 2-bromopyrimidine (or 2-chloropyrimidine, in lower yield) into pyrimidin-2-yl thiocyanate (**449**) ⟨63YZ1086⟩ and by 5-chloromethyluracil into uracil-5-ylmethyl thiocyanate ⟨66JMC97⟩. Isothiocyanates are obtained by thermal rearrangement of thiocyanates and this can occur during preparation. Thus, 4-chloro-2-ethylthiopyrimidine (**450**; R = Cl) and potassium thiocyanate in boiling acetone give 2-ethylthiopyrimidin-4-yl thiocyanate (**450**; R = SCN) but when boiling toluene is used as solvent, the product is the corresponding isothiocyanate (**450**; R = NCS) ⟨05MI21300⟩. Sodiosulfanilamide reacts with chloropyrimidines to give the substituted sulfonamide; for example, 4-chloro-6-methoxypyrimidine gives *N*-(6-methoxypyrimidin-4-yl)sulfanilamide (**451**), a type of reaction of considerable importance ⟨61JOC2764⟩.

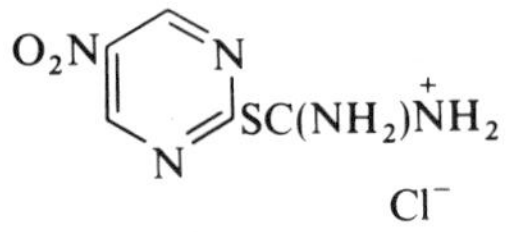

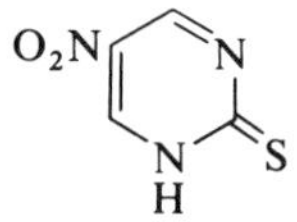

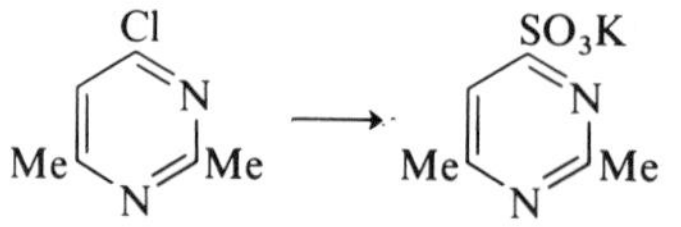

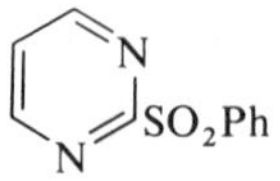

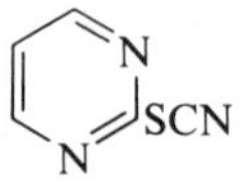

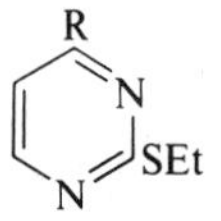

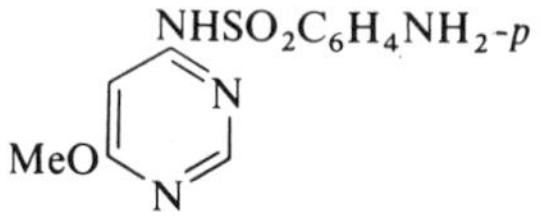

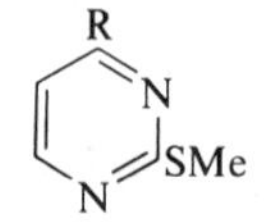

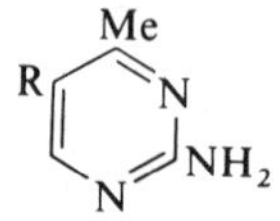

(*h*) The formation of pyrimidinecarbonitriles from halogenopyrimidines is uncommon in the 2- or 4/6-position but 4-iodo-2-methylthiopyrimidine (**452**; R = I) does give 2-methylthiopyrimidine-4-carbonitrile (**452**; R = CN) on heating with cuprous cyanide in pyridine ⟨64JHC201⟩. However, the formation of pyrimidine-5-carbonitriles in this way is reasonably common. Thus 5-bromopyrimidine gives pyrimidine-5-carbonitrile (cuprous cyanide in boiling quinoline) ⟨64MI21301⟩ and 5-bromo-4-methylpyrimidin-2-amine (**453**; R = Br) gives 2-amino-4-methylpyrimidine-5-carbonitrile (**453**; R = CN) under similar conditions ⟨67JCS(C)1928⟩. Extranuclear halogenopyrimidines also react under quite gentle conditions: 2-chloromethyl-4,6-dimethylpyrimidine and aqueous ethanolic sodium cyanide at 70–80 °C give 4,6-dimethylpyrimidin-2-ylacetonitrile in good yield ⟨80CPB3362⟩.

(*i*) Halogenopyrimidines react with active methylene groups, such as those in diethyl malonate, ethyl cyanoacetate, ketene diethylacetal, *etc.* For example, 4-chloro-6-methyl-5-nitropyrimidin-2-amine (**454**) and dimethyl sodiomalonate give dimethyl 2-amino-6-methyl-5-nitropyrimidin-4-ylmalonate (**455**) ⟨63ZOB3132⟩; 2-chloro-4,6-diphenylpyrimidine and malononitrile give 4,6-diphenylpyrimidin-2-ylmalononitrile (**456**) ⟨65IZV2087⟩; 4-chloroquinazoline and benzyl sodiocyanoacetate give the cyanoester (**457**) ⟨61JCS2689⟩; and 5-bromomethylpyrimidine (**458**) and diethyl benzyloxycarbonyl-aminomalonate (**459**) give initially, diethyl α-benzyloxycarbonylamino-α-(pyrimidin-5-ylmethyl)malonate (**460**) which can be degraded to 2-amino-3-(pyrimidin-5′-yl)propionic acid (**461**) ⟨65JHC1⟩.

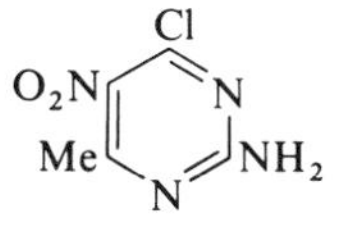

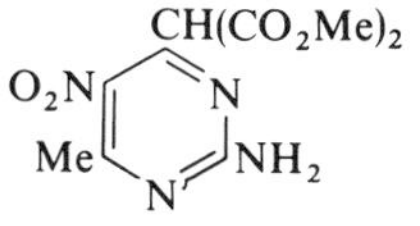

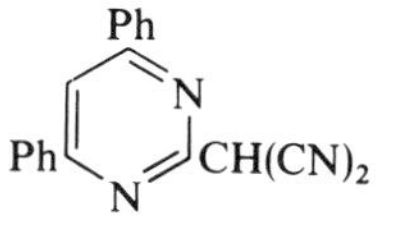

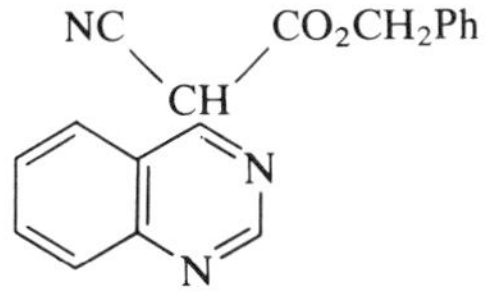

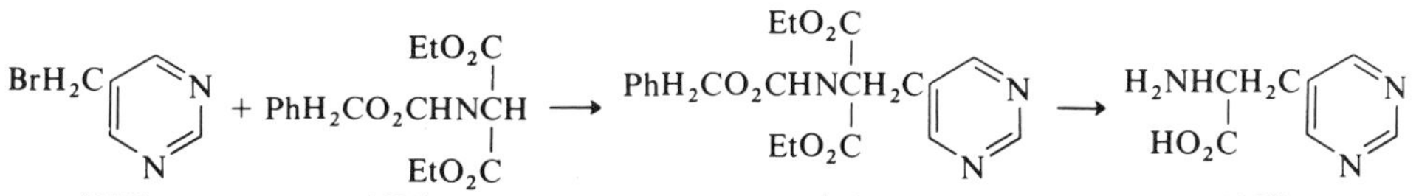

(*j*) Halogenopyrimidines undergo Ullmann, Busch and Strekowski biaryl syntheses to yield bipyrimidines. The Ullmann reaction can yield 2,2′-, 4,4′- or 5,5′-bipyrimidines by heating appropriate bromopyrimidines with copper bronze in cumene or DMF. For example, 2-bromopyrimidine gives 2,2′-bipyrimidine (**462**) in 35% yield ⟨62JOC2945⟩; 4-iodo-6-methyl-2-phenylpyrimidine gives 6,6′-dimethyl-2,2′-diphenyl-4,4′-bipyrimidine (**463**); and 5-bromo-2-phenylpyrimidine-4-carboxylic acid gives (after decarboxylation)2,2′-diphenyl-5,5′-bipyrimidine (**464**) ⟨67JCS(C)1204⟩. The Busch reaction is not always successful in the series, but heating 5-bromo-2-phenylpyrimidine in a mixture of methanolic potassium hydroxide, hydrazine hydrate and palladium/calcium carbonate does give the diphenyl-bipyrimidine (**464**) in yield comparable with that from the Ullmann reaction above ⟨67JCS(C)1204⟩. The Strekowski reaction involves treatment of a 5-bromopyrimidine with less than 1 mol of benzyllithium in THF at −70 °C to yield a 4,5′-bipyrimidine; there are at least two possible mechanisms ⟨74MI21300⟩. This synthesis has many recent examples, *e.g.* 5-bromo-*N,N*-dimethylpyrimidin-2-amine gives *N,N,N′,N′*-tetramethyl-4,5′-bipyrimi-dine-2,2′-diamine (**465**) ⟨81AJC2629⟩.

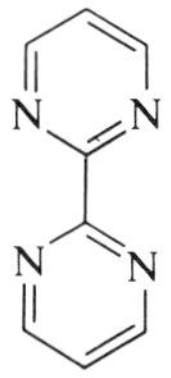

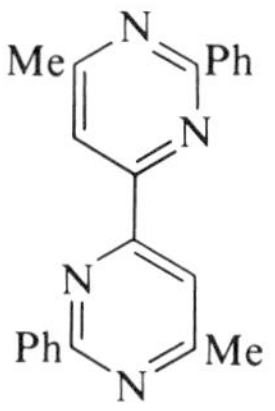

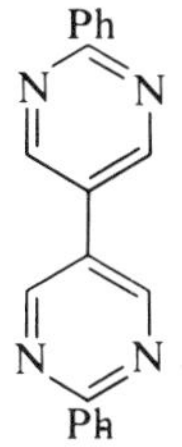

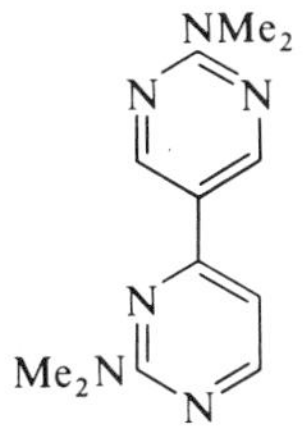

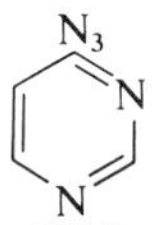

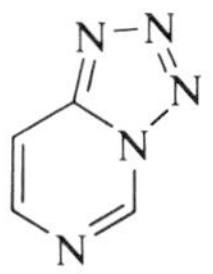

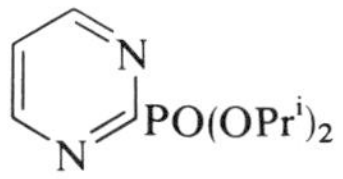

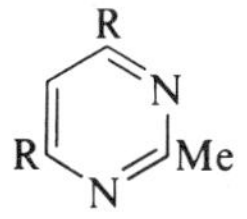

(470)        (471)        (472)        (473)

($k$) Some less common reactions of halogenopyrimidines are illustrated in the following examples. 4-Chloropyrimidine and sodium azide in DMF give 4-azidopyrimidine (466) in equilibrium with its bicyclic isomer (467) ⟨65JOC829⟩. 2-Chloropyrimidine reacts with triisopropyl phosphite, evolving isopropyl chloride to give diisopropyl pyrimidin-2-ylphosphonate (468) ⟨61JOC1895⟩. 4,6-Dichloro-2-methylpyrimidine (469; R = Cl) and ethylmagnesium bromide give 4,6-diethyl-2-methylpyrimidine (469; R = Et) ⟨52JOC1320⟩. 4-Chloro-2,6-diphenyl-5-propylpyrimidine and phenyllithium afford 2,4,6-triphenyl-5-propylpyrimidine ⟨41JCS323⟩. 5-Bromopyrimidine yields its 5-lithium analogue on treatment with butyllithium at very low temperatures (for subsequent reactions, see Section 2.13.2.2.10) but at higher temperatures, 3,4-addition of the butyllithium occurs to give (after hydrolysis) 4-butyl-5-bromo-3,4-dihydropyrimidine (470) which can be aromatized by permanganate to 5-bromo-4-butylpyrimidine (471); 5-bromo-4-(thien-2'-yl)pyrimidine can be made similarly ⟨65ACS1741⟩. 5-Bromo-2-tribromomethylpyrimidine (472; R = CBr₃) reacts with silver nitrate in acetic acid to give 5-bromopyrimidine-2-carboxylic acid (472; R = CO₂H) in good yield ⟨60MI21300⟩; likewise, 5-bromo-2-dibromomethylpyrimidine (472; R = CHBr₂) with aqueous ethanolic silver nitrate gives 5-bromopyrimidine-2-carbaldehyde (472; R = CHO) ⟨65JCS5467⟩. The bromomethyl compound (473; R = CH₂Br) can be converted into a Wittig reagent with triphenylphosphine and thence, by treatment with *p*-nitrobenzaldehyde, into the *p*-nitrostyryl derivative (473; R = CH:CHC₆H₄NO₂) ⟨66JHC324⟩.

($l$) Transhalogenation, or the replacement of one halogeno substituent by another, is of little interest as a reaction but of great preparative importance. Accordingly, it is discussed in Section 2.13.3.3.9$f$.

### 2.13.2.2.10 Reactivity of metallo derivatives

The recorded use of metallo derivatives in the pyrimidine and quinazoline series is minimal. The best described pyrimidinyllithium compounds are those derived from 5-bromopyrimidines. Their reactions are illustrated in the following examples. Pyrimidin-5-yllithium (474; R = H) reacts with solid carbon dioxide under ether to give pyrimidine-5-carboxylic acid (475; R = H) in good yield ⟨65ACS1741⟩; 4,6-dimethoxy- (474; R = OMe), 4,6-bismethylthio- (474; R = SMe) and other pyrimidin-5-yllithiums behave similarly to give the corresponding pyrimidine-5-carboxylic acids (475; R = OMe, SMe, *etc.*) ⟨65JCS5467⟩. 2,4,6-Trimethoxypyrimidin-5-yllithium (476; R = Li) reacts with DMF to give 2,4,6-trimethoxypyrimidine-5-carbaldehyde (476; R = CHO). 4,6-Dimethoxypyrimidin-5-yllithium (474; R = OMe) and sulfur give a polysulfide which can be reduced to the mercaptan (477; R = H) and isolated (by treatment with chloroacetic acid) as 2-(4',6'-dimethoxypyrimidin-5'-ylthio)acetic acid (477; R = CH₂CO₂H) ⟨65JCS5467⟩. 2,4-Diethoxypyrimidin-5-yllithium (478; R = Et) and benzaldehyde yield the secondary alcohol (479); other aromatic aldehydes behave similarly ⟨63JMC550⟩. 2,4-Dimethoxypyrimidin-5-yllithium (478; R = Me) and trimethyl borate give the corresponding dimethoxypyrimidin-5-ylboronic acid (480) ⟨64JA1869⟩. 2-Methylpyrimidin-5-yllithium and dibutyl phosphorochloridate give dibutyl 2-methylpyrimidin-5-ylphosphonate (481) in poor yield ⟨61JOC1895⟩. Such lithium derivatives may also be used to make pyrimidine-5-nucleosides ⟨66JOC2215⟩.

(474)        (475)        (476)        (477)

(478)        (479)        (480)        (481)

When the temperature is not low enough for 5-lithiation, both 2- and 5-bromopyrimidine undergo 3,4-addition by thien-2-yllithium to give, for example, the intermediate (**482**; R = Li) which gives on hydrolysis 2-bromo-4-(thien-2'-yl)-3,4-dihydropyrimidine (**482**; R = H) and on gentle permanganate oxidation, the bromothienylpyrimidine (**483**) in good yield ⟨65ACS1741⟩.

The formation of bromine-free 4,5'-bipyrimidines from 5-bromopyrimidines and butyl-lithium by the Strekowski reaction is discussed above (Section 2.13.2.2.9). It should be distinguished from a single example wherein 5-bromopyrimidine and butyllithium give 5-bromo-3,4-dihydro-4,5'-bipyrimidine (**484**) and on oxidation, 5-bromo-4,5'-bipyrimidine (**485**) ⟨65ACS(19)1741⟩.

The action of alkyllithiums and alkylmagnesium halides with functional groups on pyrimidines has been mentioned in appropriate sections on the reactivity of the substrates.

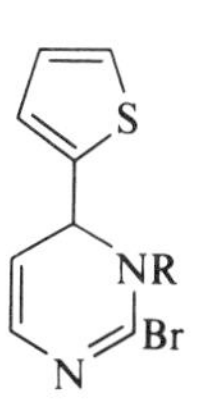

(**482**) 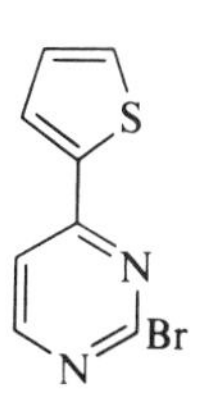  (**483**) 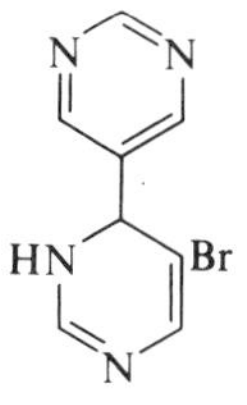  (**484**) 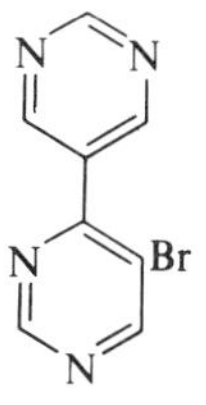  (**485**)

### 2.13.2.2.11 Reactivity of N-oxides

Although the presence of an *N*-oxide substituent modifies the reactivity of other groups, only reactions directly involving the *N*-oxide part are discussed here.

(*a*) The removal of an *N*-oxide substituent is usually done by treatment with phosphorus trichloride in chloroform. In this way, 4-methylpyrimidine *N*-oxide gives 4-methyl-pyrimidine ⟨55CPB175⟩ and 4-phenyl-6-styrylpyrimidine 1-oxide (**486**) gives the parent pyrimidine (**487**) ⟨80CPB1526⟩. An alternative method of removal, when there are no other reduction-sensitive groups present, is catalytic hydrogenation over Raney nickel or palladium. Thus is 4-ethoxy-6-β-piperidinoethylpyrimidine obtained (Raney nickel) from its 1-oxide (**488**) in >90% yield ⟨80CPB1526⟩ and 6-methylpyrimidin-4(3*H*)-one (**489**) from its *N*-oxide (using palladium) ⟨59CPB158⟩. A third (reductive) method may also be used: treatment of 4-ethoxyquinazoline 1-oxide (**490**) with sulfur dioxide gives a better yield of the parent ethoxyquinazoline than does use of phosphorus trichloride ⟨59CPB152⟩.

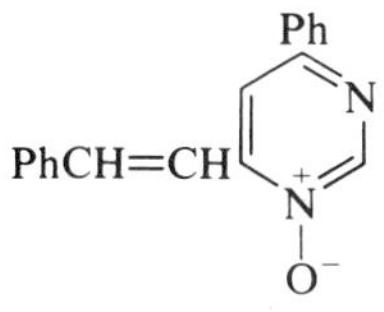

(**486**) 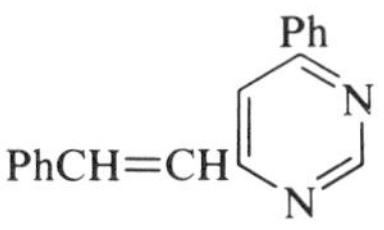  (**487**) 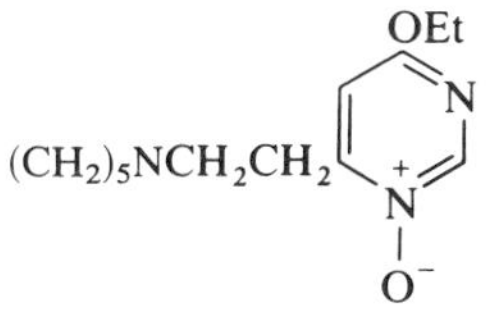  (**488**) 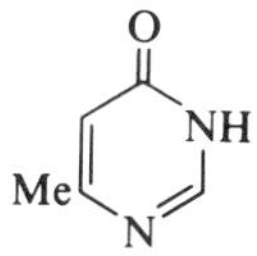  (**489**) 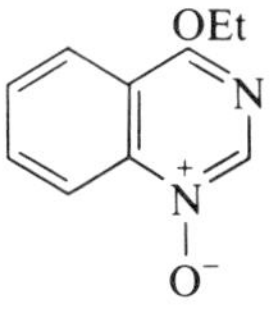  (**490**)

(*b*) Pyrimidine *N*-oxides undergo Reissert-like reactions. When 4-ethoxy-6-methyl-pyrimidine *N*-oxide (**491**; R = Et) is dissolved in aqueous potassium cyanide and then treated with benzoyl chloride followed by alkali, 4-ethoxy-6-methylpyrimidine-2-carbonitrile (**492**; R = Et) is obtained in good yield; analogues (**492**; R = Ph, Me, Bu, *etc.*) may be made similarly ⟨58CPB633⟩. If potassium carbonate is used instead of the cyanide above, 4-methoxy-6-methylpyrimidine *N*-oxide (**491**; R = Me) yields 4-methoxy-6-methyl-pyrimidin-2(1*H*)-one (**493**) ⟨55CPB175⟩. Similar rearrangements are brought about by other acylating agents. Thus, pyrimidine 1-oxide and acetic anhydride give pyrimidin-4-yl acetate (**494**) ⟨58CB2832⟩ but when the position is occupied by a methyl group, as in 4,6-dimethyl-pyrimidine 1-oxide, the product is 6-methylpyrimidin-4-ylmethyl acetate (**495**) ⟨59JCS525⟩.

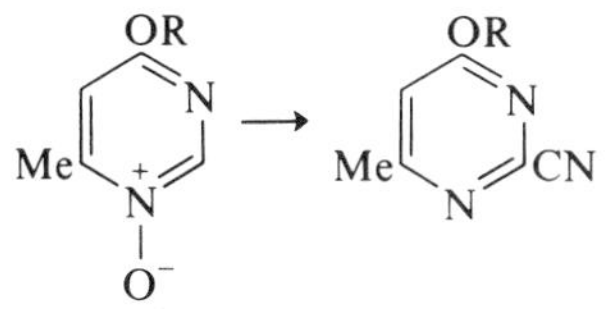

(**491**)  (**492**) 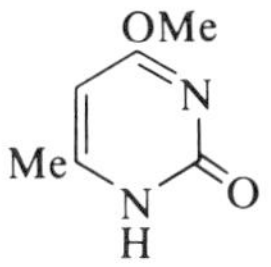  (**493**) 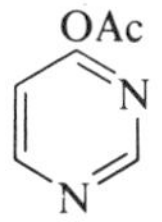  (**494**) 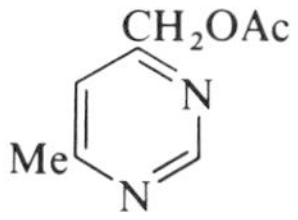  (**495**) 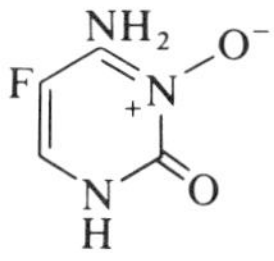  (**496**) 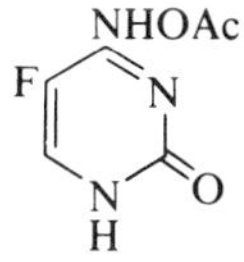  (**497**)

Likewise, 4-amino-5-fluoropyrimidin-2(1*H*)-one 3-oxide (**496**) and acetic anhydride give 4-acetoxyamino-5-fluoropyrimidin-2(1*H*)-one (**497**) ⟨68M847⟩.

(*c*) The ability of pyrimidine *N*-oxides to exist as *N*-hydroxy tautomers when appropriate substituents are present makes it possible to *O*-alkylate such compounds. Thus, 1-hydroxy-4-methoxyquinazolin-2(1*H*)-one (**498**; R = H) reacts with methyl iodide/silver hydroxide to give 1,4-dimethoxyquinazolin-2(1*H*)-one (**498**; R = Me) ⟨59CPB152⟩.

4-Ethoxy-6-methylpyrimidine 1-oxide (**499**) reacts with phenyl isocyanate to eliminate carbon dioxide and give a mixture of 4-ethoxy-6-methyl-*N*-phenylpyrimidin-2-amine (**500**) and (the derived) *N*-(4-ethoxy-6-methylpyrimidin-2-yl)-*N*,*N'*-diphenylurea (**501**); phenyl isothiocyanate reacts quite differently ⟨79CPB2642⟩.

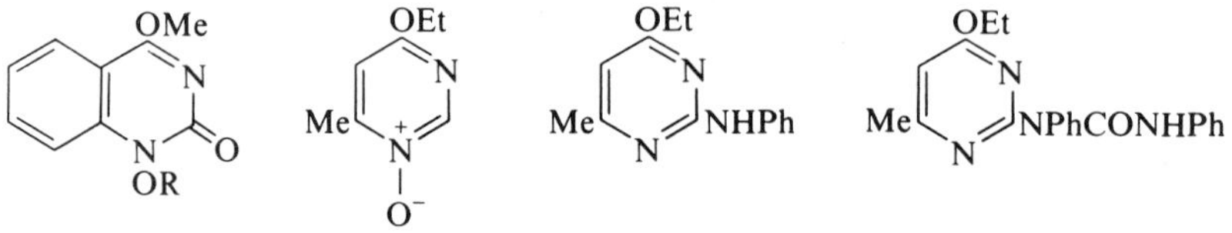

| (**498**) | (**499**) | (**500**) | (**501**) |

## 2.13.3 SYNTHESIS OF PYRIMIDINES

Despite considerable localization of $\pi$-electrons at the nitrogen atoms of pyrimidine, the ring system is still sufficiently aromatic to possess substantial stability. This is a great advantage in the primary synthesis of pyrimidines, in the synthesis of pyrimidines from the breakdown or modification of other heterocyclic systems and in the myriad of metatheses required to synthesize specifically substituted pyrimidines.

### 2.13.3.1 Primary Synthesis of the Pyrimidine Ring

The first primary synthesis of a pyrimidine from aliphatic fragments was carried out by Frankland and Kolbé in 1848. Since then, a great many quite distinct primary synthetic methods have been devised, although it is true to say that one of these (the 'Principal Synthesis') has provided upward of 80% of all known pyrimidines, either directly or indirectly.

#### 2.13.3.1.1 Primary syntheses involving formation of one bond

In many pyrimidine ring syntheses, it is possible or even desirable to isolate an intermediate ripe for ring-closure by the formation of just one bond. For example, ethyl 3-aminocrotonate (**502**) reacts with methyl isocyanate to give the ureido ester (**503**) which may be isolated and subsequently converted into 3,6-dimethyluracil (**504**) by the completion of one bond. However, viewed pragmatically, the whole synthesis involves the formation of two bonds and therefore is so classified. On such criteria, only two pyrimidine/quinazoline syntheses involve the formation of only one bond.

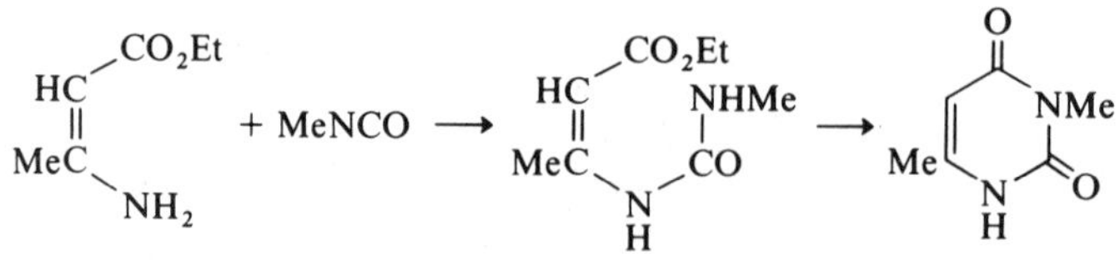

| (**502**) | (**503**) | (**504**) |

(*a*) The so-called Rinkes synthesis of uracil ⟨27RTC268⟩ is brought about by treating malediamide (**505**) with sodium hypochlorite; one of the amide groups appears to undergo the first stages of a Hofmann reaction to give the isocyanate intermediate (**506**) which then cyclizes with the other amide group to give uracil (**507**) in good yield. The reaction does not seem to have been extended within the pyrimidine series proper but there are even earlier examples of such procedures giving quinazoline-2,4(1*H*,3*H*)-dione (**509**) from phthaldiamide (**508**) and 5,6-dihydrouracil (**511**; R = H) from succindiamide (**510**; R = H) ⟨62HC(16)430⟩. The last reaction can be extended, for example, by using 2-phenylsuccin-diamide (**510**; R = Ph) to afford 6-phenyl-5,6-dihydrouracil (**511**; R = Ph).

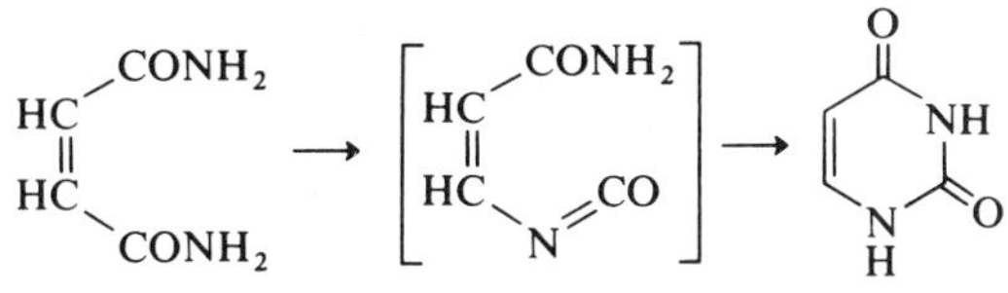

(505)   (506)   (507)

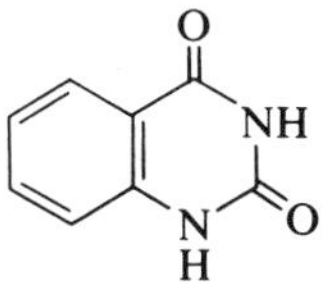

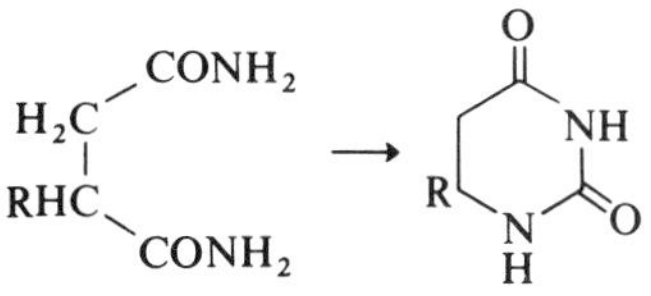

(508)   (509)   (510)   (511)

(*b*) A rare reaction is also included here: 3-acetamidocrotonamide undergoes cyclization at 200 °C to give 2,6-dimethylpyrimidin-4(3*H*)-one in good yield ⟨67YZ955⟩. Likewise, it was recorded in the last century that heating 2-*N*-methylacetamidobenzamide at 200 °C gives 1,2-dimethylquinazolin-4(1*H*)-one and this reaction is still used sometimes in several modifications ⟨67HC(24-1)99⟩.

### 2.13.3.1.2 *Primary syntheses involving formation of two bonds*

This category falls naturally into three sub-categories: syntheses from [1 + 5] atom fragments, [2 + 4] atom fragments and [3 + 3] atom fragments.

Syntheses involving [1 + 5] atom fragments may be sub-divided into two types: in the first, the one-atom fragment supplies C-2 of the final ring and in the second, it supplies N-1 or N-3. The first type includes several syntheses based on malondiamides and malon-diamidines.

(*a*) The Remfry–Hull synthesis is illustrated by the condensation of α-butylmalondiamide (**512**) and ethyl formate in ethanolic sodium ethoxide to give 5-butyl-6-hydroxypyrimidin-4(3*H*)-one (**513**) ⟨65CCC3730⟩. The diamide may be replaced by a monothiodiamide and the ester by an acid chloride: 2-ethyl-2-thiocarbamoylbutyramide (**514**) and acetyl chloride give the pyrimidinethione (**515**) ⟨64JCS3204⟩. Other variations include the use of diethyl oxalate to give with malondiamide, the pyrimidine-2-carboxylic acid (**516**) ⟨66JCS(C)226⟩ and replacement of the ester by a simple amide such as formamide which reacts, for example, with α-methoxymalondiamide to give 6-hydroxy-5-methoxypyrimidin-4(3*H*)-one (**517**) ⟨65M1677⟩. When homologues of formamide or ethyl formate are used, the reaction is slower and hence allows time for the self-condensation of malondiamide to yield the by-product (**518**) ⟨56JCS2312⟩ The older literature contains other examples of the use of one-carbon fragments with malondiamides ⟨62HC(16)82⟩.

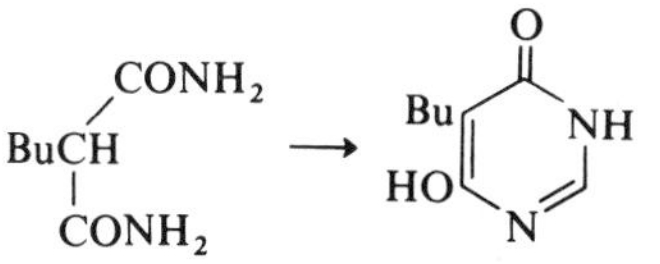

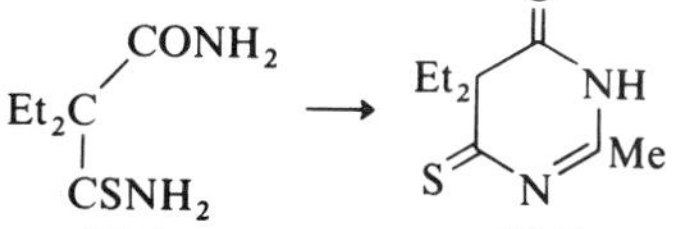

(512)   (513)   (514)   (515)

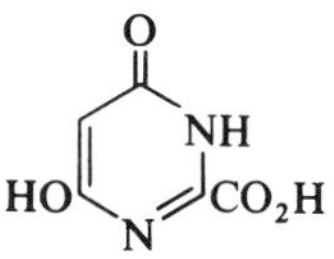

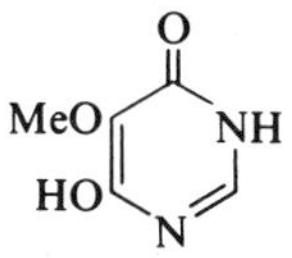

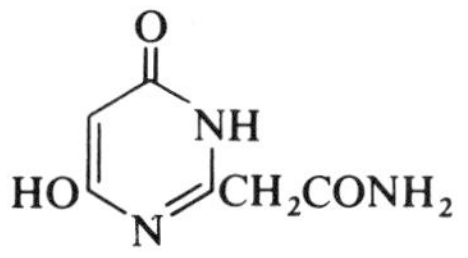

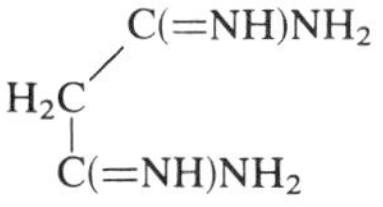

(516)   (517)   (518)   (519)

(*b*) Pyrimidine-4,6-diamines are obtained from similar condensations of malondiamidine with esters or amides. For example, malondiamidine (**519**) and ethyl benzoate give 2-phenylpyrimidine-4,6-diamine (**520**) ⟨44JCS476⟩ while the similar use of diethyl carbonate or ethyl chloroformate yields 4,6-diaminopyrimidin-2(1*H*)-one (**521**) ⟨48JA3109⟩. The use of *N*-substituted malondiamidines is seen in the condensation of symmetrical *N,N'*-diallyl-malondiamidine with ethyl formate to give *N,N'*-diallylpyrimidine-4,6-diamine (**522**) ⟨58JA2185⟩.

(**520**)          (**521**)                      (**522**)

(*c*) The use of a one-carbon fragment at a lower oxidation state leads to a dihydropyrimidine: 2-ethyl-2-phenylmalondiamide (**523**) and benzaldehyde give 5-ethyl-2,5-diphenyl-2,5-dihydropyrimidine-4,6(1*H*,3*H*)-dione (**524**) ⟨56CB2239⟩. Likewise it is possible to use a five-atom fragment, such as a propane-1,3-diamine, which has a lower oxidation state; thus, with an ester *etc.*, it gives a tetrahydropyrimidine or with an aldehyde, it gives a hexahydropyrimidine. For example, propane-1,3-diamine (**525**) and triethyl orthopropionate yield 2-ethyl-1,4,5,6-tetrahydropyrimidine (**526**) ⟨62CB1840⟩. A great many variations of this synthesis, such as the replacement of the ortho ester by an acid, ester, nitrile, amidine, *etc.* are recorded, thus making it a major route to such tetrahydropyrimidines ⟨70HC(16-S1)322⟩. It also furnishes a route to dihydroquinazolines when the diamine is replaced by an *o*-(aminomethyl)aniline: thus, *o*-benzylaminobenzylamine (**527**) and boiling formic acid give 1-benzyl-1,4-dihydroquinazoline (**528**) ⟨61JCS2697⟩; naturally, if the benzyl group in the substrate (**527**) is moved to the extracyclic amino group, a 3,4-dihydroquinazoline then results ⟨67HC(24-1)391⟩. The condensation of (monoprotonated) propane-1,3-diamine with aqueous formaldehyde gives the parent hexahydropyrimidine (**529**) and many *N*-and/or *C*-alkyl derivatives may be made similarly ⟨67AJC1643⟩; likewise, boiling an ethanolic mixture of 2-amino-4,5-dimethoxybenzylamine with *m*-nitrobenzaldehyde gives the 1,2,3,4-tetrahydroquinazoline (**530**) ⟨50JA3053⟩.

(**523**)                      (**524**)                      (**525**)                      (**526**)

(**527**)                      (**528**)          (**529**)                      (**530**)

(*d*) A variation on the use of a malondiamide above is to employ a 3-aminoacrylamide derivative, which, because of its unsaturation, is at the same overall oxidation state as malondiamide. Thus, 3-aminocinnamamide (**531**; R = H), easily obtained by hydrogenation of 3-phenylisoxazol-5-amine, reacts with benzyl chloride to give the acyl intermediate (**531**; R = COPh), which undergoes cyclization to 2,6-diphenylpyrimidin-4(3*H*)-one (**532**) ⟨54JCS665⟩. Likewise, anthranilamide with formaldehyde yields 1,2-dihydroquinazolin-4(3*H*)-one ⟨64G595⟩.

(**531**)                      (**532**)

(*e*) The Shaw synthesis and its variants are characterized by the completion of the ring with a one-nitrogen fragment and it affords access to uracil or thiouracil derivatives indicated in formula (**533**), although not all combinations are possible ⟨62HC(16)82⟩. A typical sequence involves the condensation of cyanoacetic acid with *N*-methylurethane in acetic anhydride to give *N*-cyanoacetyl-*N*-methylurethane (**534**) which is converted by triethyl orthoformate and acetic anhydride into its ethoxymethylene derivative (**535**) and thence with aniline (the one-nitrogen fragment) into the anilinomethylene analogue (**536**) followed by cyclization to 5-cyano-3-methyl-1-phenylpyrimidine-2,4(1*H*,3*H*)-dione (**537**) ⟨56JCS4118⟩. To obtain a proton or methyl group at the 5-position, the route is rather different; for example, ethyl propiolate and urethane give ethoxymethyleneacetylurethane (**538**; R = H) which reacts readily with aniline to give the intermediate (**539**; R = H) and thence 1-phenylpyrimidine-2,4(1*H*,3*H*)-dione (**540**; R = H) ⟨57JCS2363⟩. The above intermediate (**538**; R = H) reacts similarly with tri-*O*-benzoylribosylamine to give, after deacylation, uridine (**541**) ⟨58JCS2294⟩. Using the related intermediate (**538**; R = Ac) with aniline gives 5-acetyl-1-phenyluracil (**540**; R = Ac); when aniline is replaced by an amino acid, a variety of analogues result in which the 1-phenyl group is replaced by an amino acid residue ⟨61JCS3254⟩. Another convenient way to obtain 1-substituted uracils or thiouracils is illustrated in the following sequences: ethoxymethyleneacetic acid (**542**; R = OH) is converted into the acid chloride (**542**; R = Cl) and thence with potassium thiocyanate into the isothiocyanate (**543**; X = S) which reacts with methylamine to give 1-methyl-2-thiouracil (**544**; X = S, R = Me) ⟨58JCS153⟩; the same acid chloride (**542**; R = Cl) also reacts with silver cyanate to give the isocyanate (**543**; X = O) which in turn reacts with benzylamine to give 1-benzyluracil (**544**; X = O, R = CH$_2$Ph) ⟨63JCS811⟩ or with *O*-benzylhydroxylamine to give 1-benzyloxyuracil (**544**; X = O, R = OCH$_2$Ph) and thence (unambiguously) uracil 1-oxide (**544**; X = O, R = OH) ⟨64M1729⟩. This synthesis may also be used to make 1-substituted 5,6-dihydrouracils. For example, the easily available 3-bromopropionyl isocyanate (**545**) reacts with aniline to give the intermediate (**546**) which cyclizes in boiling propionic acid containing silver carbonate to give 1-phenyl-5,6-dihydrouracil (**547**) ⟨59JOC1391⟩; the route can be modified extensively to give dihydropyrimidine-2-thiones, *e.g.* (**548**).

$$R^1 = H, Me, Ph, CH_2CH_2OH, etc.$$
$$R^2 = H, Me, Ph, sugar, etc.$$
$$R^3 = H, CN, Ac, Me, etc.$$
$$R^4 = H, Me, etc.$$

(**533**)

(**534**) → (**535**) → (**536**) → (**537**)

(**538**) (**539**) (**540**) (**541**) (**542**) (**543**)

(**544**) (**545**) (**546**) (**547**) (**548**)

(*f*) A related synthesis is that in which N-3 is supplied for the cyclization of a β-acylaminovinyl alkyl (or aryl) ketone. For example, β-acetamidovinyl phenyl ketone (**550**) and ammonia at 200 °C give 2-methyl-4-phenylpyrimidine (**549**); however, yields are better if the ammonia is replaced by formamide, formamidine acetate, acetamide, *etc.*, which indicates that the mechanism is far from simple. Several analogues may be prepared in which the reagent does more than supply N-3: thus, the same ketone (**550**) with guanidine carbonate gives 4-phenylpyrimidin-2-amine (**551**) ⟨63CB1505⟩. Bischler's synthesis is seen in the cyclization of *o*-acetylphenylurethane by alcoholic ammonia to afford 4-methylquinazolin-2(1*H*)-one in good yield ⟨66JCS(C)234⟩. The so-called Grimmel synthesis is specific for 2,3-disubstituted quinazolin-4-ones. It is exemplified in the condensation of *o*-acetamidobenzoic acid with aniline and phosphorus trichloride to give 2-methyl-3-phenylquinazolin-4(3*H*)-one ⟨46JA542⟩.

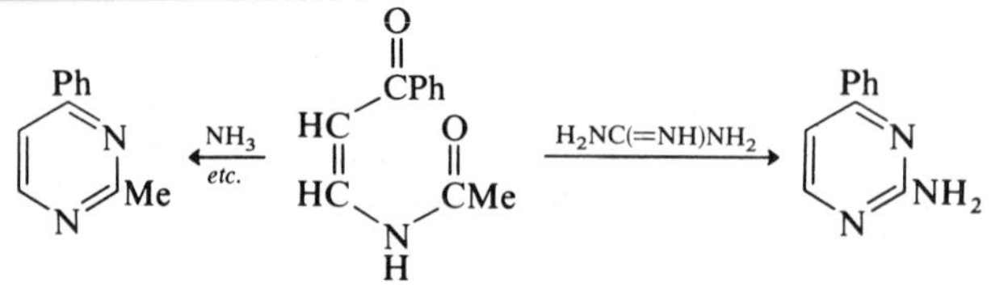

(549)    (550)    (551)

Syntheses involving [2+4] atom fragments may be subdivided into two types: in the first, the two-atom fragment supplies C-2+N-3, and in the second, it supplies C-5+C-6 of the final ring. The first type includes three distinct syntheses.

(*g*) Isocyanates and aminomethylene derivatives (such as 3-aminoacrylic acid) lead to 3-alkyluracils and related compounds. The reaction is best illustrated by its first use ⟨01LA(314)200⟩: methyl isocyanate reacts with ethyl 3-aminocrotonate (552) to give the intermediate (553) which cyclizes to 3,6-dimethyluracil (554); phenyl isocyanate reacts similarly to give 6-methyl-3-phenyluracil (555). Many other variations are possible, *e.g.* diethyl aminomethylenemalonate (556) and phenyl isocyanate give the carboxylic acid (557) ⟨34JA2754⟩. Although it has undergone some drastic modifications to produce, for example, pyrimidinethiones such as (558), the reaction has been little used of recent years ⟨70HC(16-S1)53⟩. However, it has proven useful in making dihydropyrimidines. Thus, ethyl 3-methyl-aminopropionate (559) and cyanic acid give, without isolation of the ureido intermediate, 1-methyl-5,6-dihydrouracil (560) which is converted by 5-bromination and subsequent 5,6-dehydrobromination into 1-methyluracil (561), representing its first unambiguous synthesis ⟨55JCS211⟩. The dihydropyrimidine (562) may be made by cyclization of an appropriate ureido compound ⟨66ZOR364⟩. Such reactions are widely reported for quinazolines. For example, 4-nitroanthranilic acid (563) and cyanic acid (or urea) give the ureido acid (564) which cyclizes in the acidic medium used to give 7-nitroquinazoline-2,4(1*H*,3*H*)-dione (565) ⟨48JCS1759⟩ and *o*-aminobenzaldehyde with methyl isocyanate yields 3-methylquinazolin-2(3*H*)-one (566) ⟨62JCS3129⟩.

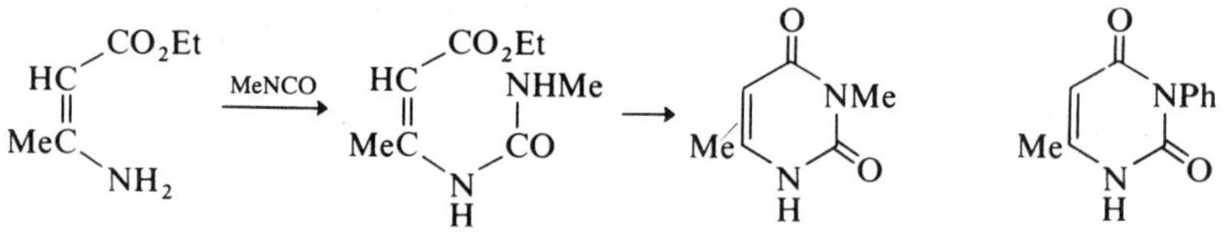

(552)    (553)    (554)    (555)

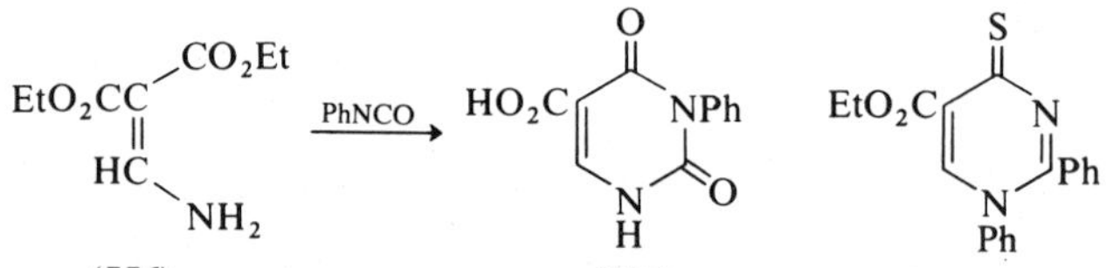

(556)    (557)    (558)

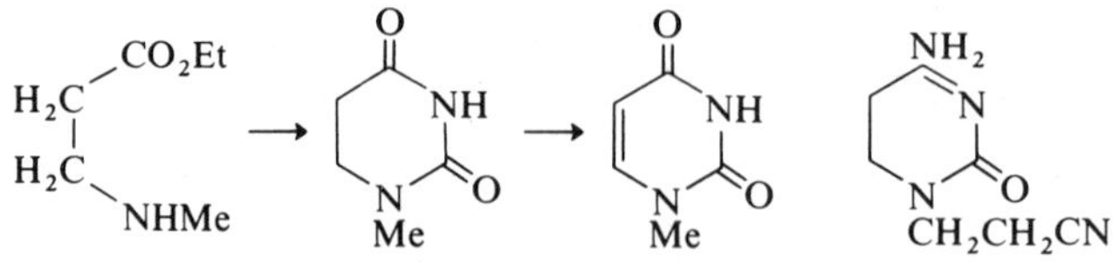

(559)    (560)    (561)    (562)

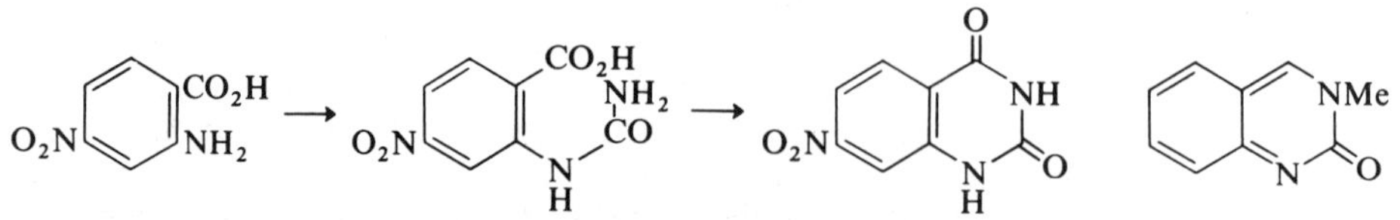

(563)    (564)    (565)    (566)

(*h*) The reaction of imino ethers or imidoyl chlorides with aminomethylene derivatives is less widely used but can afford 4-amino- or 4-hydroxy-pyrimidines fairly easily. For example, ethyl 3-aminocrotonate (**567**) reacts with ethyl benzimidate (**568**) to give 6-methyl-2-phenylpyrimidin-4(3*H*)-one (**569**) ⟨66LA(700)87⟩; if the ester is replaced by a similar nitrile, a pyrimidin-4-amine results. The reaction has been more used in the quinazoline series, as illustrated in the reaction of ammonium anthranilate with *N*-phenylbenzimidoyl chloride to yield 2,3-diphenylquinazolin-4(3*H*)-one (**570**); the mechanism is not as simple as might be supposed ⟨56JCS985⟩. It can also furnish a dihydropyrimidine; 3-amino-3-phenylpropionic acid (**571**) with ethyl benzimidate gives 2,6-diphenyl-5,6-dihydropyrimidin-4(3*H*)-one (**572**; R = Ph) ⟨66LA(696)97⟩ or with *O*-methylurea it gives the 2-amino analogue (**572**; R = NH₂) ⟨48ZOB2023⟩.

(*i*) The condensation of amides or thioamides with aminomethylene derivatives is rare in pyrimidines but better represented in the quinazolines. Ethyl 3-amino-2-cyanoacrylate (**573**) and thioacetamide yield (under alkaline conditions) ethyl 4-amino-2-methyl-pyrimidine-5-carboxylate (**574**) ⟨43JCS388⟩. The classical Niementowski synthesis of quinazolines is of this type. Thus, anthranilic acid (**575**) and formamide at 120 °C give an excellent yield of quinazolin-4(3*H*)-one (**576**) ⟨61MI21300⟩. The reaction goes well with *C*-substituted anthranilic acids but only in very poor yield with homologues of formamide; in such cases there is considerable advantage in using thioamides. For example, anthranilic acid with acetamide gives 2-methylquinazolin-4(3*H*)-one in 35% yield, but with thioacetamide the yield is *ca.* 90% ⟨62JIC368⟩. The mechanism of the Niementowski synthesis is quite complicated ⟨43JOC239⟩.

(*j*) The type of synthesis in which the two-atom fragment supplies C-5 + C-6 is uncommon but useful in preparing pyrimidine- and 5,6,7,8-tetrahydroquinazoline-2,4-diamines. Thus, dicyandiamide (**578**) with benzyl methyl ketone (**577**) yields 6-methyl-5-phenylpyrimidine-2,4-diamine (**579**), or with acetophenone it yields 6-phenylpyrimidine-2,4-diamine ⟨62JOC2708⟩. Likewise, with cyclohexanone it yields the tetrahydroquinazolinediamine (**580**) and by using *N*-substituted dicyandiamides, 2- and/or 4-alkylamino groups may be introduced ⟨65JOC1837⟩.

Syntheses involving [3 + 3] atom fragments comprise only the several versions of the Principal Synthesis. It is typified by Pinner's classical condensation of acetylacetone (**581**) with benzamidine (**582**) to give 4,6-dimethyl-2-phenylpyrimidine (**583**). The reaction is usually done under alkaline conditions, *e.g.* in ethanolic sodium ethoxide, but other solvents and neutral or acidic conditions are advantageous sometimes. The great versatility of the synthesis lies in the fact that each keto group in the three-carbon fragment can be replaced by an aldehydo, ester, nitrile or imino grouping; the amidine may be replaced by a urea, thiourea, guanidine or any of their *O*-, *S*-, or *N*-alkyl derivatives; and the three-carbon fragment can be substituted at its methylene point. Providing at least one of the fragments is symmetrical, the structure of the resulting pyrimidine will be unambiguous; if both are unsymmetrical, two pyrimidines are possible. In practice, not every combination produces the required product(s) but by a careful choice of fragments, a vast range of mono- to hexa-substituted pyrimidines can be made. Very detailed reviews of the Principal Synthesis are available: to 1960 ⟨62HC(16)31⟩ and 1960–1968 ⟨70HC(16-S1)20⟩.

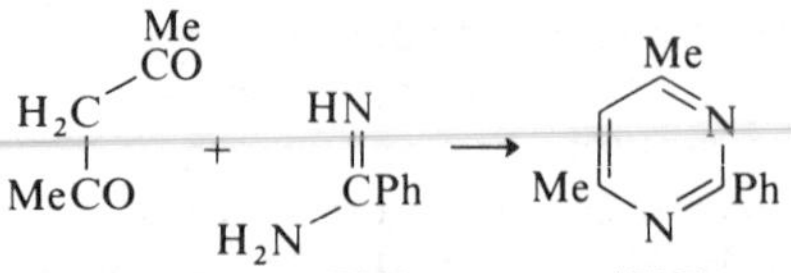

(581)    (582)    (583) 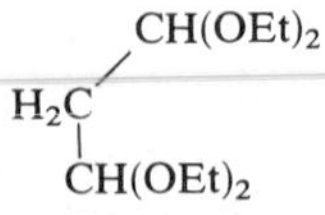    (584)

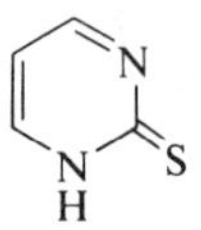

(585) 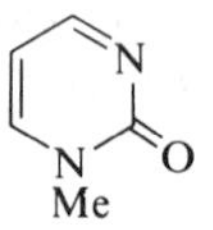    (586) 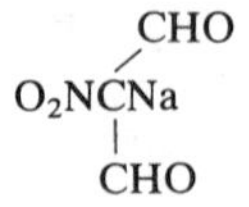    (587) 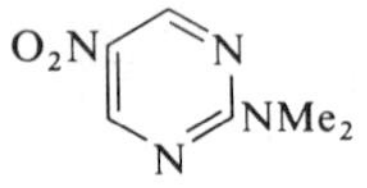    (588)

(*k*) The use of β-dialdehydes or their equivalents is seen in the condensation of 1,1,3,3-tetraethoxypropane (the tetraethyl acetal of malondialdehyde; **584**) with thiourea in alcoholic hydrochloric acid to give pyrimidine-2(1*H*)-thione (**585**) ⟨59JCS525⟩, or with *N*-methylurea under similar conditions to give 1-methylpyrimidin-2(1*H*)-one (**586**) ⟨60JA486⟩ and in the condensation of sodionitromalondialdehyde (**587**) with *N,N*-dimethylguanidine in aqueous piperidine to give *N,N*-dimethyl-5-nitropyrimidin-2-amine (**588**) ⟨67JOC3856⟩.

(*l*) The use of β-aldehydo ketones is illustrated by the reaction of 4,4-diethoxy-3-methylbutan-2-one (**589**; R = Me) with *S*-methylthiourea in ethanolic alkali to give 4,5-dimethyl-2-methylthiopyrimidine (**590**) ⟨64YZ207⟩. The related condensation of 4,4-diethoxybutan-2-one (**589**; R = H) with *N*-methylurea could give two isomeric pyrimidines because both fragments are unsymmetrical. In the event, the reaction (in ethanolic hydrochloric acid) gives a single product ⟨67JCS(C)1928⟩, shown to be 1,6-dimethylpyrimidin-2(1*H*)-one (**591**) ⟨68AJC243⟩. The fragment, ethyl ethoxymethyleneacetoacetate (**592**), could react as an aldehydo ketone, aldehydo ester or keto ester because, for practical purposes, the ethoxymethylene part is equivalent to an aldehyde. In fact, it prefers to react as the first of these: thus, with benzamidine in alcoholic alkoxide, it gives ethyl 4-methyl-2-phenyl-pyrimidine-5-carboxylate (**593**; R = Et), confirmed in structure by hydrolysis to the known carboxylic acid (**593**; R = H) ⟨62YZ462⟩.

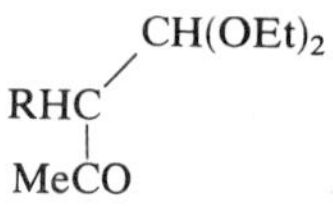

(589) 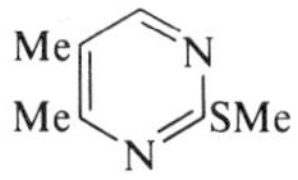    (590) 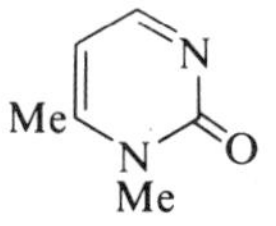    (591) 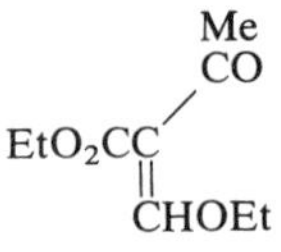    (592)

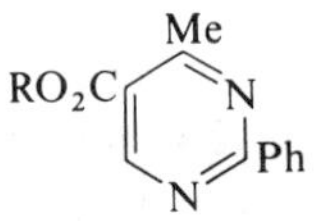

(593) 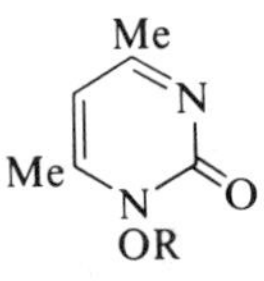    (594) 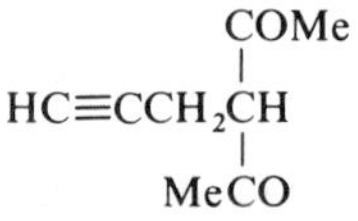    (595) 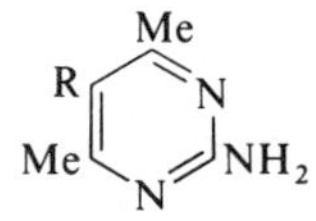    (596)

(*m*) The use of β-diketones has already been mentioned in respect of acetylacetone and benzamidine giving 4,6-dimethyl-2-phenylpyrimidine (**583**); a closer look at this reaction reveals that the yields are best in media as alkaline as consistent with the stability of benzamidine, *i.e. ca.* pH 10 ⟨51JCS3155⟩. Acetylacetone (**581**) and *N*-methoxyurea in ethanolic hydrogen chloride give 1-methoxy-4,6-dimethylpyrimidin-2(1*H*)-one (**594**; R = Me) which undergoes hydrolysis in concentrated hydrobromic acid to afford the *N*-hydroxy-pyrimidinone (**594**; R = H), an *N*-oxide tautomer ⟨67T353⟩. A warning is contained in the condensation of α-prop-2-ynylacetylacetone (**595**) with guanidine carbonate which gives a separable mixture of the expected 5-(prop-2′-ynyl)-4,6-dimethylpyrimidin-2-amine (**596**; R = CH₂C≡CH) with its prop-1-ynyl (**596**; R = C≡CMe) and allenyl (**596**; R = CH=C=CH₂) isomers, on account of prototropic changes under the strongly basic conditions of the condensation ⟨67JCS(C)1922⟩.

(*n*) The use of β-aldehydo esters is illustrated in the condensation of ethyl formylacetate (**597**) with trifluoroacetamidine in aqueous base to give 2-trifluoromethylpyrimidin-4(3*H*)-one (**598**) ⟨61JOC4504⟩; in the ambiguous condensation of ethyl 3,3-diethoxypropionate with *N*-methylthiourea which in fact gives only 3-methyl-2-thiouracil (**599**) as revealed by conversion of the product into 3-methyluracil ⟨62CPB313⟩; in the condensation of ethyl 2-ethoxymethylene-2-nitroacetate (**600**; prepared *in situ*) with *N,N′*-dimethylurea to give (in ethanolic ethoxide) 1,3-dimethyl-5-nitropyrimidine-2,4(1*H*,3*H*)-dione (**601**) ⟨63CCC2501⟩; and in the formation of isocytosine (2-aminopyrimidin-4(3*H*)-one; **602**) from formylacetic acid and guanidine sulfate in (hot) fuming sulfuric acid ⟨63MI21300⟩.

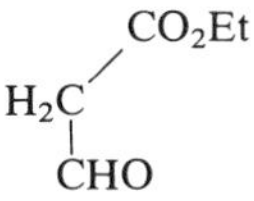

(597) 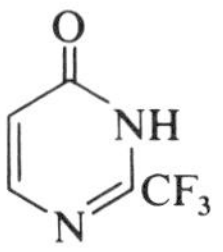    (598) 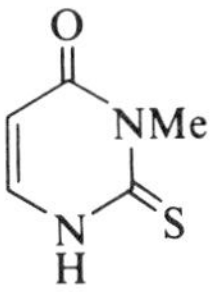    (599) 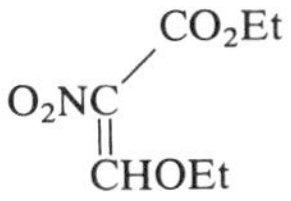    (600) 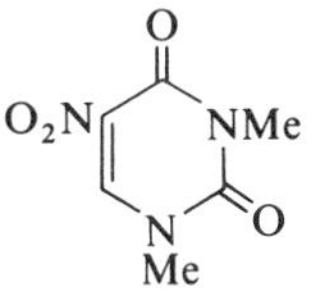    (601) 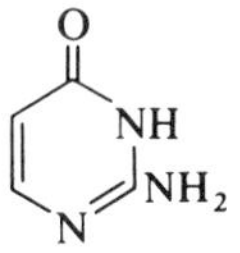    (602)

(*o*) The use of β-keto esters is seen in the behaviour of ethyl acetoacetate (**603**) with pivalamidine in methanolic sodium methoxide to give 2-*t*-butyl-6-methylpyrimidin-4(3*H*)-one (**604**) ⟨63JCS5642⟩; of ethyl 4,4-dimethoxyacetoacetate (**605**) with *N*-methylthiourea to give only 6-dimethoxymethyl-3-methyl-2-thiouracil (**606**), ripe for conversion by acid into the corresponding thiouracilcarbaldehyde ⟨64JMC337⟩; and of ethyl acetoacetate with amidinoacetamide to give the expected pyrimidinone (**607**) and the separable isomeric pyridinone (**608**) ⟨77AJC621⟩. There are several types of compound which behave as β-keto esters in the Principal Synthesis; for example, diketene (**609**) acts as ethyl acetoacetate (**603**) and the thiol-lactone (**610**) as ethyl 2-mercaptoethylacetoacetate (**611**) ⟨66CB872⟩ by giving the thiol (**612**) with guanidine.

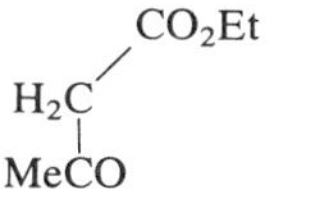

(603) 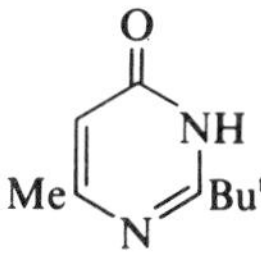    (604) 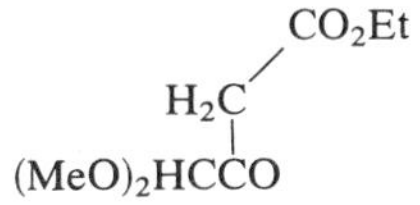    (605)

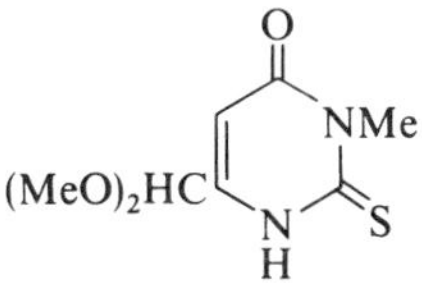

(606) 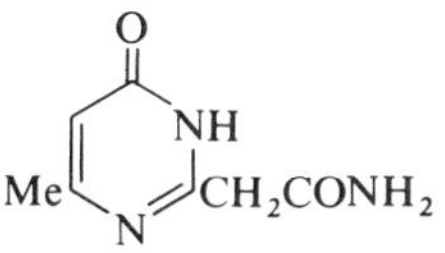    (607) 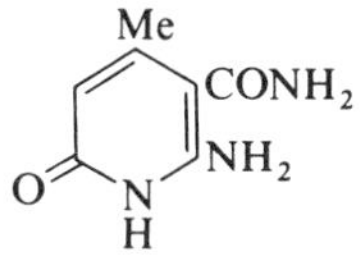    (608)

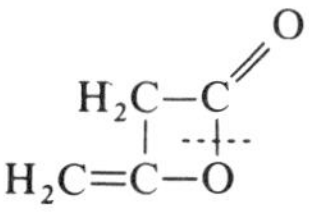

(609) 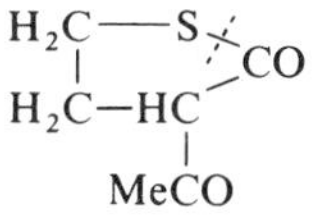    (610) 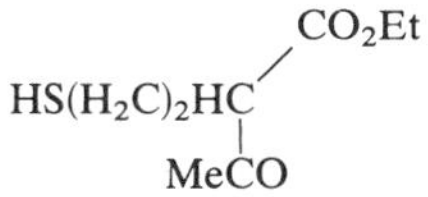    (611) 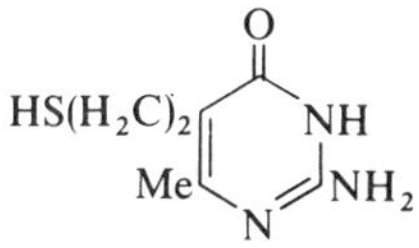    (612)

(*p*) The use of β-diesters (malonic esters) is essential in making barbiturates, of which many hundreds are known. That apart, diesters behave very predictably in the wider context of the Principal Synthesis. Thus, diethyl malonate (**613**) and 2-hydroxyacetamidine in methanolic sodium methoxide give the pyrimidinylmethanol (**614**) in good yield ⟨63ZOB2848⟩. The useful intermediate (**615**) is equivalent to the diester (**616**); accordingly, it reacts, for example, with free guanidine in DMF to give an acyclic intermediate which undergoes thermal cyclization at 150 °C to afford 2-amino-5-chloro-6-ethoxypyrimidin-4(3*H*)-one (**617**) ⟨63JOC509⟩. The condensation of ureas with α-substituted malonic esters to give barbiturates often goes rather slowly in refluxing ethanolic ethoxide so that temperatures of 105–110 °C in a sealed system are used frequently. In efforts to avoid this, condensations are sometimes done in acetic anhydride, chloroform/acetyl chloride, and even phosphoryl chloride. In the last case, potential hydroxy groups in the product may be converted into chloro substituents; for example, α-butylmalonic acid and *N,N′*-diphenylurea with phosphoryl chloride affords 5-butyl-6-chloro-1,3-diphenylpyrimidine-2,4(1*H*,3*H*)-dione (**618**), a most useful intermediate, in excellent yield ⟨60LA(638)205⟩. Malonic acid reacts immediately with *N,N′*-dicyclohexylcarbodiimide to give 1,3-dicyclohexylbarbituric acid (**619**) in 65% yield, which should be compared with the usual yield of *ca.* 5% from diphenylurea and malonyl chloride ⟨63T85⟩. The synthesis permits access to bipyrimidines, as indicated in the condensation of 2-dimethylaminopyrimidine-4-carboxamidine (**620**) with dimethyl malonate in methanolic methoxide to yield 2'-dimethylamino-6-hydroxy-2,4'-bipyrimidin-4(3*H*)-one (**621**) ⟨81AJC1353⟩.

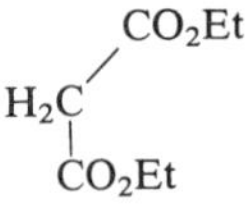

(613) 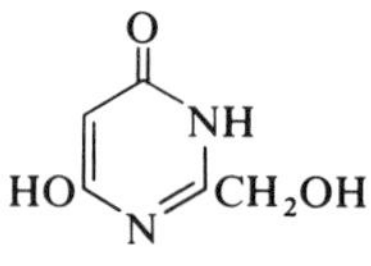   (614) 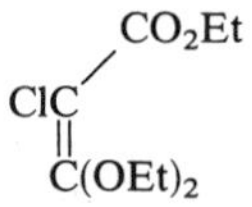   (615) 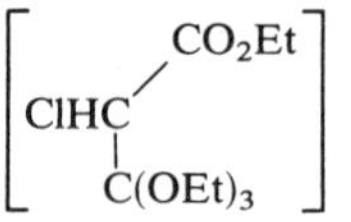   (616)

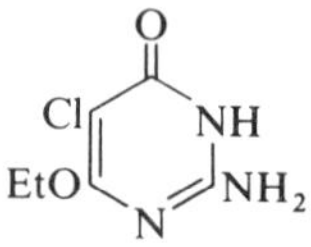

(617) 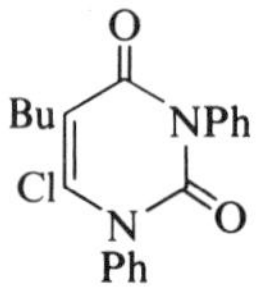   (618)

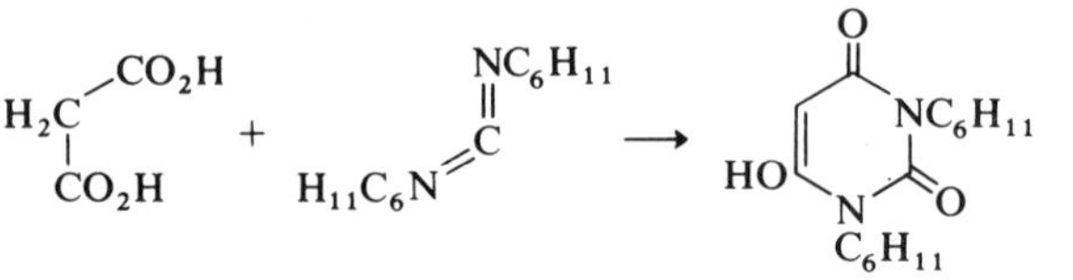

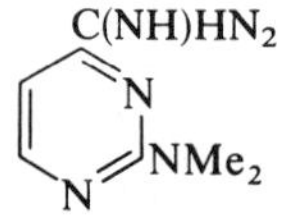

(620) 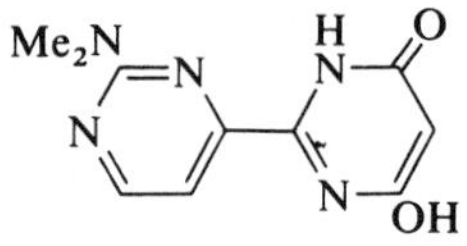   (621)

(*q*) The use of β-aldehydo nitriles gives a pyrimidin-4/6-amine. Thus 3-ethoxy-2-methoxymethylenepropionitrile (**622**) and formamidine in refluxing ethanol give 5-ethoxymethylpyrimidin-4-amine (**623**) ⟨64CPB393⟩; methoxymethylenemalononitrile (**624**) reacts with thiourea to give the thione (**625**; R = H) ⟨58MI121300⟩ or with *N*-methylthiourea to give the 3-methylated thione (**625**; R = Me) ⟨56JA5294⟩; and 2,3-diethoxyacrylonitrile (**626**) with guanidine gives 5-ethoxypyrimidine-2,4-diamine (**627**) ⟨65CB2576⟩. Although urea does not condense quite normally with aldehydo nitriles, much of the work in this area falls within the Whitehead synthesis, which leads to cytosine with a cyano, carboxy, alkoxycarbonyl, carbamoyl or nitro group in the 5-position and with or without a 3-alkyl or aryl substituent. The essential point in the synthesis is the formation of a ureidomethylene compound (**628**) from a mixture of nitrile, orthoformate and *N*-alkylurea followed by ring-closure in ethanolic ethoxide to give the cytosine (**629**). More than fifty such products are described. Typical examples are 3-methylcytosine-5-carboxylic acid (**630**; R = $CO_2H$) ⟨55JA5867⟩ which undergoes Dimroth rearrangement during decarboxylation to yield 4-methylaminopyrimidin-2(1*H*)-one (**631**) ⟨56JA5294⟩, 3-methyl-5-nitrocytosine (**630**; R = $NO_2$) ⟨59MI21300⟩, and simple 5-cyanocytosine (**632**) ⟨66CCC3990⟩.

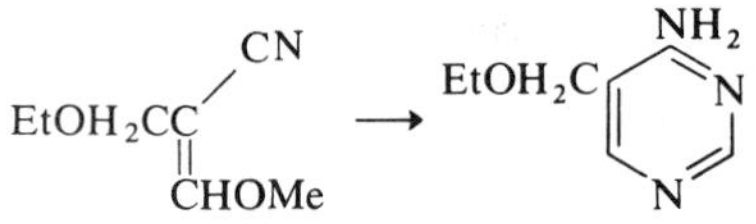

(622)   (623) 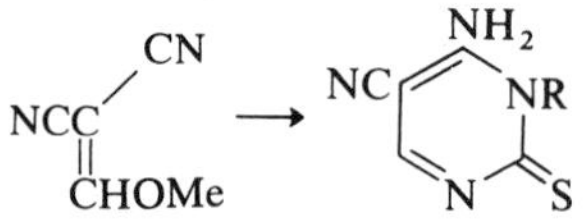   (624)   (625)

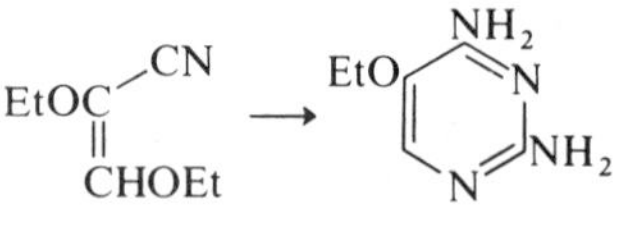

(626)   (627) 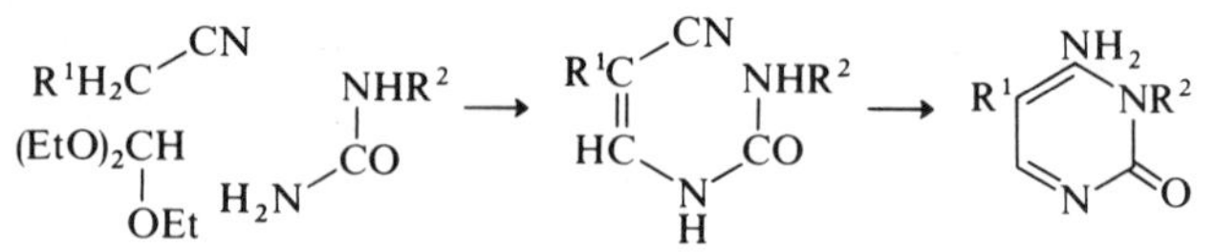   (628)   (629)

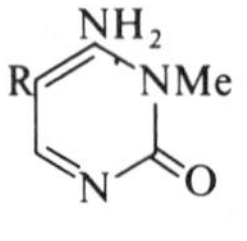

(630) 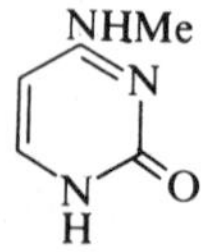   (631) 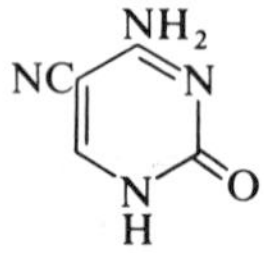   (632)

(*r*) The use of β-keto nitriles is relatively minor in the literature but straightforward examples are the condensation of 2-cyano-3-ethoxycrotononitrile (**633**) with formamidine to give 4-amino-6-methylpyrimidine-5-carbonitrile (**634**; R = H), with *S*-methylthiourea to give the thioether (**634**; R = SMe) and with thiourea to give the thione (**635**) ⟨60JA5711⟩. In fact, most examples lead to 5-arylpyrimidine-2,4-diamines of potential medicinal interest: for example, 2-*p*-chlorophenyl-3-ethoxy-4,4,4-trifluorocrotononitrile and guanidine give the diamine (**636**) ⟨65JHC162⟩.

(**633**)　(**634**)　(**635**)　(**636**)

(**637**)　(**638**)　(**639**)　(**640**)

(*s*) The use of β-ester nitriles leads to 6-aminopyrimidin-4-ones. Ethyl cyanoacetate (**637**; R = H) reacts normally with benzamidine in ethanolic ethoxide to give 6-amino-2-phenylpyrimidin-4(3*H*)-one (**638**; R = Ph), but with aliphatic amidines results are much less satisfactory ⟨63JCS3729⟩; with thiourea it gives 6-amino-2-thiouracil (**639**), but with *S*-alkylthioureas yields are very poor ⟨60JOC148⟩; and with guanidine, it gives 2,6-diaminopyrimidin-4(3*H*)-one (**638**; R = NH$_2$) ⟨52OS(32)45⟩. Similarly, ethyl isonitroso-cyanoacetate and guanidine give the nitrosopyrimidine (**640**) ⟨53MI21300⟩.

(*t*) The use of β-dinitriles is quite normal in combination with thioureas and guanidines. Thus, malononitrile (**641**; R = H) with thiourea in ethanolic ethoxide gives 4,6-diaminopyrimidine-2(1*H*)-thione (**642**; R = H) ⟨62ZOB1655⟩ or with thiosemicarbazide it gives the 1,4,6-triamino analogue (**642**; R = NH$_2$) ⟨67JOC2379⟩; likewise, malononitrile and its α-alkyl derivatives (**641**; R = Et, Pr$^i$, CH$_2$Ph but not Ph) give the corresponding 2,4,6-triaminopyrimidines (**643**) ⟨52JA3443⟩.

(**641**)　(**642**)　(**643**)　(**644**)　(**645**)　(**646**)

Malononitriles do not react successfully with ureas despite contrary claims in some patents. With amidines, unsubstituted malononitrile gives abnormal products but substituted malononitriles react normally. Thus, malononitrile (**641**; R = H) reacts with one equivalent of formamidine to give α-aminomethylenemalononitrile (**644**) which then reacts with a second molecule of formamidine to give 4-aminopyrimidine-5-carbonitrile (**645**; R = H); benzamidine gives the diphenyl derivative (**645**; R = Ph) ⟨43JCS388⟩. Likewise, trifluoroacetamidine reacts with malononitrile to give the product (**645**; R = CF$_3$) but with 'α-phenylazomalononitrile', the hydrazone tautomer of (**641**; R = N=NPh), it gives a normal product (**646**) ⟨63JMC39⟩.

(*u*) A modified Principal Synthesis may also be employed to make some di- and tetra-hydropyrimidines by using a three-carbon fragment at low oxidation state. For example, the reduced equivalents of a β-dialdehyde are a β-aldehydo alcohol (*e.g.* acrolein, **647**) or a β-dialcohol (*e.g.* 1,3-dibromopropane, **648**). While the former (**647**) is recorded several times as failing to react with urea to give a dihydropyrimidin-2-one, the latter (**648**) does react with benzamidine to give 2-phenyl-1,4,5,6-tetrahydropyrimidine (**649**) ⟨12JCS2342⟩. Likewise, the β-keto alcohol equivalent mesityl oxide (**650**) reacts with benzamidine to give 4,4,6-trimethyl-2-phenyl-3,4-dihydropyrimidine (**651**) ⟨62JOC4090⟩, and the β-ester alcohol equivalent acryloyl chloride (**652**) gives 1-phenyl-5,6-dihydrouracil (**653**) ⟨59JOC1391⟩. The β-alcohol nitrile equivalent, 2,3-diphenylacrylonitrile (**654**) does react with guanidine, but the product is 2-amino-5,6-diphenyl-5,6-dihydropyrimidin-4(3*H*)-one (**656**) rather than the expected diamine (**655**), probably on account of hydrolytic deamination at an intermediate stage ⟨56JCS1019⟩. Finally, the condensation of 2'-benzoylstyrene with benzamidine gives 2,4,6-triphenylpyrimidine rather than its 4,5-dihydro analogue because of oxidation by the excess of unsaturated intermediate ⟨52JOC461⟩.

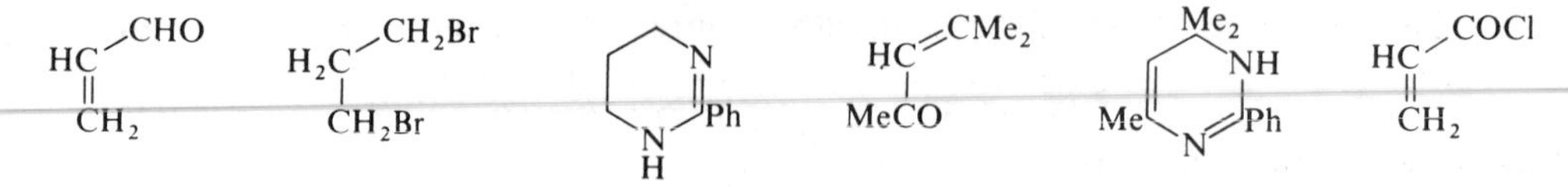

(647)  (648)  (649)  (650)  (651)  (652)

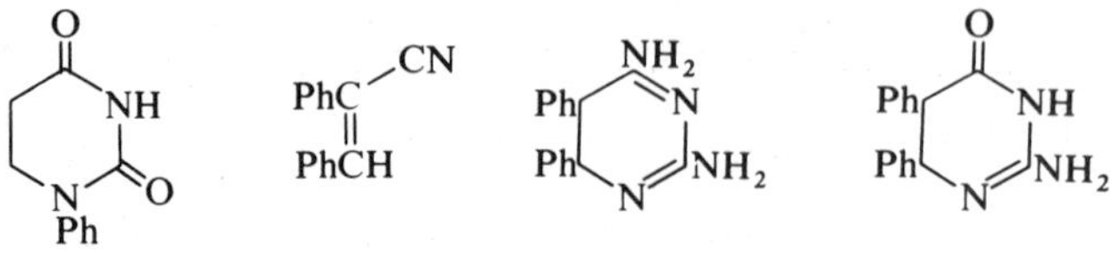

(653)  (654)  (655)  (656)

### 2.13.3.1.3 Primary syntheses involving formation of three bonds

This category embraces several otherwise unrelated syntheses, all of which involve [2+2+2] or [3+2+1] atom fragments.

(a) The Frankland–Kolbé synthesis was pioneered in 1848 and its subsequent history is fascinating ⟨62HC(16)82⟩. The synthesis involves the trimerization of three simple nitriles (657) by heating with (molten) potassium or with potassium alkoxide to give a di- or tri-alkylpyrimidin-4-amine (658) in which the 5-alkyl group has one methylene less than those in the 2- and 6-positions. This places a severe restriction on its scope but it is still used as a convenient route to 2,6-dimethylpyrimidin-4-amine (658; R = H) from acetonitrile (657; R = H) and potassium ethoxide ⟨55OSC(3)71⟩ or to 5-ethyl-2,6-dipropylpyrimidin-4-amine (658; R = Et) ⟨65JCS3357⟩. The mechanism is not understood and it may be that a 1,3,5-triazine intermediate is involved. Thus, propionitrile trimerizes when heated under pressure without any catalyst to give 2,4,6-triethyl-1,3,5-triazine (659) and this can be converted subsequently into 2,6-diethyl-5-methylpyrimidin-4-amine (658; R = Me) by alkali ⟨52JA5633⟩. Mixtures of nitriles may be used. For example, p-chlorophenylacetonitrile (1 mol) and pyridine-4-carbonitrile (2 mol) in butanolic butoxide at 115 °C give 5-p-chlorophenyl-2,6-di(pyridin-4′-yl)pyrimidin-4-amine in excellent yield ⟨67LA(704)144⟩.

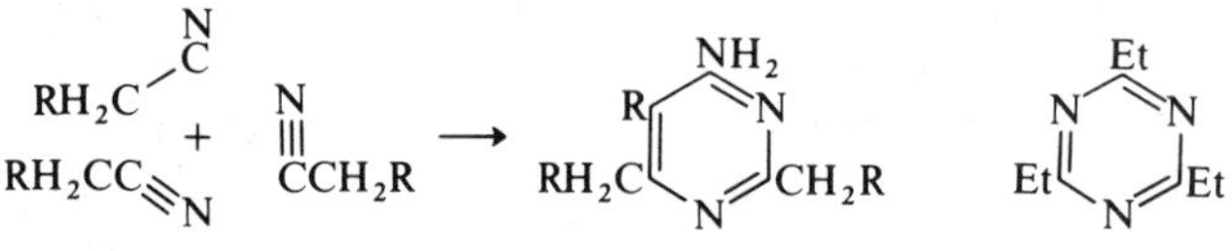

(657)  (658)  (659)

(b) The formamide/active methylene synthesis is illustrated by the condensation of acetophenone with two molecules of formamide (zinc chloride; high temperature) to give 4-phenylpyrimidine (661) via the intermediate (660). Similarly, phenylacetonitrile and formamide (under ammonia at 180 °C) give 5-phenylpyrimidin-4-amine (663), probably via the intermediate (662) ⟨45JCS347⟩. The synthesis may be used more effectively by 'combining' the molecules of formamide as trisformamidomethane, which may then furnish N-formylformamidine; thus, acetone and this reagent with toluenesulfonic acid as catalyst yield 4-methylpyrimidine (664); propanal yields 5-methylpyrimidine; and cyclohexanone yields 5,6,7,8-tetrahydroquinazoline (665) ⟨60CB1402⟩. Many other examples have been summarized ⟨70HC(16-S1)53⟩.

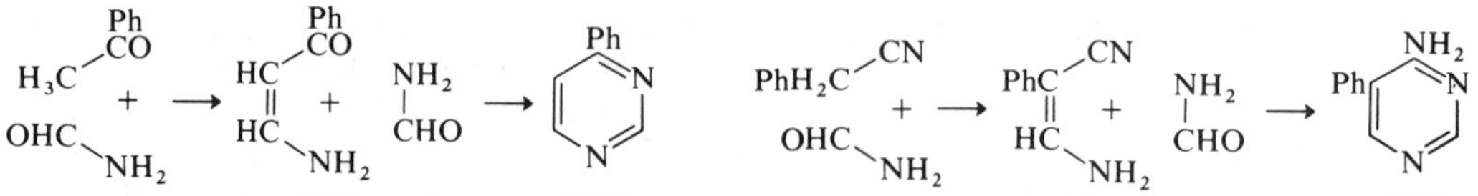

(660)  (661)  (662)  (663)

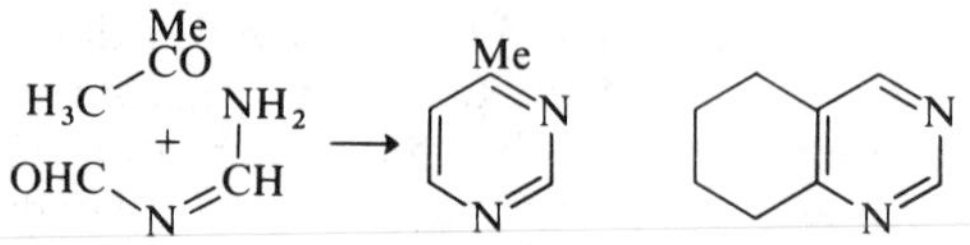

(664)  (665)

(c) The closely related formamide/β-dicarbonyl synthesis is now usually known as the Bredereck synthesis. In it, a β-dicarbonyl or equivalent compound is heated at about 200 °C in an excess of formamide. As illustration, benzoylacetone (**666**) and formamide at >220 °C give 4-methyl-6-phenylpyrimidine (**668**). The course of the reaction is indicated by heating at the lower temperature of 150 °C, when an intermediate (**667**) may be isolated; it gives the pyrimidine (**668**) on boiling with formamide ⟨57CB942⟩. The scope of the reaction seems to be parallel to that of formamidine in the Principal Synthesis: if formamidine will theoretically give a certain pyrimidine with a given dicarbonyl compound, then formamide may well do the same. A summary of substrates and resulting pyrimidines (without functional groups) is available ⟨62HC(16)82⟩. Some pyrimidines with functional groups may also be made; for example, 2-bromo-1,1,3,3-tetraethoxypropane and ethyl 1,1,3,3-tetraethoxy-propane-2-carboxylate react with formamide to give, respectively, 5-bromopyrimidine and ethyl pyrimidine-5-carboxylate ⟨62CB803⟩.

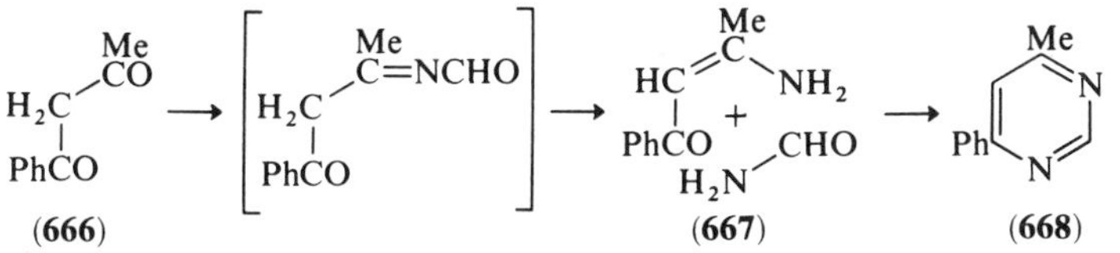

(**666**)      (**667**)      (**668**)

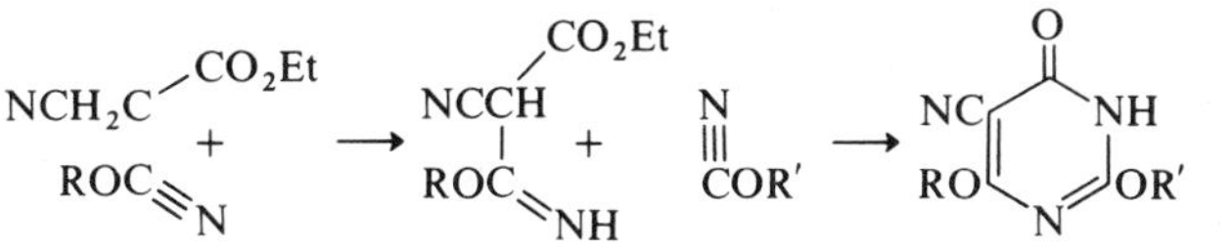

(**669**)      (**670**)

(d) The relatively little-used aryl cyanate/active methylene synthesis involves condensation of a compound containing an active methylene grouping successively with two molecules of aryl cyanate, which may be the same or different. For example, ethyl cyanoacetate and phenyl cyanate give the intermediate (**669**; R = Ph) which is then allowed to react with p-tolyl cyanate to yield 5-cyano-6-phenoxy-2-p-tolyloxypyrimidin-4(3H)-one (**670**; R = Ph, R' = $C_6H_4Me$). If malononitrile is substituted for ethyl cyanoacetate, a comparable pyrimidin-4-amine is formed ⟨65AG913⟩.

(e) The Biginelli reaction dates from 1893 and involves the condensation of a ketone having an activated but unsubstituted methylene grouping next to the carbonyl group with an aromatic aldehyde and urea or thiourea in alcoholic acid to give a dihydropyrimidin-2-one or similar thione. Two parallel mechanisms have been proposed ⟨33JA3784⟩: benzaldehyde reacts with urea (2 mol) to give the intermediate (**671**) which then has one of its urea residues displaced by ethyl acetoacetate to give the second (unisolated) intermediate (**672**) ripe for cyclization to the pyrimidinone (**673**). At the same time, ethyl acetoacetate reacts with urea to give the intermediate (**674**) which then reacts with benzaldehyde to complete the pyrimidine (**673**); thiourea gives the corresponding thione and replacement of ethyl acetoacetate by acetylacetone gives the analogous 5-acetylpyrimidinone. The use of aceto-phenone leads to the 5-unsubstituted pyrimidinones (**675**) ⟨65MI21301⟩. The reaction is now seldom used.

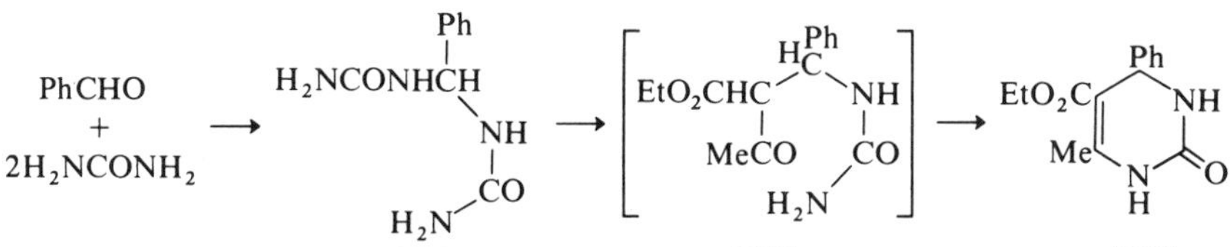

(**671**)      (**672**)      (**673**)

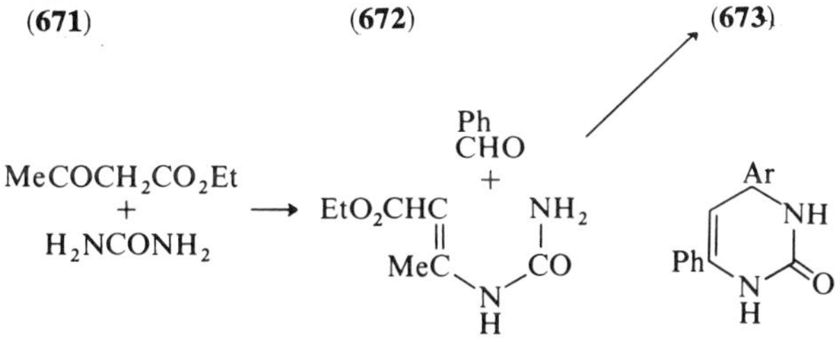

(**674**)      (**675**)

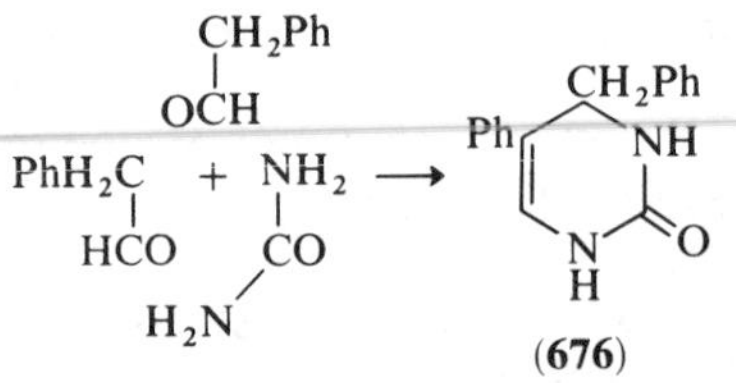

(676)

When the aromatic aldehyde is omitted from a Biginelli reaction mixture, a dihydropyrimidine is still formed. Thus, for example, phenylacetaldehyde (2 mol) and urea (1 mol) react to give 4-benzyl-5-phenyl-3,4-dihydropyrimidin-2(1*H*)-one (676) ⟨33JA3361⟩.

### 2.13.3.1.4 Primary syntheses involving formation of four or more bonds

(*a*) The synthesis of tetrahydropyrimidines from carbonyl compounds and ammonia or an amine is usually known as the acetonin synthesis. Thus, acetone reacts with liquid ammonia in the presence of calcium chloride to give a compound 'acetonin' which is now known to be 2,2,4,4,6-pentamethyl-2,3,4,5-tetrahydropyrimidine (677) ⟨47JCS1394⟩; the mechanism involves three molecules of acetone and two of ammonia, with mesityl oxide (678) and 'diacetonamin' (679) as the only definite intermediates. The reaction is capable of considerable variation, for example, in the conversion of diacetonin (679) into the thione (680); this and other aspects are summarized elsewhere ⟨70HC(16-S1)322⟩.

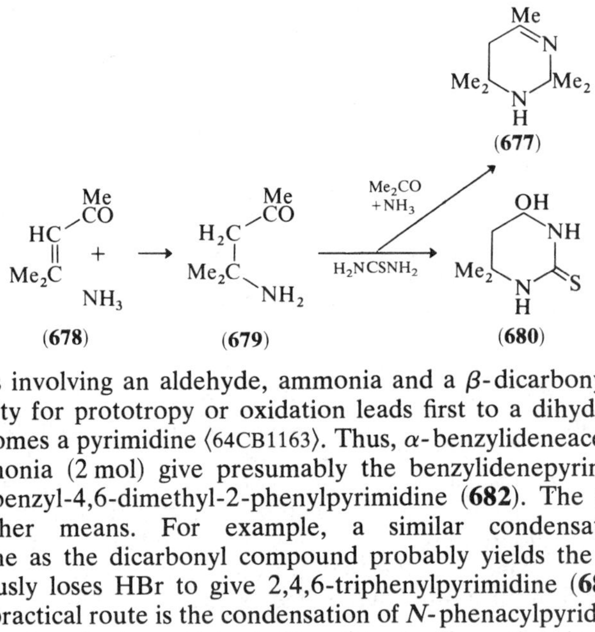

(678)          (679)          (680)

(*b*) A synthesis involving an aldehyde, ammonia and a β-dicarbonyl compound which has inbuilt capacity for prototropy or oxidation leads first to a dihydropyrimidine which subsequently becomes a pyrimidine ⟨64CB1163⟩. Thus, α-benzylideneacetylacetone, benzaldehyde and ammonia (2 mol) give presumably the benzylidenepyrimidine (681) which rearranges to 5-benzyl-4,6-dimethyl-2-phenylpyrimidine (682). The same effect can be achieved by other means. For example, a similar condensation using dibenzoylbromomethane as the dicarbonyl compound probably yields the intermediate (683) which spontaneously loses HBr to give 2,4,6-triphenylpyrimidine (684). An even more complicated but practical route is the condensation of *N*-phenacylpyridinium bromide with ammonia (2 mol) and *p*-nitrobenzaldehyde (2 mol) to give the notional tetrahydro intermediate (685) which loses the quaternary group as a pyridine salt and is further oxidized by excess of aldehyde to give 2,4-di(*p*-nitrophenyl)-6-phenylpyrimidine in >70% yield.

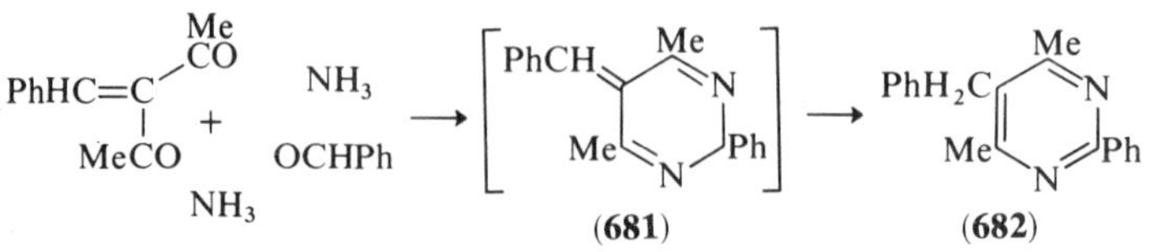

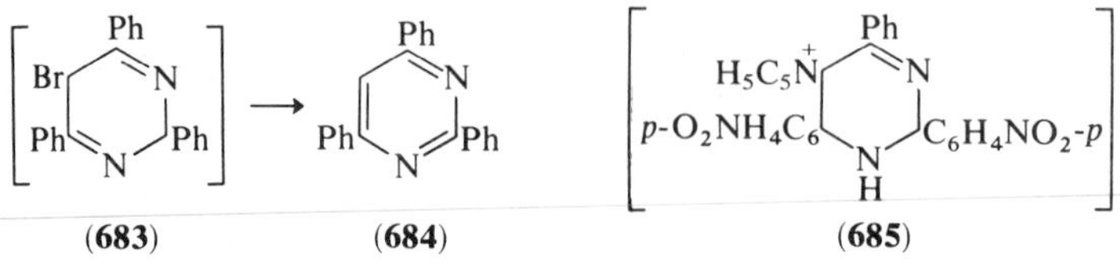

(681)          (682)

(683)          (684)          (685)

### 2.13.3.2 Syntheses of the Pyrimidine Ring from Other Heterocycles

It is possible to prepare pyrimidines from other heteromonocyclic compounds by a variety of processes, or from fused heterobicyclic systems which already contain a pyrimidine ring by preferentially degrading the unwanted second ring. In the latter case, the bicyclic system may best be made from a pyrimidine in the first place, occasionally even from the self-same pyrimidine to which it reverts on degradation. Such syntheses may be of interest but are certainly not of any utility.

#### 2.13.3.2.1 Pyrimidines from pyrroles

When the oxime of 2,4,5-triphenylpyrrol-3(2H)-one (**686**) is treated with phosphorus pentachloride in ether, two acyclic products (**687**; X = O or NOH) are formed. The first of these on heating gives rise to 2,5,6-triphenylpyrimidin-4(3H)-one (**688**) and the second on reduction with zinc/acetic acid gives the triphenylpyrimidin-4-amine (**689**); when the reaction is done in chloroform, the product (**688**) is obtained directly ⟨40G504⟩. An unrelated ring expansion occurs when 2,3,5-triphenylpyrrole is irradiated in alcoholic ammonia open to air to give 2,4,6-triphenylpyrimidine; it is evident that the second nitrogen atom is inserted between the two adjacent phenyl substituents of the pyrrole ⟨56G119⟩.

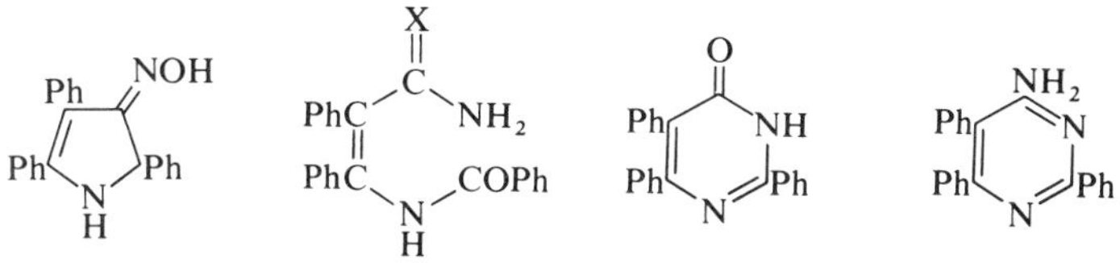

(**686**)     (**687**)     (**688**)     (**689**)

#### 2.13.3.2.2 Pyrimidines from imidazoles

Hydantoins, *i.e.* imidazole-2,4(3H,5H)-diones, sometimes arise in reactions designed to make pyrimidines. However, they can usually be converted into the desired pyrimidines, often under hydrolytic conditions. For example, diethyl oxalacetate and urea under the usual conditions do not give the expected uracilcarboxylic acid (**692**) but the hydantoin (**690**). However, this may be converted, on vigorous alkaline treatment, first into the acyclic intermediate (**691**) and thence into orotic acid (**692**) ⟨47JA674⟩. The reaction is not confined entirely to substituted orotic acids: for example, diethyl methyloxalacetate and guanidine give the imidazole (**693**) and thence the pyrimidine (**694**) ⟨65JHC162⟩. A different approach is the condensation of acetylenedicarboxylic acid and thiourea to give the thiohydantoin (**695**) which yields 2-thioorotic acid (**696**) on alkaline treatment ⟨63YZ169⟩.

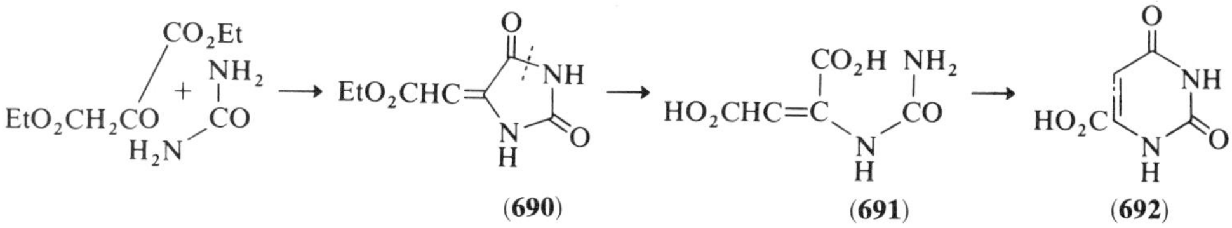

(**690**)     (**691**)     (**692**)

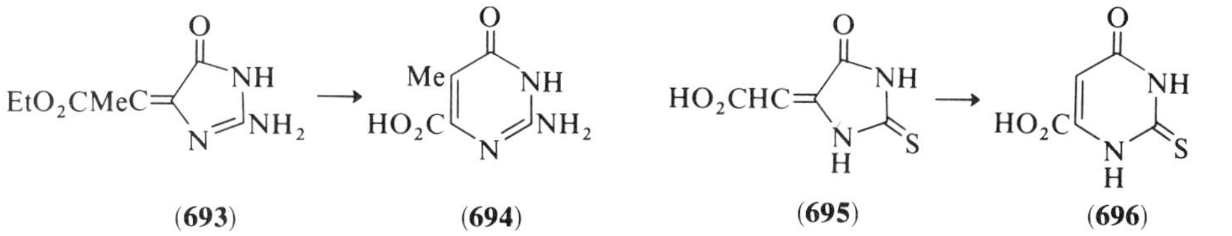

(**693**)     (**694**)     (**695**)     (**696**)

#### 2.13.3.2.3 Pyrimidines from isoxazoles and oxazoles

3,4-Dimethylisoxazol-5-amine is easily acylated to its formyl derivative (**697**) which, on catalytic hydrogenation, undergoes ring cleavage and recyclization to yield 5,6-dimethyl-pyrimidin-4(3H)-one (**698**); other acyl derivatives give analogous 2-substituted pyrimidines

⟨63AP298⟩. A similar result may be achieved by ring cleavage followed by acylation and final reclosure ⟨54JCS665⟩.

~~Treatment of the oxazole (**699**) with~~ ammonia converts it into 5-hydroxy-4,6-dimethyl-pyrimidine (**700**) and homologues may be made similarly ⟨60CB1998⟩.

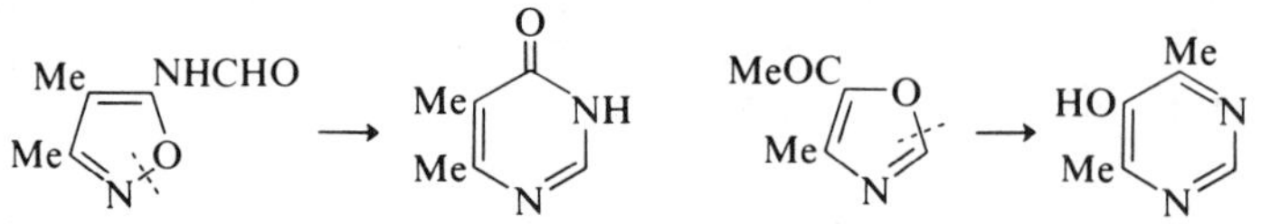

(697)          (698)          (699)          (700)

### 2.13.3.2.4 *Pyrimidines from pyridines, pyrazines and triazines*

Although not a practical method of preparing pyrimidines, certain halogenopyridines undergo amination with sodium amide in liquid ammonia to give poor to moderate yields of pyrimidines. For example, 2,6-dibromopyridine (**701**) gives 2-methylpyrimidin-4-amine (**702**) in 24% yield, along with pyridine-2,6-diamine; 2,6-dichloro- but not 2,6-difluoro-pyrimidine gives similar products and a plausible mechanism has been suggested ⟨65RTC1569⟩. Several other derivatives of 2-bromopyridine behave similarly: for instance, 2-bromo-6-phenoxypyridine (**703**) gives up to 55% of 2-methyl-4-phenoxypyrimidine (**704**) ⟨69RTC1391⟩.

(701)          (702)          (703)          (704)

The pyrolysis of pyrazine at 1000 °C/2 mmHg gives pyrimidine in 3% yield ⟨68TL3115⟩. The photolysis of pyrazine at 254 nm gives a little pyrimidine; in addition, 2-methylpyrazine gives both 4- and 5-methylpyrimidine, 2,6-dimethylpyrazine gives 4,5-dimethylpyrimidine, and 2,5-dimethylpyrazine gives both 4,6- and 2,5-dimethylpyrimidine, all in minute yield ⟨67TL3913⟩.

1,3,5-Triazine reacts with 2-ethoxycarbonylacetamidinium chloride (**705**) or with ethyl 2-ethoxycarbonylacetimidate (base; **706**) in acetonitrile to give ethyl 4-aminopyrimidine-5-carboxylate (**707**; R = NH₂) in good yield; in contrast, it reacts with the same acetimidate hydrochloride to give ethyl 4-ethoxypyrimidine-5-carboxylate (**707**; R = OEt) ⟨62JOC548⟩. This useful reaction can be modified to give a variety of 4,5-disubstituted pyrimidines, e.g. malononitrile gives 4-aminopyrimidine-5-carbonitrile, while malondiamide gives 5-carbamoylpyrimidin-4(3H)-one ⟨62JOC551⟩. In a rather different way, triazine reacts with propane-1,3-diamine to give 1,4,5,6-tetrahydropyrimidine in >60% yield ⟨62JCS527⟩.

Several triazinyl ketones isomerize to 4-acetamidopyrimidines. This is seen in the C-acylation of 2,4,6-trimethyl-1,3,5-triazine (**708**) with benzoyl chloride in the presence of sodium amide to give the ketone (**709**) which undergoes a Dimroth-like rearrangement in boiling water to afford N-(2-methyl-6-phenylpyrimidin-4-yl)acetamide (**710**): it can be seen that the acylating agent determines the identity of the 6-substituent ⟨64JHC145⟩.

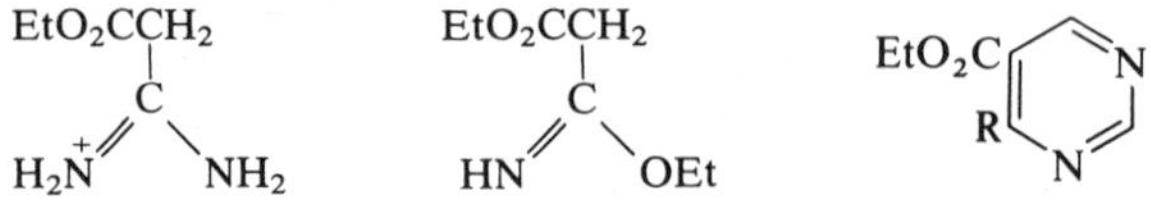

(705)          (706)          (707)

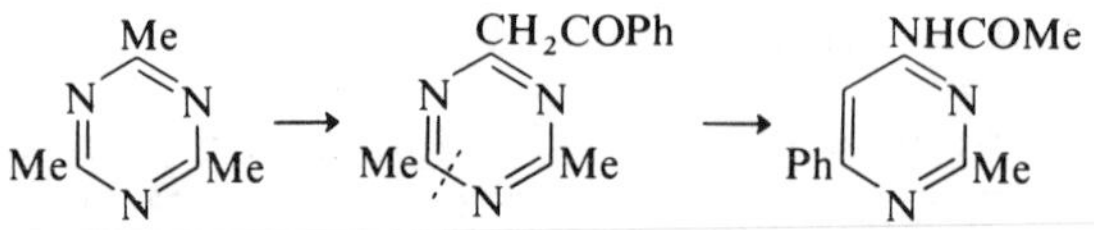

(708)          (709)          (710)

### 2.13.3.2.5 *Pyrimidines from oxazines and thiazines*

Both 1,3-oxazines and 1,3-thiazines may be converted into pyrimidines by treatment with ammonia or primary amines. Thus 2,5-diphenyl-1,3-oxazin-6-one (**711**) with ammonia or aniline yields 2,5-diphenylpyrimidin-4(3*H*)-one (**712**; R = H) or its 2,3,5-triphenyl analogue (**712**; R = Ph), respectively ⟨63N403⟩, and the 1,3-thiazine (**713**) in warm aqueous ammonia gives 6-hydroxymethyl-2-thiouracil (**714**; R = H) or with methylamine gives the 1-methyl derivative (**714**; R = Me) ⟨71AJC785⟩. It is evident that such oxazine or thiazine substrates must carry an oxo or thioxo substituent as well as two double bonds in the ring in order to yield a 'fully aromatic' pyrimidine; otherwise, a dihydropyrimidine will result. In fact, this is a useful synthesis thereof: for example, ethyl 2-anilino-1,3-thiazine-5-carboxylate undergoes rearrangement in hot formic acid to give the dihydropyrimidinethione (**715**) ⟨65JOC2290⟩.

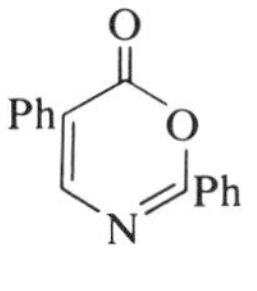

(711) 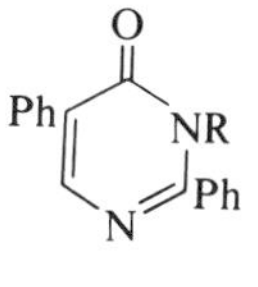    (712) 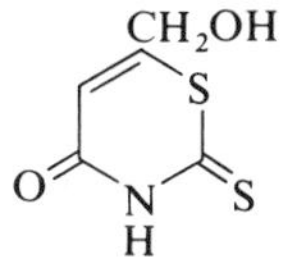    (713) 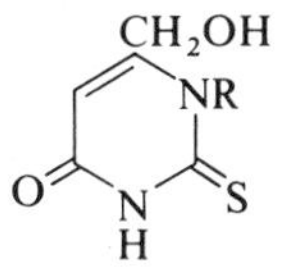    (714) 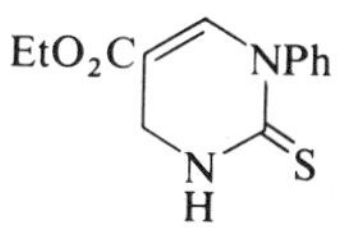    (715)

### 2.13.3.2.6 *Pyrimidines from benzofurans and other O-heterocycles*

The pyrimidine synthesis from benzofurans is very specialized, leading only to 5-(*o*-hydroxyphenyl)pyrimidines. One example from a great many is the treatment of 3-acetyl-2-ethyl-6-nitrobenzofuran (**716**) with guanidine, acetamidine, thiourea or urea to give in good yield 4-ethyl-5-(2'-hydroxy-4'-nitrophenyl)-6-methylpyrimidin-2-amine (**717**; R = NH$_2$), the 2-methyl analogue (**717**; R = Me) and the corresponding pyrimidin-2(1*H*)-thione and pyrimidinone, respectively; the reaction will be recognized as a modified Principal Synthesis ⟨73MI21300⟩. A somewhat similar result is obtained on treatment of chromones with sulfaguanidine; thus, the substrate (**718**) and sulfaguanidine in alcoholic alkoxide give the sulfonamide (**719**) in 80% yield ⟨75MI21300⟩. Single ring *O*-heterocycles behave similarly: 2,6-dimethylpyran-4-one and urea in ethanolic sodium ethoxide give 4-acetonyl-6-methyl-pyrimidin-2(1*H*)-one, albeit in poor yield ⟨74MI21301⟩.

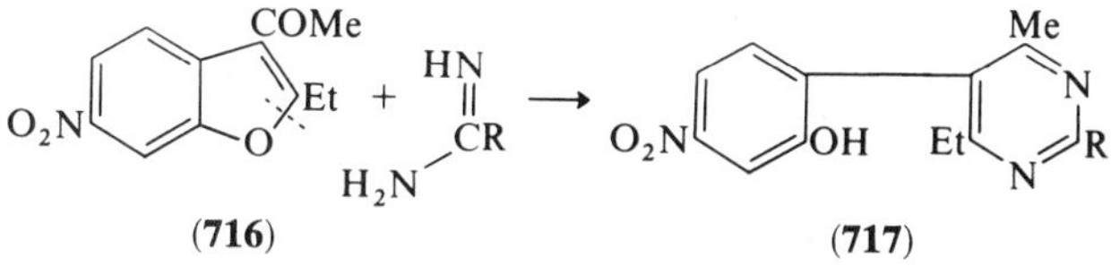

(716)    (717)

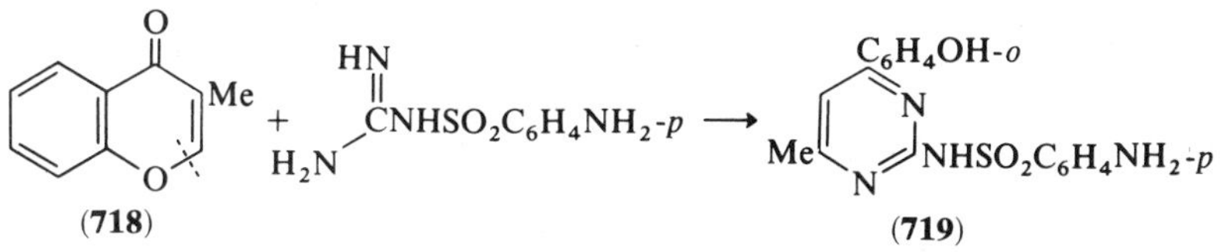

(718)    (719)

### 2.13.3.2.7 *Pyrimidines from purines and related fused systems*

Purin-2(3*H*)-one (**720**), the corresponding thione and its *S*-methyl derivative all undergo facile hydrolytic degradation to 4,5-diaminopyrimidin-2(1*H*)-one (**721**), the analogous thione or the 2-methylthio analogue, respectively; 9-methylpurine similarly gives 4-methyl-aminopyrimidin-5-amine ⟨54JCS2060⟩. However, since the above purines are best made from the pyrimidines to which they are degraded, the synthesis is worthless. In contrast, 7-methylpurine (**722**) can be made easily from aliphatic intermediates and can subsequently undergo alkaline degradation to 5-methylaminopyrimidin-4-amine (**723**), a most useful route to this compound and some analogues ⟨64LA(673)82⟩; a variation of these procedures produces 4,5-bis(alkylamino)pyrimidines ⟨64LA(673)78⟩.

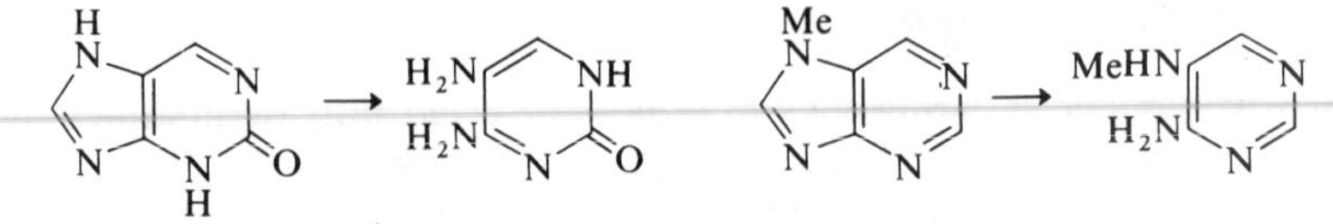

(720)          (721)               (722)          (723)

2-Phenylpyrazolo[3,4-d]pyrimidin-3(2H)-one (724) may be made easily from an available pyrazole and then degraded by catalytic hydrogenation to give 4-amino-N-phenyl-pyrimidine-5-carboxamide (725; R = NHPh) and thence by alkaline hydrolysis to the carboxylic acid (725; R = OH) ⟨62CB2796⟩; oxidative degradations are also possible ⟨61JOC451⟩. Isoxazolopyrimidines may be made from isoxazole intermediates and then degraded to pyrimidines. For illustration, hydrogenation of 3-methylisoxazolo[5,4-d]pyrimidin-4-amine (726) followed by boiling in water yields 5-acetyl-6-aminopyrimidin-4(3H)-one (727) ⟨64JOC2116⟩. Thiadiazolopyrimidines can give pyrimidines, usually those from which they are made in the first place. Of more use is the degradation of thienopyrimidines, prepared from thiophenes. For example, 5-methylthieno[2,3-d]pyrimidin-4-amine (728) undergoes desulfurization by Raney nickel to give a mixture of 5-isopropenyl- (729) and 5-isopropyl-pyrimidin-4-amine; other analogues may be made similarly ⟨67JOC2376⟩. When the thiazolo[4,5-d]pyrimidine (730), which is made from thiazole intermediates, is boiled with aqueous alkali in the absence of air, 6-anilino-5-mercaptopyrimidin-4(3H)-one (731) is formed in good yield; in the presence of air, the product is naturally the corresponding disulfide ⟨66JPR(32)26⟩.

(724)          (725)               (726)          (727)

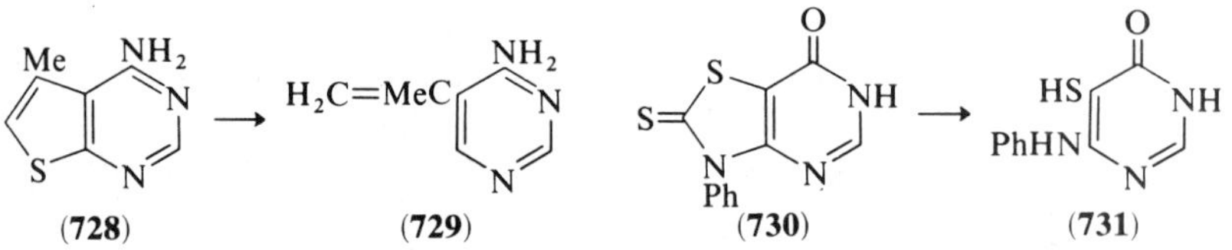

(728)          (729)               (730)          (731)

### 2.13.3.2.8 Pyrimidines from pteridines and related fused systems

Most pteridines are degraded to pyrazines and when they do yield pyrimidines, these may well be the ones from which they were made. However, some useful preparations of pyrimidines from pteridines are known. Thus, reduction of pteridin-7(8H)-one (732) and subsequent hydrolysis yields N-(4-aminopyrimidin-5-yl)glycine (733) ⟨52JCS1620⟩ and hydrolysis of 5,8-dimethylpteridine-6,7(5H,8H)-dione (734) gives N,N'-dimethyl-pyrimidine-4,5-diamine (735), not otherwise available ⟨56JCS2066⟩. Some more recent degradations are less important although of interest ⟨66JCS(C)1112⟩.

Pyrimido[4,5-d]pyrimidines may be used as pyrimidine precursors. Thus, the dihydro derivative (736) undergoes alkaline hydrolysis to the amide (737; R = PrCO) which may be deacylated in ethanolic hydrogen chloride to give 5-aminomethyl-2-propylpyrimidin-4-amine (737; R = H) ⟨64CPB393⟩; rather similarly, the pyrimidopyrimidinedione (738) reacts with amines to give, for example, 6-amino-5-benzyliminomethyl-1,3-dimethylpyrimidine-2,4(1H,3H)-dione (739; R = CH₂Ph) or the hydrazone (739; R = NH₂) ⟨74JCS(PI)1812⟩.

The oxidation of quinazoline with alkaline permanganate is still the preferred route to pyrimidine-4,5-dicarboxylic acid ⟨04CB3643⟩.

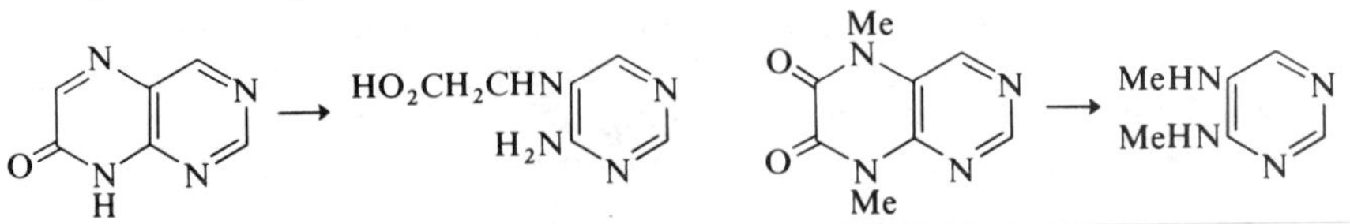

(732)          (733)               (734)          (735)

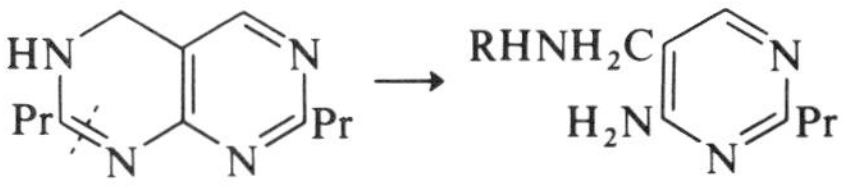

(736)      (737) 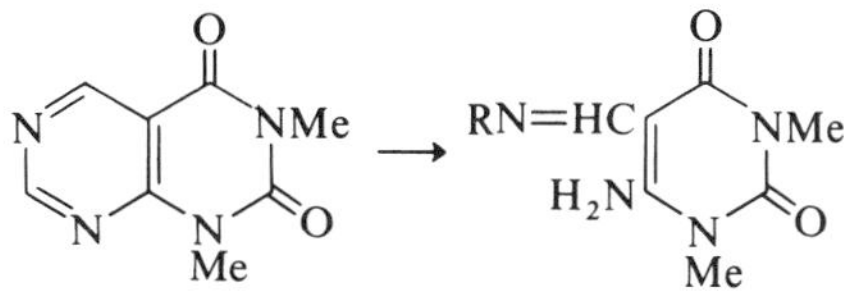      (738)      (739)

The pyrano[2,3-*d*]pyrimidine (**740**) is easily made and on alcoholysis yields the pyrimidine (**741**; R = CO$_2$Et) and thence the simple acetyl analogue (**741**; R = H); the 2-thio analogue may be made similarly ⟨64JOC219⟩.

Pyrimidines may be made also from pyrimido[5,4-*b*][1,4]thiazines, *e.g.* (**742**) ⟨62JA1904⟩ and from pyrimido[4,5-*e*][1,2,4]thiadiazines, *e.g.* (**743**) ⟨63JOC1994⟩ but the products are rather esoteric.

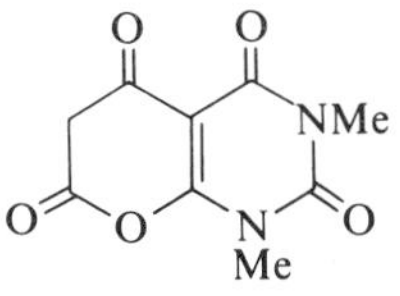

(740) 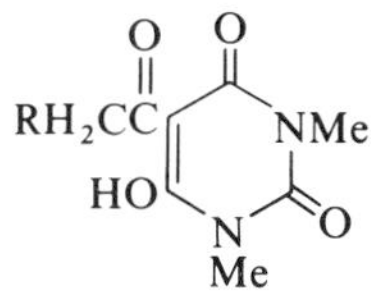      (741) 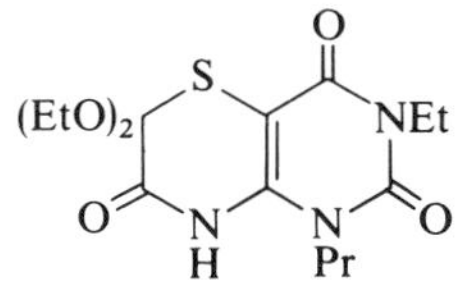      (742) 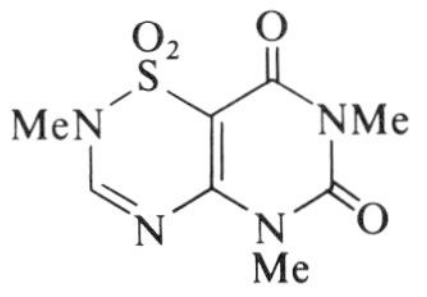      (743)

### 2.13.3.3 Preferred Synthetic Routes to Pyrimidines

This section seeks to summarize preferred routes to various types of pyrimidine. To some extent it is based on the personal experience of the present author, who is well aware of its resulting inadequacies.

#### 2.13.3.3.1 Synthesis of pyrimidine

The best way to make pyrimidine in quantity is from 1,1,3,3-tetraethoxypropane (or other such acetal of malondialdehyde) and formamide, by either a continuous ⟨58CB2832⟩ or a batch process ⟨57CB942⟩. Other practical ways to make small amounts in the laboratory are thermal decarboxylation of pyrimidine-4,6-dicarboxylic acid (**744**), prepared by oxidation of 4,6-dimethylpyrimidine ⟨59JCS525⟩, or hydrogenolysis of 2,4-dichloropyrimidine over palladium–charcoal in the presence of magnesium oxide ⟨53JCS1646⟩.

The best route to 1,4,5,6-tetrahydropyrimidine (**746**) is treatment of propane-1,3-diamine with ethyl formate to give the formyl derivative (**745**) which is heated at *ca.* 150 °C/20 mmHg to give the tetrahydropyrimidine as a distillate ⟨62JCS527⟩. It may also be made by hydrogenation of pyrimidine in aqueous hydrochloric acid over palladium or by hydrogenolysis of 2-chloro-, 2,4-dichloro-, or 4,6-dichloro-pyrimidine (**747**) in aqueous ether over palladium ⟨55JOC829⟩. Hexahydropyrimidine may be made by treatment of propane-1,3-diamine with aqueous formaldehyde at pH *ca.* 4 ⟨67AJC1643⟩.

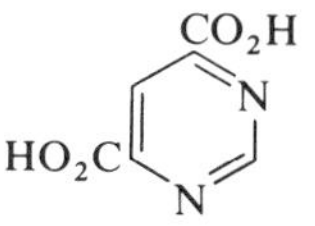

(744) 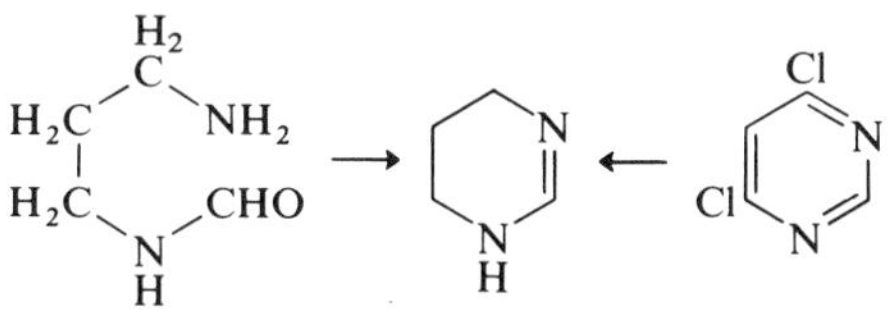      (745)      (746)      (747)

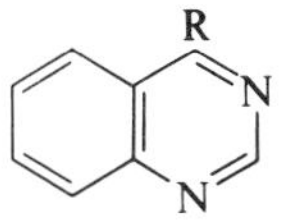

(748) 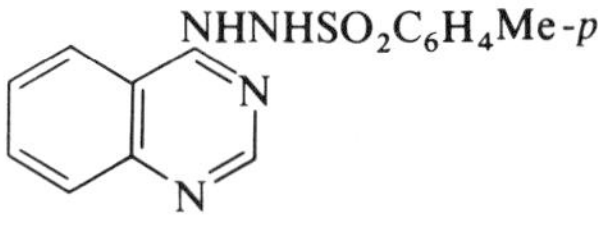      (749) 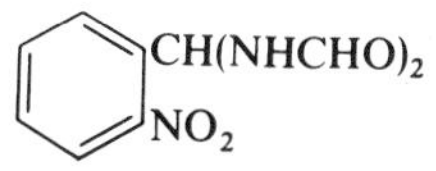      (750)

#### 2.13.3.3.2 Synthesis of quinazoline

The preferred route to quinazoline in quantity is to convert 4-chloro- (**748**; R = Cl) directly into 4-*p*-toluenesulfonylhydrazinoquinazoline (**749**) and thence by treatment with alkali in hot aqueous ethylene glycol, into quinazoline (**748**; R = H) ⟨65JCS5360⟩. Other practical routes to quinazoline are the hydrogenolysis of 4-chloroquinazoline (**748**; R = Cl) over palladium and sodium acetate in methanolic benzene ⟨61MI21300⟩ and the conversion of *o*-nitrobenzaldehyde into its bisformamido derivative (**750**) followed by reductive cyclization using zinc and acetic acid ⟨63IJC346⟩; in the latter method, the starting material poses some problems.

The least troublesome routes to 3,4-dihydro- and 1,2,3,4-tetrahydro-quinazoline are probably the reduction of quinazoline by sodium borohydride, in water for the former or in methanol for the latter. Both must be isolated as salts. The dihydroquinazoline may be formed also by reduction with LAH in ether ⟨65JHC157⟩. In contrast, 5,6,7,8-tetrahydroquinazoline is best made by primary synthesis from 2-formylcyclohexanone and formamide ⟨57CB942⟩ or from cyclohexanone and trisformamidomethane ⟨60CB1402⟩.

### 2.13.3.3.3 Synthesis of alkyl and aryl derivatives

To prepare alkylpyrimidines without other substituents, it is best to place the required alkyl group(s) in position by an appropriate primary synthesis and then, if necessary, remove unwanted functional groups. For example, 2,5-dimethylpyrimidine (**752**) could be made directly by the condensation of acetamidine with 2-ethoxymethylene- (**751**; R = OEt) ⟨77CZ305⟩ or 2-dimethylaminomethylene-acrolein (**751**, R = NMe₂) ⟨60CB1208⟩; however, for easy availability of intermediates, it might be preferable to condense acetamidine with diethyl methylmalonate to give the pyrimidinone (**754**) followed by treatment with phosphoryl chloride to give the dichloropyrimidine (**753**) which, on hydrogenolysis, gives 2,5-dimethylpyrimidine (**752**) ⟨54CI(L)786⟩. In contrast, 4,6-diphenylpyrimidine can be made directly from dibenzoylmethane and formamide ⟨57CB942⟩, albeit in only 30% yield, whereas the best indirect route *via* an hydriodic acid dechlorination of 2-chloro-4,6-diphenylpyrimidine gives an even lower overall yield and takes considerably longer ⟨73AJC443⟩.

(**751**)　　　(**752**)　　　(**753**)　　　(**754**)

(**755**)　　　(**756**)　　　(**757**)　　　(**758**)

Little general guidance on the preparation of alkylpyrimidines with functional groups present can be given here, save to point out that the method used for removing any unwanted group(s) must be compatible with the remaining groups. For example, the obvious route to 2-methyl-5-nitropyrimidine (**755**; R = H) is from its 4,6-dichloro derivative (**755**; R = Cl) which is made easily. However, the reductive removal of the chloro substituents, either directly by hydrogenolysis or indirectly by Raney nickel desulfurization of a derived mercaptothione, cannot be done without reducing the nitro group. Accordingly, an oxidative procedure must be adopted: the dichloro compound is converted into its dihydrazino analogue (**755**; R = NHNH₂) and thence by treatment with silver oxide into the required product (**755**; R = H) ⟨67JCS(C)573⟩. It is interesting that a Principal Synthesis from nitromalondialdehyde and acetamidine in the older literature, gives, not 2-methyl-5-nitropyrimidine, but an isomeric pyridine (**756**) formed by rearrangement ⟨78KGS1400⟩.

The direct or indirect introduction of alkyl/aryl groups into pyrimidines is seldom useful but replacement of a 4-chloro (or better, a 4-cyano) group using a Grignard reagent is quite successful in the quinazoline series. Thus, quinazoline-4-carbonitrile (**757**), which is easily made from quinazoline by 3,4-addition of hydrogen cyanide and subsequent oxidation of the adduct ⟨60YZ245⟩, reacts with appropriate Grignard reagents to afford the 4-alkyl- or 4-aryl-quinazolines (**758**; R = Me, Et, Prⁱ, Ph, CH₂Ph, *etc.*) in good yield ⟨62CPB1043⟩; quinazoline-2-carbonitrile does not behave similarly.

### 2.13.3.3.4 Synthesis of acyl derivatives

The preparation of pyrimidine-4-carbaldehydes without other functional groups can be done by the Principal Synthesis. For example, 1-(diethoxyacetyl)-2-dimethylaminoethylene (**759**) and formamidine give the acetal 4-diethoxymethylpyrimidine (**760**; R = H), which is treated with warm sulfuric acid to afford the aldehyde (**761**; R = H); appropriate amidines or *S*-methylthiourea give the 2-substituted aldehydes (**761**; R = Ph, Me, SMe) and the synthesis is further variable ⟨64CB3407⟩. In contrast, pyrimidine-2-carbaldehyde (**763**; R = H) is best made by an LAH reduction of methyl pyrimidine-2-carboxylate (**762**; R = H); its 4,6-dimethyl (**763**; R = Me) and 4,6-diphenyl (**763**; R = Ph) derivatives are made similarly ⟨69KGS1086⟩. A third type of synthesis is required for pyrimidine-5-carbaldehyde (**766**; R = H) and derivatives: since simple pyrimidines cannot be 5-formylated, the hydroxy-pyrimidinone (**764**; R = H) is treated with DMF and phosgene to give the intermediate quaternary salt (**764**; R = CH=N⁺Me₂) which with water gives the aldehyde (**764**; R = CHO). To obtain the simple aldehyde, it is best to carry out the original reaction in the presence of phosphoryl chloride or to treat the intermediate with phosphoryl chloride so as to isolate the dichloroaldehyde (**765**). On hydrogenolysis (palladium/magnesium oxide), this gives the aldehyde (**766**; R = H); the 2-methyl (**766**; R = Me) and 2-phenyl (**766**; R = Ph) derivatives are made similarly ⟨72LA(766)73⟩.

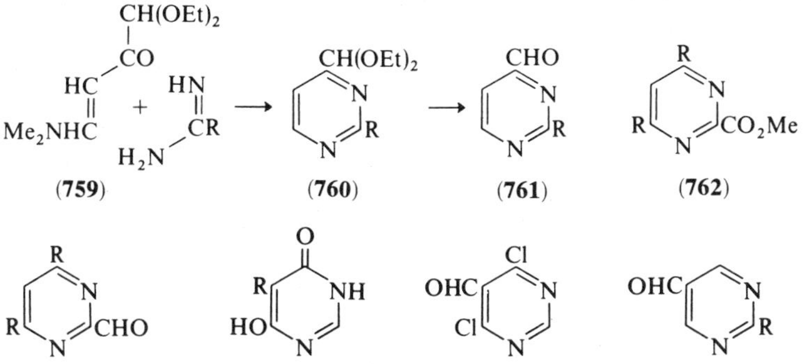

(**759**)      (**760**)      (**761**)      (**762**)

(**763**)      (**764**)      (**765**)      (**766**)

The above reactions point to several ways of making less simple pyrimidinecarbaldehydes. However, there are other practical ways too, as illustrated in the following examples. Oxidation of 6-methyl- (**767**; R = H) and 5,6-dimethyl-pyrimidine-2,4(1*H*,3*H*)-dione (**767**; R = Me) with selenium dioxide gives the pyrimidine-4-carbaldehyde (**768**; R = H) and its 5-methyl derivative (**768**; R = Me), respectively, indicating that 4/6- but not 5-methyl groups are so oxidized ⟨67JHC163⟩; 5-hydroxymethylpyrimidine-2,4(1*H*,3*H*)-dione (**769**; R = CH₂OH) may, however, be oxidized to the uracilcarbaldehyde (**769**; R = CHO) by a variety of agents ⟨66TL5253⟩ of which ammonium ceric nitrate appears to be the most satisfactory ⟨B-78MI21301⟩. 4-Aminopyrimidine-5-carbonitrile (**770**; R = CN) is hydrogenated in dilute aqueous acid over palladium to the (unisolated) imine (**770**; R = CH=NH) which undergoes hydrolysis to 4-aminopyrimidine-5-carbaldehyde (**770**; R = CHO) in good yield, and and a similar procedure can be used to make the oxodihydropyrimidinecarbaldehyde (**771**) and related compounds ⟨67CB3664⟩.

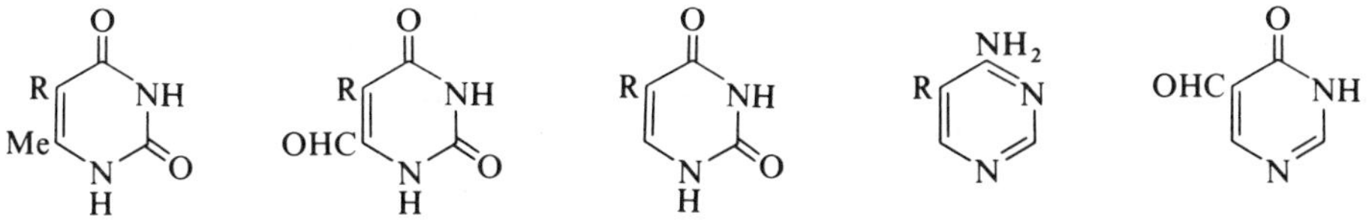

(**767**)      (**768**)      (**769**)      (**770**)      (**771**)

Almost all 5-acetyl (or higher acyl)pyrimidines must be made by primary synthesis. For example, 3-ethoxymethyleneacetylacetone (**772**) reacts as an aldehydo ketone with acetamidine to give 5-acetyl-2,4-dimethylpyrimidine (**773**; R = Me) ⟨45JA1294⟩ or with *N,N*-dimethylguanidine to give the 2-dimethylamino analogue (**773**; R = NMe₂) ⟨58JCS3742⟩. The Shaw synthesis (Section 2.13.3.1.2*e*) may also be used to produce 5-acetylpyrimidines. Some barbituric acids and analogues can be directly 5-acylated. For example, the dimethylaminodimethyluracil (**774**; R = H) reacts with boiling acetic anhydride to give the 5-acetyl derivative (**774**; R = Ac) ⟨58LA(612)173⟩. Oxidative methods are poorly represented but 5-benzyl-4,6-dimethyl-2-phenylpyrimidine (**775**) can be oxidized satisfactorily to 5-benzoyl-2-phenylpyrimidine-4,6-dicarboxylic acid (**776**; R = CO$_2$H) which is decarboxylated in boiling acetic anhydride to yield 5-benzoyl-2-phenylpyrimidine (**776**; R = H) ⟨64CB1163⟩; several secondary alcohols are also known ⟨63JMC550⟩ to undergo oxidation to 5-acyl derivatives.

Some 2- and 4-acylpyrimidines can be made from the corresponding nitrile or carboxylic acid with a lithium or Grignard reagent. For example, 4,6-dimethylpyrimidine-2-carbonitrile (**777**; R = CN) and phenylmagnesium bromide give 2-benzoyl-4,6-dimethylpyrimidine (**777**; R = COPh) ⟨56M526⟩ while 2,6-diethoxypyrimidine-4-carboxylic acid (**778**; R = CO$_2$H) and methyllithium gives the 4-acetyl analogue (**778**; R = Ac) ⟨56JA2136⟩. However, a far better method is now available in the homolytic acylation of simple pyrimidines with acyl radicals, *e.g.* (i) from pyruvic acid/silver nitrate/ammonium persulfate or (ii) from an aldehyde/ferrous sulfate/*t*-butylhydroperoxide. In this way, 2,4-dialkylpyrimidines give their 6-acyl derivatives and 4,6-dialkylpyrimidines give a mixture of 2-mono- and 2,5-di-acylated derivatives. Thus, 2-methyl-6-phenylpyrimidine by method (i) gives the 4-acetyl derivative (**779**; R = Ac); the same substrate with propionaldehyde in method (ii) gives the 4-propionyl derivative (**779**; R = COEt); and 4,6-dimethylpyrimidine (**780**) by method (i) gives a separable mixture of the 2-acetyl (**781**) and 2,5-diacetyl (**782**) derivatives, the latter in lower yield ⟨80CPB202⟩.

### 2.13.3.3.5 *Preparation of carboxy and related derivatives*

(*a*) The best route to the simple pyrimidinecarboxylic acids is oxidative. Thus, 2-methyl- is converted into 2-styryl-pyrimidine and thence by permanganate into pyrimidine-2-carboxylic acid ⟨54CI(L)786⟩. 4-Methylpyrimidine with an excess of selenium dioxide in boiling pyridine gives 80% of pyrimidine-4-carboxylic acid ⟨77JMC1312⟩, and pyrimidine-5-carbaldehyde (see Section 2.13.3.3.4) is a convenient substrate for oxidation by permanganate to pyrimidine-5-carboxylic acid ⟨72LA(766)73⟩. The oxidative route may also be used for less simple carboxylic acids, as in the oxidation of 5-hydroxymethyluracil (**783**) by chromic acid to uracil-5-carboxylic acid (**784**) ⟨60JA991⟩.

Primary synthesis has limited application in making pyrimidine-carboxylic acids or even their esters. However, some pyrimidine-4(and 5)-carboxylic acids can be effectively so made. For example, bromomucic acid (**785**) reacts as an aldehydo ketone with *S*-methyl-thiourea to give 5-bromo-2-methylthiopyrimidine-4-carboxylic acid (**786**) directly ⟨53JCS3129⟩ while the Whitehead synthesis (Section 2.13.3.1.2*q*) can give, for instance, 3-methylcytosine-5-carboxylic acid (**787**) ⟨55MI21300⟩.

In fact, most pyrimidinecarboxylic acids are made by hydrolysis of the corresponding esters, nitriles or sometimes amides, many of which can be made more easily by primary synthesis than can the acids themselves. Thus, pyrimidine-5-carboxylic acid may be made by alkaline hydrolysis of its ethyl ester ⟨62JOC2264⟩ and pyrimidin-5-ylacetic acid (**789**;

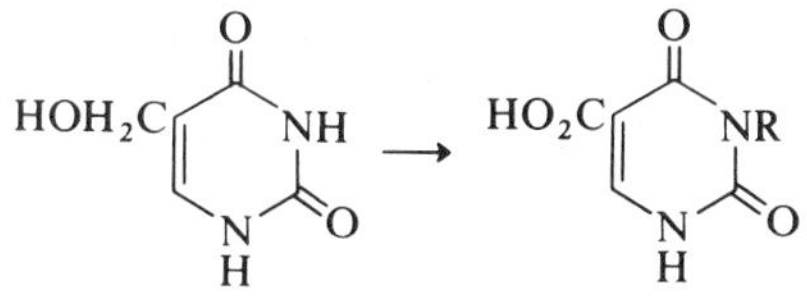

(783) (784) 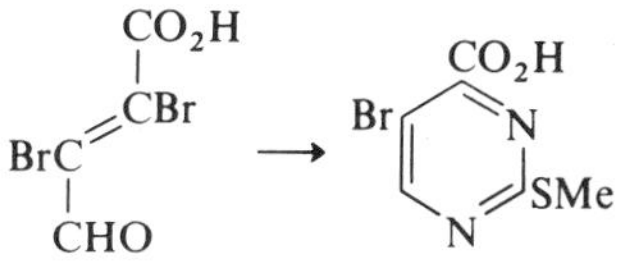 (785) (786)

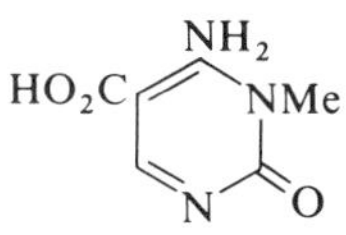

(787) 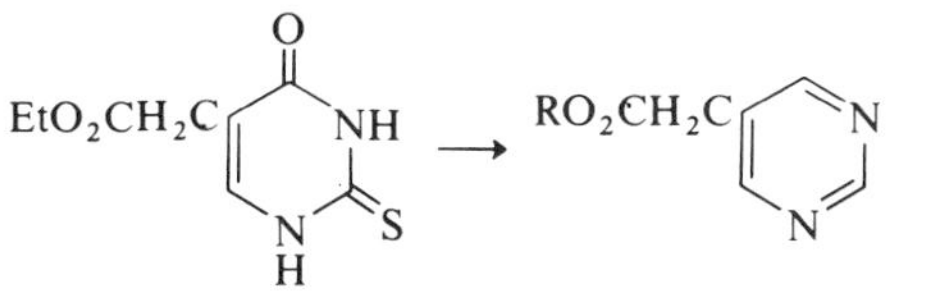 (788) (789) 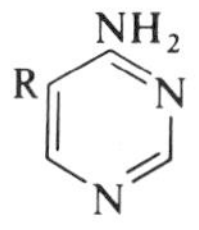 (790)

R = H) is made similarly from its ester (**789**; R = Et), itself prepared by several obvious steps (see (b) below) from the pyrimidine (**788**) which can be made by primary synthesis ⟨66AP362⟩. 4-Aminopyrimidine-5-carbonitrile (**790**; R = CN), which may be made by primary synthesis, undergoes hydrolysis in alkali to the amino acid (**790**; R = CO$_2$H); it may be made similarly from the amide (**790**; R = CONH$_2$) ⟨53JCS331⟩.

(b) The usual route to esters, simple or otherwise, is primary synthesis followed by removal or modification of other groups present. For example, condensation of diethyl 2-formylsuccinate with thiourea gives the ester (**788**) which can be desulfurized oxidatively to the pyrimidine (**791**); this is converted into the chloro ester (**792**) and then dechlorinated by hydrogenolysis to yield the simple ester (**789**; R = Et) ⟨66AP362⟩.

In addition, it may be convenient sometimes to prepare esters from carboxylic acids. For example, pyrimidine-4-carboxylic acid (**793**; R = H) gives the methyl ester (**793**; R = Me), either in methanolic hydrogen chloride (yield 86%) ⟨65JOC2398⟩ or in ethereal diazomethane (yield 75%) ⟨77JMC1312⟩. Especially when an acid must be converted into its amide, it is often better to go *via* the acid chloride than the regular ester. Thus pyrimidine-5-carboxylic acid (**794**; R = OH) and thionyl chloride give the acid chloride (**794**; R = Cl) which affords amides, e.g. (**794**; R = NEt$_2$), in >60% overall yield ⟨62JOC2264⟩. Another convenient use of an acid chloride is seen in the treatment of orotic acid (**795**; R = OH) with phosphoryl chloride/phosphorus pentachloride to give the trichlorinated pyrimidine (**796**; R = Cl) and thence by methanolic sodium methoxide to the regular ester (**796**; R = OMe), otherwise difficultly accessible ⟨62JOC3507⟩.

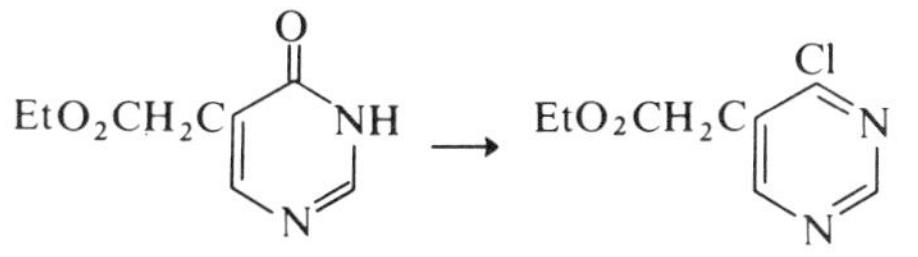

(791) (792)  (793) 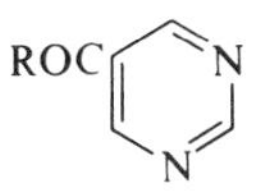 (794)

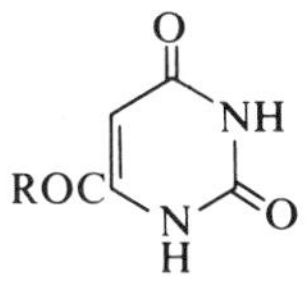

(795) 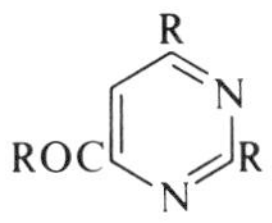 (796) 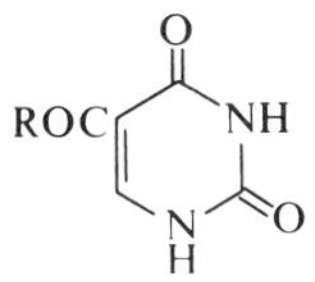 (797) 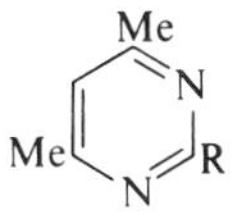 (798)

(c) The usual route to carboxamides is by aminolysis of esters (including acid chlorides). Methyl pyrimidine-4(and 5)-carboxylate both give the corresponding carboxamides by treatment with methanolic ammonia at room temperature ⟨60MI21300⟩, the esters (**795**; R = OBu) and (**797**; R = OBu) give their respective *N*-butylamides (**795**; R = NHBu) and (**797**; R = NHBu) by treatment with hot ethanolic butylamine ⟨60JOC1950⟩, and pyrimidine-5-carboxhydrazide (**794**; R = NHNH$_2$) can be made from the ethyl ester (**794**; R = OEt) with ethanolic hydrazine ⟨62CB803⟩. An example of aminolysis of an acid chloride is given in section (b) above.

A second practical route to *N*-unsubstituted amides is by the controlled hydrolysis of nitriles, which can often be made (in the 5-position) by primary synthesis or (elsewhere) by displacement of an ammonio grouping. Thus 4,6-dimethylpyrimidine-2-carbonitrile (**798**; R = CN) in warm aqueous ammonia gives the amide (**798**; R = CONH$_2$) in good yield

⟨56M526⟩, 4-amino-2-methylaminopyrimidine-5-carbonitrile gives the corresponding carboxamide by dissolution in concentrated sulfuric acid at 30 °C ⟨60JA5711⟩, and 4-aminopyrimidine-5-carbonitrile gives the 5-carboxamide by warming with alkaline hydrogen peroxide (the Radziskewski reaction) ⟨53JCS331⟩. In addition, the best route to thioamides is by heating the corresponding nitrile in ethanolic or pyridinic triethylamine saturated with hydrogen sulfide. For example, 4,6-dimethylpyrimidin-2-ylacetonitrile (**798**; R = CH$_2$CN) so gives the thioamide (**798**; R = CH$_2$CSNH$_2$) ⟨71JMC244⟩ and pyrimidine-2(4 and 5)-thiocarboxamides may be made similarly ⟨60MI21300⟩.

There are several other specialized ways to make individual amides but none is general. A random example is the self-condensation of malondiamide to give 4,6-dihydroxypyrimidin-2-ylacetamide ⟨56JCS2312⟩.

(*d*) The obvious route to pyrimidine-5-carbonitriles is primary synthesis followed by removal or modification of unwanted groups. For example, ethoxymethylenemalononitrile (**799**) and thiourea give 4-amino-2-thioxo-1,2-dihydropyrimidine-5-carbonitrile (**800**) ⟨58MI21300⟩ which can be desulfurized, *S*-alkylated, *N*-acylated, or hydrolyzed to 5-cyanocytosine *etc.*, without affecting the cyano group. The Shaw synthesis (Section 2.13.3.1.2*e*) also offers a variety of pyrimidine-5-carbonitriles. The primary synthesis of 2- or 4/6-cyano derivatives is seldom, if ever, practical.

Nitriles may also be made by displacement reactions from halogeno, ammonio, alkylsulfonyl, alkylsulfinyl or sulfo derivatives of pyrimidine but the scope of such reactions is more limited than might be expected. 4-Chloromethyl-2-isopropyl-6-methylpyrimidine (**801**; R = Cl) and aqueous alcoholic sodium cyanide give the nitrile (**801**; R = CN) ⟨80CPB3362⟩; 2-chloropyrimidine (**802**; R = Cl) with trimethylamine in benzene gives *N,N,N*-trimethylpyrimidin-2-ylammonium chloride (**802**; R = N$^+$Me$_3$) which reacts with potassium cyanide in acetamide to give pyrimidine-2-carbonitrile (**802**; R = CN) ⟨59JA905⟩; pyrimidine-2(1*H*)-thione is *S*-methylated and then oxidized to 2-methylsulfonylpyrimidine (**802**; R = SO$_2$Me) which reacts with sodium cyanide in DMF to give pyrimidine-2-carbonitrile (**802**; R = CN) ⟨67JCS(C)568⟩; and 5-bromopyrimidine and cuprous cyanide in boiling quinoline give pyrimidine-5-carbonitrile in good yield ⟨64MI21301⟩.

(**799**)          (**800**)

(**801**)          (**802**)          (**803**)

(**804**)          (**805**)          (**806**)

There are a few other routes of occasional value in making nitriles. The oximes of pyrimidinecarbaldehydes (made by nitrosation of methylpyrimidines or from the aldehyde (if available) are dehydrated to nitriles. The oxime (**803**; R = CH=NOH) in boiling acetic anhydride gives the nitrile (**803**; R = CN) ⟨67JCS(C)1172⟩; phosphoryl chloride or thionyl chloride may also be used as dehydrating agents ⟨72LA(766)73⟩. Amides may be dehydrated rather similarly and this affords a method of converting esters into nitriles (*via* amides): pyrimidine-5-carboxamide gives pyrimidine-5-carbonitrile in 77% yield by dry distillation with phosphorus pentoxide ⟨62JOC2264⟩ but it is usually better to use phosphoryl chloride with or without a solvent, as for example in the conversion of 2-methylpyrimidine-4,5-dicarboxamide (**804**) into the corresponding dinitrile by boiling in phosphoryl chloride/xylene ⟨65JHC202⟩. Quinazoline-4-carbonitrile is best made by addition of hydrogen cyanide to the 3,4-bond of quinazoline and subsequent oxidation of the adduct (**805**) with ferricyanide ⟨60YZ245⟩. The 5-aminoquinazolinone (**806**; R = NH$_2$) may be diazotized and treated with cuprous cyanide to afford the 5-cyano analogue (**806**; R = CN) ⟨45JA2112⟩ and some 5-cyanopyrimidines may be made similarly.

### 2.13.3.3.6 *Preparation of amino, nitro and related derivatives*

(*a*) Methods of making pyrimidinamines fall into three chief categories: (i) by the Principal or other primary synthesis, (ii) by nucleophilic displacement of other groups and (iii) by other means such as reduction, degradation or modification of other groups.

The Principal Synthesis may be used to obtain pyrimidines with 4/6-(primary)-amino groups, and/or a 2-(primary, secondary or tertiary)-amino group but lacking in 5- or extranuclear amino groups. For example, malononitrile and *N,N*-dimethylguanidine give 2-dimethylaminopyrimidine-4,6-diamine (**807**) ⟨51JA2864⟩. Other primary syntheses can afford pyrimidinamines too. For instance, trimerization of acetonitrile under Frankland–Kolbé conditions (Section 2.13.3.1.3*a*) is still the best way to make 2,6-dimethylpyrimidin-4-amine (**808**) ⟨55OSC(3)71⟩. In addition, they can occasionally provide 5-amino or alkylamino groups, as in the alkaline degradation of 7-ethyl-9-methylpurine-8(7*H*)-thione (**809**) to give 5-ethylamino-4-methylaminopyrimidine (**810**) in good yield ⟨67CB2280⟩. *N*-Aminopyrimidines, such as 1-aminobarbituric acid, are also made by primary synthesis ⟨79AJC153⟩.

Ammonia and amines are good nucleophiles and can displace halogeno, mercapto, alkylthio, alkylsulfonyl, alkylsulfinyl, alkoxy, sulfo and a few less important substituents, providing the substrate and product will withstand the conditions needed in the particular reaction. The best leaving groups are halogeno and alkylsulfonyl, followed by alkoxy and alkylthio, followed by mercapto; the others are of less use because they are less easily accessible on the whole. If conditions are not to be overly severe, it is specially important to choose a 'good' substrate when rather hindered amines (*e.g. s*- or *t*-butylamine, diisopropylamine *etc.*) are to be used, when a less-active position (4/6 > 2 ≫ 5) is involved, or when deactivating (*i.e.* electron-releasing) substituents are present. If more than one substrate is available, a rational choice may be made on the basis of hundreds of examples summarized elsewhere ⟨62HC(16)306, 70HC(16-S1)230⟩. The following approximate relative rates for the *n*-alkylaminolysis of some 2-substituted pyrimidines indicate the vast differences involved: 2-Cl: $3 \times 10^6$; 2-SH: —; 2-SMe: 20; 2-SO$_2$Me: $12 \times 10^6$; 2-SOMe: $6 \times 10^6$; 2-OMe: 80; 2-SO$_3$H: ?; 4,6-Me$_2$-2-Cl: $3 \times 10^5$; 4,6-Me$_2$-2-SMe: 1; 4,6-Me$_2$-2-OMe: 60; 5-Br-2-Cl: $6 \times 10^7$; 5-Br-2-SMe: 80; 5-Br-2-OMe: 400; 5-NO$_2$-2-Cl: $10^{10}$; 5-NO$_2$-2-SMe: $4 \times 10^6$; 5-NO$_2$-2-OMe: $10^8$ ⟨B-69MI21300⟩.

(807)  (808)  (809)  (810)

(811)  (812)  (813)  (814)

Because of the variation in reactivity of the same substituent when situated at the 2-, 4/6- or 5-position, considerable selectivity is possible in aminolysis of di- or trichloropyrimidines or the like. In addition, it will be realized that the insertion of each amino group deactivates any remaining leaving-group, thereby achieving selectivity even in 4,6-dichloropyrimidine, in which both chloro substituents start with equal activity. Thus, pyrimidine-2,4(1*H*,3*H*)-dithione (**812**) undergoes monoaminolysis in refluxing hexylamine to give 4-hexylaminopyrimidine-2(1*H*)-thione (**813**) from which the sulfur can be removed by Raney nickel to give *N*-hexylpyrimidin-4-amine (**811**; R = C$_6$H$_{13}$) ⟨62JMC871⟩ or modified in several other ways as required.

Because of their inherent inactivity, only the best leaving groups in the 5-position can undergo aminolysis. Thus, 5-bromo-4-methylpyrimidine (**814**; R = Br) requires aqueous ammonia at 140 °C for 50 hours to form 4-methylpyrimidin-5-amine (**814**; R = NH$_2$) in 85% yield ⟨65RTC1101⟩.

Fortunately, however, 5-aminopyrimidines can be obtained easily from reduction of 5-nitro-, 5-nitroso- or 5-phenylazo-pyrimidines. Such reduction is generally done by hydrogenation at atmospheric pressure over a Raney nickel, palladium or even a platinum catalyst but classical reducing agents may be used too. Such reductions are discussed in Section 2.13.2.2.6. Degradation of carboxamides *etc.* to amines is sometimes useful (Section 2.13.2.2.5) but, as a rule, it is best to avoid the method. Extranuclear primary amino groups can be made from nitriles: hydrogenation of the nitrile (**815**) in acidic aqueous ethanol over platinum gives the diamine (**816**) in good yield ⟨65JPS714⟩. The conversion of a primary into a secondary or tertiary amino group has a place in preparative procedures, especially if the primary amine is available easily by primary synthesis. Alkylation is complicated by the propensity of pyrimidin-2(or 4)-amines to alkylate on ring-nitrogen, possibly followed by Dimroth rearrangement (Section 2.13.2.2.6*f*): accordingly, at least in the 4/6-position, it is better to try transamination. For example, 2,6-dimethylpyrimidin-4-amine hydrochloride and a small excess of butylamine heated at 170 °C for 20 hours gives *N*-butyl-2,6-dimethylpyrimidin-4-amine (**817**) in >90% yield; many other examples are known ⟨60JA3971⟩. Cytosine (**818**; R = H) undergoes transamination with hydroxylamine at pH 7 to give 4-hydroxyaminopyrimidin-2(1*H*)-one (**818**; R = OH); the mechanism of this particular reaction involves formation of a 5,6-hydroxylamine adduct, transamination, and loss of the originally added molecule of hydroxylamine ⟨65JCS208⟩.

(**815**)          (**816**)          (**817**)          (**818**)

(**819**)          (**820**)          (**821**)          (**822**)

(*b*) The formation of nitropyrimidines does not rely much on direct primary synthesis. However, nitromalondialdehyde (**819**) does condense with urea, guanidine, *S*-alkylthiourea or aromatic amidines to give appropriate 2-substituted 5-nitropyrimidines such as the 2-methylthio (**820**; R = SMe) ⟨51JCS1218⟩ or 2-phenyl derivative (**820**; R = Ph) ⟨56JA1434⟩. Primary syntheses leading to substituted 5-nitrocytosines ⟨59MI21300⟩, to 5-nitrouracil and 5-nitro-2-thiouracil derivatives ⟨63CCC2501⟩ and to 5-nitro-2,6-diphenylpyrimidin-4(3*H*)-one ⟨68LA(716)143⟩ are useful in that some of the products cannot be obtained by other means.

Most nitropyrimidines are made by nitration with or without subsequent modification or (oxidative) removal of other groups. Almost any pyrimidine with a free 5-position, with at least one strongly electron-releasing group, and lacking groups labile under the conditions required, can be nitrated. For example, pyrimidin-2(and 4)-one are both nitrated, albeit only under extremely vigorous conditions, to give the respective 5-nitro derivatives ⟨69JHC593⟩ but pyrimidines like 6-methyluracil, cytosine *etc.* are nitrated in sulfuric/nitric acid quite rapidly at 20–25 °C. Indeed, conditions can be made sufficiently gentle to allow the presence of an alkylthio, alkoxy or even an (active) halogeno substituent: for example, the methylthiopyrimidinone (**821**; R = H) is satisfactorily nitrated by the addition of (urea-decolourized) fuming nitric acid at 0 °C ⟨65JCS3770⟩. The removal of unwanted groups is best done *via* their derived hydrazino groups, which does not involve the reductive conditions contraindicated by the presence of the nitro group. This is illustrated in the following sequence: 2-benzyl-6-hydroxypyrimidin-4(3*H*)-one is made by primary synthesis and subsequently nitrated to its 5-nitro derivative which is converted by phosphoryl chloride into the dichloro compound (**822**; R = Cl); this is converted into the dihydrazino analogue (**822**; R = NHNH$_2$) which is treated with methanolic silver oxide to give 2-benzyl-5-nitropyrimidine (**822**; R = H) ⟨67JCS(C)573⟩.

Because 5-nitrosopyrimidines are so accessible, their oxidation to the corresponding nitropyrimidines is of considerable potential use. Thus, 4,6-diaminopyrimidin-2(1*H*)-one is very easily converted into its 5-nitroso derivative which undergoes oxidation by hydrogen peroxide in trifluoroacetic acid to the 5-nitro analogue (**823**; X = O) in 97% yield; unfortunately, 4,6-diaminopyrimidine-2(1*H*)-thione, which (unlike the corresponding pyrimidinone) cannot be nitrated directly, goes through the same sequence to give the same product (**823**; X = O), rather than the desired thione (**823**; X = S) ⟨65JOC3153⟩. Similar oxidation of 5-nitrosopyrimidine-4,6-diamine gives a mixture of the same product (**823**; X = O), 5-nitropyrimidine-4,6-diamine and its 1,3-di-*N*-oxide; the ratio of products varies with conditions ⟨80AJC131⟩. There are many known pyrimidines in which nitro groups are attached to substituents and are therefore not relevant to the discussion here. 2-Nitropyrimidine has been prepared by an indirect oxidative route from the corresponding amine ⟨82JOC552⟩.

(*c*) The preparation of 5-nitrosopyrimidines is usually done by treatment of an existing pyrimidine with nitrous acid ('nitrosation'). However there are a few examples of primary synthesis, of little actual importance. In addition, a few 2-, 4/6- and extranuclear nitrosopyrimidines are known. The requirements for 5-nitrosation are outlined above (Section 2.13.2.1.1*c*). The basic technique is to add aqueous sodium nitrite to a pyrimidine substrate dissolved or suspended in dilute aqueous acid at room temperature: the coloured nitroso compound forms quite quickly. However, solvents such as ethanol or dioxane may be used if the substrate is too insoluble in water but isopentyl nitrite or the like is then used to provide nitrous acid. An example of this procedure is the formation of 6-amino-1-benzyl-5-nitroso-2-thiouracil (**824**) ⟨63JOC2304⟩. A comparison between aqueous and non-aqueous media in preparing 1,3-dimethyl-5-nitroso-6-phenylethylpyrimidine-2,4(1*H*,3*H*)-dione (**825**) indicates no significant difference in yield (*ca.* 85%) ⟨66LA(691)142⟩. When the 5-position is occupied and there is sufficient electron-release by (three) other groups, 4/6-nitrosation becomes possible to give, for example, the nitrosopyrimidine (**826**) ⟨64JCS1001⟩. *N*-Nitrosation of secondary amino substituents is always possible: thus *N*-methylpyrimidin-2-amine gives the corresponding nitrosamine (**827**) ⟨67JCS(B)273⟩ and 5-methylaminopyrimidin-4-amine gives the nitrosamine (**823**) which resists cyclization to the *v*-triazolopyrimidine ⟨66JCS(B)427⟩.

(823)    (824)    (825)    (826)

(827)    (828)    (829)

Primary synthesis of nitrosopyrimidines is seldom worthwhile but it is at least possible. Thus, isonitrosomalononitrile and *N,N*-dimethylguanidine give the nitrosotriamine (**829**) by a Principal Synthesis ⟨62JA3744⟩.

(*d*) The preparation of arylazopyrimidines is usually done by treatment of a 5-unsubstituted pyrimidine, which must bear at least one electron-releasing substituent, with a diazonium salt (see Section 2.13.2.1.1*d*). A typical example is the formation of 6-chloro-5-phenylazopyrimidine-2,4(1*H*,3*H*)-dione (**830**) without ill effect on the reasonably active chloro substituent ⟨60CB1406⟩; variations in actual procedure are possible ⟨66JPS568⟩. Since it is possible to diazotize some pyrimidin-5-amines, reverse coupling with phenols *etc.* can be done. For example, diazotized 2,4-dimethoxypyrimidin-5-amine couples with *N*-β-diethylaminoethylnaphthyl-2-amine to give the azopyrimidine (**831**) ⟨63JMC646⟩. Like nitrosation, it is possible to induce azo coupling at the 4/6-position of a pyrimidine to give, for example, 6-*p*-chlorophenylazo-5-hydroxypyrimidin-4(3*H*)-one (**832**) in 83% yield ⟨64JCS1001⟩.

Primary synthesis of arylazopyrimidines is used ⟨52JCS3448⟩. It is exemplified in the condensation of phenylazomalondiamidine with diethyl oxalate to give the azopyrimidine (**833**) ⟨66JCS(C)226⟩. Finally, 5-phenylazopyrimidine may be made by the condensation of pyrimidin-5-amine with nitrosobenzene ⟨51JCS1565⟩ but the reaction seems to have been overlooked for many years.

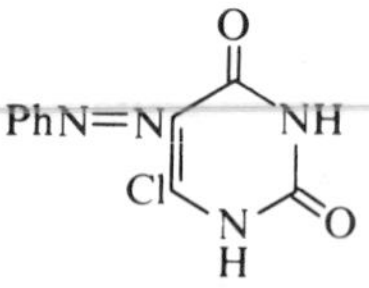

(830) 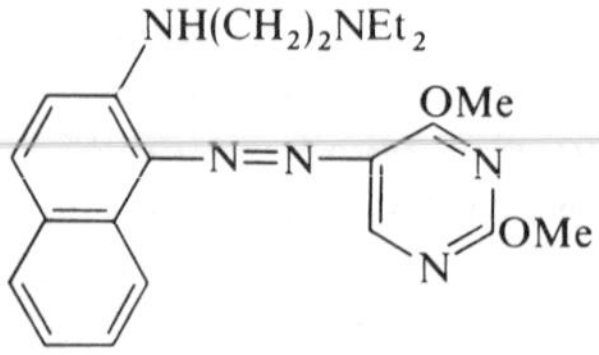            (831) 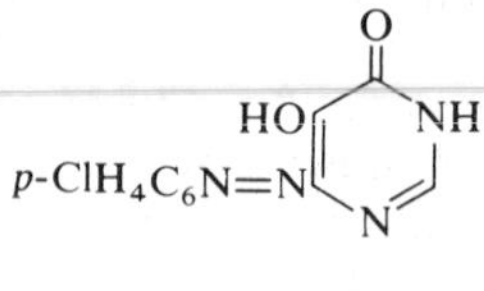            (832) 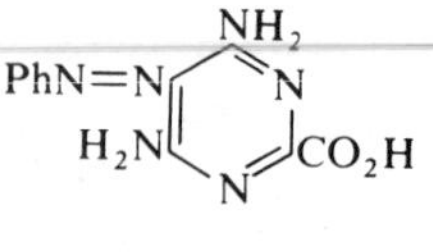            (833)

### 2.13.3.3.7 *Synthesis of hydroxy, alkoxy and related derivatives*

The preparation of 'hydroxypyrimidines' must be divided into two categories, pyrimidinones (2- or 4/6-hydroxypyrimidines) and pyrimidinols (5- or extranuclear hydroxypyrimidines) simply because the approaches are different.

(*a*) Nearly all pyrimidinones have their 'oxy' substituent(s) inserted at the primary synthesis stage, even if considerable modification of other groups is done later. The Principal Synthesis can give a variety of 2-, 4- and/or 6-oxypyrimidines and most other primary syntheses can be used to give such derivatives too. Thus pyrimidin-4(3*H*)-one (835) can be prepared directly by the condensation of ethyl 3-ethoxyacrylate (834), triethyl orthoformate and ammonia (2 mol) ⟨74S286⟩ although many chemists might prefer simply to desulfurize ⟨50MI21301⟩ 2-thiouracil (836) obtained commercially by a primary synthesis; either way, the oxy group results from primary synthesis. This is not always true. For example, cytosine (838) may be made by several routes of which two seem to be the more practical. The first ⟨70JHC527⟩ involves primary synthesis of the chlorodihydrocytosine (837) with the oxy substituent already in place, followed by dehydrochlorination to cytosine (838); the second, and possibly better, method ⟨B-68MI21301⟩ begins the 4-thiation of 2-thiouracil to give pyrimidine-2,4(1*H*,3*H*)-dithione (840) which undergoes preferential ammonolysis to 2-thiocytosine (839) and only in the last stage is the sulfur replaced by oxygen to give cytosine (838). For such cases, in which the oxy substituent is not inserted by primary synthesis, the following examples of routes to pyrimidinones may prove useful. 6-Chloro-5-nitropyrimidin-4(3*H*)-one (841; R = Cl) cannot be made by monochlorination of its hydroxy analogue (841; R = OH) but it can be made by dichlorination to (842) followed by preferential alkaline hydrolysis of an halogeno substituent ⟨64ZOB1321⟩. Because pyrimidin-2-amine is readily available commercially, it is convenient to make pyrimidin-2(1*H*)-one by hydrolysis of the amino group ⟨62JOC2945⟩ although it can indeed be made directly by primary synthesis ⟨59JCS525⟩; the same change can be done by acidic hydrolysis or by treatment of amines (primary only) with nitrous acid. The hydrolysis of an alkoxypyrimidine is useful occasionally for structural purposes, as was the case in the acid hydrolysis of 2-methoxy-*N*-methyl-5-nitropyrimidin-4-amine (843) to the known 4-methylamino-5-nitropyrimidin-2(1*H*)-one ⟨57MI21300⟩. Since neither urea nor its alkyl derivatives condense with

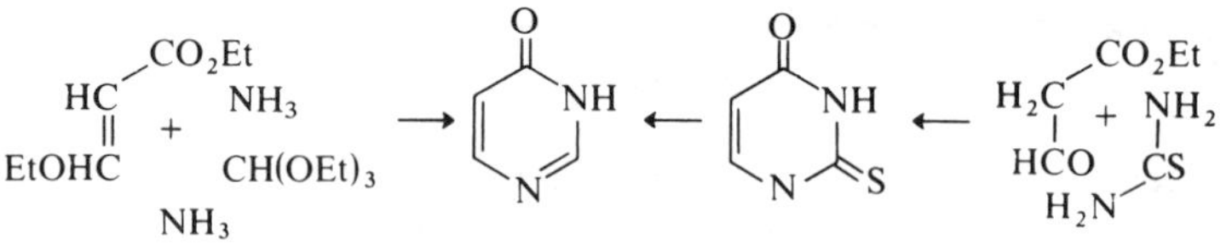

(834)            (835)            (836)

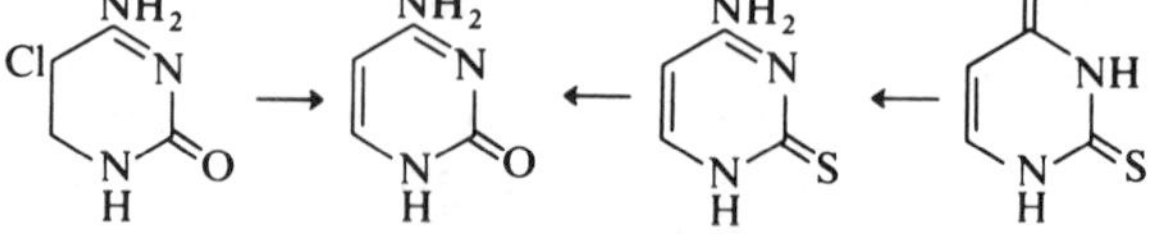

(837)            (838)            (839)            (840)

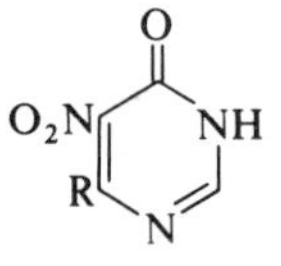

(841) 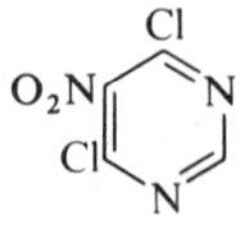            (842) 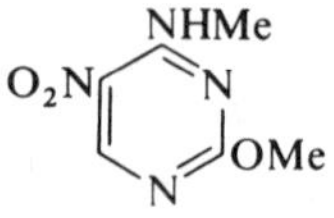            (843) 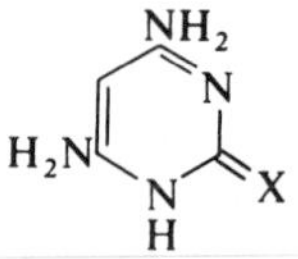            (844)

malononitrile, 4,6-diaminopyrimidin-2(1$H$)-one (**844**; X = O) is made from its 2-thioxo analogue (**844**; X = S), which is available by such a Principal Synthesis from thiourea: the conversion of pyrimidinethione to pyrimidinone is done by boiling in aqueous chloroacetic acid ⟨48JA3109⟩ or by $S$-methylation followed by acidic hydrolysis. The ability to convert a benzyloxypyrimidine into a pyrimidinone by hydrogenolysis is sometimes useful when hydrolytic conditions must be avoided, as in the conversion of 4-benzyloxy-6-fluoropyrimidin-2-amine into 2-amino-6-fluoropyrimidin-4(3$H$)-one ⟨63JMC688⟩. Examples of the hydrolysis of sulfones, sulfoxides, sulfinic acids and sulfonic acids are given in Sections 2.13.2.2.8$n$ and 2.13.2.2.8$o$.

(*b*) The preparation of pyrimidin-5-ols cannot be done directly by primary synthesis. However, primary synthesis of pyrimidin-5-yl ether is possible and the product may be converted subsequently into a pyrimidin-5-ol. For example, condensation of guanidine with ethyl 2-benzyloxy-2-formylacetate (**845**) gives 2-amino-5-benzyloxypyrimidin-4(3$H$)-one (**846**; R = CH₂Ph) which undergoes hydrogenolysis over palladium to afford the 5-hydroxy analogue (**846**; R = H) ⟨56JCS2124⟩. Similarly, 5-benzyloxypyrimidine-2,4(1$H$,3$H$)-dione (**847**; R = CH₂Ph) can be made by primary synthesis and then submitted to acidic hydrolysis to afford isobarbituric acid (**847**; R = H) ⟨64JCS1001⟩. Pyrimidines with amino, benzamido, methoxy or other groups in the 5-position may also be used as precursors for pyrimidin-5-ols ⟨63JCS5590⟩. The classical route to isobarbituric acid by reduction of nitrouracil and acidic hydrolysis of the 5-amino group ⟨25CB1685⟩ may be modified with advantage by treatment of the intermediate 5-aminouracil with nitrous acid ⟨65MI21302⟩.

Pyrimidin-5-ol is probably best made by converting 5-bromo- into 5-methoxy-pyrimidine ⟨58CB2832⟩ and then cleaving the ether with potassium hydroxide in ethylene glycol ⟨62CB803⟩. Direct hydrolysis of 5-bromopyrimidines to pyrimidin-5-ols appears to need very vigorous conditions but 5-bromo-1,3-dimethyluracil (**848**; R = Br) yields the 5-hydroxy analogue (**848**; R = OH) by simply boiling in aqueous sodium hydrogen carbonate ⟨59JA3786⟩.

The semi-direct 5-hydroxylation of pyrimidines by the Elbs persulfate reaction is summarized in Section 2.13.2.1.5$a$.

(**845**)    (**846**)    (**847**)    (**848**)

(*c*) The preparation of extranuclear hydroxypyrimidines can be done by a variety of methods. The more practical of these are summarized in the following examples. Primary synthesis: the condensation of thiourea with 2-$\beta$,$\gamma$-dihydroxypropylmalononitrile gives 4,6-diamino-5-(2′,3′-dihydroxypropyl)pyrimidine-2(1$H$)-thione (**849**) ⟨60JCS131⟩; from amino: 5-aminoethyl-2-ethylpyrimidin-4-amine may be made by primary synthesis and then converted into the hydroxymethyl analogue (**850**; R = Et) in 80% yield by treatment with nitrous acid ⟨64MI21302⟩; from halogeno: 5-bromomethyl-2-methylthiopyrimidin-4-amine undergoes hydrolysis in aqueous sodium carbonate to give the hydroxymethyl-pyrimidine (**850**; R = Me) ⟨60ZPC(322)173⟩; from ester: reduction of the ester (**851**; R = CO₂Et) with LAH gives the 6-$\beta$-hydroxyethylpyrimidine-2,4(1$H$,3$H$)-dione (**851**; R = CH₂OH) ⟨64LA(673)153⟩; from aldehydes or ketones: see Section 2.13.2.2.4$b$; hydroxymethylation: the insertion of an hydroxymethyl group at the 5-position is covered (Section 2.13.2.1.1$j$) but it can be done in many other ways too, as illustrated by the reaction of 4,6-dimethylpyrimidine-2(1$H$)-thione with ethylene chlorohydrin to give the 2-$\beta$-hydroxyethylthiopyrimidine (**852**) ⟨66IZV1613⟩.

(**849**)    (**850**)    (**851**)    (**852**)

(*d*) Most quinazolin-2(or 4)-ones are made by primary syntheses but a few may be made by metatheses ⟨67HC(24-1)69⟩. Quinazolin-5(6, 7 or 8)-ols are made either by the Riedel primary synthesis or from the corresponding methyl ethers, which are available by primary synthesis. For example, quinazolin-5-ol is best made by the Riedel technique but quinazolin-7-ol is made by cleavage of its methyl ether with aluminium chloride at 130 °C ⟨62JCS561⟩.

(e) The 2- and 4/6-alkoxypyrimidines are usually made from the corresponding pyrimidinones *via* the chloropyrimidines but the 5- or extranuclear alkoxypyrimidines are usually made by the Principal Synthesis. However, there are enough exceptions to justify some care in choosing a suitable route.

The transformation of halogeno- into alkoxy-pyrimidines is covered in Section 2.13.2.2.9c. In addition, it should be noted that 5-bromo- does give 5-methoxy-pyrimidine by vigorous treatment with methanolic methoxide at 120 °C ⟨58CB2832⟩; also that extra-nuclear halogeno substituents are extremely sensitive to alkoxide ion: 5-bromomethyl-6-methylpyrimidine-2,4(1H,3H)-dione (853; R = Br) is converted into its 5-methoxymethyl analogue (853; R = OMe) by simply refluxing in methanol ⟨64ZOB2159⟩.

The primary synthesis of alkoxypyrimidines is exemplified in the condensation of dimethyl malonate with O-methylurea in methanolic sodium methoxide at room temperature to give the 2-methoxypyrimidine (854) ⟨64M207⟩; in the condensation of diethyl phenoxymalonate with formamidine in ethanolic sodium methoxide to give the 5-phenoxypyrimidine (855) ⟨64ZOB1321⟩; and in the condensation of butyl 2,4-dimethoxyacetoacetate with thiourea to give 5-methoxy-6-methoxymethyl-2-thiouracil (856) ⟨58JA1664⟩.

(853)          (854)          (855)          (856)

Other ways of making alkoxypyrimidines include the following processes. O-Alkylation: this is seldom useful because even diazoalkanes tend to give mainly N-alkyl derivatives from pyrimidinones; however, it is applicable to pyrimidin-5-ols, to quinazolin-5(6,7 or 8)-ols and to extranuclear hydroxy compounds (Section 2.13.2.2.7d). Displacement of alkylthio, alkylsulfonyl and similar groups: this offers a route independent of hydroxy derivatives and the sequence, pyrimidinethione → alkylthiopyrimidine → alkylsulfonyl-pyrimidine → alkoxypyrimidine is quite practical, for example in making 4-propoxy-pyrimidine, for which the normal intermediate, 4-chloropyrimidine, is not easily made in quantity ⟨67JCS(C)568⟩. Transalkoxylation: the transformation of one alkoxy group into another by silver oxide in an appropriate alcohol is an occasionally useful reaction but it is confined to the 4/6-position and to additionally activated alkoxy groups ⟨70JCS(C)2661⟩; both 2- and 4-alkoxyquinazolines can undergo a similar reaction (using alcoholic sodium alkoxide) and no additional activation is needed ⟨64JMC812⟩.

(f) The N-alkylated pyrimidinones may be prepared in several ways. Many 1-alkyl-pyrimidin-2(1H)-ones and some N-alkylated pyrimidine-2,4(1H,3H)-diones are made by the Principal Synthesis or other primary synthesis, although such routes are not always unambiguous. For example, 1,1,3,3-tetraethoxy-2-methylpropane (857) and N-methylurea give 1,5-dimethylpyrimidin-2(1H)-one (858) without ambiguity ⟨68AJC243⟩ but diethyl ethoxymethylenemalonate (859) reacts as an aldehydo ester with N-methylurea to give a single product which could be either the pyrimidinone (860; R = H) or its 1-methyl isomer, but which proved to be the former on saponification and decarboxylation to the known 3-methyluracil; the same substrate (859) and N,N'-dimethylurea gives (unambiguously) the 1,3-dimethylpyrimidinone (860; R = Me) ⟨52JA4267⟩. In contrast, the primary synthesis of 1-methyl-5,6-dihydropyrimidine-2,4(1H,3H)-dione from methyl 3-(methylamino)pro-pionate and cyanic acid is indeed unambiguous and oxidation yields 1-methyluracil ⟨55JCS211⟩. The Shaw synthesis is also a source of N-alkylated pyrimidinones (Section 2.13.3.1.2e).

(857)          (858)          (859)          (860)

The most used route to N-alkylpyrimidinones is alkylation of the 'hydroxy' compound (Section 2.13.2.2.7d). A simple example is the treatment of pyrimidin-2(1H)-one (861;

R = H) with benzyl bromide in DMF to give the 1-benzyl derivative (**861**; R = CH$_2$Ph); prolonged treatment with a large excess of alkylating agent is to be avoided because of possible quaternization: pyrimidin-2(1*H*)-one, as the sodium salt, reacts with an excess of methyl iodide to give eventually the methiodide (**862**) ⟨66JOC3969⟩. Such alkylations have been reviewed in detail ⟨70HC(16-S1)269⟩.

The rearrangement of 2- or 4/6-alkoxypyrimidines, either by thermal means or by the Hilbert–Johnson reaction, also leads to *N*-alkylpyrimidinones: the former method is of little preparative value but the Hilbert–Johnson reaction is quite important. The processes are outlined in Section 2.13.2.2.7*l*.

The hydrolysis of imino- and thioxo-pyrimidines, which are alkylated on a ring-nitrogen, is a useful route to *N*-alkylpyrimidinones. For example, unlike pyrimidin-4(3*H*)-one which on methylation gives a mixture of its 1- and 3-methyl derivatives, pyrimidin-4-amine gives entirely the 1-methylated imine (**863**; X = NH) and this in alkali at 20 °C gives 1-methyl-pyrimidin-4(1*H*)-one (**863**; X = O) ⟨55JCS4035⟩. However, many pyrimidinamines are alkylated on the adjacent ring-nitrogen in which case Dimroth rearrangement usually outstrips hydrolysis in alkali. Thus, 1-methylpyrimidin-2(1*H*)-imine gives almost entirely the rearranged product, *N*-methylpyrimidin-2-amine, in alkali ⟨55JCS4035⟩ but 4-dimethylamino-1-methylpyrimidin-2(1*H*)-imine (**864**; X = NH), which rearranges much more slowly on account of electron-release by the dimethylamino group, gives mainly the pyrimidinone (**864**; X = O) at the expense of rearranged material ⟨63JCS1276⟩.

(861)  (862)  (863)  (864)  (865)  (866)

*N*-Alkylpyrimidinethiones cannot be converted into the corresponding pyrimidinones under regular hydrolytic conditions but they can be so converted oxidatively, although examples are few. Thus, 1,3,6-triphenyl-4-thiouracil (**865**; X = S) is converted by hydrogen peroxide in acetic acid into the corresponding uracil (**865**; X = O) ⟨64G606⟩ and the pyrimidinethione (**866**; X = S) into the pyrimidinone (**866**; X = O) by mercuric acetate ⟨64JOC1115⟩.

### 2.13.3.3.8 Synthesis of mercapto, alkylthio and related derivatives

(*a*) The best preparative method for mercaptopyrimidines varies from position to position. It is very easy to prepare pyrimidine-2-thiones by the Principal Synthesis using thiourea as one of the components but, although pyrimidine-4/6-thiones should be available by using dithioesters, the presparation of such starting materials is a problem and the reaction is not usable. The simplest examples from the Principal Synthesis are the condensations of 1,1,3,3-tetraethoxypropane with thiourea to give pyrimidine-2(1*H*)-thione (**867**) ⟨59JCS525⟩ or with *N*,*N*′-dimethylthiourea to give a 1,3-dimethyl-2-thioxo-1,2-dihydropyrimidinium salt (**868**) ⟨77JCS(P1)1862⟩. The Remfry–Hull synthesis can be used to make a few pyrimidine-4-thiones: thus, 2-ethyl-2-thiocarbamoylbutyramide (**869**) and acetyl chloride give the thione (**870**) ⟨64JCS3204⟩; other primary syntheses are of little use in this context.

(867)  (868)  (869)  (870)

Pyrimidine-4/6-thiones (and occasionally 2-thiones) can be made by thiation of the corresponding pyrimidinones providing that there are no sensitive groups present. The method is summarized and exemplified in Section 2.13.2.2.7*b*.

The thiolysis of halogenopyrimidines is a valuable route to pyrimidine-2(and 4/6)-thiones and to extranuclear thiols but it fails with 5-halogenopyrimidines. The process may be carried out directly with aqueous or alcoholic sodium hydrogen sulfide or indirectly *via* a thiouronium salt. The scope of each procedure is outlined in Section 2.13.2.2.9*f*.

It is sometimes convenient to change a 2- or 4/6-alkylthiopyrimidine into the corresponding thione by treatment with phosphorus pentasulfide, by ether cleavage with aluminum bromide or hydriodic acid, or by reductive cleavage (benzylthio only). For example, 3-methyl-2-methylthiopyrimidin-4(3*H*)-one (**871**) and phosphorus pentasulfide give 3-methylpyrimidine-2,4(1*H*,3*H*)-dithione (**872**) ⟨59MI21300⟩; 2-benzylthio-5-iodopyrimidin-4(3*H*)-one (**873**), in which the 2-mercapto group is so protected during iodination, gives 5-iodo-2-thiouracil (**874**) by heating with aluminum bromide in toluene ⟨54JA3666⟩; and reduction of 6-benzylthiopyrimidine-2,4,5-triamine in liquid ammonia with sodium gives the thione (**875**) ⟨64JOC3370⟩.

(**871**)          (**872**)          (**873**)          (**874**)          (**875**)          (**876**)

There are several minor routes to pyrimidinethiones from thiocyanato- or aminopyrimidines or by direct introduction of sulfur, but they are preparatively unimportant.

Special routes to the few known pyrimidine-5-thiols are illustrated in the chlorosulfonation of uracil to give 5-chlorosulfonylpyrimidine-2,4(1*H*,3*H*)-dione (**876**; R = SO$_2$Cl) which is reduced by an excess of zinc in cool sulfuric acid to give the thiol (**876**; R = SH) ⟨56JA401⟩; in the formation of 2-aminopyrimidine-5-thiol by reduction of the corresponding disulfide with dithionite ⟨64LA(675)151⟩; and, rather unsatisfactorily, by treatment of diazotized 5-aminouracil with sodium sulfide to give a dipyrimidinyl disulfide followed by reduction ⟨55JA960⟩.

The preparation of quinazoline-2(and 4)-thiones follows those of the corresponding pyrimidines ⟨67HC(24-1)270⟩ but there is at least one special primary synthesis for quinazoline-4(3*H*)-thiones, illustrated by the reaction of *o*-aminobenzonitrile with thioacetic acid at 110 °C to give 2-methylquinazoline-4(3*H*)-thione in 90% yield ⟨53JA675⟩.

(*b*) There are four methods for making alkylthiopyrimidines of which *S*-alkylation of pyrimidinethiones is by far the most important and usually the simplest (there are relatively few 5- or extranuclear alkylthiopyrimidines known). The usual technique is to shake with alkyl iodide in aqueous or alcoholic alkali but dimethylformamide dimethylacetal in toluene ⟨81AJC1729⟩ is also effective for methylations. The processes are discussed and exemplified in Section 2.13.2.2.8*b*, and an additional example is the conversion of the unstable 5-mercaptomethyluracil into its 5-methylthiomethyl and 5-ethylthiomethyl analogues, using alkyl halide ⟨66JMC97⟩.

Only 2-alkylthiopyrimidines are made by the Principal Synthesis, using *S*-alkylthiourea as one component. For example, ethoxymethylenemalononitrile (**877**) and *S*-benzylthiourea in aqueous acetone at 20 °C give 4-amino-2-benzylthiopyrimidine-5-carbonitrile (**878**) ⟨61JOC79⟩; and ethyl 2-allyl-2-formylacetate and *S*-methylthiourea in aqueous ethanolic alkali give 5-allyl-2-methylthiopyrimidin-4(3*H*)-one ⟨61JOC4425⟩.

The displacement of halogeno by alkylthio substituents is sometimes invaluable. For example, 5-methylthiopyrimidine is made by heating 5-bromopyrimidine with ethanolic sodium methanethiolate for 12 hours ⟨67JCS(C)568⟩ and 7-fluoro- gives 7-ethylthio-quinazoline likewise ⟨67JCS(B)449⟩. Another such transformation, which cannot be done *via* a mercapto intermediate because of the reducing conditions pertaining in sodium hydrogen sulfide solution, is that of 2-chloro-5-nitropyrimidin-4-amine (**879**; R = Cl) by sodium methanethiolate into the 2-methylthio analogue (**879**; R = SMe) ⟨54JCS2060⟩. The route is also useful in making arylthio derivatives which cannot be made by arylation of mercapto compounds: thus 2-chloro- gives 2-phenylthio-pyrimidine (**880**) by treatment with ethanolic sodium thiophenate at 80 °C ⟨67JCS(C)568⟩ and 2,4-dichloro-6-methylpyrimidine reacts with ethanolic potassium thiophenate to give, at 0 °C, 2-chloro-4-methyl-6-phenylthiopyrimidine (**881**; R = Cl) but at 100 °C, 4-methyl-2,6-bisphenylthiopyrimidine (**881**; R = SPh) ⟨63AP151⟩.

The direct introduction of an alkythio substituent into the 5-position of uracils and analogous pyrimidines is possible. The reagent, which is usually made quite easily *in situ*,

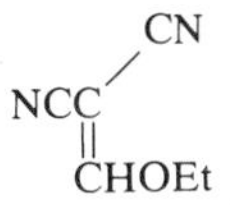

(877) 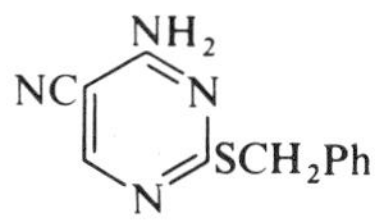    (878) 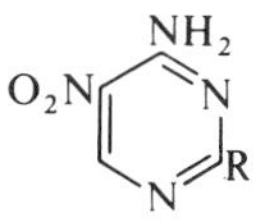    (879) 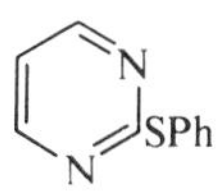    (880)

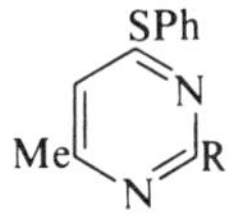

(881) 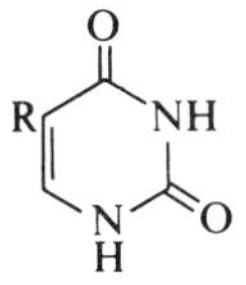    (882) 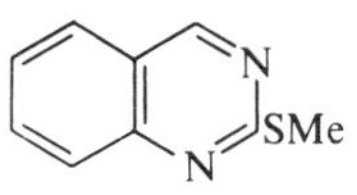    (883) 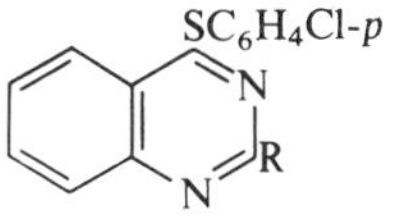    (884)

is methanesulfenyl chloride or a homologue. In this way, uracil (**882**; R = H) gives 5-methylthiopyrimidine)-2,4(1*H*,3*H*)-dione (**882**; R = SMe) in >70% yield and homologous derivatives likewise; in addition, benzeneselenenyl chloride gives the 5-phenylseleno analogue (**882**; R = SePh) ⟨79JHC567, *cf.* 70JCS(C)986⟩.

The 2- and 4-alkylthioquinazolines are formed by the second or third of the above methods as appropriate. For example, quinazoline-2(1*H*)-thione and methyl iodide/alkali give the thioether (**883**) and the 4-thioether is made similarly ⟨62JCS3129⟩; 2,4-dichloroquinazoline and sodium *p*-chlorothiophenate give 2-chloro-4-*p*-chlorophenylthio-(**884**; R = Cl) or 2,4-bis-*p*-chlorophenylthio-quinazoline (**884**; R = SC$_6$H$_4$Cl) according to conditions ⟨48JCS1766⟩.

(*c*) The preparation of dipyrimidinyl sulfides and disulfides can be done in several ways. The available routes to sulfides are illustrated in the following examples. 4-Methyl-6-phenylpyrimidine-2(1*H*)-thione (**885**) is converted into the sulfide (**886**), either by treatment with 2-chloro-4-methyl-6-phenylpyrimidine in boiling ethanol (>90% yield) or by treating the thione with copper bronze in boiling *p*-cymene (11% yield) ⟨67JCS(C)1204⟩. 5-Chloromethyl- and 5-mercaptomethyl-uracil combine (as above) to give the extranuclear sulfide (**887**) ⟨66JMC97⟩. When uracil is boiled with sulfur monochloride in formic acid, the sulfide (**888**) results in 70% yield; similar treatment with sulfur dichloride gives the same product (35%), but the reaction is confined to uracil or its 1- and/or 3-alkyl derivatives, with or without a 6-chloro, amino, hydroxy or alkyl substituent ⟨72AJC2275⟩. Occasionally, treatment of chloropyrimidines with sodium hydrogen sulfide ⟨65CCC3730⟩ or thiourea ⟨61CPB38⟩ does not give the expected thione but the corresponding sulfide; several explanations have been advanced but remain unproven. A few sulfides have been made by painstaking Principal Syntheses ⟨64LA(675)151⟩.

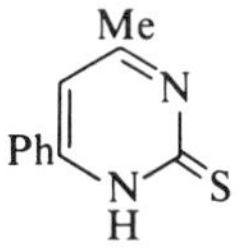

(885) 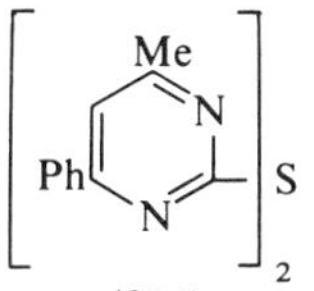    (886) 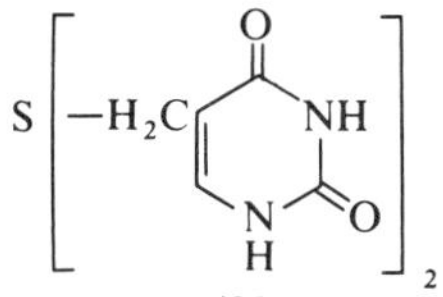    (887) 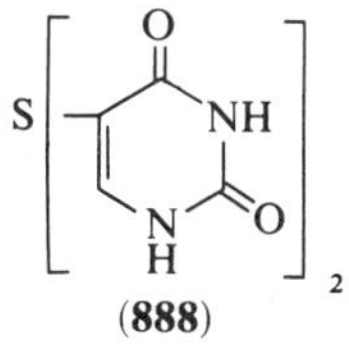    (888)

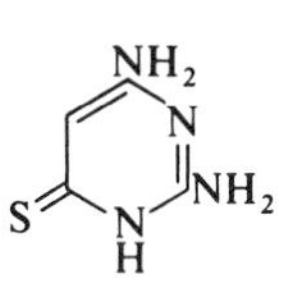

(889) 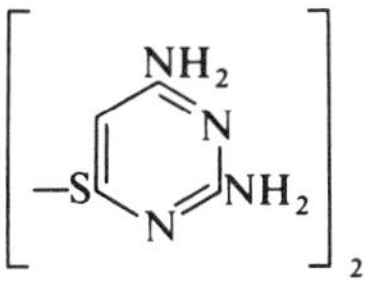    (890) 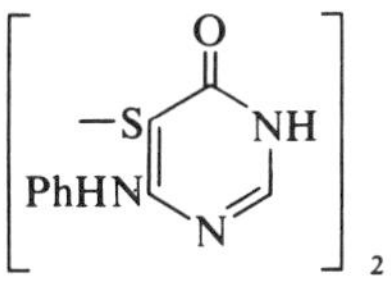    (891) 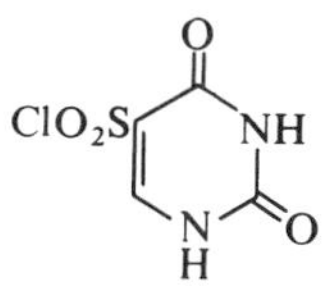    (892)

Most dipyrimidinyl disulfides are made by oxidation of the corresponding thione or thiol with bromine, iodine, air, nitrous acid, *etc.* For example, 2,6-diaminopyrimidine-4(3*H*)-thione (**889**) and iodine or nitrous acid gives the disulfide (**890**) ⟨64JMC792⟩ and 6-anilino-5-mercaptopyrimidin-4(3*H*)-one with iodine or air gives the disulfide (**891**) ⟨66JPR(32)26⟩. A few disulfides can be made by the Principal Synthesis ⟨64LA(675)151⟩, by partial reduction of uracil-5-sulfonyl chloride (**892**) ⟨56JA401⟩, by treatment of uracil-5-diazonium chloride with sodium disulfide (from sodium sulfide and sulfur) ⟨55JA960⟩, by treatment of 4-amino-5-bromouracil with boiling aqueous sodium disulfide ⟨56JCS917⟩ and by other procedures of minor importance.

(*d*) The preparation of pyrimidinesulfonic and pyrimidinesulfinic acids and their acid halides may be done in several ways, largely depending on position. The 5-sulfonation or chlorosulfonation of pyrimidines is covered in Section 2.13.2.1.1g but chlorosulfonation can also occur on aromatic rings attached to pyrimidine, especially if the 5-position is occupied: thus, 5-bromo-6-benzyluracil (**893**; R = H) reacts with chlorosulfonic acid to give the chlorosulfonyl derivative (**893**; R = SO$_2$Cl) ⟨67JMC316⟩. Quinazolin-4(3*H*)-one (**894**; R = H) chlorosulfonates at the 6-position to give the sulfonyl chloride (**894**; R = SO$_2$Cl) which yields the sulfonamide (**894**; R = SO$_2$NH$_2$) in 60% overall yield ⟨64MI21303⟩.

Oxidation of pyrimidine-2(or 4/6)-thiones (or their derived disulfides) can give sulfonic or sulfinic acids. Thus, treatment of 4,6-dimethylpyrimidine-2(1*H*)-thione with permanganate in aqueous ethanol at 20 °C gives potassium 4,6-dimethylpyrimidine-2-sulfonate (**895**) in 65% yield and other 2- and 4-sulfonates in both the pyrimidine ⟨72JCS(P1)522⟩ and quinazoline series ⟨72AJC2641⟩ are made similarly. Alkaline peroxide is probably less effective, for example, in oxidizing 4-amino-5-phenylpyrimidine-2(1*H*)-thione to the corresponding sulfonic acid ⟨62MI21300⟩. However, under controlled conditions it is the reagent of choice for converting 5,6-diamino-2-thiouracil (**896**) into the sulfinic acid (**897**) ⟨60JOC148⟩ or 6-aminopyrimidine-4(3*H*)-thione into 6-aminopyrimidine-4-sulfinic acid ⟨64JMC5⟩. Oxidation by chlorine converts pyrimidine-2(1*H*)-thione into pyrimidine-2-sulfonyl chloride (**898**; X = Cl) ⟨50JA4890⟩ and if the reaction is done in the presence of an excess of potassium hydrogen difluoride in aqueous methanol at −10 °C, the corresponding sulfonyl fluoride (**898**; X = F) is obtained in 90% yield ⟨72JCS(P1)522⟩. This and related sulfonyl fluorides are invaluable intermediates for, *inter alia*, the corresponding sulfonamides, *e.g.* (**898**; X = NH$_2$), or sulfonohydrazides, *e.g.* (**898**; X = NHNH$_2$).

(**893**)          (**894**)          (**895**)

(**896**)          (**897**)          (**898**)          (**899**)

Many pyrimidine-2(or 4/6)-sulfonic acids may be made by treatment of the corresponding chloropyrimidines with potassium sulfite. In simple cases, the acids are isolated as their salts but there are some practical difficulties in removing the last traces of inorganic salts from them; for this reason, the oxidative route is probably better. The process is illustrated in the conversion of 4-chloro-2,6-dimethylpyrimidine (**899**; R = Cl) into the potassium sulfonate (**899**; R = SO$_3$K) in >70% yield by heating in water containing fresh potassium sulfite dihydrate for 20 minutes ⟨71JCS(B)2214⟩; these conditions should be contrasted with those required for displacement of the deactivated chloro substituent in 4-chloropyrimidine-2,6-diamine, which needs heating in aqueous sodium sulfite at 145 °C for 25 hours to give sodium 2,6-diaminopyrimidine-4-sulfonate ⟨69JOC821⟩.

(*e*) Almost all pyrimidine sulfones and sulfoxides are made by oxidation of alkylthiopyrimidines; these processes are discussed and exemplified in Section 2.13.2.2.8*l*. Sulfones are occasionally made by three other methods, illustrated in the Principal Synthesis of 2-methyl-5-methylsulfonylpyrimidin-4(3*H*)-one (**901**) from ethyl 2-ethoxymethylene-2-methylsulfonylacetate (**900**) and acetamidine ⟨67MIP21300⟩; in the treatment of 2-chloropyrimidine with sodium benzenesulfinate to give 2-phenylsulfonylpyrimidine (**902**) ⟨67JCS(C)568⟩; and in the photo-Fries rearrangement of the ester, pyrimidin-4-yl methanesulfonate (**903**), into the 5-methylsulfonylpyrimidinone (**904**) in 60% yield ⟨68JCS(C)2367⟩. The

(**900**)               (**901**)               (**902**)

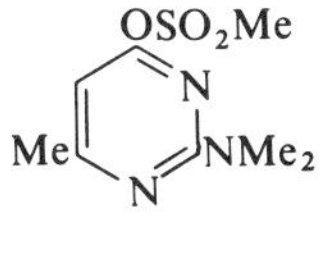

(903) 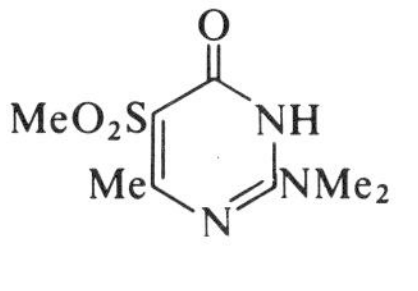  (904) 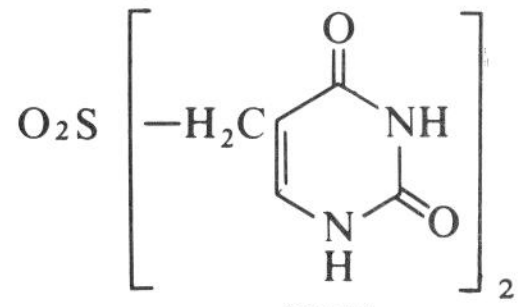  (905)

symmetrical sulfone, di(uracil-5-ylmethyl) sulfone (905) may be made by peroxide oxidation of the corresponding sulfide ⟨66JMC97⟩.

(*f*) The *N*-alkylated pyrimidinethiones may be formed by the Principal Synthesis using *N*-substituted thiourea as a component, by other primary syntheses or by thiation, usually of an *N*-alkylated pyrimidinone. These routes are exemplified in the condensation of ethoxymethyleneacetonitrile with *N*-methylthiourea to give 4-amino-1-methylpyrimidine-2(1*H*)-thione (906) of confirmed structure ⟨71JCS(C)2507⟩; in the reaction of ethyl β-anilino-crotonate with benzoyl isothiocyanate at room temperature to give the thione (907) ⟨64JOC1115⟩; and in the reaction of 3-methylpyrimidin-4(3*H*)-one with phosphorus pentasulfide in pyridine to give the corresponding thione (908) ⟨62JCS3129⟩. An additional unusual reaction is that of 4-dimethylamino-1-methyl-6-methylthiopyrimidinium iodide (909) with ethanolic sodium hydrogen sulfide to give 6-dimethylamino-3-methylpyrimidine-4(3*H*)-thione (910) ⟨65AJC199⟩.

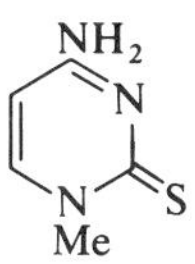

(906) 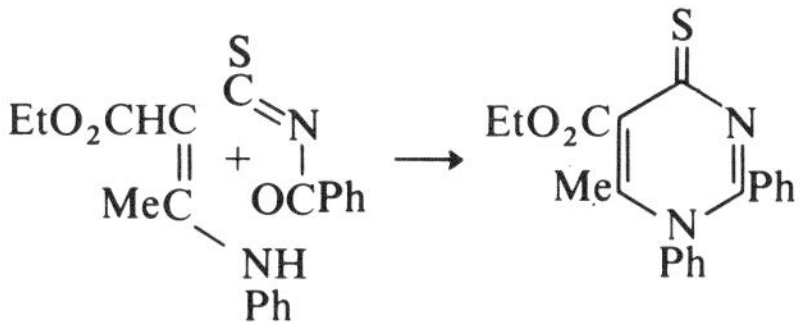  (907) 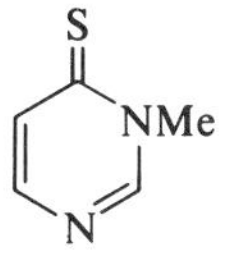  (908) 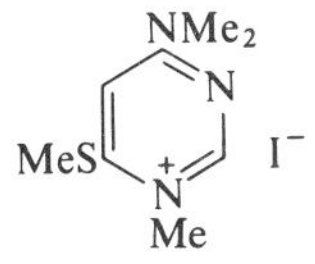  (909) 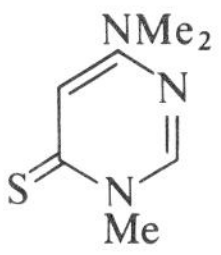  (910)

### 2.13.3.3.9 Synthesis of halogeno derivatives

(*a*) The preparation of halogeno- from 'hydoxy'-pyrimidines is the chief route to 2-, 4/6- and extranuclear chloropyrimidines and their bromo analogues. The reagent is usually the phosphoryl halide, phosphorus pentahalide, phosphorus trihalide or a mixture of two of these. For the extranuclear halogenopyrimidines, the corresponding alcohol may also be 'esterified' with thionyl chloride, hydrogen chloride, hydrogen bromide or hydrogen iodide. These regular transformations are discussed and exemplified in Section 2.13.2.2.7*a*. In addition, when the highly acidic conditions and elevated temperatures associated with the use of phosphorus halides are to be avoided, chloromethylene dimethylammonium chloride (ClCH=ŇMe₂ Cl⁻; made *in situ* from DMF and phosgene, thionyl chloride, *etc.*) may be used as an effective reagent in chloroform at room temperature. In this way, 5-benzyloxy-methyluracil (911) is converted into the dichloro analogue (912; R = Cl), which gives the dimethoxy analogue (912; R = OMe) in >70% yield overall ⟨66CCC1053⟩; however, the main use of the reagent is for such transformations in nucleoside chemistry.

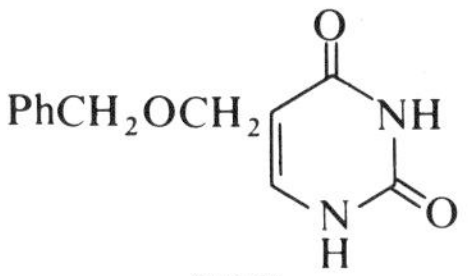

(911) 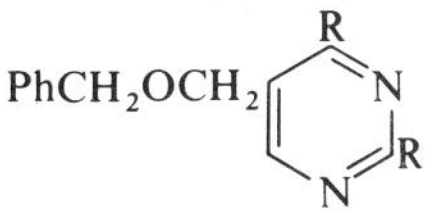  (912)

(*b*) Direct halogenation can produce 5- and extranuclear halogenopyrimidines. The reagents are usually elemental chlorine, bromine and, occasionally, fluorine under various conditions, an *N*-halogenosuccinimide, sulfuryl chloride, or iodine monochloride. Fluorination is rather special. 5-Halogenation is discussed in Section 2.13.2.1.1*e* and extranuclear halogenation in Section 2.13.2.2.2*c*.

(*c*) The conversion of amino- into halogeno-pyrimidines by diazotization and treatment with an excess of halide ion or with cuprous halide is used for pyrimidines with a chloro or bromo substituent at any position and also for 5-iodo and 2-fluoro-pyrimidines; for the last mentioned, fluoroboric acid may be used ⟨74JCS(P2)204⟩ but yields are not always good. The reaction is discussed more fully in Section 2.13.2.2.6*b*.

(*d*) It is possible to make some 5- and extranuclear halogenopyrimidines by primary syntheses. Use of the Principal Synthesis is illustrated in the condensation of ethyl fluoroformylacetate with *S*-methylthiourea to give the 5-fluoropyrimidinone (**913**) ⟨65ZOB1303⟩, of ethyl (dichloroaceto)acetate with *S*-methylthiourea to give the dichloromethylpyrimidinone (**914**) ⟨65MI21303⟩, and of chloromalondialdehyde with benzamidine to give 5-chloro-2-phenylpyrimidine ⟨51JCS2323⟩. The 5-, 6- and 7-fluoroquinazolin-4(3*H*)-ones are also made by primary synthesis ⟨67JCS(B)449⟩.

(*e*) Extranuclear halogenopyrimidines may be made by cleavage of the corresponding ethers with a hydrogen halide: 5-methoxymethyl- (**915**; R = OMe) gives 5-bromomethylpyrimidine (**915**; R = Br) with hydrobromic acid ⟨65JHC(2)1⟩ and 5-methoxymethyl-2-methylpyrimidin-4-amine (**916**; R = OMe) gives the iodomethyl analogue (**916**; R = I) with hydriodic acid ⟨64CPB558⟩.

(*f*) Transhalogenation is of great preparative importance for iodo- and fluoro-pyrimidines and to a lesser extent for bromo- and chloro-pyrimidines. Interchange is not usually attempted in the 5-position and the facile transformation of 5-bromo-6-hydroxypyrimidin-4(3*H*)-one (**917**; X = Br) into the 5-chloro analogue (**917**; X = Cl), by warming with hydrogen chloride in DMF, is all the more noteworthy for being the best route to the chloro compound ⟨64ZOB3851⟩.

(913)        (914)        (915)        (916)

(917)        (918)        (919)        (920)

Apart from 5- and some extranuclear iodopyrimidines, which are made by other routes, all iodopyrimidines are made by interchange from their chloro or occasionally their bromo analogues. Thus treatment of 2-chloro- or 2-bromo-pyrimidine with hydriodic acid at 0–5 °C gives 2-iodopyrimidine in *ca.* 50% yield; at higher temperatures, dehalogenation occurs ⟨73AJC443⟩. Treatment of 4,6-dichloropyrimidine with sodium iodide and hydriodic acid in acetone gives the 4,6-diiodo analogue ⟨62JMC1335⟩. 2,4-Dichloro-5-chloromethylpyrimidine and sodium iodide in acetone gives an excellent yield of the 5-iodomethyl analogue (**918**) without affecting the nuclear chloro substituents ⟨66LA(692)119⟩.

In view of the high price and other difficulties associated with the use of phosphoryl bromide, the transhalogenation of 2-chloro- and 2-chloro-4,6-dimethyl-pyrimidine (**919**) by treatment with inexpensive phosphorus tribromide to give their bromo analogues in >50% yield, assumes potential importance as a general method ⟨67JCS(C)1204⟩.

Some 2- and 4/6-fluoropyrimidines are best made from their chloro analogues by treatment with sulfur tetrafluoride or with potassium, cesium or silver fluoride, usually at an elevated temperature. The following are examples of fluoropyrimidines so made: 2,4,6-trifluoropyrimidine, using silver fluoride/90 °C (70%) ⟨62JOC2580⟩ or potassium fluoride/300 °C (90%) ⟨66USP3280124⟩; 2,4- and 4,6-difluoropyrimidine, using sulfur tetrafluoride/150 °C (*ca.* 70%) ⟨60JA5107⟩; 2,4-difluoro-6-methylpyrimidine, using potassium fluoride in DMF/170 °C (*ca.* 60%) ⟨67YZ1315⟩; and 4-fluoro-2-methoxypyrimidine, using cesium fluoride in *N*-methylpyrrolidinone/90 °C ⟨73MI21301⟩.

(*g*) Some 2- and 4/6-fluoropyrimidines are made by trimethylammonio displacement. Thus, the trimethylammoniopyrimidine (**920**) reacts with potassium fluoride in ethylene glycol to give 4-fluoro-2,6-dimethoxypyrimidine in good yield ⟨61JOC3392⟩; 2-isopropylpyrimidin-4-yl trimethylammonium chloride reacts at 20 °C with potassium hydrogen difluoride to give 4-fluoro-2-isopropylpyrimidine (81%) ⟨78AJC1391⟩; and trimethyl 4-phenylpyrimidin-2-ylammonium chloride reacts with potassium fluoride in DMSO at 60 °C to give 2-fluoro-4-phenylpyrimidine ⟨74RTC111⟩.

### 2.13.3.3.10 Synthesis of N-oxides and derivatives

Pyrimidine *N*-oxides may be made directly or *via* their *N*-alkoxy analogues by means of the Principal Synthesis or other primary synthesis. The alternative route is peroxide oxidation of the parent pyrimidine but this can lead to a mixture of 1- and 3-oxides if the substrate is unsymmetrical about the 2,5-axis of the molecule.

Direct primary synthesis is represented by the condensation of acetylacetone with *N*-hydroxyurea to give in poor yield the pyrimidine *N*-oxide usually written in the tautomeric form (**921**; R = H); for this product, indirect synthesis is better: condensation of acetyl-acetone with *N*-benzyloxy- or *N*-methoxy-urea gives the intermediate (**921**; R = CH₂Ph or Me), both of which give the required *N*-hydroxy compound (**921**; R = H) in excellent yield on acidic hydrolysis ⟨67T353⟩. Another (indirect) Principal Synthesis is the condensation of *N*-benzyloxyurea with diethyl ethylmalonate to give 1-benzyloxy-5-ethylbarbituric acid (**922**; R = CH₂Ph), from which hydrogenolysis over platinum gives 5-ethyl-1-hydroxybar-bituric acid (**922**; R = H) ⟨78AJC2517⟩. The *N*-oxide, 5-fluoro-1-hydroxypyrimidine-2,4(1*H*,3*H*)-dione (**924**; R = H) may be made in three ways: (i) an ambiguous Principal Synthesis from ethyl 2-fluoro-2-methoxymethyleneacetate and *N*-benzyloxyurea gives the benzyloxy intermediate (**924**; R = CH₂Ph or an isomer) which undergoes hydrogenolysis over palladium to give the *N*-hydroxy product (**924**; R = H) or its 3-hydroxy isomer; (ii) 5-fluoro-2,4-dimethoxypyrimidine (**925**) undergoes both *N*-oxidation and hydrolysis by *m*-chloroperbenzoic acid in acetic acid at 70 °C to give an *N*-hydroxy product, identical with that above; and (iii) an unambiguous primary synthesis from 3-benzyloxyamino-2-fluoroacrylamide (**923**) and oxalyl chloride which give a benzyloxy intermediate and an *N*-hydroxy product respectively, identical with those above and now of proven structure (**924**; R = CH₂Ph or H) ⟨68M847⟩.

(**921**)   (**922**)   (**923**)   (**924**)   (**925**)

(**926**)   (**927**)   (**928**)   (**929**)   (**930**)

Primary syntheses of pyrimidine 1,3-dioxide and several derivatives proceed *via* tetrahy-dropyrimidine dioxides: for example, malondialdoxime (**926**) reacts with formaldehyde to give 1-hydroxytetrahydropyrimidine 3-oxide (**927**) (or an acyclic isomeric nitrone) which is dehydrogenated by manganese dioxide to give pyrimidine 1,3-dioxide (**928**) ⟨77KGS259⟩. Quinazoline 3-oxides are usually made by primary synthesis: heating *o*-aminobenzaldoxime (**929**) with triethyl orthoformate gives the parent 3-oxide (**930**) ⟨57YZ507⟩.

The oxidative route to *N*-oxides is illustrated in the conversion of pyrimidine ⟨59JCS525⟩ and some alkyl derivatives ⟨67YZ1096⟩ into their respective *N*-oxides, by using hydrogen peroxide in glacial acetic acid as oxidant. 4-Methylpyrimidine likewise ⟨55CPB175⟩ gives a product which proved to be a separable mixture of the 1- (**931**) and 3-oxide (**932**) ⟨64TL19⟩. It must be kept in mind that during *N*-oxidation, other reactions may take place. For

(**931**)   (**932**)   (**933**)   (**934**)   (**935**)   (**936**)

example, 6-phenylthiopyrimidine-2,4-diamine and hydrogen peroxide/TFA give 6-phenyl-sulfonylpyrimidine-2,4-diamine 3-oxide (**933**) in 60% yield; likewise 6-anilino-5-nitrosopyrimidine-2,4-diamine (**934**) and peroxide/TFA afford the corresponding nitropyrimidine 3-oxide (**935**) ⟨79AJC2049⟩. Further examples of *N*-oxidation are given in Section 2.13.2.1.5*d*.

The *O*-alkyl derivatives of those *N*-oxides, which exist partly or entirely as *N*-hydroxy tautomers, may be made by primary synthesis (as above) or by alkylation. Thus, 5,5-diethyl-1-hydroxybarbituric acid (**936**; R = H) with methyl iodide/sodium ethoxide gives the 1-methoxy derivative (**936**; R = Me) or with benzenesulfonyl chloride/ethoxide it gives the alkylated derivative (**936**; R = PhSO$_2$) ⟨78AJC2517⟩.

## 2.13.4 SOME IMPORTANT PYRIMIDINES

For present purposes, pyrimidines are considered to be 'important' if they occur naturally as such or as part of a natural molecule from which the pyrimidine can be removed easily (fused pyrimidines, *e.g.* xanthine, are therefore excluded); if they are used as drugs; if they are used as agricultural chemicals; or if they have been given a generally accepted special name. Although a few pyrimidines are used as industrial chemicals for other purposes, e.g. in photography, these have assumed no great importance.

### 2.13.4.1 Naturally Occurring Pyrimidines

The subject of pyrimidine nucleosides and nucleotides cannot be discussed in an article of this nature. Because of the very rapid progress in such research, even specialized reviews become partly outdated in a few years. Most aspects of this vast field are reviewed in the following works: general introduction and broad principles ⟨B-66MI21302⟩; detailed general procedures ⟨B-71MI21300, B-66MI21303⟩; synthetic procedures and physical properties ⟨B-68MI21307, B-73MI21302, B-78MI21302⟩; and year by year progress in all aspects ⟨B-81MI21300⟩.

#### 2.13.4.1.1 *Uracil [pyrimidine-2,4(1H, 3H)-dione]*

Uracil was first isolated from the hydrolysis of herring sperm in 1900 and shortly afterwards it was obtained also from bovine thymus or spleen and from wheat germ. Its structure (**937**; X = O) was proven by synthesis in 1901. The controlled hydrolysis of such materials containing ribonucleic acids first gives nucleotides, which are phosphoric acid esters of uridine (1-β-D-ribofuranosidouracil) and usually known as uridylic acids or uridine-*x*-monophosphates. The next stage is the removal of the phosphate residue(s) to give uridine and finally this glycoside undergoes hydrolysis to uracil.

The best direct synthetic route to uracil is probably the classical procedure from malic acid and urea in concentrated sulfuric acid ⟨26JA2379⟩, despite efforts to use maleic acid, urea and polyphosphoric acid ⟨71S154⟩ or propiolic acid, urea and a little concentrated sulfuric acid ⟨77JOC2185⟩ to achieve the same result. However, the most convenient source (apart from purchase) is to convert 2-thiouracil (**937**; X = S) into uracil by boiling with aqueous chloroacetic acid ⟨52MI21300⟩ or perhaps by oxidation with DMSO in strong sulfuric acid ⟨74S491⟩.

Uracil is an excellent starting material for all manner of simple pyrimidines. For example, the sequence uracil (**937**; X = O) → pyrimidine-4,5-diamine (**942**), not only affords an invaluable substrate for pteridine and purine syntheses, but also the intermediate pyrimidines (**938–941**), each with many other synthetic uses ⟨52MI21300⟩. As well as the above nitration to 5-nitrouracil (**938**), uracil undergoes 5-halogenation to 5-mono- or 5,5-di-halogenouracil according to conditions (Section 2.13.2.1.1*e*), conversion into 2,4-dichloropyrimidine by phosphoryl chloride (Section 2.13.2.2.7*a*), alkylation to 1-alkyl- or 1,3-dialkyl-uracil according to the reagent (Section 2.13.2.2.7*d*), reduction to 5,6-dihydrouracil (**943**) (Section 2.13.2.1.5*e*), thiation by phosphorus pentasulfide to 4-thiouracil (Section 2.13.2.2.7*b*), *O*-trimethylsilylation (Section 2.13.2.2.7*e*), aminolysis to pyrimidine-2,4-diamines using appropriate phosphoramides under vigorous conditions (Section 2.13.2.2.7*h*), and the

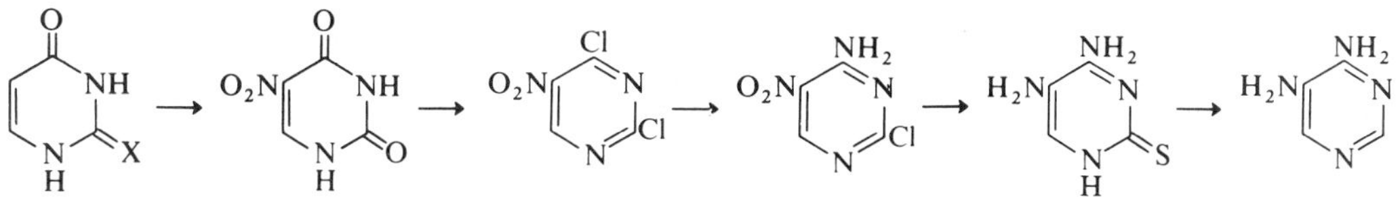

(937)   (938)   (939)   (940)   (941)   (942)

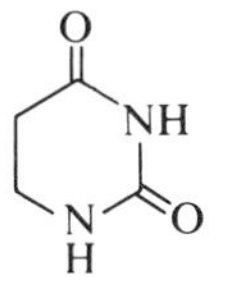

(943) 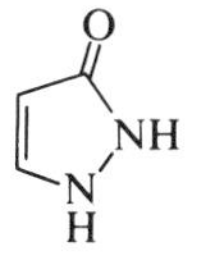   (944)

biologically important photohydration and photodimerization reactions (Section 2.13.2.1.4); also some reactions of relatively minor importance. Uracil may be detected by the non-specific Wheeler–Johnson colour test, but a better way to detect uracil, thymine or cytosine on chromatograms is to spray with phenol and sodium hypochlorite to give a bright blue spot ⟨56CI(L)1312⟩.

Uracil reacts with hydrazine to give pyrazol-3(2*H*)-one (**944**) and urea; *N*-methyl- and *N,N'*-dimethyl-hydrazine behave similarly to give the 2-methyl- and 1,2-dimethyl derivatives. The reactions of hydrazines with uridine and related nucleosides and nucleotides is well studied ⟨67JCS(C)1528⟩. The tautomerism and predominant form of uracil are discussed in Section 2.13.1.8.4.

### 2.13.4.1.2 Thymine [5-methylpyrimidine-2,4(1H,3H)-dione]

Thymine was isolated from hydrolyzates of bovine thymus or spleen in 1893, several years before uracil, but it was not made synthetically until 1901. Unlike uracil, it comes not from ribonucleic but from deoxyribonucleic acids *via* thymidine (3-D-2'-deoxyribofuranosidothymine).

The best laboratory synthesis of thymine (**947**) is probably from 3-methylmalic acid (**945**) which gives 2-formylpropionic acid (**946**; R = H) *in situ* by decarboxylation and oxidation in fuming sulfuric acid prior to condensation with urea ⟨46JA912⟩; a similar method from ethyl 2-formylpropionate (**946**; R = Et) is also described ⟨68IZV918⟩.

Having a 5-methyl group, thymine is not nitrated or halogenated normally, but with aqueous bromine it does give the dihydropyrimidine (**948**) ⟨25JBC(64)233⟩; its other reactions parallel those of uracil although its behavior on irradiation is somewhat different (Section 2.13.2.1.4).

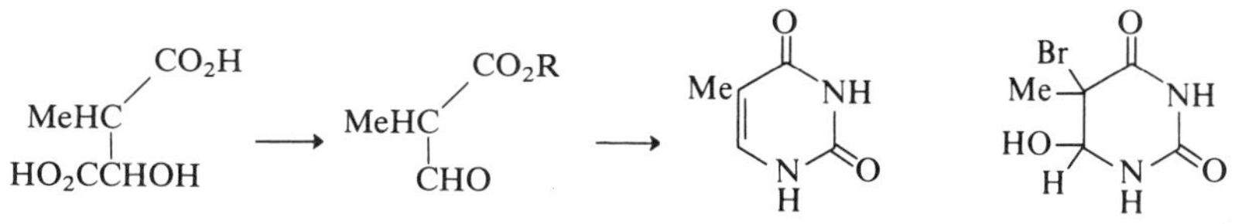

(945)   (946)   (947)   (948)

### 2.13.4.1.3 Vicine and divicine [2,6-diamino-5-hydroxypyrimidin-4(3H)-one]

Vicine was first isolated from the seeds of *Vicia sativa* in 1870 but only in 1896 was it recognized as the glycoside of a new base, 'divicine'. By 1911, this base was thought to be a pyrimidine and two leading workers suggested 5,6-diaminopyrimidine-2,4-dione and 2,5-diamino-6-hydroxypyrimidin-4-one, respectively, as its structure. Only in 1953 was the evidence re-evaluated in the light of new experimental data to show that divicine is really a third isomer, 2,6-diamino-5-hydroxypyrimidin-4(3*H*)-one (**950**; R = NH₂) and that vicine is its 5-β-D-glucopyranoside ⟨53BBA(12)462⟩.

Two interesting syntheses of the base followed in 1956: the first involves a Principal Synthesis from ethyl 2-cyano-2-(tetrahydropyran-2'-yloxy)acetate and guanidine to give the tetrahydropyranyloxypyrimidine (**949**) which undergoes gentle acidic hydrolysis to divicine (**950**; R = NH$_2$) ⟨56JCS2124⟩; the second begins with a Principal Synthesis of 2-amino-5-benzyloxypyrimidin-4(3H)-one (**951**) and its conversion into the 5-hydroxy analogue (**950**; R = H) which undergoes 6-nitrosation to the derivative (**950**; R = NO) and subsequent reduction to divicine (**950**; R = NH$_2$) ⟨56CI(L)1453, 64JCS1001⟩. However, the best synthesis is probably that from the readily available 2,6-diaminopyrimidin-4(3H)-one by an Elbs persulfate reaction ⟨73IZV2363⟩.

(949)            (950)            (951)            (952)

### 2.13.4.1.4 *Convicine and isouramil [6-amino-5-hydroxypyrimidine-2,4(1H,3H)-dione]*

Convicine was isolated from *Vicia faba* in 1881 but only in 1932 was its aglycone suspected of being 'isouramil' (**952**), a substance first synthesized in the same year and made more satisfactorily later from ethyl 2-cyano-2-(tetrahydropyran-2'-yloxy)acetate and urea followed by hydrolytic removal of the protecting group ⟨56JCS2124⟩. In 1968 isouramil was made by a third route, involving reduction of 5-hydroxy-6-nitrosopyrimidine-2,4(1H,3H)-dione, and then compared carefully with the aglycone derived from convicine: the natural base is indeed isouramil (**952**) and convicine is its 5-β-D-glucopyranoside ⟨68JCS(C)496⟩. The aglycone is one of the active substances which induce 'favism' (an acute haemolytic crisis) in some individuals who eat the favia bean.

### 2.13.4.1.5 *Cytosine [4-aminopyrimidin-2(1H)-one]*

Cytosine was isolated from hydrolysis of calf thymus in 1894 and by 1903 its structure was known and it had been synthesized from 2-ethylthiopyrimidin-4(3H)-one. The acid hydrolysis of ribonucleic acid gives nucleotides, among which are two cytidylic acids, 2'- and 3'-phosphates of cytidine; further hydrolysis gives cytidine itself, *i.e.* the 1-β-D-ribofuranoside of cytosine, and thence cytosine. The deoxyribonucleic acids likewise yield deoxyribonucleotides, including cytosine deoxyribose-5'-phosphate, from which the phosphate may be removed to give cytosine deoxyriboside and thence cytosine.

(953)        (954)        (955)        (956)        (957)        (958)        (959)

There are several good synthetic routes to cytosine. Probably the best involves conversion of uracil into 2,6-dichloropyrimidine (**953**) and then monoaminolysis into a mixture of the chloropyrimidinamines (**954**; R = Cl) and (**955**; R = Cl) which is treated with methanolic sodium methoxide to give a mixture of the methoxypyrimidinamines (**954**; R = OMe) and (**955**; R = OMe). The latter component of this mixture is easily removed by its solubility in dioxane and the remaining methoxypyrimidinamine (**954**; R = OMe) is submitted to acid hydrolysis to give cytosine (**956**; X = O); the dioxane-soluble isomer likewise gives isocytosine (**957**) ⟨57ZOB2113⟩. Another good route begins with the conversion of 2-thiouracil (**958**; X = O) into dithiouracil (**958**; X = S) and thence by aminolysis in aqueous ammonia into 2-thiocytosine (**956**; X = S) which on heating in aqueous chloroacetic acid gives cytosine (**956**; X = O) ⟨B-68MI21301⟩; the last step may also be done by warming the thiocytosine in DMSO containing a little concentrated sulfuric acid ⟨74S491⟩. There is also a well-described Principal Synthesis of cytosine, from 3,3-diethoxypropionitrile and urea, which is especially useful for making radioactive tagged material ⟨65JCS1515⟩ as well as a primary synthesis *via* the dehydrochlorination of 5-chloro-5,6-dihydrocytosine (**959**) which is of more doubtful utility ⟨70JHC527⟩.

Cytosine undergoes 5-nitration normally, but although bromination can give 5-bromocytosine, it can also give one or more 5,5-dibromodihydrocytosines, according to conditions. Acetylation of cytosine gives the 4-acetamido analogue but methylation gives several products. Cytosine undergoes an indirect transamination with hydroxylamine to give eventually 4-hydroxyaminopyrimidin-2(1*H*)-one ⟨65JCS208⟩. Although methylhydrazine and *N*,*N*'-dimethylhydrazine with cytosine give the expected 4-(1'-methylhydrazino)- and 4-(1',2'-dimethylhydrazino)-pyrimidin-2(1*H*)-one, respectively, hydrazine itself yields 4-hydrazinopyrimidin-2(1*H*)-one only at low temperatures and, according to conditions, it is mixed with substantial amounts of pyrazol-3-amine (**960**) the hydrazine (**961**) and azole (**962**) ⟨67JCS(C)1528, 65LA(686)134⟩. Cytosine with *m*-chloroperbenzoic acid gives the 3-*N*-oxide (**963**) ⟨65JOC2766⟩, elsewhere proven in structure by unambiguous synthesis. The irradiation of cytosine is not as well documented as those of uracil and thymine; leading references are given in Section 2.13.2.1.4.

(960)   (961)   (962)   (963)

### 2.13.4.1.6 5-Methylcytosine

5-Methylcytosine (**964**; X = O) was synthesized in 1901 and its isolation from hydrolyzates of tubercule bacilli was reported in 1925. However, this was later shown to be incorrect and only about 1950 was it isolated by hydrolysis of the deoxyribonucleotide fractions from thymus, wheat germ and other sources ⟨50MI21302⟩. Nucleotides and a nucleoside of 5-methylcytosine are known.

Probably the best synthesis of 5-methylcytosine is *via* its 2-thio analogue (**964**; X = S) ⟨49JBC(177)357⟩ but the following briefly-described route may be preferable: 2,4-dimethoxy-5-methylpyrimidine (**965**) with acetyl chloride gives the 1-acetyl derivative (**966**) which on aminolysis in aqueous ammonia gives 5-methylcytosine (**964**; X = O) in good overall yield ⟨68TL2171⟩.

(964)   (965)   (966)   (967)

### 2.13.4.1.7 5-Hydroxymethylcytosine

5-Hydroxymethylcytosine (**967**) was isolated only in 1952 from the T-even bacteriophages of *Escherichia coli*, in which it occurs instead of cytosine in the 2-deoxyribonucleic acid ⟨65MI21304⟩. Of several syntheses described, the most convenient is probably that beginning with ethyl 4-amino-2-methylthiopyrimidine-5-carboxylate which is reduced by LAH to 4-amino-2-methylthiopyrimidin-5-ylmethanol followed by hydrolysis to 5-hydroxymethylcytosine (**967**) ⟨B-68MI21302, B-68MI21306⟩.

### 2.13.4.1.8 Orotic acid (2,6-dioxo-1,2,3,6-tetrahydropyrimidine-4-carboxylic acid)

Orotic acid (**971**) has a chequered history. It was isolated in 1905 from the whey of cows' milk in Italy and it was subsequently synthesized in the United States in 1907. However, the workers involved were discouraged by some difference in melting points and no direct comparison of specimens was ever made. To make matters worse, the same laboratories prepared the isomeric 5-hydroxy-2-oxo-1,2-dihydropyrimidine-4-carboxylic acid and announced it as orotic acid, again without any direct comparison. Only in 1930 did a German worker actually compare directly natural and the original synthetic orotic acid, thereby showing them to be identical ⟨30CB1000⟩.

Orotic acid is the key substance in the biosynthesis of probably all naturally occurring pyrimidines. In brief, ammonia and carbon dioxide combine in the presence of adenosine triphosphate to give carbamoyl dihydrogen phosphate ($H_2NCOOPO_3H_2$) which reacts enzymically with L-aspartic acid to yield, after loss of phosphoric acid, ureidosuccinic acid (**969**), in enzymic equilibria with both 5-carboxymethylhydantoin (**968**) and dihydroorotic acid (**970**). Subsequent oxidation by dihydroorotate dehydrogenase gives orotic acid (**971**). Its utilization in making other pyrimidines is exemplified in its combination with ribose-5-phosphate-1-pyrophosphate in the presence of orotidylic acid pyrophosphorylase to give the nucleotide, orotidylic acid, which is then reversibly decarboxylated by orotidylic decarboxylase to yield uridine-5′-phosphate and so on ⟨57MI21301⟩.

There are at least eight syntheses of orotic acid in the literature. The most practical in the laboratory is that involving the condensation of diethyl oxalacetate (**972**) with S-methylthiourea to give 2-methylthio-6-oxo-1,6-dihydropyrimidine-4-carboxylic acid (**973**) which undergoes either direct acidic hydrolysis or a less smelly oxidative hydrolysis, *via* the unisolated sulfone (**974**), to afford orotic acid (**971**) ⟨B-68MI21303⟩.

Orotic acid undergoes 5-nitration, 5-bromination in hydrobromic acid with peroxide, 5,5-dibromination following decarboxylation in bromine water, esterification, methylation (rather complicated), conversion into its acid chloride (containing some anhydride) by treatment with thionyl chloride, and conversion into 2,6-dichloropyrimidine-4-carboxylic acid by phosphoryl chloride ⟨62HC(16)422⟩.

### 2.13.4.1.9 Willardiine

In 1959 a new non-protein L-$\alpha$-amino acid was isolated from the seeds of *Acacia willardiana* and later from other species of *Acacia*; it proved to be 1-$\beta$-amino-$\beta$-carboxyethyluracil (**977**) ⟨59ZPC(316)164⟩. The structure was confirmed by at least four syntheses in the next few years. The most important involves a Shaw synthesis (Section 2.13.3.1.2*e*) of the acetal (**975**) and hydrolysis to the aldyhyde (**976**) followed by a Strecker reaction (potassium cyanide, ammonia and ammonium chloride) to give DL-willardiine (**977**); after resolution, the L-isomer was identical with natural material ⟨62JCS583⟩. Although not unambiguous, a Principal Synthesis from the ureido acid (**978**) and ethyl formylacetate is the most direct route ⟨64ZOB407⟩.

### 2.13.4.1.10 Lathyrine or tingitanin

Another non-protein $\alpha$-amino acid was isolated from the seeds of *Lathyrus tingitanus* in 1961; it has been called both lathyrine and tingitanin (**982**) ⟨62MI21301⟩. Its structure rested first on spectral and ionization constant data which were later confirmed by oxidation to 2-aminopyrimidine-4-carboxylic acid and a synthesis involving condensation of 2-diacetyl-amino-4-methylpyrimidine (**979**) with diethyl oxalate to give the keto ester (**980**; X = O) followed by conversion into the oxime (**980**; X = NOH), hydrolysis to the carboxylic acid (**981**) and final stannous chloride reduction to lathyrine (**982**) ⟨65JOC115⟩. Two other syntheses follow broadly similar lines ⟨64ZOB3506⟩ and a fourth is rather different ⟨65LA(684)209⟩.

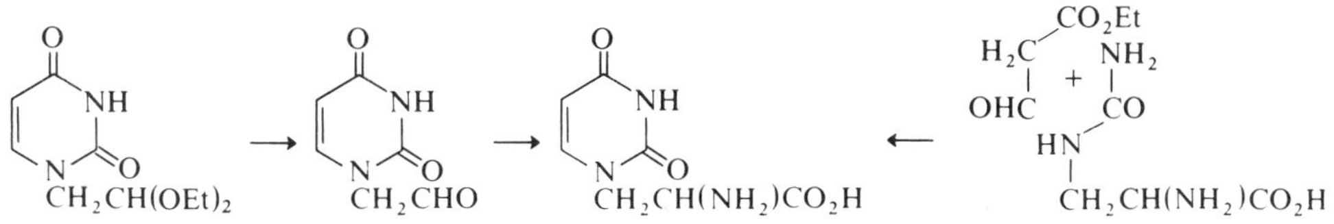

(975) → (976) → (977) ← (978)

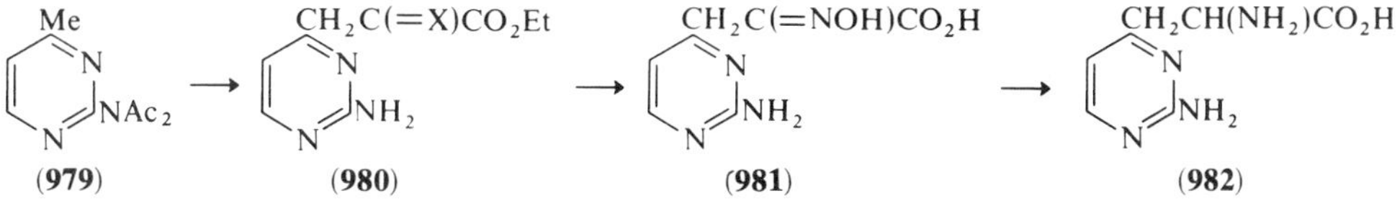

(979) → (980) → (981) → (982)

### 2.13.4.1.11 Some pyrimidine antibiotics

The simplest pyrimidine antibiotic is bacimethrin, 5-hydroxymethyl-2-methoxypyrimidin-4-amine (**985**), which was isolated in 1961 from *Bacillus megatherium* and is active against several yeasts and bacteria *in vitro* as well as against staphylococcal infections *in vivo*; it has some anticarcinoma activity in mice 〈69MI21301〉. It may be synthesized by LAH reduction of ethyl 4-amino-2-methoxypyrimidine-5-carboxylate (**984**) which may be made by primary synthesis in poor yield, or better, from the sulfone (**983**) 〈B-68MI21304〉.

The antibiotic blasticidin S was isolated from *Streptomyces griseochromogenes* about 1958 and its structure (**986**) was confirmed only in 1966 by an X-ray study 〈66BCJ1091〉. It is active against certain fungal plant pathogens and may be used to protect growing rice from 'rice blast disease'; it is toxic to animals. Of rather similar structure is the cytosine derivative, gougerotin, from *Streptomyces gougerotii*. This antibiotic was isolated in 1962 and is active against mycobacteria as well as several Gram-positive and Gram-negative bacteria. Structural studies and chemistry are reviewed 〈64MI21304〉.

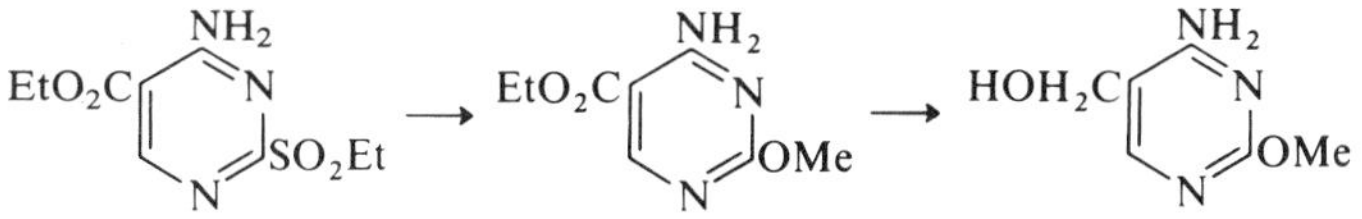

(983) → (984) → (985)

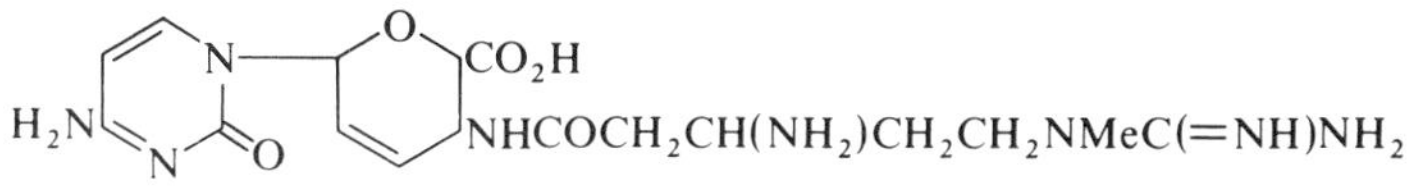

(986) 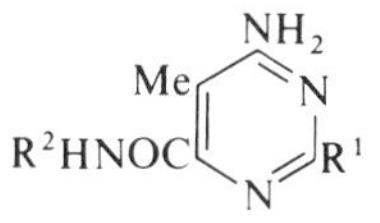  (987)

Three more antibiotics, all discovered about 1953, are also derivatives of cytosine. Amicetin, bamicetin and plicacetin may all be isolated from *Streptomyces plicatus* and all have some activity against some acid-fast and Gram-positive bacteria as well as some other microbial systems 〈69MI21301〉. Structural work in this area is fascinating 〈62JOC2991〉.

The phleomycin, bleomycin and related families are widespectrum antibiotics containing the pyrimidine (**987**); in addition, they have antineoplastic activity and bleomycin is already in clinical use for certain tumours. They were isolated about 1956 from *Streptomyces verticillus*, and in addition to the pyrimidine portion the molecules contain an amide part (R$^1$) and a complicated part (R$^2$) consisting of polypeptide, an imidazole, two sugars, a bithiazole and a polybasic side chain which can vary widely; phleomycin and bleomycin differ by only one double bond in the bithiazole section 〈78MI21303〉. The activity of such antibiotics is increased by the addition of simple heterocycles (including *inter alia* pyrimidines and fused pyrimidines) and other amplifiers 〈82MI21300〉.

### 2.13.4.1.12 Tetrodotoxin or tarichatoxin

Tetrodotoxin is one of the most powerful non-protein neurotoxins known. It occurs in the liver and ovaries of the Japanese puffer fish, *Sphoerides rubripes* and *S. phyreus*, and its lethal effects have been known for centuries, although it was isolated in crystalline form only in 1950; it also occurs in the Californian newt or salamander, *Taricha torosa* ⟨50NKK590⟩. The structure (**988**) of this complicated and labile quinazoline derivative was elucidated finally in 1964, mainly in Japan ⟨65T2059, 64PAC(9)49⟩. The total synthesis of (±)-tetrodotoxin was achieved in 1972; it had half the activity of the natural single stereoisomer (**988**) ⟨72JA9219⟩. Slight modifications of the molecule result in greatly decreased activity ⟨79AHC(24)1⟩ and simplified analogues, produced to date by elegant independent syntheses, show only a small fraction of the toxicity associated with tetrodotoxin ⟨79AJC1805⟩.

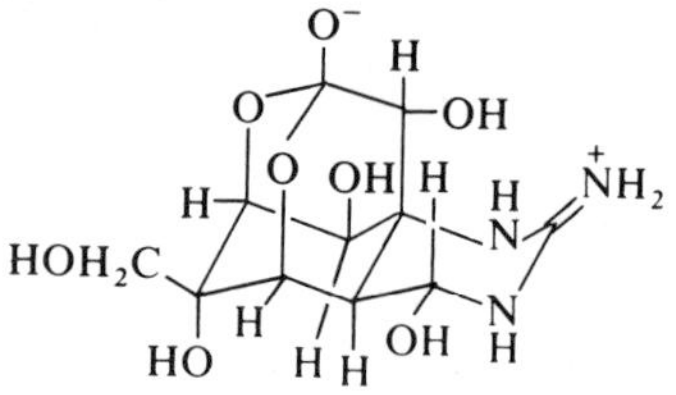

(988) 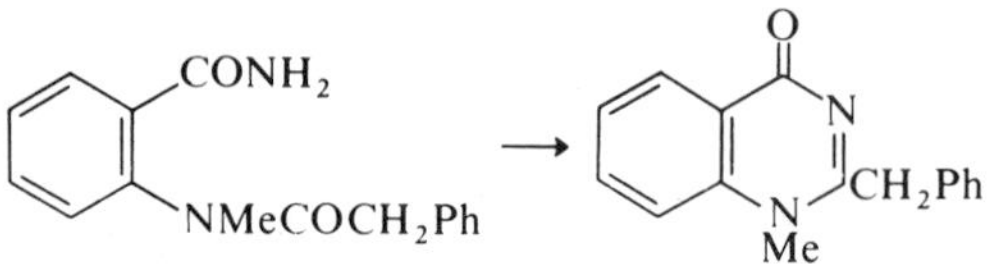            (989)            (990)

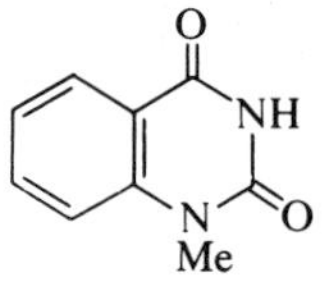

(991) 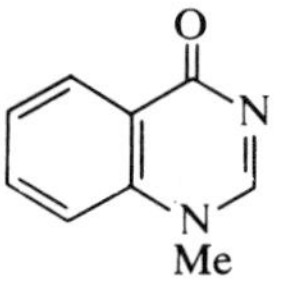            (992) 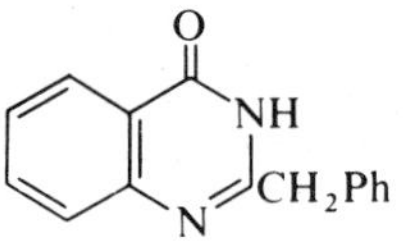            (993)

### 2.13.4.1.13 The quinazoline alkaloids

Quinazoline alkaloids are found in at least six botanical families of which the *Rutaceae* are the most important in this respect. Thus, arborine was isolated from *Glycosmis arborea* in 1952 and in the following year appeared its structure (**990**) and synthesis by thermal cyclization of the phenylacetyl derivative (**989**) of *N*-methylanthranilamide ⟨53JCS3337⟩. The same plant material yields three related alkaloids, glycosmicine (**991**), glycorine (**992**) and glycosminine (**993**) ⟨63T1011⟩.

The alkaloid vasicine was first isolated from *Adhatoda vasica* in 1925 but the structure (**996**) was elucidated only after a great amount of work, culminating in two independent syntheses in 1935 ⟨B-53MI21301⟩. It was made subsequently by a simple route from 2-aminobenzaldehyde and 4-amino-2-hydroxybutyraldehyde (**994**) followed by dehydration of the tricyclic intermediate (**995**) ⟨60TL(25)44⟩. Vasicine has bronchodilatory activity of a low order. Two related unnamed alkaloids (**997**) and (**998**) were obtained in 1965 from members of the *Araliaceae* family, *viz. Mackinlaya subulata* and *M. macrosciadia*; both had been synthesized earlier ⟨66AJC151⟩.

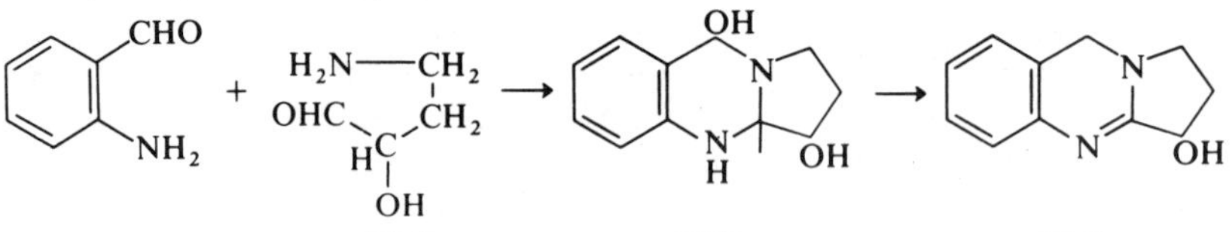

(994)            (995)            (996)

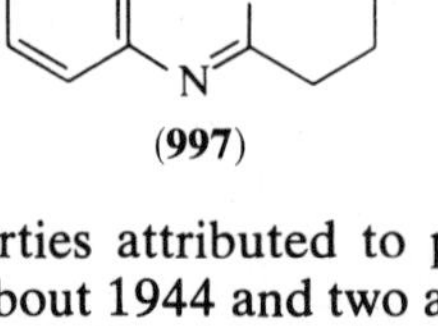

(997) 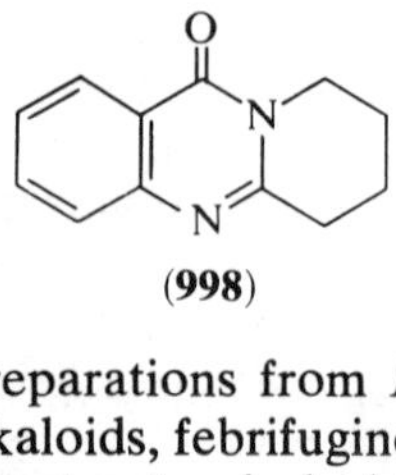            (998)

The antimalarial properties attributed to preparations from *Dichroa febrifuga* by the Chinese were confirmed about 1944 and two alkaloids, febrifugine (**999**) and isofebrifugine (**1000**), were isolated eventually. After difficult structural elucidations, syntheses of (±)-febrifugine followed: it proved to be half as active as the natural material, itself far better than quinine, but the therapeutic index was disappointingly low ⟨67HC(24-1)490⟩.

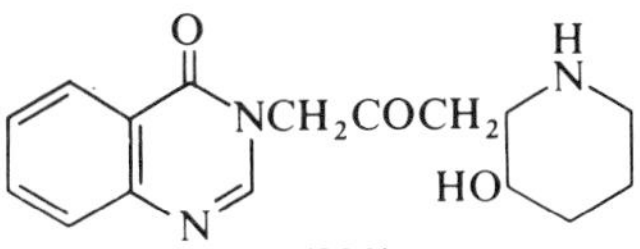

(999) 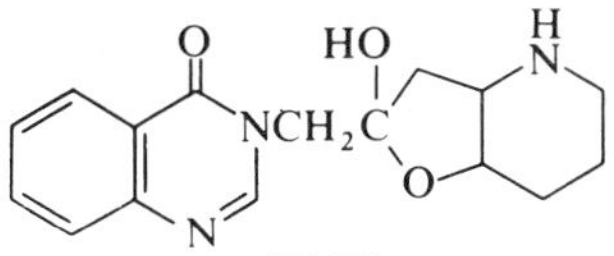     (1000)

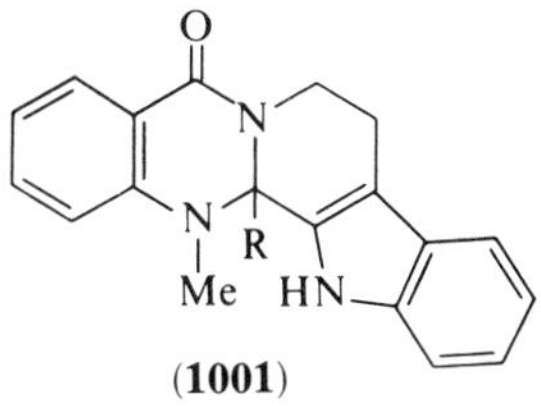

(1001) 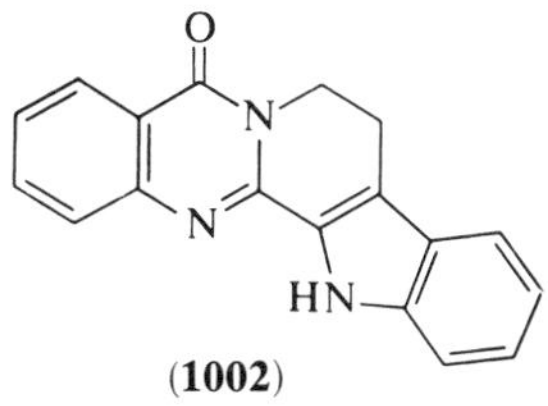     (1002)

There are several alkaloids in the structure of which are both quinazoline and indole nuclei. Evodiamine (**1001**; R = H) and rutaecarpine (**1002**) were both isolated from the seeds of *Evodia rutaecarpa* about 1916 and synthesized in 1927. They were found subsequently to occur also in *Xanthoxylum rhetsa*, which in addition yielded rhetsinine (**1001**; R = OH) ⟨59T(7)257⟩. The plant *Hortia arborea* afforded two more related alkaloids, hortiacine and hortiamine, each characterized by a methoxy group in the benzene ring of the indole portion ⟨60JA5187⟩.

### 2.13.4.1.14 Other pyrimidines from nature

Alloxan (**1003**) has been observed in the mucus associated with dysentery and it was the very first pyrimidine made synthetically when Brugnatelli oxidized uric acid in 1818. Alloxan has an interesting diabetogenic action which appears to be associated with removal of essential zinc from insulin by chelation. Such permanent diabetes may be induced in fish, dogs, cats, sheep, some birds, monkeys and other creatures, but not in man, owls or guinea-pigs; certain pyrimidines related to alloxan show some such activity.

There are many synthetic routes to alloxan. Probably the best is direct oxidation of barbituric acid (**1004**; R = H) with chromium trioxide ⟨52OS(32)6⟩ but it may be made from barbituric acid *via* its benzylidene derivative; by direct or indirect oxidation of uric acid; from 5-chlorobarbituric acid (**1004**; R = Cl) by nitration or from 5-nitrobarbituric acid (**1004**; R = NO₂) by chlorination, both *via* the intermediate (**1005**) ⟨64M1057⟩; or by permanganate oxidation of uracil (**1006**) under carefully controlled conditions ⟨73BSF1167⟩.

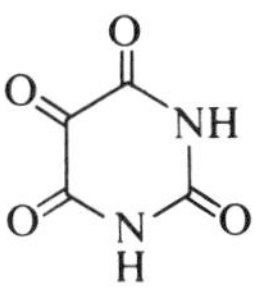

(1003) 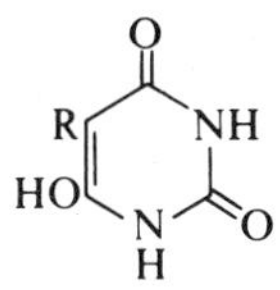     (1004) 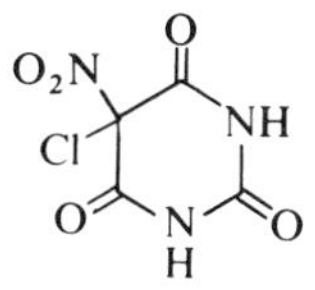     (1005) 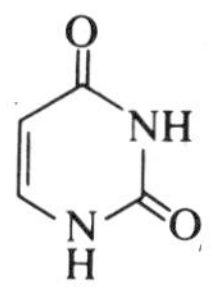     (1006)

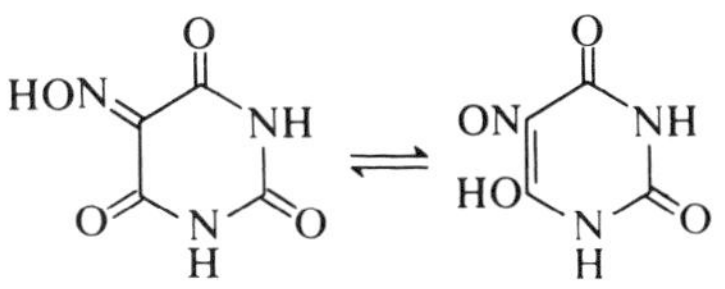

(1007) 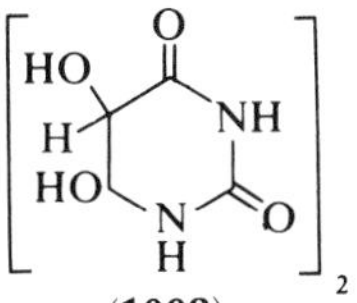     (1008) 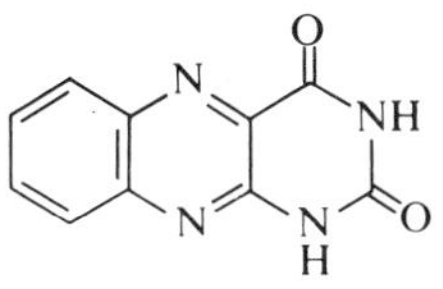     (1009)

Alloxan forms an oxime (**1007**) which is the same compound, violuric acid, as that formed by nitrosation of barbituric acid; likewise, a hydrazone and semicarbazone. Reduction of alloxan gives first alloxantin, usually formulated as (**1008**), and then dialuric acid (**1004**; R = OH); the steps are reversible on oxidation. Vigorous oxidation with nitric acid and alkaline hydrolysis both give imidazole derivatives (parabanic acid and alloxanic acid, respectively) and thence aliphatic products. Alloxan and *o*-phenylenediamine give the benzopteridine, alloxazine (**1009**) ⟨61MI21300⟩.

Toxopyrimidine (pyramin; **1010**) may be obtained from thiamine (vitamin B₁; **1011**) by acidic hydrolysis or by treatment with the thiaminase of *Bacillus aneurinolyticus*. Toxopyrimidine produces convulsions and death in rodents as do analogues, *e.g.* 2,4-dimethylpyrimidin-5-ylmethanol (**1012**) ⟨65JMC750⟩, but the effect is minimized or even inhibited by administration of vitamin B$_6$ compounds such as pyridoxamine (**1013**). Toxopyrimidine and analogues show many other and varied biological activities, apparently of no great importance ⟨69MI21301⟩. Thiamine itself is discussed in Chapter 4.19.

(**1010**)        (**1011**)        (**1012**)        (**1013**)

### 2.13.4.2 Pyrimidine Drugs

Most drugs in the pyrimidine series fall into four categories: the barbiturates, the sulfonamides, the antimicrobials and the antitumour agents. In addition there are innumerable pyrimidines with diverse biological activities, some of which are in use.

#### 2.13.4.2.1 The barbiturates

Barbituric acid, usually represented as the trione (**1014**; $R^1 = R^2 = H$), was first made about 1864 and it has no hypnotic properties. It may be made conveniently from diethyl malonate and urea in ethanolic sodium ethoxide ⟨43OSC(2)60⟩ and it does have a variety of biological properties without much practical use ⟨71MI21301⟩. The origin of the name is lost, although there are several plausible explanations associated with St. Barbara's feast day, a favourite München *Kellnerin* rejoicing in that given name, and even the Latin *barba* which is a beard or the business end of a key.

(**1014**)        (**1015**)        (**1016**)        (**1017**)

The first hypnotic barbiturate, 5,5-diethylbarbituric acid (barbital, Veronal; **1014**; $R^1 = R^2 = Et$) was made in 1904, was introduced into medicine in 1905, and is still used sometimes; the second was 5-ethyl-5-phenylbarbituric acid (phenobarbital, Luminal; **1014**; $R^1 = Et$, $R^2 = Ph$), also prepared in 1904 but used as a long-acting CNS depressant only from 1912 until the present day. Several thousand active barbiturates were made subsequently but scarcely a dozen are still used to any extent. Among these are 5-ethyl-5-(1'-methylbutyl)barbituric acid (pentobarbital, Nembutal; **1014**; $R^1 = Et$, $R^2 = CHMePr$), with action of a medium duration; 5-ethyl-5-isopentylbarbituric acid (amobarbital, Amytal; **1014**; $R^1 = Et$, $R^2 = CH_2CH_2CHMe_2$) of medium duration; 5-allyl-5-(1'-methylbutyl)barbituric acid (secobarbital, Seconal; **1014**; $R^1 = CH_2CH=CH_2$, $R^2 = CHMePr$) of medium duration; 5-(cyclohex-1'-enyl)-1,5-dimethylbarbituric acid (hexobarbital, Sombulex or Evipal; **1015**) of short duration; and 5-ethyl-5-(1'-methylbutyl)-2-thiobarbituric acid (thiopental, Pentothal; **1016**) of ultra-short duration. The whole subject is well reviewed ⟨B-75MI21301, B-75MI21304, B-70MI21302⟩.

The 'pseudo-barbiturate', 2-methyl-3-*o*-tolylquinazolin-4(3*H*)-one (methaqualone, Revonal; **1017**) has an even wider spectrum of activities than do the barbiturates proper; it appears to be quite widely used as a sedative, hypnotic, anticonvulsant, antispasmodic and local anaesthetic agent ⟨63MI21301, B-75MI21301⟩.

#### 2.13.4.2.2 The pyrimidine sulfonamides

The first pyrimidine analogues of sulfanilamide were introduced in 1942, some five years after the start of the bacterial chemotherapy revolution. Sulfadiazine (**1018**; $R^1 = R^2 = H$) and sulfamerazine (**1018**; $R^1 = Me$, $R^2 = H$) may be made by treatment of pyrimidin-2-amine or its 4-methyl derivative, either with *p*-acetamidobenzenesulfonyl chloride followed by deacetylation, or with *p*-nitrobenzenesulfonyl chloride followed by reduction; other direct condensations are also known. The chief advantage over earlier heterocyclic analogues such as sulfapyridine and sulfathiazole is good water solubility of both the drug and its bioacetylated product: this minimizes the risk of kidney blockage by precipitation and eliminates the need for an excessive water intake during therapy, even for patients with impaired renal function. Today, sulfadiazine is still widely used, although sulfamerazine is no longer used for chemotherapy of infections. A third homologue, sulfadimidine, which is the 4,6-dimethyl derivative (**1018**; $R^1 = R^2 = Me$) of sulfadiazine, is also little used although it is still found in most Pharmacopoeias. Other pyrimidine sulfonamides currently in clinical use include sulfametoxydiazine (**1019**); sulfasomidine (**1020**; $R^1 = R^2 = Me$) which is an isomer of sulfadimidine; sulfadimethoxine (**1020**; $R^1 = R^2 = OMe$); and sulfametomidine (**1020**; $R^1 = Me$, $R^2 = OMe$).

(**1018**)     (**1019**)     (**1020**)

It is interesting that sulfonamide drugs, including the above pyrimidines, are by no means superseded by antibiotics: indeed, the two groups are complementary rather than competitive and sulfonamides remain the drugs of choice in well defined areas such as some acute urinary-tract infections, cerebrospinal meningitis, or the numerous patients sensitive to penicillins; in addition, most sulfonamides have excellent therapeutic indices (margins of safety) and their side effects, if any, disappear quickly on discontinuance of the drug. Structure–activity relationships in pyrimidine sulfonamides are reviewed ⟨69MI21301⟩ and the practical pharmacology is summarized ⟨B-75MI21302⟩.

### *2.13.4.2.3 The antimicrobial pyrimidine-2,4-diamines*

Pyrimethamine (Daraprim; **1022**) inhibits the enzyme, dihydrofolate reductase (DHFR), thereby blocking the reduction of di- to tetra-hydrofolic acid, which is an essential coenzyme in nucleic acid synthesis. Pyrimethamine has a far higher affinity for protozoal than for mammalian DHFR, so that it may be used for malarial prophylaxis without adversely affecting human DHFR. It is active against both *Plasmodium falciparum* and *P. vivax*; in combination with an appropriate sulfonamide, it is effective for toxoplasmosis because the sulfonamide acts synergistically by arresting production of dihydrofolic acid from *p*-aminobenzoic acid, thus achieving a double (sequential) blockage of the folate pathway in the microorganism, which (unlike man) cannot use preformed folic acid ⟨70MI21303, 65MI21305⟩. A Principal Synthesis is used to prepare pyrimethamine (**1022**) from the enol ether (**1021**) and guanidine ⟨51JA3763⟩; many analogues are prepared similarly.

(**1021**)     (**1022**)

(**1023**)     (**1024**)

Although trimethoprim (**1024**) showed early promise as an antimalarial drug, its performance in the field was disappointing. However, it is widely used in combination with sulfamethoxazole (under the names Septrin, Bactrim, *etc.*) as a general systemic antibacterial agent against a whole variety of Gram-positive and Gram-negative organisms. The combination is effective in preventing the development of drug resistance and in the treatment of strains already resistant to single drug therapy ⟨69MI21302⟩. Trimethoprim may be made by several related Principal Syntheses: for example, condensation of 3,4,5-trimethoxy-benzaldehyde with 3-ethoxypropionitrile gives the intermediate (**1023**) which reacts with guanidine to afford trimethoprim (**1024**) ⟨63JOC1983, 64BRP957797⟩.

### 2.13.4.2.4 *Halogenopyrimidines as antitumor agents*

Uracil is used more effectively in nucleic acid synthesis within a rat hepatoma than in normal liver. This observation appears to have stimulated the synthesis of 5-fluorouracil (**1027**) as an antimetabolite mainly because the introduction of a fluorine atom involves a minimal increase in size. In the event, 5-fluorouracil did prove to have antineoplastic activity and it is now a valuable drug for treatment of tumors of the breast, colon or rectum, and to a lesser extent, gastric, hepatic, pancreatic, uterine, ovarian and bladder carcinomas. As with other drugs which interfere with DNA synthesis, the therapeutic index is quite low and great care is required during treatment ⟨69MI21301⟩.

5-Fluorouracil (**1027**) was made originally by condensation of $S$-ethylthiourea with ethyl 2-fluoro-2-formylacetate (**1025**) followed by acidic hydrolysis of the resulting ethylthiopyrimidine (**1026**) ⟨57JA4559⟩. It may be made also by direct fluorination of uracil with fluorine (55% yield), $C_{19}XeF_6$ (90%) or fluoroxytrifluoromethane (94%) or by a primary synthesis akin to the original method ⟨72JOC329, 80TL277, 67JCS(C)2206⟩.

(**1025**)     (**1026**)     (**1027**)     (**1028**)     (**1029**)

Among the 'mustards' which act as antineoplastic agents by alkylation is uramustine (uracil mustard, **1029**; R = Cl). Although not used widely at present, uramustine is available for chronic lymphocytic leukaemia, Hodgkin's disease and lymphosarcoma; it is not effective against the acute leukaemias of childhood. The drug is well tolerated in appropriate dosage; its $LD_{50}$ is <4 mg/kg in mice. Uramustine is made by treatment of 5-aminouracil (**1028**) with ethylene oxide to give the dihydroxy intermediate (**1029**; R = OH), which is treated with thionyl chloride to give the required dichloro product (**1029**; R = Cl) ⟨58JA6459⟩. Many structural variants of uramustine are known but none has achieved a place in practical chemotherapy ⟨71MI21301⟩.

### 2.13.4.2.5 *Other pyrimidine drugs*

Hyperthyroidism may be treated in several ways. One of these is interference with the synthesis of the thyroid hormones, possibly by removal of iodine. Thiourea and 'cyclic thioureas' have this effect and of such cyclic compounds, thiouracil (**1030**; R = H), its 6-alkyl derivatives (**1030**; R = Me or Pr) and thiobarbital (**1031**) are effective thyroid drugs. Today only propylthiouracil (**1030**; R = Pr) is widely used, probably because it has fewer side effects than the others ⟨71MI21302⟩. The thiouracils are made by the Principal Synthesis from a $\beta$-oxo ester (**1032**; R = H, Me, Pr, *etc.*) and thiourea ⟨45JA2197⟩; their fine structures are experimentally based ⟨64AF1004⟩.

(**1030**)          (**1031**)          (**1032**)

Cytarabine, 4-amino-1-$\beta$-D-arabinofuranosylpyrimidin-2(1$H$)-one or cytosine arabino-side (**1033**; R = H, X = NH₂), is an established drug for the treatment of acute leukaemias of childhood and adult granulocytic leukaemia. It must be given intravenously and much of the drug becomes the corresponding inactive uracil derivative *in vivo* by virtue of a deaminase in the liver; it interferes with DNA but not RNA synthesis, and it has incidental antiviral activity against herpes and vaccinia infections although contraindicated in herpes zoster types ⟨B-75MI21303⟩.

Cytarabine may be made by treatment of 2,4-dimethoxypyrimidine with 2,3,5-tri-*O*-benzyl-D-arabinofuranosyl chloride to give the 4-methoxypyrimidinone (**1033**; R = CH$_2$Ph, X = OMe) which undergoes aminolysis by ammonia to the amine (**1033**; R = CH$_2$Ph, X = NH$_2$) and subsequent hydrogenolysis to the nucleoside (**1033**; R = H, X = NH$_2$) ⟨65JOC835⟩. Another practical route involves epimerization of the glycosyl part of cytidine or cytidylic acid with phosphoric acid, followed by hydrolysis and dephosphorylation of the resulting anhydro body ⟨B-68MI21305⟩.

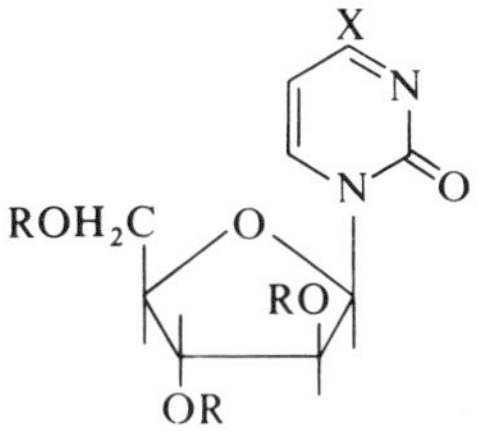

(**1033**) 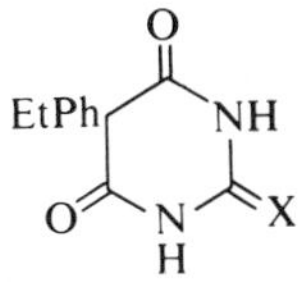　(**1034**) 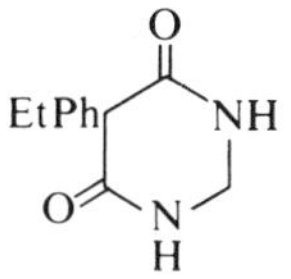　(**1035**) 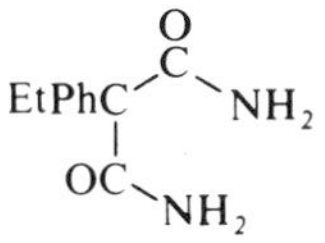　(**1036**)

The anticonvulsant primidone (**1035**) resembles phenobarbital but lacks the 2-oxo substituent. It was introduced in 1952 and has remained a valuable drug for controlling grand mal and psychomotor epilepsy. As might be expected, primidone is metabolized to yield phenobarbital (**1034**; X = O) and *C*-ethyl-*C*-phenylmalondiamide (**1036**), both of which have marked anticonvulsant properties; however, primidone does have intrinsic activity and an appropriate mixture of its metabolites has only a fraction of its activity ⟨73MI21303⟩. Primidone may be made in several ways, of which desulfurization by Raney nickel of the 2-thiobarbiturate (**1034**; X = S) or treatment of the diamide (**1036**) with formic acid (at 190 °C) seem to be the most satisfactory ⟨54JCS3263⟩.

Two related quinazolines are widely used as diuretics: quinethazone (**1037**) and metolazone (**1038**) cause the excretion of sodium ions and chloride ions in approximately equivalent amounts along with an appropriate amount of water. They are indicated for adjunctive therapy in edema associated with congestive heart failure, hepatic cirrhosis and with corticosteroid or estrogen therapy; they are used also for hypertension control. Quinethazone is the more used. Its synthesis begins with 3-chloro-6-methylacetanilide which is chlorosulfonated at the 4-position and then converted into the sulfonamide and subsequently oxidized to the benzoic acid (**1039**); deacetylation followed by fusion with propionamide gives a quinazoline, which on catalytic hydrogenation of its 1,2-bond, affords the dihydroquinazoline (**1037**) ⟨60JA2731⟩. The preparation and pharmacology of quinethazone is discussed further in a series of five papers ⟨63AF660⟩.

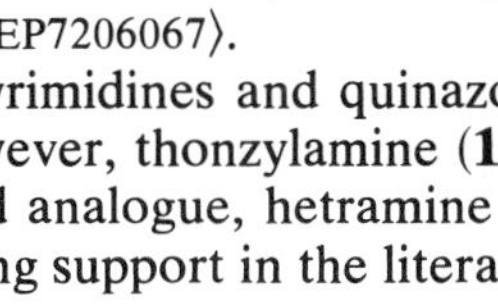

(**1037**) 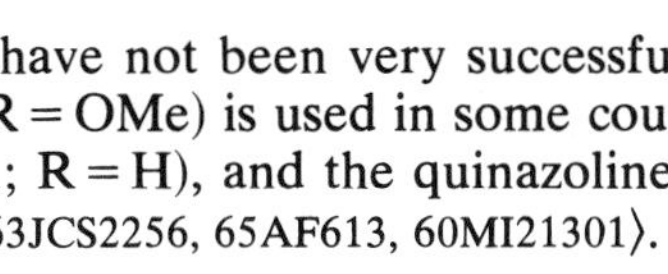　(**1038**) 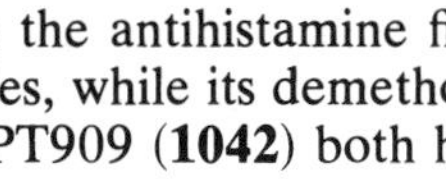　(**1039**)

Another quinazoline, prazosin (**1040**), is now a widely used antihypertensive drug, acting by vasodilation ⟨77MI21300, B-81MI21301⟩; its preparation is outlined in the patent literature ⟨72NEP7206067⟩.

Pyrimidines and quinazolines have not been very successful in the antihistamine field. However, thonzylamine (**1041**; R = OMe) is used in some countries, while its demethoxylated analogue, hetramine (**1041**; R = H), and the quinazoline HPT909 (**1042**) both have strong support in the literature ⟨63JCS2256, 65AF613, 60MI21301⟩.

(**1040**) 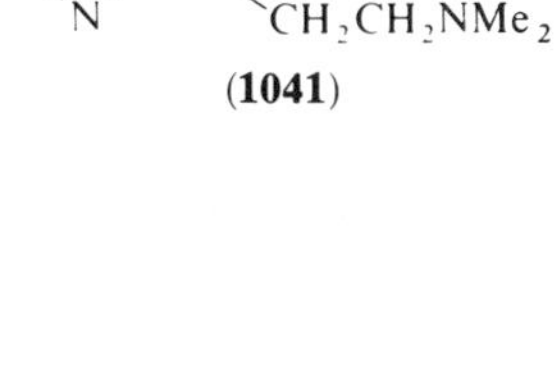　(**1041**)

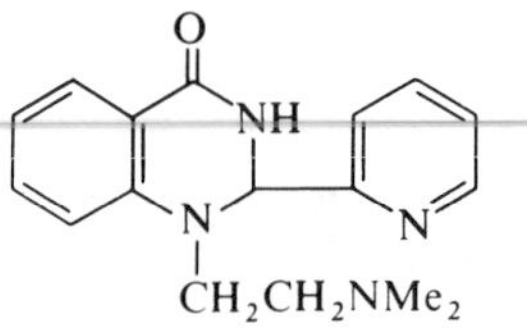

(1042) 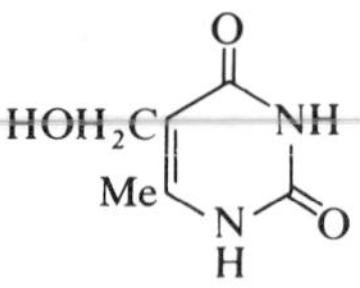                                    (1043)

5-Hydroxymethyl-6-methyluracil (**1043**) was prepared many years ago from 6-methyl-uracil and formaldehyde, or in other ways. Since 1956 it has received much attention in the USSR under the (transliterated) name pentoxyl or pentoxil. It is used in several anaemic and disease conditions. For example, a mixture of folic acid and pentoxyl quickly reduces the anaemia resulting from lead poisoning; pentoxyl stimulates the supply of serum protein after massive blood loss; it stimulates wound healing; it stimulates the immune response in typhus infection; and it potentiates the action of sulfonamides in pneumococcus infections ⟨70MI21300⟩.

### 2.13.4.3 Some Pyrimidines of Veterinary or Agricultural Interest

A great many pyrimidines have properties potentially useful in stock or crop raising. The following examples include some of the pyrimidines actually used in such areas.

#### 2.13.4.3.1 Diaveridine

Diaveridine (**1044**) is a close relative of trimethoprim (Section 2.13.4.2.3) and is made by an analogous Principal Synthesis. It is used prophylactically against coccidiosis in poultry and in combination with sulfaquinoxaline as a curative agent for the same disease; similar mixtures are also effective ⟨64MI21305⟩.

#### 2.13.4.3.2 Dimpylate

Condensation of *O,O*-diethyl chlorothiophosphate with 2-isopropyl-6-methylpyrimidin-4(3*H*)-one gives the powerful insecticide dimpylate (**1045**) ⟨57HCA1562⟩. It is used in veterinary medicine ⟨58MI21301⟩, in particular for topical application, with or without added DDT, in cases of blowfly strike in sheep.

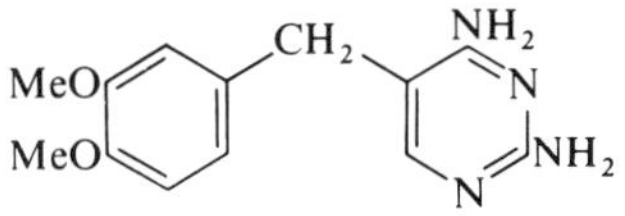

(1044) 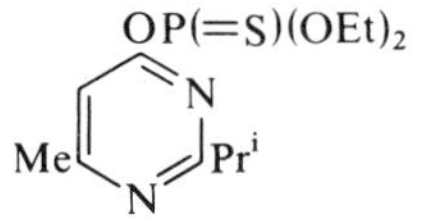                    (1045) 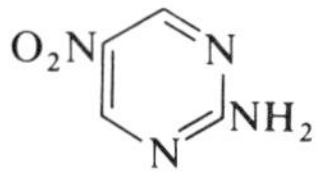                    (1046) 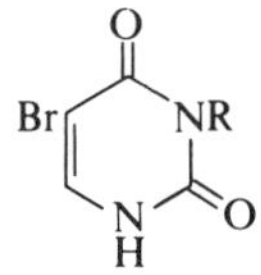                    (1047)

#### 2.13.4.3.3 'Enheptin-P'

5-Nitropyrimidin-2-amine, 'Enheptin-P' (**1046**), is effective in suppressive and curative therapy of a common protozoan infection known as enterohepatitis or blackhead in turkeys ⟨50MI21303⟩. However, it seems that less expensive analogues, *e.g.* amnitrozole (2-acetamido-5-nitrothiazole), are preferred now. The subject is reviewed ⟨58MI21302⟩.

#### 2.13.4.3.4 Bromacil, isocil, etc.

In 1962, bromacil (5-bromo-3-*s*-butyl-6-methyluracil; **1047**; R = Bu$^s$), its homologue (isocil; **1047**; R = Pr$^i$) and related *N*-alkyluracils were shown to have valuable selective phytotoxic properties and vitally no mammalian toxicity. Thus, bromacil achieves a complete kill of most unwanted broad-leaf annuals or perennials along with some grasses without affecting the food crops ⟨62MI21302⟩; it achieves this by inhibition of photosynthesis in such species ⟨64MI21301⟩. Isocil appears to be superseded now by its higher homologue; practical aspects are reviewed ⟨67MI21301⟩. Bromacil may be made by the condensation of *s*-butyl isocyanate with methyl 3-aminocrotonate to give the methyl 3-(3′-*s*-butyl-ureido)crotonate which cyclizes in alkali to 3-*s*-butyluracil and subsequent bromination gives bromacil (**1047**; R = Bu$^s$) ⟨62MI21302⟩.

### 2.13.4.4 Trivial Names for Pyrimidines

During the last 160 years, many trivial names have been used for pyrimidine and its derivatives. Some, like uracil, persist but others have now fallen out of use, save for the occasional biochemical or biological paper. Table 1 embraces most such names: WHO non-proprietary drug names, trade or proprietary names, and names of some natural products discussed in this chapter are not included because they are readily available elsewhere ⟨B-76MI21303⟩.

**Table 1** Trivial Names of Pyrimidines

| Trivial name | Proper name |
| --- | --- |
| Acetonin | 2,2,4,6,6-Pentamethyl-1,2,5,6-tetrahydropyrimidine |
| Alloxan | 2,4,5,6-Tetraoxohexahydropyrimidine |
| Aminobarbituric acid | 2-Amino-6-hydroxypyrimidin-4(3*H*)-one |
| Barbituric acid | Pyrimidine-2,4,6(1*H*,3*H*,5*H*)-trione |
| Cyanmethine | *See* Kyanmethin |
| Cytosine | 4-Aminopyrimidin-2(1*H*)-one |
| Dialuric acid | 5-Hydroxypyrimidine-2,4,6(1*H*,3*H*,5*H*)-trione |
| Diaminocytosine | 4,5,6-Triaminopyrimidin-2(1*H*)-one |
| Diaminouracil | 5,6-Diaminopyrimidine-2,4(1*H*,3*H*)-dione |
| *m*-Diazane | Hexahydropyrimidine |
| *m*-Diazine | Pyrimidine |
| Dilituric acid | 5-Nitropyrimidine-2,4,6(1*H*,3*H*,5*H*)-trione |
| Dithiouracil | Pyrimidine-2,4(1*H*,3*H*)-dithione |
| Hydrouracil | 5,6-Dihydropyrimidine-2,4(1*H*,3*H*)-dione |
| Isobarbituric acid | 5-Hydroxypyrimidine-2,4(1*H*,3*H*)-dione |
| Isocytosine | 2-Aminopyrimidin-4(3*H*)-one |
| Isodialuric acid | *See* 'Dialuric acid' (probably the same) |
| Isouracil | 5-Hydroxypyrimidin-2(1*H*)-one |
| Isouramil | 6-Amino-5-hydroxypyrimidine-2,4(1*H*,3*H*)-dione |
| Isovioluric acid | 5-Hydroxy-6-nitrosopyrimidine-2,4(1*H*,3*H*)-dione |
| Ketobarbituric acid | *See* Alloxan |
| Kyanmethin[a] | 2,6-Dimethylpyrimidin-4-amine |
| Malonylurea[b] | *See* 'Barbituric acid' |
| Methylcytosine | 4-Amino-5-methylpyrimidin-2(1*H*)-one |
| Miazine | Pyrimidine |
| Orotic acid | 2,6-Dioxo-1,2,3,6-tetrahydropyrimidine-4-carboxylic acid |
| Piperimidine | Hexahydropyrimidine |
| Pseudouracil | 6-Hydroxypyrimidin-4(3*H*)-one |
| Pseudouric acid | 5-Ureidopyrimidine-2,4,6(1*H*,3*H*,5*H*)-trione |
| Tartronyl urea | *See* 'Dialuric acid' |
| Thiobarbituric acid | 2-Thioxo-1,2-dihydropyrimidine-4,6(3*H*,5*H*)-dione |
| 2-Thiouracil | 2-Thioxo-1,2-dihydropyrimidin-4(3*H*)-one |
| 4-Thiouracil | 4-Thioxo-3,4-dihydropyrimidin-2(1*H*)-one |
| Thymine | 5-Methylpyrimidine-2,4(1*H*,3*H*)-dione |
| Uracil | Pyrimidine-2,4(1*H*,3*H*)-dione |
| Uramil | 5-Aminopyrimidine-2,4,6(1*H*,3*H*,5*H*)-trione |
| Violuric acid | 5-Nitrosopyrimidine-2,4,6(1*H*,3*H*,5*H*)-trione |

[a] 'Kyanalkine' are 2,6-dialkyl-5-(alkyl less CH$_2$)pyrimidin-4-amines: thus 'kyanisobutin' is 2,6-diisobutyl-5-isopropylpyrimidin-4-amine.

[b] To be distinguished from 'Malonurea', a trade name (?) once widely used for 5,5-diethylbarbituric acid (barbital).

# 2.14

# Pyrazines and their Benzo Derivatives

A. E. A. PORTER

*University of Stirling*

2.14.1  INTRODUCTION                                                                 157

   *2.14.1.1  Structural Considerations*                                         158
   *2.14.1.2  Proton NMR Spectroscopy*                                          159
   *2.14.1.3  $^{13}C$ NMR Spectroscopy*                                       160
   *2.14.1.4  IR Spectroscopy*                                                  161
   *2.14.1.5  UV Spectroscopy*                                                  161
   *2.14.1.6  Mass Spectrometry*                                                162
   *2.14.1.7  Physical Properties*                                              162

2.14.2  REACTIVITY                                                                   163

   *2.14.2.1  Electrophilic Substitution*                                      163
   *2.14.2.2  Nucleophilic Substitution*                                       164
   *2.14.2.3  Reactions with Radicals*                                         166
   *2.14.2.4  Side Chain Reactivity*                                           166
   *2.14.2.5  N-Oxides*                                                        168
   *2.14.2.6  Pyrazinones, Quinoxalinones and Hydroxyphenazines*               173
   *2.14.2.7  Halo Pyrazines, Quinoxalines and Phenazines*                     175
   *2.14.2.8  Amino Derivatives*                                               176
   *2.14.2.9  Reduced Pyrazines, Quinoxalines and Phenazines*                  177

2.14.3  SYNTHESIS                                                                    179

   *2.14.3.1  General Comments*                                                179
   *2.14.3.2  Type 'A' Syntheses*                                              179
   *2.14.3.3  Type 'B' Syntheses*                                              184
   *2.14.3.4  Other Synthetic Methods*                                         188

2.14.4  APPLICATIONS                                                                 191

   *2.14.4.1  Introduction*                                                    191
   *2.14.4.2  Pyrazinones, Quinoxalinones and Hydroxyphenazines*               191
   *2.14.4.3  Alkoxypyrazines*                                                 192
   *2.14.4.4  Alkylpyrazines*                                                  193
   *2.14.4.5  Chemiluminescence and Bioluminescence in Pyrazine Derivatives*   194
   *2.14.4.6  Miscellaneous Pyrazine Derivatives*                              194
   *2.14.4.7  Miscellaneous Quinoxaline Derivatives*                           195
   *2.14.4.8  Naturally Occurring Phenazine Pigments*                          195
   *2.14.4.9  Synthetic Phenazine Dyes*                                        196

## 2.14.1  INTRODUCTION

A computer search of volumes 70–95 of *Chemical Abstracts* using the keyword 'Pyrazine' resulted in more than 2600 references, and, after removal of fused pyrazine systems and cross-referencing the remaining references, this number increased to approximately 7000 in total. When the benzopyrazines quinoxaline and phenazine were added the number of references was in excess of 10 000, all of which might be considered to be relevant to a chapter devoted to pyrazines and their benzo analogues.

Clearly in the limited space available it would not be possible to include all relevant references. Generally, references to patents have been omitted since it is frequently difficult to abstract meaningful information from patent literature, and references to communications have been omitted unless the work is particularly new or experimental details are included. In addition, where several references contain similar methodology and/or results, only those which are commonly available have been selected. In surveying the literature, major reviews (*vide infra*) have been used to gather basic background information and this has been complemented by an up-to-date survey of *Chemical Abstracts* to the end of Volume 95.

### 2.14.1.1 Structural Considerations

Substitution of two carbon atoms of a benzene ring by tervalent nitrogen atoms may occur in three ways, giving rise to pyridazines (see Chapter 2.12), the pyrimidines (see Chapter 2.13) and the pyrazines, with the nitrogen atoms occupying a 1,2-, 1,3- or 1,4-disposition respectively.

In many respects pyrazines are the poor relation amongst the diazines in that although a number of reviews have been written ⟨47CRV(40)279, B-57MI21400, B-59MI21400, 72AHC(14)99⟩, no major works comparable with those written on the pyridazines ⟨73HC(28)1⟩ or the pyrimidines ⟨62HC(16)1⟩ exist. This seems strange in view of the importance of a wide variety of pyrazine derivatives with a spectrum of biological activity ranging from flavouring agents to pharmaceuticals, and represents an omission which will undoubtedly be filled in the future. In contrast, a good reference work on benzopyrazines (including quinoxalines but not phenazines) has recently appeared ⟨79HC(35)1⟩, whereas no recent survey of phenazines has been published.

In valence bond terms the pyrazine ring may be represented as a resonance hybrid of a number of canonical structures (*e.g.* **1–4**), with charge separated structures such as (**3**) contributing significantly, as evidenced by the polar character of the C=N bond in a number of reactions. The fusion of one or two benzene rings in quinoxaline (**5**) and phenazine (**6**) clearly increases the number of resonance structures which are available to these systems.

Structural parameters and interatomic distances derived from electron diffraction (**7**) ⟨77JST(42)121⟩ and X-ray diffraction (**8**) studies ⟨76AX(B)3178⟩ provide unequivocal evidence that pyrazine is planar with $D_{2h}$ symmetry. There is an increased localization of electron density in the carbon–nitrogen bonds, with carbon–carbon bonds being similar in length to those in benzene*.

Interatomic distances calculated from the detailed analysis of rotational fine structure of the UV spectrum of pyrazine are in close agreement with those observed in (**7**) and (**8**), with the calculated bond lengths for C—C of 1.395, C—N 1.341 and C—H 1.085 Å ⟨60DIS(20)4291⟩. Thermochemical data have provided a figure of 75 kJ mol$^{-1}$ for the delocalization energy of the pyrazine ring ⟨B-67MI21400⟩.

Phenazine also exhibits $D_{2h}$ symmetry and numerous reports on the X-ray structure of α-phenazine have appeared ⟨54AX129⟩. The parameters determined at 80 K are shown in

---

* *The Cambridge Crystallographic Data Centre* (CCDC), University Chemical Laboratory, Lensfield Road, Cambridge CB2 1EW, UK, has atomic coordinates for more than 30 000 structures in its database and information on any structural fragment is easily retrieved. No less than 60 structures containing the pyrazine ring were stored within this facility at the time of writing this chapter.

(**9**) and are broadly similar to those of anthracene and pyrazine. The lower symmetry of quinoxaline appears to result in increased disorder in the crystal state, which results in a somewhat lower melting point (see Table 3) and no crystal structure determination at lower temperatures appears to have been carried out.

The great increase in the number of molecular orbital calculations over the last 20 years has inevitably produced a large number of publications describing calculations of various levels of sophistication to predict electron densities (usually separated into $\sigma$ and $\pi$ contributions), bond lengths and energies, energies of electronic transitions and spectra. Such methods, when applied to pyrazine and the benzopyrazines, have produced a bewildering array of numbers which frequently vary by an order of magnitude depending upon the computational methods used and the assumptions made.

Pugmire *et al.* have published calculated electron densities for pyrazine ⟨68JA697⟩, quinoxaline ⟨69JA6381⟩ and phenazine and the calculated total electron densities ($\sigma + \pi$) are shown in (**10**), (**11**) and (**12**).

**(10)**　　　　**(11)**　　　　**(12)**

The precise numerical values of the calculated electron densities are unimportant, as the most important feature is the relative electron density; thus, the electron density at the pyrazine carbon atom is similar to that at an $\alpha$-position in pyridine and this is manifest in the comparable reactivities of these positions in the two rings. In the case of quinoxaline, electron densities at N-1 and C-2 are proportionately lower, with the highest electron density appearing at position 5(8), which is in line with the observation that electrophilic substitution occurs at this position.

The electron density at nitrogen in phenazine is intermediate between those of pyrazine and quinoxaline and the highest electron density on the carbon atoms of the benzene rings is at C-1 (with positions 4, 6 and 9 being equivalent)*.

The HOMO ($\pi$) of pyrazine has a nodal surface passing through both nitrogen atoms and predictions have been made ⟨B-70MI21400⟩ on this basis that the first ionization potential (IP) for pyrazine should be similar to those of benzene and pyridine, and this is borne out in practice, the respective values being 9.29, 9.25 and 9.26 eV. In the case of quinoxaline the situation is more complex in that the first and second IPs are closely spaced at 8.89 and 10.72 eV, and since the highest occupied $\pi$ MO and one of the non-bonding orbitals are extremely close in energy, it is uncertain from which orbital the first ionization occurs ⟨71CRV295⟩. A similar situation exists with phenazine: the IPs of 8.4 and 9.9 eV have been ascribed to ionization from the $\pi$ and non-bonding orbitals respectively ⟨67JCP(47)4863⟩, although doubt has been expressed on this assignment ⟨71ACS487⟩.

### 2.14.1.2 Proton NMR Spectroscopy

Pyrazine and its derivatives have been extensively studied by proton and $^{13}$C NMR spectroscopy and conflicting reports on the reliability of additivity rules and/or correlation of chemical shifts with calculated electron densities have appeared.

In deuterochloroform, pyrazine shows a single proton resonance at $\delta$ 8.59 ⟨72CPB2204⟩. $^3J_o$, $^4J_m$ and $^5J_p$ values between pyrazine ring protons obtained from a number of pyrazine derivatives are 2.5–3, 1.1–1.4 and 0 Hz respectively, and these values do not appear to be affected by the nature of the ring substituents. Some substituent shielding parameters are shown in Table 1.

A study of $^1$H NMR spectra of monosubstituted pyrazines in DMSO has been reported and the assigned chemical shift values of protons located *ortho* or *para* relative to the substituent correlate well with Hammett or Taft $\sigma$-substituent values. *meta*-Protons did not, however, give good correlations ⟨72JOC111⟩.

---

* The *Chemical Abstracts* numbering system of phenazine is used.

**Table 1** Substituent Shielding Parameters in Substituted Pyrazines[a]

| R | o | m | p |
|---|---|---|---|
| $CO_2Me$ | 0.74 | 0.13 | 0.19 |
| CN | 0.37 | 0.18 | 0.25 |
| OMe | $-0.35$[b] | $-0.48$ | $-0.47$ |
| Cl | $-0.04$ | $-0.2$ | $-0.08$ |
| Me | $-0.12$ | $-0.12$ | $-0.20$ |

[a] Determined in $CDCl_3$.
[b] The negative sign indicates a chemical shift value numerically smaller than that of pyrazine itself.

The use of alkylated and methoxypyrazines as potent flavouring agents has prompted several groups to examine the proton NMR spectra of alkylpyrazines in some detail ⟨72T4155, 73T3939⟩ to permit structural correlation with ring-proton/ring-proton and ring-proton/benzylic-proton coupling constants.

Generally, $^4J_o$ couplings (13) range from 0.55–0.66 Hz, $^5J_m$ couplings from 0.28–0.34 Hz (14) and $^6J_p$ (15) couplings from 0.65–0.74 Hz, with negative, positive and negative coupling constants respectively. As the 2-alkyl group increases in size, $^5J_m$ couplings remain fairly constant but $^6J_p$ couplings fall steadily with the increasing size of the alkyl group.

(13)            (14)            (15)

Proton NMR spectra of quinoxalines have been reported in a number of solvents ⟨79HC(35)1⟩, although $CCl_4$ is probably the solvent of choice. H-2(H-3) appears as a low-field singlet and the aromatic protons in the benzenoid ring appear as an AA′BB′ system with the high-field half of the spectrum, assigned to protons 6 and 7, appearing somewhat broadened due to $^6J$ couplings with H-2(H-3) ⟨65AJC707⟩.

An examination of coupling constants in a number of quinoxaline derivatives gives values in the following ranges: $J_{5,8}$ 0.3–0.8, $J_{6,8}$ 0.7–2.9, $J_{5,7}$ 1.4–2.7, $J_{2,3}$ 1.7–1.9, $J_{6,7}$ 5.0–8.3 and $J_{7,8}$ 8.4–10.3 Hz.

Phenazine gives rise to an AA′BB′ NMR spectrum with coupling constants $J_{1,2}$ 9.0, $J_{1,3}$ 1.67, $J_{1,4}$ 0 and $J_{2,3}$ 6.55 Hz ⟨66CPB419⟩. Similar coupling constants are also observed in a number of phenazines and phenazine $N$-oxides.

### 2.14.1.3 $^{13}$C NMR Spectroscopy

The proton decoupled $^{13}$C NMR spectrum of pyrazine shows a single resonance at 145 p.p.m. from TMS ⟨73JOC1313, 74OMR(6)663, 80OMR(13)172⟩. Off-resonance decoupling of a number of pyrazine derivatives has provided coupling constants of $J_{CH}$ (+)177–187, $^2J_{CH}$ (+)8–12, $^3J_{CH}$ (+)7–10 and $^4J_{CH}$ (−)1.5 Hz.

$^{13}$C Chemical shifts of pyridine and the diazines have been measured as a function of pH in aqueous solution and generally protonation at nitrogen results in deshielding of the carbon resonances by up to 10 p.p.m. ⟨73T1145⟩. The pH dependence follows classic titration curves whose inflexions yield $pK_a$ values in good agreement with those obtained by other methods.

$N$-Oxidation of pyrazines appears to result in increased shielding of the $\alpha$ and $\alpha'$ carbon resonances by 6–11 p.p.m., whereas the $\beta$ and $\beta'$ carbon atoms are deshielded by 3–4 p.p.m., a trend similar to that observed with substituted pyridines. These results have been qualitatively explained in terms of resonance polar effects ⟨80OMR(13)172⟩.

Both $^{13}$C and $^1$H NMR data are also available for a range of 2(1$H$)-pyrazinones which are structurally related to the aspergillic acids ⟨76T655⟩.

The $^{13}$C NMR spectrum of quinoxaline has been measured in $CDCl_3$ and the chemical shift values are as shown in (16) ⟨69JA6381⟩. Curiously, $^{13}$C NMR spectra of phenazine and its derivatives have been recorded in benzene solution and the chemical shift values quoted relative to benzene; however, for consistency the values in (17) are quoted relative to TMS.

## 2.14.1.4 IR Spectroscopy

A detailed analysis of the IR and Raman spectra of pyrazine and its deutero analogues has been presented ⟨64JSP(14)190⟩ and it would appear that the most useful diagnostic bands are those which appear in the ranges 1600–1575, 1550–1520, 1500–1465 and 1420–1370 cm$^{-1}$ ⟨B-69MI21400⟩. All of these bands probably have their origin in ring-skeleton vibrations or in-plane C—H bending modes. Of the four bands, those at 1550–1520 and 1600–1575 cm$^{-1}$ are usually present as weak to medium intensity absorptions. The band at 1500–1465 cm$^{-1}$ is often masked by C—H bending vibrations of alkyl substituents when present, and in substituted pyrazines vibrations between 1000 and 1050 cm$^{-1}$ have been used in the recognition of the ring substitution pattern ⟨73RTC123⟩.

IR studies on quinoxaline and its substituted derivatives have shown that eight ring stretching bands occur in the region 1620–1350 cm$^{-1}$ and these appear to be relatively insensitive to substituent changes ⟨63JCS3764⟩. Other absorptions between 1300–1000 cm$^{-1}$ have been attributed to C—H bending and ring breathing modes.

The high symmetry of phenazine ($D_{2h}$) means that many of the fundamental vibrations are not visible in the IR, and if their presence is to be confirmed Raman techniques must be used. It is interesting to note that the complexity of the Raman spectrum varies with the physical form of the sample; thus, a routine sample of phenazine shows nine lines in the solid state. In solution this increases to 17 lines, whereas a solution spectrum of ultra-purified phenazine showed 46 lines ⟨71JSP(39)536⟩. Stammer and Taurins have reported IR data on phenazine in the range 4000–650 cm$^{-1}$ ⟨63SA1625⟩. The ranges between 4000–1650 and 1400–650 were examined using a dilute solution of phenazine in $CS_2$ and the intermediate range 1650–1400 as a solid solution in KBr. Under these conditions no less than 39 absorptions were observed; however, many of these absorptions are too weak to be of diagnostic value and the stronger more diagnostic bands appear at 3065, 1515, 1432, 1362, 1112, 1029, 958, 905, 820, 752 and 745 cm$^{-1}$.

## 2.14.1.5 UV Spectroscopy

A large body of information is available on the UV spectra of pyrazine derivatives ⟨B-61MI21400, B-66MI21400⟩. Pyrazine in cyclohexane shows two maxima at 260 nm (log ε 3.75) and 328 nm (log ε 3.02), corresponding to $\pi \to \pi^*$ and $n \to \pi^*$ transitions respectively ⟨72AHC(14)99⟩. Auxochromes show similar hypsochromic and bathochromic shifts to those observed with the corresponding benzenoid derivatives.

The UV spectra of quinoxalines have been examined in several solvents. In cyclohexane, three principal absorptions are observed (Table 2). In hydroxylic solvents the vibrational fine structure disappears and in methanol or water the weak $n \to \pi^*$ transitions are obscured by the intense $\pi \to \pi^*$ transition ⟨79HC(35)1⟩.

**Table 2** Absorption Maxima of Quinoxaline in Cyclohexane ⟨B-55MI21400⟩

| $\lambda_{max}$ | log ε | Origin |
|---|---|---|
| 340 | 2.84 | $n \to \pi^*$ |
| 312 | 3.81 | $\pi \to \pi^*$ |
| 232 | 4.51 | $\pi \to \pi^*$ |

UV spectra of phenazine and its derivatives in ethanolic solution have been reported ⟨62T1095⟩. Phenazine itself shows four maxima at 250 (log ε 5.1), 350 (log ε 4.0), 364 (log ε 4.1) and 400 nm (log ε 3.3), although the transitions responsible for these maxima have not been specified. When the molecule is protonated, as in the hydrochloride salt, six maxima are observed at 251 (5.0), 257 (4.9), 366 (4.2), 375 (4.2), 383 (4.2) and 430 nm (3.2).

### 2.14.1.6 Mass Spectrometry

A feature common to the pyrazine, quinoxaline and phenazine ring systems is their remarkable stability in the mass spectrometer and in all cases with the parent heterocycles the molecular ion is the base peak. In the case of pyrazine, two major fragments are observed at $m/e$ 53 and 26, and these fragments are consistent with the fragmentation pattern shown in Scheme 1.

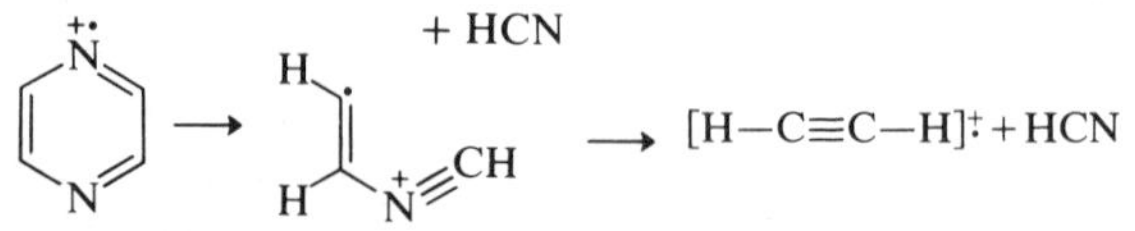

**Scheme 1**

The occurrence of a large number of pyrazines as flavouring or aroma constituents and as pheromones in extremely low concentrations has led to mass spectrometry being the method of choice for determining the gross structural details of a pyrazine nucleus. The method appears to be generally applicable and relatively specific and sensitive.

The fragmentation of quinoxaline is remarkably similar to that of pyrazine in that the molecular ion is also the base peak and the first fragmentation involves the loss of HCN ⟨70MI21401⟩. In the case of substituted quinoxalines and pyrazines the fragmentation pattern is to a certain extent governed by the nature of the substituents. Simple methyl and polymethyl derivatives usually give characteristic fragmentations which are definitive for given isomers although it is not always possible to define substitution patterns. When ethyl groups are present in the pyrazine ring then the $(M-1)^+$ peak appears as the base peak. With longer chain alkyl derivatives the McLafferty rearrangement becomes possible and usually dominates the spectrum.

In contrast to pyrazine and quinoxaline, phenazine and its derivatives behave somewhat differently in the mass spectrum. Phenazine itself shows a fragmentation in which the first steps appear to involve loss of nitrogen from the molecular ion to yield biphenylene (Scheme 2); a metastable peak supports this fragmentation ⟨66CPB426⟩. Thereafter, peaks corresponding to the typical fragmentation of biphenylene appear in the mass spectrum along with peaks corresponding to minor fragmentation pathways.

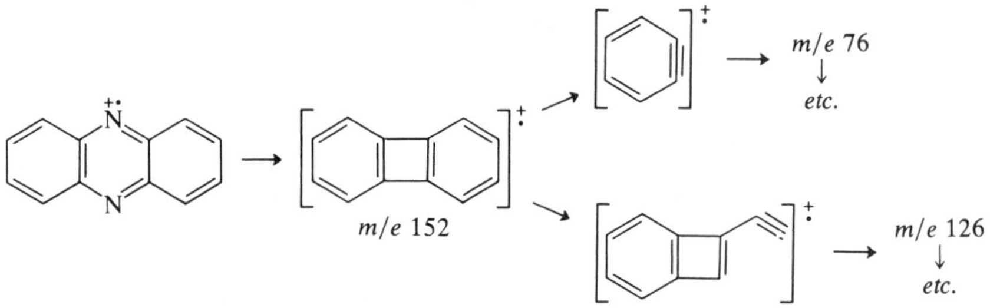

**Scheme 2**

### 2.14.1.7 Physical Properties

Table 3 lists some of the basic physical properties of pyrazine, quinoxaline and phenazine (references are given in the main text).

**Table 3** Physical Properties of Pyrazine, Quinoxaline and Phenazine

|  | *Pyrazine* | *Quinoxaline* | *Phenazine* |
|---|---|---|---|
| M.p. (°C) | 54 | 29–30 | 171 |
| B.p. (°C) | 121 | 108–111/12 mmHg | 339 |
| p$K_a$ | 0.65 (−5.8)[a] | 0.56 (−5.52) | 1.2 |
| $\rho^T$ | 1.0311 | 1.1334 | 1.33[c] |
| $n_D^T$ | 1.4953 | 1.6231 | — |
| $\mu$ (D) | 0 | 0.51[b] | 0 |

[a] Second p$K_a$ in parentheses.   [b] In benzene solution.
[c] Determined crystallographically for $\alpha$-phenazine. Phenazine exhibits polymorphism, with $\alpha$-phenazine being the common polymorph.

## 2.14.2 REACTIVITY

### 2.14.2.1 Electrophilic Substitution

The known reluctance of pyridine to take part in electrophilic substitution reactions (see Chapter 2.05) suggests that the introduction of a second azomethine nitrogen into the ring would have the effect of rendering pyrazine even less reactive towards electrophiles than pyridine, and generally this is the case. Aside from the intrinsic deactivation by the heteroatoms, the lone pairs on the ring nitrogen atoms represent a further cause of deactivation in that they can react with electrophilic species to form the conjugate acid, *e.g.* (**18**), which, being a positively charged species, is extremely reluctant to react further with electrophiles.

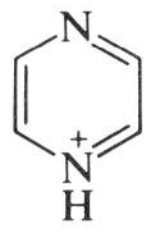

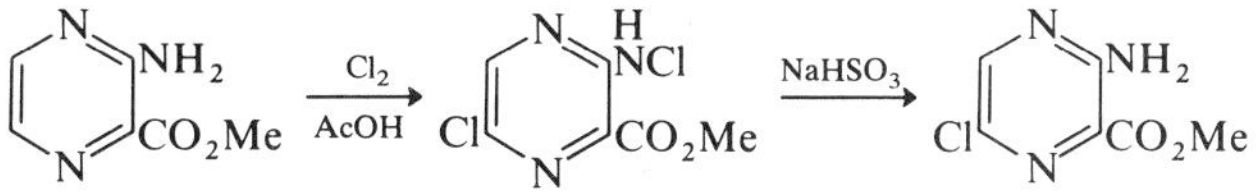

**Scheme 3**

When the ring is substituted with activating substituents ($+I$ or $+M$) then reactions which may formally be regarded as electrophilic substitution take place, but these are limited in number. Chlorination of 2-amino-3-methoxycarbonylpyrazine with chlorine in acetic acid results in ring chlorination at the 5-position but the amino group is also chlorinated (Scheme 3) ⟨B-79MI21400⟩. However, the *N*-chloro compound is readily reduced to the amine using sodium bisulphite. 2-Amino-3-bromopyrazine has been brominated similarly, using bromine in hydrobromic acid to yield 2-amino-3,5-dibromopyrazine ⟨69EGP66877⟩.

Remarkable selectivity has been observed ⟨72JCS(P1)2004⟩ in the chlorination of 2-alkyl-pyrazines with sulfuryl chloride/DMF. The reaction between this chlorinating agent and the alkylpyrazine is exothermic, but when the temperature is controlled at *ca.* 20 °C during the addition of the reagent, then yields of 2-alkyl-3-chloropyrazine of about 30% are consistently obtained. Under these conditions the pyrazine is selectively chlorinated at the 3-position and the DMF appears to play a crucial rôle since in its absence little ring chlorination is observed. When POCl$_3$/PCl$_5$ mixtures are used a greater preference for the formation of the 3-alkyl-5-chloropyrazine was observed.

Electrophilic substitution reactions of unsubstituted quinoxaline or phenazine are unusual; however, in view of the increased resonance possibilities in the transition states leading to the products one would predict that electrophilic substitution should be more facile than with pyrazine itself (*cf.* the relationship between pyridine and quinoline). In the case of quinoxaline, electron localization calculations ⟨57JCS2521⟩ indicate the highest electron density at positions 5 and 8 and substitution would be expected to occur at these positions. Nitration is only effected under forcing conditions, *e.g.* with concentrated nitric acid and oleum at 90 °C for 24 hours a 1.5% yield of 5-nitroquinoxaline (**19**) is obtained. The major product is 5,6-dinitroquinoxaline (**20**), formed in 24% yield.

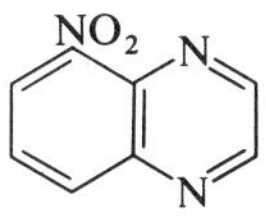

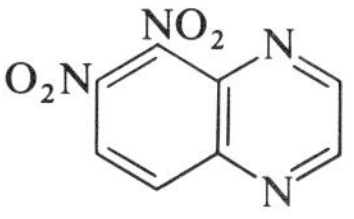

When activating substituents are present in the benzenoid ring, substitution usually becomes more facile and occurs in accordance with predictions based on simple valence bond theory. When activating substituents are present in the heterocyclic ring the situation varies depending upon reaction conditions; thus, nitration of 2(1*H*)-quinoxalinone in acetic acid yields 7-nitro-2(1*H*)-quinoxalinone (**21**) whereas nitration with mixed acid yields the 6-nitro derivative (**22**). The difference in products probably reflects a difference in the species being nitrated: neutral 2(1*H*)-quinoxalinone in acetic acid and the diprotonated species (**23**) in mixed acids.

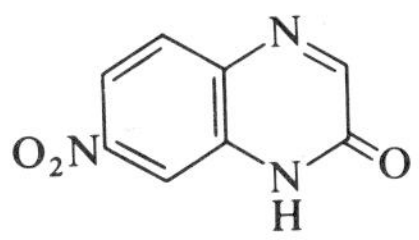

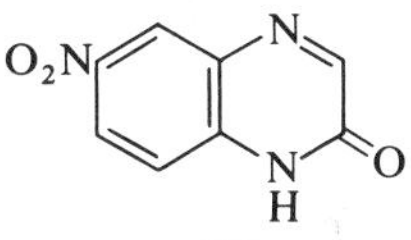

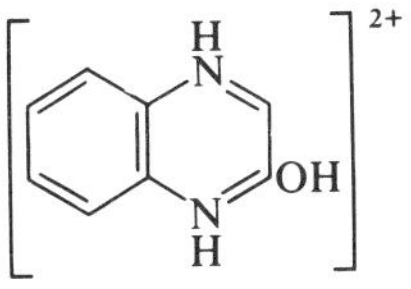

In the case of phenazine, substitution in the hetero ring is clearly not possible without complete disruption of the aromatic character of the molecule. Like pyrazine and quinoxaline, phenazine is very resistant towards the usual electrophilic reagents employed in aromatic substitution reactions and substituted phenazines are generally prepared by a modification of one of the synthetic routes employed in their construction from monocyclic precursors. However, a limited range of substitution reactions has been reported. Thus, phenazine has been chlorinated in acid solution with molecular chlorine to yield the 1-chloro, 1,4-dichloro, 1,4,6-trichloro and 1,4,6,9-tetrachloro derivatives, whose gross structures have been proven by independent synthesis ⟨53G327⟩.

Conflicting reports on the nitration of phenazine have appeared, but the situation was clarified by Albert and Duewell ⟨47MI21400⟩. The early work suggested that 1,3-dinitrophenazine could be prepared in 66% yield under standard nitration conditions; however, this proved to be a mixture of 1-nitrophenazine and 1,9-dinitrophenazine (24). As with pyrazines and quinoxalines, activating substituents in the benzenoid rings confer reactivity which is in accord with valence bond predictions; thus, nitration of 2-methoxy- or 2-hydroxyphenazine results in substitution at the 1-position.

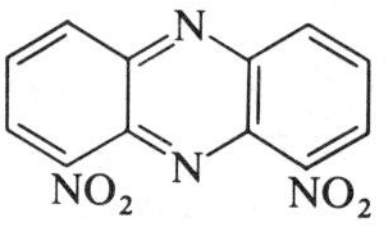

(24) 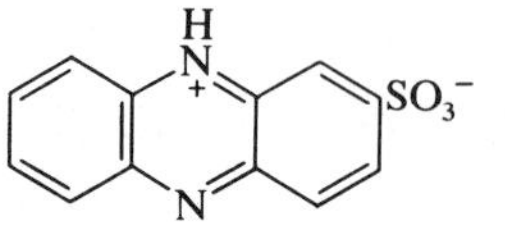    (25)

Sulfonation of phenazine has been carried out, but forcing conditions are required. Monosulfonation is best effected using oleum containing 50–70% $SO_3$ in the presence of $HgSO_4$ at 170 °C for long reaction times. Some polysulfonation occurs and residual phenazine remains after the reaction; the monosulfonic acid, which is yellow in colour, probably exists as the betaine (25). The position of sulfonation was confirmed by heating the sodium salt of the sulfonic acid with NaCN, which gave the corresponding nitrile. This was subsequently transformed into the known phenazine-2-carboxylic acid ⟨50G651⟩.

### 2.14.2.2 Nucleophilic Substitution

Broadly speaking, nucleophilic substitution may be divided into (a) the direct displacement of hydrogen and (b) the displacement of other substituents. Displacements of type (a) are rare and are typified by the Tschitschibabin reaction. Pyrazine reacts with $NaNH_2/NH_3$ to yield 2-aminopyrazine, but no yield has been quoted ⟨46USP2394963⟩. Generally, the synthesis of aminopyrazines, aminoquinoxalines and aminophenazines is more readily accomplished by alternative methods, particularly displacement of halogen from the corresponding halo derivatives, which are themselves readily available.

Phenazine reacts with benzenesulphinic acid in alcoholic hydrogen chloride to give 2-phenazinyl phenyl sulfone (26; Scheme 4), presumably by an intermediate 5,10-dihydrophenazine; this reaction is evidently a useful method of preparing 2-substituted phenazines, since the sulfone is readily displaced in substitution reactions.

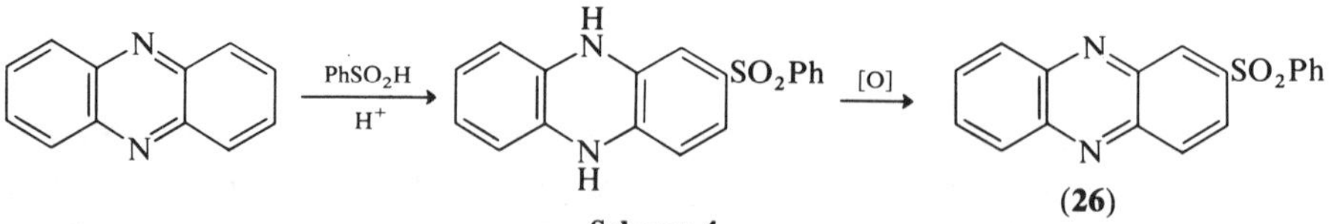

**Scheme 4**

The ease of displacement of substituents during nucleophilic substitution reactions varies depending upon the substitution pattern in the particular system. 2-Halopyrazines are much more reactive than 2-halopyridines ⟨67AJC1595⟩ and 2-haloquinoxalines are still more reactive than the halopyrazines ⟨56JCS1563⟩. In the case of halophenazines and quinoxalines substituted in the benzenoid ring, short range activation of nucleophilic substitution is not possible; however, it is known that 2-chlorophenazine reacts readily with nucleophiles. Thus, with ammonia 2-aminophenazine is formed in good yield ⟨52JA971⟩. Similarly, aqueous ethanolic KOH gives rise to 2-ethoxyphenazine in 88% yield after 3 days at reflux. In contrast, 1-chlorophenazine is considerably more resistant to displacement by nucleophiles, being stable under the above conditions. No information on the relative rates of reaction of 5- or 6-haloquinoxalines appears to be available.

In accordance with observations in halodinitrobenzene derivatives, fluoropyrazines are by far the most reactive of the halopyrazines. Fluoropyrazine undergoes facile reaction with sodium azide to give azidopyrazine (**27**), which exists in dynamic equilibrium with tetrazolo[1,5-*a*]pyrazine (**28**) ⟨66JHC435⟩.

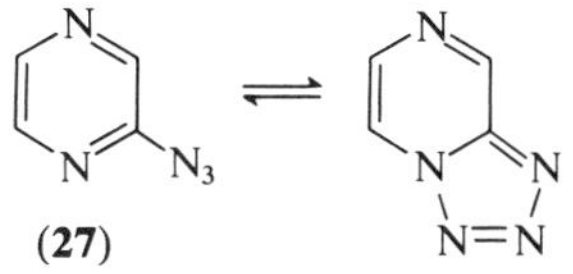

(27)        (28)

Monochloropyrazines undergo all of the expected displacement reactions with nucleophiles such as $CN^-$, ammonia and alkylamines, alkoxides and thiols; other groups such as SMe, SOMe and $SO_2Me$ also function as leaving groups in nucleophilic substitution reactions, albeit at a slower rate than the corresponding halides ⟨68JCS(B)1435⟩. One of the nitrile groups in 2,3-dicyanopyrazine appears to be very labile ⟨80JHC455⟩ and is replaced simply by heating a DMF solution with an alcohol and triethylamine as a catalyst.

Nucleophilic substitution frequently does not follow a mechanistically simple course. There is always a tendency to formulate nucleophilic substitution reactions as simple addition/elimination reactions (Scheme 5) but evidence has been presented showing that this is an oversimplification in a number of cases. Chloropyrazine reacts with sodamide or potassamide in liquid ammonia to produce a mixture of aminopyrazine, 2-cyanoimidazole and imidazole (Scheme 6) ⟨71RTC207⟩. Furthermore, when 2-chloropyrazine labelled at the 1-position with $^{15}N$ is subjected to the reaction, the $^{15}N$ is found in the amino group and not in the ring as expected on the basis of the simple addition/elimination mechanism. These results have been rationalized in terms of an Addition Nucleophilic Ring Opening Ring Closure (ANRORC) mechanism (Scheme 7) ⟨72RTC949⟩. In contrast, 2-chloroquinoxaline reacts by an addition/elimination mechanism.

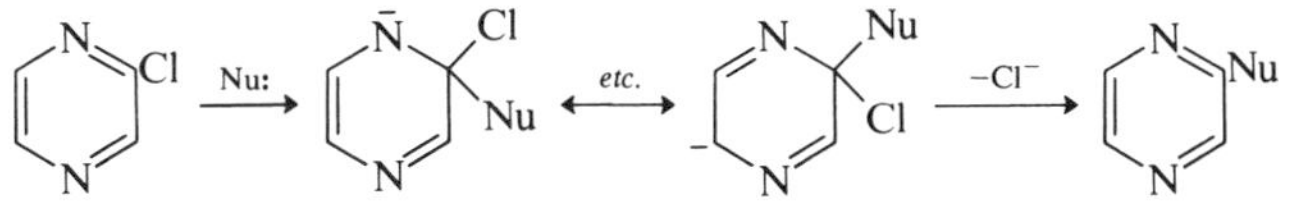

**Scheme 5**

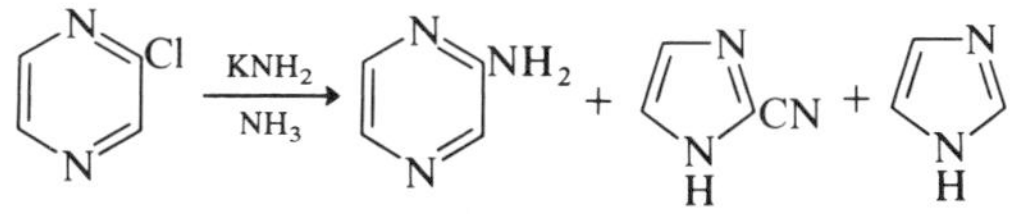

**Scheme 6**

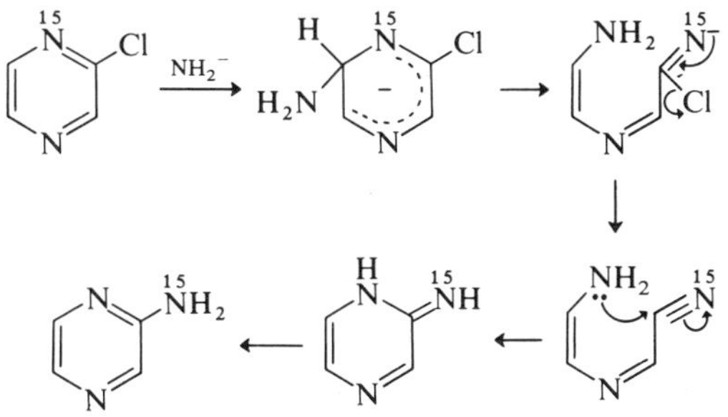

**Scheme 7**

Perhaps the most convincing evidence for nucleophilic attack at an unexpected ring position comes from the direct observation of intermediate Meisenheimer complexes in the NMR spectrum. When 2-chloro-3,6-diphenylpyrazine is treated with $KNH_2$ in liquid ammonia, the intermediate (**29**) was observed directly (Scheme 8). It was postulated that this initially formed complex rearranges to (**30**) which gives the observed product by elimination of a chloride ion ⟨73RTC708⟩.

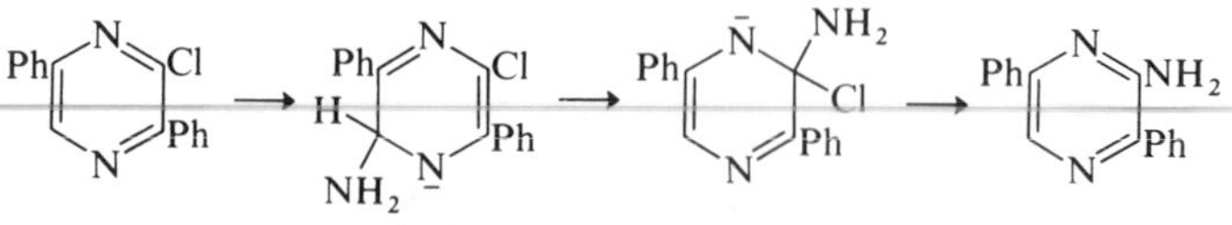

**Scheme 8**

It would appear that this type of addition may not be confined to the addition of $NH_2^-$ in liquid ammonia, since it has been observed that treatment of 2-chloro-3-dichloromethyl-pyrazine with an excess of methoxide results in the introduction of a methoxy group into the 6-position of the pyrazine ring (Scheme 9) ⟨68TL5931⟩. This reaction is best rationalized in terms of addition of the methoxide ion at the 6-position, followed by loss of chloride ion from the dichloromethyl side chain.

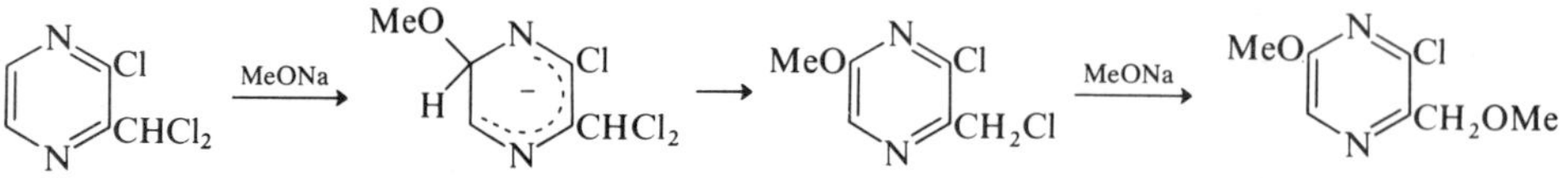

**Scheme 9**

Substitution of the pyrazine ring by electron releasing substituents reduces the reactivity of halopyrazines and more forcing conditions must invariably be employed to bring about displacement of the halogen.

### 2.14.2.3 Reactions with Radicals

Homolytic aromatic substitution of the diazines and benzodiazines appears to have been a little used phenomenon and information on such reactions in the pyrazine, quinoxaline and phenazine series is scarce, although in the case of pyrazine and quinoxaline the positions α to the ring nitrogen atoms appear to be the major site of attack ⟨72BSF1173⟩. Recent work on the substitution of pyrazine and quinoxaline with radicals generated by the oxidation of formamide ⟨70TL15⟩ indicate that the reaction has some preparative potential in that the corresponding carboxamides may be prepared in good yield; thus, pyrazine yields pyrazinecarboxamide in yields >80% (based on consumed starting material).

### 2.14.2.4 Side Chain Reactivity

Inductive and resonance stabilization of carbanions derived by proton abstraction from alkyl substituents α to the ring nitrogen in pyrazines and quinoxalines confers a degree of stability on these species comparable with that observed with enolate anions. The resultant carbanions undergo typical condensation reactions with a variety of electrophilic reagents such as aldehydes, ketones, nitriles, diazonium salts, *etc.*, which makes them of considerable preparative importance.

Methylpyrazine reacts with sodamide in liquid ammonia to generate the anion, which may be alkylated to give higher alkylpyrazines (Scheme 10) ⟨61JOC3379⟩. The alkylpyrazines have found extensive use as flavouring and aroma agents (see Section 2.14.4). Condensation reactions with esters, aldehydes and ketones are common, *e.g.* methyl benzoate yields phenacylpyrazine in 95% yield, and reactions of this type are summarized in Scheme 11.

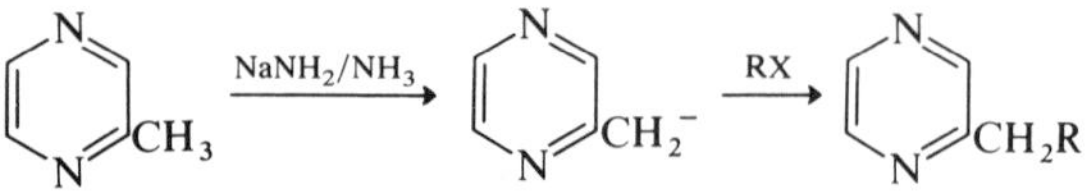

**Scheme 10**

Although most of the reactions of preparative importance involving the α-alkyl carbanions are usually carried out under controlled conditions with $NH_2^-/NH_3$ being used as the base, a number of reactions using less severe conditions are known, both in the pyrazine and quinoxaline series. In the case of alkylquinoxalines, where an increased number of resonance possibilities exist, mildly basic conditions are usually employed in condensation reactions.

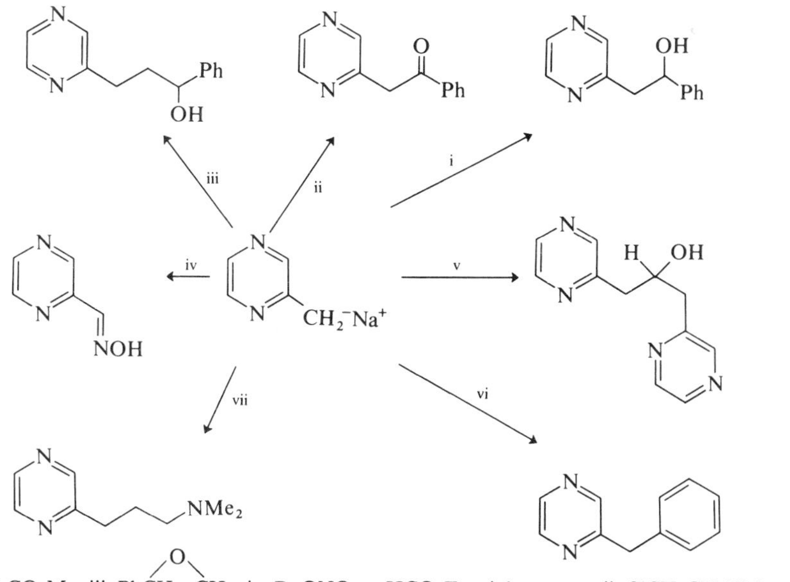

i, PhCHO; ii, PhCO₂Me; iii, PhCH—CH₂; iv, BuONO; v, HCO₂Et; vi, benzyne; vii, ClCH₂CH₂NMe₂

**Scheme 11**

2-Methylquinoxaline reacts with formaldehyde to give (**31**) and (**32**) in low yields ⟨64JGU2089⟩, and 2,3-dimethylquinoxaline reacts in a similar way. Many examples of the condensation of aromatic aldehydes with alkylquinoxalines are known and the reactions are normally carried out in refluxing acetic anhydride, not requiring the strongly basic conditions usually employed in the pyrazine series. Alkylpyrazines and alkylquinoxalines both participate in the Mannich reaction with formaldehyde and dimethylamine to yield the dialkylaminoethyl derivatives (**33**) and (**34**), which serve as useful starting materials for the corresponding vinyl derivatives.

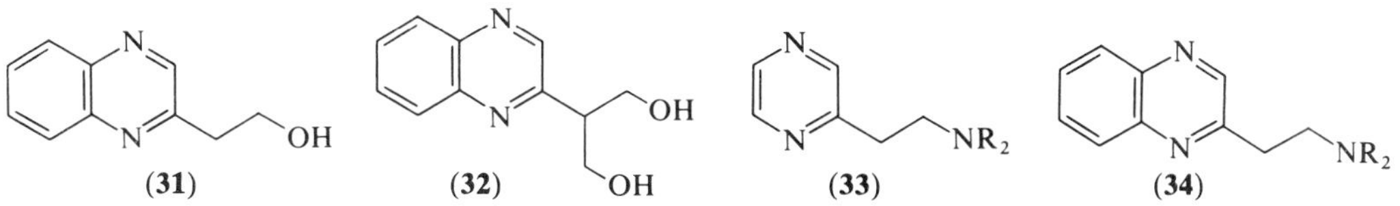

Anion formation has been accomplished using very strong bases such as butyllithium (Scheme 12) ⟨81TL1219⟩; however, this reaction is not without its complications in that normally addition of organometallic reagents to the C=N bond is facile, as illustrated by the addition of methyllithium ⟨74JOC3598⟩, ethyllithium ⟨71RTC513⟩ and phenyllithium to pyrazine derivatives. These reactions usually result in the formation of 1,2-dihydropyrazines which undergo oxidation during work up (Scheme 13). A further complicating feature in the use of strong bases is the possibility of anion formation at an unsubstituted ring position adjacent to a nitrogen atom, *e.g.* (**35**) ⟨69JHC239⟩.

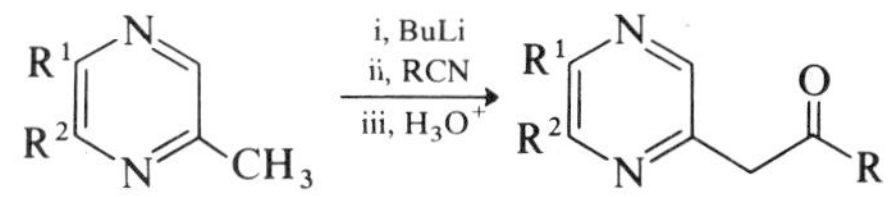

**Scheme 12**

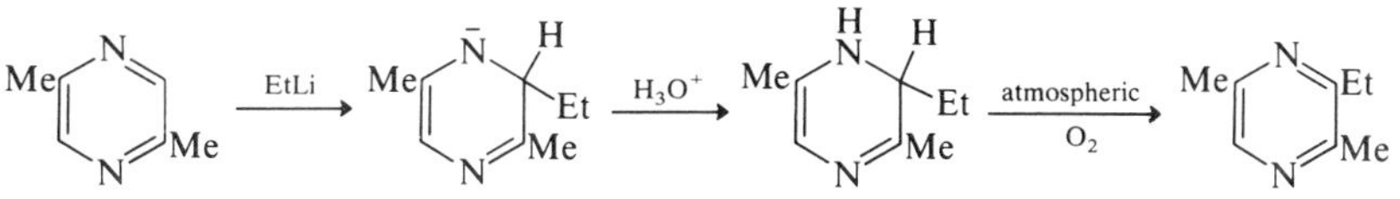

**Scheme 13**

Alkyl side chains in both pyrazines and quinoxalines are susceptible to halogenation by elemental halogens ⟨28JCS1960, 68TL5931⟩ and under radical conditions with NBS ⟨72JOC511⟩. Thus, bromination of 2-methylquinoxaline with bromine in the presence of sodium acetate yields only the tribromomethylquinoxaline (36), and chlorination of methylpyrazine with chlorine in acetic acid at 100 °C similarly gives trichloromethylpyrazine (37). Interestingly, when 2-chloro-3-methylpyrazine is similarly chlorinated the major product appears to be 2-chloro-3-dichloromethylpyrazine (38).

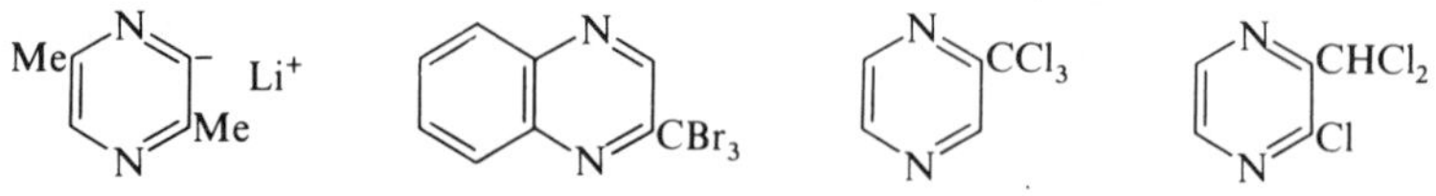

(35)            (36)            (37)            (38)

Bromination using NBS has been used to provide acetylpyrazine derivatives from the corresponding ethylpyrazines. Bromination of 2-ethyl-3-methylpyrazine gives 2-bromoethyl-3-methylpyrazine in quantitative yield; this may be oxidized using the sodium salt of 2-nitropropane or with pyridine $N$-oxide to yield 2-acetyl-3-methylpyrazine in yields of 66 and 25% respectively (Scheme 14).

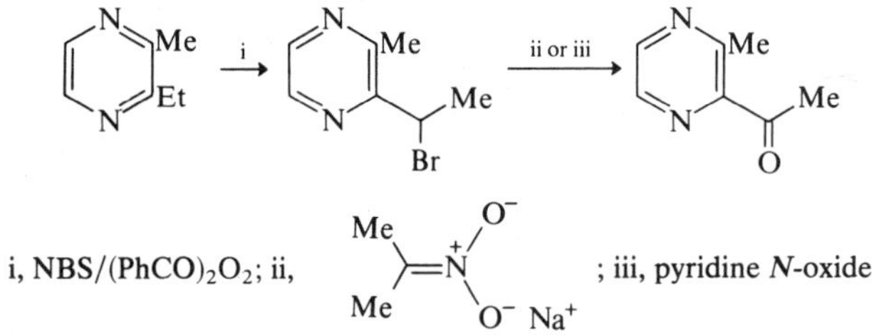

i, NBS/(PhCO)$_2$O$_2$; ii, [structure]; iii, pyridine $N$-oxide

**Scheme 14**

Direct oxidation of ethylpyrazines to the corresponding acetylpyrazines may also be carried out in favourable circumstances using hot chromic acid ⟨75JOC1178⟩. Treatment of 2-ethyl-3-alkylpyrazines with chromic acid yields the corresponding 2-acetyl-3-alkyl-pyrazines in yields of 50–70%. In the absence of the 3-alkyl substituent the yields fall dramatically to less than 10%. Acetylpyrazines are more generally prepared by the inverse addition of a Grignard reagent to a cyanopyrazine.

### 2.14.2.5 *N*-Oxides

Heterocyclic $N$-oxides hold a key position in the chemistry of the heterocycles in that they permit functional group manipulation and structural modification possibilities which are not accessible by any other methods. The preparation of the mono-$N$-oxides of pyrazine, quinoxaline and phenazine from the parent heterocycles is readily accomplished by the methods used in the preparation of their monoazine counterparts, and typical oxidizing agents such as Ac$_2$O/H$_2$O$_2$, $m$-chloroperbenzoic acid, monoperoxyphthalic acid, permaleic acid and peroxytrifluoroacetic acid are used. Standard experimental procedures by direct oxidation using H$_2$O$_2$/AcOH are available, although recently a cautionary note has appeared which indicates that, although uncommon, an explosion hazard exists ⟨77MI21400⟩. Using this method the mono-$N$-oxides of pyrazine (39), quinoxaline (40) and phenazine (41) have all been prepared in good yields.

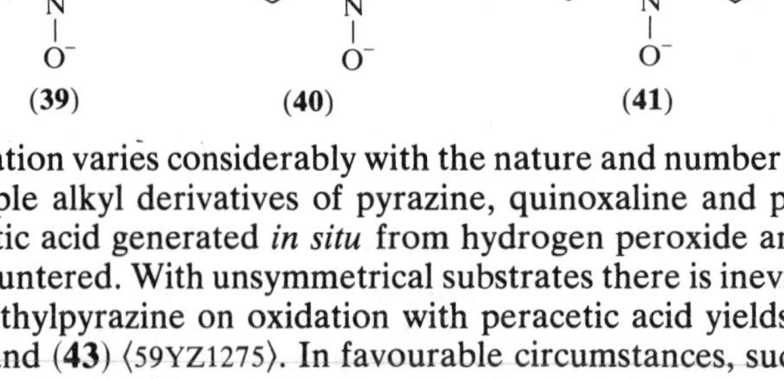

(39)            (40)            (41)

The ease of oxidation varies considerably with the nature and number of ring substituents; thus, although simple alkyl derivatives of pyrazine, quinoxaline and phenazine are easily oxidized by peracetic acid generated *in situ* from hydrogen peroxide and acetic acid, some difficulties are encountered. With unsymmetrical substrates there is inevitably the selectivity problem. Thus, methylpyrazine on oxidation with peracetic acid yields mixtures of the 1- and 4-oxides (42) and (43) ⟨59YZ1275⟩. In favourable circumstances, such product mixtures may be separated by fractional crystallization. Simple alkyl derivatives of quinoxalines are readily oxidized using acetic acid/$H_2O_2$, although it has been reported that quinoxaline itself may give rise to quinoxaline-2,3-dione (**44**) under these conditions ⟨59YZ260⟩. To form the di-*N*-oxides, protracted heating with the peracid is usually required and complete conversion is rare. It is also evident that in many cases the di-*N*-oxides undergo decomposition under the reaction conditions.

(**42**)    (**43**)    (**44**)

With pyrazines and quinoxalines, as the size of the substituent $\alpha$ to a given ring nitrogen atom increases, a greater preference for the formation of the 4-oxide is observed. This is evident in a consideration of 2-methylquinoxaline which gives a $3:2$ ratio of the 1- and 4-oxides, 2-ethylquinoxaline which gives a ratio of $23:26$ for 1- and 4-oxides, isopropylquinoxaline which gives the 4-oxide and a small quantity of the 1,4-dioxide, and 2-*t*-butylquinoxaline which gives exclusively the 4-oxide.

In the pyrazine and quinoxaline series, electron withdrawing substituents such as halogens, $CO_2H$, $CO_2R$, $CONH_2$, CN, *etc.* deactivate the ring towards *N*-oxidation. Clearly this effect will be more pronounced in the pyrazine series and quinoxalines where the deactivating substituent is bonded to the heteroaromatic ring. In quinoxalines substituted in the benzenoid ring and in phenazines, deactivating substituent effects are significantly less dominant. When the substituent is located $\alpha$ to the ring nitrogen, then it is specifically that nitrogen atom which is deactivated to the greater extent, as illustrated in the conversion of 2-chloro-3,6-dimethylpyrazine (**45**) to the 4-oxide (**46**) with peracid ⟨70JCS(C)1070⟩.

(**45**)    (**46**)    (**47**)

More recently, several groups have investigated *N*-oxidation using strong hydrogen peroxide (60–90% $H_2O_2$) in sulfuric acid ⟨71CC28, 71JHC697⟩, a reagent which forms di-*N*-oxides of the most highly deactivated pyrazine nuclei such as tetrachloropyrazine (**47**). Mixan and Pews reasoned that under these conditions the effective oxidizing agent is Caro's acid, and they have demonstrated that potassium peroxydisulfate may be used as a substitute for the high test peroxide (which is always hazardous, both in handling and in storage) with no detriment to yields ⟨77JOC1869⟩. Using this reagent some rather unusual transformations are possible. 2-Chloropyrazine or 2-chloroquinoxaline both yield the corresponding 1-oxides as the only product. Even when an excess of the reagent is used there is no evidence for 1,4-dioxide formation or the formation of the isomeric 4-oxides, and it has been argued that under normal circumstances the relative rates of oxidation of the two nitrogen atoms is governed by their relative basicities. In 2-chloropyrazine, for example, the more basic nitrogen atom is N-4. In very strong acid medium the pyrazine is protonated and it is argued that although the nucleophilicity of the monoprotonated pyrazine is reduced, the electrophilicity of peroxysulfuric acid is sufficient to effect oxidation of the unprotonated (*i.e.* less basic) nitrogen atom. This method is of particular importance, as by careful choice of reagents it is possible to exercise regioselectivity in *N*-oxidation reactions (Scheme 15).

(**48**)    **Scheme 15**

Use of Caro's acid does result in di-*N*-oxidation of 2,3-dichloropyrazine and 2,3-dichloroquinoxaline, but it is of interest to note that other heterocycles such as pyridine derivatives, pyridazines and pyrimidines do not give *N*-oxides under these conditions. It has also been observed that 2-chloropyrazine 4-oxide (**48**) is not oxidized to the di-*N*-oxide (**49**) with this reagent. No data are available on the fate of the di-*N*-oxide (**49**) in Caro's acid. Since there is precedent for deoxygenation of an *N*-oxide by electrophilic attack of a peracid on the oxygen atom of an *N*-oxide (Scheme 16) ⟨70JCS(C)1070⟩, it cannot be ruled out that (**49**) is formed during the course of the reaction and undergoes deoxygenation to the observed product.

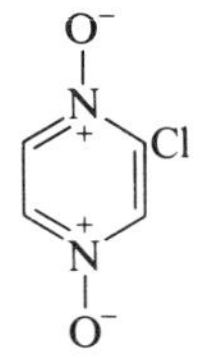

(**49**) 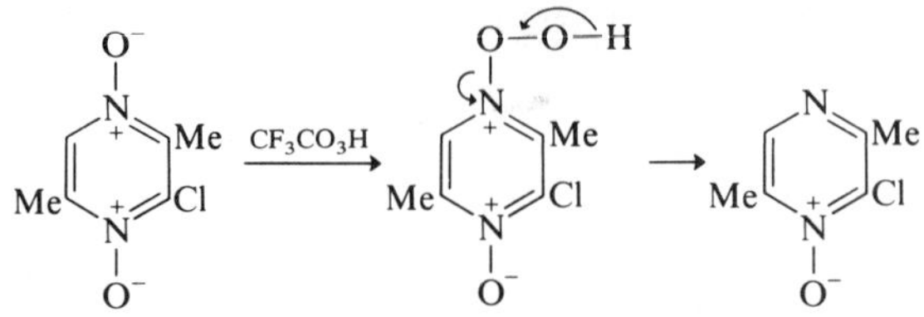                                                    Scheme 16

Cerium(IV) ammonium nitrate in methanol has been used to oxidize phenazine to the mono-*N*-oxide (**41**) in good yield ⟨75JCS(P1)1398⟩, but no other reports on the application of this reagent to the pyrazine or quinoxaline series have appeared.

Electron-donating substituents such as amino or acetamido groups α to the ring nitrogen activate the ring towards attack; thus, oxidation of 2-aminoquinoxaline with permaleic acid results in the formation of the 1-oxide (Scheme 17), although it is not possible to say whether hydrogen bonding or electronic effects are responsible for the selectivity ⟨73JCS(P1)2707⟩.

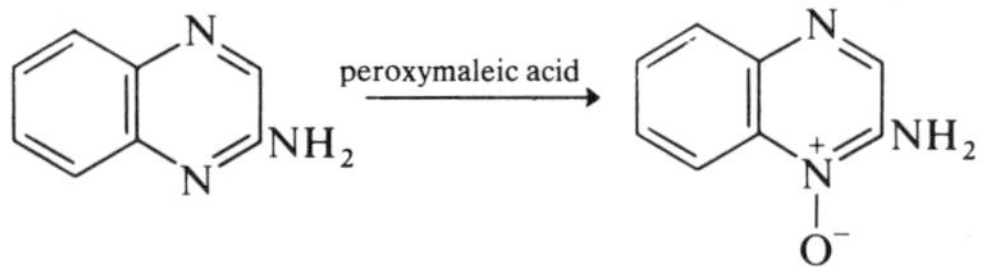

Scheme 17

Many pyrazine and quinoxaline syntheses yield mono- or di-*N*-oxides ⟨76H(4)769⟩. The condensation of α-aminooximes with 1,2-diketones results in the direct formation of pyrazine mono-*N*-oxides. The α-aminooximes themselves are not easily prepared but 2-amino-2-deoxy sugars readily form the oximes, which have been condensed with glyoxal to yield the pyrazine 4-oxides (Scheme 18) ⟨72JOC2635, 80JOC1693⟩.

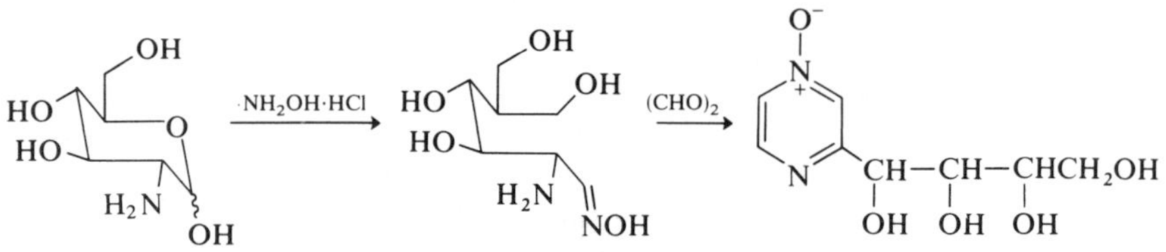

Scheme 18

Pyrazine 1,4-dioxides are available by the direct self-condensation of 1,2-hydroxyaminooximes ⟨70JOC2790⟩. 1,2-Nitrooximes are obtained by the isomerization of alkene–dinitrogen trioxide adducts, which are reduced with palladium on charcoal to the hydroxyaminooximes which undergo acid-catalyzed auto-condensation to the pyrazine 1,4-dioxides (Scheme 19).

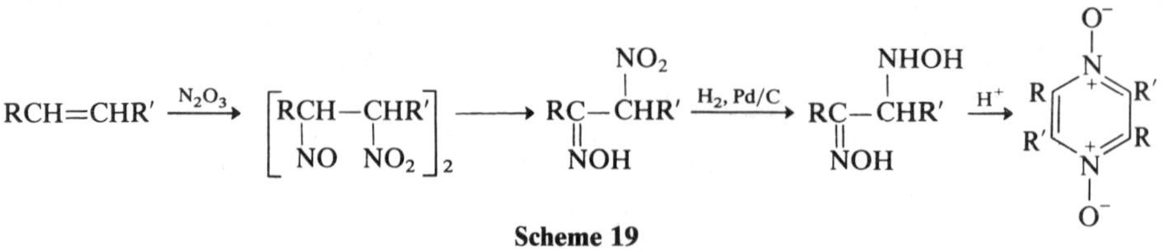

$$RCH{=}CHR' \xrightarrow{N_2O_3} \left[ \begin{matrix} RCH{-}CHR' \\ | \quad\;\; | \\ NO \;\; NO_2 \end{matrix} \right]_2 \longrightarrow \underset{NOH}{\overset{NO_2}{RC{-}CHR'}} \xrightarrow{H_2, Pd/C} \underset{NOH}{\overset{NHOH}{RC{-}CHR'}} \xrightarrow{H^+}$$

Scheme 19

Quinoxaline mono-*N*-oxides are also available by a direct synthesis from *o*-nitroaniline derivatives. Condensation of acetyl chloride derivatives with *o*-nitroaniline followed by treatment with sodium ethoxide in ethanol yields the mono-*N*-oxides in good yields (Scheme 20) ⟨64JCS2666⟩.

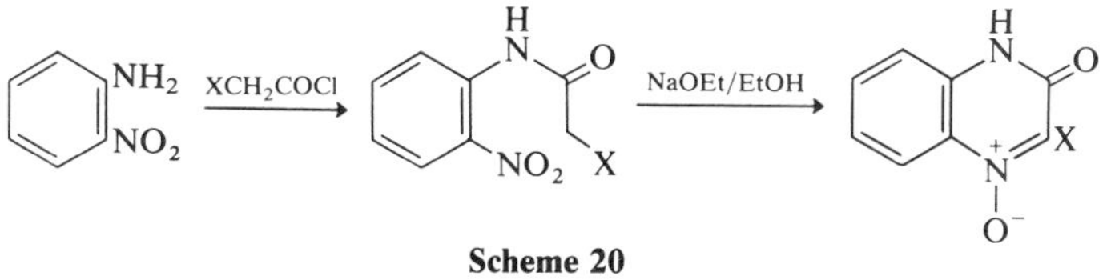

**Scheme 20**

Phenazine mono-*N*-oxides have also been prepared from nitrobenzene derivatives. Condensation of nitrobenzene with aniline using dry NaOH at 120–130 °C results in modest yields of phenazine 5-oxide, although the precise mechanism of this reaction is not well understood ⟨57HC(11)1⟩; with unsymmetrical substrates it is not possible to predict which of the isomeric *N*-oxides will be produced. Nitrosobenzene derivatives also function as a source of phenazine mono-*N*-oxides; thus, if 4-chloronitrosobenzene is treated with sulfuric acid in acetic acid at 20 °C the *N*-oxide is formed (Scheme 21).

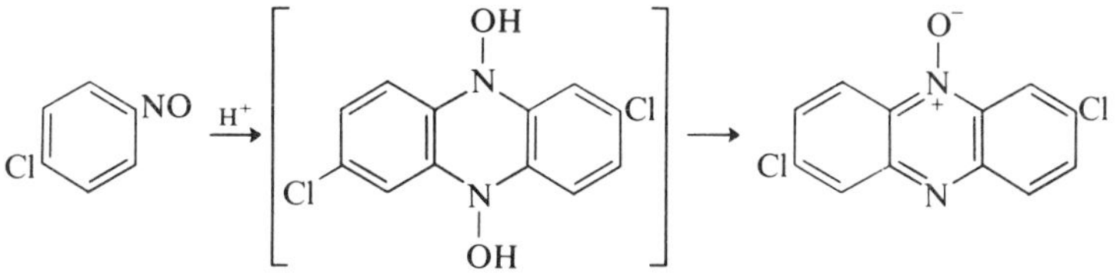

**Scheme 21**

The synthesis of the di-*N*-oxides of pyrazines, quinoxalines or phenazines may be accomplished by direct oxidation of the heterocycles. Generally it is easier to prepare phenazine dioxides in this way, since even if deactivating substituents are present they are too far removed from the ring nitrogen atoms to have any significant effect on the reaction rates and peracetic acid may generally be used. In the case of quinoxalines, when deactivating substituents are present in the benzenoid ring, similar behaviour would be expected; however, when deactivating substituents are present in the hetero ring, similar deactivation to that observed with pyrazines is normal. In these circumstances more forcing conditions become necessary; for highly deactivated systems such as the chloro or fluoro compounds, peroxytrifluoroacetic acid appears to be the oxidant of choice and gives moderate to good yields of the di-*N*-oxides.

Quinoxaline and phenazine di-*N*-oxides are also directly available by the 'Beirut reaction' (see Section 2.14.3.2).

Pyrazine and quinoxaline *N*-oxides generally undergo similar reactions to their monoazine counterparts. In the case of pyridine *N*-oxide the ring is activated both towards electrophilic and nucleophilic substitution reactions; however, pyrazine *N*-oxides are generally less susceptible to electrophilic attack and little work has been reported in this area. Nucleophilic activation generally appears to be more useful and a variety of nucleophilic substitution reactions have been exploited in the pyrazine, quinoxaline and phenazine series.

Treatment of both pyrazine 1-oxide and quinoxaline 1-oxide with POCl$_3$ results in the formation of the corresponding chlorinated derivatives (**50**) and (**51**). However, in the case of quinoxaline 1-oxide the 2-chloroquinoxaline is accompanied by 6-chloroquinoxaline (**52**) ⟨67YZ942⟩.

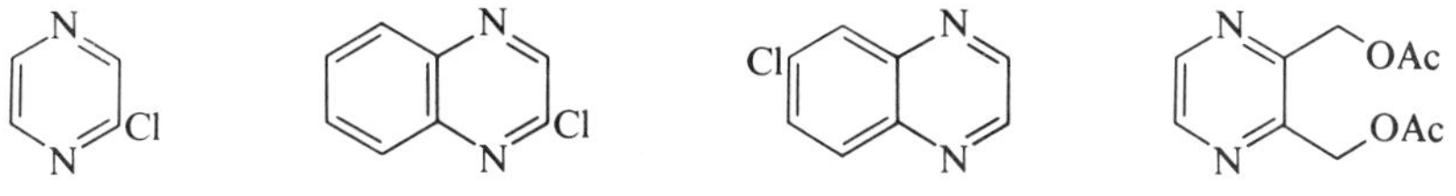

Other reactions with their counterparts in the pyridine series are also well known. Thus, 2,3-dimethylpyrazine 1,4-dioxide reacts with acetic anhydride to yield 2,3-bis(acetoxymethyl)pyrazine (**53**) in good yield ⟨72KGS1275⟩. Pyrazine 1-oxide also reacts directly with acetic anhydride to yield 2(1*H*)-pyrazinone by way of the intermediate acetate (Scheme 22). The corresponding reaction in the quinoxaline series is not so well defined and at least three products result (Scheme 23) ⟨67YZ942⟩.

In view of the known behaviour of pyrazines during nucleophilic substitution reactions, it comes as no surprise that anomalous reactions appear during nucleophilic substitution

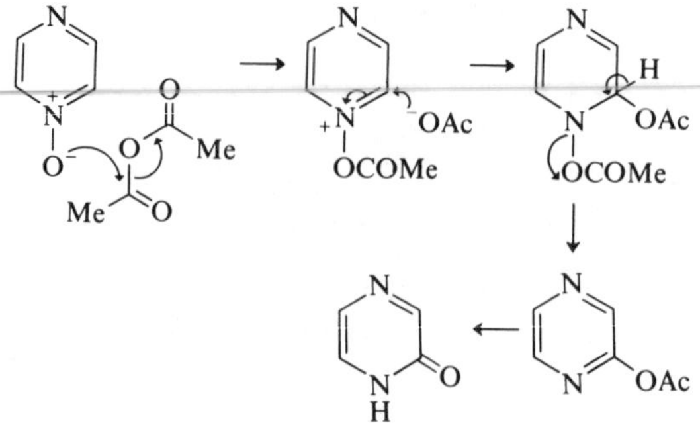

**Scheme 22**

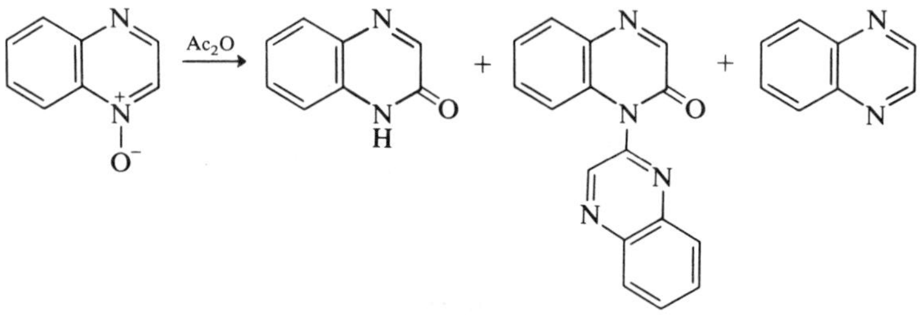

**Scheme 23**

reactions with the *N*-oxides. Treatment of 3-alkoxycarbonylpyrazine 1-oxides with phosphorus oxychloride does not result in the formation of either of the expected chloro compounds; substitution occurs $\alpha$ to the nitrogen remote from the *N*-oxide group to yield 2-chloro-6-methoxycarbonylpyrazine (**54** (Scheme 24)) ⟨69JAP6912898, 69JAP6920345⟩.

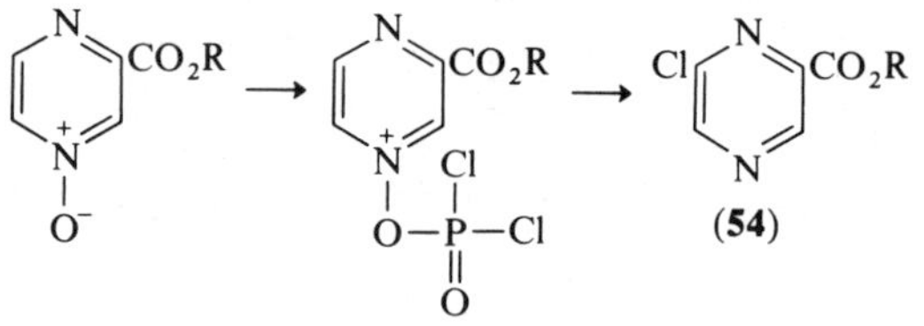

**Scheme 24**

Ring substituents show enhanced reactivity towards nucleophilic substitution, relative to the unoxidized systems, with substituents $\alpha$ to the *N*-oxide showing greater reactivity than those in the $\beta$-position. In the case of quinoxalines and phenazines the degree of labilization of a given substituent is dependent on whether the intermediate addition complex is stabilized by mesomeric interactions and this is easily predicted from valence bond considerations. 2-Chloropyrazine 1-oxide is readily converted into 2-hydroxypyrazine 1-oxide (1-hydroxy-2(1*H*)-pyrazinone) (**55**) on treatment with dilute aqueous sodium hydroxide ⟨63G339⟩, whereas both 2,3-dichloropyrazine and 3-chloropyrazine 1-oxide are stable under these conditions. This reaction is of particular importance in the preparation of pyrazine-based hydroxamic acids which have antibiotic properties.

In the case of substituted phenazine *N*-oxides some activation of substituents towards nucleophilic substitution is observed. 1-Chlorophenazine is usually very resistant to nucleophilic displacements, but the 2-isomer is more reactive and the halogen may be displaced with a number of nucleophiles. 1-Chlorophenazine 5-oxide (**56**), however, is comparable in its reactivity with 2-chlorophenazine and the chlorine atom is readily displaced in nucleophilic substitution reactions. 2-Chlorophenazine 5,10-dioxide (**57**) and 2-chlorophenazine 5-oxide both show enhanced reactivity relative to 2-chlorophenazine itself. On the basis of these observations, similar activation of 5- or 6-haloquinoxaline *N*-oxides should be observed but little information is available at the present time.

In those reactions where the *N*-oxide group assists electrophilic or nucleophilic substitution reactions, and is not lost during the reaction, it is readily removed by a variety of reductive procedures and thus facilitates the synthesis of substituted derivatives of pyrazine, quinoxaline and phenazine.

Side chain reactivity is also enhanced and is typified by the difference in reactivity of 2-methylpyrazine and 2-methylpyrazine 1,4-dioxide towards anion formation and subsequent condensation reactions. 2-Methylpyrazine undergoes condensation with benzaldehyde at 180 °C, with zinc chloride catalysis, to yield the styrylpyrazine (**58**), whereas the corresponding reaction of 2-methylpyrazine 1,4-dioxide proceeds at 25 °C under base catalysis ⟨67KGS419⟩.

### 2.14.2.6 Pyrazinones, Quinoxalinones and Hydroxyphenazines

Although some writers have found it convenient to rationalize the behaviour of these hydroxydiazines in terms of the ring retaining its heteroaromatic character with the hydroxyl substituent existing as such, *e.g.* (**59**), rather than in the tautomeric amide form (**60**), there is overwhelming evidence to suggest that in those cases where the hydroxyl group is in the same ring as the heteroatoms the molecules exist largely in the amide form. IR studies have shown that both 'hydroxypyrazines' and '2-hydroxyquinoxalines' (**61**) ⟨61JCS3983⟩ have carbonyl absorptions in the region 1690–1660 cm$^{-1}$, and it is now generally accepted that both systems exist in tautomeric equilibrium, with the position of the equilibrium being governed by substituent and solvent effects to a large extent. Hence pyrazinones and quinoxalinones are more suitable names to describe the structures of these systems.

Alkylation of pyrazinones and quinoxalinones may be carried out under a variety of conditions and it is usually observed that while *O*-alkylation may occur under conditions of kinetic control, to yield the corresponding alkoxypyrazines or alkoxyquinoxalines, under thermodynamic control the *N*-alkylated products are formed. Alkylation using trialkyloxonium fluoroborate results in exclusive *O*-alkylation, and silylation under a variety of conditions ⟨75MI21400⟩ yields specifically the *O*-silylated products. Alkylation with methyl iodide or dimethyl sulfate invariably leads to *N*-methylation.

Pyrazinones and quinoxalinones both play important roles in the chemistry of pyrazines and quinoxalines respectively, in that they are usually available by direct synthesis and serve as important starting points for halo derivatives, which in turn lead to a range of substitution products (*e.g.* see Section 2.14.3.3).

Other synthetic routes to pyrazinones and quinoxalinones are from the halo compounds, by dealkylation of ethers ⟨81JCS(P1)3111⟩ or by diazotization of the corresponding amines, but since the halo derivatives are normally derived from the hydroxy compounds, and the amines from the halo derivatives, direct synthesis seems to represent the most practical approach.

Both pyrazinones and quinoxalinones exhibit a dichotomy of behaviour in that they have the normal reactivity typical of amides but also exhibit reactions which are typical of phenols. They can be coupled with diazonium salts to produce azo dyes, and nitrated or brominated. 2(1*H*)-Quinoxalinone is nitrated (in acetic acid) ⟨61JCS1246⟩ and halogenated in the benzenoid ring at position 7; the corresponding 1-methyl compound, however, gives the 6-nitro derivative (H$_2$SO$_4$/KNO$_3$) ⟨61JCS1246⟩.

Tautomerism questions arise with hydroxyphenazines and quinoxalines which contain hydroxyl groups in the aromatic ring. Does 5-hydroxyquinoxaline (**62**) show any tendency to exist as the tautomer (**63**) or 1-hydroxyphenazine (**64**) to exist as (**65**)? Analogous tautomeric forms can also be written for 6-hydroxyquinoxaline and 2-hydroxyphenazine.

Little information is available at the present time on these systems, but UV data ⟨51JCS3204⟩ indicate that the hydroxy forms rather than the vinylogous amide forms are favoured and these systems are true 'hydroxyquinoxalines' and 'hydroxyphenazines'.

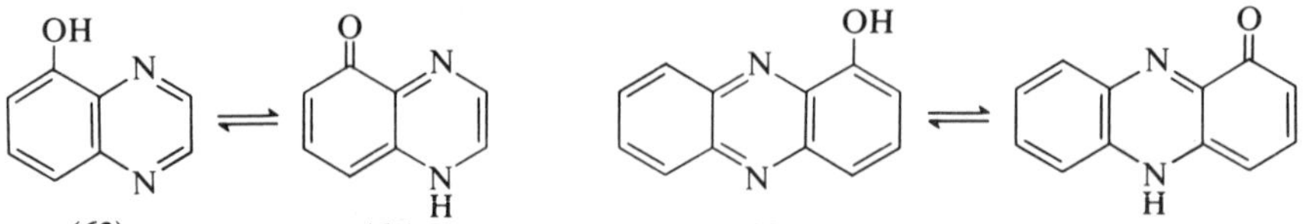

(62)     (63)     (64)     (65)

In recent years, 2,5-dioxypyrazines, 2,6-dioxypyrazines and their related derivatives have been shown to participate in a number of interesting cycloaddition reactions. In 1970 it was first demonstrated that '3-benzyl-6-methyl-2,5-dihydroxypyrazine' (66) undergoes cycloaddition reactions with dimethyl acetylenedicarboxylate to give labile bicyclic adducts which undergo a retro-Diels–Alder reaction to yield the isomeric pyridones (67) and (68) (Scheme 25) ⟨70CC1103⟩. Subsequently, it was demonstrated that strained alkenes such as norbornadiene take part in analogous reactions to give products such as (69). Several mechanistic possibilities exist for these reactions. Bearing in mind the known propensity of hydroxypyrazines to exist in amide forms, *e.g.* (60), then a [4 + 2] cycloaddition to yield the observed products can be envisaged. Indeed, blocked tautomers such as (70) have been prepared and shown to be very reactive, undergoing cycloaddition reactions with isolated double bonds and with atmospheric oxygen to yield adducts such as (71). This last reaction is particularly interesting in that it either involves the reaction of the pyrazinone (70) with oxygen in its electronic ground state, or (70) functions as a photosensitizer, generating singlet oxygen from atmospheric oxygen in the triplet ground state ⟨73JCS(P1)404⟩.

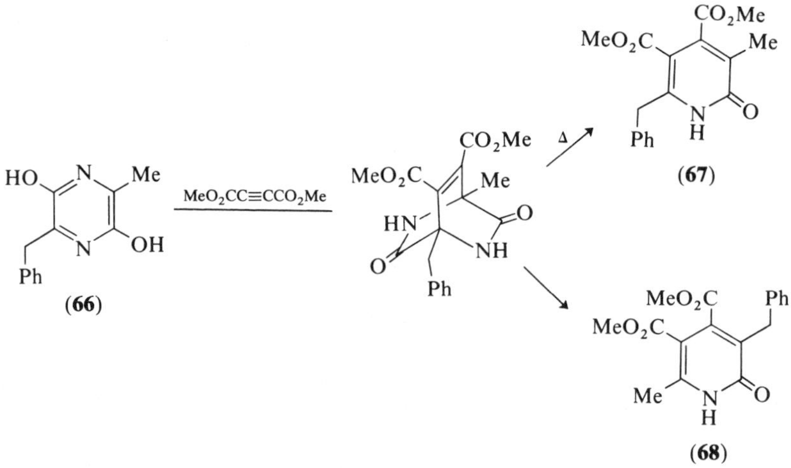

**Scheme 25**

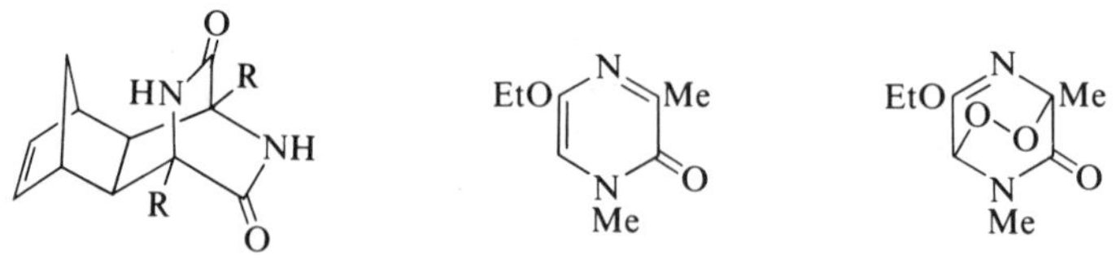

(69)     (70)     (71)

Clearly, in the case of (66) two amide tautomers (72) and (73) are possible, but if both hydroxyl protons tautomerize to the nitrogen atoms one amide bond then becomes formally cross-conjugated and its normal resonance stabilization is not developed (*cf.* 74). Indeed, part of the driving force for the reactions may come from this feature, since once the cycloaddition (of 72 or 73) has occurred the double bond shift results in an intermediate imidic acid which should rapidly tautomerize. In addition, literature precedent suggests that betaines such as (74) may also be present and clearly this opens avenues for alternative mechanistic pathways.

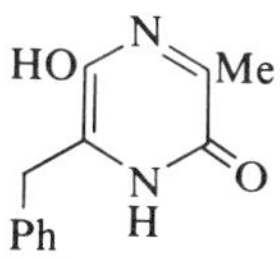

(72) 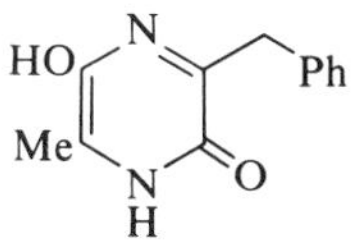          (73) 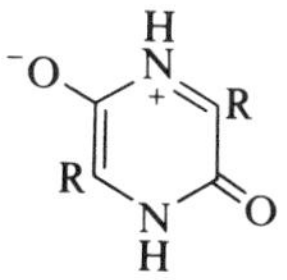          (74)

2,5-Dialkoxypyrazines and '2,5-dihydroxypyrazines' have been shown to add to singlet oxygen to yield peroxy adducts such as (75) in high yields ⟨76CC417, 79JCS(P1)1885⟩ and the reaction is believed to be important in the biosynthesis of the antibiotic bicyclomycin (76) ⟨81JCS(P1)3111⟩.

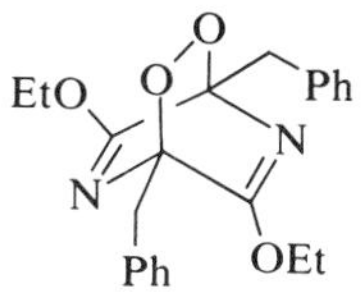

(75) 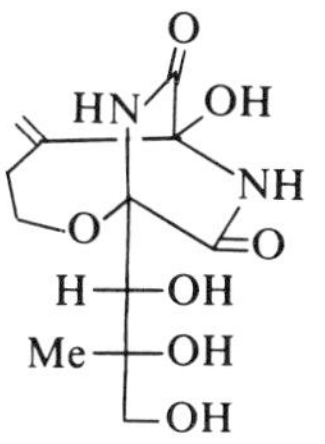          (76)

More recently, Cheeseman and coworkers have investigated cycloaddition reactions of 2,6-dioxypyrazines ⟨80JCS(P1)1603⟩. '2,6-Dihydroxy-3,5-diphenylpyrazine' (77) reacts with electron deficient dienophiles such as *N*-phenylmaleimide, diethyl maleate and diethyl fumarate (Scheme 26) to yield adducts of the 3,8-diazabicyclo[3.2.1]octane class such as (78). This reaction is believed to proceed by way of the betaine (79) and has precedent ⟨69AG(E)604⟩ in that photolysis of the bicyclic aziridine (80) generates analogous betaines which have been trapped in cycloaddition reactions.

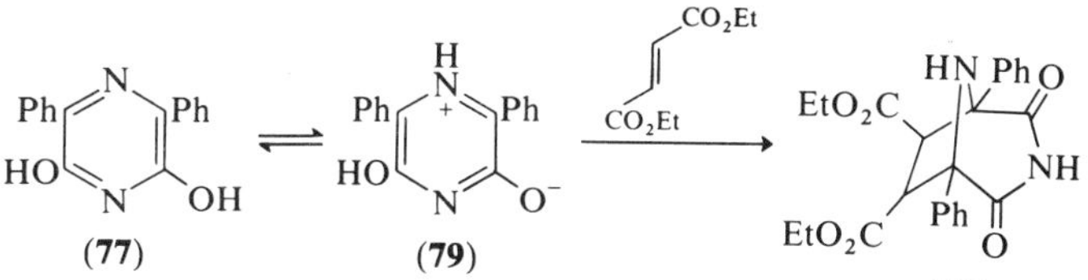

(77)          (79)          (78)          (80)

**Scheme 26**

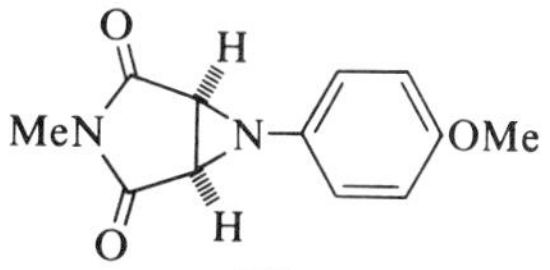

### 2.14.2.7 Halo Pyrazines, Quinoxalines and Phenazines

The classical route to halopyrazines and haloquinoxalines substituted in the heterocyclic ring, particularly the chloro and bromo derivatives, is from the corresponding oxo derivatives by treatment with phosphoryl chloride or phosphoryl bromide. Other reagents such as molecular bromine and thionyl chloride have also found extensive use. Both 2-chloroquinoxaline and 2-chloropyrazine are directly available by treatment of the corresponding oxo derivatives with thionyl chloride. The use of phosphorus pentahalides is best avoided as mixtures of products may arise (Scheme 27). Phosphorus pentahalides, particularly at elevated temperatures, bring about polyhalogenation ⟨66CI(L)1721⟩ and the synthesis of tetrachloropyrazine has been accomplished in this way (Scheme 28).

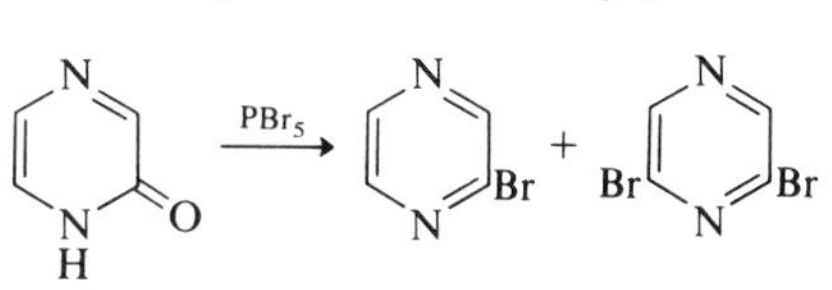

**Scheme 27**

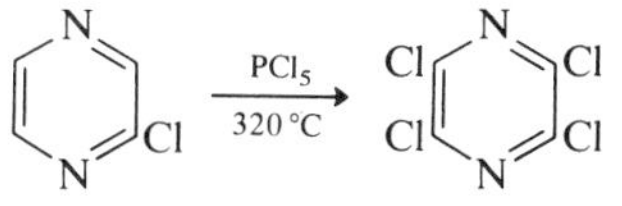

**Scheme 28**

Fluorinated and iodinated derivatives are usually prepared by halogen exchange reactions, although the Baltz–Schiemann reaction has been applied to the synthesis of 2-fluoroquin-oxaline ⟨66JHC435⟩.

Direct halogenation of quinoxaline appears to be of limited value but pyrazine may be chlorinated in the vapor phase to give monochloropyrazine at 400 °C or at lower temperatures under catalytic conditions ⟨72AHC(14)99⟩, and at higher temperatures tetra-chloropyrazine formation occurs in high yields. Mention has already been made of direct chlorination (see Section 2.14.2.1) of phenazine.

As already stated, the principal value of the halo-pyrazines, -quinoxalines and -phenazines rests in the ease of displacement of the halogen atom by a variety of nucleophiles, thus permitting functional group manipulation within a series. The reactivities of chlorodiazines towards a given nucleophile vary greatly. Thus, for reaction with ethanol at 20 °C, 2-chloropyrazine and 2-chloropyrimidine have relative rate constants of 1.5. The rates of reaction of the chlorodiazines are faster than, for example, 2-chloropyridine. In the case of the benzochlorodiazines, generalization about reactivities becomes difficult. Intuitively it would be predicted that the benzo analogs should be more reactive than the corresponding chlorodiazines in view of the greater resonance stabilization of the initial addition product, and all of the benzochlorodiazines are observed to be significantly more reactive than, for example, chloropyrazine. However, the nucleophile seems to play an important role in the determination of the relative rates of reaction of the benzochlorodiazines and prediction of the relative reactivities of the halobenzodiazines is not yet possible ⟨56JCS1563⟩.

The reactions of haloquinoxalines in which the halogen atom is bonded to the benzenoid ring have not been well studied, but by analogy with examples in the phenazine series it would seem probable that they are unlikely to be displaced with the same ease as those bonded directly to the heterocyclic ring. It is evident from the foregoing discussion that *N*-oxidation has a pronounced effect on their reactivity, and, by this means, considerable latitude in the specific functionalization of dihalo or polyhalo derivatives may be exercised.

Generally, the reactions of halopyrazines and haloquinoxalines with nucleophiles are believed to proceed by way of addition/elimination sequences, although there are clear-cut examples where this is not the case (see Section 2.14.2.2) and, consistent with a mechanism which involves bond forming, rather than bond breaking, reactions in the rate-determining step, fluoro derivatives are considerably more reactive (*ca.* $\times 10^3$) than the corresponding chloro derivatives.

Nucleophilic substitution of the chlorine atom in 2-chloropyrazine and 2-chloroquinoxa-lines has been effected with a variety of nucleophiles, including ammonia and amines, oxygen nucleophiles such as alkoxides, sodium azide, hydrazine, sulfur containing nucleophiles, cyanide, *etc.*, and reactions of this type are typical of the group (see Chapter 2.02).

2,3-Dihaloquinoxalines are extremely reactive and both halogen atoms are replaceable, on occasions explosively ⟨59RTC5⟩, whereas in the case of dihalopyrazines, and tri- or tetra-halopyrazines, there is frequently a considerable difference in reactivity of the halogen atoms. When 2,3-dichloropyrazine is treated with ammonia at 130 °C, only one chlorine atom is displaced, giving 2-amino-3-chloropyrazine ⟨66FES799⟩.

An interesting divergence of behavior is observed with 2,3,5-trichloropyrazine, which on reaction with aqueous ammonia gives 2-amino-3,6-dichloropyrazine, whereas reaction with ammonia under pressure results in the formation of 2-amino-3,5-dichloropyrazine.

### 2.14.2.8 Amino Derivatives

The progression from hydroxypyrazines/quinoxalines through the halo derivatives to the amines is a logical sequence in that, for practical purposes, this is the best method of synthesis of the amino compounds (see preceding Section). The ammonolysis proceeds most easily in the case of fluoro compounds. Fluoropyrazine reacts with aqueous ammonia at room temperature, whereas the reaction with chloropyrazine requires higher temperature and pressure.

Numerous other methods are available for the preparation of amino derivatives, and these include direct synthesis (see Section 2.14.3.2) and more traditional transformations such as the Hofmann reaction. Aminopyrazine has been prepared from pyrazinamide ⟨60G1807⟩ and 2-aminoquinoxaline from the corresponding carboxamide ⟨71JOC1158⟩. The more traditional methods of amine formation such as the reduction of nitro compounds is limited to some extent by the availability of the starting materials. No reports on the practical nitration of pyrazine have appeared; however, quinoxaline and phenazine have both been nitrated in the benzenoid ring(s) and clearly such nitro derivatives are amenable to the classical reduction methods. The reduction of azides appears to offer a useful method of preparing amines, since the azides are readily available. Shaw and coworkers have used 2,6-diazidopyrazine as a precursor of the 2,6-diamino compound ⟨80JHC11⟩, although this appears to be somewhat hazardous in view of the known sensitivity of azides, and particularly polyazides, to detonation.

In principle, aminopyrazine and 2-aminoquinoxaline are capable of existing in the form of the imino tautomers (**81**) and (**82**); however, comparison of the UV spectra of the amino, methylamino and dimethylamino derivatives indicates that in both systems the amino rather than the imino tautomer is favored ⟨60JCS242, 58JCS108⟩.

(**81**)          (**82**)

Aminopyrazines and 2-aminoquinoxalines, like their pyridine analogs, react with nitrous acid under aqueous conditions to give the 2(1*H*)-pyrazinones and 2(1*H*)-quinoxalinones. 2-Aminoquinoxalines are more readily hydrolyzed than typical heterocyclic amines and 2-amino-3-methylquinoxaline, for example, undergoes hydrolysis on heating at 100 °C with dilute sodium hydroxide ⟨59JCS1132⟩.

As might be expected from a consideration of electronic effects, an amino substituent activates pyrazines, quinoxalines and phenazines to electrophilic attack, usually at positions *ortho* and *para* to the amino group; thus, bromination of 2-aminopyrazine with bromine in acetic acid yields 2-amino-3,5-dibromopyrazine (Scheme 29).

**Scheme 29**

### 2.14.2.9 Reduced Pyrazines, Quinoxalines and Phenazines

Four dihydropyrazines, the 1,2- (**83**), 2,3- (**84**), 1,4- (**85**) and 2,5- (**86**), the two tetrahydropyrazines (**87**) and (**88**) and hexahydropyrazine or piperazine (**89**) represent the varying stages of reduction of the pyrazine ring and examples of all types are known.

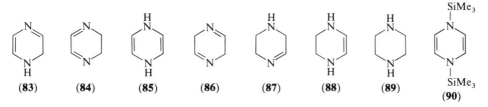

(**83**)     (**84**)     (**85**)     (**86**)     (**87**)     (**88**)     (**89**)     (**90**)

In a series of detailed studies, Armand and coworkers have examined the electrochemical reduction of pyrazines ⟨72CR(C)(275)279⟩. The first step results in the formation of 1,4-dihydropyrazines (**85**), but the reaction is not electrochemically reproducible. The 1,4-dihydropyrazine is pH sensitive and isomerizes at a pH dependent rate to the 1,2-dihydro compound (**83**). The 1,2-dihydropyrazine then appears to undergo further reduction to 1,2,3,4-tetrahydropyrazine (**88**) which is again not electrochemically reproducible. Compound (**88**) then appears to undergo isomerization to another tetrahydro derivative, presumably (**87**), prior to complete reduction to piperazine (**89**). These results have been confirmed ⟨72JA7295⟩.

1,4-Dihydropyrazines, although the initial reduction product under electrochemical conditions, appear to be unstable, as might be predicted from the fact that they contain eight

$\pi$-electrons and hence would be expected to be antiaromatic in character. Many of the earlier reports relating to 1,4-dihydropyrazines have subsequently been disproved, but the 1,4-bis(trimethylsilyl)-1,4-dihydropyrazine (**90**) has been prepared and characterized ⟨71AG(E)127⟩. Moreover, Lown and coworkers (see Section 2.14.3.3) have prepared a number of 1,4-dialkyl-1,4-dihydropyrazines and demonstrated that they undergo isomerization to 1,2-dihydropyrazines.

1,2-Dihydropyrazines are relatively stable, although they are easily oxidized. They are usually formed *via* the addition of organometallic reagents to the pyrazine ring. Similarly, 2,3-dihydropyrazines are usually easily oxidized to pyrazines and are formed during type 'A' synthesis (see Section 2.14.3.2).

2,5-Dihydropyrazines are formed by the self-condensation of $\alpha$-aminocarbonyl compounds and they are relatively stable, although again they are easily oxidized to the corresponding pyrazines. Tetrahydropyrazines are less well documented and structures such as (**87**) appear to be more stable than the enediamine (**88**).

The final reduction product of pyrazine, piperazine (**89**), is a stable compound which behaves as a typical diamine. It has found extensive use in medicinal chemistry as a linking agent and as a medicine in its own right for the treatment of helminths both in human and veterinary medicine.

The fusion of a benzene ring to pyrazine results in a considerable increase in the resistance to reduction and it is usually difficult to reduce quinoxalines beyond the tetrahydroquinoxaline state (**91**). Two possible dihydroquinoxalines, *viz.* the 1,2- (**92**) and the 1,4- (**93**), are known, and 1,4-dihydroquinoxaline appears to be appreciably more stable than 1,4-dihydropyrazine ⟨63JOC2488⟩. Electrochemical reduction appears to follow a course analogous to the reduction of pyrazine, giving the 1,4-dihydro derivative which isomerizes to the 1,2- or 3,4-dihydroquinoxaline before subsequent reduction to 1,2,3,4-tetrahydroquinoxaline (**91**). Quinoxaline itself is reduced directly to (**91**) with LiAlH$_4$ and direct synthesis of (**91**) is also possible. Tetrahydroquinoxalines in which the benzenoid ring is reduced are well known but these are usually prepared from cyclohexane derivatives (Scheme 30).

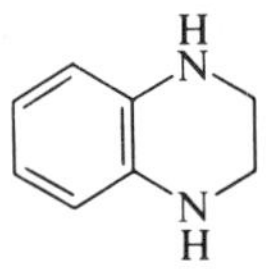

(**91**) 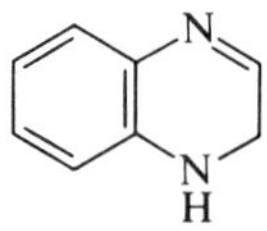                    (**92**) 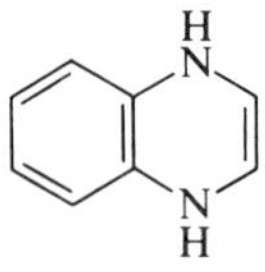                    (**93**)

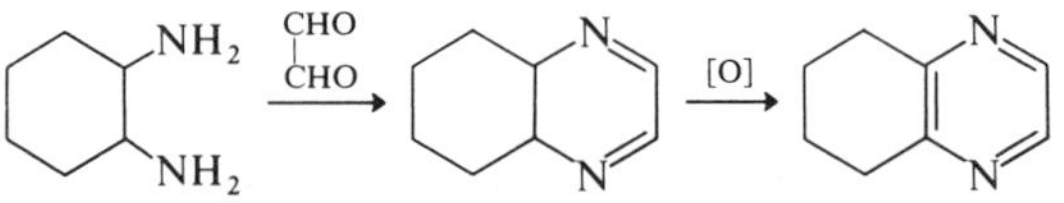

**Scheme 30**

In the case of phenazine, reduction beyond the dihydrophenazine state (**94**) is difficult and more highly saturated phenazines are usually prepared by direct synthesis. Phenazine and a number of its derivatives are reduced to (**94**) using many reducing agents; thus, reduction with lithium metal in the presence of trimethylsilyl chloride (Scheme 31) yields the dibenzo analogue of (**90**) ⟨75CB3105⟩. 1,4-Dihydrophenazines are stable compounds but they show a strong tendency to rearomatize even on attempted melting point determination if atmospheric oxygen is not excluded. More highly reduced phenazines are known, but again these are usually synthesized directly (*cf.* Scheme 30).

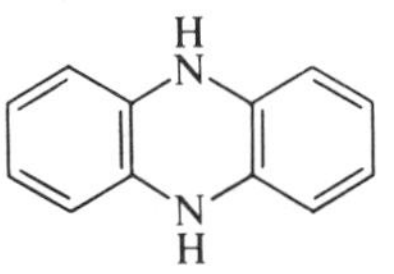

(**94**) 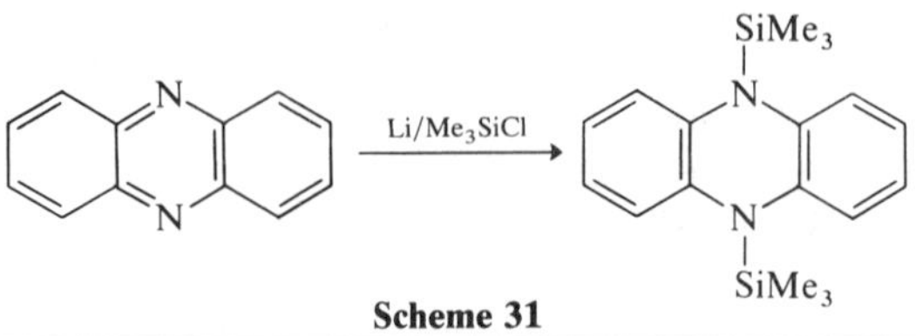                              **Scheme 31**

## 2.14.3 SYNTHESIS

### 2.14.3.1 General Comments

Cheeseman and Werstiuk have classified the synthetic methods for the construction of pyrazine rings into six categories illustrated by (A) to (F) below ⟨72AHC(14)99⟩. Clearly this classification is an over-simplification, as it seems improbable that synchronous formation of two, three or four bonds would yield pyrazines in a single step, and it is likely that many of the reactions involve 'F' as a common intermediate. Nevertheless, the classification is useful in that it indicates the origin of the various structural fragments in the pyrazine ring and it may also be applied to the synthesis of the benzo analogs.

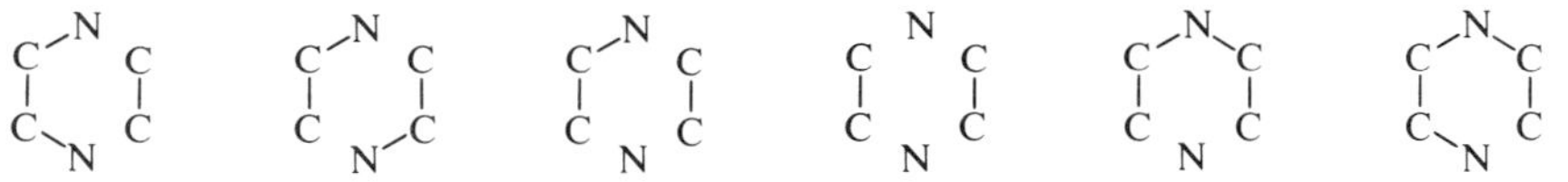

In many instances the primary reaction product is a dihydropyrazine and aromatization may be required as a final step. In addition, many pyrazines are prepared by the structural modification of a preformed pyrazine ring and hence would be classified as a reaction of the ring rather than a ring synthesis; such processes are discussed more fully in Section 2.14.2.

### 2.14.3.2 Type 'A' Syntheses

Type 'A' syntheses are certainly amongst the oldest and most general pyrazine syntheses, and usually involve the condensation of an $\alpha$-diketone with a 1,2-diaminoalkane or 1,2-diaminoarene, to give dihydropyrazines and quinoxalines respectively (Scheme 32). Type 'A' syntheses represent the most practical approach to quinoxalines. The reaction described in Scheme 32 is generally carried out in two steps: firstly the formation of the 2,3-dihydropyrazine and, secondly, oxidation to the pyrazine, although industrially it is now possible to carry out the synthesis directly by vapor phase reaction over a copper chromite catalyst. Usually, once the dihydropyrazines have been prepared they may be oxidized by air in the presence of KOH pellets ⟨67C510⟩, although more recently use of metal oxides such as MnO₂ or CuO in ethanolic KOH has resulted in a considerable improvement of both methodology and yields ⟨78MI21400⟩. The reaction has the obvious limitation that with unsymmetrical diaminoethane derivatives or unsymmetrical diketones, product mixtures arise; thus, for example, the condensation of 1,2-diaminopropane with 3-methylbutane-1,2-dione gives a 1:1 mixture of the isomeric dihydropyrazines (95) and (96) (Scheme 33) ⟨69RTC1335⟩. Oxidation of this mixture yielded the corresponding pyrazines which could not be separated by fractional distillation; however, separation was readily achieved by column chromatography over alumina.

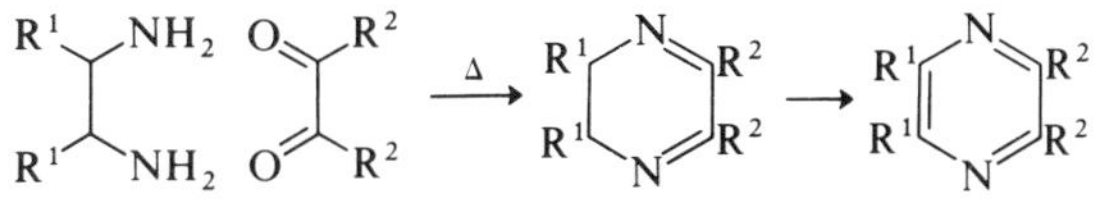

**Scheme 32**

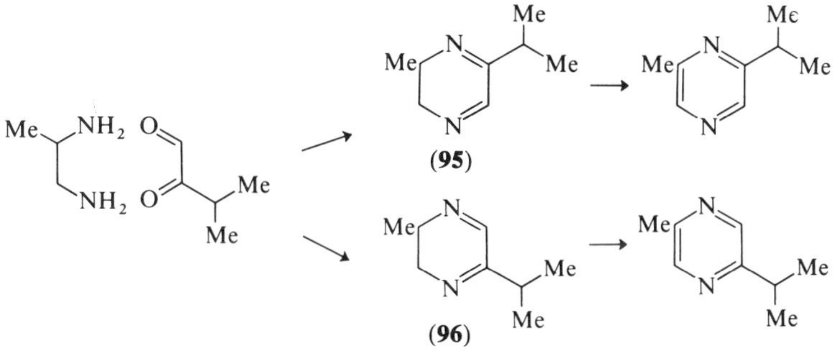

**Scheme 33**

Substitution of 1,2-diaminoethanes with *o*-phenylenediamines leads to quinoxalines and this method is perhaps the most widely used quinoxaline synthesis, since by the simple expedient of using substituted phenylenediamines and substituted α-dicarbonyl compounds virtually any substitution pattern within the quinoxaline is possible. This concept is of course directly applicable to pyrazine synthesis since, by using 1,2-diaminoethene derivatives, pyrazines are formed directly without need to recourse to a second oxidation step. Considerable progress has been made in this direction using diaminomaleonitrile as shown in Scheme 34, when highly substituted pyrazines are formed in moderate to excellent yield ⟨75CB875⟩. This reaction probably proceeds in a stepwise fashion, and when acyl cyanides are used the intermediate Schiff's bases may be isolated ⟨79JOC827⟩ and selectively hydrolyzed using $H_2O_2/Na_2MoO_4$ (Scheme 35) to amides which subsequently undergo ring closure to pyrazine derivatives.

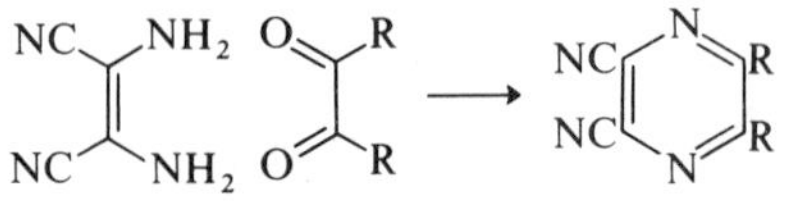

**Scheme 34**

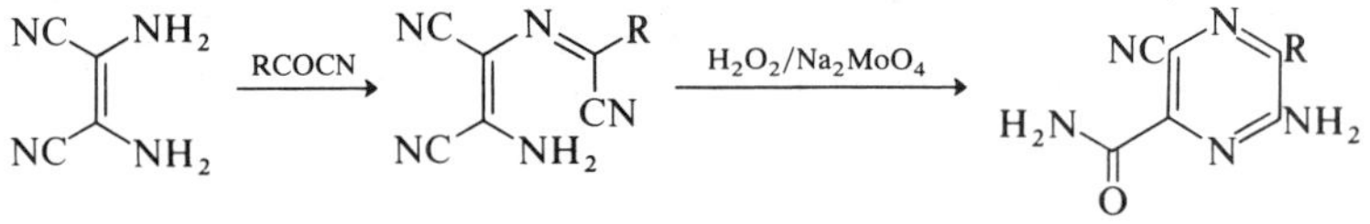

**Scheme 35**

Many variants of this procedure exist. Thus, Kano and coworkers have carried out the condensation of β-keto sulfoxides with diaminomaleonitrile (Scheme 36) ⟨78S372⟩. This reaction probably yields an intermediate dihydropyrazine which is oxidized under the reaction conditions, and it seems likely that the condensation of the carbonyl group and the amine is the first step.

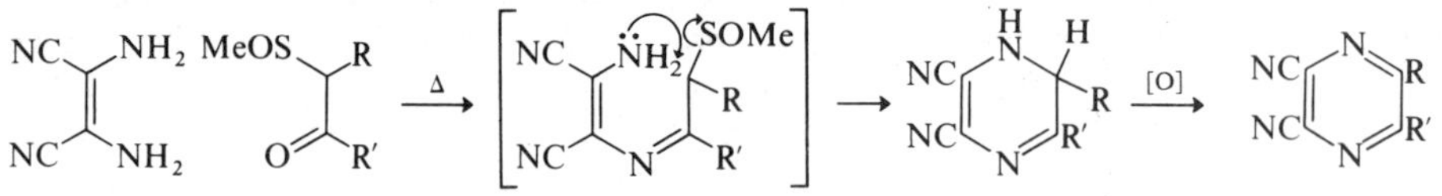

**Scheme 36**

In an approach to the synthesis of the bioluminescent substance Cypridina luciferin (**97**), Kishi *et al.* ⟨69YZ1646⟩ have utilized the condensation of 3-indolylglyoxal (Scheme 37) with aminoacetamidine to form the 2-amino-5-(3-indoloyl)pyrazine which was subsequently elaborated to (**97**). Although in principle two products might be anticipated from this condensation, only one is observed and this presumably results from the reduced reactivity of the vinylogous amide carbonyl group. The initial condensation of the more nucleophilic amidine function takes place with the more electrophilic terminal carbonyl group. A similar orientation effect has been observed in the condensation of aminomalonamidamidine with glyoxal derivatives ⟨59JA2472⟩. Both methylglyoxal and phenylglyoxal undergo condensation with aminomalonamidamidine to yield 2-amino-5-methyl- and -5-phenyl-pyrazine (Scheme 38) and once again this is readily rationalized in terms of a stepwise condensation reaction with the more nucleophilic amidine function reacting first with the more electrophilic aldehyde carbonyl group.

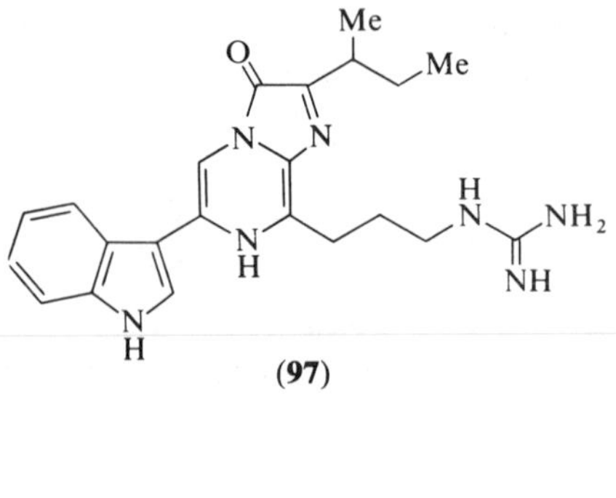

(**97**)

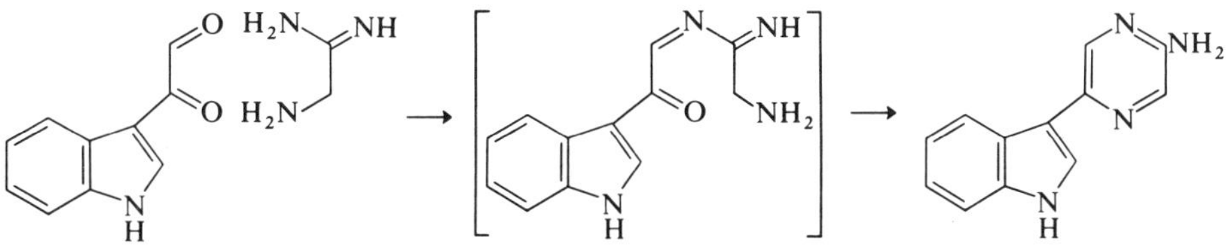

**Scheme 37**

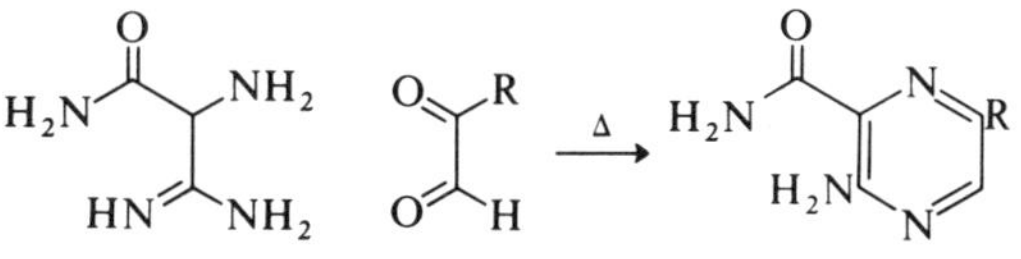

**Scheme 38**

Diaminomaleonitrile also undergoes a condensation with $\alpha$-diketone monoximes to yield pyrazine mono-$N$-oxides ⟨80JOC2485⟩ and here hydrogen cyanide is displaced (Scheme 39) to yield the aminocyanopyrazine rather than the dicyanopyrazine which is observed with $\alpha$-diketones. Mono-$N$-oxides of pyrazine also result from the condensation of $\alpha$-aminooximes and this reaction has been used predominantly in sugar series, where the oximes are readily derived from 2-amino-2-deoxy sugars (see Scheme 18) ⟨71JAP7105310⟩.

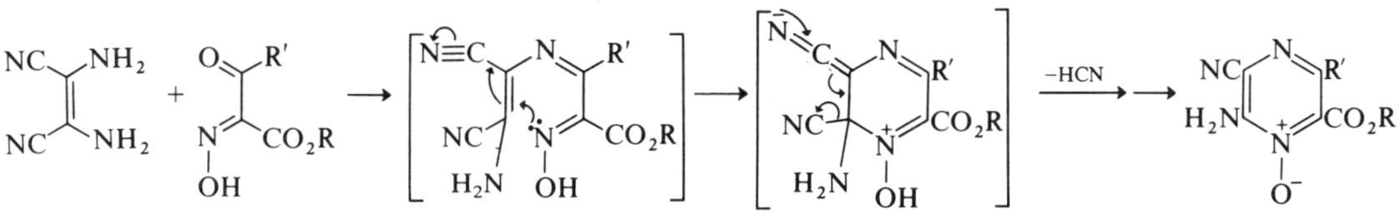

**Scheme 39**

Perhaps one of the most exciting developments in the chemistry of quinoxalines and phenazines in recent years originates from the American University of Beirut in Lebanon, where Haddadin and Issidorides first made the observation that benzofuroxans undergo reaction with a variety of alkenic substrates to produce quinoxaline di-$N$-oxides in a one-pot reaction which has subsequently become known as the 'Beirut reaction'. Many new reactions tend to fall by the wayside by virtue of the fact that they are experimentally complex or require starting materials which are inaccessible; however, in this instance the experimental conditions are straightforward and the starting benzofuroxans are conveniently prepared by hypochlorite oxidation of the corresponding $o$-nitroanilines or by pyrolysis of $o$-nitrophenyl azides.

The first Beirut reactions to be reported involved the preparation of quinoxaline di-$N$-oxides by the reaction of enamines with benzofuroxan ⟨66JOC4067⟩. The reaction of a number of enamines works well but consistently higher yields are obtained using morpholine enamines (Scheme 40); product isolation is also facilitated. Imines have also been shown to react readily, as do three-component mixtures of benzofuroxan, ketones and amines (Scheme 41); however, it is possible to rationalize all of these reactions in terms of the reaction substrate being an enamine or enolate, since tautomeric equilibria between imines and enamines and ketones/enolates is well established. Using the morpholine enamine derived from cyclohexanone, it is possible to prepare 1,2,3,4-tetrahydrophenazine 5,10-dioxide directly, and in high yield. (Scheme 42). When benzofuroxan is reacted with the enamine $Me_2C=CHNMe_2$ the 2,3-dihydroquinoxaline 1,4-dioxide (**98**) is isolated (Scheme 43); in this instance aromatization by elimination of the amine fragment is structurally precluded and in view of the existence of such dihydroquinoxaline 1,4-dioxides it has been suggested that the intermediacy of these compounds is the norm in all examples of the Beirut reaction.

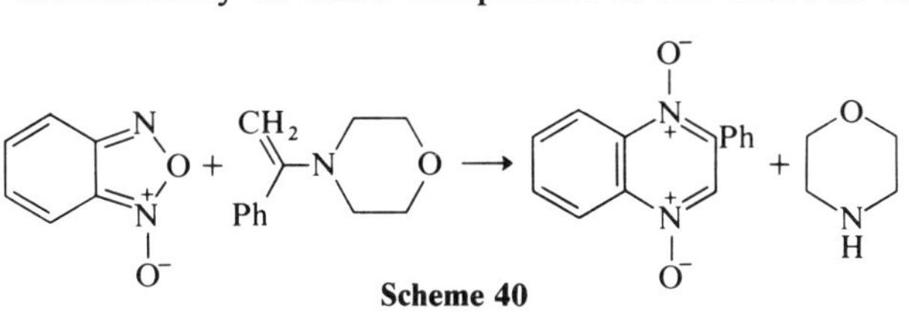

**Scheme 40**

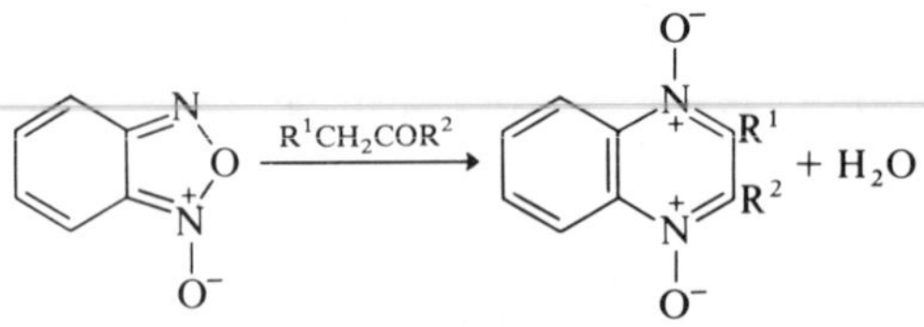

**Scheme 41**

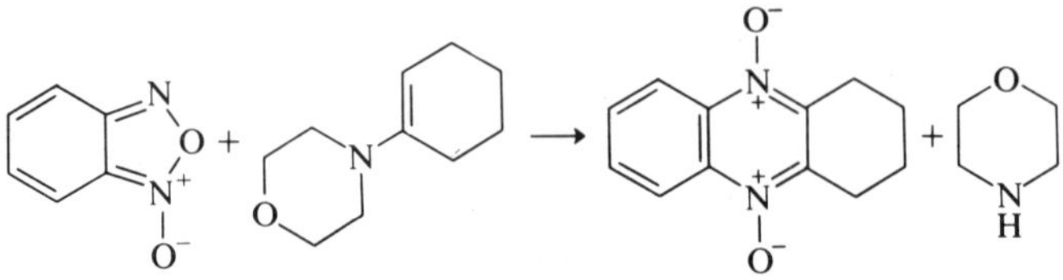

**Scheme 42**

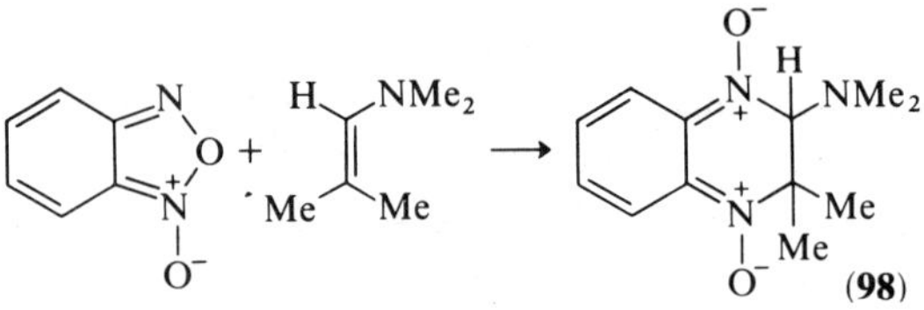

**Scheme 43**

A variety of simple carbonyl substrates react directly with benzofuroxans in the presence of bases to give quinoxaline 1,4-dioxides (Scheme 44), and in these instances the base probably functions to generate an equilibrium concentration of the enolate, which then adds to the benzofuroxan. The reaction of methyl ethyl ketone is typical; thus, treatment of a solution of the benzofuroxan with ammonia and methyl ethyl ketone in methanolic solution results in the formation of a 91% yield of 2,3-dimethylquinoxaline 1,4-dioxide. Compounds containing active methylene groups such as malononitrile and ethyl cyanoacetate also react to give high yields of the corresponding di-*N*-oxides (Scheme 45). When unsymmetrically substituted benzofuroxans are used product mixtures might be expected; however, in practice a considerable regioselectivity of cyclization may be observed. For example, in the reaction of 5-substituted benzofuroxans with benzoylacetonitrile only one regioisomer is observed (Scheme 46). Ynamines, being effectively at a higher oxidation state than enamines or enolates, yield the corresponding aminoquinoxaline 1,4-dioxides directly, without the intermediacy of the dihydroquinoxaline 1,4-dioxides (Scheme 47).

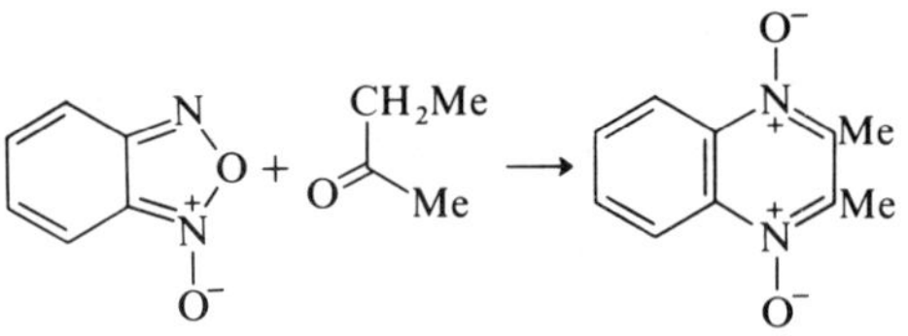

**Scheme 44**

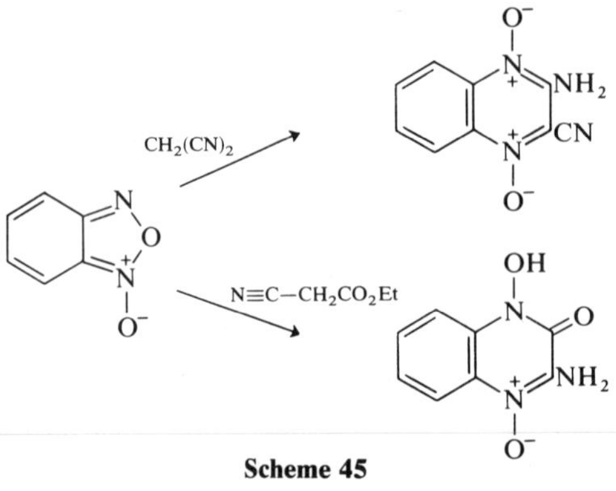

**Scheme 45**

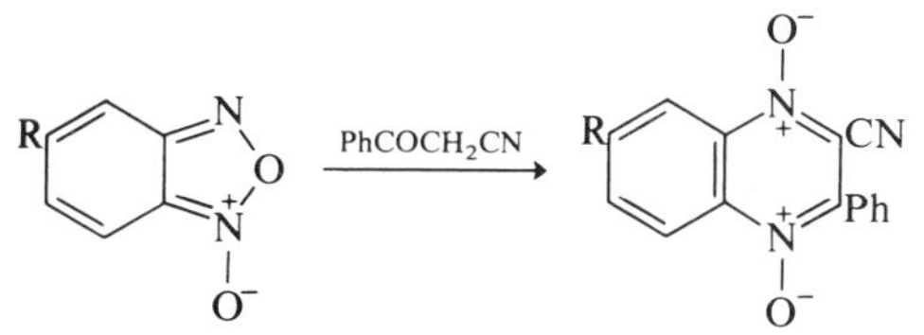

**Scheme 46**

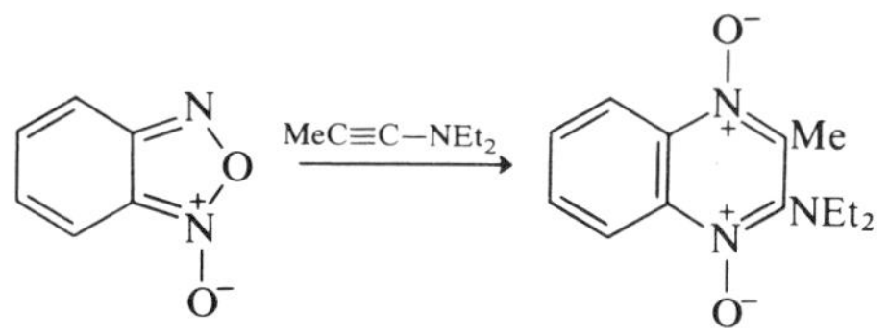

**Scheme 47**

Curiously, when phenols are used as substrates in this reaction, in the place of enolates or enamines/imines, then phenazine 5,10-dioxides are formed in high yields and under extremely mild conditions (Scheme 48), and there is little doubt that these reactions represent the method of choice for the construction of the phenazine ring system. The synthesis of 2-hydroxyphenazine 5,10-dioxide is typical and is accomplished in 97% yield simply by stirring an alkaline solution of hydroquinone and benzofuroxan at room temperature for 10 hours. The synthesis is highly specific and a large number of highly substituted phenazine derivatives have been prepared in this way. With naphthols, benzophenazine derivatives result. Polyhydric phenols have also been used in the reaction and phloroglucinol, for example, reacts to yield 1,3-dihydroxyphenazine 5,10-dioxide (**99**) in good yield. At the higher oxidation level, benzoquinone reacts with benzofuroxans in alcoholic ammonia to yield 2-hydroxyphenazine 5,10-dioxide (**100**) directly in 92% yield. Triphenylphosphine or trialkylphosphines also serve as bases in this reaction and, curiously, the products are the 1-trialkyl(aryl)phosphonium betaines (Scheme 49).

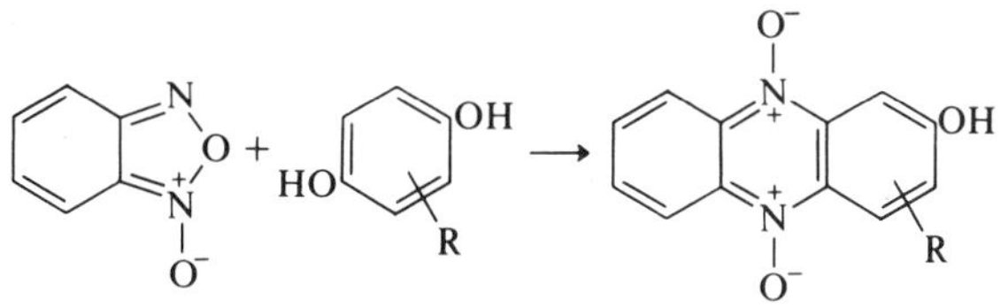

**Scheme 48**

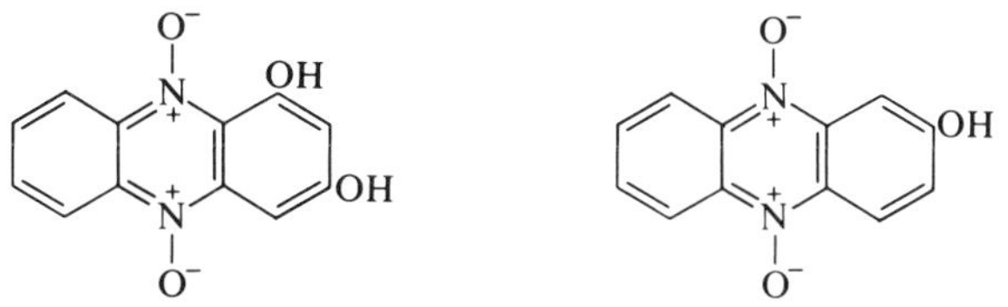

(**99**)

(**100**)

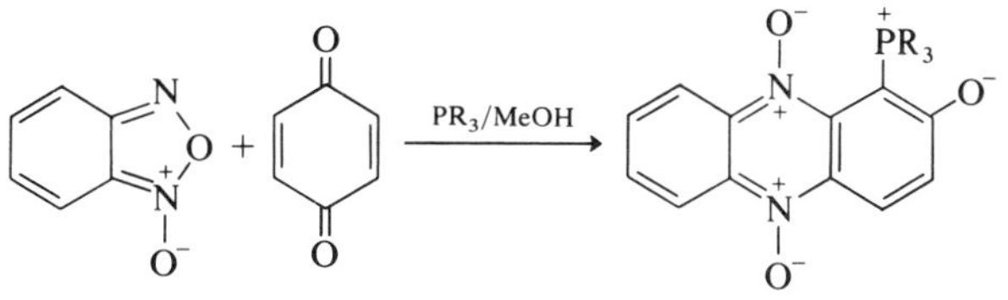

**Scheme 49**

In spite of the usefulness of the Beirut reaction, mechanistically it is not well understood. It has been suggested that the first step involves the nucleophilic attack by the enolate or the enamine at N-3 of the benzofuroxan to yield an intermediate *N*-oxide (Scheme 50) which subsequently undergoes tautomerism to an hydroxylamino derivative. This intermediate then cyclizes to the dihydroquinoxaline 1,4-dioxide. This suggestion has not been proven, and indeed there is evidence that benzofuroxan is in equilibrium with 1,2-dinitrosobenzene

(Scheme 51), so it has been suggested that the initial reaction involves the dinitrosobenzene. It does seem, however, that this may be an over-simplification, as there are documented cases where mono-*N*-oxides rather than the di-*N*-oxides are formed; for instance, the reaction of benzofuran-3(2*H*)-ones with benzofuroxan yields 3-(*o*-hydroxyphenyl)quinoxaline 1-oxide (Scheme 52). Other mechanistic possibilities may also be put forward but it seems probable that more than one pathway may be operating, particularly in view of the more recent findings on the reactions of benzofuroxans ⟨81AHC(29)251⟩.

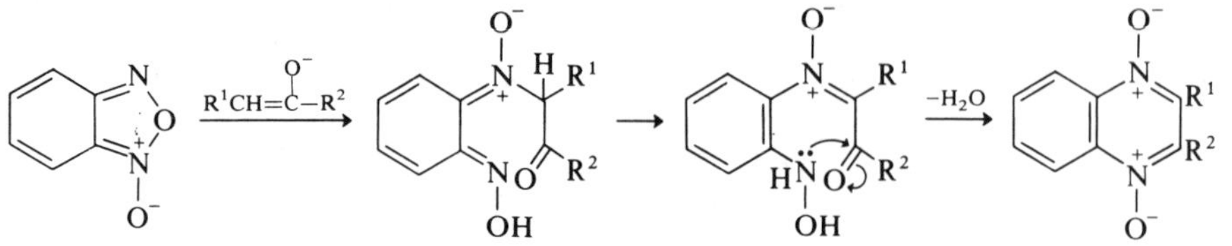

**Scheme 50**

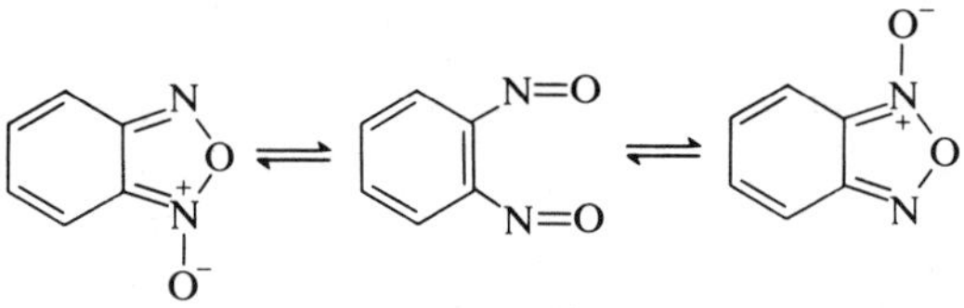

**Scheme 51**

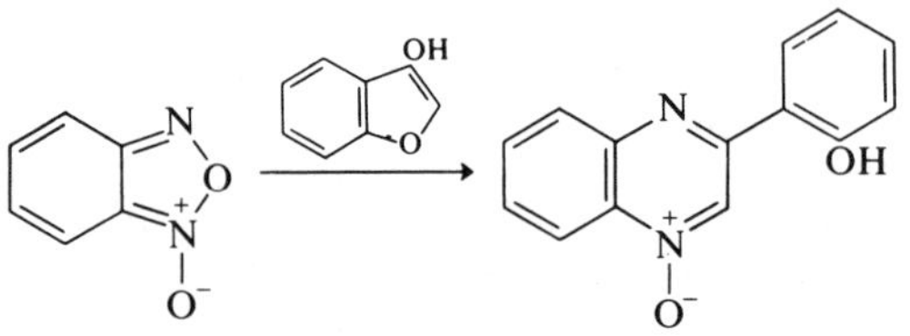

**Scheme 52**

The more traditional methods of phenazine synthesis falling into the type 'A' synthesis are altogether less satisfactory than the application of the Beirut reaction. Traditionally, Ris prepared phenazine in low yield by heating *o*-phenylenediamine and catechol in a sealed tube at 200 °C ⟨1886CB2206⟩; however, the method appears to be unsatisfactory at best and gives, in addition to phenazine, 5,10-dihydrophenazine in varying amounts (Scheme 53). Several variants of this procedure exist: *o*-benzoquinone has been used in condensation with *o*-phenylenediamine and yields as high as 35% have been reported, and 1,2,3,4-tetrahydrophenazine has been prepared by condensation of *o*-phenylenediamine with cyclohexane-1,2-dione.

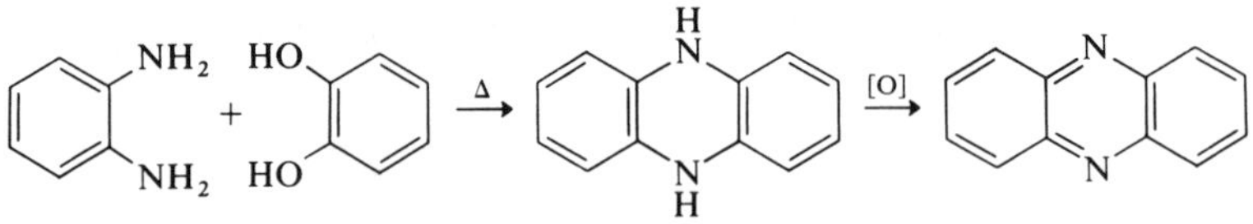

**Scheme 53**

### 2.14.3.3 Type 'B' Syntheses

Classically, type 'B' pyrazine syntheses involve self-condensation of an α-aminoacyl compound to yield a 3,6-dihydropyrazine which is subsequently oxidized to the pyrazine (Scheme 54) ⟨70CC25⟩. The aromatization usually proceeds under very mild conditions.

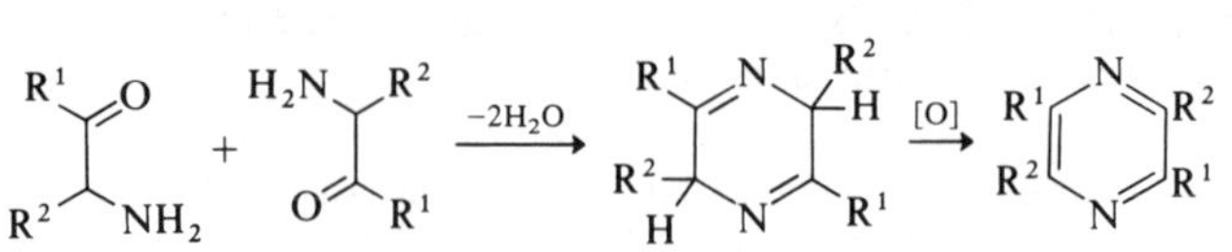

**Scheme 54**

The principal difficulty associated with this type of synthesis is in the availability of α-aminoacyl compounds, *e.g.* α-aminoaldehydes, α-aminoketones, *etc.*, and most type 'B' syntheses rely on the generation of these compounds *in situ*, where the self-condensation occurs spontaneously. A large number of research groups have addressed themselves to this problem and a variety of routes are now available.

Treatment of α-hydroxy-ketones or -aldehydes with ammonium acetate ⟨65BSF3476, 68BSF4970⟩ results in the formation of dihydropyrazines, presumably by direct amination of the hydroxyketone followed by self-condensation ⟨79AJC1281⟩. Low yields of pyrazines have been noted in the electrolysis of ketones in admixture with KI and ammonia, and again it appears probable that the α-aminoketone derived by way of the α-iodoketone is the intermediate ⟨69CI(L)237⟩.

A number of reductive procedures have found general applicability. α-Azidoketones may be reduced catalytically to the dihydropyrazines ⟨80OPP265⟩ and a direct conversion of α-azidoketones to pyrazines by treatment with triphenylphosphine in benzene (Scheme 55) has been reported to proceed in moderate to good yields ⟨69LA(727)231⟩. Similarly, α-nitroketones may be reduced to the α-aminoketones which dimerize spontaneously ⟨69USP3453279⟩. The products from this reaction are pyrazines and piperazines and an intermolecular redox reaction between the initially formed dihydropyrazines may explain their formation. Normally, if the reaction is carried out in aqueous acetic acid the pyrazine predominates, but in less polar solvents over-reduction results in extensive piperazine formation.

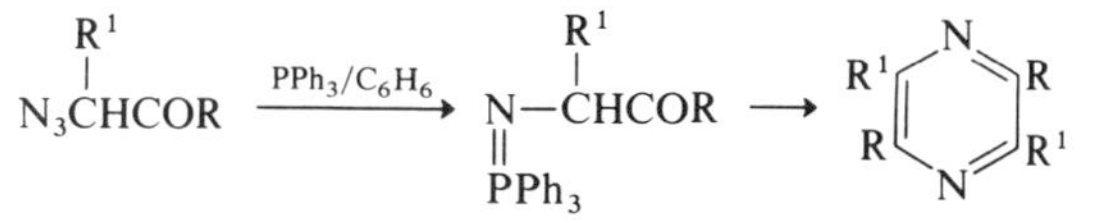

**Scheme 55**

The reduction of α-oximinoketones using Pd/C catalyst has found use in the synthesis of a number of interesting pyrazine derivatives. Oximination of β-ketoesters using NaNO₂/AcOH followed by catalytic reduction (Scheme 56) has been used to provide 2,5-dimethyl-3,6-diethoxycarbonylpyrazine ⟨73SC225⟩, whilst Evans and Mewett have used this approach in an attempted synthesis of 2,3,5,6-tetra-*t*-butylpyrazine (Scheme 57) ⟨72AJC2671⟩. The tetra-*t*-butylpyrazine was not formed by the reduction of pivalil monoxime (Scheme 57; R = Bu$^t$) or on treatment of pivaloin with ammonium acetate; however, *t*-butylglyoxime (Scheme 57; R = H) was effectively reduced to the 2,5-di-*t*-butylpyrazine which on *N*-oxidation and reaction with Bu$^t$Li yielded 2,3,5-tri-*t*-butylpyrazine. Spectroscopic studies indicated a significant twisting of the pyrazine ring and it would appear that the distortion which would be present in tetra-*t*-butylpyrazine probably precludes its synthesis.

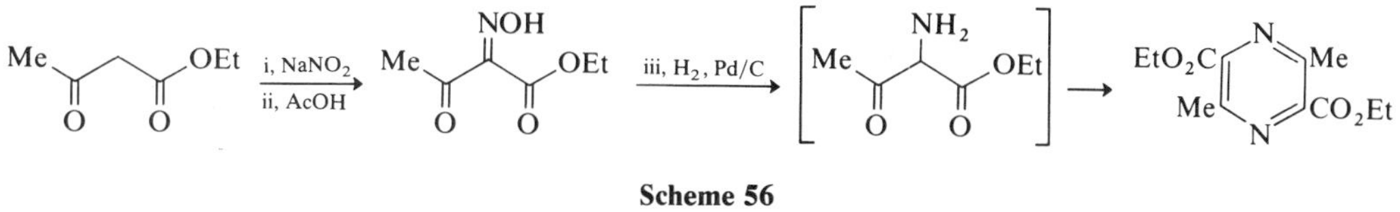

**Scheme 56**

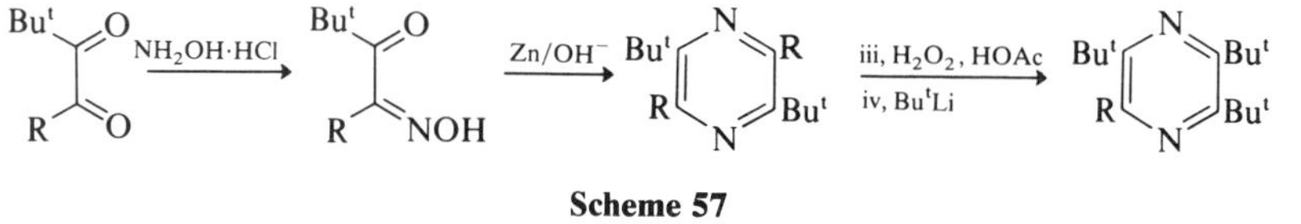

**Scheme 57**

An alternative approach to the use of α-aminoketones involves acetals ⟨72JOC221⟩ and pyrazine-2,3-diones have been synthesized by this route (Scheme 58). The acetals are readily available from the phthalimido derivatives *via* the α-chloroketones. Hemiacetals have also served as a starting point for pyrazine synthesis, although in most cases hemiacetals are too labile to be easily prepared; examples are common in the 2-amino-2-deoxy sugar series: 2-amino-2-deoxy-D-glucose for example dimerizes to the pyrazine (**101**) when generated *in situ* from the hydrochloride salt ⟨68JAP6813469⟩.

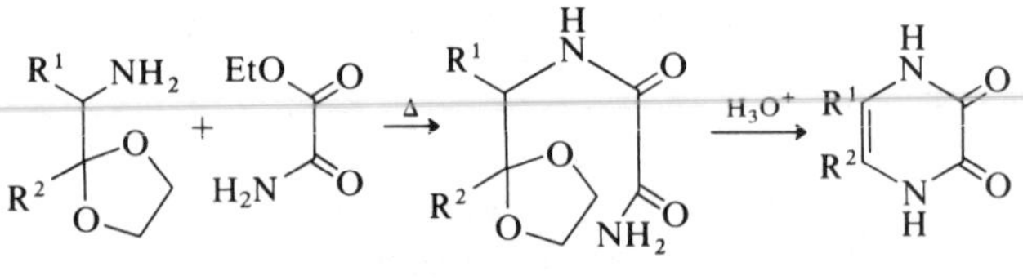

**Scheme 58**

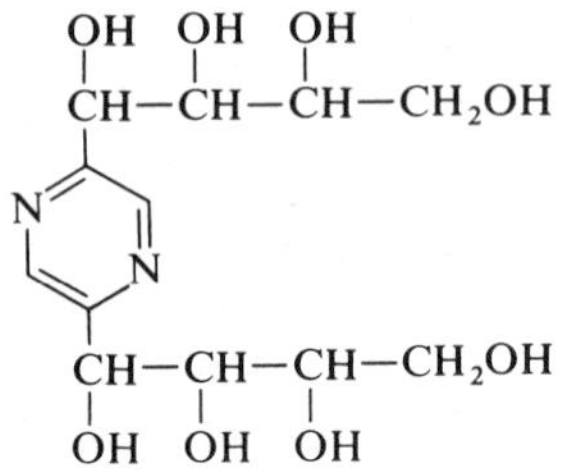

**(101)**

Other methods of generating $\alpha$-aminoketones *in situ* are common, if somewhat less general than the methods already described. 2-Nitrovinylpyrrolidine, which is readily available, yields 2,3-bis(3-aminopropyl)pyrazine on reduction and this almost certainly involves ring opening of the intermediate enamine to an $\alpha$-aminoketone which then dimerizes under the reaction conditions (Scheme 59) ⟨78TL2217⟩. Nitroethylene derivatives have also served as $\alpha$-aminoketone precursors *via* ammonolysis of the derived epoxides at elevated temperatures (Scheme 60) ⟨76S53⟩. Condensation of 1,1-disubstituted hydrazine derivatives with $\alpha$-nitro-$\beta$-ethoxyethylene derivatives has been used in the synthesis of 1,4-dialkylamino-1,4-dihydropyrazines (Scheme 61) ⟨77S136⟩.

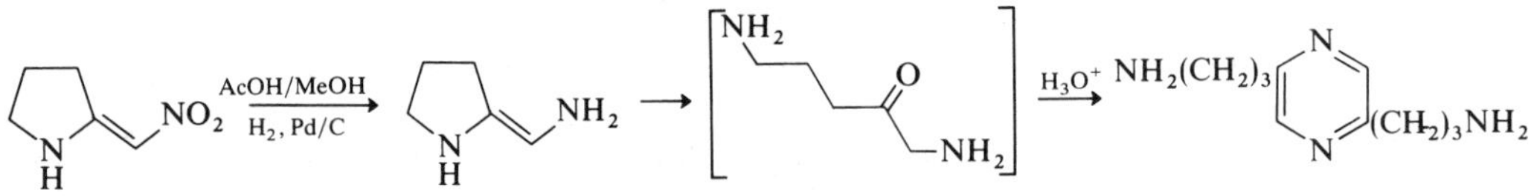

**Scheme 59**

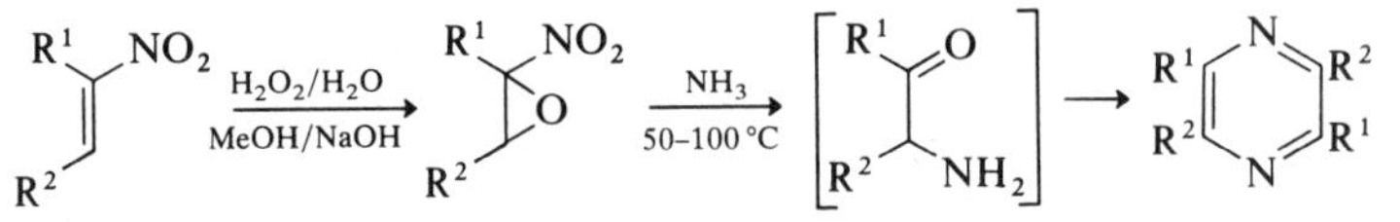

**Scheme 60**

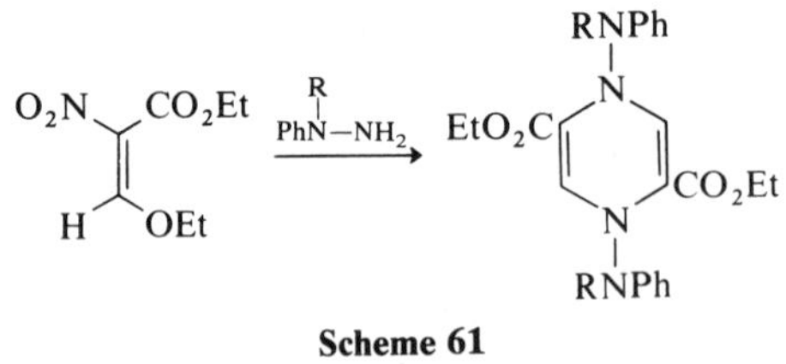

**Scheme 61**

1,4-Dialkyl-1,4-dihydropyrazines are generally unstable and are reported to undergo rearrangement to 1,2-dialkylpyrazines (Scheme 62) ⟨74TL179⟩.

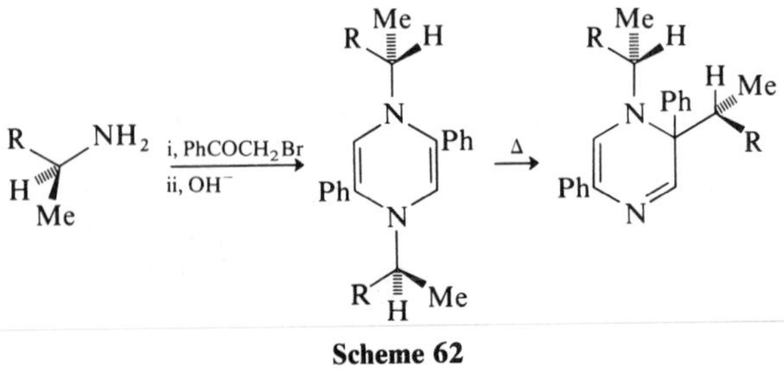

**Scheme 62**

2,5-Dioxopiperazines are amongst the most ubiquitous of natural products ⟨75FOR(32)57⟩ and they are formally derived by the cyclodimerization of $\alpha$-amino acids ⟨69CCC4000⟩ or their esters. A number of methods are available for their oxidation to the corresponding pyrazines. Treatment of 2,5-dioxopiperazines with triethyl- or trimethyl-oxonium fluoroborate followed by oxidation with DDQ, chloranil or iodine results in pyrazine formation, usually in high yields (Scheme 63) ⟨72JCS(P1)2494⟩.

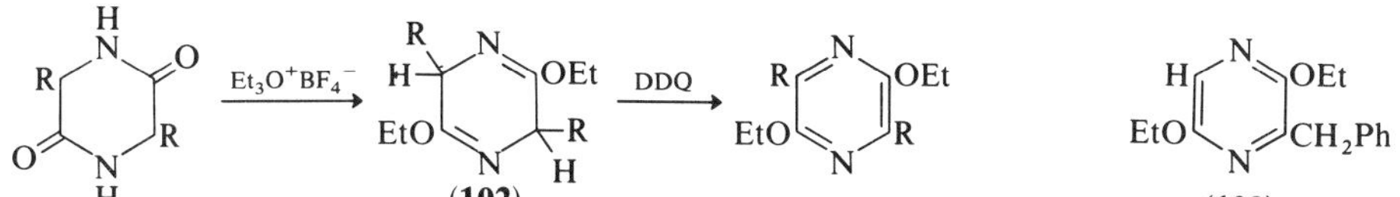

**Scheme 63**

An interesting reaction occurs when the dihydropyrazine (**102**; R = CH$_2$Ph) is pyrolyzed under vacuum. Toluene is liberated to give the monobenzylpyrazine (**103**) in high yield, presumably by a radical mechanism.

2,5-Dioxopiperazines have been converted into the corresponding dihydroxypyrazines by base catalyzed isomerization of the corresponding arylidene derivatives (Scheme 64) ⟨70JCS(C)980⟩, although this reaction appears to be limited to the synthesis of benzyl- or aryl-substituted benzylpyrazines.

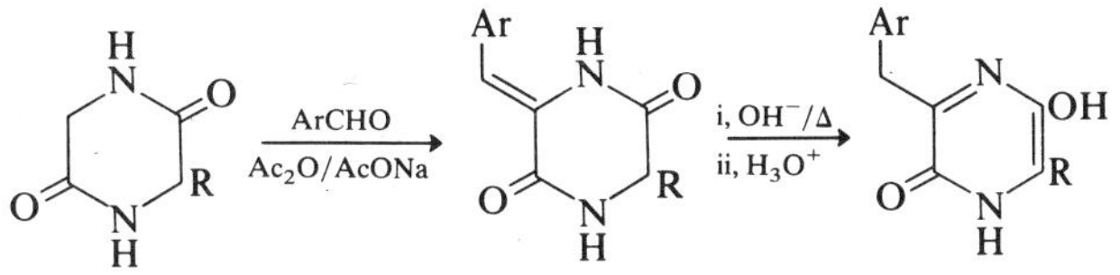

**Scheme 64**

Direct chlorination of 2,5-dioxopiperazines to produce pyrazine derivatives may be achieved using POCl$_3$ or PCl$_5$ ⟨47JCS1179, 70JCS(C)1070⟩, although the reaction is notorious for its unpredictability. Thus, chlorination of 3,6-dimethylpiperazine-2,5-dione (**104**; R = Me) with phosphorus oxychloride gives the mono- and di-chloropyrazines (Scheme 65) with the former predominating, whereas the dibenzyldioxopiperazine (**104**: R = CH$_2$Ph) gives predominantly the dichloropyrazine. These reactions are best rationalized in terms of the initial product of the reaction being the bisimidoyl halide which is then either oxidized directly to the dichloropyrazine, or undergoes isomerization and elimination under the reaction conditions.

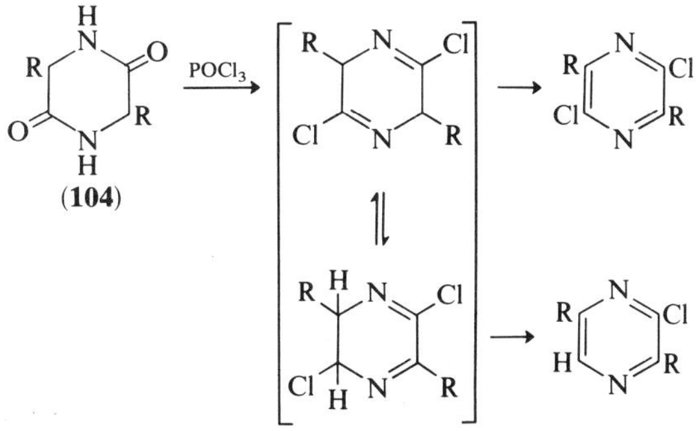

**Scheme 65**

Monooximes of $\alpha$-diketones have found applicability in the synthesis of 2-aminopyrazine 1-oxides by condensation with $\alpha$-aminonitriles, and this reaction was used by White and coworkers in an approach to the synthesis of Cypridina etioluciferamine (Scheme 66; R = 3-indoloyl) ⟨73T3761⟩. In this instance, the use of TiCl$_4$ as a catalyst was essential, since the carbonyl group in 3-acylindoles is normally deactivated and the required amine/carbonyl condensation is impractically slow. Under normal circumstances the carbonyl group in simple alkyl-substituted monooximes of $\alpha$-diketones is the more reactive site and the reaction is rapid, requiring no catalysis ⟨69LA(726)100⟩.

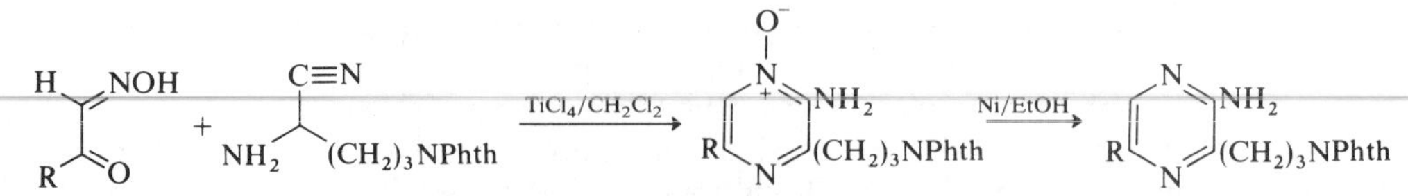

**Scheme 66**

Although at present of limited applicability, the recent observation that azirines undergo dimerization in the presence of Group VI metal carbonyls such as chromium or molybdenum hexacarbonyl ⟨75JA3541⟩ may prove to be of synthetic value. The reaction is straightforward in that simply stirring a 1:1 mixture of the azirine and Mo(CO)₆ at room temperature in dry THF is adequate to effect dimerization of the azirine. This is typified by the reaction of 2-phenylazirine (Scheme 67), which produces the pyrazine and isomeric dihydropyrazines in a ratio of 1:1. Subsequent developments have shown that yields of pyrazine and dihydropyrazines as high as 78% may be achieved ⟨77JA4330⟩. Clearly, if an oxidative work-up procedure was employed in this reaction, it would greatly simplify the isolation of the pyrazine, as the dihydropyrazines are known to be oxidized under mild conditions, and the method would be of some synthetic value. Ring opening of the azirines is not restricted to metal carbonyls; both silver catalysis ⟨76CC983⟩ and HF induced ring opening followed by dimerization ⟨80JOC5333⟩ have also been observed.

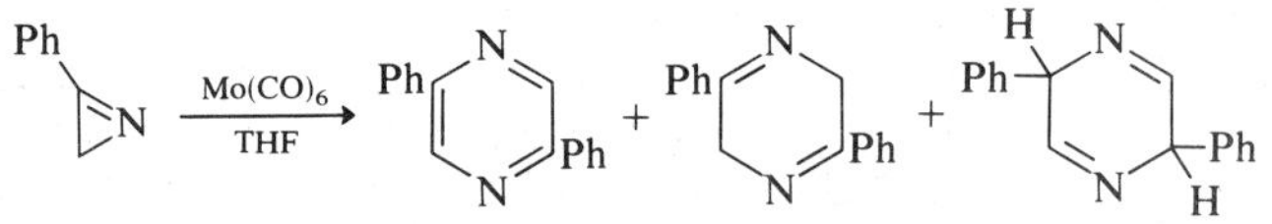

**Scheme 67**

No practical type 'B' syntheses of quinoxalines are commonly in use, largely because of the fact that type 'A' syntheses are more facile; however, some phenazine syntheses of this type are known, particularly those described in the older chemical literature. Hillemann ⟨38CB42⟩ has effected dimerization of o-bromoaniline by heating its solution in nitrobenzene with K₂CO₃ and copper powder. The reaction is believed to proceed through the intermediacy of 5,10-dihydrophenazine, but the latter has not been isolated (Scheme 68).

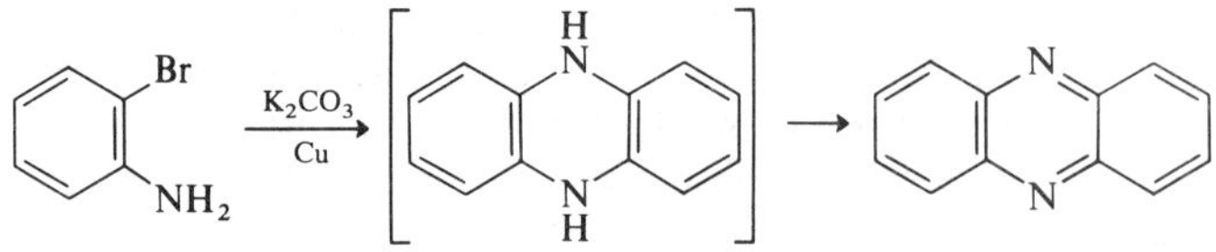

**Scheme 68**

### 2.14.3.4 Other Synthetic Methods

From the foregoing discussion it is evident that the most general methods for the synthesis of pyrazines, quinoxalines and phenazines fall into type 'A' and type 'B' categories, but other methods do exist. Although most of these are not of such general applicability, they are worthy of comment.

The reaction of $\alpha$-haloketones with ammonia formally corresponds to type 'D' synthesis, but the reaction is not simple and usually gives rise to product mixtures. This is almost certainly due to initial displacement of the halogen by ammonia to give the $\alpha$-aminoketone (Scheme 69), which can then either undergo dimerization (type 'B' synthesis) or alkylation by a second molecule of the $\alpha$-haloketone and subsequent reaction with ammonia to yield the dihydropyrazine which is oxidized *in situ* to the pyrazine. The final step in this reaction thus corresponds to a type 'E' synthesis ⟨10JCS2495⟩ or, if viewed in terms of the reaction of the initially formed $\alpha$-aminoketone with the $\alpha$-haloketone and ammonia, a type 'C' synthesis; this clearly indicates the ambiguity in attempting to apply rigid classification of this type.

The photolysis of azirines has been shown to result in dimerization to pyrazines ⟨72JA1395⟩ and although this formally corresponds to a type 'B' synthesis it involves an isolable intermediate (**105**) and does not proceed by simple dimerization (Scheme 70).

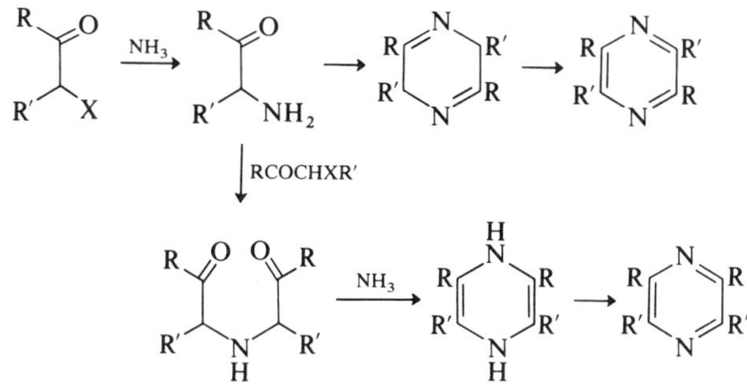

**Scheme 69**

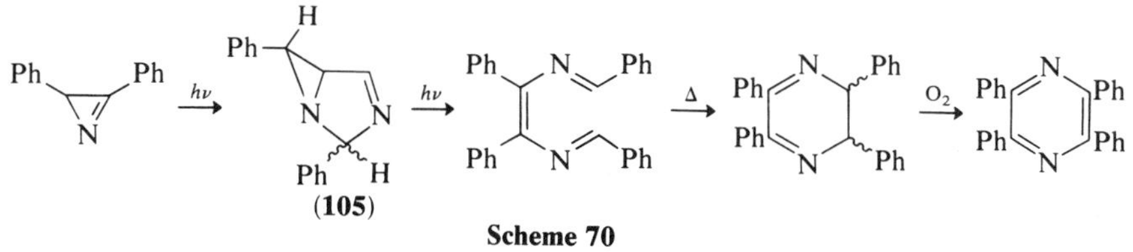

(**105**)

**Scheme 70**

Syntheses which fall into the type 'F' category are numerous, although not generally used in the construction of the pyrazine nucleus. A superior method to simple dimerization of amino acids involves the synthesis of a dipeptide ester (Scheme 71). These can be cyclized with varying ease to dioxopiperazines, which may then serve as pyrazine precursors. The principal advantage of this reaction over the simple dimerization is that it permits the synthesis of unsymmetrically substituted pyrazine derivatives ⟨71TH21400⟩. An alternative approach to the synthesis of similarly unsymmetrically substituted pyrazine derivatives has been recorded ⟨75FOR(32)57⟩: thus, treatment of 1,4-diacetylpiperazine-2,5-dione with strong base such as Bu$^t$OK or LDA followed by alkylation (Scheme 72) brings about similar results.

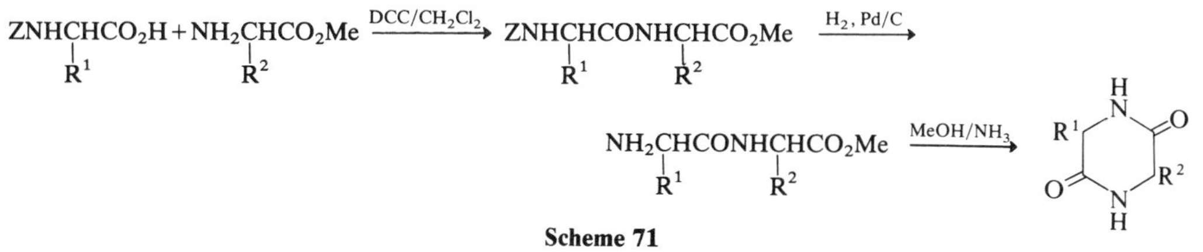

**Scheme 71**

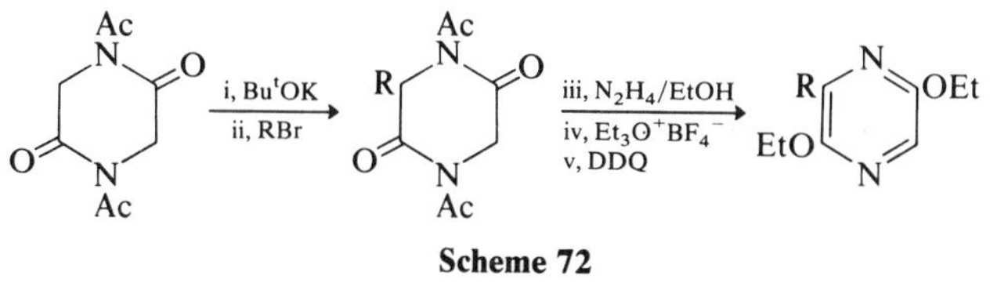

**Scheme 72**

Schöllkopf and coworkers ⟨79AG(E)863⟩ have used an approach which is conceptually similar to this in the synthesis of chiral α-amino acids. Dimerization of L-alanine produced the chiral dioxopiperazine (**106**; Scheme 73), which was readily alkylated with trimethyl-oxonium fluoroborate without racemization at the chiral centers. Lithiation of the intermediate dihydropyrazine (**107**) was effected at −70 °C using LDA or butyllithium and subsequent alkylation proceeded in a highly enantioselective fashion. Double metallation, not unexpectedly, is not observed, since dianions such as (**108**) contain eight π-electrons and would be expected to be antiaromatic. Hydrolysis of the alkylated products (**109**) with dilute aqueous hydrochloric acid proceeds smoothly to give mixtures of L-alanine and the alkylated derivatives, which are easily separated by fractional distillation of the methyl esters. During the alkylation step the *R* configuration is induced at the alkylated center and this is thought to be due to the preference of the incoming alkyl group to adopt a position *trans* relative to the methyl substituent at C-6. Enantiomeric excesses as high as 93% have been observed with this reaction sequence and it clearly has some potential.

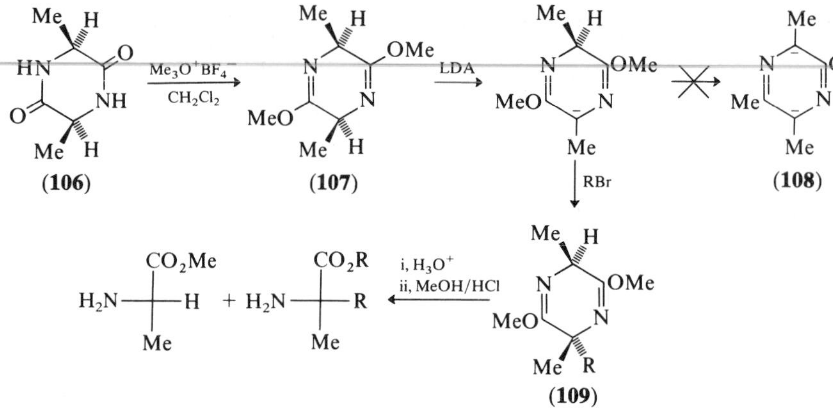

**Scheme 73**

Several examples of type 'F' syntheses have been applied to the construction of the phenazine ring system. Indeed, many of the classical syntheses of phenazines use *o*-substituted diphenylamine derivatives and effect ring closure using a reductive or oxidative procedure. This concept was first applied by Nietzki and Ernst ⟨1890CB1852⟩, who prepared phenazines from 2-aminodiphenylamines by direct oxidation using $MnO_2$, and Fischer ⟨1893CB378, 1896CB1873⟩ prepared the parent compound using lead dioxide or iron(III) ions (Scheme 74). The 2-aminodiphenylamine itself is readily available from the reaction of *o*-nitrochlorobenzene with aniline followed by reduction under standard conditions. A number of variants of this procedure have found use and many phenazine derivatives have been prepared in this way. The intermediate *o*-nitrodiphenylamine can be cyclized directly using a reductive procedure; thus, when heated with iron filings it yields phenazine in moderate yield and generally the ease with which the reaction occurs permits the use of a wide range of reducing agents such as red phosphorus, sulfur or elemental lead. More recently, $NaBH_4$ under alkaline conditions has been employed ⟨70CC1423⟩ and yields of substituted phenazines as high as 80% have been reported.

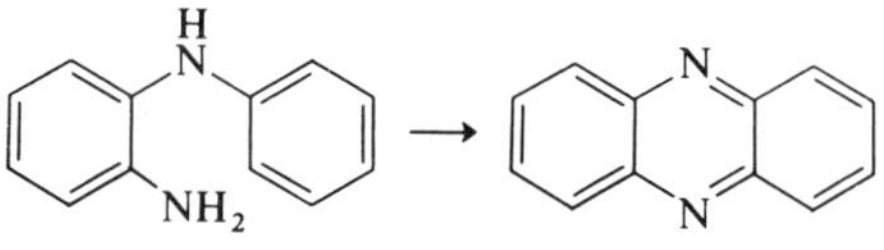

**Scheme 74**

Very good yields of phenazine derivatives have been claimed ⟨14M(35)1153⟩ by Eckert and Steiner by reduction of 2,2'-dinitrodiphenylamine with tin(II) chloride, followed by oxidation of the resulting semiquinonoid salt with $H_2O_2$ or other strong oxidizing agents; a variant involving the iron(III) chloride oxidation of 2,2'-diaminodiphenylamines also appears to give high yields of phenazine and its derivatives.

The cleavage of fused pyrazines represents an important method of synthesis of substituted pyrazines, particularly pyrazinecarboxylic acids. Pyrazine-2,3-dicarboxylic acid is usually prepared by the permanganate oxidation of either quinoxalines or phenazines. The pyrazine ring resembles the pyridine ring in its stability rather than the other diazines, pyridazine and pyrimidine. Fused systems such as pteridines may easily be converted under either acidic or basic conditions into pyrazine derivatives (Scheme 75).

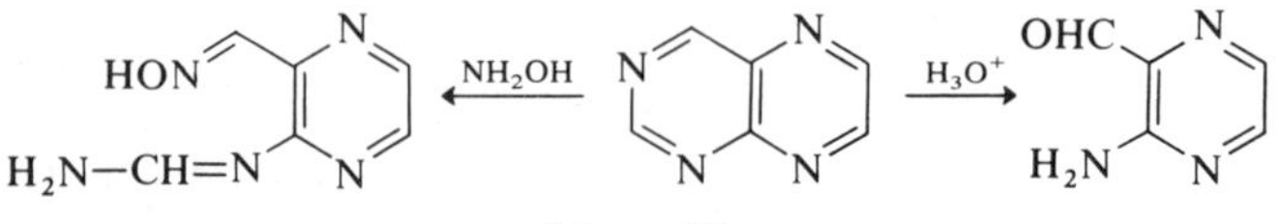

**Scheme 75**

Following the initial observation that perfluoropyridazine undergoes rearrangement to the isomeric pyrazine on irradiation with UV light ⟨69CC1200⟩, it was suggested that the reaction proceeded through a diazaprismane intermediate ⟨71CC264⟩. However, this suggestion has been refuted on the basis of the observation that 4,5-dichloro-3,6-difluoropyridazine gives rise to 2,5-dichloro-3,6-difluoropyrazine (Scheme 76) ⟨70JA7505⟩ and this result cannot be rationalized in terms of a diazaprismane intermediate. It would appear that this synthesis is of limited value as pyrazine itself, for example, has been shown to be photochemically unstable and isomerizes to pyrimidine ⟨76FRP2044534⟩.

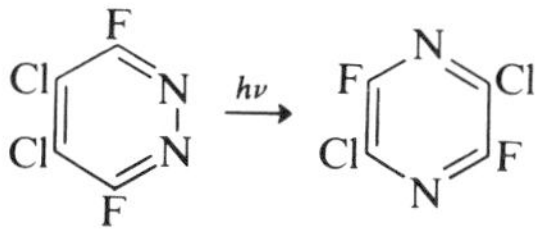

**Scheme 76**

## 2.14.4 APPLICATIONS

### 2.14.4.1 Introduction

Scarcely a single issue of *Chemical Abstracts* is published without reference to medicinal compounds containing the pyrazine or quinoxaline ring in some form, and hence it is impractical to list all applications of pyrazines, quinoxalines and phenazines. Some of the more important applications and natural products, particularly the more recent developments, are mentioned in this Section.

### 2.14.4.2 Pyrazinones, Quinoxalinones and Hydroxyphenazines

Following the discovery of penicillins, an extensive program for the screening of culture fluids and residual mycelial material commenced which resulted in the discovery of a large number of pyrazinones and related 1-hydroxy-2-pyrazinones with pronounced antibiotic character. Some examples are shown in Table 4. One of the earliest substances to be isolated, aspergillic acid (**110**; $R^1 = OH$, $R^2 = Me$, $R^3 = Et$, $R^4 = R^5 = H$, $R^6 = Pr^i$), was found to have good antibiotic properties and its structure was confirmed by total synthesis ⟨51JCS2679⟩. It appears that the hydroxamic acid function of the 1-hydroxypyrazinone is necessary for antibiotic activity as many of the simple pyrazinones are biologically inactive.

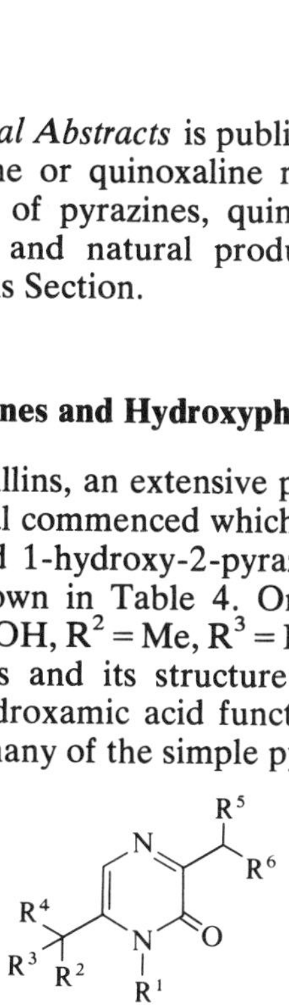

(**110**)

**Table 4** Pyrazinones with Antibiotic Character

| $R^1$ | $R^2$ | $R^3$ | $R^4$ | $R^5$ | $R^6$ |
|-------|-------|-------|--------|-------|--------|
| OH | Me | Et | H | H | $Pr^i$ |
| H | H | H | $Pr^i$ | H | $Pr^i$ |
| OH | OH | Me | Me | H | $Pr^i$ |
| OH | Me | Et | H | Me | Et |
| H | OH | Me | Et | Me | Et |

Most of the naturally-occurring pyrazine hydroxamic acids appear to be derived from valine, leucine and isoleucine, and biosynthetic studies by MacDonald and coworkers ⟨61JBC(236)512, 62JBC(237)1977, 65JBC(240)1692⟩ indicate that these amino acids are incorporated. However, it would seem that the logical intermediates, *viz.* the 2,5-dioxopiperazines such as (**111**) and (**112**), are not always incorporated. This does not rule out their intermediacy, as there may be problems such as low solubility or membrane permeability which prevent their efficient incorporation. An exception to these results was reported for pulcherrimic acid (**113**) ⟨65BJ(96)533⟩, which has been shown to be derived from cyclo-L-leu-L-leu which serves as an efficient precursor.

Mycelianamide (**114**) clearly bears a structural resemblance to pulcherrimic acid and it seems probable that both are derived from the corresponding dihydroxypyrazines (**115**)

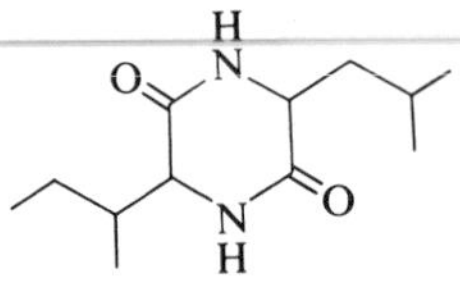

(111) 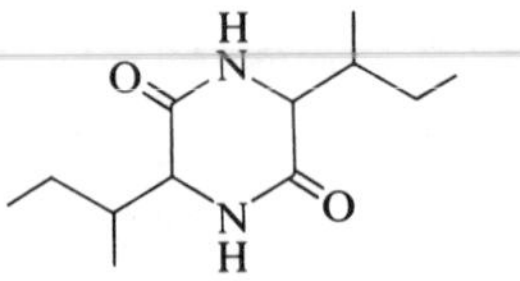          (112) 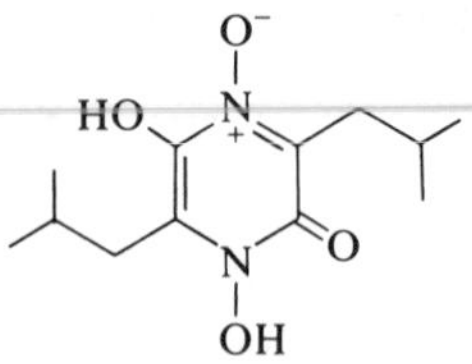          (113)

and (116) *in vivo* by way of the di-*N*-oxides (117) and (118). In these structures the 'hydroxy-*N*-oxide' tautomer is represented rather than the more probable hydroxamic acid form to illustrate the probable steps involved during their biosynthesis. Tautomerism of (118) then leads to pulcherrimic acid (113) in which only one hydroxy-*N*-oxide tautomerizes to the hydroxamic acid; the second is then effectively cross-conjugated and the isomerization cannot occur. In the case of (117), however, tautomerism results in the formation of (119) in which the increased acidity of the benzylic protons permits a rearrangement to the arylidene bishydroxamic acid found in mycelianamide. A synthetic strategy based on this concept has recently been published ⟨82JCS(P1)953⟩ and a total synthesis of mycelianamide by more classical methods has also been realized ⟨80CC1020⟩.

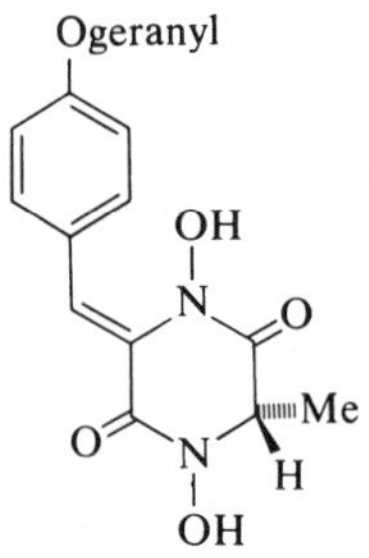

(114) 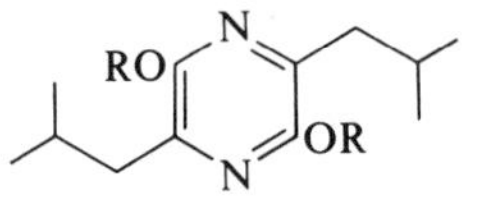          (115) 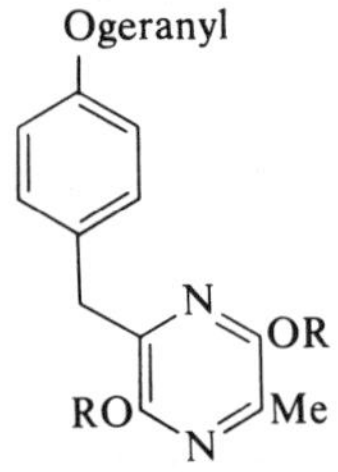          (116)

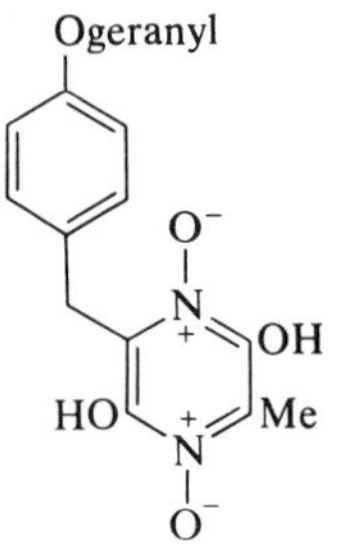

(117) 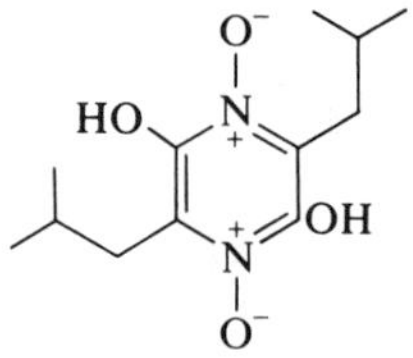          (118) 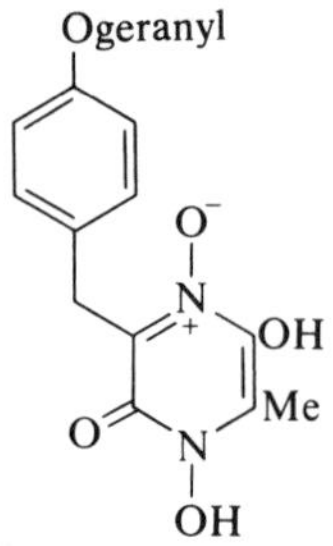          (119)

There are a number of bacterial pigments containing the phenazine ring system, and perhaps the best known of these is iodinin (120). The organism *Chromobacterium iodinium*, which was first isolated from milk, was so named because of the purple crystals with a characteristic 'coppery glint' which were produced during the late stages of its growth, and the structure of the pigment was established as (120) ⟨38JCS479⟩.

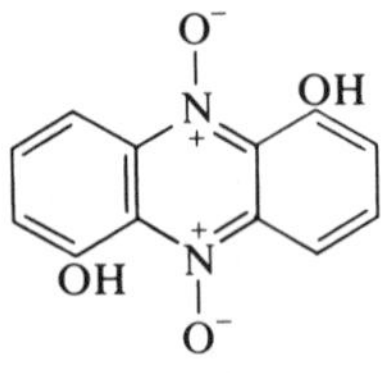

(120) 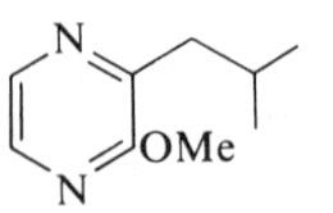          (121)

### 2.14.4.3 Alkoxypyrazines

The major component of the oil extracted from bell peppers has been shown, initially on the basis of mass spectral studies but subsequently by total synthesis, to be 2-methoxy-3-

isobutylpyrazine (**121**). The pyrazine comprises 16% of the oil which occurs as one part in $10^6$ of the weight of the plant. Subsequently, 3-isopropyl-, 3-*s*-butyl- and 3-isobutyl-2-methoxypyrazines have been isolated from green peas ⟨70CI(L)897⟩. These methoxypyrazines have extremely important organoleptic properties, being detectable at very low threshold concentrations; thus, (**121**) as a solution in water can be detected at a level of 2 parts in $10^{12}$.

### 2.14.4.4 Alkylpyrazines

During the last 20 years there has been a very large number of publications on alkyl-pyrazines of natural origin, particularly as flavor components in foodstuffs. It would appear that any foodstuff containing both carbohydrates and amino acids gives rise to alkylpyrazines during the cooking process. Pyrazines have been identified in potato chips, cooked meats, coffee, cocoa, tea, many different types of cheese, alcoholic beverages and many other sources such as cigarette smoke, and it would seem probable in most cases that the pyrazines are to a large part implicated in the characteristic aromas/flavors of the materials with which they are associated.

The number of simple alkylated pyrazines and the food products with which they are associated are enormous and a few examples are listed in Table 5. Fuller reviews on this subject are available ⟨73MI21400⟩.

**Table 5**  Food Products containing Alkylpyrazines

| Pyrazine | Source | Ref. |
|---|---|---|
| Unsubstituted | Coffee | 68MI21400 |
| 2-Methylpyrazine | Roasted barley | 69ABC1775 |
|  | Coffee | 68MI21400 |
| 2,3-Dimethylpyrazine | Boiled beef | 72MI21400 |
| 2,5-Dimethylpyrazine | Fried beef | 71MI21400 |
|  | Coffee | 68MI21400 |
| 2,6-Dimethylpyrazine | Coffee | 68MI21400 |
|  | Potato chips | 71MI21401 |
| 2,3,5-Trimethylpyrazine | Coffee | 68MI21400 |
|  | Fried beef | 71MI21400 |
| 2,3,5,6-Tetramethylpyrazine | Fried beef | 71MI21400 |
|  | Coffee | 68MI21400 |
| 2-Ethylpyrazine | Fried beef | 71MI21400 |
|  | Filberts | 72MI21401 |
| 2-Ethyl-3-methylpyrazine | Roasted barley | 71MI21402 |
| 2-Ethyl-5-methylpyrazine | Fried beef | 71MI21400 |
| 2-Ethyl-6-methylpyrazine | Coffee | 68MI21400 |
| 2,5-Diethylpyrazine | Peanuts | 71MI21403 |
| 2-Vinylpyrazine | Coffee | 67MI21401 |
| 2-Methyl-6-vinylpyrazine | Coffee | 67MI21401 |

There has been considerable speculation on the origin of the pyrazine derivatives but it is now recognized that they probably arise in the Maillard reaction ⟨12CR(154)66⟩ and model studies using glycine/glyoxal mixtures ⟨76MI21400⟩ and glucose/asparagine mixtures ⟨70MI21402⟩ have clearly demonstrated that pyrazine derivatives are produced by heating mixtures of the amino acid and sugar or sugar degradation products. Although the precise mechanism for pyrazine formation remains unclear it has been shown by isotopic labeling that the nitrogen originates from the amino acid and the carbon solely from the sugar, thus ruling out simple dimerization of amino acids followed by a reduction/oxidation sequence ⟨69MI21402⟩. It appears that ammonium ions are not involved in the transfer of nitrogen but that α-aminoketones and/or α-aminoaldehydes may be formed in a Strecker reaction and these subsequently react to yield the pyrazines.

In recent years a number of alkylpyrazines from animal sources have been discovered and generally they appear to be more abundant than those of plant origin. The ponerine ants (genus *Otontomachus*) when disturbed produce a secretion which has an odour similar to that of chocolate and a GLC analysis of the mandibular gland secretion shows that it contains a series of 3-*n*-alkyl-2,6-dimethylpyrazines ⟨73MI21401⟩. The Canadian beaver (*Castor fiber*) produces a series of 5,6,7,8-tetrahydroquinoxalines (**122a–d**) ⟨76HCA1169⟩. Pyrazines of animal and insect origin have recently been extensively reviewed ⟨80H(14)477⟩.

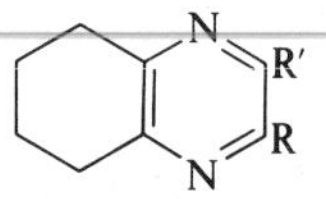

(122)  a; R = R′ = H
       b; R = Me, R′ = H
       c; R = R′ = Me
       d; R,R′ = (CH$_2$)$_4$

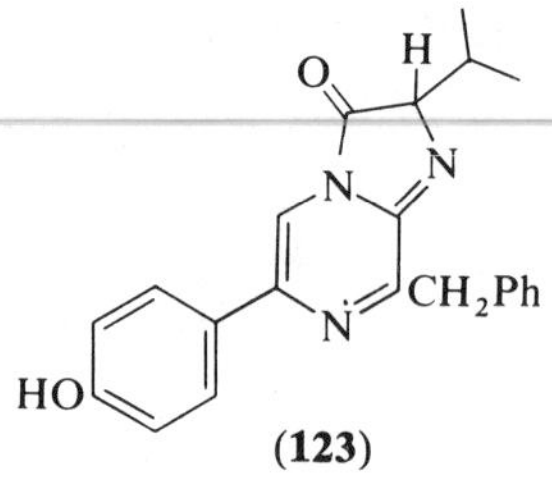

## 2.14.4.5 Chemiluminescence and Bioluminescence in Pyrazine Derivatives

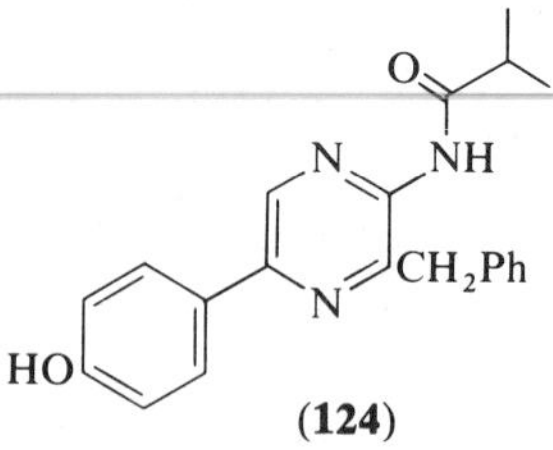

Numerous coelenterates belonging to the genera *Obdia, Mnemiopsis, Renilla, Pelagia, Campanularia, Ptilosarcus, Clytia, Lovenella,* etc. are characterized by their remarkable bioluminescence ⟨76ACR201⟩. Although there appears to be minor structural variations within the species, it is now generally recognized that simple pyrazines such as (123) are responsible for the bioluminescence during an enzyme-catalyzed oxidation, *viz.* (123) → (124). The fused pyrazines, *e.g.* (123), appear to be biosynthetically derived from a modified tripeptide precursor ⟨73CC467⟩, and McCapra and Manning have successfully produced derivatives of (123) by classical peptide synthesis (Scheme 77). Interestingly, treatment of derivatives of (123) with oxygen/base under a variety of pH conditions produces chemiluminescence with wavelengths of emitted light varying from 414 to 523 nm and it is believed that the wavelength of the emitted light during bioluminescence is probably controlled enzymatically by local pH variations within the enzyme–substrate complex. Like many chemiluminescence/bioluminescence reactions the decomposition of a peroxide (Scheme 78) is thought to result in an electronically excited species which gives off light in quantum yields ranging from 0.04–0.88. The step most likely to be responsible for the chemiluminescence is the breakdown of the dioxetane (125).

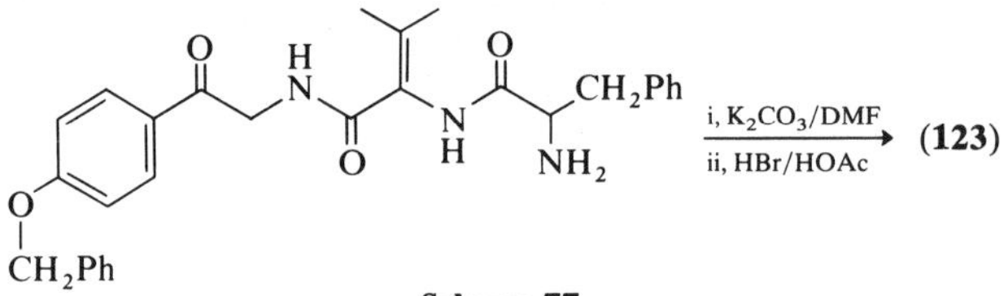

**Scheme 77**

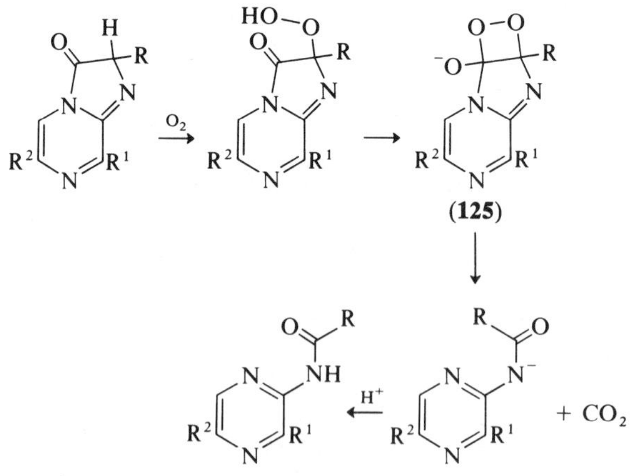

**Scheme 78**

## 2.14.4.6 Miscellaneous Pyrazine Derivatives

Before the discovery of streptomycin, pyrazinamide (126) was one of the front runners in the treatment of tuberculosis. A broad spectrum of biological activity has been associated with pyrazine derivatives, ranging from the herbicidal activity of (127) to antibiotic activity in the aspergillic acids. Antibacterial activity has also been observed in the semisynthetic penicillins (**128**) and in pyrazinium betaines (**129**). A review of the biological activities of pyrazine derivatives has been published ⟨66MI21401⟩.

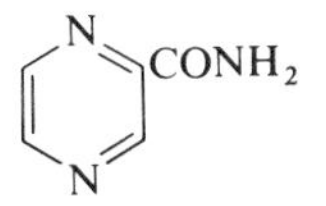

(**126**) 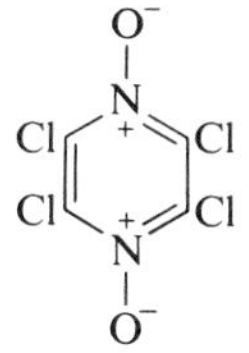     (**127**) 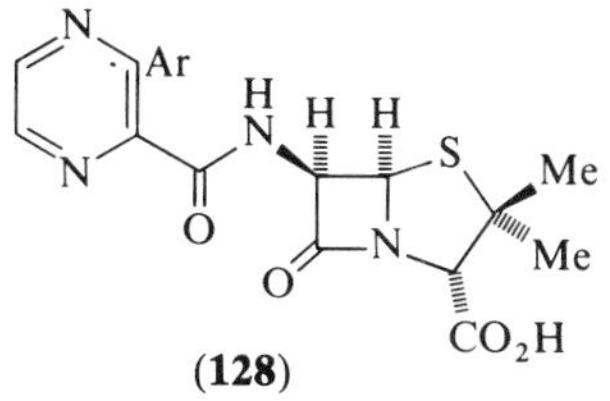     (**128**) 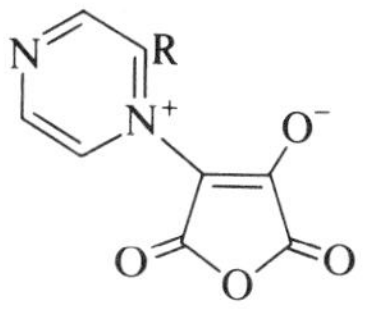     (**129**)

### 2.14.4.7 Miscellaneous Quinoxaline Derivatives

Natural products containing the quinoxaline ring are rare; however, the 1,4-dioxide of quinoxaline-2-carboxylic acid (**130**) has been isolated from cultures of *Streptomyces ambrofaciens*, and this compound and its derived esters have been shown to have antibiotic properties. It is particularly interesting to note that a large number of quinoxaline 1,4-dioxides have antibacterial activity and this activity is higher *in vivo* than *in vitro* and is increased by two orders of magnitude under anaerobic conditions. It has been suggested that the high *in vivo* activity is possibly due to the quinoxaline acting as a pro-drug with the metabolic products acting as the effective antibiotic substances ⟨75JMC637⟩, although if this is the case the active drug has not been identified.

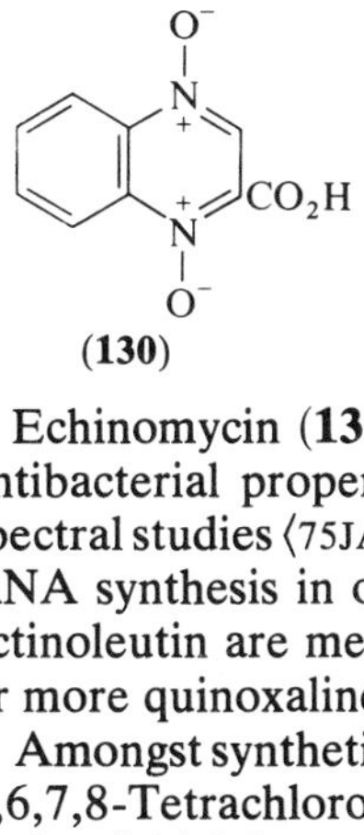

(**130**) 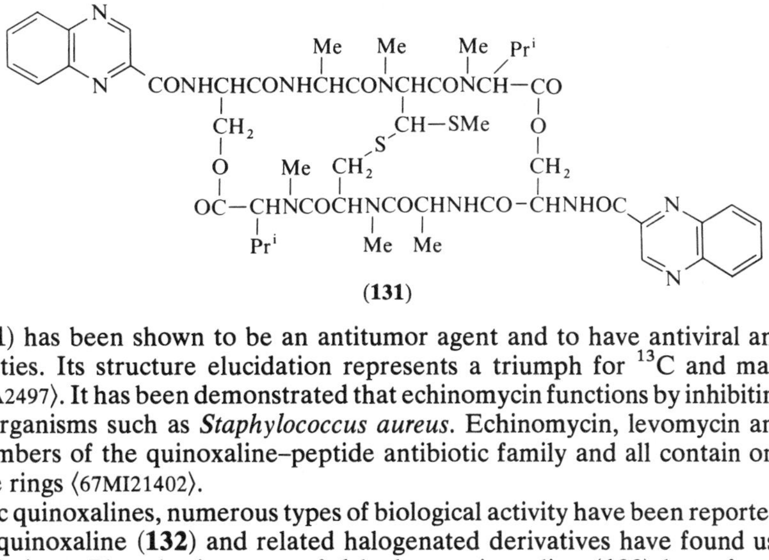     (**131**)

Echinomycin (**131**) has been shown to be an antitumor agent and to have antiviral and antibacterial properties. Its structure elucidation represents a triumph for $^{13}$C and mass spectral studies ⟨75JA2497⟩. It has been demonstrated that echinomycin functions by inhibiting RNA synthesis in organisms such as *Staphylococcus aureus*. Echinomycin, levomycin and actinoleutin are members of the quinoxaline–peptide antibiotic family and all contain one or more quinoxaline rings ⟨67MI21402⟩.

Amongst synthetic quinoxalines, numerous types of biological activity have been reported. 5,6,7,8-Tetrachloroquinoxaline (**132**) and related halogenated derivatives have found use in fungicidal formulations. Phosphoric esters of 6-hydroxyquinoxaline (**133**) have found use in insecticidal preparations, and phosphoric ester derivatives of 2-hydroxyquinoxalines, such as (**134**), function as anthelmintics.

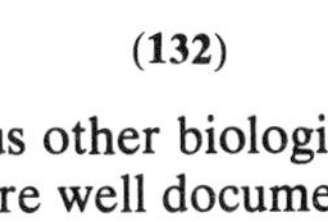

(**132**) 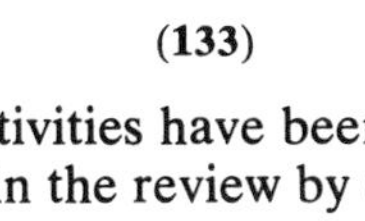     (**133**) 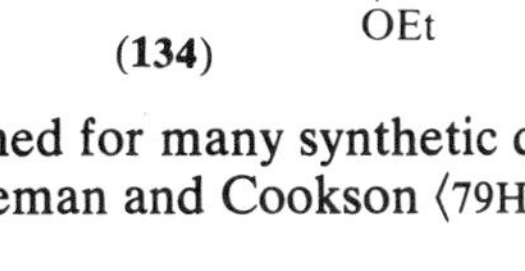     (**134**)

Numerous other biological activities have been claimed for many synthetic quinoxalines and these are well documented in the review by Cheeseman and Cookson ⟨79HC(35)1⟩.

### 2.14.4.8 Naturally Occurring Phenazine Pigments

Reference has already been made to iodinin (**120**) (1,6-dihydroxyphenazine 5,10-dioxide), the violet pigment produced by the bacterium *Chromobacterium iodinium*. Iodinin inhibits the growth of some pathogenic bacteria such as *Streptococci*, but the relatively high toxicity precludes practical clinical use.

*Pseudomonas aeruginosa* is a common organism infecting human skin, particularly sweaty areas such as the armpits where its presence may be recognized by a blue color associated with garments which have been worn during heavy exercise, particularly white cotton. It is also evident in blue pus. This blue coloration is due to the phenazine pyocyanine (**135**). In principle, pyocyanine can be represented in two forms, *viz.* the betaine (**135**) and the keto form (**136**); the dipole moment of 7.0 D demonstrates the importance of the former. It crystallizes as dark blue needles which decompose on standing to 1-hydroxyphenazine, and its structure has been confirmed by total synthesis. Thus, on treatment of phenazine with dimethyl sulfate and exposure of an aqueous solution of the resultant salt to sunlight for 24 hours, McIlwaine was able to prepare (**136**) in 45% yield ⟨37JCS1704⟩.

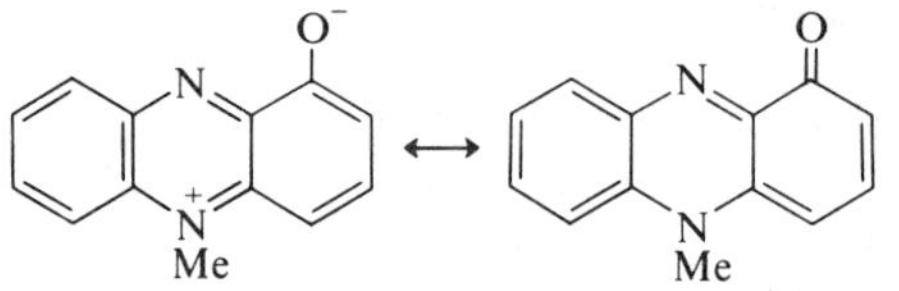

(**135**)                    (**136**) 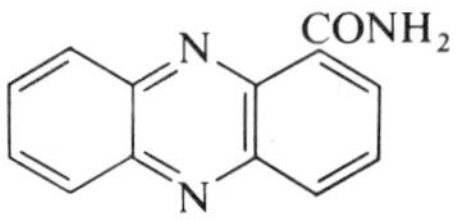                    (**137**)

Phenazine-1-carboxamide (**137**) is known as oxychlororaphine and has been isolated from cultures of *Pseudomonas chlororaphis*; it has some limited inhibitory properties, but the inhibitory action of phenazines is generally disappointing. Some phenazine derivatives have insecticidal properties; thus, phenazine itself has been found to be toxic to the clothes moth, the Hawaiian beet webworm, the rice weevil and larva of the codling moth, but under trial conditions its toxicity to plant material, as evidenced by severe burning of foliage, was found to be too high to make it of practical value.

### 2.14.4.9 Synthetic Phenazine Dyes

Historically the phenazine dyes have played an important part in the dyestuffs industry, although their use has largely been superseded by the more modern, color-fast dyes, in particular those dyes which become chemically bonded to the fibers of the materials being dyed. Amongst the earliest examples of phenazine dyes are those compounds known as the safranines. The discovery of the safranines has been attributed to Greville Williams in 1859 and they were apparently in commercial use shortly after that date, but it was not until 1886 that it was recognized that phenosafranine (**138**) was indeed a phenazine containing system.

Originally safranine was prepared by heating mixtures of aminoazotoluene and toluidine with oxidizing agents such as potassium dichromate, although it was found later that the oxidizing agent was unnecessary. It was also discovered that, simply by mixing two molecules of *p*-phenylenediamine derivatives with a monoamine (aliphatic or aromatic) under oxidizing conditions, a range of dyestuffs could be produced, and a mechanism was postulated involving the initial oxidation of the *p*-diamine to a *p*-quinonediimine followed by reaction with a second molecule of the diamine to give a 4,4′-diaminodiphenylamine intermediate (Scheme 79). Oxidation of the latter leads to the formation of indamine (**139**) which condenses with the amine to yield the safranine, phenosafranine (**138**) in the case of aniline.

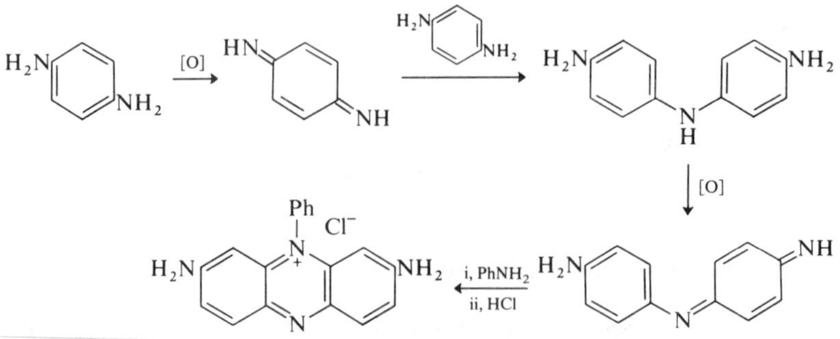

(**138**)                                    (**139**)

**Scheme 79**

Subsequent commercial development utilized 4,4'-diaminodiphenylamine as the starting material, and the whole process may be carried out *in situ* on the fabric.

By varying the structure of the starting diamine or the monoamine, a range of dyes became available, covering a spectrum of colors. Commercial safranine derivatives have found particular application in mordant dyeing; thus, the toluosafranines dyes mordanted cotton in brilliant red shades that are fast to washing, although fastness to light leaves something to be desired. Safranines have also found some use as staining agents in histology and cytology. Mauveine (**140**) which was first synthesized by W. H. Perkin, is a historically important member of the safranine class.

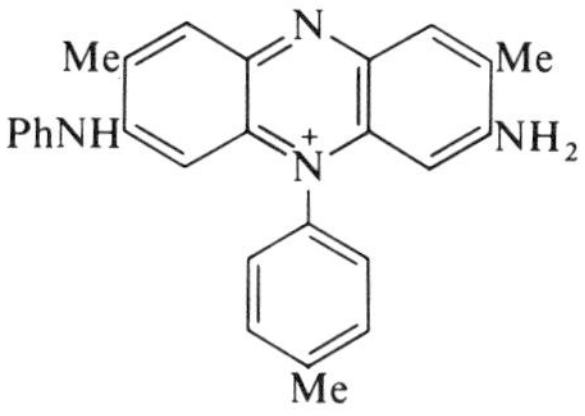

(**140**)

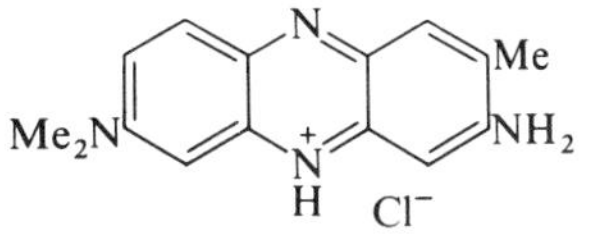

(**141**)

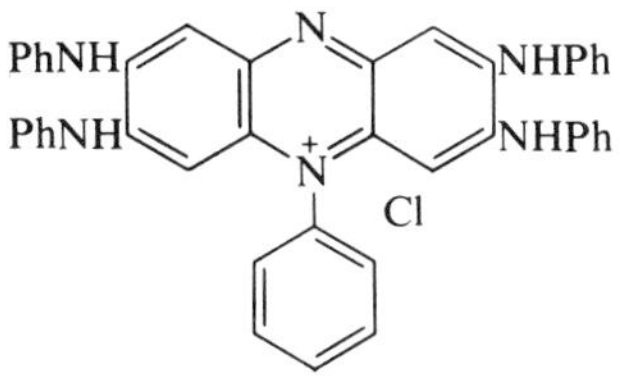

(**142**)

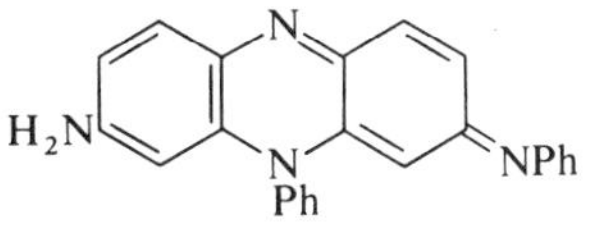

(**143**)

Numerous dyes structurally related to the safranines, such as the eurodines, *e.g.* (**141**), the indulines, *e.g.* (**142**), the nigrosines (**143**) and aniline black, a pigment of unknown structure used in the printing industry, are well known and a detailed account of their chemistry and applications has been presented ⟨57HC(11)1⟩.

# 2.15

# Pyridodiazines and their Benzo Derivatives

E. LUNT and C. G. NEWTON

*May & Baker Ltd., Dagenham*

| | | |
|---|---|---|
| 2.15.1 | PYRIDOPYRIMIDINES: INTRODUCTION | 201 |
| 2.15.2 | STRUCTURE AND PHYSICAL PROPERTIES OF PYRIDOPYRIMIDINES | 201 |
| *2.15.2.1* | *Conformational Behaviour* | 202 |
| *2.15.2.2* | *NMR Spectroscopy* | 202 |
| *2.15.2.2.1* | *$^1$H NMR spectra* | 202 |
| *2.15.2.2.2* | *$^{13}$C NMR spectra* | 202 |
| *2.15.2.3* | *IR Spectroscopy* | 204 |
| *2.15.2.4* | *UV Spectroscopy* | 204 |
| *2.15.2.5* | *MO Calculations* | 204 |
| *2.15.2.6* | *Mass Spectrometry* | 204 |
| *2.15.2.7* | *Tautomerism* | 204 |
| 2.15.3 | REACTIVITY AT RING ATOMS IN PYRIDOPYRIMIDINES | 205 |
| *2.15.3.1* | *Electrophilic Attack at Ring Carbon* | 205 |
| *2.15.3.2* | *Electrophilic Attack at Ring Nitrogen* | 206 |
| *2.15.3.3* | *Nucleophilic Attack at Ring Carbon* | 207 |
| 2.15.4 | REACTIVITY OF SUBSTITUENTS IN PYRIDOPYRIMIDINES | 209 |
| *2.15.4.1* | *Electrophilic Attack on Substituents* | 209 |
| *2.15.4.2* | *Nucleophilic Attack on Substituents* | 212 |
| *2.15.4.3* | *Nucleophilic Substitution Reactions* | 213 |
| *2.15.4.4* | *Radical Attack on Ring and Substituents and Photochemistry* | 215 |
| 2.15.5 | SYNTHESIS OF PYRIDOPYRIMIDINES | 215 |
| *2.15.5.1* | *By Formation of One Bond between Two Heteroatoms [6 + 0 (N)]* | 215 |
| *2.15.5.2* | *By Formation of One Bond adjacent to a Heteroatom [6 + 0 (α)]* | 215 |
| *2.15.5.2.1* | *From pyridine intermediates* | 215 |
| *2.15.5.2.2* | *From pyrimidine intermediates* | 217 |
| *2.15.5.2.3* | *From naphthyridines* | 218 |
| *2.15.5.2.4* | *Fused derivatives* | 218 |
| *2.15.5.3* | *Formation of One Bond β to a Heteroatom [6 + 0 (β)]* | 220 |
| *2.15.5.3.1* | *From pyridine intermediates* | 220 |
| *2.15.5.3.2* | *From pyrimidine intermediates* | 220 |
| *2.15.5.3.3* | *Fused derivatives* | 220 |
| *2.15.5.4* | *By Formation of One Bond γ to a Heteroatom [6 + 0 (γ)]* | 220 |
| *2.15.5.4.1* | *From pyridine intermediates* | 220 |
| *2.15.5.4.2* | *From pyrimidine intermediates* | 220 |
| *2.15.5.4.3* | *Fused derivatives* | 221 |
| *2.15.5.5* | *By Formation of Two Bonds from [5 + 1] Atom Fragments* | 222 |
| *2.15.5.5.1* | *From pyridine intermediates* | 222 |
| *2.15.5.5.2* | *From pyrimidine intermediates* | 224 |
| *2.15.5.5.3* | *Fused derivatives* | 224 |
| *2.15.5.6* | *By Formation of Two Bonds from [4 + 2] Atom Fragments* | 225 |
| *2.15.5.6.1* | *From pyridine intermediates* | 225 |
| *2.15.5.6.2* | *From pyrimidine intermediates* | 227 |
| *2.15.5.6.3* | *Fused derivatives* | 227 |
| *2.15.5.7* | *By Formation of Two Bonds from [3 + 3] Atom Fragments* | 228 |
| *2.15.5.7.1* | *From pyridine intermediates* | 228 |
| *2.15.5.7.2* | *From pyrimidine intermediates* | 229 |
| *2.15.5.7.3* | *Fused derivatives* | 230 |
| *2.15.5.8* | *By Formation of Three Bonds* | 231 |
| *2.15.5.9* | *By Formation of Four or More Bonds* | 232 |
| 2.15.6 | PYRIDOPYRIDAZINES: INTRODUCTION | 232 |

2.15.7   STRUCTURE AND PHYSICAL PROPERTIES OF PYRIDOPYRIDAZINES   232

  *2.15.7.1   Conformational Behaviour*   233
  *2.15.7.2   NMR Spectroscopy*   234
    *2.17.7.2.1   $^1H$ NMR spectra*   234
    *2.15.7.2.2   $^{13}C$ NMR spectra*   234
  *2.15.7.3   IR Spectroscopy*   234
  *2.15.7.4   UV Spectroscopy*   236
  *2.15.7.5   MO Calculations*   236
  *2.15.7.6   Mass Spectrometry*   237
  *2.15.7.7   Tautomerism*   237

2.15.8   REACTIVITY AT RNG ATOMS OF PYRIDOPYRIDAZINES   237

  *2.15.8.1   Electrophilic Attack at Ring Carbon*   237
  *2.15.8.2   Electrophilic Attack at Ring Nitrogen*   238
  *2.15.8.3   Nucleophilic Attack at Ring Carbon*   239

2.15.9   REACTIVITY OF SUBSTITUENTS IN PYRIDOPYRIDAZINES   240

  *2.15.9.1   Electrophilic Attack on Substituents*   240
  *2.15.9.2   Nucleophilic Attack on Substituents*   241
  *2.15.9.3   Nucleophilic Substitution Reactions*   241

2.15.10   SYNTHESIS OF PYRIDOPYRIDAZINES   242

  *2.15.10.1   By Formation of One Bond between Two Heteroatoms [6 + 0 (N)]*   242
  *2.15.10.2   By Formation of One Bond adjacent to a Heteroatom [6 + 0 (α)]*   242
    *2.15.10.2.1   From pyridine intermediates*   242
    *2.15.10.2.2   From pyridazine intermediates*   243
    *2.15.10.2.3   Fused derivatives*   244
  *2.15.10.3   By Formation of One Bond β to a Heteroatom [6 + 0 (β)]*   244
    *2.15.10.3.1   Fused derivatives*   244
  *2.15.10.4   By Formation of One Bond γ to a Heteroatom [6 + 0 (γ)]*   245
    *2.15.10.4.1   From pyridine intermediates*   245
    *2.15.10.4.2   From pyridazine intermediates*   245
    *2.15.10.4.3   Fused derivatives*   245
  *2.15.10.5   By Formation of Two bonds from [5 + 1] Atom Fragments*   245
  *2.15.10.6   By Formation of Two Bonds from [4 + 2] Atom Fragments*   246
    *2.15.10.6.1   From pyridine intermediates*   246
    *2.15.10.6.2   From pyridazine intermediates*   247
    *2.15.10.6.3   Fused derivatives*   247
  *2.15.10.7   By Formation of Two Bonds from [3 + 3] Atom Fragments*   247
    *2.15.10.7.1   From pyridazine intermediates*   247
  *2.15.10.8   By Formation of Three Bonds*   248

2.15.11   PYRIDOPYRAZINES: INTRODUCTION   248

2.15.12   STRUCTURE AND PHYSICAL PROPERTIES OF PYRIDOPYRAZINES   248

  *2.15.12.1   Conformational Behaviour*   249
  *2.15.12.2   NMR Spectroscopy*   249
    *2.15.12.2.1   $^1H$ NMR spectra*   249
    *2.15.12.2.2   $^{13}C$ NMR spectra*   249
  *2.15.12.3   IR Spectroscopy*   249
  *2.15.12.4   UV Spectroscopy*   250
  *2.15.12.5   MO Calculations*   250
  *2.15.12.6   Mass Spectrometry*   250
  *2.15.12.7   Tautomerism*   250

2.15.13   REACTIVITY AT RING ATOMS OF PYRIDOPYRAZINES   250

  *2.15.13.1   Electrophilic Attack at Ring Carbon*   250
  *2.15.13.2   Electrophilic Attack at Ring Nitrogen*   251
  *2.15.13.3   Nucleophilic Attack at Ring Carbon*   251

2.15.14   REACTIVITY OF SUBSTITUENTS IN PYRIDOPYRAZINES   252

  *2.15.14.1   Electrophilic Attack on Substituents*   252
  *2.15.14.2   Nucleophilic Attack on Substituents*   253
  *2.15.14.3   Nucleophilic Substitution Reactions*   253
  *2.15.14.4   Radical Attack on Ring and Substituents and Photochemistry*   254

2.15.15   SYNTHESIS OF PYRIDOPYRAZINES   254

  *2.15.15.1   By Formation of One Bond between Two Heteroatoms [6 + 0 (N)]*   254
  *2.15.15.2   By Formation of One Bond adjacent to a Heteroatom [6 + 0 (α)]*   254
    *2.15.15.2.1   From pyridine intermediates*   254
    *2.15.15.2.2   From pyrazine intermediates*   255
    *2.15.15.2.3   Fused derivatives*   255
  *2.15.15.3   By Formation of One Bond β to a Heteroatom [6 + 0 (β)]*   256

2.15.15.4   By Formation of One Bond γ to a Heteroatom [6 + 0 (γ)]                    256
   2.15.15.4.1   From pyridine intermediates                                  256
   2.15.15.4.2   From pyrazine intermediates                                  256
   2.15.15.4.3   Fused derivatives                                            256
2.15.15.5   By Formation of Two Bonds from [5 + 1] Atom Fragments                     257
   2.15.15.5.1   Fused derivatives                                            257
2.15.15.6   By Formation of Two Bonds from [4 + 2] Atom Fragments                     257
   2.15.15.6.1   From pyridine intermediates                                  257
   2.15.15.6.2   Fused derivatives                                            259
2.15.15.7   By Formation of Two Bonds from [3 + 3] Atom Fragments                     259
2.15.16   BIOLOGICAL ACTIVITY AND USES OF PYRIDODIAZINES                              260
   2.15.16.1   Pyridopyrimidines                                              260
   2.15.16.2   Pyridopyridazines                                              261
   2.15.16.3   Pyridopyrazines                                                261

## 2.15.1 PYRIDOPYRIMIDINES: INTRODUCTION

Of the three broad classes of compound discussed in this Chapter, the pyridopyrimidines, compounds in which a pyridine ring is attached directly to a pyrimidine ring, and their benzo fused derivatives, are by far the most completely explored, much less being known of the corresponding pyridopyridazines and pyridopyrazines. This is largely by reason of the pyridopyrimidines possessing actual or potential biological activity as isosteres of quinazolines or pteridines, or having a role in nature in the case of the benzo fused 5-deazaflavins.

The only previous general review of this class of compound is by Irwin and Wibberley ⟨69AHC(10)149⟩, but individual aspects have been the subject of reviews, and these are referred to in the appropriate Sections in the following discussion.

## 2.15.2 STRUCTURE AND PHYSICAL PROPERTIES OF PYRIDOPYRIMIDINES

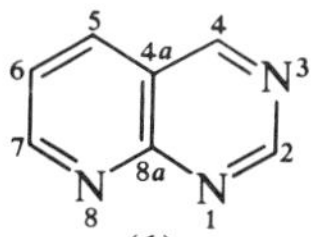

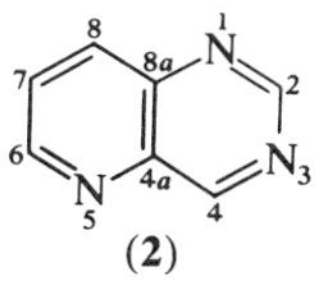

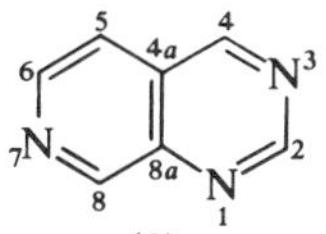

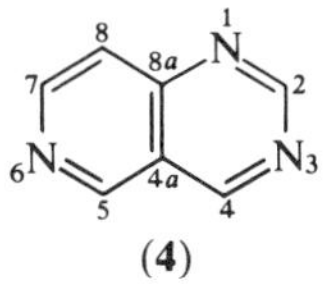

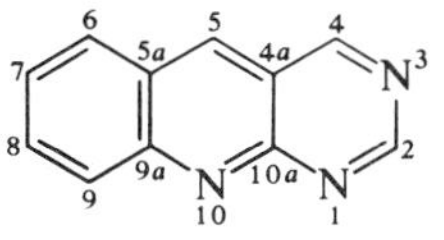

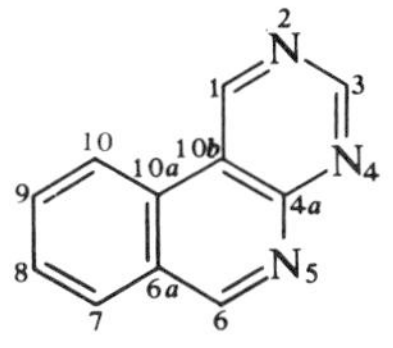

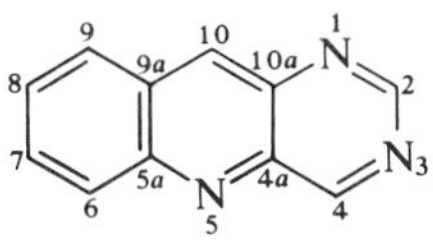

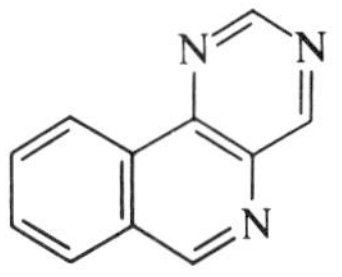

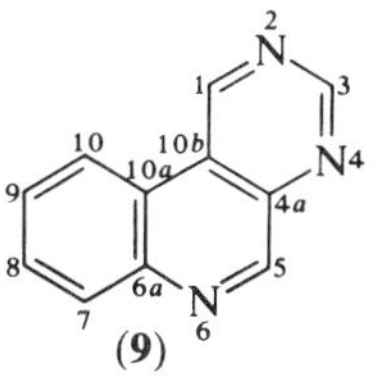

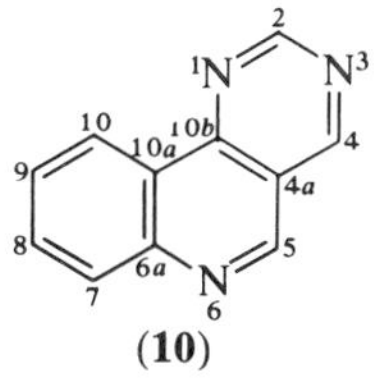

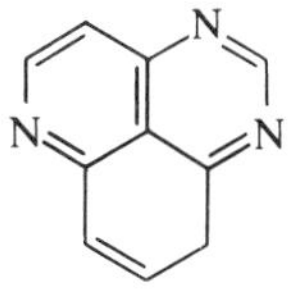

All four possible pyridopyrimidine systems, pyrido[2,3-*d*]pyrimidine (**1**), pyrido[3,2-*d*]pyrimidine (**2**), pyrido[3,4-*d*]pyrimidine (**3**) and pyrido[4,3-*d*]pyrimidine (**4**), are known, the numbering being as shown. In the older literature they may be known as 1,3,8-triazanaphthalenes, 1,3,5-triazanaphthalenes, 1,3,7-triazanaphthalenes (or copazoline) and 1,3,6-triazanaphthalenes respectively, but *Chemical Abstracts* nomenclature is used throughout this Chapter. For the reasons given above (Section 2.15.1), the pyrido[2,3-*d*]pyrimidine system is by far the best known, and the linear and angular benzo fused systems, *e.g.* (**5**) and (**6**) (numbering shown), are also known. The linear benzo fused derivatives of pyrido[3,2-*d*]pyrimidine, *e.g.* (**7**), are known but the angular system (**8**) has not been prepared. The angular benzo fused system, *e.g.* (**9**), from pyrido[3,4-*d*]pyrimidine is known, as is that, *e.g.* (**10**), from pyrido[4,3-*d*]pyrimidine. No example of a 'phenalene-type' benzo fused system, *e.g.* (**11**), is known.

### 2.15.2.1 Conformational Behaviour

Conformational behaviour of reduced pyridopyrimidines does not appear to have been systematically investigated.

### 2.15.2.2 NMR Spectroscopy

For the reasons discussed in the General Structure Chapter (2.01), in particular the non-comparability of experimental conditions, a detailed discussion of the NMR spectra of the systems is not attempted; the reader should consult the original articles referred to for the appropriate systems.

#### 2.15.2.2.1 *¹H NMR spectra*

The ¹H NMR spectra of all four parent compounds (**1**)–(**4**) have been measured and analyzed ⟨66JCS(B)750⟩. In general, after allowing for substituent effects, trends are observed in which protons on carbon atoms adjacent or in vinylogous relation to nitrogen are found at lower fields, whilst the ring protons of the corresponding pyrimidinones and pyrimidinediones, and pyridinones, show the expected marked upfield shifts. Anisotropic effects are seen, particularly at *peri*-positions, analogous to those observed in other fused six-membered nitrogen ring systems.

Amongst the more specialized studies in which ¹H NMR has been used are included a comparison with theoretical values predicted by MO calculations ⟨75MI21500⟩, studies of tautomeric ⟨64JOC219⟩ and covalent hydration ⟨81CJC2755⟩ effects in pyrido[2,3-*d*]pyrimidinones, methyl-proton lability in the same series ⟨72JCS(P1)1041, 80JA6168⟩, the electron-releasing effects of dialkylamino substituents in the pyrido[3,2-*d*]pyrimidine series ⟨74T549⟩, rotational barriers in these and related dialkylamino compounds (**12**) ⟨72JCS(P2)451⟩, and the stereochemistry of additions to the 4a-position of various 5-deazapterin derivatives such as (**13**) ⟨79MI21500⟩.

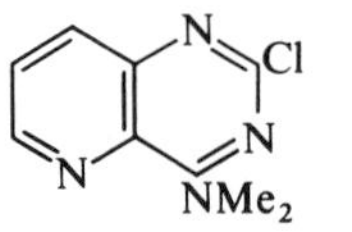

(**12**) 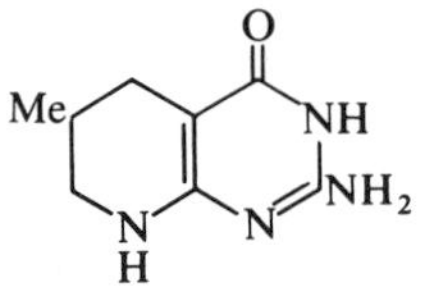   (**13**)

In addition to those cases recorded in ⟨69AHC(10)149⟩, ¹H NMR has been widely used in the solution of structural and regioisomeric problems in the field of pyridopyrimidines, their benzo fused derivatives ⟨74CC308⟩, and nucleosides. In some cases the NOE technique has proved especially useful ⟨78JOC828, 74JCS(P1)1225⟩.

The ¹H NMR data for the parent pyridopyrimidines are shown in Table 1.

#### 2.15.2.2.2 *¹³C NMR Spectra*

¹³C NMR spectra appear to have been recorded only for pyrido[3,2-*d*]- ⟨76JHC439⟩ and -[2,3-*d*]pyrimidines ⟨73T2209, 74CB2537, 77JOC221, 77JOC1919, 80JCS(P1)2645, 81JHC495⟩. The ¹³C NMR shift of the 4a-carbon has been used to follow the formation of a series of oxidation products of reduced 5-deazapterins ⟨78TL2271⟩ (see Section 2.15.3.1).

**Table 1** $^1$H NMR Data for Pyridopyrimidines

| Compound | Solvent | Chemical shift ($\delta$-values) | | | | | | Coupling constants (Hz) | Ref. |
|---|---|---|---|---|---|---|---|---|---|
| | | H-2 | H-4 | H-5 | H-6 | H-7 | H-8 | | |
| (1) | CDCl$_3$ | 9.57 | 9.58 | 8.44 | 7.72 | 9.33 | — | 8.3 (5, 6) 4.2 (6, 7) 1.9 (5, 7) | 66JCS(B)750, 69AHC(10)149 |
| (2) | CDCl$_3$ | 9.45 | 9.70 | — | 9.12 | 7.85 | 8.40 | 4.1 (6, 7) 8.3 (7, 8) 0.8 (4, 8) 1.7 (6, 8) | 66JCS(B)750, 69AHC(10)149 |
| | acetone-$d_6$ | 9.39 | 9.63 | — | 9.16 | 8.01 | 8.42 | 4.2 (6, 7) 8.3 (7, 8) 0.8 (4, 8) 1.8 (6, 8) | 66JCS(B)750 |
| | DMSO-$d_6$ | 9.40 | 9.65 | — | 9.14 | 8.06 | 8.46 | 4.1 (6, 7) 8.4 (4, 8) 0.8 (4, 8) 1.8 (6, 8) | 66JCS(B)750 |
| (3) | CDCl$_3$ | 9.57 | 9.58 | 7.79 | 8.85 | — | 9.55 | 5.8 (5, 6) 0.8 (4, 8) | 66JCS(B)750, 69AHC(10)149 |
| (4) | CDCl$_3$ | 9.54 | 9.62 | 9.47 | — | 9.00 | 7.92 | 5.8 (7, 8) 0.7 (4, 8) 0.8 (5, 8) | 66JCS(B)750, 69AHC(10)149 |
| | acetone-$d_6$ | 9.51 | 9.81 | 9.60 | — | 9.00 | 7.88 | 5.8 (7, 8) 0.7 (4, 8) 0.8 (5, 8) | 69AHC(10)149 |

### 2.15.2.3  IR Spectroscopy

The IR spectra of the parent compounds (1)–(4) in the solid state were first studied by Armarego *et al.* ⟨66SA117⟩ in a general study of the IR spectra of polyazanaphthalenes, and compared with theory. The solution IR spectrum of pyrido[3,2-*d*]pyrimidin-4(3*H*)-one was studied by Mason ⟨57JCS4874⟩, who noted the high $\nu_{C=O}$ (1745 cm$^{-1}$), indicating a possible *quasi o*-quinonoid form. Recent studies include an MO treatment ⟨77JSP(66)192⟩, whilst the Raman spectra of some 5-deazaflavin (14) analogues have been reported ⟨80BBA(623)77⟩.

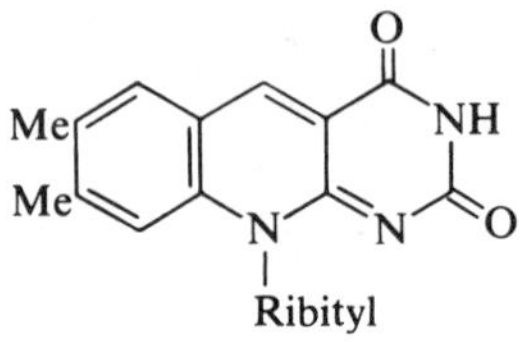

(14)

### 2.15.2.4  UV Spectroscopy

Pyridopyrimidines in general show three bands in their UV spectra due to $\pi$–$\pi^*$ transitions (typically in the 210–220, 260–280 and 290–310 nm regions in the parent compounds), together with a weak band due to an $n$–$\pi^*$ transition in the 330–350 nm region, in fair agreement with theoretical values from MO calculations ⟨65MI21500, 65MI21501, 66JSP(19)25⟩. In substituted compounds, bathochromic shifts often obscure the $n$–$\pi^*$ transitions.

UV maxima in addition to those noted in ⟨69AHC(10)149⟩ have been recorded for pyrido[2,3-*d*]pyrimidines ⟨*e.g.* 69JOC821, 81CJC2755⟩, pyrido[3,4-*d*]pyrimidines ⟨79JHC133⟩ and pyrido[4,3-*d*]pyrimidines ⟨*e.g.* 68JCS(C)2706⟩ (including benzo fused derivatives ⟨*e.g.* 78B1942⟩).

The UV spectra have been used in studies of protonation and related covalent hydration, structural assignments and tautomerism (see appropriate Sections), as well as in studies of bridgehead addition to 5-deazapterins ⟨79MI21500, 78TL2271⟩ and related 5-deazaflavin derivatives ⟨80JA1092⟩.

### 2.15.2.5  MO Calculations

In addition to the MO calculations of IR frequencies and electronic transitions referred to in Sections 2.15.2.3 and 2.15.2.4, calculations of electron densities, ionization potentials and dipole moments have been carried out on several pyridopyrimidines ⟨75MI21500, 67JCS(A)1626, 76CPB2078, 74T549, 79IJC(B)610⟩. Results are in general agreement with observed reactivity patterns. In two studies, electron densities were correlated with observed diuretic activity ⟨76CPB2078, 79MI21501⟩.

### 2.15.2.6  Mass Spectrometry

The mass spectra of pyridopyrimidines in general show many features in common with those of other related N-heterocycles, in particular quinazolines and pteridines. The pyridopyrimidines show strong molecular ions, and when breakdown of the hetero ring occurs, fragments arising from loss of CO, CN, HCN and HCNO are observed.

Specific discussions of mass spectral fragmentation patterns, in addition to those in ⟨69AHC(10)149⟩, are to be found for various pyrido[2,3-*d*]pyrimidines ⟨*e.g.* 72CPB772⟩, pyrido-[3,2-*d*]-, -[3,4-*d*]- and -[4,3-*d*]-pyrimidines ⟨*e.g.* 69JCS(C)513, 75JHC79⟩, including some benzo fused analogues ⟨*e.g.* 73T2209⟩ and *N*-oxides.

### 2.15.2.7  Tautomerism

Apart from some very early superficial work by Rydon and co-workers using a UV method ⟨56JCS1045⟩, there seems to have been little or no systematic study of the tautomerism of pyridopyrimidine derivatives. The tautomeric equilibrium in the azide–tetrazole tautomerism of the benzo fused pyrimido[4,5-*c*]quinoline (**15**) has been measured and found to favour the open-chain azide structure shown ⟨79JHC707⟩, whilst an NMR method was used in the case of the products from aminouracils and malonic acids (Section 2.15.5.7.2) ⟨64JOC219⟩.

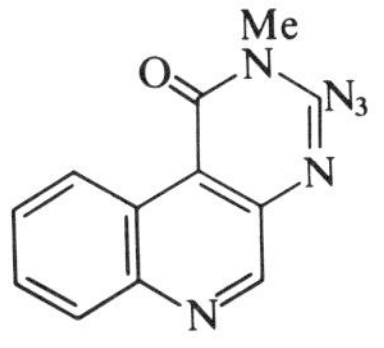

(**15**)

## 2.15.3 REACTIVITY AT RING ATOMS IN PYRIDOPYRIMIDINES

### 2.15.3.1 Electrophilic Attack at Ring Carbon

The main reaction of this type involves oxidative aromatization of various di- and poly-hydro derivatives, although this is often strictly an attack on ring hydrogen, involving such reagents as air or oxygen (either at high temperature or photolytically), high potential quinones, Pd/C or Rh/alumina, sulfur, potassium ferricyanide, chromium trioxide or diethyl azodicarboxylate.

Many of these reactions occur in the course of synthesis of fully or partly unsaturated products after initial ring closure, giving rise to more unsaturated systems, *e.g.* in the pyrido[2,3-*d*]pyrimidine series, in the preparation of piromidic and pipemidic acids (Section 2.15.4.1) and their derivatives, *e.g.* (**16a**) → (**17**) ⟨74JAP(K)7444000⟩. Examples are also found in the pyrido[3,2-*d*]pyrimidine ⟨76JHC439⟩, pyrido[3,4-*d*]pyrimidine ⟨74CR(C)(278)427⟩ and pyrido[4,3-*d*]pyrimidine ⟨66AG(E)308⟩ series.

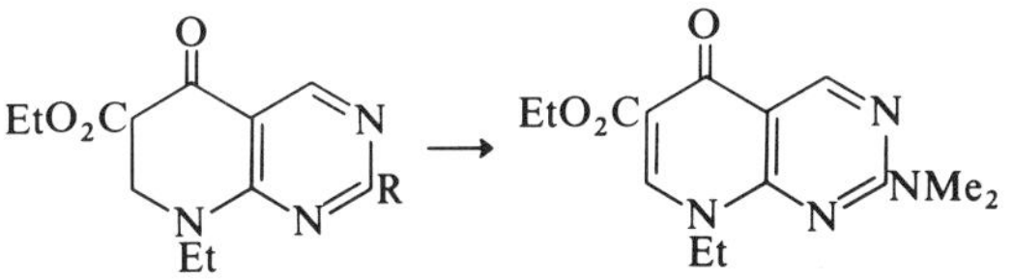

(**16**) **a**, R = NMe$_2$          (**17**)
      **b**, R = SMe

Another method of this type involving actual attack at ring carbon consists of halogenation/dehalogenation to produce the aromatic derivative, *e.g.* in the preparation of (**18**) ⟨75JHC311⟩, and many other examples in the pyrido[2,3-*d*]pyrimidine field abound in the literature. Dihydro compounds are sometimes oxidized through to ring oxo derivatives, *e.g.* with permanganate or mercury(II) acetate.

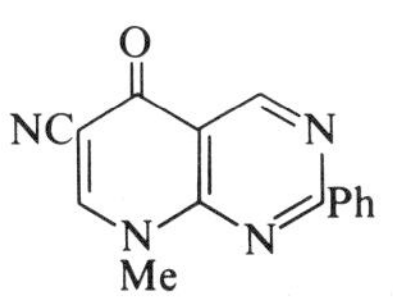

(**18**) 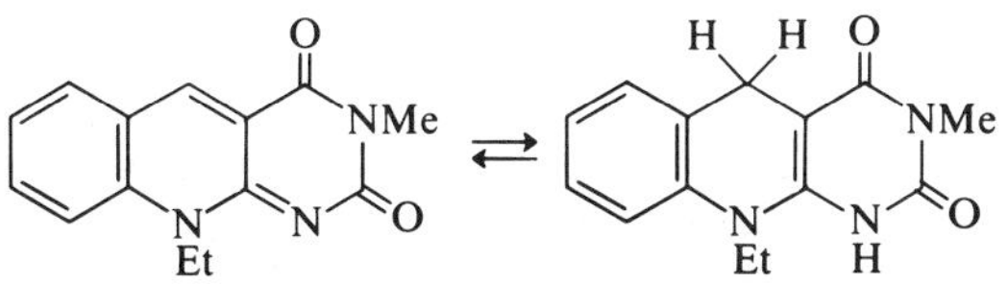                    (**19**)                    (**20**)

Oxidative substitutions at ring junction positions in various tetrahydro-5-deaza-pterins ⟨79JA6068⟩ and -flavins ⟨77JA6721⟩ have been studied, *e.g.* to give (**13**), and the oxidation–reduction reactions of 5-deazaflavins ⟨*e.g.* 78CL1177, 80CPB3514⟩ across the 1,5-positions, *e.g.* (**19**) ⇌ (**20**), are involved in their co-enzymic role in enzymic oxidations (see Section 2.15.16.1).

Enzymic oxidations at the 7-position of pyrido[2,3-*d*]pyrimidin-4-ones have also been studied ⟨75MI21501⟩, whilst formation of 6- and 7-oxygenated derivatives and of the 8-*N*-oxide have been observed in the metabolism of the pyrido[2,3-*d*]pyrimidine analogues of the antiepileptic drug methaqualone ⟨75MI21502, 74MI21500⟩.

The only genuine electrophilic aromatic substitutions have been observed in the 6-position of various 5-oxo derivatives of pyrido[2,3-*d*]pyrimidine, where reaction with phosgene (or its equivalents) gives the 6-acid chloride (**21**). Further attack to give the 6-formyl derivative (**22**) is also observed in this system as a concomitant reaction of certain cyclizations (Section 2.15.5.5.1) involving the Vilsmeier–Haack reagent. Treatment of the saturated anion of (**16b**) with alkyl iodides leads to alkylation at the 6-position to give (**23**) ⟨75USP3886159⟩, whilst a similar reaction of a 6-bromo analogue with formaldehyde led to 6-hydroxymethyl substitution ⟨80JAP(K)8038361⟩.

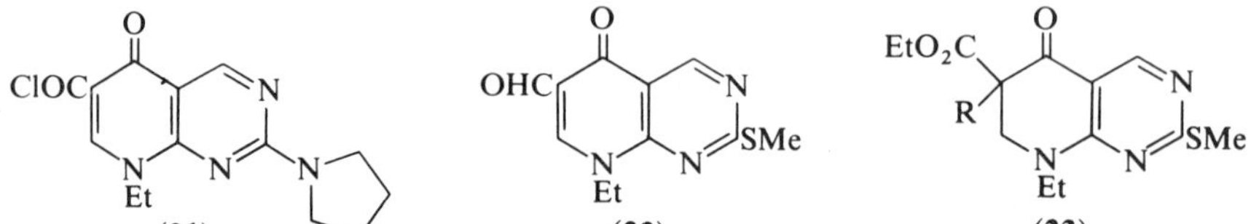

(21)    (22)    (23)

### 2.15.3.2 Electrophilic Attack at Ring Nitrogen

Protonation of parent pyridopyrimidines usually occurs at the pyrimidine nitrogen because of favourable resonance considerations and is accompanied by covalent hydration and/or ring opening, *e.g.* (**24**) → (**25**) → (**26**). This has rendered measurement of p$K_a$ values very difficult ⟨62JCS4094, 63JCS5166⟩ although 'true' p$K_a$ values have been estimated by use of a rapid-reaction technique ⟨66JCS(B)436, 67JCS(B)950⟩. Theoretical considerations have also been applied ⟨67JCS(A)1626⟩.

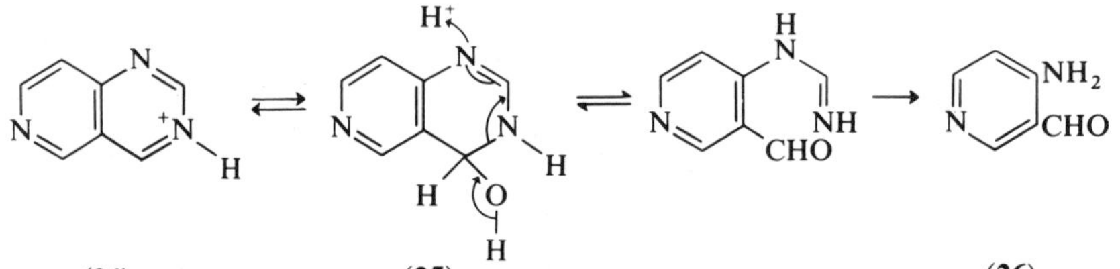

(24)    (25)    (26)

The corresponding -ones and -imines give better results ⟨*e.g.* 78KGS1666, 81CJC2755, but *cf.* 60MI21500⟩, as do the diamino compounds, *e.g.* (**27**) ⟨69JOC821⟩. The p$K_a$ values of a number of deazaflavins have been measured ⟨78B1942⟩, as has that of the important antibacterial drug pipemidic acid (Section 2.15.4.1) ⟨77CR(C)(285)431⟩.

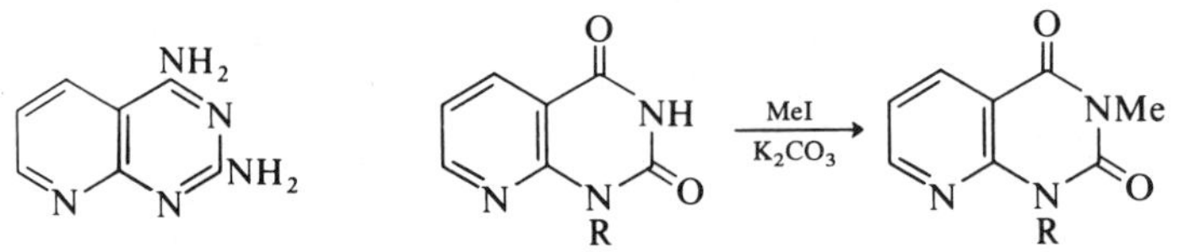

(27)    (28)    (29)

*N*-Alkylations, especially of oxo-di- and tetra-hydro derivatives, *e.g.* (**28**) → (**29**), have been carried out readily using a variety of reagents such as (usual) alkyl halide/alkali, alkyl sulfate/alkali, alkyl halide, tosylate or sulfate/NaH, trialkyloxonium fluoroborate and other Meerwein-type reagents, alcohols/DCCI, diazoalkanes, alkyl carbonates, oxalates or malonates, oxosulfonium ylides, DMF dimethyl acetal, and triethyl orthoformate/Ac$_2$O. Also used have been alkyl halide/lithium diisopropylamide and in one case benzyl chloride on the thallium derivative. In neutral conditions 8-alkylation is observed and preparation of some 8-nucleosides has also been reported ⟨78JOC828, 77JOC997, 72JOC3975, 72JOC3980⟩.

In some cases cyclic fused derivatives have resulted from alkylation reactions, *e.g.* (**30**) ⟨80YZ1187⟩, whilst rearrangements of alkyl substituents from oxygen to nitrogen or between nitrogens ⟨78JOC828, 72JOC3975⟩ have been reported. Quaternization and *N*-oxide formation usually occur at pyridine nitrogen ⟨but *cf.* 74CR(C)(278)427⟩, as does acylation in reduced derivatives, but (**31**) forms the 2-methiodide (**32**) ⟨62JCS1671⟩, and other instances of pyrimidine ring quaternization in oxo derivatives have been recorded.

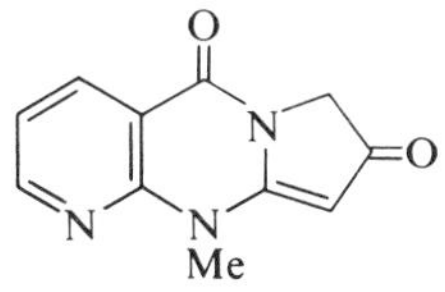

(30) 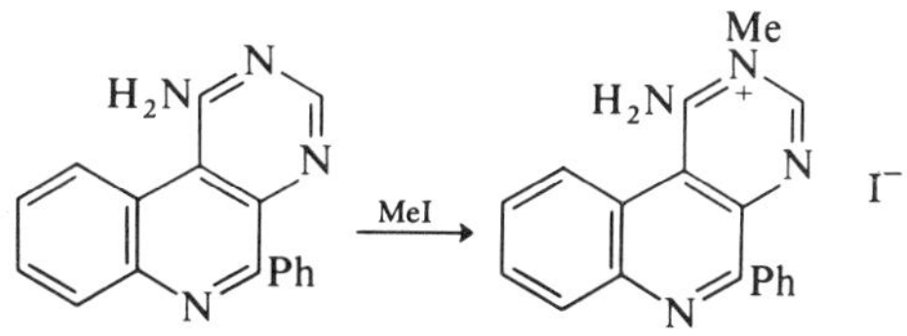          (31)          (32)

### 2.15.3.3 Nucleophilic Attack at Ring Carbon

Reduction of the ring system in pyridopyrimidines occurs fairly readily. Catalytic reduction normally leads to tetrahydro derivatives in the pyridine ring ⟨76JHC439, 56JCS1045⟩, but borohydride reduction gives either pyrimidine-ring di- or tetra-hydro derivatives (33) ⟨e.g. 72JCS(P1)353, 70GEP1961326⟩ or tetrahydropyridine derivatives, e.g. (34) ⟨72BCJ1127, 69TL1825⟩. LiAlH₄ reductions often proceed further with reductive ring opening to aminomethyl-pyridine derivatives (35) ⟨72JCS(P1)353, 72BCJ1127⟩, especially in the presence of *N*-aryl substituents ⟨71JCS(C)780⟩. 3-Alkyl quaternary salts also give tetrahydropyrimidine derivatives with NaBH₄ ⟨76JAP(K)7616693⟩.

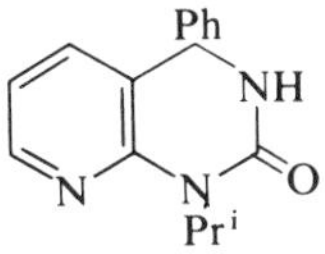

(33) 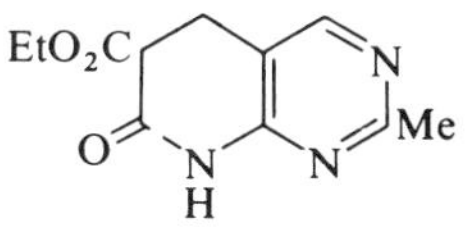          (34) 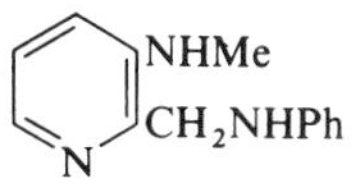          (35)

5-Deaza- and 10-deaza-flavins also undergo reduction when acting as coenzymes in oxidative systems ⟨72CC847, 80CPB3514, 80CPB3049⟩, whilst the former can also add dithionite at the 5-position to give intermediates such as (36) ⟨80JA1092⟩, from which sulfoxylate ion acts as hydride donor. They also form epoxides across the 4a,5-double bond *via* initial nucleophilic addition of hydroperoxide ⟨79JA7623⟩.

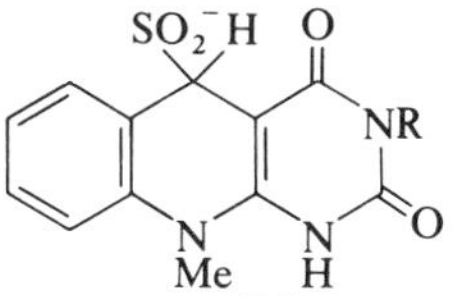

(36) 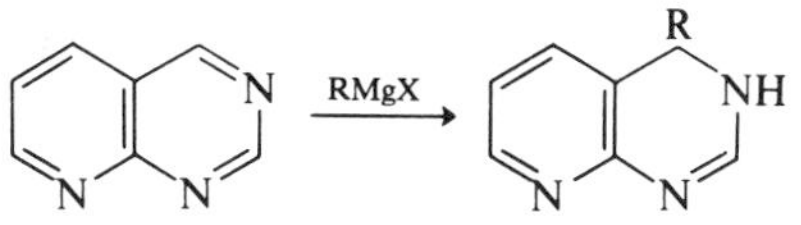          (37)

Some $E_{1/2}$ values for one-electron reductions of deazaflavins and pyridopyrimidines have been recorded ⟨79MI21502, 81CJC2755⟩. Grignard reagents add to pyrido[2,3-*d*]pyrimidine to give 4-substituted derivatives (37) ⟨70CPB1457⟩, whilst the action of alkyllithiums on the benzo fused pyrido[3,2-*d*]pyrimidine (38) gave both addition and substitution reactions leading to (39) ⟨77JHC611⟩.

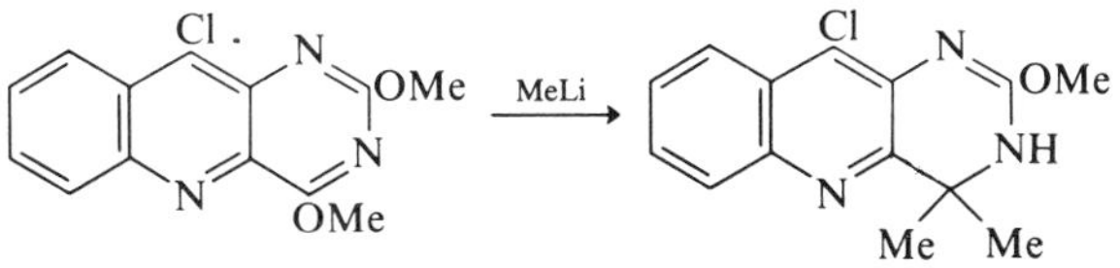

(38)          (39)

Nucleophilic attack on most pyridopyrimidines by water in acid conditions, or by various basic reagents, has been shown to lead, *via* initial nucleophilic addition to the ring, to opening of the pyrimidine ring with formation of a variety of *o*-substituted aminopyridine derivatives. Early work has been discussed in ⟨69AHC(10)149 and refs. therein⟩, and the effect on protonation equilibria has been described earlier (Section 2.15.3.2). Such ring openings are also observed as side reactions in nucleophilic substitution reactions (see Section 2.15.4.3). Pyrido[4,3-*d*]pyrimidines are most susceptible, and pyrido[3,4-*d*]pyrimidines most resistant, to ring opening, with the other two types intermediate between these, although even the pyrido[3,4-*d*]pyrimidine ring can be opened under forcing conditions, especially in its benzo fused form.

In recent years most examples of ring opening of this type have been noted in the pyrido[2,3-*d*]pyrimidine series, *e.g.* reaction of (40) with bases or concentrated H₂SO₄ gives the formylamino nitrile (41) ⟨78IJC(B)889⟩. Flavins also show similar reactions, and give 2-substituted quinoline-3-acid derivatives ⟨*e.g.* 80JCS(P1)293⟩, as do the pseudo-bases of pyrimido[4,5-*b*]quinolinium salts ⟨76JCS(P1)131⟩.

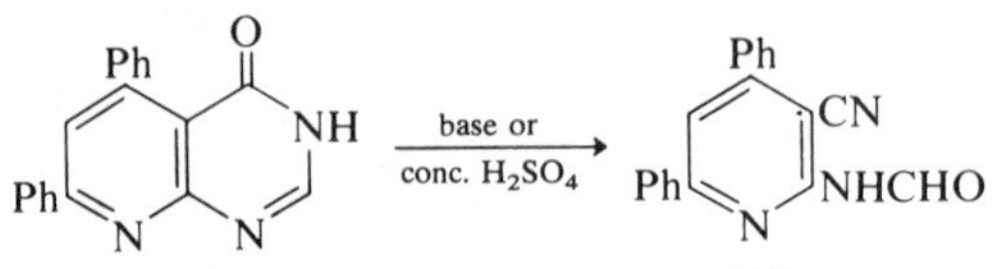

(40)          (41)

An interesting development involves the use of the ring-opening reaction as a synthetic source of 2-aminopyridine derivatives, *e.g.* 2-arylaminonicotinic acids ⟨79JAP(K)7924877⟩, and 2-aminonicotinaldehydes ⟨75JOC1438, 79JOC531⟩, and in particular an extension to a continually expanding 2-aminonicotinaldehyde synthesis, (42) → (43) → (44) and so on, leading to polymeric products as models for pyrolyzed polyacrylonitrile ⟨79MI21503⟩.

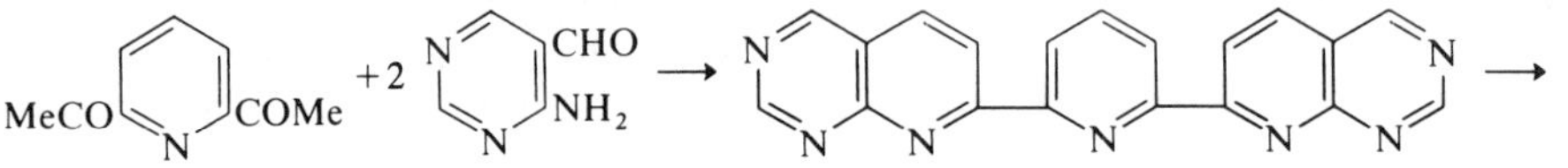

(42)                     (43)

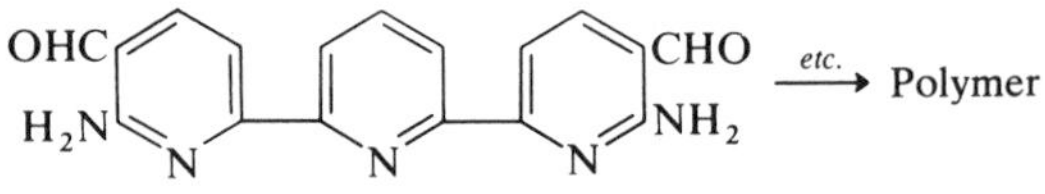

(44)

With *N*-substituted quaternary derivatives in the presence of oxidizing agents, ring opening is avoided and the classical Dekker oxidation of the pseudo-base occurs to give (45) ⟨62JCS1671⟩.

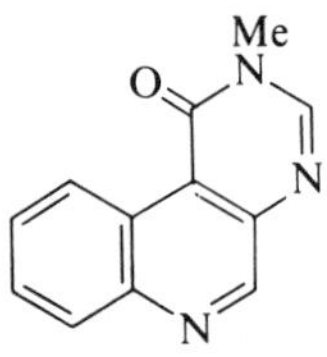

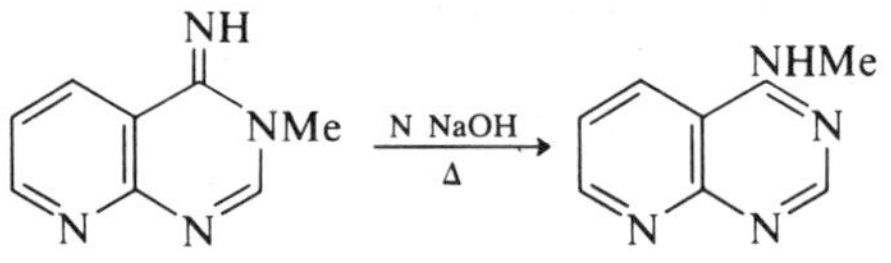

(45)                     (46)                     (47)

Another classical reaction which has been observed is the Dimroth rearrangement, *e.g.* (46) → (47) ⟨75JCS(P1)2182⟩, and related reactions lead to the transformations (48) → (49) ⟨79JHC1169⟩ and (50) → (51) ⟨76JAP(K)7636485⟩, and to formation of other similar products ⟨69MIP21500⟩.

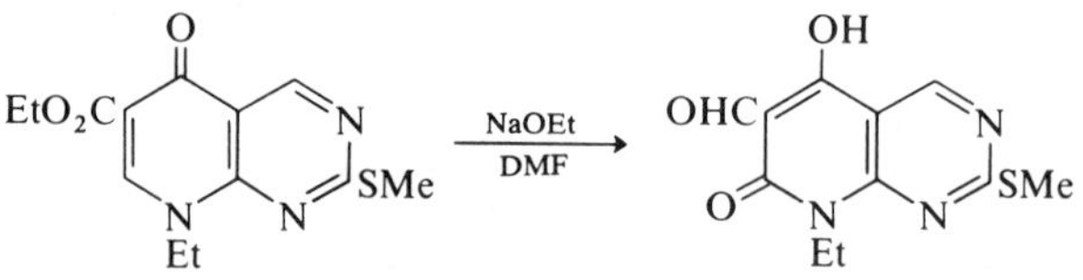

(48)                     (49)

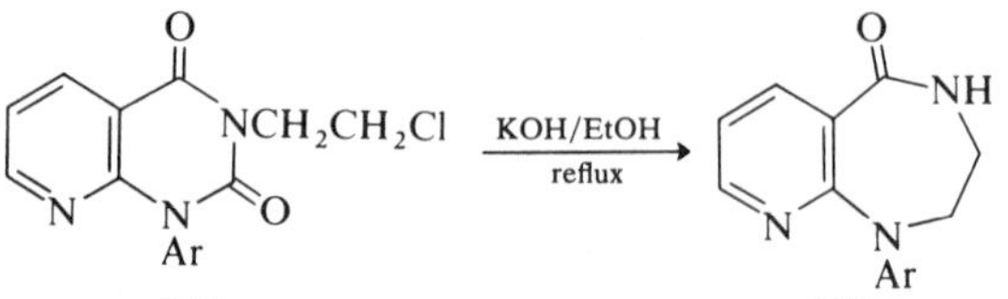

(50)                     (51)

An interesting set of transformations occurs on ring opening with NaOMe of the four isomeric series of *O*-benzenesulfonyl *N*-oxides (52a–c, 55). The initial ring-opened products (53a–c, 56) undergo the Lössen rearrangement, giving in three cases the *o*-hydrazinopyridine esters (54a–c), whilst in the pyrido[2,3-*d*]pyrimidine case the *s*-triazolo[4,3-*a*]pyridin-3(2*H*)-one (57) is isolated ⟨74JHC163⟩. Hydrolysis with alkali affords the free 3-hydroxy compounds, whilst acid hydrolysis gives simple ring-opened products ⟨72JHC1433⟩. In the

3-hydroxypyrido[2,3-*d*]pyridine series, ring opening with hydrazine, and in acid conditions, to give nicotinaldehyde oximes has been reported ⟨73CPB2643⟩.

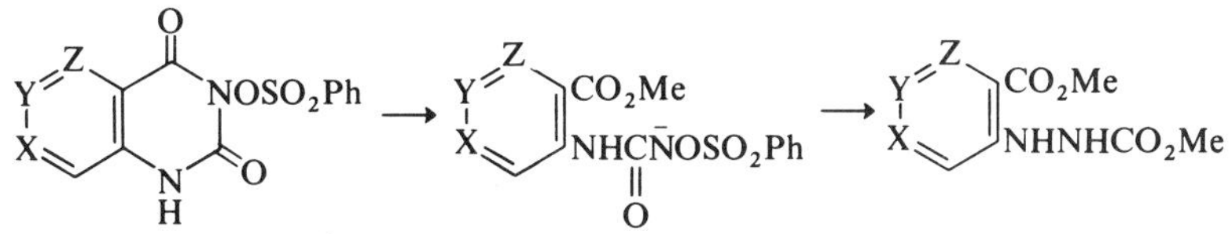

(52) **a**, X = N  Y = Z = CH
 **b**, Y = N  X = Z = CH
 **c**, Z = N  X = Y = CH

(53) a–c

(54) a–c

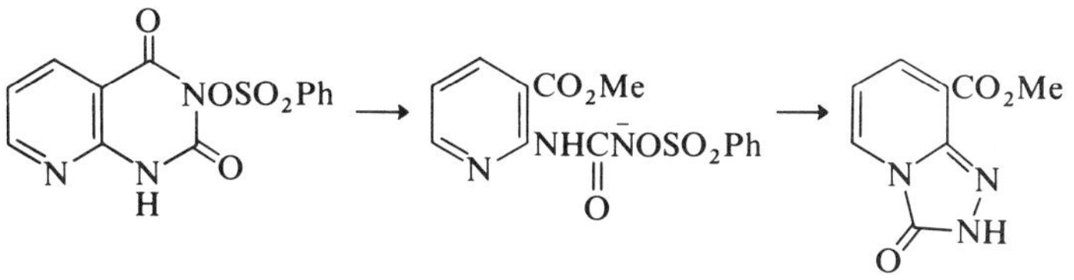

(55)  (56)  (57)

Other nucleophiles studied include the carbanions from active methylene compounds ⟨70CPB1457, 74CR(C)(278)427⟩ (which in some cases lead to ring transformations ⟨73JCS(P1)1794⟩ after ring opening, *e.g.* (58) → (59) ⟨75CPB2939⟩), cyanide ion ⟨70CPB1457⟩, and hydrazine ⟨73CPB2643⟩.

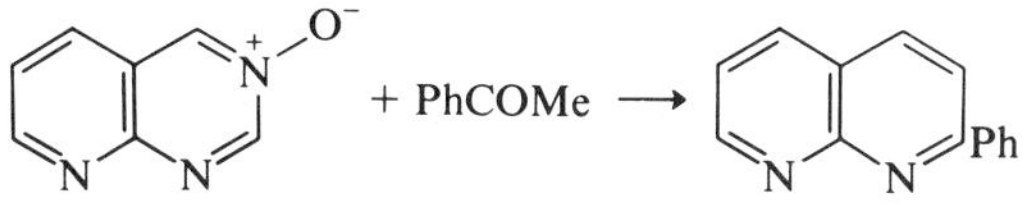

(58)  + PhCOMe ⟶  (59)

Finally, pyrido[2,3-*d*]pyrimidine 8-*N*-oxides undergo nucleophilic additions on activation by electrophilic attack at oxygen, giving, surprisingly, the 6-oxygenated derivative (61) *via* a cyclic intermediate (60) ⟨76JOC3027⟩. The corresponding 1-hydroxy compound gives stable 1-acetoxy or 1-acyloxy derivatives, but with POCl₃/PCl₅ it gives the 6-chloro derivative (62) ⟨75JOC3608⟩.

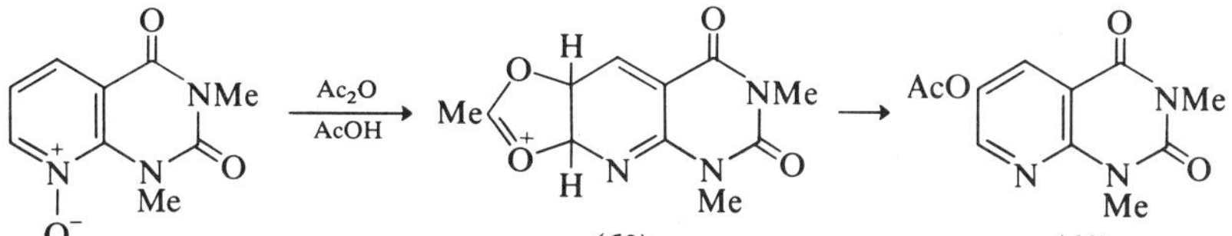

(60)  (61)

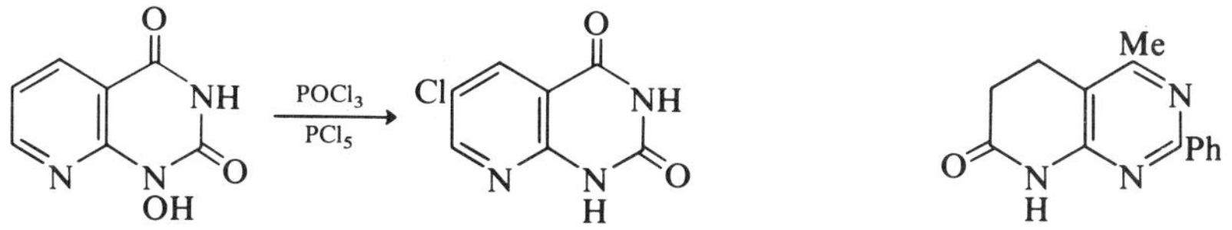

(62)  (63)

One or two cases are reported of a ring opening involving attack on a pyridinone ring in a 7-oxotetrahydropyrido[2,3-*d*]pyrimidine, *e.g.* (63) ⟨69JMC424⟩.

## 2.15.4 REACTIVITY OF SUBSTITUENTS IN PYRIDOPYRIMIDINES

### 2.15.4.1 Electrophilic Attack on Substituents

Alkyl groups attached to pyridopyrimidines adjacent to a nitrogen are 'activated', *i.e.* they are readily deprotonated and react with electrophilic reagents as their anions, or resonance stabilized equivalents, *e.g.* (64). This ready deprotonation, of course, leads to facile exchange of the alkyl protons for deuterium (Sections 2.15.2.2.1, 2.15.4.2), but, in addition, reaction with aldehydes leads to styryl or 2-furylvinyl compounds. Further oxidation of the 2-styryl compounds gives 2-aldehydes, as does a similar oxidation of the dimethylaminomethylene derivative (65), leading to the 2-acetyl derivative (66) ⟨77MI21500⟩, whilst aldehydes are also available by direct oxidation of the 2-methyl compounds with SeO₂ ⟨81JHC671, 77JHC1053⟩.

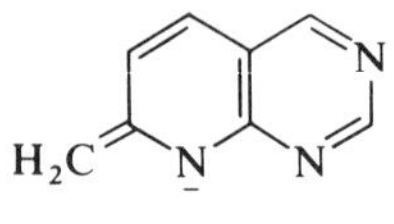

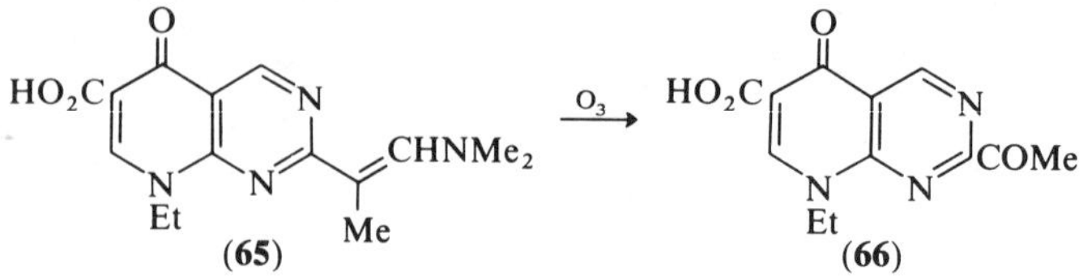

Acylation of 2-methylpyrido[2,3-*d*]pyrimidines with anhydrides gives 2-acylmethyl derivatives (67), whilst bromination to the 7-bromomethyl derivative has been reported for 7-methylpyrido[3,2-*d*]pyrimidines ⟨56JCS4433⟩ in a synthesis of potential folic acid antagonists.

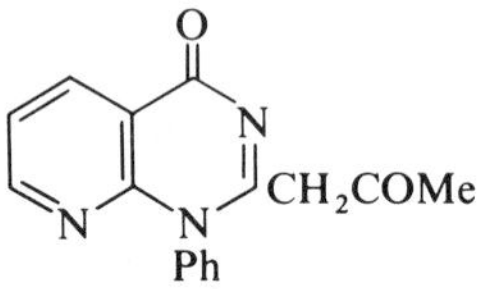

Surprisingly, there do not appear to be any reports of electrophilic aromatic substitution in the benzo-ring of benzo fused pyridopyrimidines, although some cases of substitution in isolated aromatic rings attached to carbon ⟨76MI21500, 74MI21502⟩ or nitrogen ⟨76JAP(K)7641391⟩ have been reported.

Appropriate pyrido[2,3-*d*]pyrimidin-5-ones with formyl groups in the 6-position have been oxidized to piromidic (68) and pipemidic (69) acids, or to intermediates for these, using moist silver oxide, chromium trioxide (potassium dichromate), potassium permanganate or, alternatively, sodium chlorite/hydroxylamine-*O*-sulfonic acid. 6-Acetyl groups have been similarly oxidized using sodium hypobromite in aqueous dioxane, whilst 2-acetyl groups give dimethylaminomethylene derivatives *en route* to 2-pyrazolylpyrido[2,3-*d*]pyrimidines.

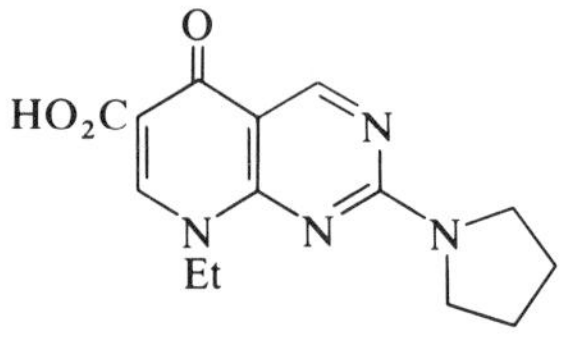

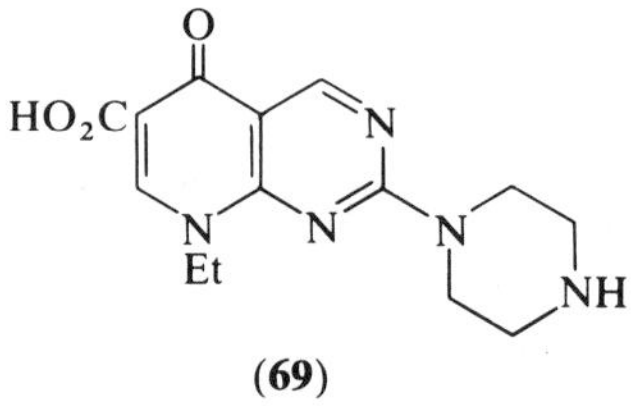

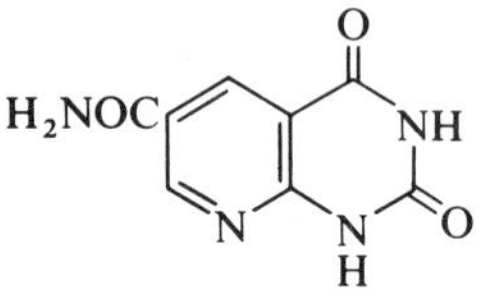

Carboxylic acid derivatives on pyridopyrimidine rings appear to undergo normal reactions with electrophilic reagents, *e.g.* the 6-amide (70) is dehydrated to the 6-nitrile with phosphorus oxychloride.

Nitrogen substituents also behave normally. For instance, amino groups undergo alkylation, acylation and silylation, react with aldehydes to form Schiff bases and isocyanates to give ureas. Amino groups in the pyrimidine ring, or in reduced pyridine rings, are converted by nitrous acid to the corresponding oxo derivative as in the purine system ⟨60JOC1368, 64JOC2674, 72BCJ1127⟩, whilst pyridine-ring amino groups undergo diazotization to salts which give coupling reactions ⟨62JOC2863⟩, are reduced to the corresponding H-compounds on boiling with ethanol, and undergo Sandmeyer reactions ⟨75JOC3608⟩.

*N*-Amino derivatives, *e.g.* (71), are converted to the corresponding NH compounds by nitrous acid and give acyl and alkoxymethylene derivatives with the usual reagents.

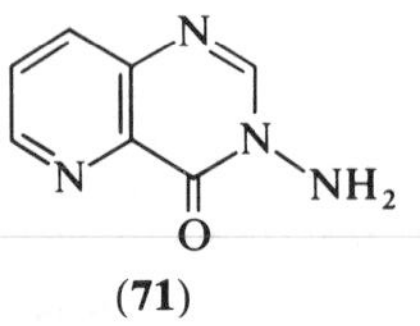

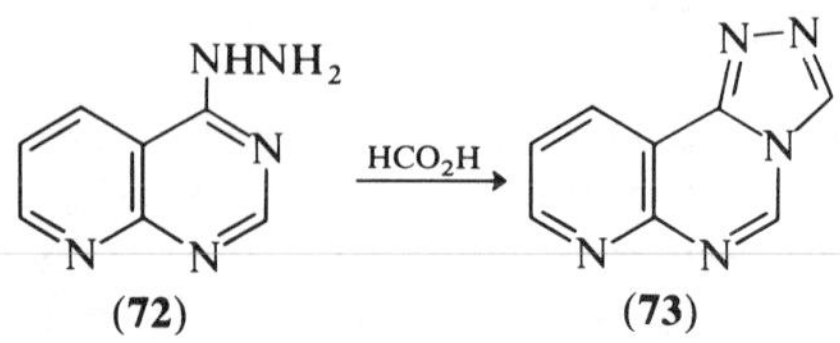

Hydrazino groups are also converted into H-compounds with mercury(II) oxide ⟨74CR(C)-(278)427⟩; in other reactions they have given hydrazones, or have been converted into pyrazoles and fused heterocyclic rings ⟨77JAP(K)7785194⟩, *e.g.* (72) → (73).

The 4-piperazinyl nitrogen in pipemidic acid (69) has been alkylated, acylated and sulfonylated in the search for enhanced antibacterial activity, whilst the 3-pyrrolidinyl position in piromidic acid (68) is hydroxylated metabolically and enzymically to (74) ⟨75MI21503⟩, from which acyloxy derivatives have been prepared.

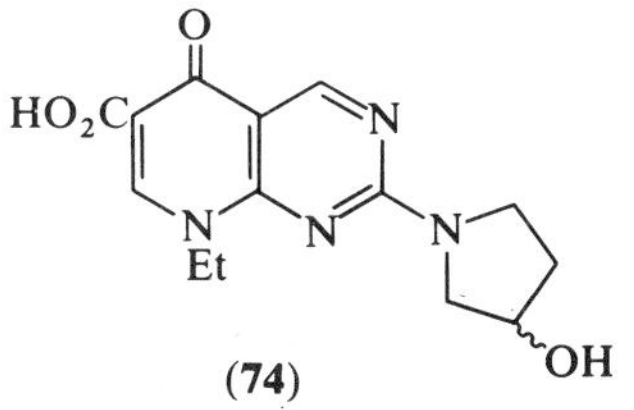

(74) 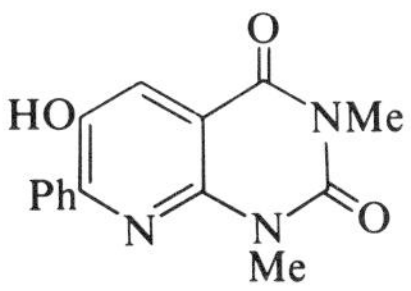          (75)

Electrophilic attack at oxygen is rare, alkylation normally occurring at nitrogen, but the 6-hydroxypyrido[2,3-*d*]pyrimidine (75) is methylated on oxygen in low yield ⟨70CB1250⟩, whilst the 6-cyano derivative (76) undergoes *O*-tosylation to (77) ⟨75JHC311⟩.

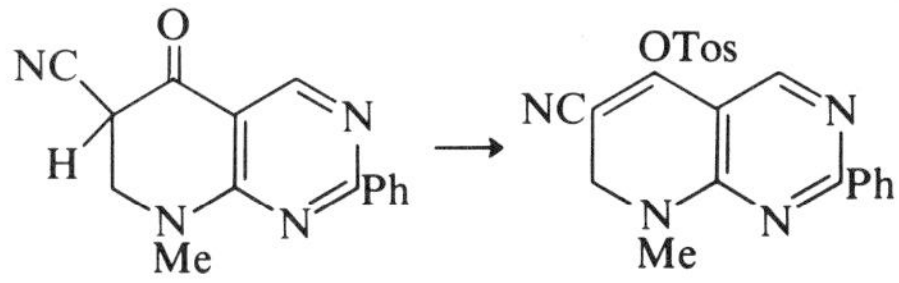

(76)          (77)

Thione groups undergo alkylation in anhydrous conditions. Examples include 2- and 4-thione groups in thiooxopyrido[2,3-*d*]pyrimidines and 2-thiones in thiooxopyrido[4,3-*d*]pyrimidines (as well as similar 2,4-dithiones), 4-thiones in thiooxopyrido[3,4-*d*]pyrimidines ⟨60JA6058⟩, 3- and 5-thiones in the tricyclic analogues (78) ⟨79JHC707⟩ and (79) ⟨75JCS(P1)2271⟩ respectively, and a 10-deazapterin-2,4-dithione ⟨74KGS554⟩.

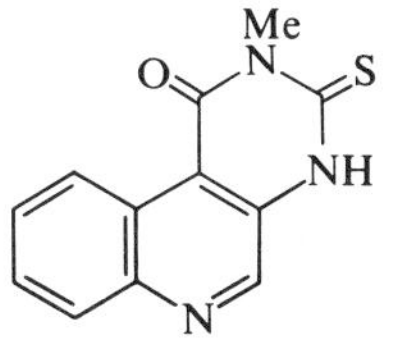

(78) 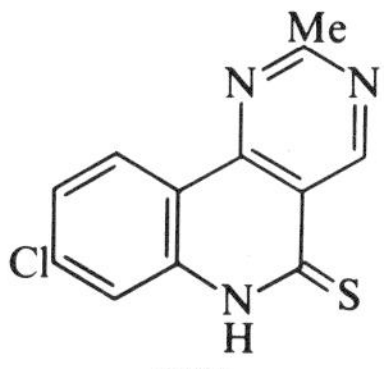          (79)

In aqueous alkaline conditions with chloroacetic acid the pyrido[4,3-*d*]pyrimidinethione (80) undergoes facile ring opening, attributed to the resonance stabilization of a delocalized covalent hydrate dianion intermediate (81) ↔ (82). Pyrido[2,3-*d*]pyrimidine-4-thiones (and also pteridines) which can exhibit similar stabilizations also undergo ring opening, but the corresponding pyrido-[3,2-*d*]- and -[3,4-*d*]-pyrimidinethiones, which cannot, give stable carboxymethylthio compounds or their oxo hydrolysis products ⟨60JA6058⟩. 7-Phenyl-pyrido[2,3-*d*]pyrimidine-2-thione is also said to give only the 2-one ⟨58JA3449⟩.

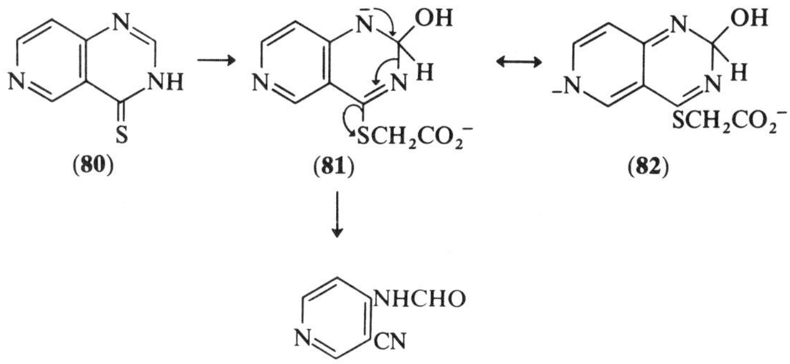

(80)          (81)          (82)

Both 2- and 7-methylthiopyrido[2,3-*d*]pyrimidines have been oxidized to sulfoxides and sulfones with hydrogen peroxide or MCPBA, as have 2-arylthio derivatives in the pyrido[3,2-*d*]pyrimidine series, whilst direct oxidation/hydrolysis of thione groups to the corresponding -ones with hydrogen peroxide or mercury(II) acetate has also been reported.

### 2.15.4.2 Nucleophilic Attack on Substituents

The kinetics of deuterium exchange on active methyl protons in pyrido[2,3-*d*]pyrimidines (5-deazalumazines) has been extensively investigated ⟨69CC290, 72JCS(P1)1041, 80JA6168⟩. Both 5- and 7-methyl groups, but not 6- and 8-, undergo exchange *via* highly delocalized anionic species similar to those previously mentioned (Section 2.15.4.1). Additional methyl substituents in the *m*-position decrease rates, but adjacent methyl groups increase the tendency to exchange.

β-Ketoalkyl groups are deacylated to alkyl groups on alkaline hydrolysis ⟨70CPB1457⟩, whilst the intermediate (**83**) is also deacylated on hydrolysis to the methoxymethyl derivative (**84**) ⟨72BCJ1127⟩. Hydroxymethyl groups, prepared by hydride reduction of esters ⟨80JOC3746⟩ or aldehydes ⟨81JHC671, 81JMC382⟩, or *via* acetoxymethyl rearrangement of α-methyl *N*-oxides ⟨B-81MI21500⟩, have been successively converted into halides (with HBr or PBr₃) and then aminomethyl derivatives, as key intermediates in syntheses of deaza analogues of folic acid, (**85**) → (**86**) → (**87**), or into methyl groups to give deazapterins ⟨81JHC671⟩.

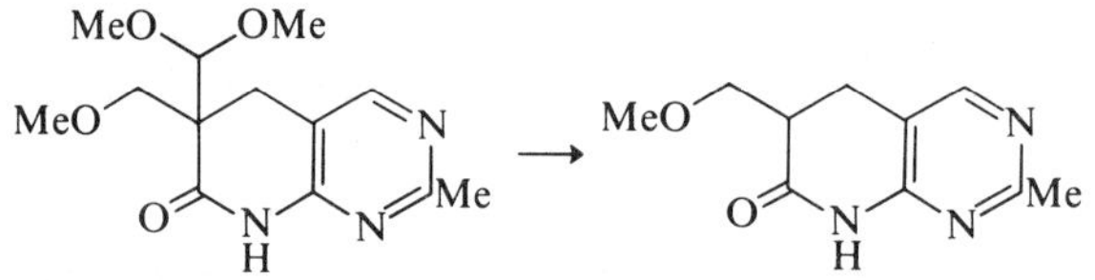

(83)        (84)

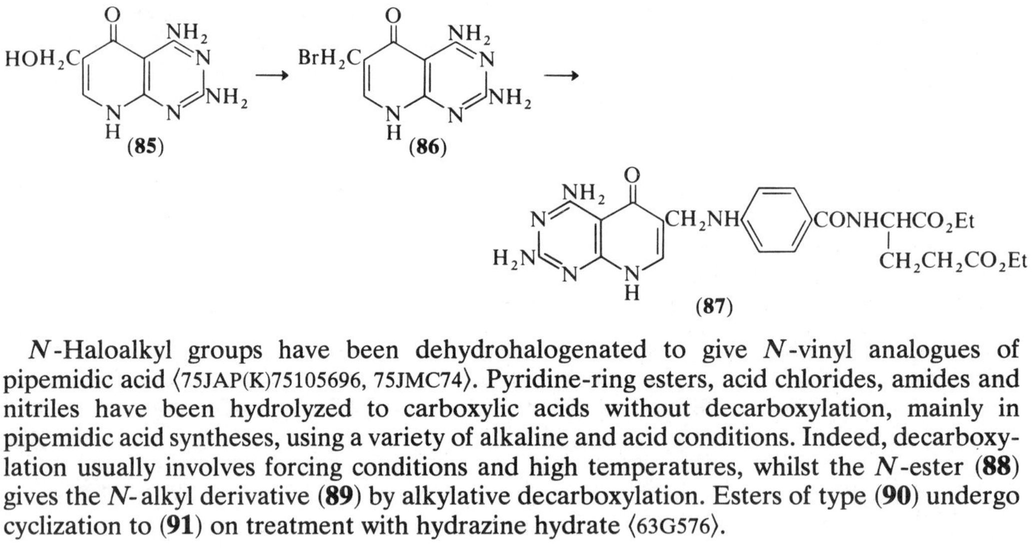

(85)        (86)

(87)

*N*-Haloalkyl groups have been dehydrohalogenated to give *N*-vinyl analogues of pipemidic acid ⟨75JAP(K)75105696, 75JMC74⟩. Pyridine-ring esters, acid chlorides, amides and nitriles have been hydrolyzed to carboxylic acids without decarboxylation, mainly in pipemidic acid syntheses, using a variety of alkaline and acid conditions. Indeed, decarboxylation usually involves forcing conditions and high temperatures, whilst the *N*-ester (**88**) gives the *N*-alkyl derivative (**89**) by alkylative decarboxylation. Esters of type (**90**) undergo cyclization to (**91**) on treatment with hydrazine hydrate ⟨63G576⟩.

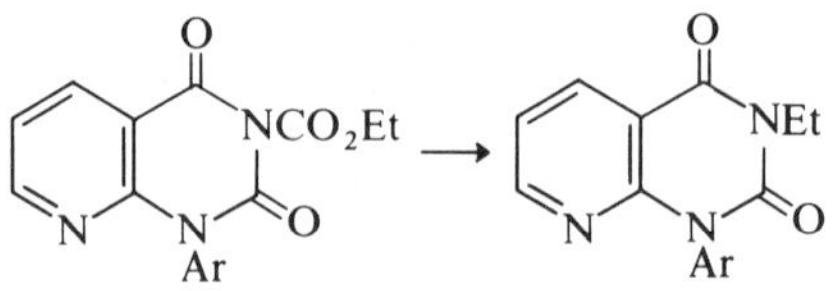

(88)        (89)

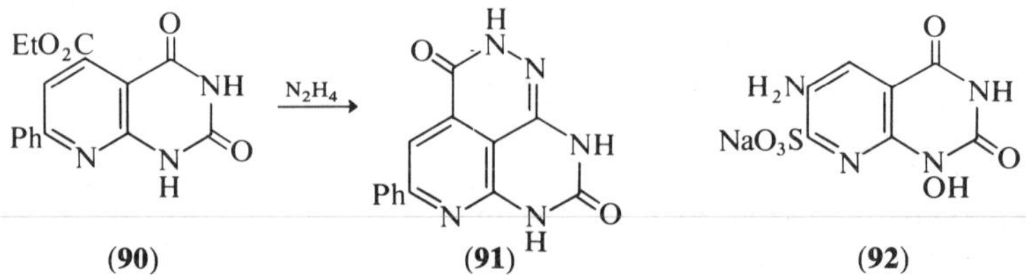

(90)        (91)        (92)

Nitro groups in the pyridine ring are reduced to amines catalytically, but side reactions can occur with dithionite, leading to, *e.g.* (**92**) ⟨75JOC3608⟩.

Amino groups $\alpha$ to nitrogen are hydrolyzed to the corresponding oxo compounds (as in the purines and pteridines) in both acid and alkaline conditions. Schiff bases are reduced to benzylamino derivatives with borohydride.

A useful reaction in early preparations of parent pyridopyrimidines was the McFadyen–Stevens decomposition of tosyl hydrazides, *e.g.* (**93**) → (**3**) ⟨62JCS4094⟩.

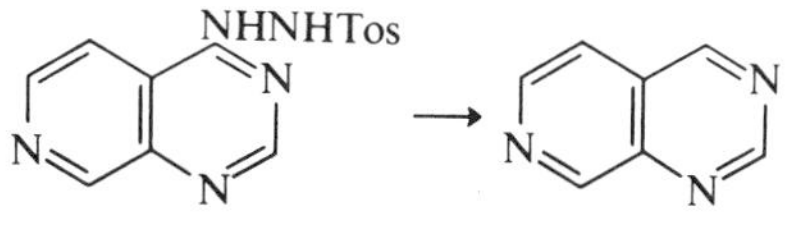

(**93**)          (**3**)

## 2.15.4.3 Nucleophilic Substitution Reactions

As in the purine and pteridine fields, nucleophilic substitution reactions, especially selective replacements, have been widely used in the manipulative chemistry of pyridopyrimidines. Oxo groups in -ones are readily replaced by halogen using the usual reagents such as phosphorus oxychloride, phosphorus oxychloride–phosphorus pentachloride, phosphorus oxychloride–tertiary base or thionyl chloride–DMF, which is useful in cases where phosphorus reagents are unsatisfactory. Monochloro derivatives are generally obtained under mild conditions, but polyhalo derivatives, *e.g.* (**94**), may require forcing conditions ⟨72JHC91⟩.

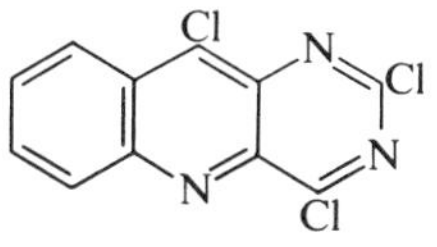

(**94**) 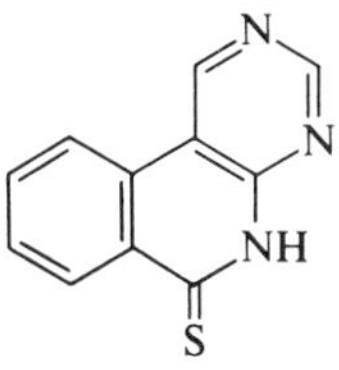          (**95**)

Where direct amination of chloro compounds has proved unsatisfactory, 4-alkoxy or 4-phenoxy intermediates have sometimes been used for reactions with amines or hydrazine.

In some cases oxo compounds have been converted directly to the corresponding thiones with phosphorus pentasulfide.

The thiones are readily desulfurized with Raney nickel to give the corresponding unsubstituted compounds in bicyclic systems in the 2-, 4- and 7-positions, and in tricyclic systems such as (**95**). The 2-methylthio derivatives may be similarly desulfurized. Thione groups in the 4-position, but not the 2-position, in pyrido-[2,3-*d*]- and -[3,2-*d*]-pyrimidines may be replaced directly with ammonia or amines.

In the 2-position, the methylthio derivatives have mostly been used, especially in the synthesis of piromidic acid, pipemidic acid and their analogues, whilst in the tricyclic series the methylthio derivative (**96**) was reacted with hydrazine to give the versatile intermediate (**97**) ⟨79JHC707⟩.

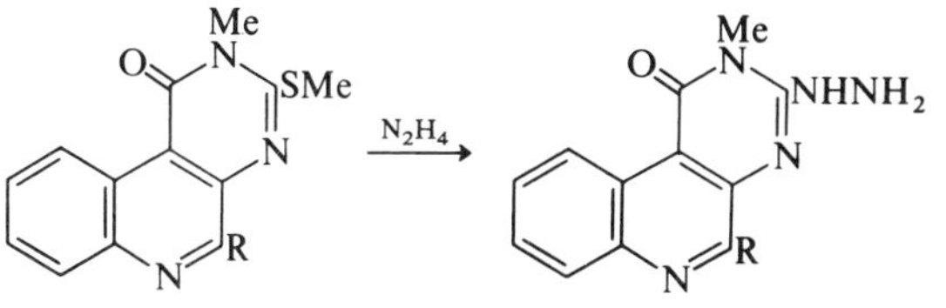

(**96**)          (**97**) 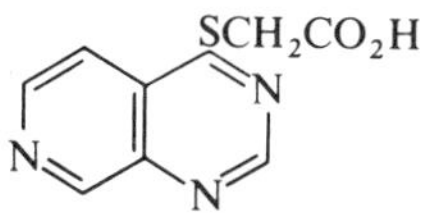          (**98**)

Alkylthio derivatives have also been converted into oxo groups, sometimes *via* the carboxymethylthio derivative in cases, *e.g.* (**98**), where ring opening (Section 2.15.4.1) does not occur.

If greater activation is desired the methyl- or aryl-thio derivatives may be oxidized to the sulfoxides/sulfones (Section 2.15.4.1), which may then be aminated ⟨64JOC2903, 80JOC3746, 77JOC997⟩.

As with the purines, the replacement of chloro groups by amino groups has been one of the major building blocks of pyridopyrimidine chemistry, being employed in a great variety of reactions too numerous to catalogue fully, particularly in the piromidic and pipemidic acid fields.

In monohalo pyrido[3,2-*d*]pyrimidines, 4-chloro groups are more readily replaced than 2-chloro, the same order holding in the 2,4-dichloro compounds ⟨56JA973, 56JCS1045, 81JHC671⟩. In the 2,4,8-trichloro compounds the reactivity order is 4>2>8, as evidenced by successive reactions with ammonia, benzyloxide ion and methylthiolate ion leading to (**99**)–(**101**) ⟨79JOC435⟩. On the other hand, in the pyrido[2,3-*d*]pyrimidines the order is 4>7>2, except with hydroxide ion, when it is 2,4>7 ⟨77JOC993⟩. The 4-chloro is readily replaced by ammonia and amines, as is a 5-chloro, but 7-chloro groups require forcing conditions. In the pyrido[4,3-*d*]pyrimidine series, 2-, 4- and 5-chloro groups are all replaceable, as are 2- and 4-chloro groups in the pyrido[3,4-*d*]pyrimidines.

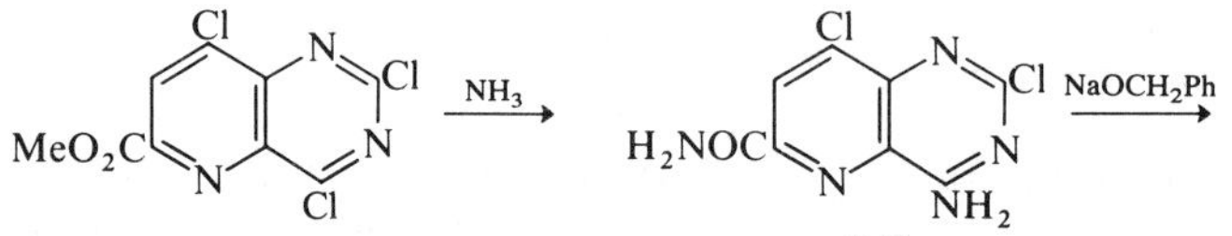

(**99**)

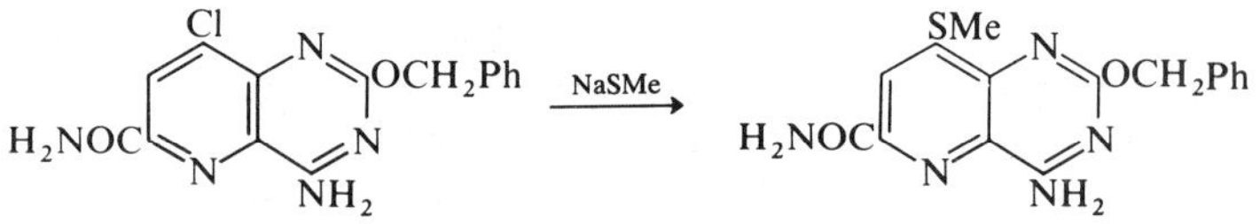

(**100**)          (**101**)

In the tricyclic series some interesting transformations have been carried out with 2,4,10-trichloropyrimido[5,4-*b*]quinoline (**94**). In early work ⟨57JCS4997⟩, formation of a 4-phenoxy-2-chloro-10-dialkylaminoalkyl derivative was claimed, but subsequent workers observed reaction only in the 2- and 4-positions with amines, alkoxide ion or hydrosulfide ⟨74KGS554, 72JHC91⟩. In the 7,8-dimethyl analogue, only reaction at the 4-position was possible, whilst the 10-chloro-2,4-dione (**102**) underwent a Dimroth rearrangement with *n*-butylamine to give the 3-butyl-10-butylamino compound (**103**) under forcing conditions ⟨77JHC611⟩.

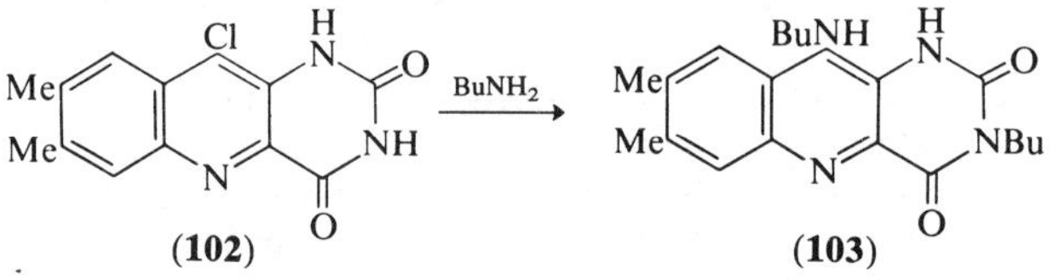

(**102**)          (**103**)

The 5-chloro in 5-deazaflavins is reactive ⟨75JHC181⟩, but not the 8-chloro, in contrast to the flavin case ⟨79LA1802⟩, and 4-chloropyrimido[4,5-*b*]quinolinium salts are readily hydrolyzed ⟨76JCS(P1)131⟩.

In the pyrimido[4,5-*c*]quinoline series the 1-chloro derivative (**104**) reacts with a variety of nucleophiles ⟨57JCS3718⟩, as does a 2-chloro in (**105**) in the isomeric pyrimido[5,4-*c*]quinoline series ⟨66MI21500⟩.

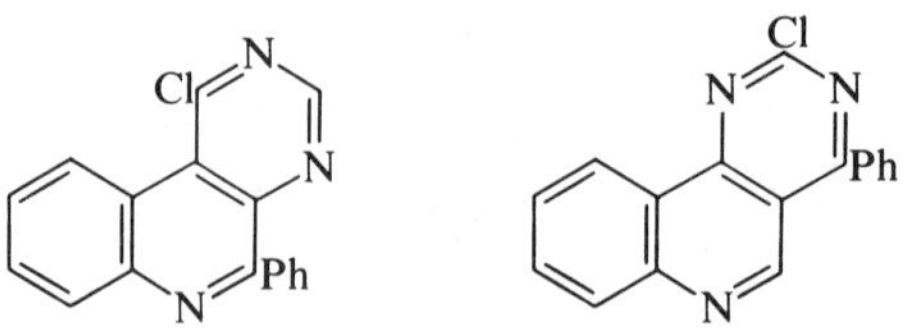

(**104**)          (**105**)

Reactive halogens in various series have been removed by catalytic hydrogenation with either platinum or palladium catalysts, and other nucleophiles which have been used in chloride displacements include hydroxide ion, alkoxides, hydrosulfide, hydrazine and toluene-*p*-sulfonylhydrazine, and trimethyl phosphite.

Reactions with carbon nucleophiles, *e.g.* from ketones ⟨73CR(C)(277)703, 74CR(C)(278)427⟩, are sometimes accompanied by deacylation *in situ* to give alkyl derivatives, *e.g.* (**106**)

⟨70CPB1457⟩, whilst with azide ion the chloro compound (**107**) underwent ring opening and reclosure to give the 2-tetrazolyl-3-aminopyridine (**108**) ⟨74CR(C)(278)1421⟩.

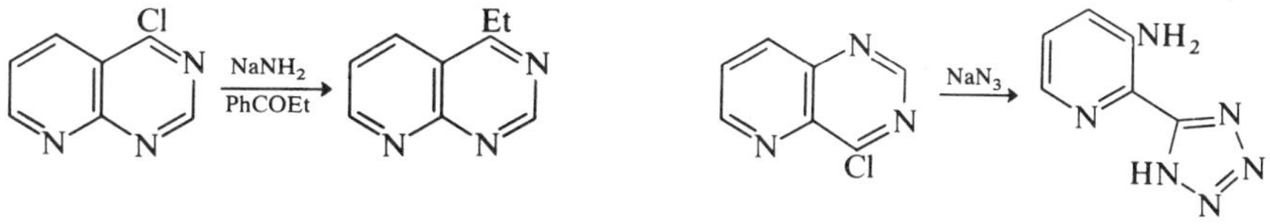

(**106**)      (**107**)      (**108**)

4-Bromopyrido[3,2-*d*]pyrimidines also give 4-amino compounds with ammonia.

## 2.15.4.4 Radical Attack on Ring and Substituents and Photochemistry

There appear to be no reports of direct radical attack on the pyridopyrimidine ring system, but radical bromination of methyl substituents in the 7-position of the pyridine ring has been utilized in the synthesis of deaza analogues of natural products ⟨62JCS4678, 79JHC133⟩.

Likewise there are no reports of systematic photochemical studies, but the pyrido[2,3-*d*]pyrimidine ring system appears relatively photostable, as photolytic removal of D-ribityl and hydroxyethyl *N*-substituents was employed in structural confirmation studies ⟨74JCS(P1)1225⟩. Photo adducts of deazaflavins with cyclohexadienes have been studied, however ⟨77ZN(B)434⟩, as have several other aspects of deazaflavin photochemistry.

## 2.15.5 SYNTHESIS OF PYRIDOPYRIMIDINES

### 2.15.5.1 By Formation of One Bond between Two Heteroatoms [6+0 (N)]

No syntheses of pyridopyrimidines by formation of bonds between two heteroatoms are possible.

### 2.15.5.2 By Formation of One Bond adjacent to a Heteroatom [6+0 (α)]

#### 2.15.5.2.1 From pyridine intermediates

One of the first syntheses of a pyridopyrimidine was of this type, by the Hofmann rearrangement of *o*-pyridinedicarboxamides, using hypohalites or lead tetraacetate ⟨02CB2831⟩. The 2,3-dicarboxamide gave only the pyrido[2,3-*d*]pyrimidine-2,4-dione with no [3,2-*d*] isomer, whilst the 3,4-dicarboxamide likewise gave only the [3,4-*d*] isomer with no [4,3-*d*] derivative ⟨68JCS(C)2756, 75USP3887550⟩. This effect has been confirmed for several other cases ⟨56JCS4433, 74JAP(K)7436700, 74MI21503, 71GEP2031230⟩, only the 6-phenyl-2,3-diamide proving an exception ⟨74MI21503⟩, but in the related Lössen rearrangement of the dihydroxamic acid a mixture of arylsulphonyl-*N*-hydroxy derivatives (**52a–c, 55**) was obtained with both systems ⟨72JHC1433⟩. Naturally, from mixed primary/secondary amides the primary amide is converted to isocyanate, determining the course of cyclization ⟨*e.g.* 76USP3960877⟩.

In a related reaction the isocyanate from azide (**109**) cyclized with aminopyridine esters *via* a ureido intermediate to give (**110**) ⟨80JHC733⟩.

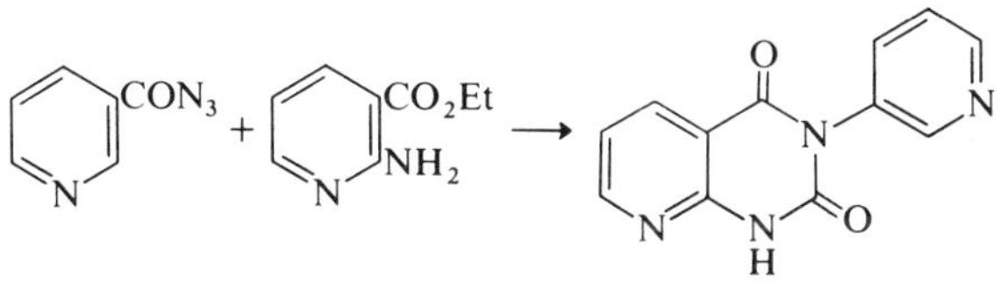

(**109**)          (**110**)

Syntheses from *o*-acylaminopyridine carboxamides, *e.g.* (**111**) → (**112**), are well known for the [2,3-*d*] ⟨*e.g.* 71GEP1963152⟩, [3,2-*d*] ⟨*e.g.* 65JCS4240⟩, [3,4-*d*] ⟨76MI21501⟩ and [4,3-*d*]

⟨67JCS(C)2613⟩ systems, and in many cases the starting materials are available *via* oxazine intermediates (*e.g.* **113**), or by *in situ* hydrolysis of a nitrile ⟨74GEP2365302⟩.

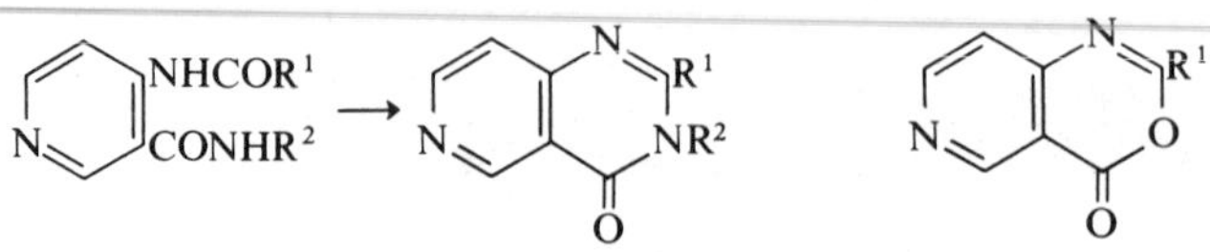

**(111)**　　　　　**(112)**　　　　　**(113)**

Although the reaction of *o*-aminopyridinecarboxylic acid derivatives with cyanic acid, isocyanates and isothiocyanates to give 2-oxo- or 2-thioxo-pyridopyrimidines (*e.g.* **114 → 115**) could be regarded as a [4 + 2] reaction, it occurs *via* a well-defined ureido intermediate and will be dealt with here. Examples using esters, amides or thioamides are found in the [2,3-*d*] ⟨*e.g.* 75JAP(K)7529599, 65CB1505, 76JAP(K)76122092⟩, [3,2-*d*] ⟨70CB82, 80JHC733⟩ and [4,3-*d*] ⟨76JAP(K)76108091⟩ series. Similarly, the use of pyridinecarboxamides with a leaving group (halide, OR, SR, SO₂R) in the *o*-position gives 1-substituted analogues ⟨*e.g.* 77JAP(K)7773897⟩. Extension to the piperideine analogues, *e.g.* (**116**), yields octahydro derivatives (**117**) ⟨76LA412⟩. The same piperideine with open-chain or cyclic imidates gives similar hexahydro derivatives (**118**) ⟨78CB2297⟩.

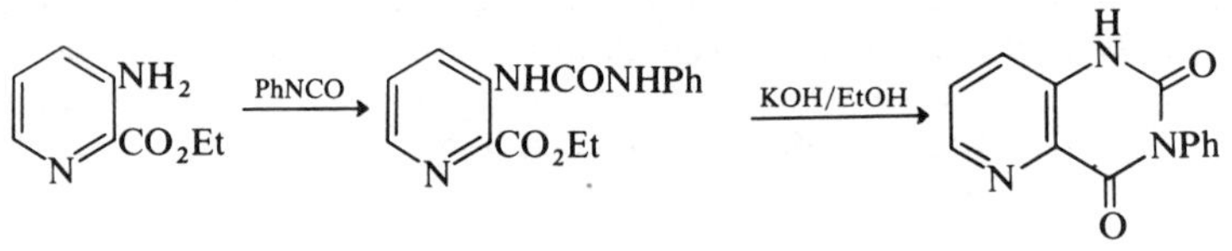

**(114)**　　　　　　　　　　　　　　　　　　**(115)**

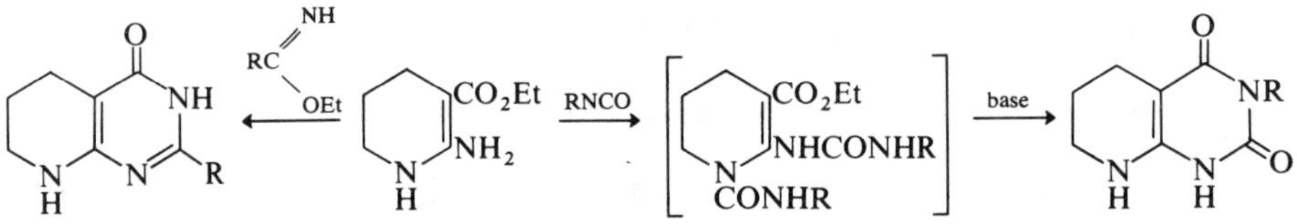

**(118)**　　　**(116)**　　　　　　　　　　　　**(117)**

The use of piperidine starting materials for preparation of perhydro derivatives is also seen in the reaction of the ethoxymethyleneaminonitrile (**119**) with sodium hydrosulfide to give (**120**) ⟨66AG(E)308⟩.

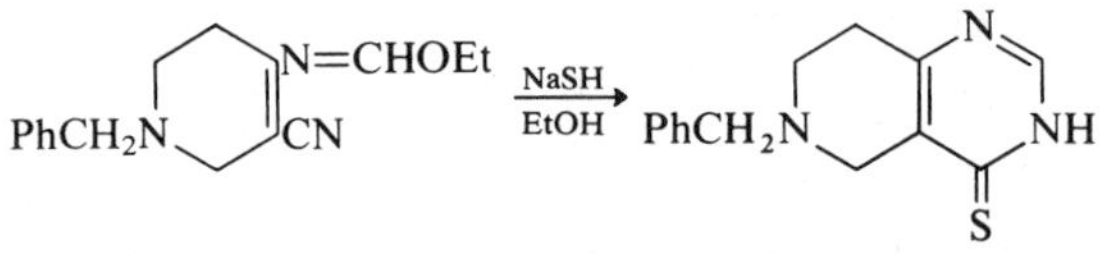

**(119)**　　　　　　　　　　　　**(120)**

Ethoxymethylene- and aminomethylene-amino intermediates are also of use for preparation of 4-ones and 4-imines, *e.g.* (**121**) → (**122**) → (**123**) from *o*-aminopyridine esters ⟨78KGS1671⟩ and nitriles ⟨*e.g.* 75JCS(P1)2182⟩.

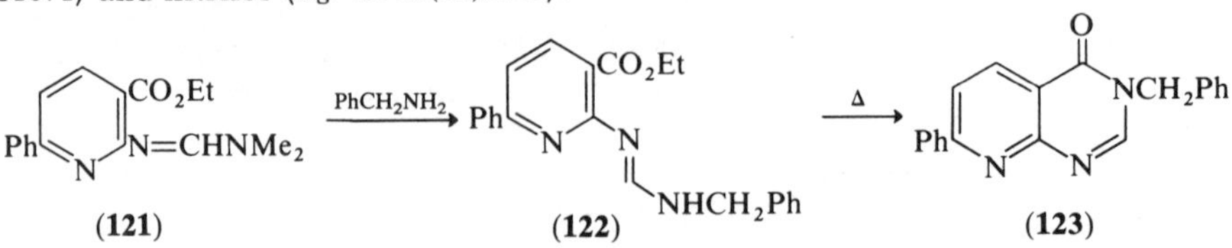

**(121)**　　　　　　　**(122)**　　　　　　　**(123)**

3-*N*-Oxides, *e.g.* (**125**), were similarly prepared using hydroxylamine to give intermediates such as (**124**) ⟨79JOC1695, 78H(9)1327⟩, and an alternative synthesis of *N*-oxides involves cyclization of 3-acylamino-2-aroylpyridine oximes (**126**) → (**127**) ⟨69MIP21500⟩.

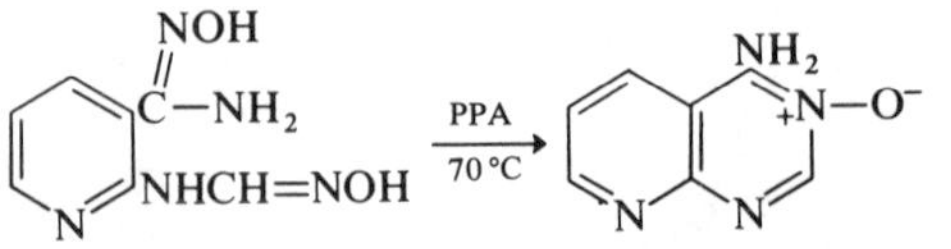

**(124)**　　　　　**(125)** 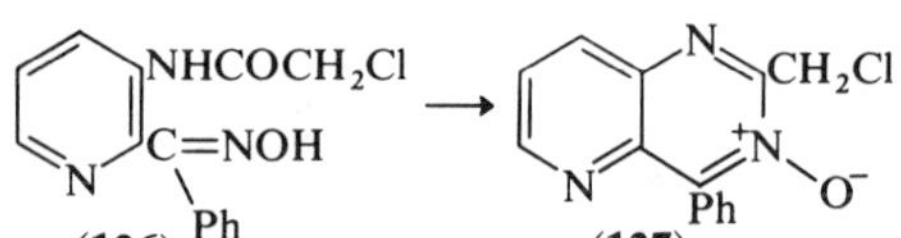　　　　　　　**(126)**　　　　　**(127)**

3,4-Dihydropyrido[2,3-*d*]pyrimidines have also been prepared from 3-acylaminomethyl-2-aminopyridines ⟨53JA656⟩.

An unusual final example of a synthesis from pyridines involves the 4-lithiotetra-chloropyridine (**128**), which with two moles of benzonitrile gave the trichlorodiphenyl-pyrido[3,4-*d*]pyrimidine (**130**) *via* the intermediate (**129**) ⟨72JCS(P1)2190⟩. The 2-lithio analogue gave the corresponding [3,2-*c*] derivative.

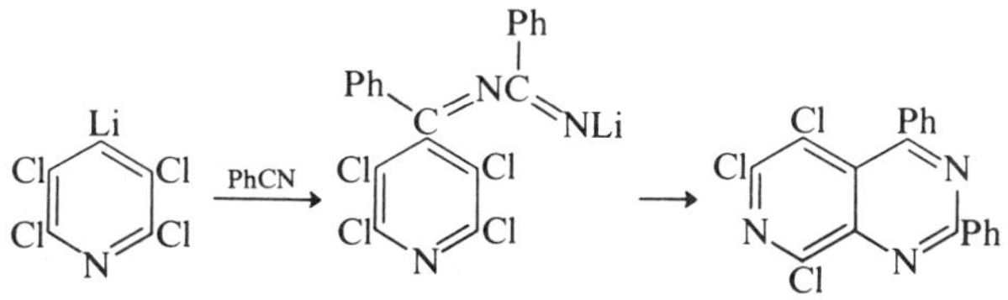

(**128**)    (**129**)    (**130**)

### 2.15.5.2.2 From pyrimidine intermediates

A major type of reaction in this class is the cyclization of 4-amino- or 4-halo-pyrimidines carrying 5-cyanoethyl or 5-ethoxycarbonylethyl groups, which cyclize to 7-amino or 7-oxo derivatives of 5,6-dihydropyrido[2,3-*d*]pyrimidine, e.g. (**131**) → (**63**). The intermediates may sometimes be prepared by reaction of 4(6)-aminopyrimidines with acrylonitrile, or even *via* a pyrimidine ring synthesis from an amidine and a cyanoacetic ester or malononitrile derivative, e.g. (**132**) → (**133**) ⟨71JOC2385, 72BCJ1127⟩.

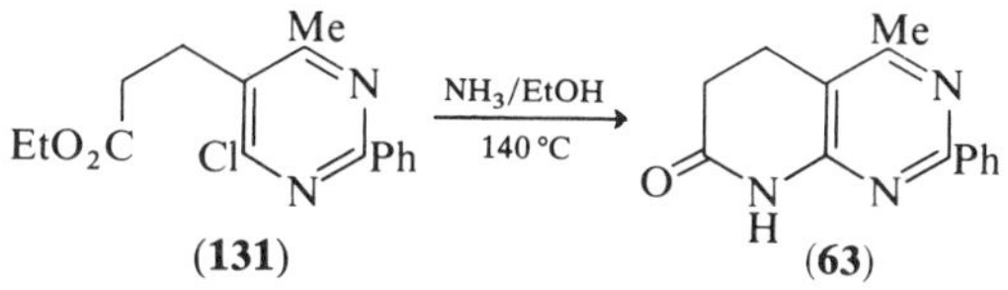

(**131**)    (**63**)

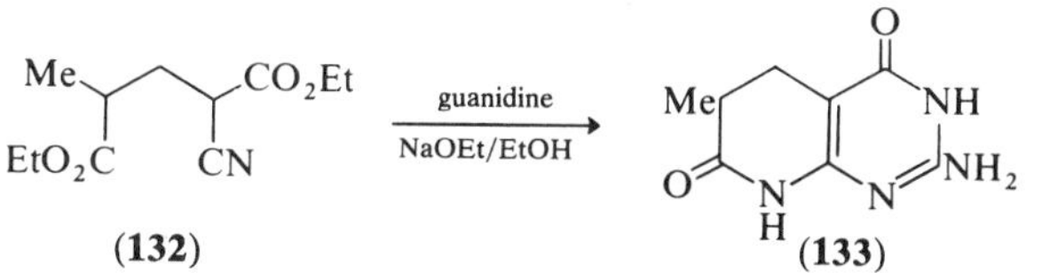

(**132**)    (**133**)

Cyclization of 5-alkylaminopropyl-4-chloropyrimidines gives the 8-substituted tetrahydro derivatives (**134**) ⟨61GEP1100030⟩. One interesting variant is the reaction of a 6-aminouracil with another uracil molecule *via* a ring opening and subsequent reclosure with elimination of dimethylurea (**135** → **136**) ⟨80H(14)407, 81JOC846⟩. Another interesting example of ring opening/reclosure involves the reaction of the pyrimido[4,5-*d*]pyrimidine (**137**) with malononitrile, with subsequent closure to (**138**) ⟨73JCS(P1)1794⟩.

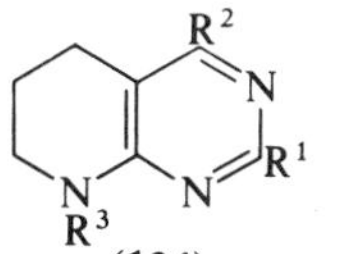

(**134**)

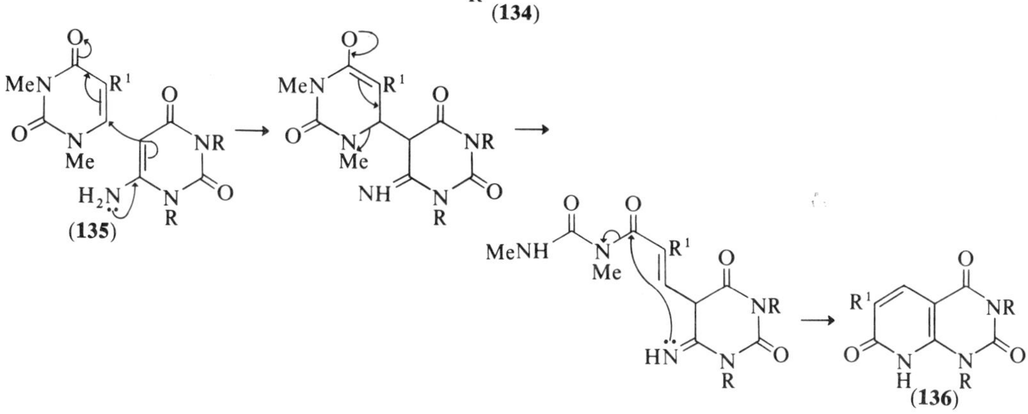

(**135**)    (**136**)

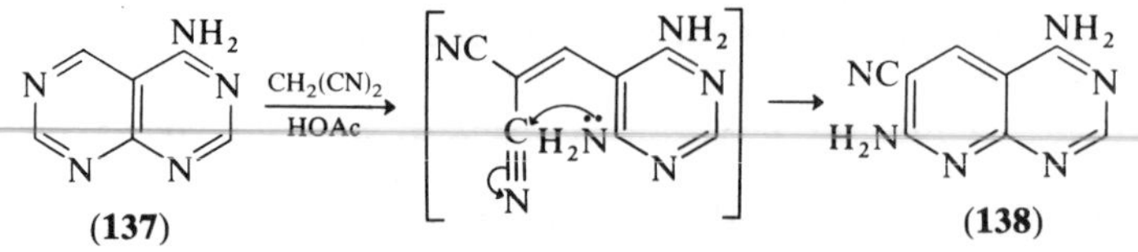

(137) → (138)

The cyclization of 6-aminouracils with three-carbon fragments such as $\alpha,\beta$-unsaturated carbonyl compounds, $\beta$-dicarbonyl compounds, acetylenic esters, *etc.*, is dealt with as a [3+3] reaction (see Section 2.15.5.7.2). Reactions with alkoxymethylenemalonates and related compounds are regarded as proceeding through [6+0 ($\gamma$)] cyclizations (see Section 2.15.5.4.2).

A reversal of the position of the substituents on the pyrimidine ring is involved in the reductive cyclization of 4-ketoalkyl-5-phenylazopyrimidines to give, *e.g.* 6-alkylpyrido[3,2-*d*]pyrimidines ⟨76JHC439⟩.

Finally, a novel synthetic route involves formation of the pyridine ring from a fused pyran intermediate, *e.g.* (139) → (140) ⟨70CB1250, 80JOC1918, 73JCS(P1)823⟩. If a pyrylium salt is used, a quaternary pyridopyrimidinium salt such as (141) is formed ⟨77KGS1484⟩.

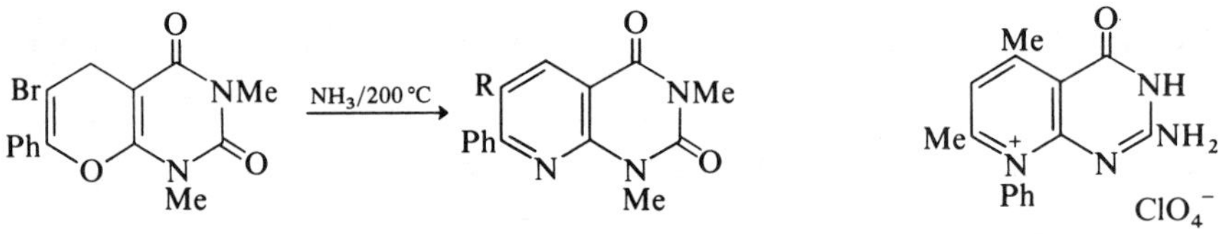

(139)    (140) R = NH₂ or OH    (141)

In a somewhat related reaction the fused lactone (142) furnishes the decahydropyrido[3,4-*d*]pyrimidine (143) during the reaction of $\alpha$-(hydantoin-5-ylidene)-$\gamma$-butyrolactone with ammonia in ethylene glycol at 200 °C ⟨71KGS1280⟩.

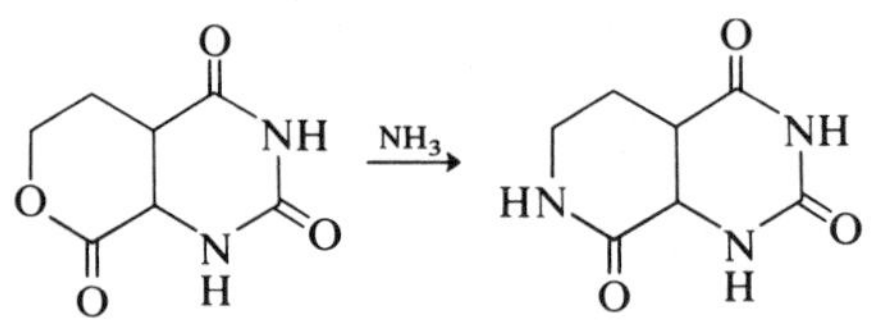

(142)    (143)

### 2.15.5.2.3 From naphthyridines

An unusual reaction, probably involving the mechanism shown, results in the partial conversion of halonaphthyridines such as (144) to pyrido-[3,2-*d*]- or -[4,3-*d*]-pyrimidines (*e.g.* 145) with KNH₂ in liquid ammonia ⟨78MI21500, 73MI21500⟩.

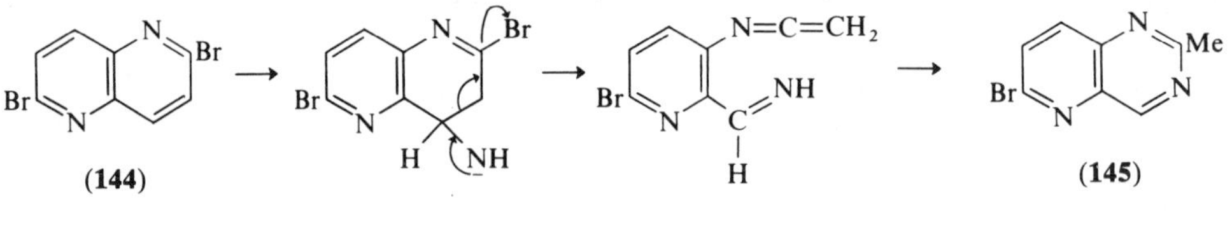

(144)    (145)

### 2.15.5.2.4 Fused derivatives

One obvious route to fused derivatives is the application of methods detailed above to bicyclic starting materials, *e.g.* with the 3-acetylaminocinchoninic acid derivative (146) ⟨25JCS1493⟩, in the cyclization of a hydroxylaminonitrile to the *N*-oxide (147) ⟨75JCS(P1)1023⟩,

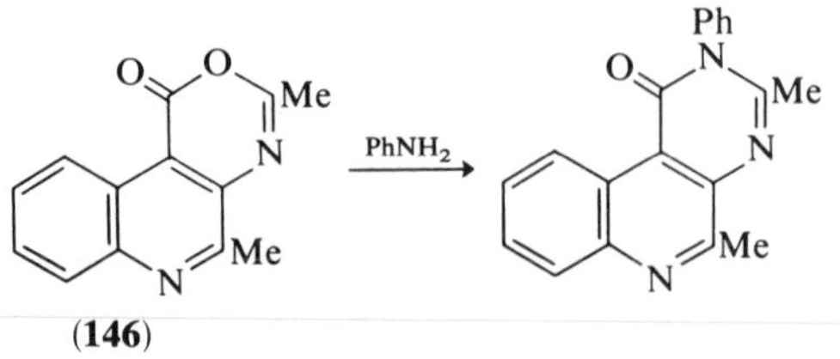

(146)

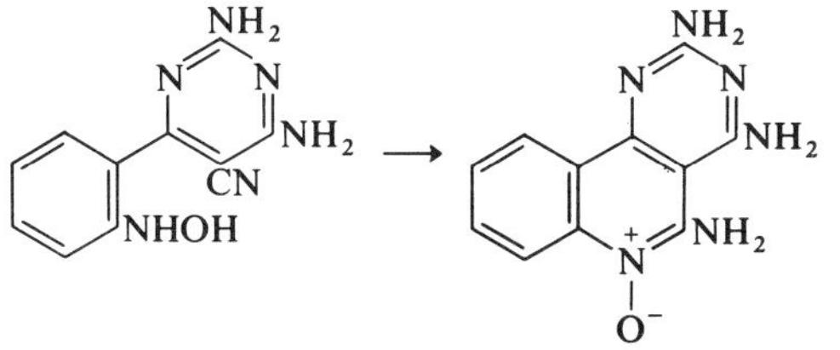

**(147)**

and in a synthesis of 3-aryl-5-deazaflavinoids from 2-amino-3-cyanotetrahydroquinolines ⟨80CPB3514⟩.

The main new type of reaction for the synthesis of fused derivatives involves cyclization to an adjacent *o*-position of an aryl ring. A simple case involves the cyclization of a 5-bromo-6-*o*-aminobenzylpyrimidine to a pyrimido[5,4-*b*]quinolinedione ⟨78AP115⟩, whilst in the pyrimido[4,5-*b*]quinoline field the reaction of the 3-lithio derivative of 2-ethoxyquinoline with benzonitrile to give (148) ⟨79E574⟩, and a similar synthesis in the pyrimido[5,4-*c*]quinoline field ⟨77CI(L)310⟩, are noteworthy. Somewhat similar intermediates are involved in the complex one-pot reaction of ethoxycarbonylmethylenetriphenylphosphorane with aryl isocyanates to give (149), which cyclizes with loss of $CO_2$ and Stevens-type rearrangement to give (150) ⟨80CB395⟩.

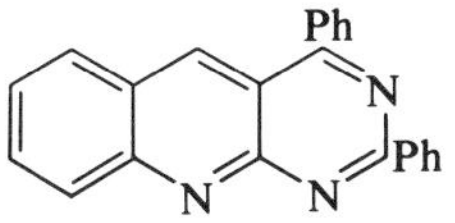

**(148)**

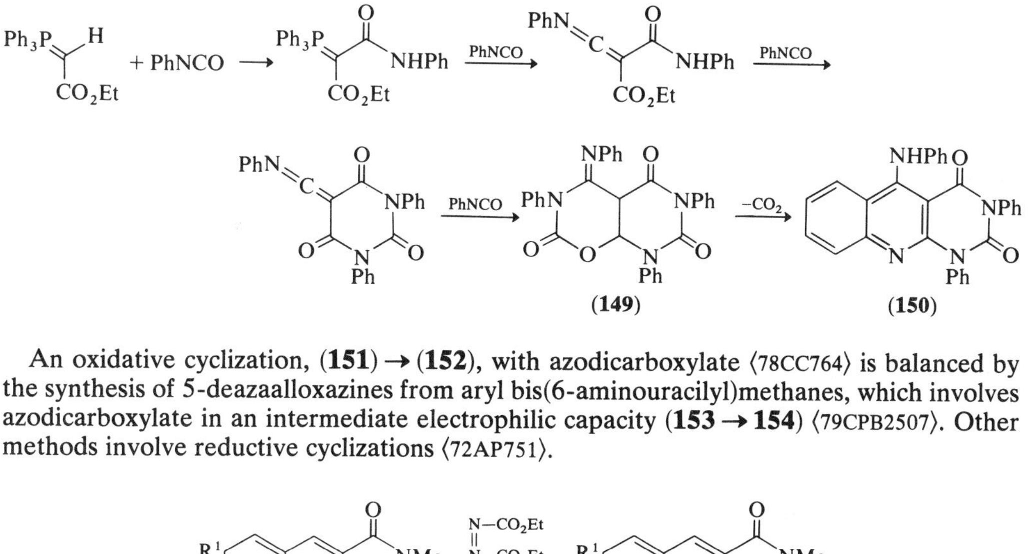

**(149)**          **(150)**

An oxidative cyclization, (151) → (152), with azodicarboxylate ⟨78CC764⟩ is balanced by the synthesis of 5-deazaalloxazines from aryl bis(6-aminouracilyl)methanes, which involves azodicarboxylate in an intermediate electrophilic capacity (153 → 154) ⟨79CPB2507⟩. Other methods involve reductive cyclizations ⟨72AP751⟩.

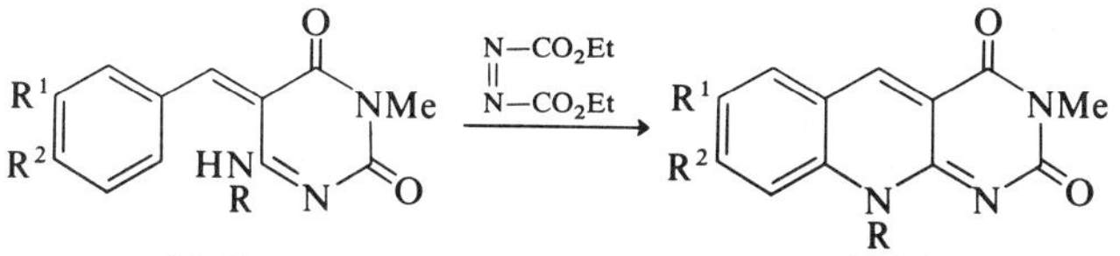

**(151)**          **(152)**

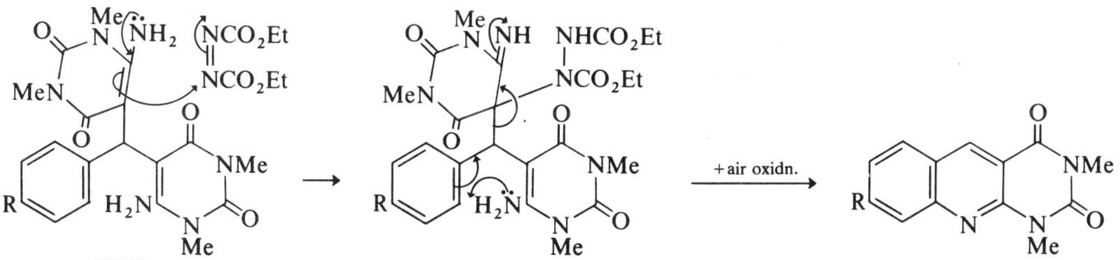

**(153)**                    **(154)**

### 2.15.5.3 By Formation of One Bond β to a Heteroatom [6+0 (β)]

#### 2.15.5.3.1 From pyridine intermediates

No examples of simple pyridopyrimidines formed in this way have been reported, although some [4+2] reactions may involve intermediates of this type.

#### 2.15.5.3.2 From pyrimidine intermediates

The only example of this type which is noteworthy is the *in situ* cyclization to (156) of the intermediate (155) from a 5-aminopyrimidine anil and DMF dimethyl acetal ⟨80S479⟩.

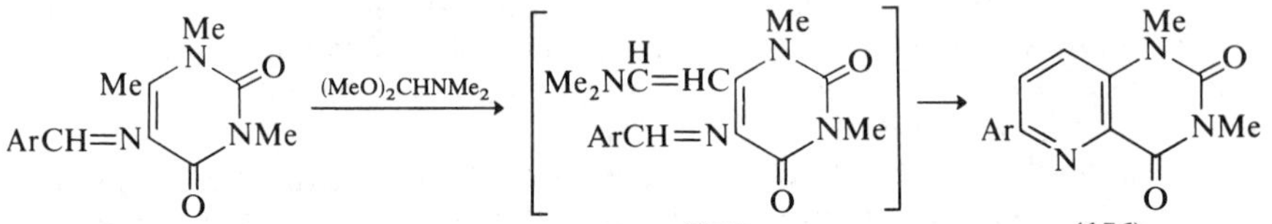

(155)                              (156)

#### 2.15.5.3.3 Fused derivatives

The preparation of benzo fused pyrido[3,2-*d*]pyrimidines has furnished the only examples of the classic reaction of this type, the Bischler–Napieralski, involving the cyclization of 5-aryl-4-acylaminopyrimidines to 6-alkylpyrimido[4,5-*c*]isoquinolines, *e.g.* (157) → (158) ⟨73YZ330⟩. As often found in this reaction, the presence of activating substituents appears necessary ⟨78CPB245⟩.

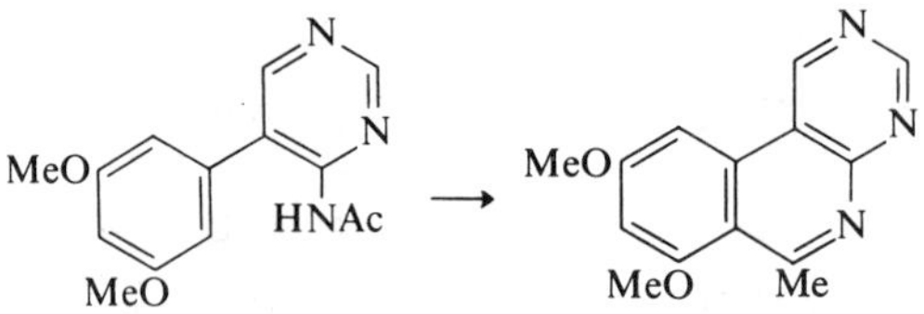

(157)                              (158)

### 2.15.5.4 By Formation of One Bond γ to a Heteroatom [6+0 (γ)]

#### 2.15.5.4.1 From pyridine intermediates

No examples exist of this class of cyclization starting from pyridine intermediates.

#### 2.15.5.4.2 From pyrimidine intermediates

Two types of cyclization predominate, one of which involves the use of 4(6)-aminopyrimidine derivatives having a free 5-position and electron donating substituents elsewhere in the molecule, often a 6-aminouracil derivative. Reaction of these with alkoxy- or dialkylamino-methylenemalonates, as in the well known Gould and Jacobs quinoline cyclization, to give an intermediate of type (159), is followed by [6+0 (γ)] cyclization on heating, usually in inert solvents, to give 5-oxopyrido[2,3-*d*]pyrimidine-6-carboxylate derivatives of type (160) ⟨70CPB1385, 71CPB1482, 67JCS(C)1745, 72JOC3980⟩. Similar cyclizations have been performed with alkoxymethylene derivatives of Meldrum's acid, a cyclic malonate

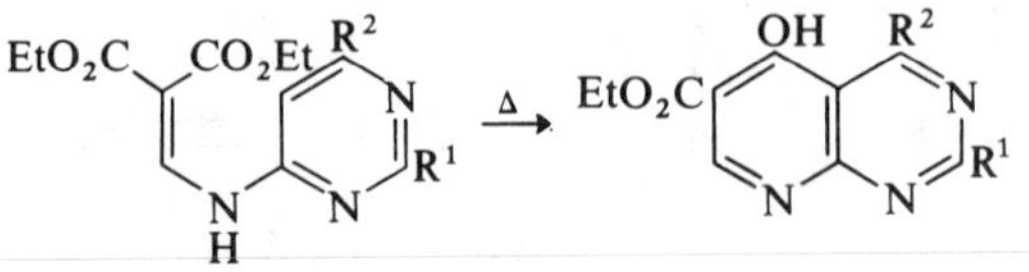

(159)                              (160)

ester (also to give **160**) ⟨81JAP(K)8199480⟩, malonodinitrile ⟨81JAP(K)81125386⟩ and cyanoacetic ester ⟨74JAP(K)7442696⟩ to give 5-oxo-6-nitriles, ethyl acetoacetate (to give 5-methyl-6-carboxylates) ⟨59AC(R)944⟩, and acetylacetone (to give 5-methyl-6-acetyl derivatives) ⟨60AC(R)505⟩. 5-Aminopyrimidines likewise give pyrido[3,2-*d*]pyrimidines ⟨67USP3320257⟩. Direct reactions of 6-aminouracils with β-keto esters are treated as [3 + 3] reactions (see Section 2.15.5.7.2), but a report ⟨76JAP(K)76139633⟩ of a [6 + 0 (γ)] cyclization to give (**162**) after preliminary formation of an intermediate of type (**161**) in such a reaction must be treated with caution, as this regioselectivity is in the opposite sense to that observed in most other cases.

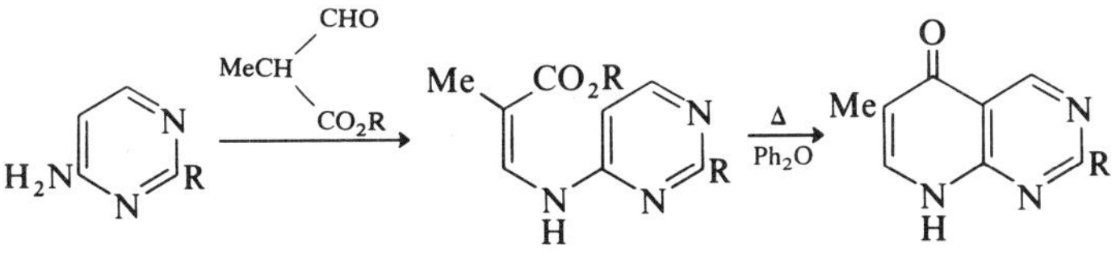

The other main reaction in this class is the Dieckmann-type cyclization of the intermediates (**163**) from 4(6)-halo-5-ethoxycarbonylpyrimidines with *N*-substituted β-alanine esters and nitriles, and related compounds, to give 5,6,7,8-tetrahydro-5-oxopyrido[2,3-*d*]pyrimidine-6-carboxylates (**164**) ⟨*e.g.* 74MI21501, 75JHC311⟩ which may be aromatized to (**165**) (see Section 2.15.3.1). A variation involves amidines, *e.g.* (**166**), which give 7-dialkylamino derivatives of type (**167**) ⟨77MIP21500⟩.

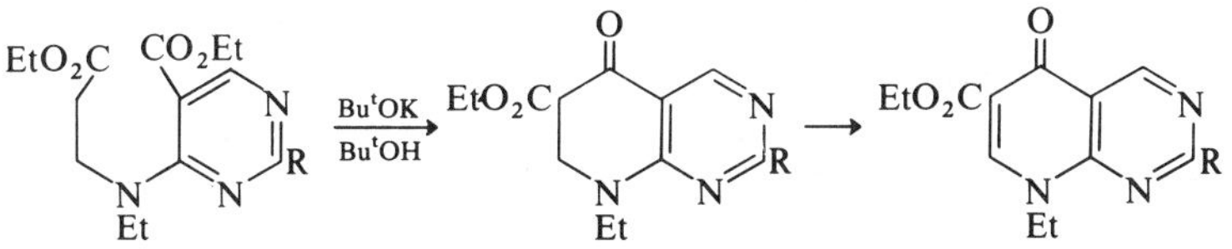

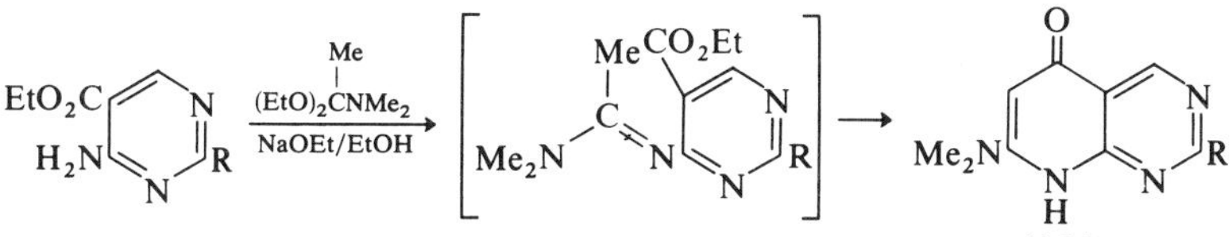

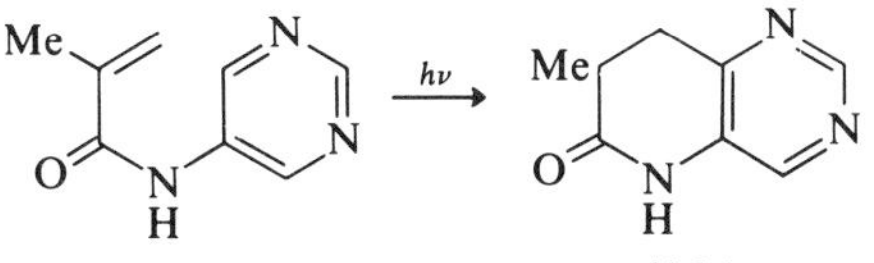

A rare type of reaction involves the photocyclization of the 5-(unsaturated acyl)aminopyrimidines (**168**) to reduced 7-methylpyrido[3,2-*d*]pyrimidin-6-ones (**169**) ⟨72CPB2264⟩. Similar reactions are known involving related 4-aminopyrimidine derivatives, leading to pyrido[2,3-*d*]pyrimidines.

### 2.15.5.4.3 Fused derivatives

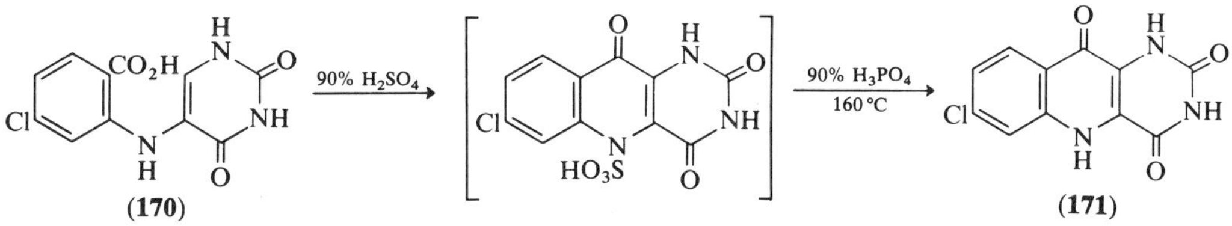

The cyclization of 5-(2-carboxyanilino)pyrimidine (**170**) with sulfuric acid/phosphoric acid leads to the 7-chloropyrimido[5,4-*b*]quinolinetriones (**171**) ⟨57JCS4997, 74KGS131⟩, the 2,4,10-trichloro compounds being obtained with phosphoryl chloride ⟨72JHC91⟩, whilst a formally similar cyclization of 4(6)-arylamino-5-ethoxycarbonylpyrimidines gives pyrimido[4,5-*b*]quinolin-5-ones ⟨80CB395⟩ (or 5-chloro analogues with phosphorus halides ⟨75JHC181⟩).

Finally, an analogous cyclization of the corresponding 5-aldehydes ⟨*e.g.* 79LA1802⟩ or their dimethylaminomethylene or other equivalents ⟨*e.g.* 78JCS(P1)716⟩ leads to the 5-unsubstituted analogues such as (**172**). The reaction may also be carried out with *in situ* formylation of 4(6)-arylaminopyrimidines using the Vilsmeier–Haack reagent ⟨76JCS(P1)1805, 76AG475⟩, as well as to produce quaternary derivatives from *N,N*-disubstituted analogues ⟨76JCS(P1)131⟩. 5-Nitriles give 5-amino-5-deazaflavins ⟨78MI21501⟩.

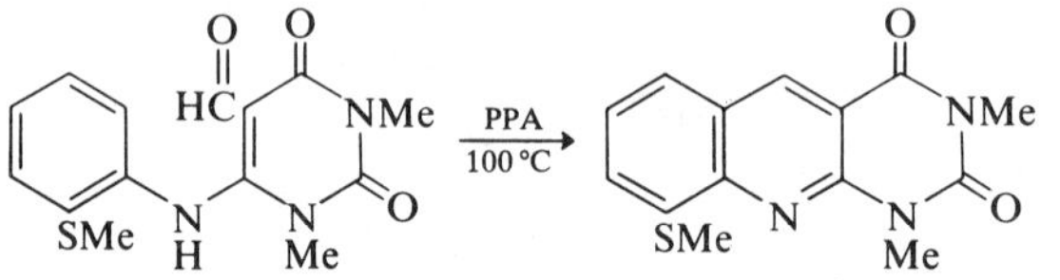

(**172**)

### 2.15.5.5 By Formation of Two Bonds from [5+1] Atom Fragments

#### 2.15.5.5.1 From pyridine intermediates

The vast majority of this type of cyclization start from pyridines. One obvious route converts *o*-acylaminopyridine acids, esters or their equivalents to pyridopyrimidines using a one-atom ammonia or amine fragment, *e.g.* (**173**) → (**174**). Examples are known mainly in the pyrido[2,3-*d*]pyrimidine field ⟨*e.g.* 73GEP2248497, 74MI21502⟩ but also in [3,4-*d*] and [4,3-*d*] cases ⟨74GEP2348111⟩. Sometimes pyridooxazine intermediates similar to (**113**) are involved, especially in the few examples with pyrido[3,2-*d*]pyrimidines ⟨65JCS4240⟩.

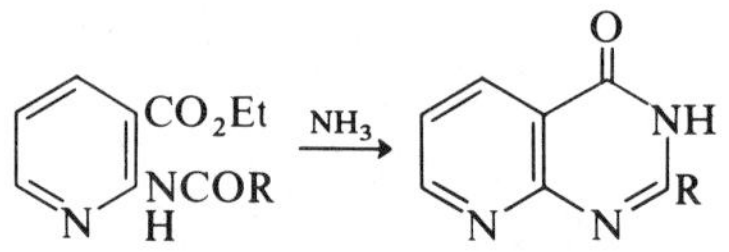

(**173**)          (**174**)

The use of hydrazine as the ring-closing fragment leads to 3-amino derivatives (**175**), either directly or *via* the 2-urethanes ⟨72MI21500, 65CB1505⟩ or dimethylaminomethylene derivatives ⟨72MI21500⟩. Hydroxylamine similarly furnishes the 3-hydroxy analogues ⟨60JCS2157⟩.

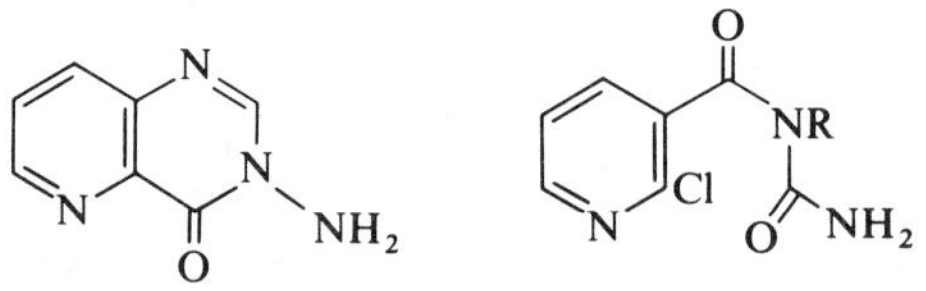

(**175**)                    (**176**)

Cyclization of *o*-formamidoaldehydes with ammonia leads to unsubstituted [2,3-*d*] and [4,3-*d*] analogues ⟨62JCS4094, 70CPB1457⟩, ethoxymethyleneaminonitriles give 4-imines ⟨75JCS(P1)2182⟩, whilst another type of aminocyclization involves the 2-halo-3-carbamoyl derivatives such as (**176**) ⟨75JAP(K)75160296⟩.

The other major type of [5+1] cyclization from pyridines involves that of *o*-aminopyridine amides or their congeners with a one-carbon fragment, usually a phosgene equivalent, giving oxo derivatives (**177**), a carboxylic acid equivalent, leading to unsubstituted, alkyl or aryl derivatives (**178**), or an aldehyde, initially giving partially reduced analogues (**179**). Ketones give *gem*-disubstituted derivatives. Apart from phosgene itself, equivalents used

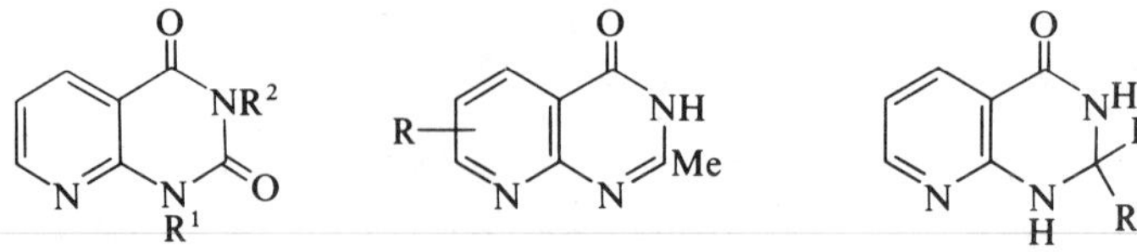

(**177**)              (**178**)              (**179**)

include carbonate esters and ethyl chloroformate or, in some cases, urea (see Section 2.15.5.6.1).

Acid moieties include formic acid itself, formates and orthoesters, formamide, DMF dimethyl acetal and ethyl diethoxyacetate, acids, acid chlorides and anhydrides, the last including a rare [3,4-*d*] derivative. With pyruvate or oxalate esters, 2-acyl or 2-ethoxycarbonyl derivatives (*e.g.* **180**) are formed.

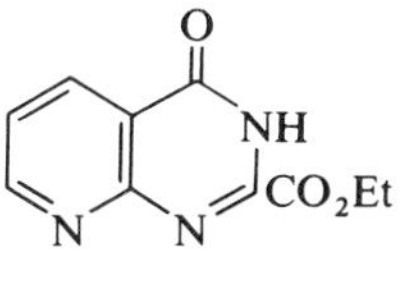

(**180**) 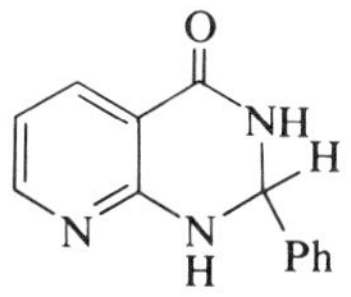          (**181**)

Aromatic aldehydes give 2-aryl-4-oxo derivatives (**181**) in the presence of concentrated sulfuric acid ⟨70EGP73039⟩, whilst pyrimidine derivatives (**182**) give octahydropyrido[4,3-*d*]pyrimidines (**183**) with formaldehyde ⟨*e.g.* 66M52⟩. A similar reaction is observed with 6-methylpyrimidinones ⟨*e.g.* 70M1415⟩.

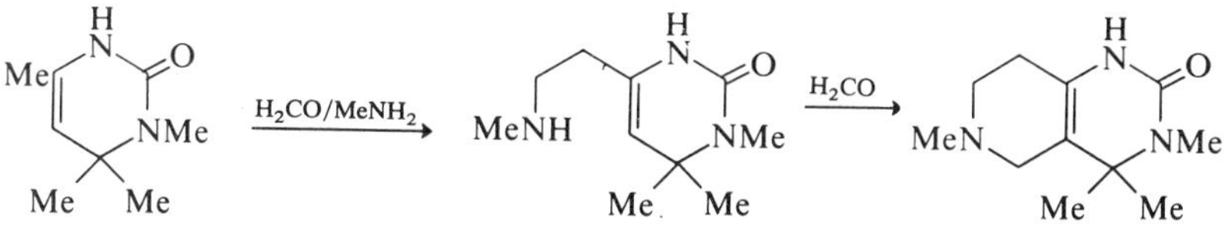

(**182**)          (**183**)

Imines (**184**), formed from nitriles with Grignard reagents ⟨*e.g.* 74JMC636⟩, amidoximes ⟨78H(9)1327, 79JOC1695⟩ (which give 4-amino 3-*N*-oxides together with 4-hydroxylamino derivatives), hydrazides, hydroxamic acids and oximes (which give 3-amino ⟨72MI21500, 78JOC393⟩, 3-hydroxy ⟨60JCS2157⟩ and 3-*N*-oxide ⟨73CPB2643⟩ derivatives respectively) have also been cyclized with a variety of the above one-carbon fragments. Amongst miscellaneous cyclizations, the Vilsmeier–Haack reagent with 2-amino-3-pyrimidyl ketones (**185**) gives the formylated product (**22**) ⟨79JAP(K)7981298/9⟩, and other unusual cyclizations involve carbon disulfide, which converts (**186**) to (**187**) *via* a 1,3-thiazine intermediate ⟨67CI(L)1452⟩, and cyanogen bromide, which with amidinopyridines (**188**) gives 2,4-diimines (**189**) ⟨78JHC877⟩.

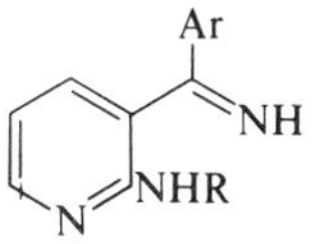

(**184**) 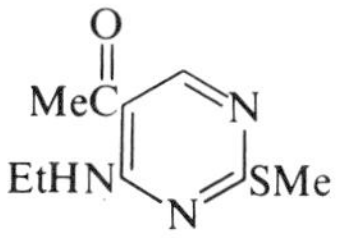          (**185**) 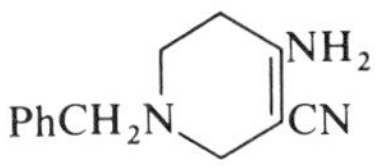          (**186**)

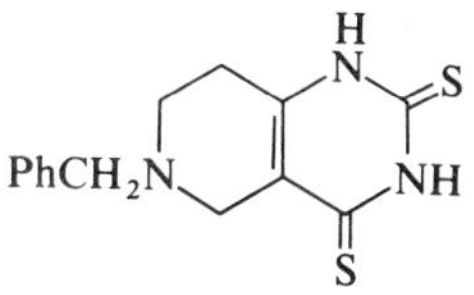

(**187**) 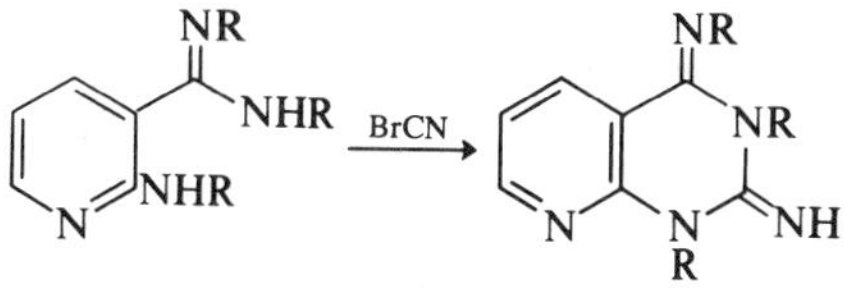          (**188**)          (**189**)

Reduced derivatives are obtained from *o*-aminomethylpyridinamines, *e.g.* (**190**) → (**191**) ⟨58MIP21500⟩, whilst the 3-oxadiazolopyridine (**192**) undergoes ring opening and reclosure to (**193**) with formic acid ⟨78JOC393⟩.

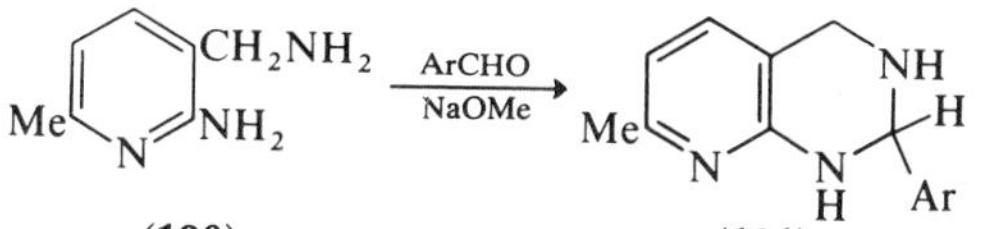

(**190**)          (**191**)

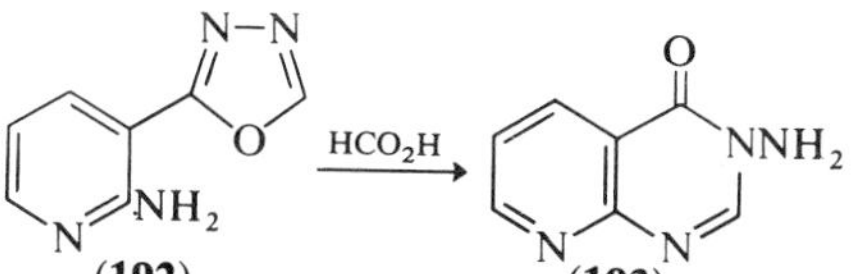

(**192**)          (**193**)

### 2.15.5.5.2 From pyrimidine intermediates

There are few examples of [5 + 1] cyclizations from pyrimidine intermediates. Two of these involve the chloropropionic ester (194), which gives the 5,6,7,8-tetrahydro-7-one (195) with ammonia ⟨59JCS1849⟩, and the cyclization of a 4-ethynylpyrimidine-5-carboxylate with ammonia to give a pyrido[4,3-*d*]pyrimid-5-one ⟨82H(19)184⟩. In a recent patent, 5-ethoxycarbonylpyrimidin-4-yl-*β*-alanine derivatives are cyclized with ammonia to pipemidic acid analogues ⟨80GEP2903850⟩. One-carbon pyrimidine [5 + 1] syntheses are included in Section 2.15.5.5.1 above.

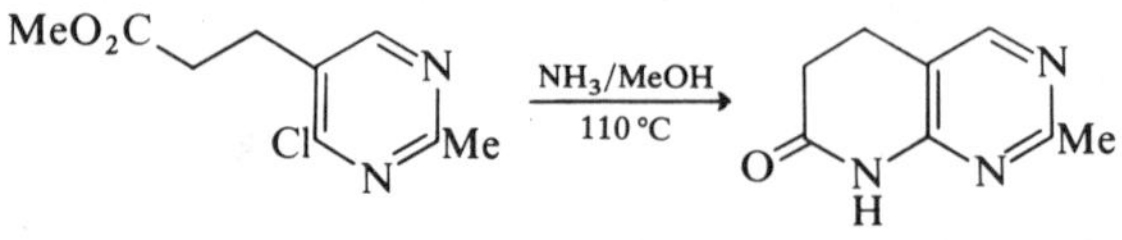

(194)                                              (195)

### 2.15.5.5.3 Fused derivatives

The simplest case gives 5-unsubstituted pyrimido[4,5-*b*]quinoline-2,4-diones (196) from 4(6)-halopyrimidine-5-carbaldehydes and aromatic amines ⟨80MI21500⟩. Other cyclizations involving closure to a substituent benzene ring include (197) → (6), the parent pyrimido[4,5-*c*]isoquinoline ⟨75CPB494⟩, and various syntheses of 5-deazaisoalloxazines from uracil derivatives, including that of the F$_{420}$ coenzyme (199) from the protected D-ribityl uracil (198) ⟨80JHC1709⟩.

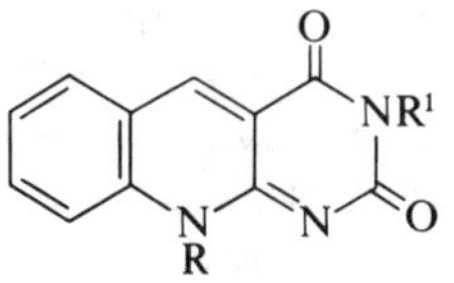

(196) 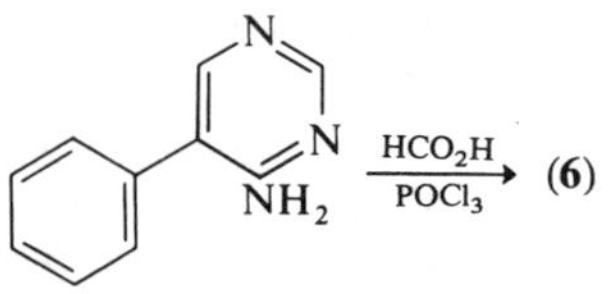                                              (197)

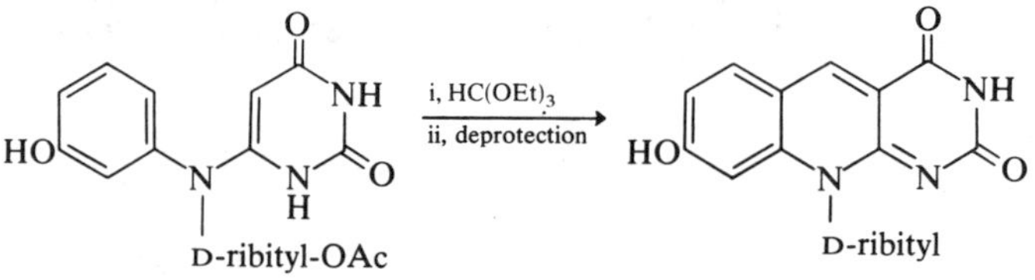

(198)                                              (199)

Other syntheses involve preformed quinoline systems in standard syntheses, including examples in the pyrimido[4,5-*b*]quinoline (*e.g.* 200 → 201) ⟨71JHC111⟩, pyrimido[4,5-*c*]quinoline (*e.g.* 202 → 203) ⟨62JCS1671⟩ and pyrimido[5,4-*c*]quinoline (*e.g.* 204 → 205) ⟨78JMC295⟩ series.

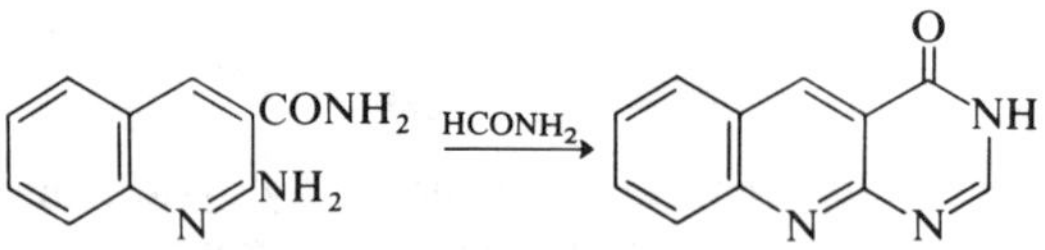

(200)                                              (201)

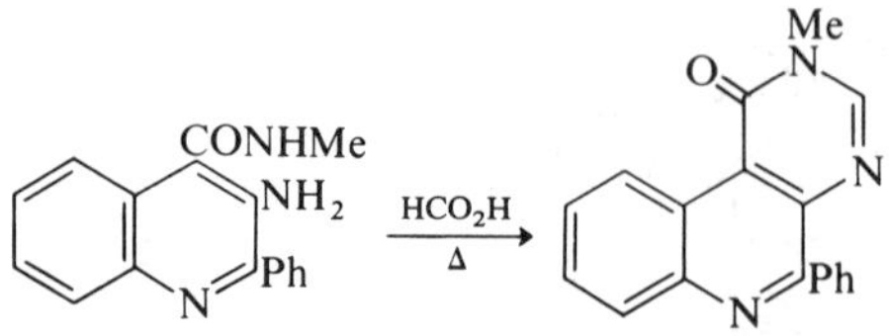

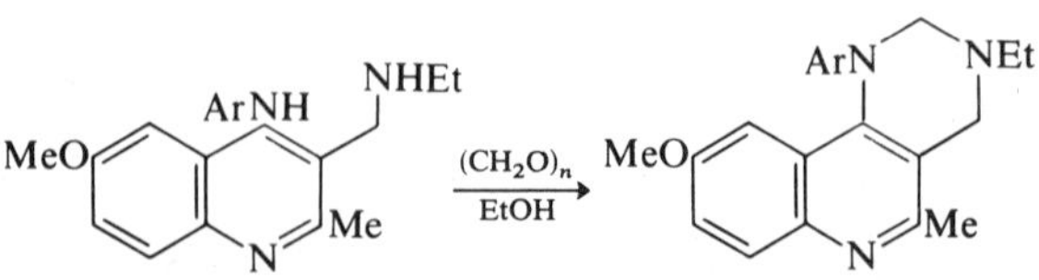

(202)            (203)            (204)            (205)

Additional pyrimido[4,5-*b*]quinoline syntheses include one involving the action of cyanogen bromide on 2-chloro-3-cyanoquinoline similar to that described for pyridines (Section 2.15.5.5.1) ⟨78JHC877⟩, and a synthesis of pyrimido[4,5-*b*]quinoline-2-carboxylic esters, important as antiallergics, by the oxalate ester route above (Section 2.15.5.5.1) ⟨*e.g.* 80JMC262, 80OPP219⟩.

The reaction of the cyanoquinoline (**206**) with $CS_2$ to give (**207**) involves a 4-amino-1,3-thiazine intermediate similar to that encountered previously (Section 2.15.5.5.1) ⟨67T891⟩.

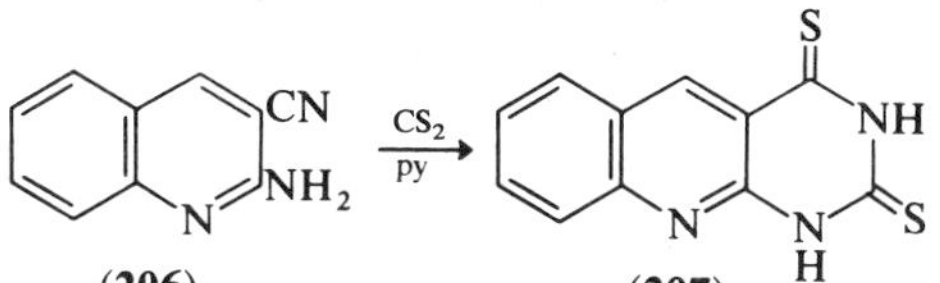

(**206**)          (**207**)

Imidazolyl-, *s*- and *v*-triazolyl-pyridinamines (*e.g.* **208**) undergo cyclization with orthoesters to give 3,4-fused pyrido-[2,3-*d*]-, -[3,2-*d*]- and -[3,4-*d*]-pyrimidines (*e.g.* **209**) ⟨74CR(C)(278)1421, 75BSF2757⟩.

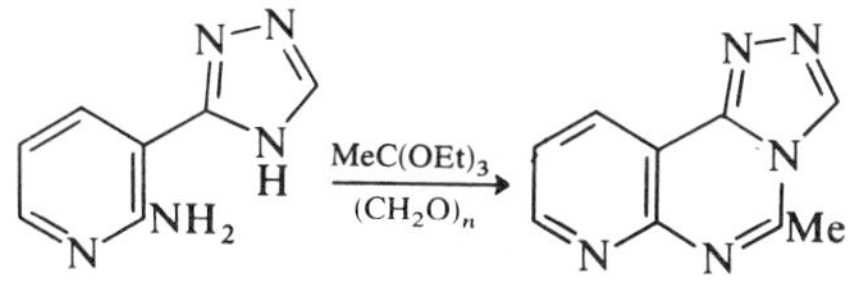

(**208**)          (**209**)

## 2.15.5.6 By Formation of Two Bonds from [4+2] Atom Fragments

### 2.15.5.6.1 From pyridine intermediates

As reactions involving isocyanates and isothiocyanates which proceed *via* identifiable intermediates in a [6+0 (α)] type cyclization have already been dealt with (Section 2.15.5.2.1), the major [4+2] type reactions involve the cyclization of *o*-aminopyridinecarboxylic acid derivatives with incorporation of the atoms of two-carbon fragments, or two of the atoms of three-carbon fragments. The simplest methods, first widely used by Hitchings' group, involve, *e.g.* 2-aminopyridine-3-carboxylic acid, its esters or amides, with urea ⟨58JA3449, 56JCS1045⟩ or urethanes and their analogues ⟨74JOC3434, 79JHC133⟩ to give 2,4-diones (*e.g.* **210** → **211**). In the case of *o*-aminoamides this may result in a [5+1] type of reaction. 2-Halonicotinamides have also been used ⟨*e.g.* 75JAP(K)75131994⟩, as have the 1,3-oxazine intermediates referred to previously (Section 2.15.5.2.1) ⟨76USP3931183⟩.

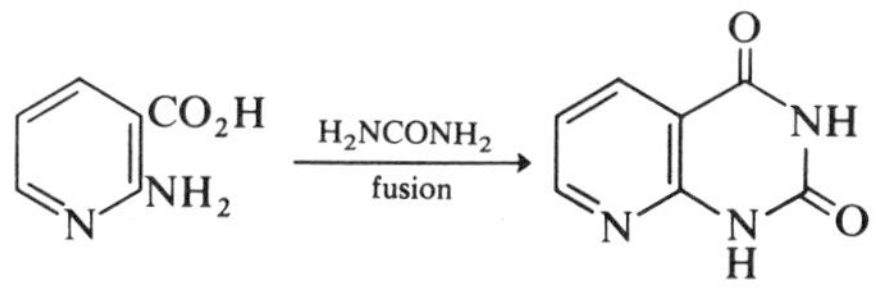

(**210**)          (**211**)

Thioureas give thioxo analogues of a variety of the above syntheses ⟨52JOC542⟩, although these thioxo products are usually prepared from isothiocyanates (Section 2.15.5.2.1). Examples are known in the pyrido-[2,3-*d*], -[3,2-*d*], -[3,4-*d*] and -[4,3-*d*]-pyrimidine fields.

*N*-Alkyl substituted ureas usually eliminate the *N*-substituted amine ⟨80JHC235⟩, but *N*-arylthioureas may give ring *N*-aryl derivatives ⟨66IJC447⟩.

The use of guanidine for cyclization gives amino substituted derivatives (*e.g.* **212**) ⟨52CB1012⟩, and in this case *o*-aminonitriles may be used to furnish diamines ⟨*e.g.* 81JOC1394⟩. An unusual reaction involving nitriles occurred during the preparation of nicotinonitrile from the amide and ammonium sulfamate, when a 60% yield of the dimeric by-product (**213**) was formed *via* the nitrile ⟨69BSB289⟩. Similar products have been obtained from

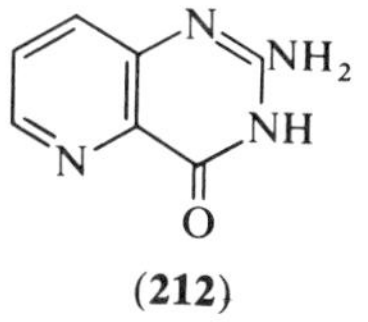

(**212**) 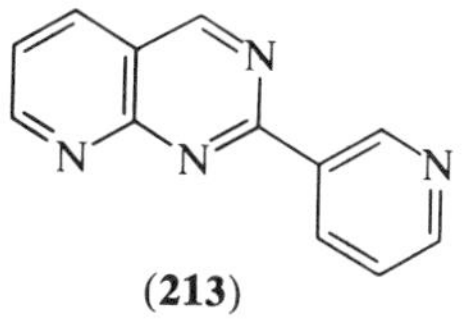          (**213**)

reaction of 2-aminonicotinonitrile with nicotinonitrile itself ⟨61JOC4967⟩, and from 2-chloronicotinonitrile and ammonia ⟨58JA427⟩.

*C*-Unsubstituted derivatives (*e.g.* **214**) are obtained with formamide ⟨*e.g.* 55JA2256⟩, *N*-methylformamide ⟨72JOC3975⟩ or formamidine ⟨77MI21501⟩ whilst acetamide gives methyl analogues ⟨23MI21500, 60JCS2157⟩. Indeed, one of the earliest syntheses of a pyridopyrimidine, a [3,4-*d*] derivative (**215**), was of this type ⟨02CB2831⟩. Benzimidates give phenyl derivatives ⟨67LA(707)250, 73JCS(P1)1794⟩.

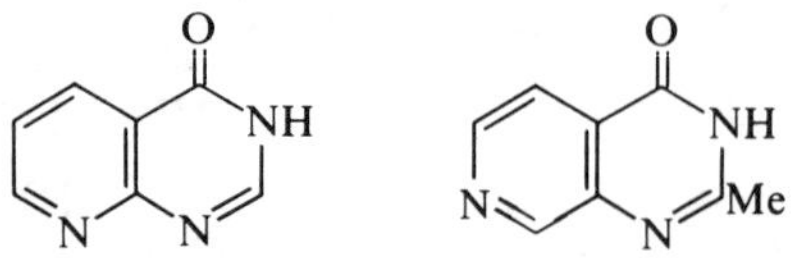

(214)          (215)

Another source of *C*-unsubstituted derivatives are *o*-aminopyridine aldehydes (*e.g.* **216 → 217**) ⟨60JCS1370⟩, and the corresponding ketones may be used to furnish *C*-aryl compounds ⟨71GEP2051013⟩.

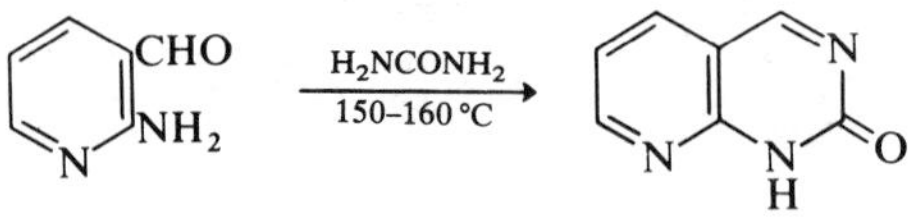

(216)          (217)

Partially reduced derivatives have been obtained from *o*-aminopyridinemethanols (**218 → 219**) ⟨75JAP(K)75157394⟩, and partially reduced pyridines (*e.g.* **220**) have also furnished this type of derivative in reactions with formamide, guanidine and imidates ⟨72JHC1113, 72JHC1123, 78CB2297⟩.

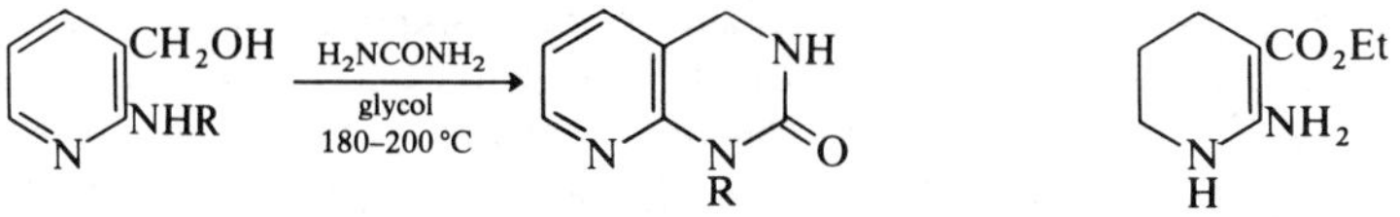

(218)          (219)          (220)

Amongst the more unusual reactions, 2,3-thiazolo fused pyrido[3,2-*d*]pyrimidines have been prepared from 3-aminopicolinic acid and 2-bromothiazoles, whilst a similar derivative resulted with allyl isothiocyanate (**221 → 222**) ⟨72IJC602⟩. Similar products are also produced in [3 + 3] reactions of 2-aminothiazoles (Section 2.15.5.7.1).

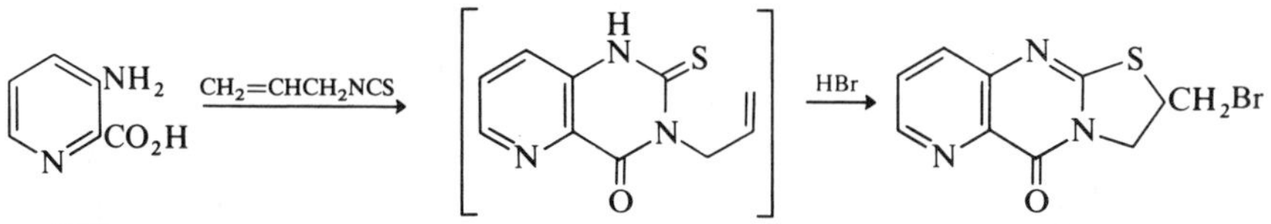

(221)          (222)

A [4 + 2] type reaction followed by a [6 + 0 (α)] cyclization is involved in the reaction of the trimethoxynitrile (**223**) with acetamidine to give (**225**) *via* an intermediate (**224**) ⟨72BCJ1127⟩.

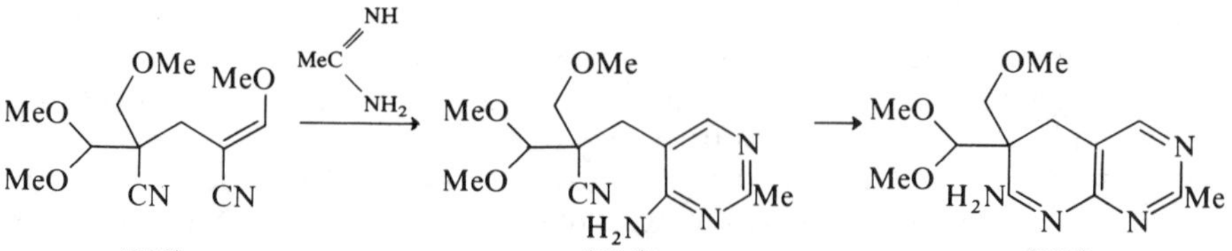

(223)          (224)          (225)

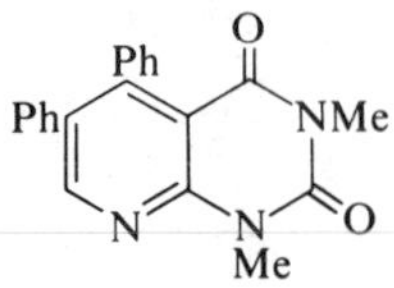

(226)

The pyrido[2,3-*d*]pyrimidine (**226**) was obtained as one of a number of products in the photoreaction of stilbene and caffeine ⟨80AG735⟩.

### 2.15.5.6.2 *From pyrimidine intermediates*

There are two main classes here. Firstly, 5-substituted 4(6)-aminopyrimidines, *e.g.* the 5-ester (**227**), are reacted with esters in the presence of sodium ⟨63CB1868⟩, or with acetals in the presence of alkoxide ⟨78KGS1549⟩, to give pyrido[2,3-*d*]pyrimidine-5,7-diones (*e.g.* **228**). Keto esters give 6-ketones (**229**) ⟨80USP4215216⟩, whilst use of aminopyrimidine nitriles gives 7-oxo-5-amino derivatives ⟨81USP4245094⟩.

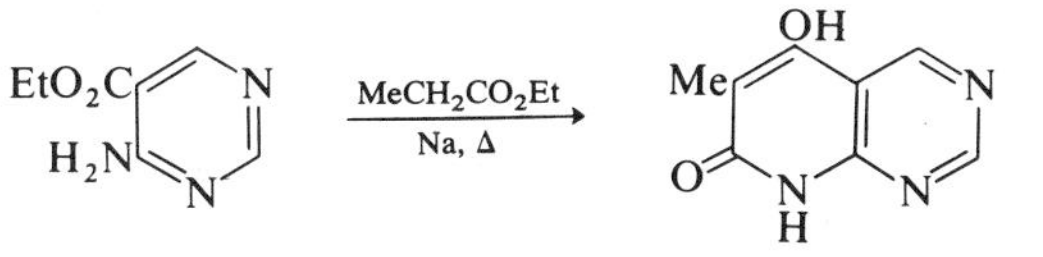

(**227**)　　(**228**) 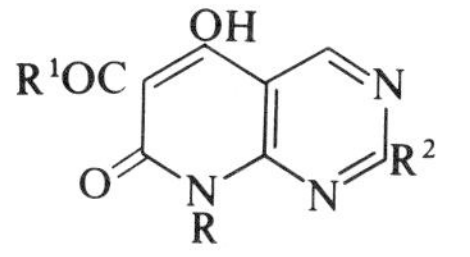　　(**229**)

With the more reactive 4-aminopyrimidine-5-carbaldehydes or their methylaminomethyl-ene equivalents, cyclization can occur with other active methylene compounds such as ketones ⟨*e.g.* 75JOC1438⟩, keto esters ⟨68CB512⟩, malonic esters ⟨72BCJ1127⟩, malonodinitrile ⟨*e.g.* 68CB512⟩, arylacetonitriles ⟨69BRP1171218, 80EUP18151⟩ and cyanoacetamides ⟨72JMC442⟩ to give a variety of 6,7-disubstituted pyrido[2,3-*d*]pyrimidines, whilst the use with 2,6-diacetylpyridine in conjunction with consecutive ring-opening reactions to prepare poly-pyridines has been mentioned previously (Section 2.15.3.3). 4-Amino-5-ketones give 5-substituted analogues ⟨64JOC2116⟩.

Other pyrimidine starting materials include 5-amino-6-methyl derivatives, which with oxalates give 6,7-dihydroxypyrido[3,2-*d*]pyrimidines (**230**) ⟨57CB738⟩, and 5-arylidenebar-bituric acid derivatives, which with malononitrile are converted into 5,6-dihydro-7-aminopyrido[2,3-*d*]pyrimidines (**231**) ⟨79RRC1191⟩.

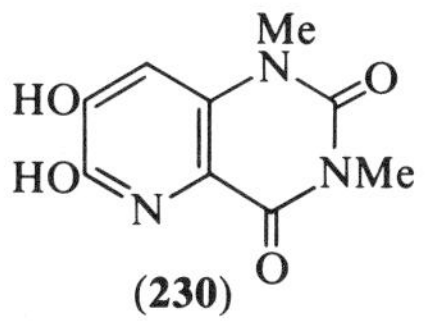

(**230**) 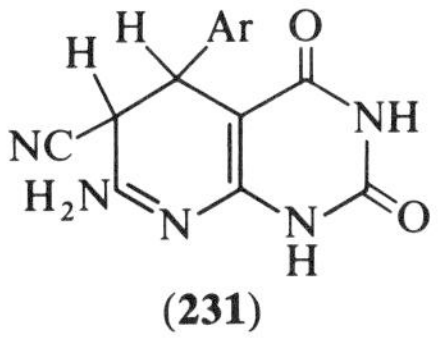　　(**231**)

### 2.15.5.6.3 *Fused derivatives*

Reactions of aminoquinoline acids with two-carbon fragments similar to those discussed in Section 2.15.5.6.1 have been used to prepare pyrimido-[4,5-*b*]- (*e.g.* **232**) ⟨58JA3449⟩, -[5,4-*b*]- ⟨77JHC1053⟩ and -[4,5-*c*]-quinolines ⟨*e.g.* 57JCS3718⟩ using urea or amides, whilst the synthesis of the pyrimido[4,5-*c*]isoquinoline derivative (**234**) from homophthalic anhy-dride and excess formamide involves similar cyclizations, *via* intermediates such as (**233**) ⟨66G1108, 74JHC1081⟩. Pyrimido[5,4-*c*]quinolines have also arisen from tetrahydroquinolines, *e.g.* (**235**) → (**236**) ⟨66MI21500⟩.

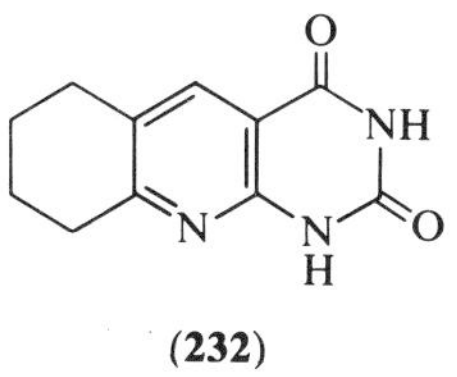

(**232**) 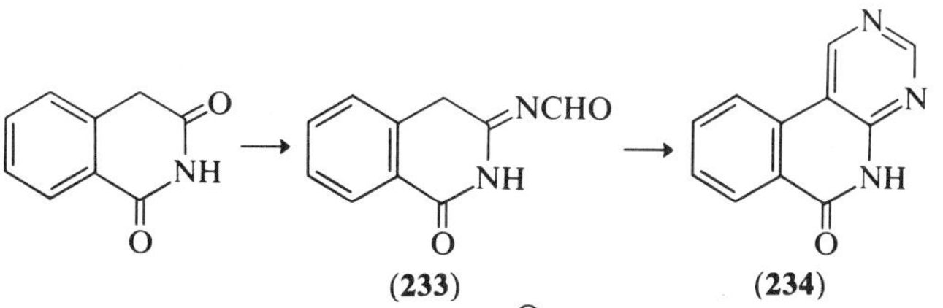　　(**233**)　　(**234**)

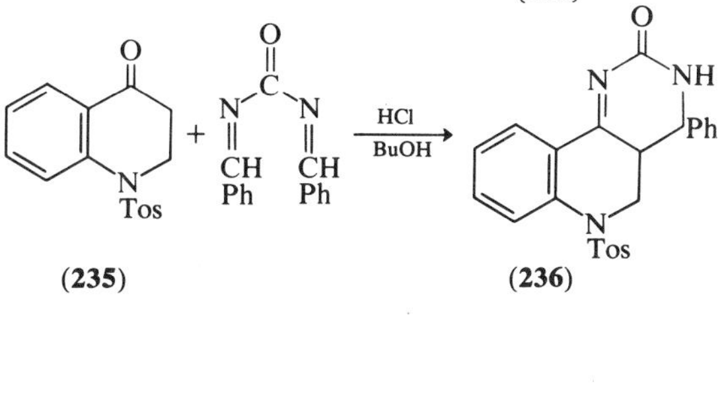

(**235**)　　(**236**)

Syntheses involving two six-membered rings have largely concerned *o*-aminobenzaldehydes or their equivalents, which give pyrimido[4,5-*b*]quinolines, *e.g.* (**237**), with 4,6-dihydroxypyrimidines ⟨47JCS726⟩. Barbituric acid has been a favourite substrate, and with 4,5-dimethyl-2-(*N*-D-ribitylamino)benzaldehyde furnished a synthesis of 10-deazariboflavin (**238**) ⟨70JHC99⟩. With a 1,3-dimethylbarbituric acid and a 2-methyl-aminobenzaldehyde, a deazaflavinium salt is obtained in acid conditions ⟨78AP196⟩.

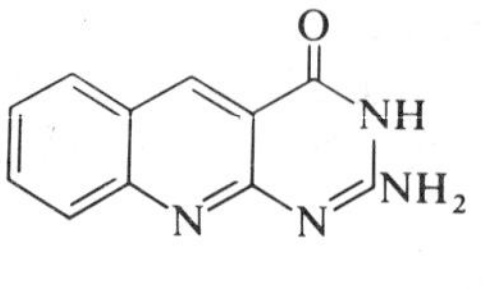

**(237)** 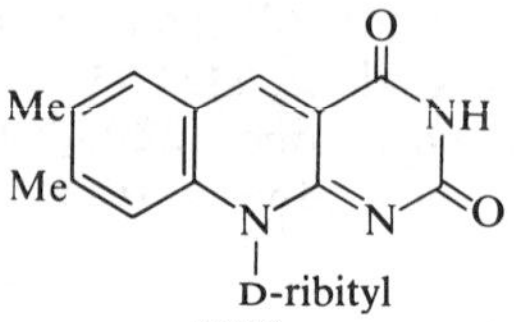 **(238)**

Finally, 4-aminopyrimidine-5-carbaldehydes with cyclic *β*-diketones, such as cyclohexane-1,3-dione, give partially reduced pyrimido[4,5-*b*]quinolones (**239**) ⟨76JOC1058⟩.

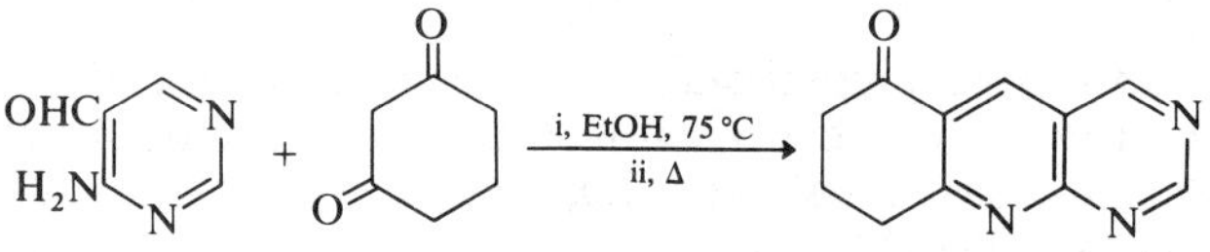

**(239)**

### 2.15.5.7 By Formation of Two Bonds from [3+3] Atom Fragments

#### 2.15.5.7.1 From pyridine intermediates

The simplest [3+3] reactions in the pyridine series involve reaction of *o*-chloropyridinecarboxylic acid derivatives with three-atom fragments such as urea, thiourea(s), amidines and guanidines, *e.g.* (**240**) → (**241**). Examples are known mainly in the pyrido-[2,3-*d*]- ⟨81JHC495, 72JMC837⟩ and -[4,3-*d*]-pyrimidine ⟨78JCS(P1)857⟩ fields. Nitriles give amino derivatives, esters give oxo analogues, whilst from a 3-acetyl-2-pyridinone the 2-oxo(thioxo)-4-methylpyrido[2,3-*d*]pyrimidines (**242**) were formed ⟨78IJC(B)332⟩. With the chloro ester and 2-amino-4-phenylthiazole the 2,3-thiazolo fused derivative (**243**) was obtained ⟨72IJC602, 80KGS1200⟩. In the pyrido[4,3-*d*]pyrimidine field also, a 4-methylthio group replaced the halogen atom ⟨78JCS(P1)857⟩, whilst partially reduced pyrido[4,3-*d*]pyrimidines are formed from piperidines ⟨81JHC327⟩. 3-Arylidenetetrahydropyridin-4-ones give 4-aryl derivatives.

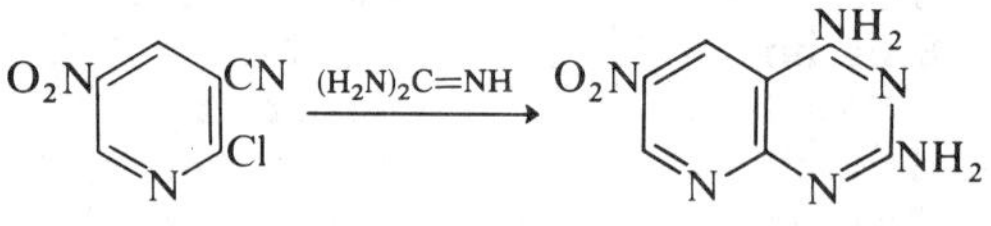

**(240)** **(241)** 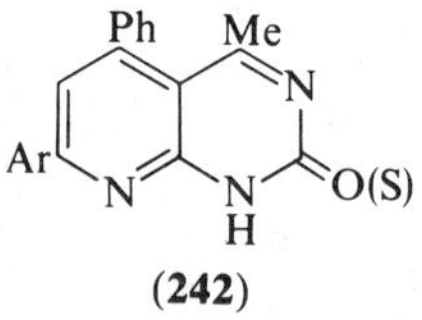 **(242)** 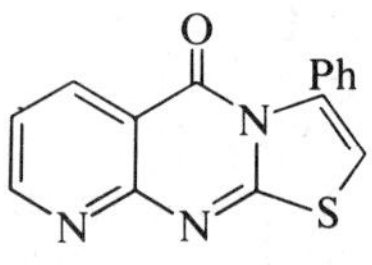 **(243)**

Similar intermediates including *o*-ethoxycarbonyl-, *o*-cyano- and *o*-dimethyl-aminomethylene-piperidones or their imines have been used to give partially reduced analogues, *e.g.* (**244**), in the [2,3-*d*], [3,4-*d*] and [4,3-*d*] series. By the use of *gem*-ethyl ethoxycarbonyl derivatives the bridgehead-alkyl reduced pyrido[2,3-*d*]pyrimidinones are obtained, *e.g.* (**245**) → (**246**) ⟨69MI21500⟩.

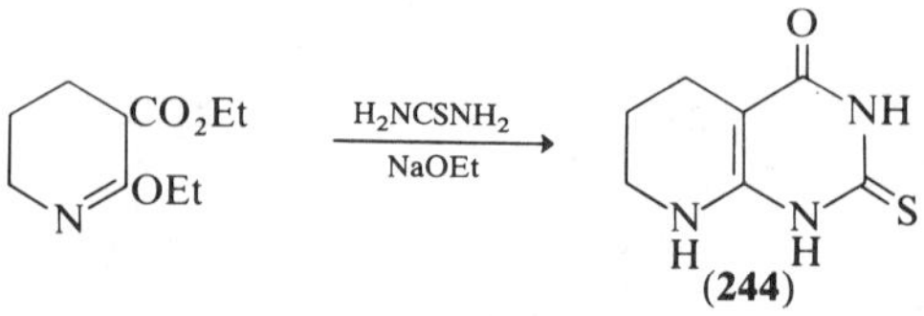

**(244)** 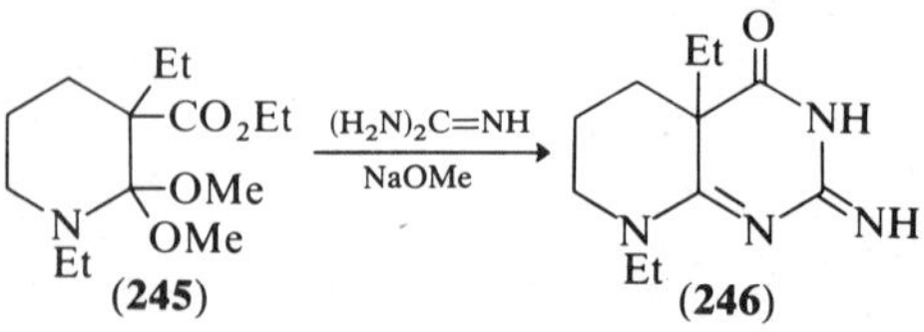 **(245)** **(246)**

An unusual reaction involving *s*-triazine (**247**) and ethyl acetoacetate with sodium ethoxide leads eventually to the pyrido[4,3-*d*]pyrimidin-5-one (**249**), possibly *via* a ring opening and Dimroth-type rearrangement of the intermediate (**248**) ⟨80JHC389⟩.

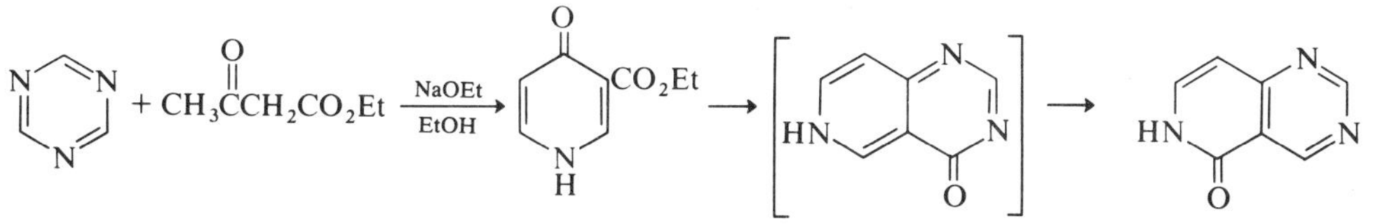

**(247)** **(248)** **(249)**

### 2.15.5.7.2 *From pyrimidine intermediates*

The principal type of reaction from pyrimidine intermediates involves the condensation of a 6-aminopyrimidine (often a uracil derivative) with a three-carbon fragment to complete the annelation to a pyrido[2,3-*d*]pyrimidine, and this has been one of the main sources of this ring system since the classical work of Hitchings and his coworkers ⟨*cf.* 58JA3449⟩. Although in some cases intermediates have been isolated or postulated, making these [6 + 0] reactions, the majority are considered here for convenience. In the original work β-dicarbonyl compounds were used as the $C_3$ fragment, but in subsequent work a wide variety of alternatives has been employed. These include β-keto aldehydes ⟨*e.g.* 74JCS(P1)1225⟩, β-dialdehydes (including nitromalonodialdehyde to give 6-nitro derivatives) ⟨*e.g.* 75JOC3608⟩, β-diketones ⟨*e.g.* 58JA3449, 56AC(R)428⟩, β-keto esters ⟨*e.g.* 80JMC327, 68JMC703⟩, and α,β-unsaturated carbonyl compounds (analogous to the Doebner–Miller quinoline cyclization) ⟨*e.g.* 76JOC3149, 74JCS(P1)1225⟩ and nitriles ⟨*e.g.* 66USP3235554, 73SC397⟩. In many cases basic catalysts are employed. Various pseudo keto aldehydes, for example the α-amino-, hydroxy-, alkoxy- and dialkylamino-methylene derivatives of aldehydes and ketones, have been used in the interests of enhanced stability ⟨*e.g.* 70M1130, 74JHC51, 74CB2537⟩, whilst the activation of these by conversion into chloroiminium derivatives such as (**250**) with phosgene has found some application ⟨*e.g.* 68JMC708⟩. Other examples of this type of reaction where the intermediate has been isolated have been dealt with in Section 2.15.5.4.2 as [6 + 0 (γ)] reactions.

$$(\overset{+}{Me_2N}=CHC=CHCl)\ Cl^-$$
$$|$$
$$R$$

**(250)**

The corresponding saturated dialkylamino ketones (Mannich bases) have also been used to provide 5,6-dihydro derivatives, which are oxidized in air to aromatic pyrido[2,3-*d*]pyrimidines ⟨78AP406⟩.

Acetylenic aldehydes and ketones or their chloroenone equivalents have been used ⟨58GEP1040040⟩, as have acetylenic esters such as DMAD ⟨*e.g.* 76JOC1095, 73CPB2014⟩.

With malonic acid as the $C_3$ fragment in the presence of acetic anhydride, 6-substituted 5,7-dihydroxy compounds are obtained ⟨64JOC219, 61M1184⟩, whilst the 6-*N*-lithio derivative of a uracil (**251**) reacted with a ketenimine to give the 7-*t*-butylamino compound (**252**) ⟨77JOC221⟩.

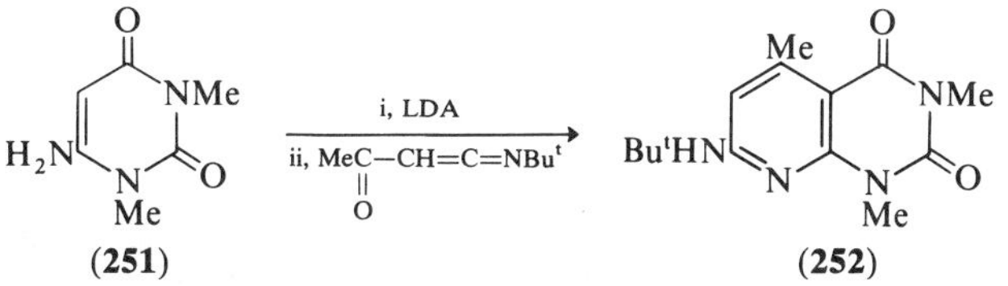

**(251)** **(252)**

A final method for the preparation of pyrido[2,3-*d*]pyrimidines which does not stem from 6-aminouracils involves the reaction of 4-chloropyrimidine-5-carboxylic acid chlorides with enamines in the presence of base to give 6,7,8-trisubstituted 5-ones (**253** → **254**)

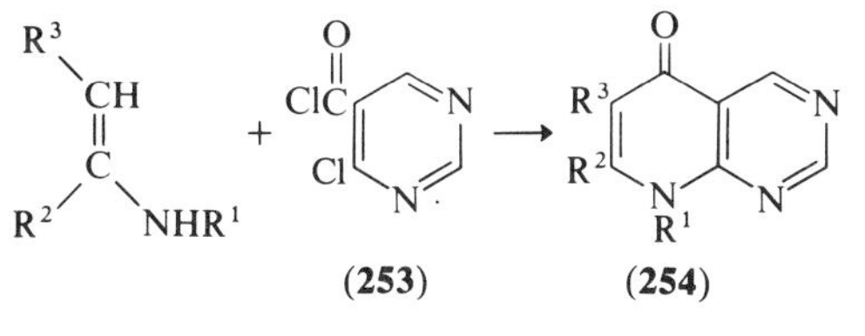

**(253)** **(254)** 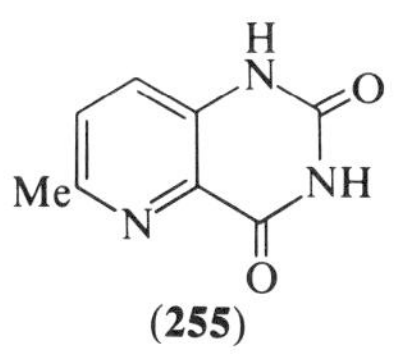 **(255)**

⟨79GEP2808070⟩. The reversed 5-chloro-4-acid chlorides give the corresponding pyrido[3,2-*d*]pyrimidinones.

The only other pyrimidine-based preparation of pyrido[3,2-*d*]pyrimidines involves reaction of 5-aminopyrimidine with crotonaldehyde to give (**255**) ⟨70JHC1219⟩.

The question of orientation of the pyrido[2,3-*d*]pyrimidines produced by the above reactions of 6-aminopyrimidines with unsymmetrical three-carbon fragments has been the subject of considerable study, as a result of which a general rule has been formulated ⟨74JCS(P1)1225⟩. In this, the most reactive position in the pyrimidine, which is the 5-position (*via* an enamine-type resonance in the anion, as distinct from the aniline NH$_2$ in the related Doebner–Miller quinoline synthesis), reacts first with the most reactive position in the C$_3$ fragment. For example, in β-keto aldehydes (RCOCH$_2$CHO) this is the aldehyde group, leading to 7-R derivatives ⟨58JA3449⟩. Esters are less reactive than ketones, so keto esters (RCOCH$_2$CO$_2$Et) give 5-R-7-ones ⟨68JMC703⟩, whilst diketo esters (RCOCH$_2$COCO$_2$Et) give 5-CO$_2$Et-7-R compounds ⟨63G576⟩, and the unsaturated ketone (RCOCH=CHCO$_2$Me) gives the 5-CO$_2$Me-7-R derivative ⟨77JOC1919⟩. DMAD gives only the 5-ethoxycarbonyl-7-one ⟨76JOC1095⟩. Most published examples are in accord with the general rule but, initially, the reactions of 6-D-ribityluracils with keto aldehydes and unsaturated ketones appeared to contradict this ⟨72JCS(P1)1041⟩. This was later found to be due to an erroneous assignment, however, which was corrected following photochemical degradation and NOE $^1$H NMR studies ⟨74JCS(P1)1225⟩.

Only in the reactions of alkoxy- and amino-methylene ketones have variations in orientation been observed. For example, as expected if the reactions proceed *via* the typical amino unsaturated ester intermediate (Section 2.15.5.4.2), reaction of 1-phenyl-3-methyl-6-aminouracil with the ethoxymethylene derivative (**256**) gives the 5- (not 7-) methyl derivative (**257**) ⟨60AC(R)505⟩. A similar orientation was observed with dimethylaminomethylene derivatives, RCOCH=CHNR$_2$, which also give 5-R compounds ⟨70M1130⟩, and ethoxymethylenepentane-2,3-dione, which is said to give 5-methyl-6-acetyl analogues ⟨59AC(R)944⟩. On the other hand, (**258**) was claimed to give the 7-*t*-butyl product (**259**) ⟨77JOC221⟩, and although other papers reported the 5-derivatives to be formed from similar cyclizations ⟨73T2209⟩, this was later corrected ⟨74CB2537⟩.

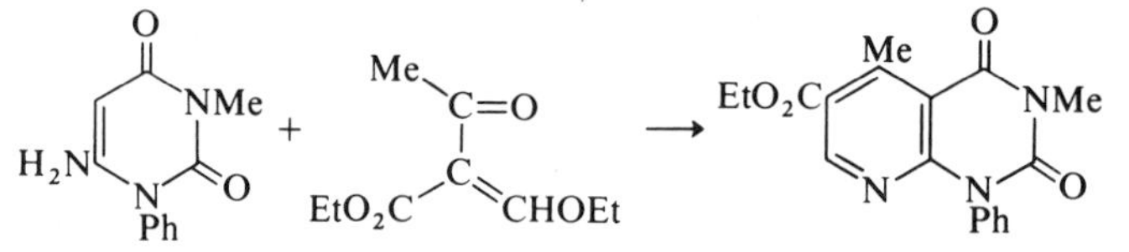

(**256**)          (**257**)

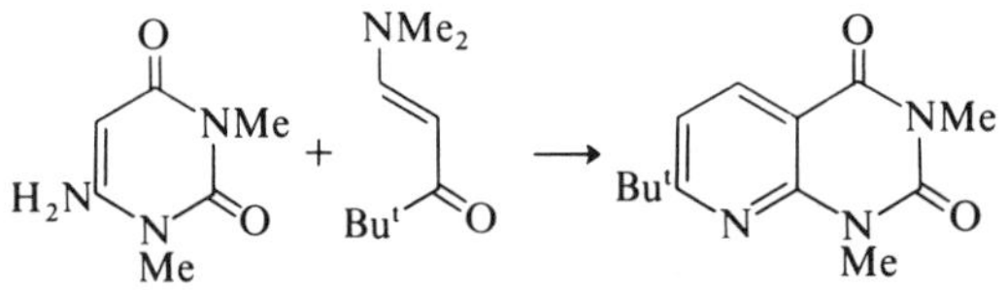

(**258**)          (**259**)

### 2.15.5.7.3 *Fused derivatives*

The synthesis of benzo fused analogues broadly follows the general methods outlined above. One method starts from *o*-quinolinonecarboxylic acid derivatives. For example, reaction of the ester (**260**) with urea, thiourea or guanidine gives pyrimido[4,5-*b*]quinolines (**261**) ⟨70MI21500⟩, and several other examples have been noted ⟨60JOC1368, 58JA3449⟩. 4-Methylthio- or 4-methoxy-3-carbonitriles and -carbaldehydes have been converted into pyrimido[5,4-*c*]quinolines ⟨78JCS(P1)857, 75JCS(P1)2271⟩, and another example from the same

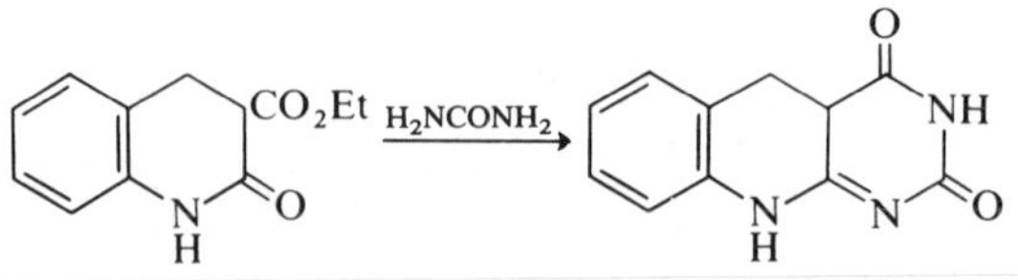

(**260**)          (**261**)

ring system gives the *N*-tosyl derivative (263) from (262) ⟨79JCR(S)268⟩. Isomeric pyrimido[4,5-*c*]isoquinolines result from 3-chloroisoquinoline-4-carbonitriles ⟨74JMC1272⟩.

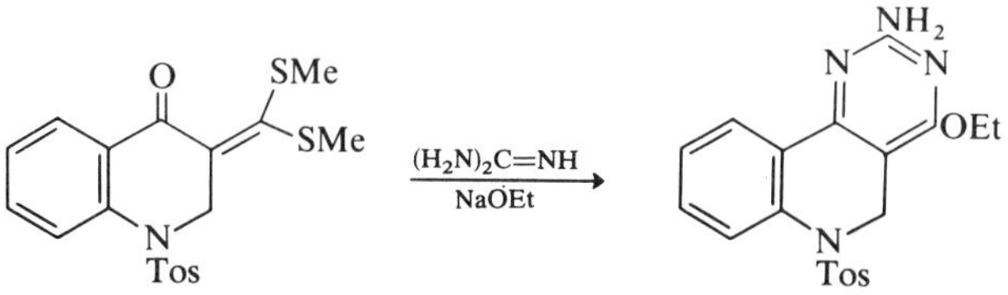

(262)          (263)

Reactions of 6-aminouracils with various 2-substituted cyclohexanones such as the aldehyde (264) give reduced pyrimido[4,5-*b*]quinolines (265) ⟨57BRP774095, 58JA3449⟩, and other cyclohexanone derivatives used include the 2-dimethylaminomethyl (Mannich) bases ⟨78AP542⟩ and the 5-benzylidenedimedones (266) formed *in situ* from dimedone and aldehydes ⟨67KGS395, *cf.* 67KGS406⟩.

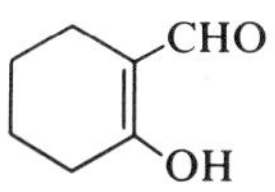

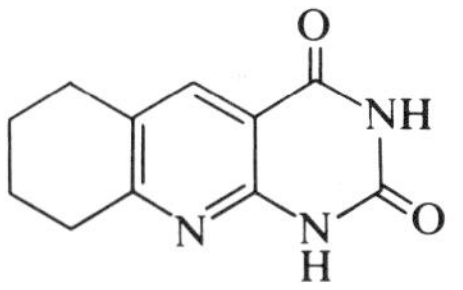

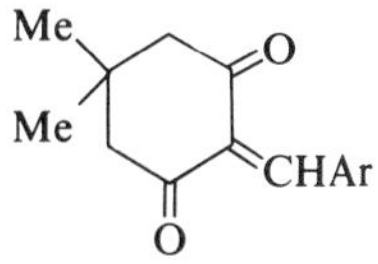

(264)          (265)          (266)

In the case of the 2-dimethylamino- (or 2-amino-)methylene derivatives, the products were at first thought to be pyrimido[4,5-*c*]isoquinolines (267), but later work with 6-(*N*-substituted amino)uracils assigned the structures of the products (268) as belonging to the isomeric pyrimido[4,5-*b*]quinoline system ⟨74CB2537⟩, in agreement with the regioselection rules above.

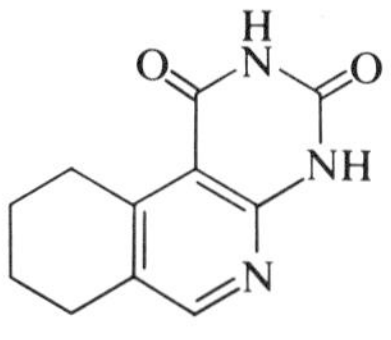

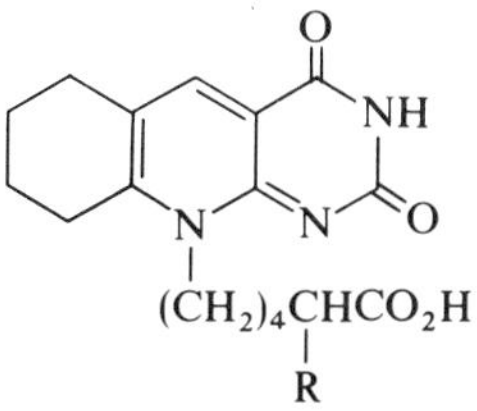

(267)          (268)

Cyclohexanone-2-carboxylic ester did give a pyrimido[4,5-*c*]isoquinoline (269) in agreement with the rules ⟨60USP2937284⟩, whilst an unusual related reaction of a 6-aminouracil with benzylideneacetone gave the diphenylpyrimido[4,5-*b*]quinoline (271) *via* a double ring closure of the dimer (270) of the unsaturated ketone ⟨76JOC3149⟩. 5-Deazaflavin analogues, *e.g.* (272), from the same system resulted from a reaction of 6-chloropyrimidine-5-carbaldehyde with *N*-alkylanilines ⟨80MI21500⟩.

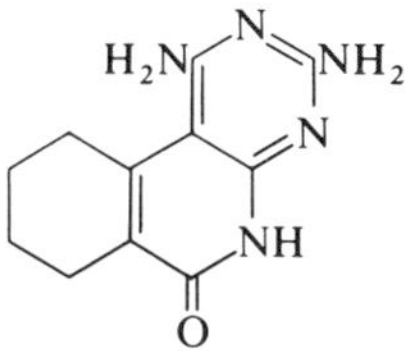

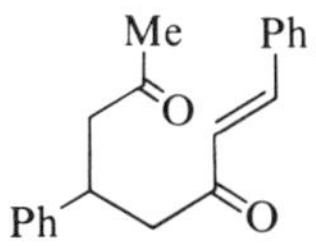

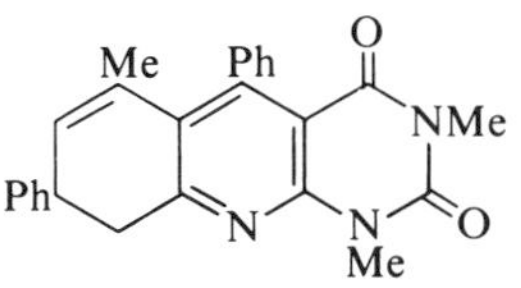

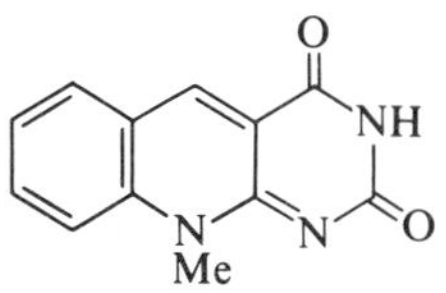

(269)          (270)          (271)          (272)

## 2.15.5.8 By Formation of Three Bonds

These constitute a miscellany of reactions which mainly involve piperidinones, which react with two two-atom fragments to form the pyrimidine ring. Thus 2-piperidinone dimethyl acetal with two moles of aryl isothiocyanate gives (273) ⟨80IJC(B)195⟩, with formamide/POCl₃ (274) is formed ⟨73BCJ2835⟩, whilst the use of benzonitrile with LDA and a piperidinone lactim ether gives (275) ⟨77JOC1808⟩. A related reaction uses 4-piperidinone with aldehydes and urea to furnish the pyrido[4,3-*d*]pyrimidine (276) ⟨78JAP(K)7856692⟩.

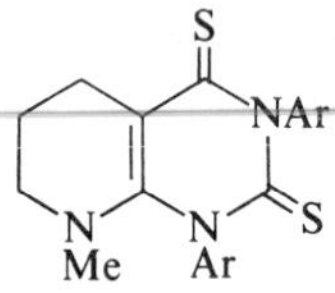

(273) 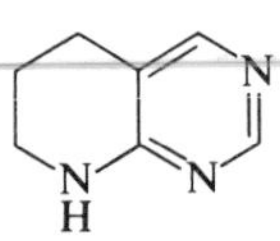              (274) 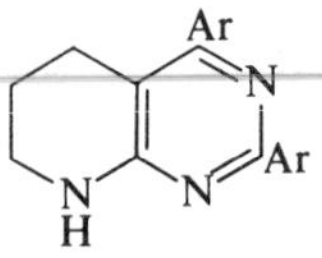              (275) 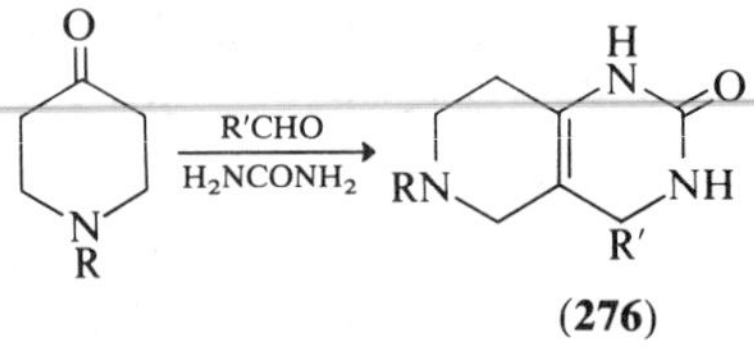                              (276)

The other main type of reaction involves 6-aminouracils, which react with β-amino-crotononitrile and aldehydes to give the 6-carbonitrile (277) ⟨72KGS422⟩, whilst from two moles of aminouracil with strong acids in DMF the back-to-back pyrimido fused molecule (278) is obtained ⟨62USP3035061⟩.

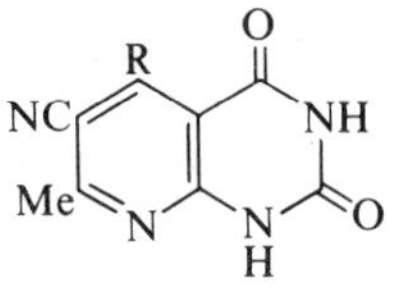

(277) 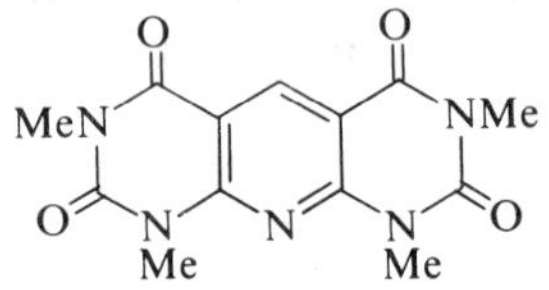                                    (278)

## 2.15.5.9 By Formation of Four or More Bonds

One of the most elegant of these involves the reaction of two moles of 2-aminopyridine with two moles of carbon dioxide under high temperature and pressure to give the 3-pyridyl-2,4-dione (279) ⟨54USP2680741⟩.

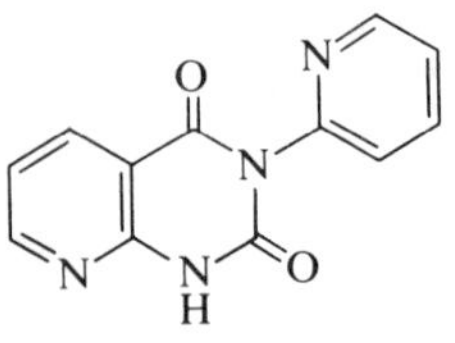

(279) 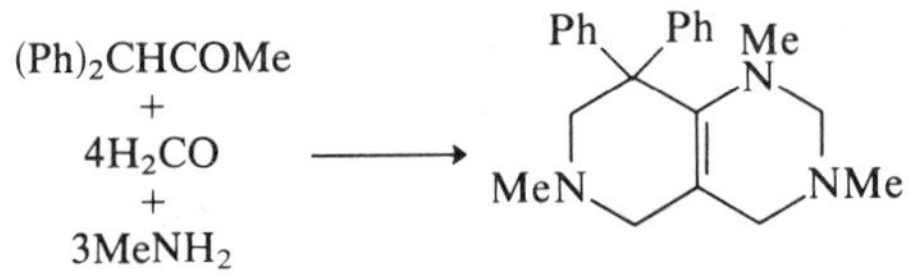                                    (280) 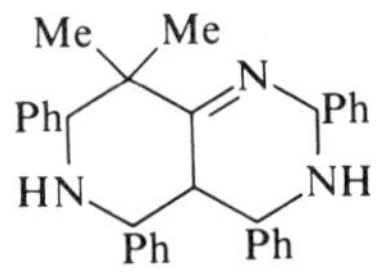              (281)

The combination of 1,1-diphenylpropanone with three moles of methylamine and four moles of formaldehyde gives (280) ⟨57BRP776335⟩, whilst in a similar reaction, *N,N'*-dibenzylidenephenylmethanediamines and isopropyl methyl ketone give the polyhydropyrido[4,3-*d*]pyrimidines (281) ⟨81S151⟩. Cyclization of 4,4,6-trimethyltetra-hydropyrimidine-2-thiones with primary amines and formaldehyde to give hexa-hydropyrido[4,3-*d*]pyrimidinethiones ⟨*e.g.* 70M1415⟩ also fall into this category.

## 2.15.6 PYRIDOPYRIDAZINES: INTRODUCTION

The field of pyridopyridazines, although potentially a rich one, has been little explored compared with that of the pyridopyrimidines. This is due partly to the difficulty of preparation of several of the member systems, to the relatively small number of different syntheses utilized for those systems which have been explored, and the relative lack of interesting biological activity, or pharmaceutical and other uses found with them.

Apart from a short section in an early book on phthalazines ⟨53HC(5)198⟩, the only previous general account of significance occurs in a chapter in a now dated volume on fused pyridazines, which covers work up to 1969 ⟨73HC(27)968⟩, and which also includes bridgehead nitrogen derivatives not relevant to this review. A later review ⟨75MI21504⟩ does exist but is in Japanese.

## 2.15.7 STRUCTURE AND PHYSICAL PROPERTIES OF
## PYRIDOPYRIDAZINES

All six isomeric pyridopyridazine systems are now known, comprising pyrido[2,3-*c*]pyridazine (282), pyrido[2,3-*d*]pyridazine (283), pyrido[3,2-*c*]pyridazine (284), pyrido[3,4-*c*]pyridazine (285), pyrido[3,4-*d*]pyridazine (286) and pyrido[4,3-*c*]pyridazine

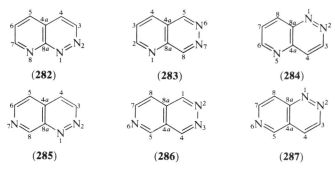

(282)        (283)        (284)

(285)        (286)        (287)

(287); the ring numbering is as shown on the structures. They have also been known at various times under a variety of other names, *e.g.* pyrido[2,3-*d*]pyridazine has been called 1,6,7-triazanaphthalene, 4,6,7-triazanaphthalene, 1,6,7-pyridopyridazine, 1,6,7-benzotriazine and 5-azaphthalazine, and corresponding names have been associated with most of the other members of the group. Only two of the four possible benzo fused pyrido[2,3-*c*]pyridazines, (288) and (289), are known, as are two of the three benzo[2,3-*d*] analogues, (290) and (291), two of the three benzo[3,2-*c*] analogues, (292) and (293), and two benzo[3,4-*c*] derivatives, (294) and (295). Of the three benzo fused pyrido[3,4-*d*]pyridazines, (296) is known, but (297), thought to be the only example of a phenalene-type benzo fused analogue, was later shown ⟨42JA2417⟩ to have the alternative ring structure (298). Finally, two of the three possible benzo fused derivatives, (299) and (300), of the pyrido[4,3-*c*]pyridazine system are exemplified. The most studied systems are the pyrido[2,3-*d*]pyridazines and pyrido[3,4-*d*]pyridazines, these being the easiest to prepare.

(288)        (289)        (290)

(291)        (292)        (293)

(294)        (295)        (296)        (297)

(298)        (299)        (300)

## 2.15.7.1 Conformational Behaviour

As with the pyridopyrimidines, conformational behaviour of reduced pyridopyridazines does not appear to have been studied.

### 2.15.7.2 NMR Spectroscopy

#### 2.15.7.2.1 *$^1$H NMR spectra*

The $^1$H NMR spectra of the parent compounds of the pyrido-[2,3-*d*]- and -[3,4-*d*]-pyridazine systems have been studied, together with those of some closely related derivatives ⟨*e.g.* 69BSF2519, 68AJC1291⟩, as have those of the [3,2-*c*] and [3,4-*c*] systems ⟨79T2027, 80KGS541⟩, and that of the 7-methylpyrido[2,3-*c*]pyridazine (**301**) has recently been described ⟨80KGS541⟩ (but not that of the parent compound, **282**). In the pyrido[4,3-*c*]pyridazine series, only the spectrum of the dihydro compound (**302**) has been recorded ⟨79T2027⟩.

(301)          (302)

In the [3,2-*c*] series the spectra of the parent compound and some closely related analogues have been compared with those predicted from MO calculations of electron densities. The results confirm the predicted occurrence of a considerable degree of bond localization in these compounds ⟨80KGS541⟩, as was previously found in their deaza (cinnoline) analogues ⟨67JCS(B)1243⟩. Other comparisons of NMR values with those expected from MO calculations for a series of azanaphthalenes have included some results with pyridopyridazines ⟨75MI21500⟩.

The barriers to rotation in a series of *N,N*-dimethylformamidine derivatives of pyrido-[2,3-*d*]- and -[3,4-*d*]-pyridazines have been measured by an NMR method ⟨78JHC1105⟩, and NMR studies have also been used in a study of tautomerism in the latter system ⟨75BSF702⟩.

The $^1$H NMR spectra of pyrido[2,3-*d*]pyridazine *N*-oxides ⟨*e.g.* 69AJC1745⟩, 1-ones ⟨69BSF3678⟩ and 1,4-diones ⟨69MI21501⟩ have been recorded and discussed. Some reduced derivatives and quaternary salts have also been studied and alkaline deuterium exchange reactions investigated ⟨77BSF919⟩.

Miscellaneous NMR measurements have been employed to elucidate structural problems in the case of various vinylpyridine-azodicarboxylate adducts (mainly in the pyrido-[3,2-*c*]-, -[3,4-*c*]- and -[4,3-*c*]-pyridazines ⟨79KGS639⟩), in the alkylthio adducts of pyrido[3,2-*c*]pyridazine ⟨78KGS1272⟩, in studies of halogen reactivity in polyhalo[2,3-*d*] derivatives (see Section 2.15.9.3) ⟨72MI21501⟩, in nucleophilic substitutions in pyrido[3,4-*d*]pyridazines ⟨75CPB2239, 75CPB2306⟩ and in various -[3,4-*c*]- ⟨76CPB1870⟩ and azolo-fused pyrido-[2,3-*d*]- and -[3,4-*d*]-pyridazines ⟨68MI21500, 69CPB2266⟩.

In the benzo fused systems only derivatives of the pyridazino[4,5-*b*]quinoline system appear to have been studied ⟨72BSF1588⟩.

$^1$H NMR data for a number of pyridopyridazines are given in Table 2.

#### 2.15.7.2.2 *$^{13}$C NMR spectra*

No systematic study of the $^{13}$C NMR spectra of pyridopyridazines has been undertaken, and only the spectra of a few isolated compounds such as the dihydro[4,3-*c*] derivative (**302**) ⟨79T2027⟩ have been recorded.

### 2.15.7.3 IR Spectroscopy

Similarly, no systematic study of the IR spectra of the pyridopyridazines has been recorded, but the spectra of the [2,3-*d*] derivatives have been discussed ⟨68AJC1291⟩. The diones of this series have also been studied ⟨69MI21501⟩, and IR used to distinguish between the structure (**303**) and the possible isomeric formulation (**304**) ⟨74JHC351⟩. The IR spectra of some of the azodicarboxylic ester adducts ⟨79T2027⟩ have been recorded, whilst in the benzo fused systems some problems with the structure of acyl derivatives in the pyridazino[4,5-*b*]quinoline series have been resolved with the help of IR spectroscopy ⟨71BSF906, 72BSF1588⟩.

**Table 2** ¹H NMR Data for Pyridopyridazines

| Compound | Solvent | Chemical shift (δ-values) | | | | | | | | Coupling constants (Hz) | Ref. |
|---|---|---|---|---|---|---|---|---|---|---|---|
| | | H-1 | H-2 | H-3 | H-4 | H-5 | H-6 | H-7 | H-8 | | |
| (282)ᵃ | CDCl₃ | — | — | 9.30 | 7.80 | 8.05 | 7.50 | — | — | 5.8 (3, 4) 8.5 (5, 6) | 80KGS541 |
| (283) | CDCl₃ | — | 9.31 | 7.88 | 8.40 | 9.63 | — | — | 9.79 | 4.5 (2, 3) 8.5 (3, 4) 1.7 (2, 4) 0.75 (4, 8) 1.5 (5, 8) | 69BSF2519 |
| | DMSO-$d_6$ | — | 6.75 | 5.45 | 6.06 | 7.20 | — | — | 7.20 | 4.5 (2, 3) 8.5 (3, 4) 1.7 (2, 4) | 69BSF2519 |
| (284) | CDCl₃ | — | — | 9.50 | 8.05 | — | 9.13 | 7.73 | 8.78 | 6.0 (3, 4) 4.0 (6, 7) 9.0 (7, 8) 1.0 (4, 8) 2.0 (6, 8) | 80KGS541 |
| | CDCl₃ | — | — | 9.55 | 8.15 | — | 9.2 | 7.8 | 8.85 | 6.0 (3, 4) 4.0 (6, 7) 8.0 (7, 8) 1.0 (4, 8) 2.0 (6, 8) | 78TL2731, 79T2027 |
| (285) | CDCl₃ | — | — | 9.35 | 7.80 | 7.60 | 8.70 | — | 9.85 | 5.8 (3, 4) 6.0 (5, 6) | 78TL2731, 79T2027 |
| (286) | CDCl₃ | 9.73 | — | — | 9.73 | 9.54 | — | 9.10 | 7.89 | 5.5 (7, 8) 0.75 (4, 8) 0.5 (5, 8) 0.5 (1, 5) | 69BSF2519, 77BSF(2)665 |
| (287)ᵇ | CDCl₃ | — | — | — | — | — | — | — | — | — | 79T2027 |

ᵃ 7-Methyl compounds: δ (7-Me) 2.85.

ᵇ Only the 1,4-dihydro-7-methyl compound was recorded: δ 2.4 (3H, s), 3.3 (2H, d, $J$ 3 Hz, H-4), 6.4 (1H, s, H-8), 6.8 (1H, t, $J$ 3 Hz, H-3), 8.0 (1H, s, H-5) and 8.4 (1H, br, exch. $D_2O$, NH).

(303)                    (304)

### 2.15.7.4 UV Spectroscopy

The UV spectra of the [3,2-c] and [3,4-c] parent compounds, and that of the dihydro[4,3-c] derivative (302), have been measured ⟨79T2027⟩, as has that of the parent [2,3-d] analogue in a study of its possible covalent hydration (see Section 2.15.8.2) ⟨68AJC1291⟩. The UV spectra of the [2,3-d]-5,8-diones have also been studied ⟨30BSF630, 33BSF151⟩.

MO calculations of electronic spectra have been performed and compared with observed values in the [2,3-d] and [3,4-d] systems ⟨69MI21502⟩. Good agreement was observed, as it was in a wider study of such calculations in a large series of azanaphthalenes, including several pyridopyridazines ⟨66JSP(19)25⟩.

The UV spectra of pyrido[2,3-d]pyridazine-1,4-diones have been recorded ⟨68MI21501⟩, whilst several of the latter were also used in chemiluminescence studies. Quinolinic and cinchomeronic hydrazides showed no chemiluminescence ⟨60NKK173, 37JPR(148)135⟩, but it was observed in the 8-hydroxy derivative (305) ⟨72YZ703⟩.

(305)

Amongst miscellaneous UV spectra recorded are those of alkylthio adducts of pyrido[3,2-c]pyridazine ⟨78KGS1272⟩, various pyrido[3,4-d]pyridazines ⟨57AC(R)728⟩ and several vinyl-pyridine–azodicarboxylic ester adducts ⟨79T2027, 79KGS639⟩.

In the benzo fused systems, tautomerism in pyridazino[4,5-b]quinolinediones has been elucidated with the help of UV methods ⟨80CPB3457⟩.

### 2.15.7.5 MO Calculations

Several interesting MO calculations and correlations have been carried out on pyridopyridazine derivatives.

Apart from the general study by Wait and Wesley on azanaphthalenes, including most of the pyridopyridazines ⟨66JSP(19)25⟩, more specialized studies have attempted to relate charge density calculations to various chemical and physical properties, with greater success in the latter case.

Comparison of the lowest electronic transition with observed values gave good agreement in all systems ⟨66JSP(19)25⟩, especially in the [2,3-d] and [3,4-d] series ⟨69MI21502, 75MI21500⟩, whilst good agreements with chemical shifts ⟨75MI21500⟩ and bond order values ⟨80KGS541⟩ from NMR observations have already been referred to. Calculation of polarographic reduction potentials also agreed well with observations ⟨78MI21502⟩.

On the chemical reactivity side, however, although reasonable agreement was observed in nucleophilic substitutions in [3,2-c] ⟨78MI21502, 79KGS403⟩ and [4,3-c] ⟨78MI21502⟩ derivatives (minor deviations being ascribed to steric hindrance to *peri* H-bonding), and in calculations of chlorine substituent reactivity in pyrido[2,3-d]pyridazines ⟨65CPB586⟩ (less so in the [3,4-d] series ⟨73HC(27)968 p. 983⟩), other comparisons were less successful. Discrepancies were observed, particularly in the pyrido[2,3-d]pyridazine series, in the case of N-oxide formation ⟨68CR(C)(266)1459, 69AJC1745⟩, N-alkylation ⟨77BSF919⟩, nucleophilic attack by reducing agents ⟨77BSF919⟩ or organometallics ⟨68CR(C)(266)1459⟩ (Section 2.15.8.3) and alkaline deuterium exchange studies ⟨77BSF919⟩.

In an interesting application, the charge density at the ring junction in [2,3-d] and [3,4-d] derivatives was related to the diuretic activity of the compounds ⟨79MI21501⟩.

Charge density values calculated by Castle for all the pyridopyridazine parent systems are tabulated in ⟨73HC(27)968 p. 989⟩.

### 2.15.7.6 Mass Spectrometry

The mass spectra of the parent pyridopyridazines have been recorded in the case of the [2,3-*d*] series (when the fragmentation proceeds by loss of $N_2$ and stepwise loss of HCN) ⟨67AJC2677, 74OMS(8)31⟩, in the [3,2-*c*] series (loss of $N_2$) ⟨79KGS1124, 79T2027⟩ and the [3,4-*d*] series ⟨74OMS(8)31⟩. Various derivatives have also been studied in the [3,2-*c*] ⟨79KGS1124, 79T2027⟩, [3,4-*c*] and [4,3-*c*] series ⟨79T2027⟩.

Structural problems in the pyrido[3,4-*c*]pyridazine series ⟨76CPB1870⟩ and in the fused pyridazino[4,5-*b*]quinoline series ⟨80CPB3457⟩ have also invoked mass spectral determinations.

### 2.15.7.7 Tautomerism

There seems to have been no systematic study of tautomerism in the pyridopyridazines, but isolated observations in the pyrido[3,4-*d*]pyridazinedione ⟨75BSF702, 69CPB2266⟩ and pyrido[2,3-*d*]pyridazinedione ⟨74JHC351⟩ series have involved methylation studies. The pyrido[2,3-*d*]pyridazine-5,8-diones are believed to be enolized at the 8-position, from metal complexation results ⟨67MI21500⟩.

## 2.15.8 REACTIVITY AT RING ATOMS OF PYRIDOPYRIDAZINES

### 2.15.8.1 Electrophilic Attack at Ring Carbon

Again, as with pyridopyrimidines, the main reaction is oxidation of di- or poly-hydro derivatives to fully aromatic structures, often merely by air or oxygen. In some cases the reagent of choice is mercury(II) oxide, whilst other reagents used include sulfur, bromine, chloranil, chromium trioxide–acetic acid, hydrogen peroxide, and potassium ferricyanide, which also caused oxidative removal of a benzyl group in the transformation (**306**) → (**307**) ⟨75CPB2239⟩.

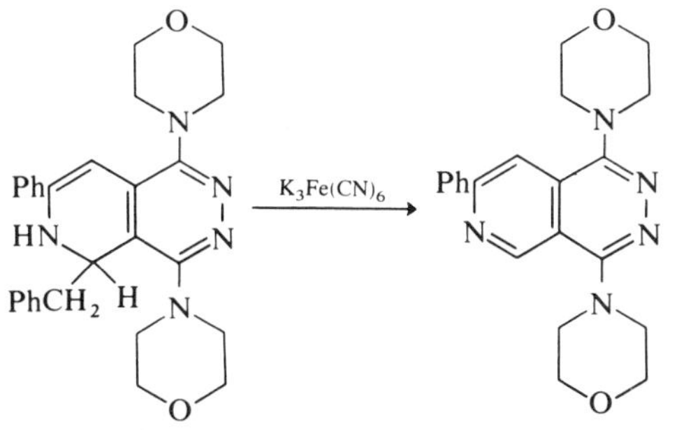

(**306**)                    (**307**)

True electrophilic substitution is very difficult in pyridopyridazines. For example, the [3,4-*d*] parent (**286**) is inert to hot 65% oleum ⟨68AJC1291⟩, and although formation of a 3-bromo derivative (**308**) was reported in the [2,3-*d*] series, it seems to have arisen by an addition–elimination reaction *via* the dibromide (**309**) ⟨69AJC1745⟩. Attempted chlorination led to ring opening. A similar effect was observed in the [3,4-*d*] system, where an 8-bromo derivative was obtained ⟨77BSF665⟩, and in *N*-oxides of the pyrido[2,3-*c*]pyridazine and fused pyridazino[3,4-*c*]isoquinoline series ⟨72JHC351⟩. The formation of (**311**) from (**310**)

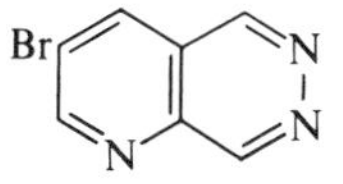

(**308**) 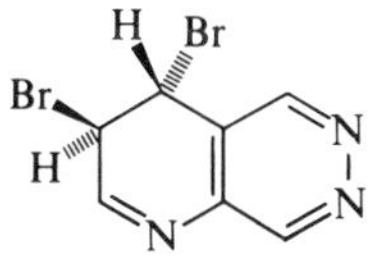            (**309**) 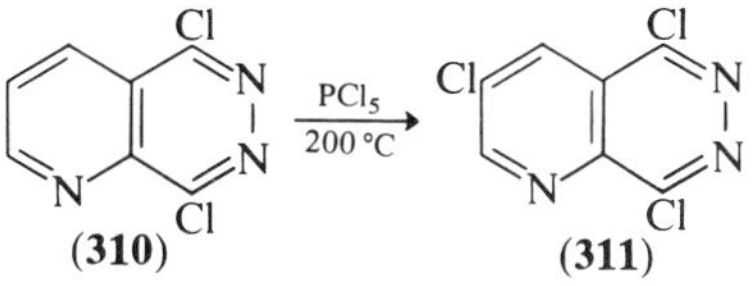            (**310**)            (**311**)

on strong heating with phosphorus pentachloride, parallelled in the [3,4-*d*] series ⟨72MI21501⟩, may have involved a similar mechanism.

Oxidative cleavage in alkaline conditions of one of the rings of pyridopyridazines has been observed to give pyridazinedicarboxylic acids, with either these acids or pyridinedicarboxylic acids being formed in acid conditions, *e.g.* on attempted nitration ⟨30BSF630, 69AJC1745⟩.

### 2.15.8.2 Electrophilic Attack at Ring Nitrogen

Protonation of pyrido[2,3-*d*]pyridazine (p$K_a$ 2.01) is believed to occur at position 1 as predicted by MO calculations ⟨68AJC1291⟩, although alkylation results (see below) cast doubt on this. Covalent hydration has not been observed in this series, nor in the slightly weaker [3,4-*d*] base (p$K_a$ 1.76), which probably protonates at N-6, though this is not certain.

Alkylation was shown for pyrido[2,3-*d*]pyridazine to give a mixture of 6- and 7-quaternary salts, whose structure was proved by reduction and subsequent oxidation to known alkylated -ones ⟨77BSF919⟩. The site of quaternization of the [3,4-*d*] analogue was at first unknown ⟨68AJC1291⟩, but was later shown to be at both the 2- and 3-positions ⟨72CR(C)(275)1383⟩. The site of alkylation of fused derivatives in pyridazino-[3,4-*c*]- ⟨59JCS6⟩ and -[4,5-*c*]-isoquinolines ⟨59JCS1⟩ and pyridazino[3,4-*c*]quinolines ⟨62JCS1671⟩ has been studied.

*N*-Alkylation appears to occur with -ones and -diones even with diazomethane (but *cf.* ⟨73GEP2322073⟩), and *N*-arylation with a reactive aromatic halide. The Mannich reaction has also been observed ⟨69MI21503⟩.

Acylation of -ones and -diones appears to occur mainly at oxygen, but in the fused pyridazino[4,5-*b*]quinoline (312) the *O*-acyl derivative (313) was formed *via* an *N,O*-diacyl derivative ⟨72BSF1588⟩. Reduced derivatives, however, are acylated at nitrogen.

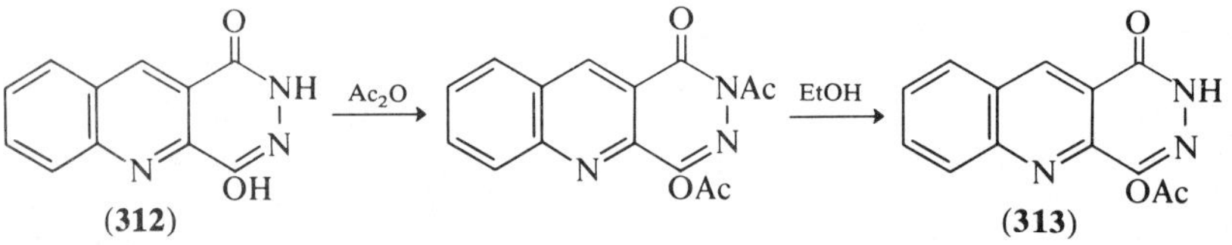

(312)   (313)

*N*-Oxide formation in the [2,3-*d*] series also occurs at the 6,7-positions ⟨69BSF2519, 69AJC1745⟩, contrary to theory, but the dialkoxy-[3,4-*d*] derivatives give the 6-*N*-oxides (314) ⟨77JAP(K)7705798⟩, whilst both mono- and di-*N*-oxides were obtained with the 5-alkoxy[2,3-*d*] compounds ⟨75H(3)381⟩ and with the fused pyridazino[3,4-*c*]quinoline (315) ⟨66JCS(C)2053⟩. Lead tetraacetate oxidation at nitrogen of the [2,3-*d*]-dione led to the quinone (316), which underwent [4 + 2] cycloaddition with butadiene (to give 317) and with a variety of other dienes ⟨70MI21501, 73JCS(P1)26⟩. Similar results were observed in the pyrido[3,4-*d*]pyridazine series ⟨73JCS(P1)26⟩.

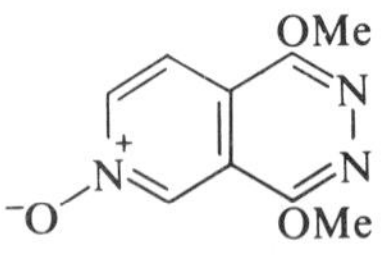

(314) 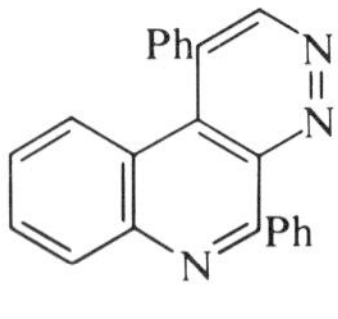   (315)

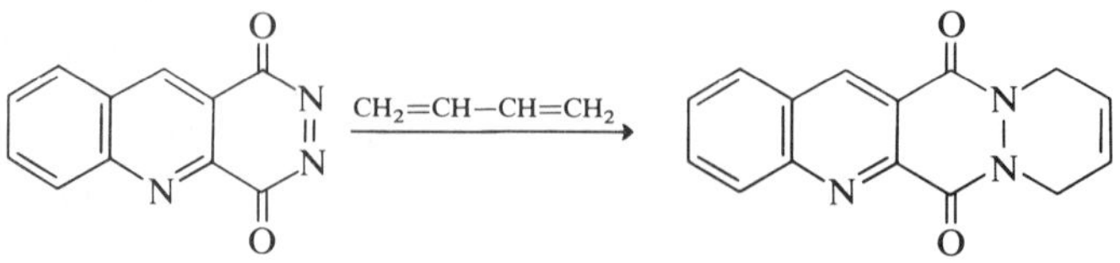

(316)   (317)

Complexation with metals has been observed with a variety of pyridopyridazinones, whilst electrophilic attack at nitrogen is involved in cyclizations to a variety of azolo and azino fused tricyclic systems, *e.g.* ⟨65CPB586, 71JOC3812⟩.

### 2.15.8.3 Nucleophilic Attack at Ring Carbon

Ring reductions in simple pyridopyridazines seem to occur mainly in the pyridazine ring, although there is one report of an unexpected catalytic hydrogenation of the pyridine ring in a [2,3-*d*]-one ⟨65CPB586⟩. Lithium aluminum hydride gives a mixture of 5,6- and 7,8-dihydro products from the [2,3-*d*] parent, instead of the 1,2-derivative predicted by MO theory ⟨77BSF919⟩, and the corresponding 4-one gives similar products ⟨67BCJ153⟩. Corresponding results with lithium aluminum hydride and methyllithium were obtained with pyrido[3,4-*d*]pyridazine ⟨77BSF665⟩, but catalytic reduction of the 5-benzyl-1,4-dimorpholino derivatives (**318**) led to the 5,6-dihydro products ⟨75CPB2239⟩, and additions of Grignard and lithium reagents also took place at the 5,6-positions ⟨75CPB2239⟩. Reduction of the [2,3-*d*] parent with zinc amalgam/acetic acid caused ring contraction to 4-azaindole ⟨75KGS1431⟩, and a similar contraction was noted in fused system (**294**) ⟨66JCS2053⟩.

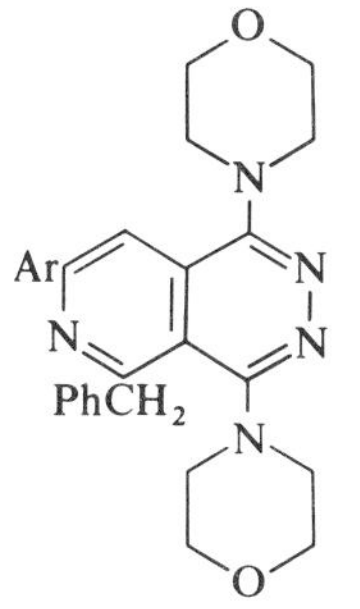

(**318**)

The 6- and 7-quaternary salts of pyrido[2,3-*d*]pyridazine are reduced by borohydride with subsequent air oxidation to 5,6- and 7,8-dihydro oxo derivatives respectively ⟨77BSF919⟩, whilst the [3,4-*d*] analogues give the corresponding 1,2- and 3,4-dihydro oxo compounds ⟨72CR(C)(275)1383⟩.

In the [2,3-*d*] system, deuterium exchange occurs very readily in acidic, neutral or dilute alkaline $D_2O$ solution in the pyridazine ring, in strong alkali in the 3- and 4-pyridine positions, and not at all in the 2-pyridine position, giving the order 8>5≫4>3 ⟨77BSF919⟩. Similarly, in the [3,4-*d*] series the pyridazine protons exchange first, giving the order 1,4≫8>5 ⟨77BSF665⟩. These results are in broad agreement with the predictions of Zoltiewicz ⟨*cf.* 69JA5501⟩ for isolated pyridine and pyridazine systems.

Direct nucleophilic addition to the ring system in the pyrido[3,2-*c*]pyridazine system occurs with soft bases in the 4-position as predicted by MO theory (see Section 2.15.7.2.5), and is observed with amines ⟨78KGS809⟩, hydrazine ⟨76KGS976⟩, thiols ⟨78KGS1272⟩ and indole ⟨78KGS1555⟩, the dihydro adducts undergoing aerial oxidation to the observed products. With hard bases, dimerization occurs *via* the anion radical ⟨79MI21504⟩. Secondary attack by reduced reagent to give (**320**) during preparation of the triazoledione adduct (**319**) (see Section 2.15.10.6.1) was reported ⟨76KGS702⟩, but this work could not be repeated ⟨79T2027⟩.

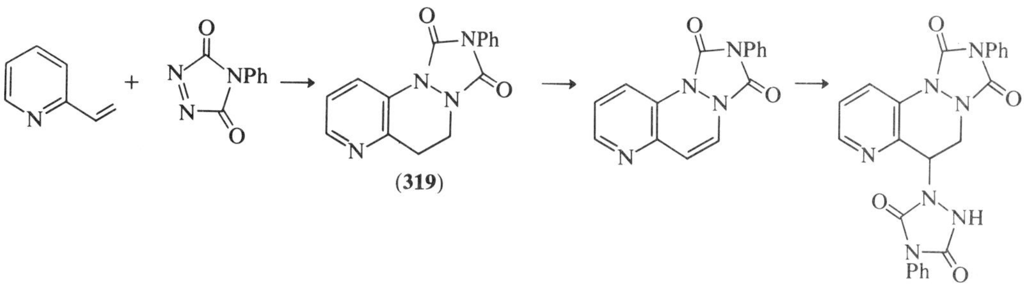

(**319**)                    (**320**)

The $E_{1/2}$ values for one-electron reductions in this series have been measured and compared with theory ⟨78MI21502⟩.

Finally, the [2,3-*d*] derivative (rather surprisingly in view of the above) is reported to undergo Tschitschibabin-type amination at the 2-position of the pyridine ring ⟨69AJC1745⟩.

Ring-opening reactions are best known in the pyridazine ring. For example, the *N*-aryl cyclic hydrazides (**321**) undergo ring opening with alkali to give the *N*-aminoimides (**322**), which recyclize in strong acid to the dione ⟨67MI21500⟩. A related reaction gives the N-benzylideneaminoimide (323) from reaction of the pyridazino[4,5-b]quinolinedione with aldehydes ⟨72BSF1588⟩. Ring opening with hydrazine can result in an apparent dearylation (324 → 325) ⟨57AC(R)728⟩, whilst the ester (326) gives the aldehyde hydrazone (327) on attempted alkaline hydrolysis ⟨66CPB1010⟩.

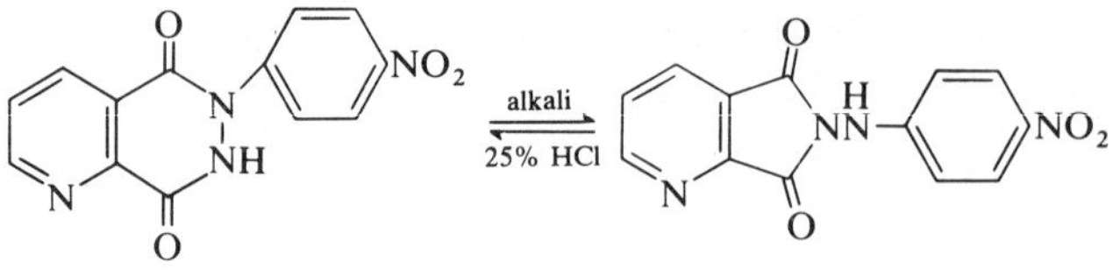

(321)                                    (322)

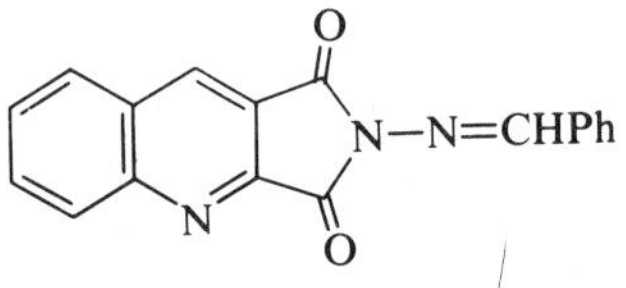

(323) 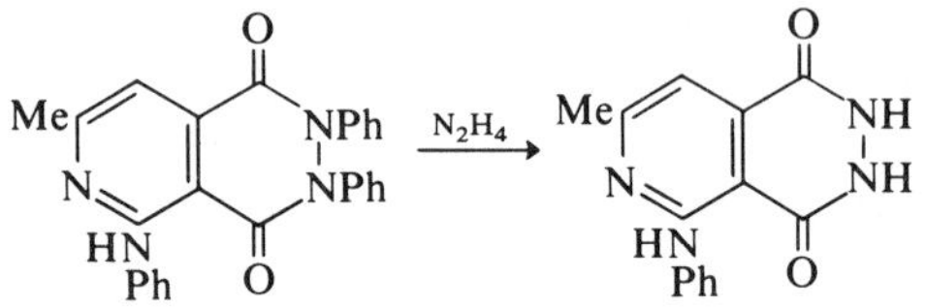                          (324)                  (325)

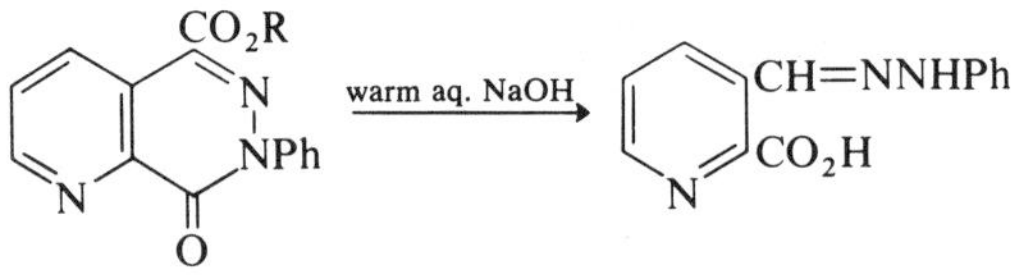

(326)              (327)

Eliminative ring-openings are illustrated by the loss of nitrogen from the pyridocinnolines (328) to give the azabiphenylene (329) ⟨75TL569⟩, and by the dry distillation of the fused betaine (330) with soda lime to give 3-amino-2-phenylquinoline (331) in studies of methylation regiospecificity ⟨62JCS1671⟩.

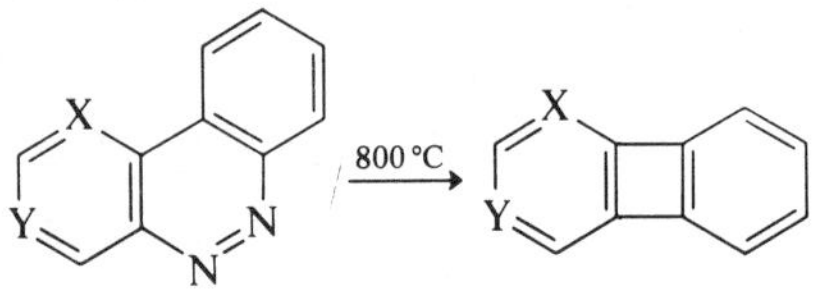

(328) a; X = N, Y = CH          (329)              (330)              (331)
      b; X = CH, Y = N

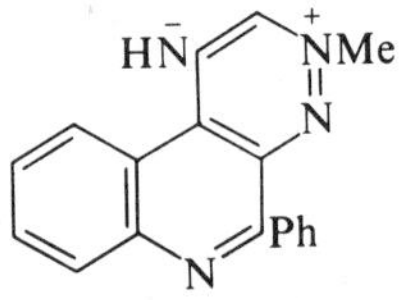

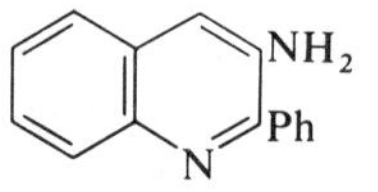

## 2.15.9 REACTIVITY OF SUBSTITUENTS IN PYRIDOPYRIDAZINES

### 2.15.9.1 Electrophilic Attack on Substituents

The only example of attack on active methyl groups in pyridopyridazine occurs in the conversion of 4-methylpyrido[3,4-c]pyridazine to the styryl derivative with benzaldehyde/zinc chloride. The styryl group was then oxidized further with permanganate to give the carboxylic acid ⟨66JCS(C)2053⟩. The fused benzene ring of pyridazino[4,5-b]quinoline was also oxidized in the same way to give the pyrido[2,3-d]pyridazine-2,3-dicarboxylic acid ⟨71BSF906⟩, but no reports of benzenoid substitution have been noted.

Acetyl ⟨79GEP2808070, 71FES1074⟩ and other groups on the pyridine ring have been oxidized to the corresponding carboxylic acids in the pyrido-[2,3-c]- and -[3,4-d]-pyridazine fields. Acid groups in pyridopyridazines behave normally on esterification ⟨66CPB1010⟩.

Amino substituents also behave normally with electrophiles, being acylated and converted to benzylidene derivatives. N-Amino groups have also been converted to benzylidene derivatives and are removed on treatment with N-nitrosodiphenylamine ⟨79JHC249⟩.

Hydrazino groups are readily acylated ⟨64MI21500, 57GEP958561⟩ or converted to hydrazones, often in the preparation of pro-drugs. On oxidation with mercury(II) oxide they are removed to give the parent ring systems ⟨69AJC1759, 68AJC1291⟩, whilst on thermal decomposition they may give the corresponding amino derivatives ⟨81ACH(108)167⟩. Their ready cyclization to azolo and azino fused pyridopyridazines has already been referred to (Section 2.15.8.2).

Removal of *N*-oxide groups by PCl₃ follows normal behaviour ⟨63JCS6073⟩, but with acetic anhydride the *N*-oxides (**332**) underwent a complex ring-opening reaction leading to (**333**), and an isomeric 8-alkoxy-6-oxide behaved similarly ⟨75H(3)381⟩.

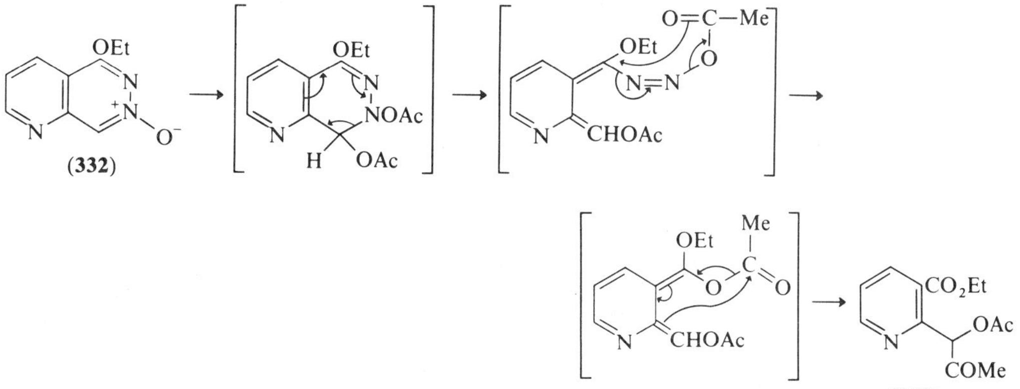

Alkylation of thioxo groups to give alkylthio derivatives occurs normally ⟨65CPB586⟩.

### 2.15.9.2 Nucleophilic Attack at Substituents

Hydroxyalkyl groups attached to pyridopyridazines have been converted to haloalkyl groups by standard methods and the products reacted with amines. Ketones have been converted to hydrazones and oximes and the resulting derivatives deprotected during syntheses.

Carboxylic esters have been converted to amides, saponified and decarboxylated in the usual way, but side reactions involving ring opening have been observed ⟨79T2027, 66CPB1010⟩. In the case of *N*-alkoxycarbonyl protecting groups from azodicarboxylate syntheses (Section 2.15.10.6.1), use of the *t*-butoxycarbonyl protecting group, removed under mild conditions with TFA, was preferred ⟨79T2027⟩.

Amino groups are formed by reduction of nitro groups in aryl substituents and behave normally ⟨62JCS1671⟩, but when attached directly to the pyridopyridazine ring they may be removed by acid hydrolysis or treatment with nitrous acid, or replaced by hydrazine.

As in the pyridopyrimidines, the MacFadyen–Stevens degradation of tosylhydrazino derivatives has been used to prepare the parent [2,3-*d*] compound ⟨63JCS6073⟩, and 1,4-bishydrazinopyrido[3,4-*d*] pyridazines are reduced to the 1,4-diamino derivatives by Raney nickel ⟨68AJC1291⟩.

### 2.15.9.3 Nucleophilic Substitution Reactions

As in the pyridopyrimidines, selective nucleophilic substitution reactions at reactive ring positions have been a fruitful source of pyridopyridazines.

The pyridopyridazinones, and especially the cyclic hydrazide-diones, have been converted to the corresponding chloro derivatives using phosphorus oxychloride or phosphorus oxychloride/phosphorus pentachloride ⟨63JCS6073⟩ (also with a pyridazino[3,4-*c*]quinoline). Both of these reagents often give relatively poor yields. Better results are obtained with phosphorus oxychloride in the presence of tertiary amines, such as a dialkylaniline (but *cf.* ⟨69CPB2266⟩) or, best, pyridine or quinoline ⟨75CPB2239, 69CPB2266⟩. *N*-Substituted diones often give only monochlorination ⟨66CPB1010⟩.

Selective substitutions have been most studied in the pyrido[3,4-*d*]pyridazine field. The 1,4-dichloro derivative (**334**) was more reactive than 1,4-dichlorophthalazine, and was expected from MO calculations to show a more reactive 4-chlorine group. In practice, however, almost equal amounts of monosubstitution products were observed on hydrolysis, although slight differences in the use of alkaline and acid conditions, or on hydrazination, have been reported ⟨69CPB2266⟩. Reaction of this 1,4-dichloro compound, or substituted derivatives, with amines under mild conditions did give the 1-halo-4-amino product ⟨69CBP2266, 75CPB2239⟩, but the diamino derivative was formed under more forcing conditions ⟨76CPB2699⟩, and similar results were obtained on methoxylation ⟨69CPB2266, 75BSF702⟩. After replacement of one chlorine, the remaining one is less reactive ⟨75BSF702⟩. With sulfur nucleophiles, however, only disubstitution was observed ⟨75CPB2239⟩.

In the 7-phenyl-1,4,5-trichloro compound (335), however, the 5-chloro was the most reactive with amines, followed by the 1-chloro and finally the 4-chloro ⟨75CPB2306⟩.

(334)        (335)        (336)

In the [2,3-*d*] series, a 5-chloro is predicted to be more reactive than an 8-chloro, and in studies of the reaction of 5,8-dichloropyrido[2,3-*d*]pyridazine (336) with alkoxides, phenoxides ⟨65CPB586⟩ or amines ⟨65CPB586, 68JHC13⟩ this proved to be the case, although an exception was noted in the latter case when aniline was the nucleophile, the 8-derivative being formed in greater amount ⟨65CPB586⟩. The results obtained by the Japanese workers on monohydrolysis in acid and alkaline conditions ⟨65CPB586⟩ have recently been reinvestigated and corrected ⟨81ACH(108)167⟩, with the 8-chloro being somewhat more reactive to alkali.

In forcing conditions with excess of reagents the 5,8-bis derivative was obtained in the above cases, with hydrazine and with sulfur nucleophiles. Other authors have also observed selective reactions in the pyrido[2,3-*d*]pyridazine series, *e.g.* ⟨69AJC1759, 67JOC1139⟩.

Many other examples are known of non-selective reactions of halo groups in pyridopyridazines with amines, alkoxides, sulfur nucleophiles such as hydrosulfide and thiolate ions, or thiourea, hydrazine(s), cyanide ion and dimethyl sulfoxide, or on catalytic reduction.

In addition to conversion to chloro compounds, -ones and diones have given thiones with phosphorus pentasulfide, whilst in some cases replacement by amine of alkoxy groups in preference to a chloro group has been observed ⟨68JHC13⟩.

Thione or alkylthio groups have also been involved in nucleophilic substitutions with hydrazine, or amines, and by desulfurization using Raney nickel or aluminum amalgam.

## 2.15.10 SYNTHESIS OF PYRIDOPYRIDAZINES

### 2.15.10.1 By Formation of One Bond between Two Heteroatoms [6+0 (N)]

Surprisingly, no fully detailed syntheses of pyridopyridazines by joining of two heteroatoms have been recorded, although a recent patent claimed the preparation of pyrido-[2,3-*c*]-, -[3,2-*c*]-, -[3,4-*c*]- and -[4,3-*c*]cinnolines by reduction of 2,2'-dinitro substituted phenylpyridines with a variety of reagents ⟨80GEP2939259⟩.

### 2.15.10.2 By Formation of One Bond adjacent to a Heteroatom [6+0 (α)]

#### 2.15.10.2.1 From pyridine intermediates

The most common reaction of this type is the cyclization of various derivatives of hydrazine and substituted hydrazines with pyridine *o*-dicarboxylic acids and related compounds. Reactions in which the acid derivative reacts directly with the hydrazine are dealt with as [4+2] reactions in Section 2.15.10.6.1.

Early experiments with pyridine-2,3-dicarboxylic mono- and bis-phenylhydrazides were unsuccessful ⟨32JIC145⟩, but later these were cyclized in acetic acid ⟨66CPB1010⟩ to give only the 7-phenylpyrido[2,3-*d*]pyridazine-5,8-dione.

The 3,4-pyridinedicarboxylic phenylhydrazide (337), however, did cyclize in one of the earliest pyrido[3,4-*d*]pyridazine syntheses ⟨1890M(11)133⟩, as did the corresponding *N*-unsubstituted analogue at a higher temperature ⟨12M(33)393, 66CPB1010⟩.

(337) 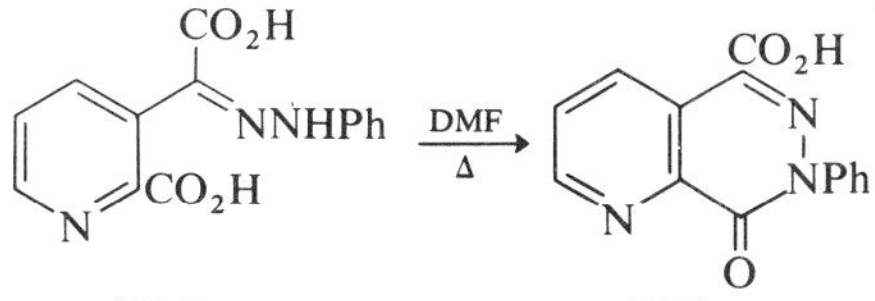　　　　　　(338)　　　　　　(339)

The hydrazones from *o*-aldehydopyridine acids ⟨66CPB1010, 56G990⟩ or nitriles ⟨80USP4223142⟩ cyclize to give 8-ones or 5-amino derivatives of the pyrido[2,3-*d*]pyridazine series, whilst the keto acid hydrazone (338) gave the 5-carboxylic acid (339) ⟨66CPB1010, 67BCJ153⟩, and a related synthesis from *o*-ethoxycarbonylamidrazones gives 5(8)-oxo-8(5)-amino derivatives ⟨81ACH(108)167⟩.

Cyclic imides ⟨75CPB2306⟩ and their *N*-alkyl ⟨59AC(R)944, 81ACH(108)167⟩ and *N*-phenyl ⟨*e.g.* 72JA8451, 76CPB2699⟩ analogues give pyrido[3,4-*d*]pyridazinones on heating with hydrazine, and similar intermediates must be involved when pyridinedicarboxylic acid anhydrides are used ⟨*e.g.* 75CPB2239, 75BSF702, 33BSF151⟩. These may include the *N*-aminoimides, *e.g.* (322), which are often in equilibrium with the pyridopyridazinones (see Section 2.15.8.3), or are readily isomerized to them ⟨*e.g.* 60G1399⟩.

The bromolactone (340) gives a pyrido[2,3-*d*]pyridazine on heating with hydrazine ⟨67BCJ153⟩, as does the lactone (341) ⟨37CB(B)2018⟩. The piperidine lactone (342) is similarly converted to a reduced pyrido[4,3-*c*]pyridazine (343) ⟨67CR(C)(264)405⟩, whilst the piperidine enamine (344) gives a similar product (345) with a haloketone semicarbazone ⟨73GEP2302383⟩. A reduced pyridopyridazine (347) is also formed by reaction of ethyl 2-bromomethylnicotinate with hydrazine, *via* the *N*-aminopyrrolidinopyridine (346) ⟨69BCJ2996⟩.

An important early method simulated the well-known Widman–Stoermer cinnoline synthesis. 3-Aminopyridine-2- or -4-alkenes such as (348) gave pyrido-[3,2-*c*]- or -[3,4-*c*]-pyridazines on diazotization and alkaline cyclization ⟨66JCS(C)2053⟩.

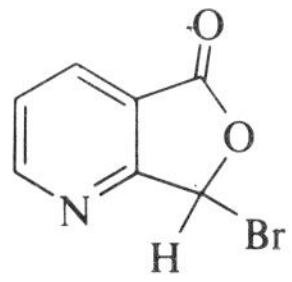

(340) 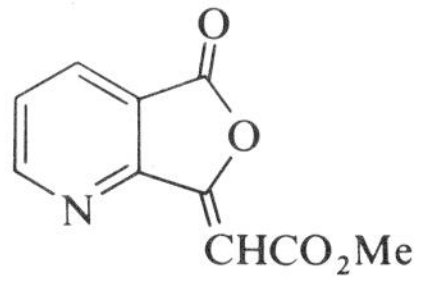　　　　(341) 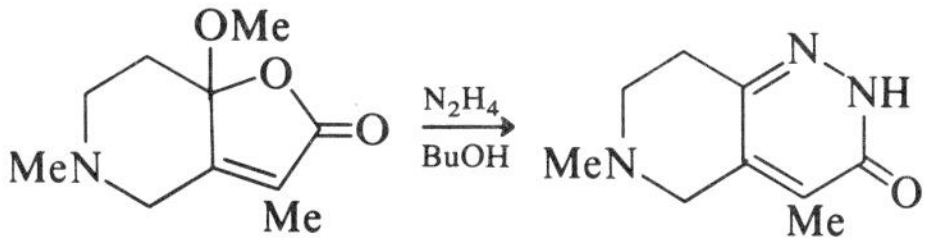　　　　(342)　　　　(343)

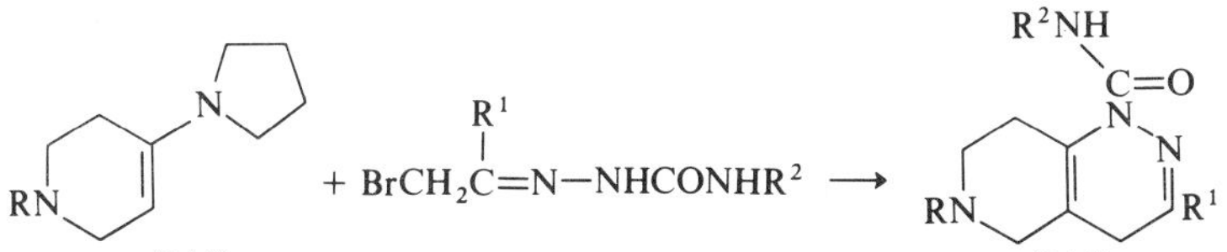

(344)　　　+ BrCH₂C=N—NHCONHR²　⟶　(345)

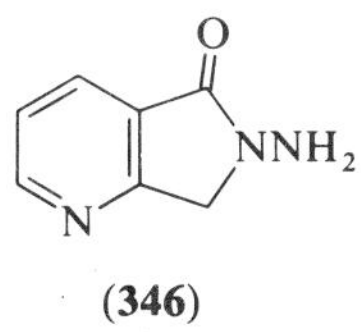

(346) 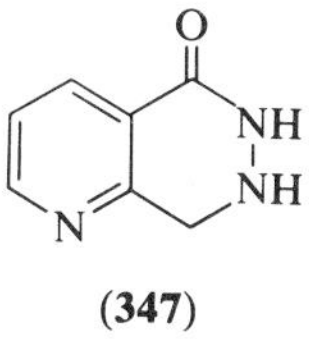　　　　(347) 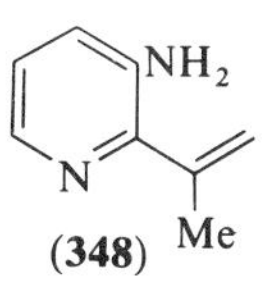　　　　(348)

### 2.15.10.2.2 From pyridazine intermediates

Few reactions of this type are recorded. The azidopyridazinone ester (349) on reduction with triethyl phosphite, hydrazine or borohydride furnished the pyrido[2,3-*c*]pyridazin-7-one (350) ⟨79JHC1559⟩, whilst an *N*-aminopyrido[2,3-*c*]pyridazine (352) resulted from the

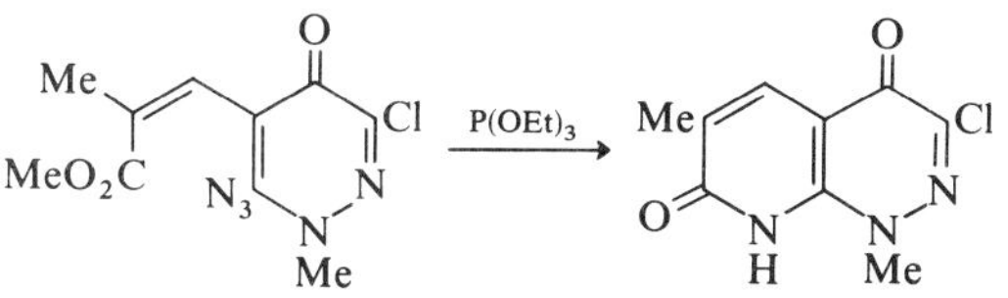

(349)　　　　　　(350)

action of hydrazine on the acrylic ester (**351**). This ester was derived from a pyrano[2,3-*d*]pyridazinone obtained by KCN/DMSO ring expansion of a furopyridazine ⟨79JHC249⟩.

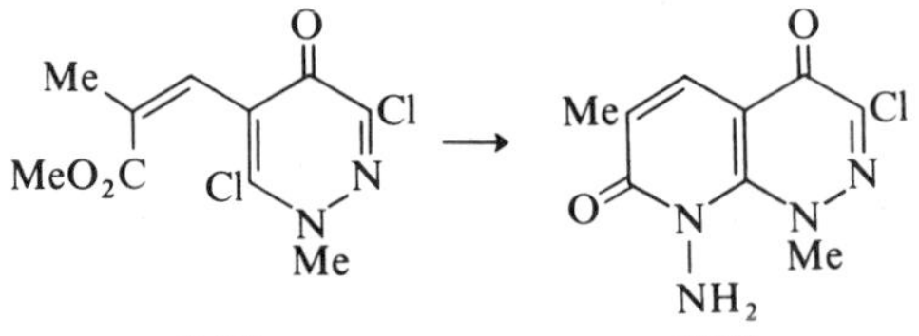

(**351**)                                    (**352**)

### 2.15.10.2.3 *Fused derivatives*

Syntheses of fused derivatives from quinoline intermediates include analogues of the Widman–Stoermer ⟨66JCS(C)2053⟩ and Borsche and Herbert ⟨75JOC3874, 57JCS3722⟩ (**353** → **354**) cinnoline syntheses in the pyridazino[3,4-*c*]quinoline series, whilst in the same ring system the 1,2,4,5-tetrazinedicarboxylic ester adduct of (**355**) gave (**356**) with loss of nitrogen and subsequent [6 + 0 (α)] cyclization ⟨76AP679⟩.

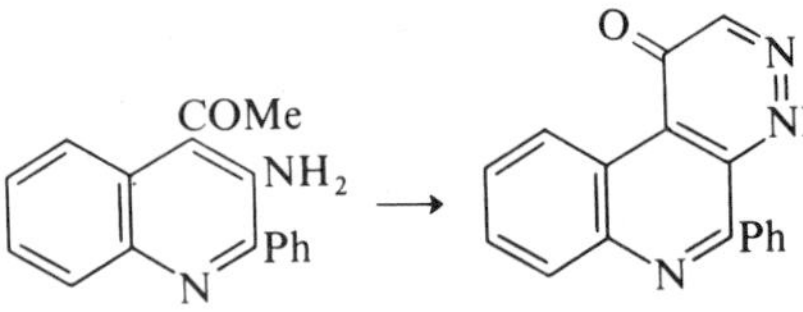

(**353**)              (**354**) 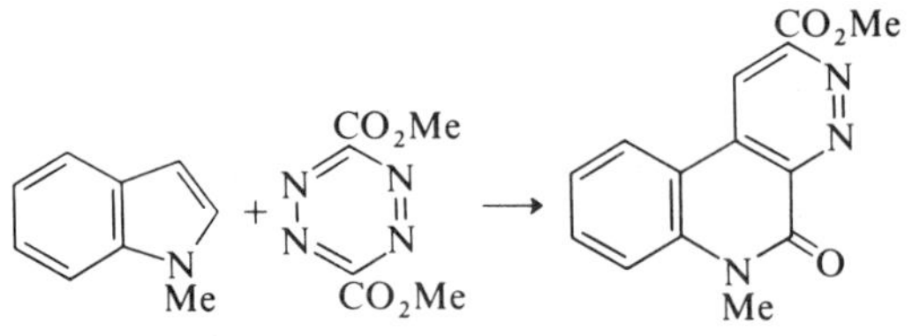                (**355**)                  (**356**)

A related diazotization reaction involving the supposed *peri*-aminolepidine (**357**) to give the unknown phenalene type system (**297**) was later shown to be incorrect due to wrong orientation of the amino substituent ⟨42JA2417⟩.

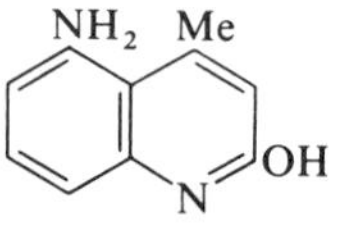

(**357**) 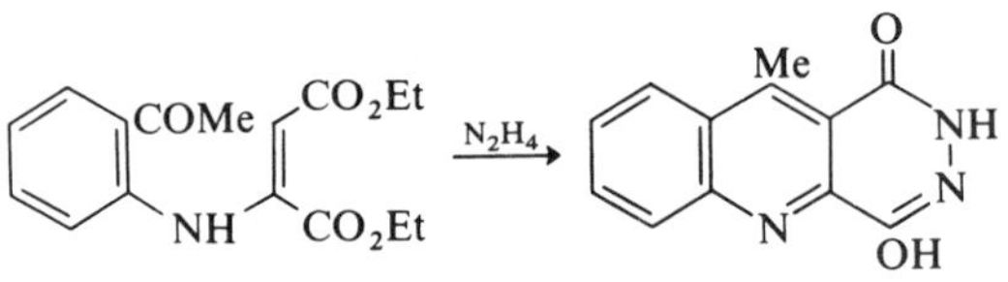                (**358**)                  (**359**) 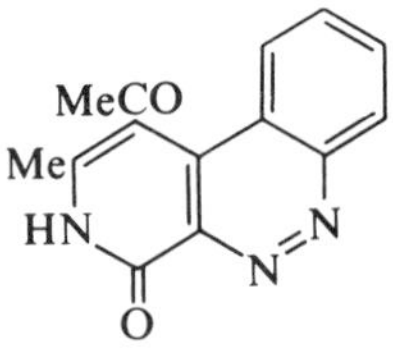                  (**360**)

Reactions involving quinoline hydrazide derivatives have been noted in the pyridazino-[4,3-*c*]- ⟨64MI21500⟩, -[4,5-*b*]- ⟨31M(58)238⟩ and -[4,5-*c*]-quinoline ⟨71CB3341⟩ series, whilst the double cyclization of (**358**) to the pyridazino[4,5-*b*]quinoline (**359**) ⟨80CPB3457⟩ and related cyclizations in the same series ⟨80H(14)267⟩ are of a basically similar type. A lone cyclization of this type from cinnoline intermediates involves the *o*-acetonylcarboxamide type formation of the pyridine ring to give the pyrido[3,4-*c*]cinnoline (**360**) ⟨76JCS(P1)592⟩.

### 2.15.10.3 By Formation of One Bond β to a Heteroatom [6 + 0 (β)]

#### 2.15.10.3.1 *Fused derivatives*

The only reactions of this type noted in the pyridopyridazine series are two classic Bischler–Napieralski type closures in benzo fused systems in which 3- or 4-benzamidopyridazines are cyclized to an adjacent phenyl group to give a 6-arylpyridazino-[3,4-*c*]- (*e.g.* **361** → **362**) or -[4,5-*c*]- isoquinoline respectively ⟨59JCS1, 59JCS6⟩.

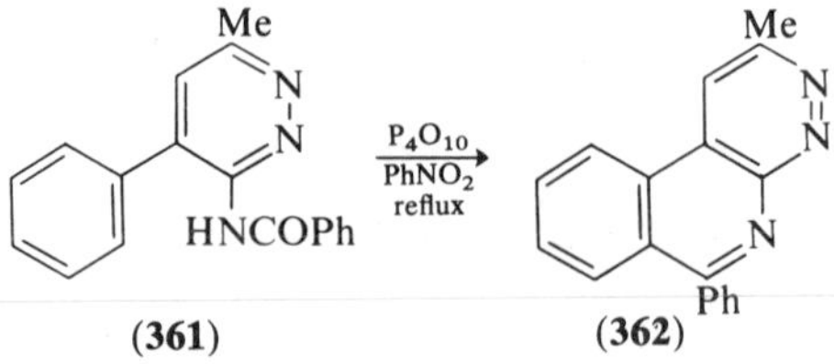

(**361**)                    (**362**)

## 2.15.10.4 By Formation of One Bond γ to a Heteroatom [6+0 (γ)]

### 2.15.10.4.1 From pyridine intermediates

The only example of a simple pyridopyridazine synthesis of this kind from a pyridine intermediate involves a variant of a well known cinnoline synthesis in which the Japp–Klingemann intermediate (**363**) gives the pyrido[3,4-*c*]pyridazine (**364**) with PPA ⟨69JHC977⟩.

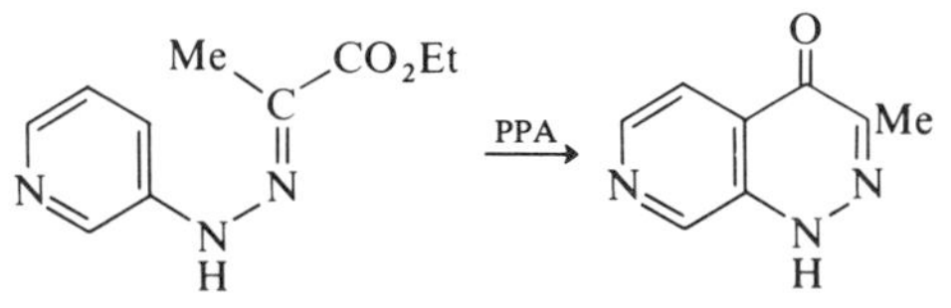

(363)　　　　(364)

### 2.15.10.4.2 From pyridazine intermediates

The only example from a pyridazine intermediate is of the Dieckmann-type used so widely in the pyridopyrimidine field (Section 2.15.5.4.2). The 3-carboxyalkylamino-4-ester (**365**) is cyclized by strong base to give the pyrido[2,3-*c*]pyridazine-6-carboxylate (**366**) ⟨77JAP(K)7733695⟩.

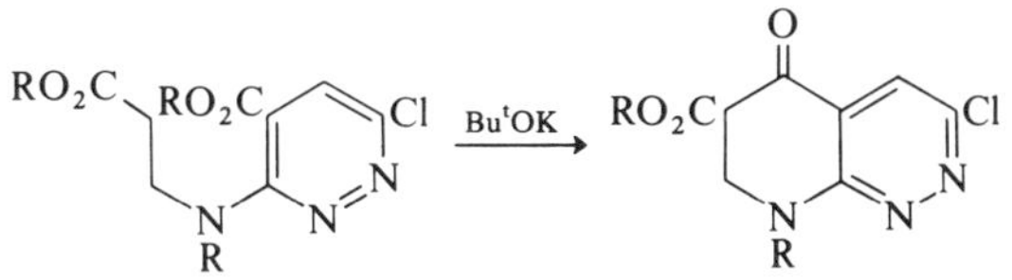

(365)　　　　(366)

### 2.15.10.4.3 Fused derivatives

Apart from an early report by Schofield and Simpson ⟨46JCS472⟩, who obtained various derivatives of the rare pyrido[4,3-*c*]cinnoline system (**300**) during the alkaline degradation of a 4-(2-pyridyl)cinnoline derivative, both the examples in this field are photochemical in nature. Firstly, 3-phenylazocinnoline on irradiation gives a mixture of pyrido-[3,2-*c*]- and -[3,4-*c*]-cinnolines (**328**), used in a synthesis of azabiphenylenes (Section 2.15.8.3) ⟨75TL569⟩, whilst the 2-phenylazo isomer gives the [2,3-*c*] analogue on irradiation in the presence of proton or Lewis acids ⟨77MI21502, 80GEP2939259⟩.

Irradiation of the 3-pyridyltetrazolium salt (**367**) similarly gives a mixture of 5,6-tetrazolo fused derivatives of these systems (**368**), which can be converted to the parent systems by various reducing agents ⟨75TL569, 56CB563⟩. Use of the 4-pyridyl isomer gave the pyrido[4,3-*c*]cinnoline anologues.

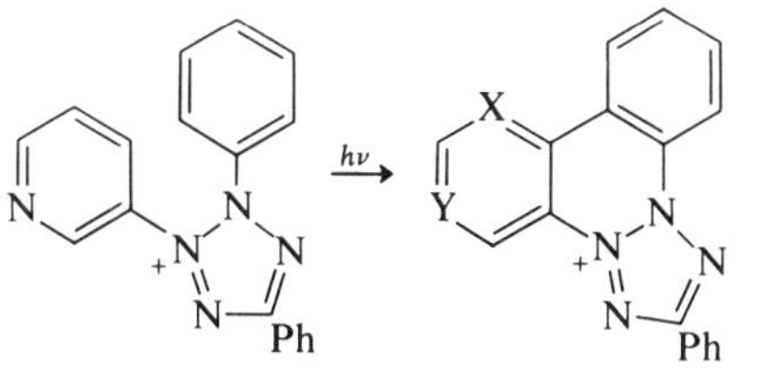

(367)　　　　(368) a; X = N, Y = CH
　　　　　　　　　　b; X = CH, Y = N

## 2.15.10.5 By Formation of Two Bonds from [5+1] Atom Fragments

No examples of this type of synthesis have been noted in the pyridopyridazine series.

**2.15.10.6  By Formation of Two Bonds from [4 + 2] Atom Fragments**

*2.15.10.6.1  From pyridine intermediates*

The vast majority of syntheses of pyrido[2,3-*d*]- and pyrido[3,4-*d*]-pyridazines fall into this category, resulting from the cyclization of various *o*-substituted pyridine derivatives (2,3- or 3,4-, respectively) with hydrazine or its congeners.

2,3-Dialdehydes gave the 5,8-unsubstituted [2,3-*d*] derivatives, including the parent compound ⟨66CR(C)(262)1335, 68AJC1291⟩, whilst other [2,3-*d*] derivatives prepared in this way include 5-, 8- and 5,8-disubstituted derivatives (from 2,3-diacylpyridines ⟨*e.g.* 69BSF4082⟩), 5,8-diones (from diacids ⟨69MI21501⟩ and diesters ⟨56JA159⟩) and 5(8)-ones from aldehydo ⟨56G990, 69BSF3678⟩ and acyl ⟨*e.g.* 56JA159, 66CPB1010, 01M(22)843⟩ acids or esters. Deuterated derivatives have also been prepared in this way, as have the corresponding *N*-oxides ⟨63JCS6073, 69CR(C)(268)1531⟩. The use of a cyanoaldehyde gave the 5-one by hydrolysis of the intermediate reactive imine ⟨73GEP2322073⟩, whilst with a 2,3-dicyanopyridine, further reaction of the diamine with hydrazine led to the 5,8-dihydrazino compound (369) ⟨55BRP732521⟩, a similar reaction being observed in the [3,4-*d*] series ⟨71JAP7129876⟩.

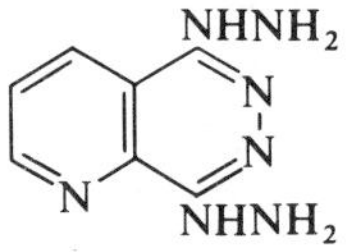

(369)

The use of an alkyl- or phenyl-hydrazine gave the 6- or 7-substituted derivatives with 2-acyl-3-acids or 3-acyl-2-acids respectively, this having constituted the first synthesis of a pyridopyridazine ⟨1893CB1501⟩. The 1,2-bisphenylhydrazine required use of the diacid chloride for successful reaction ⟨67JAP6700191⟩. In the [3,4-*d*] series, corresponding derivatives were obtained from 3,4-dialdehydes ⟨*e.g.* 69BSF2519, 69AJC1745⟩, 3,4-diacids and esters ⟨*e.g.* 60G1399, 33BSF151, 56JA159, 75CPB2239⟩, 3-acyl-4-acids (esters) ⟨*e.g.* 77H(6)547, 55JCS2685⟩ and 4-acyl-3-acids (esters) ⟨71FES1074⟩. In this series, however, the use of cyanoesters did furnish the 4-aminopyrido[3,4-*d*]pyridazines ⟨*e.g.* 69CPB2266, 42JOC286⟩, whilst the use of an *o*-methylcarbamoyl acid, either isolated or formed *in situ* by ring opening of a pyrido[2,3-*d*]pyrimidine, was also reported ⟨59AC(R)944⟩. Phenylhydrazine gave 2- or 3-phenyl derivatives ⟨*e.g.* 55JCS2685⟩, whilst in this series the reaction with 1,2-bisphenylhydrazine proceeded on the diacid ⟨57AC(R)728⟩.

The other main source of various pyridopyridazines from pyridines are the [4 + 2] cycloaddition reactions, already mentioned (Section 2.15.8.3), between vinylpyridines and azodicarboxylic esters ⟨79T2027, 79KGS639⟩ or triazolidinediones ⟨*e.g.* 78KGS651⟩. 2-Vinyl-pyridines gave reduced pyrido[3,2-*c*]pyridazines (370), 4-vinylpyridines gave [3,4-*c*] analogues, whilst 2-methyl-5-vinylpyridine furnishes a mixture of the [2,3-*c*] and [4,3-*c*] compounds. Yields are low, however, and these remain curiosities for practical synthetic purposes.

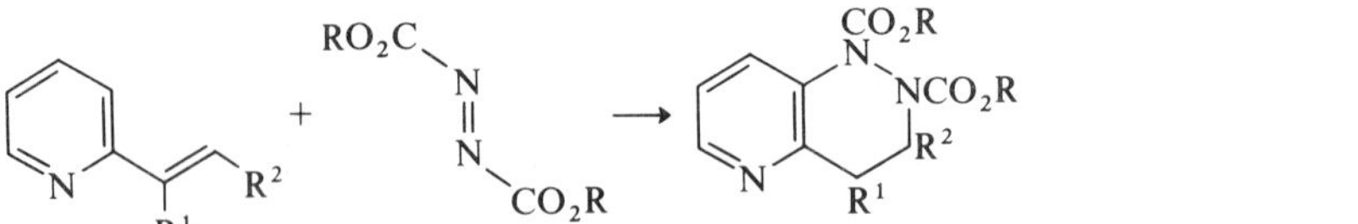

(370)

2-(1-Alkoxyvinyl)pyridines give 4-alkoxypyrido[3,2-*c*]pyridazines, but with the 1-bromovinyl analogues dehydrobromination to the fully aromatic system takes place ⟨78KGS651⟩.

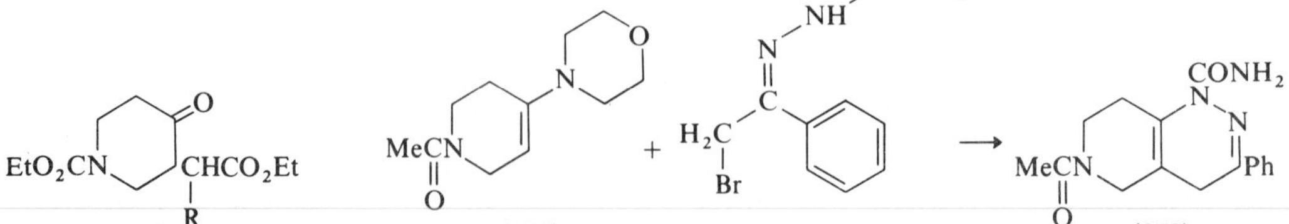

(371)          (372)          (373)

Reduced pyrido[4,3-*c*]pyridazines are obtained from piperidine derivatives such as (**371**) with hydrazine ⟨*e.g.* 79AF1835⟩, whilst another related synthesis coupled the enamine (**372**) with phenacyl bromide semicarbazone to give (**373**) ⟨74JAP(K)7488897⟩.

### 2.15.10.6.2 *From pyridazine intermediates*

The only recorded synthesis of this type from a pyridazine involves the [4 + 2] cycloaddition of the lactim ether (**374**) with 1,2,4,5-tetrazine-3,6-dicarboxylic ester, which proceeds with loss of nitrogen and methanol from the intermediate adduct to give the pyrido[2,3-*d*]pyridazine (**375**) ⟨77AP936⟩.

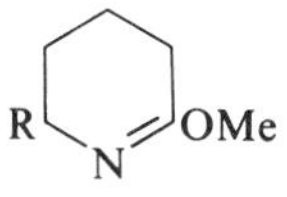

(**374**) 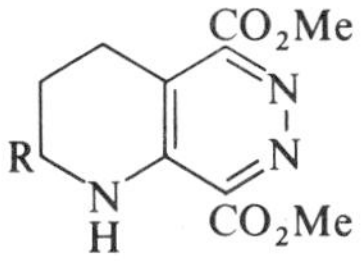　　　(**375**)

### 2.15.10.6.3 *Fused derivatives*

The majority of these syntheses are quinoline analogues of the pyridine hydrazine syntheses described above. 2,3-Quinolinedialdehydes ⟨71BSF906, 81JCS(P1)2509⟩, diketones ⟨74JOC3278, 73JOC1769⟩, aldehydo- or keto-esters ⟨71BSF906, 43LA(554)269⟩ and diesters ⟨*e.g.* 72BSF1588, 52CB204⟩ have all been reacted with hydrazine to give the appropriate pyridazino[4,5-*b*]quinolines, whilst the pyridine-2,3,5,6-tetracarboxylate gave the pyridazino-fused back-to-back derivative (**376**) ⟨71BSF906⟩. The corresponding quinoline-3,4-diesters give pyridazino[4,5-*c*]quinolinediones ⟨69ZC230, 75CR(C)(281)941⟩.

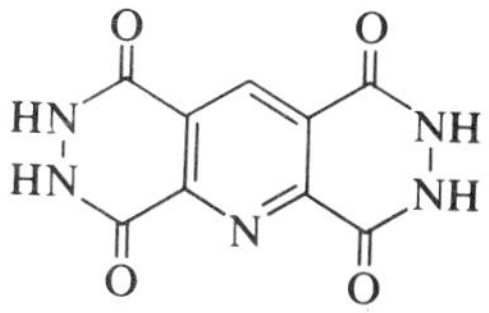

(**376**) 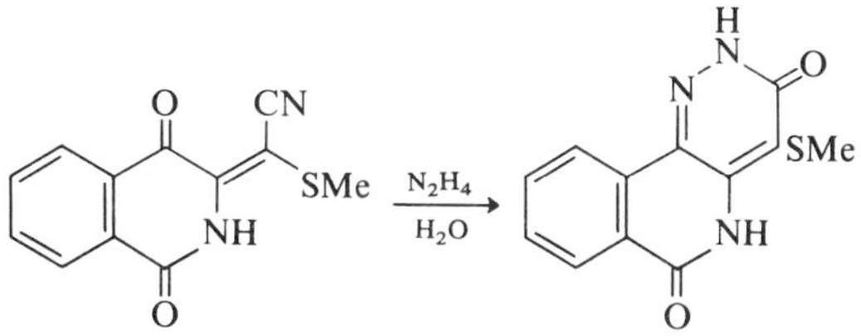　　　(**377**)　　　(**378**)

The only other synthesis of a benzo fused derivative involves the pyridazino[4,3-*c*]isoquinoline series, the isoquinoline derivative (**377**) reacting with hydrazine to give (**378**) ⟨74YZ607⟩, although an *s*-triazolo-fused pyrido[2,3-*d*]pyridazine was obtained in the reaction of 1-amino-3-iminoisoindolenine with hydrazine and formic acid ⟨56GEP951993⟩.

### 2.15.10.7 By Formation of Two Bonds from [3+3] Atom Fragments

#### 2.15.10.7.1 *From pyridazine intermediates*

The only syntheses of this type noted are analogues of corresponding [3+3] reactions in the pyridopyrimidine field.

*o*-Chloropyridazine acid chlorides, *e.g.* (**379**), react with enamines to give pyrido[3,2-*c*]pyridazin-8-ones (**380**), whilst the corresponding [2,3-*c*] analogues are obtained from the reversed isomers, and the 4,5-pyridazine derivatives give [2,3-*d*] products ⟨79GEP2808070⟩.

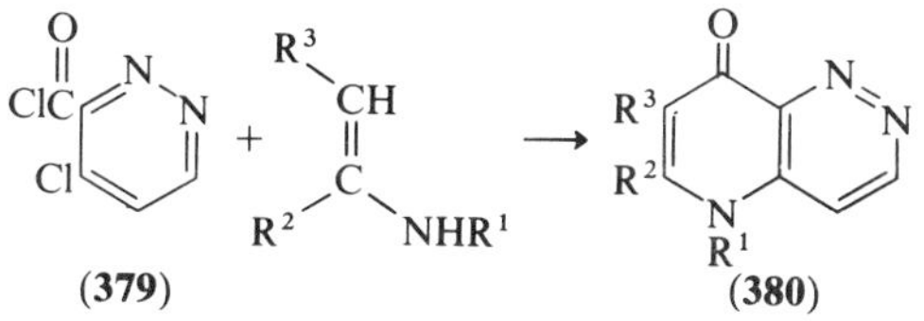

(**379**)　　　(**380**)

Finally, the activated 3-aminopyridazine *N*-oxide (**381**), mimicking 6-aminouracil, cyclizes with β-ketoesters to give the pyrido[2,3-*c*]pyridazin-7-one *N*-oxide (**382**). A reduced fused derivative in the pyridazino[3,4-c]isoquinoline series is formed if ethyl cyclohexanone-2-carboxylate is used ⟨72JHC351⟩.

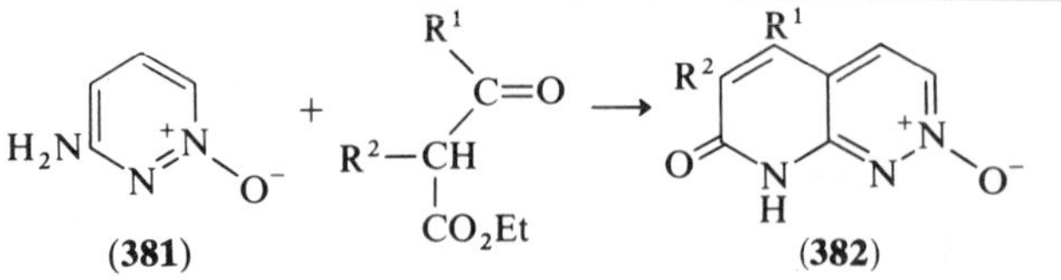

**2.15.10.8 By Formation of Three Bonds**

The only synthesis which corresponds with this description formally, though not mechanistically, resulted in formation of (**385**) from a side reaction in the attempted esterification of the pyridinone acid (**383**) with diazoethane, *via* ring expansion of the putative intermediate (**384**) ⟨76CPB1870⟩.

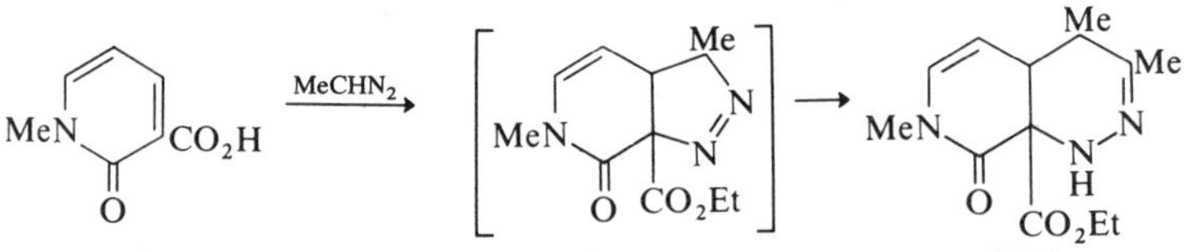

**2.15.11 PYRIDOPYRAZINES: INTRODUCTION**

Although there are only two possible pyridopyrazines, these compounds have attracted a considerable amount of attention, particularly in the study of deaza analogues of naturally occurring and/or pharmacologically active pteridines. In spite of this interest, the only full scale review of the chemistry of pyridopyridazines was in 1953 ⟨53HC(5)356⟩, although a thorough survey to 1971 appears in a Ph.D. thesis ⟨71TH21500⟩. References to reviews on various aspects of the biologically active molecules referred to above will be found in the appropriate Section below.

**2.15.12 STRUCTURE AND PHYSICAL PROPERTIES OF PYRIDOPYRAZINES**

Both the possible non-bridgehead pyridopyrazines, pyrido[2,3-b]pyrazine (**386**) and pyrido[3,4-b]pyrazine (**387**), are well known, the numbering being as shown. In the older literature they may be known as pyridino-2',3'-2,3- and -3',4'-2,3-pyrazines, as 1,4,5- and 1,4,6-triazanaphthalenes, or as 5- and 6-azaquinoxalines respectively. Some derivatives may also be referred to and numbered as deazapteridines, or as deaza derivatives of various natural products (see Section 2.15.16.3). Of the benzo fused systems, representatives of one angular (**388**) and two linear (**389**) and (**390**) derivatives of the [2,3-b] system are known, but only the linear benzo fused [3,4-b] system (**391**).

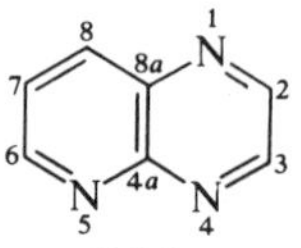

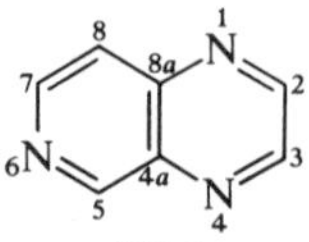

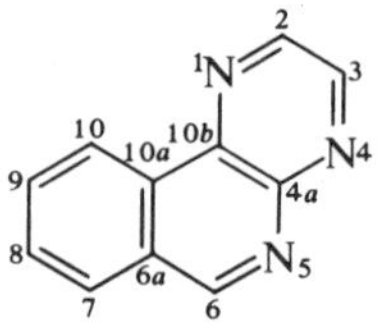

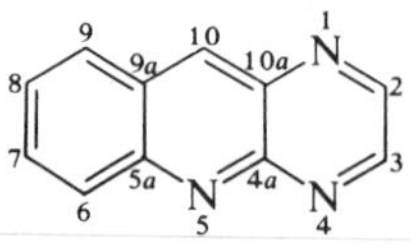

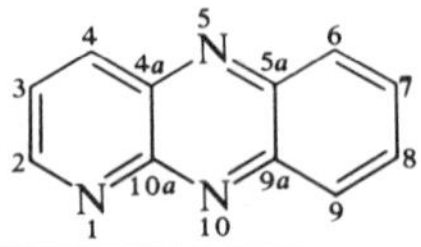

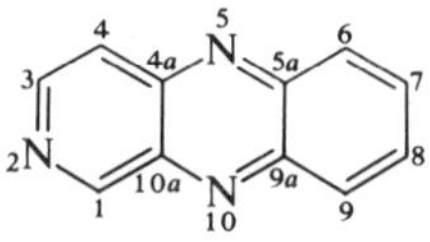

#### 2.15.12.1 Conformational behaviour

The stereochemistry and conformational behaviour of 2,3-dimethyl-1,2,3,4-tetrahydropyrido[2,3-*b*]pyrazine and some related compounds have been studied, as have the corresponding pyrido[3,4-*b*]pyrazine derivatives ⟨67JOC1378, 60JA3762⟩.

#### 2.15.12.2 NMR Spectroscopy

##### 2.15.12.2.1 ¹H NMR spectra

The ¹H NMR spectra of both the parent [2,3-*b*] and [3,4-*b*] pyridopyrazine systems have been analyzed ⟨66JCS(C)999⟩. Shift values are given in Table 3. These studies were extended to the phenomenon of covalent hydration in both systems ⟨66JCS(C)999,79JHC301⟩ (see Section 2.15.13.2), as well as the addition of other nucleophiles such as amide ion ⟨79JHC301, 79JHC305⟩.

**Table 3** ¹H NMR Data for Pyridopyrazines

| Compound | Solvent | Chemical shift (δ-values) | | | | | | Ref. |
|---|---|---|---|---|---|---|---|---|
| | | H-2 | H-3 | H-5 | H-6 | H-7 | H-8 | |
| (386) | TFA | 9.50 | 9.50 | — | 9.50 | 8.58 | 9.50 | 66JCS(C)999 |
| (387) | TFA | 9.71[a] | 9.64[a] | 10.30 | — | 9.20 | 9.00 | 66JCS(C)999 |

[a] Values interchangeable.

The europium shift NMR spectra of both parents have also been studied ⟨71OMR(3)575⟩, as has the nematic liquid crystal behaviour of pyrido[2,3-*b*]pyrazine ⟨76OMR(8)155⟩.

The ¹H NMR behaviour of methyl ⟨72T1983⟩ and phenyl ⟨78HCA2452⟩ quaternary salts of the [2,3-*b*] series has been the subject of investigation, whilst other uses in the solution of structural problems have involved 3-deazamethotrexate analogues ⟨79JMC862⟩, a series of 8-oxopyrido[2,3-*b*] pyrazine-7-carboxylic acids ⟨73MI21501⟩ (including NOE studies), and the nature of the one-electron electrochemical reduction products of pyrido[2,3-*b*]pyrazines ⟨78CB1763⟩.

The ¹H NMR spectra of some derivatives of the benzo fused pyrido[3,4-*b*]quinoxaline (1-deazaflavin) (391) system have been recorded ⟨74JCS(P1)1965⟩.

The ¹H NMR spectra of the parent pyridopyrazines are shown in Table 3.

##### 2.15.12.2.2 ¹³C NMR spectra

Relatively little work has been carried out on ¹³C NMR spectra of pyridopyrazines, but some have been utilized during studies of the covalent hydration (*q.v.*) of both parent bases ⟨66JCS(C)999, 75AG356, 79JHC301⟩ and their reaction with nucleophiles ⟨79JHC305⟩.

#### 2.15.12.3 IR spectroscopy

The IR spectra of both pyrido-[2,3-*b*]- and -[3,4-*b*]-pyrazines and some closely related compounds have been recorded and analyzed in detail ⟨*e.g.* 66SA117, 63PMH(2)1 and refs. therein⟩, and MO calculations carried out for the assignment of out-of-plane vibrations ⟨77JSP(66)192⟩. IR spectroscopic data have been used in the solution of the problem of tautomerism in pyrido-[2,3-*b*]- and -[3,4-*b*]-pyrazinones ⟨57JCS4874, 57JCS5010⟩, and in a study of the charge-transfer ⟨77MI21503⟩ and metal ion complexes ⟨75IC2378, 77MI21504⟩ of these systems.

IR spectroscopy has also been used in structural problems in 2- and 3-hydroxypyrido[3,4-*b*]pyrazines ⟨63JCS5156⟩, in 8-oxopyrido[2,3-*b*]pyrazine-7-acids ⟨73MI21501⟩ and in the pyrido[3,4-*b*]quinoxaline field ⟨74JCS(P1)1965⟩. IR spectra were recommended for the distinction of isomeric products in the Isay reaction (Section 2.15.15.6.1) ⟨71TH21500⟩; UV spectra were not satisfactory. The Raman spectra of a number of 1- and 3-deazaflavin analogues have been recorded and discussed ⟨80BBA(623)77⟩.

### 2.15.12.4 UV Spectroscopy

The UV spectra of the parent molecules and some close relatives have been measured and discussed ⟨*e.g.* 62JCS493, 62JCS3162⟩, including a discussion on electronic relaxation as a cause of diffuseness ⟨71AJC1107⟩, and circular dichroism studies have been carried out ⟨78JA4037⟩. Luminescence ⟨B-68MI21502⟩ and phosphorescence studies, especially on 3-deazaflavins ⟨76MI21502⟩, are noteworthy, whilst MO calculations of the first electronic transition have been carried out and the results compared with observations ⟨*e.g.* 66JSP(19)25, 67MI21501⟩. The spectra of charge transfer ⟨77MI21503⟩ and metal complexes ⟨*e.g.* 75IC2378, 77MI21504⟩ have been discussed.

The UV spectra of 1- and 3-deazaflavins have been measured ⟨74JCS(P1)1965, 78B1942⟩, and used in the latter instance in a study of their interaction with enzymes ⟨79B3635⟩.

The IR and UV spectra of a large number of new pyridopyrazines are recorded and discussed in ⟨71TH21500⟩.

### 2.15.12.5 MO Calculations

In addition to the calculations of electronic absorption spectra referred to in the previous section, MO calculations have been applied to circular dichroism ⟨78JA4037⟩, IR spectra ⟨77JSP(66)192⟩, interactions of non-bonding orbitals ⟨71MI21500⟩, calculation of resonance energies, ionization potentials, electron affinities, dipole moments ⟨79IJC(B)610⟩ and charge densities ⟨66JSP(19)25⟩, calculation of position of alkylation of pyrido[2,3-*b*]pyrazinones ⟨73MI21501⟩ and ligand forces in metal complexes ⟨75IC2378⟩.

### 2.15.12.6 Mass Spectrometry

No systematic study of the mass spectra of pyridopyrazines has been noted, but those of 2,3-dialkyl and 2,3-diaryl derivatives have been recorded ⟨75OMS97⟩, and mass spectrometry has been used in the elucidation of problems in the reactions of pyrido[2,3-*b*]pyrazines with amide ion (including use of $^{13}C$ and $^{15}N$ derivatives) ⟨79JHC305⟩, and of pyrido[2,3-*b*]pyrazinium salts with indoles ⟨78ZOR431⟩. The mass spectra of some 1-deazaflavins have been recorded ⟨74JCS(P1)1965⟩.

### 2.15.12.7 Tautomerism

The tautomerism of 8-hydroxypyrido[2,3-*b*]pyrazines was investigated by Mason ⟨57JCS4874⟩ using an IR method, and it was concluded that the compounds existed in the -one form ⟨*cf.* 71CR(C)(273)1529⟩.

The tautomerism of some deaza analogues of riboflavin has been studied ⟨77TL2551⟩. The tautomerism of 3-ethoxalylmethyl- and 3-acetonyl-pyrido[2,3-*b*]pyrazin-2-ones is fully discussed in ⟨71TH21500 p. 68⟩

## 2.15.13 REACTIVITY AT RING ATOMS OF PYRIDOPYRAZINES

### 2.15.13.1 Electrophilic Attack at Ring Carbon

The aromatization of hydro derivatives in the pyridopyrazine field does not appear to proceed as readily as with other pyridodiazines, although the 2,3-dimethyl-1,2,3,4-tetrahydro compounds and related derivatives aromatize spontaneously in air, as do 1,4-dihydropyrido[2,3-*b*]pyrazines. Dihydro-1-deazaflavin is also very readily oxidized by molecular oxygen with appreciable production of superoxide anion ⟨77B3586⟩. Usually, however, oxidation requires use of a chemical oxidant, and iodine, potassium permanganate in acetone, selenium dioxide, hydrogen peroxide, Pd/C dehydrogenation and bromination/dehydrobromination have all been used. The parent [3,4-*b*] derivative undergoes oxidation to the 2-oxo derivative (**393**) with alkaline permanganate *via* the covalent hydrate intermediate (**392**) ⟨63JCS5156⟩, and a similar ring oxidation is observed with the 2-oxo-[2,3-*b*] compound using potassium ferricyanide, giving the 2,3-dihydroxy derivative ⟨63JCS5737⟩. The 7-bromo-1,4-dihydro compound undergoes ready dehydrobromination ⟨78CJC1804⟩.

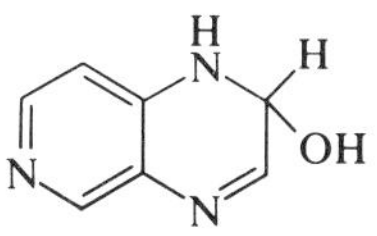

(392) 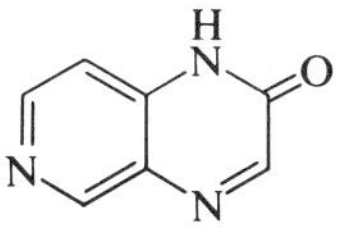     (393)

True electrophilic substitution has only been observed in some pyridine-ring amino derivatives, where bromination ⟨79JMC862, 70JHC1195⟩ and nitration (*e.g.* **394 → 395**) ⟨79JMC862⟩ have been observed. Ring substitution to give (**396**) was also surprisingly observed in an attempted NBS bromination of the nuclear methyl substituent ⟨64JOC734⟩, and 7-chloro substitution was also observed during phosphorus pentachloride chlorination of 2,3-dihydroxypyrido[2,3-*b*]pyrazine ⟨71TH21500⟩.

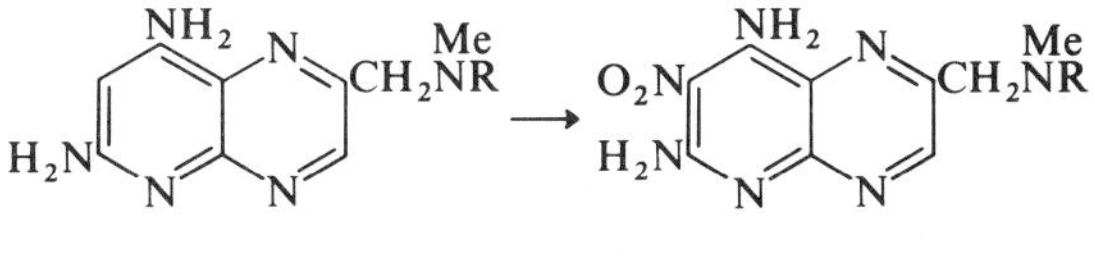

(394)     (395) 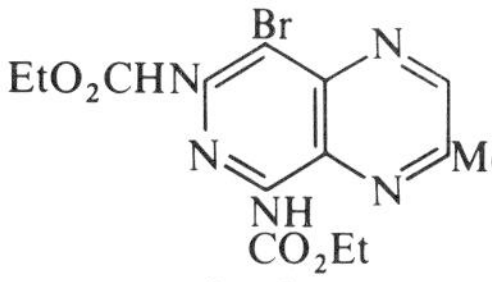     (396)

### 2.15.13.2 Electrophilic Attack at Ring Nitrogen

Protonation of pyrido[2,3-*b*]pyrazine occurs normally without covalent hydration, although the 2-hydroxy derivative did show such behaviour ⟨63JCS5737⟩. The pyrido[3,4-*b*]pyrazine parent base does show the phenomenon, although the exact structure of the covalent hydrate seemed to be in doubt between protonated (**392**) and (**397**). The issue was resolved in favour of the former by $^{13}$C NMR ⟨79JHC301, 75AG356⟩. The 3-hydroxy derivative also shows hydration effects, as does the 7-amino cation ⟨63JCS5166⟩.

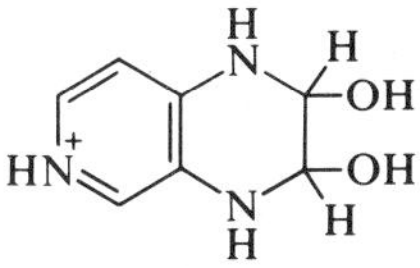

(397)

The $pK_a$ values have been measured by a rapid reaction method, as have those for a series of 5-hydroxy and 8-hydroxy analogues of oxine (*e.g.* 54JCS505, 66JCS(B)436), and for other deazaflavins and pyridopyrazines ⟨78B1942, 68JOC2393⟩.

The parent bases and several related derivatives have been converted to methiodides ⟨60USP2945037, 71TH21500⟩. The orientation was not determined but was assumed to be at pyridine nitrogen on theoretical grounds, although a later NMR study was in agreement ⟨72T1983⟩. Methylation of -ones with dimethyl sulphate or diazomethane occurs on nitrogen; for example, 2-ones are said to give 1-methyl-2-ones, 3-ones to give 4-methyl-3-ones (but *cf.* ⟨63JCS5156, 64JCS2825⟩), whilst the 7-ethoxycarbonylpyrido[2,3-*b*]pyrazin-8-ones undergo 5-methylation, in agreement with MO calculations ⟨73MI21501⟩.

*N*-Benzoylation of a 1,2,3,4-tetrahydro derivative is reported, whilst related 1,4-dialkyl derivatives resulted from borohydride reduction of the parent [2,3-*b*] compound in the presence of carboxylic acids ⟨79JHC973⟩.

Few reports of successful *N*-oxide preparation have been found ⟨48JCS1389, 71CR(C)-(273)1529⟩, whilst other papers refer to many failures in attempted *N*-oxidations, and the parent [2,3-*b*] compound gives the 6-hydroxy derivative instead of an *N*-oxide ⟨63JCS5737⟩.

On the other hand, the formation of metal complexes of both pyrido-[2,3-*b*]- and -[3,4-*b*]-pyrazines is widely reported (*e.g.* 52JCS4985, 54JCS505).

### 2.15.13.3 Nucleophilic Attack at Ring Carbon

Ring reductions in the pyridopyrazine series have been achieved with a wide variety of agents, and may lead to di- or tetra-hydro derivatives, usually in the pyrazine ring.

Aluminum or sodium amalgam reduces 3-ones to 1,2-dihydro derivatives ⟨71TH21500⟩, as does borohydride ⟨63JCS5156⟩ and catalytic hydrogenation (Pd/SrCO$_3$) ⟨71TH21500⟩. Catalytic (Pd/C) reduction of a 5-one also gave a 1,2-dihydro compound ⟨74JMC553⟩.

Catalytic hydrogenation of the parent [2,3-*b*] base gave the 1,2,3,4-tetrahydro derivative ⟨B-64MI21502⟩, and similar reductions to tetrahydro analogues have been reported with titanium(II) chloride ⟨80JHC1237⟩, zinc–acetic acid ⟨B-64MI21502⟩, sodium borohydride/TFA ⟨79JOC1719⟩ and lithium aluminum hydride ⟨73BSF3100⟩. The stereochemistry of the products in the case of the 2,3-dimethyl derivative has been shown to be *cis* ⟨60JA3762, 67JOC1378⟩. Reduction of the 1,4-di-*N*-oxides also gives 1,2,3,4-tetrahydro derivatives ⟨80M407⟩.

One-electron electrochemical reduction has been studied and redox potentials measured ⟨e.g. 78CJC1804, 78CB1763⟩, whilst preparative electrochemical reduction of [2,3-*b*] derivatives initially gives 1,4-dihydro derivatives, which may reoxidize or isomerize to 1,2-, 3,4- or 5,8-isomers ⟨78CJC1804⟩. Pyrido[3,4-*b*]pyrazines, *via* initial 1,4-reduction, give 1,2- (or 3,4-) dihydro derivatives, which on further reduction give the isolated 1,2,3,4-tetrahydro product ⟨78CJC1804⟩. Reductions of 1- and 3-deazaflavins ⟨79MI21502, 78B1942⟩ also give dihydro derivatives with H$_2$/Pt or dithionite ⟨77B3586⟩, or electrochemically.

Covalent hydration, previously discussed (Section 2.15.13.2), does not lead to ring opening in the pyridopyrazine series. Indeed, ring openings are very rare, occurring only in strongly oxidizing conditions with formation of pyrazine-2,3-dicarboxylic acids ⟨71CR(C)(273)1529⟩.

Additions of other nucleophiles to pyridopyrazines have been described in a number of cases; for instance, Grignard reagents give 2,3-dialkyl compounds (**398**) (or 6-alkyl analogues with 2,3-diaryl compounds) ⟨76BSF251⟩, and other workers have also observed 6-addition ⟨71CR(C)(273)1529⟩.

(**398**)                                                        (**399**)

Other additions have been observed with indole ⟨76KGS1146, 78ZOR431⟩, phenylhydrazines ⟨76KGS1146⟩, sulphite ion ⟨71TH21500⟩, cyanide ion/aldehydes (to give 3-benzoyl derivatives) ⟨76CPB238⟩, and amide ion (monitored by NMR spectroscopy) ⟨79JHC301⟩. The action of amide ion on 6-chloropyrido[2,3-*b*]pyrazine led to dehalogenation to the parent compound together with ring contraction to the imidazo[4,5-*b*]pyridine (**399**) ⟨79JHC305⟩ (see Section 2.15.14.3).

An unusual addition of acetoacetic acid to pyrido[2,3-*b*]pyrazin-2(1*H*)-one (**400**) to give (eventually) the 3-acetonyl derivative (**401**) was postulated ⟨71TH21500⟩ to occur *via* a cyclic transition state, and the similar addition of oxalacetic ester may occur *via* a related mechanism.

(**400**)                                                        (**401**)

4-Substituted pyrido[2,3-*d*]pyrazinium salts form cycloadducts with quinones ⟨75HCA2529⟩, similar to those formed by quinoxalinium salts.


## 2.15.14 REACTIVITY OF SUBSTITUENTS IN PYRIDOPYRAZINES

### 2.15.14.1 Electrophilic Attack on Substituents

Methyl groups in pyridopyrazines ⟨64JMC240⟩ and pyridopyrazinones ⟨71TH21500⟩ are oxidized to carboxylic acids with potassium permanganate. Aryl carbinol substituents are also very readily oxidized to benzoyl derivatives in alkaline conditions ⟨76CPB238⟩. Bromination of 2,3-dimethylpyrido[3,4-*b*]pyridazine gives the 2,3-bisbromomethyl derivative, whilst with iodine/pyridine the 2-methyl group only undergoes the Ortoleva–King reaction to give the pyridiniomethyl derivative (**402**). The related 3-phenyl salt gives the Kröhnke-type nitrone with *p*-nitrosodimethylaniline ⟨71ZC256⟩.

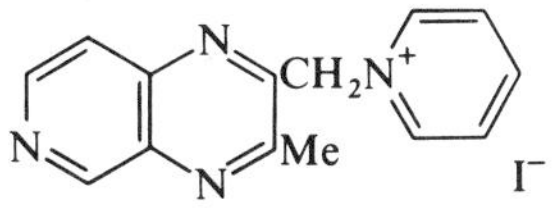

(**402**)

Both 2- and 3-methyl groups in pyrido[2,3-*b*]pyrazines are acylated by ethyl oxalate ⟨71TH21500⟩. They give (preferentially 3-) styryl derivatives with aromatic aldehydes and oximes with pentyl nitrite.

Amino groups are acetylated normally and are converted to hydroxy compounds with nitrous acid. The 2,3-diamino [2,3-*b*] derivative gives an imidazolo fused pyridopyrazine with acetic anhydride ⟨75USP3898216⟩. The oxidative removal of hydrazino groups has been used to give ring-unsubstituted derivatives ⟨79JHC305⟩.

Oxo substituents are, in a few cases, methylated with diazomethane to give methoxy derivatives in addition to the major *N*-methylation, and oxo groups in the 3-deazaflavins have been acylated.

### 2.15.14.2 Nucleophilic Attack on Substituents

Systematic studies of methyl group hydrogen exchange in pyrido[3,4-*b*]pyrazines have been carried out. Carbon acidities have been measured and compared with values from MO calculations ⟨73T3071, 79T1615⟩. The action of bases on *N*-aryl-2,3-dimethylpyrido[2,3-*b*]pyrazinium salts gives anhydro bases, *e.g.* (**403**) yields (**404**) ⟨78HCA2452, 56CB2684⟩.

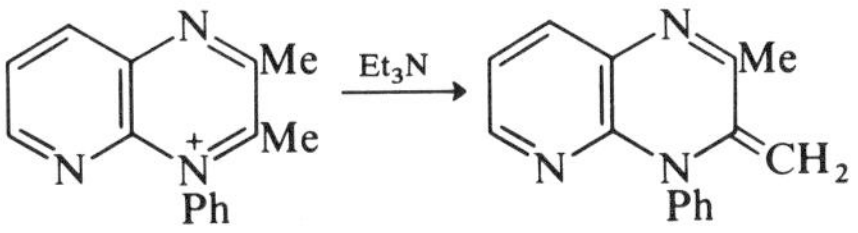

(**403**)                (**404**)

Unusual degradations have been described in the 2-position. The cyclic ketone (**405**) undergoes ring opening with alkali to give the 3-(2-carboxyphenyl)pyrido[2,3-*b*]pyrazinone (**406**) ⟨72JHC255⟩. 2-Aminomethyl [3,4-*b*] derivatives underwent extensive decomposition in alkali during an attempted synthesis of folic acid analogues ⟨71JOC2818⟩, and 2-acetonyl groups were removed by alkaline hydrolysis ⟨71TH21500⟩.

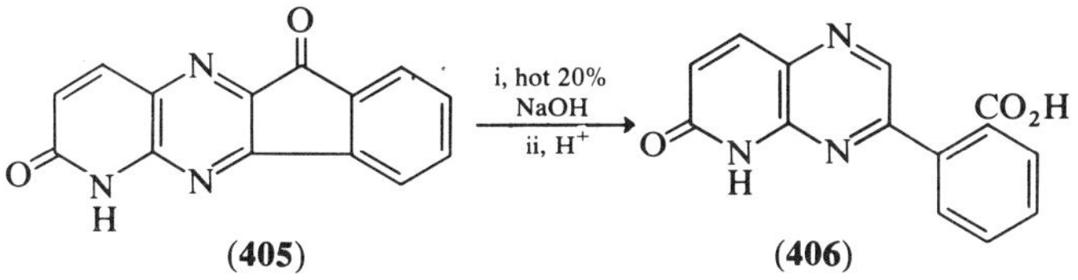

(**405**)                        (**406**)

Esters undergo hydrolysis and conversion to amides under the usual conditions, and amide side chains have also been formed from the acid and amine with DCCI. Acids have been formed from the corresponding spirohydantoins *via* ureido derivatives (Section 2.15.15.6.1), and undergo decarboxylation in the usual manner.

Nitro groups have been reduced to amino groups, whilst amino groups in the 3- and 6-positions of pyrido[2,3-*b*]pyrazines and in the 5-position of the [3,4-*b*] isomers have been hydrolyzed to the corresponding hydroxy derivatives with alkali. Protected amino groups have been liberated by hydrolysis or reduction in deazapteridine syntheses.

### 2.15.14.3 Nucleophilic Substitution Reactions

The nucleophilic substitution reactions in pyrido-[2,3-*b*]- and -[3,4-*b*]-pyridazines in general follow the usual pattern of polyaza heterocycles. Oxo groups in the 2-, 3- and 6-positions of [2,3-*b*]-ones, and in the 2- and 3-positions of [3,4-*b*]-ones have been converted to the corresponding chloro compounds with phosphorus oxychloride, phosphorus oxychloride/DMF or phosphorus oxychloride/phosphorus pentachloride ⟨71TH21500⟩. Oxo groups in the 3- and 6-positions are converted to thioxo groups with $P_2S_5$, and to bromo compounds with phosphorus oxybromide.

3-Methoxy groups undergo alkoxy exchange with potassium hydroxide in alcohols.

Thiol groups in the 2-, 3- and 6-positions have been desulfurized with Raney nickel under the usual conditions to give the corresponding unsubstituted compounds.

Halogen groups in the 2-, 3- and 6-positions in pyrido[2,3-*b*]pyrazines are replaceable in nucleophilic substitution reactions, and have been reacted with amines, hydrazine, alkoxides, thiolate ion, fluoride ion (Finkelstein reaction) or aqueous acid. There seems to be little difference in reactivity between the halogen atoms in 2,3-dichloro compounds, but in the 2,3,6-trichloro derivative the 2,3-groups were replaced by alkoxide, leaving the 6-chlorine unattacked ⟨70AP44⟩. In the pyrido[3,4-*b*]pyrazines, 5-chloro groups have been successfully removed by catalytic hydrogenation, but a 7-chloro group was resistant to reduction, hydrolysis or alcoholysis, being replaced only by ammonia under forcing conditions ⟨63JCS5156⟩.

In a series of reactions with potassium amide in liquid ammonia, 6-chloropyrido[2,3-*b*]pyrazine gave reduction and ring contraction (Section 2.15.13.3), the 6-bromo analogue underwent only reduction, whilst the 6-fluoro derivative gave only the 6-amino substitution product ⟨79JHC305⟩.

### 2.15.14.4  Radical Attack on Ring and Substituents and Photochemistry

Apart from the nuclear bromination observed (Section 2.15.13.1) in the attempted radical bromination of a side-chain methyl group leading to (**396**), which may or may not have involved radical intermediates, the only other reaction of interest in this section is a light-induced reduction of certain hydroxypyrido[3,4-*b*]pyrazines or their oxo tautomers analogous to that well-known in the pteridine field ⟨63JCS5156⟩. Related one-electron reduction products of laser photolysis experiments with 1-deazaflavins have been described ⟨79MI21502⟩.

### 2.15.15  SYNTHESIS OF PYRIDOPYRAZINES

#### 2.15.15.1  By Formation of One Bond between Two Heteroatoms [6 + 0 (N)]

No synthesis of pyridopyrazines by this type of reaction is possible.

#### 2.15.15.2  By Formation of One Bond adjacent to a Heteroatom [6 + 0 (α)]

##### 2.15.15.2.1  From pyridine intermediates

The main reaction of this type has been the reductive cyclization of nitropyridine derivatives carrying an *o*-amino ester or *o*-aminocarbonyl substituent. These cyclize *in situ via* the *o*-diamino derivative to give pyridopyrazines of known constitution, either for establishment of structure of products obtained in the ambiguous Isay synthesis (see Section 2.15.15.6.1), or in the synthesis of aza analogues of biologically active molecules.

For example, in the former case the reductive cyclization of 3-nitro-2-pyridylglycine ester (**407**) with tin(II) chloride gave the 3,4-dihydro-2-one (**408**), oxidized to compound (**400**), identical with the product of the Isay-type reaction of 2,3-diaminopyridine and ethyl glyoxylate ⟨63JCS5737⟩. Similarly, the use of 3-nitro-4-pyridylglycine ester established the structure of the isomeric 3-hydroxypyrido[3,4-*b*]pyrazine ⟨63JCS5156⟩. Mild reduction of the 3-nitro-2-pyridyl ester in alkaline solution gave the 3,4-dihydro-1-hydroxy-2-one (**409**) ⟨64MI21501⟩. Similar products have been obtained from the corresponding nitroamides ⟨75JCS(P1)1424⟩. The use of aminoketones such as (**410**) gives the 2-methyl-3,4-dihydro base (**411**) ⟨66JOC1890⟩ on catalytic hydrogenation, and variants of this last procedure with esters or ketones have been used *en route* to the synthesis of a variety of deazapteridines, *e.g.* with the 3-deazamethotrexate intermediate (**412**) ⟨71JOC2818⟩.

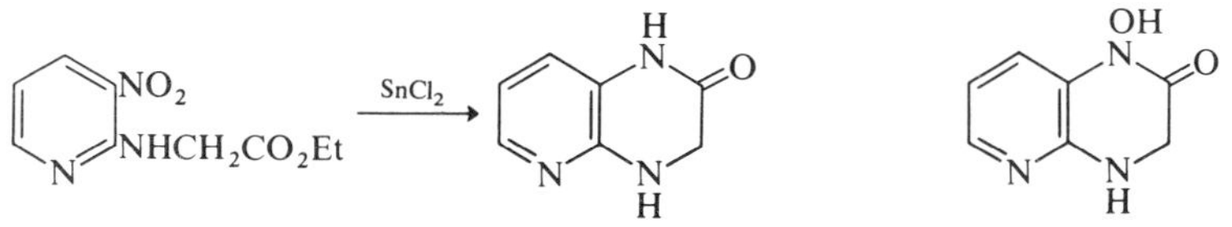

(407)　　　　　(408)　　　　　(409)

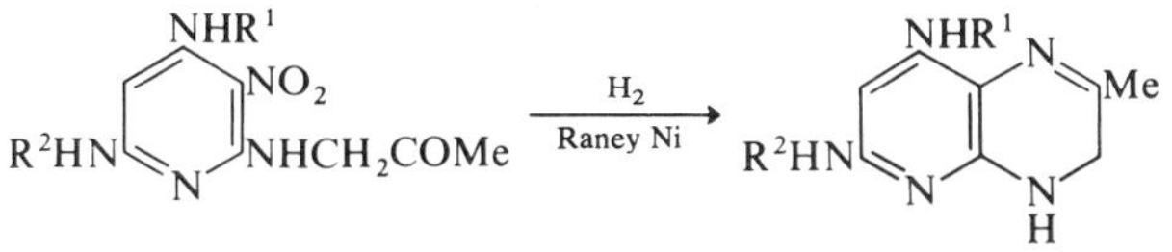

(410)　　　　　(411)

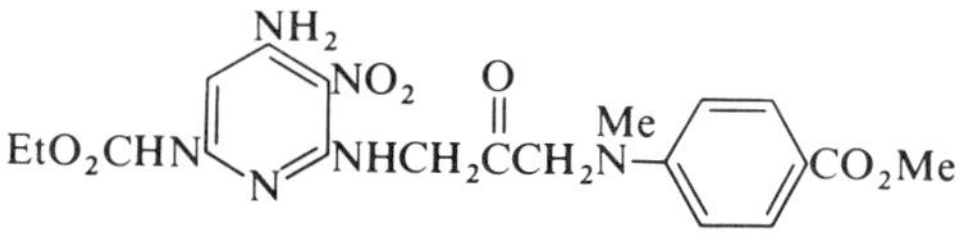

(412)

In a related reaction the amino alcohol (413) was cyclized with HBr to the 3-methyl-1,2,3,4-tetrahydro derivative (414) ⟨65JCS1558⟩, whilst the Isay reaction itself has been shown (Section 2.15.15.6.1) to probably occur, at least in some cases, *via* a [6+0 (α)] cyclization of intermediate anils.

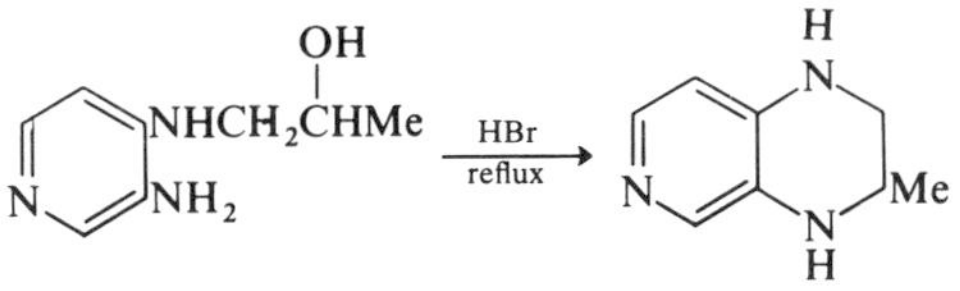

(413)　　　　　(414)

### 2.15.15.2.2 *From pyrazine intermediates*

The only reaction of this type noted involved the reaction of pteridines, *e.g.* (415), with malonodinitrile (or cyanoacetamide), *via* ring opening to (416), with final [6+0 (α)] cyclization to give the 6-amino-7-nitrile (amide) (417) ⟨73JCS(P1)1615, 73JCS(P1)1974⟩.

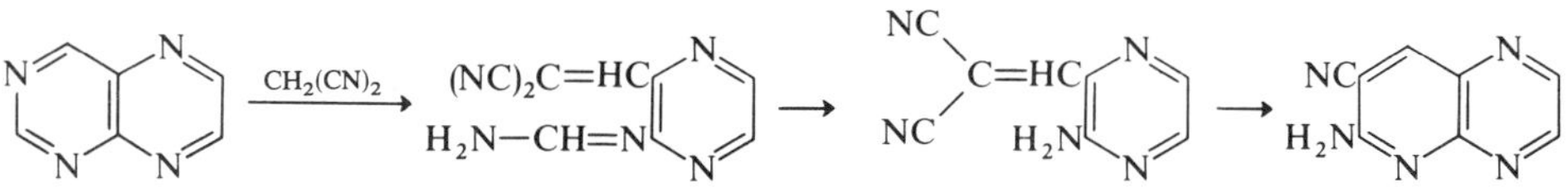

(415)　　　　　(416)　　　　　(417)

### 2.15.15.2.3 *Fused derivatives*

An early synthesis of pyrido[3,4-*b*]quinoxalines involved cyclization by strong heating of *o*-aminoanilinopyridinamine derivatives, *e.g.* (418) to give (419) ⟨49JCS2540⟩. In a related reaction, *o*-nitroanilinopyridines (420) were cyclized to pyrido-[2,3-*b*]- or -[3,4-*b*]-quinoxalines (421) by reduction with iron(II) oxalate, probably *via* a nitrene intermediate ⟨74JCS(P1)1965⟩.

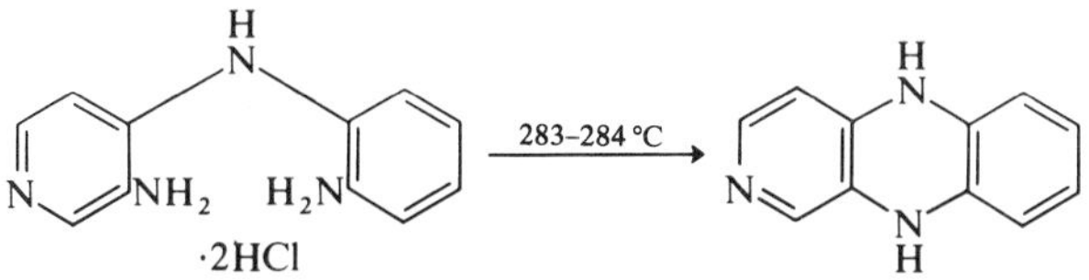

(418)　　　　　(419)

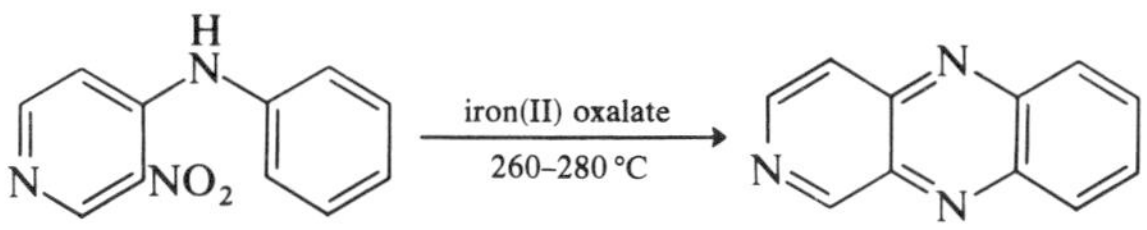

(420)　　　　　(421)

The double cyclisation of (**422**) to give the pyrazino[2,3-*b*]quinoline (**423**) must also involve a [6+0 (α)] cyclization step ⟨B-79MI21505⟩, although it might also be viewed as a [5+1] reaction.

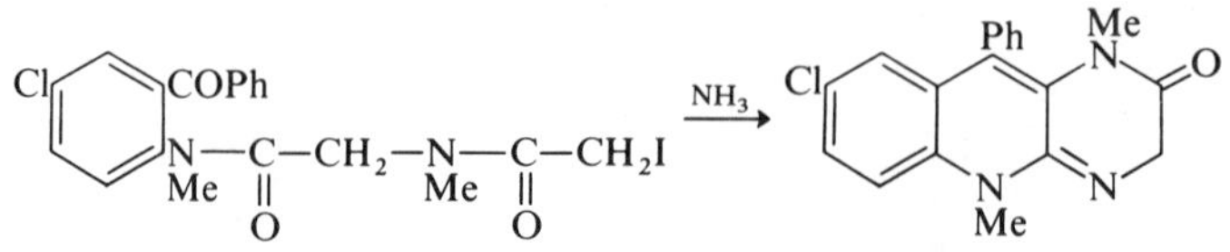

(422)                                                                          (423)

### 2.15.15.3 By Formation of One Bond β to a Heteroatom [6+0 (β)]

No syntheses of this type have been noted for pyridopyrazines.

### 2.15.15.4 By Formation of one bond γ to a Heteroatom [6+0 (γ)]

#### 2.15.15.4.1 From pyridine intermediates

There is no record of this type of synthesis in the pyridopyrazine field.

#### 2.15.15.4.2 From pyrazine intermediates

Most syntheses of this type have followed the classical Gould–Jacobs pattern (Section 2.15.5.4.2) in which 2-aminopyrazines bearing a 6-substituent give esters of 8-oxopyrido[2,3-*b*]pyrazine-7-carboxylic acids (**424**) *via* the usual intermediate ethoxymethylenemalonate adducts. In some cases the isomeric pyrazino[1,2-*a*]pyrimidines are formed in addition ⟨*e.g.* 74CPB1864⟩.

The only other type noted concerned the photocyclization of the unsaturated pyrazine amide (**425**) to (**426**) with subsequent selenium dioxide aromatization ⟨72CPB2264⟩.

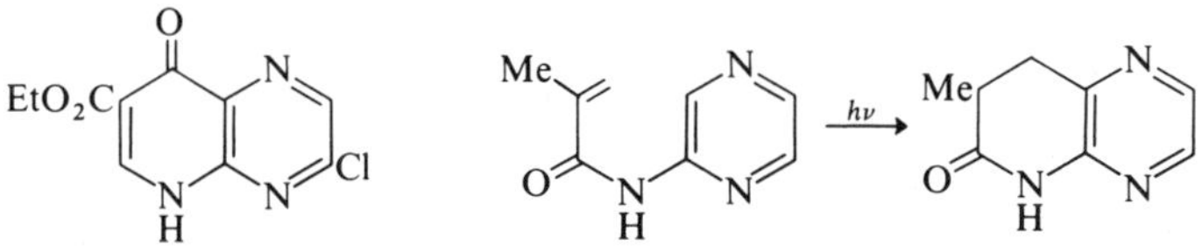

(424)                          (425)                          (426)

#### 2.15.15.4.3 Fused derivatives

The first example of this type also mimics a synthesis used in the pyridopyrimidine field, in which the quinoxaline diester (**427**) is cyclized in Dieckmann fashion to the pyrido[2,3-*b*]quinoxaline (**428**) ⟨76CR(C)(282)861⟩.

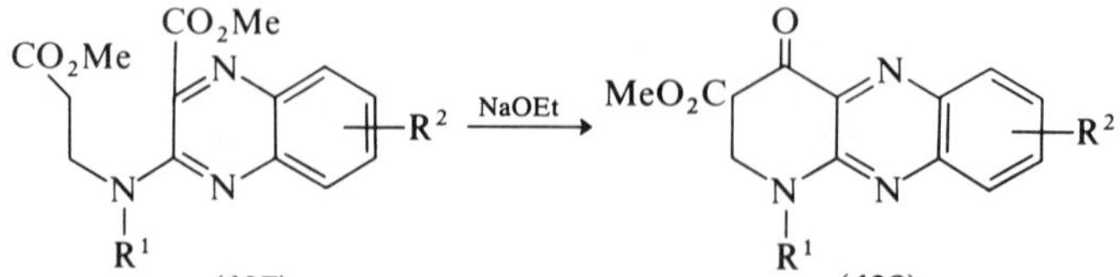

(427)                                                          (428)

The double cyclization to the pyrazino[2,3-*b*]quinoline (**423**) referred to above must also involve a [6+0 (γ)] cyclization step.

## 2.15.15.5 By Formation of Two Bonds from [5+1] Atom Fragments

### 2.15.15.5.1 Fused derivatives

No syntheses of simple pyridopyrazines by [5+1] type cyclizations are known, but a few syntheses of pyrido[3,4-*b*]quinoxalines are described.

The diester (**429**) has been cyclized with ammonia or primary amines to give the esters (**430**) ⟨72AP2⟩, whilst the related diester (**431**) has likewise been converted to (**432**). Similar reactions starting with *N*-substituted *o*-phenylenediamines have been used as the basis of a synthesis of deazariboflavins (**433**) ⟨77TL2551⟩.

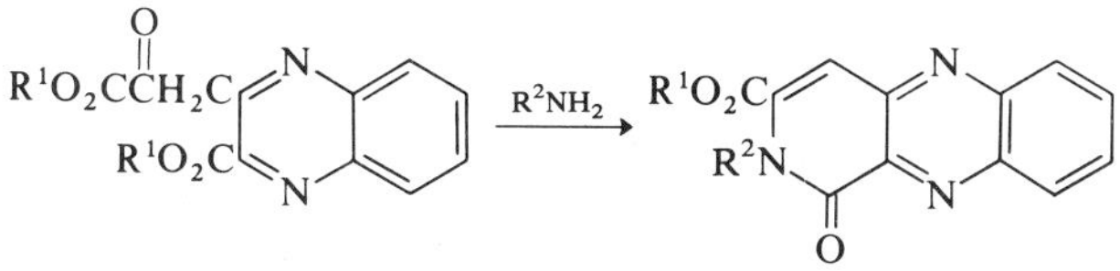

(**429**)          (**430**)

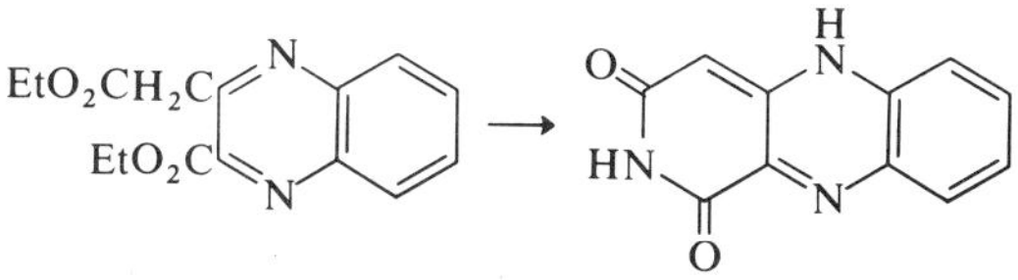

(**431**)          (**432**) 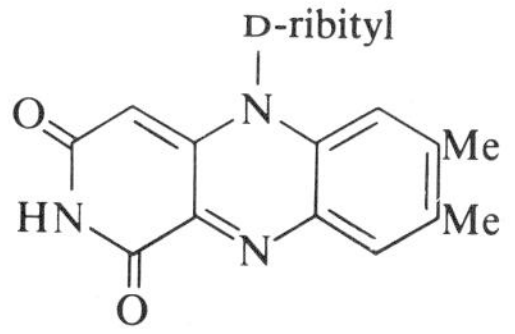          (**433**)

## 2.15.15.6 By Formation of Two Bonds from [4+2] Atom Fragments

### 2.15.15.6.1 From pyridine intermediates

By far the most important reaction for the synthesis of pyridopyrazines is the reaction of diaminopyridines with two-carbon fragments, the pyridopyrazine equivalent of the well-known Isay reaction in the pteridine field.

The two-carbon fragment has varied widely and examples noted include the dialdehyde glyoxal (giving 2,3-unsubstituted pyridopyrazines such as **386**, *e.g.* with 2,3-diamines leading to pyrido[2,3-*b*]pyrazines, with 3,4-diaminopyridines leading to pyrido[3,4-*b*]pyrazines); ketoaldehydes (giving 2(3)-monosubstituted pyridopyrazines, *e.g.* **434**), including methylglyoxal (pyruvaldehyde) (leading to both [2,3-*b*] and [3,4-*b*]), other alkyl glyoxals ([2,3-*b*] and [3,4-*b*]), aryl glyoxals ([2,3-*b*] and [3,4-*b*]) and heterocyclic glyoxals ([2,3-*b*] and [3,4-*b*]); diketones (giving 2,3-disubstituted analogues, *e.g.* **435**), including diacetyl ([2,3-*b*] and [3,4-*b*]), benzil ([2,3-*b*] and [3,4-*b*]), 2,2'-bipyridyl ([2,3-*b*]) and others ([2,3-*b*], [3,4-*b*]); aldehydo and keto acids (giving pyridopyrazinones, *e.g.* **436**), including glyoxylic acid, pyruvic acid, benzoylpyruvic acid and benzalpyruvic acid (all [2,3-*b*]); aldehydo and keto esters (similarly giving pyridopyrazinones), including glyoxylic esters, their acetates or hemiacetals ([2,3-*b*] and [3,4-*b*]), ethyl pyruvate ([2,3-*b*] and [3,4-*b*]), ethyl mesoxalate ([2,3-*b*] and [3,4-*b*]) and ethyl 2,4-dioxopentanoate ([2,3-*b*] and [3,4-*b*]); diesters (giving diones such as **437**), including ethyl oxalate ([2,3-*b*] and [3,4-*b*]); halo aldehydes such as dibromopropionaldehyde, giving dihydro derivatives which were oxidized *in situ* and used in the synthesis of (**438**), a key intermediate in a 3-deazafolic acid synthesis ⟨58JBC(231)331⟩; dihydroxytartaric acid (giving the diacids **439**) ⟨71CR(C)(273)1645⟩ ([2,3-*b*] and [3,4-*b*]); benzilic acid (giving 3,3-diphenyl [2,3-*b*] derivatives) ⟨75M1059⟩; and cyclic diketones, which give fused derivatives (see below) or, in the case of alloxan in neutral conditions, undergo ring opening to give 2- or 3-ureido derivatives, *e.g.* (**440**) and thence (**441**), *e.g.* ⟨57JCS430⟩ ([2,3-*b*]), ⟨62JCS3162⟩ ([3,4-*b*]) (*cf.* earlier papers which gave wrong structures).

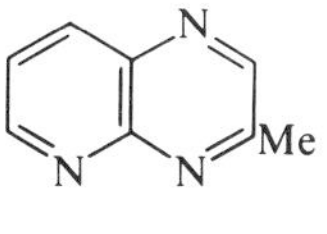

(**434**) 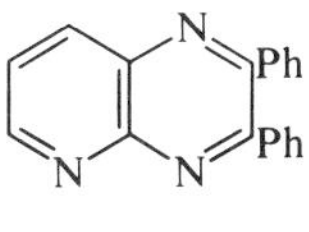          (**435**) 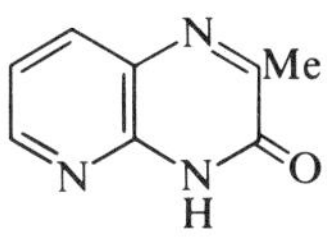          (**436**) 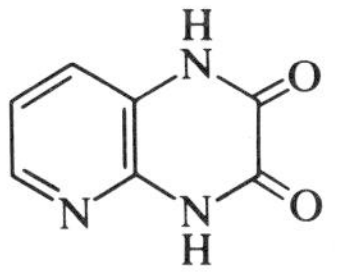          (**437**) 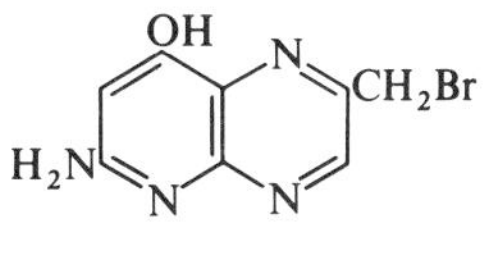          (**438**)

(439)          (440)          (441)

The use of 2-substituted amino-3-aminopyridines (or corresponding 3-substituted amino-2-amino, 3-substituted amino-4-amino and 4-substituted amino-3-amino derivatives) gives the corresponding *N*-substituted pyridopyrazinones with esters or alloxan, or with diketones gives quaternary salts such as (403) or their anhydro bases (404) ⟨*e.g.* 56CB2684, 78HCA2452⟩.

Regioselectivity in these Isay-type reactions with asymmetric substitution has previously been discussed by Albert ⟨63JCS5156⟩, and, of course, the corresponding pteridine synthesis has been dealt with in many publications or reviews too numerous to detail here (see Chapter 2.16). Montgomery ⟨68JOC2393⟩ formulated some rules by which, in general, the most reactive amino group in the pyridine (usually the 3-amino) will react preferentially with the more reactive carbonyl group in the 2-carbon fragment, and this is often found in practice in neutral or weakly acid conditions. In strongly acid conditions the 3-amino group would be preferentially protonated, and reaction of the more reactive carbonyl group in the two-atom fragment with the 2- or 4-amino group might be more favoured.

Threlfall ⟨71TH21500⟩, in a long and detailed study of this problem, in which the identity of products was established by rigorous alternative synthesis or by transformation of pyridopyrazines of established constitution, has indicated that the picture is not as comfortably clear-cut as this might suggest, and pointed out several discrepancies. For instance, the product from 3,4-diaminopyridine and methyl glyoxal, said by Albert ⟨63JCS5156⟩ to be the 3-methyl derivative, was shown to be the 2-methyl isomer, being non-identical with the authentic 3-methyl product (see below) of Hepworth and Tittensor ⟨65JCS1558⟩ The product from 2,3-diaminopyridine and glyoxylic acid was found by Threlfall to be the pyridopyrazin-2-one in neutral and weak acid solution, and only a small amount of 3-one compound was formed in strong acid. A recent patent, however, claimed the 3-hydroxy compound as the main product ⟨78USP4082845⟩. The reaction of 3,4-diaminopyridine with glyoxylate derivatives gave at least four products ⟨71TH21500⟩, a much less simple picture than previously envisaged ⟨63JCS5156, 71CR(C)(273)1645⟩. With ethyl 2,4-dioxopentanoate, 2,3-diaminopyridine furnished mainly the 2-acetonylpyrido[2,3-*b*]pyrazin-3(4*H*)-one (442) in contradiction of a previous report, whilst in the corresponding reaction with 5-bromo-2,3-diaminopyridine, increasing proportions of 3-hydroxy compounds were formed with increasing acidity, the opposite of the expected behaviour. Other discrepancies from predictions have been noted in compilation of this chapter, *e.g.* in the products of reaction of glyoxals with diaminopyridines.

(442)                    (443)

The explanation of at least some of these discrepancies may lie in an observation of Threlfall that with 5-bromo-2,3-diaminopyridine and glyoxylate derivatives in neutral or dilute acid solution, an intermediate formulated as (443) was formed from two moles of diaminopyridine and one of glyoxylate. This intermediate could then be made to yield either the 7-bromo-2-hydroxy derivative (in strong acid) or a mixture of 2- and 3-hydroxy isomers (in dilute acid). It seems that Montgomery's rules may apply only if the orientation is determined by kinetic control of formation of the first bond (followed by rapid cyclization to the second amino group), or if this initial reaction leads in any case to the thermodynamically more stable intermediate. If, on the other hand, there is effective thermodynamic control *via* an intermediate, or if an intermediate such as (443) undergoes further cyclization without splitting, then the orientation pattern may be decided by a number of factors such as the acidity of the solution (possibly acting in an opposite sense to that predicted by Montgomery), the absolute and relative $pK_a$ values of the two amino groups or of the nitrogens in any intermediate, the relative nucleophilicity of the two amino groups, possibility of covalent hydration, formation and/or breakdown of reactive groups (*e.g.* acetals) in the two-carbon fragment, *etc.* In some earlier work the relative ease of isolation of particular isomers, or misidentification of products, may have led to discrepancies, and the whole problem is in need of a systematic reinvestigation using, for example, an NMR method for identification and quantitative determination of isomers in mixtures, as was used by Dieffenbacher and von Phillipsborn in the pteridine series ⟨69HCA743⟩.

The other prominent reaction of the [4+2] type is the Timmis-type reaction which has been used in the positive identification of isomers from the Isay reaction. In this reaction a 2-amino-3-nitrosopyridine is condensed with a compound containing a —CH₂CO— or —CH₂CN grouping to give a 3-alkyl or 3-aryl ⟨71TH21500, 55JCS303⟩, 3-hydroxy ⟨55JCS303, 55JCS2032⟩ or 3-amino (*e.g.* 55JCS2032) pyrido[2,3-*b*]pyrazine, *e.g.* (**444**) → (**445**). Sodium ethoxide is usually used as catalyst, but use of pyridine as solvent/catalyst proved useful in some cases ⟨71TH21500⟩. An unusual [4+2] synthesis involved the reaction of pyridofuroxan with ketones to give the di-*N*-oxide (**446**) ⟨80M407⟩.

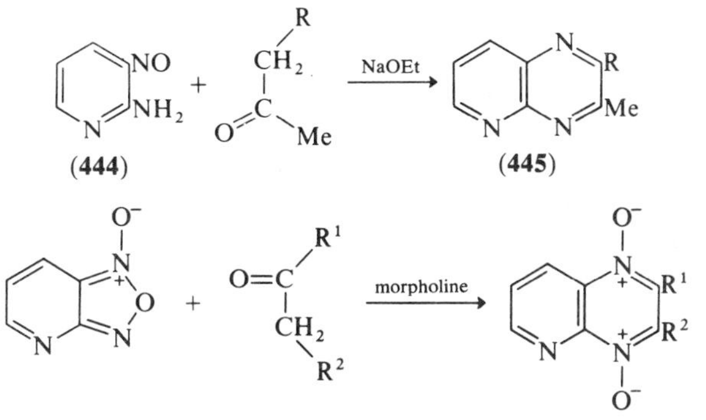

(**444**)  (**445**)

(**446**)

### 2.15.15.6.2 *Fused derivatives*

One obvious source of fused derivatives is the use of the Isay reaction with cyclic diketones or their equivalents. One of these is, or course, alloxan, which whilst giving ureidopyridopyrazines in neutral conditions, gives isoalloxazine-type products, *e.g.* (**447**), in acid ⟨*e.g.* 38CB(B)1243, 38CB(B)1323⟩. Boric acid/acetic acid has been favoured ⟨49JA1891⟩, but in some cases use of BF₃ is necessary. Other cyclic diketones used include quinones, such as phenanthraquinone, 1,2-indanedione and 1,2,3-indanetrione, and isatin. 2,4,6-Trioxopiperidine, 2,3,4-trioxopiperideine and various other related compounds also react with phenylenediamines to give pyrido-[2,3-*b*]- or -[3,4-*b*]-quinoxalines, whilst 3-hydroxy-pyridin-2-one reacted with benzofuroxan analogues by the transformation described in Section 2.15.5.6.1 above to give the corresponding pyrido[2,3-*b*]quinoxaline 5,10-di-*N*-oxides ⟨72JOC589⟩.

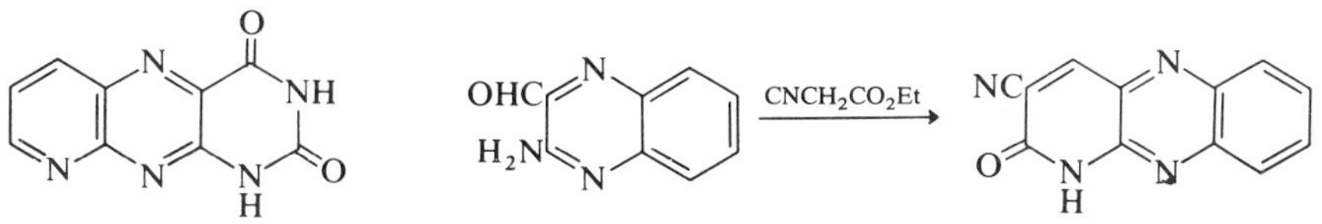

(**447**)  (**448**)  (**449**)

Finally the preformed 2-formyl-3-aminoquinoxaline (**448**) gave the pyrido[2,3-*b*]quin-oxalin-2-one (**449**) with activated esters/sodium alkoxide ⟨79ZC422⟩.

### 2.15.15.7 By Formation of Two Bonds from [3+3] Atom Fragments

The one report of a true [3+3] cyclization occurs in the only reported synthesis of the pyrazino[2,3-*c*]isoquinoline ring system. The isoquinolino fused derivative (**451**) was iso-lated from dimerization of an intermediate iminoquinone formed by air oxidation of the unstable 4-amino-2-methyltetrahydroisoquinolin-1-one (**450**) ⟨75JOC1760⟩.

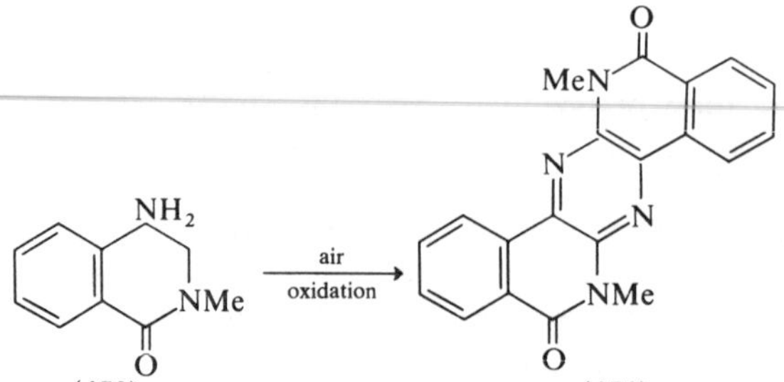

(450) → air oxidation → (451)

## 2.15.16 BIOLOGICAL ACTIVITY AND USES OF PYRIDODIAZINES

### 2.15.16.1 Pyridopyrimidines

The only naturally occurring pyridopyrimidine appears to be the co-enzyme $F_{420}$ (199), a 5-deazaflavin isolated from methanobacteria ⟨80JBC(255)1891⟩ and recently synthesized ⟨79JA4419, 80JHC1709⟩. Many deaza derivatives of natural pteridines and flavins have, however, been synthesized including 5-deazariboflavin ⟨76AG475⟩. The competitive enzyme inhibitory activity of 5-deaza- ⟨79MI21506, 79JA6068⟩ and 8-deaza-pterins ⟨81JHC671⟩ in oxidation–reduction reactions has been studied, as has that of the corresponding 5-deaza- (19) ⟨79JBC(254)12145, 79MI21500⟩ and 10-deaza-flavins (452) ⟨72CC847⟩. The former have recently been reviewed ⟨80MI21501, 80MI21502⟩. Some simple 8-ribitylpyrido[2,3-*d*]pyrimidines are inhibitors of riboflavin synthesis ⟨74JCS(P1)1225, 80JCS(P1)2645⟩.

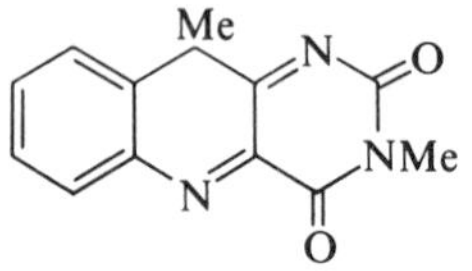

(452) 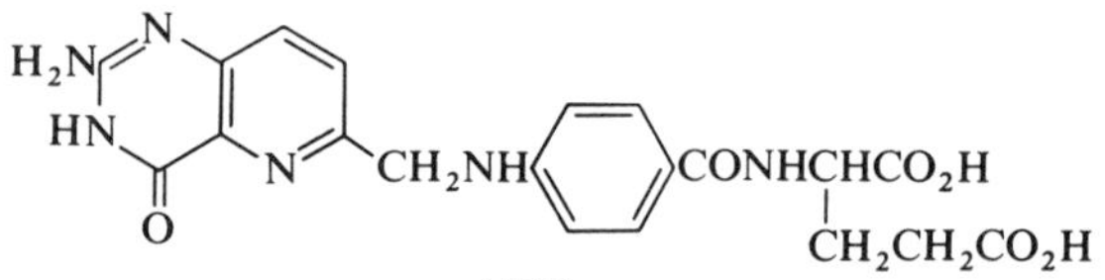                                    (453)

The pyrido[2,3-*d*]pyrimidine analogue of adenine has been prepared ⟨80JBC(255)909⟩, as have some pyrido[2,3-*d*]pyrimidine nucleosides related to the antibiotic sangivamycin ⟨77JOC997, 72JOC3980⟩. 8-Deazafolic acid (453) ⟨74JMC470, 81JOC1777, 81JMC1254⟩ has been synthesized, as has the 7-positional isomer of the 5-deaza analogue ⟨81JOC1394⟩.

5-Deazafolic acid itself has been prepared ⟨82JOC761⟩ and its antifolate properties studied ⟨81B1241⟩, and the 5-deaza-5-oxo analogues of the antifolate antitumour drugs aminopterin (454a) and methotrexate (454b) have also recently been synthesized ⟨80JOC3746⟩. Many pyrido[2,3-*d*]pyrimidines are potent dihydrofolate reductase inhibitors and are thus active as antibacterials ⟨68JMC708, 68JMC711⟩, antimalarials, antipsoriasis agents, antitumour agents (including the very active BW301U, 455 ⟨80JMC327⟩), immunosuppressants and diuretics ⟨*e.g.* 76CPB2057, 76CPB2078⟩.

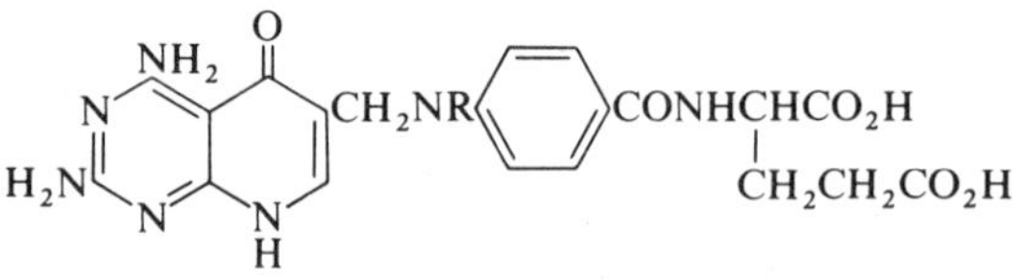

(454) a; R = H
      b; R = Me

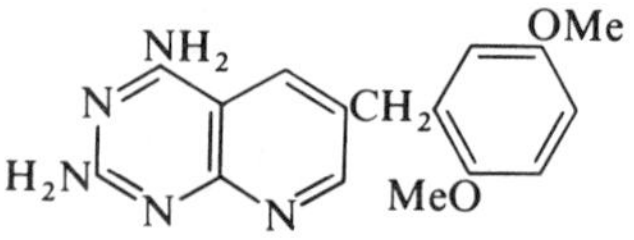

(455)

The greatest medicinal interest in pyridopyrimidines has been in another type of antibacterial derivative, the analogues of nalidixic acid, piromidic (68) ⟨71CPB1426⟩ and pipemidic (69) acids ⟨75JMC74⟩.

In addition to the foregoing, various pyridopyrimidines have been claimed as antiallergics including the interesting (456) ⟨79JMC44, 80JMC262⟩, antiinflammatories, *e.g.* (457) ⟨75AF1712⟩, hypnotics ⟨69IJC866⟩ and antiepileptics (including an active aza analogue of methaqualone, Kr-100 (458) ⟨79MI21507⟩), analgesics, CNS depressants, hypotensives, antiulcer agents, antithrombotics and platelet aggregation inhibitors (including SH869, 459), antihistamines, spasmolytics, antitussives, muscle relaxants, CNS stimulants, amoebicides,

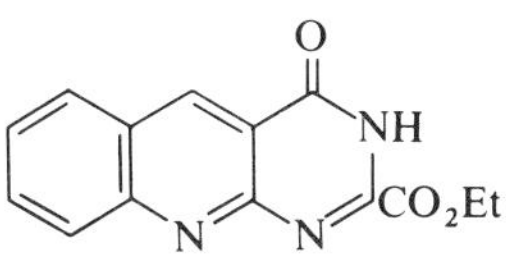

(456) 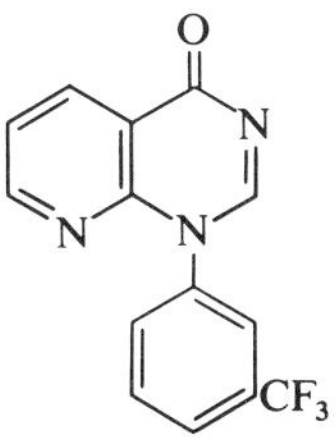   (457) 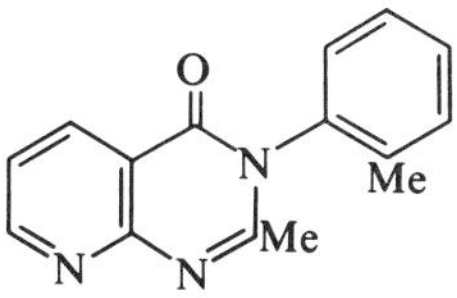   (458) 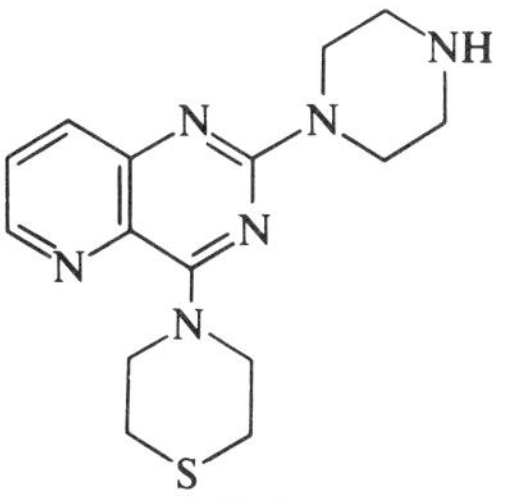   (459)

fungicides, coronary dilators, anorexics, antiphlogistics, antipyretics, neuroleptics, pyridoxal antagonists and xanthine oxidase inhibitors.

Nonmedical uses claimed for pyridopyrimidines include uses as growth promoters, cytokinins, herbicides, agricultural fungicides, coccidiostats, dyestuffs intermediates, UV absorbants and corrosion inhibitors.

### 2.15.16.2 Pyridopyridazines

No natural products or their analogues are included among the pyridopyridazines, but several interesting biologically active compounds have emerged. Some 1-chloro-4-hydrazino- and 4-chloro-1-hydrazino-pyrido[2,3-*d*]pyridazines (**460**) are very active hypotensives, whilst related dialkoxy compounds have anticonvulsant activity ⟨65CPB586⟩.

Similarly, hypotensive activity is found in other analogues of hydralazine, including the very active endralazine (**461**) ⟨79AF1835, 79AF1843⟩, a pyrido[4,3-*c*]pyridazine.

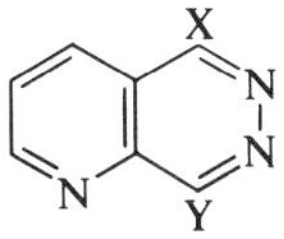

(460) **a**; X = Cl, Y = NHNH₂      (461)
      **b**; Y = Cl, X = NHNH₂

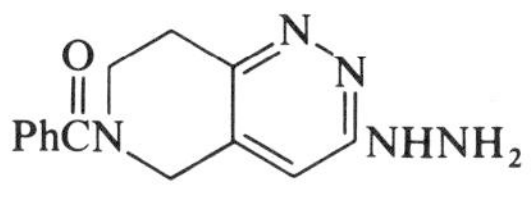

The greatest interest, however, has been shown in diuretics, culminating in the development of the drug DS-511 (**307**), a 1,4-dimorpholinopyrido[3,4-*d*]pyridazine from a large series of related heterocyclic derivatives ⟨77AF1663⟩. Its mode of action, toxicology, analysis, pharmacology and pharmacokinetics have been investigated. The relation between diuretic activity and MO charge density has already been mentioned (Section 2.15.7.2.5).

In addition, various pyridopyridazines have been claimed to have activity as antibacterials and antiseptics, antitubercular agents, analgesics, anti-inflammatories, antiallergics, tranquillizers, CNS depressants and muscle relaxants.

Nonmedical uses claimed for pyridopyridazines include fungicides, growth promotion agents, parasiticides, UV absorbants and dyestuff intermediates.

### 2.15.16.3 Pyridopyrazines

The pyridopyrazine group does not contain any actual natural products, but the considerable scope for synthesis of 1- and 3-deaza analogues of various natural pteridines or pteridine drugs has been widely exploited.

1-Deazafolic acid (**462a**) does not appear to be known, although 3-deazafolic acid (**462b**) has been synthesized ⟨58JBC(231)331⟩, as well as the $N^{10}$-methyl-1-deaza analogue (**462c**) ⟨74JMC553⟩.

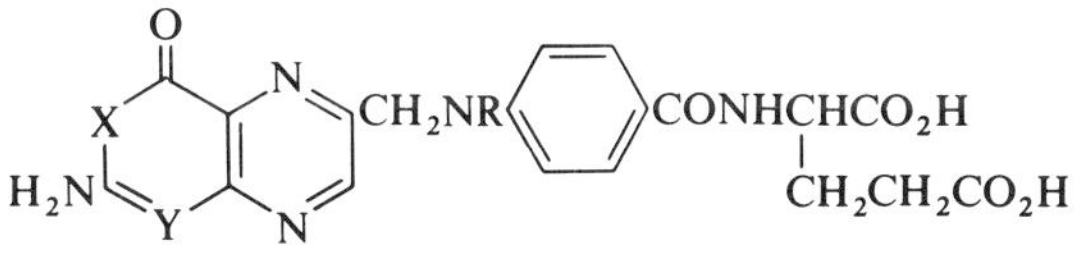

(462) **a**; R = H, X = NH, Y = CH
      **b**; R = H, X = CH₂, Y = N
      **c**; R = Me, X = NH, Y = CH

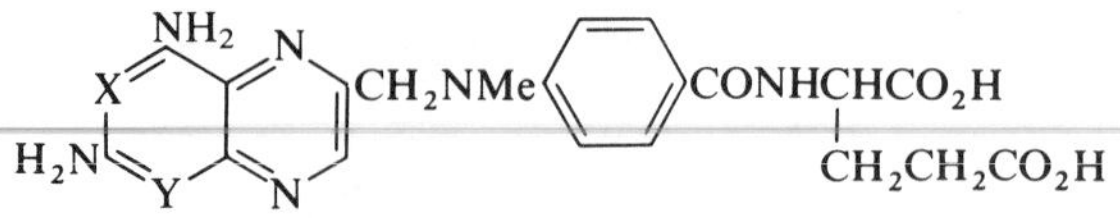

(463) a; X = N, Y = CH
b; X = CH, Y = N

The 1- and 3-deaza analogues (463a, 463b) of the DHFR-inhibitory antitumour drug methotrexate have also been investigated ⟨71JOC2818⟩.

Both 1- and 3-deazariboflavin (464, 465) have been made ⟨77TL2551⟩ and the properties of the former thoroughly investigated ⟨77B3586, 78B1942⟩. Some 3,5-dialkyl-1-deaza analogues have also been prepared ⟨81JA5494⟩.

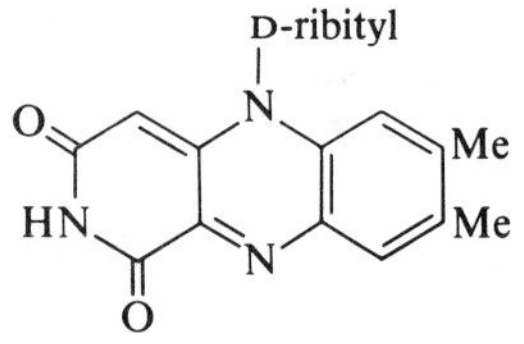

(464) 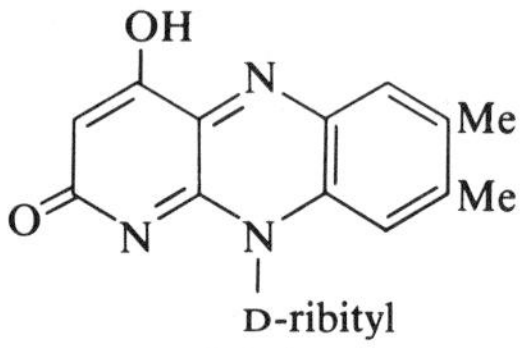　(465) 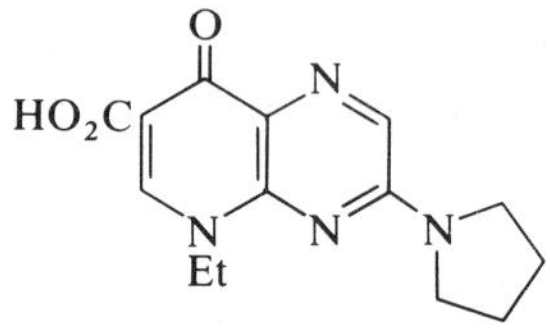　(466)

Apart from the methotrexate analogues above, and some close relatives ⟨79JMC862⟩, the only noteworthy pyridopyrazine drugs have been some antibacterial 5-substituted 8-oxopyrido[2,3-*b*]pyrazine-7-carboxylic acids (466), relatives of piromidic acid ⟨74CPB1864⟩. In addition to antibacterials, pyridopyrazines have been claimed as active as antimalarials, virucides, trichomonacides, amoebicides, anti-inflammatories, diuretics, antitumour agents, tranquillizers and CNS depressants, anxiolytics, analgesics, antiphlogistics, anorexics, antidepressants, hypnotics, antiallergics and serotonin mimetics.

Nonmedical uses envisaged include as growth promoters, indicators for copying processes, analytical complexing agents, cyanine dyes and dye-bleaching catalysts.

# 2.16

# Pteridines

W. PFLEIDERER
*Universität Konstanz*

| 2.16.1 | STRUCTURE | 264 |
| | *2.16.1.1  Introduction* | 264 |
| | *2.16.1.2  Pteridine* | 264 |
| | *2.16.1.3  Alkyl- and Aryl-pteridines* | 265 |
| | *2.16.1.4  Chloropteridines* | 266 |
| | *2.16.1.5  Aminopteridines* | 267 |
| | *2.16.1.6  Hydroxypteridines (Pteridinones)* | 271 |
| | *2.16.1.7  Thiopteridines (Pteridinethiones)* | 273 |
| | *2.16.1.8  Aminohydroxypteridines (Aminopteridinones)* | 273 |
| | *2.16.1.9  Pteridinecarboxylic Acids and Derivatives* | 276 |
| | *2.16.1.10  8-Substituted Quinonoid Pteridines* | 277 |
| | *2.16.1.11  Dihydropteridines* | 279 |
| | *2.16.1.12  Tetrahydropteridines* | 280 |
| | *2.16.1.13  Pteridine N-Oxides* | 281 |
| | *2.16.1.14  Pteridine Radicals* | 282 |
| | *2.16.1.15  Pteridine Nucleosides and Nucleotides* | 282 |
| | *2.16.1.16  Dimeric Pteridines* | 283 |
| | *2.16.1.17  Complex Pteridine Structures* | 284 |
| | *2.16.1.18  NMR Spectra* | 285 |
| | *2.16.1.19  Mass Spectra* | 285 |
| | *2.16.1.20  Polarography and Electrochemistry* | 285 |
| | *2.16.1.21  Solubility* | 286 |
| 2.16.2 | REACTIVITY | 286 |
| | *2.16.2.1  Reactivity at the Ring Carbon Atoms* | 286 |
| | *2.16.2.1.1  Electrophilic reagents* | 286 |
| | *2.16.2.1.2  Nucleophilic reagents* | 286 |
| | *2.16.2.1.3  Cine and tele substitutions* | 289 |
| | *2.16.2.1.4  Free radicals* | 290 |
| | *2.16.2.2  Reactivity of Substituents* | 291 |
| | *2.16.2.2.1  Halogen substituents* | 291 |
| | *2.16.2.2.2  N-Linked substituents* | 293 |
| | *2.16.2.2.3  O-Linked substituents* | 295 |
| | *2.16.2.2.4  S-Linked substituents* | 299 |
| | *2.16.2.2.5  C-Linked substituents* | 301 |
| | *2.16.2.3  Reactivity at the Ring Nitrogen Atoms* | 304 |
| | *2.16.2.4  Reduction* | 305 |
| | *2.16.2.5  Oxidation* | 307 |
| | *2.16.2.6  Ring Transformations* | 308 |
| | *2.16.2.7  Rearrangements* | 308 |
| 2.16.3 | SYNTHESES | 309 |
| | *2.16.3.1  Syntheses from Pyrimidines* | 309 |
| | *2.16.3.1.1  Unambiguous syntheses* | 313 |
| | *2.16.3.1.2  The Timmis reaction* | 313 |
| | *2.16.3.1.3  The Pachter reaction* | 314 |
| | *2.16.3.1.4  The Polonovski–Boon reaction* | 315 |
| | *2.16.3.1.5  Miscellaneous regioselective syntheses* | 316 |
| | *2.16.3.2  From Pyrazines* | 317 |
| | *2.16.3.2.1  The Taylor synthesis* | 318 |
| | *2.16.3.3  Transformations from Other Heterocyclic Ring Systems* | 319 |
| | *2.16.3.4  Biosynthesis* | 320 |

2.16.4  APPLICATIONS                                                      322
  2.16.4.1   *Naturally Occurring Pteridines*                             322
  2.16.4.2   *Synthetic Pteridines with Chemotherapeutic Effects*         324
  2.16.4.3   *Folic Acid and Related Derivatives*                         325

## 2.16.1 STRUCTURE

### 2.16.1.1 Introduction

Knowledge of the pteridines ⟨B-64MI21601⟩ originated in the 1890's when Hopkins ⟨1891MI21600, 42MI21600⟩ published the results of his investigations on the wing pigments of the common English brimstone butterfly and the cabbage white butterfly. At about the same time, synthetic work by Kühling ⟨1895CB1970⟩ resulted in the first laboratory preparation of a substance with the same heterocyclic nucleus, but the relationship between these results was not clearly realized until 1940 when the wing pigments, now known as 'pterins', were shown by Purrmann ⟨40LA(544)182, 40LA(546)98, 41LA(548)284⟩, after 15 years of tedious and laborious investigations in the laboratories of Schöpf in Darmstadt ⟨33LA(507)261⟩ and Wieland in Munich ⟨25CB2178, 26CB2067⟩, to be derived from the pteridine ring system. The name 'pteridine', which has its origin in the Greek word 'pteron', wing, was proposed by Wieland ⟨41LA(548)83⟩ in 1941 and has been used ever since as a trivial name for the condensed pyrazino[2,3-*d*]pyrimidine ring system (**1**), which is numbered according to the IUPAC rules. Another numbering system analogous to that of the purine system has also been used in earlier literature, creating some confusion in the nomenclature, as do also the old terms azapurine, alloxazine, and lumazine.

Most naturally occurring pteridines have an amino group in the 2-position and a 'hydroxy group' in the 4-position, thus limiting structural variations to different substituents in positions 6 and 7. The term 'pterin', which is used as a final syllable to name naturally occurring pteridine derivatives in general, should therefore be designated only to 2-amino-4-hydroxypteridine and not be used as a synonym for pteridine. If, in addition, prototropically active functions are represented in their most favoured tautomeric formula, pterin has to be written as 2-aminopteridin-4-one (**2**) and lumazine, another basic molecule of naturally occurring pteridines, in the pteridine-2,4-dione (**3**) structure. The most common butterfly wing pigments xanthopterin (**4**), isoxanthopterin (**5**), and leucopterin (**6**) can therefore also be described as 6-oxo-, 7-oxo- and 6,7-dioxo-pterins respectively.

The chemistry of pteridines was reviewed frequently during the early years of development in this field ⟨42N269, 45FOR(4)64, 47CRV(41)63, 52QR197, 54FOR(11)350⟩, but since 1964 ⟨64AG(E)114⟩ a critical survey of the experimental results has not been produced, although there have been six International Symposia on the Chemistry and Biology of Pteridines in 1952, 1954 ⟨B-54MI21600⟩, 1962 ⟨B-64MI21601⟩, 1969 ⟨B-70MI21605⟩, 1975 ⟨B-75MI21601⟩ and 1979 ⟨B-79MI21606⟩ reporting on the progress in the form of extended proceedings.

### 2.16.1.2 Pteridine

The basic pteridine nucleus (**1**) consists, according to X-ray crystallographic studies ⟨75JCS(P2)40⟩, of a more or less planar molecule with no unexpected features in so far as the bond angles and bond distances are concerned. Bond angles C—N—C with a nitrogen atom in the centre are less than 120°, as for other six-membered ring nitrogen heterocycles. The C(4)—N(3) bond, which is known to be highly reactive, is found to be slightly longer than expected, demonstrating that chemical reactivity has to involve considerations of the geometry and electronic structure of the transition state and should not be predicted from the ground-state structure alone. The bond lengths and angles of pteridine are illustrated in structure (**7**).

(**7**)

Pteridine is a yellowish substance of high solubility in all solvents ranging from light petroleum to water and is low-melting at 139.5 °C. The NMR spectrum contains two singlets which can be assigned to the protons of the pyrimidine ring and a pair of doublets from the adjacent H-6 and H-7 protons ⟨B-73NMR347⟩. Further assignment of peaks to specific protons was made by deuteration and methyl substitution ⟨65JCS623⟩, showing the proton chemical shifts in CDCl$_3$ solution ($\delta$-values in p.p.m.) to be: H-6 (9.15), H-7 (9.33) ($J_{6,7} = 1.7$ Hz), H-2 (9.65), H-4 (9.80) ⟨63JCS1773⟩ from highest to lowest field. The UV spectrum of (**1**) in *n*-hexane is characterized by three $\pi-\pi^*$ bands at 210 ($\varepsilon$ 11 000), 235 ($\varepsilon$ 2910) and 301 nm ($\varepsilon$ 7490) as well as a very weak $n-\pi^*$ band at 387 nm ($\varepsilon$ 84) ⟨55JCS2336⟩. In aqueous solution pteridine exhibits anomalous behaviour in adding a molecule of water reversibly across the 3,4-double bond to form 4-hydroxy-3,4-dihydropteridine (**9**) ⟨62JCS645⟩. At 20 °C the equilibrium ratio is 3.5:1 in favour of the 'anhydrous' pteridine, whereas its addition to acid or alkali results in a rapid and quantitative formation of the cation (**10**) and anion (**8**) respectively (equation 1). These equilibria are also responsible for the fact that (**1**), although it has no ionizable hydrogen, behaves on titration with alkali as a weak acid with p$K_a$ 11.21, and furthermore covalent hydration also explains the unexpected basic properties of (**1**) with p$K_a$ 4.79. The cation and anion finally undergo ring fission to 2-aminomethyleneamino-3-formylpyrazine in a relatively slow reaction. Kinetic studies allowed the separation of the reversible acid–base catalyzed hydration from the ring opening process ⟨63JCS2648⟩.

(1)

(**8**)     (**9**)     (**10**)

The $\pi$-electron deficiency of (**1**) is also seen in NMR spectroscopic studies. Acidifying a solution in D$_2$O with DCl to pH 2 rapidly produces the 3,4-hydrate in which the signal for H-4 has moved upfield to 6.8 p.p.m. If this acidic solution is allowed to stand at 33 °C new peaks appear and the spectrum may be interpreted as being that of an equilibrium mixture of 21% 3,4-monohydrate (**10**) and 79% 5,6,7,8-dihydrate (**11**) ⟨66JCS(C)1105⟩. This example is one of the rare cases where an intramolecular shift is seen in a time-dependent thermodynamically controlled process ⟨63JCS5151⟩, indicating that both hydrates retain one heteroaromatic ring as a molecular feature. Similarly, 1:1 and 2:1 adducts as equilibrium mixtures are formed with alcohols such as methanol, ethanol and isopropanol, whereas in the presence of sodium methoxide immediate attack takes place at the pyrazine ring to produce 6,7-dimethoxy-5,6,7,8-tetrahydropteridine ⟨71JCS(B)2423⟩.

### 2.16.1.3 Alkyl- and Aryl-pteridines

Studies on covalent hydration of N-heterocycles ⟨67AG(E)919, 76AHC(20)117⟩ have revealed the diagnostic value of alkyl substituents in structural assignments due to their steric hindrance effects in addition reactions. *C*-Methyl substituents are therefore also considered as molecular probes to solve fine-structural problems in the pteridine field. The derivatives

2-, 6- and 7-methylpteridine ⟨70JCS(C)1540⟩ behave analogously to the parent substance in forming as cations in a fast reaction the 3,4-hydrates first, then shifting to equilibrium mixtures of 3,4-mono- and 5,6,7,8-di-hydrates in amounts of 75 and 88% respectively.

A 2-methyl group increases inductively the electron density at N(3) more than at C(4) and hence increases resistance to removal of a proton, thereby stabilizing the 3,4-hydrate and causing the observed increase in the equilibrium ratio of 3,4-hydrate to 5,6,7,8-dihydrate. The methyl group of 6- and 7-methylpteridine sterically and/or inductively decreases covalent hydration at the pyrazine moiety, again favouring 3,4-hydration. In spite of the methyl groups on C(6) and C(7) respectively, the NMR integral shows that the 5,6- and 7,8-bonds are hydrated to about the same extent. On the other hand, the anhydrous cation of 4-methylpteridine gave no observable 3,4-hydrate but was converted partly into another species characterized as the 5,6,7,8-dihydrate present to 40% of the composition. 6,7-Dimethyl- and 2,6,7-trimethyl-pteridine give 3,4-hydrated cations, presumably owing to steric and inductive blocking of nucleophilic attack at C(6) and C(7) ⟨66JCS(B)1105⟩. It is therefore somewhat surprising that 4,6,7-trimethyl- and 2,4,6,7-tetramethyl-pteridine ⟨68JCS(C)2292⟩ show in the NMR spectrum the 3,4-hydrate as the predominant species, (pH 1 in DCl at 5 °C), and 5,6,7,8-dihydration could be excluded. These results show that 'steric hindrance' is not necessarily additive and a likely reason is the simplification of the structure of water aggregates in the neighbourhood of largely hydrophobic molecules ⟨45JCP(13)507⟩.

4-Trifluoromethylpteridine and its 7-methyl and 6,7-dimethyl derivatives ⟨69JCS(C)1751⟩ are, as expected, even more subject to hydration. The first two are essentially completely hydrated across the 3,4-double bond at equilibrium in neutral solution and the last is partly hydrated. On dissolution of 4-trifluoromethylpteridine in aqueous acid the 5,6,7,8-dihydrated cation is the main product initially, rearranging more slowly to the thermodynamically more stable 3,4-hydrate.

2-Phenylpteridine and its 4- and 7-monomethyl, 4,7- and 6,7-dimethyl and 4,6,7-trimethyl derivatives ⟨69JCS(C)1408⟩, as well as the corresponding 4-phenylpteridine series and its 2- and 7-methyl, 2,7- and 6,7-dimethyl and 2,6,7-trimethyl derivatives ⟨69JCS(C)1883⟩, exist as neutral molecules in aqueous solution, essentially as unhydrated species. In acid solution 2- and 4-phenylpteridine and its 4- and 2-methyl derivatives favour the 5,6,7,8-dihydrated cation state, while 7-mono- or 6,7-disubstitution shifts the equilibrium mixture towards the 3,4-monohydrates.

### 2.16.1.4 Chloropteridines

Among the simple halogenopteridines only the 2-, 4-, 6- and 7-monochloro as well as the 6,7-dichloro derivatives have been synthesized ⟨51JCS474, 52JCS1620, 52JCS4219, 54JCS3832, 64JCS1666⟩ and structurally investigated. It has generally been assumed that the dominating feature of their chemistry is the ease of replacement of the reactive chlorine atoms by a nucleophilic reagent, but in 6-chloro- (12), 7-chloro- (13), and 6,7-dichloro-pteridine (14) covalent hydration also plays an important role due to the negative inductive effect of the chlorine atoms. The spectral changes which occur on dissolving in neutral aqueous solution very closely resemble those observed when pteridine is similarly treated. In all cases partial reversible covalent hydration at the 3,4-double bond (15) is the first reaction observed and this is followed by slower ring-opening reactions. In neutral or weakly acidic solutions, hydrolysis of a chlorine atom appears to be negligible in the case of 7-chloro- and 6,7-dichloro-pteridine, but it may be more important in the 6-isomer. The cations of these three compounds are found to be almost completely hydrated. This is also seen from the $pK_a$ value, which is expected to be close to $-3$ for the anhydrous species but experimentally is found to be 3.72, 3.26 and 3.27, respectively, indicating an equilibrium $pK_a$ with a high content of the more basic hydrate form (equation 2).

$$\qquad\qquad\qquad\qquad\qquad\qquad\qquad\qquad\qquad (2)$$

(12) R = Cl, R$^1$ = H　　(15) X = OH, NH$_2$, PhCH$_2$NH, SH, PhCH$_2$S
(13) R = H, R$^1$ = Cl　　　　 R = H, Cl
(14) R = Cl, R$^1$ = Cl　　　　 R$^1$ = Cl, H

These effects can be attributed mainly to the inductive nature of the chlorine atoms, which reduces the electron density at position 4 and increases polarization of the 3,4-double bond. The dual reactivity of the chloropteridines has been further confirmed by the preparation of new adducts and substitution products. The addition reaction competes successfully, in a preparative sense, with the substitution reaction, if the latter is slowed down by a low temperature and a non-polar solvent. Compounds (**12**) and (**13**) react with dry ammonia in benzene at 5 °C to yield the 3,4-adducts (**15**), which were shown by IR spectroscopy to contain little or none of the corresponding substitution product. The adducts decompose slowly in air and almost instantaneously in water or ethanol to give the original chloropteridine and ammonia. Certain other amines behave similarly, forming adducts which can be stored for a few days at −20 °C. Treatment of (**12**) and (**13**) in acetone with hydrogen sulfide or toluene-α-thiol gives adducts of the same type.

The consequences of these interactions with nucleophiles are obviously readily explained by a rapid but reversible addition of the nucleophile to the 3,4-double bond which accompanies the normal irreversible displacement of the chlorine atoms.

2,4-Dichloro- ⟨59JA2464⟩, 2,4,6- and 2,4,7-trichloro- ⟨56JCS4621⟩ and 2,4,6,7-tetrachloropteridine ⟨41LA(548)83⟩ have not yet been studied regarding their physical properties due to their high instability and reactivity towards nucleophiles.

### 2.16.1.5 Aminopteridines

Introduction of amino groups into the 2-, 4-, 6- and 7-positions of the pteridine nucleus raises two general aspects for consideration: firstly, the electron-donating properties of this function, and secondly, its potentiality to show tautomerism with iminodihydro forms. It is expected from the strong π-electron deficiency of the unsubstituted pteridine ring that monosubstitution with an amino group is still not enough to compensate for the electron attraction of the ring N atoms stabilizing the system over the normal pH range 0–14. An interesting and complex situation is found with 2-aminopteridine (**16**), which forms a yellow aqueous solution which decolorizes on acidification, whereas that of 2-amino-4-methylpteridine remains yellow ⟨62JCS2595⟩. The latter behaviour appears normal, and the abnormality of (**16**) involves 3,4-covalent hydration of the cation to give the resonance-stabilized ion (**18**), which with its guanidinium-type stabilization and formal 3,4-dihydro structure accounts for the π-electron deficiency of the heteroaromatic resonance form (**17**; equation 3).

$$\text{(16)} \quad \rightleftharpoons \quad \text{(17)} \quad \rightleftharpoons \quad \text{(18)} \tag{3}$$

There is also evidence for stable 3,4-adducts from the X-ray analysis of 2-amino-4-ethoxy-3,4-dihydropteridinium bromide, the nucleophilic addition product of 2-aminopteridine hydrobromide and ethanol ⟨69JCS(B)489⟩. The pH values obtained by potentiometric titration of (**16**) with acid and back-titration with alkali produces a hysteresis loop, indicating an equilibrium between various molecular species such as the anhydrous neutral form and the predominantly hydrated cation. Table 1 illustrates more aspects of this anomaly. 2-Aminopteridine, paradoxically, is a stronger base than any of its methyl derivatives; each dimethyl derivative is a weaker base than either of its parent monomethyl derivatives. Thus the base strengths decrease in the order in which they are expected to increase, with only the 2-amino-4,6,7-trimethylpteridine out of order, being more basic than the 4,7-dimethyl derivative.

The basic $pK_a$ values, which have to be considered as equilibrium values, including those of anhydrous and hydrated species, reveal a destabilizing inductive effect of the 6- and 7-methyl group towards 3,4-hydrate formation, as do also the 2-methylamino and 2-dimethylamino groups for additional steric reasons. If the cation of 2-aminopteridine did not add water its $pK_a$ value would be about 1.6, arrived at by substracting from the $pK_a$ 2.6 of the essentially anhydrous 2-amino-4,7-dimethylpteridine cation 0.3 for the 7- and 0.7 for the 4-methyl group. The difference between the observed value of 4.29 and the

**Table 1** Physical Data for Aminopteridines

| Pteridine | pKa | λmax (nm) [UV absorption spectra][a] | log ε | pH | Solubility in H2O[b] 20 °C | Solubility in H2O[b] 100 °C | Melting point (°C) |
|---|---|---|---|---|---|---|---|
| 2-Amino- | — | 225, 259, 370 | 4.39, 3.81, 3.82 | 7.0 | 1350 | 100 | 275 (decomp.) |
| cation | 4.29 | 232, 302, — | 3.92, 3.87 | 2.0 | — | — | — |
| 2-Amino-6-methyl- | — | 224, [256], 376 | 4.42, [3.92], 3.84 | 7.0 | — | — | — |
| cation | 4.05 | 238, 309, — | 3.98, 3.91 | 1.0 | — | — | — |
| 2-Amino-7-methyl- | — | 228, [255], 367 | 4.35, [3.83], 3.89 | 7.0 | — | — | — |
| cation | 3.76 | 237, 306, — | 3.74, 3.98 | 1.0 | — | — | — |
| 2-Amino-6,7-dimethyl- | — | 227, [257], 367 | 4.36, [3.87], 3.88 | 7.0 | — | — | — |
| cation | 3.41 | 240, 310, — | 3.93, 3.98 | 1.0 | — | — | — |
| 2-Methylamino- | — | 229, 273, 388 | 4.36, 3.96, 3.82 | 6.5 | 320 | 35 | 219–220 |
| cation | 3.62 | 235, 304, [363] | 4.09, 3.93, [2.07] | 1.0 | — | — | — |
| 2-Dimethylamino- | — | 236, 281, 410 | 4.37, 4.02, 3.82 | 7.1 | 25 | — | 125 |
| cation | 3.03 | 238, 305, [370] | 4.17, 3.91, [2.59] | 1.0 | — | — | — |
| 2-Amino-4-methyl- | — | 225, 260, 367 | 4.44, 3.83, 3.85 | 7.0 | — | — | — |
| cation | 2.82 | 217, 307, 346 | 4.27, 3.72, 3.71 | 0.5 | — | — | — |
| 2-Amino-4,6-dimethyl- | — | 226, 258, 373 | 4.46, 3.94, 3.85 | 7.0 | — | — | 312 (dec.) |
| cation | 2.70 | 219, —, 355 | 4.40, —, 3.77 | -0.5 | — | — | — |
| 2-Amino-4,7-dimethyl- | — | 228, 257, 363 | 4.39, 3.83, 3.92 | 7.0 | — | — | 284 (dec.) |
| cation | 2.63 | 218, —, 347 | 4.27, —, 4.02 | 0.0 | — | — | — |
| 2-Amino-4,6,7-trimethyl- | — | 227, —, 366 | 3.31, —, 3.91 | 7.0 | — | — | — |
| cation | 3.03 | 219, —, 350 | 4.42, —, 4.05 | 1.0 | — | — | — |
| 4-Amino- | — | —, 244, 335 | —, 4.20, 3.82 | 7.3 | 1400 | 80 | 305 (dec.) |
| cation | 3.56 | 229, —, 324 | 4.10, —, 3.99 | 1.1 | — | — | — |

| | | | | | | | | | | | |
|---|---|---|---|---|---|---|---|---|---|---|---|
| 4-Amino-2-methyl- | — | 219 | 245 | 336 | 3.97 | 4.23 | 3.81 | 7.0 | — | — | 235 (dec.) |
| cation | 4.30 | 232 | — | 329 | 4.14 | — | 3.97 | 2.0 | — | — | |
| 4-Dimethylamino- | — | — | 241 | 362 | — | 4.15 | 3.93 | 7.1 | 60 | 4 | 165 |
| cation | 4.33 | 239 | — | 347 | 4.19 | — | 4.10 | 2.0 | — | — | — |
| 6-Amino- | — | 232 | 258 | 362 | 4.30 | 4.01 | 3.75 | 7.0 | 1500 | 110 | 300 (dec.) |
| cation | 4.15 | — | — | — | — | — | — | — | — | — | — |
| 6-Dimethylamino- | — | 231 | 279 | 399 | 4.21 | 4.17 | 3.74 | 7.0 | — | — | 212 |
| cation | 4.31 | — | — | — | — | — | — | — | — | — | — |
| 7-Amino- | — | 228 | 262 | 334 | 4.26 | 3.80 | 4.03 | 5.1 | 1400 | 170 | 320 (dec.) |
| cation | 2.96 | 217 | — | 326 | 4.25 | — | 4.00 | 0.6 | — | — | — |
| 7-Dimethylamino- | — | 240 | 279 | 364 | 4.19 | 3.98 | 4.05 | 5.1 | 6 | 2 | 204 |
| cation | 2.53 | 245 | 312 | 361 | 3.95 | 3.96 | 4.16 | 0.5 | — | — | — |
| 2,4-Diamino- | — | 225 | 254 | 363 | 4.03 | 4.31 | 3.86 | 9.0 | 3000 | 130 | 315 (dec.) |
| cation | 5.32 | 239 | 282 | 331 | 4.10 | 3.71 | 3.98 | 3.0 | — | — | — |
| 4,6-Diamino- | — | — | 263 | 375 | — | 4.18 | 3.81 | 6.7 | 24 000 | 940 | — |
| cation | 4.37 | 254 | 284 | 376 | 4.09 | 3.87 | 3.90 | 2.4 | — | — | — |
| 4,7-Diamino- | — | 233 | 257 | 343 | — | — | — | — | 5000 | 300 | — |
| cation | 4.97 | 227 | 257 | 350 | — | — | — | — | — | — | — |
| 2,4,7-Triamino- | — | 227 | 257 | 350 | 4.54 | 4.13 | 4.17 | 8.5 | 4500 | 200 | — |
| cation | 6.30 | 255 | 275 | 342 | 4.16 | 3.84 | 4.28 | 4.3 | — | — | — |
| 4,6,7-Triamino- | — | 227 | 256 | 345 | 4.33 | 4.17 | 4.11 | 8.5 | 12 500 | 450 | — |
| cation | 5.57 | 224 | 245 | 353 | 4.20 | 4.21 | 4.26 | 3.6 | — | — | — |
| 2,4,6,7-Tetramino- | — | — | — | — | — | — | — | — | 13 000 | — | >360 |
| cation | 6.86 | 235 | 305 | 360 | 3.19 | 3.88 | 4.05 | 1.0 | — | — | — |

[a] [ ] = Shoulder.
[b] Volume of water (ml) required to dissolve 1 g of heterocycle.

expected value of 1.6 suggests that the ratio of hydrated to anhydrous cation is about 500:1. The base weakening effect produced by a methyl group will lower this ratio significantly, and this can outweigh the normal base-strengthening properties.

4-Aminopteridine reveals normal behaviour in forming an anhydrous cation, presumably by protonation at N-1 giving rise to a *p*-quinonoid amidine-type structure with base-strengthening resonance. Spectral comparisons with 4-amino-2-methylpteridine, a somewhat stronger base with $pK_a$ 4.30, emphasize these conclusions. The compounds 6- and 7-aminopteridine do not show any special structural features despite the high sensitivity of the former to acid-catalyzed hydrolysis. The di-, tri- and tetra-aminopteridines fit well, in this order, into the scheme of increasing basicity counting for a formal electronic neutralization of one ring N atom by one amino group.

Studies on the tautomerism of the aminopteridines reveal a strong preference of the amino over the tautomeric iminodihydro structures, as found for most heterocyclic ring systems (equation 4) ⟨B-76MI21601⟩. A typical example is that of 4-aminopteridine (**19**) ⟨60JCS1978⟩. Its UV spectrum shows, both as neutral molecule and cation, the usual progressive bathochromic shift in the long-wavelength band with extranuclear *N*-mono- (**20**) and di-methylation (**22**) due to the stronger electron-donating power of an *N*-methyl and dimethylamino group respectively, but no shift arising from any change in the tautomeric ratio of the equilibrium-dependent molecular species. This shift is also evident in changing from the cations of 4-imino-1-methyl-1,4-dihydropteridine (**21**) to the methylimines (Table 2). Since it is reasonable to expect protonation of (**19**) to occur at the same site as methylation, the implication that N-1 is the basic centre is independently upheld by the close similarity of the cationic spectra of the amino-, methylamino- and dimethylamino-pteridines to those of the imine cation (equation 5). Noteworthy is the transannular 6,7,8-trimethyl-4-methyl-iminopteridine, with its bathochromic shift of the long-wavelength band ruling out the importance of such tautomeric forms in general.

(4)

(5)

The best indication for judging the state of the amino–iminodihydro tautomeric equilibrium is seen in the basic $pK_a$ values which differ usually by several $pK_a$ units in comparison of the most strongly basic iminopteridines with the corresponding amino derivatives. Since 4-aminopteridine shows a basic $pK_a$ of 3.56 and 4-imino-1-methyl-1,4-dihydropteridine of 9.51 it is obvious that in the unblocked free system the basic iminodihydro form is formally 'neutralized' to the less basic amino tautomer. The tautomeric ratio of $9 \times 10^5:1$ can be calculated easily from these basic $pK_a$ values because they possess corresponding cation forms. Similar studies with 7-amino-, 4,7-diamino- ⟨65JCS1175⟩ and 2,4-diamino-pteridine ⟨61JCS4413, 70CB722⟩ again establish the predominance of the amino forms in the tautomeric equilibria.

Another important correlation between structure and properties in the pteridine series is seen in the solubilities, to which insufficient attention has been paid in general. Introduction of an amino group into pteridine (**1**) lowers the solubility in all solvents despite the fact

**Table 2** Physical Data for Aminopteridines and Ring and *N*-Methylated Derivatives

| Pteridine | $pK_a$ | $\lambda_{max}$ (nm) | | | $\log \varepsilon$ | | | pH |
|---|---|---|---|---|---|---|---|---|
| 4-Amino- | — | — | 244 | 335 | — | 4.20 | 3.82 | 7.3 |
| cation | 3.56 | 229 | — | 324 | 4.10 | — | 3.99 | 1.1 |
| 4-Methylamino- | — | 226 | 248 | 352 | 4.05 | 4.06 | 3.88 | 5.7 |
| cation | 3.70 | 232 | [251] | 339 | 4.12 | [3.68] | 4.06 | 1.7 |
| 4-Dimethylamino- | — | — | 241 | 362 | — | 4.15 | 3.93 | 7.1 |
| cation | 4.33 | 239 | — | 347 | 4.19 | — | 4.10 | 2.0 |
| 1,4-Dihydro-4-imino-1-methyl-, cation | 9.51 | 233 | 333 | [350] | 4.14 | 4.00 | [3.87] | 7.4 |
| 1,4-Dihydro-4-methylimino-1-methyl-, cation | 10.34 | 233 | 344 | [354] | 4.16 | 4.07 | [4.03] | 8.3 |
| 4-Dimethylamino-1-methyl-, iodide | — | 243 | 352 | — | 4.17 | 4.11 | — | — |
| 4,8-Dihydro-6,7,8-trimethyl-4-methylimino- | — | 215 | 241 | 284 | 4.27 | 4.06 | 4.23 | |
| | | | | 383 | | | 3.68 | 8.5 |
| cation | 6.64 | 233 | 268 | 410 | 4.09 | 4.06 | 4.03 | 4.0 |
| 1,7-Dihydro-7-imino-1-methyl-, cation | 8.33 | 218 | [260] | 349 | 4.24 | [3.60] | 4.12 | 6.0 |
| 7-Amino-1,4-dihydro-4-imino-1-methyl- | 12.12 | 237 | 256 | 261 | 4.27 | 4.22 | 4.22 | |
| cation | | | 345 | [360] | | 4.15 | [4.00] | 9.8 |
| 4-Amino-2,8-dihydro-2-imino-8-methyl- | — | 241 | 288 | 313 | 4.11 | 3.93 | 3.92 | |
| | | | | [365] | | | [3.23] | 11.0 |
| cation | 8.88 | [244] | 270 | 330 | [3.73] | 4.24 | 3.52 | |
| | | | | 412 | | | 3.98 | 6.0 |

that this group almost invariably increases the solubility in water of aliphatic and aromatic substances. Monoaminopteridines are 200-fold less soluble than (**1**) and successive introduction of more amino functions decreases the solubility further (Table 1). The somewhat paradoxical insolubilizing effect of water-attracting groups is found especially with nitrogen heterocycles ⟨52JCS4219, B-69MI21601⟩, where the results are explained by the hypothesis of increased crystal-lattice forces functioning through intermolecular hydrogen-bonding. These hydrophilic groups exert even more attraction for one another than they do for the molecules of water. Evidence in support of this hypothesis comes from a study of *N*-methyl derivatives, since 2-methylaminopteridine is much more soluble and 2-dimethylaminopteridine even more so due to their reduced and missing hydrogen bonding respectively.

The tendency for pteridines with hydrogen-bonding substituents to have higher melting points is also in agreement with abnormally strong crystal-lattice forces (Table 1).

### 2.16.1.6 Hydroxypteridines (Pteridinones)

Structural considerations in hydroxypteridines are mainly concerned with the various possibilities of tautomerism of this function with the ring N atoms. Detailed studies of the tautomeric relationships have shown, as a result of UV spectrophotometric comparisons with the corresponding *O*- and *N*-methyl derivatives, that with only a few exceptions the favoured form is the lactam ⟨52JCS1620, 57CB2582, 57CB2588, 57CB2604, 57CB2617, 57CB2624, 57CB2631, 58CB1671, 59CB3190, 60MI21600, 62CB749, 62CB1605⟩. It is furthermore seen, with all four potential monohydroxypteridines, that the acidic hydrogen is always located at the adjacent ring N atom and not at one of the vinylogous positions (equation 6). The rare situation where an iminol form predominates in the tautomeric equilibrium is encountered with 1-mono- and 1,3-di-substituted 7-hydroxypteridine-2,4-diones (**23**) as well as with 7-hydroxylumazine-6-carboxylic acids (**24**). In the former case N-1 substitution forces the *peri*-located N(8)-H into the lactim configuration for steric reasons, while in (**24**) and related compounds the stabilizing effect is due to favourable intramolecular hydrogen bonding.

Considering the four potential monohydroxypteridines, pteridin-4- and -7-one ⟨56JCS3443⟩ behave normally whereas pteridin-2- and -6-one (**25**) form covalent hydrates. The reversible hydration of nitrogen heterocycles was actually discovered with pteridin-6-one ⟨52JCS1620⟩,

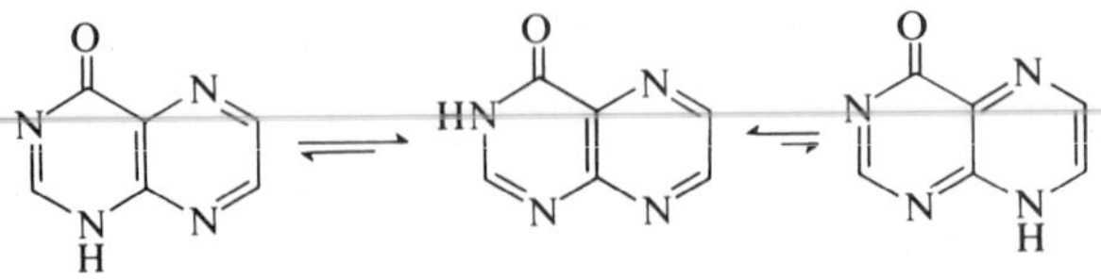

(6)

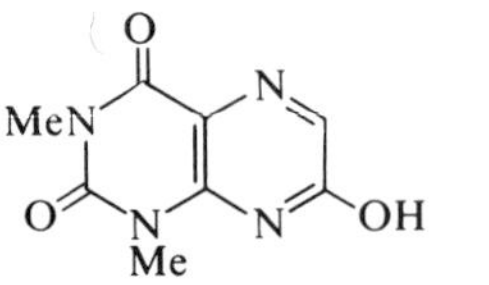

(23) 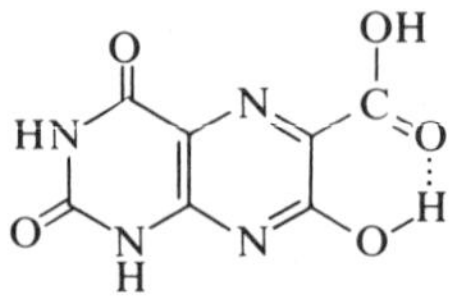                                        (24)

which shows a hysteresis curve on potentiometric titration with base and acid. The $pK_a$ values of 9.7 and 6.7 derived from these two curves are in agreement with the conclusion that the neutral form of (25) is a hydrate with covalently bound water in the 7,8-positions (26) whereas the anion (27) tends to be anhydrous (equation 7). Analogously, pteridin-2-one is a 3,4-hydrate (28) as indicated by its high $pK_a$ of 11.13, the spectrophotometric comparisons with model substances, its mild oxidation to lumazine and a marked decrease of hydrate formation in 4-methylpteridin-2-one ⟨62JCS1591⟩. The equilibrium ratio at 20 °C is found to be 320 : 1 in favour of the hydrated species ⟨62JCS2600⟩. The long-wavelength bands in the UV spectra of the neutral forms are found as low as 289 and 307 nm respectively, but anion formation is associated with large bathochromic shifts of 67 and 68 nm, suggesting that the anions are more conjugated than the neutral molecules.

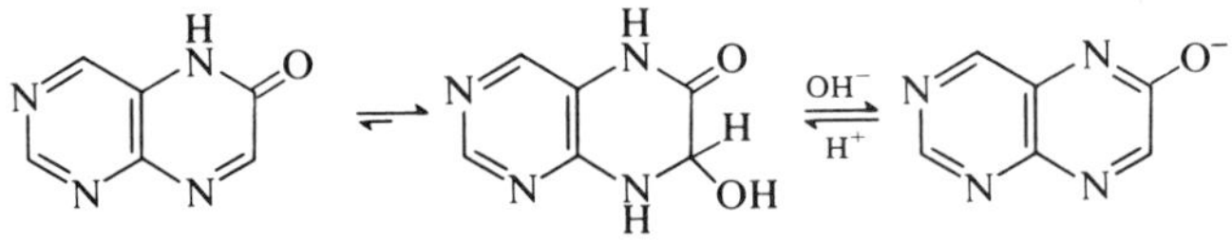

(25)                        (26)                        (27)                        (7)

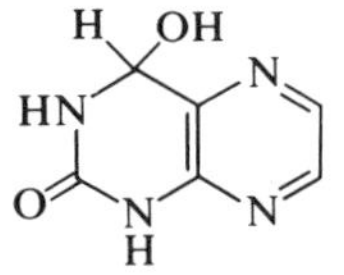

(28)

The molecular features of covalent hydration are also present in the dihydroxy series, *i.e.*, in pteridine-2,6-dione (30) and in pteridine-4,6-dione. The latter compound is hydrated only at the C(7)—N(8) double bond, whereas (30) forms two hydrated species, 7-hydroxy-7,8-dihydro- (29) and 4-hydroxy-3,4-dihydro-pteridin-2,6-dione (31) (equation 8). Structure (29) is thermodynamically the more stable substance; (31) is formed more rapidly in solution but disappears slowly with time ⟨63JCS5151⟩. Insertion of a 4-methyl group greatly reduces the extent of 3,4- in favour of 7,8-hydration by a 'blocking effect'.

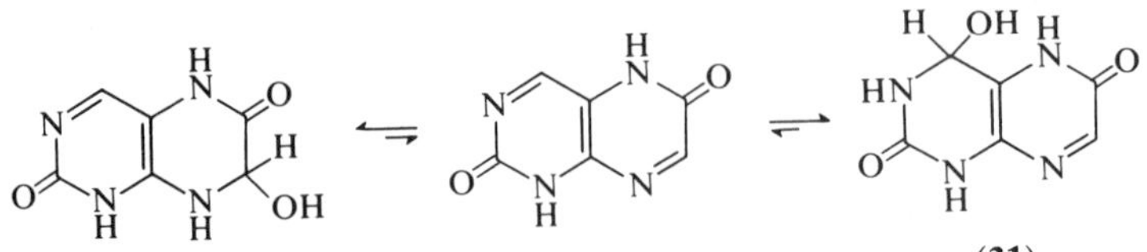

(29)                        (30)                        (31)                        (8)

The structure of lumazine has been studied more precisely by X-ray analysis ⟨72AX(B)659⟩. The crystal structure is built up of almost coplanar, hydrogen-bonded dimers of lumazine with the oxygens of the pyrimidine moiety in the keto form and the observed bond distances indicating the pyrazine ring electrons to be delocalized.

Pteridinetriones exist as anhydrous species because the $\pi$-electron deficiency is largely compensated by the electron-releasing hydroxy groups. The acidic properties of the amide functions and the sequence of ionization of the acidic protons have been determined in most polyoxopteridines by measurements of the $pK_a$ values and comparison of spectral data with the corresponding *N*-methyl derivatives as specifically blocked analogs. Since the lactam structure represents the prototype of hydrogen-bonding function, intermolecular aggregation associated with a decrease in solubility is a striking feature of this kind of pteridine derivative. As with the aminopteridines an increasing number of hydroxy groups causes a profound desolubilizing effect. Blocking of the active hydrogens by *O*- or *N*-methylation increases the solubility in water and organic solvents dramatically. The intermolecular interactions can also be seen in the remarkable variation in melting points (Table 3).

### 2.16.1.7 Thiopteridines (Pteridinethiones).

Relatively few systematic studies have been made of the thiopteridines, which in their energetically favoured thioamide tautomers have to be regarded as pteridinethiones. The monothiopteridines show in their behaviour close similarities to their hydroxy counterparts. Pteridine-6-thione forms a 7,8-hydrate and the isomeric pteridine-2-thione adds water covalently to the 3,4-double bond, while the anions are anhydrous ⟨68JCS(C)63, 65JCS27⟩. In acidic solution 2- and 4-methylthiopteridines both add two molecules of water to the pyrazine ring, giving rise to 5,6,7,8-dihydrates. Further structural information is available from extended studies with thiolumazines ⟨74CB3377, 78CB971, 79CB1499⟩, in which the 2-thioxo function seems to be the most stable; a 4-thioxo group can be displaced easily by nucleophiles, and the lumazine-6- and -7-thiones are both prone to photochemical interconversions ⟨81TL2161⟩. Other striking features of the pteridinethiones can be seen in their enhanced acidities, the lower solubilities and the red-shifted UV and visible spectra, compared with the oxo analogues.

### 2.16.1.8 Aminohydroxypteridines (Aminopteridinones)

The most important combination of substituents in the pteridine series is the 2-amino-4-hydroxy pair, which is characteristic for most naturally occurring and biologically active pteridine derivatives. 2-Aminopteridin-4-one (**2**), or 'pterin', combines in its molecular structure the features of the amino and the hydroxy pteridines in an integrated manner ⟨60CB2015⟩. Tautomeric equilibria are mainly fixed in the amino and lactam configurations which, according to their weak basic and weak acidic character respectively, very strongly interact by intermolecular hydrogen bonding, leading to very poorly soluble substances (Table 4). Other arrangements of these two substituents in the pteridine nucleus lead to isopterin (4-aminopteridin-2-one) ⟨63CB2950⟩, 6-amino-7-oxo- ⟨65JCS27⟩, 7-amino-4-oxo- ⟨65JCS1175⟩, 4-amino-7-oxo- and 4-amino-6-oxo-dihydropteridine ⟨63CB2977⟩, respectively. The last compound exists in its neutral form as a 7,8-hydrate which eliminates water on anion formation. Among the naturally occurring pteridine derivatives there is a series of 6-substituted pterins such as L-biopterin (**32**), D-neopterin (**33**), L-monapterin (**34**) and 6-hydroxymethylpterin (**35**) as well as folic acid (pteroylglutamic acid; **36**) and its derivatives.

Their physical properties closely resemble those of pterin, which has a basic p$K_a$ of 2.20 and an acidic one of 7.86 associated with N-1 protonation and a hypsochromic shift of the long-wavelength absorption band in the UV spectrum, and N-3 deprotonation effecting a bathochromic shift respectively (Table 4). The xanthopterin (**4**) and isoxanthopterin types

**Table 3**　Physical Data for Hydroxypteridines (Pteridinones)[a]

| Pteridine | pKa | λmax (nm) | log ε | pH | Solubility[b] in H₂O 20 °C | Solubility[b] in H₂O 100 °C | Melting point (°C) |
|---|---|---|---|---|---|---|---|
| 2-Hydroxy-(xH₂O) | — | 230, 307 | 3.88, 3.83 | 7.1 | 600 | 50 | 240 (dec.) |
| anion | 11.13 | 260, 375 | 3.85, 3.78 | 13.0 | — | — | — |
| 2-Hydroxy-4-methyl- | — | 231, 309, 350 | 3.53, 3.50, 2.96 | 4.5 | — | — | — |
| anion | 10.85 | 259, 373 | 3.84, 3.86 | 13.0 | — | — | — |
| 4-Hydroxy- | — | 230, 265, 310 | 3.98, 3.84, 3.82 | 5.6 | 200 | 29 | 350 |
| anion | 7.89 | 242, 333 | 4.23, 3.54, 3.79 | 10.0 | — | — | — |
| 6-Hydroxy- (xH₂O) | — | 215, 289, 356 | 4.10, 4.00, 3.60 | 5.2 | 3500 | 230 | 240 (dec.) |
| anion | 9.7 | 289, 303 | 3.90, 3.69 | 8.8 | — | — | — |
| 7-Hydroxy- | — | 227, 248, 326 | 3.79, 3.44, 3.44 | 4.0 | 900 | 76 | 230 (dec.) |
| anion | 6.41 | 226, 260, 326 | 4.27, 4.04, 4.04 | 9.0 | — | — | — |
| 2,4-Dihydroxy-(3) | — | 228, 325 | 4.03, 3.86 | 5.8 | 800 | 120 | 335 (dec.) |
| anion | 7.95 | 236, 269, 346 | 4.03, 3.98, 3.71 | 10.0 | — | — | — |
| 2,6-Dihydroxy- (xH₂O) | — | 235, 299 | 4.22, 3.79 | 3.9 | 11 000 | 4500 | — |
| anion | 11.6 | 225, 246, 412 | 4.41, 4.23 | 13.0 | — | — | — |
| 4,6-Dihydroxy- | — | 270, 356 | 3.88, 3.87 | 4.0 | 5000 | 300 | 320 (dec.) |
| anion | 6.08 | 241, 280, 359 | 4.03, 4.03, 3.38 | 7.9 | — | — | — |
| 2,7-Dihydroxy- | — | 258, 328 | 3.54, 3.83 | 3.6 | 1400 | 100 | 280 (dec.) |
| anion | 5.83 | 282, 343, 359 | 4.42, 3.99, 4.11 | 7.9 | — | — | — |
| 4,7-Dihydroxy- | — | 285, 328 | 4.38, 3.94 | 4.0 | 4000 | 600 | 350 |
| anion | 6.08 | 289, 326 | 3.98, 3.79 | 7.9 | — | — | — |
| 6,7-Dihydroxy- | — | 227, 249, 301 | 4.03, 3.71, 3.95 | 4.0 | 3000 | 290 | 350 |
| anion | 6.87 | 220, 268, 319 | 4.06, 3.71, 4.18 | 8.4 | — | — | — |
| 2,4,6-Trihydroxy- | — | 227, 251, 365 | 4.15, 4.02, 4.29 | 3.5 | 7400 | 400 | 350 |
| anion | 5.85 | 223, 265, 381 | 3.92, 4.07, 3.74 | 7.5 | — | — | — |
| 2,4,7-Trihydroxy- | — | 269, 324 | 3.92, 4.07 | 1.0 | 12 000 | 1400 | 350 |
| anion | 3.43 | 226, 274, 327 | 4.19, 3.91, 4.08 | 6.5 | — | — | — |
| 4,6,7-Trihydroxy- | — | — | — | — | 27 000 | 7000 | 350 |
| 2,4,6,7-Tetrahydroxy- | — | — | — | — | 58 000 | 7000 | 350 |
| 2-Methoxy- | — | 220, 258, 325 | 4.02, 3.45, 3.92 | 6.4 | 80 | 4 | 150 |
| 4-Methoxy- | — | 225, 258, 304 | 4.25, 3.89 | 5.7 | 80 | 9 | 195 |
| 3-Methyl-4-oxo- | — | 233, 276, 312 | 4.12, 3.57, 3.81 | 5.3 | 70 | 9 | 286 |
| 1,3-Dimethyl-2,4-dioxo- | — | 236, 331 | 4.19, 3.88 | 6.0 | 240 | — | 200 |

[a] For simplicity, structures are indicated in the hydroxy forms, wherever possible.
[b] Volume of water (ml) required to dissolve 1 g of heterocycle.

**Table 4**  Physical Data for Pterins

| Pterin | $pK_a$ | λmax (nm) | | | | log ε | | | | pH | Solubility in $H_2O$[a] 20°C |
|---|---|---|---|---|---|---|---|---|---|---|---|
| Unsubstituted (**2**) | — | 233 | 270 | — | 339 | 4.04 | 4.08 | — | 3.79 | 5.0 | 57 000 |
| cation | 2.20 | [229] | [242] | — | 314 | [3.03] | [3.94] | — | 3.89 | 0.0 | — |
| anion | 7.86 | — | 251 | — | 360 | — | 4.33 | — | 3.85 | 10.0 | — |
| 3-Methyl- | — | 240 | 274 | — | 353 | 4.17 | 4.07 | — | 3.79 | 5.0 | — |
| cation | 2.18 | 230 | — | — | 318 | 4.17 | — | — | 3.88 | 0.0 | — |
| 1-Methyl- | — | 240 | [285] | — | 327 | 4.21 | [3.63] | — | 3.93 | 5.0 | — |
| cation | 2.86 | 228 | [248] | — | 314 | 4.09 | [3.94] | — | 3.89 | 0.0 | — |
| $N^2$-Acetyl- | — | 232 | 277 | — | 325 | 4.15 | 4.13 | — | 3.89 | 5.0 | 450 |
| anion | 7.37 | — | 253 | — | 339 | — | 4.43 | — | 3.85 | 5.0 | — |
| -6-one (**4**) | — | — | 276 | [305] | 388 | — | 4.16 | [3.74] | 3.42 | 4.0 | 40 000 |
| cation | 1.6 | — | 245 | — | 355 | 4.07 | 4.07 | — | 3.82 | -3.0 | — |
| monoanion | 6.3 | 239 | 275 | — | 389 | — | 4.12 | — | 3.75 | 7.8 | — |
| dianion | 9.23 | — | 255 | — | 394 | — | 4.28 | — | 3.86 | 13.0 | — |
| 6-Amino- | — | — | 271 | — | 377 | — | — | — | — | — | — |
| cation | — | — | 260 | — | 392 | — | — | — | — | 1.0 | — |
| anion | — | — | [260] | — | 373 | 4.31 | [4.04] | — | 3.77 | 13.0 | — |
| $N^2$-Acetyl- -6-one | — | — | 267 | [278] | 340 | 4.48 | 4.00 | — | 4.14 | 7.0 | 16 500 |
| -7-one (**5**) | — | 205 | [254] | 289 | 320 | — | [3.80] | — | 4.06 | 4.0 | 200 000 |
| cation | 7.34 | 210 | 253 | 286 | 332 | 4.46 | 3.81 | — | 4.11 | -3.0 | — |
| monoanion | 10.06 | 229 | — | 280 | 339 | 4.58 | 4.05 | — | 4.14 | 8.7 | — |
| dianion | — | 221 | — | — | 336 | 4.15 | — | — | 3.89 | 13.0 | — |
| -6,7-dione (**6**) | — | 225 | 292 | 296 | 345 | 3.76 | 4.09 | [4.01] | — | 5.0 | 750 000 |
| cation | 7.56 | 245 | 288 | [320] | 341 | 4.19 | 3.94 | 4.02 | 4.03 | -4.0 | — |
| monoanion | 9.78 | 237 | 280 | 333 | 343 | 4.20 | 3.86 | — | 4.02 | 8.7 | — |
| dianion | 13.6 | 239 | 240 | 287 | 343 | 4.28 | 4.27 | 3.88 | 4.10 | 12.0 | — |
| trianion | — | 221 | 240 | 287 | 343 | 4.28 | 4.27 | 3.88 | 4.10 | 4NKOH | — |

a Volume of water required (ml) to dissolve 1 g of heterocycle.

(5), which differ structurally in the location of an additional oxo function in the 6- or 7-position, can easily be differentiated by their UV spectral behaviour. Compound (4) is long-wavelength absorbing compared to (5), as can be observed visually by the yellow colour of (4) and the colourless appearance of (5). In both cases protonation takes place at N-1, shortening the chromophoric system, as seen from the hypsochromic effect on the long-wavelength absorption band. Anion formation occurs by deprotonation first at the pyrazine moiety, followed by a second from N-3. Whereas (4) exists in its neutral form as an equilibrium mixture of 7,8-hydrate and anhydrous species ⟨62JCS2600⟩ in a ratio of about 1 : 1 and shows a successive bathochromic shift in the UV absorption on going to the mono- and di-anion respectively, (5) reveals a characteristic hypsochromic shift on deprotonation at N-8, changing to a bathochromic shift in the dianionic species ⟨61CB1⟩. Such UV shifts may help in determining ionization sequences as well as in structural assignments of new substances. From the UV spectrum of leucopterin (6) it can be concluded that the overall shape of a spectrum may be more informative than the positions of the absorption maxima ⟨61CB118⟩. Exchange of an oxo function for an amino group does not change the spectrophotometric properties very much, since the latter type corresponds quite well with the corresponding anionic species according to Jones's rule ⟨45JA2127⟩; this is experimentally borne out with 6- ⟨64JCS4769⟩ and 7-aminopterin respectively. Acylation of the 2-amino group is not only associated with hypsochromic shift of the UV absorption but also with a profound increase in solubility due to decreased intermolecular interactions.

More complicated structures bearing additional substituents in the 6- or 7-positions may change the spectral properties drastically according to their conjugative abilities and/or potential tautomerism. 6-Acetonyl- (37) and phenacyl-isoxanthopterin spectrally resemble (5), whereas the isomeric 7-acetonyl- and 7-phenacyl-xanthopterin absorb at much longer wavelength than the parent compound (4) ⟨72BCJ2829⟩. NMR spectra reveal a tautomeric structure (38) with the side chain in resonance with the nucleus *via* a vinylogous amide configuration ⟨66T3253⟩. The red butterfly pigment erythropterin (39) contains the same structural features ⟨62CB2195⟩ with a low field (¹H NMR) N—H arising by intramolecular hydrogen bonding ⟨63HCA2597⟩. The ¹H and ¹³C NMR spectra of pterins have been investigated under various conditions and by considering their different molecular species their basic structures and sites of protonation and deprotonation have been established ⟨69HCA743, 73HCA2680⟩. An X-ray analysis of xanthopterin hydrochloride proves the anhydrous nature of the cation, which is protonated at N-1 ⟨76HCA2374⟩.

### 2.16.1.9 Pteridinecarboxylic Acids and Derivatives

The high $\pi$-electron deficiency of the pteridine nucleus can be regarded as the main reason why only a few monosubstituted pteridine derivatives with electron-attracting substituents have been prepared. Since such substituents aggravate this deficiency it is not surprising that 2- ⟨70JCS(C)1540⟩ and 4-ethoxycarbonylpteridine ⟨67TL1099, 67JCS(C)1543, 68JCS(C)313⟩ are susceptible to a remarkably facile addition of two molecules of water or alcohol at the pyrazine ring. 4-Ethoxycarbonyl-6- and -7-methylpteridine are incompletely dihydrated at equilibrium in aqueous solution at 20 °C, and the 6,7-dimethyl derivative is only slightly hydrated, as expected for steric reasons. However, all the cations are dihydrated and this has made it possible to isolate the pure hydrates of these compounds by rapid neutralization followed by crystallization. The structures of the adducts were deduced from spectroscopic evidence, $pK_a$ values and elemental analysis. Hydrolysis of ethyl pteridine-4-carboxylate to the corresponding carboxylic acid led to the 5,6,7,8-dihydrated species, which could not be dehydrated without decomposition. Pteridine-4-carboxamides closely resemble the corresponding esters in their chemical and physical behaviour. The same is true for pteridine-4-carbonitrile, a compound which is difficult to synthesize ⟨71JCS(C)375⟩. A series of 2-substituted ethyl pteridine-4-carboxylates revealed 3,4-covalently hydrated species in the case of the 2-hydroxy and 4-mercapto derivatives, while the 2-chloro-,

2-ethoxy, 2-amino, 2-dimethylamino or 2-methylmercapto derivatives could be prepared both in anhydrous and 5,6,7,8-dihydrated forms, which are interconvertible ⟨68JCS(C)1124⟩. Stable pteridine-6- and -7-mono- as well as -6,7-di-carboxylic acids have only been derived from 2,4-diaminopteridine, lumazines ⟨48JA3026⟩ and pterins ⟨70LA(741)64, 78CB3790⟩. A peculiarity in this series is the overlapping basic and acidic $pK_a$ of the pterin-6- and -7-carboxylic acids, which therefore do not exist in a pure neutral form in aqueous solution.

Further stabilization can be achieved by the presence of additional electron-releasing substituents, as in 4,7-dihydroxy- ⟨53JCS74⟩ and 2-amino-7-hydroxy-pteridine-6-carboxylic acids ⟨60JA3765, 71JOC4012⟩ as well as 6-hydroxy-7-carboxy- and 7-hydroxy-6-carboxy-lumazines ⟨57CB2617, 57CB2624, 62CB749, 62CB1605⟩ and -pterins ⟨62CB1591⟩.

### 2.16.1.10 8-Substituted Quinonoid Pteridines

An interesting type of pteridine structure arises from the quaternization of 2- and/or 4-substituted pteridines at N-8 ⟨61JCS4413, 58JA6095, 63JOC1509, 60JCS1978⟩ or by condensation of 6-alkylamino- and 6-arylamino-5-aminopyrimidines with 1,2-dicarbonyl compounds (equation 9) ⟨57JCS4157, 60CB1406, 63CB2950, 66CB3503, 68CB1072, 71CB2273⟩. This substitution pattern gives rise to an entirely new $\pi$-electron distribution in the pteridine system which can be regarded as being of cross-conjugated quinonoid type. The substituted N-8 atom takes part in vinylogous amide resonance by interaction with the 2-oxo function in the lumazines and the 4-oxo group in the pterins. A characteristic feature of this chromophoric system is its red-shifted UV absorption at about 400 nm for the 8-substituted lumazines and pterins, which are yellow compounds showing an intense yellow–green fluorescence (Table 5).

Detailed structural investigations have revealed various interesting effects depending on the substituents in positions 6, 7, and 8. Protonation of the weakly basic vinylogous amide chromophore takes places at N-1 in the 8-substituted lumazines (**40**) and at N-3 in the corresponding pterins ⟨66CB3503, 68CB1072⟩, shifting the long-wavelength bands to shorter wavelengths. Anion formation is a much more complex matter due to the occurrence of nucleophilic addition of hydroxide ion to C-7. The anion of the 8-methyllumazine is a mixture of two species, the anhydrous (**41**) and hydrated anions (**42**), which are characterized by UV absorption bands at 405 and 307 nm respectively (equation 10). 3,8-Dimethyl-lumazine also has a 'basic' $pK_a$ of 10.41, describing the pseudo-base formation by nucleophilic attack at C-7 associated with a strong blue shift of the absorption band to 306 nm. 8-Substituted 6,7-diphenyl derivatives favour pseudo-base formation over simple deprotonation since the steric hindrance of the phenyl groups can be reduced at least to some extent by conversion of the $sp^2$-hybridized C-7 into an $sp^3$-carbon atom. The 8-substituted 6,7-dimethyllumazines and pterins again form mixtures of two species in alkaline pH consisting of the pseudo-base anion (**43**) and the 7-methylene (anhydro-base) (**44**,; equation 11). Their UV absorption is found at 310–315 and 365–370 nm respectively. NMR spectral evidence for the exocyclic methylene group is given by 3,8-dimethyl-7-methylene-8-phenyl-7,8-dihydropterin and 2-amino-4-methoxy-6,8-dimethyl-7-methylene-7,8-dihydropteridine, which exist in these structures as neutral molecules ⟨71CB2273⟩. There is one more additional interaction if the N-8 alkyl substituent bears a $\beta$- or $\gamma$-hydroxy group capable of intramolecular nucleophilic addition to C-7. This possibility is realized in 6,7-dimethyl–8-D-ribityllumazine, where all long-wavelength bands are missing in basic media (**45**). 1,8-Methylene-bridged lumazine analogs have also been synthesized and their properties contribute to a better understanding of this field ⟨81H(15)437⟩.

**Table 5** Physical Data for 8-Substituted Lumazines and Pterins

| Compound | $pK_a$ | $\lambda_{max}$ (nm) | | | | log $\epsilon$ | | | | pH |
|---|---|---|---|---|---|---|---|---|---|---|
| *Lumazine* | | | | | | | | | | |
| 8-Methyl- | | — | 257 | — | 392 | — | 4.21 | — | 3.97 | 6.0 |
| cation | −0.01 | 241 | 272 | — | 344 | 4.03 | 3.16 | — | 4.02 | −2.7 |
| anion | 9.89 | 230 | 280 | 307 | 405 | 4.27 | 4.10 | 3.90 | 3.04 | 13.0 |
| 3,8-Dimethyl- | | — | 259 | — | 394 | — | 4.32 | — | 3.97 | 7.0 |
| cation | 0.06 | 241 | [285] | — | 343 | 4.03 | [3.26] | — | 4.03 | −1.9 |
| anion | 10.41 | 230 | 282 | 306 | — | 4.30 | 4.13 | 3.96 | — | 13.0 |
| 8-Methyl-6,7-diphenyl- | | 266 | 289 | — | 424 | 4.19 | 4.27 | — | — | 7.0 |
| cation | 0.36 | 257 | 283 | — | 398 | 4.04 | 4.07 | — | 4.18 | −2.7 |
| anion | 9.69 | 244 | 281 | 352 | [425] | 4.27 | 4.14 | 4.05 | 4.05 | 12.0 |
| 8-(2-Hydroxyethyl)-6,7-diphenyl- | | 257 | 289 | — | 426 | 4.19 | 4.23 | — | [3.42] | 4.0 |
| cation | −0.13 | 244 | 281 | — | 398 | 4.13 | 4.12 | — | 4.08 | −2.7 |
| anion | 7.89 | 256 | [281] | 368 | — | 4.29 | [4.07] | 4.16 | 4.06 | 11.0 |
| 6,7,8-Trimethyl- | | 244 | 275 | — | 402 | 4.17 | 4.05 | — | — | 7.0 |
| cation | 0.85 | 244 | [267] | 313 | 358 | 4.04 | — | — | 4.09 | −2.7 |
| anion | 9.90 | 258 | 275 | — | 364 | 4.30 | [3.87] | 4.35 | 4.18 | 13.0 |
| 6,7-Dimethyl-8-ribityl- | | 245 | 280 | — | 407 | 4.13 | 3.95 | — | 3.78 | 6.0 |
| cation | 0.56 | 230 | 265 | [330] | 361 | 4.08 | — | — | 4.01 | −2.7 |
| anion | 8.29 | — | 276 | 313 | — | 4.13 | 4.06 | 3.91 | 4.10 | 13.0 |
| *Pterin* | | | | | | | | | | |
| 8-Methyl- | | — | 283 | [330] | 400 | — | 4.30 | [3.56] | — | 8.0 |
| cation | 5.32 | 260 | 281 | — | 386 | 4.15 | 4.06 | — | 4.00 | 3.0 |
| anion | 12.37 | 238 | 298 | 312 | 428 | 4.28 | 3.91 | 3.93 | 4.01 | 14.0 |
| 8-Methyl-6,7-diphenyl- | | — | [285] | — | 428 | — | 4.36 | — | 4.14 | 7.0 |
| cation | 5.34 | [266] | 298 | — | 422 | [4.10] | 4.27 | — | 4.18 | 1.0 |
| anion | 11.76 | 250 | [285] | 364 | — | 4.29 | [3.82] | 4.08 | — | 13.0 |
| 6,7,8-Trimethyl- | | — | 267 | 313 | 406 | — | 4.33 | 3.35 | 4.09 | 9.0 |
| cation | 6.13 | 253 | 284 | — | 395 | 4.07 | 4.12 | — | 4.11 | 3.0 |
| anion | 11.94 | 230 | — | 307 | 370 | 4.27 | — | 4.18 | 3.85 | 14.0 |

(11)

(43)          (44)          (45)

## 2.16.1.11 Dihydropteridines

Theoretically there are at least 16 different dihydropteridine structures possible, not counting those forms involving the bridgehead carbons 4a and 8a. The most important are derived by formal reduction of the 1,2-, 3,4-, 5,6- and 7,8-C=N double bonds as the thermodynamically most stable species. 3,4-Dihydropteridine, its 2- and 6-methyl derivatives, and the 4,6-dimethyl-7,8-dihydro-pteridine are so far the only simple dihydropteridines which have been obtained by unambiguous synthesis from 2-amino-3-aminomethylpyrazines ⟨70JCS(C)1540⟩ and catalytic dehalogenation of 2-chloro-4,6-dimethyl-7,8-dihydropteridine respectively ⟨62JCS2162⟩. Pteridine itself could not be reduced to a dihydro stage, but gave with lithium aluminum hydride the 5,6,7,8-tetrahydro derivative ⟨59JA2464⟩. More successful have been the selective reductions and direct syntheses of reduced amino- and hydroxy-pteridines. 2-Aminopteridine can be converted to its 3,4-dihydro form ⟨66JCS1117⟩, as can pteridin-2-one ⟨61JCS5131⟩. The 7,8-dihydro derivative of the latter compound has also been prepared, as well as some alkyl-substituted analogues ⟨62JCS2162⟩. From pteridin-4-one the 5,6- and 7,8-dihydro compounds are known and 5,6-dihydropteridin-7-one and 7,8-dihydropteridin-6-one are also available ⟨56JCS3443⟩. All these compounds, as expected from the rupture of the cyclic heteroaromatic resonance, have stronger basic and less acidic properties compared with the parent molecules (Table 6). The covalent hydrates of various pteridine derivatives as well as the different types of Michael addition products obtained with pteridine ⟨73JCS(P1)1615⟩, 6,7-dimethylpteridine ⟨73JCS(P1)1974⟩, pteridine-2-thione ⟨68JCS(C)63⟩, 2-aminopteridine ⟨66JCS(C)1117⟩, pteridin-4-one ⟨62JCS1591⟩, pteridin-6-one ⟨61JCS127, 64JCS3357⟩ and pteridin-7-one ⟨65JCS6930⟩ also belongs to this category of dihydropteridine derivatives, bearing a new C—C bond in the 4-, 6- or 7-position.

There is no easy understanding of the spectral properties of these compounds in general, which may or may not have a built-in chromophoric system responsible for a long-wavelength absorption like 7,8-dihydropteridin-4-one or a blue-shifted excitation like its 5,6-dihydro isomer. More important than the simple dihydropteridine model substances are the dihydropterins and dihydrolumazines, which are naturally occurring pteridine derivatives and reactive intermediates in redox reactions.

In 6-alkyl-substituted 7,8-dihydro-lumazines and -pterins, such as 7,8-dihydrobiopterin and 6-hydroxymethyl-7,8-dihydropterin, the most striking feature is the strong bathochromic shift of the UV spectra on monoprotonation. Cation formation takes place at N-5 ⟨66CB3008⟩, which is also the site of quaternization with alkylating agents ⟨71CB2313, 71CB3842⟩. Direct proof for the N-5 protonation has been provided by an X-ray analysis of 6-methyl-7,8-dihydropterin monohydrochloride ⟨77HCA2303⟩. The 6-acyl-7,8-dihydro-lumazine and -pterin structures also represent a special type of pteridine chromophore, since partial hydrogenation at the pyrazine moiety is again associated with a bathochromic shift of the UV maximum towards the visible region. Sepiapterin and deoxysepiapterin (46) are therefore yellow pigments from *Drosophila* eyes and can be regarded as merocyanine dyes due to the presence of a linear chromophoric system terminated by the 2-amino- and 6-carbonyl groups ⟨79CB2750⟩. There are also other cases where partial reduction of a heteroaromatic pteridine system causes a bathochromic shift in the UV spectrum, as for example in the 8-alkyl-3,4-dihydropteridin-7-ones and their 6-carboxylic acids, which are bright yellow compounds ⟨71JOC4012⟩. 6,7-Diphenyl-5,6-dihydro-pterin ⟨65HCA(48)764⟩ and -lumazine (47) ⟨68HCA1029⟩ absorb as far as 450 nm, demonstrating the great importance of the 7-phenyl group as part of the chromophore involved. The most unusual structures have been found during anaerobic photolysis of biopterin, giving rise to the formation of 5,8-dihydropterin-6-carbaldehyde (48) ⟨77TL2817⟩, and the electrochemical reduction of 6-acyllumazines to their 5,8-dihydro derivatives, respectively ⟨80H(14)1603⟩, which are red compounds showing a low extinction band in the region 475–500 nm.

**Table 6** Physical Data for Dihydropteridines

| Pteridine | $pK_a$ | UV absorption spectra | | | | | | pH |
|---|---|---|---|---|---|---|---|---|
| | | $\lambda_{max}$ (nm) | | | $\log \varepsilon$ | | | |
| 3,4-Dihydro- | — | — | — | 335 | — | — | 3.96 | 9.0 |
|   cation | 6.36 | — | — | 311 | — | — | 3.90 | 4.0 |
| 7,8-Dihydro-4,6-dimethyl- | — | 218 | 293 | — | 4.27 | 3.73 | — | 8.2 |
|   cation | 6.00 | 218 | 293 | — | 4.13 | 3.91 | — | 3.0 |
| 2-Amino-3,4-dihydro- | — | — | 282 | 339 | — | 3.89 | 3.89 | 10.0 |
|   cation | 7.73 | 245 | — | 308 | 3.65 | — | 3.89 | 6.0 |
| 3,4-Dihydro-2-hydroxy- | — | 248 | — | 317 | 3.72 | — | 3.89 | 7.0 |
|   cation | 0.0 | 254 | — | 337 | 3.78 | — | 3.85 | −2.0 |
|   anion | 12.6 | — | 281 | 343 | — | 3.98 | 3.84 | 14.0 |
| 7,8-Dihydro-2-hydroxy- | — | 223 | 290 | — | 4.35 | 3.88 | — | 7.0 |
|   cation | 3.50 | 225 | 290 | [310] | 3.86 | 3.79 | [3.75] | 1.0 |
|   anion | 12.0 | — | — | 308 | — | — | 3.95 | 14.0 |
| 5,6-Dihydro-4-hydroxy- | — | | 286 | | | 3.78 | — | 7.0 |
|   cation | 2.94 | | 258 | | | 3.74 | — | 1.0 |
|   anion | 10.29 | | 279 | | | 3.82 | — | 12.3 |
| 7,8-Dihydro-4-hydroxy- | — | 248 | — | 367 | 3.86 | — | 3.68 | 7.0 |
|   cation | 0.32 | — | 257 | 374 | — | 3.87 | 3.80 | −2.0 |
|   anion | 12.13 | — | 253 | 364 | — | 3.97 | 3.70 | 14.0 |
| 7,8-Dihydro-6-hydroxy- | — | — | 293 | — | — | 3.93 | — | 7.4 |
|   cation | 4.78 | — | 292 | — | — | 4.01 | — | 2.0 |
|   anion | 10.54 | — | — | 305 | — | — | 4.07 | 13.0 |
| 5,6-Dihydro-7-hydroxy- | — | — | 271 | 319 | — | 3.58 | 3.70 | 6.0 |
|   cation | 3.36 | 223 | 284 | 352 | 4.47 | 3.74 | 3.71 | 1.0 |
|   anion | 9.94 | 224 | — | 325 | 4.34 | — | 3.93 | 12.0 |
| 6,7,8-Trimethyl-7,8-dihydrolumazine | — | — | 280 | 313 | — | 4.20 | 3.81 | 5.0 |
|   cation | 2.80 | 236 | 273 | 351 | 4.08 | 4.33 | 3.85 | 1.0 |
|   anion | 7.24 | 230 | 280 | 320 | 4.36 | 4.12 | 3.91 | 13.0 |
| 6-Methyl-7,8-dihydropterin | — | 229 | 279 | 324 | 4.41 | 4.02 | 3.75 | 7.0 |
|   cation | 4.17 | 252 | [271] | 361 | 4.27 | [3.88] | 3.71 | 1.0 |
|   anion | 10.85 | 231 | 282 | 322 | 4.17 | 3.90 | 3.76 | 14.0 |
| Sepiapterin (**237**) | — | 212 | 266 | 417 | 4.23 | 4.24 | 4.04 | 7.0 |
|   cation | 1.27 | 233 | 283 | 400 | 4.12 | 4.07 | 3.89 | −1.0 |
|   anion | 9.95 | 266 | 313 | 438 | 4.23 | 3.38 | 4.15 | 13.0 |

Finally, a quinonoid 6,7,8-trihydropterin structure (**49**) absorbing at 303 nm plays an important role as a labile intermediate in the tetrahydrobiopterin-dependent enzymatic hydroxylation of phenylalanine ⟨67JBC(242)3934⟩.

    (46)        (47)        (48)        (49)

### 2.16.1.12 Tetrahydropteridines

Extensive reduction of (**1**) leads to 5,6,7,8-tetrahydropteridine ⟨59JA2464⟩, showing that reducing agents attack at the pyrazine rather than the pyrimidine ring of the condensed system. 1,2,3,4-Tetrahydropteridine has only been obtained by cyclization of 2-amino-3-aminomethylpyrazine with formaldehyde ⟨70JCS(C)1540⟩, and no other simple tetrahydropteridine isomers have so far been synthesized. These compounds show basic $pK_a$ values of 6.63 ⟨57JCS1⟩ and 5.62 respectively, and are thus stronger bases than the parent pteridine. They are fairly stable to oxidation and do not show any indication of covalent hydration. With the exception of 2-hydroxy- and 2-amino-pteridines, which are reduced at the 3,4-bond, all 4-mono- and the 2,4-di-substituted pteridine derivatives are converted by catalytic hydrogenation or treatment with complex hydrides into the corresponding 5,6,7,8-tetrahydro forms. The same type of compound can be obtained by intramolecular alkylation of 5-amino-4-(*N*-benzyl-$\beta$-haloethylamino)pyrimidines ⟨53JCS2234, 55JCS896, 61JCS5131⟩. 5,6,7,8-Tetrahydropteridines resemble the corresponding 4,5-diaminopyrimidines in their physical and chemical properties. The basicities of both series and their spectrophotometric behaviour are similar ⟨52JA3252, 71CB780, 71LA(747)111⟩. The presence of electron-releasing substituents at the 2- and/or 4-positions greatly influences the stability of the tetrahydro derivatives towards autoxidation ⟨59HCA1854, 64HCA2087, 65HCA816, 68MI21601⟩. 5,6,7,8-Tetrahydro-lumazines and -pterins should therefore be kept and stored as their hydrochloride salts ⟨66HCA875, 74HCA1485⟩ or even better as 5-acyl derivatives ⟨71CB2293, 71CB3842, 58HCA2170, 47JA250, 74HCA1651⟩.

Studies on the configuration and conformation of 6-methyl- and 6,7-dimethyl-5,6,7,8-tetrahydropterins ⟨75HCA1772, 76HCA2379⟩ and their N-5 acyl derivatives ⟨77HCA152, 77HCA161, 81HCA367⟩ have been performed by means of NMR, leading to the conclusion that the tetrahydropyrazine ring favours a half-chair conformation with the C-6 substituent in an equatorial position. In the 6,7-dimethyl derivatives catalytic reduction takes place in a *cis* manner, while the 7-deuterio-6-trideuteriomethyl analogues reveal the formation of an almost 1:1 mixture of the *cis*- and *trans*-isomers ⟨77JCS(P1)2529⟩. X-Ray analysis of 5,6,7-trimethyl-5,6,7,8-tetrahydropterin dihydrochloride monohydrate ⟨77HCA1926⟩ and 5-formyl-6,7-dimethyl-5,6,7,8-tetrahydropterin ⟨77HCA447⟩ prove the structures in more detail. In the cases of 5,6,7,8-tetrahydro-L-biopterin and -L-neopterin, separation into the two pure diastereomeric (6R)- and (6S)-forms could be achieved *via* the fully acylated derivatives. Enzymic reduction of 7,8-dihydropterins by dihydrofolate reductase produces the new chiral center at C-6 in the (S) configuration, as found for tetrahydrobiopterin ⟨80MI21605⟩, the natural cofactor for aromatic amino acid hydroxylases, and 6-ethyl-5,6,7,8-tetrahydropterin ⟨80CC334, 81BCJ2543⟩ as a model.

In the folic acid series, conformational analyses of 5,6,7,8-tetrahydropteroic acid, 5,6,7,8-tetrahydro-L-folic acid (50) ⟨78HCA2744⟩ and its imidazo-fused derivative (51) ⟨79HCA1340⟩ have been achieved, showing again a half-chair conformation of the tetrahydropyrazine ring with a pseudo-equatorial position of the side chain at C-6. Further structural information and assignments are derived from ¹³C ⟨74HCA2658⟩ and ¹⁵N NMR spectra ⟨78HCA2108⟩.

The chiralities at C-6 of natural 5,6,7,8-tetrahydrofolic acid and related folates, *e.g.* 5,10-methylene-, 5-methyl- and 5-formyl-5,6,7,8-tetrahydrofolic acid, from various biological systems are the same and possess the absolute configuration (S) at C-6 as deduced from an X-ray study of the ion (51) ⟨79JA6114⟩.

### 2.16.1.13 Pteridine *N*-Oxides

With four ring N atoms in the pteridine nucleus the possibilities for *N*-oxide formation are numerous. Cyclization reactions with 4-amino-5-nitrosopyrimidines lead to pteridine 5-oxides ⟨63JOC1197, 66LA(694)142⟩, while pteridine 8-oxides have been obtained in an unambiguous way from 2-aminopyrazine 1-oxide ⟨68JA2424, B-69MI21601, 73JA6407, 73JA6413, 73JA4455, 73JOC2817, 80JOC2485, 76JOC1299⟩. Direct oxidation of lumazines ⟨73CB3149⟩, pterins and 2,4-diaminopteridines ⟨73CB3175⟩ by peracids forms preferentially 8-oxides and then 5,8-dioxides.

*N*-Oxidation is very sensitive to steric effects, since 1-substituted lumazines and pterins give only 5-oxides and the presence of bulky substituents at position 7 also directs oxidation to N-5. The pteridine 5-oxide (52) and 8-oxide (53) and the 5,8-dioxide (55) contain the *N*-oxide groups as such, even when the possibility of *N*-hydroxy tautomers exists, as in (53) ⇌ (54).

No simple pteridine 1- or 3-oxides are yet known. If the *N*-atom of an amide function is formally oxidized, tautomerism favours the cyclic hydroxamic acid structure, as found for 3-hydroxypteridin-4-one ⟨55JA3927⟩, 1-hydroxylumazine ⟨64JOC408⟩ and 2,4-diamino-8-hydroxypteridin-7-ones ⟨75JOC2332⟩.

Various 4-aminopteridine 3-oxides have recently been obtained by cyclization of 2-amino-3-cyanopyrazine derivatives ⟨81H(15)293⟩.


### 2.16.1.14 Pteridine Radicals

The various dihydro- and tetrahydro-pteridines undergo one-electron reduction and oxidation to produce different types of pteridine radical. Cations of monohydropteridine radicals have been obtained in solution and according to ESR studies show a 5,8-dihydro-pteridine radical cation structure (56) ⟨67HCA411, 70MI21602⟩, where the unpaired electron is delocalized over the whole pyrazine moiety (equation 12). Analogously the hyperfine patterns of the trihydropteridine radical cations are in agreement with the computed spectra of the 5,6,7,8-tetrahydropteridine radical cation species, revealing a localization of the free electron mainly at N-5 or possibly to a lesser extent at the bridge carbons. Substitution in the pyrimidine part of the system does not affect the spin distribution to any significant extent ⟨68HCA607⟩. It is assumed that one-electron oxidation–reduction occurs in the production of intermediates in the pteridine/5,8-dihydropteridine and 6,7-dihydro-/5,6,7,8-tetrahydropteridine systems, thus permitting rapid electron transfer. Trihydropteridine radicals might be of biological importance, not of course as the strongly acidic cations but perhaps stabilized in the form of metal complexes.

Radical anions have recently been detected during electrochemical reductions of lumazines ⟨80H(14)1603⟩ and are also assumed to be reactive intermediates in the reductive acylation of 2,4-disubstituted pteridines to the corresponding 5,8-diacyl-5,8-dihydro derivatives.


### 2.16.1.15 Pteridine Nucleosides and Nucleotides

The structural relationship between the pteridine nucleus on one hand and the purine and pyrimidine systems on the other has been responsible for great efforts having been made to synthesize pteridine nucleosides as structural analogs of the nucleic acid components. The lumazine system can be glycosylated at N-1 and N-3 to give both mono- and 1,3-di-glycosyl derivatives ⟨73CB1401, 76CB3217, 77LA1217, 80CB1524⟩, while isopterins show more selective substitution at position 1 ⟨73HCA1225⟩. The NMR spectra of the 1-β-D-glucopyranosyl-lumazines and -isopterins ⟨73CB2982⟩ have revealed the syn- (57) and anti-rotamers (58) which are expected to represent the most stable conformations of nucleosides in general ⟨73CB2982⟩. Subsequently syn–anti isomeric pentofuranosyl nucleosides were also detected by NMR spectroscopic means at low temperature, showing rotational barriers of 38–54 kJ mol$^{-1}$ ⟨73CB2975⟩.

8-β-D-Ribofuranosylpteridin-7-ones (59) ⟨73CB317, 73CB1952, 74CB339, 76CB3228, 80CB1535⟩, which are formed by glycosylation *via* the silyl method or *via* cyclization reactions ⟨64JHC23, 66CB3022, 71CB770⟩, contain a very stable glycosidic linkage, whereas the corresponding 7-*O*-glycosides ⟨62CB738, 62CB1621, 66CB536⟩ are base-labile as expected.

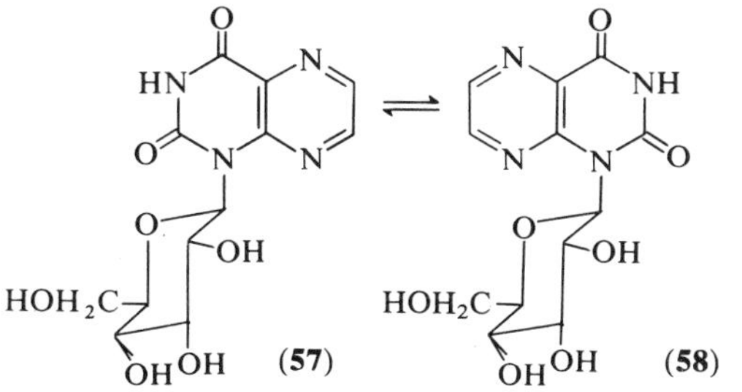

2,2'- and 2,5'-Anhydropteridine nucleosides (**60**) ⟨76CB3217, 76CB3159⟩ reveal interesting rigid structures for conformational and chiroptical studies. Various pteridine nucleotides bearing the phosphate group at the 3'- ⟨79HCA1171⟩ and 5'-position (**61**) ⟨78LA1780, 78LA1788⟩ have been synthesized as well as dinucleoside 3',5'-phosphates ⟨79HCA1179, 80LA50⟩ and trinucleoside diphosphates ⟨80LA65⟩ containing at least one pteridine nucleus as an unnatural base in the oligonucleotide chain. UV and CD studies indicate intramolecular interactions, suggesting base-stacking and helix formation.

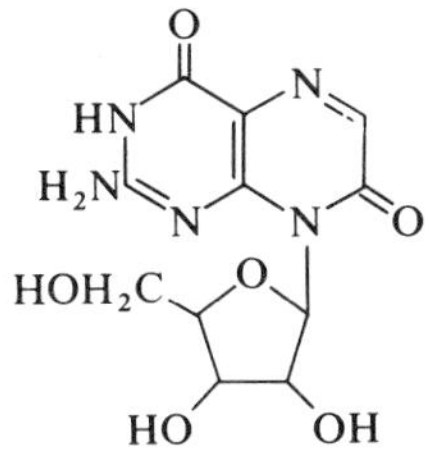

### 2.16.1.16 Dimeric Pteridines

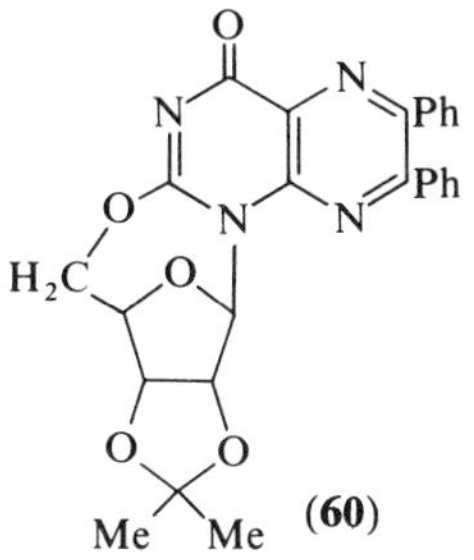

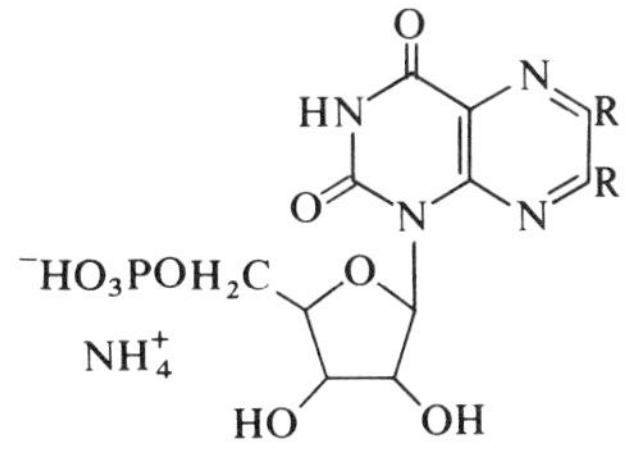

It had early been noticed during isolation and purification of butterfly pigments that higher molecular weight compounds of intense red to violet colour are formed, belonging to the class of dipteridinylmethane dyestuffs ⟨44LA(556)186, 49JA3412⟩. Pterorhodin (**62**), the most important representative of this group, was later found in the wings of tropical butterflies ⟨63ZN(B)420⟩ and in the eyes of *Ephestia* and *Ptychopoda* ⟨59ZN(B)654⟩. Its structure is fixed in a hydrogen-bonded vinylogous amide configuration connecting both xanthopterin rings *via* a methine bridge. Similar structures are assigned to the 'pteridine reds' which are derived from oxidative Michael additions of methylpterins ⟨50HCA39, 50HCA1233, 51HCA1029, 51HCA2155⟩. Michael adducts dimeric in nature are also formed between 6-methyl-7- and 7-methyl-6-pteridinone yielding 7,8-dihydro-7-(7-oxopteridin-6-ylmethyl)-7-methylpteridin-6(5*H*)-one (**63**), which could be resolved into two enantiomers ⟨64JCS3357⟩. The 7-nor derivative (**64**) is oxidized to an intensely orange-coloured compound of structure (**65**). Self-condensation of 4-methylpteridine takes place *via* position 7 and of 7-methyl- and 6,7-dimethylpteridine at position 4 in an analogous manner ⟨73JCS(P1)1974⟩.

Other types of dimer are formed from pteridin-6-one by the action of ammonia and hydrogen sulfide, giving rise to the corresponding dipteridinyl amine ⟨55JCS2690⟩ and sulfide ⟨65JCS27⟩ respectively. Two pteridine nuclei may be connected directly; on base treatment

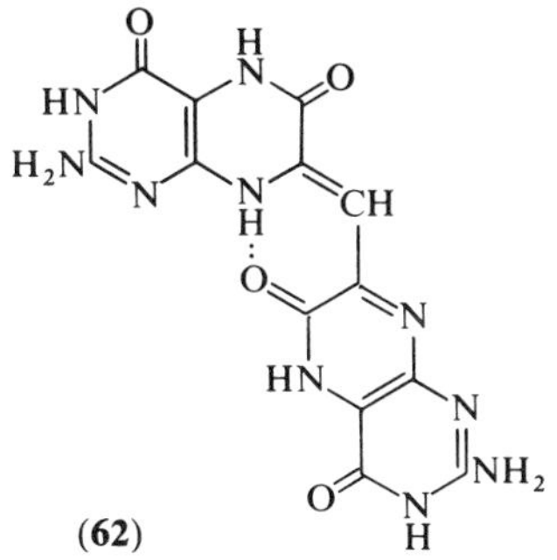

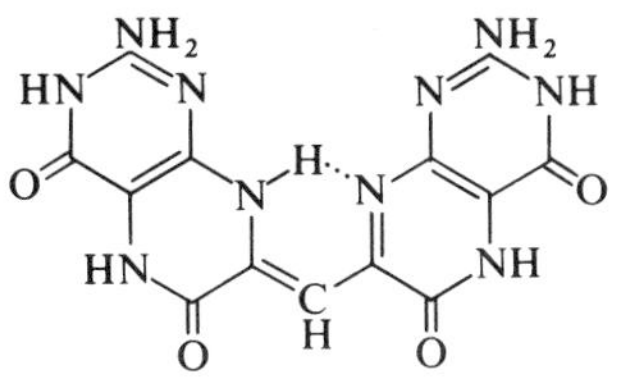

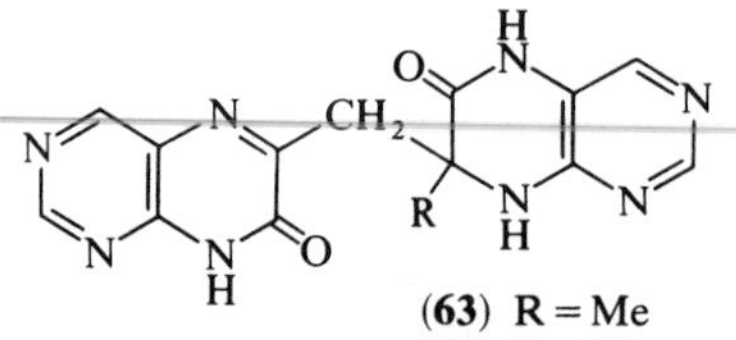

(63) R = Me
(64) R = H

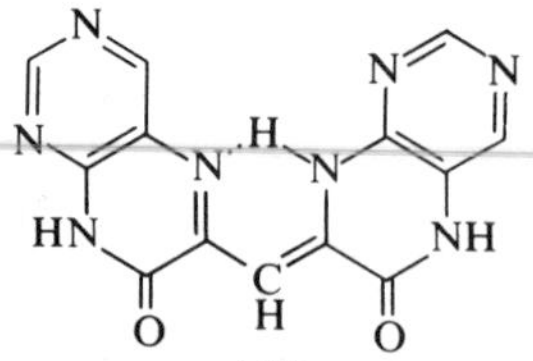

(65)

of 2- ⟨60JCS1370⟩ and 6-pteridinone ⟨55JCS2690⟩ compounds (66) and (67) are formed, and related compounds are probably formed on autoxidation of 5,6,7,8-tetrahydropterin ⟨63HCA1537⟩. Dimerization by one-electron reduction and radical recombination is observed with 1,3-dimethyllumazine, leading to 6,7′-linked dilumazinyls of structures (68) and (69). Finally, cathodic reduction of 6- and 7-acyllumazines allows dipteridinylglycol formation, presumably through the intermediary ketyls ⟨80H(14)1603⟩.

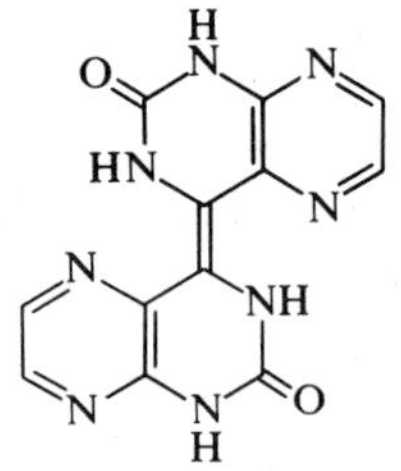

(66)

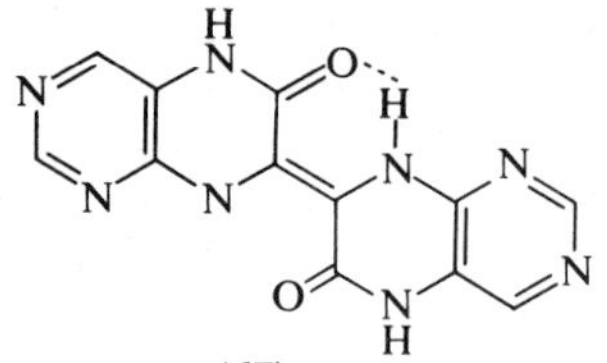

(67)

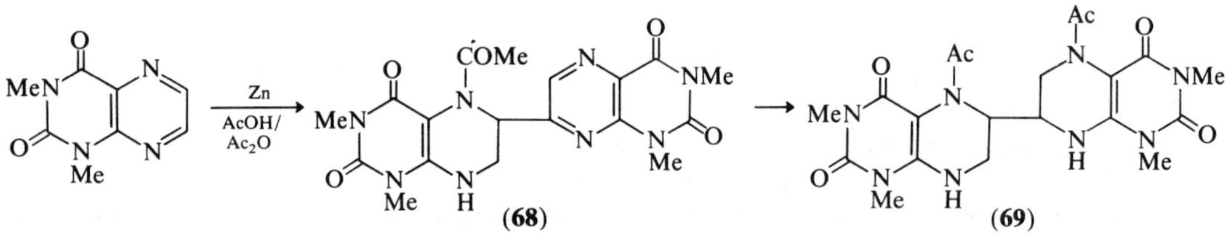

(68)

(69)

### 2.16.1.17  Complex Pteridine Structures

The pteridine nucleus may also form part of a more complex system derived by annelation of additional rings. Acrylonitrile, for example, forms pyrimido[2,1-*b*]pteridine derivatives with pterins ⟨59JA5650⟩, and intramolecular alkylation of 2-thiolumazines gives rise to thiazolo[2,3-*b*]-, thiazolo[3,2-*a*]- and thiazino[2,3-*b*]-pteridines ⟨78CB971⟩. Wieland's 'bis-alloxazine' ⟨40LA(545)209⟩ was the first representative of the pyrimido[5,4-*g*]pteridine system ⟨54JA1874⟩, which results, together with its [4,5-*g*] isomer, from oxidative self-condensations of 4,5-diaminopyrimidines ⟨53CB845, 55JA2243⟩. Isothiazolo[4,5-*g*]pteridines are valuable synthetic intermediates ⟨78JOC4154⟩, and some cycloalka[*g*]pteridines ⟨77JMC1215⟩ are active as dihydrofolate reductase inhibitors.

The most interesting structures, however, are found as natural products. Urothione (70), which was isolated from urine ⟨40ZPC(263)78, 43ZPC(277)284⟩, is derived from the thieno[3,2-*g*]pteridine system⟨69MI21605⟩, and in *Russula* species the imidazolo[4,5-*g*]-pteridine nucleus is present ⟨B-79MI21606⟩. The drosopterins, red eye pigments from *Drosophila melanogaster*, contain a pentacyclic ring system (71) ⟨78CB(111)3385⟩, and surugatoxin (72), a complex marine toxin from the Japanese ivory shell *Babylonia japonica*, can be regarded as a complicated 5-acylated 5,6,7,8-tetrahydrolumazine derivative ⟨72TL2545⟩.

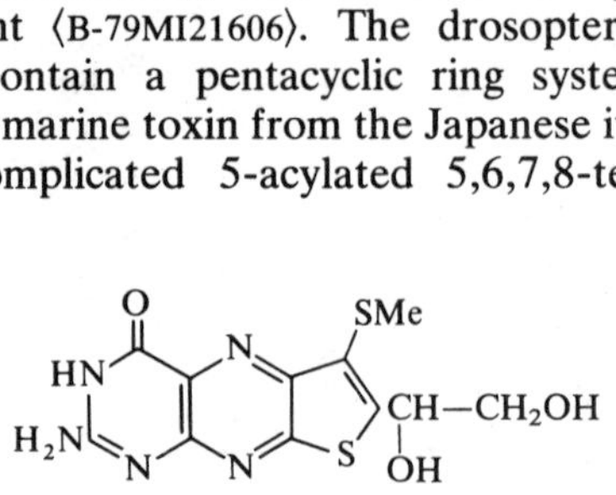

(70)

(71)

(72)

### 2.16.1.18 NMR Spectra

Proton NMR studies of pteridine (**1**) and its 2-, 4- and 7-methyl derivatives in chloroform have resulted in an assignment of the $^1$H signals ⟨63JCS1773⟩. The low solubility of pterin (**2**) and its naturally occurring derivatives in aprotic solvents does not allow easy investigation of the neutral forms, but mono- and di-cations taken in trifluoroacetic acid and fluorosulfonic acid are well characterized ⟨66HCA1355, 69HCA743⟩. Some NMR data of biologically active dihydropterins ⟨68MI21600⟩ are also available, but more information was provided by the $^{13}$C NMR spectra of pteridines ⟨73HCA2680, 74CB876, 74CB3275⟩. Assignments of simple pteridines have been extended to more complex structures such as folic acid (**36**) and related structures ⟨73B2425, 73CB3951⟩.

### 2.16.1.19 Mass Spectra

Mass spectrometry plays an important role in the structural elucidation of naturally occurring pteridines because of the small quantities of material usually available. Simple pteridine derivatives have been used as models to determine the principal fragmentation modes ⟨65JOC1844⟩, which are, however, strongly dependent on the substitution ⟨66NKK1226⟩. High resolution studies on the fragment ions of pterins show that the major processes are loss of either a CO or an HCN fragment ⟨69OMS(2)923⟩. Besides the determination of mass spectra of unprotected pteridine derivatives ⟨72TL3219⟩, acylation ⟨67NKK1320⟩ and especially trimethylsilylation ⟨70OMS(3)1365, 70MI21600⟩ reveal a substantial improvement in the spectra due to the greatly increased volatility. Mass spectral analysis of pteridines has also been carried out on the reduced analogues ⟨73JHC827⟩, on the covalent hydrates ⟨73OMS(7)737⟩, and on higher molecular weight compounds such as folic acid and methotrexate. Their determination requires methylation ⟨78MI21600⟩ or application of field desorption techniques ⟨76MI21600⟩.

### 2.16.1.20 Polarography and Electrochemistry

Since the discovery of the polarographic reducibility of naturally occurring pteridines ⟨47JA2753⟩, polarography has been used extensively for qualitative and quantitative determination of such compounds ⟨73LA1082⟩. Correlations between various structural elements and the electrochemical properties of pterins, 7,8-dihydropterins ⟨73LA1091⟩ and the 8-substituted quinonoid-type derivatives have been made ⟨73LA1099⟩. Investigations also include cyclic voltammetric studies and controlled potential electrolysis ⟨73BBA(297)285⟩, which leads in a first reversible two-electron reduction to 5,8-dihydro derivatives, which after tautomerism to the 7,8-dihydro isomers are further reduced at a more negative potential to the 5,6,7,8-tetrahydro stage. The preparative electrochemical reduction of folic acid and analogues ⟨79MI21600⟩ can be regarded as a valuable alternative to purely chemical and catalytic methods. The electrochemistry of pteridine (**1**) itself has also been studied in detail and demonstrates the complexity of this ring system ⟨75MI21600⟩. Interesting electrochemical behaviour is shown also by the 6,7-dioxotetrahydropteridines which are susceptible to a four-electron reduction to give 6-oxo-5,6,7,8- or 7-oxo-5,6,7,8-tetrahydro

derivatives ⟨78CB1763⟩. Electrochemical oxidations of pteridines have been limited to 6-
and 7-oxo and the 6,7-oxo derivatives ⟨74MI21600⟩.

### 2.16.1.21 Solubility

Structural features of organic molecules are also reflected in simple physicochemical
properties such as the solubility ⟨63PMH(1)177⟩. The solubility in water can be informative
if various derivatives of the same nucleus are compared. In nitrogen heterocycles, the
$sp^2$-hybridized nitrogen atom, because of its lone electron pair and hydrogen-bonding
interaction with protic solvents, leads to a pronounced increase in solubility in comparison
with the corresponding hydrocarbon. Pteridine is therefore quite soluble in water, one part
dissolving in 7 ml at room temperature, whereas one part of naphthalene requires 40 l.
However, the most striking solubility effect found generally with $\pi$-electron deficient
nitrogen heterocycles, and particularly with pteridines, is the decrease in solubility on
introduction of hydrophilic substitutents such as amino, hydroxy and mercapto groups. This
is quite contrary to the normal effect of increased solubility encountered in the aliphatic
and carbocyclic series. The most probable explanation for the decrease in solubility is the
presence of strong intermolecular hydrogen bonds between the molecules rather than
solvation effects. Increasing numbers of amino groups lead to a higher probability of
intermolecular interactions to give higher molecular weight linear and two-dimensional
aggregates. The solubility of the amino- (Table 1) and hydroxy-pteridines (Table 3) decreases
gradually from the mono- to the di- and tri- to the tetra-substituted derivatives. Blocking
of the hydrogen atoms of the hydrophilic substituents by the hydrophobic methyl group
results in a large increase in solubility as a consequence of reduced hydrogen bond formation.
The most striking combination of substituents is where amino and hydroxy groups are
present in the same ring, which leads to the highest degree of intermolecular association.
The solubility thus reaches its minimum in such pteridine derivatives (Table 4).

## 2.16.2 REACTIVITY

### 2.16.2.1 Reactivity at the Ring Carbon Atoms

#### 2.16.2.1.1 Electrophilic reagents

No simple electrophilic substitution, for example nitrosation, nitration, sulfonation or
halogenation of a C—H bond, has so far been recorded in the pteridine series. The strong
$\pi$-electron deficiency of this nitrogen heterocycle opposes such electrophilic attack, which
would require a high-energy transition state of low stability.

#### 2.16.2.1.2 Nucleophilic reagents

In contrast to electrophilic reagents, the highly $\pi$-deficient character of the pteridine
nucleus is responsible for its vulnerability towards nucleophilic attack by a wide variety of
reagents. The direct nucleophilic substitution of pteridine itself in a Chichibabin-type
reaction with sodamide in diethylaniline, however, was unsuccessful ⟨51JCS474⟩. Pteridin-6-
one, on the other hand, yielded pteridine-6,7-dione under the same conditions, *via* a still
unknown reaction mechanism.

More common are the nucleophilic additions at the ring carbon atoms, of which there
are many examples with oxygen, sulfur, nitrogen and carbon nucleophiles. The first example
of an adduct of a pteridine and a nucleophile was found in the covalent hydrate of
pteridin-6-one ⟨52JCS1620⟩. Covalent hydration in general, *i.e.* the addition of water to a
C=N bond in an aza heterocyclic system, was found to be a reversible reaction. The driving
force for the addition can be seen in the resulting partial compensation of the $\pi$-deficiency
of the heteroaromatic system, and quite often in a stronger stabilization of the corresponding
cation by amidinium-type resonance.

With pteridine (**1**) the covalent hydration is a complex matter since the general acid–base
catalyzed reaction provides a good example of a kinetically controlled addition to the

3,4-position (**8**) *versus* a thermodynamically favoured di-adduct formation at the pyrazine moiety leading to (**11**). The effect of substituents on the rate and site of hydration has been the subject of several studies and was reviewed recently ⟨76AHC(20)117⟩. Partial inhibition of covalent hydration by a methyl group attached to the carbon atom accepting the nucleophile is explained by steric and electronic factors. There exists also a linear free energy relationship for the covalent addition of a number of nucleophiles at an aza aromatic substrate, correlating in principle the $pK_a$ values of the conjugated acid of the adding nucleophile with the equilibrium constant for the addition reaction ⟨74JA1843⟩.

The formal hydroxylation of a C—H bond in the enzymic reaction catalyzed by xanthine oxidase can, from the mechanistic point of view, also be regarded as a nucleophilic attack leading finally to an oxidation of an —N=CH— into an amide function ⟨74JBC(249)4363⟩. 7-(*p*-Substituted-phenyl)pteridin-4-ones are converted by free and immobilized xanthine oxidase into the corresponding lumazine derivatives ⟨79RTC224⟩. A fuller study deals with the pathways and relative rates of oxidation of a large number of pteridines ⟨59BBA(33)29⟩, and shows that two series of substrates can be defined. Pteridines with a free 6-position are all converted finally into pteridine-2,4,7-trione (equation 13), whereas the presence of a preformed 6-hydroxy group induces either resistance to enzymic attack or leads to the pteridine-2,4,6,7-tetrone. From these and similar results with *N*-methylated derivatives ⟨77BBA(480)21⟩ it has been concluded that only the anhydrous species are susceptible to the xanthine oxidase reaction, since pteridin-6-one, which exists in aqueous solution as a stable 7,8-hydrate, is inactive, but its 7-methyl derivative is transformed into 7-methylpteridine-2,4,6-trione ⟨64BBR(17)461⟩.

(13)

The purely chemical analogy involving nucleophilic attack and subsequent oxidation can be achieved by hydrogen peroxide, which converts pteridin-6-one into pteridine-6,7-dione ⟨52JCS1620⟩, and xanthopterin (**4**) into leucopterin (**6**) ⟨39LA(539)179⟩. Isoxanthopterin (**5**) reacts with nitrous acid to give pteridine-2,4,6,7-tetrone ⟨44LA(555)146⟩.

The addition of primary and secondary alcohols proceeds not only in the presence of acid but also in neutral solution ⟨62JCS1591, 65JCS6930, 71JCS(B)2423⟩. Stable 1:1 mono- and 2:1 di-σ-adducts were found and characterized spectrally. *t*-Butyl alcohol does not form adducts for steric reasons.

The action of sulfur nucleophiles like sodium bisulfite and thiophenols causes even pteridines that are unreactive towards water or alcohols to undergo covalent addition reactions. Thus, pteridin-7-one smoothly adds the named *S*-nucleophiles in a 1:1 ratio to C-6 ⟨65JCS6930⟩. Similarly, pteridin-4-one (**73**) yields adducts (**74**) in a 2:1 ratio at C-6 and C-7 exclusively (equation 14), as do 4-aminopteridine and lumazine with sodium bisulfite. Xanthopterin forms a 7,8-adduct and 7,8-dihydropterin can easily be converted to sodium 5,6,7,8-tetrahydropterin-6-sulfonate ⟨66JCS(C)285⟩, which leads to pterin-6-sulfonic acid on oxidation ⟨59HCA1854⟩.

(14)

Nucleophilic addition reactions of nitrogen nucleophiles show the same reaction pattern, with simple pteridines forming 1:1 and 2:1 σ-adducts. The 1:1 adduct of ammonia across the 3,4-bond of pteridine is formed amongst other products in a dilute buffered solution ⟨74JCS(P1)357⟩, but in liquid ammonia at low temperature (−60 °C) 4-amino-3,4-dihydropteridine is the sole detectable species ⟨75RTC45⟩. Reactions with [15]N-labelled ammonia proved that no ring opening–ring closure mechanism is involved in the formation of the adducts. At room temperature, ammonia and amines attack C-6 and C-7 to form the thermodynamically most stable 2:1 adducts ⟨71JCS(C)2357, 73JCS(P1)1974⟩. The 6- and 7-chloropteridines were reported to yield 1:1 adducts to the 3,4-bond with ammonia, benzylamine and cyclohexylamine, preferably at low temperature and in apolar solvents ⟨64JCS4920⟩. Attempts to add the strongly nucleophilic amide ion to (1) failed due to decomposition.

Substituted pteridine derivatives, on the other hand, under such drastic conditions show nucleophilic displacements *via* the $S_N ANRORC$ mechanism and ring contractions to purine derivatives ⟨75CPB2678, 77H(7)205⟩. Addition of ammonia, hydrazine and primary and secondary amines to 7,8-dihydropterin (75) provides an interesting approach to 6-aminopterins (77) *via* the intermediate tetrahydro derivative (76) and subsequent oxidation (equation 15) ⟨64HCA2195⟩. Another direct amination reaction was achieved with pteridin-6-one and hydroxylamine, yielding 7-aminopteridin-6-one in 65% yield ⟨55JCS2690⟩. Furthermore, the reaction of xanthopterin-7-carboxylic acid with ammonia and subsequent oxidation by $MnO_2$ leading to 7-aminoxanthopterin provides an interesting approach to 7-substituted xanthopterins ⟨80HCA1805⟩.

$$(15)$$

The use of carbon nucleophiles in Michael-type addition reactions with pteridine and its derivatives leads to a quite complicated and divergent pattern. These reactions are strongly dependent on the nature of the carbon nucleophile and can be divided into various categories.

(a) Addition across the 3,4-bond to yield 4-substituted 3,4-dihydropteridines (79) occurs with pteridine, 2-aminopteridine, pteridin-2-one (78) and pteridine-2-thione and dimedone, barbituric acid, 2-thiobarbituric acid, acetylacetone, diethyl malonate, ethyl benzoylacetate, ethyl acetoacetate and ethyl cyanoacetate (equation 16) ⟨62JCS1591, 66JCS(C)1117, 68JCS(C)63, 73JCS(P1)1615⟩.

$$(16)$$

(b) Pteridin-4-one (73) reacts only with strong carbon nucleophiles like dimedone, barbituric acid and 2-thiobarbituric acid to form 6,7-disubstituted 5,6,7,8-tetrahydro-pteridin-4-ones ⟨73JCS(P1)2630⟩, while pteridine and its 2- and 4-methyl derivatives (80) are able to add bifunctional nucleophiles such as ethyl acetoacetate or acetylacetone to both carbon atoms of the pyrazine ring to give condensed ring systems (81; equation 17) ⟨73JCS(P1)1974, 73JCS(P1)1615⟩. Phenylation takes place exclusively at position 7 if 2-methyl-thio-4,6- and 2-methylthio-4,7-diphenylpteridine respectively are treated with phenyl-lithium ⟨77H(7)205⟩.

$$(17)$$

(c) An interesting case has been found with pteridin-7-one (83), which shows no tendency to add water or alcohols across a double bond, but nevertheless readily accepts various types of Michael reagents not only in substantially neutral solution but also at pH 2. Potentially carbanionic reagents such as acetylacetone, dimedone, ethyl acetoacetate and diethyl acetonedicarboxylate lead without exception to products isolated as 1:1 adducts with UV spectra very similar to 5,6-dihydropteridin-7-one ⟨65JCS6930⟩. Comparable batho-chromic shifts of about 50 nm for the long-wavelength band are found on passing from the cation of (83) to the cation of its 5,6-dihydro derivatives, and to those of the adducts (82; equation 18). Steric hindrance by a 6-methyl group confirms that, in these reactions, addition occurs across the 5,6-double bond. Although (83) does not react with the feebly acidic diethyl malonate in water, the desired 1:1 adduct can readily be prepared in tributylamine.

The rather low p$K_a$ values for the methylene groups of barbituric acid and its 2-thio derivative indicated a particularly easy adduct formation (**84**) which proceeds in water in excellent yield.

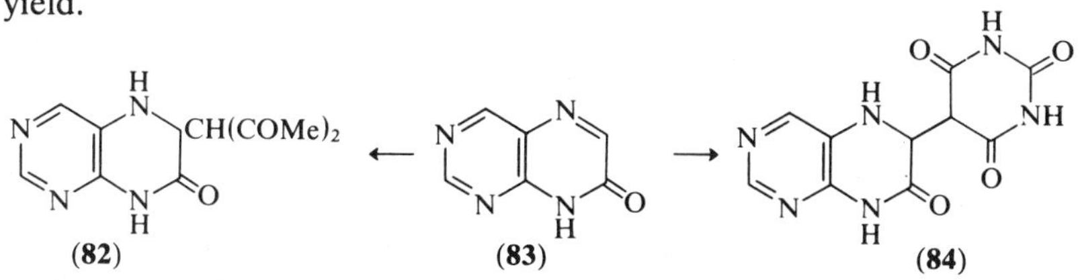

(18)

(**82**)        (**83**)        (**84**)

(d) Pteridin-6-one (**85**) is also a good substrate for the addition of carbanions in cold alkaline solution. Stable 7,8-dihydro adducts ⟨61JCS127⟩ are obtained with acetone, diethyl malonate and ethyl cyanoacetate, of which (**86**) was degraded to 7-cyanomethyl-7,8-di-hydropteridin-6-one (**87**), whereas treatment with hot 1N NaOH in air converts the adducts to pteridine-6,7-dione, thus establishing the orientation (equation 19). The presence of an electron-donating substituent at the pyrimidine moiety as seen in xanthopterin (**4**) allows also an acid-catalyzed addition to the 7,8-position, and the adduct reveals a strong tendency to be autoxidized back to the heteroaromatic oxidation level. Compound (**4**) reacted by this pathway with oxaloacetic ⟨62AG(E)115⟩ or pyruvic acid ⟨61HCA1783, 63HCA51⟩ to form erythropterin (**39**) ⟨62CB2195⟩ *via* its 7,8-dihydro derivative (**88**; equation 20).

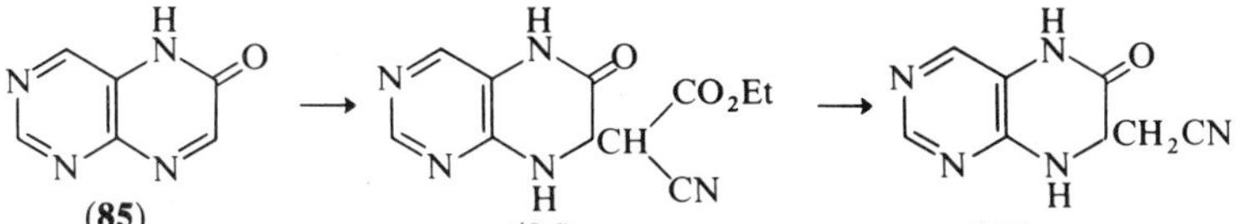

(19)

(**85**)        (**86**)        (**87**)

(20)

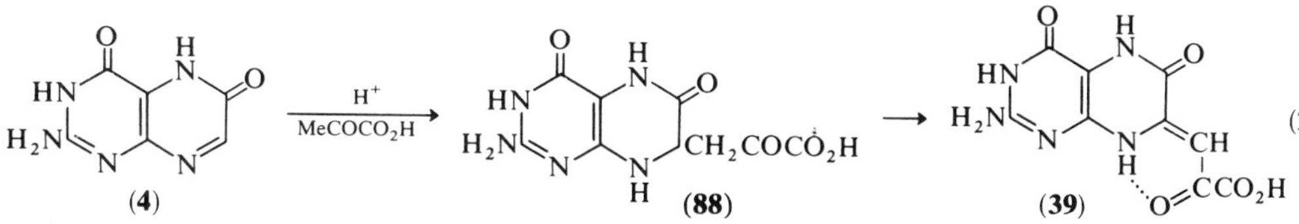

(**4**)        (**88**)        (**39**)

(e) 7,8-Dihydropterin (**75**) reacts readily with carbanions derived from Michael-type reagents to form the corresponding 5,6-adducts, which on careful oxidation give aromatic pteridines. The cyanide ion also adds to (**75**) and oxidation leads to 7,8-dihydropterin-6-carboxamide ⟨66JCS(C)285, 63HCA1181⟩. A similar mechanistic pathway may explain the formation of deoxysepiapterin (**46**) from a reduced form of pterin and α-oxobutyric acid in the presence of thiamine which produces the formal acylanion (equation 21) ⟨62MI21600⟩.

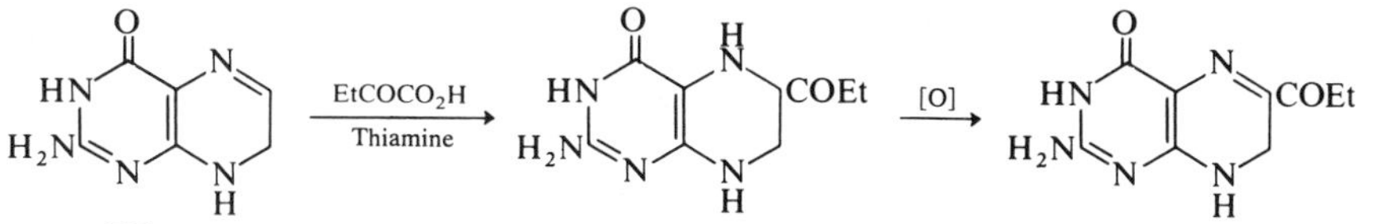

(21)

(**75**)        (**46**)

(f) The addition of carbon nucleophiles can precede ring transformation, as seen in the conversion of pteridine (**1**) by malononitrile *via* the intermediates (**89**) and (**90**) into 6-aminopyrido[2,3-*b*]pyrazine-7-carbonitrile (**91**; equation 22) ⟨73JCS(P1)1615⟩.

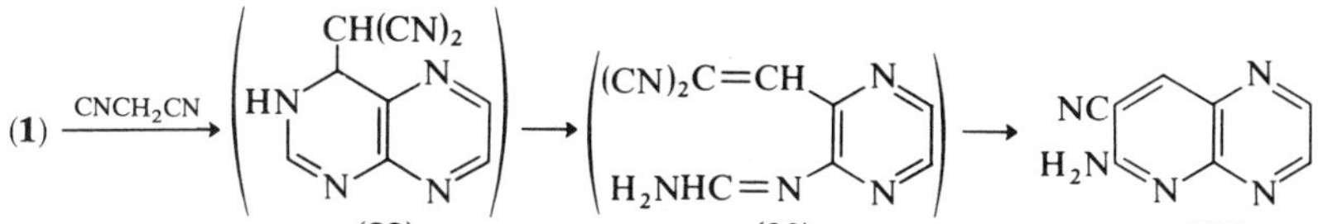

(22)

(**89**)        (**90**)        (**91**)

### 2.16.2.1.3 *Cine* and *tele* substitutions

A useful approach to the substitution of ring C—H positions lies in the activation of the heteroaromatic system by an *N*-oxide group, initiating a formal intramolecular redox reaction. 1-Methyllumazine 5-oxide reacts with acetic anhydride in a Katada rearrangement to give 6-acetoxy-1-methyllumazine ⟨73CB3149⟩, and treatment of 1,3-dimethyllumazine 5-oxide (92) with acetyl chloride yields 6-chloro-1,3-dimethyllumazine (95) almost quantitatively, indicating firstly acylation at the *N*-oxide function (93) followed by a nucleophilic attack at the adjacent (*cine*) position (94) and subsequent elimination (equation 23) ⟨82LA2135⟩.

Another interesting rearrangement starting from pterin, lumazine and 2,4-diaminopteridine 8-oxides involves a *tele* substitution, introducing the nucleophile into the 6-position. Pterin 8-oxide (96) is converted by a trifluoroacetic acid/trifluoroacetic anhydride mixture into xanthopterin (4; equation 24) ⟨73JA4455⟩. Acetyl chloride/TFA causes the formation of 6-chloropterin and 2,4-diaminopteridine 8-oxide yields 2,4-diamino-6-chloropteridine in an analogous reaction ⟨78JOC680⟩. There is, however, also the possibility of attack at C-7, since 2,4-diaminopteridine 8-oxide reacts with pyrrolidine to give 2,4-diamino-7-pyrrolidinopteridine, while with pterin 8-oxide and ethyl cyanoacetate/acetic anhydride in anhydrous HMPT an attachment of a carbon side-chain, yielding ethyl α–(pterin-7-y1)-α-cyanoacetate, can be achieved ⟨75JOC2341⟩.

### 2.16.2.1.4 Free Radicals

An interesting method for the substitution of a hydrogen atom in $\pi$-electron deficient heterocycles was reported some years ago, in the possibility of homolytic aromatic displacement ⟨74AHC(16)123⟩. The nucleophilic character of radicals and the important role of polar factors in this type of substitution are the essentials for a successful reaction with six-membered nitrogen heterocycles in general. No paper has yet been published describing homolytic substitution reactions of pteridines with nucleophilic radicals such as alkyl, carbamoyl, α-oxyalkyl and α-*N*-alkyl radicals or with amino radical cations.

Preliminary studies in this field, however, have recently been achieved by acylations of various lumazine derivatives with acyl radicals derived from aldehydes by hydrogen abstraction ⟨82UP21603⟩. 1,3-Dimethyllumazine (98) reacts preferentially at the most electron-deficient 7-position, yielding 7-acyl-1,3-dimethyllumazines (97) in good yields. 7-Alkyl substituents direct to the adjacent 6-position, as found with 1,3,7-trimethyllumazine (99) which gives the 6-acyl derivatives (100; equation 25). The compounds 6- and 7-amino-(102), methylamino- and dimethylamino-1,3-dimethyllumazines, as well as the corresponding hydroxy and methoxy derivatives (103), are very good substrates for homolytic radical substitutions, which take place especially with acyl radicals at the unsubstituted position of the pyrazine ring (101 and 104; equation 26). 2,4-Diaminopteridine reacts with propionyl radicals in excellent yield to give the 2,4-diamino-7-propionylpteridine, while the introduction of the same substituent into position 6 could be effected by a removable blocking group at C-7 such as an alkylthio function. Thus, a 2-amino-4-pentyloxy-7-propylthio-pteridine (105) is acylated at C-6 to give (106), which after desulfurization and hydrolysis leads to 6-propionylpterin (107), the synthetic precursor of the *Drosophila* pigment deoxysepiapterin (46; equation 27).

(25)

(97)     (98) R = H     (100)
           (99) R = Me

(26)

(101)     (103) R = OMe, R' = H     (104)
           (102) R = H, R' = NH$_2$

(105)     (106)     (107)  → (46)   (27)

## 2.16.2.2 Reactivity of Substituents

### 2.16.2.2.1 Halogen Substituents

It is generally agreed that in six-membered $\pi$-deficient nitrogen heterocycles there occurs a strong activation of $\alpha$- and $\gamma$-substituents due to the polarization in the ground state and pronounced stabilization of the various types of (negatively charged) transition states by the heteroatoms. The 1,3,5,8-relationship of the ring nitrogen atoms in the pteridine ring system creates two types of $\alpha$-carbon atoms, since the 2-, 4- and 7-positions are influenced not only by the adjacent but also by the $\gamma$- and vinylogous heteroatoms, whereas C-6 is restricted to activation only by N-5. These facts indicate a higher reactivity of substituents in the former positions in the sequence $7 > 4 > 2$ which parallels the energies of the different transition states of the closely related $\sigma$-complexes in nucleophilic substitution reactions.

Comparing the four simple monochloropteridines, the 7-chloro compound seems to be the most reactive as seen from its difficulty of preparation and its instability, initiated apparently by self-quaternization ⟨54JCS3832⟩. In hydrolysis studies, 2-chloropteridine was found to be by far the most stable. Activation of the chlorine atoms, however, is in all cases strong enough to allow nucleophilic displacement under mild conditions with the common O, N and S nucleophiles. Sodium methoxide yields the methoxypteridines, ammonia, amines and hydrazine the corresponding amino- and hydrazino-pteridines and sodium hydrogen sulfide and thiolates react to give pteridinethiones and alkylthiopteridines respectively ⟨52JCS1620, 54JCS3832, 64JCS4920⟩. With 4,7- ⟨65JCS1175⟩ and 6,7-dichloropteridine (107) ⟨65JCS27⟩, either stepwise or simultaneous displacement of the chlorine atoms has been achieved, attack occurring preferentially at the 7-position (equation 28). 2,4-Dichloro-6,7-dimethyl- and -6,7-diphenyl-pteridine ⟨51JA4384⟩ have not been examined for selective substitution by amines, and use of 4,6-dichloro-, 2,4,7- and 4,6,7-trichloro-pteridine ⟨56JCS4621⟩ led to complete halogen exchange with ammonia at elevated temperatures. More informative studies are available on 2,4,6,7-tetrachloropteridine (109), which can be hydrolyzed by aqueous base to 2,4-dichloropteridine-6,7-dione (108), and aminated by ammonia, presumably to 7-amino-2,4,6-trichloropteridine ⟨41LA(548)82⟩, and by liquid ammonia at −70 °C to 6,7-diamino-2,4-dichloropteridine (110) ⟨59JA2464⟩. It can be assumed that the mechanisms involve a nucleophilic substitution at the 7-position first, when the new electron-donating substituent greatly decreases the reactivity of the chlorine atoms in the pyrimidine moiety, leaving the 6-position electronically intact for further displacement by nucleophiles. More drastic conditions convert (109) to pteridine-2,4,6,7-tetrone and 2,4,6,7-tetraaminopteridine (111) respectively (equation 29).

A great variety of differently substituted pteridine derivatives have been synthesized starting from (109) or the 2,4,6,7-tetrabromo compound ⟨60USP2940972⟩, which shows a corresponding reactivity pattern. Amines at low temperature substitute only the 6- and

$$(28)$$

**(107)**

$$(29)$$

**(108)**    **(109)**    **(110)**    **(111)**

7-halogen atoms while boiling leads to 2,6,7-trisubstituted 4-halo derivatives. The last halogen atom needs very drastic conditions and temperatures up to 200 °C for displacement. Compound (**109**) has been applied as a new type of linker in reactive dyestuff chemistry to bind the colour covalently to the fibre ⟨69MI21600, 70MI21604⟩. The influence of electron-donating substituents on the reactivity of halogen atoms can be seen also in the stability of 2-amino-4-chloropteridine-6,7-dione towards hot, dilute acid and alkali. Only boiling in 6N sodium hydroxide causes hydrolysis to leucopterin (**6**). 7-Chloropterin ⟨62CB1621⟩ is also fairly stable towards hydrolysis and requires prolonged boiling in sodium methoxide for displacement of the chlorine atom. 2,4-Di- and 2,4,7-tri-substituted 6-halopteridines are also very stable and can easily be prepared from pyrazine precursors by base-catalyzed cyclizations without touching the halogen atom ⟨68JMC322⟩. Nucleophilic substitution may also depend on additional factors, as 6-chloropterin and 2,4-diamino-6-chloropteridine both react smoothly with arenethiols and alkanethiolates to give a series of 6-*S*-substituted pteridine derivatives, while all attempts to displace the chlorine by primary amines failed, even in the presence of reducing agents ⟨78JOC680, 81JMC1001⟩. Secondary amines like *N*-methylarylmethylamines, piperidines and pyrrolidines, however, do react under defined reaction conditions ⟨81JMC140⟩. The influence of only one electron-donating group on the reactivity of the 6-chlorine atom is expected to be less and explains the general susceptibility of 6-chloropteridin-4-one and its 3-methyl derivative towards nucleophilic displacements by alkoxides, thiolates and amines ⟨67JMC899⟩.

The different reactivities of halogen atoms also becomes obvious in the lumazine series ⟨82UP21600⟩. 7-Chloro-1,3-dimethyllumazine (**112**), which can be regarded as a vinylogous acid chloride due to the mesomeric influence of the 4-carbonyl group on position 7, reacts with a wide variety of nucleophiles almost spontaneously at room temperature (equation 30), whereas the corresponding 6-chloro isomer (**117**) requires prolonged heating, especially with amines. 6,7-Dichloro-1,3-dimethyllumazine (**114**) is an excellent substrate for stepwise displacement of the chlorine atoms, always substituting first at C-7, to form 7-amino-6-chloro-1,3-dimethyllumazine (**115**) with ammonia, and the corresponding 6-substituted derivative (**116**) after subsequent treatment with various O, N, S and C nucleophiles. The hardness and softness of the reactive centres and nucleophiles seems to play an important role in the choice of the reaction conditions, since (**114**) reacts with piperidine to form 1,3-dimethyl-6,7-dipiperidinolumazine only after 8 h boiling whereas the conversion into 1,3-dimethyllumazine-6,7-dithione (**113**) proceeded by treatment with sodium sulfide at room temperature (equation 31). Displacement reactions with carbanions are achieved with some difficulty even at the reactive C-7 position and, starting from (**117**), further activation by quaternization at N-5 (**118**) is required to bring the 6-chlorine atom into the reaction

$$(30)$$

**(112)**

$$(31)$$

**(114)**    **(115)**    **(116)**

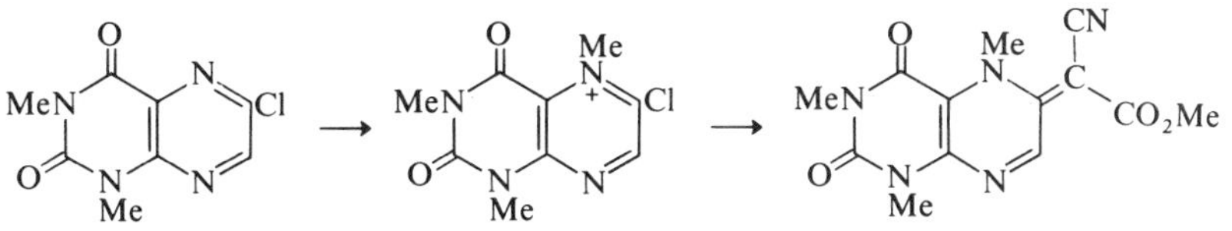

$$(32)$$

$$(33)$$

to form (119; equation 32). On the other hand, both centres of (114) may react if an intramolecular displacement is involved, as in the reaction with dibenzoylmethane which with base-catalysis yields (120; equation 33).

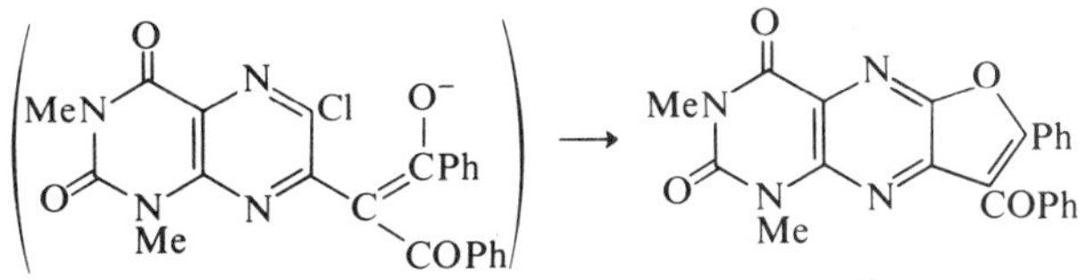

Reactions with strongly basic nucleophiles such as potassium amide in liquid ammonia may prove much more complex than direct substitution. 2-Chloro-4,6,7-triphenylpteridine reacts under these conditions *via* an $S_N ANRORC$ mechanism to form 2-amino-4,6,7-triphenylpteridine and the dechlorinated analogue 〈78TL2021〉. The attack of the nucleophile exclusively at C-4 is thereby in good accord with the general observation that the presence of a chloro substituent on a carbon position adjacent to a ring nitrogen activates the position *meta* to the chlorine atom for amide attack.

Replacement of chlorine by hydrogen had been effected previously by hydriodic acid, which converts 2-amino-4-chloropteridine-6,7-dione into 2-aminopteridine-6,7-dione 〈40LA(543)209〉 and removes the chlorine analogously from the 2- 〈44LA(555)146〉 and 7-positions of other pteridines 〈48JA14〉. 2-Chloro-7,8-dihydropteridin-6-one is reduced to 7,8-dihydropteridin-6-one by hydrogen iodide and red phosphorus at 160 °C and the 4-chloro isomer behaves similarly 〈51JCS96〉. Catalytic dehalogenations are feasible in principle, especially in the presence of bases, but they encounter the difficulty of further reduction to dihydro and tetrahydro stages which usually necessitates an oxidative work-up 〈78JOC680〉. An effective alternative includes nucleophilic displacement of the chlorine atom by hydrazine followed by oxidation by copper(II) ions, as exemplified by the sequence 7-chloropterin (121)→7-hydrazinopterin (122)→pterin (2) (equation 34).

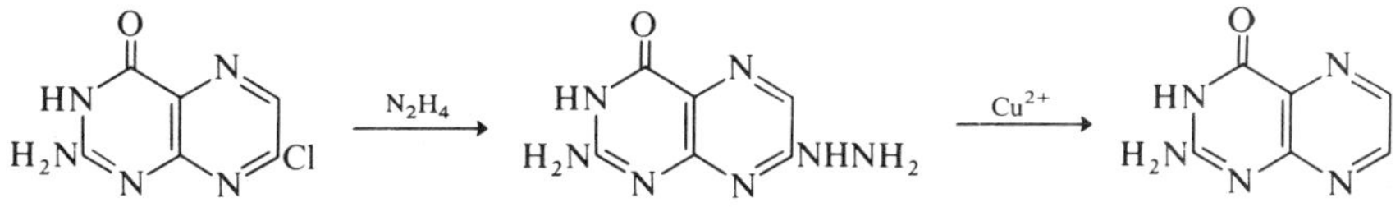

$$(34)$$

Very little information is available about other halopteridines such as bromo, fluoro and iodo derivatives, indicating that there is little call for their preparation since, for example, 6-bromo- and 6-chloro-1,3-dimethyllumazine react very similarly 〈82UP21600〉.

### 2.16.2.2.2 *N-Linked Substituents*

The reactivity of the amino groups at the pteridine nucleus depends very much upon their position. All amino groups form part of amidine or guanidine systems and therefore do not behave like benzenoid amino functions which can usually be diazotized. The 4-, 6- and 7-amino groups are in general subject to hydrolysis by acid and alkali, whereas the 2-amino group is more stable under these conditions but is often more susceptible to removal by nitrous acid.

7-Aminopteridine is the most sensitive to acid hydrolysis, and 6-amino- and 6-dimethyl-amino-pteridine are also hydrolyzed, even by cold 0.01N hydrochloric acid, too rapidly for accurate determination of the cation form 〈52JCS1620〉. 2-Amino- and 4-amino-pteridine are not readily attacked by 1N HCl at 20 °C but at 100 °C the former compound is destroyed and the latter converted into pteridin-4-one 〈51JCS474〉. 2,4-Diaminopteridine can be hydrolyzed by refluxing in 6N HCl for 30 minutes to 2-aminopteridin-4-one (pterin; 2) and after

60 h pteridine-2,4-dione (lumazine; **3**) is obtained. Isopterin (4-aminopteridin-2-one) similarly gives lumazine in 20 min ⟨49JA2538⟩.

Alkaline hydrolysis is apt to be gentler and is therefore often preferred as a preparative method. Refluxing 4-aminopteridine in 0.01N NaOH for 3 min yields pteridin-4-one exclusively but 4-amino-6,7-diethylpteridine requires longer heating. The 2-amino isomer is rapidly destroyed by boiling 2.5N NaOH, but 2-amino-6,7-diethylpteridine is unaffected after 2 h. This is consistent with other evidence that electron-donating groups make the acid and alkaline hydrolysis more difficult in the pteridine series. 2,4-Diaminopteridine and its 6- and 7-methyl derivatives are hydrolyzed to the corresponding pterins in excellent yield by boiling in 1N NaOH ⟨49JA1753⟩, analogous to the conversion of 4-amino, 4-piperidyl- and 4-dimethylamino-folic acid into folic acid ⟨50JA1914⟩. 2,4-Diaminopteridine 8-oxide leads to pterin 8-oxide and 2,4-diamino-7-(1-pyrrolidino)pteridine gave isoxanthopterin (**5**) after prolonged heating ⟨75JOC2341⟩. The 2-amino groups show high stability in these systems against alkaline hydrolysis because of anion formation, while the 1-methylpterin derivatives are hydrolyzed relatively easily to the corresponding 1-methyl-lumazines ⟨58JA6095⟩. 3-Methylpterins on the other hand are stabilized by Dimroth rearrangement to 2-methylaminopteridin-4-ones. Thus, in case of hydrolysis by either acid or alkali, 4- and 7-amino groups are always more reactive than 2-amino groups. The nature of the substrate as well as the reaction conditions may also play an important role in product formation. 4-Amino-8-benzyl-6-methylpteridine-2,7-dione (**124**) is hydrolyzed by acid to 8-benzyl-6-methylpteridine-2,4,7-trione (**123**), but treatment with alkali results in attack at the amide function of the pyrazine ring. This initiates a rearrangement *via* a ring opening–ring closure mechanism to 4-benzylamino-6-methylpteridine-2,7-dione (**125**), which resists further hydrolysis due to dianion formation and electrostatic repulsion (equation 35) ⟨63CB2964⟩.

(35)

The slow acid hydrolysis of 2-amino groups can sometimes be greatly accelerated by the addition of sodium nitrite ⟨51CB801⟩. Pterin gives lumazine in this way in 5 min, but 2- and 4-aminopteridine as well as xanthopterin decompose ⟨39LA(539)128⟩. Rhizopterin ⟨47JA2753⟩ and leucopterin ⟨40LA(544)163⟩ can be deaminated by nitrous acid which is, however, without effect on 4-aminopteridine-2,6,7-trione and 2,4-diaminopteridine ⟨49JA2538⟩. Mild hydrolysis can be achieved enzymatically in special cases since adenosine deaminase catalyzes the conversion of 4-aminopteridine into pteridin-4-one ⟨73B392⟩. Kinetic similarities between this reaction and the hydration of pteridine suggests that in the deamination water attacks the substrate directly and that the transition state in this reaction is reached before a tetrahedral intermediate is fully developed. An enzyme deaminating 2-amino group was found in the silkworm *Bombyx mori*, and converts sepiapterin into sepialumazine ⟨67E116, 71MI21602, 80MI21604⟩.

Replacement reactions of amino groups by nucleophiles other than water work only with primary and secondary amines, and all efforts at direct conversion into chloro- or thio-pteridines have been unsuccessful. During replacement reactions of 4-amino-6,7-dimethylpteridine-2-thione it was noticed that at elevated temperatures and strenuous conditions not only does exchange of the thione function occur but also that of the 4-amino group ⟨51JA4384⟩ leading to 2,4-bis-(substituted amino) pteridine derivatives. Hence a ring opening–ring closure mechanism has been postulated for the displacement of the 2-substituent while the more reactive 4-amino group may be replaced by a direct nucleophilic substitution mechanism. Thus, 2,4-diamino-6,7-diphenylpteridine (**126**) when heated under reflux with benzylamine gave 2-amino-4-benzylamino-6,7-diphenylpteridine (**127**) in 67% yield, but addition of a small amount of hydrochloric acid to the reaction mixture led to the formation of the 2,4-bis-(benzylamino) derivative (**128**) in 91% yield (equation 36) ⟨52JA1648⟩. 2-Hydroxyethylamine, piperidine, morpholine, 3-dimethylaminopropylamine, 3-diethylaminopropylamine and 3-isopropylaminopropylamine reacted similarly. With 7-amino-1,3-dimethyllumazine an easy displacement of the amino group by other amines proceeds on heating with the corresponding amine or substituted ammonium acetate.

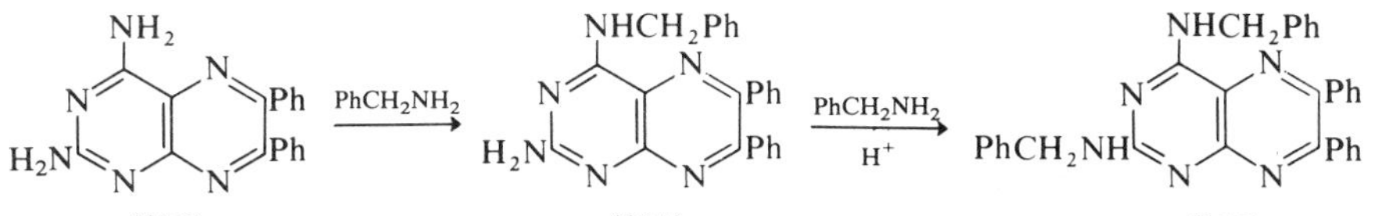

(36)

Aminopteridines are also subject to acylation, which is usually performed by boiling in the anhydrides. Acylated derivatives of 2-, 4-, 6- and 7-aminopteridines are obtained in good yield by refluxing in acetic anhydride for a few minutes. 2,4-Diaminopteridine ⟨65JCS1530, 70CB722⟩ and its 6- and 7-methyl derivatives form 2,4-bis(acetamido) derivatives, but 2,4-diamino-6,7-diphenylpteridine gives only a monoacetyl compound, of undetermined orientation, after 16 h refluxing. Addition of sulfuric acid facilitates the diacetylation in this case. Pterin (2) and xanthopterin (4) are acetylated relatively easily at the 2-amino group, but *N*-benzoylation with (4) requires 200 °C with benzoic anhydride. Leucopterin (6) cannot be substituted by acetic anhydride alone or with pyridine; here again sulfuric acid catalysis overcomes the difficulty ⟨33LA(507)226⟩. Acetamidopteridines are in the main lower-melting compounds, more soluble in all solvents than the corresponding amino derivatives. Hydrolysis by acid and base usually proceeds under mild conditions and in special cases like 6-acetoxymethyl-2,4-bis(acetamido)pteridine (130) cleavage of the amide function takes place even in alcohols to give 6-acetoxymethyl-2-acetamido-4-amino- and 6-acetoxymethyl-2,4-diamino-pteridine (129) ⟨80CB1514⟩. Boiling in water converts (130) into pterin derivatives (131) and (132), removing the entire acetamido group by nucleophilic attack at C-4 (equation 37). This type of reaction is not unique, since 2-acetamido-1,6,7-trimethylpteridin-4-one yields 1,6,7-trimethyllumazine on acid treatment ⟨61JOC2129⟩.

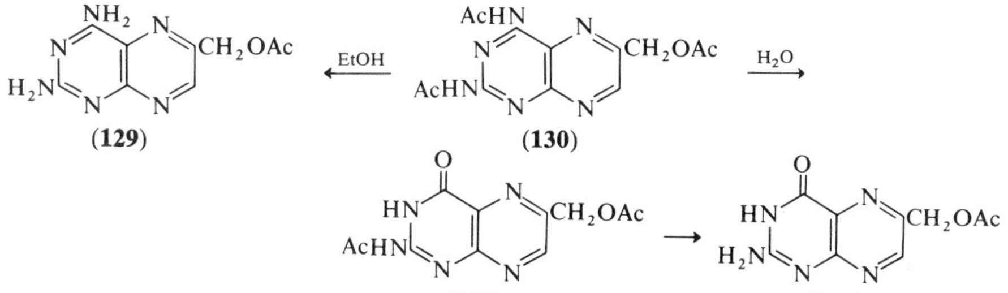

(37)

An interesting behavior of the $N^2N^2$-disubstituted pterins was found during oxidation of euglenapterin ⟨80AG474⟩ with potassium permanganate. Not only is the carbon side-chain in the 6-position oxidized in the usual manner to the carboxyl group, but also the *N*-alkyl groups are oxidatively removed, leading to pterin-6-carboxylic acid. Analogously, $N^2,N^2,7$-trimethylpterin (134) can be oxidized to pterin-7-carboxylic acid (135) and $N^2,N^2$-dibenzylpterin is easily deblocked to pterin (2). The unusual reactivity of the $N^2$-alkyl substituents is further reflected in the photochemical demethylation of (134) to $N^2,7$-dimethylpterin (133; equation 38) ⟨81TH21600⟩.

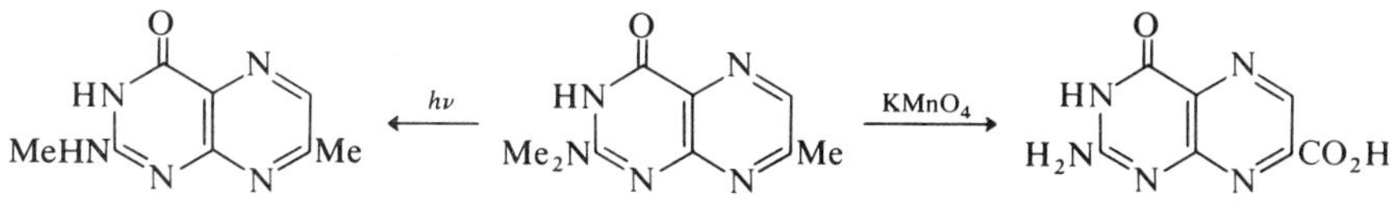

(38)

### 2.16.2.2.3 *O-Linked Substituents*

When hydroxypteridines are considered, it must be borne in mind that these compounds exist principally in the pteridinone forms, containing thermodynamically stable amide functions, and consequently have low reactivity. Their stability towards acid and alkali correlates well with the number of electron-donating groups which apparently redress the deficit of $\pi$-electrons located at the ring nitrogen atoms. Quantitative correlations can be seen in the decomposition studies of various pteridinones (Table 7). These results are consistent with the number of the oxy functions and their site at the pteridine nucleus. The

**Table 7**   Decomposition of Pteridines by Excess of Acid or Alkali at 110 °C

| Pteridine[a] | Decomposition (%) in 1 h by | | |
|---|---|---|---|
|  | *1N H$_2$SO$_4$* | *1N NaOH* | *10N NaOH* |
| Unsubstituted | 74 | 53 | 57 |
| 2-Hydroxy- | 55 | 7 | 89 |
| 4-Hydroxy- | 60 | 57 | 94 |
| 6-Hydroxy- | 2 | 94 | 100 |
| 7-Hydroxy- | 52 | 31 | 76 |
| 2,4-Dihydroxy- | 6 | — | 4 |
| 6,7-Dihydroxy- | 7 | — | 12 |
| 4,6,7-Trihydroxy- | — | — | 4 |
| 2,4,6,7-Tetrahydroxy- | — | — | 6 |

[a] For simplicity, structures are indicated in the hydroxy forms.

somewhat higher percentage of decomposition of the tetra- over the tri-oxopteridine in alkali is attributed not to simple hydrolysis but apparently to a base-catalyzed oxidative degradation.

Phosphorus halides, especially POCl$_3$, are very effective reagents for the conversion of oxo into chloro groups in this series, as illustrated in the high yielding preparation of 6-(**117**) and 7-chloro-1,3-dimethyllumazine (**112**) from the corresponding trioxo derivatives ⟨82LA2135⟩. It is found that addition of chloride ions sometimes accelerates the reaction. The oxo groups of the pyrazine moiety seem to be more reactive in these displacements, since 1- and 3-methylpteridine-2,4,7-trione can selectively be converted into 1- and 3-methyl-7-chlorolumazine by treatment with POCl$_3$ at 90 °C ⟨83TH21600⟩. Similarly, 2-methylthiopteridine-4,7-dione (**136**) forms first 7-chloro-2-methylthiopteridin-4-one (**137**) and then on prolonged heating in POCl$_3$ at 100 °C the 4,7-dichloro-2-methylthio compound (**138**; equation 39). In a number of cases only a mixture of POCl$_3$ and PCl$_5$ is successful, whereby full or partial substitution takes place. Pteridin-6-one ⟨52JCS1620⟩, pteridine-2,4,6,7-tetrone ⟨41LA(548)82⟩, 6,7-diphenyllumazine ⟨51JA4384⟩ and 6,7-diphenylpterin ⟨49JA892⟩ are fully chlorinated, but with leucopterin the 2-amino-4-chloropteridine-6,7-dione ⟨33LA(507)226⟩, with 4-aminopteridine-2,6,7-trione the 4-amino-2-chloropteridine-6,7-dione and with isoxanthopterin ⟨62CB1621⟩ and its 6-carboxylic acid ⟨48JA14⟩ the corresponding 7-chloro derivatives are obtained in these reactions. There are also pteridinones which are destroyed by the vigorously reactive POCl$_3$/PCl$_5$ mixture.

$$(39)$$

(**136**)                    (**137**)                    (**138**)

The direct conversion of a lactam function into an amino group has been little studied in the pteridine series. Zav'yalov's method ⟨69IZV655, 69IZV2857, 70IZV904, 70IZV953, 70IZV1198⟩ using phenylphosphordiamidates works quite well with 6- and 7-hydroxy-1,3-dimethyllumazines, to give the corresponding amino, methylamino and dimethylamino derivatives (equation 40) ⟨82LA2135⟩. On the other hand, 6,7-dimethyl-2-thiolumazine is transformed into 2,4-bis(methylamino)-6,7-dimethylpteridine by alcoholic methylamine at 190 °C for 18 h ⟨51JA4384⟩.

$$(40)$$

(**139**)                    (**140**)                    (**141**)

The 5,6-amide function has been found to be unusually reactive in the 2,4-disubstituted pteridin-6-one series, forming the corresponding 6-alkoxy derivatives, *e.g.* (**141**), with alcoholic HCl directly ⟨62CB755, 70CB735⟩.

Direct sulfurizations of simple monooxopteridines have not been successful, but the more stable lumazines and 2-thiolumazines are good substrates for P$_4$S$_{10}$ treatment, leading preferentially to C-4 substitution ⟨74CB3377⟩. With 7-hydroxy-1,3-dimethyllumazines it was found that the various amide groups can selectively be substituted in the order C-7, C-4, C-2 by $P_4S_{10}$ treatment (equation 41) ⟨79CB1499⟩. 2,4-Diamino- and 2-amino-4-alkoxy-pteridin-7-ones react analogously to the corresponding 7-thiones, and 1,3-dimethyl-6-oxodihydrolumazine also undergoes a highly selective reaction to give 1,3-dimethyl-6-thioxo-5,6-dihydrolumazine.

$$(41)$$

Among the alkylation reactions of pteridinones, methylations have attracted most attention. Pteridin-7-one reacts with diazomethane to give 8-methylpteridin-7-one with only a trace of 7-methoxypteridine ⟨52JCS1620⟩. Isopterin can be methylated in a base-catalyzed reaction by methyl iodide to 4-amino-1-methylpteridin-2-one ⟨61JCS4413⟩ and lumazine (**3**) forms its 1,3-dimethyl derivative if an excess of base, which affords ring cleavage, is avoided. Lumazine-6-carboxylic acid forms a mixture of the 1,3- and 3,8-dimethyl derivatives, and pterin-6-carboxylic acid (**142**) gives rise to three products which were identified as the 1- (**144**) and 3-monomethyl- (**143**) and the 3,8-dimethyl-pterinium-6-carboxylate zwitterion (**145**; equation 42). The isomeric pterin-7-carboxylic acid, on the other hand, is methylated only at N-1 and N-3 and no alkylation on the pyrazine ring could be detected, probably because of steric hindrance by the carboxyl group ⟨62JOC892⟩.

$$(42)$$

The methylation of 6- and 7-phenyl- and 6,7-diphenyl-pterin is also influenced by steric effects ⟨63JOC1509⟩. 7-Hydroxy-1,3-dimethyllumazine gives 7-methoxy-1,3-dimethyl-lumazine with diazomethane as well as with dimethyl sulfate and base ⟨57CB2588⟩ and methylation of 8-methyl-7-oxodihydrolumazines is sterically directed to O-2 rather than to N-1 ⟨58CB1671⟩. Careful methylation of leucopterin (**6**) by dimethyl sulfate in weakly alkaline medium gives the 3,5,8-trimethyl derivative (**146**) as the main reaction product, along with a small amount of 6-methoxy-3,8-dimethylisoxanthopterin (**147**) ⟨61CB118⟩. Both compounds could also be detected after diazomethane treatment of (**6**) in methanol (equation 43) ⟨41LA(547)180⟩.

$$(43)$$

Pteridines containing amide functions react easily with silylating agents such as trimethyl-silylacetamide, bis(trimethylsilyl)acetamide, trimethylsilyl chloride or hexamethyldisilazane to form the corresponding trimethylsilyloxypteridine derivatives. Lumazine and its derivatives are converted into the 2,4-bis(trimethylsilyloxy)pteridines ⟨73CB1401⟩, as are pteridin-6- and -7-ones ⟨73CB317⟩. Amino groups are monosubstituted in this reaction as with isoxan-thopterin (**5**), which gives 4,7-bis(trimethylsilyloxy)-2-trimethylsilylaminopteridine (**148**) ⟨73CB1952⟩ and a tetrakis(trimethylsilyl) derivative is formed from leucopterin (**6**). The silyl derivatives are thermostable compounds which can be distilled in high vacuum, providing an alternative purification method for high-melting, less soluble pteridines. They are also extremely valuable starting materials, especially for pteridine nucleoside synthesis (**149**) *via* the Hilbert–Johnson–Birkofer procedure (equation 44) ⟨73CB317, 73CB1401, 73CB1952, 73HCA1225, 74CB339, 77LA1217⟩. Glycosidation of amide groups in the pteridine series using heavy metal methods leads usually to *O*-glycoside formation ⟨62CB738, 66CB536⟩ but in some unusual cases, like 6,7-diphenyl-2-thiolumazine, *N*-substitution can be achieved with a mercury salt and sugar halides ⟨78CB2571⟩.

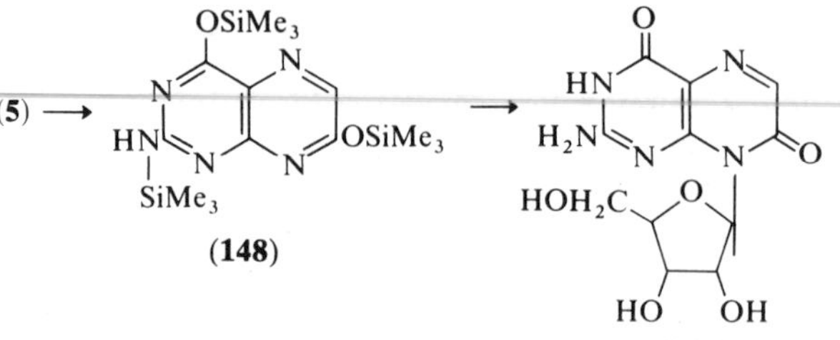

$$(5) \longrightarrow \quad (148) \longrightarrow \quad (149) \tag{44}$$

The replacement of an oxo function by a hydrogen atom has been accomplished in several ways. One route introduces a chloro group, which is then removed by hydrogen iodide or by catalytic reduction. The direct reduction of an amide group can be achieved in pteridine-6,7-diones with sodium amalgam, yielding 7,8-dihydropteridin-6-ones. Leucopterin (**6**) with sodium amalgam forms 7,8-dihydroxanthopterin (**150**) but electrolytic reduction in 75% sulfuric acid produces isoxanthopterin (**5**) ⟨44LA(555)146⟩; this formal amide reduction has been studied by pH-dependent electrolysis ⟨78CB1763⟩. Pteridine-6,7-diones are electrochemically active at the two adjacent oxo functions which in a four-electron process forms the 6,7-dihydroxy-5,8-dihydropteridine of enediol structure, followed by tautomerism to position 7, water elimination and further reduction; most 2- and 4-monosubstituted and 2,4-disubstituted derivatives provide 7,8-dihydropteridin-6-ones as relatively stable end products. Leucopterins and 6,7-dioxotetrahydrolumazines, on the other hand, show pH dependence; in alkaline medium the 6-oxo-5,6,7,8-tetrahydro derivatives are formed but in acid the reduction provides 7-oxo-7,8-dihydro-pterins and -lumazines respectively. This reaction again consumes four electrons to give the endiol (**149**), then proton tautomerism to the 6-position, water elimination and subsequent reduction, generating the 7-oxo-5,6,7,8-tetrahydro derivatives which are, however, very sensitive to oxidation and on work-up are autoxidized at the 5,6-position (equation 45). With 8-methylpteridine-6,7-diones the reduction process consists only of a 2-electron wave and stops at the stage of pseudo-base formation (**152**), as exemplified in equation (46).

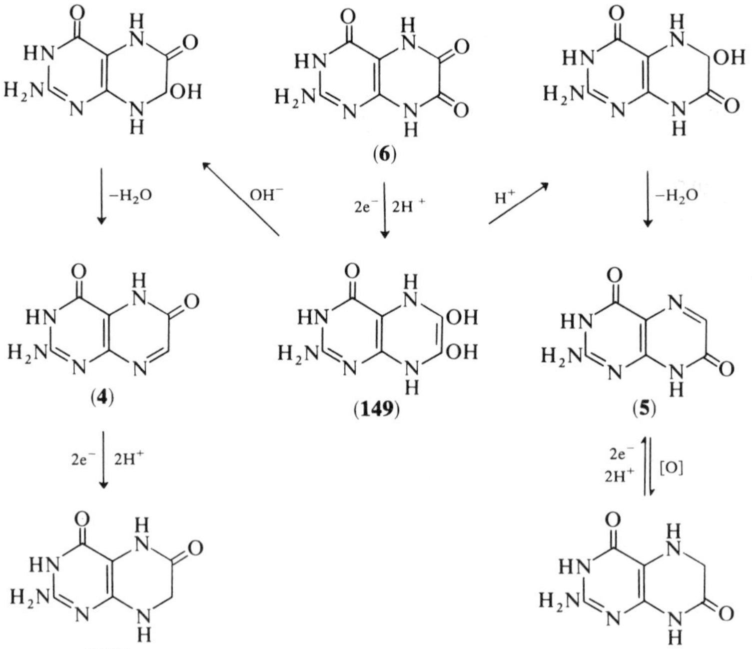

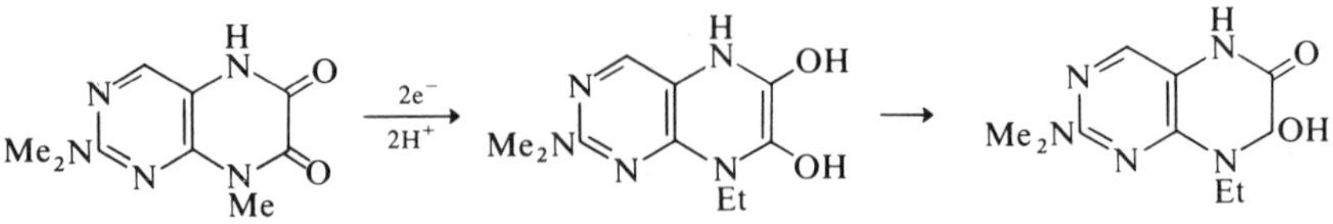

$$(45)$$

$$(46)$$

There is also the possibility of removing the 2-oxo group by ring cleavage and subsequent recyclization. Lumazine can be hydrolyzed by strong alkali to 2-aminopyrazine-3-carboxylic acid (**153**) which is converted first into the amide (**154**) and then cyclized by ethyl orthoformate into pteridine-4-one (**155**; equation 47) ⟨51JCS474⟩.

$$ \text{(3)} \quad \longrightarrow \quad \text{(153)} \quad \longrightarrow \quad \text{(154)} \quad \longrightarrow \quad \text{(155)} \tag{47} $$

Alkoxypteridines can be classed as iminoesters and are therefore subject to easy hydrolysis and nucleophilic substitution. Since they are often inaccessible their use as intermediates is limited ⟨61CB12, 61CB2708⟩. The monomethoxypteridines are easily hydrolyzed by sodium hydroxide to the corresponding pteridinones. Additional electron-donating groups show an expected stabilizing effect. In 2,4-dimethoxypterin-7-ones, alkoxy displacement by benzyl oxide proceeds in high yield to the corresponding 2,4-bis(benzyloxy) derivatives which then can be debenzylated catalytically over palladium/charcoal ⟨80CB1535⟩.

### 2.16.2.2.4 S-Linked Substituents

Pteridinethiols are potentially interesting synthetic intermediates due to the versatile reactivity of the thioamide function. Alkylation takes place on sulfur forming the corresponding alkylthio derivatives, as was shown by 1945 with 2-thiolumazines ⟨45BSF(12)78⟩. Similarly, 1,3-dimethyllumazine-7-thiol (**159**) ⟨79CB1499⟩ and 2-amino-4-pentyloxypteridine-7-thione give the 7-methylthio analogues on treatment with methyl iodide in alkaline medium. 4-Thiolumazines (**156**) give the 4-methylthiopteridin-2-ones in high yield and 2,4-dithiolumazines form the 2,4-bis(methylthio)pteridine derivatives.

The alkylthio group is replaceable by nucleophiles. The positions 7 and 4 react under mild conditions in that order; the 2-alkylthio functions require more drastic treatment. Conversion of 1-methyl-4-methylthiopteridin-2-one (**157**) into the 4-methylamino derivative (**158**) can be achieved by stirring with methylamine at room temperature (equation 48). The reactivity of an alkylthio group can often be further enhanced by oxidation to the corresponding sulfoxide and sulfone. Thus, reaction of 1,3-dimethyl-7-methylthiolumazine (**160**) with *m*-chloroperbenzoic acid yields 7-methylsulfinyl- (**161**) and 7-methylsulfonyl-1,3-dimethyllumazine (**162**; equation 49) ⟨82UP21601⟩. 4-Amino-2-methylthio-7-phenylpteridine and 4-amino-2-methylthio-8-methylpteridin-7-one are oxidized to the 2-methylsulfonyl derivatives, which are highly reactive towards various nucleophiles ⟨B-79MI21607⟩. Oxidation of 2-methylthio-7-phenylpteridin-4-one forms such a reactive sulfone that immediate hydrolysis or substitution by alkoxy groups takes place. 8-Methyl-2-methylthiopteridine-4,7-dione (**163**) behaves anomalously on oxidation, leading to additional hydroxylation in the 6-position (**165**) without detectable formation of the intermediate methylsulfonyl derivative (**164**; equation 50). 6-Alkyl- and 6-aryl-thiopterins and the corresponding 2,4-diaminopteridines are very unreactive and could not be forced to give nucleophilic substitution reactions ⟨78JOC680⟩. 6-Methylsulfonylpterin and 2,4,7-triamino-6-methylsulfonylpteridine ⟨73JHC133⟩ reacted only with S nucleophiles and no displacement of the methylsulfonyl group was observed upon treatment with amines.

$$ \text{(156)} \quad \longrightarrow \quad \text{(157)} \quad \xrightarrow{\text{MeNH}_2} \quad \text{(158)} \tag{48} $$

$$ \text{(159)} \quad \longrightarrow \quad \text{(160)} \quad \longrightarrow \quad \text{(161)} \quad \longrightarrow \quad \text{(162)} \tag{49} $$

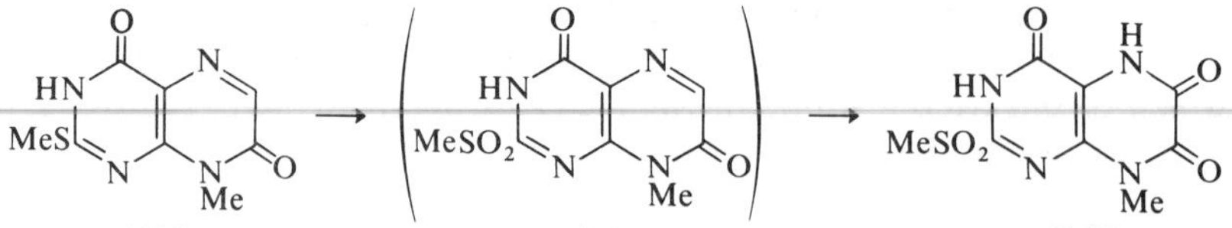

(50)

The sulfur atom can be used to initiate C—C bond formation. 2-Thio- and 4-thio-6,7-diphenyllumazine (166) react with phenacyl halides to give the phenacylthio derivatives (167), which on heating in DMF in the presence of triphenylphosphine extrude sulfur to form the benzoylmethyl derivative (168) in its tautomeric vinylogous amide form (169; equation 51).

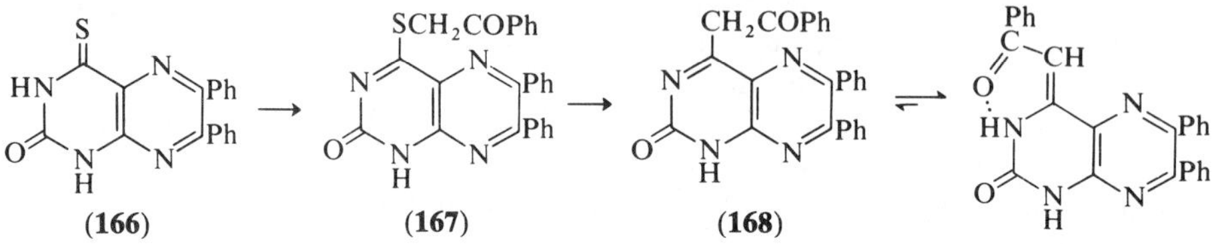

(51)

The thioamide function frequently shows high reactivity, as in 1- and 3-substituted 4-thiolumazines, which undergo direct nucleophilic substitution with amines under mild conditions. 3-Mono- and 1,3-disubstituted 4-thiolumazines react, for example, with methylamine to give the corresponding 4-methylamino (170) and 4-methylimino derivatives (171; equation 52). 2-Thiolumazines are much less reactive and require prolonged heating with high boiling amines if a reaction takes place at all. To overcome these difficulties an oxidative substitution process has been developed ⟨83TH21601⟩. Heating 6,7-diphenyl-2-thiolumazine (172) with morpholine in the presence of hydrogen peroxide gives 2-morpholino-6,7-diphenylpteridin-4-one (173) in 85% yield after a few minutes (equation 53). This procedure also works with 6,7-diphenyl-4-thiolumazine, which is converted by benzylamine and hydrogen peroxide after 1 min boiling into 4-benzylamino-6,7-diphenylpteridin-2-one, whereas the same reaction with benzylamine and acetic acid ⟨69KGS908⟩ or in the presence of mercury(II) oxide ⟨53JA1904⟩ needed refluxing for 1.5 and 5 h, respectively. These results suggest that the activation of the thioamide group proceeds *via* a sulfinic or sulfonic acid intermediate. Compounds of these structures have recently been synthesized from 2-thiolumazine and its 6- and 7-substituted derivatives (175). Hydrogen peroxide leads to 4-oxodihydropteridine-2-sulfinic acids (174), whereas potassium permanganate oxidizes to the 2-sulfonic acids (176), which were isolated as potassium salts (equation 54) ⟨83TH21601⟩. Both functional groups are highly reactive towards nucleophiles, as shown in their easy displacement in boiling amines, for example (178), or in their hydrolysis with acid or alkali to lumazines (179). The free sulfinic acids are very unstable and undergo desulfination by anhydrous acids to the corresponding C—H derivatives (177). This oxidative desulfurization can also be achieved directly from (175) by hydrogen peroxide treatment in anhydrous formic acid. Furthermore, pteridinethiones are light-sensitive and photochemically active. Photooxidation of 6- and 7-mercapto-1,3-dimethyllumazine (181) is easily achieved, giving various reaction products depending on the molecular form of the excited state; disulfides (180), sulfinic (182) and sulfonic acids (183) are obtained (equation 55) ⟨81TL2161⟩.

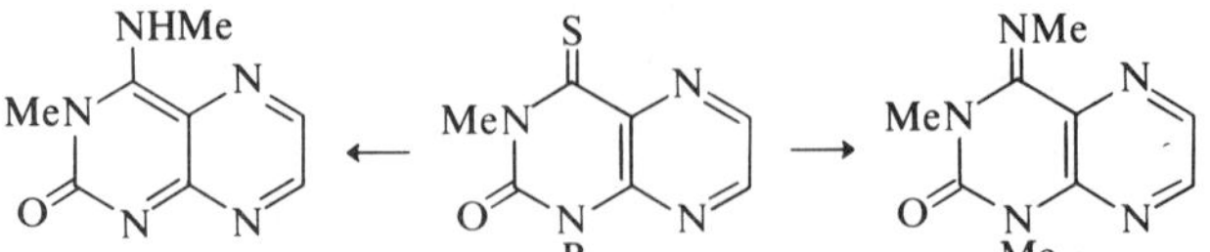

(52)

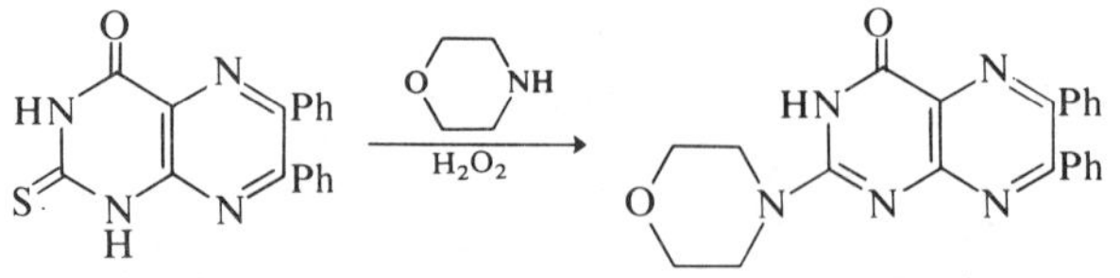

R = H, Me

(53)

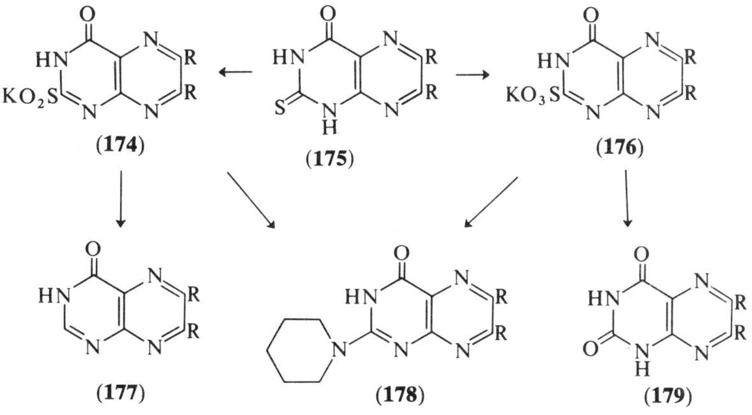

(54)

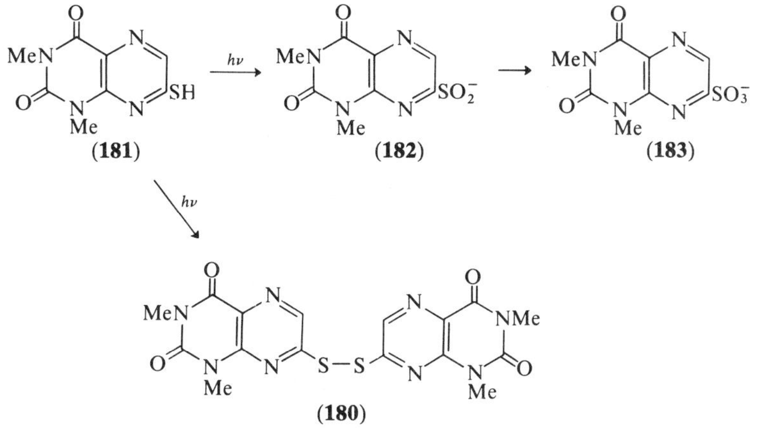

(55)

Attempts to replace a thiol group by a hydrogen atom using Raney nickel is generally unsuccessful in the pteridine series despite the fact that this method works in high yield with thiopyrimidines and thiopurines. Pteridinethiones are either unchanged under these conditions or are completely destroyed ⟨78JOC4154⟩. A rare positive result has been reported with urothione (**70**), which was smoothly desulfurized with Raney nickel to pterin-6-propionic acid ⟨69MI21615⟩. Free amino and oxy groups may specially be responsible for the difficulties due to strong interaction with the Raney nickel. Alkyl substitution at these functions suppresses complex formation with the metal and gives better results. The method of choice, however, is the use of less reactive Raney metals such as Raney cobalt, or even better Raney copper, which was successful with 2-amino-4-pentyloxy-6-propionylpteridine-7-thione (**184**) and its 7-methylthio derivative (**186**), forming 2-amino-4-pentyloxy-6-propionylpteridine (**185**) in 44 and 71% yield (equation 56) ⟨81TH21601⟩.

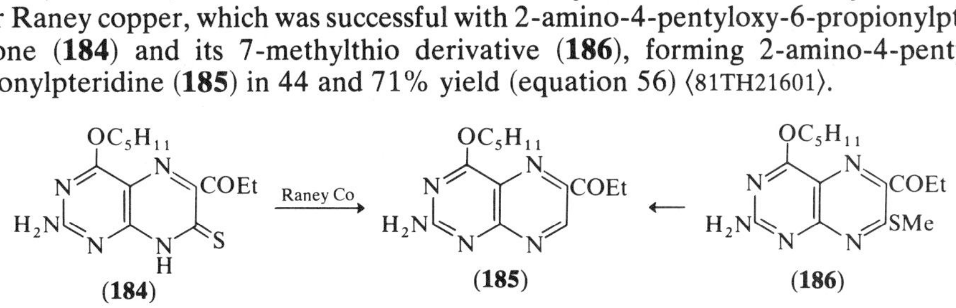

(56)

### 2.16.2.2.5 *C-Linked substituents*

As a result of the π-deficiency of the pteridine nucleus, alkyl pteridines are activated in the α-positions. The common reactions based on C—H acidity are found with a wide variety of compounds. Bromination of 6- and 7-methyl groups leads to mono- and di-substitution; selective formation of the monobromomethyl derivatives has not yet been achieved satisfactorily. 6-Methylisoxanthopterin is claimed to give the 6-bromomethyl derivative with bromine in acetic and sulfuric acids at 100 °C for 2 min ⟨50ZN(B)132⟩ and with 1,7-dimethyl-lumazine a 90% yield of the 7-bromomethyl derivative ⟨60CB2668⟩ is obtained after 4 h heating in acetic acid. 1,6,7-Trimethyllumazine (**187**) reacts with two equivalents of bromine in acetic acid to form 6,7-bis(bromomethyl)-1-methyllumazine (**188**). Treatment with pyridine leads to the corresponding pyridinio-methyl salts (**189**), which form nitrones (**190**) on reaction with nitrosodimethylaniline (equation 57). 2,4-Diamino-6-bromomethyl-pteridine (**191**), which was prepared from the corresponding 6-hydroxymethyl derivative, is the first example of a halide converted into a pteridine ylide (**192**) and used in a Wittig reaction to prepare (**193**) (equation 58) ⟨80JMC320⟩. Compound (**191**) is highly reactive and is a valuable intermediate in the preparation of aminofolic acid and methotrexate ⟨77JOC208⟩ as well as analogues by displacement of the bromine by O, S and N nucleophiles ⟨79JHC537⟩.

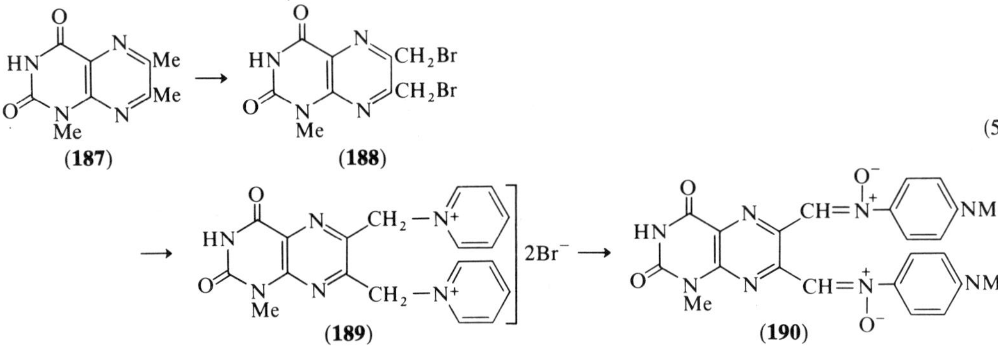

(57)

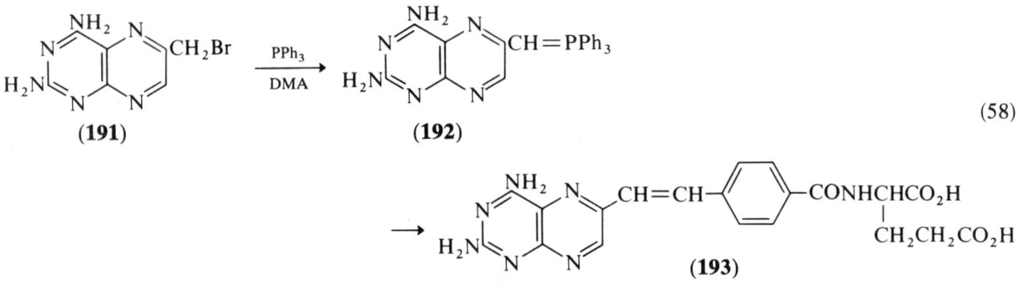

(58)

(**193**)

Excess of bromine converts each methylpteridine compound into the dibromomethyl derivative which on hydrolysis gives good yields of the corresponding aldehyde. An interesting variation of the reaction conditions was found in the treatment of the C-methylpteridines with POBr₃, which leads to the same mono- and di-bromomethyl derivatives. 6-Methylpterin reacts with more difficulty and with an excess of bromine in hydrobromic acid forms the 6-dibromomethylpterin in only 45% yield ⟨50JA4630⟩, while considerable amounts of the monobromomethyl derivative are formed with bromine in an autoclave ⟨48JA27⟩. Chlorination can be achieved to a small extent with sulfuryl chloride when benzoyl peroxide is used as catalyst. Additional electron-donating groups at the pteridine nucleus usually do not decrease the reactivity of C-alkyl substituents. 6-Methylpteridine-4,7-dione ⟨53JCS74⟩ and 7-hydroxy-1,3,6-trimethyllumazine (**194**) ⟨56CB641⟩ form the dibromomethyl derivatives (**195**) which on hydrolysis give the corresponding aldehydes (**196**; equation 59).

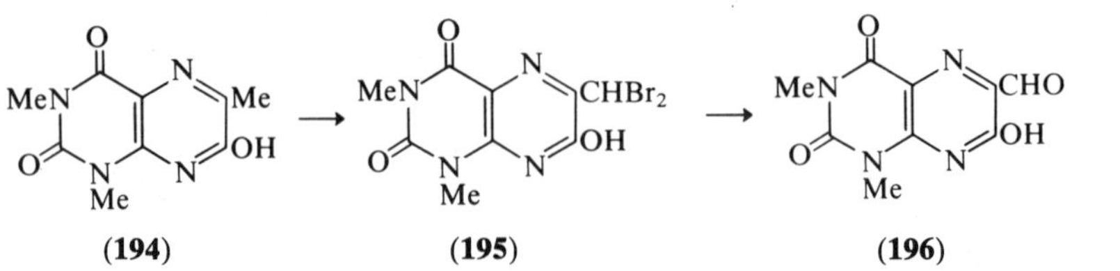

(59)

Various 6- and 7-methyl- and 6,7-dimethyl-pteridines bearing either oxo or amino groups in the 2- and 4-positions can be oxidized to the corresponding carboxylic acids by alkaline potassium permanganate on heating. Various lumazine and pterin mono- and di-carboxylic acids have been prepared in this way ⟨48JA3026, 78CB3790⟩.

The α-ionization of 7-methylpteridines can also be utilized in aldol-type condensation reactions. 7-Methyl-pterin and -lumazine and 2,4-diaminopteridine condense readily in aqueous base with aromatic aldehydes to afford 7-alkylidenepteridines ⟨77JOC2951⟩. A Claisen condensation requires the protection of the acidic hydrogens of the amide bonds, as in the conversion of 3,5,7-trimethylxanthopterin (**197**) into 3,5-dimethylerythropterin methyl ester (**198**; equation 60) ⟨62CB2195⟩. Acid-catalyzed Michael-type addition reactions also proceed easily, forming deeply coloured dipteridylmethines on subsequent oxidation, with pterorhodine (**62**) and its structural analogs as the most prominent examples ⟨49JA3412⟩. The 6- and 7-methylpterins form the pteridine reds in a similar manner ⟨50HCA1233, 51HCA1029, 51HCA2155⟩. Strong activation is also found in 6,7,8-trimethyllumazine (**200**) showing an unusual general acid–general base catalysis on deuterium exchange at the 7-methyl group ⟨72JCS(P2)376⟩. Similarly, the rate of oxidation of (**200**) with permanganate is more rapid in buffered than in unbuffered solutions ⟨74CJC3884⟩, yielding in a multistep process 6,8-dimethyl-7-oxodihydrolumazine (**199**) ⟨66JCS(C)1065⟩. Use of potassium *trans*-1,2-diaminocyclohexanetetraacetatomanganate(III), an exclusively one-electron oxidant, produces (**199**) and formaldehyde at pH 6.0 under oxygen, but under nitrogen oxidative dimerization to (**201**) occurs with one and to (**202**) with two equivalents of the oxidant (equation 61) ⟨74CJC3879⟩. The activation is also present in pteridine *N*-oxides, since 3,6,7-trimethyllumazine 5,8-dioxide on heating in acetic anhydride undergoes a Boekelheide rearrangement ⟨54JA1286⟩, first to 7-acetoxymethyl-3,6-dimethyllumazine 5-oxide and then on prolonged boiling to 6,7-diacetoxymethyl-3-methyllumazine ⟨73CB3149⟩.

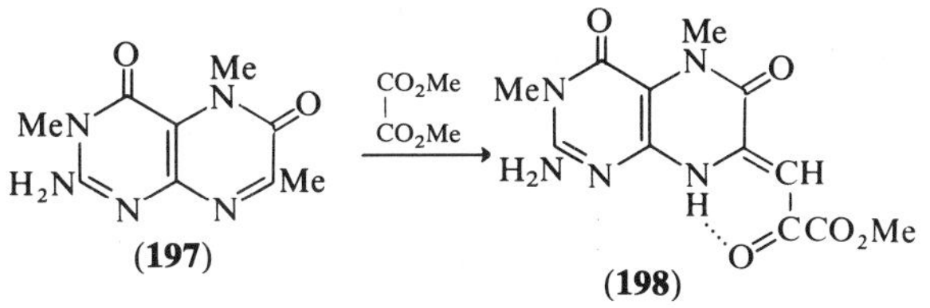

(60)

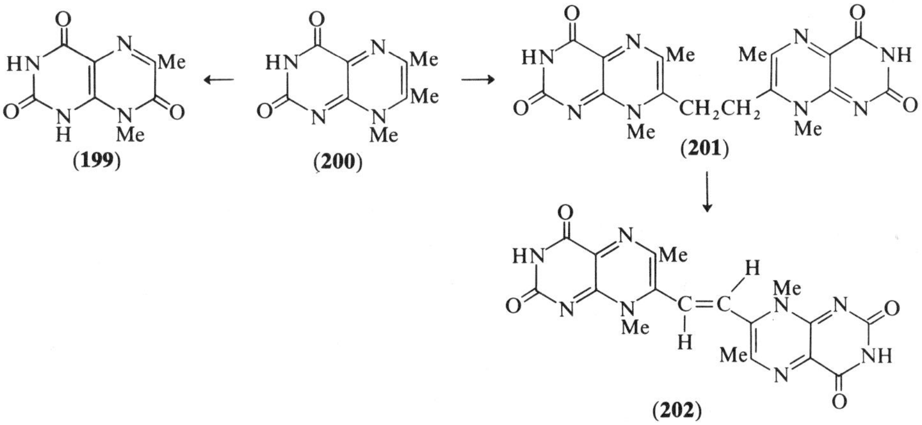

(61)

Among the functionalized methyl groups, alkoxycarbonylmethylpteridines show enhanced reactivity and on hydrolysis behave like β-ketocarboxylic acids in decarboxylating easily ⟨53JCS74⟩. 7-Acetonylxanthopterin (**203**) functions as a 1,3-dicarbonyl compound analogue, reacting with strong aqueous base (5N/NaOH) in a reverse Claisen cleavage to yield 7-methylxanthopterin (**204**) and in sodium bicarbonate with nucleophilic attack at C-7 to form leucopterin (**6**). On hydrogen peroxide treatment in alkaline medium, oxidative scission occurs to give xanthopterin-7-carboxylic acid (**205**; equation 62) ⟨82UP21602⟩.

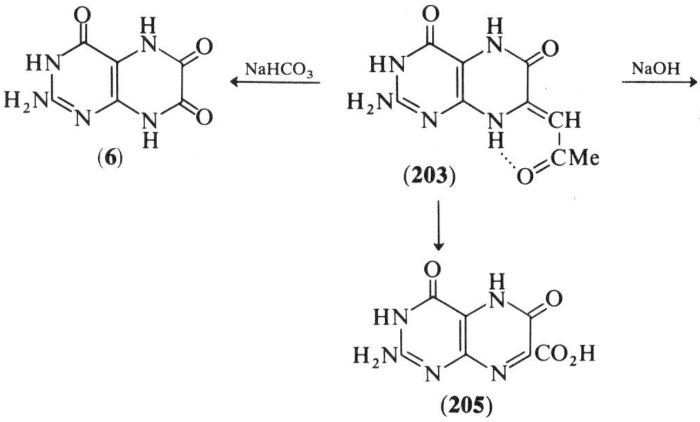

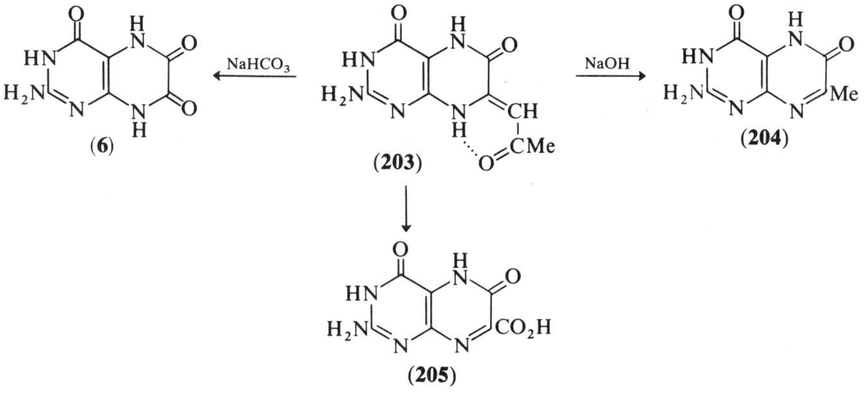

(62)

More elaborate modifications of the carbon side-chains have sometimes been tried in the synthesis of natural products, as for example urothione (**70**) ⟨69MI21606⟩. Polyhydroxyalkyl side-chains are found in various naturally occurring pteridines, but no systematic investigations have been made to modify these potentially versatile substituents. Dehydrations, reductions and formal substitution reactions have been observed on sodium dithionite treatment of D-erythroneopterin ⟨73BCJ939⟩. Cleavage of a 1,2-diol structure by periodate leads to the corresponding aldehyde and oxidation by permanganate gives the pteridinecarboxylic acid. 6-Hydroxymethyl-pterin and -2,4-diaminopteridine are valuable synthetic intermediates ⟨64JOC3610, 80CB1514⟩ on account of their easy conversion into the bromomethyl derivatives ⟨74JHC279, 78USP4080325⟩.

Pteridine-carbaldehydes show the chemical behavior of their aromatic counterparts and readily form phenylhydrazones, oximes, anils ⟨50JA4630⟩ and aldol-type products ⟨56CB641⟩. Pterin-6- (**206**) and -7-carbaldehyde and their $N^2$-acetyl derivatives form anils with *p*-aminobenzoic acid ⟨71JOC860, 81JOC1394⟩ and *p*-aminobenzoylglutamic acids ⟨55JA6365, 75HCA1374⟩ which are intermediates in the synthesis of pteroic and folic acids, since reduction of the azomethine function proceeds readily with arenethiols, dimethylamine–borane and sodium borohydride. The aldehyde (**206**) undergoes the Cannizzaro reaction, giving pterin-6-carboxylic acid and 6-hydroxymethylpterin in good yield ⟨50JA4630⟩. The reaction with Wittig reagents in DMF is a useful route to 6-alkylidenepteridine derivatives (**207**; equation 63).

$$(206) \quad + \quad PhCH=PPh_3 \quad \longrightarrow \quad (207) \tag{63}$$

Pteridine-6- and -7-carboxylic acids are usually very stable if they are derived from pterin or lumazine, which possess electron-donating substituents on the pyrimidine ring. In these cases a carbon side-chain can be oxidized to the carboxylic acid group. The presence of an additional oxo group in the pyrazine ring does not usually allow permanganate oxidation without oxidative degradation of the nucleus. Xanthopterin-7- and isoxanthopterin-6-carboxylic acid ⟨41LA(548)284, 62CB1591⟩ and the lumazine analogues ⟨57CB2617, 57CB2624, 62CB749, 62CB1605⟩ are therefore generally obtained by direct synthesis and not by oxidation processes. 2,4-Diaminopteridine-, pterin- and lumazine-6,7-dicarboxylic acids ⟨48JA3026⟩ have been known for a long time, but they are of little synthetic use and have only recently been studied in any detail ⟨78CB3790⟩. Decarboxylation requires quite high temperatures: lumazine-7-carboxylic acid is obtained from lumazine-6,7-dicarboxylic acid in boiling quinoline, while the pterin-6,7-dicarboxylic acid hydrochloride yields as the main product pterin-6-carboxylic acid. The presence of an adjacent oxo function usually stabilizes the carboxyl group, which can, however, be removed more readily if the pyrazine ring is further reduced to form a β-keto acid-type molecule ⟨52JA3877, 71JOC4012⟩. The pteridinecarboxylic acids can be esterified by the Fischer method or by diazomethane, and the corresponding amides can be produced from the esters ⟨62CB749⟩ or directly by various condensations, including the Timmis reaction ⟨63JOC1187, 68JMC542⟩.

Of the classical Hofmann, Curtius, Lossen and Schmidt degradations, only a rare example of the first is known, hypobromite converting 4,7-diamino-2-phenyl-6-pteridinecarboxamide (**208**) into 8-amino-2,3-dihydro-6-phenyl-1*H*-imidazo[4,5-*g*]pteridin-2-one (**209**; equation 64) ⟨63JOC1203⟩.

$$(208) \quad \longrightarrow \quad (209) \tag{64}$$

### 2.16.2.3 Reactivity at the Ring Nitrogen Atoms

Among the substitution reactions involving the ring nitrogen atoms of the pteridine nucleus, alkylations of amide functions are preeminent. Under base-catalyzed conditions it is usually the nitrogen atom adjacent to the carbonyl function which is substituted electrophilically in an $S_N2$ mechanism, while quaternizations of neutral species take place at vinylogous or even transannular ring nitrogen positions. Pterin (**3**) and its 6,7-dimethyl derivative react with methyl iodide to form 8-methylpterins, and analogously 2,4-diamino-pteridine can be substituted at N-1 and N-8 ⟨61JCS4413⟩. 7-Aminopteridine (**210**) as a more simple case reacts at positions 1 (**211**) and 3 (**212**), but 4,7-diaminopteridine yielded only the 1-methyl derivative for steric reasons. Treatment of 3-methyl-6-phenylpterin with dimethyl sulfate/glacial acetic acid/DMF yields 63% of the 3,8-dimethyl and only 13% of the 1,3-dimethyl derivative, while the isomeric 3-methyl-7-phenylpterin leads to exclusive formation of 1,3-dimethyl-2-imino-4-oxo-7-phenyltetrahydropteridine ⟨63JOC1509⟩. 7-Amino-1,3-dimethyllumazine is quaternized at the less hindered N-5 position, which is also, for electronic reasons, the site of attack in 7,8-dihydrolumazines ⟨71CB3842⟩ and 7,8-dihydropterins ⟨71CB2313⟩.

$$(65)$$

Cyanoethylation of pteridinones with acrylonitrile in pyridine/water provides a parallel to the alkaline alkylations, with substitution at the nitrogen atoms of the lactam groups ⟨62JOC1366, 61JOC2364⟩.

By direct oxidation with $H_2O_2/H^+$, various types of $N$-oxide can be obtained ⟨65AG(E)1075⟩. Lumazine, pterin and 2,4-diaminopteridine (**213**) behave similarly and form in the first step the corresponding 8-oxides (**214**) ⟨73CB3149, 73CB3175⟩, which react further with an excess of hydrogen peroxide and under more forcing conditions to give the 5,8-dioxides (**55**; equation 66). On the other hand, $N$-oxidation is very sensitive to the steric effects of adjacent or *peri* located substituents. 1-Methyl-lumazines and -pterins give exclusively the 5-oxides (**52**) and no further reaction at the N-8 atom can be achieved. Moreover, in 7-*t*-butyl- and 7-phenyl-pteridines the N-5 atom is attacked selectively and in the highly hindered 1,3-dimethyl-6-phenyllumazine this same position is oxidized, as the 1-methyl group does not allow electrophilic reaction at N-8. There is a close relationship between the reactivity of 7-amino- and 7-alkoxy-1,3-dimethyllumazines towards proton-ation, quaternization and $N$-oxidation ⟨73CB3203⟩, which all take place at the N-5 atom.

$$(66)$$

## 2.16.2.4 Reduction

The discovery of naturally occurring dihydro- and tetrahydro-pteridines has directed some attention to the chemistry of hydrogenated pteridines, which show a wide range of stability depending on the nature and site of the various substituents. The chemical and physical properties of this class of compound can best be understood by considering first 5,6,7,8-tetrahydropteridine, which is produced either by reduction of (**1**) with lithium aluminum hydride or by stepwise reductive dehalogenation of 2,4,6,7-tetrachloro-pteridine (**109**) *via* 2,4-dichloro-5,6,7,8-tetrahydropteridine ⟨59JA2464⟩. This behavior illus-trates the greater susceptibility of the pyrazine ⟨47CR(40)279⟩ over the pyrimidine ring ⟨51JCS2323⟩ in reduction processes. In contrast to pteridine (**1**), which is unstable towards acid and base hydrolysis, the 5,6,7,8-tetrahydro derivative survives even boiling in 1N HCl or 1N NaOH. All attempts to reoxidize the latter compound to (**1**) have so far failed. If electron-donating substituents are introduced in sequence into the 2- and 4-positions of dihydro- and tetrahydro-pteridines, the stability gives way successively to sensitivity to oxidation, which is enhanced with increase in the electron donor properties of the sub-stituents. Although the alkyl derivatives of 5,6,7,8-tetrahydropteridine are chemically stable ⟨53JCS2234, 54JCS4109, 55JCS896, 57JCS1⟩ and the 2- and 4-monosubstituted hydroxy and amino derivatives of dihydro- and tetrahydro-pteridines can be isolated ⟨61JCS5131, 62JCS2162⟩ and are stable as solids, 2,4-disubstituted 5,6,7,8-tetrahydropteridines are very sensitive to autoxidation in their neutral and anion forms, while the protonated cationic species resist attack of oxygen much better ⟨58HCA2170, 59HCA1854, 66HCA875, 71LA(747)111⟩, and acylation at N-5 inhibits autoxidation completely.

If 5,6,7,8-tetrahydropterin is oxidized by air the 7,8-dihydro compound (215) is formed, *via* several other thermodynamically less stable dihydro stages ⟨66JCS(C)285⟩. This is actually the only known simple dihydropterin out of the five theoretically possible isomers (215–219). While there is vinylogous amide resonance stabilization in (215) the *o*- (218) and *p*-quinonoid (216) and the cross-conjugated (219) and isolated dihydro structures (217) are energetically on a higher level and prone to easy oxidation or tautomerism. As intermediates, however, they play a substantial role in chemical reactions and biochemical interconversions ⟨78MI21602⟩. The 5,8-dihydropteridine structure contains an 8-electron antiaromatic π-system which is formed in electrochemical reductions ⟨77MI21600, 80H(14)1603⟩, as detected by subsequent acylation to 5,8-diacetyl-5,8-dihydro-lumazines and -pterins ⟨77AX(B)2911⟩. Electron-attracting substituents stabilize the 5,8-dihydro forms to some extent under anaerobic conditions, as seen, for example, in 6-formyl-5,8-dihydropterin (220), which is a photodegradation product of biopterin (32) and neopterin (33); (equation 68) ⟨77TL2817⟩. Oxidations of tetrahydro-pterins and -lumazines proceed by complicated mechanisms ⟨74CI(L)233⟩ to give at first 6*H*-7,8-dihydro derivatives (218) which are stabilized by prototropy to the 7,8-dihydro isomers (215) ⟨70MI21601⟩. The very labile quinonoid 6*H*-7,8-dihydropterins ⟨64JBC(239)332⟩ have been detected as intermediates in various enzymatic hydroxylation processes, in which 5,6,7,8-tetrahydrobiopterin (221) functions as a cofactor in the interconversion of phenylalanine to tyrosine ⟨67JBC(242)3934, 68MI21601⟩, tyrosine to dopa ⟨65MI21602⟩ and tryptophan to 5-hydroxytryptophan ⟨66JBC(241)192⟩ by shuttling with the 6*H*-7,8-dihydrobiopterin (222) in a strictly thermodynamically reversible redox reaction ⟨70MI21603, 72AG1088⟩. Reduction of pterins with zinc in alkaline medium ⟨69JCS(C)928, 66CB3008⟩ or sodium dithionite ⟨68MI21600, 73BCJ939⟩ leads to 7,8-dihydro derivatives which show interesting interconversions, in the case of 7,8-dihydrobiopterin (223) involving also the carbon side-chain ⟨79CB2750⟩.

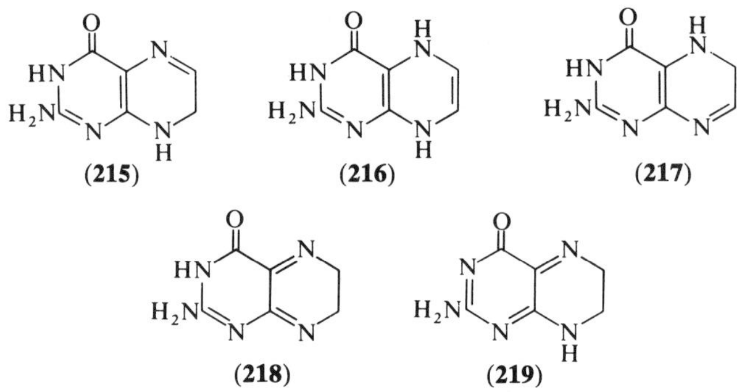

(67)

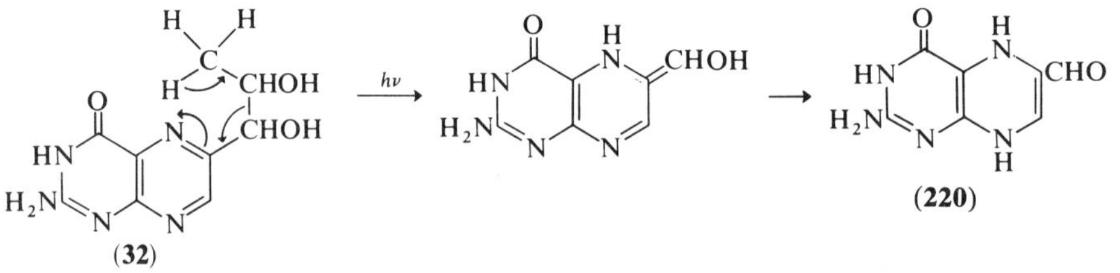

(215)     (216)     (217)

(218)     (219)

(68)

(32)     (220)

(221)     (222)     (223)

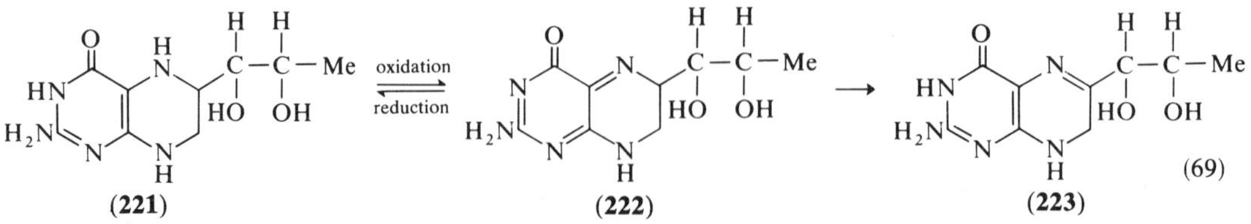

(69)

Little information is available about 5,6-dihydropteridines, of which various 6,7-diphenyl-5,6-dihydropterins ⟨65HCA764, 69HCA306⟩ and -lumazines ⟨68HCA1029, 70HCA789⟩ have been synthesized and characterized. As noticed already ⟨51BSF521⟩, this type of compound isomerizes in an acid-catalyzed reaction to the 7,8-dihydro derivative ⟨77HCA922⟩ or oxidizes to the heteroaromatic compound. Another isomerization has been detected with 8-alkyl-7-oxo-3,4-dihydropteridine-6-carboxylic acids and esters in trifluoroacetic acid solution, which are converted into the 5,6-dihydro isomers ⟨71JOC4012⟩. Reduction, with various reagents, of all mono- and poly-oxopteridines has also been achieved and leads to the corresponding dihydro and tetrahydro derivatives ⟨62JCS2162⟩.

Investigations of the mechanism of catalytic reduction of various 6- and 7-substituted pterins reveal a dependence on the site of the substituent as well as on the pH of the medium. 7-Methyl- and 6,7-diphenylpterin react in acidic solution, first at the 5,6-double bond followed by a 1,2-hydrogen shift leading to the thermodynamically more stable 7,8-dihydropterins ⟨78HCA2246, 80HCA395⟩. In neutral or weakly acidic medium 6- and 7-methylpterins are reduced in the first step directly to 7,8-dihydro derivatives, which on further hydrogenation yield the 5,6,7,8-tetrahydropterins ⟨80HCA1754⟩.

Catalytic reduction of folic acid to 5,6,7,8-tetrahydrofolic acid (**225**) proceeds fast in trifluoroacetic acid ⟨66HCA875⟩, but a modified method using chemical reductants leads with sodium dithionite to 7,8-dihydrofolic acid (**224**). Further treatment with sodium borohydride gives (**225**) which has been converted into 5-formyl-(6*R*,*S*)-5,6,7,8-tetrahydro-L-folic acid (leucovorin) (**226**) by reaction with methyl formate (equation 70) ⟨80HCA2554⟩.

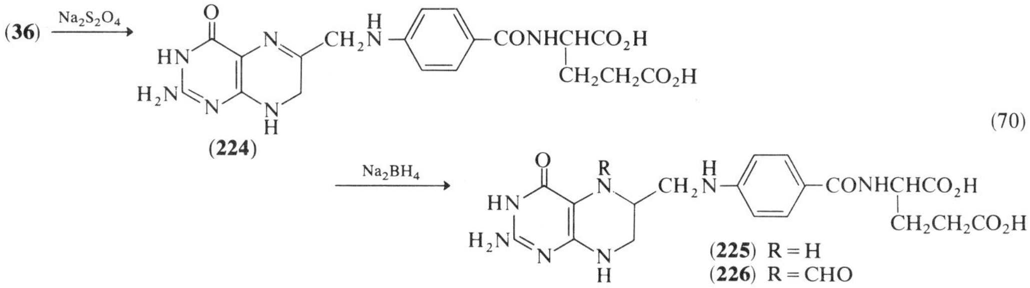

(70)

## 2.16.2.5 Oxidation

Oxidations in the pteridine series comprise (i) replacement of hydrogen by hydroxyl, (ii) glycol formation at the central C=C bond (iii) the removal of hydrogen atoms from dihydro and tetrahydro derivatives.

Various types of oxidizing agents are able to convert an azomethine function directly into an amide group, as described in Section 2.16.2.1.2. Easy oxidation is not always associated with covalent hydrate formation since 7-hydroxy-1,3-dimethyllumazine reacts quickly with hydrogen peroxide in formic acid to give 1,3-dimethyl-6,7-dioxotetrahydrolumazine ⟨73CB3203⟩. Leucopterin (**6**) can, like uric acid, be oxidized further by chlorine water to a glycol (**227**) by attack at the junction of the two rings. Chlorine in methanol gives the corresponding 4a,8a-dimethoxy derivative (**228**; equation 71) ⟨33LA(507)226, 37LA(530)152⟩.

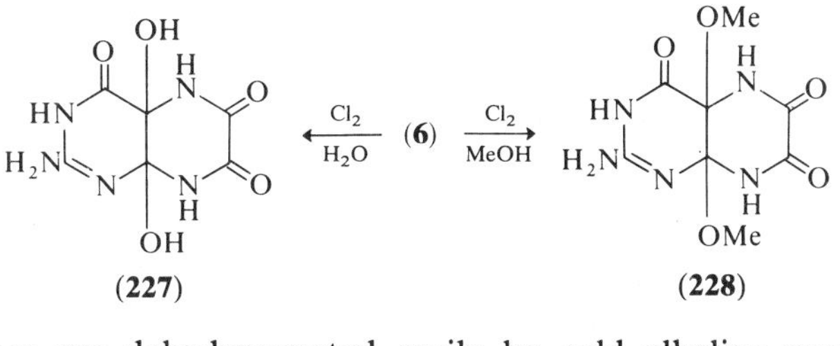

(71)

Dihydropteridines are dehydrogenated easily by cold alkaline permanganate, which appears to be the best general reagent for this purpose, although derivatives with several electron-donating substituents can be oxidized by milder reagents such as oxygen in alkaline solution, hydrogen peroxide, alkaline silver nitrate, methylene blue, sodium hypobromite, chloramine-T, benzoquinone and dichlorophenol/indophenol (Tillman's reagent). 7,8-Dihydropteridin-6-one, 5,6-dihydropteridin-7-one ⟨52JCS1620⟩ and 7,8-dihydroxanthopterin ⟨49JA741⟩ are oxidized in excellent yields, but autoxidation of the last compound to (**4**) by ammonia catalysis seems to be an interesting alternative ⟨74CB785⟩. In cases of sensitive dihydro- and tetrahydro-pteridine derivatives, as obtained for example during oxidative condensation reactions of biopterin, the right combination of dehydrogenating reagents results in moderate to good yields. Iodine, iron(III) chloride or a mixture of potassium ferricyanide, potassium iodide and hydrogen peroxide at pH 2–3 have been used with varying success ⟨79BCJ181⟩.

The complex mechanism of autoxidation of 5,6,7,8-tetrahydro-pterins and -lumazines, involving radical intermediates ⟨67HCA2222, 74CI(L)233, 74JCS(P2)80⟩ shows strong pH dependence, leading to heteroaromatic analogs in acidic medium and to the 7,8-dihydro derivatives in neutral solution ⟨71LA(747)111, 67HCA1492, 65HCA816⟩. Substitution at N-8 does not alter the scheme in general but 8-substituted quinonoid-type compounds are obtained in acid ⟨71CB2293⟩. Oxidation can also effect a more drastic structural change, as seen from the autoxidative rearrangement of 5-methyl-6,7-diphenyl-5,6,7,8-tetrahydro-pterin (**229**) and -lumazine to 2-amino-8-methyl-4,9-dioxo-*cis*-6,7-diphenyl-6,7,8,9-tetrahydro-4*H*-pyrazino[1,2-*a*][1,3,5]triazine (**231**) and the 2-oxo analogue respectively ⟨75T533, 75T541⟩ in alkaline medium. The mechanism of rearrangement may involve an intermediate 4a-peroxy derivative (**230**; equation 72). In the cases of 1,3,5,6,7- (**233**) and 1,3,6,7,8-pentamethyl-5,6,7,8-tetrahydrolumazine (**234**), ring contraction takes place to give the corresponding C(8a)- (**232**) and C(4a)-spirohydantoin derivative (**235**) respectively (equation 73) ⟨76T2303⟩. On changing the medium, N(5)-demethylation in (**233**) may compete with the formation of spirohydantoins and a ring opening of the peroxide in a Criegee-like rearrangement could occur as an alternative reaction. These results have led recently to a reinvestigation of the oxidation of 5-methyltetrahydrofolate by hydrogen peroxide at pH 6 which has shown that a rearrangement to 2-amino-8-methyl-4,9-dioxo-6,7,8,9-tetrahydro-4*H*-pyrazino[1,2-*a*][1,3,5]triazin-7-ylmethyl *p*-aminobenzoylglutamate also takes place ⟨81BBR(101)1259⟩.

(72)

(73)

(**233**) R = Me, R¹ = H
(**234**) R = H, R¹ = Me

## 2.16.2.6 Ring transformations

Even at the fully aromatic oxidation level the pteridine nucleus can undergo ring transformation reactions. Ring contractions involving the pyrazine moiety and leading to purines are the most commonly observed transformations. 2-Methylthio-4,6,7-triphenylpteridine is converted by potassium amide in liquid ammonia at −33 °C in part into 2-methylthio-6,8-diphenylpurine, which is actually the major product on similar treatment of 4,6- and 4,7-diphenyl-2-methylthiopteridine ⟨75CPB2678, 77H(7)205⟩. Boiling 7-hydroxy-1,3,6-trimethyllumazine 5-oxide in acetic anhydride results in the formation of 1,3-dimethyluric acid ⟨73CB3203⟩ and treatment of 1,3,6,7-tetramethyllumazine with zinc in acetic acid/acetic anhydride leads to 1,3,8-trimethylxanthine ⟨72RTC1137⟩. In an enzymatic degradation, xanthopterin (**4**) ⟨62B1161⟩ and isoxanthopterin (**5**) ⟨63JBC(238)1116⟩ are transformed by the bacterium *Alcaligenes faecilis* into xanthine-8-carboxylic acid *via* pteridine-2,4,6,7-tetrone ⟨64JBC(239)4272⟩.

## 2.16.2.7 Rearrangements

The high chemical stability of pterins towards aqueous base is due to anion formation suppressing nucleophilic attack at a ring carbon atom by electrostatic repulsion. Substitution at N-3, however, forms an alkali-sensitive molecule which can react by Dimroth rearrangement to give the corresponding 2-(substituted amino)pterin ⟨58JA6095⟩. This reaction works with alkyl, aryl and carboxyl groups in the 6- and 7-positions ⟨63JOC1509⟩ and also includes 3,8-disubstituted derivatives, as seen from the conversion of 3,8-dimethylpterin-6-carboxylic acid (**145**) into the $N^2$,8-dimethyl isomer (**236**; equation 74) ⟨62JOC892⟩. 3,8-Dimethylisoxanthopterin and its 6-carboxylic acid show the same rearrangement of the pyrimidine moiety of the molecule ⟨67MI21602⟩.

Another unusual rearrangement is performed by *Bacillus subtilis* during the catabolism of sepiapterin (**237**), in converting the whole side-chain with subsequent oxidation of the pyrazine ring into 6-(1-carboxyethoxy)pterin (**238**; equation 75).

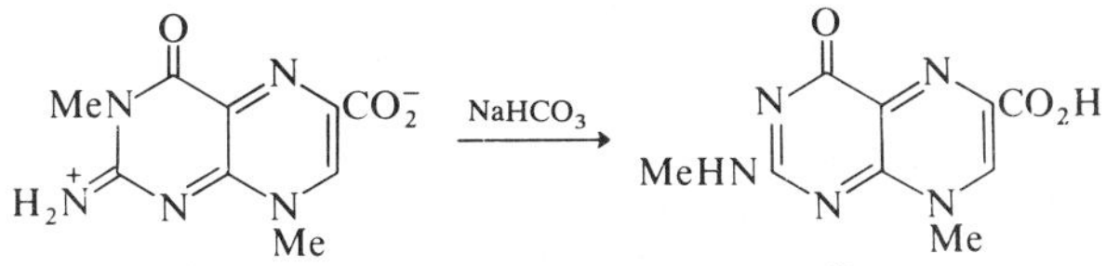

$$ (74) $$

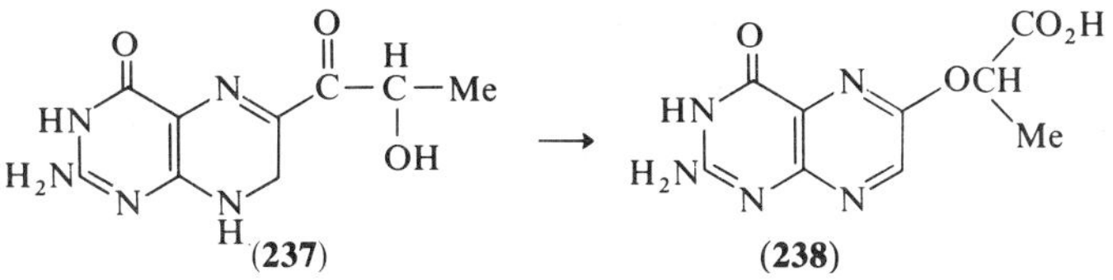

$$ (75) $$

## 2.16.3 SYNTHESES

Examination of the pyrazino[2,3-*d*]pyrimidine structure of pteridines reveals two principal pathways for the synthesis of this ring system, namely fusion of a pyrazine ring to a pyrimidine derivative, and annelation of a pyrimidine ring to a suitably substituted pyrazine derivative (equation 76). Since pyrimidines are more easily accessible the former pathway is of major importance. Less important methods include degradations of more complex substances and ring transformations of structurally related bicyclic nitrogen heterocycles.

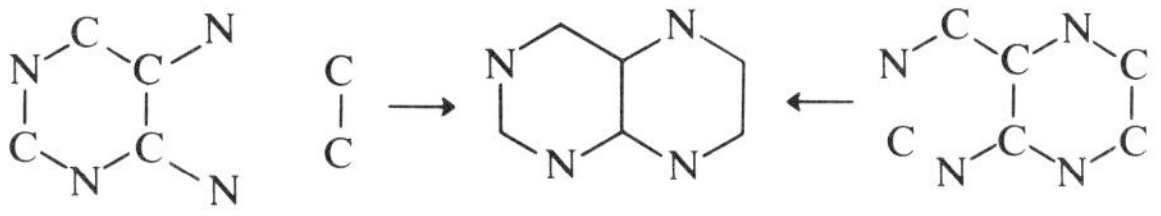

$$ (76) $$

### 2.16.3.1 Syntheses from Pyrimidines

The most straightforward synthesis of a pteridine is the Gabriel–Isay reaction ⟨52QR197⟩ involving the condensation of a 5,6-diaminopyrimidine with a 1,2-dicarbonyl compound, as in Gabriel's preparation of 4-methyl- ⟨01CB1234⟩ and Isay's preparation of 6,7-diphenyl-pteridine ⟨06CB250⟩ by this route at the beginning of this century (equation 77). A fuller study of the reaction was published two years later ⟨08CB3957⟩ and it has since become the synthetic method most frequently used. Analogous condensation reactions with 5-amino-6-(monosubstituted amino)pyrimidines lead to N(8)-substituted pteridines possessing a cross-conjugated $\pi$-electron system of long wavelength absorption (equation 78) ⟨66CB3503, 68CB1072, 71CB2273⟩. Syntheses with symmetrical 1,2-dicarbonyl compounds (glyoxal, diketones, oxalic acid derivatives) present no problems, but unsymmetrical dicarbonyl components such as $\alpha$-ketoaldehydes, $\alpha$-ketoacids, *etc.* give rise to mixtures of substituted pteridines isomeric at positions 6 and 7. Since separation and purification of such mixtures involves losses of time and yield and is often extremely difficult, syntheses which selectively produce a single isomer are sought.

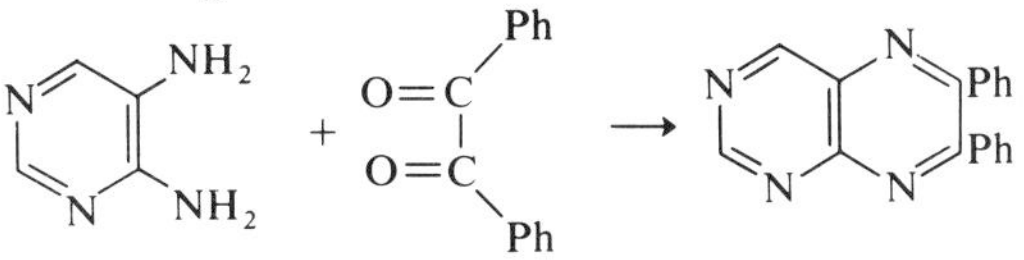

$$ (77) $$

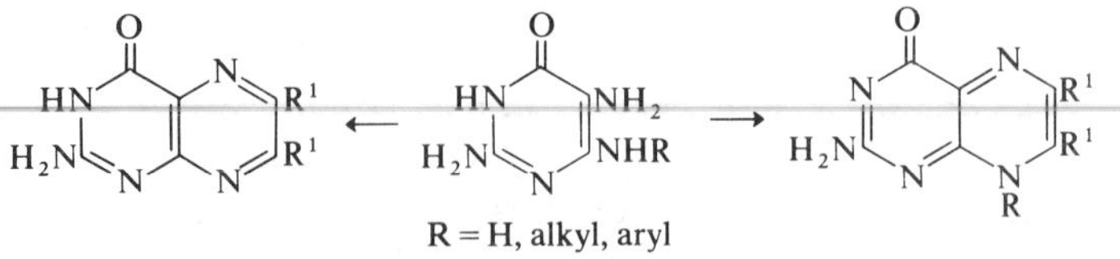

$$R = H, \text{alkyl, aryl} \tag{78}$$

The principle of pH-dependent condensation, first recognized by Purrmann ⟨41LA(548)284⟩ and later applied generally ⟨50JA78⟩, offers the possibility of influencing the direction of the reaction using the differences in the basicity and nucleophilic reactivity of the amino groups in the 5- and 6-positions. Particularly in condensations with 5,6-diaminouracil (**240**) ⟨57CB2588, 57CB2604, 57CB2617, 57CB2624⟩, 2,5,6-triamino-6-pyrimidinone (**251**), 2,4,5,6-tetraaminopyrimidine (**248**) and their corresponding substituted analogs with α-keto-acids and -esters, neutral or weakly acidic media as well as organic solvents ⟨61CB2708⟩ favour the formation of pteridin-7-ones (**241**), whereas in a strongly acidic medium protonation of the more basic 5-amino group leads to the opposite orientation of the substituents, to give pteridin-6-ones (**239**; equation 79). Condensations with ethyl glyoxylate hemiacetal proceed particularly smoothly if the alkyl 5-azomethinecarboxylate (**243**) is isolated first and then cyclized under mild basic conditions ⟨57CB2588, 59CB3190, 61CB1⟩. In the direct condensation reactions much less specific behavior is found depending to a large extent on the nature of the pyrimidine substituents. 5,6-Diaminopyrimidin-4-one, for example, does not show a pH-dependence and forms pteridine-4,6-dione at pH 7.5 and 0 almost quantitatively ⟨53JCS74⟩, whereas with 5,6-diaminopyrimidine the 6- and 7-oxo ratio can be influenced in the usual manner ⟨52JCS1620⟩. The most direct way to obtain 2,4-disubstituted 7-methylpteridin-6-ones (**239**; R = Me) is *via* the corresponding 7-acetonyl derivatives (**242**), obtained in pure form from 5,6-diaminopyrimidines and 2,4-dioxopentanoic esters in acidic medium ⟨57CB2604, 69CB4032⟩ followed by hydrolysis of the side-chain by alkali. Another regiospecific condensation takes place between 2,5,6-triamino-4-isopropyl-oxypyrimidine and 4-substituted 2-trifluoromethylpseudooxazolones in glacial acetic acid, yielding 7-substituted xanthopterins after subsequent alkaline hydrolysis of the 4-isopropyl-oxy group ⟨64CB3456⟩.

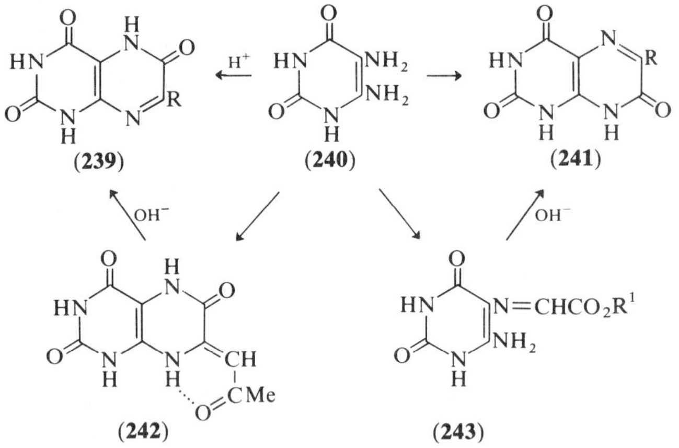

$$\tag{79}$$

7-Oxopteridine-6-carboxylic acids can also be synthesized in a quite regiospecific manner using alkyl mesoxalates at neutral pH ⟨57CB2617, 62CB1591, 61CB2708⟩ or alloxan in dilute aqueous alkali ⟨59JA2474, 60JA3765⟩. It is difficult to prepare the isomeric 6-oxopteridine-7-carboxylic acids from the same starting materials; changes in pH affect the isomer ratio only slightly in the direction of the 6-oxo derivatives. A synthetic approach to this series of compounds was found in the reaction of the free 5,6-diaminopyrimidine bases with 1,3-dimethylalloxan (**244**) leading to 6-oxopteridine-7-*N*-methylcarboxamides (**245**), which can be hydrolyzed by alkali to the corresponding acid (**246**) (equation 80) ⟨57CB2624, 62CB749⟩.

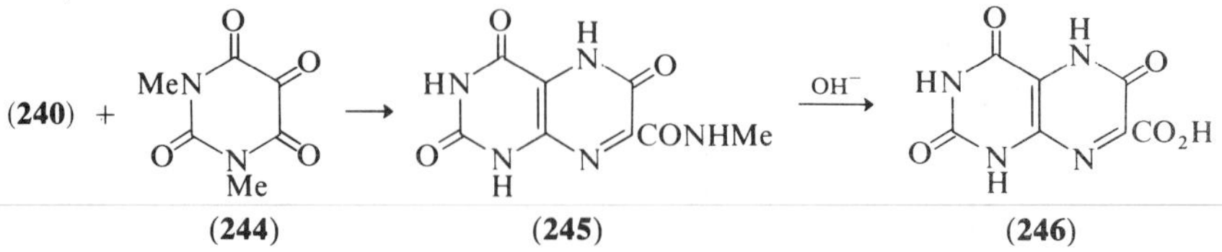

$$\tag{80}$$

When the unsymmetrical dicarbonyl compound is neither an acid nor an ester the most successful attempts to influence orientation have involved the use of aldehyde- and ketone-binding reagents such as hydrazine or sodium hydrogen sulfite, which tends to direct an alkyl group into the 6-position. 2,4,5,6-Tetraaminopyrimidine (248) and methylglyoxal give 2,4-diamino-7-methylpteridine (247) in weakly acidic medium, but in aqueous sodium sulfite or bisulfite only a trace of this material was produced and the 6-methyl isomer (249) predominated (equation 81) ⟨49JA1753⟩. Again, 2,5,6-triaminopyrimidin-4-one (251) reacts at the 5-position with methylglyoxal in weakly alkaline medium to give the corresponding Schiff's base, which cyclizes to 7-methylpterin in high yield in boiling dilute acetic acid. On the other hand, the isomeric 6-methylpterin ⟨70LA(741)64⟩ is formed predominantly if methylglyoxal is first allowed to react with hydrazine ⟨49JCS2077⟩ in a 1:6 ratio. Similarly, (251) and glucosone give the 6- or 7-(tetrahydroxybutyl) analogue depending on whether hydrazine is present or absent ⟨49JCS79⟩. In general, $\alpha$-ketoaldehydes form 7-substituted pteridines with the pH having little influence on the orientation, as can be seen from the reaction of (240) with methylglyoxal to give 7-methyllumazine and (251) with phenyl- and hydroxymethyl-glyoxal to give 7-phenyl- (250) and 7-hydroxymethyl-pterin ⟨45JA802⟩ respectively. However, the purity of the reaction products always has to be checked, since simple chromatography often does not differentiate between the 6- and 7-isomers. Besides carbonyl binding reagents, chemically modified ketones may help to solve the orientation problem. 1,1-Dichloracetone and $\omega,\omega$-dichloracetophenone react with (251) in an acetate buffer to give 6-methyl- and 6-phenyl-pterin (252); (equation 82) ⟨52JCS2144⟩.

(81)

(82)

Another general synthesis of pteridines consists of the condensation of 5,6-diaminopyrimidines with $\alpha$-substituted carbonyl compounds, especially of the aldehydo- and keto-alcohol type. The mechanisms of these reactions are complex and product formation often depends on the conditions. The most common reaction products with (251) are 7-alkyl-5,6-dihydropterins, which are oxidized by air to the fully aromatic oxidation level. In this way (251) gives 7-methylpterin with hydroxyacetone ⟨49JCS2077⟩ and also with dihydroxyacetone, by elimination of water from the side-chain. The tetramine (248) behaves somewhat differently in yielding 2,4-diamino-6-methylpteridine (249) with dihydroxyacetone and sodium sulfite and the 7-hydroxymethyl derivative with hydroxymethyl-glyoxal under similar conditions ⟨65JCS1530⟩. More detailed studies on the synthesis of 6-hydroxymethylpteridines from various 2,4-disubstituted 5,6-diaminopyrimidines and dihydroxyacetone revealed the necessity for very special reaction conditions, namely high concentration of sodium acetate, addition of cysteine and adjustment of the pH conditions ⟨64JOC3610⟩. With 2,4,5,6-tetraaminopyrimidine (248), however, 20% of (249) is still formed in addition to 2,4-diamino-6-hydroxymethylpteridine ⟨74JHC279⟩, but further improvement can be effected by more rigorous pH control ⟨78USP4080325⟩, exchange of sodium acetate for ammonium chloride, and oxidation of the intermediate dihydro stage by pure oxygen ⟨80CB1514⟩.

Analogous reactions of (251) with sugar hexoses take place simultaneously by both mechanisms, each giving a mixture of the same 7-tri- and 7-tetra-hydroxybutylpterin. In the presence of hydrazine, glucose and fructose give the 6-tetrahydroxybutyl isomer ⟨49JCS79, 49JCS2077⟩ and this has been attributed not only to a dehydrogenation of the sugar to an osone but also to an orientation effect. The same regioselectivity has been found in the condensations of (251) with *p*-tolyl-D-isoglucosamine ⟨49CB25⟩, *p*-tolyl-D-isogalactosamine and *p*-tolyl-L-arabinosamine ⟨56CB2904⟩ in the presence of hydrazine. The striking complexity of these reactions is revealed in a more detailed study which showed a great variety of reaction products ⟨64CB1002⟩. Nevertheless, a regiospecific synthesis of D-mona- and L-neo-pterin from (251) and 1-benzylamino-1-deoxy-D-xylulose and -L-ribulose was developed ⟨68HCA1495⟩. Another improvement climaxed in the condensation of 2,5,6-triaminopyrimidin-4-one dihydrochloride with the phenylhydrazones of L-xylose and D-arabinose (253) respectively, to form D-neo- (33) and L-mona-pterin (34) in high yields ⟨70HCA1202⟩ *via* a series of intermediates (equation 83). The biologically important biopterin (32) was first synthesized from (251) and rhamnotetrose ⟨56JA5868⟩, but the reaction always led to a mixture of the 6- and 7-isomers ⟨58HCA108, 58ZPC(311)79, 62ZPC(329)291⟩ and ever since that time new approaches to this naturally occurring pterin have been developed ⟨72HCA574, 75BCJ3767⟩. The best method uses the benzylphenylhydrazone of 3,4-di-*O*-acetyl-5-deoxy-L-arabinose for the condensation with (251) dihydrochloride in the usual manner and subsequent oxidation of the tetra- and di-hydro intermediates ⟨77HCA211⟩.

(83)

Aromatic keto alcohols of the benzoin type give pairs of stable isomeric dihydropteridines. 6,7-Diphenyl-7,8-dihydropteridines are formed in the presence of acetic acid and the corresponding 5,6-dihydro isomers under neutral reaction conditions ⟨48BSF963, 51BSF423, 51BSF428, 51BSF521, 65HCA764, 68HCA1029⟩. Complex reaction pathways may also occur with α-halo carbonyl compounds, because an adjacent hydroxy function can react instead of a 5- or 6-amino group, forming pyrimido[4,5-*b*][1,4]oxazines ⟨70JCS(C)437⟩ instead of dihydropteridines. With 1,3-dichloroacetone or α,β-dichloropropionaldehyde, dehydrohalogenation from the side-chain is also observed, leading to 6- and 7-methylpteridines in preference to the chloromethyl derivatives. A useful direct condensation reaction of (251) to 6-chloromethylpterin, which is a desirable intermediate for the synthesis of folic acid (36), has not yet been achieved. However, it has been noticed that the condensation of (251) with 1-(2-bromo-2-formylethyl)pyridinium iodide is regioselective, resulting in formation of 6-pterinylmethylpyridinium iodide ⟨48JA23⟩. On this basis a useful chemical synthesis of (36) can be realized starting from (251) and 1-bromo-2,2-diethoxypropionaldehyde and leading in the first step to the 6-(diethoxymethyl)pterin (255) as the hydrogen peroxide oxidation product of the intermediate 5,6-dihydro derivative (254). Acetylation of the amino group followed by hydrolysis of the acetal function to the $N^2$-acetylpterin-6-carbaldehyde (256), condensation with *p*-aminobenzoylglutamic acid, reduction of the Schiff's base (257), and basic hydrolysis of the acetyl group finally yields (36) in a respectable overall yield ⟨55JA6365⟩ (equation 84). Compound (256) is also the starting material for the synthesis of pteroic acid, which was obtained from the condensation with ethyl *p*-aminobenzoate, subsequent reduction by dimethylamine–borane and basic hydrolysis ⟨71JOC860⟩.

A new, versatile and selective synthesis of 6- and 7-substituted pteridines was reported by Rosowsky ⟨73JOC2073⟩. β-Keto sulfoxides, which can be viewed as latent α-keto aldehydes, react with (251) to give 6-substituted pterins, and the use of α-keto aldehyde hemithioacetals leads in a regiospecific synthesis to the isomeric 7-substituted pterins (equation 85).

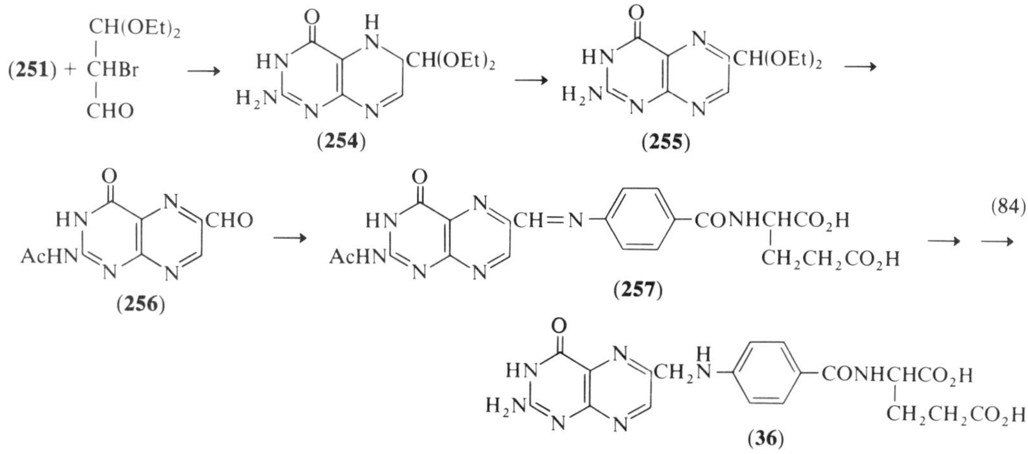

(251) + [structure with CH(OEt)₂, CHBr, CHO] → (254) → (255) →

(256) → (257) → (84)

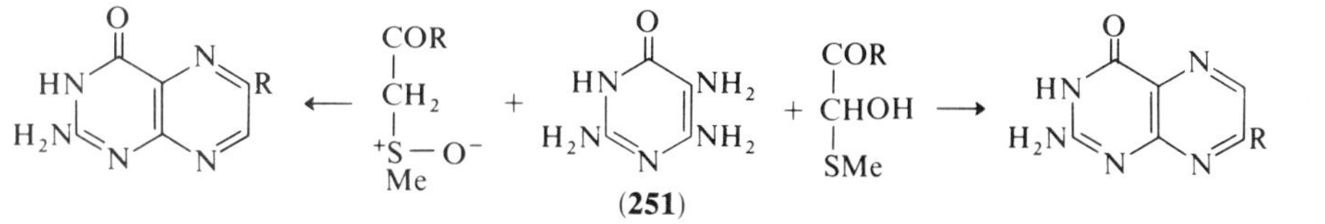

(36)

(251) → (85)

## 2.16.3.1.1 Unambiguous syntheses

Since the structures of the Gabriel–Isay condensation products of 5,6-diaminopyrimidines with unsymmetrical 1,2-dicarbonyl or α-substituted monocarbonyl compounds are always ambiguous, the synthesis of 6- and 7-substituted pteridines by an unambiguous approach was and still is a necessity and an important challenge.

## 2.16.3.1.2 The Timmis Reaction

In 1949 Timmis discovered that ketomethylene structures will react with 6-amino-5-nitrosopyrimidines ⟨49MI21600⟩, providing a new general and unequivocal route to 6- and 7-substituted pteridines respectively. These condensation reactions are not restricted to ketones and aldehydes (equation 86) ⟨54JCS2881, 54JCS2895, 56JCS213⟩ but work also with nitriles ⟨68JMC549⟩, esters, and acyl halides possessing an activated methylene group adjacent to these functions (equation 87) ⟨54JCS2887, 55JCS2036, 55JCS2038, 64MI21602⟩. With highly activated methylene compounds no special base catalysis is necessary for the condensation reactions, since 2,6-diamino-4-isopropyloxy-5-nitrosopyrimidine (259) reacts with methyl cyanoacetate to give 2-amino-6-cyano-4-isopropyloxypteridin-7-one (258) and with malononitrile to give 2,7-diamino-6-cyano-4-isopropyloxypteridine (260) (equation 88) ⟨61CB2708⟩. Treatment of 6-amino-1,3-dimethyl-5-nitrosouracil with phenacylpyridinium halides in pyridine and in the presence of aqueous sodium hydroxide yields the corresponding 7-aryl-1,3-dimethyllumazin-6-ones ⟨77H(6)1907⟩.

The addition of phosphonate carbanions to 6-amino-5-nitrosopyrimidines may be regarded as an extension of the Timmis principle and it proceeds under mild conditions and in

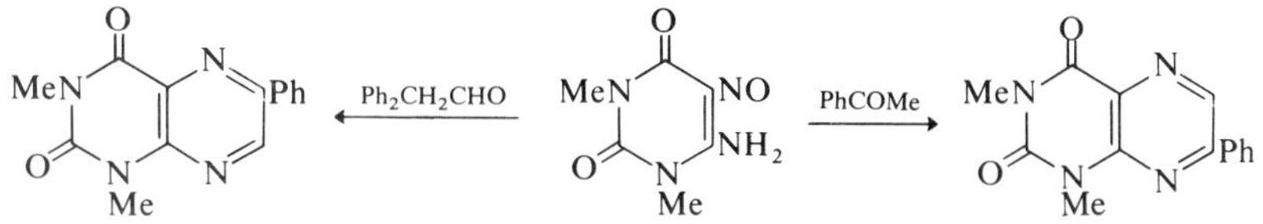

[structure] ←Ph₂CH₂CHO— [structure] —PhCOMe→ [structure] (86)

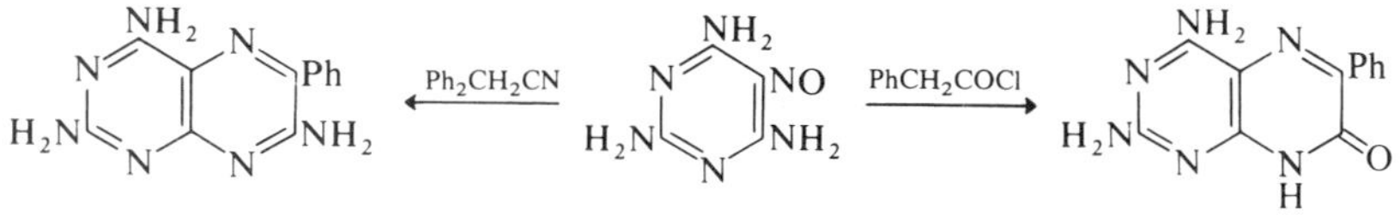

[structure] ←Ph₂CH₂CN— [structure] —PhCH₂COCl→ [structure] (87)

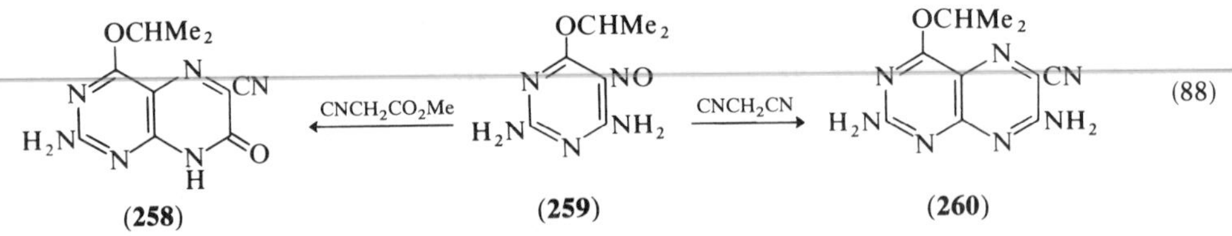

$$(88)$$

Structures (258), (259), (260) with reagents CNCH$_2$CO$_2$Me and CNCH$_2$CN

(258)   (259)   (260)

high yield ⟨70CC1371, 71CC189⟩. 4,6-Diamino-5-nitroso-2-phenylpyrimidine (262) and the anion of triethyl phosphonoacetate gave a 90% yield of 4-amino-2-phenyl-pteridin-7-one (261) in THF. Analogously, a variety of other phosphonate anions, readily prepared by reaction of α-bromo-esters, -nitriles or -ketones with triethyl phosphite followed by addition of base, form pteridin-7-ones, 7-amino- (263) and 7-alkyl- or aryl-pteridines respectively (equation 89). In the same vein, 7-substituted 1,3-dimethyllumazines have been obtained from 6-amino-1,3-dimethyl-5-nitrosouracil and phenacylidenetriphenylphosphoranes ⟨76CC588⟩. Moreover, the reactions of dimethyl acetylenedicarboxylate with 6-amino-5-nitroso-, 6-amino-5-phenylazo- and 6-hydrazino-5-nitroso-1,3-dimethyluracil offer synthetic routes to the pteridine nucleus, leading to 6,7-bis(methoxycarbonyl)-1,3-dimethyl-lumazine ⟨81H(15)757⟩.

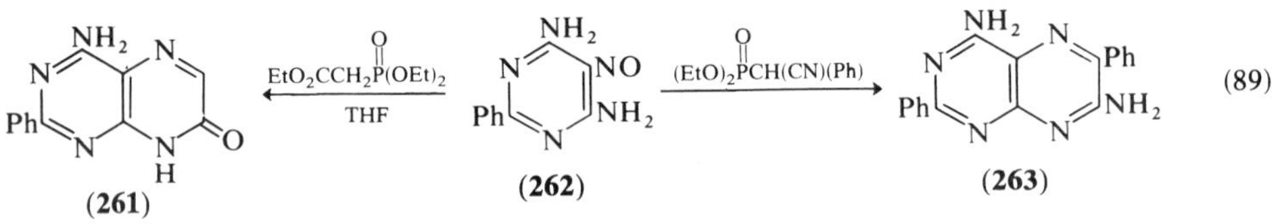

$$(89)$$

Structures (261), (262), (263) with reagents EtO$_2$CCH$_2$P(OEt)$_2$/THF and (EtO)$_2$PCH(CN)(Ph)

(261)   (262)   (263)

### 2.16.3.1.3 The Pachter reaction

The versatility of 5-nitrosopyrimidines in pteridine syntheses was noticed by Pachter ⟨64MI21603⟩ during modification of the Timmis condensation between (262) and benzyl methyl ketone; simple condensation leads to 4-amino-7-methyl-2,6-diphenylpteridine (264) but in the presence of cyanide ion 4,7-diamino-2,6-diphenylpteridine (265) is formed (equation 90). The mechanism of this reaction is still uncertain ⟨63JOC1187⟩; it may involve an oxidation of an intermediate hydroxylamine derivative, nitrone formation similar to the Kröhnke reaction, or nucleophilic addition of the cyanide ion to the Schiff's base function (266) followed by cyclization to a 7-amino-5,6-dihydropteridine derivative (267), oxidation to a quinonoid-type product (268) and loss of the acyl group (equation 91). Extension of these principles to α-aryl- and α-alkyl-acetoacetonitriles omits the oxidation step and gives higher yields, and forms 6-alkyl-7-aminopteridines, which cannot be obtained directly from simple aliphatic ketones.

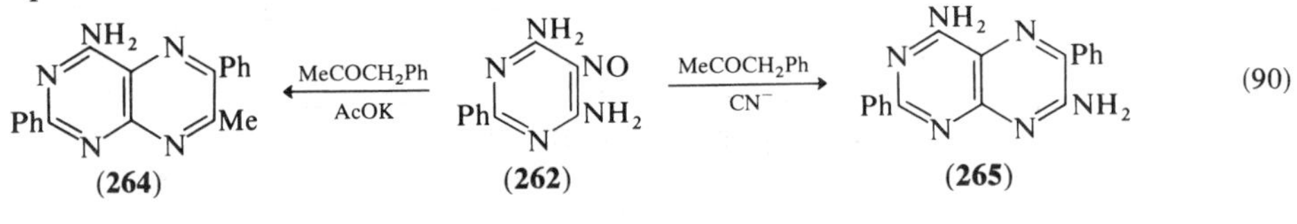

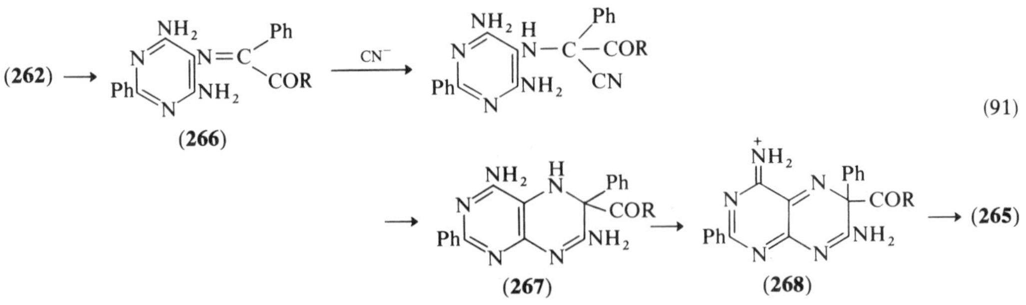

$$(90)$$

Structures (264), (262), (265) with reagents MeCOCH$_2$Ph/AcOK and MeCOCH$_2$Ph/CN$^-$

(264)   (262)   (265)

$$(91)$$

$(262) \rightarrow$ Structure (266) with reagent CN$^-$ → intermediate structure → Structures (267), (268) → (265)

(266)

(267)   (268)

4-Amino-5-nitrosopyrimidines also condense with benzoylacetonitrile, phenacyl-pyridinium bromide and acetonylpyridinium chloride in the presence of sodium cyanide to produce 7-amino-6-pteridinyl ketones ⟨63JOC1197⟩. Pteridine syntheses from pyridinium salts are not limited to the preparation of pteridyl ketones since pyridinium acetamide reacts with (**262**) in the presence of sodium cyanide to give 4,7-diamino-2-phenyl-6-pteridinecarboxamide (**269**); with pyridinium benzyl cyanide in the presence of sodium acetate, 4,7-diamino-2,6-diphenylpteridine 5-oxide (**270**) is obtained (equation 92).

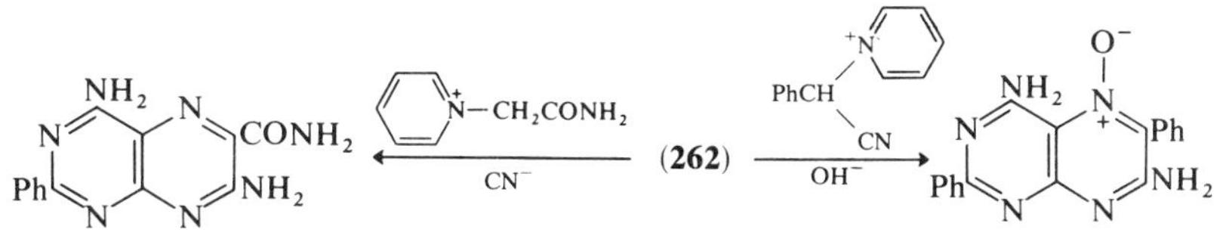

(92)

An alternative method of synthesis is the Blicke–Pachter method ⟨54JA2798, 63JOC1191⟩ involving the formation of aminonitriles from 5,6-diaminopyrimidines, aldehydes, and hydrogen cyanide followed by base-catalyzed cyclization with sodium methoxide and oxidation with hydrogen peroxide, to produce 7-aminopteridines (equation 93). The reaction works with formaldehyde (R=H), to form 6-unsubstituted amines, and also with heteroaromatic aldehydes to give 7-amino-6-heteroarylpteridines ⟨68JMC560⟩. α,β-Unsaturated aldehydes are reduced intramolecularly in this process with formation of 6-phenethyl- and 6-styryl-7-aminopteridines starting from cinnamaldehyde and phenylpropargylaldehyde respectively (equation 94) ⟨63JOC1191⟩.

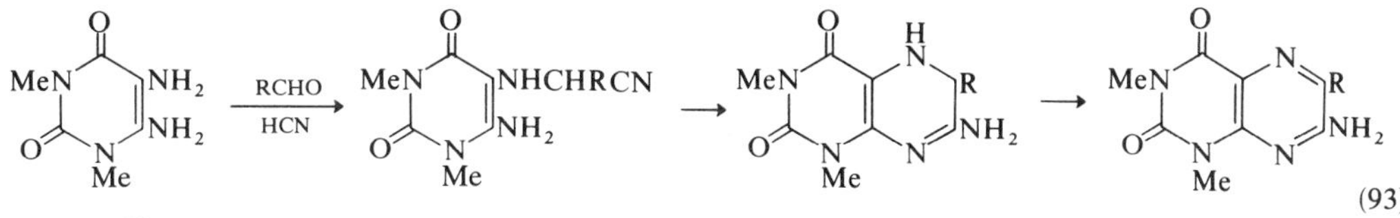

(93)

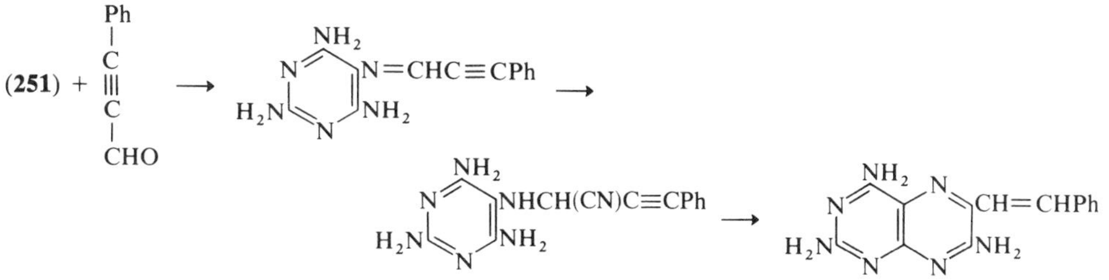

(94)

### 2.16.3.1.4 The Polonovski–Boon reaction

Another unambiguous approach to 6- and 7-substituted pteridines was found by Polonovski ⟨50CR(230)392⟩ and Boon ⟨51JCS96⟩ in the nucleophilic substitution of 6-chloro-5-nitropyrimidines with α-aminocarbonyl compounds, which on reduction cyclize to 7,8-dihydropteridines. 7,7-Dimethyl-7,8-dihydroxanthopterin (**273**) was thus obtained from 2-amino-6-chloro-5-nitropyrimidin-4-one (**271**) and ethyl α-aminoisobutyrate on catalytic or chemical reduction of the intermediate (**272**; equation 95) ⟨64JCS4769⟩. 6-Chloro-5-phenylazopyrimidines react similarly with α-aminoketones to yield 6-substituted 7,8-dihydropteridines ⟨51JCS1497⟩; the carbonyl function sometimes has to be protected to avoid self-condensation ⟨67JHC12, 67JHC124⟩. More complex molecules such as L-neopterin ⟨63MI21600⟩ and biopterin (**32**) have been synthesized by this route. Compound (**271**) reacts with 1-amino-1,5-dideoxy-L-erythropentulose to produce (**274**); reduction leads to 7,8-dihydrobiopterin (**223**), which on manganese dioxide oxidation yields biopterin (**32**; equation 96) ⟨69JCS(C)928⟩. Nature uses a similar approach in the biosynthesis of the pteridine

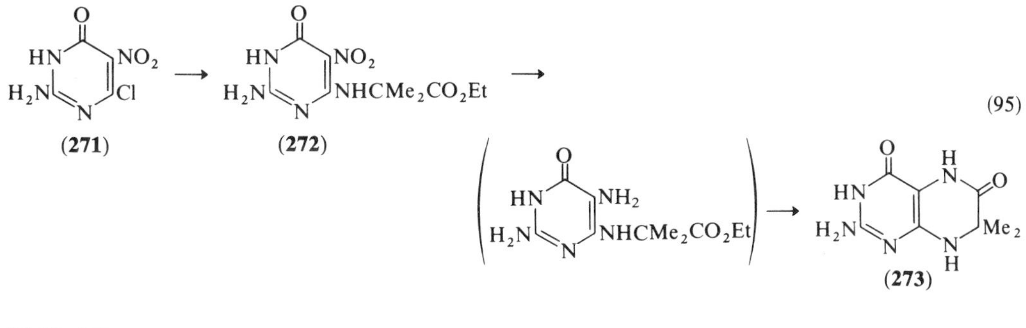

(95)

ring system starting from guanosine 5′-triphosphate (GTP). After cleavage of the imidazole ring an Amadori rearrangement of the 6-ribosylamino function to the corresponding 1-deoxyribulosylamino derivative takes place, followed by condensation with the adjacent 5-amino group to form 7,8-dihydro-D-neopterin 3′-triphosphate ⟨B-69MI21602, B-75MI21602⟩.

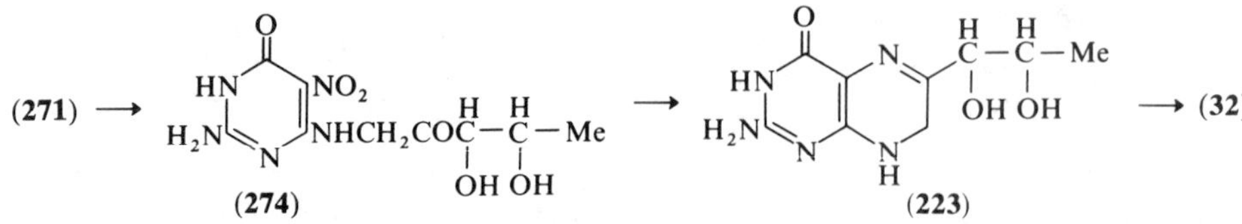

$$(271) \rightarrow (274) \rightarrow (223) \rightarrow (32) \quad (96)$$

Another approach uses the reaction of 6-chloro-5-nitropyrimidines with $\alpha$-phenyl-substituted amidines followed by base-catalyzed cyclization to pteridine 5-oxides, which can be reduced further by sodium dithionite to the heteroaromatic analogues (equation 97) ⟨79JOC1700⟩. Acylation of 6-amino-5-nitropyrimidines with cyanoacetyl chloride yields 6-(2-cyanoacetamino)-5-nitropyrimidines (**276**), which can be cyclized by base to 5-hydroxypteridine-6,7-diones (**275**) or 6-cyano-7-oxo-7,8-dihydropteridine 5-oxides (**277**), precursors of pteridine-6,7-diones (**278**; equation 98) ⟨75CC819⟩.

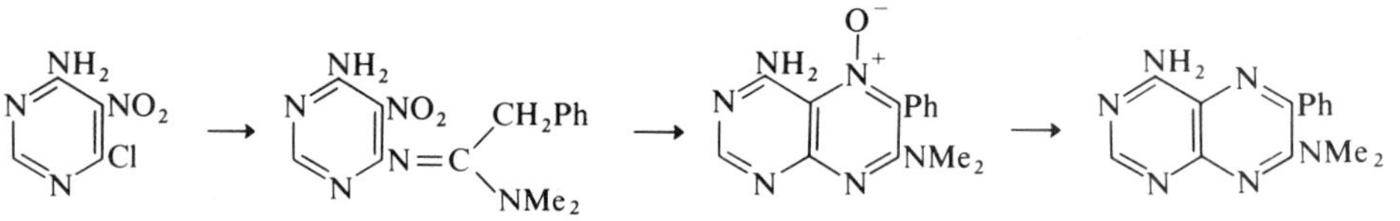

$$(97)$$

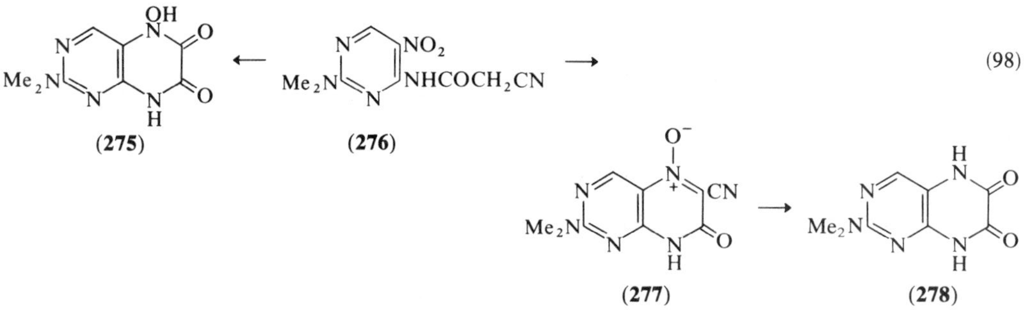

$$(98)$$

A kind of modification of the Polonovski–Boon synthesis is the reaction of 5,6-dihalopyrimidines with ethylenediamine derivatives. Depending on the bulkiness of the amino substituents a more or less regiospecific condensation may proceed ⟨71CB780⟩, as shown recently in the reaction of 5-bromo-6-chloro-1,3-dimethyluracil (**279**) with 2-methyl-amino-*n*-propylamine to form 1,3,5,6-tetramethyl-5,6,7,8-tetrahydrolumazine (**280**; equation 99) ⟨80BCJ3385⟩.

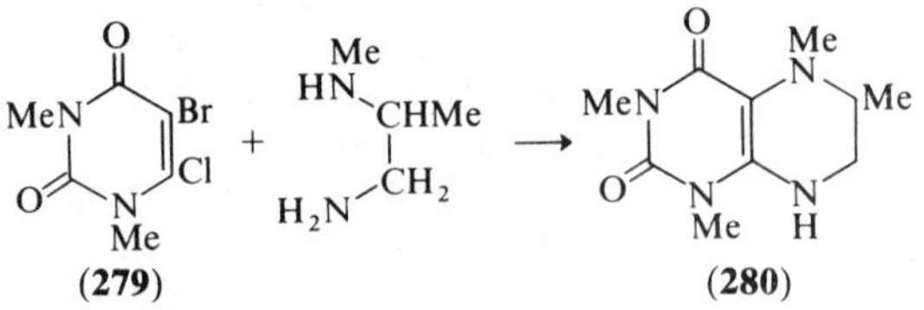

$$(279) + \rightarrow (280) \quad (99)$$

### 2.16.3.1.5 *Miscellaneous Regioselective Syntheses*

An unequivocal synthesis of 6-arylpteridines can be achieved by intramolecular cyclization of diazahexatrienes ⟨77JCS(P1)1336⟩. Treatment of 6-amino-5-benzyl-ideneaminopyrimidines with an excesss of triethyl orthoformate in DMF affords the 6-ethoxymethyleneamino derivatives, which undergo thermal cyclization by valence isomeriz-ation and aromatization by elimination of ethanol to give 6-arylpteridines (equation 100). Analogous treatment with dimethylformamide diethyl acetal gives the 6-dimethyl-aminomethyleneamino analogues, which on heating in tetramethylene sulfone cyclize in the same manner. 1,3-Dimethyl-6-($\alpha$-methylbenzylidenehydrazino)uracils react with

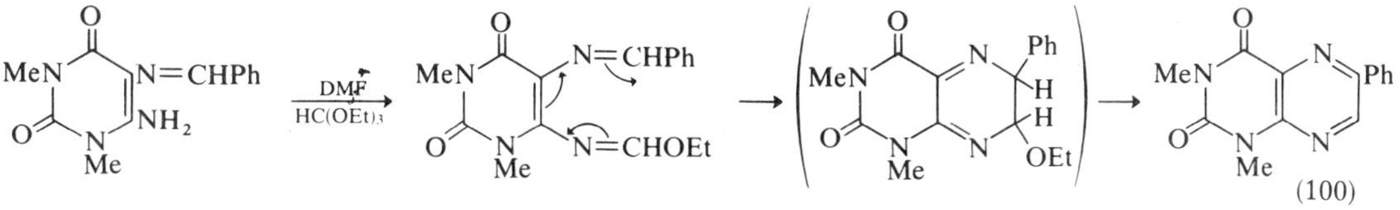

(100)

sodium nitrite in acetic acid followed by dithionite in formic acid to give 6-aryl-1,3-dimethyllumazines ⟨77H(6)693⟩.

The reaction of 6-amino-5-(1,2-diethoxycarbonylhydrazino)pyrimidines with enamines represents another convenient method for the preparation of pteridines. Fusion of 5-(1,2-diethoxycarbonylhydrazino)-2,4,6-triaminopyrimidine (**281**) with an excess of morpholinocyclohexene leads to 2,4-diaminotetrahydrobenzo[*g*]pteridine, and with the morpholinoenamine (**282**) from 17β-hydroxy-5α-androstan-3-one regioselective condensation to the fused pteridine (**283**) takes place in almost quantitative yield (equation 101) ⟨71CC83⟩. 6-Amino-5-nitroso- and 6-amino-5-phenylazo-pyrimidines react similarly, imitating the Timmis-type reaction ⟨72CPB1428⟩.

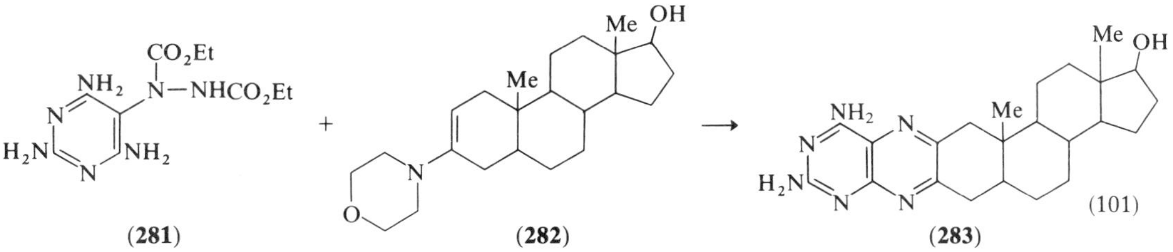

(101)

A novel type of ring closure is the reaction of 6-amino-5-dichloroacetylaminopyrimidines (**285**) with sulfur and morpholine under the conditions of a Kindler reaction ⟨B-64MI21605⟩. 7-Morpholinopteridin-6-ones (**287**) are formed, either *via* thiooxamide derivatives (**286**) or *via* corresponding 7,8-dihydropteridines (**284**; equation 102). Chloral hydrate also reacts with 2-substituted 5,6-diaminopyrimidin-4-ones to form pteridin-6-ones ⟨56JCS3311, 64JCS565⟩ by a so far unknown mechanism.

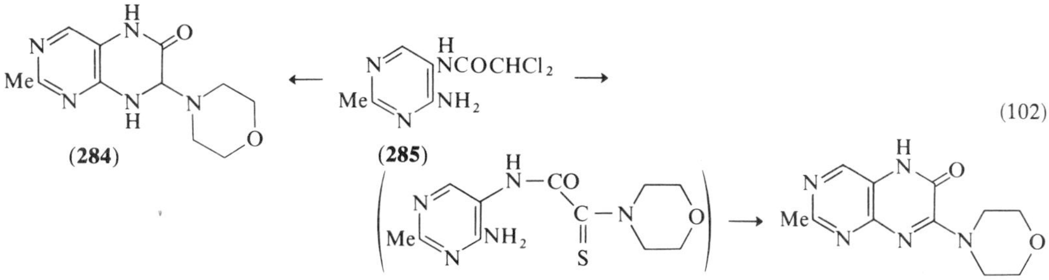

(102)

An unusual approach to the lumazine nucleus was found in the photochemical transformation of 6-azido-1,3-dimethyluracil (**289**) with various amino compounds ⟨78JA7661⟩. Irradiation of (**289**) in the presence of ethyl α-amino acid esters forms 7-substituted 7,8-dihydrolumazin-6-ones (**288**), and with α-aminoketones 6-substituted 7,8-dihydrolumazines (**290**) are formed (equation 103).

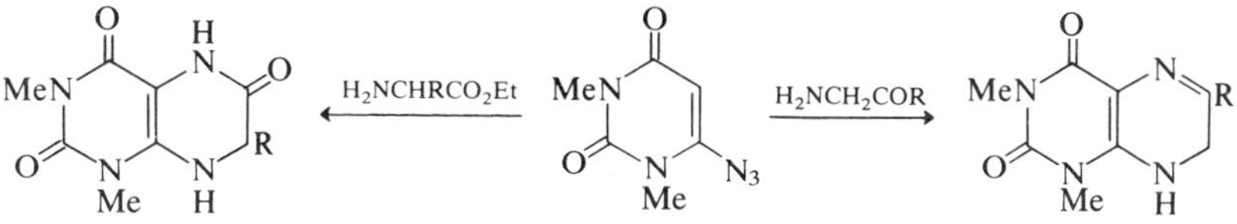

(103)

## 2.16.3.2 From Pyrazines

Despite the fact that one of the first pteridine syntheses was based on an intramolecular Hofmann carboxamide degradation of pyrazine-2,3-dicarboxamide by action of potassium hypobromite and leads to lumazine (equation 104), ⟨07CB4857⟩, pyrazine derivatives in general have not often been used because of availability problems. The reaction of alkyl

2-chloropyrazine-3-carboxylates with guanidinium salts leads to pterins ⟨55JCS1379⟩ and cyclization of 2-aminopyrazine-3-carbohydrazide and -3-hydroxamic acid gives 3-amino- ⟨59JA2479, 62LA(660)98⟩ and 3-hydroxypteridin-4-ones respectively ⟨55JA3927⟩. Pteridine-4-one and -4-thione are formed similarly, by heating 2-aminopyrazine-3-carboxamide or -3-carbothiamide with ethyl orthoformate and acetic anhydride ⟨51JCS474⟩, and the 2-substituted analogues are obtained with homologous orthoesters, amidines or acid anhydrides ⟨79JCS(P1)1574⟩. The synthesis of 3,4-dihydro and 4-unsubstituted pteridine derivatives can be achieved from 2-amino-3-aminomethylpyrazine ⟨70JCS(C)1540⟩ and 3-aminopyrazine-2-carbaldehyde ⟨71JCS(C)2357⟩ respectively (equation 105).

(104)

(105)

Aminolytic or hydrolytic ring cleavage of suitable pteridine derivatives provides a good source of 2-aminopyrazine-3-carboxamides ⟨B-54MI21601⟩, which on recyclization result in new pteridines (equation 106). This procedure is quite general and can be varied in many ways ⟨53JA1904, 56JA210⟩. On the other hand, 2-aminopyrazine-3-carboxamides are also available by reductive cleavage of the N—N bond in 3-oxo-2,3-dihydropyrazolo[3,4-b]pyrazines ⟨58JA421⟩.

(106)

### 2.16.3.2.1 The Taylor synthesis

A versatile new route to pteridines was found by Taylor ⟨68JA2424, B-70MI21607⟩ in the reaction of an oximino-aldehyde or -ketone with ethyl α-aminocyanoacetate or aminomalononitrile, to give 2-amino-3-ethoxycarbonyl- or -3-cyano-pyrazine 1-oxides which cyclize with guanidine to pterin 8-oxides or 2,4-diaminopteridine 8-oxides respectively (equation 107) ⟨73JA6407, 73JA6413⟩. Reduction of these 8-oxides to the heteroaromatic forms as well as the corresponding 7,8-dihydropteridine derivatives provides a wide variety of compounds. This new approach also solves the orientation problem of the 6- and/or 7-substituents, since their structure is already determined in the α-oximinocarbonyl component. The reaction of α-ketoaldoximes with ethyl α-aminocyanoacetate is general and leads in a direct route to 6-substituted pteridines. Double oximation of acetone to α,α'-dioximinoacetone and analogous condensation gives 2-amino-3-ethoxycarbonyl-5-oximinomethylpyrazine 1-oxide, which on reaction with guanidine cyclizes to 6-oximinomethylpterin 8-oxide. This compound curiously could not be hydrolyzed directly to the corresponding 6-carbaldehyde, but conversion to pterin-6-carbaldehyde (206), a potent inhibitor of xanthine oxidase, could then be accomplished *via* the two-step sequence of sodium sulfite reduction followed by iodine oxidation ⟨69LA(726)100⟩. Two other routes, involving 2-amino-3-cyano-5-chloromethyl- ⟨73JOC2817⟩ and 2-amino-3-cyano-5-(dimethoxymethyl)-pyrazine 1-oxide ⟨81JOC1394⟩ respectively, lead also to the aldehyde (206), which functions as a key intermediate for the preparation of pteroic acid, folic acid and various derivatives ⟨78JOC736⟩. The pyrazine 1-oxide approach has also been used for the synthesis of biopterin (32) ⟨74JA6781⟩, by reaction of 5-deoxy-2-keto-L-ribose oxime with benzyl α-aminocyanoacetate in the usual manner. Further, the oxime cyclization route is capable of still another extension, leading to the unequivocal synthesis of a series of isomeric 7-substituted pteridines ⟨76JOC1299⟩. Addition of nitrosyl chloride to acrolein gives α-oximino-β-chloropropionaldehyde which condenses with α-aminomalononitrile to yield 2-amino-3-cyano-6-chloromethylpyrazine 1-oxide (**291**), a versatile intermediate for various interconversions, as for example to 2,4-diamino-7-styrylpteridine (**292**; equation 108). The universality of the new method ⟨B-75MI21603⟩ is again illustrated by the synthesis of 8-hydroxypteridin-7-ones (**293**), so-called pteridine hydroxamic acids ⟨75JOC2332⟩, or the more complex total synthesis of the naturally occurring asperopterin B (**294**) ⟨75JOC2336⟩.

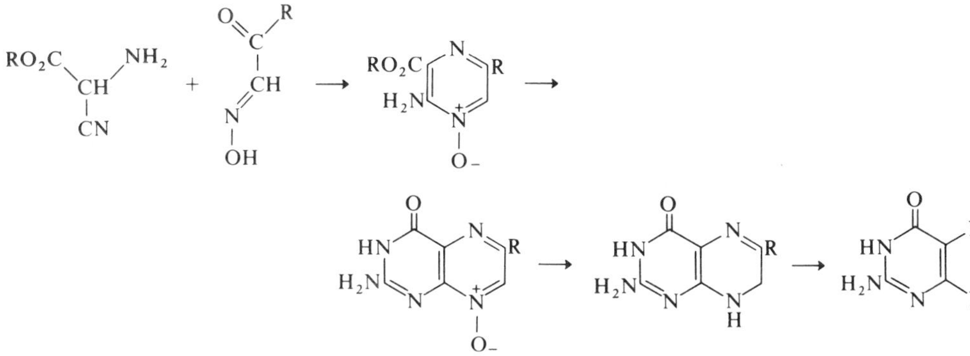

(107)

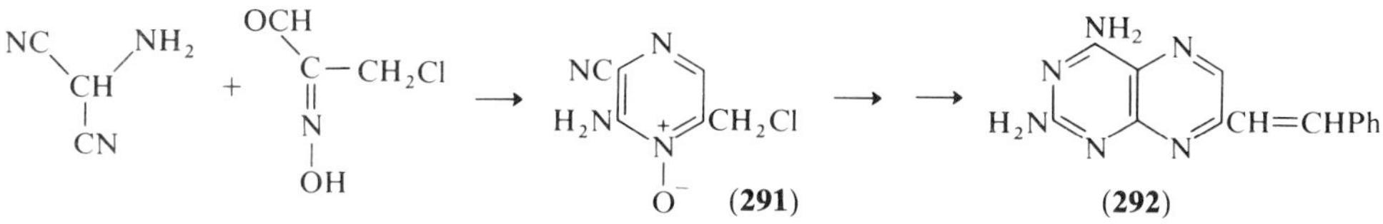

(108)

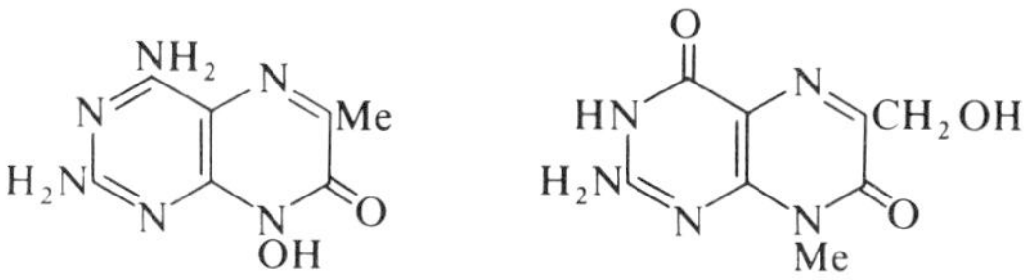

(**293**)       (**294**)

## 2.16.3.3 Transformations from Other Heterocyclic Ring Systems

Various condensed heterocyclic ring systems which contain the masked 5,6-diaminopyrimidine or 2-aminopyrazine-3-carboxylic acid structural element may act as potential pteridine synthons. Some acid-labile purines can be converted into pteridines through hydrolytic ring-opening of the imidazole moiety and subsequent condensation with a 1,2-dicarbonyl compound (equation 109) ⟨57BJ(65)124⟩. More stable purine derivatives require more drastic reaction conditions and, as with uric acid, lead to a complex mixture of compounds in low yield ⟨59CB2468⟩. On the other hand, the imidazole ring can be labilized by quaternization to facilitate nucleophilic attack at C-8 and ring-opening. Thus, 9-methyl-guanine was alkylated by ethyl bromoacetate to 7-ethoxycarbonylmethyl-9-methyl-guaninium bromide (**295**), which on treatment first with base and then with acid reacted *via* (**296**) to give 8-methylisoxanthopterin (**297**; equation 110) ⟨73CB1389⟩. In an analogous conversion, isoxanthopterin-8-riboside resulted from guanosine ⟨74CB575⟩. A similar ring transformation starting from $N^6$-benzoyl-9-benzyl-7-phenacyladeninium bromide gave 4-benzylamino-7-phenylpteridine ⟨81H(15)895⟩.

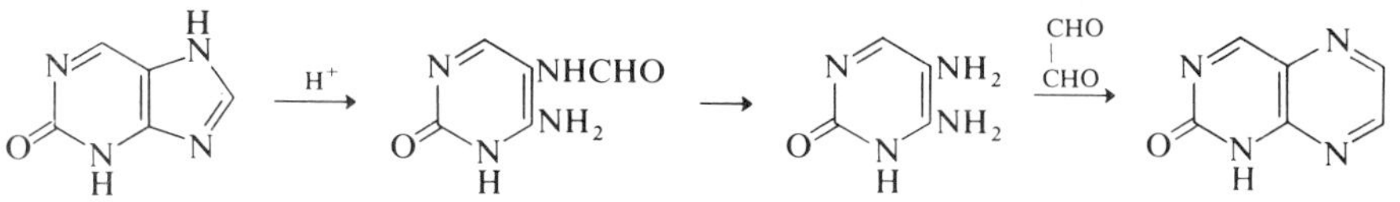

(109)

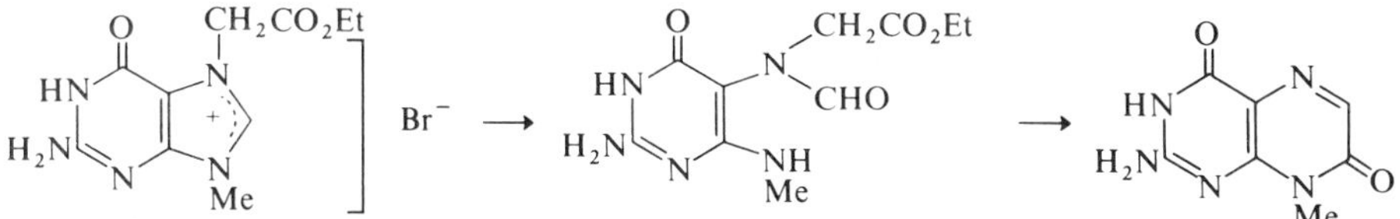

(110)

(**295**)       (**296**)       (**297**)

7-Aminofurazano[3,4-*d*]pyrimidines represent latent 4,5,6-triaminopyrimidines due to the easy reductive cleavage of the N—O—N linkage. Reduction of 7-benzoylmethylamino-5-phenylfurazano[3,4-*d*]pyrimidine (**298**) followed by acid-catalyzed cyclization and oxidation leads to an unequivocal synthesis of 4-amino-2,6-diphenylpteridine (**299**; equation 111) ⟨79JOC302⟩. Besides the reductive ring-opening of 3-oxo-2,3-dihydropyrazolo[2,3-*d*]pyrazines to 2-aminopyrazine-3-carboxamides and subsequent ring-closures to differently substituted pteridines ⟨56JA5451, 58JA421⟩, 2-substituted pyrazino[2,3-*d*][1,3]oxazin-4-ones are valuable intermediates for the preparation of 2-mono- and 2,3-di-substituted pteridin-4-ones by treatment with ammonia and primary amines respectively (equation 112) ⟨72TL3359⟩.

$$(111)$$

$$(112)$$

The degradation of more complex substances can be regarded as another route to pteridine derivatives. Already in 1895 'tolualloxazine' was oxidized by alkaline permanganate to lumazine-6,7-dicarboxylic acid, and further heating led in a stepwise decarboxylation to lumazine (**3**) ⟨1895CB1970⟩.


### 2.16.3.4 Biosynthesis

Investigations of the biosynthesis of the pteridine nucleus have been initiated by applying radioactive labelled precursor techniques to insects and bacteria ⟨B-69MI421602, 72AG1088⟩. When *Xenopus* larvae ⟨56ZN(B)82⟩ were shown to transform purines into pteridines, and a possible biogenetic relationship between purines and riboflavin was suggested ⟨54JBC(208)513⟩, more discussion of this structurally related interconversion followed ⟨58JA739, 58JA951, 58LA(619)70, 58BJ(68)40, 60JA217⟩. Feeding of [1-$^{14}$C]glucose to *Drosophila* larvae led to incorporation of the label in positions 6 and 7 of the pteridine ring ⟨59HCA2254, 61HCA1480⟩, whereas analogous studies with butterflies pointed to a purine nucleotide-pteridine transformation ⟨55AG328, 61AG402, 63ZN(B)757⟩ with a guanosine derivative as the most likely direct natural precursor. The ring enlargement proceeds in guanosine 5'-triphosphate (**300**) with loss of the C-8 atom of the purine nucleus (**301**) and subsequent intramolecular condensation of the ribose moiety after the Amadori-type rearrangement to a ribulose structure (**302**) ⟨67JBC(242)565⟩, to give 7,8-dihydro-D-neopterin 3'-triphosphate (**303**; equation 113) ⟨66JBC(241)2220⟩.

This interesting conversion of a five- into a six-membered heterocyclic ring was proven by the isolation of the enzyme GTP-cyclohydrolase from *E. coli* ⟨71MI21600⟩ and a similar one from *Lactobacillus platarum* ⟨B-71MI21601⟩ which catalyzes the reaction (**300**) → (**303**). Dephosphorylation leads to 7,8-dihydro-D-neopterin (**304**), which is then cleaved in the side-chain to 6-hydroxymethyl-7,8-dihydropterin (**305**), the direct precursor of 7,8-dihydropteroic acid and 7,8-dihydrofolic acid (**224**). The alcohol (**305**) requires ATP and Mg$^{2+}$ for the condensation with *p*-aminobenzoic and *p*-aminobenzoylglutamic acid, indicating pyrophosphate formation to (**306**) prior to the substitution step.

The various other pterin derivatives found in nature are the result of complex multistep metabolisms of mainly unknown detail. There is an epimerase which converts (**303**) into its L-*threo*-isomer ⟨B-75MI21602⟩, and a chicken kidney preparation catalyzes the conversion of the same substrate to dihydrobiopterin (**223**) ⟨81JBC(256)2963⟩. The proposed metabolic pathway thus produces 6-pyruvyl-7,8-dihydropterin (**307**) and sepiapterin (**237**) (equation 114). The triphosphate (**303**) is also found as precursor in the biosynthesis of the drosopterin eye pigments, showing that 6-acetyl-2-amino-4-oxo-3,4,7,8,9-tetrahydropyrimido[4,5-*b*]-[1,4]diazepine (**308**) functions as an intermediate in the build-up of the complex drosopterin structure (**71**) ⟨81JBC(256)10399⟩. 6,7-Dimethyl-8-D-ribityllumazine (**311**), the biogenetic

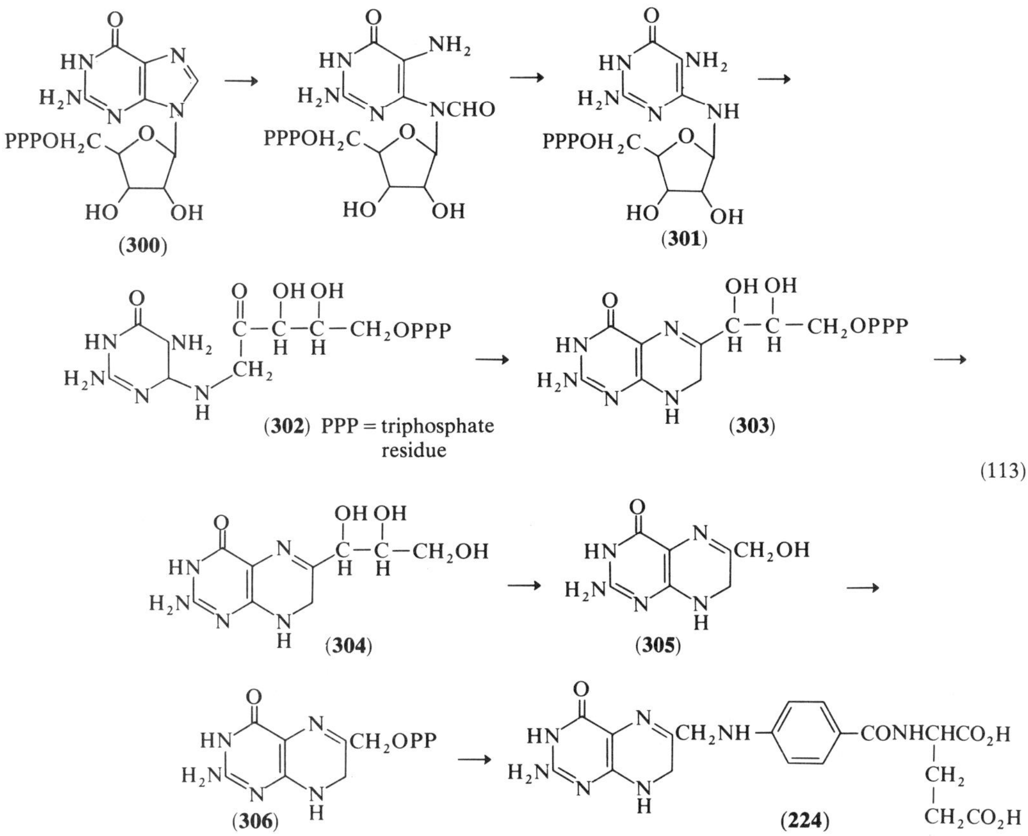

(300) → (301) →

(302) PPP = triphosphate residue   (303) →

(113)

(304)   (305) →

(306)   →   (224)

precursor of riboflavin (312) ⟨61MI21600⟩, is also derived from a guanosine derivative, possibly again from GTP (300) or GMP and passing through 2,5-diamino-6-ribityl-aminopyrimidin-4-one (309) and 5-amino-6-ribitylaminouracil (310) respectively (equation 115) ⟨70JBC(245)4647⟩. A similar pathway is postulated for the biosynthesis of 6-(3-indolyl)- (313) and 6-*p*-hydroxyphenyl-8-D-ribityllumazin-7-ones (314) in *Achromobacter petrophilum* ⟨B-70MI21606⟩ and *Pseudomonas ovalis* ⟨71BCJ(44)1869⟩ through condensation of (310) with the tryptophan- and tyrosine-derived 3-indolyl- and *p*-hydroxyphenyl-glyoxylic acids. A direct introduction of the carbon side-chain is assumed in the biosynthesis of erythropterin (39), since xanthopterin (4) and 7,8-dihydroxanthopterin (150) are directly incorporated in the butterflies *Colias eurytheme* and *C. croceus* ⟨67JBC(242)565⟩ and is also found in eggs of *Oncopeltus fasciatus* in the presence of oxalylacetic acid.

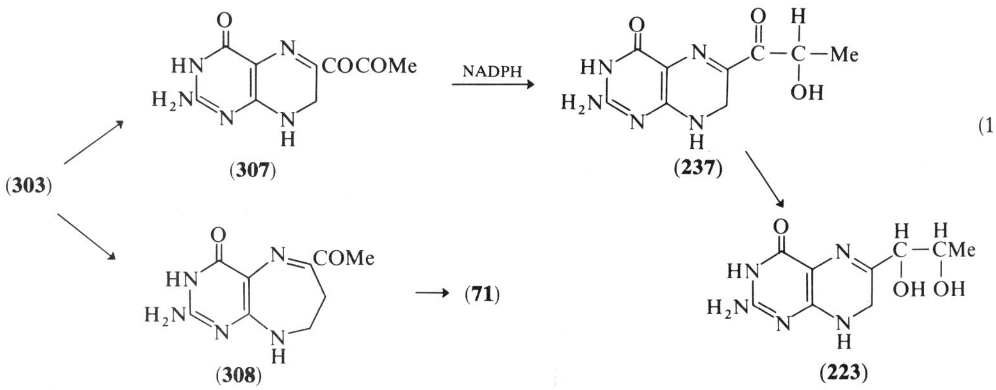

(303)

(307) →(NADPH)→ (237)

(308)  → (71)

(223)

(114)

Studies on the catabolism of pteridines revealed a degradation of tetrahydrobiopterin (221) and tetrahydroneopterin by rat-liver homogenases *in vitro* ⟨69BBA(184)386, 69BBA(184)589, 71BBA(230)117⟩ as well as *in vivo* ⟨71BBA(237)365⟩ to simple lumazine derivatives, including a cleavage of the carbon side-chain in the 6-position, to 7,8-dihydropterin (75), its deamination to 7,8-dihydrolumazine, and further hydroxylations and oxidations to lumazine-6- and -7-one and -6,7-dione respectively ⟨72AG1088⟩. In an analogous sequence

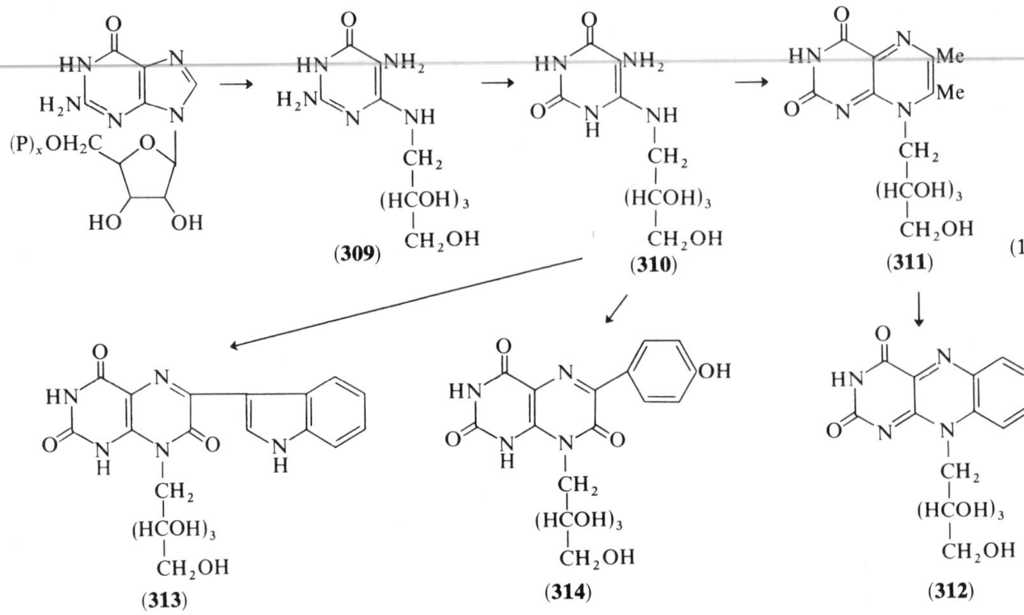

(115)

of reactions, pterin (**2**), xanthopterin (**4**), 7,8-dihydroxanthropterin (**150**), isoxanthopterin (**5**) and leucopterin (**6**) are formed if the pterin-deaminase is missing (equation 116). It has furthermore been shown in a variety of insects ⟨70MI21609⟩ that no further degradation of the pteridin-7-ones and -6,7-diones occurs, marking these derivatives as the end-products of pterin and lumazine metabolism. So far only in the microorganism *Alcaligenes faecalis* ⟨63JBC(238)1116, 64JBC(239)332⟩ has a ring-contraction of lumazine-6,7–dione into xanthine-8-carboxylic acid been observed.

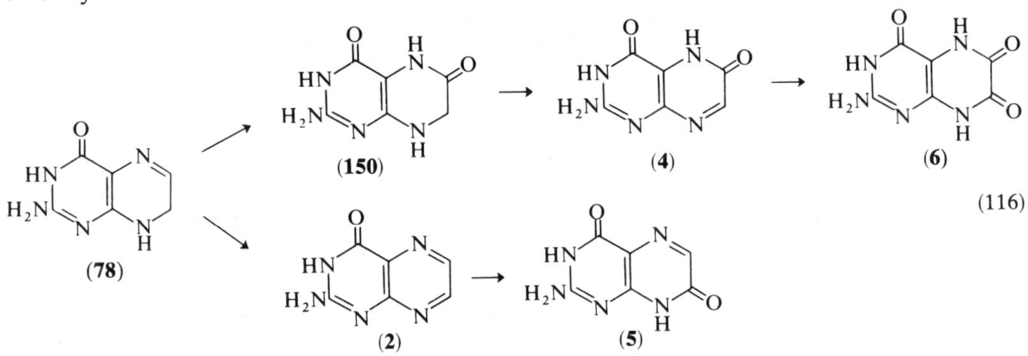

(116)

## 2.16.4 APPLICATIONS

The general interest in the pteridines is due to their widespread occurrence in both the animal and plant kingdoms, implying potential biological activity and drug-type properties in structural analogues.

### 2.16.4.1 Naturally Occurring Pteridines

Pteridines can often be recognized directly in natural materials as pigments in the wings and eyes of insects, and in the skin of fish, amphibia and reptiles ⟨64AG(E)114, 56MI21600, B-65MI21604, B-69MI21603⟩. Moreover, with the exception of the 5,6,7,8-tetrahydro forms, they usually possess a very characteristic fluorescence which allows them readily to be detected and identified by chromatographic means. Despite the fact that more sophisticated analytical methods and advanced chromatographic techniques are now available, investigations of those natural products are still limited by their occurrence in normally very low concentrations in the natural material. It must therefore be considered fortunate that the three best-known butterfly pigments, *viz.* xanthopterin (**4**), isoxanthopterin (**5**) and leucopterin (**6**), occur in relatively large quantities in the wings of these insects. Hence, no better starting material could be found for the initial studies on the isolation and structure elucidation of natural pteridines.

All compounds of this heterocyclic ring system so far found in nature are derivatives of pterin (**2**) and lumazine (**3**) carrying different substituents in the 6- and/or 7-positions. The most common representatives of these series are listed in Tables 8 and 9.

Among the more simple pterin derivatives, euglenapterin ⟨80AG474⟩ reveals the most interesting structure due to the fact that it is the first naturally-occurring pterin which is further modified at the 2-amino substituent to a dimethylamino group. The biological significance of this methylation is, however, not yet known. There is in general very little information available about the physiological and biochemical activity of the natural pteridines so far isolated, suggesting that these mainly oxidized derivatives are either end-products of a metabolic pathway or artefacts from the isolation procedure. It is believed that the *in vivo* activity of pteridines is mainly associated with the dihydro and tetrahydro levels, as revealed by the cofactor activity of tetrahydrobiopterin (**221**) in biological hydroxylations ⟨70MI21603⟩, in mitochondrial electron transfer ⟨72MI21603⟩ and in photosynthesis ⟨69PNA(63)1311⟩. Tetrahydrobiopterin activates phenylalanine hydroxylase for the interconversion of phenylalanine into tyrosine and is itself oxidized during this hydroxylation to the quinonoid-type 6*H*-7,8-dihydrobiopterin (**315**). In the redox shuttle NADPH is also involved in catalysis of the back-reaction to (**221**; equation 117). A cofactor activity of

**Table 8** Naturally Occurring Pterin Derivatives

| *Trivial name* | *Systematic or semisystematic name* | *Structure* |
|---|---|---|
| Leucopterin | 2-Aminopteridine-4,6,7-trione | (**6**) |
| Xanthopterin | 2-Aminopteridine-4,6-dione | (**4**) |
| Dihydroxanthopterin | 2-Amino-7,8-dihydropteridine-4,6-dione | (**150**) |
| Chrysopterin | 2-Amino-7-methylpteridine-4,6-dione | — |
| Ekapterin | 7-Xanthopteryllactic acid | — |
| Lepidopterin | 7-Xanthopteryl-$\alpha$-aminoacrylic acid | — |
| Erythropterin | 7-Xanthopterylpyruvic acid | (**39**) |
| Pterorhodin | Di(7-xanthopteryl)methane | (**62**) |
| Isoxanthopterin | 2-Aminopteridine-4,7-dione | (**5**) |
| Cyprino-purple B | Isoxanthopterin-6-carboxylic acid | — |
| Ichthyopterin | 6-(1′,2′-Dihydroxypropyl)isoxanthopterin | — |
| Cyprino-purple $C_1$ | 6-(1′-Hydroxy-2′-acetoxypropyl)isoxanthopterin | — |
| Cyprino-purple $C_2$ | 6-(1′-Acetoxy-2′-hydroxypropyl)isoxanthopterin | — |
| Asperopterin B | 6-Hydroxymethyl-8-methylisoxanthopterin | (**294**) |
| Asperopterin A | 8-Methyl-6-($\beta$-D-ribosyloxymethyl)isoxanthopterin | — |
| | 6-(1′-Hydroxypropyl)-8-methylisoxanthopterin | |
| Pterin | 2-Aminopteridin-4-one | (**2**) |
| Ranachrome-3 | 6-Hydroxymethylpterin | (**35**) |
| Pterincarboxylic acid | Pterin-6-carboxylic acid | (**142**) |
| Biopterin | 6-(L-*erythro*-1′,2′-Dihydroxypropyl)pterin | (**32**) |
| Biopterin glucoside | Biopterin 1′-$\alpha$-D-glucoside | — |
| Bufochrome | 6-(L-*erythro*-1′,2′,3′-Trihydroxypropyl)pterin | — |
| Neopterin | 6-(D-*erythro*-1′,2′,3′-Trihydroxypropyl)pterin | (**33**) |
| Neopterin phosphate | Neopterin 3′-phosphate | — |
| Neopterin cyclophosphate | Neopterin 2′,3′-cyclic phosphate | — |
| | Neopterin 3′-$\beta$-D-glucuronide | |
| | Neopterin 3′-triphosphate | |
| Monapterin | 6-(L-*threo*-1′,2′,3′-Trihydroxypropyl)pterin | (**34**) |
| Ciliapterin | 6-(L-*threo*-1′,2′-Dihydroxypropyl)pterin | — |
| Sepiapterin | 6-Lactoyl-7,8-dihydropterin | (**237**) |
| Deoxysepiapterin | 6-Propionyl-7,8-dihydropterin | (**46**) |
| | 6-Hydroxymethyl-7,8-dihydropterin | (**305**) |
| | 7,8-Dihydrobiopterin | (**223**) |
| | 7,8-Dihydroneopterin 3′-triphosphate | (**303**) |
| Hynobius-blue | 5,6,7,8-Tetrahydrobiopterin | (**221**) |
| Euglenapterin | 2-Dimethylamino-6-(L-*threo*-1′,2′,3′-trihydroxypropyl)-pteridin-4-one | — |
| Urothione | 7-Amino-2-(1′,2′-dihydroxyethyl)-3-methylthio-5-oxo-5,6-dihydrothieno[3,2-*g*]pteridine | (**70**) |
| Drosopterin, isodrosopterin, neodrosopterin, aurodrosopterin | — | (**71**) |

**Table 9** Naturally Occurring Lumazine Derivatives

| Trivial name | Systematic or semisystematic name | Structure |
|---|---|---|
| Lumazine | Pteridine-2,4-dione | (3) |
| | 6-Hydroxymethyllumazine | |
| Lumazinecarboxylic acid | Lumazine-6-carboxylic acid | — |
| Violapterin | | |
| (isoxantholumazine) | Pteridine-2,4,7-trione | — |
| Ribolumazine | | |
| (compound G) | 6,7-Dimethyl-8-D-ribityllumazine | (301) |
| Compound V | 6-Methyl-8-D-ribitylpteridine-2,4,7-trione | — |
| Photolumazine C | 8-D-Ribitylpteridine-2,4,7-trione | — |
| Putidolumazine | 6-(2-Carboxyethyl)-8-D-ribitylpteridine-2,4,7-trione | — |
| Photolumazine A | 6-(L-1′,2′-Dihydroxyethyl)-8-D-ribitylpteridine-2,4,7-trione | — |
| Photolumazine B | 6-Hydroxymethyl-8-D-ribitylpteridine-2,4,7-trione | — |
| — | 6-(3-Indolyl)-8-D-ribitylpteridine-2,4,7-trione | (313) |
| — | 6-p-Hydroxyphenyl-8-D-ribitylpteridine-2,4,7-trione | (314) |
| Luciopterin | 8-Methylpteridine-2,4,7-trione | — |
| Sepialumazine | 6-Lactoyl-7,8-dihydrolumazine | — |
| Leucettidine | 6-(1′-Hydroxypropyl)-1-methyllumazine | — |
| Surugatoxin | — | (72) |

tetrahydropterins is further found in other monooxygenase reactions, as in the oxidation of higher alkyl ethers of glycerol to the corresponding fatty acids ⟨64JBC(239)4081⟩, the $17\alpha$-hydroxylation of progesterone ⟨64MI21600⟩, the transformation of tyrosine into 3,4-dihydroxyphenylalanine ⟨64JBC(239)2910, 64BBR(17)177⟩, the hydroxylation of cinnamic acid to p-coumarinic acid ⟨65P161⟩ and that of tryptophan in the brain ⟨65MI21601⟩, as well as the conversion of 8,11,14-eicosatrienoic acid into prostaglandins. The natural cofactor promotes growth of *Crithidia fasciculata*, as does biopterin (**32**) ⟨B-64MI21606⟩.

Patients with congenital errors causing phenylketonuria and related variants ⟨79MI21605⟩ show defects in biopterin metabolism ⟨78MI21601, 78MI21603, 79MI21602⟩ due to deficiencies in phenylalanine-4-hydroxylase, dihydrobiopterin synthetase ⟨79MI21604⟩ and dihydropteridine reductase ⟨80MI21603⟩. Since tetrahydrobiopterin shows some interesting chemotherapeutic effects in this respect, there is hope for the more successful treatment of these diseases and related mental disorders in the future.

Excretion of pteridines in man has led to the isolation of biopterin ⟨55JA3167, 56JA5871⟩ and neopterin ⟨67MI21601⟩ from human urine. Quantitative determinations ⟨72JBC(247)4549⟩ of these urinary pterins for diagnostic use have only recently been performed ⟨82MI21600, 82MI21601, 80MI21602⟩, recognizing the neopterin/creatinine ratio as characteristic. On the basis of confidence limits on the normal values for each sex and age group this method can be used to assess neoplasias; pteridine excretion is correlated with cell proliferation ⟨80N610⟩ and functions therefore as an indicator. Rapid or malignant proliferation is expressed in higher neopterin levels ⟨79ZPC(360)1957, 81MI21600, 82MI21601⟩ and in the urine of Ehrlich ascites tumor-bearing mice increased excretion of 7,8-dihydrolumazin-6-one is found ⟨79MI21603⟩.

## 2.16.4.2 Synthetic Pteridines with Chemotherapeutic Effects

The discovery of folic acid as a vitamin and the recognition of its versatile modes of action in biological systems focused early attention on simple synthetic pteridines with antifolic acid activity in bacteria ⟨47JBC(170)747⟩. 6,7-Disubstituted 2,4-diaminopteridines revealed growth-inhibiting activity ⟨49JA892⟩. They possess a broad spectrum of biological activity as dihydrofolate reductase inhibitors ⟨B-69MI21604, 69JMC662, 77JMC1215⟩, antitumor agents ⟨B-66MI21600⟩, antimalarials ⟨67JMC431⟩ and especially diuretics ⟨B-64MI21606, B-70MI21608⟩.

Out of many hundreds of pteridines tested, 2,4,7-triamino-6-phenylpteridine (316), 4,7-diamino-2-phenylpteridine-6-carboxamide, 4-amino-7-(2-methoxyethylamino)-2-phenylpteridinecarbox-*N*-(2-methoxyethyl)amide (317) ⟨69JPS(58)867⟩ and 2,4-diamino-6,7-dimethylpteridine showed the most interesting diuretic activity as a result, at least in part, of antagonism effects of mineralocorticoids on electrolytes. Compound (316) is used under the trade name 'Triamterene' as a potent diuretic drug exhibiting a strong natriuretic activity with simultaneous retention of potassium ⟨65MI21603⟩. Recently it has been found that the main metabolites of (316) (hydroxytriamterene and hydroxytriamterene sulfate ester) ⟨76AF533, 76AF2125⟩ are also biologically active. For 2,4-diamino-6,7-diisopropylpteridine, a vibriostatic activity ⟨72MI21602⟩ has been reported, and 6-adamantyl-2,4-diaminopteridine ⟨72JMC1331⟩ shows inhibition of bacterial growth. Furthermore, a large number of 6,7,7-trisubstituted 7,8-dihydropterins has been synthesized ⟨70BRP1303171⟩ since 6-hydroxymethyl-7,7-dimethyl- (318) and 6,7,7-trimethyl-7,8-dihydropterin have bacteriostatic activity, being particularly effective against *Clostridium perfringens* and *Dermatophilus dermatonomous*.

(316)     (317)     (318)

### 2.16.4.3 Folic Acid and Related Derivatives

The name 'folic acid' was coined in 1944 to designate a substance, present in leaves and mammalian organs, which was able to stimulate the growth of the bacterium *Streptococcus faecalis* R ⟨44JA267⟩. In 1946 a substance of similar biological properties was isolated from liver and was shown by degradation and synthesis to be pteroylglutamic acid (36). At about the same time pteroyltriglutamic acid was found in culture liquors of certain *Corynebacteria* ⟨48JA1⟩ but revealed very little growth-promoting effect on *S. faecalis* R. Pteroylhepta-glutamic acid was obtained from yeast ⟨46JA1392⟩ and showed almost no bacteria-stimulating activity. A substance known as rhizopterin is present in *Rhizopus nigrans* and its structure was elucidated as 10-formylpteroic acid ⟨47JA2753⟩. Finally, the 'leucovorum factor' or 'leucovorin' was isolated and identified as 5-formyl-5,6,7,8-tetrahydrofolic acid (226) ⟨51JA3067⟩. Their structural relationship, the occurrence of folate derivatives in nature, their biological significance in the metabolism of purines, pyrimidines, amino acids and proteins and the effects of folate deficiency on cells and organisms are manifold and have been summarized in detail ⟨B-69MI21604⟩. Metabolic interactions work on the tetrahydrofolate (225) (THF) level; however, many enzymic reactions require not THF itself but formyl, methyl and other derivatives as substrates. Their interconversions provide a broad spectrum of 'active' one-carbon fragments including 5,10-methylene-THF (319), 5,10-methinyl-THF (320), 10-formyl-THF (321), 5-formyl-THF (324), 5-formimino-THF (323), and 5-methyl-THF (322) (equation 118).

Mechanistic aspects of the action of folate-requiring enzymes involve one-carbon unit transfer at the oxidation level of formaldehyde, formate and methyl ⟨78ACR314, 80MI21600⟩ and are exemplified in pyrimidine and purine biosynthesis. A more complex mechanism has to be suggested for the methyl transfer from 5-methyl-THF (322) to homocysteine, since this transmethylation reaction is cobalamine-dependent to form methionine in *E. coli*.

Since folate derivatives occupy key positions in cellular metabolism, analogues of folic acid are of general pharmacological interest; this is not only because of their extensive use in chemotherapy of cancer and of other diseases but because they are also useful tools for the elucidation of the biochemical pathways. The effects of folate analogues on bacteria and yeasts were recognized early and the most potent inhibitors of growth were found to be in 4-amino-4-deoxyfolic acid (aminopterin) (325) ⟨61JOC3351⟩, 4-amino-4-deoxy-10-methylfolic acid (amethopterin, methotrexate) (326) ⟨60MI21600⟩ and reduced derivatives

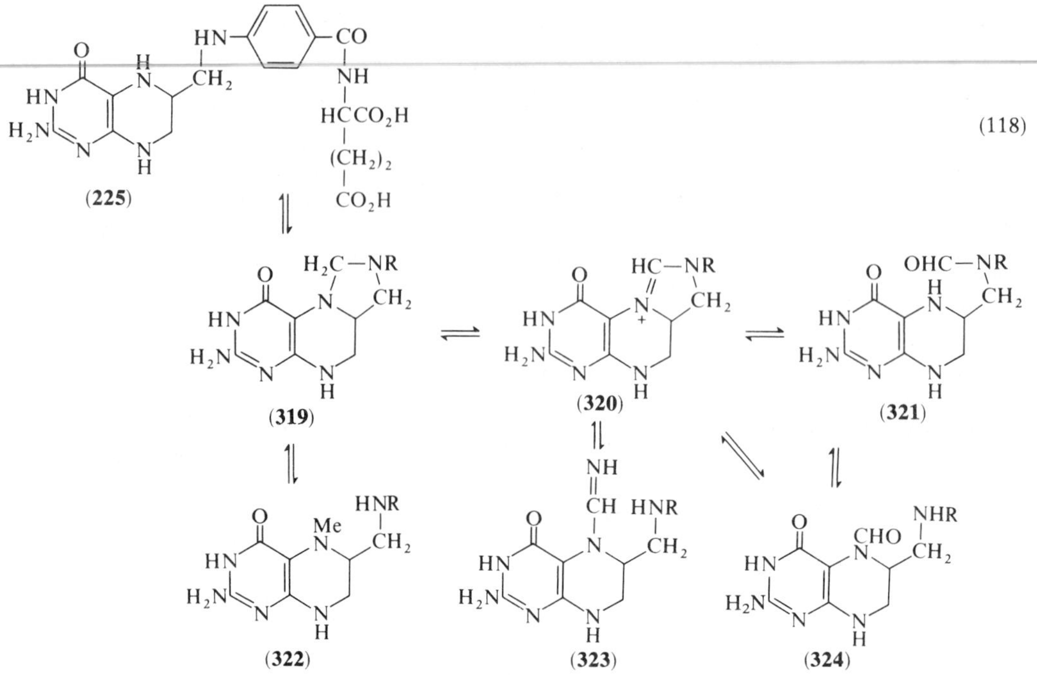

(225)                                                                                    (118)

(319)          (320)          (321)

(322)          (323)          (324)

of these compounds. Compounds (325), (326) and other folate derivatives are extremely toxic to rats, mice, dogs and chickens, whereas 3',5'-dichloroamethopterin is much less toxic than (326) when given in repeated daily doses. Clinical manifestations of intoxication by (325), (326) and 4-amino-4-deoxypteroylaspartate in rats, mice and dogs include progressive weight loss, anorexia, diarrhoea, progressive depression and terminal collapse and coma. Gastrointestinal and haematological changes are particularly marked. One of the first intestinal effects is the inhibition of mitosis in Crypt cells. Later, swelling and cytoplasmic vacuolation of the epithelial lining occurs and is followed by desquamation of epithelium. In general, aminopterin, methotrexate and other folate analogues produce similar effects in man. Gastrointestinal toxicity results in abdominal pains, nausea and vomiting, diarrhoea, and if toxicity is severe, haemorrhagic diarrhoea. Mitotic activity in Crypt cells is decreased and effects on the bone marrow and peripheral circulation are substantial. The marrow and peripheral blood return to normal in a week to a month after discontinuation of drug administration.

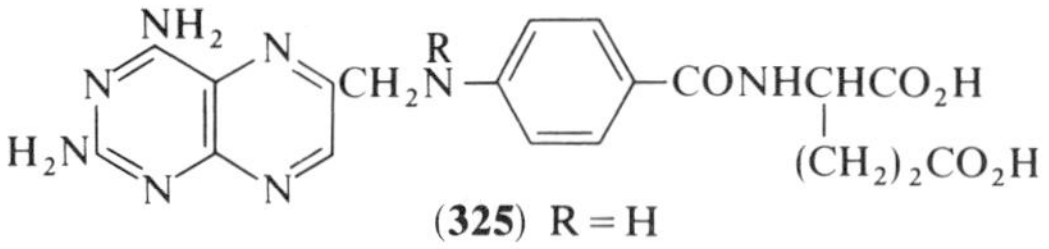

(325) R = H
(326) R = Me

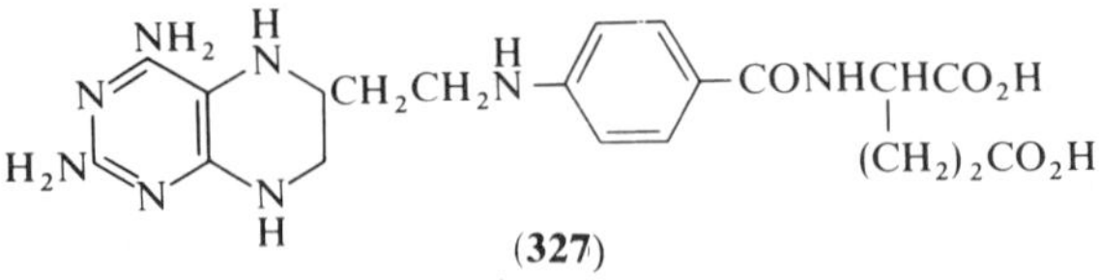

(327)

In view of the well-documented inhibition of dihydrofolate reductase by aminopterin (325), methotrexate (326) and related compounds it is generally accepted that this inhibitory effect constitutes the primary metabolic action of folate analogues and results in a block in the conversion of folate and dihydrofolate (DHF) to THF and its derivatives. As a consequence of this block, tissues become deficient in the THF derivatives, and this deficiency has many consequences similar to those resulting from nutritional folate deficiency. The crucial effect, however, is a depression of thymidylate synthesis with a consequent failure in DNA synthesis and arrest of cell division that has lethal results in rapidly proliferating tissues such as intestinal mucosa and bone marrow (B-69MI21604, B-69MI21605).

Aminopterin and methotrexate have also been tested for their effect on many transplantable tumors in rats, mice, birds and rabbits. L1210, an acute lymphoid leukaemia, has been extensively used in studies of the antitumor action of these analogues. Administration of 5-formyl-THF (**324**) to tumor-bearing animals prior to, or simultaneous with, the administration of a folate analogue, protects them against toxic effects but also reverses the antitumor effects of the analogue ⟨B-69MI21605⟩. The therapeutic effectiveness of methotrexate against L1210 can also be increased by administration of the analogue in combination with various other drugs ⟨71ANY(186)423⟩. Tetrahydrohomofolate (**327**) ⟨64JA308⟩ increased the survival time of mice inoculated with an antifolate-resistant variant of L1210, but decreased the prolongation of survival time caused by (**326**).

In the treatment of human neoplastic diseases methotrexate has largely supplanted aminopterin in chemotherapy, due to the better therapeutic index of the former in experimental animals, although this superiority over (**325**) has not been conclusively demonstrated in man.

The cure of choriocarcinoma has been the most spectacular achievement recorded for chemotherapy with folate analogues ⟨65MI21600⟩. Long-term remissions have also been produced by therapy with folate analogues in patients with acute leukemia ⟨65MI21606⟩. The use of combinations of drugs in cyclic therapy is hopeful for the chemotherapy of acute lymphocytic leukemia of childhood ⟨67MI21600⟩ and of adult acute myelogenous leukemia ⟨65MI21605⟩. For reasons not yet well understood, most solid human tumors are resistant to therapy with folate analogues.

The development of methotrexate (MTX) analogues with potentially superior clinical properties and whose sole mode of action is inhibition of dihydrofolate reductase (DHFR) is generally held to have little promise ⟨B-74MI21601⟩. However, the design of folate analogues which act as substrates for DHFR, producing 'spurious' coenzymes which inhibit other enzymes in the folate cycle, remains an attractive rational strategy ⟨79MI21601⟩ since its initial proposal ⟨61MI21601⟩. The potent inhibition of thymidylate synthetase and the antitumor activity of several reduced folate, methotrexate and quinazoline analogues support this strategy. The chemical syntheses and biological evaluation of prodrug derivatives of methotrexate include various classes of compounds such as diesters ⟨80MI21601⟩, bis(amides) ⟨77JMC925⟩, $\alpha$- and $\gamma$-glutamyl conjugates ⟨78JMC170⟩, monoesters ⟨78JMC380⟩, hydrazides and hydroxamic acids ⟨81JMC559⟩. Among homofolate analogues, aza analogues of folic acid ⟨79JMC874⟩ and 5,11-methinyltetrahydrohomofolate ⟨81JMC1086⟩ have recently been synthesized. More severe structural changes have been realized by the synthesis of isofolic acid (**328**) ⟨74JMC223⟩ and isoaminopterin (**329**) ⟨74JMC1268⟩ as well as 10-thio- (**330**) ⟨75JOC1745⟩ and 10-oxa-folic acid (**331**) ⟨76JMC825⟩ and 11-oxahomoaminopterin ⟨81JMC1068⟩.

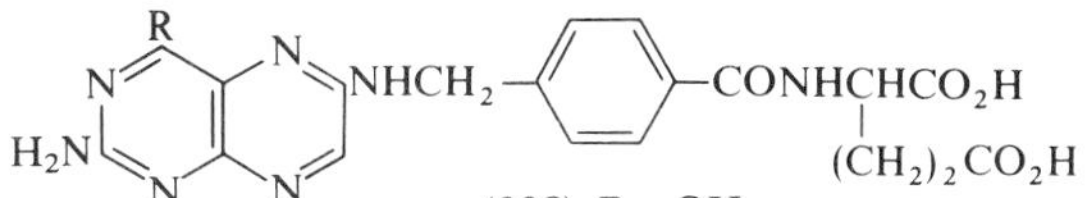

(**328**) R = OH
(**329**) R = NH$_2$

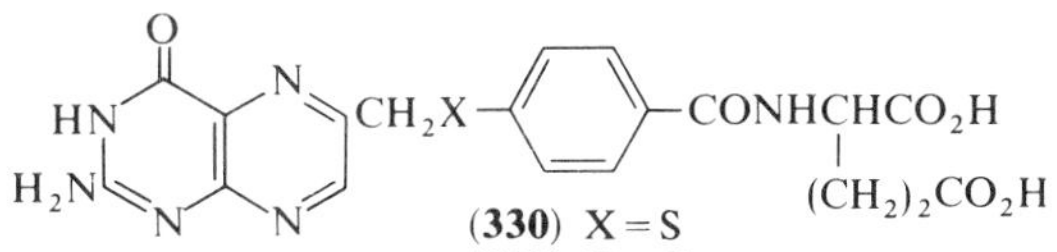

(**330**) X = S
(**331**) X = O

# 2.17

# Other Diazinodiazines

R. N. CASTLE
*University of South Florida*

and

S. D. PHILLIPS
*Olin Research Center, New Haven*

| | | |
|---|---|---|
| 2.17.1 | INTRODUCTION | 330 |
| 2.17.2 | STRUCTURE | 331 |
| *2.17.2.1* | *Pyridazino[1,2-a]pyridazine* | 331 |
| *2.17.2.2* | *Pyridazino[3,4-c]pyridazine* | 332 |
| *2.17.2.3* | *Pyridazino[4,3-c]pyridazine* | 332 |
| *2.17.2.4* | *Pyridazino[4,5-c]pyridazine* | 332 |
| *2.17.2.5* | *Pyridazino[4,5-d]pyridazine* | 332 |
| *2.17.2.6* | *Pyridazino[1,2-b]pyridazine* | 333 |
| *2.17.2.7* | *Pyrimido[4,5-c]pyridazine* | 335 |
| *2.17.2.8* | *Pyrimido[5,4-c]pyridazine* | 336 |
| *2.17.2.9* | *Pyrimido[4,5-d]pyridazine* | 336 |
| *2.17.2.10* | *Pyrazino[1,2-b]pyridazine* | 337 |
| *2.17.2.11* | *Pyrazino[2,3-c]pyridazine* | 337 |
| *2.17.2.12* | *Pyrazino[2,3-d]pyridazine* | 337 |
| *2.17.2.13* | *Pyrimido[1,2-a]pyrimidine* | 337 |
| *2.17.2.14* | *Pyrimido[1,6-a]pyrimidine* | 338 |
| *2.17.2.15* | *Pyrimido[1,6-c]pyrimidine* | 338 |
| *2.17.2.16* | *Pyrimido[4,5-d]pyrimidine* | 338 |
| *2.17.2.17* | *Pyrimido[5,4-d]pyrimidine* | 339 |
| *2.17.2.18* | *Pyrazino[1,2-a]pyrimidine* | 339 |
| *2.17.2.19* | *Pyrazino[1,2-c]pyrimidine* | 340 |
| *2.17.2.20* | *Pyrazino[1,2-a]pyrazine* | 340 |
| *2.17.2.21* | *Pyrazino[2,3-b]pyrazine* | 340 |
| 2.17.3 | REACTIVITY | 341 |
| *2.17.3.1* | *Pyridazino[1,2-a]pyridazine* | 341 |
| *2.17.3.2* | *Pyridazino[3,4-c]pyridazine* | 341 |
| *2.17.3.3* | *Pyridazino[4,3-c]pyridazine* | 342 |
| *2.17.3.4* | *Pyridazino[4,5-c]pyridazine* | 342 |
| *2.17.3.5* | *Pyridazino[4,5-d]pyridazine* | 342 |
| *2.17.3.6* | *Pyrimido[1,2-b]pyridazine* | 343 |
| *2.17.3.7* | *Pyrimido[4,5-c]pyridazine* | 344 |
| *2.17.3.8* | *Pyrimido[5,4-c]pyridazine* | 345 |
| *2.17.3.9* | *Pyrimido[4,5-d]pyridazine* | 345 |
| *2.17.3.10* | *1H-Pyrazino[1,2-b]pyridazine* | 347 |
| *2.17.3.11* | *Pyrazino[2,3-c]pyridazine* | 347 |
| *2.17.3.12* | *Pyrazino[2,3-d]pyridazine* | 347 |
| *2.17.3.13* | *Pyrimido[1,2-a]pyrimidine* | 348 |
| *2.17.3.14* | *Pyrimido[1,6-a]pyrimidine* | 349 |
| *2.17.3.15* | *Pyrimido[1,6-c]pyrimidine* | 349 |
| *2.17.3.16* | *Pyrimido[4,5-d]pyrimidine* | 349 |
| *2.17.3.17* | *Pyrimido[5,4-d]pyrimidine* | 350 |
| *2.17.3.18* | *Pyrazino[1,2-a]pyrimidine* | 350 |
| *2.17.3.19* | *Pyrazino[1,2-c]pyrimidine* | 351 |
| *2.17.3.20* | *Pyrazino[1,2-a]pyrazine* | 351 |
| *2.17.3.21* | *Pyrazino[2,3-b]pyrazine* | 351 |

2.17.4   SYNTHESIS                                                              351
    *2.17.4.1   Pyridazino[1,2-a]pyridazine*                                    351
    *2.17.4.2   Pyridazino[3,4-c]pyridazine*                                    352
    *2.17.4.3   Pyridazino[4,3-c]pyridazine*                                    352
    *2.17.4.4   Pyridazino[4,5-c]pyridazine*                                    353
    *2.17.4.5   Pyridazino[4,5-d]pyridazine*                                    353
    *2.17.4.6   Pyrimido[1,2-b]pyridazine*                                      354
    *2.17.4.7   Pyrimido[4,5-c]pyridazine*                                      356
    *2.17.4.8   Pyrimido[5,4-c]pyridazine*                                      357
    *2.17.4.9   Pyrimido[4,5-d]pyridazine*                                      358
    *2.17.4.10  Pyrazino[1,2-b]pyridazine*                                      359
    *2.17.4.11  Pyrazino[2,3-c]pyridazine*                                      359
    *2.17.4.12  Pyrazino[2,3-d]pyridazine*                                      359
    *2.17.4.13  Pyrimido[1,2-a]pyrimidine*                                      360
    *2.17.4.14  Pyrimido[1,6-a]pyrimidine*                                      362
    *2.17.4.15  Pyrimido[1,6-c]pyrimidine*                                      362
    *2.17.4.16  Pyrimido[4,5-d]pyrimidine*                                      363
    *2.17.4.17  Pyrimido[5,4-d]pyrimidine*                                      364
    *2.17.4.18  Pyrazino[1,2-a]pyrimidine*                                      365
    *2.17.4.19  Pyrazino[1,2-c]pyrimidine*                                      366
    *2.17.4.20  Pyrazino[1,2-a]pyrazine*                                        366
    *2.17.4.21  Pyrazino[2,3-b]pyrazine*                                        367

2.17.5   APPLICATIONS                                                          367
    *2.17.5.1   Pyridazino[1,2-a]pyridazine*                                    367
    *2.17.5.2   Pyridazino[4,5-c]pyridazine*                                    367
    *2.17.5.3   Pyrimido[4,5-c]pyridazine*                                      367
    *2.17.5.4   Pyrimido[4,5-d]pyridazine*                                      368
    *2.17.5.5   Pyrimido[4,5-d]pyrimidine*                                      368
    *2.17.5.6   Pyrimido[5,4-d]pyrimidine*                                      368
    *2.17.5.7   Pyrazino[2,3-b]pyrazine*                                        368

## 2.17.1   INTRODUCTION

Twenty-four diazinodiazine ring systems are possible. Of these, 22 have been reported in the literature. There are six pyridazinopyridazines, namely, pyridazino[1,2-*a*]pyridazine (**1**), pyridazino[2,3-*b*]pyridazine (**2**) (9a*H* shown), pyridazino[3,4-*c*]pyridazine (**3**), pyridazino[4,3-*c*]pyridazine (**4**), pyridazino[4,5-*c*]pyridazine (**5**) and pyridazino[4,5-*d*]-pyridazine (**6**). Examples of all of these ring systems are known except (**2**).

(**1**) [1,2-*a*]          (**2**) [2,3-*b*]          (**3**) [3,4-*c*]

(**4**) [4,3-*c*]          (**5**) [4,5-*c*]          (**6**) [4,5-*d*]

There are five pyrimidopyridazines: pyrimido[1,2-*b*]pyridazine (**7**) (9a*H* shown), pyrimido[3,4-*b*]pyridazine (**8**) (9a*H* shown), pyrimido[4,5-*c*]pyridazine (**9**), pyrimido-[5,4-*c*]pyridazine (**10**) and pyrimido[4,5-*d*]pyridazine (**11**). Of these, all are known except (**8**).

(**7**) [1,2-*b*]     (**8**) [3,4-*b*]     (**9**) [4,5-*c*]     (**10**) [5,4-*c*]     (**11**) [4,5-*d*]

There are three pyrazinopyridazines, namely pyrazino[1,2-*b*]pyridazine (**12**) (4a*H* shown), pyrazino[2,3-*c*]pyridazine (**13**) and pyrazino[2,3-*d*]pyridazine (**14**), and all ring systems are known ⟨73HC(27)1012⟩.

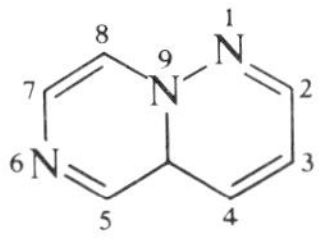

**(12)** [1,2-*b*] 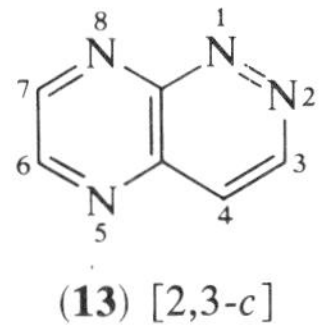         **(13)** [2,3-*c*] 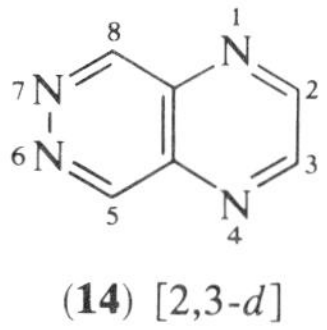         **(14)** [2,3-*d*]

Of the five possible pyrimidopyrimidines, all are known. They are pyrimido-[1,2-*a*]pyrimidine **(15)** (9a*H* shown), pyrimido[1,6-*a*]pyrimidine **(16)** (9a*H* shown), pyrimido[1,6-*c*]pyrimidine **(17)** (9a*H* shown), pyrimido[4,5-*d*]pyrimidine **(18)** and pyrimido[5,4-*d*]pyrimidine **(19)**.

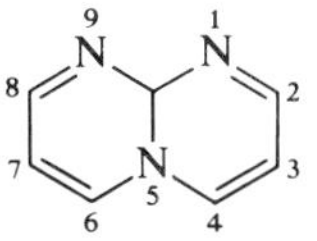

**(15)** [1,2-*a*] 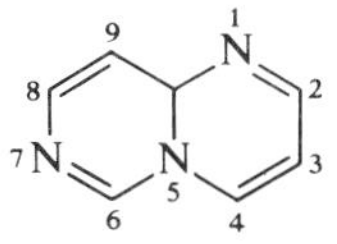    **(16)** [1,6-*a*] 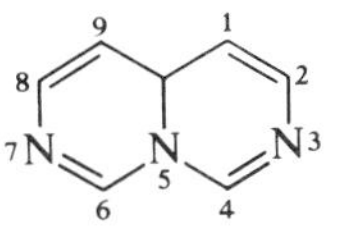    **(17)** [1,6-*c*] 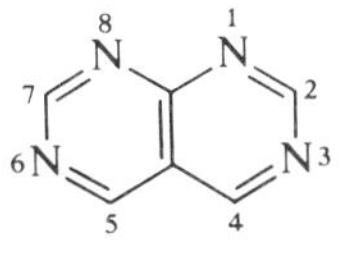    **(18)** [4,5-*d*] 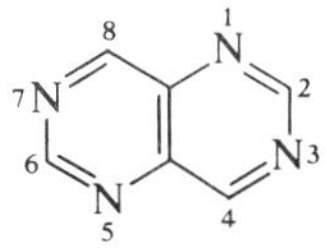    **(19)** [5,4-*d*]

All three of the possible pyrazinopyrimidines are known, namely pyrazino[1,2-*a*]-pyrimidine **(20)** (9a*H* shown), pyrazino[1,2-*c*]pyrimidine **(21)** (9a*H* shown) and pyrazino[2,3-*d*]pyrimidine (pteridine); the last group is treated in Chapter 2.16.

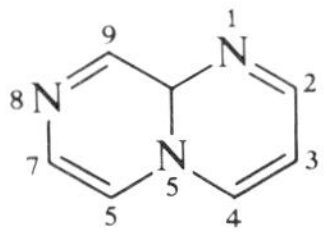

**(20)** [1,2-*a*] 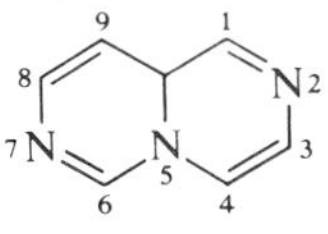         **(21)** [1,2-*c*]

Both pyrazinopyrazines are known and these are pyrazino[1,2-*a*]pyrazine **(22)** (9a*H* shown) and pyrazino[2,3-*b*]pyrazine **(23)** ⟨79HC(35)568⟩.

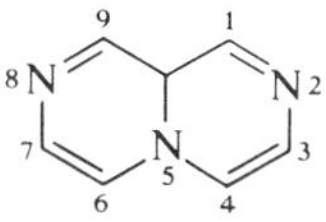

**(22)** [1,2-*a*] 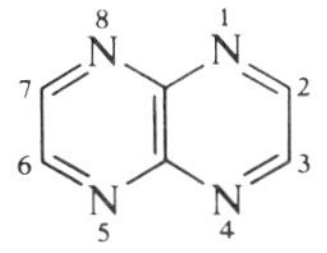         **(23)** [2,3-*b*]

## 2.17.2 STRUCTURE

### 2.17.2.1 Pyridazino[1,2-*a*]pyridazine

The structure of pyridazino[1,2-*a*]pyridazine-1,4,5,8-tetrone has been determined by X-ray crystallography. These studies reveal that although each nitrogen atom in the ring is trigonal $sp^2$ hybridized, the molecule is by no means planar. The bond lengths also suggest that the molecule is not aromatic, although some electron delocalization in the O=C—N—C=O system is observed. The packing structure is layered with the molecules lying in sheets. Interactions between carbon and oxygen are prevalent between the layers ⟨77AX(B)2464⟩.

¹H NMR has been used to study the thermodynamics of the conformational inversion of various pyridazino[1,2-*a*]pyridazines. In the case of 1,2,3,4,6,9-hexahydro-7,8-dimethyl-pyridazino[1,2-*a*]pyridazine-1,4-dione, analysis of the signals from the C-3 and C-6 methyl-ene protons at −60 °C shows that the energy barrier to the conformational inversion must be less than 43.5 kJ mol⁻¹ ⟨66T3477⟩.

A variable temperature ¹³C NMR conformational study of several reduced pyridazino[1,2-*a*]pyridazines has been reported. It is known that the relative energies of the conformations (diequatorial alkyl groups (*ee*) and axial, equatorial alkyl groups (*ae*)) for hexahydropyridazines varies widely and is dependent on substitution (Scheme 1). The *ee* conformation is favored when the alkyl groups are linked in a six-membered ring. For example, by ¹³C NMR techniques, only the *ee* was detected for octahydropyridazino-[1,2-*a*]pyridazine **(24)**. The *ae* conformation is favored by the presence of α-methyl groups and unsaturation. This fact is demonstrated in the ¹³C NMR analysis of the hexahy-dropyridazino[1,2-*a*]pyridazine **(25)** where only the *ae* form was detected. The ¹³C NMR data were used to calculate various thermodynamic parameters for these pyridazino-[1,2-a]pyridazines. Based on these activation parameter conformational changes, it was demonstrated that both nitrogen inversion and ring reversal must play a role in the conformational transformations of these compounds ⟨78JA4004⟩.

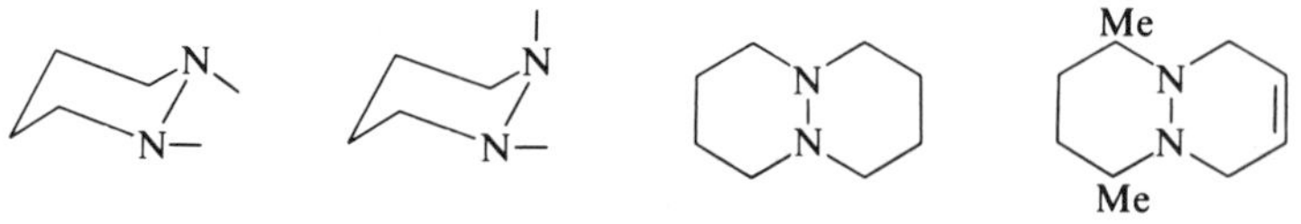

ee              ae              (24)              (25)

**Scheme 1**

The conformational data as reported using $^{13}$C NMR techniques are also supported by photoelectron spectra ⟨74JA6987⟩ and cyclic voltammetry ⟨76JA5269⟩. Using cyclic voltammetry, $K_{eq}$ for the *ee–ae* equilibrium can be measured directly when the activation barrier separating the two is greater than 46 kJ mol$^{-1}$ ⟨78JA4012⟩. ESR spectroscopy has also been used to determine the conformation of octahydropyridazino[1,2-a]pyridazine ⟨74JA2916⟩.

The molecular symmetry of octahydropyridazino[1,2-a]pyridazine (24) has been analyzed using band numbers, frequency coincidences and Raman depolarization factors in the IR and Raman spectra. A centrosymmetric *trans* conformation was confirmed for this compound ⟨76SA(A)157⟩. IR spectral data also indicate that the carbonyl absorption for pyridazino[1,2-a]pyridazine-1,4-diones (26) occurs at *ca.* 1625 cm$^{-1}$ ⟨72TL1885⟩.

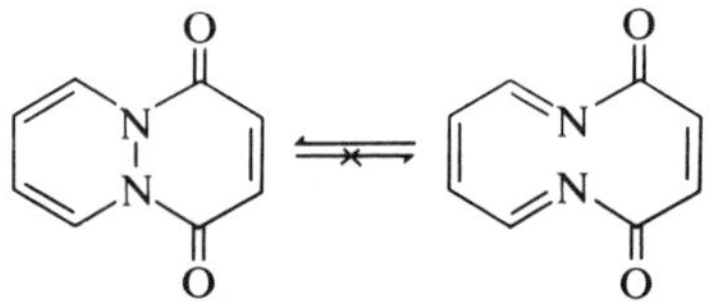

(26)

The UV spectrum for the aromatic compound (26) in methanol gives $\lambda_{max}$ at 214 nm ($\varepsilon$ 13 100), 411 (6500 sh), 437 (7300) and 467 (5700), thus indicating the presence of a $\pi$-conjugated ring system. This spectrum in conjunction with the $^1$H NMR spectrum reveals that this compound exists as an aromatic ten $\pi$-electron system rather than a monocyclic diazaannulenedione ⟨72TL1885⟩.

The mass spectra of various pyridazino[1,2-a]pyridazines have been reported ⟨74JOC47⟩. Hückel molecular orbital calculations have been reported for this ring system ⟨66JSP(19)25⟩.

### 2.17.2.2 Pyridazino[3,4-c]pyridazine

Hückel MO calculations have also been made for this ring system ⟨66JSP(19)25⟩.

### 2.17.2.3 Pyridazino[4,3-c]pyridazine

An X-ray crystal structure determination has been made for the only reported derivative of this ring system ⟨80JHC617⟩. Hückel MO calculations have also been made ⟨66JSP(19)25⟩.

### 2.17.2.4 Pyridazino[4,5-c]pyridazine

The $^1$H NMR spectrum for the parent unsubstituted heterocycle gives absorptions at $\delta$ 8.17 (H-4), 9.78 (H-5), 9.88 (H-3) and 10.20 (H-8), $J_{3,4} = 6$ Hz ⟨73JHC1081⟩. $^1$H NMR, IR and UV spectral data have also been reported for several pyridazino[4,5-c]pyridazine derivatives ⟨67JHC393⟩. Hückel MO calculations have been made for this ring system ⟨66JSP(19)25⟩.

### 2.17.2.5 Pyridazino[4,5-d]pyridazine

The crystal structure of the parent unsubstituted ring system has been determined using X-ray crystallography. These studies reveal that the molecule is planar and that each molecule is in contact with eight more in the crystal. An interaction between the lone pair of electrons on the nitrogen atoms and CH groups of adjacent molecules is suggested. This crystal feature would result in a special attraction between neighboring molecules, and might account for the unusually high melting point of pyridazino[4,5-*d*]pyridazine ⟨69AX(B)2231⟩. The X-ray crystal structure data for 2,6-dimethyl-4,8-dichloro-2*H*,6*H*-pyridazino[4,5-*d*]pyridazine-1,5-dione ⟨72AX(B)1173⟩ and 1,4,5,8-tetramethoxy-pyridazino[4,5-*d*]pyridazine ⟨72AX(B)1178⟩ have also been reported. These data reveal that the former compound exists in a very compact crystal structure, accounting for its extremely high melting point. The latter compound was shown to be roughly planar, with molecular dimensions which very nearly approximate those found for the parent pyridazino-[4,5-*d*]pyridazine.

The $^{1}$H NMR spectrum of 1-hydroxypyridazino[4,5-*d*]pyridazine shows two singlets at δ 8.60 and 9.86 and a broad signal near δ 13.6 (integrated ratio 1:2:1). These results suggest that the oxo form is the predominant tautomer for this compound, since the peaks can be assigned to H-4, H-5 + H-8, and NH absorptions, respectively. Other 1-hydroxy-pyridazino[4,5-*d*]pyridazine derivatives were also shown to exist predominantly in the oxo form by analogous $^{1}$H NMR data. In the case of 2,5,8-trimethyl-2*H*,3*H*-pyridazino-[4,5-*d*]pyridazine-1,4-dione (**27**), two singlets at δ 2.93 and 2.99 are observed for the 5- and 8-methyl groups, respectively. Such a difference in chemical shifts seems to indicate that tautomer (**28**) is predominant, since compound (**29**) exhibits a singlet at δ 2.99 for the same groups. Analogous chemical shift data for other 5,8-dimethylpyridazino[4,5-*d*]-pyridazine-1,4-diones have been used to determine predominant tautomeric structures ⟨72G169⟩.

(27)            (28)            (29)

Another interesting feature in the $^{1}$H NMR spectra of various pyridazino[4,5-*d*]-pyridazines is long-range coupling between protons on the ring. Coupling is reported to occur between H-4 and H-8, and also between H-5 and H-8. In these cases, doublets are observed with coupling constants of $J_{4,8} < 1$ ⟨72G169⟩ and $J_{5,8} = 1.5-2.0$ Hz ⟨79M365⟩.

The IR spectra of several hydroxypyridazino[4,5-*d*]pyridazines have also been used to determine predominant tautomeric structures. These data confirm the $^{1}$H NMR data which have already been discussed ⟨72G169⟩. The IR spectrum of the parent unsubstituted heterocycle reveals a high molecular symmetry, showing a limited number of sharp absorptions at 3065, 3020, 1530, 1445, 1308, 1290, 1250, 1170, 953, 892, 680, 668 and 478 cm$^{-1}$ ⟨67CC1006⟩.

UV spectroscopy has been used to study tautomerism in hydroxypyridazino[4,5-*d*]-pyridazines ⟨68JCS(C)2857⟩. The spectrum of 1-hydroxypyridazino[4,5-*d*]pyridazine ($\lambda_{max}$ 243, 251 and 304 nm) and of 5-methyl-1-hydroxypyridazino[4,5-*d*]pyridazine ($\lambda_{max}$ 253, 261, 295sh and 317 nm) are very similar. However, these spectra are sufficiently different from that of 1-methoxy-5,8-dimethylpyridazino[4,5-*d*]pyridazine to support the oxo structure as the predominant tautomeric form for these two compounds ⟨72G169⟩.

Hückel MO calculations have also been reported for this ring system ⟨66JSP(19)25⟩.

### 2.17.2.6 Pyrimido[1,2-*b*]pyridazine

An extensive $^{1}$H NMR study of pyrimido[1,2-*b*]pyridazin-2-ones and pyrimido[1,2-*b*]-pyridaziniums has appeared in the literature. Data for some of these compounds are compiled in Table 1. From these data it can be seen that a long range coupling constant ($J_{4,9}$) is apparent. In addition, only the 4-methyl derivatives of these compounds display a small coupling constant between the 4-methyl group and H-3. No such interactions with H-3 are observed with 2-methyl derivatives. $^{1}$H NMR data have further demonstrated that methanol and 3-ethoxycarbonylpyrimido[1,2-*b*]pyridazin-4-one (**30**) exist in equilibrium with the adduct (**31**) in deuteriochloroform solution ⟨71JOC2457⟩.

**Table 1**    ¹H NMR Data for Pyrimido[1,2-b]pyridazines[a]

| | H-2 | H-3 | H-4 | H-7 | H-8 | H-9 | $J_{2,3}$ | $J_{3,4}$ | $J_{2,4}$ | $J_{7,8}$ | $J_{8,9}$ | $J_{7,9}$ | $J_{4,9}$ | 4-Me | $J_{3,4\text{-Me}}$ |
|---|---|---|---|---|---|---|---|---|---|---|---|---|---|---|---|
| 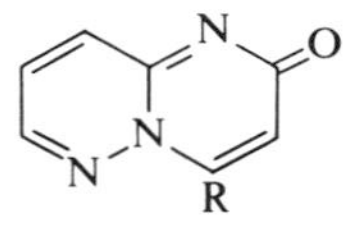 R = H | — | 3.36 (d) | 1.78 (d) | 1.32 (dd) | 2.48 (dd) | 2.10 (dd) | — | 6.3 | — | 4.9 | 8.9 | 2.1 | 0.1 | — | — |
| R = Me | — | 3.43 (q) | — | 1.38 (dd) | 2.46 (dd) | 2.12 (dd) | — | — | — | 3.9 | 9.0 | 2.0 | — | 7.51 (d) | 0.8 |
| 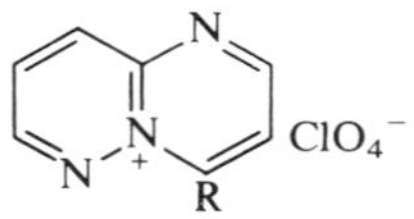 R = H | 0.21 (dd) | 1.52 (dd) | 0.0 (ddd) | 0.34 (dd) | 1.48 (dd) | 0.97 (ddd) | 4.5 | 6.7 | 1.8 | 4.5 | 9.4 | 1.8 | 0.9 | — | — |
| R = Me | 0.45 (d) | 1.54 (dd) | — | 0.42 (dd) | 1.66 (dd) | 1.11 (dd) | 4.2 | — | — | 4.3 | 9.0 | 1.8 | — | 6.96 | 0.8 |

[a] Chemical shifts in p.p.m.; coupling constants in Hz.

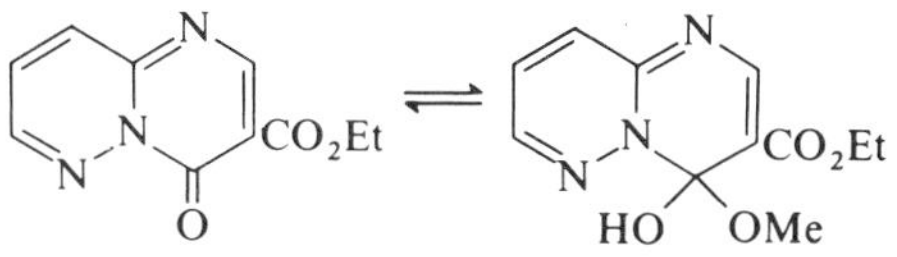

(30)　　　　　　　　(31)

In the IR spectra, the pyrimido[1,2-*b*]pyridazin-2-one carbonyl absorption occurs at a longer wavelength than the isomeric 4-one carbonyl absorption. Thus, the compound (32) exhibits a C=O absorption at 1643 cm⁻¹ whereas the isomeric (33) exhibits a C=O absorption at 1708 cm⁻¹ ⟨71JOC3506⟩.

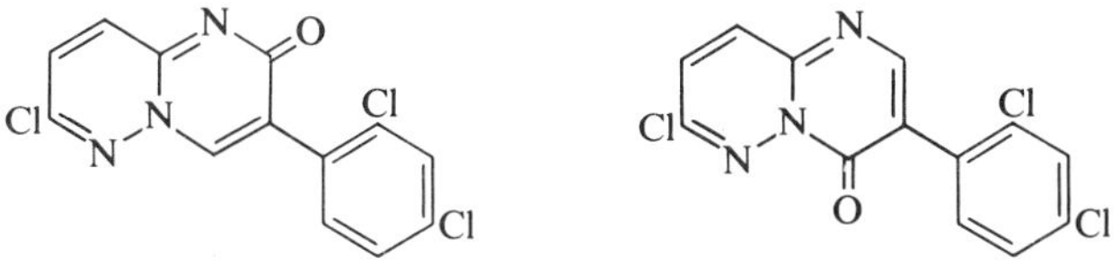

(32)　　　　　　　　(33)

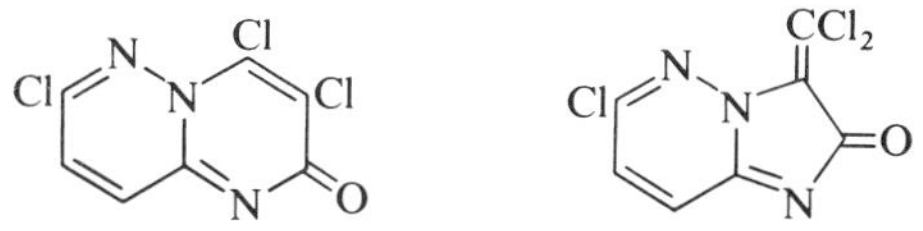

(34)　　　　　　　　(35)

Mass spectrometry was used to differentiate between the two products (34) and (35) obtained from the same reaction run at different temperatures. The prominent ion *m/e* 94 (C₂Cl₂) from (35) was not observed in the mass spectrum of (34). Possibly due to its high degree of symmetry, dichloroacetylene (ClC≡CCl) may leave the ionization chamber without apparent ionization ⟨71JOC3506⟩.

MO calculations for this ring system have also been reported ⟨71JOC2457⟩.

### 2.17.2.7 Pyrimido[4,5-*c*]pyridazine

¹H NMR data have been used to determine the structure of certain unusual products obtained in the reaction of chlorohydroxypyrimido[4,5-*c*]pyridazines with phosphorus oxychloride and *N,N*-dimethylaniline ⟨71CPB1849⟩. ¹H NMR has also been used in the structure determination of several 7-aminopyrimido[4,5-*c*]pyridazin-5(1*H*)-ones. Structures for the phenyl compounds (36) and (37) can be assigned based on deshielding of the *ortho* protons on the phenyl substituent in the 3-phenyl isomer only. This deshielding effect can be explained by diamagnetic anisotropy, which is allowed by the more coplanar geometry of the 3-substituted isomer. Nonplanar geometry is required for the 4-substituted isomer because of interaction with the 5-oxo substituent. Further structural information can be obtained from the relative chemical shifts of the vinyl protons of the pyridazine moieties of (36) and (37). The vinyl signal for the 3-phenyl isomer appears significantly downfield from that of the 4-phenyl isomer. Similar observations can be made for other 3- and 4-aryl substituted pyrimido[4,5-*c*]pyridazines ⟨82JOC674⟩.

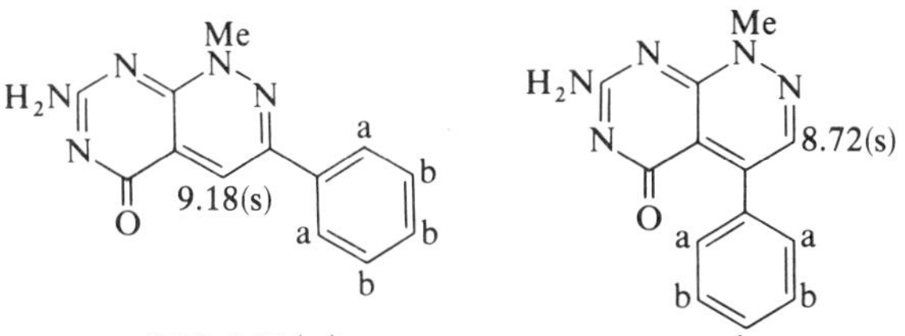

a = 8.00–8.22 (m)
b = 7.58–7.77 (m)
(36)

a = b = 7.60 (s)
(37)

¹³C NMR spectroscopy has also been used in structural assignments of 7-aminopyrimido[4,5-*c*]pyridazin-5(1*H*)-ones. Generally, C-3, adjacent to the ring nitrogen

atom, is shifted to lower field than C-4. These absorptions appear as doublets in the proton-coupled spectra of the 3- and 4-phenyl substituted isomers ⟨82JOC674⟩.

IR spectral data have been used to determine the structure of 1,2-dihydro derivatives of this ring system ⟨78JHC781⟩. UV analysis of compounds of the type (36) and (37) reveal that the 3-phenyl isomer absorbs at a longer wavelength than the 4-phenyl isomer. This phenomenon may be explained by the extended conjugation through the attached benzene ring in the 3-phenyl substituted isomer. This conjugation effect would be minimized in the 4-phenyl isomer due to interaction with the 5-oxo substituent ⟨82JOC674⟩.

The $pK_a$ values for certain pyrimido[4,5-c]pyridazines have been determined ⟨58LA(615)48⟩. Mass spectral fragmentations for several derivatives of this ring system have also been reported ⟨82JOC674⟩, as well as Hückel MO calculations ⟨66JSP(19)25⟩.

### 2.17.2.8 Pyrimido[5,4-c]pyridazine

The $^1$H NMR spectra for several 6-amino-2-phenylpyrimido[5,4-c]pyridazin-8-ones have been reported. A common feature in these compounds is the appearance of a pair of doublets at $\delta$ 9.2 (H-3) and 7.4 (H-4) ($J_{3,4} = 7.1$–7.6 Hz). These data are contrasted with data obtained for several corresponding pyrrolo[3,2-d]pyrimidines ⟨78JOC2536⟩.

The IR and UV spectral data for several derivatives of this ring system have been reported ⟨68JHC523⟩. One interesting physical property which is common to several pyrimido-[5,4-c]pyridazines is a strong fluorescence. This property facilitates the visual detection and identification of these intermediates, and is particularly useful in silica gel plate chromatography ⟨78JOC2536⟩. Hückel MO calculations have also been reported for this ring system ⟨66JSP(19)25⟩.

### 2.17.2.9 Pyrimido[4,5-d]pyridazine

$^1$H NMR spectroscopy has been used to distinguish between the two isomeric products obtained as a mixture in the reaction of 5,8-dichloro-2-phenylpyrimido[4,5-d]pyridazine with aliphatic amines. In these reactions the isomeric products are separated by column chromatography and then the chemical shift of the H-4 proton is noted. This proton is shifted to lower field (*ca.* $\delta$ 10.2) for the 5-amino substituted products (39), when compared with the chemical shift (*ca.* $\delta$ 9.9) for the 8-amino substituted products (38) ⟨72CPB1522⟩. Chemical shifts for several pyrimido[4,5-d]pyridazin-5-ones have also been reported (Table 2) ⟨76BSF1549⟩.

(38)          (39)

**Table 2** $^1$H NMR Data for Pyrimido[4,5-d]pyridazin-5-ones[a]

| Substitution | 2-Me | 8-Me | 2-NH$_2$ | 6-Ph | 2-H | 4-H | 8-H | NH | Solvent |
|---|---|---|---|---|---|---|---|---|---|
| R$^1$ = Me, R$^2$ = R$^3$ = H | 2.36 | — | — | — | — | 9.55 | 8.38 | 13.07 | DMSO |
| R$^1$ = R$^2$ = H, R$^3$ = Me | — | 2.66 | — | — | 9.83 | 9.70 | — | 11.00 | CDCl$_3$ |
| R$^1$ = R$^3$ = Me, R$^2$ = H | 2.96 | 2.65 | — | — | — | 9.66 | — | 11.10 | CDCl$_3$ |
| R$^1$ = NH$_2$, R$^2$ = H, R$^3$ = Me | — | 2.40 | 7.80 | — | — | 9.30 | — | 12.60 | DMSO |
| R$^1$ = R$^3$ = H, R$^2$ = Ph | — | — | — | 7.52 (m) | 9.82 | 9.62 | 8.50 | — | CDCl$_3$ |
| R$^1$ = R$^3$ = Me, R$^2$ = Ph | 2.98 | 2.72 | — | 7.61 (m) | — | 9.81 | — | — | CDCl$_3$ |
| R$^1$ = NH$_2$, R$^2$ = Ph, R$^3$ = Me | — | 2.56 | 7.45 | 7.60 (m) | — | 9.50 | — | — | CDCl$_3$ |

[a] Chemical shifts in p.p.m.

IR spectral data for several pyrimido[4,5-*d*]pyridazines have been reported ⟨68JHC845⟩. A comparison of UV data for several 8-amino substituted 5-chloro-2-phenylpyrimido-[4,5-*d*]pyridazines and 5-amino substituted 8-chloro-2-phenylpyrimido[4,5-*d*]pyridazines has been made. The p$K_a$ values for these compounds have also been reported (Table 3) ⟨76BSF1549⟩. Hückel MO calculations have been made ⟨66JSP(19)25⟩.

**Table 3** UV Data for Isomeric Pyrimido[4,5-*d*]pyrimidines

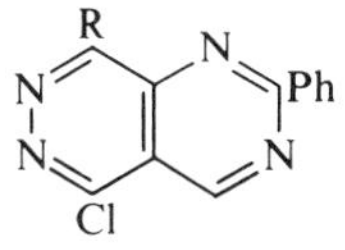

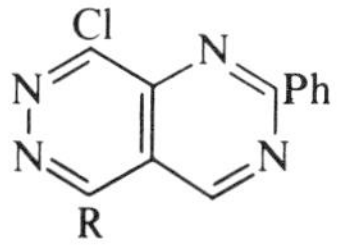

| R | $\lambda_{max}(EtOH)$ (nm) ($\varepsilon \times 10^4$) | p$K_a$ | $\lambda_{max}(EtOH)$ (nm) ($\varepsilon \times 10^4$) | p$K_a$ |
|---|---|---|---|---|
| Pr$^i$NH | 288 (2.0), 363 (0.7) | 2.6 | 278 (2.46), 378 (0.54) | 2.9 |
| BuNH | 285 (2.02), 365 (0.7) | 2.8 | 279 (2.63), 378 (0.52) | 2.9 |
| Bu$^t$NH | 287 (2.05), 362 (0.75) | 3.4 | 288 (2.5), 377 (0.52) | 3.6 |
| PhCH$_2$NH | 283 (2.06), 365 (0.71) | 3.4 | 279 (2.79), 372 (0.54) | 3.5 |
| HOC$_2$H$_4$NH | 284 (2.10), 360 (0.65) | 2.8 | 278 (2.62), 370 (0.52) | 2.8 |

### 2.17.2.10 Pyrazino[1,2-*b*]pyridazine

$^1$H NMR and mass spectra were used to identify the structure and conformation of a pyrazino[1,2-*b*]pyridazine isolated as a degradation product of monomycin D$_1$ and H$_1$. This is the only compound reported in the literature which is a derivative of this ring system ⟨71JCS(C)526⟩.

### 2.17.2.11 Pyrazino[2,3-*c*]pyridazine

Only very limited physical data have been reported for derivatives of this ring system ⟨81AJC1361⟩. Hückel MO calculations have been reported ⟨66JSP(19)25⟩.

### 2.17.2.12 Pyrazino[2,3-*d*]pyridazine

The $^1$H NMR spectrum of 2-methylpyrazino[2,3-*d*]pyridazine reveals a singlet at δ 9.76 for both the H-5 and H-8 protons. The H-2 proton absorbs at δ 9.17. The $^1$H NMR spectrum for 5,8-dichloropyrazino[2,3-*d*]pyridazine exhibits a singlet at δ 9.42 for both the H-2 and H-3 protons. Spectral data for several other 5,8-disubstituted derivatives of this ring system all give a singlet for the two H-2 and H-3 protons at δ 9.1–9.4 ⟨66JHC512⟩.

ESR spectral data for the parent unsubstituted heterocycle have been recorded. These are in good agreement with values obtained by MO calculations ⟨71JA5850⟩.

The IR spectrum of the parent pyrazino[2,3-*d*]pyridazine shows absorptions at 3055 (m), 2965 (w), 1570 (w), 1445 (w), 1425 (s), 1325 (m), 1300 (w), 1290 (m), 1175 (m), 1100 (w), 1022 (s), 968 (s), 935 (m), 854 (w), 793 (w), 637 (w), 630 (w), 548 (w) and 532 (w) cm$^{-1}$ ⟨66JHC512⟩. IR data for 5-bromopyrazino[2,3-*d*]pyridazin-8-one reveal a strong carbonyl absorption at 1710 cm$^{-1}$, indicating the predominance of the oxo tautomer ⟨68JHC53⟩.

The UV spectrum for the parent pyrazino[2,3-*d*]pyridazine shows absorptions at $\lambda_{max}$ 219 ($\varepsilon$ 13 300), 284 (1330) and 295 (980). Hückel MO calculations have also been reported ⟨66JSP(19)25⟩.

### 2.17.2.13 Pyrimido[1,2-*a*]pyrimidine

The $^1$H NMR spectrum of the hydrobromide salt of 2*H*-pyrimido[1,2-*a*]pyrimidin-2-one reveals absorptions at δ 6.85 (d, H-3), 8.10–7.84 (m, H-7), 8.40 (d, H-4) and 9.76–9.40

(m, H-6 and H-8). A characteristic feature of this spectrum is the AB pattern shown by the protons at positions 3 and 4. The resonance for the proton at position 4 also exhibits secondary splitting due to coupling with the *peri* hydrogens. The $^1$H NMR spectrum for the hydrochloride salt of the 3,4-dihydro derivative of 2$H$-pyrimido[1,2-$a$]pyrimidin-2-one gives absorptions at $\delta$ 3.44 (t, CH$_2$), 5.10 (t, CH$_2$), 7.76 (q, aromatic), 8.96 (q, aromatic) and 9.20 (q, aromatic) (71JOC604).

IR data for pyrimido[1,2-$a$]pyrimidin-4-ones (73CJC2650) and tetrahydropyrimido-[1,2-$a$]pyrimidin-4-ones have been reported (73LA103). UV data for pyrimido[1,2-$a$]-pyrimidin-2-ones have also been reported (71JOC604).

The p$K_a$ value for 2,3,6,7,8,9-hexahydro-4$H$-pyrimido[1,2-$a$]pyrimidine has been determined to be 11.22 using potentiometric titration methods. This value is nearly identical for the p$K_a$ values reported for the corresponding imidazo[1,2-$a$]pyrimidine and imidazo-[1,2-$a$][1,3]diazepine heterocycles (62CJC1160).

### 2.17.2.14 Pyrimido[1,6-$a$]pyrimidine

IR and UV spectral data have been reported for 6$H$-1,2,3,4-tetrahydropyrimido-[1,6-$a$]pyrimidine-4,6-dione (64JOC1762). Hückel MO calculations have also been made for this ring system (66JSP(19)25).

### 2.17.2.15 Pyrimido[1,6-$c$]pyrimidine

No extensive physical structural data for compounds containing this ring system have been reported.

### 2.17.2.16 Pyrimido[4,5-$d$]pyrimidine

The $^1$H NMR spectral data for several 3-aryl substituted 1,2,3,4-tetrahydropyrimido-[4,5-$d$]pyrimidines have been reported. In these compounds the N-1 proton absorbs at $\delta$ 6.2–6.5, whereas the N-3 proton absorbs at $\delta$ 2.0–2.2. The two methylene protons in the ring at the 4-position absorb at $\delta$ 3.8–3.9 and the proton on C-2 appears at $\delta$ 5.4–5.8 (69JOC2760). $^1$H NMR data for substituents on aromatic pyrimido[4,5-$d$]pyrimidines have also been reported (74CPB1765).

IR spectral data for 5-aminopyrimido[4,5-$d$]pyrimidine-2,4-diones indicate a broad NH absorption at 3360 cm$^{-1}$ for the amino group and absorptions at *ca.* 1700 and 1650 cm$^{-1}$

**Table 4**   UV Data for Pyrimido[4,5-$d$]pyrimidines

| R$^1$ | R$^2$ | R$^3$ | R$^4$ | *Solvent* | $\lambda_{max}$(nm)(log $\varepsilon$) | *Ref.* |
|---|---|---|---|---|---|---|
| NH$_2$ | H | NH$_2$ | H | pH 7 | 244 (4.47), 314 (3.81) | 60JA5711 |
| H | H | NH$_2$ | H | 0.1N HCl | 232 (3.86), 302 (3.70) | 60JA5711 |
| Ph | H | NH$_2$ | H | 0.1N HCl | 263 (4.37), 324 (4.28) | 60JA5711 |
| NH$_2$ | NH$_2$ | Ph | H | 1N HCl | 251 (4.30), 276 (4.11) | 68JMC568 |
| NH$_2$ | NH$_2$ | H | Ph | pH 7 | 261 (4.53), 333 (3.98) | 60JA5711 |
| NH$_2$ | NH$_2$ | Me | NH$_2$ | 1N HCl | 226 (4.60), 234 (sh), (4.57), 300 (4.07) | 68JMC568 |
| NH$_2$ | NH$_2$ | H | NH$_2$NH | pH 7 | 236 (4.60), 323 (4.14) | 60JA5711 |
| Ph | NH$_2$ | H | EtS | pH 7 | 271 (4.62), 334 (4.15) | 60JA5711 |
| H | OH | H | Me$_2$N | pH 7 | 262 (4.52), 280 (4.17), 336 (3.50) | 60JA5711 |
| H | OH | H | H | pH 7 | 286 (3.75), 315 (3.54) | 58JOC1451 |
| Me | OH | Me | NH$_2$ | pH 7 | 246 (4.67), 306 (3.74) | 60JA5711 |
| Me | OH | H | MeS | 0.1N NaOH | 260 (4.39), 318 (3.00) | 60JA5711 |
| Me | SH | H | MeS | 0.1N NaOH | 241 (4.25), 269 (4.21) | 60JA5711 |
| H | SH | H | H | pH 7 | 254 (4.05), 382 (3.97) | 58JOC1451 |
| H | MeNH | H | EtS | pH 7 | 269 (4.45), 337 (4.10) | 60JA5711 |

for the two carbonyl groups in the ring. If these compounds are unsubstituted at N-3, an additional NH absorption is observed at 3280 cm$^{-1}$ ⟨79H(12)503⟩.

A great amount of UV spectral data for a wide variety of pyrimido[4,5-*d*]pyrimidines has been reported. Some of these data for selected derivatives of this ring system are shown in Table 4.

The p$K_a$ values of pyrimido[4,5-*d*]pyrimidin-4-one and pyrimido[4,5-*d*]pyrimidine-4-thione have been determined to be 7.20 and 6.23, respectively. When compared with the p$K_a$ values of the corresponding purine and pteridine analogs, it can be seen that the pyrimido[4,5-*d*]pyrimidines are more acidic. The greater acidity of these compounds can probably be attributed to their lack of ability to form strong internal hydrogen bonds ⟨58JOC1451⟩.

The mass spectrum of compound (**40**) exhibits a molecular ion peak at *m/e* 412 and other prominent fragment ions at *m/e* 370, 369, 264 (base peak) and 147. The first two of these fragment ions correspond to $M^+ - COCH_2$ and $M^+ - COCH_3$, respectively. The proposed structures for the last two fragment ions are shown in Scheme 2 ⟨74CPB305⟩.

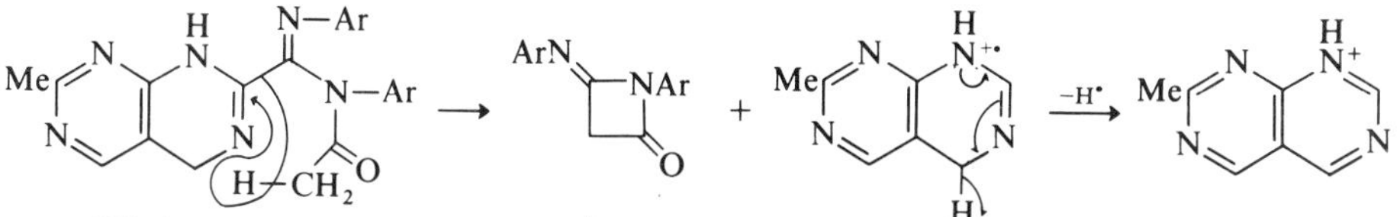

**Scheme 2**

Hückel MO calculations have also been reported for this ring system ⟨66JSP(19)25⟩.

## 2.17.2.17 Pyrimido[5,4-*d*]pyrimidine

The X-ray crystal structure for the disodium salt of pyrimido[5,4-*d*]pyrimidine-2,4,6,8-tetrone tetrahydrate has been determined. This molecule was shown to be planar, but the bond distances reported indicate that it is only partially aromatic. The bond distance C(4)—C(4a) (1.47 Å) is significantly greater than the normal bond distance of aromatic compounds, and is approximately equal to that of a $C(sp^2)$—$C(sp^2)$ single bond. The C—C and C—O bond distances for this compound bear a close resemblance to similar bond distances obtained for barbituric acid derivatives. In the molecular packing structure, each sodium atom is surrounded by six oxygen atoms in the shape of an octahedron. Three of these oxygen atoms originate from water, and three originate from the title heterocycle. The water molecules are arranged in such a manner as to form donor–acceptor hydrogen bonds with the oxygen and nitrogen atoms in the heterocyclic ring system ⟨66JCS(A)639⟩.

$^1$H NMR and IR data have been reported for 3,6-disubstituted 4-aminopyrimido-[5,4-*d*]pyrimidine-2,8-diones. The $^1$H NMR spectra of these compounds exhibit the presence of three NH protons which are exchangeable with deuterium oxide. One of these arises from the external nitrogen (*ca.* δ 11.3), and two result from the NH groups in the ring (*ca.* δ 8.0 and 7.8) ⟨78JOC3231⟩.

The pyrimido[5,4-*d*]pyrimidinetrione (**41**) was shown to exist in the trione form by UV spectral studies. The p$K_a$ values for this compound and several of its methylated derivatives have also been reported ⟨62LA(651)112⟩.

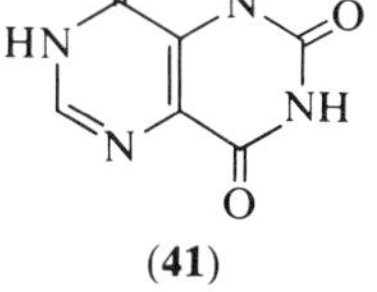

(**41**)

## 2.17.2.18 Pyrazino[1,2-*a*]pyrimidine

The $^1$H NMR spectrum of 2-hydroxy-3-phenyl-4*H*-pyrazino[1,2-*a*]pyrimidin-4-one has been reported. The IR spectrum of this compound exhibits a very broad absorption at

2300–3200 cm$^{-1}$, indicating an intramolecular bonded OH or NH$^+$. The carbonyl region of the spectrum has five peaks at 1760, 1735, 1680, 1660 and 1600 cm$^{-1}$. The IR spectral properties of this compound are similar to those of malonyl-$\alpha$-aminopyridine, both compounds exhibiting five peaks in the carbonyl region. The UV spectrum of the conjugate acid of 2-hydroxy-3-phenyl-4$H$-pyrazino[1,2-$a$]pyrimidin-4-one (obtained in 5N sulfuric acid) exhibits $\lambda_{max}$ at 231 ($\varepsilon$ 23 700), 267 (8780), 346 (4110) and 379 nm (4090). Again, a similarity with the UV spectrum of malonyl-$\alpha$-aminopyridine is noted ⟨68JMC1045⟩.

### 2.17.2.19 Pyrazino[1,2-$c$]pyrimidine

IR and $^1$H NMR data have been reported for 7-ethyl-2-methyl-1,2,3,6,7,8,9,9a-octahydropyrazino[1,2-$c$]pyrimidin-6-one ⟨71JMC929⟩.

### 2.17.2.20 Pyrazino[1,2-$a$]pyrazine

The $^1$H NMR and IR spectra of certain octahydropyrazino[1,2-$a$]pyrazines have been reported ⟨77JHC307⟩.

### 2.17.2.21 Pyrazino[2,3-$b$]pyrazine

The X-ray crystal structure for the parent unsubstituted heterocycle has been reported ⟨71MI21700⟩. Variable temperature proton and pulsed proton noise-decoupled Fourier transform $^{13}$C NMR have been used to determine the inversion barriers and conformations of *cis*- and *trans*-1,4,5,8-tetramethyldecahydropyrazino[2,3-$b$]pyrazine. Although several conformers for both the *cis* and *trans* isomer are possible, only two are significantly populated for each isomer. Using $^{13}$C NMR data and the Eyring equation, the activation energy for ring inversion was determined to be 48.9 kJ mol$^{-1}$ for the *cis* isomer (246 K) and 38.0 kJ mol$^{-1}$ for the *trans* isomer (202 K). Analogous results were obtained from $^1$H NMR spectral data ⟨76JCS(P2)1564⟩.

$^1$H NMR has also been used to study the covalent hydration of the cations of pyrazino-[2,3-$b$]pyrazines in aqueous solution. The spectrum of the anhydrous pyrazino-[2,3-$b$]pyrazine cation consists of one sharp singlet at $\delta$ 9.49 due to the time averaged equivalence of the four protons on the heterocyclic ring. In aqueous solution a covalently hydrated cation of this parent heterocycle is formed, exhibiting two sharp singlets of equal intensity at $\delta$ 7.52 and 5.37. This spectrum is incompatible with the monohydrate (**42**) but is consistent with the dihydrate (**43**). By analogy, it was shown that the cations of both 2-methyl- and 2,3-dimethyl-pyrazino[2,3-$b$]pyrazine exist as dihydrates in aqueous solution ⟨66JCS(C)999⟩. These data refute the earlier claim that cations of pyrazino[2,3-$b$]pyrazine exist as monohydrated species in aqueous solution. This earlier study was based on IR and UV spectral data ⟨63JCS4304⟩.

(42)          (43)

Visible and UV spectral data have been reported for the parent unsubstituted heterocycle ⟨62JCS493⟩. The apparent p$K_a$ values of several derivatives of this system were determined using spectrophotometric methods. The parent heterocycle has a p$K_a$ of 2.51. Values of 1.14, −0.25, −0.53 and −0.02 were obtained for the p$K_a$ of the 2-methyl, 2,3-dimethyl, 2,3,6-trimethyl and 2,3,6,7-tetramethyl derivatives of this ring system, respectively ⟨63JCS4304⟩. ESR data for the parent heterocycle have also been reported ⟨71JA5850⟩.

## 2.17.3 REACTIVITY

### 2.17.3.1 Pyridazino[1,2-*a*]pyridazine

Pyridazino[1,2-*a*]pyridazine derivatives are easily reduced with either hydrogen over a catalyst or with metal hydrides. Reaction of the pyridazino[1,2-*a*]pyridazine-1,4,6,9-tetrone (44) with hydrogen over platinum oxide gives the hydrogenated product (45) ⟨66JOC1311⟩. This product can be further reduced with lithium aluminum hydride to give the octahydropyridazino[1,2-*a*]pyridazine (24) ⟨67JA4875⟩.

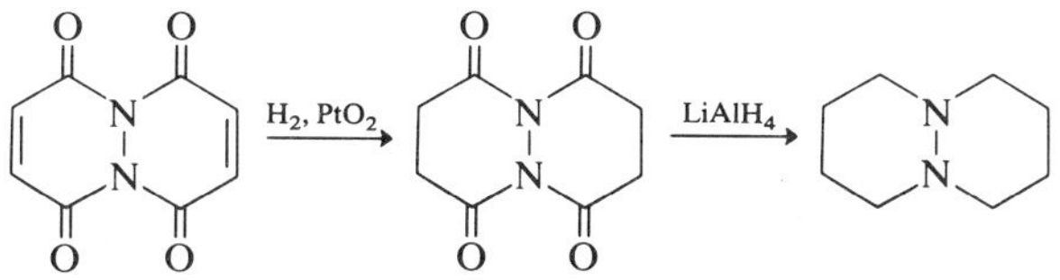

6,9-Dihydropyridazino[1,2-*a*]pyridazine-1,4-dione (46) is readily brominated in methylene chloride at −78 °C to give the dibromide (47) ⟨72TL1885⟩. The tetrabromide (48) can be formed by treatment of the starting material with two equivalents of bromine in chloroform at room temperature ⟨75JOC47⟩. Debromination of the dibromide (47) with 1,5-diazabicyclo[4.3.0]non-5-ene in refluxing benzene yields pyridazino[1,2-*a*]pyridazine-1,4-dione. Debromination of the reduced tetrabromo compound (49) occurs with lithium aluminum hydride to give 1,4,6,9-tetrahydropyridazino[1,2-*a*]pyridazine (50).

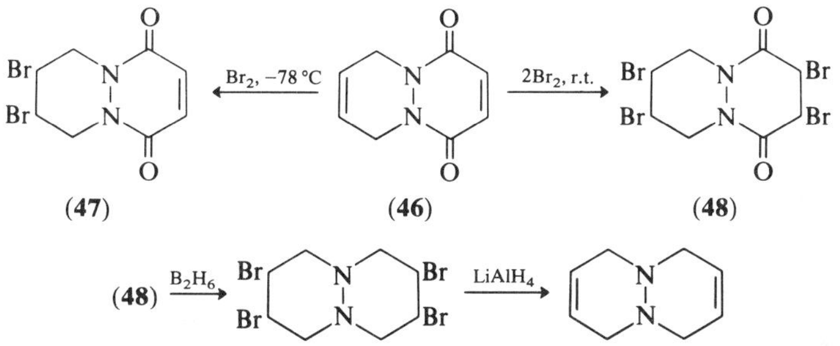

The perhydrotetrabromo compound (49) undergoes an interesting rearrangement to the pyrroles (51a) and (51b) on reaction with methyllithium in ether ⟨75JOC47⟩. In this reaction it was shown that pyrrole (51b) arises as a secondary product from (51a). Reaction of 1,4,6,9-tetrahydropyridazino[1,2-*a*]pyridazine with methyllithium in ether also gives rise to the rearranged product (51b).

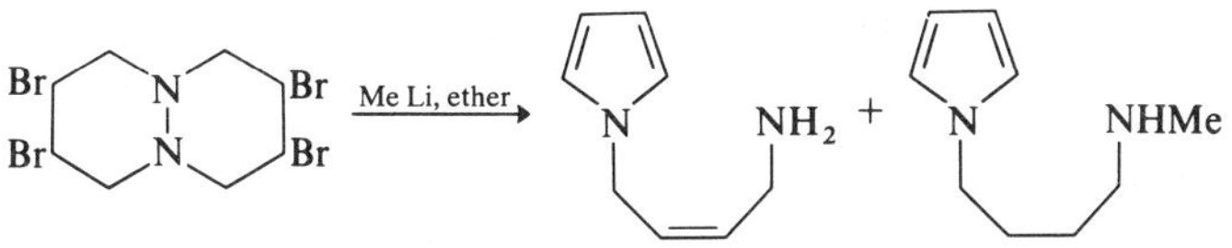

One final rearrangement of note occurs with the reduced pyridazino[1,2-*a*]pyridazine-1,4,6,9-tetrone (45). This compound gives *N,N'*-bisuccinimide (52) either on heating in a sealed tube or on reaction with succinoyl chloride in refluxing dioxane ⟨67JA4875⟩.

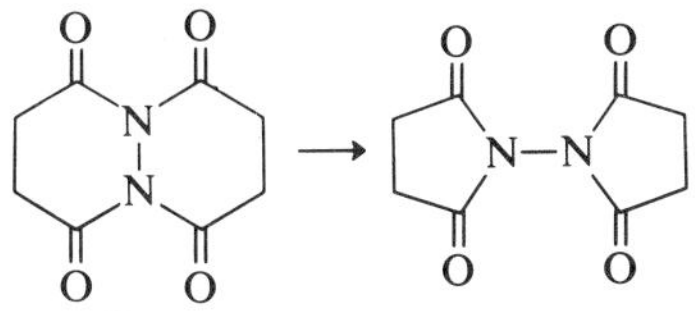

### 2.17.3.2 Pyridazino[3,4-*c*]pyridazine

No reactions involving derivatives of this ring system have been reported.

### 2.17.3.3  Pyridazino[4,3-*c*]pyridazine

Only one derivative of this ring system has been reported. No reactions involving this heterocycle are known.

### 2.17.3.4  Pyridazino[4,5-*c*]pyridazine

Only a few reactions involving this ring system are known. 3-Phenylpyridazino-[4,5-*c*]pyridazine-5,8-dione (**53**) is reported to undergo chlorination on treatment with phosphorus oxychloride to give the dichloro product (**54**) ⟨72YZ1327⟩. However, it is also reported that pyridazino[4,5-*c*]pyridazine-5,8-dione (**55**) and the 3-methyl derivative of this compound both fail to undergo chlorination using the usual procedures ⟨67JHC393⟩.

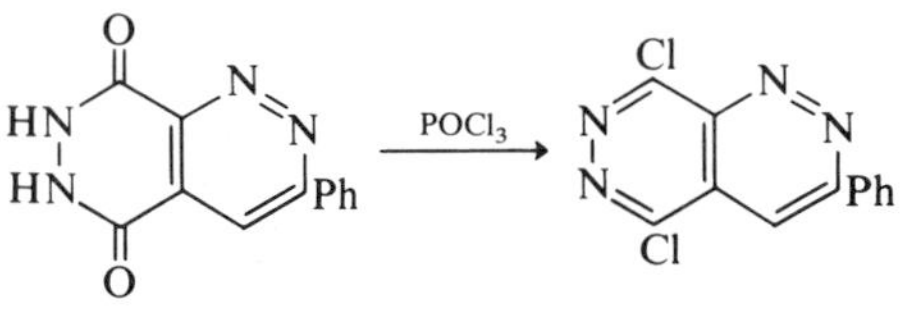

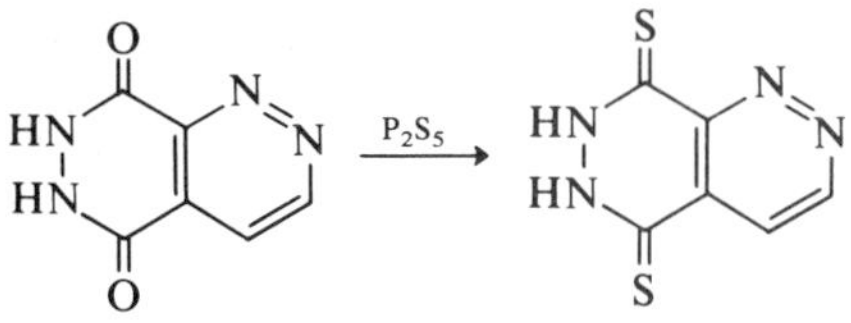

5,8-Dichloro derivatives of this ring system undergo displacement with amines ⟨72YZ1327⟩. The dithione (**56**) is readily formed from the dione (**55**) by reaction with phosphorus pentasulfide. This reaction was found to be sensitive to reaction conditions. A good yield of (**56**) is obtained when phosphorus pentasulfide is added portionwise to a refluxing solution of the starting dione (**55**) in pyridine ⟨67JHC393⟩.

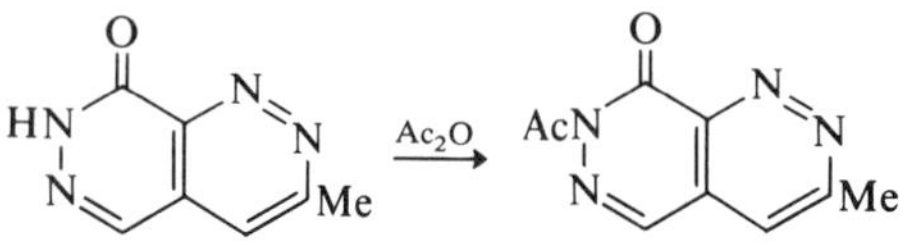

Reaction of compound (**57**) with acetic anhydride is reported to give the *N*- rather than the *O*-acetylated product ⟨73JHC1081⟩. The dihydro compound (**58**) undergoes aromatization on reaction with *p*-chloranil in refluxing xylene ⟨75JCS(P1)1326⟩.

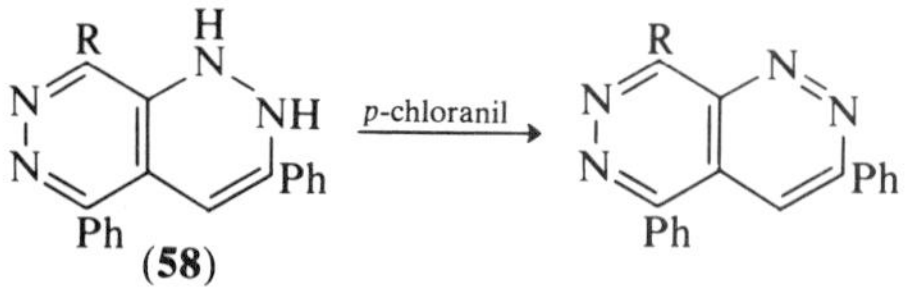

### 2.17.3.5  Pyridazino[4,5-*d*]pyridazine

Derivatives of this ring system undergo a variety of nucleophilic displacements. Chloro substituted pyridazino[4,5-*d*]pyridazines are readily prepared from the corresponding hydroxy compounds with phosphorus oxychloride or phosphorus oxychloride–phosphorus pentachloride mixtures. These chloro substituted heterocycles undergo nucleophilic displacement with sodium methoxide or sodium ethanethiolate to give the corresponding methoxy or ethylthio derivatives, respectively. The chlorine atom is also easily dechlorinated catalytically with palladium on carbon. Hydrolysis of tetrachloropyridazino[4,5-*d*]-pyridazine (**59**) in base gives a mixture of the isomeric products (**60 and 61**) ⟨72JCS(P1)953⟩.

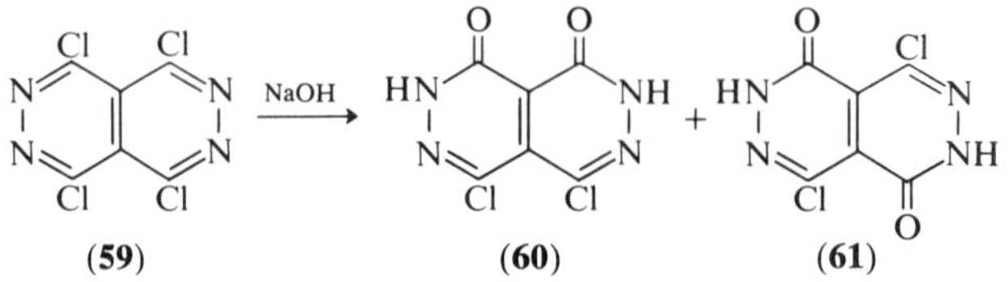

1,4-Dithiopyridazino[4,5-*d*]pyridazine (**63**) is readily prepared by reacting the corresponding dioxo heterocycle (**62**) with phosphorus pentasulfide in pyridine. Methylation of the thiol group occurs in methyl iodide and base to give the bis(methylthio) product (**64**). Methylthiopyridazino[4,5-*d*]pyridazines undergo nucleophilic substitution when reacted with amines and alkoxides, yielding amino and alkoxy derivatives, respectively ⟨67JHC491⟩.

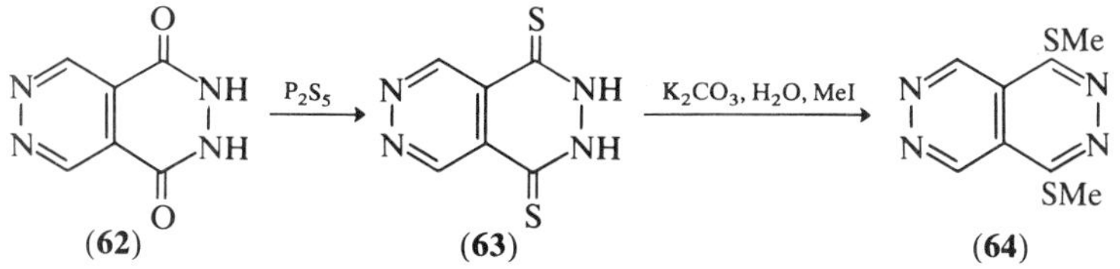

The methylation of hydroxy-, dihydroxy- and tetrahydroxy-pyridazino[4,5-*d*]pyridazines with diazomethane has been studied extensively. These reactions give a mixture of both *N*-methyl and *O*-methyl substituted products. For example, when 1,4-dihydroxy-5,8-dimethylpyridazino[4,5-*d*]pyridazine (65) is allowed to react with diazomethane, a mixture of methylated products is obtained ⟨75JHC95⟩.

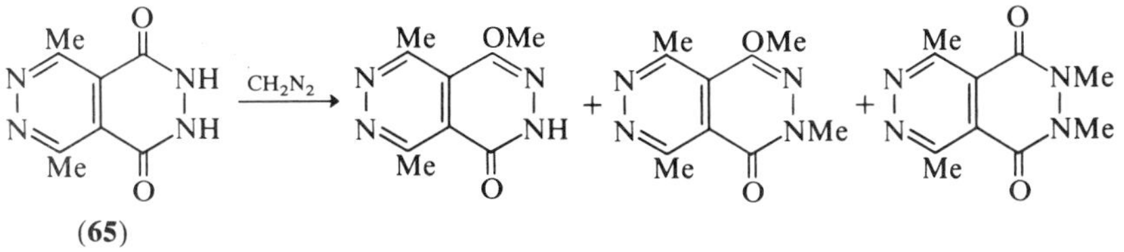

Derivatives of this ring system undergo Diels–Alder type additions. The Diels–Alder adduct (67) is formed by reacting (65) with lead tetraacetate and butadiene. This adduct is formed *via* the intermediate (66) ⟨75JHC95⟩.

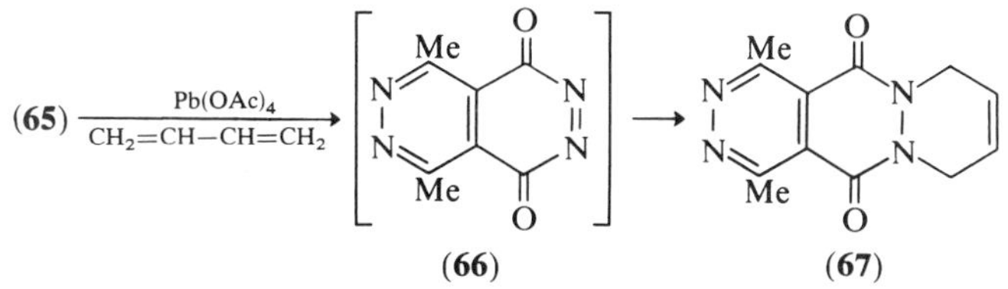

Nitration of 1,4-diaminopyridazino[4,5-*d*]pyridazine (68) resulted in the isolation of the nitramine (69). This nitro group could not be rearranged on to the other heterocyclic ring ⟨68JHC53⟩.

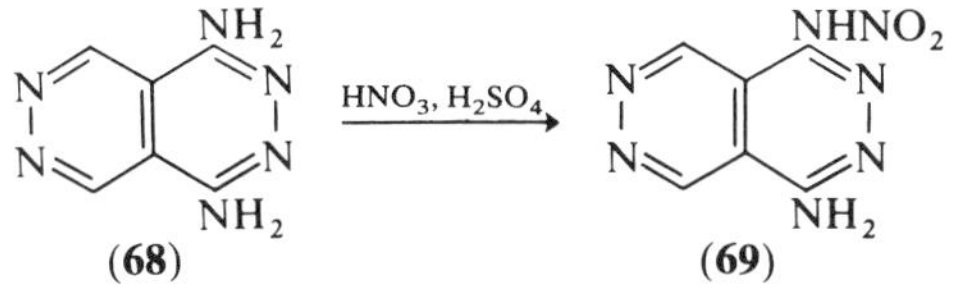

### 2.17.3.6 Pyrimido[1,2-*b*]pyridazine

Chlorinated 2*H*-pyrimido[1,2-*b*]pyridazin-2-ones are known to undergo a variety of nucleophilic displacement reactions. Displacement of the chlorine atom in both the 4- and 7-positions of these compounds with methylamine and sodium methanethiolate is facile. Displacement of the chlorine atom in the 3-position does not occur. Therefore, 3,4,7-trichloro-2*H*-pyrimido[1,2-*b*]pyridazin-2-one (34) reacts with methanethiol in sodium methoxide to give 3-chloro-4,7-bis(methylthio)-2*H*-pyrimido[1,2-*b*]pyridazin-2-one (70) ⟨71JOC3506⟩.

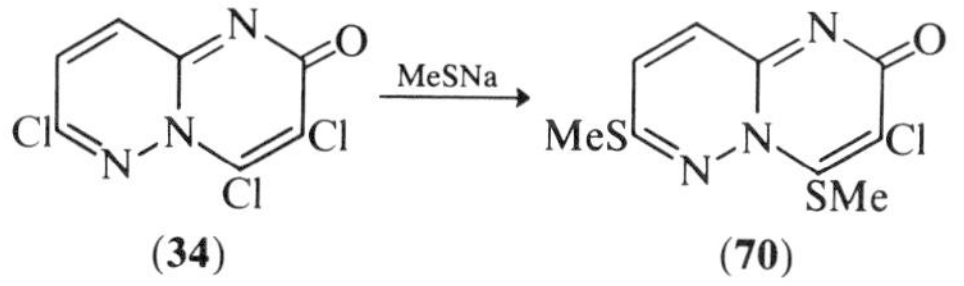

4-Phenyl-2*H*-pyrimido[1,2-*b*]pyridazin-2-one (71) is cleaved by reaction with potassium isopropoxide in refluxing isopropyl alcohol to give the pyrimidinone (72) ⟨72M1591⟩.

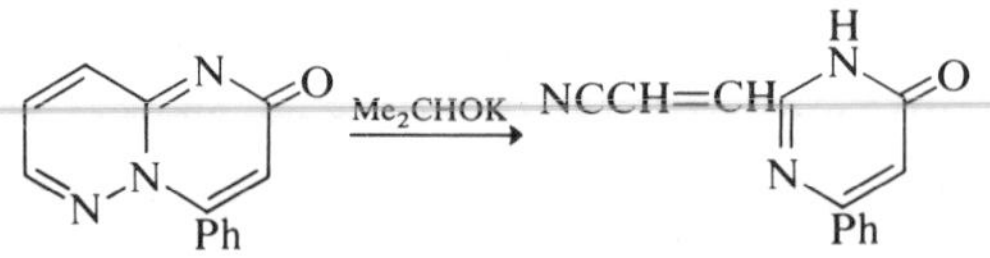

(71)                                          (72)

### 2.17.3.7 Pyrimido[4,5-*c*]pyridazine

The unusual formation of *N,N*-dimethylaminophenyl substituted pyrimido[4,5-*c*]-pyridazines (**74**) by the reaction of the oxo compound (**73**) with phosphorus oxychloride and *N,N*-dimethylaniline has been reported ⟨71CPB1849⟩. The chlorination of other oxo substituted pyrimido[4,5-*c*]pyridazines with phosphorus oxychloride has been reported to be unsuccessful. Chloro derivatives of this heterocyclic ring undergo nucleophilic displacement with amines and hydrazine to give the corresponding amino and hydrazino substituted products. The catalytic dechlorination of these chloro substituted heterocycles has also been reported ⟨68JHC523⟩.

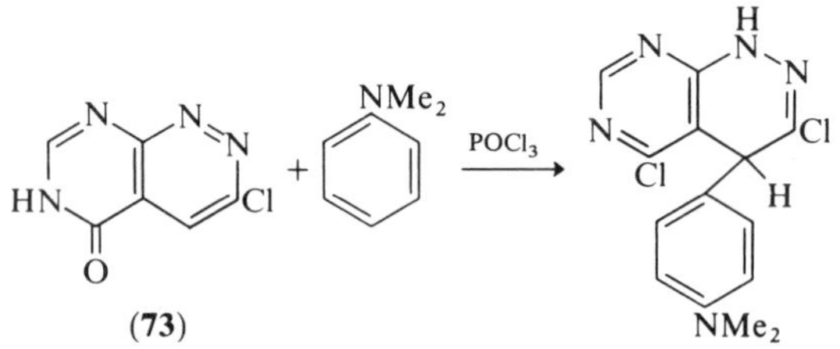

(73)                                          (74)

Oxidation of the hydroxy heterocycle (**75**) with benzoyl peroxide in the presence of triethylamine gives the corresponding oxo derivative (**76**). However, treatment of (**75**) with potassium permanganate under basic conditions gives the pyrazolo[3,4-*d*]pyrimidine (**77**). The proposed mechanism for this rearrangement involves bond migration in the intermediate hydroxy compound followed by dehydration and decarbonylation of the resulting α-ketocarboxylic acid ⟨74TL3893⟩. The air oxidation of other dihydropyrimido[4,5-*c*]-pyridazines has also been reported ⟨82JOC674⟩.

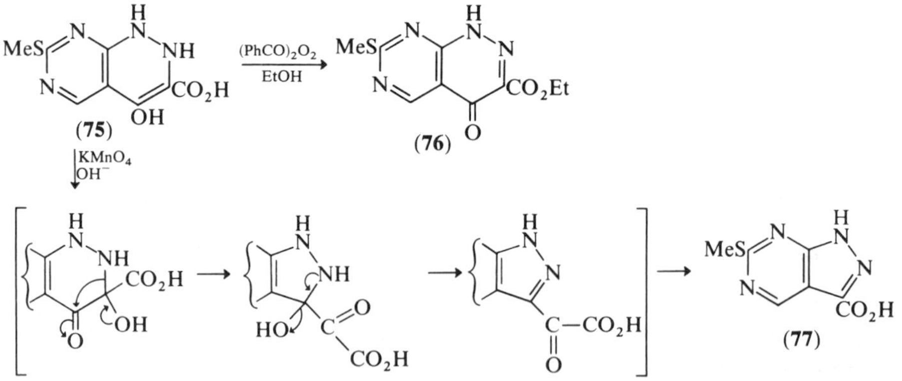

The acetylation of compound (**78**) has been reported under a variety of conditions. Reaction of (**78**) with acetic anhydride at 50 °C gives the diacetyl product (**79**). However, when the acetylation is carried out at 120 °C the cyclized product (**80**) is obtained. This latter product can also be obtained by reacting the diacetyl intermediate (**79**) with acetic anhydride at 120 °C. Hydrolysis of (**80**) in 1N aqueous sodium hydroxide at 80 °C regenerates (**79**) ⟨76CPB2637⟩.

The pyrimidine ring portion of pyrimido[4,5-*c*]pyridazines undergoes cleavage under basic conditions. Pyrimido[4,5-*c*]pyridazine-5,7-dione is cleaved in 50% aqueous sodium hydroxide to give 3-aminopyridazine-4-carboxylic acid following neutralization with hydrochloric acid ⟨68JHC523⟩. If this cleavage reaction is carried out with 20% aqueous sodium hydroxide, followed by neutralization with acetic acid, it is possible to isolate 3-aminopyridazine-4-carboxamides ⟨71CPB1849⟩.

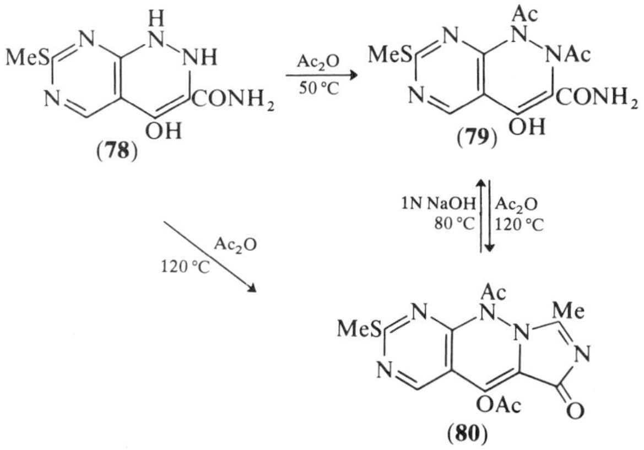

## 2.17.3.8 Pyrimido[5,4-c]pyridazine

Only a few reactions involving pyrimido[5,4-c]pyridazines have been reported in the literature. Pyrimido[5,4-c]pyridazine-8-thione can be obtained from the corresponding oxo compound by reaction with phosphorus pentasulfide. Pyrimido[5,4-c]pyridazin-8-one could not be chlorinated with phosphorus oxychloride or phosphorus oxychloride–phosphorus pentachloride mixtures. This compound is hydrolytically cleaved in base to give 4-aminopyridazine-2-carboxylic acid ⟨68JHC523⟩. Pyrimido[5,4-c]pyridazin-8-ones (**81**) undergo rearrangement on hydrogenation over palladium on carbon in glacial acetic acid or on reduction with Raney nickel in ethanol to give 5H-pyrrolo[3,2-d]pyrimidines (**82**) ⟨78JOC2536⟩. This reductive ring contraction necessarily involves three steps: (1) cleavage of the N—N bond by hydrogenolysis; (2) a hydrogenating step; and (3) ring closure with concomitant liberation of aniline. Each of these steps involves a number of possible intermediates and the exact order of these events is not known.

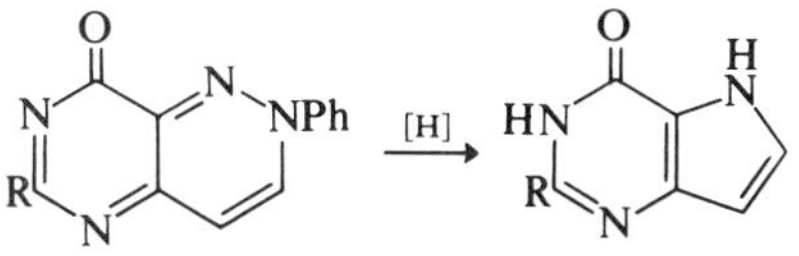

## 2.17.3.9 Pyrimido[4,5-d]pyridazines

Pyrimido[4,5-d]pyridazines undergo a variety of nucleophilic substitutions. 2-Aryl-pyrimido[4,5-d]pyridazine-5,8-diones (**83**) undergo chlorination readily in a mixture of phosphorus oxychloride and phosphorus pentachloride to give the corresponding dichloro compounds (**84**) in exceptionally high yields ⟨72CPB1513⟩. Pyrimido[4,5-d]pyridazine-2,4-dione (**85**) undergoes nucleophilic displacement with phosphorus pentasulfide in pyridine. If the reaction mixture is refluxed for 1 hour, the thione (**86**) is obtained. A reflux time of 19 hours results in the formation of the dithione (**87**) ⟨68JHC845⟩.

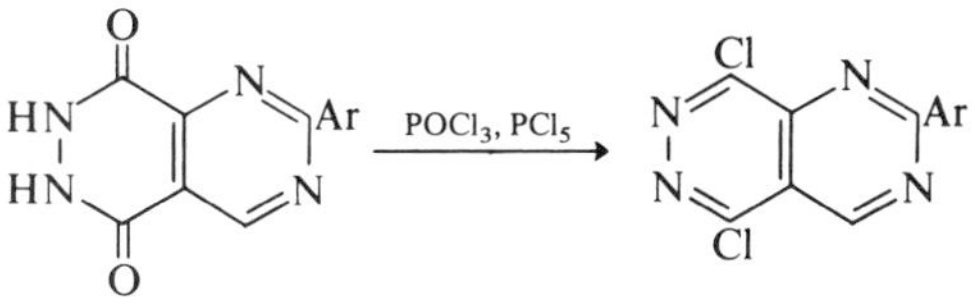

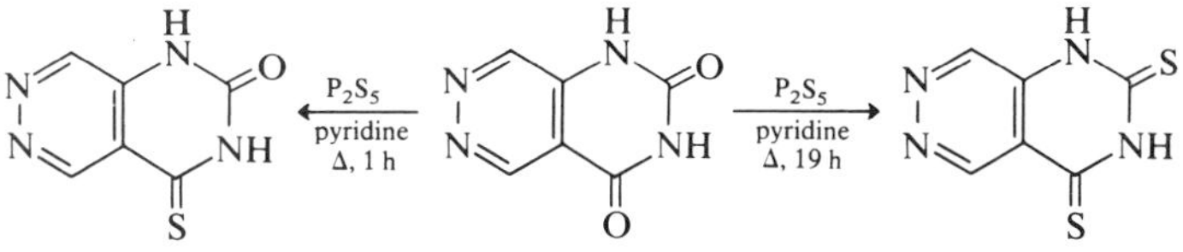

5,8-Dichloropyrimido[4,5-*d*]pyridazines react readily with amines, potassium hydro-sulfide, sodium azide ⟨72CPB1528⟩ and alkoxides ⟨72CPB1522⟩ to give the corresponding disubstituted amines, thiones, azides and alkoxides, respectively. Hydrolysis of the starting compounds (**88**) with 1% sodium hydroxide yields a mixture of the chloro products (**89**) and (**90**) ⟨72YZ1312⟩.

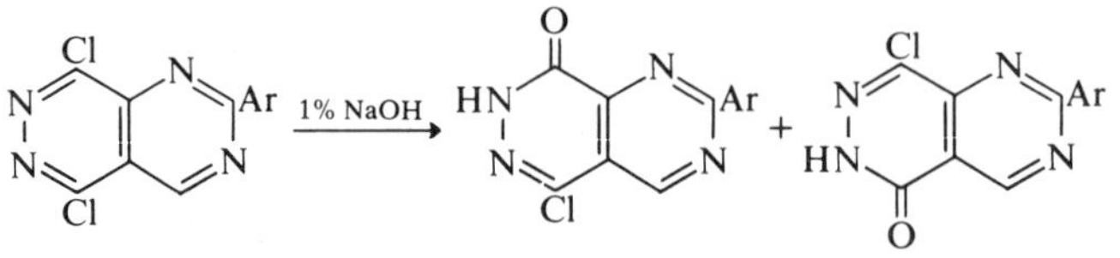

(**88**)                    (**89**)                    (**90**)

Pyrimido[4,5-*d*]pyridazinethiones undergo nucleophilic displacement with amines ⟨70JHC209⟩ and chlorine gas to give the corresponding amino and chloro compounds, respectively. These thiones are also readily alkylated with alkyl halides to give alkylthio products ⟨72CPB1528⟩.

3,4-Dihydro-2,5,8-trisubstituted pyrimido[4,5-*d*]pyridazines are readily alkylated at N-2 with either alkyl halides or alkyl sulfates and sodium hydride in DMF ⟨72YZ1316⟩. If this reaction is carried out with benzyl bromide and sodium hydride, the 1,4-bisbenzyl product can be obtained ⟨73YZ1043⟩. Alkylation of (**91**) with ethylmagnesium bromide gives the dialkylated product (**92**) ⟨75CPB1488⟩.

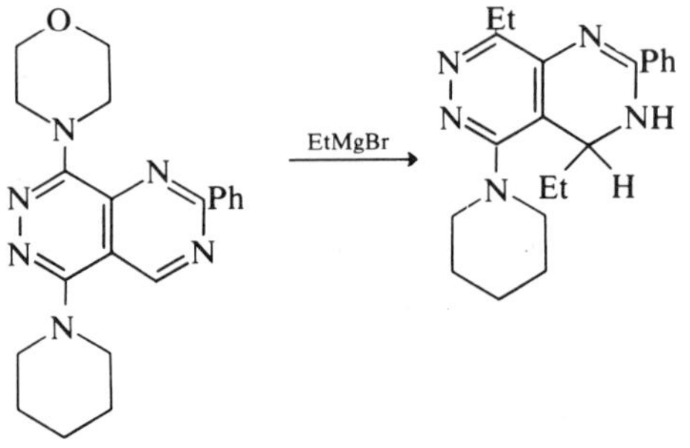

(**91**)                              (**92**)

Treatment of the hydrazonium salt (**93**) with 10% hydrochloric acid gives the rearranged product (**94**). The proposed mechanism for this ring contraction is shown ⟨72CPB1513⟩.

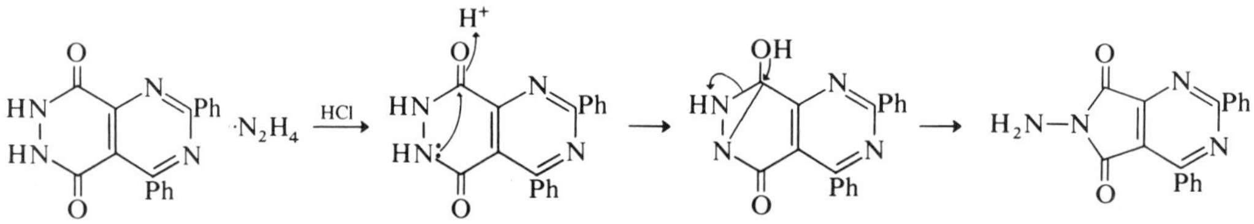

(**93**)                                                                      (**94**)

Pyrimido[4,5-*d*]pyridazine derivatives are readily cleaved under both acidic and basic conditions. Reaction of 2-phenyl-5,8-dimorpholinopyrimido[4,5-*d*]pyridazine (**95**) with 10% hydrochloric acid gives a mixture of four isolated products: 2-phenyl-8-morpholinopyrimido[4,5-*d*]pyridazin-5(6*H*)-one (**96**), 4-hydroxy-6-morpholinopyridazin-3(2*H*)-one (**97**), 4-amino-5-formyl-3,6-dimorpholinopyridazine (**98**) and benzoic acid (**99**).

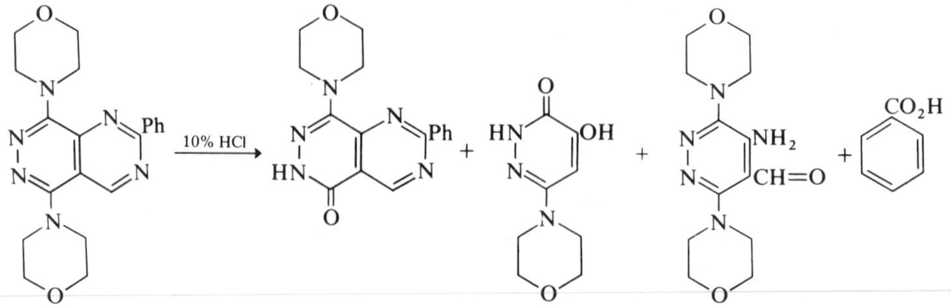

(**95**)                    (**96**)              (**97**)              (**98**)              (**99**)

A mechanism has been proposed to account for the formation of each of these products 〈72YZ1312〉. Pyrimido[4,5-*d*]pyridazine-2,4-dione (**85**) is cleaved by 10% aqueous sodium hydroxide to give 5-hydroxypyridazine-4-carboxylic acid (**100**). Cleavage with concentrated ammonium hydroxide also occurs, yielding 5-aminopyridazine-4-carboxylic acid (**101**) 〈68JHC845〉.

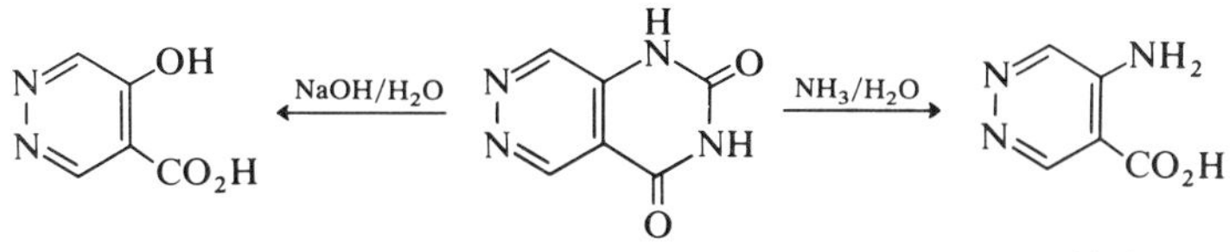

(**100**)　　　　　　　　(**85**)　　　　　　　　(**101**)

### 2.17.3.10　1*H*-Pyrazino[1,2-*b*]pyridazine

Only one derivative of this ring system is reported in the literature 〈71JCS(C)528〉. No reactions of this compound are reported.

### 2.17.3.11　Pyrazino[2,3-*c*]pyridazine

Only reactions involving the preparation of this ring system have been reported.

### 2.17.3.12　Pyrazino[2,3-*d*]pyridazine

Pyrazino[2,3-*d*]pyridazines undergo nucleophilic displacements readily. Pyrazino-[2,3-*d*]pyridazine-5,8-dione can be chlorinated in a mixture of phosphorus oxychloride and phosphorus pentachloride to give 5,8-dichloropyrazino[2,3-*d*]pyridazine. The corresponding dibromo product is prepared by reaction of the starting dione with phosphorus oxybromide and bromine. However, all attempts to chlorinate pyrazino[2,3-*d*]pyridazin-5-one failed 〈69JHC93〉.

Chloro substituted pyrazino[2,3-*d*]pyridazines react readily with amine and alkoxide nucleophiles to give the corresponding amino and alkoxy substituted products 〈66JHC512〉. 5,8-Dichloropyrazino[2,3-*d*]pyridazine (**102**) can be hydrolyzed in 2% aqueous sodium hydroxide to yield 5-chloropyrazino[2,3-*d*]pyridazin-8-one (**103**). Displacement of only one halogen substituent is also observed in the basic hydrolysis of the 5,8-dibromo analog 〈68JHC53〉.

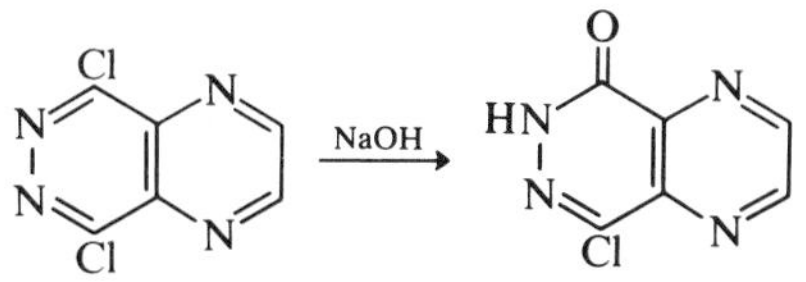

(**102**)　　　　　　　　(**103**)

Pyrazino[2,3-*d*]pyridazine-5,8-dione (**104**) can be dithiated with phosphorus pentasulfide to give the unisolated intermediate (**105**), which yields the bis(benzylthio) product (**106**) on reaction with benzyl chloride. Acetylation and tosylation of this starting dione (**104**) produce the monoacetyl and the monotosyl products, respectively 〈66JHC512〉.

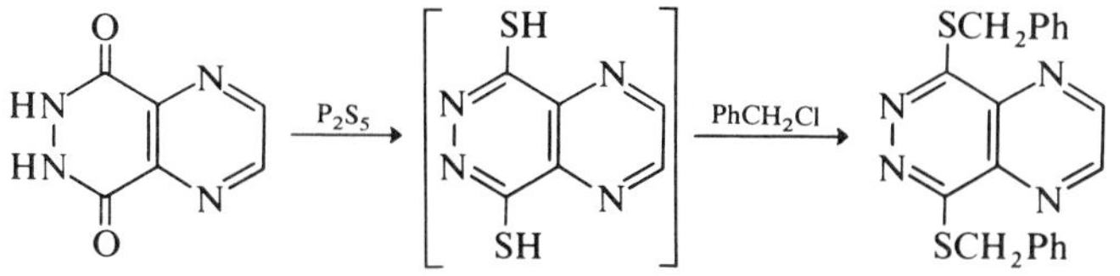

(**104**)　　　　　(**105**)　　　　　(**106**)

5,8-Diaminopyrazino[2,3-*d*]pyridazine yields bis(picrylamino)pyrazino[2,3-*d*]-pyridazine when reacted with picryl fluoride in dry DMSO 〈69JHC255〉. This diamino starting material (**107**) also reacts with benzoyl chloride and triethylamine in chloroform to give the perbenzoylated product (**108**) 〈78JHC1451〉.

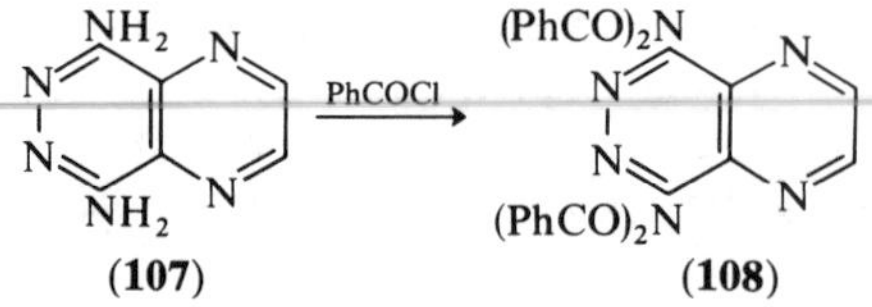

(107)  →[PhCOCl]  (108)

The reaction of the diamino compound (107) under several different nitration conditions led to the formation of 5-nitroamino-8-aminopyrazino[2,3-*d*]pyridazine exclusively ⟨68JHC53⟩. 5,8-Dimorpholino substituted pyrazino[2,3-*d*]pyridazines are known to react with Grignard reagents to give 3-substituted 3,4-dihydro derivatives ⟨75CPB1488⟩. The dimorpholino heterocycles are also known to react photochemically with alcohols and cyclic ethers in the presence of photosensitizers to give 3-substituted 3,4-dihydro derivatives (Scheme 3) ⟨75CPB1500⟩.

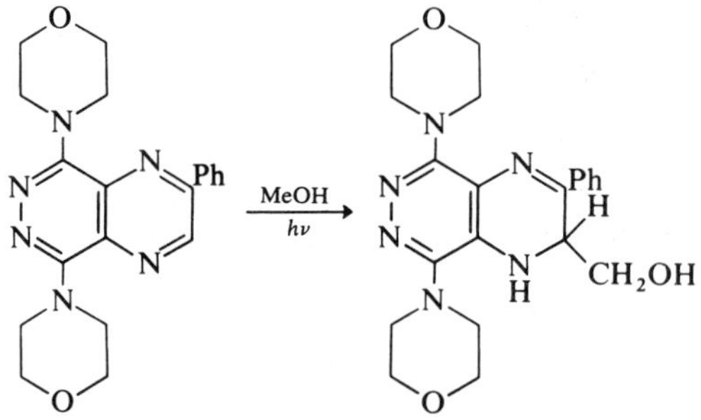

**Scheme 3**

Catalytic hydrogenation of the chloro substituted starting material (103) over palladium on carbon gives the dechlorinated and hydrogenated product (109) ⟨68JHC53⟩. 3,4-Dihydropyrazino[2,3-*d*]pyridazines can be dehydrogenated either with 2,3-dichloro-5,6-dicyanoquinone or potassium ferricyanide ⟨75CPB1505⟩.

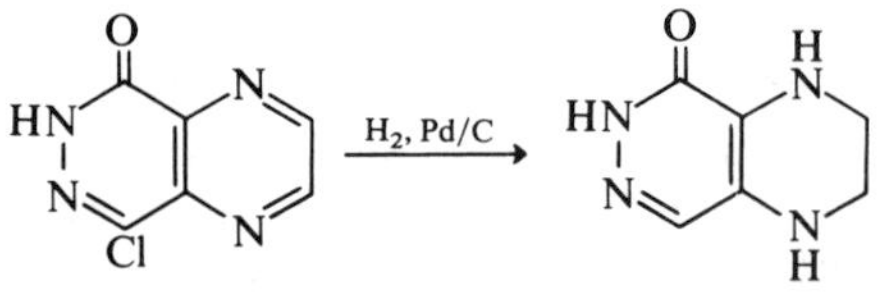

(103)  →[H₂, Pd/C]  (109)

### 2.17.3.13 Pyrimido[1,2-*a*]pyrimidine

Only a few reactions involving this ring system have been reported. Dipolar pyrimido-[1,2-*a*]pyrimidines undergo cycloadditions, followed by cleavage, to give pyrido-[1,2-*a*]pyrimidines. Reaction of the heterocyclic betaine (110) with either dimethyl acetylenedicarboxylate, dicyanoacetylene or dibenzoylacetylene gives the pyrido-[1,2-*a*]pyrimidin-6-ones (112). These rearranged products are presumably formed from the intermediate adduct (111). In these reactions the starting betaines can be viewed as containing 'masked' 1,4-dipoles. Alkenic dipolarophiles such as *N*-phenylmaleimide, maleic anhydride and tetracyanoethylene are reported not to form cycloaddition products with the betaine ⟨73JOC3485⟩.

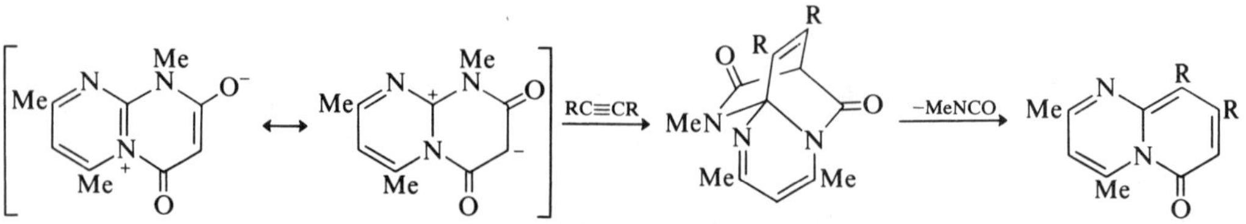

(110)  →[RC≡CR]  (111)  →[−MeNCO]  (112)

Pyrimido[1,2-*a*]pyrimidine derivatives are reported to undergo ring cleavage under a variety of reaction conditions. Reaction of the starting heterocycle (113) with guanidine at room temperature leads to the formation of the ring opened product (**114**). This same starting material (**113**) reacts on heating in boiling water to give the pyrimidinylpropionic acid (**115**) ⟨68JOC3354⟩. Analogous acid–base ring cleavage reactions have also been reported ⟨61JOC1891⟩.

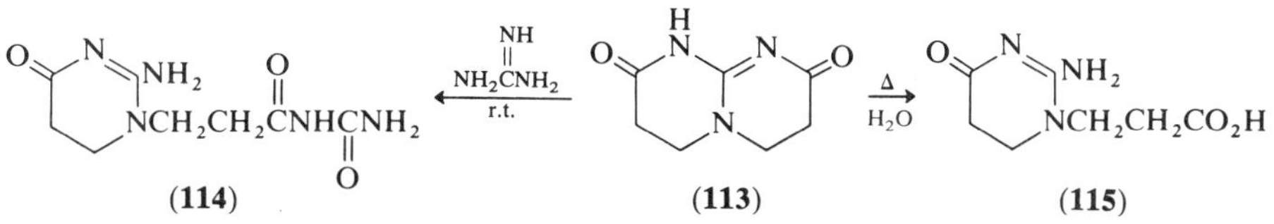

(**114**)    (**113**)    (**115**)

### 2.17.3.14 Pyrimido[1,6-*a*]pyrimidine

Only very few reactions involving derivatives of this ring system have been reported. The starting material (**116**) undergoes ring cleavage in boiling water to give the acid (**117**). Reaction of the same starting heterocycle with ethanolic hydrogen chloride gives the ester (**118**) which is isolated as the hydrochloride salt. These two ring cleavage reactions can be reversed to yield the starting material again, by reaction of the products with refluxing acetic anhydride and sodium bicarbonate, respectively ⟨64JOC1762⟩.

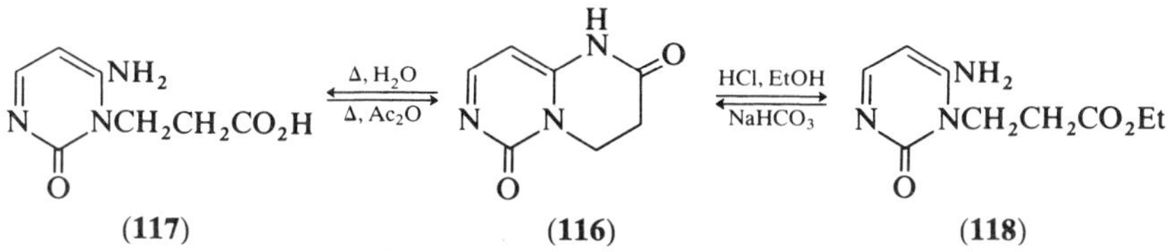

(**117**)    (**116**)    (**118**)

### 2.17.3.15 Pyrimido[1,6-*c*]pyrimidine

No reactions involving derivatives of this ring system have been reported.

### 2.17.3.16 Pyrimido[4,5-*d*]pyrimidines

Pyrimido[4,5-*d*]pyrimidine derivatives undergo a wide variety of nucleophilic displacement reactions. Pyrimido[4,5-*d*]pyrimidin-4-ones are readily chlorinated in either phosphorus oxychloride ⟨62JOC4211⟩ or a mixture of phosphorus pentachloride and thionyl chloride ⟨74JMC451⟩ to give the corresponding 4-chloropyrimido[4,5-*d*]pyrimidines. These chloro compounds are generally not purified but are isolated crude and used for subsequent nucleophilic displacements.

4-Chloropyrimido[4,5-*d*]pyrimidines react with primary and secondary amines to give the corresponding 4-amino compounds ⟨74JMC451⟩. The 4-chloro substituent of the 2,4-dichloropyrimido[4,5-*d*]pyrimidine-5,7-dione (**119**) can be selectively displaced with ethylamine at 20 °C to give the corresponding 2-chloro-4-aminopyrimido[4,5-*d*]pyrimidine-5,7-dione (**120**) ⟨74LA2066⟩. Displacement of the 2-chloro substituent occurs at 80 °C. 4-Chloropyrimido[4,5-*d*]pyrimidines also react with thiourea to give the corresponding thiones ⟨62JOC4211⟩.

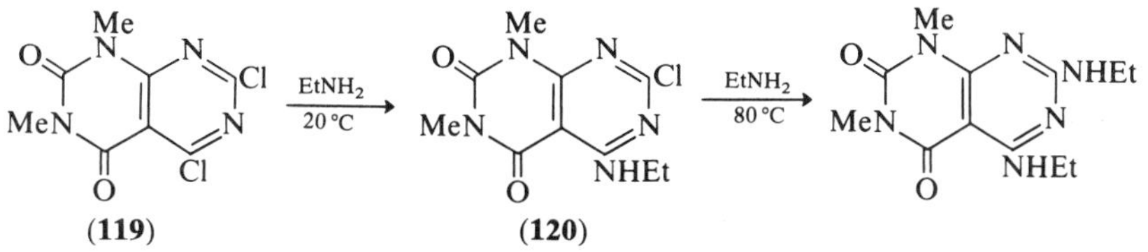

(**119**)    (**120**)

Pyrimido[4,5-*d*]pyrimidine-4-thiones are also formed by reaction of pyrimido[4,5-*d*]-pyrimidin-4-ones with phosphorus pentasulfide. Methylation of these thiones with dimethyl sulfate or methyl iodide gives the corresponding methylthio derivatives ⟨60JA5711⟩. 5-Amino-1,3-dimethyl-7-methylthiopyrimido[4,5-*d*]pyrimidine-2,4(1*H*,3*H*)-dione (**121**) undergoes

Raney nickel dethiation, affording 5-amino-1,3-dimethylpyrimido[4,5-*d*]pyrimidine-2,4(1*H*,3*H*)-dione (**122**) ⟨79H503⟩. Alkylthiopyrimido[4,5-*d*]pyrimidines are also known to undergo nucleophilic displacement with alkylamines to give the corresponding alkylamino compounds ⟨60JA5711⟩. 4-Aminopyrimido[4,5-*d*]pyrimidine (**124**) is conveniently prepared by reaction of 4-methylthiopyrimido[4,5-*d*]pyrimidine (**123**) with formamidine ⟨60JA3138⟩. All of these nucleophilic displacements of alkylthio derivatives are known to occur with the corresponding thiones.

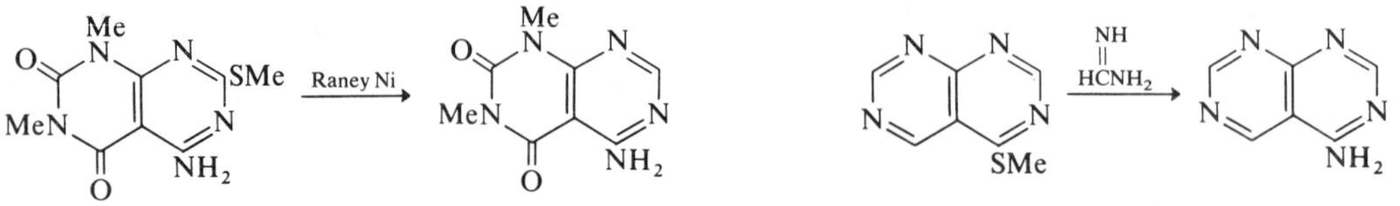

Pyrimido[4,5-*d*]pyrimidines are cleaved by both organic and inorganic bases. Reaction of the pyrimidopyrimidine (**125**) with ethylamine or methylamine in a sealed tube at 100 °C gives the ring cleaved product (**126**) ⟨74JCS(P1)1812⟩. 4-Methylthiopyrimido[4,5-*d*]-pyrimidine (**127**) is readily cleaved with sodium carbonate at 20 °C to give the pyrimidine (**128**) ⟨60JA6058⟩.

4-Aminopyrimido[4,5-*d*]pyrimidine (**124**) undergoes an interesting rearrangement to the pyridopyrimidine (**129**) when condensed with malononitrile in glacial acetic acid ⟨73JCS(P1)1794⟩.

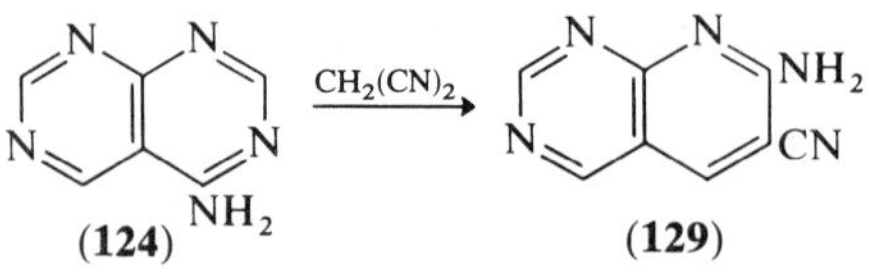

### 2.17.3.17 Pyrimido[5,4-*d*]pyrimidines

Pyrimido[5,4-*d*]pyrimidines are known to undergo a variety of nucleophilic substitutions. Chloro derivatives of this ring system are readily prepared by the reaction of hydroxy-pyrimido[5,4-*d*]pyrimidines with a mixture of phosphorus pentachloride and phosphorus oxychloride. Using this method, 2,4,6,8-tetrachloro and 2,4,8-trichloro-pyrimido[5,4-*d*]-pyrimidine can be prepared in excellent yield from the corresponding tetrahydroxy and trihydroxy substituted starting materials. These chloro substituted pyrimido[5,4-*d*]-pyrimidines are known to undergo ready nucleophilic diplacement with alkoxides, amines and iodide. As illustrated by the reaction shown, nucleophilic substitution of chlorine by amines at positions 4 and 8 occurs more readily than that at positions 2 and 6. This same trend is observed in the displacement of chlorine with iodide anion. However, reaction of the tetrachloro compound (**130**) with sodium methoxide gives the tetramethoxy product (**131**) ⟨60LA(631)147, 66JMC610⟩.

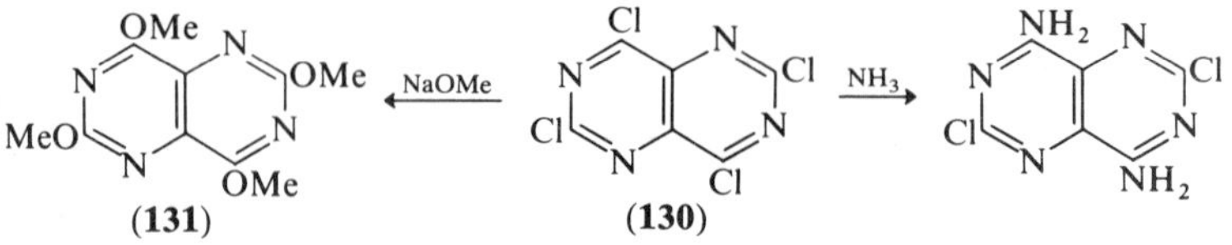

### 2.17.3.18 Pyrazino[1,2-*a*]pyrimidine

Only very few reactions of derivatives of this ring system have been reported. The iminopyrazino[1,2-*a*]pyrimidine (**132**) reacts with acetic anhydride to give the diacetylated product (**133**). 2-Hydroxypyrazino[1,2-*a*]pyrimidines undergo base catalyzed *O*-alkylation on reaction with tertiary aminoalkyl chlorides ⟨68JMC1045⟩.

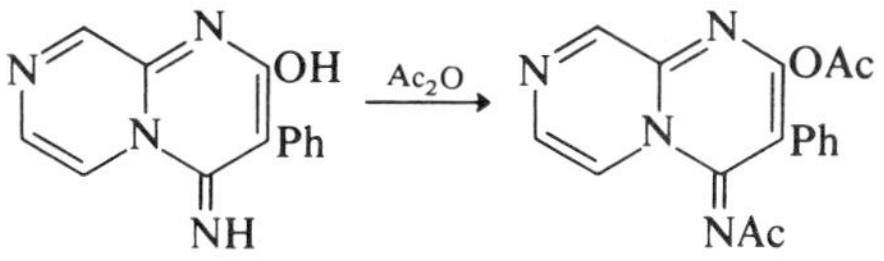

(**132**)          (**133**)

### 2.17.3.19 Pyrazino[1,2-*c*]pyrimidine

Several reactions of 2,7-disubstituted octahydropyrazino[1,2-*c*]pyrimidin-6-ones have been reported. These compounds react with such reagents as isocyanates, isothiocyanates and ethyl chloroformate to give the expected products with substitution occurring at the N-2 position on the heterocyclic ring. This position on the ring is also easily alkylated with alkyl halides. Reaction of the dibenzyl heterocycle (**134**) with hydrogen in the presence of palladium on carbon gives hydrogenolysis of only the N-2 benzyl group ⟨75JMC913⟩.

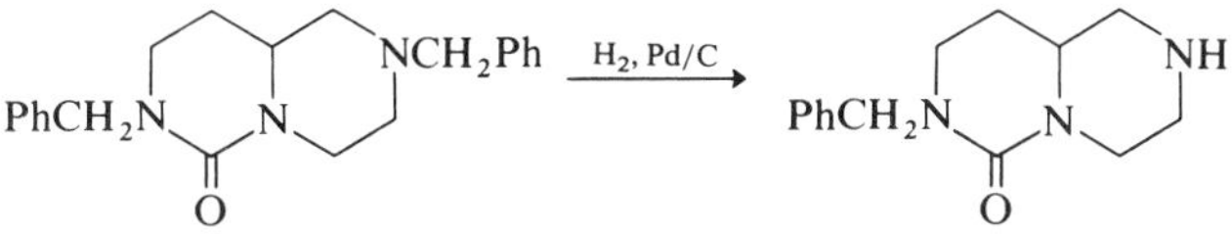

(**134**)

### 2.17.3.20 Pyrazino[1,2-*a*]pyrazine

The reduction of oxopyrazino[1,2-*a*]pyrazines is reported to occur readily with lithium aluminum hydride ⟨77JHC307⟩. The dioxo starting material (**135**) reacts with LAH to give the octahydropyrazino[1,2-*a*]pyrazine (**136**) in good yield. This product was isolated as the tripicrate ⟨62AP121⟩.

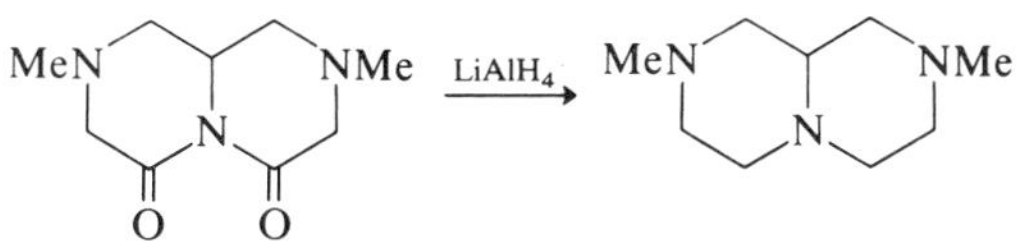

(**135**)          (**136**)

### 2.17.3.21 Pyrazino[2,3-*b*]pyrazine

Pyrazino[2,3-*b*]pyrazines are known to undergo covalent hydration ⟨63JCS4304⟩. In aqueous solution the cation of the parent pyrazino[2,3-*b*]pyrazine (**137**) exists as the covalently dihydrated form (**43**) ⟨66JCS(C)999⟩.

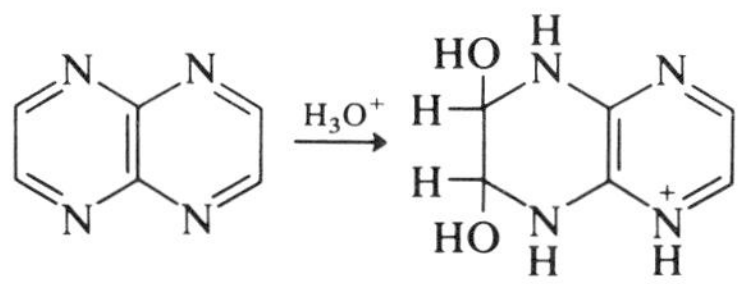

(**137**)          (**43**)

## 2.17.4 SYNTHESIS

### 2.17.4.1 Pyridazino[1,2-*a*]pyridazine

Only a few synthetic pathways to the pyridazino[1,2-*a*]pyridazine ring system are known. Most of these involve condensation of pyridazine derivatives which are unsubstituted at positions 1 and 2 with diacid chlorides ⟨67JA4875⟩, γ-ketoacid chlorides ⟨69JOC2720⟩ and cyclic anhydrides ⟨61JOC559⟩. For example, the cyclic hydrazide (**138**) condenses with succinoyl chloride to give the perhydropyridazino[1,2-*a*]pyridazine-1,4,6,9-tetrone (**45**).

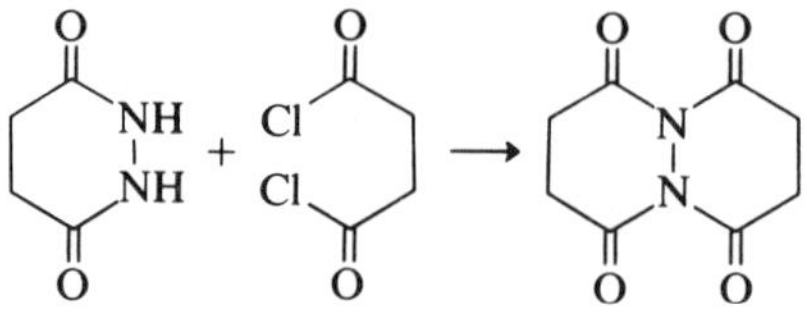

(**138**)          (**45**)

Another pathway which is often used in the preparation of this ring system involves the Diels–Alder reaction. 3,6-Pyridazinedione (**139**) is known to be exceptionally reactive toward dienes. This dienophile readily condenses with butadiene, 2,3-dimethylbutadiene and coumalic acid to give the respective Diels–Alder adducts (**140**). 3,6-Pyridazinedione is also reported to decompose with evolution of nitrogen gas at temperatures below 0 °C, giving the pyridazino[1,2-*a*]pyridazine-1,4,6,9-tetrone (**44**) as the major product ⟨62JA966⟩.

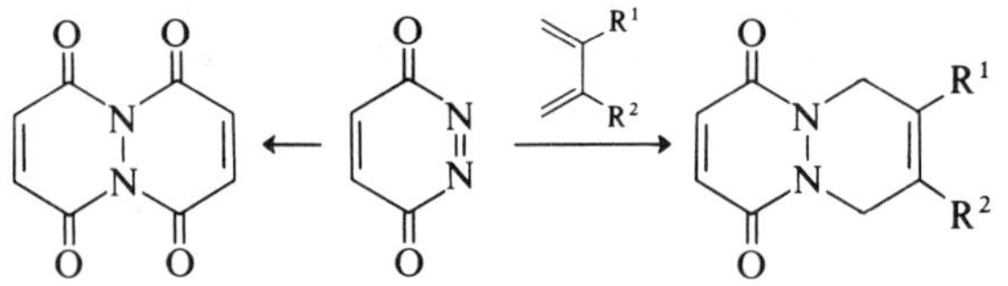

(**44**)          (**139**)          (**140**)

### 2.17.4.2  Pyridazino[3,4-*c*]pyridazine

Only one unrefuted synthesis of this ring system has been reported in the literature ⟨76MIP21700⟩. A mixture of (**142**) and the desired ring system (**143**) are reported as the products from the reaction of the anhydride (**141**) with an excess of hydrazine hydrate in water at reflux temperature for 3.5 hours. If the reaction is carried out using equimolar amounts of the starting anhydride and hydrazine hydrate in aqueous ethanol for 5 days, the formation of the pyridazinone (**144**) is reported.

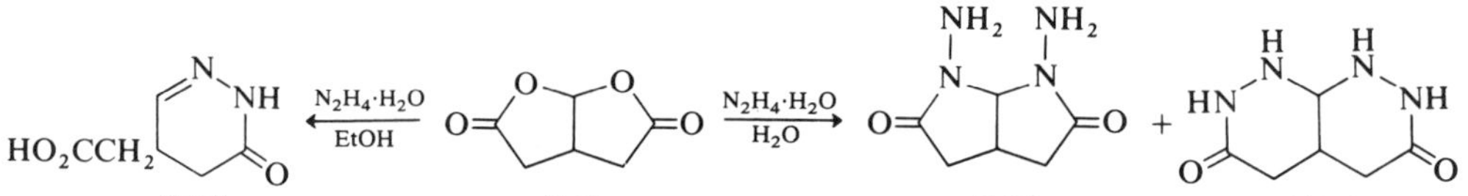

(**144**)          (**141**)          (**142**)          (**143**)

It was reported that 7-chloro-3-phenyl-4*H*-pyridazino[6,1-*c*]-1,2,4-triazine underwent reaction with hydrazine to give the unexpected product (**146**) ⟨69AC(R)552⟩. However, it was later established that the assigned structure for this reaction product was incorrect, and the correct hydrazino structure (**147**) for this product was confirmed by an independent synthesis ⟨70S180⟩.

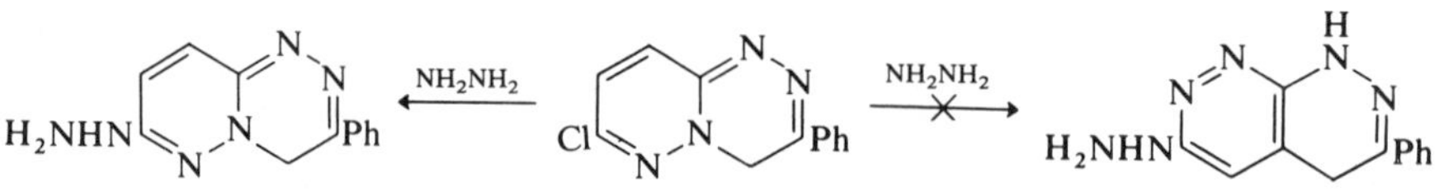

(**147**)          (**145**)          (**146**)

### 2.17.4.3  Pyridazino[4,3-*c*]pyridazine

Only one example of a derivative of this ring system has been reported. This compound, 2,6-dimethyl-3,4-dioxo-2,3,4,6,7,8-hexahydropyridazino[4,3-*c*]pyridazine    4-(*p*-bromophenylhydrazone) (**149**), was obtained in the reaction of the lactone (**148**) with methylhydrazine. A proposed mechanism for this unusual reaction is shown ⟨80JHC617⟩.

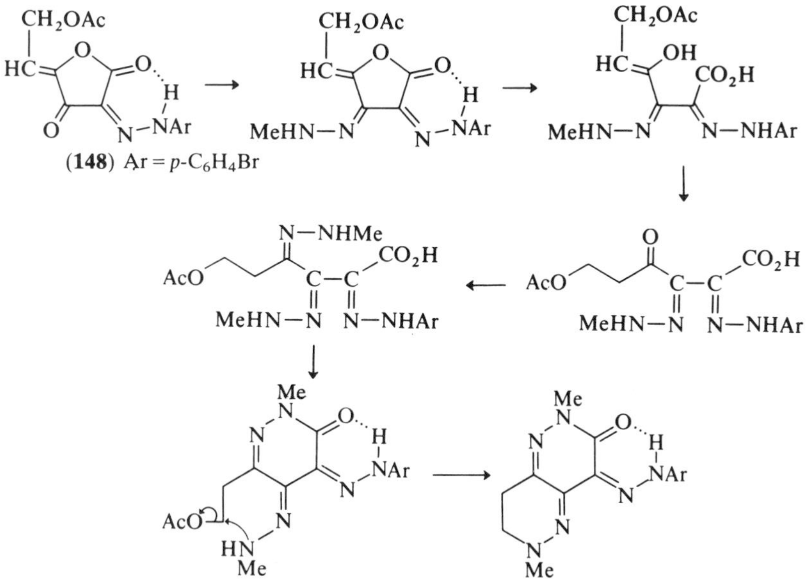

(149)

## 2.17.4.4 Pyridazino[4,5-*c*]pyridazine

Compounds containing this ring system are prepared by the reaction of 3,4-disubstituted pyridazines with hydrazine. 6-Methylpyridazine-3,4-dicarbaldehyde (150) condenses with hydrazine to give 3-methylpyridazino[4,5-*c*]pyridazine (151) ⟨73JHC1081⟩. 5-Oxo and 5,8-dioxo substituted pyridazino[4,5-*c*]pyridazines are prepared in like manner by condensing ethyl 3-formylpyridazine-4-carboxylates and diethyl pyridazine-3,4-dicarboxylates, respectively, with hydrazine. Pyridazine-3,4-dicarbonitrile (152) also reacts with hydrazine to give the diamino heterocycle (153) ⟨67JHC393⟩.

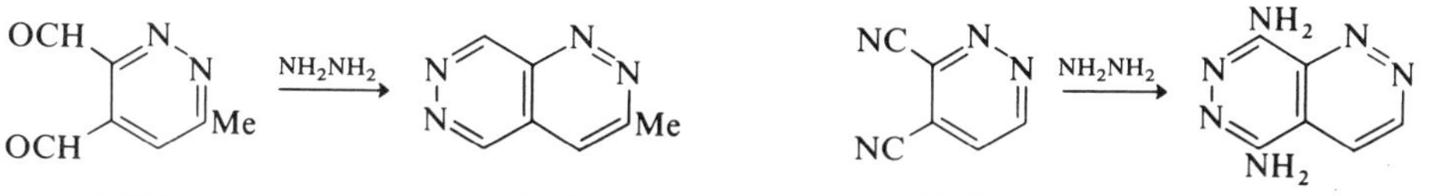

One final preparation of this ring system involves the novel transformation of the dihydrofurans (154). Reaction of these compounds with hydrazine affords tetrahydropyridazino[4,5-*c*]pyridazines (155). A possible mechanism for this reaction is shown ⟨75JCS(P1)1326⟩.

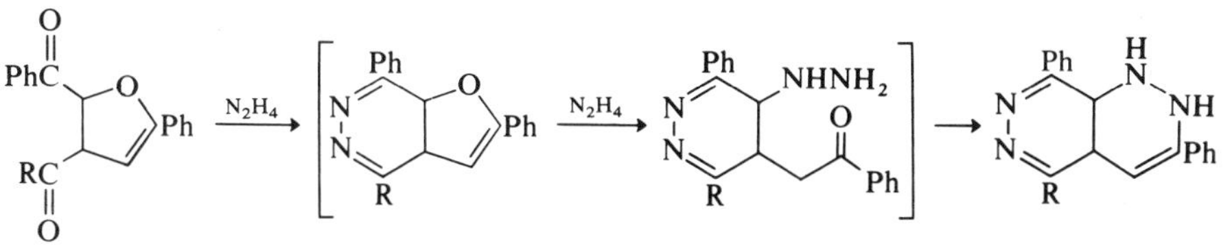

## 2.17.4.5 Pyridazino[4,5-*d*]pyridazine

All preparations of this ring system involve the condensation of hydrazine and substituted hydrazines with functionally disubstituted pyridazines. The unsubstituted ring compound (6) is prepared by reducing diethyl pyridazine-4,5-dicarboxylate with LAH to give the dialdehyde (156). This intermediate product is not isolated but is reacted immediately with hydrazine to give the desired ring system (6) ⟨67CC1006⟩.

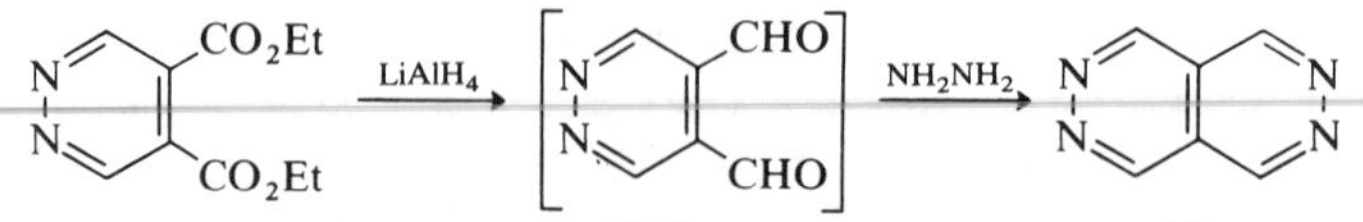

(156)                              (6)

1,4-Diaminopyridazino[4,5-*d*]pyridazine is prepared by the condensation of 4,5-dicyanopyridazine with hydrazine ⟨68JHC53⟩. In like manner, condensation of the pyridazine-4,5-dicarboxylic esters (**157**) with hydrazines gives the corresponding pyridazino[4,5-*d*]-pyridazine-1,4-(2*H*,3*H*)-diones (**158**). These compounds can also be prepared by condensing pyridazine-4,5-dicarboxylic acid anhydrides with hydrazines ⟨75JHC95⟩.

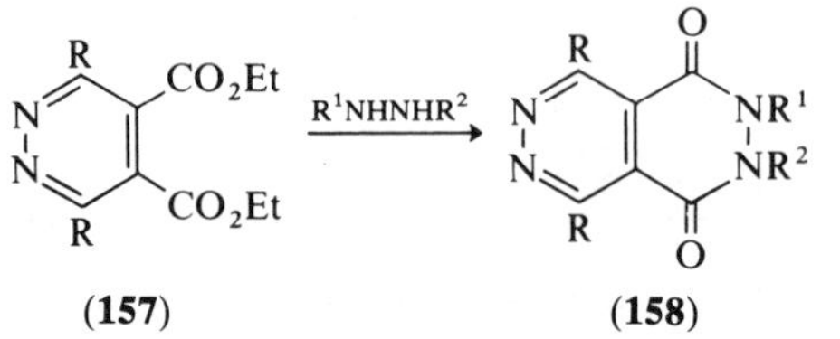

(157)                              (158)

1,4,5,8-Tetrahydroxypyridazino[4,5-*d*]pyridazine is prepared by condensing tetraethyl ethylenetetracarboxylate with hydrazine ⟨72JCS(P1)953⟩. Condensation of hydrazine with ethyl 5-acyl-4-pyridazinecarboxylates (**159**) gives pyridazino[4,5-*d*]pyridazin-1(2*H*)-ones (**160**) bearing an alkyl or aryl substituent at the 4-position. Pyridazines (**159**) are prepared by the homolytic acylation of ethyl 4-pyridazinecarboxylate ⟨79M365⟩.

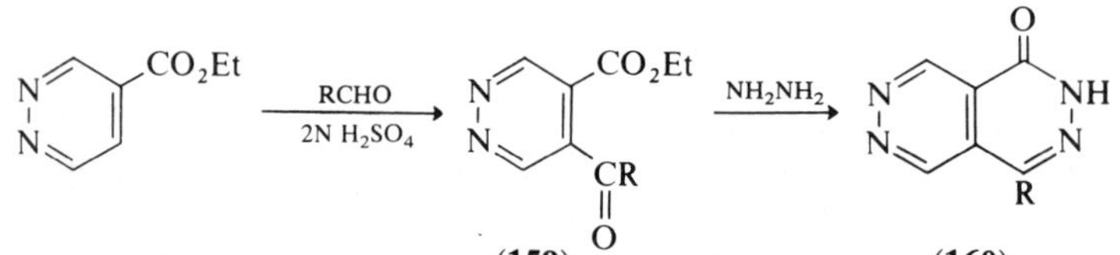

(159)                              (160)

An early report of the preparation of 4a,8a-dihydro-1,4-dihydroxy-5,8-dimethyl-pyridazino[4,5-*d*]pyridazine by the reaction of diethyl diacetylsuccinate with hydrazine ⟨04CB91⟩ was later shown to be incorrect ⟨56JA159⟩. The structure of this compound was shown to be 3,3'-dihydroxy-5-methyl-4,4'-bipyrazolyl.

### 2.17.4.6  Pyrimido[1,2-*b*]pyridazine

2*H*-Pyrimido[1,2-*b*]pyridazin-2-ones are prepared by the reaction of 3-aminopyridazines with a variety of condensing reagents. Reaction of 3-amino-6-chloropyridazine (**161**) with 3-chloroacrylic acids (and acid chlorides) (**162**) gives the corresponding intermediate acrylamides (**163**). Cyclization of these intermediates in refluxing xylene yields 7-chloro-2*H*-pyrimido[1,2-*b*]pyridazin-2-ones (**164**). In the case of the reaction of the pyridazine (**161**) with β,2,4-trichloroatropoyl chloride (**165**), the intermediate acrylamide was not isolated, but rather the cyclized 2*H*-pyrimido[1,2-*b*]pyridazin-2-one (**32**) was isolated as the major product. A small amount of 4*H*-pyrimido[1,2-*b*]pyridazin-4-one (**166**) was also isolated ⟨71JOC3506⟩.

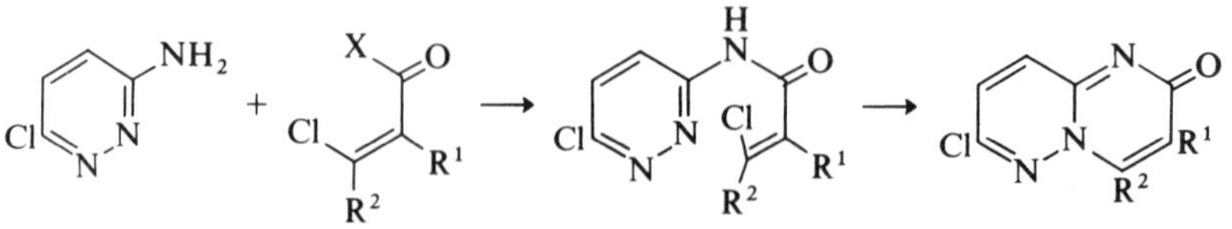

(161)           (162)              (163)              (164)

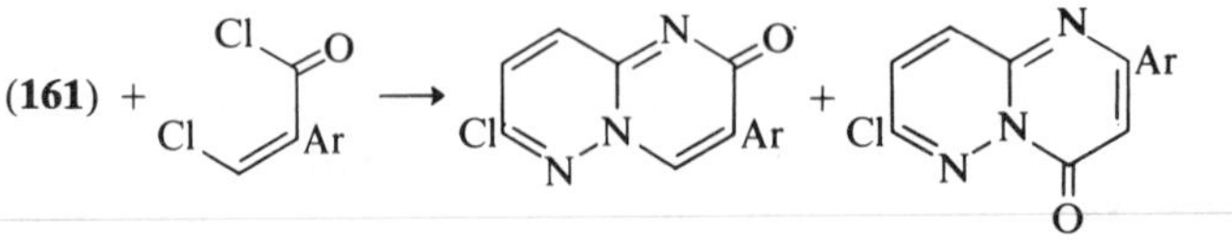

(165) Ar = 2,4-Cl₂C₆H₃       (32) 70%        (166) 2%

When cyclization of the acrylamide (**167**) was attempted in refluxing 1,2,4-trichloroben-zene, 6-chloro-3-dichloromethylene-3*H*-imidazo[1,2-*b*]pyridazin-2-one (**35**) was formed instead of the expected pyrimido[1,2-*b*]pyridazin-4-one (**34**). The latter compound could be obtained by conducting the cyclization in refluxing xylene. Since (**34**) and (**35**) are in practice not thermally interconvertible (at 210 °C), it can be assumed that (**34**) is not a precursor in the formation of (**35**) ⟨71JOC3506⟩.

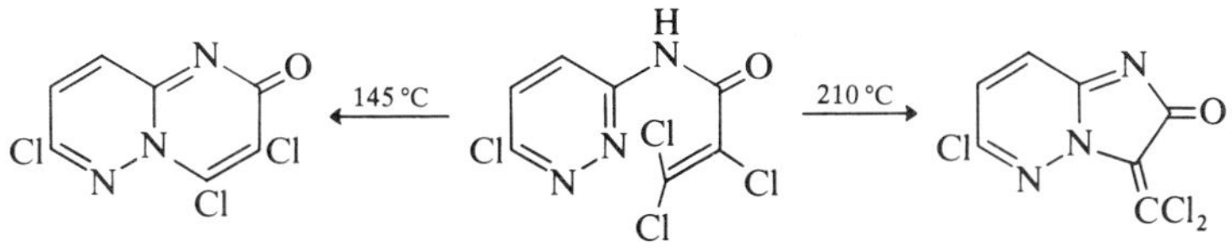

(**34**)          (**167**)          (**35**)

3-Aminopyridazines can also be condensed with $\beta$-keto esters in the presence of polyphos-phoric acid to give 2*H*-pyrimido[1,2-*b*]pyridazin-2-ones. 3-Ethoxycarbonylpyrimido-[1,2-*b*]pyridazin-4-ones (**169**) are obtained by the reaction of 3-aminopyridazines (**168**) with diethyl ethoxymethylenemalonate in the presence of polyphosphoric acid ⟨71JOC2457⟩.

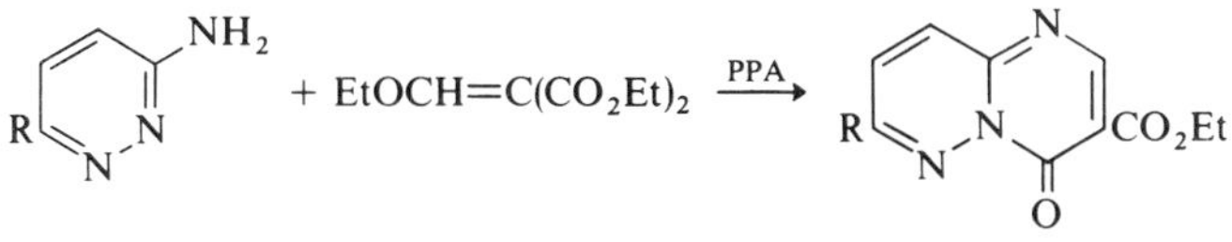

(**168**)          (**169**)

The fully aromatic and unsubstituted pyrimido[1,2-*b*]pyridazinium perchlorate (**171**) has been prepared by the condensation of 3-aminopyridazine (**170**) with 1,1,3,3-tetraethoxypropane in the presence of polyphosphoric acid. The product was precipitated from the reaction mixture by treatment with perchloric acid and ice. Several alkyl, aryl and halogenated derivatives of this ring system have also been prepared ⟨71JOC2457⟩.

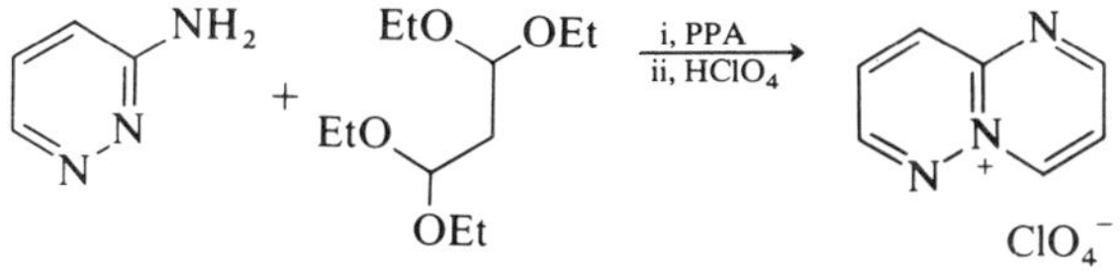

(**170**)          (**171**)

The mesomeric betaine pyrimido[1,2-*b*]pyridazine-2,4-diones (**173**) have been prepared from 2-aminopyridazines (**172**) by reaction with bis(2,4,6-trichlorophenyl) methylmalonate. These heterocycles are extremely stable molecules. They have high melting points and are stable to nucleophilic attack by benzylamine in refluxing chloroform, ethanol or acetonitrile for 7 days ⟨76JPS(65)1505⟩.

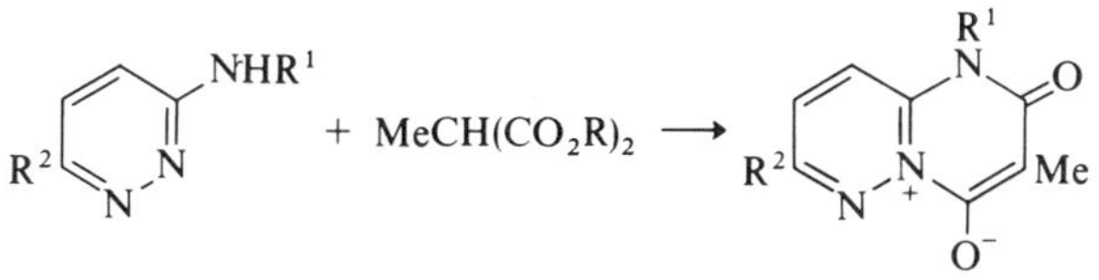

(**172**)          (**173**)

One final interesting preparation of this ring system involves the reaction of the tetrazolo[1,5-*b*]pyridazine (**174**) with polyphosphoric acid to give 7-azido-4*H*-pyrimido[1,2-*b*]pyridazin-4-one (**175**). This reaction involves azido–tetrazolo valence isomerization, which has been studied in other heterocyclic systems ⟨71JHC1055⟩.

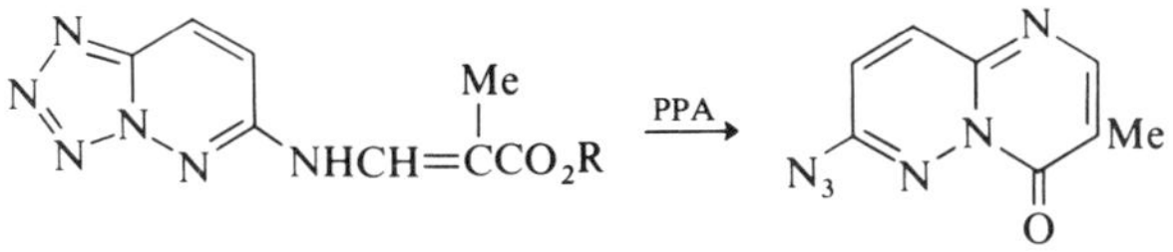

(**174**)          (**175**)

### 2.17.4.7 Pyrimido[4,5-c]pyridazine

The majority of pyrimido[4,5-c]pyridazines have been prepared from pyrimidine precursors. The chloropyrimidines (**176**) give the desired heterocyclic ring (**177**) on reaction with hydrazine ⟨72BSF1483⟩. Hydrazine also reacts with ethyl α-diazo-β-oxo-5-(4-chloro-2-methylthiopyrimidine)propionate (**178**) to give the pyrimido[4,5-c]pyridazine-3-carboxamide (**78**). A mechanism for this interesting reaction has been proposed as shown, on the basis of the detection of hydrogen azide in the reaction mixture. There is no precedent for the reaction of the α-carbon of α-diazo-β-oxopropionates with nucleophiles under basic conditions ⟨76CPB2637⟩.

The diazopropionate (**178**) also undergoes reaction with triphenylphosphine to give the product (**179**). This product was shown to be formed by way of the intermediate phosphazene and hydrazone as shown ⟨78CPB14⟩.

Hydrazinopyrimidines can also serve as starting materials for the preparation of pyrimido[4,5-c]pyridazines. Pyrimido[4,5-c]pyridazine-5,7-diones (**181**) are prepared in nearly quantitative yield by reacting the hydrazinouracils (**180**) with aqueous glyoxal ⟨75JHC1221⟩. In an analogous reaction, hydrazinopyrimidines react readily with α-keto esters to give pyrimido[4,5-c]pyridazines with an oxo substituent in the 4-position ⟨78JOC4844⟩.

If this type of condensation is carried out with hydrazinoisocytosines (**182**) and α,γ-diketo esters, 1H-pyrimido[4,5-c]-1,2-diazepines (**183**) are formed as the major products. Only a small amount of pyrimido[4,5-c]pyridazines (**184**) can be isolated from these reactions ⟨82JOC667⟩.

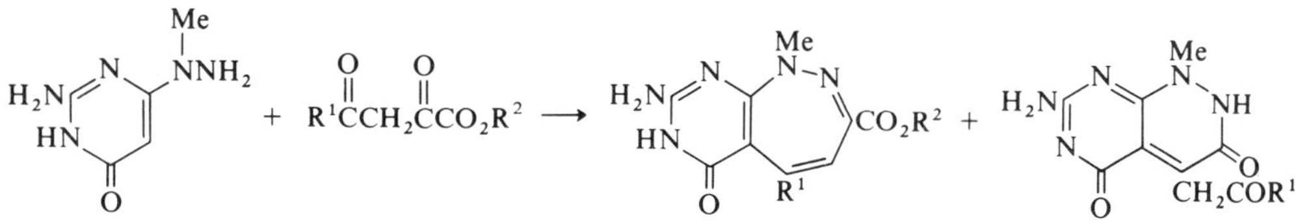

**(182)** + R$^1$CCH$_2$CCO$_2$R$^2$ → **(183)** + **(184)**

6-Hydrazino-1,3-dimethyluracil (**185**) reacts with phenacyl bromides in DMF to give the corresponding 3-aryl-6,8-dimethylpyrimido[4,5-*c*]pyridazine-5,7-diones (**186**). 6-Benzyl-idenehydrazino-1,3-dimethyluracils are also known to react with DMF dimethyl acetal to give pyrimido[4,5-*c*]pyridazines ⟨78JHC781⟩.

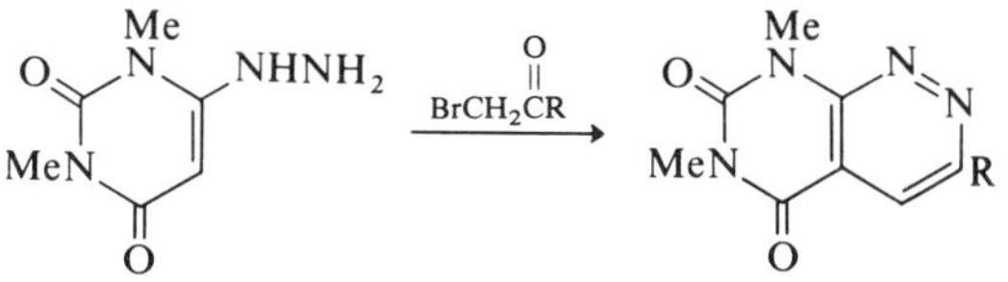

**(185)** → **(186)**

Pyridazine-3,4-dicarboxamide (**187**) undergoes the Hofmann reaction followed by ring closure to give pyrimido[4,5-*c*]pyridazine-5,7-dione (**188**) as the major product. In an analogous type of ring closure, 3-amino-6-methylpyridazine-4-carboxamide reacts with ethyl orthoformate to give 3-methylpyrimido[4,5-*c*]pyridazin-5-one ⟨68JHC523⟩.

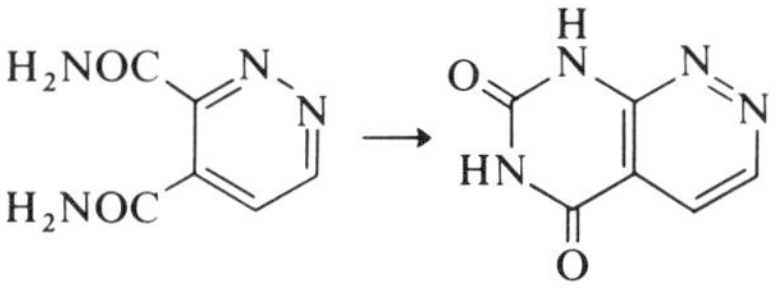

**(187)** → **(188)**

### 2.17.4.8 Pyrimido[5,4-*c*]pyridazine

Pyrimido[5,4-*c*]pyridazines have been prepared by three major routes, namely: (1) the Hofmann reaction on pyridazine-4,5-dicarboxamides; (2) condensation of 4-aminopyridazine-3-carboxamide or 4-amino-3-cyanopyridazine with various reagents; and (3) reaction of 6-methyl-5-phenylazopyrimidines with *t*-butoxybis(dimethylamino)methane. The first of these routes is not of practical synthetic value since the desired ring system is obtained as the minor product along with the isomeric pyrimido[4,5-*c*]pyridazines.

3-Methylpyridazine-4,5-dicarboxamide (**189**) gives a mixture of the two products (**190**) and (**191**), along with an unidentified product, when reacted with potassium hydroxide in a Hofmann type reaction ⟨68JHC523⟩.

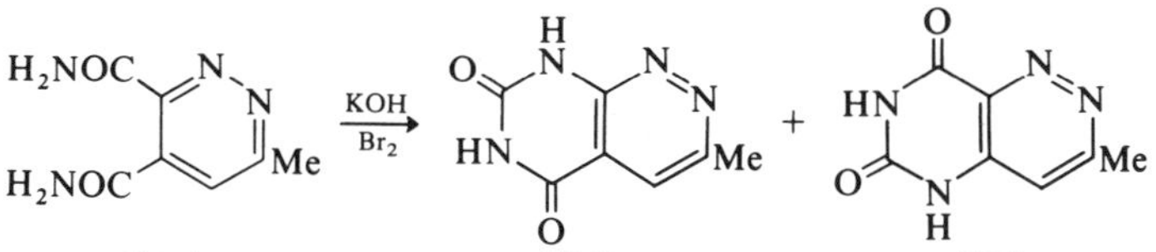

**(189)** → **(190)** + **(191)**

4-Aminopyridazine-3-carboxamide condenses with urea and ethyl orthoformate to give pyrimido[5,4-*c*]pyridazine-6,8-dione and pyrimido[5,4-*c*]pyridazin-8-one, respectively. In like manner, 8-aminopyrimido[5,4-*c*]pyridazine is formed by the reaction of 4-amino-3-cyanopyridazine with formamide ⟨68JHC523⟩.

Treatment of 6-methyl-5-phenylazouracil (**192**) with the formylating reagent *t*-butoxybis(dimethylamino)methane (BBDM) in DMF affords the pyrimido[5,4-*c*]-pyridazinedione (**193**). This reaction probably occurs by the mechanism shown. Reaction of the pyrimidinethione (**194**) under these conditions gives 6-(*N,N*-dimethylamino)-2-phenylpyrimido[5,4-*c*]pyridazin-8(2*H*)-one (**195**) ⟨78JOC2536⟩.

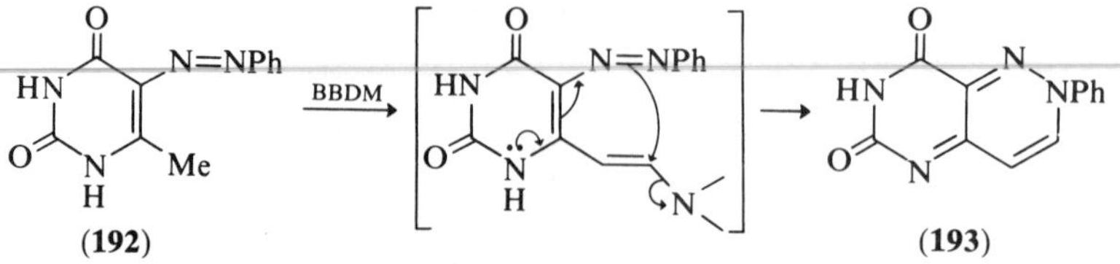

**(192)** → BBDM → **(193)** → **(193)**

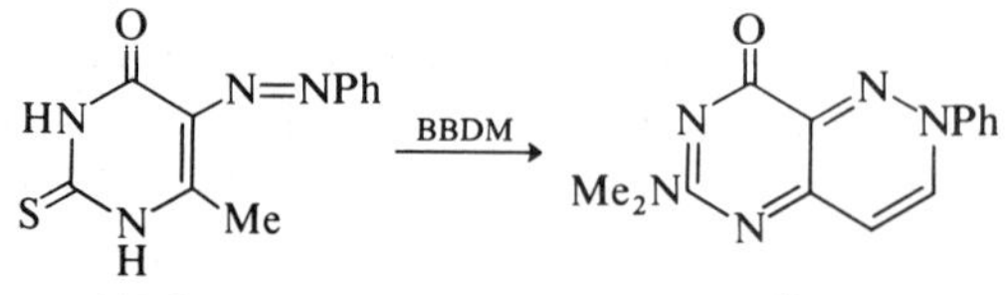

**(194)** → BBDM → **(195)**

### 2.17.4.9  Pyrimido[4,5-*d*]pyridazine

Most pyrimido[4,5-*d*]pyridazines are prepared from 4,5-disubstituted pyridazine precursors. The versatile intermediate pyrimido[4,5-*d*]pyridazine-2,4-dione (**85**) can readily be formed in high yield from pyridazine-4,5-dicarboxamide (**196**) by reaction with sodium hypobromite in a Hofmann type reaction 〈68JHC53〉.

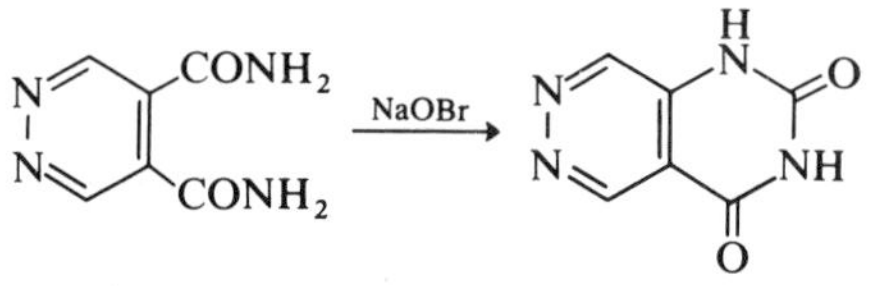

**(196)** → NaOBr → **(85)**

Ethyl 5-aminopyridazine-4-carboxylate (**197**) is a useful intermediate for the preparation of a variety of pyrimido[4,5-*d*]pyridazines. *N*-Acetylation with acetic anhydride followed by reaction with ethanolic ammonia at room temperature gives 2-methylpyrimido-[4,5-*d*]pyridazin-4-one (**198**) in high yield. Treatment of (**197**) with ethanolic ammonia with heating under pressure gives the intermediate carboxamide (**199**), which can be ring closed to the product (**200**) on reaction with ethyl orthoformate. 2-Aminopyrimido[4,5-*d*]-pyridazin-4-one (**201**) can be prepared in one step from compound (**197**) by fusion with guanidine carbonate 〈68JHC845〉. The carboxamide (**199**) is also known to react with 1,3-diphenylguanidine and sodium hydride in anhydrous THF to give 2-anilinopyrimido-[4,5-*d*]pyridazin-4(3*H*)-one 〈77JCS(P1)1020〉.

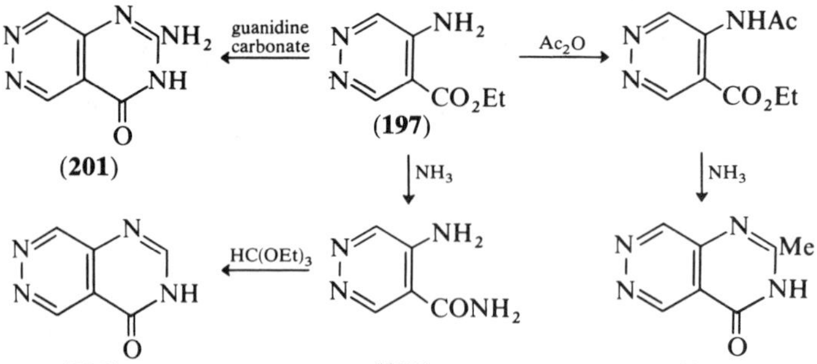

**(201)**    guanidine carbonate ←    **(197)**    Ac₂O →    

**(200)**    ← HC(OEt)₃    **(199)**    NH₃    **(198)**

Diethyl pyrimidine-4,5-dicarboxylates react readily with hydrazine to give pyrimido-[4,5-*d*]pyridazine-5,8-diones 〈72CPB1513〉. In a similar manner the acyl pyrimidines (**202**) react with hydrazines to give the products (**203**) 〈76BSF1549〉. Hydrazines also react with

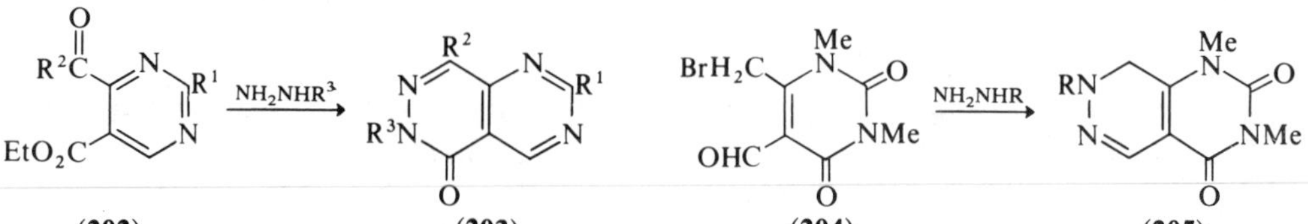

**(202)**    NH₂NHR³ →    **(203)**        **(204)**    NH₂NHR →    **(205)**

6-bromomethyl-1,3-dimethyl-5-formyluracil (**204**) to give the pyrimido[4,5-*d*]pyridazine-2,4-diones (**205**). However, when acetylhydrazines are reacted with the starting uracil (**204**), the ring opened acetylhydrazones are formed ⟨78S463⟩.

### 2.17.4.10 1*H*-Pyrazino[1,2-*b*]pyridazine

Only one derivative of this ring system is reported in the literature, namely octahydro-1*H*-pyrazino[1,2-*b*]pyridazine-5,8-dione (**206**) ⟨71JCS(C)528⟩. This compound is isolated in both the acid and base hydrolysis of the monamycins, a mixture of cyclohexadepsipeptide antibiotics.

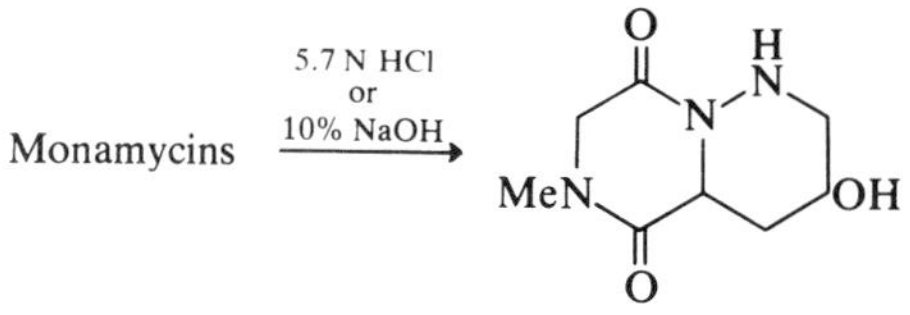

(**206**)

### 2.17.4.11 Pyrazino[2,3-*c*]pyridazine

Several 5,6,7,8-tetrahydropyrazino[2,3-*c*]pyridazines and a few derivatives of the fully aromatic ring system have been reported. 3,4,6-Trichloropyridazine (**208**) reacts with the ethylenediamine (**207**) to give the tetrahydro compound (**209**). Methyl chloride and hydrogen chloride are noted as by-products in this reaction. With the appropriately substituted ethylenediamines, the 8-ethyl and other 5-alkyl derivatives of this ring system can be prepared ⟨58AG5⟩. The fully aromatic ring system is formed by the condensation of 3,4-diaminopyridazine (**210**) with benzil, biacetyl and phenylglyoxal to give 6,7-diphenyl-, 6,7-dimethyl- and 6(7)-phenyl-pyrazino[2,3-*c*]pyridazine (**211**), respectively ⟨81AJC1361⟩.

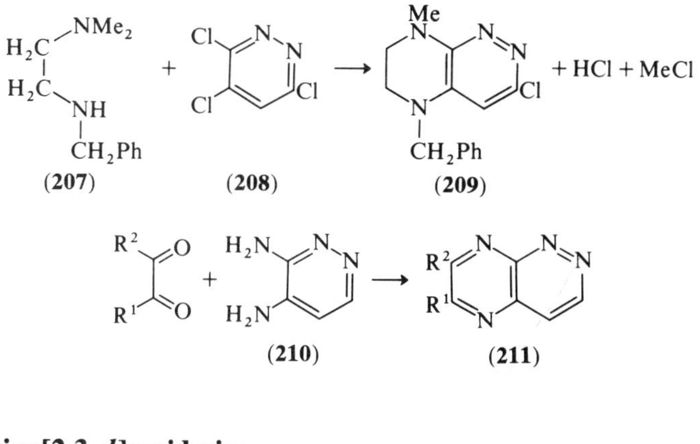

(**207**)     (**208**)     (**209**)

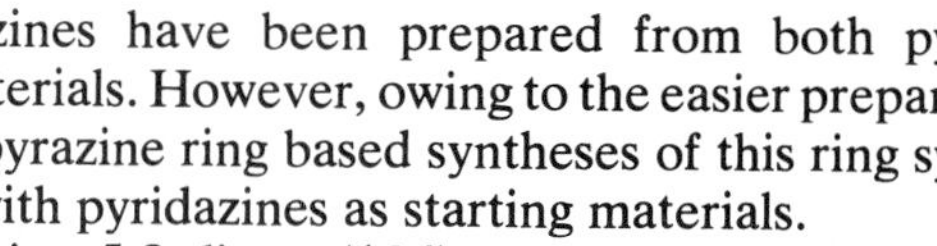

(**210**)     (**211**)

### 2.17.4.12 Pyrazino[2,3-*d*]pyridazine

Pyrazino[2,3-*d*]pyridazines have been prepared from both pyrazine and pyridazine derivatives as starting materials. However, owing to the easier preparation of the appropriate pyrazine intermediates, pyrazine ring based syntheses of this ring system are more popular than those which begin with pyridazines as starting materials.

Pyrazino[2,3-*d*]pyridazine-5,8-dione (**104**) can be prepared from either pyrazine-2,3-dicarboxylic acid anhydride (**212**) or dimethyl pyrazine-2,3-dicarboxylate (**213**) by reaction with hydrazine hydrochloride or hydrazine hydrate, respectively. The preparation based

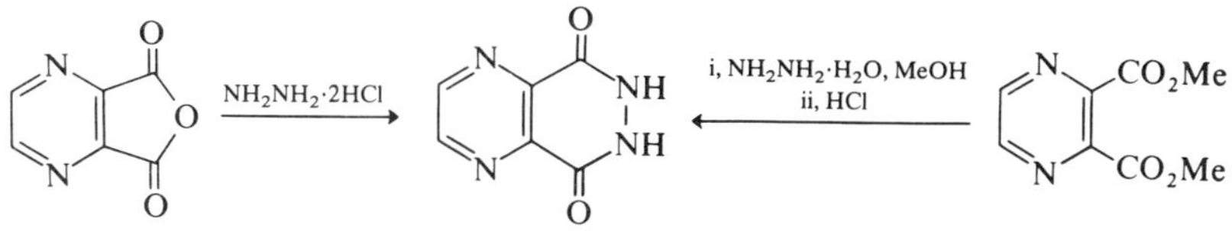

(**212**)     (**104**)     (**213**)

on the diester (**213**) gives a higher yield of the desired product. In like manner, reaction of pyrazine-2,3-dicarbonitrile (**214**) with hydrazine gives the diamino product (**215**) ⟨66JHC512⟩.

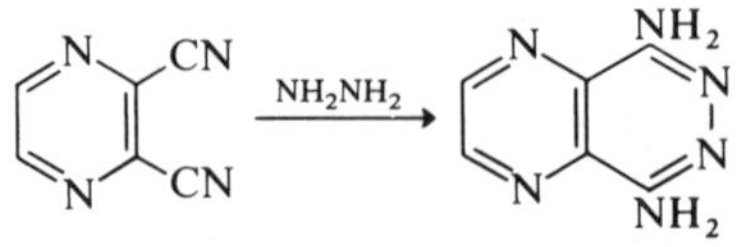

(**214**)                                    (**215**)

Unsubstituted pyrazino[2,3-*d*]pyridazine (**14**) is prepared by the reaction of 4,5-diaminopyridazine (**216**) with glyoxal. Reaction of this starting pyridazine with pyruvaldehyde produces the methyl substituted product (**217**) ⟨66JHC512⟩. Similarly, pyrazino-[2,3-*d*]pyridazin-5-one (**218**) can be prepared by the reaction of 4,5-diamino-3-pyridazinone with glyoxal ⟨69JHC93⟩.

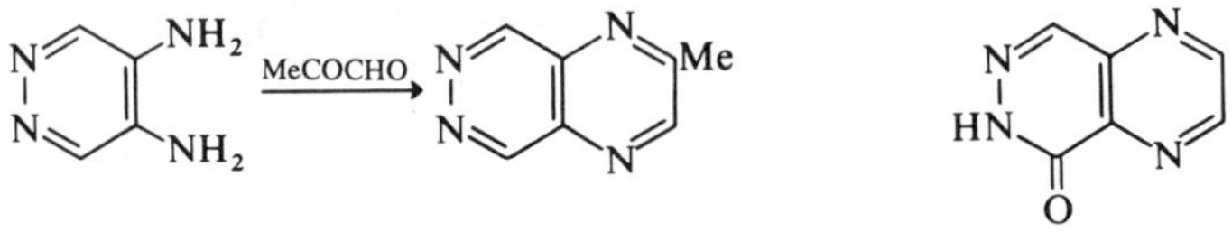

(**216**)                  (**217**)                  (**218**)

## 2.17.4.13 Pyrimido[1,2-*a*]pyrimidine

Most derivatives of this ring system are prepared from 2-aminopyrimidines. Ethoxymethylenemalonic ester (**220**) is a versatile synthon which can be reacted with 2-aminopyrimidines to give pyrimido[1,2-*a*]pyrimidine precursors. Reaction of 2-aminopyrimidine (**219**) with this ester gives the aminomethylenemalonic ester (**221**), which can be thermally cyclized to the heterocyclic ester (**222**). The enamine analogous to (**221**) which is derived from 2-amino-4,6-dimethylpyrimidine cannot be cyclized under these conditions. Presumably this cyclization fails due to steric hindrance of the methyl groups ⟨72JMC1203⟩.

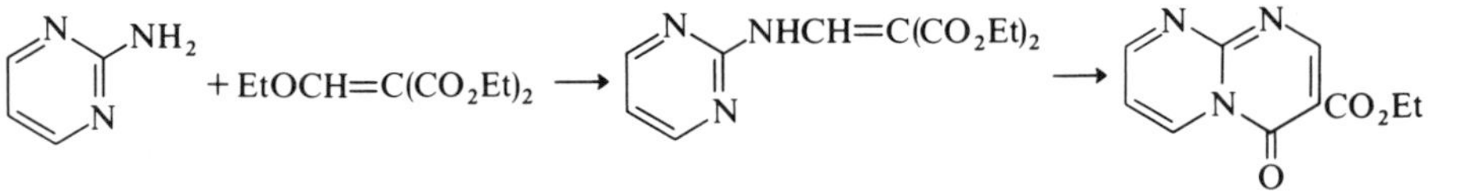

(**219**)          (**220**)                  (**221**)                  (**222**)

2-Aminopyrimidine (**219**) also reacts readily with 3-chloropropionyl chloride in chloroform–pyridine solution to give the amide (**223**). If this amide is heated in DMSO, ring closure occurs giving 3,4-dihydro-2*H*-pyrimido[1,2-*a*]pyrimidin-2-one hydrochloride (**224**) in near quantitative yield. Reaction of the starting pyrimidine (**219**) with acryloyl chloride gives the amide (**225**), which can be cyclized by heating in *m*-xylene to give the product (**226**). The intermediate acryloylaminopyrimidine (**225**) can also be prepared by reaction of the amide (**223**) with aqueous potassium carbonate ⟨71JOC604⟩.

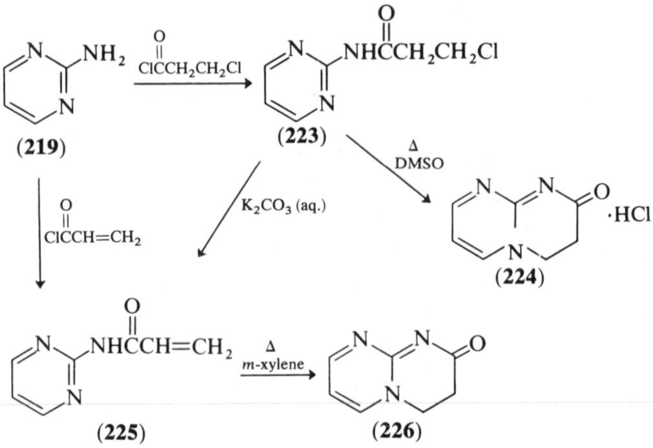

(**219**)                  (**223**)                          (**224**)

(**225**)                          (**226**)

The preparation of 4*H*-3-(2-hydroxyethylthio)-2-methylpyrimido[1,2-*a*]pyrimidin-4-one (**228**) involves an interesting ring opening of the 5,6-dihydro-1,4-oxathiin ring system. Reaction of 2-aminopyrimidine (**219**) with 5,6-dihydro-2-methyl-1,4-oxathiin-3-carbonyl chloride (**227**) gives the desired product (**228**) ⟨73CJC2650⟩. In a similar manner, 3-hydroxyethylpyrimido[1,2-*a*]pyrimidin-4-ones are formed *via* ring opening of the tetrahydrofuran ring system ⟨73LA103⟩.

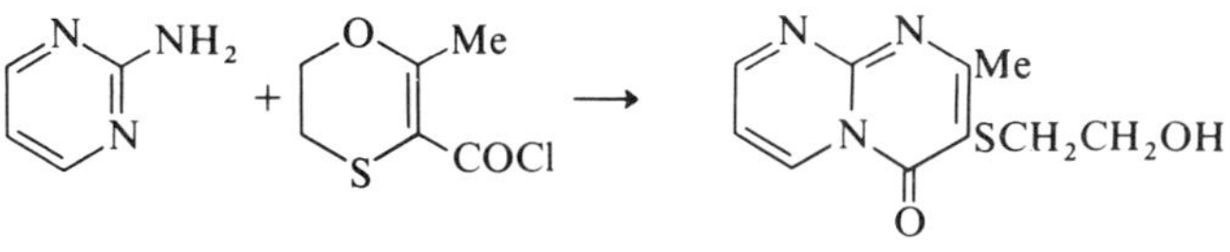

(**219**)        (**227**)                    (**228**)

Reaction of 2-amino-4-pyrimidinone (**229**) with an excess of acrylonitrile gives the intermediate (**230**). This intermediate can be cyclized to the product (**231**) in excellent yield by heating in glacial acetic acid ⟨61JOC1891⟩. Similarly, 3,4-dihydro-6,8-dimethyl-pyrimido[1,2-*a*]pyrimidin-2-one hydrobromide (**233**) is prepared by the reaction of 2-amino-4,6-dimethylpyrimidine (**232**) with 3-bromopropionic acid. This reaction also proceeds *via* a pyrimidinylpropionic acid intermediate ⟨61JOC1891⟩.

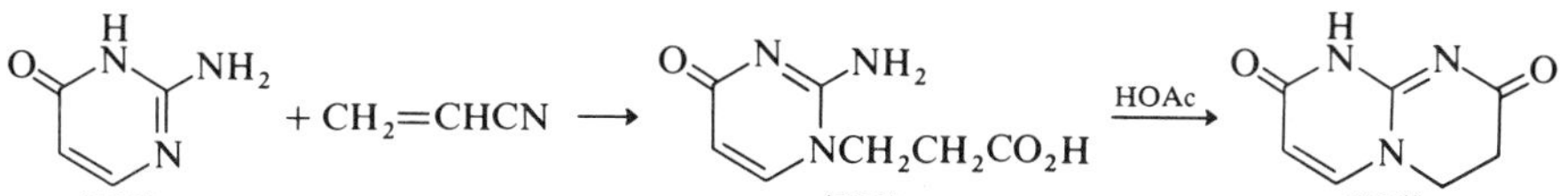

(**229**)                    (**230**)                    (**231**)

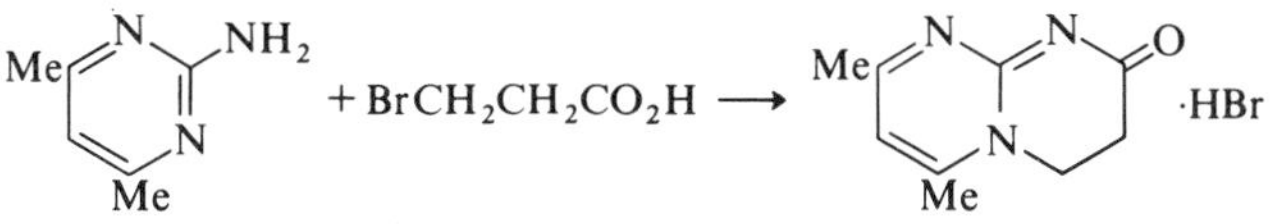

(**232**)                    (**233**)

The heteroaromatic betaine (**110**) is formed in the reaction of 4,6-dimethyl-2-methyl-aminopyrimidine (**234**) with carbon suboxide ⟨73JOC3485⟩.

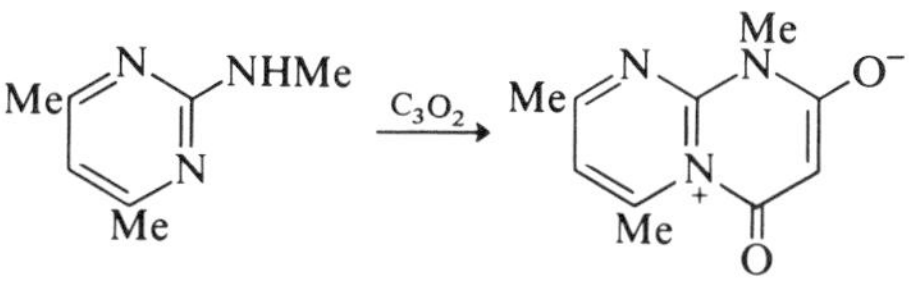

(**234**)                    (**110**)

One interesting preparation of pyrimido[1,2-*a*]pyrimidines from sulfimides has been reported. Reaction of the sulfimides (**235**) with diphenylcyclopropenone gives the cyclized products (**236**) in good yield. The starting sulfimides can be readily prepared by reaction of the corresponding 2-aminopyrimidines with *t*-butyl hypochlorite and dimethyl sulfide. This reaction is postulated to proceed *via* a ketene intermediate ⟨75JCS(P1)1969⟩.

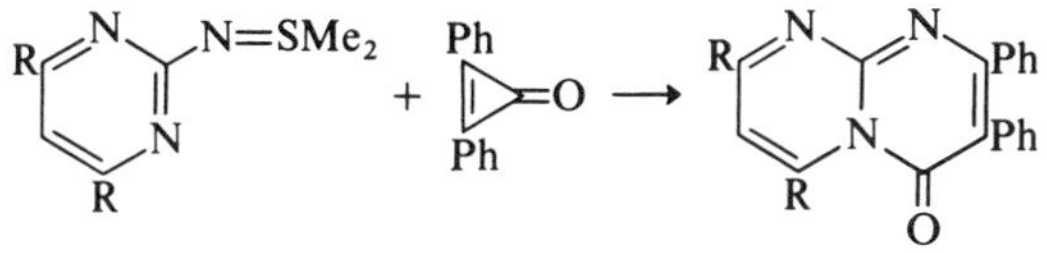

(**235**)                    (**236**)

A one step synthesis of the diimine (**237**) has been reported. This proceeds by the reaction of guanidine with acrylonitrile in DMF to give the product in good yield ⟨68JOC3354⟩.

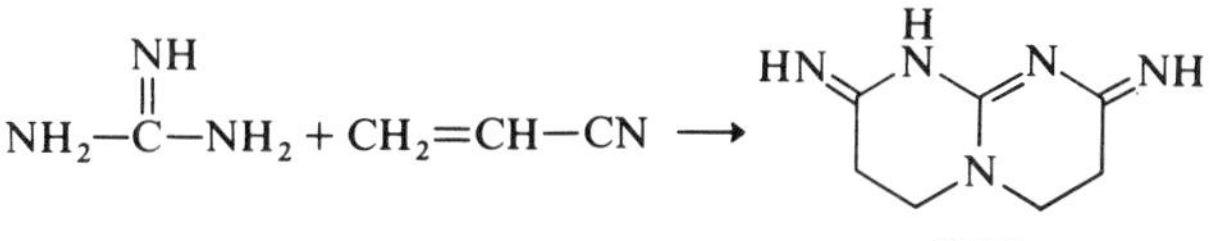

(**237**)

## 2.17.4.14 Pyrimido[1,6-*a*]pyrimidine

Pyrimido[1,6-*a*]pyrimidines are conveniently prepared from functionally 4-substituted pyrimidine precursors. Specifically, 4-chloro-, 4-amino- and 4-methylthio-pyrimidines serve as useful intermediates for the synthesis of derivatives of this ring system.

2,4-Dichloro-5-fluoropyrimidine (**238**) reacts with 3-aminopropanol in the presence of triethylamine to give the intermediate hydroxypropylamino substituted pyrimidine (**239**). Reaction of this compound with thionyl chloride results in the desired ring closure to the product (**240**), which is isolated as the hydrochloride salt ⟨65USP3320256⟩.

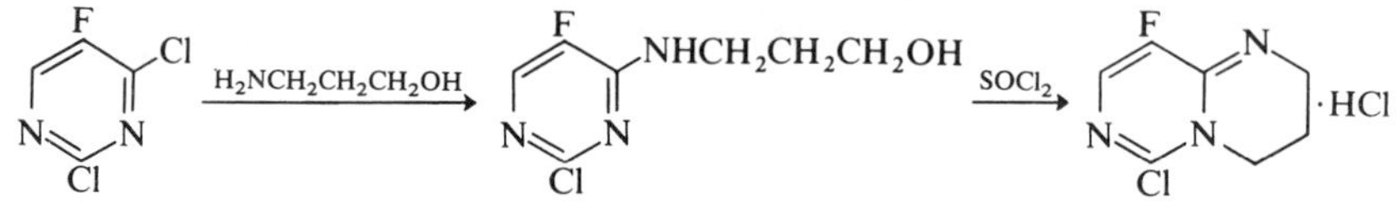

(**238**)                    (**239**)                    (**240**)

4,6-Diaminopyrimidine (**241**) reacts readily with diethyl ethoxymethylenemalonate under fusion conditions to give the intermediate (**242**). Following the acetylation of the amino group on this pyrimidine intermediate in acetic anhydride, thermal cyclization yields the condensed heterocycle (**243**) ⟨72JOC3980⟩. Reaction of the aminopyrimidine (**244**) with the benzyl substituted malonic ester (**245**) gives the desired product (**246**) in good yield. Other 4-aminopyrimidines are reported to react in like manner ⟨61M1184⟩.

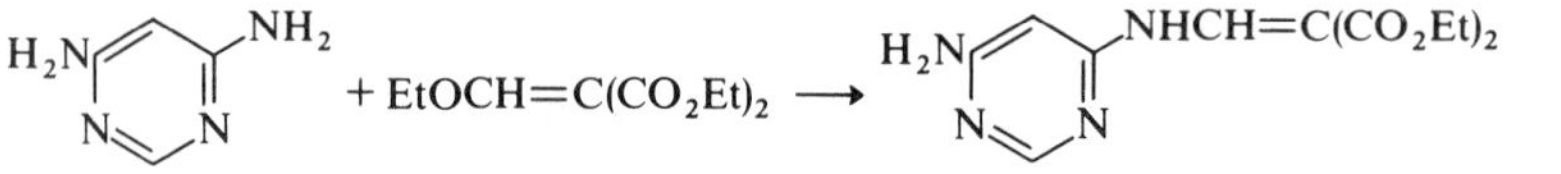

(**241**)                    (**242**)

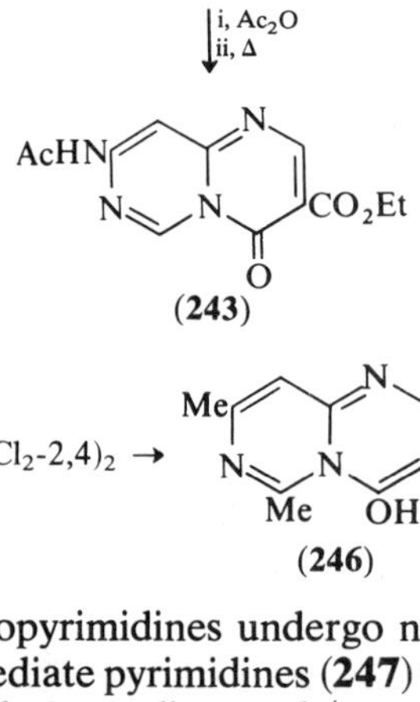

(**243**)

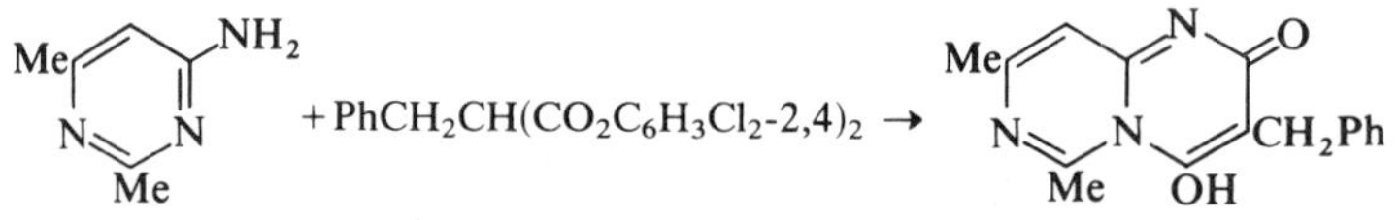

(**244**)                    (**245**)                    (**246**)

4-Methylthiopyrimidines as well as 4-chloropyrimidines undergo nucleophilic displacement with 3-aminopropanol to give the intermediate pyrimidines (**247**) which can be reacted to yield the desired ring system, as has already been discussed ⟨63JA4024⟩. These methylthiopyrimidines also react with amino acids to give pyrimidinylamino acids. *N*-(1*H*-2-Oxo-4-pyrimidinyl)-β-alanine cyclizes with rearrangement on reaction with refluxing acetic anhydride to give 1*H*-1,2,3,4-tetrahydropyrimido[1,6-*a*]pyrimidine-2,6-dione in good yield. A mechanism for this novel rearrangement has been proposed ⟨64JOC1762⟩.

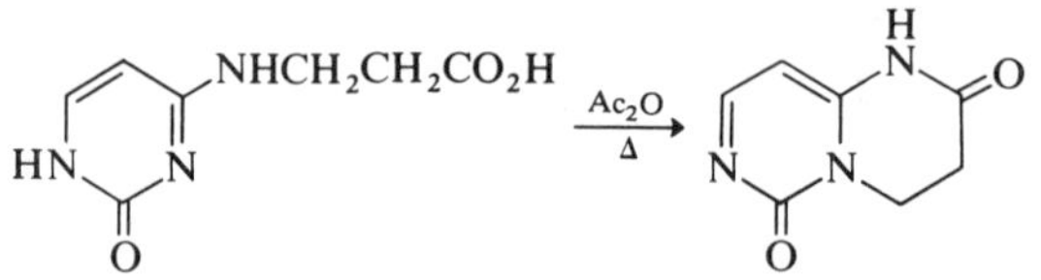

(**247**)                    (**116**)

## 2.17.4.15 Pyrimido[1,6-*c*]pyrimidine

Only one preparation of this ring system has been reported. This preparation involves the interesting reaction of compound (**248**) with benzylamine hydrochloride and formaldehyde to give the cyclized product (**249**) ⟨70M1824⟩.

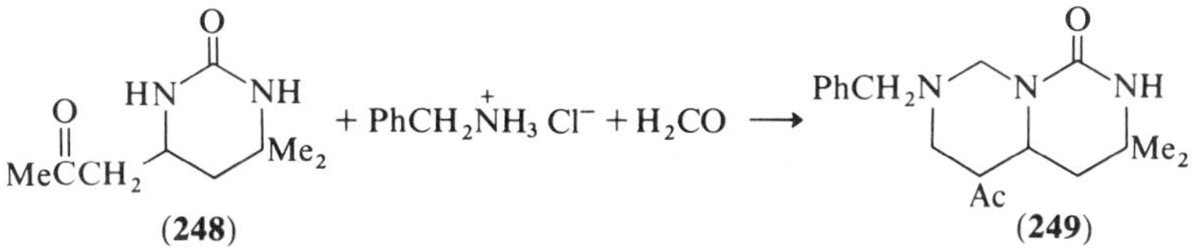

(248)       (249)

### 2.17.4.16 Pyrimido[4,5-*d*]pyrimidine

The most popular preparations of pyrimido[4,5-*d*]pyrimidines are based on cyanoethylene or cyanomethylene derivatives. In many of these preparations the appropriate acid chloride ($R^1$ = alkyl, aryl) is reacted with malononitrile to give an enol (250), which is readily alkylated to the enol ether (251). This can be condensed with guanidine or a variety of amidines ($R^2$ = H, alkyl, aryl) to yield the key intermediate 4-amino-5-cyanopyrimidines (252) ⟨81S955⟩. Reaction with a second mole of guanidine or amidine is a convenient preparation of a variety of pyrimido[4,5-*d*]pyrimidines. Under the appropriate conditions the final ring system can be obtained from the enol ether in one step ⟨68JMC568⟩.

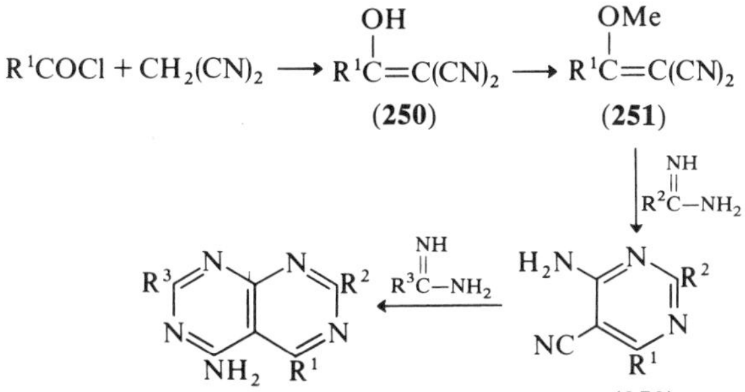

(250)       (251)

(252)

The 5,7-dioxopyrimido[4,5-*d*]pyrimidine-5,7-dione (254) can be obtained by fusion of the amide (253) with urea ⟨68JMC568⟩. This amide is prepared by hydrolysis of the corresponding cyanopyrimidine in concentrated sulfuric acid. 4-Aminopyrimidine-5-carboxamides can also be condensed with alkyl and aryl esters to give pyrimido[4,5-*d*]pyrimidin-4-ones ⟨75JHC1311⟩.

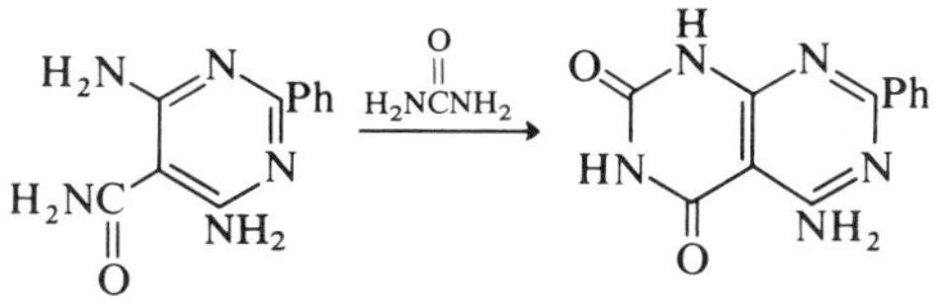

(253)       (254)

6-Aminouracils have been used extensively in the preparation of tetrasubstituted pyrimido[4,5-*d*]pyrimidines with functional groups at positions 2, 4, 5 and 7. 4-Thioxopyrimido[4,5-*d*]pyrimidine-5,7-diones (257) are formed in the thermal cyclization

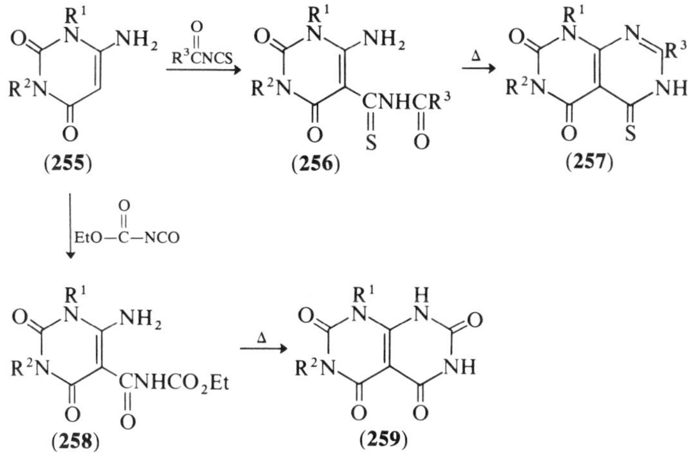

(255)       (256)       (257)

(258)       (259)

of the *N*-acylthiocarboxamides (**256**). These intermediates are obtained by electrophilic attack of *N*-acylisothiocyanates on the appropriate 6-aminouracils (**255**) ⟨74LA2019⟩. Reaction of the starting uracils with ethyl isocyanatoformate results in the formation of the intermediate *N*-(ethoxycarbonyl)carboxamidouracils (**258**). Again, these intermediates can be thermally cyclized to give pyrimido[4,5-*d*]pyrimidine-2,4,5,7-tetrones (**259**) ⟨70JHC243⟩.

5-Amino-7-methylthiopyrimido[4,5-*d*]pyrimidine-2,4-diones (**261**) can be prepared in one step from 6-aminouracils (**255**) using the versatile intermediate dimethyl cyanoimidodithiocarbonate (**260**). Condensation of the uracils (**255**) with (**260**) in the presence of potassium carbonate in DMF gives the desired products (**261**) ⟨79H503⟩.

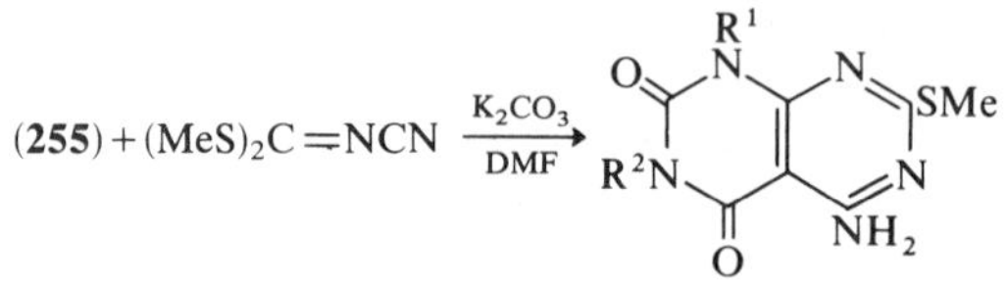

$$(255) + (MeS)_2C{=}NCN \xrightarrow[\text{DMF}]{K_2CO_3}$$

(**260**)                     (**261**)

4-Chloro-5-acylpyrimidines have also been used as intermediates in the preparation of pyrimido[4,5-*d*]pyrimidines by reaction with amidines in the presence of sodium ethoxide (Scheme 4) ⟨73CB3524⟩.

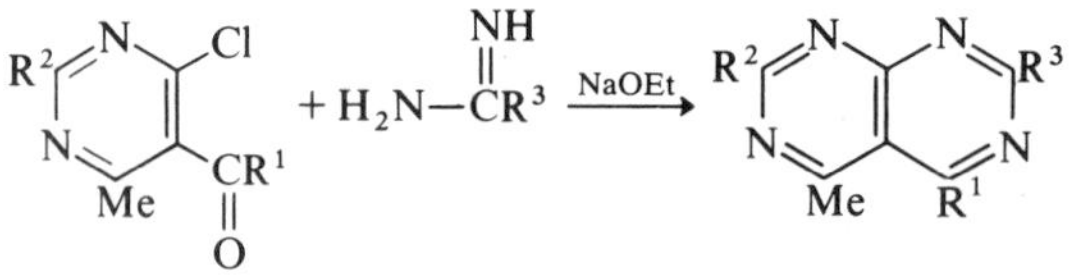

$$\text{(diagram)} + H_2N{-}\overset{\overset{NH}{\|}}{C}R^3 \xrightarrow{NaOEt} \text{(diagram)}$$

**Scheme 4**

1,3-Dimethyl-5-nitrosopyrrolo[2,3-*d*]pyrimidine-2,4(1*H*,3*H*)-diones (**262**) react with tosyl chloride in DMF in a Beckmann type rearrangement to give 1,3-dimethylpyrimido-[4,5-*d*]pyrimidine-2,4(1*H*,3*H*)-diones (**263**) ⟨73CPB473⟩. This is a novel example of the conversion of a nitrosopyrrole derivative into a pyrimidine by the Beckmann rearrangement. Reaction of the starting pyrrolopyrimidines (**262**) with phosphorus oxychloride in DMF (Vilsmeier–Haack conditions) also led to the rearranged products (**263**).

$$\text{(diagram)} \xrightarrow[\text{POCl}_3,\text{ DMF}]{\text{TsCl, DMF or}} \text{(diagram)}$$

(**262**)                     (**263**)

### 2.17.4.17 Pyrimido[5,4-*d*]pyrimidine

Pyrimido[5,4-*d*]pyrimidines are readily prepared by the condensation of 5-amino-6-carboxypyrimidine-2,4-diones (**264**) with either formamide or urea. Reaction of the starting pyrimidine (**264**) with formamide gives the pyrimido[5,4-*d*]pyrimidine-4,6,8-trione (**265**). The corresponding tetrone (**266**) results from fusion of (**264**) with urea ⟨51LA(572)217⟩. This same reaction can be carried out with the starting pyrimidine and methyl substituted ureas or methyl and phenyl substituted formamidines to give 7-methyl and 7-phenyl substituted pyrimido[5,4-*d*]pyrimidine products ⟨60LA(633)158⟩. In a similar manner, reaction of methyl 2,5,6-triaminopyrimidine-4-carboxylate with benzamidine furnishes 2-phenyl-4-hydroxy-6,8-diaminopyrimido[5,4-*d*]pyrimidine ⟨68JMC568⟩.

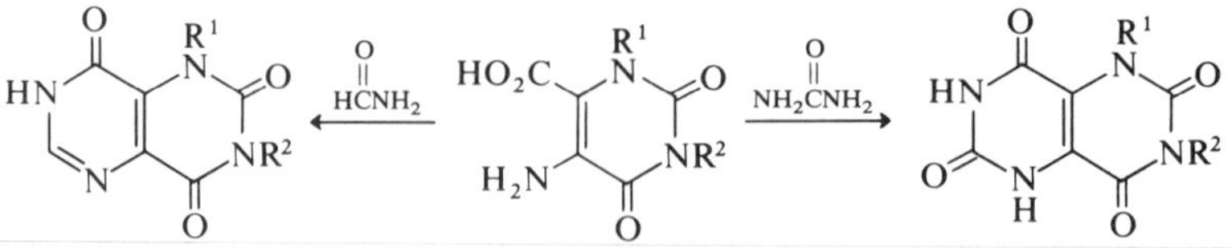

(**265**)                     (**264**)                     (**266**)

A recently reported novel synthesis of this ring system begins with diaminomaleonitrile (**267**), the tetramer of hydrogen cyanide. Reaction of this starting material with alkyl or aryl isocyanates gives the urea derivatives (**268**). These urea intermediates react with alkyl and aryl aldehydes in the presence of triethylamine to give 3,6-disubstituted 4-iminopyrimido[5,4-*d*]pyrimidine-2,8-diones (**271**). A proposed mechanism for this cyclization is shown. The preparation of the desired products can also be achieved by the reaction of the urea intermediates (**268**) with ethyl acylacetates. This condensation apparently proceeds by elimination of ethyl acetate. Ketones also react with the ureas (**268**), but in this case the corresponding intermediate imines (**270**) are isolated as products ⟨78JOC3231⟩.

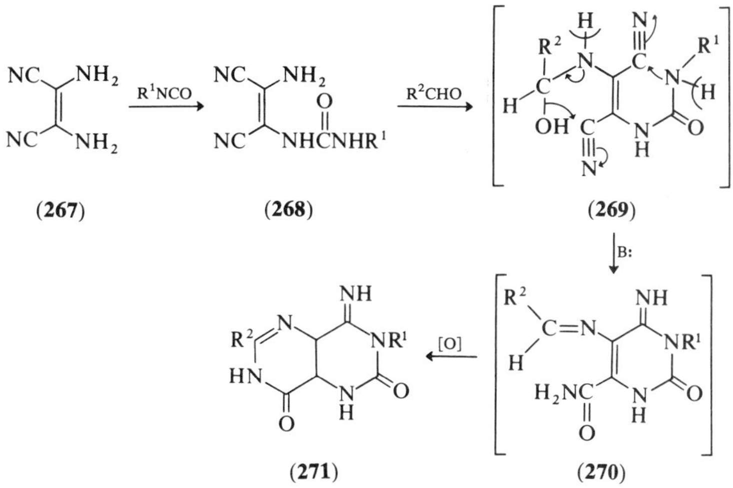

In a similar reaction it is also reported that 4,8-diaminopyrimido[5,4-*d*]pyrimidine is formed in the reaction of hydrogen cyanide with liquid ammonia. This reaction also presumably proceeds by way of the hydrogen cyanide tetramer, diaminomaleonitrile ⟨69BCJ1454⟩.

One final interesting preparation of pyrimido[5,4-*d*]pyrimidines involves a ring transformation of pyrazolo[4,3-*d*]pyrimidines. Reaction of the *N*-oxides (**272**) with sodium ethoxide in refluxing ethanol leads to the isolation of 1,3-dimethyl-6-substituted pyrimido[5,4-*d*]pyrimidine-2,4-diones (**273**) in good yield. A mechanism has been proposed for this transformation ⟨78TL2295⟩.

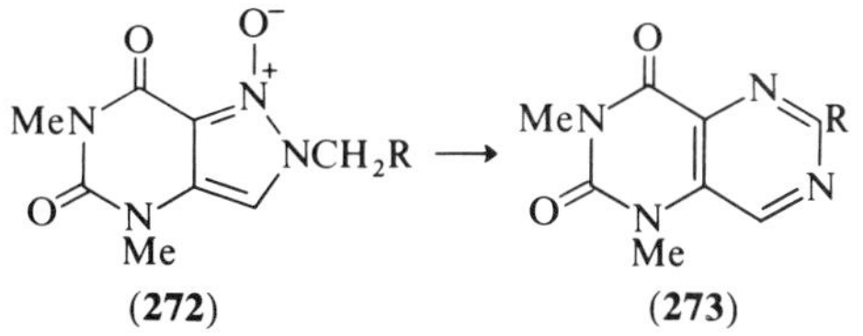

### 2.17.4.18 Pyrazino[1,2-*a*]pyrimidine

Pyrazino[1,2-*a*]pyrimidines are most commonly prepared from 2-aminopyrazines. Reaction of 2-aminopyrazine (**274**) with ethyl ethoxymethylenemalonate followed by heating in

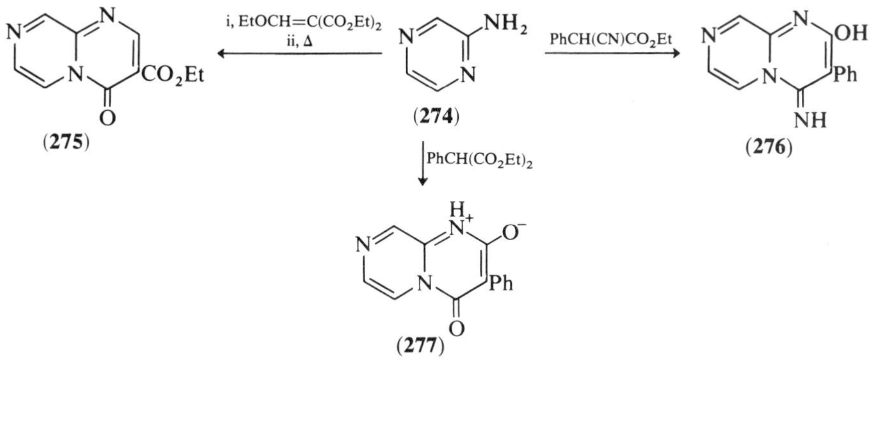

Dowtherm gives 3-ethoxycarbonyl-4*H*-pyrazino[1,2-*a*]pyrimidin-4-one (**275**) in good yield. Ethyl phenylcyanoacetate also condenses with the starting pyrazine to give 2-hydroxy-4-imino-3-phenyl-4*H*-pyrazino[1,2-*a*]pyrimidine (**276**). Condensation of (**274**) with ethyl phenylmalonate gives zwitterion (**277**) ⟨68JMC1045⟩.

Pyrazino[1,2-*a*]pyrimidines can also be prepared from piperazin-2-ones. Catalytic hydrogenation of 4-benzyl-1-(2-cyanoethyl)piperazin-2-one (**279**) over Raney nickel gives the cyclized product (**280**) in near quantitative yield. The cyanoethyl piperazine can be prepared from the corresponding piperazine (**278**) by reaction with acrylonitrile ⟨73BCJ3612⟩.

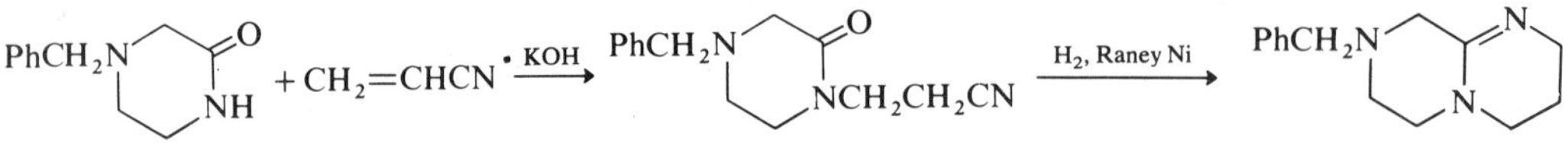

(**278**)                                          (**279**)                                          (**280**)

### 2.17.4.19 Pyrazino[1,2-*c*]pyrimidine

Pyrazino[1,2-*c*]pyrimidines are most commonly prepared from the intermediate 1-alkyl-3-(2-alkylaminoethyl)piperazines (**281**). This intermediate is readily prepared in three steps from either benzyloxycarbonylsarcosylaspartate ⟨71JMC929⟩ or from *N*-benzylethylenediamine and diethyl fumarate ⟨75JMC913⟩. Condensation of the piperazine intermediate (**281**) with ethyl chloroformate at pH 3–3.5 gives the monocarbamates (**282**). Cyclization of these carbamates by treatment with sodium ethoxide gives the desired ring system (**283**) ⟨75JMC913⟩.

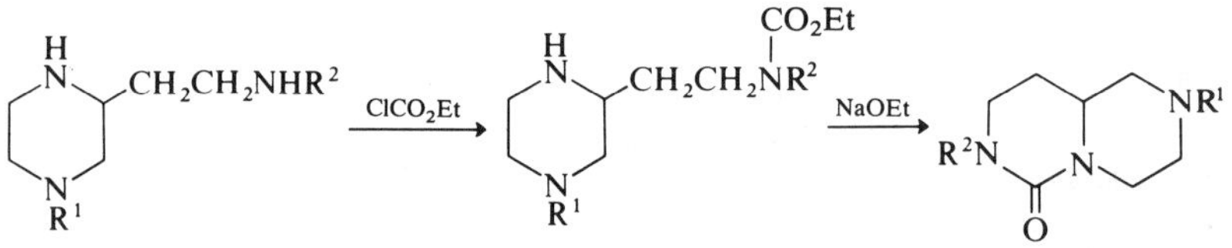

(**281**)                                          (**282**)                                          (**283**)

The preparation of derivatives of this ring system has also been achieved by reaction of substituted piperazines with urea. Reaction of the starting piperazinone (**284**) with urea under fusion conditions gives the desired ring system (**285**) ⟨66CPB194⟩.

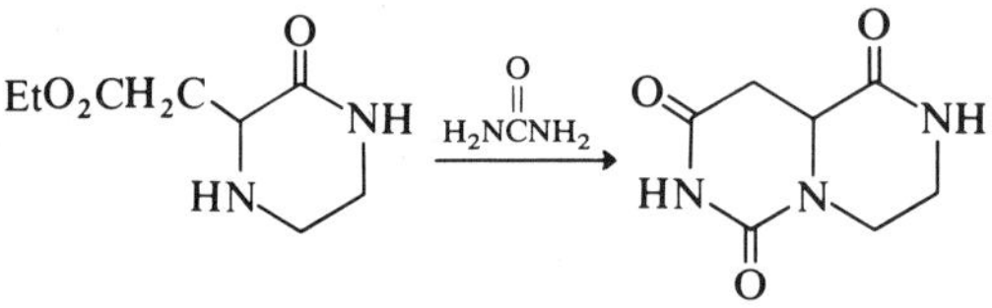

(**284**)                                          (**285**)

### 2.17.4.20 Pyrazino[1,2-*a*]pyrazine

The first reported preparation of this ring system involves an interesting two-step preparation starting with the dinitrophenylhydrazone (**287**). Reaction of this hydrazone with the amino ester (**286**) gives the intermediate hydrazone (**288**) in near quantitative yield. The intermediate can be cyclized to the ring closed product (**135**) by reaction with Raney nickel ⟨62AP121⟩.

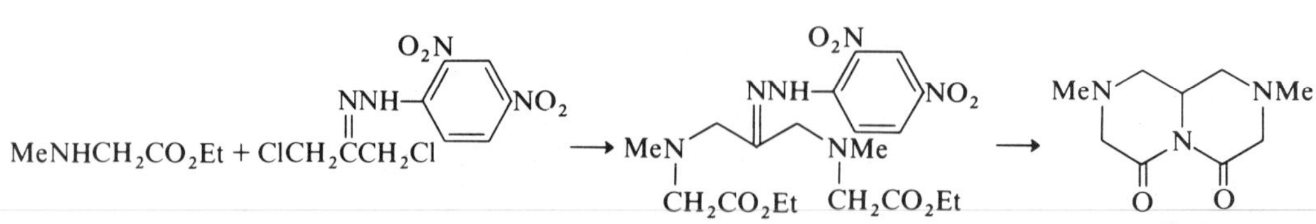

(**286**)                    (**287**)                              (**288**)                          (**135**)

The title ring system is also prepared by the reaction of aziridine with piperazinecarboxylic esters. Ethyl 4-methyl-2-piperazinecarboxylate (**289**) reacts with aziridine in ethanolic hydrogen chloride to give the cyclized product (**290**) in moderate yield ⟨77JHC307⟩.

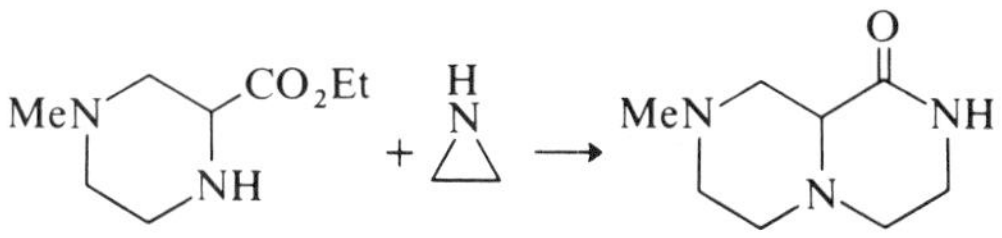

(**289**)                                    (**290**)

### 2.17.4.21 Pyrazino[2,3-*b*]pyrazine

Pyrazino[2,3-*b*]pyrazines are most readily prepared by the reaction of 2,3-diaminopyrazines with glyoxal. 2,3-Dimethylpyrazino[2,3-*b*]pyrazine (**292**) is prepared from glyoxal and 2,3-diamino-5,6-dimethylpyrazine (**291**), and from diacetyl and 2,3-diaminopyrazine ⟨48JA1257⟩.

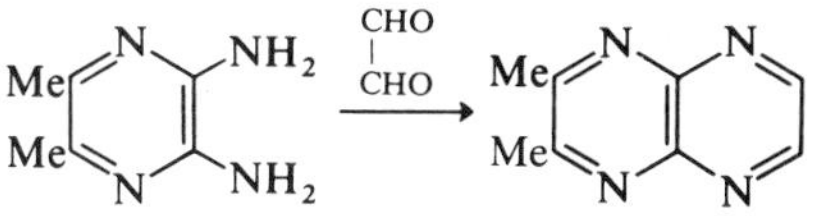

(**291**)                (**292**)

Glyoxal can also be condensed with *N,N'*-dimethylethane-1,2-diamine (**293**) to give the perhydropyrazino[2,3-*b*]pyrazine (**294**). This reaction leads to both *cis* and *trans* isomers of the product ⟨76JCS(P2)1564⟩.

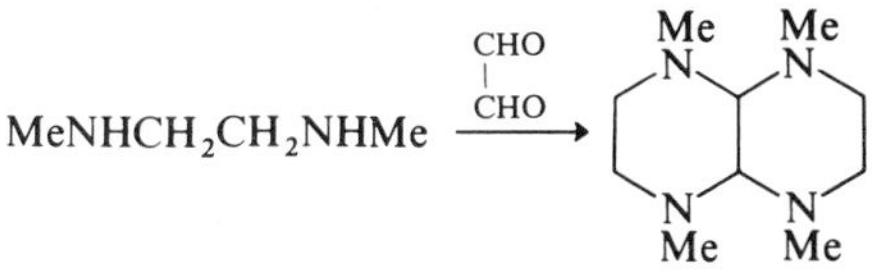

(**293**)                (**294**)

### 2.17.5 APPLICATIONS

Most of the applications of members of the diazinodiazines are modelled after the biologically active pteridines; however, applications in the diazinodiazines are limited. Most applications are only suggested from research data and are not actually practiced.

### 2.17.5.1 Pyridazino[1,2-*a*]pyridazine

1-(*p*-Methoxyphenyl)-1,2,3,4,6,7,8,9-octahydropyridazino[1,2-*a*]pyridazine has been shown to possess antidepressant properties ⟨72USP3657239⟩.

### 2.17.5.2 Pyridazino[4,5-*c*]pyridazine

5,8-Bis(morpholino)- and 5,8-bis(piperidino)-3-phenylpyridazino[4,5-*c*]pyridazine exhibit diuretic activity ⟨72YZ1327⟩.

### 2.17.5.3 Pyrimido[4,5-*c*]pyridazine

1,6-Dimethyl-1,5,6,7-tetrahydropyrimido[4,5-*c*]pyridazine-5,7-dione (4-deazatoxoflavin) inhibited the growth of *Pseudomonas* 568 and was also demonstrated to bind to herring sperm DNA ⟨75JHC1221⟩.

### 2.17.5.4 Pyrimido[4,5-*d*]pyridazine

5,8-Bis(morpholino)-2-phenylpyrimido[4,5-*d*]pyridazine and related compounds in this ring system have been found to possess diuretic activity ⟨76CPB2057⟩.

### 2.17.5.5 Pyrimido[4,5-*d*]pyrimidine

2,5-Dialkyl- and 2,5,7-trialkyl-pyrimido[4,5-*d*]pyrimidin-4(3*H*)-ones have been found to possess pre- and post-emergence herbicidal activity ⟨74USP3830812⟩. 2,4,7-Triamino-5-phenylpyrimido[4,5-*d*]pyrimidine has diuretic activity similar to triamterene (6-phenyl-2,4,7-pteridinetriamine) ⟨68JMC573⟩.

### 2.17.5.6 Pyrimido[5,4-*d*]pyrimidine

Dipyridamole (2,6-bis(diethanolamino)-4,8-dipiperidinopyrimido[5,4-*d*]pyrimidine) (295) is marketed as a coronary vasodilator ⟨59BRP807826⟩. It is useful for long-term therapy in patients suffering from chronic angina pectoris. This drug is chemically unrelated to digitalis or the nitrate esters. At recommended dosages adverse reactions are minimal and transient; thus the drug is fairly safe. The $LD_{50}$ in rats is 8.4 g kg$^{-1}$ orally and 208 mg kg$^{-1}$ i.v. It may require two or three months of continuous therapy before a satisfactory clinical response is evident.

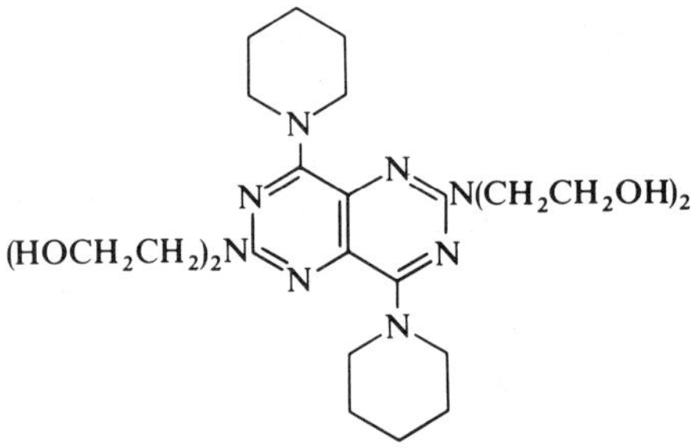

(295)

### 2.17.5.7 Pyrazino[2,3-*b*]pyrazine

Derivatives of decahydropyrazino[2,3-*b*]pyrazine are vulcanization accelerators for thiodiethanol copolymer rubbers ⟨80USP4218559⟩.

# 2.18

# 1,2,3-Triazines and their Benzo Derivatives

## H. NEUNHOEFFER
*Technische Hochschule Darmstadt*

| | | |
|---|---|---|
| 2.18.1 | INTRODUCTION | 369 |
| 2.18.2 | STRUCTURE AND SPECTRA | 370 |
| 2.18.3 | REACTIVITY AND REACTIONS | 374 |
| 2.18.4 | SYNTHESES | 381 |
| 2.18.5 | APPLICATIONS AND IMPORTANT COMPOUNDS | 384 |

## 2.18.1 INTRODUCTION

Of the three possible triazine systems the 1,2,3-triazines are by far the least studied class. A number of reviews on the chemistry of 1,2,3-triazines have been published: the first by Erickson in 1956 ⟨56HC(10)1⟩, covering the literature through *Chemical Abstracts* 1950, a second by Horwitz in 1961 ⟨B-61MI21800⟩, another by Kobylecki and McKillop in 1976 ⟨76AHC(19)215⟩ and finally a comprehensive review in 1978 ⟨78HC(33)3⟩, covering the literature through *Chemical Abstracts* 1974. Some aspects of 1,2,3-triazine chemistry were summarized in 1974 ⟨74MI21800⟩ and in 1976 ⟨76MI21800⟩.

The parent compound of the 1,2,3-triazine series has structure (**1**) and was prepared for the first time at the end of 1981 ⟨81CC1174⟩. In addition to the name 1,2,3-triazine, *v*-triazine or *β*-triazine can also be found in the older literature.

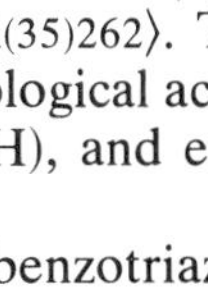

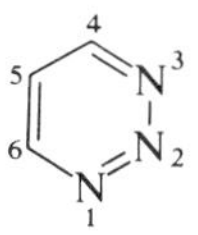

(**1**) 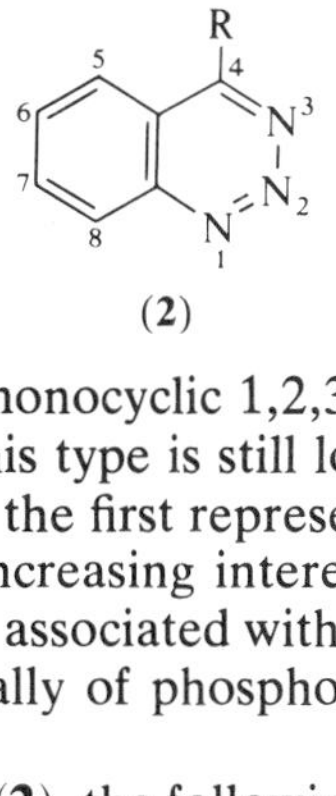    (**2**) 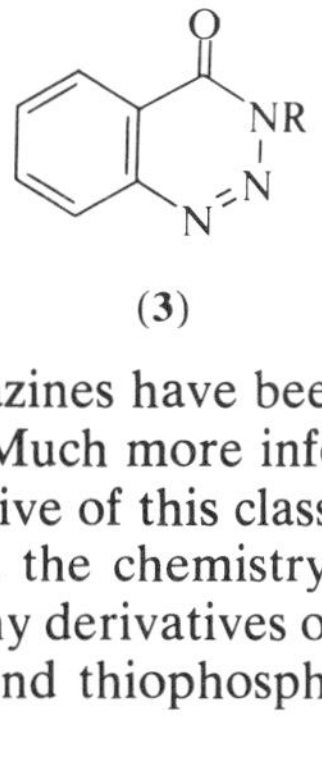    (**3**)

Only a few papers dealing with monocyclic 1,2,3-triazines have been published, and the number of known compounds of this type is still low. Much more information is available on the 1,2,3-benzotriazines (**2**) and the first representative of this class was prepared in the last century ⟨1887JPR(35)262⟩. The increasing interest in the chemistry of (**2**) is a result of the wide range of biological activity associated with many derivatives of 1,2,3-benzotriazin-4-(3*H*)-one (**3**; R=H), and especially of phosphoric and thiophosphoric acid derivatives of (**3**).

Besides the 1,2,3-benzotriazines (**2**), the following 1,2,3-triazine systems condensed with carbocycles are known: naphtho[2,3-*d*][1,2,3]triazines (**4**), 1*H*-naphtho[1,8-*de*][1,2,3]-triazines (**5**) and 1*H*-acenaphtho[5,6-*de*][1,2,3]triazines (**6**).

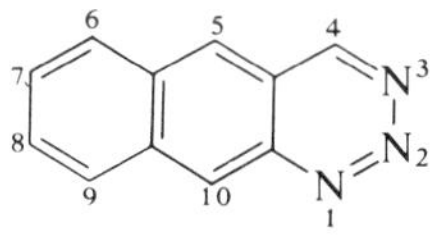

(**4**) 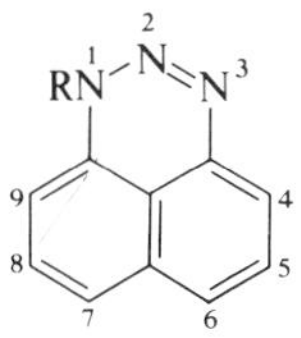    (**5**) 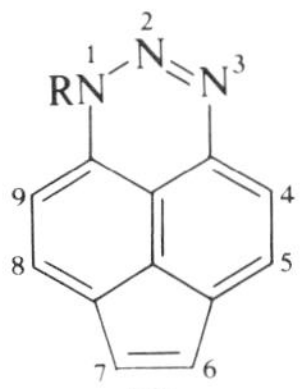    (**6**)

369

## 2.18.2 STRUCTURE AND SPECTRA

Only a small number of monocyclic 1,2,3-triazines are known, so knowledge of the structure of these compounds is poor. The available data on the parent compound (**1**) and its methyl derivatives do not permit an evaluation of the degree of electron delocalization and aromaticity of (**1**) or its alkyl derivatives. Attempts at a crystallographic analysis of (**1**) were unsuccessful as the crystal was unstable under X-ray irradiation.

The only monocyclic 1,2,3-triazine of which the structure has been determined by X-ray crystallographic analysis is 4,5,6-tris(4-methoxyphenyl)-1,2,3-triazine 〈72CB3704〉. The results show that the 1,2,3-triazine ring is planar, as one expects for a molecule with some degree of electron delocalization. The experimental bond distances and angles are given in Table 1. They are in reasonable agreement with calculated values.

**Table 1** Experimental and Calculated Bond Distances and Experimental Bond Angles in Tris(4-methoxyphenyl)-1,2,3-triazine[a]

| | | Bond distance (Å) | | | |
|---|---|---|---|---|---|
| | | | Calculated | | |
| Bond | Experimental | PPP[b] | SPO[b] | Experimental bond angle (°) | |
| $N^1-N^2$ | 1.319 (7) | 1.302 | 1.302 | $C^6-N^1-N^2$ | 120.2 (4) |
| $N^2-N^3$ | 1.314 (6) | | | $N^1-N^2-N^3$ | 122.8 (4) |
| $N^3-C^4$ | 1.368 (6) | 1.328 | 1.331 | $N^2-N^3-C^4$ | 119.3 (4) |
| $C^4-C^5$ | 1.406 (6) | 1.399 | 1.399 | $N^3-C^4-C^5$ | 120.6 (4) |
| $C^5-C^6$ | 1.388 (6) | | | $C^4-C^5-C^6$ | 116.2 (4) |
| $C^6-N^1$ | 1.363 (7) | 1.328 | 1.331 | $C^5-C^6-N^1$ | 120.4 (4) |

[a] The standard deviations in the last figure are given in parentheses.
[b] 〈66JCP(44)759〉.

Data on the structures of monocyclic dihydro- or hexahydro-1,2,3-triazines, on 1,2,3-triazine *N*-oxides and 1,2,3-triazinones are not yet available. From studies on the one-electron reduction of the tetrahydro-1,2,3-triazinium salts (**7**) it was concluded that the heterocyclic ring is flexible and not planar 〈80LA285〉. No detailed information on the structure of 3-benzyl-1,5-diphenyl-1,2-dihydro-1,2,3-triazine-4,6(3*H*,5*H*)-dione (**8**) or of the 6-hydroxy-4-oxo-1,4-dihydro-1,2,3-triazinium hydroxide inner salts (**9**) seems to be available.

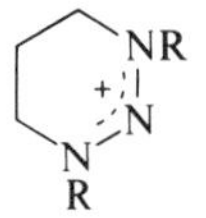

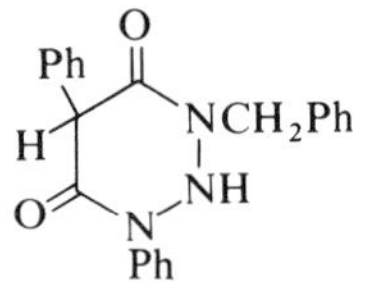

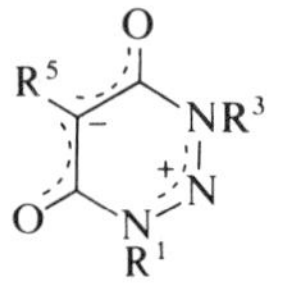

Hjortas has published an X-ray crystallographic analysis of the 1,2,3-benzotriazin-4-one (**10**) which shows that of the four possible tautomeric forms (**10a–10d**) only the 3*H* form (**10a**) is present in the crystalline state 〈73AX(B)1916〉. The N1—N2 distance is found to be 1.274 Å. Grabowski 〈60MI21800〉 has published some crystallographic data on 3-phenyl-1,2,3-benzotriazin-4-one (**11**). The structures of anhydro-2-methyl-4-(2-nitrophenyl)imino-1,2,3-benzotriazinium hydroxide (**12**; Ar = 2-O$_2$NC$_6$H$_4$) 〈78AX(B)2514〉 and of 2-(2,4-dibromophenyl)-4-oxo-1,2,3-benzotriazin-2-ium-3-ide (**13**; Ar = C$_6$H$_3$Br$_2$) 〈76AX(B)2240〉 were determined by X-ray crystallographic analysis. The bond distances and angles for the heterocyclic ring are as shown in Figure 1. Data are also available on the 1-oxide of (**13**; Ar = Ph) 〈75AX(B)626〉.

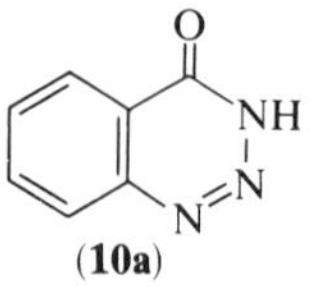

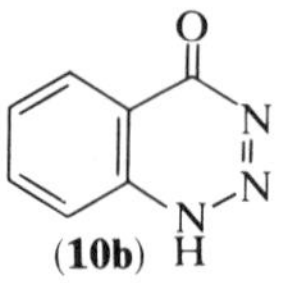

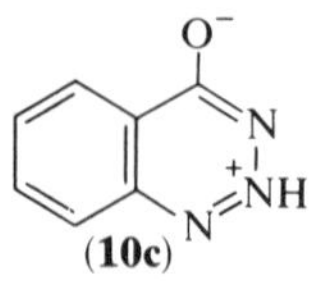

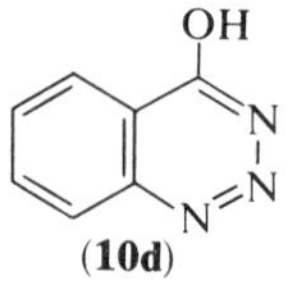

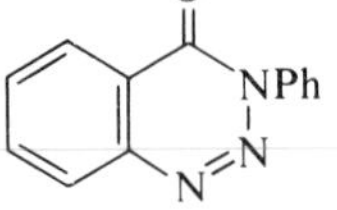

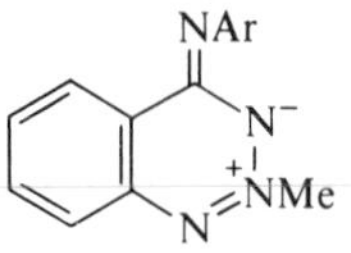

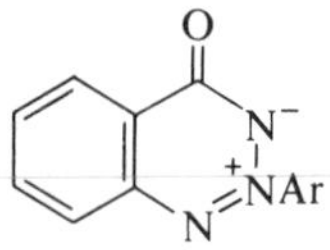

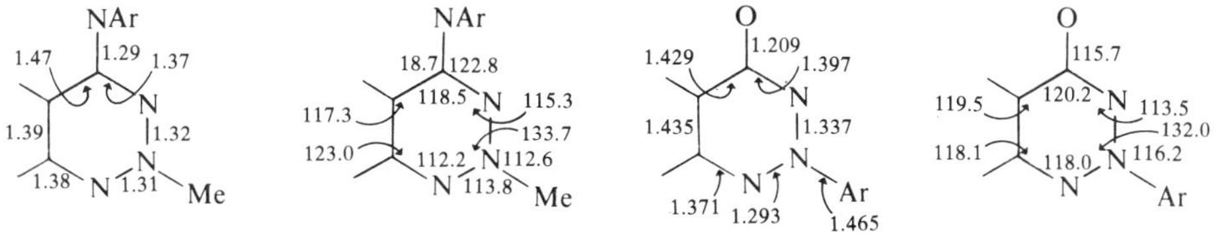

(12) Ar = 2-O$_2$NC$_6$H$_4$          (13) Ar = 2,4-Br$_2$C$_6$H$_3$

**Figure 1**  Bond lengths (Å) and angles (°) in compounds (12) and (13)

The $^1$H NMR spectra of most known 1,2,3-triazines have been published. The following spectrum is reported for the parent 1,2,3-triazine (1): $\delta = 9.06$ (2H, d, $J = 6.0$ Hz) and 7.45 (1H, t, $J = 6.0$ Hz) ⟨81CC1174⟩. For 4-methyl-1,2,3-triazine (14) the following values were reported: $\delta = 8.92$ (1H, d, $J = 6$ Hz), 7.33 (1H, d, $J = 6$ Hz) and 2.70 (3H, s) ⟨81CC1174⟩. In the $^1$H NMR spectrum of trimethyl-1,2,3-triazine (15) two signals at $\delta = 7.34$ and 7.68 with relative intensities of 2 : 1 were observed ⟨72JOC1051⟩, and the spectrum of tris(dimethyl-amino)-1,2,3-triazine (16) shows two signals at $\delta = 7.10$ and 7.25 ⟨73AG918⟩. The chemical shifts and the coupling constant of 6.0 Hz found in the spectrum of (1) are in the same region as in other six-membered heterocyclic aza systems.

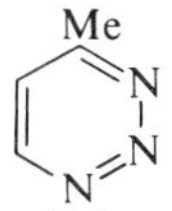

(14) 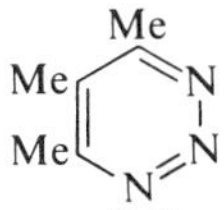          (15) 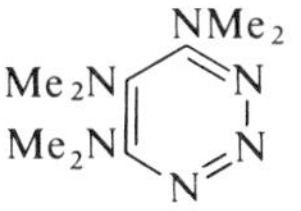          (16)

The formation of an *N*-oxide group influences the shifts of all protons of the heterocyclic ring, as is shown for 4-methyl-1,2,3-triazine 3-oxide (17) and 2-oxide (18). For (17): $\delta = 8.43$ (1H, d, $J = 5.0$ Hz), 7.44 (1H, d, $J = 5.0$ Hz) and 2.52 (3H, s); for (18) $\delta = 8.50$ (1H, d, $J = 6.0$ Hz), 6.79 (1H, d, $J = 6.0$ Hz) and 2.53 (3H, s) ⟨80CC1182⟩.

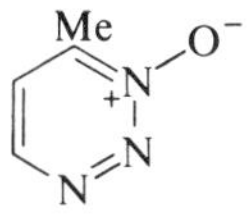

(17) 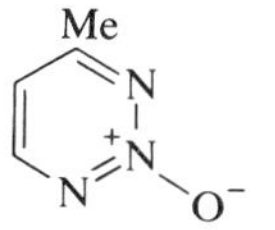          (18) 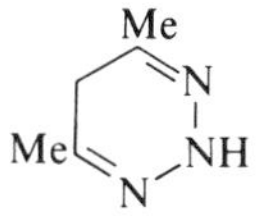          (19)

The $^1$H NMR spectrum of 4,6-dimethyldihydro-1,2,3-triazine (19) clearly shows that only the 2,5-dihydro form is present, and no evidence for the presence of the 1,4-dihydro form could be found. The spectrum of (19) is as follows: $\delta = 8.35$ (1H, br. s), 2.45 (2H, s) and 2.00 (6H, s) ⟨80CC1182⟩.

The following $^{13}$C NMR spectra have been published for 1,2,3-triazine (1) and its 4-methyl derivative (14). For (1): (CDCl$_3$) $\delta = 117.9$ (C-5, $J_{CH} = 175.0$ Hz) and 149.7 p.p.m. (C-4,6, $J_{CH} = 187.5$ Hz); for (14): $\delta = 21.4$ (CH$_3$, $J_{CH} = 129.8$ Hz), 117.8 (C-5, $J_{CH} = 172.1$ Hz), 148.8 (C-6, $J_{CH} = 184.3$ Hz) and 159.7 p.p.m. (C-4, s) ⟨81CC1174⟩.

Nitrogen NMR spectra of monocyclic 1,2,3-triazines have not yet been published, but chemical shifts for 1,2,3-triazines ⟨72BAP91⟩ and 1,2,3-triazine *N*-oxides ⟨76SA(A)345⟩ have been calculated.

In the $^1$H NMR spectrum of 1,2,3-benzotriazine (20) two signals were observed: a 1H doublet at $\delta = 9.85$ and a 4H multiplet at 8.14–8.55 ⟨75JCS(P1)31⟩. The NH proton of 1,2,3-benzotriazin-4(3*H*)-one (21) gives rise to a signal at $\delta = 11.00$ while the four aromatic protons appear as a multiplet at 7.60–8.10 ⟨72JOC196⟩.

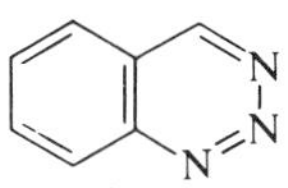

(20) 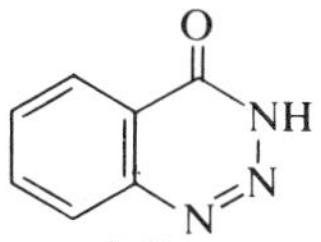          (21)

Infrared spectral data for many monocyclic and condensed 1,2,3-triazines have been published but no detailed studies of the influence of substituents on the vibrational modes of the ring skeleton have been made. Four IR spectral bands were reported for 1,2,3-triazine (1) and its 4-methyl derivative (14) ⟨81CC1174⟩. For (1): $\nu_{max}$(KBr) = 3045 (m), 1565 (s), 1440 (m) and 1385 cm$^{-1}$ (w); for (14): $\nu_{max}$(KBr) = 3055 (m), 1550 (s), 1412 (m) and 1380 cm$^{-1}$ (w). No IR spectra are available of monocyclic 1,2,3-triazine *N*-oxides or dihydro-1,2,3-triazines. Incomplete data on the IR spectra of tetrahydro-1,2,3-triazinium salts (7)

⟨79CB445⟩ and of the dihydro-1,2,3-triazine-4,6(3$H$,5$H$)-dione (**8**) have appeared ⟨78CB2173⟩.

1,2,3-Benzotriazine (**20**) shows the following IR bands: $\nu_{max}$(KBr) = 1473, 1383, 877, 800 and 750 cm$^{-1}$ ⟨75JCS(P1)31⟩. In the IR spectra of 1,2,3-benzotriazin-4-ones an intense band between 1700 and 1667 cm$^{-1}$ is observed, showing that the hydroxy tautomer (**10d**) can be excluded. Similarly, it was shown that in 1,2,3-benzotriazine-4-thiones (**22**) the thione form predominates in the solid state. The IR spectra of 4-anilino-1,2,3-benzotriazines (**23**) show a strong absorption band at 1145 ± 10 cm$^{-1}$ which is absent in the 3-aryl-4-imino isomers (**24**) and presents a useful means of distinction between these isomers ⟨70JCS(C)765⟩.

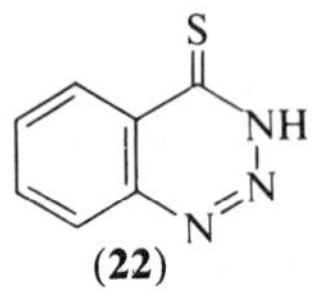

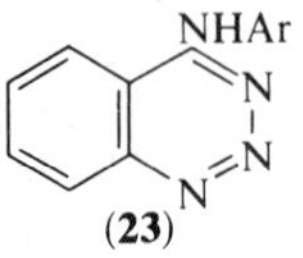

The UV spectra of monocyclic and condensed 1,2,3-triazines are well documented. The parent 1,2,3-triazine (**1**) shows the following UV spectrum: $\lambda_{max}$(EtOH) = 325 (sh), 268 (log$_{10}$ ε 2.935) and 232 nm (sh), while for the 4-methyl derivative (**14**) bands at 313 (sh), 286 (2.71) and 288 nm (sh) are reported ⟨81CC1174⟩. (The last band for (**14**) at 288 nm is probably located at 238 nm). The UV spectra of 1,2,3-triazine $N$-oxides and dihydro-1,2,3-triazines have not yet been published. The 1,3-bis(4-methylphenyl)-3,4,5,6-tetrahydro-1,2,3-triazinium perchlorate (**7**; R = 4-MeC$_6$H$_4$) has two absorption bands at 348 (4.20) and 248 nm (3.79) ⟨79CB445⟩.

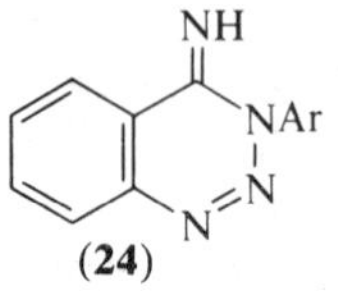

Ultraviolet data have been obtained of many different classes of benzotriazine derivatives. Representative examples of a number of these are given in Table 2.

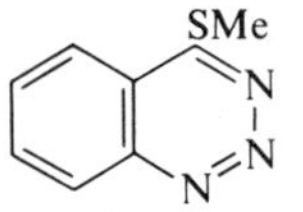

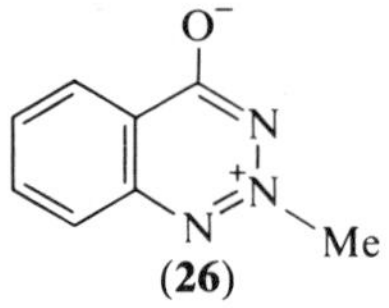

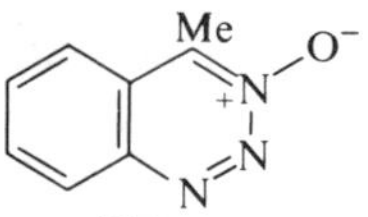

**Table 2**    Representative UV Spectra of Benzotriazine Derivatives

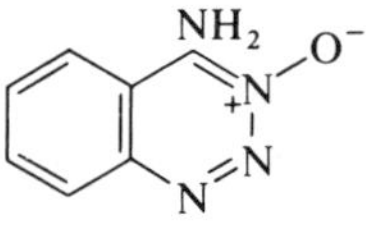

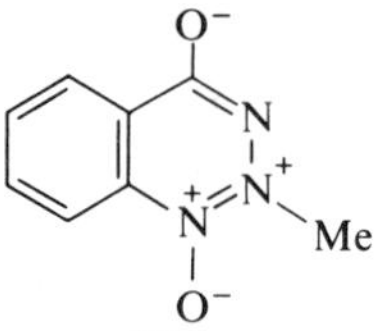

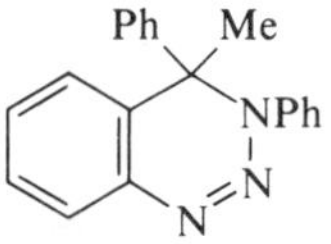

| Compound | $\lambda_{max}$ (nm) (log$_{10}$ ε) | Ref. |
| --- | --- | --- |
| (**2**; R = Me) | 275 (2.83), 227 (4.0), 207 (3.58) | 75JCS(P1)31 |
| (**2**; R = Ph) | 293 (2.95), 232 (4.1), 206 (3.86) | 72JCS(P1)1315 |
| (**3**; R = CH$_3$) | 316 (sh, 3.61), 300 (sh, 3.79), 285 (3.89), 252 (sh, 3.69), 225 (4.35), 214 (4.31) | 74 JOC2710 |
| (**21**) | 307 (3.42), 296 (3.63), 278 (3.80), 250 (3.71), 224 (4.28), 211 (4.15) | 74JOC2710 |
| (**23**; Ar = Ph) | 333 (4.30), 273 (3.76) | 70JCS(C)765 |
| (**24**; Ar = Ph) | 318 (3.77), 307 (3.76), 268 (3.96), 260 (3.97) | 70JCS(C)765 |
| (**25**) | 316 (3.92), 261 (3.65) | 70JCS(C)765 |
| (**26**) | 335 (3.86), 276 (sh, 3.43), 266 (sh, 2.53), 250 (sh, 3.82), 238 (4.00), 231 (4.01), 212 (4.02) | 74JOC2710 |
| (**27**) | 395 (3.08), 310 (3.83), 302 (3.91) | 73JA2390 |
| (**28**) | 390 (3.27), 325 (3.65), 258 (sh, 3.97), 232 (4.36) | 58CI(L)1234 |
| (**29**) | 384 (3.76), 278 (3.89), 226 (4.29) | 74JOC2710 |
| (**30**) | 317, 268, 259, 252 | 71TL3117 |

Mass spectrometry has been used to determine the structures of isomeric monocyclic 1,2,3-triazines and their $N$-oxides. The general fragmentation pattern of monocyclic 1,2,3-triazines shows peaks for M$^+$ − N$_2$, and for nitriles and alkynes, which indicates a fragmentation as shown in Scheme 1 ⟨72CB3695, 79CB1514⟩. According to this fragmentation pattern the following mass spectrum was observed for the unsubstituted 1,2,3-triazine (**1**): 81 ($M^+$, 47%), 53 ($M^+ - N_2$, 69%), 27 (HCN, 13%) and 26 ($C_2H_2$, 100%); and for the 4-methyl-1,2,3-triazine (**14**): 95 ($M^+$, 13%), 67 ($M^+ - N_2$, 26%), 40 ($C_3H_4$, 100%), 27 (HCN, 18%) and 26 ($C_2H_2$, 22%). No peak for $MeCN^+$ was reported ⟨81CC1174⟩.

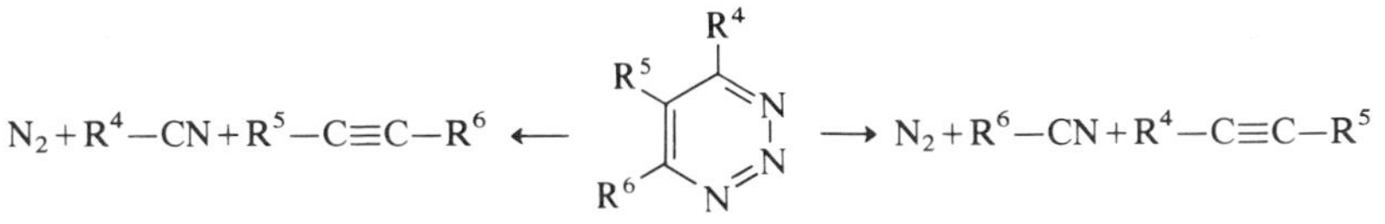

$$N_2 + R^4 - CN + R^5 - C \equiv C - R^6 \longleftarrow \quad \longrightarrow N_2 + R^6 - CN + R^4 - C \equiv C - R^5$$

**Scheme 1**

Mass spectra have been used to determine the site of oxidation of monocyclic 1,2,3-triazines. In 1,2,3-triazine 1(3)-oxides (*e.g.* **17**) peaks for $M^+ - N_2$ and for $(RCNO)^{+}$ were observed while these peaks were absent or very weak in 1,2,3-triazine 2-oxides (*e.g.* **18**) ⟨72CB3695, 81CC1174⟩. The mass spectra of dihydro-1,2,3-triazines have not yet been reported. As one might expect, in the mass spectrum of 1-benzyl-3,5-diphenyl-1,2-dihydro-1,2,3-triazine-4,6(3*H*,5*H*)-dione (**8**) no $M^+ - N_2$ peak is observed; the following spectrum was recorded: 357 ($M^+$, 6%), 238 (13%), 210 (30%), 183 (5%), 118 (100%), 105 (6%), 91 (100%), 77 (23%), 65 (7%) and 44 (20%) ⟨78CB2173⟩.

The following peaks were reported for 1,2,3-benzotriazine (**20**): 131 ($M^+$), 103 ($M^+ - N_2$) and 76 ($C_6H_4$) ⟨75JCS(P1)31⟩. The mass spectrometric fragmentation of 1,2,3-benzotriazin-4-ones has been studied by various groups. As with all other 1,2,3-triazines the fragmentation starts with loss of nitrogen (Scheme 2). Labelling with $^{15}N$ shows that the nitrogen in the 3-position remains in the molecule. The loss of nitrogen is followed by loss of carbon monoxide. There is controversy over whether the first fragmentation product is a benzazetinone (**31**) or not ⟨69OMS(2)355, 69T5869⟩.

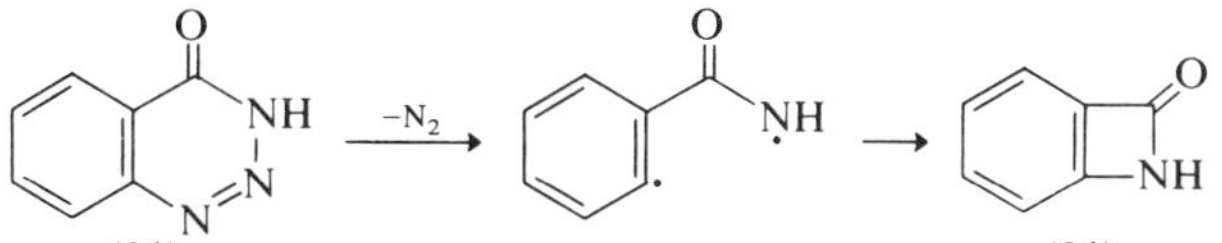

**Scheme 2**

The mass spectra of 4-anilino-1,2,3-benzotriazines (**23**), 3-aryl-4-imino-1,2,3-benzotriazines (**24**) and 2-alkyl-4-arylimino-1,2,3-benzotriazinium betaines (**32**) have been extensively studied ⟨70JCS(C)1238⟩. The fragmentation of (**23**) and (**24**) starts with loss of nitrogen, while for (**32**) the loss of an *ortho* substituent in the aryl group (Ar) or the elimination of ($RN_2$) is observed. 1,2,3-Benzotriazinium betaine 1-oxides (**33**) undergo fragmentation in the mass spectrometer to give prominent ions at $M^+ - 16$, $M^+ - 28$, $M^+ - 44$ and $M^+ - 72$ ⟨74JOC2710⟩.

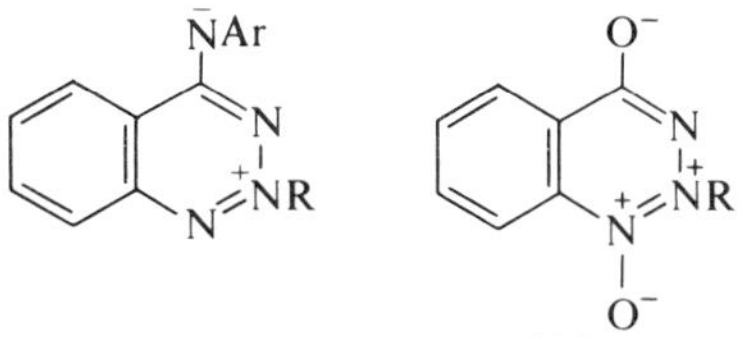

A large number of theoretical calculations on the 1,2,3-triazine system have been published. References to these papers can be found in the 1978 review ⟨78HC(33)53⟩. Since then only a few further papers have been published ⟨76SA(A)345, 76MI21800, 77IJC(B)168⟩.

The thermodynamic stability of 1,2,3-triazines and 1,2,3-benzotriazines depends very much on the substituents bound to the heterocyclic ring. Further details will be discussed in the next section on the reactivity and reactions of 1,2,3-triazines.

The photoelectron spectra of 1,2,3-triazines and 1,2,3-benzotriazines have not yet been published.

## 2.18.3 REACTIVITY AND REACTIONS

Most known monocyclic 1,2,3-triazines and 1,2,3-benzotriazines are stable at room temperature. No detailed study of the stability of monocyclic 1,2,3-triazines towards water, aqueous acids or bases has been published, but one can assume from the reaction conditions used in the preparation of monocyclic 1,2,3-triazines, by oxidation of *N*-aminopyrazoles, that they are stable to water, aqueous acids or bases at room temperature, at least for a short time. Treatment of triaryl-1,2,3-triazines with aqueous hydrochloric acid at higher temperatures leads to hydrolysis of the ring and formation of 1,3-dicarbonyl compounds (Scheme 3) ⟨60TL(13)19, 76UP21800⟩.

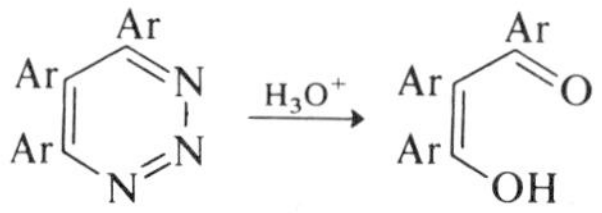

**Scheme 3**

1,2,3-Benzotriazine, and especially its conjugate acid, reacts very easily with nucleophiles such as water or acetic acid to give derivatives of 2-aminobenzaldehyde ⟨75JCS(P1)31, 71CC828⟩. 4-Alkyl- or 4-aryl-substituted 1,2,3-benzotriazines are more stable towards attack of water or acids. 2-Aminobenzophenone (100%) is obtained when 4-phenyl-1,2,3-benzotriazine is heated under reflux for 10 min in 10% aqueous ethanol containing concentrated sulfuric acid ⟨75JCS(P1)31⟩.

1,2,3-Benzotriazin-4-ones (3) are slightly acidic compounds which are easily soluble in aqueous or alcoholic bases; addition of acid to these solutions reprecipitates them unchanged. A number of 1,2,3-benzotriazin-4-ones are soluble in concentrated hydrochloric acid and are reprecipitated on addition of water. Heating 1,2,3-benzotriazin-4-ones (3) in acidic media affords several products, depending on the structure of the starting material and reaction conditions used. As with other 1,2,3-benzotriazines, most isolated compounds should be formed by a reversible ring scission between N-2 and N-3, especially in strongly acidic media, yielding the diazonium ion (34; Scheme 4) which can be transformed into the isolated products ⟨63JCS3539, 78HC(33)53⟩.

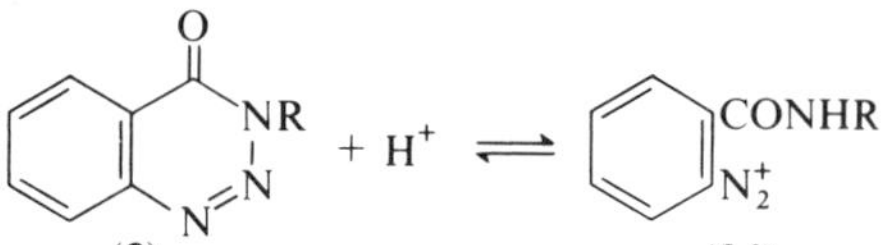

**Scheme 4**

3-Amino-1,2,3-benzotriazin-4-ones (35) are transformed by acid primarily into the diazonium ions (36) which then give anthraniloyl azides (37) ⟨65T2191, 27JPR(116)9, 25JPR(111)36⟩.

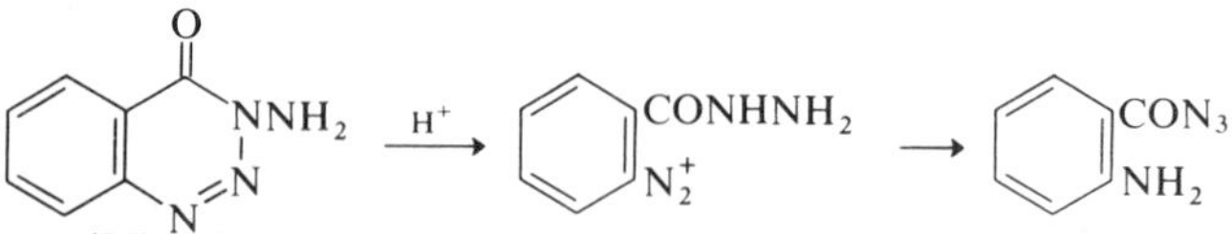

1,2,3-Benzotriazinium betaines (13) and (26) are stable toward acids. They are slightly basic compounds and form salts with acids which are easily hydrolyzed. The corresponding 1-oxides (33) behave similarly, being, for instance, souble in concentrated hydrochloric acid from which they are reprecipitated on dilution with water ⟨27JCS323⟩. 4-Amino-1,2,3-benzotriazines (38) and 4-imino-1,2,3-benzotriazines (39) behave as masked diazonium compounds (40) and the products of acidic treatment of these compounds are best explained as reaction products of the diazonium salts ⟨78HC(33)78⟩. Treatment of 1,2,3-benzotriazine

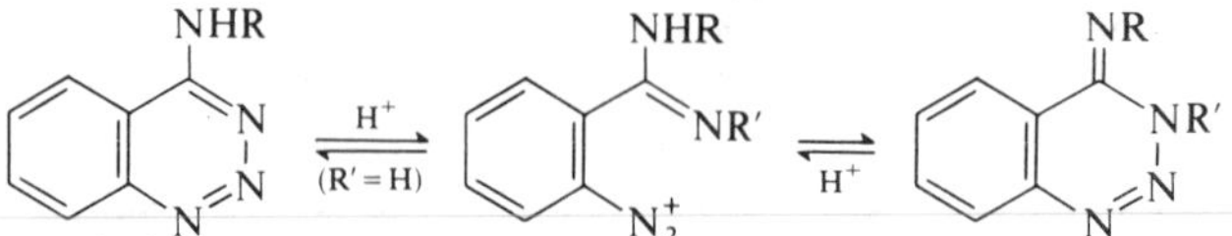

3-oxides (**41**) with dilute mineral acids affords 2-azidobenzaldehydes or 2-azidophenyl ketones (**42**) ⟨59JOC963, 27CB1736, 01CB1309, 73JA2390⟩.

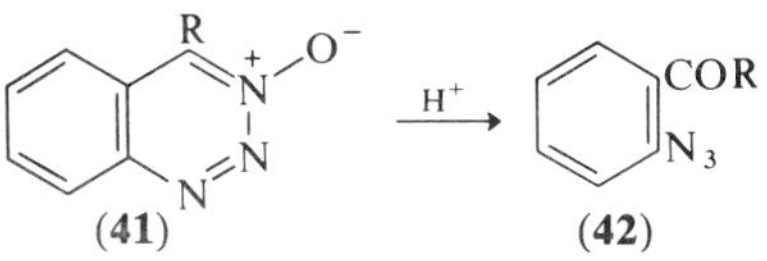

The reaction of dihydro-1,2,3-benzotriazines (**43**) in acidic media are also best explained by intermediate formation of the diazonium ion (**44**), which reacts with nucleophiles as do simple aromatic diazonium ions ⟨78HC(33)94⟩.

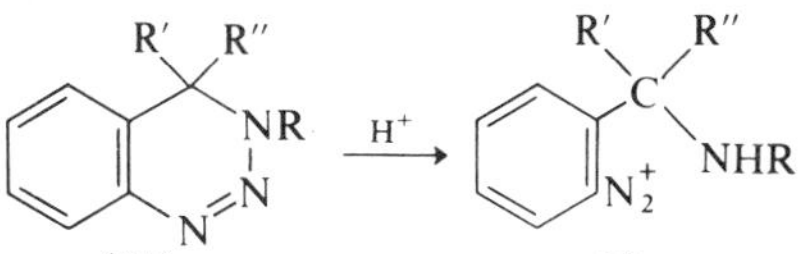

4-Phenyl-1,2,3-benzotriazine (**2**; R = Ph) reacts with hydrazine to give 2-aminobenzophenone hydrazone ⟨75JCS(P1)31⟩. Alkaline degradation of 1,2,3-benzotriazin-4-ones (**3**) affords triazenes (**45**) or anthranilic acids ⟨1896JPR(53)210, 01JPR(63)241⟩. 3-Amino-1,2,3-benzotriazin-4-ones (**35**) are attacked by hydroxide ion at the carbonyl group, leading to tetrazene-like intermediates (**46**) which may either lose nitrogen, affording 2-hydrazinobenzoates (**47**), or an amide ion, forming 2-azidobenzoate (**48**) ⟨65T2191⟩. 2-Aryl-1,2,3-benzotriazinium betaines (**13**) react with hydroxide ion at C-4 to give the triazenes (**49**) ⟨27JCS323⟩. Reaction of the 2-methyl derivative (**26**) with 2.5N NaOH is reported to give 1,2,3-benzotriazin-4-one (**10**) and anthranilic acid ⟨68MI21800⟩.

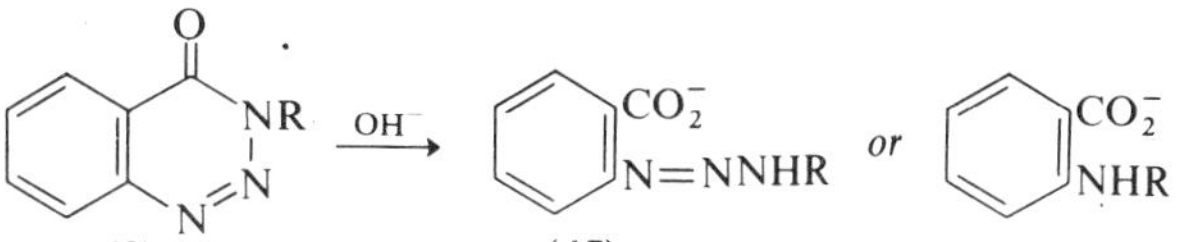

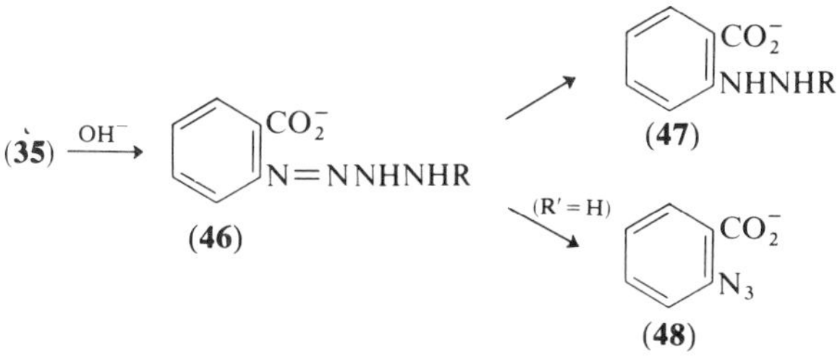

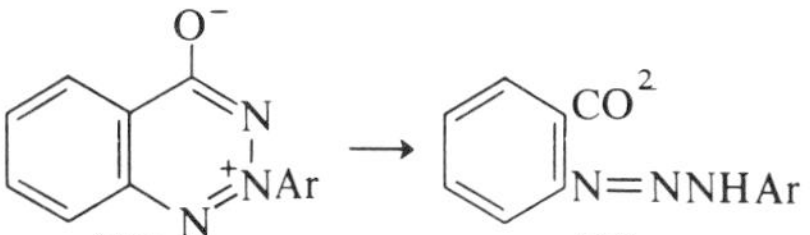

4-Imino-1,2,3-benzotriazines (**24**) react with sodium ethoxide to give the triazene ion (**50**), which can be trapped by acids ⟨70JCS(C)2284⟩. Treatment of 1,2,3-benzotriazine 3-oxides (**41**) with bases affords 2-azidobenzaldehyde or 2-azidophenyl ketones (**42**) which can be transformed further into anthranils (**51**) ⟨78HC(33)86⟩. 4-Amino-1,2,3-benzotriazine 3-oxides (**28**) with bases are reported to yield 2-aminobenzonitriles or 2-aminobenzaldehydes and salicylaldehydes ⟨59JOC272, 1898JPR(58)333⟩. 1,2,3-Benzotriazinium betaine 1-oxides (**33**) are readily decomposed by alcoholic potash or ammonia; the structures of the products have not yet been completely identified ⟨27JCS323⟩.

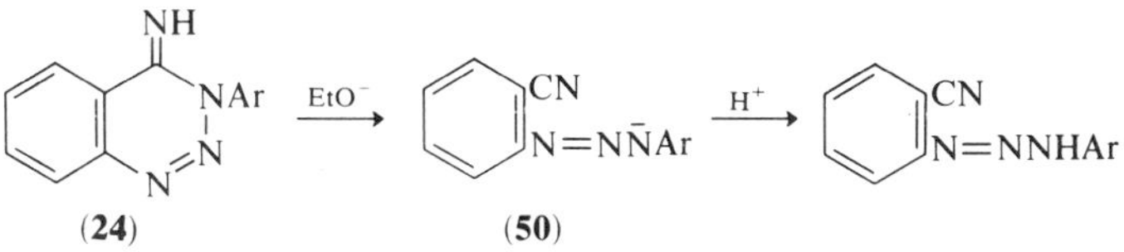

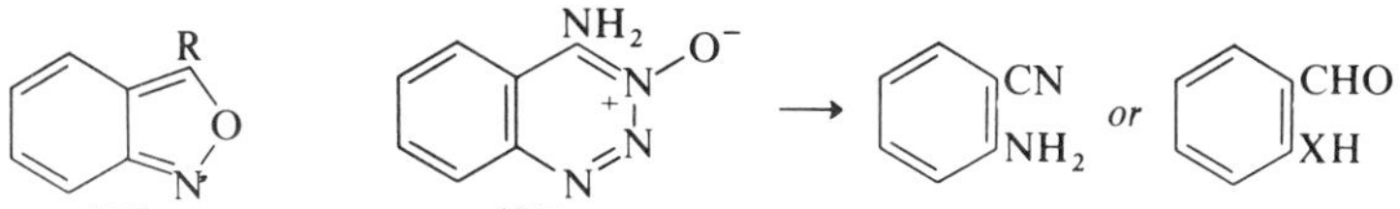

Alkylation and acylation of 1,2,3-triazines have not yet been studied extensively. Protonation of monocyclic 1,2,3-triazines usually takes place at N-2. The tetrafluoroborate salt
of 5-chloro-4,6-bis(dimethylamino)-1,2,3-triazine could be isolated. Similarly, reaction of
1,2,3-triazines with methyl iodide affords 2-methyl-1,2,3-triazinium iodides ⟨79CB1514⟩.
2-Substituted 1,2,3-triazinium salts were also obtained by reaction with chloroformate,
picryl chloride and chlorobis(dialkylamino)cyclopropenylium perchlorates. Alkylation and
acylation of 1,2,3-benzotriazin-4-ones have been thoroughly investigated. In most cases
substitution takes place at N-3, and only a few examples have been published which lead
to *O*-substituted products ⟨78HC(33)54⟩. Alkylation of 1,2,3-benzotriazine-4-thiones gives
mostly 4-alkylthio-1,2,3-benzotriazines ⟨78HC(33)68⟩.

Alkylation of 4-arylamino-1,2,3-benzotriazines (23) with alkyl iodides in ethanol affords
the 2-alkyl-4-arylamino-1,2,3-benzotriazinium iodides (52) which on basification form the
deep red 2-alkyl-4-arylimino-1,2,3-benzotriazinium betaines (32). Alkylation in the presence of sodium ethoxide affords a mixture of (32) and 3-alkyl-4-arylimino-1,2,3-
benzotriazines (53) ⟨70JCS(C)2289⟩.

Oxidation of monocyclic 1,2,3-triazines with peracid affords mainly 1(3)-oxides together
with smaller amounts of the 2-oxides ⟨80CC1182, 73UP21800⟩. The *N*-oxides can be reduced
by trivalent phosphorus compounds ⟨73UP21800⟩.

Oxidation of 3-amino-1,2,3-benzotriazin-4-ones (35) with lead tetraacetate leads to the
formation of the nitrenes (54) which loses either one molecule of nitrogen to give the
indazolones (55) or two molecules of nitrogen to give the benzocyclopropenones (56). The
suggested mechanisms were supported by $^{15}$N labelling experiments ⟨71JCS(C)981, 69CC221⟩.
The benzocyclopropenones could be trapped with pentafluoroacetone ⟨80CC585⟩, and the
indazolones by tetracyclone ⟨71JCS(C)981⟩.

1,2,3-Benzotriazinium betaines (13) are not affected by peracids. Oxidation of (13) with
potassium permanganate led to the degradation of the heterocyclic ring and 2-nitrobenzoic
acids were isolated ⟨27JCS323⟩. 2-Substituted 1,2,3-benzotriazinium betaine 1-oxides (33)
yield 2-nitrobenzaldehydes on oxidation with potassium permanganate.

Reduction of monocyclic 1,2,3-triazines with lithium aluminum hydride affords 2,5-
dihydro-1,2,3-triazines ⟨80CC1182, 73UP21800⟩. Reaction of triphenyl-1,2,3-triazine with zinc
in acetic acid or with hydrogen over palladium gives 3,4,5-triphenylpyrazole ⟨60TL(13)19⟩.
Reduction of 1,2,3-benzotriazin-4-ones (3) with Raney nickel in 95% ethanol at 60 °C
is claimed to give 1,2-dihydro-1,2,3-benzotriazin-4-ones (57) ⟨66MIP21800⟩. Electrochemical reduction of (3; R = H) is reported to give indazolinone (58) ⟨76BSF433⟩.
Reduction of (3) over palladium on charcoal in acetic acid led to the isolation of 5,6,7,8-
tetrahydro-1,2,3-benzotriazin-4-ones (59) in yields between 21 and 52% ⟨76S717⟩. 1,2,3-
Benzotriazin-4-ones are stable toward sodium amalgam and Adams catalyst. Raney nickel
and hydrazine in ethanol, tin(II) chloride in hydrochloric acid, boiling titanium(III) chloride,
and zinc and ammonia, all lead to the destruction of the heterocyclic ring of (3). Reduction
of 3-amino-1,2,3-benzotriazin-4-ones (35) with zinc in acetic acid yields 1,2,3-benzotriazin-
4-ones (10) ⟨25JPR(111)36⟩.

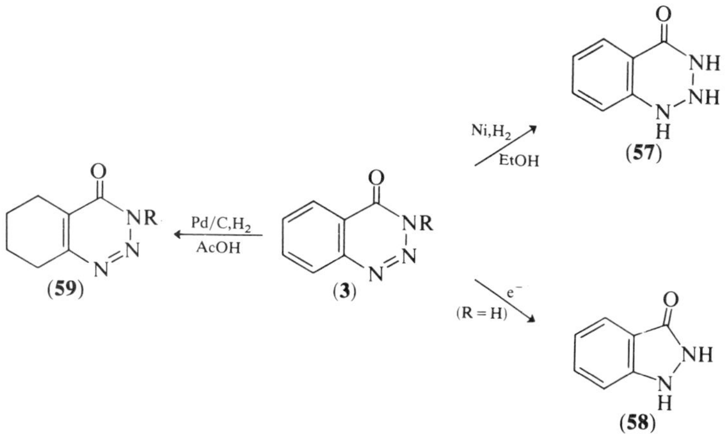

2-Arylbenzotriazinium betaines (**13**) give ring-opened products on reduction with tin and hydrochloric acid ⟨27JCS323, 72JOC1587⟩. Similarly, 4-amino-1,2,3-benzotriazines (**23**), 4-imino-1,2,3-benzotriazines (**24**) and 2-alkyl-4-imino-1,2,3-benzotriazinium betaines (**32**) all provide ring-opened products on reduction ⟨70JCS(C)2308, 70JCS(C)2289, 64JCS3663⟩, as do 1,2,3-benzotriazine 3-oxides (**41**) and 3,4-dihydro-1,2,3-benzotriazines (**43**) ⟨78HC(33)86⟩. 1,2,3-Benzotriazinium betaine 1-oxides (**33**) are deoxygenated by boiling ethanol or tin(II) chloride to the betaines (**13**) ⟨30JCS157, 30JCS843, 31JCS2787, 31JCS2792, 35JCS1005⟩. 1,3-Diaryl-tetrahydro-1,2,3-triazinium salts (**7**; R = Ar) are electrochemically reduced to the radicals (**60**) ⟨80LA285⟩.

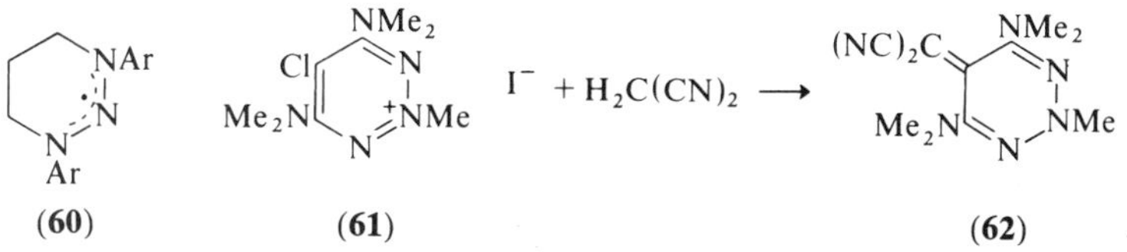

Only a few exchange reactions of substituents directly bound to the heterocyclic ring have been reported. Gompper has studied the nucleophilic substitution of bromo- and chloro-1,2,3-triazines and observed replacement of bromine or chlorine with sodium ethoxide, sodium ethanethiolate and amines. In most cases yields are quantitative. With the trihalo compound, first the 4-mono- then the 4,6-di-substituted derivative is obtained ⟨79CB1529⟩. Reaction of 5-chloro-2-methyl-4,6-bis(dimethylamino)-1,2,3-triazinium iodide (**61**) with malononitrile affords compound (**62**). Compounds of the general structure (**63**) are hydrolyzed to 1,2,3-triazin-5(2*H*)-ones (**64**) ⟨79CB1535⟩.

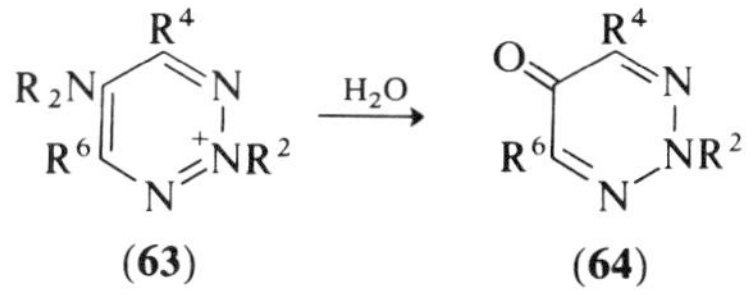

1,2,3-Benzotriazin-4-ones (**3**) afford 1,2,3-benzotriazine-4-thiones (**22**) in good yields on treatment with phosphorus pentasulfide ⟨68AP923, 68MI21800, 64BEP641818⟩. Similarly, 2-alkyl-1,2,3-benzotriazinium betaines (**26**) were transformed into their thio analogs by treatment with phosphorus pentasulfide in pyridine ⟨68MI21800⟩. 1,2,3-Benzotriazine-4-thiones (**22**) and their *S*-alkylated derivatives react with amines, hydrazines and hydroxyl-amine to give the 4-amino-, 4-hydrazino- and 4-hydroxylamino-1,2,3-benzotriazines ⟨61JCS4930, 71JHC785, 59JOC272, 69JHC779⟩. The hydrazino group in (**65**) can be transformed into the azide group (**66**), by reaction with nitrous acid, which in turn can be replaced by thioglycolate ⟨71JHC785⟩.

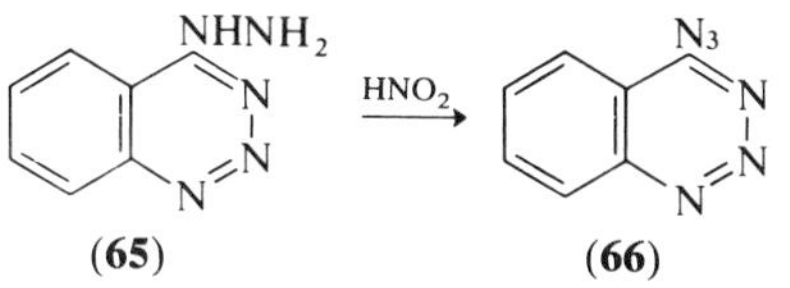

Reaction of 3-substituted 1,2,3-benzotriazin-4-ones with Grignard reagents led to opening of the heterocyclic ring ⟨60JOC1501⟩. 4-Methyl- and 4-phenyl-1,2,3-benzotriazine 3-oxide (41; R = Me, Ph) with Grignard reagents give complex reaction mixtures from which it is claimed that 1,4- (67) and 3,4-dihydro-1,2,3-benzotriazines (68) were isolated ⟨71TL3117⟩.

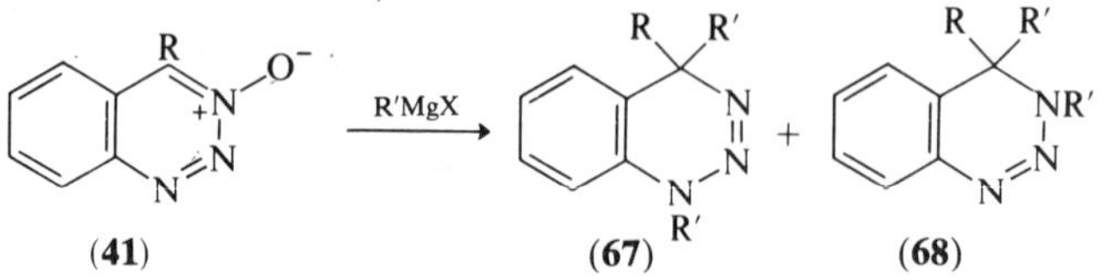

(41)                    (67)           (68)

Reactions which can be considered as cycloadditions are rare in the 1,2,3-triazine field. 4-Substituted 1,2,3-benzotriazines (2) react with diphenylcyclopropenone to give one, two or three of the following products: pyrazolo[1,2-*a*][1,2,3]benzotriazin-1-ones (69), pyrazolo[1,2-*a*][1,2,3]benzotriazin-3-ones (70) or pyrazolo[2,3-*a*]quinazolines (71) ⟨80CC808⟩.

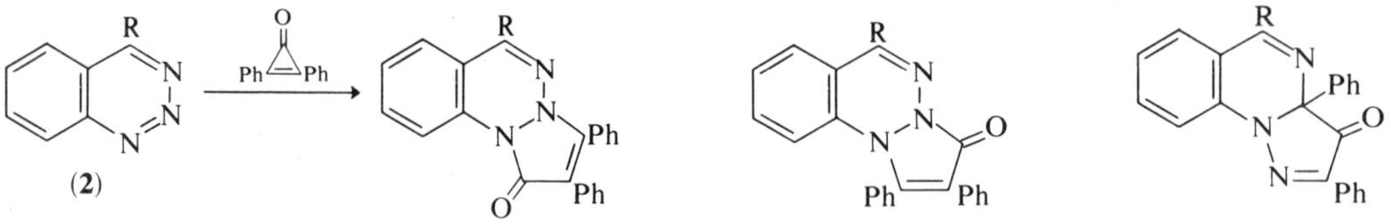

(2)              (69)              (70)              (71)

2-Methylnaphtho[1,8-*de*][1,2,3]triazine (72) reacts with dimethyl acetylenedicarboxylate in refluxing *o*-dichlorobenzene *via* a thermally allowed 1,11-dipolar cycloaddition to give the dihydroacenaphtho[5,6-*de*][1,2,3]triazine (73) which was dehydrogenated under the reaction conditions to (74) ⟨72CC1281⟩.

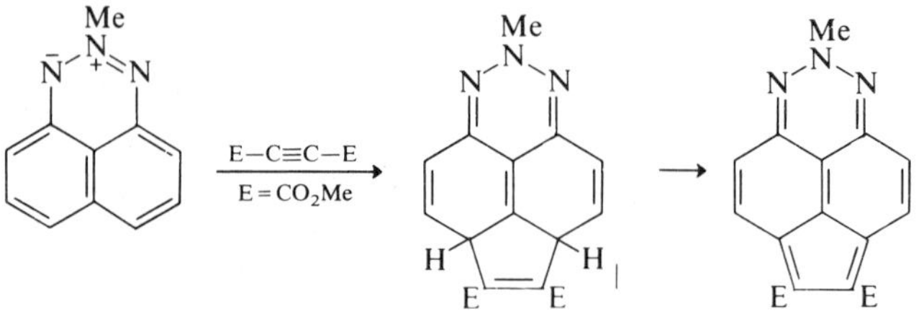

(72)                    (73)                (74)

The most carefully studied reactions of 1,2,3-triazines and 1,2,3-benzotriazines are their thermolysis (pyrolysis) and photolysis, as in all Kekulé structures these compounds have an N=N double bond. This gives the chemist the hope that nitrogen can be eliminated easily and azacyclobutadienes (azetes) (75) or benzazetes (76) can be formed. According to this idea flash vacuum pyrolysis of 4,5,6-tris(dimethylamino)-1,2,3-triazine (16) led to the isolation of a red unstable compound which is formulated as tris(dimethylamino)azete (75a; $R^4 = R^5 = R^6 = Me_2N$) ⟨73AG918, 73AG920⟩. Vapor phase pyrolysis of 1,2,3-benzotriazines led to benzazetes (76) only under special circumstances and in a few cases ⟨73CC19, 75JCS(P1)31, 75JCS(P1)41, 75JCS(P1)45, 76TL4647⟩. Pyrolysis of 4,5,6-trimethyl- or 4,5,6-triphenyl-1,2,3-triazine gave mainly nitrogen, a nitrile and an alkyne; no azetes could be detected ⟨60TL(13)19, 72JOC1051, 75JCS(P1)45⟩. Heating the 4,5,6-triaryl-1,2,3-triazines (77) to 250 °C yielded mainly the indenone imines (78) ⟨73TL219⟩.

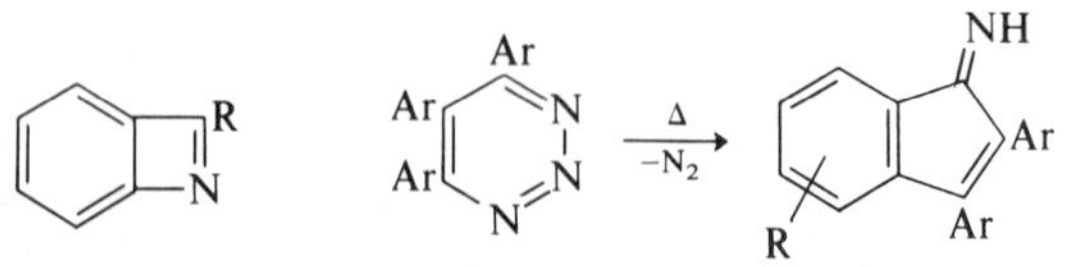

(76)              (77)              (78)

No products could be identified when 3-alkyl- or 3-aryl-4-methylene-3,4-dihydro-1,2,3-benzotriazines were refluxed in ethanol or *p*-xylene ⟨75CJC3714⟩. Flash vacuum pyrolysis of 3-adamantyl-1,2,3-benzotriazin-4-one (79) led to the isolation of 1-adamantylbenzazet-2(1*H*)-one (80) ⟨73JCS(P1)868⟩. In most other cases the isolation of the benzazetone failed. When 1,2,3-benzotriazin-4-one (21) is heated in diethylene glycol dimethyl ether, quinazolino[3,2-*c*][1,2,3]benzotriazin-8-one (**81**) is isolated in 70% yield ⟨70JCS(C)2070⟩. Heating the same compound in 1-methylnaphthalene at 250 °C yielded 2-(2-aminophenyl)-3,1-benzoxazin-4-one (**82**) ⟨73JCS(P1)1169, 68JCS(C)2730, 66TL3465⟩. The formation of these compounds is rationalized by intermediate formation of the ketene (**83**).

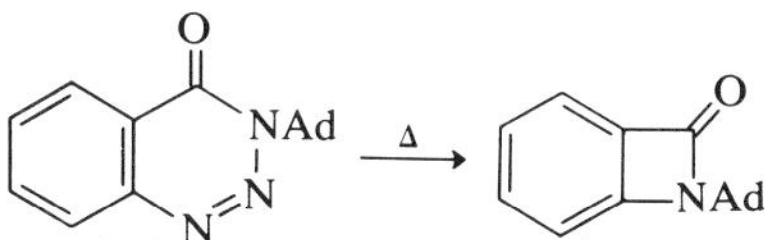

Ad = 1-Adamantyl

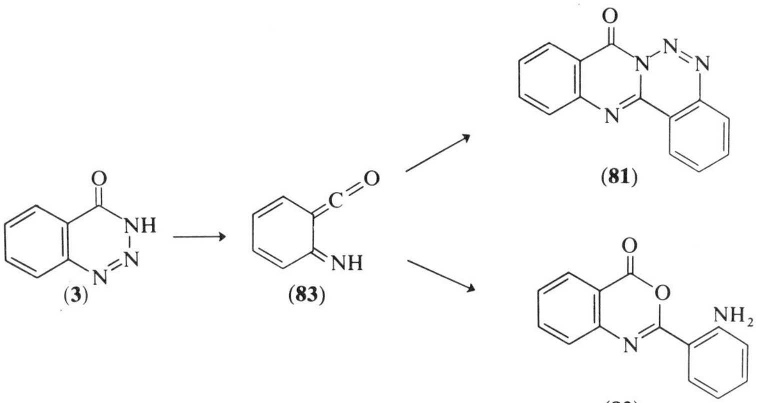

3-Aryl-1,2,3-benzotriazin-4-ones (**84**) in boiling 1-methylnaphthalene, or as solids, undergo thermal decomposition and rearrangement to 9-acridinones (**85**) and phenanthridin-6-ones (**86**), but in paraffin oil at 300 °C only benzanilides are formed ⟨76LA946, 62CI(L)1332, 68JCS(C)1028, 79JCS(P1)2203⟩. 3-(1-Naphthyl)-1,2,3-benzotriazin-4-one in hot paraffin oil at 300 °C gave a mixture of benz[*c*]acridinone (**87**) and benzo[*c*]phenanthridinone (**88**) ⟨79JCS(P1)2203⟩. 3-Alkenyl-1,2,3-benzotriazin-4-ones (**89**) undergo thermal decomposition to give 3-substituted quinolin-4(1*H*)-ones (**90**) in practicable yield ⟨79JCS(P1)2203⟩.

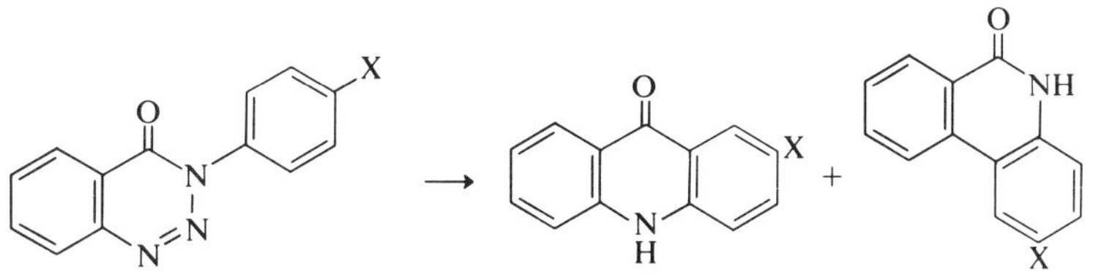

Thermal decomposition of 3-arylidenamino- (**91**) and 3-imidoyl-1,2,3-benzotriazin-4-ones (**93**) in solution gives 2-aryl- (**92**) and 2,3-diaryl-quinazolin-4(1*H*)-ones (**94**) respectively ⟨80JCS(P1)633, 75S187, 75S709⟩. In the latter decomposition phenanthridinones (**95**) are also formed in minor yields. Mechanisms to account for the isolated products were discussed. 3-Amino-1,2,3-benzotriazin-4-ones (**96**) in boiling 1-methylnaphthalene yield in all cases 2-substituted indazolin-3-ones (**97**) ⟨80JCR(S)246⟩. A mechanism for the formation of these products is suggested. Vapor phase pyrolysis of (**96**; R = H) led to the isolation of indazolin-3-one (**97**; R = H) in 80% yield ⟨73JCS(P1)868⟩. Thermolysis of 3-hydroxy-1,2,3-benzotriazin-4-one (**98**) gives 3-(2-aminobenzoyloxy)-1,2,3-benzotriazin-4-one (**99**) ⟨73TL4547⟩. A mechanism for its formation is presented.

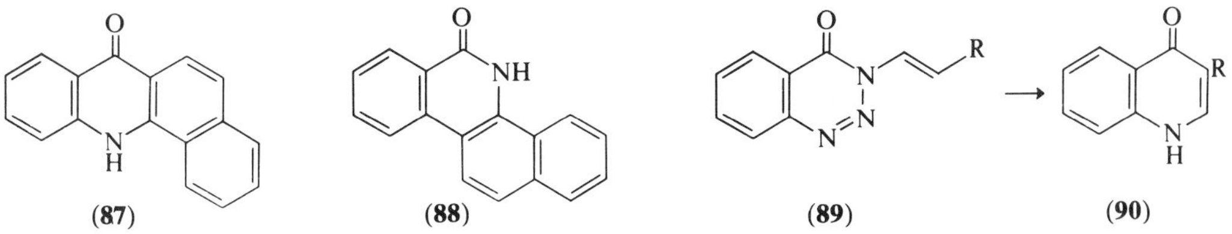

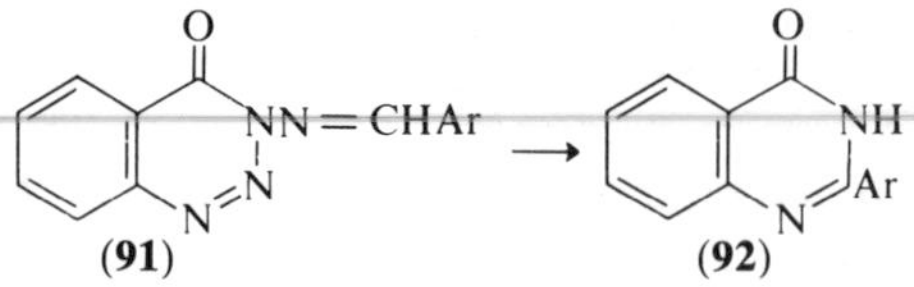

(91)  →  (92)

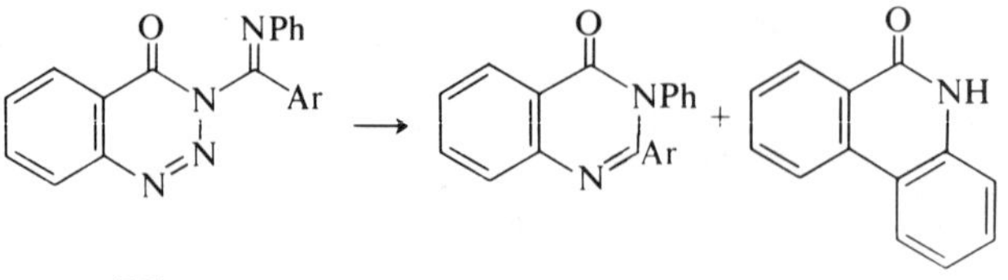

(93)  →  (94)  +  (95)

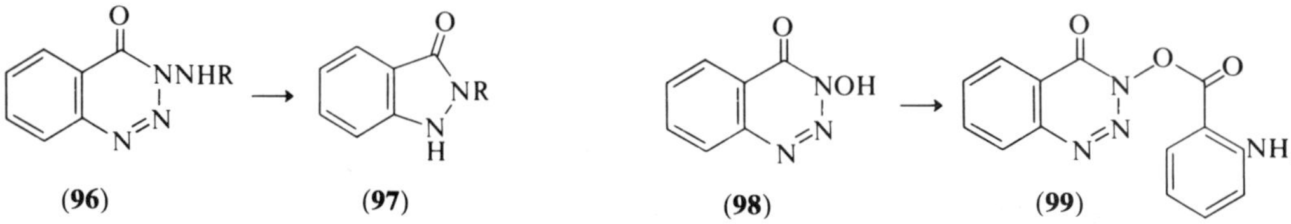

(96)  →  (97)     (98)  →  (99)

Pyrolysis of 4-phenyl-1,2,3-benzotriazine 3-oxide (100) at 420 °C gave 3-phenylbenz-isoxazole (101), 3-phenylindazole (102) and acridin-9-one (103). However, the pyrolysis of 4-phenylnaphtho[2,3-*d*][1,2,3]triazine (104) at 470 °C gave the orange 2-phenyl-naphth[2,3-*b*]azete (105) which is quite stable at room temperature ⟨75JCS(P1)45⟩.

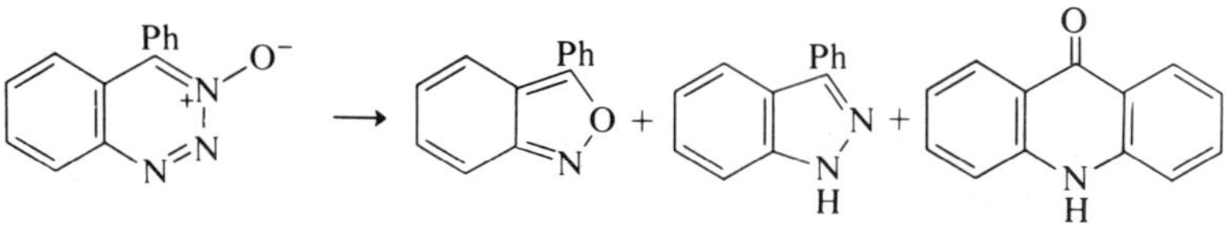

(100)  →  (101)  +  (102)  +  (103)

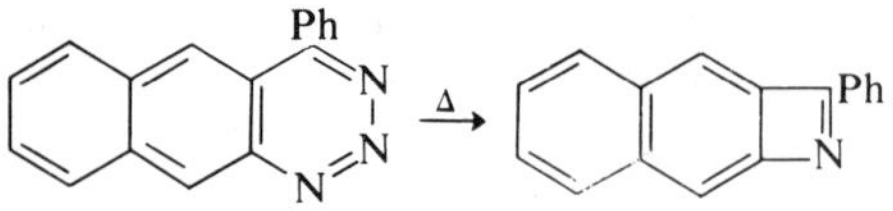

(104)  →  (105)

The irradiation of monocyclic 1,2,3-triazines has been studied by different groups ⟨60TL(13)19, 72JOC1051, 73JOC176, 74JOC940⟩. In most cases nitrogen, a nitrile and an alkyne are formed. Trimethyl-1,2,3-triazine did not give an azete on photolysis in a matrix ⟨80LA798⟩. The fused triazine (106) cleaves quantitatively to give the strain-free alkynic nitrile (107) ⟨77H(8)319⟩. Photolysis of 4-phenyl- or 4-*t*-butyl-1,2,3-benzotriazine (2; R = Ph, Bu$^t$) yielded dimers (108) of the expected benzazetes ⟨75JCS(P1)45, 76CC411, 76TL4647⟩. The 2-phenylbenz-azete is stable up to −40 °C.

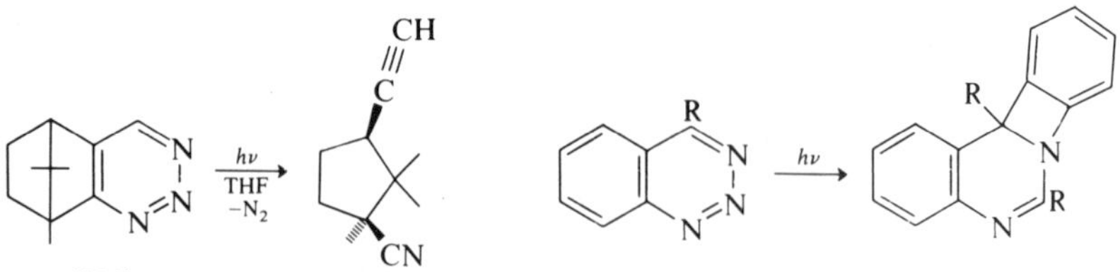

(106)  →  (107)     (2)  →  (108)

The photolysis of 1,2,3-benzotriazine-4-ones has been carefully examined by Ege and his group. 3-Unsubstituted and 3-alkyl-1,2,3-benzotriazin-4-ones seem to be stable towards irradiation with UV light. 3-Aryl-1,2,3-benzotriazin-4-ones (84) lose nitrogen and form benzazetinones (109) which could not be isolated but were detected spectroscopically ⟨76LA946, 78HC(33)58⟩.

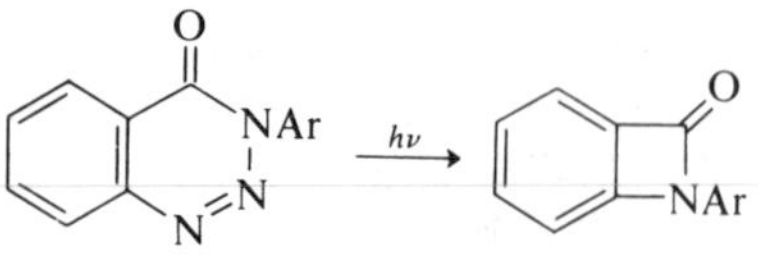

(84)  →  (109)

Indazolin-3-one (**58**) is formed by irradiation of 3-amino-1,2,3-benzotriazin-4-one (**35**) in acetonitrile ⟨73JCS(P1)868⟩. Photolysis of 4-alkyl-1,2,3-benzotriazine 3-oxides (**41**; R = alkyl) yields mainly the 3-alkylanthranils (**51**), while from the 4-aryl compounds (**100**) both 3-arylindazoles (**102**) and the anthranils (**101**) are isolated ⟨73JA2390⟩. 2-Substituted 1,2,3-benzotriazinium betaines (**13**) rearrange almost quantitatively to the corresponding 3-substituted 1,2,3-benzotriazin-4(3*H*)-ones (**84**) on irradiation in acetonitrile ⟨79JCS(P1)1199⟩. 1-Phenylbenzazetine (**111**), benzalaniline and 5,6-dihydrophenanthridine (**112**) were obtained from the photolysis of 3-phenyl-3,4-dihydro-1,2,3-benzotriazine (**110**) ⟨66JA1580⟩, while 3-amino- (**113a**) and 3-phenyl-naphtho[2,3-*d*][1,2,3]triazin-4-one (**113b**) gave the corresponding naphth[2,3-*b*]azet-2(1*H*)-one (**114a**, **114b**) ⟨68AG316, 73JCS(P1)868⟩.

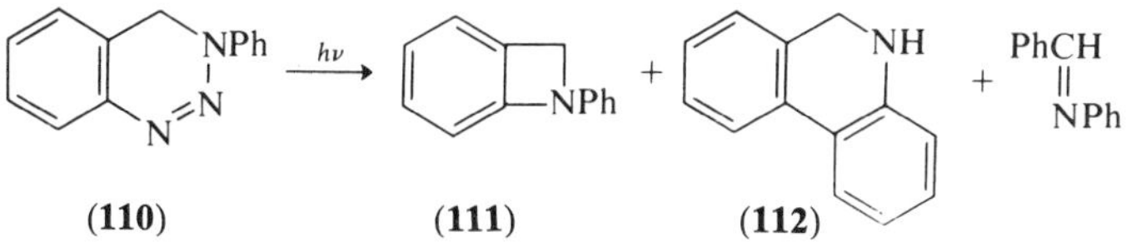

(**110**)    (**111**)    (**112**)

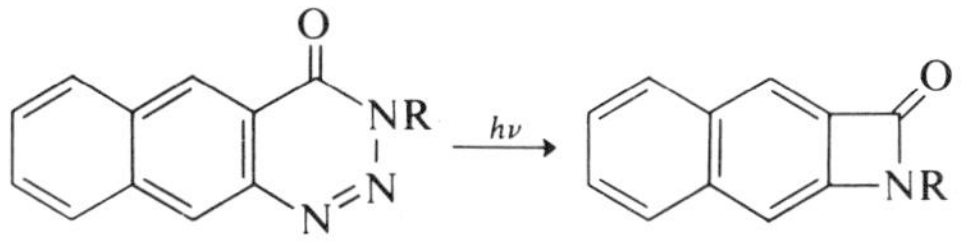

(**113a**) R = NH₂
(**113b**) R = Ph          (**114**)

## 2.18.4 SYNTHESES

The most general method for the synthesis of monocyclic 1,2,3-triazines and of 1,2,3-benzotriazines (**2**) seems to be the oxidation of *N*-aminopyrazoles (**115**) and *N*-aminoindazoles (**116, 117**) with LTA or nickel peroxide ⟨81CC1174, 80CC1182, 75JCS(P1)31, 71CC828⟩. Reaction of 3-phenyl-1-aminoindazole (**116**; R = Ph) with butyllithium and tosyl azide is reported to yield 4-phenyl-1,2,3-benzotriazine (**2**; R = Ph) ⟨72JCS(P1)1315⟩. 1-Amino- (**118**) and 3-amino-quinazolin-2-ones (**119**) can also be oxidized to 1,2,3-benzotriazines (**2**) ⟨75JCS(P1)31⟩. *C*-Amino compounds can also give triazines; oxidation of 3-aminoindazole (**120**) with hydrogen peroxide, permanganate, persulfate or dichromate led to the isolation of 1,2,3-benzotriazin-4-one (**10**) ⟨1898CB2636, 1899LA(305)289⟩.

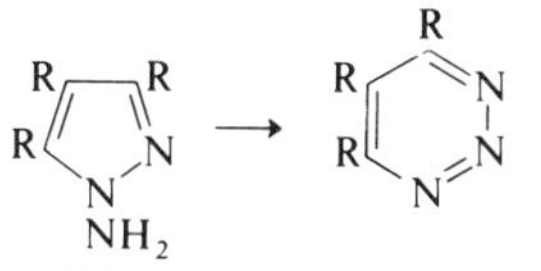

(**115**)

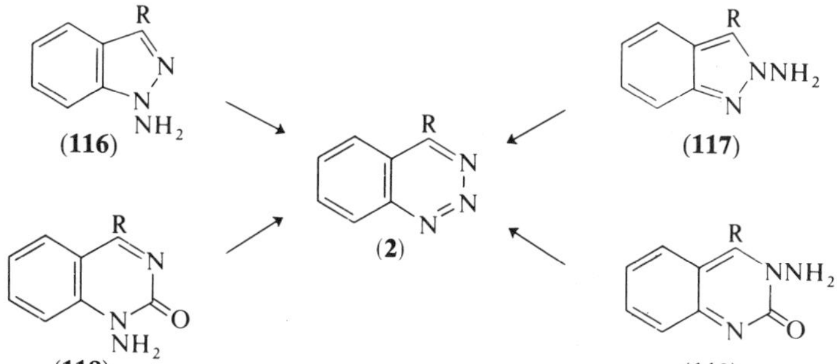

(**116**)    (**117**)
(**2**)
(**118**)    (**119**)

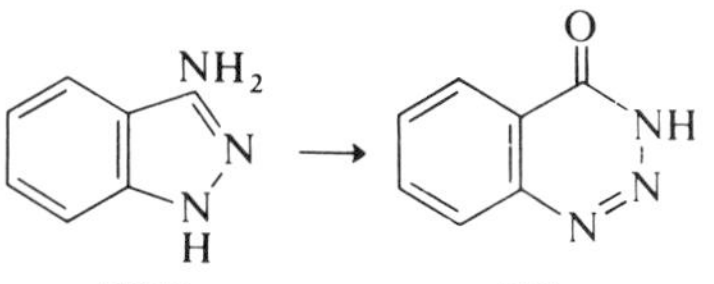

(**120**)    (**10**)

The rearrangement of cyclopropenyl azides (**121**) is the method most used for the synthesis of monocyclic 1,2,3-triazines (**1**) ⟨60TL(13)19, 72CB3695, 72JOC1051, 73AG918, 73ZN(B)535, 72TL3293, 73JOC3149, 79CB1514⟩. Tribromo- (**123**; X = Br) and trichloro-1,2,3-triazine (**123**: X = Cl) have been obtained by reaction of trimethylsilyl azide with tetrabromo- and tetrachloro-cyclopropene (**122**; X = Br, Cl) respectively ⟨79CB1529⟩. Mass spectrometric fragmentation of the 1,2,3-triazines obtained by rearrangement of isomeric mixtures of cyclopropenyl azides (**121**) always reveals a single isomer in which the substituent with the highest electron-donating power is located at the 5-position ⟨72CB3695, 79CB1514⟩. Attempts to prepare 1,2,3-benzotriazines (**2**) by rearrangement of benzocyclopropenyl azides have not been reported. Diphenylchloroazirine (**124**) and diazomethane yielded 4,5-diphenyl-1,2,3-triazine (**125**) in 10–20% yield ⟨81TL2909⟩. However, no reaction was observed between (**124**) and ethyl diazoacetate or diphenyldiazomethane.

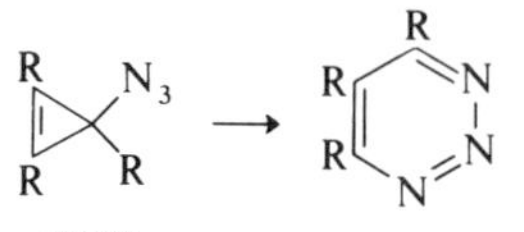

(121)

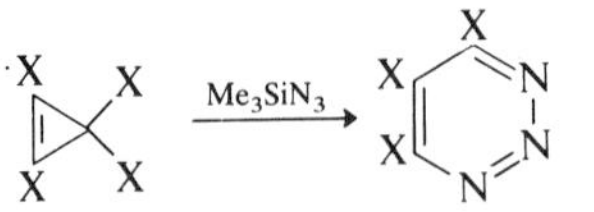

(122)   (123) 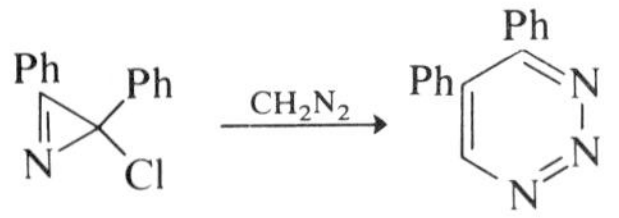   (124)   (125)

For the synthesis of the 1,2,3-benzotriazine system a ring closure by the [5 + 1] atom fragment method is very often used, where the one-atom fragment is mostly N-2 of the heterocyclic ring. The following 1,2,3-benzotriazines have been prepared by this synthetic principle from the indicated starting materials by reaction with nitrous acid: 4-methoxyphenyl-1,2,3-benzotriazine (**2**; Ar = $C_6H_4OMe$) from 2-(1-imino-2-methoxyphenylene)aniline (**126**) ⟨53JCS716⟩, 1,2,3-benzotriazin-4-ones (**3**) from 2-aminobenzamides (**127**; X = O) ⟨78HC(33)17, 75JHC199⟩, 1,2,3-benzotriazine-4-thiones (**128**; X = S) from 2-aminothiobenzamides (**127**; X = S) ⟨71JHC785, 59JOC272, 09CB3710⟩, 3-amino-1,2,4-benzotriazin-4-ones (**128**; R = NH₂, NHR'; X = O) from 2-aminobenzohydrazides ⟨78HC(33)17⟩, 3-hydroxy-1,2,3-benzotriazin-4-ones (**98**) from 2-aminobenzohydroxamic acids (**127**; X = O, R = OH) ⟨78HC(33)17⟩, 1,2,3-benzotriazine 3-oxides (**41**) from 2-aminophenyl-aldoximes or -ketoximes (**129**; R = H, alkyl, aryl) ⟨78HC(33)84⟩, 4-amino-1,2,3-benzotriazines (**38**) or 4-imino-1,2,3-benzotriazines (**39**) from 2-aminobenzamidines (**130**) ⟨61JCS4930, 70JCS(C)2308, 64JCS3663, 24MI21800⟩ and 4-amino-1,2,3-benzotriazine 3-oxides (**28**) from 2-aminobenzamidoximes (**129**; R = NH₂) ⟨61JCS4930⟩. 1,2,3-Benzotriazin-4-one (**3**) is also obtained by reaction of 2-aminobenzohydrazide (**127**; R = NH₂) with excess of nitrous acid ⟨27JPR(116)9⟩ and from isatoic diamide (**131**) and nitrous acid ⟨47G308⟩.

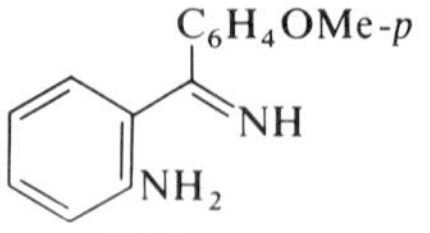

(126) 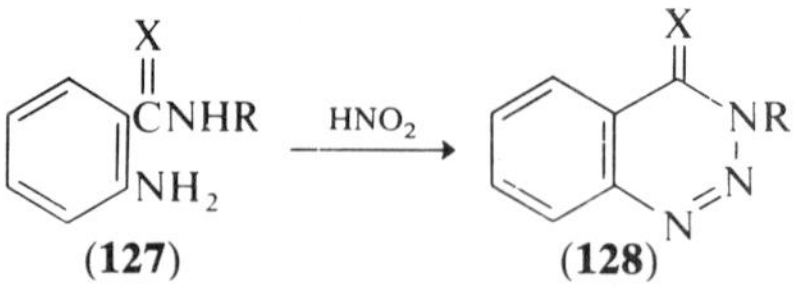   (127)   (128)

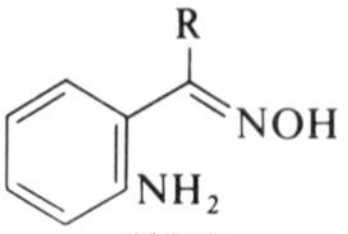

(129) 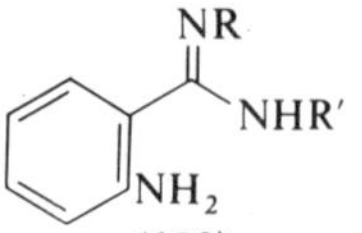   (130)    (131)

Dihydro-1,2,3-benzotriazines (**132**) have been obtained by reaction of 2-aminobenzyl-amines (**133**) and of 2-aminobenzylhydrazines (**134**) with nitrous acid ⟨78HC(33)91, 75JHC1155⟩.

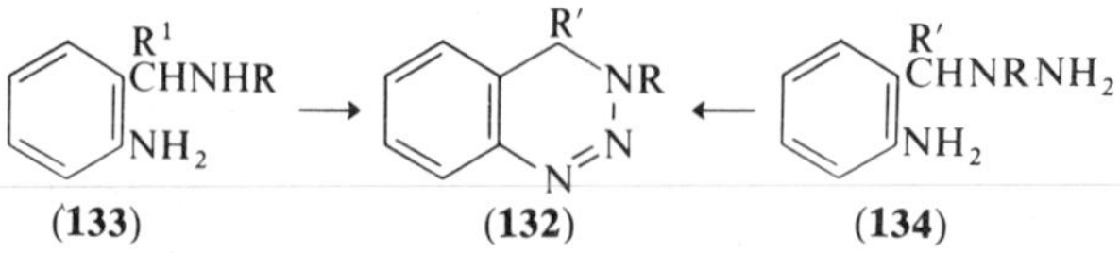

(133)   (132)   (134)

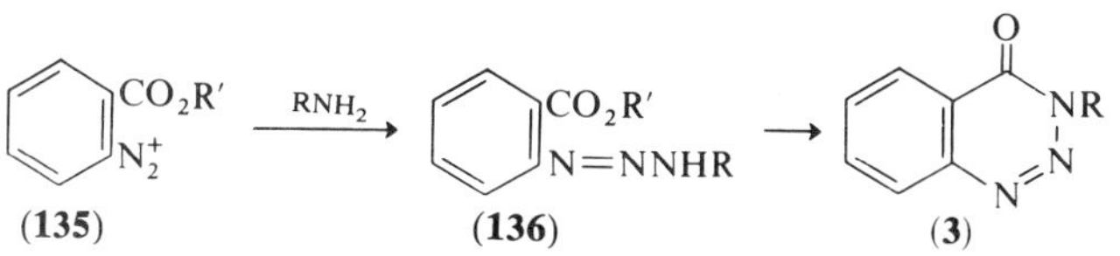

In some cases N-3 is the one-atom fragment in the synthesis of 1,2,3-benzotriazines. Diazotization of anthranilates affords the diazonium compounds (135) which react with amines to give the triazenes (136) which cyclize to the 1,2,3-benzotriazin-4-ones (3) ⟨78HC(33)18, 78TL5041⟩. Similarly, 2-aminobenzonitriles can be diazotized and the diazonium compounds (137) react *via* the triazenes (138) to yield 4-imino-1,2,3-benzotriazines (139) or the rearranged amino compounds (38) on boiling in aqueous ethanol, in ethanol/piperidine, or in aqueous acids ⟨78HC(33)72⟩. Benzenediazonium-2-carboxylate, when treated with a variety of compounds such as chlorosulfonyl isocyanate, eventually gives the triazine (10) or its substituted derivatives (3) ⟨72JOC196, 72BCJ3504⟩.

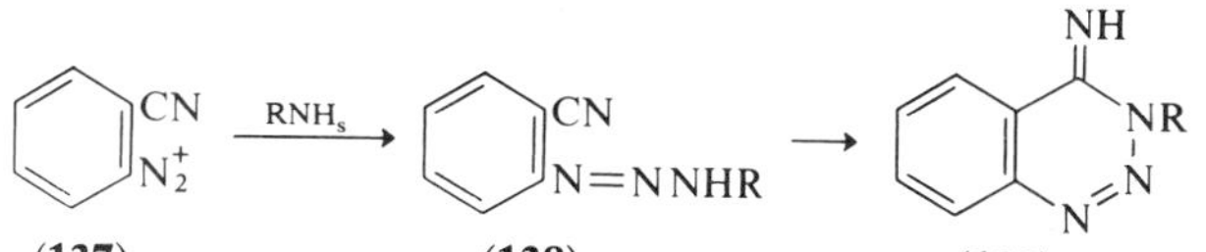

The [5 + 1] atom fragment method has also been used indirectly for the synthesis of one monocyclic 1,2,3-triazine. The 5-aminoimidazo-4-carboxamidine (140) was cyclized with nitrous acid to give the 4-aminoimidazo[4,5-*d*][1,2,3]triazine (141) which gave 4,5-diamino-6-cyclopentylamino-1,2,3-triazine (142) together with other products on hydrolysis ⟨72JMC182⟩.

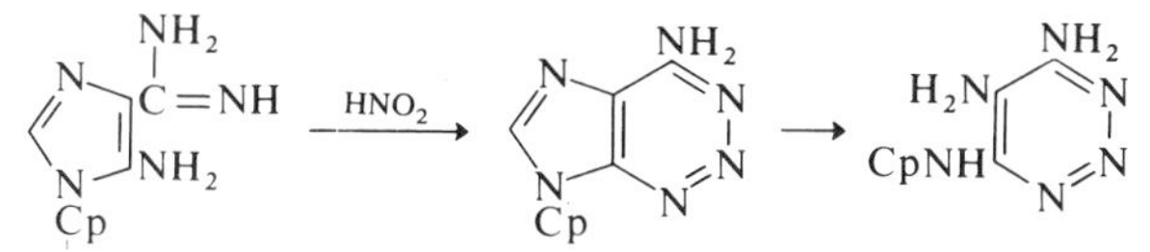

(140) Cp = Cyclopentyl   (141)   (142)

Oxidation of 2-azidophenyl ketone hydrazones (143) affords the 2-azidophenyldiazoalkanes (144) which can be cyclized thermally to 1,2,3-benzotriazines (2) ⟨75JCS(P1)31, 71CC828⟩. Similarly, 2-aminophenyl ketone hydrazones (145) gives 1,2,3-benzotriazines (2) on oxidation with LTA ⟨75JCS(P1)31, 71CC828⟩.

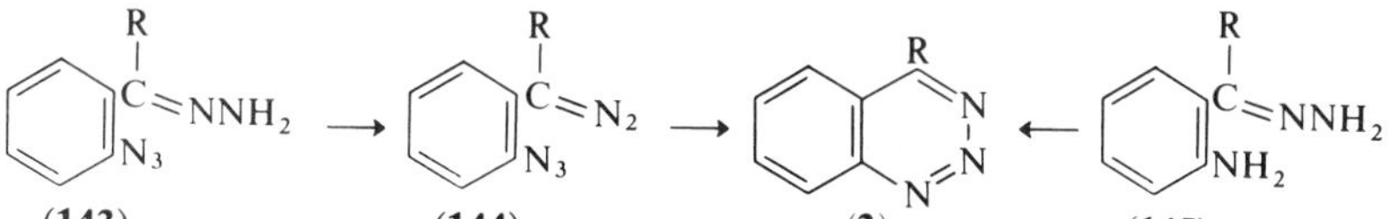

(143)   (144)   (2)   (145)

Reaction of 2-nitrobenzaldehyde arylhydrazones (146) with a halogen, followed by treatment with base, gives compounds which, after a long and controversial history, were finally shown to be 2-aryl-4-oxido-1,2,3-benzotriazinium betaine 1-oxides (33; R = Ar) ⟨72JOC1587, 72JOC1592, 74JOC2710⟩. Oxidation of the methylhydrazone of 2-nitrobenzaldehyde with LTA afforded the 2-methyl compound (33; R = Me), and other hydrazones (146) similarly formed the betaines (33; R = Ar) directly ⟨74JOC2710⟩. When the bromo derivative (147; X = Br) is irradiated with UV light, the 3-aryl-1,2,3-benzotriazin-4-ones (3; R = Ar) are produced ⟨78S382⟩.

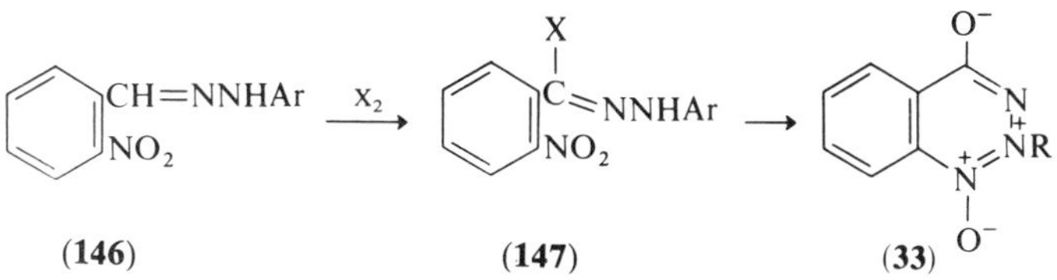

(146)   (147)   (33)

Electrochemical oxidation of 2-hydroxylaminobenzohydrazide (148) affords (10), and similar treatment of 1-(2-hydroxylaminobenzyl)-1-phenylhydrazine (149) yields 3-phenyl-3,4-dihydro-1,2,3-benzotriazine (150) ⟨76BSF433⟩.

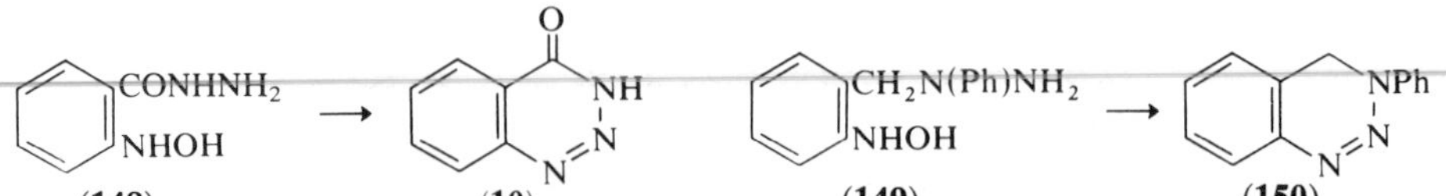

Cyclization of the triazenes (151) has been used for the synthesis of the 1,3-disubstituted 3,4,5,6-tetrahydro-1,2,3-triazinium salts (7) ⟨79CB445⟩. Another monocyclic 1,2,3-triazine system, the mesomeric betaine (9), is formed by reaction of the triazenes (152) with chloroformylketenes (153) ⟨78CB2173⟩. The thermal rearrangement of 3-arylazo-2,1-benzisoxazoles (154) in *o*-dichlorobenzene led to the 3-aryl-1,2,3-benzotriazin-4-ones (84) ⟨79TL4687⟩. In another apparent rearrangement, the *p*-dimethylaminophenyl derivative (84; Ar = C₆H₄NMe₂) is produced in the reaction of the azohydrazide (155) with nitrous acid, presumably *via* the corresponding acid azide ⟨72JOC1592⟩.

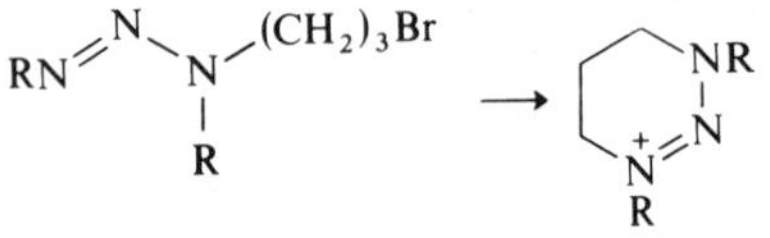

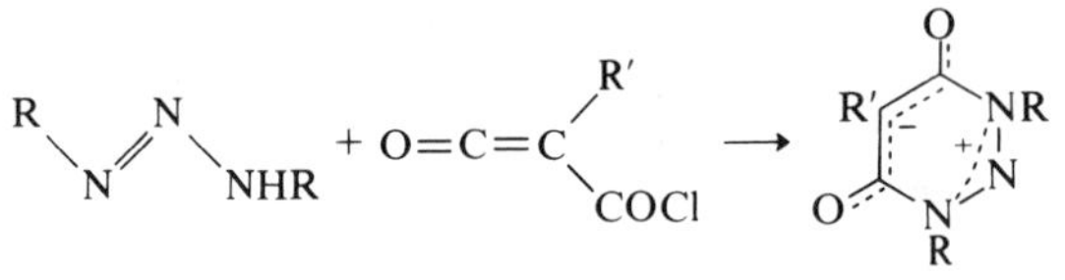

## 2.18.5 APPLICATIONS AND IMPORTANT COMPOUNDS

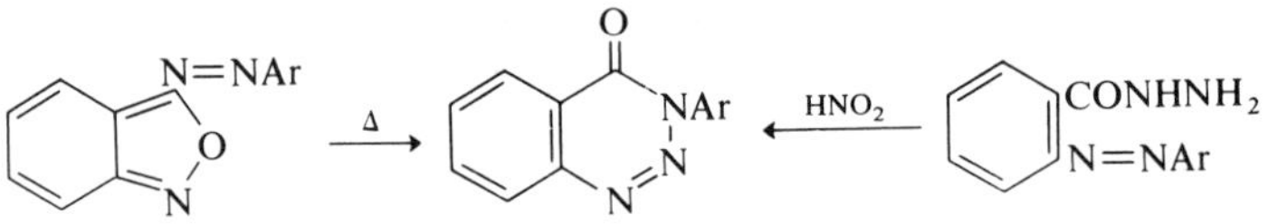

So far no compound containing the 1,2,3-triazine system has been isolated from natural sources. The most important compound in the 1,2,3-benzotriazine series seems to be the plant protection agent (156) (Bayer 17147, Guthion, Azinphos methyl). There are a very large number of papers and patents dealing with the chemical and biochemical properties of this compound; the references up to *Chemical Abstracts*, volume 81, can be found in the earlier monograph ⟨78HC(33)165⟩.

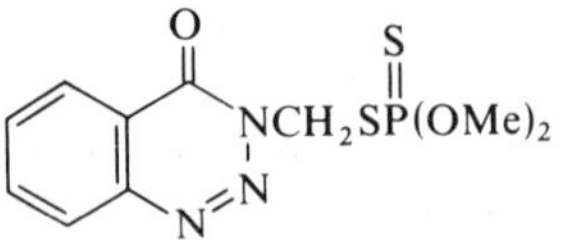

(156)

The use of 1-*t*-butyl-3-ethyl-2-methylhexahydro-1,2,3-triazine as a corrosion inhibitor for steel is mentioned in a number of Russian publications ⟨78HC(33)13⟩, but no further information on this compound could be found in the literature. For a number of other compounds containing the 1,2,3-triazine system, claims for biochemical or technical applications have been made, but these seem to have been mostly for the purposes of obtaining patents on the compounds involved, and it appears that no significant uses are yet known.

# 2.19

# 1,2,4-Triazines and their Benzo Derivatives

H. NEUNHOEFFER

*Technische Hochschule Darmstadt*

| | | |
|---|---|---|
| 2.19.1 | INTRODUCTION | 385 |
| 2.19.2 | STRUCTURE AND SPECTRA | 386 |
| | *2.19.2.1 Structure* | 386 |
| | *2.19.2.2 Tautomeric Studies* | 389 |
| | *2.19.2.3 NMR Spectra* | 393 |
| | *2.19.2.4 IR Spectra* | 395 |
| | *2.19.2.5 UV Spectra* | 395 |
| | *2.19.2.6 Mass Spectra* | 396 |
| | *2.19.2.7 Photoelectron Spectra* | 398 |
| | *2.19.2.8 Other Spectral and Thermodynamic Properties* | 398 |
| 2.19.3 | REACTIONS AND REACTIVITY | 399 |
| 2.19.4 | SYNTHESES | 430 |
| | *2.19.4.1 [4+2] Atom Combinations* | 430 |
| | *2.19.4.1.1 Combination A [N(1)N(2)C(3)N(4)+C(5)C(6)]* | 430 |
| | *2.19.4.1.2 Combination B [C(3)N(4)C(5)C(6)+N(1)N(2)]* | 437 |
| | *2.19.4.1.3 Combination C [N(4)C(5)C(6)N(1)+N(2)C(3)]* | 440 |
| | *2.19.4.1.4 Combination D [C(5)C(6)N(1)N(2)+C(3)N(4)]* | 441 |
| | *2.19.4.1.5 Combination E [C(6)N(1)N(2)C(3)+N(4)C(5)]* | 442 |
| | *2.19.4.2 [5+1] Atom Combinations* | 443 |
| | *2.19.4.2.1 C(3) as the one-atom fragment* | 443 |
| | *2.19.4.2.2 N(4) as the one-atom fragment* | 446 |
| | *2.19.4.3 [6+0] Atom Combinations (Unimolecular Cyclizations)* | 446 |
| | *2.19.4.4 [3+3] Atom Combinations* | 452 |
| | *2.19.4.5 Synthesis from More than Two Fragments* | 453 |
| 2.19.5 | APPLICATIONS AND IMPORTANT COMPOUNDS | 455 |

## 2.19.1 INTRODUCTION

1,2,4-Triazines are well-known compounds and a wide variety of synthetic methods for the preparation of substituted derivatives are available. Compounds containing the 1,2,4-triazine moiety are found in natural materials and some of these show biological activity. A large number of synthetic 1,2,4-triazines also have biological activity and have been used for various purposes. The number of publications dealing with 1,2,4-triazines is high, particularly on account of their biochemical properties. Very recently 1,2,4-triazines have been shown to be intermediates for the synthesis of other nitrogen-containing heterocycles and this observation has aroused further interest in the chemistry of this ring system.

A number of reviews on the chemistry of 1,2,4-triazines have been published. The first, in 1956, covered the literature through *Chemical Abstracts* 1950 ⟨56HC(10)44⟩. A few years later a second was published by Horwitz ⟨B-61MI21900⟩. Hadaček and Slouka reviewed the chemistry of 1,2,4-triazines in three volumes in 1965, 1966 and 1970 ⟨65MI21900, 66MI21900, 70MI21900⟩. In 1971 a short account was published by Jones and Kershaw ⟨71MI21900⟩. A comprehensive review covering the whole literature on 1,2,4-triazines through *Chemical Abstracts* 1974 appeared in 1978 ⟨78HC(33)189⟩. This review listed 2316 references, and since then about 1900 additional publications have been summarized in *Chemical Abstracts*.

Besides the reviews which have been mentioned, a number of accounts of special aspects of the chemistry or biochemistry of 1,2,4-triazines have been published, but it seems to us that it would be beyond the scope of this article to list all of these.

The parent compound of the 1,2,4-triazine series has structure (**1**) and was prepared for the first time by Paudler and Barton in 1966 〈66JOC1720〉. In *Chemical Abstracts* (**1**) is numbered as indicated. Besides the name 1,2,4-triazines, *as*-triazines, $\alpha$-triazines and isotriazines can also be found in the (older) literature.

Two Kekulé structures (**1a** and **1b**) can be drawn for this molecule. Theoretical calculations suggest that structure (**1a**) gives a higher contribution to the ground state of the molecule. This is supported by X-ray crystallographic structure determinations of simple 1,2,4-triazines.

(1a)          (1b)

In contrast to the 1,2,3-triazines, the monocyclic 1,2,4-triazines are well known, but there is less information on 1,2,4-benzotriazines (**2**). Besides 1,2,4-benzotriazines a large number of 1,2,4-triazines condensed with carbocycles are known, such as cyclobuta-, cyclopenta-, cyclohepta-, cyclooct-, naphtho[1,2-*e*]- (**3**), naphtho[2,1-*e*]- (**4**) and naphtho[2,3-*e*]-1,2,4-triazines (**5**).

(2)          (3)          (4)          (5)

A number of 1,2,4-triazines are of interest owing to their biological activity. 1,2,4-Triazine-3,5-diones (**6**) represent aza analogues of pyrimidine nucleic acid bases, a number of natural antibiotics are derivatives of pyrimido[5,4-*e*][1,2,4]triazine (**7**), and 4-amino-6-*t*-butyl-3-methylthio-1,2,4-triazin-5-one (**8**) and 4-amino-3-methyl-6-phenyl-1,2,4-triazin-5-one (**9**) are used as herbicides.

(6)          (7)          (8)          (9)

## 2.19.2 STRUCTURE AND SPECTRA

### 2.19.2.1 Structure

Because of the great interest in 1,2,4-triazine chemistry, many detailed data on the structure of various 1,2,4-triazines have been published. Until now about 15 X-ray crystallographic analyses of compounds containing the 1,2,4-triazine moiety can be found in the literature. The simplest 1,2,4-triazine of which the structure has been determined by X-ray crystallography is the 5-(4-chlorophenyl) derivative (**10**) 〈74JHC743〉. The dimensions obtained for the 1,2,4-triazine ring of this compound are not expected to be significantly different from those of the parent 1,2,4-triazine (**1**). A comparison of the observed bond distances with those of similar azabenzenes shows that the Kekulé structure of the 1,2,4-triazine with a single bond between N-1 and N-2 (**10a**) more closely represents the ground tate of this ring system than the one with a double bond between N-1 and N-2 (**10b**).

Further compounds for which X-ray crystallographic analyses have been published are 5[2-(dimethylamino)propenyl]-6-methyl-3-phenyl-1,2,4-triazine (**11**) 〈73LA1970〉, 1,2,4-triazine-3,5-dione (6-azauracil; **12**) 〈74AX(B)1430〉, 6-methyl-1,2,4-triazine-3,5-dione (6-azathymidine; **13**) 〈75AX(B)2519〉, 2-(*β*-D-ribofuranosyl)-1,2,4-triazine-3.5-dione (6-azauridine; **14**) 〈73MI21900〉, 2-(*β*-D-ribofuranosyl)-6-methyl-1,2,4-triazine-3,5-dione

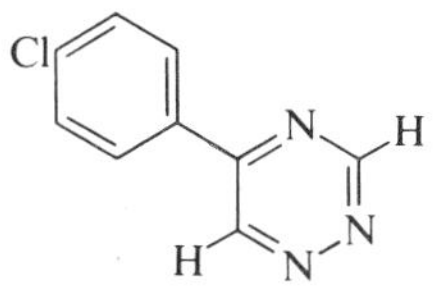

(10a) 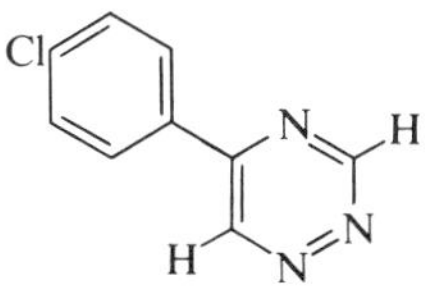  (10b) 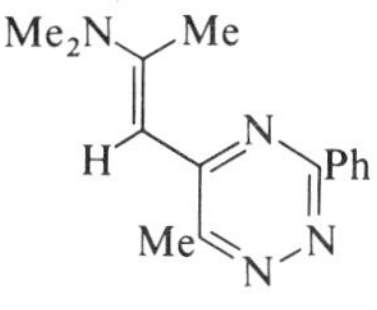  (11)

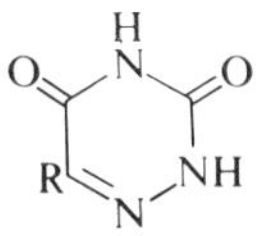

(12) R = H
(13) R = Me

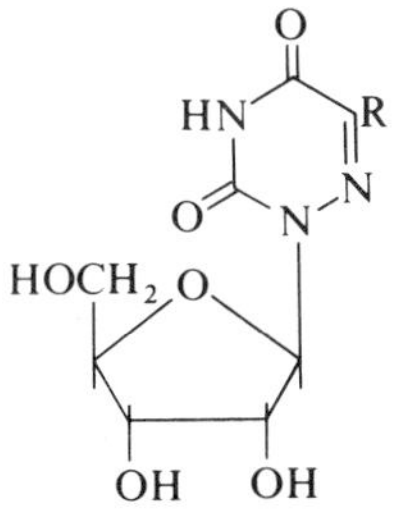

(14) R = H
(15) R = Me

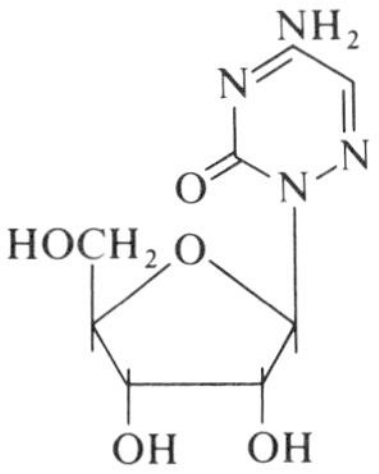

(16)

(6-azathymine; **15**) ⟨75AX(B)2519⟩, 5-amino-2-(β-D-ribofuranosyl)-1,2,4-triazin-3-one (6-azacytidine; **16**) ⟨74JA1239⟩, 6-amino-3-methyl-1,2,4-triazin-5-one monohydrate (**17**) ⟨77CL1231⟩, 6-methyl-3-thioxo-1,2,4-triazin-5-one (**18**) ⟨78TL4431⟩, 2-methyl-3-methylthio-5,6-diphenyl-2,5-dihydro-1,2,4-triazine (**19**) ⟨81BCJ41⟩, the betaine (**20**) ⟨77LA1421⟩, 3-methylnaphtho[1,2-*e*][1,2,4]triazin-2(3*H*)-one 1-oxide (**21**) ⟨78JCS(P1)789⟩ and imidazo[1,2-*b*][1,2,4]triazine (**22**) ⟨77AX(B)274⟩.

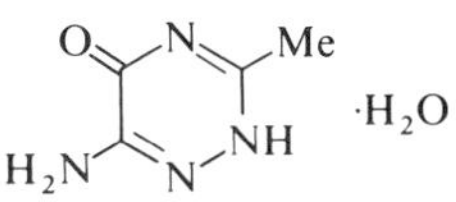

(17a) 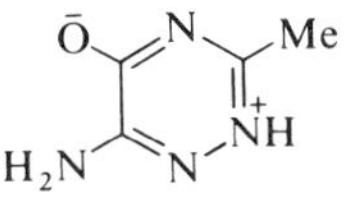  (17b) 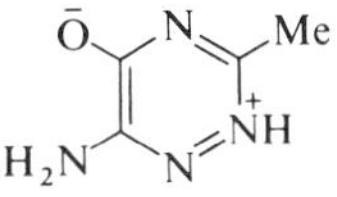  (17c)

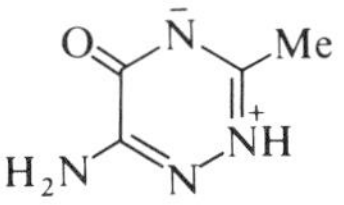

(17d) 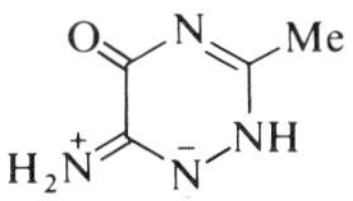  (17e) 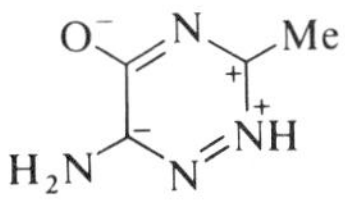  (17f)

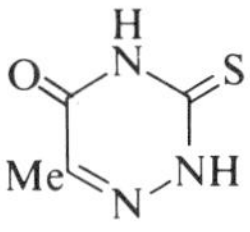

(18) 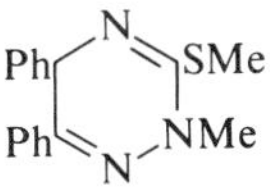  (19) 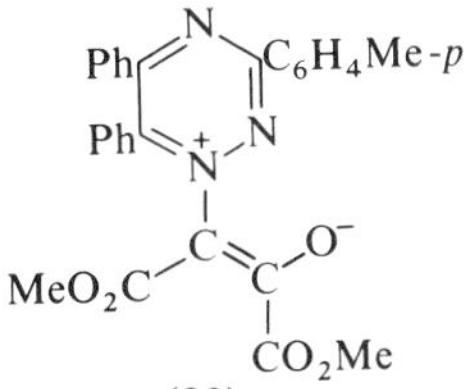  (20)

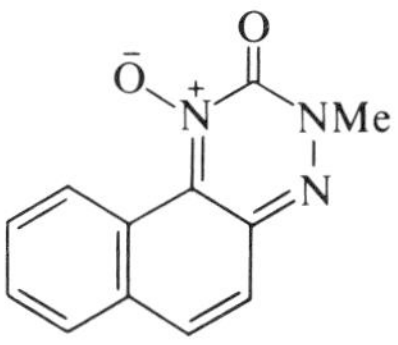

(21) 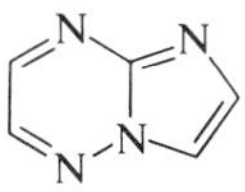  (22)

This enumeration gives only a selection of the published data. In Table 1 the experimental bond distances and bond angles are given and compared with calculated values.

The structure determination of (**10**) and (**11**) has shown that the molecules are planar, as one expects for a system with some degree of electron delocalization. The bond lengths in (**10**) and (**11**) are very similar and in good agreement with those calculated by semi-empirical methods ⟨66JCP(44)759⟩.

The C(4a)—N(8) bond length in imidazo[1,2-*b*][1,2,4]triazine (**22**) is distinctly longer than in other 1,2,4-triazines, showing that this bond is much more a single bond than in the other compounds.

**Table 1** Experimental Bond Distances and Angles of 1,2,4-Triazines and Calculated Values

| Bond/angle | (10) | (11) | (20) | (22) | Calc. by PPD | Calc. by SPO | (16) | (21) | (19) |
|---|---|---|---|---|---|---|---|---|---|
| N(1)—N(2) | 1.335 | 1.35 | 1.351 | 1.350 | 1.335 | 1.348 | 1.356 | 1.338 | 1.394 |
| N(2)—C(3) | 1.314 | 1.33 | 1.323 | 1.404 | 1.310 | 1.304 | 1.389 | 1.361 | 1.383 |
| C(3)—N(4) | 1.339 | 1.34 | 1.348 | 1.341 | 1.361 | 1.357 | 1.357 | 1.448 | 1.269 |
| N(4)—C(5) | 1.317 | 1.34 | 1.317 | 1.313 | 1.304 | 1.298 | 1.323 | 1.354 | 1.472 |
| C(5)—C(6) | 1.401 | 1.44 | 1.432 | 1.408 | 1.443 | 1.460 | 1.454 | 1.427 | 1.502 |
| C(6)—N(1) | 1.317 | 1.31 | 1.338 | 1.312 | 1.297 | 1.289 | 1.289 | 1.320 | 1.286 |
| C(3)—subst. | 1.01 | — | — | 1.320 | — | — | 1.246 | 1.206 | 1.769 |
| C(5)—subst. | 1.480 | — | — | 0.99 | — | — | 1.328 | 1.452 | 1.523 |
| C(6)—subst. | 0.97 | — | — | 1.00 | — | — | — | 1.439 | 1.470 |
| C(6)N(1)N(2) | 118.8 | 120.0 | 124.4 | 112.8 | — | — | 117.6 | 116.8 | 116.1 |
| N(1)N(2)C(3) | 117.1 | 117.0 | 115.4 | 124.1 | — | — | 123.0 | 126.6 | 116.9 |
| N(2)C(3)N(4) | 127.2 | 126.0 | 125.0 | 121.0 | — | — | 119.0 | 114.3 | 124.0 |
| C(3)N(4)C(5) | 115.9 | 117.0 | 119.0 | 115.3 | — | — | 118.5 | 121.7 | 114.4 |
| N(4)C(5)C(6) | 118.4 | 117.0 | 119.5 | 122.7 | — | — | 120.7 | 116.3 | 110.1 |
| C(5)C(6)N(1) | 122.5 | 122.0 | 116.5 | 124.1 | — | — | 120.0 | 124.1 | 120.5 |

| Bond/angle | (12) | (13) | (14) | (15) | (17) | Calc. for (17) (18) |
|---|---|---|---|---|---|---|
| N(1)—N(2) | 1.291 | 1.270 | 1.372/1.369 | 1.336 | 1.374 | 1.35/1.36 |
| N(2)—C(3) | 1.456 | 1.473 | 1.386/1.378 | 1.369 | 1.312 | 1.36 |
| C(3)—N(4) | 1.359 | 1.366 | 1.390/1.381 | 1.382 | 1.332 | 1.37/1.40 |
| N(4)—C(5) | 1.378 | 1.387 | 1.373/1.381 | 1.366 | 1.348 | 1.37/1.39 |
| C(5)—C(6) | 1.366 | 1.346 | 1.469/1.467 | 1.472 | 1.484 | 1.45/1.49 |
| C(6)—N(1) | 1.351 | 1.367 | 1.286/1.291 | 1.290 | 1.306 | 1.29/1.30 |
| N(2)—subst. | — | — | 1.462/1.473 | 1.473 | — | — |
| C(3)—subst. | 1.224 | 1.240 | 1.209/1.217 | 1.226 | 1.498 | 1.62/1.71 |
| C(5)—subst. | 1.224 | 1.240 | 1.224/1.216 | 1.220 | 1.239 | 1.17/1.23 |
| C(6)—subst. | — | — | — | 1.500 | 1.337 | 1.45/1.48 |
| C(6)N(1)N(2) | — | — | 118.6/118.6 | 119.0 | — | 116/119 |
| N(1)N(2)C(3) | — | — | 124.5/124.2 | 124.5 | — | 125/126 |
| N(2)C(3)N(4) | — | — | 113.6/114.1 | 115.0 | — | 114/121 |
| C(3)N(4)C(5) | — | — | 126.3/126.2 | 124.7 | — | 117/125 |
| N(4)C(5)C(6) | — | — | 112.7/112.2 | 114.3 | — | 114/118 |
| C(5)C(6)N(1) | — | — | 123.9/124.0 | 122.4 | — | 122 |

The heterocyclic ring in 1,2,4-triazine-3,5-dione (**12**) and its 6-methyl derivative (**13**) is planar. The data clearly show that in the crystal only the dione structure is present; no indication of any other tautomeric structure was found. The methyl group in the 6-position of (**13**) has an influence on the bond lengths of the two carbonyl groups. While in (**12**) the two carbonyl groups are equal (1.224 Å), in (**13**) two different distances were found: C(3)=O = 1.213 Å and C(5)=O = 1.242 Å. In the ribosides (**14**) and (**15**) both carbonyl groups are nearly the same length.

The data obtained for 6-amino-3-methyl-1,2,4-triazin-5-one (**17**) show the surprising result that the non-ionic structure (**17a**) contributes little to the resonance while the ionic structures (**17b–f**) have to be considered in the crystalline state, which is probably due to the stabilization of these structures by hydrogen bonding or other crystal fields acting on this molecule.

In 2-methyl-3-methylthio-5,6-diphenyl-2,5-dihydro-1,2,4-triazine (**19**) the heterocyclic ring is folded at N-2 and C-5 to take a boat form with a dihedral angle between the two planes of 146.5°. The 6-phenyl ring is parallel to the plane to which it is attached, while the 3-methylthio group lies in the other plane.

Katritzky has published a few papers on the conformation of 1,2,4-trimethylhexahydro- (**23**) and 1,2,3,4-tetramethylhexahydro-1,2,4-triazine (**24**). From studies on the photoelectron spectra of (**23**) it was proposed that this compound exists in the gas phase predominantly as the *eae*-conformer (**25a**) ⟨80JCS(P2)91⟩. In solution it was found by $^1$H and $^{13}$C NMR spectroscopy that the predominant conformer is *eae*, as in the gas phase, with as minor components the *eea*- (**25b**) and *eee*-conformers (**25c**) which are 'frozen' out at −15 °C, and the *aee*-conformer (**25d**) which is 'frozen' out at −95 °C. The tetramethyl compound (**24**) exists in solution predominantly as the *eaee*-conformer (**26a**), while the minor components *aeea* (**26b**) and *eeae* (**26c**) were 'frozen' out at −90 °C and 15 °C respectively ⟨77TL3803, 79JCS(P2)984⟩.

(**23**) R = H
(**24**) R = Me

(**25a**)   (**25b**)   (**25c**)   (**25d**)

(**26a**)   (**26b**)   (**26c**)

## 2.19.2.2 Tautomeric Studies

UV spectra have been used to determine the predominant tautomeric forms of 1,2,4-triazinones, 1,2,4-triazinethiones and amino-1,2,4-triazines. In most cases it has been shown that 1,2,4-triazines with oxygen or sulfur substituents exist predominantly or exclusively in the oxo or thioxo form if this is possible, while 1,2,4-triazines with a nitrogen substituent form predominantly or exclusively the amino, hydrazino or hydroxylamino tautomer. A few exceptions to this rule are mentioned later. 1,2,4-Triazin-3-ones exist predominantly in the 2*H*-form, as shown by comparison of the following UV spectra: 5,6-diphenyl-1,2,4-triazin-3-one (**27**): 335sh ($\varepsilon$ = 4170), 295sh (6550), and 252 nm (14 880); 2-methyl-5,6-diphenyl-1,2,4-triazin-3-one (**28**): 338 (4970), 290 (6160) and 254 nm (14 300); 4-methyl-5,6-diphenyl-1,2,4-triazin-3-one (**29**): 292 (10 720), 230sh (10 920) and 215 nm (15 430); and 3-methoxy-5,6-diphenyl-1,2,4-triazine (**30**): 328 (8000) and 257 nm (13 000) ⟨71JOC3921, 66JOC3914⟩.

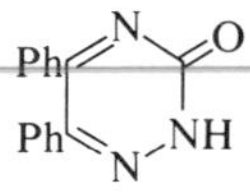

(27) 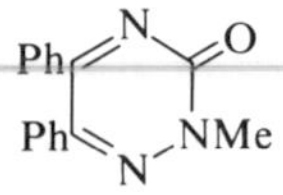        (28) 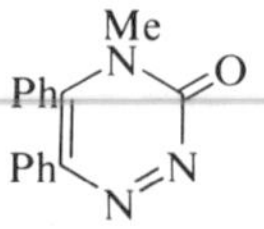        (29) 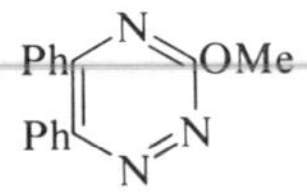        (30)

In a similar way, 1,2,4-triazin-5-ones (**31**) were found to exist in solution predominantly in the 2*H*-form, but in the solid state the 5-hydroxy tautomers (**32**) predominate, at least in the 3-phenyl and 6-phenyl derivatives ⟨78HC(33)189, p. 250, 74T3171⟩.

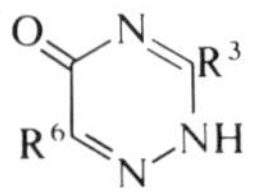

(31) 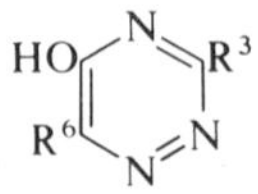                    (32)

1,2,4-Triazine-3,5-diones (**33**) show an absorption maximum in the UV at 261 nm (ethanol, dioxane) or 259 nm ($\varepsilon = 5100$) (H$_2$O; pH = 3) ⟨61CCC2155⟩. Methylation at N-2 shifts the maximum to 273 nm while methylation at N-4 causes a hypsochromic shift (258 nm). Dimethylation leads to a maximum which is given by additive contributions of the individual methyl group effects. Similar results were obtained for 1,2,4-triazine-3-thiones (**34**) and the 3,5-dithiones (**35**).

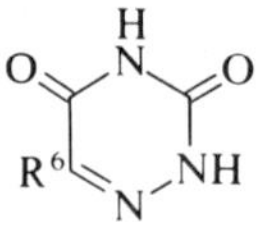

(33) 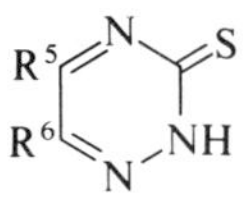            (34) 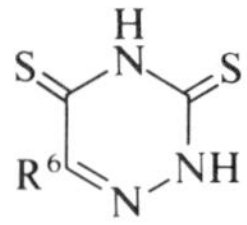            (35)

The unsubstituted 5-thioxo-1,2,4-triazin-3-one (**36**) shows an UV spectrum which is similar to those of the 2-methyl, 4-methyl and 2,4-dimethyl derivatives, implying that the 2*H*,4*H*-tautomeric form is the predominant one ⟨62CCC1886⟩. 5-Methylmercapto-1,2,4-triazin-3-one (**37**) and its 2-methyl derivative (**38**) show completely different spectra. Similar results have been obtained for 3-thioxo-1,2,4-triazin-5-ones ⟨61CCC986, 68CCC2962, 62CCC1886⟩.

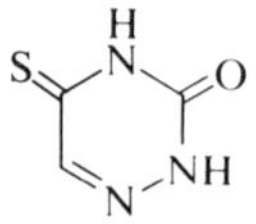

(36) 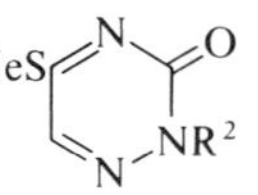            (37) R$^2$ = H            (39)
               (38) R$^2$ = Me

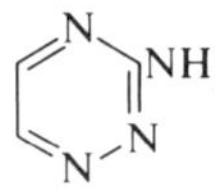

3-Amino-1,2,4-triazine (**39**) has two absorption maxima at 394 ($\varepsilon = 505$) and 310 nm (2730). The first band is attributed to an $n \rightarrow \pi^*$ transition and the second to a $\pi \rightarrow \pi^*$ ⟨59JCS1247⟩.

$^1$H NMR spectroscopy shows that 5-alkyl-1,2,4-triazin-3-ones (**40**) with a proton at C-1 of the alkyl group can occur in two tautomeric forms: the alkyl form (**40a**) and the structure (**40b**) with an alkylidene group. The ration of the two tautomers depends on the solvent, and in a few cases both have been isolated ⟨71JOC3921, 72BSF4637, 70CR(C)(270)1042⟩. 5-Alkyl-4-methyl-1,2,4-triazin-3-ones (**41**) form predominantly the 5-alkylidene tautomers. This is explained by the fact that structures with a formal N=N double bond are energetically unfavourable; they are destabilized and so tend to be avoided ⟨67JMC883⟩. This destabilization of structures containing an N=N double bond is also found in other areas of heterocyclic chemistry.

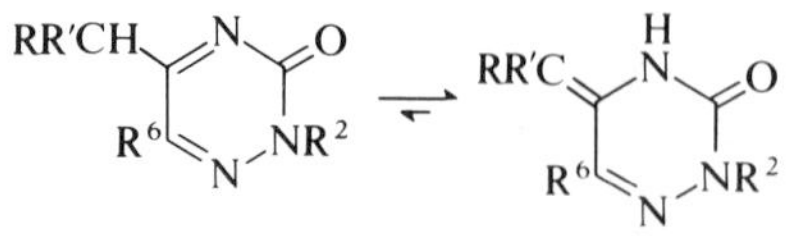

(40a)            (40b) 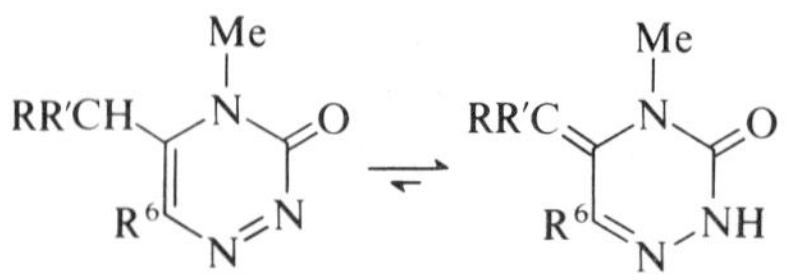            (41a)            (41b)

As with 5-alkyl-1,2,4-triazin-3-ones, so also 5-alkyl-1,2,4-triazine-3-thiones (**42**) display proton tautomerism between the alkyl group and a ring nitrogen, as shown by $^1$H NMR spectroscopy ⟨68TL2747⟩.

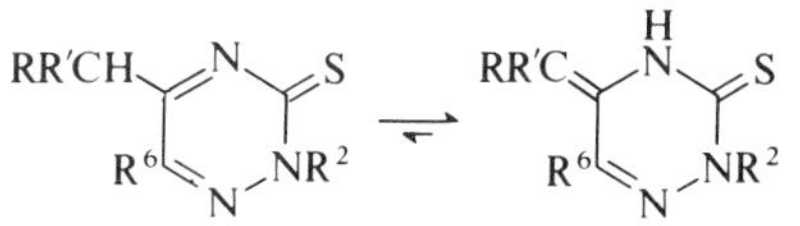

(42a)    (42b)

5-Amino-4-methyl-1,2,4-triazin-3-ones (43a) were shown to exist predominantly in the 5-imino forms (43b), due to the destabilization of the amino tautomer (43a) by the formal N=N double bond ⟨64CCC1394, 65JCS5230⟩. On the other hand, for 3-amino-4,6-dimethyl-1,2,4-triazin-5-one (44a) the predominance of the amino tautomer was proven and no imino tautomer (44b) could be detected.

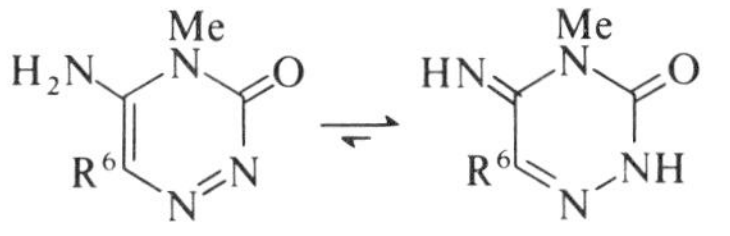

(43a)    (43b) 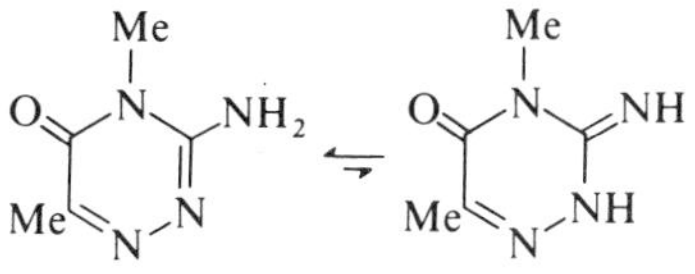    (44a)    (44b)

3-(Dicyanomethyl)-6-methyl-1,2,4-triazin-5-one (45a) is formulated as the methylene tautomer (45b) ⟨67CB2585⟩. 2-Anilino-2-hydrazino-1-nitroethylene (46) reacts with biacetyl to give a compound which is formulated as the dimethylene structure (47) ⟨77JPR149⟩.

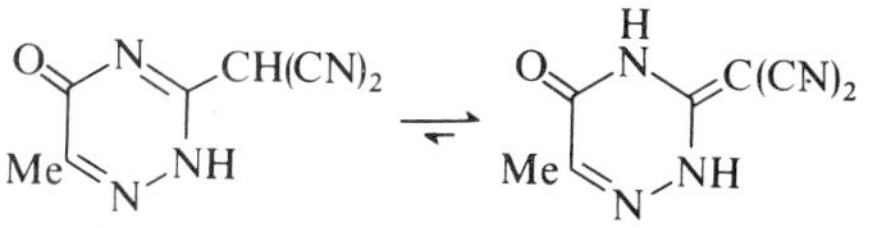

(45a)    (45b)

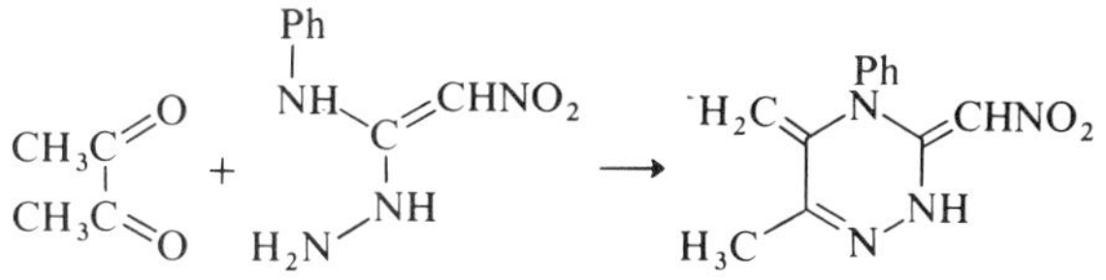

(46)    (47)

As was mentioned already, 3- and 6-phenyl substituted 1,2,4-triazin-5-ones exist in the solid state as the 5-hydroxy tautomers, as was shown by IR spectroscopy ⟨74T3171⟩. From UV spectroscopic studies it was suggested that the zwitterionic 2-methyldihydro-1,2,4-triazinium-6-olates (48a) are better formulated as such, rather than as the tautomeric alternatives (48b, 48c) ⟨78JHC1271⟩.

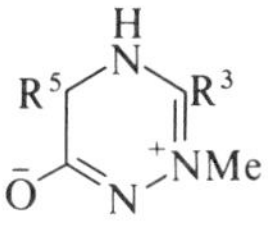

(48a) 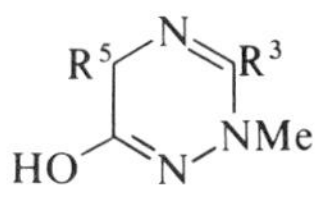    (48b) 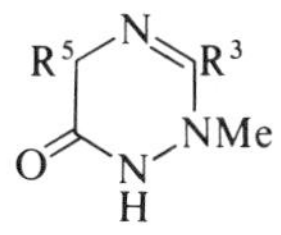    (48c)

The tetrahydro-1,2,4-triazine 4-oxides (49a) in tetrachloromethane exist in equilibrium with the open-chain structures (49b) ⟨77ZOR2617⟩, and equilibria between triazinones (50a) and the open-chain form (50b) were also observed ⟨80ZOR2297⟩. Ring–chain tautomerism was also found with the aldehydes (51a⇌51b); here the triazine ring is present throughout ⟨77CB1492⟩.

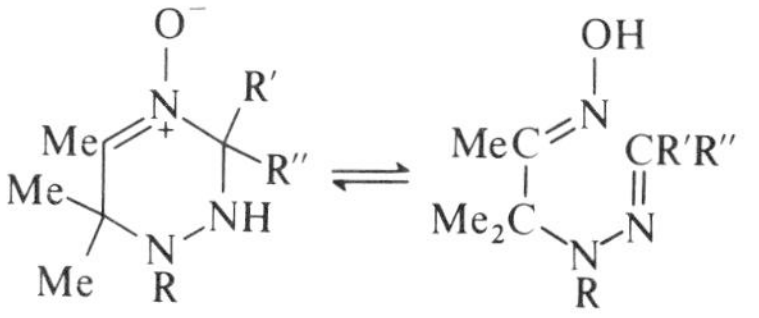

(49a)    (49b) 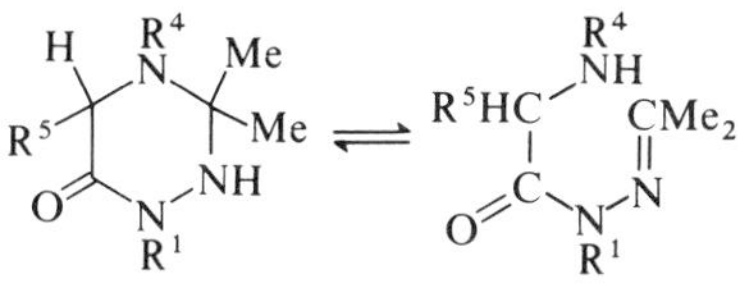    (50a)    (50b)

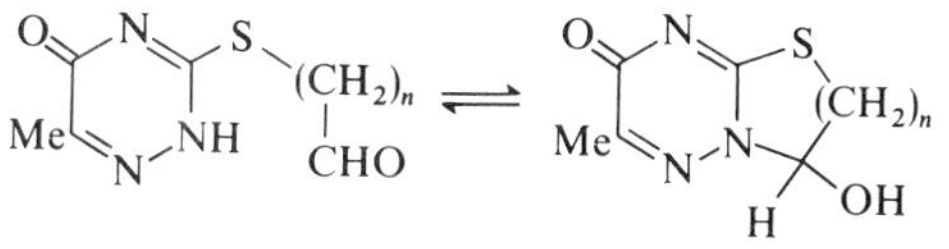

(51a)    (51b)

Azido-1,2,4-triazines may exist either as azido or as tetrazolo fused tautomers. Therefore, for 3-azido-1,2,4-triazines three structures (**52a–c**) have to be considered. As several groups have reported, 3-azido-1,2,4-triazines spontaneously cyclize to give the tetrazolo[1,5-*b*]-[1,2,4]triazines (**52b**) if cyclization to N-2 is possible. No cyclization to N-4 was observed. The structure of the bicyclic compound was proven by X-ray crystallographic analysis ⟨76JOC2860⟩. While alkyl-, dialkyl- or 5-aryl-3-azido-1,2,4-triazines do not show any azide peak in the IR spectra (Nujol mull), traces of the 3-azido-5,6-diphenyl-1,2,4-triazine (**53**) can be detected in THF, DMSO or trifluoroacetic acid solutions. This azido tautomer could be trapped by cycloaddition with 1-morpholinocyclohexene, giving the triazoline (**54**) ⟨72JCS(P1)1221⟩. 3-Azido-2-methyl-1,2,4-triazin-5-ones (**55**) did not show any azide IR absorption in the solid state. In deuterochloroform the ${}^1$H NMR spectra showed that a mixture of the two valence tautomers (**55a**, **55b**) was present, but in DMSO-$d_6$ only the tetrazolo form (**55b**) was detected ⟨77JOC1866⟩.

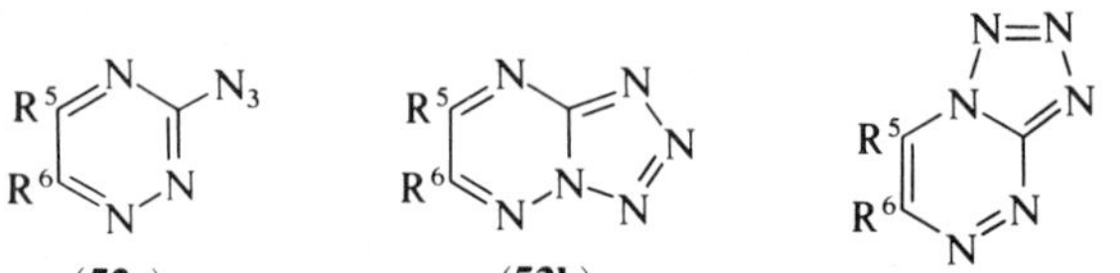

(52a)          (52b)          (52c)

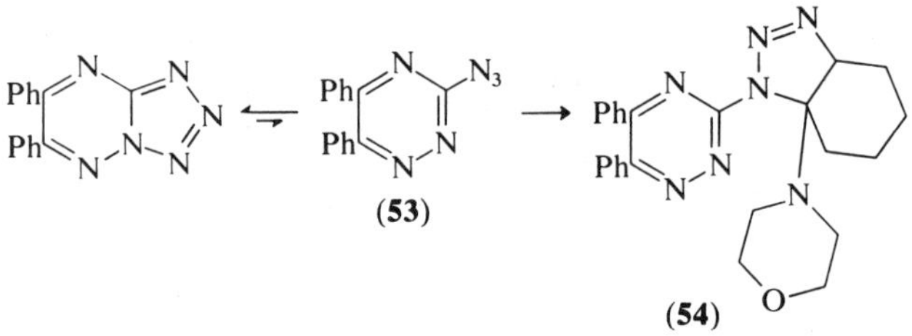

(53)          (54)

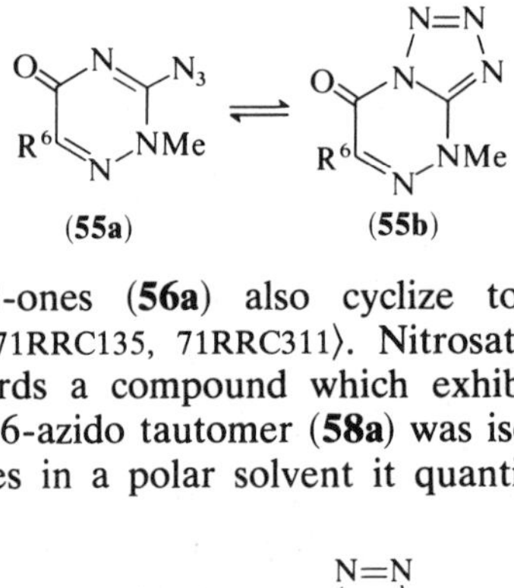

(55a)          (55b)

5-Azido-2*H*-1,2,4-triazin-3-ones (**56a**) also cyclize to give the tetrazolo[5,1-*d*]-[1,2,4]triazin-3-ones (**56b**) ⟨71RRC135, 71RRC311⟩. Nitrosation of 3-amino-6-hydrazino-1,2,4-triazin-5-one (**57**) affords a compound which exhibits an intense absorption at 2140 cm${}^{-1}$, showing that the 6-azido tautomer (**58a**) was isolated. When this compound was stirred for a few minutes in a polar solvent it quantitatively formed the tetrazole tautomer (**58b**) ⟨79JHC555⟩.

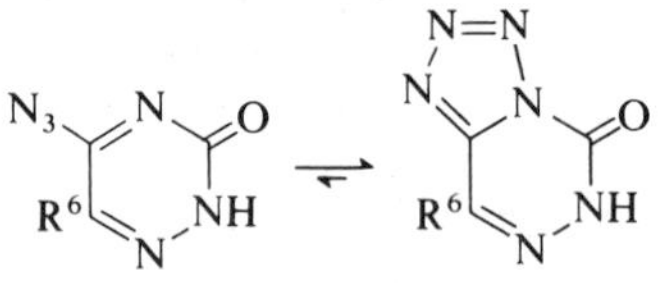

(56a)          (56b)

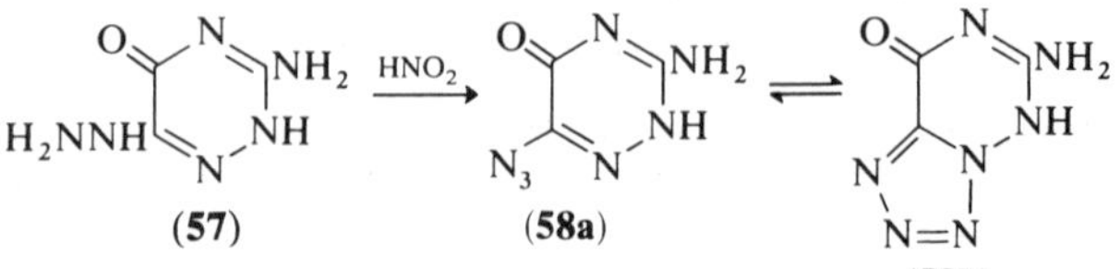

(57)          (58a)          (58b)

*N*-Oxidation of 3-azido-1,2,4-triazines at either N-1 or N-2 destabilizes the bicyclic compounds so that only the monocyclic azide tautomers (**59**, **60**) are observed ⟨77JHC1221⟩. 3-Azido-1,2,4-benzotriazine was also found to exist as the azido tautomer (**61b**) in solution, but in the crystalline state it favoured the tetrazolo[5,1-*c*][1,2,4]benzotriazine structure (**61a**) ⟨73JHC575⟩. In DMSO the third isomer, the tetrazolo[1,5-*b*]-fused compound (**61c**), was detected in small amount (10%) by ${}^{13}$C NMR spectroscopy ⟨79JOC1823⟩. Solvent and substituent effects on this equilibrium have also been studied ⟨82JOC3886⟩; in some cases the linear tetrazole isomer is the predominant one.

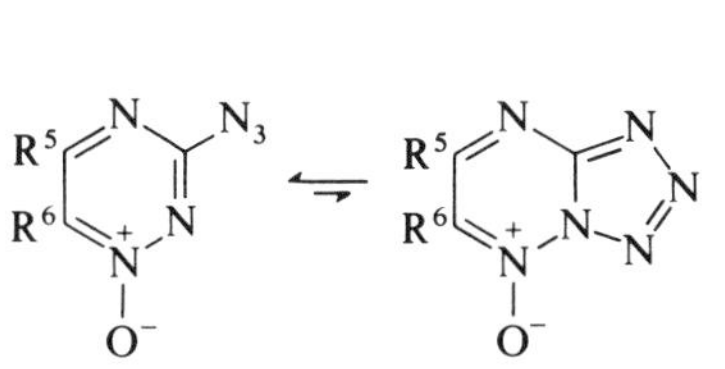

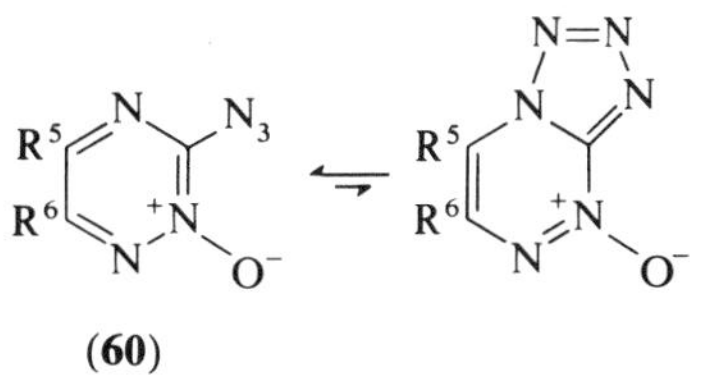

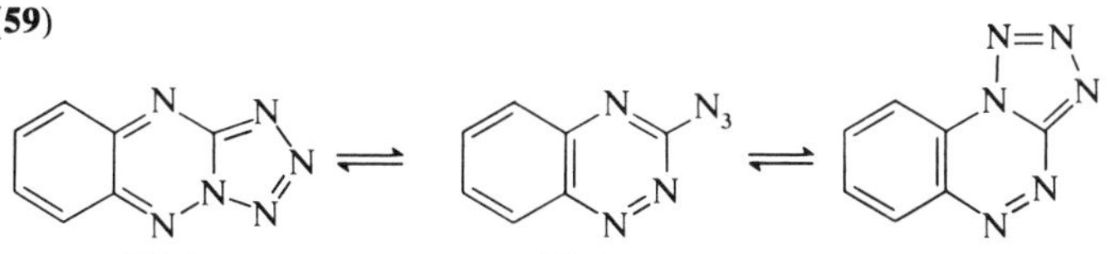

### 2.19.2.3 NMR Spectra

NMR spectra of 1,2,4-triazines are well documented, especially the $^1$H and $^{13}$C spectra. The parent compound (1) shows three signals in the $^1$H NMR spectrum, as expected, the shifts depending on the solvent as shown in Table 2. In liquid ammonia a $\sigma$-complex is formed. The signal for H-6 is a doublet of doublets, while the signals for H-3 and H-5 are simple doublets. Since to our knowledge coupling between H-3 and H-5 is never observed in simple 1,2,4-triazines, $^1$H NMR spectroscopy provides a simple method to distinguish between isomeric derivatives.

**Table 2** $^1$H NMR Shifts ($\delta$ values) of 1,2,4-Triazine (1) in Different Solvents and Predicted Chemical Shifts

| Solvent | H-3 | H-5 | H-6 | Ref. |
|---------|-----|-----|-----|------|
| CDCl$_3$ | 9.88 | 8.84 | 9.48 | 66JOC1720 |
| | 9.73 | 8.70 | 9.34 | 78RTC273 |
| CCl$_4$ | 9.63 | 8.53 | 9.24 | 69TL3147 |
| CD$_3$OD | 9.86 | 8.93 | 9.52 | 69TL3147 |
| DMSO-$d_6$ | 9.75 | 8.88 | 9.42 | 69TL3147 |
| C$_6$D$_6$ | 9.45 | 7.92 | 8.68 | 69TL3147 |
| NH$_3$ | 7.40 | 4.15 | 6.65 | 78RTC273 |
| Predicted shifts | 9.74 | 8.74 | 8.98 | 68CC1028 |

The influence of *N*-oxidation on the $^1$H NMR spectra of 1,2,4-triazines has been studied by various groups. Comparison of the $^1$H NMR spectrum of 1,2,4-triazine (1), the 1-oxide (62) and the 2-oxide (63) (Table 3) shows that on oxidation at N-1 the proton at the 6-position becomes more shielded by about 1.4 p.p.m., the proton at the 5-position by about 0.3 p.p.m., and that at the 6-position by about 0.9 p.p.m. Oxidation at N-2 leads to a shielding of H-3 by about 1.1 p.p.m., of H-5 by about 0.8 p.p.m., and of H-6 by about 1.1 p.p.m. ⟨71JOC787, 77JOC546⟩. The signals for the two heterocyclic protons in 6-methyl-1,2,4-triazine 4-oxide (64) were observed at $\delta = 9.28$ (H-3) and 8.22 (H-5) ⟨71LA(750)12⟩. It is noteworthy that in 6-substituted 1,2,4-triazine 4-oxides the H-3 and H-5 signals are doublets, while in 6-substituted 1,2,4-triazines they are singlets.

**Table 3** Influence of *N*-Oxidation on the $^1$H NMR Shifts ($\delta$ values) of 1,2,4-Triazines

| 1,2,4-Triazine | H-3 | H-5 | H-6 | Ref. |
|----------------|-----|-----|-----|------|
| 1,2,4-Triazine | 9.88 | 8.84 | 9.48 | 66JOC1720 |
| 1-oxide | 9.00 | 8.57 | 8.05 | 71JOC787 |
| 2-oxide | 8.82 | 8.00 | 8.42 | 77JOC546 |
| 3-Methoxy-1,2,4-triazine | — | 8.56 | 9.00 | 78RCT273 |
| 1-oxide | — | 8.37 | 7.83 | 71JOC787 |
| 2-oxide | — | 7.76 | 8.12 | 77JOC546 |
| 6-Methyl-1,2,4-triazine | 9.55 | 8.55 | — | 71LA(750)12 |
| 4-oxide | 9.28 | 8.22 | — | 71LA(750)12 |

The structures (62), (63), (64) are shown.

Only a few data on the $^1$H NMR spectra of dihydro-1,2,4-triazines are available. The proton in the 5-position of 3,5,6-triphenyl-4,5-dihydro-1,2,4-triazine (65) is observed at $\delta = 5.93$ (DMSO-$d_6$) ⟨72CJC1581⟩. The CH$_2$ group in 6-duryl-3,4-diphenyl-4,5-dihydro-1,2,4-triazine (66) gives a signal at $\delta = 4.00$ (CCl$_4$) which is shifted to 4.86 in trifluoroacetic acid. In 1-alkyl-4,5-dihydro-1,2,4-triazinium salts (67) the CH$_2$ group in the 5-position absorbs between $\delta = 4.72$ and 5.24 ⟨74ZOR2429⟩. In 2,4,6-triaryl-4,5-dihydro-1,2,4-triazin-3(2$H$)-ones (68) the peak for the CH$_2$ group at position 5 is found between $\delta = 4.96$ and 5.02 ⟨78LA2033⟩. Similarly, 2,4,6-triaryl-2,3,4,5-tetrahydro-1,2,4-triazines (69) show the expected two methylene group signals in the region $\delta = 4.2$–4.6 ⟨78LA2033⟩.

The structures (65), (66), (67), (68), (69) are shown.

The following $^1$H NMR spectrum is reported for 1,2,4-trimethylhexahydro-1,2,4-triazine (23) at 34 °C: $\delta = 2.16$ (s; 4-Me), 2.46 (s; 1-Me), 2.51 (s; 2-Me), 2.74 (bm; CH$_2$CH$_2$) and 3.42 (s; CH$_2$). A study of the effect of temperature on the spectrum of this compound and also of the tetramethyl analogue (24) provided information on their conformational equilibria ⟨79JCS(P2)984⟩.

The $^{13}$C NMR spectra of the parent 1,2,4-triazine (1) and its three monomethyl derivatives, of two dimethyl derivatives and the trimethyl compound, were recorded by Braun and Frey ⟨75OMR(7)194⟩. The parent 1,2,4-triazine (1) shows the following absorptions and coupling constants: $\delta = 158.1$ (C-3), 149.6 (C-5) and 150.8 p.p.m. (C-6); 207.1 Hz (C$^3$–H$^3$), 9.1 Hz (C$^3$–H$^5$), 1.3 Hz (C$^3$–H$^6$), 188.0 Hz (C$^5$–H$^5$), 7.5 Hz (C$^5$–H$^3$), 9.0 Hz (C$^5$–H$^6$), 187.5 Hz (C$^6$–H$^6$), 9.5 Hz (C$^6$–H$^5$) and 2.0 Hz (C$^6$–H$^3$). Substitution of the heterocyclic ring with methyl groups shifts the signals to lower fields as is shown for the 3,5,6-trimethyl-1,2,4-triazine: $\delta = 164.7$ (C-3), 158.3 (C-5) and 155.0 (C-6). 1,2,4-Triazine 2-oxide (63) has signals for the three carbon atoms at $\delta = 132$ (C-3), 143 (C-5) and 143 (C-6) ⟨77JOC546⟩. To compare the influence of oxidation at N-1 and N-2 on the $^{13}$C spectra of 1,2,4-triazines the values are given for 3-methoxy-1,2,4-triazine (70) and its 1-oxide (71) and 2-oxide (72): (70): $\delta = 166.1$ (C-3), 151.6 (C-5) and 145.5 p.p.m. (C-6) ⟨78RTC273⟩; (71): $\delta = 166.5$ (C-3), 154 (C-5) and 124.5 p.p.m. (C-6) ⟨77JOC546⟩; (72): $\delta = 152.5$ (C-3), 130 (C-5) and 135.5 p.p.m. (C-6) ⟨77JOC546⟩. From these data it follows that oxidation at N-1 has a large influence on C-6 and a small influence on C-3 and C-5 while oxidation at N-2 affects all three carbon atoms, but especially C-3 and C-5. Comparison of the $^{13}$C spectra of the unsubstituted 1,2,4-triazine (1) and its 2-oxide (63) gives different results since the largest influence is on C-3 and the effects on C-5 and C-6 are smaller.

The structures (70), (71), (72), (73) are shown.

Comparison of the $^{13}$C NMR spectra of a 1,2,4-triazine and its 4-oxide shows that there is a large influence on C-3 ($-10.4$ p.p.m.) and C-5 ($-15.8$ p.p.m.) while the effect on C-6 is smaller (3.0 p.p.m.) ⟨79JHC1389⟩.

The $^{13}$C NMR spectrum of 1,2,4-triazine-3,5-dione (10a) shows three signals at 149.5 (C-3), 158.1 (C-5) and 137.1 p.p.m. (C-6) ⟨79OMR(12)612⟩. The 2-($\beta$-D-ribofuranosyl)-1,2,4-triazine-3,5-dione (11a) has very similar values ⟨73JA4761⟩.

The shifts of the $^{13}$C NMR signals of 3,5,6-trichloro-1,2,4-triazine (**73**) were compared with calculated values obtained by first and second order SCS methods: $\delta = 160.8$ (C-3; 157.7, 163.4), 157.0 (C-5; 155.8, 153.0) and 154.7 (C-6; 140.8, 151.9) ⟨80OMR(13)363⟩.

The $^{15}$N–$^{13}$C coupling constants were measured for $^{15}$N 4-labelled 3-methyl- and 3,5,6-trimethyl-1,2,4-triazine ⟨76OMR(8)273⟩. They are in the same range as found for other azines.

Witanowski has shown that $^{14}$N NMR spectroscopy can be used for the location of an N-oxide position in 1,2,4-triazines ⟨80OMR(14)305⟩.

### 2.19.2.4 IR spectra

IR spectral data for most known 1,2,4-triazines have been published. The absorption of the 1,2,4-triazines in the IR region are those expected for this system. The IR spectrum of the parent compound (**1**) shows three absorption bands for the C—H stretching vibrations at 3090, 3060 and 3030 cm$^{-1}$, five bands for C=N and C=C stretching vibrations at 1560, 1529, 1435, 1380 and 1295 cm$^{-1}$, three for the C—H in-plane deformations at 1163, 1135 and 1113 cm$^{-1}$, two for the characteristic ring skeleton vibrations at 1050 and 995 cm$^{-1}$ and three bands for the C—H out-of-plane deformation vibrations at 851, 768 and 713 cm$^{-1}$ ⟨68CB3952⟩. These values are in good agreement with similar bands for pyridine, pyridazine, pyrimidine and pyrazine. Alkyl and aryl derivatives of 1,2,4-triazine show similar bands in their IR spectra, with additional bands from the substituents.

1,2,4-Triazin-3-ones, *e.g.* (**27**), show strong carbonyl absorption at *ca.* 1685 cm$^{-1}$ which is observed both in KBr and in chloroform ⟨62CJC1053⟩, and a weak peak for the N—H stretching vibration at 3450 (KBr) or 3350 cm$^{-1}$ (CHCl$_3$). The C=O stretching vibration for 1,2,4-triazin-5-ones (**31**) occurs at 1640–1670 cm$^{-1}$ for the 2*H*-form and 1675–1690 cm$^{-1}$ for the 4*H*-form. The N—H stretching vibration is observed at *ca.* 3170 cm$^{-1}$ ⟨78HC(33)189, pp. 251, 256⟩. 1,2,4-Triazin-6-ones have a strong carbonyl absorption at 1665 cm$^{-1}$ ⟨70JPR669⟩.

1,2,4-Triazine-3,5-diones (**33**) show two N—H stretching vibrations at 3378 and 3374 cm$^{-1}$ which were assigned to H-4 and H-2 according to comparisons with the 2- and 4-methyl derivatives. Also two bands in the carbonyl region were observed with 1,2,4-triazine-3,5-diones at 1730 and 1700 cm$^{-1}$. The first band is shifted to 1723–1720 cm$^{-1}$ if N-2 is alkylated, while the second band is shifted to 1687–1680 cm$^{-1}$ when N-4 is alkylated. IR spectroscopy can therefore be used to determine the position of substituents in 1,2,4-triazine-3,5-diones ⟨61CCC1680⟩.

1,2,4-Triazine-3-thiones show no S—H band in the IR spectrum; therefore the thiol tautomer can be excluded.

Further information on the IR spectra of 1,2,4-triazines can be found in the 1978 review ⟨78HC(33)189⟩.

### 2.19.2.5 UV Spectra

As mentioned already, UV spectra have been used to determine tautomeric structures in the 1,2,4-triazine series. Unsubstituted 1,2,4-triazine (**1**) has two absorption bands in methanol at 374 ($\varepsilon = 400$) and 247.8 nm (3020). In 0.1 N HCl/methanol only one band is observed at 230 nm ($\varepsilon$ 4460) while in 0.1 N NaOH/methanol bands at 382 (380) and 249.5 nm (3430) were recorded ⟨68CB3952⟩. In the gas phase this compound has several bands in the region between 420 and 380 nm ⟨75TH21900⟩. Alkyl-substituted 1,2,4-triazines show similar electronic spectra.

The electronic spectra of 1,2,4-triazin-3-ones and 1,2,4-triazine-3,5-diones were given at the beginning of Section 2.19.2.2. 1,2,4-Triazine-3,5-dithiones (**74**) have two absorption maxima around 320 and 280 nm ⟨62CCC1886⟩. 3-Amino-1,2,4-triazine (**39**) has two absorption bands at 394 ($\varepsilon = 505$) and 310 nm (2370); the first is attributed to an $n \rightarrow \pi^*$ band and the second to a $\pi \rightarrow \pi^*$ band ⟨59JCS1240⟩.

(**74**)

For 1,2,4-triazine 4-oxides a band at *ca.* 270 nm and an inflection in the 300 nm region were observed ⟨71LA(750)12⟩. Further information on the UV spectra of the different 1,2,4-triazines can be found in the 1978 review ⟨78HC(33)189⟩.

### 2.19.2.6 Mass Spectra

The mass spectra of 1,2,4-triazines have been studied extensively since they can be used for structure determination. The fragmentation pattern depends on the structure (aromatic, quinoid, alicyclic) of the 1,2,4-triazine system and on the substituents; no general underlying system is observed. The mass spectrum of unsubstituted 1,2,4-triazine shows peaks at 81, 53, 52, 51, 40, 39, 38, 28, 27, 26 and 25 mass units, being accounted for in part by the fragmentation pattern of Scheme 1. This was confirmed by the mass spectrum of 1,2,4-triazine-$d_3$ ⟨67JHC224⟩. [4-$^{15}$N]-3-Methyl-1,2,4-triazine shows a similar fragmentation pattern ⟨72LA(760)102⟩. From these data it follows that the fragmentation starts with loss of nitrogen.

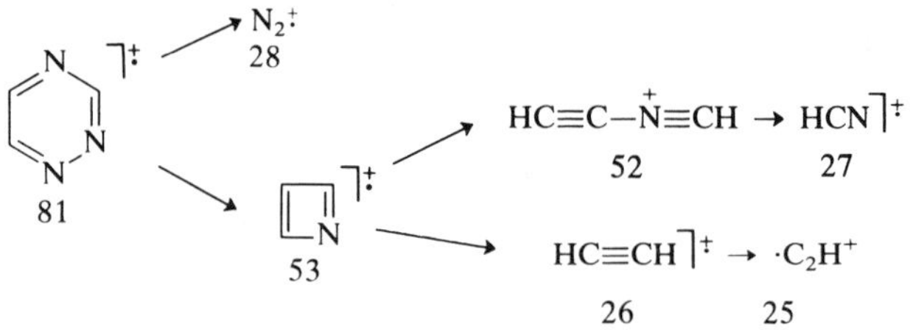

**Scheme 1**

The suggested pattern for 5,6-diphenyl-1,2,4-triazin-3-one (**27**) is quite different from that of the parent compound (**1**). The first fragmentation step is not the loss of nitrogen but loss of an NCO moiety as shown in Scheme 2 ⟨71OMS(5)1085⟩. Similarly, in the mass spectrum of 3,6-diphenyl-1,2,4-triazin-5-one (**31**; R$^3$ = R$^6$ = Ph), the first step is not the elimination of nitrogen but of benzonitrile. From the observed data the fragmentation pattern of Scheme 3 is suggested ⟨70JPR669⟩. On the other hand, 3,5-diphenyl-1,2,4-triazin-6-one (**75**) starts with loss of nitrogen as was found for simple 1,2,4-triazines. The main fragmentation pattern is given in Scheme 4 ⟨70JPR669⟩.

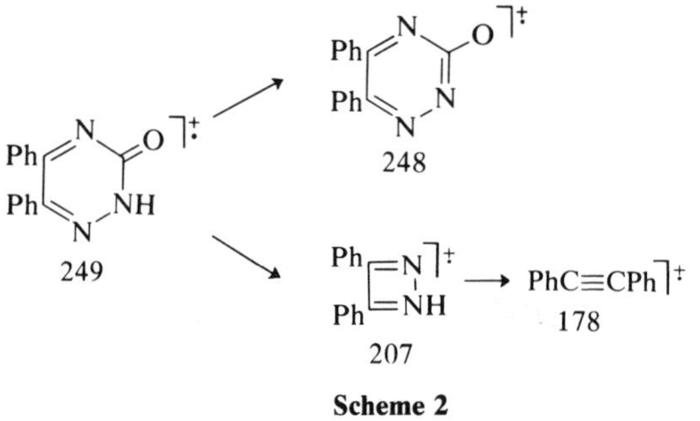

**Scheme 2**

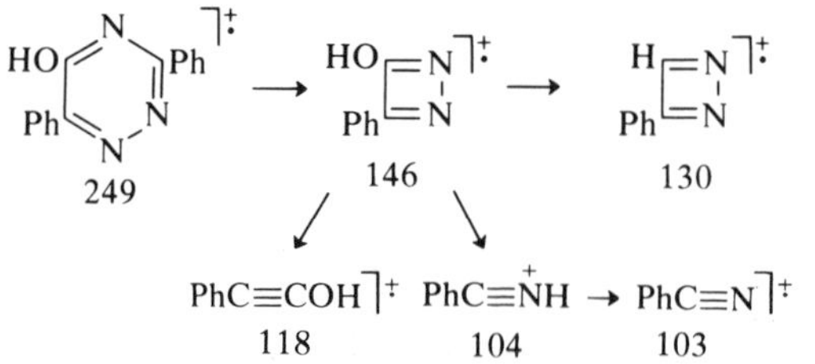

**Scheme 3**

Extensive studies on 1,2,4-triazines substituted with oxygen or sulfur in the 3- and 5-positions have shown that five distinct fragmentation patterns can be observed upon electron impact, but none involve initial loss of nitrogen. The following systems were studied: 1,2,4-triazine-3,5-diones (**33**), 1,2,4-triazine-3,5-dithiones (**74**), 3-thioxo-1,2,4-

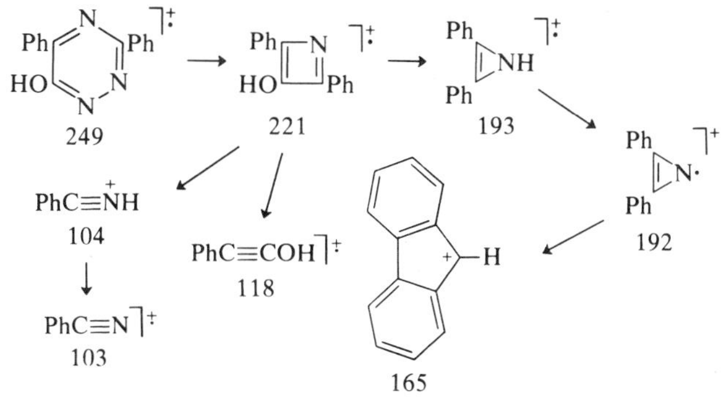

**Scheme 4**

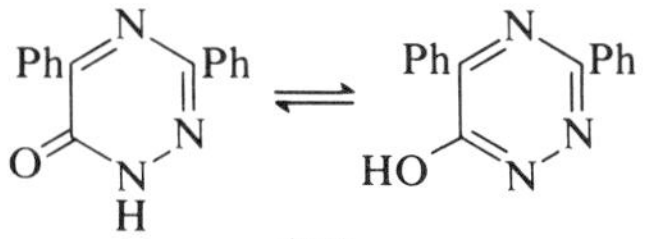

(75)

triazin-5-ones (**76**), 3-methoxy-1,2,4-triazin-5-ones (**77**), 3-methylthio-1,2,4-triazin-5-ones (**78**), 3-methylthio-1,2,4-triazine-5-thiones (**79**), 5-methoxy-3-methylthio-1,2,4-triazines (**80**) and 3,5-bis(methylthio)-1,2,4-triazines (**81**) ⟨76OMS1002⟩.

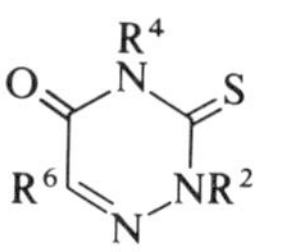

| (76) | (77) | (78) X = O<br>(79) X = S | (80) X = O<br>(81) X = S | (82) |

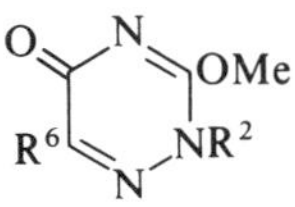

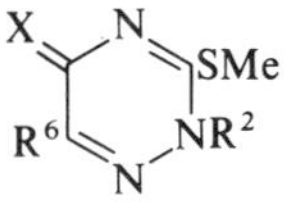

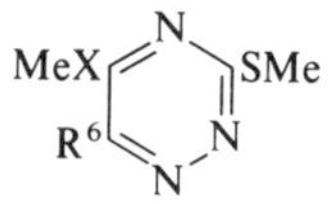

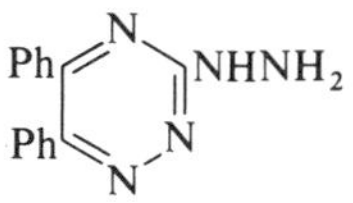

The observed five fragmentation patterns start with rupture of the N(1)—N(2) and C(5)—C(6) bonds and loss of a nitrile (A), with rupture of the N(2)—C(3) and C(5)—C(6) bonds (B), with the rupture of the N(1)—N(2) and C(3)—N(4) bonds (C), with the rupture of the N(2)—C(3) and N(4)—C(5) bonds (D) or with the rupture of the N(1)—C(6) and N(4)—C(5) bonds (E), as shown in Scheme 5.

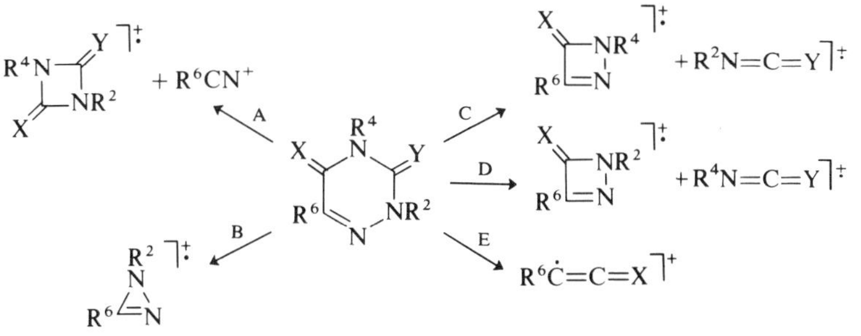

**Scheme 5**

Also in the mass spectrum of 3-hydrazino-5,6-diphenyl-1,2,4-triazine (**82**) the elimination of nitrogen from the molecular ion could not be detected. The dominant process in this case is the fragmentation of the molecular ion into a highly delocalized diphenylacetylene radical ion at 178 mass units ⟨71OMS(5)1085⟩.

Paudler and his group have studied the fragmentation of 1,2,4-triazine 1-oxides and 2-oxides. The mass spectrum of the unsubstituted 1,2,4-triazine 2-oxide has an intense mass peak, which is the base peak, and others at $(M-16)^+$ and $(M-30)^+$. Comparison of the mass spectra of 1,2,4-triazine 1-oxides and 2-oxides has shown that the patterns are very similar, with the relative peak intensities varying with the type of substituent present. For 1,2,4-triazine 1-oxides without a substituent in the 6-position the fragmentation starts with loss of oxygen, nitric oxide or nitrous oxide, and the same was found with the 2-oxides. A substituent in the 6-position of 1,2,4-triazine 1-oxides leads to elimination of OH, as does a substituent in the 3-position of 1,2,4-triazine 2-oxides, as proven by deuterium labelling ⟨71JHC317, 77JHC1389⟩.

The mass spectra of 1,2,4-triazine 4-oxides show very intense molecular ion peaks but very weak peaks for $(M - 16)^+$ ⟨71LA(750)12⟩.

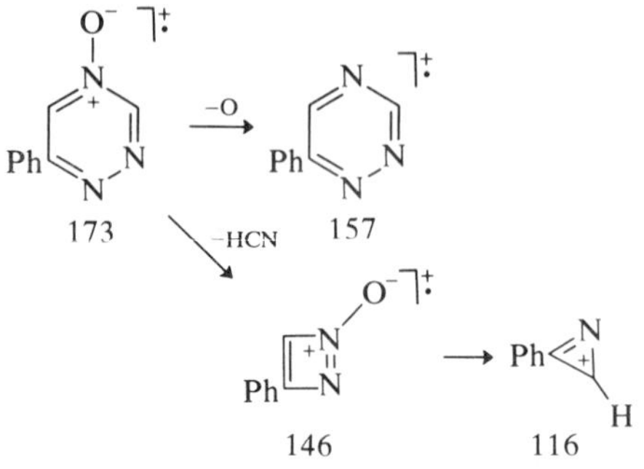

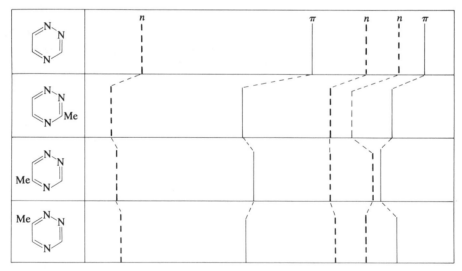

**Scheme 6**

### 2.19.2.7  Photoelectron Spectra

Photoelectron spectra of unsubstituted 1,2,4-triazine, its three monomethyl derivatives, two dimethyl derivatives and the trimethyl derivative have been measured by Gleiter ⟨77MI21900⟩. The He(I) photoelectron spectrum of the parent 1,2,4-triazine (**1**) is characterized by a single band at 9.6 eV followed by a complicated band system of several overlapping peaks between 11 and 13 eV. A further composite band is found at 15 eV. Methyl substitution seems to help in the resolution of the peaks. The experimental PE spectra were interpreted by comparison with the result of MO calculations and empirical correlation procedures. In Figure 1 the correlation of the first five bands of the 1,2,4-triazine (**1**) and its three methyl derivatives are given.

**Figure 1**

The PE spectrum of 1,2,4-trimethylhexahydro-1,2,4-triazine (**25**) was measured, in an investigation of the predominant conformer of this compound in the gas phase ⟨80JCS(P2)91⟩.

### 2.19.2.8  Other Spectral and Thermodynamic Properties

For many 1,2,4-triazines of varied structure the $pK_a$ values were determined. Some of these values are collected in Table 4.

The heats of sublimation were determined for 6-azauridine (**14**), 6-azathymine (**15**) and 6-azacytosine (**16**) in order to elucidate the effect of substituents on the intermolecular energies. For 6-azauridine (**14**) a value of 130.9 kJ mol$^{-1}$ was determined by the Knudsen method ⟨74MI21900, 75MI21900⟩.

**Table 4** $pK_a$ Values of 1,2,4-Triazines

| Compound | $pK_a$ | Ref. |
|---|---|---|
| 1,2,4-Triazine-3,5-dione | 7.00, 12.9 | 68CCC2962 |
| 2-Methyl-1,2,4-triazine-3,5-dione | 6.99 | 61CCC2155 |
| 4-Methyl-1,2,4-triazine-3,5-dione | 9.25 | 61CCC2155 |
| 6-Methyl-1,2,4-triazine-3,5-dione | 7.42 | 68CCC2962 |
| 2,6-Dimethyl-1,2,4-triazine-3,5-dione | 7.49 | 62CCC716 |
| 4,6-Dimethyl-1,2,4-triazine-3,5-dione | 10.1 | 62CCC716 |
| 2-Ethyl-1,2,4-triazine-3,5-dione | 7.07 | 62CCC716 |
| 4-Ethyl-1,2,4-triazine-3,5-dione | 9.59 | 62CCC716 |
| 6-Benzyl-1,2,4-triazine-3,5-dione | 7.31 | 62CCC716 |
| 6-Hydroxy-1,2,4-triazine-3,5-dione | 2.95 | 61JOC1118 |
| 1,2,4-Triazine-3,5-dithione | 5.66 | 61CCC986 |
| 2-Methyl-1,2,4-triazine-3,5-dithione | 5.76 | 61CCC986 |
| 4-Methyl-1,2,4-triazine-3,5-dithione | 7.37 | 61CCC986 |
| 3-Amino-1,2,4-triazine | 3.09 | 54JA1451 |
| 5-Thioxo-1,2,4-triazin-3-one | 6.33 | 61CCC986 |
| 2-Methyl-5-thioxo-1,2,4-triazin-3-one | 6.25 | 61CCC986 |
| 4-Methyl-5-thioxo-1,2,4-triazin-3-one | 8.57 | 61CCC986 |
| 5-Methylthio-1,2,4-triazin-3-one | 9.18 | 61CCC986 |
| 3-Thioxo-1,2,4-triazin-5-one | 5.98 | 61CCC986 |
| 2-Methyl-3-thioxo-1,2,4-triazin-5-one | 6.24 | 61CCC986 |
| 4-Methyl-3-thioxo-1,2,4-triazin-5-one | 8.12 | 61CCC986 |
| 6-Methyl-3-thioxo-1,2,4-triazin-5-one | 6.39 | 68CCC2962 |
| 3-Methylthio-1,2,4-triazin-5-one | 5.94 | 61CCC986 |
| 3-Amino-1,2,4-triazin-5-one | 1.55 | 66CCC1864 |
| 3-Amino-6-methyl-1,2,4-triazin-5-one | 2.19 | 66CCC1864 |
| 3-Amino-2,6-dimethyl-1,2,4-triazin-5-one | 1.90 | 66CCC1864 |
| 3-Amino-4,6-dimethyl-1,2,4-triazin-5-one | 4.52 | 66CCC1864 |
| 3-Hydrazino-1,2,4-triazin-5-one | 2.73, 8.40 | 68CCC2962 |
| 3-Hydrazino-6-methyl-1,2,4-triazin-5-one | 2.81, 8.94 | 68CCC2962 |
| 6-Benzyl-3-hydrazino-1,2,4-triazin-5-one | 2.82, 8.80 | 68CCC2962 |
| 6-Bromo-1,2,4-triazine-3,5-dione | 6.00 | 62CCC716 |
| Ethyl 3,5-dioxo-1,2,4-triazine-6-carboxylate | 6.34 | 62CCC716 |
| 1,6-Dihydro-1,2,4-triazine-3,5-dione | 10.3 | 62CCC716 |
| 2-Methyl-1,6-dihydro-1,2,4-triazine-3,5-dione | 10.6 | 62CCC716 |
| 4-Methyl-1,6-dihydro-1,2,4-triazine-3,5-dione | >11 | 62CCC716 |

Irradiation of a single crystal of deuterated and undeuterated 6-methyl-1,2,4-triazine-3,5-dione (**13**) with γ- or X-rays afforded radicals, the EPR spectra of which were recorded by Horak and Schoffa ⟨71MI21901⟩. The 19-line spectrum of the deuterated radical indicates interaction of the unpaired electron with two equivalent protons, and two non-equivalent nitrogen atoms. The EPR spectra of metastable triplet states of 6-methyl-1,2,4-triazine-3,5-dione (**12**) in ethylene glycol/water glass at −196 °C were recorded by Shulman and Rahn ⟨66JCP(45)2940⟩. It was found that the lowest triplet state is a $\pi \rightarrow \pi^*$ state.

## 2.19.3 REACTIONS AND REACTIVITY

Most known 1,2,4-triazines and 1,2,4-benzotriazines are yellow, orange or white compounds, usually crystalline at room temperature. They are mostly stable at room temperature and can be distilled under reduced pressure or recrystallized without decomposition.

The most reactive position of the 1,2,4-triazine ring is position 5, where nucleophiles can attack very easily. Very often nucleophilic attack at position 5 is followed by an electrophilic attack at 4 (or 2).

Unsubstituted 1,2,4-triazine (**1**) and its lower alkyl derivatives can be stored at temperatures below 0 °C, but in time they become dark. After a long time in the refrigerator the parent compound (**1**) became an amber-like solid of constitution $(C_3H_3N_3 \cdot H_2O)_n$. The structure of this compound was not elucidated ⟨70UP21900⟩. Paudler has shown that addition of water to a solution of 1,2,4-triazine (**1**) in trifluoroacetic acid rapidly generates a new species, the amount of which is directly proportional to the quantity of water present. The $^1H$ NMR spectrum was interpreted in terms of structure (**83a**), formed by covalent addition of water to the N(4)—C(5) bond of (**1**), although (**83b**), by N(2)—C(5) addition, seems an alternative possibility. Addition of base to the acidic solution regenerated the triazine quantitatively, and attempts to isolate the hydrated species failed ⟨70JHC767⟩. Addition of water, alcohols and thiols to C-5 of 1,2,4-triazines has been observed for 1,2,4-triazin-3-ones (**84**), 1,2,4-triazine-3-thiones (**85**) and 3-amino-1,2,4-triazines (**86**); adducts have structures of type (**87**) or (**88**). (The site of the added proton is questionable. Since in most cases the original authors have drawn the 4*H*-forms, we have retained them here.) For references the 1978 review should be consulted ⟨78HC(33)189⟩.

3-Amino-1,2,4-triazines (**86**) also add sulfurous acid to give the covalent addition products (**89**).

6-Amino-1,2,4-triazine-5-carboxylates (**90**) add water or alcohols to C-3 yielding compounds (**91**), some of which can be isolated ⟨71JOC2974⟩.

Photochemical hydration of the C(6)—N(1) bond of 1,2,4-triazines is reported for 1,2,4-triazine-3,5-diones (**92**) and 5-amino-1,2,4-triazin-3-ones (**94**), yielding compounds (**93**) and (**95**) respectively ⟨78HC(33)189, pp. 328, 426⟩.

Paudler observed alcohol addition to the N(1)—C(6) bond of 1,2,4-triazine 2-oxides unsubstituted at the 6-position (**96**) in the presence of hydrogen chloride. Elimination of water followed, leading to 6-alkoxy-1,2,4-triazines (**97**) ⟨77JOC3489⟩.

The addition of methoxide ion to 3-methylthio-1,2,4-triazine-5-ones (**78**) leading to (**98**) and then to (**99**) is reported by Piskala ⟨75CCC2340⟩. Hydrazine adds to the C=O double bond of 3-thioxo-1,6-dihydro-1,2,4-triazin-5-ones (**100**) ⟨70BSF1606⟩.

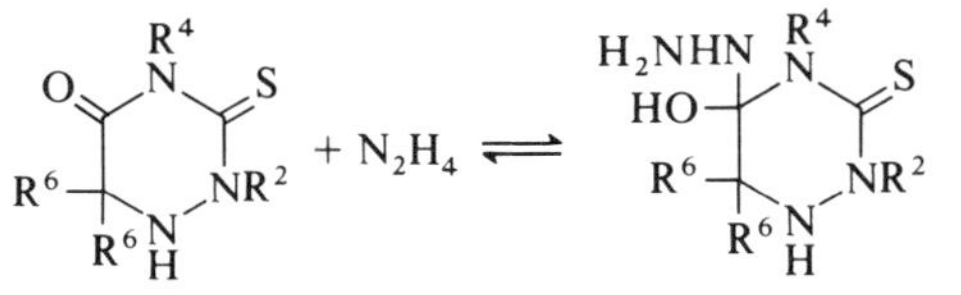

(100)

A number of 1,2,4-triazines such as the parent compound, the 3-methylthio, 3-methoxy and 3-amino-6-bromo derivatives add ammonia in liquid ammonia at −44 °C to the N(4)—C(5) bond to give adducts of type (101), as was shown by $^1$H and $^{13}$C NMR spectroscopy. No reaction was observed with 3-methoxy-1,2,4-triazine 1-oxide (71), other 3-amino-1,2,4-triazines, or compounds with a substituent in the 5-position. As was shown using $^{15}$N-enriched ammonia, an equilibrium between the covalent addition products (101) and the open-chain isomers (102) does not occur ⟨78RTC273⟩.

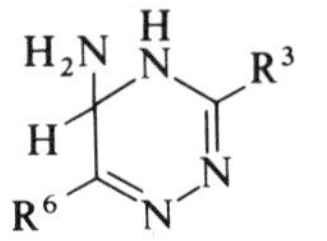

(101) 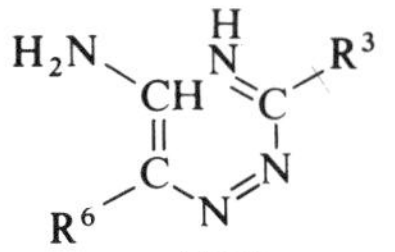          (102)

1,2,4-Triazine and some of its 3-methoxy, 3-methylthio or 3-amino derivatives (103) with the 5-position unsubstituted react with potassium cyanide to afford two products, the i-triazinyls (105) and the 1,2,4-triazine-5-carboxamides (104). These are proposed to be formed *via* a cyanide adduct ⟨73JHC343, 74JHC43⟩. The bi-1,2,4-triazinyls (105) were also isolated when the triazines (103) were treated with sodium methoxide or potassium in liquid ammonia. It is suggested that the methoxide-catalyzed dimerization proceeds *via* an anionic intermediate (106), while the reaction with potassium in liquid ammonia occurs *via* a free radical process.

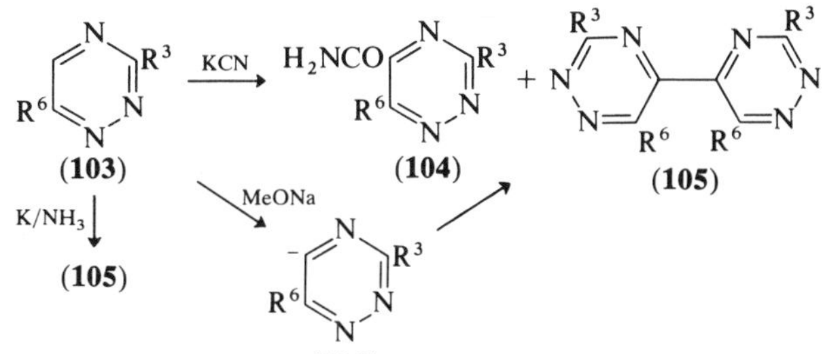

(103)          (104)          (105)

(105)          (106)

H–D exchange has been observed in neutral or basic media with 1,2,4-triazines unsubstituted at the 3-position (107) and with 5-methyl-1,2,4-triazines. Two different mechanisms have been discussed for the H–D exchange at the 3-position. In neutral media only those 1,2,4-triazines undergo H–D exchange which show covalent addition of water to the N(4)—C(5) bond (see Scheme 7). In alkaline media a carbanionic intermediate (108) has been postulated ⟨73T2495⟩.

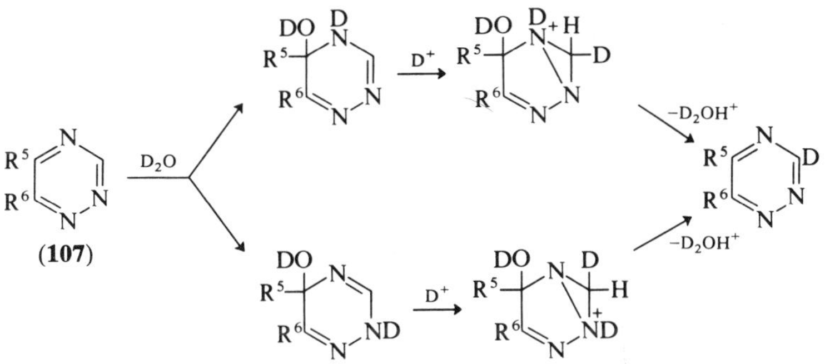

(107)

**Scheme 7**

In methyl-1,2,4-triazines (109) the 5-methyl groups shows rapid H–D exchange in basic media while the other methyl groups show very slow exchange ⟨74UP21900⟩.

(107)                    (108)

(109)

In most 1,2,4-triazines the heterocyclic ring is stable toward acids but longer treatment with concentrated acids at higher temperatures leads to its destruction. Many 1,2,4-triazines are soluble in acids and are recovered unchanged on addition of water or bases. Salts of the triazines with mineral acids have been isolated.

The salts of simple 1,2,4-triazines are prepared by addition of dry acids to a solution of the triazine in an organic solvent. Most 1,2,4-triazines with aryl substituents form deep red solutions in concentrated sulfuric acid, from which they are reprecipitated unchanged on addition of water. Addition of concentrated hydrochloric acid to a solution of 5,6-diphenyl-1,2,4-triazine in sulfuric acid leads to the precipitation of the crystalline hydrochloride ⟨54CB1540⟩.

1,2,4-Triazine-3-thiones also form red solutions in concentrated sulfuric or hydrochloric acid, and are reprecipitated unchanged on addition of water ⟨03CB4126, 53MI21900⟩.

Treatment of 4-amino-2,3,4,5-tetrahydro-1,2,4-triazines (110) with hydrochloric acid led to the isolation of 2,3-dihydro-1,2,4-triazines (111) ⟨36JPR(144)273⟩.

(110)                    (111)

Very often treatment of 1,2,4-triazines with hetero substituents in the 3-, 5- or 6-position with aqueous mineral acids leads to the replacement of the substituent by an oxo group. Transformations of this type have been reported for the thioxo group, for methylthio and methoxy groups, for amino, hydrazino and hydroxylamino groups, and for chloro or bromo substituents. If there is more than one hetero substituent bound to the triazine ring then usually the substituent in the 5-position is hydrolyzed first. This is consistent with the greatest susceptibility of the 5-position to nucleophilic attack. An exception to this rule seems to be the 3-methylthio-1,2,4-triazine-5-thione in which the methylthio group is hydrolyzed first. Sometimes the leaving group can be modified to produce a preference for displacement at another site, as, for instance, by oxidation of a methylthio group to methylsulfonyl in the 3,5-bis(methylthio) compound, which is hydrolyzed first at position 3 ⟨69BSF3670⟩. After position 5, the next site of nucleophilic attack seems to depend on the nature of the reagent. Neutral nucleophiles tend to attack the 3-position while anionic species apparently prefer the 6-position. It is thus possible to isolate 6-chloro-1,2,4-triazine-3,5-dione (112) from acidic hydrolysis of 3,5,6-trichloro-1,2,4-triazine (73) ⟨61JOC1118⟩. Reaction of (73) with two moles of sodium methoxide affords a mixture of 6-chloro-3,5-dimethoxy- (113) and 3-chloro-5,6-dimethoxy-1,2,4-triazine (114) ⟨76CB1113, 75CCC2680⟩.

(112)            (73)            (113)            (114)

Complete destruction of the heterocyclic ring of 1,2,4-triazines on prolonged treatment with aqueous acids is reported for 1,2,4-triazin-3-ones (84), 1,2,4-triazine-3-thiones (85), 3-thioxo-1,2,4-triazine-5,6-diones (115), 1,2,4-triazine *N*-oxides, and 1,2,4-benzotriazin-3-ones (116). This is a fairly general observation in 1,2,4-triazine systems.

The 1,2,4-triazine ring is even less stable toward bases than toward acids. Although many with acidic functionality can be dissolved in bases and reprecipitated unchanged many others are destroyed, especially on prolonged treatment. The half-life of the parent 1,2,4-

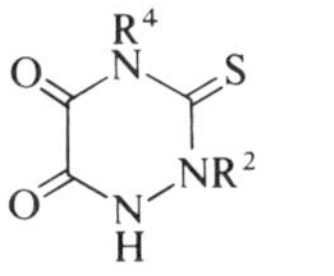

(115) 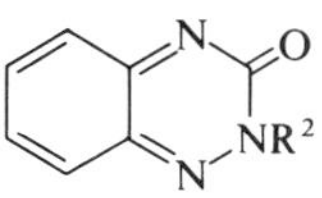          (116)

triazine (1) in 0.5N aqueous sodium hydroxide is 4.25 h and that of the 3-methyl compound is 168 h. The hydrolysis products were not identified ⟨75TH21900⟩.

1,2,4-Triazin-3-ones (117) are stable toward bases, dissolving because they are weak acids, and being reprecipitated on acidification. A number of 1,2,4-triazin-3-ones (117), 1,2,4-triazine-3-thiones (118) and thioxo-1,2,4-triazinones (*e.g.* 74; $R^2$ and/or $R^4$ = H) can be titrated with bases in the presence of phenolphthalein.

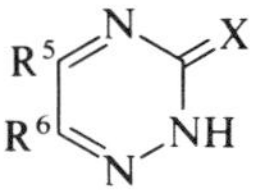

(117) X = O
(118) X = S

4-Alkyl-substituted 1,2,4-triazine-3,5-diones (119) can be hydrolyzed by bases to semi-carbazones of α-ketocarboxylic acids (120) ⟨78HC(33)189, p. 326⟩. Clearly nucleophilic attack of hydroxide ion at position 5 brings this reaction about. 3-Thioxo-1,2,4-triazine-5,6-diones react with bases to afford 5-thioxo-1,2,4-triazoline-3-carboxylic acids ⟨70BSF1599, 70BSF1590, 68CR(C)(267)1726, 68CR(C)(267)904⟩. This is illustrated for the 2,4-dimethyl compound (121). Again, one can assume nucleophilic attack of a hydroxide ion at the 5-position of the 1,2,4-triazine ring and recyclization of the open-chain compound (122) to the isolated triazole (123).

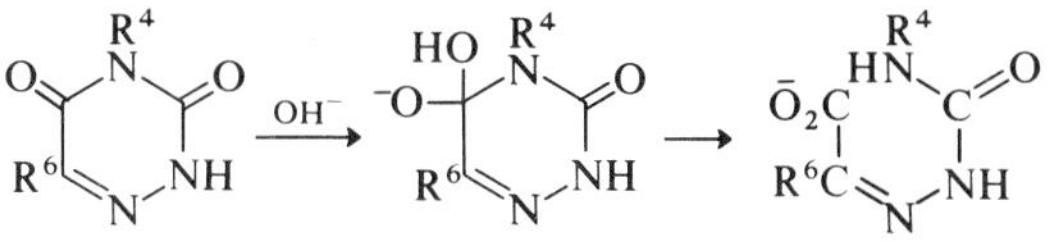

(119)          (120)

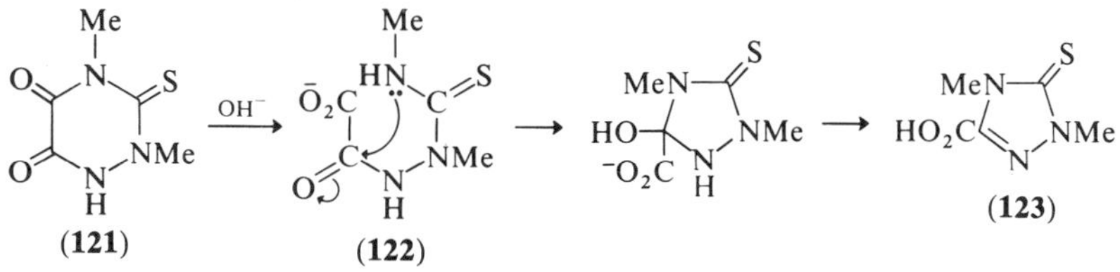

(121)          (122)          (123)

1,2,4-Triazine 4-oxides (124) can be hydrolyzed by bases yielding 2-acylhydrazono oximes (125) ⟨76LA153⟩. Here the initial nucleophilic attack evidently occurs at the 3-position.

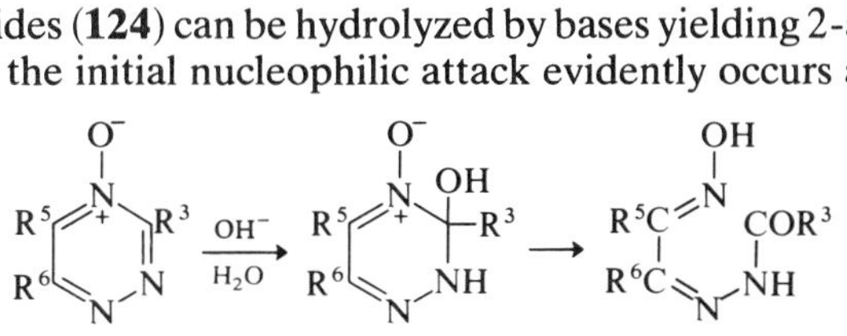

(124)          (125)

Treatment of 1-benzoyl-6-methyl-1,6-dihydro-1,2,4-triazine-3,5-dione (126) with bases led to the 5-phenyl-1,2,4-triazolin-3-one (127) ⟨03JA386⟩.

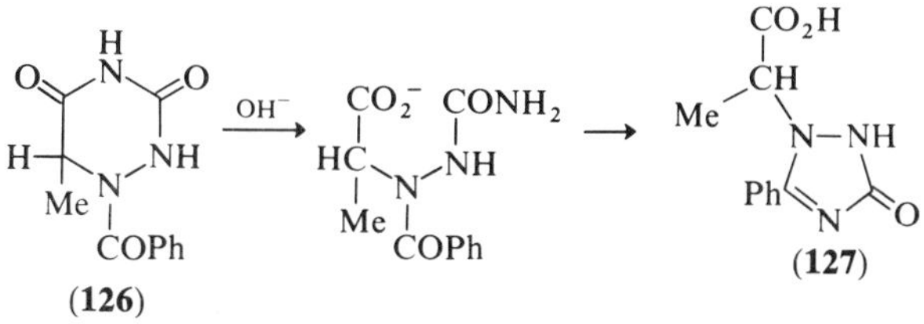

(126)          (127)

Heating 1,2,4-benzotriazines (128) with bases gives aniline, nitrogen and a carboxylic acid ⟨60G1113⟩. 2-Substituted 1,2,4-benzotriazin-3-ones (116) with potassium hydroxide in ethanol afford the ring-opened products (**129**). Aniline and ammonia in ethanol cause a similar ring cleavage with formation of the ureas (**130**) ⟨1899CB2959⟩.

Treatment of 1,2,4-benzotriazin-3-one 1-oxide (**131**) with sodium hydroxide afforded benzotriazole (**132a**). Similarly, 3-amino-1,2,4-benzotriazine 1-oxide (**133**) with a base gave benzotriazole or 1-acetylbenzotriazole (**132b**) while 3-amino-1,2,4-benzotriazine 2-oxide (**134**) was converted into benzotriazole and benzotriazole-1-carboxamine (**132c**) ⟨62JOC185⟩.

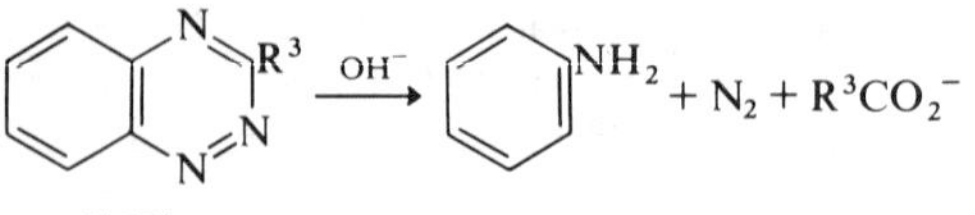

(**128**)

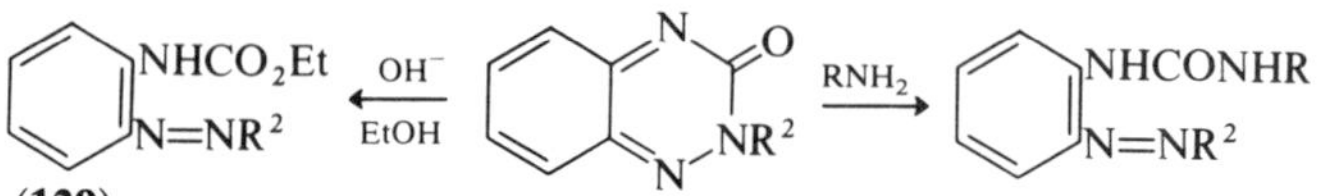

(**129**)    (**116**)    (**130**)

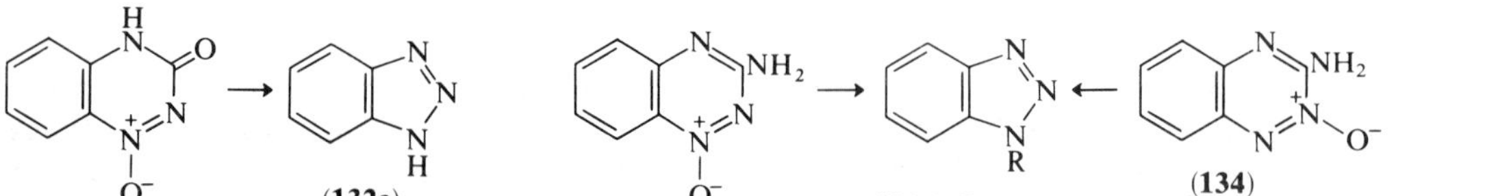

(**131**)    (**132a**)    (**133**)    (**132a**) R = H
                                       (**132b**) R = COMe
                                       (**132c**) R = CONH₂    (**134**)

3-Amino-1,2,4-triazines (**86**) react with phenylhydrazine to afford 2-phenyl-1,2,3-triazoles (**135**) ⟨60MI21900⟩. 3-Thioxo-1,2,4-triazin-5-ones (**76**) are transformed into 1,2,4-triazole-3-thiones (**136**) by treatment with sodium methoxide ⟨72BSF1511⟩. Treatment of compound (**137**) with acetic anhydride in the presence of sulfuric acid led to the isolation of the pyrazole (**138**) ⟨10JA1499⟩.

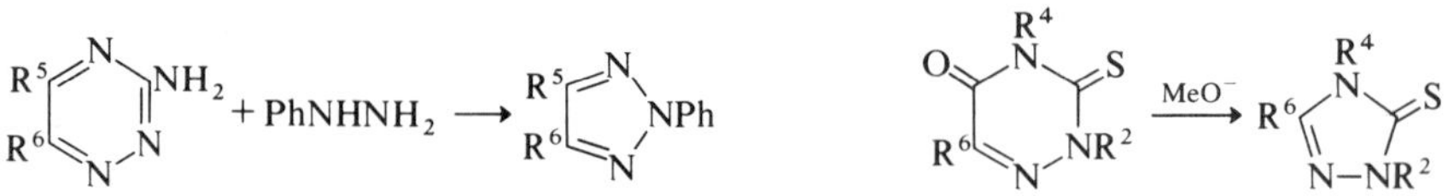

(**86**)         (**135**)         (**76**)         (**136**)

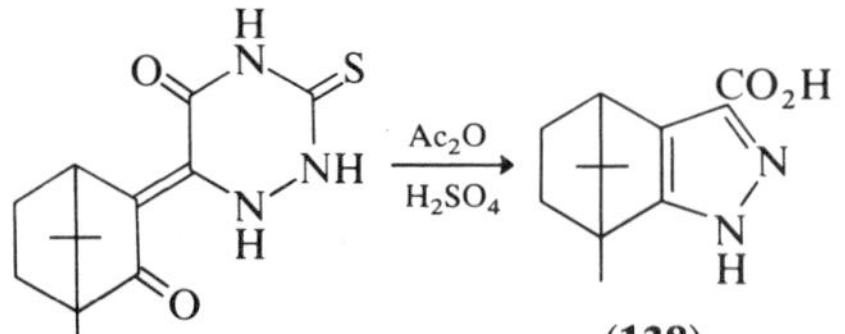

(**137**)         (**138**)

1,6-Dihydro-1,2,4-triazine-3,5(2*H*,4*H*)-dione (**139**) rearranges with aldehydes, with which they form *N*-aminoimidazoledione derivatives (**140**) ⟨68CCC2087⟩.

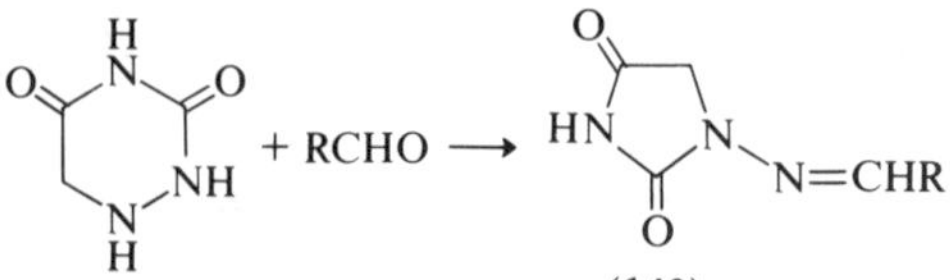

(**139**)         (**140**)

The reaction of 1,2,4-triazin-3-ones (**141**) with hydroxylamine-*O*-sulfonic acid, chloroamine and sodium hypochlorite has been studied by Rees and coworkers. In the first case imidazolinones (**142**) were obtained, while in the other cases 1,2,3-triazoles (**143**) were isolated ⟨73JCS(P1)545⟩. Similarly, treatment of 1,2,4-benzotriazin-3-one (**144**) with chloroamine or lead tetraacetate led to the isolation of benzotriazole (**146a**) or 1-acetyl-benzotriazole (**146b**), while (**144**) was converted into benzimidazolinone (**145**) by hydroxylamine-*O*-sulfonic acid ⟨73JCS(P1)545⟩.

(142) (141) (143)

(145) (144) (146a) R = H (146b) R = Ac

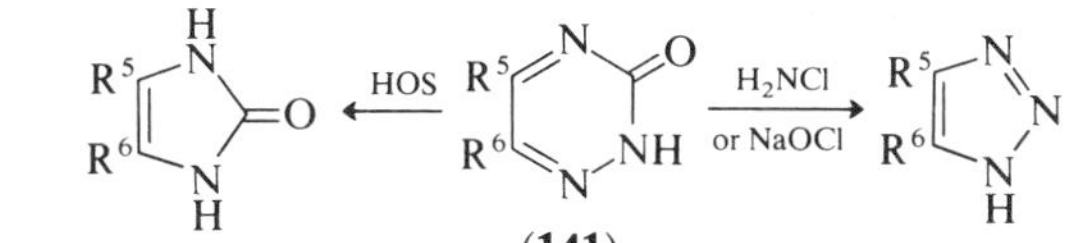

A considerable number of studies have been made of the reactions of various 1,2,4-triazine derivatives with Grignard reagents. 3,5,6-Triphenyl-1,2,4-triazine (**147**) with phenylmagnesium bromide is reported to give 2,4,6-triphenyl-1,3,5-triazine (**148**) and 2,4,4,5-tetraphenyl-4*H*-imidazole (**149**) ⟨71JPR699⟩.

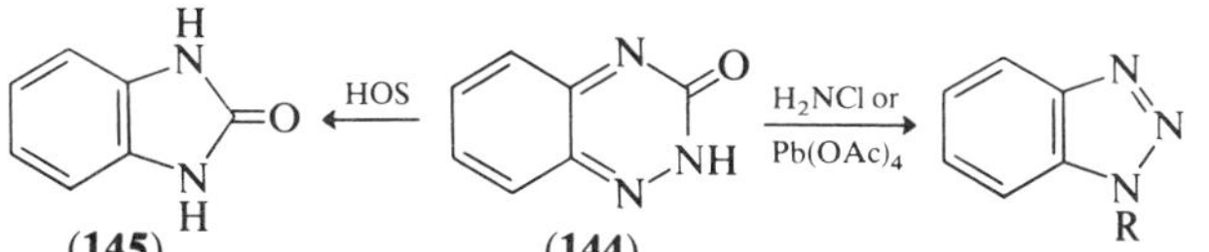

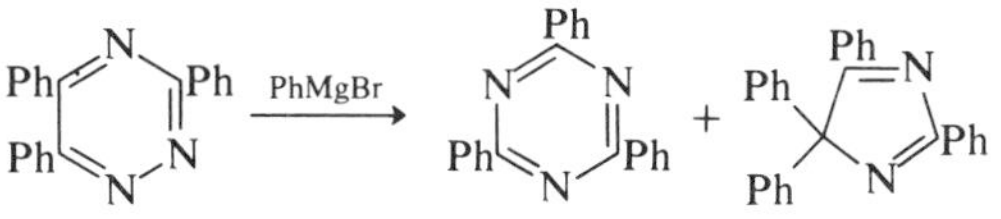

(147) (148) (149)

Reaction of 1,2,4-triazin-3-ones (**84**) with Grignard reagents affords mainly 4,5-dihydro-1,2,4-triazin-3-ones (**150**), but the isolation of 1,2,4-triazines or imidazole derivatives has also been reported ⟨78HC(33)189, p. 247⟩.

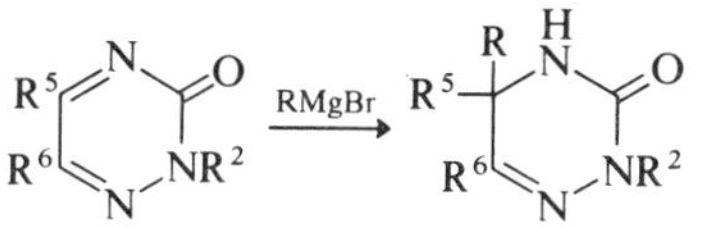

(84) (150)

Reaction of 1,2,4-triazin-5-ones (**151**, **152**) or 5-methoxy-1,2,4-triazines (**156**) with Grignard reagents afford, depending on the structure of the 1,2,4-triazine used, 1,2,4-triazines (**153**), 3,4-dihydro- (**154**), or 1,6-dihydro-1,2,4-triazin-5-ones (**155**) ⟨73JHC559⟩.

(151) (153) (154a)

(152) (155a) (154b) (156) (155b)

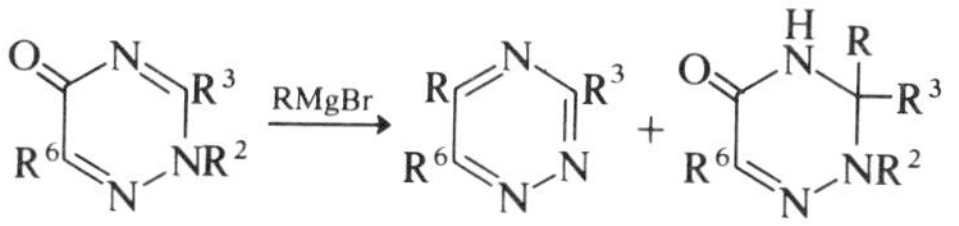

From the reaction of 6-phenyl-1,2,4-triazine-3,5-dione (**157**) with phenylmagnesium bromide, 5,6-diphenyl-1,2,4-triazin-3-one (**27**), 3,5,6-triphenyl-1,2,4-triazine (**147**), 3,3,5,6-tetraphenyl-2,3-dihydro-1,2,4-triazine (**158**), 2,4,6-triphenyl-1,3,5-triazine (**148**) and 2,4,4,5-tetraphenylimidazole (**149**) were isolated ⟨71JPR699⟩. From the results obtained in the reaction of 3,5,6-triphenyl-1,2,4-triazine (**147**) with phenylmagnesium bromide it can be supposed that the last two compounds were formed *via* the intermediate formation of (**147**).

(157) (27) (158)

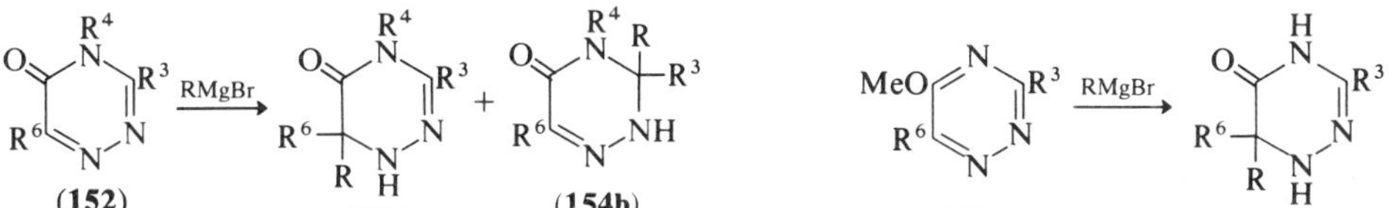

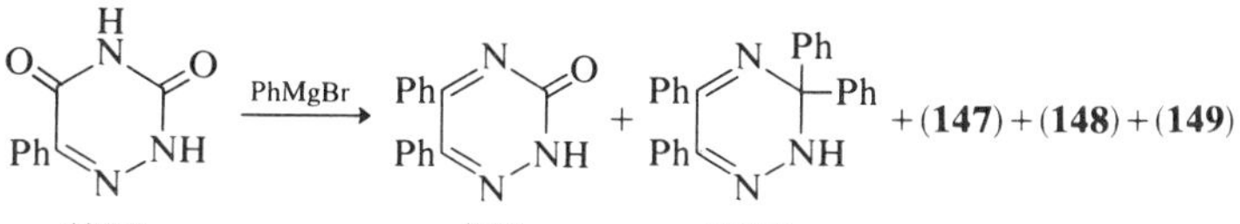

1,2,4-Triazine-3-thiones (**159**) gave 4,5-dihydro-1,2,4-triazine-3-thiones (**160**) on treatment with Grignard reagents ⟨78HC(33)189, p. 346⟩. 3-Chloro-5,6-diphenyl-1,2,4-triazine (**161**) afforded 5,6-diphenyl-3-*p*-tolyl-1,2,4-triazine (**162**) with *p*-tolylmagnesium bromide, while the reaction of (**161**) with alkylmagnesium bromides gave 4,5-dihydro-1,2,4-triazin-3-ones (**163**) after work-up ⟨70LA(733)177⟩.

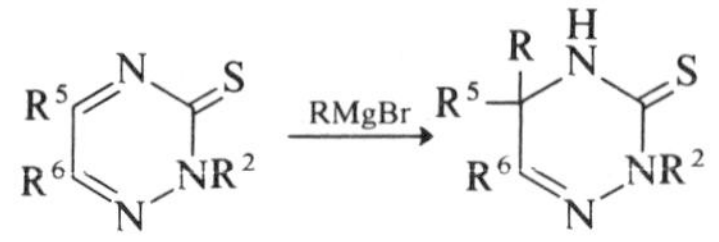

(**159**)                    (**160**)

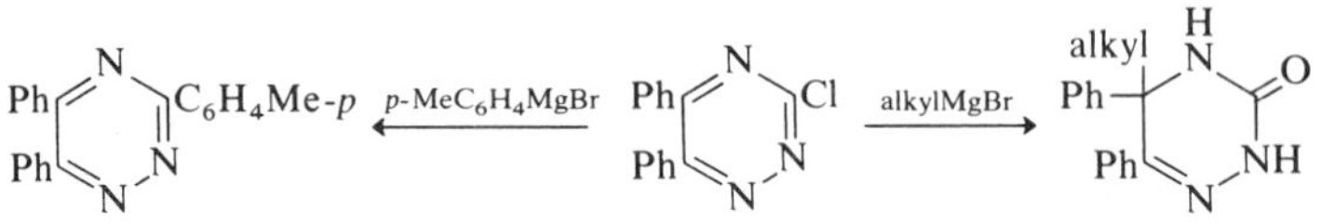

(**162**)                    (**161**)                    (**163**)

Phenylmagnesium bromide converts 3-amino-6-phenyl-1,2,4-triazin-5-ones (**164**) into 3-amino-5,6-diphenyl-1,2,4-triazines (**165**) ⟨73JPR221⟩.

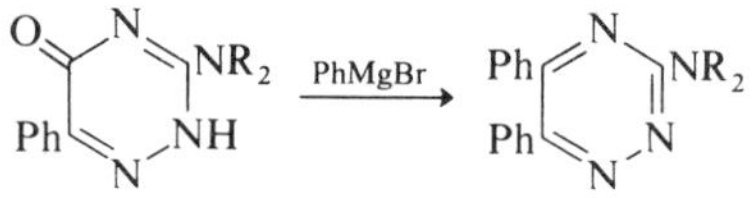

(**164**)                    (**165**)

2,4,6-Trimethyl-5-oxo-1,2,4-triazinium methosulfate (**166**) afforded 2,3,4,6-tetramethyl-3,4-dihydro-1,2,4-triazin-5(2*H*)-one (**167**) when treated with methylmagnesium bromide ⟨74BSF999⟩. 2-*p*-Tolyl-1,2,4-benzotriazin-3-one (**168**) was converted into 4a-phenyl-2-*p*-tolyl-4,4a-dihydro-1,2,4-benzotriazin-3-one (**169**) by phenylmagnesium bromide ⟨63JOC3519⟩. However, the organometallic residue was not incorporated when 1,2,4-benzotriazine 1-oxides (**170**) reacted with Grignard reagents; 1,2,4-benzotriazines (**2**) and benzimidazoles (**171**) were isolated ⟨73CC622⟩.

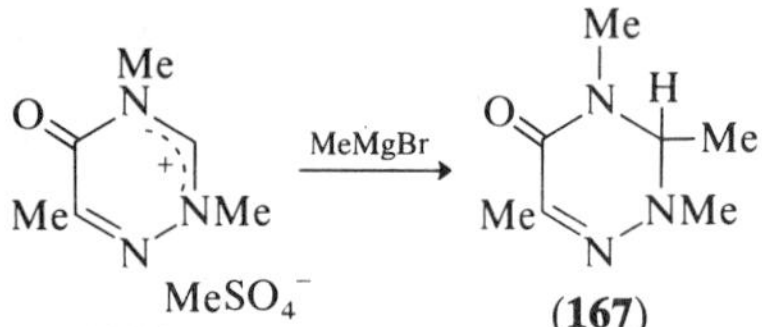

(**166**)                    (**167**) 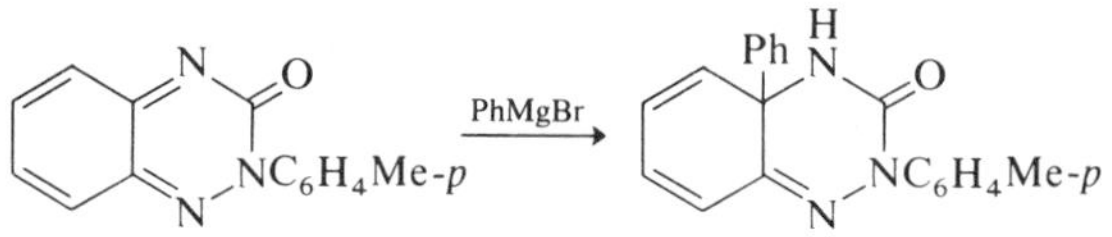                    (**168**)                    (**169**)

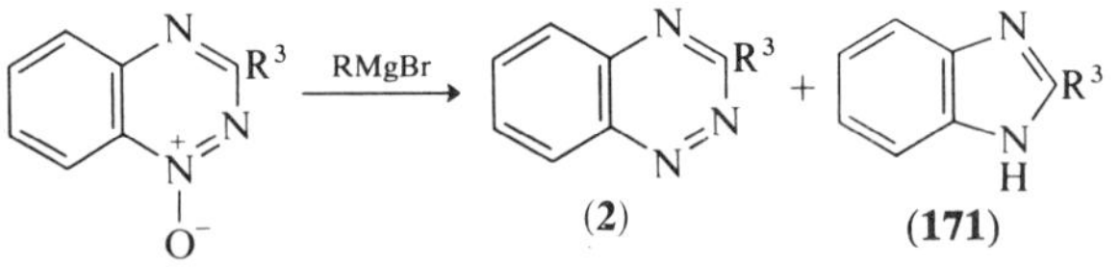

(**170**)                              (**2**)              (**171**)

Alkylation and acylation of the different 1,2,4-triazine systems have also been extensively studied. Alkylation of 1,2,4-triazines (**172**) with methyl iodide gave mainly 1-methyl-1,2,4-triazinium iodides (**173**), which are red, and in a few cases the colourless 2-methyl isomers (**174**). The best results were obtained when nitromethane was used as the solvent for the alkylation. The formation of (**173**) or (**174**) depends on the substituents on the 1,2,4-triazine ring ⟨63JCS1628⟩.

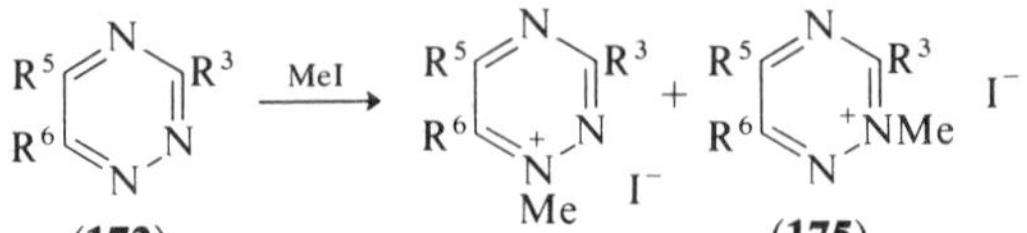

(**172**)                    (**173**)              (**175**)

When 1,2,4-triazin-3-ones (**117**) are alkylated the products are mainly the 2-alkyl-1,2,4-triazin-3-ones (**175a**), along with the 3-alkoxy-1,2,4-triazines (**176a**) in minor amounts ⟨78HC(33)189, p. 245⟩. 2-Substituted 1,2,4-triazin-3-ones (**175b**) were also obtained when

(117) reacted with formaldehyde and amines ⟨70LA(733)177⟩. Acylation of (117) affords *N*(2)-acylated (175c) and *O*-acylated products (176b). Alkylation of 1,2,4-triazin-5-ones (31) with methyl iodide gives mainly 2-methyl-1,2,4-triazin-5-ones (177) and smaller amounts of the 4-methyl isomers (178). Different results were obtained when the 5-ones (31) were methylated with diazomethane: three methylation products, (177), (178) and the 5-methoxy-1,2,4-triazines (156), were isolated in different ratios depending on the solvent used. Further alkylation of (177) or (178) leads always to the formation of 2,4-dimethyl-1,2,4-triazinium iodides (179) ⟨78HC(33)189, p. 257⟩.

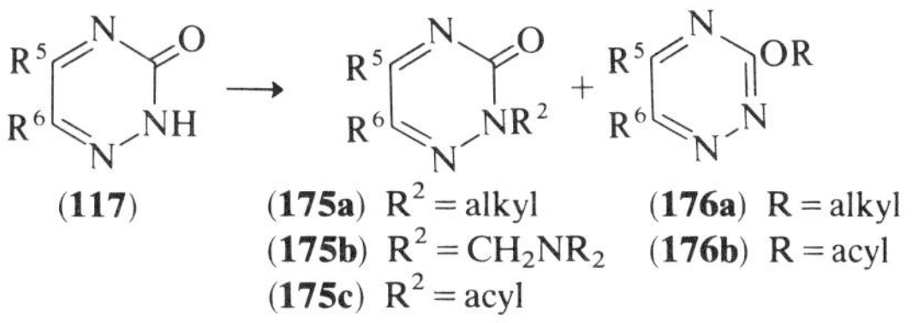

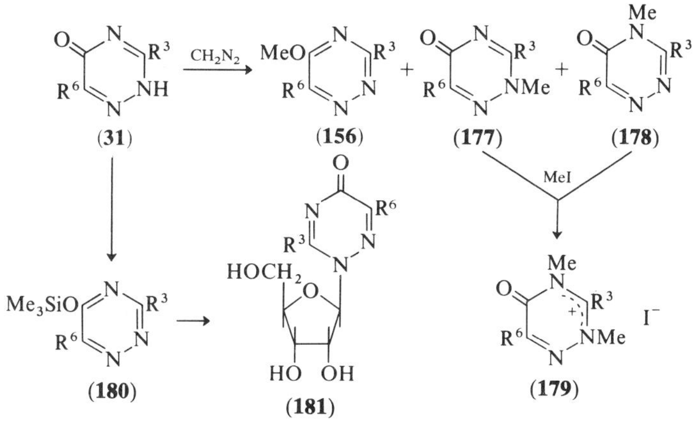

**(117)**  **(175a)** $R^2$ = alkyl  **(176a)** R = alkyl
**(175b)** $R^2$ = CH$_2$NR$_2$  **(176b)** R = acyl
**(175c)** $R^2$ = acyl

**(31)**  **(156)**  **(177)**  **(178)**

**(180)**  **(181)**  **(179)**

1,2,4-Triazin-5-ones (31) give the 5-trimethylsilyloxy-1,2,4-triazines (180) on treatment with trimethylchlorosilane in the presence of hexamethyldisilazane. The silyloxy compounds (180) were used for the preparation of N-2 glycosides (181) of the 1,2,4-triazin-5-ones ⟨74JOC3668⟩.

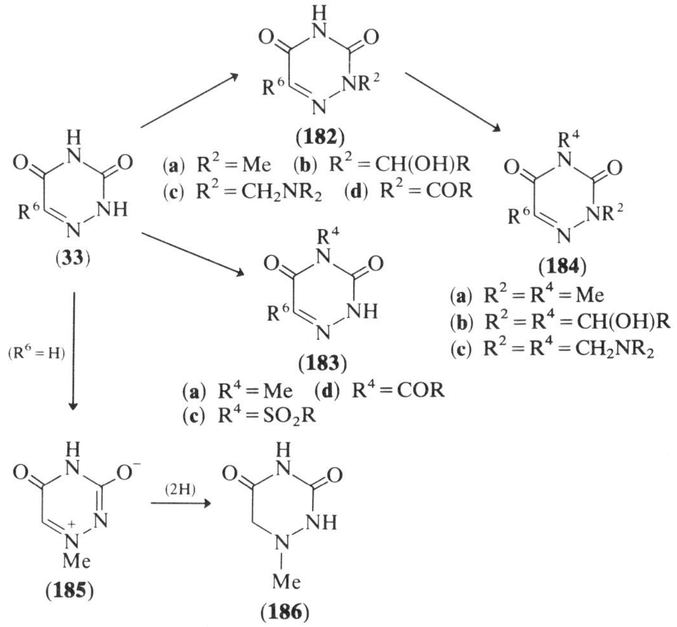

**(182)**
(a) $R^2$ = Me  (b) $R^2$ = CH(OH)R
(c) $R^2$ = CH$_2$NR$_2$  (d) $R^2$ = COR

**(33)**  **(184)**
(a) $R^2$ = $R^4$ = Me
(b) $R^2$ = $R^4$ = CH(OH)R
(c) $R^2$ = $R^4$ = CH$_2$NR$_2$

($R^6$ = H)  **(183)**
(a) $R^4$ = Me  (d) $R^4$ = COR
(c) $R^4$ = SO$_2$R

**(185)**  **(186)**

Alkylation of 1,2,4-triazine-3,5-diones (33) with methyl iodide begins at N-2 (182a), while dimethyl sulfate or diazomethane convert (33) initially into the 4-methyl derivatives (183a). In all cases 2,4-dimethyl-1,2,4-triazine-3,5-diones (184a) were isolated on further alkylation ⟨78HC(33)189, p. 327⟩.

Alkylation of 1,2,4-triazine-3,5-dione (**33**; R = H) with dimethyl sulfate at 135–145 °C without a base afforded 1-methyl-5-oxo-1,2,4-triazin-3-olate (**185**), the structure of which was proved by hydrogenation to be the dihydro compound (**186**) ⟨73CCC934⟩.

Reaction of (**33**) with formaldehyde, acetaldehyde or formaldehyde and amines also leads to N(2)-substituted (**182b, c**) and 2,4-disubstituted products (**184b, c**).

When 1,2,4-triazine-3,5-diones (**33**) are acylated, the 2-acyl derivatives (**182d**) are the principal products, and in only one case was a 4-acylated compound (**183d**) obtained. However, sulfonyl chlorides give predominantly the 4-sulfonyl derivatives (**183e**).

All 1,2,4-triazines containing a thioxo group either in the 3-, 5- or 6-position are alkylated by methyl iodide or dimethyl sulfate at the sulfur if this is possible. This observation is reported for 1,2,4-triazine-3-thiones, -5-thiones, -3,5-dithiones, -3,5,6-trithiones, 3-thioxo-1,2,4-triazin-5-ones, 5-thioxo-1,2,4-triazin-3-ones, 5-amino-1,2,4-triazine-3-thiones, 6-thioxo-1,2,4-triazine-3,5-diones, 5,6-dithioxo-1,2,4-triazin-3-ones, 3-thioxo-1,2,4-triazine-5,6-diones, 3,5-dithioxo-1,2,4-triazine-6-carboxylates, 4-amino-3-thioxo-1,2,4-triazin-5-ones, 3-thioxo-1,6-dihydro-1,2,4-triazin-5-ones, tetrahydro-1,2,4-triazine-3-thiones and 1,2,4-benzotriazine-3-thiones ⟨78HC(33)189⟩.

Both 3-thioxo-1,2,4-triazin-5-ones (**187**) and 5-thioxo-1,2,4-triazin-3-ones (**188**) are first alkylated by alkyl iodides or dimethyl sulfate at the sulfur, and then with further reagent a second alkylation occurs at N-2, to form compounds (**189**) and (**190**) respectively. The 6-alkylthio-5-thioxo-1,2,4-triazin-3-ones (**191**) afford the 5,6-bis(alkylthio) compounds (**192**) as the dialkylated products. 3-Thioxo-1,2,4-triazine-5,6-dione (**193**) yields as the dialkylated product the 6-alkoxy-3-alkylthio-1,2,4-triazin-5-one (**194**). 3,5-Dithioxo-1,2,4-triazine-6-carboxylates (**195**) are alkylated at both sulfur atoms to give 3,5-bis(alkylthio)-1,2,4-triazine-6-carboxylates (**196**).

Sometimes different results are obtained when diazomethane is used as the alkylating agent. For instance, 1,2,4-triazine-3,5-dithiones (**35**) are methylated by diazomethane at nitrogen, leading to the 2-methyl- (**197**) and 4-methyl-3,5-dithione (**198**) derivatives ⟨62CCC1898⟩. 5-Thioxo-1,2,4-triazin-3-ones and 3-thioxo-1,2,4-triazin-5-ones are methylated by diazomethane at sulfur or at the 4-nitrogen depending on the solvent used. Controversial results have been published on the alkylation of 3-amino-1,2,4-triazines. Alkylation of the 5,6-diphenyl compound (**199**) with ethyl iodide is reported to yield 2-ethyl-3-imino-5,6-diphenyl-1,2,4-triazine (**200a**) ⟨63AF3⟩. Alkylation of (**199**) with bromo- or chloro-acetic acid also affords 2-substituted 3-imino-1,2,4-triazines (**200b**) ⟨79KGS1561⟩. However, alkylation of 3-amino-1,2,4-triazines (**86**) with dimethyl sulfate is reported to give 1-alkyl-1,2,4-triazinium salts (**201**) ⟨64MI21900⟩, and an Italian group has reported the isolation of a 1 : 1 mixture of 1-methyl- (**201**) and 4-methyl-3-amino-1,2,4-triazinium salts (**202**) when 3-amino-1,2,4-triazines (**86**) were alkylated with methyl iodide ⟨64MI21900⟩. Alkylation of 3-amino-1,2,4-triazin-5(2H)-ones affords mainly the 2-alkylated products.

Acylation of 3-amino-1,2,4-triazines (**86**) always leads to 3-acylamino-1,2,4-triazines (**203**). Similarly, 3-amino-1,2,4-triazin-5-ones, 5-amino-1,2,4-triazin-3-ones, 6-amino-1,2,4-triazin-5-ones and 6-amino-3-ethoxy-1,2,4-triazine-5-carboxylate esters give the respective acylamino derivatives.

Alkylation and acylation of 6-halo-1,2,4-triazine-3,5-diones yields 2- and 4-alkylated and -acylated derivatives. With trimethylchlorosilane they afford the 6-halo-3,5-bis(trimethylsilyloxy)-1,2,4-triazines. 1,6-Dihydro-1,2,4-triazine-3,5-diones (204) are acylated at N-1 and N-2 and the diacylated product (205) can then be methylated by diazomethane at N-4 to afford compound (206) ⟨63CCC2527⟩.

1,2,4-Benzotriazin-3-one (144) forms 3-methoxy-1,2,4-benzotriazine (207) when an ethereal solution of diazomethane is added slowly to (144), but when (144) is added to an excess of diazomethane the 2-methyl (208) and 4-methyl (209) derivatives are produced ⟨50MI21900⟩.

The product of methylation of 1,2,4-benzotriazin-3-one 1-oxide (210) with methyl iodide was identified as 4-methyl-1,2,4-benzotriazin-3-one 1-oxide (211) ⟨59JOC813⟩. Like the amines in the monocyclic series mentioned above, 3-amino-1,2,4-benzotriazine 1-oxides (133) are acylated on the amino group ⟨54JA3551⟩.

1-Aryl-1,4-dihydro-1,2,4-benzotriazines (212) are alkylated at both free nitrogens yielding 2-alkyl- (213) and 4-alkyl-1-aryldihydro-1,2,4-benzotriazines (214) ⟨80CB1205⟩.

Reaction of 5,6-diphenyl-1,2,4-triazin-3-one (215) or 5,6-diphenyl-1,2,4-triazine-3-thione (216) with benzene under Friedel–Crafts conditions led to the isolation of 5,5,6-triphenyl-4,5-dihydro-1,2,4-triazin-3-one (217) and -3-thione (218) respectively ⟨76IJC(B)273⟩. Reacting under similar conditions (C₆H₆/AlCl₃) the 6-styryl compounds (219–221) are converted into the corresponding 6-(2,2-diphenylethyl) derivatives, *e.g.* (222) from (219) ⟨75IJC1098⟩.

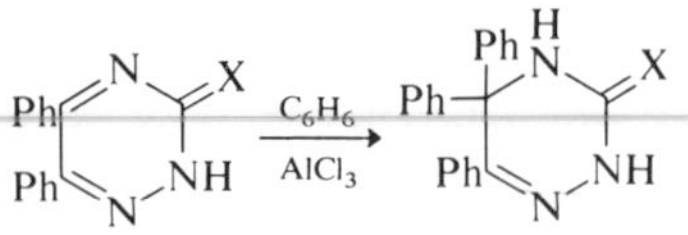

(215) X = O
(216) X = S

(217) X = O
(218) X = S

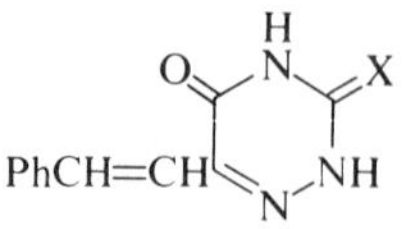

(219) X = O
(220) X = S

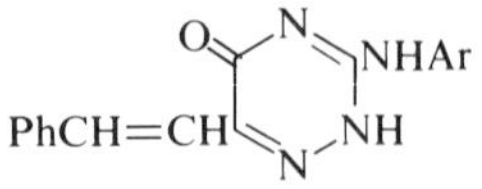

(221)

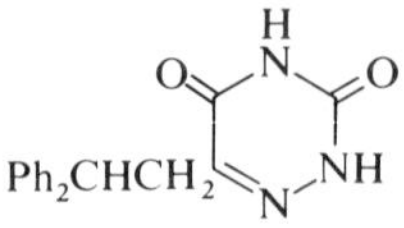

(222)

Oxidation of 1,2,4-triazines may take several courses. Thus, oxidation at a nitrogen of the heterocyclic ring, leading to 1,2,4-triazine *N*-oxides, oxidation of the heterocyclic ring, preferentially in the 5-position, affording 1,2,4-triazin-5-ones or 1,2,4-triazine-5,6-diones, oxidation of a methyl group to a carboxyl group, of a benzyl group to a benzoyl group and of dihydro-1,2,4-triazines to 1,2,4-triazines, have been reported. Furthermore, oxidation of thioxo-1,2,4-triazines to disulfides, of alkylmercapto groups to alkylsulfonyl groups, and of thioxo groups to sulfonic acids are known.

1,2,4-Triazines (173) without a substituent in the 5-position are oxidized by peracids to 1,2,4-triazin-5-ones (31), or 1,2,4-triazine-5,6-diones (223) if the 6-position is also unsubstituted ⟨72LA(758)111⟩. If the 5-position is substituted, 1,2,4-triazine 1-oxides (224) are isolated as the major products and 1,2,4-triazine 2-oxides (225) as the minor products ⟨71JOC787⟩.

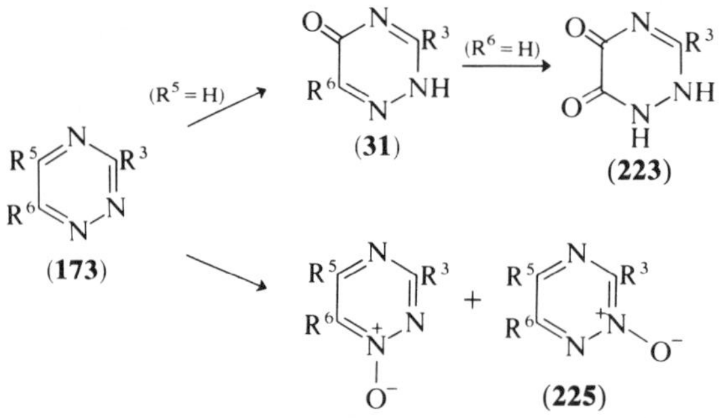

(173)

(31)

(223)

(224)

(225)

A methyl group in the 5-position can be oxidized by potassium permanganate to a carboxyl group, while one in the 3- or 6-position is converted into an oxo group under the same reaction conditions, leading to 1,2,4-triazin-3- or -6-ones. Oxidation of 5,6-dimethyl-1,2,4-triazines (226) with potassium permanganate was reported to yield 6-oxo-1,2,4-triazine-5-carboxylic acids (227) ⟨58CB422, 57CB481⟩.

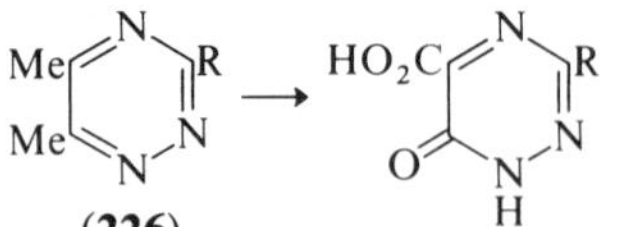

(226)

(227)

Oxidation of 3-methoxy- or 3-phenoxy-1,2,4-triazines (228) with perbenzoic acid converted them into the appropriate 1-oxides (229) ⟨71JOC787⟩. Peracid treatment of 1,2,4-triazin-3-ones with an unsubstituted 5-position (84; $R^5 = H$) gave 1,2,4-triazine-3,5-diones (182) ⟨69JHC403⟩, while 5-substituted 1,2,4-triazin-3-ones (84) afforded the appropriate 1-oxides (230) ⟨66JOC3914⟩. 3-Amino-1,2,4-triazines with an unsubstituted (86a) or a monosubstituted amino group (86b) were oxidized by peracids to give the 2-oxides (231), while the 3-dialkylamino compounds (86c) afforded the 1-oxides (232) ⟨77JOC546⟩.

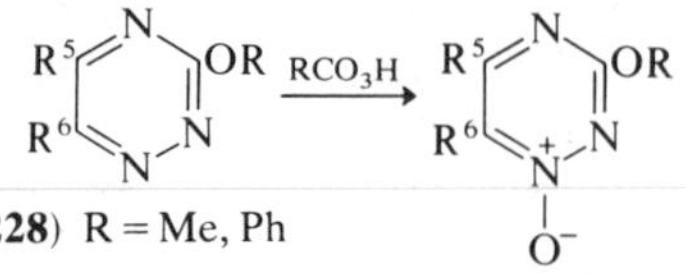

(228) R = Me, Ph

(229)

Oxidation of 3-amino-1,2,4-triazine (**39**) with peracetic acid led to 3-amino-1,2,4-triazin-5-one (**233**) ⟨79JHC555⟩ while oxidation with permanganate gave the 3,5-dione (**234**) ⟨79JHC1649⟩. When 3-amino-5,6-dimethyl-1,2,4-triazine (**235**) was treated in basic or acidic medium with potassium permanganate the only isolated product was the 3-amino-6-methyl-1,2,4-triazin-5-one (**237**) and not the 5,6-dione (**238**) as was reported earlier ⟨63MI21900⟩. The same product (**237**) was obtained when 3-amino-6-methyl-1,2,4-triazine (**236**) was oxidized with potassium permanganate, chromic acid or hydrogen peroxide in acetic acid ⟨58CCC1588⟩.

Mild oxidation of thioxo-1,2,4-triazines with iodine affords the disulfides, as with 3-thioxo-1,2,4-triazines (**34**) leading to (**239**), for 3-thioxo-1,2,4-triazin-5- and -6-ones, 3-thioxo-1,6-dihydro-1,2,4-triazin-5-ones and 1,2,4-benzotriazine-3-thione 1-oxides.

1,2,4-Triazine-3-thiones (**34**) with hydrogen peroxide formed 1,2,4-triazin-3-ones (**117**) while oxidation with potassium permanganate led to the isolation of 1,2,4-triazine-3-sulfonic acids (**240**) in a few cases, and in general also 1,2,4-triazin-3-ones (**117**) were produced ⟨78HC(33)189, p. 346⟩.

Oxidation of 3,5-bis(methylthio)-1,2,4-triazines (**81**) with potassium permanganate afforded the 3-monosulfones (**241**) which could be hydrolyzed to 5-methylthio-1,2,4-triazin-3-ones (**242**) ⟨69BSF3670⟩. 6-Benzyl-3-thioxo-1,2,4-triazin-5-one (**243**) with potassium permanganate afforded the 6-benzoyl-3-sulfonic acid (**244**) ⟨67BSF2551⟩. Similarly, 5-amino-1,2,4-triazine-3-thione (**245**) gave the sulfonic acid (**246**) ⟨56JA1938⟩. Oxidation of 6-alkylthio-1,2,4-triazine-3,5-diones (**247**) with chlorine was used to prepare the 6-alkylsulfonyl compounds (**248**) ⟨70RRC1121⟩, while reaction of the 6-mercapto compounds (**247**; R = H) affords the 6-sulfonyl chlorides (**249**).

(**81**) (**241**) (**242**)

(**243**) (**244**)

(**245**) (**246**)

(**249**) (**247**) (**248**)

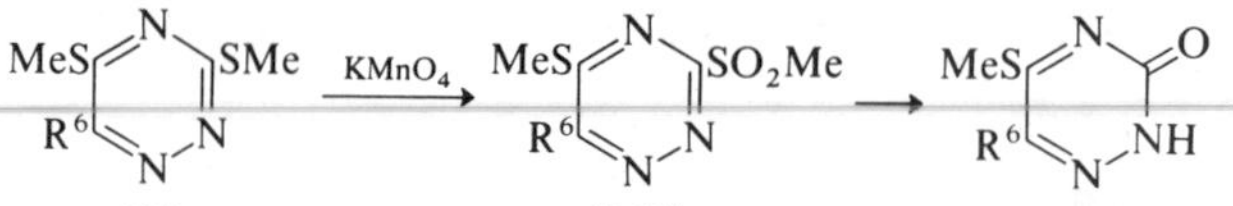

5,6-Disubstituted 1,2,4-triazine 4-oxides (**124**) with peracids give the corresponding 1,4-dioxides (**250**). If the 5-position is unsubstituted the 5-oxo 4-oxides (**251**) are formed, and these are also the products of oxidation by potassium permanganate ⟨76LA153⟩.

(**251**) (**124**) (**250**)

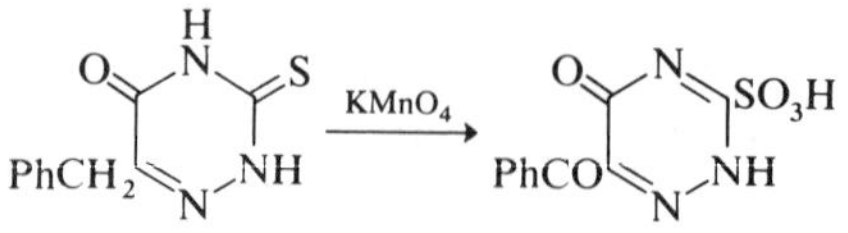

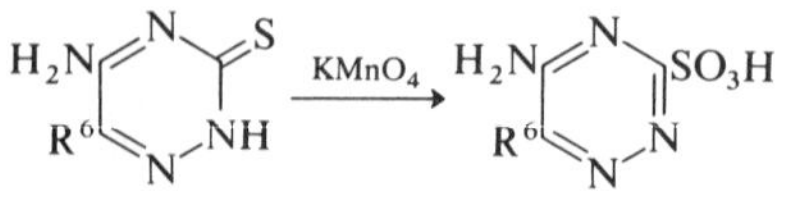

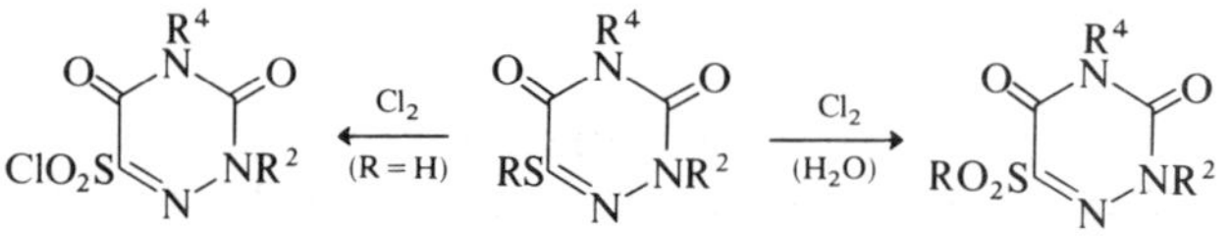

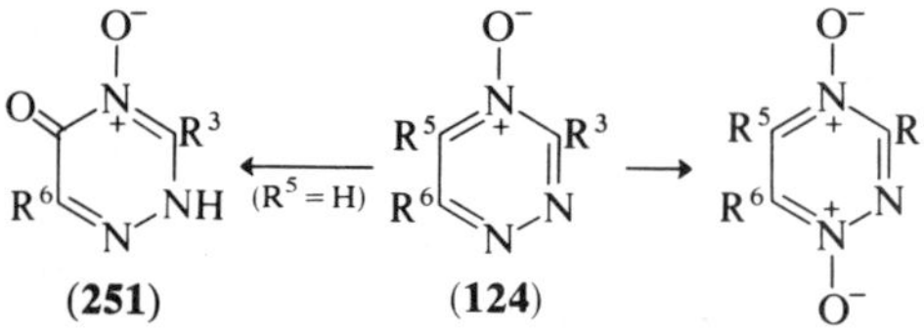

A wide variety of dihydro-1,2,4-triazines have been oxidized to the corresponding aromatic compounds. *p*-Benzoquinone seems to be one of the best reagents for this purpose. The oxidation of 4-hydroxy-2,3,4,5-tetrahydro-1,2,4-triazines (**252**) with lead dioxide to the 1,2,4-triazine 4-oxides (**124**) is reported ⟨73KGS134⟩. Peracid oxidation of 1,2,4-benzotriazines (**128**) affords the 1-oxides (**253**) or 2-oxides (**254**) depending on the reaction conditions used ⟨57JCS3186⟩. 3-Chloro- and 3-ethoxy-1,2,4-benzotriazine were oxidized by peracid to the appropriate 1-oxides (**253**; R = Cl, OEt) ⟨69CB3818⟩. 3-Amino-1,2,4-benzotriazines (**255**) were oxidized primarily to the 2-oxides (**256**), but prolonged oxidation at 50 °C affords the 1,4-dioxides (**257**). Oxidation of the 1-oxides (**258**) also affords the 1,4-dioxides (**257**) ⟨70JCS(B)911⟩.

(**252**) (**128**) (**253**) (**254**)

(**255**) (**256**) (**257**) (**258**)

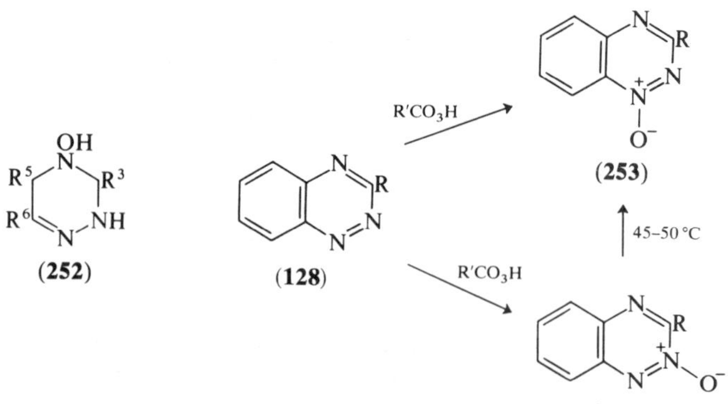

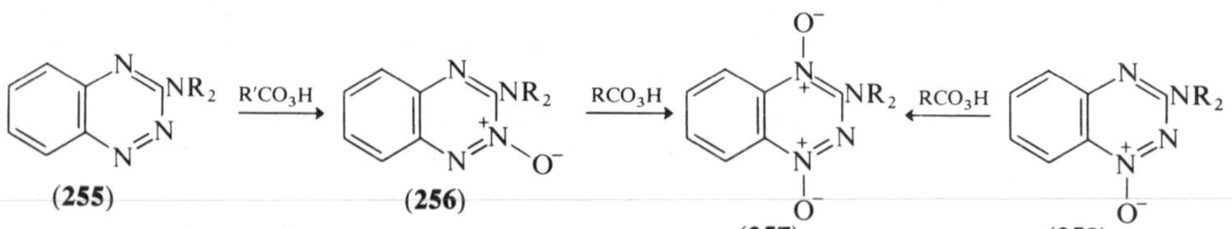

3-Phenyl-1,2,4-benzotriazine 2-oxide (**254**; R = Ph) has been shown to rearrange to the 1-oxide (**253**) on prolonged heating at 45–50 °C ⟨57JCS3186⟩.

1-Aryl-1,4-dihydro-1,2,4-benzotriazines (**212**) are readily oxidized in basic solution, even by air, to the stable, deeply coloured and crystalline 1-aryl-1,2,4-benzotriazinyl radicals (**259**) ⟨68TL2701, 69CB1848, 74CC485, 80CB1205⟩.

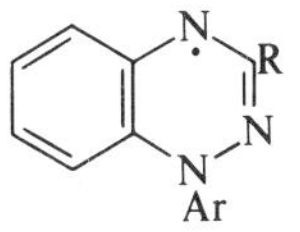

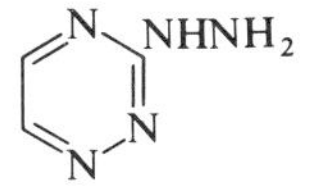

Finally, oxidation of hydrazino-1,2,4-triazines leads to the replacement of the hydrazino group by hydrogen. This was used for the preparation of the parent 1,2,4-triazine (**1**), 3-hydrazino-1,2,4-triazine (**260**) being oxidized by manganese dioxide ⟨70JHC767⟩. Oxidation of the following hydrazines has also been reported in the literature: 5-hydrazino-1,2,4-triazines, 3-hydrazino-1,2,4-triazin-5-ones, 3-hydrazino-1,2,4-triazine 1-oxides, and 3-hydrazino-1,2,4-benzotriazine 1-oxide; mercury(II) oxide or copper ions are alternative oxidizing agents. When 5,6-dimethyl-3-(3-pyridyl)-1,2,4-triazine (**261**) was oxidized with hydrogen peroxide in acetic acid the triazine ring was lost: the product isolated was nicotinamide *N*-oxide (**262**) ⟨60AK(15)387⟩.

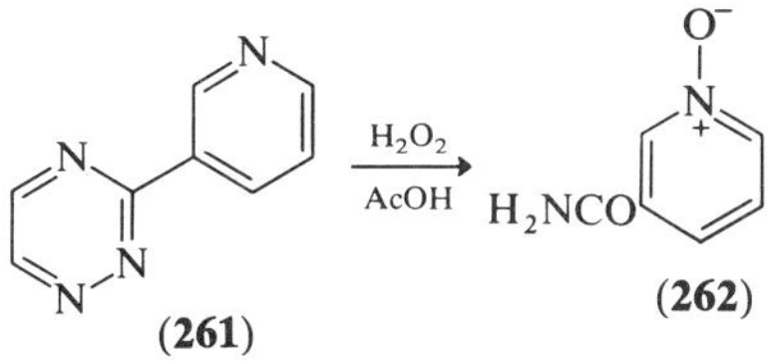

The reduction of 1,2,4-triazines with a variety of reducing agents has been published but no systematic studies have been made, and sometimes different structures for the same product are found in the literature.

Reduction with zinc and acetic acid affords 1,2-dihydro-1,2,4-triazines (**263**) and imidazoles (**264**). The latter are secondary products, formed from (**263**) ⟨57T103, 59CB2481, 63JCS1628⟩. The compounds obtained by electrochemical reduction of 1,2,4-triazines have been shown to be 1,2-dihydro- (**263**) and 4,5-dihydro-1,2,4-triazines (**265**) ⟨72CL1185, 72CJC1581⟩. Reduction of 1-methyl-3,5,6-triphenyl-1,2,4-triazinium iodide (**266**) with zinc and acetic acid affords 2,4,5-triphenylimidazole (**267**) and methylamine ⟨63JCS1628⟩.

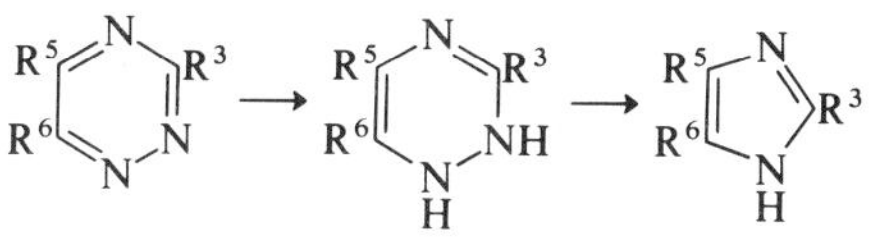

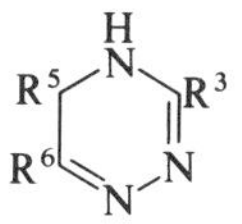

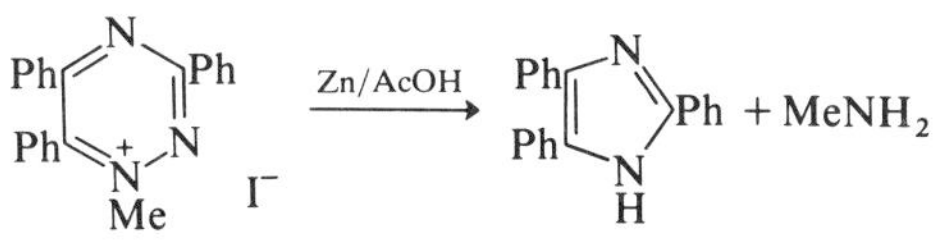

Reduction of 1,2,4-triazin-3-ones (**84**) with Raney nickel, zinc and acetic acid, lithium aluminum hydride, sodium borohydride, titanium(III) chloride, *p*-toluenethiol, hydrogen and a palladium catalyst, or electrochemically, produces 4,5-dihydro-1,2,4-triazin-3-ones (**268**) ⟨78HC(33)189, p. 246, 80JHC1237⟩, which may be further reduced to 1,4,5,6-tetrahydro-1,2,4-triazin-3-ones (**269**). 1,2,4-Triazin-3-ones (**84**) with hydriodic acid and phosphorus yielded imidazoles ⟨05LA(339)243⟩. 3-Alkoxy-1,2,4-triazines (**126**) and sodium borohydride gave the 2,5-dihydro derivatives (**270**) ⟨80JOC4594⟩.

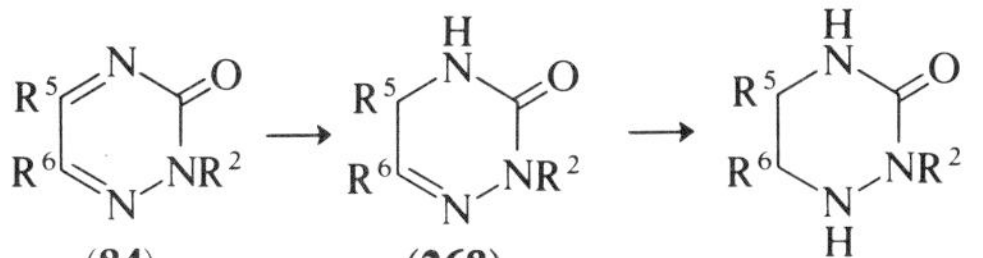

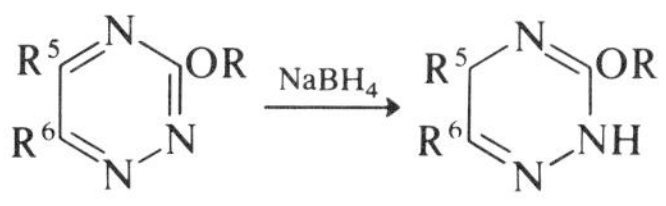

6-Methyl-1,2,4-triazin-5-one (**271**) with lithium aluminum hydride formed the 1,2,5,6-tetrahydro-1,2,4-triazine (**272**), and reduction of the 2,6-dimethyl-1,2,4-triazin-5-one (**273**) and 4,6-dimethyl-1,2,4-triazin-5-one (**274**) led to the dimethyl-3,4-dihydro-1,2,4-triazin-5(2*H*)-ones (**275**) and (**276**), respectively ⟨73JHC559⟩.

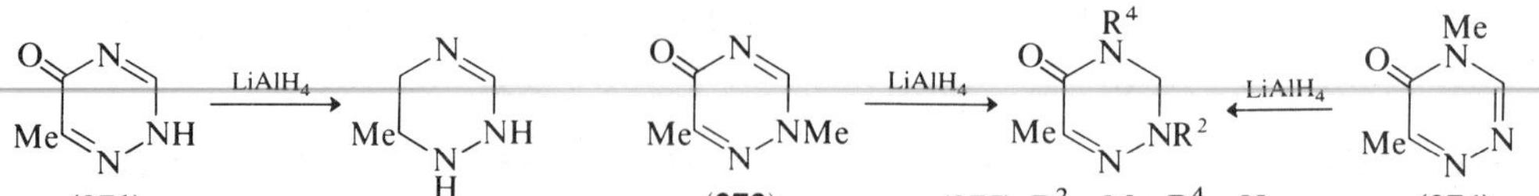

(271)     (272)     (273)     (275) $R^2 = Me, R^4 = H$     (274)
(276) $R^2 = H, R^4 = Me$

1,2,4-Triazine-3,5-diones (**6**) with hydrogen and a catalyst, or electrochemically, afforded the dihydro-1,2,4-triazine-3,5-diones (**277**); reduction with zinc and acetic acid led to the isolation of (**277**) and ring-opened compounds, while sodium amalgam converted (**6**) into ring-opened compounds only ⟨78HC(33)189, p. 328⟩.

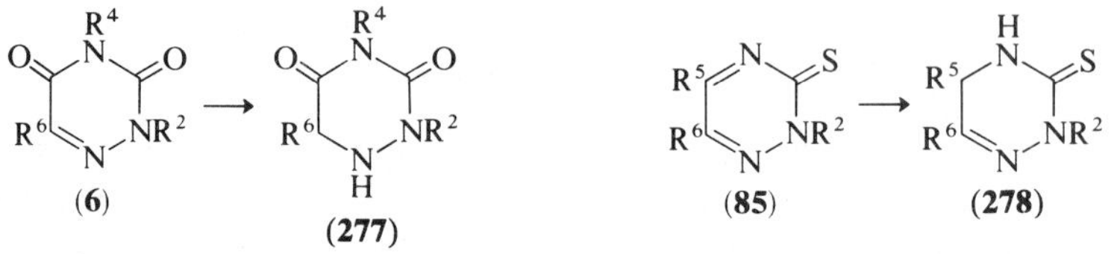

(6)       (277)       (85)       (278)

Reduction of 1,2,4-triazine-3-thiones (**85**) occurs at the N(4)—C(5) bond, yielding 4,5-dihydro-1,2,4-triazine-3-thiones (**278**) ⟨78HC(33)189, p. 346⟩.

The 3-methylthiodihydro-1,2,4-triazines obtained from the reduction of 3-methylthio-1,2,4-triazines (**279**) with sodium borohydride were formulated by Mansour ⟨74ZN(B)792⟩ as the 4,5-dihydro derivatives (**280**) and by Sasaki ⟨80JOC4587⟩ as the 2,5-dihydro compounds (**281**). It is not certain which of these tautomeric forms is predominant, but it has been shown by X-ray crystallography that the product of methylation of the reduced 5,6-diphenyl compound (**279**; $R^5 = R^6 = Ph$) is the 2-methyl-3-methylthio-5,6-diphenyl-2,5-dihydro-1,2,4-triazine (**19**).

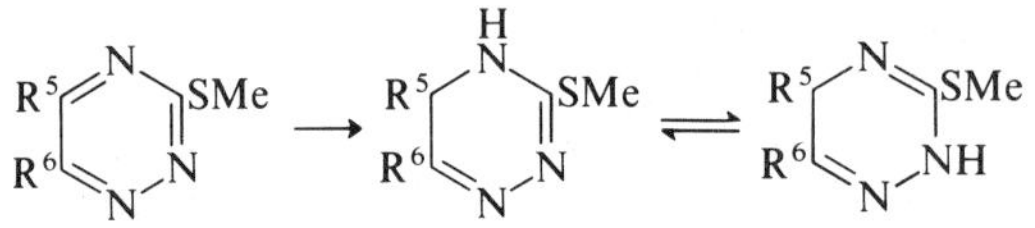

(279)       (280)       (281)

2-Substituted 3-thioxo-1,2,4-triazin-5-ones (**76**; $R^4 = H$) are stable toward sodium amalgam, but the 4-substituted derivatives (**76**; $R^2 = H$) are reduced to the 1,6-dihydro compounds (**282**). Hydrogenation of (**76**; $R^2 = H$) affords ring-opened products. 3-Methyl-thio-1,2,4-triazin-5-ones (**78**) are reduced by sodium amalgam to 3-methylthio-3,4-dihydro-1,2,4-triazines (**283**) while sodium borohydride affords 3,4-dihydro-1,2,4-triazin-5-ones (**284**) ⟨81JHC1053⟩. Hydrogenation of (**78**) converts them into the 1,6-dihydro compounds (**285**) ⟨78HC(33)189, p. 467⟩.

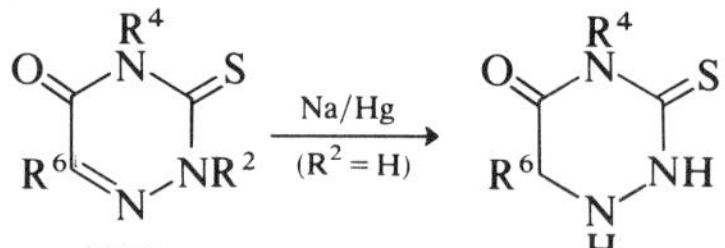

(76)       (282)

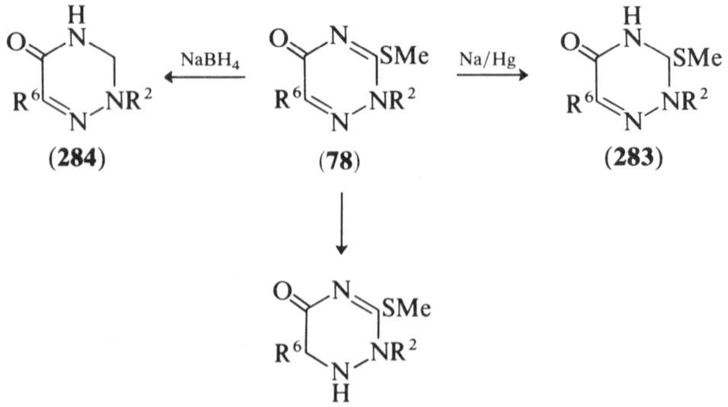

(284)       (78)       (283)

(285)

5-Methylthio-1,2,4-triazin-3-ones (**286**) are reduced by sodium borohydride to 4,5-dihydro-1,2,4-triazin-3-ones (**287**) ⟨81JHC631⟩.

Desulfurization of 1,6-dihydro-1,2,4-triazine-3,5-dithiones (**288**) leads either to tetrahydro-1,2,4-triazin-3-ones (**289**) or to tetrahydro-1,2,4-triazines (**290**), depending on the conditions ⟨70BSF1606⟩.

Reduction of 4-amino-1,2,4-triazin-5-ones (**291**) affords compounds which were formulated (rather improbably) as 4-amino-5-hydroxy-3,4-dihydro-1,2,4-triazines (**292**) ⟨75GEP2346936⟩.

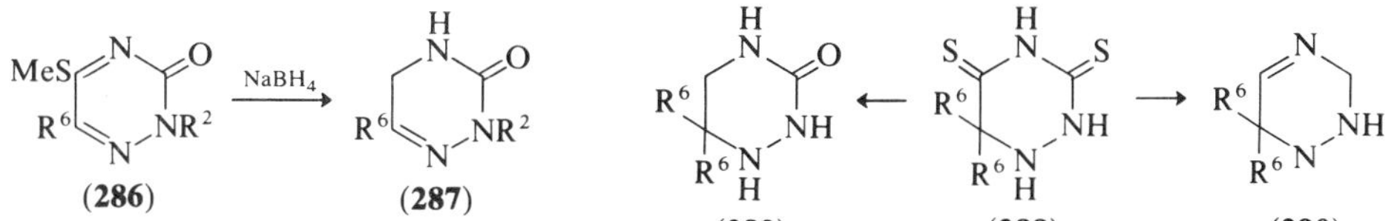

(**286**)  (**287**)  (**289**)  (**288**)  (**290**)

(**291**)  (**292**)

Hydrogenation of 5,6-diphenyl-1,2,4-triazine-3-carboxylic acid or its ethyl ester (**293**) led to the pyrazole-3-carboxylic acid or ester (**294**) respectively ⟨59CB564⟩.

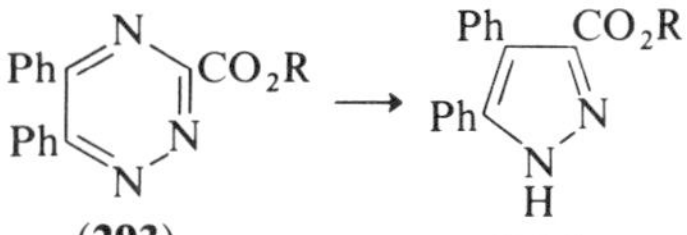

(**293**) → (**294**)

1,2,4-Triazine $N$-oxides can be deoxygenated by trivalent phosphorus compounds or by sodium hydrogen sulfite. Reduction of 1,2,4-triazin-3-one $N$-oxides by zinc and acetic acid or by catalytic hydrogenation affords dihydro-1,2,4-triazin-3-ones ⟨73JOC3277, 66JOC3914⟩. Electrochemical reduction of 3,5,6-triphenyl-1,4-dihydro-1,2,4-triazine (**295**) converts it into the *cis*-1,4,5,6-tetrahydro compound (**296**) ⟨72CJC1581⟩. Zinc and acetic acid reduces 2,3-dihydro-1,2,4-triazines (**297**) to the 2,3,4,5-tetrahydro derivatives (**298**) ⟨36JPR(144)273⟩. Reduction of 1,2,4-triazin-3-ones (**84**) or 4,5-dihydro-1,2,4-triazin-3-ones (**268**) either catalytically or electrochemically was used for the synthesis of tetrahydro-1,2,4-triazin-3-ones (**269**) ⟨78HC(33)189, p. 643⟩. Similar reduction of the 3-methoxy analogues (**176**) or (**270**) (R = Me) yields tetrahydro-3-methoxy-1,2,4-triazines ⟨55BSF1171⟩. Dihydro-1,2,4-triazine-3-thiones (**278**) can be reduced to the tetrahydro compounds (**299**) ⟨72CJC1581⟩, and 4-aminotetrahydro-1,2,4-triazines (**300**) similarly give the 4-aminohexahydro-1,2,4-triazines (**301**) ⟨60LA(635)82, 67CB3101⟩.

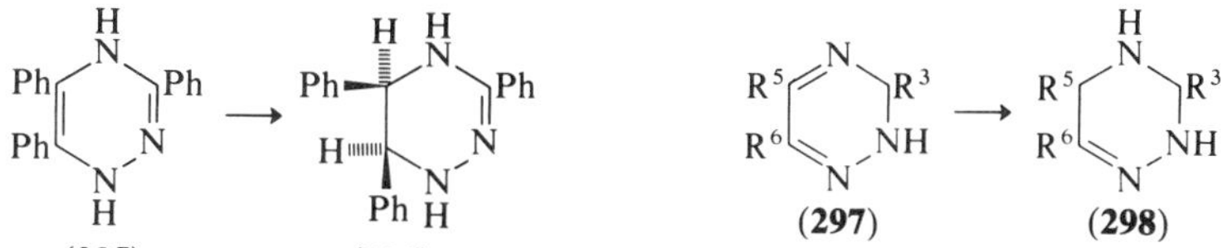

(**295**) → (**296**)    (**297**) → (**298**)

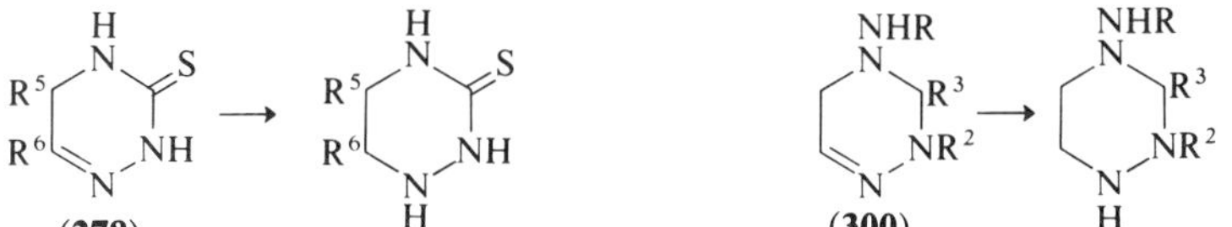

(**278**) → (**299**)    (**300**) → (**301**)

1,2-Dihydro-1,2,4-benzotriazines (**302**) have been isolated from the reduction of 1,2,4-benzotriazines (**2**) by sodium dithionite ⟨67JCS(C)1297, 67JCS(C)2658⟩ or sodium borohydride ⟨76S459⟩. 1,2,4-Benzotriazin-3-ones (**116**) afford 1,4-dihydro-1,2,4-benzotriazin-3-ones (**303**) on reduction, and the reduction products of 1,2,4-benzotriazine-3-carboxylates (**304**) were formulated as the 1,4-dihydro compounds (**305**). Reduction of the various 1,2,4-benzotriazine $N$-oxides affords 1,2,4-benzotriazines or dihydrobenzotriazines, the product ratio depending on the conditions used.

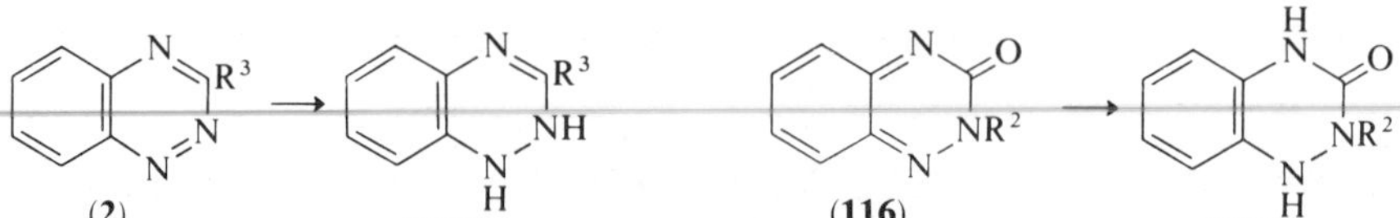

(2) → (302) (116) → (303)

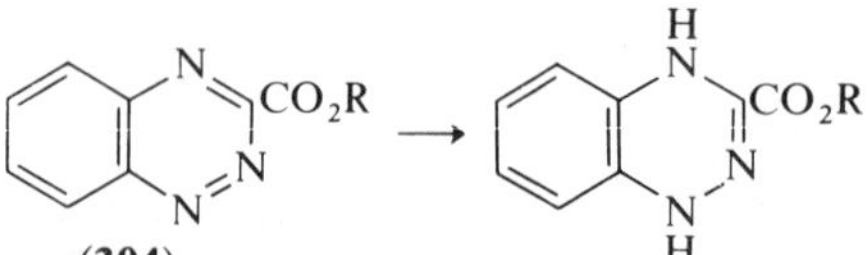

(304) → (305)

Substituent exchange in the 1,2,4-triazine ring is an extensively studied reaction, as in other systems. As earlier mentioned, the introduction of a hydrogen atom into the 1,2,4-triazine ring can be achieved by oxidation of hydrazino-1,2,4-triazines (306), as was shown for 3-hydrazino and 5-hydrazino compounds. Another method is by the decarboxylation of carboxylic acids (307). This reaction was used by Paudler and Barton for the preparation of the parent 1,2,4-triazine (1) by decarboxylation of the 3-carboxylic acid ⟨66JOC1720⟩, but decarboxylation of 5- and 6-carboxylic acids are also well-documented reactions.

(306) (307)

The reduction of chloro compounds (308), mainly by catalytic means, has been used for the synthesis of a variety of 1,2,4-triazines. For instance, the methoxy- and dimethoxy-triazines (309–311) have been made from the trichloro compound (73) by treatment with methoxide, and hydrodechlorination of the methoxy–halo compounds ⟨66CB1113⟩.

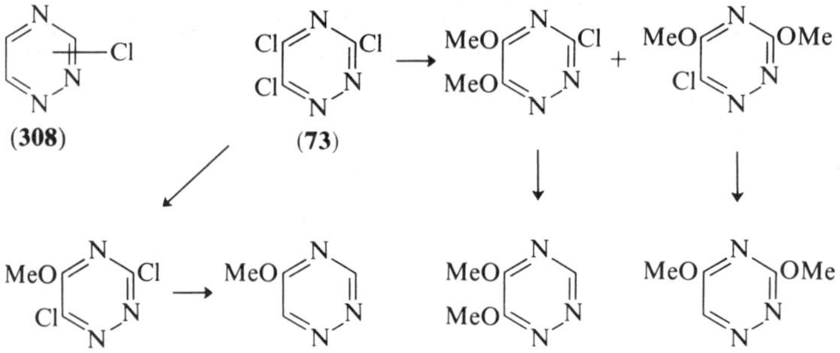

(308) (73)

(309) (310) (311)

For the debromination of 6-bromo-1,2,4-triazine-3,5-diones (312), both sodium in ammonia and butyllithium have been used ⟨58JA976, 67RRC913⟩. Other methods of obtaining an unsubstituted ring position include desulfurization with Raney nickel of 3-thioxo-1,2,4-triazin-5-ones (187) to the triazin-5-ones (313) ⟨73BSF2126⟩, and treatment of the 3-sulfonylhydrazino-1,2,4-triazines (314) with a base to afford the compounds (107) ⟨55MI21900⟩.

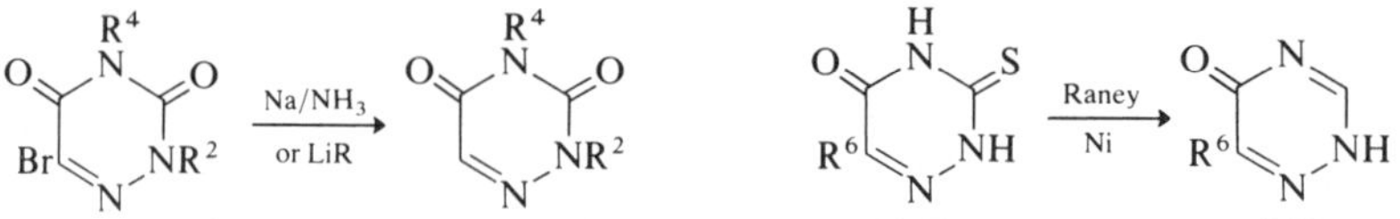

(312) (187) (313)

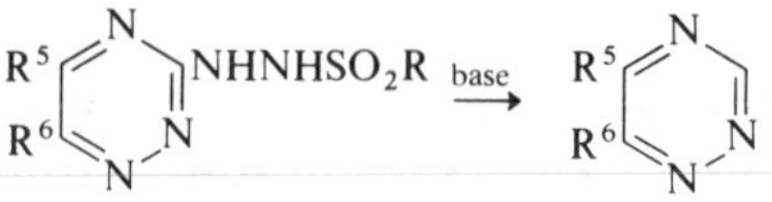

(314) (107)

On the other hand, a proton directly bound to the 1,2,4-triazine ring can be replaced by halogen. This is an electrophilic substitution reaction at carbon, and, as expected for such a heavily aza-substituted ring system, it needs considerable activation by electron-donating substituents. The halogenation reaction is best known in bromination at the 6-position. From the published data it seems that either an amino group in the 3-position or an oxo group in the 5-position is necessary. The formation of 6-halo-substituted compounds has been reported for 1,2,4-triazin-5-ones, 1,2,4-triazine-3,5-diones, 3-amino-1,2,4-triazines and 3-amino-1,2,4-triazin-5-ones.

1,2,4-Triazine 1-oxides (315) with an amino or methoxy group in the 3-position can also be brominated in the 6-position to yield 6-bromo-1,2,4-triazine 1-oxides (316) ⟨77JOC3498⟩. The reaction of 3-amino-6-methyl-1,2,4-triazine (317) with bromine has been reported to afford the 5-bromo product (318) ⟨71MI21902⟩. We believe, however, that the starting material in this reaction was in fact the 5-methyl compound (320), and the product should therefore be assigned the 3-amino-6-bromo-5-methyl-1,2,4-triazine structure (321). Nitration has also been observed with an aminotriazine. As with the bromination described above, the reported reaction is the formation of the 3-amino-6-methyl-5-nitro compound (319), from (317) ⟨63MI21900⟩, but we believe that both starting material and product structures are misassigned, and that the reaction in fact observed was (320)→(322).

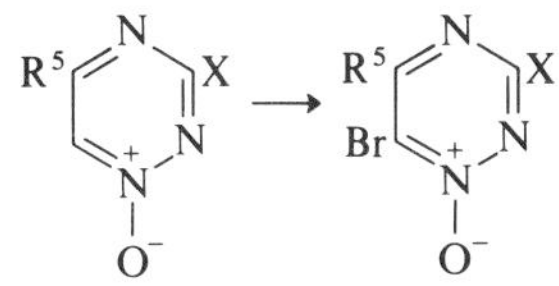

(315) X = OMe, NH$_2$    (316)

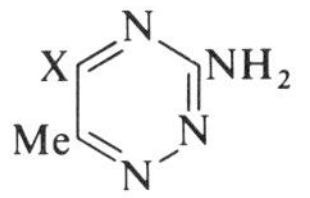

(317) X = H
(318) X = Br
(319) X = NO$_2$

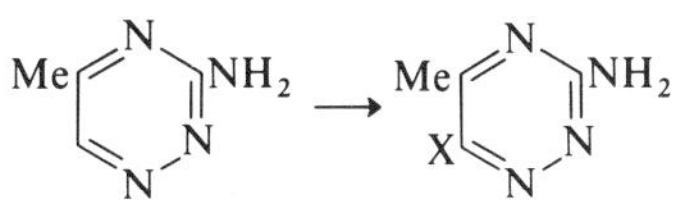

(320)

(321) X = Br
(322) X = NO$_2$

The transformation of one hetero substituent into another is a reaction frequently observed in the 1,2,4-triazine series. Oxo group transformations are indicated in Scheme 8. They can be transformed into thioxo groups by phosphorus pentasulfide or Lawesson's reagent. Pyridine seems to be the best solvent for the phosphorus pentasulfide reaction and a small amount of water (0.1% or less) seems to improve results. An oxygen in the 5-position is more easily replaced than one in the 3- or 6-position. Transformation of the oxo (hydroxy) group into a chloro substituent is achieved by phosphorus oxychloride, phosphorus pentachloride, or thionyl chloride and DMF. In this reaction also the oxygen is most reactive in the 5-position.

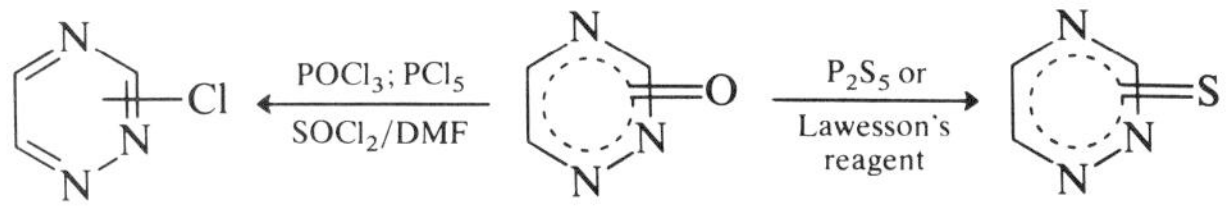

**Scheme 8**

Alkoxy-1,2,4-triazines can be transformed into amino-, hydrazino- or hydroxylamino-1,2,4-triazines.

A halogen directly attached to the 1,2,4-triazine ring is very reactive and can be replaced by most other nucleophiles as outlined in Scheme 9. The reactivity toward neutral nucleophiles decreases from the 5-position to the 3- and 6-positions, while that towards anionic nucleophiles decreases from the 5-position to the 6- and 3-positions. Among nucleophiles which displace halogen atoms are water, alcohols, amines, hydrazine, hydroxylamine, thiols and hydrogen sulfide. It is also reported that the bromine in 6-bromo-1,2,4-triazine-3,5-dione can be replaced by fluorine (323) or by a cyano group (324).

Thioxo and alkylmercapto groups are easily replaced by other groups. The sulfur in the 5-position of the 1,2,4-triazine ring is the most reactive and is replaced initially. Water, alcohols, amines, hydrazine and hydroxylamine can replace the thioxo or alkylmercapto group. Scheme 10 summarizes these reactions. Since the preparation of 3-thioxo-1,2,4-triazines is very easy and proceeds in high yield, these replacement reactions at the 3-position are of much preparative value.

Amino groups are relatively stable toward exchange by other nucleophiles. Hydrolysis, leading to 1,2,4-triazinones, is the reaction most often observed. The transformation of an amino group into a chloro substituent by nitrous acid in the presence of chloride ions has also been reported (Scheme 11).

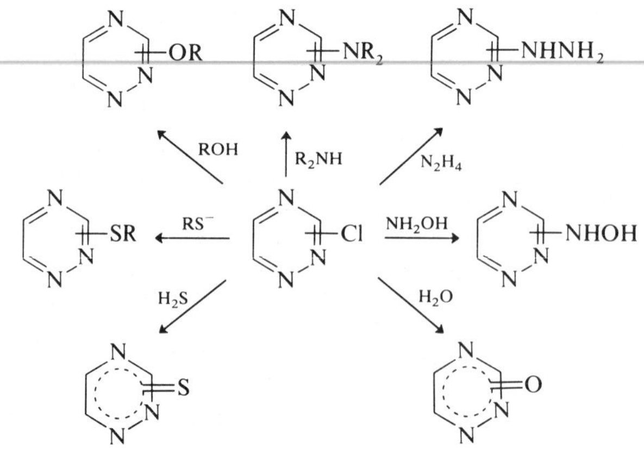

**Scheme 9**

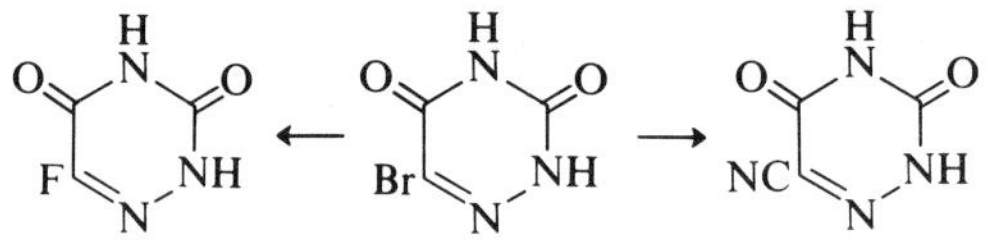

(323)    (324)

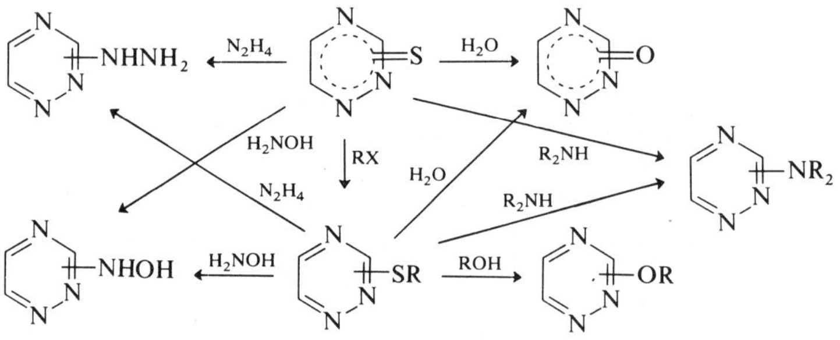

**Scheme 10**

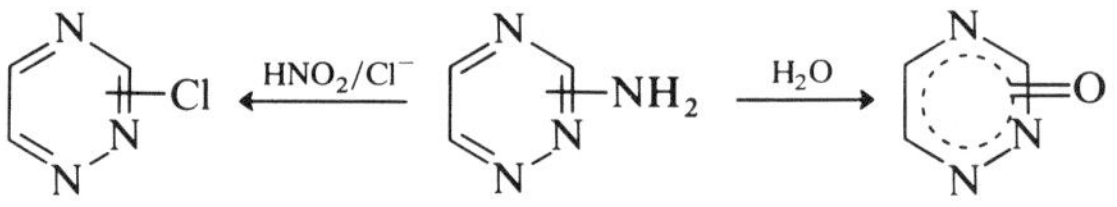

**Scheme 11**

Further published transformations of functional groups include the formation of an azido group from a hydrazino with nitrous acid, reaction of a carboxyl group with sulfur tetrafluoride to give a trifluoromethyl, formation of an amino group from a carboxyhydrazido group *via* the azidocarbonyl group and its Curtius degradation, and substitution of chlorine by a methyl group by reaction with methyltriphenylphosphonium bromide. A carboxamide group can be transformed into a nitrile, and this can react with hydroxylamine to afford the amidoxime.

Amino groups react very easily with aldehydes or ketones, and with aldehydes in the presence of amines, they can be acylated by the usual acylating agents, and they react with amidacetals, Vilsmeier reagents and nitroso compounds (Scheme 12). As mentioned earlier, alkylation leads mainly to $N(2)$-alkylated products. The hydrazino group reacts in the same way as the amino group with aldehydes or ketones, with acyl chlorides or carboxylic anhydrides, with sulfonyl chlorides, ortho esters, carbon disulfide and with nitrous acid. The last three reactions have mainly been used for the synthesis of condensed 1,2,4-triazines.

3,5,6-Trichloro-1,2,4-triazine (**73**) reacts with potassium fluoride to give the trifluoro compound (**325**). This reacts with hexafluoropropene to give 5-heptafluoroisopropyl-3,6-difluoro- (**326**), 3,5-bis(heptafluoroisopropyl)-6-fluoro- (**327**) and a small amount of 5,6-bis(heptafluoroisopropyl)-3-fluoro-1,2,4-triazine (**328**) and finally to the

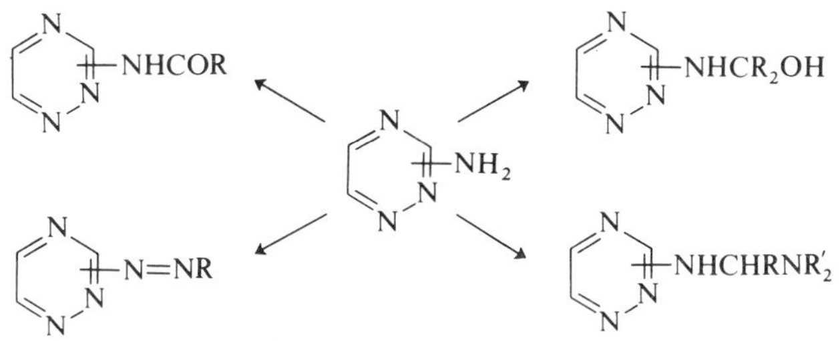

**Scheme 12**

tris(heptafluoroisopropyl) compound (**329**). In this reaction, therefore, the order of reactivity of the fluorines is in the sequence 5 > 3 > 6. In contrast, when 3,5,6-trifluoro-1,2,4-triazine (**325**) was treated with sodium bis(trifluoromethyl)amide in the presence of cesium fluoride, the main products isolated were the 3,6-di- (**330**) and 3,5,6-tri-substituted compound (**331**). Only small amounts of the 6- (**332**) and the 5-monosubstituted compound (**333**) were detected. This means that the 5-position is the least reactive toward nucleophilic attack by the amide anion in this case ⟨80JCS(P1)2254⟩.

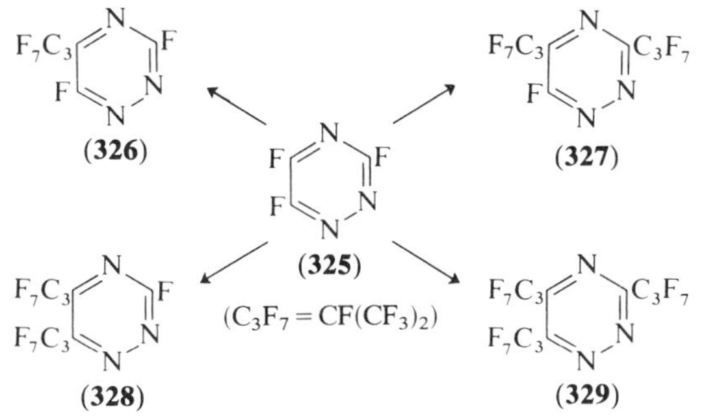

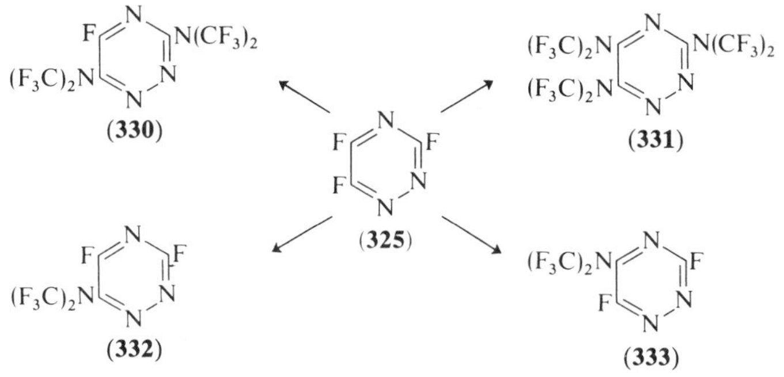

Alkyl substituents in the 5-position of 1,2,4-triazines are very reactive. As was mentioned earlier, the protons of a 5-methyl group can easily be exchanged by deuterium. With phosphorus pentachloride a 5-benzyl group (**334**; R = Ph) gives a dichlorobenzyl derivative (**335**) ⟨77CCC2182⟩. *N*-Bromosuccinimide forms a 5-bromoalkyl group (**336**) ⟨77UP21900⟩. 5-Methyl-1,2,4-triazines (**334**; R = H) react also with ketene-*O,N*-acetals (**337**) to afford the dimethylaminopropenyl compounds (**338**). The reaction starts with deprotonation of the methyl group by the acetal followed by combination of the two ions (**339**, **340**) formed. Elimination of alcohol from the adduct (**341**) yields the isolated product (**338**) ⟨73LA1963⟩.

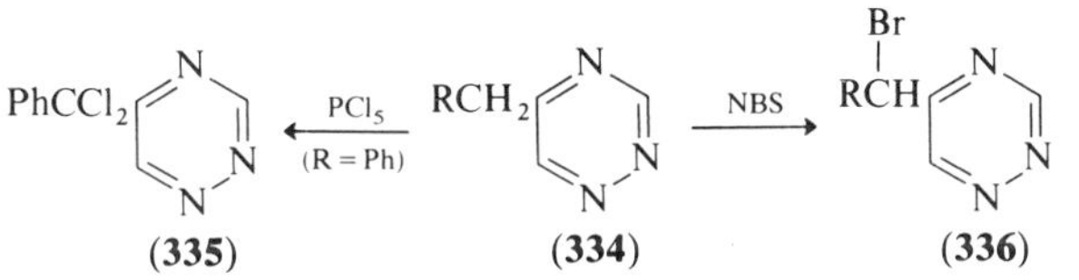

Reaction of 4-aryl-3-methylthio-1,2,4-triazin-5-ones (**342**) with hydrazine led to 4-amino-3-arylamino-1,2,4-triazin-5-ones (**343**) instead of the expected 4-aryl-3-hydrazino derivatives (**344**). This is explained by nucleophilic attack of the hydrazine at the 5-position, ring opening and subsequent reclosure. The same result was obtained when 4-aryl-3-thioxo-

(337)) (339) (340) (338) (341)

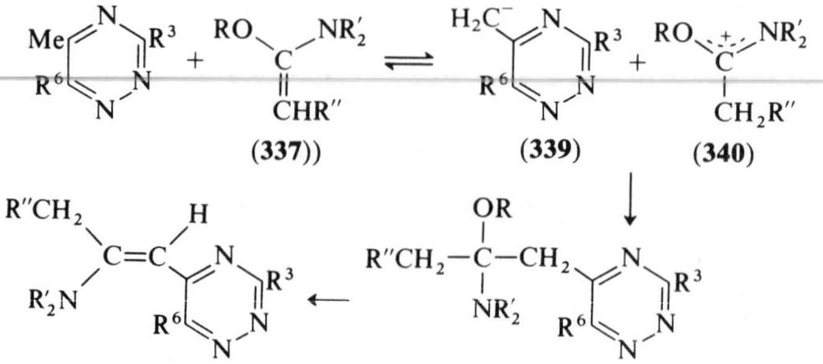

1,2,4-triazin-5-ones (**345**) reacted with hydrazine, while 4-aryl-1,2,4-triazine-3,5-diones (**346**) did not react ⟨80JHC1733⟩. When no 4-aryl group was present the rearrangement was not observed. Tetrahydro-1,2,4-triazin-3-ones with a 2-alkyl group (**347**) are rearranged by sodium amide and sodium hydride to the 1-alkylaminoimidazolin-2-ones (**348**) ⟨71FES580⟩.

(342) (343) (344) (346) (345)

(347) (348)

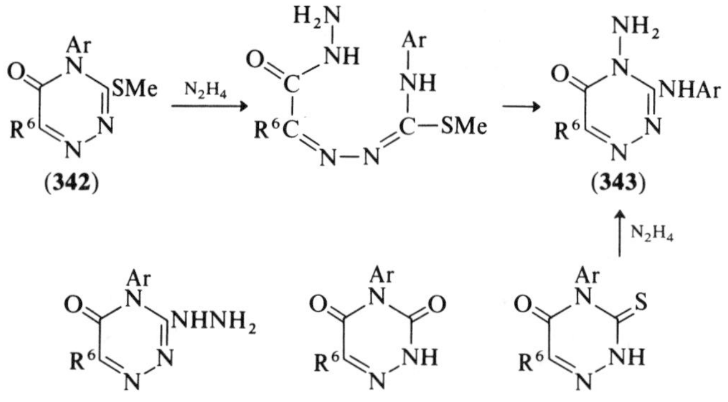

Most 1,2,4-triazines are very stable thermally. Only a few data on their thermolytic behavior are therefore to be found in the literature. We observed a few years ago that triphenyl-1,2,4-triazine (**349**) can be refluxed in an open tube at 550 °C without extensive decomposition. When the tris(dimethylamino) compound (**350**) was heated at 650 °C the colorless starting material was isolated ⟨76UP21900⟩.

(349) (350)

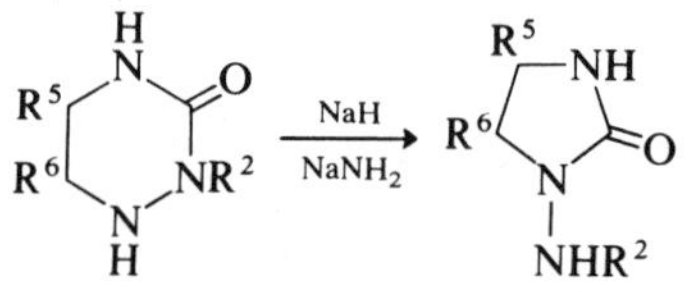

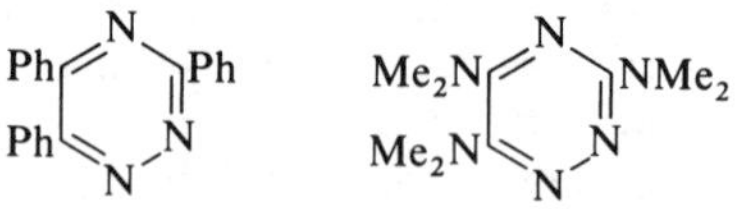

Pyrolysis of trichloro-1,2,4-triazine (**73**) at 660 °C is reported to afford trichloroacrylonitrile and tetrachloroethylene, besides nitrogen. The formation of the nitrile is explained by intermediate formation of trichloroazete (**351**), ring opening and migration of chlorine from C-2 to C-4 ⟨79JCS(P1)1978⟩.

(73) (351)

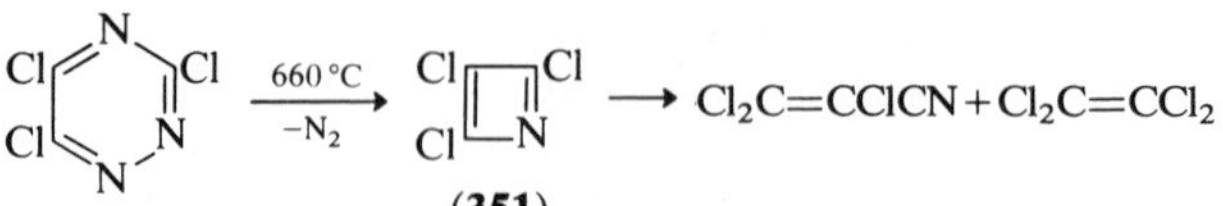

Pyrolysis of tris(heptafluoroisopropyl)-1,2,4-triazine (**329**) affords bis(heptafluoroisopropyl)acetylene and perfluorobutyronitrile, besides nitrogen ⟨79JCS(P1)1978, 80JCS(P1)2254⟩. The thermal fragmentation of this compound is analogous to the mass spectral fragmentation of many other 1,2,4-triazines.

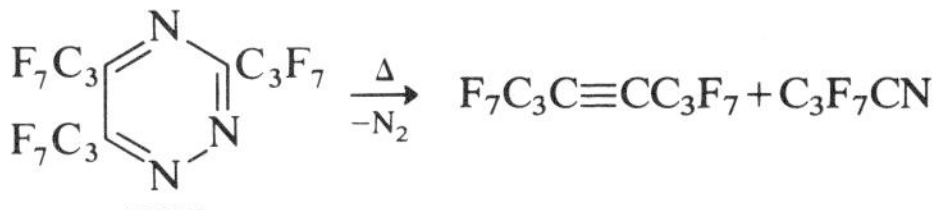

(329)

Heating 3-methoxy-1,2,4-triazine (**70**) for 2 hours to 200 °C led to the isolation of 2-methyl-1,2,4-triazin-3-one (**352**) ⟨53CR(235)1310⟩. When 6-methoxy-5-dimethylamino-1,2,4-triazine (**353**) was heated for a longer time at higher temperatures, 5-dimethylamino-1,2,4-triazin-6-one (**354**) was isolated ⟨76CB1113⟩.

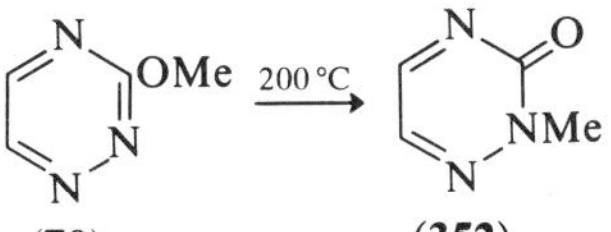

(70)          (352) 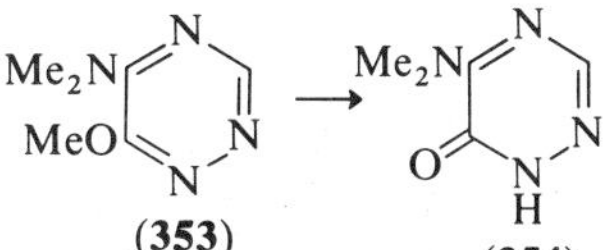          (353)          (354)

There are also not very many data on the photolysis of 1,2,4-triazines. Many 1,2,4-triazines are not only stable to irradiation but they increase the photolytic stability of other systems. Meier tried to prepare trimethylazete (**356**) by matrix photolysis of trimethyl-1,2,4-triazine (**355**) but observed only the formation of 2-butyne, acetonitrile and nitrogen ⟨80LA798⟩. Photolysis of tris(heptafluoroisopropyl)-1,2,4-triazine (**329**) afforded tris(heptafluoroisopropyl)-1,3,5-triazine (**357**), nitrogen, perfluorobutyronitrile and bis(heptafluoroisopropyl)acetylene. The last three products are the same as in the pyrolysis of (**329**) ⟨80JCS(P1)2254⟩.

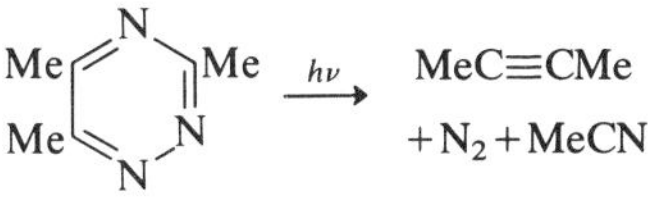

(355) 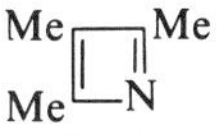          (356)

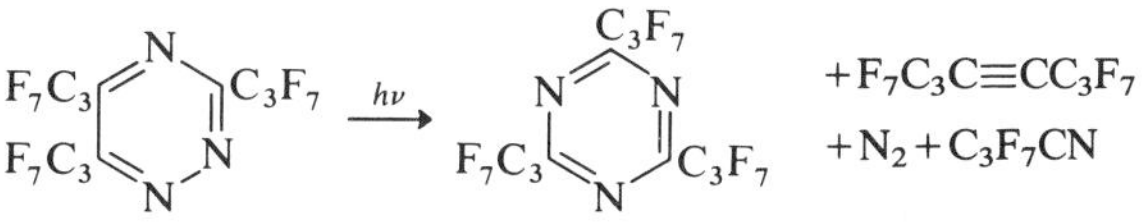

(329)          (357)

Irradiation of 5,6-diphenyl-3-*p*-tolyl-1,2,4-triazine (**358**) in methanol with UV light afforded three products, 2,4-diphenyl-6-*p*-tolyl-1,3,5-triazine (**359**), 5,6-diphenyl-3-*p*-tolyl-2,5-dihydro-1,2,4-triazine (**360**) and 4,5-diphenyl-2-*p*-tolylimidazole (**361**) ⟨72UP21900⟩.

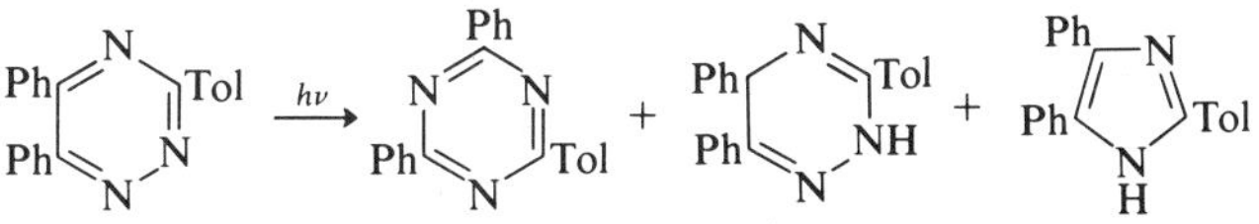

(358)          (359)          (360)          (361)

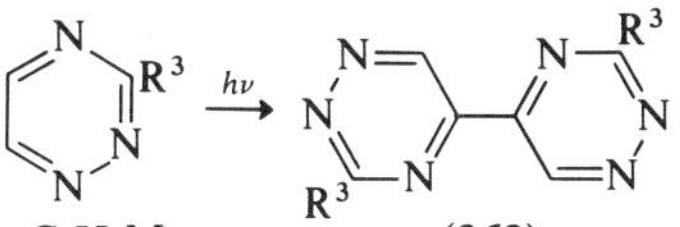

R³ = Me, C₆H₄Me-*p*          (363)

(362)

Photolysis of 3-methyl- or 3-*p*-tolyl-1,2,4-triazine (**362**) in methanol yielded the 5,5′-dimerization products (**363**) ⟨72UP21900⟩.

Irradiation of 1,2,4-triazine 4-oxides (**124**) led to the deoxygenated products, and also, when position 5 was unsubstituted, to 1,2,4-triazoles (**364**) ⟨76LA153⟩. 4-Amino-1,2,4-triazin-5-ones (**365**) are photochemically deaminated. This has been demonstrated, in particular, with the 6-*t*-butyl-3-methylthio compound (**8**) which is used as a herbicide ⟨71MI21902⟩.

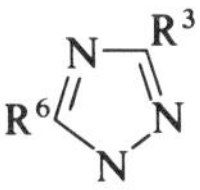

(364) 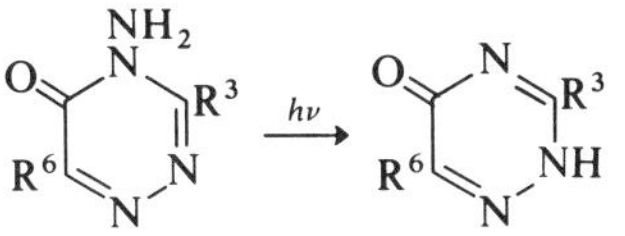          (365)          (31)

1,2,4-Triazines are reactive electron-deficient dienes in Diels–Alder reactions with inverse electron demand. They react with alkenes, strained double bonds, electron-rich and electron-deficient alkynes and C=N double bonds. In most cases it is found that the dienophile addition occurs across the 3- and 6-positions of the triazine ring, but ynamines can also add across the 2- and 5-positions. The reactions are still under active theoretical and practical investigation.

The trichloro compound (73) is the only 1,2,4-triazine which has so far been found to react with simple alkenes such as ethylene and *cis*-2-butene. As with other [4 + 2] cycloaddition reactions of 1,2,4-triazines the following mechanism is suggested: addition of the alkene to the 3- and 6-positions to give the bicyclic product (366) which is not isolated, as nitrogen is very quickly lost *via* a retro-Diels–Alder reaction and the dihydropyridine (367) is formed. To explain the isolated products, 2,6-dichloropyridine (368a) and 2,6-dichloro-3,4-dimethylpyridine (368b), one may assume a 1,5-sigmatropic shift of hydrogen followed by elimination of hydrogen chloride from (369) ⟨79CC658⟩.

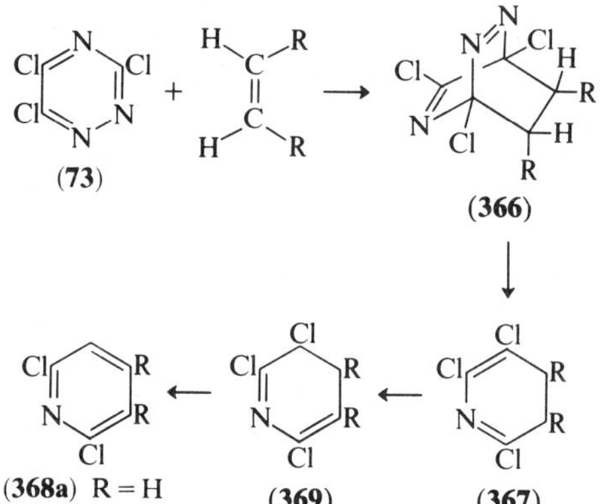

(73)  (366)

(368a) R = H
(368b) R = Me

(369)   (367)

Condensed pyridines were obtained when 1,2,4-triazines reacted with cyclic alkenes. Trichloro-1,2,4-triazine (73) afforded the condensed pyridines (370a, b) in high yield with cyclopentene and (Z)-cyclooctene, respectively. In the reaction with cyclopentene the bis adduct (371) was isolated in small amount but the bis adduct (372a) is the sole reaction product when (73) is treated with norbornene. The more reactive trifluoro-1,2,4-triazine (325) affords bis adducts (371b, c; 372b) with cyclopentene and (Z)-cyclooctene as well as with norbornene ⟨79CC658⟩.

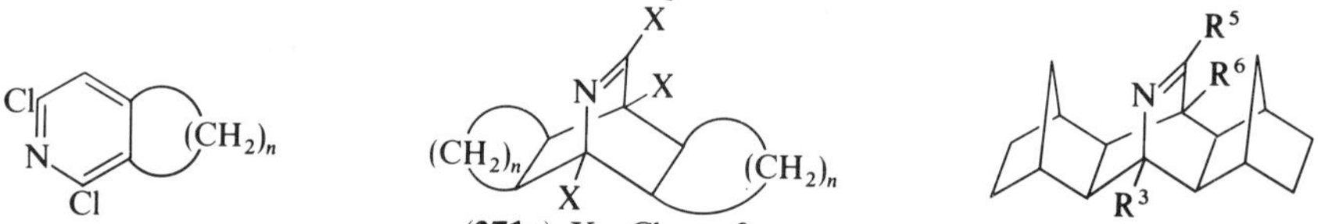

(370a) $n = 3$
(370b) $n = 6$

(371a) X = Cl, $n = 3$
(371b) X = F, $n = 3$
(371c) X = F, $n = 6$

(372a) $R^n$ = Cl
(372b) $R^n$ = F

Besides the trihalo compounds (73) and (325), many other 1,2,4-triazines have been found to react with cyclopentene, cyclohexene, cycloheptene and cyclooctene to give condensed pyridines (373). In most cases *p*-benzoquinone was added to oxidize the initially-formed dihydropyridines (374) ⟨69TL5171⟩.

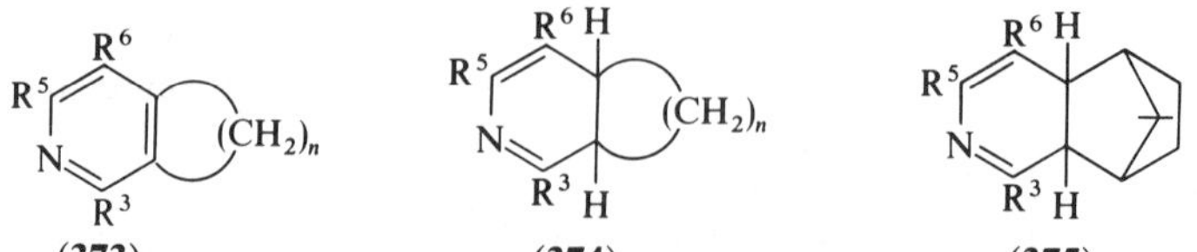

(373)    (374)    (375)

Other 1,2,4-triazines react with norbornene to give either the condensed pyridines (375) and/or the bis adducts (372). The ratio of the two products depends on the reaction conditions and the relative proportions of reactants ⟨69TL5171⟩.

Reaction of (325) with norbornadiene affords 2,3,6-trifluoropyridine (376a), the formation of which is explained by elimination of cyclopentadiene from the dihydropyridine

(**377**) (**376a**) R″ = F

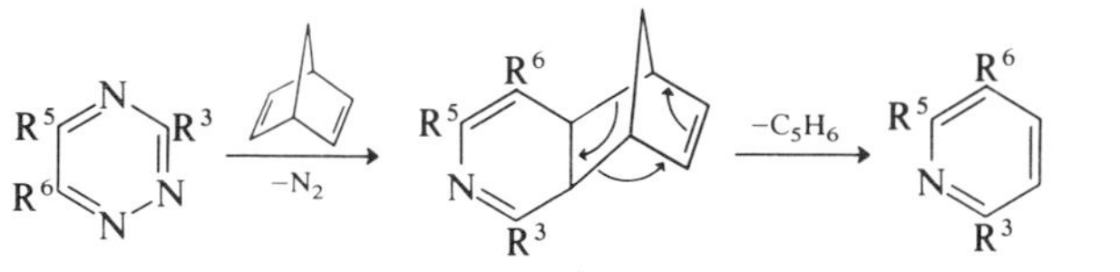

(**377**). Other 1,2,4-triazines behave similarly, to give pyridines (**376**) with norbornadiene ⟨79CC658⟩.

Cyclopropenes react very easily with 1,2,4-triazines. The azanorcaradienes (**378**), or their valence tautomers, the 4*H*-azepines (**379**), were isolated *via* the tricyclic intermediates (**380**) and loss of nitrogen. The equilibrium between (**378**) and (**379**) depends on the substituents. The 4*H*-azepines rearrange to the 3*H*-compounds (**381**) ⟨80TL595, 72JA2770⟩. The parent 1,2,4-triazine (**1**) also reacts with cyclopropene, but the product seems to be a trimer of the azanorcaradiene, of uncertain structure ⟨70UP21901⟩.

(**380**) (**378**) (**381**) (**379**)

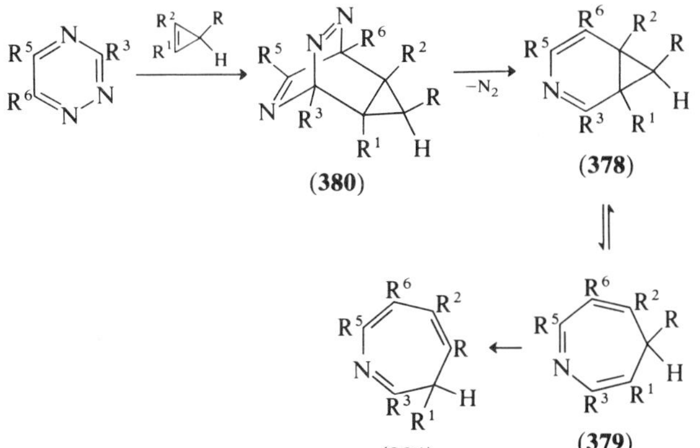

The cyclohepta[*e*][1,2,4]triazinone (**382**) reacted with cyclopropene to give compound (**383**), which showed no tendency to tautomerize to the azaheptalene derivatives (**383a**) or (**383b**) ⟨81UP21900⟩. Reaction of (**382**) with 1-methylcyclopropene affords the condensed norcaradiene (**384**), which tautomerizes very slowly, particularly in the presence of silica gel, to the methylene derivative (**385**). The azanorcaradiene (**384**) reacts with a second molecule of 1-methylcyclopropene to form the polycyclic compound (**386**).

(**382**) (**383**) R = H
(**384**) R = Me
(**383b**) (**383a**)
(**385**) (**386**)

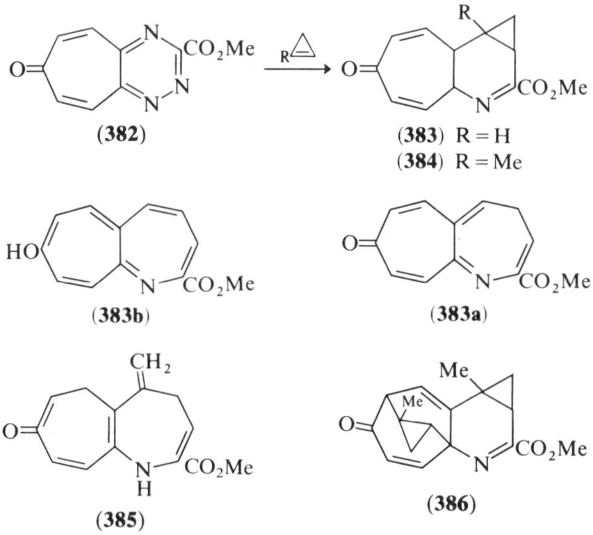

Reaction of 1,2,4-triazines (**387**) with the spirocyclopropene (**388**) did not afford the expected spiroazanorcaradienes but instead gave cyclopenta[*e*]pyrido[2,1-*f*]-[1,2,4]triazines, (**389**) as shown by X-ray crystallography ⟨75LA1445⟩. This result is explained by attack of N-1 at the C=C double bond of the cyclopropene to afford the dipole (**389**) which cyclizes. Elimination of hydrogen chloride then leads to the isolated products (**390**).

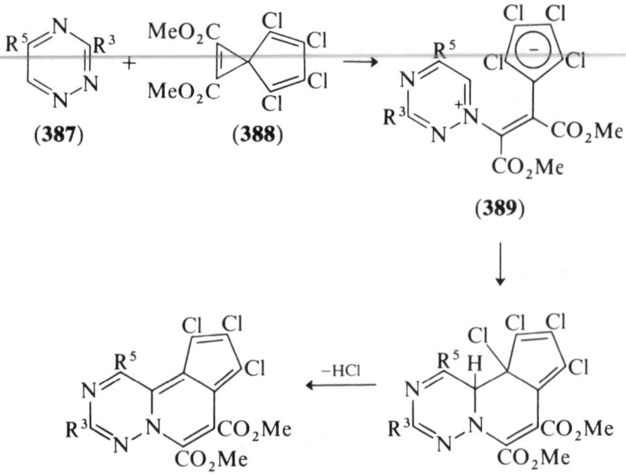

(387)  (388)  (389)

(390)  −HCl

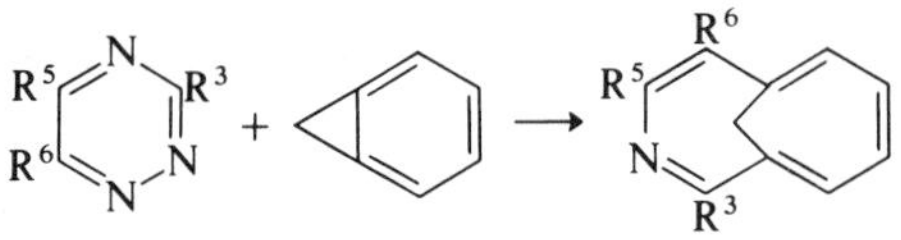

(391)

Recently it has been shown that benzocyclopropene also reacts with 1,2,4-triazines: the products are the azamethano[10]annulenes (**391**) ⟨80TL7⟩. As Dauben showed, running this reaction at high pressure (15–50 kbar) not only gives better results, but also forces 1,2,4-triazines without an activating substituent into reaction.

1,2,4-Triazines react with tricyclo[4.2.2.0$^{2,5}$]deca-3,7,9-triene-7,8-dicarboxylate esters (**392**) in refluxing toluene to give the azocines (**393**) *via* the intermediates shown (**394**, **395**) ⟨70TL1837, 72AJC865⟩. For the synthesis of azocines (**393**), aryl groups in the 5- and 6-positions ($R^5$, $R^6$) of the triazines are necessary. Starting with 1,2,4-triazines with alkyl groups or a proton in the 5- and/or 6-positions the polycyclic compounds (**396**) were isolated, probably from reaction of (**392**) with the azabicyclo[4.2.0]octatrienes (**395**), valence tautomers of the azocines. Heating (**396**) to 150 °C under reduced pressure leads to elimination of phthalate and isolation of derivatives (**397**) of the (CH)$_{11}$N system ⟨77UP21901⟩. With the tricyclodiene esters (**398**), which are reduced analogues of (**392**), the condensed azocines (**399**) and the polycyclic compounds (**400**) result, presumably by the route indicated. Here dihydrophthalate is not eliminated readily, unlike in the above examples ⟨70TL1837, 72AJC865⟩.

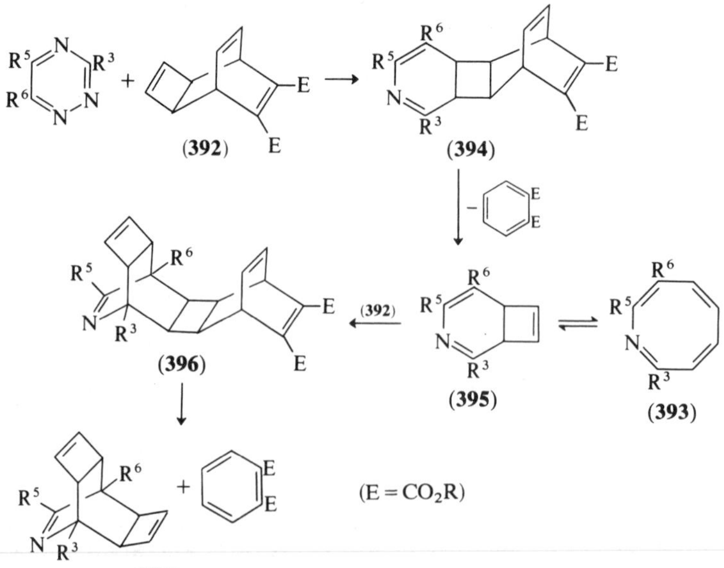

(392)  (394)

(396)  (392)  (395)  (393)

(397)  (E = CO$_2$R)

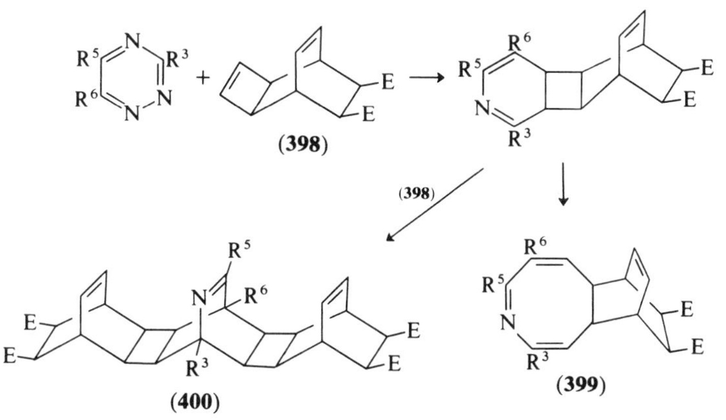

(398) (400) (399)

1,2,4-Triazines react also with electron-rich dienophiles such as ethyl vinyl ether (**401**; R = Et) or vinyl acetate (**401**; R = Ac) in boiling dioxane to yield the pyridine derivatives (**376**). After the usual [4 + 2] cycloaddition and nitrogen elimination from the bicyclic compound (**402**), the dihydropyridines (**403**) eliminate ethanol or acetic acid to give the aromatic pyridines (**376**). The dienophiles (**401**) can therefore be used as alkyne equivalents ⟨69TL5171⟩.

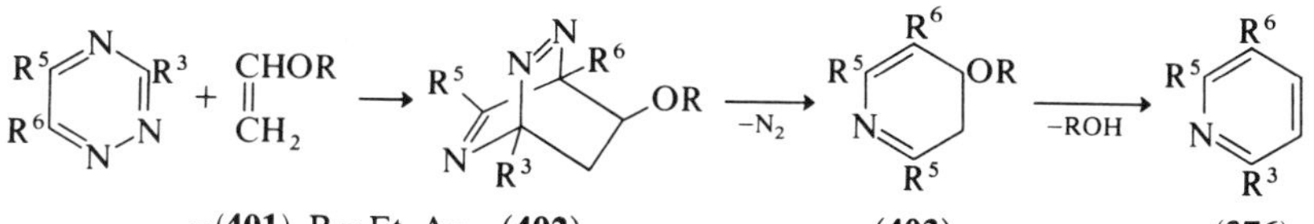

(**401**) R = Et, Ac (**402**) (**403**) (**376**)

Enamines are even more reactive dienophiles. 1-Dialkylaminocyclopentenes (**404a**) for instance, can be used for the synthesis of cyclopenta[*c*]pyridines (**405a**) ⟨69TL5171, 81JOC2179⟩. The corresponding cyclohexenes (**404b**) and cycloheptenes (**404c**) afford the cyclohexa[*c*]pyridines (**405b**) and cyclohepta[*c*]pyridines (**405c**). Enamines of aldehydes and acyclic ketones can also be used as dienophiles.

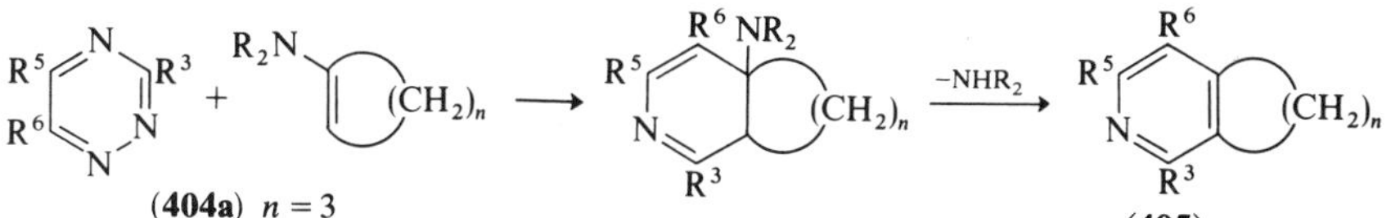

(**404a**) $n = 3$
(**404b**) $n = 4$
(**404c**) $n = 5$
(**405**)

At this point the relative orientations of the dienophile and the diene must be considered. Since both the 1,2,4-triazines and the enamines are unsymmetrical, two orientations in the transition state have to be discussed, A and B of Scheme 13.

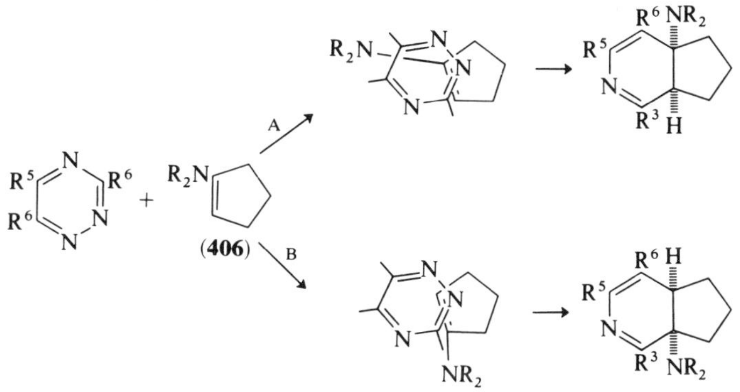

**Scheme 13**

Detailed studies of Sauer and Neunhoeffer have shown that secondary orbital interactions dominate the orientation if these interactions are possible. As a result of these, in the transition state the amino group of the enamines is always on the same side as substituents with π-electrons on the triazine. Only in this orientation can secondary orbital interactions occur between the lone pair of the amino group and the π-electrons of the substituent.

We illustrate this by two examples. Reaction of the 6-phenyl-1,2,4-triazine (**407**) with the enamine (**406**) follows orientation A since secondary orbital interactions between the *n*-electrons of the amino group and the π-electrons of the phenyl ring are possible. The dihydropyridine (**408**) can eliminate the amine to form the cyclopenta[*c*]pyridine (**409**) since the *cis* orientated proton at C-3 can shift to the nitrogen to form the 1,4-dihydropyridine (**410**) from which the amine is eliminated.

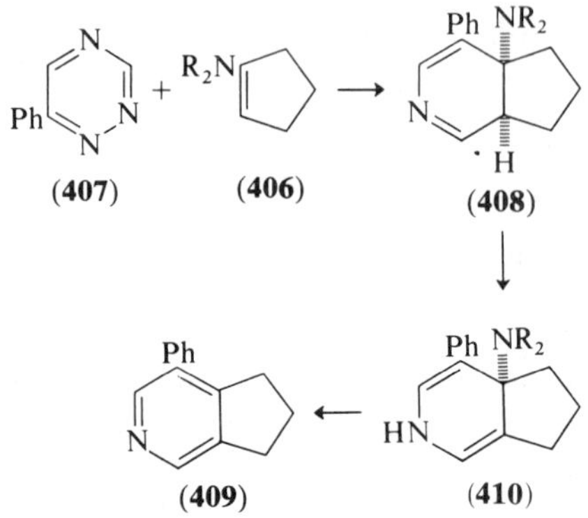

When 3-phenyl-1,2,4-triazine (**411**) reacts with the enamine (**406**), orientation B is followed. In this case the 3,4-dihydropyridine (**412**) is formed and is isolated. Elimination of the amine directly is unfavourable, because of the *cis* orientation, and transfer of the proton to the pyridine nitrogen is impossible. Oxidation to the *N*-oxide (**413**) and Cope elimination affords the condensed pyridine (**414**) ⟨78UP21900⟩.

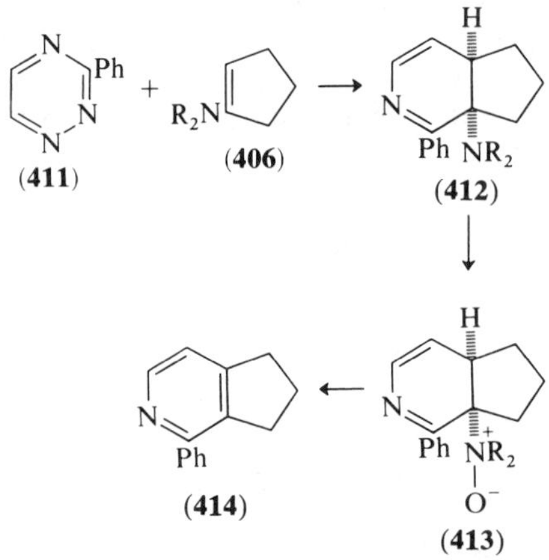

Orientation B applies only in those cases where the 1,2,4-triazines have a substituent with π-electrons in the 3-position.

The dominant influence of secondary orbital interactions is observed in the reaction of the cyclopenta[*e*][1,2,4]triazine ester (**415**) with the two enamines (**416**) and (**417**). In the first case the fused pyridine (**418**) was obtained directly; in the latter the dihydropyridine (**419**) was isolated, and aromatized, by *N*-oxidation and Cope elimination, to (**420**) ⟨82CB2807⟩.

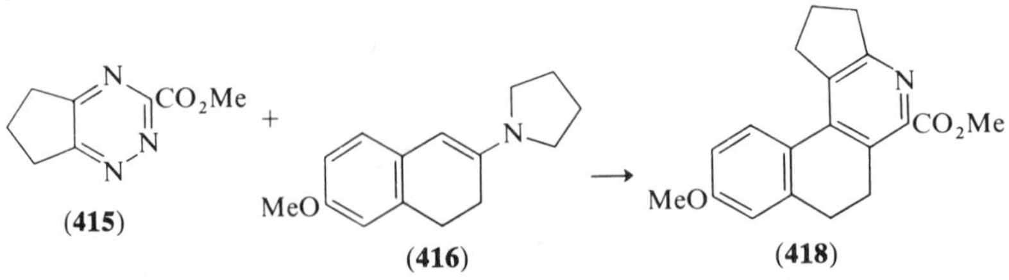

The reaction of 1,2,4-triazines with ketene derivatives such as the *O,O*-acetals (**421a**), *O,N*-acetals (**421b**), *N,N*-aminals (**421c**) and *N,S*-aminals (**421d**) has been studied extensively. In all cases pyridines (**422**) were isolated (Scheme 14). These reactions follow the usual cycloaddition route. Again it is found that the orientation is dominated by secondary

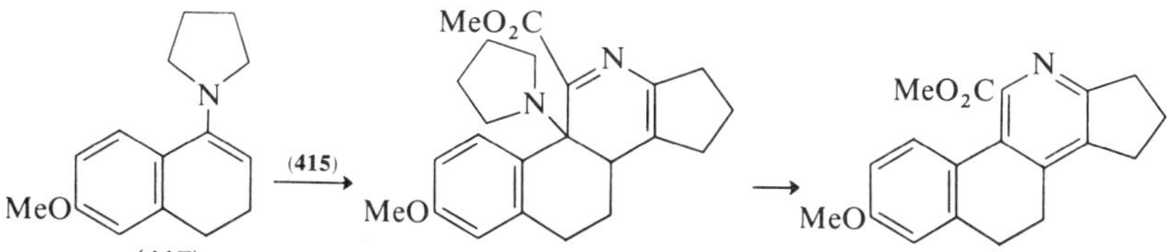

**(417)** → **(419)** → **(420)**

orbital interactions, leading either to the '*p*-substituted' (**422a**) or the '*m*-substituted' pyridines (**422b**). As Sauer observed, the ketene-*N,N*-aminals (**421c**) give mainly the orientation opposite to the other ketene derivatives ⟨72LA(758)120, 75TL2897⟩. The effect of solvent on these reactions has been studied ⟨75TL2897, 74UP21900⟩.

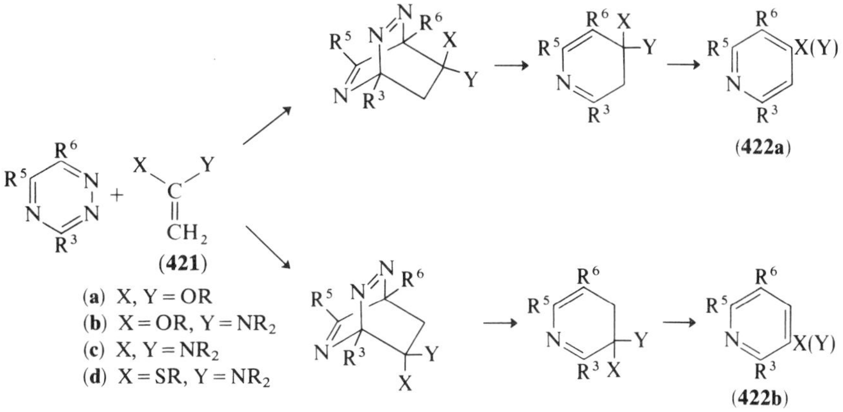

(**421**)

(a) X, Y = OR
(b) X = OR, Y = NR$_2$
(c) X, Y = NR$_2$
(d) X = SR, Y = NR$_2$

**Scheme 14**

Another well-studied cycloaddition of 1,2,4-triazines is the reaction with ynamines (**423**). In this the dienophile often attacks the 1,2,4-triazines across the 2- and 5- rather than the 3- and 6-positions. This can perhaps be due to the transition state of the cycloaddition with ynamines being more polar than that in the cycloaddition with alkenes, and a partial negative charge in the 1,2,4-triazine ring is better stabilized at a nitrogen (N-2) than at a carbon (C-6). The products isolated from these reactions are pyrimidines (**424**). It was shown by using $^{15}$N-labelled 3-methyl-1,2,4-triazine that the reaction is in fact a [4 + 2] cycloaddition to N-2 and C-5 and not a [2 + 2] cycloaddition to the N(4)—C(5) bond ⟨72LA(758)125⟩.

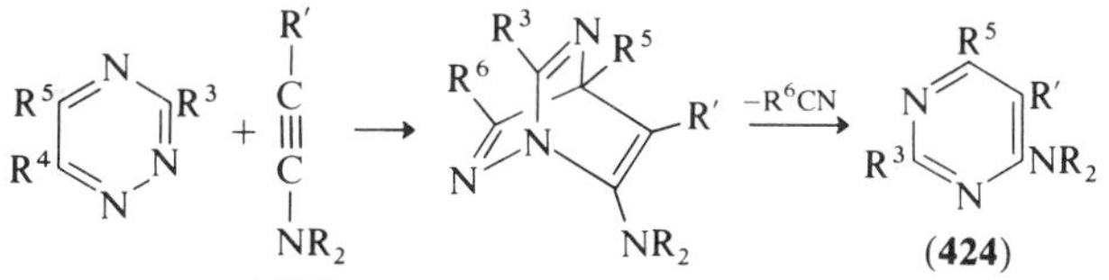

(**423**) → (**424**)

The attack of the ynamine at the triazine 5-position is observed particularly in those cases where that position is unsubstituted. If the 5-position is substituted attack of the dienophile at the 3-position predominates ⟨70TL3355, 70TL3357, 72LA(758)125, 75TL2901⟩.

Neunhoeffer and Lehmann have shown that it is possible to reverse the diene character of the 1,2,4-triazine ring by introducing alkoxy or dialkylamino groups into the ring. Alkoxy-, dialkoxy- and dialkylamino-1,2,4-triazines are therefore less reactive toward ynamines but they still react with these dienophiles. Bis(dialkylamino)-, trialkoxy- and tris(dialkylamino)-1,2,4-triazines (**425**) behave as electron-rich dienes and give cycloaddition reactions with acetylenedicarboxylate (**426**) but not with ynamines. Compounds (**425**) and (**426**) afford the 2,4-bis(dialkylamino)pyrimidine-5,6-dicarboxylates (**427**) ⟨77LA1413⟩.

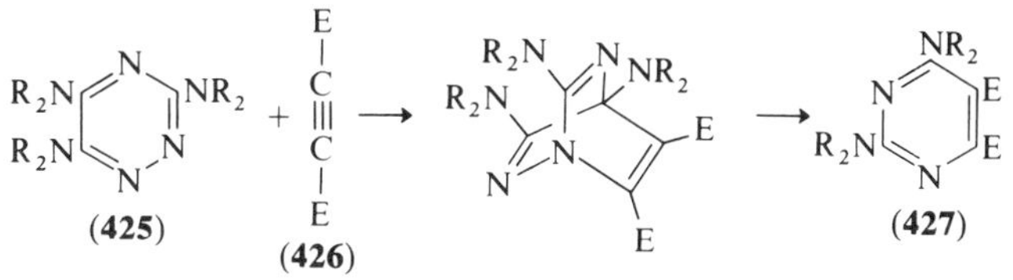

(**425**) + (**426**) → (**427**)

1,2,4-Triazines with aryl substituents gave complex reaction mixtures with acetylene-dicarboxylate. From the reaction of 5,6-diphenyl-3-*p*-tolyl-1,2,4-triazine (**358**) with (**426**) the following two products were isolated: 4,4a-diphenyl-2-*p*-tolylpyrido[2,1-*f*]-[1,2,4]triazine-5,6,7,8-tetracarboxylate (**428**) and 1,2-bis(alkoxycarbonyl)-2-(5,6-diphenyl-2-*p*-tolyl-1,2,4-triazin-1-io-1-yl)vinylate (**429**) ⟨77LA1421⟩.

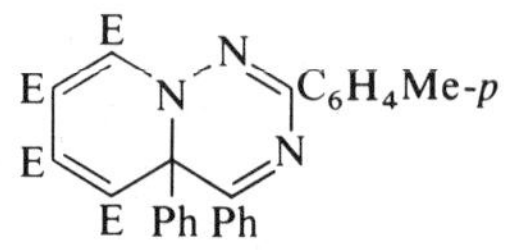

(**428**) 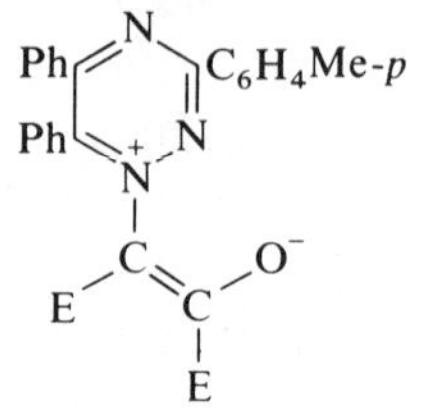                                              (**429**)

1,2,4-Triazine 4-oxides (**430**) react with 1-ethoxy-*N*,*N*-dimethylvinylamines (**431**) (ketene-*O*,*N*-acetals) in two different ways. 1,2,4-Triazine 4-oxides with an unsubstituted 5-position and without a methyl group in the 6-position (**430**) react with (**431**) by 1,3-dipolar cycloaddition to give the isoxazolo-1,2,4-triazines (**432**) which could not be isolated. 5-Methyl-1,2,4-triazines (**433**) were obtained in those cases where R′ = H while 1,2,4-triazin-5-yl-acetamides (**434**) were isolated in the other cases. 1,2,4-Triazine 4-oxides with a methyl group in the 5- (**435**) or 6-position (**438**) react with the dienophiles (**436**) to give dimethylaminopropenyl-1,2,4-triazines (**437**, **439**) with a strong preference for the reaction at the methyl group in the 5-position. The mechanism of these reactions is probably exactly analogous to that described earlier with the ketene-*O*,*N*-acetals (**337**) and 5-methyltriazines without the *N*-oxide group. No reaction was observed between 1,2,4-triazine 1-oxides and ketene-*O*,*N*-acetals ⟨78CB240⟩.

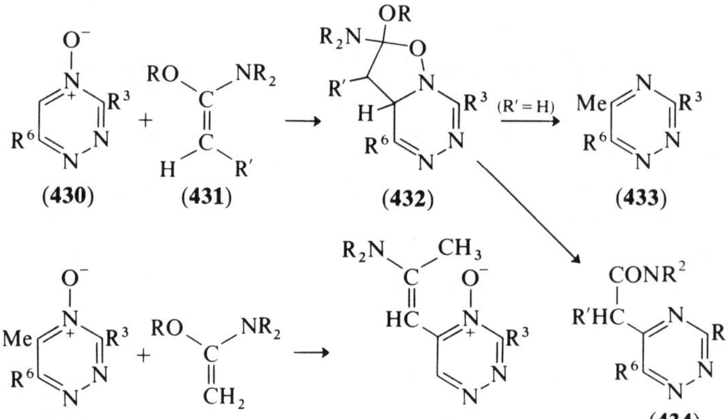

(**430**)        (**431**)              (**432**)              (**433**)

(**435**)        (**436**)              (**437**)              (**434**)

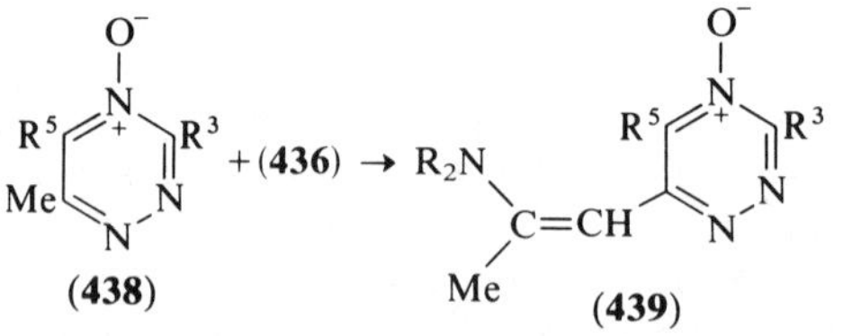

(**438**)                    (**439**)

1,2,4-Triazine 4-oxides (**440**) react with ynamines (**423**) to give the same pyrimidine derivatives (**424**) which are formed from the reaction of the appropriate 1,2,4-triazine. It was shown that the first step in these reactions is the reduction of the *N*-oxide (**440**) by the ynamine ⟨78CB240⟩.

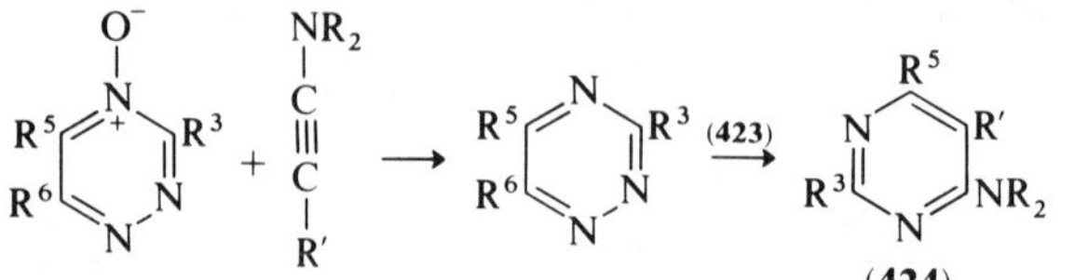

(**440**)        (**423**)                                    (**424**)

A few cases of [4 + 2] cycloaddition reactions of 1,2,4-triazines with C=N double bonds have been reported. Reaction of 1,2,4-triazines (441) with benzamidine (442) in boiling toluene led to the isolation of the 1,3,5-triazines (443). It may be supposed that here the dienophile adds to the 2- and 5-positions of the 1,2,4-triazine, as in the reaction with ynamines, then a nitrile ($R^6CN$) is eliminated and aromatization follows by loss of ammonia. In one case, the initially formed 1,3,5-triazine (443) reacted with a second molecule of benzamidine by [4 + 2] cycloaddition and elimination of $R^3CN$ and ammonia, resulting finally in triphenyl-1,3,5-triazine (444). This is a known reaction in the 1,3,5-triazine series (see Chapter 2.20) ⟨81TL1393⟩.

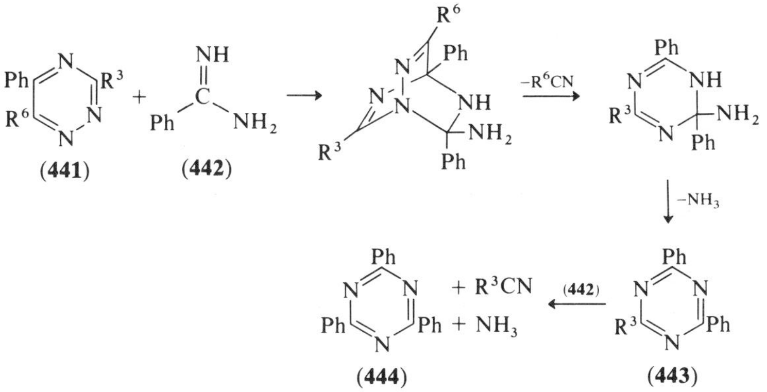

Dimethyl 3-cyano-1,2,4-triazine-5,6-dicarboxylate (445) reacts with formaldehyde dimethylhydrazone (446) *via* a [4 + 2] cycloaddition to give the dihydropyrimidine (447) in very low yield ⟨82CZ100⟩.

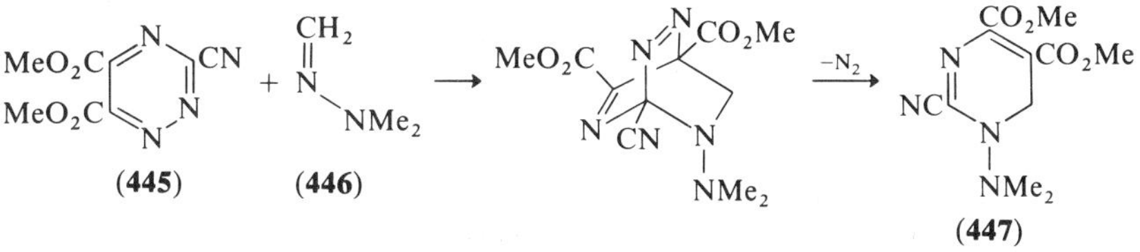

Besides [4 + 2] cycloadditions, a number of [2 + 2] cycloaddition reactions of 1,2,4-triazines with alkenes are known. All cases involve photochemical addition of the alkene to the N(1)—C(6) double bond of 1,2,4-triazine-3,5-diones (92). The first step is the formation of the azeto[2,1-*f*][1,2,4]triazinediones (448), which can be isolated in good yield by careful chromatography. Hydrolysis of the photoaddition products (448) affords the 1,6-dihydrodiones (449), which can be oxidized to 1,2,4-triazine-3,5-diones (450) if the 6-position of the starting material is unsubstituted ⟨74JHC453, 74JHC917, 72CC1144, 74JA4879⟩. 1,2,4-Triazine-3,5-diones (92) react also with dibromomaleimides to yield the cycloaddition products (451) ⟨80AG1066⟩.

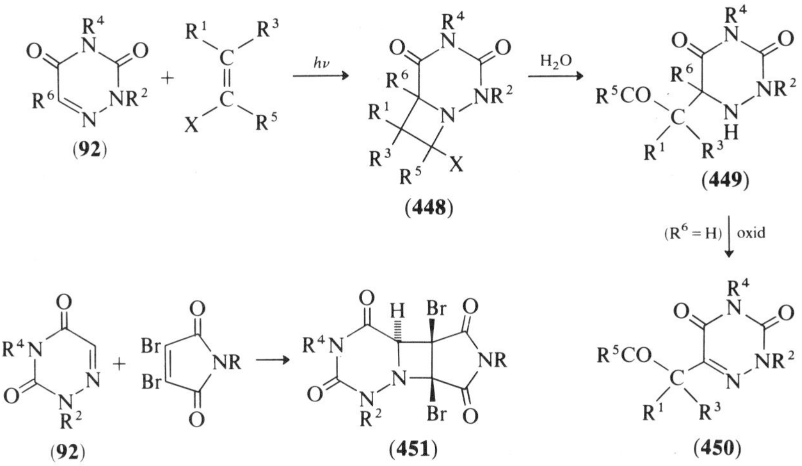

## 2.19.4 SYNTHESES

### 2.19.4.1 [4+2] Atom Combinations

The most frequently-used methods of forming six-membered rings, and the ones with the greatest applicability to the synthesis of 1,2,4-triazines, are those which proceed from [4+2] atom fragments. Examples of five of the six possible [4+2] atom fragment combinations for 1,2,4-triazines are published. These are the following: N(1)N(2)C(3)N(4)+ C(5)C(6) (A), C(3)N(4)C(5)C(6)+N(1)N(2) (B), N(4)C(5)C(6)N(1)+N(2)C(3) (C), C(5)C(6)N(1)N(2)+C(3)N(4) (D) and C(6)N(1)N(2)C(3)+N(4)C(5) (E). We have not been able to find an example of the last possible combination: N(2)C(3)N(4)C(5)+C(6)N(1) (F) (Scheme 15). Of the five known methods, the first (A) is the most often used, followed by method (B).

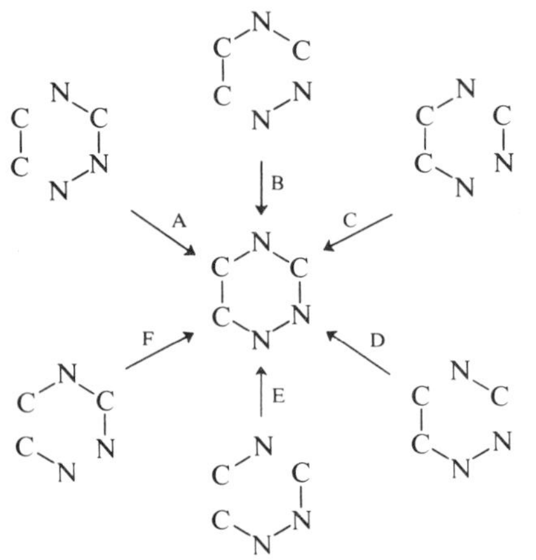

**Scheme 15**

### 2.19.4.1.1 Combination A [N(1)N(2)C(3)N(4)+C(5)C(6)]

The reaction of 1,2-dicarbonyl compounds (**452**) with amidrazones (**453**) is the best method for the synthesis of alkyl, aryl or hetaryl substituted 1,2,4-triazines ⟨78HC(33)189, p. 195⟩. No limitation of this synthetic principle is reported, except that it is, of course, preferable for the dione (**452**) to be symmetrical. The best reaction procedure is to add the dicarbonyl compound to a solution of the free amidrazone, or amidrazonium salt in the presence of one mole of base, and allow a reaction time of about 12 h. Since the first step of this reaction, *i.e.* condensation of the hydrazono group with one carbonyl group, is fast, while the second, *i.e.* condensation of the amide group with the other carbonyl, is slow, the intermediate (**454**) has been isolated in a few cases. This method has been used also for the synthesis of compounds containing more than one 1,2,4-triazine nucleus, and for the parent 1,2,4-triazine (**1**) ⟨68CB3952⟩.

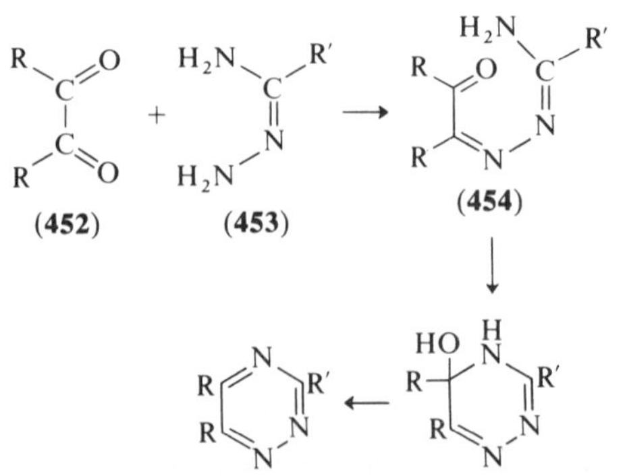

The reaction with unsymmetrical 1,2-dicarbonyl compounds often leads to a mixture of isomeric triazines. If a monosubstituted glyoxal (**455**) is used, the 5-substituted 1,2,4-triazines (**456**) predominate.

(455) (453) (456) (458) (457)

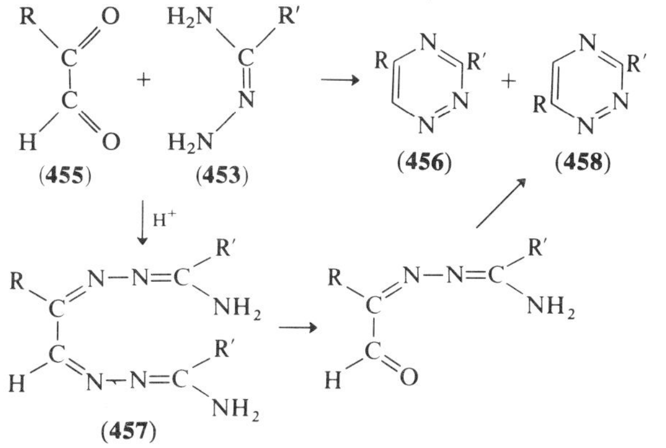

The reaction of amidrazones (**453**) with dicarbonyl compounds (**452, 455**) in acidic media may lead to the 'bis(amidrazones)' (**457**). These can be transformed into the 1,2,4-triazines by acidic hydrolysis. When a monosubstituted glyoxal (**455**) was used as the 1,2-dicarbonyl compound, under these conditions the 6-substituted 1,2,4-triazines (**458**) were formed predominantly ⟨72LA(760)88⟩.

When semicarbazide (**459a**), thiosemicarbazide (**459b**), selenosemicarbazide (**459c**) or aminoguanidine (**460a**) are used instead of the amidrazones, the corresponding 1,2,4-triazin-3-ones (**461a**) ⟨78HC(33)189, p. 230⟩, 3-thiones (**461b**) ⟨78HC(33)189, p. 335⟩, 3-selenones (**461c**) ⟨78HC(33)189, p. 357⟩ or the 3-amino-1,2,4-triazines (**462a**) ⟨78HC(33)189, p. 358⟩ are prepared. Similarly, *N*(2)- and *N*(4)-substituted semicarbazides and thiosemicarbazides can be used for the synthesis of 2- (**463**) or 4-substituted 1,2,4-triazin-3-ones (**464a**) and 3-thiones (**464b**) and use of substituted aminoguanidines (**460c**; Y = NHR) leads to 3-(substituted amino)triazines (**462c**). Reaction of (*S*)-methylthiosemicarbazide (**460b**) with (**452**) affords 3-methylthio-1,2,4-triazines (**462b**).

The primarily formed 'hydrazones' (**465, 466**) can in most cases be isolated, and cyclized in a second step, but this is not necessary, as direct cyclization is usually achieved without any problem. Intermediates of type (**467**) could be isolated, but in only a few cases.

(452) (459a) X = O (459b) X = S (459c) X = Se (465) (467) (461a) X = O (461b) X = S (461c) X = Se

(463) (464)

(**a**) X = O
(**b**) X = S
(**c**) X = Se

(455) + (460) (466) (462)

(**a**) Y = $NH_2$
(**b**) Y = SMe
(**c**) Y = NHR

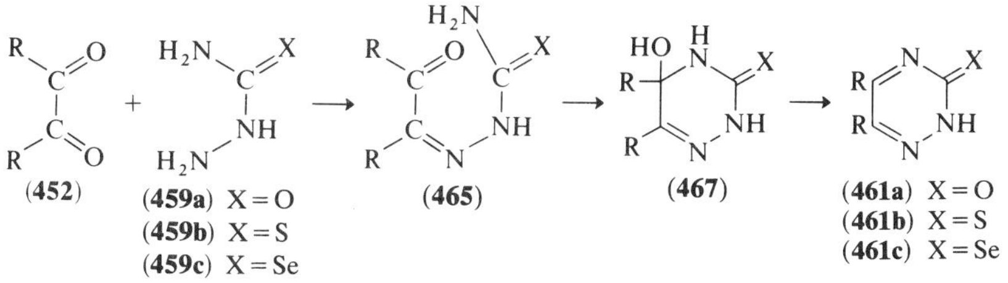

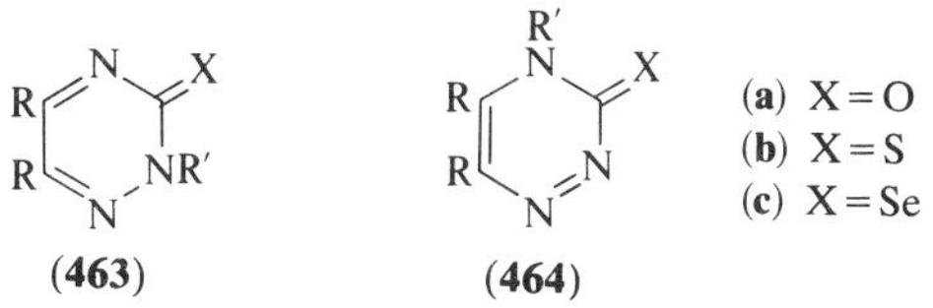

Semicarbazides (**459a**) and thiosemicarbazides (**459b**) also react with 1,2-dicarbonyl monooximes, leading to 1,2,4-triazin-3-ones and 3-thiones (**461a, b**). This method is of preparative value for the synthesis of 6-substituted 1,2,4-triazin-3-ones, which are formed only in minor amounts when a monosubstituted glyoxal reacts with semicarbazide. For instance, phenylglyoxal (**455**; R = Ph) and semicarbazide afford mainly 5-phenyl-1,2,4-triazin-3-one (**468**), while phenylglyoxal aldoxime (isonitrosoacetophenone) (**469**) reacts *via* the semicarbazone (**470**) to give the 6-phenyl compound (**471**).

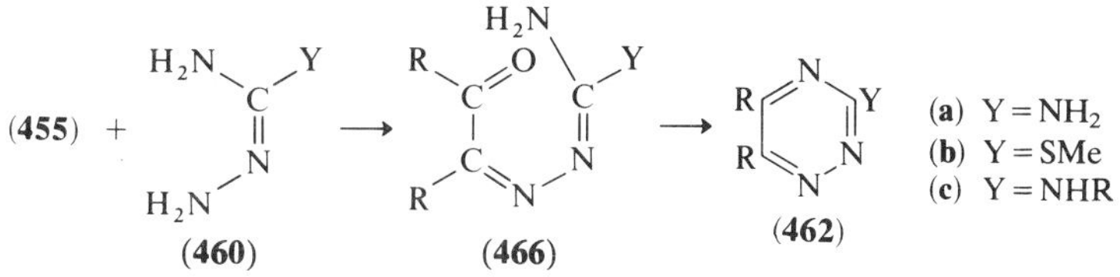

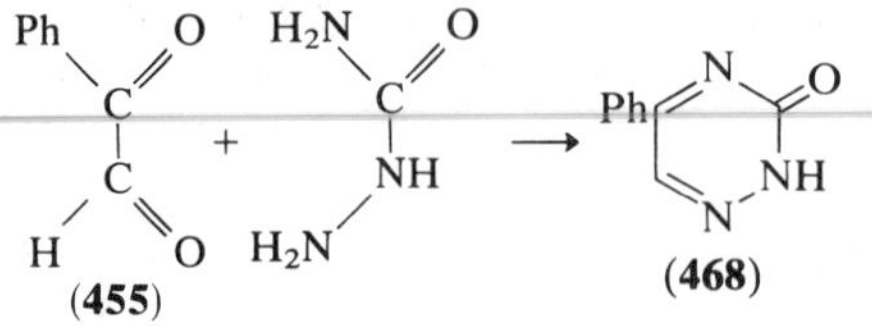

(455)   (468)

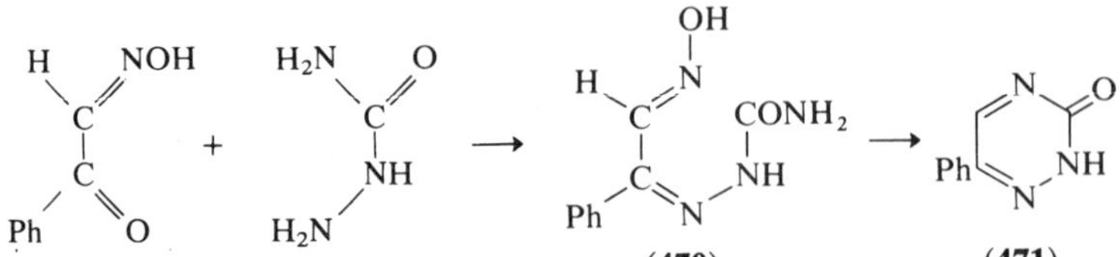

(469)   (470)   (471)

The reaction of monosubstituted glyoxals (455) with aminoguanidine (460a) has been extensively studied. Either 5- or 6-substituted 3-amino-1,2,4-triazines can be obtained depending on the reaction conditions. *N*-Alkyl-substituted aminoguanidines with 1,2-dicarbonyl compounds condense to give 3-alkylamino-1,2,4-triazines, while aminoguanidine (460a) and phenylglyoxal aldoxime (469) afford the 3-amino-6-phenyl compound. Dibromoketones (472) can be used instead of 1,2-dicarbonyl compounds, reacting with aminoguanidine to give 6-substituted 3-amino-1,2,4-triazines (473). Reaction of bis(arylidene) or bis(alkylidene) acetone (474) with aminoguanidine gives the hydrazones (475); these are cyclized to give the 4,5-dihydro-1,2,4-triazines (476), from which elimination of RMe affords the 5-unsubstituted 3-aminotriazines (477) ⟨78HC(33)189, p. 360⟩. Reaction of diaminoguanidine (478) with 1,2-dicarbonyl compounds (452) was used for the synthesis of 3-hydrazino-1,2,4-triazines (479) ⟨78HC(33)189, p. 396⟩.

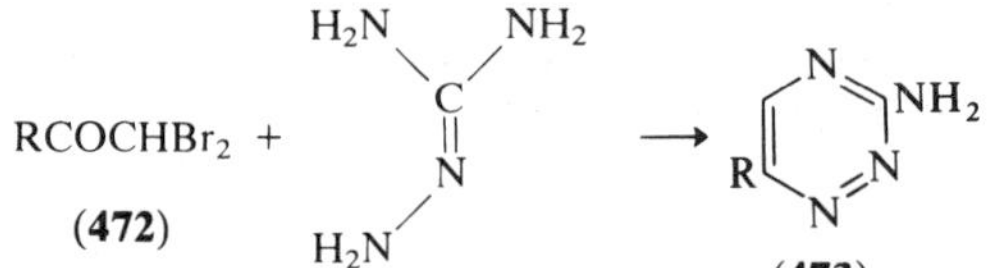

(472)   (473)

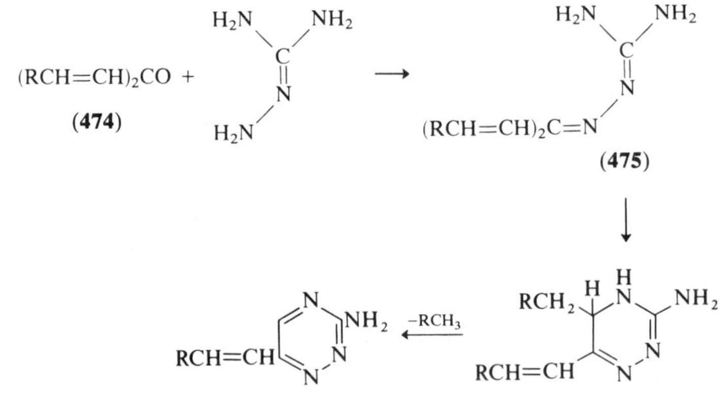

(474)   (475)

(477)   (476)

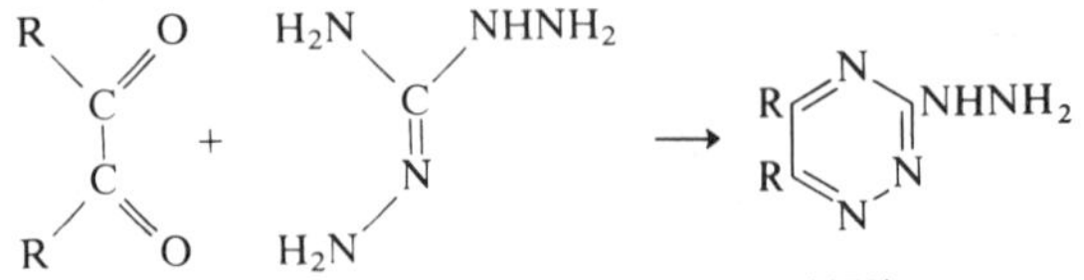

(452)   (478)   (479)

Using $\alpha$-ketocarboxylic acids or their derivatives (480) instead of 1,2-dicarbonyl compounds is the general method for the synthesis of 1,2,4-triazin-5-ones. Compounds (480) react with amidrazones (453), semicarbazide (459a), thiosemicarbazide (459b), selenosemicarbazide (459c), aminoguanidine (460a) or diaminoguanidine (478) to yield 1,2,4-triazin-5-ones (481) ⟨78HC(33)189, p. 248⟩, 1,2,4-triazine-3,5-diones (482a) ⟨78HC(33)189, p. 264⟩, 3-thioxo-1,2,4-triazin-5-ones (482b) ⟨78HC(33)189, p. 430⟩, 3-selenoxo-1,2,4-triazin-5-ones (482c) ⟨78HC(33)189, p. 470⟩, 3-amino-1,2,4-triazin-5-ones (483)

⟨78HC(33)189, p. 475⟩ and 3-hydrazino-1,2,4-triazin-5-ones (**484**) ⟨78HC(33)189, p. 494⟩, respectively. It is not necessary to isolate the initially formed hydrazones, but these can be cyclized by heating in DMF (Scheme 16).

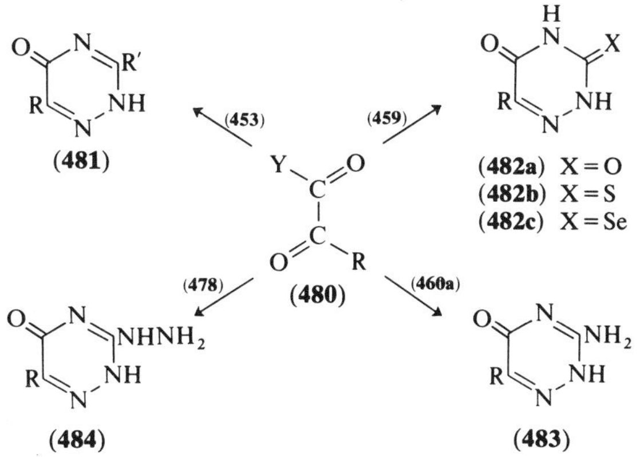

**Scheme 16**

The most fully studied reaction of this type is that of 2-keto acids (**480**; Y = OH) with thiosemicarbazide (**459b**) and its derivatives. In most cases the initial hydrazone was isolated and cyclized in a second step under basic conditions. A wide range of substituted 3-thioxo-1,2,4-triazin-5-ones have been prepared by this method: *viz.* the *N*(2)-, *N*(4)- and *S*-monosubstituted, and also *N*(2),*S*- and *N*(4),*S*-disubstituted derivatives. To prevent hydrolysis of the thioxo or methylthio group, the reaction should be carried out in a nonaqueous solvent, if vigorous reaction conditions are necessary ⟨78HC(33)189, p. 430⟩.

Reaction of acylcyanides (**485**) with amidrazones (**453**), semicarbazide (**459a**), thiosemicarbazide (**459b**) or aminoguanidine (**460a**) is used for the synthesis of 5-amino-1,2,4-triazines (**486**) ⟨78HC(33)189, p. 387⟩, the 5-amino-3-ones (**487a**) ⟨78HC(33)189, p. 416⟩, and 3-thiones (**487b**) ⟨78HC(33)189, p. 470⟩ and 3,5-diamino-1,2,4-triazines (**488**) ⟨78HC(33)189, p. 391⟩, respectively (Scheme 17). The amino group in the 5-position can be hydrolyzed.

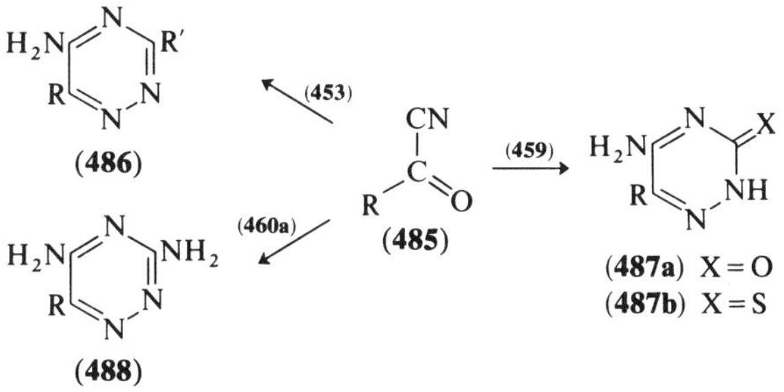

**Scheme 17**

The reaction of 1,2-dicarbonyl compounds with amidrazones, semicarbazide, thiosemicarbazide and aminoguanidine has also been used for the synthesis of condensed 1,2,4-triazines. Cyclohexane-, cycloheptane- and cyclooctane-1,2-diones (**489**) with amidrazones (**453**) afford the cycloalkatriazines (**490**) ⟨78HC(33)189, p. 661⟩.

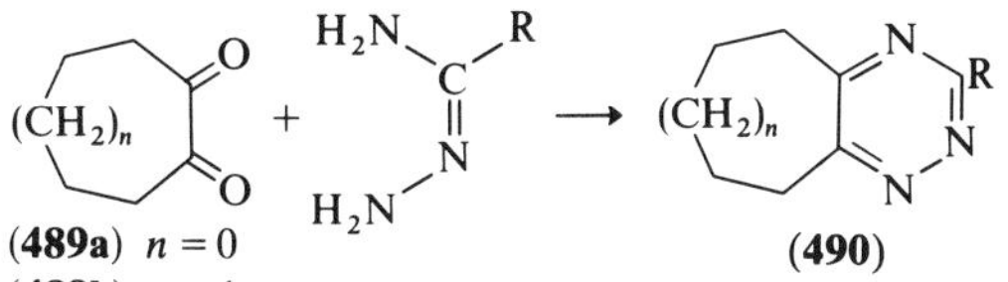

*o*-Benzoquinone (**491**) reacts with formamidrazone (**492**) to give 1,2,4-benzotriazine (**493**) in very low yield ⟨68CB3952⟩. Similarly, naphtho[2,1-*e*]- (**494**) and naphtho[1,2-*e*]-1,2,4-triazine (**495**) are both obtained in low yields from 1,2-naphthoquinone (**496**) and formamidrazone.

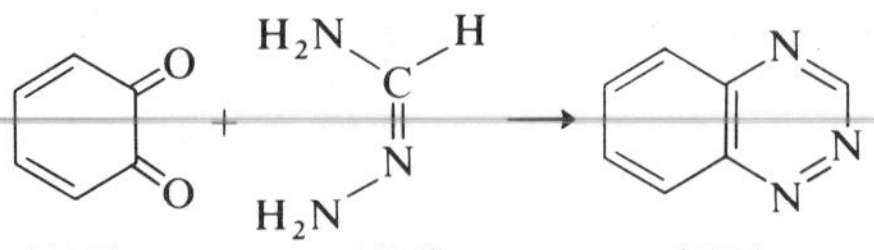

**Scheme 18**

(491)    (492)    (493)

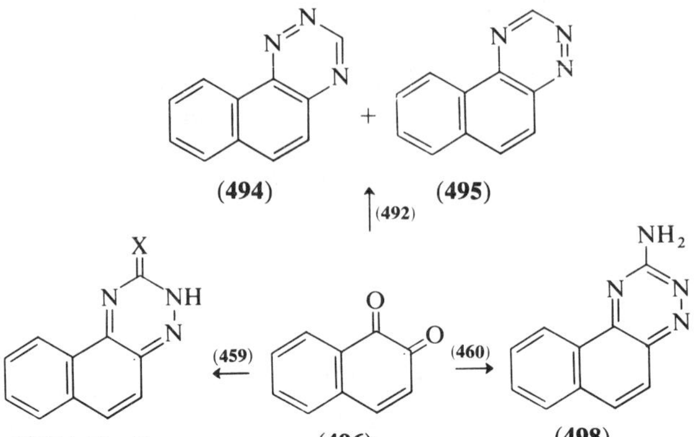

(494)    +    (495)

(497a) X = O
(497b) X = S

(496)

(498)

Better yields were obtained from 1,2-naphthoquinone (**496**) using semicarbazide (**459a**), thiosemicarbazide (**459b**) or aminoguanidine (**460a**); compounds (**497**) and (**498**) are formed, apparently regiospecifically (Scheme 18) ⟨78HC(33)189, p. 725⟩.

Reaction of 1,2-naphthoquinone 1-oxime (**499**) with 4-phenylthiosemicarbazide (**500**) yielded either 2-anilinonaphtho[1,2-*e*][1,2,4]triazine (**501**) or the corresponding 1-oxide (**502**) depending on the reaction conditions ⟨78HC(33)189, p. 727⟩.

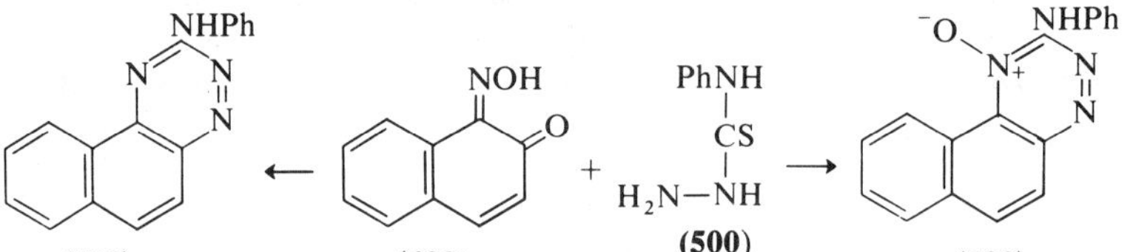

(501)    (499)    (500)    (502)

In a similar way, reactions of phenanthroquinone, chrysenoquinone, ninhydrin, acenaphthoquinone or phenalene-1,2,3-trione with amidrazones (**453**), semicarbazide (**459a**), thiosemicarbazide (**459b**) or aminoguanidine (**460a**) have been used for the synthesis of phenanthro[9,10-*e*][1,2,4]triazines (**503**), chryseno[5,6-*e*][1,2,4]triazines (**504**), indeno[1,2-*e*][1,2,4]triazines (**505**), acenaphtho[1,2-*e*][1,2,4]triazines (**506**) and phenaleno[1,2-*e*][1,2,4]triazines (**507**), respectively ⟨78HC(33)189, pp. 740–748⟩.

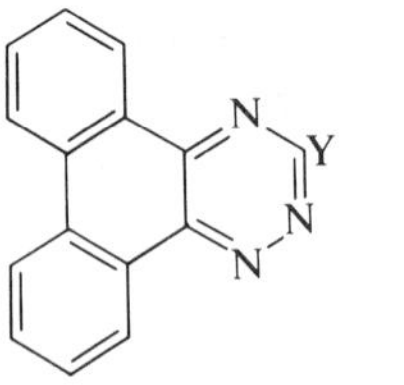

(503) Y = R, OH, SH, NH$_2$

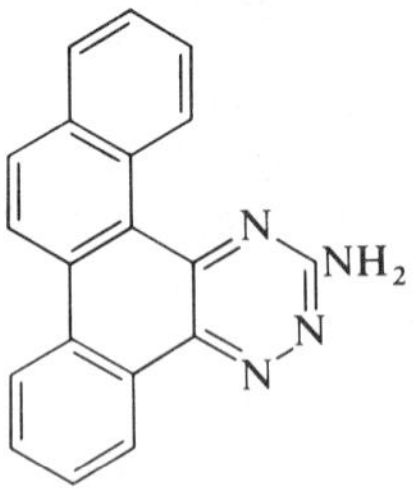

(504)

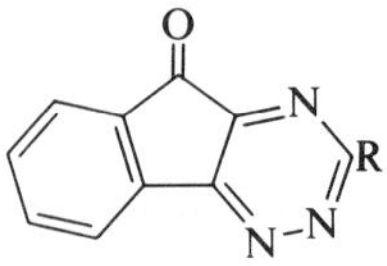

(505)

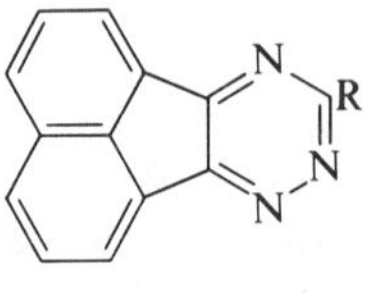

(506)

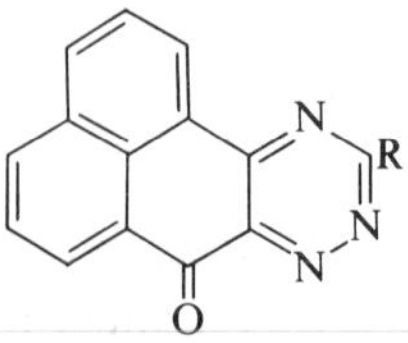

(507)

The reaction of α-hydroxy (**508a**), α-methoxy (**508b**), α-halo (**508c**), α-amino (**508d**) or α,β-unsaturated ketones (**509**) with semicarbazide (**459a**) is used for the synthesis of dihydro-1,2,4-triazin-3-ones (**510, 512**) ⟨78HC(33)189, p. 585⟩. Similarly, α-hydroxy (**508a**) or α-bromo ketones (**508c**) with thiosemicarbazide (**459b**) have been shown to give the corresponding 3-thiones (**510b**) ⟨78HC(33)189, p. 595⟩. The reaction of α-halo ketones (**508c**) or α,β-unsaturated ketones (**509**) with aminoguanidine (**460a**) gives the 3-aminodihydrotriazines (**511, 513**). When the dihydrotriazines (**513**) are treated with a base, RMe may be eliminated; *cf.* the conversion (**476**) → (**477**) mentioned earlier, to give the 5-unsubstituted 3-amino compound (**514**). Reaction of α-aminoketones (**508d**) with thiosemicarbazide (**459b**) yields a mixture of 3-amino- (**511**) and 3-thioxodihydro-1,2,4-triazines (**510b**) ⟨78HC(33)189, p. 600⟩. The use of α-amino ketones (**508d**) in these reactions is mentioned here because of its relationship to the other α-substituted ketones. It is, however, apparent that its reactions form a [3+3] atom fragment sequence rather than a [4+2], since the amino group of the α-amino ketone and not the amino group of semicarbazide, thiosemicarbazide or aminoguanidine becomes N-4 of the 1,2,4-triazine ring.

(**510a**) X = O
(**510b**) X = S
(**508a**) Y = OH
(**508b**) Y = OMe
(**508c**) Y = Hal
(**508d**) Y = $NH_2$
(**511**)

(**512**) (**509**) (**513**) (**514**)

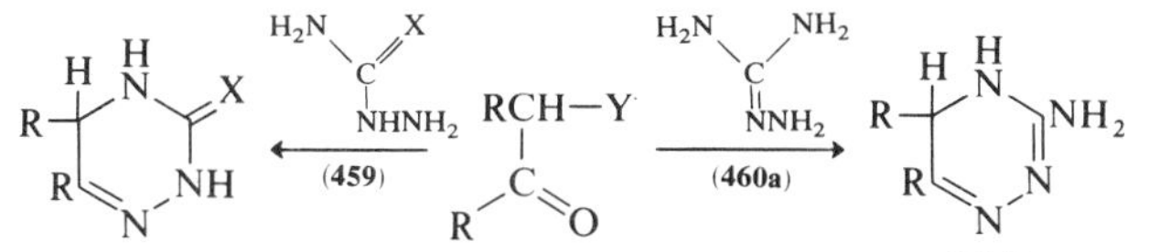

The reaction of 1,3-dicarbonyl compounds (**515**) with azodicarboxamidine (**516**) affords 1-amidino-3-amino-1,2-dihydro-1,2,4-triazines (**517**) ⟨73AP697, 73AP801⟩.

(**515**) (**516**) (**517**)

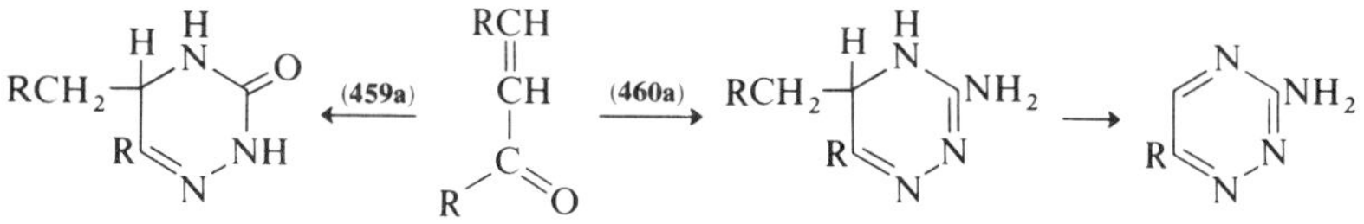

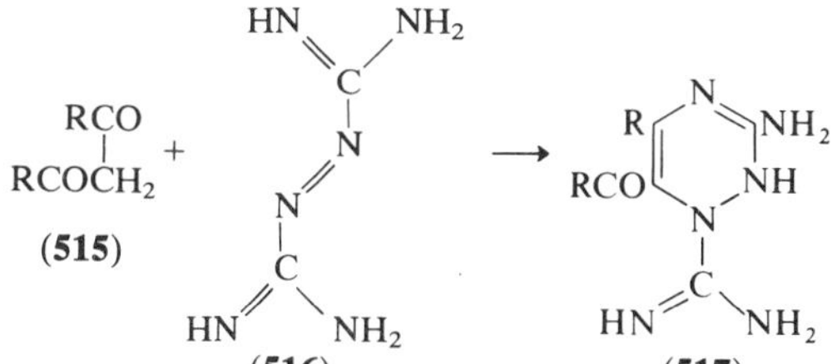

Some interesting syntheses originate with phenacyl halides (**518**). With diaminoguanidine (**478**) they react to give the 3-hydrazinodihydrotriazines (**519**) ⟨68CB29⟩. When the bromides (**520**) were treated with *S*-methylthiosemicarbazide (**460b**), 5-aryl-3-methylthio-1,2,4-triazines (**521**) were isolated, probably being formed by the route outlined in Scheme 19 ⟨78BCJ1846⟩.

(**518**) (**478**) (**519**)

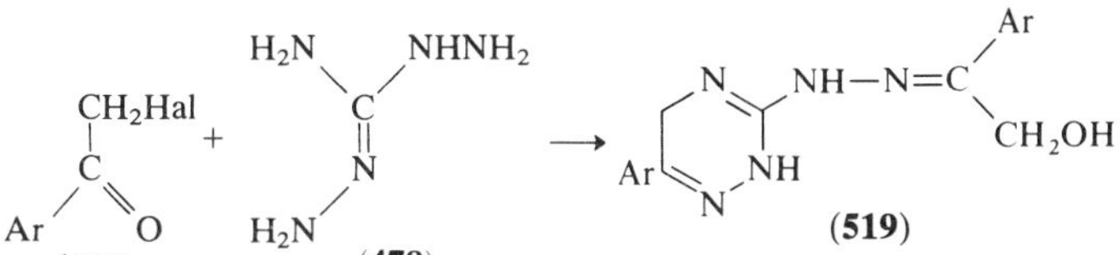

From the reaction of aminoguanidine with 2-bromoisobutyraldehyde dimethylacetal (**522**) a product was isolated which was assigned structure (**523**) or a tautomer thereof ⟨70BSF1606⟩. Semicarbazide (**459a**) reacts with the glyoxal–bisulfite addition product (**524**) to give a product formulated as structure (**525**) ⟨28JA2731⟩.

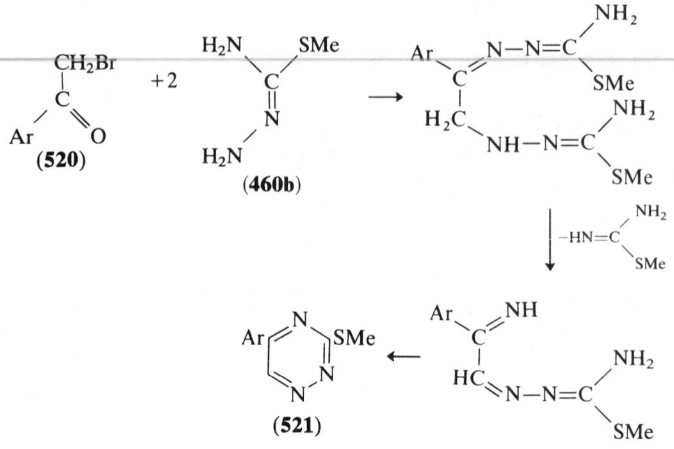

**Scheme 19**

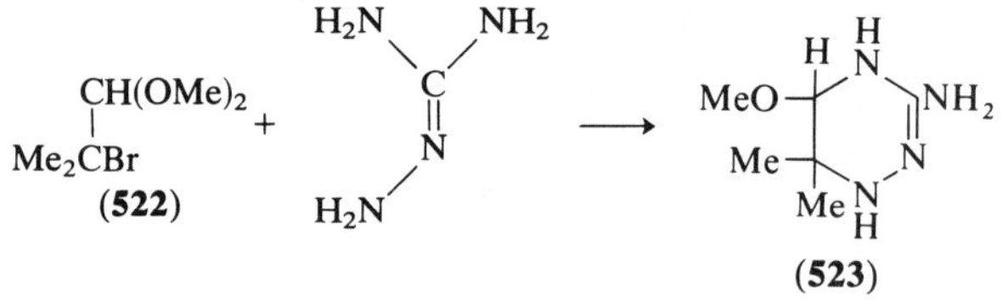

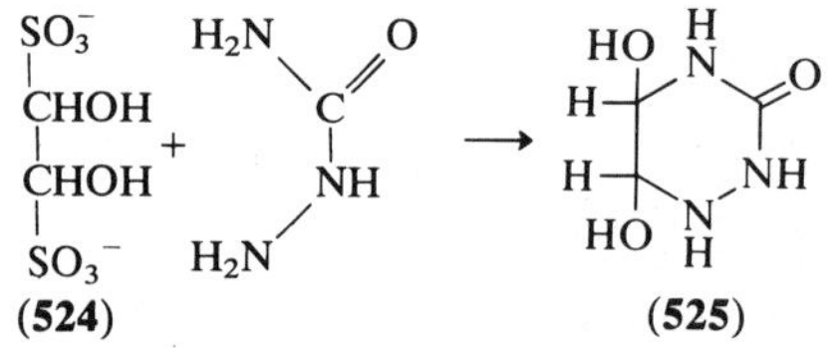

Reaction of hydrazidines (**526**) with α-ketocarboxylic acids or their derivatives is a straightforward route to 4-amino-1,2,4-triazin-5-ones (**527**) ⟨76LA2206, 77GEP2556835⟩. Similarly, carbonohydrazide (**528**) or thiocarbonohydrazide (**529**) and α-ketocarboxylic acids or their derivatives afford the 4-amino-3,5-diones (**530a**) and 4-amino-3-thioxo-1,2,4-triazin-5-ones (**530b**), respectively. Diaminoguanidine (**526**; R = NH₂) and α-ketocarboxylates yield 3,4-diamino-1,2,4-triazin-5-ones (**527**). 4,5-Diamino-1,2,4-triazin-3-ones (**533a**) and the 3-thiones (**533b**) were obtained by treatment of α-iminonitriles (**532**) with carbonohydrazide (**528**) or thiocarbonohydrazide (**529**), respectively ⟨78HC(33)189, pp. 563–573⟩.

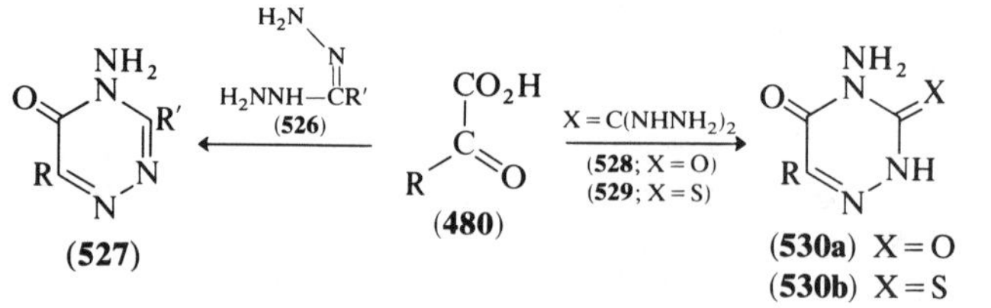

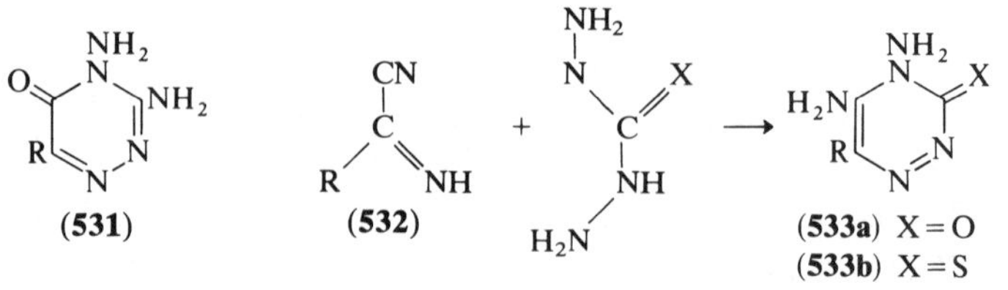

4-Amino-1,2,4-triazine-5,6-diones (**534**), 4-amino-4,5-dihydro-1,2,4-triazines (**535**), 4-amino-5-imino-1,2,4-triazines (**536**) and 4,6-diamino-1,2,4-triazin-5-ones (**537**) were prepared by reaction of hydrazidines (**526**) with oxalates, α-bromo ketones, acyl cyanides (**485**) and thioxamidates, respectively ⟨80UP21900⟩.

**(534)** **(535)** **(536)** **(537)**

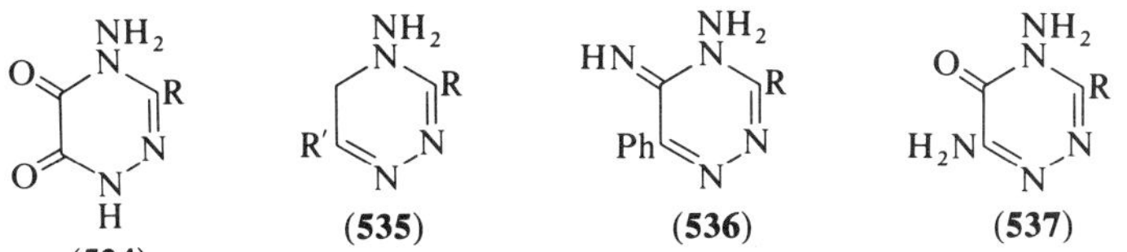

The mixed chloroacetic–acetic anhydride (**538**) with semicarbazones of aliphatic aldehydes or ketones (**539**) affords compounds (**540**) which cyclize to the 1,2,4-triazinium salts (**541**) ⟨73ZOR834⟩.

**(538)** **(539)** **(540)** **(541)**

**(542)** **(543)**

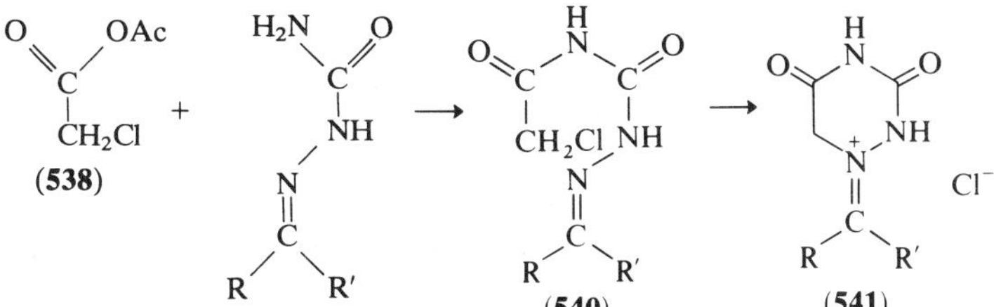

The silatriazolinethione (**542**) reacts with oxalyl chloride to give the triazine derivative (**543**) ⟨78ZC336⟩.

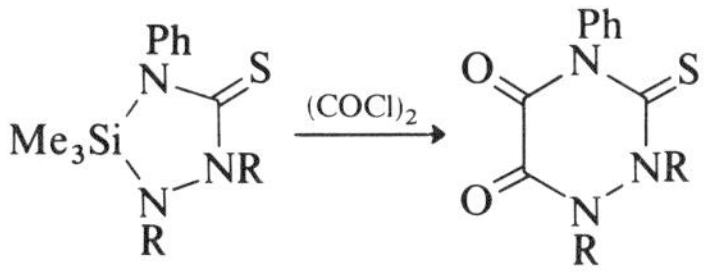

Dimethyl acetylenedicarboxylate (**544**) and the formazan (**545**) afford an addition product (**546**) and also the triazine diester (**547**) ⟨70KGS1704⟩.

**(544)** **(545)** **(546)** **(547)**

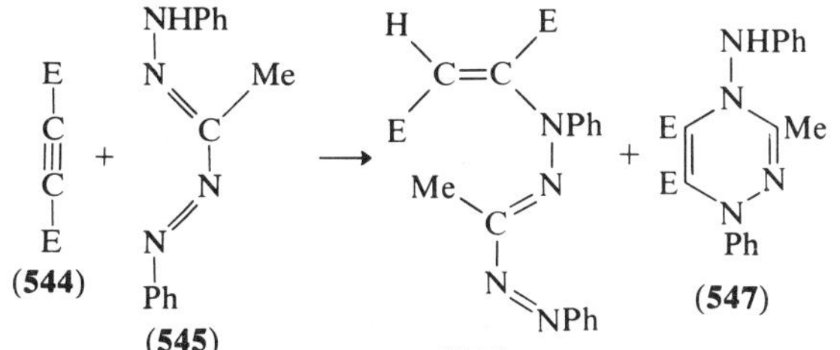

### *2.19.4.1.2 Combination B [C(3)N(4)C(5)C(6)+N(1)N(2)]*

The combination of C(3)N(4)C(5)C(6) with N(1)N(2) is another very often used method for synthesis of 1,2,4-triazines. In many cases it is a question of appropriateness whether one takes the C(3)N(4)C(5)C(6) fragment as the starting material or considers it as a combination of [3+1] atom fragments. For instance, $\alpha$-acylamino and $\alpha$-thioacylamino ketones (**548**; X=O,S) react with hydrazine to give dihydro-1,2,4-triazines (**549**) which can be oxidized to the 1,2,4-triazines ⟨78HC(33)189, p. 197⟩. Since $\alpha$-acylamino ketones are usually prepared by acylation of $\alpha$-amino ketones (**508d**) one may also call the synthesis of 1,2,4-triazines by this method a [3+2+1] atom fragment sequence. By using urethanes instead of acylamino compounds, dihydro-1,2,4-triazin-3-ones can be prepared, as illustrated with the aminocamphor derivatives (**550** → **551**) ⟨30HCA444⟩.

**(508d)** **(548)** **(549)**

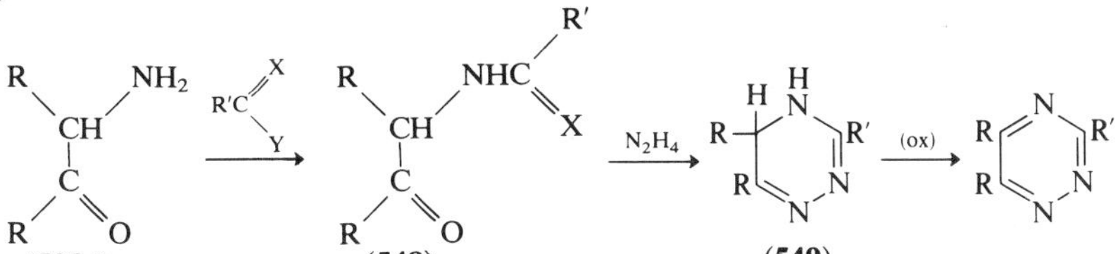

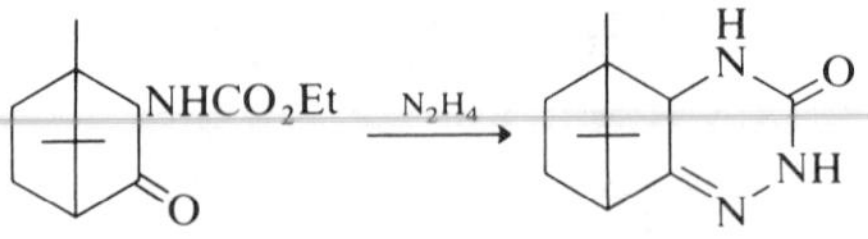

(550) (551)

Reaction of the ketonitrones (552) with hydrazine affords 4-hydroxytetrahydrotriazines (553) which can be oxidized to the 1,2,4-triazine 4-oxides (554) ⟨73KGS134⟩.

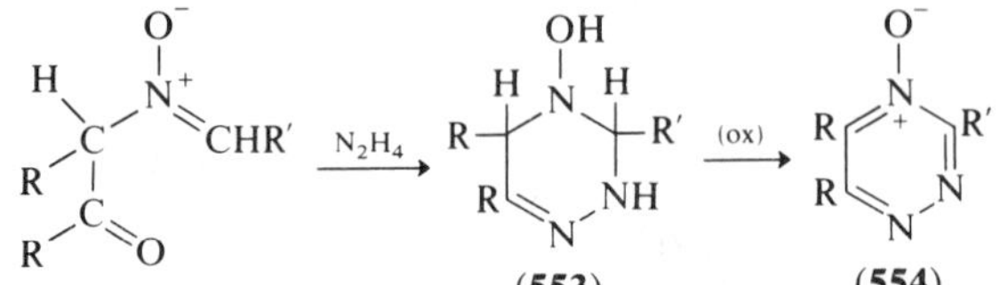

(552) (553) (554)

Tetrahydro-1,2,4-triazine-3-thiones (556) can be obtained by reaction of ketoisothiocyanates (555) with hydrazines. In cases where $R^1$ is hydrogen, water is eliminated and dihydrotriazine-3-thiones (557) are isolated ⟨78HC(33)189, p. 596, 76CB154⟩. An enol acetate has been used instead of the ketones (555), in the preparation of the 4,5-dihydro-1,2,4-triazine-3-thione (559) from (558) ⟨79JCR(S)240⟩. Reaction of α-isothiocyanato esters (560) with hydrazines led to the 3-thioxo-dihydro-1,2,4-triazin-6-ones (561) ⟨80AP77⟩.

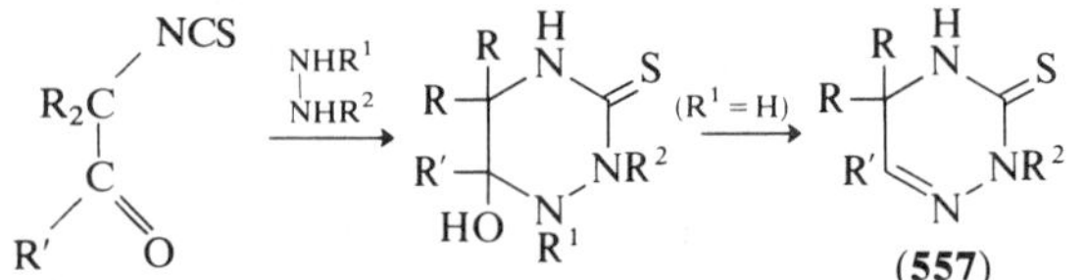

(555) (556) (557)

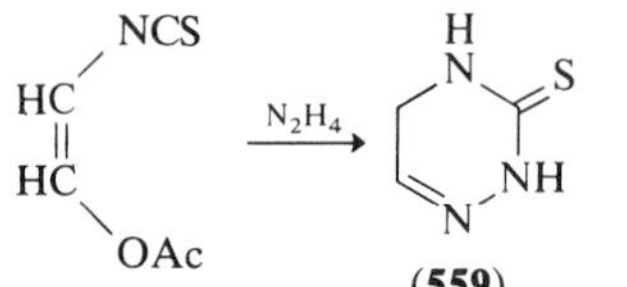

(558) (559) 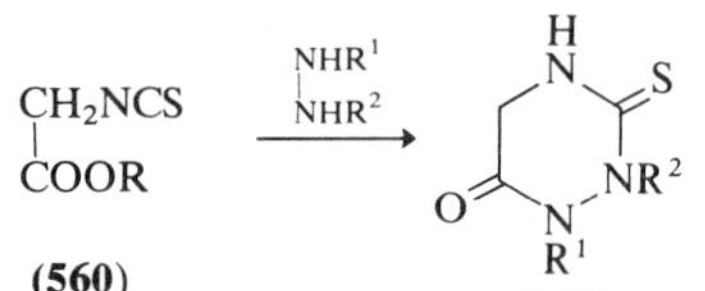 (560) (561)

Similarly, hydrazine or substituted hydrazines effected the conversion of α-imidoyl esters (562) into dihydro-1,2,4-triazin-6-ones (563) ⟨78JHC1271⟩, of α-(alkoxycarbonylamino)-carboxylates and -thiocarboxylates (564) into the 3,6-diones (565; X = O) and 6-thioxo-3-ones (565; X = S) ⟨78HC(33)189, p. 620⟩, of the α-isocyano esters (566) into 3-unsubstituted dihydro-1,2,4-triazin-6-ones (567) ⟨80JHC1621⟩, and of N-(cyanomethyl)-imidates (568) into 6-amino-1,2-dihydro-1,2,4-triazines (569) ⟨61JCS4845, 74BSF1453⟩. When the N-(2-bromoethyl)oxamidate (570) was treated with hydrazine the tetrahydro-1,2,4-triazine-3-carbohydrazide (571) was isolated ⟨63MI21901⟩. The reaction of N-chlorocarbonyl-N-(2-chloroethyl)amines (572) with hydrazines provides a synthesis of tetrahydro-1,2,4-triazin-3-ones (573); reaction with 1,1-dimethylhydrazine affords the 1,2,4-triazinium salts (574) ⟨62JOC2270, 71FES580⟩.

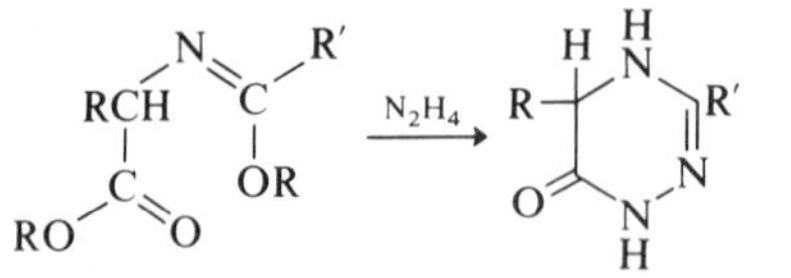

(562) (563) 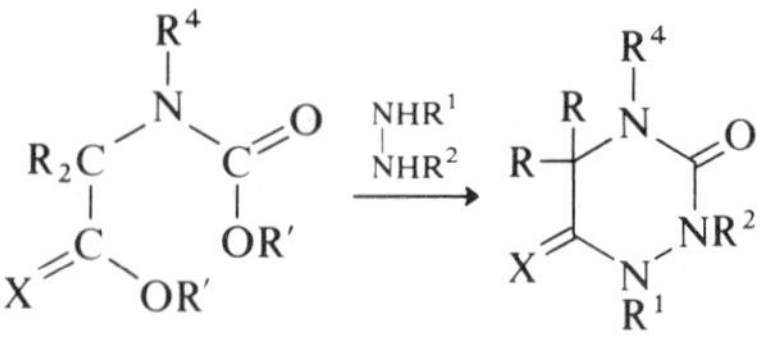 (564) (X = O or S) (565)

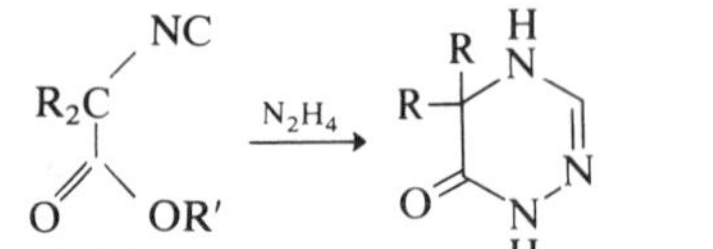

(566) (567) 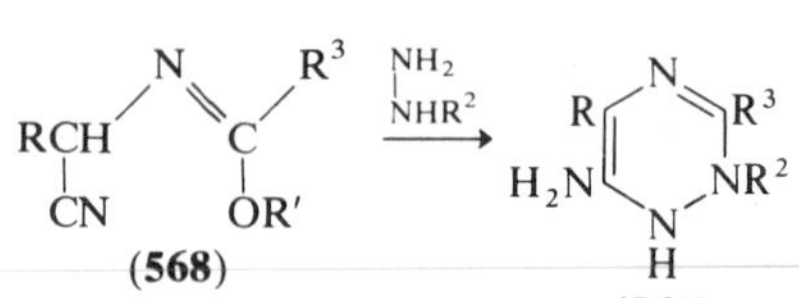 (568) (569)

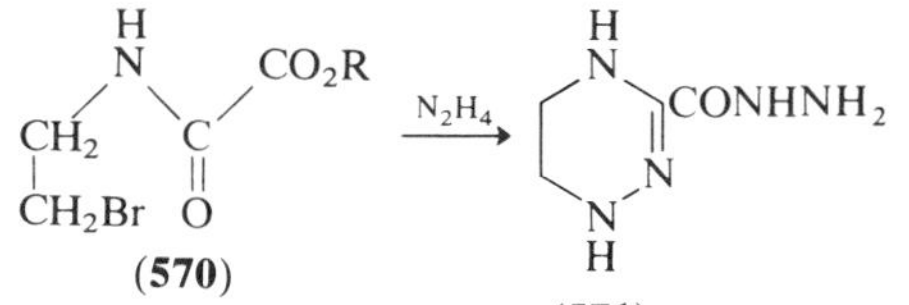

(570) (571)

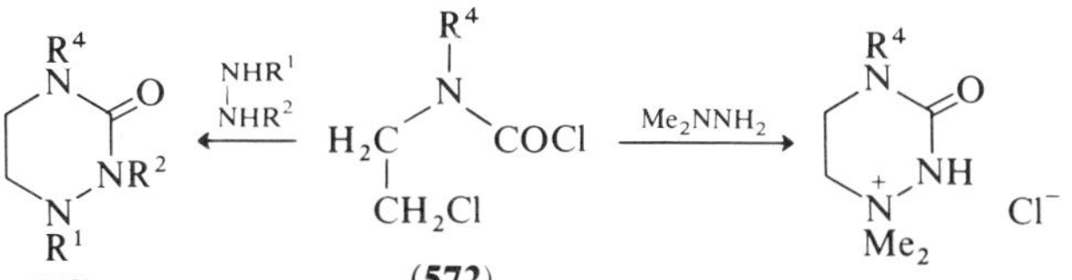

(573) (572) (574)

1,2-Dihydro-1,2,4-benzotriazines (**576**) are obtained from the reaction of *N*-(2-chloro-phenyl)imidates (**575**) with hydrazines. They can be oxidized to 1-alkyl-1,2,4-benzotriazinium salts (**577**) ⟨67MI21900⟩.

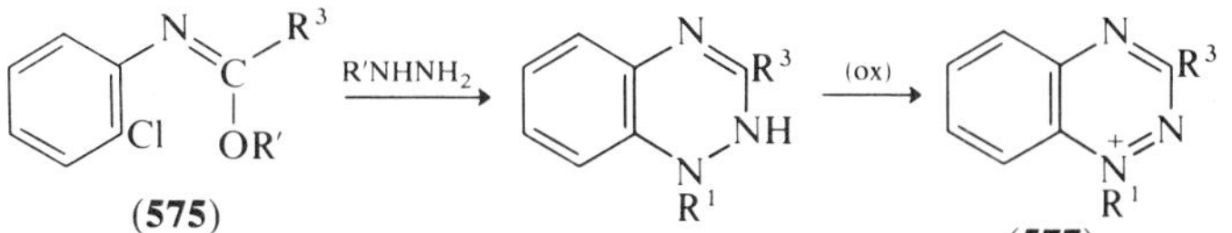

(575) (576) (577)

Another principle for the introduction of N(1)N(2) is the reaction of diazonium salts either with activated methylene groups or with aromatic compounds. *N*-(Cyanoacetyl)urethanes (**578**) react with aromatic diazonium salts to give the arylhydrazones (**579**) which can be cyclized to 2-aryl-6-cyano-1,2,4-triazine-3,5-diones (**580**) ⟨78HC(33)189, p. 537⟩. Similarly, the azo compounds (**582**), formed by coupling of the urethanes (**581**) with aryldiazonium salts, can be cyclized to 1,2,4-benzotriazine-3,6-diones (**583**) ⟨77CCC3449⟩. Coupling of 1,3-bis(cyanamido)benzene with aryldiazonium salts provides a synthesis of 2-aryl-6-cyanamido-1,2,4-benzotriazin-3-imines (**584**) and 2-cyanamidonaphthalene similarly gives 2-arylnaphtho[2,1-*e*][1,2,4]triazin-3-imines (**585**) ⟨08MI21900⟩.

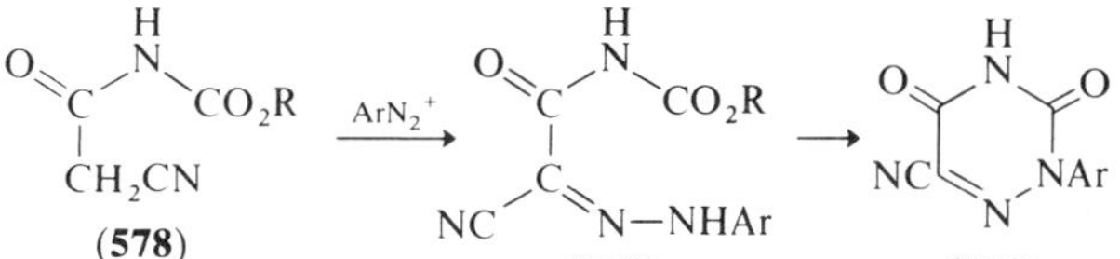

(578) (579) (580)

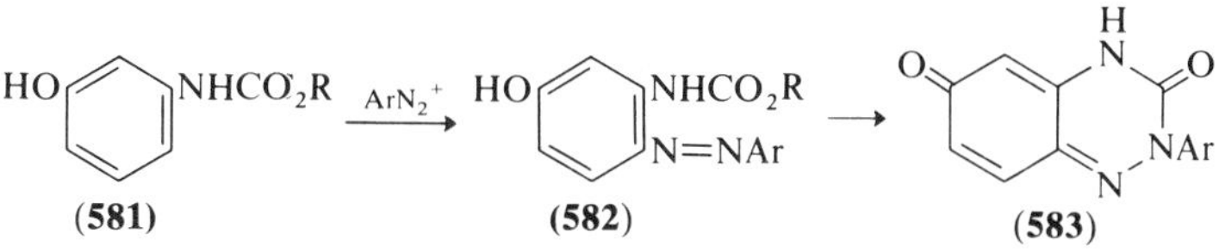

(581) (582) (583)

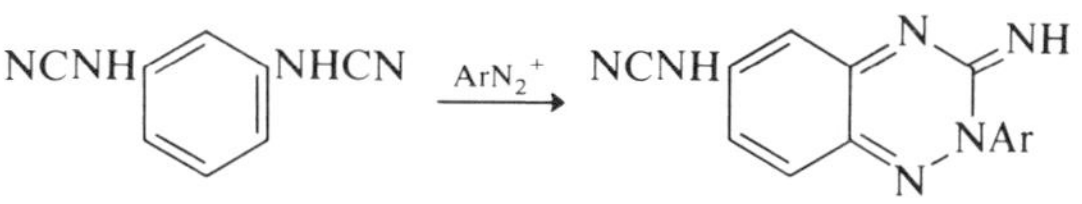

(584)

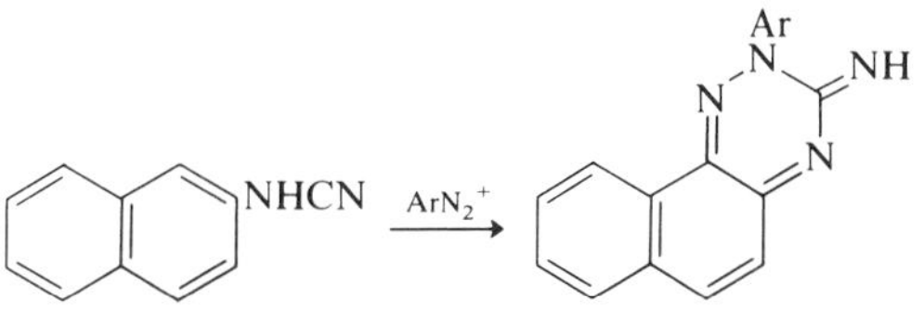

(585)

Frequently, five-membered heterocyclic systems have provided the C(3)N(4)C(5)C(6) fragment in 1,2,4-triazine syntheses. For instance, the oxazol-5-one (**586**) reacts with hydrazines to give, *via* open-chain intermediates (**587**), the 1,2,4-triazine-3,6-diones (**588**)

**(586)** **(587)** **(588)**

**(589)** **(590)**

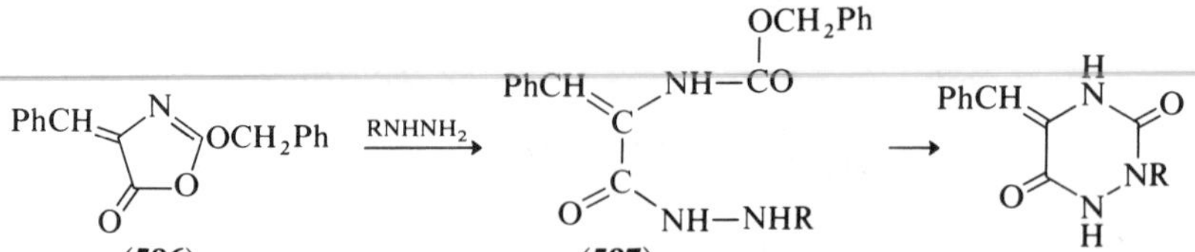

⟨80JCS(P1)858⟩ and the rhodanines (**589**) with hydrazine afford 3-thioxo-1,2,4-triazin-6-ones (**590**) ⟨50JCS1892, 58JCS4588⟩.

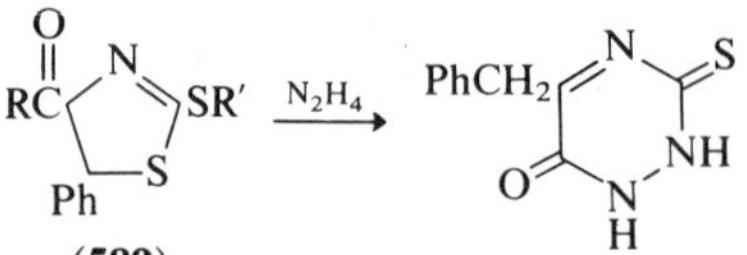

Reaction of the oxazolium salts (**591**) with methylhydrazine led to the isolation of 6-hydroxytetrahydro-1,2,4-triazines (**592**) ⟨74ZOR2429⟩, while the oxazolium or thiazolium salts (**593**; X = O,S) with hydrazine gave dihydro-1,2,4-triazines (**594**) ⟨78HC(33)189, p. 576⟩. 2-Amino-3-phenacyloxadiazolium salts (**595**) and hydrazine afforded the 3,4-diamino compounds (**596**) ⟨71LA(749)125⟩.

**(591)** **(592)**

**(593)** **(594)** **(595)** **(596)**

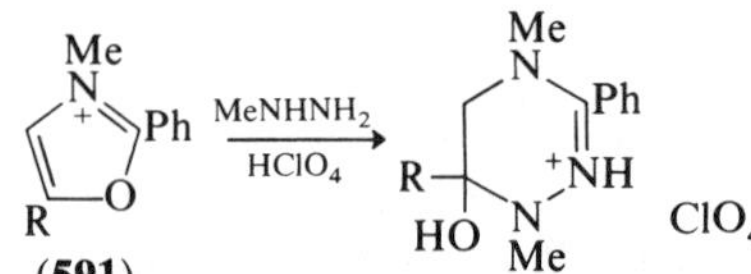

There is one case in which a seven-membered ring has provided the four-atom fragment in this synthetic combination. Reaction of the 1,4-oxazepinone (**597**) with hydrazine provided 3-(diphenylmethyl)-6-phenyl-4,5-dihydro-1,2,4-triazine (**598**) ⟨73JOC3466⟩.

**(597)** **(598)** **(599)** **(600)**

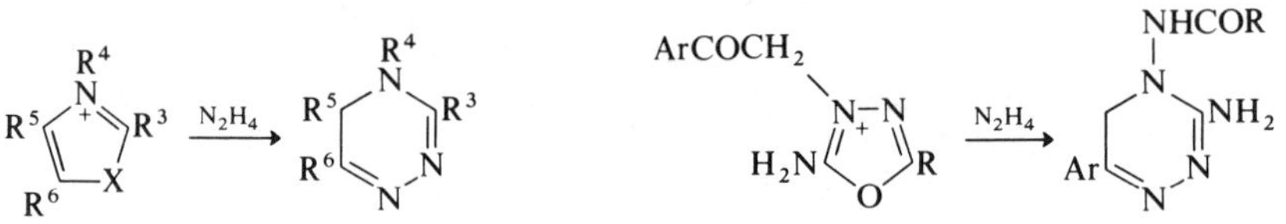

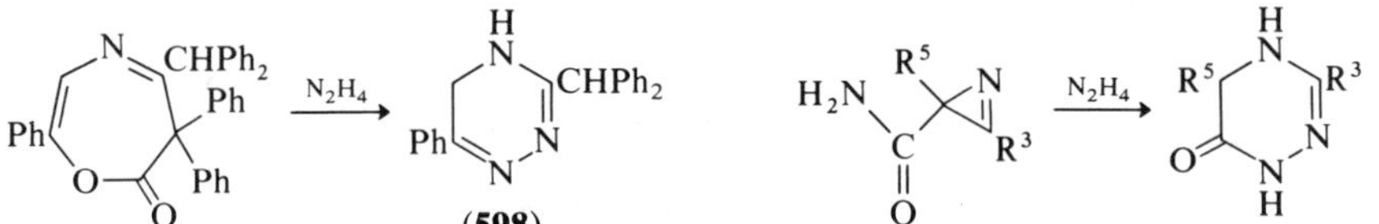

Azirine-3-carboxamides with a second substituent in the 3-position (**599**) were used by Nishikawa and Sato for the synthesis of dihydro-1,2,4-triazin-6-ones (**600**) ⟨71JCS(C)2648⟩.

### *2.19.4.1.3 Combination C [N(4)C(5)C(6)N(1) + N(2)C(3)]*

Only a few examples of the formation of 1,2,4-triazines by this fragment combination have been published, all cases being 1,2,4-benzotriazine syntheses. 2-Nitroaniline reacts with cyanamides to give 2-nitrophenylguanidines (**601**) which cyclize under basic conditions to 3-amino-1,2,4-benzotriazine 1-oxides (**602**) ⟨78HC(33)189, p. 699⟩. With benzoyl isothiocyanate it affords *N*-(2-nitrophenyl)-*N'*-benzoylthiourea (**603**), which cyclizes to 1,2,4-benzotriazine-3-thione 1-oxide (**604**) ⟨80MI21900⟩.

The reaction of benzofurazan oxide (**605**) with amines has been studied by Haddadin *et al.* ⟨79T681⟩. With diethylamine, beside other products, 3-methyl-1,2,4-benzotriazine (**606**) and its 4-oxide (**607**) were isolated, the production of which was explained by the intermediate formation of (**608**) and (**609**).

(601) (602) (603) (604)

(605) (606) (607) (608) (609)

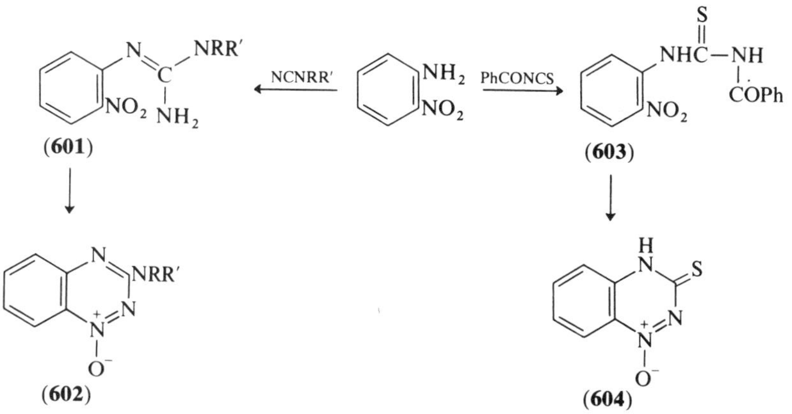

#### 2.19.4.1.4 Combination D [C(5)C(6)N(1)N(2) + C(3)N(4)]

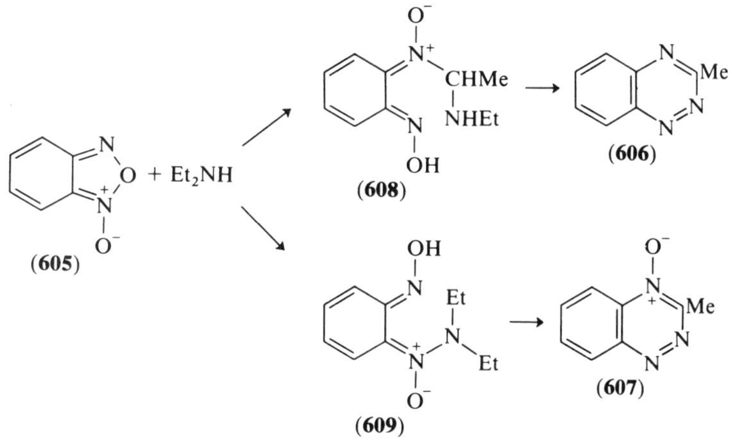

Benzil monohydrazone (**610**) and urea are reported to form 5,6-diphenyl-1,2,4-triazin-3-one (**611**) ⟨71MI21903⟩, and α-hydrazonocarboxylic acids (**612**) can be cyclized with imidates to 1,2,4-triazin-5-ones (**613**) ⟨74T3171⟩. Both these reactions can also be treated as [2+2+2] syntheses, since the hydrazones (**610**) and (**612**) are prepared from the appropriate ketones and hydrazine.

(610) (611) (612) 613)

(614)

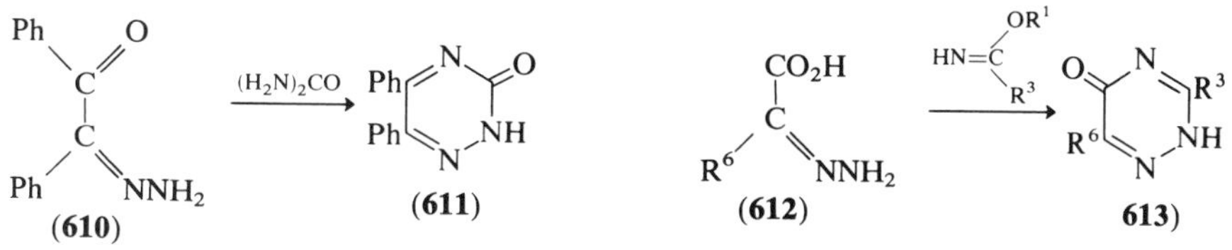

With α-hydrazinocarboxylates (**613**) instead of hydrazonocarboxylic acids, imidates give the dihydro-1,2,4-triazin-5-ones (**614**) ⟨1895CB1223, 64LA(676)121⟩.

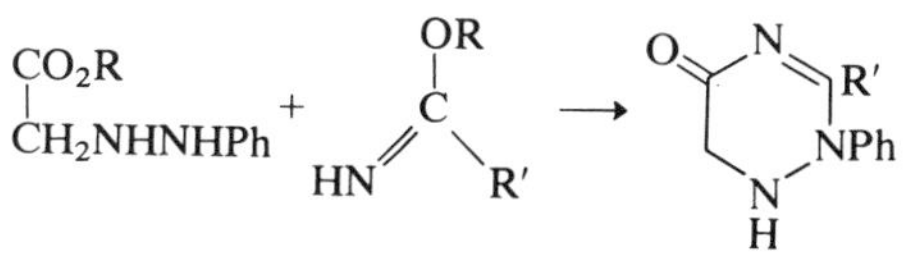

2-Hydrazinoethanols (**615**) react with nitriles to give amidrazones (**616**) which can be cyclized ($H_2SO_4$) to the tetrahydro-1,2,4-triazines (**617**) ⟨66JMC881⟩.

(615) (616) (617)

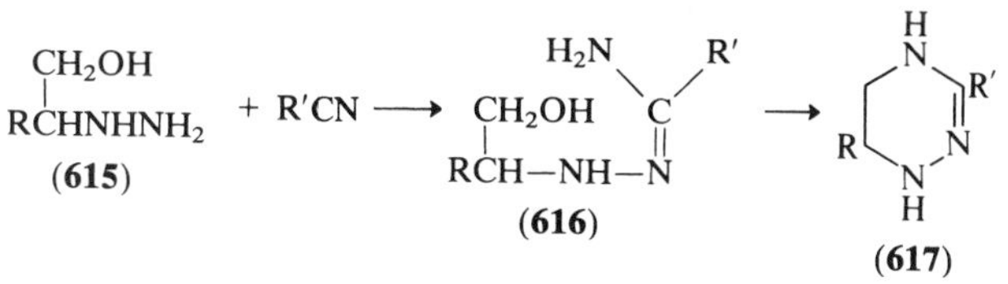

5,6-Diamino-1,2,4-triazin-3-one (**620**) was prepared by reaction of cyanoformamidrazone (**618**) with potassium cyanate to give the semicarbazone (**619**) which cyclized in acidic medium ⟨61JOC3783⟩. In a similar reaction sequence, hydrazonomalononitriles (**621**) with isocyanates or isothiocyanates gave, *via* the intermediate semicarbazones (**622a**) or thiosemicarbazones (**622b**), the 6-cyano-5-imino-1,2,4-triazin-3-ones (**623a**) and -3-thiones (**623b**) respectively ⟨80ZN(B)485, 78HCA1175⟩.

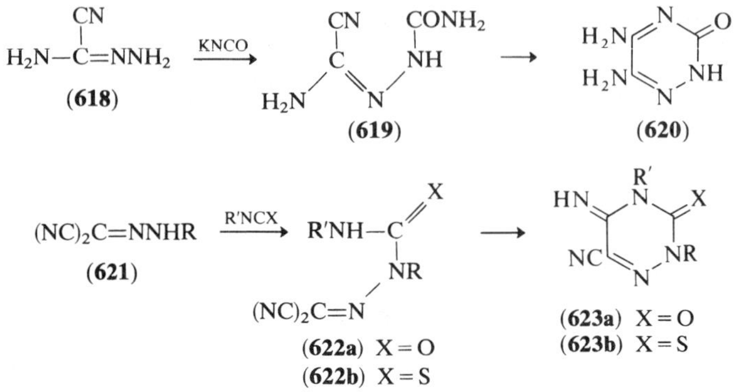


(**618**) (**619**) (**620**)

(**621**) (**622a**) X = O (**622b**) X = S (**623a**) X = O (**623b**) X = S

### *2.19.4.1.5 Combination E [C(6)N(1)N(2)C(3)+N(4)C(5)]*

1,2,4,5-Tetrazines (**624**) are reactive dienes in Diels–Alder reactions with inverse electron demand. They react with both C—C and C—N multiple bonds. Cycloaddition of (**624**) with imidates thus affords 1,2,4-triazines (**625**) which are formed *via* the bicyclic intermediates (**626**) and the dihydro-1,2,4-triazines (**627**) ⟨69JHC497⟩. Further studies have been made on the limitations of this reaction.

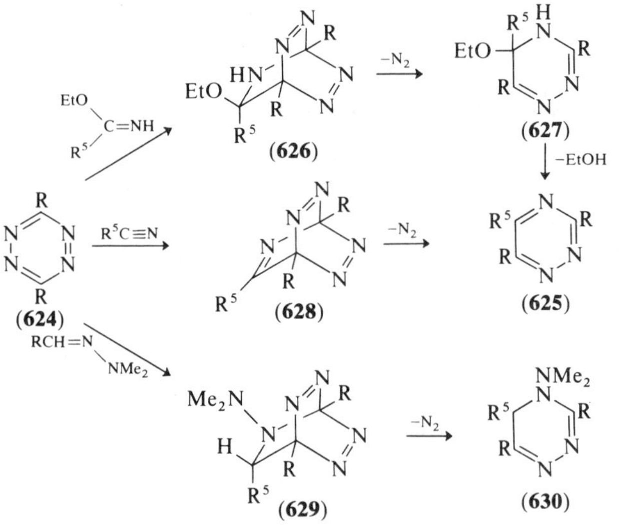


(**624**) (**626**) (**627**) (**628**) (**625**) (**629**) (**630**)

**Scheme 20**

By reacting tetrazines (**624**) with cyanamides the bicyclic intermediates (**628**) lead directly to 5-amino-1,2,4-triazines (**625**; $R^5 = NR_2$) by elimination of nitrogen ⟨79CZ230⟩.

1,2,4,5-Tetrazines (**624**) react also with the C=N double bond of aliphatic aldehyde dimethylhydrazones. Elimination of nitrogen from the adducts (**629**) affords 4-aminodihydro-1,2,4-triazines (**630**). The elimination of amine from (**630**) to give (**625**) has not been reported (Scheme 20) ⟨79AP452, 81AP376⟩.

Aromatic aldazines when treated in boiling toluene with potassium *t*-butoxide, furnished 3,5,6-triaryl-1,2,4-triazines (**632**) and their 2,5-dihydro (**633**) and 1,2,5,6-tetrahydro derivatives (**634**), besides triazoles, which were the major products ⟨76JCS(P1)207⟩. The formation of the triazines is best explained by a [4+2] cycloaddition or a two-step process *via* the carbanion (**631**) and elimination of benzalimine to give (**633**), which can be oxidized to (**632**; Scheme 21). The formation of (**634**) is still in doubt.

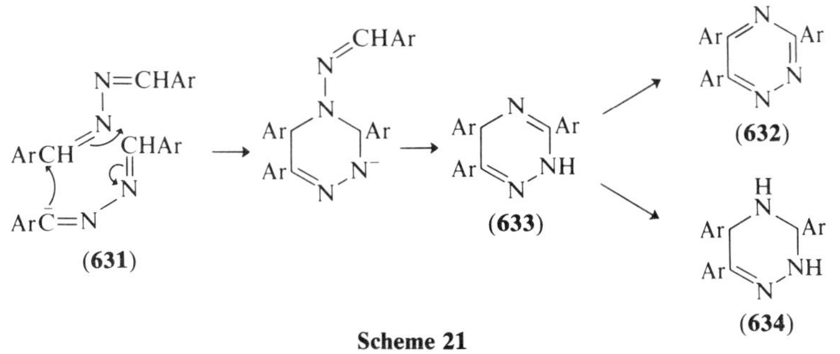

**Scheme 21**

## 2.19.4.2 [5 + 1] Atom Combintations

Of the six possible [5 + 1] atom fragment combinations for synthesizing 1,2,4-triazines, only two are found in the literature; in these C(3) and N(4) are the one-atom fragments.

Many of the methods discussed in this section can also be treated as combinations of more than two fragments. For instance, in the cyclization of α-hydrazono oximes with orthocarboxylates described in the next section, the α-hydrazono oximes (635) are treated as a five-atom fragment, but since these compounds are usually prepared from ketones by nitrosation and reaction of the oximino ketones with hydrazine, the entire reaction sequence may be described as a [2 + 1 + 2 + 1] combination.

### 2.19.4.2.1 C(3) as the one-atom fragment

The cyclization of α-hydrazono oximes (635) with orthocarboxylates is an excellent way of preparing 1,2,4-triazine 4-oxides (636) ⟨73TL1429, 71LA(750)12, 77LA1713⟩. Using aldehydes or ketones instead of orthoesters provides 2,3-dihydro derivatives (637; Scheme 22) ⟨75S794⟩. Cyclization of α-hydrazonocarboxamides (638) with carbonic acid derivatives such as ethyl chloroformate affords 1,2,4-triazine-3,5-diones (639) ⟨68M1808⟩. The α-hydrazonocarboxamides (638) can be prepared by the reaction of diazonium salts with α-activated acetamides such as cyanoacetamide as shown.

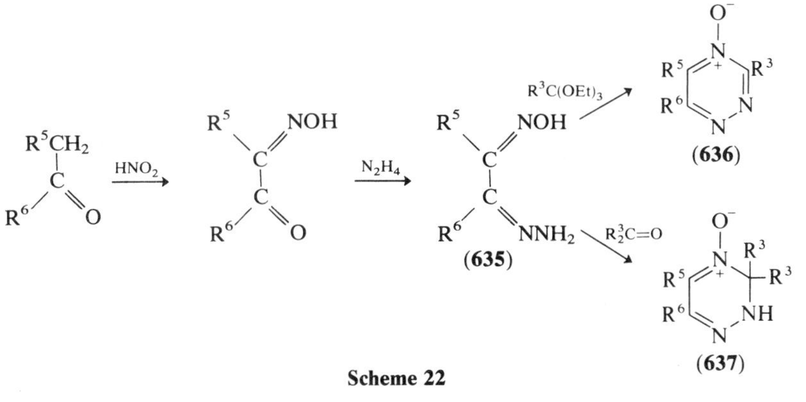

**Scheme 22**

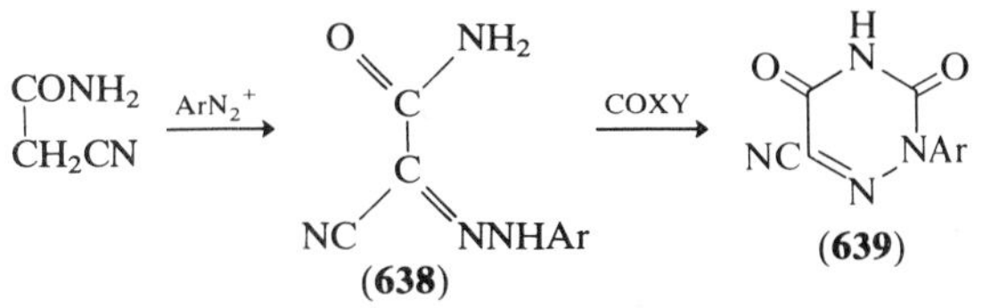

β-Aminohydrazines (640) are useful intermediates for the synthesis of reduced 1,2,4-triazines. With nitriles, carboxylic acids, imidates, thioimidates or orthocarboxylates, they form 1,4,5,6-tetrahydro-1,2,4-triazines (641) ⟨78HC(33)189, p. 629⟩, with carbon disulfide they give the 3-thiones (642) ⟨78HC(33)189, p. 648⟩, and with aldehydes or ketones hexahydro-1,2,4-triazines (643) are formed ⟨78HC(33)189, p. 657⟩.

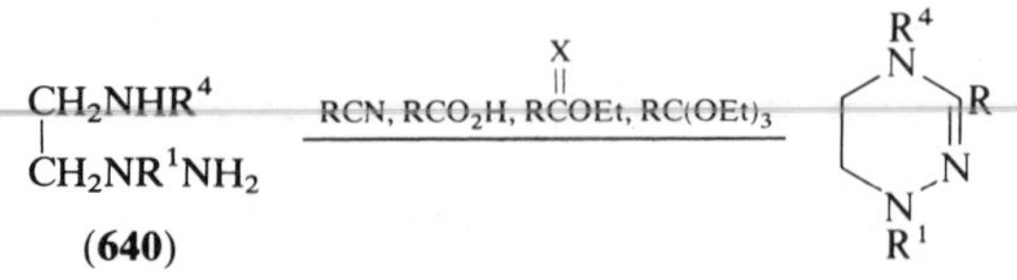

$$\text{(640)} \quad \begin{array}{c} \text{CH}_2\text{NHR}^4 \\ | \\ \text{CH}_2\text{NR}^1\text{NH}_2 \end{array} \xrightarrow[\text{RCN, RCO}_2\text{H, RCOEt, RC(OEt)}_3]{\overset{X}{\overset{||}{}}} \text{(641)}$$

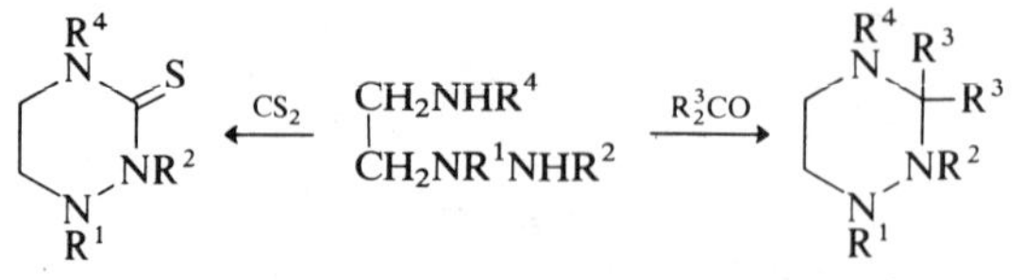

$$\text{(642)} \xleftarrow{\text{CS}_2} \begin{array}{c} \text{CH}_2\text{NHR}^4 \\ | \\ \text{CH}_2\text{NR}^1\text{NHR}^2 \end{array} \xrightarrow{\text{R}_2^3\text{CO}} \text{(643)}$$

The use of $\alpha$-aminohydrazones (**644**) instead of $\alpha$-aminohydrazines as the five-atom fragment affords 2,3,4,5-tetrahydro-1,2,4-triazines (**645**) with aldehydes ⟨78LA2033, 79JCS(P2)984⟩, 4,5-dihydro-1,2,4-triazin-3-ones (**646a**) with phosgene ⟨78LA2033⟩ and the corresponding 3-thiones (**646b**) with carbon disulfide ⟨78HC(33)189, p. 596⟩. Another group of versatile five-atom building blocks are the $\alpha$-aminocarbohydrazides (**647**). Cyclization of (**647**) with carboxylic acids affords 4,5-dihydro-1,2,4-triazin-6-ones (**648**) ⟨78HC(33)189, p. 605⟩, with thiophosgene they give 4,5-dihydro-3-thioxo-1,2,4-triazin-6-ones (**649**) ⟨78HC(33)189, p. 621⟩, while reaction with *t*-butyl isocyanide led to the isolation of 3-amino-2,3,4,5-tetrahydro-1,2,4-triazin-6-ones (**650**) (Scheme 23) ⟨77JOM(131)121⟩.

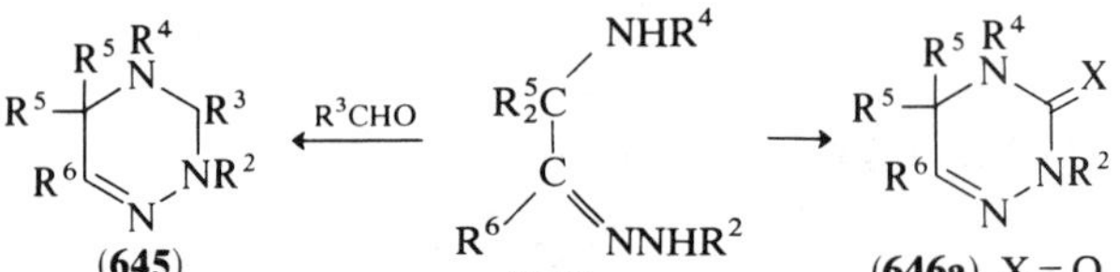

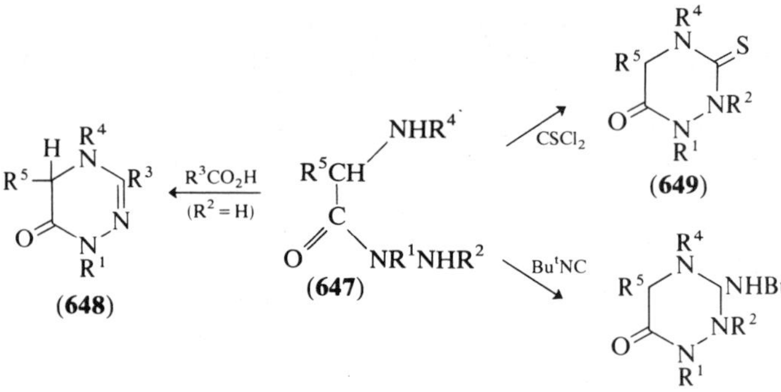

**Scheme 23**

$\alpha$-Hydrazinocarboxamides (**651**) have been cyclized with phosgene and thiophosgene to yield 1,6-dihydro-1,2,4-triazine-3,5-diones (**652a**) and the 3-thioxo-5-ones (**652b**), respectively ⟨1898LA(301)55, p. 68, 60MI21901⟩.

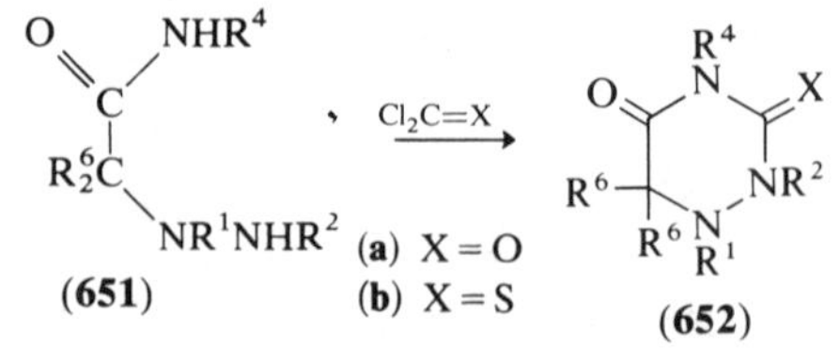

$$\begin{array}{c} \text{(651)} \end{array} \xrightarrow{\text{Cl}_2\text{C=X}} \begin{array}{c} \text{(a) X = O} \\ \text{(b) X = S} \end{array} \text{(652)}$$

The cyclization of $\alpha$-hydrazinohydrazones (**653**) with ketones has been used for the synthesis of 4-amino-2,3,4,5-tetrahydro-1,2,4-triazines (**654**), and their reaction with phosgene affords 4-amino-4,5-dihydro-1,2,4-triazin-3-ones (**655**) ⟨78HC(33)189, pp. 608, 656⟩. 3-Thioxo-1,2,4-triazine-5,6-dione (**657**) was obtained when oxamohydrazide (**656**) reacted with thiophosgene ⟨76ACS(B)71⟩.

The [5 + 1] atom fragment method is frequently used for the synthesis of 1,2,4-benzotriazines. 2-Aminophenylhydrazine is reported to be cyclized with urea or xanthates

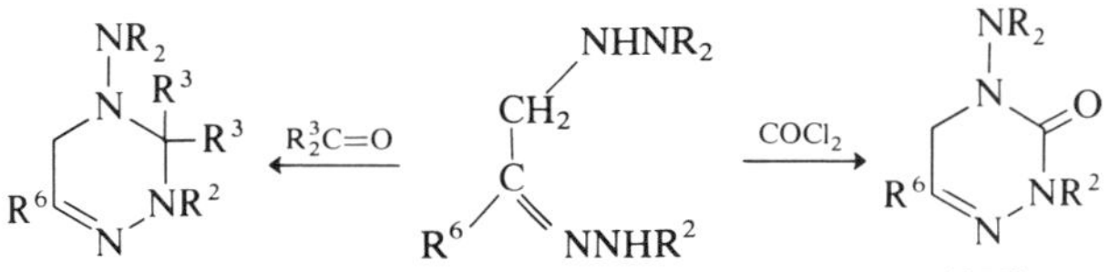

(654) (653) (655)

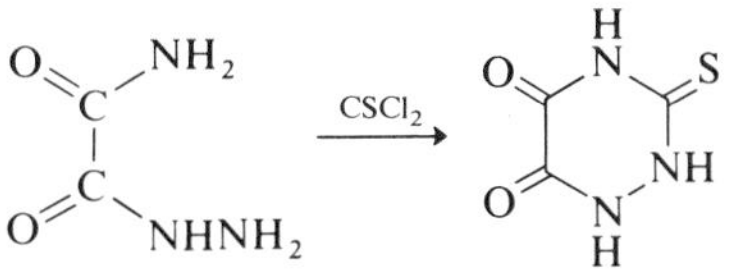

(656) (657)

to yield dihydro-1,2,4-benzotriazin-3-one (**658a**) or the 3-thione (**658b**), respectively
⟨25JIC84⟩. 1,2,4-Benzotriazin-3-ones (**660**) were obtained when 2-aminophenylazo compounds (**659**) were condensed with phosgene or ethyl chloroformate ⟨80CCC1379⟩. Fischer
has shown that the reaction of (**659**) with aldehydes affords 1-aminobenzimidazoles (**661**)
and not the 2,3-dihydro-1,2,4-benzotriazines (**662**) ⟨24JPR(107)16, 22JPR(104)102⟩.

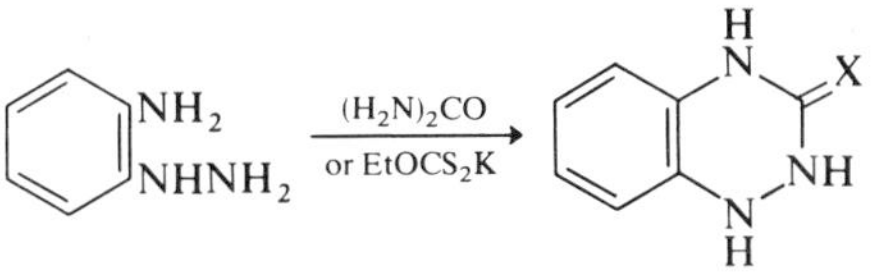

(658a) X = O
(658b) X = S

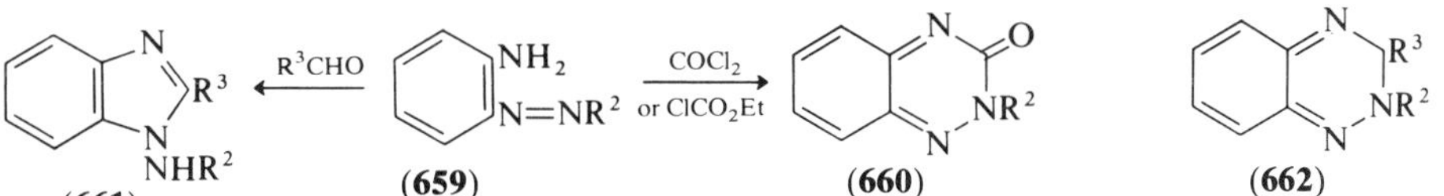

(661) (659) (660) (662)

The same methods used for 1,2,4-benzotriazines have also been applied to the synthesis
of naphtho-1,2,4-triazines. 1-Amino-2-arylazonaphthalenes (**663**) with phosgene yield
naphtho[1,2-*e*][1,2,4]-triazin-2-ones (**664**) ⟨05MI21900⟩, and the isomeric 2-amino-1-
arylazonaphthalenes (**665**) with phosgene or isocyanates similarly gave naphtho[2,1-*e*]-
[1,2,4]triazin-3-ones (**666**) ⟨78HC(33)189, p. 732⟩. 3-Iminonaphtho[2,1-*e*][1,2,4]triazines
(**667, 668**) were obtained when (**665**) reacted with cyanogen bromide ⟨08MI21900⟩ or with
isothiocyanates and mercury(II) oxide (Scheme 24).

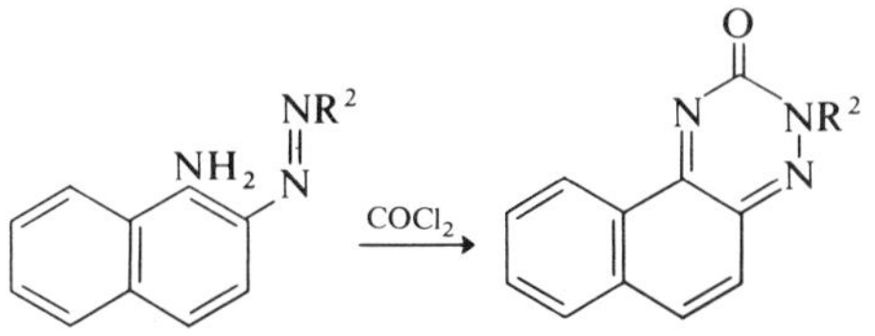

(663) (664)

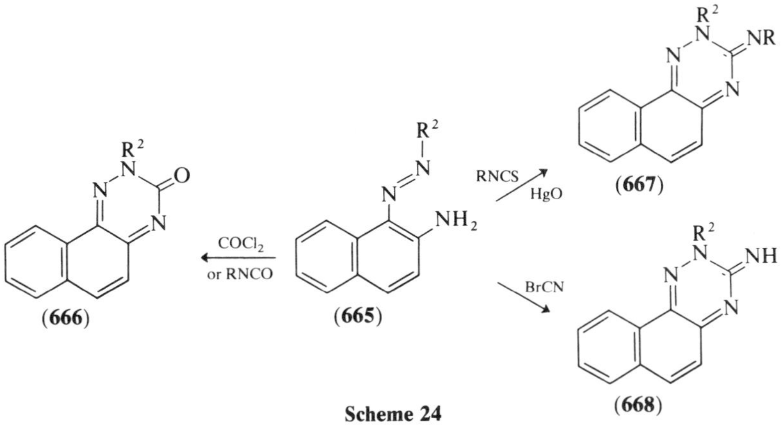

(667)

(666) (665)

(668)

**Scheme 24**

### 2.19.4.2.2 N(4) as the one-atom fragment

α-Acylhydrazono ketones (**669**) react with ammonia to give 1,2,4-triazines (**670**), either directly or by conversion first into the α-chloroazines (**671**; Scheme 25) ⟨78HC(33)189, p. 196⟩. Since the acylhydrazono ketones (**669**) are obtained from 1,2-diketones and acylhydrazines this reaction sequence can also be regarded as a [3 + 2 + 1] combination synthesis.

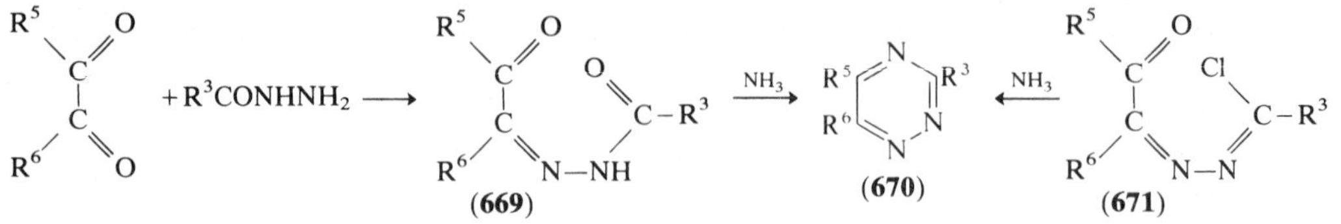

**Scheme 25**

4-Amino-1,2,4-triazin-5-ones (**674**) have been prepared by the reaction of α-acyl-hydrazonocarboxylates (**672**) or the chloro derivatives (**673**) with hydrazine. The chloroazines (**673**) with hydroxylamine afford 4-hydroxy-1,2,4-triazin-5-ones (**675**; Scheme 26) ⟨78HC(33)189, p. 563⟩.

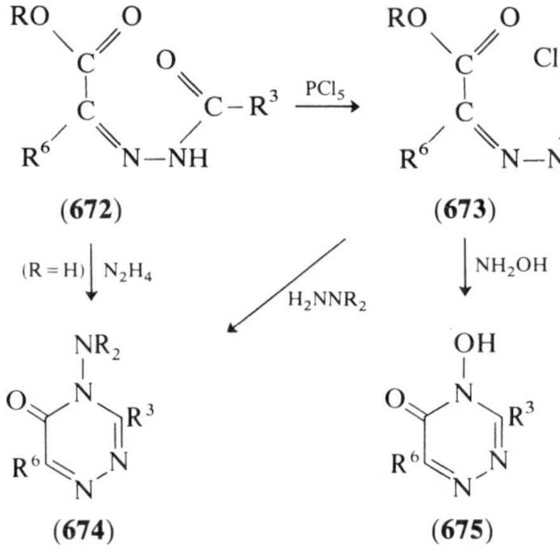

**Scheme 26**

5-Amino-1,2,4-triazines (**678**) can be prepared from the chloronitriles (**677**) and ammonia ⟨78HC(33)189, p. 249⟩ or from the nitrosopyrazoles (**676**) by treatment with phosphorus pentachloride and then ammonia ⟨58T(3)209⟩.

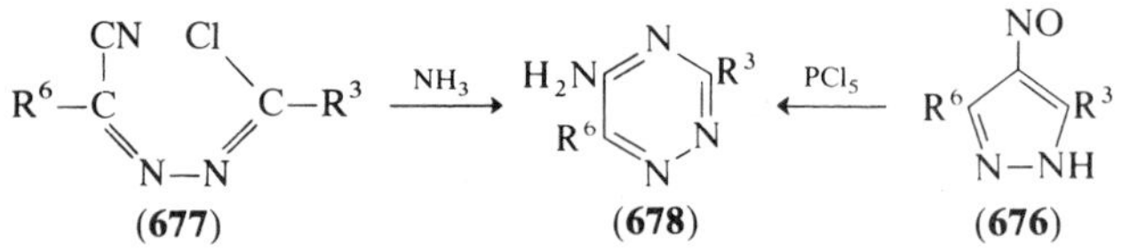

### 2.19.4.3 [6 + 0] Atom Combinations (Unimolecular Cyclizations)

Many of the reactions discussed in this section can be treated as syntheses of 1,2,4-triazines from more than one fragment.

Photolysis of 1,6-diazido-2,5-diphenyl-3,4-diaza-2,4-hexadiene (**679**) afforded 3,6-diphenyl-1,2,4-triazine (**680**) in 21% yield ⟨72CL1185⟩.

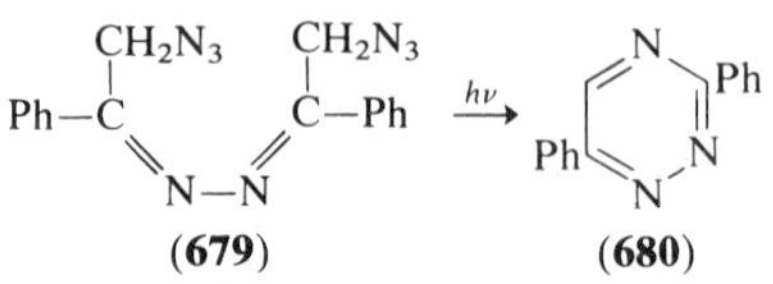

Hydrolysis in acid of compounds assigned the α-cyanoazoacetophenone structures (**681**) affords arylglyoxal semicarbazones (**682**). These can be cyclized in hot sodium hydroxide to 5-aryl-1,2,4-triazin-3-ones (**683**) ⟨02LA(325)129⟩.

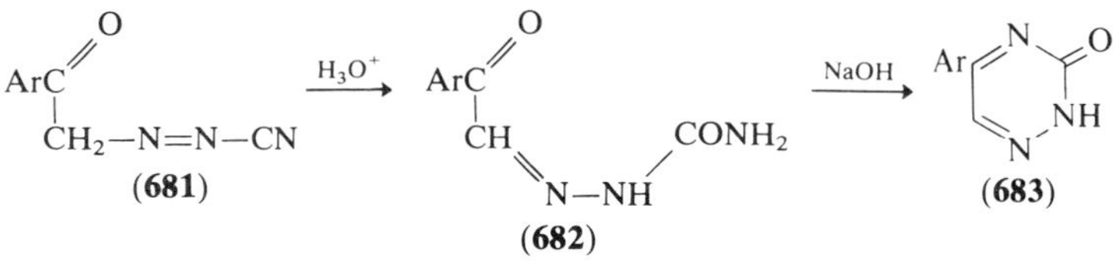

(681) (682) (683)

Cyclization of α-(acylamino)acrylohydrazides (**684**) with sodium hydroxide or acetic acid/acetic anhydride is the most commonly used method for preparing 1,2,4-triazin-6-ones (**685**) ⟨78HC(33)189, p. 258⟩. But since the acrylohydrazides (**684**) are usually made by hydrazinolysis of azlactones (*cf.* also **586** → **588**, Section 2.19.4.1.4), this method can also be treated as a [4+2] atom fragment method.

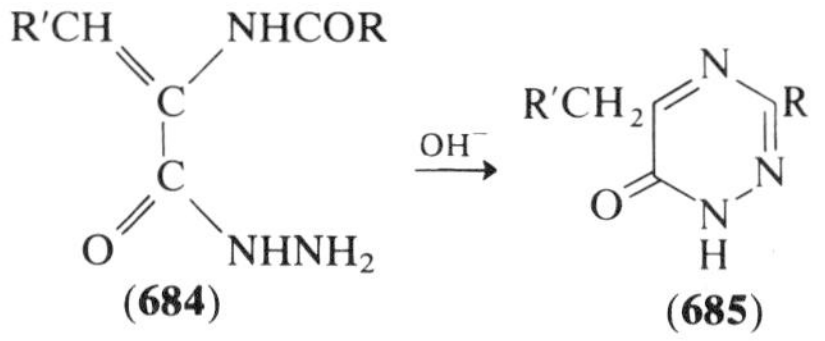

(684) (685)

Oxidation of methyl ketone guanylhydrazones (**686**) with selenium dioxide affords the corresponding monosubstituted glyoxal derivatives (**687**) which cyclize to the 5-unsubstituted 3-amino-1,2,4-triazines (**473**) ⟨78HC(33)189, p. 360⟩. We have already noted the cyclization of bis(alkylidene)- or bis(arylidene)-acetone guanylhydrazones (**475**) in the synthesis of 6-(vinyl-substituted) 3-amino-1,2,4-triazines (**477**) (Section 2.19.4.1.1).

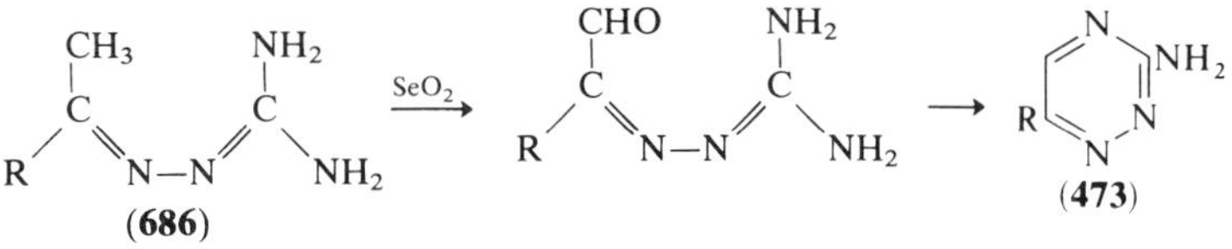

(686) (473)

Treatment of the formazans (**688**) with potassium hydroxide affords 6-arylazo-1,2,4-triazine-3,5-diones (**689**) ⟨27JCS521⟩. Basic or thermal cyclization of the hydrazonourethanes (**690**) gives the 6-cyano-1,2,4-triazine-3,5-diones (**691**) which can be hydrolyzed to the carboxylic acids (**692**), and then decarboxylated to the 6-unsubstituted triazinediones ⟨78HC(33)189, p. 537⟩. Basic cyclization of the hydrazonomalonic diamides (**693**) leads to the same 3,5-dioxo-1,2,4-triazine-6-carboxylic acids (**692**) while their thermal cyclization affords the corresponding carboxamides (**694**) ⟨78HC(33)189, p. 537⟩. Alloxan semicarbazone (**695**) with sodium hydroxide forms 3,5-dioxo-1,2,4-triazine-6-carboxylic acid (**692**; R = H), with cleavage of the pyrimidine ring (Scheme 27) ⟨69JPR438, 64CB5⟩.

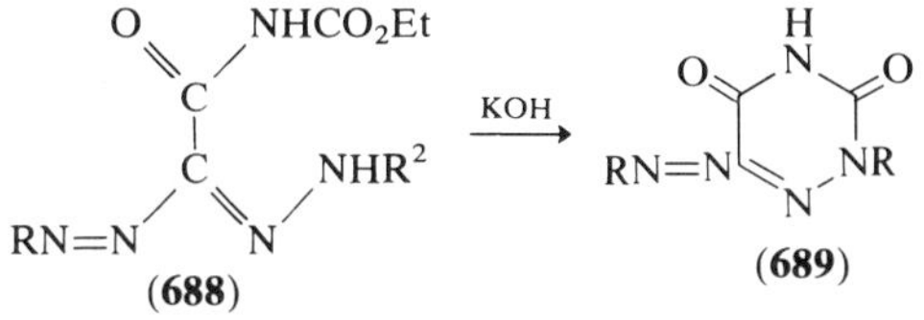

(688) (689)

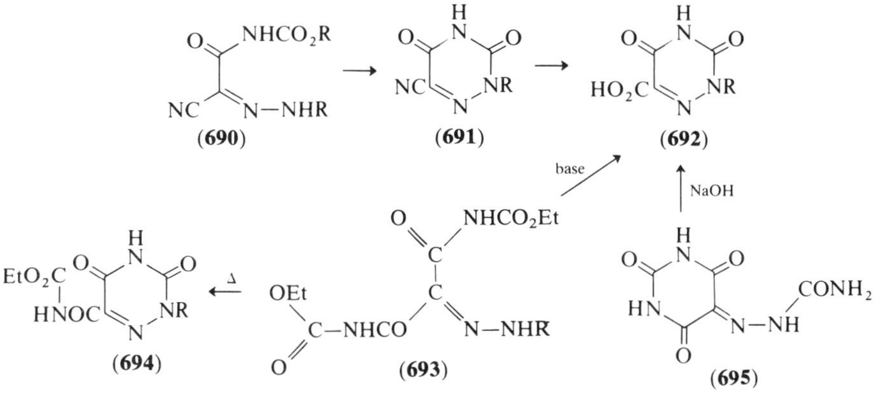

(690) (691) (692) (693) (694) (695)

**Scheme 27**

A compound formulated as 5-methyl-1,4-dihydro-1,2,4-triazin-3(2*H*)-one (**696**) was obtained by sodium hydroxide treatment of compound (**697**). Reacting (**697**) with thallium ethoxide led to the isolation of diethyl 5-methyl-3-oxo-1,2,3,4-tetrahydro-1,2,4-triazine-1,6-dicarboxylate (**698**) ⟨70JOC3792⟩. Since compound (**697**) is formed by reaction of aminocrotonic ester with diethyl azodicarboxylate, these reactions can also be treated as [3 + 3] atom component preparations.

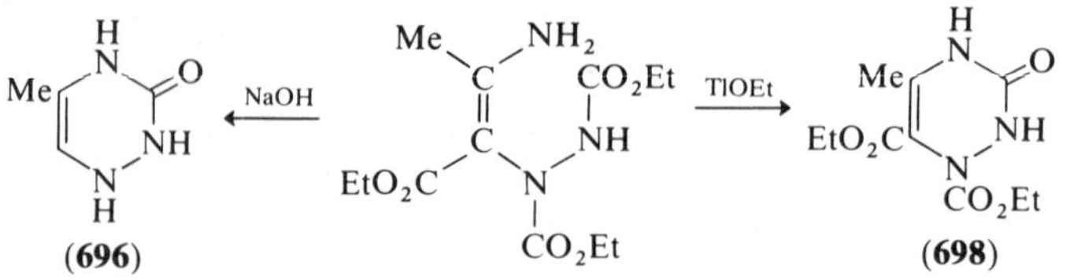

The semicarbazone of phenacylhydrazine (**699**) yielded 4-amino-6-phenyl-4,5-dihydro-1,2,4-triazin-3-one (**700**) on treatment with base ⟨36JPR(144)273⟩.

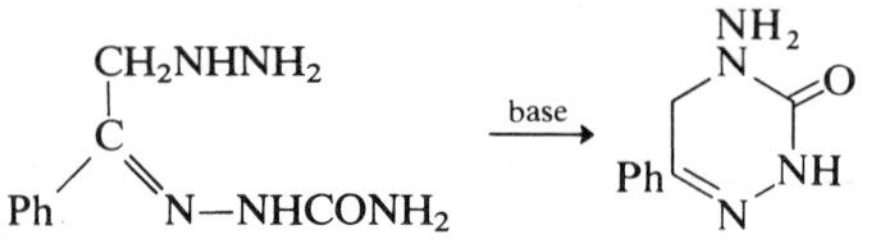

Cyclization of α-semicarbazidocarboxylic acid derivatives (**701**) is a method very frequently used for the synthesis of dihydro-1,2,4-triazine-3,5-diones (**702**) ⟨78HC(33)189, p. 613⟩. The same compounds (**702**) are obtained by cyclization of the α-hydrazinoureides (**703**) ⟨1899AP346⟩. Using thiosemicarbazidocarboxylic acid derivatives (**704**) instead of (**701**), 3-thioxo-1,6-dihydro-1,2,4-triazin-5-ones (**705**) can be prepared. Treatment of α-semicarbazidonitriles (**706**) with hydrogen chloride is another route to (**705**) ⟨78HC(33)189, p. 621⟩.

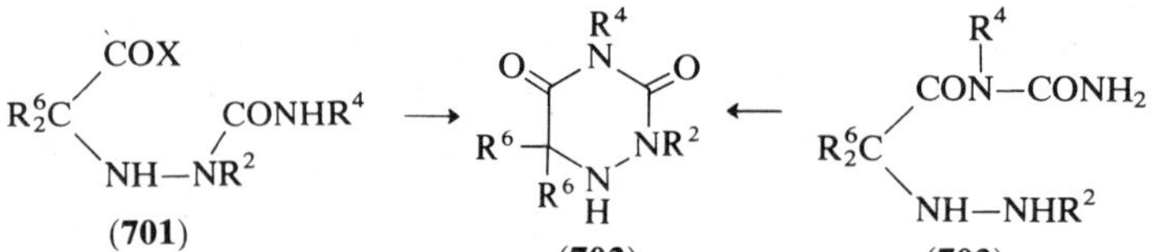

3-Thioxo-4,5-dihydro-1,2,4-triazin-6-ones (**708**) are reported to be prepared by treatment of α-(4-thiosemicarbazido)acetates (**707**) with sodium ethoxide ⟨64CB994⟩; there is, however, some doubt about the structure of these products ⟨68CCC2087⟩.

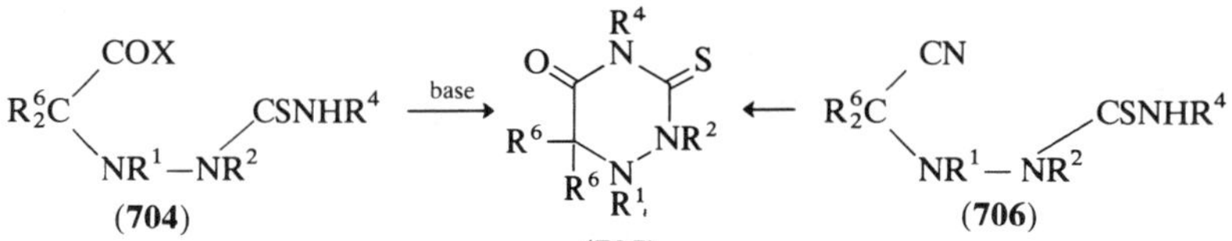

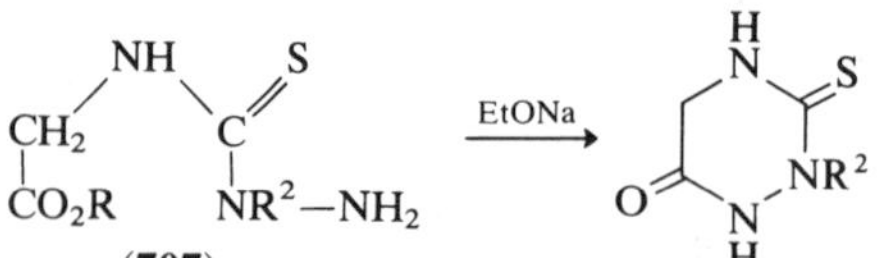

The *N*-substituted 4-amino-3,5-dione (**710**) was obtained when the carbonohydrazide derivative (**709**) was treated with a base ⟨03CB3877⟩. In another base-induced cyclization, 1-(2-chloroethyl)-2-(dimethylaminomethyl)-1-methyl hydrazine (**711**) afforded the 1,4,4-trimethylhexahydro-1,2,4-triazinium salts (**712**) ⟨73CB3540⟩.

8-Thioxo-6,7,9-triazaspiro[4.5]decan-10-one (**714a**) and 3-thioxo-1,2,4-triazaspiro[5.5]undecan-5-one (**714b**) were obtained when the 1-(1-thiosemicarbazido)cycloalkanecarboxamides (**713**) were treated with sodium ethoxide ⟨63FRP1324339⟩. 2-Amino-1-(benzoylhydrazono)cycloheptatrienes (**715**) and 2-amino-1-(ethoxycarbonyl-

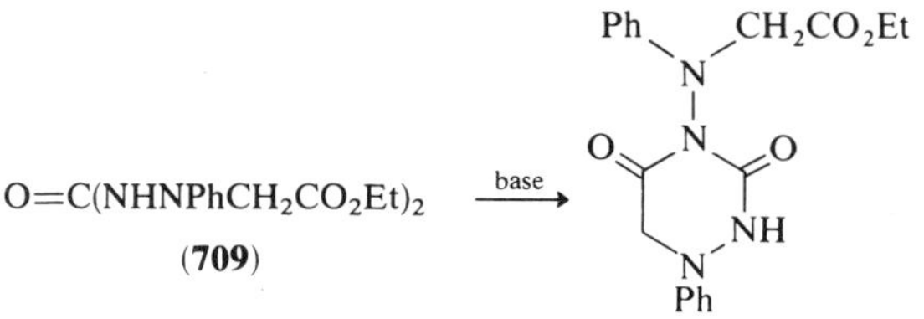

(709) → (710)

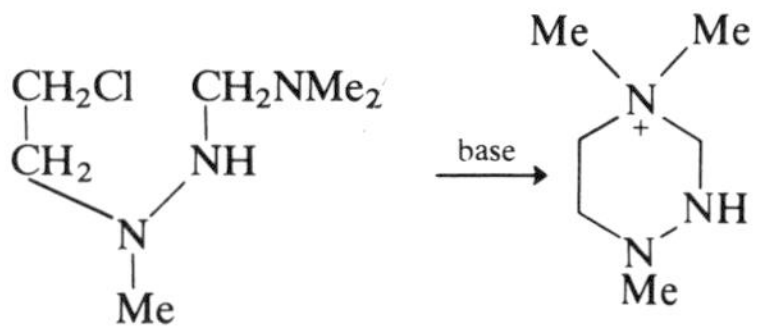

(711) → (712)

hydrazono)cycloheptatrienes (716) can be cyclized to 3-phenyl-4*H*-cyclohepta[*e*]-[1,2,4]triazines (717) and 4*H*-cyclohepta[*e*][1,2,4]triazin-3(2*H*)-ones (718), respectively ⟨78HC(33)189, p. 662⟩.

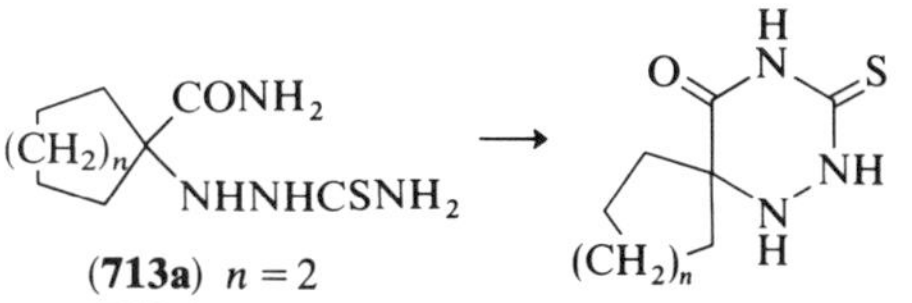

(713a) n = 2
(713b) n = 3

(714a) n = 2
(714b) n = 3

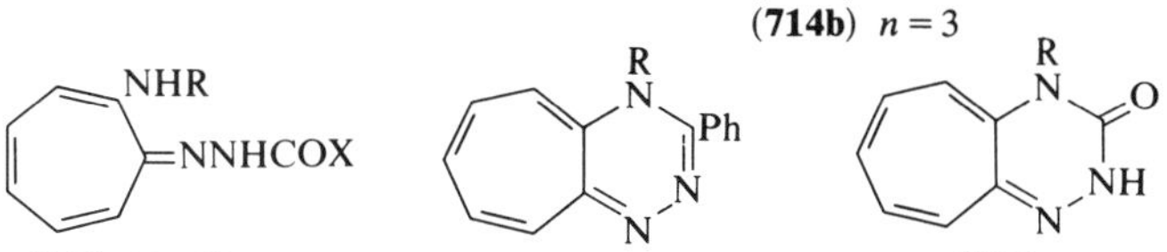

(715) X = Ph
(716) X = OEt

(717)

(718)

The six-atom cyclization is commonly encountered in 1,2,4-benzotriazine synthesis. In most cases the reactions are real [6+0] atom fragment methods since, although reactive intermediates may be involved, the precursors usually have all six atoms present in the molecule. As with the monocyclic compounds, it is usually a C—N bond which is formed in the cyclization but some syntheses take place with N—N bond formation.

2-(2-Nitrophenyl)hydrazides (719) give 1,2-dihydro-1,2,4-benzotriazines (720) when the nitro group is reduced with sodium amalgam in ethanol. In most cases the initial dihydro compounds (720) are not isolated but are oxidized by potassium ferricyanide to the aromatic 1,2,4-benzotriazines (721) ⟨78HC(33)189, p. 666⟩. Similarly, reduction of the nitrohydrazones (722) (or of the tautomeric azo compounds 723) affords 1,2,4-benzotriazines (721) ⟨78HC(33)189, p. 666⟩. 2-Aminophenylhydrazides (724) are cyclized and oxidized to (721) when treated with hydrochloric acid and sodium *m*-nitrobenzenesulfonate (Scheme 28) ⟨73JCS(P1)842⟩.

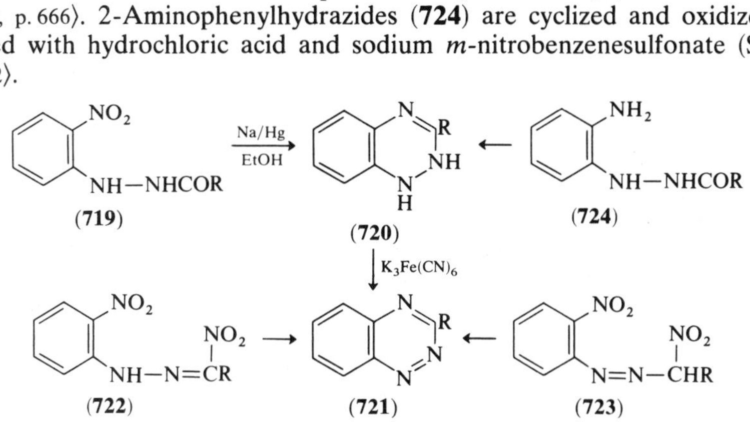

(719) → (720) ← (724)

(722) → (721) ← (723)

**Scheme 28**

A very commonly used method of 1,2,4-benzotriazine synthesis is the cyclization of formazans (725) by treatment with sulfuric acid in acetic acid. When mixed formazans are used, Jerchel and Woticky have shown that both possible benzotriazines (**726**) are formed ⟨57LA(605)191⟩. 3-Acyl-1,2,4-benzotriazines (**727**) are similarly produced from the acylformazans (**728**) in sulfuric acid ⟨78HC(33)189, p. 688⟩. The unsubstituted compound (**493**) was isolated when the formazan ester (**728**; R = OEt) was cyclized, ester hydrolysis and decarboxylation also occurring under the reaction conditions ⟨02JPR(65)123, 1892CB3201⟩.

(**725**) (**726**)

(**493**) (**728**) (**727**)

(**729**) (**730**)

(**731**) (**732**)

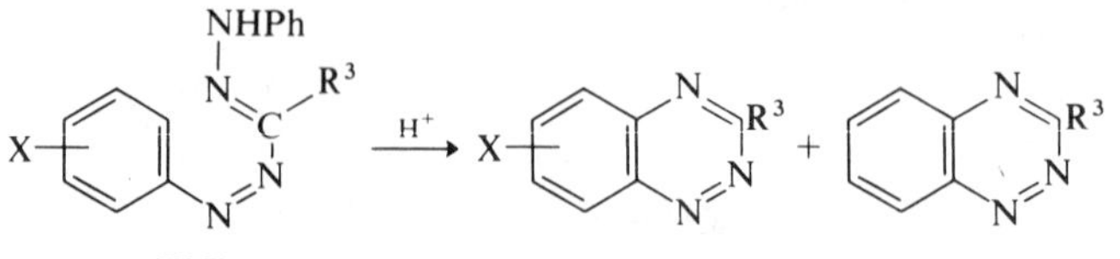

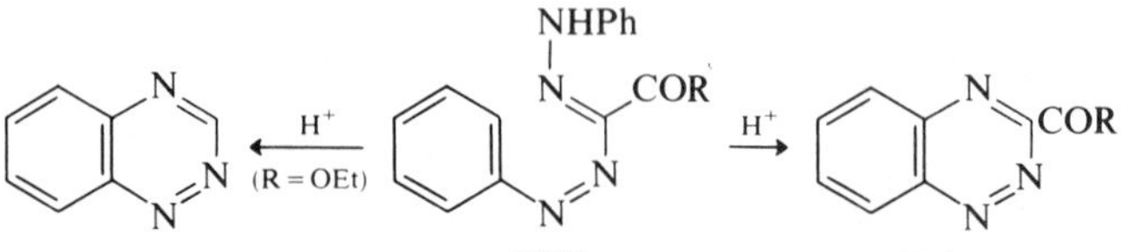

Cyclization of the 2-acylaminoazobenzenes (**729**) leads to 2-aryl-1,2,4-benzotriazinium salts (**730**) ⟨74GEP2241259⟩. When the azo compound (**731**) was heated without a solvent to 200 °C 3-amino-1,2,4-benzotriazine (**732**) was isolated ⟨27CB2598⟩; this involves N(1)—C(8a) [N(1)—C(6)] bond formation, unusual in 1,2,4-benzotriazine syntheses.

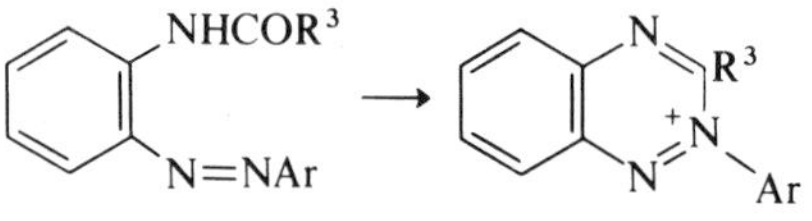

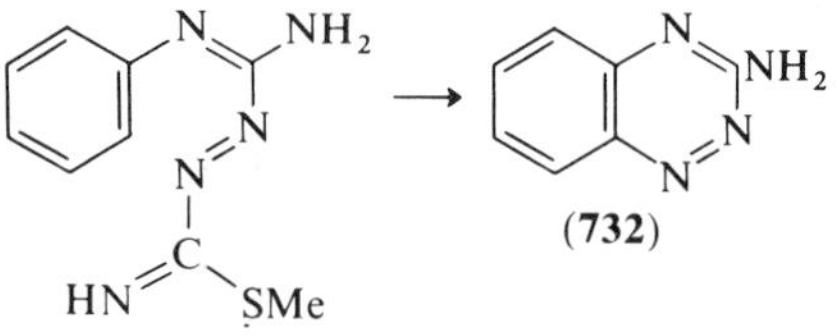

Reduction of 2-nitrophenyloxaloamidrazonates (**733**) with iron and hydrochloric acid was used for the synthesis of 1,2-dihydro-1,2,4-benzotriazine-3-carboxylates (**734**), which can be oxidized to the fully aromatic compounds (**735**) ⟨56G484⟩.

(**733**) (**734**) (**735**)

(**a**) X = R
(**b**) X = OH
(**c**) X = SH
(**d**) X = $NH_2$

(**736**) (**737**)

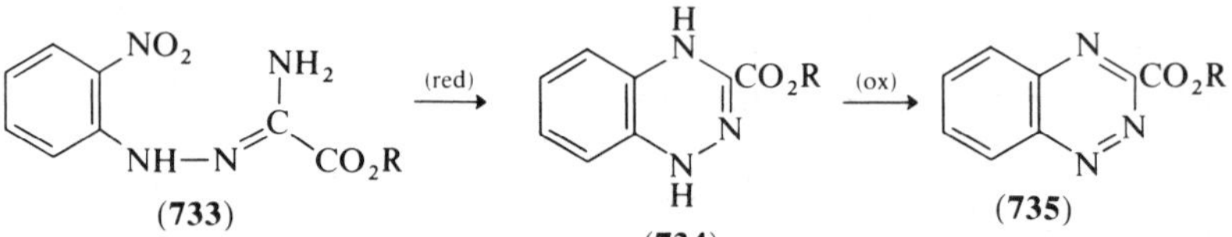

2-Nitrophenylamidines (**736a**) with base afford 1,2,4-benzotriazine 1-oxides (**737a**) ⟨78HC(33)189, p. 690⟩. Similar treatment of the 2-nitrophenyl-ureas (**736b**), -thioureas (**736c**) and -guanidines (**736d**) yields the 3-hydroxy- (**737b**), 3-mercapto- (**737c**) and 4-amino-1,2,4-benzotriazine 1-oxides (**737d**) or their tautomers ⟨78HC(33)189, pp. 695, 698, 699⟩.

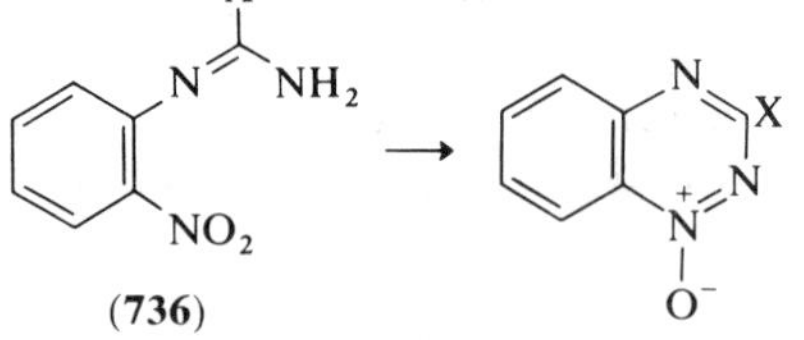

2-Aminophenyl-hydrazides (**738a**), -semicarbazides (**738b**) and -thiosemicarbazides (**738c**) can be cyclized to give 1,2-dihydro-1,2,4-benzotriazines (**739a**), 3-hydroxy-dihydro-1,2,4-benzotriazines (**739b**) and dihydro-1,2,4-benzotriazine-3-thiols (**739c**) or their tautomers ⟨78HC(33)189, p. 715⟩. Very often the starting materials (**738**) were not isolated but generated as intermediates by reduction of the appropriate nitro derivatives (**740**).

1,2-Diaminoimidazoles (**741**) and 1,2-diaminobenzimidazoles (**743**) can be oxidized by manganese dioxide or lead tetraacetate to give a useful synthesis of 3-amino-1,2,4-triazines (**742**) and -benzotriazines (**744**) ⟨76SC457, 76TL903, 77JOC542, 78JOC2693⟩.

(**740a**) X = O, Y = R
(**740b**) X = NH, Y = OH
(**740c**) X = NH, Y = SH
(**738**) (**739**)

(**741**) (**742**)

(**743**) (**744**)

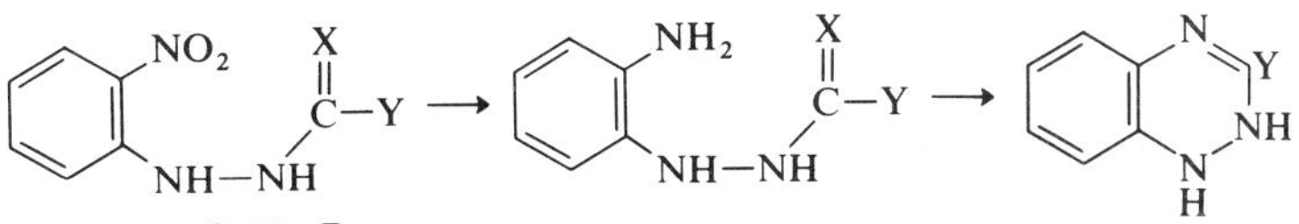

Catalytic reduction of the 2-nitrophenylhydrazones (**745**) and subsequent air oxidation of the tetrahydro-1,2,4-benzotriazines (**746**) which are formed is used for the synthesis of 3,4-dihydro-1,2,4-benzotriazines (**747**) ⟨80FES715⟩. In another cyclization of an *o*-nitro group, which somewhat resembles others (*e.g.* **736** → **737**) described earlier, treatment of 1,1-dialkyl-2-(2-nitrophenyl)hydrazines (**748**) with acids affords 2-alkyl-1,2,4-benzotriazinium salts (**749**) ⟨77JCS(P1)478⟩. The 2-nitrophenylbromohydrazone (**750**) with sodium ethoxide and ethyl cyanoacetate forms 6-bromo-3-phenyl-1,2,4-benzotriazine (**751**); a mechanism for this interesting transformation, in which the cyanoacetate plays no part, has been proposed ⟨79JHC33⟩.

(**745**) (**746**) (**747**)

(**748**) (**749**)

(**750**) (**751**)

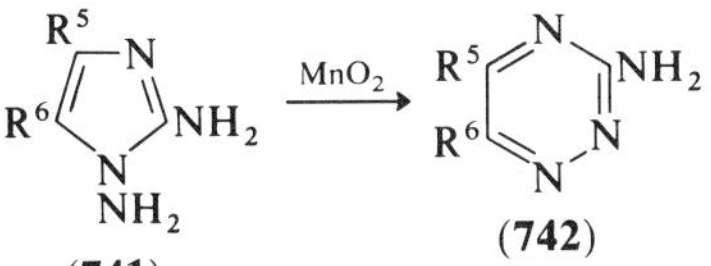

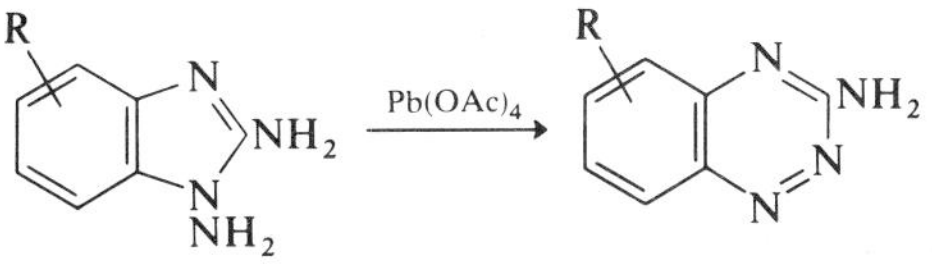

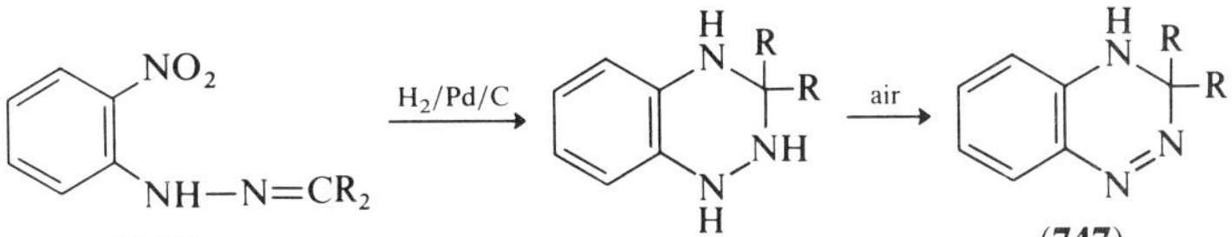

6-Amino-1,2,4-triazin-5-ones (**753**) were obtained when the imidazolones (**752**) were treated with acid ⟨75ZN(B)603⟩. Reaction of 3-methyl-1-phenyl-4-isonitrosopyrazol-5-one (**754**) with the β-lactone (**755**) afforded the 1,2,4-triazine-5,6-dione (**756**), which was the first 5,6-dione in this series to be isolated ⟨69ZOR2039⟩. Another conversion of a five-membered heterocycle into a 1,2,4-triazine is the thermal or photochemical rearrangement of the 5-hydrazinoisoxazoles (**757**); compounds (**758**) are formed, along with pyrazole derivatives ⟨77JCS(P1)971, 78TL4439⟩.

(**752**) (**753**)

(**754**) (**755**) (**756**)

(**757**) (**758**)

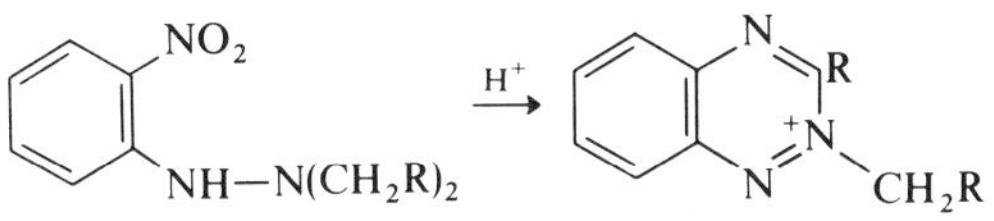

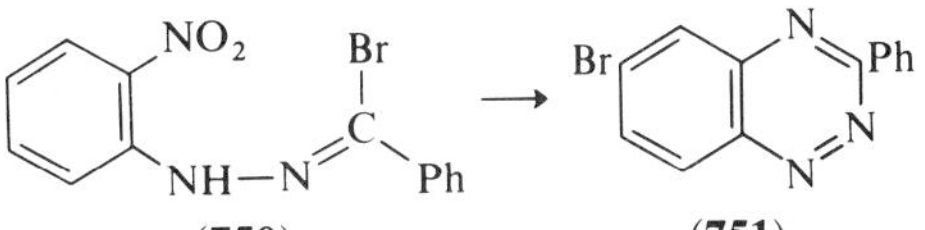

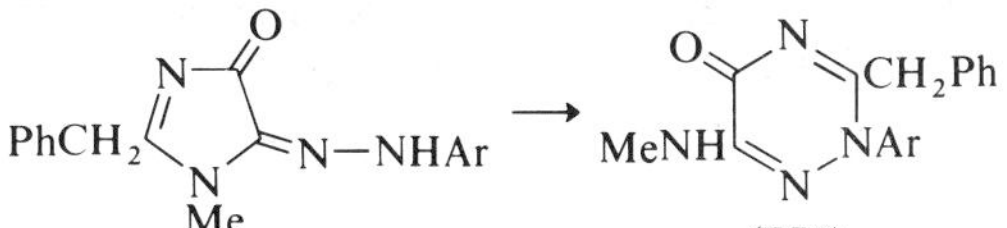

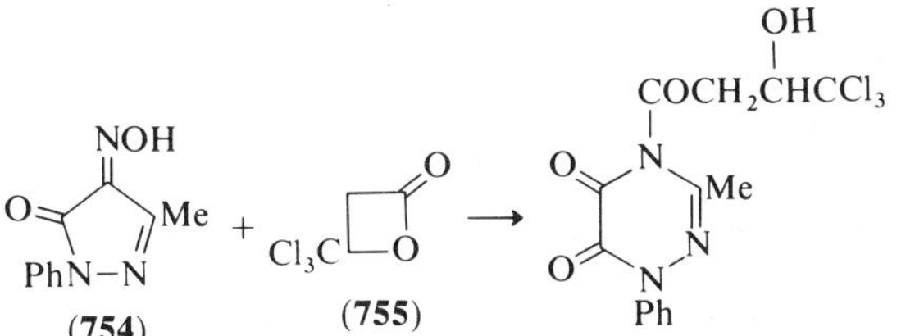

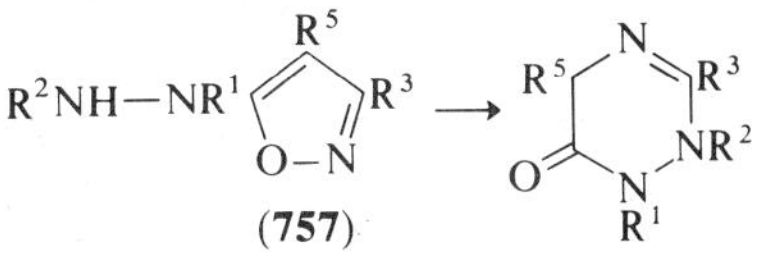

In a further [6 + 0] cyclization reaction, heating the acylhydrazonocarbohydrazides (**759**) for 25 h in boiling ethanol with sodium acetate affords 4-amino-1,2,4-triazin-5-ones (**527**) ⟨79GEP2366215⟩. Intermediates in reactions mentioned in earlier sections, which could in principle be isolated and cyclized in a separate operation, include compounds (**454**), (**457**), (**465**), (**466**), (**470**) and (**475**), among others. The 1,2-bis(semicarbazone) (**760**) is cyclized by aqueous acid to (**461**), and the corresponding bicondensation products of 1,2-diketones and amidrazones behave similarly. These are, however, more appropriately considered as [4 + 2] atom syntheses (see Section 2.19.4.1.1).

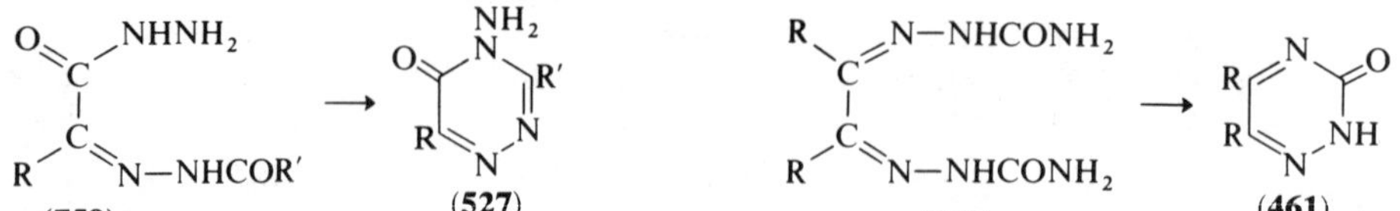

(**759**)  (**527**)  (**760**)  (**461**)

### 2.19.4.4 [3+3] Atom Combinations

Only a few reactions are known which can be treated as synthetic methods for 1,2,4-triazines from two three-atom fragments.

The reaction of $\alpha$-amino ketones (**508d**) with thiosemicarbazides (**761**) affords 3-amino-4,5-dihydro-1,2,4-triazines (**762**) and the dihydrotriazine-3-thiones (**763**) ⟨57MI21900⟩. 2,2-Dimethyl-3-dimethylamino-2*H*-azirine (**764**) reacts with benzohydrazides (**765**; R = Ar) *via* the open-chain intermediates (**766**) to give 3-aryl-6-dimethylamino-2,5-dihydro-1,2,4-triazines (**767**; R = Ar). With ethyl carbazidate (**765**; R = OEt) the corresponding triazin-3-one (**768**) is formed ⟨78HCA2419, 78C332⟩.

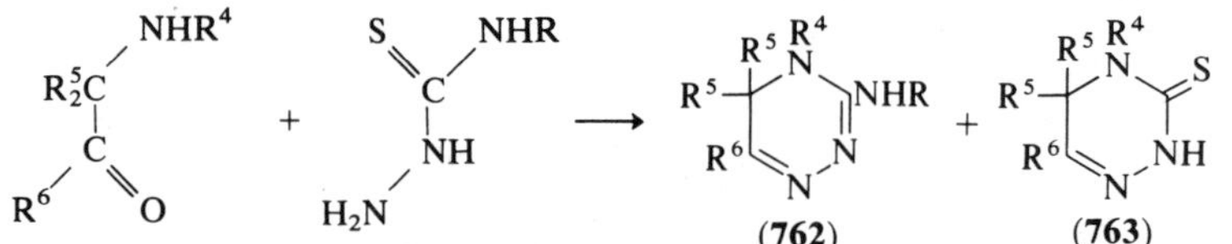

(**761**)  (**762**)  (**763**)

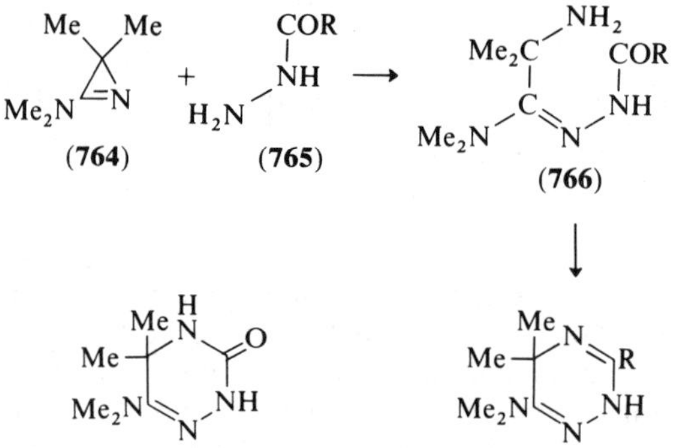

(**764**)  (**765**)  (**766**)

(**768**)  (**767**)

*o*-Phenylenediamine (**769**) has been reported to react with methyl $\alpha$-phenyldithiocarbazinate (**770**) to give 1-phenyl-1,4-dihydro-1,2,4-benzotriazin-3-one (**771**) ⟨28JIC163⟩. From this result it appears that an amino group of the phenylenediamine is replaced, which is an unusual reaction.

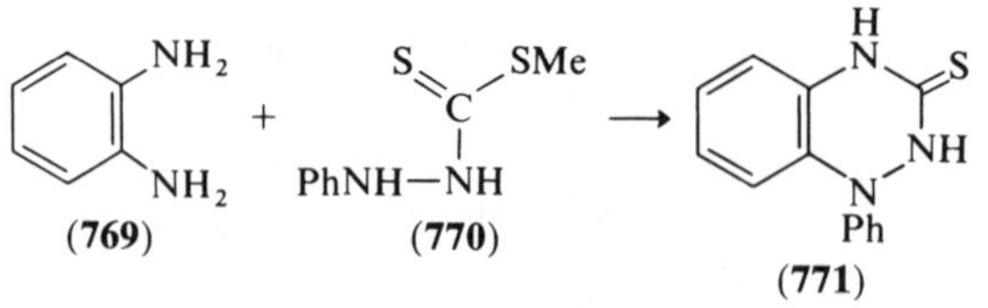

(**769**)  (**770**)  (**771**)

1,2-Naphthoquinone 1-oxime (**772**) reacts with aminoguanidine (**773**) and similar compounds to give 2-aminonaphtho[1,2-*e*][1,2,4]triazine 1-oxides (**774**). 1-Hydroxy-naphtho[1,2-*e*][1,2,4]triazin-2-ones can be prepared in an analogous reaction ⟨78HC(33)189, p. 728⟩.

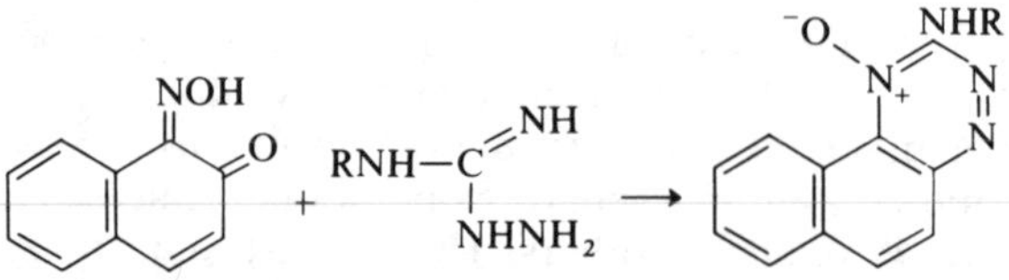

(**772**)  (**773**)  (**774**)

2-Aminonaphthalene (**775**) reacts with diethyl azodicarboxylate to give the trisubstituted hydrazine (**776**) which can be cyclized in wet piperidine to 1,4-dihydro-3-oxonaphtho-[2,1-*e*][1,2,4]triazine-1-carboxylate (**777**) ⟨21CB213⟩.

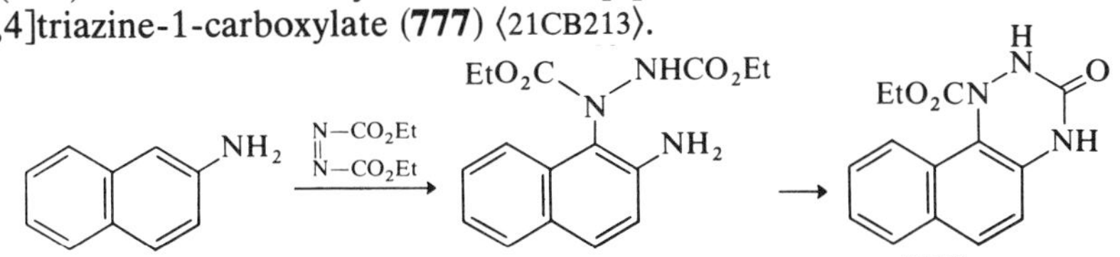

(**775**) (**776**) (**777**)

In a number of cases nitrilimines (**778**) have been used as a three-atom synthetic fragment. Reaction of the spiroaziridine (**779**) with nitrilimines (**778**) led to the 1,4,5,6-tetrahydro-1,2,4-triazines (**780**) while with the spiroazirine (**781**) the rearranged cyclobuta[*e*]-[1,2,4]triazines (**782**) were formed ⟨79BCJ3654⟩. 8-Azaheptafulvalenes (troponimines) (**783**) react with the nitrilimine (**784**) to give a mixture of the two isomeric cyclohepta[*e*]-[1,2,4]triazines (**785** and **786**) ⟨80T935⟩. 1,2-Dihydro-1,2,4-benzotriazines (**788**) were isolated when the nitrilimine (**784**) was reacted with the sulfimines (**787**) ⟨74CC485⟩.

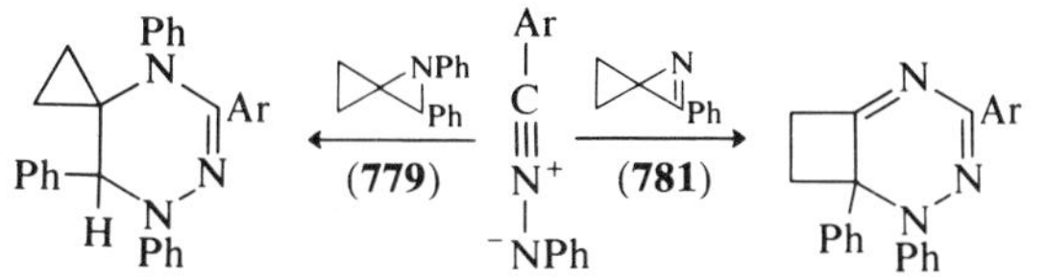

(**780**) (**778**) (**782**)

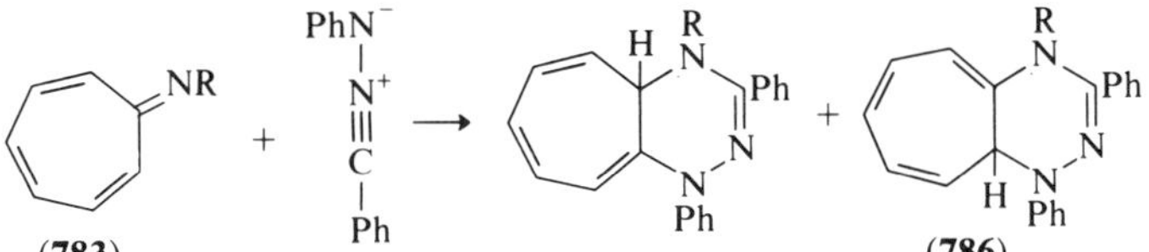

(**783**) (**784**) (**785**) (**786**)

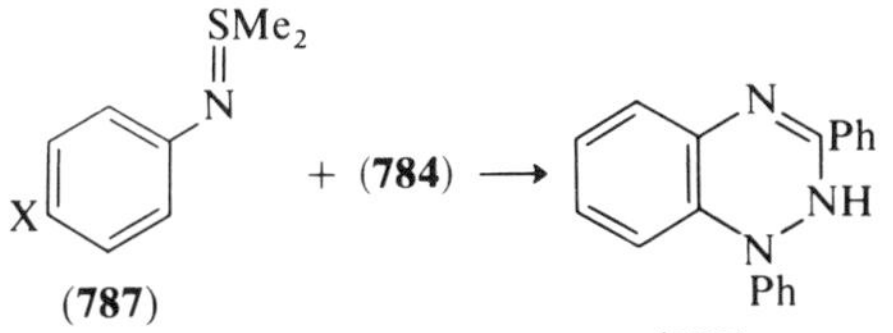

(**787**) (**788**)

### 2.19.4.5 Synthesis from More than Two Fragments

In the previous chapters we mentioned a number of reactions which can be treated as methods for the preparation of 1,2,4-triazines from more than two fragments since one of the starting materials is prepared from two or three compounds. Here we present some further reactions.

Interaction of phenacyl bromides (**789**) with acylhydrazines (**790**) affords 1,2,4-triazines (**791**), the formation of which is rationalized as shown in Scheme 29. In this mechanism the phenacyl bromide is a two-atom fragment, one acylhydrazine a three atom fragment, while the second acylhydrazine provides N(4) as a one-atom fragment ⟨71TL2315, 77T1043⟩.

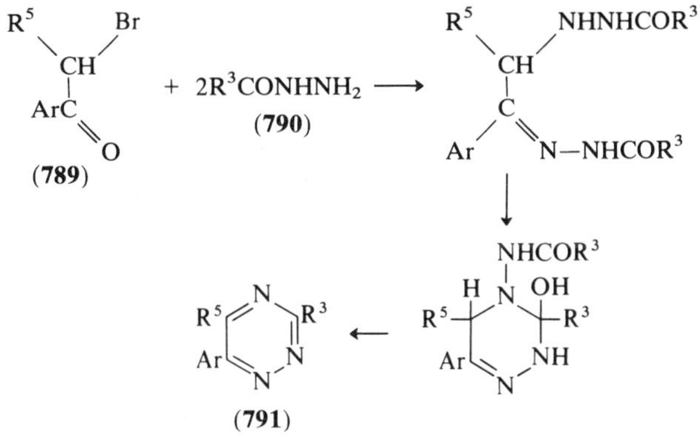

(**789**) (**790**) (**791**)

**Scheme 29**

$\alpha$-Ketocarboxylates react with acylhydrazines to give the acylhydrazones (**672**) which can be cyclized with hydrazine or hydroxylamine to yield 4-amino- and 4-hydroxy-1,2,4-triazin-5-ones (**674**, **675**); see Scheme 26 ⟨78HC(33)189, p. 563⟩.

$\alpha$-Amino ketones (**792**) and azirines (**793**) react as three-atom fragments with hydrazine to give the $\alpha$-aminohydrazones (**794**) which are cyclized by acyl chlorides to give 4,5-dihydro-1,2,4-triazines (**795**), and by aldehydes to form 2,3,4,5-tetrahydro-1,2,4-triazines (**796**; Scheme 30) ⟨78HC(33)189, p. 575, 78KGS342⟩.

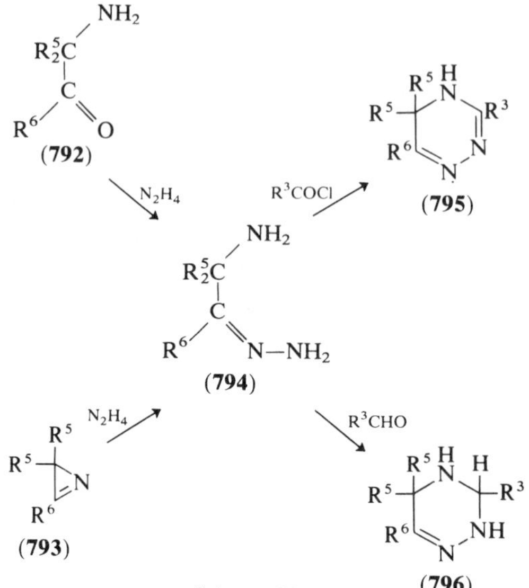

**Scheme 30**

When the hydroxyliminoenamines (**797**) were treated with hydrazine, 2,3-dihydro-1,2,4-triazine 4-oxides (**799**) were obtained. The intermediates are the $\alpha$-hydrazono oximes (**798**) which as indicated above (*cf.* **635** → **637**, Section 2.19.4.2.1) can also be cyclized with ketones to afford the 2,3-dihydro derivatives (**800**; Scheme 31) ⟨75S794, 76GEP2527490⟩.

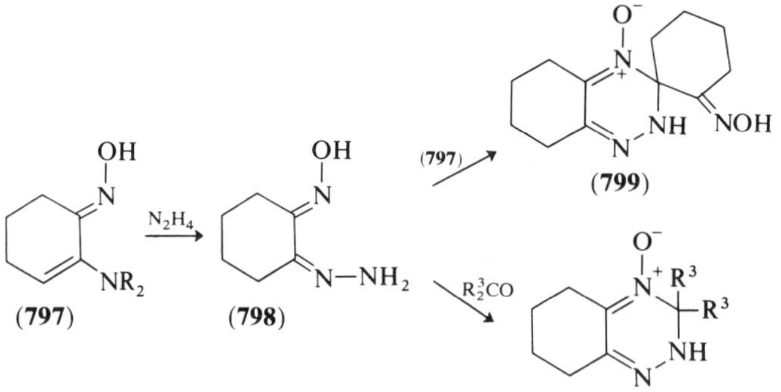

**Scheme 31**

When the compound (**801**) was heated for 1 h at 125–133 °C, purple–red crystals were isolated to which structure (**802**) was assigned ⟨64CB566⟩.

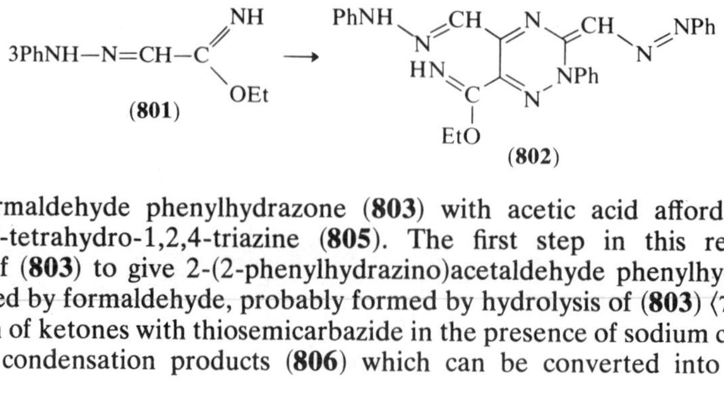

Treating formaldehyde phenylhydrazone (**803**) with acetic acid affords 4-anilino-2-phenyl-2,3,4,5-tetrahydro-1,2,4-triazine (**805**). The first step in this reaction is the dimerization of (**803**) to give 2-(2-phenylhydrazino)acetaldehyde phenylhydrazone (**804**) which is cyclized by formaldehyde, probably formed by hydrolysis of (**803**) ⟨73AJC1297⟩.

The reaction of ketones with thiosemicarbazide in the presence of sodium cyanide affords Strecker-type condensation products (**806**) which can be converted into 3-thioxo-1,6-

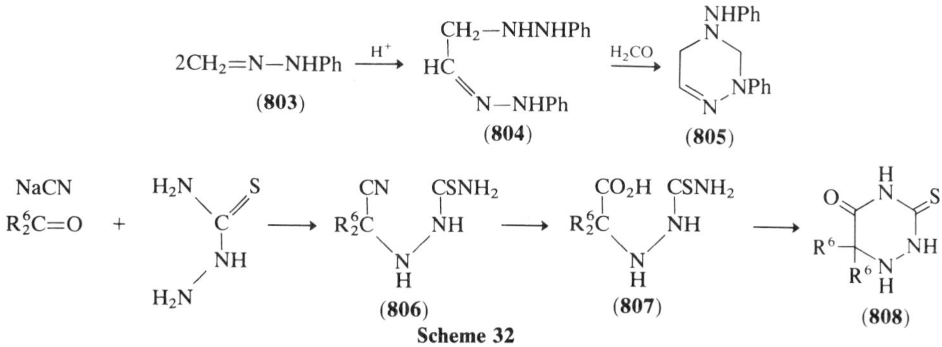

**Scheme 32**

dihydro-1,2,4-triazin-5-ones (**808**) by hydrolysis of the nitrile group and cyclization of the thiosemicarbazido–acetic acids (**807**; Scheme 32) ⟨76JCS(P1)83⟩.

2-Hydroxy-1-nitrohydrazones (**809**) react with primary amines and formaldehyde to give 2,3,4,5-tetrahydro-1,2,4-triazines (**810**) ⟨64MI21901⟩, and similar compounds (**812**) can be obtained from the reaction of aldehyde hydrazones (**811**) with primary amines and aldehydes ⟨74KGS425, 75KGS1290, 74UKZ1220⟩.

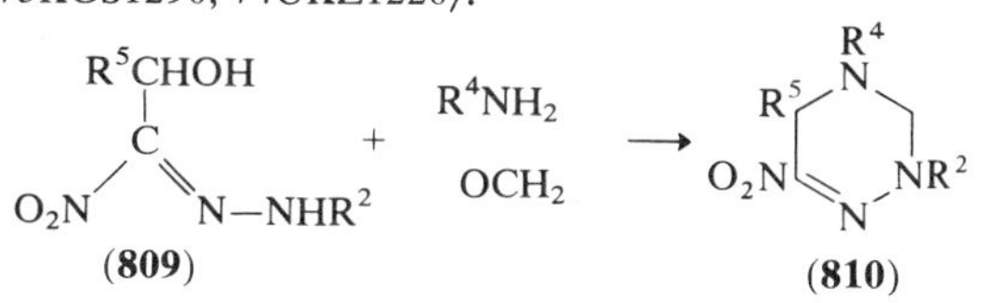

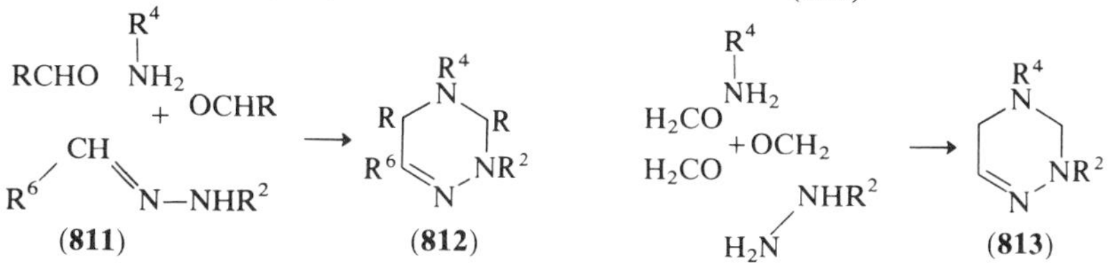

A third method for the preparation of 2,3,4,5-tetrahydro-1,2,4-triazines is to react hydrazines with primary amines and formaldehyde ⟨78HC(33)189, p. 630⟩. The first step here is probably the formation of a formaldehyde hydrazone (**811**; $R^6 = H$) which reacts with another molecule of formaldehyde to give compound (**813**), the structure of which is similar to (**809**).

## 2.19.5 APPLICATIONS AND IMPORTANT COMPOUNDS

A large number of suggested uses for compounds containing the 1,2,4-triazine ring have been recorded in the literature. It seems to us, however, that only a few are of any significance. 4-Amino-1,2,4-triazin-5-ones are biochemically highly active substances and their use as herbicides is claimed by different groups ⟨78HC(33)189, p. 1001⟩. Two compounds of this group are widely used as herbicides, *viz.* 4-amino-6-*t*-butyl-3-methylthio-1,2,4-triazin-5-one (**8**; sencor, metribuzin, BAY 94337) and 4-amino-3-methyl-6-phenyl-1,2,4-triazin-5-one (**9**; Goltix). Another group of biochemically active 1,2,4-triazines are the 6-(5-nitrofuryl)-substituted 1,2,4-triazines, especially 3-amino-6-[(5-nitro-2-furyl)vinyl]-[1,2,4]triazine (**814**; $R^5 = H$: panfuran). These compounds have been tested for their pharmacological, antibacterial and tuberculostatic activity ⟨78HC(33)189, p. 1001⟩.

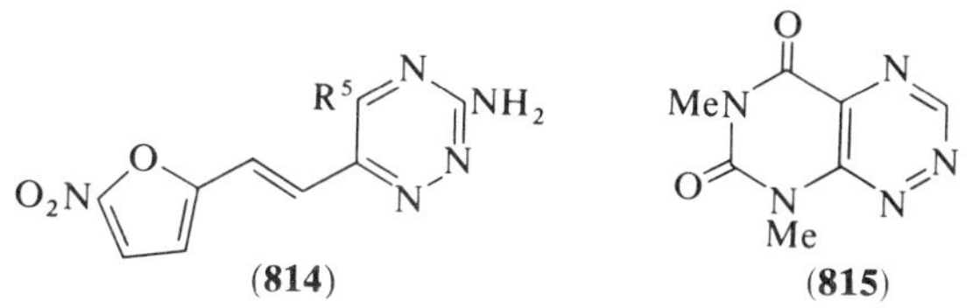

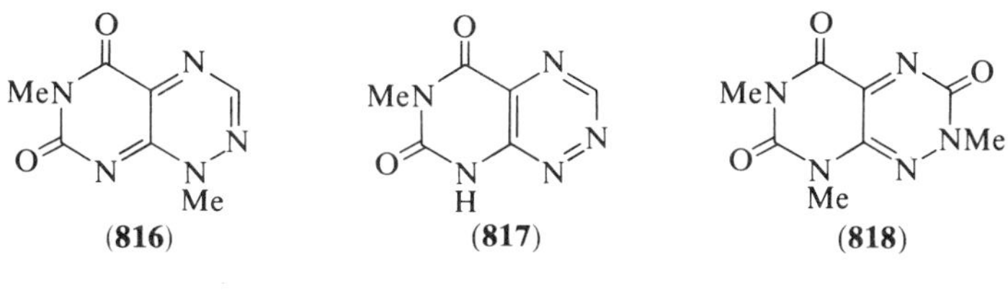

*N*-Methyl derivatives of pyrimido[5,4*e*][1,2,4]triazine-5,7-dione are the naturally occurring antibiotics fervenulin (planomycin; **815**), toxoflavin (xanthothricin; **816**) and reumycin (**817**). The related antibiotic MSD-92 is 2,6,8-trimethylpyrimido[5,4-*e*]-[1,2,4]triazine-3,5,7-trione (**818**) ⟨78HC(33)189, p. 808⟩.

The biochemical properties of 1,2,4-triazine-3,5-dione (6-azauracil; **12**), 6-methyl-1,2,4-triazine-3,5-dione (6-azathymine; **13**), 2-(*β*-D-ribofuranosyl)-1,2,4-triazine-3,5-dione (6-azauridine; **14**) and 2-(*β*-D-deoxyribofuranosyl)-6-methyl-1,2,4-triazine-3,5-dione (6-azathymidine; **819**) have been extensively studied and various uses proposed ⟨78HC(33)189, p. 1002⟩.

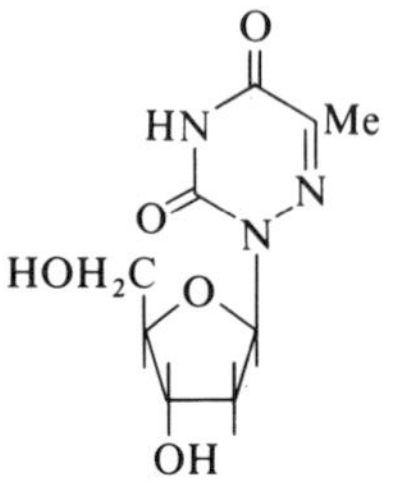

(**819**) 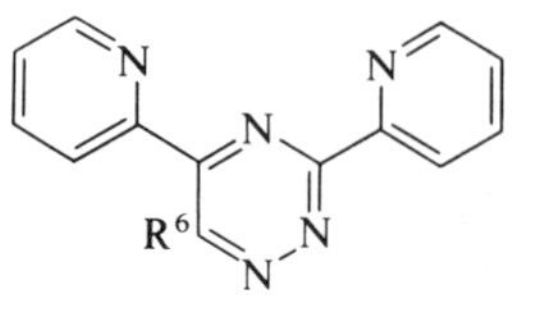      (**820**) 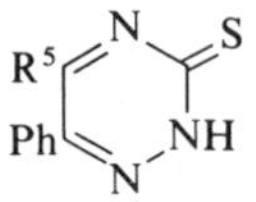      (**821a**) R⁵ = H   (**821b**) R⁵ = Ph

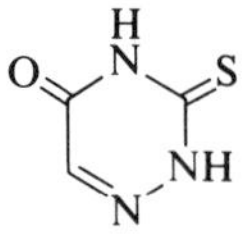

(**822**) 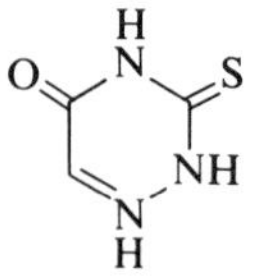      (**823**) 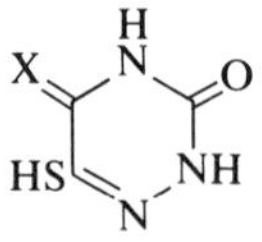      (**824a**) X = O   (**824b**) X = S      (**825**)

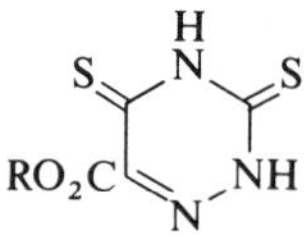

Many 1,2,4-triazines form complexes with metal ions and can be used for their determination. Thus, 3- and/or 5-(2-pyridyl)-substituted 1,2,4-triazines (*e.g.* **820**) can be used for the determination of iron(II), cobalt(II), nickel(II), zinc(II) and copper(I) ions, thallium and palladium ions can be analyzed by 6-phenyl- (**821a**) and 5,6-diphenyl-1,2,4-triazine-3-thione (**821b**), while osmium can be determined by 3-thioxo-1,2,4-triazin-5-one (**822**), 3-thioxo-dihydro-1,2,4-triazin-5-one (**823**), 6-mercapto-1,2,4-triazine-3,5-dione (**824a**), 6-mercapto-5-thioxo-1,2,4-triazin-3-one (**824b**) and 3,5-dithioxo-1,2,4-triazine-6-carboxylates (**825**) ⟨78HC(33)189, p. 1004⟩.

Pyrimido[4',5':5,6][1,2,4]triazino[4,3-*e*]indazoles (**826**), and derivatives of 1,2,4-benzotriazine-7-carboxylic acid 1-oxide (**827**), can be used as dyes ⟨78HC(33)189, p. 1004⟩.

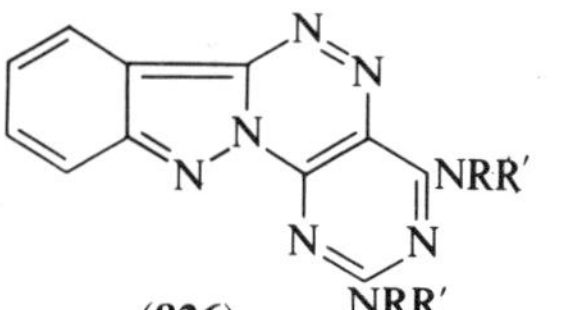

(**826**) 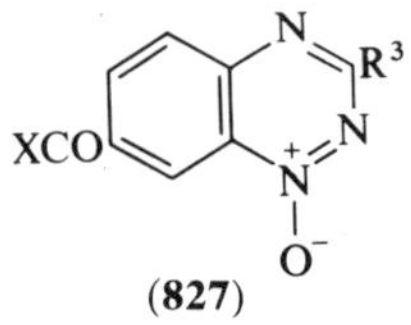      (**827**)

# 2.20

# 1,3,5-Triazines

J. M. E. QUIRKE

*University of Durham* and *Florida International University*

| | | |
|---|---|---|
| 2.20.1 | INTRODUCTION | 459 |
| | *2.20.1.1  1,3,5-Triazine* | 459 |
| | *2.20.1.2  Cyanuric Acid* | 460 |
| | *2.20.1.3  Cyanuric Chloride* | 460 |
| | *2.20.1.4  Melamine* | 460 |
| 2.20.2 | STRUCTURE | 460 |
| | *2.20.2.1  Molecular Dimensions of 1,3,5-Triazines* | 461 |
| | *2.20.2.2  Molecular Spectra* | 462 |
| | *2.20.2.2.1  NMR spectra* | 462 |
| | *2.20.2.2.2  UV and related spectra* | 464 |
| | *2.20.2.2.3  IR spectra* | 464 |
| | *2.20.2.2.4  Mass spectra* | 465 |
| | *2.20.2.2.5  Photoelectron spectra* | 466 |
| | *2.20.2.3  Aromaticity* | 466 |
| | *2.20.2.4  Shape and Conformation of Saturated and Partially Saturated 1,3,5-Triazines* | 466 |
| | *2.20.2.5  Tautomerism* | 467 |
| | *2.20.2.6  Structure of the Polyazacycl[3.3.3]azines* | 468 |
| 2.20.3 | REACTIVITY | 469 |
| | *2.20.3.1  Reactions of 1,3,5-Triazine* | 469 |
| | *2.20.3.1.1  Reactions with electrophiles* | 469 |
| | *2.20.3.1.2  Reactions with nucleophiles* | 470 |
| | *2.20.3.1.3  Cycloaddition reactions* | 471 |
| | *2.20.3.2  Reactions of Alkyl-1,3,5-triazines* | 472 |
| | *2.20.3.2.1  Reactions of the 1,3,5-triazine ring* | 472 |
| | *2.20.3.2.2  Reactions of the alkyl substituents* | 472 |
| | *2.20.3.3  Reactions of Aryl-1,3,5-triazines* | 473 |
| | *2.20.3.3.1  Reactions with nucleophiles* | 473 |
| | *2.20.3.4  Reactions of 1,3,5-Triazine Aldehydes* | 474 |
| | *2.20.3.5  Reactions of Ketoalkyl-1,3,5-triazines* | 474 |
| | *2.20.3.6  Reactions of 1,3,5-Triazinecarboxylic Acids and Related Compounds* | 474 |
| | *2.20.3.7  Reactions of 1,3,5-Triazinecarbonitriles* | 474 |
| | *2.20.3.8  Reactions of Amino-1,3,5-triazines* | 475 |
| | *2.20.3.8.1  Reactions of the 1,3,5-triazine ring* | 475 |
| | *2.20.3.8.2  Reactions of the amino group* | 475 |
| | *2.20.3.8.3  Reactions of alkylamino-1,3,5-triazines* | 476 |
| | *2.20.3.8.4  Rearrangements of amino-1,3,5-triazines* | 476 |
| | *2.20.3.8.5  Reactions of bis(1,3,5-triazinyl)hydrazines and related compounds* | 476 |
| | *2.20.3.9  Reactions of Azido-1,3,5-triazines* | 477 |
| | *2.20.3.10  Reactions of 1,3,5-Triazines Bearing O-Linked Substituents* | 477 |
| | *2.20.3.10.1  Reactions of oxo-1,3,5-triazines* | 477 |
| | *2.20.3.10.2  Reactions of isocyanurates and triazine-2,4-diones* | 478 |
| | *2.20.3.10.3  Reactions of alkyl cyanurates* | 480 |
| | *2.20.3.10.4  Reactions of aryl cyanurates* | 481 |
| | *2.20.3.11  Reactions of 1,3,5-Triazines Bearing S-Linked Substituents* | 481 |
| | *2.20.3.11.1  Reactions of 1,3,5-triazinethiones* | 481 |
| | *2.20.3.11.2  Reactions of thiocyanurates and isothiocyanurates* | 482 |
| | *2.20.3.12  Reactions of Halogenated 1,3,5-Triazines* | 482 |
| | *2.20.3.12.1  Nucleophilic substitution of cyanuric chloride* | 482 |
| | *2.20.3.12.2  Cyanuric chloride as a halogenating and deoxygenating agent* | 485 |
| | *2.20.3.12.3  Reactions of fluoro-1,3,5-triazines* | 486 |

2.20.3.13   *Reactions of Hexahydro-1,3,5-triazines*                                          486
2.20.3.14   *Reactions of Hexamethylenetetramine*                                             487
2.20.3.15   *Reactions of Fused 1,3,5-Triazine Systems*                                       488
   2.20.3.15.1   *Rearrangements of fused 1,3,5-triazines*                      488
   2.20.3.15.2   *Reactions of triazacycl[3.3.3]azine and related compounds*     489
2.20.3.16   *Reactions of 1,3,5-Triazine Betaines*                                            489

2.20.4   SYNTHESES                                                                            490

2.20.4.1   *Synthesis of 1,3,5-Triazines by Formation of One Carbon–Nitrogen Bond*           490
   2.20.4.1.1   *Synthesis from urea, thiourea and their derivatives*           490
   2.20.4.1.2   *Synthesis of fused rings*                                       491
2.20.4.2   *Synthesis of 1,3,5-Triazines by Formation of Two Carbon–Nitrogen Bonds from [5 + 1]*
   *Atom Fragments*                                                             492
   2.20.4.2.1   *Synthesis using biguanides as the 5-atom fragment*             492
   2.20.4.2.2   *Synthesis from other 5-atom fragments*                          493
   2.20.4.2.3   *Synthesis of fused ring systems*                                494
2.20.4.3   *Synthesis of 1,3,5-Triazines by Formation of Two Carbon–Nitrogen Bonds from [2 + 4]*
   *Atom Fragments*                                                             495
   2.20.4.3.1   *Synthesis from N-cyanoamidine and related compounds*            495
   2.20.4.3.2   *Synthesis from dicyandiamide*                                   498
   2.20.4.3.3   *Synthesis using other four-atom components*                     499
   2.20.4.3.4   *Synthesis of fused ring systems*                                500
2.20.4.4   *Synthesis of 1,3,5-Triazines by Formation of Two Carbon–Nitrogen Bonds from [3 + 3]*
   *Atom Fragments*                                                             500
   2.20.4.4.1   *Synthesis using amidines and related compounds*                 500
   2.20.4.4.2   *Synthesis of fused ring systems*                                502
2.20.4.5   *Synthesis of Symmetrical 1,3,5-Triazines by Formation of Three Carbon–Nitrogen Bonds*  503
   2.20.4.5.1   *Synthesis from nitriles*                                        503
   2.20.4.5.2   *Synthesis from imino derivatives*                               506
   2.20.4.5.3   *Synthesis of symmetrical isocyanurates from isocyanates*        507
   2.20.4.5.4   *Synthesis of 1,3,5-triazine derivatives from the reaction of ammonia with aldehydes*  508
   2.20.4.5.5   *Synthesis of melamine*                                          509
   2.20.4.5.6   *Synthesis of fused ring systems*                                509
2.20.4.6   *Synthesis of Unsymmetrical 1,3,5-Triazines by Formation of Three Carbon–Nitrogen*
   *Bonds*                                                                      510
   2.20.4.6.1   *Synthesis by cotrimerization reactions*                         510
   2.20.4.6.2   *Synthesis from amidines, isocyanates, aldehydes and ketones*    512
   2.20.4.6.3   *Synthesis of fused ring systems*                                517
2.20.4.7   *Synthesis of 1,3,5-Triazines by Rearrangement or Cleavage of Other Heterocyclic Systems*  517
   2.20.4.7.1   *Synthesis by formation of one carbon–nitrogen bond*            517
   2.20.4.7.2   *Synthesis by formation of two carbon–nitrogen bonds*           518
   2.20.4.7.3   *Synthesis by formation of three carbon–nitrogen bonds*         520
   2.20.4.7.4   *Synthesis of fused ring systems*                                521
2.20.4.8   *Synthesis of Polyaza[3.3.3]cyclazines*                                            521
2.20.4.9   *An Assessment of the Application of the Synthetic Routes in the Synthesis of 1,3,5-*
   *Triazines*                                                                  522
   2.20.4.9.1   *Synthesis of 1,3,5-triazine*                                    522
   2.20.4.9.2   *Synthesis of monosubstituted 1,3,5-triazines*                  522
   2.20.4.9.3   *Synthesis of 2,4-disubstituted 1,3,5-triazines bearing a single type of substituent*  522
   2.20.4.9.4   *Synthesis of 2,4-disubstituted 1,3,5-triazines bearing two different substituents*  522
   2.20.4.9.5   *Synthesis of 2,4,6-trisubstituted 1,3,5-triazines*             523
   2.20.4.9.6   *Synthesis of other 1,3,5-triazine derivatives*                 523

2.20.5   OCCURRENCE OF 1,3,5-TRIAZINES                                                        524

2.20.6   APPLICATIONS                                                                         524

2.20.6.1   *Applications of 1,3,5-Triazine*                                                   524
2.20.6.2   *Applications of Aryl-1,3,5-triazines*                                             525
2.20.6.3   *Applications of Amino-1,3,5-triazines*                                            525
2.20.6.4   *Applications of Cyanurates and Isocyanurates*                                     525
2.20.6.5   *Physiological Importance of 5-Azacytidine and Related Compounds*                  526
2.20.6.6   *Applications of Alkylthio-1,3,5-triazines*                                        526
2.20.6.7   *Applications of Halogenated 1,3,5-Triazines*                                      526
   2.20.6.7.1   *Synthetic applications of cyanuric chloride*                    526
   2.20.6.7.2   *1,3,5-Triazine dyes and optical brighteners*                   526
   2.20.6.7.3   *Miscellaneous applications of cyanuric chloride*               527
   2.20.6.7.4   *Applications of fluoro- and fluoroalkyl-1,3,5-triazines*       527
2.20.6.8   *1,3,5-Triazine Herbicides*                                                        528
2.20.6.9   *Applications of 1,2-Dihydro-1,3,5-triazines*                                      529
2.20.6.10   *Applications of Organometallic 1,3,5-Triazines*                                  529
2.20.6.11   *Applications of Hexahydro-1,3,5-triazine and Hexamethylenetetramine*             529

## 2.20.1 INTRODUCTION

The 1,3,5-triazines are amongst the oldest known organic molecules. Originally they were called the symmetric triazines, usually abbreviated to *s*- or *sym*-triazines. The numbering follows the usual convention of beginning at the heteroatom as shown for the parent compound 1,3,5-triazine (**1**). Rather non-systematic nomenclature is prevalent even in the current literature, because some of the compounds have been known for at least 150 years. The non-systematic names of some of the more important 1,3,5-triazines are listed in Table 1. The terms melamine, cyanuric acid and cyanuric chloride will be used throughout this chapter, and the term triazine will refer to 1,3,5-triazines only. In addition to the above names, 2,4,6-trialkoxy-1,3,5-triazines (**2**) are called cyanurates. Similarly, 1,3,5-trialkyl-1,3,5-triazines (**3**) are called isocyanurates.

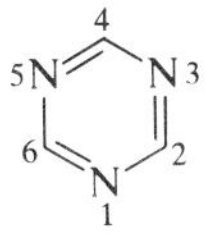

**Table 1** Common Names of Some Important 1,3,5-Triazines:

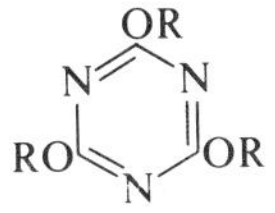

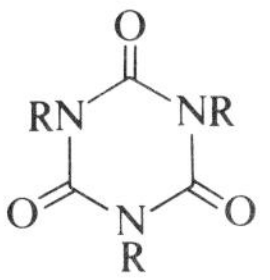

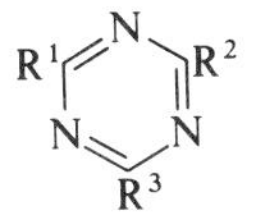

| Name | $R^1$ | $R^2$ | $R^3$ |
|---|---|---|---|
| Cyanuric chloride | Cl | Cl | Cl |
| Cyanuric acid[a] | OH | OH | OH |
| Melamine | NH$_2$ | NH$_2$ | NH$_2$ |
| Ammeline[a] | OH | NH$_2$ | NH$_2$ |
| Ammelide[a] | OH | OH | NH$_2$ |
| Acetoguanamine | Me | NH$_2$ | NH$_2$ |
| Acetoguanide[a] | Me | NH$_2$ | OH |
| Acetoguanamide[a] | Me | OH | OH |

[a] Exist mainly in keto forms (see Section 2.20.2)

There have been many developments in the chemistry of the 1,3,5-triazines since the comprehensive reviews of Smolin and Rapoport ⟨59HC(13)1⟩ and Modest ⟨B-61MI22000, p.627⟩. It is impossible to include all the vast range of 1,3,5-triazines reported in the literature: 138 pages of the Compound Index in the Ninth Collective Index of *Chemical Abstracts* are dedicated to this system. This review will concentrate on the more important recent developments of 1,3,5-triazine chemistry, and reference will be made to previous reviews, or typical papers, to illustrate the better-known aspects of the subject.

The 1,3,5-triazines are an unusual class of compound. Although they include some of the oldest known organic compounds, the parent compound 1,3,5-triazine was not correctly identified until 1954 (see Section 2.20.1.1). 1,3,5-Triazine is atypical of its class, being highly reactive and undergoing ring cleavage very easily. Cyanuric chloride remains the most important starting material for the synthesis of 1,3,5-triazines; the majority of 1,3,5-triazine herbicides, for example, are prepared from this compound.

Before embarking on a discussion of recent developments in 1,3,5-triazine chemistry, it is appropriate to look at the 1,3,5-triazines in their historical context. Thus, the original syntheses and characterization of four of the more important compounds, 1,3,5-triazine, cyanuric chloride, cyanuric acid and melamine, will be discussed briefly.

### 2.20.1.1 1,3,5-Triazine

In 1895, Nef treated hydrogen cyanide with ethanol in an ether solution saturated with hydrogen chloride. The resultant salt was treated with base to yield the imidate (**4**) which formed 1,3,5-triazine on distillation (equation 1) ⟨1895LA(287)333⟩. The yield was low (10%).

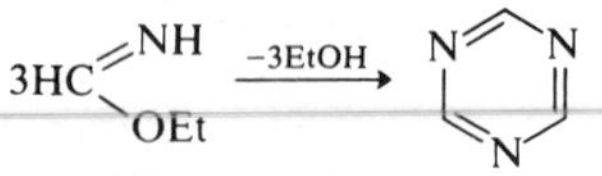

$$3HC \underset{OEt}{\overset{NH}{<}} \xrightarrow{-3EtOH} \text{(triazine)} \qquad (1)$$

(4)

On the basis of a single cryoscopic molecular weight determination, the compound was tentatively assigned as the dimeric species C=NCH=NH (5) (the cryoscopic determination in benzene gave a value of 64, rather than 54 for $C_2H_2N_2$, but the discrepancy was attributed to poor quality benzene). It was not until 1954 that Grundmann and Kreutzberger ⟨54JA632⟩ proved that the compound was the trimer (1) by repeating the cryoscopic determination in benzene. These workers employed a modification of Nef's original method, in which hydrogen chloride and hydrogen cyanide reacted to form 1,3,5-triazine *via* the intermediate (6; Scheme 1).

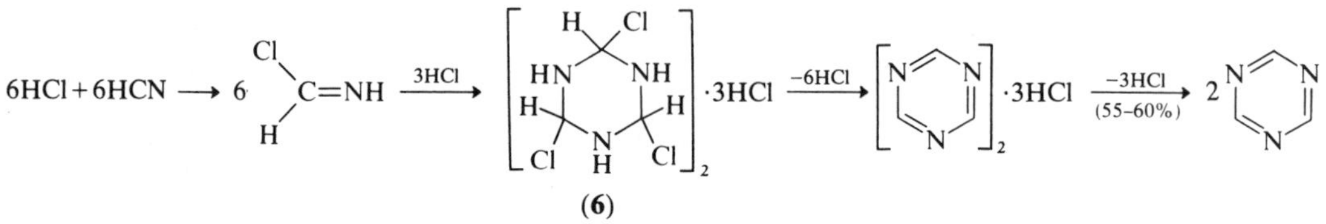

$$6HCl + 6HCN \longrightarrow 6 \underset{H}{\overset{Cl}{<}} C=NH \xrightarrow{3HCl} \left[ \text{(6)} \right] \cdot 3HCl \xrightarrow{-6HCl} \left[ \text{(triazine)} \right]_2 \cdot 3HCl \xrightarrow[(55-60\%)]{-3HCl} 2 \text{ (triazine)}$$

(6)

**Scheme 1**

### 2.20.1.2 Cyanuric acid

In 1776, Scheele first prepared cyanuric acid by pyrolysis of uric acid (7) ⟨B-1793MI22000⟩. In 1830, the first correct analysis was obtained by Liebig and Wohler ⟨1835MI22000⟩, but the molecular formula was incorrectly assigned as $C_6H_6N_6O_6$. Dreschel deduced the correct molecular formula, $C_3H_3N_3O_3$, in 1875 ⟨1875JPR(11)289⟩. Cyanuric acid does not have many industrial applications itself, but many of its derivatives are valuable monomers (see Section 2.20.5).

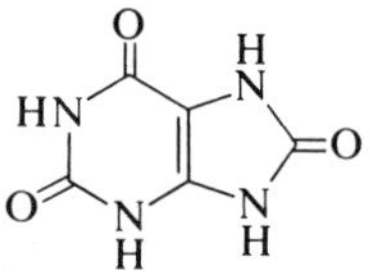

(7)

### 2.20.1.3 Cyanuric chloride

Cyanuric chloride was first prepared by the trimerization of cyanogen chloride (CNCl) in sunlight ⟨1828MI22000⟩. Its composition was determined by Liebig ⟨1829MI22000⟩. Initially, it was thought that Serullas had prepared an isomer of cyanogen chloride rather than the trimer.

### 2.20.1.4 Melamine

Melamine was first prepared in 1834 by Liebig by fusing potassium thiocyanate with twice its weight of ammonium chloride ⟨1834LA(10)17⟩. The product was, in fact, predominantly melamine thiocyanate, but treatment with alkali yields the free compound. The commercial potential of melamine in polymer syntheses was not realized until a century after its discovery.

## 2.20.2 STRUCTURE

There have been significant advances in the study of the structure of the 1,3,5-triazines since the last major review ⟨62RCR712⟩. Perhaps the most interesting developments have been in the study of the conformations of the hexahydro-1,3,5-triazines (see Section 2.20.2.4).

### 2.20.2.1 Molecular Dimensions of 1,3,5-Triazines

There have been several studies on the molecular dimensions of the 1,3,5-triazines. The compounds follow the general trend of other heterocyclic systems; the CNC bond angles are less than 120°, and thus the ring is distorted from a regular hexagon (see Chapter 2.01). The bond lengths of several symmetrical 1,3,5-triazines are summarized in Table 2.

**Table 2** Molecular Dimensions of Symmetrical 1,3,5-Triazines (10)

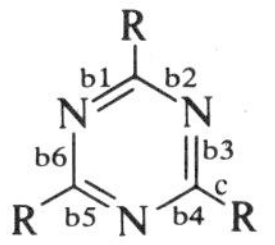

(10)

| R | Bond lengths (Å) | Bond angles (°) | | Method | Ref. |
|---|---|---|---|---|---|
| | | CNC | NCN | | |
| H | b1, b2, b3, b4, b5, b6 = 1.319; c = 0.998 | 113.2 | 126.8 | X[b] | 72JCS(P2)642 |
| Cl | b1, b2, b3, b4, b5, b6 = 1,33; c = 1.68 | 115 | —[c] | E[a] | 72JCS(P2)642 |
| Ar[d] | b1, b4 = 1.31; b2, b5 = 1.33; b3, b6 = 1.30 | —[c] | —[c] | X[b] | 65SA1563 |
| $NH_2$ | b1, b4, b6 = 1.35; b2, b3 = 1.34; c = 1.37 | 116–117 | 123–125 | X[b] | 72JCS(P2)642 |
| $NMe_2$ | b1, b2, b3, b4, b5, b6 = 1.345; c = 1.366 | 112.7 | 127.3 | X[b] | 72JCS(P2)642 |
| $N_3$ | b1, b3, b5 = 1.38; b2, b4, b6 = 1.31; c = 1.38 | 113 | 127 | X[b] | 35MI22000 |
| $NHNH_2$ | b1 = 1.355; b2, b3, b5, b6 = 1.336–1.339; b4 = 1.362; c = 1.343–1.362 | 112.7–113.8 | 126.3–127.4 | X[b] | 76AX(B)2101 |

[a] E = Electron diffraction.　[b] X = X-ray crystallography.　[c] Not determined.　[d] Ar = $4\text{-}ClC_6H_4$.

The presence of substituents has relatively little effect on the bond lengths and angles of the ring. 2,4,6-Trihydrazino-1,3,5-triazine molecules are hydrogen bonded, and the shortest hydrogen bond is particularly close, 2.83 Å. The ring C—N bond lengths resemble those of melamine (Table 2), and the C—N bond length of the substituent (bond c in **10**) indicates a possible resonance contribution of the type (**8**). There is no evidence of the imido tautomer (**9**). X-Ray analyses of cyanuric acid have confirmed the predominance of the triketo form of the molecule. The C—N and C—O bond lengths are 1.371 and 1.223 Å respectively and the N—H bond length is 0.9 Å. The CNC and NCN bond angles are 124.6 and 115.4° respectively 〈64MI22000〉. The molecule is hydrogen bonded and the bond lengths are 2.772–2.798 Å. Usanmaz has reported the structure of phenyl isocyanurate 〈79AX(B)1117〉. The C—O and C—N bond lengths are 1.191 and 1.392 Å respectively, and the CNC and NCN bond angles are 125.1 and 114.9° respectively. The benzene rings make an angle of 72.9° with the triazine ring. The triazine (**11**) has also been analyzed 〈77JHC857〉. The structure will not be discussed in detail, but the heterocyclic ring exists in the boat form. The C—S bond length is 1.659 Å.

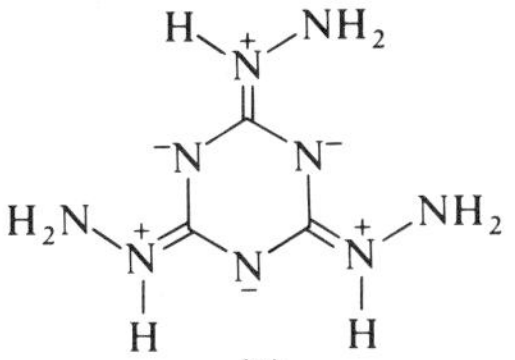

(8)

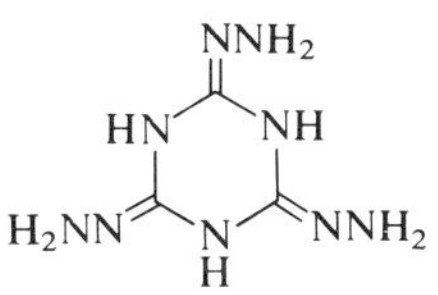

(9)

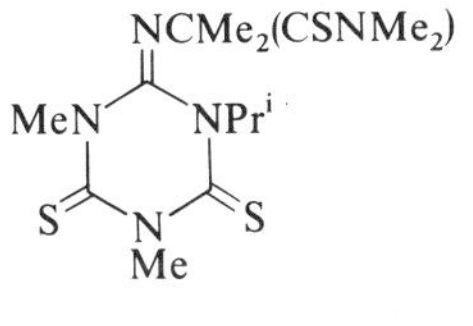

(11)

The dihydrotriazines are of interest because of their antimalarial activity, and their potential as cancer chemotherapeutic drugs (see Section 2.20.5.10). The X-ray crystallographic analysis of the antimalarial drug 'cycloguanil' (**12**) (as the hydrochloride salt) has been reported ⟨78CC188⟩, and the bond lengths are shown. The triazine ring atoms N(1), N(3), N(5), C(2) and C(4) are coplanar within ±0.5 Å and the phenyl group makes an angle of 80° with the ring. The bond angles were not quoted. The structure of 'Baker's antifol' (13) has also been determined (see Section 2.20.5.10). The ring atoms N(1), N(3), N(5), C(2) and C(4) are nearly coplanar ⟨79AX(B)2113⟩. The most important structural feature of 'Baker's antifol' is that the molecule adopts a linearly extended conformation.

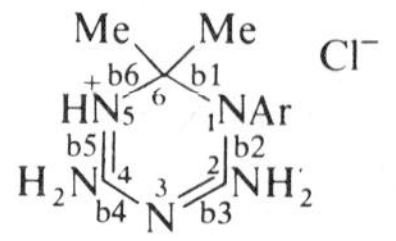

(12) Ar = 4-ClC₆H₄, b1 = 1.493, b2 = 1.345, b3 = 1.348, b4 = 1.333, b5 = 1.342, b6 = 1.464, N—Ar = 1.446, C—NH₂ = 1.326 Å

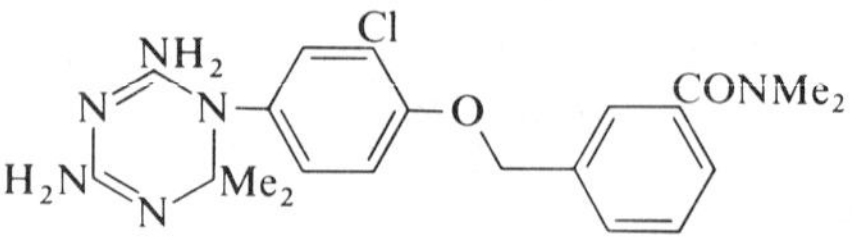

(13)

There have been relatively few reports on the hexahydro-1,3,5-triazines. Lund analyzed (14) and found that the C—C and C—N bond lengths were normal (1.53 and 1.47 Å respectively). The NCN and CNC bond angles were 114.6 and 109.0° respectively. The ring was in the chair form with the methyl groups bonded equatorially to the ring carbon ⟨58ACS1768⟩. Choi et al. have investigated the structure of 1,3,5-triacetylhexahydro-1,3,5-triazine (15). The bond lengths vary slightly — from 1.446 to 1.467 Å — and the bond angles vary from 109.6 to 111.5°. The three acetamide groups are essentially planar, and the ring is mainly in the chair form ⟨75AX(B)2934⟩.

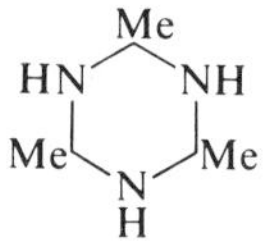

(14) 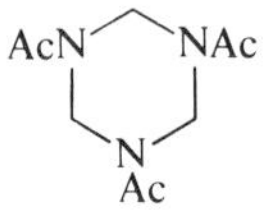          (15)

## 2.20.2.2 Molecular Spectra

### 2.20.2.2.1 NMR spectra

The $^1$H NMR spectra of the 1,3,5-triazines are quite simple. The ring protons give sharp lines which are not broadened by nitrogen quadrupolar relaxations ⟨65BSB119⟩. The chemical shifts of the ring protons are 1 to 2 p.p.m. downfield from benzene protons, presumably

**Table 3**  $^1$H NMR Chemical Shifts of 1,3,5-Triazines

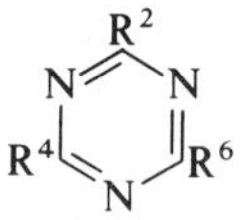

| $R^2$ | $R^4$ | $R^6$ | Solvent | Chemical shifts δ (assignment) | Ref. |
|-------|-------|-------|---------|-------------------------------|------|
| H | H | H | CCl₄ | 9.18 | 65BSB119 |
| H | H | H | CDCl₃ | 9.25 | 65BSB119 |
| Me | Me | H | CCl₄ | 2.54 (Me), 8.80 (6-H) | 65BSB119 |
| MeO | MeO | H | CCl₄ | 4.01 (Me), 8.51 (6-H) | 65BSB119 |
| Et₂N | Et₂N | H | CCl₄ | 1.19 (NCH₂CH₃), 3.54 (NCH₂Me), 8.19 (6-H) | 65BSB119 |
| Ph | Ph | H | CCl₄ | 7.41 (H$_{meta}$ + H$_{para}$), 8.85 (H$_{ortho}$), 9.07 (6-H) | 65BSB119 |
| MeS | MeS | MeS | CDCl₃ | 2.53 (SMe) | 66JCS(C)909 |
| CH₂Cl | Ph | Cl | CDCl₃ | 4.6 (CH₂Cl), 7.5 and 8.45 (Ph) | 81AJC623 |
| MeO | MeO | —CH=C(Me*)(OH) | — | 2.1 (Me*), 4.1 (MeO), 5.5 (H of R⁶), 13.07 (OH of R⁶) | 74JHC317 |

because of the effect of the ring nitrogens (see Chapter 2.01), and the presence of electron releasing substituents leads to slight upfield shifts. Typical chemical shifts of 1,3,5-triazine derivatives are given in Table 3.

The spectra of the isocyanurates are summarized in Table 4. The hexahydro-1,3,5-triazines are discussed in Section 2.20.2.4. There are few reports on the $^{13}$C NMR of the 1,3,5-triazines, and in several instances only a part of the spectrum is quoted. Examples of the $^{13}$C NMR chemical shifts are shown in Table 5. The ring carbons are deshielded because they each lie between two nitrogens. The $^{13}$C–$^1$H coupling constants for 1,3,5-triazine are 207.2 and 8.0 Hz for C(2)–H(2) and C(2)–H(4) interactions respectively. The $^{13}$C–$^1$H coupling constant for the interaction between methyl protons and the adjacent ring carbon is 6.9 Hz ⟨75OMR(7)194⟩. Both $^{13}$C NMR and $^1$H NMR spectra confirm that 'hydroxy' and 'mercapto' 1,3,5-triazine occur mainly in the keto forms. Thus the $^{13}$C NMR spectrum of (16) contains peaks at 143.6 and 177.8 δ assigned to C=O and C=S respectively ⟨81CB2075⟩.

(16)

**Table 4** $^1$H NMR Chemical Shifts of Isocyanurates and Related Compounds

| $R^1$ | $R^3$ | $R^5$ | X | Y | Z | Solvent | Chemical shifts δ (assignment) | Ref. |
|---|---|---|---|---|---|---|---|---|
| Ph | Ph | Allyl | O | O | O | (CD$_3$)$_2$SO | 4.43 (CH$_2$CH=CH$_2$), 5.08–6.18 (CH$_2$CH=CH$_2$), 7.43 (Ph) | 70JHC725 |
| Ph | Ph | Bu$^n$ | O | O | O | (CD$_3$)$_2$SO | 3.87 (CH$_2$CH$_2$CH$_2$Me), 0.90–1.52 (CH$_2$CH$_2$CH$_2$CH$_3$), 7.43 (Ph) | 70JHC725 |
| Allyl | Allyl | Me | O | O | O | (CD$_3$)$_2$SO | 3.20 (Me), 4.38 (CH$_2$CH=CH$_2$), 5.17–5.87 (CH$_2$CH=CH$_2$) | 70JHC725 |
| Et | Et | Et | O | O | O | CCl$_4$ | 3.95 (CH$_2$Me), 1.24 (CH$_2$CH$_3$) | 65JCS6858 |
| Me | Me | Me | O | O | O | CCl$_4$ | 3.44 (Me) | 65JCS6858 |
| Me | Me | Me | S | S | S | CDCl$_3$ | 4.19 (Me) | 66JCS(C)909 |
| CH$_2$Ph | H | H | S | S | O | A$^a$ | 6.01 (CH$_2$Ph), 13.06 (NH)$^b$ | 81CB2075 |
| Pr$^i$ | H | CH$_2$CN | O | O | O | CDCl$_3$ | 1.34 (CHMe$_2$), 4.40 (CH$_2$CN), 4.88 (CHMe$_2$), 11.88 (NH) | 76TH22000 |
| Pr$^i$ | H | H | O | O | O | CDCl$_3$ | 1.45 (CH(CH$_3$)$_2$), 4.95 (CHMe$_2$), 11.20 (NH) | 76TH22000 |

$^a$ A = (CD$_3$)$_2$SO/(CD$_3$)$_2$CO (3:1).　$^b$ No value for phenyl given.

**Table 5** $^{13}$C NMR Chemical Shifts of 1,3,5-Triazines in CDCl$_3$

| | Compound | | | | | Chemical shifts (δ) | | | | | | |
|---|---|---|---|---|---|---|---|---|---|---|---|---|
| $R^2$ | $R^4$ | $R^6$ | C-2 | C-4 | C-6 | $C_s^d$ | $C_o^d$ | $C_m^d$ | $C_p^d$ | CN | CONH$_2$ | Ref. |
| H | H | H | 166.1 | 166.1 | 166.1 | — | — | — | — | — | — | 75OMR(7)194 |
| Me | Me | Me | 175.9 | 175.9 | 175.9 | 25.4 | — | — | — | — | — | 75OMR(7)194 |
| Ph | Ph | Ph | 171.7 | 171.7 | 171.7 | 136.4 | 128.6 | 129.1 | 132.5 | — | — | 75OMR(7)194 |
| Ph | Ph | CONH$_2$$^a$ | 166.4$^b$ | 171.7 | 171.7 | 134.7$^c$ | 128.9$^c$ | | 133.3 | — | 163.6$^b$ | 78JHC1055 |
| Ph | Ph | CN | 153.6 | 172.6 | 172.6 | 133.9 | 129.0$^c$ | 129.4$^c$ | 134.1 | 114.9 | — | 78JHC1055 |

$^a$ In (CD$_3$)$_2$SO.　$^{b,c}$ Assignments may be interchanged.
$^d$ C$_s$ = ring attached C; C$_o$, C$_m$, C$_p$ = ortho, meta and para C's respectively.

$^{14}$N NMR is likely to become a most valuable spectrometric technique. The spectrum of 1,3,5-triazine in dioxane has been reported ⟨71T3129⟩, and the chemical shift ($\delta_N = 98$, relative to MeNO$_2$) reflects the high $\pi$-electron density at the ring nitrogens. The $^{19}$F NMR spectrum of perfluoro-2,4-diisopropyl-1,3,5-triazine has been described. The chemical shifts (upfield from CFCl$_3$) are as follows: 31.5 (F), 73.7 (CF($CF_3$)$_2$) and 183.1 ($CF$(CF$_3$)$_2$) ⟨77JCS(P1)1605⟩.

### 2.20.2.2.2 UV and related spectra

The UV spectra of the 1,3,5-triazines have been reviewed previously ⟨62RCR712⟩. The UV absorption of 1,3,5-triazines shows two bands at 272 and 222 assigned as $n-\pi^*$ and $\pi-\pi^*$ transitions respectively. The spectra of the parent compound and other symmetrical 1,3,5-triazines are given in Table 6. The spectrum of cyanuric acid is markedly dependent on the pH; it undergoes a hypsochromic shift with decreasing pH ⟨64SA397⟩. The compound absorbs in the far UV region, but there is a very weak band ($\varepsilon = 0.2$) at *ca.* 280 nm.

**Table 6**   UV Spectra of Symmetrical 1,3,5-Triazines ⟨62RCR712⟩

| Substituent | Absorbance $\lambda_{max}(\varepsilon)$ | Solvent |
|:---:|:---:|:---:|
| H | 272 (891), 222 (151) | Cyclohexane |
| Me | 264 (701), 227 (289) | Cyclohexane |
| Et | 259 (696) | MeOH |
| MeO | 258 (521), 187 (42 000) | H$_2$O |
| NH$_2$ | 236 (weak), 208 (51 000) | H$_2$O |
| Et$_2$N | 230 (weak), 210 (55 000) | H$_2$O |
| OH | 280 (0.2), 215 (10 500) | H$_2$O |
| Cl | 256 (600), 196.5 (45 000) | H$_2$O |

The presence of substituents which can conjugate with the triazine ring produces the expected bathochromic shift and the band may be very intense; for example, the 4,6-dichloro-2-(4-nitrophenyl) derivative shows an intense band ($\varepsilon = 24\ 700$) at 279 nm ⟨61RTC158⟩. Mono- and di-amino-1,3,5-triazines show a significant band at *ca.* 260 nm; 2-amino-1,3,5-triazine absorbs at 261 nm ($\varepsilon = 1960$).

### 2.20.2.2.3 IR spectra

1,3,5-Triazine has $D_{3h}$ symmetry. The IR spectrum and assignment of the fundamental modes of 1,3,5-triazine and its fully deuterated analogue have been reported ⟨61SA155⟩, and the data are summarized in Table 7. Three of these frequencies persist in a variety of symmetrical 1,3,5-triazines although the band at *ca.* 675 cm$^{-1}$ is often absent. Perfluorophenyl and some partially fluorinated alkoxy derivatives do not contain the band at 675 cm$^{-1}$ either ⟨65SA663⟩.

IR spectroscopy has proved to be particularly valuable in the study of tautomeric forms of the triazines. Thus, cyanuric acid was shown to exist in the keto form by the presence

**Table 7**   Fundamental Frequencies (cm$^{-1}$) of Symmetrical 1,3,5-Triazines ⟨62RCR712⟩

| Substituent | $E^I$ | $E^I$ | $A_2^{II}$ | $E^I$ |
|:---:|:---:|:---:|:---:|:---:|
| | | *Symmetry of vibration* | | |
| H | 1555 | 1410 | 735 | 675 |
| D | 1530 | 1284 | 667 | 577 |
| Me | 1560 | 1410 | 775 | — |
| Et | 1560 | 1387 | 785 | — |
| Cl | 1510 | 1268 | 735 | 705 |
| MeO | 1565 | 1390 | 813 | 730 |
| NH$_2$ | 1534 | 1211 | 810 | — |
| ND$_2$ | 1530 | 1410 | 828 | — |

of peaks at 1710 cm$^{-1}$ and at 3210 and 3060 cm$^{-1}$ ⟨52JA3545⟩. Similarly, the spectrum of trithiocyanuric acid shows only two very weak bands in the 2450–2600 cm$^{-1}$ region, corresponding to S—H stretching vibrations, but strong N—H stretching bands between 2900 and 3250 cm$^{-1}$, thereby indicating that the compound exists mainly in the thioamide form. In addition the bands at 1115, 1330 and 1540 cm$^{-1}$ were attributed to the N—C=S system. By contrast, melamine appears to exist in the amino rather than the imino form, as shown by the strong amine absorbance at 3200–3600 cm$^{-1}$ and confirmed by a Raman study ⟨66JCP(45)3155⟩.

In general, the 1,3,5-triazine substituents absorb in the expected region of the spectrum; thus, 1,3,5-triazine aldehydes are characterized by a sharp carbonyl band at *ca.* 1720 cm$^{-1}$. Dichloro-1,3,5-triazines and cyanuric chloride give a diagnostic peak at 840 cm$^{-1}$ due to a complex C—Cl stretch involving part of the triazine ring ⟨67JCS(B)123⟩. The IR spectra of the isocyanurates show bands characteristic of the ring at *ca.* 1690 and 1430 cm$^{-1}$ ⟨70JHC725⟩. The spectra of the hexahydro-1,3,5-triazines are discussed below (see Section 2.20.2.3). The IR spectra of 35 1,3,5-triazine derivatives have been published and are a valuable reference ⟨58JA803⟩, and a similar study of chloro-1,3,5-triazines is also useful ⟨61SA600⟩.

### 2.20.2.2.4 *Mass spectra*

1,3,5-Triazine shows the molecular ion (*m/e* 81) as base peak, as would be expected for an aromatic compound. The major fragments (*m/e* 54 and 27) are formed by the stepwise loss of two molecules of hydrogen cyanide. The 2-alkoxycarbonyl derivative (**17**) readily losses carbon dioxide and ethylene, probably *via a* cyclic fragmentation process (Scheme 2). Ethylamino-1,3,5-triazines lose ethylene in a similar way. Amino-1,3,5-triazines lose NHCN and NH$_2$CN fragments and the imino tautomer may make an important contribution in the pathway. The presence of a substituted aryl ring modifies the pathway considerably. Usually *meta* or *para* substituents are lost easily, followed by cleavage of the triazine ring; *ortho* substituents may undergo rearrangements, as in the example shown in Scheme 3 ⟨70OMS(3)863⟩.

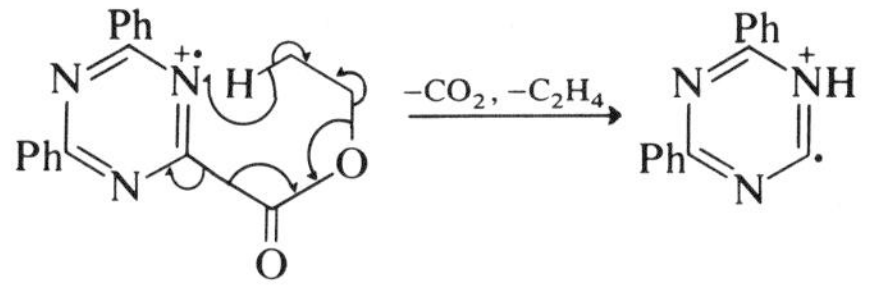

(**17**) *m/e* 305      *m/e* 223

**Scheme 2**

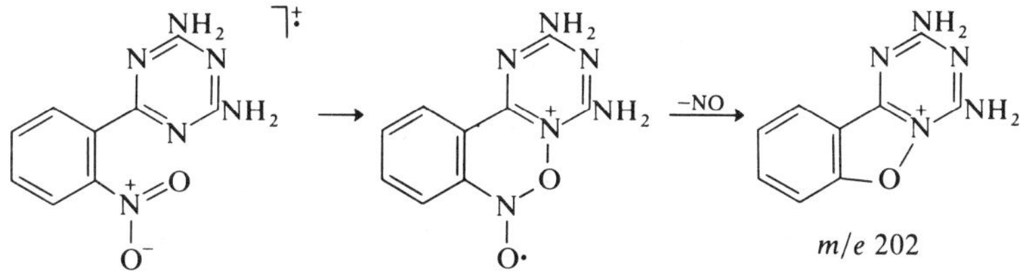

**Scheme 3**

The spectra of cyanuric chloride and other chloro-1,3,5-triazines are characterized by either loss of a chlorine atom, which is important for 2-amino-4,6-dichloro-1,3,5-triazine and similar compounds, or ring cleavage with expulsion of cyanogen chloride, which is important for cyanuric chloride ⟨70OMS(3)219⟩. The GLC–MS analyses of the alkyl cyanurates (2,4,6-trialkoxy-1,3,5-triazines) have been investigated ⟨78OMS43⟩. The molecular ions decrease in intensity with increasing molecular weight. In general, *m/e* 130, *i.e.* (**18**) or its tautomer, is the base peak. The major fragments are attributed to the loss of alkyl side chains; for example, the major fragment ions for *n*-butyl cyanurate are: *m/e* 297 (M$^+$), 242

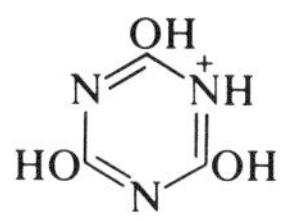

(**18**)

$(M-C_4H_7)$, 186 $[M-(C_4H_7+C_4H_8)]$, 130 (base peak). The mass spectra of the corresponding isocyanurates (1,3,5-trialkyl-1,3,5-triazine-2,4,6-triones) show similar fragmentation pathways.

The hexahydro-1,3,5-triazines are difficult to analyze because the samples may decompose prior to electron impact ⟨76OMS1221⟩. The spectra show weak molecular ions, and the $M-1^+$ ion is usually more intense. The alkyl groups may cleave in a variety of ways; one pathway is shown in Scheme 4 ⟨66HCA1439⟩. There have been studies of 1,3,5-trinitrohexahydro-1,3,5-triazine using high resolution and isotopic labelling, but the spectra did not show sufficient metastable ions to confirm the fragmentations ⟨70OMS(3)13⟩.

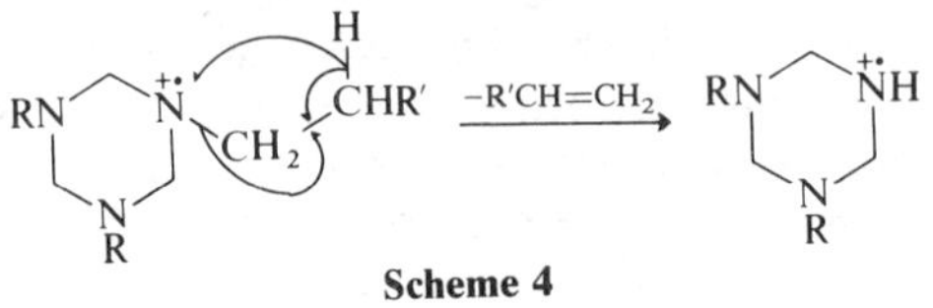

**Scheme 4**

Beynon *et al.* have reported that the energy release during the decomposition of metastable peaks in the mass spectrum of 1,3,5-triazine was 187 μeV, a lower value than for either pyridine or pyrimidine ⟨71OMS(5)229⟩.

### 2.20.2.2.5 Photoelectron spectra

The photoelectron spectrum of 1,3,5-triazine shows five major bands. The ionization potentials and peak assignments were recorded ⟨72HCA255⟩. The bands at 10.37 and 13.21 eV were assigned as being ionizations from lone pair molecular orbitals, whilst those at 11.67 and 14.67 eV were due to π-orbital ionizations, and the band at 14.88 eV was assigned as σ-orbital ionization. An analysis of 1,3,5-triazine by ESCA gave very similar results, although there was no report of the σ-orbital ionization ⟨72JA1466⟩.

### 2.20.2.3 Aromaticity

X-Ray data for 1,3,5-triazine demonstrate the aromatic character of the ring, but the structure is not a regular hexagon (see Section 2.20.2.1). Coulson and Looyenga ⟨65JCS6592⟩ have calculated the bond angles and have shown that the replacement of CH by N in aromatic rings leads to a perturbation which changes all the angles of the ring by varying amounts. The calculations were in good agreement with experimental values.

The resonance energy of the triazine ring has been calculated using the Pariser–Parr–Pople-type SCF–LCAO–MO treatment and a value of 10.903 eV was obtained ⟨66JCP(44)759⟩. The delocalization energy was calculated by the valence bond method and a value of 121 kJ mol$^{-1}$ was obtained (compared with 172 kJ mol$^{-1}$ for benzene) ⟨46JCS670⟩. The electron affinity of 1,3,5-triazine has been measured as 0.45 eV by electron transmission spectroscopy ⟨75JCP(62)1747⟩. The value is higher than for pyridine or the diazines as is expected.

1,3,5-Triazine continues the trend of decreasing aromaticity observed on going from pyridine to the diazines. The polarity of the C=N bonds increases as more π-electrons are partially located on the ring nitrogens, thereby decreasing delocalization energy. The charge distribution is shown below (**19**). The polar canonical form (**20**) makes a major contribution to the structure. These properties are reflected in the reactivity of 1,3,5-triazine towards nucleophiles. 1,3,5-Triazine is completely decomposed by cold water, whilst pyrimidine is unstable in hot alkali (see Section 2.20.3.1).

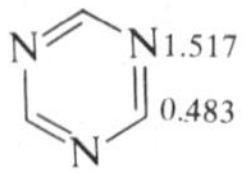

**(19)** 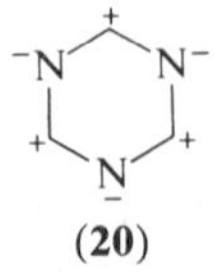　　　　**(20)**

### 2.20.2.4 Shape and Conformation of Saturated and Partially Saturated 1,3,5-Triazines

The 1,3,5-trisubstituted 1,3,5-hexahydrotriazines have been the subject of considerable research in recent years. 1,3,5-Trimethyl-1,3,5-hexahydrotriazine exists in the diequatorial monoaxial form (**21**) ⟨70JCS(B)135⟩. The compound was studied by variable temperature [1]H NMR. On reducing the temperature to −59 °C the methylene protons (δ 3.05) split into an AB quartet whilst the methyl group (δ 2.19) remained as a singlet. This result implies slow ring inversion and fast nitrogen inversion ⟨74JA1591⟩. Subsequently, variable temperature [13]C NMR was used and the signals for both the ring carbon and methyl carbon split at low temperatures (−116 and −130 °C respectively). The energies of activation for ring inversion and nitrogen inversion were calculated as 53.5 and 30.1 kJ mol$^{-1}$ respectively ⟨78JCS(P2)377⟩, which are in good agreement with previous studies ⟨67JCS(B)387⟩. It is likely that the monoaxial conformer is preferred because it relieves apparent *syn*-axial lone pair–lone pair repulsions. The diaxial conformer would result in serious *syn*-axial methyl–methyl repulsions. The barriers to inversion of both the ring and nitrogen decrease with increasing size of substituent. Thus, it was not possible to measure the barrier for nitrogen inversion for 1,3,5-tri-*t*-butyl-1,3,5-hexahydrotriazine by the [13]C NMR technique. 1,3,5-Trimethoxy-1,3,5-hexahydrotriazine and the corresponding 1,3,5-trismethanesulphonyl derivative also exist in the diequatorial monoaxial form ⟨77G363, 78T3413⟩.

(21)          (22)

The 2,4,6-trialkyl-1,3,5-hexahydrotriazines are rather less interesting structurally. The chair form is preferred, and the alkyl groups are equatorial ⟨58ACS1768, 73JOC3288⟩. Colebrook *et al.* ⟨76CJC3757⟩ have investigated the 1,2-dihydro-1,3,5-triazines (**22**) by [1]H NMR and have shown that there are substantial barriers to internal rotation about the aryl C—N bonds. For most of the 2,2-dimethyl compounds the barriers are too high to be measured by [1]H NMR lineshape analysis. It is clear from the variable temperature [1]H NMR study that the barrier to internal rotation is greater than 80 kJ mol$^{-1}$.

### 2.20.2.5 Tautomerism

The tautomerism of 1,3,5-triazines has been reviewed previously ⟨B-76MI22000, pp.138,152,168⟩. [1]H NMR and IR studies have shown that cyanuric acid exists mainly in the oxo form. Although the cyanurates and isocyanurates are the two major derivatives, there is no doubt that compounds with both types of functional group present in the same molecule are possible (*e.g.* Scheme 36). Trithiocyanuric acid exists predominantly in the thioamido form. In contrast, melamine exists in the amino form. β-Oxoalkyl-1,3,5-triazines exist mainly in the enol form, and thus they undergo 'ene' type reactions with dienophiles.

The structures of 5-azacytosine and related compounds are of interest because of their biological importance (see Section 2.20.5.6). 1-Methyl-5-azacytosine exists in the amino-oxo form (**23**). 5-Azauracil (1,3,5-triazine-2,4-dione) is of particular interest. IR spectra indicate that it exists in the dioxo form in the solid, but [1]H NMR studies have been interpreted to show that it exists in the monoenolic form in solution. The spectra showed a non-exchangeable sharp singlet at 8.18 δ (H*, **24**) ⟨76OMR(8)224⟩. Derivatives of 5-azacytosine and 5-azauracil are covalently hydrated. Thus 5-azauridine exists entirely in the crystal form as (**25**) ⟨76MI22000, p.139⟩.

(23)          (24)          (25)

Weber *et al.* have reported that the condensation product (**26**) of the reaction between methyl cyanoacetate and 6-chloro-2,4-dimethoxy-1,3,5-triazine exists as two tautomeric forms in rapid equilibrium. The ester is enolized readily (A). The second exchange involves the breaking of an intramolecular hydrogen bond (B) (Scheme 5) ⟨78RTC107⟩.

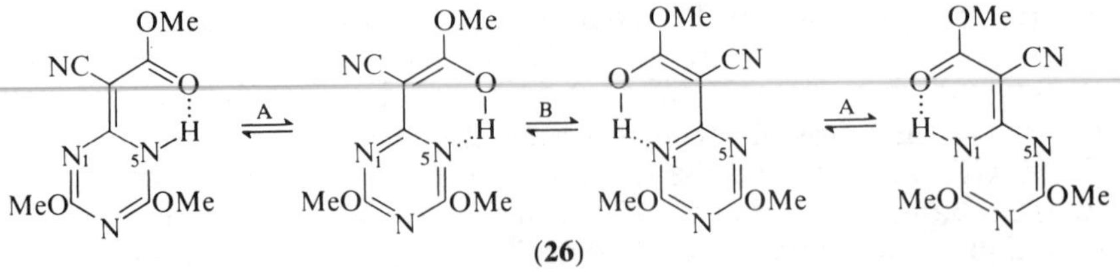

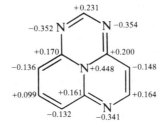

(26)

**Scheme 5**

### 2.20.2.6 Structure of the Polyazacycl[3.3.3]azines

1,3,6-Triazacycl[3.3.3]azine (**27**) and its derivatives possess aromatic properties; the compounds have high melting points and undergo electrophilic substitution reactions (see Section 2.20.3.15.2). Ceder and Anderson ⟨72ACS596⟩ have carried out MO calculations on (**27**) using the Hückel II program. The resonance energy was calculated as 4.3726 $\beta$, which is in good agreement with other cycl[3.3.3]azine systems. The values of the charge densities (Figure 1) and free valencies (Figure 2) indicate that electrophilic and radical substitution occur preferentially at C-4, C-7 and C-9 positions in (**27**), which is indeed the case (see Section 2.20.3.15.2). The charge density of the central nitrogen indicates that there is a delocalization of electrons from the central nitrogen atom. The bond orders (Figure 3) show that the system is highly delocalized.

**Figure 1**   Charge densities of 1,3,6-triazacycl[3.3.3]azine

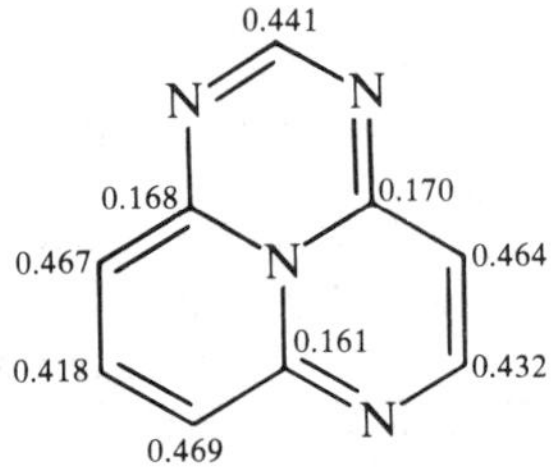

**Figure 2**   Free valencies of 1,3,6-triazacycl[3.3.3]azine

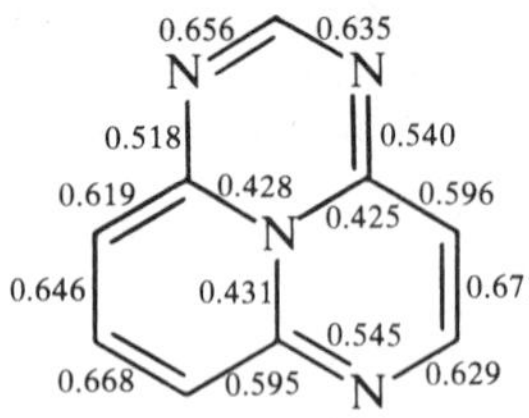

**Figure 3**   Bond orders of 1,3,6-triazacycl[3.3.3]azine

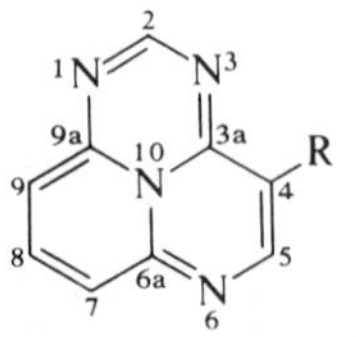

(**27**) R = H
(**28**) R = CN

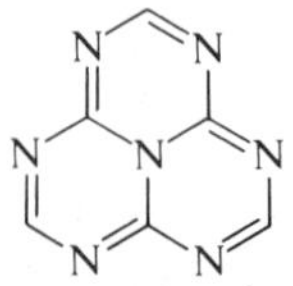

(**29**)

The IR spectrum of (27) has not been reported fully, although Ceder and Anderson have stated that it lacks NH absorptions. The UV spectrum contains peaks at 223, 234 ($\varepsilon = 11\,220$), 343, 362, 372, 378, 494, 531, 580 ($\varepsilon = 570$) and 632 nm. The long wavelength absorptions at 490–670 nm are responsible for the blue color. The fine structure and presence of seven distinct bands between 220–400 nm is indicative of an aromatic system. The chemical shifts of the ring protons of the 4-cyano derivative (28) lie in the range 6.6–8 $\delta$ ⟨72ACS596⟩. The mass spectrum of (27) shows $m/e$ 170 (molecular ion) as base peak, and major fragments at $m/e$ 143 and 116 from successive losses of HCN. Doubly-charged ions are also observed ⟨72ACS611⟩.

Recently, Leonard and coworkers ⟨82JA5497⟩ reported the synthesis and spectral and structural features of the hexaazacyclazine (29), the parent ring system of 'melem' (see Section 2.20.3.8).

## 2.20.3 REACTIVITY

There are considerable differences between the reactions of the parent compound and its derivatives. 1,3,5-Triazine is very labile, and undergoes ring cleavage with a wide variety of nucleophiles, whilst the substituted 1,3,5-triazines undergo ring opening reactions much less readily. In general, the role of the triazine ring in the chemistry of 1,3,5-triazine derivatives is limited to its effect on charge distribution, and many of the reactions simply reflect the chemistry of the ring substituents.

### 2.20.3.1 Reactions of 1,3,5-Triazine

The reactions of 1,3,5-triazine have been reviewed thoroughly ⟨63MI22000,63AG(E)309⟩. The compound is thermally stable as would be expected because of its aromatic character. It decomposes to form hydrogen cyanide on heating above 600 °C. In general, its properties reflect the significant contribution of the dipolar canonical form (20) to the structure. Triazine is a catalyst poison, and therefore it cannot be hydrogenated.

#### 2.20.3.1.1 Reactions with electrophiles

1,3,5-Triazine is resistant to electrophilic substitution. Chlorination requires vigorous conditions, and yields are low; bromination is a more efficient process (Scheme 6). The reagents employed in the attempted sulfonation or nitration preferentially hydrolyze the ring ⟨63AG(E)309⟩. Recently, Korolev and Mal'tseva have reported that 1,3,5-triazine is protonated and hydrated to form the cation shown in equation (2) ⟨75ZOR2613⟩.

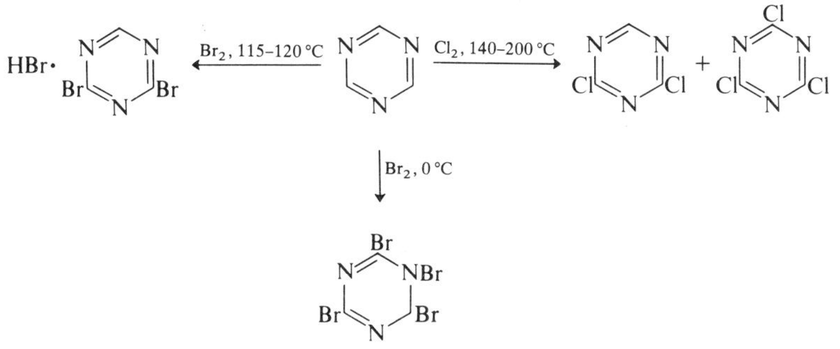

**Scheme 6**

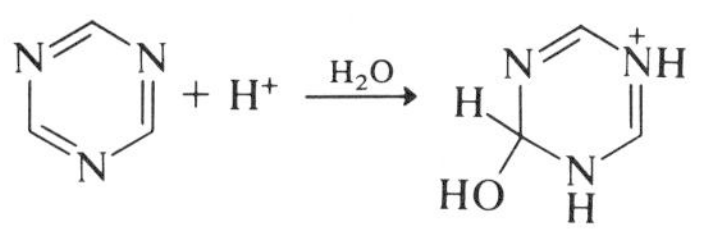

(2)

### 2.20.3.1.2 Reactions with nucleophiles

The characteristic reaction is ring cleavage. The reaction between 1,3,5-triazine and a primary amine (Scheme 7) is typical. Further examples of reactions with nucleophiles are shown in Scheme 8 ⟨63AG(E)309⟩.

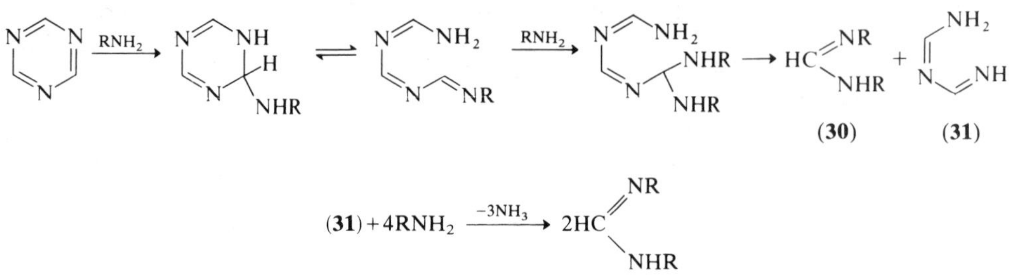

$$(31)+4RNH_2 \xrightarrow{-3NH_3} 2HC\overset{NR}{\underset{NHR}{\diagdown}}$$

**Scheme 7**

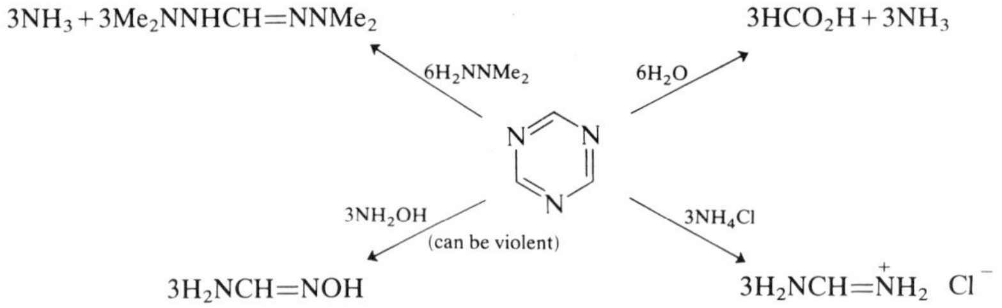

**Scheme 8**

A variety of heterocycles can be prepared from the reaction of bifunctional amines and related compounds with 1,3,5-triazine (equation 3). Similarly, 1,2,4-triazole is formed from hydrazine hydrochloride (Scheme 9) ⟨63AG(E)309⟩.

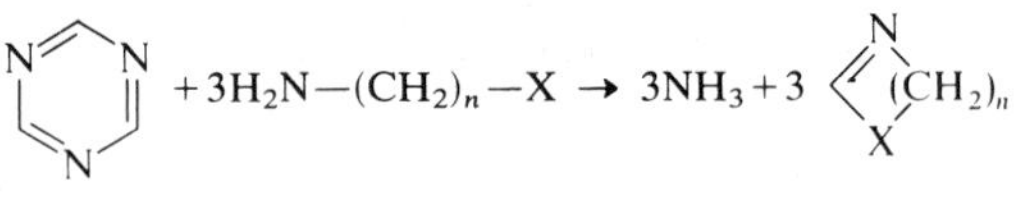

$$+3H_2N-(CH_2)_n-X \rightarrow 3NH_3+3 \underset{X}{\diagup}(CH_2)_n \qquad (3)$$

$$n = 1-3; X = NHR, OH, SH$$

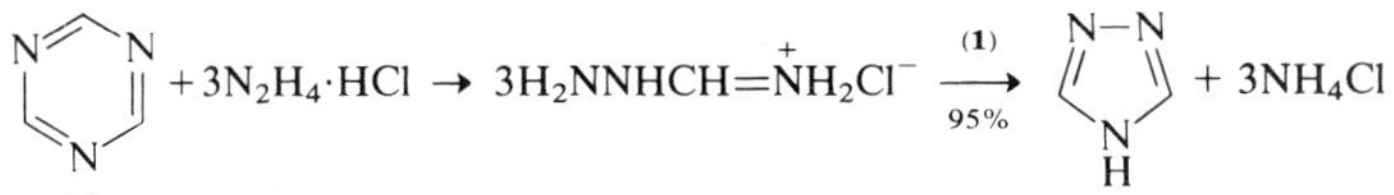

$$(1) +3N_2H_4\cdot HCl \rightarrow 3H_2NNHCH=\overset{+}{N}H_2Cl^- \xrightarrow[95\%]{(1)} \underset{H}{\diagup}\overset{N-N}{\diagdown} + 3NH_4Cl$$

**Scheme 9**

1,3,5-Triazine reacts with imidates *via* (32) to provide a simple route to 2-alkyl (or aryl) 1,3,5-triazines ⟨61JOC2784⟩. Amidines behave similarly, and the 2,4-dialkyl-1,3,5-triazines are formed by using an excess of the appropriate amidine (Scheme 10) ⟨59JA1470⟩.

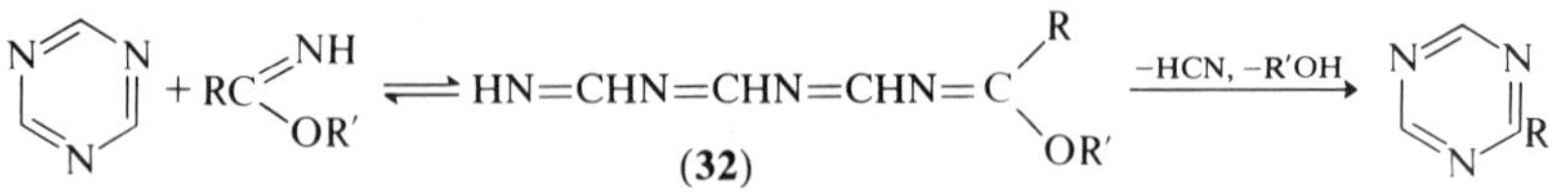

$$+ RC\overset{NH}{\underset{OR'}{\diagdown}} \rightleftharpoons HN=CHN=CHN=CHN=C\overset{R}{\underset{OR'}{\diagup}} \xrightarrow{-HCN, -R'OH}$$

(32)

R = alkyl or aryl

**Scheme 10**

Pyridines and 4,5-disubstituted pyrimidines can be synthesized from 1,3,5-triazine with compounds containing a reactive methylene group (Schemes 11 and 12) ⟨63AG(E)309⟩.

Several dehydro Mannich bases have been prepared in a similar way. The reaction between 1,3,5-triazine, morpholine, and diethyl malonate is a typical example (Scheme 13) ⟨76T2603⟩.

Kurabayashi and Grundmann have reported the preparation of the 1,2,4-oxadiazoles (34) from 1,3,5-triazine and aryl nitrile oxides in the presence of boron trifluoride. The mechanism has not been fully elucidated, but it is most likely that the initial stage is the formation of the complex (33) ⟨78BCJ1484⟩.

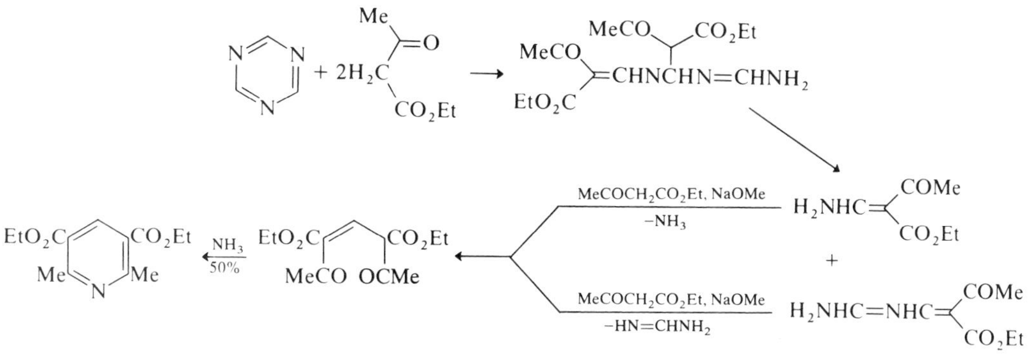

**Scheme 11**

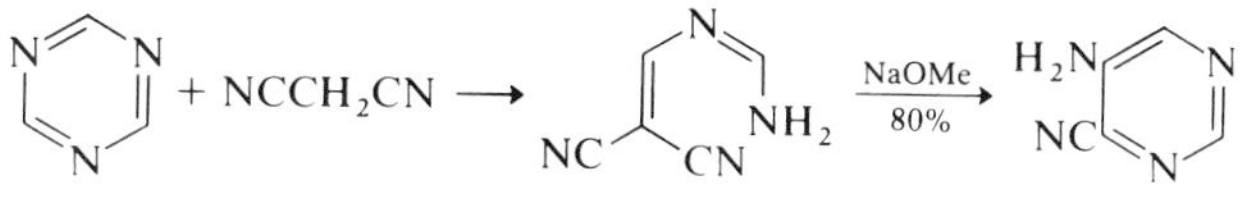

**Scheme 12**

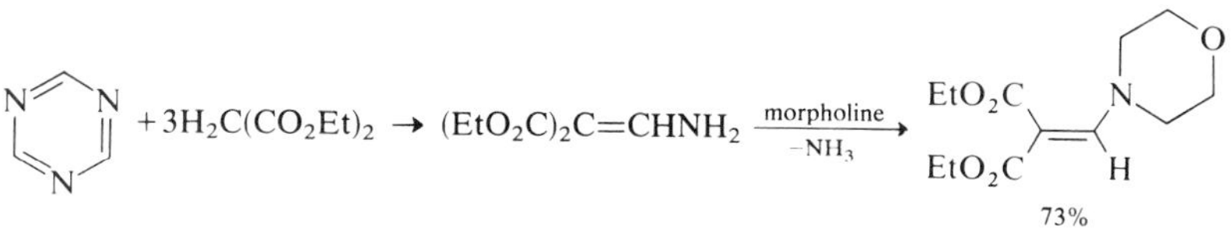

**Scheme 13**

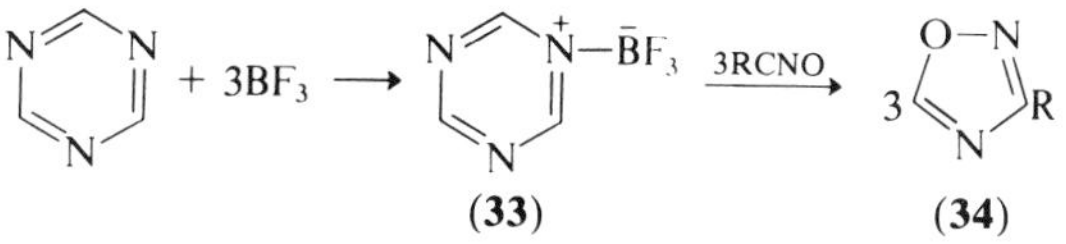

**Scheme 14**

1,3,5-Triazine may be used in the Gatterman aldehyde synthesis in place of hydrogen cyanide (Scheme 15) ⟨67AG(E)940⟩. On treatment of 1,3,5-triazine with aryl Grignard reagents, the aryl aldehyde is formed (Scheme 16) ⟨63MI22000⟩.

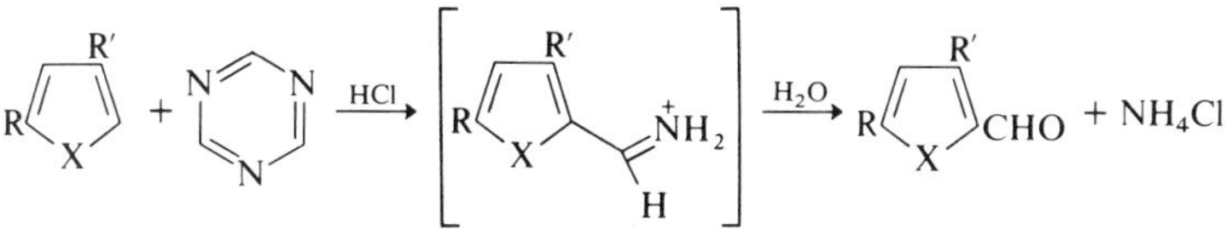

R,R' = Me or H, X = O, NH or NMe

**Scheme 15**

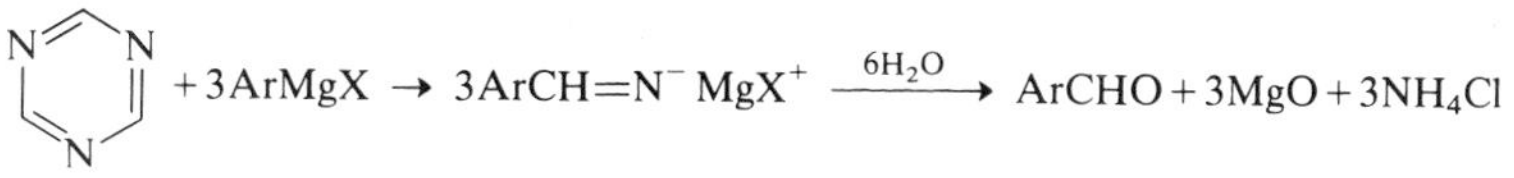

$$\text{(1,3,5-triazine)} + 3\text{ArMgX} \rightarrow 3\text{ArCH}=\text{N}^-\text{MgX}^+ \xrightarrow{6\text{H}_2\text{O}} \text{ArCHO} + 3\text{MgO} + 3\text{NH}_4\text{Cl}$$

**Scheme 16**

### 2.20.3.1.3 Cycloaddition reactions

1,3,5-Triazine and its derivatives react with dienophiles by a [4 + 2] cycloaddition reaction (equation 4) ⟨75CB3877⟩. The 1,2-dihydro-1,3,5-triazines (**35**) can also undergo Diels–Alder reactions with potent dienophiles (equation 5) ⟨77KGS122⟩.

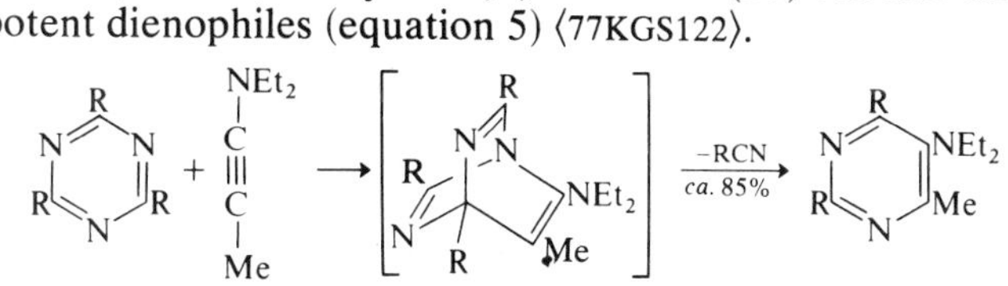

(4)

R = H, Me, CO₂Et, *etc.*

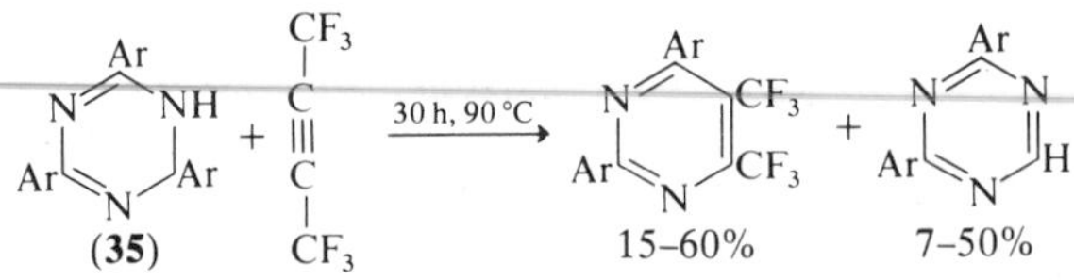

$$\text{(35)} + \underset{\overset{|}{CF_3}}{\overset{CF_3}{\underset{|}{C}}} \overset{30\,h,\,90\,°C}{\longrightarrow} \quad \underset{15\text{–}60\%}{} + \underset{7\text{–}50\%}{} \tag{5}$$

## 2.20.3.2 Reactions of Alkyl-1,3,5-triazines

There have not been many reports on the reactions of alkyl-1,3,5-triazines because these compounds are difficult to synthesize (see Section 2.20.4). Most of the work reported has been centered on the symmetrical 2,4,6-trialkyl derivatives.

### 2.20.3.2.1 Reactions of the 1,3,5-triazine ring

The triazine ring is markedly more stable than in the parent compound, but it may be opened under severe conditions (equations 6, 7, 8) ⟨59HC(13)1, p.147⟩.

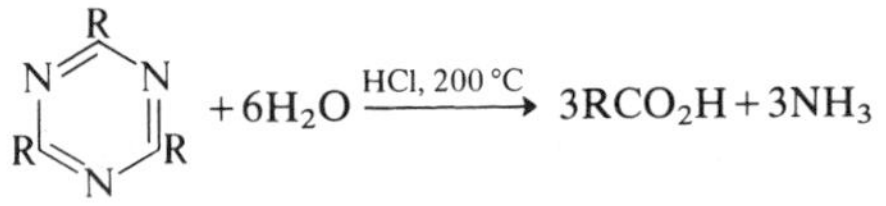

$$\text{[R-triazine]} + 6H_2O \overset{HCl,\,200\,°C}{\longrightarrow} 3RCO_2H + 3NH_3 \tag{6}$$

$$\text{[Et-triazine]} \overset{\text{ammoniacal MeOH}}{\underset{8500\,\text{atm},\,150\,°C}{\longrightarrow}} \text{[Me, NH}_2\text{-triazine]} \tag{7}$$

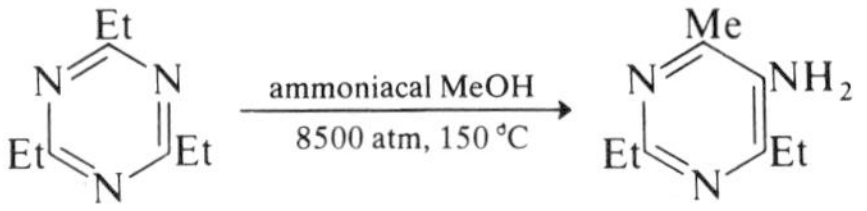

$$\text{[R-triazine]} \overset{Zn/MeCO_2H}{\longrightarrow} NH_3 + \text{[R-imidazole]} \tag{8}$$

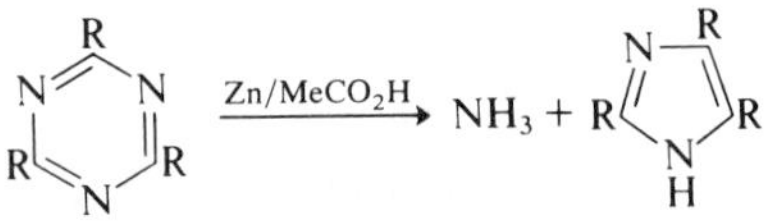

R = alkyl or aryl

### 2.20.3.2.2 Reactions of the alkyl substituents

Each alkyl group is α to two ring nitrogen atoms and therefore is reactive (see Chapter 2.02). Deprotonation of alkyl groups in basic conditions is a facile process. Typical examples of such reactions are shown in Schemes 17 and 18 and equation (9) ⟨59HC(13)1, p.168, 64JHC128, 64JHC145, 64JOC678, 72RRC2043⟩.

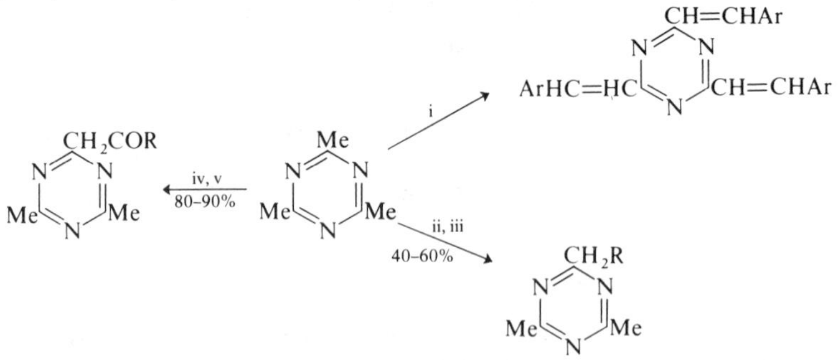

i, ArCHO, NaOH; ii, KNH₂, liquid NH₃; iii, RBr; iv, KNH₂, liquid NH₃; v, RCO₂R′

**Scheme 17**

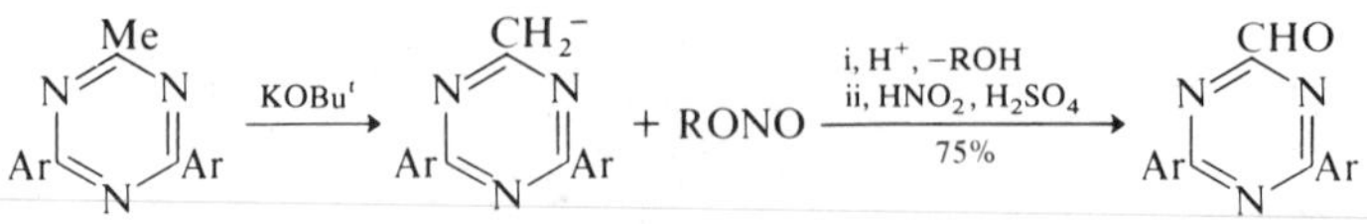

**Scheme 18**

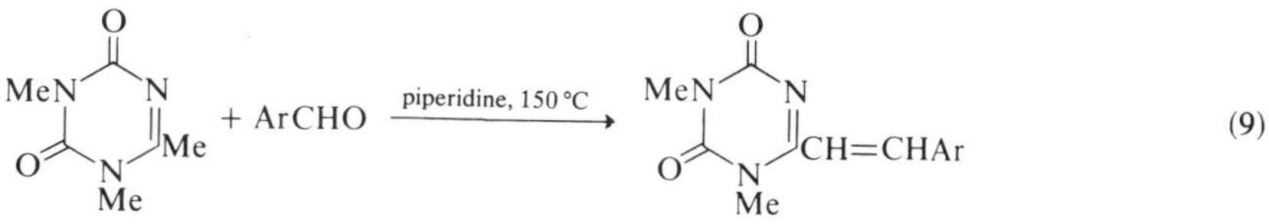

$$\text{(9)}$$

The $\alpha$-hydrogens of 2,4,6-trialkyl-1,3,5-triazines are readily chlorinated or brominated in the presence of acid. The reaction is believed to occur *via* an ionic mechanism. The final products using an excess of chlorine or bromine are strongly dependent on the reaction conditions used (Scheme 19) ⟨64JOC1527⟩. 2-Alkyl-1,2-dihydro-1,3,5-triazines (36) may be rearranged in good yields to pyrimidines (Scheme 20) ⟨79TL1241⟩.

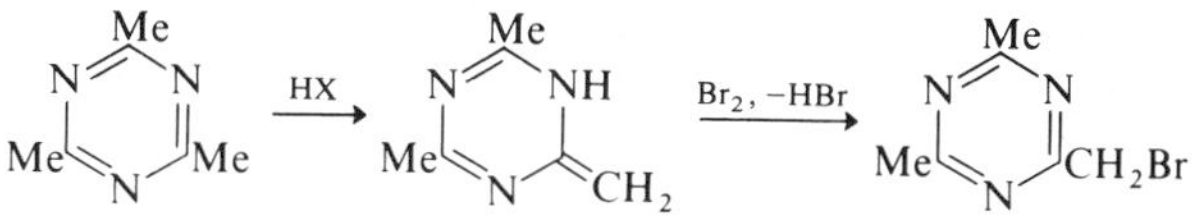

**Scheme 19**

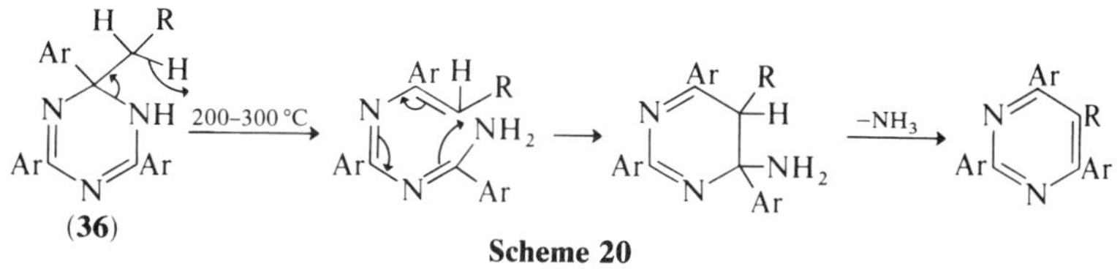

**Scheme 20**

### 2.20.3.3 Reactions of Aryl-1,3,5-triazines

Triaryl-1,3,5-triazines are more readily prepared than are the alkyl derivatives (see Section 2.20.4). In general, the compounds are stable, and 2,4,6-triphenyl-1,3,5-triazine, for example, can be distilled at *ca.* 350 °C without decomposition. These compounds undergo similar ring-opening reactions to the 2,4,6-trialkyl-1,3,5-triazines, but the reaction requires even more vigorous conditions (see Section 2.20.3.2) ⟨59HC(13)1, p.178⟩.

It is possible to effect electrophilic substitution of the aryl ring. The tri-*p*-tolyl compound, for instance, can be nitrated to give 2,4,6-tris(4-methyl-3-nitrophenyl)-1,3,5-triazine ⟨59HC(13)1, p.178⟩.

#### 2.20.3.3.1 Reactions with nucleophiles

2,4-Diphenyl-1,3,5-triazine undergoes a Chichibabin-type amination reaction, the first example of this reaction to be reported for the 1,3,5-triazines. The 2-methylthio derivative (37; X = MeS) reacts exclusively *via* an $S_N$(ANRORC) mechanism rather than the $S_N$AE mechanism (Scheme 21) ⟨76RTC113⟩.

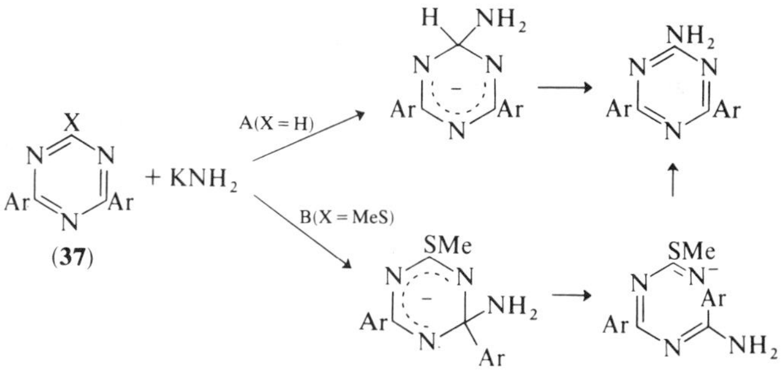

A, $S_N$(AE); B, $S_N$(ANRORC)

**Scheme 21**

2-Phenyl-1,3,5-triazine has fewer sterically hindered positions of attack, and therefore can undergo amination by an $S_N$(ANRORC) process. Indeed, Simig and van der Plas investigated the reaction by $^1$H NMR and showed that it does occur partially (*ca.* 55%) by this mechanism ⟨76RTC125⟩.

### 2.20.3.4  Reactions of 1,3,5-Triazine Aldehydes

These compounds are crystalline solids which readily form stable hydrates, and polymerize on standing; the starting material is regenerated on heating the polymer ⟨58JA5547⟩. 1,3,5-Triazine aldehydes undergo the typical reactions of the aldehyde group although oxidation to the corresponding carboxylic acid requires vigorous conditions ⟨61JOC957⟩.

### 2.20.3.5  Reactions of Ketoalkyl-1,3,5-triazines

Studies using $^1$H NMR and IR spectroscopy ⟨73BSF2039, 70JHC987⟩ have shown that the compounds exist mainly in the enol form. They do exhibit some ketonic properties, *e.g.* they form oximes ⟨63JOC2933⟩. The ketoalkyl-1,3,5-triazines are readily hydrolyzed, but the products are markedly dependent on the conditions employed (Scheme 22). The pyrimidine (**38**) is formed quantitatively by an $S_N$(ANRORC) mechanism on refluxing the triazine in distilled water ⟨64JHC145⟩. The ketoalkyl-1,3,5-triazines undergo an 'ene' reaction with dienophiles (equation 10) ⟨74JHC317⟩.

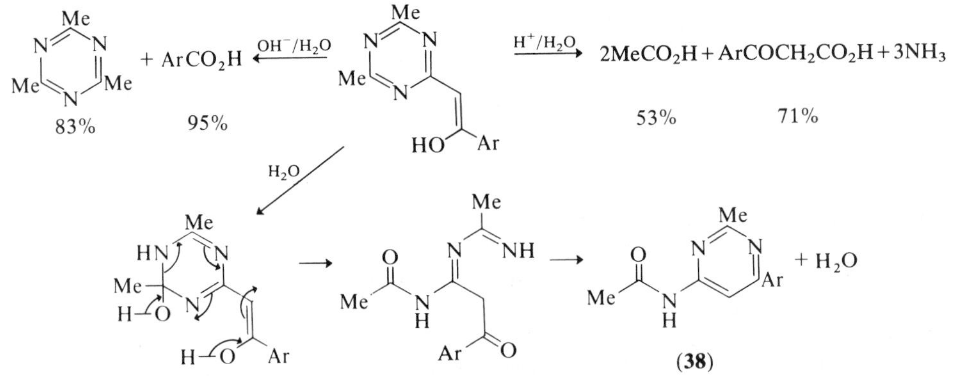

**Scheme 22**

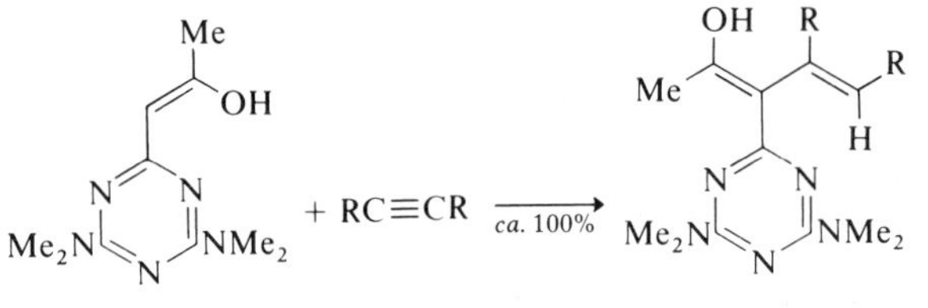

$$(10)$$

### 2.20.3.6  Reactions of 1,3,5-Triazinecarboxylic Acids and Related Compounds

The properties of the carboxylic acids are quite predictable. Esterification is a facile process, and the compounds are decarboxylated on heating. The 1,3,5-triazine ring is readily ruptured on treatment with warm water (equation 11) ⟨59HC(13)1, p.169⟩. The corresponding esters are hydrolyzed to the acid by cold alkali solution, but the ring is cleaved if the ester is heated with base.

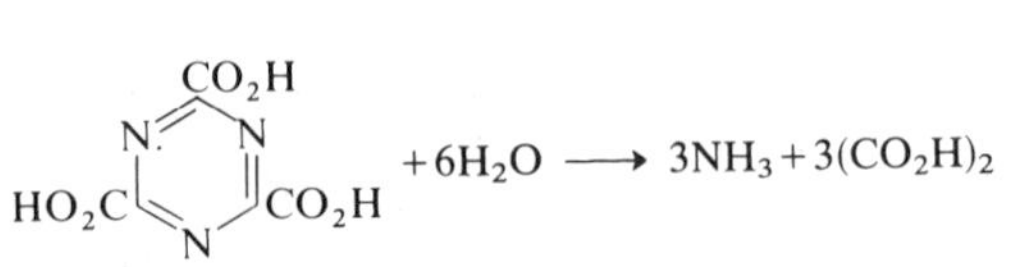

$$(11)$$

### 2.20.3.7  Reactions of 1,3,5-Triazinecarbonitriles

These compounds are thermally stable. 2,4,6-Tricyano-1,3,5-triazine reacts rapidly with water to form cyanuric acid and hydrogen cyanide (equation 12) ⟨59HC(13)1, p.170⟩.

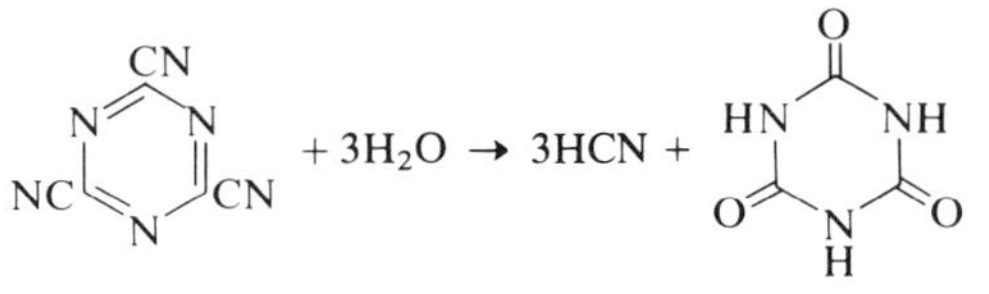

$$+ 3H_2O \rightarrow 3HCN + \text{[structure]} \tag{12}$$

## 2.20.3.8 Reactions of Amino-1,3,5-triazines

The amino-1,3,5-triazines are amongst the most important of the 1,3,5-triazines. Melamine is a particularly valuable industrial compound (see Section 2.20.5.4 and Chapter 1.11). Many of the reactions are well known ⟨58CRV131, 59HC(13)1, pp.217, 269 and 309⟩.

### 2.20.3.8.1 Reactions of the 1,3,5-triazine ring

The triazine ring is cleaved under very vigorous conditions (Scheme 23). Novak and Dobas have reported that the first and second protonations of 2-substituted 4,6-diamino-1,3,5-triazines occur on the ring nitrogens ⟨76CCC3378⟩. Similarly, dialkyl sulfates react with melamine at the ring nitrogens (Scheme 23).

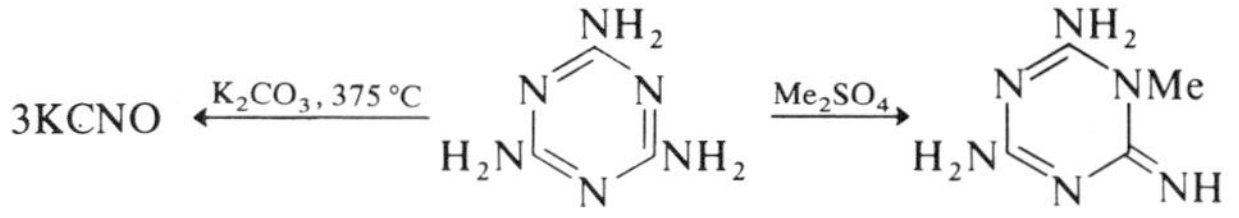

**Scheme 23**

Melamine undergoes self-condensations to form a variety of polycyclic compounds, depending on the reaction conditions (Scheme 24) ⟨58CRV131⟩.

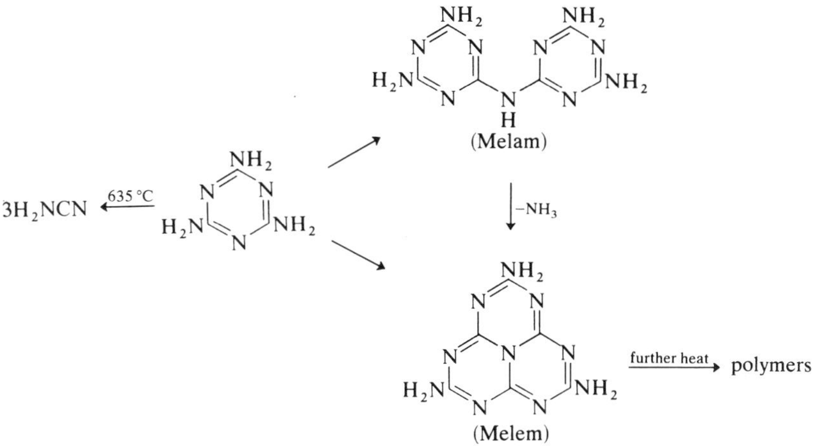

**Scheme 24**

### 2.20.3.8.2 Reactions of the amino group

In some of its reactions the amino substituent resembles an amido group. Thus amino-1,3,5-triazines are not acylated by acid chlorides, although the reaction may be effected by heating the 1,3,5-triazine with acid anhydrides.

It is possible to displace the amino groups. Thus melamine is hydrolyzed to cyanuric acid by sequential nucleophilic substitution of the amino groups. Similarly, the amino groups are transaminated with alkyl or aryl amines at high temperature ⟨58CRV131⟩. The amino-1,3,5-triazines are also attacked by electrophiles. Melamine is partly nitrated with nitric acid at low temperatures (equation 13). (At higher temperatures cyanuric acid is formed.) It is also possible to halogenate melamine to form triazines bearing from one to six halogens depending on the reaction conditions ⟨59HC(13)1, p.330⟩.

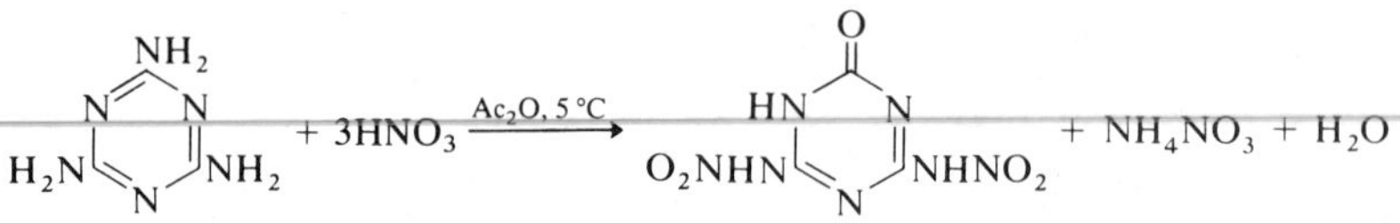

$$
\text{(13)}
$$

The reaction of melamine with formaldehyde is a useful one, as the initial product (**39**) forms a resin on heating. Such condensates are very important polymers (see Section 2.20.6.3 and Chapter 1.11). Melamine and other amino-1,3,5-triazines form salts with aqueous acids. In addition, the amino-1,3,5-triazines form potassium and silver salts ⟨58CRV131⟩.

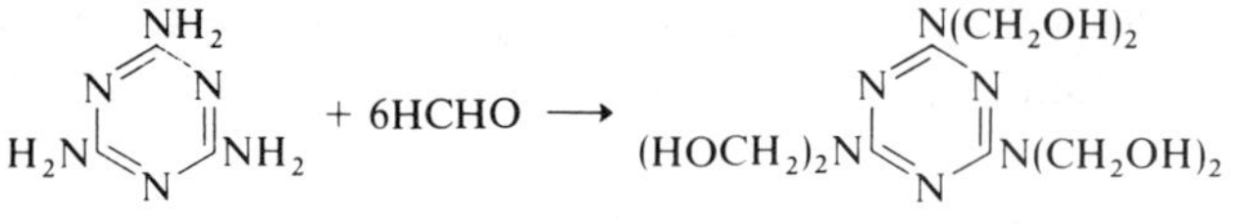

$$
\text{(14)}
$$

### 2.20.3.8.3 Reactions of alkylamino-1,3,5-triazines

2,4-Bis(alkylamino)-6-chloro-1,3,5-triazines are important herbicides which undergo dealkylation as the major detoxication process in animals. *In vitro* studies have shown that thermal dealkylation occurs *via* a Chugaev-like mechanism (equation 15). The dealkylation is also effected by sonication (irradiation with ultrasonic waves), but the mechanism has not been elucidated ⟨71MI22000, 77JCS(P1)1257⟩. It is also possible to dealkylate such 1,3,5-triazines by using hydroxyl radicals ⟨71MI22001⟩.

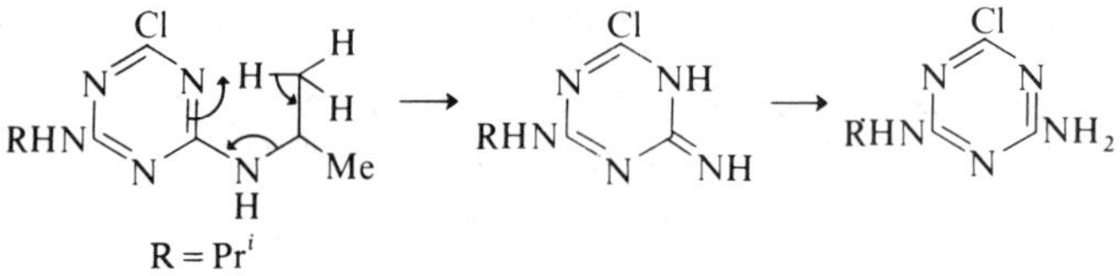

$$
\text{(15)}
$$

### 2.20.3.8.4 Rearrangements of amino-1,3,5-triazines

Tsunoda *et al.* ⟨73BCJ3499⟩ have reported a novel reverse Smiles rearrangement of *N,N*-bis-(1,3,5-triazinyl)-2-aminophenols (Scheme 25). 2,4,6-Tris(1-aziridinyl)-1,3,5-triazine is rearranged by acid into the tetracyclic compound (**40**; equation 16) ⟨55JA5922⟩.

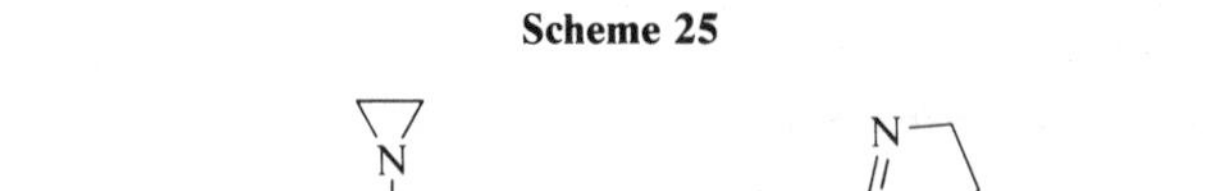

**Scheme 25**

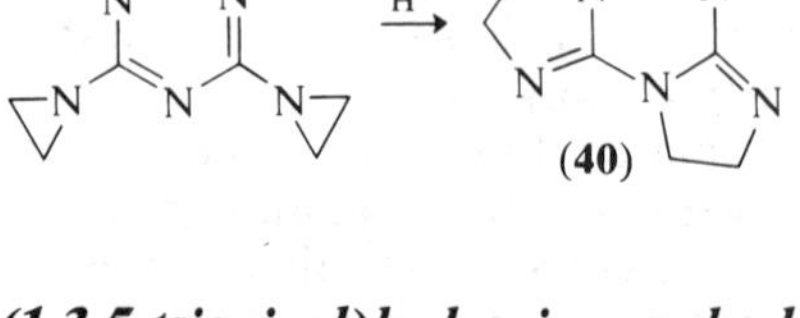

$$
\text{(16)}
$$

### 2.20.3.8.5 Reactions of bis(1,3,5-triazinyl)hydrazines and related compounds

Bis(1,3,5-triazinyl)hydrazines (**41**) are oxidized to the corresponding azo derivatives which are very reactive dienophiles, as illustrated in Scheme 26 ⟨76JHC829⟩.

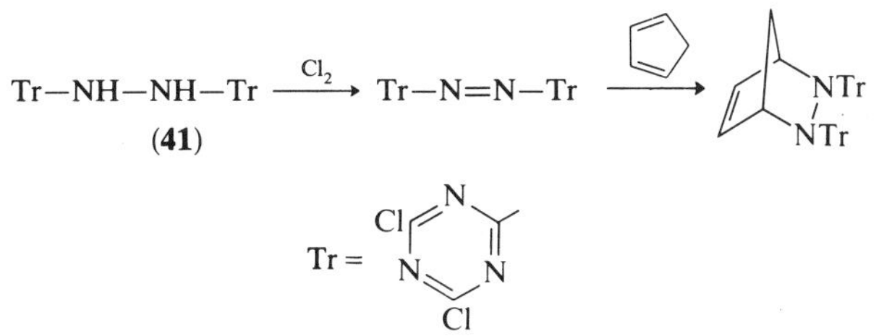

**Scheme 26**

## 2.20.3.9 Reactions of Azido-1,3,5-triazines

2,4,6-Triazido-1,3,5-triazine is highly unstable and detonates very easily ⟨59HC(13)1, p.300⟩. Recent studies by Matsui and coworkers have shown that singlet nitrenes are formed on irradiation of monoazide derivatives. These nitrenes can either form cycloadducts with the solvent or they can undergo intersystem crossing to form the triplet nitrene, and subsequently produce amino-1,3,5-triazines. Typical examples are shown in Schemes 27 and 28 ⟨75BCJ3309, 78JOC1361⟩. The triazinylnitrenes are inert towards nitro compounds. Thus, the photolysis of 2-azido-4,6-dimethoxy-1,3,5-triazine in nitromethane gives rise to products derived from the combination of two molecules of the starting material ⟨79BCJ1231⟩.

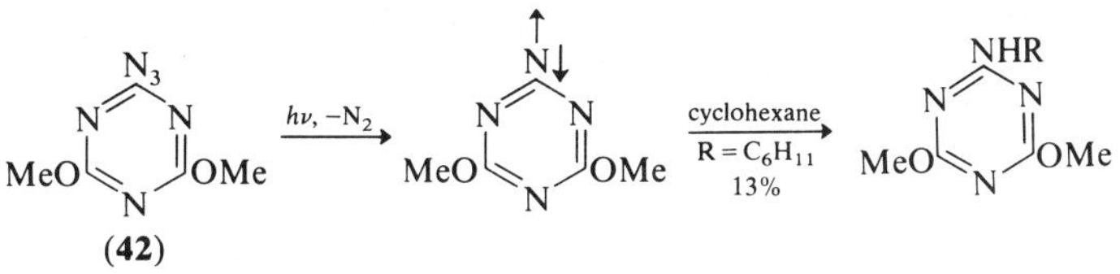

**Scheme 27**

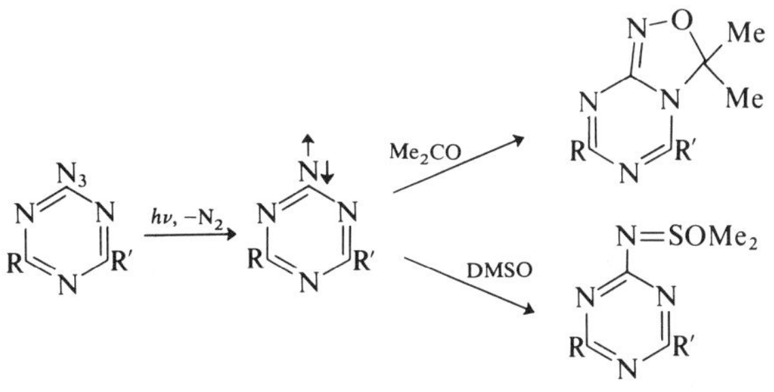

R = naphthyl, R′ = OMe

**Scheme 28**

## 2.20.3.10 Reactions of 1,3,5-Triazines Bearing *O*-Linked Substituents

The chemistry of the hydroxy-1,3,5-triazines, the cyanurates and isocyanurates are closely interlinked; thus it is most convenient to discuss their reactivities in one section.

### 2.20.3.10.1 Reactions of oxo-1,3,5-triazines

The reactions of these compounds have been reviewed previously ⟨59HC(13)1, p. 17, B-79MI22001⟩. Cyanuric acid will be used to illustrate the reactions of the oxo-1,3,5-triazines. It is thermally stable (it is unaffected by heating in a sealed tube to 500 °C) and is not readily hydrolyzed. The oxo group undergoes nucleophilic substitutions under forcing conditions (Scheme 29).

Cyanuric acid forms a variety of salts, most of which are not of major significance, but the trisilver salts are used to prepare tribromoisocyanurate and alkyl isocyanurates (Scheme 30) ⟨67M1613⟩.

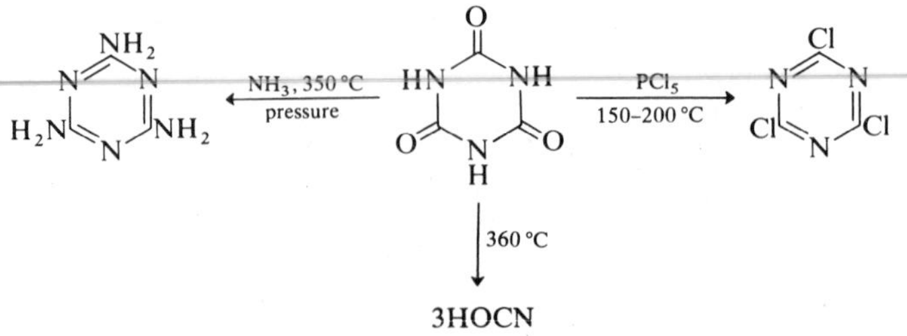

**Scheme 29**

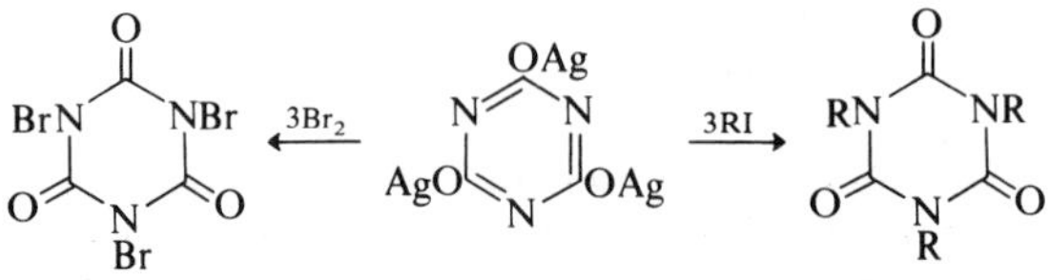

**Scheme 30**

Cyanuric acid exists in two tautomeric forms corresponding to keto–enol tautomerism in carbonyl compounds. The keto form predominates, and most of the reactions of cyanuric acid have their counterparts in the chemistry of the cyclic imides. Many of the reactions involve the replacement of all three imido hydrogens (Scheme 31). Usually, the reaction cannot be controlled to produce the mono- or di-substituted isocyanurates specifically, but there are exceptions, *e.g.* the reaction between cyanuric acid and aziridine (Scheme 31) ⟨B-79MI22001, 63JOC85, 63AHC(2)245⟩.

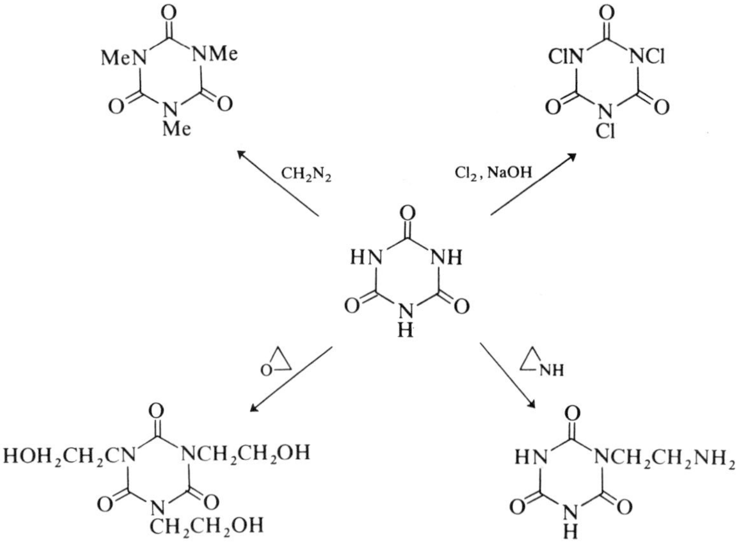

**Scheme 31**

## 2.20.3.10.2 Reactions of isocyanurates and triazine-2,4-diones

The reactions of the isocyanurates have been reviewed previously ⟨59HC(13)1, p.389⟩. The following discussion will concentrate on three facets of isocyanurate chemistry: ring cleavage, reactions with Grignard reagents and rearrangements.

Trialkyl isocyanurates are hydrolyzed in base to the corresponding trialkyl biuret ⟨1876CB1008⟩. Argabright and Phillips ⟨70JHC999⟩ have shown that the site of ring cleavage is dependent on the nature of the substituents by studying the hydrolysis of disubstituted isocyanurates. The 1,5-disubstituted biurets (**43**) are formed specifically, and the reaction is favoured by the presence of electronegative substituents. The mechanism is shown in Scheme 32. Aqueous alkaline hydrolysis of tris(2-hydroxyethyl)isocyanurate (**44**) produces ethanolamine (equation 17) ⟨61JCS3148⟩.

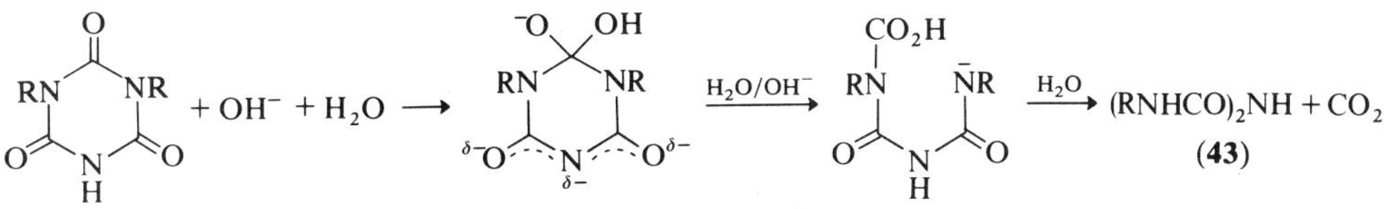

**Scheme 32**

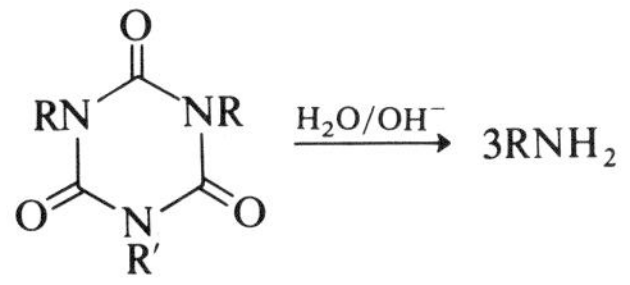

(17)

(**44**)  R = HOCH$_2$CH$_2$

Isocyanurates react with Grignard reagents to form the alkene derivatives (**45**) which can undergo further reactions as shown in Scheme 33 ⟨75CR(C)(281)563, 77CR(C)(285)321⟩. A novel ring transformation of the 1,3,5-triazine-2,4-dione (**46**) to form the pyrimidine (**47**) has been reported (Scheme 34) ⟨79JOC3982⟩. 2-Hydroxyethyl isocyanurate is thermally decomposed to form 2-oxazolidone (equation 18) ⟨60JOC1944⟩.

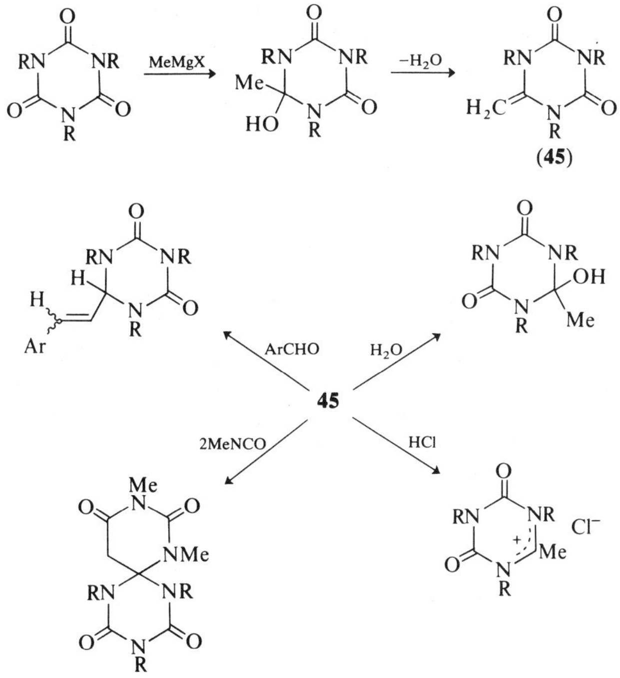

**Scheme 33**

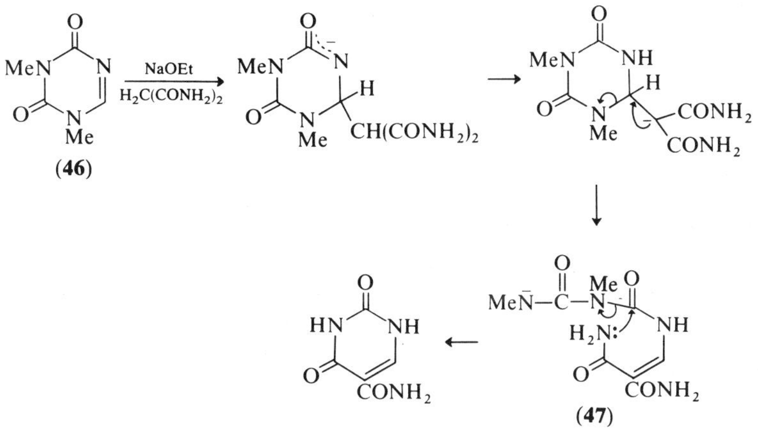

**Scheme 34**

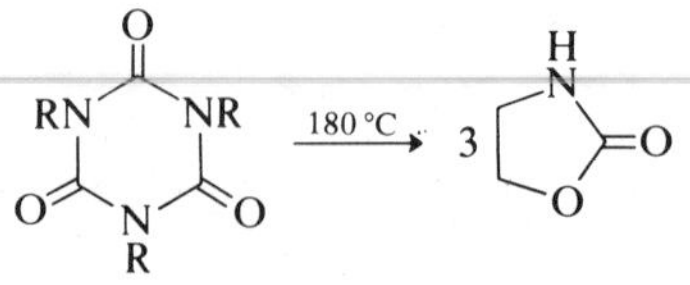

$$\text{(18)}$$

R = CH_2CH_2OH

### 2.20.3.10.3 Reactions of alkyl cyanurates

The alkyl cyanurates are thermally stable, although they decompose on passing over red hot platinum metal to yield cyanic acid and alkenes. (Trimethyl cyanurate is exceptional; it decomposes to methyl isocyanate.)

It many ways the alkyl cyanurates behave as typical esters. They are hydrolyzed to cyanuric acid and the appropriate alcohol, and readily undergo transesterification reactions. Cyanurates react with ammonia to form melamine under severe conditions ⟨59HC(13)1, p.17⟩. The reaction with amines usually gives mixed products, but triallyl cyanurates react to form the monosubstituted products in excellent yields (Scheme 35) ⟨75S182, 75S184⟩.

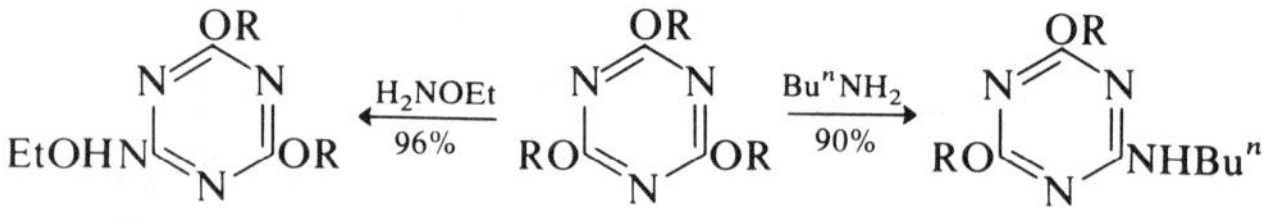

R = CH_2—CH=CH_2

**Scheme 35**

The alkyl cyanurates can isomerize to form isocyanurates on heating ⟨1886CB2061⟩. This is often the preferred synthetic route to isocyanurates. Tosato and coworkers have studied the mechanism of the isomerization and have shown that it proceeds in a stepwise manner (Scheme 36) ⟨68JHC533⟩. The proposed mechanism is complex, and involves the generation of methyl cations ⟨79JCS(P2)1371⟩.

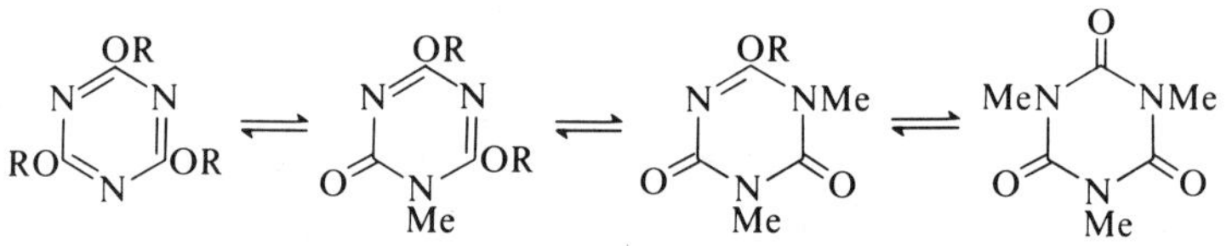

**Scheme 36**

There have been several other rearrangements of cyanurates reported. Dovlatyan and Dovlatyan have described the thermal rearrangements of the 2-chloroethoxy-1,3,5-triazine (**48**) and related compounds to form fused 1,3,5-triazines as shown in equations (19) and (20) ⟨79KGS124, 80KGS411⟩. The propargyl ether (**49**) undergoes a Claisen rearrangement and cyclization (equation 21) ⟨80TL4731⟩.

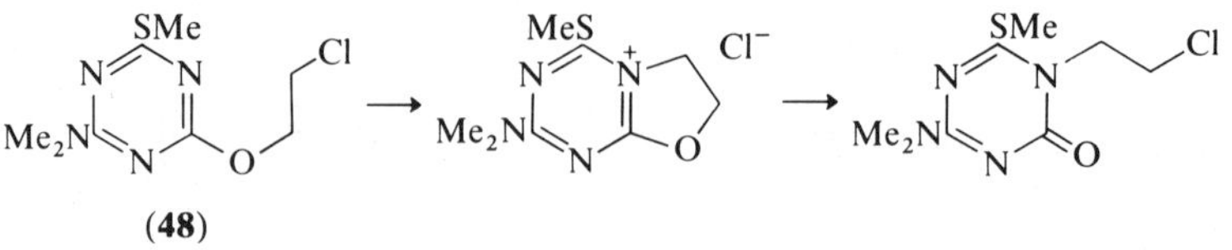

$$\text{(19)}$$

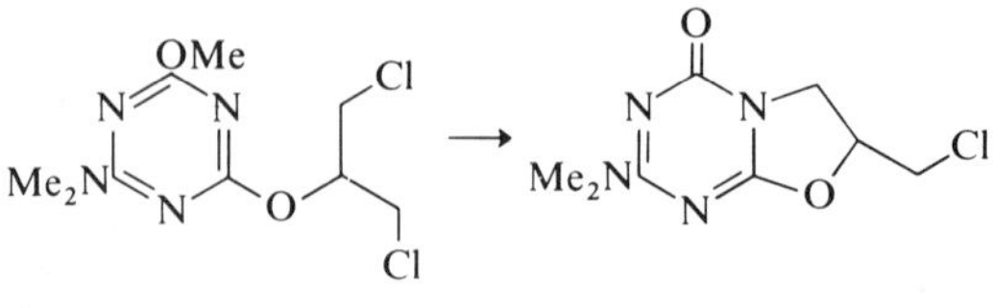

(**48**)

$$\text{(20)}$$

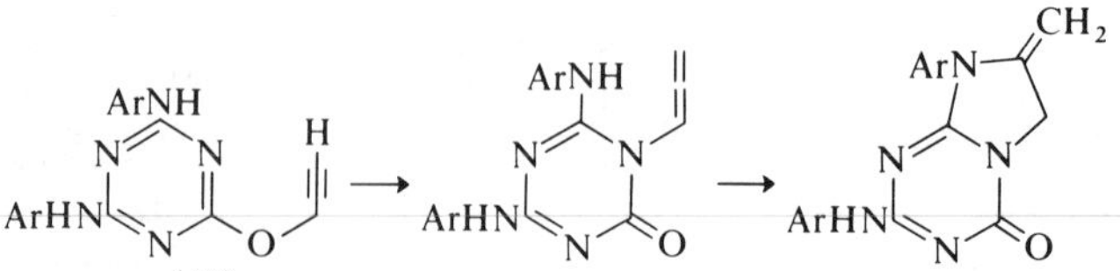

$$\text{(21)}$$

(**49**)

### 2.20.3.10.4 *Reactions of aryl cyanurates*

These compounds are more stable than the alkyl derivatives. They do not undergo thermal isomerization, and hydrolysis requires very severe conditions. The aryl cyanurates react smoothly with amines to yield 2,4,6-triamino-1,3,5-triazines. It is possible to effect electrophilic substitution of the aryl rings ⟨59HC(13)1, p.17⟩. Aryl cyanurates are hydrogenated to form cyanuric acid and the arene (equation 22) ⟨74RTC204⟩.

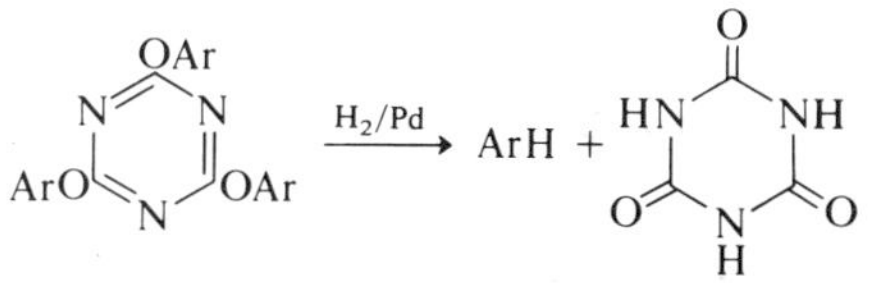

$$\text{(22)}$$

The 2-aminophenoxy triazine (50) readily undergoes the Smiles rearrangement in base (equation 23) ⟨70JHC981⟩, or photochemically ⟨70TL1467⟩. Budziarek reported a similar reaction with the naphthyl derivative (51; Scheme 37) ⟨71JCS(C)74⟩.

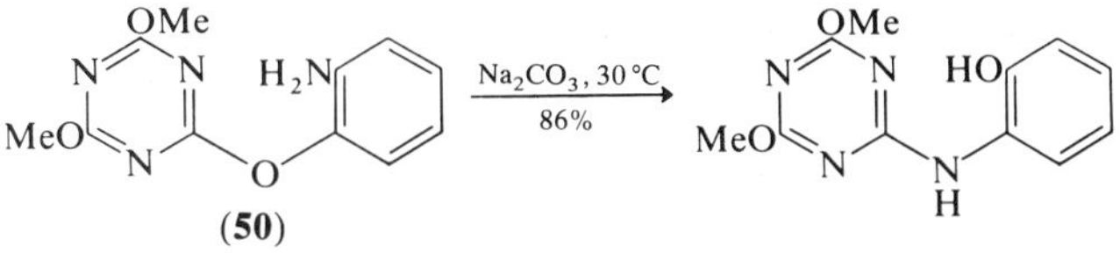

$$\text{(23)}$$

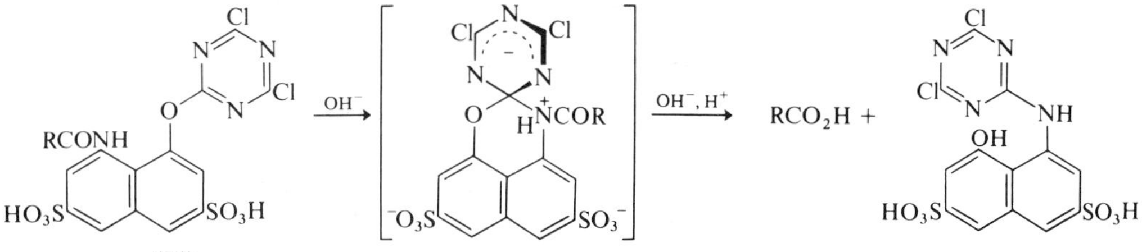

**Scheme 37**

### 2.20.3.11 Reactions of 1,3,5-Triazines bearing *S*-Linked Substituents

The chemistry of the mercapto-1,3,5-triazines, thiocyanurates and isothiocyanurates will be discussed in this section.

### 2.20.3.11.1 *Reactions of 1,3,5-triazinethiones*

In general, the thiones are not very reactive. Trithiocyanuric acid decomposes on heating above 200 °C, and is hydrolyzed to cyanuric acid and hydrogen sulfide in hydrochloric acid at 200 °C. Melamine is produced in 40% yield on heating trithiocyanuric acid with ammonia under pressure ⟨59HC(13)1, p.105⟩. Garmaise has reported the formation of the disulfide derivatives (52) ⟨66CJC1801⟩.

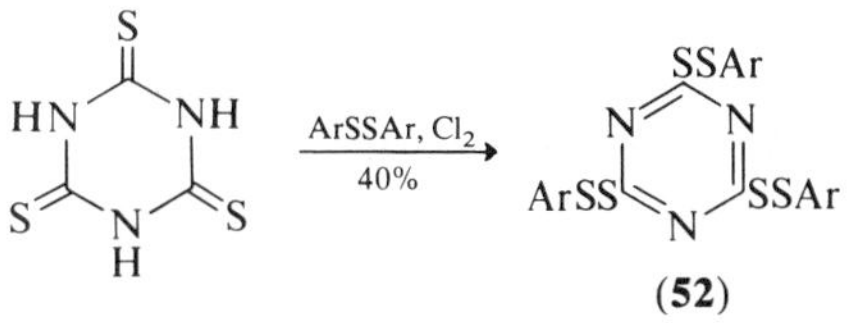

$$\text{(24)}$$

$$Ar = 4\text{-}ClC_6H_4$$

The triazinethiones form sulfur-bridged oligomers by reaction with chloro-1,3,5-triazines, and are oxidized to yield the dimeric species (53) ⟨66JHC137⟩.

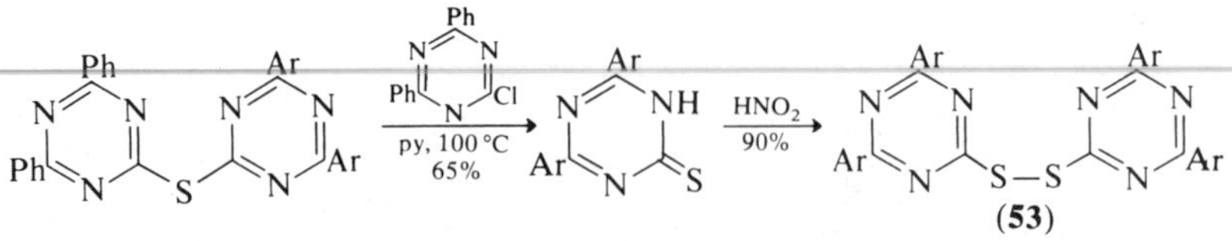

**Scheme 38**

### 2.20.3.11.2 Reactions of thiocyanurates and isothiocyanurates

The thiocyanurates are readily hydrolyzed in acid to cyanuric acid and a thiol. Trimethyl trithiocyanurate reacts with sodium sulfide to give the trisodium salts. Melamine is formed on treatment of the thiocyanurates with aqueous ammonia at high temperatures; the alkylthio groups are replaced sequentially ⟨59HC(13)1, p.111⟩, and can be removed using Raney nickel (equation 25) ⟨B-61MI22000, p.641⟩.

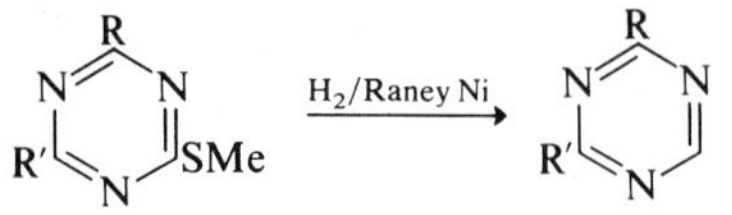

In contrast to the cyanurates, the trithiocyanurates are thermally stable. In fact, the trithioisocyanurates isomerize on heating to thiocyanurates. Tosato has shown the reaction occurs sequentially (equation 26) ⟨79JCS(P2)1371⟩. Shiohima *et al.* have reported a novel dimerization of the 1,3,5-triazine derivative (**54**) ⟨73BCJ2549⟩.

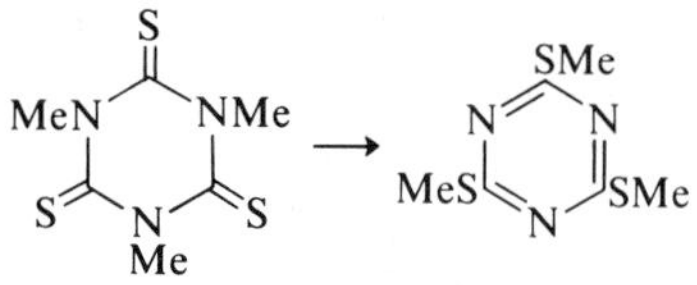

### 2.20.3.12 Reactions of Halogenated 1,3,5-Triazines

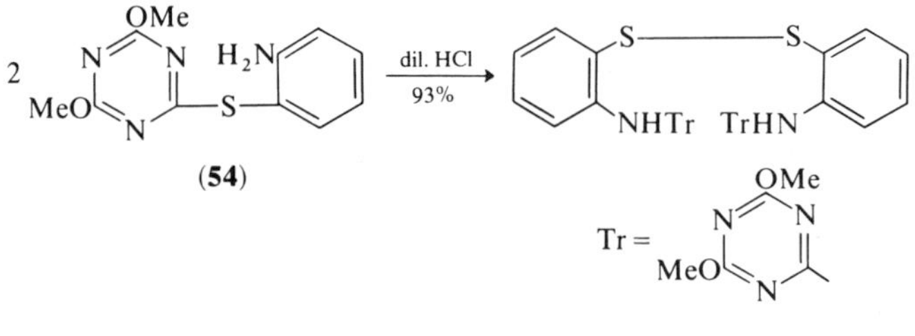

Cyanuric chloride and other chloro-1,3,5-triazines are very important starting materials for the synthesis of many of the 1,3,5-triazines. In general, the fluoro-1,3,5-triazines resemble the chloro derivatives.

### 2.20.3.12.1 Nucleophilic substitution of cyanuric chloride

The chemistry of cyanuric chloride has been reviewed previously ⟨59HC(13)1, p.48, 61MI22000, p.666, 64RCR92⟩. It may be considered as the nitrogen analogue of an acid chloride. The chlorine atoms may often be replaced sequentially, and it is this property that makes cyanuric chloride so valuable. The substitutions are usually increasingly more difficult to carry out in the presence of electron-donating substituents, which impede the attack of other nucleophiles. Clearly, electron-withdrawing groups activate the carbon–chlorine bonds. Typical reaction conditions for the displacement of the chlorine atoms are given in Table 8. Yields are usually good. There can be major variations from the conditions given in the table, *e.g.* two chlorine atoms are displaced by arylaminosulfonic acids at 0 °C, whilst deactivated amines are completely inert ⟨64RCR92⟩.

**Table 8** Synthesis of 1,3,5-Triazines from Cyanuric Chloride

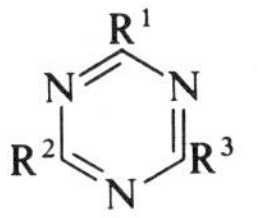

| | Triazine product | | | Temperature | |
|---|---|---|---|---|---|
| $R^1$ | $R^2$ | $R^3$ | *'Typical' conditions* | (°C) | *Ref.* |
| ArO | Cl | Cl | ArOH, NaOH | 0–5 | B-61MI22000, p. 666 |
| ArO | ArO | Cl | ArOH, NaOH | 15–20 | B-61MI22000, p. 666 |
| ArO | ArO | ArO | ArONa | 170–210 | B-61MI22000, p. 666 |
| RO | Cl | Cl | ROH, NaHCO$_3$ | 30 | B-61MI22000, p. 666 |
| RO | RO | Cl | ROH, NaHCO$_3$ | 65 | B-61MI22000, p. 666 |
| RO | RO | RO | ROH, NaOH | 25–30 | 59HC(13)1, p. 48 |
| HO | HO | Cl | H$_2$O, NaOH | 100 | B-61MI22000, p. 666 |
| HO | HO | HO | H$_2$O, NaOH | 125 | B-61MI22000, p. 666 |
| HS | HS | HS | Na$_2$S, HCl | 40 | 59HC(13)1, p. 48 |
| RS | Cl | Cl | RSH | −25 to 0 | 64RCR92 |
| RS | RS | RS | i, NaHS, MeOH; ii, RI, NaOH, MeI | 20 | 56JOC641 |
| ArS | ArS | ArS | ArSNa, MeOH | 20 | 59HC(13)1, p. 48 |
| CHN$_2$ | Cl | Cl | CH$_2$N$_2$, Et$_2$O | 20 | B-61MI22000, p. 666 |
| N$_3$ | N$_3$ | N$_3$ | NaN$_3$, H$_2$O | 0–5 | 59HC(13)1, p. 48 |
| R$_2$N | Cl | Cl | R$_2$NH, H$_2$O | 0 | B-61MI22000, p. 666 |
| R$_2$N | R$_2$N | Cl | R$_2$NH, H$_2$O | 30–50 | B-61MI22000, p. 666 |
| R$_2$N | R$_2$N | R$_2$N | R$_2$NH, H$_2$O | 90–100 | B-61MI22000, p. 666 |
| ▷N | ▷N | ▷N | ▷NH, H$_2$O | 0 | 59HC(13)1, p. 48 |
| (Me$_3$Si)$_2$N | (Me$_3$Si)$_2$N | (Me$_3$Si)$_2$N | (Me$_3$Si)$_2$NLi, THF | 60 | 65JCS5452 |
| Ph$_2$P | Ph$_2$P | Ph$_2$P | Ph$_2$PH, C$_6$H$_6$ | 80 | 64JCS1020 |
| (RO)$_2$P=O | (RO)$_2$P=O | (RO)$_2$P=O | P(OR)$_3$ | 60 | 57JOC444 |
| CH=C(OEt)$_2$ | Cl | Cl | H$_2$C=C(OEt)$_2$, Et$_2$O | 30 | 61JOC4705 |
| F | F | F | NaF, sulfolane | 243–248 | 81AHC(28)1 |

The mechanism of the reaction has been studied extensively, and has been shown to vary with the reaction conditions ⟨64AHC(3)285⟩. Cyanuric chloride is insensitive to both acid catalysis and autocatalysis, but the 2,4-dichloro (**55**) and 2-chloro derivatives (**56**) exhibit both acid catalysis and autocatalysis on solvolysis in ethanol–acetone solutions.

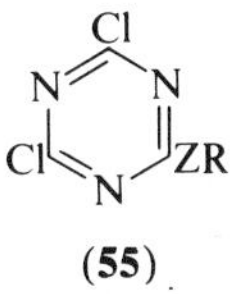

(**55**) 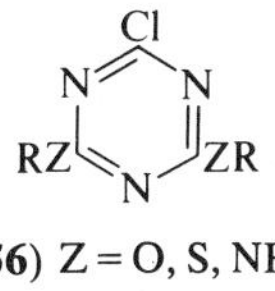          (**56**) Z = O, S, NH

Zollinger ⟨61AG125⟩ investigated the reaction of cyanuric chloride with aniline in benzene, and showed that it is catalyzed by both acid and base. The reaction proceeds *via* the complexes (**57**) and (**58**) in acidic and basic conditions respectively (Scheme 39) ⟨61HCA812⟩. The preparation of 2,4,6-triamino-1,3,5-triazines from the mono-chloro derivative (**59**) is readily effected in dipolar aprotic solvents, but there is no reaction under the same conditions in acetone–water ⟨73BSF2112⟩.

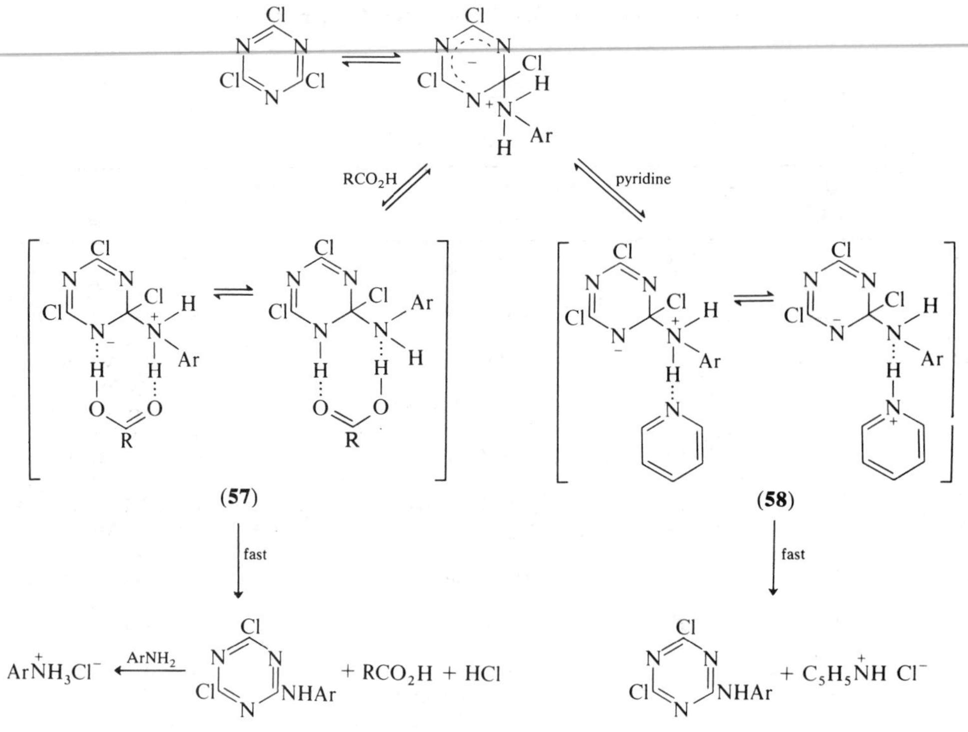

**Scheme 39**

$$(28)$$

**(59)** $R^1 = C_6H_{11}$, $R^2 = H$

Loew and Weis ⟨76JHC829⟩ reported that two equivalents of cyanuric chloride react with hydrazine to form (**60**). There have been several reports of the reactions of cyanuric chloride with a variety of heterocycles, as illustrated in Scheme 40 ⟨73JCS(P2)2075, 75T1879⟩.

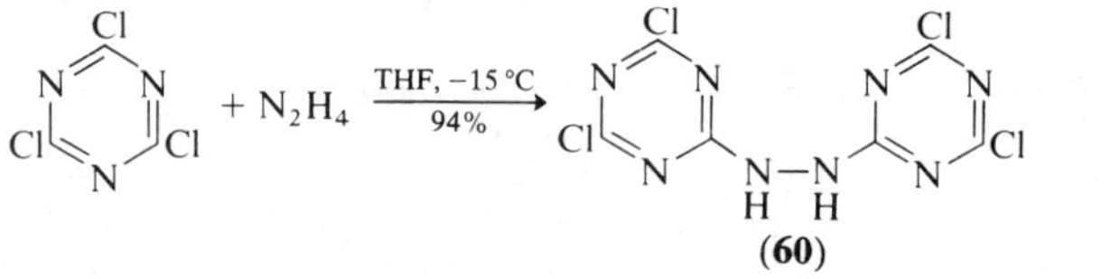

$$(29)$$

**(60)**

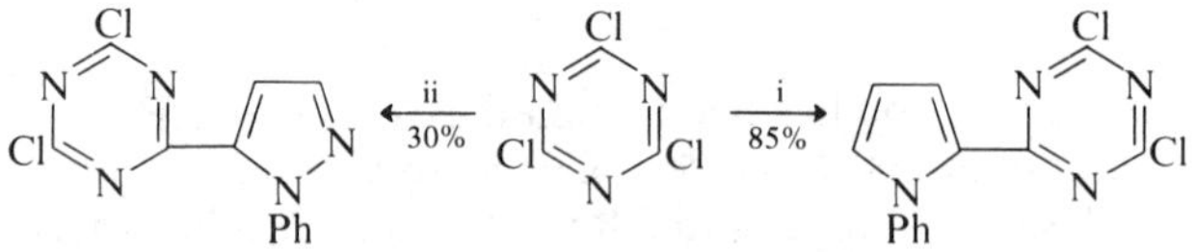

i, *N*-phenylpyrrole, 80 °C; ii, 1-phenylpyrazole, Bu$^n$Li, −10 °C

**Scheme 40**

In addition to the substitutions described above, it is possible to prepare mono- or di-alkyl-1,3,5-triazines using Grignard reagents (Scheme 41) ⟨50HCA1365⟩. The synthesis of the dialkyl derivative must be effected under mild conditions to prevent the attack of the triazine double bonds by the Grignard reagent to form (**62**). The product (**61**) is unstable, and undergoes self-condensation on standing ⟨65JOC702⟩. Cyanuric chloride undergoes the Friedel–Crafts and Wurtz–Fittig reactions to form the triaryl-1,3,5-triazines ⟨64RCR92⟩. Unsubstituted positions may be generated by reductive elimination of the chlorines using lithium aluminum hydride ⟨59HC(13)1, p.48⟩ or *via* the methylthio derivative (see Section 2.20.3.11.2). An interesting exchange of halogens between cyanuric chloride and cyanuric fluoride has been promoted by antimony pentafluoride (equation 30) ⟨81AHC(28)1⟩.

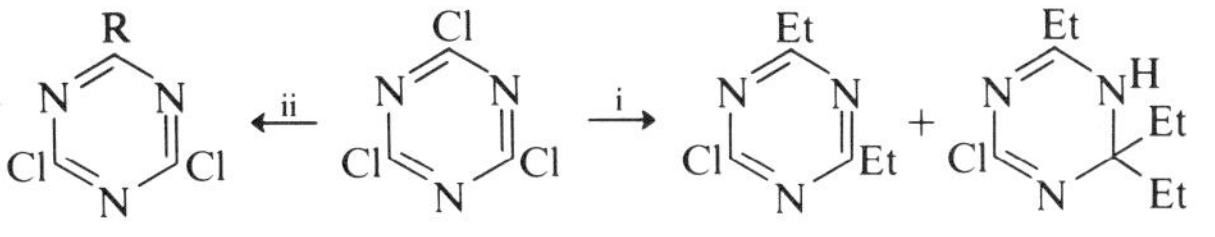

(61) 90%    (62)

i, 3.5 equivalents EtMgBr, Et$_2$O–CH$_2$Cl$_2$(1:1), −15 °C; ii, 1.5 equivalents RMgBr, Et$_2$O, 0 °C

**Scheme 41**

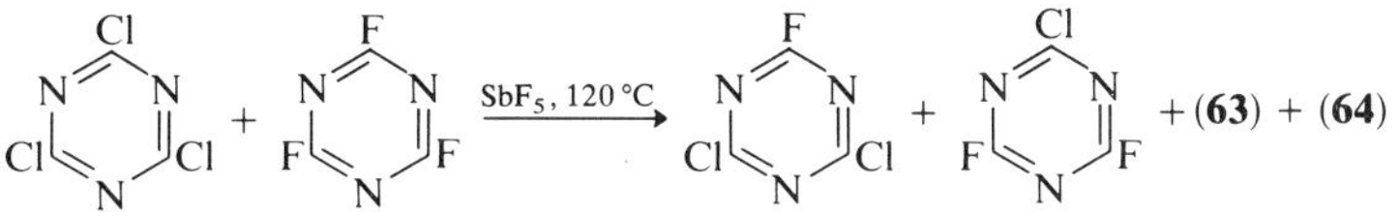

(63)    (64)    (30)

Budziarek and Hampson ⟨71JCS(C)1167⟩ reported that the reaction between cyanuric chloride and 3-hydroxy-2-naphthanilide occurs in base with an *O*-triazine to *N*-triazine rearrangement (Scheme 42). A stable triazine peroxide (66) has been synthesized from (65) and hydrogen peroxide ⟨77CI(L)232⟩, and (65) forms a novel adduct with the cyclopentadienyl anion (Scheme 43) ⟨68HCA249⟩.

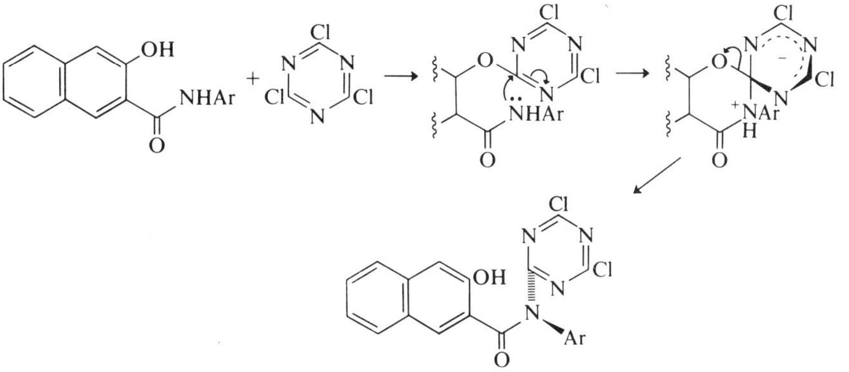

**Scheme 42**

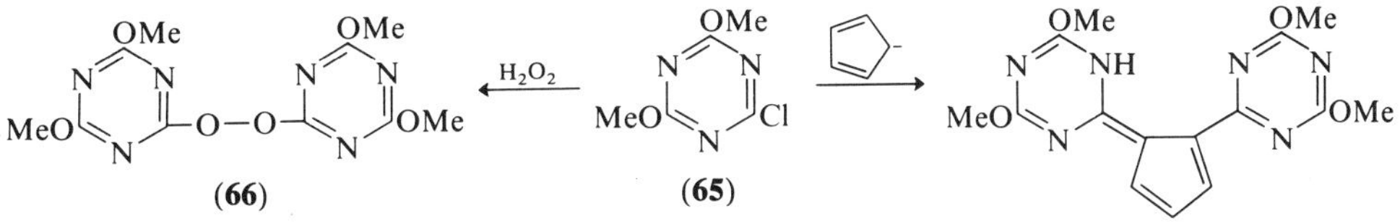

(66)    (65)

**Scheme 43**

### 2.20.3.12.2 *Cyanuric chloride as a halogenating and deoxygenating agent*

Recently, it has been shown that acyl chlorides may be prepared from cyanuric chloride and carboxylic acids ⟨79TL3037⟩. Cyanuric chloride is also valuable for effecting the formation of macrocyclic lactones (Scheme 44) ⟨80TL1893⟩. Similarly, aryl oximes may be converted to

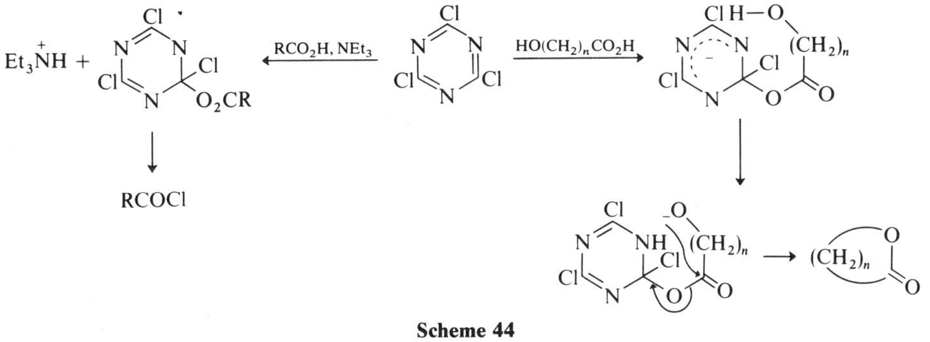

**Scheme 44**

the corresponding cyanides ⟨72CC1226⟩, and carbodiimides are prepared from the appropriate thiourea ⟨80CB79⟩. Sandler has reported the synthesis of alkyl halides; the reaction occurs without isomerization (Scheme 45) ⟨70JOC3967⟩.

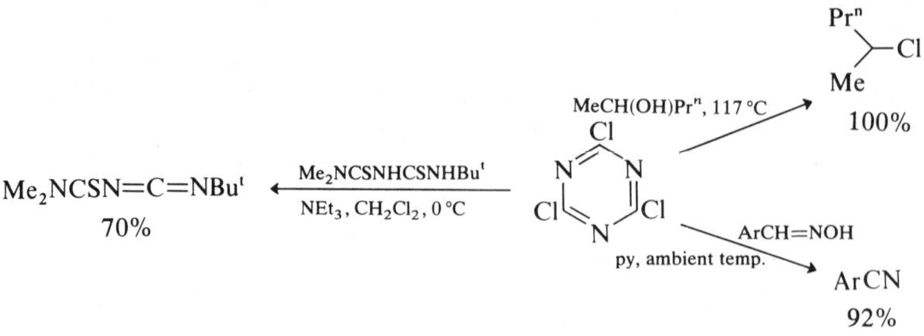

**Scheme 45**

In addition to undergoing Friedel–Crafts arylations, cyanuric chloride has been used in the Friedel–Crafts synthesis of (**67**) from pyridine (equation 31) ⟨68ZOB1368⟩.

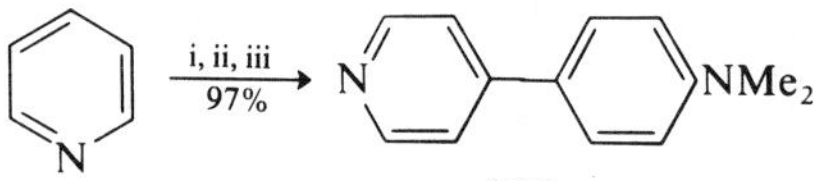

$$\text{(31)}$$

(**67**)

i, cyanuric chloride, 70 °C; ii, PhNMe$_2$, AlCl$_3$, ambient temp.; iii, 100 °C

### 2.20.3.12.3 Reactions of fluoro-1,3,5-triazines

In general, the chemistry of fluoro-1,3,5-triazines resembles that of the chloro derivatives, and therefore will not be discussed in detail. There is evidence that cyanuric fluoride is less reactive than cyanuric chloride in reaction with aniline ⟨76CCC3378⟩. Chambers *et al.* have reported the isolation of a stable anion σ-complex from the reaction of cyanuric fluoride and cesium fluoride (equation 32) ⟨77JCS(P1)1605⟩. Such species are believed to be intermediates in the formation of (**68**) from cyanuric fluoride and perfluoropropene (equation 33). Olah *et al.* have shown that cyanuric fluoride deoxygenates sulfoxides efficiently (equation 34) ⟨80S221⟩.

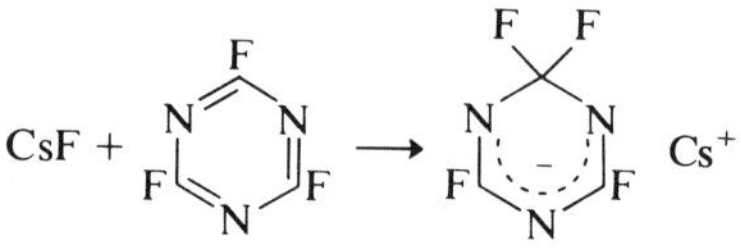

$$\text{(32)}$$

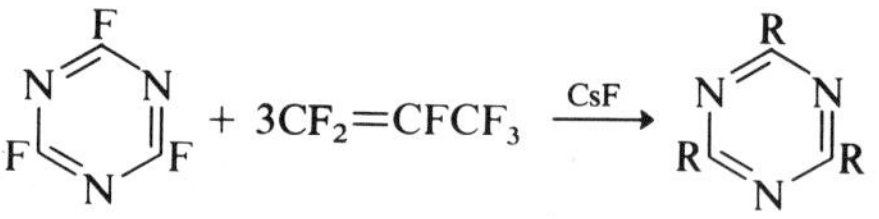

$$\text{(33)}$$

(**68**) R = (CF$_3$)$_2$CF

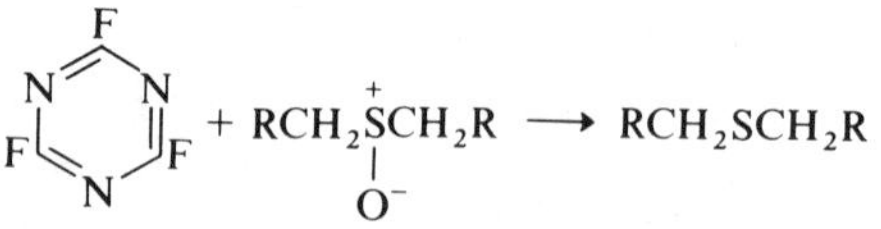

$$\text{(34)}$$

### 2.20.3.13 Reactions of Hexahydro-1,3,5-triazines

1,3,5-Trialkylhexahydro-1,3,5-triazines are readily hydrolyzed in acid, and they may also decompose to the parent amine and formaldehyde on heating strongly with water. They

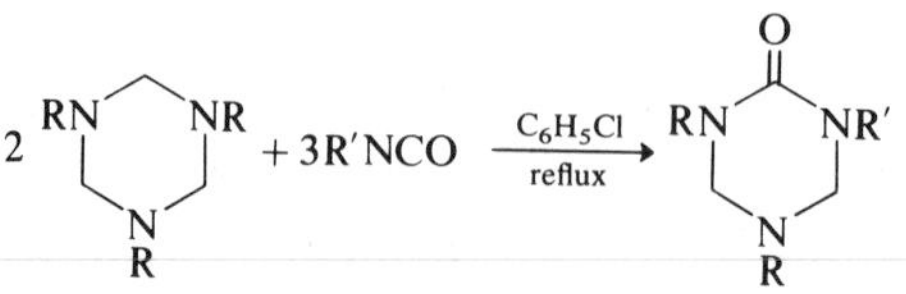

$$\text{(35)}$$

(**69**)

form quaternary salts with alkyl iodides, and are reduced by zinc and hydrochloric acid to the corresponding methylamine. In general, the 1,3,5-triarylhexahydro-1,3,5-triazines display similar reactivity to the alkyl derivatives ⟨59HC(13)1, p.473⟩. Hexahydro-1,3,5-triazines react with aryl isocyanates to form the triazinones (**69**) in high yield ⟨68JHC211⟩.

1,3,5-Trialkylhexahydro-1,3,5-triazines have also been used to prepare α-*N*-alkyl-acetamidomethylated carbonyl compounds (equation 36) ⟨81CPB1156⟩ and as aminomethyl-ating agents ⟨71JHC597, 71JHC605, 71JHC611⟩. In contrast to conventional Mannich reagents, these reagents favour *para* substitution of phenols (Scheme 46). *N*-Chloromethylcar-boxamides may also be prepared conveniently from hexahydro-1,3,5-triazines (Scheme 47) ⟨79S810⟩.

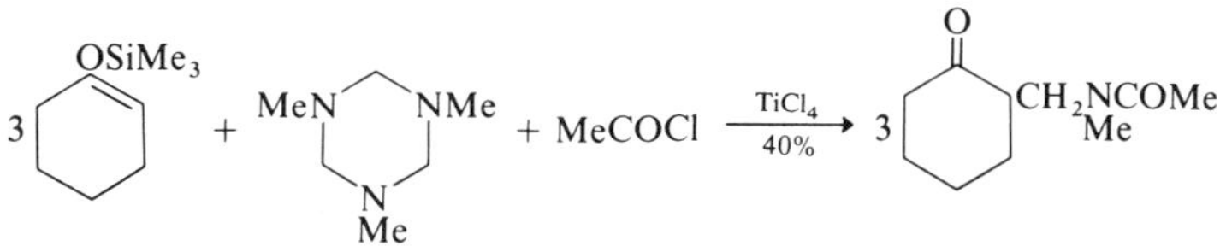

(36)

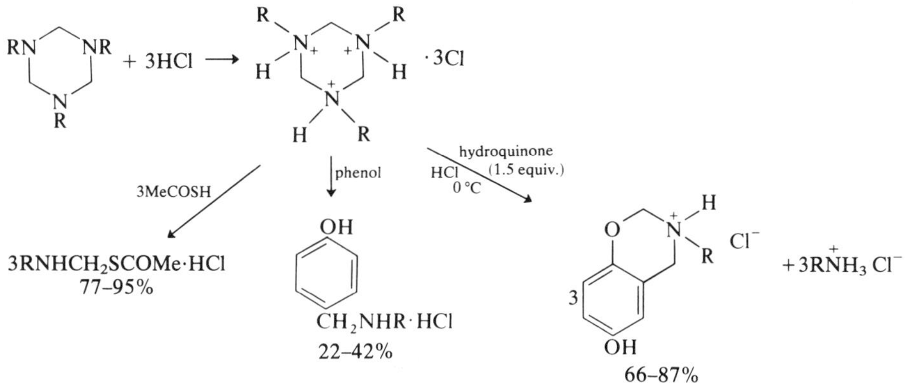

**Scheme 46**

**Scheme 47**

### 2.20.3.14 Reactions of Hexamethylenetetramine

The chemistry of hexamethylenetetramine, or hexamine (**70**), has been reviewed thoroughly ⟨B-61MI22000, p. 688, 59HC(13)1, p.545, 79S161⟩. The discussion below summarizes these reviews. Hexamethylenetetramine has a symmetrical adamantane-like structure, and is quite stable. It is hydrolyzed by hot mineral acid to formaldehyde and the appropriate

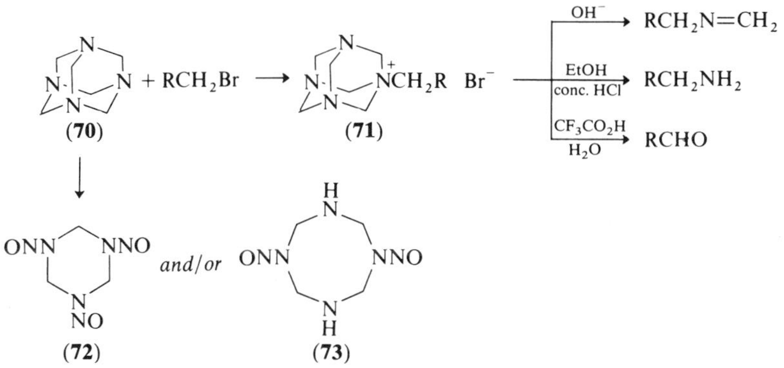

**Scheme 48**

ammonium salt, but in cold dilute organic or mineral acid it forms salts. It is reduced to ammonia and formaldehyde in zinc and mineral acid. Quaternary salts (71) are formed by treatment of hexamine with alkyl halides. These salts are valuable synthetic reagents which may be used to prepare amines, imines and aldehydes (Scheme 48) and to effect ring closure.

Hexamine undergoes degradative nitrosation to form either 1,3,5-trinitrosohexahydro-1,3,5-triazine or 1,5-dinitrosooctahydro-1,3,5,7-tetrazocine, depending on the reaction conditions. At pH 1, (72) is formed exclusively, but between pH 3 and 6 only (73) is formed. It reacts with nitric acid and acetic anhydride in a similar way.

### 2.20.3.15  Reactions of Fused 1,3,5-Triazine Systems

There have been a number of syntheses of a wide range of fused 1,3,5-triazine systems (see Section 2.20.4), but the chemistry of most of these compounds has yet to be fully investigated. The following discussion is limited to an outline of novel rearrangements of some fused 1,3,5-triazines and the reactions of cycl[3.3.3]azine derivatives.

#### 2.20.3.15.1  Rearrangements of fused 1,3,5-triazines

The pyrido[1,2-*a*][1,3,5]triazine (74) reacts with cyanamide to form (75; equation 37) ⟨75JHC407⟩. 1,3,5-Triazino[1,2-*c*][1,2,3]benzotriazines (76) are transformed into (77) on treatment with sodium azide and acetic acid ⟨77JCS(P1)103⟩. The product loses nitrogen on heating to form a fused triazinoindazole (equation 38).

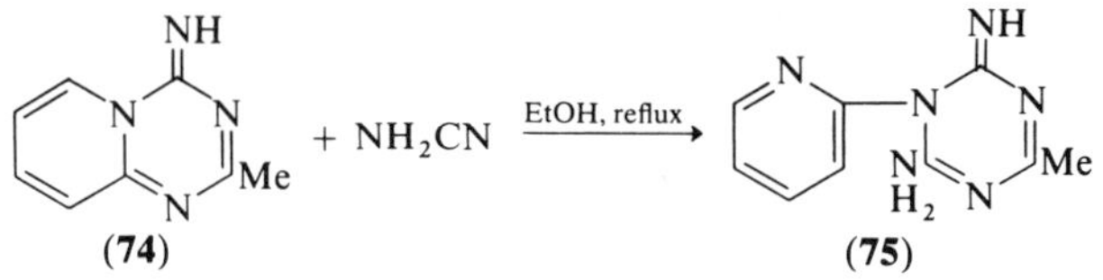

(37)

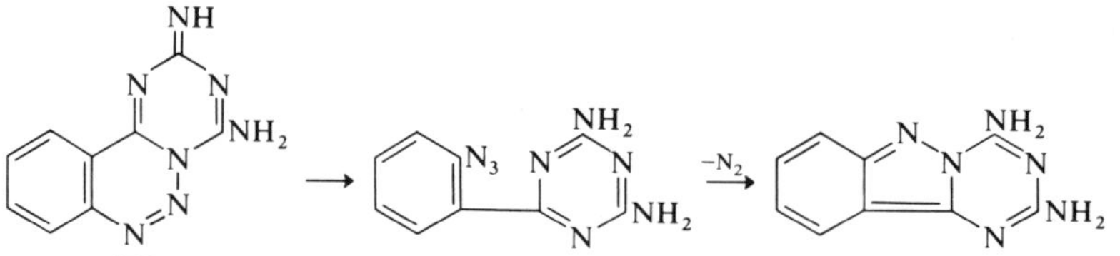

(38)

The *s*-triazolo[4,3-*a*][1,3,5]triazine (78) undergoes a Dimroth-type rearrangement on heating with excess dimethylamine (Scheme 49) ⟨73JHC231⟩. Kobe *et al.* have reported a

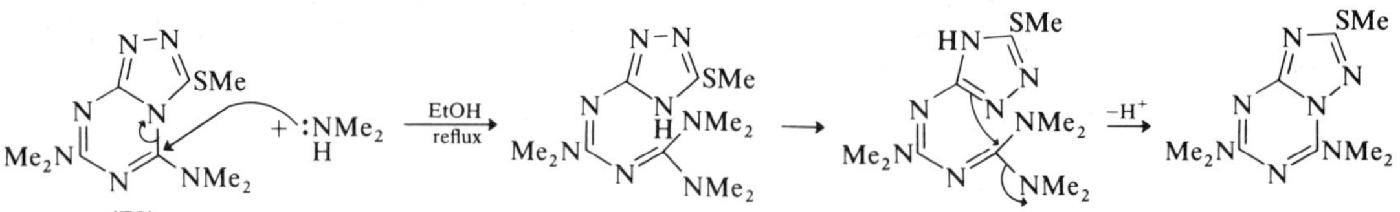

**Scheme 49**

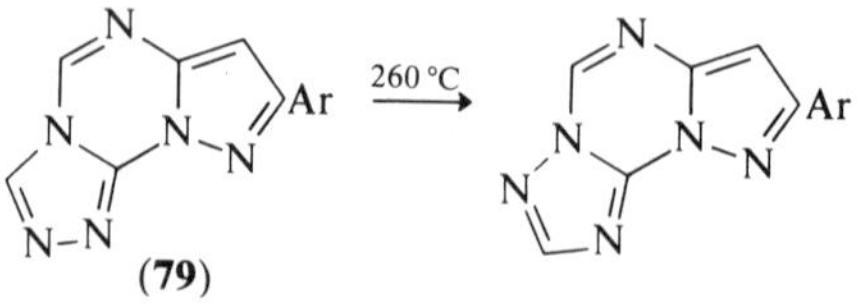

(39)

similar thermal rearrangement of (79; equation 39) ⟨74JHC991⟩. The fused 1,3,5-triazine system (29) is very stable, although its derivatives can undergo reactions similar to the corresponding substituted 1,3,5-triazine ⟨59HC(13)1, p. 423⟩.

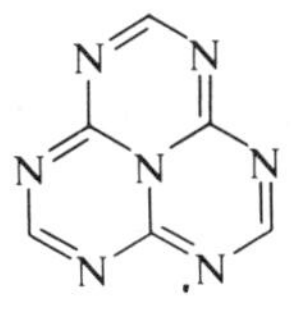

(29) 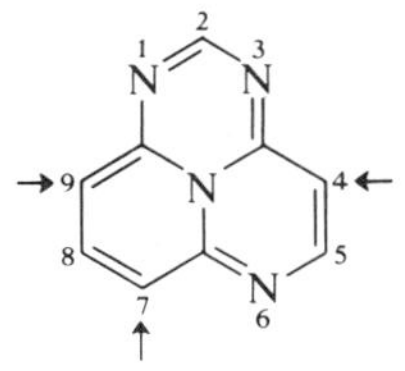    (80)

### 2.20.3.15.2 *Reactions of triazacycl[3.3.3]azine and related compounds*

Most of the reports in the literature discuss electrophilic substitution of 1,3,6-triazacycl[3.3.3]azine (**80**) and the 2-methyl, 4-cyano and 4-ethoxycarbonyl derivatives. Electrophilic bromination occurs preferentially at the 4-position (if available), and subsequently at the 9- and 7-positions. These data support the electron charge density calculations (see Section 2.20.2) ⟨73ACS3264⟩. Nitrations are carried out using copper(II) nitrate and acetic anhydride. The central nitrogen is completely non-basic, and in the triazacyclazines protonation occurs initially at position 6 (**80**) ⟨77ACS(B)239⟩. Ceder and Vernmark have reported that piperidine reacts with (**81**) *via* the aryne intermediate; the *A–E cine*-substitution mechanism is an attractive alternative (equation 40).

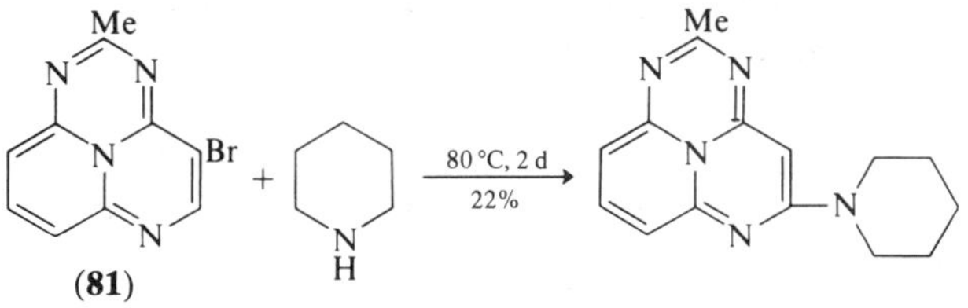

(40)

The chemistry of the ring substituents remains largely unknown. Shaw *et al.* have shown that the methyl group has similar properties to those of an alkyl-1,3,5-triazine ⟨81JHC75⟩. The 4-cyano group may be removed by heating with polyphosphoric acid at 200 °C ⟨73ACS3259⟩.

### 2.20.3.16 Reactions of 1,3,5-Triazine Betaines

Coburn and Bhoosan reported the first synthesis of 1,3-disubstituted 1,3,5-triazine-4,6-dione (**82**) and the corresponding thione derivative. The dione (**82**) is thermally stable, but forms ring-opened products on exposure to moisture. The reaction of (**82**) with dimethyl acetylenedicarboxylate gives the 1 : 1 adduct rather than the expected cycloaddition products (Scheme 50) ⟨75JHC187⟩. The fused system (**83**) reacts with ethylamine to produce the 1,3,5-triazine (**84**; equation 41) ⟨81JCS(P1)331⟩.

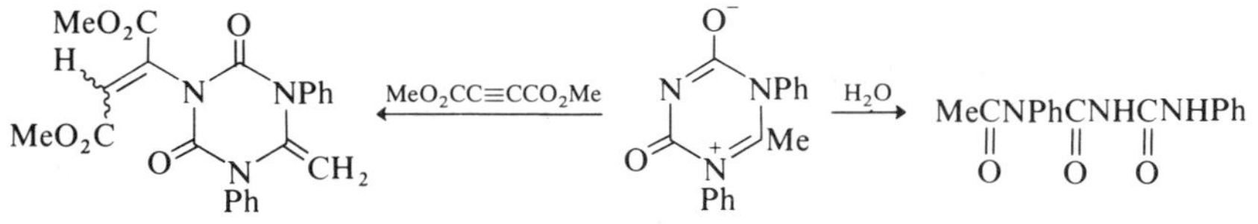

**Scheme 50**

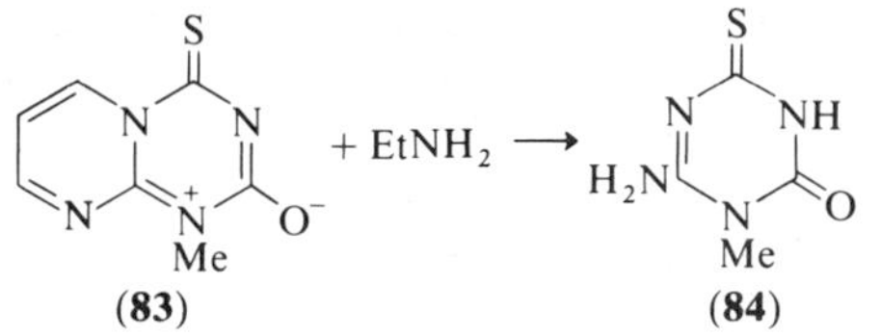

(41)

## 2.20.4 SYNTHESES

The fundamental synthetic strategies remain largely unchanged since the reviews of Smolin and Rapoport ⟨59HC(13)1⟩ and Modest ⟨B-61MI22000⟩. The major synthetic problem continues to be the preparation of 1,3,5-triazines bearing a different substituent on each carbon, but the use of *N*-cyanoamidines as starting materials has been a significant development. Cyanuric chloride is still the most valuable starting material for many important triazines (see Section 2.20.3.12). Discussion of this method will be confined to an intercomparison of the available routes (see Section 2.20.4.9).

### 2.20.4.1 Synthesis of 1,3,5-Triazines by Formation of One Carbon–Nitrogen Bond

There are few examples of this type of synthetic route. (It is more appropriate to discuss Pinner type syntheses in Section 2.20.4.6.) The best known examples involve derivatives of urea or thiourea.

#### 2.20.4.1.1 Synthesis from urea, thiourea and their derivatives

1,3,5-Triazinedione is synthesized from formylbiuret (**85**; Scheme 51) ⟨60AG836⟩. The latter compound has been postulated as an intermediate in the reaction of biuret with ethyl formate ⟨61CCC2519⟩. The synthesis is also effected by treatment of *N,N'*-dicarbamoylformamidine (**86**) with sodium ethoxide (Scheme 51).

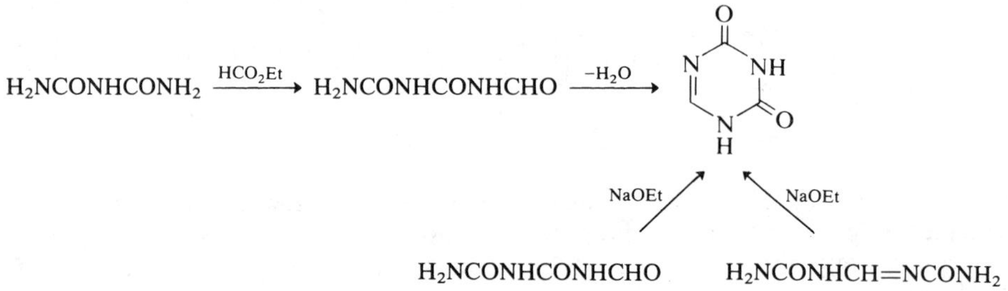

$$H_2NCONHCONH_2 \xrightarrow{HCO_2Et} H_2NCONHCONHCHO \xrightarrow{-H_2O}$$

H₂NCONHCONHCHO          H₂NCONHCH=NCONH₂
        (**85**)                              (**86**)

NaOEt                    NaOEt

**Scheme 51**

Hexahydro-1,3,5-triazine-2,4-dione and 4-thioxo-tetrahydro-1,3,5-triazin-2-one may be synthesized in an analogous manner (equations 42 and 43 respectively) ⟨73ZC408⟩.

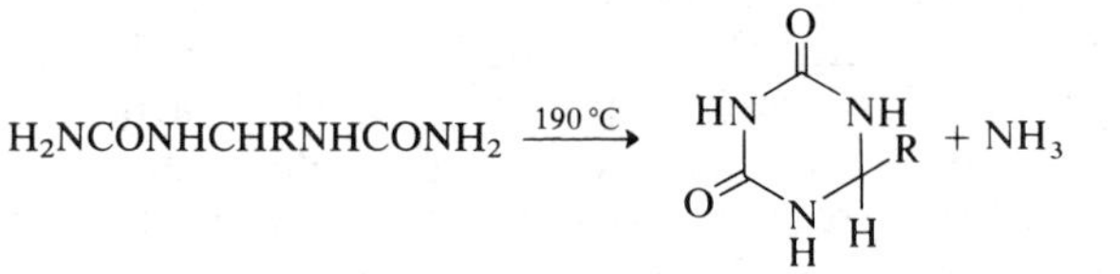

$$H_2NCONHCHRNHCONH_2 \xrightarrow{190\,°C} \quad + NH_3 \tag{42}$$

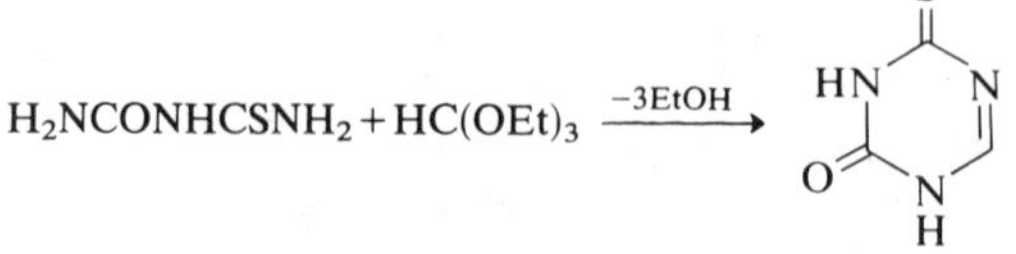

$$H_2NCONHCSNH_2 + HC(OEt)_3 \xrightarrow{-3EtOH} \tag{43}$$

5-Alkyl-hexahydro-1,3,5-triazin-2-ones are prepared by the reaction between urea, formaldehyde and primary amines ⟨73S243⟩. The intermediate polycondensate (**87**) undergoes intramolecular cyclization on heating to yield the desired product (Scheme 52). By using substituted ureas (or thioureas) and/or different aldehydes, a wide variety of *N*-substituted 5-alkyl-hexahydro-1,3,5-triazin-2-ones (or 2-thiones) may be prepared.

Diisothiocyanates react with aryl amines to form the 1,3,5-triazine *via* the thiadiazinone intermediate (**88**; Scheme 53) ⟨81CB2075⟩. The reaction also works for benzylic amines.

Dicyanoguanidines (**89**) are also valuable starting materials. The synthesis of 1,3,5-triazines by the reaction of hydrogen chloride and dicyanoguanidines at 5 °C (equation 44) was the first specific application of this cyclization to yield a heterocyclic compound

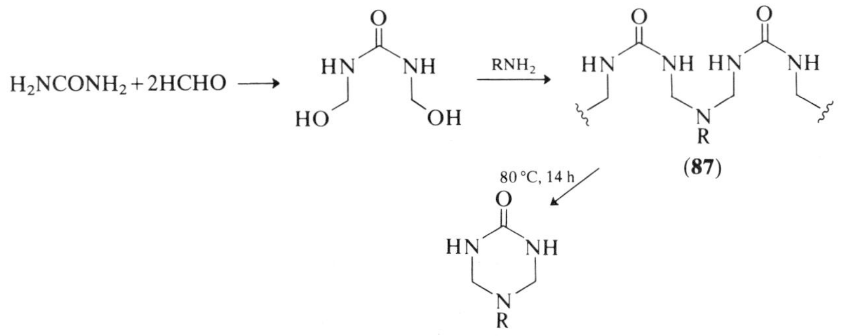

(87)

Scheme 52

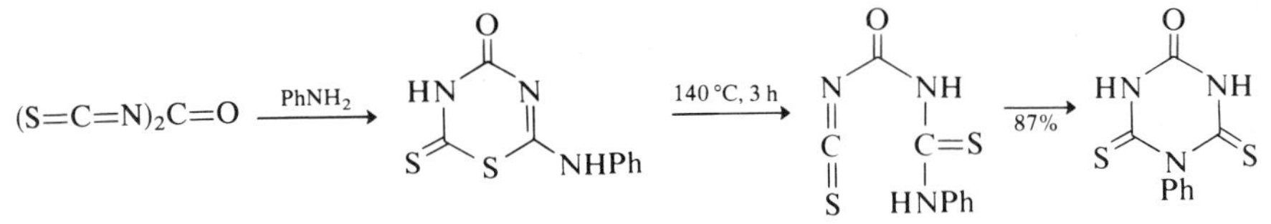

(88)

Scheme 53

⟨53USP2630433⟩. Heating dicyanoguanidines with excess aqueous hydrochloric acid gives the corresponding 1,3,5-triazin-2-one (equation 45) ⟨53USP2653937⟩ and 4,6-diamino-1,3,5-triazine-2-thione may be prepared similarly (equation 46) ⟨54USP2688016⟩.

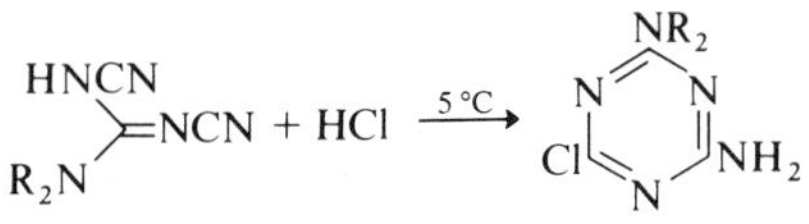

(89)

(44)

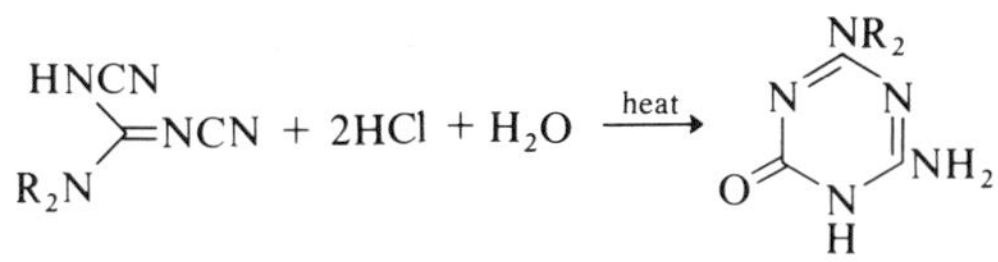

(45)

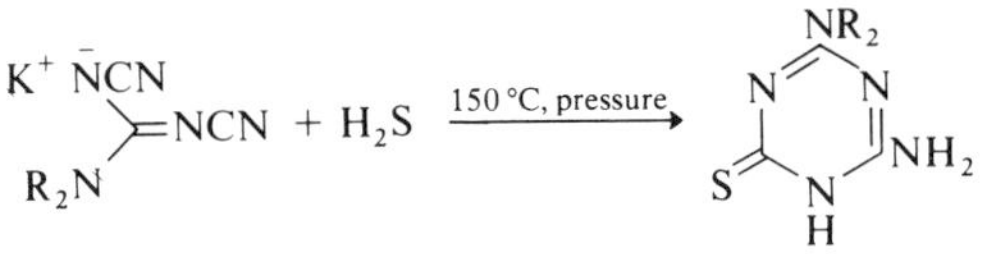

(46)

### *2.20.4.1.2 Synthesis of fused rings*

Fused 1,3,5-triazine systems can also be prepared from urea or thiourea derivatives. The syntheses of (**90**; equation 47) ⟨74JHC199⟩ and (**91**; equation 48) ⟨76JHC589⟩ illustrate the method.

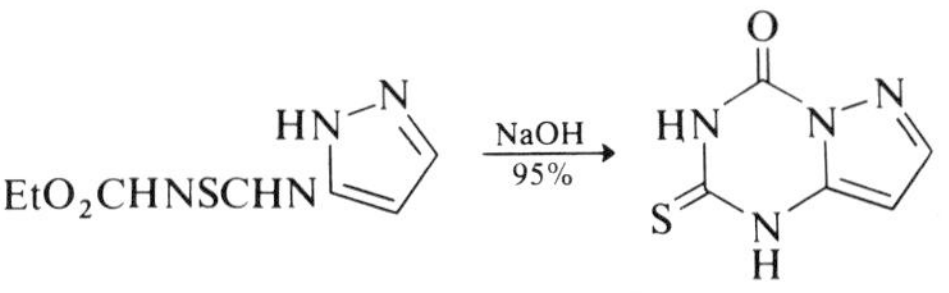

(90)

(47)

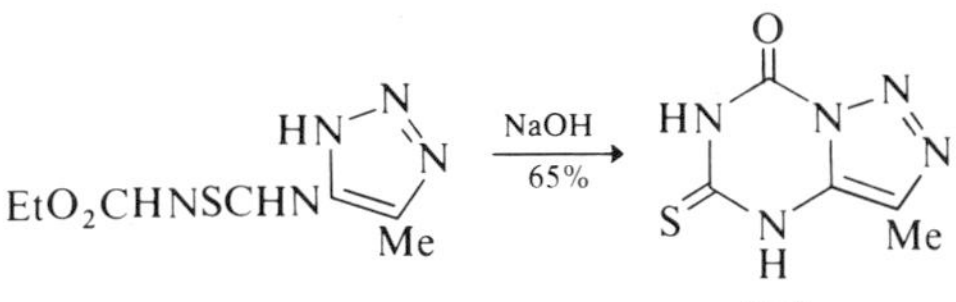

(91)

(48)

### 2.20.4.2 Synthesis of 1,3,5-Triazines by Formation of Two Carbon–Nitrogen Bonds from [5+1] Atom Fragments

There are many examples of such syntheses. Usually, the five-atom component has nitrogens at each end. The biguanides are the most important starting materials, but there are also a number of other valuable reagents.

#### 2.20.4.2.1 Synthesis using biguanides as the five-atom fragment

Biguanide (**92**; R = H) and its derivatives react with a variety of carboxylic acid derivatives including lactones, amides, orthoesters, esters, acid anhydrides and acid chlorides to produce a wide range of 6-substituted 2,4-diamino-1,3,5-triazines ⟨59HC(13)1, p. 228⟩. Many of these triazines exhibit marked diuretic activity. The synthetic route was discovered by Bamberger and coworkers who prepared the 1,3,5-triazine (**93**) whilst attempting to dry a chloroform solution of piperidinylbiguanide over potassium hydroxide (equation 49) ⟨1892CB525, 1892CB534⟩. The syntheses are effected in basic or neutral conditions, and yields are usually good. Typical examples are given in Table 9.

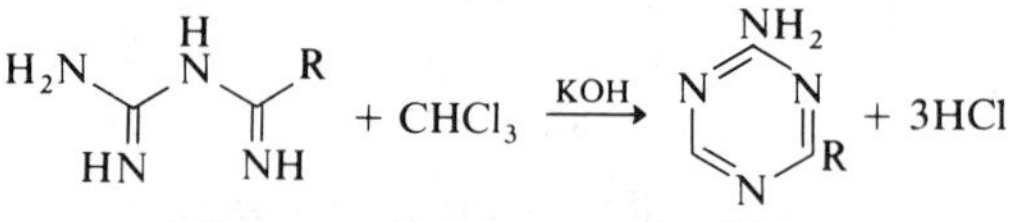

$$\text{(92)} + CHCl_3 \xrightarrow{KOH} \text{(93)} + 3HCl \tag{49}$$

(**92**)  R = piperidine

(**93**)

**Table 9**  Synthesis of 2,4-Diamino-1,3,5-Triazines from Biguanides

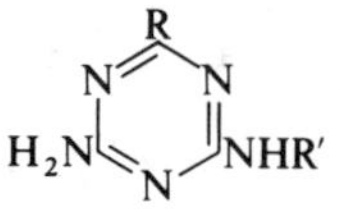

| Other starting reagent | Conditions | 1,3,5-Triazine[a] R | Yield (%) | Ref. |
|---|---|---|---|---|
| $Ac_2O$ | NaOH, dioxane, r.t. | Me | 61 | 59HC(13)1, p. 242 |
| $C_{17}H_{31}COCl$ | $Na_2CO_3$, toluene, 110 °C, 16 h | $C_{17}H_{31}$ | 90 | 59HC(13)1, p. 242 |
| $H_2C=CHCOCl$ | NaOH, 0 °C, 1 h | $CH=CH_2$[b] | 37 | 59HC(13)1, p. 242 |
| $(CO_2Et)_2$ | i, MeOH; ii, EtOH, 78 °C | $CO_2Et$[b] | 92 | 59HC(13)1, p. 242 |
| PhCOCl | NaOH | Ph | 83 | 59HC(13)1, p. 242 |
| 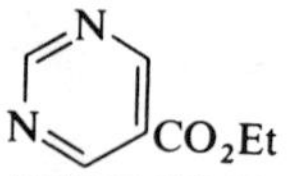 (pyrazine-$CO_2Et$) | MeOH | 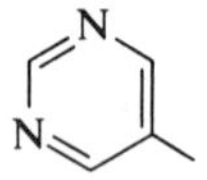 (pyrazinyl) | 90 | 59HC(13)1, p. 242 |
| $NCCH_2CO_2Et$ | MeOH, 60 °C, 5 h | $CH_2CN$ | 85 | 62CPB1215 |
| $CHF_2CO_2Et$ | MeOH, 60 °C, 5 h | $CHF_2$ | 70 | 65ZOB1156 |
| $C_2F_4$ | DMF, 60–70 °C | $CHF_2$ | — | 66KGS122 |
| $C_3F_7CO_2Me$ | i, 160 °C; ii, r.t., 24 h | $C_3F_7$ | 75 | 59JOC1809 |
| 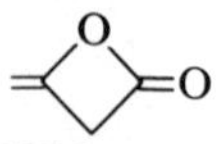 (diketene, O=⟨⟩=O) | $Me_2CO$, r.t., 2 h | $CH_2COMe$ | 44 | 73S536 |
| $HCO_2Et$ | NaOMe, r.t., 18 h | H | 52 | 57JA5064 |
| $CS_2$ | KOH, EtOH, 78 °C, 14 h | SH | 13 | 68JMC1167 |
| $ClCH_2CO_2Et$ | MeOH, 3 h | $CH_2Cl$ | 81 | 57JA941 |
| 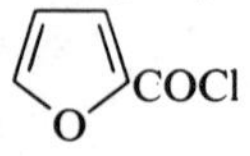 (furyl-COCl) | NaOH, $Me_2CO$, 0 °C to r.t., 2 h | 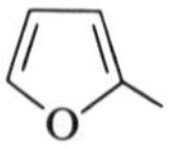 (furyl) | 88 | 73JMC1305 |

[a] R′ = H unless otherwise indicated.   [b] R$^1$ = Ph.

The reaction probably occurs *via* the acylation of the biguanide followed by cyclization (Scheme 54). The mechanism of the reaction between ethyl oxalate and cyclohexylbiguanide has been investigated. An intermediate claimed to have structure (**94**) was isolated, and subsequently converted into the 1,3,5-triazine (Scheme 55) ⟨57JA5064⟩.

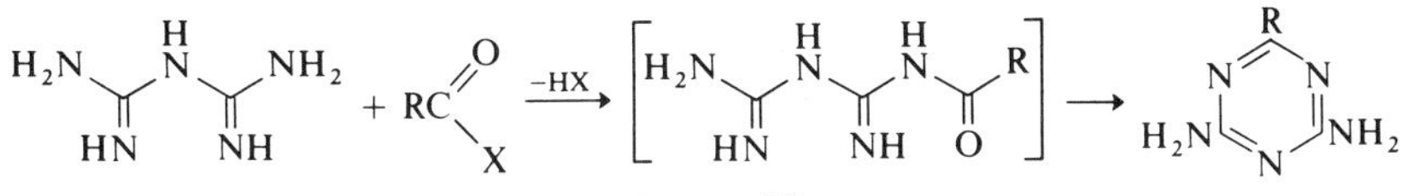

**Scheme 54**

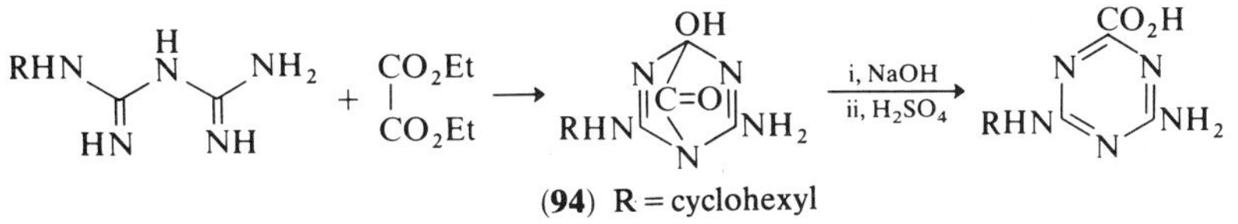

(**94**) R = cyclohexyl

**Scheme 55**

Ethylene carbonate (**95**) reacts with arylbiguanides *via* the intermediate (**96**) to form (**97**) (Scheme 56) ⟨74BCJ2893⟩, and the corresponding thio derivative can be prepared in a similar manner.

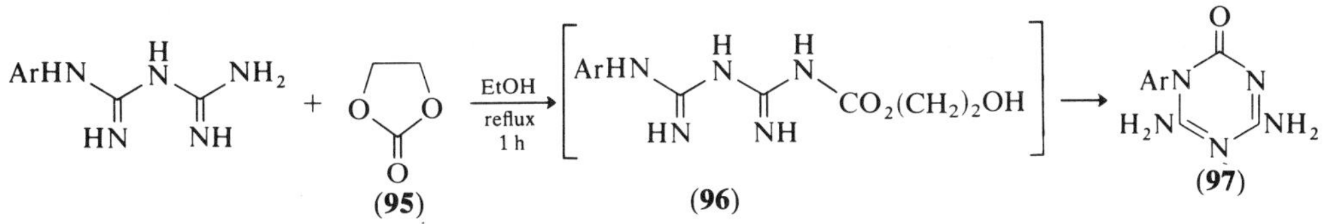

**Scheme 56**

Biguanides react with carbodiimides to form melamine derivatives in 60–70% yields (equation 50) ⟨64JCS3459⟩, and with isothiocyanates to form the 1,3,5-triazinethione (**98**; equation 51) ⟨65JCS6296⟩.

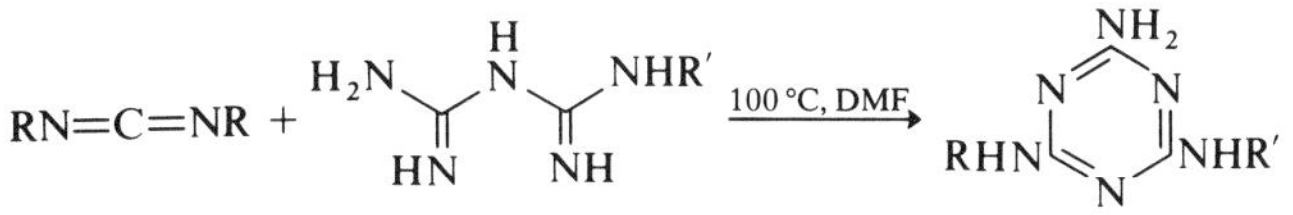

(50)

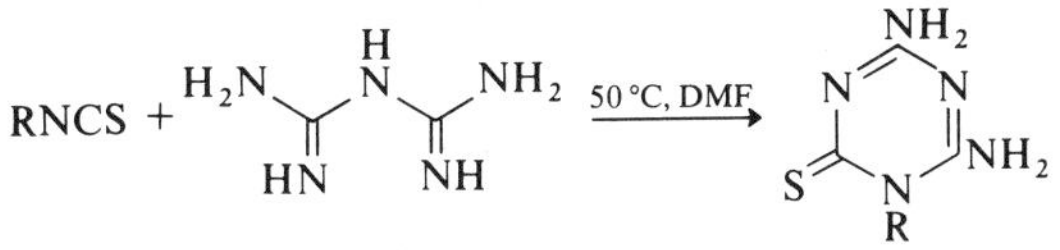

(51)

When aldehydes or ketones are used instead of the carboxylic acid derivatives, the corresponding 1,2-dihydro-1,3,5-triazines are formed (equation 52). The reaction is acid catalyzed ⟨61MI22000, p. 697⟩.

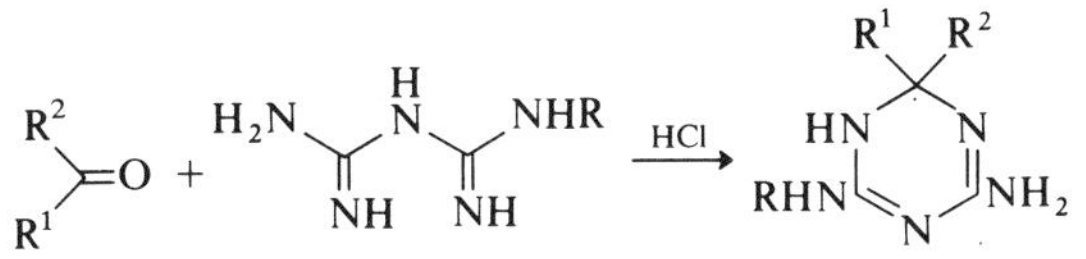

(52)

### 2.20.4.2.2 Synthesis from other five-atom fragments

Monohydroxyphenyl-1,3,5-triazines are of industrial importance as stabilizers in polymers (Section 2.20.5.3), and they may be synthesized from (**99**) and methanol in the presence of sodium methoxide (Scheme 57) ⟨72HCA1566⟩.

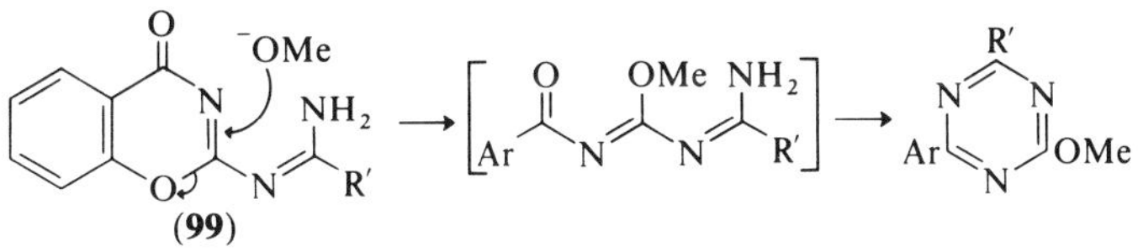

(**99**)

i, MeOH, reflux, 5 min; R′ = alkyl, diamino, SH, Ar = 2-HOC$_6$H$_4$

**Scheme 57**

Aryl cyanates react with ammonia to form the bis(iminoethers) (**100**). These compounds form 1,3,5-triazines by reaction with a further equivalent of the cyanate (Scheme 58) ⟨67AG(E)206⟩. *N*-Amidoamidines condense with ethyl oxalate to form 2-amino-1,3,5-triazines (equation 53) ⟨71CPB1789⟩.

$$2ArOCN + NH_3 \longrightarrow \text{(100)} \xrightarrow[\text{H}_2\text{O, 30–40 °C}]{ArOCN, (NH_4)_2CO_3}$$

**(100)**

**Scheme 58**

$$(53)$$

Recently, Richter and Ulrich devised an efficient synthesis of 2,4-bis(arylimino)-1,3,5-trialkylhexahydrotriazinones (**102**) from the reaction of 2-arylimino-1,3-diaryldiazetidin-4-ones, phosgene and an aryl amine. The intermediate 1:1 adduct (**101**) has been partially characterized, and the proposed mechanism is shown in Scheme 59 ⟨81JOC3011⟩.

**(101)**                    **(102)**

**Scheme 59**

Shaw has reported a synthesis of 2,6-diamino-4-methyl-1,3,5-triazine 1-oxide from potassium dicyanoacetamidine and hydroxylamine (equation 54) ⟨62JOC3890⟩.

$$(54)$$

### 2.20.4.2.3 Synthesis of fused ring systems

There are not many examples of syntheses of fused 1,3,5-triazines from [5 + 1] atom cyclizations. The basic synthetic strategy is similar to the syntheses of 1,3,5-triazines from biguanides, but in this case the five-atom component incorporates a heterocyclic substituent. The examples in equations (55)–(57) illustrate the approach ⟨74JHC991, 75HCA761, 80JHC1121⟩.

$$+ HC(OEt)_3 \xrightarrow[67\%]{\text{reflux, 3 h}} \qquad (55)$$

$$+ MeCO_2CH(OEt)_2 \xrightarrow[55\%]{\text{reflux, 5 min}} \qquad (56)$$

$$+ MeC(OEt)_3 \xrightarrow[91\%]{\text{dioxane reflux, 2 h}} \qquad (57)$$

### 2.20.4.3 Synthesis of 1,3,5-Triazines by Formation of Two Carbon–Nitrogen Bonds from [2+4] Atom Fragments

There are many examples of such syntheses. The development of $N$-cyanoamidines as a starting material is a major advance in the synthesis of 1,3,5-triazines bearing a different substituent on each carbon. Dicyandiamide continues to be a valuable starting material for the synthesis of 2,4-diamino-1,3,5-triazines, although it has been superseded by urea as the starting material in the industrial synthesis of melamine.

#### 2.20.4.3.1 Synthesis from N-cyanoamidine and related compounds

The original studies on the potential of $N$-cyanoamidines (**103**) as precursors of 1,3,5-triazines concentrated on condensations with nitriles, imidates and amidines. The reaction produced triazines, but yields were moderate (15–50%) ⟨63JOC1812⟩. The condensation of $N$-cyanoamidines with chloromethyleniminium salts (**105**), which are prepared *in situ* from amides (**104**; Scheme 60), is a marked improvement, and it provides an efficient, convenient route to many 1,3,5-triazines (Table 10) ⟨80S841, 81AJC623⟩, although the reaction is of limited value for the synthesis of 2,4-dialkyl-1,3,5-triazines. Usually the chloro-1,3,5-triazines (**106**; $R^5$ = Cl) are the major products, and the corresponding amino derivatives are formed in trace (*ca.* 5%) amounts; however, if $N$-arylamides (**104**; $R^2$ = Ph) are used, the amino derivative may become the major product (Table 10). Clearly the chloro-1,3,5-triazines are valuable in their own right as it is possible to carry out numerous substitutions of the chlorine atom (see Section 2.20.3.12). Harris has tentatively proposed a mechanism (path A, Scheme 61) but two other possibilities cannot be discounted (Scheme 61, paths B and C) ⟨81AJC623⟩.

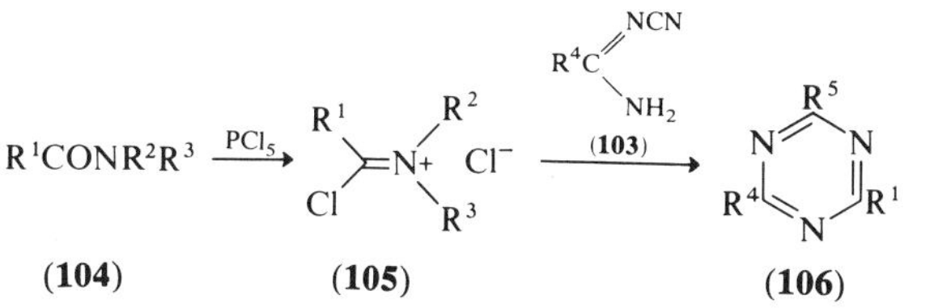

**Scheme 60**

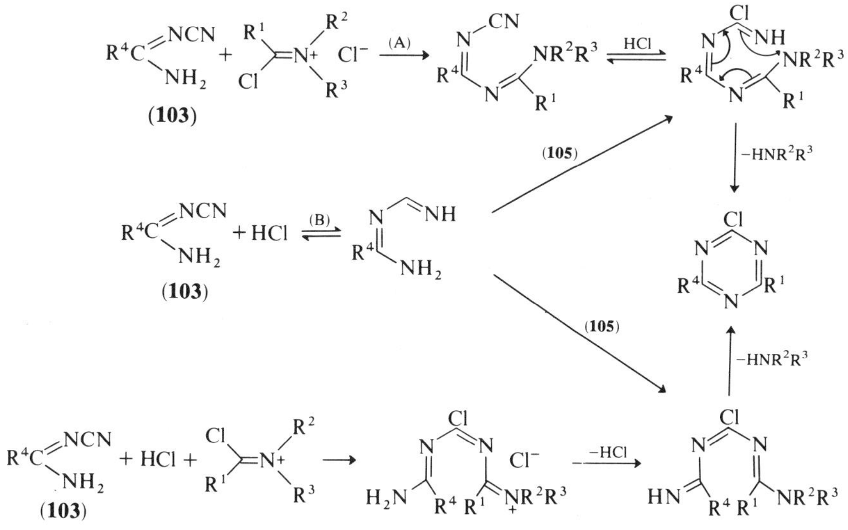

**Scheme 61**

The condensation of $N$-cyanoamidines with thiocarbamate esters (**107**) provides routes to 1,3,5-triazine thioethers (**108**; Scheme 62), and dimethylamino-1,3,5-triazines (**109**; Scheme 63). In addition, novel chloromethylenimino salts (*e.g.* **110**; $R^6$ = OPh) may be prepared *in situ* (Scheme 63). Typical examples are shown in Table 11.

**Table 10** Typical Examples of 1,3,5-Triazines (106) Synthesized from *N*-Cyanoamidines (103) and Chloromethyleneiminium salts (105)

*Starting material*

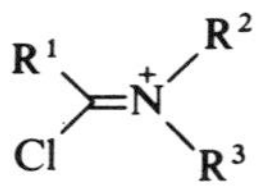

(105)

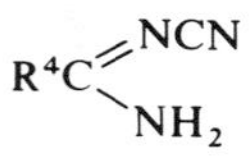

$R^4C{=}NCN$ — $NH_2$

(103)

*1,3,5-Triazine*

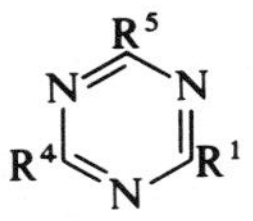

(106)

| $R^1$ | $R^2$ | $R^3$ | $R^4$ | Reaction conditions | $R^1$ | $R^4$ | $R^5$ | Yield (%) | Ref. |
|---|---|---|---|---|---|---|---|---|---|
| Ph | Me | Me | Ph | POCl$_3$, MeCN, reflux 30 min | Ph | Ph | Cl | 70 | 80S841 |
| H | Me | Me | Ph | POCl$_3$, MeCN, ambient 15 min | H | Ph | Cl | 56 | 80S841 |
| Me | Me | Me | Ph | POCl$_3$, C$_6$H$_6$, ambient 18 h | Me | Ph | Cl | 81 | 80S841 |
| Ph | Me | Me | CH$_2$Cl | POCl$_3$, MeCN, reflux 1 h | Ph | CH$_2$Cl | Cl | 88 | 80S841 |
| MeS | Me | Ph | Ph | POCl$_3$, reflux 30 min | MeS | Ph | Cl | 56 | 80S841 |
| Me | H | Ph | Ph | POCl$_3$, MeCN, reflux 1 h | Ph | Me | NHPh | 84 | 80S841 |
| Ph | Me | Me | Et | POCl$_3$, MeCN, reflux 1 h | Ph | Et | Cl | 64 | 81AJC623 |
| C$_{11}$H$_{23}$ | C$_6$H$_{11}$ | H | Ph | POCl$_3$, MeCN, reflux 30 min | C$_{11}$H$_{23}$ | Ph | Cl | 87 | 81AJC623 |
| Ph | C$_6$H$_{11}$ | H | CCl$_3$ | PCl$_5$, 150 °C 15 min | Ph | CCl$_3$ | Cl | 83 | 81AJC623 |
| Me | Me | Me | CCl$_3$ | POCl$_3$, MeCN, reflux 1 h | Me | CCl$_3$ | Cl | 73 | 81AJC623 |

**Table 11** Typical Examples of 1,3,5-Triazines (**106**) Synthesized from *N*-Cyanoamidines (**103**) and Dithiocarbamate-derived Chloromethyleneiminium salts (**105**) ⟨81AJC623⟩

*Starting material* — *1,3,5-Triazines*

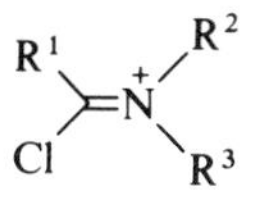

(**105**)

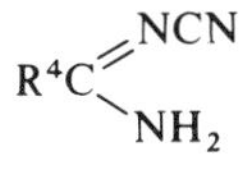

(**103**)

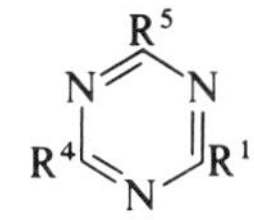

(**106**)

| $R^1$ | $R^2$ | $R^3$ | $R^4$ | *Reaction conditions* | $R^1$ | $R^4$ | $R^5$ | *Yield* (%) |
|---|---|---|---|---|---|---|---|---|
| EtS | Me | Me | Ph | i, $COCl_2$; ii, MeCN, reflux, 30 min | EtS | Ph | Cl | 52 |
| MeS | Me | Me | Me | Not given | MeS | Me | Cl | 66 |
| MeS | Me | Ph | Ph | i, $COCl_2$; ii, $POCl_3$, 100 °C, 60 min | MeS | Ph | $NMe_2$ | 37 |
| Cl | Me | Me | $CH_2Cl$ | MeCN, r.t. 12 h | Cl | $CH_2Cl$ | $NMe_2$ | 58 |
| $CCl_3$ | Me | Me | $CCl_3$ | MeCN, reflux, 30 min | Cl | $CCl_3$ | $NMe_2$ | 77 |
| Me | Me | OPh[a] | Ph | MeCN, reflux 1.5 h | $NMe_2$ | Ph | OPh | 53 |

[a] The phenoxy group was introduced *in situ* (Scheme 63).

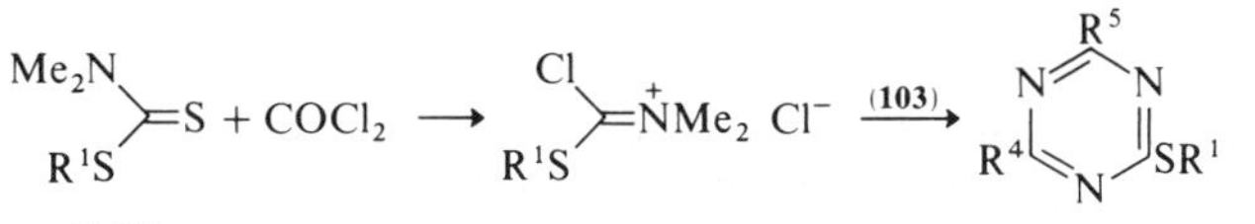

(**107**) (**108**)

**Scheme 62**

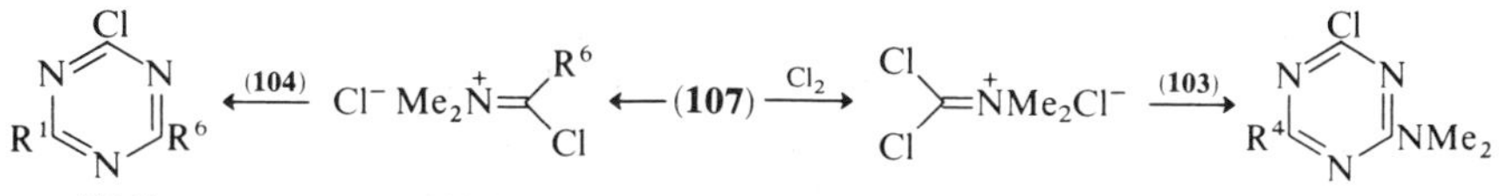

(**111**) (**110**) (**109**)

**Scheme 63**

*N*-Cyanocarbamimides (**112**) may be condensed with chloromethyleniminium salts in a similar way (equation 58). Yields are good to excellent, and reaction conditions are mild (Table 12) ⟨81S907⟩. The synthesis provides an easy route to 4-alkoxy-2-chloro-1,3,5-triazines (**113**) and the corresponding mercapto derivatives.

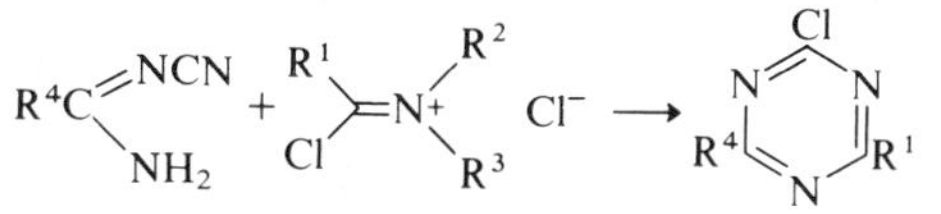

(**112**) (**113**) (58)

**Table 12** Typical Examples of 2-Chloro-1.3.5-triazines (**113**) from *N*-Cyanocarbamides (**112**) and Chloromethyleneiminium salts (**105**) ⟨81S907⟩

*Starting material* — *Product*

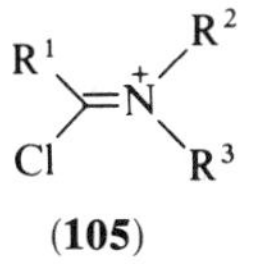

(**105**)

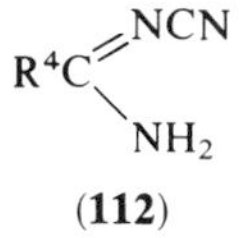

(**112**)

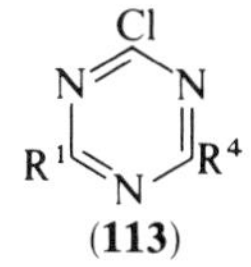

(**113**)

| $R^1$ | $R^2$ | $R^3$ | $R^4$ | Conditions | $R^1$ | $R^4$ | Yield (%) |
|---|---|---|---|---|---|---|---|
| H | Me | Me | OPh | $CH_2Cl_2$, 1 h, 25 °C | H | OPh | 73 |
| $ClCH_2$ | H | $C_6H_{11}$ | OPh | MeCN, 18 h, 25 °C | $ClCH_2$ | OPh | 74 |
| 2-Furyl | $-(CH_2)_4-$ | | OEt | MeCN, 1 h, reflux | 2-Furyl | OEt | 44 |
| Ph | Me | Me | SMe | MeCN, 1 h, reflux | Ph | SMe | 85 |
| H | Me | Me | Cl | $CH_2Cl_2$, 18 h, 25 °C | H | Cl | 69 |

Aryl cyanates react with cyanamides to form (**114**), which reacts with a further equivalent of aryl cyanate to form the 2,4-bis(aryloxy)-1,3,5-triazine (**115**) in strongly alkaline conditions (Scheme 64) ⟨67AG(E)206⟩.

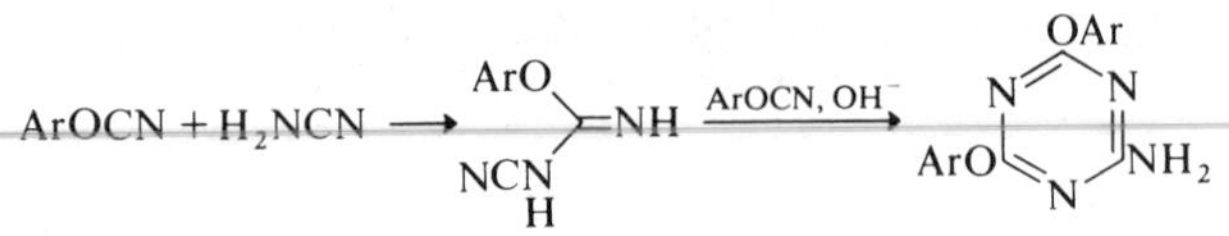

$$\text{ArOCN} + \text{H}_2\text{NCN} \longrightarrow \underset{\substack{| \\ \text{NCN} \\ | \\ \text{H} \\ \textbf{(114)}}}{\overset{\text{ArO}}{\diagdown}}{C}{=}\text{NH} \xrightarrow{\text{ArOCN, OH}^-} \quad \textbf{(115)}$$

**Scheme 64**

### 2.20.4.3.2 Synthesis from dicyandiamide

Dicyandiamide (**116**) reacts with a variety of two-atom components to form 2,4-diamino-1,3,5-triazines. The route, which bears some resemblance to the syntheses from biguanides (see Section 2.20.4.2.1), was discovered by Ostrogovich ⟨11MI22000⟩. He found that 1,3,5-triazines were prepared efficiently on heating alkyl or aryl nitriles with dicyandiamide (equation 59).

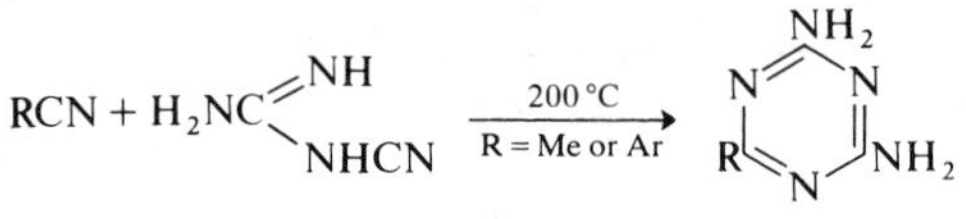

$$\text{RCN} + \text{H}_2\text{NC}\underset{\text{NHCN}}{\overset{\text{NH}}{\diagdown\diagup}} \xrightarrow[\text{R = Me or Ar}]{200\,^\circ\text{C}} \qquad\qquad\qquad (59)$$

$$\textbf{(116)}$$

Subsequently, it was shown that the reaction was catalyzed by base ⟨48M(79)106⟩. Dicyandiamide reacts with nitriles (the most valuable method), amidines, cyanamides, ammonia, cyanates, thiocyanates, carboxylic acids and anhydrides to yield 1,3,5-triazines (Table 13). This synthetic route has been reviewed thoroughly several times ⟨59HC(13)1, p. 219, 61MI22000, p. 650, 73ZC408⟩. The base-catalyzed reaction of dicyandiamide with alkyl or aryl nitriles (Scheme 65) proceeds *via* the imino ether anion and the rate determining step is solvent dependent. In DMSO the formation of the imino ether is rate determining, but in 2-methoxyethanol the reaction between the anion and dicyandiamide controls the rate ⟨66T157⟩.

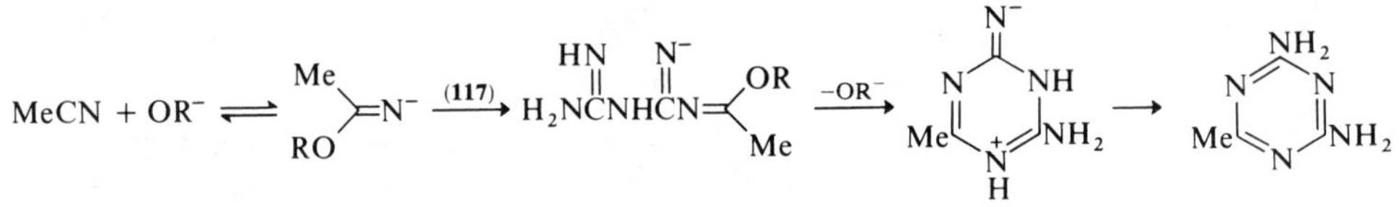

$$\text{MeCN} + \text{OR}^- \rightleftharpoons \underset{\text{RO}}{\overset{\text{Me}}{\diagdown\diagup}}{C}{=}\text{N}^- \xrightarrow{\textbf{(117)}} \text{H}_2\text{NCNHCN}{=}\underset{\text{Me}}{\overset{\text{OR}}{\diagup\diagdown}} \xrightarrow{-\text{OR}^-} \longrightarrow$$

**Scheme 65**

**Table 13**  Examples of 6R-2,4-Diamino-1,3,5-triazines Synthesized from Dicyandiamides

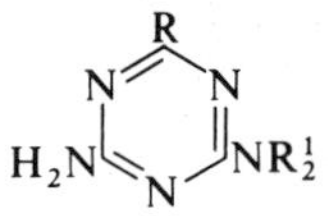

$$\underset{\text{H}_2\text{N}\diagdown_N\diagup\text{NR}_2^1}{\overset{\text{R}}{\diagup_N\diagdown}}$$

| Other starting reagent | Conditions | Product[a] R | Yield (%) | Ref. |
|---|---|---|---|---|
| MeCN | NaOH, 1 h, 100 °C | Me | 96.5 | 59HC(13)1, p. 244 |
| CN(CH₂)₂CN | NaOH, 1 h, 100 °C | (CH₂)₂CN | 87 | 59HC(13)1, p. 244 |
| Cl₃CCN | NaOH, 1 h, 100 °C | CCl₃ | 72 | 59HC(13)1, p. 244 |
| PhCN | Methyl cellosolve, KOH, 5 h, reflux | Ph | 93 | 59HC(13)1, p. 244 |
| Bu$^t$C(NH₂)=... CN | Bu$^n$OH, KOH, 2 h, reflux | NHBu$^t$ | 82 | B-61MI22000, p. 650 |
| H₂NCONH₂ | Butyl cellosolve, KCNO, 5–6 h, 155 °C | OH[c] | 90 | B-61MI22000, p. 650 |
| NaSCN | HCl, 5.5 h, 100 °C | SH[b] | 94 | 59JA5663 |
| Me₂NCN | PrOH, KOH, 90 °C | NMe₂ | 81.5 | 58CRV131 |
| CH₂=CHO-(CH₂)₂N(Me)CN | PrOH, KOH, 6.5 h, 80 °C | NMe(CH₂)₂OCH=CH₂ | 93 | 58JOC1032 |
| Bu$^n$OCH(Me)OCH₂CN | PrOH, KOH, 16 h, 80 °C | CH₂OCHMeOBu$^t$ | 93 | 58JOC724 |
| HCONH₂ | 5 h, 165 °C | NH₂ | 73 | 61CB1883 |

[a] R$^1$ = H unless indicated otherwise.   [b] R$^1$ = C₄H₉.   [c] Or keto tautomer.

Melamine may be prepared easily and in high yield by heating dicyandiamide to fusion (equation 60) ⟨40MI22000⟩. Melamine was prepared industrially by the same basic method, but the reaction was carried out under pressure and in the presence of ammonia ⟨B-79MI22000⟩.

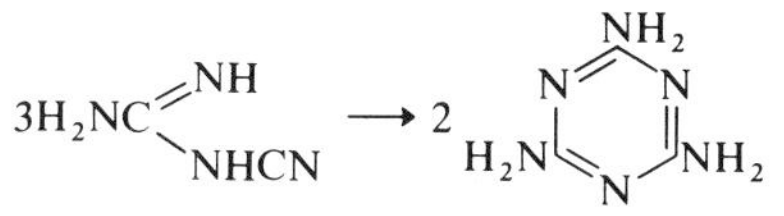

$$3H_2NC\overset{NH}{\underset{NHCN}{\big\langle}} \longrightarrow 2 \quad \text{(melamine)} \tag{60}$$

Dicyandiamide reacts with thioacetamides *via* the intermediate (**117**) to form the 1,3,5-triazinethione (**118**) in low yield (Scheme 66). This is one of the few examples of a 1,3,5-triazine synthesis in which the 2,4-diamino product is *not* formed ⟨74AJC2627⟩.

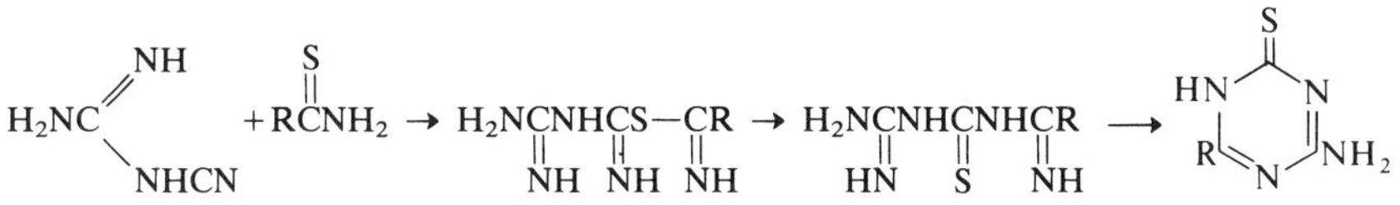

(**117**) R = Me or Ph                    (**118**)

**Scheme 66**

The reaction of dicyandiamide with thiosemicarbazides provides an efficient route to the 1-amino-1,3,5-triazines (**119**) ⟨76AJC1051⟩.

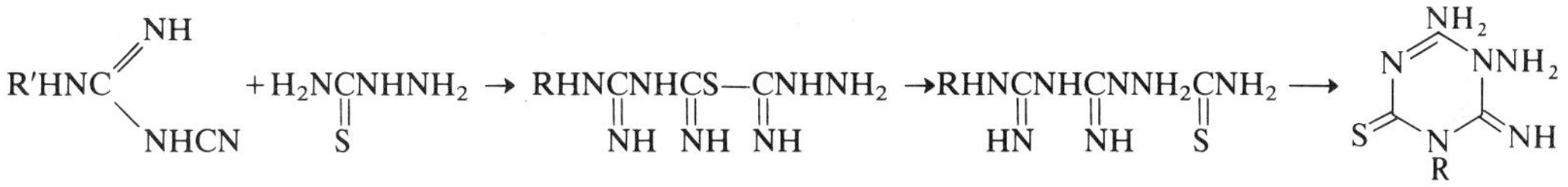

**Scheme 67**                    (**119**)

The method may be extended by using a variety of substituted dicyandiamides; 1,3,5-triazinethiones (**120**) are produced by the reaction of dicyandiamides with isothiocyanates ⟨69IJC20⟩.

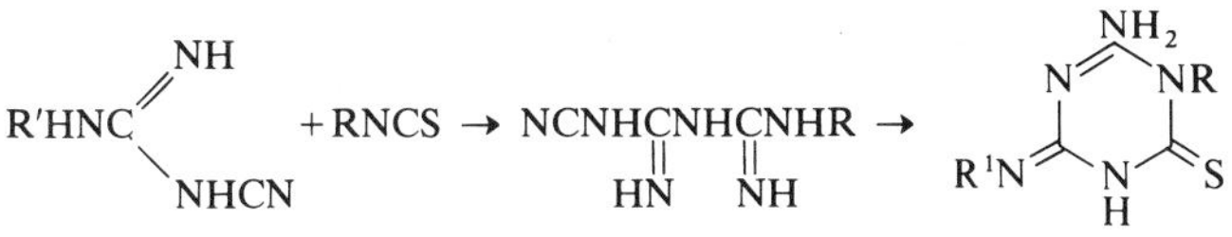

**Scheme 68**                    (**120**)

### 2.20.4.3.3 Synthesis using other four-atom components

1,3,5-Triazine is synthesized on heating triformamidomethane above its melting point ⟨63CB3260⟩. The crucial intermediate is formylamidine (**121**; Scheme 69) ⟨63AG(E)655⟩. 2-Amino-, 2-alkylamino- and 2-arylamino-1,3,5-triazines may be synthesized in a similar way (equation 61).

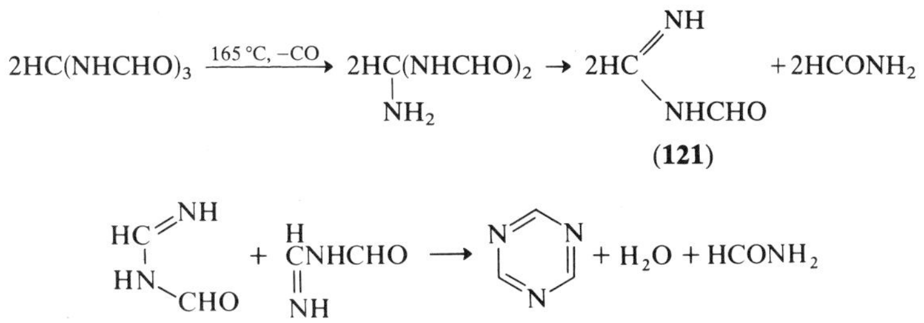

$$2HC(NHCHO)_3 \xrightarrow{165\,°C,\,-CO} 2HC(NHCHO)_2 \longrightarrow 2HC\overset{NH}{\underset{NHCHO}{\big\langle}} + 2HCONH_2$$

(**121**)

**Scheme 69**

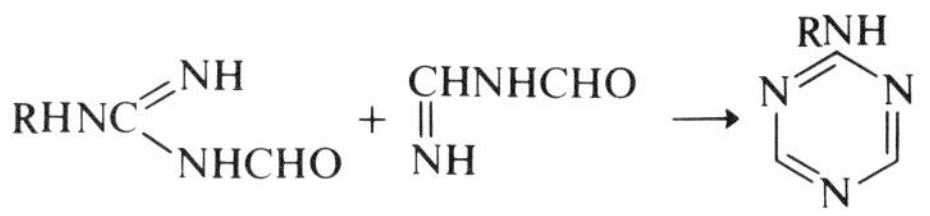

$$RHNC\overset{NH}{\underset{NHCHO}{\big\langle}} + \overset{CHNHCHO}{\underset{NH}{\big\|}} \longrightarrow \quad \text{(triazine)} \tag{61}$$

R = H, 80%; R = alkyl, 39–63%; R = aryl, 60–90%

Burger *et al.* have reported a synthesis of the novel 1,3,5-triazine (**124**) by the reaction of the cyanamide (**122**) and the triene (**123**) (equation 62) ⟨79CZ264⟩.

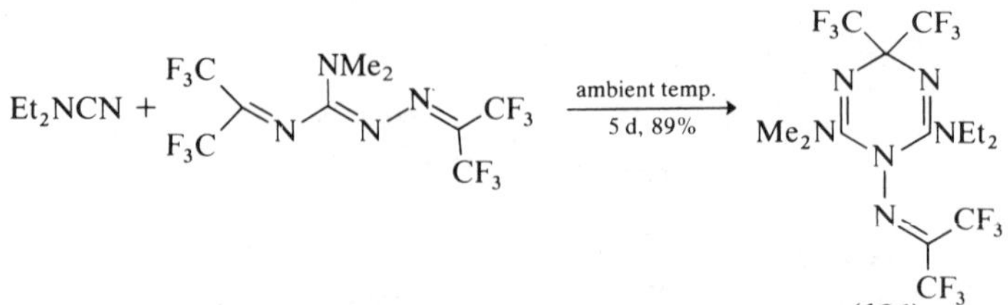

$$\text{Et}_2\text{NCN} + \quad \text{(123)} \quad \xrightarrow[\text{5 d, 89\%}]{\text{ambient temp.}} \quad \text{(124)} \tag{62}$$

(**122**)            (**123**)                              (**124**)

### 2.20.4.3.4 Synthesis of fused ring systems

2,4-Diamino-1,3,5-triazino[1,2-*a*]benzimidazole (**125**) has been prepared from 2-aminobenzimidazole and dicyandiamide (equation 63) ⟨78CB3007⟩.

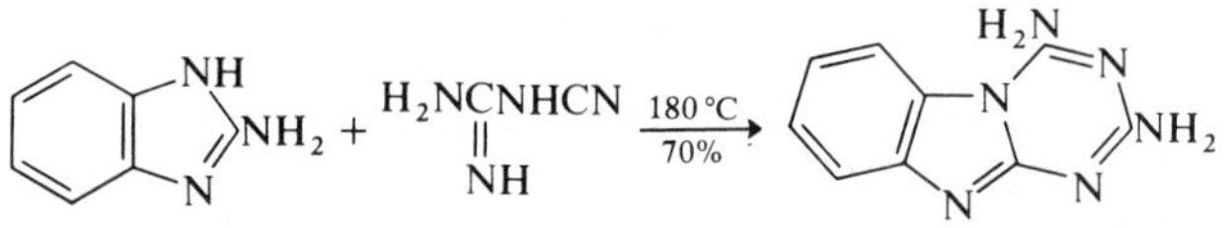

$$\xrightarrow[\text{70\%}]{180\,^\circ\text{C}} \tag{63}$$

(**125**)

Pyrolysis of 2*H*-[1,2,4]oxadiazolo[2,3-*a*]pyridine-2-thione (**126**; X = CH) yields the 1,3,5-triazine (**128**; X = CH). The reaction is thought to proceed *via* the isocyanate (**127**; X = CH). The pyridazine analogue (**126**; X = N) behaves in a similar manner ⟨80CPB3570⟩. A similar fused system (**130**) has been synthesized from diphenylcarbodiimide and (**129**) (equation 65) ⟨79ZC59⟩.

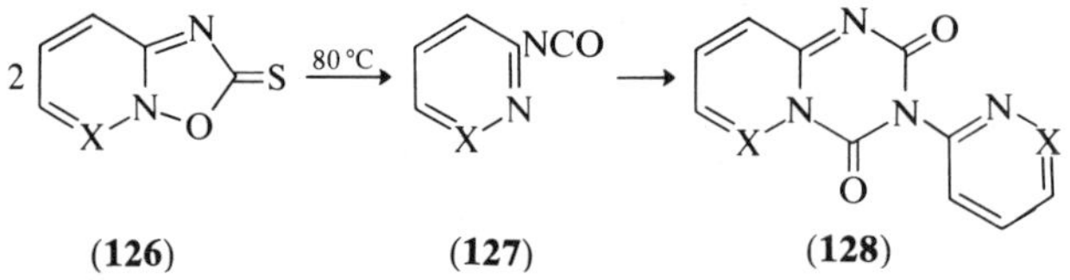

$$2 \quad \xrightarrow{80\,^\circ\text{C}} \quad \longrightarrow \tag{64}$$

(**126**)              (**127**)              (**128**)

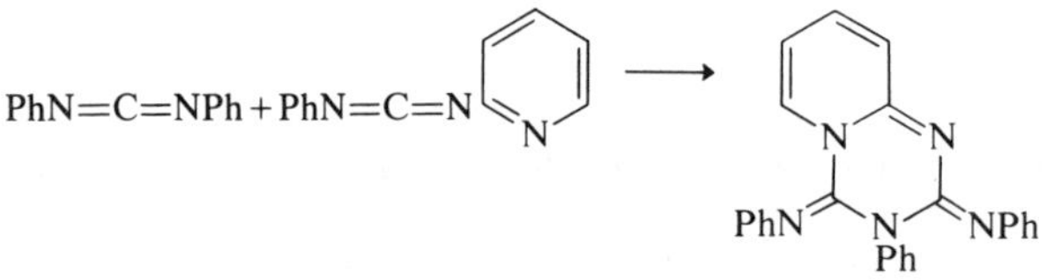

$$\text{PhN}{=}\text{C}{=}\text{NPh} + \text{PhN}{=}\text{C}{=}\text{N} \quad \longrightarrow \tag{65}$$

(**129**)                          (**130**)

### 2.20.4.4 Synthesis of 1,3,5-Triazines by Formation of Two Carbon–Nitrogen Bonds from [3 + 3] Atom Fragments

Nearly all such syntheses are based on the use of amidines as one of the starting reagents. The role of amidines in the synthesis of 1,3,5-triazines has been reviewed recently by Gautier *et al.* ⟨B-75MI22000⟩.

### 2.20.4.4.1 Synthesis using amidines and related compounds

Amidines react with derivatives of cyanic esters to give a variety of 1,3,5-triazines with an aryloxy substituent. The method has been reviewed by Grigat ⟨72AG(E)949⟩ and typical examples are shown in Scheme 70.

The reaction between amidines and polychloroimines (**131**) is a valuable route to chloro-1,3,5-triazines with a variety of alkyl or aryl substituents (equation 66) ⟨66AG(E)960a⟩.

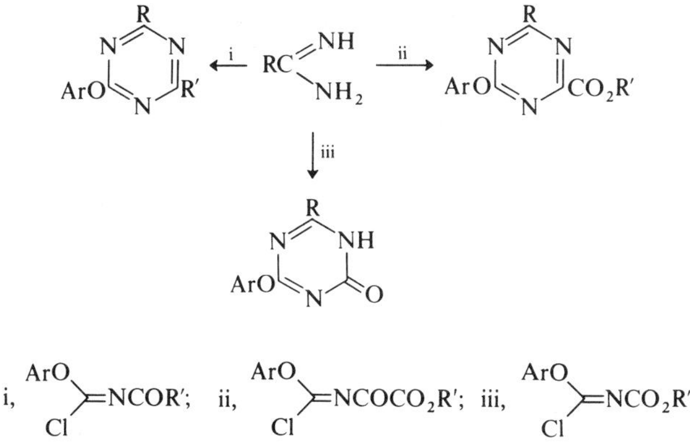

**Scheme 70**

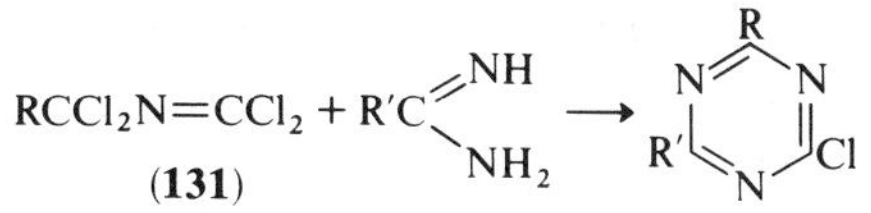

$$RCCl_2N{=}CCl_2 + R'C\overset{NH}{\underset{NH_2}{\diagup}} \longrightarrow \quad (66)$$

$$\textbf{(131)}$$

R = CCl$_3$, or aryl, R' = H, allyl or aryl

1,3,5-Triazin-2-ones may be prepared in almost quantitative yields from the reaction of amidines and an *N*-(α-chloroalkylidene)carbamoyl chloride (**132**; equation 67) ⟨66AG(E)960b⟩.

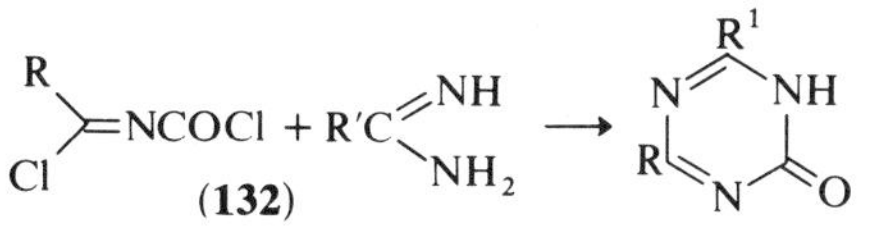

$$\overset{R}{\underset{Cl}{\diagdown}}{=}NCOCl + R'C\overset{NH}{\underset{NH_2}{\diagup}} \longrightarrow \quad (67)$$

$$\textbf{(132)}$$

R = CCl$_3$ or aryl; R' = H, alkyl or aryl

1,3,5-Triazinethiones are formed in good yields by the condensation of isothiocyanates and amidines. The reaction between ethoxycarbonyl isothiocyanate and benzamidine is used to illustrate the method (equation 68) ⟨71CB1606⟩.

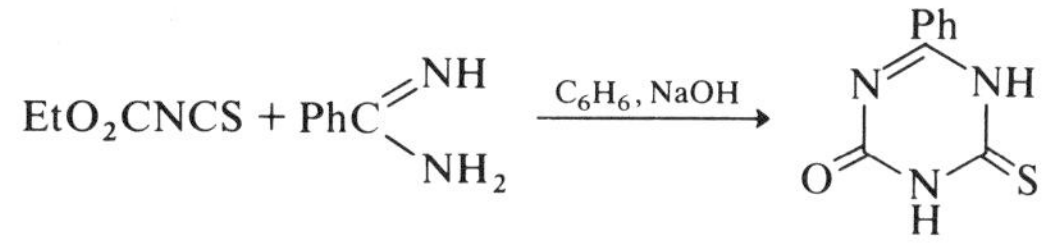

$$EtO_2CNCS + PhC\overset{NH}{\underset{NH_2}{\diagup}} \xrightarrow{C_6H_6, NaOH} \quad (68)$$

The reaction between amidines and *N*-chlorocarbonyl isocyanate gives the ionic compounds (**133**; Scheme 71). If the amidine is replaced by a urea, the 1,3-disubstituted isocyanurate (**134**) is formed in very good yield. Even potentially labile substituents, *e.g.* *t*-butyl, survive this reaction intact. Thioureas also may be used; they form salts (**135**) which

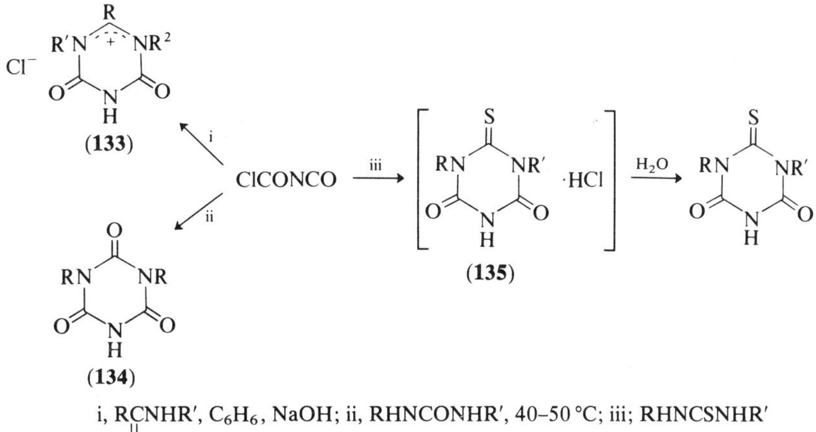

i, RCNHR', C$_6$H$_6$, NaOH; ii, RHNCONHR', 40–50 °C; iii; RHNCSNHR'
$\overset{\parallel}{\underset{R^2N}{}}$

**Scheme 71**

give the expected product on hydrolysis (Scheme 71) ⟨77AG(E)743⟩. Recently, dimethyl *N*-cyanodithioimidocarbonate (136) has proved to be a valuable synthon for 1,3,5-triazine preparations. It has been used, for example, in the syntheses of 2,4-dimethylthio-1,3,5-triazines (equation 69) and 1,3,5-triazinedithiones (Scheme 72) ⟨75JHC37⟩.

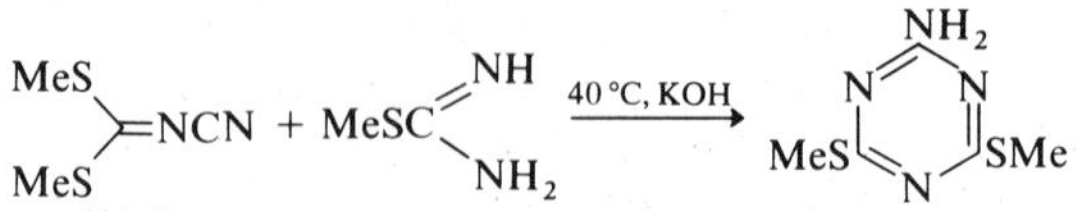

(69)

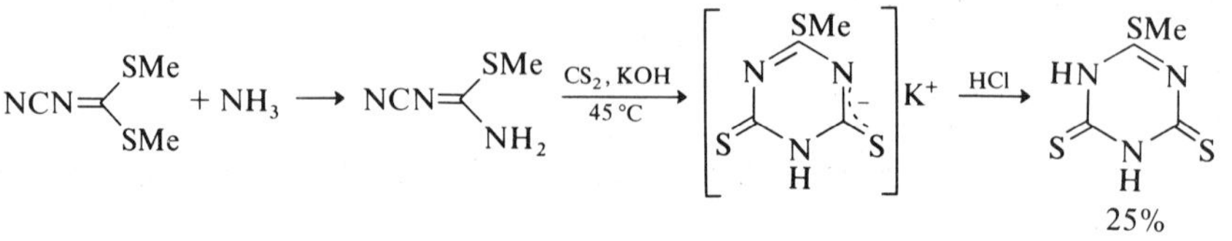

**Scheme 72**

Mesomeric 1,3,5-triazine-dione, -one-thione and -dithione betaines may be synthesized from *N,N'*-disubstituted acetamidines (Scheme 73) ⟨75JHC187⟩. Huffman and Schaefer have reported the synthesis of the triazine 1-oxide (137) in 45% yield from methyl *N*-cyanoacetimidate and benzamidoxime (equation 70) ⟨63JOC1816⟩.

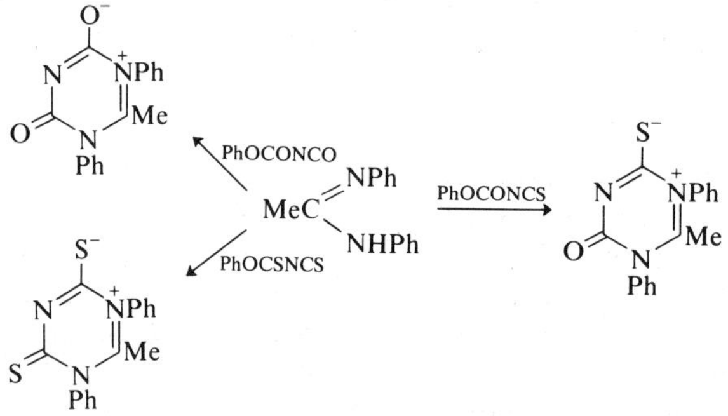

**Scheme 73**

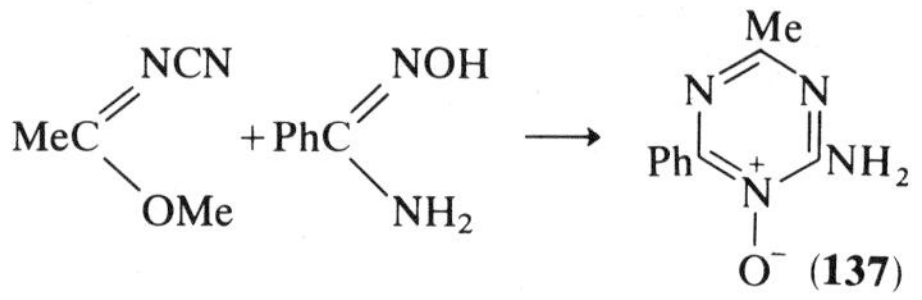

(70)

### 2.20.4.4.2 Syntheses of fused ring systems

Most of the syntheses of fused triazines have been based on the use of 2-amino-substituted nitrogen heterocyclic systems. These compounds can react with the other reagents to give the desired product by formation of two carbon–nitrogen bonds. There has been very little published on the mechanism. The examples shown in equations (71)–(73) and Scheme 74 illustrate the value of the method, and include an example of the synthesis of a fused mesomeric 1,3,5-triazine betaine ⟨74JOC1819, 75JHC407, 80JHC1121, 81JCS(P1)331⟩.

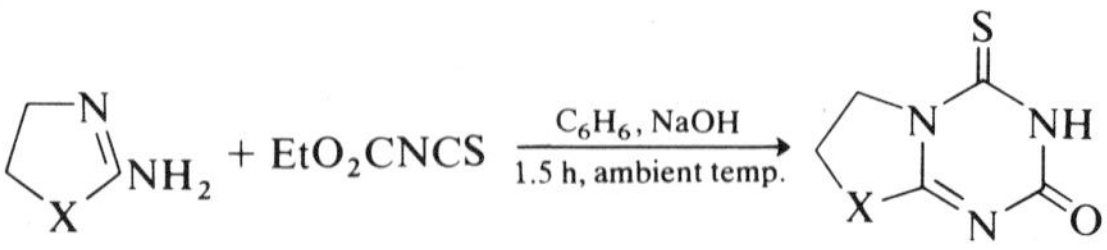

(71)

$$X = S, 65\% ; X = Se, 83\%$$

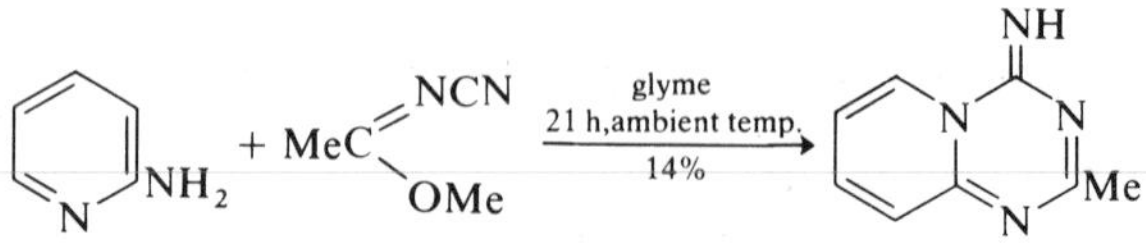

(72)

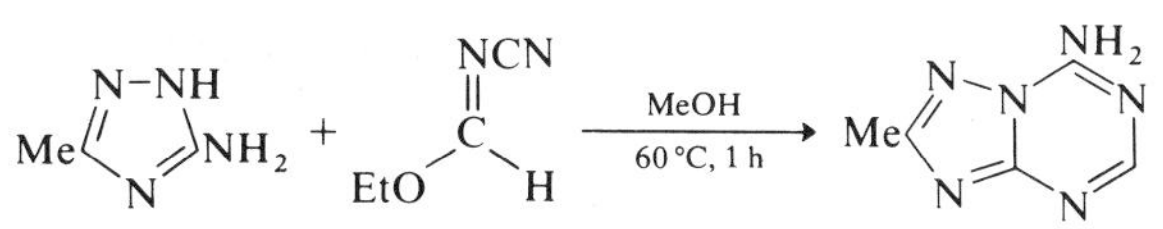

(73)

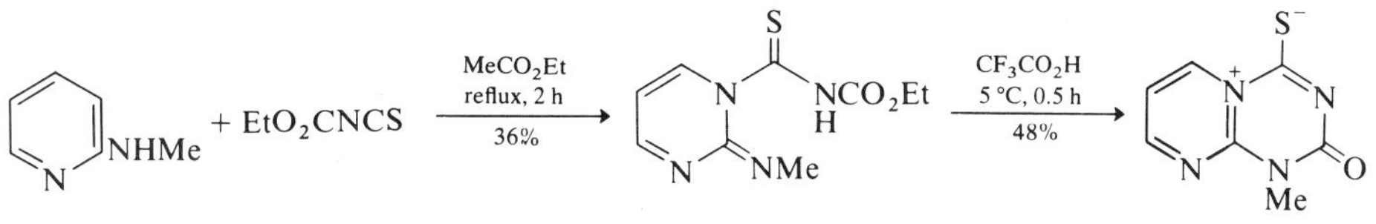

**Scheme 74**

### 2.20.4.5 Synthesis of Symmetrical 1,3,5-Triazines by Formation of Three Carbon–Nitrogen Bonds

Probably the cyclotrimerization of nitriles is the best known route to 1,3,5-triazines. The reaction has the obvious limitation that it is of value for preparing the symmetrical derivatives only. Nevertheless, many important triazines, such as cyanuric chloride, are made in this way. There are a number of other cyclotrimerization reactions which are also useful, in particular the trimerization of imidates. An easy route to 1,3,5-triazine from ammonium acetate has been developed.

#### 2.20.4.5.1 Synthesis from nitriles

This method has been reviewed recently ⟨78RCR975⟩. Alkyl cyanides are difficult to trimerize, needing both very high pressures and temperatures (Table 14). It is one of the few organic reactions which requires pressures above 1000 atm. In contrast to other nitriles, the alkyl cyanides will not trimerize in acid conditions; thus in the presence of hydrogen chloride, alkyl cyanides form $N$-substituted amidines ⟨73BCJ292⟩.

**Table 14** Formation of Symmetrical 1,3,5-Triazines $(RCN)_3$ by Cyclotrimerization of Alkyl and Aryl Cyanides and Aryl Cyanates

| R | Catalyst | Pressure $(10^3 \text{ atm})$ | Temp. (°C) | Solvent | Time (h) | Yield (%) | Ref. |
|---|---|---|---|---|---|---|---|
| Me | — | 7.5 | 60 | MeOH | 24 | 39 | 52JA5633 |
| Et | — | 7.5 | 70 | MeOH | 24 | 36 | 52JA5633 |
| Bu$^n$ | — | 7.5 | 150 | MeOH | 18 | 35 | 52JA5633 |
| Bu$^n$ | — | 7.5 | 100 | MeOH | 18 | 7 | 52JA5633 |
| cyclo-Pr | — | 8.5 | 200 | EtOH | 10 | 90 | 75T619 |
| CCl$_3$ | BF$_3$–Et$_2$O/HCl | — | −10 to 25 | — | 12 | 94 | 69BCJ2924 |
| (CH$_2$)$_3$OH | HCl | — | 0 | EtOH | 8 | 73 | 75BRP1399345 |
| CO$_2$Et | HCl | — | 25 | — | 300 | 96 | 19CB656 |
| CF$_3$ | — | 0.07 | 300 | — | 16 | 31 | 57JOC698 |
| C$_2$F$_5$ | — | — | 300 | — | 120 | 48 | 57JOC698 |
| Ph | HCl | — | 100 | — | — | trace | 73BCJ306 |
| Ph | AlCl$_3$ | — | 150 | — | 6 | — | 58JA1442 |
| Ph | PCl$_5$/HCl | — | 100 | — | 24 | 98 | 73BCJ306 |
| 2-MeC$_6$H$_4$ | H$_3$PO$_4$ | 37.5 | 350 | — | 0.26 | 100 | 58JA1442 |
| 4-MeC$_6$H$_4$ | PCl$_5$/HCl | — | 100 | — | 90 | 60 | 73BCJ306 |
| 3-MeC$_6$H$_4$ | — | 50 | 545 | — | 0.11 | 100 | 58JA1442 |
| 3-CNC$_6$H$_4$ | ZnCl$_2$ | — | 360 | — | 0.71 | 100 | 74USP3839331 |
| OPh | NaOEt, ZnCl$_2$ | — | 200 | — | 0.1 | 85 | 64TL2829 |
| 3-MeC$_6$H$_4$O | PCl$_5$ | — | 20 | CCl$_4$ | ~2 | 95 | 69CB2508 |
| 4-OMeC$_6$H$_4$O | NaOEt, ZnCl$_2$ | — | 200 | — | 0.1 | 98 | 64TL2829 |
| 4-ClC$_6$H$_4$O | PCl$_5$ | — | 20 | CCl$_4$ | ~2 | 35 | 69CB2508 |

Alkyl cyanides with electron withdrawing substituents are more reactive and cyclotrimerize under both acidic and basic conditions as well as at high pressure (Table 14). Complexes of Lewis acids and hydrogen halides are particularly valuable catalysts. Two mechanisms for the acid-catalyzed cyclotrimerization have been postulated. Grundmann *et al.* suggest the reaction proceeds *via* the adduct (**138**) followed by addition of a third cyanide (Scheme 75) ⟨52LA(577)77⟩, whilst Zil'berman ⟨62RCR615⟩ has proposed that the reaction occurs *via* a cyclic electron transfer involving both free cyanides and complexes with hydrogen chloride (Scheme 76).

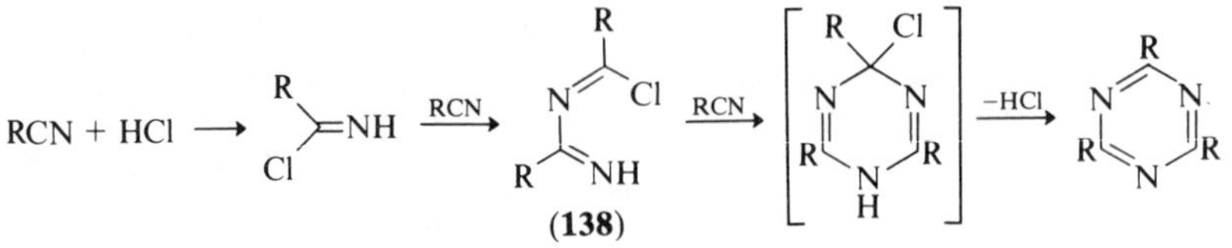

**Scheme 75**

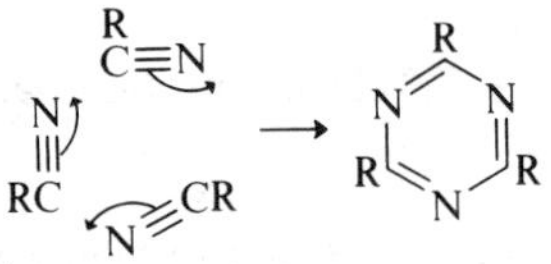

**Scheme 76**

Base-catalyzed trimerizations are facile also; for example, trifluoromethyl cyanide trimerizes in the presence of ammonia, presumably through the formation of the amidine intermediate (**139**; Scheme 77) ⟨67JOC231⟩. Similarly, perfluoro-*n*-propyl cyanide forms the 1,3,5-triazine in the presence of sodium methoxide, probably *via* the imidate (**140**; Scheme 78) ⟨52JA5633⟩.

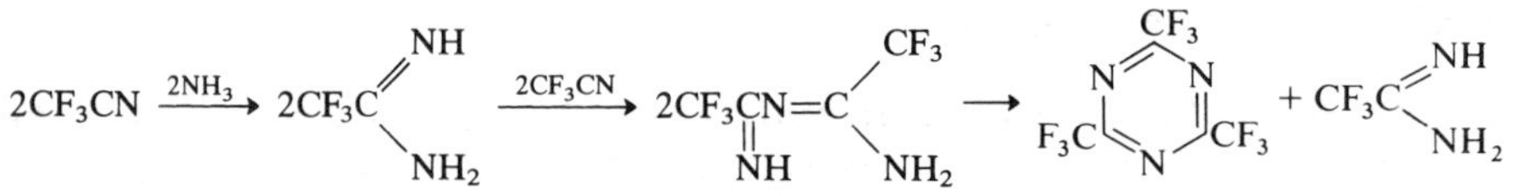

**Scheme 77**

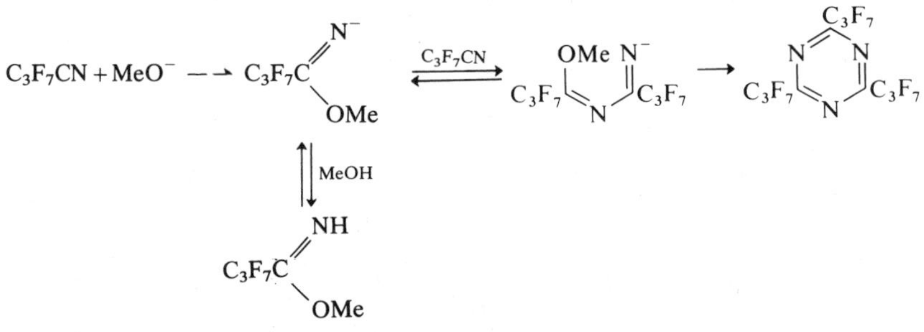

**(140)**    **Scheme 78**

Aryl cyanides are of similar activity to alkyl cyanides containing electron-withdrawing groups. A combination of hydrogen chloride and a Lewis acid is particularly valuable for the acid-catalyzed cyclization. Lewis acids alone are inefficient (Table 14), and hydrogen chloride alone forms the stable adduct (**141**). The presence of the Lewis acid, such as PCl₅, decreases the nucleophilicity of the chloride counter ion, permitting the reaction to go to completion ⟨73BCJ306⟩.

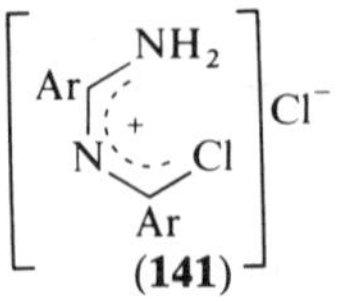

**(141)**

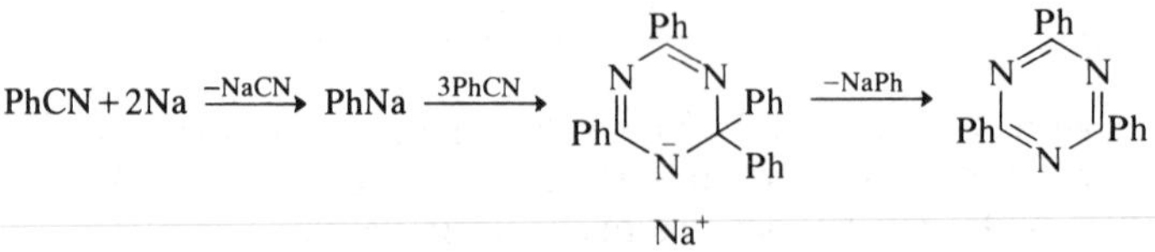

**Scheme 79**

Phenylsodium is formed initially in the reaction between benzonitrile and sodium (Scheme 79) ⟨59JOC208⟩. Trimerizations of aryl or heterocyclic nitriles catalyzed by amines or alkoxides resemble the reactions of the perfluoronitriles described above (Schemes 77 and 78 respectively). Kurabayashi *et al.* ⟨71BCJ3413⟩ studied the trimerization of benzonitrile under pressure in methanol, and have shown that the rate determining step is the reaction of the benzimino ether with benzonitrile. Steric factors exert an important influence in the trimerizations of aryl cyanides. The cyclizations of *ortho*-substituted aryl cyanides need more severe conditions than either the corresponding *meta* or *para* derivatives (Table 14).

Cyclotrimerization of nitriles with heteroatomic substituents are also facile and important reactions. Melamine is formed on heating cyanamide above its melting point (equation 74) ⟨59HC(13)1, p. 309⟩. Substituted cyanamides react to give either the expected 1,3,5-triazine or the isomer (**142**). The 1,3,5-triazine is the preferred product at high temperatures, under acid catalysis, and for cyanamides with bulky substituents, whilst the isomer is favoured by basic conditions and low temperatures (Scheme 80). The mechanism of formation of (**142**) has been proposed (Scheme 81) ⟨78RCR975⟩.

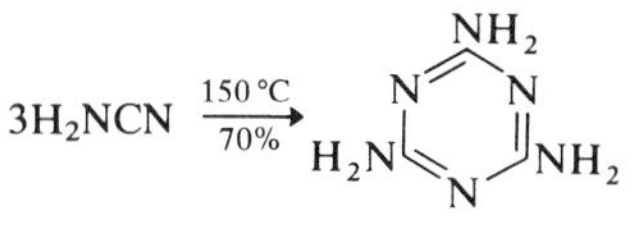

$$3H_2NCN \xrightarrow[70\%]{150\,^{\circ}C} \quad (74)$$

**Scheme 80**

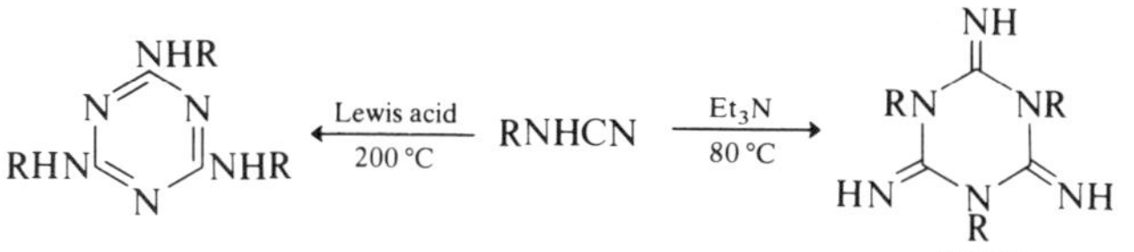

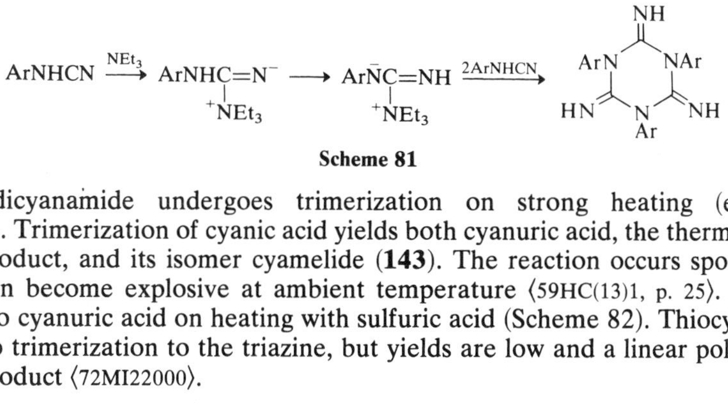

**Scheme 81**

Sodium dicyanamide undergoes trimerization on strong heating (equation 75) ⟨22LA(427)26⟩. Trimerization of cyanic acid yields both cyanuric acid, the thermodynamically preferred product, and its isomer cyamelide (**143**). The reaction occurs spontaneously at 0 °C, and can become explosive at ambient temperature ⟨59HC(13)1, p. 25⟩. Cyamelide is isomerized to cyanuric acid on heating with sulfuric acid (Scheme 82). Thiocyanic acid will also undergo trimerization to the triazine, but yields are low and a linear polymer (**144**) is the major product ⟨72MI22000⟩.

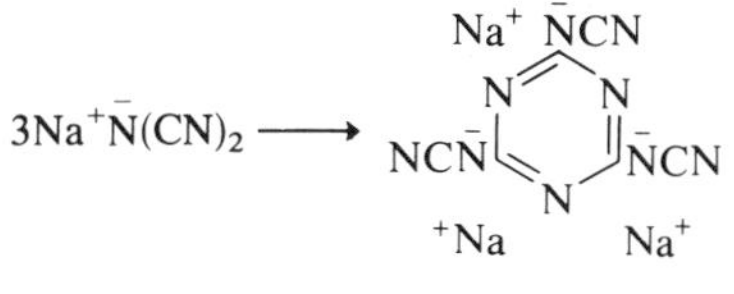

$$3Na^+N(CN)_2 \longrightarrow \quad (75)$$

**Scheme 82**

$$\left[\begin{array}{c} -C-NH- \\ \| \\ S \end{array}\right]_n$$

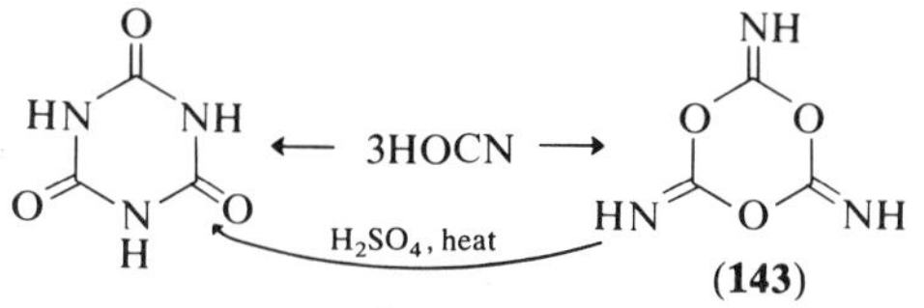

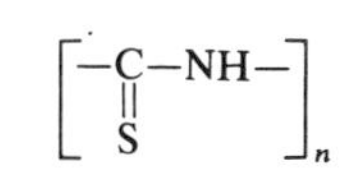

(**144**)

Cyanuric halides are readily synthesized from the cyanogen halides. Cyanuric fluoride is formed at room temperature, whilst cyanuric chloride and cyanuric bromide are prepared from their respective cyanogen halides in the presence of a wide range of catalysts including acids, metal oxides, activated charcoal and zeolites. There have been few mechanistic details published on the acid-catalyzed trimerizations. The trimerization of cyanogen chloride using crystalline molecular sieves is a first-order reaction with an activation energy of 63.4 kJ mol$^{-1}$ ⟨72ZC293⟩. The trimerization of cyanogen chloride in the gas phase on activated charcoal is probably the most useful industrial route to cyanuric chloride. The reaction apparently occurs owing to the presence of a graphitic phase on which the cyanuric chloride is spatially orientated to facilitate trimerization ⟨71DOK(199)146⟩. In addition to forming cyanuric chloride, cyanogen chloride can also form the tetramer (equation 76).

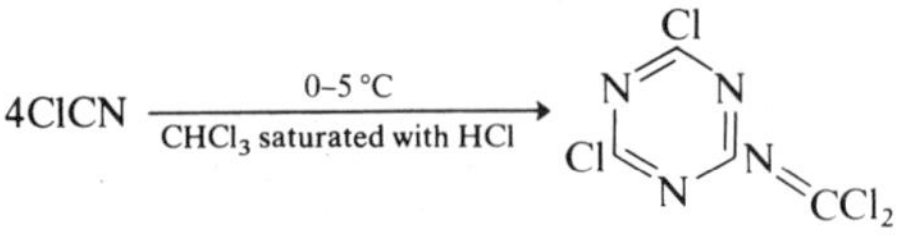

$$4\text{ClCN} \xrightarrow[\text{CHCl}_3 \text{ saturated with HCl}]{0-5\,°\text{C}}$$ (76)

Aryl cyanates (cyanic esters), which can be prepared *in situ* ⟨67AG(E)206⟩, form 1,3,5-triazines under either acid or base catalysis (Scheme 83) ⟨77RCR278, 78RCR975⟩. The acid-catalyzed reactions are not properly understood, but it seems that the mechanism is dependent on the precise nature of the Lewis acid used. Yields are usually good (Table 14). The reaction of aryl cyanates with ketoximines yields 1,3,5-triazines (**146**) *via* the intermediate (**145**) (Scheme 84) ⟨66CB2361⟩. Alkyl cyanates are very reactive, and isomerize readily; therefore they are unsuitable starting materials for 1,3,5-triazine synthesis.

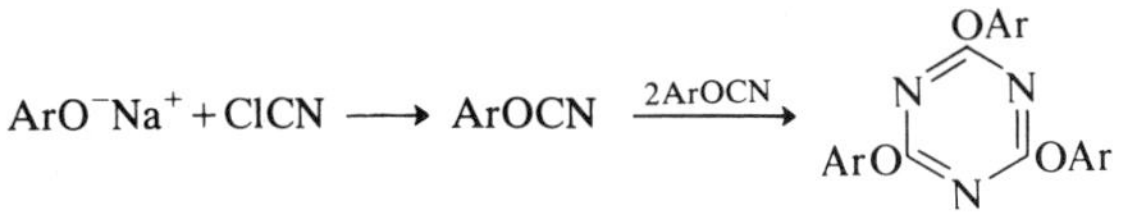

$$\text{ArO}^-\text{Na}^+ + \text{ClCN} \longrightarrow \text{ArOCN} \xrightarrow{2\text{ArOCN}}$$

**Scheme 83**

$$\text{ArOCN} + \text{HON}{=}\!\!\begin{smallmatrix}R\\R'\end{smallmatrix} \longrightarrow \text{ArOCON}{=}\!\!\begin{smallmatrix}R\\R'\end{smallmatrix} \xrightarrow[37\%]{2\text{HON}{=}}$$

(**145**)      (**146**)

$$X = \text{O}-\text{N}{=}\!\!\begin{smallmatrix}R\\R'\end{smallmatrix}$$

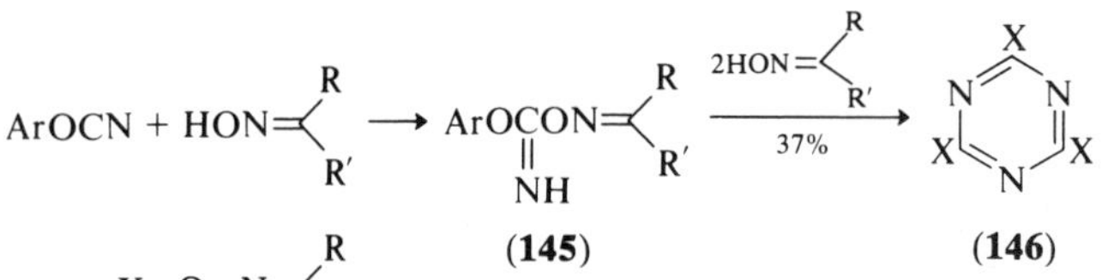

**Scheme 84**

The preparation of 1,3,5-triazine 1,3,5-trioxides by trimerization of a nitrile oxide was reported ⟨09CB803⟩; however, the structure of the product has been reassigned as a polymer ⟨65LA(687)191⟩.

### 2.20.4.5.2 Synthesis from imino derivatives

A variety of imino derivatives has been used in the synthesis of 1,3,5-triazine (Scheme 85) ⟨B-61MI22000, p. 680⟩. The trimerization of the imidates is the most valuable of these routes to 1,3,5-triazine. Imidates can be considered as activated nitriles, and cyclotrimerize more readily than do the corresponding nitriles. The symmetrical 2,4,6-trialkyl-1,3,5-triazines are easily formed (Table 15), although large alkyl substituents may give rise to steric hindrance ⟨61JOC2778⟩. The mechanism is shown in Scheme 86.

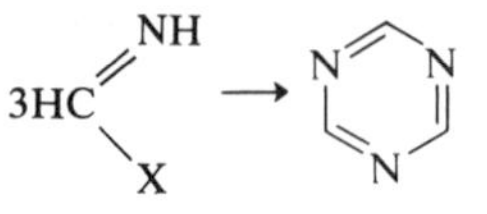

$$3\text{HC}{\underset{X}{\overset{NH}{\diagup}}} \longrightarrow$$

$$X = \text{OEt, Cl, NH}_2, \text{SH, OCH}_2\text{Ph}$$

**Scheme 85**

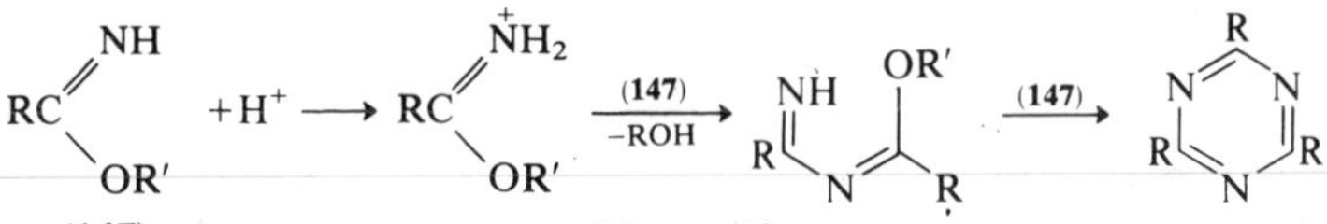

$$\text{RC}{\underset{\text{OR}'}{\overset{\text{NH}}{\diagdown}}} + \text{H}^+ \longrightarrow \text{RC}{\underset{\text{OR}'}{\overset{\overset{+}{\text{NH}}_2}{\diagdown}}} \xrightarrow[-\text{ROH}]{(147)} \xrightarrow{(147)}$$

(**147**)      **Scheme 86**

**Table 15** Syntheses of Symmetrical 1,3,5-Triazines from Imidates ⟨61JOC2778⟩

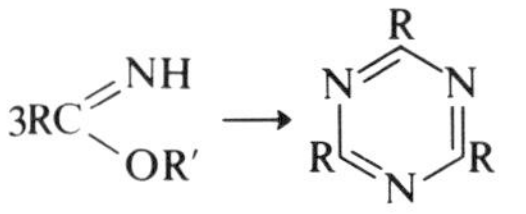

$$3RC\overset{NH}{\underset{OR'}{\diagup}} \longrightarrow \text{(triazine)}$$

| Imidate | | | Conditions | | Product | |
|---|---|---|---|---|---|---|
| R | R' | Temp. (°C) | Catalyst | Time (h) | R | Yield (%) |
| Me | Et | r.t. | MeCO$_2$H | 16 | Me | 85 |
| Et | Et | r.t. | MeCO$_2$H | 24 | Et | 74 |
| Pr$^i$ | Et | r.t. | MeCO$_2$H | 24 | Pr$^i$ | 68 |
| 3-NO$_2$C$_6$H$_4$ | Et | 70–80 | MeCO$_2$H | 4 | 3-NO$_2$C$_6$H$_4$ | 79 |
| 3-NO$_2$C$_6$H$_4$ | Me | 70–90 | MeCO$_2$H | 2.5 | 3-NO$_2$C$_6$H$_4$ | 75 |

Primary alkyl enamines undergo spontaneous trimerization to form 1,3,5-hexahydrotriazines *via* the imino derivatives (**148**; equation 77) ⟨80T2497⟩.

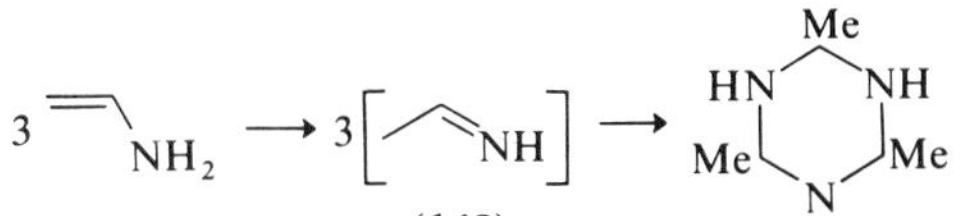

$$3 \overset{}{\underset{NH_2}{\diagup}} \longrightarrow 3\left[\overset{}{\underset{NH}{\diagup}}\right] \longrightarrow \text{(148)} \tag{77}$$

Flynn and Michl have recently succeeded in synthesizing ethylene iminocarbonate (**149**), which is converted into the trimer (**150**) at room temperature ⟨74JOC3442⟩.

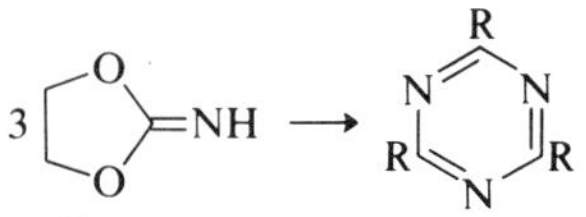

$$3 \text{ (149)} \longrightarrow \text{ (150)} \tag{78}$$

(**149**)    (**150**)

R = OCH$_2$CH$_2$OH

Substituted amidines are of limited use in the synthesis of symmetric 1,3,5-triazines. Only formamidine or amidines bearing strongly electron withdrawing groups react readily. The proposed mechanism ⟨59JA1466⟩ resembles that of the trimerization of imidates (Scheme 87).

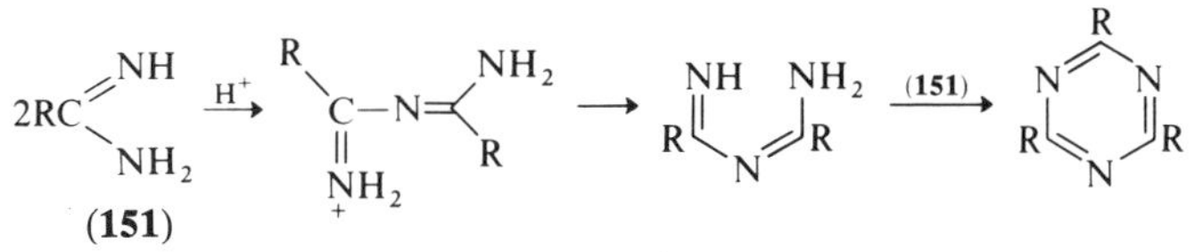

$$2RC\overset{NH}{\underset{NH_2}{\diagup}} \xrightarrow{H^+} \cdots \xrightarrow{\text{(151)}} \text{(triazine)}$$

(**151**)

**Scheme 87**

Bredereck and coworkers ⟨79S690⟩ have discovered a new route to 1,3,5-triazine which has many advantages over the methods of Grundmann and Kreutzberger (see Section 2.20.1.1) and from the imino ether ⟨61JOC2778⟩. Treatment of formamidinium acetate with triethyl orthoformate produces 1,3,5-triazine in high yield (81%). Formamidine acetate may be prepared *in situ*, and the reaction is effected with only a slight drop in yield (Scheme 88). Thus it is possible to prepare 1,3,5-triazine efficiently from readily available compounds, without the use of hydrogen cyanide.

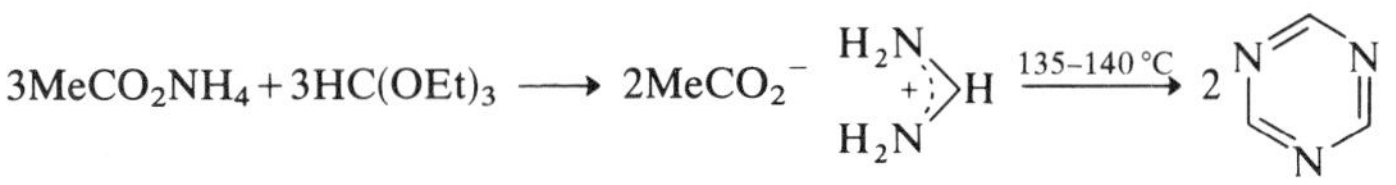

$$3MeCO_2NH_4 + 3HC(OEt)_3 \longrightarrow 2MeCO_2^- \quad \overset{H_2N}{\underset{H_2N}{\diagup}}\overset{+}{\underset{}{\diagdown}}H \xrightarrow{135-140\,°C} 2 \text{ (triazine)}$$

**Scheme 88**

### 2.20.4.5.3 Synthesis of symmetrical isocyanurates from isocyanates

Symmetrical isocyanurates are readily synthesized from isocyanates in the presence of a variety of catalysts including tertiary amines, phosphines, sodium acetate and sodium methoxide in anhydrous conditions. The method has been reviewed thoroughly ⟨B-71MI22002,

57CRV47, 48CRV(43)203⟩. The probable reaction mechanism, using a tertiary amine catalyst, involves a series of polar additions (Scheme 89).

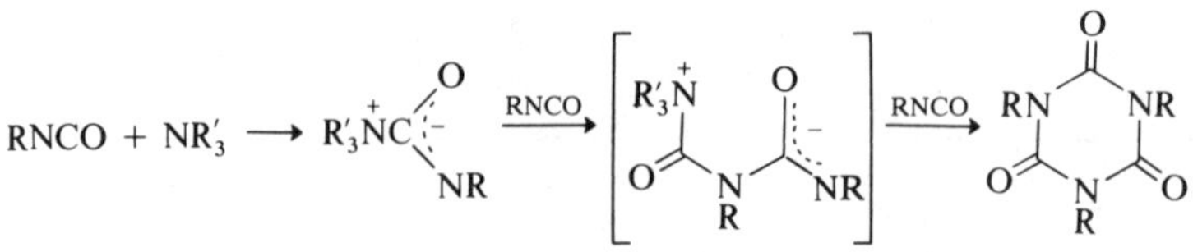

$$RNCO + NR_3' \longrightarrow R_3'\overset{+}{N}C\overset{O}{\underset{NR}{\overset{..}{C}}^{-}} \xrightarrow{RNCO} \left[ \begin{array}{c} R_3'\overset{+}{N} \quad O \\ O\overset{..}{\underset{N}{\overset{}{\parallel}}}\overset{}{\underset{R}{N}}NR \end{array} \right] \xrightarrow{RNCO} \begin{array}{c} O \\ RN\overset{}{\underset{}{\parallel}}NR \\ O\overset{}{\underset{N}{\parallel}}\overset{}{\underset{R}{}}O \end{array}$$

**Scheme 89**

Although the direct trimerization of alkyl isocyanates is facile, it is preferable to prepare these compounds *in situ* because the isocyanates are both expensive and poisonous (equation 79) ⟨65BCJ1586⟩.

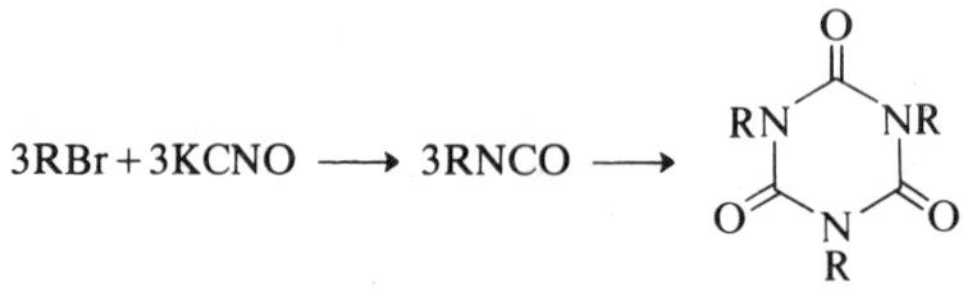

$$3RBr + 3KCNO \longrightarrow 3RNCO \longrightarrow \begin{array}{c} O \\ RN\overset{}{\underset{}{\parallel}}NR \\ O\overset{}{\underset{N}{\parallel}}\overset{}{\underset{R}{}}O \end{array} \qquad (79)$$

The preparation of a variety of isocyanurates from isocyanates is summarized in Table 16.

**Table 16**   Synthesis of Symmetrical Isocyanurates from Isocyanates (RNCO)

| Starting materials | Conditions[a,b] | Product R | Yield (%) | Ref. |
|---|---|---|---|---|
| MeI | KCNO, DMF, 134 °C, 1 h; 2 h, 100–110 °C | Me | 90 | 65BCJ1586 |
| n-C$_7$H$_{15}$Br | KCNO, DMF, 134 °C, 1 h; 3 h, 140–150 °C | C$_7$H$_{15}$$^n$ | 75 | 65BCJ1586 |
| Bu$^s$Br | KOCN, DMF, 134 °C, 1 h; 3 h, 140–150 °C | C$_4$H$_9$$^s$ | 75 | 65BCJ1586 |
| PhCH$_2$Cl | KOCN, DMF, 134 °C, 1 h; 3 h, 140–145 °C | PhCH$_2$ | 83 | 65BCJ1586 |
| MeOCH$_2$CH$_2$Br | KOCN, DMF, 135 °C, 3 h | MeOCH$_2$CH$_2$ | 55 | 65BCJ1586 |
| H$_2$C=CHCH$_2$NCO | PhCO$_2$Na, DMF, 100–110 °C, 6 h | H$_2$C=CHCH$_2$ | 97 | 66BCJ1922 |
| H$_2$C=CHNCO | Et$_3$P, −78 to 0 °C, 2 h | H$_2$C=CH | 55 | 61JOC3334 |
| F$_3$CSCl | AgOCN, C$_6$H$_6$, 80 °C, 8–10 h | Cl$_3$CS | 90 | 65CB3353 |
| MeCONCO | Et$_3$N, −68 °C, 6 h | MeCO | 41 | 72BSF251 |
| PhNCO | EC, NMM, 125 °C, 25 h | Ph | 98.8 | 56JA4911 |
| 2-MeC$_6$H$_4$NCO | EC, NMM, 125 °C, 20 h | 2-MeC$_6$H$_4$ | 85 | 56JA4911 |
| 4-NO$_2$C$_6$H$_4$NCO | EC, NMM, 125 °C, 1 h | 4-NO$_2$C$_6$H$_4$ | 80 | 56JA4911 |

[a] EC = PhNHCO$_2$Et.   [b] NMM = N-Methylmorpholine.

### 2.20.4.5.4 Synthesis of 1,3,5-triazine derivatives from the reaction of ammonia with aldehydes

The reaction between aldehydes and ammonia leads to compounds usually called 'aldehyde ammonias' (equation 80). The first reported compound ('acetaldehyde ammonia') was synthesized by Liebig ⟨1835LA(14)133⟩. Nielsen *et al.* ⟨73JOC3288⟩ have shown that most of these compounds are hexahydro-1,3,5-triazines, which are usually isolated as the trihydrates (Table 17). When aldehydes containing a branched alkyl substituent are used, the free hexahydro-1,3,5-triazines may be formed (Table 17). The kinetics of the reaction between acetaldehyde and ammonia have been studied by $^1$H NMR spectroscopy ⟨73JOC2931⟩, and shown to be rather complex. It is beyond the scope of this review to discuss the reaction fully.

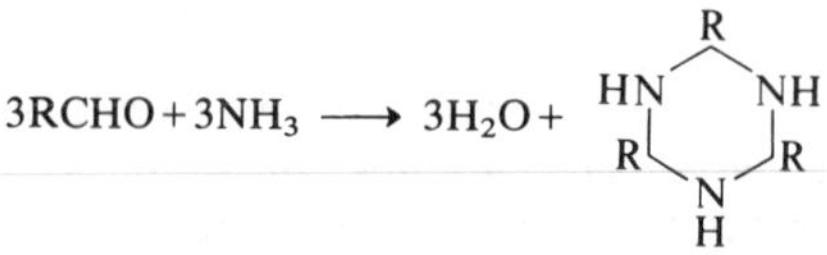

$$3RCHO + 3NH_3 \longrightarrow 3H_2O + \begin{array}{c} R \\ HN\overset{}{\underset{}{}}NH \\ R\overset{}{\underset{N}{}}R \\ H \end{array} \qquad (80)$$

**Table 17** Typical Examples of 1,3,5-Hexahydrotriazines (RCHCNH)₃ Synthesized from Aldehydes and Ammonia

| | | Products (%) | | |
| --- | --- | --- | --- | --- |
| *R* | *Conditions*[a,b] | $(RCHNH)_3 \cdot 3H_2O$ | $(RCHNH)_3$ | *Ref.* |
| Me | A | 75 | — | 73JOC3288 |
| Et | i, A; ii, Drierite | — | 89 | 73JOC3288 |
| Pr | A, 8 d | 67 | — | 73JOC3288 |
| Pr$^i$ | A, 1 d | — | 92 | 73JOC3288 |
| PhCHMe | B, 3 d | — | 79 | 74JOC1349 |

[a] A = 4 equivalents 15M aq. NH₃, 0–5 °C, stored at 0 °C.
[b] B = 1.2 equivalents 9M aq. NH₃, 5–7 °C, stored at −15 °C.

Formaldehyde reacts with amines to yield the 1,3,5-trisubstituted hexahydro-1,3,5-triazines and with nitriles to form the acyl derivatives (Scheme 90) ⟨59HC(13)1, p. 486, B-61MI22000, p. 658⟩. Hexamethylenetetramine (**70**) is prepared by heating an ammoniacal solution of formaldehyde ⟨B-61MI22000, p. 688⟩.

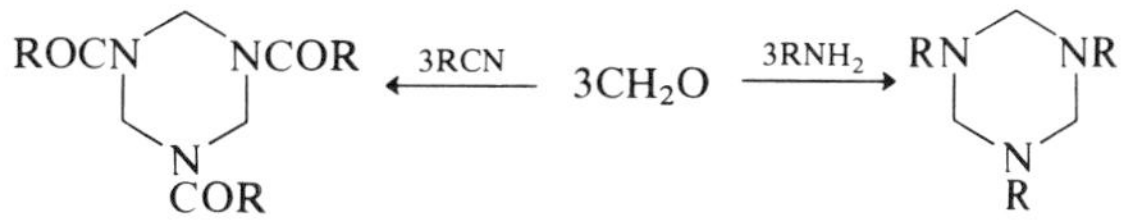

**Scheme 90**

Acetaldehyde undergoes electrochemical oxidation in methanolic ammonia to give 2,4,6-trimethyl-1,3,5-triazine in better than 20% yield (Scheme 91). The reaction works very poorly with aromatic aldehydes ⟨76CB1346⟩.

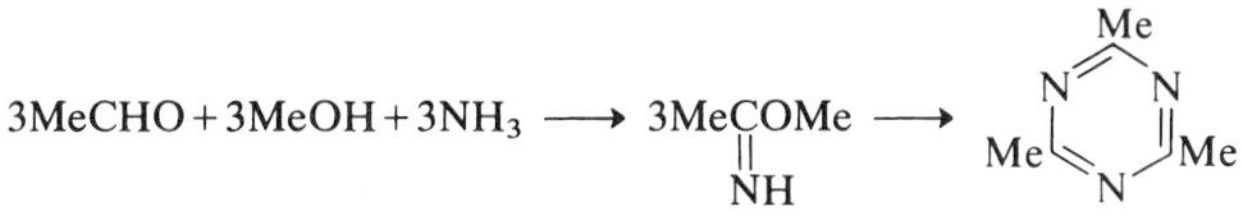

$$3MeCHO + 3MeOH + 3NH_3 \longrightarrow 3MeCOMe(NH) \longrightarrow$$

**Scheme 91**

### 2.20.4.5.5 Synthesis of melamine

The condensation of urea is the preferred industrial preparation of melamine. The details of the process have been reviewed recently ⟨B-79MI22000⟩.

### 2.20.4.5.6 Synthesis of fused ring systems

$C,C,N^\alpha$-Triaryl-$N^\beta$-cyanoazomethine imines (**152**) are trimerized on heating in solution to yield (**153**). The reaction is believed to occur *via* a succession of three 1,3-dipolar cycloadditions, the third being intramolecular ⟨80AG(E)906⟩.

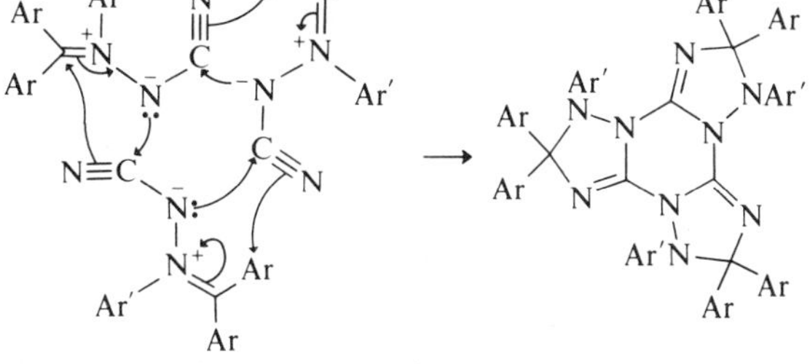

(**152**)　　　　(**153**)

**Scheme 92**

The trimerization of imidazole derivatives is the best known route to fused 1,3,5-triazines by formation of three carbon–nitrogen bonds. A typical example is shown in equation 81 ⟨75BCJ956⟩. Takeuchi *et al.* investigated the trimerization of 2-fluoroimidazoles, and found all the imidazoles tested were successfully trimerized (equation 82). When unsymmetrical imidazoles (**154**; $R^1 \neq R^2$) are used, isomeric mixtures of 1,3,5-triazines are formed ⟨79JOC4243⟩.

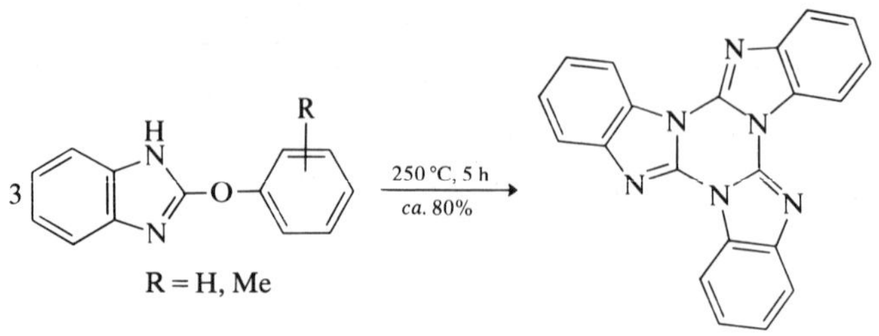

$$\text{(81)}$$

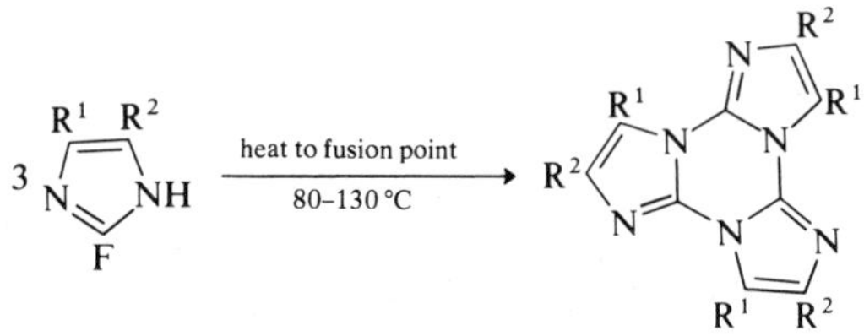

$$\text{(82)}$$

(**154**) $R^1, R^2$ = H, alkyl, $(CH_2)_2NHCOCF_3$, $(CH_2)_2NHCO_2Et$

*N*-Chloropiperidine is trimerized on heating with alcoholic potassium hydroxide. The initial stage is the loss of hydrogen chloride to give the intermediate (**155**), and mixed products are produced (Scheme 93) ⟨59HC(13)1, p. 446⟩. Recently, Warning and Mitzlaff have found a selective synthesis of $\alpha$-tripiperideine from 1-formyl-2-methoxypiperidine (equation 83) ⟨79TL1565⟩.

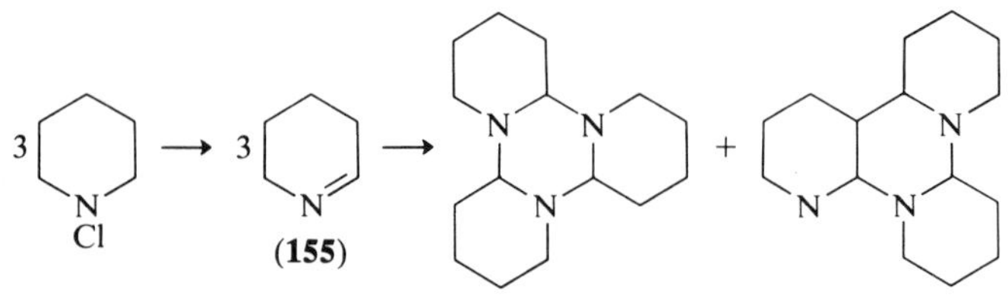

**Scheme 93**

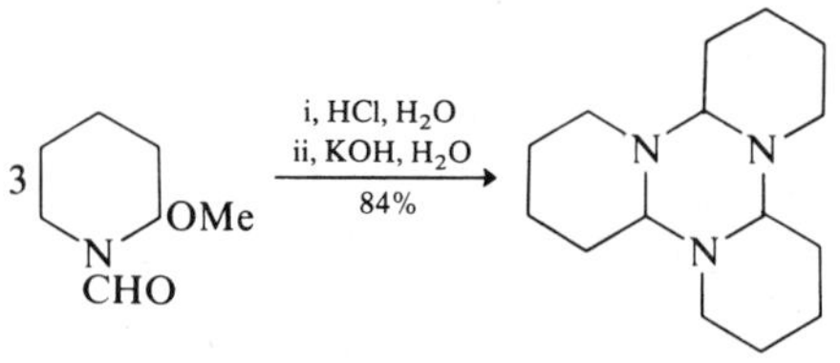

$$\text{(83)}$$

### 2.20.4.6 Synthesis of Unsymmetrical 1,3,5-Triazines by Formation of Three Carbon–Nitrogen Bonds

There are very many examples of such routes, and therefore discussion will be confined to illustrative examples. In general, the starting reagents have already been encountered in the previous sections and mechanisms are similar to those described already.

#### 2.20.4.6.1 Synthesis by cotrimerization reactions

In theory, the cotrimerization of two or three nitriles should give mixtures of four or ten 1,3,5-triazines respectively; however, reactions may be carried out selectively if two nitriles of different reactivities are used. Thus the reaction between two equivalents of trichloromethyl cyanide and one equivalent of acetonitrile gives just one product ⟨74JOC2591⟩. Trichloromethyl cyanide readily forms the adduct (**156**), whilst acetonitrile does not, hence the selectivity is ensured (Scheme 94). Clearly, this route can only be applied to the synthesis of a very limited number of 1,3,5-triazines. Cotrimerization of amidines is only effective when formamidine or amidines containing electron withdrawing substituents, such as 2,2,2-trichloroacetamidine, are used ⟨59JA1466⟩. Cotrimerization of imidates is also of limited value, as is shown by the reaction in equation (84) ⟨62JOC3362⟩.

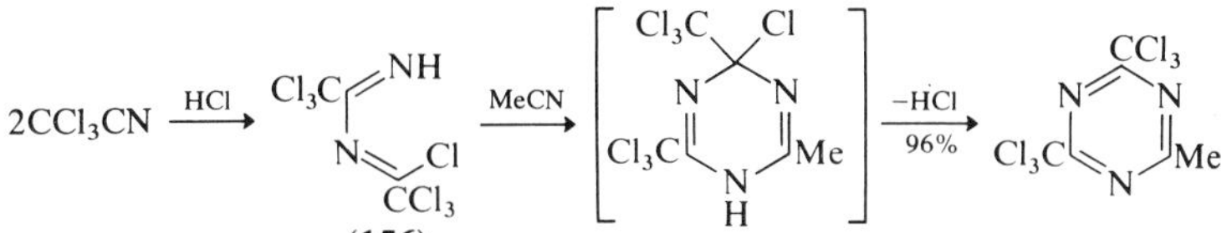

**Scheme 94**

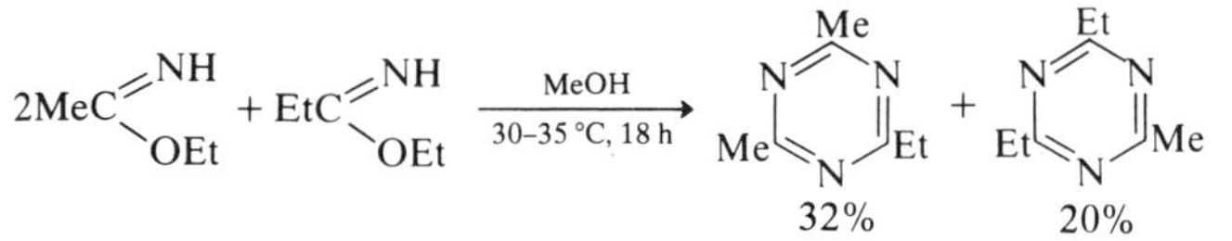

(84)

In general, mixed trimerization of isocyanates also gives mixtures ⟨61JOC3334⟩; however, it is possible to prepare isocyanurates of the type (157) from arenesulfonyl isocyanates in the presence of 1,2-dimethylimidazole as catalyst ⟨76JOC3409⟩. The mechanism is similar to that of the trimerization of isocyanates (Scheme 95).

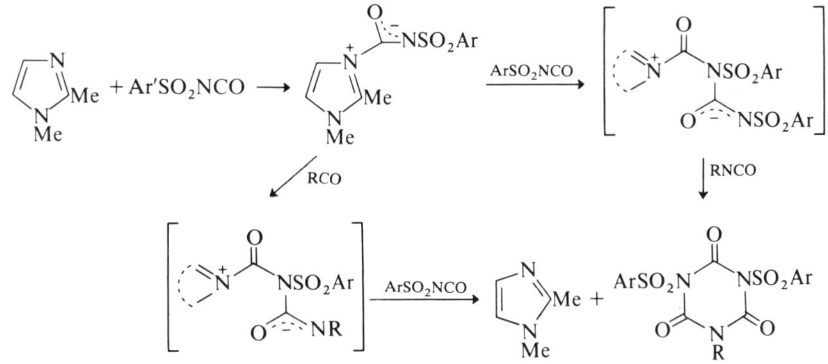

**Scheme 95**

The inherent problems of cotrimerization may be circumvented by using the reaction of inorganic cyanates with organic isocyanates to form the salt (158), which can react with an alkyl halide to give the isocyanurate (159; Scheme 96) ⟨70JHC725⟩. The reaction does give the symmetrical isocyanurate as a by-product but it is readily separated from the salt. The route can be applied to the synthesis of many isocyanurates.

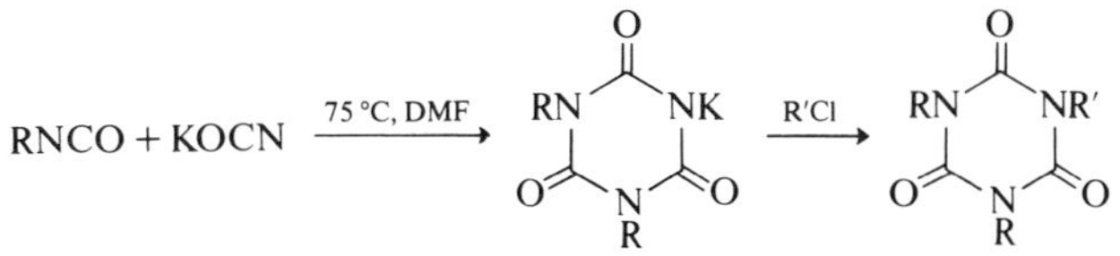

$R = Ph, CH=CHCH_2$; $R' = PhCH_2, C_8H_{17}^n, Bu^n, Bu^s, Me$

**Scheme 96**

Bloodworth and Davies devised a similar route ⟨65JCS6858⟩. By reaction of a trialkyltin alkoxide with an isocyanate the intermediate 1:1 adduct (160; Scheme 97) may be isolated,

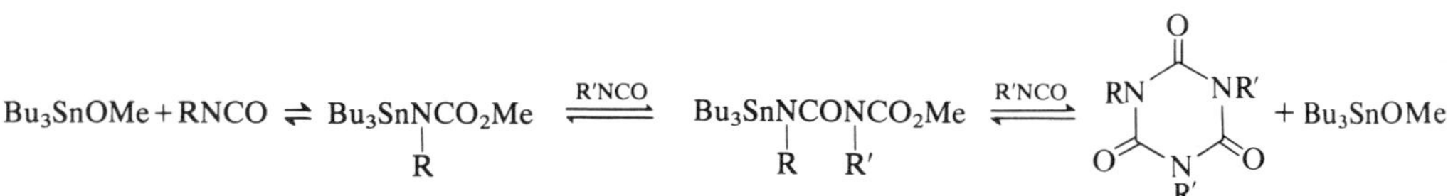

**Scheme 97**

and subsequently reacted with two equivalents of another isocyanate to yield the desired
product. The isocyanurate may take several days to crystallize out.

### 2.20.4.6.2 Synthesis from amidines, isocyanates, aldehydes and ketones

The value of amidines, nitriles and isocyanates in triazine syntheses has been demonstrated
already. The reactions described below are largely modifications of the routes discussed in
the previous sub-sections. Aryl amidines react with carbonyl chloride derivatives to give
2,4-diaryl-1,3,5-triazines. This valuable synthetic route was discovered by Pinner
⟨1890CB2912⟩. The intermediate (**161**; Scheme 98) undergoes cyclization on heating above
the melting point ⟨56JA2447⟩. The reaction fails when alkyl amidines are used, but 2,4-
dimethyl-1,3,5-triazine and 2,4-dimethyl-1,3,5-triazinone may be prepared by a
modification of the Pinner synthesis (Scheme 99) ⟨56JA2447⟩.

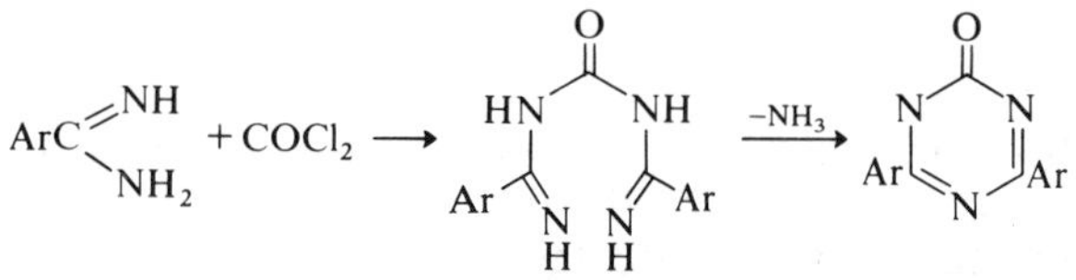

**Scheme 98**

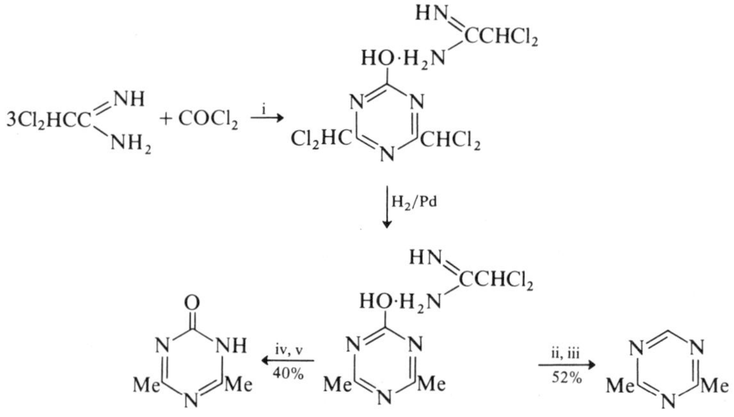

i, NaOH, toluene, 0 °C; ii, POCl₃; iii, H₂/Pd; iv, HCl; v, NaOH

**Scheme 99**

Amidines can react with a wide range of carboxylic acid derivatives; the reactions work
on similar principles to the original Pinner synthesis. Typical examples are shown in
equations (85)–(89) ⟨B-61MI22000, p. 662, 72HCA1566, 72GEP2145174, 63CB3265, 68BCJ1368⟩.

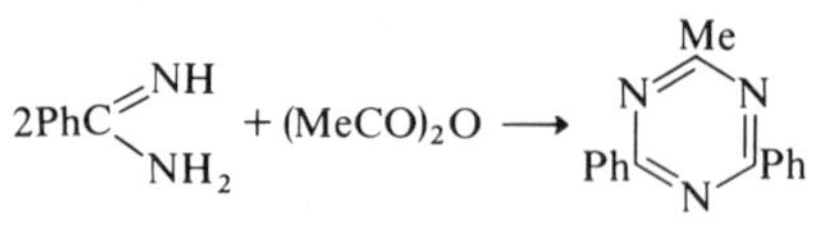

(85)

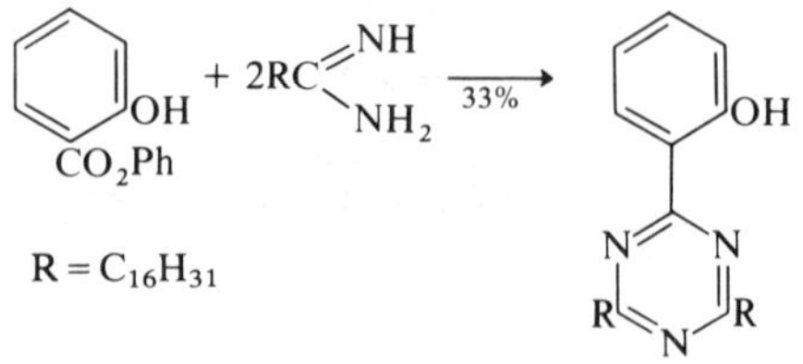

$R = C_{16}H_{31}$

(86)

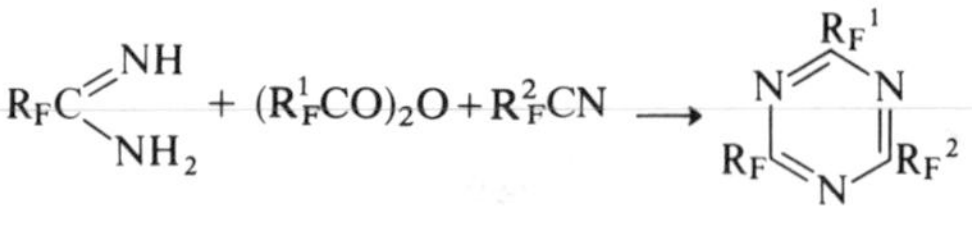

(87)

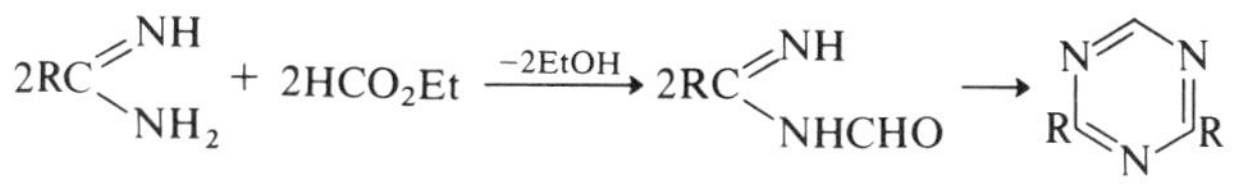

$$2RC\overset{NH}{\underset{NH_2}{\diagup}} + 2HCO_2Et \xrightarrow{-2EtOH} 2RC\overset{NH}{\underset{NHCHO}{\diagup}} \longrightarrow \quad (88)$$

$$ArC\overset{NH}{\underset{NH_2}{\diagup}} + R\overset{OH}{\underset{H}{\overset{|}{-}}}CN \longrightarrow \quad (89)$$

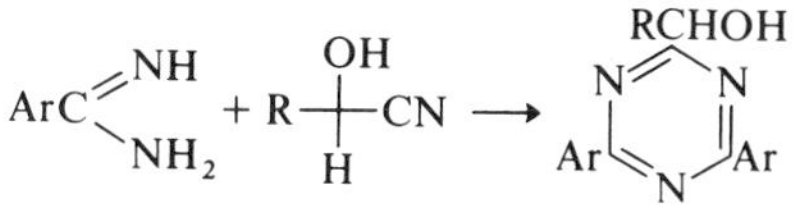

The reaction between amidine salts and acetimidates or propionimidates provides a valuable route to unsymmetrical 1,3,5-triazines bearing two methyl or two ethyl groups respectively (equation 90). Yields vary, and mixtures can be formed (Table 18). The reaction does not work well for amidines with large alkyl substituents ⟨62JOC3608⟩.

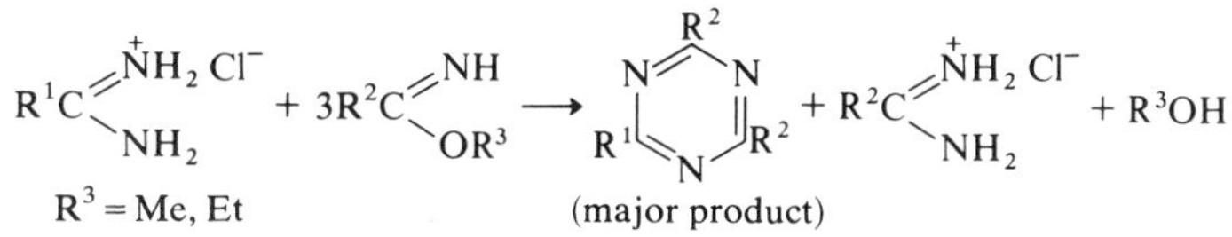

$$R^1C\overset{\overset{+}{N}H_2\ Cl^-}{\underset{NH_2}{\diagup}} + 3R^2C\overset{NH}{\underset{OR^3}{\diagup}} \longrightarrow \quad + R^2C\overset{\overset{+}{N}H_2\ Cl^-}{\underset{NH_2}{\diagup}} + R^3OH \quad (90)$$

$$R^3 = Me, Et \qquad\qquad (major\ product)$$

**Table 18** Synthesis of 1,3,5-Triazines from the Reaction of Imidates and Amidines ⟨62JOC3608⟩

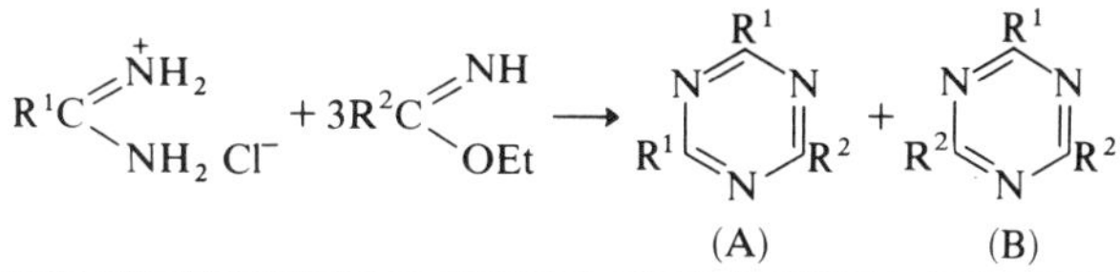

$$R^1C\overset{\overset{+}{N}H_2}{\underset{NH_2\ Cl^-}{\diagup}} + 3R^2C\overset{NH}{\underset{OEt}{\diagup}} \longrightarrow \quad (A) \qquad (B)$$

| Reactants | | | Products (%) | |
|---|---|---|---|---|
| $R^1$ | $R^2$ | Conditions | A | B |
| Ph | Me | i, 1 h, 60 °C; ii, 18 h, r.t. | 55 | — |
| 2-C₅H₄N | Me | i, 0.5 h, 50 °C; ii, 4 d, r.t. | 81 | — |
| MeO | Me | i, 0.5 h, 80 °C; ii, 1.5 h, 80–25 °C | 73 | — |
| MeS | Et | i, 1 h, 40 °C; ii, 18 h, r.t. | 56 | — |
| (EtO)₂CH | Me | 2 d, r.t. | 45 | 20 |
| CHCl₂ | Me | 4 d, r.t. | 40 | 11 |

Bader has described the first preparative route to 1,3,5-triazines containing three different alkyl or aryl substituents ⟨65JOC702⟩. The initial step is the formation of an acyl imidate (**162**) followed by condensation with an amidine to give the product (Scheme 100). Although mixtures are formed, the method is valuable because such triazines are otherwise very difficult to prepare. Typical examples are given in Table 19.

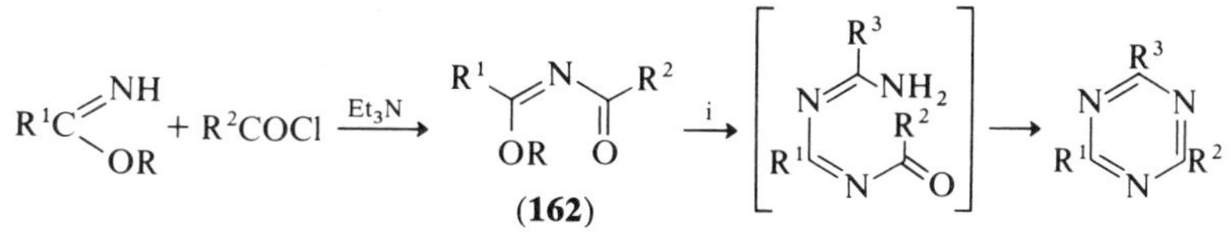

$$R^1C\overset{NH}{\underset{OR}{\diagup}} + R^2COCl \xrightarrow{Et_3N} \underset{OR\ \ O}{R^1\diagdown N\diagup R^2} \xrightarrow{i} \left[ \quad \right] \longrightarrow \quad$$

**(162)**

**Scheme 100**

2-(4-Pyridyl)-1,3,5-triazine may be prepared from 4-cyanopyridine and formamidine *via* the imidate (**163**; Scheme 101) ⟨80JHC333⟩.

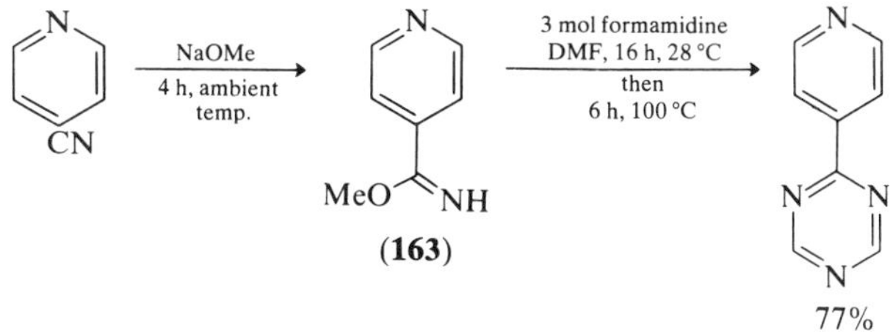

$$\text{(pyridine-CN)} \xrightarrow[\substack{4\ h,\ ambient \\ temp.}]{NaOMe} \underset{MeO\diagdown\ NH}{\text{(163)}} \xrightarrow[\substack{then \\ 6\ h,\ 100\,°C}]{\substack{3\ mol\ formamidine \\ DMF,\ 16\ h,\ 28\,°C}} \quad 77\%$$

**(163)**

**Scheme 101**

**Table 19** Synthesis of 1,3,5-Triazines from the Reaction of Acylimidates and Amidines ⟨65JOC702⟩

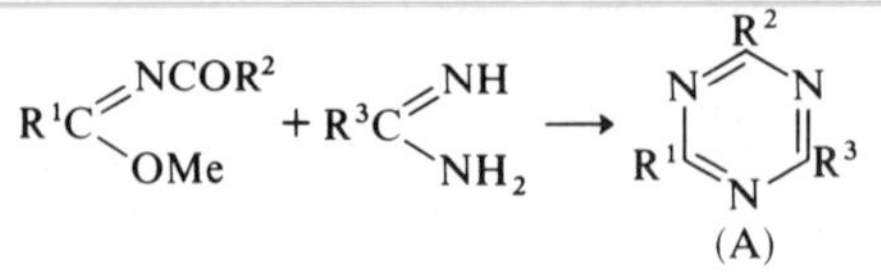

$$R^1C\begin{matrix}NCOR^2\\\\OMe\end{matrix} + R^3C\begin{matrix}NH\\\\NH_2\end{matrix} \longrightarrow \underset{(A)}{R^1\underset{N}{\overset{R^2}{\bigcirc}}R^3}$$

| | Reactants | | Conditions | | A (% yield) | Triazine products + By-products | | | (% yield) |
|---|---|---|---|---|---|---|---|---|---|
| $R^1$ | $R^2$ | $R^3$ | Solvent | Time (h) | | $R^1$ | $R^2$ | $R^3$ | |
| Et | PhCH=CH$_2$ | Me | Bu$^t$OH | 4 | (53) | — | — | — | — |
| Et | Et | MeO | MeOH | 3 | (38.6) + | Et | Et | Et | (7.6) |
| | | | | | | NH$_2$ | Et | OMe | (33.8) |
| | | | | | | Et | OMe | OMe | (6.3) |
| Ph | Et | Me | a | — | (47.6) +Ph | | Me | Me | (2.2) |
| Ph | Et | MeS | a | — | (25) +Ph | | Et | OMe | (6.5) |

$^a$ Mixture of MeOH and Et$_2$O (1:2.5).

Aryl cyanides react with the anion of urea (or thiourea) to give the 2,4-diaryl-1,3,5-triazinone (**164**) in good to excellent yields except when *o*-chloroaryl isocyanates are used ⟨76JHC917⟩.

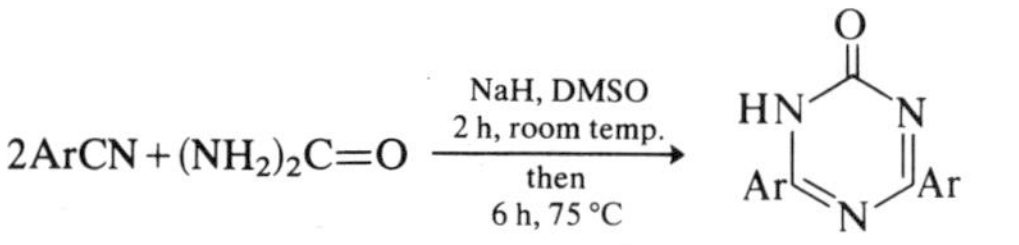

$$2ArCN + (NH_2)_2C=O \xrightarrow[\substack{\text{then}\\\text{6 h, 75 °C}}]{\substack{\text{NaH, DMSO}\\\text{2 h, room temp.}}} \underset{(164)}{\text{HN}\overset{O}{\underset{Ar}{\bigcirc}}Ar} \qquad (91)$$

The reaction of benzonitrile with carbonyl chloride in the presence of hydrogen chloride gives 2-chloro-4,6-diphenyl-1,3,5-triazine in *ca.* 80% yields on prolonged heating at 100 °C. Experiments using 4-methoxybenzonitrile and other derivatives failed. The proposed mechanism of the reaction (Scheme 102) is supported by isotopic labelling experiments ⟨73S189⟩.

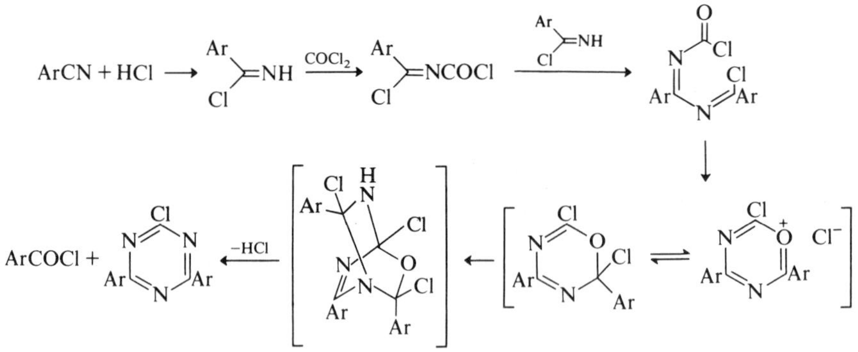

**Scheme 102**

The isocyanates play a most important role in the syntheses of 1,3,5-triazine-2,4-diones. Two equivalents of the isocyanate react with an equivalent of another reagent, usually an amidine or imine, to yield the desired products. Once again the reaction occurs *via* a dipolar intermediate (**165**; Scheme 103) (see Sections 2.20.4.5.3 and 2.20.4.6.1).

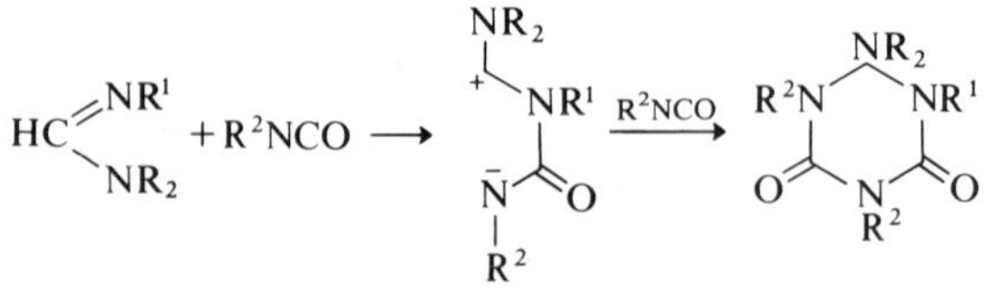

**Scheme 103**

A typical example of the reaction of isocyanates with formamidines is shown in equation (92). Alkyl isocyanates can react in a similar way, but there can be mixed products formed under certain circumstances, as in the reaction of methyl isocyanate with *N,N*- dialkyl-*N'*- arylformamidines (**166**). The predicted product (**167**; equation 93) is formed except when *N,N'*-dimethyl-*N*-arylformamidines are used. In the latter case a mixture of (**168**) and (**169**) is formed (equation 94). The mechanism of the formation of (**169**) is shown in Scheme 104 ⟨73HCA776⟩.

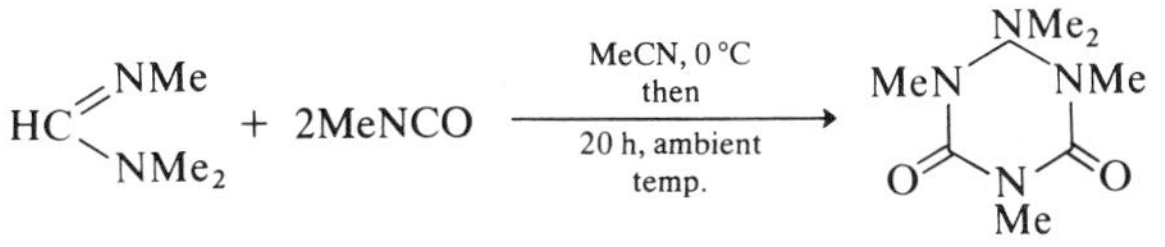

$$\text{HC}\overset{\text{NMe}}{\underset{\text{NMe}_2}{\big\langle}} + 2\text{MeNCO} \xrightarrow[\substack{20\text{ h, ambient}\\ \text{temp.}}]{\substack{\text{MeCN, 0 °C}\\ \text{then}}} \qquad (92)$$

$$\text{(166)} \quad \text{HC}\overset{\text{NAr}}{\underset{\text{NR}_2}{\big\langle}} + 2\text{MeNCO} \xrightarrow[20\text{ h, ambient temp.}]{0\text{ °C then}} \qquad (93)$$

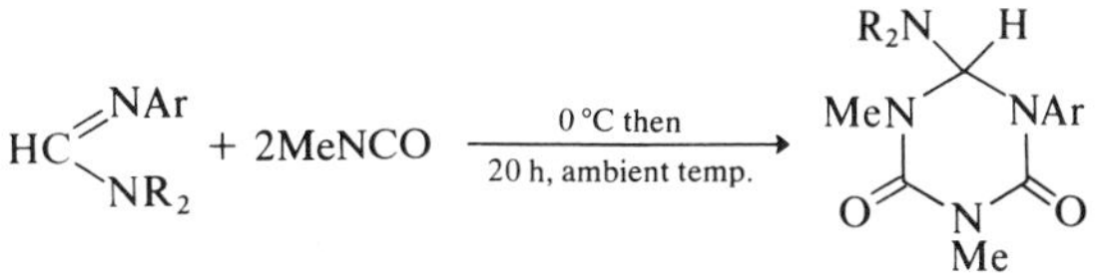

(**167**) Ar = 4-ClC$_6$H$_4$, *etc.*; R = Et, Pr, *etc.*

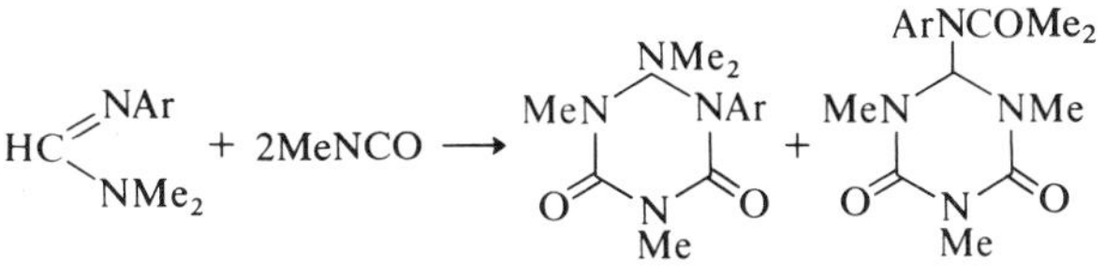

$$\text{HC}\overset{\text{NAr}}{\underset{\text{NMe}_2}{\big\langle}} + 2\text{MeNCO} \longrightarrow \text{(168)} + \text{(169)} \qquad (94)$$

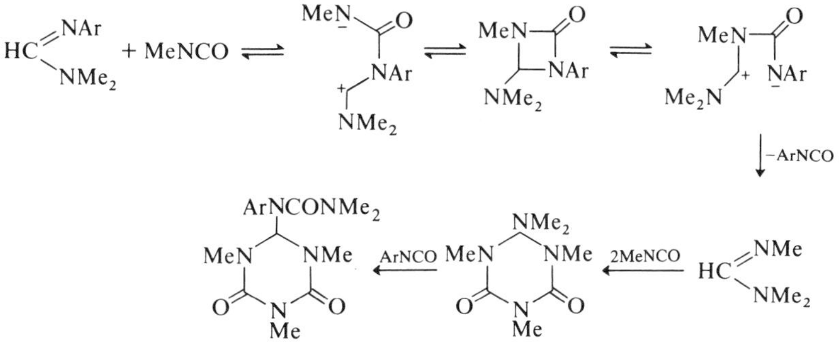

Scheme 104

Kantlehner *et al.* ⟨78LA512⟩ have reported a novel synthesis of triazine-2,4-diones from (**170**) and methyl isocyanate (equation 95).

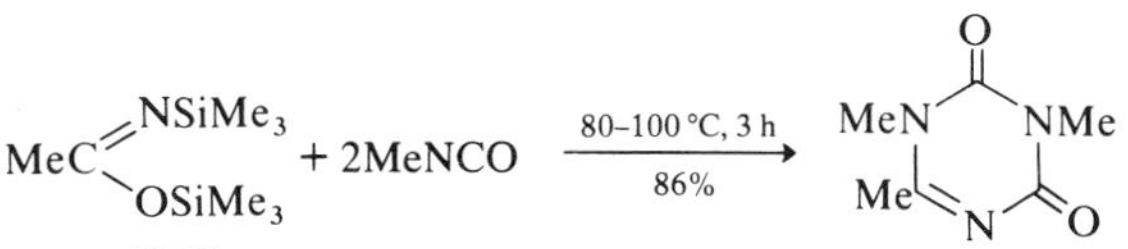

$$\text{MeC}\overset{\text{NSiMe}_3}{\underset{\text{OSiMe}_3}{\big\langle}} + 2\text{MeNCO} \xrightarrow[86\%]{80–100\text{ °C, 3 h}} \qquad (95)$$

(**170**)

Isocyanates (one equivalent) react smoothly with azomethines (two equivalents) in the presence of Lewis acids to form 1,3,5-triazin-2-ones in moderate to excellent yields (40–100%) (equation 96). Isothiocyanates can be used in an analogous way ⟨61JOC767⟩.

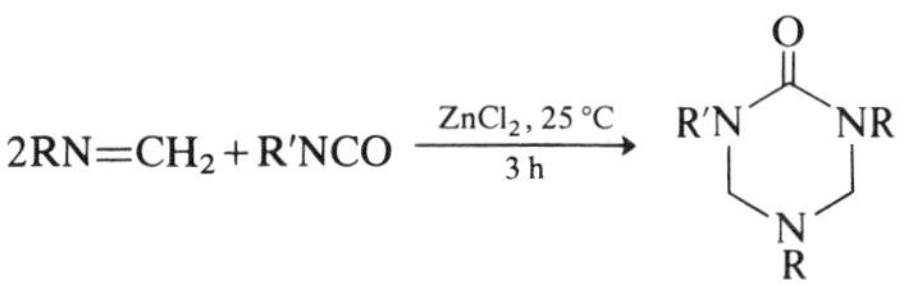

$$2\text{RN}{=}\text{CH}_2 + \text{R'NCO} \xrightarrow[3\text{ h}]{\text{ZnCl}_2,\ 25\text{ °C}} \qquad (96)$$

R = Me, Et, Bu$^t$, *etc.*; R' = H, Bu$^n$, Ph, *etc.*

When two equivalents of an isocyanate are treated with one equivalent of an enamine, a 1,3,5-triazine-2,4-dione is formed. The novel spiro compound (**171**) may be prepared by this route. The spiro derivative is isomerized on heating ⟨69CB931⟩.

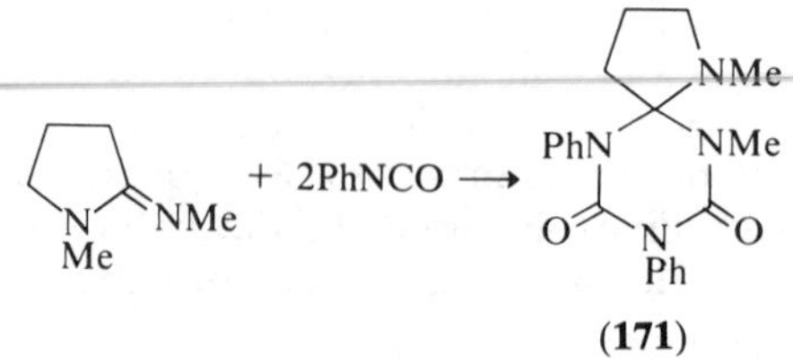

(97)

**(171)**

Chlorosulfonyl isocyanate is a valuable starting reagent, which reacts with most anils in a similar way. The products *e.g.* (**172**), are readily dechlorosulfonylated (Scheme 105) ⟨77JCS(P1)47⟩.

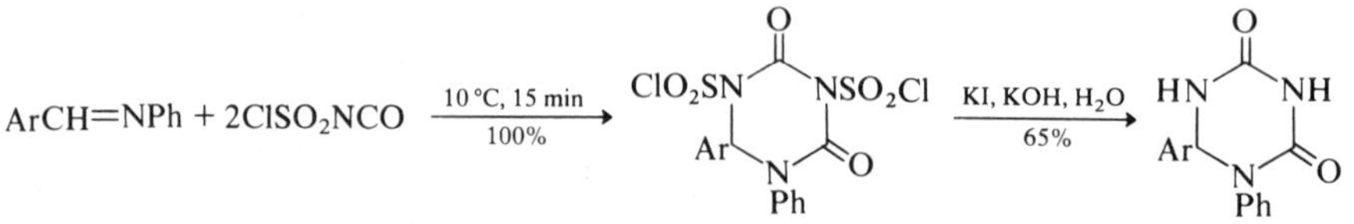

$$\text{ArCH=NPh} + 2\text{ClSO}_2\text{NCO} \xrightarrow[100\%]{10\,°C,\,15\,min}$$

(**172**) Ar = 4-Me$_2$NC$_6$H$_4$

**Scheme 105**

Aryl isocyanates undergo a novel cycloaddition with *N*-benzylidenealuminum amides (**173**) *via* a series of dipolar intermediates to yield the triazines (**174**; Scheme 106) ⟨78LA1111⟩.

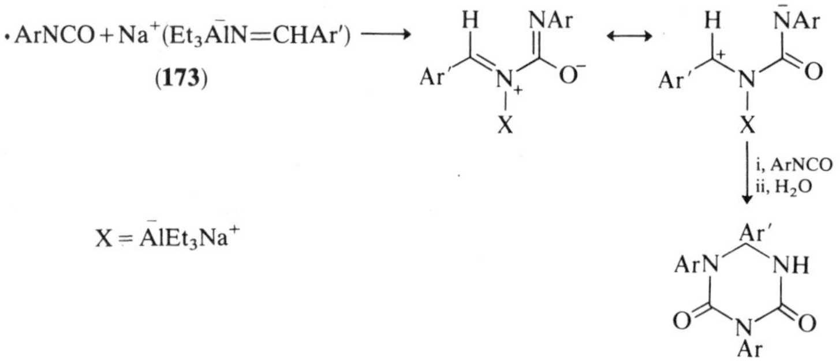

$$\cdot \text{ArNCO} + \text{Na}^+(\text{Et}_3\bar{\text{A}}\text{lN=CHAr}') \longrightarrow$$

(**173**)

X = $\bar{\text{A}}$lEt$_3$Na$^+$

i, ArNCO
ii, H$_2$O

**Scheme 106**          (**174**)

Aldehydes and ketones are valuable starting materials for the preparation of 1,2-dihydro-1,3,5-triazines. The two-component reaction of aryl biguanides and an aldehyde or ketone has been mentioned earlier (see Section 2.20.4.2.2). In addition, three-component syntheses by the direct condensation of an aldehyde or ketone and a variety of other reagents has been reported. The strategy of the route is illustrated by the examples in equations (98)–(102) ⟨59JOC573, 61JOC767, 60JOC147, 65JOC930, 75JHC37⟩.

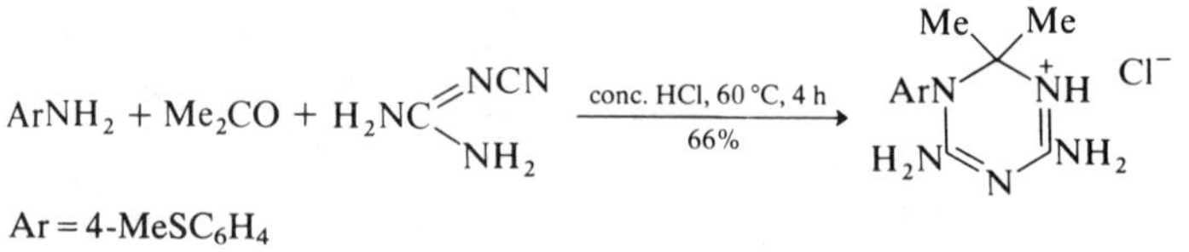

$$\text{ArNH}_2 + \text{Me}_2\text{CO} + \text{H}_2\text{NC}\begin{smallmatrix}\diagup\text{NCN}\\\diagdown\text{NH}_2\end{smallmatrix} \xrightarrow[66\%]{\text{conc. HCl, }60\,°C,\,4\,h}$$

Ar = 4-MeSC$_6$H$_4$

(98)

$$\text{RNHCONHR} + 2\text{H}_2\text{CO} + \text{RNH}_2 \xrightarrow[57\%]{100\,°C,\,12\,h}$$

R = Bu$^n$

(99)

$$\text{Bu}^n\text{NH}_2 + 2\text{H}_2\text{CO} + \text{H}_2\text{NC}\begin{smallmatrix}\diagup\text{NCN}\\\diagdown\text{NH}_2\end{smallmatrix} \xrightarrow[80\%]{54\,°C,\,1\,h}$$

(100)

$$\text{R}^1\text{C}\begin{smallmatrix}\diagup\text{NH}\\\diagdown\text{NH}_2\end{smallmatrix} + \text{R}^2\text{C}\begin{smallmatrix}\diagup\text{NH}\\\diagdown\text{OR}^3\end{smallmatrix} + \text{R}^4\text{R}^5\text{C=O} \longrightarrow$$

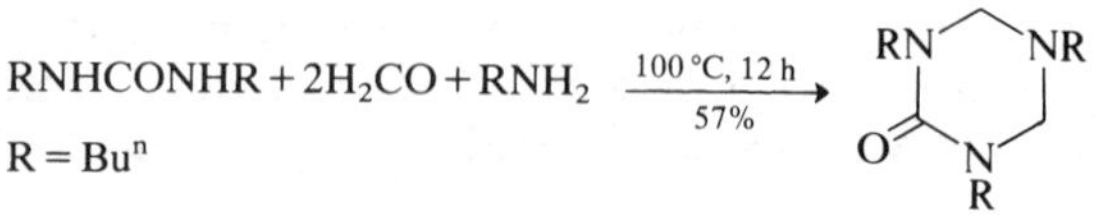

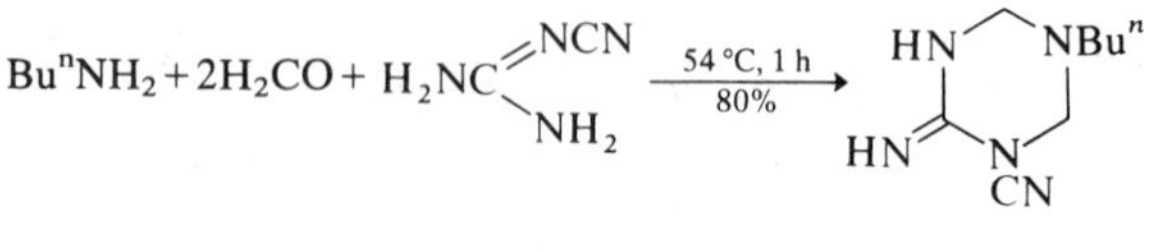

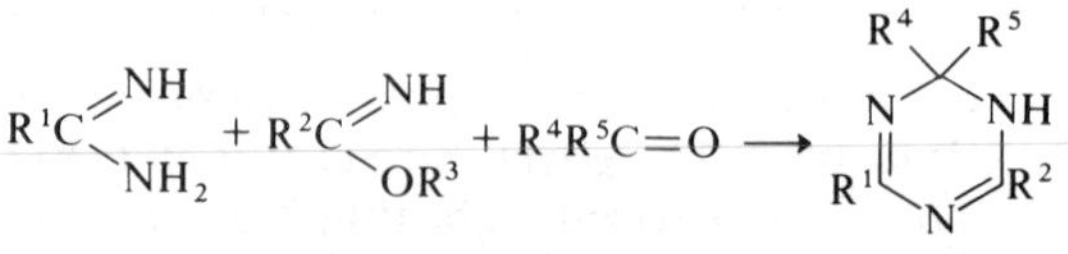

(101)

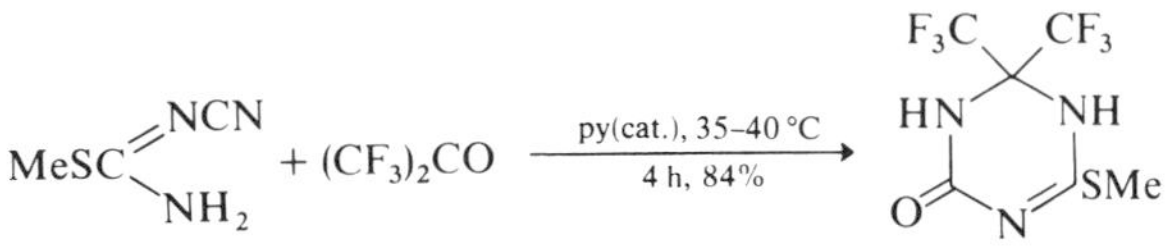

$$\text{(102)}$$

### 2.20.4.6.3 Synthesis of fused ring systems

There are a considerable number of such syntheses in the literature. Most of the syntheses are based on the reaction of two equivalents of isocyanate with a nitrogen heterocycle. The two examples given are typical of the route (equation 103 and Scheme 107). The mechanism of formation of the 1,3,4-oxadiazolo[3,2-*a*][1,3,5]triazine-5,7-diones (**175**) is included as it is of interest ⟨74T221, 74JCS(P1)1786⟩.

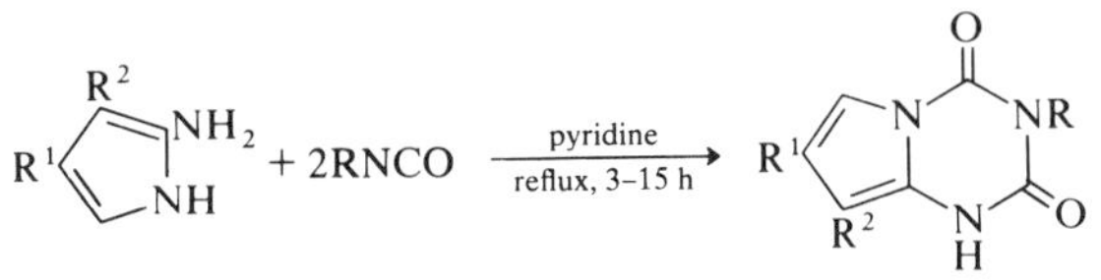

$$\text{(103)}$$

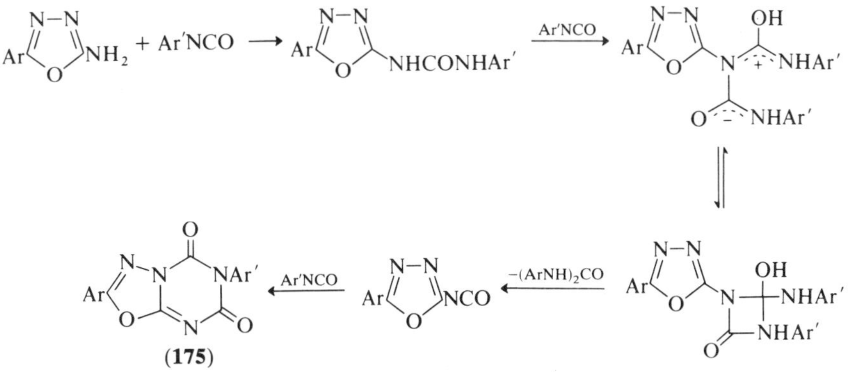

**Scheme 107**

The mixed trimerization of heterocyclic compounds is another potentially valuable route to fused systems. The reaction of isoquinoline and 2-chloro-4,5-dihydroimidazole is a recent example of this approach (equation 104) ⟨81S154⟩.

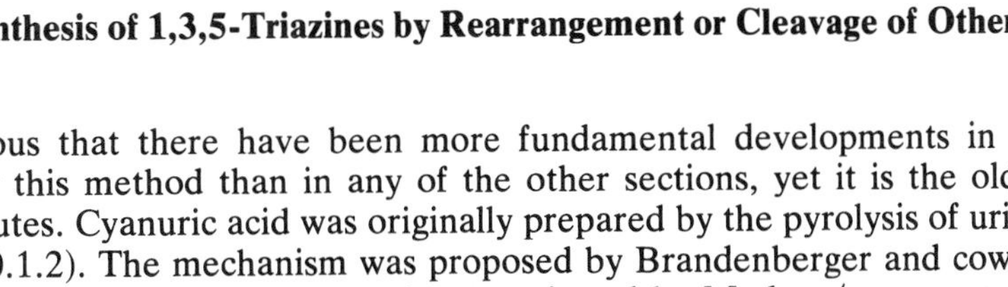

$$\text{(104)}$$

### 2.20.4.7 Synthesis of 1,3,5-Triazines by Rearrangement or Cleavage of Other Heterocyclic Systems

It is curious that there have been more fundamental developments in 1,3,5-triazine synthesis by this method than in any of the other sections, yet it is the oldest of all the synthetic routes. Cyanuric acid was originally prepared by the pyrolysis of uric acid (**7**) (see Section 2.20.1.2). The mechanism was proposed by Brandenberger and coworkers on the basis of isotopic labelling studies, and was reviewed by Modest ⟨B-61MI22000, p. 706⟩. The rearrangement of pyrimidines to 1,3,5-triazines is the most important advance in this area.

### 2.20.4.7.1 Synthesis by formation of one carbon–nitrogen bond

There are very few examples of such syntheses. The 1,3,5-triazinedione (**176**) is formed by the thermal rearrangement of 6-ethoxycarbonyl-3-phenyl-1,3,5-triazabicyclo-

[3.1.0]hexane-2,4-dione ⟨75JA5611⟩. Tetrahydropteridines (**177**) undergo an autoxidative rearrangement to the 1,3,5-triazines (**178**) ⟨75T533⟩. The proposed mechanism is shown in Scheme 108.

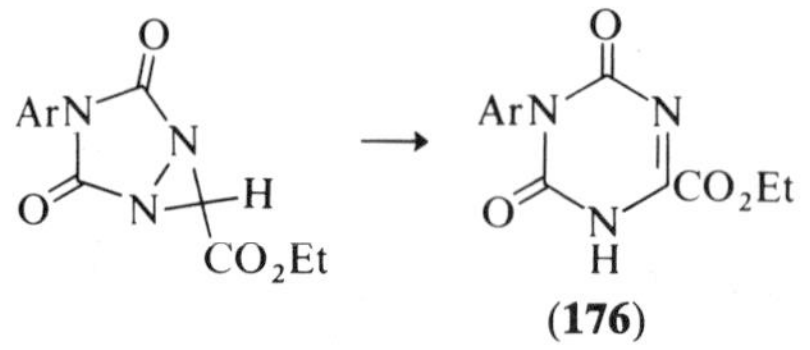

(105)

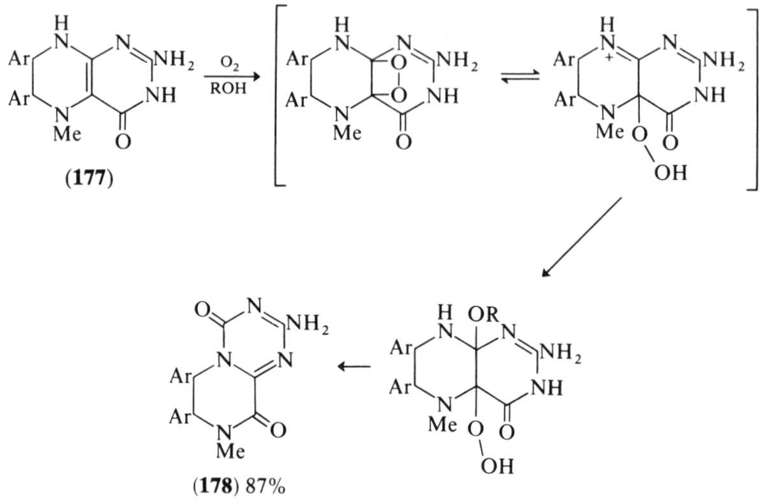

(**176**)

(**177**)

(**178**) 87%

**Scheme 108**

### 2.20.4.7.2 Synthesis by formation of two carbon–nitrogen bonds

The most important route is the conversion of pyrimidines into 1,3,5-triazines. The first one-step transformation was effected by Taylor and Jefford ⟨62JA3744⟩ by heating the pyrimidine (**179**) with benzenesulfonyl chloride in pyridine (equation 106). The reaction may be considered as an example of an abnormal Beckmann rearrangement. The mechanism of the reaction of the 4-aminopyrimidine (**180**) is probably dependent on the nature of the 2-substituent (**180**, R). If R is an electron-releasing moiety, pathway B seems more likely (Scheme 109). The 4-hydroxypyrimidine (**179**; R' = OH) behaves similarly. Many 2-cyano-1,3,5-triazines may be synthesized by this method.

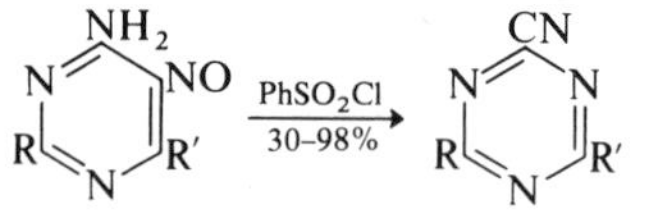

(106)

(**179**)  R = SMe, NMe₂, Ph; R' = NH₂, OH

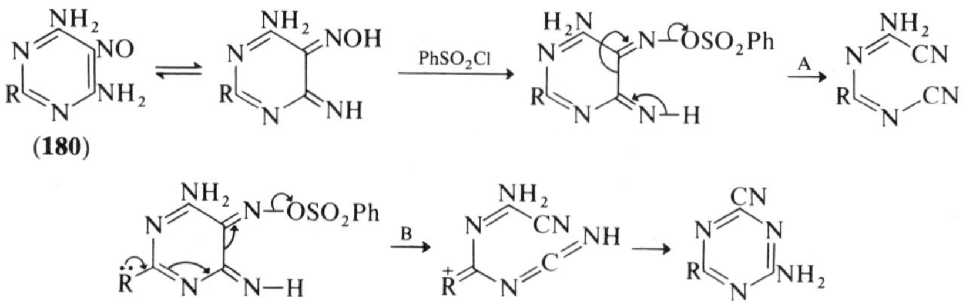

(**180**)

**Scheme 109**

Van der Plas and coworkers have devised a synthesis of 2-substituted 4-methyl-1,3,5-triazines (**182**) by treatment of 2-substituted 4-chloropyrimidines with potassium amide in liquid ammonia (Table 20). The transformation of 4-chloro-2-dimethylaminopyrimidine has been studied by $^{13}$C NMR spectroscopy, which indicates that the acetylene anion (181) is an intermediate in the pathway (Scheme 110) ⟨78JOC2682⟩.

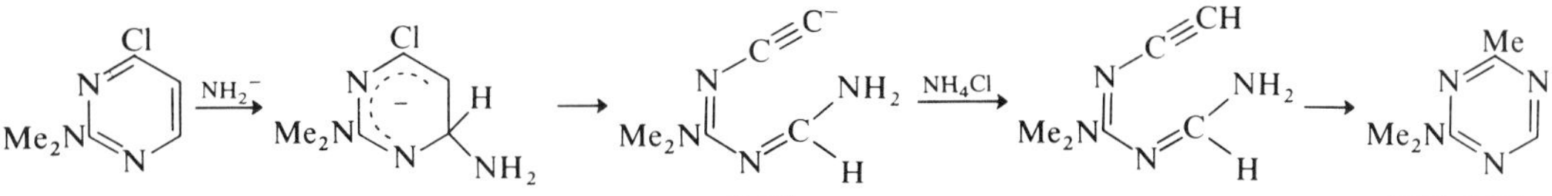

**Scheme 110**

**Table 20** Synthesis of 1,3,5-Triazines from 2-Substituted 4-Chloropyrimidines

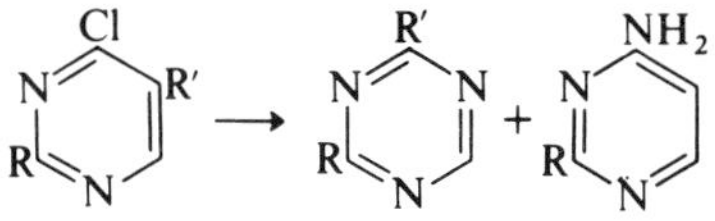

(193)

| Reactant | | | Product (193) | | | |
| R | R' | Conditions[a] | R | R' | Yield (%) | Ref. |
| --- | --- | --- | --- | --- | --- | --- |
| Me | H | i, ii | Me | Me | 25–30 | 66RTC1101 |
| Et | H | i, ii | Et | Me | 20–25 | 66RTC1101 |
| Ph | H, | i, ii | Ph | Me | 44 | 66RTC1101 |
| Me | Me | i, ii | Me | Et | 14–15 | 66RTC1101 |
| C$_5$H$_{10}$N | H | i, ii | C$_5$H$_{10}$N | Me | 80 | 69RTC1156 |
| 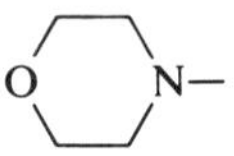 O⌁N— | H | i, ii | 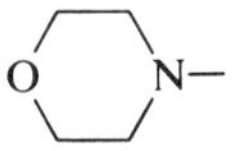 O⌁N— | Me | 85 | 69RTC1156 |
| β-C$_{10}$H$_7$ | H | i, ii | β-C$_{10}$H$_7$ | Me | 63 | 69RTC426 |
| MeNPh | H | i, ii | MeNPh | Me | 65 | 69RTC426 |
| NMe$_2$ | H | i, ii | NMe$_2$ | Me | 75 | 69RTC426 |

[a] i, KNH$_2$/liq. NH$_3$, −33 °C, 2 h; ii, NH$_4$Cl.

There have also been interesting developments in the conversion of 1,2,4-triazines into 1,3,5-triazines. Chambers *et al.* reported the photolytic rearrangement of perfluorotriisopropyl-1,2,4-triazine in low yield. The 1,3,5-triazine may be formed *via* the triazabenzvalene derivative (183) (Scheme 111) ⟨79JCS(P1)1978⟩.

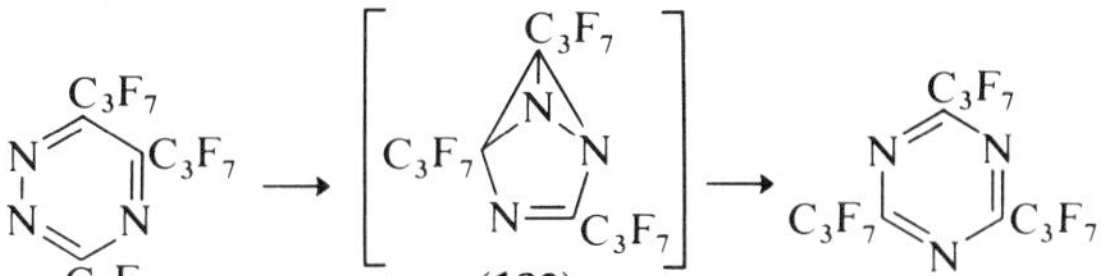

**Scheme 111**

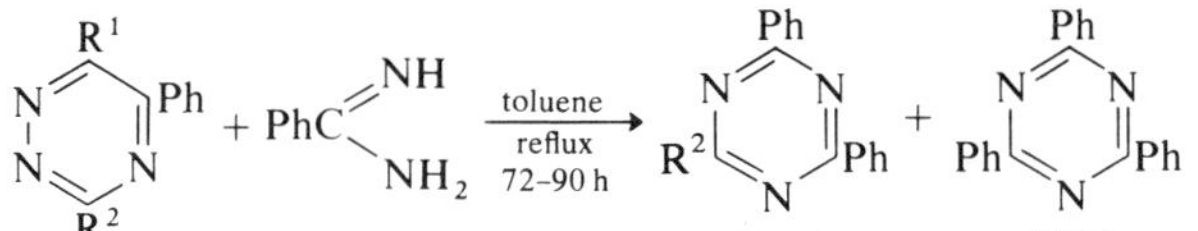

(107)

R$^1$ = Ph or pyridyl; R$^2$ = Ph, Me or 2-pyridyl

Recently, it has been shown that some 1,2,4-triazines react with aryl amidines to form the 1,3,5-triazines (184) and (185) (equation 107) ⟨81TL1393⟩. It is proposed that the 1,3,5-triazine (184) is formed by a Diels–Alder reaction which produces the adduct (186) with subsequent cleavage of the N—N single bond, loss of the cyanide R$^1$CN and of ammonia to give the desired product (Scheme 112); 2,4,6-triphenyl-1,3,5-triazine is formed by an analogous Diels–Alder reaction of (184) with excess benzamidine (Scheme 113). Support for the hypothesis was obtained when 2-(2-pyridyl)-4,6-diphenyl-1,3,5-triazine (184; R$^2$ = 2-pyridyl) reacted with benzamidine and the predicted product (185) was formed in 66%

yield. These are the first examples of Diels–Alder reactions with inverse electron demand occurring between an open chain carbon to nitrogen double bond and a 1,2,4- or a 1,3,5-triazine system.

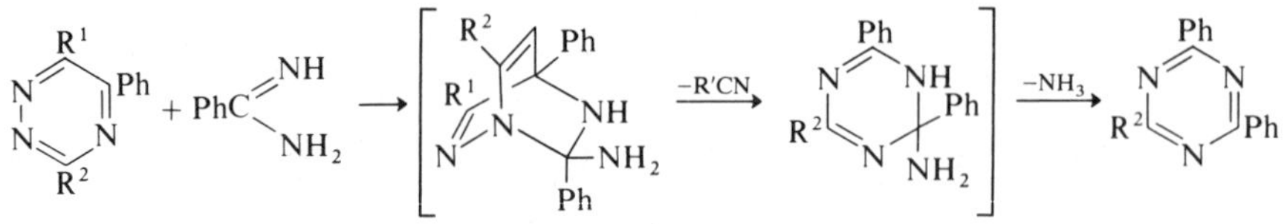

**Scheme 112**

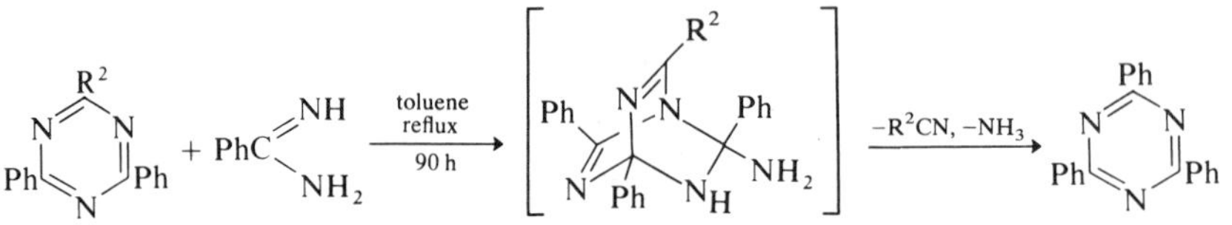

**Scheme 113**

4,6-Diaryl-1,2,3,5-oxathiadiazine 2,2-dioxides (**187**) are valuable starting reagents which are readily prepared by treatment of the aryl cyanide with sulfur trioxide. Amidines, imidates and trichloroacetonitrile react with (**187**) to give a wide range of 4,6-diaryl-1,3,5-triazines in 50–90% yields under acidic conditions (equation 108) ⟨63CB2070⟩.

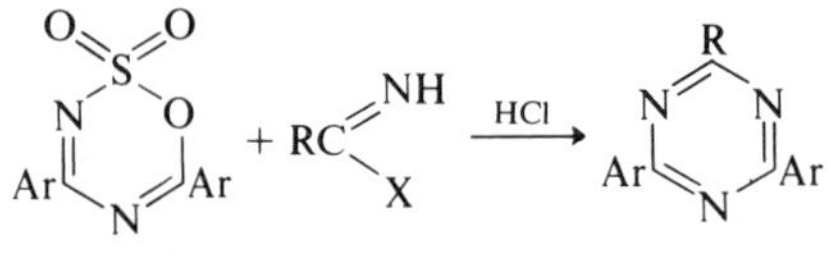

(108)

R = Me, CCl$_3$, NH$_2$, OH, *etc.*; Ar = Ph, 4-MeC$_6$H$_4$; X = NH$_2$ or alkoxy

1,3,5-Triazinediones may be prepared from triazolinediones in 50–70% yields (equation 109) ⟨77JCR(M)2826⟩. The ring oxygen atom of 4-oxotetrahydro-1,3,5-oxadiazines can be exchanged for nitrogen by a transureidoalkylation mechanism to produce 5-alkyl-2-oxohexahydro-1,3,5-triazines (equation 110) ⟨73S243⟩.

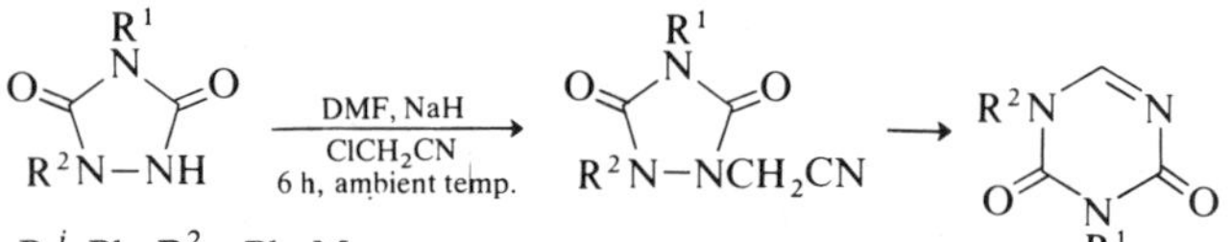

(109)

R$^1$ = Pr$^i$, Ph, R$^2$ = Ph, Me

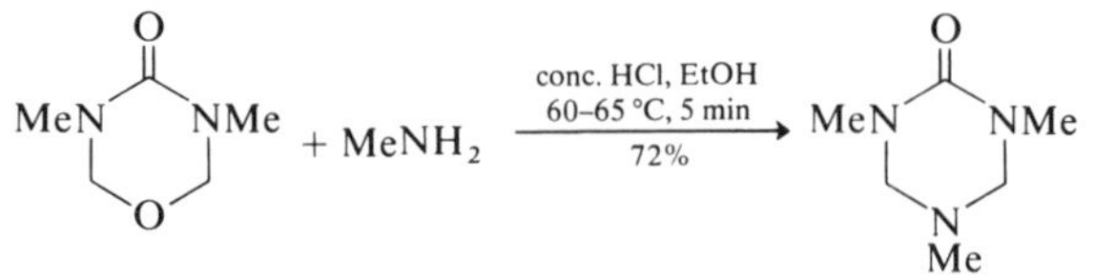

(110)

### 2.20.4.7.3 Synthesis by formation of three carbon–nitrogen bonds

There are few examples of this type of synthesis. Typically, an azirine reacts with isocyanates or isothiocyanates to yield the triazine *via* a series of dipolar intermediates. The products may be contaminated with the isocyanate trimer; the example of Scheme 114 below is typical. The novel spiro compound (**188**) is formed by a similar route (Scheme 115) ⟨79HCA1429, 81LA264⟩.

Fitzsimmons *et al.* have shown that alkoxyphosphazenes (**189**) may be converted into 2,4,6-triphenyl-1,3,5-triazines by reaction with benzoyl chloride. The mechanism is not properly elucidated (equation 111) ⟨62MI22000⟩.

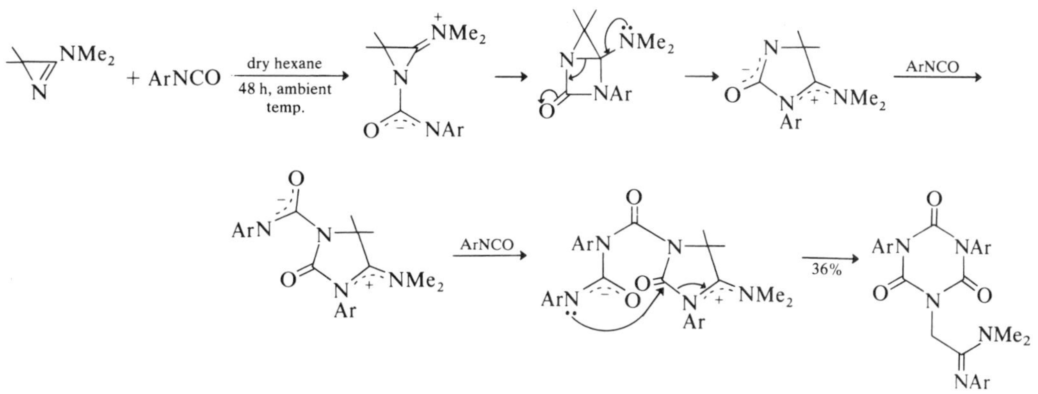

**Scheme 114**

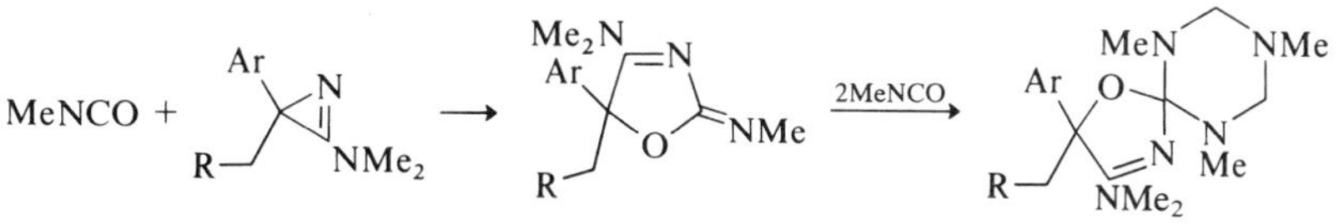

**(188)**

**Scheme 115**

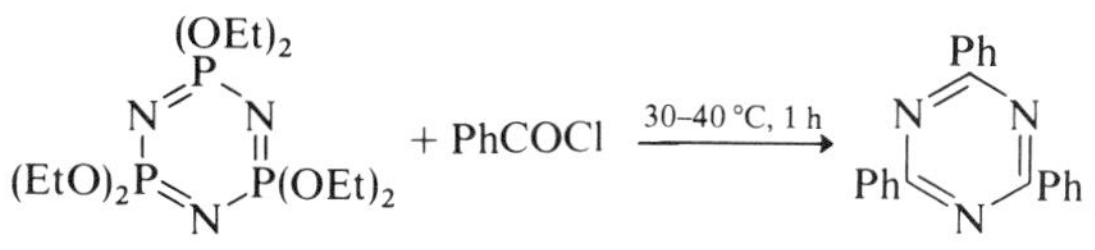

**(189)**

(111)

#### *2.20.4.7.4 Synthesis of fused ring systems*

Imidazo[1,5-*a*][1,3,5]triazinones (**191**) have been prepared from pyrimidines such as (**190**) by cleavage of the C(4)–C(5) bond and cyclization to the desired product (Scheme 116) ⟨79JOC1740⟩. The allyl derivative (**192**) is formed in the same way (equation 112) ⟨81JOC3681⟩.

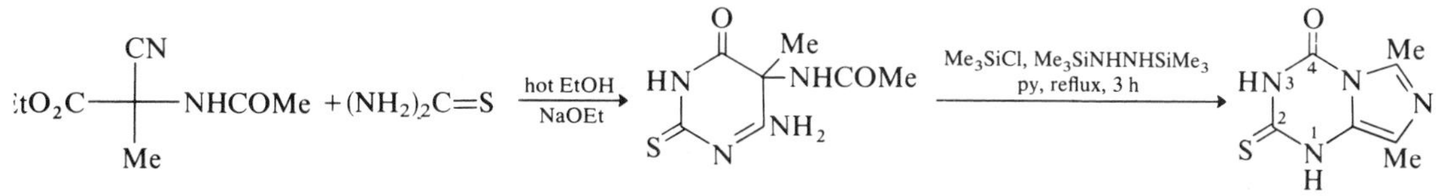

(**190**) 54%

(**191**) 83%

**Scheme 116**

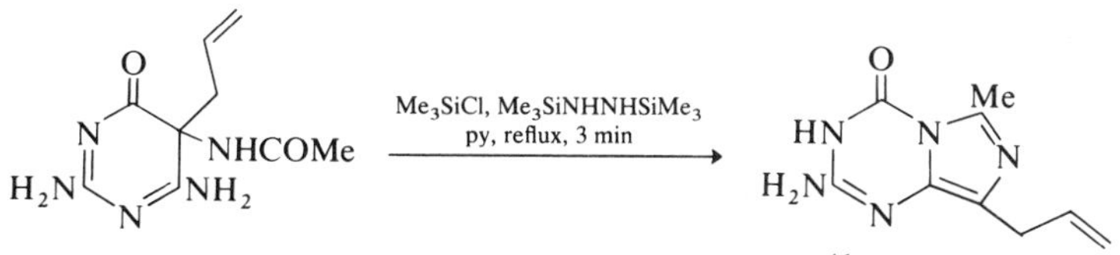

(**192**) 95%

(112)

### 2.20.4.8 Synthesis of Polyaza[3.3.3]cyclazines

The polyaza[3.3.3]cyclazines have become of increasing interest. Ceder, Shaw and their coworkers have been predominant in this area. It is not possible to discuss the available syntheses in any detail but the three examples of Schemes 117–119 illustrate the strategies in the preparation of these compounds ⟨77JHC679, 72ACS596, 74JHC627⟩.

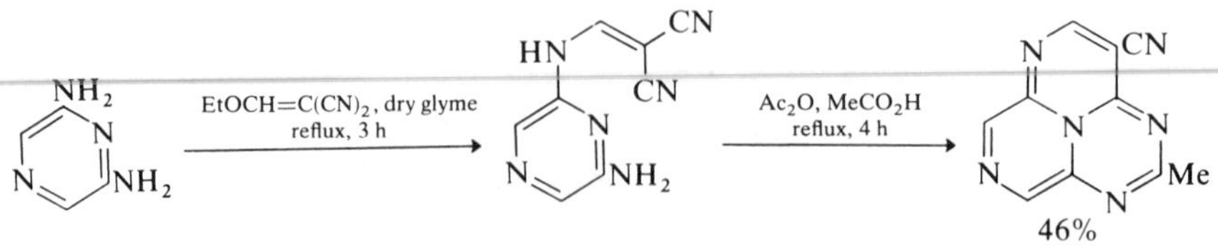

**Scheme 117**

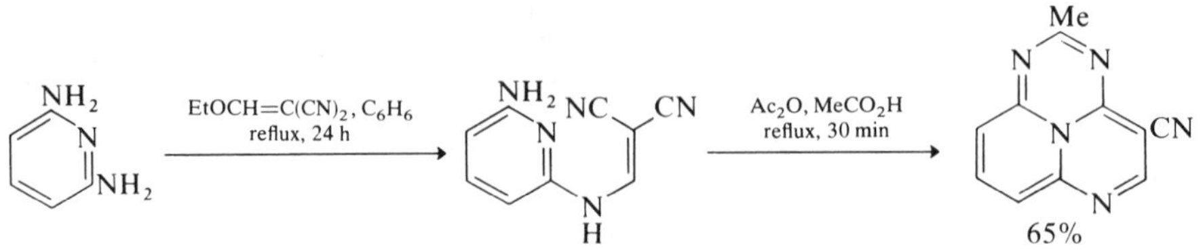

**Scheme 118**

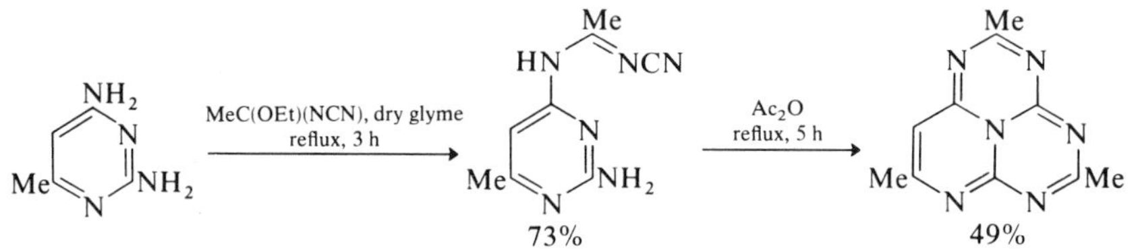

**Scheme 119**

### 2.20.4.9 An Assessment of the Application of the Synthetic Routes in the Synthesis of 1,3,5-Triazines

This final sub-section outlines how the application of the routes described in the earlier sub-sections may best be applied in 1,3,5-triazine syntheses.

#### 2.20.4.9.1 Synthesis of 1,3,5-triazine

The best route is the synthesis from ammonium acetate and triethyl orthoformate (see Section 2.20.4.5.2).

#### 2.20.4.9.2 Synthesis of monosubstituted 1,3,5-triazines

The 2-alkyl and 2-aryl derivatives are best prepared by the reaction between 1,3,5-triazine and an imidate or amidine (see Section 2.20.3.1.2). 2-Amino-1,3,5-triazines may be prepared efficiently by: (a) treatment of 1,3,5-triazine with 1-amidinopyrazole (see Section 2.20.3.1); (b) condensation of triformidomethane and a formylguanidine (see Section 2.20.4.3.3). 2-Aryloxy derivatives are produced by hydrogenation of the corresponding dichloro-1,3,5-triazine derivative (see Section 2.20.3.12).

#### 2.20.4.9.3 Synthesis of 2,4-disubstituted 1,3,5-triazines bearing a single type of substituent

The preferred routes to several of these compounds are summarized in Table 21.

#### 2.20.4.9.4 Synthesis of 2,4-disubstituted 1,3,5-triazines bearing two different substituents

There are too many compounds to discuss efficient routes to all of them. The guidelines below are useful in determining the method of choice: (a) 1,3,5-Triazines bearing heteroatomic substituents are often formed from cyanuric chloride. (b) 1,3,5-Triazines bearing 2-amino substituents may be formed efficiently from biguanides (see Section

**Table 21** Preferred Syntheses of Disubstituted 1,3,5-Triazines (A)

(A)

| R | Preferred method(s) | Section |
|---|---|---|
| Me | i, 2ClCH$_2$C(=NH)(NH$_2$) + (a) COCl$_2$, (b) H$_2$/Pd | 2.20.4.6.2 |
| | ii, 2-Chloro-4-methylpyrimidine + KNH$_2$ | 2.20.4.7.1 |
| Alkyl[a] | i, s-Triazine + amidine | 2.20.3.1 |
| | ii, 2RC(=NH)(NH$_2$) + HCO$_2$Et[b] | 2.20.4.6.2 |
| Aryl | 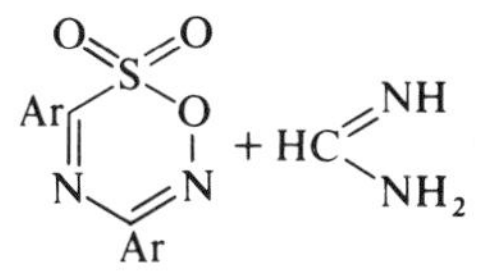 ArSO$_2$ triazinone + HC(=NH)(NH$_2$) | 2.20.4.7.1 |
| OH | i, C$_3$N$_3$Cl$_3$ + (a) 2MeO$^-$, (b) H$_2$/Pd, (c) H$^+$ | 2.20.3.12 |
| | ii, (H$_2$NCO)$_2$NH + HCO$_2$Et | 2.20.4.1.1 |
| SH | (H$_2$NCS)$_2$NH + HCO$_2$Et | 2.20.4.1.1 |
| NH$_2$ | i, H$_2$NC(=NH)(NH$_2$) + HCN | 2.20.4.3.4 |
| | ii, C$_3$N$_3$Cl$_3$ + (a) NH$_3$ + (b) H$_2$/Pd | 2.20.3 |

[a] Including Me.  [b] Trisubstituted 1,3,5-triazines may be prepared by using other carbonyl substituents.

2.20.4.2.1). (c) Aryl cyanates are valuable starting materials for preparation of aryloxy derivatives (see Sections 2.20.4.3.1, 2.20.4.4.1 and 2.20.4.5.1). (d) The reaction between *N*-cyanoamidine and the appropriate chloromethyleniminium salt is a valuable and quite general method (see Section 2.20.4.3.1). (e) There is no general synthetic route to 1,3,5-triazines with two different alkyl or aryl substituents; however, 2-alkyl(or aryl)-4-methyl-1,3,5-triazines may be prepared from the appropriate 4-chloropyrimidine (see Section 2.20.4.7.1).

### 2.20.4.9.5 Synthesis of 2,4,6-trisubstituted 1,3,5-triazines

The routes to symmetrical derivatives are summarized in Table 22.

1,3,5-Triazines containing two different substituents are formed by a variety of methods. The guidelines described above are applicable to these compounds also (see Section 2.20.4.9.4). The condensation of imidates with amidines provides a valuable route to 6-substituted 2,4-dialkyl-1,3,5-triazines (see Section 2.20.4.6.2).

1,3,5-Triazines with three different substituents are formed by variations on the above methods. There is no efficient route to trialkyl or triaryl derivatives, although the reaction of acylimidates and amidines has met with some success (see Section 2.20.4.6.2). In general the best routes available are as follows. (a) Substitution reactions of cyanuric chloride. (b) The condensation of *N*-cyanoamidine with chloromethyliminium salts. (c) 2-Aryl-1,3,5-triazin-2-ones may be efficiently prepared by ring closure reactions on *N*-(α-chloroalkylidene)carbamoyl chloride and amidines (see Section 2.20.4.4.1).

### 2.20.4.9.6 Synthesis of other 1,3,5-triazine derivatives

1,2-Dihydro-1,3,5-triazines are prepared by the reaction of aldehydes or ketones with a variety of aliphatic nitrogen compounds, including biguanides and dicyandiamides.

**Table 22**  Preferred Synthetic Routes to Symmetrical 1,3,5-Triazines

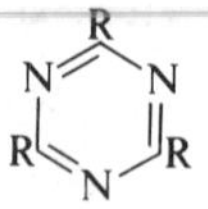

| R | Starting Material | Section |
|---|---|---|
| Alkyl | i, Imidate | 2.20.4.5.1 |
|  | ii, Alkyl nitrile | 2.20.4.5.1 |
| Aryl | Aryl nitrile | 2.20.4.5.1 |
| NH$_2$ | Dicyandiamide | 2.20.4.5.1 |
| Cl | Cyanogen chloride | 2.20.4.5.1 |
| OH | Cyanuric chloride + MeCO$_2$H | 2.20.3.12 |
| Alkoxy | Cyanuric chloride + NaOH + ROH | 2.20.3.12 |
| SH | Cyanuric chloride + Na$_2$S + HCl | 2.20.3.12 |
| SMe | Cyanuric chloride + NaSH + MeI | 2.20.3.12 |
| Aryloxy | i, Cyanuric chloride + NaOH + ArOH | 2.20.3.12 |
|  | ii, Aryl cyanate | 2.20.4.5.1 |

Alternatively, the reaction of isocyanates and enamines provides an efficient route. 1,3,5-Hexahydrotriazines are prepared efficiently by the condensation of aldehydes and ammonia or amines (see Section 2.20.4.6.2). Isocyanates are very valuable for the preparation of 1,3,5-triazin-2-ones, -2,4-diones and -2,4,6-triones (see Section 2.20.4.6.2). The isocyanurates or 1,3,5-triazine-2,4,6-triones may also be synthesized from the corresponding cyanurates.

Fused ring systems are usually formed by reaction of a 2-amino nitrogen heterocycle with an appropriate reagent to generate the 1,3,5-triazine rings.

## 2.20.5  OCCURRENCE OF 1,3,5-TRIAZINES

The only report of 1,3,5-triazines in nature was the detection of cyanuric acid in soil humus ⟨59HC(13)1, p. 24⟩. Hayatsu and coworkers have detected melamine and 4,6-diamino-1,3,5-triazin-2-one in the Orgueil meteorite ⟨64MI22001, 73MI22000⟩. Melamine has also been isolated from the Murchison meteorite ⟨75MI22001⟩. Clearly the origin of these extraterrestrial compounds is unknown, but it would not be surprising if they were formed *via* the trimerization of cyanides.

## 2.20.6  APPLICATIONS

The uses of 1,3,5-triazine and its derivatives will be discussed according to the nature of the substituents. The fibre reactive triazine dyes are discussed in Section 2.20.6.7. It is more convenient to describe the properties of all the triazine herbicides in a separate section.

### 2.20.6.1  Applications of 1,3,5-Triazine

Although 1,3,5-triazine is of limited use in the synthesis of triazine derivatives, it can be used to prepare a variety of other heterocyclic systems, including 4,5-disubstituted pyrimidines containing electron-withdrawing substituents (see Section 2.20.3.1.3). It is becoming of increasing value in the synthesis of dehydro Mannich bases. The preparation of 5-aminomethylenebarbituric acids (**193**) which exhibit activity against lymphoma may prove of particular importance (equation 113) ⟨78AF1684⟩. 1,3,5-Triazine is a good alternative to hydrogen cyanide in the Gattermann reaction.

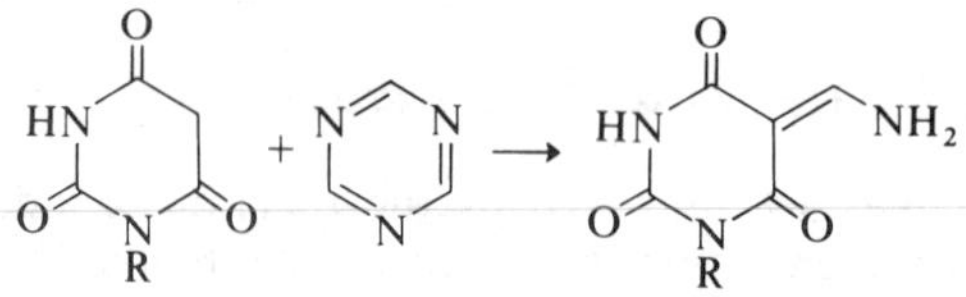

(113)

(**193**)

1,3,5-Triazine is a good solvent for solvent-refined coals. This may prove to be exceptionally important in spectrometric studies on the structures of coals ⟨81MI22000⟩.

## 2.20.6.2 Applications of Aryl-1,3,5-triazines

Aryl-1,3,5-triazines themselves are not very useful, but the 2-hydroxyphenyl derivatives are valuable as UV absorbers and stabilizers for polymers ⟨69MI22000⟩.

## 2.20.6.3 Applications of Amino-1,3,5-triazines

Melamine and its polymers have uses in many industrial fields. The commercial importance of the compounds is shown by the fact that $34 \times 10^6$ kg of melamine was produced in the USA in 1970. The chemistry of the melamine resins has been discussed already (see Chapter 1.11). The major uses of the resins are in the formation of high-pressure laminates for home furniture, and as moldings for crockery. In addition the resins are used in finishing textiles, to improve crease resistance, and as coatings for wet strength paper.

2,4-Diamino-6-diethoxyphosphinyl-1,3,5-triazine (DAPT) (**194**) and the corresponding 6-(3,3,3-tribromopropyl) derivative (**195**) are valuable fire retardants ⟨B-80MI22000⟩. 2,4-Diamino-1,3,5-triazines also exhibit diuretic activity, *i.e.* they increase the rate of urine formation. The 4-chlorophenyl derivative chloroazanil (**196**) is the most important of the triazine diuretics. Although it has never been fully established in clinical use, it was the first compound studied which interfered with the mineralocorticoid-regulated potassium secretion ⟨59MI22000⟩. A variety of 2,4-diamino-1,3,5-triazines containing 6-cycloalkyl or 6-aryl substituents (*e.g.* **197** and **198**) have promise as anti-inflammatory agents ⟨73JMC1305⟩. Menazon (**199**) exhibits important aphicidal activity ⟨61CI(L)630⟩. The 1,3,5-triazine herbicides are discussed separately (see Section 2.20.6.8).

(**194**) R = (EtO)₂PO  (**195**) R = CBr₃CH₂CH₂  
(**196**) R = 4-ClC₆H₄  (**197**) R = cyclohexyl  (**198**) R = 4-FC₆H₄

(**199**) R = (EtO)₂PSCH₂ (with S doubly bonded to P)

The trimethyl-1,3,5-triazinylammonium salts have been used as starting reagents for the synthesis of other 1,3,5-triazines by nucleophilic substitution of the trimethylamine group.

## 2.20.6.4 Applications of Cyanurates and Isocyanurates

Triallyl cyanurate is prepared commercially by the reaction of cyanuric chloride with allyl alcohol. It is used as a minor comonomer with a range of monomers and preformed polymers, imparting heat resistance, solvent resistance, adhesion and strength to the polymers. It is particularly valuable for the preparation of high temperature electrical insulation components. It is also used in curing fluoro polymers such as Viton. Triallyl isocyanurate may be used similarly ⟨B-78MI22000⟩. The use of cyanurates and isocyanurates in polymer chemistry has been reviewed thoroughly ⟨77RCR278⟩. 1,3,5-Trichloroisocyanurate exhibits the reactions typical of the nitrogen–chlorine bond. It has been used as a chlorinating agent in, for example, the preparation of chlorobenzene (equation 114) ⟨70JOC719⟩. The major use of the compound is as a disinfectant in swimming pools ⟨74MI22000⟩.

(114)

### 2.20.6.5 Physiological Importance of 5-Azacytidine and Related Compounds

5-Azacytidine (**200**), the nitrogen analogue of cytidine, exhibits cancerostatic, bacterio-static and mutagenic properties ⟨64E202⟩. It is incorporated into both RNA and DNA, and it disrupts protein synthesis probably through its incorporation into RNA ⟨67BBA(145)771⟩. It has proved particularly effective against myelogeneous leukaemia ⟨73MI22001, 74MI22001⟩, but the full clinical use of the drug has been limited by its facile hydrolysis. The dihydro derivative (**201**) is more stable than the parent compound, and shows potential as an antitumour drug ⟨77JMC806⟩.

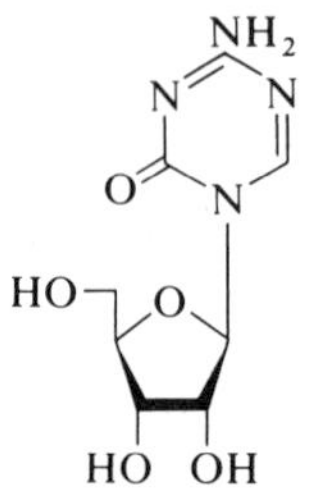

(**200**) 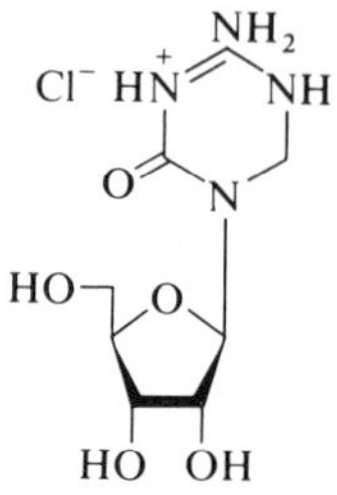                    (**201**)

### 2.20.6.6 Applications of Alkylthio-1,3,5-triazines

2,4-Diamino-6-thio-1,3,5-triazines are potentially valuable as crosslinking agents in polymers and as accelerators in vulcanization processes ⟨70MI22000⟩.

### 2.20.6.7 Applications of Halogenated 1,3,5-Triazines

Cyanuric chloride is the most important of these compounds. The uses of chloro-1,3,5-triazines derived from cyanuric chloride will be discussed within the same section. The uses of tris-*N*-chloroisocyanuric acid were mentioned in Section 2.20.6.4.

#### 2.20.6.7.1 Synthetic applications of cyanuric chloride

Cyanuric chloride is used to synthesize all of the more important 1,3,5-triazine herbicides and many of the fibre-reactive dyes. The syntheses involve successive nucleophilic displace-ments of two or three of the chlorine atoms (see Section 2.20.3.12). The triazine herbicides will be discussed later (see Section 2.20.6.8).

In addition to being the most valuable starting material for the 1,3,5-triazines, cyanuric chloride has been used for the preparation of alkyl halides, acid chlorides, peptides, dicarbodiimides and macrocyclic lactones (see Section 2.20.3.12.2).

#### 2.20.6.7.2 1,3,5-Triazine dyes and optical brighteners

Cyanuric chloride is the precursor of many of the fibre-reactive dyes. It is beyond the scope of this review to discuss the chemistry of the dyes in depth, but there are many excellent reviews on the subject ⟨68MI22000, B-70MI22001, B-72MI22001, 74MI22002, B-77MI22000, B-79MI22002⟩. The fibre-reactive dyes are prepared by nucleophilic displacement of one or

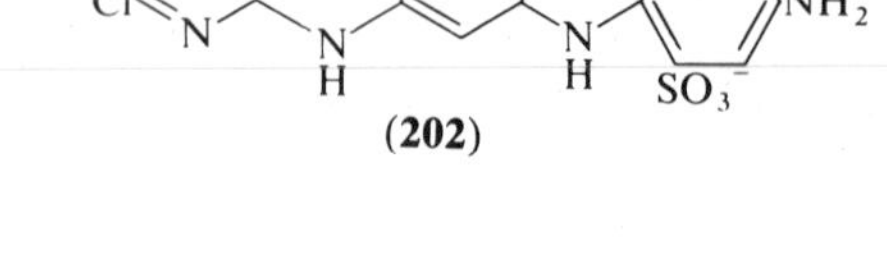

(**202**)                                        (**203**)

two chlorines with a dye or dyes. Then the product is bound on to the textile by displacement of the third chlorine with the hydroxy group in cellulose fibre or with amino functions in polyamides, silk and wool. These dyes were developed initially by ICI in 1956 and CIBA in 1957, and manufactured under the trade names Procion and Cibacron respectively. Cibacron blue 3GA (**202**) and Procion scarlet MXG (**203**) are typical examples of reactive dyes.

Cyanuric chloride is also used for the synthesis of optical brighteners for the whitening of paper, textile finishing and brightening plastics and paints. Derivatives of flavic acid, *e.g.* (**204**), are particularly valuable for brightening cotton and paper whilst coumarin derivatives (**205**) are valuable plastics brighteners ⟨75AG(E)665⟩.

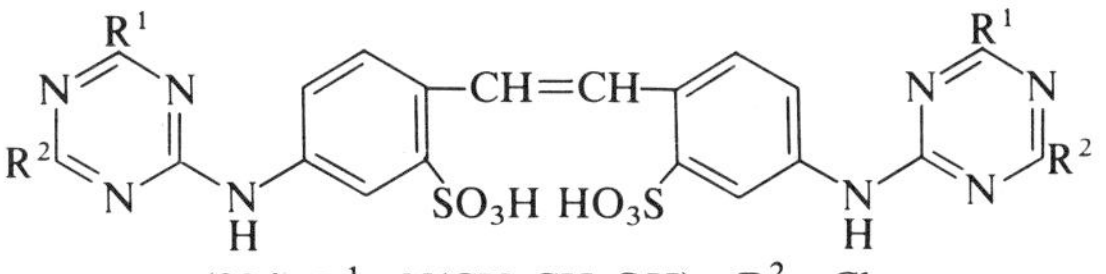

(**204**) $R^1 = N(CH_2CH_2OH)_2$, $R^2 = Cl$ 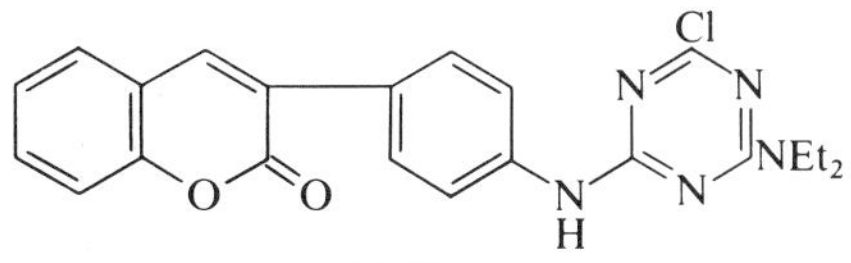 (**205**)

(**206**) $R^1 = N(CH_2CH_2OH)_2$, $R^2 = HN$⟨⟩I, $HO_2C$

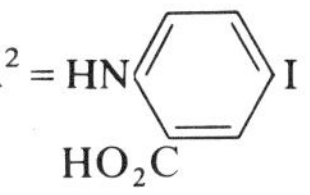

### 2.20.6.7.3 Miscellaneous applications of cyanuric chloride

Madelmont and Veyre have reported the preparation of (**206**) from cyanuric chloride ⟨75MI22002⟩. The compound is valuable in the dynamic exploration of the lymphatic system, when it is prepared using radioactive isotopes of iodine. 2,4-Dichloro-1,3,5-triazines containing a secondary amino group with a vinylic substituent have promise in binding dental resins to biological tissues. The chlorines react with the dentine tooth enamel whilst the vinyl moiety bonds to the resin ⟨75GEP2630745⟩.

The chloro-1,3,5-triazines have potential in chromatography. 2,4-Dichloro-6-methoxy-1,3,5-triazine has been used to prepare affinity chromatography adsorbents (*e.g.* Scheme 120). These compounds are valuable for the purification of enzymes, including chymotrypsin ⟨77JCS(P1)2189⟩. In dextran conjugates with Cibacron blue F3GA (**202**) the dye is responsible for the interaction with enzymes. The role of triazine dyes in protein purification has been reviewed recently ⟨B-79MI22003⟩. The optically active 1,3,5-triazine (**207**) is used as a gas chromatographic packing for the separation of (+) and (−) isomers of amino acids ⟨79JAP(K)7976587⟩.

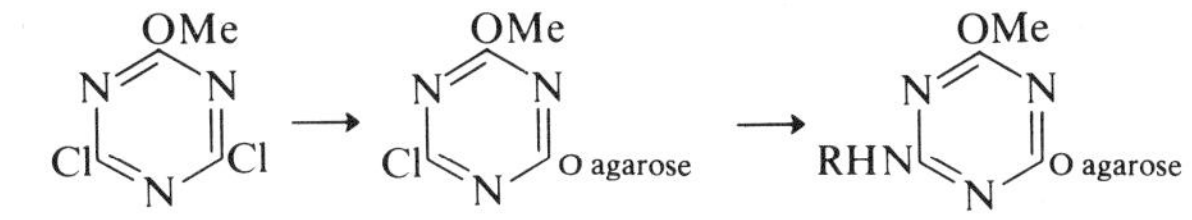

$R = (CH_2)_3NH(CH_2)_3NH_2$

**Scheme 120**

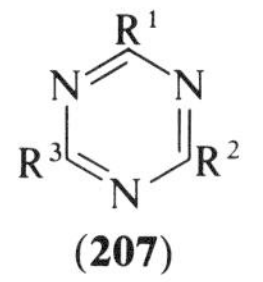

$R^1 = HN(CH_2)_4CH(CONHBu^t)NHCO(CH_2)_{10}Me$;
$R^2 = R^1$, OEt; $R^3 = R^1$, OEt, Cl

(**207**)

### 2.20.6.7.4 Applications of fluoro- and fluoroalkyl-1,3,5-triazines

2,4,6-Perfluoroalkyl-1,3,5-triazines are very stable to oxidation and heat. They have potential as lubricating fluids for use in aerospace work ⟨B-74MI22003⟩. Their polymeric products are stable elastomers with excellent rubber-like properties for use as electrical wire insulation and seals for hydraulic, lubricating and fuel systems of aircraft ⟨B-80MI22001⟩. Perfluoro-2,4,6-tri-*n*-heptyl-1,3,5-triazine is a valuable reference standard for precise mass measurement up to $m/e$ 1600. It does not yield ions which interface with hydrocarbon ions, and it is a liquid ⟨68MI22001⟩.

Cyanuric fluoride has similar properties to cyanuric chloride. Fibre reactive dyes based on cyanuric fluoride are the subject of a great deal of effort, and they may prove very valuable in the future ⟨B-79MI22004⟩.

### 2.20.6.8 1,3,5-Triazine Herbicides

A wide range of 1,3,5-triazines exhibit selective herbicidal properties. The structures of some of the more important compounds are listed in Table 23. Simazine and atrazine are the most important of the triazine herbicides. The chemistry, mode of action and degradation of these compounds have been reviewed thoroughly ⟨B-60MI22000, B-70MI22002, B-75MI22003, B-76MI22001, 81MI22001⟩. The following discussion merely summarizes the more important aspects of these reviews.

**Table 23**   1,3,5-Triazine Herbicides

| $R^2$ | $R^4$ | $R^6$ | Name | Co. |
|---|---|---|---|---|
| Cl | NHEt | NHEt | Simazine | Ciba |
| Cl | NHEt | NHPr$^i$ | Atrazine | Ciba |
| Cl | NHEt | NHBu$^s$ | Sebuthylazine | Ciba |
| Cl | NHEt | NHBu$^t$ | Terbuthylazine | Ciba |
| Cl | NHEt | NHEt$_2$ | Trietazine | Ciba |
| Cl | NHEt | NHCMe$_2$CN | Cyanazine | Shell |
| Cl | NHPr$^i$ | NEt$_2$ | Ipazine | Ciba |
| Cl | NHPr$^i$ | NH(CH$_2$)$_3$OMe | Metoprotazine | Ciba |
| Cl | NHPr$^i$ | HN◁ | Cyprazine | Gulf |
| Cl | NEt$_2$ | NEt$_2$ | Chlorazine | Ciba |
| Cl | NPr$^i_2$ | NPr$^i_2$ | Siprazine | Ciba |
| MeO | NHEt | NHEt | Simeton | Ciba |
| MeO | NHEt | NHBu$^s$ | Secbumeton | Ciba |
| MeO | NHEt | NHBu$^t$ | Terbumeton | Ciba |
| MeO | NHEt | NHCHMe$_2$ | Atraton | Ciba |
| MeO | NHPr$^i$ | NHPr$^i$ | Prometon | Ciba |
| MeO | NH(CH$_2$)$_3$OMe | NH(CH$_2$)$_3$OMe | Methometon | Ciba |
| MeS | NHMe | NHPr$^i$ | Desmetryne | Ciba |
| MeS | NHEt | NHEt | Simetryne | Ciba |
| MeS | NHEt | NHPr$^i$ | Ametryne | Ciba |
| MeS | NHEt | NHPr$^i$ | Terbutryne | Ciba |
| MeS | NHPr$^i$ | NHPr$^i$ | Prometryne | Ciba |
| MeS | NHPr$^i$ | NH(CH$_2$)$_3$OMe | Methoprotryne | Ciba |
| MeS | NHPr$^i$ | N$_3$ | Aziprotryne | Ciba |
| EtS | NHPr$^i$ | NHPr$^i$ | Dipropetryn | Ciba |
| MeS | NH(CH$_2$)$_3$OMe | NH(CH$_2$)$_3$OMe | Lambast | Ciba |

The triazine herbicides cause striking physiological effects on plants. They act as inhibitors of photosynthesis by interrupting the light-driven flow of electrons from water to NADP. The compounds drastically inhibit the Hill reaction, *i.e.* the evolution of oxygen from water in the presence of chloroplasts and a suitable electron acceptor. The site of action of the inhibitor molecule seems to be at the water-splitting of the photosystem. The triazine herbicides can also affect the nucleic acid metabolism, and even act as plant-growth regulators.

The selectivity of the 1,3,5-triazines towards plants is primarily determined by the pathway and the rate of detoxication of these compounds in a given plant. Non-enzymatic hydrolysis of chloro-1,3,5-triazines is catalyzed by cyclic hydroxamates such as 2,4-dihydroxy-7-methoxy-1,4-benzooxazin-3(2$H$)-one, but the degree of tolerance to the triazines does not coincide with the content of these compounds. The chloro-1,3,5-triazines can be completely detoxified by conjugation with glutathione, a reaction catalyzed by the enzyme glutathione-*S*-transferase. Unlike the benzoxazinone system, the enzyme is active only in resistant plants such as sorghum, maize and sugar cane. In addition, oxidative degradation reactions such as *N*-dealkylation play a significant role in 1,3,5-triazine selectivity. These reactions are particularly important in plants which are partially susceptible to the herbicides, for example atrazine is degraded to 2-amino-6-chloro-4-isopropylamino-1,3,5-triazine in pea plants. The dealkylation reactions contribute appreciably to selectivity in resistant plants such as maize and sorghum.

### 2.20.6.9 Applications of 1,2-Dihydro-1,3,5-triazines

4,6-Diamino-2,2-dimethyl-1,2-dihydro-1,3,5-triazine derivatives (Baker triazines) are becoming of increasing importance as pharmaceuticals. Many of these compounds are inhibitors of dihydrofolate reductase, and some have shown activity against leukemia. Currently the Baker triazine 'antifol' (**208**) is undergoing clinical trials as a drug in cancer chemotherapy ⟨73JMC209, 80JMC1248⟩. Kim *et al.* have shown that the activity of the 1,2-dihydro-1,3,5-triazines is interpretable using quantitative structure–activity relationships (QSAR) ⟨80JMC1248⟩. The presence of 2-substituents (at * in **208**) lowers activity. Hopfinger has predicted that (**209**) will be highly active from QSAR studies ⟨80JA7196⟩.

4,6-Diamino-1-(4-chlorophenyl)-2,2-dimethyl-1,2-dihydro-1,3,5-triazine is the active antimalarial metabolite of the British antimalarial Paludrine (or proguanil) (**210**) ⟨B-61MI22000, p. 717⟩.

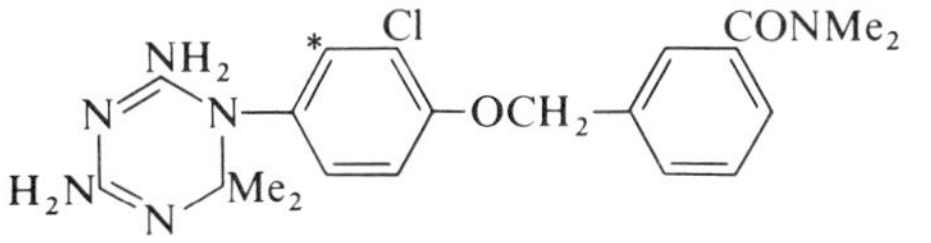

(**208**) 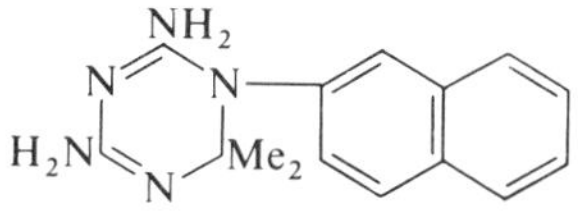        (**209**) 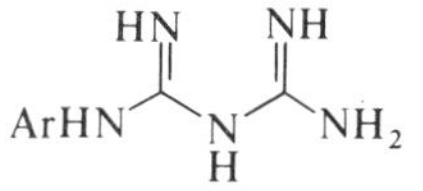        (**210**) Ar = 4-ClC₆H₄

### 2.20.6.10 Applications of Organometallic 1,3,5-Triazines

Melarsen (**211**) is active against both early and late stages of African trypanosomiasis. The antimony derivative (**212**) is used similarly ⟨B-61MI22000, p. 715⟩. 2,4,6-Tripyridyl-1,3,5-triazine forms a purple bis(triazinyl) complex with iron(III) ions which provides a spectrophotometric method for estimating iron in wines ⟨59MI22001⟩.

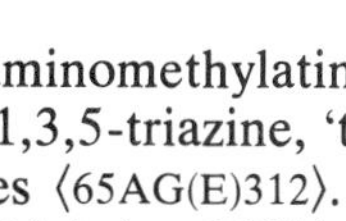

(**211**)  Ar = 4-H₂O₃AsC₆H₄
(**212**)  Ar = 4-H₂O₃SbC₆H₄

### 2.20.6.11 Applications of Hexahydro-1,3,5-triazine and Hexamethylenetetramine

These compounds are valuable as aminomethylating reagents (see Section 2.20.3.11). In addition, 1,3,5-triacryloylhexahydro-1,3,5-triazine, 'triacryloformal' (**213**), forms the basis of the Basazol range of reactive dyes ⟨65AG(E)312⟩. The schematic reaction is shown in Scheme 118. The colour yield is very high (*ca.* 90%).

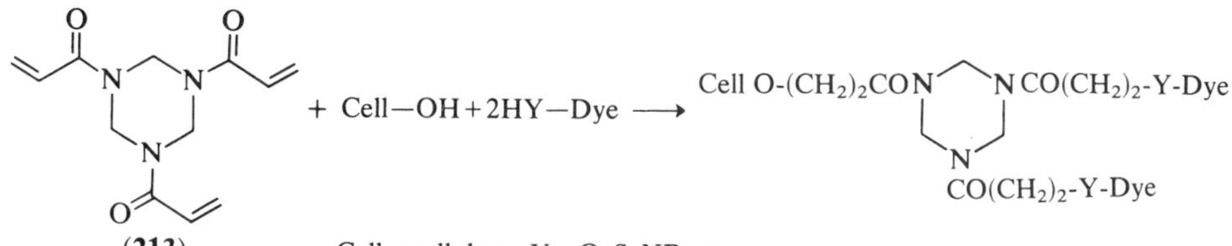

(**213**)        Cell = cellulose; Y = O, S, NR, *etc.*

**Scheme 121**

Hexamethylenetetramine (**70**) has been widely used as a urinary antiseptic under the names Methenamine or Urotropin ⟨B-61MI22000, p. 712⟩. The synthetic applications have already been discussed (see Section 2.20.3.14).

1,3,5-Trinitro-hexahydro-1,3,5-triazine (**214**), which is better known as RDX or hexogen, is an important explosive. It is less toxic than TNT and it may be handled without physiological effects if appropriate precautions are taken to ensure cleanliness of operations. On detonation, mainly gaseous products of low molecular weight are formed, without the intermediate formation of solids. RDX is stable even at relatively high temperatures; it has been stored at 85 °C for ten months without appreciable deterioration. In the USA it is mainly used in 'Composition B', a mixture of RDX, TNT and wax. The properties of RDX have been reviewed recently ⟨B-80MI22002⟩.

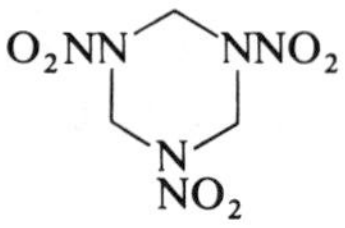

$$O_2NN \qquad NNO_2$$

$$\underset{NO_2}{N}$$

(**214**)

# 2.21

# Tetrazines and Pentazines

H. NEUNHOEFFER

*Technische Hochschule Darmstadt*

| | |
|---|---|
| 2.21.1  1,2,3,4-TETRAZINES | 531 |
| *2.21.1.1  Structure and Spectra* | 531 |
| *2.21.1.2  Reactions* | 533 |
| *2.21.1.3  Syntheses* | 534 |
| 2.21.2  1,2,3,5-TETRAZINES | 535 |
| 2.21.3  1,2,4,5-TETRAZINES | 536 |
| *2.21.3.1  Introduction* | 536 |
| *2.21.3.2  Structure and Spectra* | 537 |
| *2.21.3.3  Reactions* | 543 |
| *2.21.3.4  Syntheses* | 555 |
| *2.21.3.4.1  Synthesis by the [6 + 0] atom fragment method* | 556 |
| *2.21.3.4.2  Synthesis by the [5 + 1] atom fragment method* | 557 |
| *2.21.3.4.3  Synthesis by the [4 + 2] atom fragment method* | 560 |
| *2.21.3.4.4  Synthesis by the [3 + 3] atom fragment method* | 563 |
| *2.21.3.4.5  Synthesis by the [2 + 1 + 2 + 1] atom fragment method* | 569 |
| *2.21.3.4.6  Synthesis by the [3 + 2 + 1] atom fragment method* | 571 |
| *2.21.3.5  Applications and Important Uses* | 572 |
| 2.21.4  PENTAZINES | 572 |

## 2.21.1  1,2,3,4-TETRAZINES

### 2.21.1.1  Structure and Spectra

The 1,2,3,4-tetrazines have structure (**1**) and are numbered as indicated. Besides the name 1,2,3,4-tetrazine, *v*-tetrazine and osotetrazine can also be found in the literature. Two Kekulé structures can be drawn for 1,2,3,4-tetrazine (**1a** and **1b**), but one may assume that structure (**1a**) gives a higher contribution to the ground state than (**1b**) since in (**1a**) only one N=N double bond is present.

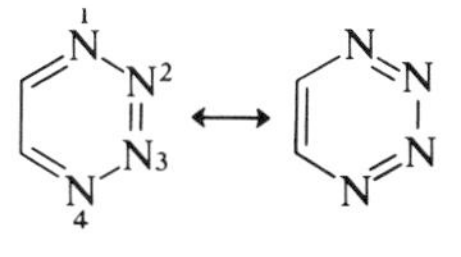

(**1a**)     (**1b**)

At the time of writing, no aromatic 1,2,3,4-tetrazine has been prepared. In the older literature a number of publications can be found describing the synthesis of 2,3-dihydro-1,2,3,4-tetrazines (**2**), but, as was shown later, none of these compounds had the proposed structure. The older literature was reviewed by Wystrach in 1956 ⟨56HC(10)138⟩ and in 1967 ⟨B-67MI22100⟩, and finally in 1978 by Wiley ⟨78HC(33)1287⟩ who covered the whole literature through *Chemical Abstracts* 1974. In this last review the syntheses of the so-called 2,3-dihydro-1,2,3,4-tetrazines (**2**) and the determination of their correct structures were covered, and therefore we do not discuss this problem here. The first compound having four adjacent nitrogen atoms in a six-membered ring was prepared in 1972 by Nelsen and

Fibinger ⟨72JA8497⟩. They isolated 1,4-dimethyl-1,4,5,6-tetrahydro-1,2,3,4-tetrazine (**4**) in less than 5% yield from the oxidation of $N^1,N^{1'}$-dimethyl-1,2-dihydrazinoethane (**3**) with sodium hypochlorite in the presence of sodium hydroxide.

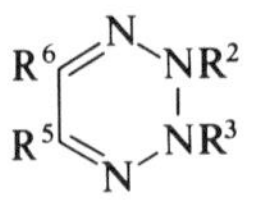

(2) 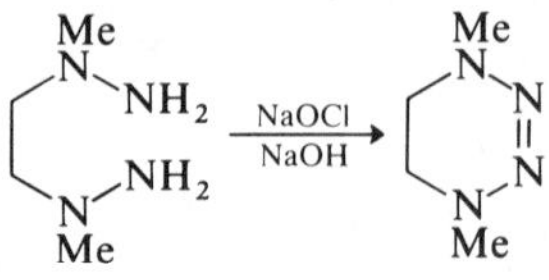                    (3)                    (4)

In the following years further 1,4,5,6-tetrahydro-1,2,3,4-tetrazines, their 2-oxides (**5**) and 2,3,4,5-tetrahydro-1,2,3,4-tetrazines (**7**) have been prepared.

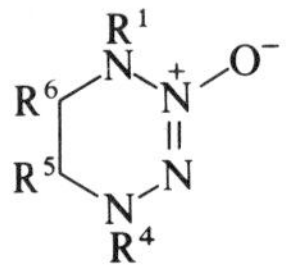

(5) 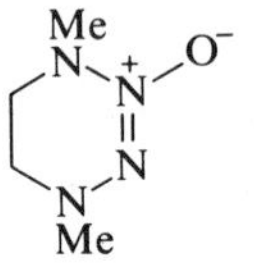                    (6) 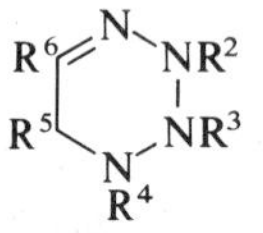                    (7)

Only about 20 compounds with the tetrahydro-1,2,3,4-tetrazine structure are known, but two X-ray crystallographic analyses have been reported. The structure of 1,4-dimethyl-1,4,5,6-tetrahydro-1,2,3,4-tetrazine 2-oxide (**6**) is a twisted boat conformation with non-planar terminal nitrogen atoms, which reflects the repulsion in the four-atom six-electron N—N=N—N system. The bond distances and angles are given in Figure 1 ⟨78HCA1622⟩. A similar study of dimethyl 2-phenyl-2,3,4,4a,5,6,7,8-octahydrobenzo-1,2,3,4-tetrazine-3,4-dicarboxylate (**8**), an example of a 2,3,4,5-tetrahydro derivative, shows a distorted half-chair conformation. The bond distances and angles for the heterocyclic ring of this compound are given in Figure 1 ⟨79AG757⟩.

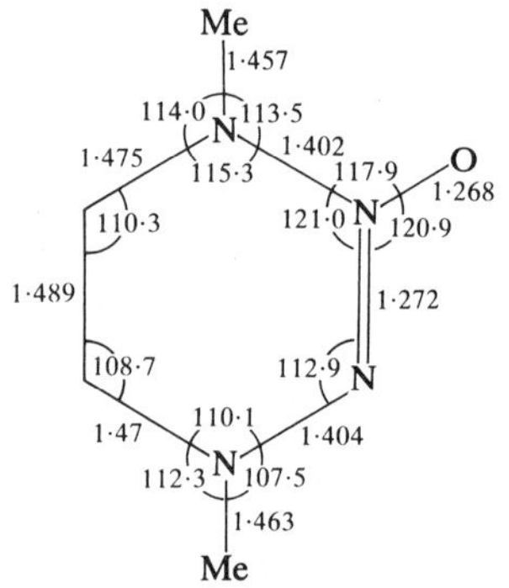

(6) 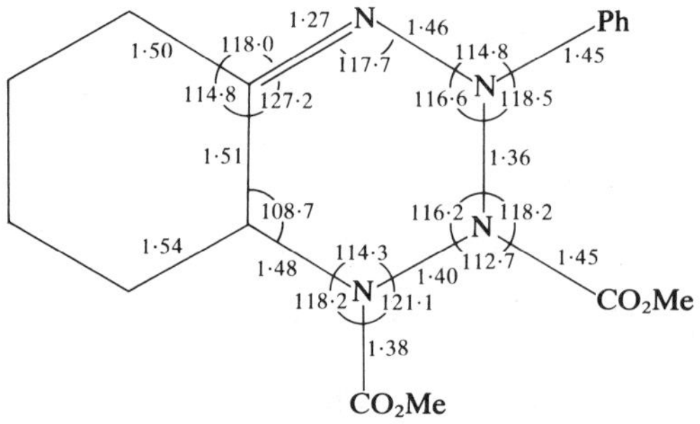                    (8)

**Figure 1**

UV, $^1$H NMR and mass spectra of most of the known tetrahydro-1,2,3,4-tetrazines have been published.

1,4-Dimethyl-1,4,5,6-tetrahydro-1,2,3,4-tetrazine (**4**) has two absorption bands in the UV region at 268 ($\varepsilon = 5580$) and 225 nm ($\varepsilon = 2970$). In the $^1$H NMR spectrum of (**4**) in CDCl$_3$ two signals in the ratio 2:3 were observed at $\delta = 3.12$ and 3.04. The mass spectrum of (**4**) shows the molecular ion peak, and fragments at $m/e$ 71, 43 and 42 ⟨72JA8497⟩. Similar spectra have been obtained for other 1,4,5,6-tetrahydro-1,2,3,4-tetrazines ⟨73CL51, 73CB3097, 73AG504, 78HCA1622⟩.

The photoelectron spectrum of (**4**) has four bands at 8.03, 9.0, 10.20 and 11.37 eV. These were assigned on the basis of semiempirical MO calculations ⟨75T1415⟩.

IR spectral details have been published for compounds (**4**) and (**6**) ⟨78HCA1622⟩.

The *N*-oxide (**6**) shows a single maximum at 268 nm (log $\varepsilon$ 4.01) (methanol) in the UV spectrum. The NMR spectra (CDCl$_3$) are as follows. $^1$H: $\delta = 2.92$ (s, 3H), 3.08 (s, 3H), 3.05–3.60 (m, 4H); $^{13}$C: $\delta = 38.98$, 44.00, 49.19 and 52.69 p.p.m. ⟨78HCA1622⟩.

The $pK_B$ values determined for (**4**) and (**6**) are (**4**): $pK_{B1} = 8.50$, $pK_{B2} = 8.75$; and (**6**): $pK_{B1} = 11.3$ and $pK_{B2} = 12.9$ ⟨78HCA1622⟩.

The following spectra were recorded for 2,6-dimethyl-3,4-dibenzoyl-2,3,4,5-tetrahydro-1,2,3,4-tetrazine (9): $^1$H NMR (CDCl$_3$): $\delta = 1.90$ (s, 3H), 2.79 (s, 3H), 4.20 and 4.80 (AB system, 2H); $^{13}$C NMR (CDCl$_3$): $\delta = 22.0$, 42.0, 43.9, 152.9, 172.1 and 173.6; IR (KBr): 1637, 1665 and 1685 cm$^{-1}$ ⟨79AG757⟩.

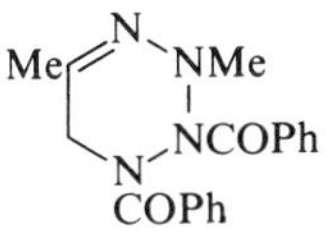

(9) 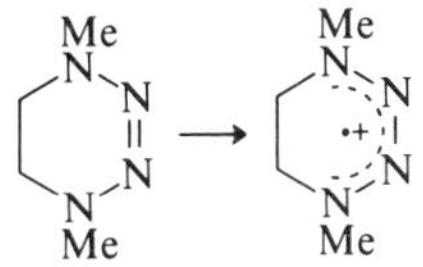         (4)    (10)

On electrolytic oxidation of compound (4) the radical cation (10) was formed, which showed the following ESR hyperfine splitting constants: 17.5 G for the methylene hydrogen atoms, 10.5 G for the methyl hydrogen atoms and 3.17 G for the N(2)—N(3) nitrogen atoms ⟨72JA8497⟩. A large number of theoretical calculations on the 1,2,3,4-tetrazine system have been published. Most of them can be found in the 1978 review ⟨78HC(33)1287⟩. Further references are: ⟨72JSP(43)477, 74TL253, 74JCS(P2)420, 74RRC859, 72MI22100, 76SA(A)345, 75PAC(44)767 and 79MI22102⟩.

## 2.21.1.2 Reactions

At the time of writing, only a few reactions of the 1,4,5,6-tetrahydro-1,2,3,4-tetrazines (11) and their 2-oxides (5) have been reported. The 2-oxides (5) can be reduced by trimethyl phosphite or lithium aluminum hydride to give the deoxygenated compounds (11) ⟨73AG504,

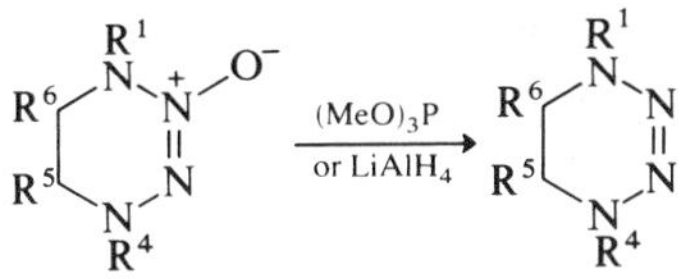

(5)       (11)

74GEP2327545, 78HCA1622⟩. Thermolysis and photolysis of 1,4-dimethyl-1,4,5,6-tetrahydro-1,2,3,4-tetrazine (4) affords 1,3,5-trimethylhexahydro-1,3,5-triazine (12) ⟨72JA8497, 78HCA1622⟩. Oxidation of (4) with potassium permanganate affords 1,4-dimethyl-1,2,3,4-tetrazine-5,6-dione (13) in low yield. Treatment of (4) with chromium pentacarbonyl–THF complex gave 1,3-dimethylimidazolin-2-one (14; Scheme 1). Compound (4) forms complexes with silver nitrate and mercury(II) oxide ⟨78HCA1622⟩.

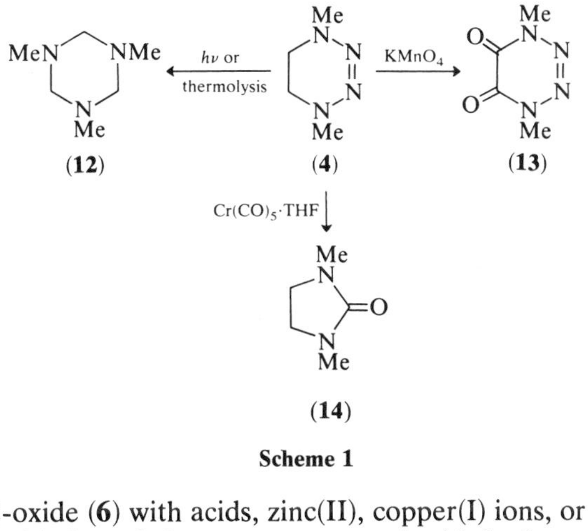

(12)     (4)     (13)

(14)

**Scheme 1**

Treatment of the 2-oxide (6) with acids, zinc(II), copper(I) ions, or methyl iodide affords the tetrazene (15). Oxidation of the same compound with hydrogen peroxide led to the bis(nitrosamine) (16); manganese dioxide oxidizes (6) to 1,4-dimethyl-1,6-dihydro-1,2,3,4-tetrazin-5(4H)-one 2-oxide (17), while on oxidation with potassium permanganate a compound was isolated which was formulated as 1,4-dimethyltetrazol-5-one 2-oxide (18; Scheme 2) ⟨78HCA1622⟩.

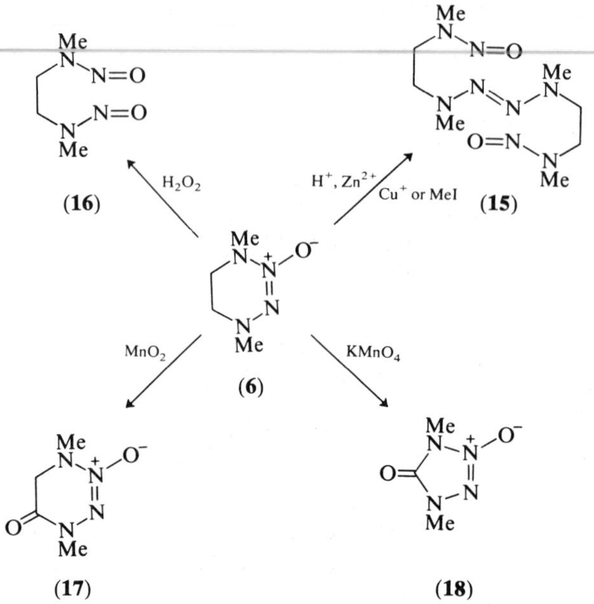

**Scheme 2**

### 2.21.1.3 Syntheses

So far there are only four known methods for preparing tetrahydro-1,2,3,4-tetrazines. The method which gives the best yields of the 1,4,5,6-tetrahydro compounds is that from α-lithiated *N*-alkylnitrosamines (**19**). These compounds decompose at −73 °C to yield the 1,4,5,6-tetrahydrotetrazine 2-oxides (**5**) which can be reduced by trimethyl phosphite or lithium aluminum hydride to the 1,4,5,6-tetrahydro-1,2,3,4-tetrazines (**11**) ⟨73AG504, 74GEP2327545, 78HCA1622⟩.

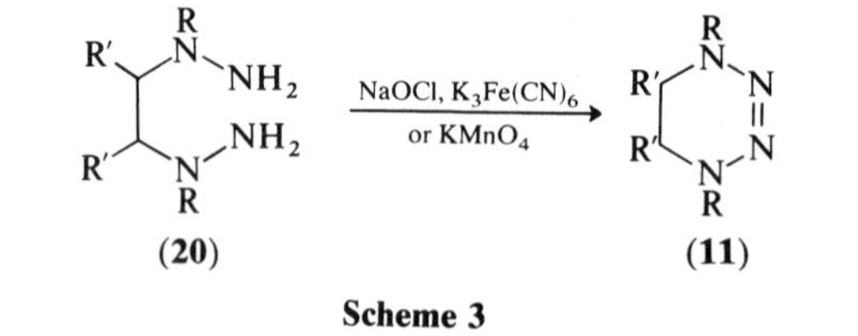

**Scheme 3**

Compounds (**11**) were also prepared in low yield by oxidation of the bishydrazine (**20**) by sodium hypochlorite ⟨72JA8497⟩, potassium ferricyanide or potassium permanganate (Scheme 3) ⟨73CB3097⟩. The 1,4-dibenzyl-5,6-diphenyl compound (**22**) was prepared from 1,1,4,4-tetrabenzyltetrazene (**21**) by oxidative cyclization using thionyl chloride ⟨73CL51⟩.

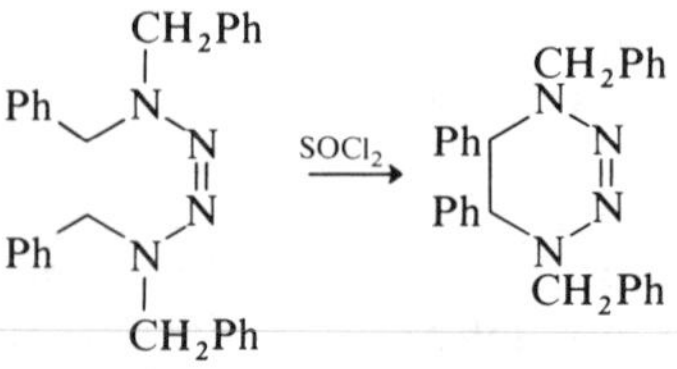

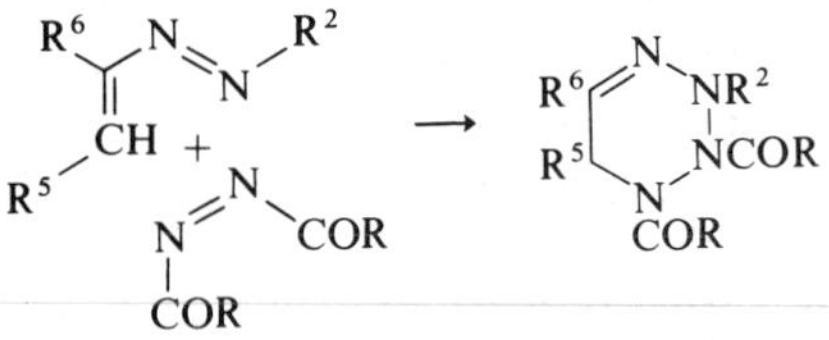

2,3,4,5-Tetrahydro-1,2,3,4-tetrazines (**23**) are made by cycloaddition of electron-deficient azo compounds to vinylazo compounds ⟨79AG757⟩.

## 2.21.2 1,2,3,5-TETRAZINES

Of the three possible tetrazine systems the 1,2,3,5-tetrazines (**24**), also called *as-*tetrazines, are by far the least studied class. No compound with this structure was mentioned in the last review ⟨78HC(33)1296⟩, which covered the literature through *Chemical Abstracts* 1974. The parent compound has structure (**24**) and is numbered as indicated.

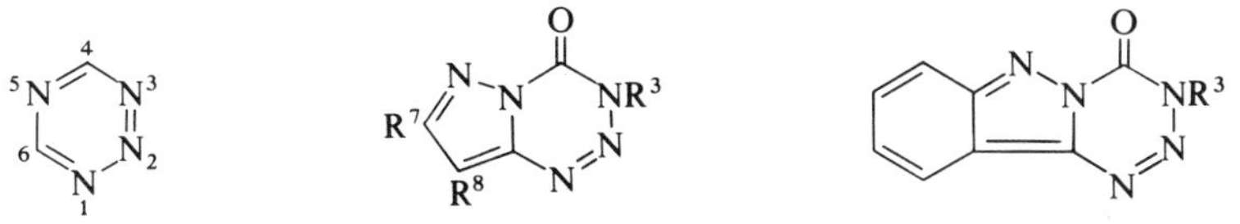

(**24**)          (**25**)          (**26**)

Since 1974 two publications have appeared dealing with the preparation of monocyclic 1,2,3,5-tetrazines ⟨76BCJ1339, 79LA870⟩. Beside these two articles, Ege and his group have published the synthesis of condensed 1,2,3,5-tetrazines such as pyrazolo[5,1-*d* ]-[1,2,3,5]tetrazin-4-ones (**25**) and 1,2,3,5-tetrazino[5,4-*b* ]indazol-4-ones (**26**) ⟨79TL4253⟩.

When cyanamide was electrochemically oxidized at a platinum anode in 2M potassium hydroxide, and the anolyte was neutralized with hydrochloric acid or carbon dioxide, several products were isolated, and to three of these were assigned 1,2,3,5-tetrazine structures. These were the 4-amino-6-cyanamino- (**27**), 4-amino-6-cyanoguanidino- (**28**) and 4-ureido-6-cyanoguanidino-1,2,3,5-tetrazines (**29**) ⟨76BCJ1339⟩. The structures were consistent with the IR spectral data and elemental analyses, but it would be desirable to have unambiguous evidence on these compounds, particularly as their method of synthesis is not an obvious one for forming the connection of three adjacent nitrogen atoms.

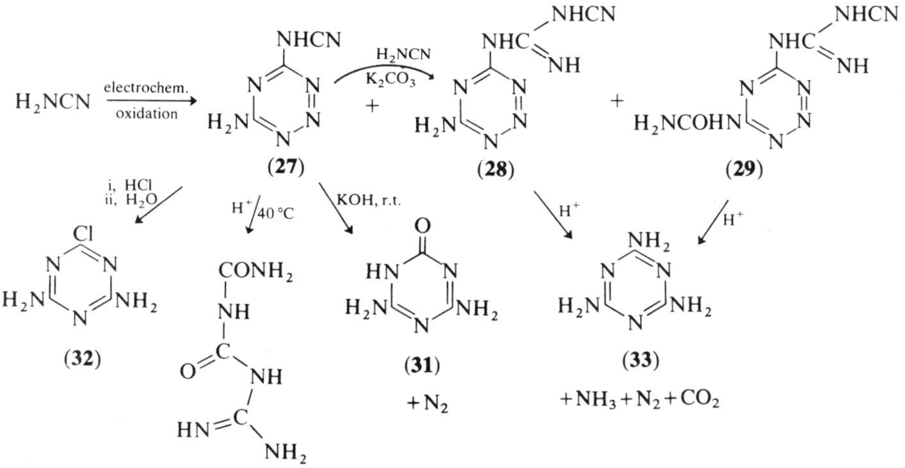

**Scheme 4**

Treatment of (**27**) with cyanamide in aqueous potassium carbonate affords (**28**). Amidinobiuret (**30**) and nitrogen were formed when (**27**) was warmed at 40 °C in dilute sulfuric or hydrochloric acid. After 24 h in aqueous potassium hydroxide, (**27**) was transformed into ammeline (2,4-diamino-1,3,5-triazin-6-one; **31**) and nitrogen.

When hydrogen chloride was introduced into a solution of the potassium salt of (**27**) in dry acetone, and the resulting precipitate treated with water, 2,4-diamino-6-chloro-1,3,5-triazine (**32**) was isolated. Compound (**28**) decomposed to melamine (**33**), ammonia, nitrogen and carbon dioxide when heated in dilute mineral acid, and the same compounds were obtained from (**29**) in the same way (Scheme 4).

3-Substituted 3-chlorodiazirines (**34**) react with bis(trimethylsilyl)mercury (**35**) under irradiation to give 2-trimethylsilyl-1,2,3,5-tetrazinyl radicals (**36**), most likely *via* a diazirinyl radical. The tetrazinyl radicals (**36**) are in equilibrium with the corresponding dimers and are thermolabile. They decompose on attempted isolation. The analogous 2-trimethylgermyl radical (**37**) was prepared similarly ⟨79LA870⟩.

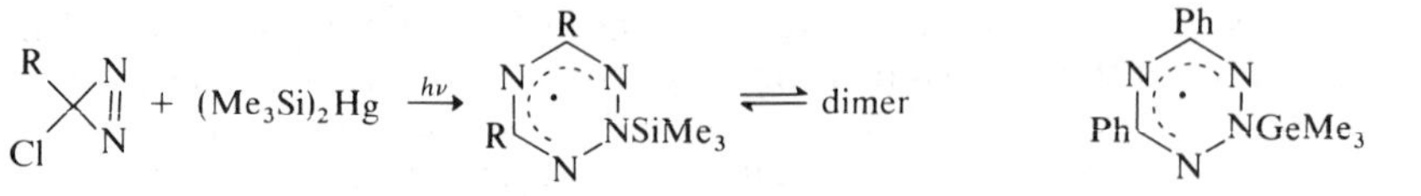

Although only a few 1,2,3,5-tetrazines have been prepared there are many published theoretical calculations on this system. The references up to 1974 are collected in the 1978 review ⟨78HC(33)1296⟩; further calculations appeared in the following articles: ⟨74TL253, 74JCS(P2)420, 74MI22103, 76SA(A)345, 75PAC(44)767⟩.

### 2.21.3  1,2,4,5-TETRAZINES

#### 2.21.3.1  Introduction

Of the three possible tetrazine systems the 1,2,4,5-tetrazines, also called *s*-tetrazines or *sym*-tetrazines, are by far the best known class. The parent compound of the 1,2,4,5-tetrazine series has structure (**38**) and is numbered as indicated. It was prepared for the first time by Hantzsch and Lehmann in 1900 ⟨00CB3668⟩. The two Kekulé structures for (**38**) are degenerate, as was confirmed by X-ray crystallographic analysis.

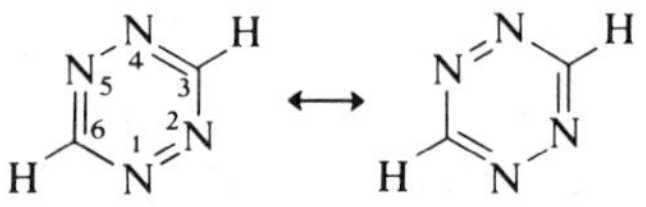

There are many methods known for the synthesis of 1,2,4,5-tetrazines (**39**) but most afford the desired compounds only in low yield. There has been great interest in these compounds by physical organic chemists on account of their physical and spectroscopic properties, while preparative organic chemists have been interested in their high reactivity as dienes in cycloaddition reactions.

A number of excellent reviews on the 1,2,4,5-tetrazines (**39**) have been published. The first, by Wiley in 1956 ⟨56HC(10)179⟩, covered the literature through *Chemical Abstracts* 1950; a second, by Wystrach, appeared eleven years later ⟨B-67MI22101⟩, and finally in 1978 Wiley comprehensively reviewed the whole literature through *Chemical Abstracts* 1978 ⟨78HC(33)1075⟩. Since that time about 350 publications dealing with physical, chemical or biochemical aspects of the 1,2,4,5-tetrazines have been covered by *Chemical Abstracts*.

In addition to the aromatic 1,2,4,5-tetrazines (**39**), dihydro-1,2,4,5-tetrazines, formulated as 1,2- (**40**), 1,4- (**41**), 1,6- (**42**) or 3,6-dihydro-1,2,4,5-tetrazines (**43**), 1,2,3,4-tetrahydro-1,2,4,5-tetrazines (**44**) and hexahydro-1,2,4,5-tetrazines (**45**) are known. The 1,4-dihydro-1,2,4,5-tetrazine-3,6(2*H*,5*H*)-dione (**46**) or its tautomer, 3,6-dihydroxy-1,4-dihydro-1,2,4,5-tetrazine (**47**), is called *p*-urazine. Many preparations of this compound have been claimed but all substances isolated had another structure. Finally Neugebauer ⟨82LA387⟩ was able to isolate a compound with the correct structure.

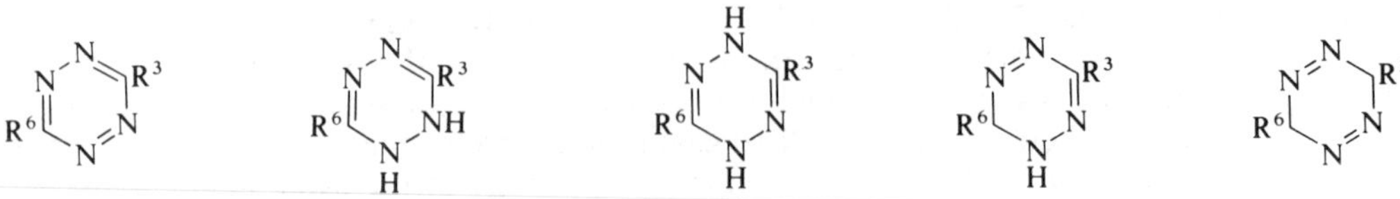

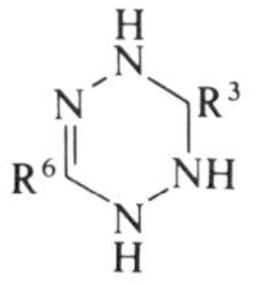

(44) 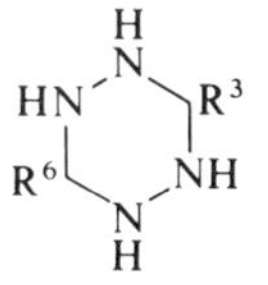　　　　　　(45) 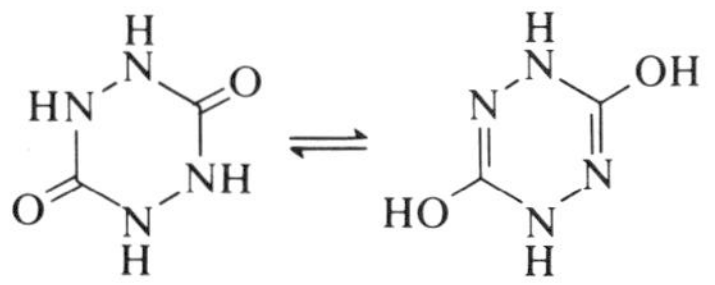　　　　　　(46)　⇌　(47)

Another class of 1,2,4,5-tetrazine compounds which is well known and intensively studied is the stable radicals verdazyls (**48**), also called 1,2,4,5-tetrazin-1(2*H*)-yls. Benzo derivatives of the 1,2,4,5-tetrazine system are not possible since in all cases the second ring has to be condensed to the 1,2,4,5-tetrazine ring either by two nitrogen atoms as in (**49**), or by one carbon and one nitrogen atom as in (**50**).

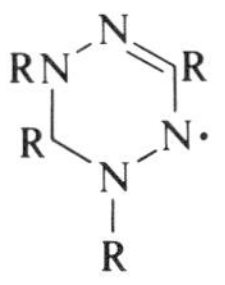

(48) 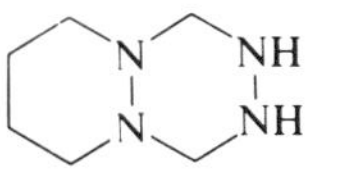　　　　　　(49) 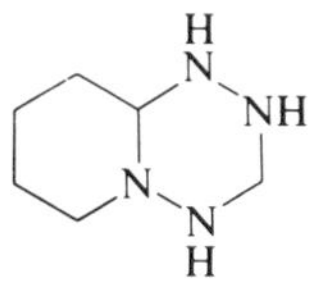　　　　　　(50)

## 2.21.3.2 Structure and Spectra

The number of known 1,2,4,5-tetrazines, their hydro derivatives and of verdazyls is tremendous. Extensive studies on the structure, and especially on the physical properties, of these compounds have been made.

The structure of the parent 1,2,4,5-tetrazine (**38**) was determined by Bertinolli *et al.* ⟨56AX510⟩ by X-ray crystallographic analysis. The bond distances found were 1.334 Å for the C—N bonds and 1.321 Å for the N—N bonds. The angles at the nitrogen atoms were found to be 115° 57′ and those at the carbon atoms are 127° 22′ (Figure 2). The distances between the nearest atoms belonging to different molecules in the lattice range between 3.27 and 3.67 Å for CH···N contacts, between 3.41 and 3.83 Å for N···N contacts and between 3.74 and 3.79 Å for CH···CH contacts.

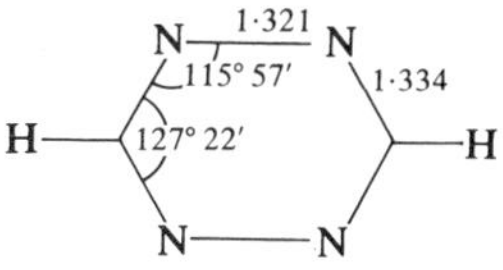

(38) 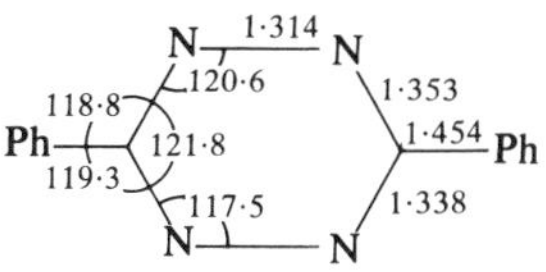　　　　　　(51)

**Figure 2**

The X-ray crystallographic analysis of 3,6-diphenyl-1,2,4,5-tetrazine (**51**) was published in 1972 ⟨72AX(B)739⟩. The observed bond distances and angles are in the same range as were found for the parent compound (**38**; Figure 2). In both cases the heterocyclic ring is planar as expected for a molecule with some degree of electron delocalization.

Dihydro-1,2,4,5-tetrazines have been formulated as 1,2-dihydro- (**40**), 1,4-dihydro- (**41**), 1,6-dihydro- (**42**) and 3,6-dihydro-1,2,4,5-tetrazines (**43**). It seems to us that there is still a great confusion over the structure of these substances, especially the 1,2-dihydro- and 1,4-dihydro-1,2,4,5-tetrazines which are mixed up, and the same compound is often formulated as both structures. In most cases the dihydro structure is presented, which would be the first reaction product, or authors have formulated their compounds in the dihydro structure which seemed to be the most accepted at that time. An X-ray crystallographic analysis of the parent dihydro-1,2,4,5-tetrazine has shown that this compound has the 1,4-dihydro structure (**41**; R = H), at least in the solid state ⟨83CB2261⟩, and the same applies to the 3,6-di(2-pyridyl)dihydro-1,2,4,5-tetrazine (**52**) ⟨76AX(B)1467⟩. For the tetrazine ring in (**52**) a boat-like conformation was found, with the two pyridine rings each inclined by 20° and the pyridine N-atoms in the *anti* conformation. Torsion between adjacent tetrazine rings is minimized by two intramolecular N—H···N hydrogen bonds. The bond distances and angles are given in Figure 3. It should be mentioned that the structure found in the crystal is not necessarily the same as in solution, but Huisgen and Chae have suggested the 1,4-dihydro structure for 3,6-diphenyldihydro-1,2,4,5-tetrazine, also on the basis of spectroscopic studies in solution ⟨65TH22100⟩. No X-ray crystallographic analysis of a tetrahydro-1,2,4,5-tetrazine (**44**) has to our knowledge been published so far.

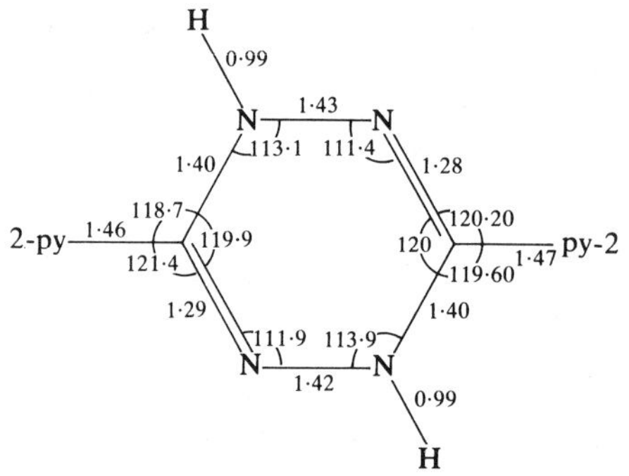

**(52)**

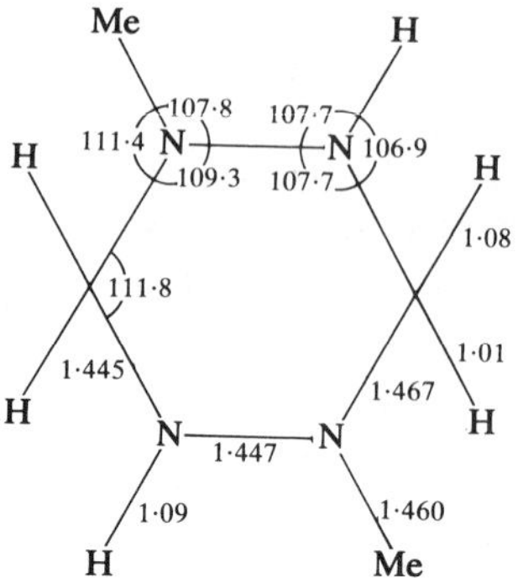

**(53)**

**Figure 3**

The structure of hexahydro-1,2,4,5-tetrazines has been studied both by X-ray crystallography and by NMR and photoelectron spectroscopy. The results obtained by the different methods agree very well. All monocyclic hexahydro-1,2,4,5-tetrazines are chair-shaped in the crystal and in solution.

X-ray crystallographic analysis of 1,4-dimethylhexahydro-1,2,4,5-tetrazine (**53**) has shown that the molecule is centrosymmetric and chair-shaped; the two *N*-methyl groups are equatorial and the two *N*-hydrogens are axial. Hydrogen bonding of the two axial hydrogens to the nearest methyl-carrying nitrogens in the molecules above or below is observed in the crystal ⟨70CC446, 75JCS(P2)270⟩. The observed bond distances and angles are given in Figure 3.

1,2,4,5-Tetrabenzylhexahydro-1,2,4,5-tetrazine (**54**) is also a centrosymmetric molecule in the chair conformation. Two of the *N*-benzyl groups are equatorial and the other two are axial (Figure 4) ⟨78AX(B)933⟩. The same applies to the *N*-methyl groups of compound (**55**), as shown by $^1$H and $^{13}$C NMR ⟨74JA576⟩ and confirmed by photoelectron spectroscopy ⟨75CB1557⟩.

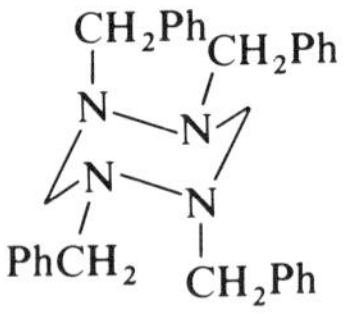

**(54)**

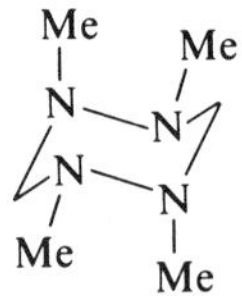

**(55)**

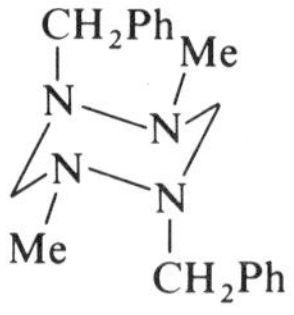

**(56)**

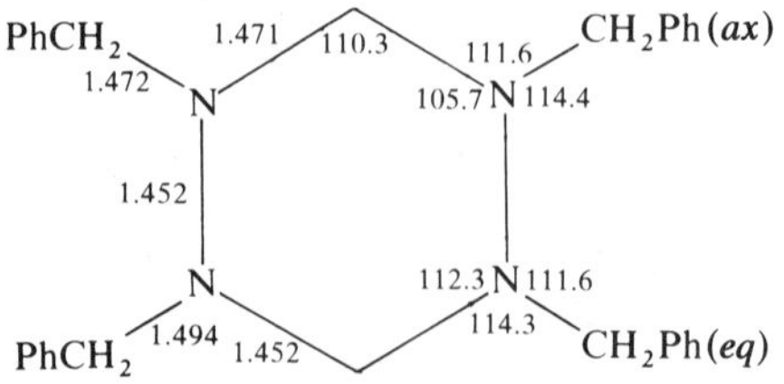

**(54)**

**Figure 4**

X-ray crystallography of 1,4-dibenzyl-2,5-dimethylhexahydro-1,2,4,5-tetrazine (**56**) again reveals the molecule in a chair conformation, with the two methyl groups equatorial

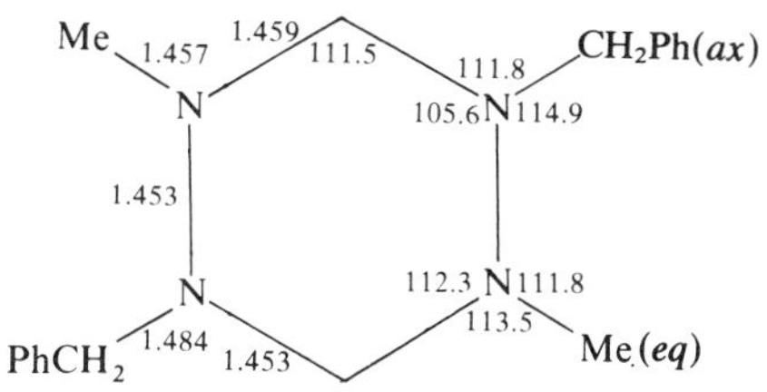

**(56)**

**Figure 5**

and the two larger benzyl groups axial ⟨80JCS(P2)1733⟩. The bond distances and angles of (**56**) are given in Figure 5.

Variable temperature $^{13}$C NMR spectroscopy has shown that the 1,2,3,4,5,6-hexamethyl-hexahydro-1,2,4,5-tetrazine (**57**) is the *trans* isomer with the two *C*-methyl groups equatorial and the *N*-methyl groups symmetrically diaxial and diequatorial. The observed dynamic $^{13}$C NMR effects are consistent with (**57a**) and (**57b**) as the major conformation and probably (**57c**) as the minor one. The first coalescence represents the 'freezing' out of (**57c**) while the (**57a**) ⇌ (**57b**) interconversion remains fast. The nitrogen inversion barrier was found to be 32.2 kJ mol$^{-1}$ ⟨79JCS(P2)981⟩. The He(I) photoelectron spectrum of (**57**) confirmed the NMR data ⟨80JCS(P2)91⟩.

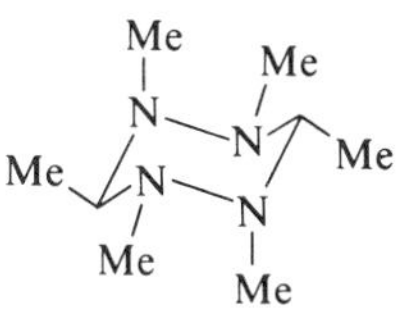

**(57a)** 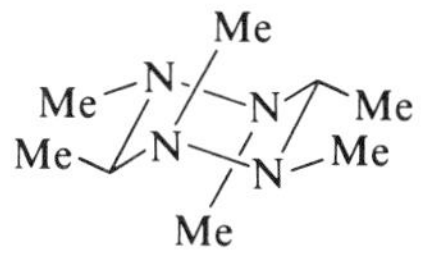          **(57b)** 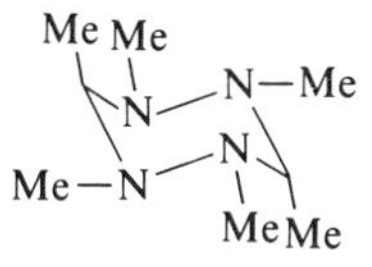          **(57c)**

Study of the conformation of the fused 1,2,4,5-tetrazines (**58**) and (**59**) has shown that all rings are in the chair conformation. In (**58**) three substituents are equatorial and one is axial leading to the conformation (**58a**) while in (**59**) all substituents are equatorial, resulting in conformation (**59a**) ⟨80JCS(P2)1733⟩. Both heterocyclic rings are in the chair conformation in compound (**60**) also, but here all four hydrogen bonds are axial (**60a**), as was found by photoelectron spectroscopy ⟨75CB1557⟩ and by IR and Raman spectroscopy ⟨76SA(A)157⟩.

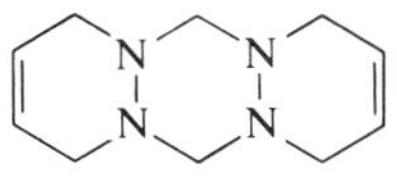

**(58)** 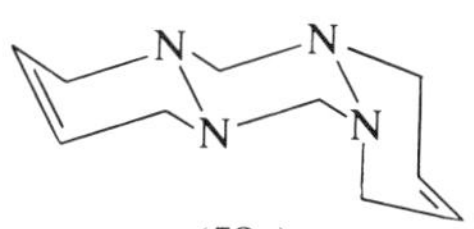          **(58a)** 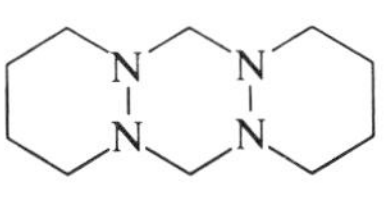          **(59)**

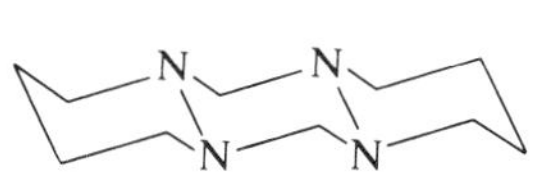

**(59a)** 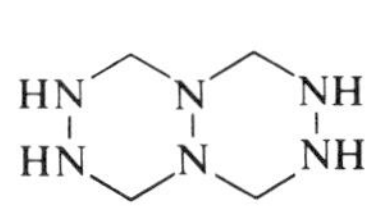          **(60)** 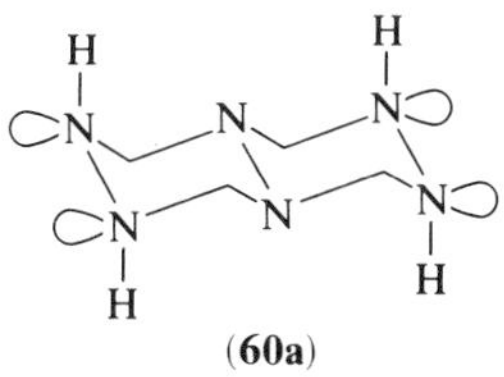          **(60a)**

There are also a number of X-ray crystallographic studies of verdazyls (3-hydro-1,2,4,5-tetrazinyls; **48**) published. The heterocyclic ring has an asymmetric boat form with four coplanar nitrogen atoms. In 1,3,5-triphenylverdazyl (**61**), C-3 is displaced from the plane of the nitrogen atoms by 43° and C-6 by 9° ⟨73AX(B)96⟩. Similar results were obtained for 3-methyl-1,3,5-triphenylverdazyl (**62**), 1,3,5-tri(*p*-tolyl)verdazyl (**63**) and 1-phenyl-3,5-di(*p*-tolyl)verdazyl (**64**). The bond lengths and angles for (**61–64**) are given in Figure 6 ⟨73AX(B)96, 75BCJ819, 75BCJ825, 80BCJ2671⟩. The arrangement of the molecules in the crystal suggests a weak exchange interaction between unpaired electrons ⟨75BCJ819⟩ or a fairly strong antiferromagnetic exchange interaction between unpaired electrons ⟨75BCJ825⟩.

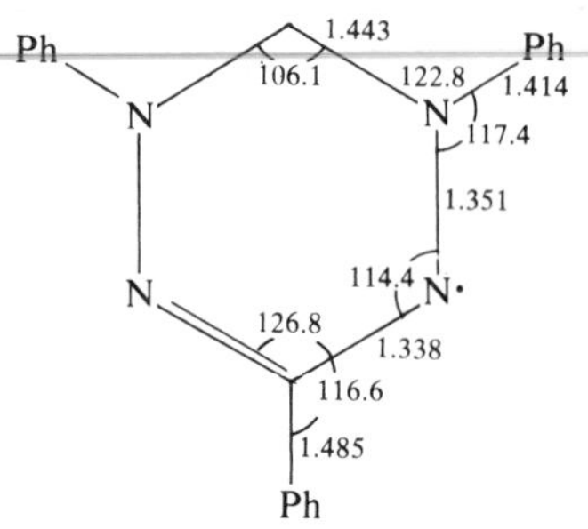

**Figure 6**

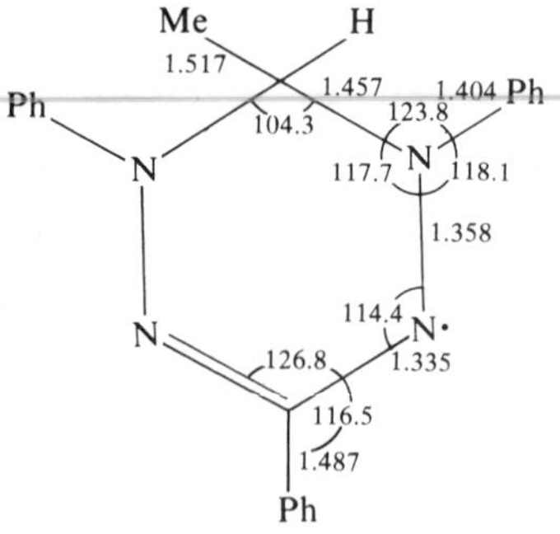

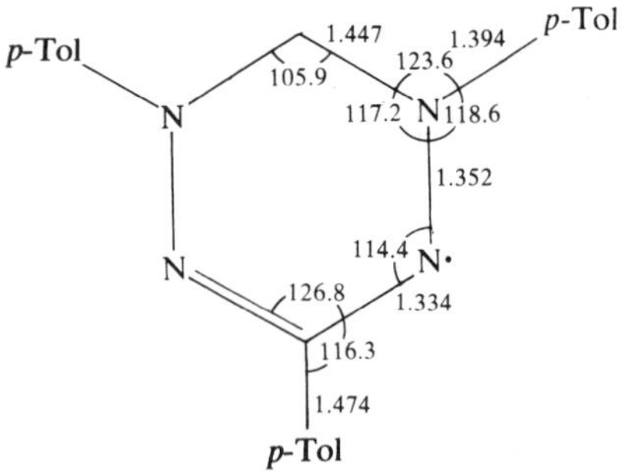

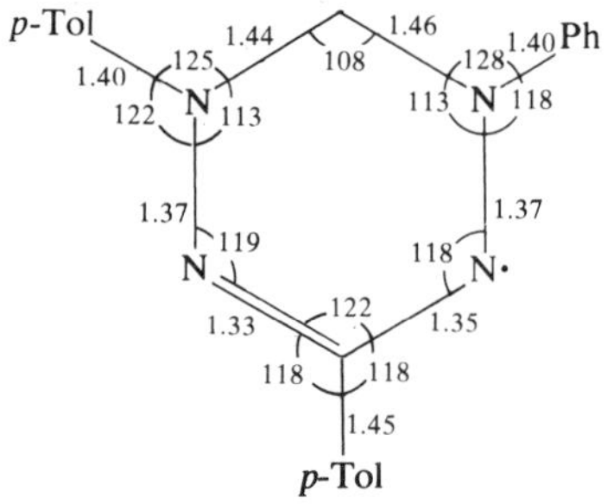

ESR data indicate that the 1,3,5-triphenylverdazyl (**61**) exists in solution as the chair and/or boat conformer. Lowering the temperature does not affect the conformational equilibrium noticeably ⟨80MI22100⟩. Soviet authors have reported the X-ray crystallographic analysis of 1,3,5-triphenyl-1,6-dihydro-1,2,4,5-tetrazinium bromide ⟨79MI22102⟩.

The X-ray crystallographic data of 1,2,4,5-tetrazines are summarized in Table 1.

The spectra most fully studied of the 1,2,4,5-tetrazines are the electronic spectra. All the aromatic 1,2,4,5-tetrazines are strongly coloured: they are red, violet, red-violet or bluish red. These spectra are of great interest to physicists and physical chemists, being well documented and compared with MO calculations. The absorption in the visible region is found between 570 and 520 nm with an absorptivity between $10^2$ and $10^3$. This absorption is due to an $n \to \pi^*$ transition, of energy *ca.* 2.5 eV. The absorption in the UV region is usually observed between 300 and 250 nm, with absorptivity between $2 \times 10^3$ and $4 \times 10^3$. The absorption in the UV region is due to a $\pi \to \pi^*$ transition. Most MO calculations agree well with experimental data.

The following electronic spectrum is published for the parent 1,2,4,5-tetrazine (**38**): 583 ($\varepsilon$ 22), 559 (610), 539 (830), 517 (620), 497 (540), 481 (140), 461 (45), 335 (sh) and 252 nm (2150) in cyclohexane, and 510 (360), 418 (sh), 308 (sh) and 255 (2850) in methanol ⟨B-67MI22102⟩. The resonance energy of 1,2,4,5-tetrazine (**38**) has been calculated from visible absorption data in the gas phase which led to a value of 84 kJ mol$^{-1}$ ⟨46JCS670⟩. 3,6-Dimethyl-1,2,4,5-tetrazine has two absorption maxima in ethanol at 538 ($\varepsilon$ 560) and 274 nm (3620) ⟨66TL5067⟩. Similar electronic spectra have been recorded for other 3,6-dialkyl-1,2,4,5-tetrazines. For 3-phenyl-1,2,4,5-tetrazine the absorption maximum in the visible region is observed at 540 nm ($\varepsilon$ 560) ⟨58JA3155⟩, and that of 3,6-diphenyl-1,2,4,5-tetrazine (in ether) at 554 nm ($\varepsilon$ 480) ⟨54JA427⟩. The maximum for the 3,6-dicarboxylic acid was found at 513 nm (180) in water and at 521 nm (460) in dioxane, and for the dimethyl ester (in ether) at 526 nm (580). 3,6-Diamino-1,2,4,5-tetrazine has its absorption maximum in water at 428 nm (1340), while in dioxane two maxima were observed, at 428 (1980) and 528 nm (590) ⟨54JA428⟩.

**Table 1** Bond Distances (Å) and Angles (°) in 1,2,4,5-Tetrazines, Dihydro-1,2,4,5-tetrazines, Hexahydro-1,2,4,5-tetrazines and Verdazyls

| Bond | (38)[a] | (51)[b] | (52)[c] | (53)[d] | (54)[e] | (56)[f] | (61)[g] | (62)[h] | (63)[i] | (64)[j] |
|---|---|---|---|---|---|---|---|---|---|---|
| N(1)—N(2) | 1.321 | 1.314 | 1.42 | 1.447 | 1.452 | 1.453 | 1.351 | 1.358 | 1.352 | 1.37 |
| N(2)—C(3) | 1.334 | 1.353 | 1.29 | 1.445 | 1.471 | 1.459 | 1.338 | 1.335 | 1.334 | 1.35 |
| C(3)—N(4) | — | 1.338 | 1.40 | 1.467 | 1.452 | 1.453 | — | — | — | 1.33 |
| N(4)—N(5) | — | — | 1.43 | — | — | — | — | — | — | 1.37 |
| N(5)—C(6) | — | — | 1.28 | — | — | — | — | — | — | 1.44 |
| C(6)—N(1) | — | — | 1.40 | — | — | — | 1.443 | 1.457 | 1.447 | 1.46 |
| N(1)—subst. | — | — | 0.99 | 1.460 | 1.494 | 1.484 | 1.414 | 1.404 | 1.394 | 1.40 |
| N(2)—subst. | — | — | — | 0.99 | 1.472 | 1.457 | — | — | — | — |
| C(3)—subst. | — | 1.454 | 1.46 | 1.01/1.08 | — | — | 1.485 | 1.487 | 1.474 | 1.45 |
| N(4)—subst. | — | — | 0.99 | — | — | — | — | — | — | — |
| N(5)—subst. | — | — | — | — | — | — | — | — | — | 1.40 |
| C(6)—subst. | — | — | 1.47 | — | — | — | — | 1.517 | — | — |

| Angle | (38)[a] | (51)[b] | (52)[c] | (53)[d] | (54)[e] | (56)[f] | (61)[g] | (62)[h] | (63)[i] | (64)[j] |
|---|---|---|---|---|---|---|---|---|---|---|
| N(1)—N(2)—C(3) | 115°57′ | 117.5 | 111.9 | 107.7 | 112.3 | 112.3 | 114.4 | 114.4 | 114.4 | 118 |
| N(2)—C(3)—N(4) | 127°22′ | 121.8 | 119.9 | 111.8 | 105.7 | 105.6 | 126.8 | 126.8 | 126.8 | 122 |
| C(3)—N(4)—N(5) | — | 120.6 | 113.1 | 109.3 | 110.3 | 111.5 | — | — | — | 119 |
| N(4)—N(5)—C(6) | — | — | 111.4 | — | — | — | — | — | — | 113 |
| N(5)—C(6)—N(1) | — | — | 120.1 | — | — | — | 106.1 | 104.3 | 105.9 | 108 |
| C(6)—N(1)—N(2) | — | — | 113.9 | — | — | — | 117.8 | 117.7 | 117.2 | 113 |
| C(6)—N(1)—subst. | — | — | — | 111.4 | 111.6 | 111.8 | 122.8 | 123.8 | 123.6 | 128 |
| N(2)—N(1)—subst. | — | — | — | 107.8 | 114.4 | 114.9 | 117.4 | 118.1 | 118.6 | 118 |
| N(1)—N(2)—subst. | — | — | — | 107.7 | 111.6 | 111.8 | — | — | — | — |
| C(3)—N(2)—subst. | — | — | — | 106.1 | 114.3 | 113.5 | — | — | — | — |
| N(2)—C(3)—subst. | — | 119.3 | 121.4 | — | — | — | 116.6 | 116.5 | 116.8 | 118 |
| N(4)—C(3)—subst. | — | 118.8 | 118.7 | — | — | — | — | — | — | 119 |
| N(4)—N(5)—subst. | — | — | — | — | — | — | — | — | — | 122 |
| C(6)—N(5)—subst. | — | — | — | — | — | — | — | — | — | 125 |
| N(5)—C(6)—subst. | — | — | — | 120.2 | — | — | — | — | — | — |
| N(1)—C(6)—subst. | — | — | — | 119.6 | — | — | — | — | — | — |

[a] ⟨56AX510⟩. [b] ⟨72AX(B)739⟩. [c] ⟨76AX(B)1467⟩. [d] ⟨75JCS(P2)270⟩. [e] ⟨78AX(B)993⟩. [f] ⟨80JCS(P2)1733⟩. [g] ⟨73AX(B)96⟩. [h] ⟨75BCJ825⟩. [i] ⟨75BCJ819⟩. [j] ⟨80BCJ2671⟩.

Dihydro-, tetrahydro- and hexahydro-1,2,4,5-tetrazines are less coloured than the aromatic 1,2,4,5-tetrazines. In most cases these substances are yellow, yellow-orange or colourless. Most dihydro-1,2,4,5-tetrazines have two absorption maxima but their positions depend on the substituents on the ring, and on the pattern of hydrogenation. Bands in the visible (*ca.* 430 nm; $\varepsilon$ 400–600) are found with the 1,6-dihydro compounds (**42**); 1,4-dihydrotetrazines (**41**) absorb at somewhat shorter wavelength (*ca.* 300 nm, $\varepsilon$ 100) in ethanol ⟨72HCA1404⟩.

3,6-Dimethyl-1,2,3,4-tetrahydro-1,2,4,5-tetrazine (**65**) shows only end absorption, while 3,6-dimethyl-1,2,3,6-tetrahydro-1,2,4,5-tetrazine (**66**) has a maximum at 392 nm (153) and a shoulder at 260 nm (100) ⟨72HCA1404⟩.

(**65**)          (**66**)          (**67**)

Hexahydro-1,2,4,5-tetrazines without an aromatic substituent show only end absorption in the UV region. The verdazyls (**48**) are strongly coloured compounds as one expects for radicals; they are green, blue-black or black. In the visible region a strong absorption is observed between 750 and 670 nm with an absorptivity between $10^3$ and $10^4$. Depending on the substituents, a second absorption maximum can be found in the visible region, its absorptivity being around $10^4$. The position of the maximum in the UV region depends very much on the substituents on the 1,2,4,5-tetrazine ring. In triarylverdazyls it is found around 280 nm and in monoalkyldiarylverdazyls at about 350 nm. The absorptivity of these maxima is $>10^4$. 1,3,5-Triphenylverdazyl (**61**) has the following spectrum in methanol: 710 (log $\varepsilon$ 3.64), 416 (3.88), 320–310 (broad), 284 (4.39), 244 nm (4.15) ⟨67ZN(B)105⟩.

The IR spectra of 1,2,4,5-tetrazines are as expected for this system. For the parent compound (**38**) the IR spectrum at $-182$ °C and at 25 °C has been documented ⟨71SA(A)747⟩. The spectra of 3,6-dimethyl-1,2,4,5-tetrazine (**39**; $R^3 = R^6 = Me$) ⟨70HCA251⟩, 3,6-dimethyl-1,6-dihydro-1,2,4,5-tetrazine (**42**; $R^3 = R^6 = Me$) ⟨71HCA1922⟩, 3,6-dimethylhexahydro-1,2,4,5-tetrazine (**45**; $R^3 = R^6 = Me$) and 1,3,4,6-tetramethylhexahydro-1,2,4,5-tetrazine (**67**) ⟨70HCA251⟩ are recorded in the references cited.

The following NMR spectral details have been reported: for 1,2,4,5-tetrazine (**38**): $^1$H: $\delta = 11.05$ (acetone-$d_6$) ⟨76ZN(B)1489⟩, $^{13}$C: 161.2 p.p.m.; 3-methyl-1,2,4,5-tetrazine: $^1$H: $\delta = 10.27$ and 3.10 ⟨78JHC445⟩; 3,6-dimethyl-1,2,4,5-tetrazine: $^1$H: $\delta = 2.97$ ⟨70HCA251⟩, $^{13}$C: 166.6 and 20.8 p.p.m.; 3,6-dimethyl-1,6-dihydro-1,2,4,5-tetrazine (**66**): $^1$H: $\delta = 2.00$, 2.47 and 6.30; 3,6-dimethylhexahydro-1,2,4,5-tetrazine (**45**; $R^3 = R^6 = Me$) (NaOD): $^1$H: $\delta = 1.05$ and 3.78; 1,3,4,6-tetramethylhexahydro-1,2,4,5-tetrazine (**67**) (NaOD): $^1$H: $\delta = 1.15$, 2.41 and 3.40; 1,2,3,4,5,6-hexamethylhexahydro-1,2,4,5-tetrazine (**57**) (NaOD): $^1$H: $\delta = 1.12$, 2.42 and 4.27 ⟨70HCA251⟩; 1,4-dimethylhexahydro-1,2,4,5-tetrazine (**53**): $^1$H: $\delta = 3.60$ (CH$_2$), 2.90 (NH) and 2.38 (Me) ⟨69JA2443⟩. A listing of NMR spectral data of 1,2,4,5-tetrazines covering literature to June 1976 is available ⟨77MI22101⟩. A compilation of the $^1$H paramagnetic shifts and coupling constants of verdazyls has also appeared ⟨74T2841⟩.

The mass spectra of 1,2,4,5-tetrazines are simple. The major fragmentation mode of dialkyltetrazines is initiated by loss of nitrogen from the molecular ion, followed by a simple cleavage, with formation of ions with a nitrile structure (a) and by simultaneous rearrangement of one hydrogen atom leading to $(a+H)^+$ ions and in some cases $(a-H)^+$ ions (Scheme 5) ⟨67ACS(B)2855⟩. The spectra of the parent 1,2,4,5-tetrazine and its 3,6-dimethyl derivative show these general features ⟨70HCA251⟩. In the spectra of 3,6-diaryl-1,2,4,5-tetrazines peaks at mass unit $(M+2)^+$ are observed, which are explained by the formation of the 3,6-diaryldihydro-1,2,4,5-tetrazines in the ion source ⟨68TL3929⟩. Mass spectra have also been recorded for 3,6-dimethyl-1,4-dihydro- (**41**; $R^3 = R^6 = Me$), 1,3,4,6-tetramethyl-1,4-dihydro-, 3,6-dimethyl-1,2,3,4-tetrahydro- (**65**), and 3,6-dimethyl-1,2,3,6-tetrahydro-1,2,4,5-tetrazine (**66**) ⟨72HCA1404⟩.

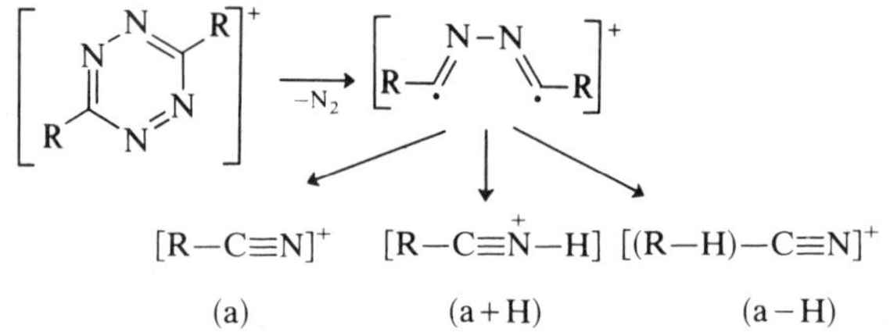

$$[R-C{\equiv}N]^+ \qquad [R-C{\equiv}\overset{+}{N}-H] \qquad [(R-H)-C{\equiv}N]^+$$

(a)                          (a + H)                       (a − H)

**Scheme 5**

The ESR spectra of verdazyls have been extensively studied. In general a nine-line spectrum is observed, with extensive hyperfine splitting. The most fully studied compound seems to be the 1,3,5-triphenyl derivative (**61**) for which the ESR spectrum was recorded by Kuhn and Trischmann. The ratio of line heights for this radical was found to be $1:4:10:16:19:16:10:4:1$ ⟨64M457⟩. The influence of the substituents at N-1, C-3, N-5 and C-6 was studied by Neugebauer and his group. The g factors are in the range of 2.0030–2.0036. The hyperfine splitting constants reported are 4.9 to 6.5 G for $a_{N(1,5)}$ and 5.9 to 6.0 G for $a_{N(2,4)}$.

A number of compounds with two or three independent verdazyl rings have also been prepared and their spectra studied. In those cases where no interaction of the unpaired electrons is possible, the normal nine-line spectrum is observed; in the other cases an interaction needed to be considered.

Photoelectron spectra have been published for simple 1,2,4,5-tetrazines (**39**) ⟨72HCA255⟩ and for hexahydro-1,2,4,5-tetrazines (**45**). In the latter cases the spectra formed part of a conformational study. The following values were recorded for 1,2,4,5-tetrazine (**38**): 9.72, 12.05, 12.78, 13.36, 13.50 and 15.84 eV; 3,6-dimethyl-1,2,4,5-tetrazine: 9.08, 10.72, 11.15, 11.98, 12.66 and 13.50 eV ⟨72HCA255⟩; 1,2,4,5-tetramethylhexahydro-1,2,4,5-tetrazine: 7.90, 8.45, 9.00 and 10.37 eV ⟨75CB1557⟩; 1,4-dibenzyl-2,5-dimethylhexahydro-

1,2,4,5-tetrazine: 7.71, 8.34, 9.02–9.16 and 10.22 eV; 1,2,4,5-tetrabenzylhexahydro-1,2,4,5-tetrazine: 7.44, 8.09 and 9.07 eV; 1,2,3,4,5,6-hexamethylhexahydro-1,2,4,5-tetrazine: 7.63, 8.08, 8.75, 9.43 and 10.00 eV ⟨80JCS(P2)91⟩; hexaazadecalin (**60**): 8.65, 9.65, 10.75 and 11.05 eV ⟨75CB1557⟩; compound (**51**): 7.73, 8.32, 9.16–9.40 and 10.06 eV ⟨80JCS(P2)91⟩.

The fluorescence and phosphorescence of 1,2,4,5-tetrazines have also been studied. The fluorescence lifetime of 1,2,4,5-tetrazine (**38**) was found to be $450 \pm 55$ ps at 300 K in benzene, corresponding to a quantum yield of $1.1 \times 10^{-3}$ ⟨76JA5443⟩ and $1.5 \pm 0.2$ ns at 1.8 K in benzene with a quantum yield of $1.8 \times 10^{-3}$, and in the gas phase $6.0 \pm 0.3$ ns ⟨74MI22101⟩. The phosphorescence lifetime of (**38**) was $(96.8 \pm 2.1) \times 10^{-6}$ s ⟨74MI22102⟩. The fluorescence lifetime of 3,6-dimethyl-1,2,4,5-tetrazine in benzene is $(1.5 \pm 0.2) \times 10^{-9}$ s ⟨74MI22101⟩ and its phosphorescence lifetime $(85.2 \pm 3.4) \times 10^{-6}$ s ⟨77JA3923⟩. The fluorescence and phosphorescence lifetimes of 3-phenyl-1,2,4,5-tetrazine were $(14.3 \pm 1.3) \times 10^{-9}$ s and $(59.6 \pm 1.3) \times 10^{-6}$ s, respectively ⟨77JA3923⟩.

Van der Waals complexes between 1,2,4,5-tetrazine (**38**) and a number of light gases (He, Ar, H₂) were observed and characterized by laser spectroscopic studies of free supersonic expansion of (**38**) in the carrier gas. The observed complexes are of the form X · (**38**) or X₂ · (**38**), where X is He, Ar or H₂. The spectra are consistent with the gas in both types of complexes being bound on or near the out of plane $C_{2v}$ axis on top of and/or below the 1,2,4,5-tetrazine ring. For the He and H₂ complexes, analysis of the rotational structure indicates that the van der Waals bond length is ~3.3 Å ⟨78JCP(68)2487, 79JCP(71)4757⟩.

### 2.21.3.3 Reactions

Most 1,2,4,5-tetrazines are unstable both to acids and bases. Acidic hydrolysis affords nitrogen and/or hydrazine and either carboxylic acids or aldehydes. It seems that the products formed during hydrolysis have not in all cases been detected ⟨78HC(33)1075, p. 1093⟩.

Hydrolysis of 1,6-diphenyl-1,2,4,5-tetrazinium 3-sulfide (**68**) led to the isolation of 2-amino-5-phenyl-1,3,4-thiadiazole (**69**) ⟨67CC1045⟩.

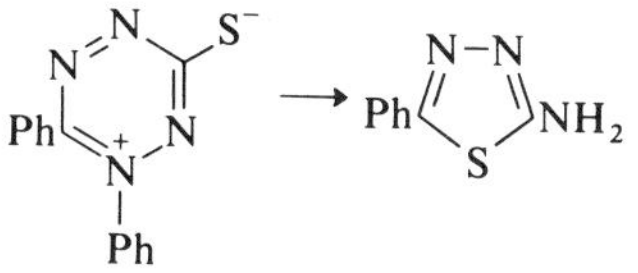

Dihydro-1,2,4,5-tetrazines, formulated as the 1,2-dihydro tautomers (**40**), rearrange under the influence of acids to 4-amino-1,2,4-triazoles (**70**) ⟨78HC(33)1075, p. 1151⟩. Further acidic hydrolysis of (**70**) resulted in the formation of various products, such as diacylhydrazines, hydrazine, carboxylic acids and 1,3,4-oxadiazoles (**71**) ⟨78HC(33)1075, p. 1151⟩.

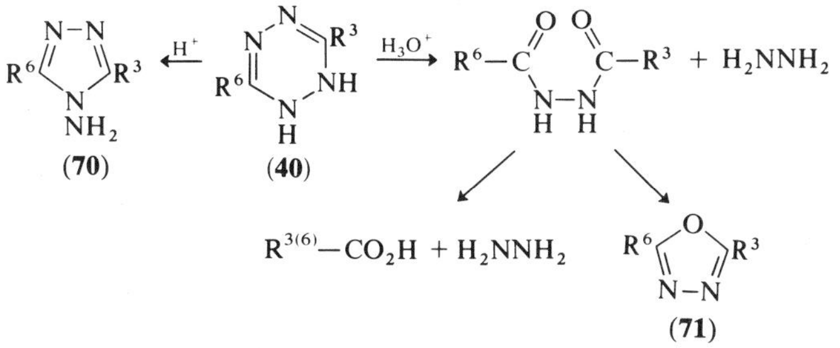

1,4-Diphenyl-1,4-dihydro-1,2,4,5-tetrazine (**72**) afforded phenylhydrazine and formic acid in aqueous acid ⟨50JCS3389⟩. When the same compound was treated with ethanolic hydrogen chloride, 1,5-diphenylformazan (**73**) was isolated ⟨65CB1476⟩.

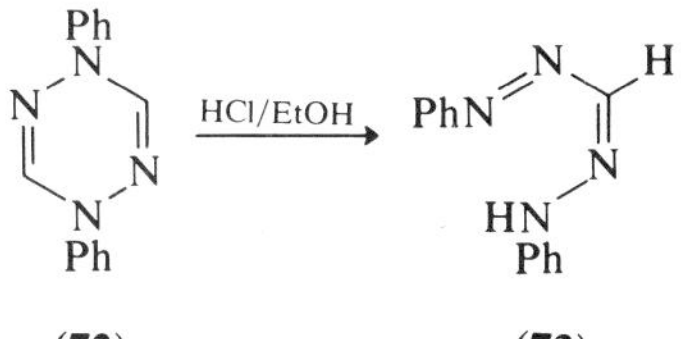

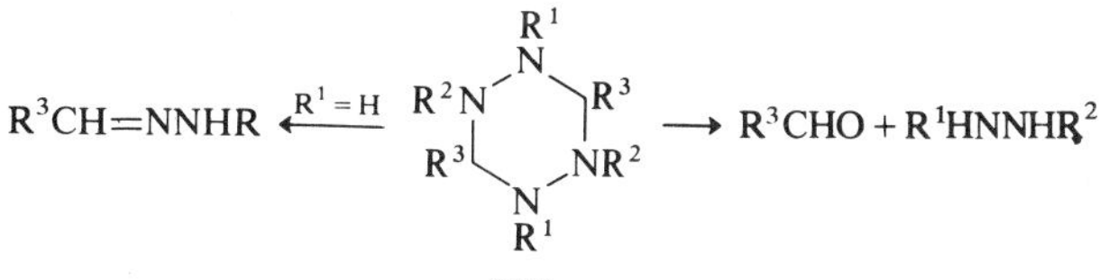

Hexahydro-1,2,4,5-tetrazines (**74**) can also be hydrolyzed by aqueous acids; the products isolated were aldehyde hydrazines or aldehydes and hydrazine ⟨70HCA251, 78HC(33)1075, p. 1213⟩.

Treatment of verdazyls (**48**) with mineral acids resulted in disproportionation to 1,2,3,4-tetrahydro-1,2,4,5-tetrazines (**75**) and 1,6-dihydro-1,2,4,5-tetrazinium salts (**76**). Here one molecule of the verdazyl is reduced to (**75**) and the other is oxidized to (**76**). The mechanism of this reaction has been studied by Polumbrik and his group ⟨72ZOR1925⟩. Heating 3-phenyl-1,2-dihydro-1,2,4,5-tetrazine-6(5*H*)-thione (**77**) in 2N hydrochloric acid led to the isolation of 3-phenyl-1,2,4-triazole-5-thione (**78**) ⟨77KGS1564⟩.

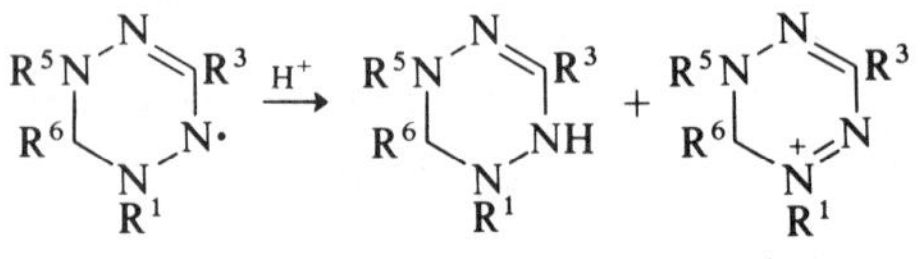

(**48**)        (**75**)        (**76**) 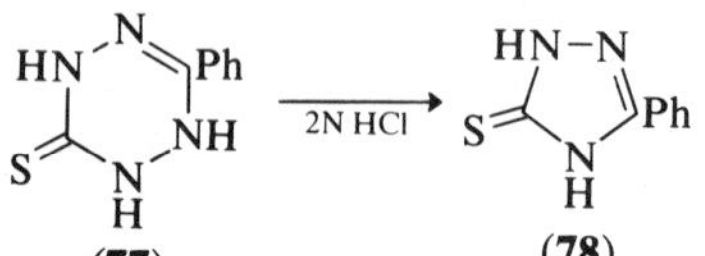                    (**77**)            (**78**)

Treatment of 3,6-diphenyl-1,2-di(phenylsulfonyl)-1,2-dihydro-1,2,4,5-tetrazine (**79**) with concentrated sulfuric acid affords 3,6-diphenyl-1,4-dihydro-1,2,4,5-tetrazine (**80**) in good yield ⟨79BCJ483⟩.

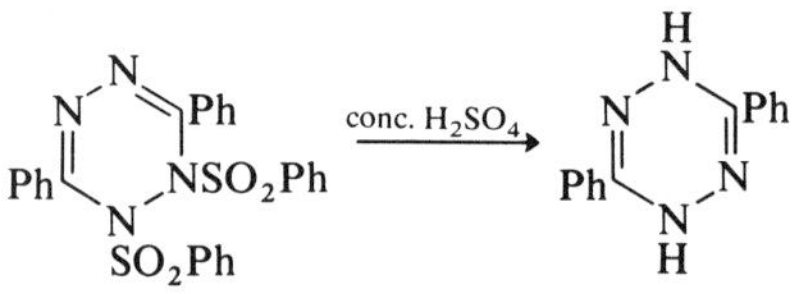

(**79**)                              (**80**)

Basic hydrolysis of 3-phenyl-1,2,4,5-tetrazine (**81**) led to the isolation of benzalazine (**82**) in 80% yield ⟨71KGS708⟩. In most other cases studied the isolated product was an aldehyde acylhydrazone (**83**), indicating that during the basic hydrolysis one carbon atom of the tetrazine ring is reduced from the acid oxidation level to an aldehyde level ⟨78HC(33)1075, p. 1094⟩. A mechanism for the reaction has been proposed by Libmann and Slack ⟨56JCS2253⟩. Benzalazine (**82**) was also isolated when 3-azido-6-phenyl-1,2,4,5-tetrazine (**84**) was hydrolyzed by base ⟨71KGS711⟩.

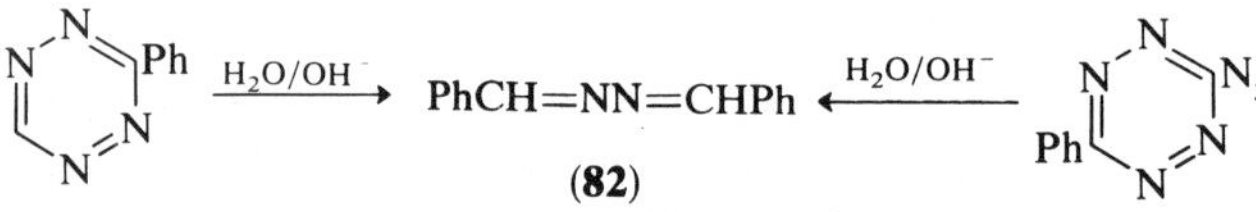

(**81**)                    (**82**)                    (**84**)

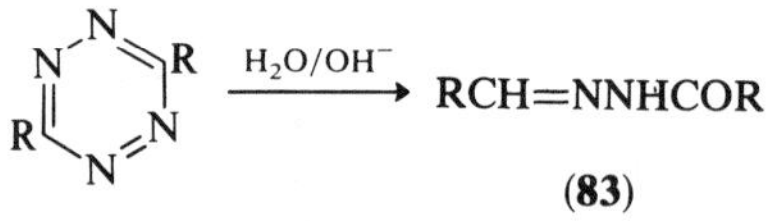

(**83**)

3,6-Dibenzyl-1,2,4,5-tetrazine (**85**) with potassium hydroxide in methanol afforded 3-benzyl-7-methoxy-6-phenylimidazo[1,2-*b*][1,2,4,5]tetrazine (**86**), the structure of which was determined by X-ray crystallography ⟨79JCS(P1)333⟩.

Basic hydrolysis of 1,4-disubstituted 1,4-dihydro-1,2,4,5-tetrazines (**87**) resulted in the formation of 2,4-disubstituted 3-imino-1,2,4-triazoles (**88**) ⟨78HC(33)1075, p. 1152⟩, and aqueous base converted 3,6-disubstituted dihydro-1,2,4,5-tetrazines (**89**) into 3,5-disubstituted 1,2,4-triazoles (**90**) ⟨56JCS2253⟩.

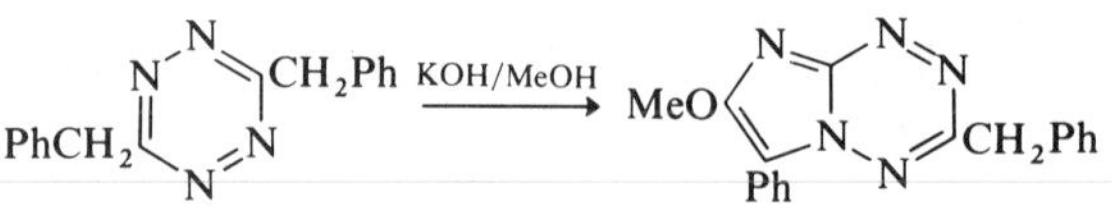

(**85**)              (**86**) 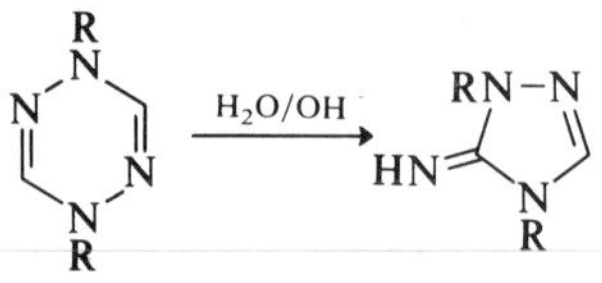                    (**87**)            (**88**)

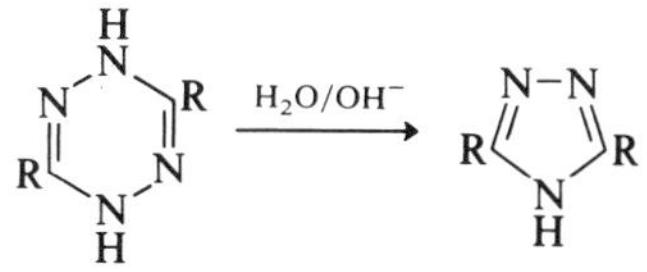

(89)       (90)

1,6-Dihydro-1,2,4,5-tetrazine-3,6-dicarboxylic acid and its derivatives (**91**) when heated with a base rearrange to the 1,2-dihydro tautomers (**92**) ⟨06CB3776, 09CB3270⟩ which were isomerized on further base treatment to 1,2,4-triazoles (**93**). With compounds (**89**) and (**92**) there arises the tautomerism problem alluded to earlier, and also with (**89**) the question whether the difference in behaviour observed between it and (**87**) is a result of the unsubstituted nitrogens in (**89**) or the unsubstituted carbons in (**87**). In connection with the tautomerism, we recall that the compound (**89**) used was found to be the 1,4-dihydro tautomer in the crystal.

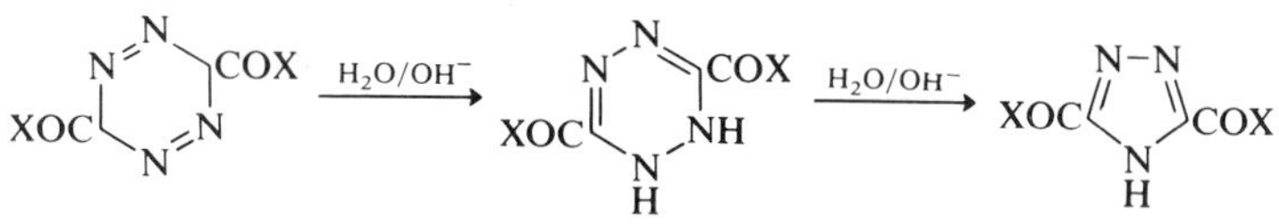

(91)       (92)       (93)

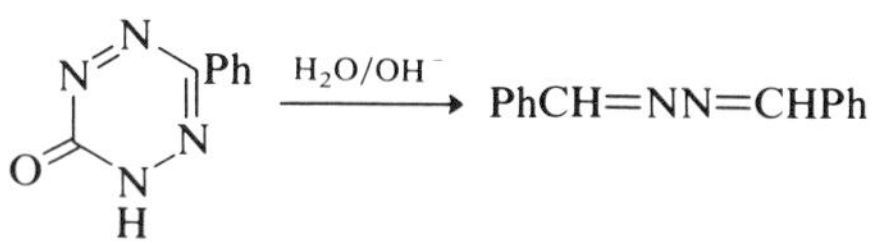

(94)       (82)

Treatment of 3-phenyl-1,6-dihydro-1,2,4,5-tetrazin-6-one (**94**) with base led to the isolation of benzalazine (**82**) ⟨69KGS566⟩. Alkaline hydrolysis of hexahydro-1,2,4,5-tetrazines (**74**) gave an aldehyde hydrazone or an aldehyde and hydrazine as with the acidic hydrolysis ⟨78HC(33)1075, p. 1213⟩.

The alkaline cleavage of 1,6-dihydro-1,2,4,5-tetrazinium salts (**95**) was studied by Soviet chemists. The kinetic data they obtained indicate that the cation (**96**) is formed as an intermediate, which reacts with hydroxide ion in the rate-determining step ⟨76MI22100⟩.

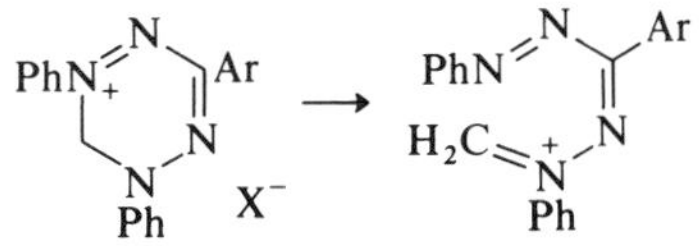

(95)       (96)

1,2,4,5-Tetrazines (**39**) are readily attacked by nucleophiles such as ammonia or hydrazine. If the tetrazine is unsubstituted at position 6, 6-amino adducts (**97**) can be oxidized by potassium permanganate to give the 6-amino-1,2,4,5-tetrazines (**98**) ⟨81JHC123, 81JOC5102, 81JOC3805⟩.

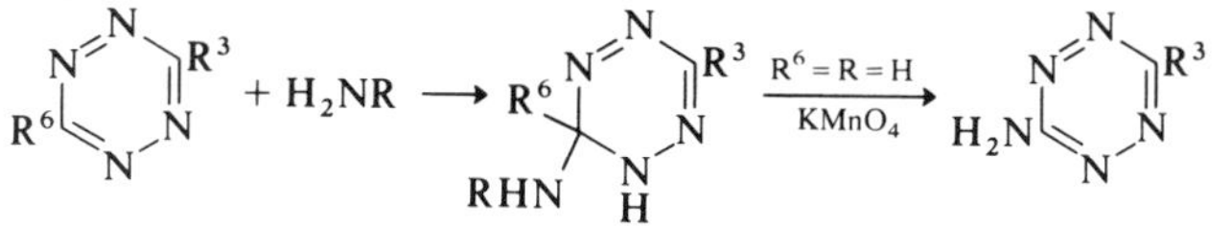

(39)       (97)       (98)

When 3-amino-6-methyl-1,2,4,5-tetrazine (**99**) was treated with hydrazine, 3-hydrazino-6-methyl-1,2,4,5-tetrazine (**100**) was isolated. It was shown by using labelled hydrazine that this reaction proceeds not only by a nucleophilic attack of the hydrazine at C-3 ($A_E$ mechanism) giving (**100a**) but also by a nucleophilic attack of the hydrazine at C-6, ring opening to (**101**), reclosure and incorporation of the labelled hydrazine into the heterocyclic ring ($S_N$(ANRORC) mechanism), forming (**100b**; Scheme 6) ⟨78JHC445⟩.

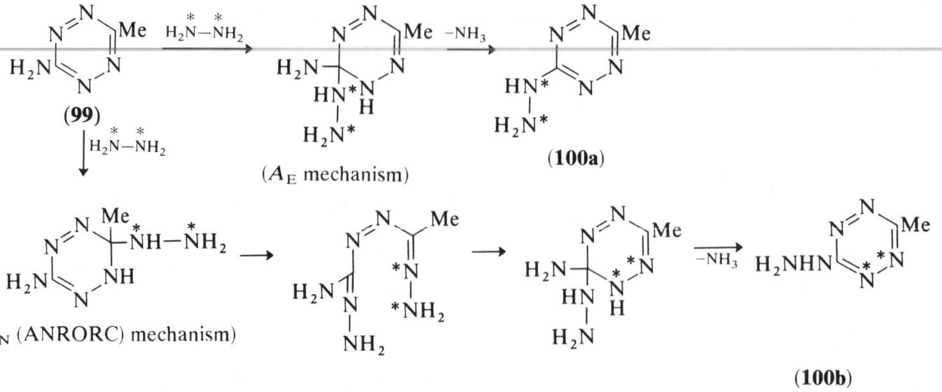

**(99)**

*(A_E mechanism)*

**(100a)**

*(S_N (ANRORC) mechanism)*

**(100b)**

**Scheme 6**

The formation of 1,2,4,5-tetrazine $N$-oxides (**102**) by oxidation of 1,2,4,5-tetrazines (**39**) has not yet been reported. Oxidation with peracetic acid resulted in the formation of 2,5-disubstituted 1,3,4-oxadiazoles (**103**) ⟨62JOC1463, 73JCS(P1)335⟩.

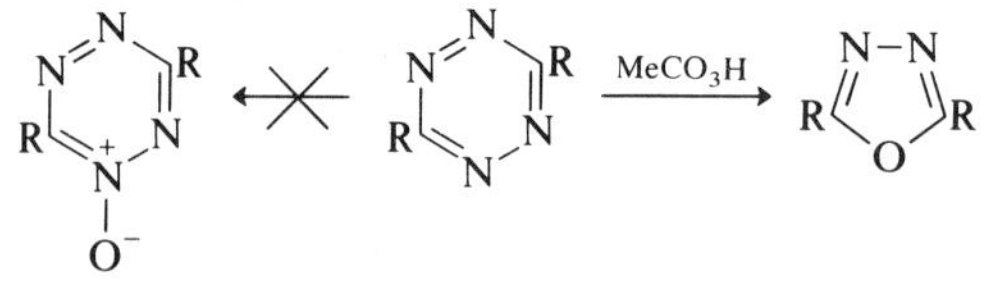

**(102)**            **(39)**            **(103)**

Dihydro-1,2,4,5-tetrazines with unsubstituted nitrogens are easily oxidized to 1,2,4,5-tetrazines. This oxidation is part of most preparative methods for the synthesis of the tetrazines. As oxidizing agents nitrous acid, nitric acid, oxygen, halogens, iron(III) chloride, hydrogen peroxide and lead tetraacetate have been used ⟨78HC(33)1075, p. 1077⟩.

1,4-Disubstituted 1,4-dihydro-1,2,4,5-tetrazines (**104**) can be electrochemically oxidized to give stable radical cations (**105**). In these oxidations the substituent on the nitrogens has more influence than that on the carbon atoms ⟨78IZV2499⟩. Tetrahydro-1,2,4,5-tetrazines (**106**) are oxidized either to dihydro-1,2,4,5-tetrazines (**107, 108**) or to verdazyls (**109**) depending on the substituents ⟨78HC(33)1075, p. 1184⟩. Hexahydro-1,2,4,5-tetrazines (**110**) can also be oxidized; the products are either 1,2,3,4-tetrahydro (**106**) or dihydro compounds (**107, 108**), again depending on the substituents ⟨78HC(33)1075, p. 1214, 70HCA251, 71HCA1922, 72HCA1404⟩.

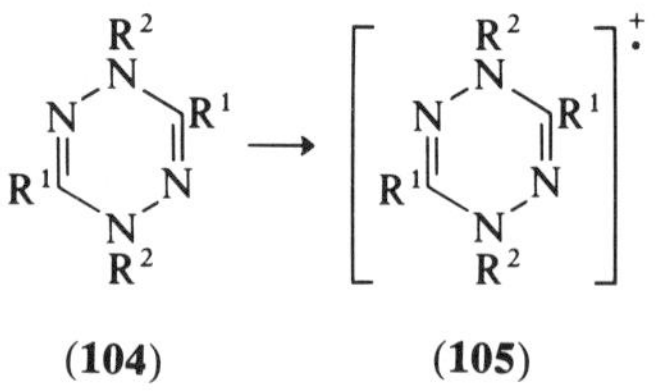

**(104)**            **(105)**

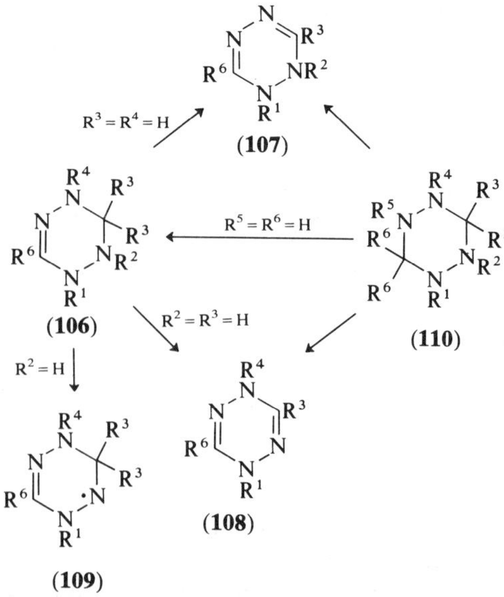

**(107)**

**(106)**

**(110)**

**(108)**

**(109)**

1,3,5-Triphenylverdazyl reacts with oxygen to give a complex with a strong contribution from a peroxide structure and some contribution from a charge transfer complex ⟨77ZOB2396⟩. If the verdazyls (**111**) are oxidized by oxygen in the presence of activated charcoal, $N$-formylformazans (**112**) are obtained ⟨67M726⟩. Electrochemical oxidation of

(**111**) led to the isolation of 1,6-dihydro-1,2,4,5-tetrazinium salts (**113**) (verdazylium salts), as did the oxidation of (**111**) with halogens or metal salts such as those of copper(II) or mercury(II) ⟨64M457, 66M517, 66M525, 66M1280⟩.

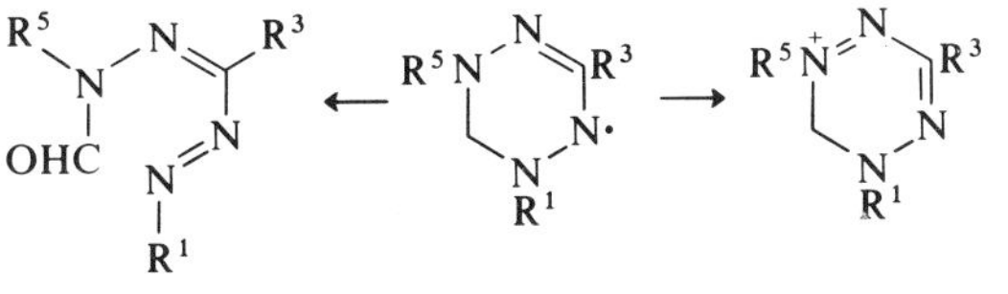

(**112**)          (**111**)          (**113**)

The reaction of verdazyls (**111**) with diacyl peroxides, such as benzoyl peroxide, dilauroyl peroxide, perbenzoic acid and substituted perbenzoic acids has received considerable attention ⟨76TL1893, 78ZOR2471, 78MI22101⟩. One mole of the radical always consumes three moles of the peroxide. The first product is the verdazylium salt (**113**) which is then transformed into the *N*-formylformazan (**112**).

Verdazyls (**111**) can also transfer an electron to *o*-quinones to give the verdazylium cation (**113**) and a semiquinone anion (**114**) ⟨80IZV2785⟩, or to tetranitromethane to give the cation (**113**) and the tetranitromethane anion radical (**115**) ⟨74MI22100⟩. Rate constants and activation parameters for the electron transfer from triphenylverdazyl to tetracyanoethylene have been determined by Soviet chemists ⟨79ZOR2344⟩.

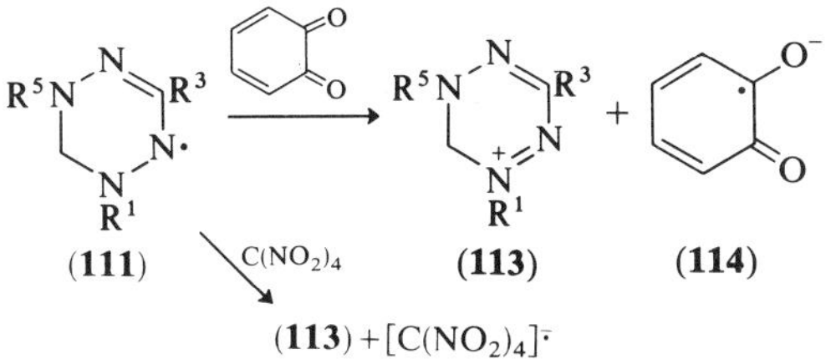

(**111**)          (**113**)          (**114**)

(**113**) + [C(NO₂)₄]⁻

(**115**)

1,2,4,5-Tetrazines can be reduced very easily; very often they can be used as oxidizing agents for dihydro aromatic compounds. For instance, in cycloaddition reactions of 1,2,4,5-tetrazines (**39**) with alkenes, dihydropyridazines (**116**) are formed, which can be oxidized by excess of the tetrazine (**39**) to yield the pyridazine (**117**) and the dihydrotetrazine (**41**). Other reducing agents which have been used are hydrogen sulfide, sodium dithionite and methanol under the influence of light ⟨78HC(33)1075, p. 1094⟩. Reduction of 1,2,4,5-tetrazines (**39**) with sodium borohydride affords 1,6-dihydro-1,2,4,5-tetrazines (**42**), as was shown by ¹H NMR spectroscopy (Scheme 7). From the NMR spectra of derivatives (**42**) in which

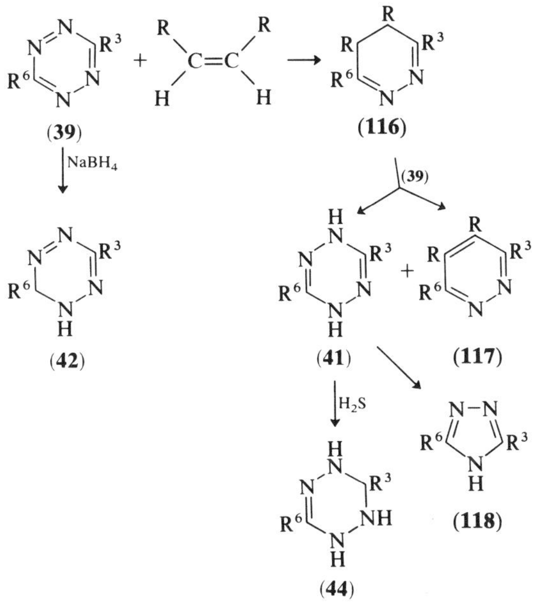

(**39**)          (**116**)

(**42**)          (**41**)          (**117**)

(**44**)          (**118**)

**Scheme 7**

$R^6 = H$ it was also concluded that these compounds are homoaromatic. One of the two protons at C-6 is oriented above and the other away from the plane of the ring atoms 1–5, thus resulting in a large difference in their chemical shifts ⟨81JOC2138⟩.

Dihydro-1,2,4,5-tetrazines (41) can be reduced by hydrogen sulfide to tetrahydro-1,2,4,5-tetrazines (44) ⟨72HCA1404⟩, while with zinc in acetic acid, sodium in ethanol or diimide they afford 1,2,4-triazoles (118; Scheme 7) ⟨78HC(33)1075, p. 1154⟩.

Verdazyls (119) can very easily be reduced to 1,2,3,4-tetrahydro-1,2,4,5-tetrazines (120). This can be achieved catalytically or by using dihydropyridines, hydrazobenzene, phenyl-hydrazine, hydrazine, hydroquinone, 2-phenylimidazoline, dihydro-1,2,4,5-tetrazines, thiophenol, phosphines and protic acids ⟨78HC(33)1075, p. 1244, 76UKZ510, 76UKZ724, 77AJC221, 75MI22100, 77MI22100, 77MI22101⟩.

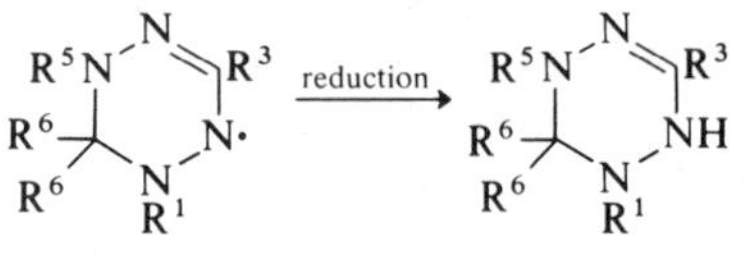

(119)                    (120)

1,2,4,5-Tetrazines (39) form nitriles and nitrogen on thermal degradation ⟨59JA4342⟩. Pyrolysis of *N*-unsubstituted dihydro-1,2,4,5-tetrazines (41) is reported to lead to the tetrazines (39) but this reaction was probably an oxidation since oxygen was not excluded. 3,6-Dimethyl-1,4-dihydro-1,2,4,5-tetrazine (121) rearranges to 4-amino-3,5-dimethyl-1,2,4-triazole (122) on heating at its melting point ⟨15CB1614⟩. 3,6-Diphenyl-1,4-dihydro-1,2,4,5-tetrazine (80) affords 2,4,6-triphenyl-1,3,5-triazine (123), 3,5-diphenyl-1,2,4-triazole (124), benzonitrile and ammonia on heating without a solvent to 190 °C ⟨62LA(654)146⟩. Acetaldehyde azine (126) and nitrogen were isolated from the pyrolysis of 3,6-dimethyl-1,6-dihydro-1,2,4,5-tetrazine (125) ⟨72HCA1404⟩.

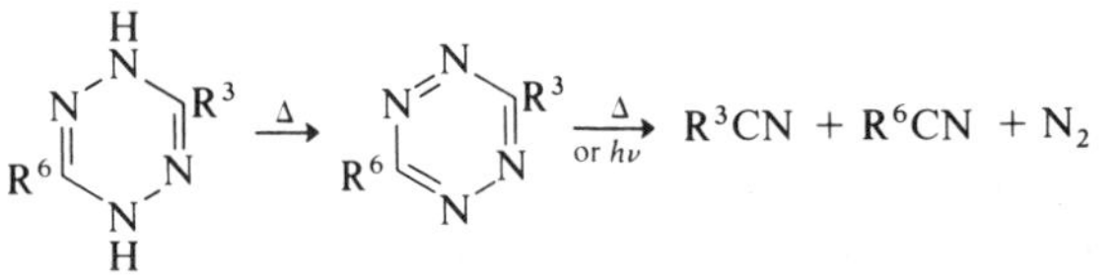

(41)            (39) 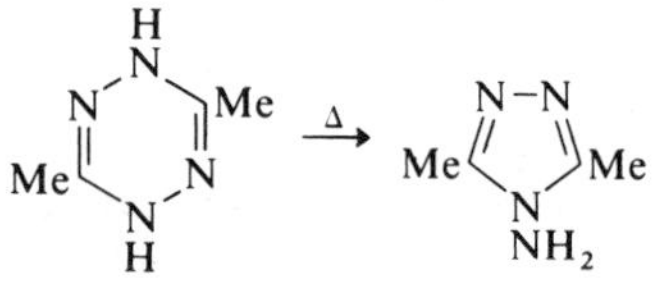                                    (121)            (122)

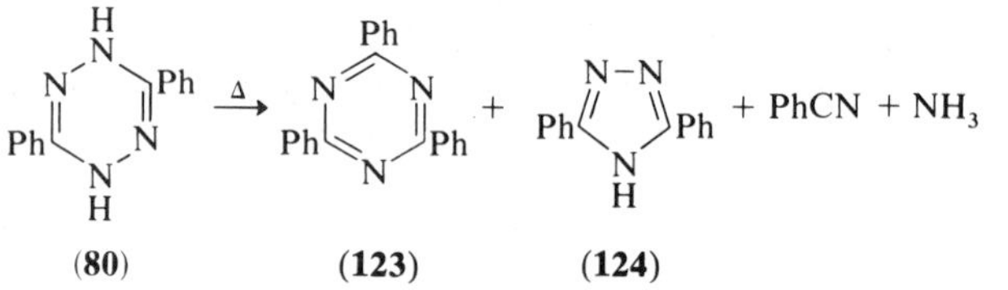

(80)            (123)            (124)

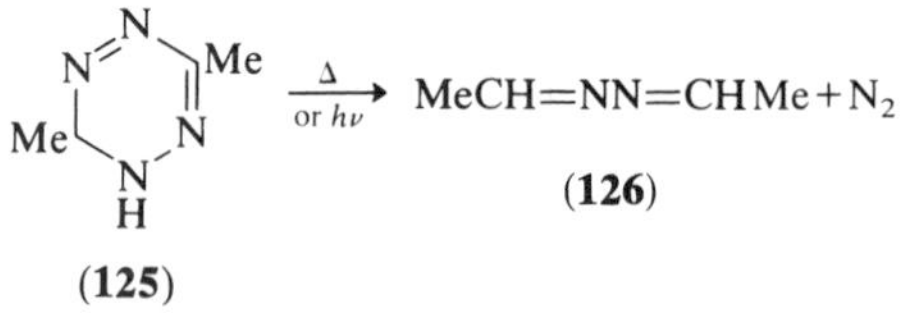

(125)

Thermolysis of 3,6-diphenyl-1,4-di(phenylsulfonyl)-1,4-dihydro-1,2,4,5-tetrazine (127) in boiling toluene gives benzenesulfonic anhydride (128), phenyl benzenethiosulfonate (129), small amounts of diphenyl disulfide (130), 3,6-diphenyl-1,2,4,5-tetrazine (51) and a rearrangement product, 3,6-diphenyl-1,2-di(phenylsulfonyl)-1,2-dihydro-1,2,4,5-tetrazine (79) ⟨79BCJ483⟩. 3,6-Disubstituted hexahydro-1,2,4,5-tetrazines (132) afforded aldehyde hydrazones (133) when heated at their melting point ⟨63AG1204⟩.

Boiling the 1,3,5-trisubstituted verdazyls (134) for five hours in acetone afforded 1,3-disubstituted 5-amino-1,2,4-triazoles (135, 136) ⟨78KGS1137⟩ and tetrahydro-1,2,4,5-tetrazines (137) ⟨64M457, 72CB549⟩; at higher temperatures, besides the triazoles (135, 136), 1,3-diphenyl-1,2,4-triazole (138) and aniline were obtained ⟨72CB549⟩.

Heating the tetrazolyl-substituted verdazyl (139) in boiling acetone afforded the tricyclic system (140) ⟨77KGS557⟩.

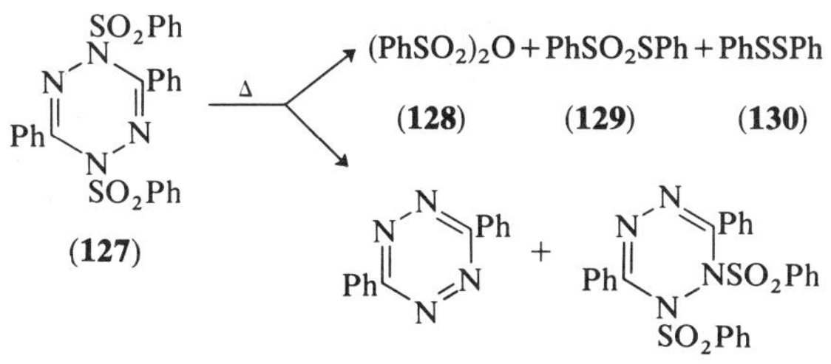

Photolysis of 1,2,4,5-tetrazines (**39**) affords nitriles and nitrogen ⟨70T2619⟩. This reaction has been studied very fully. The results obtained are consistent with a mechanism wherein the chromophore first absorbs a photon, then decays to an intermediate species which under certain conditions requires absorption of a second photon to achieve decomposition to products ⟨78JA3242⟩. The photodissociation of unsubstituted 1,2,4,5-tetrazine was found to be exothermic by more than $418\,\text{kJ mol}^{-1}$ ⟨78MI22100⟩. By using isotopically selective laser-induced photodecomposition of 1,2,4,5-tetrazine, a weighable quantity of $[^{13}\text{C}]$-1,2,4,5-tetrazine and of $[^{15}\text{N}]$-1,2,4,5-tetrazine was produced in high purity. The enrichment factor was in excess of 1000 ⟨79MI22101, 79MI22102⟩.

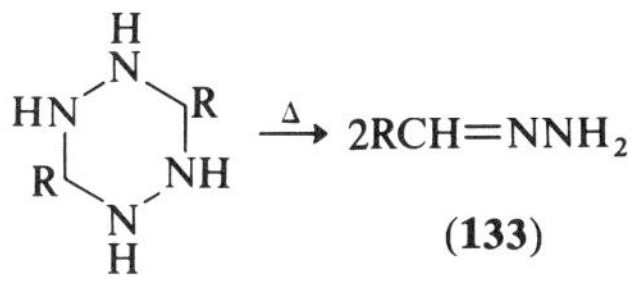

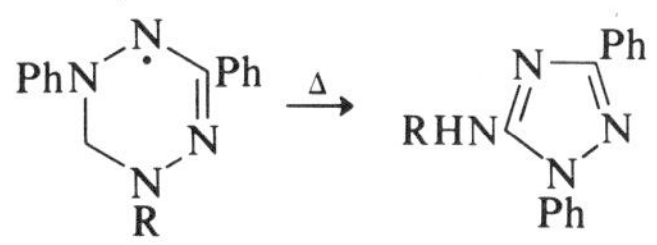

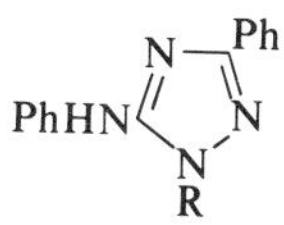

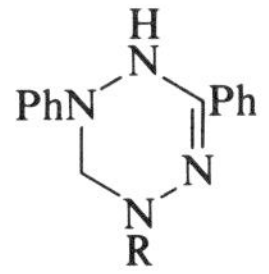

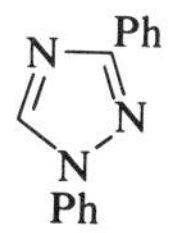

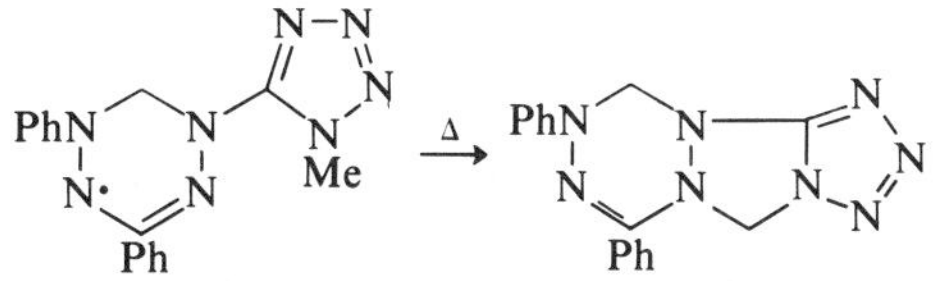

Different results were reported for the photolysis of dihydro-1,2,4,5-tetrazines. 3,6-Diphenyl-1,4-dihydro-1,2,4,5-tetrazine (**80**) afforded 3,5-diphenyl-1,2,4-triazole (**124**) on photolysis ⟨70T2619, 69JOC199⟩, while the 3,6-dimethyl-1,6-dihydro compound (**125**) yielded acetaldehyde azine (**126**) and nitrogen on photolysis, as in the thermolysis reaction ⟨72HCA1404⟩.

No photolyses of hexahydro-1,2,4,5-tetrazines have been published. The kinetics of the photodecomposition of 1,3,5-triphenylverdazyl (**61**) was studied by Soviet authors in various hydrocarbons. The rate depends on the solvent. The excited radical is consumed by H extraction from the solvent and by dissociation at the N—N bond, forming the product 1,3-diphenyl-1,2,4-triazole (**138**), which catalyzes the decomposition of the verdazyl radical ⟨74IZV2204⟩. The dipole moment of (**61**) in the excited state is $9\pm1\,\text{D}$; in the ground state it is 2.94 D.

Both mononuclear $[\text{M(CO)}_5(\text{C}_2\text{H}_2\text{N}_4)]$ and binuclear complexes $[\{\text{M(CO)}_5\}_2(\text{C}_2\text{H}_2\text{N}_4)]$ are formed when unsubstituted tetrazine $(\text{C}_2\text{H}_2\text{N}_4)$ is added to a photoirradiated solution of the metal hexacarbonyl $[\text{M(CO)}_6]$ (M = Cr, Mo, W) in ethanol. For the mononuclear complexes two signals were obtained in the $^1\text{H}$ NMR spectra while in the binuclear complexes only one signal was found ⟨76ZN(B)1489⟩.

Reaction of verdazyls (**119**) with organolithium compounds, organoaluminum compounds or Grignard reagents proceeds through a bimolecular homolytic substitution process to give 1,2,3,4-tetrahydro-1,2,4,5-tetrazines (**141**) in yields up to 50% ⟨76BCJ253, 75BCJ3765, 76BCJ1715, 75CL19⟩.

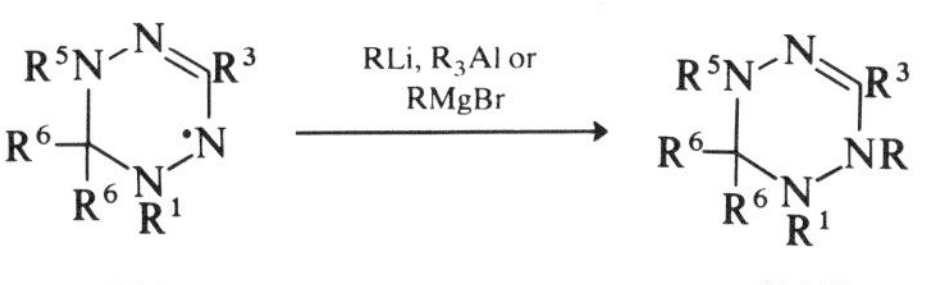

A reaction in the 1,2,4,5-tetrazine series which has received a great deal of attention is the cycloaddition reaction with alkenes and alkynes. The 1,2,4,5-tetrazines (**39**) act as the diene while the alkene or the alkyne is the dienophile. From the kinetics of these reactions they have been classified as Diels–Alder reactions with inverse electron demand. Besides alkenes and alkynes, compounds with C=N double bonds such as azomethines or azirines, and even nitriles, can react with 1,2,4,5-tetrazines. The first observations on this mode of reaction were made by Lindsay and Carboni, who used fluoroalkyl tetrazines ⟨57USP2817662, 62USP3022305, 59JA4342⟩. The cycloaddition was subsequently investigated by many groups, and these studies showed that the reactions of most 1,2,4,5-tetrazines with alkenes are particularly rapid and proceed in high yield. Nenitzescu suggested the use of dimethyl 1,2,4,5-tetrazine-3,6-dicarboxylate for the titrimetric determination of alkenes ⟨62CB2248⟩. Because of their high reactivity, 1,2,4,5-tetrazines can also be used for the detection of unstable compounds which have a reactive double or triple bond.

The first product of the reaction of 1,2,4,5-tetrazines (**39**) with alkenes (**142**) is the bicyclic compound (**143**), which has never been isolated since it loses nitrogen immediately by a retro-Diels–Alder reaction to give 4,5-dihydropyridazine (**144**). These compounds have been isolated in a few cases but usually a product is obtained which is formed from (**144**) in one of four different ways:

1. Tautomerization of (**144**) to yield the 1,4-dihydropyridazine (**145**).
2. Oxidation of (**144**) to give the pyridazine (**146**). For this reaction it is necessary that the alkene has protons at each end of the double bond. Very often an excess of the tetrazine acts as the oxidizing agent.
3. Elimination of H—X if the alkene used had a leaving group at one carbon atom of the double bond and a proton at the other carbon atom, leading to pyridazines (**146**).
4. Further cycloaddition of the 4,5-dihydropyridazine (**144**) with excess of the alkene to yield the isolable 2,3-diazabicyclo[2.2.2]oct-2-ene (**147**; Scheme 8).

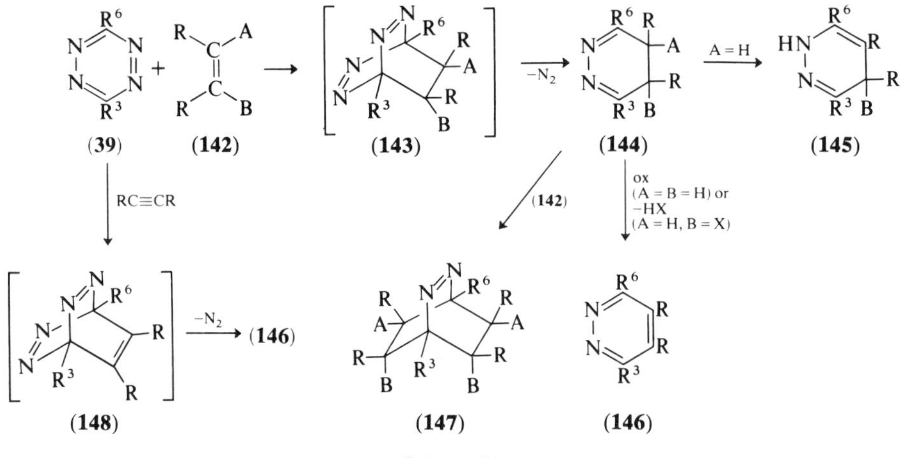

**Scheme 8**

By reacting 1,2,4,5-tetrazines with alkynes the initially-formed 2,3,5,6-tetraazabicyclo[2.2.2]oct-2,5,7-triene (**148**) decomposes without isolation to give the pyridazine (**146**) by elimination of nitrogen. In most cases the reaction with alkynes requires higher temperatures and longer reaction times than with alkenes (Scheme 8).

The number of 1,2,4,5-tetrazines used in these cycloaddition reactions is limited. The most frequently used are the 3,6-dicarboxylates, the 3,6-diphenyl, 3,6-dimethyl, 3,6-di(2-pyridyl) and bis(perfluoroalkyl) compounds; a few others have also been used. The most reactive substances seem to be the 1,2,4,5-tetrazinedicarboxylates, possessing the extra electron-withdrawing effect of the two carboxylate groups.

In the following we discuss a few of these cycloadditions. A summary of these reactions by Sauer and Sustmann is soon to be published.

1,2,4,5-Tetrazines (**39**) react with cyclopropenes (**149**) *via* the unisolated tricyclic intermediates (**150**) to form the diazanorcaradienes (**151**) by loss of nitrogen. Three reactions of the diazanorcaradienes have been observed: (1) opening of the ring to give the 5*H*-diazepines (**152**); (2) isomerization by the 'walk' mechanism to the diazanorcaradienes (**153**), which may be transformed into the 4*H*-diazepines (**154**); (3) reaction with another molecule of the cyclopropene, to afford the tetracyclic compounds (**155**) which can be transformed into homotropilidenes (**156**), the bis(homo-Dewar benzenes) (**157**) or the bis(homo-benzenes) (**158**) ⟨79TL1299, 78TL3906, 77TL4393, 76TL4321, 79BSB905, 72JA2770, 70JA3787, 70TL1617, 66TL4979⟩.

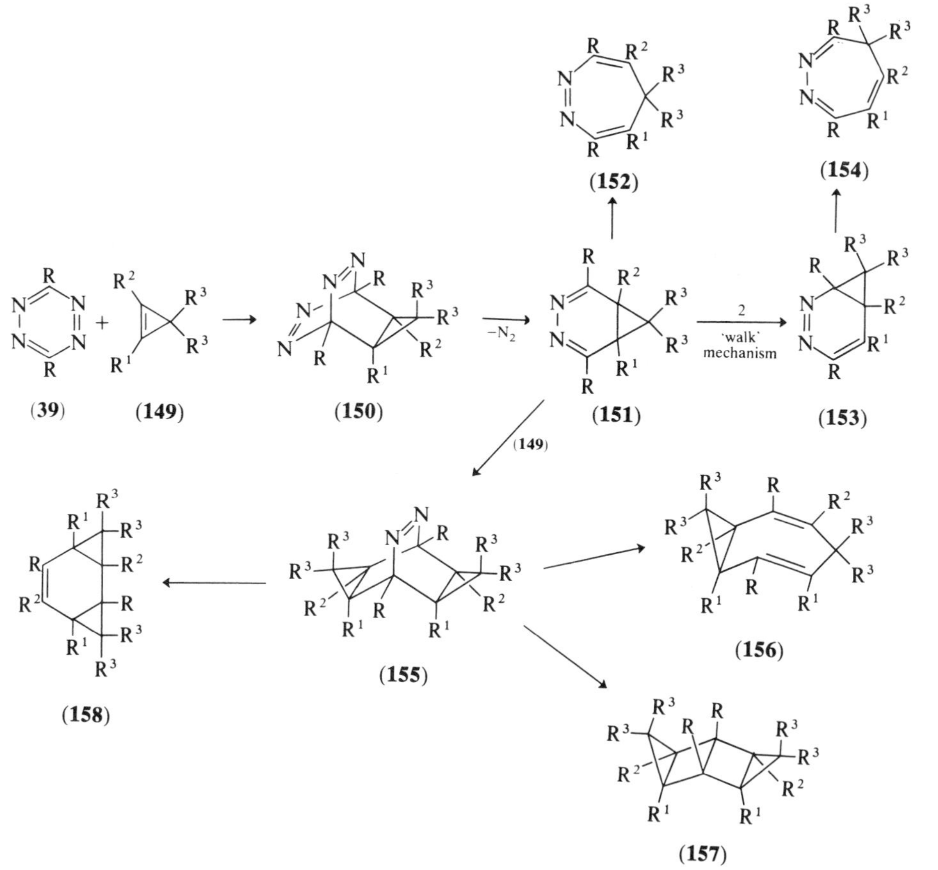

**Scheme 9**

Reaction of 1,2,4,5-tetrazines (**39**) with the dimethylbicyclopropenyl (**159**) proceeds *via* diazanorcaradienes (**160**) and diazasnoutenes (**161**) to the 2,4,6,8-tetrasubstituted semibull-valenes (**162**). In a few instances the intermediate diazanorcaradienes (**160**) could be detected ⟨80AG464⟩. Cyclobutadiene (**163**) also reacts with 1,2,4,5-tetrazines. In all cases studied, two molecules of (**163**) reacted with one of (**39**) to afford the tetracyclic compound (**164**). Analogously, benzocyclobutadiene (**165**) reacts to yield (**166**) ⟨69TL4509, 71JA7179, 76AG447⟩.

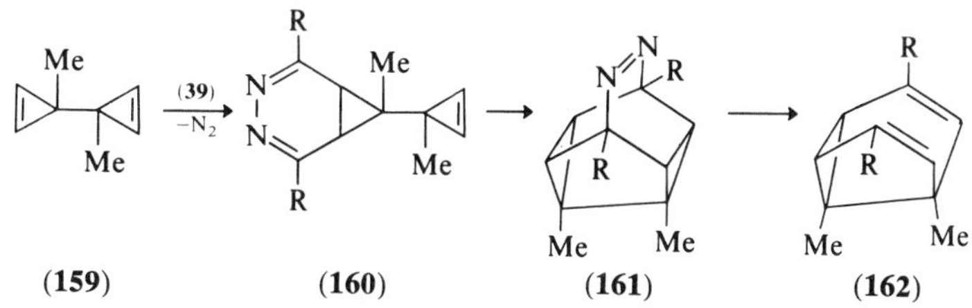

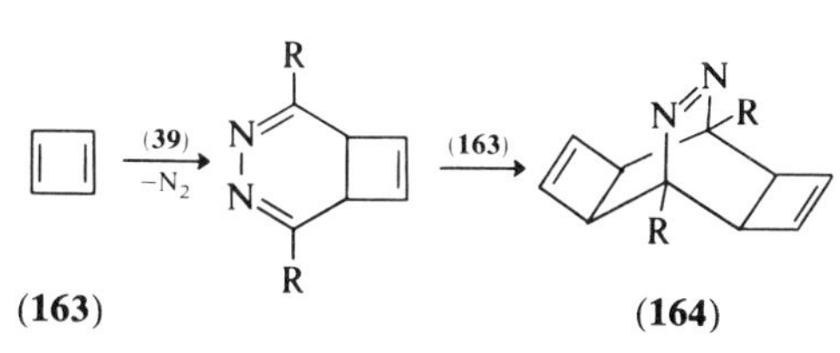

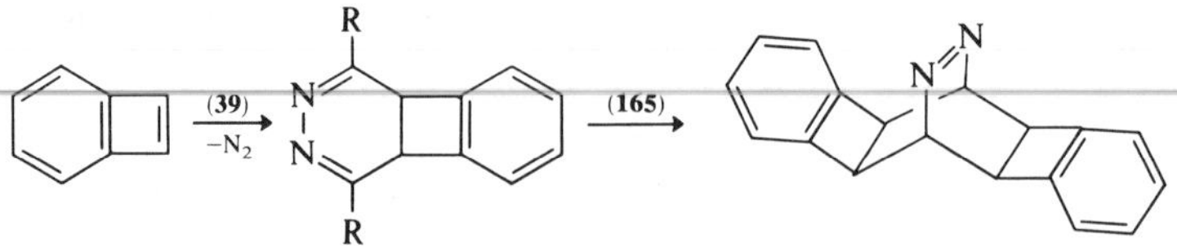

(165)                                                                    (166)

The reaction of 1,2,4,5-tetrazines (**39**) with 1*H*-azepines, oxepin, bicyclo[6.1.0]nona-2,4,6-triene (**167**) and its 9-aza- (**168**) and 9-oxa analogues (**169**), thiophene, 1-methyl-imidazole, benzofuran, benzothiophene, *N*-methylindole (**170**) and the pyrrolopyridine (**171**) have been investigated, and the products (**172**)–(**182**), respectively, were isolated ⟨78AP786, 78AP728, 76AP679, 76BCJ1725, 76CC313, 75AG842, 75CZ292, 74RTC321⟩.

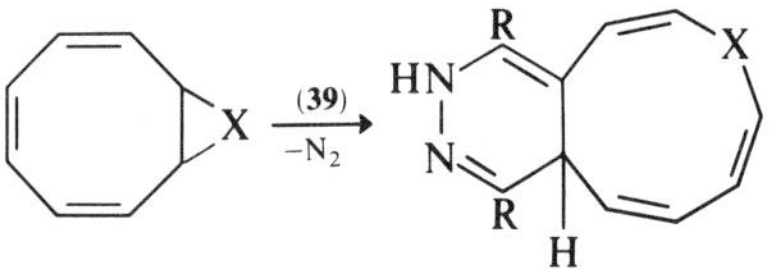

(167) X = CH₂        (174) X = CH₂                    (170) X = CH          (181) X = CH
(168) X = NR         (175) X = NR                     (171) X = N           (182) X = N
(169) X = O          (176) X = O

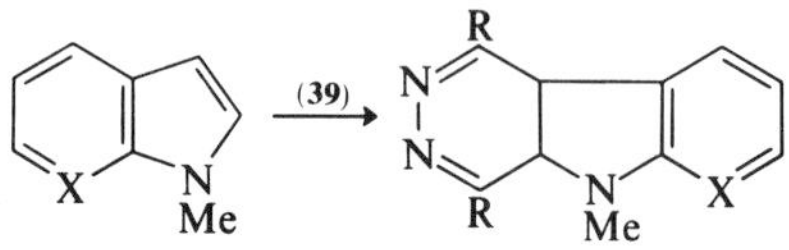

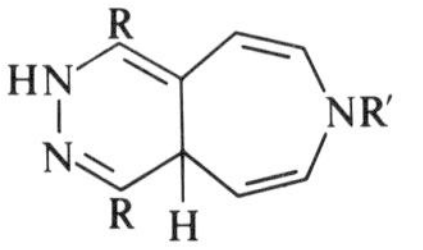

(172) 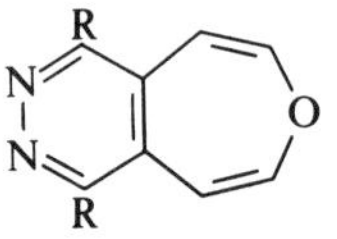                (173) 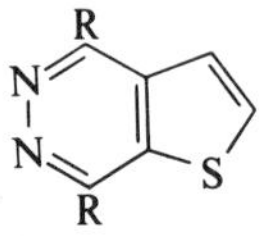                (177) 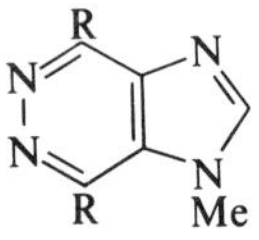                (178)

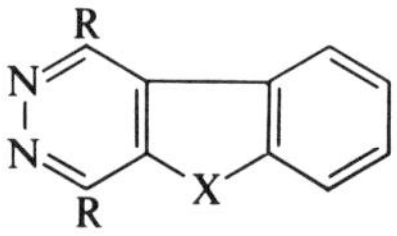

(179) X = O
(180) X = S

1,4-Epoxy-1,4-dihydronaphthalene (**183**) reacted with 3,6-di(2-pyridyl)-1,2,4,5-tetrazine (**184**) to afford the azine (**185**) which rapidly decomposed at room temperature to give isobenzofuran (**186**) and 3,6-di(2-pyridyl)pyridazine (**187**) ⟨71JA2346⟩. This method of removing the elements of acetylene to generate highly reactive *o*-quinonoid systems has been applied also to the synthesis of isoindole and substituted isoindoles (**189**) from (**188**). The reaction was extremely rapid in chloroform and no evidence for an intermediate analogous to (**185**) could be obtained ⟨72TL4295⟩. Starting with (**190**) the isobenzofulvene (**191**) was obtained in good yield ⟨72BCJ1999⟩.

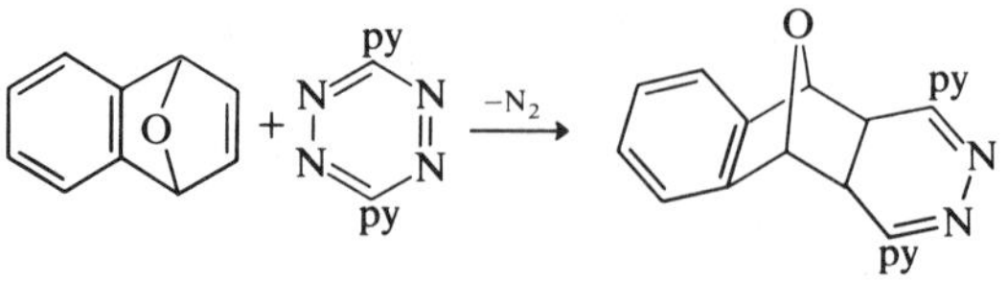

(183)            (184)                    (185)

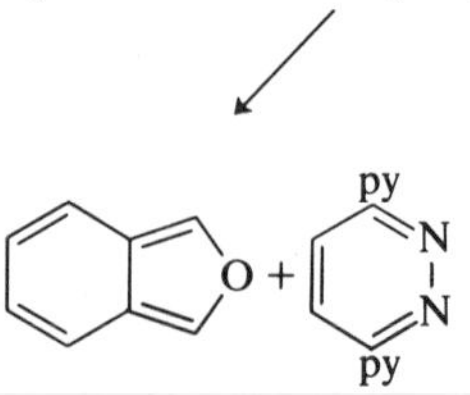

(186)            (187)

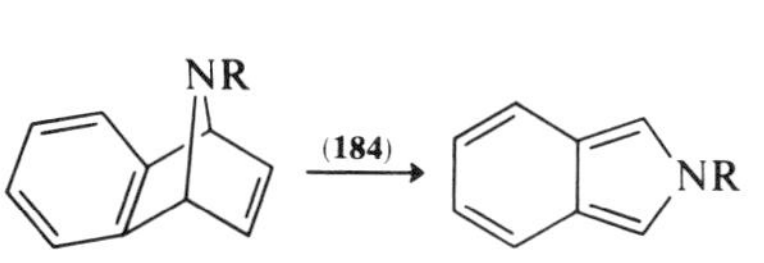

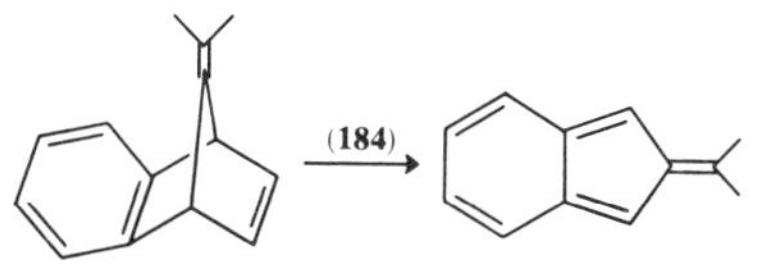

(188)  (189)  (190)  (191)

Reaction of benzyne with arsabenzene affords 1,4-etheno-1,4-dihydro-1-arsanaphthalene (192) from which an etheno bridge can be removed with 3,6-di(2-pyridyl)-1,2,4,5-tetrazine (184) to afford, *via* the polycyclic intermediate (193), the arsanaphthalene (194) ⟨79CC880⟩.

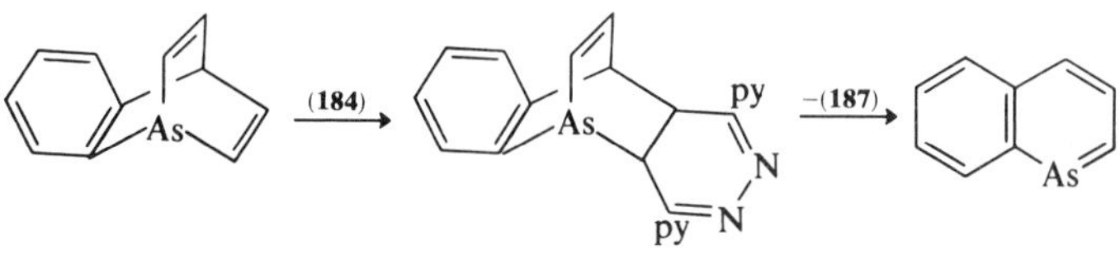

(192)  (193)  (194)

The cycloaddition of 3,6-diphenyl-1,2,4,5-tetrazine (51) with 6-dimethylaminofulvene (195) has been studied by Sasaki *et al.* ⟨75JOC1201⟩ and by Neunhoeffer and Bachmann ⟨79LA675⟩. The Japanese group claim to have isolated 4,7-diphenyl-5,6-diazaazulene (196), produced *via* a [6+4] cycloaddition and dimethylamine elimination, but Neunhoeffer and Bachmann found that the reaction proceeds by a [4+2] cycloaddition, then loss of nitrogen and dehydrogenation, to form 5-dimethylaminomethylene-1,4-diphenyl-5*H*-cyclopenta[*d*]pyridazine (197).

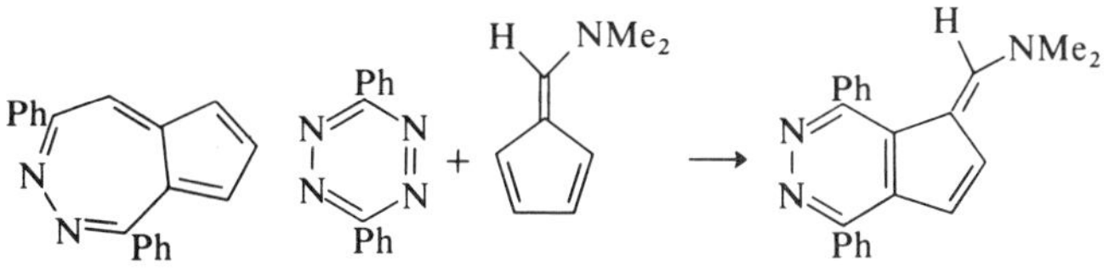

(196)  (51)  (195)  (197)

In 1969 the reaction of 1,2,4,5-tetrazines (39) with imidates (198) was reported. The formation of the 1,2,4-triazine products (199) is explained by a [4+2] cycloaddition of the tetrazine with the C=N double bond of the imidate to give the bicyclic intermediate (200) which eliminates nitrogen and alcohol ⟨69JHC497⟩. Further studies on this reaction seem to

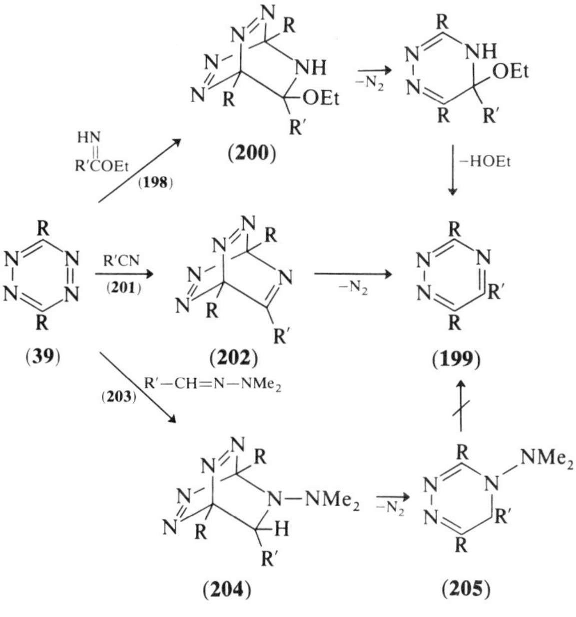

**Scheme 10**

reveal its limitations ⟨77AP936⟩. By reacting 1,2,4,5-tetrazines (**39**) with cyanides (**201**) the bicyclic intermediates (**202**) are formed, which lead directly to the 1,2,4-triazines (**199**) by loss of nitrogen ⟨79CZ230⟩. The tetrazines (**39**) also react with the C=N double bond of dimethylhydrazones (**203**; R = alkyl or β-styryl). The bicyclic intermediate (**204**) loses nitrogen to give 4-amino-4,5-dihydro-1,2,4-triazines (**205**); aromatization with loss of dimethylamine was not observed (Scheme 10) ⟨79AP452, 81AP376⟩.

Azirines (**206**) behave rather similarly to cyclopropenes with 1,2,4,5-tetrazines. The tricyclic intermediate (**207**) eliminates nitrogen to give the triazanorcaradiene (**208**) which may form the 5H-1,2,4-triazepines (**209**) by valence tautomerization. None of these two compounds have so far been isolated but a large variety of heterocyclic substances have been obtained, which were formed from (**209**) by isomerization or elimination reactions. The nature of the products isolated appears to be strongly dependent upon both the azirine (**206**) and 1,2,4,5-tetrazine (**39**) chosen and the reaction time. The following compounds have been isolated: 2H-1,2,4-triazepines (**210**), 6H-1,2,4-triazepines (**211**), pyrimidines (**212**), pyrazoles (**213**) and aziridinylpyrazoles (**214**; Scheme 11) ⟨75S483, 75BCJ2605, 75JHC183, 74CC782, 74TL2303⟩.

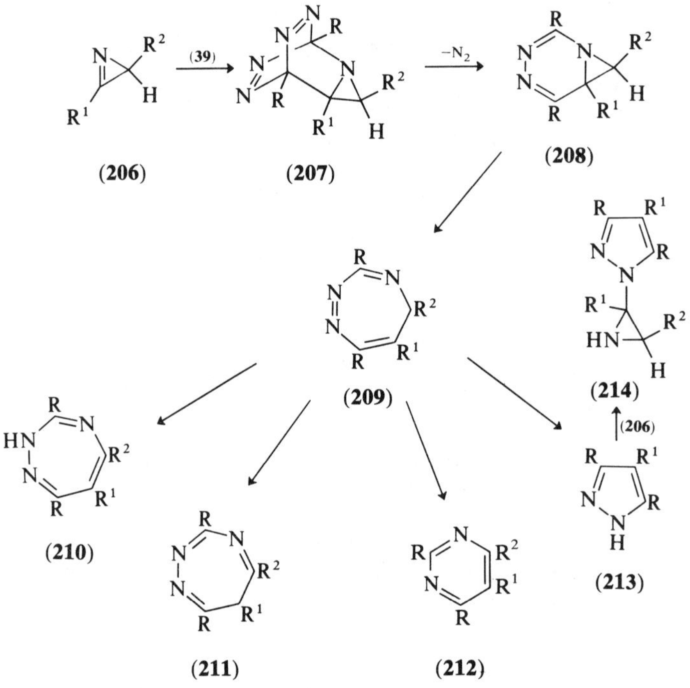

**Scheme 11**

3,6-Di(2-pyridyl)-1,2,4,5-tetrazine (**184**) has been used to trap 2-phenylbenzazete (**215**). The product isolated was the 1,2,4-benzotriazocine (**216**) ⟨75JCS(P1)45⟩.

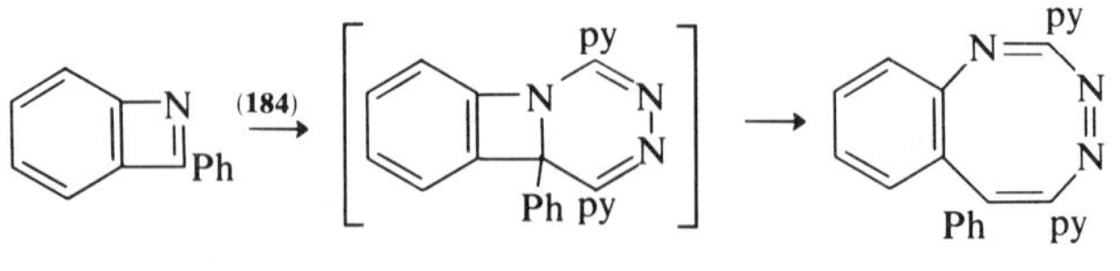

Reaction of 1,2,4,5-tetrazines (**39**) with *N*-thionitrosodimethylamine (**217**) and *N*-sulfinylanilines (**218**) led to the isolation of 4-dimethylamino-4H-1,2,4-triazoles (**219**) and 4-aryl-4H-1,2,4-triazoles (**220**), respectively. The formation of these products was explained by a [4 + 2] cycloaddition to give bicyclic intermediates, then elimination of nitrogen to yield the 1,2,4,5-thiatriazines, which form the products by extrusion of sulfur and sulfur oxide, respectively (Scheme 12) ⟨79CZ230, 77AP269⟩.

Substituents bound to the 1,2,4,5-tetrazine ring behave very much as expected. Heterosubstituents can be exchanged by nucleophilic substitution as is shown in Scheme

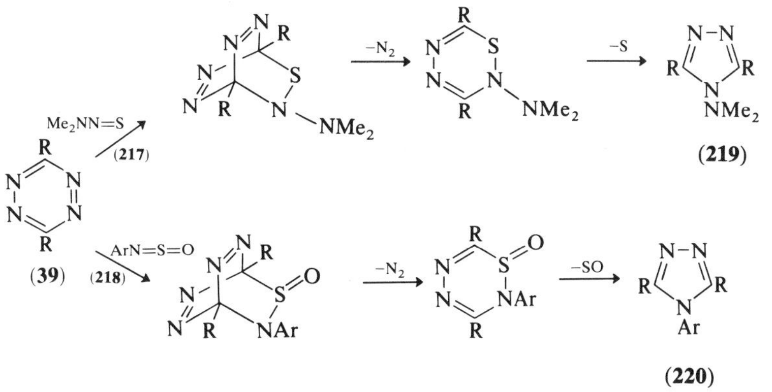

**Scheme 12**

13. Furthermore, the hydrazino group can be replaced by a proton *via* oxidation or can be transformed into the azido group 〈78HC(33)1075, p. 1101, 80JHC501, 79MIP22100, 78JHC445, 78DOK(241)366, 76USP3951941, 77JHC587, 77KGS1564〉.

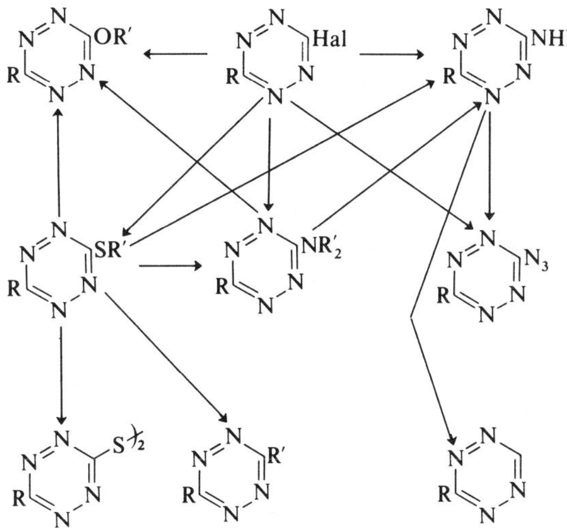

**Scheme 13**

Decarboxylation of 1,2,4,5-tetrazine-3,6-dicarboxylic acid (**221**) and the dihydrodicarboxylic acid (**222**) was used to prepare the parent 1,2,4,5-tetrazine (**38**) and the dihydro-1,2,4,5-tetrazine (**223**), respectively 〈78HC(33)1075, p. 1078, 1888JPR(38)531, 00CB58〉.

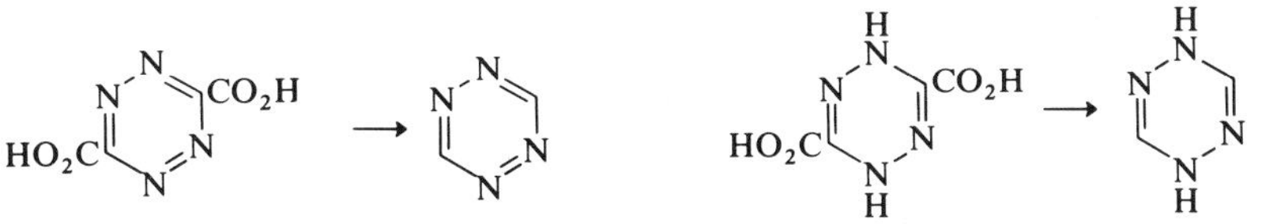

### 2.21.3.4 Syntheses

Most methods for the preparation of 1,2,4,5-tetrazines are basically derived from [1 + 2 + 1 + 2] atom fragments. Starting from these four components the intermediates (**224**)–(**226**) can be prepared, which then can be used for the preparation of the 1,2,4,5-tetrazines (**228**) by a [6 + 0] method (**227**), a [5 + 1] method (**225**), a [4 + 2] method (**224**), or a [3 + 3] atom fragment method (**226**; Scheme 14).

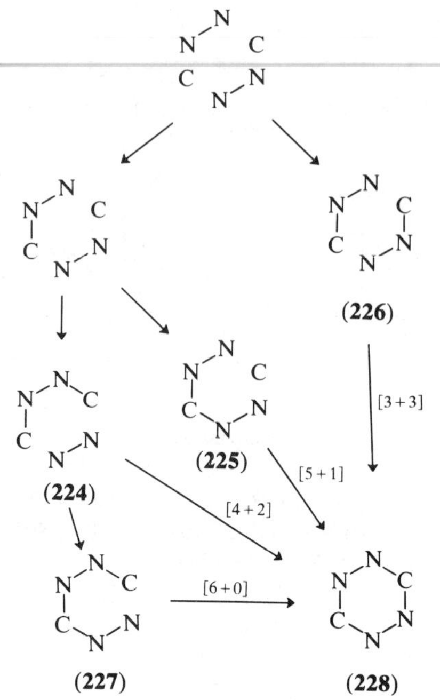

**Scheme 14**

### 2.21.3.4.1 *Synthesis by the [6+0] atom fragment method*

Heating 4-amino-3-azido-4*H*-1,2,4-triazoles (**229**) in chlorobenzene to 110 °C gives the 3-amino-1,2,4,5-tetrazines (**232**). This reaction is explained by the intermediate formation of the nitrene (**230**) which opens to the formazane (**231**) and subsequent ring closure by nucleophilic attack of the hydrazono group on the nitrile function ⟨66TL5369⟩.

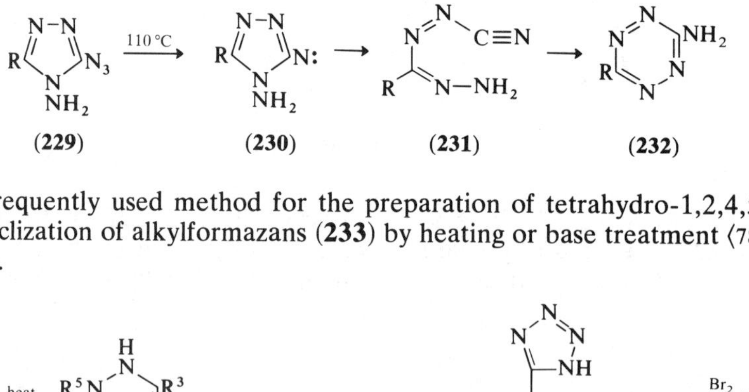

(**229**) (**230**) (**231**) (**232**)

The most frequently used method for the preparation of tetrahydro-1,2,4,5-tetrazines (**234**) is the cyclization of alkylformazans (**233**) by heating or base treatment ⟨78HC(33)1075, p. 1173, 75T555⟩.

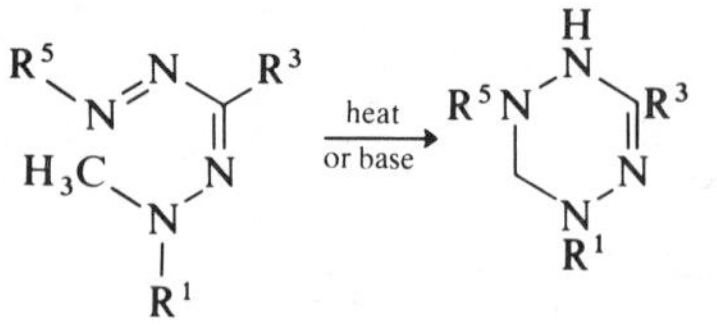

(**233**) (**234**)

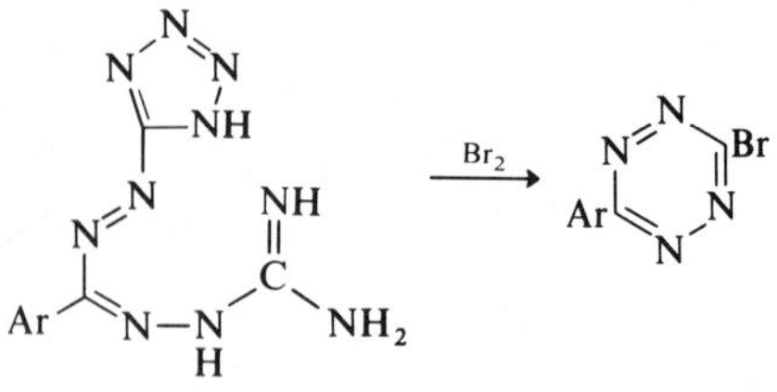

(**235**) (**236**)

The *C*-aryl-*N*-guanyl-*N'*-(5-tetrazolyl)formazans (**235**) react with bromine in acetic acid to afford 6-aryl-3-bromo-1,2,4,5-tetrazines (**236**) ⟨58JA3155⟩.

The benzimidoyl thiocarbodihydrazide (**237**) was cyclized by Soviet chemists to yield 3-phenyl-1,4-dihydro-1,2,4,5-triazine-6(5*H*)-thione (**238**). But since (**237**) was prepared by reaction of ethyl benzimidate and thiocarbodihydrazide this reaction can also be classified as a synthesis by a [5+1] atom fragment method ⟨77KGS564⟩. When the perfluoroalkyl hydrazidines (**239**) were heated with anhydrous iron(III) chloride, cyclization occurred with concurrent dehydration and oxidation forming 3,6-bis(perfluoroalkyl)-1,2,4,5-tetrazines (**240**) ⟨66JOC781⟩.

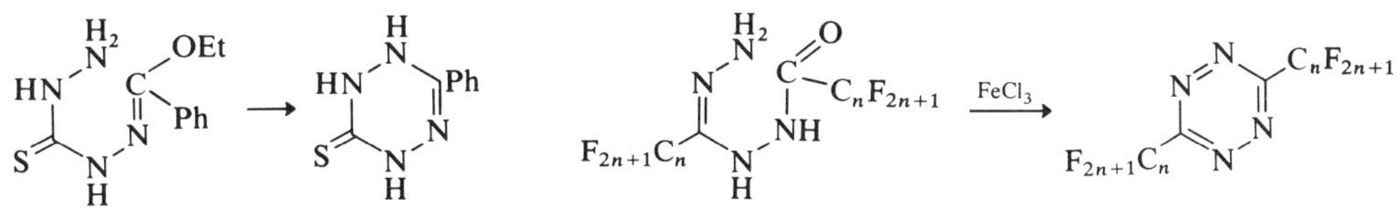

Interaction of 1,3,5-triazine (**241**) with hydrazine affords 1,2-diformylhydrazine dihydrazone (**242**) which is a light-sensitive compound which turns pink, especially in the presence of air. The color is attributed to the formation of 1,2,4,5-tetrazine (**38**) but due to its high volatility the tetrazine could not be isolated. Heating (**242**) for two hours with acetic anhydride again affords an unisolated red, volatile product formulated as (**38**), and 1,2-diacetyl-1,2-dihydro-1,2,4,5-tetrazine (**243**). Since compound (**242**) is prepared from hydrazine and (**241**) this preparation is in the final analysis a [2 + 1 + 2 + 1] method ⟨57JA2839⟩.

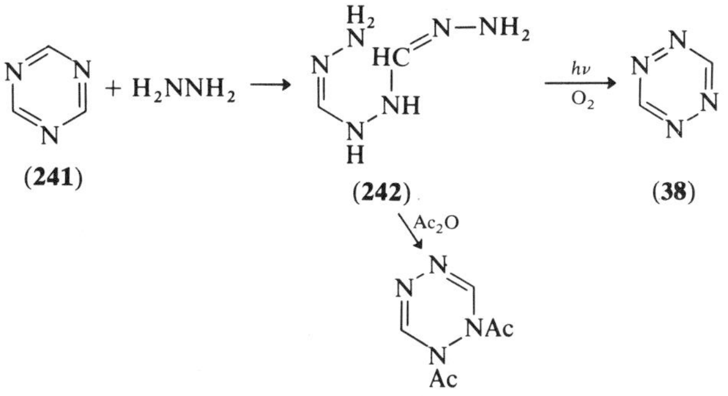

Heating compound (**244**) for several minutes with copper(II) sulfate in pyridine led to the isolation of 3,6-diphenyl-1,2,4,5-tetrazine (**51**). The mechanism of this reaction seems to be cyclization of (**244**) to give the intermediate (**245**) which eliminates prussic acid, affording the dihydrotetrazine (**80**), which is oxidized by copper(II) sulfate ⟨60AC(R)277⟩.

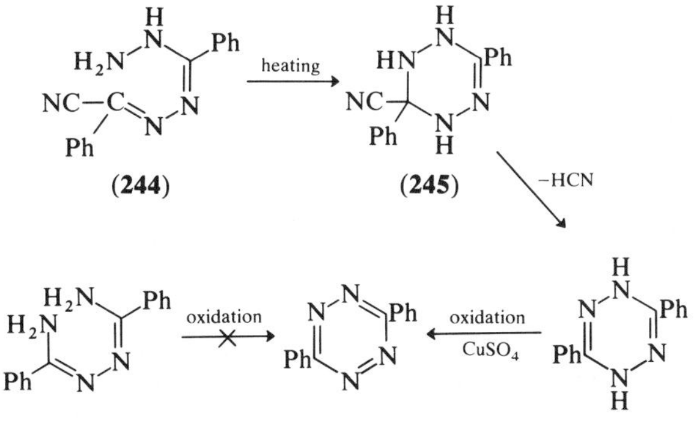

Although Wuyts and Lacourts ⟨36BSB685⟩ have reported the formation of 3,6-diphenyl-1,2,4,5-tetrazine (**51**) on oxidation of the azine (**246**), this was later shown by Lutz ⟨B-67MI22101⟩ to be incorrect; the isolated 1,2,4,5-tetrazine (**51**) was formed by oxidation of the dihydrotetrazine (**80**), an impurity in the starting azine (**246**).

### 2.21.3.4.2 Synthesis by the [5 + 1] atom fragment method

As mentioned in the preceding section, alkylformazans (**249**) can be cyclized by heat or with a base to tetrahydro-1,2,4,5-tetrazines (**250**). The alkylformazans (**249**) are prepared by alkylation of formazans (**248**). The transformation can be carried out without isolation of the intermediate (**249**), at room temperature in the presence of barium oxide or barium hydroxide octahydrate in DMF ⟨78HC(33)1075, p. 1173⟩. Very often the tetrahydro-1,2,4,5-tetrazine (**250**) is oxidized to yield verdazyls (**251**), but these can be reduced back to the tetrahydrotetrazines (**250**). Because the formazans are synthesized from aldehyde arylhydrazones (**247**) and diazonium salts, the substituents at the nitrogens are limited to aryl or hetaryl groups (Scheme 15).

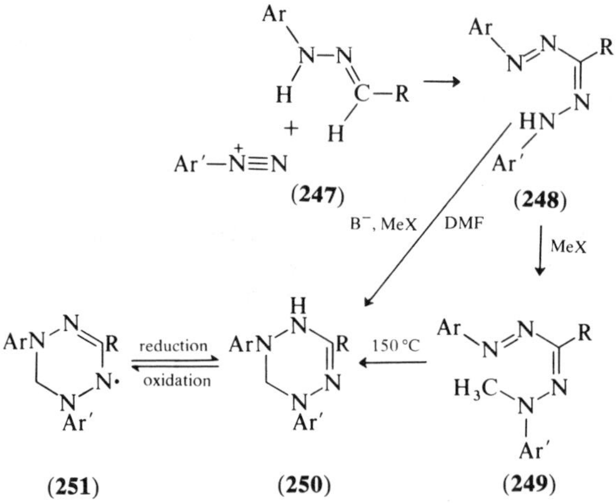

**Scheme 15**

Another method for the preparation of tetrahydro-1,2,4,5-tetrazines is by the cyclization of formazans (**248**) with an aldehyde in the presence of acid to yield the verdazylium cation (**252**), which is reduced under basic conditions to give the verdazyl (**253**) and further to the tetrahydro-1,2,4,5-tetrazines (**254**) ⟨78HC(33)1075, p. 1174, 74T2841, 80CB2049, 78KGS991, 78KGS1137⟩.

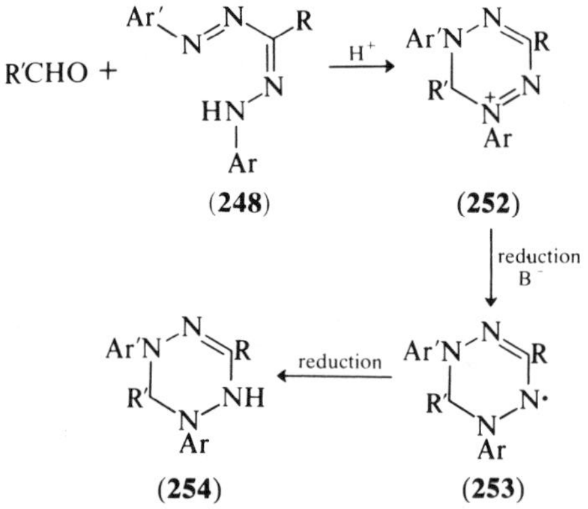

Besides formazans, other compounds of the hydrazidine type can be cyclized to the 1,2,4,5-tetrazine system. Neunhoeffer and Degen ⟨75LA1120⟩ reported the cyclization of hydrazidines (**255**) with orthocarboxylates or imidates to give 4-amino-1,2,4-triazoles, but under certain reaction conditions dihydro-1,2,4,5-tetrazines (**256**) were also formed which could be oxidized to afford the 1,2,4,5-tetrazines (**257**) ⟨75UP22100⟩. A variant on this method was published by the same authors. Reaction of (**255**) with mucochloric acid (**258**) or phthalaldehydic acid (**259**) led to the condensed 1,2,4,5-tetrazines (**260**, **261**) which could be transformed into the 1,2,4,5-tetrazines (**262**, **263**) by oxidation with potassium permanganate in methanol (Scheme 16) ⟨75CB3509, 79CB1981⟩.

Cyclization of the disubstituted hydrazidines (**264**) with phosgene and carbon disulfide was used to prepare dihydro-1,2,4,5-tetrazin-6-ones (**265**) ⟨76CZ496⟩ and dihydro-1,2,4,5-tetrazine-6-thiones (**266**) respectively ⟨73CZ565⟩.

Thiobenzoylation of the methylthioformhydrazidine (**267**) affords dihydromethylthio-1,2,4,5-tetrazines (**268**) which can be oxidized by bromine to the aromatic compounds (**269**) ⟨75JCS(P1)1787, 79JHC881⟩. Similarly, thiocarbodihydrazide (**270**) with potassium

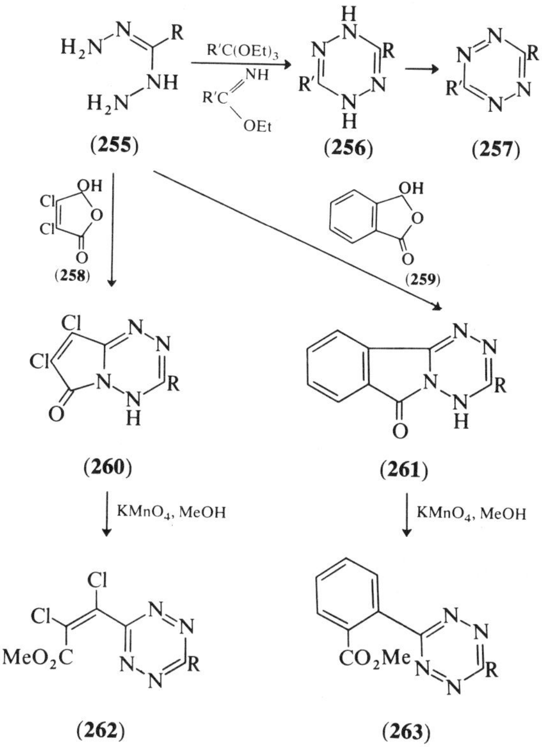

**Scheme 16**

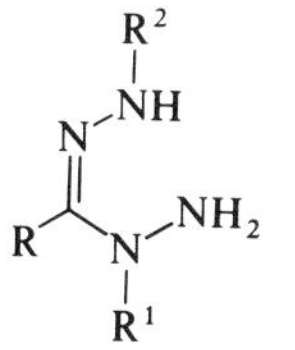

**(264)**

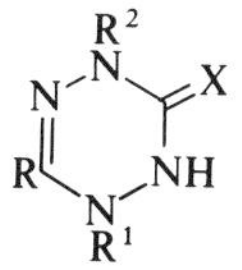

(**265**) X = O
(**266**) X = S

dithioisonicotinate (**271**) gave 3-(4-pyridyl)-1,2-dihydro-1,2,4,5-tetrazine-6-thione (**272**) (or its mercapto tautomer) ⟨54CB825, 56GEP953801⟩, and the 3,6-dithione (**273**) was obtained by reaction of (**270**) with bis(carboxymethyl)trithiocarbonate (**274**) in basic solution at room temperature ⟨61ACS1575, 80JHC501⟩.

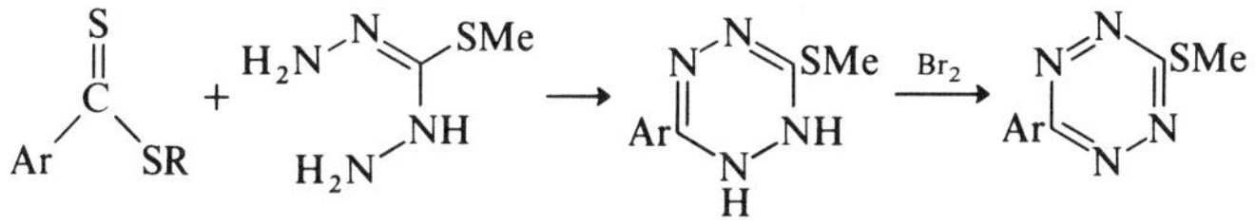

(**267**) (**268**) (**269**)

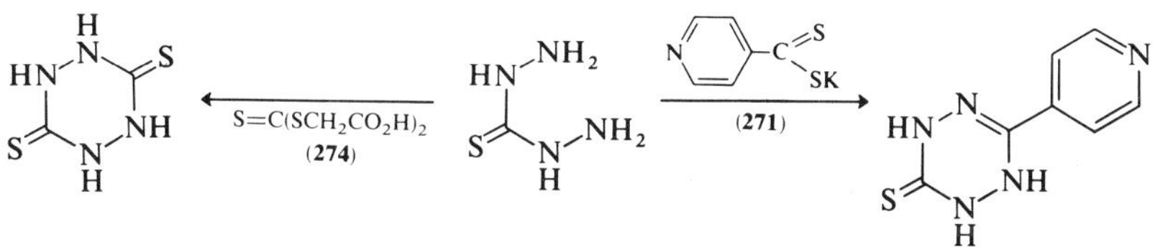

(**273**) (**270**) (**272**)

Like the thione (**270**), carbodihydrazide (**275**) and its derivatives can also be used as five-atom fragments for the preparation of tetrazines. Cyclization of (**275**) with orthoformate afforded a compound which was formulated as 1,4-dihydro-1,2,4,5-tetrazin-6(5*H*)-one

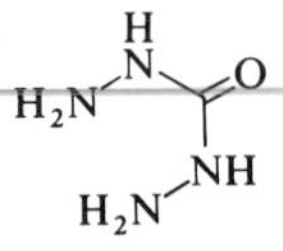

(275) 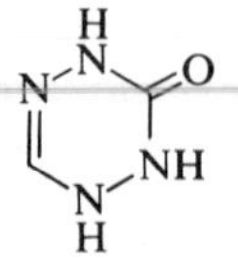                    (276) 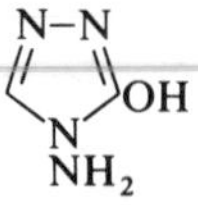                    (277)

(**276**) ⟨07CB2093⟩; however, Stollé proposed the alternative structure (**277**) ⟨07JPR(75)416⟩ and this, or a tautomer thereof, is likely to be correct.

Reaction of 1-phenylcarbodihydrazide (**278**) with carbon disulfide in ethanol in the presence of base was used for the synthesis of 1-phenyldihydro-6-thioxo-1,2,4,5-tetrazin-3-one (**279**) ⟨30JIC933⟩.

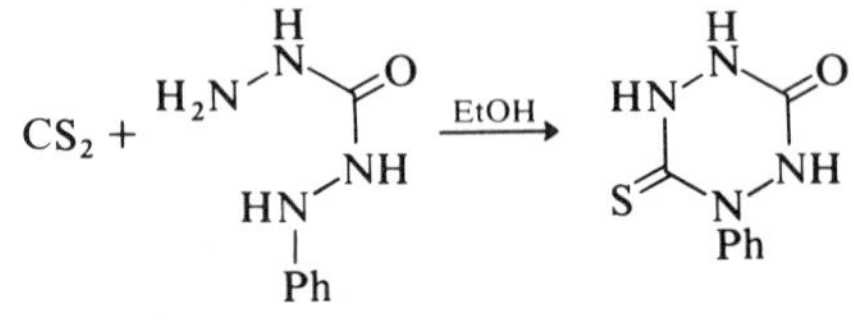

(278)                    (279)

Dimethylcarbodihydrazide (**280**) reacts with aldehydes to afford 1,5-dimethyltetrahydro-1,2,4,5-tetrazin-6-ones (**281**), which can be oxidized by silver oxide, potassium ferricyanide or lead dioxide to yield radicals (**282**) related to the verdazyls; these can be transformed into tetrahydro-1,2,4,5-tetrazines (**283**) by hydrogenation over palladium ⟨80AG766⟩.

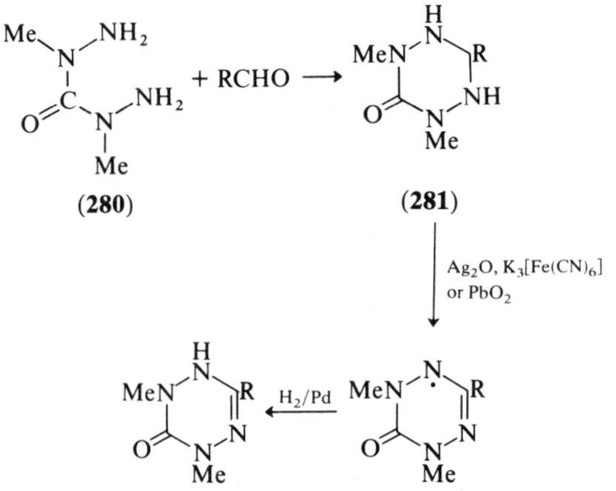

(280)                    (281)

Ag$_2$O, K$_3$[Fe(CN)$_6$]
or PbO$_2$

(283)                    (282)

Kuhn *et al.* found that triphenylverdazyl (**61**) was produced by reaction of 2,3,5-triphenyl-tetrazolium chloride (**284**) with diazomethane ⟨66M846⟩.

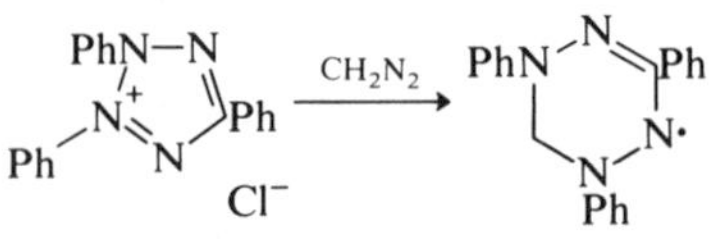

(284)                    (61)

### 2.21.3.4.3 Synthesis by the [4 + 2] atom fragment method

Compounds of the 1,3,4-oxadiazole or 1,3,4-thiadiazole series react with hydrazine or substituted hydrazines to afford 1,2,4,5-tetrazines or 4-amino-1,2,4-triazoles, depending on the reaction conditions used.

2,5-Bis(perfluoroalkyl)-1,3,4-oxadiazoles (**285**) are attacked readily by hydrazine with ring opening and formation of the perfluoroacylated hydrazidines (**286**). Concurrent dehydration and oxidation of these intermediates with iron(III) chloride produces the corresponding 3,6-bis(perfluoroalkyl)-1,2,4,5-tetrazines (**287**). Ring closure of (**286**) without oxidation produces the corresponding 3,5-bis(perfluoroalkyl)-4-amino-1,2,4-triazoles ⟨66JOC781⟩.

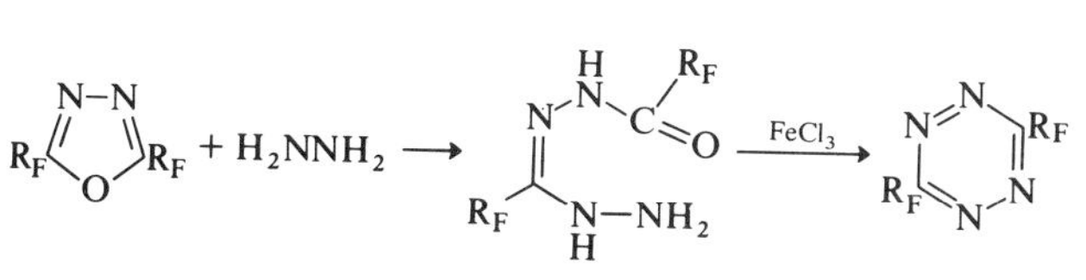

**(285)**         **(286)**         **(287)**

The 1,3,4-oxadiazolium salts **(288)** react with hydrazines to give the dihydro-1,2,4,5-tetrazines **(290)** and 4-amino-1,2,4-triazoles ⟨76KGS629, 74ZOR377, 71DOK(200)134, 71MIP22100⟩. This reaction probably proceeds *via* the intermediates **(289)**. Similarly, 2-methylmercapto-1,3,4-oxadiazolium salts **(291)** and hydrazines afford the 1,4-dihydro-1,2,4,5-tetrazin-6(5*H*)-ones **(292)** ⟨76CZ496⟩.

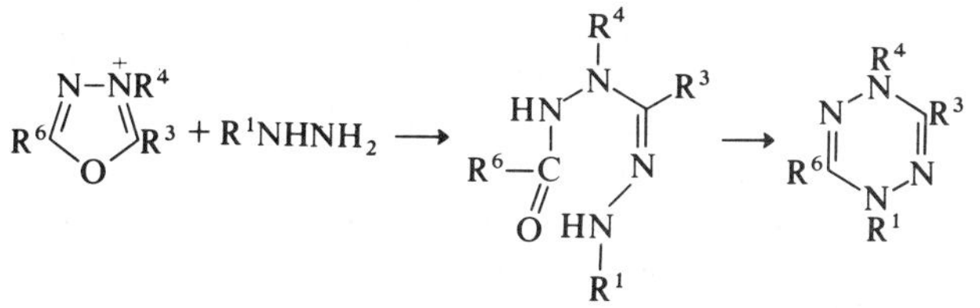

**(288)**         **(289)**         **(290)**

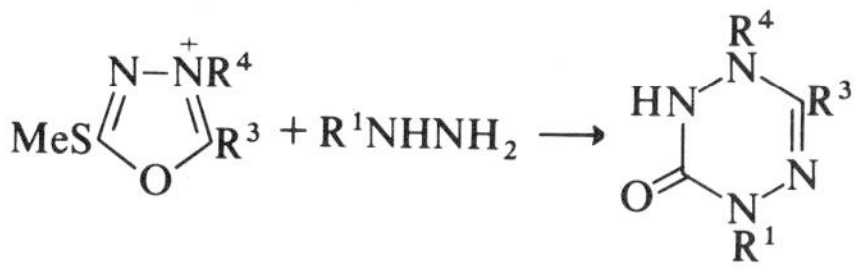

**(291)**         **(292)**

Artemov and Shvaika treated a number of 2-aryl-1,3,4-oxadiazole-5(4*H*)-thiones **(293)** with methylhydrazine (R = Me) and obtained a mixture of 1-methyl-1,4-dihydro-1,2,4,5-tetrazine-6(5*H*)-thiones **(294)** and 4-methylamino-1,2,4-triazole-3-thiones **(295**; R = Me) ⟨71KGS905⟩. A similar reaction was reported by König and his coworkers, who treated 5-(4-pyridyl)-1,3,4-oxadiazole-2-thione **(293**; Ar = py) with hydrazine (R = H) and obtained a compound which was either the 1,2,4,5-tetrazine-6-thione **(294**; R = H) or the 4-amino-1,2,4-triazole-3-thione **(295**; R = H) ⟨54CB825, 56GEP953801⟩.

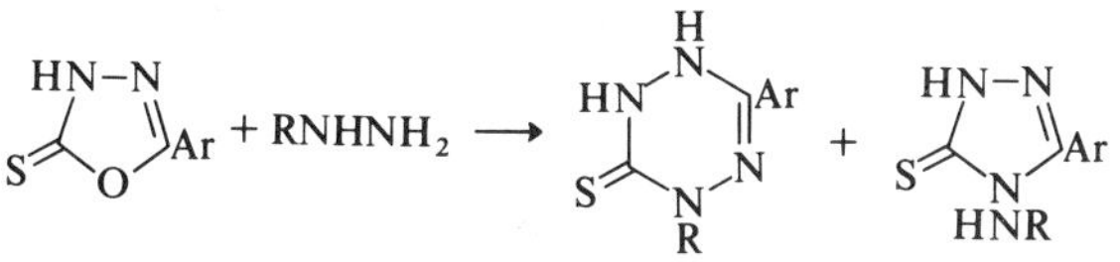

**(293)**         **(294)**         **(295)**

Like the oxadiazolium salts **(288)**, 1,3,4-thiadiazolium salts **(296)** can also react with hydrazines to afford 1,4-disubstituted 1,4-dihydro-1,2,4,5-tetrazines **(290)**. By this method, 1,4-dihydro-1,2,4,5-tetrazines with four different substituents can be prepared; this seems to be the only method for preparation of such compounds, and proceeds in yields of 70–99%. The 4-amino-1,2,4-triazoles were isolated as side-products ⟨74ZOR377, 71DOK(200)134, 71MIP22100⟩.

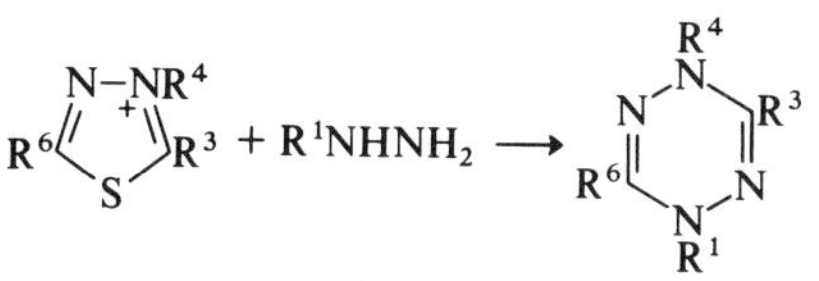

**(296)**      **(290)** 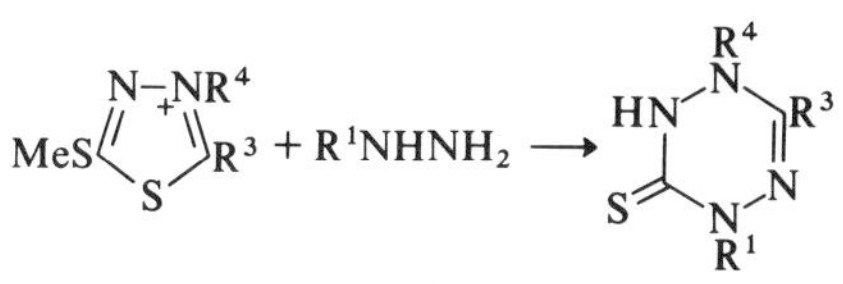         **(297)**      **(298)**

3-Methylthio-1,3,4-thiadiazolium salts **(297)** were treated with hydrazines in the ratio 1 : 2 to afford 1,4-dihydro-1,2,4,5-tetrazine-6(5*H*)-thiones **(298)**. In most cases the acyclic intermediate could not be isolated ⟨73CZ565, 73CZ566, 01CB2311⟩.

Interaction of the mesoionic thiadiazolium salt **(299**; Ar = Ph) with hydrazine led to the formation of 1,4-dihydro-6-hydrazino-1,2,4,5-tetrazine **(300)** ⟨76KGS713⟩. The reaction of salts **(299)** with azodicarboxylate esters was at one time thought to provide the mesomeric

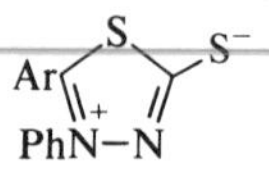

(299) 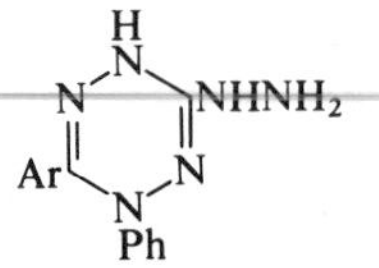    (300) 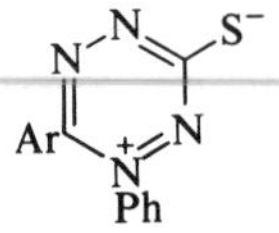    (301) 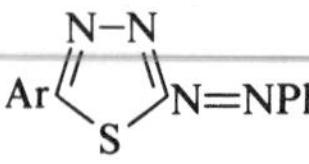    (302)

dipolar structures (**301**), but later it was shown that the products were in fact the phenyl-azothiadiazoles (**302**) ⟨71CC837, 72CC1300, 73CZ566⟩.

A very frequently used preparation of dihydro-1,2,4,5-tetrazines is the cyclization of 1,4-dichloroazines (**303**) with hydrazines ⟨76JOC3392, 75USP3860588, 75USP3860589, 79EUP5912, 78HC(33)1075, pp. 1117, 1122⟩. Dichloroazines (**303**) are prepared by the following methods: reaction of 1,2-diacylhydrazines (**304**) with phosphorus pentachloride; chlorination of aldehyde 2-acylhydrazones (**305**); and chlorination of aldehyde azines (**306**). In the chlorination of aldehyde 2-acylhydrazones (**305**) it was found that best results are obtained when the acyl group is first transformed into the chloroimidoyl group by reaction with thionyl chloride and then the azomethine group chlorinated ⟨76JOC3392⟩. In the cases where $R^1 = R^2 = H$ (when hydrazine is used for the cyclization reaction) the dihydro-1,2,4,5-tetrazines (**107**) can be oxidized to 1,2,4,5-tetrazines (**39**) by the usual methods. The direct cyclization of 1,2-diacylhydrazines (**304**) with hydrazines to yield the dihydro-1,2,4,5-tetrazines (**107**) is also reported. Because of the two-step preparation of the starting material (**305**) this method can be used for the synthesis of unsymmetrically substituted tetrazines (**39**; $R^3 \neq R^6$; Scheme 17).

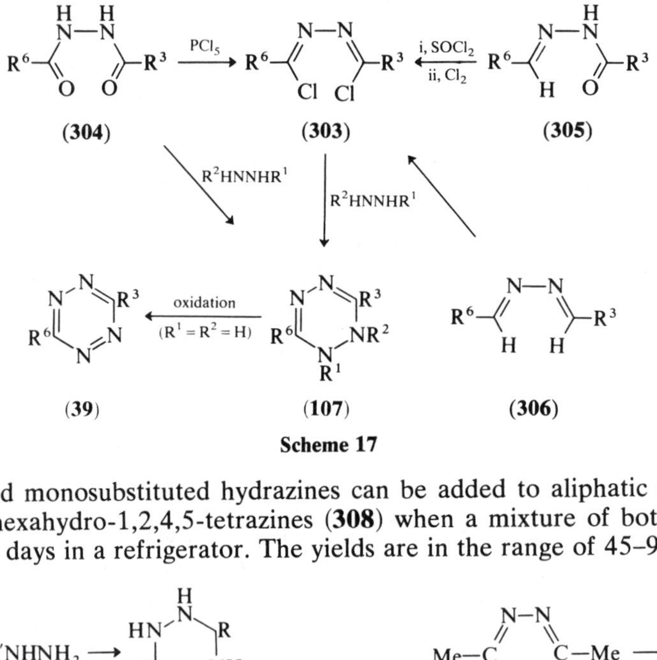

**Scheme 17**

Hydrazine and monosubstituted hydrazines can be added to aliphatic aldehyde azines (**307**) to yield hexahydro-1,2,4,5-tetrazines (**308**) when a mixture of both compounds is kept for several days in a refrigerator. The yields are in the range of 45–95% ⟨70HCA251⟩.

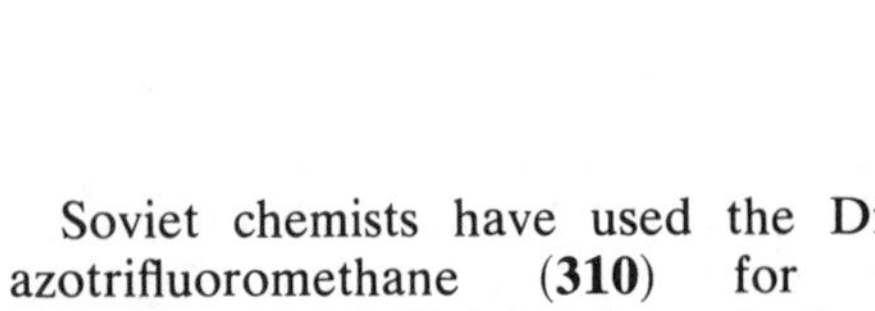

(307)    (308) 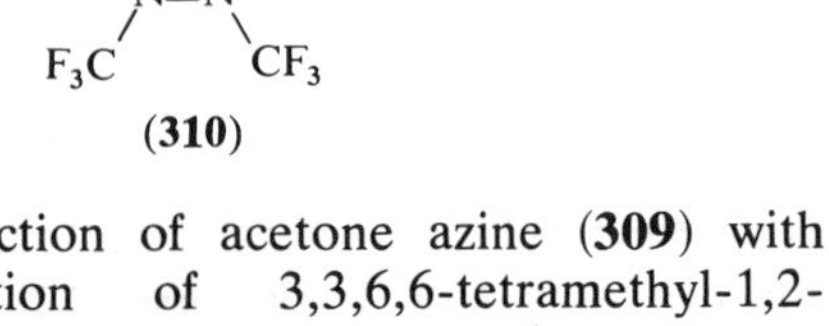    (309)    (311)

+

(310)

Soviet chemists have used the Diels–Alder reaction of acetone azine (**309**) with azotrifluoromethane (**310**) for the preparation of 3,3,6,6-tetramethyl-1,2-bis(trifluoromethyl)-1,2,3,6-tetrahydro-1,2,4,5-tetrazine (**311**) ⟨62DOK(142)354⟩.

This by no means exhausts the reported syntheses of 1,2,4,5-tetrazines by [4+2] atom cyclizations, but other examples are of less general application or value as synthetic methods.

### 2.21.3.4.4 *Synthesis by the [3 + 3] atom fragment method*

A [3 + 3] atom fragment method for the preparation of dihydro-1,2,4,5-tetrazines is the dimerization of diazo compounds under the influence of bases. In 1900 Hantzsch and Lehmann observed the formation of a new compound when ethyl diazoacetate (**312**) was treated with sodium hydroxide ⟨00CB3668⟩. The structure which they proposed originally was incorrect; it was shown later that the product isolated was the disodium salt of 1,2-dihydro-1,2,4,5-tetrazine-3,6-dicarboxylic acid (**313**) (or the 1,4-dihydro compound) ⟨78HC(33)1075, p. 1078⟩.

$$\text{N}_2\text{CHCO}_2\text{Et} \xrightarrow{\text{NaOH}}$$

(**312**)  (**313**)

The use of ammonia or amines instead of sodium hydroxide for the dimerization of ethyl diazoacetate leads to the formation of dihydro-1,2,4,5-tetrazine-3,6-dicarboxamides. The products isolated were formulated either as 1,2-dihydro-1,2,4,5-tetrazine derivatives (**314**), as 1,6-dihydro-1,2,4,5-tetrazine derivatives (**315**), or as mixtures of both. The dimerization is usually run in alcoholic solution at temperatures from cold to 100 °C ⟨78HC(33)1075, pp. 1157, 1158, 76G1⟩.

(**314**)  (**315**)  (**316**)

The reaction of ethyl diazoacetate (**312**) with sodium or potassium ethoxide was studied by Curtius and his coworkers ⟨08CB3140⟩. They isolated a compound for which they proposed structure (**316**), which is a metal salt of diethyl 1,6-dihydro-1,2,4,5-tetrazine-3,6-dicarboxylate complexed with one mole of the ethoxide used.

Treatment of diazoacetophenone (**317**) with sodium methoxide in the presence of methanol gave, in addition to many other products, a compound which was formulated as 3,6-dibenzoyl-1,2-dihydro-1,2,4,5-tetrazine (**318**) ⟨77CJC145⟩. Similarly, the phosphorus-substituted diazomethane (**319**) could be dimerized by potassium hydroxide to yield the 3,6-bis(diphenylphosphoryl)-1,4-dihydro-1,2,4,5-tetrazine (**320**) as its monopotassium salt ⟨76LA225⟩.

$$\text{PhCOCHN}_2 \xrightarrow[\text{MeOH}]{\text{MeONa}}$$

$$\text{Ph}_2\text{P(O)CHN}_2 \xrightarrow{\text{KOH}}$$

(**317**)  (**318**)  (**319**)  (**320**)

Another method for the preparation of the 1,2,4,5-tetrazine system from a diazo compound was reported by Staudinger and Meyer ⟨19HCA619⟩. They reacted diaryldiazomethanes (**321**) with triethylphosphine and isolated the phosphazines (**322**). In moist benzene or chloroform these compounds were transformed into 3,3,6,6-tetraaryl-1,2,3,6-tetrahydro-1,2,4,5-tetrazines (**323**). This result was only obtained when triethylphosphine was used. The intermediate formation of hydrazones seems unlikely, since these compounds are stable and do not dimerize. A reaction which has a certain similarity to the above was reported by Merrill and Shechter ⟨75TL4527⟩. They obtained 3,6-diphenyl-1,4-dihydro-1,2,4,5-tetrazine (**80**) when the phosphazine (**324**) was hydrolyzed.

$$\overset{\text{Ar}}{\underset{\text{Ar}}{>}}\text{C}=\text{N}_2 \xrightarrow{\text{Et}_3\text{P}} \overset{\text{Ar}}{\underset{\text{Ar}}{>}}\text{C}=\text{NN}=\text{PEt}_3 \xrightarrow{\text{H}_2\text{O}}$$

(**321**)  (**322**)  (**323**)

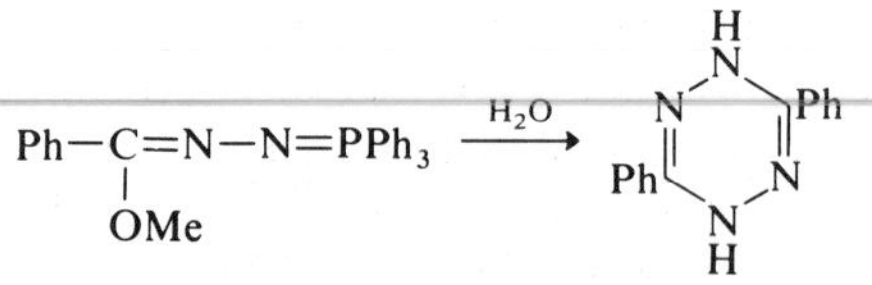

(324)                          (80)

Many reactions used for the preparation of 1,4-dihydro-1,2,4,5-tetrazines can be formulated *via* a nitrilimine intermediate (325) which dimerizes to yield the 1,4-dihydro-1,2,4,5-tetrazine (104). Although the intermediacy of the nitrilimine is not accepted by all authors, the following preparations of the tetrazine (104) can be formulated as proceeding *via* such species: photolysis of sydnones (326), tetrazoles (327) or 1,2,3-triazoles (328) ⟨79CB1635, 73JOC3627, 72JHC87, 78HC(33)1075, p. 1078⟩; thermolysis of tetrazoles (327) ⟨78HC(33)1075, p. 1079⟩; reaction of tetrazoles (329) with tosyl chloride ⟨61CB1555⟩; elimination of hydrogen halide from hydrazidoyl halides (330), mainly with triethylamine; elimination of pyridine from the pyridinium derivatives (331) ⟨77JHC1089, 76ZOR1676, 78HC(33)1075, p. 1126⟩; photolysis of the hydrazidoyl chlorides (330; Hal = Cl) ⟨75MI22101⟩; elimination (KOH) of sulfinate from the sulfonyl-substituted hydrazones (332) ⟨78JHC515⟩.

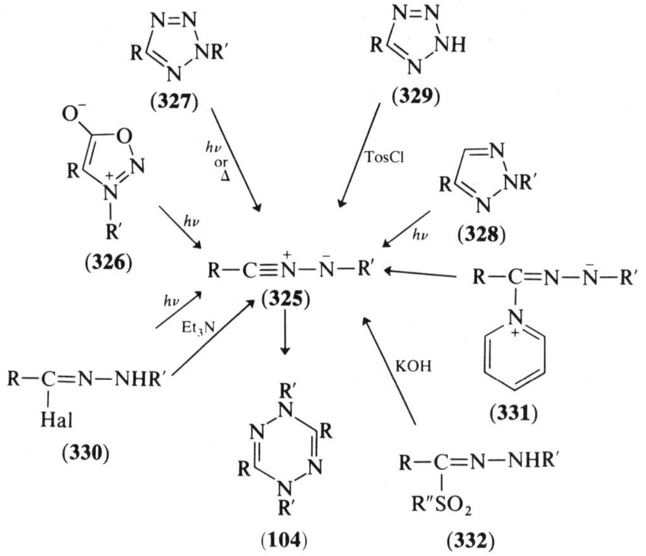

Scheme 18

The reaction of phenacyltriphenylarsonium bromides (333) with aromatic diazonium salts in methanol containing sodium acetate provides a simple method for synthesizing 3,6-diaroyl-1,4-dihydro-1,2,4,5-tetrazines (334) in good yield ⟨78JOM(155)293⟩. The reaction has been postulated to proceed *via* the nitrilimine intermediate (335), whose formation can be explained by the sequence shown in Scheme 19.

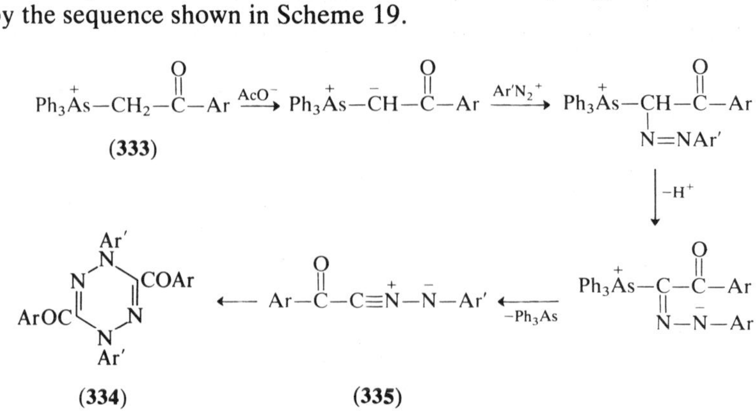

(333)

(334)                          (335)

Scheme 19

Electrochemical oxidation of hydrazidoyl halides (**330**) also affords 1,4-dihydro-1,2,4,5-tetrazines (**104**). A nitrilimine intermediate is not suggested for this reaction. The main process is the dehydrodimerization of the initially formed hydrazonyl radical, while a concurrent side-reaction leads to the 1,4-dihydro-1,2,4,5-tetrazines (**104**), which are transformed into the corresponding cation radicals (**336**) on further oxidation ⟨77IZV393, B-75MI22102⟩.

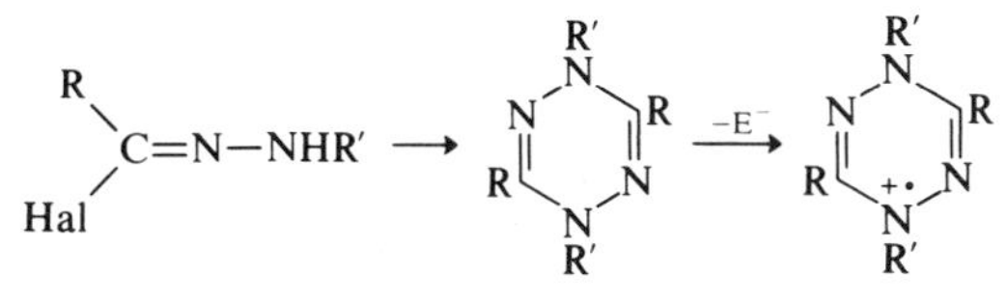

(**330**)          (**104**)          (**336**)

Activated hydrazine derivatives of carboxylic acids (**337, 338**) can dimerize to afford dihydro-1,2,4,5-tetrazines (**89**) which can be oxidized to the tetrazines (**39**) but which can also rearrange to 4-amino-1,2,4-triazoles (**339**). In many cases where the isolated products have been formulated as dihydro-1,2,4,5-tetrazines (**89**) it has either been shown, or it can be assumed, that in fact 4-amino-1,2,4-triazoles (**339**) were obtained. Very often the formation of dihydro-1,2,4,5-tetrazines (**89**) is a side-reaction in the synthesis of activated hydrazine derivatives of carboxylic acids, such as amidrazones (**337**; Y = NH₂), hydrazidines (**337**; Y = NHNH₂), thiohydrazides (**338**; X = S) and so on and not the attempted reaction. Therefore the number of effective synthetic processes for dihydrotetrazines (**89**) or tetrazines (**39**) by this synthetic principle is not very high.

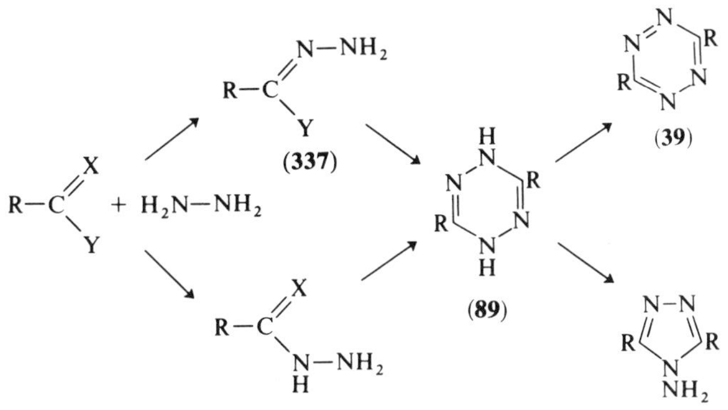

(**337**)          (**39**)          (**89**)          (**338**)          (**339**)

Many of the reactions discussed in this section also belong to the [2+1+2+1] atom fragment method, such as the reaction of nitriles (**340**) with hydrazine for a long time on a steam bath, which affords either dihydro-1,2,4,5-tetrazines (**89**) or 1,2,4,5-tetrazines (**39**) ⟨79H(12)745, 73AJC389, 78HC(33)1075, p. 1122⟩. The first step of the reaction should be the formation of the amidrazones (**341**), which dimerize to give the dihydro tetrazines (**89**); these can be oxidized under the reaction conditions. The dimerization of isolated amidrazones (**341**) has also been reported ⟨65JOC318, 62ZOB3394, 1897CB1871, 1894CB984⟩.

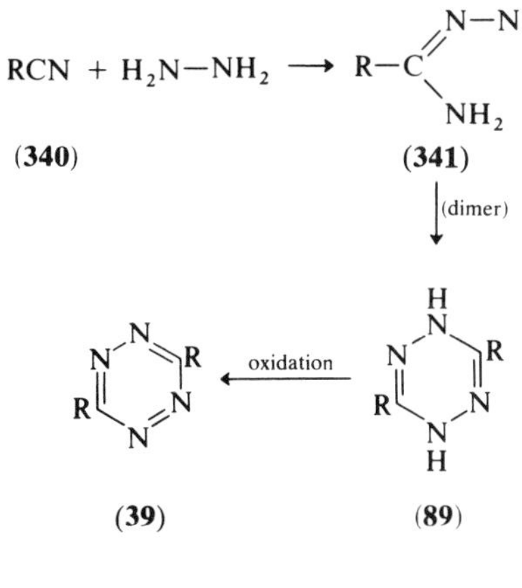

RCN + H₂N—NH₂ → R—C(=N—NH₂)(NH₂)

(**340**)          (**341**)

(dimer)

(**39**)   ←oxidation—   (**89**)

The most commonly used compounds of the type (**337/8**) are the thiohydrazides (**342**) or the thiohydrazoic esters (**343**) ⟨78HC(33)1075, pp. 1081, 1102, 1120, 1159, 79H(12)745, 79ACS(A)137⟩. Since the thiohydrazoic esters are more reactive than the thiohydrazides, the preparation of dihydrotetrazines (**89**) from (**342**) is activated by the presence of an alkylating agent ⟨61ACS1124, 67ACS1984, 67ACS2855⟩. The thiohydrazides (**342**) are also intermediates in the reaction of thioamides (**344**), thiocarboxylates (**345**) and dithiocarboxylates (**346**) with hydrazine ⟨68ZC335, 67RTC907, 55CR(241)1783⟩. The substituent R can also be an amino or substituted amino group, but in these cases not all compounds isolated are 1,2,4,5-tetrazines (**39**) as 4-amino-1,2,4-triazoles (**339**) may be formed (Scheme 20) ⟨27BSF637⟩.

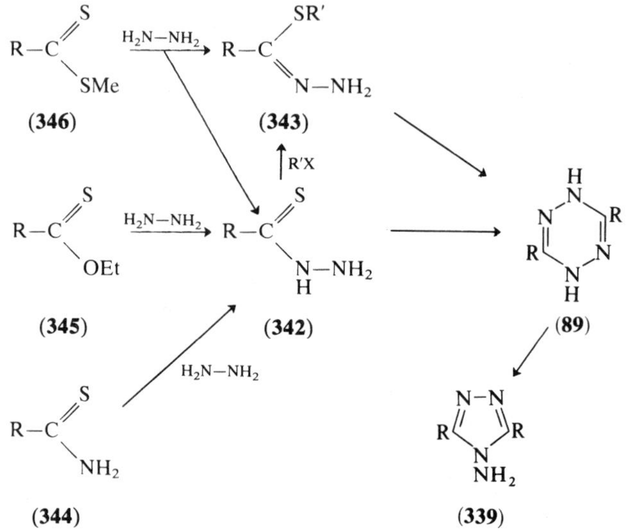

**Scheme 20**

When the imidazoline-4-thione (**347**) was treated with hydrazine the 1,2,4,5-tetrazine (**348**) was obtained, probably *via* the intermediate thiohydrazide (**349**) ⟨72LA(761)95⟩. Reaction of the alkynic thioacetate (**350**) with hydrazine affords 3,6-dipentyl-1,4-dihydro-1,2,4,5-tetrazine (**351**), the formation of which can be rationalized by the intermediate formation of (**352**) ⟨67RTC907⟩.

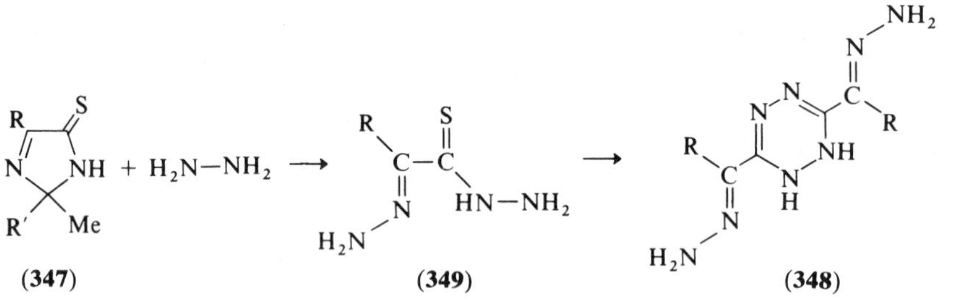

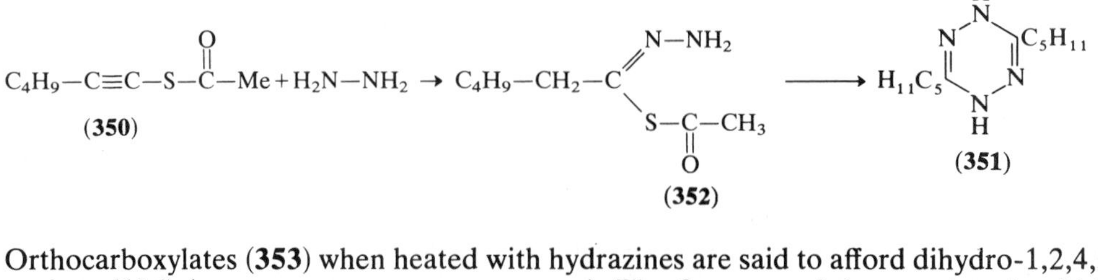

Orthocarboxylates (**353**) when heated with hydrazines are said to afford dihydro-1,2,4,5-tetrazines (**104**) ⟨66M1195, 64JMC814, 03JPR(68)464⟩. The first step of the reaction should be the formation of the hydrazidic ester (**354**). The thermal cyclization of aliphatic hydrazides (**355**) is also claimed to yield 1,4-dihydro-1,2,4,5-tetrazines (**104**) but, as was found later, at least in the cases where R' is a hydrogen, 4-amino-1,2,4-triazoles (339) were isolated ⟨78HC(33)1075, p. 1120⟩. Heating hydrazides of aromatic acids (355; R = aryl) with phenylphosphorus tetrachloride ⟨74ZOR124, 76MIP22100⟩ or thionyl chloride ⟨80KGS1130⟩ affords 1,2,4,5-tetrazines (39). The hydrazidoyl chlorides (356) may be formulated as intermediates; these are known starting materials for the synthesis of (104). In most cases a nitrilimine has been proposed as an intermediate in the transformation of (356) into (104). Interaction of *N,N*-diacetylaniline (357) with hydrazine led to the isolation of a compound the structure of which was given as 3,6-dimethyl-1,4-dihydro-1,2,4,5-tetrazine (358) ⟨00JCS1185⟩. To rationalize this result one has to assume acetohydrazide as an intermediate, but this would suggest that the product has the 4-amino-1,2,4-triazole structure (359).

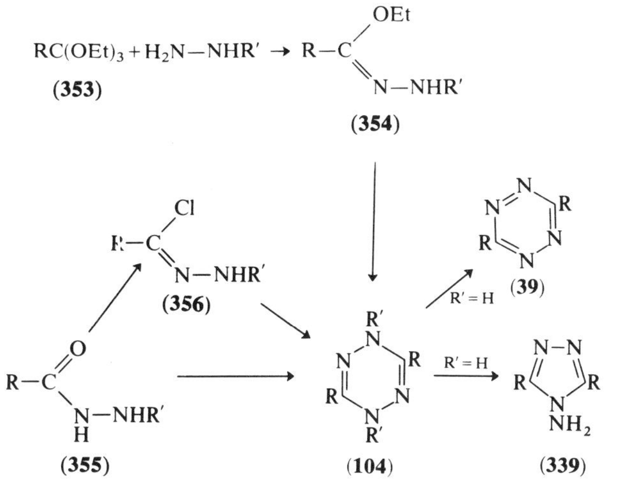

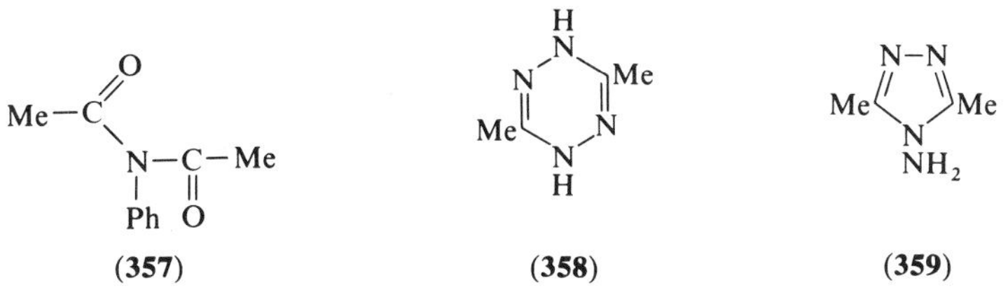

Perfluoropropene (360) with hydrazine at 130 °C produced 3,6-bis(1,2,2,2-tetrafluoroethyl)-1,4-dihydro-1,2,4,5-tetrazine (361), presumably *via* the hydrazidoyl

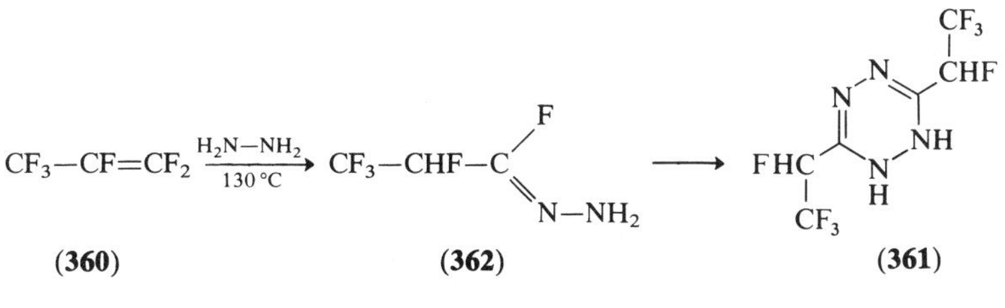

fluoride (362) ⟨57USP2817662⟩. An intermediate *N*-phenylamidrazone (363) no doubt arises in the reaction of the imidoyl chloride (364) with hydrazine ⟨14JPR(89)310⟩ and of the hydrazidoyl chloride (365) with aniline ⟨14CB1132⟩. In both cases 3,6-diaryl-1,2,4,5-tetrazines (366) were isolated, formed by oxidation of the dihydro-1,2,4,5-tetrazines (367).

The 3,6-dipyrazolyl-1,4-dihydro-1,2,4,5-tetrazine (368) was obtained when acetylacetone was treated with diaminoguanidine ⟨70JCS(C)2510, 57AG506⟩. The dipyrazolylhydrazone (369) was suggested as intermediate, but the pyrazolyl formhydrazidine (370) is also a possibility.

Hexahydro-1,2,4,5-tetrazines (371) were obtained when methyl hydrazines (372) were oxidized, either with lead tetraacetate ⟨69TL4449, 77AP764⟩ or with oxygen in ether or

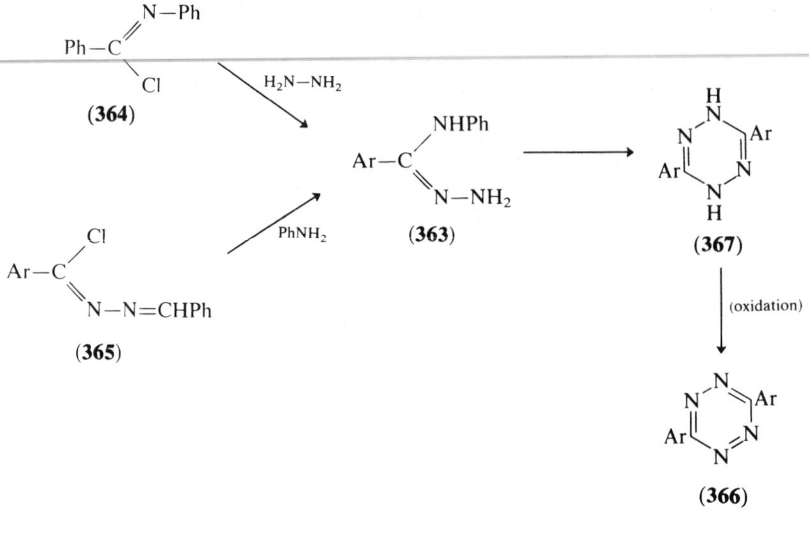

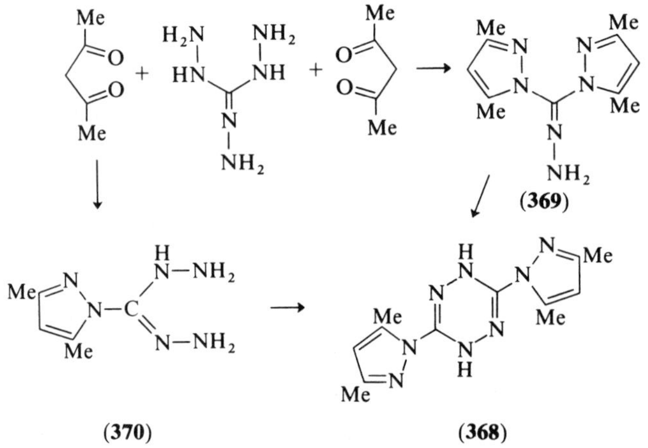

(364)  (365)  (363)  (367)  (366)

(369)  (370)  (368)

cyclohexane at 20–30 °C ⟨81IC426⟩. The azomethinimine (**373**) was suggested as an intermediate.

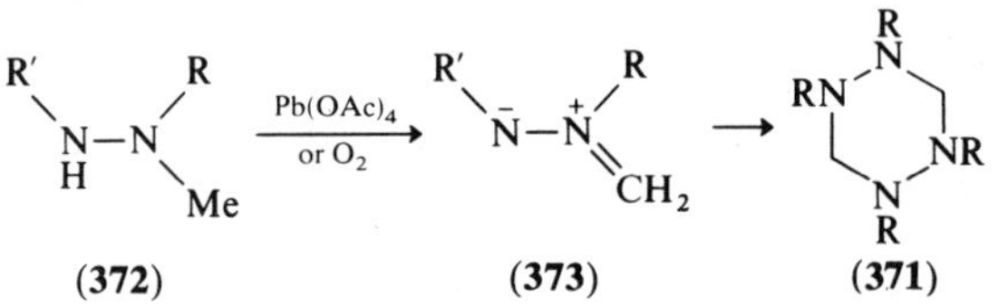

(372)  (373)  (371)

A number of aliphatic hydrazones (**374**) are slowly transformed into the solid hexahydro-1,2,4,5-tetrazines (**375**) when left at room temperature. Addition of acetic acid/acetate buffer (pH = 5.5) to the hydrazone causes almost instantaneous conversion of the hydrazone into the hexahydrotetrazine. This acceleration of the rate of dimerization suggests the intermediacy of a protonated hydrazone (**376**) ⟨77TL3155, 73ACS779, 72TL949, 71AJC1859, 70TL2199, 63AG1204⟩. In acidic media there is an equilibrium between the hexahydrotetrazines and the hydrazones ⟨70HCA251⟩. The hydrazones (**374**) are also intermediates in the preparation of the hexahydrotetrazines from aldehydes and hydrazines.

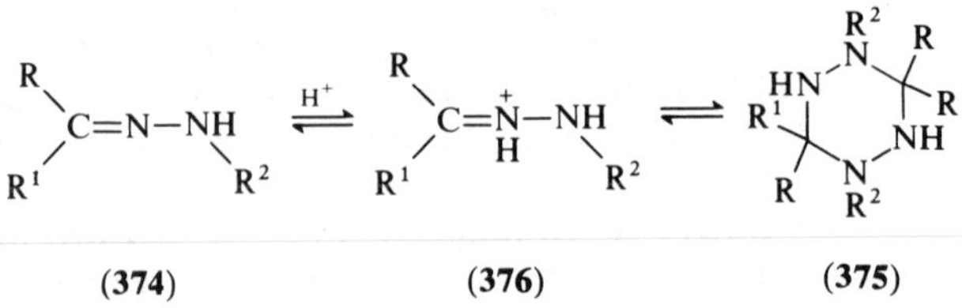

(374)  (376)  (375)

### 2.21.3.4.5 *Synthesis by the [2+1+2+1] atom fragment method*

Reaction of aldehydes with hydrazine, or mono- or 1,2-disubstituted hydrazines is the method used almost exclusively for the preparation of hexahydro-1,2,4,5-tetrazines (**377**). All types of hydrazines have been used: hydrazine itself, monoalkyl-, monoaryl-, dialkyl-, diaryl-, alkylaryl-, alkylacyl- or arylacyl-hydrazines. The aldehydes used were mainly aliphatic aldehydes. An excellent review by Wiley of the literature on this reaction up to 1974 is available ⟨78HC(33)1075, p. 1190⟩.

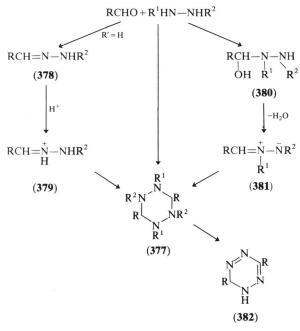

**Scheme 21**

The mechanism of these reactions has been studied by several authors. The first product of the reaction between aldehydes and hydrazine or monosubstituted hydrazines is the hydrazone (**378**), which can be isolated in a few cases. The dimerization of the hydrazones proceeds at room temperature and is catalyzed by acids. As mentioned above, protonation affords the intermediate (**379**) which dimerizes to yield (**377**; $R^1 = H$).

The reaction of aldehydes with disubstituted hydrazines gives, *via* the addition product (**380**), the azomethine imine intermediate (**381**), which dimerizes to the hexasubstituted hexahydrotetrazine (**377**; Scheme 21).

Reaction of aldehydes with hydrazine followed by catalytic oxidation of the hexahydro-1,2,4,5-tetrazines (**377**; $R^1 = R^2 = H$) has been used for the preparation of 1,6-dihydro-1,2,4,5-tetrazines (**382**) ⟨66TL5067, 70HCA251, 71HCA1922, 72HCA1404⟩.

In cases where formaldehyde is used as the aldehyde component and hydrazine or monosubstituted hydrazines are the second component, further reactions can occur such as hydroxymethylation of the hexahydrotetrazine, or the formation of condensed systems.

Compounds such as methylene chloride or methylene bromide ⟨73ACS779, 72ACS1258, 63USP3086016⟩ or tris(chloromethyl)amine ⟨74LA1851⟩, which can be treated as formaldehyde precursors or substitutes, have also been used for the preparation of hexahydro-1,2,4,5-tetrazines (**377**).

As already mentioned in the section on 1,2,4,5-tetrazine preparations by the [3+3] atom fragment method, the reaction of carboxylic acid derivatives with hydrazine or monosubstituted hydrazines is an excellent method for the synthesis of dihydro-1,2,4,5-tetrazines (**89**). For the one-carbon carboxylic acid fragments the following have been used: carboxylate esters ⟨61CCC2871, 49USP2475440⟩; orthocarboxylates ⟨66M1195, 64JMC814, 03JPR(68)464⟩; imidoyl esters ⟨78HC(33)1075, p. 1114, 76BSF621⟩; amidines ⟨80LA1448, 73JCS(P1)335, 70JCS(C)719⟩; amides ⟨70JAP7019295, 70GEP1950392⟩; thiono esters ⟨68ZC335⟩; thioamides ⟨78HC(33)1075, p. 1114⟩; dithiocarboxylates ⟨36BSB685, 67RTC907⟩; nitriles ⟨77HC(33)1075, p. 1122, 73AJC389, 75MI22103⟩; imidoyl chlorides ⟨14JPR(89)310⟩; selenocarboxylates ⟨78JHC1113⟩; selenoamides ⟨78JHC1113⟩ and Vilsmeier reagents ⟨78ZOR622⟩.

To obtain unsymmetrically substituted dihydro-1,2,4,5-tetrazines (**385**) a mixture of imidates (**383**) and amidines (**384**) was reacted with hydrazine ⟨79MI22103, 76JMC1404, 75USP3904614, 75USP3863010, 75JHC1143⟩.

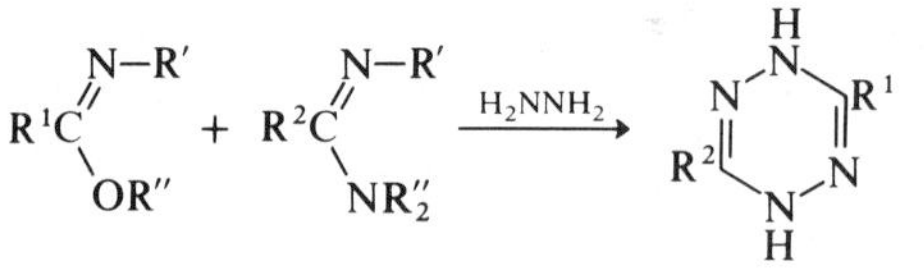

$$\text{(383)} \quad \text{(384)} \quad \xrightarrow{\text{H}_2\text{NNH}_2} \quad \text{(385)}$$

Monosubstituted hydrazines (**386**) are said to afford dihydro-1,2,4,5-tetrazine-3,6(1*H*,4*H*)-diones (**387**) with phosgene or chloroformate, and the diimino-1,2,4,5-tetrazines (**388**) with chlorocyanogen ⟨1892G(22II)99, 60USP2964524, 66JCS(C)2031⟩, but there is some doubt about the structures assigned to these products.

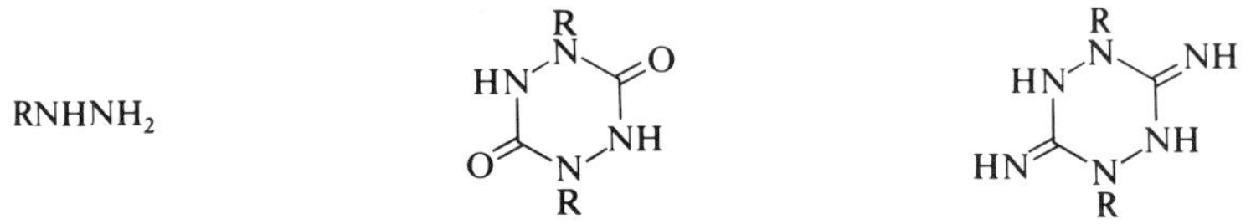

RNHNH₂

(**386**)          (**387**)          (**388**)

*N*-Phenylcarbimidedichloride (**389**) reacts with methylhydrazine in THF at −10 °C to give 3,6-dianilino-1,4-dimethyl-1,4-dihydro-1,2,4,5-tetrazine (**390**) in 55% yield ⟨81AP94⟩.

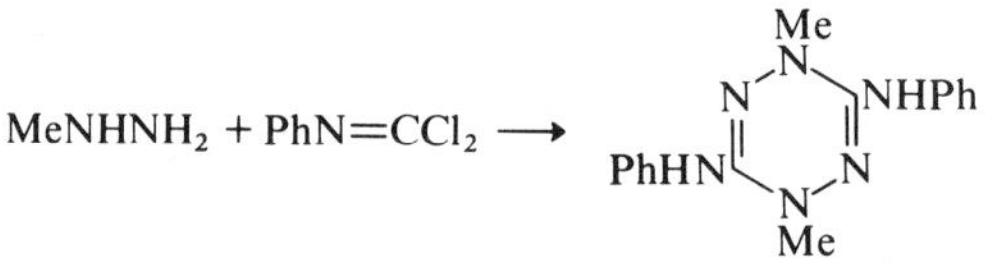

$$\text{MeNHNH}_2 + \text{PhN}{=}\text{CCl}_2 \longrightarrow$$

(**389**)          (**390**)

Treatment of the thiazinedithione (**391**) with hydrazine results in the formation of the dihydro-1,2,4,5-tetrazine (**392**) and the 1,2,4-triazolethione (**393**). It was suggested that the thiohydrazide (**394**) is an intermediate in the formation of (**392**) ⟨80JCR(S)148⟩.

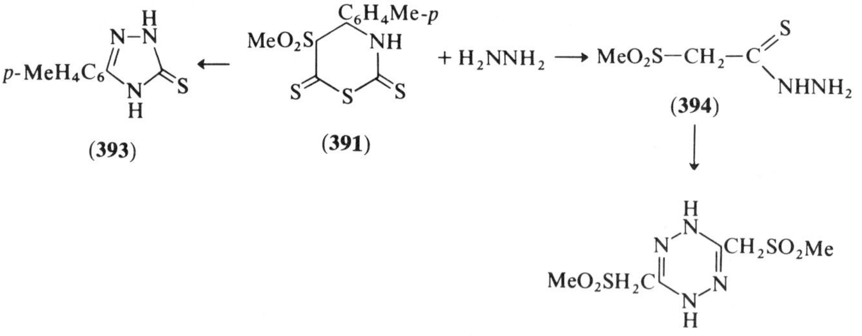

$$\text{(391)} + \text{H}_2\text{NNH}_2 \longrightarrow \text{MeO}_2\text{S}{-}\text{CH}_2{-}\text{C} \quad \text{(394)}$$

(**393**)          (**391**)

(**392**)

When the phenacylpyridinium bromides (**395**) were treated with diazonium salts (**396**) 1,4-dihydro-1,2,4,5-tetrazines (**397**) were obtained ⟨79JPR519, 79IJC(B)362⟩.

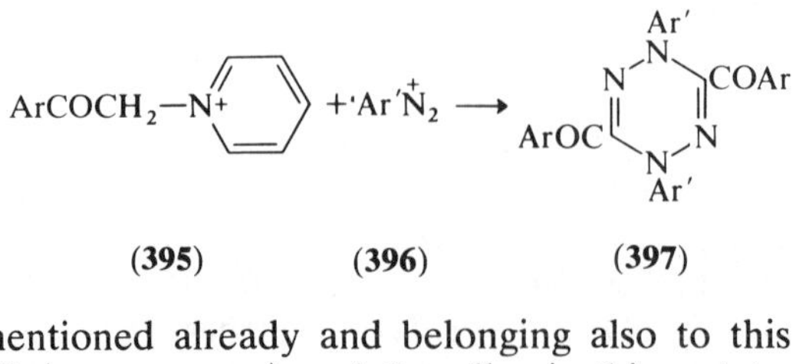

$$\text{ArCOCH}_2{-}\text{N} \quad + \text{'Ar'}\overset{+}{\text{N}}_2 \longrightarrow$$

(**395**)          (**396**)          (**397**)

Further reactions mentioned already and belonging also to this section are those of perfluoropropene (**360**) ⟨57USP2817662⟩ and the alkynic thioacetate (**350**) with hydrazine ⟨67RTC907⟩, which afford the dihydrotetrazines (**361**) and (**351**), respectively.

### 2.21.3.4.6 Synthesis by the [3+2+1] atom fragment method

The most frequently used method of this type is the reaction of hydrazones (247) with diazonium compounds to afford formazans (248). These can be cyclized with carbonyl compounds in the presence of acid to yield verdazylium cations (252) which are reduced under basic conditions to give the verdazyls (253) which can be further reduced to the tetrahydro-1,2,4,5-tetrazines (254). The formazans (248) can be alkylated and the alkylformazans (249) cyclized by heating or treatment with a base to form the tetrahydrotetrazines (250). It is not necessary to isolate the alkylated compounds (249); the complete reaction (248–250) can be carried out at room temperature in the presence of barium oxide or hydroxide in DMF. Oxidation of (250) provides a synthesis of verdazyls (251) (see Section 2.21.3.4.2).

Heating the hydrazides (398) with thionyl chloride in diglyme affords the chlorohydrazones (399) which yield formazans (400) on reaction with pentafluorophenylhydrazine. The formazans (400) undergo cyclocondensation with formaldehyde to give the verdazyls (401) which can be hydrogenated to the tetrahydro-1,2,4,5-tetrazines (402; Scheme 22) ⟨80KGS1130⟩.

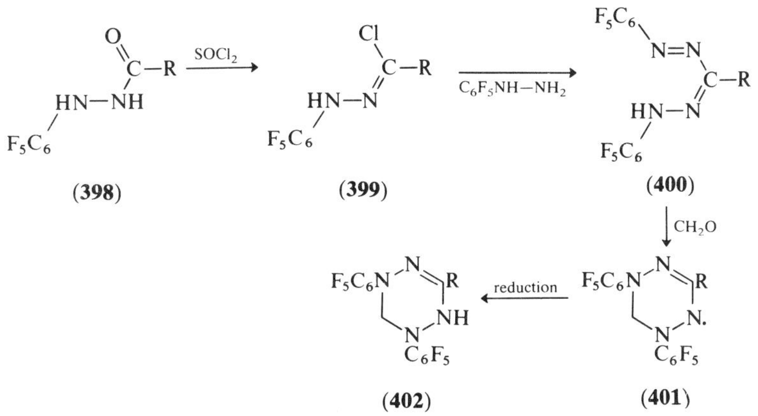

**Scheme 22**

When the nitroaldehyde arylhydrazones (403) were heated with aldehydes in boiling ethanol the nitrotetrahydro-1,2,4,5-tetrazines (404) were isolated in 31–58% yields ⟨77UKZ1192⟩. Clearly, here the hydrazones (403) provide the three- and two-atom fragments while the aldehyde furnishes the one-atom fragment. Treatment of the bromohydrazones (405) with hydrazine affords the $N^4$-arylhydrazidines (406) which can be cyclized with dimethyl acetylenedicarboxylate in refluxing THF to yield 6-alkyl- or 6-aryl-3-methoxycarbonyl-3-methoxycarbonylmethyl-1,2,3,4-tetrahydro-1,2,4,5-tetrazines (407) ⟨77BCJ953⟩.

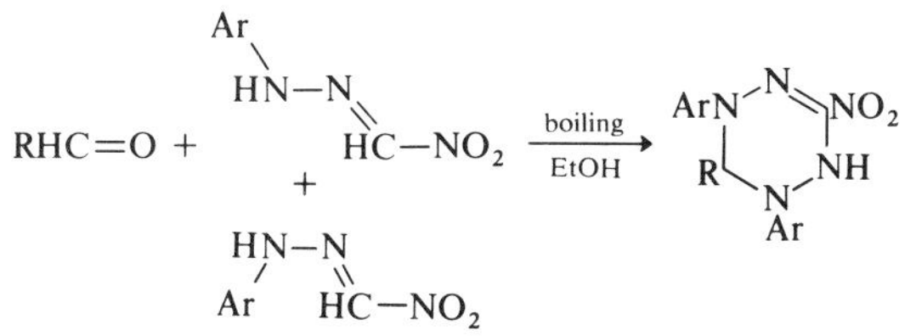

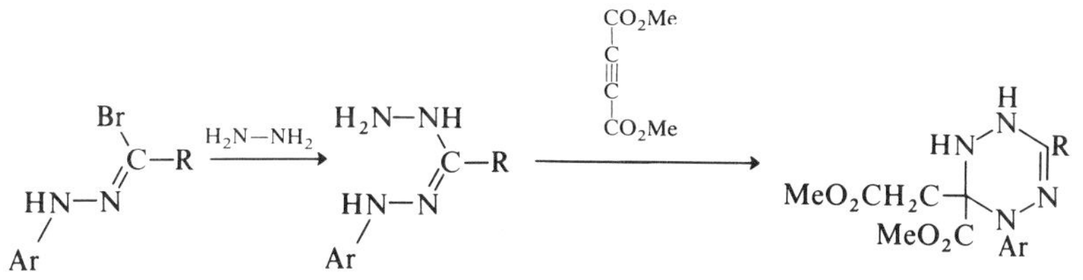

### 2.21.3.5 Applications and Important Uses

A large number of suggested uses for compounds containing the 1,2,4,5-tetrazine skeleton have been recorded in the literature. It seems that most claims must be taken with a pinch of salt and it seems doubtful that any is of significant value.

The following uses have been claimed for compounds containing the 1,2,4,5-tetrazine skeleton: as herbicides, insecticides, bactericides, as pharmaceuticals, as explosives and rocket propellants, as photographic dyes and additives to photographic layers, as reducing agents, as inhibitors of steel corrosion and as fuel additives.

On the other hand, as an earlier Section (2.21.3.3) has detailed, the usefulness of 1,2,4,5-tetrazines in synthesis, for the preparation of pyridazines and 1,2,4-triazines and for the removal of ethene bridges, is manifest.

### 2.21.4 PENTAZINES

No compound containing five nitrogen atoms in one six-membered ring is known. The only pentazine structure published ⟨26JCS113⟩ was later shown to be a tetrazole derivative ⟨62T1001⟩. The parent compound has structure (408) and is numbered as indicated.

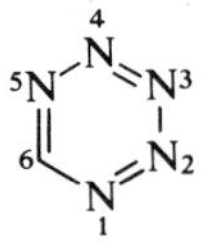

(408)

Although no pentazines have been prepared, a number of theoretical calculations have been published to predict some of the properties, mostly molecular orbital, of this system. The references up to 1974 were compiled in the review published in 1978 ⟨78HC(33)1296⟩. Further publications are: ⟨75PAC(44)767, 77IJC(B)168, 77ACH(92)65⟩. Most calculations have shown that the pentazine ring should be too unstable to be isolated.

Attempts to synthesize a condensed pentazine derivative by reaction of 6-chloro-1,2,4,5-tetrazino[3,2-a]isoindole (409) with sodium azide led to the isolation of 3-methyl-6-(2-cyanophenyl)-1,2,4,5-tetrazine (412) rather than the azide (410) or its valence tautomer (411) which contains a pentazine structure (Scheme 23) ⟨82UP22100⟩.

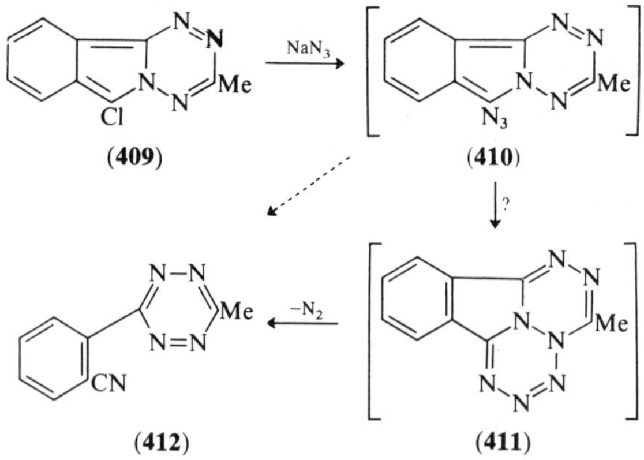

**Scheme 23**

# 2.22

# Pyrans and Fused Pyrans: (i) Structure

P. J. BROGDEN, C. D. GABBUTT and J. D. HEPWORTH
*Preston Polytechnic*

| | | |
|---|---|---|
| 2.22.1 | INTRODUCTION: NOMENCLATURE | 574 |
| 2.22.2 | THEORETICAL METHODS | 575 |
| 2.22.3 | NMR SPECTROSCOPY | 576 |
| | *2.22.3.1  $^1H$ NMR Spectra* | 576 |
| | *2.22.3.1.1  Pyrans, their benzologues and reduced derivatives* | 576 |
| | *2.22.3.1.2  Pyranones, their benzologues and reduced derivatives* | 580 |
| | *2.22.3.1.3  Flavonoids* | 584 |
| | *2.22.3.1.4  Pyrylium salts* | 585 |
| | *2.22.3.2  $^{13}C$ NMR Spectra* | 585 |
| | *2.22.3.2.1  Pyrans, their benzologues and reduced derivatives* | 585 |
| | *2.22.3.2.2  Pyranones, their benzologues and reduced derivatives* | 587 |
| | *2.22.3.2.3  Flavonoids* | 591 |
| | *2.22.3.2.4  Pyrylium salts* | 592 |
| | *2.22.3.3  $^{17}O$ NMR Spectra* | 592 |
| | *2.22.3.4  Other Nuclei* | 593 |
| 2.22.4 | IR SPECTROSCOPY | 593 |
| | *2.22.4.1  Pyrans, their Benzologues and Reduced Derivatives* | 593 |
| | *2.22.4.2  Pyranones, their Benzologues and Reduced Derivatives* | 595 |
| | *2.22.4.3  Flavonoids* | 597 |
| | *2.22.4.4  Pyrylium Salts* | 598 |
| 2.22.5 | UV SPECTROSCOPY | 598 |
| | *2.22.5.1  Pyrans, their Benzologues and Reduced Derivatives* | 598 |
| | *2.22.5.2  Pyranones, their Benzologues and Reduced Derivatives* | 599 |
| | *2.22.5.3  Flavonoids* | 601 |
| | *2.22.5.4  Pyrylium Salts* | 603 |
| 2.22.6 | MASS SPECTROMETRY | 603 |
| | *2.22.6.1  Pyrans, their Benzologues and Reduced Derivatives* | 603 |
| | *2.22.6.2  Pyranones, their Benzologues and Reduced Derivatives* | 607 |
| | *2.22.6.3  Pyrylium Salts* | 619 |
| 2.22.7 | MOLECULAR STRUCTURE AND CONFORMATION | 620 |
| | *2.22.7.1  X-Ray Determinations* | 620 |
| | *2.22.7.2  Microwave Spectroscopy* | 625 |
| | *2.22.7.3  Dipole Moments* | 626 |
| | *2.22.7.4  Conformations* | 628 |
| | *2.22.7.4.1  Reduced pyrans and their benzologues* | 628 |
| 2.22.8 | AROMATICITY | 632 |
| | *2.22.8.1  Pyranones and their Benzologues* | 632 |
| | *2.22.8.1.1  Pyran-2-one and coumarin* | 632 |
| | *2.22.8.1.2  Pyran-4-one and chromone* | 637 |
| | *2.22.8.1.3  Pyranopyrandiones* | 639 |
| | *2.22.8.2  Pyrylium Salts and their Benzologues* | 640 |
| 2.22.9 | TAUTOMERISM | 641 |
| | *2.22.9.1  Prototropy* | 641 |
| | *2.22.9.1.1  Prototropy involving annular carbon atoms* | 641 |
| | *2.22.9.1.2  Prototropy involving functional groups* | 642 |
| | *2.22.9.2  Ring–Chain Tautomerism* | 644 |
| 2.22.10 | BETAINES AND OTHER UNUSUAL STRUCTURES | 644 |
| | *2.22.10.1  Betaines* | 644 |
| | *2.22.10.2  Other Systems* | 645 |

## 2.22.1 INTRODUCTION: NOMENCLATURE

Six-membered oxygen heterocycles constitute a group of compounds which occur widely throughout the plant kingdom. Add to that the importance of some of their number as drugs and a general propensity towards biological activity and it is plain to see why these heterocycles have been extensively studied. Their reactions and syntheses are discussed in the following two chapters, which include a brief survey of the major literature in this area. Many of these works include structural aspects of oxygen heterocycles, the subject of the present chapter, and hence reference should be made to these texts; the few more specialized reviews are mentioned in individual sections. Generally, however, structural information is widely but thinly dispersed throughout the chemical literature and it is quite impossible to give comprehensive coverage. Where appropriate, tables of data have been included in an attempt to alleviate this problem.

A variety of names has been used for the benzologues of pyran and its derivatives, many of which have been described in two treatises on oxygen heterocyclic compounds ⟨77HC(31)1, 81HC(36)1⟩. It is of value to indicate the nomenclature which has been adopted in this and the two following chapters, since it is not always considered appropriate to use the more cumbersome and less familiar systematic names.

All of the compounds discussed are based on three molecules: 2*H*-pyran (**1**), 4*H*-pyran (**2**) and the pyrylium cation (**3**). Names which have been used for the benzologue (**4**) of 2*H*-pyran include 2*H*-1-benzopyran, benzo-α-pyran, chrom-3-ene and 2*H*-chromene. A similar situation exists for the corresponding derivative (**5**) of 4*H*-pyran. The unambiguous and simplest name chromene is used in the present work. The benzologue (**6**) of pyrylium is known both as benzopyrylium and chromylium; the former name is preferred here. Higher benzologues are referred to as naphthopyrans, such as 2*H*-naphtho[1,2-*b*]pyran (**7**), but the names xanthene and xanthylium are used for (**8**) and (**9**).

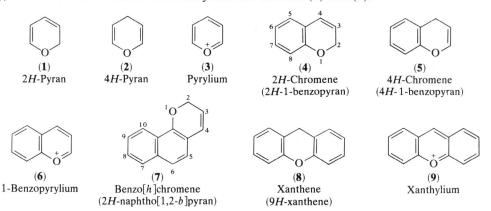

|              |              |            |                           |                           |
|:------------:|:------------:|:----------:|:-------------------------:|:-------------------------:|
| (**1**)      | (**2**)      | (**3**)    | (**4**)                   | (**5**)                   |
| 2*H*-Pyran   | 4*H*-Pyran   | Pyrylium   | 2*H*-Chromene             | 4*H*-Chromene             |
|              |              |            | (2*H*-1-benzopyran)       | (4*H*-1-benzopyran)       |

|                  |                           |               |              |
|:----------------:|:-------------------------:|:-------------:|:------------:|
| (**6**)          | (**7**)                   | (**8**)       | (**9**)      |
| 1-Benzopyrylium  | Benzo[*h*]chromene        | Xanthene      | Xanthylium   |
|                  | (2*H*-naphtho[1,2-*b*]pyran) | (9*H*-xanthene) |            |

The trivial names flavene (**10**), isoflavene (**11**) and flavylium (**12**) are in widespread use and are retained here. However, these compounds will usually be considered as the phenyl derivatives of 2*H*-chromene or benzopyrylium rather than in separate sections.

|                           |                           |                           |
|:-------------------------:|:-------------------------:|:-------------------------:|
| (**10**)                  | (**11**)                  | (**12**)                  |
| Flavene                   | Isoflavene                | Flavylium                 |
| (2-phenyl-2*H*-1-benzopyran) | (3-phenyl-2*H*-1-benzopyran) | (2-phenyl-1-benzopyrylium) |

The partially reduced pyrans (**13**) and (**14**) are named from 2*H*-pyran. The benzologue (**15**) is called chroman, whilst flavan is used for the 2-phenyl derivative (**16**).

|                              |                              |                                  |                              |
|:----------------------------:|:----------------------------:|:--------------------------------:|:----------------------------:|
|                              |                              |                                  | (**16**)                     |
|                              |                              |                                  | Flavan                       |
|                              |                              |                                  | (2-phenyl-3,4-dihydro-2*H*-  |
|                              |                              | (**15**)                         | 1-benzopyran)                |
|                              |                              | Chroman                          |                              |
| (**13**)                     | (**14**)                     | (3,4-dihydro-2*H*-1-benzopyran)  |                              |
| 3,4-Dihydro-2*H*-pyran       | 5,6-Dihydro-2*H*-pyran       |                                  |                              |

The ketones derived from the pyrans are known as pyranones (**17**) and (**18**), although the name pyrone has enjoyed much usage. Trivial rather than systematic names are in general use for the benzologues (**19**), (**20**) and (**21**) and these, too, are used throughout. The 2-phenyl derivative of chromone is known as flavone (**22**).

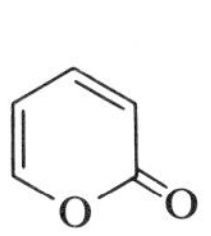

**(17)**
Pyran-2-one
(2*H*-pyran-2-one)

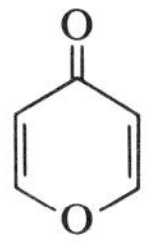

**(18)**
Pyran-4-one
(4*H*-pyran-4-one)

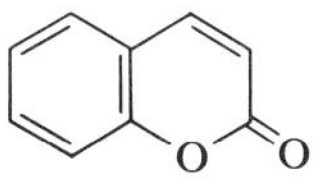

**(19)**
Coumarin
(2*H*-1-benzopyran-2-one)

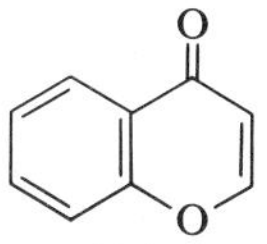

**(20)**
Chromone
(4*H*-1-benzopyran-4-one)

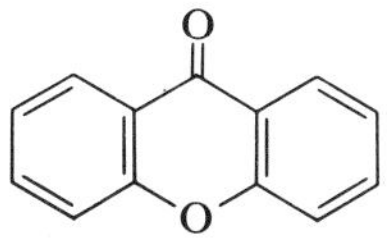

**(21)**
Xanthone
(9*H*-xanthen-9-one)

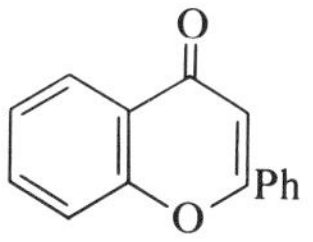

**(22)**
Flavone
(2-phenyl-4*H*-1-benzopyran-4-one)

The bicyclic molecules (**23**), (**24**) and (**25**) in which the heterocyclic ring is saturated are named as shown.

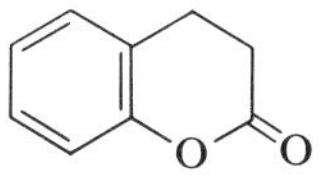

**(23)**
Dihydrocoumarin
(3,4-dihydro-2*H*-1-benzopyran-2-one)

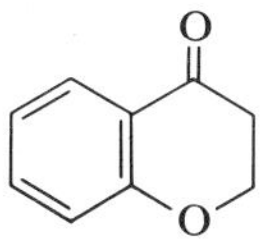

**(24)**
Chroman-4-one
(2,3-dihydro-4*H*-1-benzopyran-4-one)

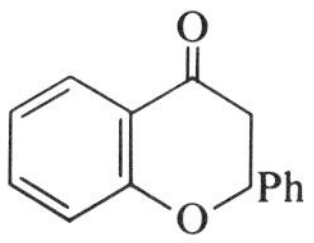

**(25)**
Flavanone
(2-phenyl-2,3-dihydro-4*H*-1-benzopyran-4-one)

The names used for some less common oxygen heterocycles are shown below (**26–29**) together with the systematic names. The list is not intended to be comprehensive and the names used for other examples will be apparent from the text.

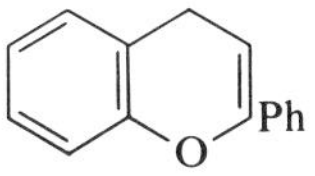

**(26)**
4*H*-Flavene
(2-phenyl-4*H*-1-benzopyran)

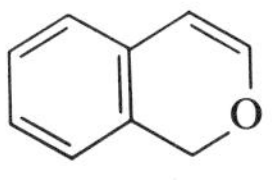

**(27)**
Isochromene
(1*H*-2-chromene,
1*H*-2-benzopyran)

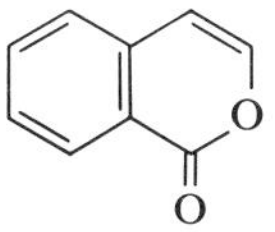

**(28)**
Isocoumarin
(1*H*-2-benzopyran-1-one)

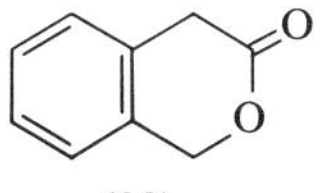

**(29)**
Isochroman-3-one
(1,4-dihydro-3*H*-2-benzo-pyran-3-one)

## 2.22.2 THEORETICAL METHODS

The application of theoretical methods to the question of aromaticity is discussed in Section 2.22.8.

Much interest has been shown in the $\pi$-electronic structures of pyranones and their benzologues. A review of sulfur-containing pyranones includes a survey of the theoretical studies on pyran-4-one and the 4-thione and the corresponding 2-isomers ⟨67AHC(8)219⟩. Various chemical properties have been explained on the basis of HMO calculations ⟨65CCC3016, 67MI22200⟩. Good agreement was found between the experimental values for $\pi \rightarrow \pi^*$ transition energies of pyran-2-ones and coumarins and those derived from SCF–CI calculations using the PPP method, but this approach was not as successful for pyran-4-ones ⟨72CPB677⟩. Changes in the $\pi$-electron distribution on going from the ground state to the lowest singlet and triplet states have been calculated.

Electron densities and bond orders for pyran-2-one, pyran-4-one and eight benzologues or dibenzologues have been calculated using the simple Hückel, the Hückel autocoherent and the semi-empirical Pople methods ⟨74BSF538⟩. Similar properties have been obtained for flavone and the electronic absorption spectra of some substituted flavones have been examined in detail ⟨74CHE1218⟩. Excitation is accompanied by a transfer of charge from the components of the heterocyclic ring to the benzene rings. Calculated $\pi$-bond orders and charge densities for a range of pyranopyrandiones have been correlated with IR, NMR and empirical resonance energies ⟨73MI22200⟩.

$\pi$-Electron densities and SCF orbital energies for chromone and some derivatives calculated by the PPP method have been correlated with $^1$H shifts and polarographic half-wave potentials ⟨78IJC(A)531⟩.

$\pi$-Electron energies of coumarin and various isomers have been calculated by a one-dimensional topology-based method, though it is recognized that such an approach to a multifunctional species can provide only a crude estimate of chemical reactivity ⟨82JHC625⟩.

The photochemical and photobiological behaviour of coumarins is well documented. It is considered that the lowest excited triplet state plays an important role in such processes and information about the nature of the $T_1$ state of coumarin has been derived from magnetic and spectroscopic studies ⟨76JA3460⟩.

An MO study of triplet coumarin has indicated that the mobile bond order for the 3,4-double bond decreases significantly on excitation ⟨73MI22201⟩. Furthermore, the electron density at C-3 is decreased but is increased at C-4 and hence an explanation of the orientational specificity of the photoaddition of coumarin to 1,1-diethoxyethylene is available (equation 1) ⟨66TL1419⟩. The reactivity of the 3- and 4-positions of triplet coumarin is also substantiated by the high spin-densities calculated for these positions ⟨70JPC4234⟩.

$$\text{[structure]} + CH_2{=}C(OEt)_2 \xrightarrow{h\nu} \text{[structure]} \qquad (1)$$

HMO indices for coumarin overestimate the reactivity of the 4-position ⟨74BSF538⟩, whereas SCF–MO free valences correctly predict that C-3 is the most reactive site towards homolytic substitution ⟨79JHC97⟩.

Carbonyl stretching frequencies for pyran-2-one, pyran-4-one, coumarin, isocoumarin and chromone derived from the bond orders obtained using EHT agree well with the experimental data ⟨78IJC(A)64⟩.

Roumanian workers have carried out extensive calculations on pyrylium salts and this work along with that by other groups has been reviewed ⟨82AHC(S2)211⟩. It is especially noteworthy that the valence electron distribution in the pyrylium cation indicates a slight negative charge on oxygen. Most of the positive charge of the cation is absorbed by the hydrogen atoms and the resonance structures based on the polarization of the $\pi$-electrons thus provide an inadequate picture of pyrylium ⟨75JCS(P2)841⟩. Calculations also indicate a higher positive charge at C-2 and C-6 than at C-4 ⟨70ACS3417⟩.

Charge densities have been calculated by CNDO/2 and EHMO for xanthene (**8**) and dibenzoxanthene and the derived cations ⟨75JA5472⟩. The $^{13}$C chemical shifts correlate well with the CNDO/2 charge densities and indicate extensive delocalization of the charge in the xanthylium species.


## 2.22.3 NMR SPECTROSCOPY

### 2.22.3.1 $^1$H NMR Spectra

#### 2.22.3.1.1 Pyrans, their benzologues and reduced derivatives

Although 2*H*-pyran itself has not yet been isolated, the $^1$H spectra of a number of derivatives which are stabilized by disubstitution at C-2 have been studied. The data for the 2,2,4,6-tetramethyl derivative are shown in Figure 1 and other 2*H*-pyran $^1$H spectra are found in ⟨72BSF707, 81CCC748⟩.

An area of particular interest is the valence isomerization of *cis*-dienones to 2*H*-pyrans. The $^1$H spectrum of the 2*H*-pyran (**30**), prepared by irradiation of *trans*-$\beta$-ionone, showed a series of weak signals in addition to the expected pyran spectrum ⟨66JA619⟩. These were assigned to *cis*-$\beta$-ionone (**31**) which exists in equilibrium with the pyran (Figure 2). Integration of the AB patterns of H-3 and H-4 for the two compounds allowed the equilibrium concentrations and hence the equilibrium constant for the isomerization to be determined.

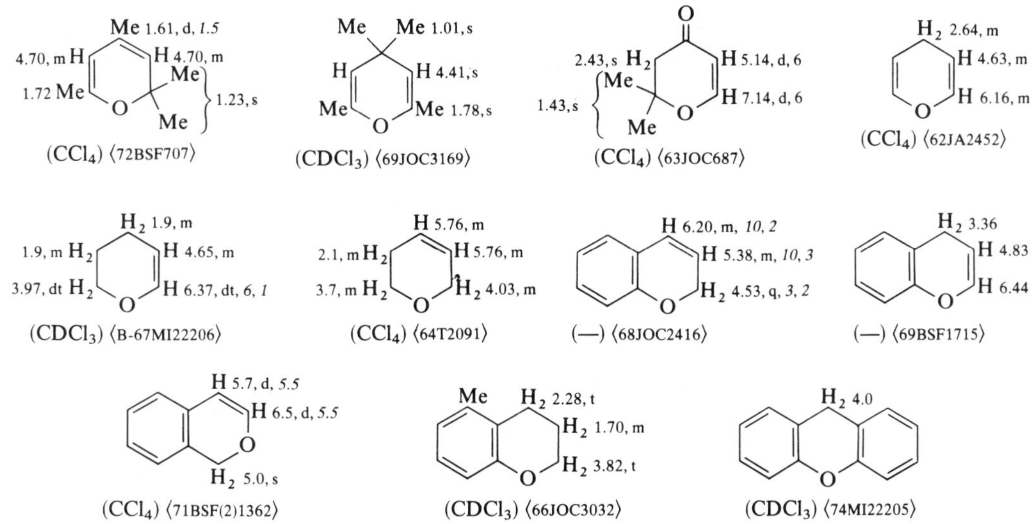

Me 1.61, d, *1.5*
4.70, m H  H 4.70, m
1.72 Me  Me
O  Me  } 1.23, s
(CCl₄) ⟨72BSF707⟩

Me  Me 1.01, s
H  H 4.41, s
Me  Me 1.78, s
O
(CDCl₃) ⟨69JOC3169⟩

2.43, s H₂  H 5.14, d, 6
(Me  H 7.14, d, 6
1.43, s  Me  O
(CCl₄) ⟨63JOC687⟩

H₂ 2.64, m
H 4.63, m
O  H 6.16, m
(CCl₄) ⟨62JA2452⟩

H₂ 1.9, m
1.9, m H₂  H 4.65, m
3.97, dt H₂  O  H 6.37, dt, 6, *1*
(CDCl₃) ⟨B-67MI22206⟩

H 5.76, m
2.1, m H₂  H 5.76, m
3.7, m H₂  O  H₂ 4.03, m
(CCl₄) ⟨64T2091⟩

H 6.20, m, *10, 2*
H 5.38, m, *10, 3*
O  H₂ 4.53, q, *3, 2*
(—) ⟨68JOC2416⟩

H₂ 3.36
H 4.83
O  H 6.44
(—) ⟨69BSF1715⟩

H 5.7, d, *5.5*
H 6.5, d, *5.5*
O
H₂ 5.0, s
(CCl₄) ⟨71BSF(2)1362⟩

Me  H₂ 2.28, t
H₂ 1.70, m
O  H₂ 3.82, t
(CDCl₃) ⟨66JOC3032⟩

H₂ 4.0
O
(CDCl₃) ⟨74MI22205⟩

**Figure 1**  $^1$H NMR spectra of representative nonaromatic pyrans and fused pyrans

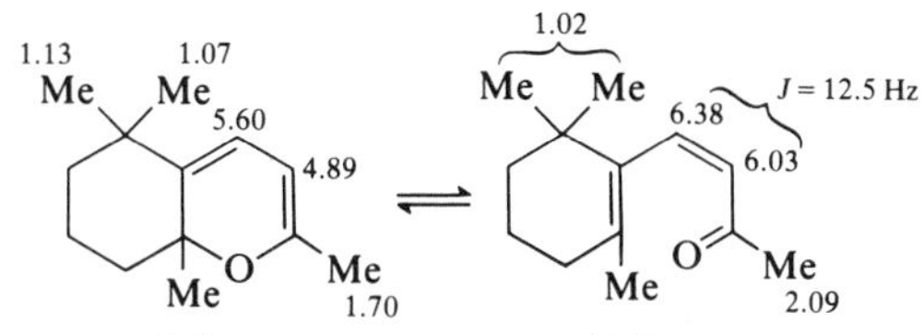

1.13  1.07  1.02
Me  Me  Me  Me  *J* = 12.5 Hz
5.60  6.38
4.89  6.03
O  Me  O  Me
Me  Me  1.70  Me  Me  2.09
(30)  (31)

**Figure 2**  $^1$H Chemical shift data (p.p.m.) for *cis*-β-ionone (**31**) and its valence isomer (**30**)

The rigidity of the bicyclic 2*H*-pyran is apparent from the appearance of separate signals for the two methyl groups at C-5.

NMR was used to distinguish 2,2,4,6-tetramethyl-2*H*-pyran from the isomeric exocyclic methylene compound (**32**) into which it is converted on treatment with acid ⟨64BSF1492⟩.

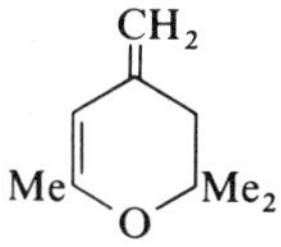

CH₂
Me  O  Me₂
(32)

Both dimethyl 2,4-diacetylpent-2-enedioate and 1,1,3,3-tetraacetylpropene exist in the cyclic forms which are substituted 2*H*-pyrans ⟨79JCS(P1)478⟩. The $^1$H NMR spectra do not show a low field signal for a chelated OH group expected in the acyclic structure. The influence of solvent polarity, increasing temperature and base-catalysis on the acetyl resonances suggests an equilibrating system of cyclic and acyclic structures (Scheme 1).

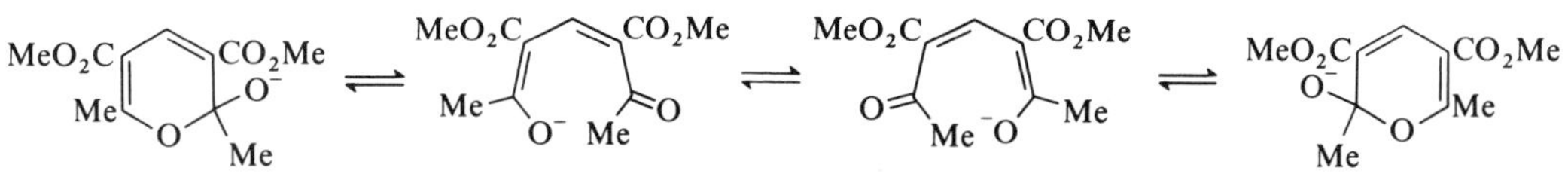

MeO₂C  CO₂Me
Me  O⁻
O  Me

MeO₂C  CO₂Me
Me  O⁻  Me
O

MeO₂C  CO₂Me
O  Me  O  Me

MeO₂C  CO₂Me
O⁻  Me
O
Me

**Scheme 1**

Two groups have reported the spectrum of 4*H*-pyran (**2**) ⟨62JA2452, 65MI22200⟩, the latter study establishing all the couplings in the system. The deshielding effect of the heteroatom is manifest in the chemical shift of H-2.

Substitution in 4*H*-pyrans has little effect on the coupling constants, but it is worth noting that the homoallylic coupling between H-5 and the 6-methyl group is clearly resolved when X is COMe or CO₂Me in (**33**), but that when X is CO₂H the proton signal appears as a broad singlet ⟨69JOC3169⟩.

The downfield shifts of the methyl protons in (**34**) on protonation or quaternization indicate an increase in the positive nature of C-2 and C-6, whilst the corresponding effects

(33)        (34)

on H-3 and H-5 are considered to arise from a combination of through-bond and anisotropic effects ⟨81JCS(P2)303⟩.

Chemical shift data for two 4*H*-pyrans are shown in Figure 1 and other examples may be found in references ⟨62JA2452, 72BSF707, 78TL2995, 81JCS(P2)303⟩ and elsewhere.

Dihydropyrans are the heterocyclic analogues of cyclohexene, for which a half-chair conformation is preferred. Conformational analysis of this heterocyclic system is more complicated than that of simple alicyclic systems because of the non-equivalence of each ring position and the occurrence of non-bonded interactions associated with the presence of the double bond and the heteroatom. However, the allylic and homoallylic coupling constants that characterize the conformations, such as (35) and (36), of 5,6-dihydro-2*H*-pyrans are close to the corresponding *J* values for cyclohexenes and it may be concluded that a half-chair conformation is also preferred for these heterocycles ⟨80CHE571⟩. Evidence from variable temperature NMR shows that *trans*-2-alkoxy-5,6-dihydro-2*H*-pyran-6-carboxylic acids exist exclusively in conformation (37) ⟨70OMR(2)55⟩. The coupling constants for H-6 with the two protons at C-5 in the two half-chair conformations (38) and (39) of some 6-substituted 5,6-dihydro-2*H*-pyrans have been used to examine the conformational equilibria ⟨72OMR(4)537⟩. In the favoured structure, H-6 is axial. A configurational and conformational study of epimeric esters of 2,5-dimethyl-5,6-dihydro-2*H*-pyrans (40) involving spin–spin decoupling and shift reagent techniques in addition to variable temperature work has shown that the compounds exist in half-chair like conformations ⟨72JOC3997⟩. A comparison of the observed coupling constants with those calculated from the Karplus equation suggests the need for a modification to that equation for many systems.

(35)        (36)        (37)

(38)        (39)        (40)

Chemical shift data have been reported for 2-alkoxy- and 2-phenoxy-3,4-dihydro-2*H*-pyrans ⟨75JOC2234, 78JOC667⟩ and for the analogous 6-alkoxy derivatives ⟨82TL603⟩. Most of the compounds which are monosubstituted at C-2 exist predominantly (*ca.* 80%) in the conformation in which the anomeric proton is equatorial (41; X=Y=H) as predicted by the anomeric effect ⟨55CI(L)1102⟩. However, 2-methoxy-4-methyl-3,4-dihydro-2*H*-pyran exists as a *cis–trans* mixture of diastereoisomers at the anomeric 2-position (41 and 42; R=Y=Me; X=H). A similar situation obtains for the 2-ethoxy-3-methyl analogue. One conformer is detected by NMR for the 2-methoxy-2-methyl derivatives and the preferred conformation is considered to be that in which the methoxy group is axial, requiring the methyl group to occupy an equatorial position (41; X=Me; Y=H).

(41)        (42)

Reactions of unsaturated esters with electron-rich alkenes have been reported to yield only cyclobutane derivatives. However, NMR examination of the products has indicated the formation of substituted 3,4-dihydro-2*H*-pyrans. The most informative feature of the spectra is the C-2 proton coupling constants of *ca.* 3 Hz with the two different protons at

C-3, since coupling between the vicinal protons in a cyclobutane structure should be greater than 6 Hz ⟨82TL603⟩.

The structures of the fused dihydropyrans Edulan I (**43**) and II (**44**) were conclusively established with the aid of lanthanide shift reagents ⟨75JCS(P1)1736⟩. The protons at C-8 undergo the largest shifts confirming their proximity to the heteroatom. The chemical shifts of the various protons in the two epimers are shown in Figure 3.

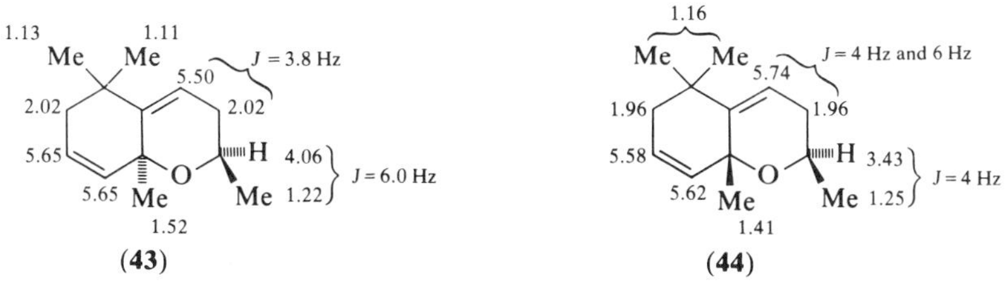

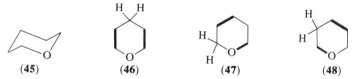

**Figure 3** ¹H Chemical shift data (p.p.m.) for edulan I (**43**) and edulan II (**44**)

¹H NMR spectral data for the 3,4- and 5,6-dihydro-2*H*-pyrans are shown in Figure 1, and details for other substituted derivatives can be found in ⟨78JOC667, 79JOC21⟩ (3,4-dihydro), ⟨79S41, 79S743, 80CHE571⟩ (5,6-dihydro), and elsewhere.

Tetrahydropyran (**45**) is similar to cyclohexane and piperidine in that the most stable conformation is the chair form. The chemical shift difference between axial and equatorial protons at C-4, $\delta_{ae}(\gamma)$, is 0.32 p.p.m. and is influenced chiefly by the anisotropy of the C(2)—C(3) and C(5)—C(6) bonds which are $\beta$ to the 4-methylene group (**46**); the equatorial proton resonates at lower field than the axial proton ⟨75CRV611⟩. The situation is more complicated for the $\alpha$-protons because of the proximity of the oxygen atom ⟨76JOC1380, 77JA5689⟩. The properties of the C—C and C—O bonds $\beta$ to the 2-CH₂ protons are expected to determine $\delta_{ae}(\alpha)$ (**47**). The combined diamagnetic anisotropies of the C—C and C—O bonds appear to have very nearly the same effect as the two C—C bonds in cyclohexane since the value of 0.50 p.p.m. for $\delta_{ae}(\alpha)$ is very close to that for the alicyclic compound (0.48 p.p.m.) ⟨B-69MI22200⟩. $\delta_{ae}(\beta)$ is determined by identical bonds (**48**) to those influencing $\delta_{ae}(\alpha)$ and a similar value is anticipated. However, $\delta_{ae}(\beta)$ is −0.074 p.p.m. (H-3$_{eq}$ upfield of H-3$_{ax}$) and it is proposed that the C—O bond exerts different shielding effects when viewed from opposite ends. Thus, the fragment CH₂OC in (**47**) deshields H-2$_{ax}$, whereas OCCH₂ in (**48**) shields H-3$_{ax}$.

NMR has been used to show that the anomeric effect operates in 2-bromo- and 2-chloro-tetrahydropyrans, which exist predominantly with the halogen atom axial ⟨66JOC544⟩, whilst the equatorial preference of a 4-halogen atom is less pronounced than in cyclohexane and decreases with decreasing electronegativity ⟨78SA(A)297⟩.

A detailed study of the coupling constants of tetrahydropyran and comparison with cyclohexane data indicated that all the vicinal *J* values change when an oxygen atom replaces a methylene group (Table 1) ⟨76JOC1380⟩. Coupling constant data have enabled the geometrical isomers of 3,4-dibromo-2-ethoxytetrahydropyran to be distinguished ⟨71JOC3633⟩. For both isomers, $J_{3,4}$ is *ca.* 10 Hz, indicative of a *trans* diaxial orientation; the two bromine atoms are therefore *trans* and equatorial. However, $J_{2,3}$ is 7.8 Hz for the isomer in which the ethoxy group and the 3-bromine atom are *trans* and equatorial, but is 2.8 Hz when the ethoxy group is axial. A similar approach has been applied to

**Table 1** ¹H NMR Data[a] for Deuterated Analogues of Cyclohexane and Tetrahydropyran

| Compound | $J_{2a,3e}$ | $J_{2e,3e}$ | $J_{2a,3a}$ | $J_{2e,3a}$ | $\Delta\nu_{2e,2a}$ | $\Delta\nu_{3e,3a}$ | Ref. |
|---|---|---|---|---|---|---|---|
| Cyclohexane-$d_8$ | 3.65 | 2.96 | 13.12 | 3.65 | 0.479 | 0.479 | 68JA6543 |
| Tetrahydropyran-$d_4$ or -$d_6$ | 1.9 | 1.5 | 12.4 | 4.5 | 0.527 | 0.074 | 76JOC1380 |

[a] Hz or p.p.m.

stereochemical assignments of 2-, 3- and 4-amino-substituted tetrahydropyrans ⟨72JPS963, 75CR(C)(280)1525, 80JOC4352⟩.

NMR spectroscopy has proved of great value in the structural identification of both natural and synthetic chromenes. Much information has been accumulated and it is impractical to attempt even a partial listing of references here: the specialist review literature should be consulted.

In addition to the aromatic multiplet, the $^1$H spectrum of 2*H*-chromene (**4**) exhibits a quartet at $\delta$ 4.53 (2-CH$_2$) and peaks at 5.38 (3-H) and 6.20 (4-H) (Figure 1). The last peaks are particularly significant, appearing in a variety of multiplet structures, the nature of which depends on the substitution pattern. However, the benzylic proton always appears downfield of H-3. Alkyl-substitution at C-2 causes an upfield shift of both alkenic protons, but disubstitution results in shifts to lower field. In the widely occurring 2,2-dimethylchrom-3-enes, the *gem*-dimethyl groups resonate as a singlet at *ca.* $\delta$ 1.5. The protons at C-3 and C-4 now appear as two characteristic doublets, with $J_{3,4} = 10$ Hz.

In general, monosubstitution in the benzenoid ring results in small downfield shifts of the heteroring protons. The aromatic multiplet is characteristic of the appropriate trisubstituted benzene derivative, a feature which is of value in structural assignments to natural products containing the chromene nucleus. Long range coupling is observed if the 8-position is unsubstituted; $J_{4,8}$ is *ca.* 0.6 Hz. Further information in this area has accrued from the application of solvent effects ⟨67TL4201, 68ACS352⟩. Some data are available on naphthopyrans ⟨68JOC2416, 71AJC2347⟩.

Rather less information is available on 4*H*-chromenes. In the parent (**5**), the alkenic protons absorb at $\delta$ 6.44 and 4.83 (Figure 1). The former has been reported as a doublet ⟨62JA813⟩ and as a sextet ⟨69BSF1715⟩ and the signal from 3-H as a multiplet and a sextet. The coupling constant $J_{2,3}$ of *ca.* 6 Hz is smaller than $J_{3,4}$ for 2*H*-chromenes.

The spectra of chromans are more complex than those of chromenes. The heterocyclic ring adopts a half-chair conformation and consequently substituents at the 2-, 3-, and 4-positions may assume axial or equatorial orientations. However, in many cases signals from the methylene groups at C-2, C-3 and C-4 appear as triplets suggesting that the ring is flexible. A similar situation obtains for 2,2-dimethylchromans. The coupling constant $J_{3,4}$ is smaller in the chromans than in the chromenes in keeping with the increase in the length of the C(3)—C(4) bond.

Substitution at C-3 or C-4 imposes restrictions on the flexibility of the chroman ring. For example, it has been established that the 3,4-dihalogenochromans which result from the addition of halogen across the double bond of a chromene exist in the *trans* diaxial conformation (**49**). However, the most probable configuration for the corresponding 2,2-dimethyl compounds is considered to be *trans* diequatorial (**50**), which avoids 1,3-diaxial interaction between a 2-methyl group and a 4-halogeno substituent. A detailed discussion of this topic is available ⟨81HC(36)169⟩.

(**49**)                                    (**50**)

NMR data have been of value in the assignment of configuration to some chromanamines ⟨72BSF696, 73JCS(P2)227, 76MI22200⟩ and to 3-substituted chroman-4-ols ⟨70BCJ442, 70CB2768, 70JCS(C)1006⟩.

It is possible to distinguish axial and equatorial hydroxy groups in chroman-4-ols on the basis of their chemical shifts in DMSO ⟨79BCJ2163⟩. The pseudoaxial OH proton resonates at a significantly higher field; the different behaviour is attributed to solute–solvent hydrogen bonding.

### 2.22.3.1.2 *Pyranones, their benzologues and reduced derivatives*

The spectrum of pyran-2-one (**17**) consists of two complex multiplets of equal intensity centred at $\delta$ 6.4 (H-3 and H-5) and 7.6 (H-4 and H-6). Despite the complexity, PMR is described as a reliable technique for establishing the position of substituents, since their nature has little effect on the coupling constants of pyranone ring protons ⟨69JHC1, 69JOC2239⟩. Figure 4 contains the spectral parameters for pyran-2-one, and other data on substituted derivatives can be found in ⟨68JHC275, 69JHC1, 69JOC2239⟩ and elsewhere.

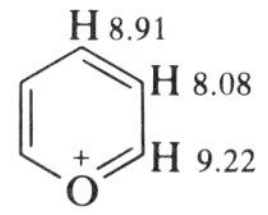
H 8.91
H 8.08
H 9.22
O

($CF_3CO_2D$) ⟨73OMR(5)251⟩

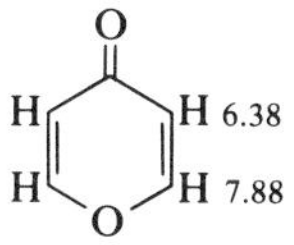
O
H
H 6.38
H
H 7.88
O

$J_{2,3}$ 5.96; $J_{3,5}$ 2.68 ⟨65SA1277⟩

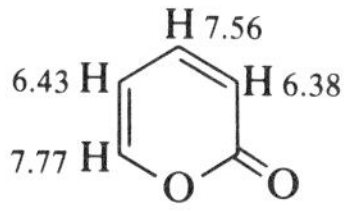
H 7.56
6.43 H
H 6.38
7.77 H
O
O

$J_{3,4}$ 9.4; $J_{4,5}$ 6.3; $J_{5,6}$ 5.0 ($CDCl_3$) ⟨69JHC1⟩

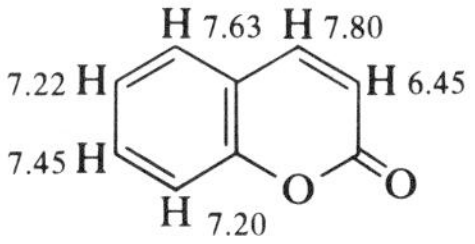
H 7.63 H 7.80
7.22 H
H 6.45
7.45 H
H 7.20
O
O

$J_{3,4}$ 9.8; $J_{5,6}$ 8.5 ($CDCl_3$) ⟨64AJC1305⟩

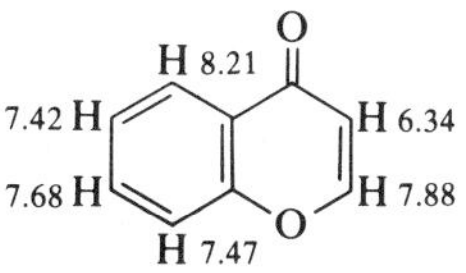
O
H 8.21
7.42 H
H 6.34
7.68 H
H 7.88
H 7.47
O

$J_{2,3}$ 6; $J_{5,6}$ 8; $J_{6,7}$ 7; $J_{7,8}$ 8.4 ($CDCl_3$) ⟨64SA871⟩

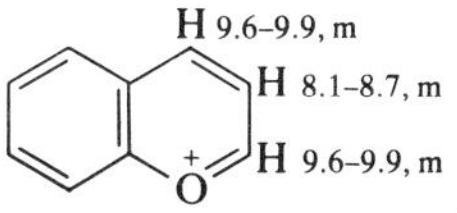
H 9.6–9.9, m
H 8.1–8.7, m
H 9.6–9.9, m
O

($CF_3CO_2H$) ⟨81JHC1325⟩

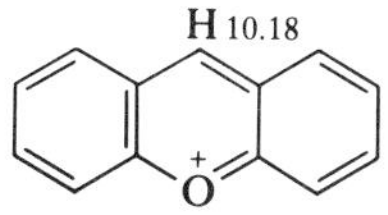
H 10.18
O

($D_2SO_4$) ⟨75JA5472⟩

**Figure 4** $^1$H NMR spectra of representative aromatic pyrans and fused pyrans

The $^1$H spectrum of dehydroacetic acid indicates its structure to be the fully enolized form (**51**) rather than that originally proposed ⟨64JCS5200⟩. However, the technique is not always so successful and difficulties arise in the distinction of the isomeric species 4-alkoxypyran-2-ones and 2-alkoxypyran-4-ones ⟨62JOC3715, 63JCS4483, 67JCS(C)413, 68JHC275, 69JHC13⟩.

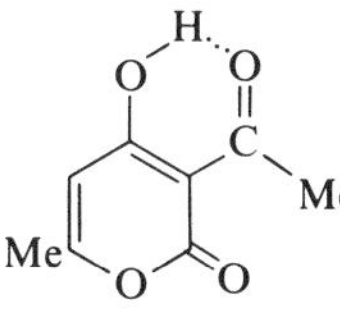
H
O
O
C
Me
Me
O
O

(**51**)

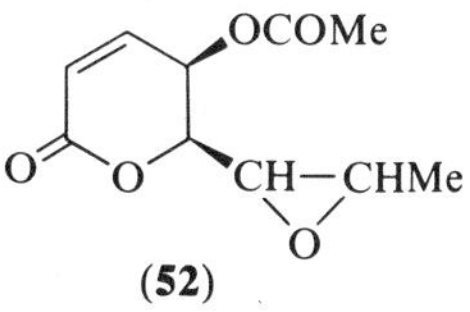
OCOMe
O
O
CH—CHMe
O

(**52**)

In addition to the coupling ($J = 10$ Hz) between H-3 and H-4 in 5,6-disubstituted derivatives of 5,6-dihydropyran-2-one (**52**), long range coupling of *ca.* 2 Hz between H-4 and H-6 is observed ⟨69JHC1⟩. This can be accounted for if the 5- and 6-substituents are *trans* diaxial, since H-4, C-4, C-5, C-6 and H-6 are then arranged in a near planar 'W' arrangement, for which four-bond coupling has been observed ⟨62JA1594⟩.

The assignment of *cis* and *trans* configurations in pairs of stereoisomers (**53**) and (**54**) was based on the values of $J_{5,6}$ and $J_{4,6}$ ⟨79MI22200⟩. The *trans* isomers, with a pseudoequatorial H-6, were identified by large $J_{5,6}$ and small $J_{4,6}$ values.

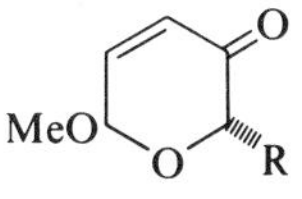
O
MeO
O
R

(**53**)

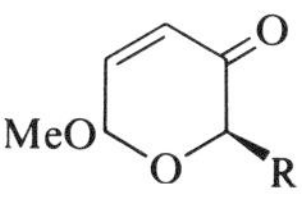
O
MeO
O
R

(**54**)

Conformational studies of substituted tetrahydropyran-2-ones (valerolactones) have utilized NMR spectroscopy ⟨67TL5119⟩.

Much of the PMR work on pyran-4-ones has been concerned with the aromatic nature of the system and several reports of aromatic character are based on the downfield shifts relative to non-aromatic dihydro analogues ⟨68T923⟩. The more acidic the solvent used in the spectral studies, the further downfield the signals are shifted. The protonated form (**55**) is clearly more aromatic and hence this shift is not unexpected, but whether this is a result of ring current is uncertain.

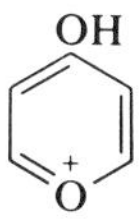
OH
O

(**55**)

Pyran-4-one can be considered as an $A_2B_2$ system and as such the spectrum consists of two multiplets (Figure 4). By comparison with the chemical shift of substituted pyran-4-ones, it was possible to assign the upfield doublet to H-3 and H-5. All the couplings of pyran-4-one have been assigned ⟨64SA871, 67SA(A)55⟩.

Spectral data for other pyran-4-ones can be found in ⟨64JOC2678, 64JOC2682, 67JCS(C)828, 68JHC275, 76JA7733⟩ and elsewhere.

The chemical shifts of H-3 and H-4 in coumarin (**19**) are similar to those in pyran-2-one (Figure 4). The values are closely related to the shifts of the $\alpha$ and $\beta$ protons of *o*-coumaric acid (**56**), implying that the heteroring has little or no aromatic character ⟨62PIA(A)(56)71⟩. The signal for H-3 appears at higher field than that from H-4 and distinction between 3- and 4-substituted isomers is usually possible. Coupling between the alkenic protons is typical of a *cis*-alkene ($J_{3,4} = 9.8$ Hz) and the pair of doublets is a characteristic feature of the spectra of coumarins.

(**56**)

The calculated electron densities for the aromatic ring correlate well with the observed chemical shifts, H-5 and H-7 occurring downfield of H-6 and H-8. All the aromatic proton coupling constants are as expected for a benzenoid ring, but in addition long range couplings are observed between H-4 and H-8 ($J$ *ca.* 0.6 Hz) ⟨67JCS(C)2000⟩. Later work presents values for other inter-ring couplings ⟨68AJC2445⟩.

Substitution in the aromatic ring has a small influence on the signals from H-3 and H-4. For instance, an electron-releasing substituent at C-5 or C-7 shifts the H-3 signal upfield, whereas a C-5 substituent deshields the proton at C-4. The chemical shifts of substituents at C-3 and C-4 are also influenced by substitution in the benzenoid ring. Thus, the shift of a 4-methyl group depends on both the nature and position of substituents in the benzene ring ⟨72MI22200⟩ and similar behaviour is reported for coumarins containing both methoxy and methyl groups ⟨78T1221⟩. The H-3 vinyl proton couples with a C-4 methyl group, exhibiting a $J$ value of 1.1–1.5 Hz. The elucidation of the structure of a novel coumarin may be achieved with the help of a set of rules which relate spectra and structure ⟨72MI22201⟩.

The value of solvent-induced shifts in structural assignments has been demonstrated notably in work on methoxycoumarins. The ASIS effect, $\delta(CDCl_3) - \delta(C_6D_6)$, is greatest for a proton at C-4 and least at C-8 ⟨66T3301⟩. Methoxy groups are also shielded in benzene compared with chloroform provided there is at least one *ortho* proton. Again the effect is most pronounced at C-4, whilst 5- and 7-methoxy groups are affected more than those at C-6 and C-8. This behaviour is consistent with conjugation of a 4-, 5- or 7-methoxy substituent with the carbonyl group which reduces the electron density at the oxygen atom of the methoxy group, thereby facilitating interaction with benzene.

Lanthanide shift reagents have also proved of value in the elucidation of the substitution patterns of coumarins ⟨74CC632⟩. Eu(fod)$_3$ complexes with the carbonyl oxygen atom and the resulting shifts for a given substituent are in the order $3 \gg 4 \approx 8 > 5 \approx 7 > 6$. By assuming that Eu(fod)$_3$ bonds to coumarin in the plane of the ring and from the direction of H-3 it is possible to correlate the lanthanide-induced shifts with substituent positions ⟨78JCS(P2)391⟩.

A PMR study of warfarin (**57**) and phenprocoumon (**58**), oral anticoagulants, has revealed that the chemical shift of the benzylic proton can be used to indicate conformational preferences of the substituent in some 3-substituted 4-hydroxycoumarins ⟨78JMC231⟩.

(**57**) R = CH$_2$COMe
(**58**) R = Et

The several different dimers which are formed on prolonged exposure of coumarin to light have been identified by NMR ⟨66CB625⟩ and spectral parameters for a number of isocoumarins (**28**) have been reported ⟨67BSF2224⟩.

The conformational aspects of *cis*- and *trans*-4-alkyl-3-phenyl-3,4-dihydrocoumarins have been studied with the conclusion that for both of the isomers (**59**) and (**60**) an axial disposition of the 3-phenyl group is preferred ⟨75MI22200⟩.

(59)　　　　　　　　(60)

The chromone system (**20**) is very similar to that of coumarin and their NMR spectra closely resemble each other (Figure 4). Thus, the chemical shifts of H-2 and H-3 are largely unaffected by annelation of the benzenoid ring onto pyran-4-one, indicating little or no aromaticity in the heteroring of chromone. However, the proximity of the carbonyl group to the aromatic ring results in downfield shifts of all of the benzenoid protons in chromone relative to those in coumarin. This effect is greatest at C-5 and is sufficiently strong to separate this signal from those of the other aromatic protons. Coupling of H-5 with both H-6 and H-7 is observable.

A detailed $^1$H NMR study of chromone has been published ⟨64SA871⟩ and the chemical shifts agree well with those predicted from electron density calculations. The position of substituents can frequently be determined from NMR spectra and this technique has been widely used in the structural elucidation of naturally occurring chromones ⟨70JCS(C)2230, 70JCS(C)2609, 71JCS(C)1482, 73JCS(P1)2781, 76JHC211⟩. 5-Hydroxychromones have been particularly well studied and two consequences of hydrogen bonding between the 5-OH and 4-C=O groups have been noted: the resonances of H-2 and H-3 occur at lower field and the 5-OH signal is well downfield ($\delta$ 12–13). A similar effect on the heteroring protons is observed for 8-hydroxychromones, suggesting weak chelation of the heteroatom with the 8-OH group.

A methyl group is commonly encountered at C-2 in chromones. Its effect is to shift the signal from H-3 upfield ⟨67IJC93⟩.

Spectral data for substituted coumarins and chromones are widely spread in the literature, and references should be sought from the review literature.

The spectrum of chroman-4-one was reported in 1966 ⟨66CB3076⟩, although the assignments of H-7 and H-8 were incorrect; the error has been remedied ⟨68JHC133⟩. The two methylene groups appear as triplets at $\delta$ 2.80 (3-CH$_2$) and 4.55 (2-CH$_2$). Studies on 3-bromo-, 2-iodo- and 3-methyl-chroman-4-ones indicated that the halogen atom was axial in the predominant conformation ⟨68JHC745⟩. A 3-*t*-butyl group preferentially occupies an axial position, whereas 3-methyl, 2-isopropyl and 2-*t*-butyl groups prefer equatorial conformations ⟨73BCJ1839⟩.

The only coupling constant that can be obtained for the heteroring protons of chroman-4-ones is ($J + J'$), where $J = \frac{1}{2}(J_{aa} + J_{ee})$ and $J' = \frac{1}{2}(J_{ae} + J_{ea})$ assuming that the ring is a rapidly inverting half-chair form. A value of 13.0 Hz is obtained ⟨68JHC745⟩. The aromatic protons show *ortho* and *meta* coupling ⟨68JHC133⟩.

The NMR spectra of various substituted chroman-4-ones occur in papers largely concerned with their synthesis. The spectra of some rotenoids such as (**61**), which contain the chromanone unit, have been analyzed and their structures elucidated ⟨62JCS775, 66JCS(C)542, 72JOC1636⟩.

(61)　　　　　　　　(62)

Chroman-3-one (**62**) shows singlets for 2-CH$_2$ ($\delta$ 4.32) and 4-CH$_2$ (3.51). The H-5 signal is no longer separated from the other aromatic protons and a complex multiplet is observed ⟨70JHC197, 81JHC1123⟩.

From the point of view of NMR spectroscopy, xanthones may be considered as *o*-disubstituted benzenes. The spectrum of the parent compound (**21**) reveals deshielding of H-1 and H-8 by the carbonyl group ⟨65T1833⟩. A complete analysis of xanthone and oxygenated analogues has been carried out based on the spectra in various solvents ⟨70JCS(B)603⟩. There is good agreement between the observed shifts of H-1 and those calculated using substituent shielding effects in benzene derivatives ⟨75JOC2088⟩. The value of aromatic solvent induced shifts in structural elucidation of xanthones has been assessed ⟨67JCS(C)2500, 72JCS(P1)1382⟩. The effects of lanthanide shift reagents have been studied ⟨75JCS(P1)1563⟩.

A review of the PMR spectra of phytoxanthones contains an extensive list of spectral data for oxygenated xanthones ⟨79H(12)421⟩.

### 2.22.3.1.3 Flavonoids

Although the PMR spectra of most of the six-membered oxygen heterocycles have already been covered, their 2- and 3-aryl derivatives have been studied in such detail that a separate treatment is considered to be worthwhile. However, the literature on flavonoids is vast as befits such an important group of compounds and only a basic discussion can be given here. Reference to texts devoted to flavonoid chemistry ⟨B-70MI22200, B-75MI22201, B-82MI22200⟩ will give access to major references, whilst the review by Batterham ⟨B-73NMR385⟩ contains an invaluable guide to the literature up to 1970 classified according to substitution patterns. An earlier paper contains a useful bibliography and summary of the applications of NMR to flavonoids ⟨66BSF2405⟩.

Early work on the PMR spectra of flavonoids was hindered by their lack of solubility in $CDCl_3$ and $CCl_4$. Progress was made following the introduction of DMSO-$d_6$ ⟨64AJC428⟩, but the most significant advance arose from the conversion of flavonoids into their more soluble trimethylsilyl ethers ⟨64TL513, 65P177⟩.

The two aromatic rings exert little or no influence on the spectral characteristics of each other. Ring A (63) is usually 5,7-disubstituted, in which case H-6 and H-8 appear as *meta*-split doublets, or 7-substituted. In the latter instances, H-5, which is strongly deshielded by the function at C-4, is *ortho*-split ($J = 9$ Hz) and H-6 appears as a doublet of doublets.

(63)

The substitution pattern for ring B is rather more variable but the usual aromatic couplings are observed. The appearance of the signals from ring C protons varies both with the location of the aryl ring B and with the oxidation state of the flavonoid.

The chemical shift of the single proton of the heterocyclic ring of flavones (22) and isoflavones (64) provides a means of distinguishing the two systems. The H-3 flavone proton ($\delta$ 6.0–6.5) appears upfield of the isoflavone proton H-2 ($\delta$ 7.7–8.0) which is deshielded by the heteroatom and in DMSO is shifted further downfield ($\delta$ 8.5–8.7) ⟨67JHC61⟩.

(64)

The spectra of reduced flavonoids, the flavanones (25) and their 3-hydroxy derivatives, are more complex. H-2 of flavanones couples with the non-equivalent protons at C-3 ($J_{trans} = 11$ Hz; $J_{cis} = 5$ Hz). The protons at C-3 interact with each other ($J = 17$ Hz) and with H-2. In the dihydroflavonols, H-2 and H-3 are disposed *trans* diaxially to each other and $J = 11$ Hz.

The distinction between flav-2-enes (26) and flav-3-enes (10) is almost a formality using NMR spectroscopy ⟨68AJC2059⟩. In the former, the protons at C-4 are magnetically equivalent and appear as a doublet ($\delta$ *ca*. 3.5), whilst H-3 is observed as a triplet ($\delta$ 5.3). On the other hand, the heterocyclic ring protons of flav-3-enes form an ABX system, with H-2 and H-3 absorbing at $\delta$ 5.7 and H-4 at 6.5. Analysis of the coupling constants indicates that the 2-aryl group is equatorially disposed. The same paper provides data on a range of flavans, isoflavans and various derivatives.

Some attention has been paid to the NMR of flavylium salts (**12**) ⟨65T3697⟩ and to the glycosides of various flavonoids ⟨65JOC4346, 68JOC1571⟩.

Benzene-induced shifts of methoxy protons have proved of value in the structural elucidation of various flavonoids ⟨68JCS(C)2477⟩. Additionally, information on the position of hydroxy groups follows from ASIS shown by the trimethylsilyl ethers of the flavonoids ⟨72P409⟩. Lanthanide shift reagents also contribute usefully to structural assignments, notably with biflavones ⟨75JCS(P1)1563⟩.

### *2.22.3.1.4 Pyrylium salts*

As expected of an aromatic system containing a strongly electron-withdrawing substituent, the NMR of the pyrylium cation shows that all of the protons are strongly deshielded. H-2 and H-6 are furthest downfield, whilst H-4 is downfield of H-3 and H-5 in accord with resonance theory ⟨73OMR(5)251⟩. $^1$H spectral parameters for the heterocyclic ring in pyrylium, 1-benzopyrylium and xanthylium salts are shown in Figure 4. They are rather sensitive to changes in solvent.

Comparison of the chemical shifts for a series of 2,6-di- and 2,4,6-tri-substituted pyrylium salts with the values for alkylbenzenes indicates that any reduction in ring current arising from replacement of a CH group by the heteroatom is more than offset by the deshielding effect of the charged oxygen atom ⟨64JCS1646, 76RRC101⟩.

Much of the NMR work on pyrylium compounds has been carried out on the perchlorate salts, but measurements on the tetrafluoroborate and hexachloroantimonates show that the nature of the anion has little effect on the chemical shifts of the ring protons. A paramagnetic anion, such as $FeCl_4^-$, may cause line broadening.

## 2.22.3.2 $^{13}$C NMR Spectra

### *2.22.3.2.1 Pyrans, their benzologues and reduced derivatives*

Presumably because of their relative inaccessibility, there is little information available on the $^{13}$C spectra of 2*H*-pyran, 4*H*-pyran or their simple derivatives. Some examples are shown in Figure 5.

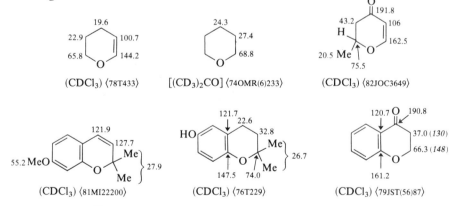

**Figure 5**  $^{13}$C NMR spectra of representative nonaromatic pyrans and fused pyrans

The reaction of pyrylium salts with sodium methoxide was shown to give 2*H*-pyrans in part by virtue of the lack of symmetry of the $^{13}$C spectra ⟨79H(12)775⟩. The ring carbon atoms of 2-amino-2*H*-pyrans are shielded relative to the analogous 2-methoxy compounds. Phenyl groups attached at C-4 of 2*H*-pyrans are deshielded at C-1′ in keeping with the non-aromatic nature of these pyrans ⟨81CS(18)256⟩.

The $^{13}$C data for a series of 4-aryliminopyrans (**65**) and their salts (**67**) have been recorded and the majority of the signals have been unambiguously assigned ⟨81JCS(P2)303, 81JCR(S)200⟩. The large downfield shifts of C-2, C-4 and C-6 observed on changing the solvent from CDCl₃ to the more polar CD₃CN are indicative of an increase in the positive nature of these atoms and are attributed to a greater contribution from the resonance structure (**66**).

(65)            (66)            (67)            (68)

Still larger deshieldings are observed in the salts in accord with a significant contribution from (68) to the overall structure.

In contrast, a useful amount of information has accumulated on reduced pyrans, generally as a result of investigations into stereochemical features of these compounds.

The $^{13}$C spectra of 2-methoxy-5,6-dihydro-2$H$-pyran (69; R = H) and eight 6-substituted derivatives have been recorded ⟨74MI22200⟩. The carbon atoms of the double bond resonate in the 126–130 p.p.m. range, with the signal from C-3 occurring slightly downfield of that from C-4. The proximity to the heteroatom causes C-2 and C-6 to appear downfield of C-5, whilst the additional deshielding effect of the methoxy group at C-2 shifts the signal from C-2 still further downfield.

(69)                    (70)

The C-2 and C-6 signals of the *trans* isomer (69) always appear at higher field than those of the corresponding *cis* isomer (70), attributable to the 1,3-shielding effect of an axial substituent. It is also possible to estimate the position of the conformational equilibrium in the *cis* isomers from the chemical shift difference of C-6 in the *cis* and *trans* compounds.

Configurational assignments for 5,6-dihydro-2$H$-pyrans have also been made from the magnitude of the geminal $^{13}$C–H coupling constants ⟨73TL1037⟩ and from the sensitivity of the chemical shift to steric hindrance ⟨B-72MI22202⟩.

Comparison of the chemical shift of the $\beta$-carbon atom of the vinyl group in 6-methyl-3,4-dihydro-2$H$-pyran (71) with that in 2-methylenetetrahydropyran (72) indicates that p–$\pi$ conjugation is greater in the former compound, which is accordingly the more stable ⟨78T433⟩.

(71)            (72)

Although the $^{13}$C spectrum of tetrahydropyran was reported in 1965 ⟨65JPC3925⟩, almost a decade passed before this saturated system was investigated by this technique.

Tetrahydropyran exists preferentially in the chair form and hence the chemical shifts of the carbon atoms relative to that of cyclohexane ($\delta$ 27.7) reflect the electronic effects of the oxygen heteroatom. Examination of the data in Figure 5 reveals a pronounced deshielding of C-2 and a small upfield shift of C-4; C-3 is almost unaffected.

From a study of a range of 2-substituted tetrahydropyrans it was concluded that the effects of an electronegative substituent were to deshield C-2 by about 30 p.p.m. and C-3 to a much reduced extent (*ca.* 4 p.p.m.), but to cause small upfield shifts in the signals from C-4, C-5 and C-6 ⟨74OMR(6)233⟩. The influence of a 2-substituent on the chemical shift of C-4 can be used for quantitative conformational analysis provided suitable reference compounds are available. A comparatively large shielding of the C-4 signal indicates a preference of the 2-substituent for the axial position.

The chemical shifts of a range of *cis* and *trans* isomeric pairs of 2-alkoxytetrahydropyrans (73) and (74) show a consistent upfield shift of the three carbon atoms attached to oxygen in the axial isomers. The effect is greatest at C-6 and smallest at the alkoxy carbon atom ⟨81JOC4948⟩.

(73)                    (74)

$^{13}$C chemical shift differences have also been employed for the assignment of configuration of a 4-amino substituent; an axial NH$_2$ group exerts a shielding of 4–5 p.p.m. ⟨80JOC4352⟩.

$^{13}$C data indicate that the axial conformer of 4-halogenotetrahydropyrans is stabilized relative to that in the corresponding cyclohexanes. The effect is greatest for the 4-fluoro derivative and is indicative of an electrostatic interaction between the heteroatom and the slightly positive C-4 atom ⟨78SA(A)297⟩.

The influence of an oxygen atom relative to other heteroatoms on the $^{13}$C shielding of the three γ-carbon atoms in 3,3-dimethylheterocyclohexanes (**75**) has been examined ⟨81OMR(17)270⟩.

**(75)**             **(76)**

Information relating to the $^{13}$C spectra of chromenes is sparse, being restricted to some 2,2-disubstituted derivatives (*cf.* Figure 5). Substituents in the aromatic ring of chromenes affect the electron density at the double bond and an attempt has been made to relate the antijuvenile hormone activity of the precocenes (**76**; R = H or OMe) to the electron density at C-3 and C-4, using $^{13}$C chemical shifts as a measure of the latter property. Whilst C-4 is essentially unaffected by substitution, it appears that a $^{13}$C shift of C-3 at less than 130 p.p.m. is a necessary condition for activity ⟨81MI22200⟩.

Little information is available on chromans *per se*, but since 6-hydroxy-2,2-dimethyl-chromans (**77**; R = Me) are model compounds of the tocopherols (**77**; R = phytyl), their $^{13}$C spectra have been thoroughly investigated ⟨76T229⟩. In most cases, unambiguous assignments have been made with the aid of off-resonance decoupling, shift reagent techniques and deuterium labelling. The two oxygenated carbon atoms (C-6 and C-8a) were distinguished by their behaviour towards Eu(fod)$_3$; the induced shift at the phenolic carbon was significantly larger than that at the ether carbon.

**(77)**            **(78)**            **(79)**

The $^{13}$C spectra indicate a preference for a pseudoaxial orientation of the hydroxy group in chroman-4-ol and *trans*-2-methylchroman-4-ol (**78**). The *cis* isomer (**79**) has a pseudoequatorial hydroxy group and the resonances for C-2, C-3 and C-4 appear at lower field than those of the *trans* compound ⟨77JCS(P1)217⟩. A similar situation obtains for the related flavanols.

### 2.22.3.2.2 *Pyranones, their benzologues and reduced derivatives*

A considerable volume of work has been carried out on the $^{13}$C spectra of pyranones and their derivatives, largely because the structural information which accrues can be extrapolated to the many naturally occurring molecules which incorporate these systems.

A study of the $^{13}$C spectra of pyran-2-one and a number of its simple derivatives indicated that the C-4 and C-6 resonances occur downfield of those from C-3 and C-5, consistent with localization of positive charge at the former positions ⟨74JOC1935⟩. A comparison of the spectra of (**80**; R = H or OMe) shows that, whilst both C-3 and C-5 are deshielded by the introduction of a 4-methoxy group, the shift of the former (25 p.p.m.) is much larger than that of the latter (3 p.p.m.). It is postulated that the degree of double bond character between C-4 and C-5 is much less than that between C-4 and C-3. The carbonyl C-2 resonance is virtually unaffected by substituents and it is claimed that the assignment of

**(80)**

any substituted carbon in pyran-2-ones is straightforward. However, a later report ⟨76TL1311⟩ advises caution in making these assignments until a firm substitution rule is obtained for the chemical shifts of the pyran-2-one ring.

Reference to Table 2 enables the effects of substitution of a methoxycarbonyl group at various positions of pyran-2-one to be assessed. The general picture is that the substituent causes large deshieldings of the $\beta$- and $\gamma$-carbon atoms of the substituted double bond system by the resonance effect ⟨80OMR(13)244⟩.

**Table 2**  $^{13}$C NMR Data for some Pyran-2-ones

| R | Solvent | C-2 | C-3 | Chemical shift (p.p.m.) ($^1J_{CH}$, Hz) C-4 | C-5 | C-6 | R | Ref. |
|---|---|---|---|---|---|---|---|---|
| H | (CD$_3$)$_2$CO | 162.0 | 116.7 (170) | 144.3 (163) | 106.8 (173) | 153.3 (200) | — | 74JOC1935 |
| 4-Me | CDCl$_3$ | 161.8 | 113.7 (169) | 156.1 | 109.3 (169) | 151.1 (199) | 21.1 | 74JOC1935 |
| 5-Me | CDCl$_3$ | 161.2 | 115.7 (171) | 146.5 (162) | 114.7 | 148.0 (197) | 14.4 | 74JOC1935 |
| 6-Me | (CD$_3$)$_2$CO | 162.0 | 112.6 | 144.1 | 103.4 | 162.9 | 19.8 | 76JOC2777 |
| 3-CO$_2$Me | CDCl$_3$ | 157.3 | 118.0 | 148.6 | 105.9 | 156.6 | 163.8 (C=O) 52.6 (OMe) | 80OMR(13)244 |
| 6-CO$_2$Me | CDCl$_3$ | 159.7 | 121.0 | 141.9 | 110.0 | 149.6 | 159.9 (C=O) 53.0 (OMe) | 80OMR(13)244 |
| 4-OMe-2-Me | (CD$_3$)$_2$CO | 162.1 | 87.3 | 171.4 | 100.3 | 164.6 | 19.7 (Me) 55.9 (OMe) | 76JOC2777 |
| 3,6-H$_2$-6-Me | CDCl$_3$ | — | 29.78 | 121.37 | 79.56 | 128.01 | 21.77 (Me) | 79SC889 |

$^{13}$C–$^1$H coupling constants have been obtained for a number of pyran-2-ones. A value of 200 Hz is typical for $J_{CH}$ of carbon atoms attached to oxygen in aromatic heterocycles ⟨B-72MI22203⟩. $J_{CH}$ values for C-3 and C-5 are similar and are usually larger than those for C-4. The five different carbon atoms all couple with each of the four non-equivalent protons and 16 long-range couplings are observed. The magnitude and signs of all of these couplings have been reported ⟨75CJC1980⟩.

Although some data were obtained from a study of $^{13}$C satellites in the $^1$H spectra of some pyran-4-ones ⟨64SA871, 67SA(A)55⟩, information on this system is derived mainly from an investigation concerned primarily with kojic acid (**81**) ⟨76JOC2777⟩. The data are collected in Table 3. The spectra in acid solution provide some evidence for a degree of aromaticity of pyran-4-ones despite the small changes in chemical shift and coupling constants.

(**81**)

**Table 3**  $^{13}$C NMR Data for some Pyran-4-ones

| R | Solvent | Chemical shift (p.p.m.) ($^1J_{CH}$, Hz) C-2 | C-3 | C-4 | C-5 | C-6 | R | Ref. |
|---|---|---|---|---|---|---|---|---|
| 2-CHCl$_2$ | CDCl$_3$ | 160.6 | 113.4 | 177.6 | 116.6 | 155.8 | 41.2 | 82S500 |
| 5-OH | (CD$_3$)$_2$SO | 155.4 | 114.2 (167) | 172.9 | 146.4 | 139.9 (198) | — | 76JOC2777 |
| 5-OH-2-CH$_2$OH | (CD$_3$)$_2$SO | 167.8 | 109.6 (166.5) | 173.5 | 145.4 | 139.0 (197.5) | 59.3 | 76JOC2777 |
| 5-OH-2-CH$_2$Cl | (CD$_3$)$_2$SO | 161.5 | 113 | 173.4 | 145.8 | 139.8 | 41.2 | 76JOC2777 |
| 3-OH-2-Me | (CD$_3$)$_2$SO | 154.3 (199.3) | 113.2 (167.6) | 172.1 | 142.6 | 148.9 | 13.8 | 76JOC2777 |
| 3-OH-2-CH$_2$OH | (CD$_3$)$_2$SO | 154.9 (197.5) | 113.3 (168) | 173.1 | 142.4 | 150.1 | 54.9 | 76JOC2777 |
| 2-OMe-6-Me | CDCl$_3$ | 167.4 | 89.9 | 181.7 | 112.7 | 161.5 | — | 79CJC1451 |
| 2,3-H$_2$-2-Me | CDCl$_3$ | 162.5 | 106 | 191.8 | 43.2 | 75.5 | 20.5 | 82JOC3649 |

Coupling constants for pyran-4-ones are readily derived by a first-order analysis. The $^1J_{CH}$ value for C-2 is characteristic of a carbon atom bonded to a hetero-oxygen atom, whereas the value for C-3 is considerably smaller (*ca.* 165 Hz). The two-bond coupling constants are generally less than 9 Hz and appear to be reduced by substitution at C-2. Although the $^3J$ couplings *via* the ring oxygen (*ca.* 8 Hz) and through the carbonyl C-4 (*ca.* 6 Hz) are quite large, they are, generally speaking, smaller than the corresponding values for pyran-2-ones ⟨80OMR(13)244⟩.

Analysis of the $^{13}$C chemical shifts of C-2 and C-6 in a series of 2-substituted 6-methoxy-3,6-dihydropyran-3-ones (**82**) allows the assignment of configuration to the isomeric pairs. C-2 is shielded by *ca.* 5 p.p.m. and C-6 by *ca.* 2.5 p.p.m. in the *trans* isomer compared with the corresponding signals of the *cis* compound ⟨79MI22200⟩. The upfield shift is associated with the 1,3-diaxial disposition of the substituent and the proton on the γ-carbon atom. This γ-effect has also been noted in the dihydropyran ring system ⟨74MI22200⟩.

MeO⟋⟍O⟍R
**(82)**

A paper devoted to heterocyclic analogues of cyclohexanones includes $^{13}$C data for some 2,6-diphenyltetrahydropyran-4-ones ⟨79JOC471⟩. The deshielding effect on the α-carbon atom is greatest when the heteroatom is oxygen.

There is a substantial body of data concerned with the $^{13}$C spectra of coumarins. The chemical shifts and coupling constants for coumarin itself (see Figure 6) and for numerous substituted coumarins have been reported ⟨75JOC1175, 75T2587, 76JMR(21)241, 77JOC1337, 77T899, 79OMR(12)284, 81T2021⟩. Long range coupling constants have proved of value in identifying some very close resonance signals ⟨77JOC1337⟩. The $^{13}$C chemical shifts for C-2, C-3 and C-4 in coumarin are at 159.6, 115.7 and 142.7 p.p.m. respectively and correspond closely to those for pyran-2-one (Figure 6). A good correlation between $^{13}$C shifts and charge densities, calculated by the CNDO/2 method, has been established ⟨75JOC1175⟩. Chemical shifts of the pyranone ring system in coumarin and 6- and 7-substituted derivatives have been correlated with Hammett constants ⟨79JCS(P2)435⟩. A good fit is found for C-3 with $\sigma^+$ and of C-2 and C-4 with $\sigma$. The ring junction carbons (4a and 8a) correlate with $\sigma_p^+$, but not with $\sigma_m$ or $\sigma_m^+$.

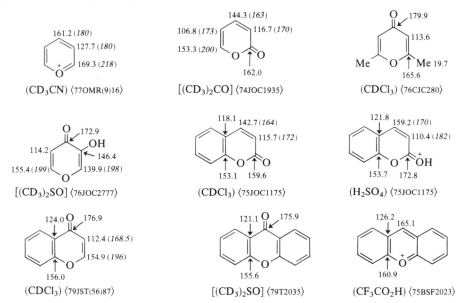

**Figure 6** $^{13}$C NMR spectra of representative aromatic pyrans and fused pyrans

Although most substituted coumarins give rise to signals which agree with those calculated by the normal additivity of substituent chemical shifts, deviations from the predicted values are observed for C-6 and C-7 and for the carbon atoms of the substituents in 6,7-dimethylcoumarin ⟨81T2021⟩. These *ortho* proximity effects have been noted in aromatic and other heteroaromatic systems. The vast majority of naturally occurring coumarins are oxygenated in the benzenoid ring. The presence of a hydroxy or methoxy group shifts the signal from the substituted carbon atom downfield by about 30 p.p.m. Upfield shifts are observed for the carbon atoms *ortho* and *para* to the substituent ⟨75T2719⟩.

$^1J_{CH}$ for C-3 is frequently larger than the other $^1J_{CH}$ values, a feature which has proved useful in structure elucidation amongst the coumarins. Typically, $^2J_{CH}$ values are no larger than 4 Hz, three-bond coupling constants are in the 4–10 Hz range and $^4J_{CH}$ values are less than 2 Hz.

The $^{13}C$ chemical shifts indicate that in concentrated sulfuric acid coumarin is protonated at the carbonyl oxygen (see Figure 6). Thus, C-2 is deshielded by 13.2 p.p.m. and C-4 by 16.5 p.p.m., whilst C-3 is shielded by 5.3 p.p.m., implying significant contributions from the resonance forms (**83**) and (**84**; Scheme 2) in which C-2 and C-4 carry a high degree of charge density. In (**84**), both vinylic oxygens are able to exert a $\beta$ shielding effect on C-3, and it would appear that contributions from this structure are important. The presence of the benzopyrylium ion (**85**) was discounted because the C-8a chemical shift remains virtually constant; $\delta(CHCl_3)$ 153.1, $\delta(H_2SO_4)$ 153.7, $\Delta\delta = 0.6$. This observation also excludes the possibility of protonation on the endocyclic oxygen ⟨75JOC1175⟩.

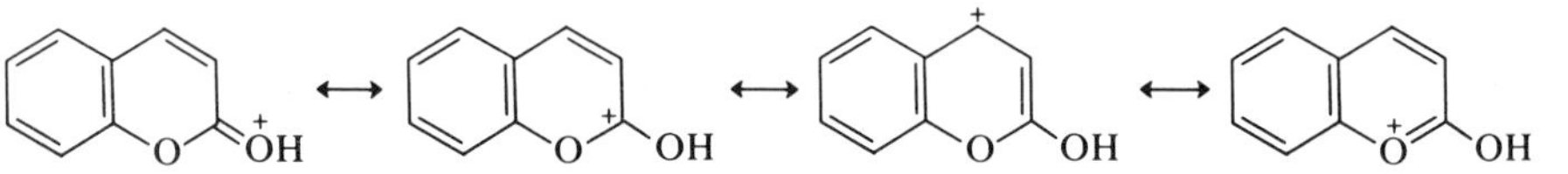

(**83**)                    (**84**)                    (**85**)

**Scheme 2**

A number of general features can be recognized in the spectra of chromones. The signal from the carbonyl carbon atom is always at lowest field and is essentially unaffected by substitution in the system. Of the remaining signals, that from C-2 is at a lower field and that associated with C-3 at higher field than all other C—H signals. Substitution at C-2 or C-3 has a marked influence on the resonances of these carbon atoms. For example, both methyl and phenyl groups induce downfield shifts of the carbon to which they are attached, but upfield shifts of the adjacent carbon atoms ⟨79JST(56)87⟩.

The resonances of the ring junction carbon atoms (4a and 8a) are not affected to any appreciable extent by substitution in either ring. The signals associated with the benzenoid carbon atoms are unaffected by changes in substituents in the heterocyclic ring, though they are perturbed by substitution in the aromatic ring. Conversely, the pyranone ring carbon shifts are insensitive to substituents at C-6 or C-7 ⟨81JCS(P1)2557⟩.

The values of $^1J_{CH}$ couplings for C-2 and C-3 correspond to those observed for pyran-4-one. Two-bond coupling constants for these atoms are typically 6–7 Hz, whilst $^3J_m$ couplings of 9 Hz have been observed for C-8a and of 4–5 Hz for C-4a. The couplings within the aromatic ring are similar to those observed for related compounds such as coumarins ⟨79JST(56)87⟩.

Reference to $^{13}C$ spectral data of a range of chromones can be found ⟨79JST(56)87, 79S889, 80CJC1211, 81JCS(P1)2557⟩.

The $^{13}C$ resonances of xanthone (**21**) have been assigned on the basis of shift analysis of the structurally related compounds anthrone and xanthene ⟨77P735⟩ and by correlation with the 'component molecules' diphenyl ether and benzophenone ⟨78T1837⟩. Delocalization of the non-bonding electron pairs at oxygen, leading to an increase in aromaticity of the pyranone ring, accounts for the deviation of the chemical shifts of C-4a, C-8a and the carbonyl carbon atom from the calculated values.

The spectra of xanthones are characterized by the carbonyl resonance at low field which is generally unaffected by substituents (*cf.* Figure 6). However, a 1-hydroxy group is able to form a hydrogen bond with the carbonyl group with the result that the carbonyl carbon atom is deshielded.

Much of the interest in xanthones centres around the naturally occurring polysubstituted derivatives and a number of studies have sought to evaluate substituent increments for xanthones ⟨77JCS(P1)2158, 78T1837, 79T2035⟩. This work has culminated in the introduction of a computer program and a set of additivity rules which allow a rapid identification of unknown polyhydroxyxanthones ⟨80T3273⟩.

The $^{13}C$ spectra of chroman-4-ones are similar to those of the corresponding chromone, although the C-2 and C-3 resonances occur at much higher field, as expected of methylene carbon atoms (see Figure 5). Both atoms are deshielded by substitution of a methyl group at either C-2 or C-3 ⟨77BCJ2789⟩. The heteroatom exerts its influence in two notable ways. A large $^1J_{CH}$ coupling is shown by C-2 ⟨79JST(56)87⟩ and, since the substituent parameter of an oxygen atom is negative for a γ-carbon, the carbonyl carbon signal is upfield of that for the corresponding atom in tetral-1-one, the homocyclic analogue of chroman-4-one.

### 2.22.3.2.3 *Flavonoids*

In keeping with the rapid development of $^{13}$C NMR spectroscopy, numerous publications containing $^{13}$C NMR data of flavonoids have appeared since *ca.* 1975. In many cases, the information is specific to an individual compound and it is not possible to cover all such work in this account. However, several studies which are more general in approach are discussed below and should direct the reader to data for each flavonoid system.

It is apparent from work published to date that the chemical shifts of C-2, C-3 and C-4 offer a means of distinguishing the different classes of flavonoids. A report on the electronic structure of parent flavonoid compounds compares observed chemical shifts with the values calculated from electron-density measurements with quite good agreement ⟨B-77MI22200⟩.

The response of $^{13}$C chemical shifts to the acetylation of phenolic groups of flavonoids is helpful in making structural assignments ⟨76JCS(P1)2475⟩. The signals from the carbon atoms *ortho* and *para* to the hydroxy group are shifted downfield, whilst that from the carbon atom bearing the substituent is moved to higher field. The shifts arising from methylation of these groups tend to be the opposite of those resulting from acetylation, but are somewhat variable ⟨78T1389⟩. The value of $^{13}$C–$^1$H coupling constants in structural elucidation has also been noted ⟨76JOC1881, 76TL1799, 77OMR(9)179, 78MI22200⟩.

The application of the principle of the additivity of substituent effects to flavonoids has been critically appraised ⟨76JCS(P1)2475⟩ and in general has been found to be reliable ⟨75JOC1120, 76T2607⟩. Of course, when the substituent can interact electronically or sterically with the flavonoid molecule, deviations from the predicted chemical shifts are observed. For instance, hydrogen bonding of a 5-hydroxy substituent to the carbonyl group at C-4 shifts the C-5 and C-4 signals further downfield than calculated.

The spectra of flavone (**22**) and isoflavone (**64**) show a close resemblance to that of chromone ⟨79JST(56)87⟩. The effect of the phenyl group is significant only at C-2 and C-3, when a downfield shift occurs at the substitution site but an upfield shift at the adjacent carbon atom. As a consequence, C-2 resonates at lower field in flavones than in isoflavones, whereas C-3 is at higher field. Two additional features help to distinguish flavones and isoflavones. In the latter system, the signal for C-4 appears as a doublet (*J* = 6 Hz) as a result of coupling with H-2 and a large $^1J_{CH}$ (*ca.* 200 Hz) is observed for C-2. A study which includes a compilation of $^{13}$C data for isoflavones ⟨78JCS(P1)666⟩ concludes that for both flavones and isoflavones the carbon atoms of the fused benzenoid ring are insulated from changes in the remainder of the molecule.

Consideration of the chemical shifts of the skeletal carbon atoms of a large number of flavone derivatives in DMSO enabled the parameters and correction factors necessary for calculating the carbon shifts to be deduced ⟨80CPB708⟩. The data were applied to the structural elucidation of polysubstituted flavones. Similarly, the background data accumulated for a range of hydroxylated flavones and flavonols have proved of value in the interpretation of the spectra of dihydroflavonoids and flavonoid glycosides. The sites of methylation, acylation and glycosylation are readily distinguished by $^{13}$C NMR spectroscopy ⟨78T1389⟩. Some general relationships between substitution patterns in isoflavones and chemical shifts which are of potential in the identification of naturally occurring isoflavones have been established ⟨80CJC1211⟩.

The effect of reducing the 2,3-double bond in flavones and isoflavones is manifest in a downfield shift of the C-4 signal to *ca.* 190 p.p.m. The spectra of flavanone (**25**) and isoflavanone (**86**) are similar to that of chroman-4-one ⟨77BCJ2789⟩. Overall downfield shifts of C-2 and C-3 are observed in both of the phenylchromanones relative to the parent molecule. The characteristic feature of the flavanone system is the highfield resonance of C-3 at 40–45 p.p.m. whilst the C-2 and C-3 signals in isoflavanones (*ca.* 70 and 50 p.p.m., respectively) complete the differentiation of the two systems ⟨76TL1799, 76JCS(P1)2475, 76T2607, 78P1363⟩.

(86)                    (87)

Other sources of $^{13}$C chemical shift data for flavonoids include those on flavonoid glycosides ⟨79ACS(B)119, 81P1977⟩, procyanidins ⟨77JCS(P1)1628, 80JCS(P1)2278, 82JCS(P1)1217⟩, rotenoids (61) ⟨75JCS(P1)1497⟩ and pterocarpans (87) ⟨77T1735⟩.

### 2.22.3.2.4 *Pyrylium salts*

The chemical shifts of C-2, C-4 and C-6 of pyrylium salts are the lowest reported for six-membered rings with a single heteroatom, a feature which is in keeping with the severe perturbation of the $\pi$-cloud brought about by the positively charged oxygen atom. Indeed, INDO MO calculations of charge density in the isoelectronic pyrylium, pyridinium, pyridine and benzene systems indicated that the downfield shifts of the $\alpha$ and $\gamma$ carbon atoms of pyrylium arise predominantly from changes in charge density associated with the heteroatom ⟨77OMR(9)16⟩.

In general, alkyl groups shift the resonances of the pyrylium ring carbons to lower field, although the chemical shifts no longer correlate with the calculated charge densities. Phenyl substituents at the 2-, 4- or 6-positions interact with the pyrylium system and the resonances of C-2', C-4' and C-6' are shifted to lower field relative to the analogous benzenoid compound, biphenyl. The deshielding effect is reduced when a substituent adjacent to the phenyl group decreases the coplanarity of the two ring systems and is of course much less apparent for *m*-phenyl substituents. When the phenyl ring is held in a coplanar arrangement by ring formation, the positive charge is efficiently delocalized to C-2', which accordingly is shifted downfield.

A $^{13}$C NMR study of the reaction of amines with pyrylium salts includes assignments of the resonances of a range of substituted pyrylium salts ⟨81CS(18)256⟩.

The $^{13}$C chemical shifts of 1-arylisobenzopyrylium salts (88) are as expected when a positively charged oxygen atom replaces the carbon atom at the 2-position of the corresponding naphthalene derivative ⟨76OMR(8)324⟩. Substitution in the 1-aryl ring has no appreciable effect on the chemical shifts of the carbon atoms of the heterocyclic system. Mesomeric interaction would be prevented if the aryl ring is perpendicular to the plane of the heterocyclic system as molecular models suggest.

(88)

$^{13}$C data for pyrylium and xanthylium ions are summarized in Figure 6.

### 2.22.3.3 $^{17}$O NMR Spectra

As yet this relatively new development of NMR spectroscopy has not been applied to any appreciable extent in the study of six-membered oxygen heterocycles. $^{17}$O NMR of organic molecules is handicapped by the low natural abundance of this isotope (0.037%) and line broadening due to $^{17}$O quadrupole relaxation. However, a report of the $^{17}$O spectra of 2-alkoxytetrahydropyrans concerned with the anomeric effect reveals a role that this method of analysis may play in structural elucidation of oxygen heterocycles ⟨81JOC4948⟩. Where both *cis* and *trans* isomers were available, both the ring and alkoxy oxygen atoms were shielded more in the *trans* compound, in which the alkoxy substituent is predominantly axial. Conformational and configurational assignments of 2-alkoxytetrahydropyrans are conveniently made using $^{17}$O chemical shifts and the anomeric effect in more complicated systems could well be investigated by this technique.

#### 2.22.3.4 Other Nuclei

The $^{19}$F NMR spectra of several highly fluorinated derivatives of 2*H*-pyran ⟨81MI22201, 83JCS(P1)1239⟩ and 4*H*-pyran ⟨83JCS(P1)1235⟩ have been reported, as have those of some fluorinated 2,2-dimethyl-2*H*-chromenes ⟨80JHC1377⟩. The structures of some pyran-2-ones derived from the reaction between phenylacetylene and fluorinated allenes have been ascertained with the aid of their $^{19}$F spectra ⟨77IZV2517⟩.

The $^{15}$N chemical shift for the iminopyran (**75**) is consistent with a build-up of negative charge at the nitrogen atom and with a lower bond order than in *N*-benzylidene-4-nitroaniline. Salt formation generally brings about an upfield shift of the nitrogen resonance by *ca.* 135 p.p.m. ⟨81JCR(M)1648⟩. Values for $J_{N,H}$ and $J_{N,C}$ are helpful in assessing the geometry of the molecules.

### 2.22.4 IR SPECTROSCOPY

Although a number of papers are devoted exclusively to IR and Raman studies of oxygen heterocycles, the present trend is to incorporate such spectroscopic data with other aspects of their chemistry. Much of the earlier work has been summarized ⟨63PMH(2)242, 71PMH(4)339⟩ and many of the papers referred to in Chapters 2.23 and 2.24 include IR data which it has not been possible to cover in the present account.

#### 2.22.4.1 Pyrans, their Benzologues and Reduced Derivatives

Relatively little information is available on the IR of simple derivatives of 2*H*-pyrans, doubtless because of their instability. The highest C=C stretching frequency occurs near 1650 cm$^{-1}$ ⟨64LA(678)183⟩, although its position is influenced by substituents. Thus, 2,2,4,6-tetramethyl-2*H*-pyran absorbs at 1666 cm$^{-1}$ ⟨64BSF1492⟩, whereas a series of 2-alkyltetrachloropyrans shows this absorption between 1648 and 1635 cm$^{-1}$ ⟨75LA240⟩. Both groups show a second C=C vibration at 1610–1600 cm$^{-1}$ and the position of the C—O band is reported between 1080 and 1070 cm$^{-1}$. Fusion of an alicyclic ring has little effect, (**89**) exhibiting peaks at 1650 and 1598 cm$^{-1}$ ⟨57JA2318⟩.

(**89**)          (**90**)

Esters of 2,4-diacetylpent-2-enedioate exist predominantly in the cyclic form (**90**). The lack of absorption at 1690–1665 cm$^{-1}$ confirms the absence of an *α*-unsaturated acetyl and a chelated ester. Bands at 1707 and 1636 cm$^{-1}$ are ascribable to the unsaturated ester and the C=C stretch of the 2*H*-pyran ⟨79JCS(P1)464⟩.

Rather more data have been reported on 4*H*-pyrans, including the spectrum of the parent compound (1700 and 1660 cm$^{-1}$, C=C; 1280–1260 cm$^{-1}$, C—O) ⟨62JA2452⟩. Studies of various derivatives revealed that this class of compound is characterized by absorption around 1720 and 1670 cm$^{-1}$ ⟨69JOC3169⟩, whilst a later account lists strong bands at 1680–1660 cm$^{-1}$ (C=C) and 1270–1250 and 1160–1135 cm$^{-1}$ (enolic C—O) as characteristic of 4*H*-pyrans ⟨78JHC57⟩. Again, the position of the highest C=C stretching frequency is variable, occurring between 1728 and 1690 cm$^{-1}$ for a series of 2-amino-4*H*-pyrans ⟨76BSF987⟩.

It is therefore possible to make at least a tentative distinction between 2*H*- and 4*H*-pyrans on the basis of their IR spectra. For instance, 2-hydroxy-2,4,6-triphenyl-2*H*-pyran shows peaks at 1610, 1585 and 1575 cm$^{-1}$ associated with the pyran double bonds and the phenyl groups, whilst 2,4,6-triphenyl-4*H*-pyran absorbs at 1640, 1585 and 1575 cm$^{-1}$ ⟨69TL2195⟩. 2,4,4,6-Tetramethyl-4*H*-pyran, isomeric with the 2*H*-pyran above, exhibits peaks at 1724 and 1667 cm$^{-1}$, providing further evidence that the highest C=C stretching frequency occurs at higher wavenumbers in 4*H*-pyrans than in 2*H*-pyrans.

The C=C stretching frequencies of 3,4-dihydro-2*H*-pyran (13) and 5,6-dihydro-2*H*-pyran (14) occur at 1630 cm$^{-1}$ ⟨58MI22200⟩ and 1640 cm$^{-1}$ ⟨58CB1589⟩, respectively. In the former series, the influence of substituents at a saturated carbon is small and somewhat variable, but a methyl or methoxy substituent at C-5 or C-6 raises the frequency by *ca.* 40 cm$^{-1}$ ⟨55JA4571, 78JOC667, 81HCA1247⟩.

Frequencies have been assigned to the ring stretching vibrations and methylene bending modes of tetrahydropyran between 1500 and 800 cm$^{-1}$ ⟨50JA4397⟩. The spectrum of the fully reduced pyran has been reexamined together with the spectra of related deuterated compounds and assignments have been extended beyond those previously listed ⟨70JCP(53)376⟩. The spectrum of tetrahydropyran-4-ol has been reported ⟨58CB1589⟩ and subsequently assignments of most of the observed bands in the spectra of all of the monohydroxy and of the 2-hydroxymethyl derivatives of tetrahydropyran have been made ⟨60JCS4565⟩. This latter work also includes the Raman spectra of tetrahydropyran and the 4-hydroxy compound.

Detailed assignments have been reported for 2*H*-chromenes ⟨66MI22200, 77HCA215⟩. The C=C stretch of the pyran ring double bond at about 1640 cm$^{-1}$ and the C—O stretch of the aromatic ether in the range 1260–1200 cm$^{-1}$ are important features, though the positions of both bands are affected by substituents especially when they are present at C-4. An electron-donating group shifts the C=C absorption to higher frequency, whilst halogen substitution at either C-3 or C-4 results in a lower value.

The characteristic absorptions of the *gem*-dimethyl group at 1380–1360 cm$^{-1}$, the aromatic ether (1270 and 1120 cm$^{-1}$) and a band at 900 cm$^{-1}$ are important for the detection of the 2,2-dimethylchromene entity in natural products.

Spirobi(2*H*-chromenes) (91), their 3,3′-trimethylene bridged analogues (92) and naphthopyran derivatives (93) show absorptions at 1670–1635 cm$^{-1}$, 1245–1210 cm$^{-1}$ and 975–915 cm$^{-1}$ for the C=C stretch, the ether C—O linkage and the spiro C—O stretch, respectively ⟨66ZN(B)291, 71T811, 77HCA215⟩.

(91)                              (92)                              (93)

The stretching frequency of the double bond in the heterocyclic ring of 4*H*-chromene occurs at 1665 cm$^{-1}$, higher than the corresponding vibration in 2*H*-chromene, and the same observation applies to the C—O stretch at 1273 and 1050 cm$^{-1}$. The IR spectra of a number of dialkyl derivatives of 4*H*-chromene are available ⟨70JOC2282, 72TL4453⟩, in addition to that of the parent compound ⟨62JA813⟩.

The C=C and C—O stretching absorptions in 1*H*-2-chromenes (27) occur at 1650–1620 and 1070–1040 cm$^{-1}$, respectively ⟨72HCA10⟩.

Chromans exhibit a C—O stretching vibration at 1260–1215 cm$^{-1}$ ⟨64CB682⟩. A detailed analysis of the spectrum of 2,2-dimethylchroman is reported ⟨68JCS(C)1837⟩ and information on variously substituted chromans is available ⟨81HC(36)1⟩.

The main feature of the spectra of 3- and 4-chromanols is the hydroxyl absorption band which occurs in the range 3625–3200 cm$^{-1}$. Some derivatives show two peaks in this region, whilst others exhibit just a single band ⟨70BSF1139, 70JOC2282⟩.

It has been concluded that some secondary chroman-4-ols adopt a half-chair conformation in which a quasi-axial OH group, which absorbs at *ca.* 3618 cm$^{-1}$, is predominant over a quasi-equatorial hydroxyl, which shows two peaks (3622 and 3600 cm$^{-1}$) ⟨74BCJ509⟩. A study of more complex tertiary chromanols has indicated that the *cis* epimer (94) exhibits peaks at 3619 cm$^{-1}$ (free OH group) and 3604 cm$^{-1}$ (bonded OH). The *trans* epimer (95) shows only one band (3620 cm$^{-1}$) which is also associated with a hydrogen bonded hydroxyl group, though now in a quasi-axial orientation ⟨81JCS(P2)944⟩.

(94)                              (95)

The spectra of *cis*- and *trans*-chroman-3,4-diols show two and three bands, respectively, in the OH stretching region ⟨68BSF4203⟩.

The effects of substitution in the aromatic ring on the C=C and C—O stretching frequencies (1587 and 1250 cm$^{-1}$, respectively) in chroman-6-ols have been discussed ⟨59JCS3362⟩. The IR spectra of a number of tocopherols have been reported ⟨48JBC(173)439, 50JBC(187)83, 56MI22200⟩ and a detailed discussion is available ⟨81HC(36)66⟩.

## 2.22.4.2 Pyranones, their Benzologues and Reduced Derivatives

Pyran-2-ones are characterized by absorption at 1730–1704 cm$^{-1}$ associated with the carbonyl group ⟨62BCJ1323⟩ and this may be accompanied by a second less intense band at higher frequency (1770–1740 cm$^{-1}$) ⟨59CJC2007⟩. The carbonyl stretching frequency is sensitive to substitution ⟨54JA3642, 56JA2393, 57JOC1257, 58CB2849⟩. A red shift is caused by an α-bromine atom, but the presence of a 3-hydroxy group brings about a more significant shift of the C=O stretch to lower wavenumber (1685 cm$^{-1}$). The appearance of a broad band at 3200 cm$^{-1}$ establishes that 3-hydroxypyran-2-one exists in a hydrogen bonded enolic form (**96**) rather than as the tautomeric pyran-2,3-dione.

(**96**)

Two carbonyl bands are present in the spectra of the 5- and 6-formyl derivatives of pyran-2-one and it appears that the aldehyde carbonyl absorption occurs at 1765–1745 cm$^{-1}$. This unusually high stretching frequency has been attributed to a tightening of the bond by the considerable positive nature of the partially aromatic pyranone ring ⟨68JHC275⟩.

A spectroscopic analysis of a range of xanthyrones (**97**) assigned the pyran ring carbonyl absorptions in the range 1775–1745 cm$^{-1}$, with ester and acetyl groups attached to the ring at 1730–1700 cm$^{-1}$ and 1690–1680 cm$^{-1}$, respectively ⟨79JCS(P1)478⟩.

(**97**)

Two bands, which are separated by *ca.* 80 cm$^{-1}$, feature in the C=C stretching region (1667–1540 cm$^{-1}$) of many pyran-2-ones ⟨61JCS4490⟩.

Pyran-4-one shows four absorption bands in the 1700–1600 cm$^{-1}$ region, two of which are of carbonyl stretching character. Although early workers assigned the intense band of highest frequency to the carbonyl group, it is now clear that in 2,6-dimethylpyran-4-one the C=O stretching frequency is lower than that of a C=C vibrational mode; a similar situation obtains with the parent compound ⟨61SA64⟩. Protonation ⟨63CJC505⟩ and complexation ⟨61CJC1184⟩ studies support this assignment. The C=O absorption at 1639 cm$^{-1}$ is shifted to *ca.* 1490 cm$^{-1}$ when the pyranone is complexed with various Lewis acids, its exact position being dependent on the anion.

The spectrum of 4-pyranone-3,5-$d_2$ allowed analysis of the nature of the splitting observed in the 1667 cm$^{-1}$ peak of pyran-4-one. The splitting was attributed to a slight anharmonic Fermi resonance involving the out-of-plane deformation mode ascribed to H-3 and H-5 which appears at 851 cm$^{-1}$ ⟨59CJC2007⟩. This assignment is confirmed since there is no appreciable absorption between 900–800 cm$^{-1}$ in the spectrum of the deuterated derivative, which also exhibits an unsplit peak at 1648 cm$^{-1}$ ⟨64JOC2678⟩.

Amongst the derivatives of pyran-4-ones which have been studied are 3,5-dibenzyl ⟨57JA156⟩, 3,5-diacyl ⟨59JOC1804⟩, 3-formyl ⟨61JOC1028⟩ and 2,3-fused alicyclic compounds ⟨62JCS1857⟩. The last work includes data on pyran-4-oximes, which are characterized by strong absorption in the 1681–1656 cm$^{-1}$ region, assigned to the C=N stretch.

Although the structures of a series of 6-substituted 4-methoxypyran-2-ones have been established by chemical means, their alternative formulation as derivatives of 2-methoxy-pyran-4-ones was mainly excluded on the basis of an absorption at 1733 cm$^{-1}$ attributable to the pyran-2-one system ⟨59JA2427⟩. This report concludes that pyran-2-ones exhibit their first carbonyl band at *ca.* 1725 cm$^{-1}$, whereas pyran-4-ones absorb at *ca.* 1667 cm$^{-1}$. Similarly, absorption at 1724 cm$^{-1}$ indicated a pyran-2-one structure for yangonin (**98**), the major constituent of kava resin, and this has been confirmed by an unambiguous synthesis. Pseudoyangonin (**99**) absorbs at 1667 cm$^{-1}$, characteristic of a pyran-4-one ⟨60JCS502⟩. Comparison of the spectra of some constituents of the rosewoods with that of yangonin established them as pyran-2-ones. The products from the reaction of malonyl chloride with some 1,3-diketones were identified in part by IR spectroscopy.

(**98**)  (**99**)

The different C=O stretching frequencies of the two systems is a reflection of the stronger basicity of the pyran-4-one. The effects of substitution on this absorption band are smaller in pyran-4-ones, and this feature has been attributed to a more significant contribution of the aromatic structure (**100**) to the hybrid structure than the contribution of (**101**) to the structure of pyran-2-one ⟨B-77MI22201⟩.

(**100**)  (**101**)

The carbonyl absorption band in simple coumarins is found in the range 1710–1695 cm$^{-1}$ in chloroform ⟨57CB1519⟩, but at 1730 cm$^{-1}$ in Nujol ⟨60M774⟩. However, a range of 1740–1670 cm$^{-1}$ is more realistic and illustrates the considerable influence of substituents. Other significant bands occur around 1600 cm$^{-1}$ and in the ranges 1270–1210 and 1145–1065 cm$^{-1}$, the latter peaks being assigned to the C—O stretching frequencies ⟨76RRC1207⟩.

The effect of substituents on the C=O stretch is variable, halogen and alkyl groups tending to raise the frequency somewhat, whilst hydroxy groups lead to a significant lowering (1700–1650 cm$^{-1}$). The latter effect is most pronounced for substitution at the 4-position and is considered to be a result of intermolecular hydrogen bonding ⟨64T2859⟩. The similar, though much smaller, influence of a 4-methoxy group ($\nu_{C=O}$ 1700 cm$^{-1}$) has been attributed to the conjugative properties of the substituent which reduce the double bond character of the carbonyl group ⟨78T1221⟩.

The C=O stretching frequency of a number of 4-hydroxy- and 4-alkoxy-coumarins, including several anticoagulant drugs, has been identified by isotopic replacement of the carbonyl carbon atom by $^{13}$C as the highest frequency band in the 1750–1550 cm$^{-1}$ region. Introduction of the isotopic atom causes a reduction in the C=O frequency of *ca.* 30 cm$^{-1}$ ⟨82JHC475⟩.

The carbonyl stretching band is at *ca.* 1650 cm$^{-1}$ in simple chromones ⟨61JCS798⟩, though this and other absorptions vary with the solvent ⟨70BSB89⟩. The effect of a hydroxyl group on the carbonyl absorption in 2-methylchromone has been investigated ⟨69SA(A)1067⟩, but a more significant observation in this study is the much enhanced shift in 2-hydroxychromone ($\nu_{C=O}$ 1705 cm$^{-1}$) attributed to tautomerism to a coumarin.

A comprehensive study of some 5-hydroxy- and 5-alkoxy-chromones has been reported ⟨69T5819⟩. The use of $^{2}$H and $^{18}$O labelling has revealed that the splitting of the carbonyl absorption of 5-hydroxychromone into two bands at 1665 and 1630 cm$^{-1}$ is a result of Fermi resonance with a mode derived from 3-H ⟨69T5839⟩. Absorption bands for mono- and di-hydroxy- and methoxy-chromones have been assigned in the 3560–2400 and 1670–1580 cm$^{-1}$ regions, the former range being complex as a result of overlap of OH and CH bands ⟨76JHC211⟩.

IR data are presented and analyzed in other reports of chromone derivatives ⟨70JCS(C)2230, 70JCS(C)2609, 74T3563⟩ and in the monograph by Ellis ⟨77HC(31)1⟩.

The IR spectra of chroman-4-ones are typical of aromatic acyclic ketones, showing strong carbonyl absorption at *ca.* 1680 cm$^{-1}$, although the frequency varies with the position and nature of substituents. An extensive compilation is provided in the review of chroman-4-ones ⟨77HC(31)215⟩. The proposition that 3-halogen atoms occupy mainly axial positions in chroman-4-ones is supported by IR studies, the observed C=O stretch at 1670 cm$^{-1}$ agreeing well with the calculated value of 1675 cm$^{-1}$ ⟨68JHC745, 70JHC187⟩.

The carbonyl stretch in xanthone (**21**) at 1660 cm$^{-1}$ ⟨58JCS294⟩ can shift to as low a frequency as 1400 cm$^{-1}$ according to protonation studies. The order of complexing ability of cations towards xanthone suggests that steric factors are of significance ⟨63CJC522⟩. Hydroxyxanthones show characteristic hydroxyl (3300 cm$^{-1}$) and carbonyl absorption bands ⟨67JCS(C)2500⟩.

IR data on other systems containing a pyranone nucleus are collated in Table 4.

**Table 4**  Typical Carbonyl Stretching Frequencies of some Reduced and Fused Pyranones

| Compound | $\nu_{C=O}$ (cm$^{-1}$) | Ref. |
|---|---|---|
| 5,6-Dihydro-2*H*-pyran-2-ones | 1735–1708 | 81HCA1247, 59AG523 |
| 3,6-Dihydro-2*H*-pyran-2-ones | 1740–1725 | 59AG523 |
| 3,4-Dihydro-2*H*-pyran-2-ones | 1775–1770 | 59AG523 |
| 2*H*-Pyran-3(6*H*)-ones | 1693–1685 | 82JOC3054 |
| Tetrahydropyran-2-ones | 1740–1730 | 59AG523, 57JOC1257 |
| Tetrahydropyran-3-ones | 1740–1725 | 77NJC79 |
| Tetrahydropyran-4-one | 1724 | 58CB1589 |
| Dihydrocoumarins | 1775–1720 | 74JCS(P1)569 |
| Isocoumarins | 1760–1700 | 77JOC1329, 72MI22206 |
| Dihydroisocoumarins | 1720–1715 | 77JOC1329 |
| Chroman-3-ones | 1740–1730 | 70JHC197, 64JGU2699 |
| Isochroman-4-one | 1700 | 71BSF1351 |

### 2.22.4.3  Flavonoids

The characteristic carbonyl stretching frequencies for various flavonoids are presented in Table 5.

**Table 5**  Carbonyl Stretching Frequencies of Flavonoids

| Flavonoid | $\nu_{C=O}$ (cm$^{-1}$) | Ref. |
|---|---|---|
| Flavones | 1660–1640 | 53JA1622, 58JOC93, 59FOR1 |
| Flavonols | 1658–1652 | 58JOC93, 59FOR(17)1, 60AC(R)875 |
| Flavanones | 1680 | 53JA1622, 58JOC93 |
| Isoflavones | 1620 | 63HCA49 |
| Isoflavanones | 1661 | 59JOC1655 |
| Flavan-3-ones | 1735–1730 | 80JCS(P1)1025 |

The absorption band in flavanones is lowered to *ca.* 1665 cm$^{-1}$ by a 3′- or 4′-hydroxy group and to *ca.* 1655 cm$^{-1}$ by a 3-OH substituent as a result of intermolecular hydrogen bonding. Interaction of hydroxy groups at C-5 or C-7 can shift the absorption to as low as 1620 cm$^{-1}$. By contrast, acetoxy groups have little effect ⟨53JA1622⟩. The stereochemistry of 3-bromoflavanones, 3-bromoflavan-4-ols and flavan-3,4-diols has been assigned on the basis of IR data ⟨57JIC753⟩. A detailed analysis of the major modes of flavanones includes a discussion of substituent effects ⟨75M333⟩.

Considerable attention has been paid to the hydroxy derivatives of flavones with particular reference to the ability to hydrogen bond and its consequences on the IR spectra ⟨57MI22200, 62JOC381, 75MI22202⟩. The combined effect of a hydroxy group at C-3 or C-4′ and of a hydroxy or methoxy group at C-5 on the carbonyl absorption is of interest, since it is lower in the presence of a 5-methoxy group than a 5-hydroxy substituent. It appears that an intramolecular hydrogen bond between the 5-OH and carbonyl groups, which has little influence on the absorptions of the two groups, sterically prevents other, more effective, forms of hydrogen bonding. When a 5-methoxy group is present, this intramolecular hydrogen bond is absent but other hydrogen bonds are formed which bring about a reduction in the carbonyl stretching frequency ⟨63SA2099, 65BSF779⟩.

### 2.22.4.4 Pyrylium Salts

Pyrylium salts have not been extensively investigated by IR techniques, but vibrational modes for pyrylium, 2-*D*-pyrylium, 3-*D*-pyrylium and pyrylium-$d_5$ perchlorates have been assigned and compared with similar data for related pyridine compounds ⟨72SA(A)1001⟩. A close similarity in the spectra is noted, but the C—H and C—D stretching frequencies occur at appreciably higher wavenumbers in the pyrylium derivatives. This effect is attributed to the greater electronegativity of the oxygen atom and to the positive charge associated with the pyrylium ring.

The IR spectrum of the 2-benzopyrylium species has been reported ⟨64ACH(40)217⟩.

## 2.22.5 UV SPECTROSCOPY

Reference to two chapters on the UV spectra of heterocycles reveals that, in contrast to the multitudinous information on nitrogen heterocycles, rather limited data are available on oxygen heterocycles ⟨63PMH(2)1, 71PMH(3)67⟩. However, characteristic spectra are reported for certain classes of these compounds which may be helpful in confirming or disproving postulated structures.

### 2.22.5.1 Pyrans, their Benzologues and Reduced Derivatives

Absorption at 240–250 nm by 4*H*-pyrans has been noted ⟨61CB1784⟩, although this has been disputed and in general 4*H*-pyrans are characterized by a weak shoulder at *ca*. 225 nm ⟨69JOC3169⟩. Introduction of an ethoxycarbonyl or acetyl group at C-3 causes a shift of the maximum to 270 and 284 nm, respectively, which are further shifted to 285 and 296 nm by a second of these substituents. These maxima are considerably different from those of β-alkoxy α,β-unsaturated esters and ketones indicative of additional conjugation with the second double bond of the pyran.

The equilibrium between 2*H*-pyrans and *cis*-dienones is much in favour of the latter species and consequently UV data on these pyrans are scarce. However, steric destabilization of the dienone valence isomer results in the existence of some highly substituted 2*H*-pyrans either alone or as a mixture with the dienone. The UV spectra of such compounds are similar to those of *s*-*cis*-dienes. For example, 2,2,4,6-tetramethyl-2*H*-pyran absorbs at 221 and 278 nm, and *cis*-β-ionone, which exists as an equilibrium mixture (Figure 2), at 208 and 253 nm ⟨71JOC1977⟩.

The vacuum UV spectrum of the dihydropyran (**13**) in solution shows a broad band ($\lambda_{max}$ 195 nm) ⟨48JCP(17)466⟩, but that of the vapour of tetrahydropyran consists of two systems of closely spaced narrow bands, one beginning at 193 nm and the other, of higher intensity, at 184 nm ⟨51JA4865⟩.

The photoelectron spectrum of tetrahydropyran has been recorded ⟨72JA5599⟩, and analyzed along with the spectra of 3,4-dihydro-2*H*-pyran and 4*H*-pyran ⟨78HCA1388⟩.

2*H*-Chromene is variously reported to exhibit maxima at 264 and 307 nm in ethanol ⟨63CPB1042⟩, 266.5 and 314 nm in hexane ⟨64T1185⟩, and 263 nm in ethanol ⟨62JA813⟩. Both absorptions may be benzenoid (B) bands or the peak near 260 nm may be a conjugation (K) band; both are associated with $\pi \rightarrow \pi^*$ transitions.

The UV spectra of a range of simple derivatives of 2*H*-chromene show maxima in the ranges 220–240, 260–280 and 300–340 nm; in most cases log ε lies between 3.0 and 4.5 ⟨63CPB1042⟩. A few 2*H*-chromenes with hydroxyl and ketone groups on the aromatic ring absorb at longer wavelength. Spectral data are available on naphthopyrans ⟨63JA1178, 68JOC2416, 71DOK(196)640⟩. Spirobi(2*H*-chromenes) (**102**) are essentially two separate 2*H*-chromene units in spectroscopic behaviour and as such absorb at similar wavelengths to the parent system, namely in the ranges 255–260 and 290–320 nm ⟨69LA(722)162⟩. The spectra of chromenes spiro-annelated at C-2 with an indoline moiety (**103**) are closely related, but show a small shift to the red and an additional longer wavelength band ⟨70JA1289⟩.

Spectral information on 4*H*-chromenes, in which the double bond of the pyran ring is not conjugated with the aromatic ring, is scanty. Absorption occurs at 275–285 nm, as with many di- or poly-substituted phenols, and a shorter wavelength absorption is also present ⟨60CB1025⟩.

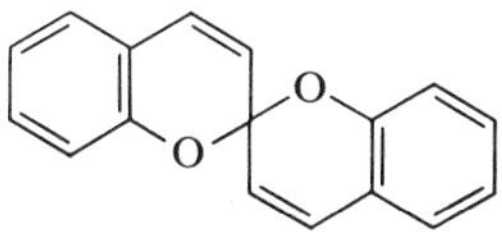

**(102)** 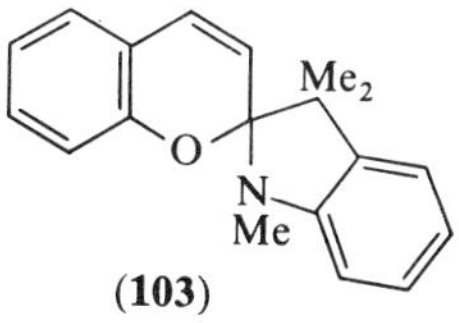          **(103)**

Xanthene shows maxima at 250, 278 and 292 nm ⟨51BSF693, 53JA3333⟩ and in xanthydrol (**104**) the latter peaks show a red shift but the first moves towards the blue. The ESR spectra of the radical anion of xanthene and several of its derivatives have been measured. It is suggested that the ion is not symmetrical but that the electron is localized on one of the aromatic rings and jumps rapidly between the two sites ⟨75JCS(P1)1652⟩.

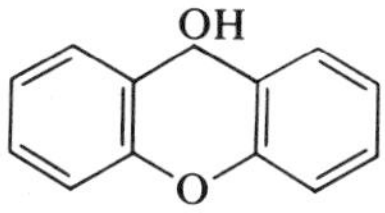

**(104)**

Aromatic ethers show at least two peaks in the 260–290 nm region. This is generally the case with chromans, the longer wavelength band being the more intense. The effect of substituents is not very pronounced unless they contain non-bonding electrons which can interact with the $\pi$-electrons of the aromatic ring (Table 6).

**Table 6**  UV Data for some Chromans

| Substituent in chroman | $\lambda_{max}$(nm)($\varepsilon_{max}$) | Solvent | Ref. |
|---|---|---|---|
| None | 270 (3.18) 276 (3.33) 284 (3.40) | Hexane | 67JCS(B)859 |
| 8-Me | 274 (3.28) 278 (3.26) 283 (3.34) | Hexane | 67JCS(B)859 |
| 2,2-Me$_2$ | 276 (3.34) 284 (3.33) | Hexane | 60JCS602 |
| 2,2-Me$_2$-6-OH | 295 (3.33) 306.5 (2.31) | Hexane | 68ACS3160 |
| 7-NO$_2$ | 237 (4.30) 279 (3.28) 332 (3.24) | Ethanol | 73JHC623 |
| 2-Ph | 227 (3.82) 245 (3.52) 274 (3.21) 282 (3.18) | Ethanol | 66ACS1561 |

The UV photoelectron spectra of several chromans have been compared with other cyclic and acyclic ethers. The chroman spectra are clearly distinguishable from those of the other ethers and the differences have been interpreted in terms of the decreased conjugative effects of oxygen and increased hyperconjugative effects of the 4-methylene group ⟨76T167⟩.

### 2.22.5.2 Pyranones, their Benzologues and Reduced Derivatives

UV analysis has been employed in the structural elucidation of derivatives of pyran-2-one ⟨54JA3642, 56JA2393, 61NKK932, 68JHC275⟩. UV and IR correlations have been used to assign structures to 4-methoxy-6-methylpyran-2-one (**105**) ($\lambda_{max}$ 280 nm) and 2-methoxy-6-methylpyran-4-one (**106**) ($\lambda_{max}$ 240 nm) ⟨58T(4)36, 60JCS502⟩. The assignments proved critical in deciding the structures of naturally occurring pyran-2-ones.

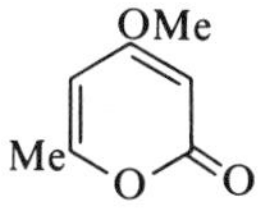

**(105)** 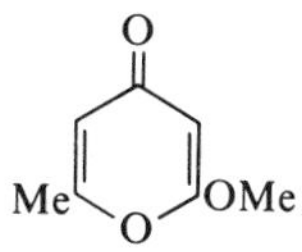          **(106)**

The UV spectrum of pyran-2-one itself shows end absorption at 216 nm and a maximum at 289 nm ⟨51M662, 57G243⟩. By contrast, pyran-4-one absorbs at 246 and 260 nm ⟨49HCA1752, 55AC(R)128⟩. It is thus possible to distinguish the two pyranones, provided both isomers are available; the one with the longer wavelength maximum is the pyran-2-one ⟨59JA2427⟩. Provided no additional chromophore is present, a UV analysis of the type described by Polish workers ⟨52MI22200⟩ is usually sufficient to differentiate between the 2- and 4-ketones.

An extensive study ⟨75JCS(F2)1812⟩ of the electronic absorption and emission spectra of pyran-4-ones has been carried out. The absorption spectrum of pyran-4-one shows the characteristic strong bands between 200 and 300 nm, as well as much weaker transitions to longer wavelengths. The magnitude of the extinction coefficients and the direction of the solvent effect (blue shifting in ethanol) are both consistent with an $n \rightarrow \pi^*$ assignment to these longer wavelength transitions. The strong transitions are red shifted in ethanol, which together with their intensity indicates that these are $\pi \rightarrow \pi^*$ transitions. In ethanol, pyran-4-one exhibits very weak fluorescence with $\lambda_{max}$ *ca.* 360 nm at room temperature, the fluorescence intensity being slightly increased at 77 K. Methyl and phenyl derivatives show similar weak fluorescence in the 360–395 nm region in ethanol. Pyran-4-ones have also been observed to phosphoresce in rigid media at 77 K. The phosphorescence spectrum of pyran-4-one shows vibrational structure with a progression having spacing *ca.* 750 cm$^{-1}$, possibly due to a C—H out-of-plane vibration.

An extensive range of coumarins has been analyzed by UV spectroscopy and it has been established that coumarins show a prominent minimum at $244 \pm 4$ nm and a second minimum at $300 \pm 5$ nm, with a principal maximum at $275 \pm 4$ nm and a second peak at $315 \pm 8$ nm ⟨56JOC1415⟩. In view of later examples, these limits are somewhat narrow. The principal minimum and maximum are generally sharp and well defined, whilst the subsidiary features are usually broad. It is pertinent to note that the major peak of coumarins occurs at longer wavelength than that of chromones, an observation which may be attributed to the linear conjugation associated with coumarins as opposed to the cross-conjugated system in chromones. A similar situation obtains with the corresponding pyranone isomers.

The UV spectra of many coumarins have been studied in different solvents with a view to correlating the effects of the nature and position of substituents ⟨74MI22201⟩. The spectral parameters do indeed show a dependence on these properties and several general rules have been proposed to enable the spectral features of a substituted coumarin to be predicted. Not unexpectedly, such predictions are not exact and a study of some disubstituted derivatives of coumarin has shown that the rules are not always obeyed ⟨78T1221⟩. The compounds which do not adhere to the rules possess a 4-methoxy group and the spectra are characterized by two double absorption bands in the ranges 267–296 and 303–322 nm. This behaviour has been attributed to the additional cross-conjugation provided by the methoxy substituent.

Both the absorption and emission maxima of some 3-phenylcoumarins can be correlated by additive substituent rules, which can be used to predict these parameters for a given compound in this class ⟨77H(7)933⟩.

The changes in the spectral parameters of hydroxycoumarins brought about by the addition of acids or bases are of value in structural elucidation ⟨60JOC2183, 67JCS(C)2545⟩. For example, both 5- and 7-hydroxycoumarins show red shifts and increases in intensity upon the addition of base, whereas for the 6- and 8-substituted compounds the bathochromic shift is accompanied by a reduction in intensity ⟨69JCS(C)526⟩. Several other additives have proved useful in structure elucidation, notably sodium acetate ⟨66IJC120⟩ and aluminum chloride ⟨69MI22201⟩.

A substantial listing of UV spectra of coumarins is available ⟨71PMH(3)190⟩ and examples of the identification of natural products based on the coumarin ring system by UV analysis, amongst other methods, have been reported ⟨58JA3686, 69MI22202⟩.

The UV spectrum of isocoumarin shows maxima at 228, 239, 253, 261 and 318 nm in ethanol ⟨60NKK654⟩.

In view of the importance of coumarins as fluorescent brightening agents (see Section 2.24.5), tunable dye lasers ⟨80MI22200⟩ and in various biological applications ⟨B-69MI22202, 76H(5)839⟩, an account of their spectral properties would be incomplete without some mention of their fluorescence spectra. A detailed account of fluorescence spectroscopy is available ⟨74PMH(6)147⟩, but, in brief, fluorescence is the emission of energy from the lowest excited singlet state following excitation. The emitted fluorescence has lower energy than the absorbed radiation in accord with Stokes' Law.

Although coumarin is unusual in that it fluoresces under mechanical stress (tribofluorescence) ⟨74JA4690⟩, it has only a low fluorescence quantum yield. However, the presence of electron-releasing groups at the 4-, 5-, 6- or 7-positions shifts the emission to longer wavelength and increases the intensity. Coumarins containing amino or hydroxy groups at C-7 are strongly fluorescent. An electron-withdrawing group at C-3 further increases the fluorescence intensity and the shift to the red. It has been proposed that the fluorescence is associated with the $^1(\pi, \pi^*)$ state partially localized on the heterocyclic ring ⟨70JPC4234⟩.

The fluorescence properties of coumarins show a dependence on solvent polarity and pH. For example, the emission from 7-amino-4-methylcoumarin shifts to the red as the polarity of the solvent increases ⟨74NKK1744⟩. This behaviour is indicative of polar character in the excited state, presumably associated with a contribution from the canonical form (**107**).

(**107**)    (**108a**)    (**108b**)    (**108c**)

The excited states (**108a–108c**) have been proposed as the species responsible for the fluorescence of 7-hydroxy-4-methylcoumarin at various acidities ⟨74JA4699⟩. In strongly alkaline solution, the coumarin is hydrolyzed to an *o*-hydroxycinnamic acid which emits at 500 nm ⟨70MI22201⟩. However, a detailed study of several 7-hydroxycoumarin derivatives indicated that complex fluorescence spectra were obtained when aqueous solvents were used ⟨77JCS(P2)262⟩.

The spectra of chromone and many of its derivatives consist of four bands centred around 205, 225, 240 and 300 nm. The major absorption peaks at 240 and 300 nm are separated by a minimum at *ca.* 270 nm. In contrast to coumarins, chromones do not show a second minimum.

A study of the spectrum of chromone in a range of solvents of differing polarity suggests that the three higher wavelength bands arise from $\pi \to \pi^*$ transitions, since they are red-shifted as the solvent polarity increases ⟨72SA(A)707⟩. The low wavelength band undergoes a small hypsochromic shift and is probably an $n \to \sigma^*$ transition involving the lone-pair electrons on the heteroatom.

In general, electron-attracting groups at C-2 shift the $\pi \to \pi^*$ bands to the red, whilst a methyl group causes a small blue shift. Introduction of $NHCO_2Et$ at this position produces pronounced changes in the spectrum which have been associated with a charge transfer species (**109**).

(**109**)

Chromones with hydroxy or methoxy groups in the aromatic ring show high intensity bands at 240–260 and 290–360 nm. Both bands are red-shifted relative to chromone and in sodium methoxide solution these shifts are enhanced as a result of ionization of the phenolic groups ⟨76JHC211⟩.

Chroman-4-one shows maxima at *ca.* 260 and 330 nm in ethanol ⟨35BSF1381⟩. The peaks are blue-shifted in 3-methylpentane ⟨70CJC3928⟩. Weakly electron-donating groups have little effect on the spectra ⟨73BSB705⟩, but hydroxy or methoxy groups bring about a significant bathochromic shift of the higher wavelength band ⟨50G750, 58BSB22⟩. In the presence of sodium acetate, the peaks shift to the red if a 7-hydroxy group is present. A compilation of UV spectra of some chroman-4-ones is available ⟨77HC(31)215⟩.

Absorption at 274.5 nm is reported for chroman-3-one ⟨64JGU2699⟩.

The UV spectra of most simple xanthones consist of three or four strong bands between 230 and 340 nm ⟨58JCS4234, 59T(6)315, 60JCS191⟩. Naturally occurring xanthones are oxygenated to varying extents and this is marked by a red shift of the highest wavelength band or the appearance of a further less intense band towards the red which accounts for the yellow colour of these xanthones. The spectra show a significant dependence on the pattern of oxygenation, a feature which, together with the additional changes brought about by varying the basicity of the solvent, allows substitution patterns to be elucidated ⟨61CRV591, 66T1777, 67JCS(C)785⟩. The absorption spectra of some phytoxanthones have been documented ⟨79H(12)269⟩.

### 2.22.5.3 Flavonoids

Flavonoid chemistry has benefited greatly from the application of UV techniques. This is partly a consequence of the need for only a small amount of pure material to obtain a

UV spectrum. However, it is more significant that the quantity of structural information gained can be enhanced by measuring the spectrum in the presence of different additives which react specifically at various sites in the flavonoid molecule. Valuable reviews have appeared on UV spectroscopy of flavonoids ⟨B-62MI22200, B-70MI22200⟩.

The UV spectra of most flavonoids consist of two major absorption bands. That at 300–400 nm (band I) is considered to arise from the cinnamoyl system (ring B), whilst the maximum in the region 240–285 nm (band II) is associated with ring A, the benzoyl moiety (**110**) ⟨B-67MI22201⟩.

**(110)**                    **(111)**

It is often possible to distinguish the various types of flavonoids by examination, particularly of the band I absorption. Thus, flavones (**22**) absorb at 300–350 nm, whilst in flavonols (**111**) this peak occurs in the region 350–385 nm. The general effect of hydroxy groups is bathochromic, although these effects are quite specific. Substitution in ring B leads to a red shift of band I, but leaves band II much the same. Conversely, band II is affected by substitution in ring A.

In the spectra of isoflavones (**64**) and flavanones (**25**), in which rings A and B are not conjugated, band I frequently appears only as a shoulder on band II. Not surprisingly, introduction of oxygen substituents in ring B has little effect on the spectra, whereas substitution in ring A shifts band II to the red.

The spectra of both anthocyanidins and anthocyanins are well documented ⟨B-62MI22200, B-63MI22200⟩. Anthocyanidins show a peak in the range 270–280 nm, but the spectra are marked by a very pronounced red shift of band I to 460–550 nm. Methylation or glycosylation of hydroxy groups in either of the benzenoid rings brings about a hypsochromic shift of the longer wavelength band, whilst an increase in oxygenation of the 2-phenyl ring causes it to red shift. An example of particular value is provided by the shift in $\lambda_{max}$ shown by the three major colouring components of plants, pelargonidin (**112**), cyanidin (**113**) and delphinidin (**114**).

| (**112**) $R^1 = R^2 = H$ | $\lambda_{max}$ 520 nm |
| (**113**) $R^1 = OH; R^2 = H$ | $\lambda_{max}$ 553 nm |
| (**114**) $R^1 = R^2 = OH$ | $\lambda_{max}$ 564 nm |

The information which accrues from the determination of the spectra following the addition of various reagents to alcoholic solutions of flavonoids is quite impressive. Addition of sodium methoxide generally ionizes all hydroxy groups in the flavonoid molecule and shifts of the absorption bands to longer wavelength result. On the other hand, addition of the less basic sodium acetate leads to the ionization of only the more acidic hydroxy groups, though again red shifts of both bands are seen. The value of aluminum chloride in this area centres on its ability to chelate with 1,2-diols and 1,2-hydroxy ketones (**115**). The outcome of this reaction is a red shift of one or both bands in the spectrum. A detailed survey of spectral shifts induced by various reagents is included in a text on flavonoids ⟨B-75MI22201⟩, and other pertinent references include ⟨75P1605, 77MI22202, 78MI22201, 81P1181⟩.

**(115)**

Fluorescence spectra of a range of flavones have been reported ⟨80P2443⟩. The excited state proton transfer in 3-hydroxyflavone has been found to exhibit a solvent dependence ⟨81JA6916, 82JA4146⟩, whilst the fluorescence of flavone has been shown to be pH dependent ⟨81JCS(P2)1443⟩.

### 2.22.5.4 Pyrylium Salts

Information on the UV spectra of pyrylium salts has been reviewed ⟨82AHC(S2)173⟩. The parent perchlorate absorbs at 270 nm ⟨60IZV2064⟩, whilst introduction of methyl groups at the 2-, 4- and 6-positions gives rise to maxima at 230 and 285 nm ⟨60T(9)163⟩. The presence of phenyl groups shifts the absorption into the visible region and pyrylium perchlorates containing two or three phenyl groups are yellow. The nature of the anion influences the colour of the compounds, the iodides notably exhibiting much deeper colours. This feature has been attributed to a charge transfer band, the position of which is intermediate between that of tropylium and pyridinium halides ⟨64T119⟩.

Benzopyrylium perchlorate exhibits maxima at 241, 326 and 349 nm ⟨81JHC1325⟩. Substitution at C-8 results in bathochromic shifts of the lowest and highest absorptions.

Many aryl-substituted pyrylium salts are intensely fluorescent. It is possible to predict these spectral properties by a consideration of the shape of the molecule, the nature of substituents and the length of the $\pi$-electron system ⟨75MI22203⟩. Pyrylium salts have been used as Q-switches for neodymium and ruby lasers in acetonitrile ⟨68MI22200⟩.

## 2.22.6 MASS SPECTROMETRY

A number of review articles deal with electron impact mass spectra of oxygen heterocycles ⟨69BSF4545, B-71MS139, B-74MI22202⟩, but little use appears to have been made of techniques such as chemical ionization or negative ion spectrometry.

Electron impact mass spectral data for conjugated, reduced and higher flavonoids have been summarized ⟨B-75MI22201⟩ and a review of electron impact, chemical ionization and field desorption mass spectra of flavones, flavonols, isoflavones and flavanones has been published ⟨B-80MI22201⟩.

### 2.22.6.1 Pyrans, their Benzologues and Reduced Derivatives

There have been no mass spectral studies of monocyclic 2*H*-pyrans. However, the benzologue, 2*H*-chromene (4) and its derivatives have received considerable attention. In general, the fragmentation of chromenes proceeds with loss of one of the C-2 substituents from the molecular ion. The pathway proposed for chromene itself ⟨64T1185⟩ is presented in Scheme 3. In this case, elimination of a hydrogen radical from [M]$^{+}$ leads to the formation of the benzopyrylium ion (4a), which appears at *m/e* 131 as the base peak. Decomposition of (4a) occurs with hydrogen migration by expulsion of carbon monoxide to give (4b), which decomposes by loss of acetylene. The high stability of (4a) is reflected by the low intensity of the daughter ion (4b).

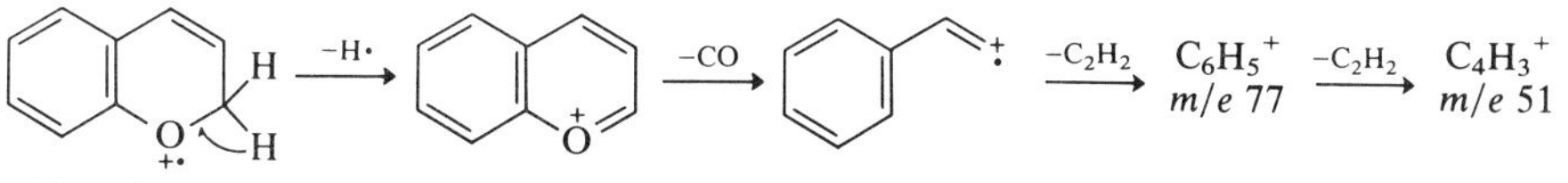

(4) *m/e* 132    (4a) *m/e* 131 (100%)    (4b) *m/e* 103 (8%)

**Scheme 3**

The fragmentation of 2-methyl-2*H*-chromene follows an analogous pathway to that described for (4). $\alpha$-Cleavage results in loss of a methyl radical giving the benzopyrylium ion (4a) as the base peak and subsequent fragmentations are identical to those depicted in Scheme 3. The alternative loss of a hydrogen radical from the molecular ion is not observed ⟨72TL4503⟩.

Both modes of $\alpha$-cleavage have been observed for flav-3-ene (10), which eliminates both hydrogen and phenyl radicals from the molecular ion ⟨67JCS(C)1933⟩. A further pathway is operative for 2-(*o*-hydroxyphenyl)chromene, the spectrum of which shows an ion at *m/e* 118 in addition to the expected fragments. This ion is thought to possess a benzofuran structure. Its formation has been rationalized by ring closure of the molecular ion with concomitant retro-Diels–Alder fragmentation ⟨67JCS(C)1933⟩.

The mass spectra of numerous 2,2-dimethylchromenes have been studied. In general, the molecular ion is not of great abundance, whilst the base peak invariably corresponds to a benzopyrylium ion ⟨64AJC975, 64T1185⟩. In the case of bis(2,2-dimethylchromenes) such as (116) the base peak corresponds to $[M-Me]^+$ ions ⟨69OMS(2)965⟩. The spiran (117) displays the molecular ion with fragment ions corresponding to $[M-Me]^+$, 3-methyl-chromone and chromone ⟨77HC(31)25⟩.

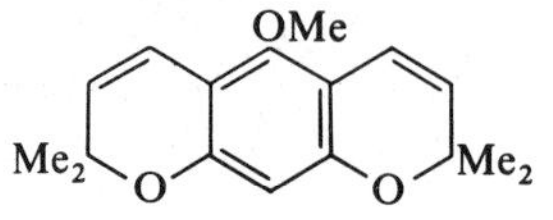

(116) 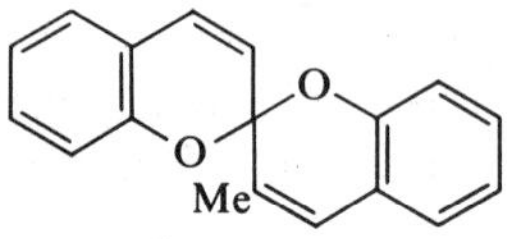          (117)

There appear to be no mass spectral studies of the isomeric 1*H*-2-chromenes (27).

There is a dearth of information regarding the mass spectra of 4*H*-pyran (2) and its benzologue (5), although numerous analogues of both have been prepared. The only 4*H*-pyran for which information is available is the dibenzologue, 9*H*-xanthene (8). The most facile fragmentation pathway involves loss of a hydrogen radical from the molecular ion to give a highly stable xanthylium ion. This subsequently yields a fragment at *m/e* 152 by elimination of a CHO radical. A study of xanthene-9-$d_2$ suggested that the xanthylium structure for the $[M-1]^+$ fragment may be inappropriate and a rearrangement involving the participation of tropylium ions has been invoked ⟨69OMS(2)829⟩.

Extensive deuterium labelling and measurements of metastable ion transitions aided elucidation of the fragmentation pathways for tetrahydropyran ⟨68OMS(1)403⟩. These are presented in Scheme 4.

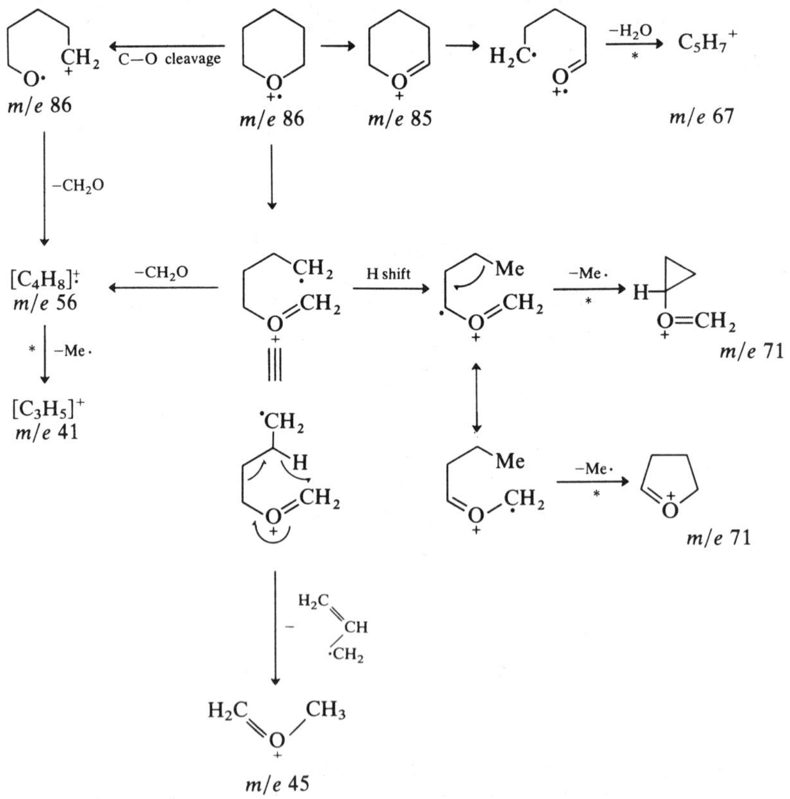

**Scheme 4**

Fragmentation of 2-alkyltetrahydropyrans is dominated by $\alpha$-cleavage with loss of the substituent, resulting in the formation of an oxonium ion.

Alkyl tetrahydropyranyl ethers (**118**; R = Et or Bu) behave similarly ⟨68JOC2266⟩. In both cases, the base peak occurs at *m/e* 85, resulting from loss of an RO· radical from [M]⁺. Pathways which involve ring cleavage of the molecular ion are also operative and appear to be of greater importance for the ethyl ether (**118**; R = Et).

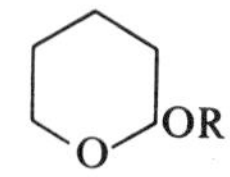

(**118**)

The mass spectral fragmentation of 3-hydroxytetrahydropyran has been studied in some detail ⟨72T1881⟩, whereas information on the 4-hydroxy compound appears to be lacking. In many respects, its behaviour resembles that of simple cyclanols such as cyclopentanol and cyclohexanol.

3-Hydroxytetrahydropyrans fragment with the formation of an $[M-CH_2OH]^+$ ion (*m/e* 71). The eliminated ·CH₂OH radical has been shown by deuterium labelling to originate from combined loss of the 3-OH and C-2 methylene groups.

The nature of $C_4H_7O^+$ ions derived from various tetrahydropyran derivatives has been investigated ⟨81T781⟩.

The electron impact mass spectral fragmentation of 3,4-dihydro-2*H*-pyran (**13**) parallels that of cyclohexene. Decomposition of the molecular ion occurs by the routes depicted in Scheme 5 ⟨63IJC20, 65T1855⟩. Retro-Diels–Alder (RDA) cleavage follows the two pathways, A and B. The former results in charge retention on the ethylene fragment (*m/e* 28), whilst the latter gives ionized acrolein (**13a**) (*m/e* 56). The base peak at *m/e* 55 is thought to be due to the hydrocarbon fragment (**13c**), although (**13b**) may also contribute and the situation here remains unclear. In the case of dihydropyrans bearing either an ether or carbonyl (ketone or ester) group at C-2, the course of the fragmentation is markedly dependent upon the nature of the substituent ⟨82OMS327⟩. Thus, the ethers (**119**; R = H, Me; R' = Me, Et, Pr, Bu) decompose by two pathways whereby α-cleavage generates an oxonium ion and RDA fission produces a vinyl ether fragment.

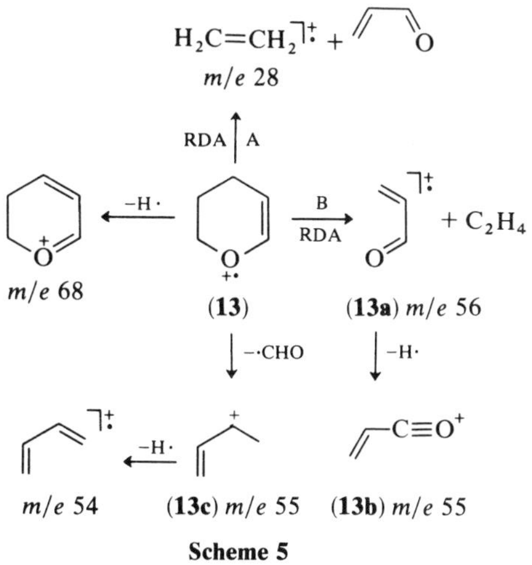

Scheme 5

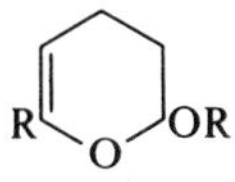

(**119**) 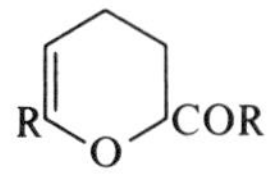       (**120**)

On the other hand, in the carbonyl compounds (**120**; R = H, Me; R' = Me, Et, OMe, OEt, *etc.*), α-cleavage occurs whilst the RDA pathway is completely suppressed. These differences in behaviour have been rationalized on the basis of thermodynamic considerations.

5,6-Dihydro-2*H*-pyran (**14**) fragments by the RDA pathway with charge retention on the diene fragment (**14a**). Hydrogen radical abstraction from the [M]⁺ ion produces (**14b**), which subsequently loses ethylene to give (**14c**; Scheme 6). Cleavage of the molecular ion by other pathways affords the fragments (**14d–14f**) ⟨70IZV1184⟩. Substituents may markedly alter the course of fragmentation.

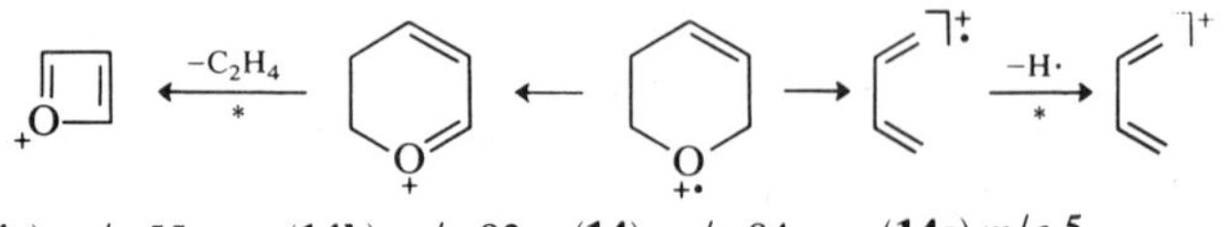

(**14c**) *m/e* 55        (**14b**) *m/e* 83   (**14**) *m/e* 84     (**14a**) *m/e* 5

**Scheme 6**

(**14d**)            (**14e**)          (**14f**)

In the case of the *gem* diester (**121**), the major pathway results in aromatization of the dihydropyran ring with production of the pyrylium species (**121a**). This is derived from the $[M-CO_2Et]^+$ ion with subsequent loss of carbon monoxide and elimination of ethanol. It was shown that RDA fragmentation of (**121**), a minor pathway only, is a thermally induced process ⟨72CJC1539⟩.

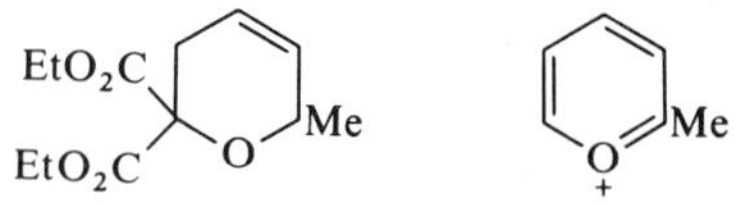

(**121**)                (**121a**)

Mass spectra of chroman (**15**), the benzologue of 3,4-dihydro-2*H*-pyran, and its analogues have been intensively studied and reviewed ⟨69BSF4545, B-74MI22202⟩. Fragmentation upon electron impact occurs by the pathways shown in Scheme 7 ⟨64T1185⟩. The molecular ion appears as the base peak. Retro-Diels–Alder cleavage occurs by path A giving (**15a**), which yields the benzene radical cation by loss of carbon monoxide. Hydrogen transfer in the RDA cleavage is also operative (path B) and this generates (**15c**). Loss of a hydrogen radical by $\alpha$-cleavage (path C) produces the oxonium ion (**15d**), *m/e* 133. An ion (**15e**) at *m/e* 119 corresponds to methyl radical loss from (**15**) (path D). The alternative possibility that this $[M-Me]^+$ ion arises by methylene extrusion from (**15d**) has been ruled out ⟨70OMS(3)409⟩.

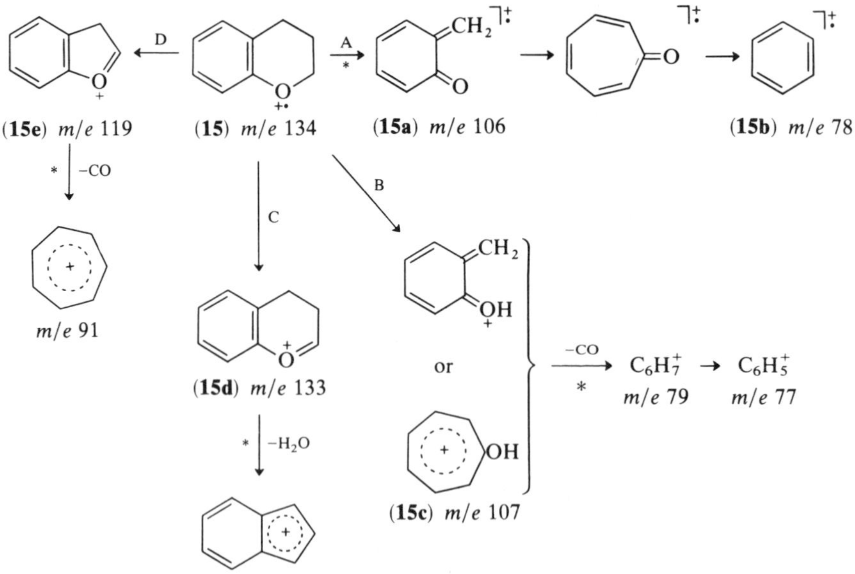

(**15e**) *m/e* 119   (**15**) *m/e* 134   (**15a**) *m/e* 106                                      (**15b**) *m/e* 78

*m/e* 91

(**15d**) *m/e* 133            or

                                                              $\xrightarrow[*]{-CO}$ $C_6H_7^+$ → $C_6H_5^+$
                                                                      *m/e* 79    *m/e* 77

(**15c**) *m/e* 107

*m/e* 115

**Scheme 7**

The mechanistic routes to (**15e**) have been investigated. Labelling experiments revealed that the expelled methyl radical originates separately from C-2, C-3 and C-4 of the chroman ring by three competing pathways ⟨70OMS(3)753⟩. A similar observation has been made for

2,2-dimethyl- and 2,2-diphenyl-chroman, which show weak $[M-C_2H_5]^+$ and $[M-C_7H_7]^+$ peaks, respectively.

Apart from the molecular ion $[M]^{\ddagger}$ $m/e$ 162, the only prominent peaks in the mass spectrum of 2,2-dimethylchroman arise from $[M-Me]^+$ $m/e$ 147, and the $[RDA+H]^+$ ion (15c), which gives the base peak at $m/e$ 107 ⟨64T1185⟩.

Electron impact mass spectra of some natural products which possess a 2,2-dimethylchroman skeleton have been reported ⟨69OMS(2)965, 70OMS(3)941⟩. The mass spectra of the isomeric $\alpha$-, $\beta$-, and $\gamma$-tocopherols have been reported ⟨72MI22204⟩.

The principal ions from chroman-4-ol correspond to $[M-H]^+$, $[M-OH]^+$, and $[M-H_2O]^+$. RDA expulsion of ethylene from the molecular ion occurs producing an abundant peak at $m/e$ 122, which is surpassed in intensity by the derived $[RDA-H]^+$ ion, $m/e$ 121 ⟨64T1185⟩.

A study of the mass spectra of some 2,2-dimethylchroman-4-ols has been undertaken ⟨78OMS653⟩. Deuterium labelling established that the chroman C-4 hydrogen is eliminated in formation of the $[RDA-H]^+$ ion.

Little information is available for the isomeric chroman-3-ol. However, from a comparative study with the 4-ol, it was shown that the carbonium ion centre at C-4 is 20 times more stable than that of C-3. This conclusion was reached by measuring the relative abundances of the $[M-OH]^+$ ions derived from both chromanols ⟨72IJC924⟩. Differences in the spectra of 3- and 4-chromanols have been summarized ⟨77HC(31)146⟩.

### 2.22.6.2 Pyranones, their Benzologues and Reduced Derivatives

The mass spectrum of pyran-2-one ⟨65JA3022⟩ displays three prominent peaks. The molecular ion (17) is of high abundance, whilst carbon monoxide loss produces an ion at $m/e$ 68, corresponding to $[C_4H_4O]^{\ddagger}$ which is of greater intensity. The cyclopropenium ion (17b) appears as the base peak. Fragmentation pathways are given in Scheme 8. A further decomposition mode of (17) involves loss of a hydrogen radical to produce (17c) from which the direct elimination of two CO molecules occurs in a stepwise manner to furnish (17b). The structure of the $[C_4H_4O]^{\ddagger}$ fragment has been the subject of much debate. Since its decomposition pathways are similar to those of furan, it seems reasonable to assign the furan structure (17a) to this ion. Numerous deuterium labelling experiments and work based on metastable ion peak characteristics suggested that the furan structure was inappropriate (see ⟨B-75MI22204⟩ for a summary of these studies). However, collisional activation mass spectral studies have demonstrated that the non-fragmenting $[M-CO]^{\ddagger}$ ions derived from pyran-2-one consist predominantly of furan radical cations.

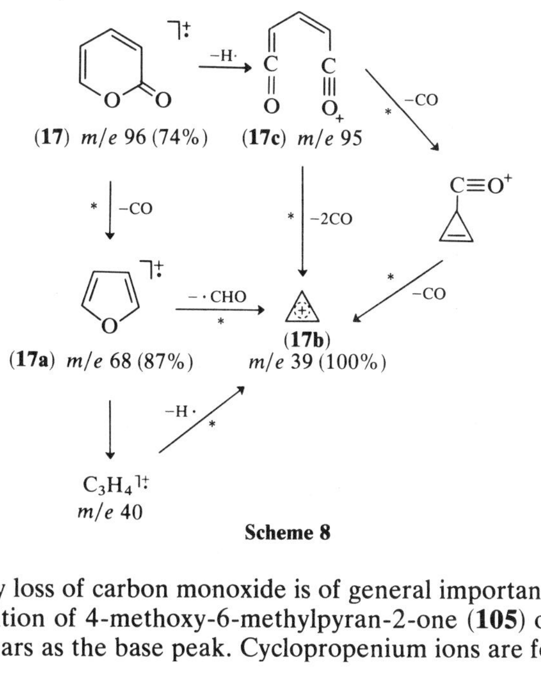

(17) $m/e$ 96 (74%)   (17c) $m/e$ 95

(17a) $m/e$ 68 (87%)   (17b) $m/e$ 39 (100%)

$C_3H_4^{\ddagger}$
$m/e$ 40

**Scheme 8**

Fragmentation by loss of carbon monoxide is of general importance for pyran-2-ones.
In the decomposition of 4-methoxy-6-methylpyran-2-one (105) on electron impact, the $[M-CO]^{\ddagger}$ ion appears as the base peak. Cyclopropenium ions are formed from $[M-CO]^{\ddagger}$

by expulsion of formyl and acetyl radicals. Cleavage of the molecular ion with loss of the C-6 methyl group occurs giving an acyclic fragment at $m/e$ 125. Breakdown of $[M-CO]^{+}$ also generates ions at $m/e$ 97 and 43. The latter corresponds to an acetyl cation, a fragment often observed in the spectra of 2-methylfurans ⟨65T1855⟩.

Some excellent comprehensive reviews on the mass spectrometry of coumarins have appeared ⟨B-71MS139, B-74MI22202, 75RCR603⟩.

On electron impact under mass spectral conditions, coumarin generates a highly abundant molecular ion (**19**). Decomposition by extrusion of a CO molecule represents the only facile pathway and the resulting $[M-CO]^{+}$ ion gives the base peak at $m/e$ 118. Subsequent loss of carbon monoxide and then a hydrogen radical produces ions of unknown structure at $m/e$ 90 and 89; the latter may also arise by direct loss of a formyl (CHO) radical from $[M-CO]^{+}$. In addition, there are two very minor peaks at $m/e$ 63 and 51 which arise by cleavage of the $[C_7H_5]^{+}$ ion ($m/e$ 89; Scheme 9) ⟨63IZV2215, 64AJC975⟩.

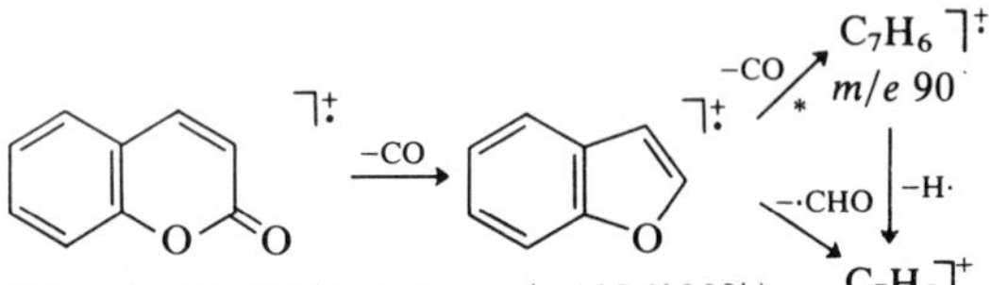

(**19**) $m/e$ 146 (98%)  (**19a**) $m/e$ 118 (100%)

**Scheme 9**

Since the spectra of coumarin and benzofuran show peaks below $m/e$ 118 of the same $m/e$ ratios and comparable intensities, the benzofuran structure (**19a**) for the $[M-CO]^{+}$ ion has been proposed. However, the validity of this type of comparison has been frequently questioned, and arguments against ⟨67JA5954, 68AJC997⟩ and in favour ⟨69JA5202⟩ of a benzofuran structure have been advanced. Despite the uncertainty, formulation of $[M-CO]^{+}$ fragments as benzofuran radical cations is of convenience and has found widespread practice. As with pyran-2-one, carbon monoxide loss represents an important fragmentation pathway for coumarins generally, although in some instances substituents may render other breakdown modes more favourable.

Mass spectra of some methyl-substituted coumarins have been reported ⟨71OMS(5)249⟩. The fragmentation of 4-methylcoumarin proceeds with expulsion of carbon monoxide giving the 3-methylbenzofuran radical cation. Subsequent loss of a hydrogen radical occurs producing the highly abundant benzopyrylium species at $m/e$ 131. A peak at $m/e$ 131 is also of prominence in the spectrum of 6-methylcoumarin.

Methyl radical loss from [3-methylbenzofuran]$^{+}$ gives a minor peak at $m/e$ 77. In contrast, expulsion of a methyl radical from the $[M-CO]^{+}$ ion derived from 7-methoxy-4-methyl-coumarin is facile. Deuterium labelling experiments confirmed that the methoxy methyl group is preferentially eliminated ⟨66JCS(C)1712⟩.

In coumarins which bear large alkyl side chains, such as osthol (**122**), the primary fragmentation does not involve carbon monoxide loss. The dominant processes are side chain cleavage, methyl radical elimination from the side chain, and loss of a methoxyl radical. The mass spectra of this and related naturally occurring coumarins have been intensively studied and reviewed ⟨75RCR603⟩.

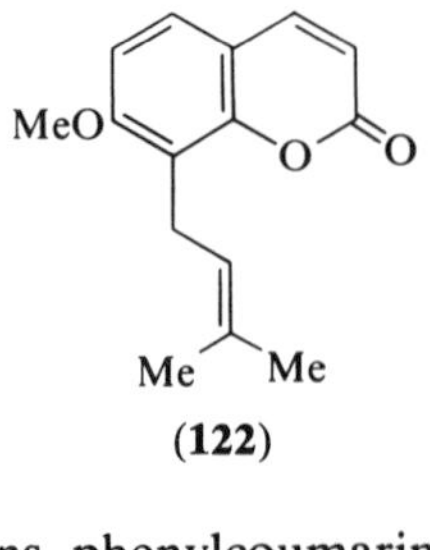

(**122**)

As with the simple alkylcoumarins, phenylcoumarins readily eliminate carbon monoxide from the molecular ion. Mass spectral fragmentation of 4-phenylcoumarin follows the route given in Scheme 10. The molecular ion (**122a**) appears as the base peak. Formyl radical elimination from (**122b**) produces the fluorenyl cation (**122c**), $m/e$ 165 ⟨63TL891⟩.

In the case of the 7-methoxy analogue, expulsion of a methyl radical from the $[M-CO]^{+}$ ion (**123**) is strongly favoured and yields the quinonoid ion (**123a**) ⟨72IJC19⟩.

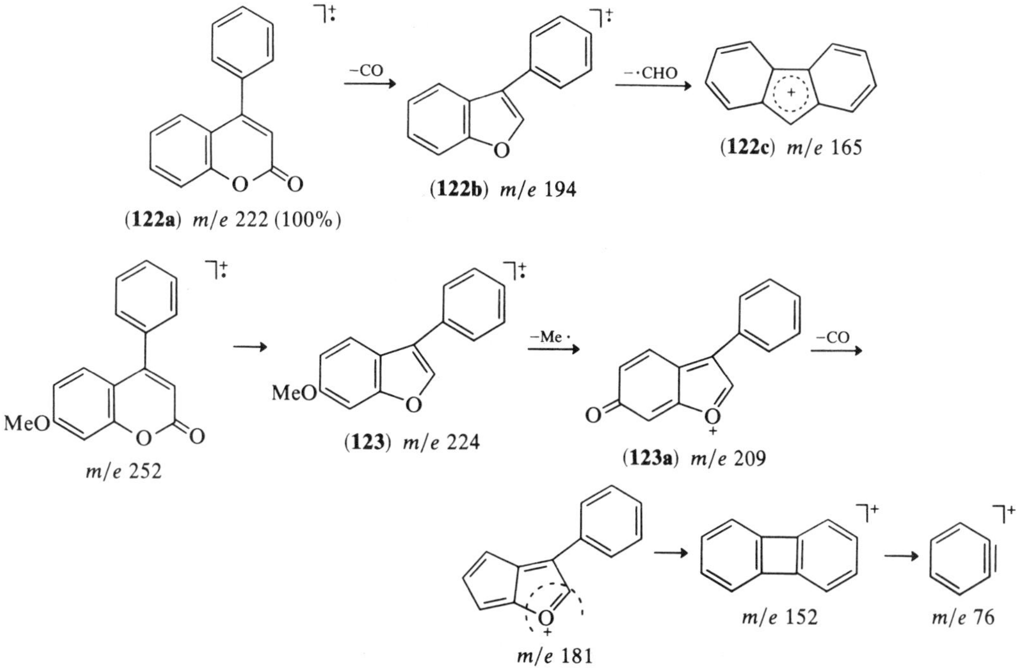

(**122a**) *m/e* 222 (100%)    (**122b**) *m/e* 194    (**122c**) *m/e* 165

(**123**) *m/e* 224    (**123a**) *m/e* 209

*m/e* 252

*m/e* 181    *m/e* 152    *m/e* 76

**Scheme 10**

Mass spectra of hydroxy- and alkoxy-coumarins have been very intensively studied. The decomposition sequence of 3-hydroxycoumarin is initiated by carbon monoxide loss from the molecular ion giving a 2-hydroxybenzofuran ion. Subsequent fragmentation occurs by two major pathways, involving a further loss of CO and expulsion of a formyl radical. The former leads to the base peak, and thence by another loss of CO to give the abundant benzene radical cation at *m/e* 78. The other main pathway gives a benzoyl cation which leads to the phenonium ion at *m/e* 77 ⟨77IJC(B)816⟩.

Fragmentation of 4-hydroxycoumarin (**124**) occurs in a completely different manner to that observed for the 3-isomer. Carbon monoxide expulsion from the molecular ion is virtually suppressed, whilst the base peak at *m/e* 120 corresponds to loss of a ketene (C₂H₂O) fragment from the molecular ion. It was proposed that fission of the heterocyclic ring occurs by a retro-Diels–Alder (RDA) reaction, a mode which has been rationalized by assuming participation of the tautomeric chromandione molecular ion (**124a**; Scheme 11). The ion at *m/e* 121 presumably arises from a hydrogen transfer reaction in the course of RDA cleavage. An intense fragment at *m/e* 92 arises by loss of CO from the [RDA]⁺· ion ⟨71OMS(5)249⟩. Subsequent work has demonstrated the enolic nature of the molecular ions derived from 4-hydroxycoumarins. Since variation of the inlet temperature did not alter the spectral characteristics, the molecular ion would appear to exist as a single tautomer. Support for the enol structure was also obtained from deuterium labelling experiments ⟨72JCS(P1)1924⟩.

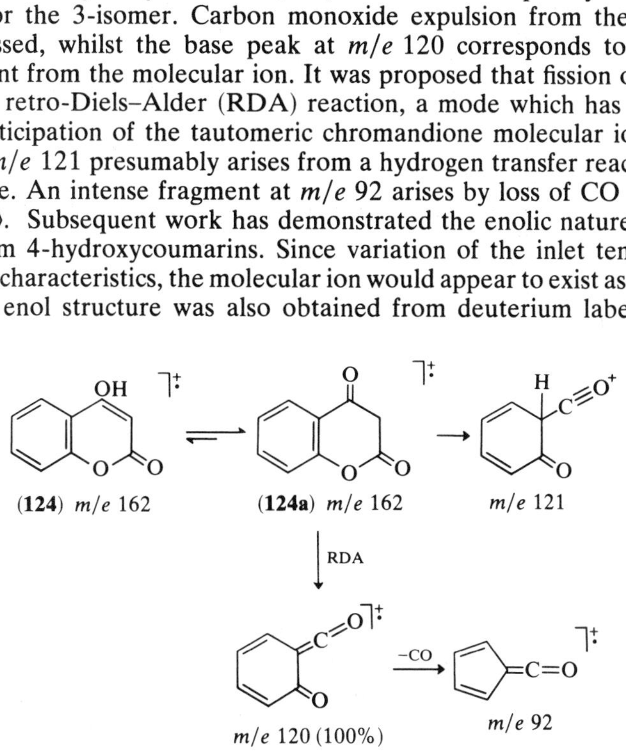

(**124**) *m/e* 162    (**124a**) *m/e* 162    *m/e* 121

*m/e* 120 (100%)    *m/e* 92

**Scheme 11**

The fragmentation patterns of 4-hydroxycoumarins which possess a carbonyl function (as an ester or an acyl group) at C-3 have been studied in considerable detail ⟨66JCS(C)1712⟩.

Coumarins which possess a hydroxy group in the carbocyclic ring fragment by processes analogous to those observed in the parent compound. However, the $[M-CO]^{+}$ ion derived from 7-hydroxycoumarin decomposes with two further losses of carbon monoxide. By analogy, fragmentation of 7-methoxycoumarin is dominated by methyl radical loss from the $[M-CO]^{+}$ ion. This is a highly favoured pathway since a stabilized quinonoid species results. Subsequent decomposition occurs by a further two losses of carbon monoxide ⟨64AJC975⟩. In the case of 6,7-dimethoxycoumarin, loss of a methyl radical occurs from both the molecular ion and $[M-CO]^{+}$. Deuterium labelling experiments confirmed that the C-6 methyl is preferentially eliminated. The reason for this appears to be the formation of the quinonoid ions (**125**) and (**126**) ⟨64AJC975, 65JOC955⟩.

(**125**)  (**126**)

Mass spectra of numerous other simple coumarins have been studied, including formyl- and keto-coumarins ⟨66JCS(C)1712⟩, halogenocoumarins ⟨71ZOR388⟩ and aminocoumarins ⟨80ZOB940⟩.

An immense amount of mass spectral data is available on more complex coumarins. Reviews have appeared on the mass spectrometry of furocoumarins and pyranocoumarins ⟨B-71MS139, 75RCR603⟩, many analogues of which occur naturally. The pyranocoumarin (**127**) is of interest since it fragments with four successive losses of carbon monoxide ⟨66JCS(C)1712⟩.

By contrast, mass spectra of isocoumarins have been little studied.

(**127**)

The isocoumarin (**128**) gives a weak molecular ion peak at $m/e$ 280. A prominent peak at $m/e$ 262 corresponds to an $[M-H_2O]^{+}$ fragment. Its formation is rationalized in Scheme 12 ⟨65BSF3025⟩.

(**128**) $m/e$ 280

$m/e$ 262  **Scheme 12**

The fragmentation pathways of pyran-4-one which were elucidated with the aid of $^{2}$H- and $^{18}$O-labelled compounds ⟨64JA3833⟩ are presented in Scheme 13. The molecular ion is the base peak. In contrast to pyran-2-one, breakdown *via* the formation of $[M-CO]^{+}$ ions constitutes only a minor pathway. The prominent feature of the spectrum is the formation of an $[M-C_2H_2]^{+}$ ion (**18a**), which arises by a retro-Diels–Alder (RDA) cleavage of the molecular ion. The formylketene radical cation (**18a**) thus formed fragments further by two pathways which give rise to ions (**18b**) and (**18c**). As with pyran-2-one, the structure of the $[M-CO]^{+}$ ion has been questioned. Early work based on metastable peak characteristics suggested that the furan structure (**18d**) was inappropriate ⟨67JA5954⟩, but a comparative

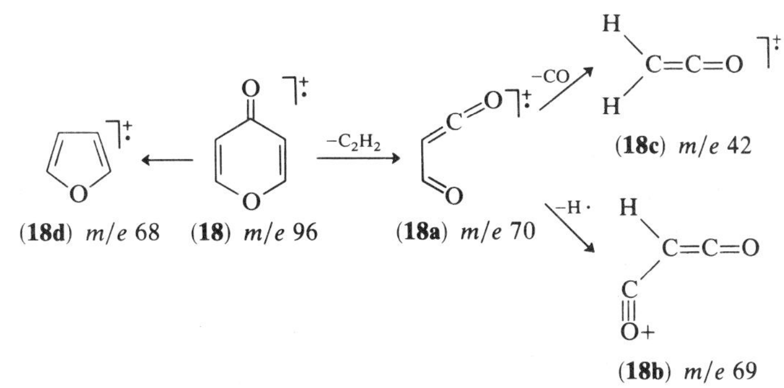

**Scheme 13**

study of the collisional activation spectra of 2- and 4-pyranones and of furan revealed that the $[C_4H_4O]^{\ddagger}$ ions do indeed possess the furan structure ⟨79JA4973⟩.

In general, substituted pyran-4-ones fragment in a similar fashion to the parent compound. This is exemplified by the behaviour of 2,6-dimethyl- and 3,5-dimethyl-pyran-4-ones (129) and (130) ⟨64JA3833⟩. The mass spectrum of (129) has been studied in detail and metastable peaks have been observed for most steps ⟨67MI22202⟩. A characteristic feature is the cleavage of the pyranone ring *via* a retro-Diels–Alder collapse. Elimination of carbon monoxide from the molecular ion constitutes the major reaction pathway, in contrast with that of pyran-4-one itself, and generates a furan ion at $m/e$ 96, from which abstraction of a hydrogen atom gives an oxonium ion.

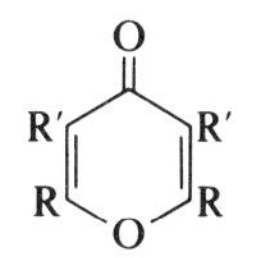

(129) R = Me, R′ = H
(130) R = H, R′ = Me

The RDA cleavage of (129) consists of expulsion of neutral methylacetylene with the formation of the acetylketene radical cation at $m/e$ 84; loss of a methyl radical gives the abundant ketene ion (18b). The isomeric pyran-4-one (130) fragments similarly. In this case, however, RDA cleavage of the pyranone ring is followed by loss of CO, producing methylketene ions, $m/e$ 56. Carbon monoxide loss from the molecular ion also occurs but to a lesser extent than for (129).

The decomposition mode of 6-substituted 2-methoxypyran-4-ones is governed by the nature of the substituent ⟨65T1855⟩. The fragmentation of 2-methoxy-6-methylpyran-4-one (106) occurs by RDA cleavage and by loss of CO. RDA expulsion of methylacetylene constitutes a minor pathway, giving rise to a ketene radical cation at $m/e$ 100 (8%). Carbon monoxide loss generates a furan radical cation, which decomposes *via* methyl radical loss and carbon monoxide expulsion to give a $[C_4H_5O]^{\ddagger}$ ion, $m/e$ 69, which is thought to possess an oxete structure.

In the mass spectrum of the 6-phenyl analogue, fragmentation by loss of CO is of less significance, and here the RDA pathway is prominent. The most abundant ions are due to a phenylacetylene radical cation as the base peak and the derived phenonium ion at $m/e$ 77 (21%). The driving force for the RDA fragmentation appears to result from efficient charge stabilization on the phenylacetylene fragment ⟨65T1855⟩.

In all the foregoing examples, the prominent fragmentation of pyran-4-ones involves a retro-Diels–Alder (RDA) cleavage of the molecular ion. However, atypical behaviour is sometimes displayed, particularly when C-3 bears an oxygen function, which can cause other more complex pathways to supervene. Some examples are given below.

The major pathways for the fragmentation of kojic acid (81, 5-hydroxy-2-hydroxymethyl-pyran-4-one), are shown in Scheme 14; support for each route was provided by the appearance of metastable ion peaks ⟨67MI22203⟩. An RDA cleavage followed by loss of a ·CH₂OH radical produces ion (81a), $m/e$ 69, the structure of which was substantiated by deuteration experiments. The ion at $m/e$ 97 arises by extrusion of CO from the molecular ion and loss of HO· from the side chain; structures (81b) and (81c) were proposed. Decomposition of [M]⁺ occurs to give ethylene and an HC≡O⁺ fragment. The initial stage involves loss of a ·CHO radical from the hydroxymethyl substituent, a process which has also been observed for benzyl alcohol ⟨B-67MI22204⟩, giving the oxonium ion (**81d**). Further decomposition (**81d → 81e**) occurs by ring opening and elimination of CO. Subsequent hydrogen migration, ring closure and displacement of carbon monoxide generates the oxetonium ion (**81f**) which readily cleaves. The peak at $m/e$ 39 is thought to arise from a cyclopropenium ion, though the mechanism of its formation remains unclear.

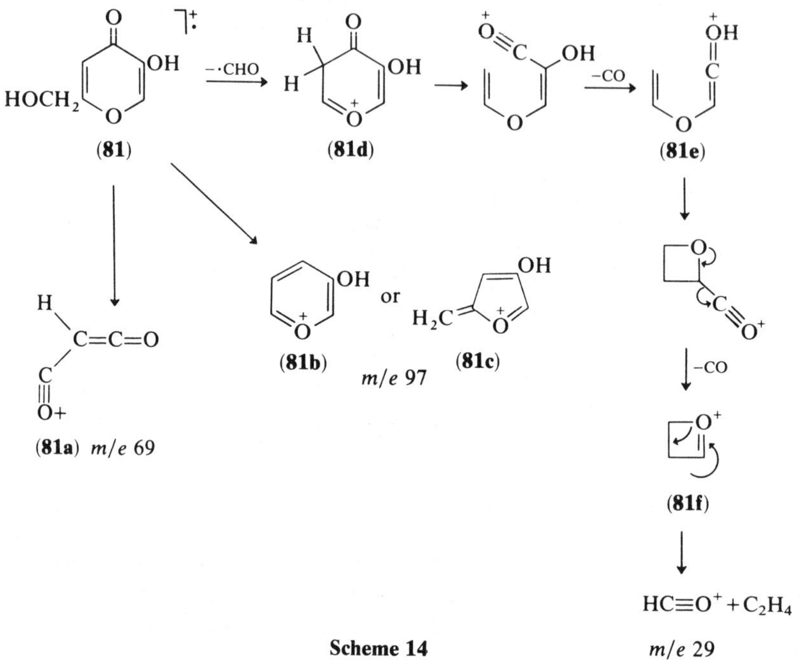

**Scheme 14**

The spectra of methyl maltol (**131**, 3-methoxy-2-methylpyran-4-one) and methyl allomaltol (**132**, 5-methoxy-2-methylpyran-4-one) are of considerable interest, since both these compounds fragment by pathways which originate from a primary rearrangement of the molecular ion. In contrast to other pyran-4-ones there are no fragments which correspond to loss of carbon monoxide or RDA cleavage of the molecular ion. Use was made of the methoxy (OCD$_3$) labelled compounds in elucidating the structures of the major daughter ions ⟨78OMS296⟩. The decomposition modes of (**131**) and (**132**) are summarized in Scheme 15. An unusual feature is the elimination of water from the molecular ion, a process which does not occur in the 3-hydroxy compound (**81**).

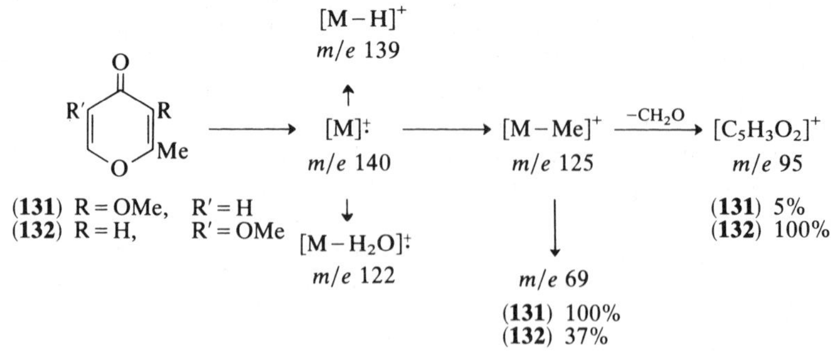

**Scheme 15**

The complex and diverse fragmentation pathways of methyl maltol and methyl allomaltol prompted a study of the mass spectral characteristics of other methoxypyran-4-ones ⟨80OMS31⟩.

A detailed examination of the spectrum of (133) revealed a strong similarity to that of methyl allomaltol (132). It was found that production of a rearranged molecular ion *via* a 1,5-hydrogen shift from the methoxy to the carbonyl group is the initial step with subsequent loss of $H_2O$, $CH_2O$, and R· (*i.e.* $CH_2OH$). The base peak appears at $m/e$ 95 and corresponds to loss of $CH_2O$ and R· fragments. Peaks for $[M - CH_2O]^{\ddagger}$ and $[M - R]^+$ ($m/e$ 125) ions are also present. Most of the compounds studied showed similar behaviour and the fragmentation pattern is largely independent of the nature of the substituents. However the introduction of a cyano group (as in 134 and 135) causes a profound change in the observed pathways, with the most intense fragments resulting from RDA cleavage.

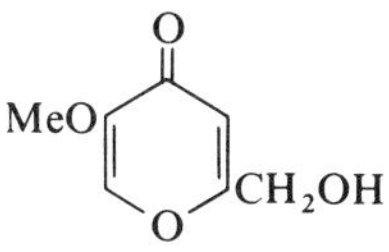

(133) 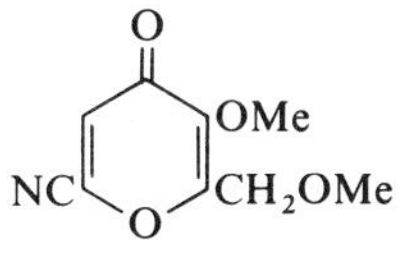          (134) 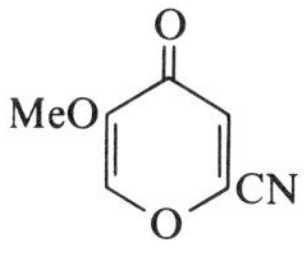          (135)

Guidelines for the interpretation of the mass spectra of substituted 3-methoxypyran-4-ones have been formulated ⟨80OMS31⟩.

The mass spectral fragmentation of the monobenzologue of pyran-4-one, chromone (20), has been reported ⟨65T1855, 65MI22201, 69BSF4545⟩.

Chromone yields an abundant molecular ion which fragments by two pathways involving either expulsion of carbon monoxide or ring cleavage by a retro-Diels–Alder (RDA) reaction. In contrast to pyran-4-one itself, the former process predominates. The fragmentation pathways are outlined in Scheme 16. The base peak is due to the molecular ion (20a) and appears at $m/e$ 146. Loss of acetylene by the RDA pathway occurs giving initially (20b), $m/e$ 120, which subsequently decomposes by further losses of CO by the route shown. The formation of the $[M - CO]^{\ddagger}$ ion by extrusion of the carbonyl group from (20a) parallels the behaviour of the isomeric system coumarin. This ion is thought to possess the benzofuran structure (20c), although other structures cannot be ruled out. Decomposition of (20c) occurs to give ions at $m/e$ 90 and 89, which correspond to loss of CO and then a hydrogen radical. Analogous behaviour was observed for benzofuran itself ⟨64T1185⟩.

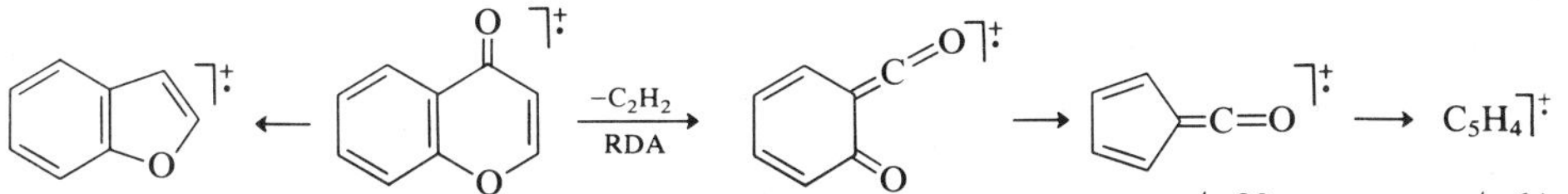

(20c) $m/e$ 118     (20a) $m/e$ 146     (20b) $m/e$ 120     $m/e$ 92     $m/e$ 64

**Scheme 16**

The presence of substituents may divert the course of fragmentation from that described above.

A mass spectral study of 2-methyl-, 3-methyl- and 2,3-dimethyl-chromone (136), (137) and (138) has been reported ⟨79OMS345⟩. In each case the molecular ion appears as the base peak, together with ions which correspond to $[M - CO]^{\ddagger}$, $[M - CHO]^{\ddagger}$, $[RDA]^{\ddagger}$, $[RDA + H]^+$ and $[RDA - CO]^{\ddagger}$. Metastable peaks confirmed that the formation of $[M - CHO]^{\ddagger}$ occurs in two steps from $[M]^{\ddagger}$. The reaction pathway for (136) and (138) is given in equation (2). In compound (137), $[M - CHO]^{\ddagger}$ is an abundant fragment ion (60%, *cf.* 35% for 136). That its generation occurs by more than one route is suggested not only by its high abundance, but also by the appearance of appropriate metastable ion peaks; two pathways are operative (Scheme 17).

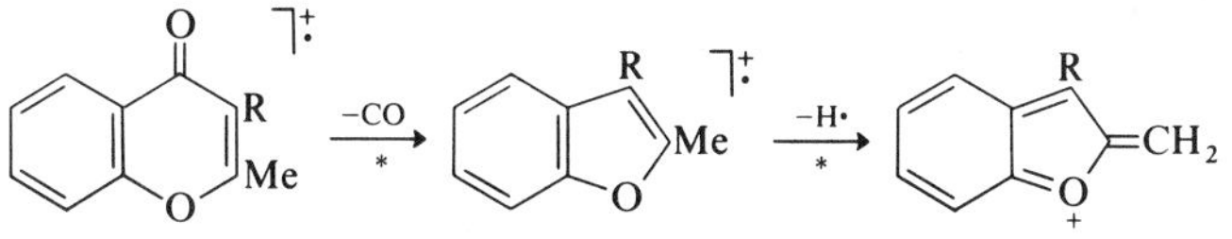

(136) R = H
(138) R = Me

(2)

Breakdown *via* RDA cleavage with the formation of $[RDA]^{\ddagger}$ ions constitutes an important pathway for chromones (20) and (136). Interestingly in (137) and (138), the intensities of the $[RDA]^{\ddagger}$ ion peaks decrease, whilst those of the $[RDA + H]^+$ ion increase. The latter arise by a hydrogen transfer reaction prior to ring cleavage. Relative peak intensities of

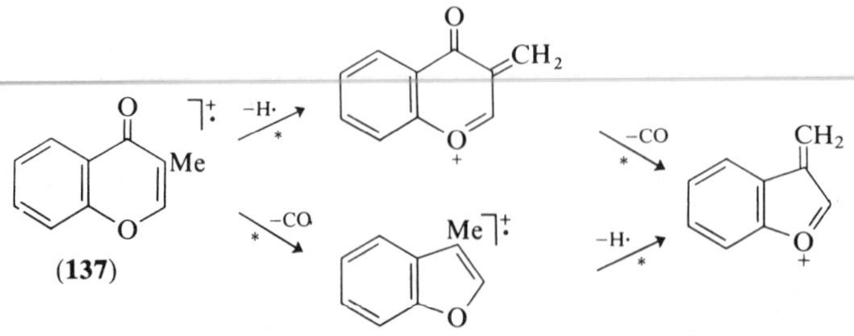

**Scheme 17**

the two ions for the chromones (**20**, **136–138**) are presented in Table 7. In the dimethyl compound (**138**) the [RDA + H]$^+$ species is prominent whilst all other fragment ions appear as minor peaks. Measurements using a low ionization potential (20 eV) demonstrate that hydrogen transfer in the RDA fragmentation of these chromones is energetically facile. Deuterium labelling experiments and investigation of metastable ions using the ion-kinetic energy defocusing technique amply demonstrate that two mechanisms for the formation of [RDA + H]$^+$ ions from (**138**) are operative (Scheme 18).

**Table 7**   Relative Peak Intensities of [RDA]$^{\ddagger}$ and [RDA + H]$^+$ Ions in the 70 eV Spectra of Some Chromones ⟨79OMS345⟩

| Compound | [RDA]$^{\ddagger}$ $m/e$ 120 | [RDA + H]$^+$ $m/e$ 121 |
|---|---|---|
| Chromone | 97 | 3 |
| 2-Methylchromone | 93 | 7 |
| 3-Methylchromone | 78 | 22 |
| 2,3-Dimethylchromone | 21 | 79 |

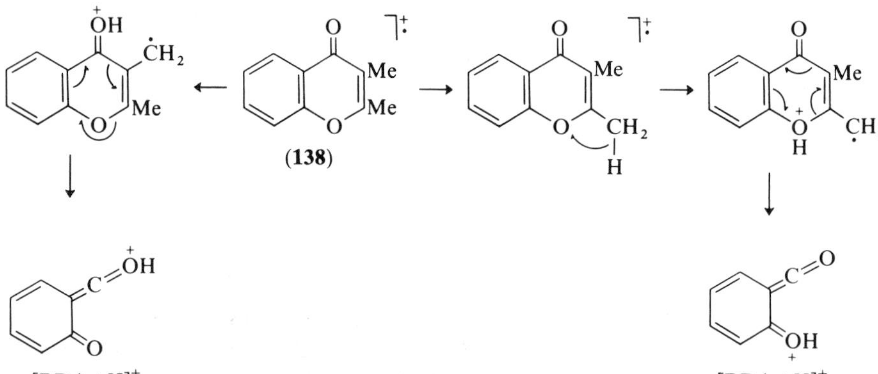

[RDA + H]$^+$                **Scheme 18**                [RDA + H]$^+$

Introduction of a phenyl substituent into the pyran ring in chromone provides flavone (2-phenylchromone, **22**) and the isomeric 3-phenyl compound isoflavone (**64**). The high stability of the flavones is reflected by the intensity of their molecular ions, which often appear as the base peak. Flavone itself fragments by an RDA pathway in an analogous manner to chromone, giving an [RDA]$^{\ddagger}$ ion, $m/e$ 120, and ionized phenylacetylene, $m/e$ 102. As with chromone, an [RDA − CO]$^{\ddagger}$ fragment, $m/e$ 92 is also observed. Carbon monoxide elimination from the molecular ion gives the phenylbenzofuran ion (**22a**) at $m/e$ 194 and the derived doubly charged ion at $m/2e$ 97. Further breakdown of (**22a**) occurs giving a benzoyl ion (**22b**) as a minor fragment (equation 3). An (M − 1) ion is also present in the spectrum, although its structure remains unresolved ⟨64AJC975⟩. Halogenated flavones fragment similarly, thus further substantiating the validity of the pathways outlined above ⟨66MI22201⟩.

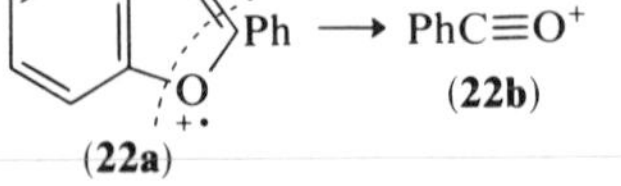

(3)

$m/e$ 194   $m/2e$ 97

The chemical ionization mass spectra of flavone and various congeneric compounds have been reported ⟨73T4083⟩. Methane was used as the reactant gas. However the usefulness of this technique appears to be limited, since few diagnostic fragments were observed. The +E spectrum of the flavone molecular anion has been described ⟨76AJC115⟩. +E spectra result from high energy ion molecule reactions in the analyzer regions of the mass spectrometer, whereby non-decomposing negative ions undergo a charge inversion reaction $M^- \rightarrow M^+$. The peaks in the spectrum result from the decomposing molecular cation produced from the charge inversion reaction. Thus, flavone exhibits peaks characteristic of $[M-(H_2O+H_2O)]^{\ddagger}$ ions together with a prominent $[M-CO_2]^{\ddagger}$ fragment. The structures of these fragments are unknown. The electron impact spectra of isoflavones strongly resemble those of flavones and once again cleavage by a retro-Diels–Alder pathway is prominent. As with flavone, the substitution pattern determines the nature of the charged fragment. The fragmentation pathways of isoflavone (64) are given in Scheme 19. The base peak is due to the (M−1) ion (64a) which, owing to its high stability, does not fragment further ⟨66BSF2892⟩. The formation of a doubly charged molecular ion is a notable feature of isoflavones in general. The RDA pathway generates the ions at *m/e* 120 and *m/e* 102. In contrast to flavone, the $[M-CO]^{\ddagger}$ peak is very weak and indeed is not observed in the spectra of (139) or (140). Isoflavones with 2'-substituents (*i.e.* in ring C) undergo fragmentation, often with loss of the substituent, generating an oxonium species analogous to (64a) ⟨77OMS51⟩.

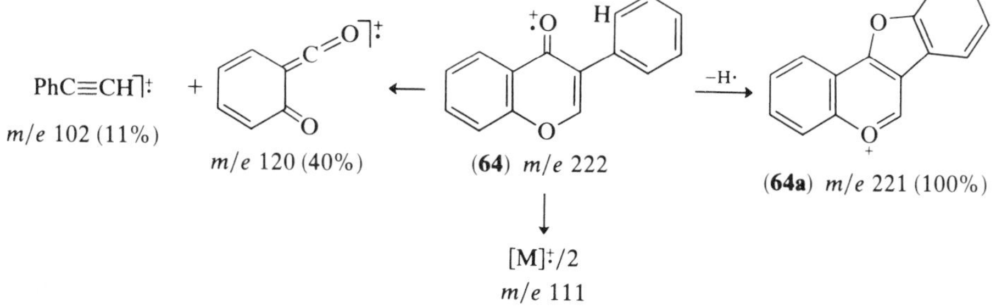

PhC≡CH]$^{\ddagger}$

*m/e* 102 (11%)

*m/e* 120 (40%)

(64) *m/e* 222

−H·

(64a) *m/e* 221 (100%)

[M]$^{\ddagger}$/2

*m/e* 111

**Scheme 19**

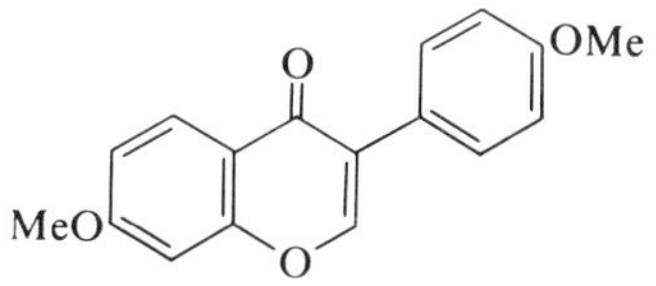

(139)

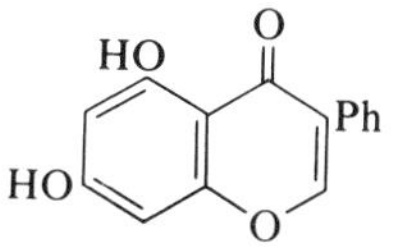

(140)

An intensive study of the mass spectral decomposition of analogues of ethyl chromone-2-carboxylate has been undertaken ⟨71OMS(5)857⟩. Virtually all of the compounds studied gave the molecular ion as the base peak and the important fragment ions corresponded to $[M-C_2H_4]^{\ddagger}$, $[M-OEt]^+$, $[M-C_2H_4-CO]^{\ddagger}$ and the tropylium ion. The mass spectral fragmentation of chromone-3-carbaldehyde exhibits ions due to loss of CO from the aldehyde moiety and an H· radical. The $[M-CO]^{\ddagger}$ ion probably possesses a chromone structure, since it decomposes in an identical fashion to (20) ⟨74T3553⟩. The pyranobenzoxazoles (141) and (142) show only breakdown by loss of NO, $NO_2$ and $C_2H_4$ from the nitro and ester groups, respectively. Fragments which result from rupture of the pyranone and oxazole rings are of low intensity. There is no evidence of a retro-Diels–Alder cleavage ⟨71JCS(C)1482⟩. In complete contrast, the pyranobenzothiazoles (143) and (144) display an

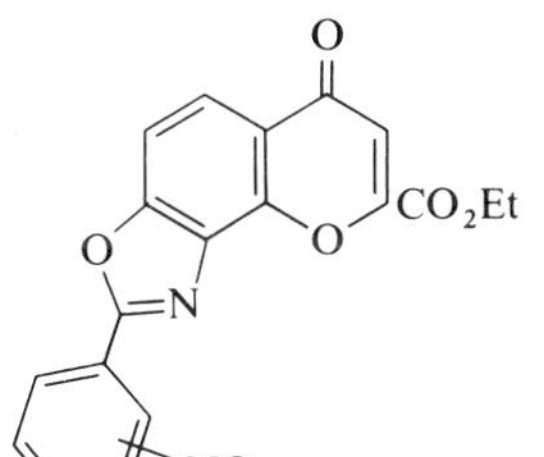

(141)

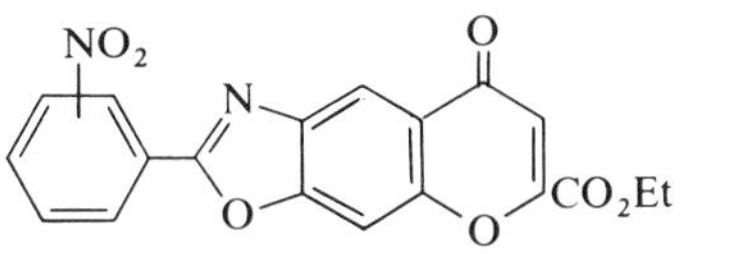

(142)

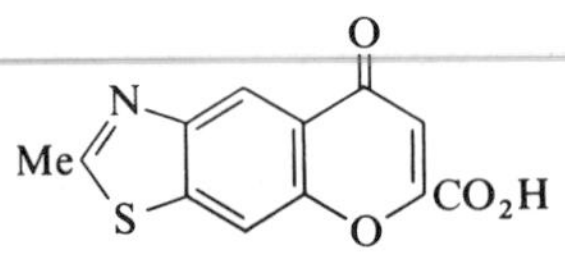

(143) 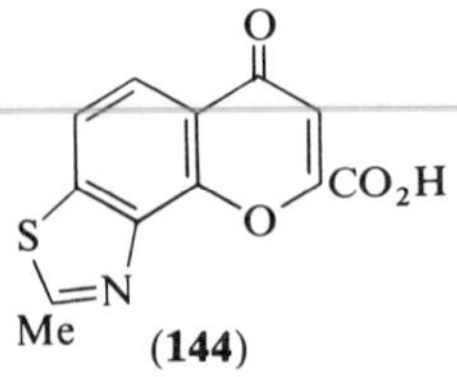                                   (144)

[RDA]$^{\ddagger}$ ion as the base peak at $m/e$ 191 corresponding to loss of a propiolic acid fragment ⟨70JCS(C)1553⟩.

Xanthone (21) is extremely stable and, as might be expected, the molecular ion appears as the base peak. RDA fragmentation does not take place and decomposition of [M]$^{\ddagger}$ occurs by extrusion of the carbonyl group as carbon monoxide giving the dibenzofuran ion (21a; Scheme 20) which fragments further giving ions at $m/e$ 140 and 139 ⟨64AJC975⟩. Xanthones bearing hydroxy, nitro or amino groups at C-1 or C-2 fragment in an identical manner ⟨74MI22203⟩. However, the decomposition mode of the monomethoxy analogues is dependent upon the position of the substituent. The 1-methoxy compound exhibits an *ortho* effect; its spectrum has been interpreted by means of deuterium and $^{18}$O labelling and use of metastable ion defocusing techniques ⟨73OMS(7)667⟩.

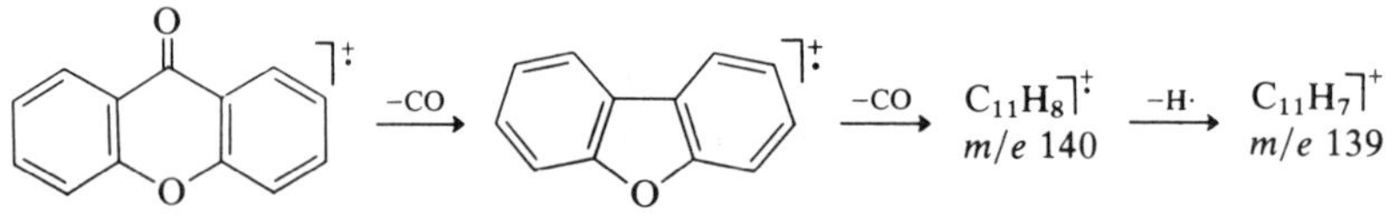

$m/e$ 196 (100%)          (21a) $m/e$ 168

**Scheme 20**

The mass spectra of some naturally occurring xanthones have been briefly reviewed ⟨B-74MI22202⟩.

The mass spectra of tetrahydropyran-2-one (145; R = H) and related lactones have been reviewed ⟨B-71MS179⟩. A generalized fragmentation pathway is outlined in Scheme 21. Expulsion of R· radicals from the molecular ion generates the oxonium ion at $m/e$ 99. Subsequent decomposition occurs by loss of CO and CHO radicals giving peaks at $m/e$ 71 and 70, respectively. Additionally δ-lactones often display an intense peak at $m/e$ 42 which has been ascribed to a ketene fragment ⟨65MI22202, 65ACS370⟩.

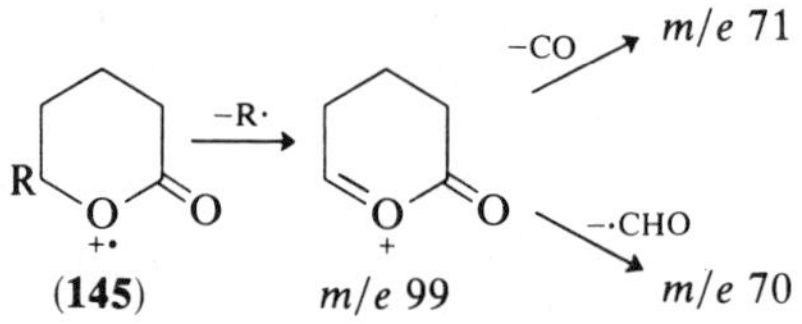

(145)          $m/e$ 99          $m/e$ 70

**Scheme 21**

The elimination of $CO_2$ from the molecular ion may also be of importance in certain monosubstituted δ-lactones. A further pathway common to all these systems is the elimination of the ring oxygen and the adjacent carbon atom with its substituents as a neutral carbonyl molecule ⟨68OMS(1)279⟩. A highly detailed investigation of the mass spectral fragmentation of tetrahydropyran-3-one (146) has been reported ⟨74JOC279⟩. Deuterium labelling and metastable defocusing experiments enabled the fragmentation pathways to be elucidated. The principal ions from (146) have $m/e$ ratios of 100 (the molecular ion), 71, 70, 55, 45, 42 (base peak) and 41; mechanisms for their formation are given in Scheme 22. Decomposition of the molecular ion occurs by direct expulsion of a CHO radical giving (146a), $m/e$ 71, or by rupture of the 2,3-bond. This latter process gives ion (146b), in which the charge is retained by the ring oxygen atom, and (146c), where the charge resides on the carbonyl oxygen. A hydrogen transfer process converts (146b) to (146c), $m/e$ 45. The ion (146d) loses formaldehyde to give (146e), $m/e$ 70, which decomposes by the pathways shown giving (146f–146g) and (146h) and (146i).

A similar fragmentation is observed for 6-methyltetrahydropyran-3-one ⟨74JOC279⟩. However, in contrast to the isomeric δ-lactones, expulsion of a methyl radical from the molecular ion yields only a minor fragment.

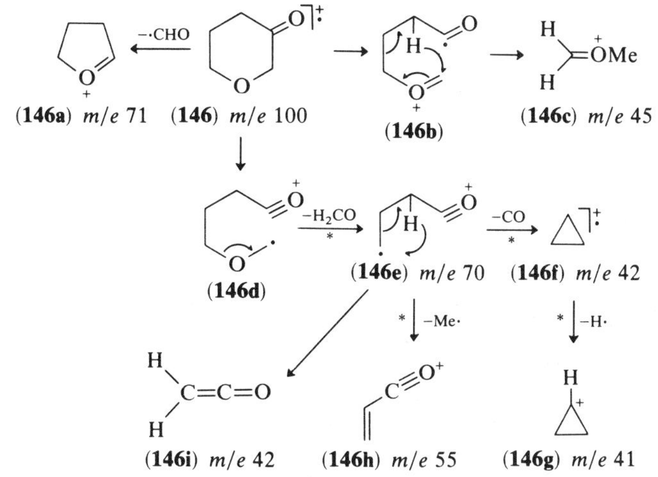

**Scheme 22**

There appears to be no detailed information regarding the mass spectral behaviour of tetrahydropyran-4-ones.

There is little information on the mass spectra of monocyclic dihydropyranones.

Some dihydropyran-2-ones are natural products and mass spectrometry has proved to be a useful tool in structure elucidation. For example, massoilactone (**147**), which is secreted by formicine ants, was characterized by its mass spectrum. On electron impact, the alkyl group is readily cleaved generating an ion at $m/e$ 97 ⟨68AJC2819⟩. This process parallels the behaviour of other simple lactones.

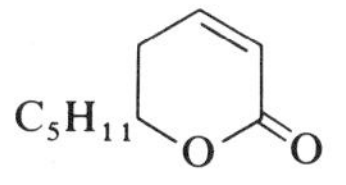

**(147)**

There is a similarity in mass spectral behaviour between dihydrocoumarin (**23**) and coumarin itself, since both decompose by two successive losses of carbon monoxide from the molecular ion. However, unlike coumarin, the formation of an [RDA]$^{\ddagger}$ ion at $m/e$ 106 by elimination of ethylene also occurs ⟨64AJC975⟩.

Chroman-3-ones are not readily accessible and consequently there are few mass spectral data available. The major peaks in the spectrum of the parent compound appear at $m/e$ 148 (M$^+$; 53%) and at 92, 91 and 89 ⟨73BSB283⟩. The structures of the fragment ions and the pathways responsible for their formation have not been established.

There have been numerous studies of the electron impact spectra of chroman-4-ones. In the parent compound (**24**), decomposition of the molecular ion occurs *via* a retrodiene fragmentation, giving the [RDA]$^{\ddagger}$ ion (**24a**), $m/e$ 120, as the base peak. The only other significant fragment appears at $m/e$ 92 and arises by elimination of carbon monoxide from (**24a**), generating ion (**24b**). Further decomposition of (**24b**) occurs giving (**24c**) by loss of CO (Scheme 23). In addition, there is a minor peak (9%) at $m/e$ 121, which corresponds to an [RDA+H]$^+$ ion, formed by hydrogen transfer prior to ring cleavage ⟨64T1185⟩. Chromanones with substituents in the aromatic ring have been reported to fragment in an analogous manner. The relative stabilities of a range of methyl- and chloro-substituted chroman-4-ones to electron impact fragmentation has been investigated ⟨73ZOR1748⟩.

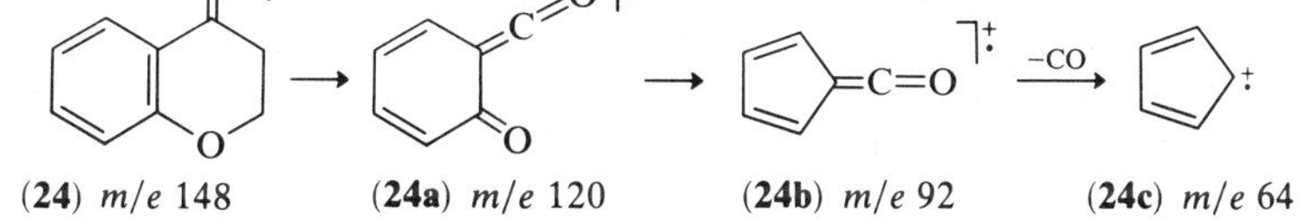

**Scheme 23**

The abundance of the [RDA+H]$^+$ ion is markedly greater in chroman-4-ones with *gem*-dimethyl groups at C-2. Spectra of 2,2-dimethylchroman-4-ones with aromatic methoxy substituents have been studied ⟨78OMS653⟩. All showed intense peaks corresponding to $[M]^{+}$, $[M-Me]^{+}$, $[RDA]^{+}$ and $[RDA+H]^{+}$ ions. Deuterium labelling experiments confirmed that the hydrogen atom in the $[RDA+H]^{+}$ fragment originates from the C-2 methyl groups rather than from C-3. The abundance of the $[RDA+H]^{+}$ ions is considerably greater than that of the corresponding $[RDA]^{+}$ fragments. However, this situation is reversed when a methoxy group is present at C-5. Measurements on 7-methoxy-2,2-dimethyl-chroman-4-one (**148**) at 20 eV confirmed that the formation of $[RDA+H]^{+}$ ions is a low energy process.

(**148**)

The $[M-Me]^{+}$ ion arises by loss of one of the C-2 methyl groups, a pathway which is generally prominent in 2,2-dimethylchroman-4-ones ⟨73MI22202⟩.

The flavanones constitute an important class of natural products. As a result, their mass spectra have been studied extensively and have been comprehensively reviewed ⟨B-74MI22202, B-75MI22201⟩. Flavanone (**25**) fragments predominantly by a retro-Diels–Alder pathway giving the characteristic $[RDA]^{+}$ ion at $m/e$ 120 as the base peak, together with ionized phenylethylene ($m/e$ 104). An intense $(M-1)$ peak has been ascribed to the 4-hydroxyflavylium ion. The mass spectrum of (**25**) and the 2'-hydroxychalcone (**149**) are virtually identical, since both possess common metastable ion characteristics and the major ions in each case have the same $\Delta H_f$ values. From a detailed comparative study of deuterium labelled analogues of (**25**) and (**149**) ⟨72OMS(6)1333⟩, it was concluded that a ring–chain tautomeric equilibrium exists between a flavanone and a chalcone type molecular ion (equation 4). The appearance of an $[M-OH]^{+}$ ion in the spectrum of (**25**) results from loss of an ·OH radical from the ring-opened chalcone tautomer. In the mass spectrum of (**149**), the base peak is due to an $[M-Ph]^{+}$ ion. Its formation by an energetically unfavourable vinylic cleavage of (**149**) is difficult to envisage, but loss of Ph· from the flavanone tautomer will be facile.

(4)

A characteristic of the mass spectrum of isoflavanone (3-phenylchroman-4-one, **86**) is its simplicity. Here again, fragmentation of the molecular ion occurs to give an $[RDA]^{+}$ ion as the base peak which decomposes by loss of CO. Hydrogen transfer occurs to give an ion corresponding to $[RDA+H]^{+}$ ⟨66BSF2892⟩.

An investigation established a relationship between the intensity of the $[M-CO_2]^{+}$ ion in the spectra of substituted isochroman-3-ones (**29**) and the yield of the corresponding benzocyclobutene (**29a**), obtained from flash vacuum pyrolysis of (**29**; equation 5) ⟨77JOC2989⟩.

(5)

The rotenoids are a class of natural products which are based on the chromanochromanone system (**150**), the fragmentation of which proceeds by the routes shown in Scheme 24 ⟨66BSF2892⟩. Rupture of ring B occurs by a retro-Diels–Alder reaction giving (**150a**) and (**150b**). Although either fragment may carry the charge it is preferentially retained by (**150b**), which gives rise to the base peak. Loss of a hydrogen radical from the latter produces the highly stable benzopyrylium ion (**150c**). An abundant fragment at $m/e$ 121, (**150d**), is formed by RDA cleavage of ring B with hydrogen transfer. Similar pathways have been observed for other rotenoids ⟨63JCS5949, 67T4741⟩.

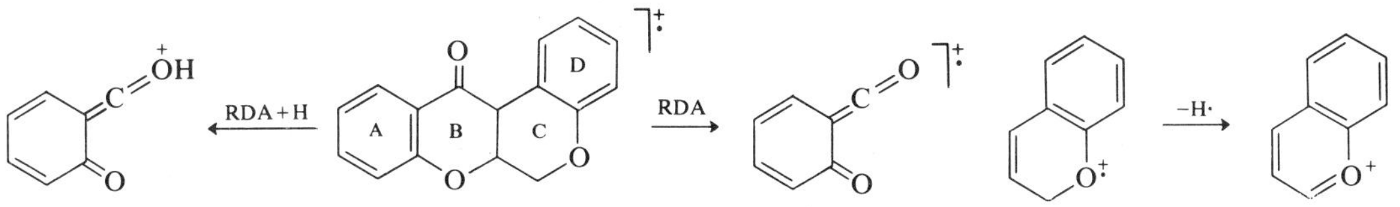

(**150d**) *m/e* 121 (**150**) (**150a**) (**150b**) *m/e* 132 (**150c**) *m/e* 131

**Scheme 24**

### 2.22.6.3 Pyrylium Salts

There have been few mass spectral studies of pyrylium salts, possibly as a result of their low volatility. The examples reported so far have been concerned with the behaviour of alkyl- and aryl-substituted compounds under electron impact conditions. Information regarding the parent compound appears to be lacking. In the initial work, use was made of 2,4,6-trisubstituted pyrylium salts. The halides (iodides and bromides) were examined because of their enhanced volatility over the corresponding perchlorates and tetrafluoroborates ⟨71OMS(5)87⟩.

For compound (**151**; R = Me or Ph), the base peak arises from an $[M-HI]^{+\cdot}$ ion. The molecular ion from the parent compound is not observed. Metastable ion peaks aided the elucidation of the fragmentation pathways which are outlined in Scheme 25. The $[M-HI]^{+\cdot}$ ion (**151a**) may possess either a methylene pyran or an oxepin structure. Further decomposition of this ion occurs by loss of a hydrogen radical. Expulsion of a methyl radical from (**151a**) generates (**151b**) which decomposes as shown (Scheme 25).

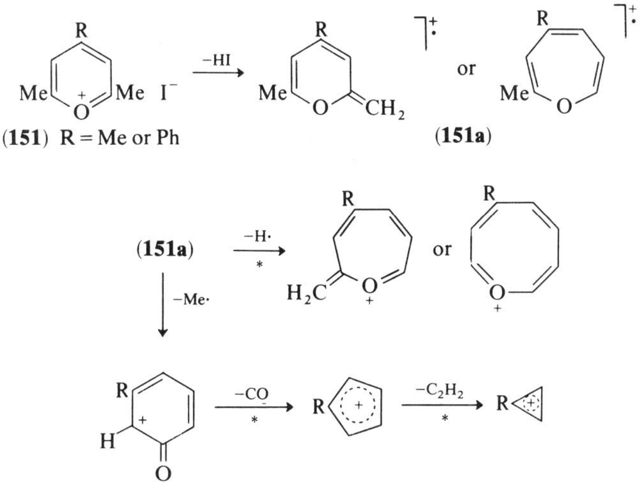

(**151**) R = Me or Ph (**151a**) (**151b**)

**Scheme 25**

An ion at *m/e* 43 due to the acetyl cation ($MeCO^+$) is a prominent low mass fragment from pyrylium salts which carry C-2 or C-6 methyl groups. Elimination of a C-4 phenyl substituent is never observed. In the pyrylium salt (**152**), fragments which correspond to formal loss of both methyl and phenyl substituents from the abundant $[M-HI]^{+\cdot}$ ion are apparent. Here the C-2 phenyl may be eliminated as a $PhCO^+$ fragment (*m/e* 105) or as a phenyl radical by a mechanism analogous to the formation of (**151b**) from (**151a**). Methyl radical expulsion from $[M-HI]^{+\cdot}$ of (**152**) is thought to occur from the rearranged ion (**152a**; Scheme 26) giving ultimately (**152b**) at *m/e* 231.

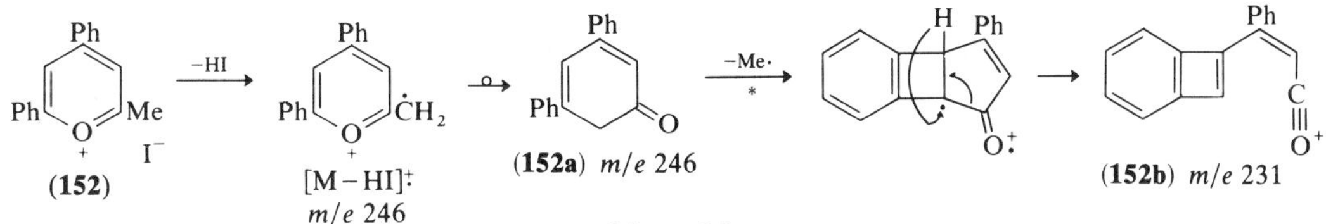

(**152**) $[M-HI]^{+\cdot}$ *m/e* 246 (**152a**) *m/e* 246 (**152b**) *m/e* 231

**Scheme 26**

Measurement of ionization potentials has shown that 2,4,6-triphenylpyrylium salts undergo a thermal reduction in the mass spectrometer giving the corresponding free radicals. Steric properties preclude dimerization ⟨74OMS(9)80⟩. This type of reduction appears to be largely independent of the nature of the anion; the bromide and iodide salts behave identically. However, in the case of the tetrafluoroborate salt, adduct formation between a cation and a fluoride ion gave a minor peak. Anomalous behaviour is displayed by the perchlorates: the anion effects oxidation of the cation upon evaporation giving a base peak which corresponds to $[M+O-H]^+$.

The appearance potential measurements indicate that thermal elimination of HI from (153) gives the corresponding methylenepyran (154) ⟨74OMS(9)80⟩.

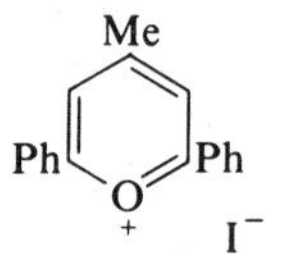

(153) 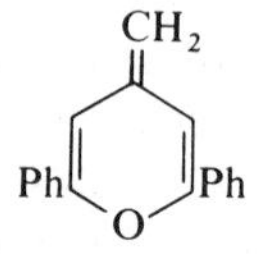        (154)

The only pyrylium benzologue for which mass spectral data are available is the flavylium ion (12) ⟨69AK(30)393⟩. On electron impact, (12) forms a highly stable molecular ion as the base peak which decomposes by two pathways (Scheme 27). Pathway A generates the $[M-H]^+$ ion (12a). There is some evidence to suggest that loss of CO from (12a) gives the phenanthrene radical ion (12b) ($m/e$ 178). Subsequent ejection of an acetylene molecule generates the biphenylene ion (12c).

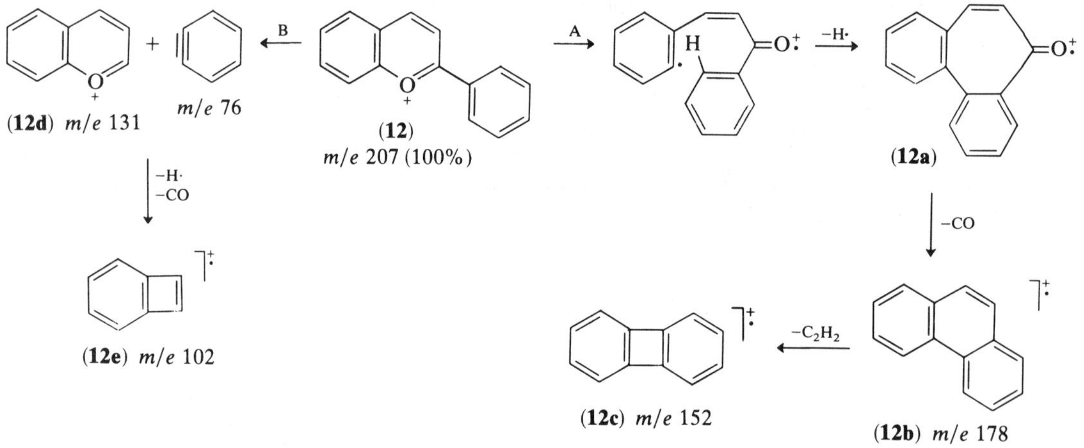

Scheme 27

The alternative fragmentation by pathway B results in loss of the phenyl substituent giving the benzopyrylium species (12d), which decomposes further to give (12e).

Mass spectra of xanthylium salts have not been reported. However, elimination of a hydrogen radical from the molecular ion of 9H-xanthene gives a xanthylium cation as base peak which fragments further by loss of a ·CHO radical (see Section 2.22.6.1).

## 2.22.7 MOLECULAR STRUCTURE AND CONFORMATION

### 2.22.7.1 X-Ray Determinations

A short chapter on X-ray diffraction studies of heterocyclic compounds provides background information to the technique ⟨62PMH(1)161⟩, whilst a comprehensive list of molecular dimensions derived from X-ray investigations prior to 1970 includes a number of six-membered oxygen heterocycles ⟨72PMH(5)1⟩.

There are few examples of X-ray studies of pyrans. The product from the reaction of 2,6-diphenylpyrylium perchlorate with water in the presence of bases was shown to be the methylenepyran (155) by a combination of spectroscopic and crystallographic measurements ⟨78T2131⟩.

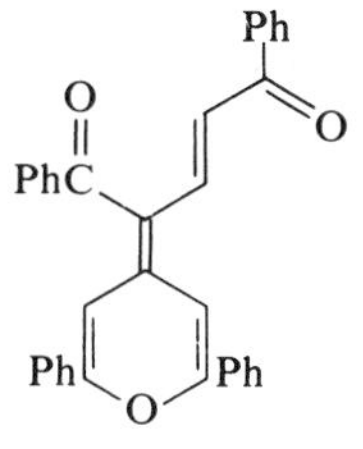

**(155)**

Crystal structures of 2,6-dimethyl-4-(*p*-nitrophenylimino)pyran (**65**; X = 4-NO$_2$) and its hydrobromide and methobromide salts have been determined ⟨81JCS(P2)303⟩. The 4-phenyl ring is rotated out of the plane of the pyran ring to avoid interaction between H-3 and H-2′. The length of the C=N bond increases along the series free base < hydrobromide < methobromide, and this feature is accompanied by changes in the parameters of the pyran ring. The C(2)—C(3) bond increases in length, but both the O—C(2) and C(3)—C(4) bonds decrease. These variations are interpreted in terms of a greater contribution from a pyrylium canonical form such as (**68**) to the overall structure of the methobromide.

Molecular dimensions have been reported for the 2,2-dimethylchromene bromouliginosin B (**156**), a derivative of an antibiotic from *Hypericum uliginosum* ⟨68JA4723⟩. Crystallographic data are also available for the pyranobenzopyran dibromoeriostoic acid (**157**) ⟨67AX(B)120⟩.

The steric interactions in the region of the central double bond in polymorphic dixanthylenes (**158**) may be relieved by deviations from coplanarity. Two situations can be envisaged, involving either folding of the molecule or rotation about the double bond. X-Ray studies indicate that in the ground state the molecule adopts a folded geometry in which the central rings are boat shaped and the two tricyclic halves are folded in opposite directions ⟨63JCS308⟩. However, in the coloured form of dixanthylene, it is considered that the two halves are planar, but are twisted about the central double bond ⟨73JA6177, 76JCS(P2)438⟩. $^1$H NMR studies confirm the folded nature of 2,2′-disubstituted dixanthylenes in solution, but the low energy barrier for thermal *E,Z* isomerization is indicative of a high energy content of the folded conformation ⟨79JA665⟩.

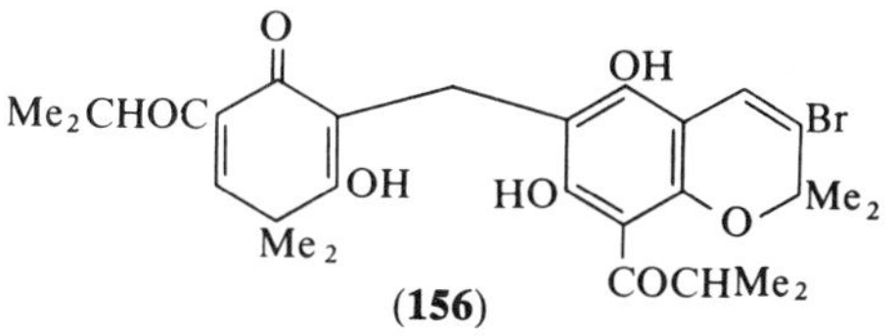

**(156)** 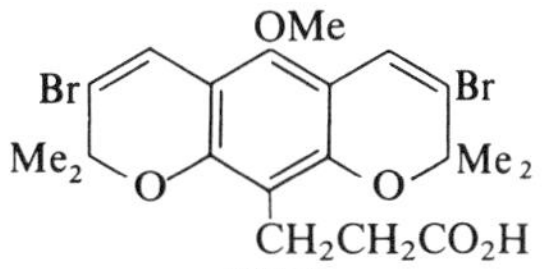          **(157)** 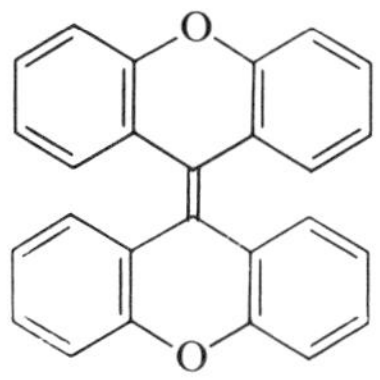          **(158)**

Interest in reduced pyrans centres around tetrahydropyrans and chromans. Many structural determinations have been reported for pyranoses and various derivatives ⟨72PMH(5)1⟩.

X-Ray investigations of a series of 2-aryloxytetrahydropyrans in which the substituent is axial revealed that the bond to the exocyclic oxygen atom is lengthened according to the electronegativity of that oxygen. A concomitant decrease in the endocyclic C—O bond is observed ⟨79AX(B)242⟩. These variations in bond lengths are related to the rates of hydrolysis of the acetal to the tetrahydropyranol, depending linearly on the p$K_a$ of the leaving group ⟨79CC288⟩.

The molecular structure of gaseous tetrahydropyran has been determined by electron diffraction; the dimensions are presented in Figure 7 ⟨79ACS(A)225⟩. The molecule exists in the chair form with $C_s$ symmetry consistent with the conclusions based on NMR, rotational and vibrational spectra. The torsional angles suggest that the heteroatom causes no flattening of the ring relative to that of cyclohexane.

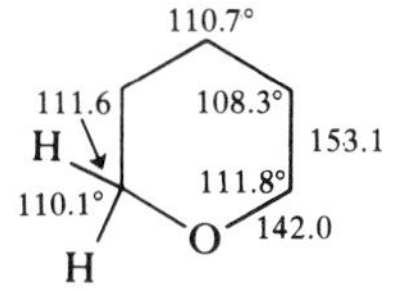

**Figure 7** Bond lengths (pm) and bond angles (°) for tetrahydropyran

Although no crystallographic studies of the tocopherols have been made, molecular dimensions have been calculated for both 2,2,5,7,8-pentamethylchroman-6-ol and α-tocopherol ⟨81HC(36)63⟩. These results are based on the agreement between the X-ray data for 4-(*p*-hydroxyphenyl)-2,2,4-trimethylchroman (Dianin's compound, **159**) ⟨70JA3749⟩ and the computer-produced values for a strain-free model.

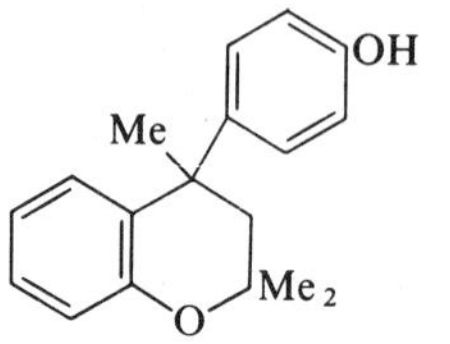

**(159)** 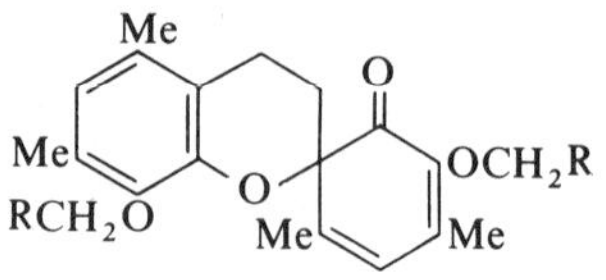          **(160)**

The conformation and dimensions of the substituted spirocyclic chromans (**160**) have been ascertained. The heterocyclic ring adopts a C-2, C-3 half-chair form, whilst the spiro-annelated cyclohexadienone ring is a C-1′ envelope ⟨80TL4973, 81AX(B)1620, 82AX(B)1001⟩.

The structure of 8-bromotetra-*O*-methyl-(+)-catechin (**161**) has been determined. The conformation of the heterocyclic ring in the crystal is between a C-2, C-3 half-chair and a C-2 sofa arrangement. The 2*R*,7*R* absolute configuration of (+)-catechin is confirmed ⟨78CC695⟩.

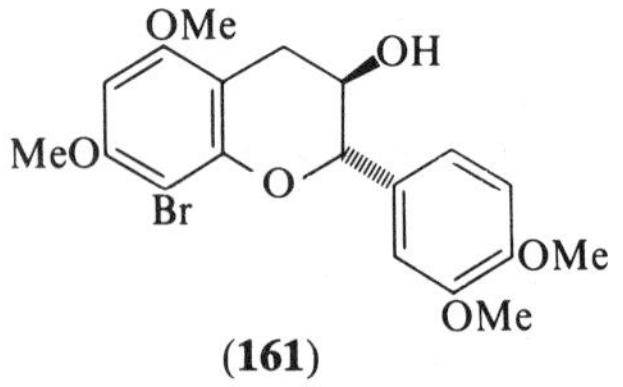

**(161)** 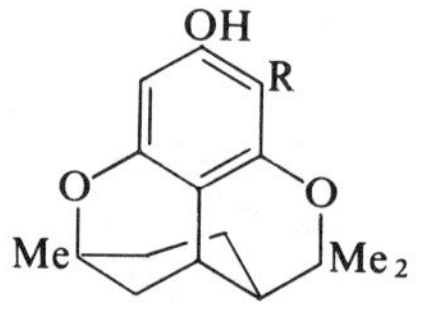          **(162)** 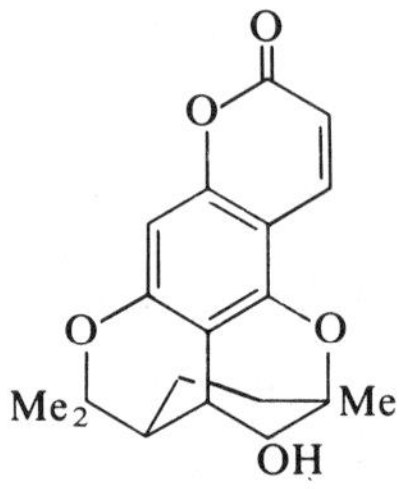          **(163)**

X-Ray studies indicate that in the strained molecules acetylcitran (**162**; R = COMe) and formylcitran (**162**; R = CHO), the aromatic ring shows considerable deviation from planarity and assumes a shallow boat form ⟨77JCS(P1)2393⟩. The isoprenylated coumarin bruceol (**163**) also contains the citran unit and crystallographic data are available both for it and deoxy-bruceol ⟨77JCS(P1)2402⟩.

Of the simple pyran-4-ones whose structures have been studied by X-ray crystallography, detailed molecular dimensions are available for 2,6-dimethylpyran-4-one ⟨65ACS217⟩ and the corresponding thione ⟨56BSB213⟩. The pyranone ring of 2,6-bis(diethylamino)-3,5-diphenylpyran-4-one is slightly non-planar and the substituents themselves show large deviations from coplanarity with the ring ⟨76AX(B)915⟩.

Molecules of 2-hydroxymethyl-5-phenacyloxypyran-4-one (phenacyl kojate, pak, **164**) are linked through a hydrogen bond between the hydroxymethyl and carbonyl groups ⟨76CJC2723⟩. Pak forms complexes with alkali metal halides in one of two stoichiometries, MX(pak)$_2$ or MX(pak), depending on the sum of the radii of the anion and cation. Crystal structures of these complexes have been reported ⟨75JCS(D)1071⟩.

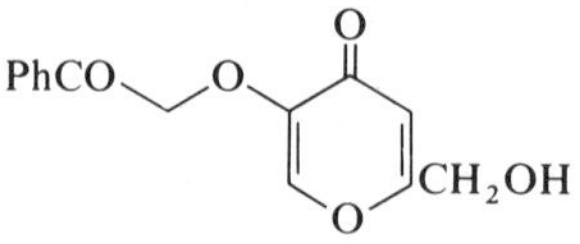

**(164)**

Several groups have investigated the structure of coumarin ⟨73K720, 74AX(B)1351, 76MI22201⟩ and the bond lengths and angles are depicted in Figure 8. The orthorhombic crystals consist of nearly planar coumarin molecules held together mainly by van der Waals forces.

Both direct and heavy-atom methods have been used to determine structures of substituted coumarins. Examples include 4-hydroxycoumarin and its 3-bromo derivative ⟨65AX927, 66AX646⟩ and the 7,8- and 6,7-dihydroxycoumarins, daphnetin and esculetin, respectively ⟨76AX(B)946, 77AX(B)283⟩. In the last compounds, hydrogen bonds link the molecules in the crystal. The stacking distances are comparable to those for other substituted coumarins ⟨75AX(B)1287⟩. A study of the structure of 7-hydroxy-4-methylcoumarin contains

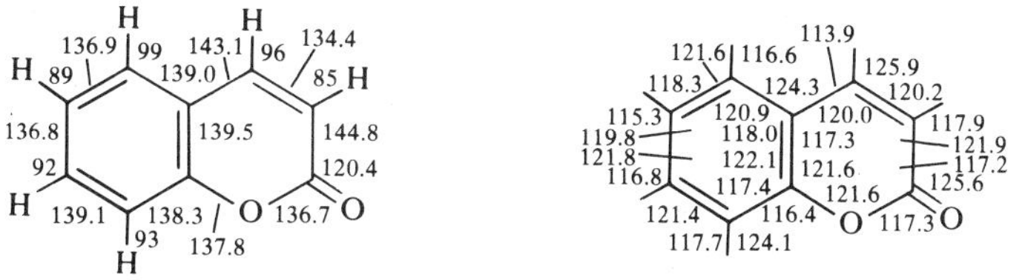

**Figure 8** Bond lengths (pm) and bond angles (°) for coumarin.

a useful compilation of the molecular geometry of the pyranone ring in a variety of coumarins ⟨75AX(B)1287⟩. Crystal data are available on pyranocoumarins ⟨70AX(B)2022⟩ and furocoumarins ⟨72AX(B)2485⟩.

Both racemic and enantiomeric warfarin (**165**; R = Ph) are hemiketals in the crystal, with the hydroxy group *trans* to the phenyl ring ⟨75AX(B)954, 76MI22201⟩. Crystal structures for related pyranobenzopyranones (**165**; R = Me) have been determined ⟨79JOC798⟩.

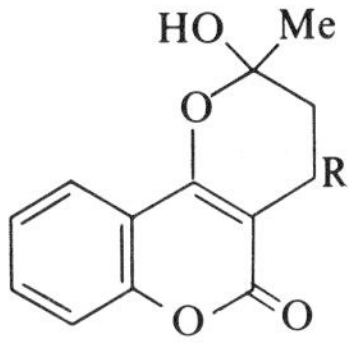

(**165**)

Bond angles and lengths for 3-bromoflavone have been determined ⟨82AX(B)983⟩. Whilst the two rings of the chromone system are essentially coplanar, the 2-phenyl group is twisted by 45.9° relative to the rest of the molecule; apparently the steric interaction between the bromine atom at C-3 and H-2′ outweighs the resonance stabilization which would be associated with a completely planar structure.

7,8-Benzoflavone (**166**) is non-planar with a torsion angle of 23° between the phenyl group and the remainder of the molecule, in keeping with the proximity of H-3 and H-2′ ⟨80MI22202⟩. However, 5,6-benzoflavone (**167**) is approximately planar despite similar steric features. It is postulated that interaction between the carbonyl group and H-10 increases the contribution of the canonical form (**168**), which leads to a more planar structure than would otherwise be expected.

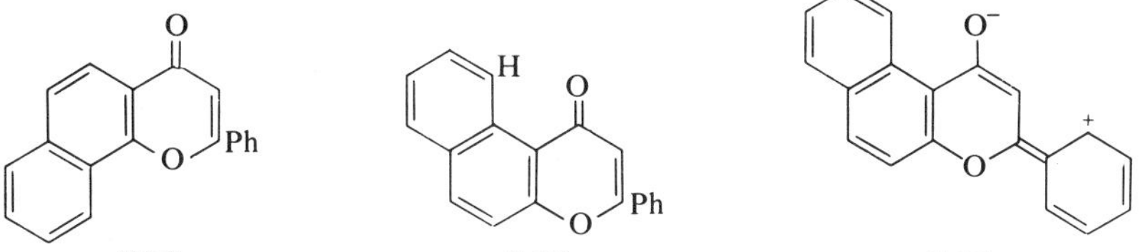

(**166**)          (**167**)          (**168**)

In the bischromone (**169**), which is valuable in the treatment of asthma, the two chromone moieties are inclined at about 53° to each other. Furthermore, the carboxyl groups lie in the plane of the chromone rings, whilst the central hydroxyl group is directed away from the chromone ring planes ⟨71JPS1458, 74MI22204⟩.

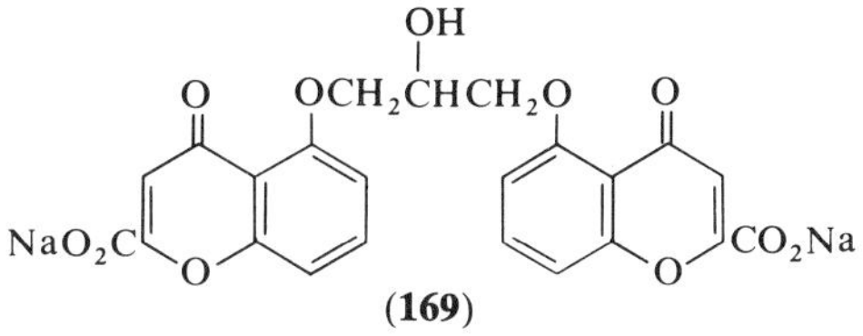

(**169**)

Rubrafusarin (**170**), a colouring component of the fungus *Fusarium culmorum*, was shown to be a linear naphthopyranone rather than a xanthone by X-ray crystallography and various spectroscopic techniques ⟨62AX451, 62AX1060⟩.

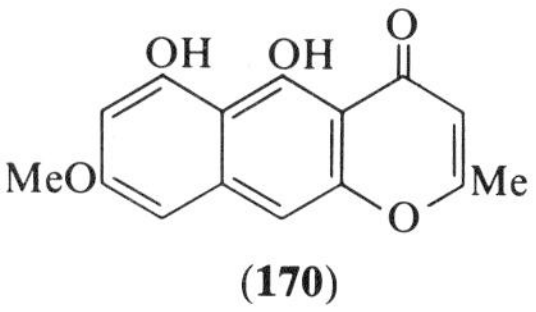

(**170**)

The molecular dimensions for 2-hydroxy-1,3,4,7-tetramethoxyxanthone have been reported ⟨69T1975⟩.

Crystallographic determinations on reduced pyranones include those on iridomyrmecin (**171**) ⟨64AX472⟩ and the pyran-3-one (**172**) ⟨82JOC3054⟩.

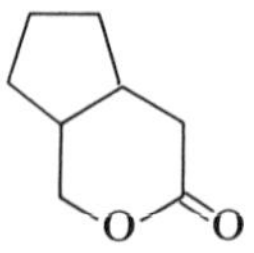

**(171)** 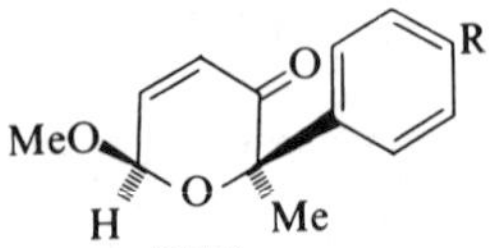          **(172)**

The phenyl ring plane in *cis*-3-bromoflavanone is at 42.2° with respect to the plane of the fused benzenoid ring. The pyranone ring adopts a 1,2-diplanar ('sofa') conformation with the phenyl group equatorial and the bromine atom axial ⟨82AX(B)981⟩. In the crystal of 4′-bromoflavanone the 4-bromophenyl ring is equatorially disposed and is rotated so that its plane is almost perpendicular to the mean plane of the rest of the molecule. The pyranone ring is in the half-chair conformation ⟨74AX(B)154⟩.

The crystal structure of (*S*)-(−)-6-bromo-5,7-dihydroxy-8-methyl-2-phenyl-2,3-dihydro-4*H*-1-benzopyran-4-one (**173**) has been determined ⟨82AJC1851⟩ and is consistent with the *S*-configuration proposed for (−)-cryptostrobin ⟨62MI22201⟩.

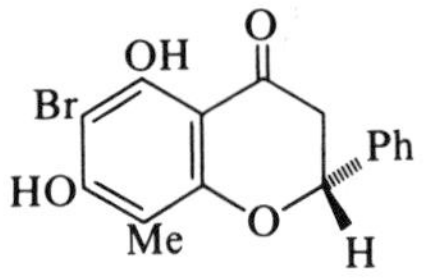

**(173)**

The rings in the reduced chromanone (**174**) are *cis*-fused and the pyranone ring is chair-shaped. The axial 2-hydroxy group is intermolecularly hydrogen bonded to the carbonyl group ⟨81JCS(P1)1096⟩.

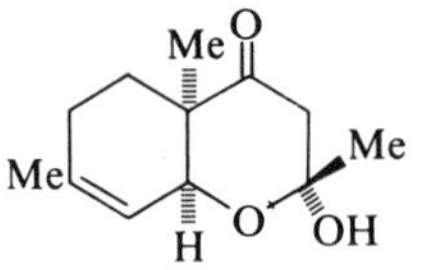

**(174)** 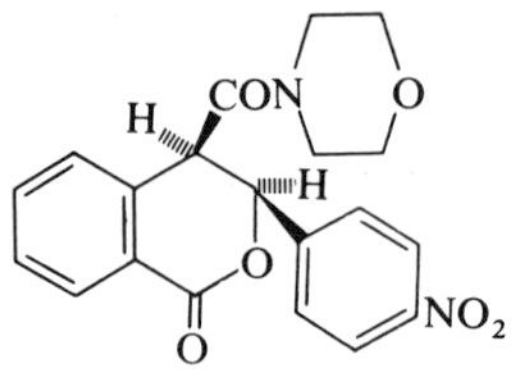          **(175)**

X-Ray diffraction shows that the *p*-nitrophenyl group is equatorial and the morpholinocarbonyl moiety axial in the isochroman-1-one *cis*-3,4-dihydro-4-morpholinocarbonyl-3-*p*-nitrophenyl-1*H*-2-benzopyran-1-one (**175**) ⟨78JCS(P1)1351⟩. The pyranone ring has a distorted twist conformation.

Structural details are also available for a number of reduced pyranopyranones.

A revised structure for the secoiridoid xylomollin (**176**) has been proposed on the basis of ¹H coupling constant data and the revised *trans*-fused arrangement was confirmed by crystallographic analysis ⟨78JA7079⟩. Both rings are in a chair conformation and the bridge-head hydrogen atoms are disposed *trans* to each other.

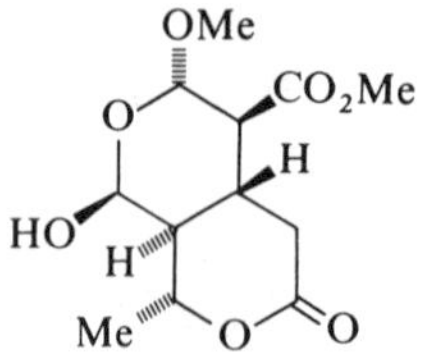

**(176)**

X-Ray studies of 8′-bromorotenone indicate that (−)-rotenone (**177**) adopts a bent geometry ⟨75CC850⟩ and confirm the stereochemistry as 6a*S*,12a*S*,5′*R* ⟨61JCS2843⟩.

The stereochemistries of papuanic and isopapuanic acids have been assigned on the basis of the structure and configuration of the pyranopyrandione (**178**) ⟨68JOC4191⟩.

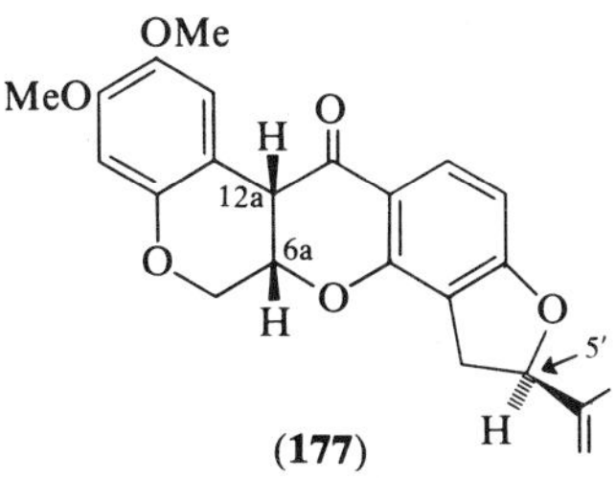

(177) 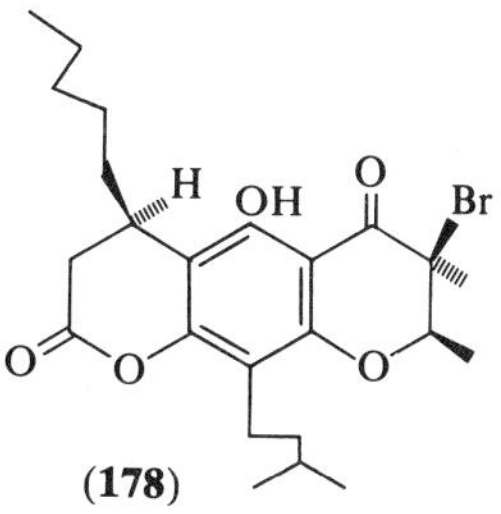                    (178)

Crystallographic data on pyrylium salts are not extensive. The 3-acetyl-2,4,6-trimethyl-pyrylium ion is planar and the dimensions are given in Figure 9 ⟨74CC614⟩. Similar geometry for the pyrylium ring of 6-(2-hydroxyprop-1-enyl)-2,4-dimethylpyrylium chloride has been reported ⟨75CC284⟩. However, the phenyl rings at C-2, C-4 and C-6 are tilted by 10.4°, 18.0° and 2.3°, respectively, from the plane of the heterocyclic ring in 2,4,6-triphenyl-pyrylium 1,1,3,3-tetracyanopropenide ⟨74BCJ832⟩. A crystal structure determination on 2,4,6-triphenylpyrylium tetrachloroferrate has also been performed ⟨76MI22203⟩.

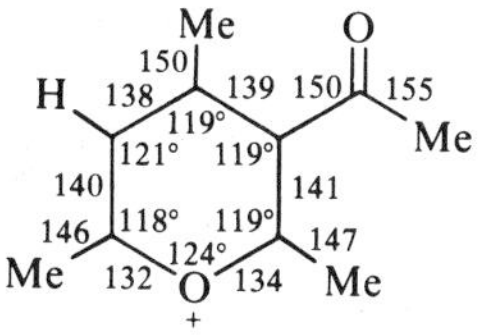

**Figure 9** Bond lengths (pm) and bond angle (°) for the 3-acetyl-2,4,6-trimethylpyrylium cation

The two features of interest in the structure of 6-(2-hydroxyprop-1-enyl)-2,4-dimethyl-pyrylium cation (**179**) are the enolic nature of the 6-substituent and the shortness of the C(6)—C(1′) bond. The latter is considered to indicate a significant contribution from the methylenepyran structure (**180**) ⟨75CC284⟩.

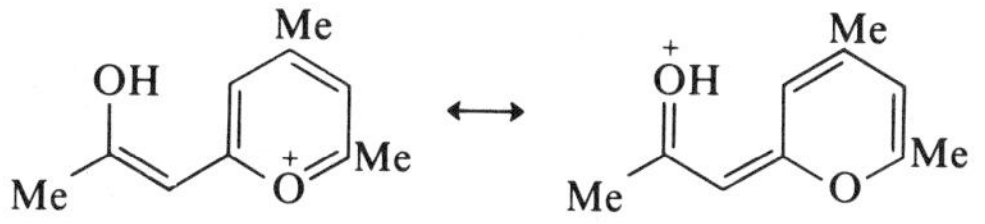

(179)                    (180)

X-Ray data for the dibenzoxanthenium salt (**181**) indicate that, whilst each individual ring is planar, the molecule as a whole is distorted from coplanarity ⟨73CSC91⟩.

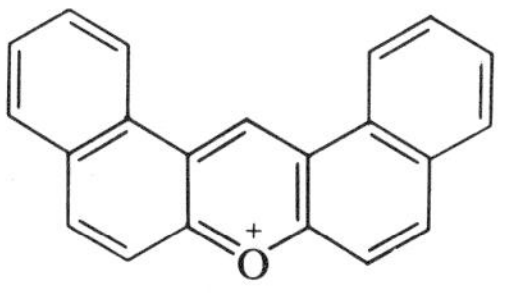

(181)

### 2.22.7.2 Microwave Spectroscopy

It has been concluded from a study of the microwave spectrum of 5,6-dihydro-2*H*-pyran that the ring skeleton is twisted such that the heteroatom and C-2 are on opposite sides of the plane formed by the other four carbon atoms ⟨74JCP(60)3987⟩. Interatomic distances and bond angles are given in Figure 10.

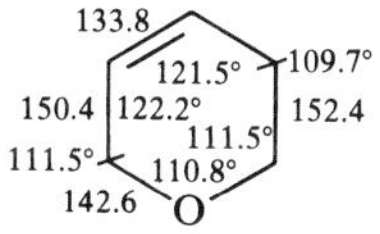

**Figure 10** Bond lengths (pm) and bond angles (°) for 5,6-dihydro-2*H*-pyran

The microwave spectra of 2-aminomethyltetrahydropyran (**182**; X = NH₂) and tetrahy-dropyran-2-ylmethanol (**182**; X = OH) indicate that both molecules exist predominantly in

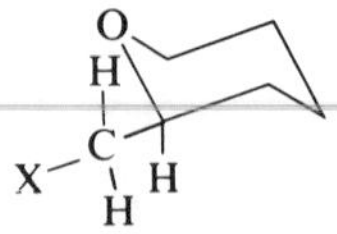

**(182)**

a chair form in which the equatorially disposed substituent adopts a synclinal conformation which allows internal hydrogen bonding ⟨79JA4499⟩.

Microwave spectra have been reported for both pyran-2-one and pyran-4-one ⟨73JA2766⟩. The structure of the latter has been corrected in a study of pyran-4-one and various sulfur analogues ⟨81JCS(F2)79⟩. The small negative values of the inertial defects indicate that pyran-4-one and pyran-4-thione are planar molecules. It is of interest to note that replacement of the exocyclic oxygen atom by sulfur results in a decrease in the C(3)—C(4) and C(2)—O(1) bond lengths, but lengthens the C(2)—C(3) double bond. The bonding in the thione is therefore more delocalized than in pyran-4-one.

The molecular dimensions of these pyranones are given in Figure 11.

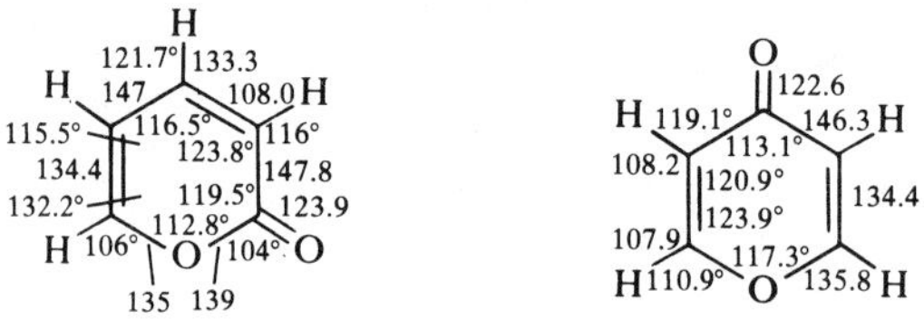

**Figure 11**   Bond lengths (pm) and bond angles (°) for pyran-2-one and pyran-4-one

### 2.22.7.3 Dipole Moments

Whilst the measurement of dipole moments is worthwhile in its own right, it is particularly valuable for the estimation of interactions between component parts of a molecule and for the determination of the stereochemistry of compounds. Useful surveys of the techniques of determination of dipole moments and their value in organic chemistry are available ⟨63PMH(1)189, B-75MI22205⟩. A compilation of dipole moments of oxygen heterocycles is presented in Table 8.

**Table 8**   Dipole Moments of Six-membered Oxygen Heterocycles

| Compound | Solvent | Temperature (°C) | Dipole moment (D) | Ref. |
|---|---|---|---|---|
| 3,4-Dihydro-2*H*-pyran | $C_6H_6$ | 20 | 1.39 | 68LA(712)201 |
| Tetrahydropyran | $C_6H_6$ | 25 | 1.55 | 59JCS3521 |
|  | $C_6H_{12}$ | 25 | 1.44 | 71PMH(4)13 |
| 3-Bromotetrahydropyran | $C_6H_6$ | 25 | 2.25 | 80JPR429 |
| 3-Chlorotetrahydropyran | $C_6H_6$ | 25 | 2.02 | 80JPR429 |
| 4-Bromotetrahydropyran | $C_6H_6$ | 25 | 1.64 | 80JPR429 |
| 4-Chlorotetrahydropyran | $C_6H_6$ | 25 | 1.59 | 80JPR429 |
| 2-Methoxytetrahydropyran | $C_6H_6$ | 25 | 1.23 | 69T3365 |
| 2-Ethoxytetrahydropyran | $C_6H_6$ | 25 | 1.24 | 69T3365 |
| 2-*n*-Propoxytetrahydropyran | $C_6H_6$ | 25 | 1.22 | 69T3365 |
| 2-*i*-Propoxytetrahydropyran | $C_6H_6$ | 25 | 1.40 | 69T3365 |
| 2-*n*-Butoxytetrahydropyran | $C_6H_6$ | 25 | 1.28 | 69T3365 |
| 2-*t*-Butoxytetrahydropyran | $C_6H_6$ | 25 | 1.51 | 69T3365 |
| 4,6-Dimethylpyran-2-one | Dioxane | 25 | 5.6 | 76JHC609 |
| 3-Acetyl-4-hydroxy-6-methylpyran-2-one | $C_6H_6$ | 25 | 2.83 | 37JCS1088 |
| Pyran-4-one | $C_6H_6$ | 20 | 3.73 | 52AC(R)673 |
| 2,6-Dimethylpyran-4-one | $C_6H_6$ | 25 | 4.65 | 37JCS1088 |
|  | $C_6H_6$ | 20 | 4.05 | 33JCS87 |
|  | Dioxane | 25 | 4.9 | 76JHC609 |
|  | — | 20 | 4.48 | 36PIA(A)(4)687 |
| 3,5-Diacetyl-2,6-dimethylpyran-4-one | $C_6H_6$ | 25 | 4.06 | 37JCS1088 |
| Pyran-4-thione | $C_6H_6$ | 20 | 4.08 | 52AC(R)673 |
| 2,6-Diphenylpyran-4-one | $C_6H_6$ | 20 | 3.82 | 33JCS87 |
|  | $C_6H_6$ | 20 | 4.74 | 54AC(R)430 |

**Table 8** *continued*

| Compound | Solvent | Temperature (°C) | Dipole moment (D) | Ref. |
|---|---|---|---|---|
| 3,5-Dimethyl-2,6-diphenyltetrahydropyran-4-one | $C_6H_6$ | 25 | 1.80 | 37JCS1088 |
| Coumarin | $C_6H_6$ | 25 | 4.48 | 37JCS1088 |
| | | 20 | 4.51 | 36PIA(A)(4)687 |
| 3-Phenylcoumarin | $C_6H_6$ | 25 | 4.30 | 37JCS1088 |
| Flavone | $C_6H_6$ | — | 4.07 | 70MI22202 |
| Isoflavone | $C_6H_6$ | — | 4.24 | B-77MI22200 |
| Xanthene | $C_6H_6$ | 30 | 1.43 | 72MI22205 |
| | $C_6H_6$ | 28 | 1.28 | 36MI22200 |
| | $CCl_4$ | 25 | 1.09 | 69JCS(B)980 |
| | $C_6H_6$ | 25 | 1.14 | 69JCS(B)980 |
| 2,7-Dibromoxanthene | Decalin | 30 | 0 | 72MI22205 |
| Xanthone | $CCl_4$ | 25 | 2.93 | 71JCS(B)82 |
| | $C_6H_6$ | 25 | 3.11 | 37JCS196 |
| | $C_6H_6$ | 25 | 2.95 | 71JCS(B)82 |
| 2,7-Dibromoxanthone | $C_6H_6$ | 30 | 4.18 | 72MI22205 |
| | $C_6H_6$ | 25 | 4.10 | 37JCS196 |
| 2,4-Dinitroxanthone | $C_6H_6$ | 25 | 2.98 | 37JCS196 |
| 2,7-Dinitroxanthone | $C_6H_6$ | 25 | 5.72 | 37JCS196 |

Disparity between the observed moment and that calculated by vector addition of the individual group moments may indicate that the actual structure is not truly represented by the simple molecular diagram. There are several instances of the application of this technique to structural investigations in the field of *O*-heterocyclic chemistry, not the least of which is the question of the aromaticity of pyranones and their benzologues.

The dipole moments of pyran-4-one and the corresponding thione have been determined from Stark effects ⟨81JCS(F2)79⟩ and the values compare favourably with those obtained from dielectric measurements ⟨52AC(R)673⟩. Their magnitudes (*ca.* 4 D) are higher than calculated, suggesting some contribution from polar canonical forms to the actual structure of the molecule. The implication to the aromaticity of pyranones is discussed in more detail in Section 2.22.8.1.

The observed dipole moments of a number of pyranopyrandiones also differ from the values calculated by vector addition. However, CNDO/2 calculated moments show a much closer agreement with the experimental values, as indeed does the value for pyran-4-one itself ⟨76JHC609⟩.

The dipole moment of xanthone (**21**) (2.95 D) is considerably higher than that predicted by vector addition (1.85 D) of the moments of benzophenone (3.0 D) and diphenyl ether (1.15 D). Whilst this may be attributed to contributions from dipolar structures such as (**183**), it was concluded that there is but a small contribution from such species ⟨37JCS196⟩.

(**183**)

Whereas the Kerr constant for xanthone suggests that the molecule is virtually planar in solution ⟨71JCS(B)82⟩, it is inferred from a study of the dipole moments of the parent ketone and its 2,7-dibromo derivative that these molecules are not planar ⟨72MI22205⟩.

In the case of xanthene (**8**), the moment (1.14 D) is in agreement with the value of 1.15 D for diphenyl ether. On the basis of the Kerr effect, it was postulated that in solution the preferred conformation of xanthene is a folded arrangement in which the dihedral angle between the two aromatic planes is $160 \pm 6°$ ⟨69JCS(B)980⟩. However, the zero moment observed for 2,7-dibromoxanthene is considered to be supportive evidence for a planar xanthene molecule, arising from the equal but opposite directional properties of the resultant moment (1.5 D) of the two C—Br bonds and the moment of xanthene ⟨72MI22205⟩. It should be noted that the latter work is based on a moment of 1.43 D for xanthene, which is substantially higher than other reported values.

A dipole moment study of the conformational equilibrium in some halogenotetrahydropyrans has been carried out ⟨80JPR429⟩. Comparison of the observed moments with those calculated by CNDO/2 and by vector addition for the pure conformers indicated that a halogen in the 3-position is preferably equatorially disposed. On the other hand, an axial arrangement is favoured by a 4-halogen atom.

The greater stability of the axial conformer of 2-alkoxytetrahydropyrans, ascribed to the anomeric effect, has been confirmed by a combined NMR and dipole moment investigation ⟨69T3365⟩. Additionally, out of the six conformers which in principle can arise from rotational isomerism about the exocyclic C—OR bond, two were shown to predominate in the equilibrium mixture.

Dipole moments of flavone (**22**) and isoflavone (**64**) have been calculated as a function of the angle by which the phenyl ring is rotated out of the plane of the chromone moiety ⟨B-77MI22200⟩. Correlation of the observed and calculated moments occurs when the ring is turned through less than 30° in the case of flavone, but through approximately 50° for the other flavonoid, and it thus appears that neither molecule is planar.

### 2.22.7.4 Conformations

There is a considerable amount of information on the conformation of six-membered oxygen heterocycles and several texts include useful reviews ⟨B-71MI22200, B-77SH(2)78, B-80MI22203⟩.

#### 2.22.7.4.1 Reduced pyrans and their benzologues

The chair conformation has been established as the prevalent structure for tetrahydropyran by microwave spectroscopy; no rotational spectrum for the boat form was observed in the range −40 °C to +30 °C ⟨69CJC1289⟩. This conformational assignment is supported by NMR studies, which indicated that replacement of a methylene group in cyclohexane by an oxygen heteroatom does not significantly affect the conformational mobility ⟨67JCS(B)1203⟩. The vibrational spectra have been assigned on the basis of a chair conformation of $C_s$ symmetry ⟨67SA(A)391, 75SA(A)339⟩, and electron diffraction data confirm this structure for gaseous tetrahydropyran ⟨79ACS(A)225⟩.

The extent of deformation from the chair conformation of cyclohexane has been investigated by the $R$-value method ⟨67JA1836⟩. The $R$-value is based on vicinal coupling constants and is defined by the expression:

$$R = \frac{J_{trans}}{J_{cis}} = \frac{\frac{1}{2}(J_{aa} + J_{ee})}{\frac{1}{2}(J_{ae} + J_{ea})}$$

An $R$-value for tetrahydropyran of 1.91, obtained from averaged coupling at ambient temperature, was interpreted in terms of a flattening of the ring relative to cyclohexane ($R = 2.16$). However, the use of individual coupling constants derived from deuterium decoupled $^1$H NMR spectra of the deuterated tetrahydropyrans (**184**) and (**185**) (Table 1) leads to an $R$-value of 2.17 for tetrahydropyran ⟨76JOC1380⟩. The coupling constants show a stereodependence on the electronegativity effects of the oxygen, as a result of which the changes in $J_{cis}$ and $J_{trans}$ relative to cyclohexane cancel each other.

(**184**)       (**185**)

The free energy of activation for ring inversion in tetrahydropyran (Scheme 28) is 43.0 kJ mol$^{-1}$ at 212 K by the coalescence temperature method ⟨73JA4634⟩, virtually identical with that for cyclohexane (42.3 kJ mol$^{-1}$) ⟨67JA760⟩. Similar values have been obtained using alternative methods ⟨66JOC3429, 67JCS(B)1203, 67JA5921⟩. It thus appears that the introduction of the heteroatom has little effect on the conformational mobility of cyclohexane. On the basis of a torsional barrier of 10.5 kJ mol$^{-1}$ for dimethyl ether ⟨73JA4634⟩, it is suggested that the transition state for inversion is the half-chair conformation (**186**).

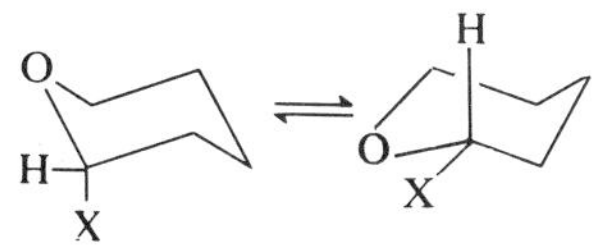

**Scheme 28**

**(186)**

Substituents in the 3- and 4-positions of tetrahydropyran are subject to the normal interactions encountered in cyclohexanes, although 1,3-interactions now involve a lone pair of electrons on the heteroatom.

However a substituent adjacent to the oxygen atom may exhibit a preference for axial stereochemistry, providing a manifestation of the anomeric effect ⟨B-63MI22201⟩. Such behaviour is particularly pronounced with electron-rich substituents such as alkoxy, hydroxy and halogens. The position of the equilibrium (Scheme 28) not only varies with the nature of the substituent, but is also affected by the presence and relative stereochemistry of other substituents and by the polarity of the solvent. By virtue of its greater dipole moment, the equatorial isomer is stabilized relative to the axial in solvents of high dielectric constant ⟨55CI(L)1102⟩ and the anomeric effect is consequently greater in non-polar solvents.

NMR spectroscopy has been successfully applied to the determination of conformational preferences of various 2-alkoxytetrahydropyrans ⟨68JOC2572⟩, deuterated 2-methoxytetrahydropyran ⟨69CJC4427⟩ and 2-chlorotetrahydropyran ⟨66JOC544⟩. Low temperature $^{13}$C NMR studies have provided information on the conformational energies of non-polar substituents, such as 2-alkyl, 2-alkenyl, 2-alkynyl and 2-methoxycarbonyl ⟨82JA3635⟩, whilst information is also available for 2,6-diphenyltetrahydropyrans ⟨75M229, 80T3565⟩. Configurational and conformational assignments of 2-alkoxytetrahydropyrans have been made on the basis of $^{17}$O chemical shifts and this method appears to be especially attractive ⟨81JOC4948⟩.

A combination of NMR and dipole moment measurements has proved useful for 2-alkoxytetrahydropyrans ⟨69T3365, 76JA6477⟩, 2-chlorotetrahydropyran ⟨69JCS(B)855⟩, 2-aminotetrahydropyran ⟨82JCS(P2)249⟩ and 2-dialkylamino-3-(2-tetrahydropyranyl)tetrahydropyrans ⟨81RRC253⟩.

A $^{35}$Cl nuclear quadrupole resonance study of 2,2,3,3-tetrachlorotetrahydropyran indicated an anomeric shift between the chlorine atoms in the 2-position. The lower frequency was assigned to the axially oriented halogen ⟨77CC19⟩.

Quantitatively, the anomeric effect is defined as the difference in conformational free energy for the process shown in Scheme 28 and the corresponding process in cyclohexane. Conformational energies for a range of substituents are available ⟨82JA3635⟩.

A discussion of the anomeric effect is not appropriate here, especially in view of the various reviews ⟨67QR364, 69AG(E)157, 72AG(E)739, B-79MI22201⟩ and papers ⟨71JCS(B)136, 73JST(16)357, 74T1717, 75JA4056, 78JA373, 81JOC4948, 82JA3635⟩ which relate to the topic. However, it is worth noting that the current interpretation of the anomeric effect is based on dipole–dipole interactions, gauche effects, and overlap of the oxygen *p*-type lone pair with adjacent antibonding orbitals ⟨69MI22203⟩.

Analysis of the far IR-spectra of 3,4-dihydro-2*H*-pyran (**13**) ⟨72JCP(57)2572⟩ and 5,6-dihydro-2*H*-pyran (**14**) ⟨81JST(71)97⟩ indicates that for both molecules the most stable conformation is a half-chair form. The barrier to planarity is greater for the former compound. These preferred structures are in accord with the half-chair conformation established for cyclohexene and its derivatives. The conformational mobility of cyclohexene is greater than that of the 3,4-dihydropyran. The increased stabilization of the pyran has been attributed to delocalization of the π-electrons of the alkenic carbon atoms and the oxygen lone-pairs ⟨69TL4713⟩.

Raman spectroscopy confirms that the dominant form of 3,4-dihydro-2*H*-pyran at room temperature is the half-chair ⟨74JCP(60)3098⟩ and an angle of twist of 23° is indicated. A pulsed IR laser study of the half-chair to boat interconversion of this dihydropyran has shown that provided at least 20 kJ mol$^{-1}$ is absorbed a transient UV absorption at 237 nm is observed which is assigned to the boat form ⟨80JA6407⟩.

An anomeric effect is observed in 3,4-dihydro-2*H*-pyrans. For example, a 2-alkoxy group preferentially occupies an axial position ⟨71DOK(196)367⟩. Indeed, a study of the NMR spectra of some 2-alkoxy-3,4-dihydro-2*H*-pyrans and their 4-methyl derivatives established that the anomeric effect was more important in the unsaturated heterocycles than in the corresponding tetrahydropyrans ⟨72BSF1077⟩. The axial preference of an alkoxy group is even more accentuated when the double bond is associated with a fused benzenoid ring, as in the 2-alkoxychromans. It is also of interest to note that the role of the polarity of the solvent on the conformational equilibrium is less important than for the saturated analogues.

A study of the coupling constants between the hydrogen atoms at C-2 and C-3 in *trans*-2-methoxy-4-methyl-3,4-dihydro-2*H*-pyran established that the conformer in which the alkoxy group is axial (**187**) predominates in the equilibrium mixture. However, the *cis* isomer exists preferentially as (**188**) despite the anomeric effect, since there is no 1,3-diaxial interaction between the 2-methoxy and 4-methyl groups ⟨72BSF1077, 78JOC667⟩.

(**187**)                    (**188**)

Little or no bias towards one particular conformer is shown by substituted 4-alkyl- or 4-aryl-5,6-dihydro-2*H*-pyrans except when the equatorial preference of a bulky substituent at C-6 dominates the situation ⟨80CHE571⟩.

An anomeric effect is also shown by the hydroxy group in the dihydropyran ring of the anticoagulant drug warfarin (**165**; R = Ph). The related *trans*-methylketal (**165**; R = Me) exists as the all-staggered half-chair conformation in which the alkoxy group is axial. However, the *cis* isomer interconverts between the two half-chair conformations ⟨79JOC798⟩.

Studies of the conformations of chromans include those on 3,4-dihalogenochromans (**189**; X = Br) and 3-halogenochroman-4-ols (**189**; X = OH) ⟨70JCS(C)1006⟩, which have been summarized ⟨81HC(36)169⟩ together with the related 2,2-dimethyl derivatives ⟨65JCS5049⟩.

(**189**)

Conformational free energies have been determined for similarly substituted flavans ⟨66T621⟩. Preferred conformations for some 3-chromanamines have been established ⟨73JCS(P2)227, 76MI22200⟩. Conformational assignments have been made to chroman-4-ols containing bulky substituents at C-3 ⟨73BCJ1839⟩ and to chroman-4-ol itself ⟨74BCJ2607⟩ and its 4-methyl derivative ⟨81JCS(P2)944⟩. In all cases, a half-chair conformation appears to be favoured and there is evidence for an anomeric effect in chromans ⟨72BSF696⟩.

It is of interest to note that the barrier to ring inversion in the 1,8-bridged naphthalene (**190**) (26.3 kJ mol$^{-1}$) is considerably lower than that for tetrahydropyran. This has been attributed to the fact that only the heteroatom is out of the plane imposed on the system by the naphthalene framework ⟨81JCS(P2)741⟩. The transition state for the inversion process is calculated to be planar (Scheme 29) and the barrier to inversion is considered to arise mainly from bond angle deformation.

(**190**)                    **Scheme 29**

The conformational equilibria for a number of *cis*- and *trans*-1-oxadecalins, which may be regarded as fully reduced chromans, have been evaluated from NMR data. The *trans*-fused system (**191**) is thermodynamically favoured ⟨74JOC2040⟩.

The preferred conformations of various reduced xanthenes have been reported ⟨69BSF2490, 71BSF3006, 71BSF3010⟩.

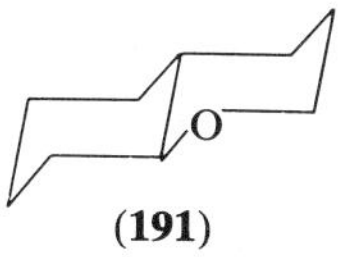

**(191)**

The 'Atlas of Stereochemistry' includes material on isochromans, flavanoids and rotenoids ⟨B-78MI22202⟩ and an extensive discussion of the stereochemistry of flavan-3,4-diols is available ⟨B-75MI22201⟩.

Configurations have been assigned to the four racemates of the flavan-3,4-diol diacetate (**192**) on the basis of their NMR spectra ⟨62JCS3858⟩. The heterocyclic ring is half-chair and the 2-aryl group is equatorial. A similar approach was used to investigate the relative configurations of 3,4-diphenylisochromans ⟨65JOC2035⟩, whilst the absolute configurations of all four stereoisomers of 7-methoxy-2,2-dimethylchroman-3,4-diol, a metabolite of precocene I, have been assigned from their NMR and circular dichroism spectra ⟨82TL1655, 82TL2693⟩. Analysis of the $^{13}$C spectra of chroman-4-ol and various methyl and phenyl derivatives indicates a preference for a pseudoaxial orientation of the hydroxy group ⟨77JCS(P1)217⟩ and confirms the configurational relationships established for the flavanols based on the $^{1}$H spectra ⟨63AJC107⟩.

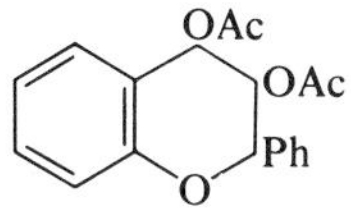

**(192)**

The CD spectra of *cis* and *trans* 4-substituted flavans are best explained by assuming a sofa conformation for the heterocyclic ring, with C-2 out of plane, rather than by a half-chair conformation. A similar conformation has been used in a discussion of the CD spectra of naturally occurring flavanols ⟨71T5459⟩.

In the CD spectrum of (+)-flavanone (**193**), the $n \rightarrow \pi^*$ band is negative proving its absolute configuration to be 2*R* ⟨73T909⟩. The majority of naturally occurring 3-hydroxy-2-phenylchroman-4-ones have the configuration 2*R*,3*R*. This area has been briefly discussed ⟨B-75MI22201⟩.

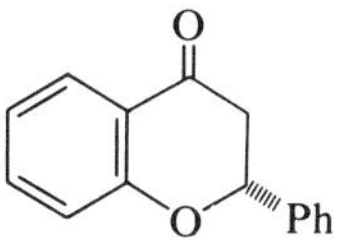

**(193)**

Half-chair conformations have been established for the *cis*- and *trans*-tetrahydropyran-2-ones on the basis of IR, NMR and CD data ⟨74JOC3890⟩. In the case of the *cis* isomer, conformation (**194**) is preferred in order to avoid the 1,3-diaxial interactions of the 3- and 5-methyl groups, whilst a somewhat flattened conformation (**195**) is predicted for the *trans* compound.

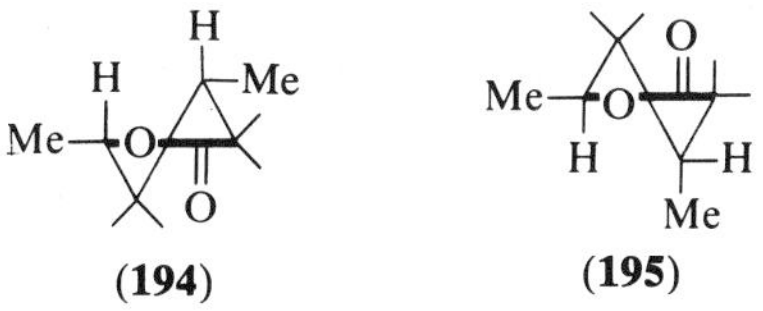

**(194)**      **(195)**

The conformation (**196**) is expected for *trans*-5,6-dimethylpyran-2-one, whilst the two conformations (**197**) and (**198**) probably contribute equally to the *cis* isomer ⟨82CJC29⟩.

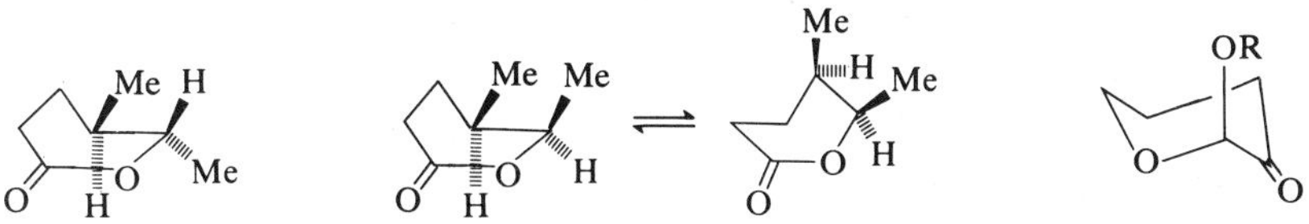

**(196)**      **(197)**      **(198)**      **(199)**

The very marked anomeric effect shown by 2-alkoxytetrahydropyran-3-ones (**199**) has been interpreted in terms of stabilizing orbital interactions ⟨77NJC79⟩. The influence of the heteroatom and the carbonyl group on the stabilization of the conformer in which the alkoxy group is axial is apparent.

A study of the influence of the heteroatom on the $^{13}$C shifts of some 1-hetera-2,6-diarylcyclohexan-4-ones includes a brief consideration of tetrahydropyran-4-ones ⟨79JOC471⟩.

The sofa or 1,2-diplanar conformation in which five adjacent atoms lie in one plane is predicted for dihydropyranones ⟨66MI22202⟩.

The conformational equilibria of a range of *cis* and *trans* 2-substituted 6-methoxy-3,6-dihydropyran-3-ones have been investigated, using the equatorial preference of a *t*-butyl group at C-2 as a basis for the interpretation of the NMR data ⟨79MI22200⟩. Thus, (200; R = CMe$_3$) and (201; R = CMe$_3$) are expected to be the exclusive conformations. In general, for the *trans* isomers, the anomeric effect at C-6 and the equatorial preference of the 2-substituent work towards the same end and conformation (201) is very predominant. However, steric interactions and conformational effects favour different conformations of the *cis* compound and consequently both conformers (200) and (202) are present in the equilibrium mixture.

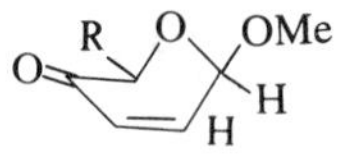

(200) 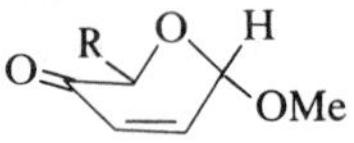          (201) 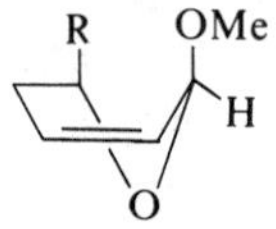          (202)

The two diastereoisomers of the dihydropyran-3-one (203) also exhibit different conformational preferences in solution ⟨82JOC3054⟩. Although the conformer with the aryl group pseudoaxially oriented is predominant for the *cis* isomer, both conformers contribute significantly to the conformational equilibrium associated with the *trans* isomer.

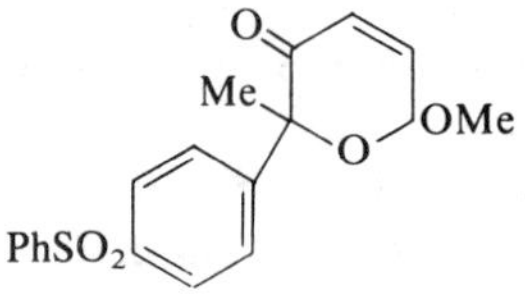

(203)

## 2.22.8 AROMATICITY

### 2.22.8.1 Pyranones and their Benzologues

#### 2.22.8.1.1 Pyran-2-one and coumarin

The pyran-2-one ring system (17) is a potentially aromatic species. The extent to which the system develops aromaticity is dependent upon the importance of contributions from the pyrylium-2-olate structure (101; Scheme 30).

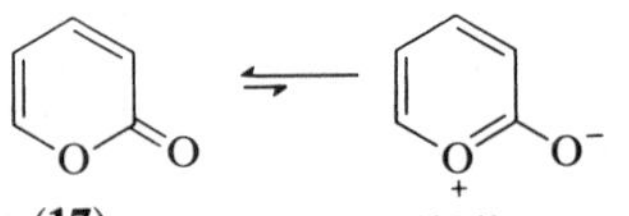

(17)                (101)

Scheme 30

From the viewpoint of chemical reactivity, pyran-2-ones behave as enol lactones, which suggests that these molecules do not possess significant aromatic character.

Although apparently undergoing electrophilic substitution, the reaction of pyran-2-one with bromine involves addition to the double bond and subsequent elimination of hydrogen bromide (Scheme 31) ⟨69JOC2239⟩. In general electrophilic substitution of pyran-2-ones is a difficult process and forcing conditions are usually required.

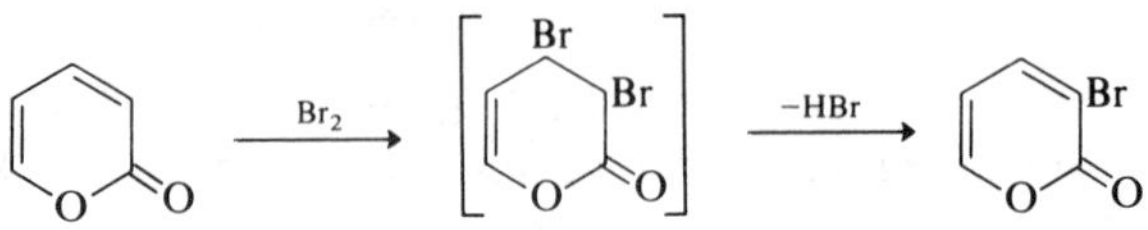

Scheme 31

The facile cleavage of the ring by nucleophiles (Scheme 32) is typical of a lactone rather than an aromatic system ⟨62CI(L)968, 62CI(L)1829⟩ and contrasts greatly with the difficulty of electrophilic substitution.

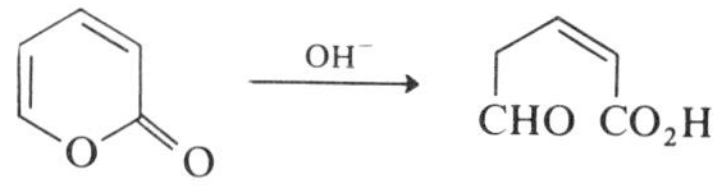

**Scheme 32**

Cycloadditions may be considered typical reactions of dienes. The diene-like nature of pyran-2-one is clearly illustrated by its reaction with maleic anhydride to give the adduct (**204**; Scheme 33) ⟨31LA(490)257⟩. Further details of these and other reactions are given in Chapter 2.23.

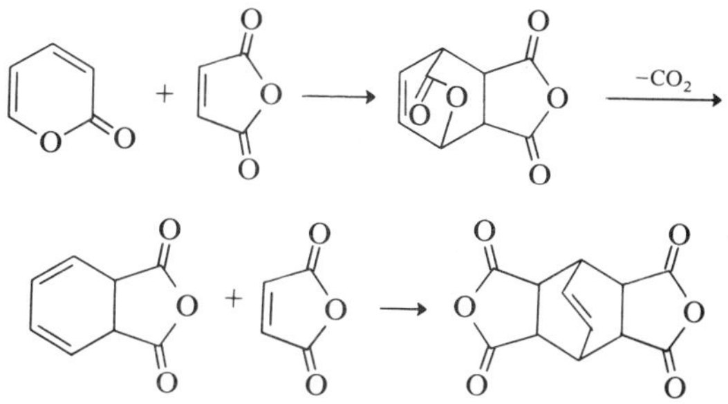

(**204**)

**Scheme 33**

The various methods available for assessing aromaticity have been reviewed ⟨B-75MI22206⟩. Those techniques which have been applied to heteroaromatic compounds are the subject of a separate review ⟨74AHC(17)255⟩, which includes a discussion of the evidence for aromaticity in pyrylium salts and 2- and 4-pyranones and their benzologues. There have been numerous theoretical studies on the pyranones. The MO delocalization energy (DE) of pyran-2-one has been calculated using a number of techniques of varying sophistication. An early MO–LCAO treatment ⟨62CCC1242⟩ indicated considerable aromaticity for both 2- and 4-pyranone; the DE values were calculated to be 2.896 $\beta$ and 2.868 $\beta$, respectively (*cf.* benzene 2 $\beta$). However, different parameters were used in each case, and thus comparison of the absolute values cannot be made. Further work also revealed that the orders of DE for 2- and 4-pyranone, calculated by the HMO method, are parameter dependent ⟨64T831⟩. Calculation of DE by means of an iterative procedure ($\omega, \beta$), gave significantly lower values which were considered to be more realistic than those calculated using the Wheland–Pauling method ⟨69CR(C)(269)298⟩. These results suggest that 2- and 4-pyranone are equally aromatic. However, later work based on the additivity of $\pi$-bond energies in non-aromatic model compounds concludes that neither pyranone is aromatic ⟨76JA2750⟩. The aromaticity indices ($A_1$), introduced by Julg and François ⟨67MI22205⟩ and which are unity for benzene and pyridine, are indicative of considerable aromatic character for both pyranones (Table 9).

**Table 9** Delocalization Energies ($\beta$), Specific Delocalization Energies ($\beta$) and Aromaticity Indices for Pyranones

|  | *Method* | *Pyran-2-one* | *Pyran-4-one* | *Ref.* |
|---|---|---|---|---|
| Delocalization energy DE | $\omega\beta$ | 1.595 | 1.598 | 69CR(C)(269)298 |
|  | Wheland–Pauling | 1.732 | 1.618 | 69CR(C)(269)298 |
| Specific delocalization energy DE$_{sp}$ | $\omega\beta$ | 0.199 | 0.200 | 69CR(C)(269)298 |
|  | Wheland–Pauling | 0.218 | 0.217 | 69CR(C)(269)298 |
|  | Hess and Shard | 0.031 | 0.042 | 76JA2750 |
|  | Aihara | 0.010 | 0.010 | 76JA2750 |
| Aromaticity indices $A_1$ | $\omega\beta$ | 0.79 | 0.72 | 69CR(C)(269)298, 67MI22205 |
|  | Wheland–Pauling | 0.96 | 0.94 | 69CR(C)(269)298 |

Determination of the values of $DE_{sp}$ for 2- and 4-pyranones and a number of their benzologues has been performed using three different MO techniques, namely a simple HMO method, the more sophisticated autocoherent HMO $\omega'\omega''\beta$ method, and one of Pople's semi-empirical MO methods ⟨74BSF538⟩. $DE_{sp}$ values for the simple pyranones calculated using these methods are given in Table 9. The results obtained from both the $\omega'\omega''\beta$ and Pople's method give good qualitative agreement for the order of $DE_{sp}$, whereas application of the HMO method produces a different relative order. From the HMO values, the magnitude of $DE_{sp}$ is shown to decrease in the order: tricyclic pyranones > bicyclic pyranones > 2- and 4-pyranone.

This trend is consistent with the intuitive idea that the aromaticity of a system (and hence the $DE_{sp}$) is augmented by the presence of fused benzene rings. However, the other two methods suggest that the order is tricyclic pyranones > 2- and 4-pyranone > bicyclic pyranones, which indicates that 2- and 4-pyranones possess greater $DE_{sp}$ values, and are therefore more aromatic, than their monobenzologues. The original authors suggest that since the $\omega'\omega''\beta$ and Pople's methods are much more elaborate than the simple HMO technique, the results from the former have greater credibility ⟨74BSF545⟩. Despite this it was stated that the problem is worthy of further consideration, and may be best approached theoretically by means of one of the methods which consider participation of all the valence electrons. In the same study, use was made of MO calculated bond lengths to determine Julg aromaticity indices, by an approach which was considerably refined over that reported previously. The results from all three MO methods indicated that pyran-2-one is considerably more aromatic than pyran-4-one. Thus from a theoretical viewpoint at least, it would appear that pyran-2-one possesses significant aromaticity.

Studies of molecular spectra (UV, IR and NMR) are potentially useful in providing information about aromaticity. Further information may be obtained from magnetic susceptibilities and dipole moment measurements. The UV spectra of pyran-2-one and some of its analogues have been reported ⟨71PMH(3)67⟩, and are consistent with the enol lactone structure (**17**).

The position of the carbonyl group absorption in the IR spectrum is dependent upon the C—O bond order and therefore should provide a means of establishing the importance of the betaine structure (**101**). The carbonyl stretching frequency in numerous pyran-2-ones occurs in the range 1720–1740 cm$^{-1}$, which is typical for lactones. The betaine structure (**101**) would absorb at a much lower frequency ⟨63PMH(2)252⟩. The carbonyl absorption maxima for pyran-2-one and some saturated 'non-aromatic' model compounds are given in Figure 12. MO calculations employing Extended Hückel Theory (EHT) indicated a high $\pi$-bond order for the carbonyl group in pyran-2-one and some coumarins. Carbonyl bond stretching frequencies calculated using MO derived bond orders are in good agreement with those obtained experimentally ⟨78IJC(A)64⟩; details are presented in Table 10. From IR studies it thus appears that pyran-2-ones do not possess betaine character.

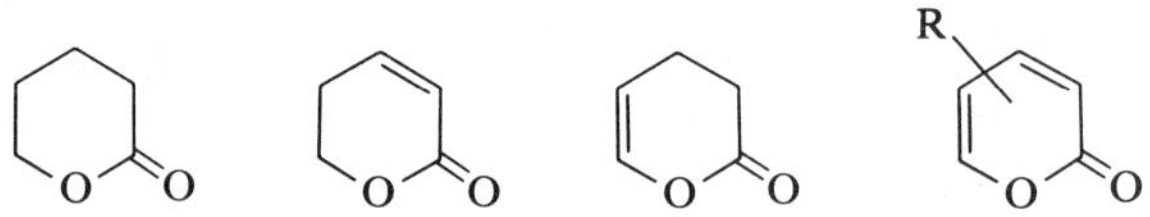

$\nu_{max}$ (cm$^{-1}$) (CCl$_4$)   1740–1730        1710        1775        1720–1740

**Figure 12**   IR carbonyl frequencies for some pyran-2-ones

**Table 10**   EHT Calculated Carbonyl $\pi$-Bond Orders ($P_{CO}$) and IR Stretching Frequencies ⟨78IJC(A)64⟩

| Compound | $P_{CO}$ | $\nu_{C=O}$ (cm$^{-1}$) Calc. | Obs. |
|---|---|---|---|
| Pyran-2-one | 0.927 | 1711 | 1721 |
| Coumarin | 0.935 | 1728 | 1725 |
| Isocoumarin | 0.943 | 1744 | 1736 |
| 3-Methylcoumarin | 0.932 | 1722 | 1730 |
| Pyran-4-one | 0.901 | 1658 | 1660 |
| Chromone | 0.908 | 1972 | 1665 |

The $^1$H NMR chemical shifts for pyran-2-one have been reported ⟨69JHC1⟩ and are given in Figure 4. There is no evidence of a significant diamagnetic ring current in this molecule, although the alternating pattern of chemical shifts suggests there may be some degree of charge delocalization. The upfield shifts induced by electron-releasing groups parallels the behaviour shown by classical aromatic systems. Despite this, there is evidence to suggest that there is considerable bond localization in pyran-2-ones. In 4-methylpyran-2-one (**205**), the four-bond coupling between H-3 and the 4-Me group, and the absence of a similar coupling between H-5 and 4-Me, indicates the presence of the 3,4-double bond. A direct relationship between the vicinal coupling constant and bond order has been established. Measurement of this coupling constant should provide an indication of the delocalization, and hence aromaticity, of the system. The method is not completely reliable, since similarities with non-aromatic model compounds have been observed. The $^{13}$C chemical shifts show the same alternating pattern observed with the $^1$H resonances; C-4 and C-6 absorb at lower field than C-3 and C-5 (Figure 6) ⟨74JOC1935⟩. Charge distribution is also indicated by the chemical shifts of methyl substituents. Thus, 4-methyl and 6-methyl carbons appear at 20 and 21 p.p.m. respectively, but the 5-methyl carbon resonates at 14 p.p.m. The $^{13}$C NMR of pyran-2-ones provides evidence of a high degree of double bond localization in these molecules. A particularly striking example is provided by 4-methoxy-6-methylpyran-2-one (**105**) in which the methoxy group causes an upfield shift of C-3 of 30 p.p.m., whereas C-5 shows an upfield shift of only 3 p.p.m. ⟨74JOC1935⟩.

$$^4J_{Me-H(3)} = 1\ Hz$$
$$^4J_{Me-H(5)} = 0\ Hz$$

(**205**)

Although single bond carbon-13 proton coupling constants ($^1J_{CH}$) show a dependency on the geometry of the ring, a value of 200 Hz is considered typical for a carbon attached to oxygen in a heteroaromatic system ⟨B-72MI22203⟩. Values for 2- and 4-pyranone and furan are presented in Table 11.

**Table 11** Single Bond $^{13}$C–$^1$H Coupling Constants for the Pyranones and Furan ⟨74JOC1935⟩

| Compound | C-2 | C-3 | $^1J(^{13}C-H)$ (Hz) C-4 | C-5 | C-6 |
|---|---|---|---|---|---|
| Pyran-2-one | — | 170 | 163 | 173 | 200 |
| Pyran-4-one | 200 | 175 | — | — | — |
| Furan | 201 | 175 | — | — | — |

The usefulness of magnetic susceptibility and diamagnetic susceptibility exaltation measurements as a means of assessing aromaticity has been outlined elsewhere ⟨68JA811, 74AHC(17)255, B-75MI22206⟩.

The molecular magnetic susceptibility anisotropies for the pyranones have been determined by microwave techniques ⟨71JA5591, 73JA2766⟩. Values for the parameter $\Delta\chi$, which represents the out-of-plane *minus* the average in-plane molecular magnetic susceptibilities, were obtained for benzene, furan, 2- and 4-pyranone, and tropone. $\Delta\chi$ may be separated into local and non-local contributions with the aid of known local group contributions for non-aromatic molecules. The results are presented in Table 12.

**Table 12** $\Delta\chi$ Values[a] for some Monocyclic Compounds ⟨73JA2766⟩

| Compound | $\Delta\chi_{expl.}$ | $\Delta\chi_{local}$ | $\Delta\chi_{non-local}$ |
|---|---|---|---|
| Benzene | −59.7 | −26.4 ± 2.4 | −33.3 ± 2.4 |
| Furan | −38.7 ± 0.5 | −15.6 ± 3.1 | −23.1 ± 3.6 |
| Pyran-2-one | −24.8 ± 1.3 | −26.5 ± 4.2 | — |
| Pyran-4-one | −22.9 ± 1.7 | −26.5 ± 4.2 | — |
| Tropone | −36.6 ± 4.5 | −37.3 ± 3.5 | — |

[a] Units of $10^{-13}\ mol\ G^2\ J^{-1}$.

Molecules which are aromatic and thus possess substantial $\pi$-electron delocalization exhibit large non-local contributions to their magnetic susceptibilities. By the absence of these contributions, 2- and 4-pyranones can be classed as non-aromatic.

Complete aromaticity in coumarin (**19**), isocoumarin (**28**) and 3*H*-2-benzopyran-3-one (**206**) can only be realized if the O—CO function contributes two electrons to form a $10\pi$-electron system, which would necessarily possess a betaine structure.

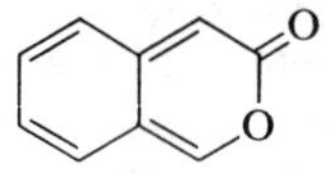

**(206)**

The available evidence suggests that the heterocyclic ring in compounds (**19**), (**28**) and (**206**) is not markedly aromatic. Much of the chemistry of coumarins can be rationalized in terms of a benzo-fused enol lactone structure.

The chemical reactivity of coumarin to a large extent resembles that of pyran-2-one. Electrophilic substitution occurs preferentially in the carbocyclic ring at C-6. Substitution at C-3 can also occur when more vigorous conditions are employed. Coumarin is readily attacked by nucleophiles, giving rise to a variety of ring-opened products.

The empirical resonance energies for the pyranone benzologues have not been determined.

Calculation of the specific delocalization energy for coumarin, isocoumarin and chromone (**20**) has been performed using various MO methods. Data are presented in Table 13.

**Table 13** Specific Delocalization Energies for Pyranones and some Benzologues ⟨74BSF545⟩

| Compound | Specific delocalization energy ($\beta$) | | |
| | HMO | $\omega'\omega''\beta$ | Pople |
| --- | --- | --- | --- |
| Pyran-2-one | 0.216 | 0.485 | 0.464 |
| Coumarin | 0.297 | 0.426 | 0.421 |
| Isocoumarin | 0.294 | 0.446 | 0.656 |
| Benzo[*c*]coumarin | 0.338 | 0.865 | 0.763 |
| Benzo[*h*]coumarin | 0.330 | 0.830 | 0.727 |
| Pyran-4-one | 0.210 | 0.477 | 0.428 |
| Chromone | 0.294 | 0.456 | 0.401 |
| Xanthone | 0.336 | 0.882 | 0.750 |
| Benzo[*g*]chromone | 0.324 | 0.873 | 0.708 |
| Benzo[*h*]chromone | 0.327 | 0.849 | 0.712 |

The topological resonance energies (TRE) of coumarin and some isomers have been obtained ⟨82JHC625⟩. TRE values and the corresponding resonance energies per electron, expressed as percentages with respect to benzene as a standard, are given in Table 14. The results indicate that coumarin and isocoumarin have comparable aromaticities, whilst that of chromone is somewhat higher. The low TRE of (**206**) indicates the lack of aromaticity in this system, in keeping with a classical *o*-quinonoid structure. The instability of this compound has precluded its synthesis ⟨70JCS(C)536⟩. However the 1,4-diphenyl analogue has been obtained, and its chemistry closely resembles that of other reactive *o*-quinonoids ⟨70JCS(C)530⟩.

**Table 14** Topological Resonance Energies (TRE) for some Benzopyranones ⟨82JHC625⟩

| Compound | TRE | TRE(PE)[a] |
| --- | --- | --- |
| Coumarin | 0.244 | 44.68 |
| Isocoumarin | 0.230 | 42.23 |
| Chromone | 0.256 | 46.90 |
| 3*H*-2-Benzopyran-3-one | 0.090 | 16.56 |

[a] TRE(PE) topological resonance energy per electron. The values are expressed as percentages of the value for benzene.

The IR spectra of coumarin and isocoumarin show carbonyl frequencies characteristic of lactones. There is no evidence to suggest that contributions from betaines (**207**) and (**208**) are significant. Moreover, EHT MO calculations predict high $\pi$-bond orders for the carbonyl groups in these molecules. IR data and C=O $\pi$-bond orders are presented in Table 10.

(**207**)     (**208**)

In the $^1$H NMR spectrum of coumarin, the H-3 and H-4 protons absorb at $\delta$ 6.45 and 7.80, respectively. The magnitude of the vicinal coupling constant $J_{3,4}$ is comparable to that of *cis*-alkenes.

The $^{13}$C chemical shifts for C-2, C-3 and C-4 in coumarin are at 159.6, 115.7 and 142.7 p.p.m. respectively, and correspond closely to those for pyran-2-one. A good correlation between $^{13}$C shifts and charge densities, calculated by the CNDO/2 method, has been established ⟨75JOC1175⟩.

As with pyran-2-one, the NMR spectra of coumarins indicate some degree of charge delocalization. However, the aromaticity of the heterocyclic ring is only marginal.

### 2.22.8.1.2 *Pyran-4-one and chromone*

The early work on aromaticity of pyran-4-one has been reviewed ⟨68T923⟩.

The pyran-4-one ring (**18**) undergoes substitution at C-3 with the usual electrophilic reagents (see Section 2.23.8.3). Ring cleavage occurs in the presence of base. However, pyran-4-one is devoid of normal ketonic properties. The Diels–Alder reaction has been advocated as a means of assessing the aromaticity in pyran-4-ones and congeneric compounds ⟨75JHC785⟩. If pyran-4-one possesses the localized structure (**18**), then it should exhibit dienophilic properties like those of 1,4-benzoquinone. Reaction of 2,3-dimethyl-buta-1,3-diene with pyran-4-one failed to produce any of the adduct (**209**). The 4*H*-thiopyran-4-one (**210**) was also inert, although the sulfone (**211**) reacted readily. From these observations, it was concluded that both (**18**) and (**210**) have substantial $\pi$-electron delocalization, consistent with the betaine structure (**212**). The sulfone (**211**) behaves as a dienophile since there is no possibility of heteroatom participation to give an aromatic 6$\pi$-electron system.

(**209**)     (**210**)     (**211**)     (**212**) X = O, S

The basicity of pyran-4-one, which is significantly greater than that of simple aromatic and aliphatic ketones, is apparent from its p$K_a$ of 0.1 and has been rationalized in terms of a betaine structure (**212**; X = O). Benzo-fusion increases the basicity: the p$K_a$ of chromone (**20**) is 2.0 ⟨65ZOB1707⟩. As a result of their basicity, both pyran-4-one and chromone have a great propensity to form salts with acids ⟨63CJC505⟩.

There appear to be no data on the heats of combustion of simple pyran-4-ones and therefore ERE values are not available.

The MO delocalization energy (DE) calculated using a simple LCAO approach indicated that pyran-4-one (DE, 2.868$\beta$) possesses greater aromaticity than benzene (DE 2$\beta$) ⟨62CCC1242⟩. Subsequent more refined calculations gave substantially lower values, but still indicated considerable aromaticity for pyran-4-one. Values of DE, DE$_{sp}$ and aromaticity indices for pyran-4-one and chromone which were obtained using various MO methods are presented in Table 9.

Calculated $\pi$-electron densities and orbital energies, obtained for chromone using a semi-empirical PPP method, were shown to correlate well with $^1$H NMR chemical shifts and polarographic half-wave potentials ⟨78IJC(A)531⟩.

The UV spectra of pyran-4-ones are unremarkable and resemble those of cyclic enones and quinones.

Studies using IR spectroscopy have been undertaken in attempts to assess the carbonyl group bond order, and hence the extent of $\pi$-electron conjugation in the pyran-4-one ring system ⟨73MI22203⟩. The carbonyl stretching frequency in pyran-4-one itself occurs in the region 1661–1678 cm$^{-1}$ ⟨62BCJ1323, 59CJC2007⟩. Benzo-fusion lowers the frequency and chromone absorbs in the region 1650–1665 cm$^{-1}$ ⟨63PMH(2)254, 63PMH(2)257⟩ (*cf. ca.* 1730 cm$^{-1}$ for pyran-2-ones). At first sight, it might appear that the carbonyl group has considerable single-bond character consistent with the betaine structure (**212**; X = O). However, comparison with cyclohexadienones, which are not potentially aromatic, reveals that $\nu$(CO) in these systems is around 1650 cm$^{-1}$. Hence it seems unnecessary to invoke a betaine structure to explain the low carbonyl frequency. Moreover, in the hydrochloride (**213**) the protonated carbonyl group still appears to have significant double bond character. The CO absorption is at 1500 cm$^{-1}$, whilst $\nu$(CO) in phenols is much lower and usually occurs at 1140–1230 cm$^{-1}$. MO calculations have established that, like pyran-2-one, pyran-4-one and chromone have high carbonyl group $\pi$-bond orders. Satisfactory correlations between calculated and experimental carbonyl group frequencies were observed ⟨74BSF545⟩.

(213)

The EHT MO calculated bond orders and IR carbonyl frequencies for pyran-4-one and chromone are given in Table 10.

If pyran-4-one is a 6$\pi$-electron aromatic system, it must adopt the betaine structure (**212**; X = O), which would exhibit a large dipole moment. The observed moment for (**18**) is 3.7 D. A dipole moment of 1.9 D is obtained by vector summation of bond and group moments. Clearly, if mesomeric effects were absent, the difference between the calculated and observed moments would be very small. The magnitude of the observed dipole moment suggests there is considerable $\pi$-electron delocalization in pyran-4-one. However, the calculated value for the betaine structure (**212**; X = O) is 21 D ⟨73MI22203⟩ and thus it appears that the extent of delocalization is, after all, quite small.

In general the application of dipole moments to derive information about aromaticity is beset with difficulties ⟨B-75MI22206⟩.

The proton chemical shifts and coupling constants for some pyran-4-ones are given in Figures 1 and 4. The chemical shifts observed for pyran-4-one suggest the presence of a diamagnetic ring current. Use was made of $^1$H chemical shifts to estimate the magnitude of the ring current in 2,6-dimethylpyran-4-one (**214**) relative to that in benzene. The dihydro compound (**215**) was used as a model to estimate the diamagnetic anisotropic effect of the ring current in (**214**). It was concluded that the pyranone (**214**) and benzene have comparable ring currents ⟨68T923⟩. However, since the size of the coupling constant $J_{2,3}$ is virtually identical in pyran-4-one and the dihydro compound (**215**) (see Figure 1), it would appear that (**214**) lacks significant aromaticity.

(214)              (215)

Additional evidence for a ring current in pyran-4-one is provided by the observation that the ring protons are deshielded by 1 p.p.m. compared to those in 2,3-dihydropyran-4-one.

Replacement of the oxygen heteroatom by sulfur induces downfield shifts of the ring protons which suggest increased ring currents and greater aromaticity in the sulfur compounds (Figure 13) ⟨65JPC1⟩.

The $^{13}$C NMR spectra resemble both $\alpha,\beta$-unsaturated ketones and to a lesser degree heteroaromatic systems. Data are given in Figure 6. The pyran-4-one chemical shifts follow

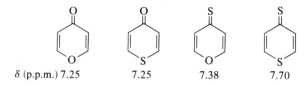

δ (p.p.m.) 7.25        7.25        7.38        7.70

**Figure 13** Average chemical shifts for some pyran-4-ones and thiones and related compounds ⟨65JPC1⟩

the same pattern as those of pyran-2-one, though the carbonyl carbon (C-4) is at much lower field. The single bond $^{13}$C–H coupling constants for pyran-4-one and furan are comparable, and are given in Table 11. From a $^{13}$C NMR study of kojic acid (**81**) it was concluded that pyran-4-one possesses some degree of aromaticity ⟨76JOC2777⟩.

The magnetic susceptibility anisotropy of pyran-4-one was determined by microwave molecular Zeeman effect studies. Data for 2- and 4-pyranone are given in Table 12. Since pyran-4-one exhibits a negligible non-local effect it was suggested that this molecule is best regarded as being non-aromatic ⟨73JA2766⟩.

Precise and complete structures of pyran-4-one and some analogues have been determined by microwave spectroscopy ⟨81JCS(F2)79⟩ and are summarized in Figure 11. Comparison of these molecular dimensions with the corresponding data for non-aromatic compounds such as cyclopentadiene or acrolein supports the conclusion reached in the earlier Zeeman studies ⟨73JA2766⟩ that the pyran-4-one ring is largely non-aromatic. There is, however, evidence for a small degree of delocalization. The bonding appears to be more delocalized in pyran-4-thione in view of the shortening of the C(3)—C(4) bond and increase in length of the C(2)—C(3) linkage. Hence the thione is more aromatic than pyran-4-one itself.

### 2.22.8.1.3 Pyranopyrandiones

Consideration of all the possible fusions between two 2- and/or 4-pyranone rings reveals that there are twenty possible pyranopyrandiones. These compounds are not readily accessible and comparatively few have been synthesized. Once again all are potentially aromatic.

The synthesis and properties of 7-methylpyrano[4,3-*b*]pyran-2,5-dione (**216**; R = H) have been reported ⟨68JCS(C)543⟩. Aromaticity can be realized if the betaine structures (**217a–c**) are adopted. The system, *cf.* (**217c**), will be iso-π-electronic with naphthalene. Measurement of the heat of combustion enabled the empirical resonance energy (ERE) of (**216**) to be calculated. An exceedingly high value of 406 kJ mol$^{-1}$ (*cf.* benzene 163 kJ mol$^{-1}$) was obtained. Bearing in mind the lability and lack of aromaticity in simple pyran-2-ones, it is questionable if this ERE is meaningful. Some of the difficulties associated with obtaining ERE values for heterocycles have been described ⟨74AHC(17)255⟩.

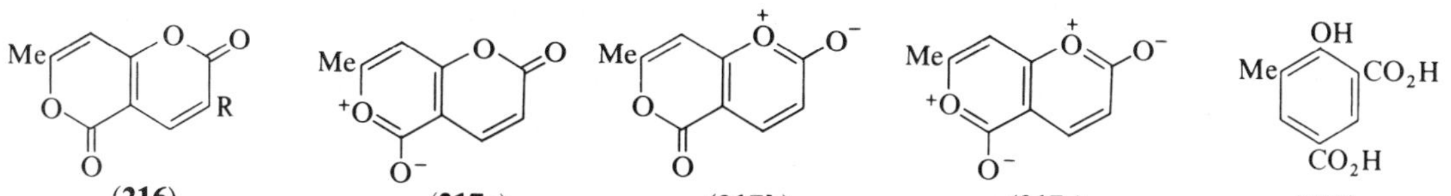

| (216) | (217a) | (217b) | (217c) | (218) |

In marked contrast to pyran-2-one, electrophilic substitution in (**216**) occurs readily at C-3, giving the monosubstituted compounds (**216**; R = NO$_2$, Br, SO$_3$H). As might be expected, (**216**) is susceptible to nucleophilic attack. Heating with 10% aqueous sodium hydroxide results in the formation of α-coccinic acid (**218**).

The $^1$H NMR spectrum of (**216**) is consistent with substantial charge delocalization, and the ring protons absorb in the regions characteristic for aromatic systems. Data for (**216**) and related compounds are given in Figure 14. However, the magnitude of $J_{3,4}$ (9.6 Hz) suggests a high π-bond order, typical of a *cis*-alkene.

The carbonyl group frequencies in the IR spectra of pyranopyrandiones are characteristic of simple lactones (Figure 14).

An MO study of all twenty pyranopyrandiones has been reported ⟨73MI22200⟩. Use was made of HMO calculations together with the ω technique. Calculations were also performed on some methyl-substituted pyranopyrandiones and afforded values for the delocalization energy, π-bond orders and π-charge densities. Empirical resonance energies were obtained

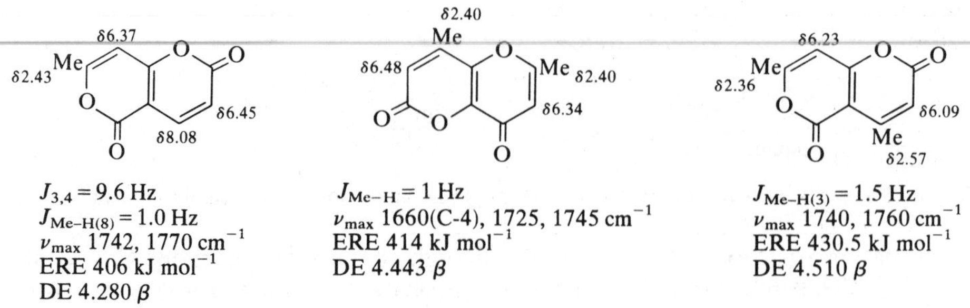
δ6.37
δ2.43
Me
O
O
O
O
δ6.45
δ8.08
δ2.40
Me
δ6.48
O
Me δ2.40
δ6.34
O
O
O
δ6.23
δ2.36
Me
O
O
O
O
δ6.09
Me
δ2.57
$J_{3,4} = 9.6$ Hz
$J_{Me-H(8)} = 1.0$ Hz
$\nu_{max}$ 1742, 1770 $cm^{-1}$
ERE 406 kJ $mol^{-1}$
DE 4.280 $\beta$
$J_{Me-H} = 1$ Hz
$\nu_{max}$ 1660(C-4), 1725, 1745 $cm^{-1}$
ERE 414 kJ $mol^{-1}$
DE 4.443 $\beta$
$J_{Me-H(3)} = 1.5$ Hz
$\nu_{max}$ 1740, 1760 $cm^{-1}$
ERE 430.5 kJ $mol^{-1}$
DE 4.510 $\beta$

**Figure 14** $^1$H NMR and IR data and resonance energies for some pyranopyrandiones ⟨73MI22200⟩

for (**216**) and two related compounds, both of which gave extremely high values. All the compounds were calculated to have large delocalization energies of $4\beta$. The results are summarized in Figure 14.

In order to obtain further insight into the $\pi$-electron distribution in these molecules and the possible involvement of betaine structures, a dipole moment study was undertaken ⟨76JHC609⟩. However, the results did not enable any firm conclusions to be drawn.

### 2.22.8.2 Pyrylium Salts and their Benzologues

The substitution of a CH unit in benzene by $O^+$ (the oxonia group) gives rise to the pyrylium cation (**3**). Since this ring still possesses 6 $\pi$-electrons, it may be expected to exhibit aromatic properties. As the oxygen is primarily tricovalent, the pyrylium ring may be formally regarded as a cyclic oxonium ion. However its enhanced stability relative to aliphatic and alicyclic oxonium salts is doubtless due to its aromatic nature.

From the point of view of reactivity, pyrylium salts are not typically aromatic. As might be expected, the highly electronegative heteroatom has a large perturbing effect on the distribution of $\pi$-electron density. Hence the 2-, 4- and 6-positions are rendered susceptible to nucleophilic attack, as can be inferred by consideration of resonance forms.

Attack by nucleophiles occurs readily, especially at C-2, and further reactions, usually involving ring opening, lead to a variety of products. A detailed discussion is given in Chapter 2.23. In spite of this, the aromatic nature of pyrylium salts is clearly exhibited by their spectral properties. Thus the $^1$H NMR spectrum for pyrylium perchlorate shows low field absorptions in the region $\delta$ 8.02–9.22 ⟨73OMR(5)251⟩, indicating the presence of a substantial diamagnetic ring current. Calculations of $\pi$-electron density using the HMO method reveal that there is considerable delocalization of the positive charge. The electron density decreases in the order C-3 > C-4 > C-2, which is to be expected from simple resonance considerations. Correlation of proton chemical shifts and $\pi$-electron densities shows good qualitative agreement ⟨66RRC1193⟩, increasing downfield shifts occurring with decreasing electron density.

Total charge densities for the pyrylium cation have been calculated by the INDO MO method and show an increase of positive charge at the 2-, 4- and 6-positions. A linear relationship with $^{13}$C chemical shifts was shown to exist ⟨77OMR(9)16⟩.

The aromatic character of pyrylium salts is also suggested by their UV spectra. A comparison can be drawn with the pyridinium ion, since this system is iso-$\pi$-electronic with pyrylium and belongs to the same symmetry point group ($C_{2v}$) ⟨71PMH(3)67⟩. Thus, the energies of their respective molecular orbitals will be comparable and this is reflected in the similarity of their UV maxima.

Benzo-fusion to a pyrylium ring can give rise to two isomers, namely the 1- and 2-benzopyrylium ions (**6**) and (**219**) both of which are formally aromatic $10\pi$-electron systems.

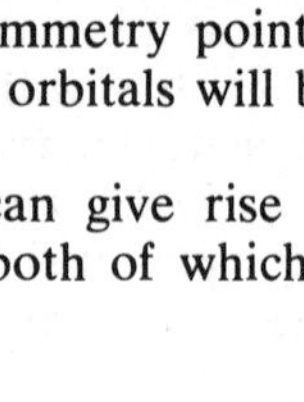
O+

(**219**)

Calculations on the distribution of charge densities in the 1-benzopyrylium ion have been carried out using the CNDO/2 method ⟨70ACS2745⟩. The results indicate that whilst the positive charge is distributed over the whole of the conjugated system, C-2 and C-4 carry a substantial fraction of the total charge density and as such will be the preferred sites for nucleophilic attack. This process occurs in the parent compound exclusively at C-2 with concomitant ring opening. However, attack at C-8a is never observed since this would result in disruption of the aromatic benzene $\pi$-system. As with the pyrylium ion, the $^1$H NMR spectrum of (6) indicates the presence of a ring current and shows absorptions in the range $\delta$ 9.60–9.90 p.p.m. ⟨81JHC1325⟩. Additional support for the aromaticity of the 1-benzopyrylium ion is provided by its UV spectrum which shows similarities to that of the quinolinium ion ⟨71PMH(3)67⟩.

The 2-benzopyrylium ion (219) displays a similar reactivity pattern to that of the 1-isomer. A $^{13}$C NMR study of a series of 1-aryl-2-benzopyrylium salts indicated that there is a high degree of charge density in both heterocyclic and carbocyclic rings. It was concluded that the $^{13}$C shifts are in accord with those of naphthalenes in which C-2 has been substituted by an oxonia group, thus substantiating the evidence for the aromaticity of this ring system ⟨76OMR(8)324⟩.

Of the higher benzologues of pyrylium, the only one of any significance is the xanthylium ion (9). The chemistry of this ion differs significantly from that of pyrylium and its simple benzologues. Although the ring system is susceptible to nucleophilic attack, this occurs exclusively at C-9 rather than at the carbon atoms adjacent to the heteroatom. Consequently, ring cleavage reactions are unknown. This would suggest that of the many possible resonance forms for the xanthylium ion, only the carbonium ion structure (220; Scheme 34) is of importance. It is evident that the properties of xanthyliums resemble those of carbonium ions derived from di- and tri-phenylmethanes rather than those of simple pyrylium salts.

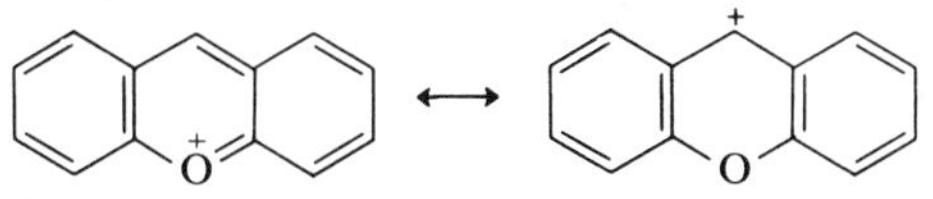

**Scheme 34**

Although the reactivity of the xanthylium ion is consistent with a high degree of charge density localized at C-9, the positive charge is nevertheless delocalized over the entire ring system. Evidence for this is provided by a comparison of the $^{13}$C NMR chemical shifts of xanthylium salts ⟨75BSF2023, 75JA5472⟩ with those of di- and tri-phenylmethyl carbonium ions ⟨75JA5472, 69JA5801⟩. In the latter, the central carbon carries a high degree of the total charge density as indicated by $^{13}$C shifts of 202.2 and 212.7 p.p.m. for the diphenyl- and triphenyl-methyl carbonium ions, respectively. In contrast, C-9 in the xanthylium cation absorbs at 165.1 p.p.m. suggesting that there is considerable charge delocalization in this cation, consistent with its planarity and great stability. CNDO/2 calculations support this view.

## 2.22.9 TAUTOMERISM

A treatise on tautomerism in heterocycles ⟨76AHC(S1)1⟩ reviews the work in this area on pyrans and fused pyrans.

### 2.22.9.1 Prototropy

#### 2.22.9.1.1 Prototropy involving annular carbon atoms

Annular prototropy in pyrans involves migration of protons between different sites in the ring. The equilibrium situation is depicted in equation (6). The systems in which this

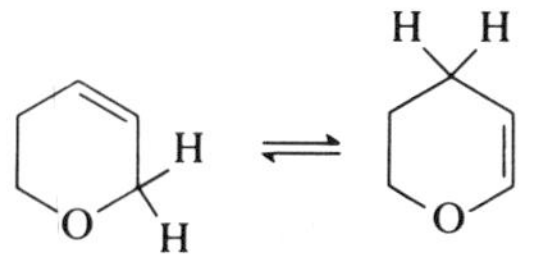

(6)

type of tautomerism could be realised are the 2*H*- and 4*H*-pyrans (**1**) and (**2**) and their respective benzologues (**4**) and (**5**).

4*H*-Pyran has been prepared ⟨62AG(E)699⟩ and many derivatives are also known, but there is no evidence to suggest that tautomerism between it and (the unknown) 2*H*-pyran takes place. The benzologues (**4**) and (**5**) have been obtained, and again it appears that interconversion between these isomers does not occur.

### 2.22.9.1.2 *Prototropy involving functional groups*

#### (i) *Potential methyl groups*

Infrared and NMR studies revealed the presence of a thermodynamic equilibrium between the 4-alkyl-2*H*-chromenes (**221**; R = H or Me) and the corresponding chromans (**222**) ⟨68BSF4203⟩.

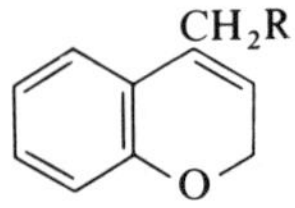

(**221**) R = H or Me 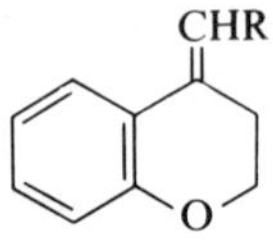          (**222**) R = H or Me

Both pyrans (**223**) and (**224**) have been isolated; their $^1$H NMR spectra show that they exist as such in solution and do not undergo interconversion ⟨64BSF1492⟩. The thermodynamics for the isomerization of (**225**) → (**226**) in solution have been studied; $\Delta G°$ for the process was determined to be $-21.8$ kJ mol$^{-1}$ in favour of (**226**) ⟨74ACS(B)1234⟩.

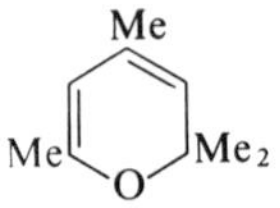

(**223**) 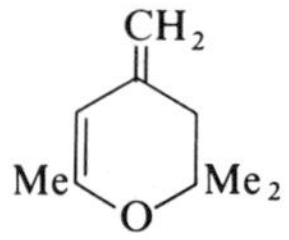          (**224**) 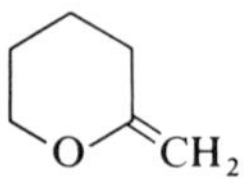          (**225**) 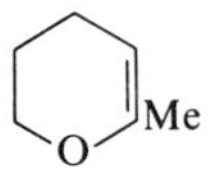          (**226**)

#### (ii) *Potential hydroxy groups*

Tautomerism in hydroxypyranones has been extensively studied and all the important investigations in this field have been reviewed ⟨63AHC(1)339, 76AHC(S1)116⟩. Of the three possible hydroxypyranone tautomers (**227**), (**228**) and (**229**), it has been observed that in most cases the 4-hydroxy tautomer (**227**) is of greatest stability.

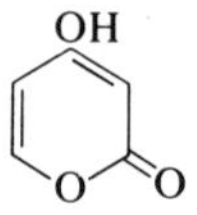

(**227**) 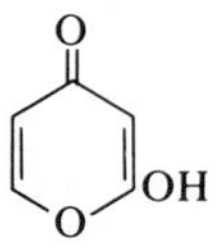          (**228**) 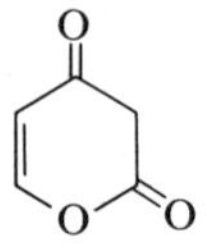          (**229**)

From measurements of the IR carbonyl frequencies, it was inferred that the 4-hydroxy-pyran-2-one (**230**) and the 2-hydroxypyran-4-one (**231**) exist as such in the solid state. In ethanol solution, however, (**231**) gives the tautomer (**232**) ⟨59JA2427⟩.

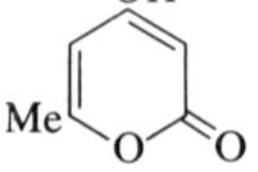

(**230**) 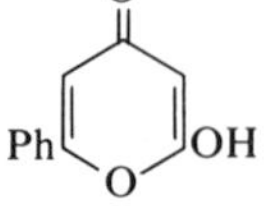          (**231**) 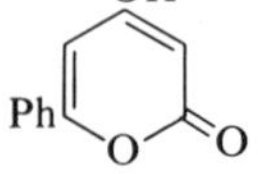          (**232**)

Spectral investigations have revealed that the ester (**233**) possesses the 4-hydroxypyran-2-one structure both in CDCl$_3$ solution and the solid state ⟨62JCS2606⟩. A 4-hydroxypyranone structure is the predominant form of the 3,5-dimethyl compound (**234**) ⟨68AC(R)664⟩.

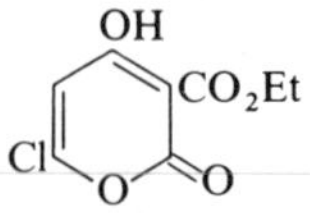

(**233**) 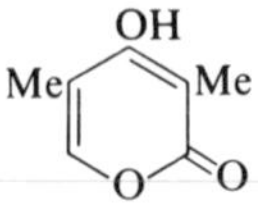          (**234**)

IR studies reveal that tetrahydropyran-2,4,6-trione (235) exists in equilibrium with the 4-hydroxydione (236). Similarly the 3-acetyl analogue is a tautomeric mixture of (237) and (238) ⟨71JCS(C)2721⟩.

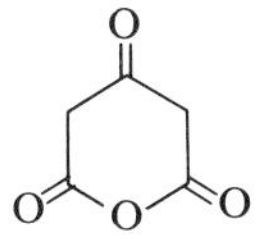

(235) 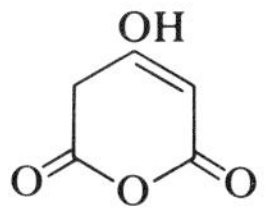        (236) 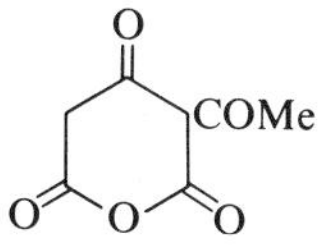        (237) 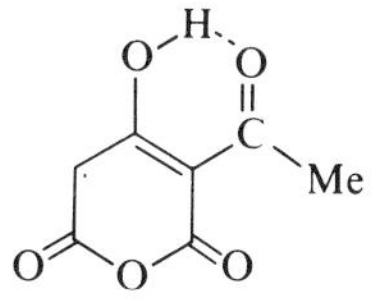        (238)

The existence of tautomers and rotamers of dehydroacetic acid (239) and derived Schiff's bases has been investigated by $^1$H, $^{13}$C and $^{15}$N NMR spectroscopy ⟨82JCS(P2)513⟩. The tautomeric form (239) proposed earlier ⟨61AK(17)523⟩ was confirmed and its existence as the rotamer (240) was established by variable temperature studies.

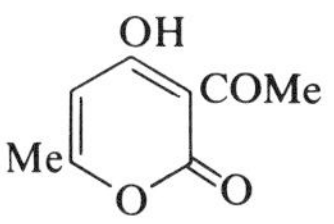

(239) 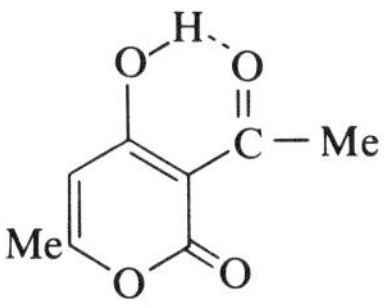        (240)

Prototropy in hydroxybenzopyranones has been investigated. The hydroxychromone (241) does not undergo tautomerism to the dione (242; equation 7). The NMR spectrum of a CDCl$_3$ solution shows the methyl signal at $\delta$ 2.50 and since this appears as a singlet, it would seem that the contribution of (242) is not significant. Solid state and chloroform solution IR spectra also support this conclusion ⟨71MI22201⟩.

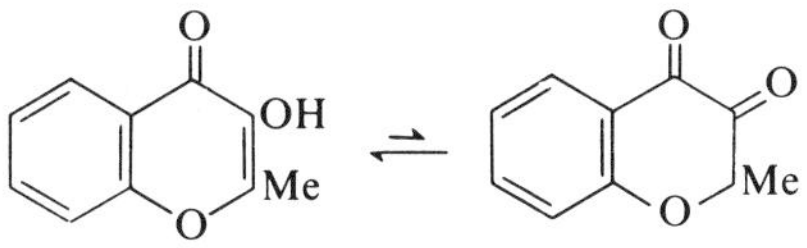

(241)        (242)        (7)

The tautomerism of 4-hydroxycoumarins has been extensively studied ⟨63AHC(1)339⟩. From recent IR studies, it appears that 4-hydroxycoumarin (243) exists as this tautomer in chloroform, dioxan and DMSO solutions. Evidence supporting the presence of the 2-hydroxychromone tautomer (244) was not obtained. However, (244) is present in anhydrous, solid state 4-hydroxycoumarin ⟨82JHC385⟩. From the $^1$H NMR spectrum of (243) in CD$_3$OD solution, the presence of the chroman-2,4-dione tautomer (245) has been claimed ⟨76MI22202⟩.

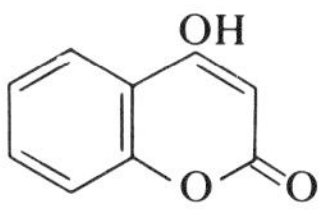

(243) 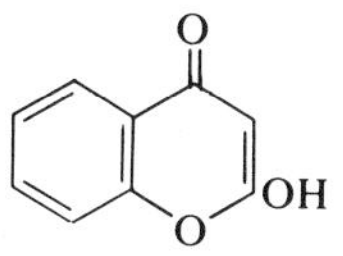        (244) 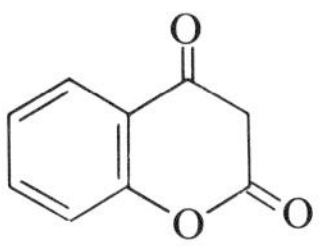        (245)

The structure of 3-arylazo-4-hydroxycoumarin (246) has been investigated ⟨78IJC(B)295⟩. The authors stated that IR and UV spectral data indicate (246) to exist as such in the solid state and in solution; the presence of the other possible tautomers (247) and (248) was discounted.

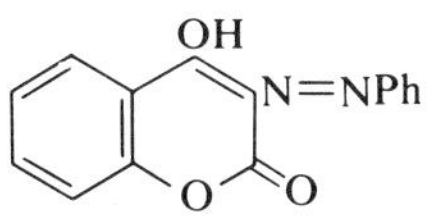

(246) 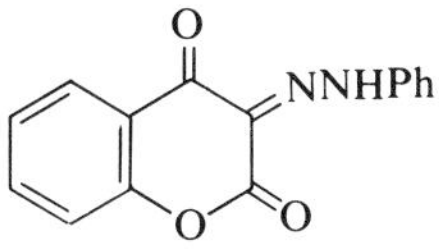        (247) 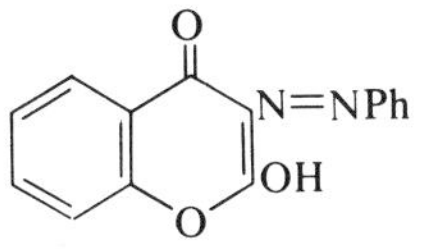        (248)

An investigation into the phototautomerism of 7-hydroxy-4-methylcoumarin has been undertaken ⟨74JA4699⟩.

(iii) *Potential amino groups*

Aminopyranones and related compounds have not been extensively investigated.

2-Aminochromone (**249**) is potentially tautomeric with 2-imino-4-hydroxycoumarin (**250**). From their IR spectra, 2-acylaminochromones (**251**; R = OEt or Me) were shown to exist in the form shown, since very intense absorptions at 1635 cm$^{-1}$, characteristic of a chromone carbonyl group, were observed ⟨67KGS782⟩.

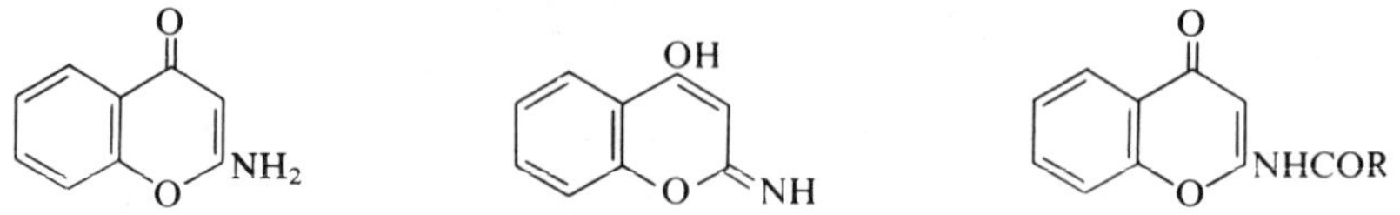

(**249**)          (**250**)          (**251**)

### 2.22.9.2 Ring–Chain Tautomerism

It has been demonstrated that 2-alkylaminotetrahydropyrans (**252**; R = H or alkyl, R′ = alkyl) exist in a tautomeric equilibrium with the hydroxyimines (**253**) ⟨79KGS1317⟩. The situation is depicted in equation (8). In the absence of solvent, the IR spectra of all the compounds studied exhibit intense bands corresponding to C=N and O—H absorptions (due to **253**). There is a marked decrease of these bands in the solution spectra, since (**252**) is predominant. The signals observed in the NMR spectrum arise from ring and chain forms. It was shown that the equilibrium concentration of tautomers is dependent upon the substituents present.

$$ \rightleftharpoons \quad HOCH_2CH_2CH_2CH_2C=NR' \qquad (8) $$

(**252**)                (**253**)

The $^1$H NMR spectrum of the dienone *cis-β*-ionone (**31**) reveals that this compound exists in an equilibrium mixture with its valence isomer the 2*H*-pyran (**30**; Figure 2). The two forms interconvert readily and thus separation is not possible ⟨66JA619⟩.

A ring–chain tautomeric equilibrium between the pyranopyrandione (**254**) and the 4-pyranonecarboxylic acid (**255**; equation 9) in CDCl$_3$ solution has been investigated by $^1$H NMR spectroscopy ⟨73AC(R)291⟩. The equilibrium constant $K_T$ was calculated as 1.7 at 298 K and $\Delta H°$ for the equilibrium reaction as 16.2 kJ mol$^{-1}$.

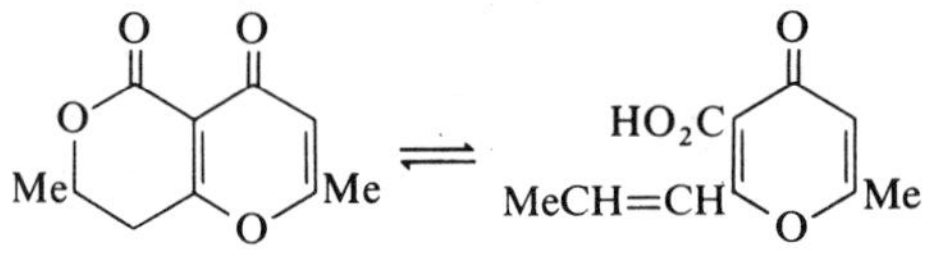

(**254**)                (**255**)          (9)

In chloroform solution the 2-hydroxyflavanones (**256**; X = Br, OH, PhCO$_2$) are in equilibrium with the chain forms (**257**; equation 10). However, in the solid state the flavanone form is preferred ⟨68BCJ2798, 69BCJ3345⟩.

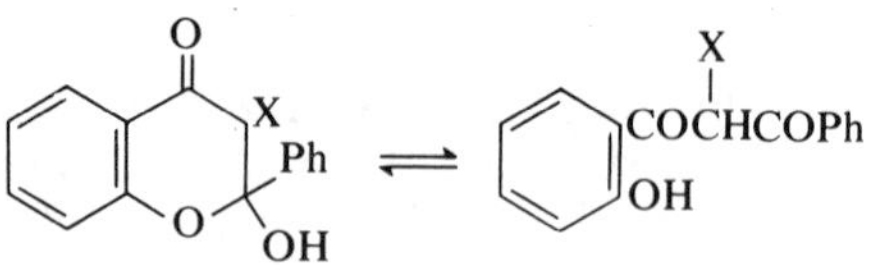

(**256**)                (**257**)          (10)

## 2.22.10 BETAINES AND OTHER UNUSUAL STRUCTURES

### 2.22.10.1 Betaines

The most important pyran betaines are the heteroaromatic pyrylium-3-olate (**258**) and 2-benzopyrylium-4-olate (**259**) systems. Numerous analogues of both are known, and it has been shown that these have a high propensity to participate in cycloadditions ⟨72JOC3838⟩,

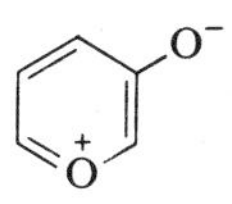

**(258)** 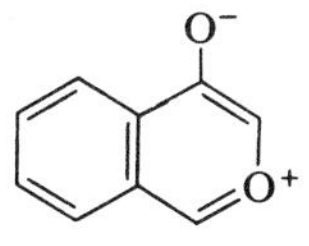    **(259)**

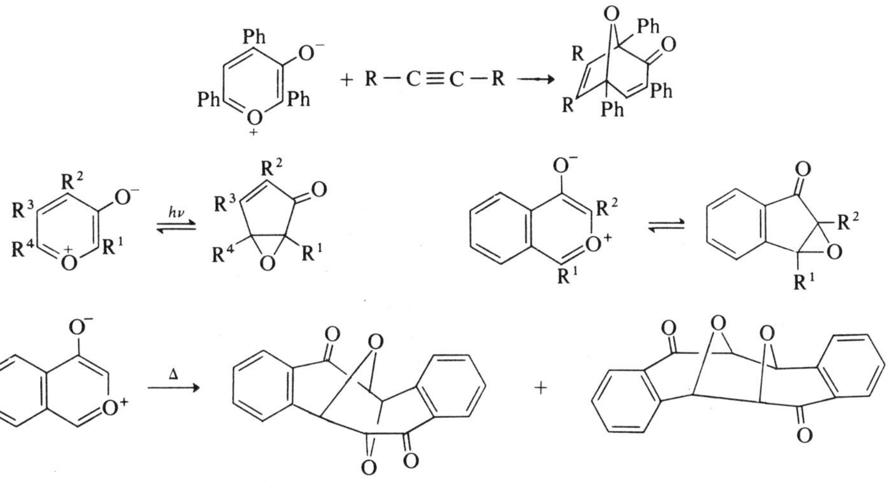

**Scheme 35**

dimerizations and rearrangements ⟨76ACS(B)619⟩. In addition, valence tautomerism of these compounds is initiated by irradiation with UV light ⟨63JA3529, 64JA3814⟩ (Scheme 35).

## 2.22.10.2 Other Systems

The mixed fulvalene (**260**; R = Me or Ph) has been obtained ⟨71TL4799⟩. Dipole moment measurements indicated that the negative pole resides in the phenalene ring. Polarization in this direction will be strongly favoured since in (**261**) both rings possess aromatic character.

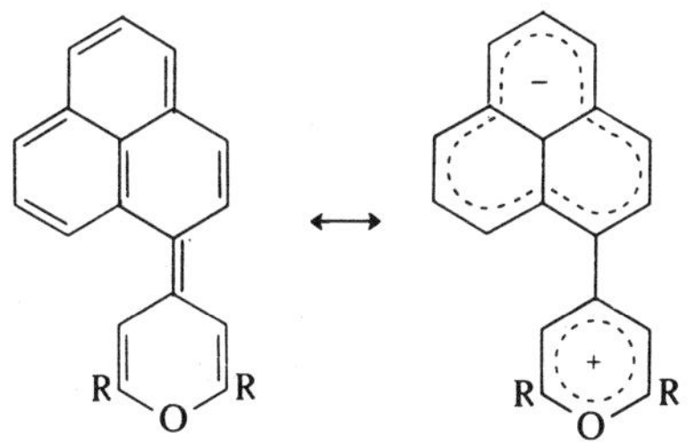

**(260)**    **(261)**

# 2.23

# Pyrans and Fused Pyrans: (ii) Reactivity

G. P. ELLIS

*University of Wales Institute of Science and Technology*

| | | |
|---|---|---|
| 2.23.1 | INTRODUCTION | 648 |
| 2.23.2 | PYRYLIUM SALTS: REACTIVITY AT RING ATOMS | 649 |
| | *2.23.2.1 Photochemical Reactions* | 649 |
| | *2.23.2.2 Electrophiles* | 649 |
| | *2.23.2.3 Nucleophiles* | 652 |
| | *2.23.2.3.1 Introduction* | 652 |
| | *2.23.2.3.2 Nucleophilic addition* | 652 |
| | *2.23.2.3.3 Formation of a new six-membered N-ring* | 654 |
| | *2.23.2.3.4 Formation of a new benzene ring* | 656 |
| | *2.23.2.3.5 Formation of other rings* | 659 |
| | *2.23.2.3.6 Reduction* | 661 |
| 2.23.3 | PYRYLIUM SALTS: REACTIVITY OF SUBSTITUENTS | 662 |
| | *2.23.3.1 Introduction* | 662 |
| | *2.23.3.2 C-Linked Substituents* | 662 |
| | *2.23.3.3 O-Linked Substiuents* | 664 |
| | *2.23.3.4 S-Linked Substituents* | 665 |
| 2.23.4 | 2H- AND 4H-PYRANS AND THEIR BENZO DERIVATIVES: REACTIVITY AT RING ATOMS | 665 |
| | *2.23.4.1 Introduction* | 665 |
| | *2.23.4.2 Photochemical and Thermal Reactions* | 666 |
| | *2.23.4.3 Electrophiles* | 667 |
| | *2.23.4.3.1 Electrophilic substitution and addition* | 667 |
| | *2.23.4.3.2 Oxidation* | 669 |
| | *2.23.4.3.3 Carbenes* | 671 |
| | *2.23.4.4 Nucleophiles* | 672 |
| | *2.23.4.4.1 Bases* | 672 |
| | *2.23.4.4.2 Reduction* | 673 |
| | *2.23.4.5 Other Reactions* | 673 |
| 2.23.5 | 2H- AND 4H-PYRANS AND THEIR BENZO DERIVATIVES: REACTIVITY OF SUBSTITUENTS | 674 |
| | *2.23.5.1 Introduction* | 674 |
| | *2.23.5.2 Fused Benzene Rings* | 674 |
| | *2.23.5.3 C-Linked Substituents* | 674 |
| | *2.23.5.4 O-Linked Substituents* | 675 |
| 2.23.6 | PYRAN-2-ONES AND THEIR BENZO DERIVATIVES: REACTIVITY AT RING ATOMS | 675 |
| | *2.23.6.1 Introduction* | 675 |
| | *2.23.6.2 Thermal and Photochemical Reactions* | 677 |
| | *2.23.6.3 Electrophiles* | 679 |
| | *2.23.6.3.1 Addition and substitution* | 679 |
| | *2.23.6.3.2 Oxidation* | 681 |
| | *2.23.6.4 Nucleophiles* | 681 |
| | *2.23.6.4.1 C-Nucleophiles* | 681 |
| | *2.23.6.4.2 N-Nucleophiles* | 683 |
| | *2.23.6.4.3 O-Nucleophiles* | 685 |
| | *2.23.6.4.4 Reduction* | 686 |
| | *2.23.6.5 Radicals* | 687 |
| | *2.23.6.6 Other Reactions* | 688 |

2.23.7  PYRAN-2-ONES AND THEIR BENZO DERIVATIVES: REACTIVITY
        OF SUBSTITUENTS                                                    689

   *2.23.7.1   Introduction*                                              689
   *2.23.7.2   Fused Benzene Rings*                                       689
   *2.23.7.3   C-Linked Substituents*                                     689
   *2.23.7.4   N-Linked Substituents*                                     691
   *2.23.7.5   O-Linked Substituents*                                     692

2.23.8  PYRAN-4-ONES AND THEIR BENZO DERIVATIVES: REACTIVITY
        AT RING ATOMS                                                     692

   *2.23.8.1   Introduction*                                              692
   *2.23.8.2   Thermal and Photochemical Reactions*                       693
   *2.23.8.3   Electrophiles*                                             696
      *2.23.8.3.1   Addition and substitution*                           696
      *2.23.8.3.2   Oxidation*                                            698
   *2.23.8.4   Nucleophiles*                                              698
      *2.23.8.4.1   C-Nucleophiles*                                       698
      *2.23.8.4.2   N-Nucleophiles*                                       700
      *2.23.8.4.3   O-Nucleophiles*                                       703
      *2.23.8.4.4   Reduction*                                            704
   *2.23.8.5   Radicals*                                                  705
   *2.23.8.6   Other Reagents*                                            705

2.23.9  PYRAN-4-ONES AND THEIR BENZO DERIVATIVES: REACTIVITY
        OF SUBSTITUENTS                                                   707

   *2.23.9.1   Introduction*                                              707
   *2.23.9.2   Fused Benzene Rings*                                       707
   *2.23.9.3   C-Linked Substituents*                                     708
   *2.23.9.4   N-Linked Substituents*                                     714
   *2.23.9.5   O-linked Substituents*                                     715
   *2.23.9.6   S-Linked Substituents*                                     717
   *2.23.9.7   Halogens*                                                  717

2.23.10  REDUCED PYRANS AND PYRANONES AND THEIR BENZO DERIVATIVES:
         REACTIVITY AT RING ATOMS                                         718

   *2.23.10.1   Introduction*                                             718
   *2.23.10.2   Thermal and Photochemical Reactions*                      719
   *2.23.10.3   Electrophiles*                                            722
      *2.23.10.3.1   Addition and substitution*                          722
      *2.23.10.3.2   Oxidation*                                           724
      *2.23.10.3.3   Carbenes*                                            725
   *2.23.10.4   Nucleophiles*                                             726
      *2.23.10.4.1   C-Nucleophiles*                                      726
      *2.23.10.4.2   N-Nucleophiles*                                      726
      *2.23.10.4.3   O-Nucleophiles*                                      728
      *2.23.10.4.4   Reduction*                                           729
   *2.23.10.5   Other Reactions*                                          730

2.23.11  REDUCED PYRANS AND PYRANONES AND THEIR BENZO DERIVATIVES:
         REACTIVITY OF SUBSTITUENTS                                       731

   *2.23.11.1   Introduction*                                             731
   *2.23.11.2   Fused Benzene rings*                                      731
   *2.23.11.3   C-Linked Substituents*                                    732
   *2.23.11.4   N-Linked Substituents*                                    734
   *2.23.11.5   O-Linked Substituents*                                    734
   *2.23.11.6   Halogens*                                                 735

## 2.23.1  INTRODUCTION

Of the classes of compounds discussed in this chapter the chromones (**1**; R ≠ Ph), including flavones and isoflavones (**1**; R = 2-Ph and R = 3-Ph), and coumarins (**2**) are the most numerous. Many of these compounds are found in plants and there is a wide variety of structures amongst them. There appears not to be a single well-defined biological role for

(1)                    (2)

This chapter is arranged so that compounds which have the maximum number of double bonds in the pyran or pyranone ring are discussed first and those with fewest or no double bonds last. Compounds which contain two oxygen heterocyclic rings are not included; they were reviewed in 1979 ⟨B-79MI22304⟩ and subsequently ⟨B-80MI22301, B-81MI22301, B-82MI22302⟩.

## 2.23.2 PYRYLIUM SALTS: REACTIVITY AT RING ATOMS

### 2.23.2.1 Photochemical Reactions

Irradiation of 2,4,6-trimethylpyrylium perchlorate in water gives a rearranged pyrylium salt (4) which is cleaved to the 5-oxohexenal (3). Replacement of the 4-methyl by 4-*t*-butyl alters the course of the photolysis to give the cyclopentene (5) as the main product, together with small amounts of several others ⟨72CC1240, 73JA2406⟩.

Several rearrangement products (*e.g.* pyran-2-ones and 2-acylfurans) are obtained when substituted 4-hydroxypyrylium sulfates are irradiated in acid solution. Mechanisms involving many intermediate species as, for example, in the irradiation of (6) and (7) have been advanced to explain the observations ⟨73JA7914, 79JA7510⟩.

### 2.23.2.2 Electrophiles

The positive charge on the pyrylium ring discourages attack by electrophiles and 2,4,6-triphenylpyrylium perchlorate (8) is nitrated in the benzene rings to give the *m*- and *p*-oriented product (9). This result suggests that, under the strongly acidic conditions of this reaction, the positive charge is largely on the ring oxygen atom and thus has a greater electronic effect on the 2- and 6- than on the 4-phenyl ring. The corresponding flavylium, 1-phenyl-2-benzopyrylium and 9-phenylxanthylium salts give the 3'-nitro derivatives. Under more drastic nitrating conditions, the pyrylium salts break down to nitrobenzoic acids.

When 2,4,6-trimethylpyrylium perchlorate is heated with [O-²H]acetic acid, the 3- and 5-hydrogens are deuterated. This substitution is unlikely to occur by direct electrophilic displacement in the cyclic compound but may proceed through an acyclic dienone formed by prior addition of acetate ion at C-2 ⟨69RRC247⟩.

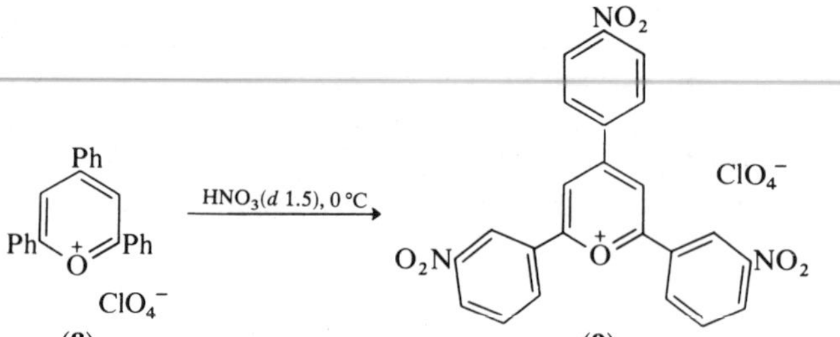

Oxidizing conditions tend to break down pyrylium salts and the products, which are usually isolated in low yield, vary with the oxidant and the salt. 2,4,6-Trialkylpyrylium salts, such as (10), are oxidized by hydrogen peroxide to 2-acylfurans (11) in a process which may be initiated by an addition at C-2 ⟨60CB599⟩. More recently, a similar oxidation of isoflavylium perchlorate (12) has given high yields of 2-phenylbenzofuran ⟨79TL1109⟩. Hydrogen peroxide cleaves the pyran ring of 2,3-diphenylbenzopyrylium perchlorate, for example, at C-2 and C-4 ⟨31JPR(131)1⟩ but the flavylium salt (13), on reaction with peroxyphthalic acid, gives a low yield of the flavone (14) ⟨42HCA1138⟩.

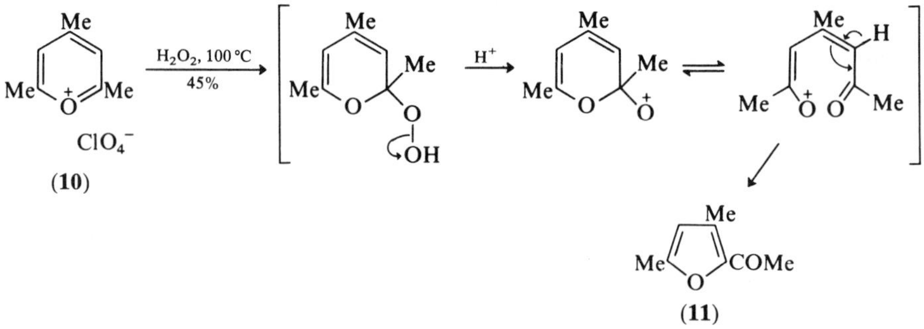

Oxidation of pyrylium salts to pyrones or vinylogous pyrones, such as (17), (18) or (21), is achieved in a two-step process in good to excellent yields, for instance, through the pyranylidene derivatives (16) or (20) ⟨78CHE1067⟩. The pyranylidenes may be regarded as oxidized or dehydrogenated derivatives of pyrylium salts and those salts which are unsubstituted at C-4 are amenable to this conversion; 2,6-diphenylpyrylium perchlorate (15) can give either the dipyranylidene (16) ⟨79JOC4456⟩ or the monocyclic product (17) whose structure was confirmed by X-ray analysis ⟨78T2131⟩.

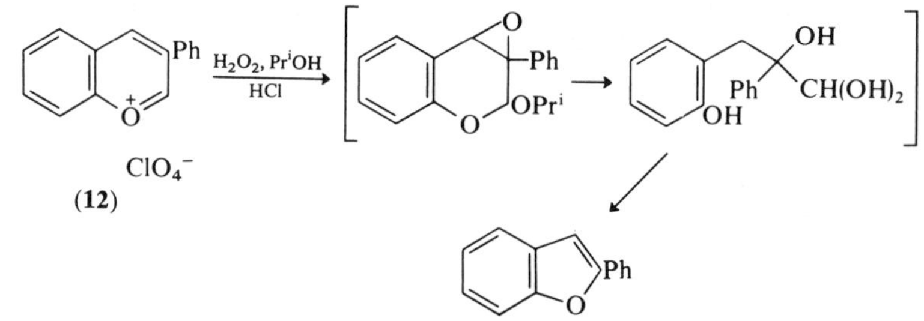

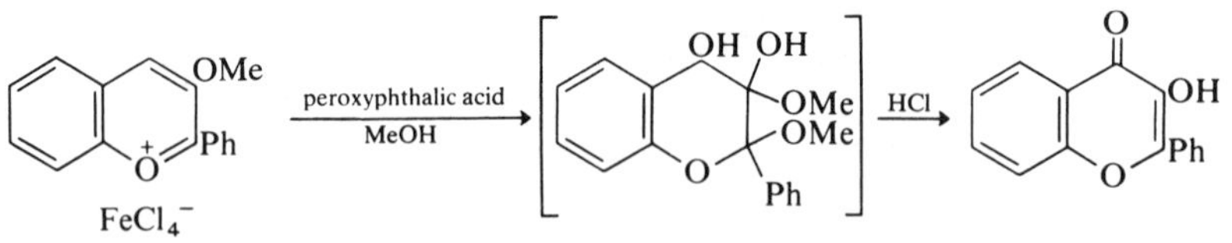

Anthocyanins and their aglycons (flavylium salts) are degraded into a mixture of compounds with alkali and air, or hydrogen peroxide; these reactions have been widely used in the determination of structure ⟨B-62MI22300, B-62MI22301, B-62MI22302⟩. Many flavylium salts have a 3-substituent (OH, OMe or Me) and such compounds are converted by peroxide into esters such as (22) by a process which resembles a Baeyer–Villiger oxidation ⟨64JOC2602⟩, but in the absence of a 3-substituent, the product is a dihydroxyketone (23) which, on warming with acid, is cyclized to a 3-hydroxypyrylium salt (24). 4'-Methoxyflavylium chloride is degraded into a mixture of two propanols (25) when subjected to hydroboration ⟨73AJC819⟩.

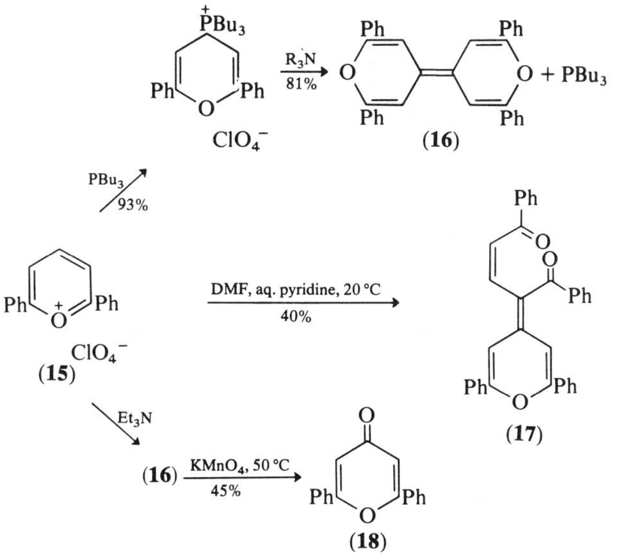

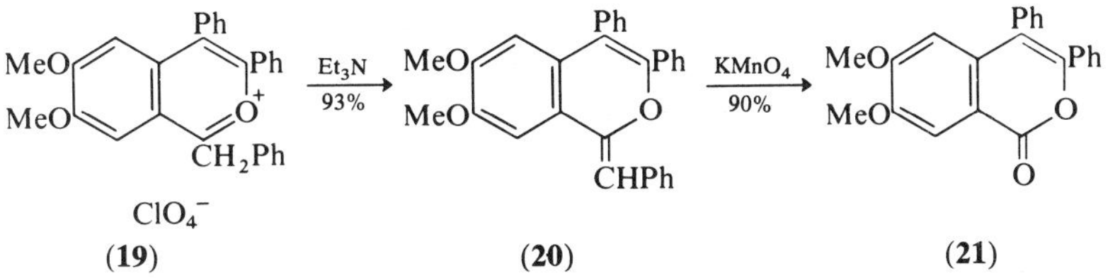

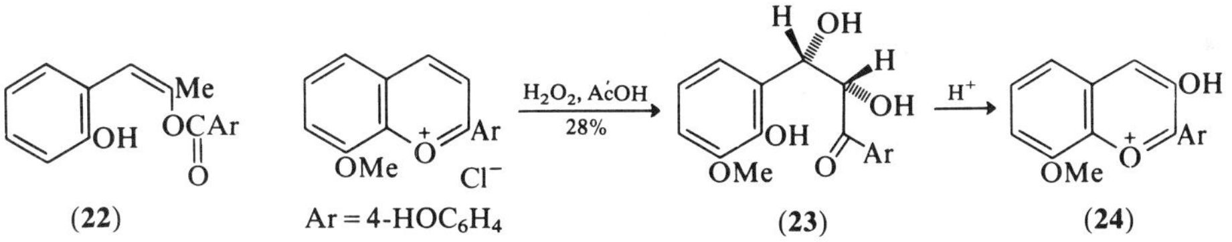

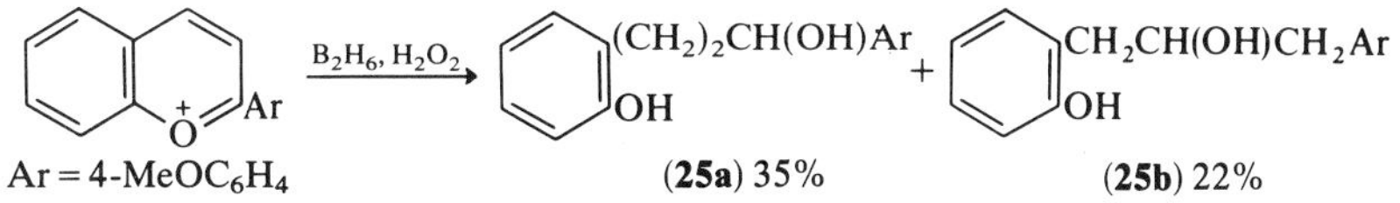

Ozonization of pyrylium salts followed by the usual peroxide or zinc–ethanol treatment cleaves the ring at the α-position to give oxo acids such as (26) ⟨51JOC1064⟩ or salicylaldehydes (27) ⟨40JA2711⟩.

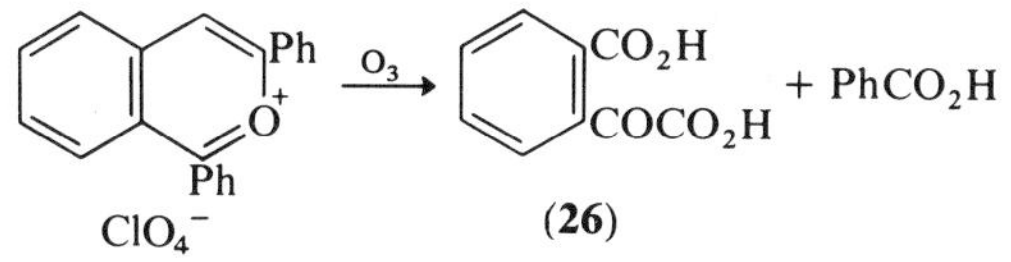

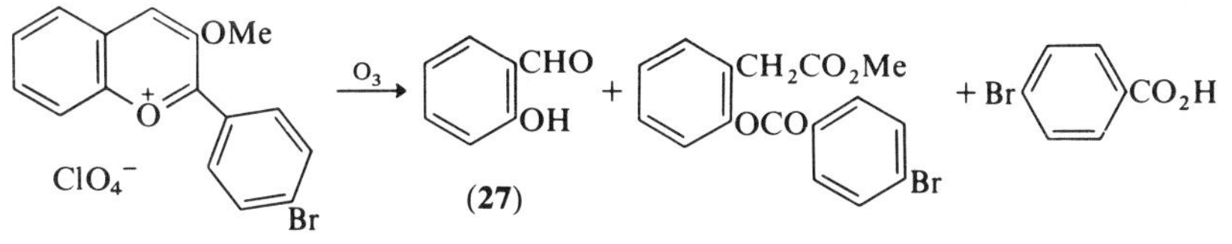

### 2.23.2.3 Nucleophiles

#### 2.23.2.3.1 Introduction

Pyrylium salts react readily with many nucleophiles either by addition or, when a suitable potential leaving group is present, by substitution. These reactions are the most important and useful conversions of pyrylium salts ⟨B-79MI22300, B-71MI22300⟩. Reduction of the salts is also discussed in this section. Although the positive charge is now usually written on the ring oxygen atom, the chemical behaviour of these compounds shows that resonance forms (28)–(31), in which the charge is at C-2, C-4 or C-6, also contribute to the overall reactivity of the molecule. In the 1-benzopyrylium ion there are only two other possible positions for the positive charge; in the xanthylium ion (32), there is only one.

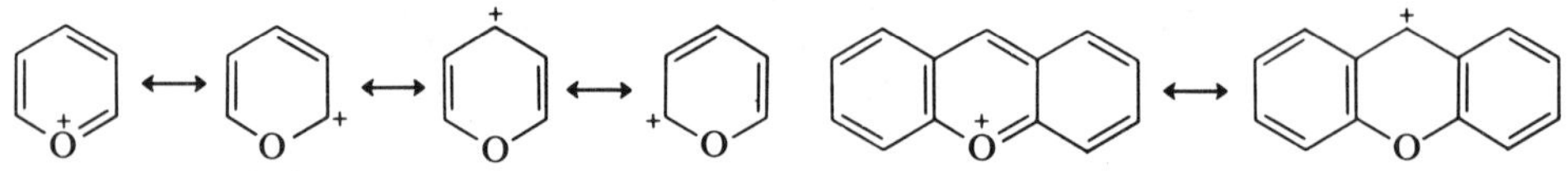

#### 2.23.2.3.2 Nucleophilic addition

Pyrylium salts vary considerably in their reactivity towards nucleophiles, the unsubstituted salt being the most reactive; for example, it adds hydroxyl ion on contact with water at ambient temperature to form 2-hydroxypyran (33; R = H) which is in equilibrium with the ring-opened enedione (34; R = H). Addition of alkali promotes the formation of the dienone. The presence of alkyl or aryl groups on the ring increases stability so that 2,4,6-triphenylpyrylium perchlorate is little affected by water but adds hydroxide ion to give the enedione (34; R = Ph) or its acyclic tautomer. Degradation of this by alkali is possible (especially when R = H or alkyl) but until that happens, acidification with strong acid reverses the process and regenerates the salt. Benzopyrylium salts are similarly attacked by aqueous alkali to give the chromen-2-ol (e.g. 35) but more drastic treatment with hot strong alkali degrades the salt; for example, peonidin (36) gives phloroglucinol and vanillic acid. This and other classical methods of degradation of flavylium salts are well documented ⟨B-62MI22300, B-62MI22301, B-62MI22302, B-63MI22300⟩.

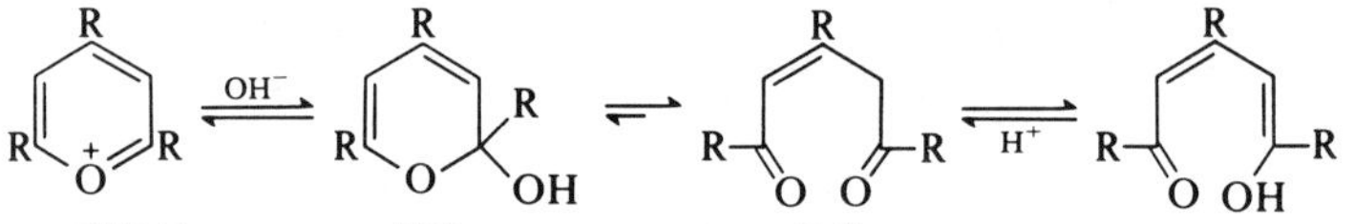

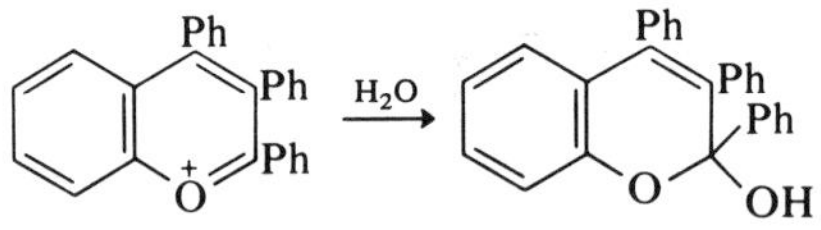

The distribution of charge in the resonance forms (28)–(31) suggests that nucleophiles may attack at C-2, C-4 or C-6 (or at C-2 or C-4 in 1-benzopyrylium cations, and at C-1, C-3 or C-4a in 2-benzopyrylium ions) but they most commonly add at C-2; for example, attack by cyanide ion gives a 2H-pyran (37) which exists partly or wholly as the acyclic isomer (38). Steric and electronic effects in the reactants probably have a role in determining the course of the reaction of trisubstituted pyrylium salts with nucleophiles. A mixture of both 2H- and 4H-pyrans is sometimes produced, for example, from methoxide ion and 2,4,6-triphenylpyrylium perchlorate (39); no acyclic product was detected in this reaction

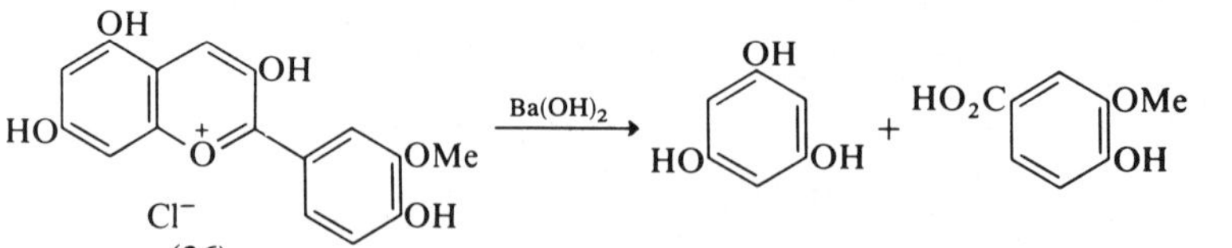

⟨80JOC5160⟩. Benzylmagnesium chloride is exceptional as it consistently attacks at C-4 even when this is a ring-fusion atom as in 1,3-dimethyl-5,6,7,8-tetrahydro-2-benzopyrylium perchlorate (**40**) ⟨64LA(678)183⟩.

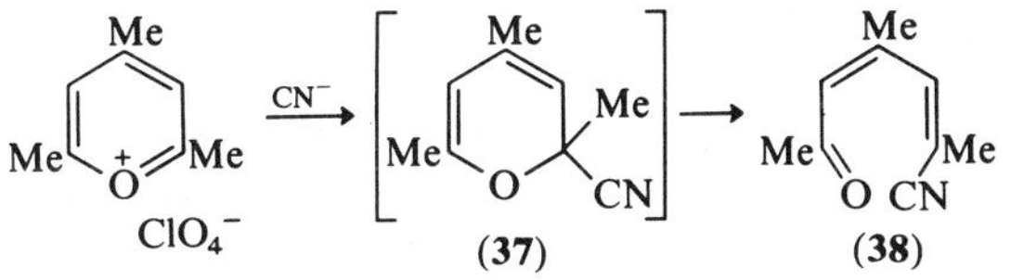

2,6-Disubstituted pyrylium salts, such as (**41**), are attacked at C-4 only ⟨74JOU2015⟩ and in the presence of perchloric acid a new pyrylium salt (**42**) is formed. Flavylium (**43**) and xanthylium (**44**) salts also react at the γ-position, for instance, with CH-acidic reagents such as pentane-2,4-dione, malonic acid derivatives or aromatic electron-rich compounds like *N,N*-dimethylaniline, 1,3-dimethoxybenzene or *N*-methylindole ⟨59CB46, 74CHE1019⟩; some examples are shown in Scheme 1. 2,4-Disubstituted pyrylium cations, *e.g.* (**45**), react at C-6 ⟨80JOC5160⟩.

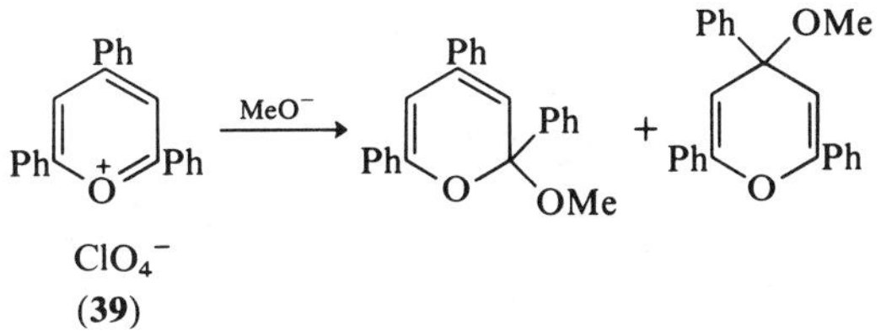

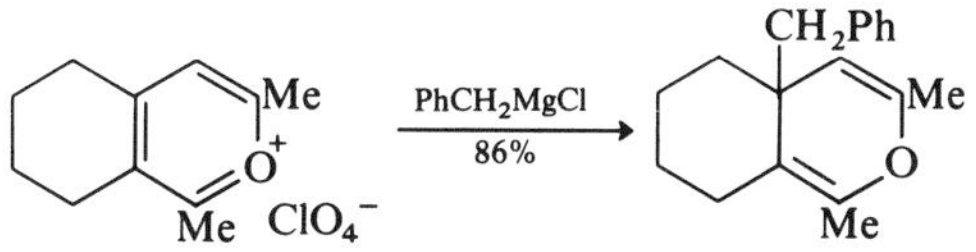

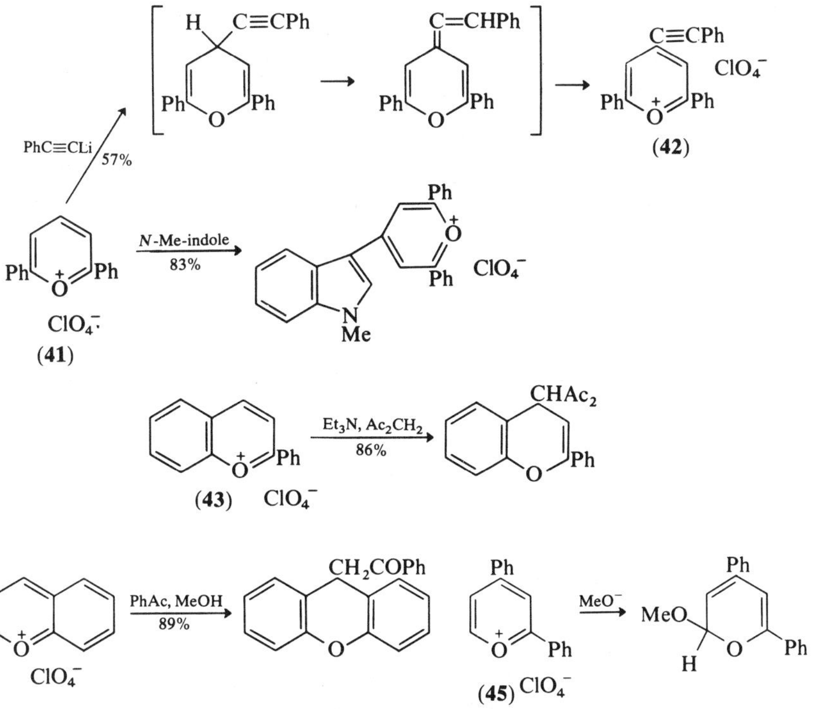

**Scheme 1**

There is evidence that the 2*H*-pyrans are thermodynamically more stable than their 4*H*-isomers. For example, the addition of methoxide ion to 2,6-diphenyl- and 4-methoxy-2,6-diphenyl-pyrylium perchlorates (**46**; R = H and R = OMe) was followed by NMR spectroscopy. The formation of the 4*H*-pyran (**47**; R = H or OMe) is rapid and at −30 °C

(**47**) is the only product, but at room temperature the dienone (**49**; R = H), presumably formed *via* the 2*H*-pyran (**48**; R = H), begins to form while the 4*H*-pyran (**47**; R = H) disappears. The dimethoxy-4*H*-pyran (**47**; R = OMe) shows no tendency to change into the dienone (**49**; R = OMe) ⟨78JOC4112⟩.

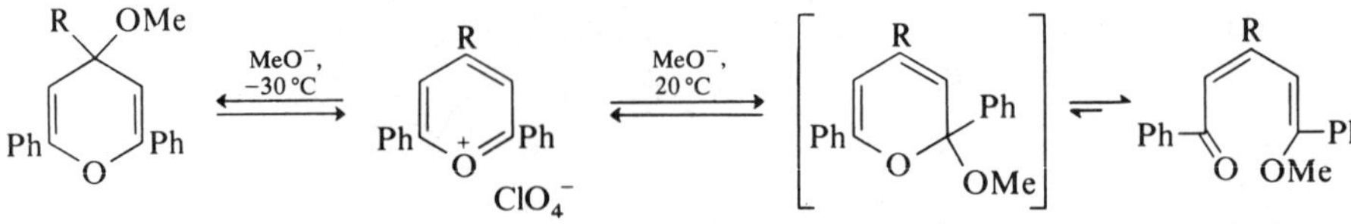

The site of addition of organometallic compounds (for examples, see Scheme 1) to pyrylium salts is dependent on the bulk of the alkyl group R of RMgX and on that of the substituents already present on the ring, but when the latter factor is unimportant, attack occurs at C-2 and the products tautomerize to the dienones. The reaction of methylmagnesium iodide with a number of pyrylium salts has been studied (Scheme 2) ⟨72BSF707⟩. Xanthylium salts react only at C-9 and this suggests that the contribution of the alternative ion (**50**) is small.

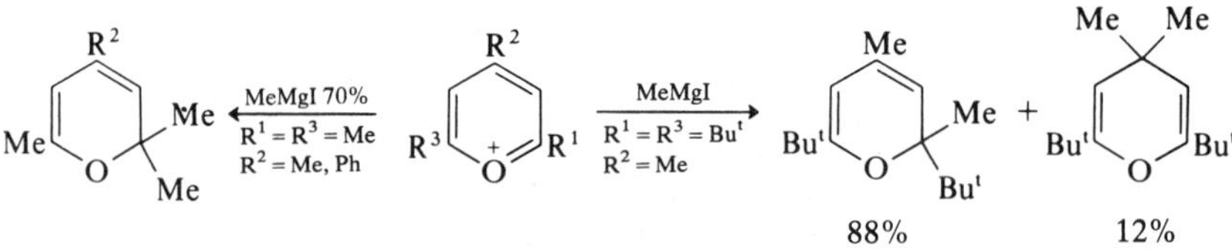

**Scheme 2**

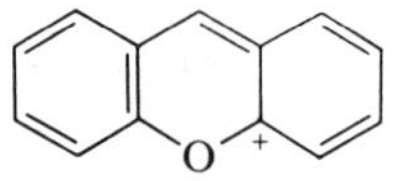

In the reaction of a Grignard reagent with a pyrylium salt which carries a displaceable substituent such as a 2- or 4-alkoxy group (see Section 2.23.3.4), the nucleophile can either add at C-2 (or C-4) or displace the alkoxy group. Addition at C-2 occurs when 4-ethoxybenzopyrylium perchlorate (**51**) is treated with one mole of Grignard reagent while ring opening results with two moles ⟨81ZOR880⟩. 4-Methoxy-2,6-dimethylpyrylium perchlorate (**52**) exhibits both types of reaction with methylamine. Use of equimolar amounts of reactants gives the pyridinium salt (*via* addition at C-2, see Section 2.23.2.3.3) while an excess of amine results in addition and replacement of the methoxy group ⟨46JCS117⟩.

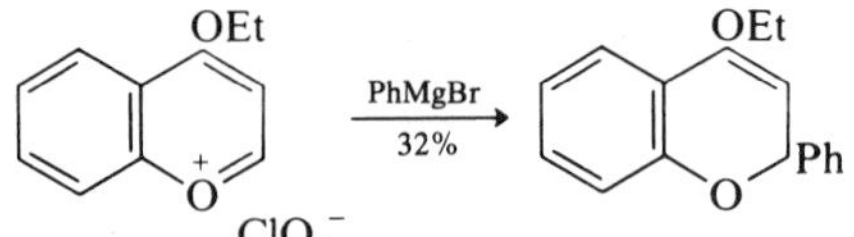

### 2.23.2.3.3 Formation of a new six-membered N-ring

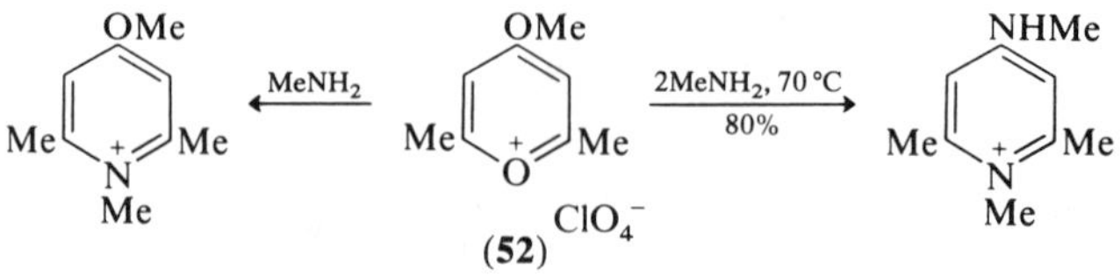

Some nucleophiles open the pyrylium ring to give intermediates which are capable of spontaneous or acid-induced cyclization to a new ring. The conversion of pyrylium salts into pyridines is the most thoroughly studied example and was first described in 1911 ⟨B-64MI22300⟩. Ammonia, ammonium acetate or carbonate react with pyrylium salts such as (**53**) under mild conditions to give high yields of pyridine derivatives. Primary amines, hydrazines, acylhydrazines, ureas, semicarbazide, hydroxylamine and amino acids react similarly to produce pyridinium compounds ⟨B-71MI22300, 79CHE265, 79JCS(P1)426, 79JCS(P1)446⟩. Scheme 3 shows some typical examples of this reaction, which has been applied to the synthesis of compounds of potential biological interest ⟨80RRC1505⟩.

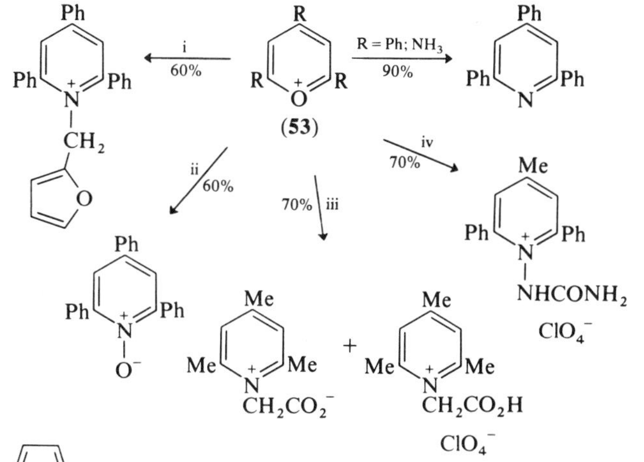

i, R = Ph; CH₂NH₂; ii, R = Ph; NH₂OH, AcOH; iii, R = Me; H₂NCH₂CO₂H;

iv, R = Me; H₂NCONHNH₂

**Scheme 3**

The nucleophile attacks at C-2 of the salt (**54**) and the ring opens in a fast base-catalyzed reaction to give the 5-aminodienone (**55**). Its cyclization to the pyridinium salt (**56**) is a slow acid-catalyzed step ⟨80T1643, 80AG(E)306⟩. When secondary amines react with a pyrylium salt (**54**), the reaction usually terminates at the dienone but a 2-methyl group allows the formation of a benzene ring (Section 2.23.2.3.4). A recent report describes the reaction of piperidine with 2,4,6-triphenylpyrylium perchlorate to give a high yield of the 2*H*-pyran (**57**) ⟨81ZC282⟩. 1-Benzopyrylium cations do not react with ammonia to give aminodienones but 2-benzopyrylium compounds, for example (**58**), are converted into isoquinolines ⟨75CHE18⟩ and 5,6,7,8-tetrahydro-1-benzopyrylium perchlorates, such as (**59**), yield the tetrahydroquinolines (**60**) ⟨B-64MI22300⟩.

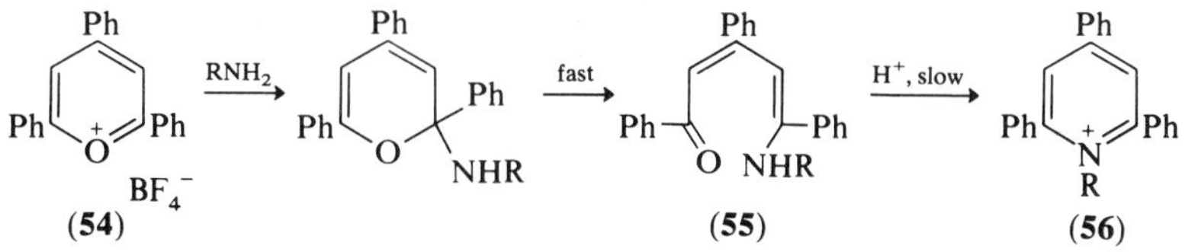

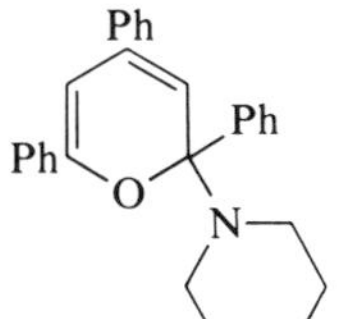

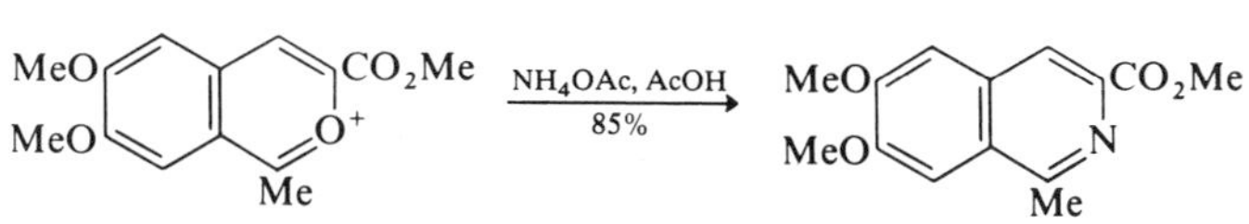

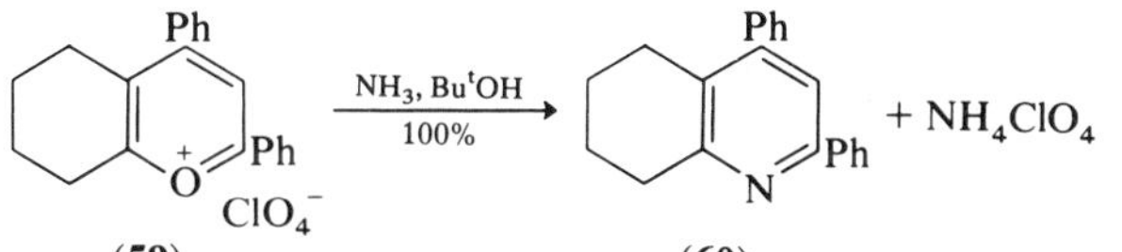

An interesting variation on the above conversion of pyrylium into pyridinum salts is the reaction of the former with cyanamide in which a 3-acyl-2-aminopyridine (**61**) is formed in 75% yield, probably by the mechanism shown ⟨71JCS(C)3873⟩.

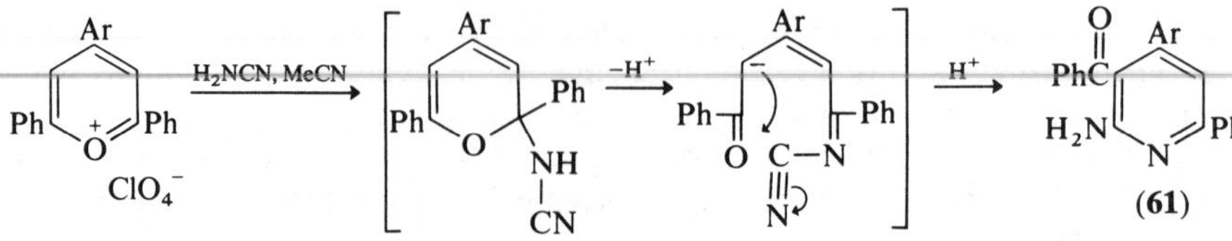

1,4-Disubstituted pyridinium salts are not readily prepared but a new route extends the use of pyrylium salts for this purpose. 2,6-Diethoxycarbonyl-4-phenylpyrylium tetrafluoroborate (62) reacts with primary amines to give the expected pyridinium salt (63), which on heating with *t*-butylamine is de-ethoxycarboxylated to the desired pyridinium salt in high overall yield ⟨81S959⟩.

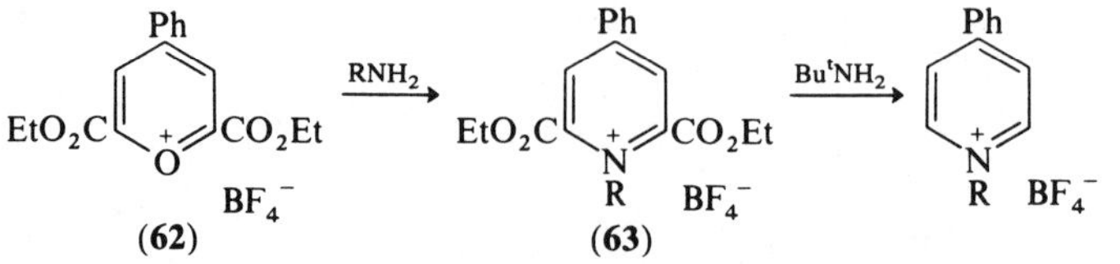

The CH$_2$—N$^+$ bond of *N*-substituted pyridinium salts such as (64) is susceptible to attack by a wide variety of nucleophiles. Thus, primary amines may be converted into tertiary amines, halides, thiocyanates, nitriles, *N*,*N'*-diarylcarbodiimides, *N*,*N*-dimethyl-dithiocarbamates, thiobenzoates and isocyanates among others, usually in good yields (Scheme 4). This work has been conveniently summarized ⟨80T679⟩. Although much of the work on pyrylium salts has been concerned with the perchlorate and tetrafluoroborate anions, the sensitivity of the former to heat the difficulty in obtaining accurate elemental analytical figures for the latter have stimulated the search for their replacement, for example by trifluoromethanesulfonates, naphthalenesulfonates, fluorosulfonates and trifluoroacetates ⟨80JCR(S)310⟩. Replacement of one or more of the 2-, 4- and 6-substituents on the pyrylium cation by heterocyclic rings has also been investigated ⟨80JCR(S)312⟩.

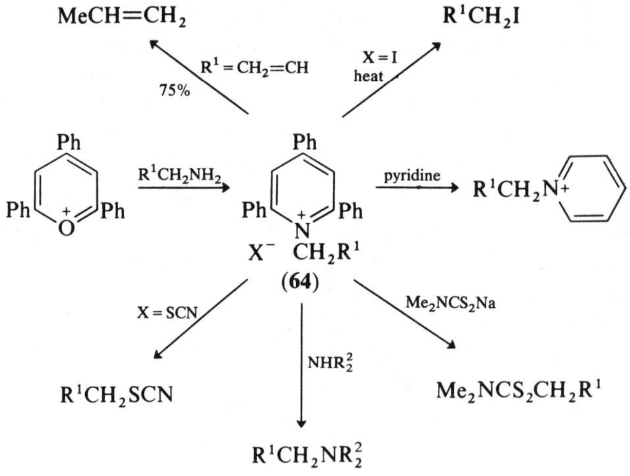

**Scheme 4**

Recently, pyrylium salts have been treated with various nitrogenous compounds, for example guanidine, aminoguanidine, isothiourea or sulfanilylguanidines, to form 4,6-diphenylpyrimidines (65) and (66). In this way *N*-methylisothiourea and 2,4,6-triphenyl-pyrylium perchlorate (39) give 2-methylthio-4,6-diphenylpyrimidine (65) ⟨80CHE574, 79CHE265, 75CHE240⟩.

When the amino compound contains another suitably placed reactive atom, as in 2-aminobenzimidazole (67), a new heterocyclic system (68) is produced together with some of the expected product (69) ⟨75CHE1025⟩.

### 2.23.2.3.4 *Formation of a new benzene ring*

Pyrylium cations may be converted into benzene rings by a long-established reaction ⟨14LA(407)332⟩ with carbanions; this can be a valuable synthetic route. Such reactions may

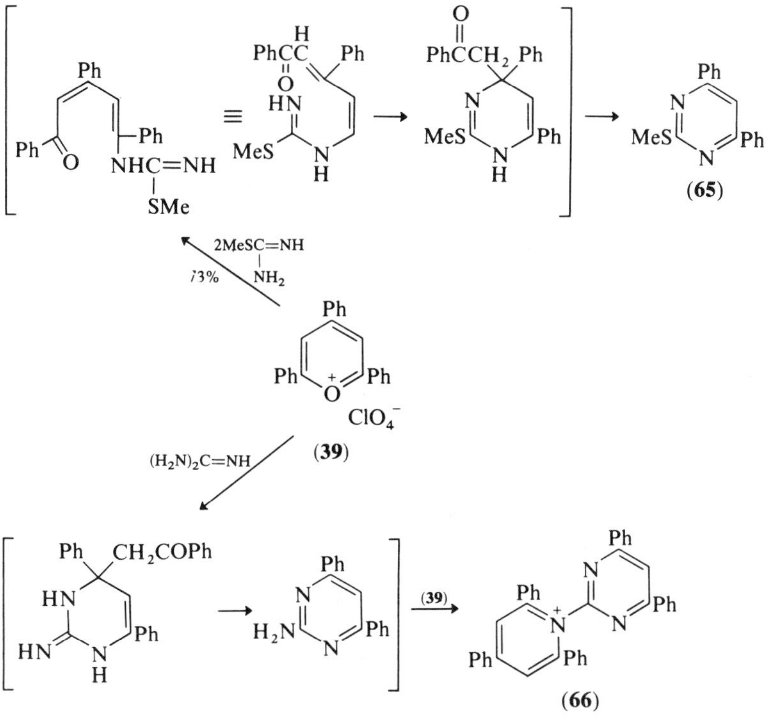

(65)

(66)

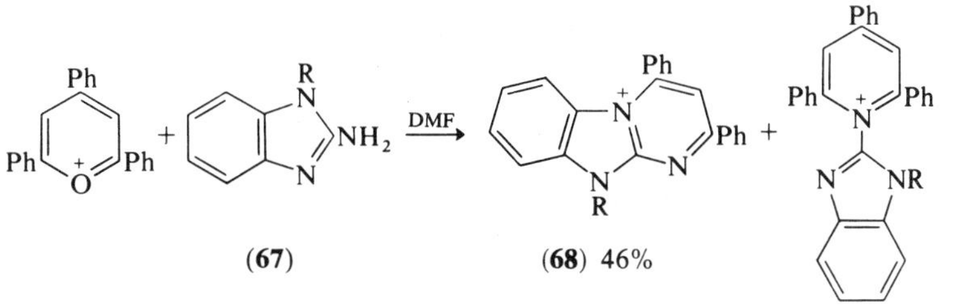

(67)   (68) 46%   (69) 32%

be divided into two types according to the source of the carbon atom which replaces the oxonium oxygen atom. Firstly pyrylium salts (**70**) which contain a 2-RCH$_2$ substituent (where R is H or alkyl) are converted into phenols (**72**) in which the $\alpha$-CH$_2$ is incorporated in the ring but a 6-aryl group diverts the reaction along another path. Addition of hydroxide ion at C-2 and ring opening to the dienone (**71**) is followed by cyclization. It provides a convenient synthesis of compounds such as prehnitenol 〈80SC195〉. When the nucleophile is a secondary amine, the pyrylium salt reacts at ambient temperature to produce *N,N*-disubstituted anilines 〈27CB716〉 such as (**73**).

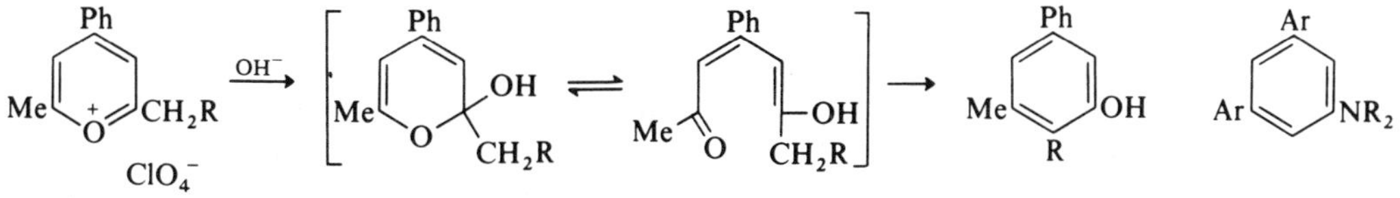

(70)   (71)   (72)   (73)

In the second type, the sixth carbon atom of the benzene ring is provided by a reagent which contains a methylene group activated by an adjacent electron-attracting function, for example nitro, carbonyl, cyano, triphenylphosphine or *S*-methylide. Table 1 shows examples of these reactions which were first described in 1956 〈56AG519〉 and have been widely studied. The reaction of nitromethane and phenylnitromethane can take one of several courses depending on the amount of base, and thus give the different products shown in Scheme 5 for nitromethane 〈69AG(E)370, 66CB3040〉 and Scheme 6 for phenyl-nitromethane 〈66CB399〉. The latter reaction can give a mixture of the phenol and nitrobenzene but by isolating the intermediate nitrodienol (**74**) and subjecting it to thermolysis or

**Table 1** Conversion of Pyrylium Salts into Benzenoid Compounds

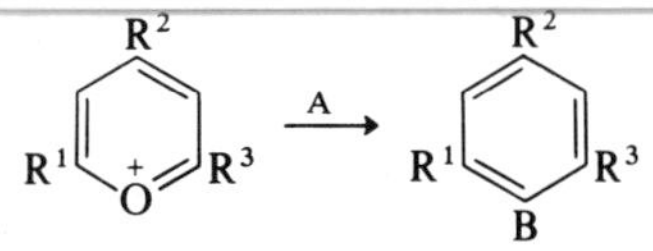

| $R^1$ | $R^2$ | $R^3$ | A | B | Yield (%) | Ref. |
|-------|-------|-------|---|---|-----------|------|
| Me | Me | Me | MeNO$_2$ | NO$_2$ | 72 | 57CB1668 |
| Me | Me | Me | CH$_2$=SMe$_2$ | OH[a] | 52 | 81H(16)17 |
| Me | Me | Me | PhMgBr | [b] | 81 | 61CB1796 |
| Me | Me | Me | CH$_2$(CN)$_2$ | [c] | 75 | 61CB1796 |
| Me | MeO | MeO | (MeO)$_2$PO EtO$_2$CCH$^-$ | CO$_2$Et | 65 | 75CC675 |
| Me | Et | Me | NaOH | OH | 40 | 59LA(625)74 |
| Ph | Ph | Ph | PhAc | Ac | 70 | 59CB2042 |
| Ph | Ph | Ph | CH$_2$Ac$_2$ | Ac | 70 | 59CB2042 |
| Ph | Ph | Ph | Ph$_3$P=CH$^-$ EtO$_2$C | CO$_2$Et | 59 | 62AG(E)511 |
| Ph | Ph | Ph | CH$_2$CN CO$_2$Et | CN | 81 | 59CB2042 |
| Ph | Ph | Ph | MeNO$_2$[d] | NO$_2$ | 85 | 57CB1634 |
| Ph | Ph | Ph | MeNO$_2$[e] | OH | 36 | 69AG(E)370 |
| Ph | Ph | Ph | i, PhCH$_2$MgCl ii, base | Ph | 51 | 64LA(678)183 |
| Ph | Ph | Ph | [f] | NHCOPh | 14 | 72JCS(P1)1142 |
| Ph | Ph | Ph | CH$_2$(CN)$_2$ | [g] | 73 | 59CB2046 |
| Ph | 4-MeC$_6$H$_4$ | Ph | PhCH$_2$NO$_2$ | [h] | 52 | 66CB399 |
| | | | | [i] | 51 | 66CB399 |
| 4-MeC$_6$H$_4$ | 4-MeC$_6$H$_4$ | Me | NHMe$_2$ | NMe$_2$ | — | 27CB716 |
| [j] | Ph | Ph | MeNO$_2$ | NO$_2$ | — | B-64MI22300 |

[a] Main product; 2,3,5-Me$_3$C$_6$H$_2$OH also formed (33% yield). [b] Product is 3,5-Me$_2$C$_6$H$_3$Ph. [c] Product is 3,5-Me$_2$C$_6$H$_3$CH(CN)$_2$. [d] 2 mol of base. [e] 1 mol of base. [f] 2-Phenyloxazol-5-one. [g] Product is 2-NH$_2$-3-PhCO-4,6-Ph$_2$-benzonitrile. [h] Product is 6-*p*-MeC$_6$H$_4$-2,3,4-Ph$_3$C$_6$HOH. [i] Product is 6-*p*-MeC$_6$H$_4$-2,3,4-Ph$_3$C$_6$HNO$_2$. [j] 2,3-(CH$_2$)$_4$ homologue of reactant and product.

basic conditions, good yields of the phenol or the nitrobenzene may be obtained. The absence of an absorption at 1680–1770 cm$^{-1}$ in the IR spectrum of (**74**) excludes the possibility that it has the dienone (**75**) structure.

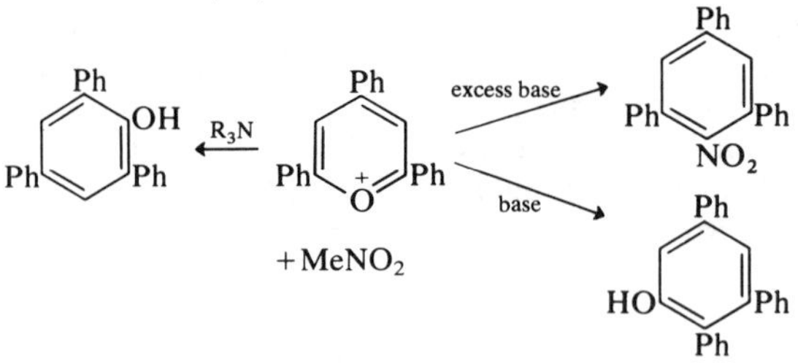

**Scheme 5**

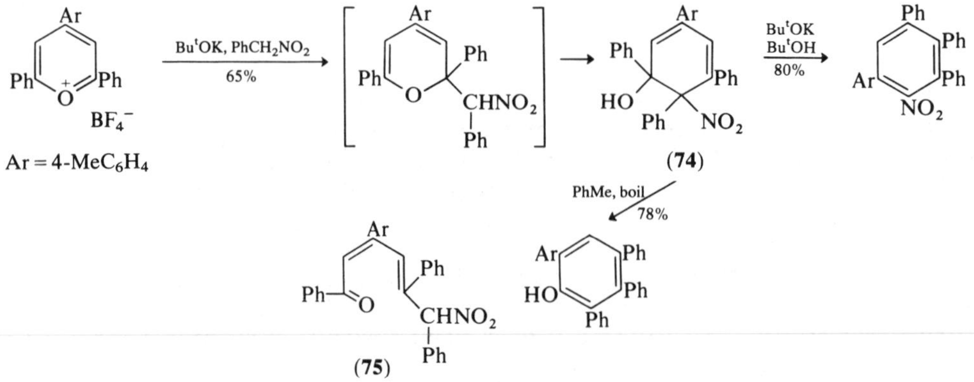

**Scheme 6**

Under the basic reaction conditions, fragments of the carbonyl compounds are lost; for example, the acetyl group of 3-oxobutanoate and pentane-2,4-dione are cleaved. Examples are given in Table 1. 1-Benzopyrylium salts undergo this kind of reaction only when the carbocyclic ring is reduced as, for example, in the reaction of 2,4-diphenyl-5,6,7,8-tetrahydrobenzopyrylium tetrafluoroborate (**76**) with nitromethane. The product may be dehydrogenated to give a nitronaphthalene ⟨B-64MI22300⟩. The reaction of nitromethane is believed to proceed through an acyclic valence isomer (**77**) and the sixth carbon atom of the new ring is that of nitromethane.

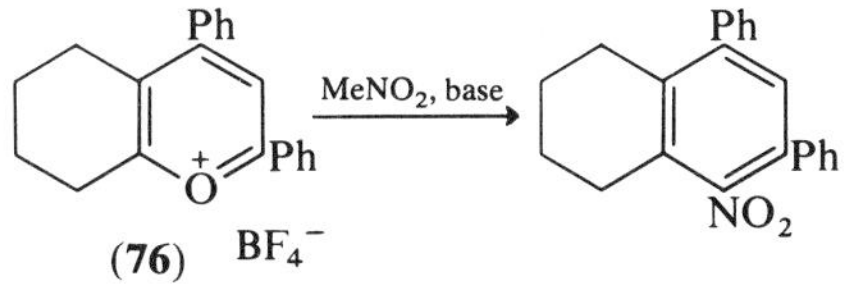

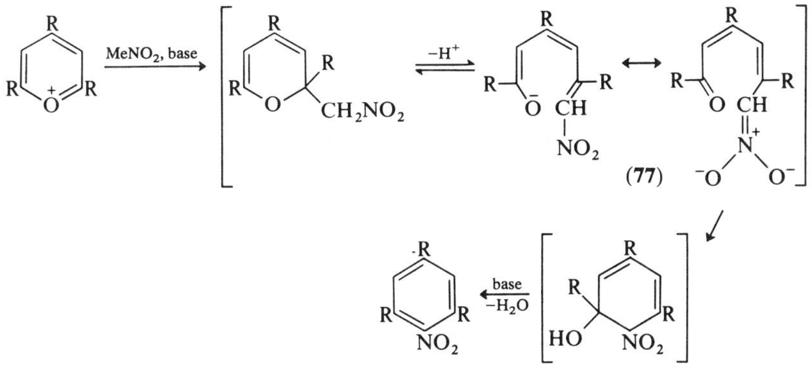

### 2.23.2.3.5 *Formation of other rings*

Pyrylium salts, especially the 2,4,6-triaryl substituted, react with a wide variety of nucleophiles some of which lead to the formation of rings other than benzene or pyridine. Sodium sulphide, for instance, attacks the pyrylium ring at C-2 to produce the blue-coloured dienone (**78**) which then cyclizes on treatment with acid to the thiopyrylium salt (**79**) ⟨56HCA207⟩.

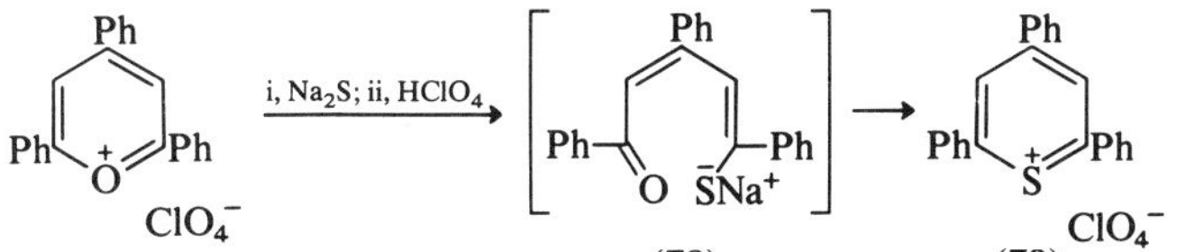

A phosphorus atom replaces oxygen when the pyrylium salt is treated with one of several nucleophilic phosphines; for example, tris(trimethylsilyl)phosphine gives 2,4,6-triphenylphosphorin (**80**) ⟨67AG(E)458⟩. The earliest use of 2,4,6-tri-*t*-butylpyrylium tetrafluoroborate (**81**) was its conversion to the corresponding phosphabenzene (**82**) ⟨68AG(E)461⟩.

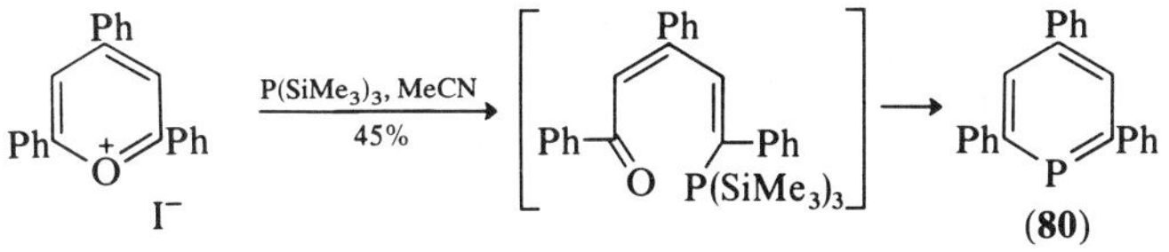

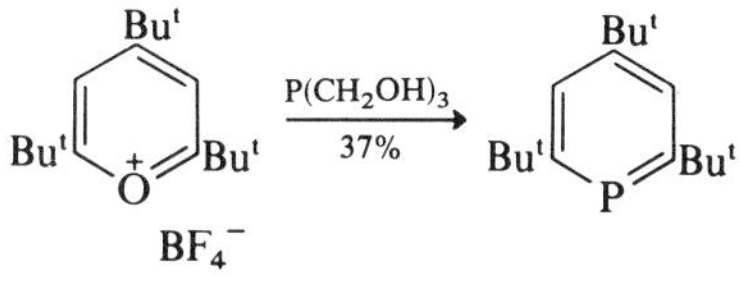

The formation of furans from pyrylium salts under oxidative conditions has been mentioned in Section 2.23.2.2. Ring opening and recyclization of 4-chloromethyl-2,6-dimethylpyrylium perchlorate (83) by aqueous alkali and DMF leads to 2-methylfuran-4-ylacetone (84) ⟨77CHE918⟩. When 2,3,4,5-tetraphenylpyrylium perbromide is hydrolyzed to the dienone and then subjected to bromination–dehydrobromination, 2-benzoyl-3,4,5-triphenylfuran (85) is formed.

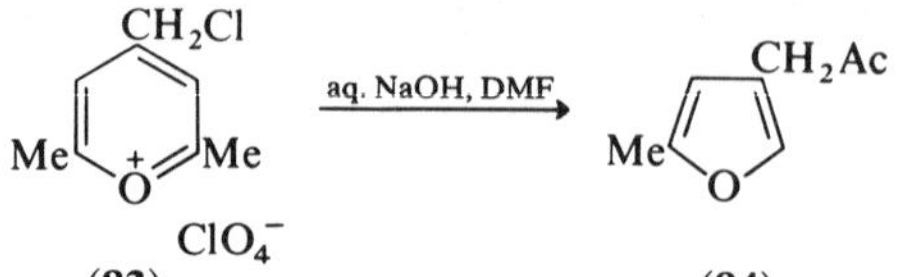

Triphenylpyrylium perchlorate reacts with sodium nitrite in an alcohol to give a mixture of cyclic and acyclic products; when the reaction is conducted in acetonitrile, the main product is 3,5-dibenzoyl-4-phenylisoxazole (86) ⟨75ACS(B)285⟩ which is a minor product under different conditions. Hydroxylamine and hydrazines add to the triphenylpyrylium salts to give acyclic products which cyclize again to give isoxazoles (87) and pyrazoles (88) respectively ⟨68T5059, 70T739⟩, but under some conditions, 4*H*-1,2-diazepines (89) are the main products ⟨80CJC494⟩. An oxazepine (90) is obtained when 2,3,5,6-tetraphenylpyrylium perchlorate is treated at low temperature with sodium azide and the intermediate azide and azirine are pyrolyzed ⟨75CR(C)(280)37⟩.

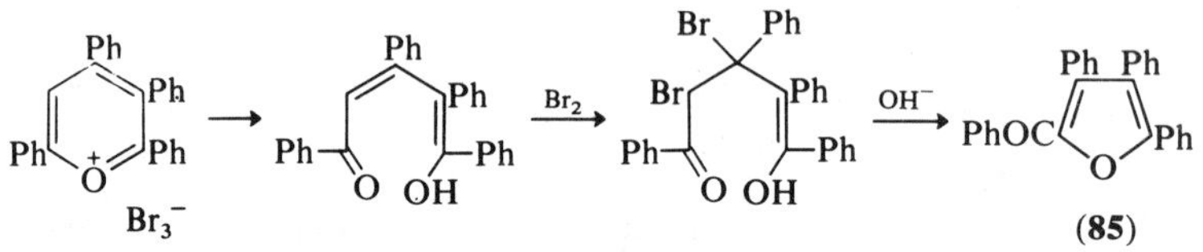

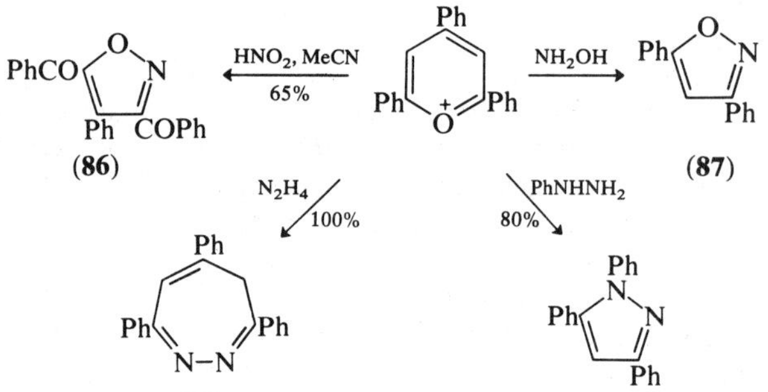

Some pyrylium salts undergo a combined ring expansion and annelation when treated at room temperature with the sodium or preferably lithium salt of cyclopentadiene ⟨58LA(618)140⟩; thus, 2,4,6-trimethylpyrylium perchlorate reacts to form 4,6,8-trimethylazulene (91) which has recently been treated again with a pyrylium salt to form 4-(4,6,8-trimethylazulen-1-yl)-2,6-diphenylpyrylium perchlorate (92). The latter behaved as expected with ammonium acetate and gave the pyridine derivative (93) ⟨80CHE807⟩. In another conversion of a pyrylium salt (two molecules) into an azulene derivative (94), methylenetriphenylphosphorane was the reagent and its methylene carbon becomes a part of the seven-membered ring while a C-2 carbon atom becomes an exocyclic carbonyl group ⟨60AG778⟩.

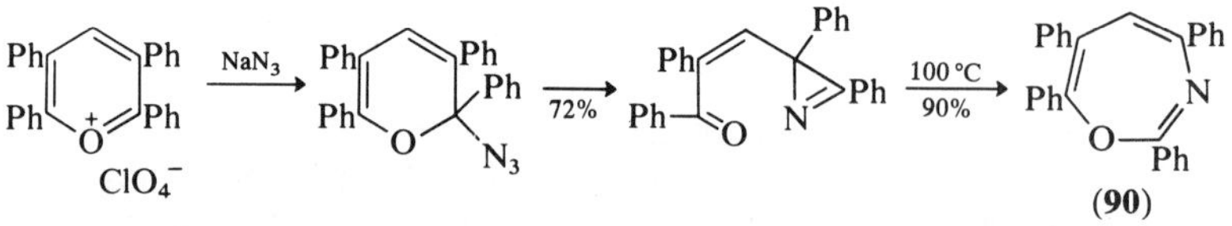

An interesting reversible conversion of the bipyrylium salt (95) into a spiro compound (96) is effected in high yield and under mild conditions, namely, aqueous sodium acetate at ambient temperature. The spiran (96) is reactive, for example, towards nucleophiles such as hydroxylamine (R = OH) or phenylhydrazine (R = PhNH) ⟨80CHE345⟩.

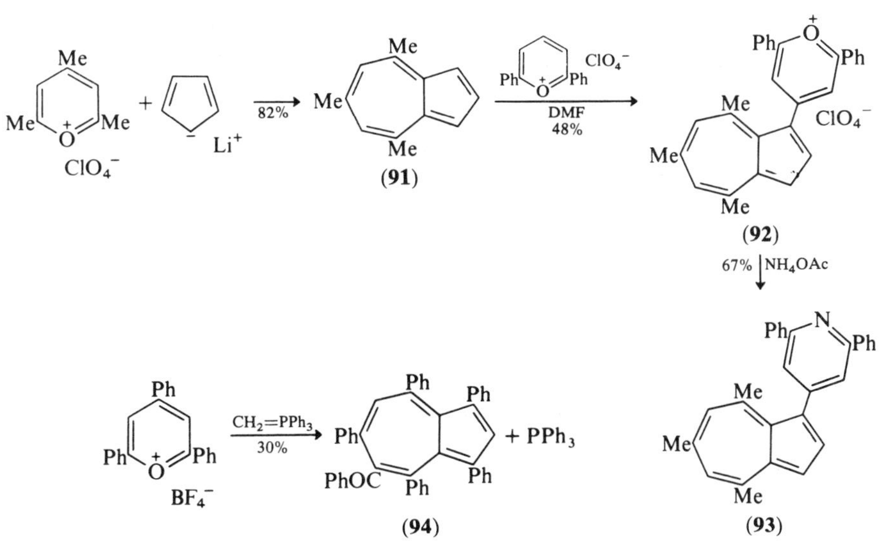

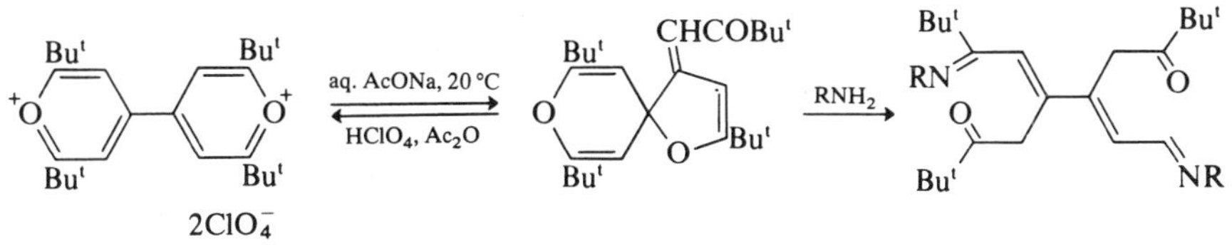

(91)

(92)

(93)

(94)

(95)

(96)

### 2.23.2.3.6 Reduction

2,4,6-Trisubstituted pyrylium salts are reduced by chemical and by catalytic methods but the two do not usually yield the same products. Reduction with zinc and water or by the dipotassium salt of cyclooctatetraene gives a dimeric product ⟨61JOC2260⟩, while sodium borohydride produces a mixture of acyclic (major) and cyclic products which result from attack at C-2 and C-4 followed by ring opening (Scheme 7). Spectral evidence is available for the formation of the 2*H*-pyran (**97**) which is rapidly cleaved at 25 °C to the dienone (**98**) in a reaction of first order kinetics ⟨72JOC3036⟩. The ratio of 4*H*-pyran to dienone varies according to the nature of the substituents on the pyrylium ring; for instance, for the trimethyl salt, the ratio is 1 : 7 but 2,6-di-*t*-butyl-4-methylpyrylium perchlorate is reduced solely to the dienone ⟨72BSF2510⟩. When 2,4,6-triphenylpyrylium cations are electrochemically reduced in the presence of alkyl halides, they give the 4-alkyltriphenylpyran ⟨80MI22300⟩.

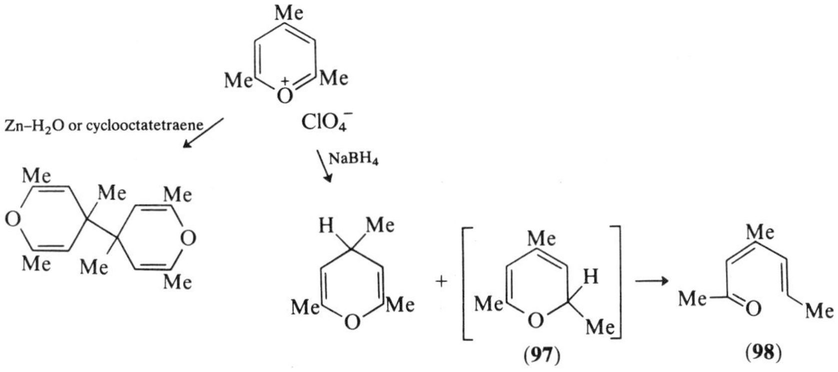

(97)

(98)

**Scheme 7**

Benzopyrylium salts are reduced by sodium borohydride to the 4*H*-chromene but LAH yields the 2*H*-isomer. The reduction products of flavylium salts vary with the conditions used; even the method of isolating the product has to be carefully controlled since, for example, 4*H*-flavenes may be isomerized by contact with acid to 2*H*-flavenes. Flavylium perchlorate in acetonitrile at 35 °C is reduced under nitrogen in 85% yield by zinc and in the absence of water to the bis(4*H*-flavene) (**99**) but on heating the same mixture in air, flavone and a polymeric material are produced ⟨67JOC3772⟩. Reduction of the salt with sodium borohydride at 18 °C gives either a bisflavan (**100**) or a 4*H*-flavene (**101**) according to the solvent employed ⟨67JOC3616⟩. A more recent report shows that high yields of 3-alkyl-2*H*-chromenes and 2*H*-isoflavene are obtained by a similar reduction in THF at ambient temperature, but when 3-ethylbenzopyrylium perchlorate is treated with hot formic acid–pyridine, a 1 : 3 mixture of 4*H*- and 2*H*-chromene is obtained ⟨77BSF1187⟩.

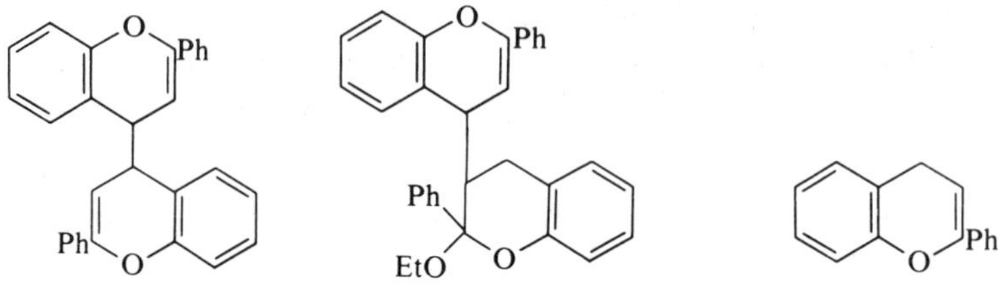

(**99**)                                 (**100**)                                 (**101**)

Potassium borohydride in THF reduces 3-methoxyflavylium perchlorate to the 2*H*-flavene in high yield; similarly, flavylium salts give a flavene or flavan ⟨58JCS4302, 72P3491, 76BSF1967⟩. Two stereoisomeric bis(4*H*-flavenes) (**103**) and (**104**) are obtained by treating the flavylium chloride (**102**; Ar = 4-HOC₆H₄) with zinc and ethanol; the reaction is reversed by benzoquinone ⟨68T2801⟩.

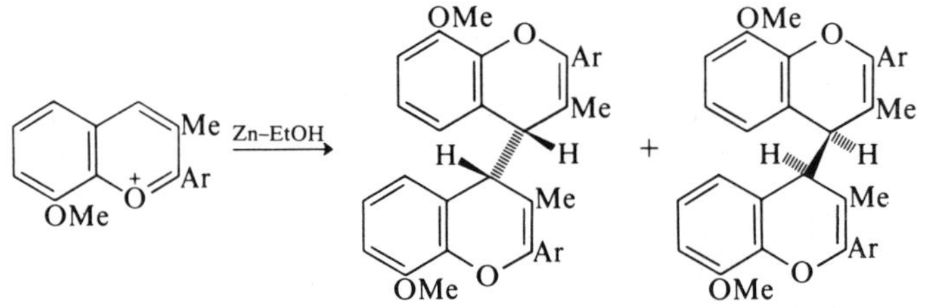

(**102**) Ar = 4-HOC₆H₄                (**103**)                                 (**104**)

In view of these and similar reports, it is clear that the reduction products of pyrylium and flavylium salts depend to a considerable extent on the nature of any substituents present, the reducing agent and the precise conditions used.


## 2.23.3 PYRYLIUM SALTS: REACTIVITY OF SUBSTITUENTS

### 2.23.3.1 Introduction

The pyrylium ring has a marked effect on substituents which are attached to it but the effect varies considerably according to the reagent and the position of the substituent. The positive character of the pyrylium ring of benzopyrylium salts inhibits attack by electrophiles on both rings but the ease with which hydrogen atoms of a 2- or 4-methyl group may be removed is increased. Similarly, nucleophilic substitution of a good leaving group attached at C-2 or C-4 of a pyrylium ring is readily achieved.

Nucleophiles add to the pyrylium ring to give 2*H*- or 4*H*-pyrans together with some ring-cleaved products which may be capable of being cyclized again to give a different ring. This type of reaction enables pyrylium salts to be employed in the synthesis of a variety of non-pyran compounds.


### 2.23.3.2 C-Linked Substituents

Alkyl groups RCH₂, where R is a hydrogen or an alkyl or aryl group, attached to the 2- or 4-positions of pyrylium rings are more reactive than those of the corresponding pyran or pyridine. The methylene hydrogen atoms are acidic, that is they are easily abstracted with consequent formation of an anionic centre, for example in (**105**). The neutral molecule

(**106**) is a resonance form of (**105**) and the stability of such species is increased by electron-withdrawing substituents or conjugation.

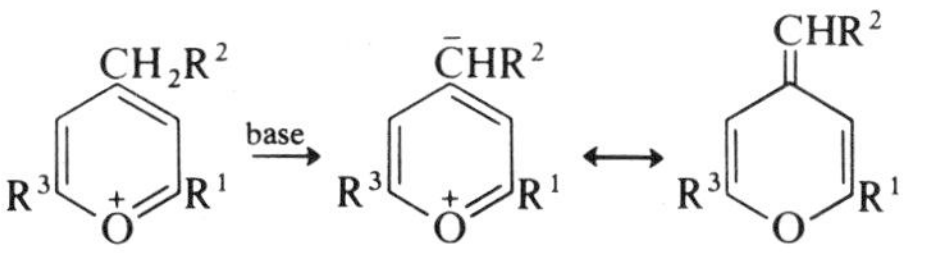

(**105**)    (**106**)

In contrast to the preferential attack on ring atoms at C-2 by nucleophiles (Section 2.23.2.3.2), a substituent at C-4 is more susceptible (than if it were at C-2) to attack by an electrophile such as a carbonyl compound. Thus, 2,4,6-trimethylpyrylium perchlorate (**107**) yields the coloured 4-styryl derivative (**109**) as the only product of reaction with a benzaldehyde ⟨67JOU1809, 71T3503⟩ but a 2-methyl group in cations such as 4,6-diphenyl-2-methylpyrylium (**108**) also reacts; reaction at a second methyl group in (**107**) requires drastic conditions ⟨72BSF3173⟩.

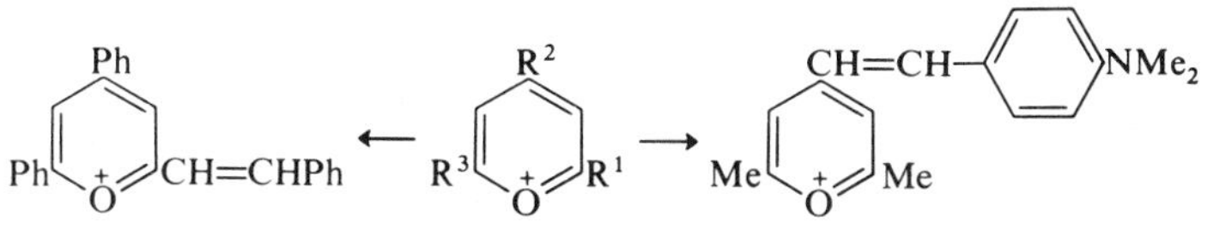

(**107**) $R^1 = R^2 = R^3 = Me$        (**109**)
(**108**) $R^1 = Me$, $R^2 = R^3 = Ph$

Condensation occurs in a basic, neutral or acidic medium and high yields are often obtained at moderate temperatures. An electron-releasing group in the aryl ring(s) attached to the pyrylium ion decreases the reactivity of the methylene group ⟨66JGU1724⟩. Similar reactions occur in the benzopyrylium and xanthylium series and the products sometimes have potential as dyes.

The α- and γ-methylene groups of pyrylium salts condense with pyranones and benzopyranones when the salts are heated in acetic anhydride or phosphorus oxychloride. Some of the resulting compounds are potentially interesting as dyes, for example the tripyran (**110**) ⟨68JOC4418⟩. In hot acetic anhydride equimolar amounts of the salt (**111**) and flavone condense to give a green-coloured salt (**112**) ⟨78AP236⟩.

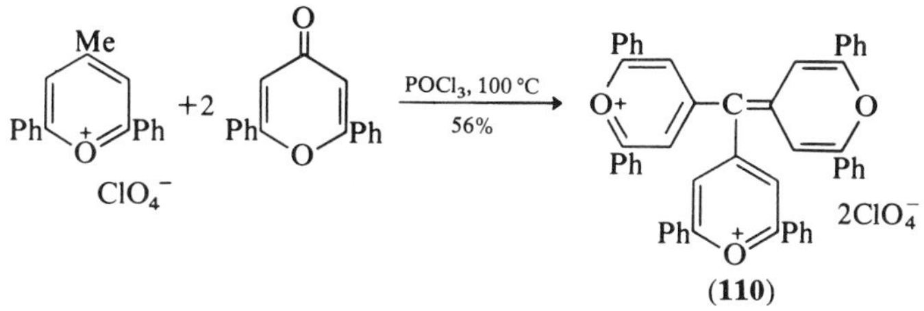

(**110**)

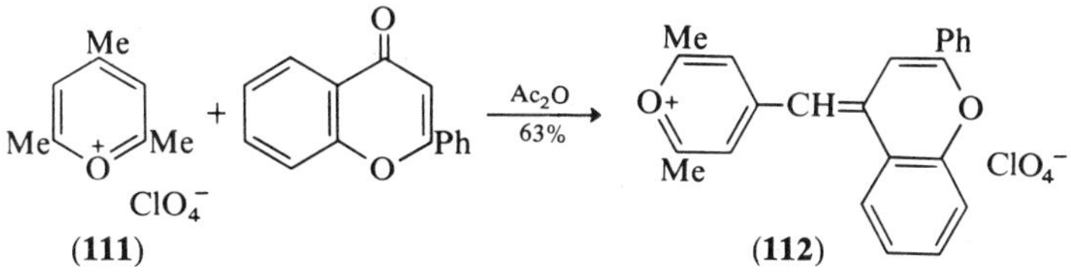

(**111**)        (**112**)

Potential dyes in which the pyran rings are separated by three carbon atoms are formed by reaction of a 4-methylpyrylium salt (**113**) with DMF–acetic anhydride and heating the resultant 4-formylmethylenepyran (**114**) with another molecule of 4-methylpyrylium salt. The product (**115**) absorbs strongly at 598 nm ($\varepsilon$ 267 000 mol$^{-1}$ cm$^{-1}$). The relationship between colour and structure in such compounds has been discussed in detail ⟨73T795⟩.

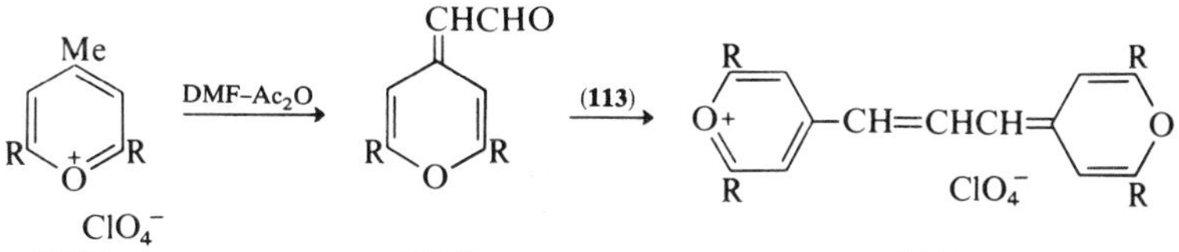

(**113**)        (**114**)        (**115**)

Methylene groups of pyrylium cations also react under mild conditions with nitrosobenzenes in the presence of acetic anhydride and an alkali metal salt, which plays an essential role in the formation of Schiff's bases (**116**). The latter are hydrolyzed to the carbonyl compound (**117**) ⟨71BSF3603⟩. It is possible to form potential dyes by treatment of pyrylium salts with ortho esters such as triethyl orthoformate. The reactants are heated together in acetic anhydride or pyridine and trimethine dyes of type (**118**) and (**119**) are formed ⟨70JHC1395, 71JOC600, 59CB2309⟩.

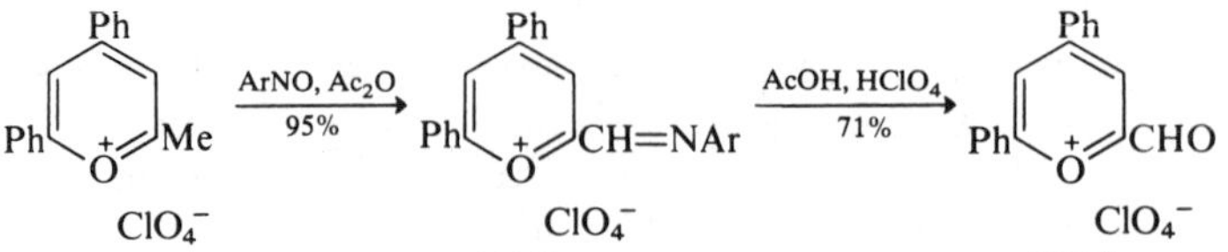

(**116**)    (**117**)

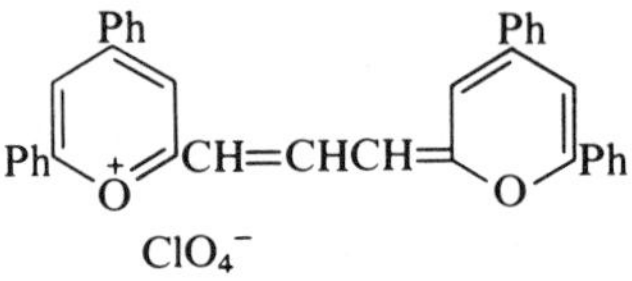

(**118**) 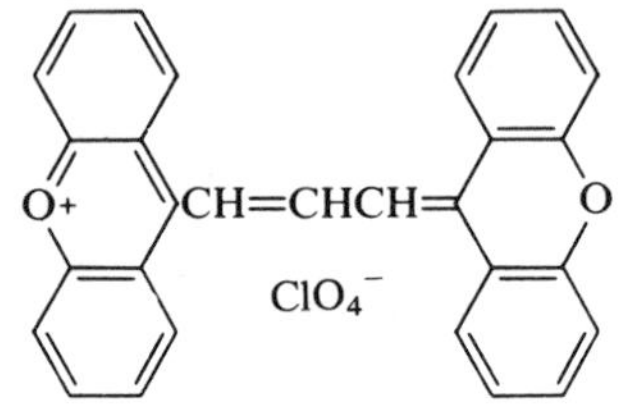    (**119**)

The 2- and 4-methyl groups of pyrylium salts undergo several other reactions ⟨79H(12)51⟩; for example, with acyl halides they form dimeric products (**122**) ⟨44CB529⟩, with Mannich reagents they give the normal bases (**120**) ⟨65JOU983⟩ and on nitration under vigorous conditions, they yield nitromethylpyrylium salts (**121**) ⟨64T483⟩.

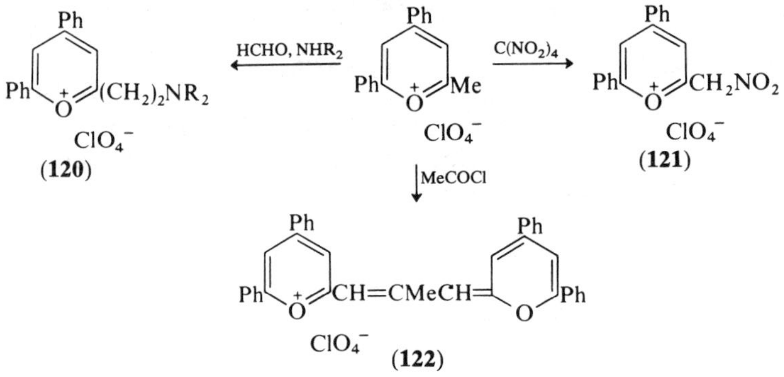

(**120**)    (**121**)

(**122**)

### 2.23.3.3. *O*-Linked Substituents

Pyrylium salts containing a substituent at C-2 or C-4 which forms a good leaving group, react with nucleophiles to give 2*H*- or 4*H*-pyrans; one such group is alkoxy. Water acts as a nucleophile in the convenient synthesis of chromones (such as 3-hydroxychromone) from benzopyrylium salts ⟨78JCR(S)47⟩; a 4-alkoxy group may also

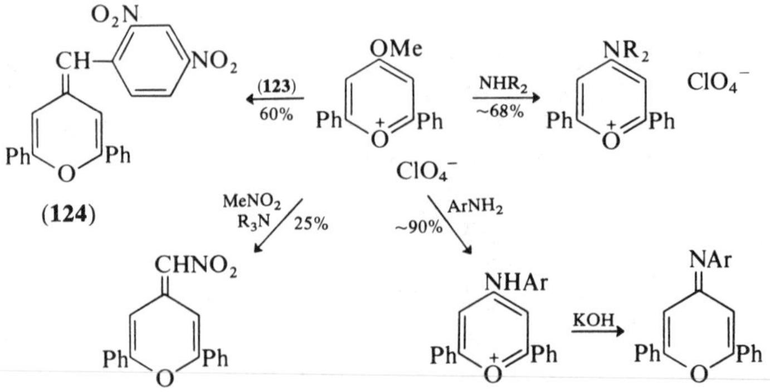

(**124**)

**Scheme 8**

be displaced by primary or secondary amines, hydrogen sulfide or active CH groups (Scheme 8), for example 2,4-dinitrotoluene (**123**) gives 4-(2,4-dinitrobenzylidene)-2,6-diphenyl-pyran (**124**) ⟨73JOC2834, 74JHC1065, 72JHC783⟩.

Sometimes, an alkoxy group may be unaffected by a nucleophile which reacts at another site in the molecule. An example is given in Section 2.23.2.3.2.

### 2.23.3.4 *S*-Linked Substituents

Alkylthio groups at C-4 are amongst the best leaving groups and are useful precursors for several 4-substituted pyrylium salts, for example 2,6-dimethyl-4-piperidinopyrylium iodide (**125**) ⟨55JOC448⟩.

(**125**)

## 2.23.4 2*H*- AND 4*H*-PYRANS AND THEIR BENZO DERIVATIVES: REACTIVITY AT RING ATOMS

### 2.23.4.1 Introduction

The simple pyrans are rather unstable compounds and of little biological or industrial significance but some of their benzo derivatives are of considerable interest. Many 2*H*-1-benzopyrans (chromenes) are found in plants, for example evodionol (**126**), lapachenole (**127**), lonchocarpin (**128**), rottlerine (**129**) and edulan (**130**). A few of the constituents of marijuana (hashish) belong to this class, for example cannabinol (**131**).

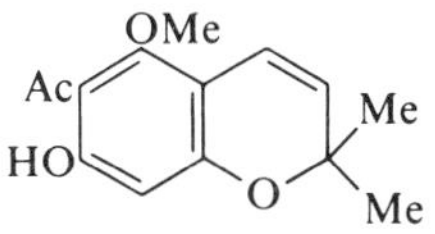

(**126**) 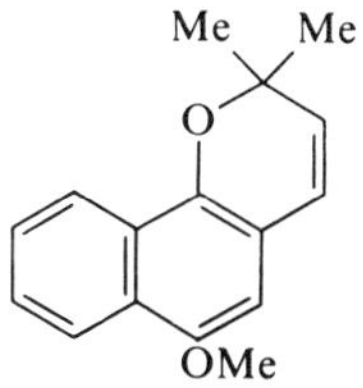    (**127**) 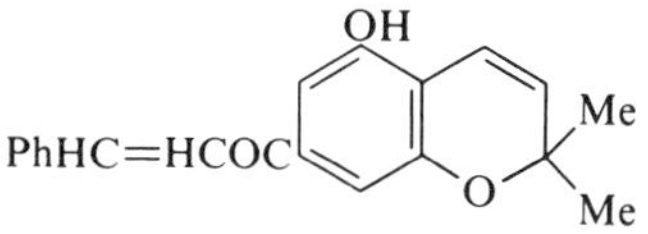    (**128**)

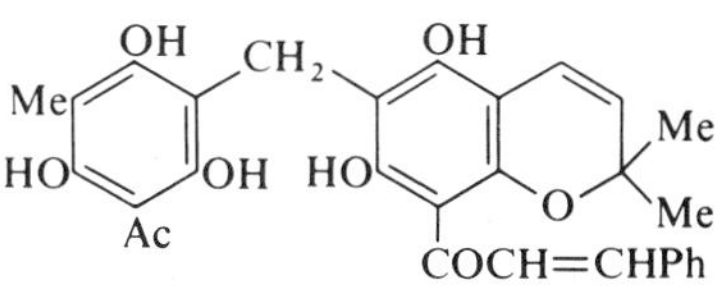

(**129**) 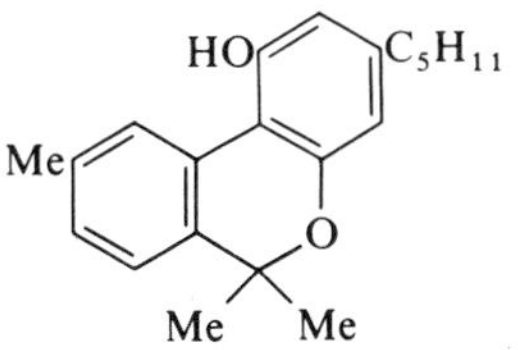    (**130**) 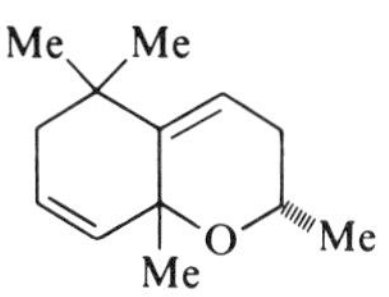    (**131**)

Recently, the biological role of precocene I (**132**) and II (**133**) and their conversion into the epoxides has been studied. Precocene II suppresses juvenile hormone ⟨79MI22301, 79CC920⟩ and this has stimulated interest in synthetic analogues ⟨80JHC1377⟩.

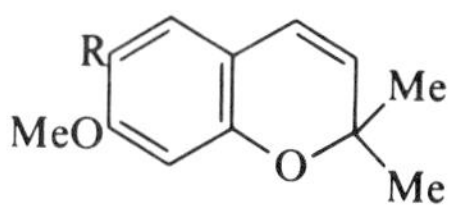

(**132**) R = H
(**133**) R = MeO          (**134**) 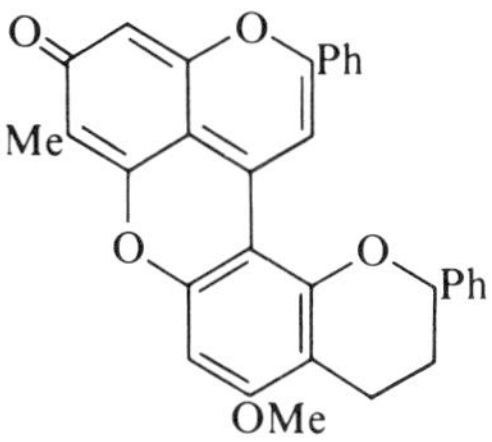          (**135**)

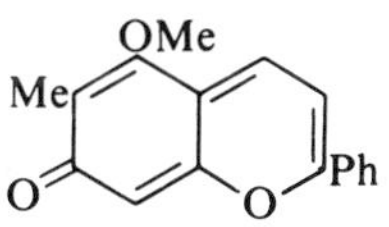

Some natural dyes are coming back into use and amongst these is one called Dragon's Blood (C.I. Natural Red) which consists of the flavenes dracorhodin (**134**) and dracorubin (**135**).

In recent years, many spirochromenes such as (**136**) and (**137**) have been studied because of their photochromic or thermochromic properties, while derivatives of xanthene are useful as dyes and indicators, for example pyronine G (**138**), fluorescein (**139**) and rhodamine 6G (**140**).

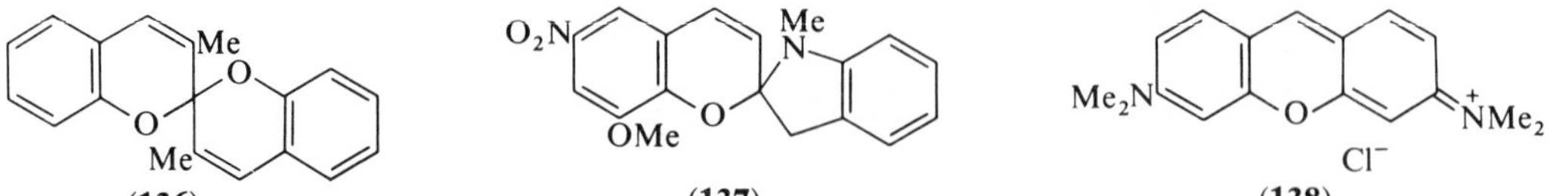

(**136**)          (**137**)          (**138**)

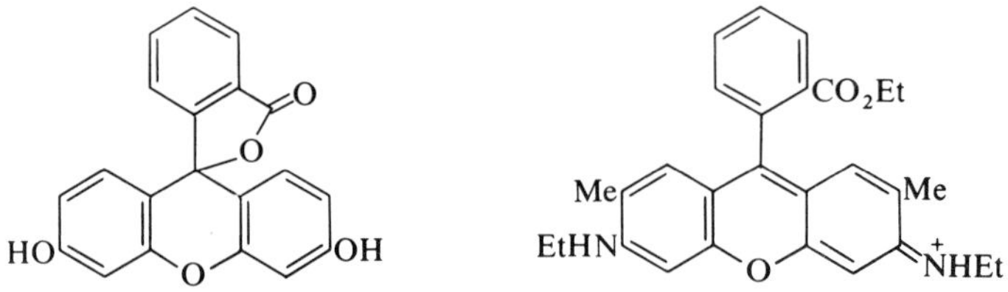

(**139**)          (**140**) Cl⁻

Reviews of 2*H*-1-chromenes ⟨75AHC(18)159⟩, of 2*H*- and 4*H*-1-chromenes ⟨77HC(31)11⟩, of pyrans, chromenes, flavenes and xanthenes ⟨B-77MI22301⟩ and of their naturally occurring members ⟨B-63MI22301⟩ are available.

### 2.23.4.2 Photochemical and Thermal Reactions

Thermal ring-opening of the 2*H*-pyran (**141**) has been studied spectroscopically, by GLC and by reduction of the acyclic ketone with LAH. The rate of ring cleavage increases with temperature for this pyran and for its bicyclic analogue (**142**) which, on reduction, gives *cis*-β-ionol (**143**) ⟨72JOC2992⟩.

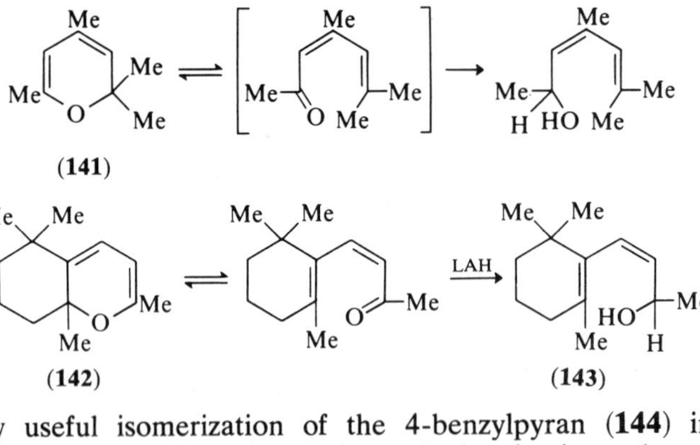

(**141**)

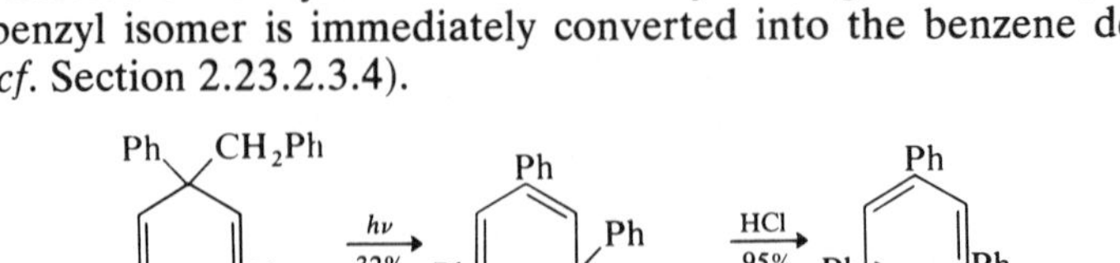

(**142**)          (**143**)

A synthetically useful isomerization of the 4-benzylpyran (**144**) into its 2-benzyl isomer (**145**) is effected either by UV irradiation or by heating in a solvent, but in the latter process the 2-benzyl isomer is immediately converted into the benzene derivative (**146**) ⟨64LA(678)183⟩ (*cf.* Section 2.23.2.3.4).

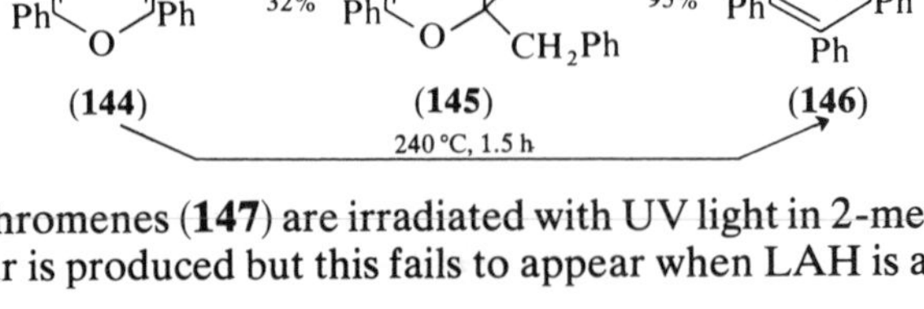

(**144**)          (**145**)          (**146**)
                   240 °C, 1.5 h

When some 2*H*-chromenes (**147**) are irradiated with UV light in 2-methyl-THF at −75 °C, an intense red colour is produced but this fails to appear when LAH is added to the reaction

mixture. As in the case of the pyrans mentioned above, the ring-opened valence isomer
(**148**) is reduced to the phenol (**149**) or (**150**) and the colour is, therefore, due to the
quinonoid photoproduct (**148**) which, in the absence of LAH, reverts to the chromene
⟨67JPC4045⟩.

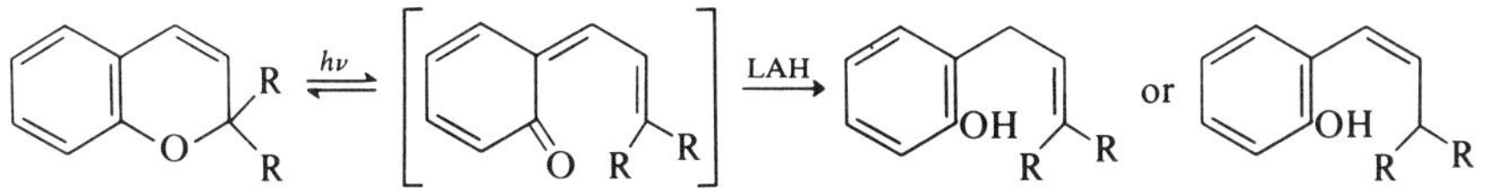

Irradiation of 2,2-dimethyl-2*H*-chromene (**151**) in methanol causes ring cleavage followed by addition of methanol. Two isomeric phenols (**152**) and (**153**) are formed in the
ratio of 7:3 when a Pyrex filter is used but a Corex filter led to (**153**) and three other
products. When the irradiation was conducted in the presence of xanthone as sensitizer,
(**153**) was the only product (90% yield) ⟨75JOC1142⟩.

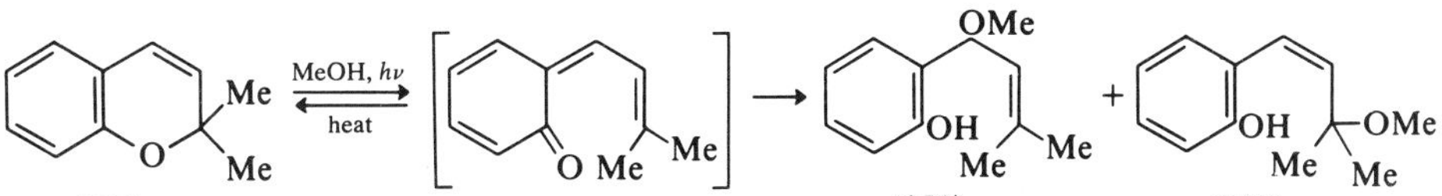

Xanthene is photostable in the absence of oxygen, but in its presence it is converted into
xanthone ⟨66BCJ1694⟩. When oxygen is bubbled through a solution of xanthene in sulfuric
acid at 25 °C, two molecules of xanthyl ion (**154**) are formed per mole of oxygen in a
reaction which is first order with respect to xanthene and to hydrogen ion activity. The
hydroperoxy cation $HO_2^+$ is probably an intermediary ⟨69JA5237⟩. The xanthyl ion can also
be formed from xanthene in a radical reaction ⟨66RTC899⟩.

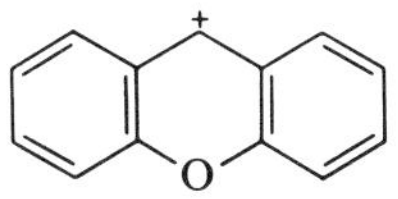

## 2.23.4.3 Electrophiles

### 2.23.4.3.1 Electrophilic substitution and addition

4*H*-Pyran decomposes in contact with moist air or acids but its tetrasubstituted derivatives,
*e.g.* (**155**), are stable and are usefully converted into naphthalenes (**156**) by treatment with
perchloric acid. A 4-benzyl or 4-naphthylmethyl substituent is essential. Pyran ring opening
occurs and is followed by cyclization and elimination of a simple ketone which is formed
from the 2- or 6-substituent. Alkyl groups at C-2 or C-6 are converted into the ketone
more easily than a phenyl group ⟨64LA(678)202⟩.

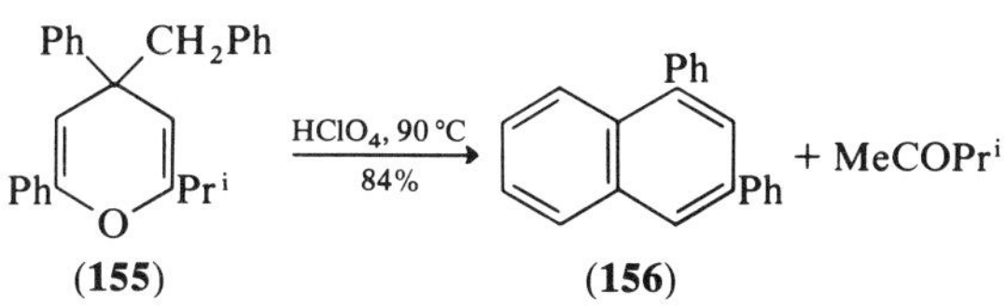

Chromenes, *e.g.* (**157**), which have a 2-hydrogen atom are converted by strong acids into
the pyrylium salt (**158**) and the stereoisomeric chromans (**159**) (the latter pair in equal
amounts), but a more efficient conversion of a 2*H*-chromene (and also of a 4*H*-pyran) into
the pyrylium salt, *e.g.* (**160**), is effected by addition of triphenylmethyl perchlorate in an
inert solvent ⟨67AC(R)1045⟩.

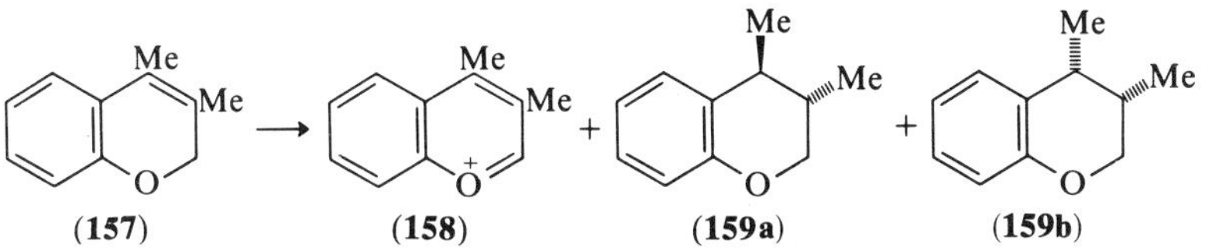

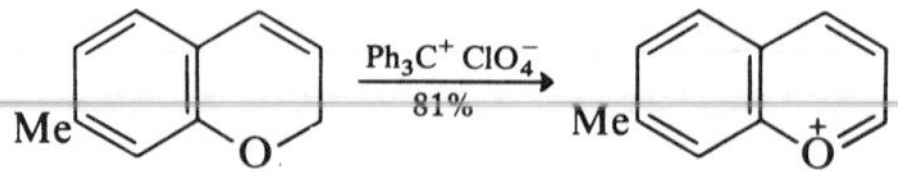

**(160)**

The benzochromene (2*H*-naphtho[1,2-*b*]pyran) lapachenole (**161**), on treatment with acid, gives a dimer (**162**). A similar product is obtained when coumarin is treated with methylmagnesium iodide followed by acidification ⟨68T1981⟩.

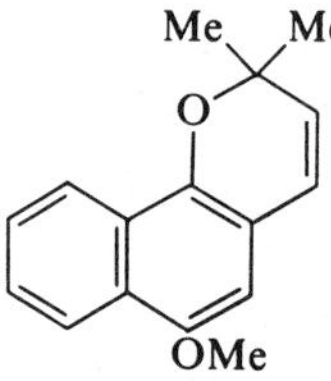

**(161)** 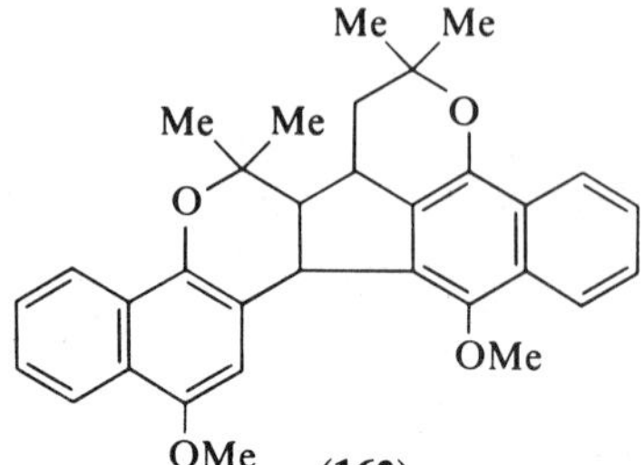                                          **(162)**

2*H*-Chromenes add chlorine and bromine at ambient or lower temperature to give the 3,4-dihalides but a 7-methoxy substituent appears to inhibit the addition ⟨66JCS(C)2013⟩ while a spontaneous dehydrobromination occurs with some dibromides so that the 3-bromo-2*H*-chromene is isolated. Hypobromous acid reacts with 2,2-dimethyl-2*H*-chromene to yield 4-bromo-2,2-dimethylchroman-3-ol, and similarly with 2,2-dimethyl-3*H*-naphtho[2,1-*b*]pyran (**163**) ⟨60JCS3094⟩. 2*H*-Flavene adds bromine at room temperature to give 2,3-*cis*-3,4-*trans*-3,4-dibromoflavan (**164**) as the main product, together with a small amount of the 2,3-*trans*-3,4-*trans*-isomer (**165**) ⟨67T341⟩.

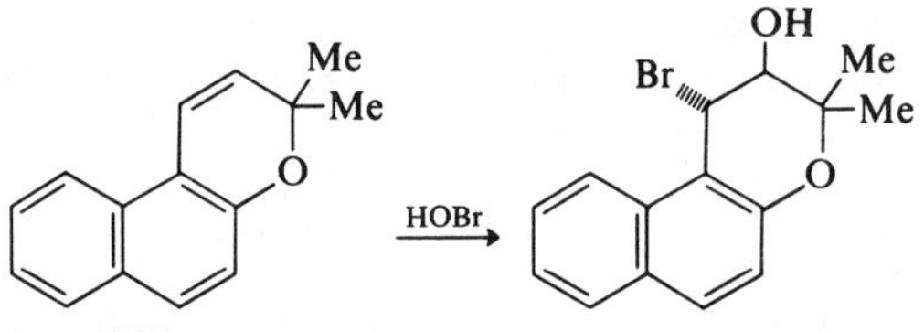

**(163)**

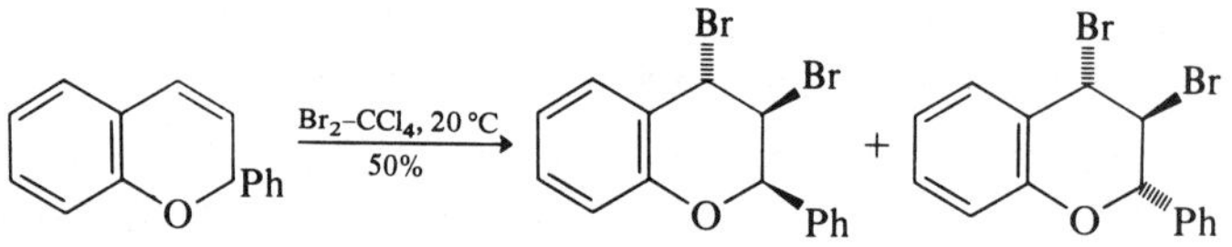

**(164)**          **(165)**

Acid-catalyzed addition of water and alcohols to 4*H*-chromenes gives the expected products as predicted by Markovnikov's rule ⟨56JCS4785⟩; an anti-Markovnikov addition of methanol followed by the reintroduction of a double bond in the alternative position gives an overall effect of substitution of hydrogen by methoxy and this is effected by treating methyl 2*H*-chromene-3-carboxylate (**166**) with triphenylmethyl perchlorate and addition of methanol to the resulting benzopyrylium salt (**167**) ⟨72CR(C)(274)650⟩.

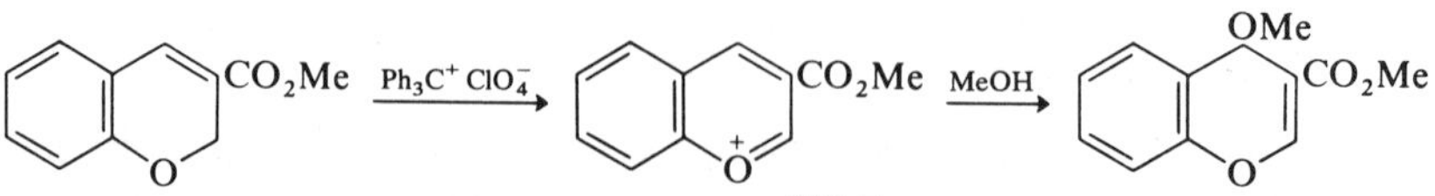

**(166)**                    **(167)**   $ClO_4^-$

The chromene ring of spirobenzopyrans such as (**168**) and (**169**) is opened by treatment with mineral acid and closed again by alkali — a reaction of potential importance in the thermochromic or photochromic application of such compounds ⟨69T5995, 74JOU1516⟩.

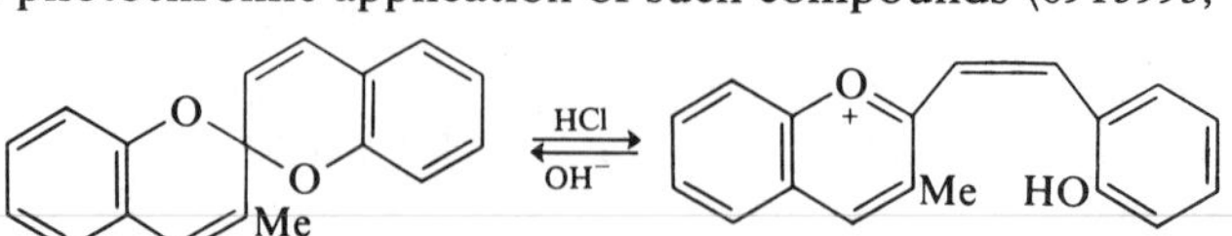

**(168)**                                      $Cl^-$

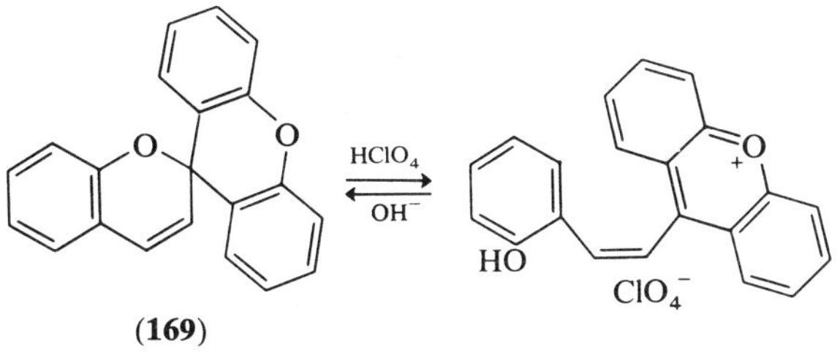

**(169)**

### 2.23.4.3.2 Oxidation

A variety of reagents attack the double bond or the methylene group of chromenes to give products of differing degrees of oxidation, and by careful choice of reagent and conditions this type of reaction can be a useful route to many compounds.

Hydroboration of 2*H*-chromene followed by oxidation gives a mixture of chroman-4-ol (55%) and the isomeric 3-ol (3%) but 4*H*-chromene yields chroman-3-ol only. Under similar conditions, 4'-methoxy-4*H*-flavene suffers cleavage of the pyran ring, the products being the isomeric diphenylpropanols (**25a**) and (**25b**) in yields of 44% and 12%, respectively ⟨73AJC819⟩. The isomeric 2*H*-flavene, however, retains its pyran ring and gives *trans*-flavan-4-ol (**170**) ⟨73AJC809⟩. Markovnikov addition of water or methanol to 2,4,4-trimethyl-4*H*-chromene (**171**) proceeds normally ⟨56JCS4785⟩.

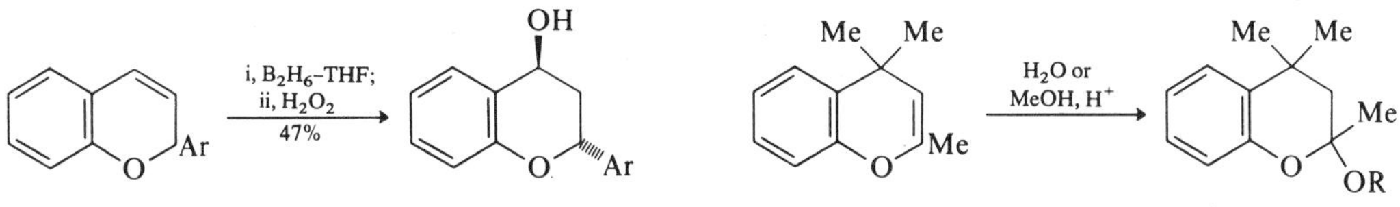

Ar = 4-MeOC$_6$H$_4$        (**170**)      (**171**)      R = H or Me

Amongst the other oxidizing agents which give specific products on reaction with 2*H*-chromenes are osmium tetroxide–hydrogen peroxide, osmium tetroxide–potassium periodate ⟨67BSF1164⟩, ozone ⟨68T497⟩, alkaline permanganate, perbenzoic acid ⟨68YZ816⟩, thallium trinitrate ⟨79CB3879⟩ and hydrogen peroxide–formic acid ⟨68BSF4203⟩. Scheme 9 shows typical products that are obtainable from 2,2-dimethyl-2*H*-chromene.

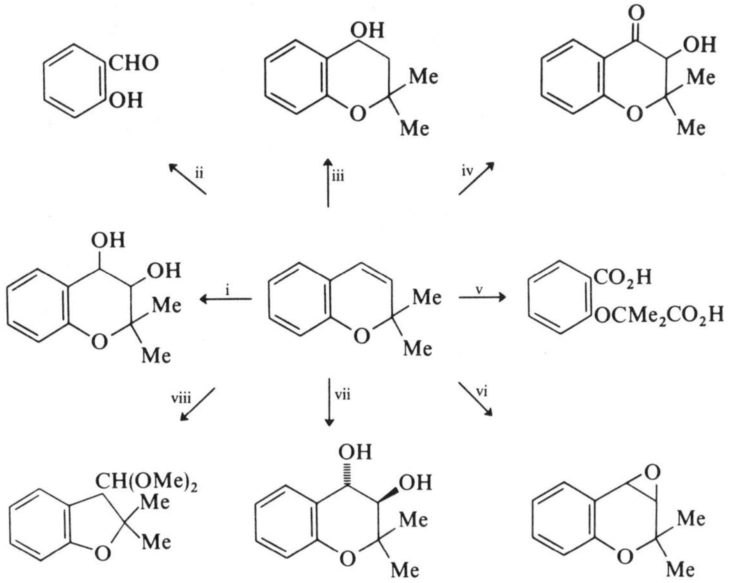

i, OsO$_4$–H$_2$O$_2$; ii, O$_3$, Zn; iii, B$_2$H$_6$, H$_2$O$_2$; iv, OsO$_4$–NaIO$_4$;
v, KMnO$_4$, OH$^-$, 100 °C; vi, PhCO$_3$H; vii, H$_2$O$_2$, HCO$_2$H; viii, Tl(NO$_3$)$_3$, MeOH

**Scheme 9**

In addition to the two types of products formed by osmium tetroxide in Scheme 9 is its action on 5,10-diacetoxy-2-methyl-2*H*-naphtho[2,3-*b*]pyran (**172**) in conjunction with sodium chlorate at room temperature to give a mixture of stereoisomeric chromanones (**173**) ⟨78CB1285⟩.

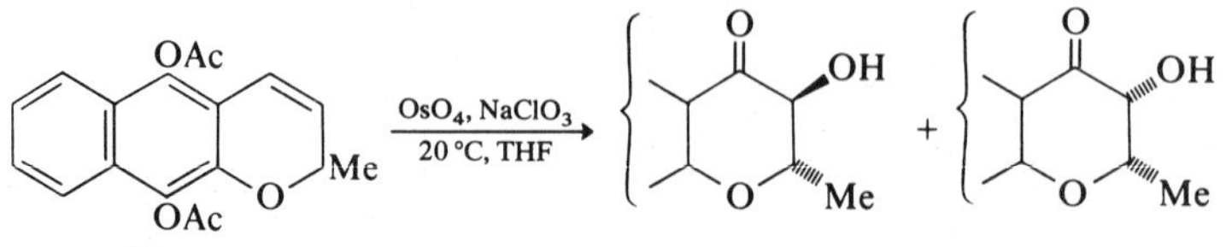

The use of a quaternary ammonium hydroxide as a phase transfer catalyst provides an efficient conversion of xanthene (**174**) to xanthone under mild conditions ⟨77TL2117⟩, while the use of permanganate under milder conditions than those shown in Scheme 9 enables flavenes such as (**175**) and (**176**) to be converted into flavones ⟨77BCJ3298, 78BCJ1175⟩. The phase transfer oxidation requires strongly alkaline conditions which may render it less suitable for the oxidation of flavenes and chromenes. Xanthene has also been converted into xanthone by means of permanganate or chromic acid.

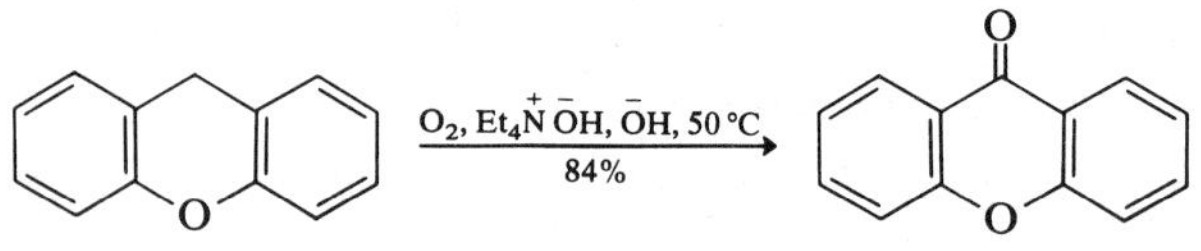

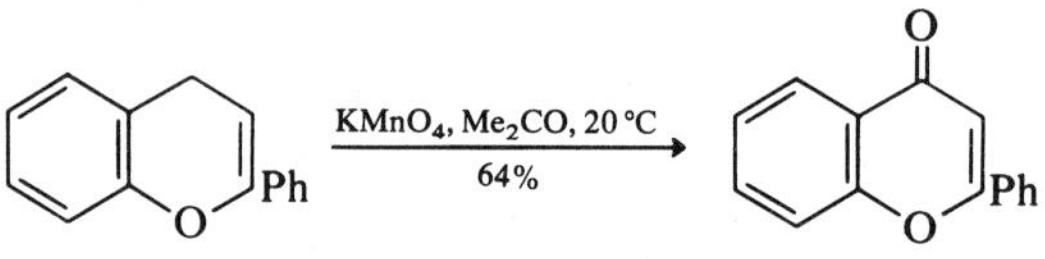

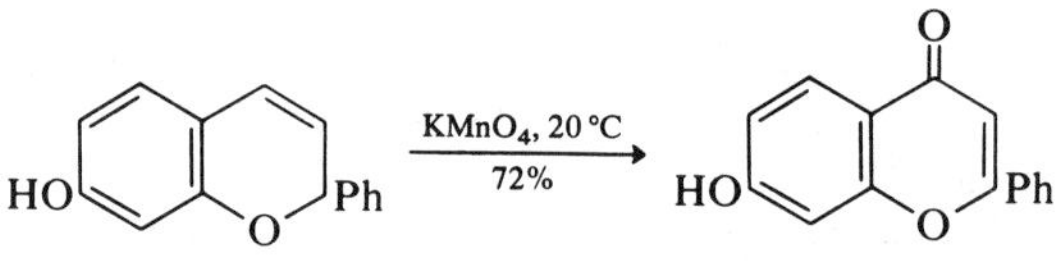

An arylidene group in the 2-benzopyran (**177**) may be oxidatively degraded to a carbonyl using permanganate ⟨78CHE1067⟩. Benzylidene derivatives may themselves be prepared by dehydrogenation of pyrans which contain the requisite atoms at C-4 of 4*H*-pyrans such as (**178**) or (**179**) ⟨59CB46, 64LA(678)183⟩.

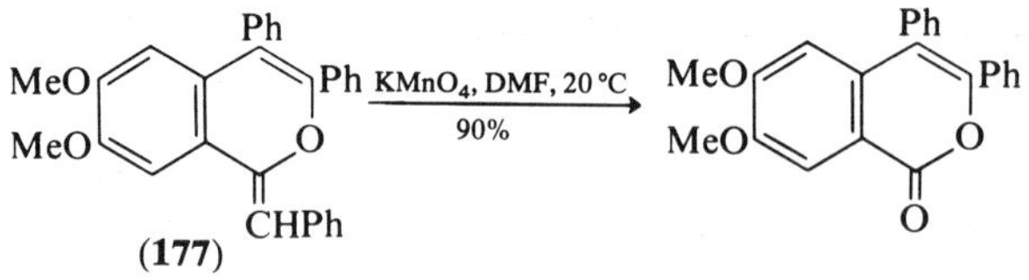

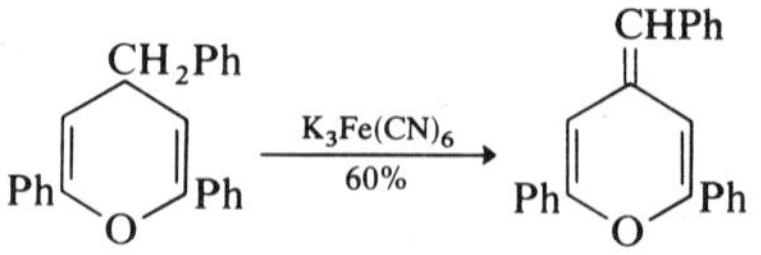

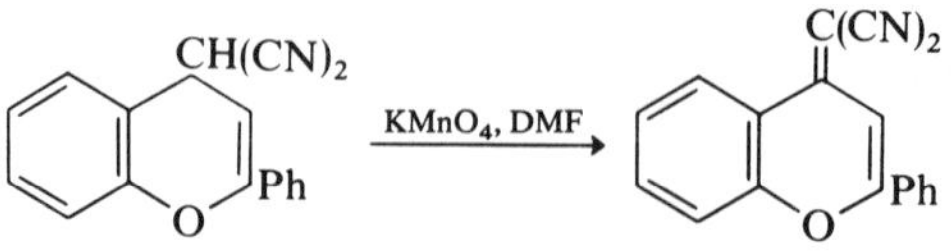

When 2,2-dimethyl-2*H*-chromene is irradiated in the presence of a quinone such as phenanthraquinone (**180**), the latter adds across the chromene 3,4-double bond with the formation of the adduct (**181**) ⟨65CB3102⟩. Similarly, xanthene reduces 1,4-naphthoquinone to the dihydroxynaphthalene and is itself oxidized to 9,9'-bixanthyl (**182**) when the two compounds are irradiated in benzene at 0–5 °C ⟨74BCJ1960⟩.

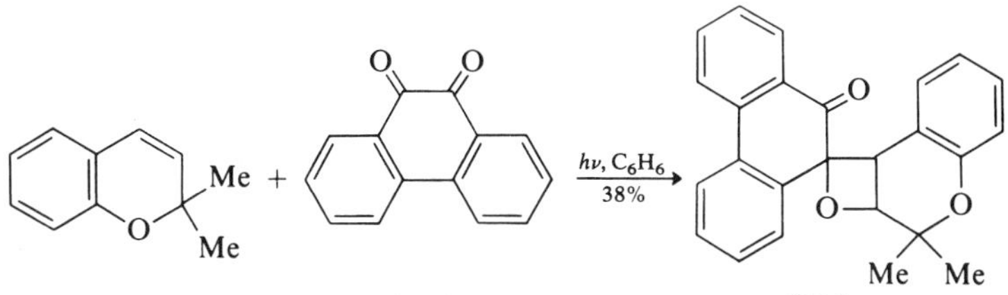

(**180**)        (**181**)

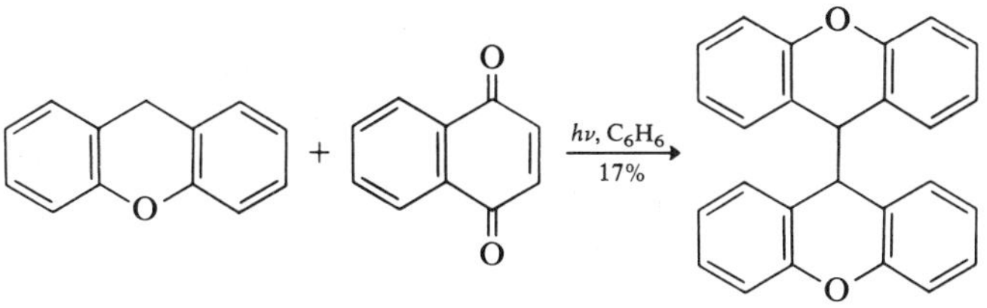

(**182**)

In an effort to elucidate the mechanism of oxidative phosphorylation, 6-hydroxy-2,2,5-trimethyl-2*H*-naphtho[1,2-*b*]pyran (**183**) was treated with silver(II) oxide. The product (**184**) was a stable quinone hemiketal ⟨71JOC4045⟩.

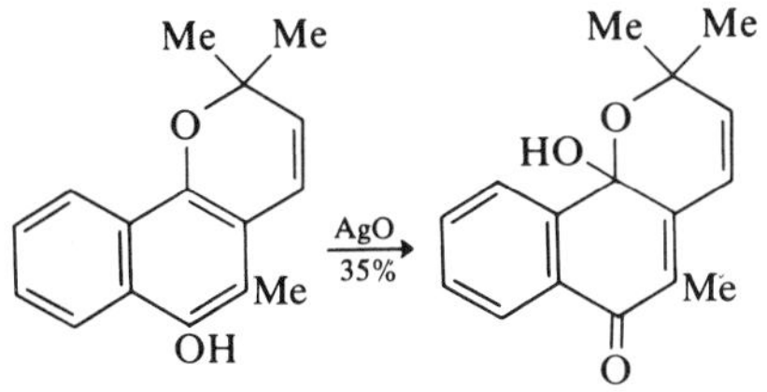

(**183**)        (**184**)

### 2.23.4.3.3 Carbenes

The addition of carbenes to double bonds of chromenes gives good yields of cyclopropa-chromenes such as (**185**) and (**188**). Addition to 2*H*-chromenes takes place mainly on the less hindered side when a 2-substituent is present to give the *trans* products ⟨71HCA306, 62JA813, 66CB2351⟩. Compounds of type (**185**) and (**186**) undergo decomposition or rearrangement at temperatures above 210 °C. The dihalides (**185**) are dehalogenated by sodium and methanol, and the product thus derived from (**185b**) undergoes a thermal rearrangement to 4-ethoxy-2-methyl-2*H*-chromene (**187**) ⟨71BSF2557⟩.

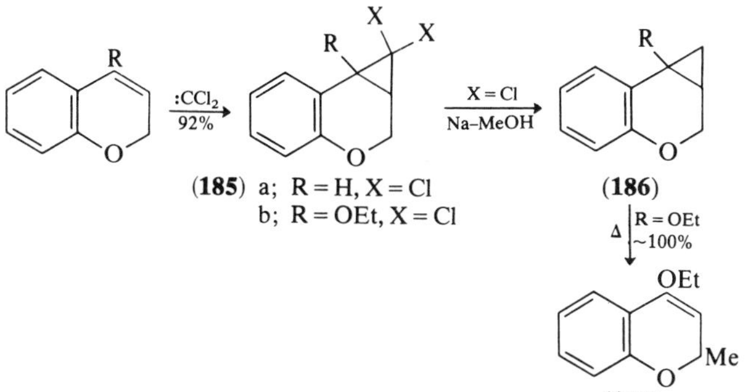

(**185**) a; R = H, X = Cl        (**186**)
         b; R = OEt, X = Cl

(**187**)

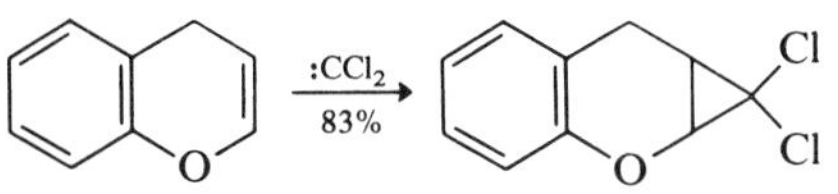

(**188**)

## 2.23.4.4 Nucleophiles

The addition of water and alcohols across the double bond of a chromene has been mentioned in Sections 2.23.4.2 and 2.23.4.3.1 because the reactions are initiated by irradiation and an electrophile.

### 2.23.4.4.1 Bases

Aqueous alkali normally opens the pyran ring of chromenes to give more than one product as a result of rearrangement or degradation of the opened ring. A phenol is usually produced together with an aldehyde and/or ketone such as acetone, depending on the structure of the chromene. For example, the chromene (**189**) is converted by heating with aqueous alkali into 6-methoxyresacetophenone (**190**) and 3-methylbut-2-enal which may then be degraded further to acetaldehyde and acetone ⟨64T1317⟩. Magnesium in THF degrades 2*H*-chromene (**191**) to a mixture of *o*-propenylphenols, which arise on acidification of the primary cyclic product ⟨69JOM(18)249⟩.

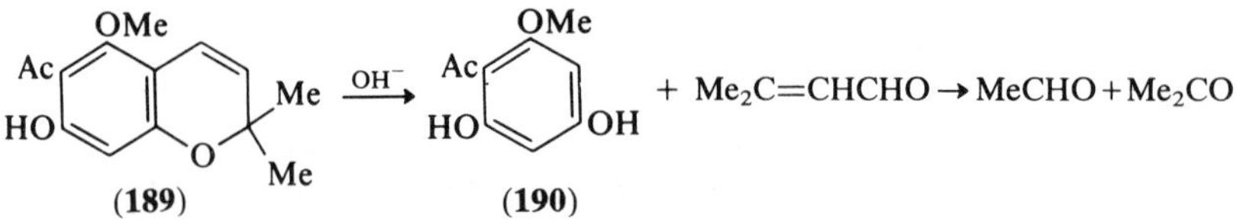

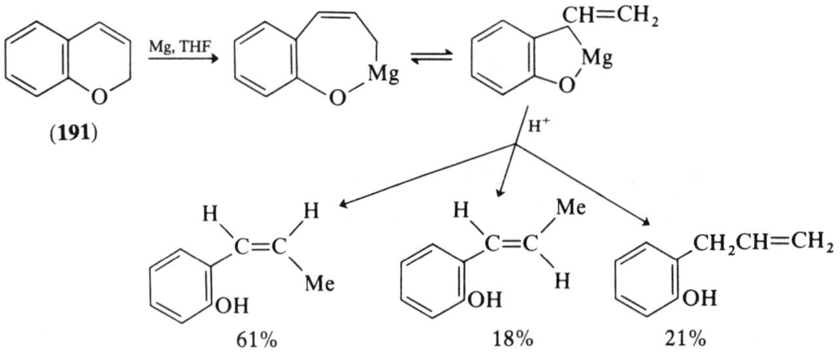

Nucleophilic attack on spirochromenes such as (**192**) is accompanied by opening of one or both rings at the common carbon atom; for example, water, hydrogen sulfide or sodium borohydride degrades the spirans in this way ⟨78JHC1439⟩.

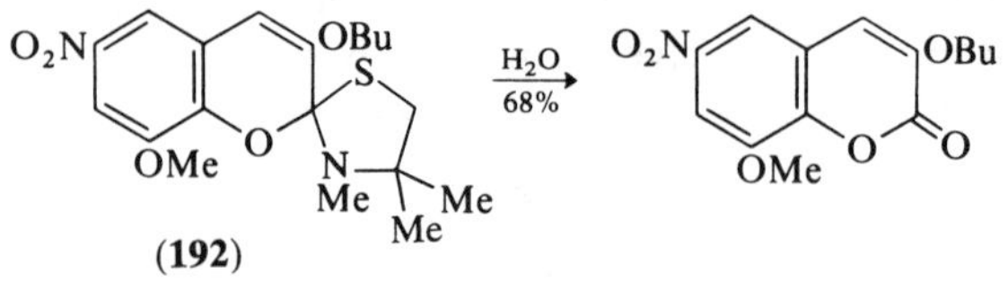

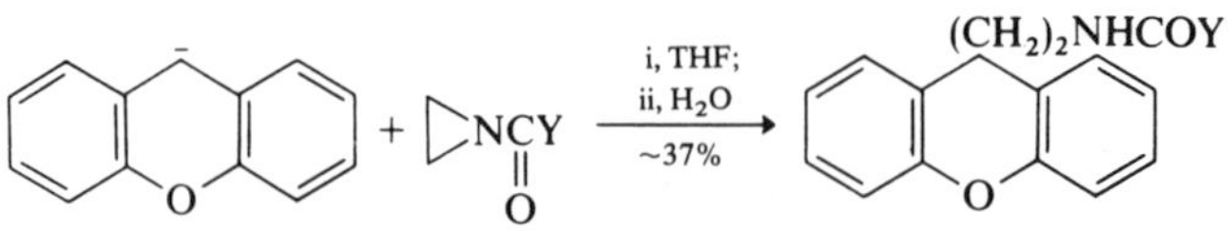

Strong bases abstract a proton from xanthene and the resulting anion (**193**) reacts with aziridines to give 9-β-aminoethyl and related derivatives (**194**) ⟨79AP133⟩.

Many pyrans are less stable or less accessible than pyrylium salts or pyranones but some suitably substituted 4*H*-pyrans can be used to prepare pyridines. Ethyl 6-amino-5-cyano-2,4-diphenyl-4*H*-pyran-3-carboxylate (**196**) on heating with ammonium acetate gives the corresponding pyridine-3-carboxylate (**195**), but when the ester group in (**196**) is replaced by a cyano group, a dihydropyridine (**197**) is obtained by a different and more complex reaction ⟨81JHC309⟩.

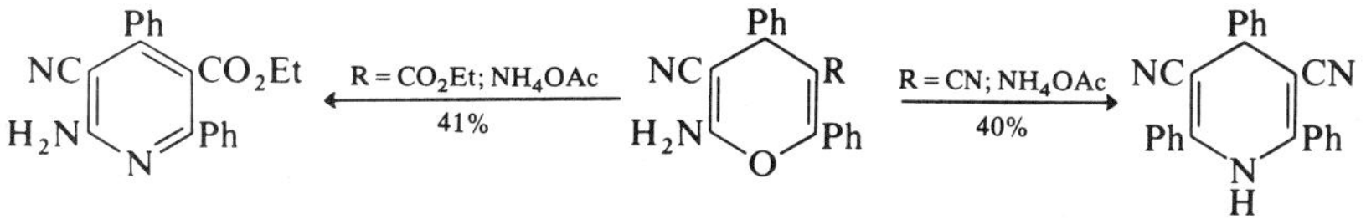

(195)    (196)    (197)

### 2.23.4.4.2 Reduction

The double bond of chromenes is readily hydrogenated catalytically to give the chroman. The choice of solvent can sometimes be important; for example, ethyl 3-hydroxy-7-methoxy-2*H*-chromene-4-carboxylate (**198**) gives the expected chroman (**199**) in ethanol but the 3-substituent is lost when the reaction is conducted in acetic acid ⟨36JCS419⟩. The two substituents in (**199**) are probaly *cis*, as are the methyl groups in chroman (**200**) ⟨68T949⟩. Hydrogenation of 4-bromo-2,2-dimethyl-2*H*-chromene is accompanied by dehalogenation ⟨60JCS3094⟩.

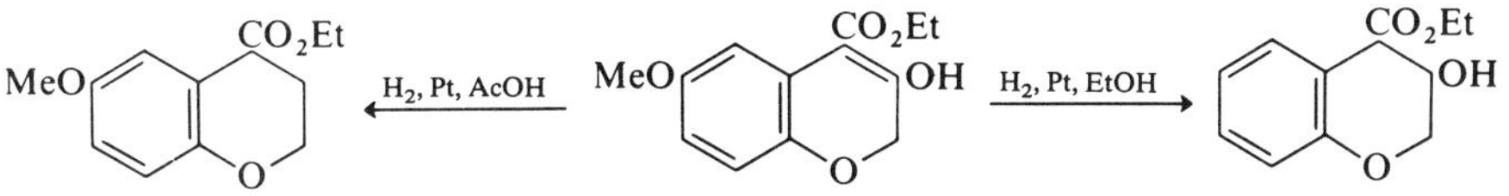

(198)    (199)

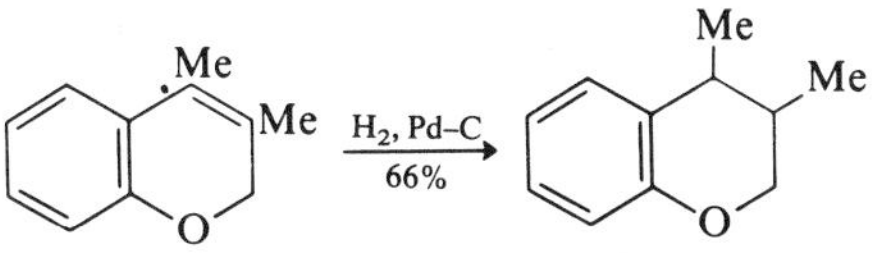

(200)

Stereochemical differences are observed between the products of hydrogenation and those of LAH reduction of some chromenes, for example 2-methyl-3-nitro-2*H*-chromene, as shown in Scheme 10 ⟨73JCS(P2)227⟩.

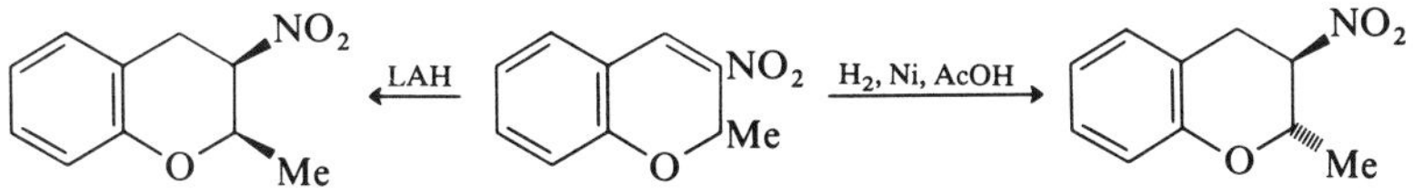

**Scheme 10**

The ether linkage is cleaved by reducing agents such as metal–ammonia; for instance, lithium–ammonia opens the pyran ring of 2,2-dimethyl-2*H*-chromene (**201**) ⟨69AJC1923, 74TL4315⟩.

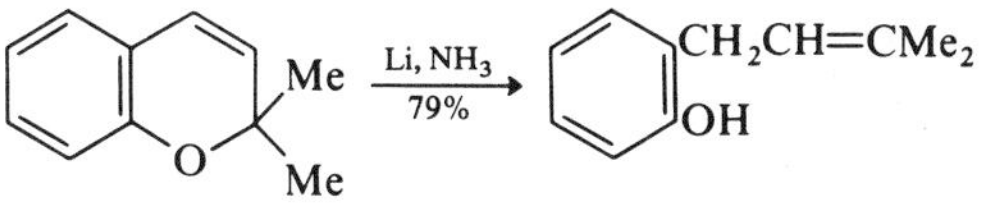

(201)

### 2.23.4.5 Other Reactions

Although chromenes are not sufficiently reactive, 2*H*-naphtho[1,2-*b*]pyrans (**202**) add reactive dienophiles such as dimethyl acetylenedicarboxylate in a Diels–Alder reaction.

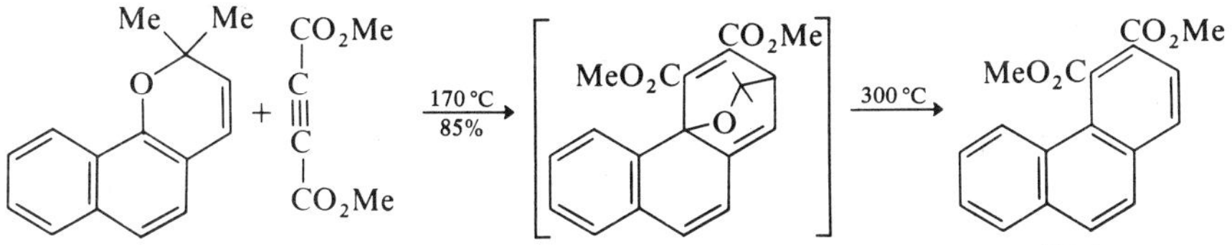

(202)    (203)

The adduct is not isolated but heated to 300°C to give the phenanthrene diester (**203**) in high yield ⟨63TL1267⟩.

Dipolar molecules such as nitrile oxides and nitrilimines add across the C-3, C-4 double bond of 2*H*-chromene to give fused tricyclic heterocycles like (**204**) or (**205**) ⟨81BCJ217⟩.

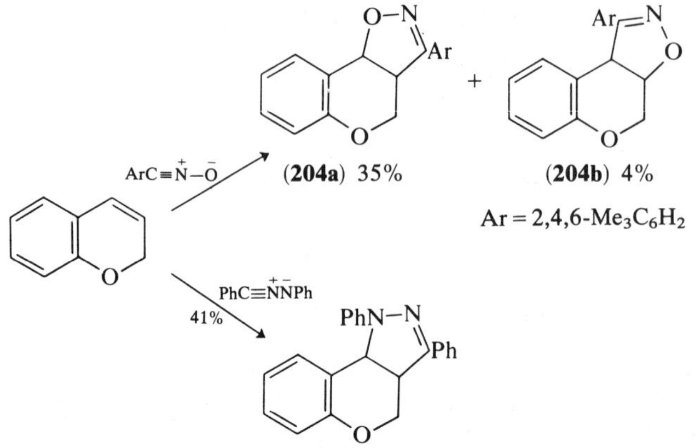

(**204a**) 35%          (**204b**) 4%

Ar = 2,4,6-Me$_3$C$_6$H$_2$

ArC≡$\overset{+}{N}$–$\overset{-}{O}$

PhC≡$\overset{+}{N}$$\overset{-}{N}$Ph

41%

(**205**)

## 2.23.5 2*H*- AND 4*H*-PYRANS AND THEIR BENZO DERIVATIVES: REACTIVITY OF SUBSTITUENTS

### 2.23.5.1 Introduction

Substituents attached to pyran behave much as if they were attached to the unsaturated side-chain of styrene. A brief selection of some typical reactions is therefore given.

### 2.23.5.2 Fused Benzene Rings

The 2*H*-chromene part of the naturally occurring lonchocarpin (**206**) remains unaltered when the compound is subjected to oxidation. This helped to determine the structure of lonchocarpin ⟨53MI22300⟩.

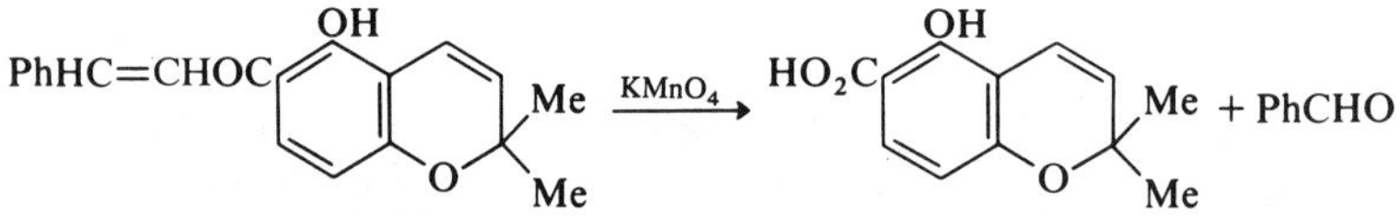

$\xrightarrow{\text{KMnO}_4}$          + PhCHO

(**206**)

Electrophiles attack xanthene at C-2 and sometimes also at C-7 when conditions are sufficiently vigorous. Nitric acid simultaneously oxidizes xanthene to the xanthone and thus gives 2-nitroxanthone. Friedel–Crafts acetylation of xanthene gives either 2-acetyl- or 2,7-diacetylxanthene according to the temperature (0 or 18 °C) and the proportion of acetyl chloride ⟨74BSF2963⟩.

### 2.23.5.3 *C*-Linked Substituents

A convenient synthesis of lonchocarpin (**206**) and similar compounds is the Knoevenagel condensation of a 6-acetylchromene (**207**) with benzaldehyde ⟨71JCS(C)796, 71JCS(C)811⟩. Carbonyl groups present in chromenes as aldehydes or ketones form oximes and 2,4-dinitrophenylhydrazones ⟨77HC(31)11⟩.

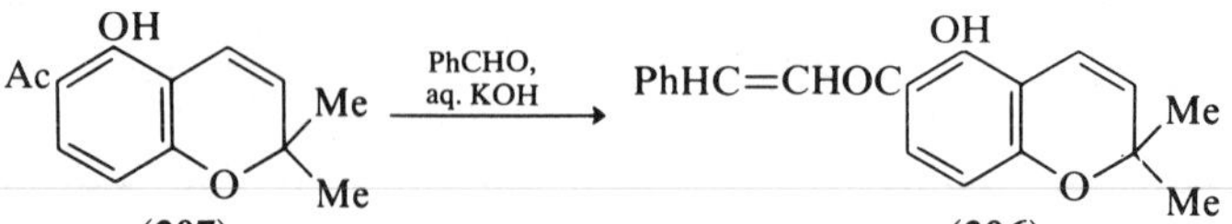

$\xrightarrow[\text{aq. KOH}]{\text{PhCHO,}}$

(**207**)          (**206**)

When 2*H*-chromene-3-carboxamide (**208**) is treated with LAH, the amide and the pyran double bond are reduced ⟨68BRP1043857⟩. The ester group of the chromene (**209**) is similarly reduced by diborane during hydroboration ⟨73SC231⟩.

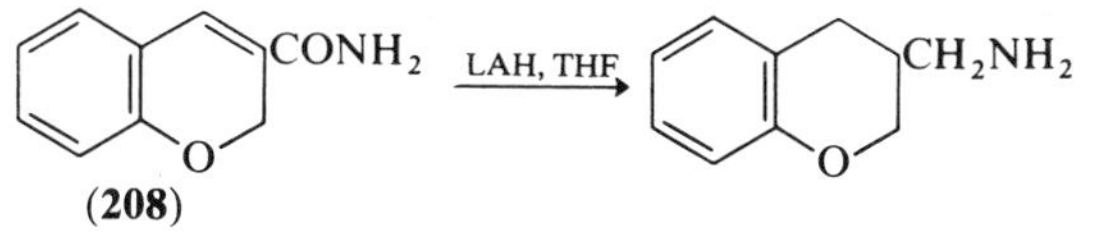

(**208**)

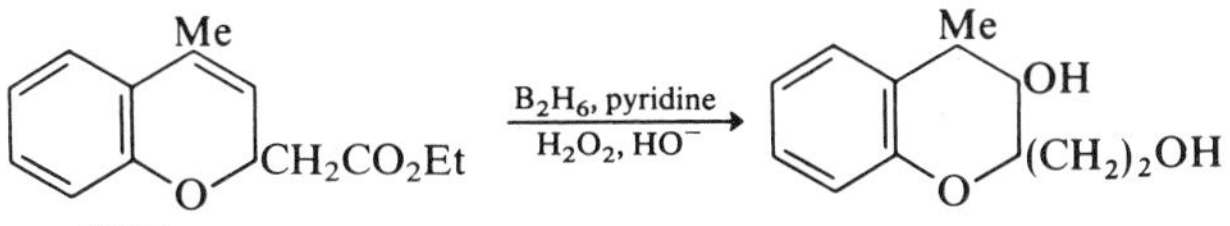

(**209**)

### 2.23.5.4 *O*-Linked Substituents

Hydroxyl groups in chromenes behave normally; for example, they are acylated or methylated by the usual reagents. A natural 2*H*-chromene isoevodionol (**210**) is methylated by dimethyl sulfate to the dimethyl ether (**211**) but its inertness to diazomethane shows hydrogen bonding of the hydroxyl group ⟨70TL3945⟩. Such observations are useful in determination of structure.

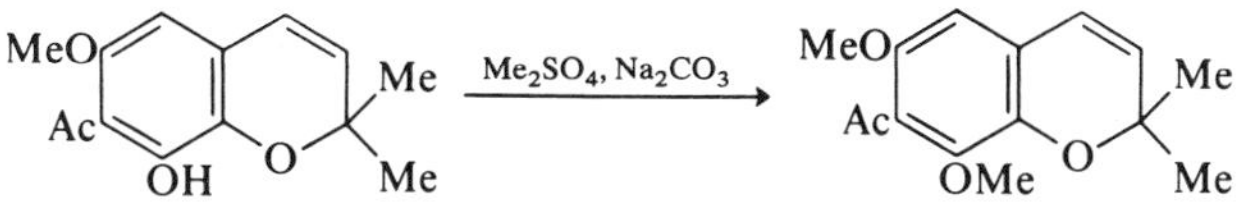

(**210**)     (**211**)

Among the less easily predicted reactions of phenols and 2*H*-chromenes is the simultaneous addition–cyclization of citral (**213**) and 5-pentylresorcinol (**212**) on heating in pyridine to give a 26% yield of a tetracyclic product called citrylidene-cannabis (**215**) together with the 2*H*-chromene, cannabichromen (**214**) (in 15% yield). Under drastic conditions, the chromene may be converted into citrylidene-cannabis and is photocyclized in acetone to cannabicyclol (**216**). The latter and cannabichromen are constituents of the cannabis plant ⟨71JCS(C)796⟩. From this work, an efficient method was developed of synthesizing 2,2-dimethyl-2*H*-chromenes using the acetal of either 3-methylbut-2-enal or citral ⟨71JCS(C)811⟩.

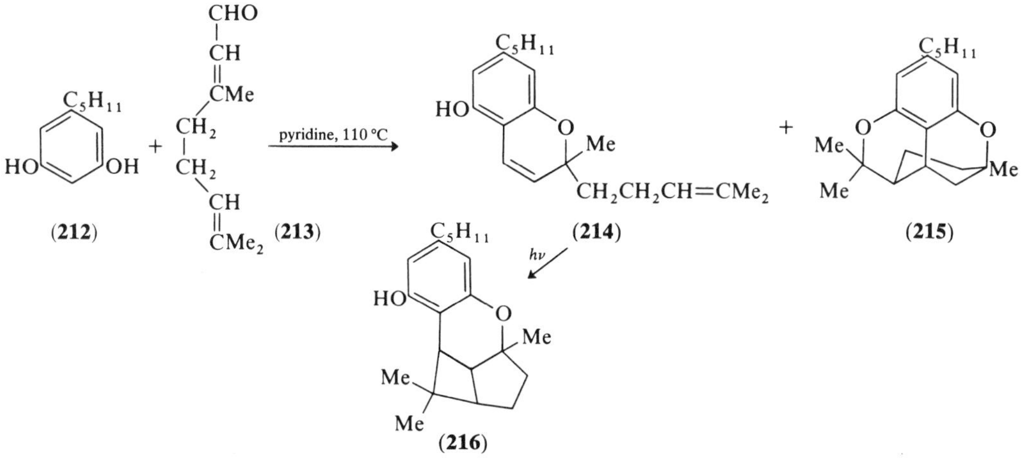

(**212**)     (**213**)     (**214**)     (**215**)

(**216**)

### 2.23.6 PYRAN-2-ONES AND THEIR BENZO DERIVATIVES: REACTIVITY AT RING ATOMS

#### 2.23.6.1 Introduction

Pyran-2-ones are unsaturated lactones which also show some aromaticity. Their properties are therefore a mixture of those of lactones, 1,3-dienes and an aromatic ring. Reaction conditions and reagents control the selectivity of these three types of reactions and it is possible for addition to a double bond to be rapidly followed by elimination so that the product which is isolated appears formally to have been the result of a simple aromatic substitution. Most of this section is devoted to pyran-2-ones and coumarins but some reactions of isocoumarins (1*H*-2-benzopyran-1-ones) are also mentioned. The reactions of these compounds have been reviewed ⟨64CRV229⟩.

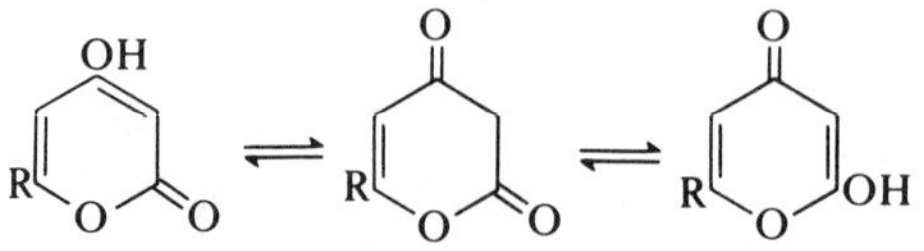

(217a)            (217b)            (217c)

Many pyran-2-ones have a 4-hydroxy group; such compounds are capable of tautomerism as shown by formulae (217a)–(217c). The formula of individual derivatives of this kind may be determined spectroscopically and by measuring or comparing the basicity of the compounds. Pyran-4-ones are more basic than their pyran-2-one isomers and form stable salts with strong acids. Pyran-2-ones, unlike their 4-one isomers, act as dienes in Diels–Alder reactions (Section 2.23.6.6). The UV absorption of methyl ethers of (217) of known structures are useful guides and the IR absorption of the carbonyl group differs in the 2- and 4-ones; the typical lactone absorption in pyran-2-ones is about 1735 cm$^{-1}$ and that of pyran-4-ones is about 1670 cm$^{-1}$. Methylation of 4-hydroxypyran-2-ones may occur at the 2- or 4-oxygen atoms. The reaction of 4-hydroxy-6-methylpyran-2-one (217a, or its tautomers; R = Me) with dimethyl sulfate gives a monomethyl ether which absorbs at 1736 and 1722 cm$^{-1}$ and does not form a salt but reacts with diethyl acetylenedicarboxylate to form a Diels–Alder adduct which loses carbon dioxide on heating to give 5-methoxy-3-methylphthalic acid (219). It is, therefore, a pyran-2-one (218). Methylation with diazomethane gives the same monomethyl ether together with an isomer (220) which forms a stable hydrochloride, does not react as a diene and its C=O group absorbs at 1692 and 1677 cm$^{-1}$. Naturally occurring pyranones have been assigned structures on similar criteria ⟨60JCS502, 59JA2427⟩.

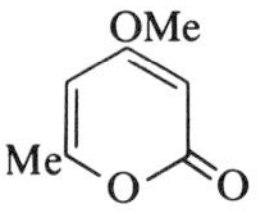

(218) 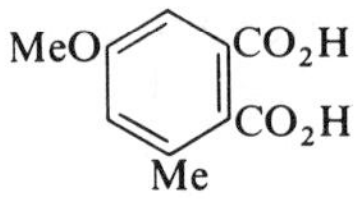                 (219) 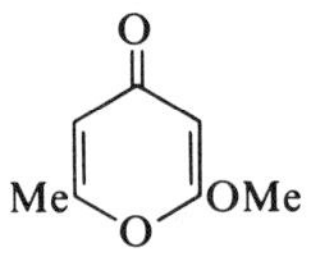                 (220)

Many coumarins are found in plants and a comprehensive review is available ⟨78FOR(35)199⟩. Coumarin has a pleasant taste and odour and was used for flavouring until its toxic action on the liver was discovered. Among other coumarins which have toxic effects on mammals are a group called aflatoxins, *e.g.* aflatoxin G$_1$, (221), which are secondary metabolites of fungi of the *Aspergillus* genus. Some of these compounds have been present in animal feeds and caused deaths of the animals. In contrast, other coumarin derivatives,

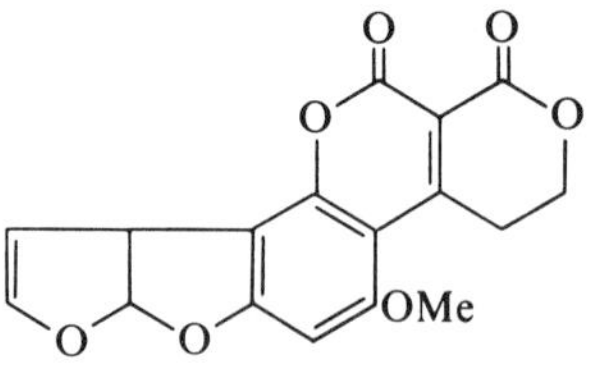

(221) 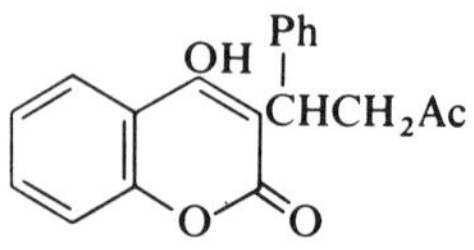                              (222)

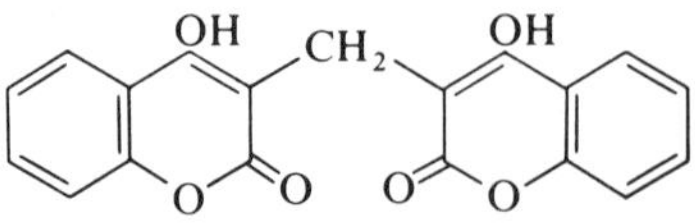

(223) 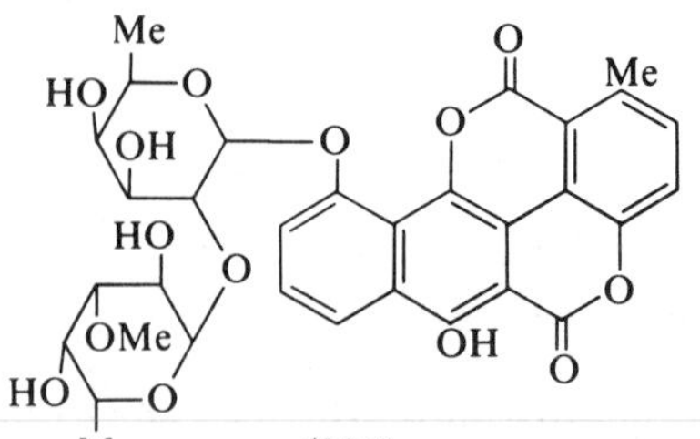                              (224)

*e.g.* warfarin (**222**) and dicoumarol (**223**), are valuable drugs used in humans to minimize the tendency for blood to clot; in larger doses, the compounds are rodenticides, causing death by excessive permeability of the blood vessels ⟨74MI22300⟩. Chartreusin (**224**) is an antibiotic of unusual structure ⟨82MI22301⟩.

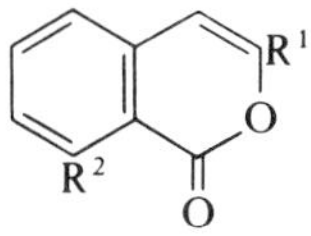

(**225**) R$^1$ = CH$_2$C≡CMe, R$^2$ = H
(**226**) R$^1$ = Me, R$^2$ = OH

Isocoumarins are less numerous than coumarins in nature. The acetylenic isocoumarin capillarin (**225**) is present in at least two *Artemisia* species and a fungus, *Marasmius ramealis*, produces 8-hydroxy-3-methylisocoumarin (**226**).

4-Arylcoumarins and 4-arylchromans are important members of a class of compounds which were first found in plants in 1953 and are collectively called neoflavanoids. Their chemistry has been reviewed ⟨B-75MI22301⟩.

## 2.23.6.2 Thermal and Photochemical Reactions

Pyran-2-one and its alkyl derivatives are reactive when irradiated and give a variety of interesting compounds, some of which are difficult to obtain otherwise. The presence or absence of oxygen and the solvent used are important factors in deciding the course of the reaction. When pyran-2-one is irradiated, it is converted at low temperature (≈8 K) into the ketene (**229**) but, as the temperature is raised, a competing reaction leads to the lactone (**227**). Further irradiation results in decarboxylation and the formation of cyclobutadiene (**228**) ⟨73JA614⟩. In another study, different intermediates have been shown to be present; for example irradiation of 4,6-dimethylpyran-2-one (**230**) in methanol gives two inseparable and reactive lactones (**231**) and (**232**). Mild treatment of these with acid gives the esters (**233**)–(**236**) whose structures were confirmed by synthesis. There is no evidence in this work ⟨73JA247⟩ for ketene formation, but other workers have obtained spectral evidence of its presence during irradiation of pyran-2-one in solid argon at 20 K with 313 nm light ⟨73JA248⟩.

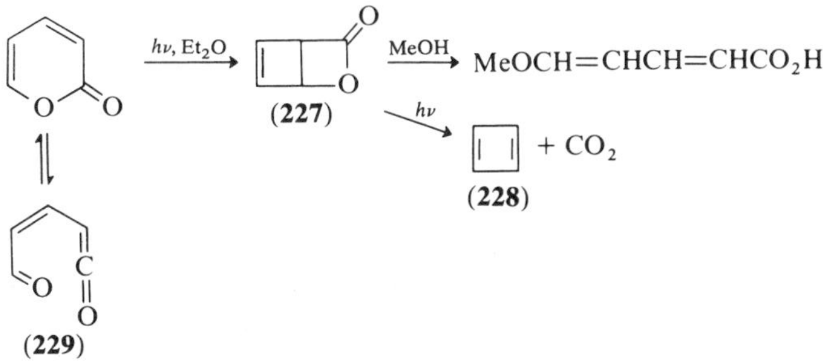

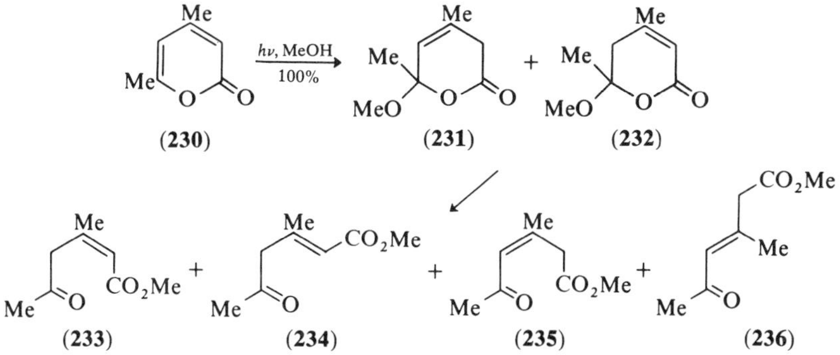

The effect of solvent and sensitizer on the course of photolysis of pyran-2-one is evident from Scheme 11. When methanol is present, either the ring is cleaved and an ester (**237**) is formed, or, with a sensitizer, dimers (**238**) are produced in equal amounts. The bicyclic lactone (**227**) and the ester (**237**) are believed to be formed *via* singlet excited states of the pyranone while the dimers (which are photostable) result from the triplet state ⟨68TL5279⟩.

A symmetrical dimer (**239**) is formed when 4,6-dimethylpyran-2-one (**230**) is irradiated in benzene and in the presence of a sensitizer (benzophenone). Following this experiment, the authors irradiated an alkylene-$\alpha$,$\omega$-bispyranone (**240**) which gives (in the presence of a sensitizer) one or both stereoisomers (**241**) and (**242**) in which the pyran rings are fused together, the length of the alkylene chain having an effect on the ratio of isomers formed ⟨72TL2247⟩.

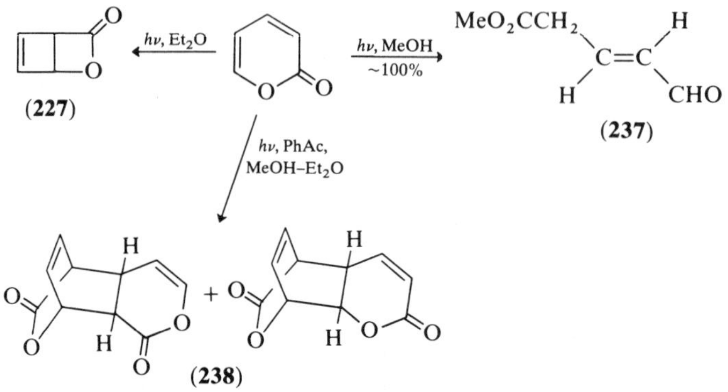

**Scheme 11**

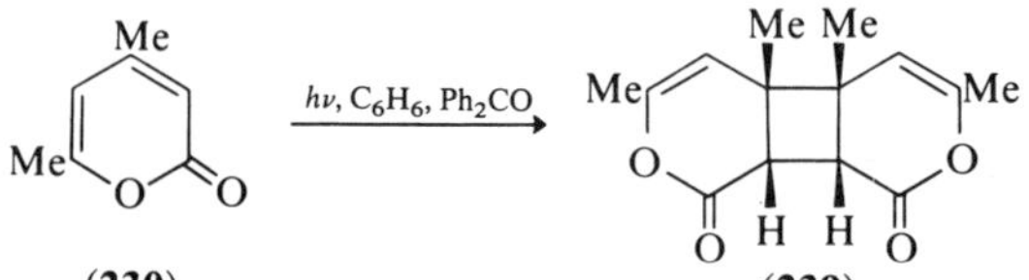

When pyran-2-ones (and a sensitizer) are dissolved in 1,2-dichloroethane and oxygen is bubbled through while the solution is irradiated, a high yield of an endoperoxide, *e.g.* (**243**), is formed. Such compounds are described as having hyperenergetic properties; when warmed, they exhibit chemiluminescence and lose carbon dioxide ⟨79JA5692⟩.

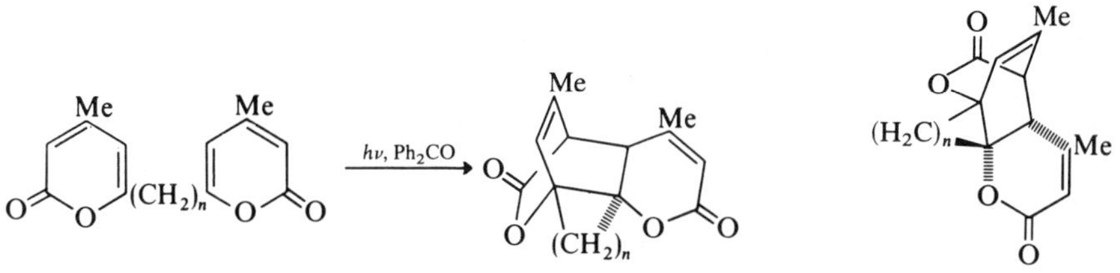

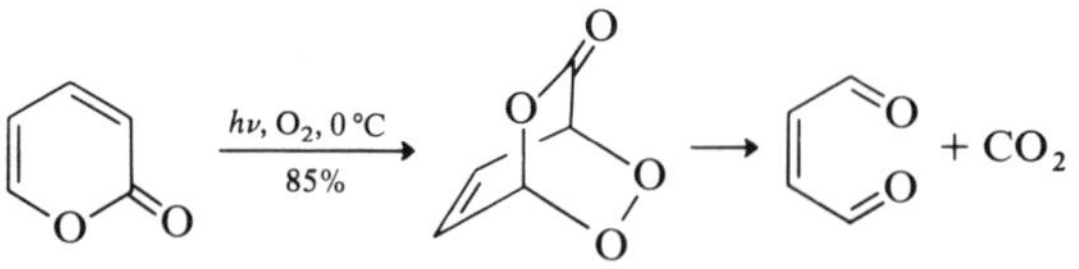

The photodimerization of coumarin has been studied in several solvents and their nature has an effect on this rather complex reaction. In a polar medium such as methanol, the only product is the *cis* head-to-head isomer (**244**) but in acetonitrile, this is accompanied by the *trans* head-to-head dimer (**245**), which becomes the main product of the reaction in non-polar solvents like benzene or dioxan. Small amounts of the head-to-tail isomers (**246**) are also formed in non-polar solvents ⟨66JA5415⟩. In some solvents, the presence of benzophenone as sensitizer is essential and it also alters the relative proportion of dimeric isomers ⟨64JA3103, 66CB625⟩. Photolysis of osthole (**247**) in the presence of acetophenone gives a mixture of two quite different dimers (**248**) and (**249**) ⟨80MI22302⟩, representing addition at the exocyclic and endocyclic double bond respectively.

**(244)** **(245)** **(246)**

**(247)** **(248)** **(249)**

Pn = $CH_2CH{=}CMe_2$

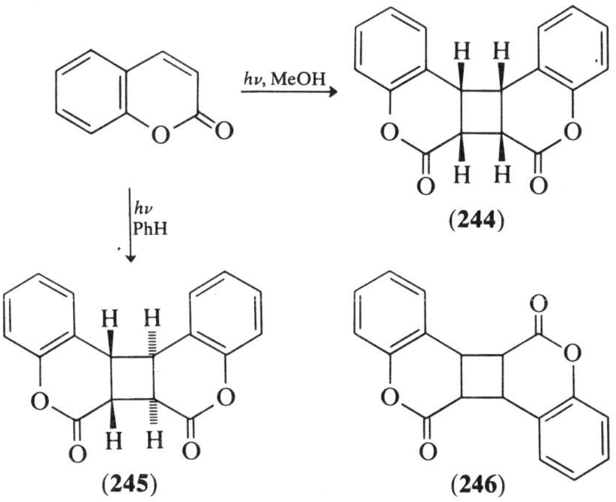

### 2.23.6.3 Electrophiles

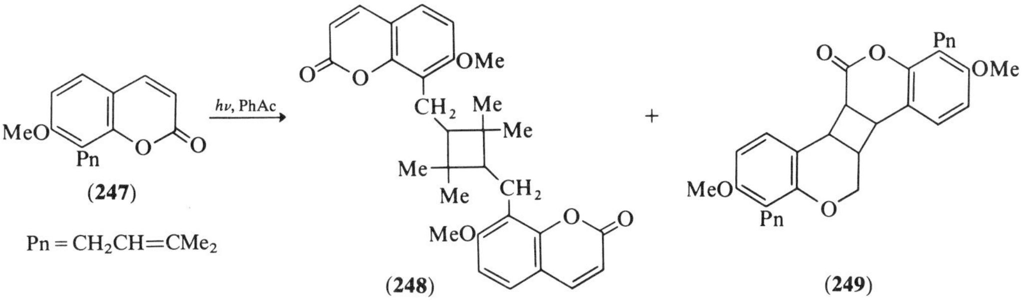

#### *2.23.6.3.1 Addition and substitution*

Electrophiles either add to a double bond or directly substitute the pyran-2-one ring. Bromine, for example, adds to the parent compound at 60 °C to give 3,4,5,6-tetrabromo-3,4,5,6-tetrahydropyran-2-one (**253**), but in boiling carbon tetrachloride it yields 3-bromopyran-2-one (**251**). Irradiation of the pyranone with bromine in 1,2-dichloroethane at −78 °C gives a high yield of the dibromide (**250**). A careful study suggests that in the reaction at 60 °C, one or possibly two dibromides (**252**) are formed first, and also that the reluctance of pyran-2-one to undergo typical electrophilic bromination may be due to the inability of the σ-complex to lose a proton. It is likely that 3-bromopyran-2-one is formed *via* the dibromide (**252**). Chlorine behaves similarly ⟨69JOC2239⟩. Pyridinium hydrobromide perbromide brominates methyl coumalate (**254**) very effectively at C-3 ⟨73JCS(P1)1130⟩.

Sulfuryl chloride, which is usually a source of chlorine radicals, reacts with 6-phenylpyran-2-one to give the 3,5-dichloro derivative ⟨63ZOR3434⟩.

**(250)** **(251)** **(252)** **(253)**

**(254)**

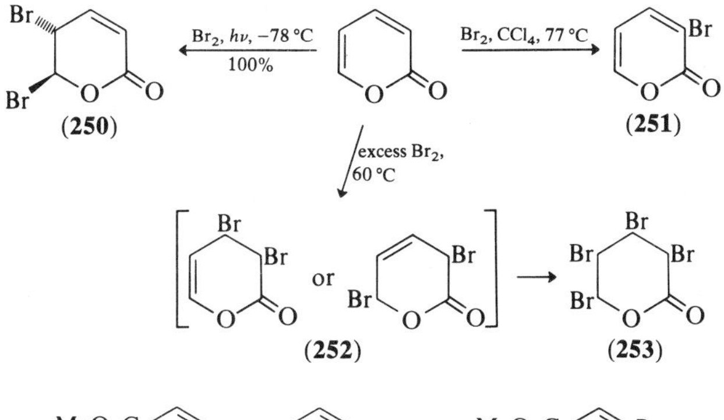

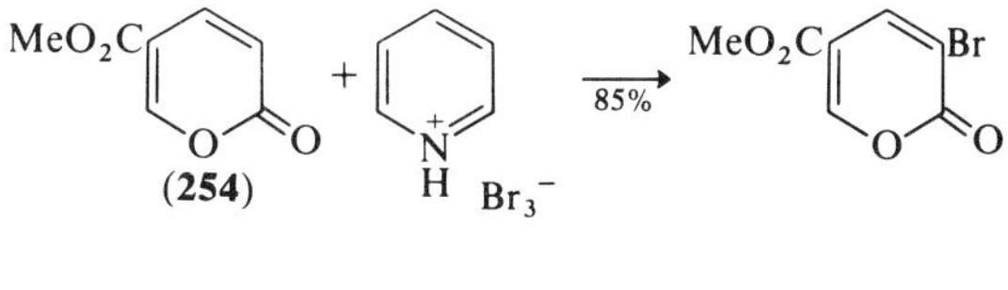

The nitration of 6-phenylpyran-2-one (**255**) gives an insight into the comparative reactivity of the two rings towards the nitronium ion. With 94% nitric acid or mixed nitric and sulfuric acids (which may protonate the pyranone), the 4-nitrophenyl isomer is formed but 67% nitric acid yields the 3-nitro isomer ⟨66JOU1113⟩. Nitration of pyran-2-one with nitronium tetrafluoroborate gives a moderate yield of 5-nitropyran-2-one (**257**) in a two-stage reaction which probably involves a 2-nitrate ester (**256**) ⟨69JHC313⟩. The nucleophilicity of the carbonyl oxygen atom is also apparent in its comparatively facile methylation by trimethyl fluoroborate and the formation of the pyrylium salt (**258**) ⟨75CC675⟩.

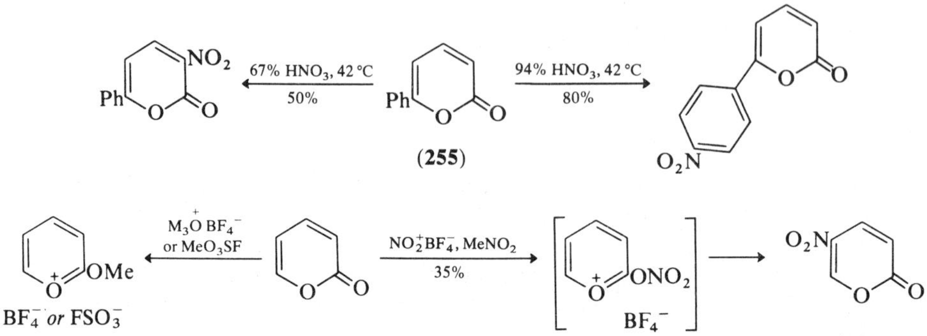

Coumarin absorbs one molecule of bromine to form the 3,4-dibromide, which readily eliminates hydrogen bromide to form 3-bromocoumarin. 3-Chloroisocoumarin adds chlorine to form 3,3,4-trichloro-3,4-dihydroisocoumarin in high yield ⟨73JOU2160⟩. The presence of a 4-hydroxy group directs electrophiles to C-3, for example, in the formation of 3-acyl-4-hydroxycoumarin ⟨53MI22301⟩ and 4-hydroxy-3-nitrocoumarin ⟨78JCR(S)47⟩. Sulfonation of coumarin with chlorosulfonic acid at 100 °C yields the 6-sulfonyl derivative, but at 130–140 °C a second substituent is introduced to give 3,6-disulfonyl chloride. The balance of reactivities with electrophiles is reversed by the presence of a 6-nitro group and sulfonation then proceeds at C-3 ⟨57JIC35⟩.

Heating some pyran-2-ones, including coumarins e.g. (**259**), with aqueous acid opens the ring and the presence of other functional groups sometimes allows recyclization to a different ring, a phenomenon which is reminiscent of the pyrylium salts (Section 2.23.2) ⟨70T5255⟩.

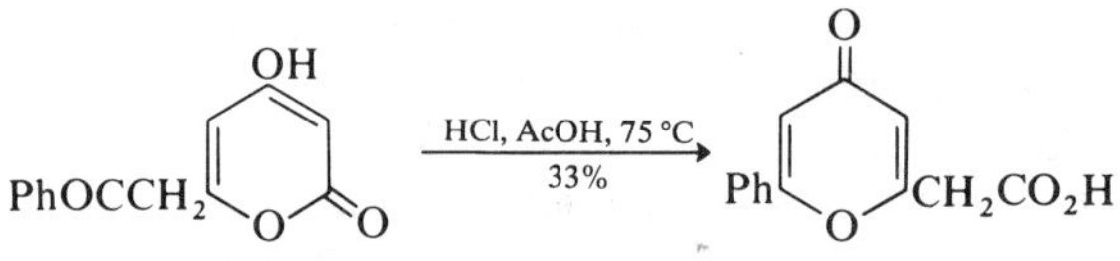

Chloromethylation of pyran-2-ones, such as 5,6-dimethylpyran-2-one (**260**), and coumarins proceeds at C-3. The related Mannich reaction on 4-hydroxycoumarin yields the 3-aminomethyl derivative ⟨53JA1883⟩ while 3-hydroxycoumarin (**261**) is aminomethylated at C-4 ⟨69JMC531⟩.

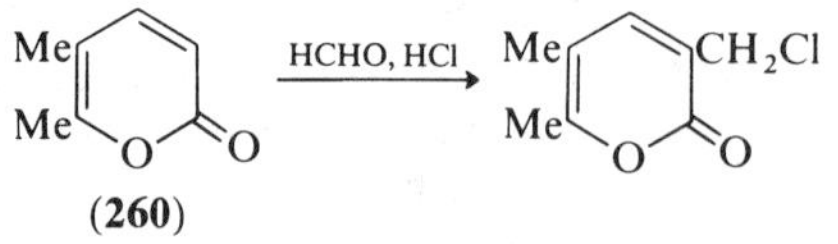

An electrophilic substitution of potential synthetic value but one which is rarely utilized is the introduction of a mono- or di-acetoxymethyl group by the action of manganese(III) acetate on a coumarin. This ring system is more amenable than many to this reagent and a variety of products may be obtained by altering the ratio of reactants (Scheme 12) ⟨79BCJ2386⟩.

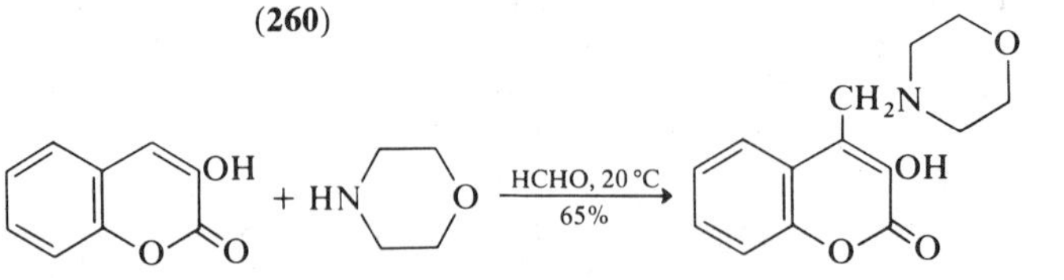

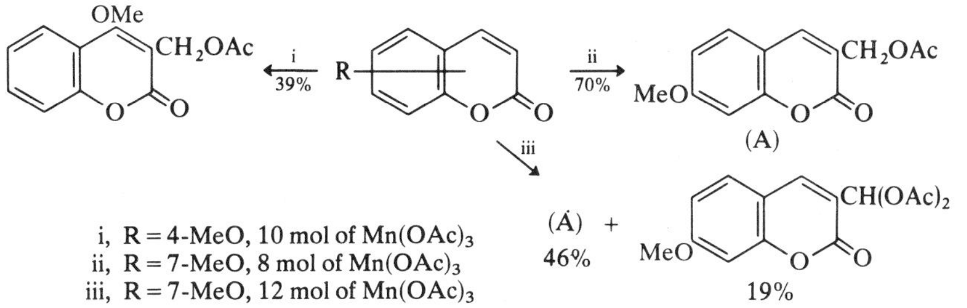
OMe
CH2OAc
i
39%
R
ii
70%
MeO
CH2OAc
(A)
iii
(A) +
46%
MeO
CH(OAc)2
19%
i, R = 4-MeO, 10 mol of Mn(OAc)3
ii, R = 7-MeO, 8 mol of Mn(OAc)3
iii, R = 7-MeO, 12 mol of Mn(OAc)3

**Scheme 12**

### *2.23.6.3.2 Oxidation*

Photooxidation of pyran-2-ones has been mentioned (Section 2.23.6.1). Potassium permanganate cleaves the pyranone ring of 3-nitro-6-phenylpyran-2-one to benzoic acid ⟨66JOU1113⟩. Coumarin, being an ester, is relatively stable to oxidizing agents; for instance, chromic acid is without effect on it, but isocoumarins are degraded by many oxidizing agents to keto acids such as (**262**). Capillarin (**225**) on ozonolysis yields phthalic acid.

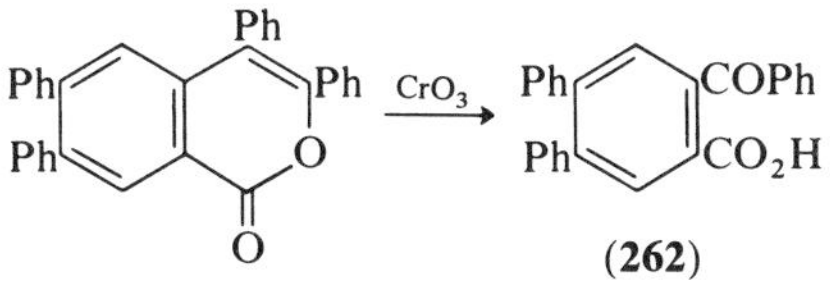
Ph
Ph
Ph
Ph
CrO3
Ph
Ph
COPh
CO2H

(**262**)

## 2.23.6.4 Nucleophiles

Several kinds of nucleophiles react with pyran-2-ones and coumarins; some of these reactions involve ring opening and, occasionally, recyclization into another ring. A nucleophile (Nu) which cleaves the ring attacks to break one of the bonds of the ring oxygen atom as shown in Scheme 13.

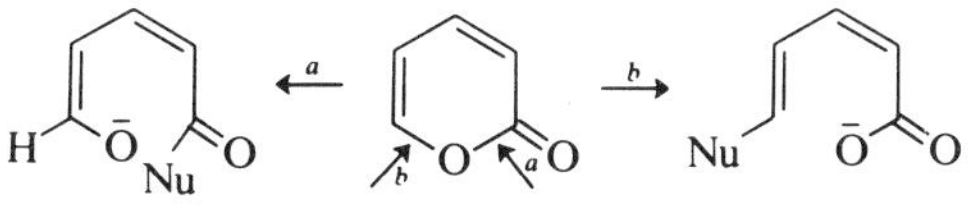
a
b
H
Nu
Nu

**Scheme 13**

### *2.23.6.4.1 C-Nucleophiles*

The most widely used and synthetically useful nucleophilic reactions of pyran-2-ones are those of organometallic compounds containing magnesium, lithium or zinc. The products obtained depend on the molar ratio of reagents and on the substituents already present on the ring. With two moles of reagent, a 2,2-disubstituted pyran, *e.g.* (**263**), or chromene is formed ⟨61CB1794, 77HC(31)11, 79MI22302⟩ but with a larger excess (and provided a methyl or methylene group is available at C-2 or C-6) the pyran ring is cleaved to a dienone (**264**) which reacts with another molecule of reagent. The methyl or methylene group then becomes a member of the newly formed benzene ring in (**265**) ⟨61CB1796⟩. When one mole of Grignard reagent reacts with an isocoumarin such as (**266**) and perchloric acid is added, a 2-benzopyrylium salt (**267**) is obtained in high yield ⟨78CHE1067⟩.

Conversion of a pyran-2-one into a benzene ring may be effected by means of a Reformatsky reagent, *e.g.* $BrZnCH_2CO_2Et$, or the corresponding lithium compound ($LiCH_2CO_2Et$). The latter reagent has the advantage that any alkoxy groups present are unaffected by the dealkylating action of zinc bromide; the reaction of 8-methoxy-3-methylisocoumarin (**268**) is a good example ⟨78JHC1535⟩.

The nucleophilic carbon of diazomethane attacks the ring of pyran-2-ones, especially when electron-withdrawing substituents are present; for example, methyl 2-oxopyran-5-carboxylate (methyl coumalate, **269**) is methylated successively at C-6 and C-4

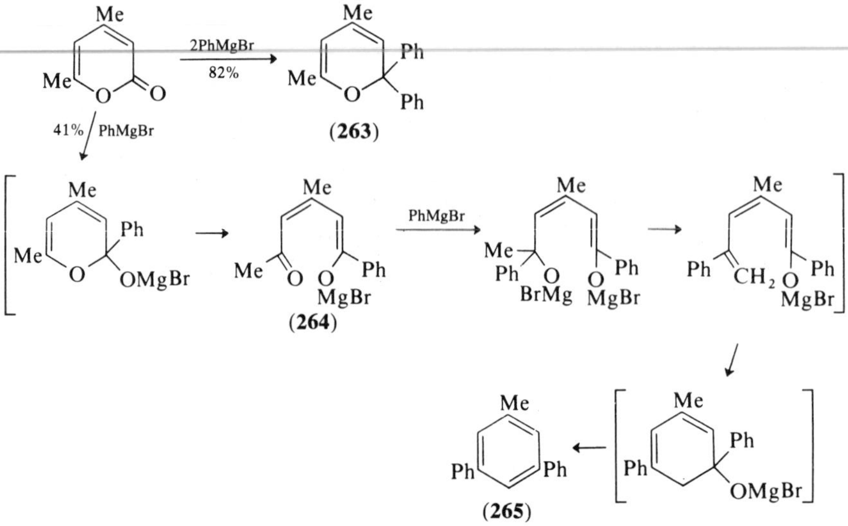

⟨41JOC577, 63MI22302⟩; 3-cyano- and 3-nitrocoumarin (**270**) are readily converted into their 4-methyl homologues ⟨70M1123⟩ but coumarin is transformed into the pyrazolone (**271**) ⟨41JA2017⟩.

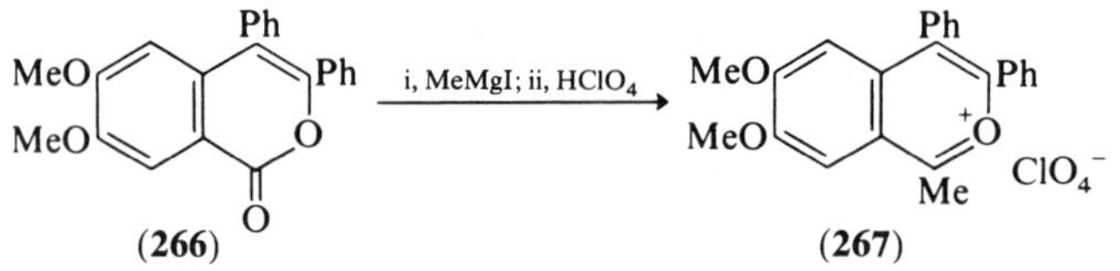

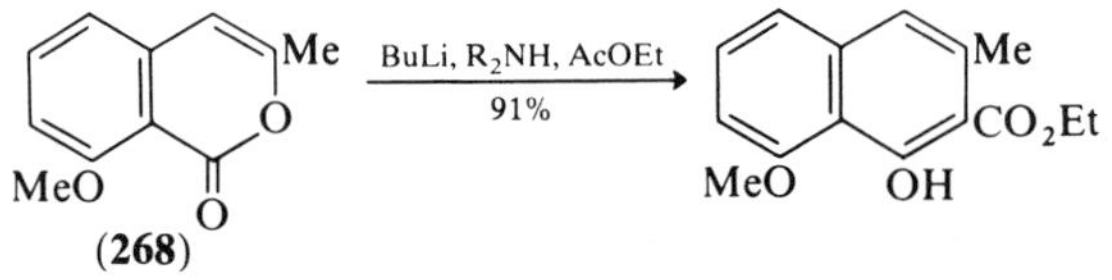

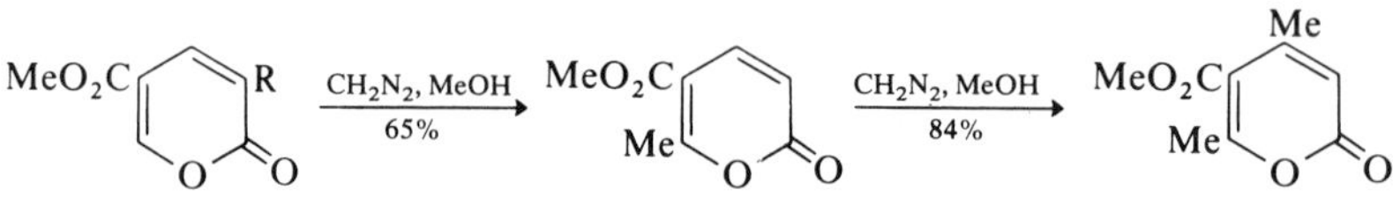

R = CN, 92%
R = NO$_2$, 59%

Carbanions generated *in situ* by strong bases cleave pyran-2-one rings but acidification may result in the formation of a new ring, as, for example, by reaction of 4-methoxycoumarin with the methanesulfinylmethide ion (**272**), which results in the formation of a chromone (**273**) ⟨78S208⟩.

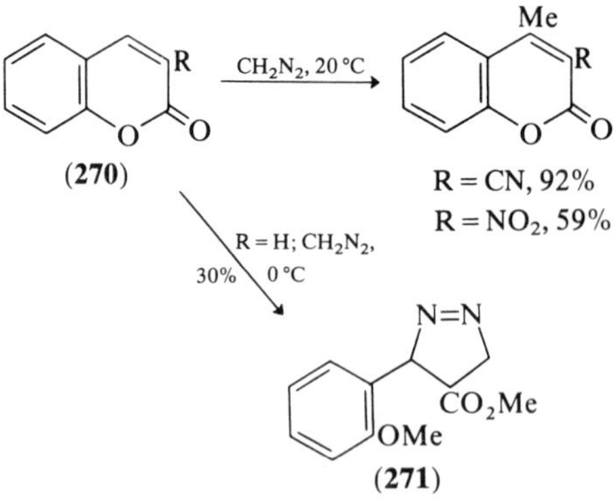

The attack of a cyanide ion at the 1,6-bond of a pyran-2-one is stereospecific and for some compounds stereochemical control can be exerted by suitable choice of reaction conditions. Ring opening occurs under mild conditions to give a 5-cyanohexa-2,4-dienoic acid (**274**) ⟨65JOC203⟩.

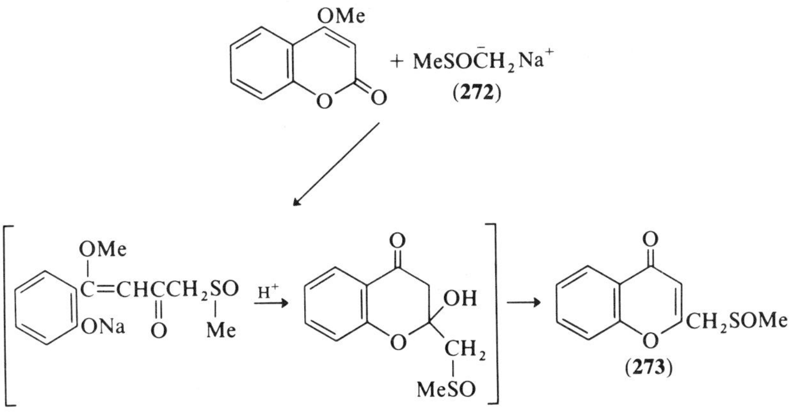

(272)

(273)

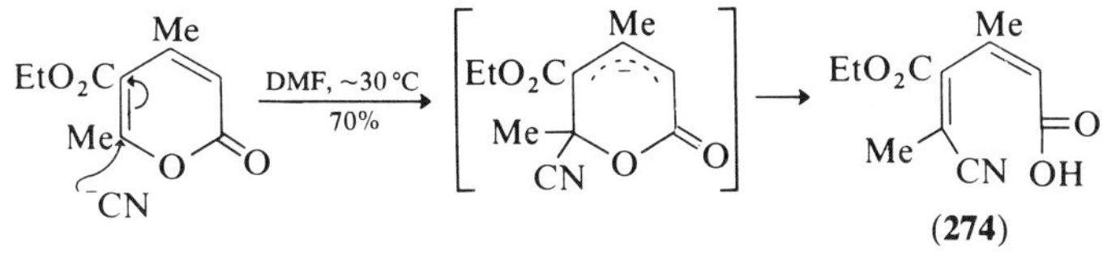

(274)

### 2.23.6.4.2 N-Nucleophiles

Ammonia and amines open pyran-2-one rings and the acyclic products may cyclize again on acidification to pyridones or a benzene ring, a reaction reminiscent of those of pyrylium salts (Section 2.23.2.3). Under mild conditions, unsaturated amino acids such as (275) are formed but at higher temperatures and longer reaction times, a pyridone (276) or benzamide (277) is formed. The probable course of benzamide formation is as shown in Scheme 14 ⟨74JOU852⟩.

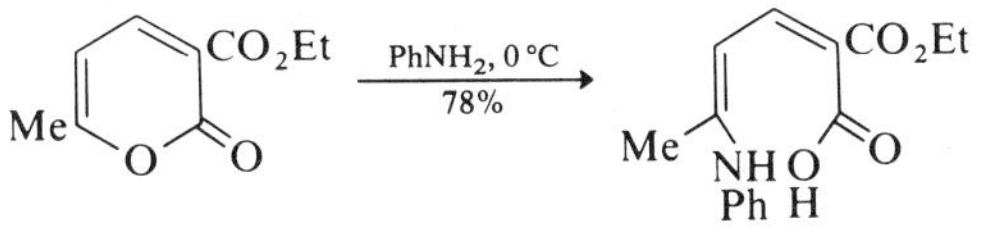

(275)

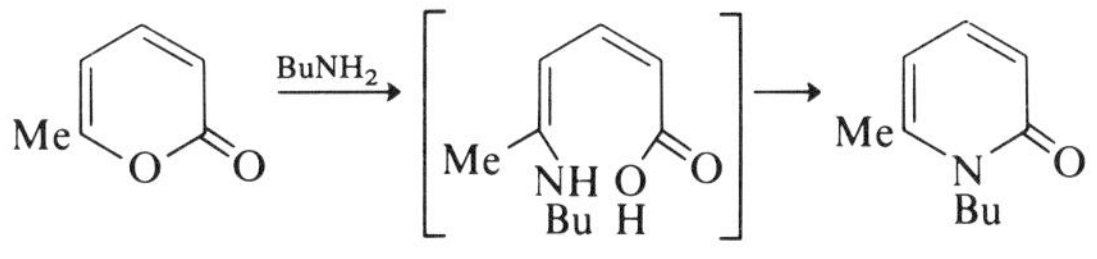

(276)

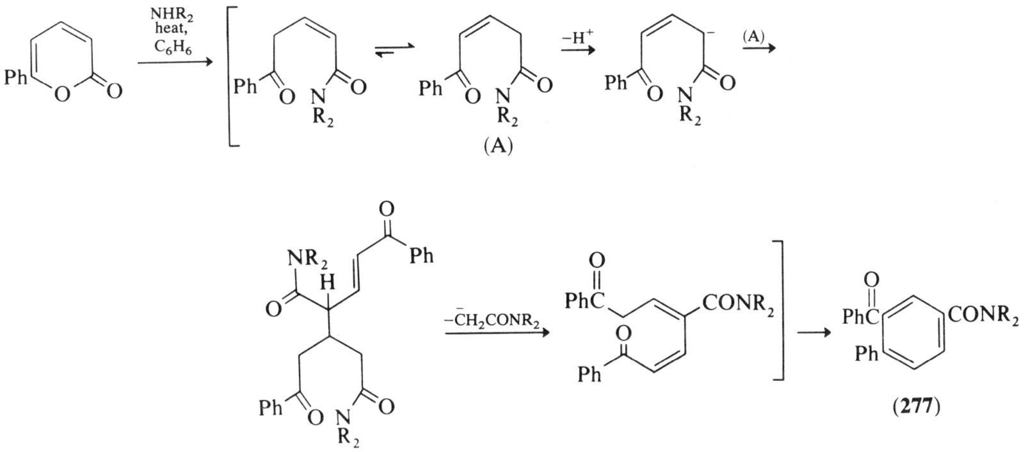

(A)

(277)

**Scheme 14**

LDA abstracts a proton from C-3 of pyran-2-ones and the resulting anion reacts with deuterium oxide, trimethylsilyl chloride and carbon dioxide as shown in Scheme 15 ⟨80CC1224, 80CC1227⟩.

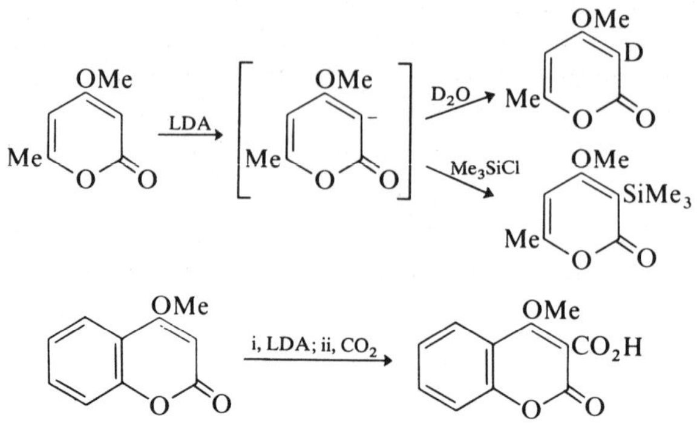

**Scheme 15**

Hydroxylamine converts pyran-2-ones into pyridine $N$-oxides by attack at the 1,2-bond; for example, the pyran-2-one (**278**), obtained by heating 3-acetyl-4-hydroxy-6-methyl-pyran-2-one (dehydroacetic acid) and DMF dimethyl acetal in xylene, reacts with two moles of hydroxylamine under mild conditions to give the pyrano[2,3-$b$]pyridine $N$-oxide (**279**) ⟨77JHC931⟩. Hydrazine converts pyran-2-ones into 1-amino-2-pyridones ⟨78JHC759⟩.

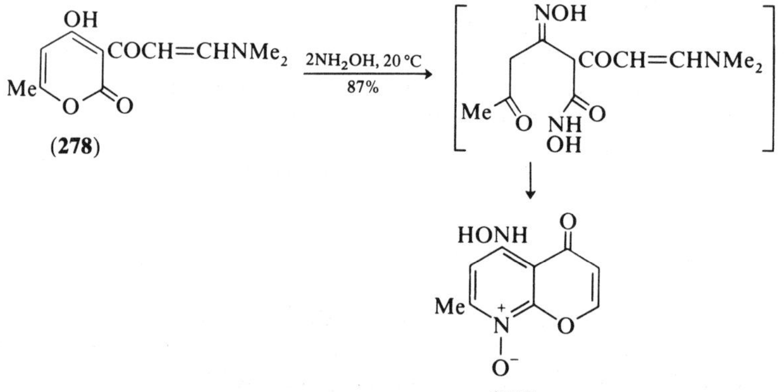

Aliphatic primary and secondary amines attack the 1,2-bond of coumarins to give various derivatives depending on the reaction conditions and substituents on the substrate. In addition to the expected amides (**280**) from coumarin itself, some $\beta$-amino amides (**281**) are obtained from 4-hydroxycoumarin. In addition, the latter reacts with secondary aliphatic and primary aromatic amines to yield substitution products such as 4-piperidinocoumarin

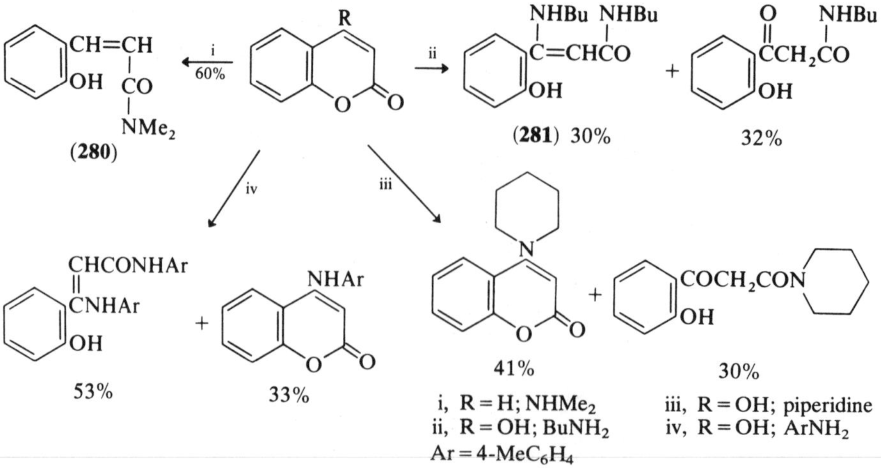

i, R = H; NHMe$_2$  
ii, R = OH; BuNH$_2$  
Ar = 4-MeC$_6$H$_4$  
iii, R = OH; piperidine  
iv, R = OH; ArNH$_2$

**Scheme 16**

(Scheme 16). 3-Bromocoumarin reacts with amines to give substitution and degradation products and also good yields of benzofuran-2-carboxamides ⟨81JHC105, 77MI22302, 65JOU1936⟩.

Isocoumarins react with aqueous ammonia or alkylamines to give good yields of isoquinolin-1-ones; for example 3-(4-methoxyphenyl)-1,2-dihydroisoquinolin-1-one (**282**). A 7-nitro group in the isocoumarin may be reduced by treatment with hydrazine and Raney nickel to the corresponding 7-amino derivative (**284**), but with hydrazine alone a 2,3-benzodiazepin-1-one (**283**) is formed ⟨68JCS(C)2205, 57JGU2342, 81CPB249⟩.

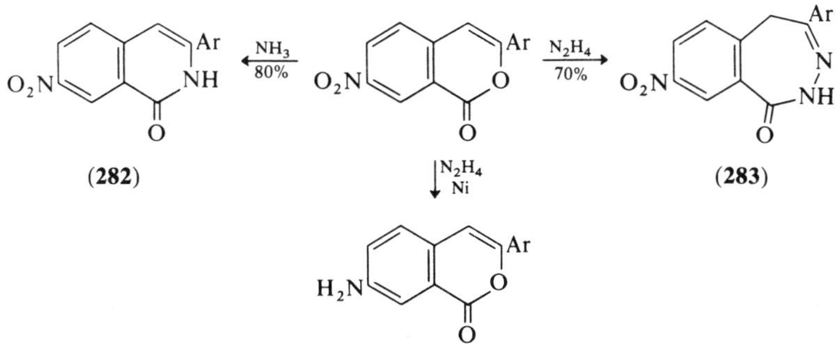

Pyran-2-ones and coumarins react with urea, thiourea or guanidine to form pyrimidines; for example, dehydroacetic acid (**285**) gives 2-amino-4-methyl-6-(2-oxopropyl)pyrimidine (**286**) ⟨73MI22300, 77BSF369⟩.

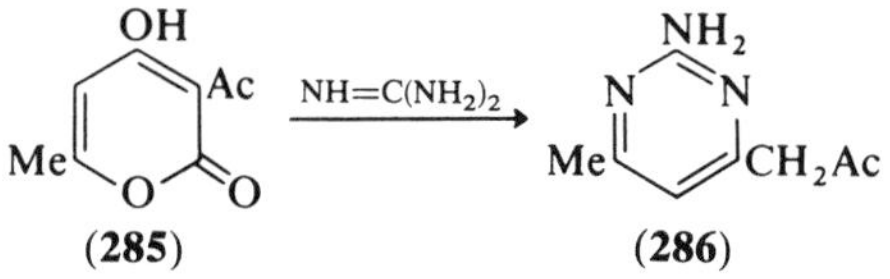

### 2.23.6.4.3 O-Nucleophiles

Aqueous alkali hydrolyzes lactones and the products, *e.g.* (**287**), are frequently unstable or recyclize, depending on other substituents present. Coumarin is hydrolyzed by dilute alkali first to the yellow *cis* acid (coumarinic acid) salt (**288**) which recyclizes to coumarin on acidification but when heated with alkali isomerizes to the *trans* acid (coumaric acid) salt (**289**). When it is desirable to identify the hydrolytic product of such a reaction it is better to incorporate a methylating agent so that the reverse reaction cannot then occur. Hot aqueous alkali converts methyl 3-bromocoumalate (**290**) into furan-2,4-dicarboxylic acid ⟨73JCS(P1)1130⟩.

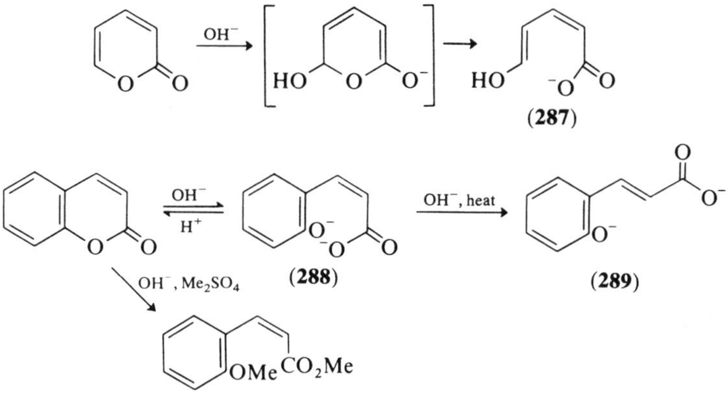

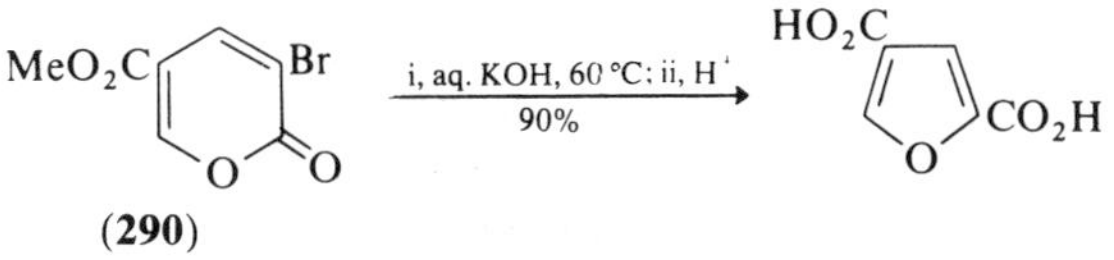

Similarly, a substituted coumarin is converted into a benzofuran-2-carboxylic acid (**291**), a reaction which is useful in the determination of structure ⟨73JCS(P1)2781⟩.

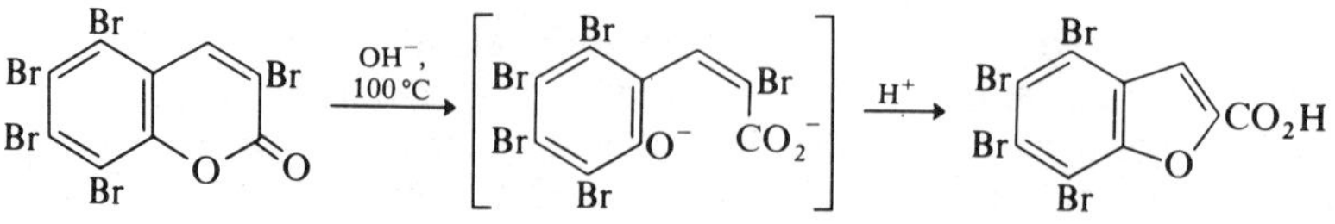

Ethanolic alkali, on the other hand, cleaves the pyran ring and may degrade the resulting chain if the temperature is high. For example, 4-methoxyparacotoin (**292**), a constituent of a South American rosewood essential oil, gives different breakdown products according to the temperature of reaction ⟨57JA4507⟩.

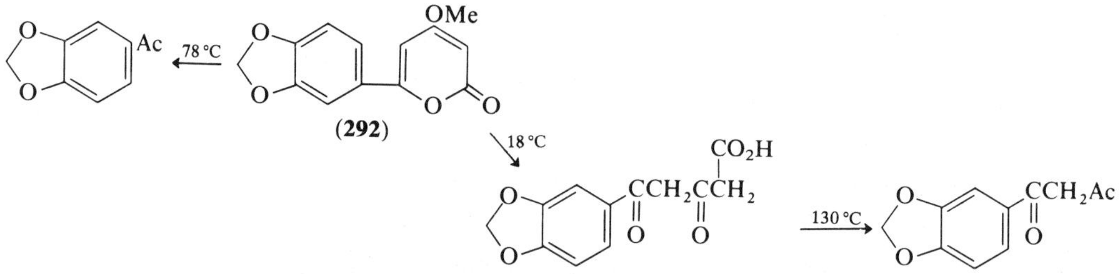

The effect of other substituents on the products of alkaline hydrolysis is well illustrated in the conversion of 4-chloromethylcoumarin (**293**) to the benzofuran-3-acetic acid (**294**) in high yield ⟨79CHE815⟩, of 4-hydroxy-6-phenacylpyran-2-one (**259**) into the benzophenone (**295**) ⟨70T5255⟩ and of 5-amino-3-methylisocoumarin (**297**) (formed from the nitro compound **296**) into the indole (**298**) ⟨76JCS(P1)1073, 81CPB249⟩.

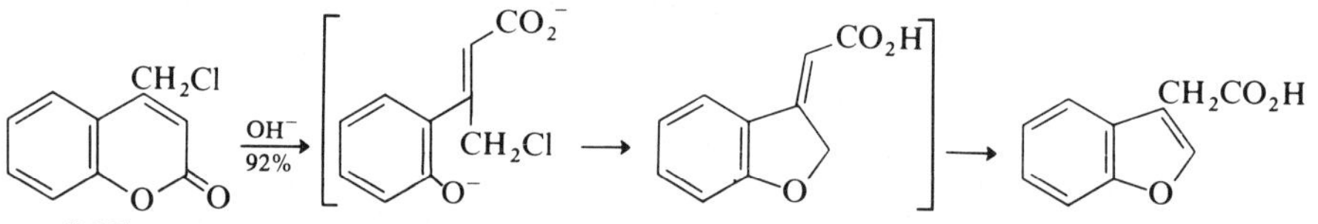

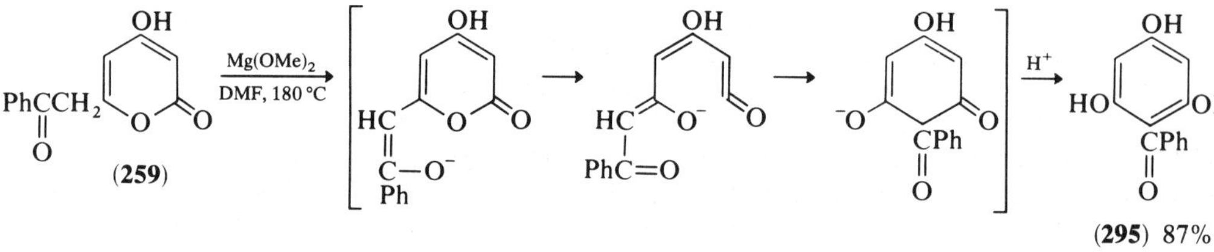

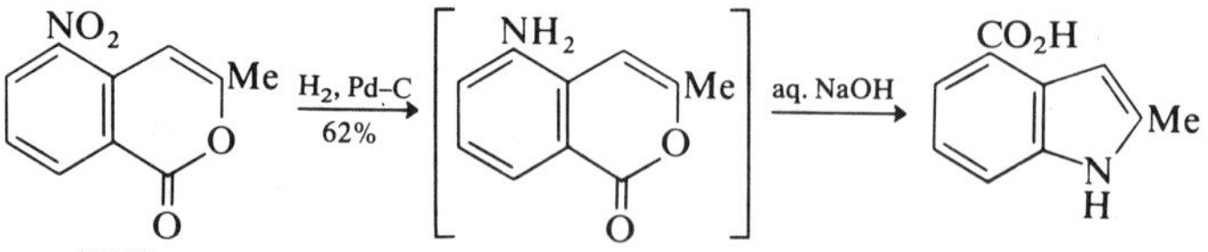

Fusion of isocoumarins (**299**) with strong alkali gives a benzoic acid ⟨57JGU2342⟩.

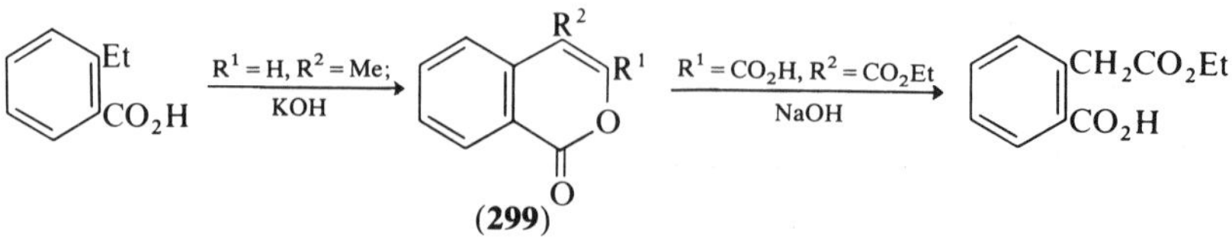

### 2.23.6.4.4 *Reduction*

Stepwise catalytic reduction of pyran-2-ones to the dihydro (**300**) and tetrahydro (**301**) pyran-2-ones is achieved by suitable choice of catalyst and reaction conditions (Scheme 17). Anhydrous copper(II) sulfate simultaneously dehydrates the alcohol ⟨78JHC1153⟩.

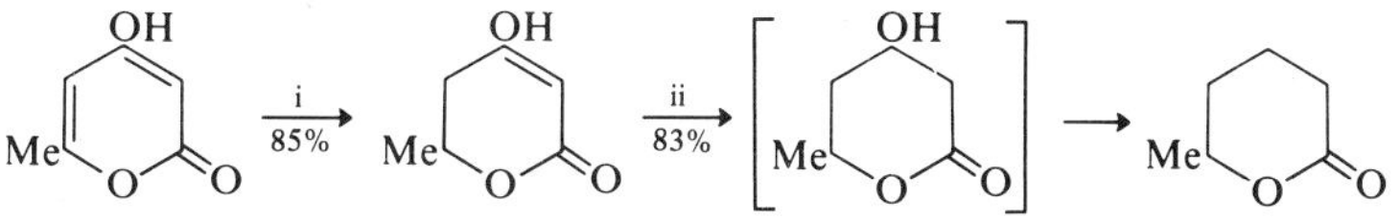

**(300)**                                                                        **(301)**

i, H$_2$, Ni or Pd–C, 5 kg cm$^{-3}$, 20 °C; ii, H$_2$, Pd–C, CuSO$_4$(anhydrous), 20 kg cm$^{-3}$, 75 °C

**Scheme 17**

LAH attacks pyran-2-ones, such as (**303**), at C-6, opens the ring and forms 3-methylhexa-2,4-dienoic (3-methylsorbic) acid (**302**). Sodium borohydride is more selective and is the reagent of choice for reducing ethyl isodehydroacetate (**303**; R = CO$_2$Et) to give, after alkali hydrolysis of the ester, the dicarboxylic acid (**304**) ⟨62CI(L)268⟩.

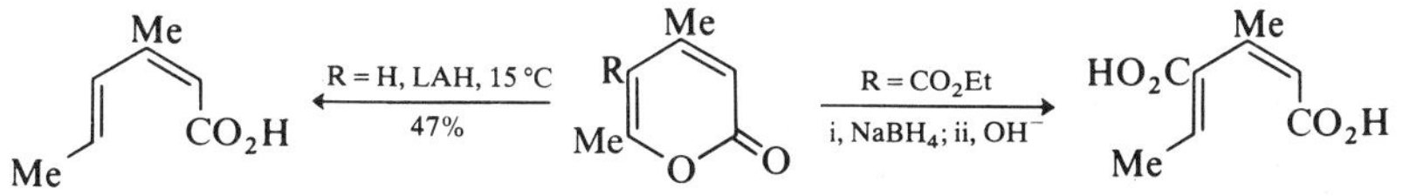

**(302)**                                    **(303)**                                    **(304)**

Many methods have been described for the reduction of coumarins ⟨B-77MI22301⟩ but undesired byproducts are frequently formed. A thorough study of the catalytic reduction of several coumarins has recently shown that careful choice of conditions can lead to high yields of either the 3,4-dihydrocoumarin (**306**) or the ring opened product (**305**) which can be cyclized efficiently (Scheme 18) ⟨80JHC1597⟩.

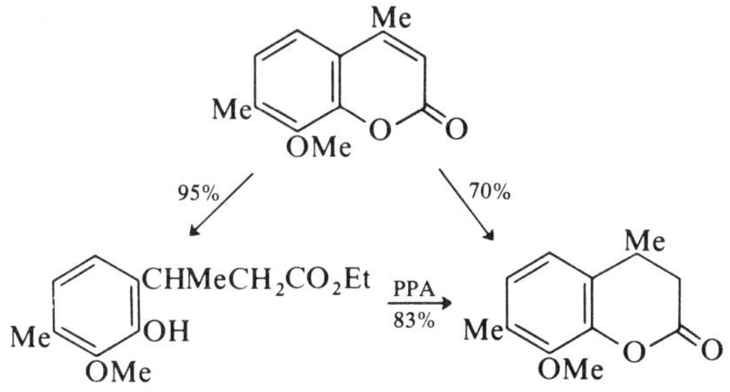

**(305)**                                    **(306)**

i, H$_2$, Pd–C, 150 °C, 1500 p.s.i., EtOH; ii, H$_2$, Pd–C, 100 °C, 90 p.s.i.

**Scheme 18**

Diborane followed by hydrogen peroxide has two effects on coumarins: the carbonyl group is reduced to methylene and the elements of water are added across the 3,4-double bond in an anti-Markovnikov manner but the overall yield is usually low (Scheme 19) ⟨70JOC2282⟩.

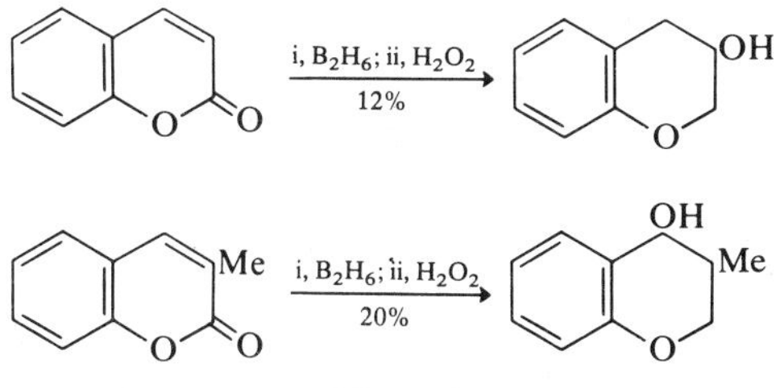

**Scheme 19**

Catalytic reduction of isocoumarins leads to 3,4-dihydroisocoumarins but LAH opens the pyranone ring to form phenethyl alcohols.

## 2.23.6.5 Radicals

The 3-position of coumarin is the most susceptible to attack by radicals and reaction with 1,3-diphenyltriazene gives 3-phenylcoumarin ⟨79JHC97⟩. The reaction of sulfuryl chloride with pyran-2-ones is described in Section 2.23.6.3.1. Coumarin (**307**) is arylated at C-3 by treatment with 4-nitrobenzenediazonium chloride under Meerwein reaction conditions. Variations of solvent and pH show that acetone and a buffered pH of 2–4 give the best result ⟨55JA3401⟩. 3-Phenylation of coumarin is achieved also by heating 4-hydroxy-coumarin in chloroform with benzoyl peroxide ⟨78IJC(B)292⟩.

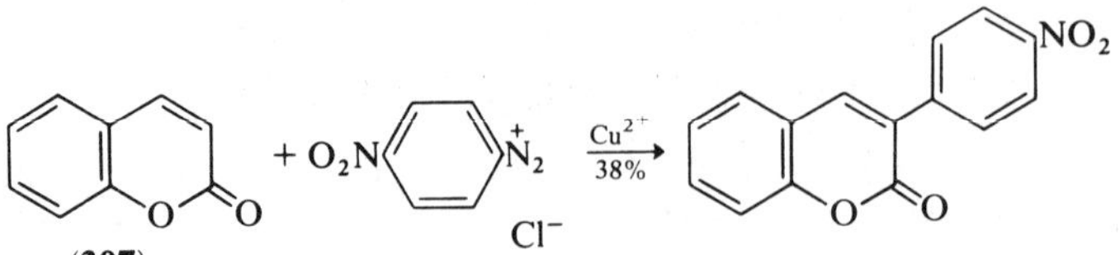

### 2.23.6.6 Other Reactions

Perhaps the most useful reaction of pyran-2-ones is their ability to act as dienes in a Diels–Alder reaction ⟨74RCR851⟩, a principle which was recognized in 1931 by the discoverers of this reaction. Pyran-2-one reacts with maleic anhydride to give the expected adduct ⟨31LA(490)257⟩ which is the *endo* isomer (**308**) ⟨70JGU1402⟩. Some adducts are convertible *in situ* by prolonged heating and loss of carbon dioxide into benzenoid compounds, for example into phthalic acid derivatives (**309**) or hemimellitic acid esters (**310**) ⟨75TL2389, 79S987, 64JOC2534⟩.

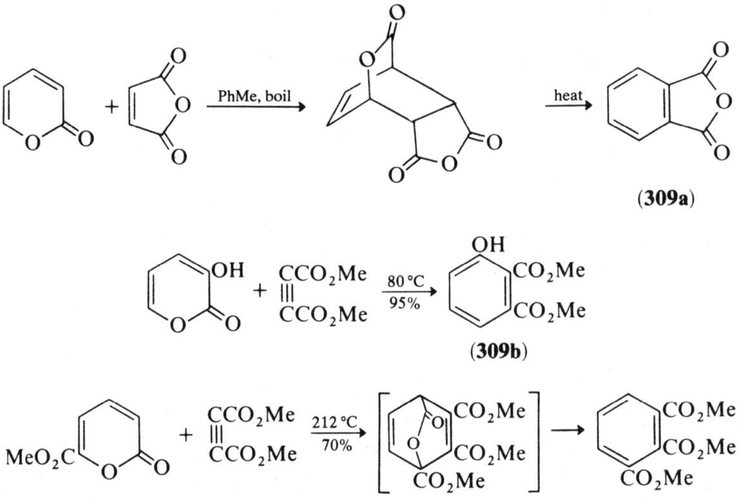

Two interesting variations on the role of pyran-2-one as a synthon are worthy of mention. Irradiation of the compound in methanol containing acetophenone as sensitizer gives equal amounts of the photostable dimers (**311**) and (**312**). In this reaction the pyranone acts as both diene and dienophile and the dimers are believed to be formed *via* triplet excited states of the pyranone ⟨68TL5279⟩.

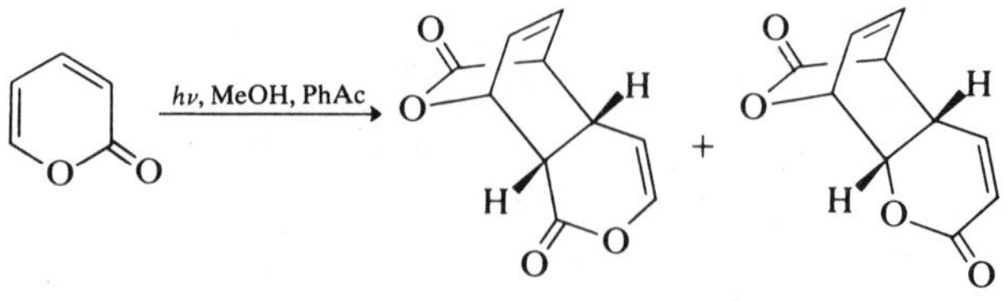

The second variant is the reaction of a pyran-2-one (methyl coumalate, **313**) as dienophile with a cyclopentadiene (**314**). The balance between the dienic and dienophilic capabilities of pyran-2-one is tipped in favour of the latter by the electron-withdrawing properties of the carboxylic ester group and this fine balance enhances the potential of these compounds as synthons ⟨72CC388⟩.

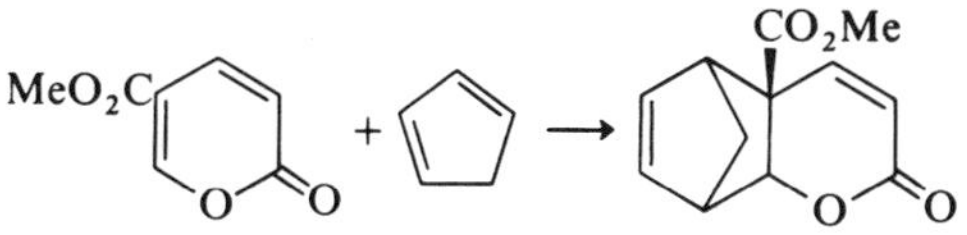

(313)    (314)

## 2.23.7 PYRAN-2-ONES AND THEIR BENZO DERIVATIVES: REACTIVITY OF SUBSTITUENTS

### 2.23.7.1 Introduction

A methyl group in the 4- or 6-position of pyran-2-ones is reactive and is readily halogenated and condensed with aldehydes but the 4-methyl group of coumarins is much less reactive. Tautomerism of 4-hydroxypyran-2-ones leads to the formation of 4-alkoxy- or 2,4-dialkoxy derivatives. Some of the pyran-2-ones found in nature contain a benzene ring, for example paracotoin (**315**), hispidin (**316**) and yangonin (**317**). Some hydroxy-coumarins and hydroxyisocoumarins are present in plants as their glucosides and the aglycones are often fluorescent. Umbelliferone (**318**), osthenol (**319**) and the microbiologically produced reticulol (**320**) are typical examples.

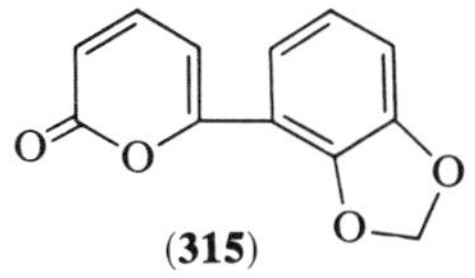

(**315**)

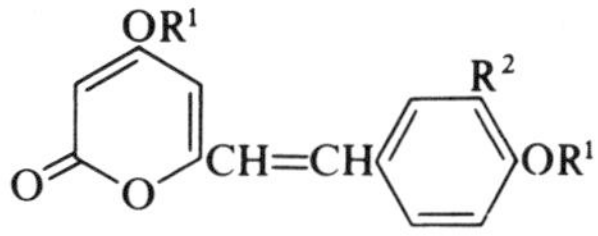

(**316**) $R^1 = H$, $R^2 = OH$
(**317**) $R^1 = Me$, $R^2 = H$

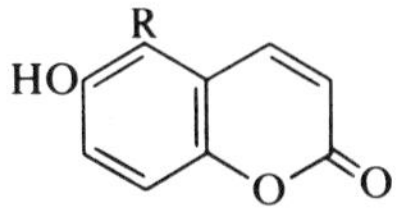

(**318**) $R = H$
(**319**) $R = CH_2CH=CMe_2$

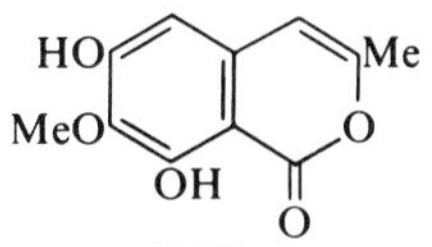

(**320**)

### 2.23.7.2 Fused Benzene Rings

Nitration of coumarin gives the 6-nitro derivative and a small amount of 8-nitrocoumarin. Similarly, Friedel–Crafts acylation and sulfonation give 6-substitution but in the presence of a hydroxyl group in the benzene ring, the incoming group usually enters a position *ortho* to it.

### 2.23.7.3 *C*-Linked Substituents

A methyl group at C-4 or C-6 of pyran-2-ones (like those at C-2 and C-6 of pyrylium salts, Section 2.23.3.2) is more reactive than that at C-3 or C-5 and readily forms an anion which attacks, for example, a carbonyl group. Thus, 4- or 6-styrylpyran-2-ones (**321**) are conveniently prepared.

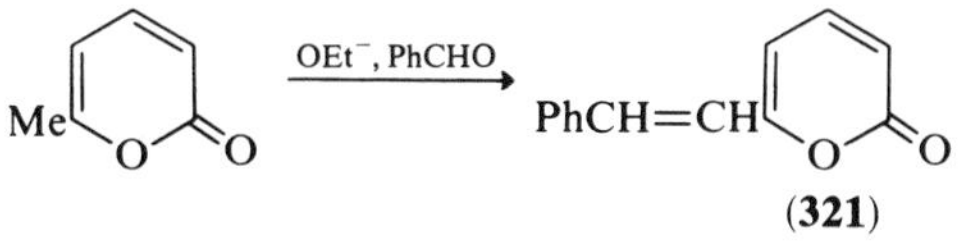

(**321**)

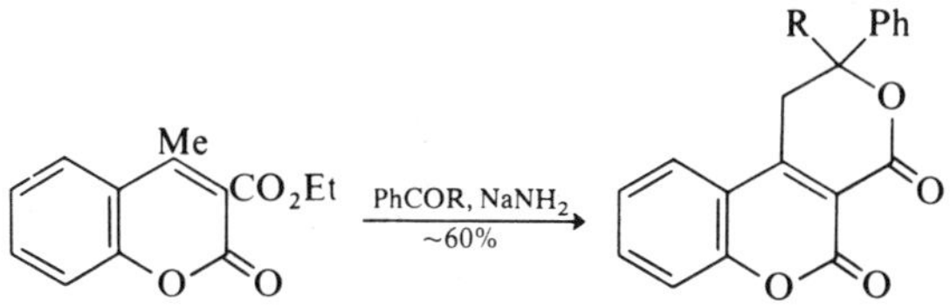

(**322**) $R = H$ or Me

The corresponding group of 4-methylcoumarin is not as reactive and normally fails to form an anion ⟨33JCS616⟩, but recently it was reported that the methyl group of ethyl 4-methylcoumarin-3-carboxylate (322) condenses with carbonyl groups and forms a third ring ⟨79S732⟩.

The acetyl group of dehydroacetic acid (323) behaves normally towards benzeldehydes and the resulting double bond can be selectively hydrogenated, as also may the acetyl group of dehydroacetic acid itself ⟨56JA3201⟩. Sodium cyanoborohydride shows a similar selectivity in reducing the 3-acyl-4-hydroxycoumarin (324; R = Et) ⟨80JOC4606⟩. The acetyl group of dehydroacetic acid condenses with DMF dimethyl acetal in the absence of base to give the enamine (325) in high yield ⟨77JHC931⟩.

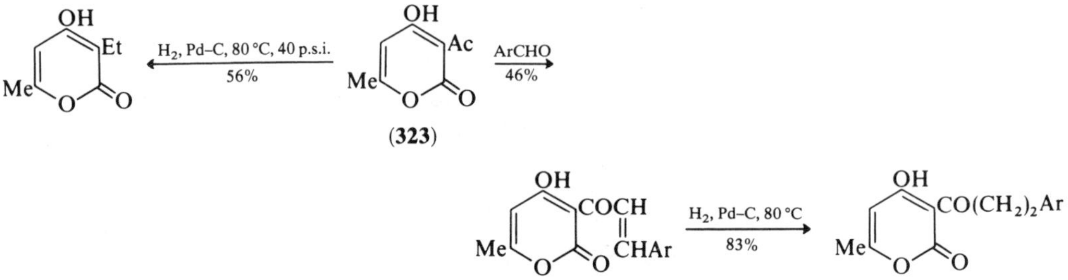

(323)

Ar = 3, 4-(MeO)$_2$C$_6$H$_3$

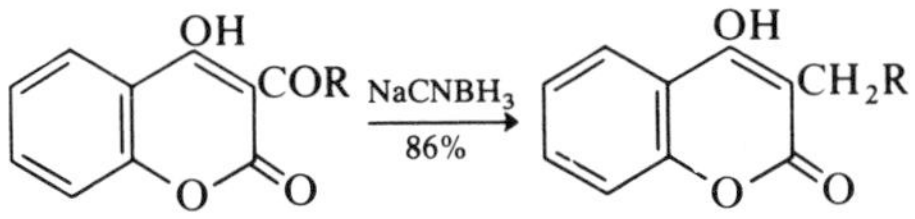

(324)

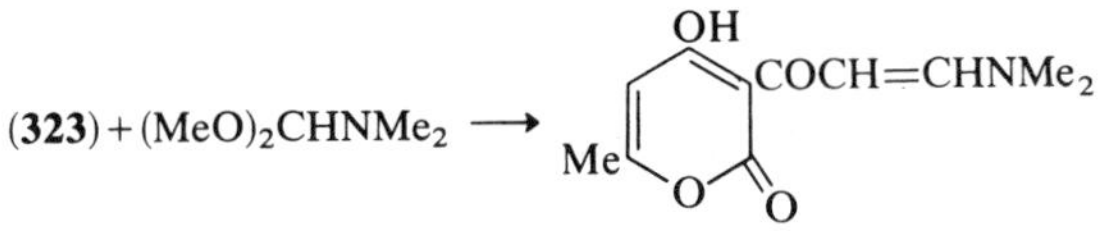

(323) + (MeO)$_2$CHNMe$_2$ ⟶

(325)

Direct oxidation of a methyl group attached to a pyran or benzopyran ring is often difficult as the ring is easily oxidized and opened, but selenium dioxide exhibits useful selectivity, for example, in oxidizing the methyl group of 5,6,7-trimethoxy-3-methyl-isocoumarin (326) in high yield ⟨79IJC(B)642⟩. The aldehyde (327) is oxidized further by peracetic acid and the carboxylic acid (328) loses carbon dioxide on heating.

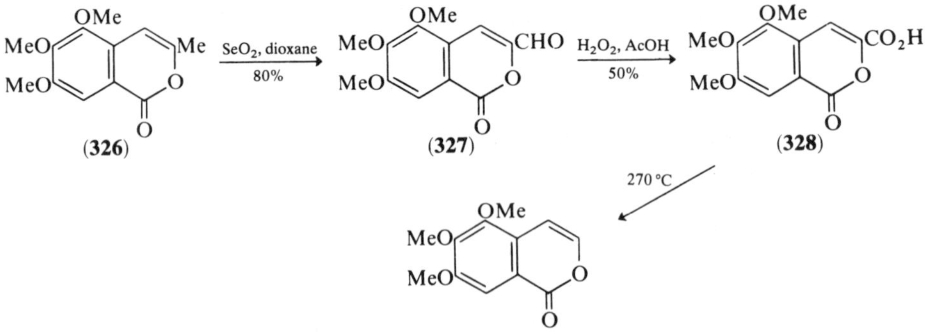

(326)	(327)	(328)

270 °C

Bromination of a methyl group in a pyran-2-one is a useful first step to a number of derivatives, for example 4-methoxy-6-(1-propenyl)pyran-2-one (329) ⟨74JOC3615⟩. NBS monobrominates methyl groups only on the benzene ring of coumarins and nuclear bromination has not been observed ⟨47CR937⟩.

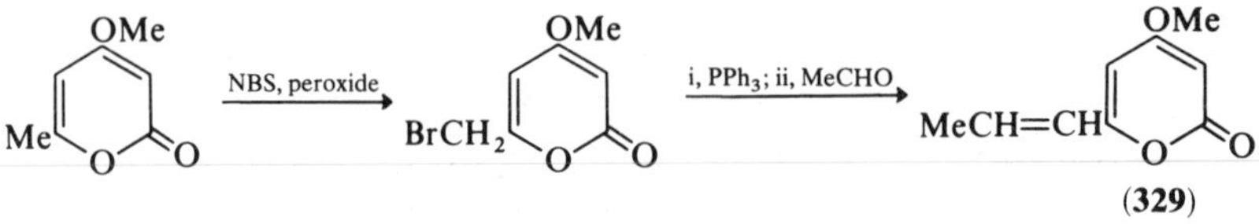

(329)

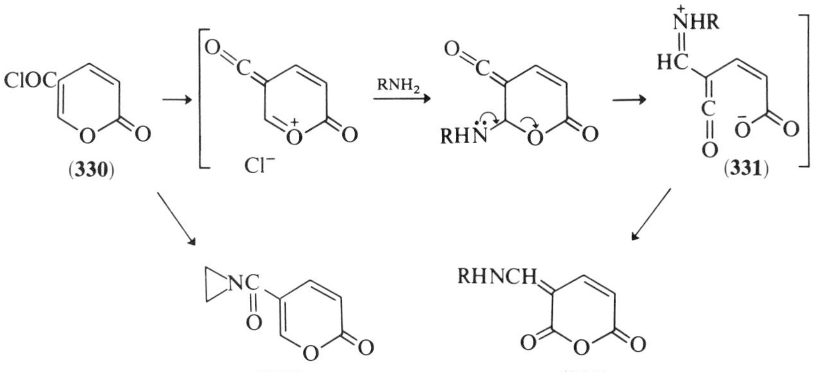

(330)  (331)

(333)  (332)

Coumalyl chloride (**330**) reacts normally with ethyleneimine to give the carboxamide (**333**), but primary and other secondary amines such as benzylamine, *t*-butylamine, diethylamine and pyrrolidine give unsaturated anhydrides (**332**), possibly through an acyclic intermediate (**331**). Aniline gives a mixture of both kinds of product ⟨78JOC4415⟩.

The two ester groups of diethyl isocoumarin-3,4-dicarboxylate (**334**) exhibit a curious lability to treatment with hydrochloric acid in that the 4-carboxyl group is lost on refluxing with 10% aqueous acid while 37% acid at 70 °C hydrolyzes the 3-carboxylate group only (Scheme 20) ⟨57JGU2342⟩.

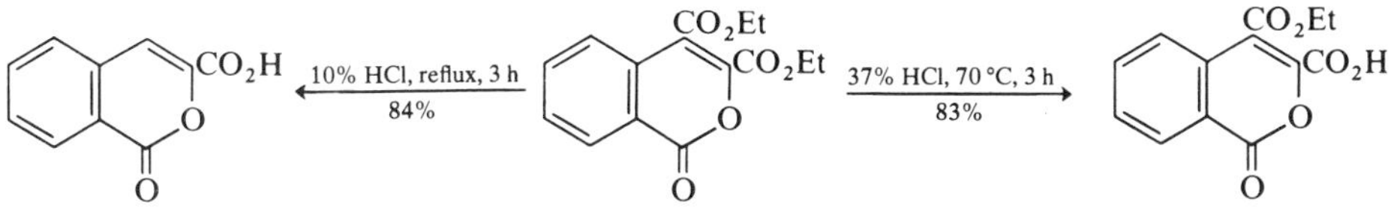

(334)

**Scheme 20**

The ester group of ethyl 4-hydroxycoumarin-3-carboxylate (**335**) reacts with aniline to give the amide (**336**) which was also obtained by the action of phenyl isocyanate on 4-hydroxycoumarin (**337**) ⟨67CJC767⟩.

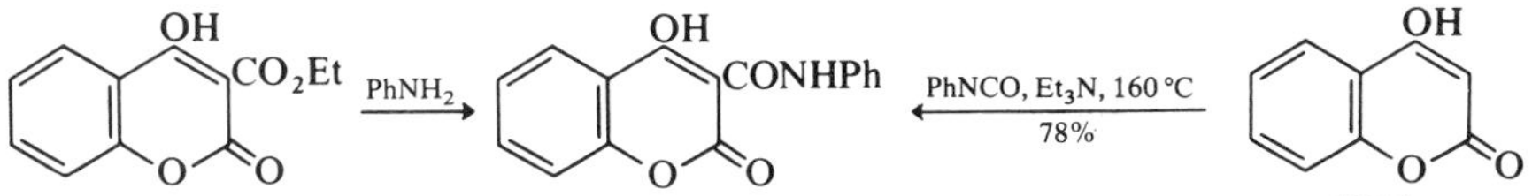

(335)  (336)  (337)

### 2.23.7.4 *N*-Linked Substituents

The 3,4-double bond of coumarins and isocoumarins is easily reduced and selective reduction of the nitro group of ethyl 6-nitrocoumarin-3-carboxylate (**338**; R = OEt) or its diethylamide (**338**; R = NEt$_2$) by catalytic hydrogenation using a variety of catalysts is not possible ⟨49JA3602⟩ unless the reaction medium is acidic so that the amine produced is converted into its salt ⟨68CPB2093⟩. Selective reduction of a nitro group is also possible by iron–acetic acid ⟨42JA825⟩ or aqueous titanium(III) chloride ⟨81CPB249⟩, but hydrazine hydrate–Raney nickel reduces the nitro group and converts isocoumarins such as (**339**) into isoquinolinones (**340**) ⟨68JCS(C)2205⟩.

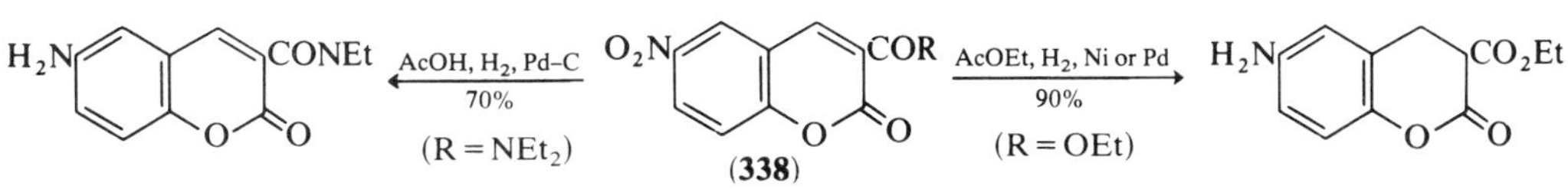

(338)

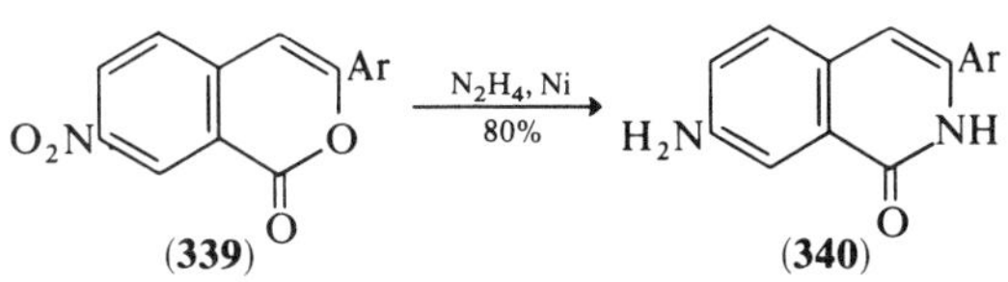

(339)  (340)

### 2.23.7.5 *O*-Linked Substituents

Methylation of a hydroxypyran-2-one often gives either a mixture of products or an undesired product ⟨67CC577⟩; for example, methylation of 4-hydroxy-6-methylpyran-2-one (**342**) with dimethyl sulfate gives the pyrylium salt (**341**). More selective reagents are therefore desirable; one of these is dimethyl sulfate–HMPT–sodium hydride which methylates the hydroxyl group at room temperature ⟨78S144⟩. The same authors have devised an efficient procedure for methylating alkali-sensitive hydroxyl groups under neutral conditions by means of a betaine (**343**), which is generated *in situ* from diethyl azoformate and triphenylphosphine, and a readily available alcohol. This reaction probably proceeds *via* the enolate (**344**) and the cation (**345**). 4-Hydroxycoumarins are similarly alkylated in 66–94% yield with primary and secondary alcohols ⟨79JCR(S)110⟩.

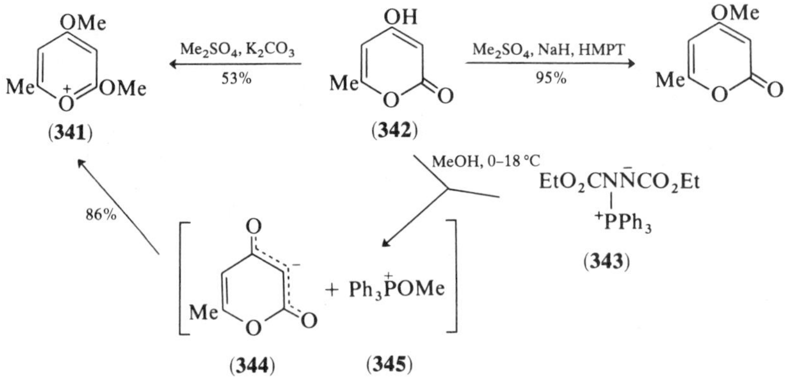

Replacement of a 4-hydroxy group of coumarin by a primary amine is described in Section 2.23.6.4.2.


## 2.23.8 PYRAN-4-ONES AND THEIR BENZO DERIVATIVES: REACTIVITY AT RING ATOMS

### 2.23.8.1 Introduction

Several of the pyran-4-ones found in nature are acidic by virtue of carboxylic acid or hydroxyl groups. Chelidonic (**346**), meconic (**347**), comenic (**348**) and pyromeconic (**349**) acids and maltol (**350**) are of plant origin but kojic acid (**351**) is produced by microorganisms, for example *Aspergillus oryzae*. Several compounds of this type are used to enhance the flavor of foods. The pyran-4-ones are more basic than their isomeric pyran-2-ones and this results in the formation of more stable salts with acids such as perchloric acid.

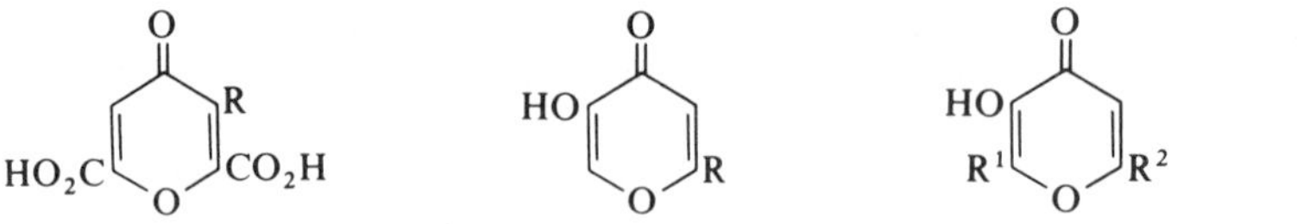

(**346**) R = H
(**347**) R = OH

(**348**) R = CO$_2$H
(**349**) R = H

(**350**) R$^1$ = Me, R$^2$ = H
(**351**) R$^1$ = H, R$^2$ = CH$_2$OH

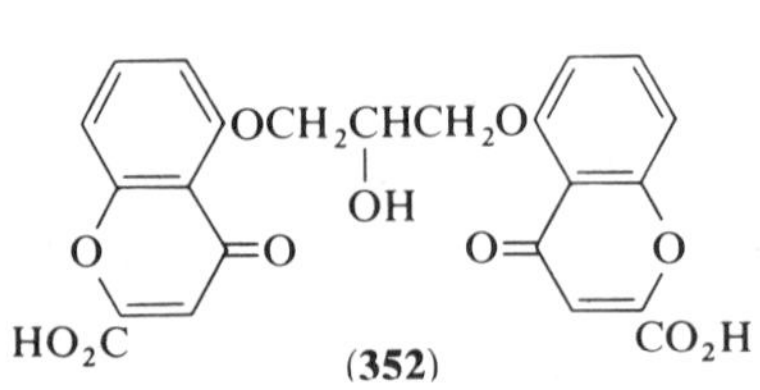

(**352**)

In 1962, about 15 chromones were known to occur in plants but by 1975 this number had grown to 55 and has continued to grow. A few chromones have been known for many years to have pharmacological activity and the discovery in 1962 of cromoglycic acid (**352**) as the first drug to have the ability to prevent asthmatic attacks stimulated a great deal of medicinal chemical research ⟨70MI22302⟩.

The flavonoids, derivatives of flavone (**353**) and of isoflavone (**354**), are widely distributed among plants of many kinds. They provide colour (from pale yellow to orange) in flowers and their potential as dyes is being revived now that awareness of the toxicity of some synthetic dyes is increasing, for example morin (**355**; C.I. Natural Red) is one of many possible hydroxylated flavones with dyeing properties.

(**353**) R = 2-Ph    (**355**)    (**356**)
(**354**) R = 3-Ph

Xanthone (**356**) and its derivatives occur less widely in nature; they are mostly hydroxyxanthones and are colored yellow.

The chemistry of chromones has been comprehensively reviewed up to about 1974 ⟨77HC(31)557⟩; naturally occurring pyran-4-ones, chromones, flavones, isoflavones and xanthones have also been described ⟨B-63MI22301⟩ and the flavones and isoflavones are discussed in separate chapters of a book ⟨B-75MI22300⟩. Pyran-4-ones and their mono- and di-benzo derivatives are covered in Rodd's 'Chemistry of Carbon Compounds' ⟨B-77MI22301⟩; useful details of individual compounds are provided. A review of chromone-2- and -3-carboxylic acids and their 3- and 2-aryl derivatives was published in 1973 ⟨73MI22306⟩. A comprehensive survey of naturally occurring flavones, flavanones and flavanols appeared recently ⟨81P869⟩.

### 2.23.8.2 Thermal and Photochemical Reactions

Irradiation of 2,6-dimethylpyran-4-one (**357**) in ethanol or benzene or in the solid state yields the cage-like molecule (**358**) from which the pyran-4-one is regenerated by hydrochloric acid ⟨63JA1208⟩. When the photodimerization is conducted in water but in the absence of air, a small amount of 4,5-dimethylfuran-2-aldehyde is also formed ⟨63JA2956⟩. Pyran-4-one is photoisomerized to pyran-2-one (**359**); the intermediates postulated resemble those suggested in the irradiation of pyrylium salts (Section 2.23.2.1) ⟨79JA7521⟩.

(**357**) $\xrightarrow[30\%]{h\nu}$ (**358**)

→ (**359**)

Photolysis of other substituted pyran-4-ones gives a variety of products depending on the substituents and the solvent. Reaction is usually more facile in methanol but trifluoroacetic acid, trifluoroethanol, and water have also been used. Maltol (**360**) gives the dioxan (**361**) or (**362**) which, on hydrogenation, forms 2-hydroxy-3-methylcyclopent-2-en-1-one (**363**), a natural flavoring ingredient of coffee ⟨72TL4655⟩. Highly substituted pyran-4-ones such as (**364**) are also photolyzed mainly to cyclopentenones ⟨76TL1939⟩ but when phenyl groups are attached at C-3 and C-5, as in (**365**), a rearranged pyran-2-one (**366**) is obtained. Monophenyl pyranones (**365**; R = H), however, are photostable in acetonitrile even in the presence of a sensitizer such as acetophenone ⟨73JA463⟩.

Photoisomerization of this type in acetonitrile is not quenched by dienes and the process, therefore, proceeds through the $\pi,\pi^*$ singlet but the yield is dependent on the type of substitution. For example, the 3,5-di-*p*-tolyl homologue of (**365**) gives a better yield (77%)

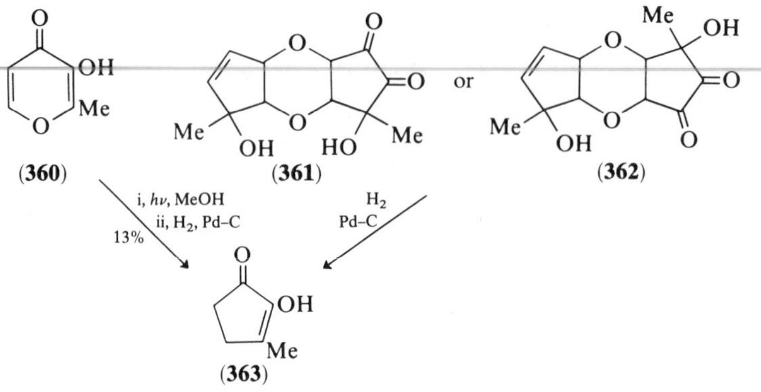

**(360)**   **(361)**   or   **(362)**

i, *hv*, MeOH
ii, H$_2$, Pd–C
13%

H$_2$
Pd–C

**(363)**

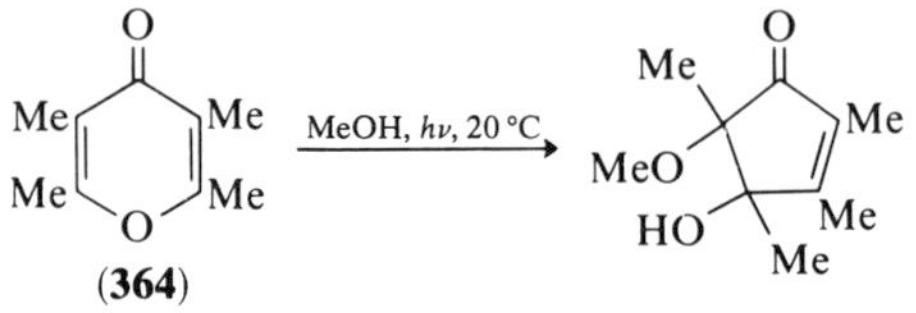

**(364)**   MeOH, *hv*, 20 °C

R   Ph   R = Ph, *hv*, MeCN   Me   Ph

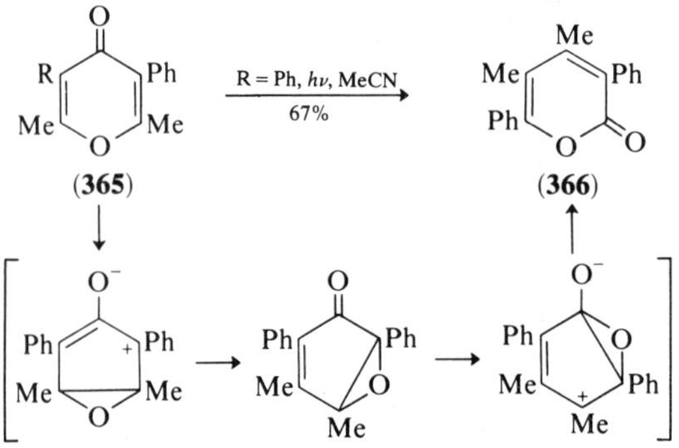

**(365)**   67%   **(366)**

than the corresponding di-*p*-bromophenyl compound (10%) ⟨78JOC2138⟩. When the 3- and 5-substituents are different, two isomeric pyran-2-ones are formed (Scheme 21) ⟨78JOC2144⟩.

Attempts to rationalize these and other photochemical reactions have been based on the nature of the lowest excited state or the stereochemistry of substituents. A wider-ranging explanation is based on stabilization of the common intermediate zwitterion (**367**) by either electron-releasing substituents at C-3 and C-5 or polar solvents ⟨79JA7521⟩.

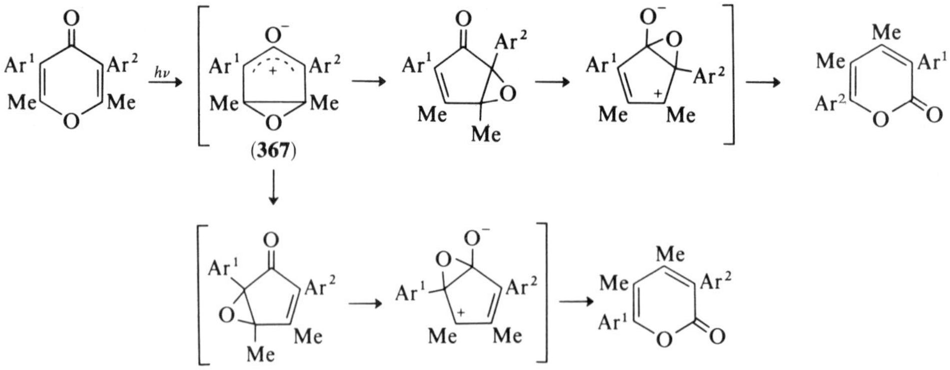

Ar$^1$   Ar$^2$   *hv*   **(367)**

**Scheme 21**

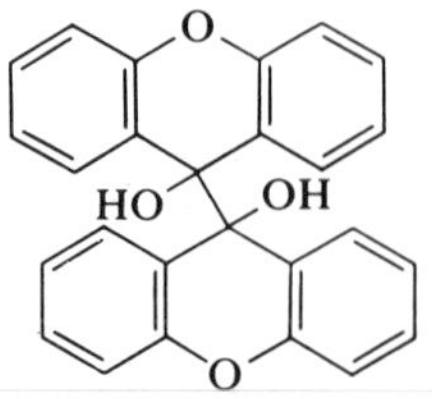

**(368)**

When xanthone is irradiated in methanol or benzene it is reduced to xanthene, but in cyclohexane 9-cyclohexylxanthydrol is formed. In hexane, the dimeric 9,9'-dihydroxydixanthene (**368**) is produced ⟨66BCJ1694⟩. Irradiation of xanthone using a nitrogen laser shows a strong solvent dependence ⟨80JA7747⟩.

Simple chromones are less susceptible to photolysis but 3,5,7-trimethoxy-2-methylchromone (**369**) is converted into a dimer which may have structure (**370**) ⟨73TL5073⟩. The 3-benzylchromone (**371**) is cyclized to the tetracycline analogue (**372**) ⟨72JCS(P1)1103⟩.

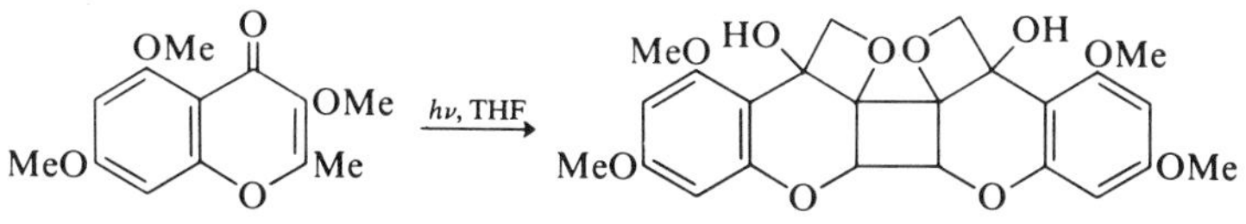

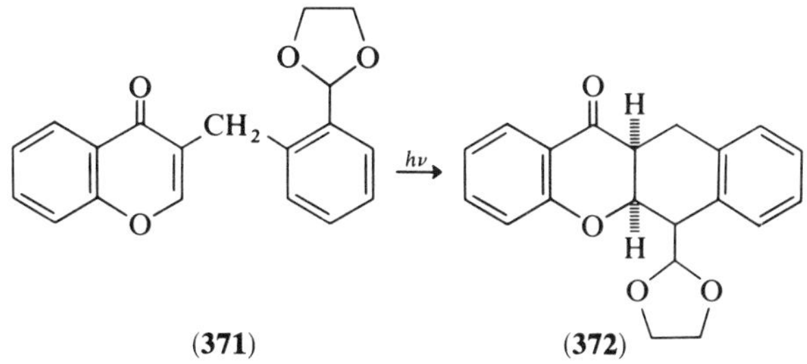

One of the most interesting developments in the photochemistry of chromones is the behavior of the colorless 3-aroylchromones. These compounds on irradiation produce orange colors which fade slowly in the dark or in a nonpolar solvent but rapidly in a polar solvent. A reversible photoenolization occurs, for example, with 3-benzoyl-2-benzylchromone (**373**; R = H) or 2-benzhydryl-3-benzoylchromone (**373**; R = Ph) which is converted into the colored enol (**374**); this, on further treatment in benzene with very intense visible and a weaker UV radiation while oxygen is bubbled through, gives a compound of probable structure (**375**; R = H or Ph), $\lambda_{max}$ 440 nm, together with minor amounts of others of known structure. These compounds have been studied for their potential value as photochromic agents ⟨65JA5424, 69USP3444212⟩.

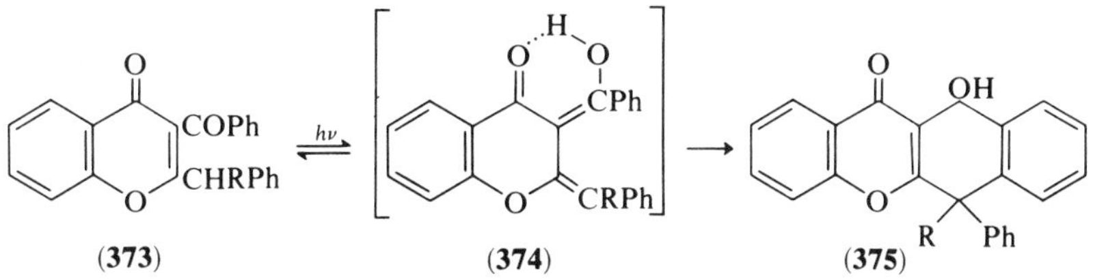

Flavones which contain a 3-hydroxy group are photolabile. Flavon-3-ol is converted into the benzofuranone (**376**) in high yield when irradiated in isopropanol in a Pyrex vessel ⟨71TL1539⟩ but photosensitized oxygenation opens the pyranone ring of quercitin (**377**; R = H) or its tetramethyl ether (**377**; R = Me), and probably through a peroxide such as (**378**), gives a good yield of the phenolic ester (**379**). Again 3-methoxyflavones are relatively stable to photooxidation ⟨70T435⟩.

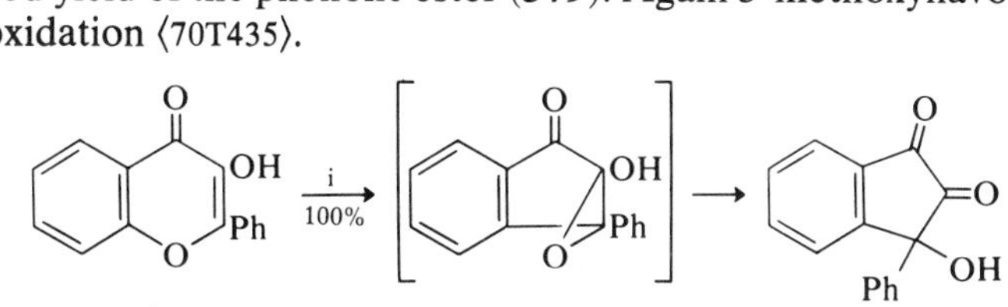

i, *hν*, Pr$^i$OH, PhH, Pyrex filter

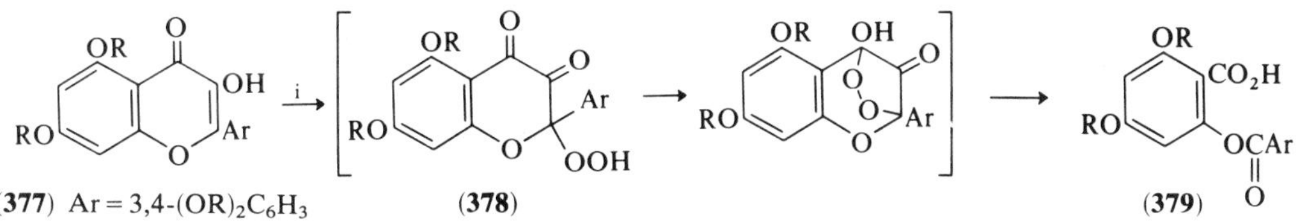

(**377**) Ar = 3,4-(OR)$_2$C$_6$H$_3$

i, *hν*, MeOH, Rose Bengal, O$_2$, Pyrex filter

Several flavonoids have given mixtures of products on irradiation; for example, the 3-phenylflavone (**380**) in methanol is converted into the isocoumarins (**381**) and (**382**); the latter is the major product and is derived from the former ⟨75CC241⟩. Quercitin pentamethyl ether (**383**) in methanol and in the absence of oxygen yields four tetracyclic products. In deoxygenated benzene, the reaction rate is doubled but only (**384**) and (**385**) are formed. This photocyclization is not inhibited by triplet quenchers and the primary photointermediate is in the triplet state ⟨67JA6213⟩.

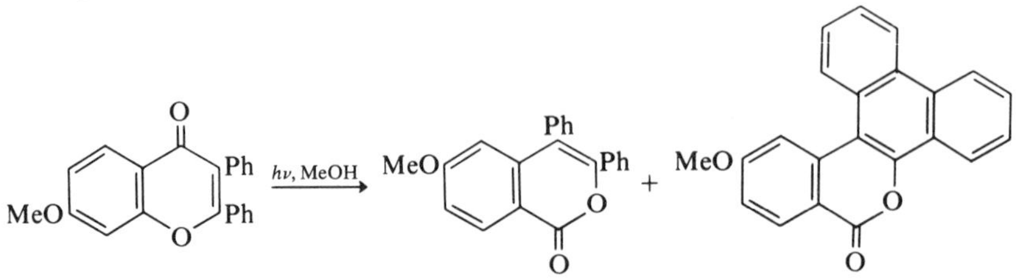

(**380**)　　　(**381**) 10%　　　(**382**) 42%

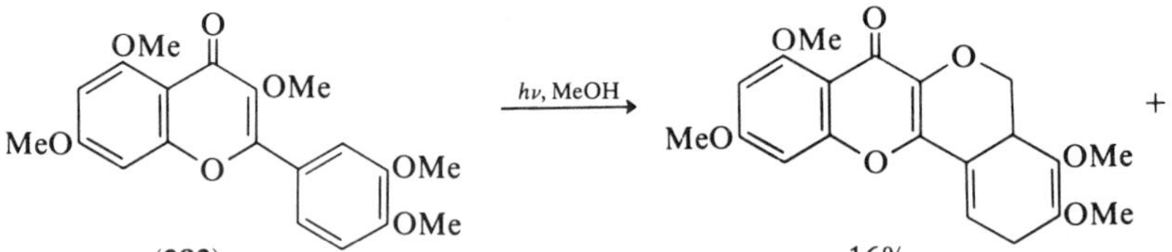

(**383**)　　　16%

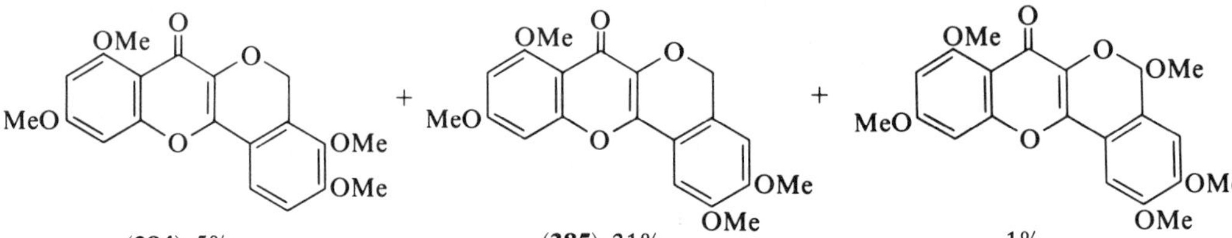

(**384**) 5%　　　(**385**) 31%　　　1%

## 2.23.8.3 Electrophiles

### 2.23.8.3.1 Addition and substitution

Pyran-4-ones are slightly more basic than their 2-one isomers and form salts with strong acids; these can be isolated and are best represented by the resonance forms (**386**) and (**387**). Methyl iodide adds similarly to form a pyrylium salt (**388**). Deuteration of pyran-4-ones at C-3 and C-5 is acid catalyzed and probably proceeds through the acyclic trione (**389**) (*cf.* pyrylium salts, Section 2.23.2.2) ⟨64JOC2678⟩.

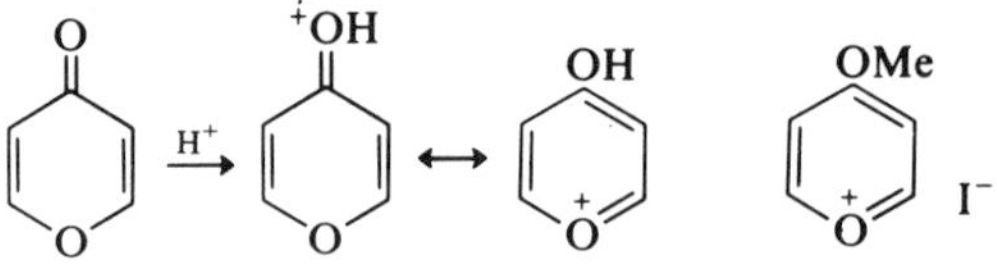

(**386**)　　　(**387**)　　　(**388**)

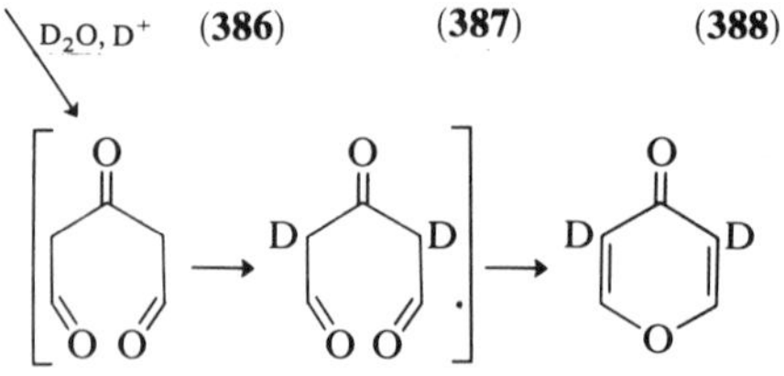

(**389**)

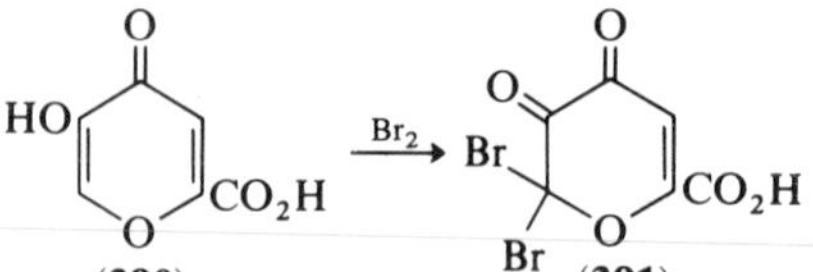

(**390**)　　　(**391**)

Bromination of pyran-4-ones occurs at C-3 or C-5 (or both) but comenic acid (**390**) gives the dibromide (**391**). Claims in the literature that kojic acid and acyl chlorides under Friedel–Crafts conditions give high yields of ketones are erroneous. The products are actually *O*-acyl derivatives ⟨72JOC1444⟩. Bromine in warm carbon disulfide adds across the 2,3-double bond of chromone to give an adduct which reacts with a secondary amine to form 3-bromochromone. The pyran ring of chromones and flavones is unreactive towards electrophiles but an electron-releasing substituent at C-2 or C-3 activates the adjacent vacant position, for example the carbamate ester (**392**) is nitrated, brominated and aminomethylated (by the Mannich reaction) at C-3 ⟨73MI22301⟩. 3-Hydroxychromone (**393**) is readily brominated, nitrated and aminated at C-2, a position which is usually depleted of electrons ⟨78JCR(S)47, 75AP385⟩. 2-Methyl- and 2-styryl-chromone are chloromethylated at C-3 in 58% and 86% yield respectively ⟨80JCR(S)159⟩.

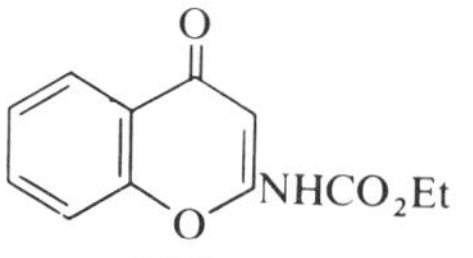

(**392**) 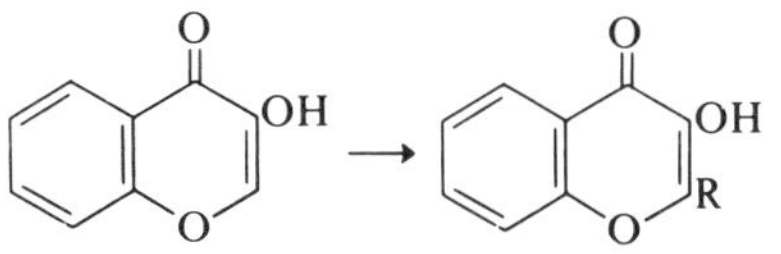                    (**393**)                    R = Br, NO$_2$ or NR$_2$

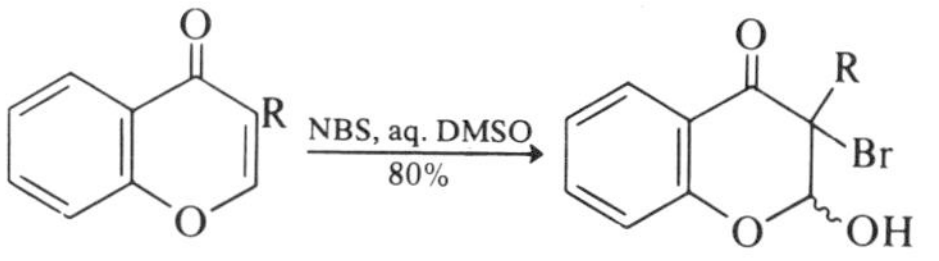

(**394**)  R = H or Br                    R = H or Br

The elements of hypobromous acid are added across the 2,3-double bond when chromones (**394**) are treated with NBS in aqueous DMSO ⟨75JHC981⟩.

Electrophilic reagents are usually either strongly acidic (nitric–sulfuric acid mixture or sulfuric acid) or produce strong acids during reaction (halogens) and are likely to protonate the pyran ring and thus inhibit further attack on it by the electrophile. Aminomethylation under Mannich reaction conditions is often achieved under less acidic conditions and this may explain the formation of 3-aminomethylchromone (**395**) salts. A methyl substituent at C-2 inhibits the reaction, but chloromethylation does occur ⟨80JCR(S)159⟩. When a Mannich base, *e.g.* (**396**), is heated with acetic anhydride, a methylenebischromone (**397**) is formed ⟨74FES247⟩.

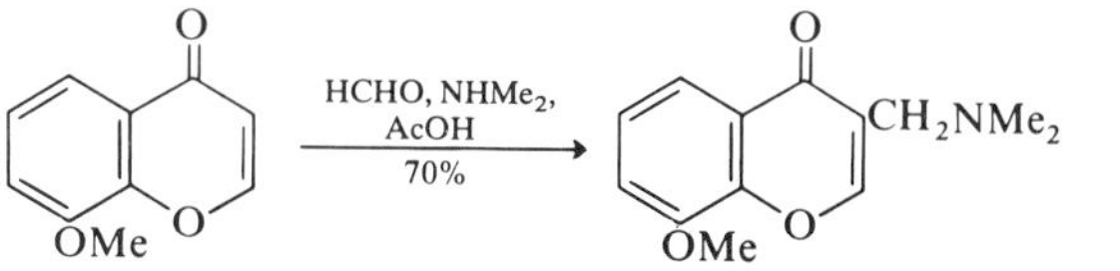

(**395**)

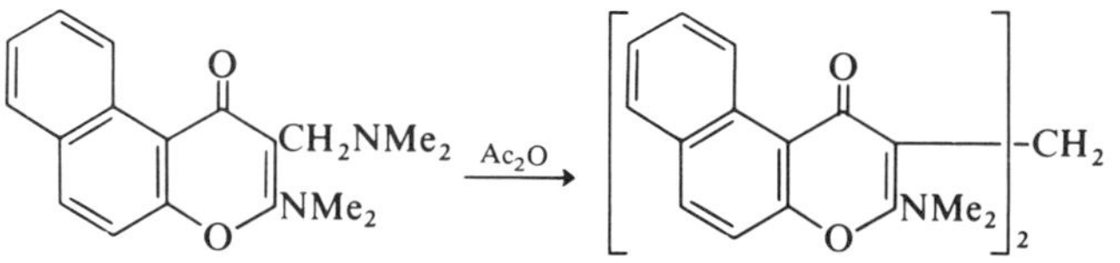

(**396**)                    (**397**)

An interesting and occasionally useful rearrangement may be induced in chromones, flavones, isoflavones and xanthones by heating a methoxylated or hydroxylated compound with hydriodic acid. This rearrangement is known by the names of two Austrian chemists, Wessely and Moser, who discovered the conversion while investigating the chemistry of the flavone scutellarein (**399**). This was formed not only by simple demethylation of the 6-methyl ether (**398**) but also by demethylation of the dimethyl ether (**400**) with hydriodic acid. The latter conversion, the Wessely–Moser rearrangement, proceeds by ring opening to the diketone (**401**) which then cyclizes preferentially to scutellarein rather than return to the 5,7,8-trihydroxy pattern ⟨30M(56)97⟩.

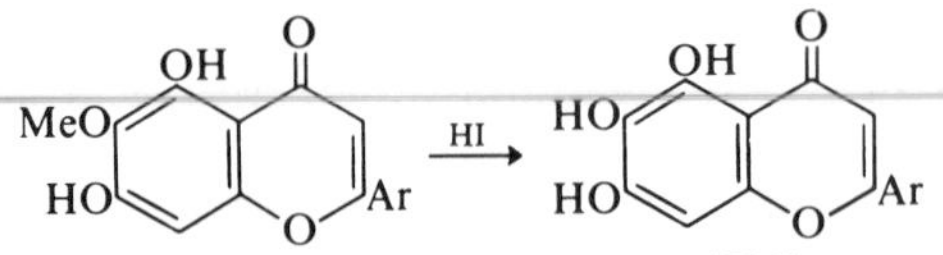

(398)         (399)

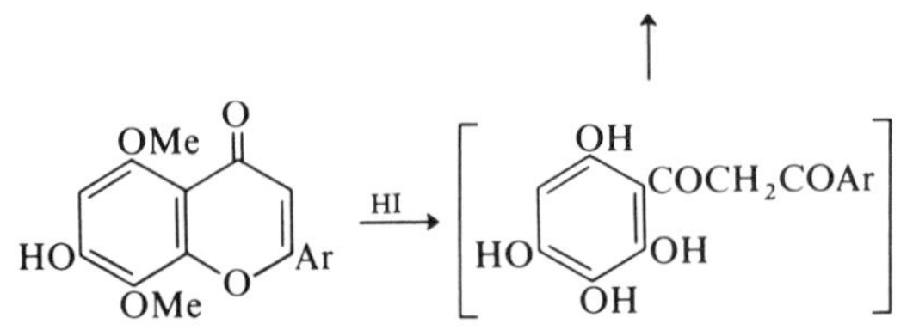

(400) Ar = 4-HOC$_6$H$_4$       (401)

### 2.23.8.3.2 Oxidation

Chromones are readily oxidized by permanganate or dichromate with opening of the pyran ring and the formation of a salicylic acid (402). Flavones and isoflavones are also degradatively oxidized; for example oxidation of munetone (403) with alkaline hydrogen peroxide yields 2-methoxybenzoic acid and 4-hydroxy-2-isopropylbenzofuran-5-carboxylic (isotubaic) acid.

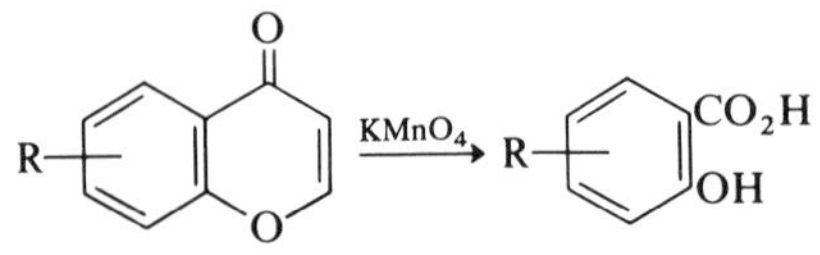

(402)

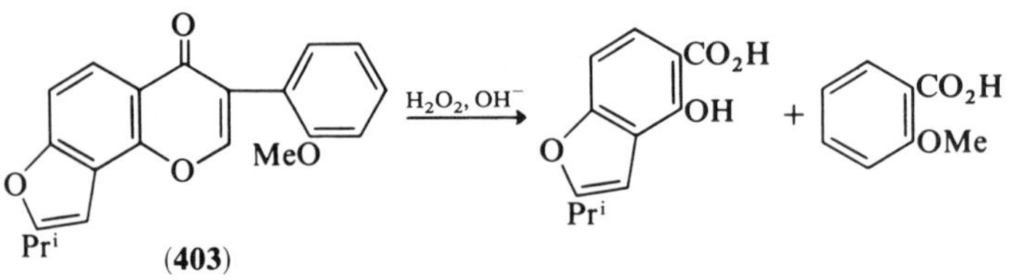

(403)

### 2.23.8.4 Nucleophiles

Pyran-4-ones and their monobenzo derivatives are easily attacked and degraded by nucleophiles of all kinds. Because of the ease with which such rings are opened, it is often difficult to prepare oximes and hydrazones in the usual way.

### 2.23.8.4.1 C-Nucleophiles

Reaction of equimolar amounts of a pyran-4-one and a Grignard reagent gives a pyran-4-ol which is usually converted by acid into the more stable pyrylium salt, e.g. (404), but a two-fold excess or more of Grignard reagent gives a 4,4-dialkyl- or -diaryl-pyran (405) in moderate yield ⟨61LA(648)114, 61CB1784⟩.

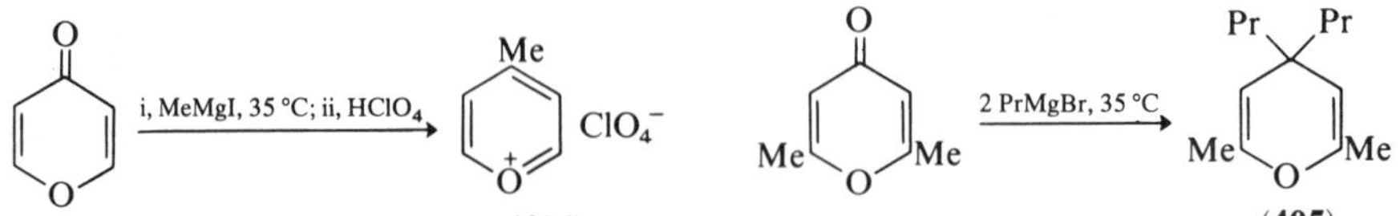

(404)                       (405)

An anion derived from a ketone such as acetophenone cleaves the 1,2-bond of pyran-4-one; the resulting dione (406) is cyclized by mineral acid to 4-hydroxybenzophenone (407).

An interesting conversion of 2,6-dimethylpyran-4-one (**408**) into 7-methyl-5-phenylchromone (**410**) is achieved by reaction with the anion of acetophenone and cyclization of the 2-hydroxyacetophenone (**409**) ⟨77AP744⟩.

(**406**) (**407**)

(**408**) (**409**) (**410**)

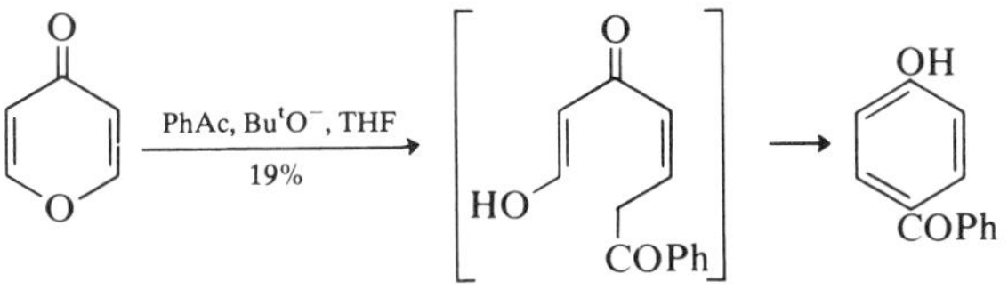

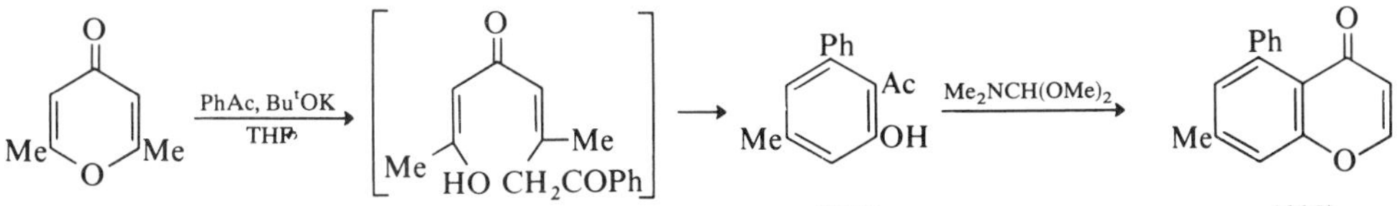

Compounds containing an active methylene group condense with the carbonyl function of pyran-4-ones and the resulting pyranylidene reacts with primary amines to form a pyridine derivative (**411**) ⟨68JOC4418⟩ but when the active methylene compound contains a nitro group, the pyranylidene (**412**) undergoes a simultaneous rearrangement and replacement of oxygen by nitrogen to form the 3-nitropyridone (**413**) ⟨74JOC989⟩. Active methylene in the presence of DBN or DBU cleaves the pyran ring of chromones at room temperature and in the case of 3-bromochromone (**414**) a furan (**415**) is formed ⟨79JOC3988⟩.

(**411**)

(**412**) (**413**)

(**414**) (**415**)

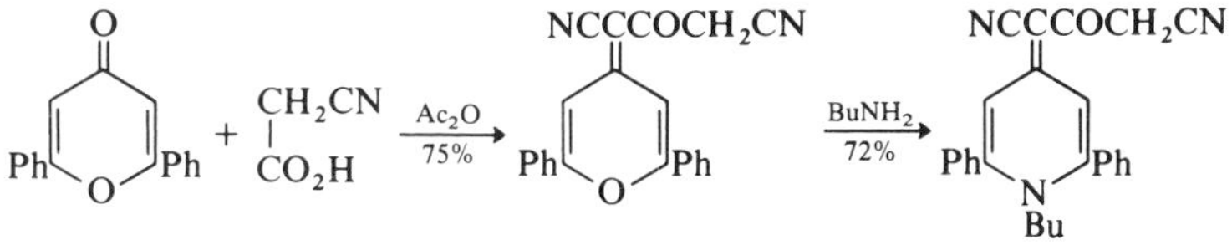

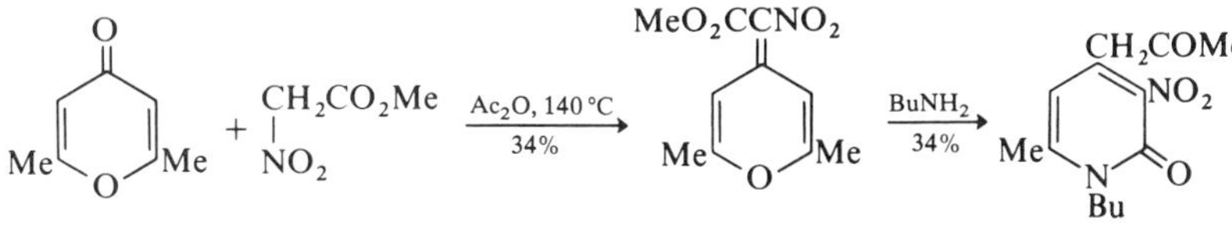

Grignard reagents attack the carbonyl group of chromones such as 3,7-dimethoxy-2-methylchromone (**416**); the product obtained after acid treatment is the benzopyrylium salt (**417**). Recently a benzopyran-4-ol was isolated from 2-methylchromone ⟨75ACH(84)319⟩. A Grignard reaction on the isoflavone (**418**) in the presence of copper(I) chloride ruptured the pyran ring and left the carbonyl group unchanged ⟨79IJC(B)182⟩.

(**416**) (**417**)

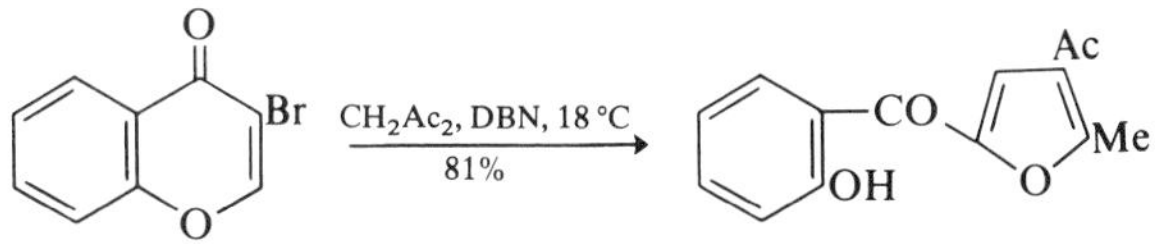

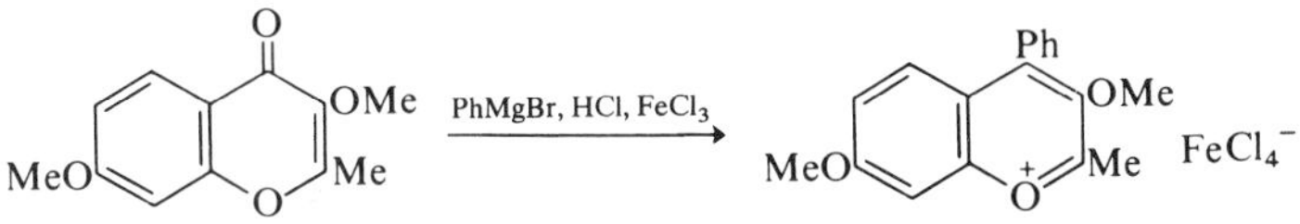

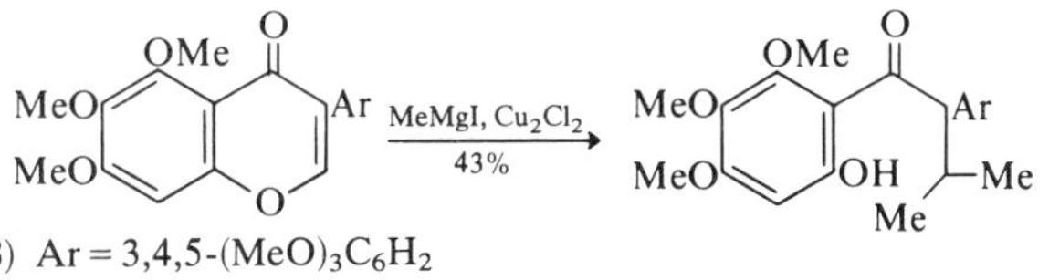

(**418**) Ar = 3,4,5-$(MeO)_3C_6H_2$

Diazomethane reacts with 3-nitrochromone (**419**) under mild conditions to give the cyclopropabenzopyran (**420**; R = H) and a homologous dimethylcyclopropane (**420**; R = Me) is obtained from 2-diazopropane. These compounds undergo useful reactions with opening of the small ring ⟨80JCS(P1)2049⟩.

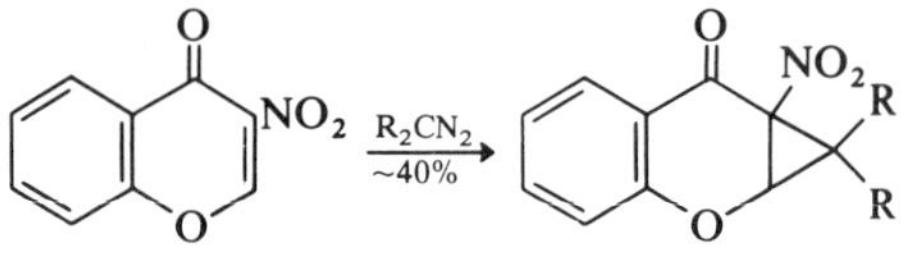

(**419**)        (**420**)

### 2.23.8.4.2 *N-Nucleophiles*

The reaction of N-containing nucleophiles such as hydroxylamine, hydrazine and its derivatives, with pyran-4-ones and their benzo derivatives has been studied intensively. The early workers assumed that oximes and hydrazones were formed in these reactions, but it was later realized that under normal conditions the pyrans rarely yield the expected derivatives because of the susceptibility of the hetero ring to the nucleophile. This is especially true when the pH of the medium is above 7; those reactions which proceed under acidic conditions usually lead to the expected derivative ⟨77T3183⟩.

The reaction of amines with pyran-4-ones usually results in ring opening, but this may be followed by cyclization to pyridine derivatives. For example dehydroacetic acid (**421**) on heating with butylamine and subsequent treatment with perchloric acid gives a pyridinium salt (**422**) in good yield; an excess of the amine first cleaves the pyranone ring ⟨74MI22301⟩. Kojic acid and ammonia give 4,5-dihydroxy-2-hydroxymethylpyridine. Weaker primary or secondary amines, *e.g.* 2- or 4-nitroaniline, are less likely to open the ring and, in the presence of phosphoryl chloride and followed by a strong acid, a high yield of a pyrylium salt (**423**) is obtained. Methanolic alkali converts this into the azomethine (**424**) ⟨74JHC1065⟩.

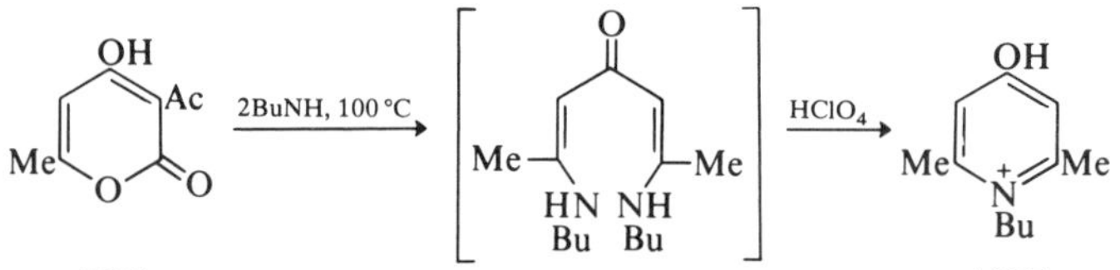

(**421**)                              (**422**)

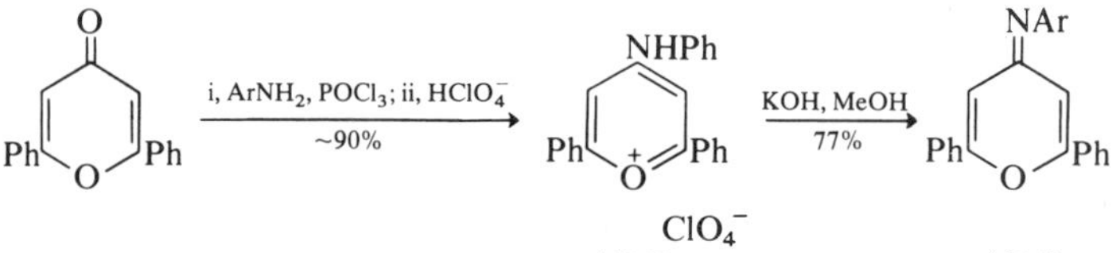

(**423**)                              (**424**)

The pyran ring of chromones (**425**) is cleaved by reaction with primary or secondary amines to yield enamines (**426**) which may be recyclized by treatment with acid ⟨73MI22305, 72CHE(8)416, 65JCS3610⟩.

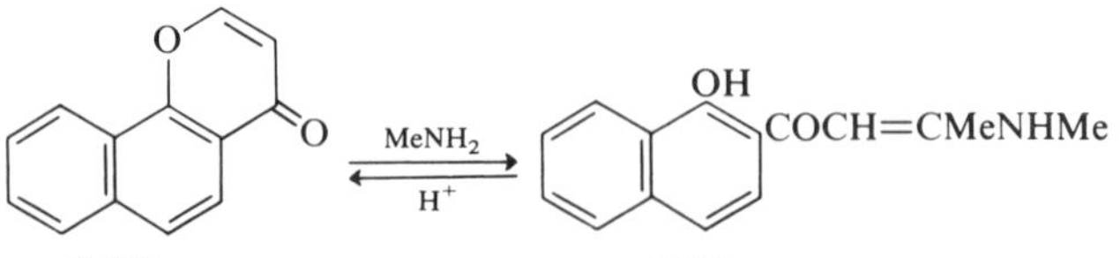

(**425**)                              (**426**)

Reaction of pyran-4-ones with hydroxylamine or hydrazine causes ring opening followed usually by cyclization to a different ring; substituents on the pyran-4-one often play a part in determining the course of the cyclization (Scheme 22) ⟨79AP591, 62CJC2146⟩. Kojic acid (**427**) is converted into pyrazoles by an excess of phenylhydrazine ⟨74MI22302⟩.

Under normal reaction conditions (heating the reactants with or without a solvent in an alkaline medium) the pyranone ring of chromones, flavones and isoflavones is opened at

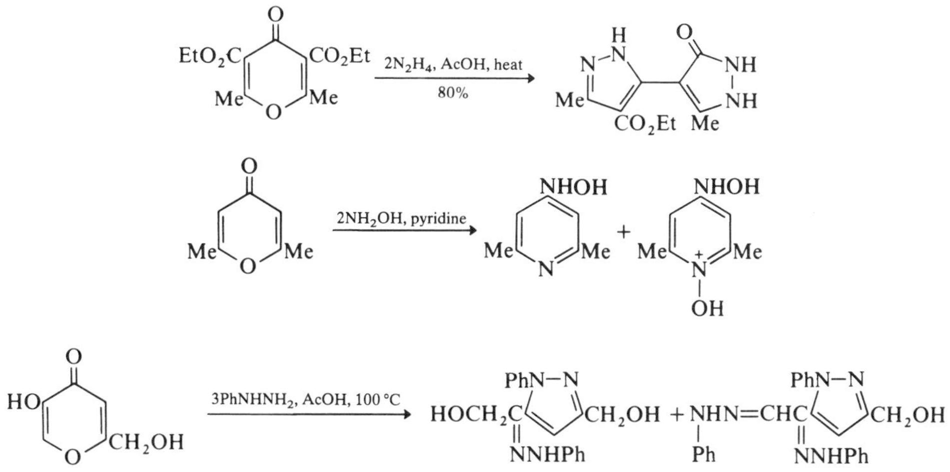

(**427**)

**Scheme 22**

the 1,2-bond by hydroxylamine and hydrazine; the initial product cyclizes to give either an isoxazole (**428**) or a pyrazole (**429**) ⟨77T3183, 78ACH(97)69, 78ACH(98)457⟩.

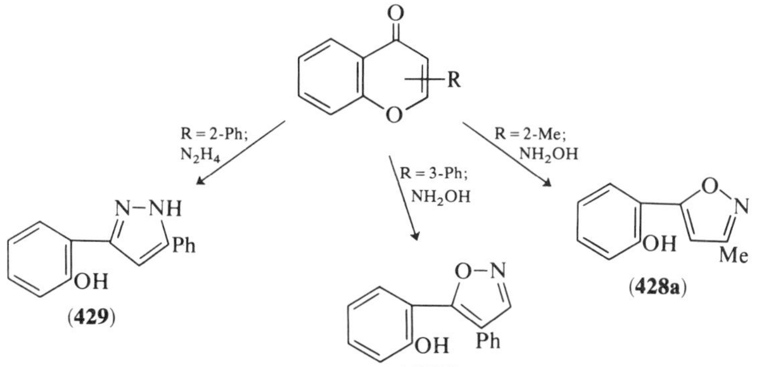

(**429**) (**428b**) (**428a**)

When chromone and hydroxylamine hydrochloride are heated together in ethanol, a mixture of isomeric isoxazoles (**430**) is obtained, (**430a**) being the major product. The two isomers can be distinguished by mass spectrometry. When the reactants are brought together under anhydrous conditions, chromone oxime (**431**) is obtained; its structure is supported by $^{13}C$ NMR ⟨77JOC1356⟩.

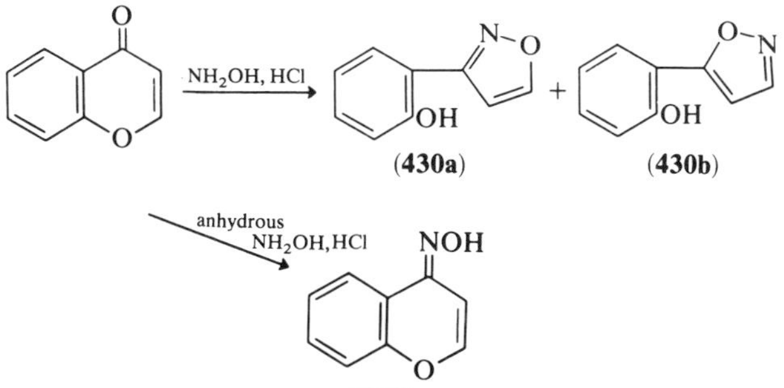

(**430a**) (**430b**)

(**431**)

Phenylhydrazine also causes ring cleavage of chromones, but under some conditions the phenylhydrazone is formed (Scheme 23) ⟨73MI22302, 77MI22303⟩.

Flavone reacts with hydroxylamine in anhydrous pyridine to give two products, the oxime (**432**) and isoxazole (**433**). The structure of the latter was confirmed by mass spectrometry. It appears from this work that 3-(2-hydroxyaryl)isoxazoles of type (**430a**) are not produced in the reaction of flavones with hydroxylamine ⟨B-77MI22304⟩. On the other hand, flavone and tosylhydrazine in hot ethanolic hydrochloric acid yield only the tosylhydrazone (**434**) ⟨65TL2269⟩.

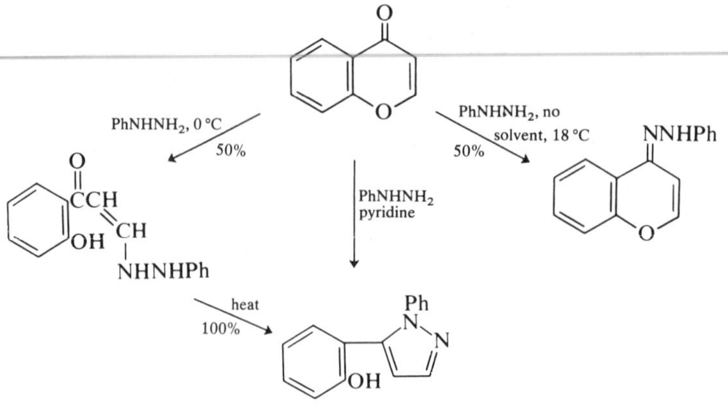

**Scheme 23**

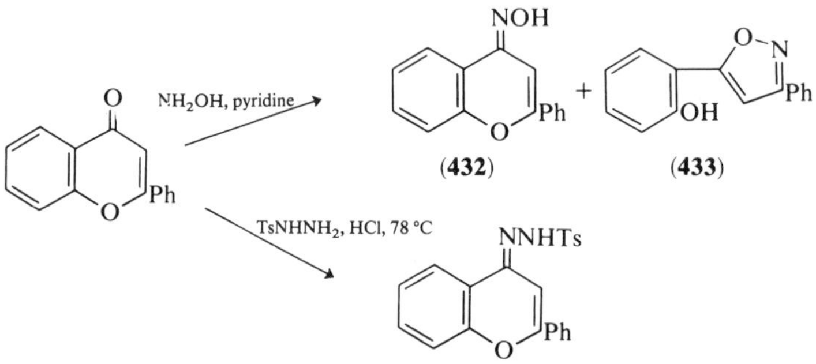

(432)            (433)

(434)

Xanthone is unreactive towards hydrazine and phenylhydrazine. The oxime is obtained by reaction of xanthione (xanthene-9-thione) with hydroxylamine, or from xanthone and hydroxylamine in pyridine. When the oxime is heated in water with phenylhydrazine, the phenylhydrazone is formed. In acid solution, xanthone reacts normally with 2,4-dinitro-phenylhydrazine but xanthone-1-carboxylic acid (435) gives the pyridazinone (436), possibly *via* the hydrazone ⟨57JCS1922⟩. When the oxime is heated with phosphorus pentachloride it undergoes a Beckmann rearrangement to give the amide (437) ⟨70MI22300⟩.

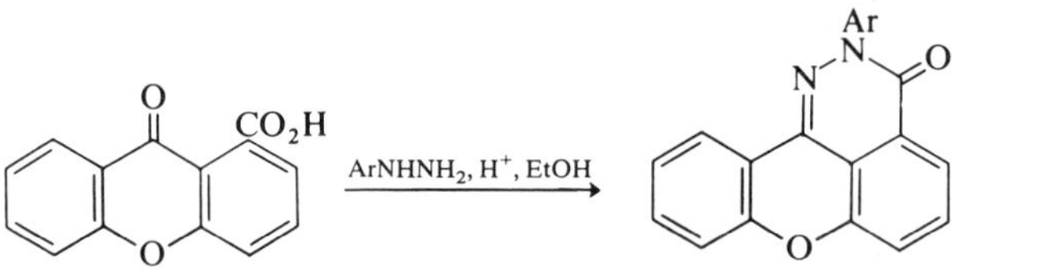

(435)                                    (436) Ar = 2,4-(NO₂)₂C₆H₃

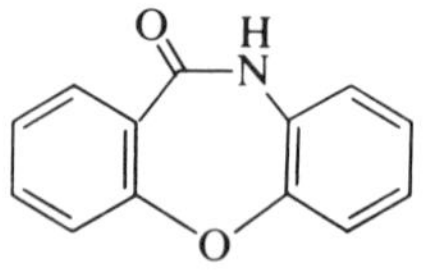

(437)

Replacement of the carbonyl oxygen atom by sulfur may be effected by heating with phosphorus pentasulfide ⟨49JCS2142⟩, boron sulfide or silicon disulfide which gives high yields under mild conditions, as, for example, in the synthesis of 7-methoxy-2-methylchromene-4-thione (438) ⟨69JCS(C)2192⟩. Such compounds are more easily converted into their oximes or hydrazones than the oxygen compounds.

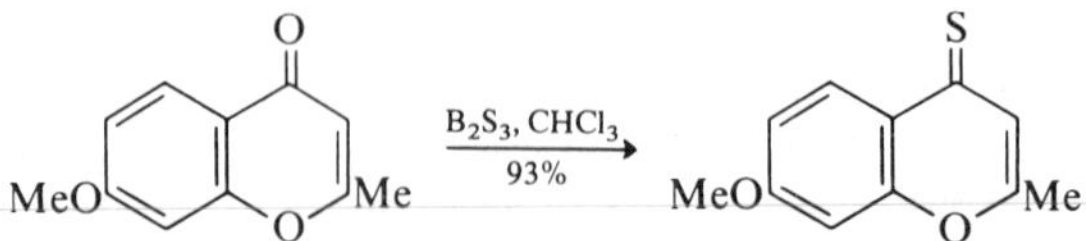

(438)

Pyran-4-ones, chromones and flavones are converted into pyrimidines, usually under base catalysis, by reaction with compounds which contain the grouping N—C—N; urea, thiourea, guanidine, aminoguanidine, acetamidine and dicyandiamide are examples of this type. Scheme 24 shows some typical examples ⟨77BSF369, 81JHC619⟩.

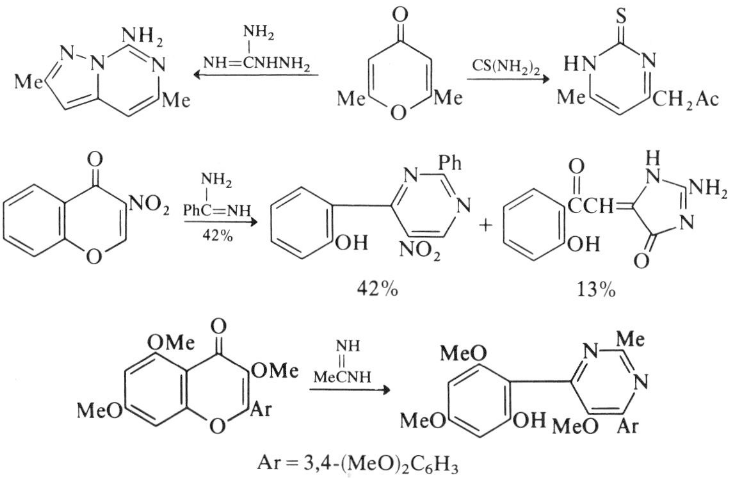

**Scheme 24**

### 2.23.8.4.3 O-Nucleophiles

Pyran-4-ones and their monobenzo derivatives undergo ring opening when treated with aqueous alkali. This reaction has been widely used in the investigation of structure of chromones, flavones and isoflavones but xanthones are little affected. The nucleophile attacks the 1,2-bond in a rate-determining step whose rate constant varies with the electron density in the ground state at C-2. A kinetic study of the reaction with several chromones supports the mechanism shown in Scheme 25 ⟨79ACH(101)73⟩.

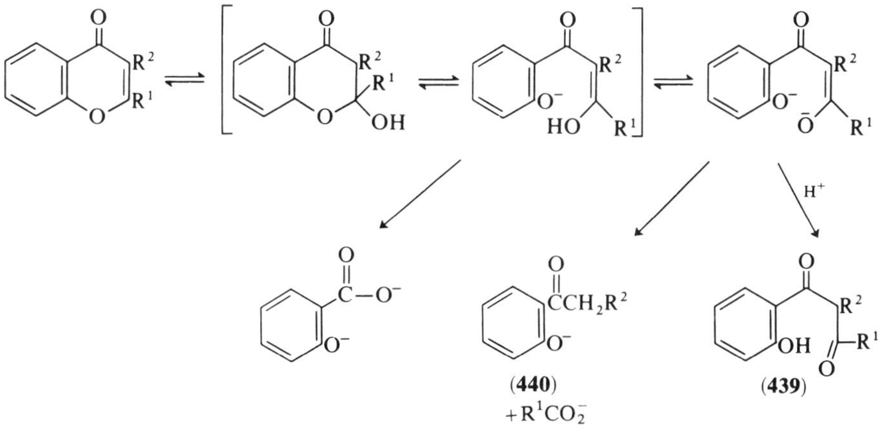

**Scheme 25**

After ring cleavage, the product, under favorable conditions, may be the 1,3-diketone (**439**) but more drastic conditions may result in the formation of an *o*-hydroxyketone (**440**), salicylic acid or phenol. The absence of oxygen frequently has an effect on the course of the reaction. Reaction of 2,6-dimethylpyran-4-one with barium hydroxide gives heptane-2,4,6-trione, and chelidonic acid (**441**) yields 2,4,6-trioxoheptanedioic acid (**442**), which may be further degraded. Ethyl 6,8-dibromo-7-hydroxychromone-2-carboxylate (**443**) is degraded to 3,5-dibromo-2,4-dihydroxyacetophenone on boiling with aqueous sodium hydroxide ⟨70JCS(C)2609⟩; an isoflavone (**444**) gives the deoxybenzoin (**446**) on heating with dilute ethanolic alkali, but fusion with potassium hydroxide degrades it to phenol and a phenylacetic acid (**445**).

(**441**) (**442**)

(**443**)

(**445**) (**444**) (**446**)

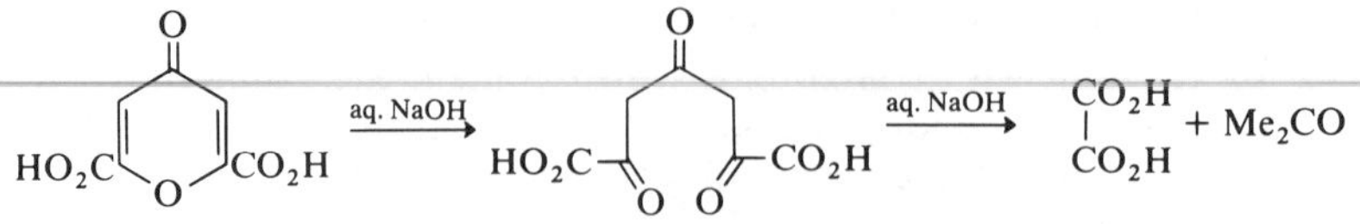

Sodium alkoxide may open a pyran-4-one ring; for example, 2-methylchromone (**447**) gives a 50% yield of a dimeric product (**448**) which is reconverted in 73% yield into 2-methylchromone on treatment with acid ⟨61CB660⟩. In a related reaction in which oxygen is bubbled through a solution of a flavone (**449**) and *t*-butoxide ion, a high yield of an aroylsalicylic acid (**450**) is formed under mild conditions ⟨79JCS(P1)2511⟩.

(**447**) (**448**)

(**449**) Ar = 4-$HOC_6H_4$

(**450**)

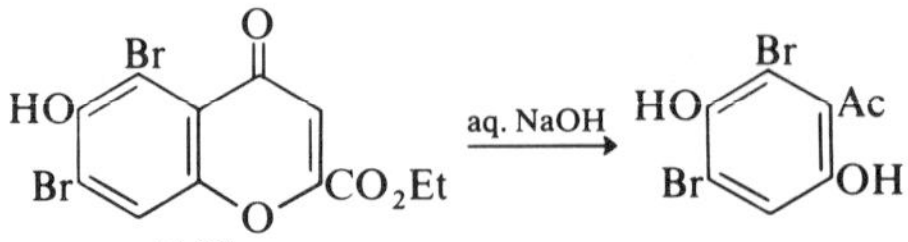

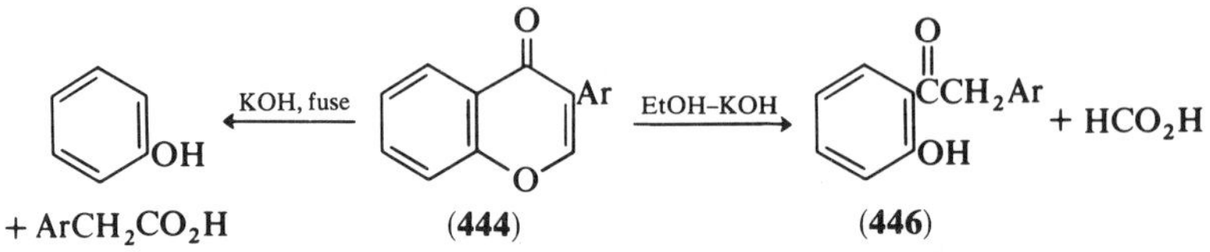

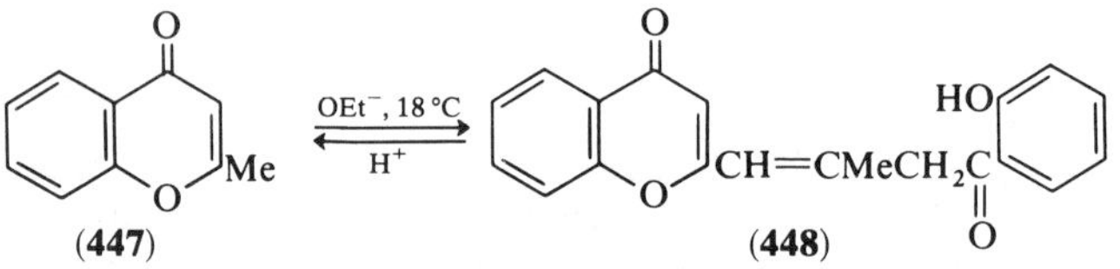

The presence of certain functions or leaving groups in a side chain can alter the course of the reaction (see Section 2.23.9.3).

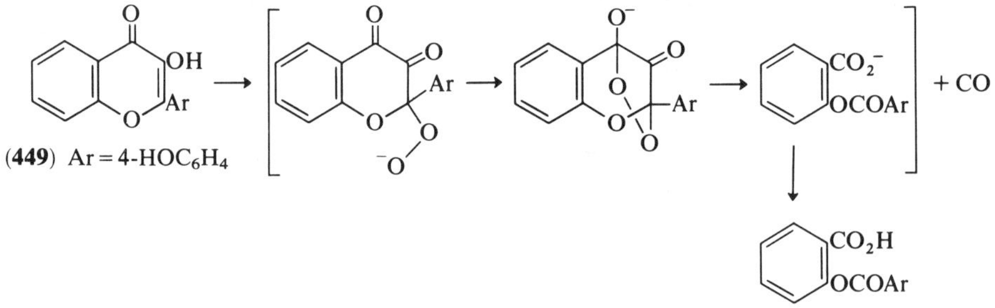

### *2.23.8.4.4 Reduction*

Reduction of pyran-4-ones can take one of several courses. Catalytic hydrogenation can be controlled so as to effect partial or complete reduction; it is thus possible to obtain 2,3-dihydropyran-4-ones, pyran-4-ols or tetrahydropyran-4-ones. Reduction of the carbonyl group usually requires pressure. LAH gives pyran-4-ol while sodium amalgam reduces the C—C double bonds. Several products may be obtained when chromones and the flavonoids are reduced catalytically, and for a particular substrate the most favorable set of conditions to produce a particular product has to be determined ⟨77HC(31)557⟩. (See also Section 2.23.9.3.) 6-Chlorochromone-2-carboxylic acid (**451**) is unaffected by hydrogenation at low pressure with Raney nickel or at high pressure and temperature over copper chromate, but at low pressure with palladium–charcoal in acetic acid the 2,3-double bond, the carbonyl and chlorine functions are lost ⟨71JMC758⟩. The reduction of ethyl 6-phenylchromone-2-carboxylate (**452**) can give several products and thorough investigations have shown that a remarkable degree of selectivity is possible. Scheme 26 shows some of the products obtained ⟨75JMC934⟩.

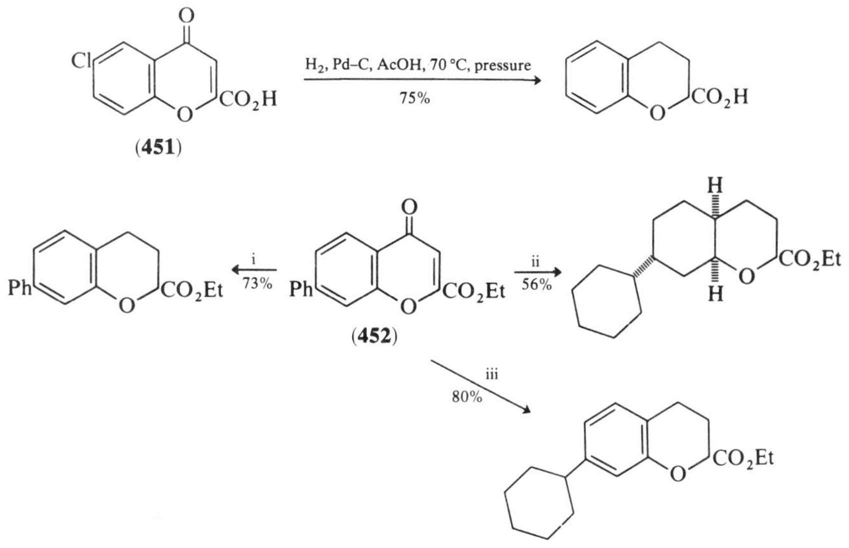

i, H$_2$, Pd–C, AcOH, 60 °C, 40 p.s.i., 4 h; ii, H$_2$, Pd–C, AcOH, 90 °C, 50 p.s.i. 16 h; iii, H$_2$, Pd–C, AcOH, 90 °C, 48 p.s.i., 8 h

**Scheme 26**

Metal hydrides usually reduce the carbonyl to alcohol but LAH converts xanthone into a dixanthyl ether (**454**) or xanthene (**453**) according to the conditions. The latter product is also obtained by the Wolff–Kishner–Huang–Minlon reduction. Metal–liquid ammonia opens the pyran-4-one ring of chromones and flavones to form dihydrochalcones and other compounds ⟨79JOC1494⟩.

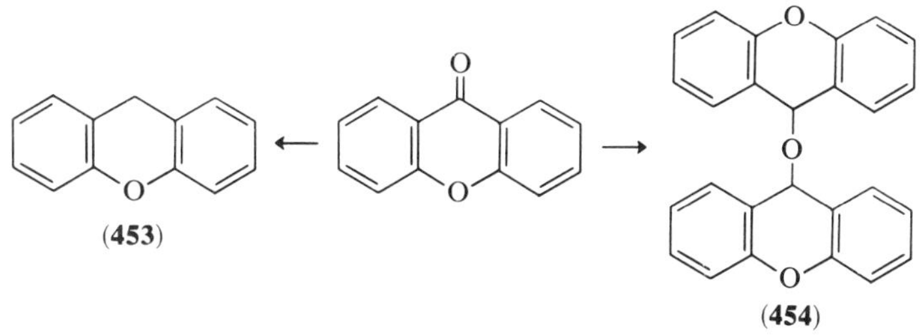

## 2.23.8.5 Radicals

Few interactions of radicals with the pyran-4-one family are recorded. Sulfuryl chloride (usually a source of chlorine radicals) converts flavone into the 2,3,3-trichloro derivative (**455**) which on catalytic hydrogenolysis in the presence of a base gives 3-chloroflavone ⟨70CC380⟩. Chromone reacts with sulfuryl chloride to give 2,3-dichlorochromanone (**456**) ⟨67CHE624⟩.

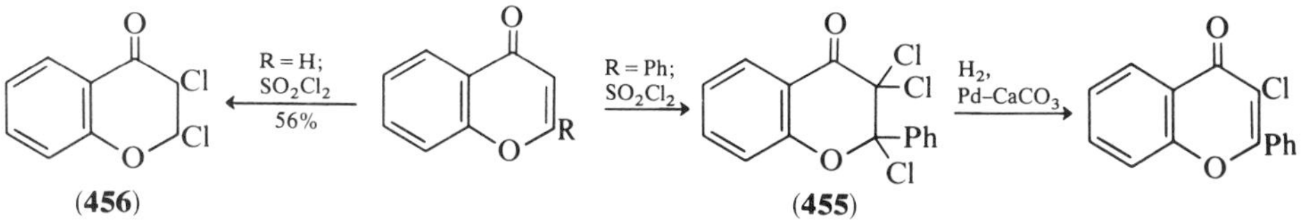

## 2.23.8.6 Other Reagents

Mono- and bi-cyclic pyran-4-ones undergo photoaddition with unsaturated organic compounds ⟨73BCJ690, 66TL1419⟩. When chromone is irradiated with an alkene or alkyne, addition occurs across the 2,3-bond and a mixture of products is obtained. 2,3-Dimethylbut-2-ene and chromone undergo photoaddition to give four compounds (457)–(460) in the ratio of 13:5.5:2.6:2 ⟨69JA4494⟩. Photoaddition of cyclopentene to chromone yields only two adducts, the major product being the cyclobutane corresponding to (457). Replacement of cyclopentene by but-2-yne gives cyclobutene (461) as the sole product. From these and similar experiments in which a sensitizer was shown to have no effect on the reaction products or their relative amounts, it appears that the excited chromone singlet quickly undergoes intersystem crossing to the triplet. This $\pi-\pi^*$ triplet reacts with the alkene, probably to give a diradical ⟨69JA4494⟩.

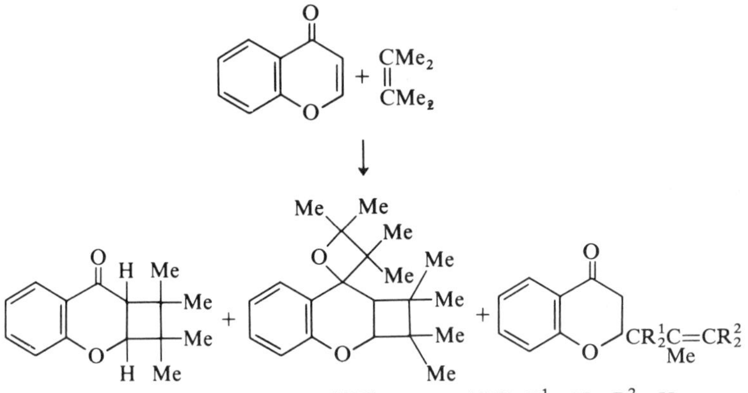

(457)    (458)    (459) $R^1 = Me$, $R^2 = H$
(460) $R^1 = H$, $R^2 = Me$

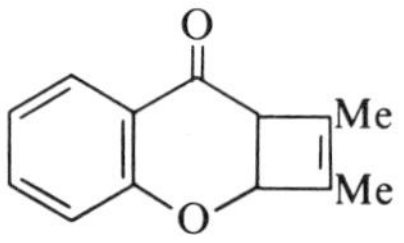

(461)

Xanthene reacts differently with alkynes to give mostly the 9-xanthylidene derivative (462) and a small amount of its cyclized analogue (463) ⟨73RTC845⟩. The spiran (464) is formed in 15% yield when xanthone and 1,3-diacetylimidazol-2-one or its homologues are irradiated in acetone ⟨67CB3961⟩.

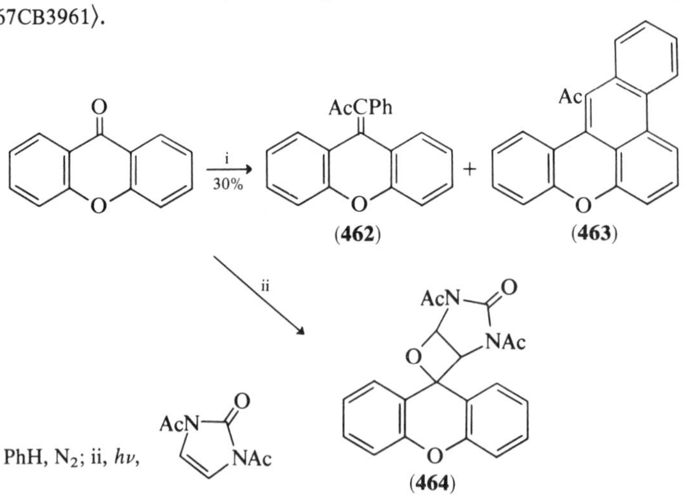

(462)    (463)

i, PhC≡CMe, hν, PhH, N₂; ii, hν,    (464)

A number of chromones react with thionyl chloride on prolonged boiling to form 4,4-dichloro-4*H*-chromene (465) which condenses with arylamines to give Schiff's bases (466) and reverts to the chromone on hydrolysis ⟨67CHE624⟩.

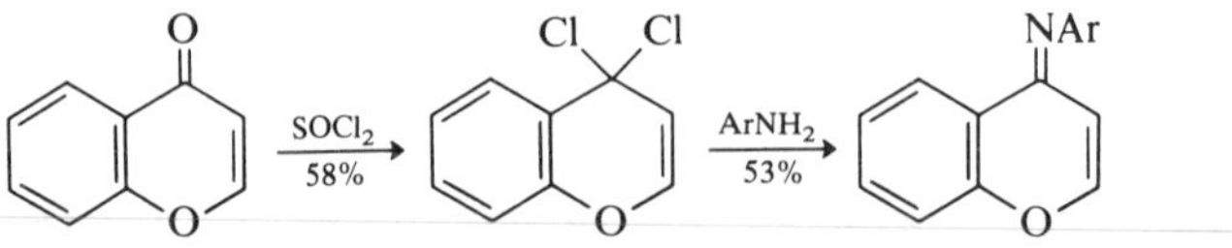

(465)    (466) $Ar = 4\text{-}NO_2C_6H_4$

## 2.23.9 PYRAN-4-ONES AND THEIR BENZO DERIVATIVES: REACTIVITY OF SUBSTITUENTS

### 2.23.9.1 Introduction

The pyran-4-one ring (whether annulated to benzene or not) exerts an effect on some attached groups, for example, on the ease with which protons are detached from methyl groups and on the $pK_a$ of some carboxylic acids. Annulation of a benzene ring to the pyran-4-one increases the latter's stability to electrophiles and the effect is even more marked when two benzene rings are fused to the pyran, as in xanthone.

In all 5-hydroxy-1-benzopyranones, hydrogen bonding exists between the hydroxy and the carbonyl groups. This affects the reactivity of the hydroxyl group. Such compounds are also capable of chelating metal ions such as copper(II), beryllium(II), aluminum(III), tin(IV), titanium(IV), uranium(VI), iron(III), and palladium(II). These complexes are ether-soluble but those formed with thorium(II), cobalt(II) and nickel(II) are not ⟨73M122303, 71MI22301⟩. Similarly, 5-hydroxyflavones have been shown to be useful reagents for colorimetric analysis of metals ⟨78BCJ2425⟩. 3-Hydroxychromone also forms complexes with metal ions; the stability constants of some of these have been determined and their application to spectrophotometric problems is described ⟨69MI22300⟩. 2-Hydroxychromones are tautomeric with 4-hydroxycoumarins.

In addition to the chromone carboxylic acids mentioned in Section 2.23.8.1, a number of other chromones have shown promising activity as antiallergic compounds, for example 3-hydroxymethyl-8-methoxychromone ⟨76MI22300⟩, chromone-3-carboxylic acid ⟨74CPB2959⟩ and *N*-tetrazolyl chromonecarboxamides ⟨78JMC1120⟩.

### 2.23.9.2 Fused Benzene Rings

In chromones, flavonoids and xanthones, the carbonyl group is adjacent to the benzene ring and depresses the reactivity of the latter towards electrophiles. However, some electrophiles successfully enter this ring but acylation and alkylation under Friedel–Crafts conditions, chloromethylation, aminomethylation, sulfonation and halogenation (with one exception, see below) are without effect on the benzene ring. Nitration of chromone gives good yields of 6-nitrochromone but flavone yields a mixture of 2'-, 3'- and 4'-nitro derivatives, the 2-phenyl ring evidently being more reactive than the fused ring. Xanthone is nitrated at the 2- and then 7-positions; when the reaction time is extended to 4 days, 2,4,7-trinitroxanthone is obtained ⟨53JCS1348⟩. In contrast, it is not possible to introduce two nitro groups into the benzene ring of chromone.

The exception mentioned above to the inertness of the benzene ring of chromone to halogenation is its bromination at room temperature by dibromoisocyanuric acid (1,3-dibromo-1,3,5-triazine-2,4,6-trione, DBI). This reagent in sulfuric acid is a source of strongly electrophilic bromine and can introduce up to four bromine atoms into each chromone molecule according to the molar ratio of reagent to substrate (Scheme 27). 6-Nitrochromone was similarly converted into 3,8-dibromo-6-nitrochromone ⟨73JCS(P1)2781⟩.

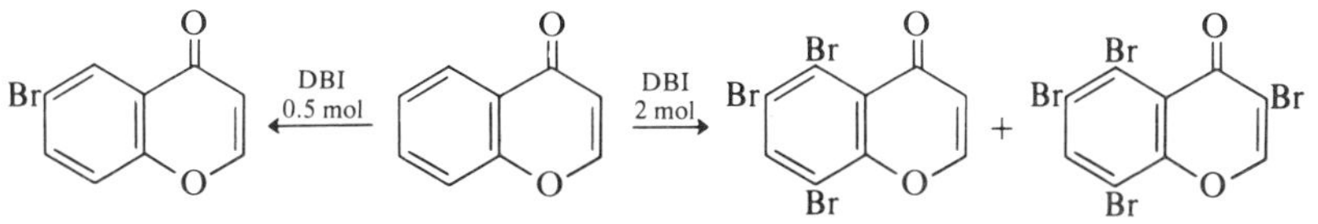

**Scheme 27**

The presence of an electron-releasing substituent facilitates electrophilic attack, so that, for example, 7-hydroxy-2-methylchromone undergoes bromination by bromine, acylation under Friedel–Crafts conditions and Mannich reaction at C-8. Similarly, a Fries or Claisen rearrangement is possible on 7-acyloxy or -allyloxy compounds, respectively. However, the balance of reactivity is narrow, as is shown by a Mannich reaction on 8-methoxychromone; the product is the 3-aminomethyl derivative, the methoxy group not being sufficiently activating at C-5 or C-7 ⟨77FES635⟩. Scheme 28 summarizes some reactions of 6-hydroxy-2-methylchromone and its derivatives ⟨70JCS(C)2230⟩. Hydroxyl groups in flavone exert a similar directing effect ⟨78CHE497⟩. Chromone is resistant to attack by chlorosulfonic or sulfuric acid but 2,3-dimethylchromone and chlorosulfonic acid give the 6-sulfonyl chloride ⟨56JOC1104⟩ and 7-methoxychromone is chlorosulfonated at C-8 ⟨81UP22300⟩.

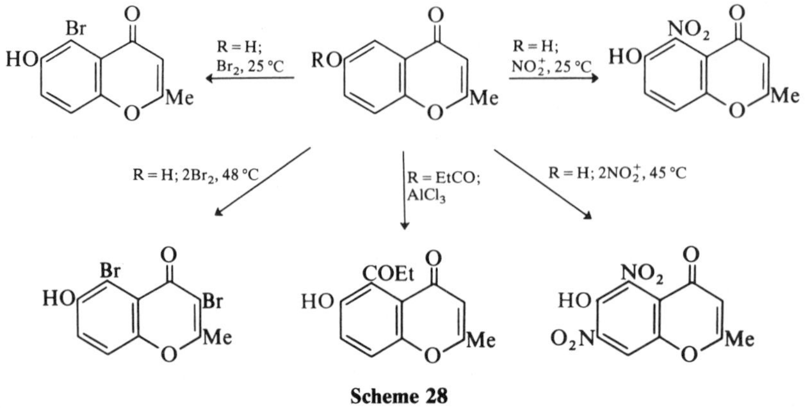

**Scheme 28**

Chloromethylation of suitably activated chromones yields useful intermediates for further synthesis, for example, in the preparation of methyl, aminomethyl, cyanomethyl or carboxaldehyde derivatives. A methoxy group in the benzene ring is sufficiently electron-releasing to promote this reaction, for instance 7-methoxy-2-methylchromone is converted into the 8-chloromethyl analogue ⟨62JIC507⟩.

### 2.23.9.3 *C*-Linked Substituents

The varying electron density at different carbon atoms of the pyran-4-one ring is reflected in the different reactivities of methyl groups. In the presence of a base, a 2-methyl group condenses with a benzaldehyde to give the 2-styryl derivative (**467**). Larger alkyl groups at C-2 or a ring, *e.g.* (**468**), behave similarly towards a variety of benzaldehydes; a styryl derivative (**469**) of naphtho[1,2-*b*]pyran-4-thione forms violet or yellow crystals which exhibit different colours in transmitted and reflected light ⟨57JA6020⟩. 2-Styryl-3-unsubstituted chromones, *e.g.* (**470**), undergo Diels–Alder reactions with dienophiles such as maleic anhydride ⟨75ACH(84)319⟩.

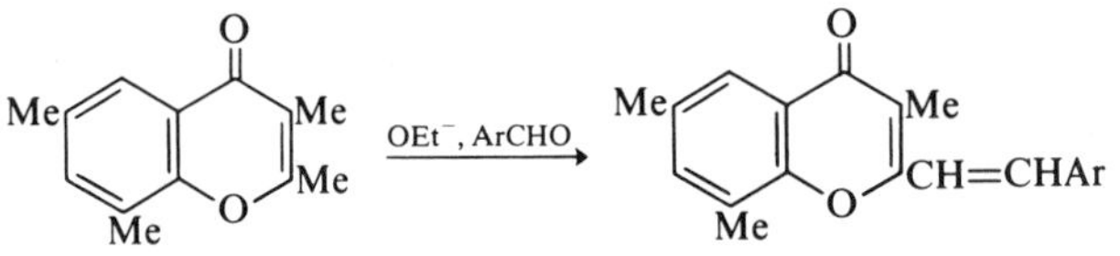

(**467**)

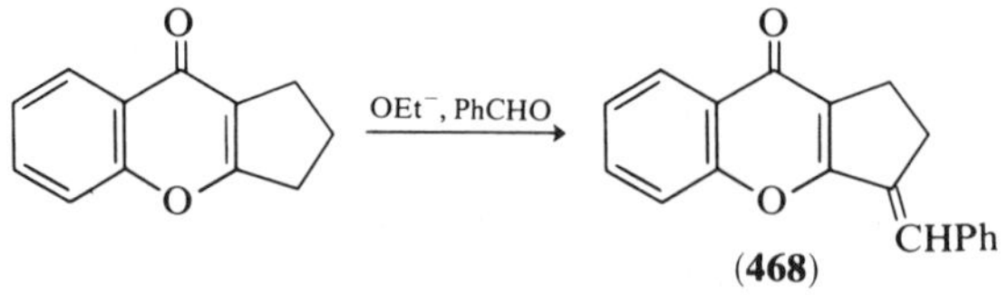

(**468**)

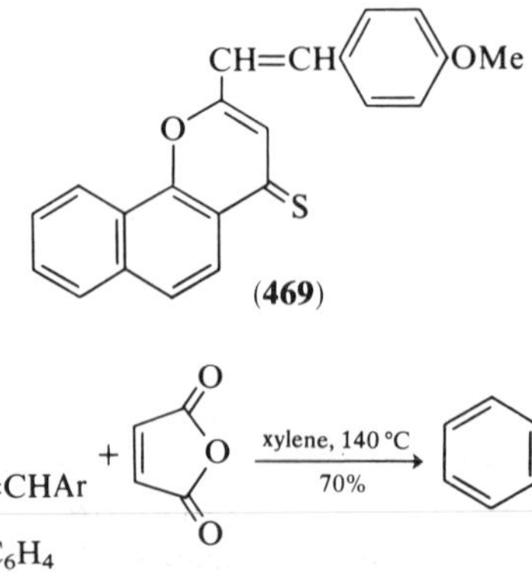

(**469**)

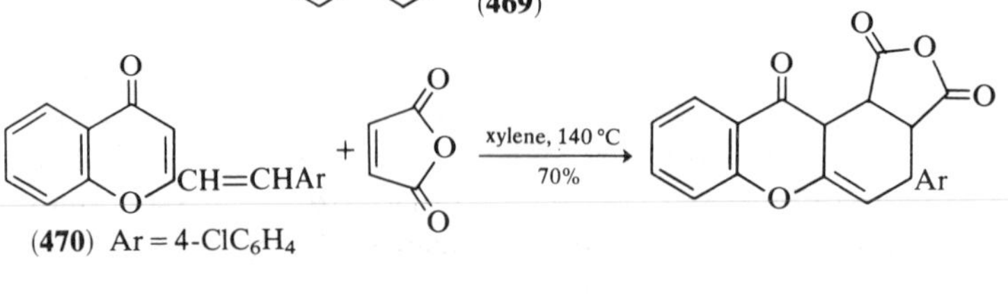

(**470**) Ar = 4-ClC₆H₄

A 2-methyl group of pyran-4-ones, *e.g.* (**471**), and chromones is susceptible to halogenation with either bromine or NBS and advantage has been taken of this in the synthesis of 3-methylchromone-2-carboxaldehyde (**474**) from 2,3-dimethylchromone (**472**) through the dibromide (**473**) ⟨80JCR(S)159⟩.

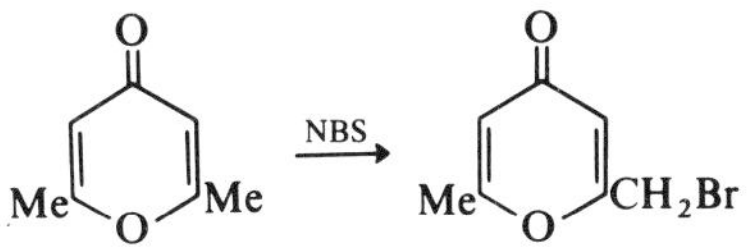

(**471**)

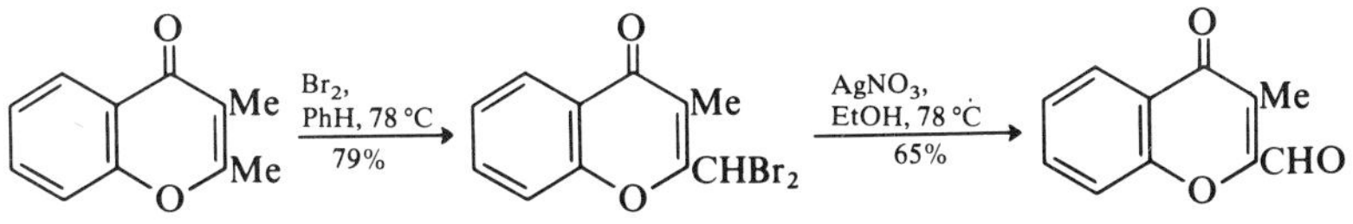

(**472**)              (**473**)

In the absence of a 2-alkyl group, other methyl groups may become chlorinated on prolonged reaction with sulfuryl chloride but yields are usually low; for example, 6-methylflavone (**475**) is chlorinated both at C-3 and in the methyl group ⟨71T4837⟩.

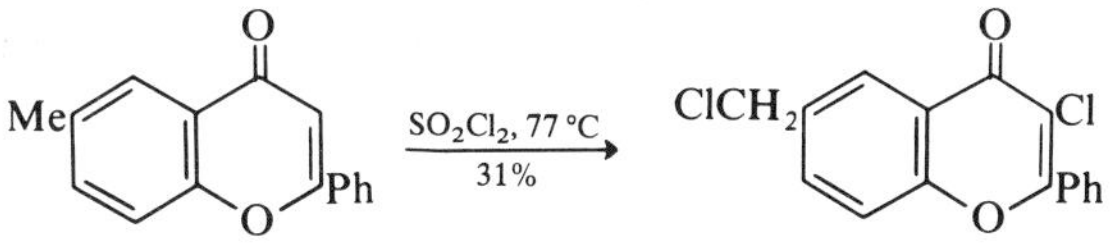

(**475**)

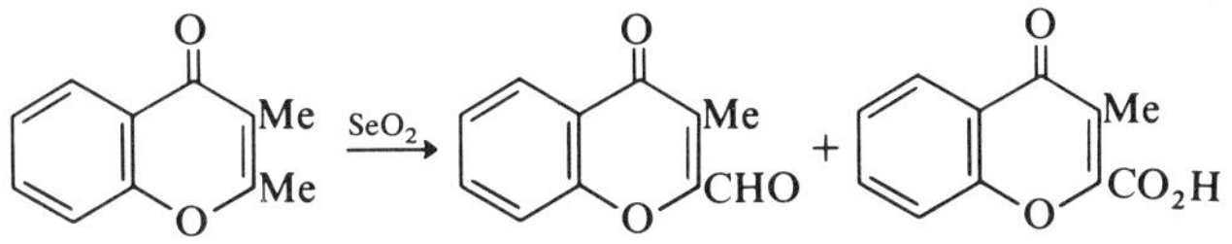

(**472**)      57%        16%

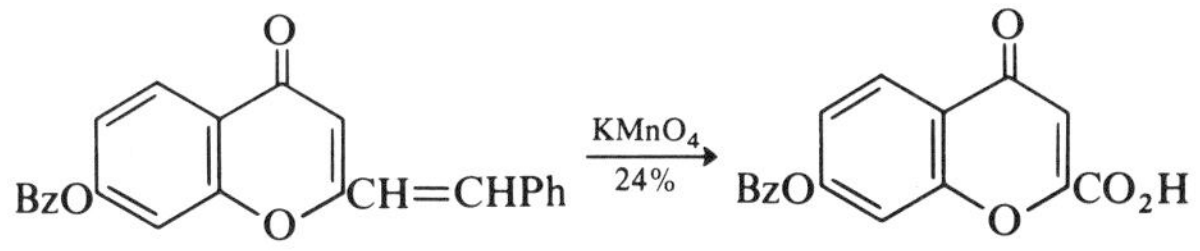

(**476**)

Chromones and flavones are susceptible to oxidative degradation (Section 2.23.8.3.2) and it is therefore difficult to convert a methyl group attached to the ring system to a carboxyl. Selenium dioxide oxidizes 2-methylchromone to the carboxylic acid in low yield, but a reasonable yield of the aldehyde is obtained when this reagent is applied to 2,3-dimethylchromone (**472**) ⟨52HCA1168⟩. Oxidation of a double bond in a side chain by means of permanganate in pyridine gives low to moderate yields of the carboxylic acid (**476**); ozonolysis of the same group gives the aldehyde in low yield ⟨58T(2)203⟩.

Conversion of an aldehyde into a carboxyl group is effected with chromium trioxide–sulfuric acid in moderate yield. This is a synthetically valuable conversion because of the availability of chromone-3-carboxaldehydes (**477**). Photooxidation of the aldehyde (**477**; R = H, Me, Cl, OMe or OAc) in the presence of NBS gives high yields of 3-carboxylic acids ⟨80SC889⟩.

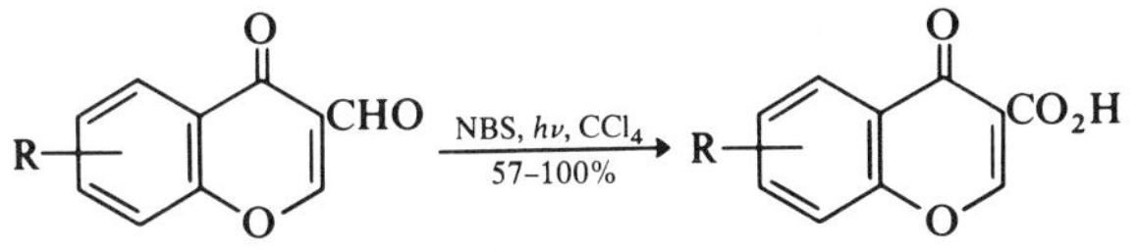

(**477**)

The action of gaseous ammonia on diethyl chromone-2,6-dicarboxylate (**478**) is a good example of the different reactivity between identical substituents in the same molecule. The 2-carboxylate function is more reactive even than an ester group in a side chain such as —OCH$_2$CO$_2$Et, which is unaffected by ammonia. This is consistent with the low electron density of C-2. 2-Carboxamides, such as (**479**), need drastic conditions for their dehydration

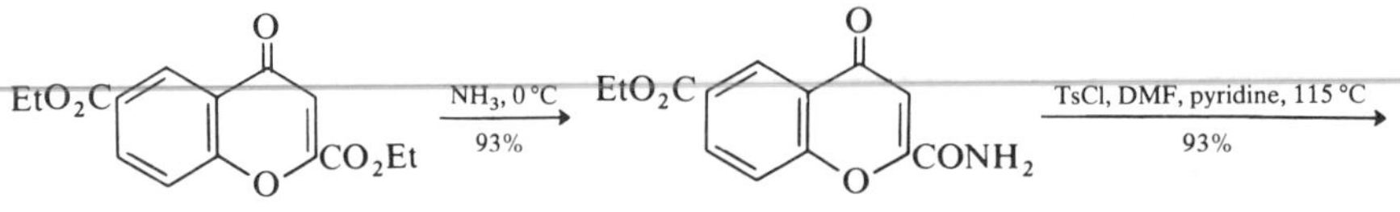

(478)                                              (479) 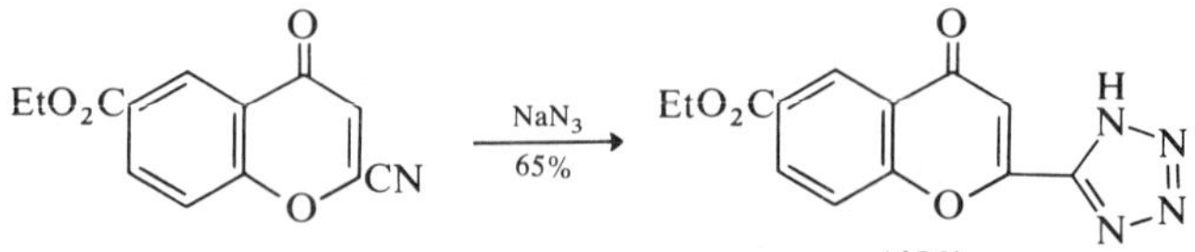                                              (480)

to nitriles. Several of the commonly used reagents, for example thionyl chloride, phosphoryl chloride and phosphorus pentoxide, have no effect, but heating with toluene-4-sulfonyl chloride, pyridine and DMF for 8 h gives good yields of nitriles ⟨72JCS(P1)779⟩. These are readily converted into tetrazoles by reaction with either sodium azide and ammonium chloride ⟨72JMC865⟩ or sodium azide and aluminum chloride in THF ⟨79JMC290⟩. The p$K_a$ of compounds such as the tetrazole (480) is comparable with that of the 2-carboxylic acids, namely, about 2.9 ⟨70MI22301, 72JMC865⟩.

Nitriles may be prepared by dehydration of aldoximes; an attempt to effect a Beckmann rearrangement on the oxime (481) of 5-methoxy-4-oxopyran-2-carboxaldehyde gives the nitrile (482) but under milder conditions only the chloride (483) and the chloronitrile (484) are obtained ⟨75JHC219⟩.

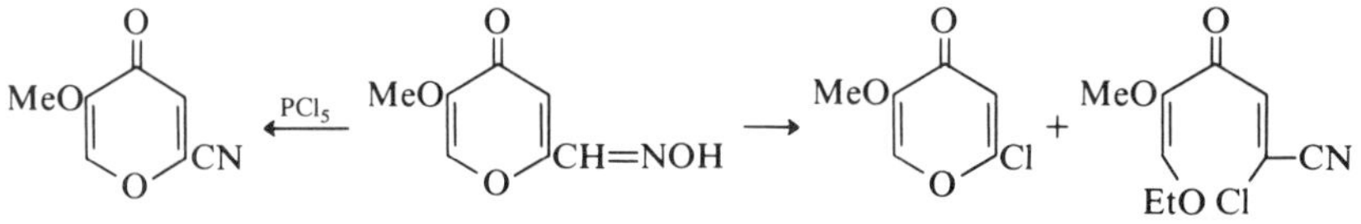

(482)                    (481)                    (483)        (484)

The carbonyl and 2,3-double bond of pyran-4-ones and their monobenzo derivatives are reducible and therefore attempts to reduce functional groups such as carboxaldehyde, carboxylate ester, carboxylic acid chloride or nitrile need careful selection of reagents. The reducibility of a particular group varies with the nature of other substituents present and it is not possible to generalize about the conditions required for each reaction.

Carboxaldehyde groups attached to the benzene ring may be reduced catalytically or with sodium borohydride to a primary alcohol ⟨60JOC1097, 74T3553⟩ and with zinc–hydrochloric acid to the methyl group ⟨53JA4992⟩. Methyl chromone-2-carboxylate is reduced selectively at the ester group by sodium borohydride–methanol to give the primary alcohol in 53% yield ⟨79S889⟩. In the absence of the ester group, the chromone (485) is reduced by borohydride in boiling ethanol to the chromanol (486) ⟨62BCJ1329⟩; LAH in THF at −80 °C reduces only the 2,3-double bond ⟨65LA(685)167⟩.

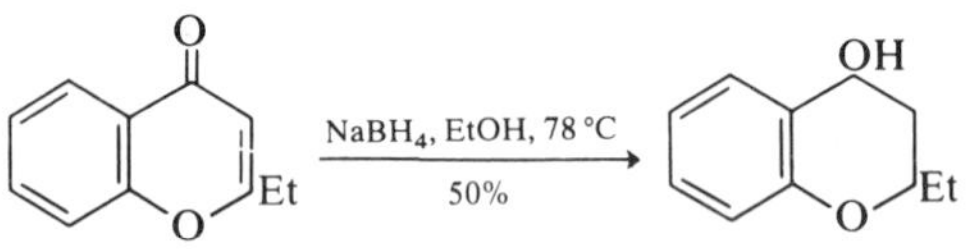

(485)                              (486)

Decarboxylation is sometimes useful in structure investigations and in assignment of NMR signals. Chelidonic acid (488; R = H), which is widely distributed in plants, is doubly decarboxylated to pyran-4-one (487) on heating with copper powder. The monoester is decarboxylated to comanic acid ester (489).

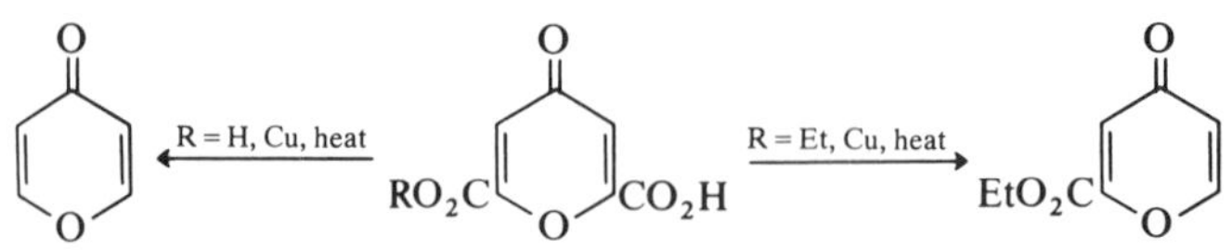

(487)                    (488)                    (489)

When meconic acid (490) is partly esterified and heated, it loses carbon dioxide and forms comenic acid (491) which is itself decarboxylated (as is meconic acid) to pyromeconic acid (**492**). Chromone was first synthesized by decarboxylation of the 2-carboxylic acid; the 3-carboxylic acid is also readily decarboxylated by heating. Sometimes quinoline and copper bronze, charcoal or copper powder facilitate the loss of carbon dioxide at or near the melting point. A carboxyl group on the benzene ring is eliminated on heating, for example that of 7-hydroxy-2-methylchromone-5-carboxylic acid, but selective monodecarboxylation at C-2 of the diacid (**493**) is possible in good yield ⟨75TH22300⟩.

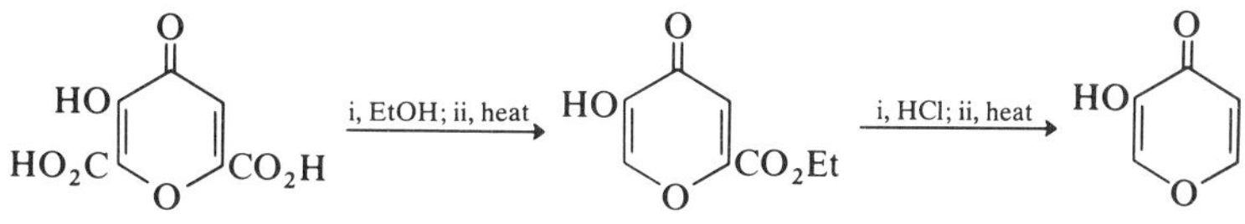

(**490**)     (**491**)     (**492**)

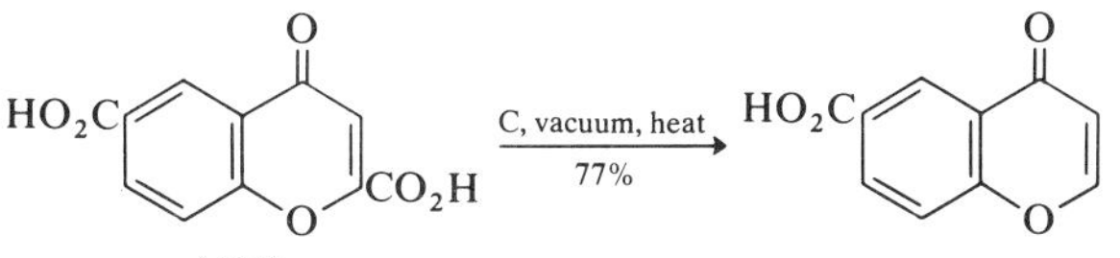

(**493**)

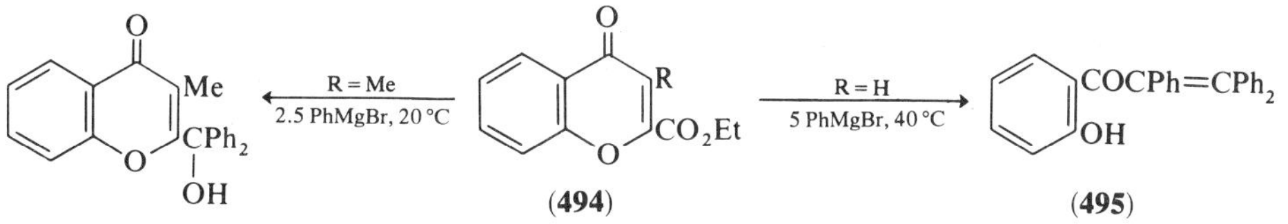

(**494**)     (**495**)

The carbonyl group of chromones is attacked by Grignard reagents (Section 2.23.8.4.1) but the carboxylate group of ethyl chromone-2-carboxylate is attacked preferentially by a Grignard reagent (2.5 molar ratio); a five molar excess of reagent caused ring opening to the hydroxyketone (**495**). Prolonged reaction of ethyl 3-methylchromone-2-carboxylate (**494**; R = Me) with the reagent at room temperature gives the tertiary alcohol as the sole product ⟨73ACS2020⟩.

Chromone-2-carbonyl chloride (**496**) reacts readily with a Grignard reagent to yield a ketone (**497**) without the need to add iron(III) chloride or to maintain the temperature at −70 °C ⟨81JCS(P1)2552⟩ as is customary in order to suppress the formation of *t*-alcohol.

The preparation of chromone-2-carbonyl chlorides from the acid and thionyl chloride often leads to the trichloride (**498**) as a byproduct ⟨61JGU523⟩ but this is suppressed when a few drops of DMF are added ⟨72JMC865⟩ or phosphorus pentachloride and cyclohexane are used ⟨73BSF2392⟩. When the trichloride is treated with water, an amine or acid, it rearranges to 4-chlorocoumarin (**499**) ⟨63JGU1806⟩.

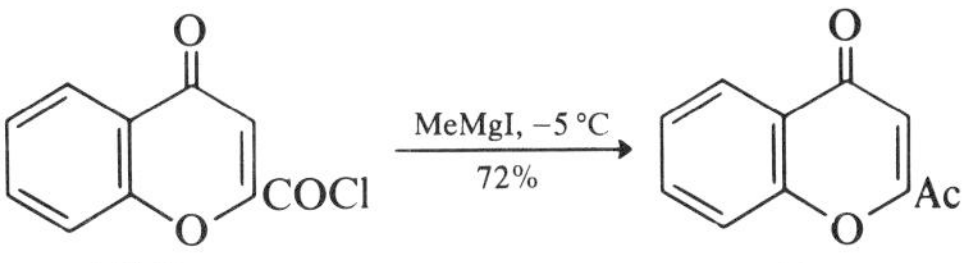

(**496**)     (**497**)

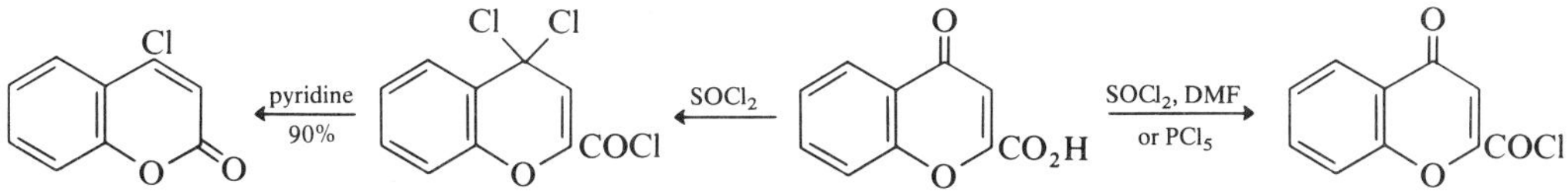

(**499**)     (**498**)

Aldehyde groups attached to pyran-4-one rings undergo Knoevenagel condensation but strongly basic conditions are to be avoided if the ring is to survive. Comenaldehyde methyl ether (**500**) condenses with ethyl cyanoacetate in acetic acid containing glycine ⟨79JHC1281⟩.

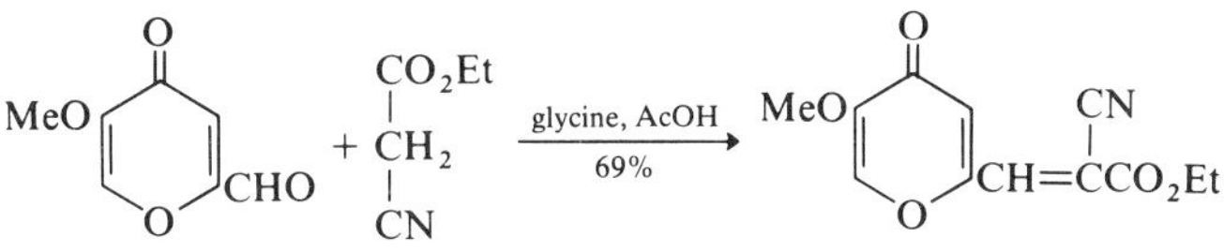

(**500**)

The reactions of chromone-3-carboxaldehyde (**501**) have been thoroughly investigated; some of them are of considerable synthetic interest. Its oxime (**502**) (prepared in acid medium, see Section 2.23.8.4.2) gives the expected nitrile (**503**) on dehydration but when it (or the nitrile) is heated with aqueous alkali, ring opening and reclosure occurs to produce the amino aldehyde (**504**) in high yield. This compound is cyclized by isocyanates and other reagents ⟨76LA1659⟩. On heating with diphenylketene, the aldehyde (**501**) forms a new pyran-2-one ring ⟨79CB1791⟩.

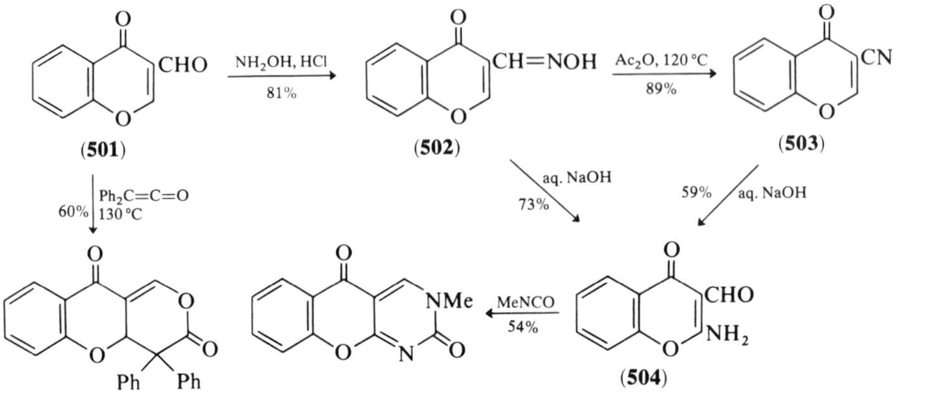

Chromone-3-carboxaldehyde forms Schiff's bases or anils (**505**) with arylamines; the aryliminomethyl group exerts an unexpected influence on the pyran ring: a nucleophile adds across the 2,3-double bond rather than open the ring as it does with simpler chromones. Thus, the Schiff's base (**505**) reacts with another molecule of arylamine under mild conditions to form the chromanone (**506**) ⟨79S337⟩. Hydrogen bonding is believed to be the stabilizing feature of the adduct because the second molecule of arylamine is eliminated when hydrogen bonding is prevented by removal of the hydrogen. When the arylamine (**507**) has a second amino group in the *ortho* position, the dihydrotetraaza[14]annulene (**508**) is obtained in high yield; it forms the expected complexes with Ni, Cu and Zn ⟨82HCA275⟩.

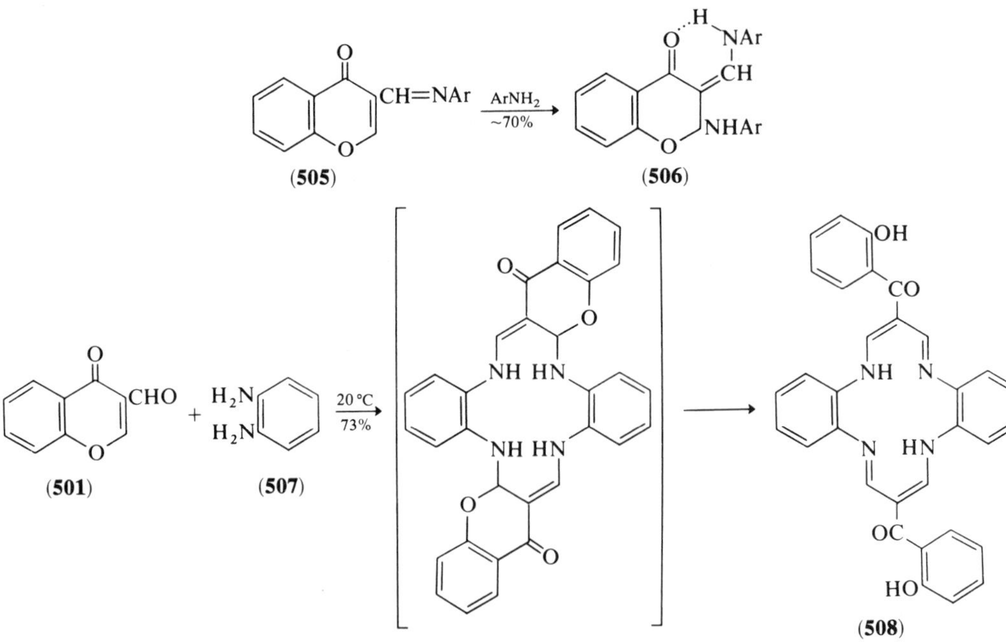

Although the pyran carbonyl group does not readily form oximes and semicarbazones, it does take part in several cyclizations with nitrogenous reagents. For example, chromone-3-carboxaldehyde gives the benzopyrano[4,3-*d*]pyrimidine (**509**) with amidines ⟨76S274, 76LA1663⟩ and the benzopyrano[4,3-*b*]pyridine (**510**) with enamines ⟨78S691⟩; chromone-3-carboxylic acid and phenylhydrazine yield the benzopyrano[4,3-*c*]pyrazol-4-one (**511**) ⟨78S779⟩.

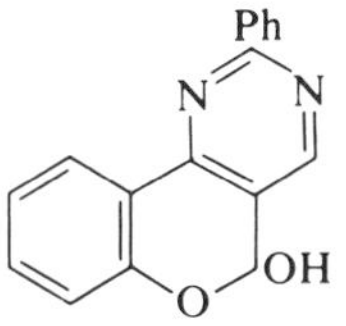

(509) 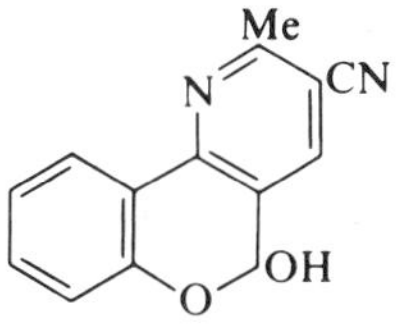          (510) 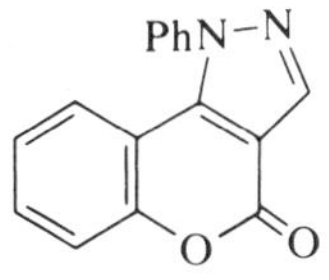          (511)

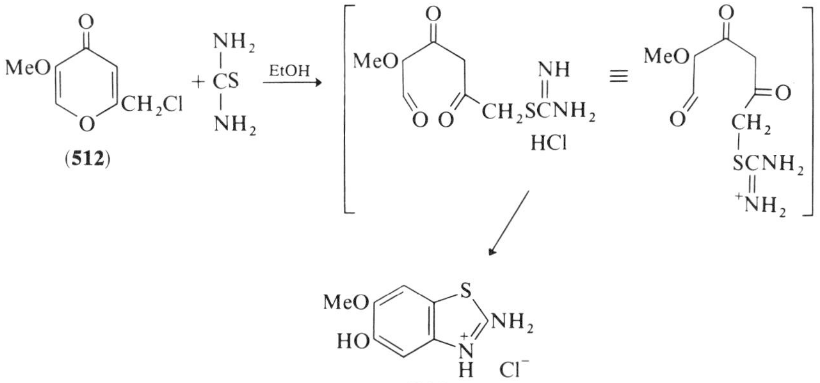

(512)          (513)

An unexpected reaction occurred when an attempt was made to convert 2-chloromethyl-5-methoxypyran-4-one (**512**) into its isothiuronium salt. Opening of the pyran ring and recyclization to a benzene ring followed by thiazole ring formation accounts for the formation of 2-amino-5-hydroxy-6-methoxybenzothiazole hydrochloride (**513**) ⟨80JHC817⟩.

The pyranone ring of chromones is opened by nucleophiles (Section 2.23.8.4.3) and the presence of certain side chains at C-3 can affect the course of this reaction. For example, the pyranone ring of 3-bromoacetylchromone (**514**) is opened by hot aqueous alkali and displacement of bromide ion produces the dihydrobenzoxepin-4-carboxaldehyde (**515**) ⟨77S61⟩.

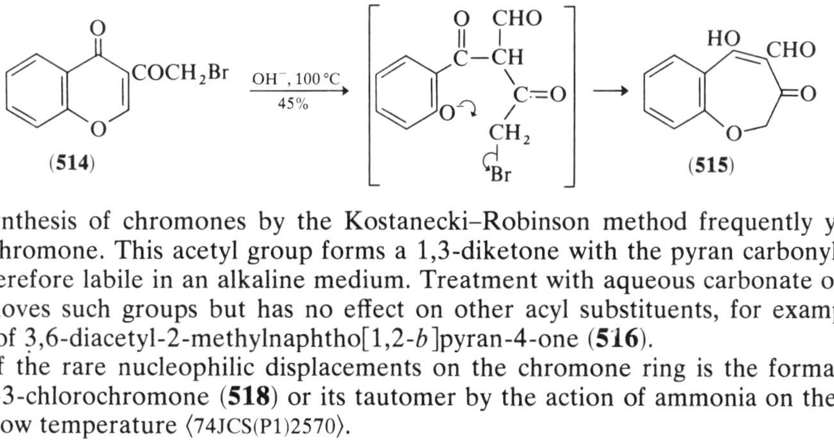

(514)          (515)

The synthesis of chromones by the Kostanecki–Robinson method frequently yields a 3-acetylchromone. This acetyl group forms a 1,3-diketone with the pyran carbonyl group and is therefore labile in an alkaline medium. Treatment with aqueous carbonate or other base removes such groups but has no effect on other acyl substituents, for example the 6-acetyl of 3,6-diacetyl-2-methylnaphtho[1,2-*b*]pyran-4-one (**516**).

One of the rare nucleophilic displacements on the chromone ring is the formation of 2-amino-3-chlorochromone (**518**) or its tautomer by the action of ammonia on the nitrile (**517**) at low temperature ⟨74JCS(P1)2570⟩.

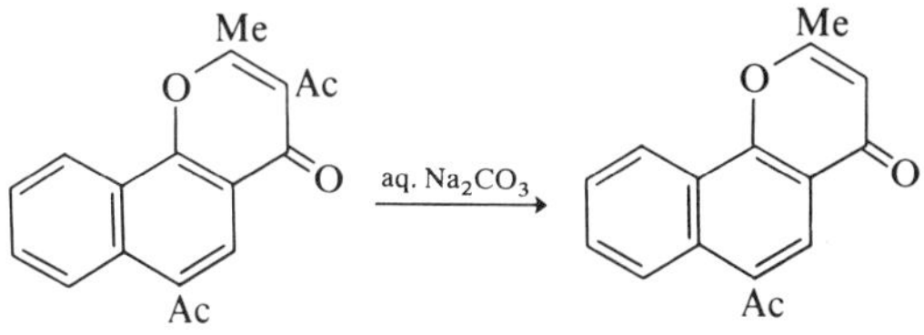

(516) 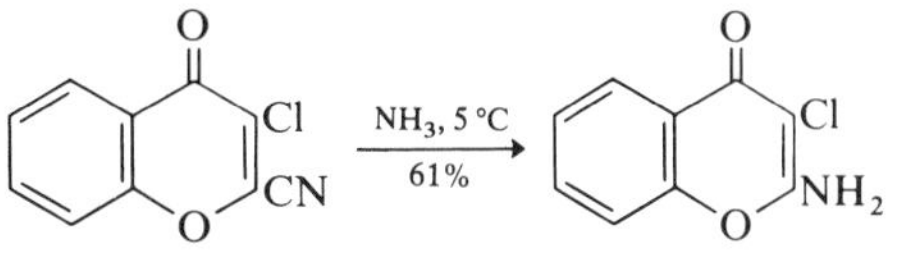          (517)          (518)

A number of natural flavonoids, coumarins and chromones have a prenyl (3-methylbut-2-enyl) side chain which is capable of cyclization on to an oxygen atom to form a pyran ring ⟨77HC(31)633⟩. Prenylation is effected by reaction with either 3-methylbut-2-enyl (prenyl) bromide or 2-methylbut-3-en-2-ol in the presence of a base or a Lewis acid. When this reaction is applied to flavonoids, the prenyl group has a choice of two benzene rings and mixtures of products are often obtained. A recent paper demonstrates the effect of a 7-benzyloxy or 7-benzoyloxy group on the nature of the product. In both compounds (**519**; R = PhCO) and (**519**; R = PhCH₂), the prenyl group enters the 2-(4-hydroxyphenyl) ring only. Moreover, the 7-benzyloxyflavone is converted directly into the pyran derivative (**522**) whereas the 7-benzoyloxyflavone gives the 3′-prenyl compound (**520**). Removal of the protecting groups gives neobavaisoflavone (**521**) and isoneobavaisoflavone (**523**) ⟨78BCJ2398⟩.

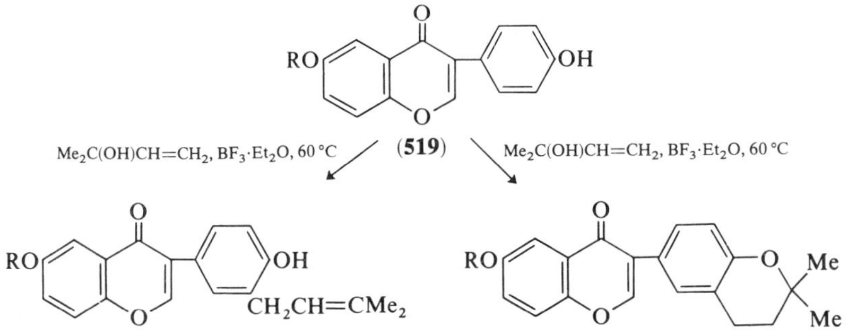

Some reactions of the hydroxymethyl group of kojic acid are discussed in Section 2.23.9.5.

### 2.23.9.4 *N*-Linked Substituents

Amino groups attached to the chromone ring system would not be expected to have identical properties since only one ring is fully aromatic. The degree of aromaticity of the pyran-4-one ring is lower than that of benzene, as the 2,3-double bond does take part in some addition reactions, as has been mentioned. The aromatic character of the pyran-4-one ring is supported by the ability of the 3-amino group to be diazotized and converted into the hydroxyl group ⟨78JCR(S)47⟩. There is, on the other hand, doubt about the nature of 2-aminochromone and its derivatives; these are tautomeric and on hydrolysis give coumarins. This suggests that they exist at least to some extent in the imino form (**524**) but spectroscopic and chemical evidence, while not conclusive, strongly support the NH₂ form for the diacetyl product (**526**) prepared from 2-amino-3-chlorochromone (**518**) or its tautomer ⟨74JCS(P1)2570⟩. The lower nucleophilicity of the 2-amino group is evident in other reactions and is consistent with its being (like 4-aminocoumarin) a vinylogous amide ⟨81JHC697⟩.

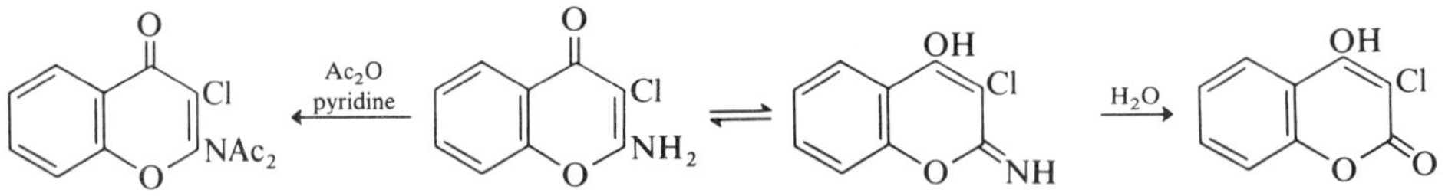

A nitrochromone has three easily reducible functions and, for its reduction, conditions and reagents which have little or no effect on the carbonyl or the 2,3-double bond should be chosen. Nitrochromones are reduced by tin–hydrochloric acid, zinc–ammonium chloride, iron–acetic acid, iron–hydrochloric acid or sodium dithionite. It may be easier to control the severity of the conditions in catalytic hydrogenation. Scheme 29 shows that with proper choice of conditions (temperature, pressure, solvent, catalyst), it is often possible to optimize the yield of the desired product (**527**). Extending the reaction time from about 30 min to 2.5 h increased the yield of the chromanone (**528**) and none of the hydroxylamine (**529**) was then detected ⟨70JCS(C)2230⟩.

Hydrogenation of a nitro group in the presence of other suitable side-chain functions can lead to the spontaneous formation of a new ring, for example the nitro ketone (**530**) on reduction with Raney nickel at atmospheric pressure and room temperature gives the dihydropyrano[3,2-*g*]benzoxazine (**531**). However, the 2,3-double bond of the isomeric nitro ketone (**532**) is retained on hydrogenation under pressure in the presence of palladium–charcoal to give the pyrano[2,3-*f*][1,4]benzoxazine (**533**) ⟨65JCS7348, 71JCS(C)2079⟩.

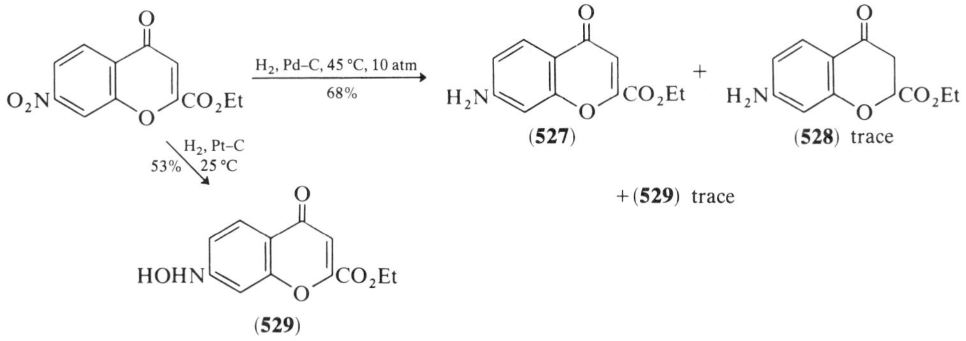

**Scheme 29**

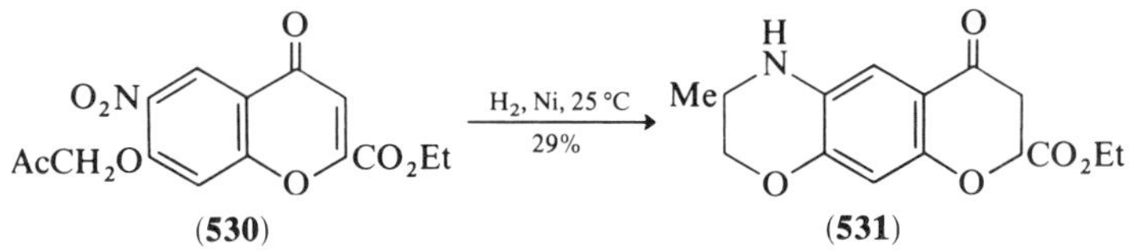

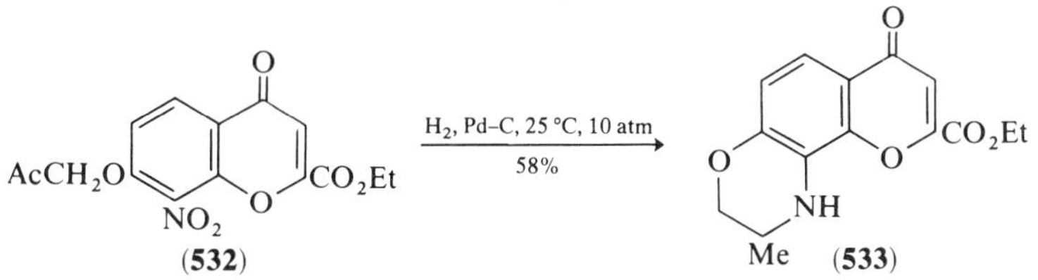

## 2.23.9.5 *O*-Linked Substituents

Most naturally occurring pyran-4-ones, chromones and flavonoids contain one or more oxygen-linked groups. Alkylation and acylation of hydroxyl groups and dealkylation of methoxyl groups has, therefore, received much attention.

Kojic acid has two hydroxyl groups and their reactions have been well studied as befits an important natural product. Until recently it was difficult to obtain the individual monoacyl

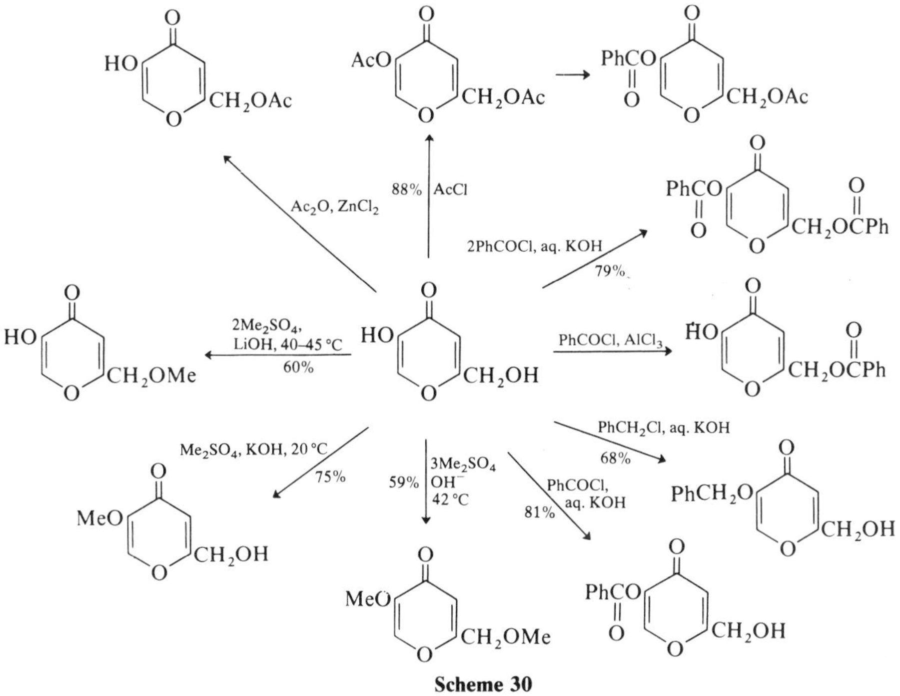

**Scheme 30**

and monoalkyl derivatives separately, but selective acylations and alkylations are now possible and are shown in Scheme 30 ⟨72JOC1444, 78JOC2842, 80BCJ569⟩.

Hydrogen bonding between a 5-hydroxy group and the pyran carbonyl group in chromones and flavonoids has an appreciable effect on the ease with which this substituent is alkylated or acylated. Therefore, other hydroxyl groups which may be present in the molecule react more easily; for example, diazomethane methylates the 3-hydroxyl group of 3,5-dihydroxy-2-methylchromone in 65% yield. When it is desired to alkylate only a 5-hydroxyl group, it is usual to protect other hydroxyl groups in the molecule, for instance by tosylation, acetylation or benzylation. The conditions for alkylating the 5-hydroxyl group may then be applied; they are usually more drastic or more lengthy than for other phenolic groups, for example the chromone (**534**) was 7-*O*-allylated by heating for a few hours with allyl bromide and potassium carbonate in acetone but heating for 24 h was required to methylate the 5-hydroxyl group with dimethyl sulfate ⟨53MI22302⟩. The synthesis of 5-methoxy-3,4',6,7-tetrahydroxyflavone (**537**) shows selective methylation of a 5-hydroxyl group. 3,4',5,6,7-Pentahydroxyflavone (**535**) on treatment with just over four molecular proportions of acetic anhydride in pyridine at 20 °C gives the tetraacetate (**536**) which is methylated and deacetylated to give vogeletin (**537**) ⟨66CB2430⟩.

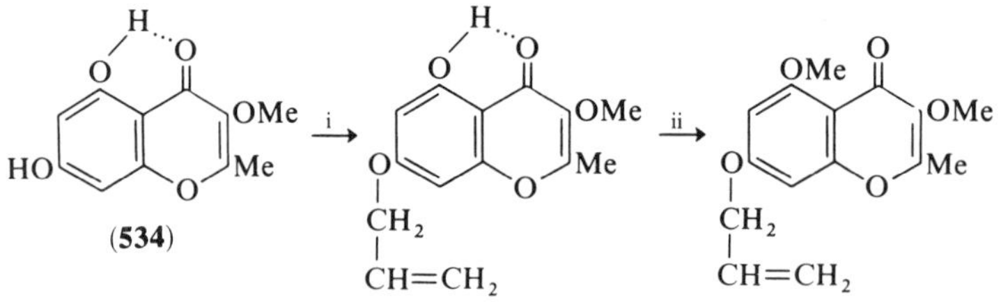

i, CH$_2$=CHCH$_2$Br, K$_2$CO$_3$, Me$_2$CO, 6 h; ii, Me$_2$SO$_4$, K$_2$CO$_3$, Me$_2$CO, 24 h

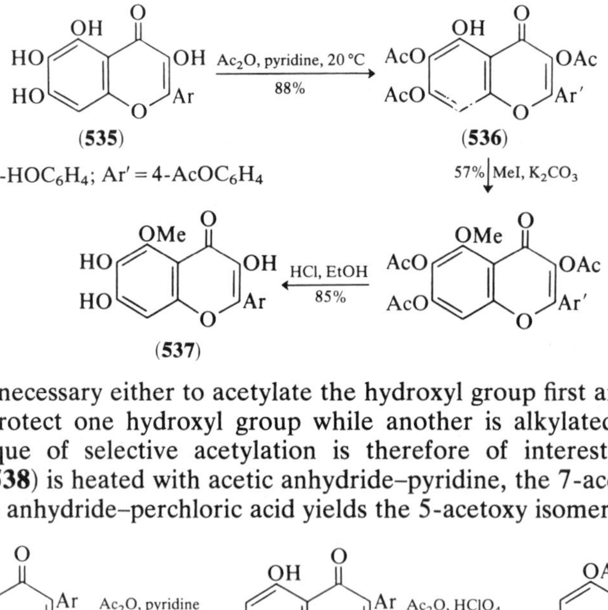

Ar = 4-HOC$_6$H$_4$; Ar' = 4-AcOC$_6$H$_4$

It is sometimes necessary either to acetylate the hydroxyl group first and then methylate this ester or to protect one hydroxyl group while another is alkylated, for instance. A promising technique of selective acetylation is therefore of interest: when 5,7-dihydroxyisoflavone (**538**) is heated with acetic anhydride–pyridine, the 7-acetoxy derivative is formed, but acetic anhydride–perchloric acid yields the 5-acetoxy isomer ⟨79CC264⟩.

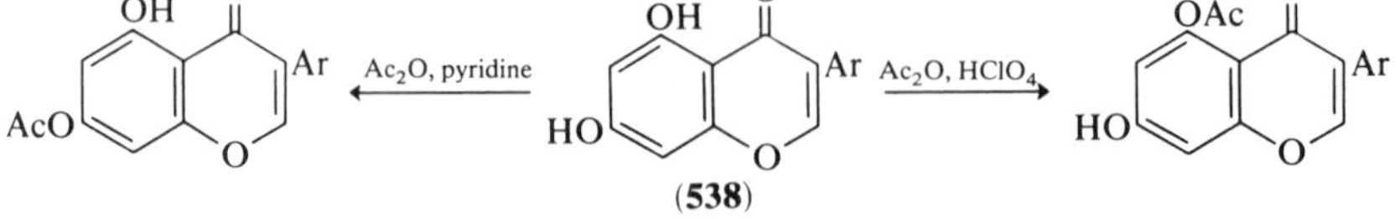

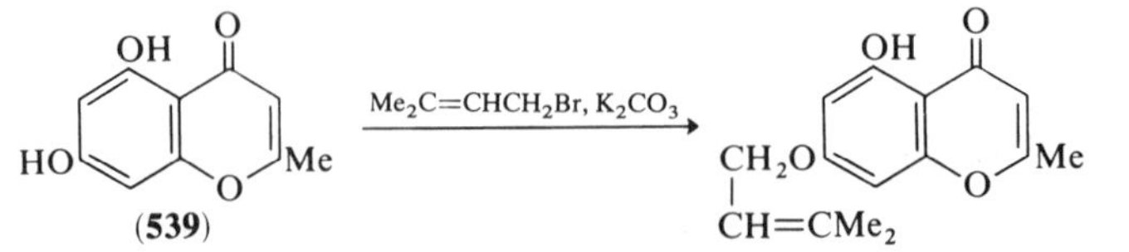

In the *O*-prenylation of hydroxychromones or hydroxyflavonoids, *C*-prenylation often proceeds simultaneously but when 5,7-dihydroxy-2-methylchromone (**539**) is treated with

prenyl (3-methylbut-2-enyl) bromide and potassium carbonate in acetone, *O*-alkenylation predominates ⟨51HCA186⟩. Using sodium methoxide as base, or 2-methylbut-3-en-2-ol and boron trifluoride etherate often alters the ratio of *C*- to *O*-alkenylated products or even the position at which substitution occurs ⟨71IJC1322⟩.

Several reagents are available for *O*-dealkylation; the long established hydriodic acid is still occasionally useful but the possibility of causing a Wessely–Moser rearrangement (Section 2.23.8.3.1) (an example is given in Scheme 31) with this or other mineral acid must be considered ⟨64CPB307⟩. Reagents which are unlikely to cause simultaneous rearrangement are aluminum chloride, magnesium iodide or boron trichloride. It is easier to demethylate a 5-methoxy group than any other. In a recent synthesis of tabularin (**540**), 7-*O*-benzyl and 5-*O*-methyl groups were simultaneously removed by boron trichloride–dichloromethane at 0 °C ⟨78T1593⟩.

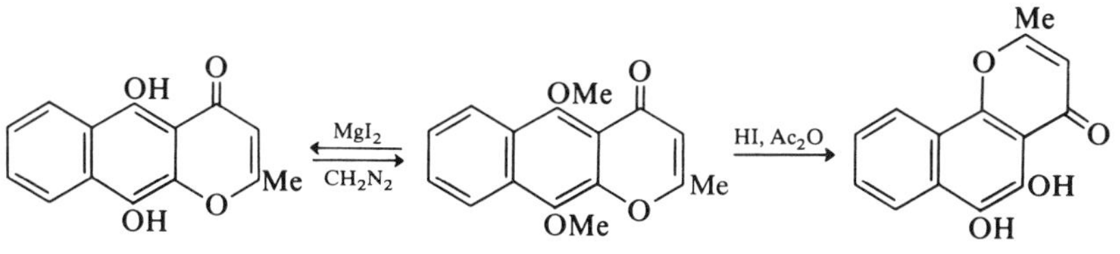

**Scheme 31**

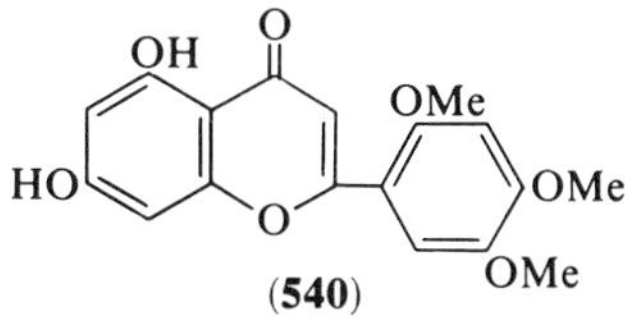

(**540**)

### 2.23.9.6 *S*-Linked Substituents

A recent report states that a 2-sulfinyl group serves as a good leaving group in the introduction of an amino function into the chromone ring system. 2-Ethylsulfinyl-5,8-dimethoxychromone (**541**) reacts at 20 °C with primary or secondary alkylamines to give the 2-alkylamino-5,8-dimethoxychromones (**542**) ⟨81JHC679⟩.

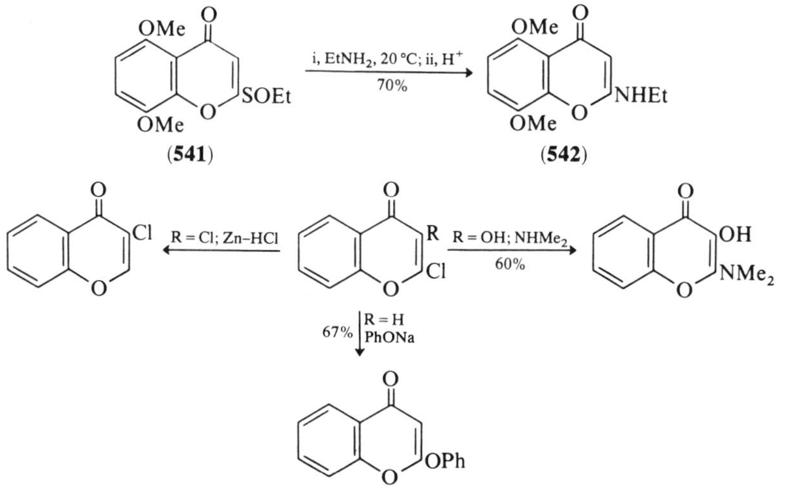

**Scheme 32**

### 2.23.9.7 Halogens

Replacement of halogens attached to a pyran-4-one ring has only recently been studied. The lability of a 2-chloro atom in the chromone ring has been known since 1960. It is readily displaced by nucleophiles and reductively displaced by hydrogen (Scheme 32) ⟨60CR(250)2819, 75AP385, 77H(6)1581⟩.

A halogen at C-3 is less reactive than one at C-2 but it may be displaced by nucleophiles. When 3-bromo-2-methylchromone is warmed with piperidine, the 3-piperidino derivative is obtained, but it may not be by direct displacement ⟨52JA3999⟩. More recently, it was found that ethyl 3-chlorochromone-2-carboxylate (543) reacts rapidly and at 0 °C with ammonia to give the amino carboxamide (544). It is difficult to believe that the ring opens and closes again under these conditions and the observation that a 3-bromo atom is more easily substituted suggests that a nucleophilic substitution mechanism is involved ⟨74JCS(P1)2570, 83JCS(P1)1705⟩.

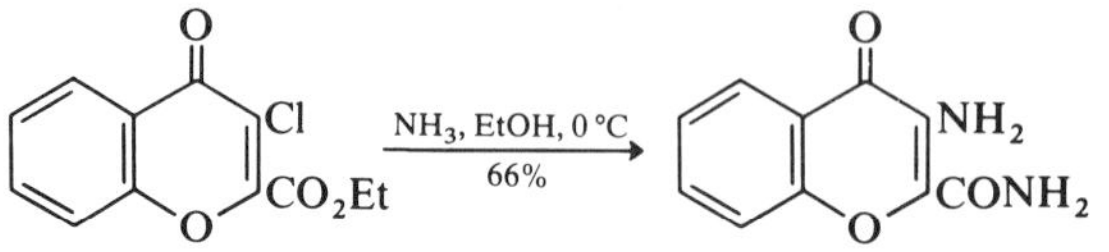

A bromine atom on the benzene ring of chromones takes part in the Rosenmund–von Braun reaction to give moderately good yields of the nitrile, for example ethyl 8-cyano-6-methylchromone-2-carboxylate (545) ⟨72JCS(P1)779⟩. Iodochromones and iodoflavones undergo the Ullmann reaction to form bis-compounds, some of which occur naturally or are related to such biflavonoids ⟨B-75MI22302, 72JMC583⟩.

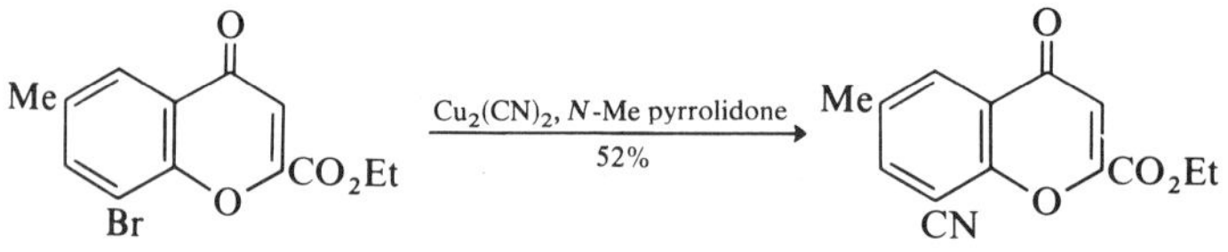

## 2.23.10 REDUCED PYRANS AND PYRANONES AND THEIR BENZO DERIVATIVES: REACTIVITY AT RING ATOMS

### 2.23.10.1 Introduction

Compounds which contain a pyran or pyranone ring with or without one double bond are very numerous. Many of these occur naturally and some have biological importance; for example, parasorbic acid (546) is present in the berries of mountain ash, rosellinic acid (8-hydroxy-2-methyl-4-oxochroman-6-carboxylic acid) is a plant growth inhibitor produced by a bacterium, vitamin E (547) is an antioxidant, hesperitin (548) is found as its 7-rutinoside in citrus fruits, catechin (549) is one of many such compounds present in the woody parts of trees, $\Delta^1$-tetrahydrocannabinol (550) is a constituent of marijuana, and rotenone (551) is a natural insecticide ⟨63FOR275, 66MI22300⟩. Several complex antibiotics contain a tetrahydropyran ring ⟨B-80MI22302, B-81MI22301, B-82MI22302⟩.

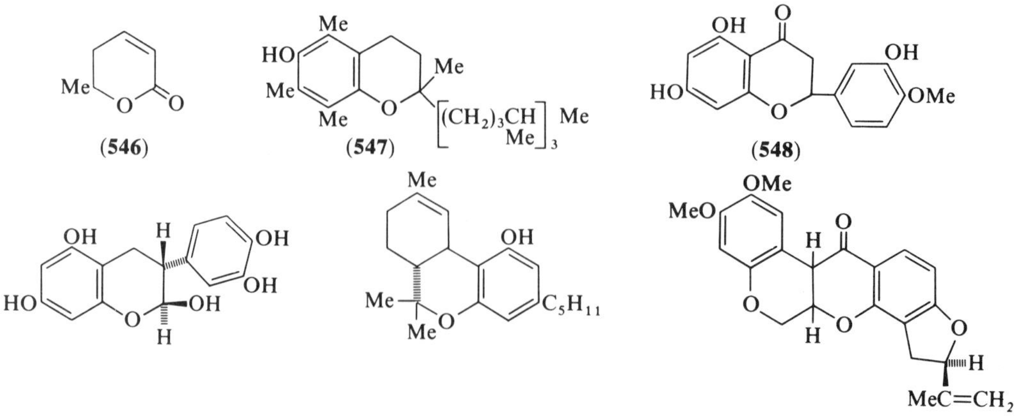

Several pyran-4-ones, *e.g.* maltol, are used as flavouring agents and 6-hydroxy-2,5,7, 8-tetramethylchroman-2-carboxylic acid is a potent synthetic antioxidant ⟨75MI22303⟩.

2,3-Dihydropyran reacts with hydroxyl (including phenolic) groups to form the 2-tetrahydropyranyl ether, which is alkali-stable but acid-labile and useful as a protecting group ⟨B-81MI22300⟩.

Most of the carbonyl compounds discussed in this section are either pyran-2- or -4-ones but a small number of pyran-3-ones and chroman-3-ones are known ⟨77HC(31)193⟩. Tetrahydropyran-2-ones are lactones of 5- (or δ-) hydroxycarboxylic acids and are not discussed in detail.

### 2.23.10.2 Thermal and Photochemical Reactions

Irradiation of 2-alkoxytetrahydropyrans (553) in non-aqueous solvents converts them into the lactone (554), the yields being higher when propargyl or allyl groups are present. A 6-methoxy substituent alters the course of the photolysis to give an open chain ester (552) ⟨77BSF101⟩.

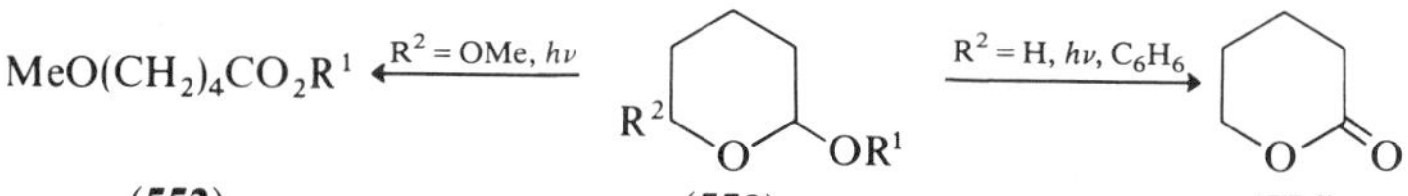

$$MeO(CH_2)_4CO_2R^1 \xleftarrow{R^2 = OMe,\ h\nu}$$

(552)   (553)   (554)

When 3,4-dihydro-2*H*-pyran (555) is given pulses of a laser beam, it decomposes by a retro-Diels–Alder reaction into acrolein and ethylene ⟨78JA6111⟩. 4-Methyl-5,6-dihydro-2*H*-pyran (555a) when irradiated in methanol through which oxygen is passed gives a mixture of four cyclic products on treatment with sodium borohydride ⟨79JCS(P1)1806⟩. Pyrolysis of the dihydropyran (555) at 350 °C yields butadiene.

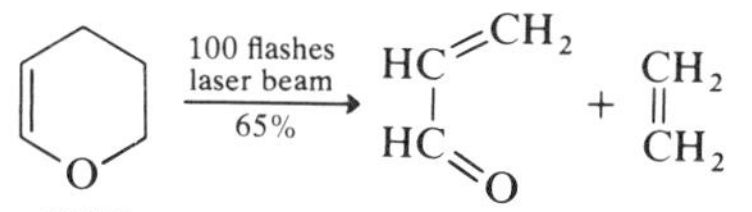

(555)

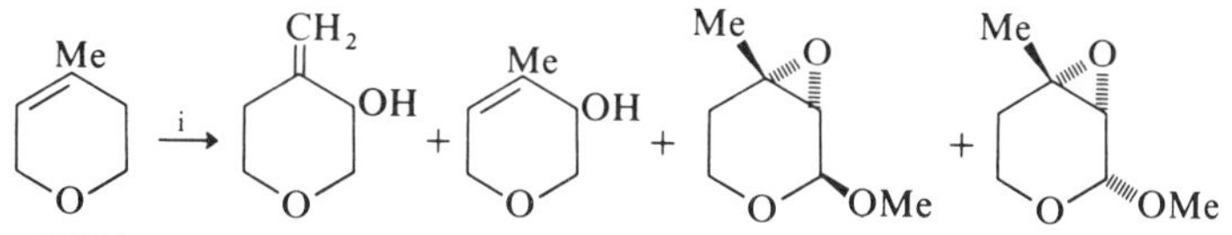

(555a)

i, *hν*, O₂, MeOH, Rose Bengal, 20 °C, then NaBH₄

6-Aryl-3,4-dihydro-2*H*-pyrans (556) are slowly oxidized by oxygen in the dark, but incandescent light greatly accelerates the process, especially when a small amount of haematoporphyrin is present as a sensitizer. 3-Formylpropyl 4-methoxybenzoate (558) was the only product and is probably formed *via* the dioxetan (557) rather than a hydroperoxide (558a) of the kind which is formed in the photolysis of dihydropyran ⟨71JCS(C)784⟩.

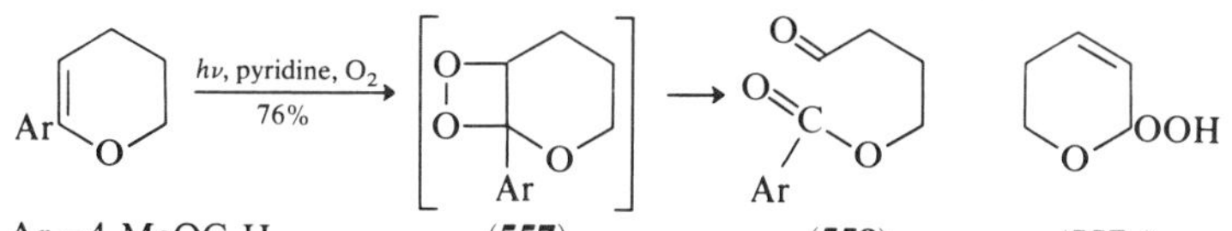

(556) Ar = 4-MeOC₆H₄   (557)   (558)   (557a)

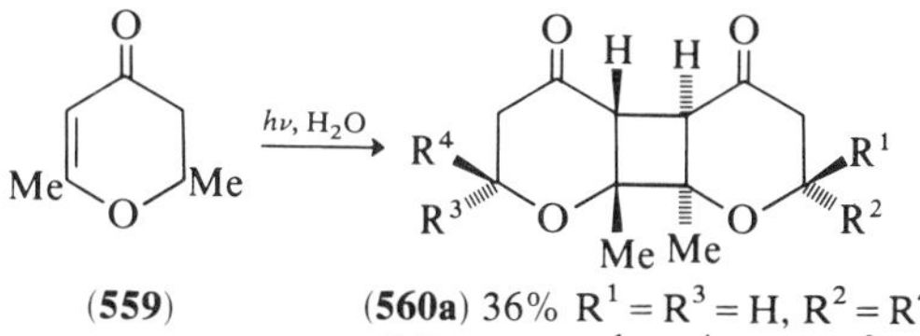

(559)   (560a) 36% R¹ = R³ = H, R² = R⁴ = Me
(560b) 60% R¹ = R⁴ = Me, R² = R³ = H
(560c) 4% R¹ = R³ = Me, R² = R⁴ = H

Irradiation of reduced pyran-4-ones frequently gives dimeric products which sometimes revert to the monomer. 2,6-Dimethyl-2,3-dihydropyran-4-one (559) in water gives a 96% yield of a mixture of three dimers (560) in a reaction which is reversible ⟨73CJC1267⟩.

3,3-Dimethylpyran-2,4-dione (**561**) yields only one photodimer (**562**) ⟨73T1317⟩, and the dipyran (**563**) suffers photodegradation on irradiation ⟨80JHC45⟩. Increasing substitution appears to bring greater photostability to reduced pyran-4-ones; 2,3-dihydro-2,2,5-trimethylpyran-4-one (**564**) is only slightly affected by irradiation as only a small amount of the dimer (**565**) is formed ⟨77BSF911⟩.

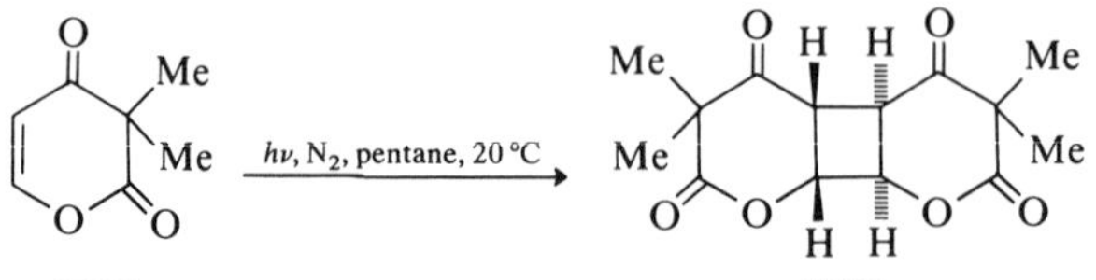

(**561**) (**562**)

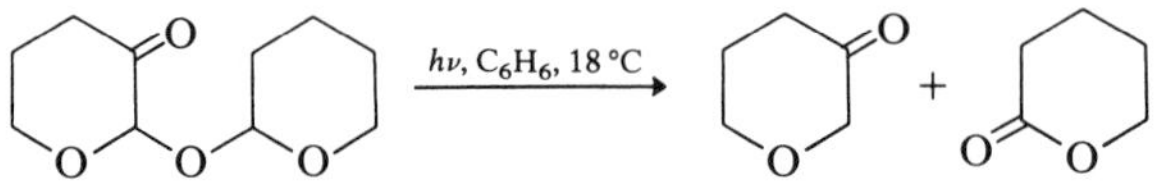

(**563**)

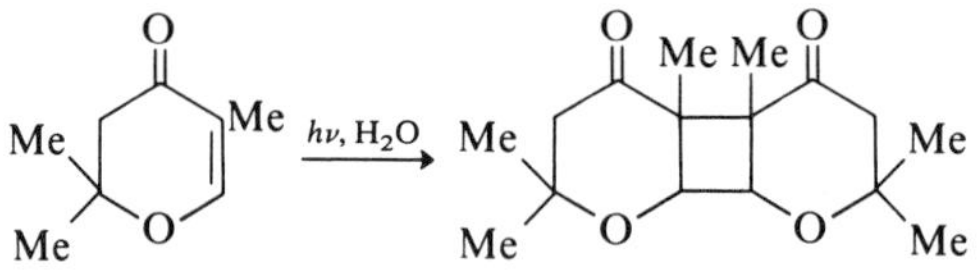

(**564**) (**565**)

α-Tocopherol (vitamin E; **566**; $R = C_{16}H_{33}$) is an antioxidant and its sensitivity to photooxidation is of considerable interest. When irradiated with light from an incandescent lamp in methanol in the presence of oxygen or nitrogen at 25 °C, there is virtually no reaction, but when a dye (proflavin) is present, photolysis occurs and four products are obtained, one of which (**567**) consists of four stereoisomers (Scheme 33) ⟨72JA866⟩.

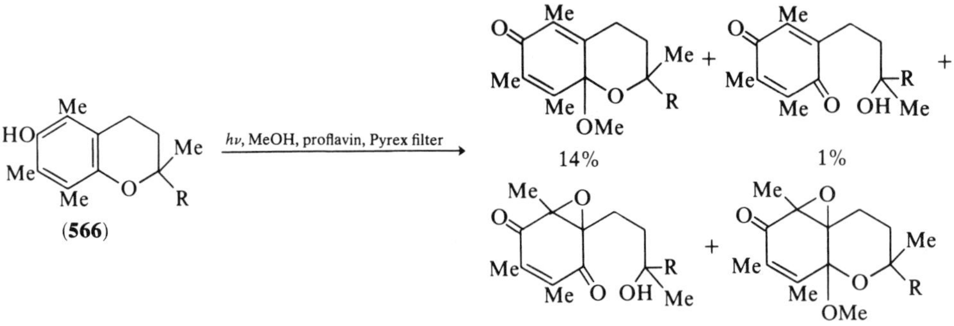

(**566**) 14% 1% 6% (**567**) 34%

**Scheme 33**

Photolysis of flavan-3-ols gives, after methylation, non-pyran products (**568**) and (**569**), but 2,3-*trans*-3-hydroxyflavan-4-ones give the *cis*-isomer (**570**, 22%) and a small amount of a rearranged product 3′,4′,7-trimethoxyisoflavone (**571**), which is formed *via* the spirodienone (**572**) ⟨77JCS(P1)125⟩. Flavanone (**573**), however, behaves differently when irradiated in benzene; 2′-hydroxychalcone (**574**) is formed by ring opening and the other product 3,4-dihydro-4-phenylcoumarin (**575**) is the result of a Fries-like rearrangement with the mechanism shown ⟨72CC451⟩.

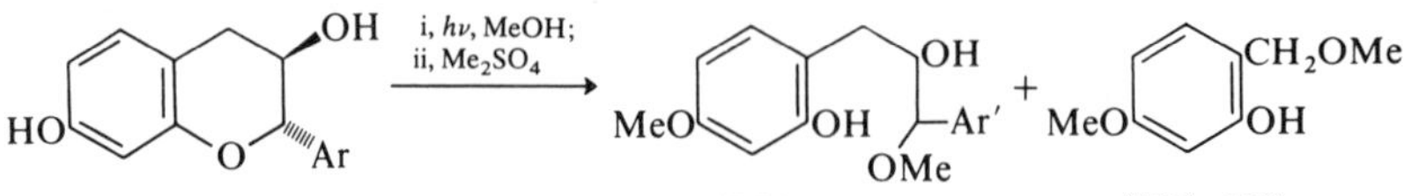

(**568**) 40% (**569**) 28%

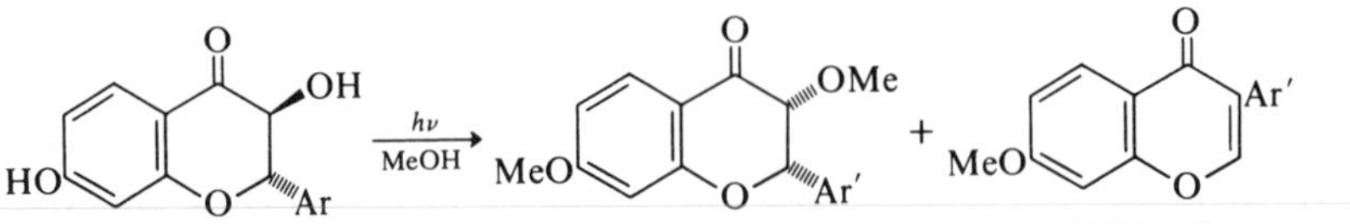

(**570**) 22% (**571**) 8%

Ar = 3,4-$(HO)_2C_6H_3$
Ar′ = 3,4-$(MeO)_2C_6H_3$

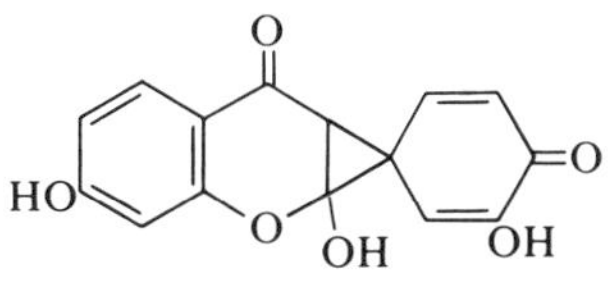

**(572)**

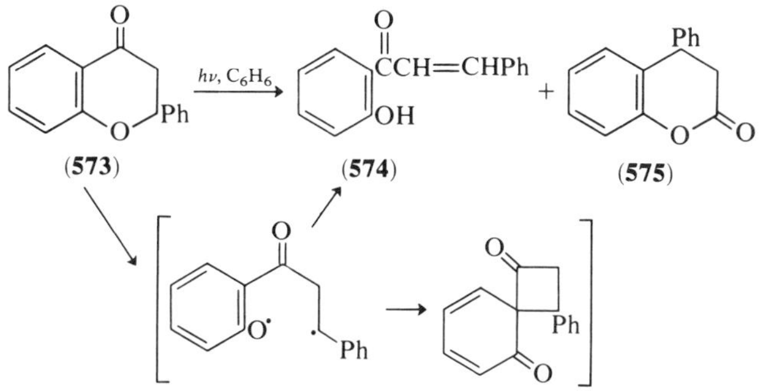

**(573)** $\xrightarrow{h\nu,\ C_6H_6}$ **(574)** + **(575)**

Irradiation of several members of the cannabinoid family has provided useful routes to some derivatives. Cannabichromene **(576)** is converted into cannabicyclol **(577)** when it is irradiated in *t*-butyl alcohol–acetone ⟨71JCS(C)796⟩ and tetrahydrocannabinolic acid **(578)** is aromatized to cannabinolic acid **(579)** ⟨70CPB1327⟩.

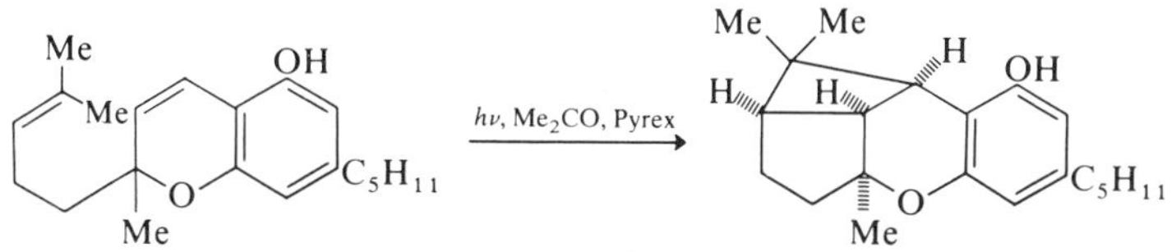

**(576)** $\xrightarrow{h\nu,\ Me_2CO,\ Pyrex}$ **(577)**

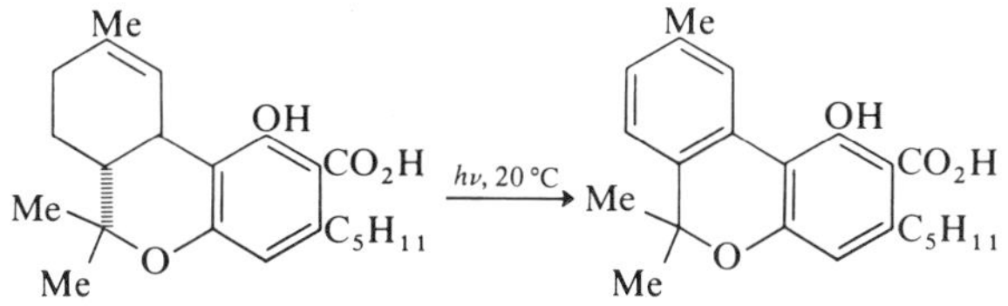

**(578)** $\xrightarrow{h\nu,\ 20\ °C}$ **(579)**

Thermal reactions are important in cannabis chemistry; for example, cannabichromenic acid **(580)** loses carbon dioxide on heating to give cannabichromene **(581)** ⟨68CPB1157⟩. Such reactions together with NMR spectroscopy have contributed to the determination of structure of many new cannabinoid compounds and are important also because many of the acids are biologically inactive but may be converted into active cannabinoids on decarboxylation, which may occur during the drying of marijuana or even during prolonged storage in a warm place. It also occurs during GLC examination under normal conditions (but not on esters). Another change which may occur when $\Delta^1$-tetrahydrocannabinol **(582)** is heated is its isomerization into the $\Delta^6$-isomer **(583)** and the importance of this during cannabis smoking is uncertain but it occurs *in vitro* on heating at 230 °C. $\Delta^1$-Tetrahydrocannabinol **(582)** is converted into the biologically inactive cannabinol **(584)** during smoking.

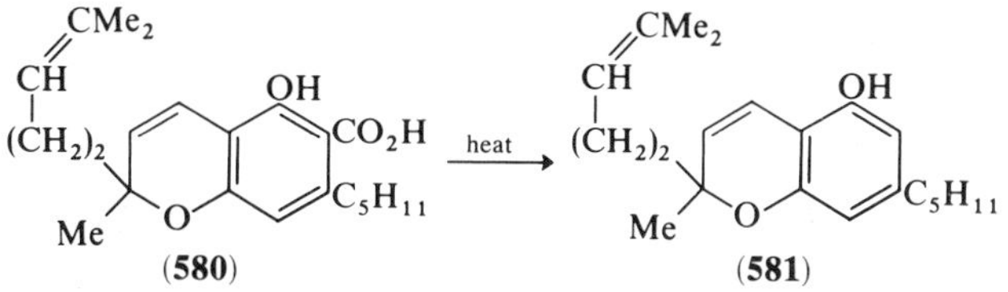

**(580)** $\xrightarrow{heat}$ **(581)**

**(584)** $\xleftarrow{heat}$ **(582)** $\xrightarrow{heat}$ **(583)**

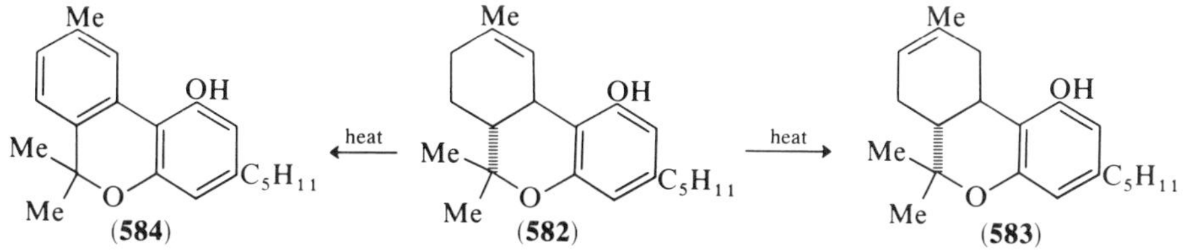

A reaction which involves both ring atoms and substituent is of considerable interest because the substrate is of a naturally occurring class of compounds. 3-Arylidene-chromanones (585) or homoisoflavanones, when heated with nickel in xylene isomerize to the 3-benzylchromone (586) ⟨74IJC281⟩; a number of homoisoflavanones occur in the bulbs of *Eucomis bicolor*, for example, eucomin (585; R = OH, Ar = 4-MeOC₆H₄). The stereochemistry about the double bond of compound (585; R = MeO) is altered from (*E*) to (*Z*) by irradiation at 300–400 nm ⟨81FOR(40)105⟩.

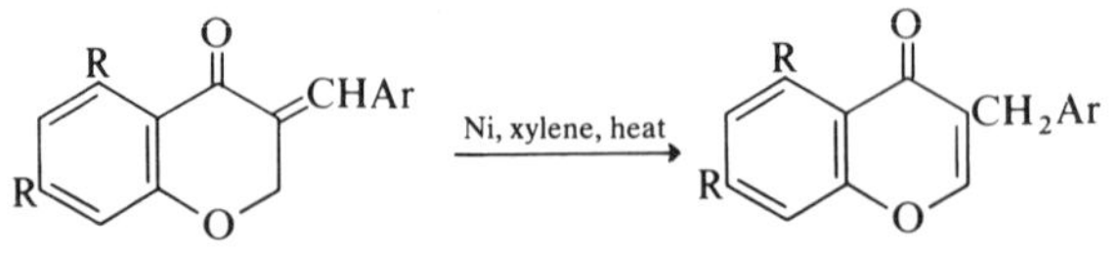

(585)     (586)

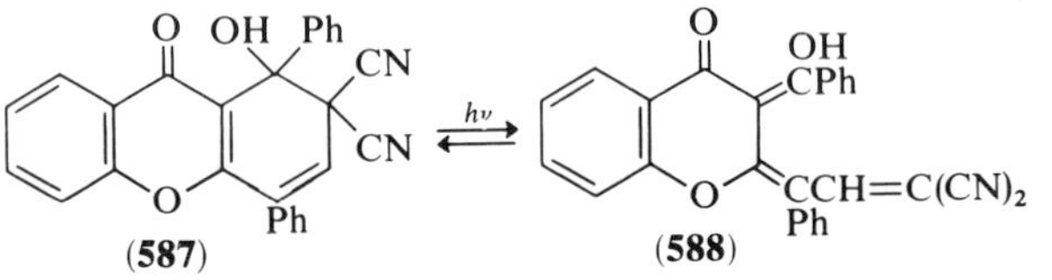

(587)     (588)

The photochromic behaviour of the reduced xanthone (587) when irradiated has been studied. The yellow or orange colour thus produced is due to ring opening to the chromanone (588). This colour fades gradually as the reverse reaction proceeds ⟨69JOC2407⟩.

When chroman is heated on alumina at 250–350 °C, 2-methyl-2,3-dihydrobenzo[*b*]furan is formed, the amount of product increasing with temperature. At the higher end of this temperature range, 2-methyl-, 2-ethyl- and 2-propyl-phenols are formed ⟨71MI22302⟩. At 300–400 °C in the presence of activated carbon, some 2-methylbenzo[*b*]furan (589) is also produced ⟨75CHE278⟩.

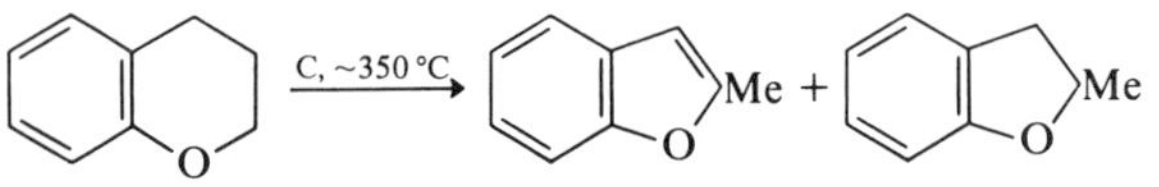

(589)

Few pyran-3-ones are known but 6-acetoxy-2,6-dihydropyran-3-one (590) dimerizes at room temperature in the presence of a base to give the diketone (591); on heating this at 140 °C, it isomerizes to dimer (592). This has a pair of each functional group but, as it is possible to react one of these selectively, the compound is valuable as a synthon ⟨80JOC3361⟩.

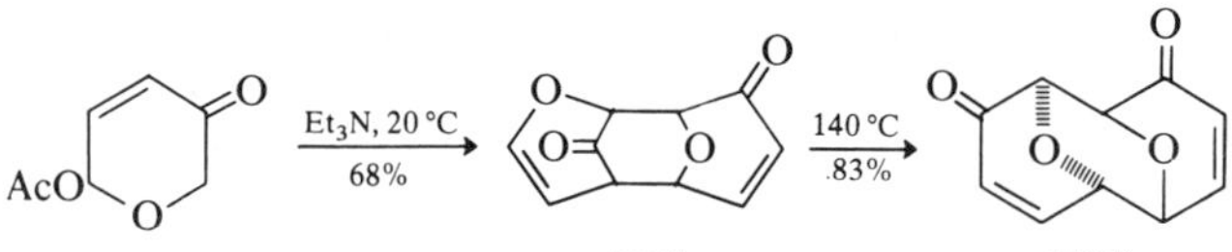

(590)     (591)     (592)

### 2.23.10.3 Electrophiles

#### 2.23.10.3.1 Addition and substitution

Halogens and hydrogen halides add to the double bond of most dihydropyrans; for example, 3,4-dihydro-2*H*-pyran (593) gives 2,3-dichlorotetrahydropyran ⟨76OS(55)62⟩. Chromanones show some of the properties of cyclohexanone. Halogenation gives 3-mono- or 3,3-di-halogeno derivatives, condensation with benzaldehydes yields 3-benzylidene derivatives, and the Mannich reaction proceeds normally at C-3. In a related reaction, formaldehyde in acid solution forms a spiran (594) ⟨79BCJ1169⟩.

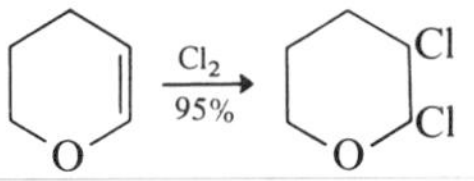

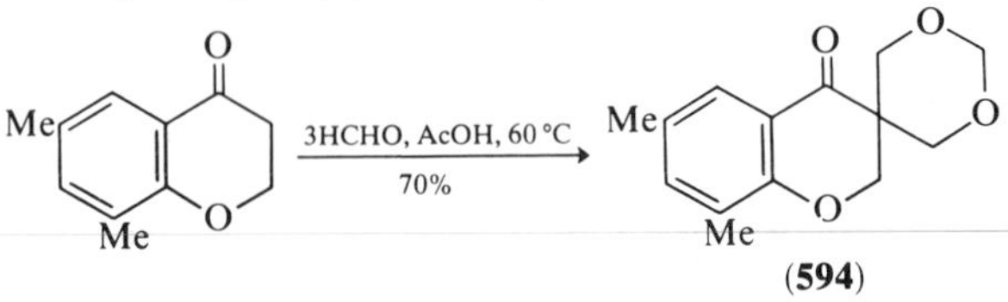

(593)     (594)

The carbanion formed when chromanone is treated with a base such as sodium methoxide reacts with ethyl formate to give the 3-hydroxymethylene derivative (**595**) and such compounds have been used as intermediates ⟨72JHC1341, 70IJC203⟩. When treated with isopentyl nitrite, chromanones usually give the 3-isonitroso derivative but this reaction occasionally behaves anomalously without obvious reason. For example, 7-methoxychromanone is readily converted into its 3-isonitroso derivative (**596**), but with 7-hydroxychromanone the reaction fails ⟨77HC(31)207⟩.

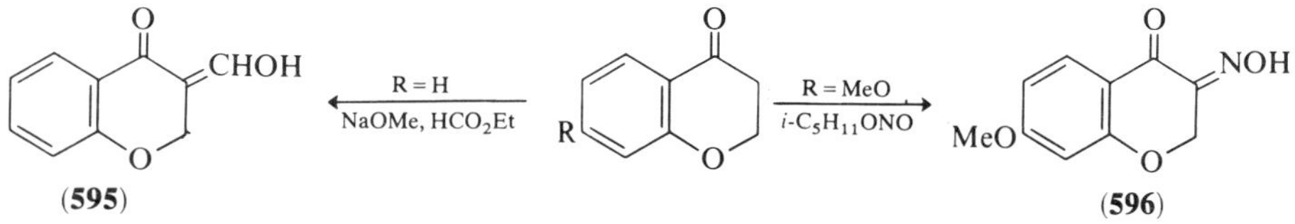

(**595**)    (**596**)

4-Hydroxy-6-methyl-5,6-dihydropyran-2-one (**597**) is brominated by NBS at C-3 ⟨78JHC1153⟩. More than one product is often obtained in this type of reaction, for example from 2,3-dihydro-4*H*-pyran (**593**), but in acetic acid this reaction gives 2-acetoxy-3-bromotetrahydropyran (**599**) ⟨58JOC1128⟩. 2,3-Dihydro-4*H*-pyran reacts normally with hydroboration reagents to give tetrahydropyran-3-ol in 80% yield ⟨70JOC2282⟩.

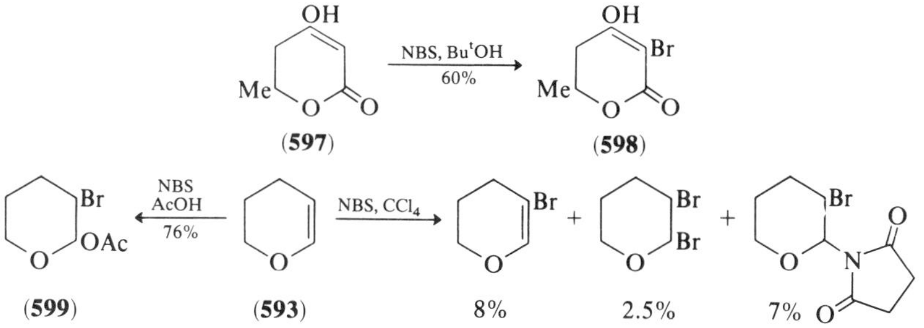

(**597**)    (**598**)

(**599**)    (**593**)    8%    2.5%    7%

Tetrahydropyran-4-ones, for example diethyl 6-methyl-4-oxotetrahydropyran-3,3-dicarboxylate (**600**), are brominated at C-3 ⟨77JCS(P1)1647⟩. Chromanones are readily halogenated at C-3 (see above); when they are treated with phosphorus pentachloride, the reaction does not stop at monochlorination; the carbonyl group is attacked and 3,4-dichloro-2*H*-chromene (**601**) is formed, an uncommon conversion of a chromanone into a chromene ⟨79TL3901⟩. Chroman yields 4- and 6-bromides when treated with NBS–benzoyl peroxide.

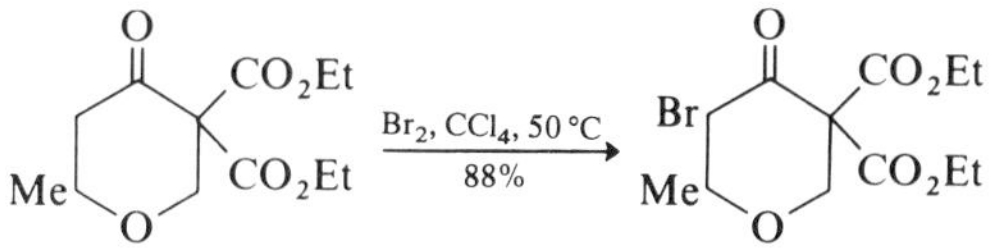

(**600**)

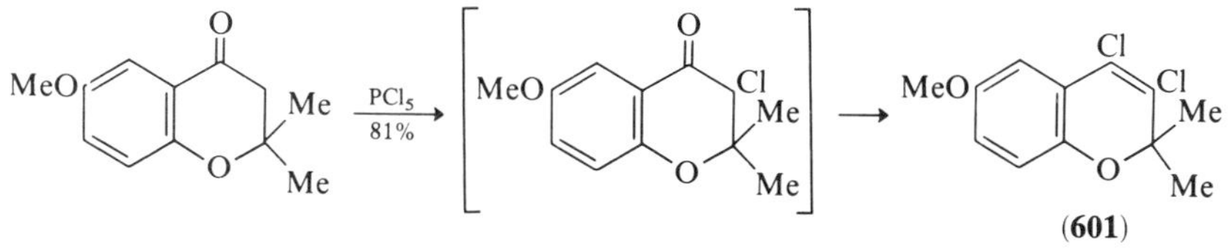

(**601**)

The pyran-3-one (**602**) is readily halogenated in the ring and the resulting chloride (**603**) loses the halogen on heating to yield maltol (**604**).

Hot mineral acids open dihydropyran rings and recyclization to a benzene ring is feasible when a suitable substituent is present, as for example in 6-ethyl-3-vinyl-2,3-dihydro-4*H*-pyran (**605**) ⟨74T1015⟩.

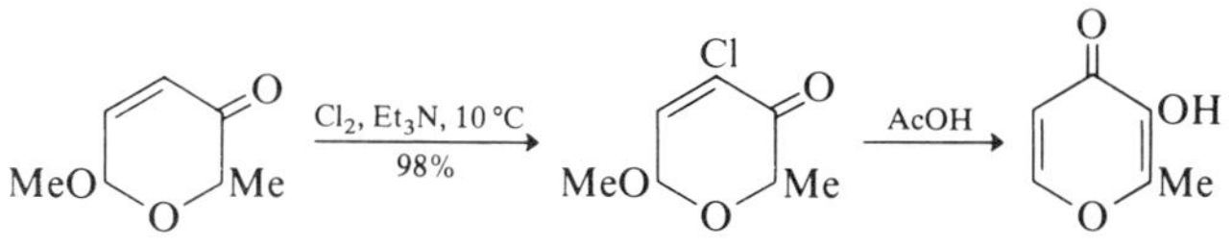

(**602**)    (**603**)    (**604**)

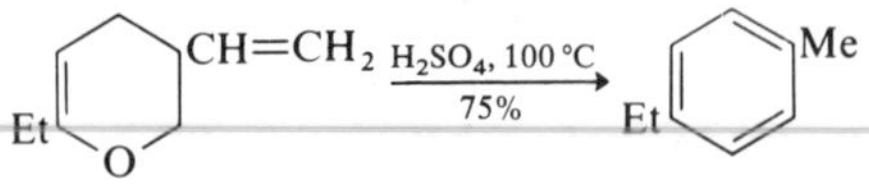

(605)

### 2.23.10.3.2 Oxidation

2,3-Dihydro-4*H*-pyrans are oxidized by several reagents as shown in Scheme 34. The hydroperoxide obtained by ozonolysis may be converted into the ester or aldehyde ⟨62JOC4498⟩.

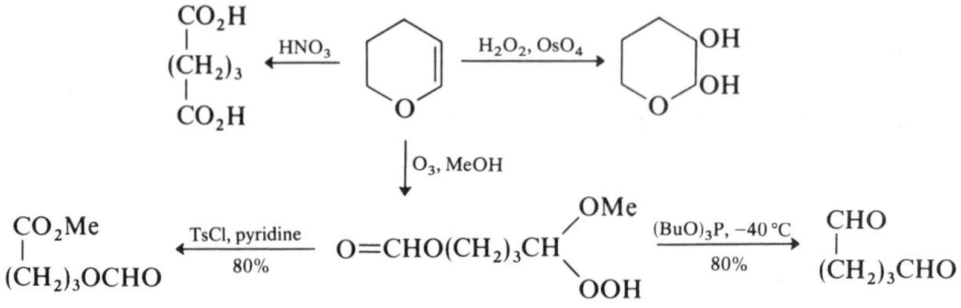

**Scheme 34**

Dehydrogenation of tetrahydropyran-4-ones to the dihydro analogues is effected by heating at 130 °C in DMSO ⟨77JCS(P1)1647⟩. 5,6-Dihydropyran-2-one (606) is efficiently converted into pyran-2-one in one step by radical bromination and dehydrobromination (70% yield) ⟨77OS(56)49⟩. Chromanones are stable to many oxidizing agents but may be dehydrogenated to chromones, for example 2,3-dihydronaphtho[2,1-*b*]pyran-1-one (607) to the naphthopyranone ⟨60CI(L)1192⟩. Flavanones are converted into flavones by bromination–dehydrobromination, by iodine–acetic acid followed by dehydroiodination with potassium acetate, or by selenium dioxide as in the synthesis of the biflavone (608) ⟨70AP428⟩.

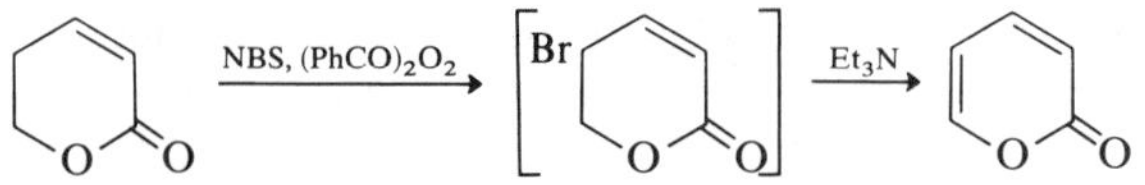

(606)

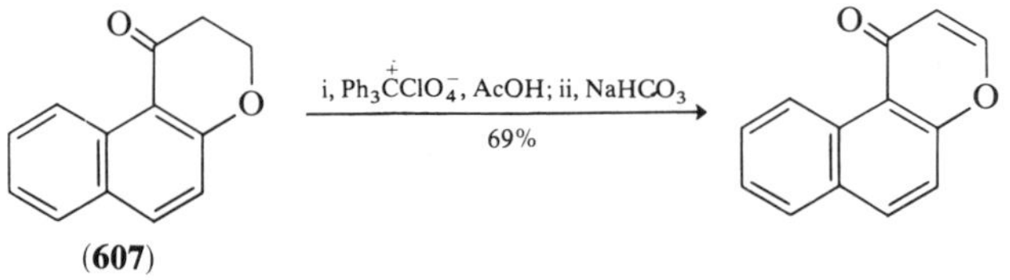

(607)

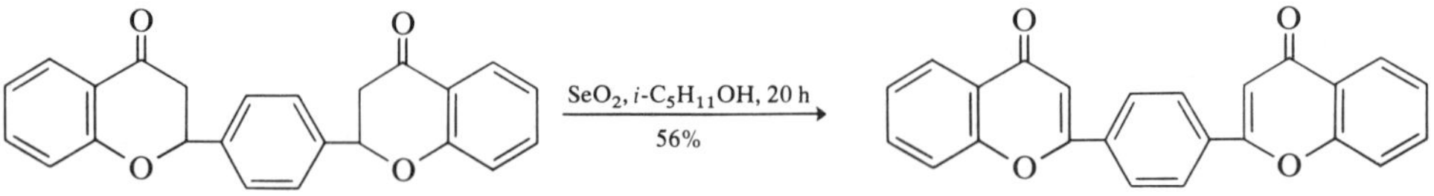

(608)

Oxidation of chromans has been extensively studied, especially of those which resemble the tocopherols which act as antioxidants in nature and in food. The behaviour of α-tocopherol (vitamin E, **547**) and many simpler model compounds under oxidizing conditions varies with the oxidizing agent. Among the many products formed are the *o*- and *p*-quinones (609) and (610), the spiran (611) and dimer (612). A detailed account is available ⟨81HC(36)59⟩.

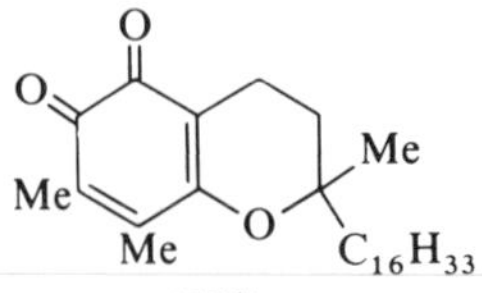

(609) 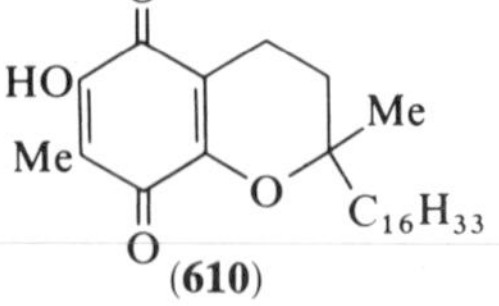                    (610)

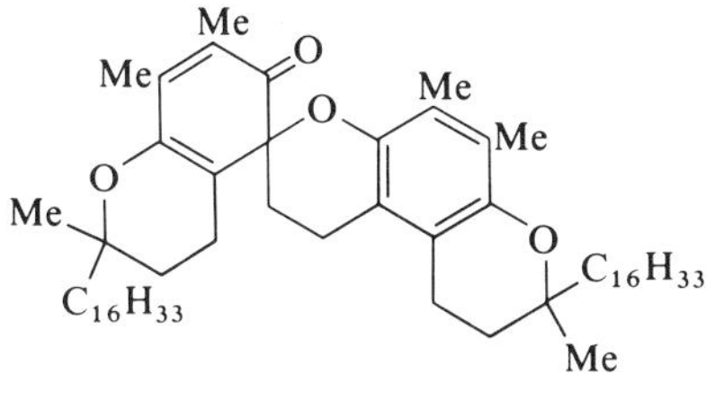

(**611**) 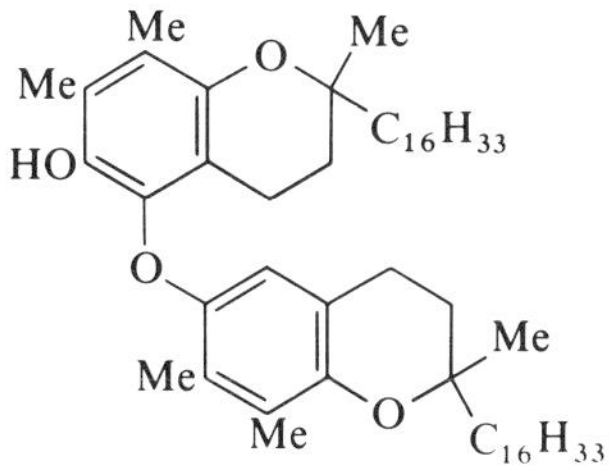                                    (**612**)

A synthetic antioxidant, 6-hydroxy-2,5,7,8-tetramethylchroman-2-carboxylic acid (**613**) is oxidized by iron(III) chloride or alkaline potassium ferricyanide to the quinone (**614**) and lactone (**615**) ⟨75MI22303⟩.

Pyran-3-ones are less numerous than the 2- and 4-isomers. The epoxidation of 6-ethoxy-2-methyl-6*H*-pyran-3-one (**616**) is achieved at low temperature and the product is rearranged with ion exchange resin to maltol (**617**) ⟨76TL1363⟩.

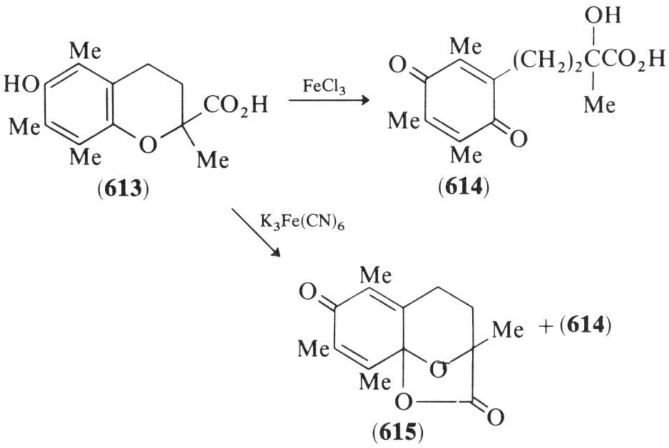

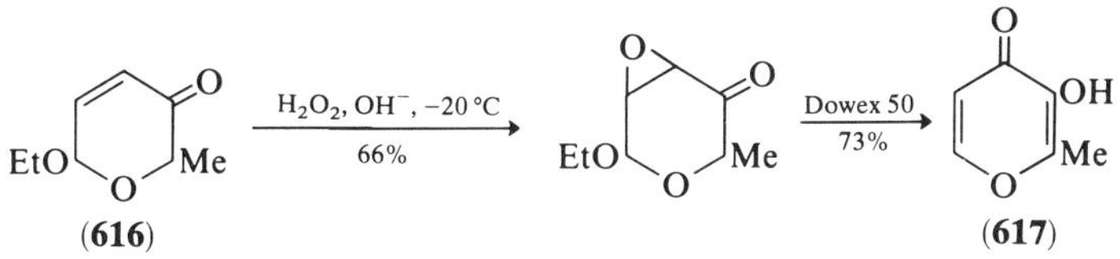

(**613**)                    (**614**)

(**615**)

(**616**)                    (**617**)

### 2.23.10.3.3 Carbenes

Carbenes add to the double bond of dihydropyrans, as they do to benzopyrans, to form cyclopropapyrans. For example, dichlorocarbene reacts with 2,3-dihydro-4*H*-pyran (**593**) to give the bicyclic pyran (**618**) which on heating at reduced pressure with quinoline undergoes ring expansion to the 3-chlorooxepine (**620**); the two chlorine atoms are removed by sodium–ammonia ⟨60JA4085⟩. The reaction of dihalocarbenes with pyrans and benzopyrans has been reviewed ⟨63OR(13)55⟩.

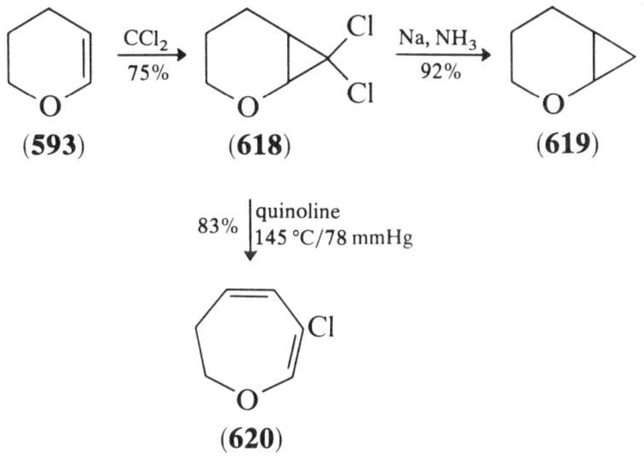

(**593**)          (**618**)          (**619**)

(**620**)

### 2.23.10.4 Nucleophiles

#### 2.23.10.4.1 C-Nucleophiles

The carbonyl group of tetrahydropyran-4-ones, chroman-3-ones and chroman-4-ones behaves like an aliphatic carbonyl function towards organometallic compounds. With ethyl 2-chloropropionate and sodium, 2,2-dimethyltetrahydropyran-4-one gives the spiran (621) ⟨79MI22303⟩. Reformatsky-type reagents attack 2-*t*-aminotetrahydropyrans to give 7-hydroxyheptanoates or the corresponding nitriles (622; R = CO₂Et or CN) in good yield ⟨77BSF337⟩.

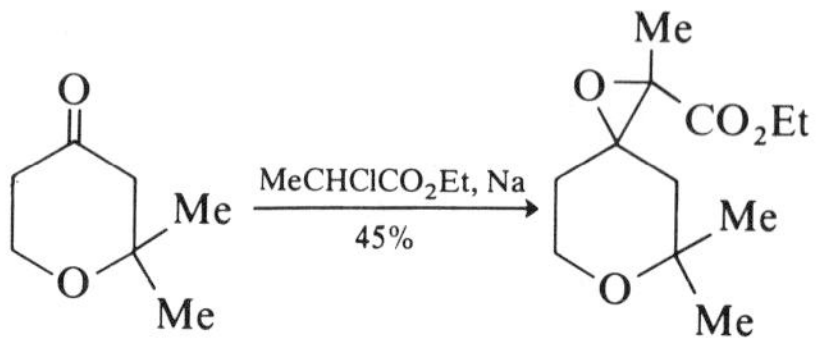

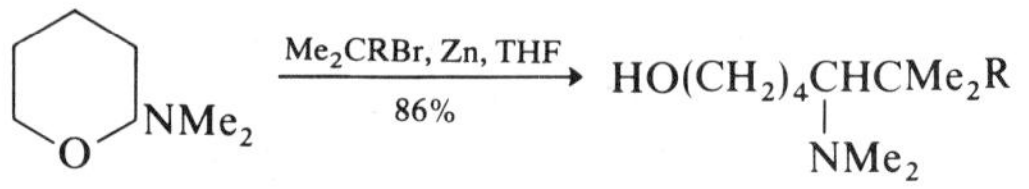

Chromanone and its isomer isochroman-4-one (623) add ethyl diazoacetate at the carbonyl group and the resulting compound undergoes ring enlargement on treatment with mineral acid. The benzoxepin-4(3H)-one (624) is obtained on heating with acid ⟨79JCR(S)142⟩.

The anion of 2,2-dimethyl-5-hydroxy-7-pentylchromanone (625) adds but-3-en-2-one to give a 1:1 mixture of cannabichromanone (626) and the spiran (627) both of which are devoid of action on the central nervous system ⟨81H(16)1899⟩.

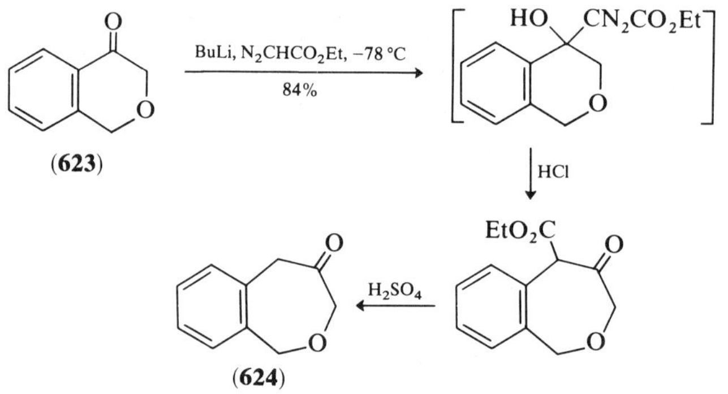

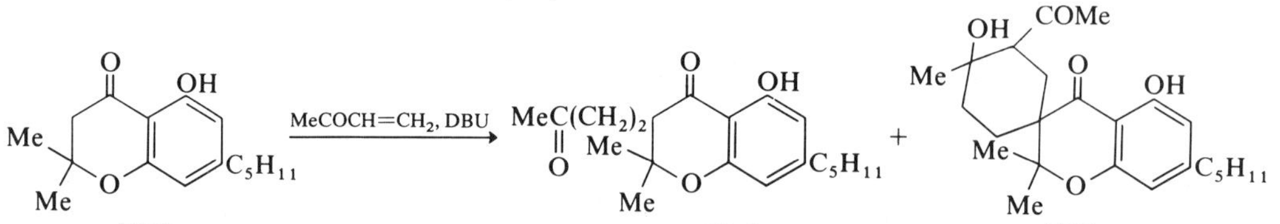

#### 2.23.10.4.2 N-Nucleophiles

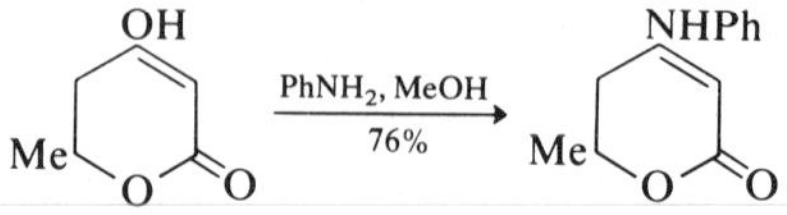

The reaction of ammonia and amines with pyrylium salts and pyranones to form pyridines has a parallel in the dihydropyran-2-ones. 4-Hydroxy-6-methyl-5,6-dihydropyran-2-one (628) on heating with aniline gives a good yield of the dihydropyrone (629) ⟨78JHC1153, 81JHC543⟩ but the presence of another reactive group in the pyran-2-one can give rise to more than one product. Thus, ethyl 5,6,6-trimethyl-2-oxopyran-4-carboxylate (631) and benzylamine at room temperature give the carboxamide (630), but at 140 °C with two moles of the amine the pyridone (632) is the main product ⟨78CHE368⟩. Radioactive

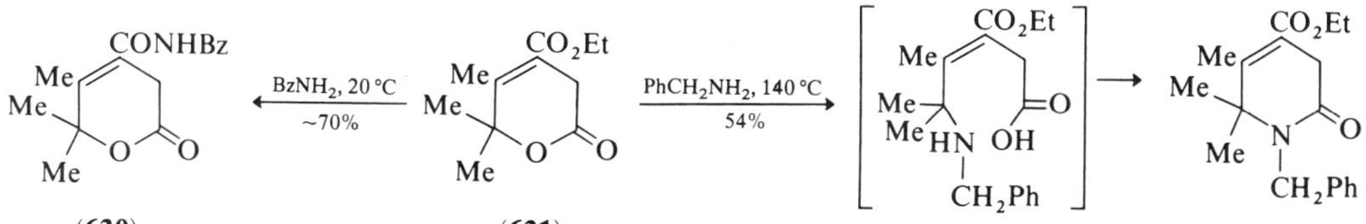

(630) (631) (632)

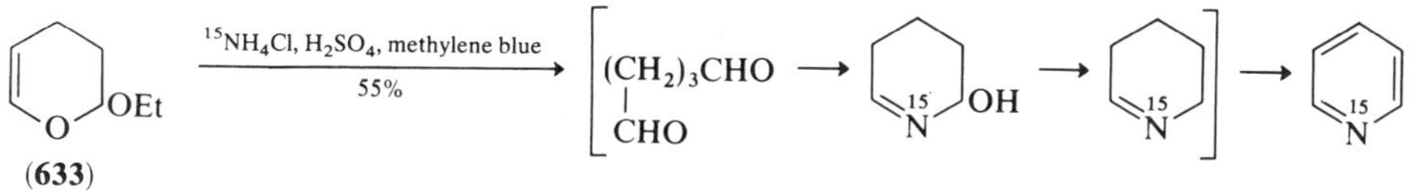

(633)

ammonium chloride converts 2-ethoxy-3,4-dihydro-2*H*-pyran (**633**) into labelled pyridine in the presence of methylene blue as dehydrogenating agent ⟨74MI22303⟩.

Hydrazines react with 2,3-dihydropyran-4-ones such as (**634**) with consequent ring opening and recyclization to a pyrazole; the conditions are mild and the yield is high ⟨78S900⟩. The dihydropyran-2-one (**635**) behaved less predictably and gave, on heating with hydrazine hydrate, the pyridone (**636**) with loss of a nitrogen atom ⟨79JHC1⟩. Phenylhydrazine needs rather more drastic conditions to give the pyrazole (**638**), and since one of its nitrogens is less nucleophilic the phenylhydrazone (**637**) is formed at room temperature, but is converted into the pyrazole (**638**) on heating in ethanolic acetic acid. A trace of acetic acid in the reaction mixture promotes the direct formation of the pyrazole ⟨79JHC657⟩.

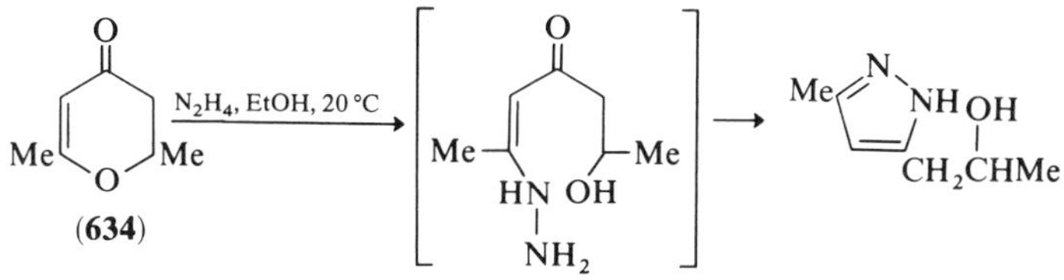

(634)

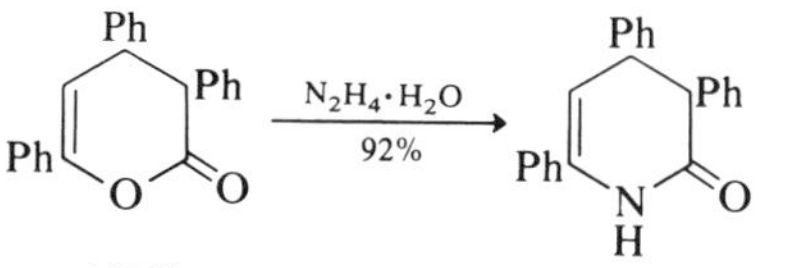

(635) (636)

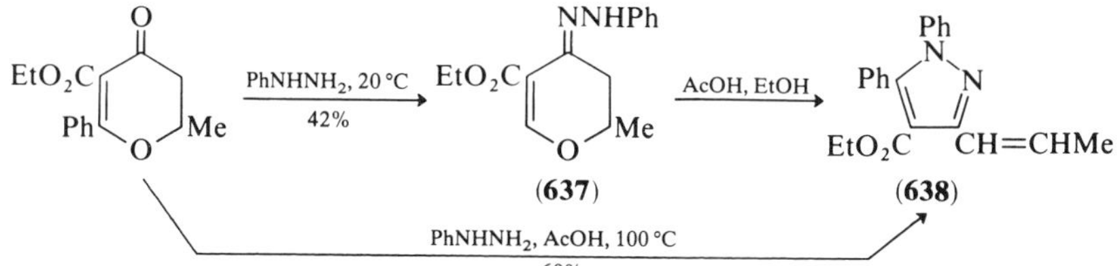

(637) (638)

Chromanones form oximes, hydrazones and semicarbazones normally but flavanone (**639**) can give one of three products with hydrazine according to the conditions (Scheme 35)

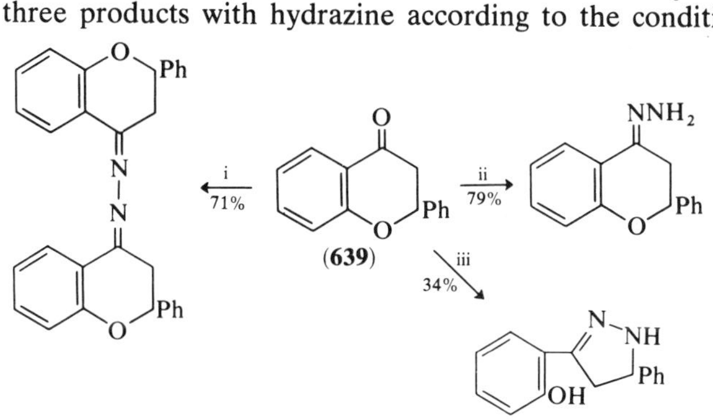

(639)

i, $N_2H_4 \cdot H_2O$(4.6 mol), pyridine, 100 °C; ii, $N_2H_4 \cdot H_2O$ (4 mol), EtOH, 22 °C; iii, $N_2H_4 \cdot H_2O$ (18 mol), EtOH, 100 °C

**Scheme 35**

⟨65T3037⟩. 3-Aroylflavanones (640) offer a choice of sites for attack by hydroxylamine, but the endocyclic carbonyl is preferentially attacked to give the isoxazoline (641) ⟨79IJC(B)510⟩. Oximes (642) of flavanones rearrange in acid solution to the isoxazoline (643) ⟨80H(14)1319⟩.

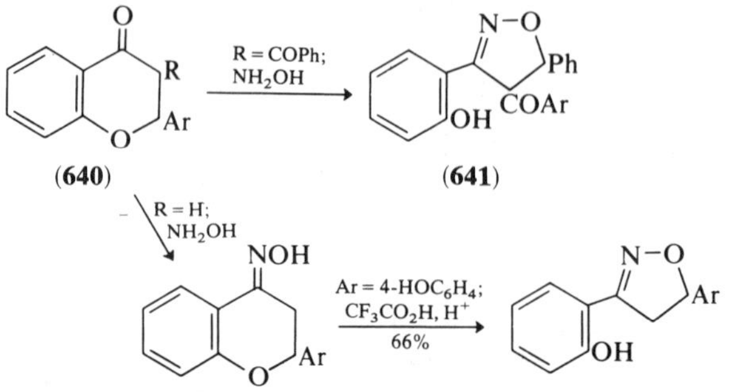

(640)                    (641)

(642)                    (643)

The Schmidt reaction is conducted in strong acid and the essential step is the attack by the nucleophilic hydrazoic acid, $HN-\overset{+}{N}\equiv N$ on the carbonyl group. Chromanone is converted into 1,4-benzoxazepin-5-one (644) but a few chromanones give the 1,5-benzoxazepin-4-one (645) while others produce both types of oxazepines ⟨77HC(31)207⟩.

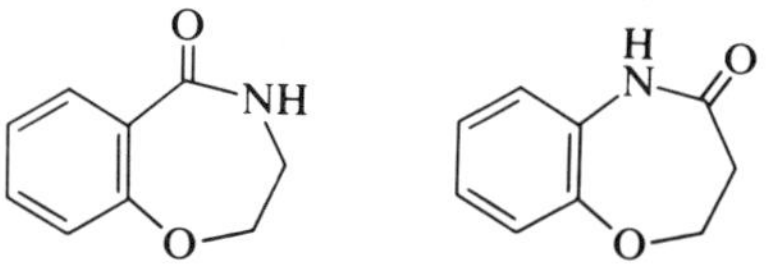

(644)                    (645)

### 2.23.10.4.3 O-Nucleophiles

Susceptibility to attack by nucleophiles such as the hydroxyl ion decreases in general in the order benzopyran-4-one > benzopyran > 2,3-dihydrobenzopyran-4-one > 2,3-dihydrobenzopyran but substituents present may affect this sequence. Tetrahydropyrans are more resistant than pyran-4-ones to aqueous alkali. Flavanones are readily converted into the chalcones (646) by a trace of alkali and so are difficult to obtain in a pure state.

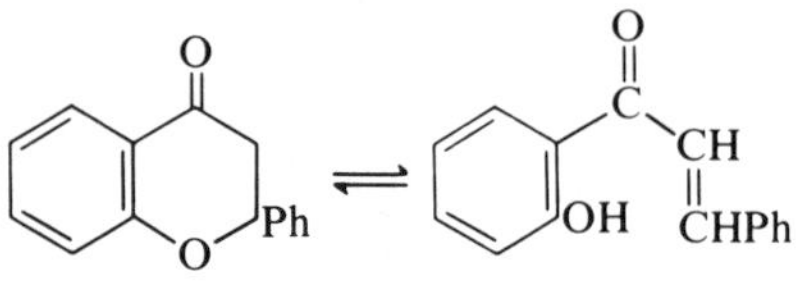

(646)

Chromanones may or may not be attacked by aqueous or ethanolic alkali depending on the substituents. 7-Methoxychromanone is quite stable to alkali but 7-methoxy-2,2-dimethylchromanone (647) is degraded, probably to an acrylophenone (648), which may itself be cyclized to the chromanone by treatment successively with aqueous sodium hydroxide and hydrochloric acid.

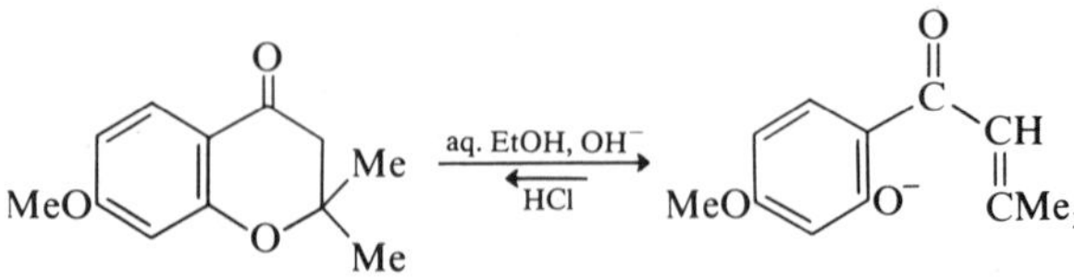

(647)                    (648)

When a homoisoflavanone, for example 3-benzylidenechromanone (649), is heated with a base in DMF, two products are obtained. The main reaction is the migration of the exocyclic double bond to form 3-(3'-hydroxy-4'-methoxybenzyl)chromone (650); a skeletal rearrangement through a ring opening reaction accounts for the formation of the other minor product, 3'-hydroxy-4'-methoxy-3-methylflavone (651). Labelling with $^{14}C$ showed

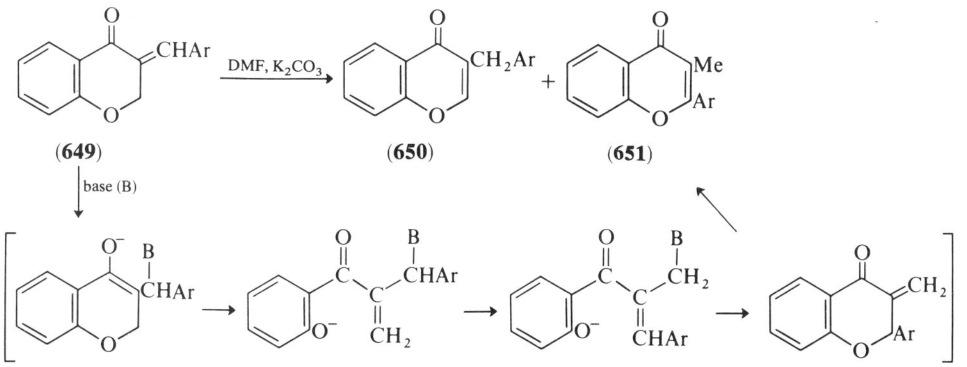

**Scheme 36**

that the methylene carbon atom of the benzylidenechromanone becomes C-2 of the flavone and the mechanism shown in Scheme 36 is in accord with the observations ⟨79JCR(S)137⟩.

3-Hydroxyflavanones (**653**) may react with alkali in several ways (Scheme 37) and a mixture of products is sometimes formed. On warming with alkali in the absence of oxygen, dehydrogenation to 3-hydroxyflavone (**654**) is often accompanied by loss of hydroxyl group to form a flavanone (**655**). Ring contraction may occur with or without benzilic acid rearrangement to give a benzofuran-2-one (**656**) or a benzofuran-3-one (**652**) ⟨54ACS734⟩.

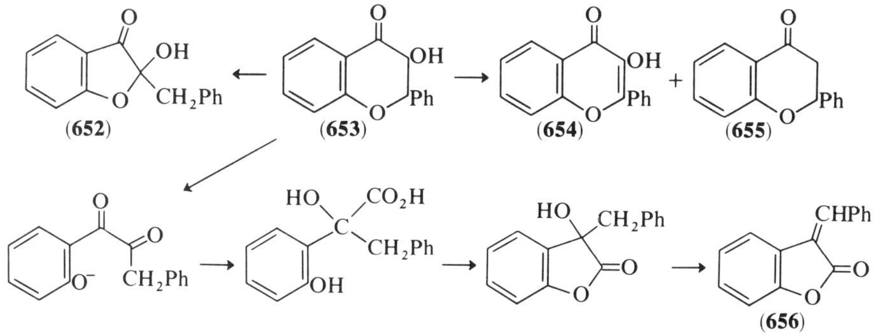

**Scheme 37**

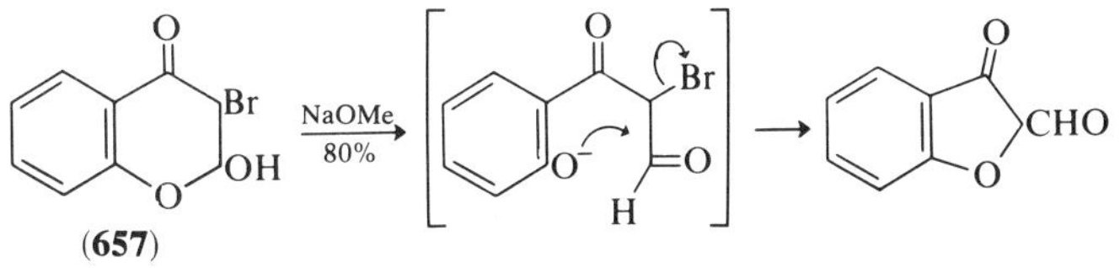

The reaction of 3-bromo-2-hydroxychromanone (**657**) with sodium methoxide also results in pyran ring cleavage followed by the formation of a furanone ring ⟨75JHC981⟩.

### 2.23.10.4.4 Reduction

The carbonyl group of di- and tetra-hydropyran-4-ones, *e.g.* (**658**), is reduced to the alcohol by lithium tri-*s*-butylaluminum hydride, the ratio of conformers being about 3:1 in favour of the *trans* isomer. Reduction of 2-methyltetrahydropyran-3-one (**659**) similarly gives a predominance of *trans*-2-methyltetrahydropyran-3-ol ⟨80JOC919⟩.

Several chroman-3- and -4-ones have been catalytically reduced to the alcohol, for example, ethyl 6,7-dimethoxy-3-oxochroman-4-carboxylate (**660**) ⟨58BCJ267⟩ and 3-acetamido-2-methylchromanone (**661**), the latter to the chroman in two stages ⟨73JCS(P2)227⟩.

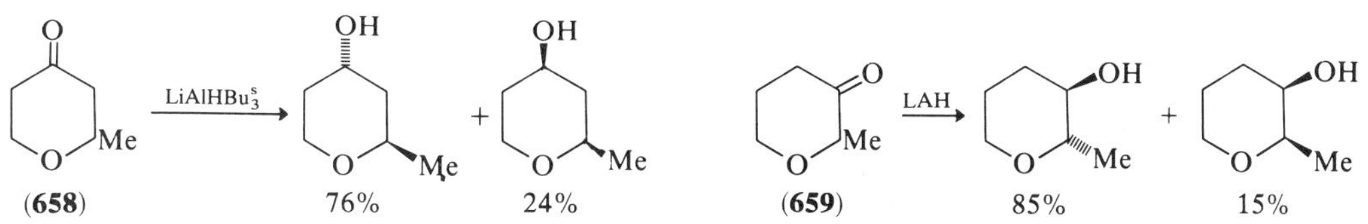

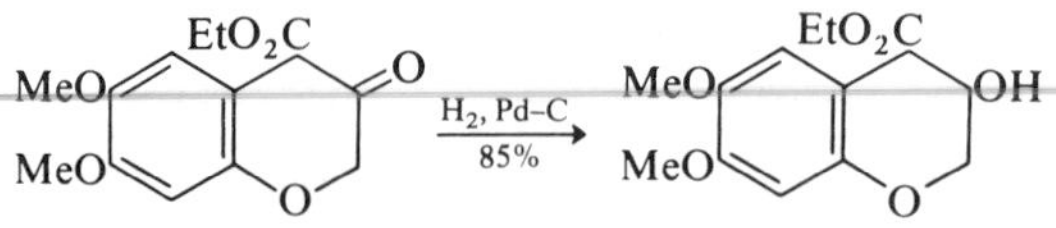

The double bond of 4-hydroxy-6-methyl-5,6-dihydropyran-2-one (**662**) requires pressure to reduce it catalytically and the presence of anhydrous copper(II) sulfate promotes simultaneous dehydration to 6-methyl-3,4,5,6-tetrahydropyran-2-one (**663**) ⟨78JHC1153⟩.

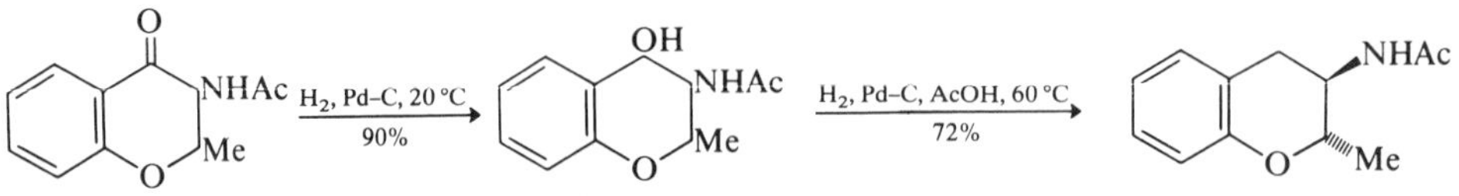

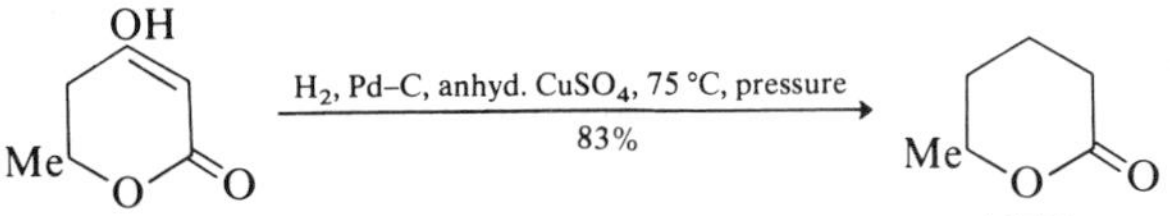

Chromanones and flavanones are reduced to the chromanols or the chromans according to the reagent used or the severity of the conditions. Sodium borohydride gives a good yield of the alcohol, but in the presence of boron trifluoride, the borohydride gives mainly the chroman. Similarly, borohydride–aluminum chloride yields the chroman as the major product together with some chromanol. Meerwein–Ponndorf reduction gives the chromanol. *cis*-2-Methylchroman-4-ol is formed when the chromanone is reduced catalytically, by borohydride or LAH but the Meerwein–Ponndorf method gives a mixture of isomers. 3-Methylchromanone is reduced by LAH or aluminum isopropoxide–isopropanol mainly to *trans*-3-methylchroman-4-ol but catalytic hydrogenation gives a mixture, the predominant isomer depending on the catalyst used.

Both chemical and catalytic methods are available to convert chromanones to chromans, and flavones to flavans. Clemmensen's method is the most commonly used and usually gives good yields, but diborane is sometimes preferable. The Wolff–Kishner method appears to degrade the pyran ring and is rarely used on chromanones. The conditions of catalytic reduction have to be rather more drastic in order to produce chromans and flavans but much depends on the activity of the catalyst and on the solvent used. If the chromanol is first formed, more severe conditions are usually required to complete the reduction ⟨77HC(31)207⟩.

### 2.23.10.5 Other Reactions

Dihydropyrans and tetrahydropyrans add on alkenes when irradiated but some of the products are unstable. 3,3-Dimethylpyran-2,4-dione (**664**) and an unsymmetrical diene such as 2-methylprop-1-ene (**665**) give regiospecifically a 1 : 1 mixture of *cis*- and *trans*-fused pyran-4-one (**666**) ⟨73T1317⟩. Tetrahydropyran (**667**) does not undergo photoaddition unless a sensitizer such as benzophenone is present, but then it reacts with diethyl maleate (**668**) ⟨67T3193⟩.

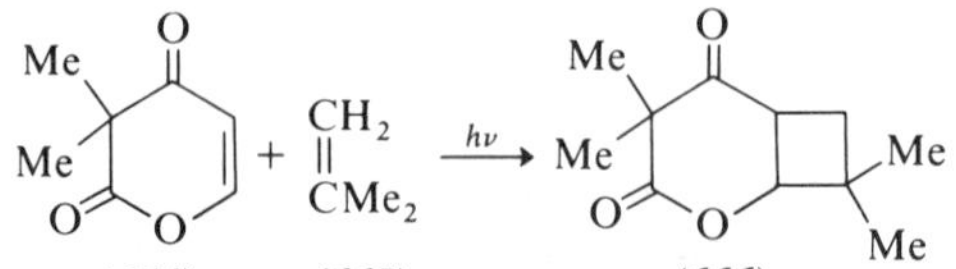

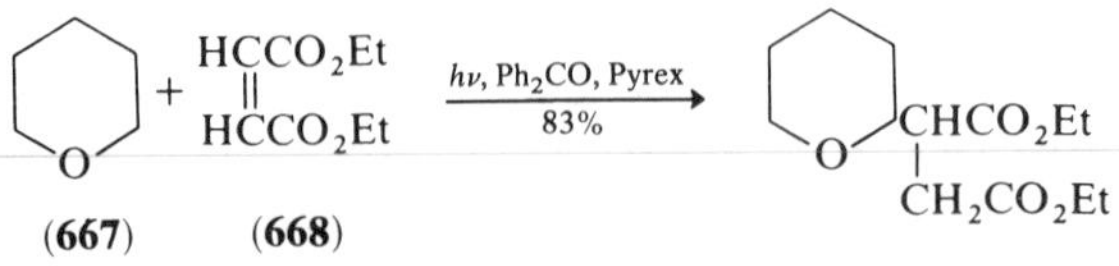

The double bond of a dihydropyran in conjunction with an exocyclic double bond conjugated with it acts as a diene in Diels–Alder reactions. In this way, the dihydrochroman (**669**) has been synthesized and may be dehydrogenated to the chroman-5,6-dicarboxylate (**670**) ⟨76CR(C)(282)357⟩. Another Diels–Alder reaction which has been recently demonstrated is that with 6-acetoxy-2,6-dihydropyran-3-one (**671**), which is a precursor of the pyrylium zwitterion (**673**). When heated with propenal, it gives the adduct (**672**) which consists of a mixture of 4:1 of the *exo*:*endo* isomers, but all reactions are regiospecific ⟨80JOC3361⟩.

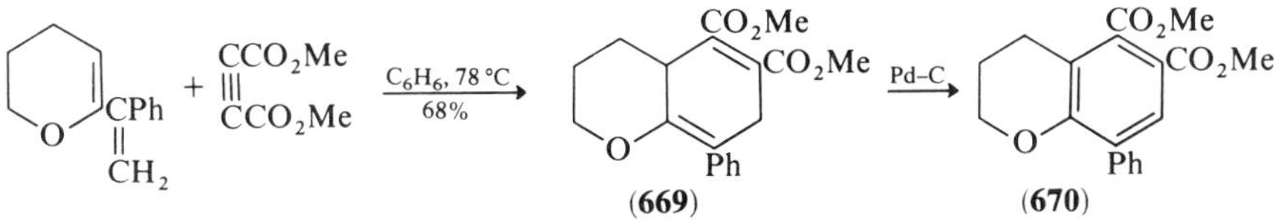

## 2.23.11 REDUCED PYRANS AND PYRANONES AND THEIR BENZO DERIVATIVES: REACTIVITY OF SUBSTITUENTS

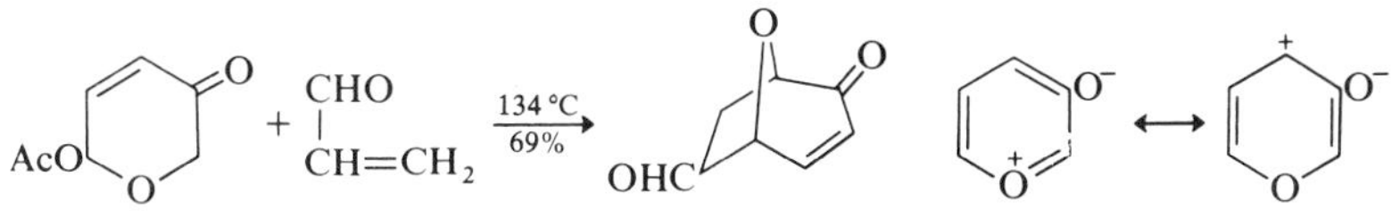

### 2.23.11.1 Introduction

Individual functional groups attached to a partially or fully reduced pyran ring behave much as expected of their aliphatic equivalents but there is often a quantitative difference in their reactivity which enables selective reactions to be carried out on polysubstituted compounds. Many examples of this are known in the tocopherol series which are the most important members of the chroman family. Since they are known by trivial names, these are shown with their structures (**674**). The most important tocopherol is natural vitamin E or $\alpha$-tocopherol; the four natural tocopherols have *R*-configuration at each of their asymmetric centres at C-2, C-4' and C-8'.

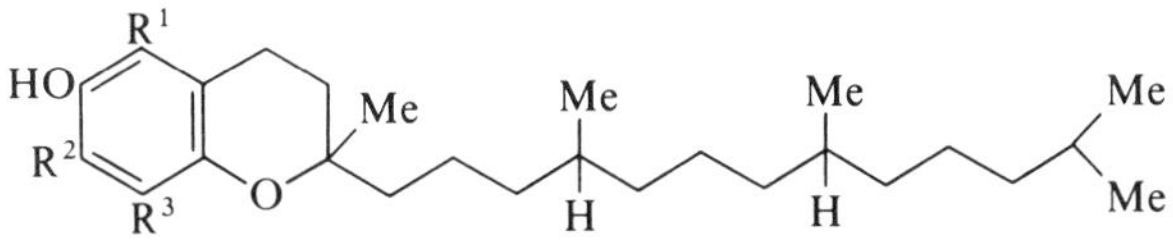

| $R^1$ | $R^2$ | $R^3$ | |
|------|------|------|------|
| Me | Me | Me | $\alpha$-tocopherol |
| Me | H | Me | $\beta$-tocopherol |
| H | Me | Me | $\gamma$-tocopherol |
| H | H | Me | $\delta$-tocopherol |

### 2.23.11.2 Fused Benzene Rings

The benzene ring of fused dihydropyrans is more easily attacked by electrophiles than that of the corresponding pyran. However, the activating effect of the carbonyl group on the 3-position of chromanones and flavanones ensures that halogenation normally occurs at that position preferentially. Thus, 3-mono- and 3,3-di-bromination is readily achieved with appropriate molar proportions of bromine but bromination of 2,3-dihydronaphtho[2,1-*b*]pyran-1-one (**675**) in acetic acid gave the 7-bromo derivative (**676**) ⟨48JA599⟩.

Bromination, chlorination, nitration and Friedel–Crafts acylation of chroman give the 6-substituted derivatives in good yields by standard methods. When 6,8-dimethyl-chromanone (**677**) is warmed with formaldehyde in sulfuric acid, hydroxymethylation at C-5 occurs ⟨78BCJ1874⟩.

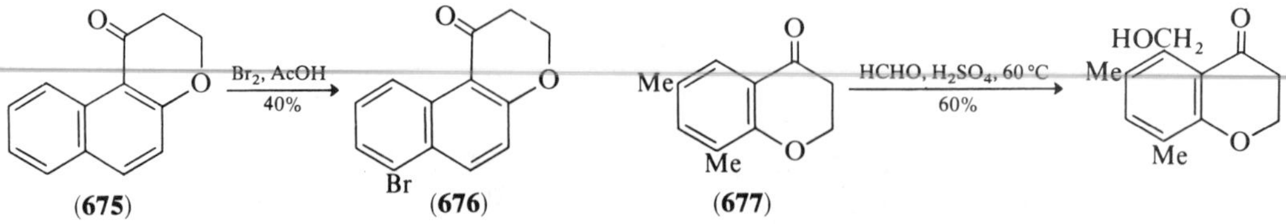

(675)            (676)            (677)

The directing effect of a substituent in a chroman is sometimes different from that in chromone; for example, bromination of the 7-methoxy compounds yields the 8-bromo-chromone not the 6-bromochroman, and nitration of 8-hydroxychromone gives mainly the 7-nitro derivative while the corresponding chroman is nitrated at C-6 ⟨65MI22300⟩. A 7- or 8-methoxy-chroman or -chromanone (678) reacts in the Vilsmeier–Haack formylation to give the 6-carboxaldehyde; 8-hydroxychroman is acylated with acetic acid–boron trifluoride largely at C-7 but the same reagent converts 8-methoxychroman into the 6-acetyl derivative ⟨81HC(36)251⟩.

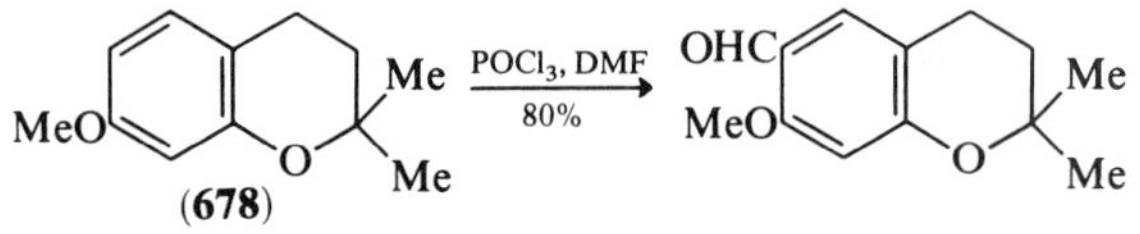

(678)

### 2.23.11.3  *C*-Linked Substituents

Halogenation of polymethylchromans with a variety of reagents shows that a 5-methyl group is the most susceptible. For example, chlorine, thionyl chloride, sulfuryl chloride, phosphorus pentachloride and bromine produce this type of reaction and the product, *e.g.* (679), is useful as an intermediate in the synthesis of compounds such as bischromans (680) or pyridinium salts (681) ⟨74MI22304⟩. When vitamin E (682; $R^1 = R^2 = Me$) is heated with trimethylamine oxide, the 5- or 7-methyl group is oxidized to the aldehyde group but $\gamma$-tocopherol (682; $R^1 = H$, $R^2 = Me$) is formylated at C-5 ⟨74ABC2545⟩.

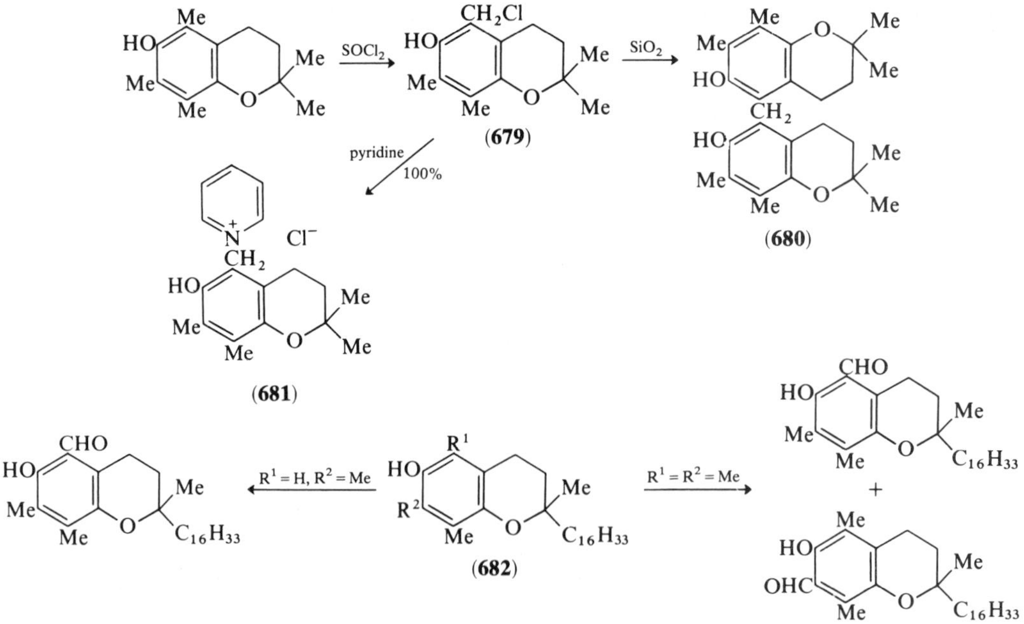

(679)

(681)

(680)

(682)

The formation of a third ring by a PPA-catalyzed intramolecular cyclization of the malononitrile derivative (683) gives a single product (684), in contrast to the mixture obtained from a similar reaction on the thiochromanone ⟨76JHC123⟩.

5,6-Dihydropyran-2-carbonitrile (685) is converted into the ketone (686) by a Grignard reagent, and the ring is stable to a Wittig reaction which converts the carbonyl into a methylene group ⟨76CR(C)(282)357⟩. The tetrahydropyran ring also survives a Reformatsky reaction on 2,2-dimethyltetrahydropyran-4-carboxaldehyde (687) ⟨B-79MI22305⟩.

(683) (684)

(685) (686) (687)

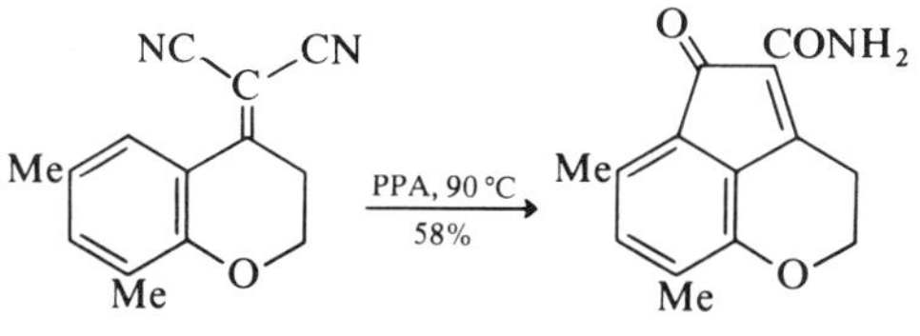

An unusual oxidative breakdown of a *p*-tolyl group to a carboxylic acid (**688**) is accomplished in moderate yield without degrading the chroman system ⟨51JCS76⟩ and the stability of this system to oxidation (*cf.* chromone, Section 2.23.9.3) is further illustrated by the conversion of an ethyl to a carboxyl group in the formation of 2,4,4-trimethylchroman-2,7-dicarboxylic acid (**689**) ⟨57JCS3060⟩.

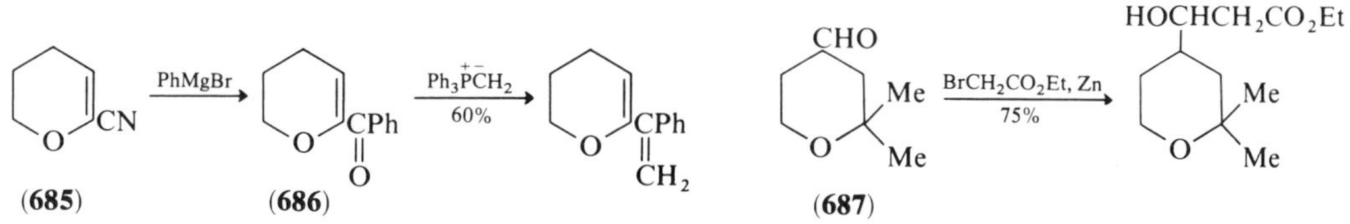

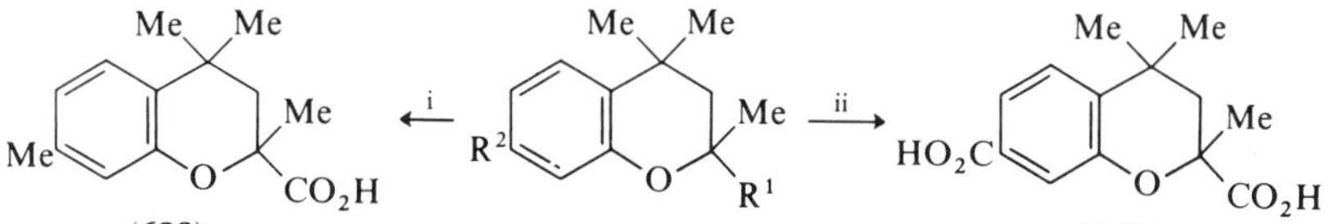

(688) (689)

i, $R^1 = 2\text{-HO-4-MeC}_6H_3$, $R^2 = Me$; $KMnO_4$, $Me_2CO$; ii, $R^1 = CO_2H$, $R^2 = Et$; $KMnO_4$, aq. $Na_2CO_3$, 100 °C

The chroman ring system is stable to organometallic reagents, for example in the formation of the tertiary alcohol (**690**) in high yield ⟨63HCA650⟩, and to the usual interconversion of carboxylic acid, acyl chloride, carboxamide and nitrile.

(690)

(691) (692)

(693) (694)

(695) (696)

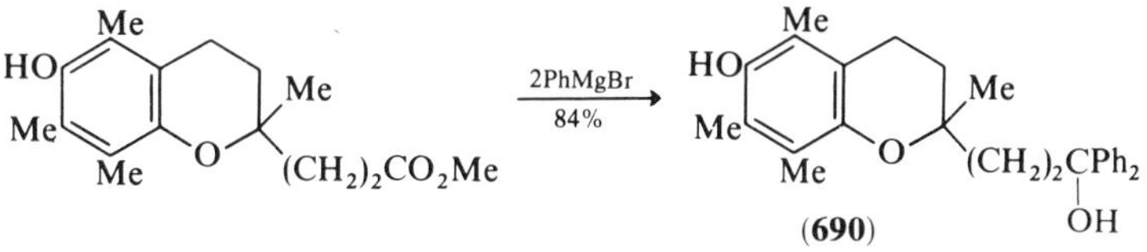

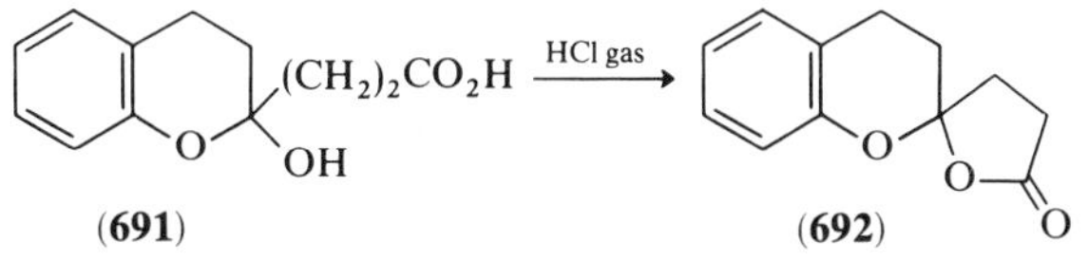

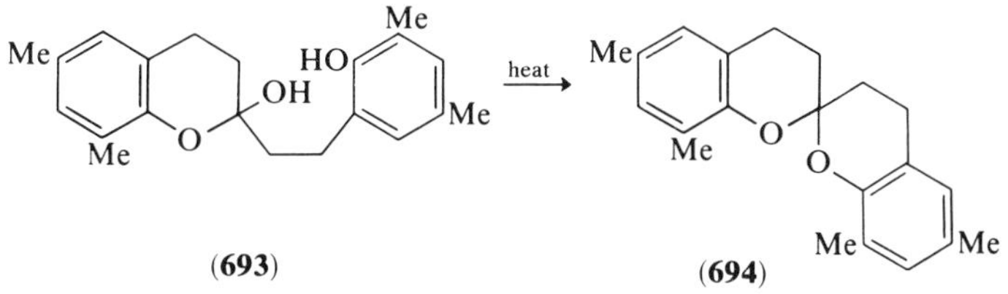

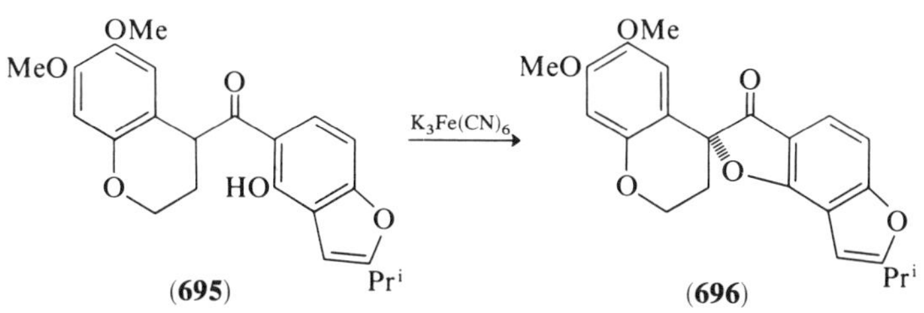

Several chromans have been converted into spiro compounds; for example when the chroman-2-ol (**691**) is treated with acid, the spirofuranone (**692**) is formed ⟨65M220⟩. Heating the chroman-2-ol (**693**) above its melting point converts it into 6,6',8,8'-tetramethyl-2,2'-spirobichroman (**694**) ⟨50JA3009⟩. The 4-acylchroman (**695**) is converted into a spirorotenoid (**696**) by mild oxidation, and a number of other compounds have been similarly prepared ⟨62JCS775⟩.

### 2.23.11.4 _N_-Linked Substituents

Nitro groups attached to the benzene ring of chromans or chromanones may be reduced catalytically to give the amines but 3-aminochromanones (**698**) are usually obtained from the oxime (**697**) of the chromanone by the Neber rearrangement. The free bases are rather unstable ⟨69JMC277⟩.

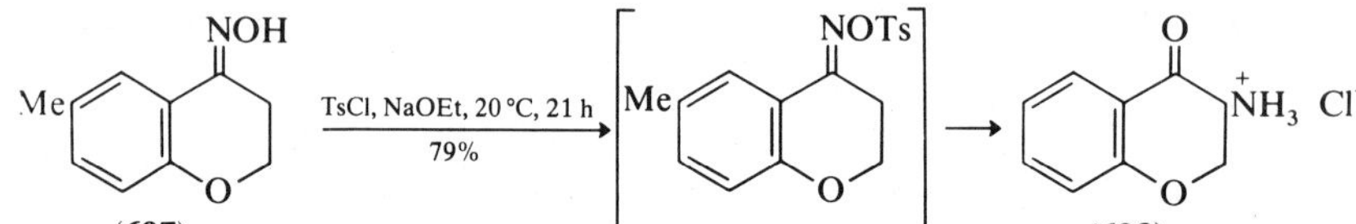

A 2-amino group in chroman is more labile than its isomers and is hydrolyzed to the alcohol by acids; for example both the amines (**699**) and (**700**) give the corresponding alcohols by treatment with nitrous acid and 50% hydrochloric acid, respectively. 6-Amino-chromans are of interest because of their chemical and biological resemblance to the tocopherols. The tocopheramines (**701**; $R^1$, $R^2$, $R^3$ = H or Me) show antioxidant and other properties of the corresponding phenols and are no more toxic. They may be obtained by catalytic or chemical reduction of the nitrochromans ⟨81HC(36)189⟩.

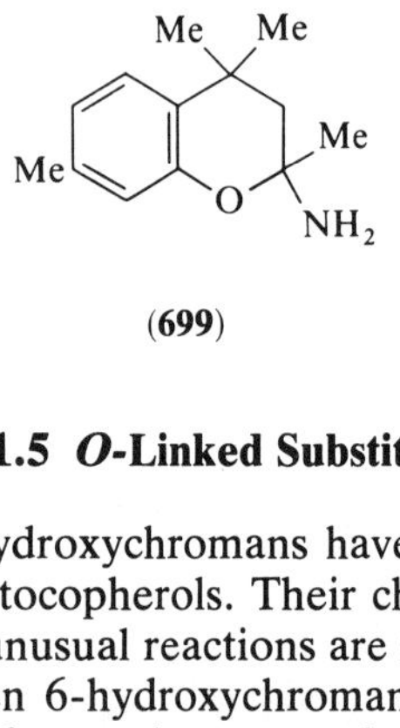

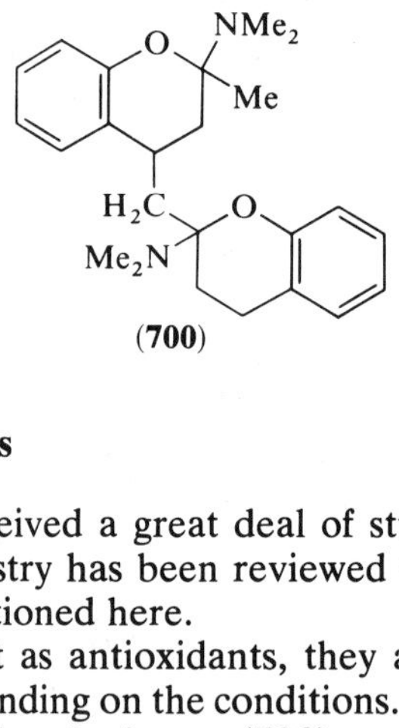

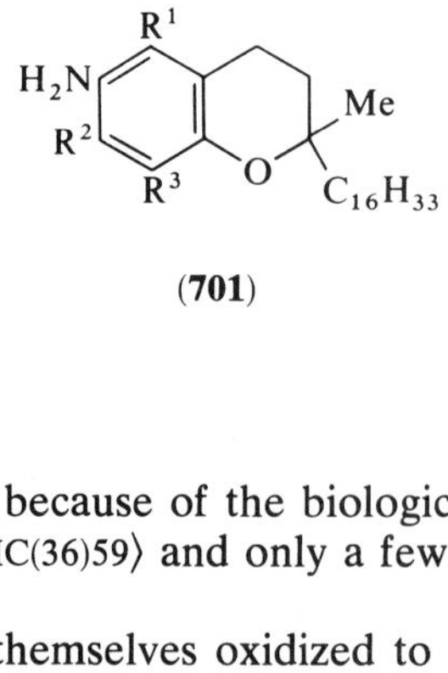

### 2.23.11.5 _O_-Linked Substituents

6-Hydroxychromans have received a great deal of study because of the biological role of the tocopherols. Their chemistry has been reviewed ⟨81HC(36)59⟩ and only a few of the more unusual reactions are mentioned here.

When 6-hydroxychromans act as antioxidants, they are themselves oxidized to one or more of several compounds depending on the conditions. Amongst the commonest products from α-tocopherol are α-tocopherylquinone (**702**), α-tocored (**703**), the 5-benzoyloxy-methyl derivative (**704**) and the spiro dimer (**705**).

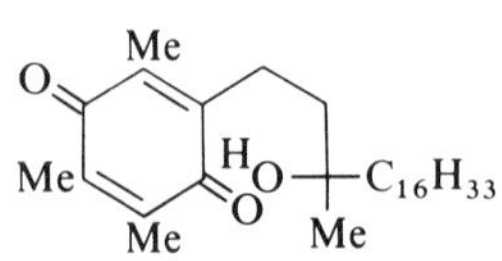

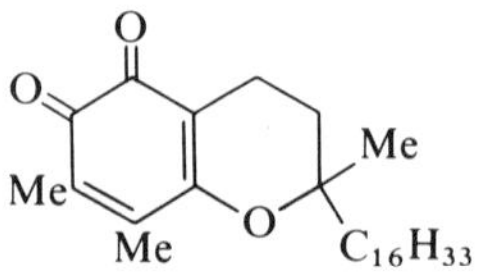

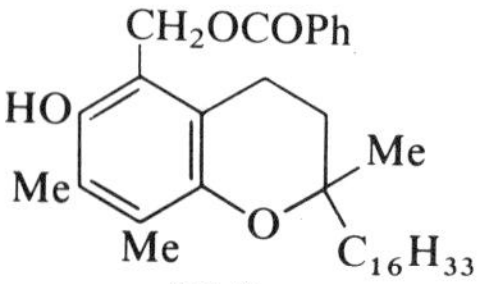

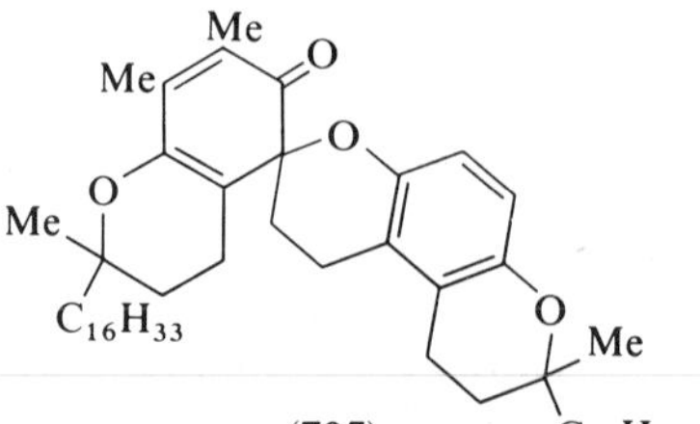

An interesting conversion of a 3-hydroxyflavanone (**706**) into a 3-aminoflavone (**707**) is effected by mesylation followed by reaction with sodium azide under mild conditions ⟨B-77MI22305⟩.

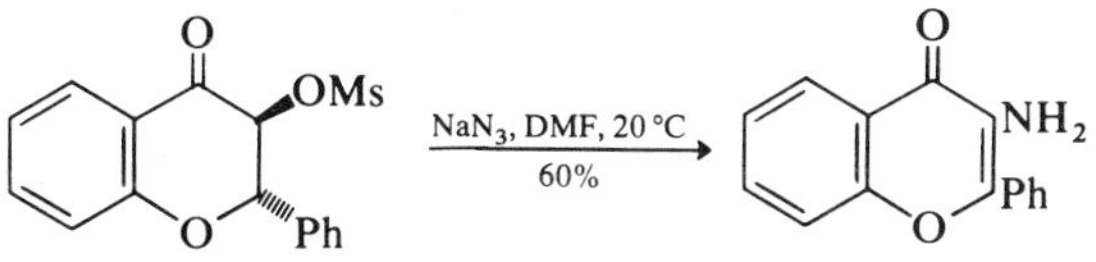

Halogenation of α-tocopherol gives the 5-halogenomethyl derivative and 6-hydroxy-chroman is converted into 5-bromo-6-hydroxychroman ⟨71ACS94⟩. The pyran ring of 2,2-dimethylchromans is stable to many reagents but boron trichloride degrades the dimethoxychroman (**708**) to the pyrogallol derivative (**709**) ⟨74CC318⟩. 3,4-Epoxychromans, obtained from 2*H*-chromenes or from 3-bromo-4-hydroxychroman (**710**), are useful intermediates in the synthesis of many chromans carrying substituents in the pyran ring as Scheme 38 shows ⟨81T2613, 68YZ816⟩. 2,3-Epoxychromanones are also useful intermediates ⟨79T2883⟩. 4-Hydroxychromans such as (**710**) are readily oxidized to the chromanones ⟨66JCS(C)2013⟩.

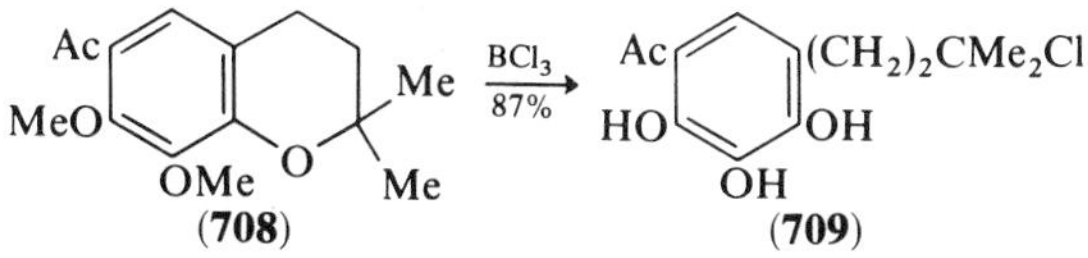

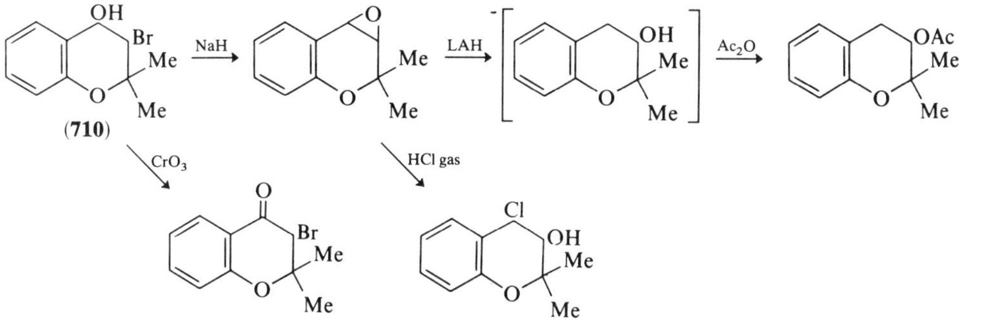

**Scheme 38**

An alkoxy group at C-1 of isochroman (**711**) is displaced by nucleophiles such as phenols or alcohols and a variety of interesting compounds have thus been obtained (Scheme 39) ⟨80CPB2967⟩.

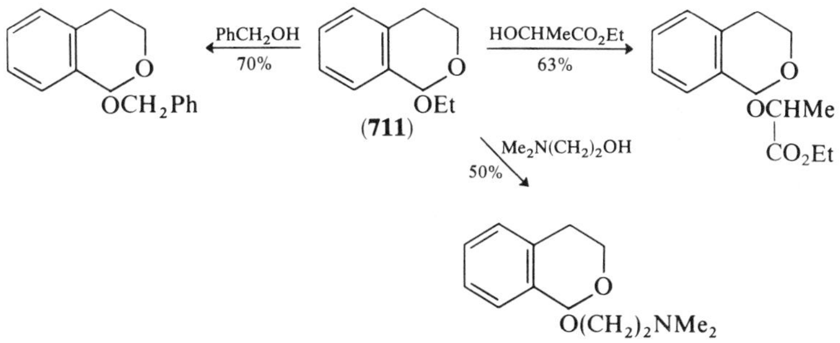

**Scheme 39**

### 2.23.11.6 Halogens

Halogens attached to pyran or pyranone rings have a reactivity comparable with that of alkyl halides but there is some difference in reactivity; for example the 2-halogen of 2,3-dichlorotetrahydropyran (**712**) is the more reactive ⟨76CR(C)(282)357⟩ and is preferentially eliminated in the reaction of 5,6-dibromo-5,6-dihydropyran-2-one (**713**) with a base

⟨69JOC2239⟩. In 3,4-dihalogenochromans, hydrolysis and methanolysis affect the 4-halogen only, as illustrated by the dibromo compound (**714**); strong alkali converts the bromide (**715**) into the epoxide (**716**), a useful intermediate ⟨81T2613, 81HC(36)161⟩.

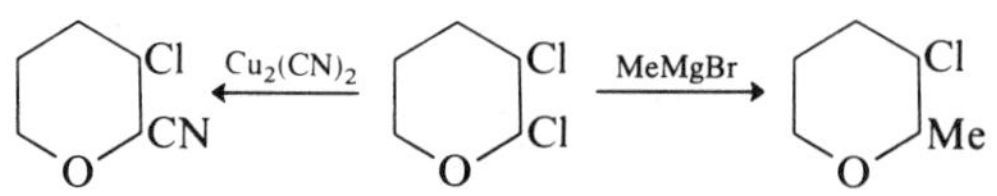

(**712**)

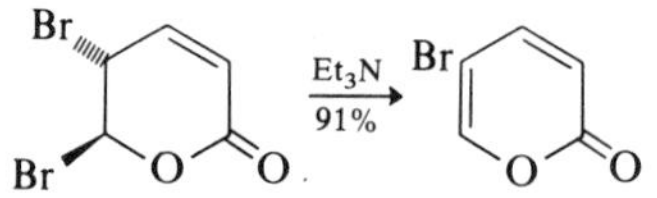

(**713**)

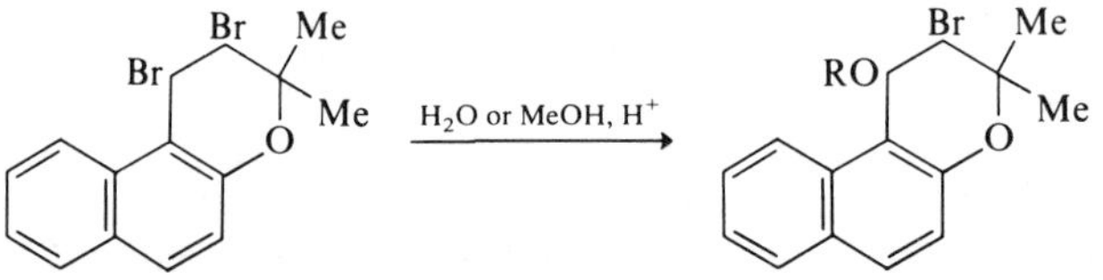

(**714**)                                                    R = H or Me

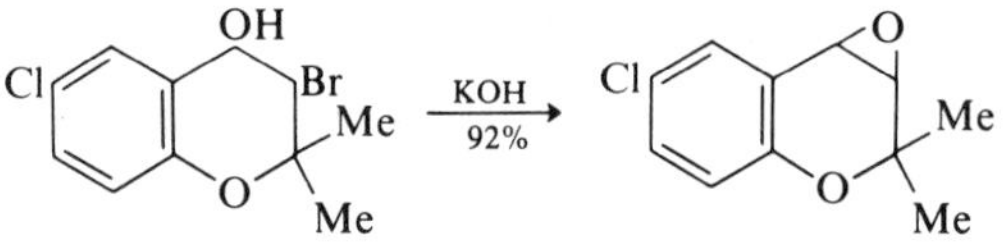

(**715**)                          (**716**)

# 2.24

# Pyrans and Fused Pyrans: (iii) Synthesis and Applications

J. D. HEPWORTH

*Preston Polytechnic*

| | | |
|---|---|---|
| 2.24.1 | INTRODUCTION | 738 |
| 2.24.2 | PYRANS AND FUSED PYRANS | 738 |
| *2.24.2.1* | *2H-Pyrans* | 739 |
| *2.24.2.1.1* | *Formation of one bond* | 739 |
| *2.24.2.1.2* | *From a preformed heterocyclic ring* | 740 |
| *2.24.2.2* | *2H-Chromenes (2H-1-Benzopyrans)* | 741 |
| *2.24.2.2.1* | *Formation of one bond* | 741 |
| *2.24.2.2.2* | *From other heterocycles* | 753 |
| *2.24.2.2.3* | *From a preformed heterocyclic ring* | 754 |
| *2.24.2.3* | *4H-Pyrans* | 756 |
| *2.24.2.3.1* | *Formation of one bond* | 757 |
| *2.24.2.3.2* | *Formation of two bonds* | 760 |
| *2.24.2.3.3* | *From other heterocycles* | 760 |
| *2.24.2.3.4* | *From a preformed heterocyclic ring* | 761 |
| *2.24.2.4* | *4H-Chromenes (4H-1-Benzopyrans)* | 763 |
| *2.24.2.4.1* | *Formation of one bond* | 763 |
| *2.24.2.4.2* | *Formation of two bonds* | 764 |
| *2.24.2.4.3* | *From other heterocycles* | 764 |
| *2.24.2.4.4* | *From a preformed heterocyclic ring* | 764 |
| *2.24.2.5* | *1H-Isochromenes (1H-2-Benzopyrans)* | 765 |
| *2.24.2.5.1* | *Formation of one bond* | 766 |
| *2.24.2.5.2* | *From other heterocycles* | 767 |
| *2.24.2.5.3* | *From a preformed heterocyclic ring* | 767 |
| *2.24.2.6* | *Xanthenes (Dibenzo[b,e]pyrans)* | 767 |
| *2.24.2.6.1* | *Formation of one bond* | 767 |
| *2.24.2.6.2* | *Formation of two bonds* | 767 |
| *2.24.2.6.3* | *From a preformed heterocycle* | 768 |
| *2.24.2.7* | *Reduced Pyrans* | 769 |
| *2.24.2.7.1* | *Dihydropyrans* | 769 |
| *2.24.2.7.2* | *Tetrahydropyrans* | 774 |
| *2.24.2.7.3* | *Chromans (dihydrobenzopyrans)* | 778 |
| *2.24.2.7.4* | *Isochromans (dihydro-2-benzopyrans)* | 787 |
| 2.24.3 | PYRANONES AND FUSED PYRANONES | 789 |
| *2.24.3.1* | *Pyran-2-ones (2-Pyrones)* | 789 |
| *2.24.3.1.1* | *Formation of one bond* | 789 |
| *2.24.3.1.2* | *Formation of two bonds* | 796 |
| *2.24.3.1.3* | *From other heterocycles* | 797 |
| *2.24.3.1.4* | *From a preformed heterocycle* | 798 |
| *2.24.3.2* | *Coumarins (2H-1-Benzopyran-2-ones)* | 799 |
| *2.24.3.2.1* | *Formation of one bond* | 799 |
| *2.24.3.2.2* | *From other heterocycles* | 809 |
| *2.24.3.2.3* | *From a preformed heterocycle* | 809 |
| *2.24.3.3* | *Pyran-4-ones* | 810 |
| *2.24.3.3.1* | *Formation of one bond* | 811 |
| *2.24.3.3.2* | *Formation of two bonds* | 814 |
| *2.24.3.3.3* | *From other heterocycles* | 814 |
| *2.24.3.3.4* | *From a preformed heterocyclic ring* | 815 |
| *2.24.3.4* | *Chromones (4H-1-Benzopyran-4-ones)* | 816 |
| *2.24.3.4.1* | *Formation of one bond* | 816 |
| *2.24.3.4.2* | *From other heterocycles* | 827 |
| *2.24.3.4.3* | *From a preformed heterocycle* | 828 |

|  |  |  |
|---|---|---|
| 2.24.3.5 | *Isocoumarins (1H-2-Benzopyran-1-ones)* | 830 |
| 2.24.3.5.1 | *Formation of one bond* | 830 |
| 2.24.3.5.2 | *From other heterocycles* | 834 |
| 2.24.3.5.3 | *From a preformed heterocycle* | 834 |
| 2.24.3.6 | *Xanthones* | 835 |
| 2.24.3.6.1 | *Formation of one bond* | 835 |
| 2.24.3.6.2 | *From a preformed heterocyclic ring* | 839 |
| 2.24.3.7 | *Reduced Pyranones* | 841 |
| 2.24.3.7.1 | *Dihydropyran-2-ones* | 841 |
| 2.24.3.7.2 | *Dihydropyran-3-ones* | 843 |
| 2.24.3.7.3 | *Dihydropyran-4-ones* | 844 |
| 2.24.3.7.4 | *Tetrahydropyran-2-ones and tetrahydropyran-4-ones* | 845 |
| 2.24.3.7.5 | *Dihydrocoumarins* | 848 |
| 2.24.3.7.6 | *Chromanones* | 848 |
| 2.24.3.7.7 | *Isochromanones* | 857 |
| 2.24.4 | PYRYLIUM SALTS | 860 |
| 2.24.4.1 | *Formation of One Bond* | 861 |
| 2.24.4.1.1 | *Adjacent to the heteroatom* | 861 |
| 2.24.4.1.2 | *β to the heteroatom* | 871 |
| 2.24.4.2 | *From Other Heterocycles* | 872 |
| 2.24.4.2.1 | *From pyridine* | 872 |
| 2.24.4.3 | *From a Preformed Heterocyclic Ring* | 872 |
| 2.24.4.3.1 | *From pyranones and benzopyranones* | 872 |
| 2.24.4.3.2 | *From pyrans and benzopyrans* | 873 |
| 2.24.4.3.3 | *From chromans* | 874 |
| 2.24.5 | APPLICATIONS | 874 |

## 2.24.1  INTRODUCTION

The presence of a six-membered oxygen heterocyclic ring in a range of naturally occurring compounds provided the stimulus for the development of synthetic routes to these compounds. Many of the methods which evolved in the closing years of the 19th century and the earlier decades of this century are still of value today. These routes have been supplemented by modern approaches and consequently there is often a choice of methods available for the synthesis of a particular oxygen heterocyclic compound.

The literature covering oxygen heterocycles is quite extensive and it is worthwhile mentioning those contributions which give comprehensive coverage of these systems and to which the reader should refer. The earlier work on monocyclic oxygen heterocycles is discussed in Volume 1 and that on their benzologues in Volume 2 of Elderfield's treatise on heterocyclic compounds ⟨B-50MI22400, B-51MI22400⟩, whilst more complex systems are reviewed in Volume 7 ⟨B-61MI22400⟩. Almost a whole volume of Rodd's 'Chemistry of Carbon Compounds' is devoted to six-membered oxygen heterocyclic compounds and it provides an excellent account of the whole field ⟨B-77MI22400⟩. Comprehensive surveys of narrower areas include those covering chromenes, chromanones, chromones and chromans ⟨77HC(31), 81HC(36)⟩ and a detailed and outstanding account of naturally occurring oxygen heterocycles ⟨B-63MI22400⟩, parts of which have been updated ⟨B-73MI22400⟩. A number of books are concerned exclusively with flavonoids ⟨B-62MI22400, B-75MI22400⟩, whilst the published proceedings of bioflavonoid symposia provide access to current work ⟨B-82MI22400⟩.

## 2.24.2  PYRANS AND FUSED PYRANS

Simple pyrans are unstable compounds as is to be expected of a dienolic ether system. Thus, 2*H*-pyran (**1**) has not yet been synthesized, whilst the preparation of 4*H*-pyran (**2**) was only achieved in 1962. Introduction of unsaturated groups lends some stability, but a

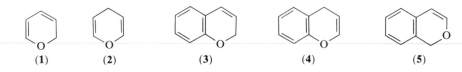

(**1**)         (**2**)         (**3**)         (**4**)         (**5**)

much more significant stabilization results from the fusion of aromatic rings. 2*H*-1-Benzopyrans (chrom-3-enes; **3**) are widespread in the higher plants and enjoy an extensive chemistry. Two other benzologues are considered, namely 4*H*-1-benzopyran (**4**) and 1*H*-2-benzopyran (isochromene; **5**).

### 2.24.2.1 2*H*-Pyrans

The literature on derivatives of 2*H*-pyran is relatively sparse, reflecting the overall instability of this system. The heterocycle undergoes facile electrocyclic ring opening to the dienone (**6**) and generally only 2,2-disubstituted derivatives are stable under normal conditions. However, it has been suggested that it is steric destabilization of the dienone, rather than a specific substitution pattern, which is essential for the existence of the pyran ⟨71JOC1977⟩.

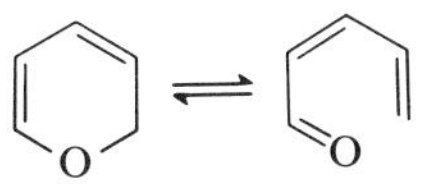

(**6**)

#### 2.24.2.1.1 *Formation of one bond*

(i) *Adjacent to the heteroatom*

Several groups have sought to displace the dienone–2*H*-pyran equilibrium in favour of the heterocycle. Thus, irradiation of the dienone (**7**) produced the *trans* isomer (**8**) together with an equilibrium mixture of the pyran (**9**) and its ring-opened isomer, which consisted mainly of (**9**) ⟨71JOC1988⟩.

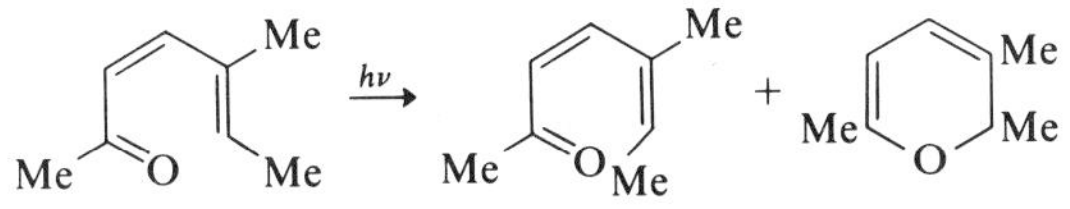

(**7**)          (**8**)          (**9**)

Brief irradiation of β-ionone (**10**) using a high pressure mercury lamp with a Pyrex filter afforded the fused pyran (**11**) in good yield and on a preparative scale; without the filter the dienone (**12**) was produced in quantitative yield ⟨76JCS(P1)532⟩.

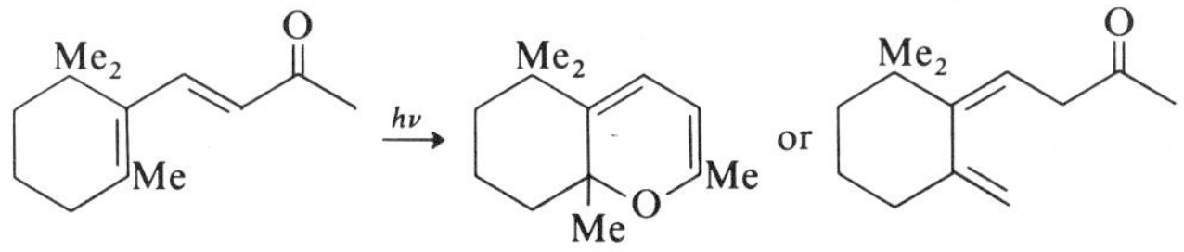

(**10**)          (**11**)          (**12**)

3,4,5,6-Tetrahydro-2-vinylbenzaldehyde (**13**) has been synthesized from 2-*s*-butoxymethylenecyclohexanone ⟨70HCA485⟩. Above 70 °C the aldehyde equilibrates with its valence isomer, the fused 2*H*-pyran (**14**). However, the ring-opened form is the thermodynamically favoured isomer. The pyran itself is unstable but has been trapped as the adduct (**15**) with TCNE (Scheme 1).

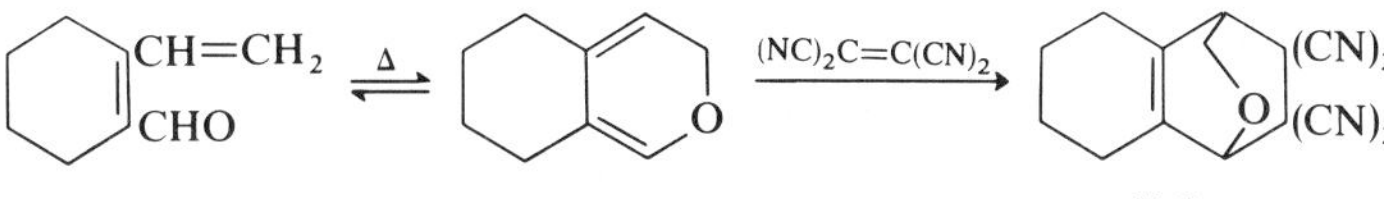

(**13**)          (**14**)          (**15**)

**Scheme 1**

The thermal ring opening of the cyclobutene (**16**) also leads to a 2*H*-pyran (**17**) and this too has been trapped as a Diels–Alder adduct ⟨69TL4987⟩.

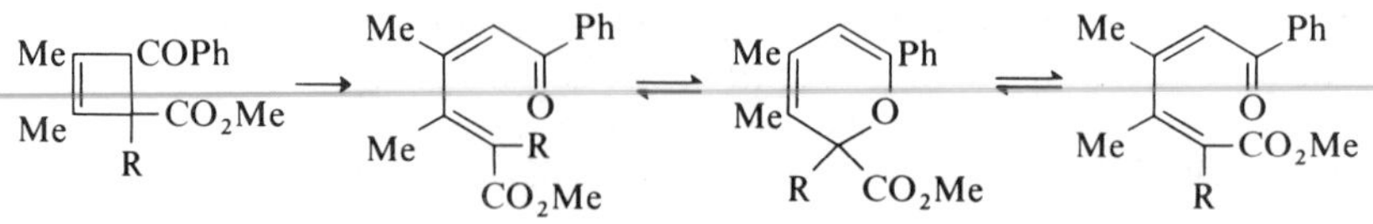

**(16)**　　　　　　　　　　　　　　**(17)**

The ring closure of a number of tetrachloropentadienols to 2*H*-pyrans (**18**) has been effected by treatment with sodium in boiling benzene ⟨75LA240⟩. The initially formed alkoxide undergoes an intramolecular nucleophilic substitution. Although the pyrans are not very stable at room temperature, rearranging to 4*H*-pyrans, they are obtained in most acceptable amounts. The pentadienols are prepared through the reaction of a Grignard reagent with an acid chloride at −60 °C, though separation from other dienols is necessary (Scheme 2).

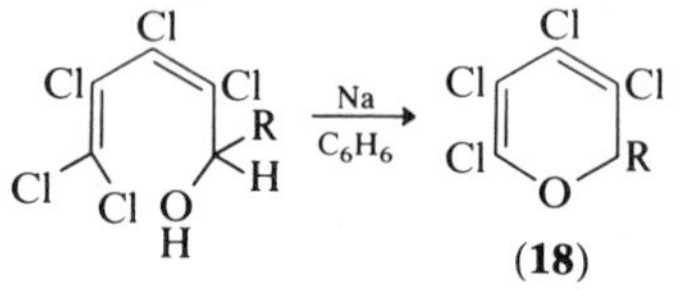

**(18)**

**Scheme 2**

In the presence of 'freeze-dried' potassium fluoride, perfluoro-2-methylpent-2-ene reacts with activated methylene compounds to yield pyrans ⟨81MI22400⟩. The fluoride ion abstracts a proton from the methylene group and subsequent condensation of the carbanion with the perfluoroalkene affords a dienone (**19**) which ring closes to the pyran (**20**; Scheme 3). In the case of pentane-2,4-dione a divinyl ether (**21**) is also formed. This product is considered to arise from reaction of the alkene at the oxygen of the enolate ion.

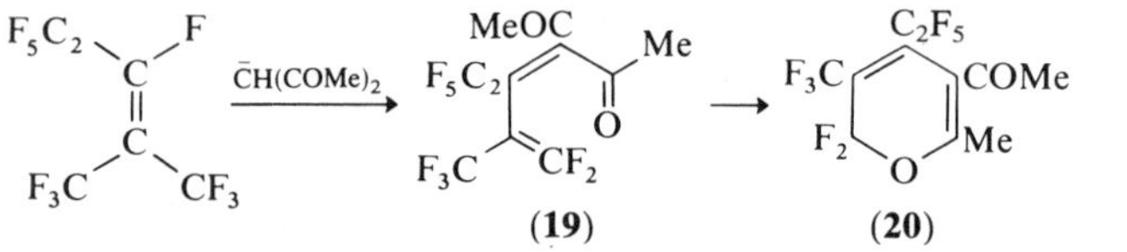

**(19)**　　　　**(20)**

**Scheme 3**

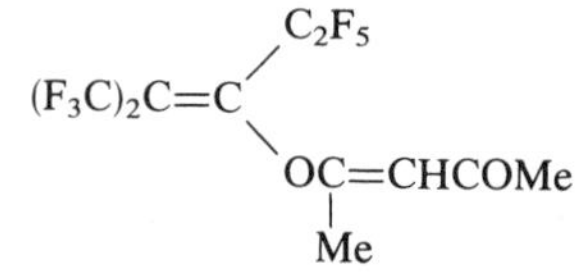

**(21)**

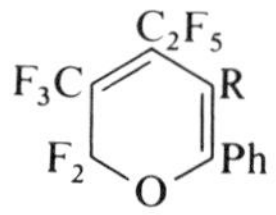

**(22)** R = CN
**(23)** R = COCF₃
**(24)** R = COC₂F₅

Ethyl acetoacetate similarly gave two products in approximately equal quantities. However, benzoylacetonitrile, benzoylmethyl trifluoromethyl ketone and the analogous pentafluoroethyl ketone afforded the pyrans (**22**), (**23**) and (**24**) exclusively and in high yield.

The cyclocondensation of benzoylacetonitrile to 3,5-dicyano-2-cyanomethyl-2,4,6-triphenyl-2*H*-pyran is best effected in a mixture of xylene and acetic acid, with ammonium acetate as catalyst and continuous separation of water ⟨81CCC748⟩. The reaction has been extended to a range of *p*-substituted derivatives of the ketonitrile. The pyrans are high melting solids and show surprising stability. However, the stabilizing effect of cyano groups has been noted for other non-aromatic heterocycles ⟨78JHC57⟩. Whilst no intermediate species have been isolated from this reaction, it seems reasonable to assume that the product arises from two aldol condensations each accompanied by dehydration. The final step may involve an electrocyclic ring closure or an internal Michael addition.

### 2.24.2.1.2 From a preformed heterocyclic ring

(i) *From pyranols*

Dehydration of mannitol with formic acid yields 2-methylpyran ⟨20LA(422)133⟩.

(ii) *From pyrylium salts*

The 2- and 4-positions of pyrylium salts are readily attacked by nucleophiles (see Chapter 2.23) and in the absence of a good leaving group these reactions give rise to 2*H*- and 4*H*-pyrans, respectively. The 2-adduct is thermodynamically favoured by virtue of the conjugated double bonds, although this isomer often undergoes ring opening to the valence tautomer. A $^{13}$C NMR study of the reaction of methoxide ion with 2,4,6-triarylpyrylium salts has established that the 2-methoxy-2*H*-pyran (**25**; R = Me) is the sole product. No evidence was found for either the ring-opened valence bond isomer or the 4*H*-pyran ⟨79H(12)775⟩.

Suitable substitution at C-2 and C-4 thus causes a shift in the pyran–dienone equilibrium towards the heterocyclic system. Generally, whilst the adducts can be isolated, they tend to decompose during normal purification methods ⟨80JOC5160⟩. However, provided substituents are additionally present in the 3- or the 3- and 5-positions, the compounds are sufficiently stable to be isolated in a pure state ⟨81ZC260⟩. The pyran results when a suspension of the pyrylium salt is heated with an equimolar quantity of triethylamine, which acts as the proton acceptor, or simply on treatment with sodium alkoxide.

The reaction of pyrylium salts with Grignard reagents or organolithium compounds leads to the pseudobase by attack at C-2 ⟨59CB2042⟩. A similar reaction occurs with the anions derived from malononitrile and diethyl malonate, although the pyran derivatives were not always isolated ⟨59CB2046⟩.

The influence of pH on the reaction of 2,4,6-triphenylpyrylium salts is significant. In neutral conditions the sole product is the ring-opened enedione (**26**), but at high pH the 2-hydroxypyran (**25**; R = H) is formed (Table 1) ⟨69TL2195⟩. Reaction of this pyran with Grignard reagents affords an equimolar mixture of the substituted 2*H*- and 4*H*-pyrans (**27**) and (**28**), whilst 2,4,6-triphenyl-2*H*-pyran results on reduction with potassium borohydride.

| | | | Ph | R |
| --- | --- | --- | --- | --- |
| (25) | (26) | (27) | (28) |

**Table 1**  Influence of pH on the Basic Hydrolysis of 2,4,6-Triphenylpyrylium Perchlorate

| pH | 7.5 | 10.35 | 11.35 | 11.80 | 12.48 | 14 |
| --- | --- | --- | --- | --- | --- | --- |
| (**26**) (%) | 100 | 100 | 80 | 63 | 16 | 0 |
| (**25**; R = H) (%) | 0 | 0 | 20 | 37 | 84 | 100 |

### (iii) *From pyranones*

The reaction of Grignard reagents with pyran-2-ones at low temperature leads to 2,2-disubstituted 2*H*-pyrans ⟨61CB1784⟩. These appear to exist solely in the pyran form since they do not form 2,4-dinitrophenylhydrazones and their UV spectra are similar to those of *s-cis*-dienes.

## 2.24.2.2  2*H*-Chromenes (2*H*-1-Benzopyrans)

There has long been an interest in the chemistry of 2*H*-chromenes or chrom-3-enes as a consequence of their widespread natural occurrence ⟨B-63MI22400, B-73MI22400⟩. The discovery that some methoxychromenes possess insecticidal properties stimulated activity in this area which led to the development of new and improved syntheses of chrom-3-enes. The reader is referred to two comprehensive surveys of chromene chemistry ⟨75AHC(18)159, 77HC(31)11⟩.

### 2.24.2.2.1  Formation of one bond

#### (i)  *Adjacent to a heteroatom*

The photochemical ring opening of chrom-3-enes (**29**) has been shown to produce *o*-quinoneallides (**30**). These have been trapped as the phenols (**31**) and (**32**) by their *in situ* reduction with lithium aluminum hydride ⟨67JPC4045⟩. The reverse process, thermal electrocyclic ring closure of the allide, is a facile reaction and hence any synthesis of an *o*-quinoneallide is effectively a synthesis of a chrom-3-ene. Many chromene syntheses are considered to proceed through the intermediacy of a quinoneallide and all of these are treated as syntheses involving the formation of one bond adjacent to the heteroatom (Scheme 4). Quinone methides have been reviewed ⟨64QR347⟩.

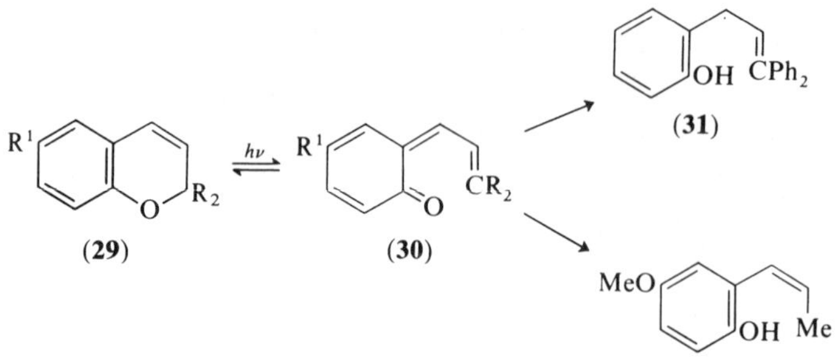

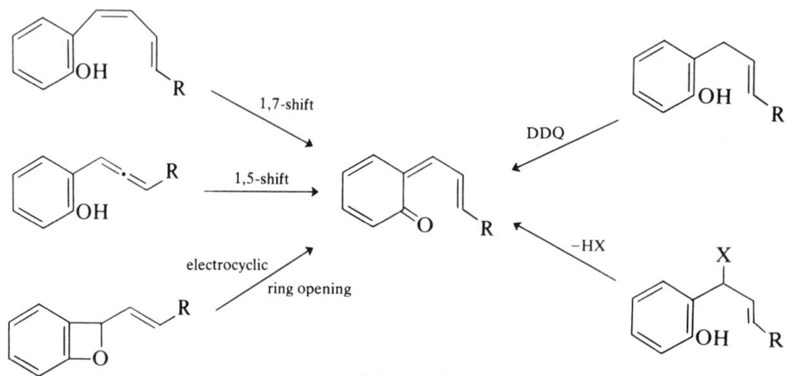

**Scheme 4**

A particularly useful chromene synthesis involves the thermal rearrangement in an inert solvent of aryl propargyl ethers (**33**) ⟨62CPB926, 63CPB1042⟩ which are conveniently prepared from a phenol and a chloroalkyne. An indication of the ease of cyclization is apparent from the presence of chromene in the ethers prepared in this manner.

The reaction is considered to proceed *via* a Claisen-like [3,3]-sigmatropic rearrangement which is followed by a [1,5]-sigmatropic shift. An electrocyclic rearrangement completes the process. Supporting evidence for this mechanism includes the internal trapping of the allenic intermediate (**34**) as the tricyclooctenone derivative (**35**) and the observation that the allenylphenol also rearranges thermally to the chromene ⟨68HCA1510⟩.

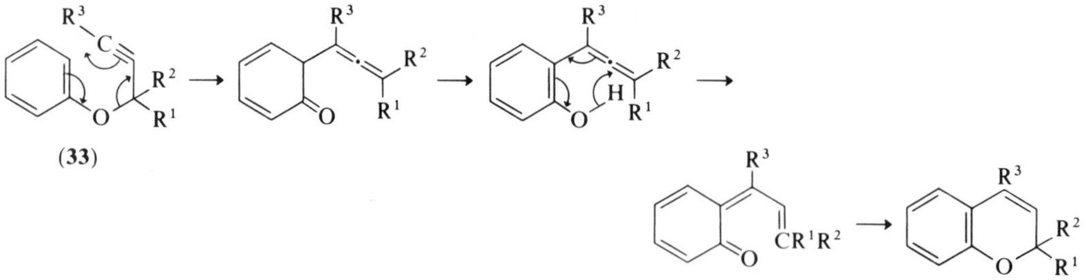

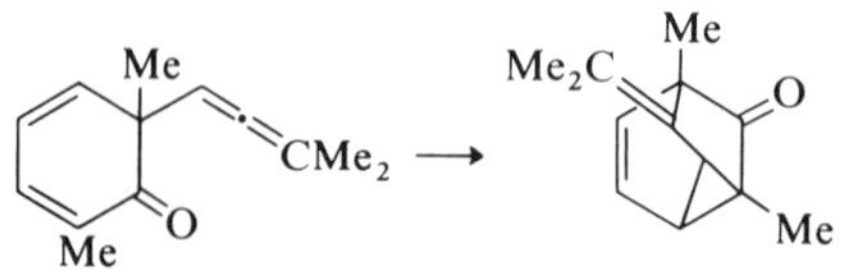

Mercury(II) oxide and acetic acid effect the cyclization of 1,4-diaryloxybut-2-ynes to 4-aryloxymethylchromenes. The transformation was attributed to cyclization of the butanone which resulted from hydration of the alkyne ⟨72JHC489⟩. However, it has since been shown that similar butanones do not cyclize to chromenes under the cyclization conditions ⟨78JOC3856⟩. Instead, a mechanism is proposed which involves a charge-induced Claisen rearrangement which is triggered by π-complex formation between the metal ion and the triple bond. The metal ion also participates in the conversion of the allenic phenol to the chromene. Catalysis by silver ions is particularly efficient, doubling the rate of cyclization at room temperature ⟨73HCA2981⟩.

A similar catalytic action was observed in the cyclization of the hexa-2,4-diyne (37), obtained by a Glaser oxidative coupling of the propargyl ether (36), to the 4,4′-bichromene (38) ⟨73TL5003⟩. In the original work, *N,N*-diethylaniline was used as the solvent, but *o*-dichlorobenzene and cumene are equally suitable. In fact, whilst 3-(4-nitrophenoxy)propyne decomposed in the basic solvent, successful cyclization to the chromene was achieved in dichlorobenzene. Although it has been recognized that a basic solvent is preferred in a true Claisen rearrangement, it appears that this is not a prerequisite for the propargyl ether → chromene transformation. The involvement of solvent in the mechanism has been postulated ⟨79MI22400⟩ with the conclusion that ethers derived from phenols which are more acidic than *p*-acetylphenol yield chromenes in a non-basic solvent, but may also form benzofurans in diethylaniline. It has been shown that the use of potassium carbonate in sulfolane also leads to the benzofuran, even from phenols of relatively low acidity ⟨73HCA1457⟩.

(36)            (37)            (38)

The reaction was initially utilized to synthesize naphthopyrans and a range of chromenes substituted in the aryl ring. Later developments have included the introduction of a substituent at C-1 of the alkyne ⟨67T1893⟩ enabling 4-substituted chromenes to be prepared, and the use of 3-substituted alkynes, which leads to 2-substituted derivatives of chrom-3-ene. It is also possible to produce 2,2-disubstituted chromenes in which the substituents may be identical or different ⟨78UP22400⟩. In view of the natural occurrence of the 2,2-dimethylchromene system and of the insecticidal properties of 7-methoxy- and 6,7-dimethoxy-2,2-dimethylchromenes, this approach has found wide application ⟨71AJC2347⟩. Yields vary over a wide range, depending especially upon the nature of 3-substituents in the propargyl ether, but also on the substituents in the aromatic ring. When the alkyne is unsubstituted, yields of up to 50% result, but yields of the order of 80–90% are reported for 2,2-dimethylchromene derivatives. This increase in yield, which may be a result of a decreased tendency to polymerize, is accompanied by an increase in the rate of cyclization. It has been proposed ⟨72JOC841⟩ that the latter arises from a conformational effect. When $R^1 = R^2 = H$, as in (39), the methylene group is nearer to the benzene ring than is the ethynyl moiety, a conformation which is unfavourable for cyclization. However, the more bulky methyl groups will take up a position remote from the aromatic ring and in (40) the alkynic bond is in a position to react. Nevertheless, it has been pointed out that neither conformation appears to have the correct geometry for the rearrangement.

(39)            (40)

The syntheses of xanthyletin (41), lonchocarpin (42), lapachenole (43) and cyclopiloselloidin (44) illustrate the application of this route to natural product chemistry ⟨71AJC2347, 78PIA(A)247, 80CB261, 81M119⟩.

Substituents in the *p*-position of the aromatic ring have little effect on the yield of chromene, but it is reported that electron-releasing groups in the *m*-position lead to increased yields whilst electron-withdrawing groups give much lower amounts of product. Such data are explained if it is assumed that the Claisen rearrangement is the rate determining step.

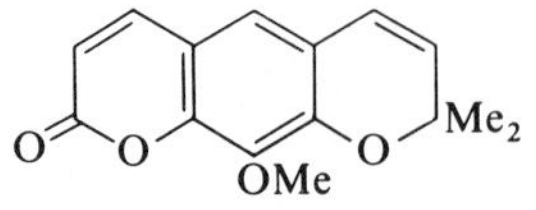

(41)

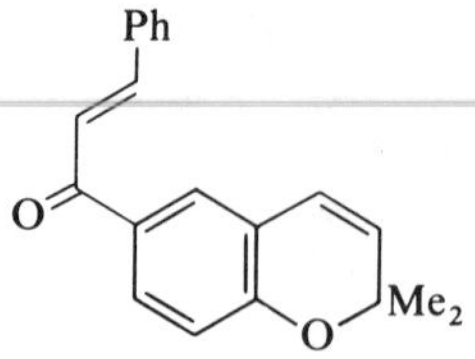

(42)

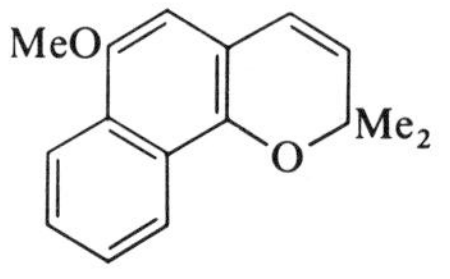

(43)

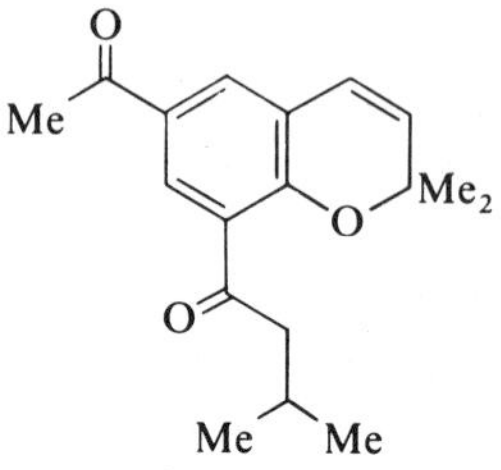

(44)

Electron-withdrawing *m*-substituents will decrease the electron density at the aromatic terminus of the [3,3]-rearrangement, thereby retarding the reaction. This allows polymerization of the terminal alkyne to compete successfully with the cyclization. However, the presence of a *m*-substituent has a much more significant effect; the cyclization of *m*-substituted aryl propargyl ethers can lead to two isomeric products (**46a** and **b**).

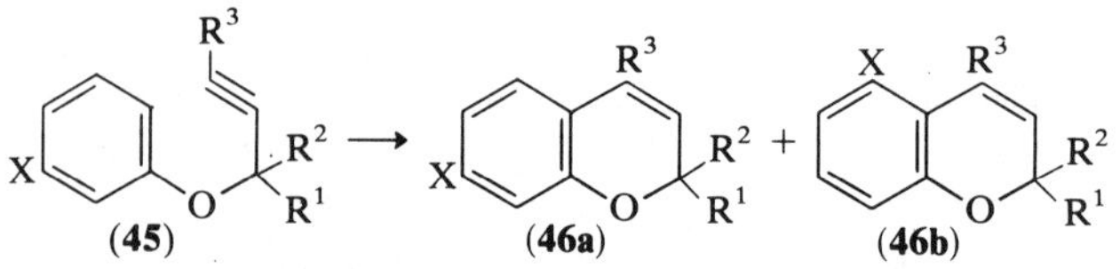

Two groups have reported that the cyclization proceeds in a regiospecific manner, leading only to the 7-substituted chromene ⟨71AJC2347, 63CPB1042⟩. However, other workers have found that both the 5- and 7-substituted isomers result from a variety of simple *m*-substituted aryl propargyl ethers and that in many cases the former isomer predominates ⟨74JOC881, 78UP22400⟩. These results are presented in Table 2. In many of the cases cited the isomers were separated on a preparative scale and their structures were elucidated by NMR spectroscopy, confirmation being achieved in some instances by unequivocal synthesis.

**Table 2**  Isomer Distribution in the Cyclization of *m*-Substituted Phenyl Propargyl Ethers

| | Substituents in ether (45) | | | Total yield (%) | Isomer distribution | | Ref. |
| X | $R^1$ | $R^2$ | $R^3$ | | (46a) | (46b) | |
|---|---|---|---|---|---|---|---|
| OMe | H | H | H | 51 | 46 | 54 | 74JOC881 |
| Me | H | H | H | 89 | 53 | 47 | 74JOC881 |
| OCOMe | H | H | H | 51 | 43 | 57 | 74JOC881 |
| NEt$_2$ | H | H | H | 20 | — | 100 | 74JOC881 |
| OCOPh | H | H | H | 63 | — | 100 | 79H(12)451 |
| OMe | H | H | Me | 86 | 53 | 47 | 74JOC881 |
| Me | H | H | Me | 92 | 64 | 36 | 74JOC881 |
| OCOMe | H | H | Me | 96 | 54 | 46 | 74JOC881 |
| COMe | H | H | Me | 75 | 45 | 55 | 74JOC881 |
| OMe | Me | Me | H | 84 | 51 | 48 | 78UP22400 |
| Me | Me | Me | H | 53 | 39 | 60 | 78UP22400 |
| Cl | Me | Me | H | 72 | 30 | 70 | 78UP22400 |
| Br | Me | Me | H | 70 | 30 | 70 | 78UP22400 |
| NO$_2$ | Me | Me | H | 92 | 30 | 70 | 78UP22400 |

Although the isomeric chromenes (**48**) and (**49**) obtained from the isoflavanone (**47**) and 3-chloro-3-methylbut-1-yne could not be separated by chromatography, a chemical separation was achieved. Ring contraction of the 7-methoxy isomer (**49**) to the benzofuran (**50**) occurred on reaction with thallium(III) nitrate in methanol; the 5-methoxy compound was unaffected. The resulting mixture was readily separated ⟨81G211⟩.

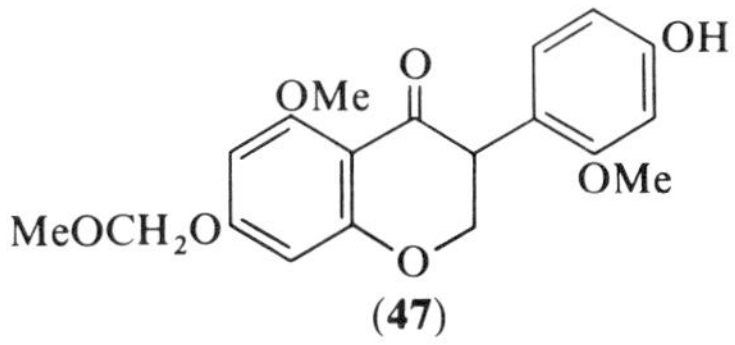

(47)

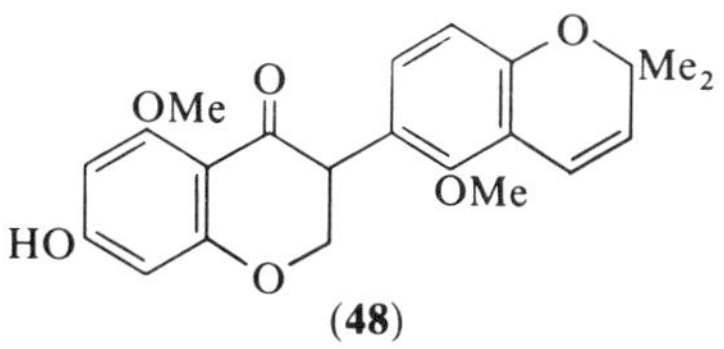

(48)

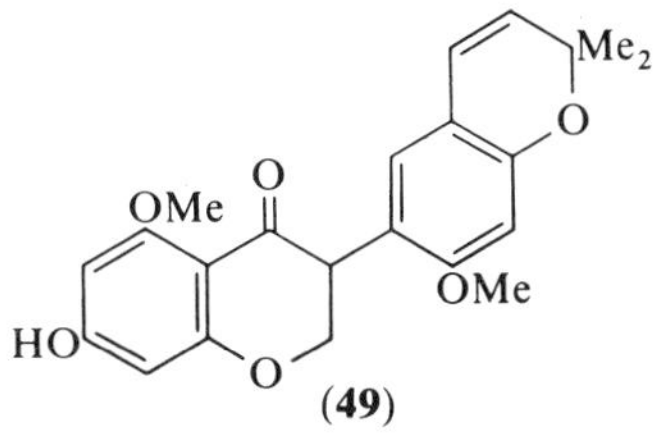

(49)

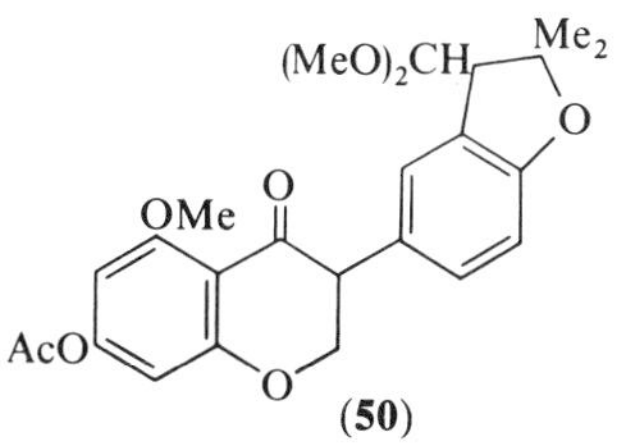

(50)

Regioselectivity in the cyclization of aryl propargyl ethers is clearly subject to subtle interactions, since minor changes can alter the isomeric composition of the product. For instance, whereas 2,4-dihydroxyacetophenone yields only the 5-hydroxychromene (52; R=H), acetylation of the intermediate propargyl ether (51) and subsequent cyclization gives a 2:1 mixture of the 5- and 7-substituted chromenes (52; R = Ac and 53; Scheme 5) ⟨72CB863⟩. The regioselectivity has been attributed to the influence of steric effects in either the initial Claisen rearrangement or the subsequent enolization of the dienone ⟨74JOC881⟩.

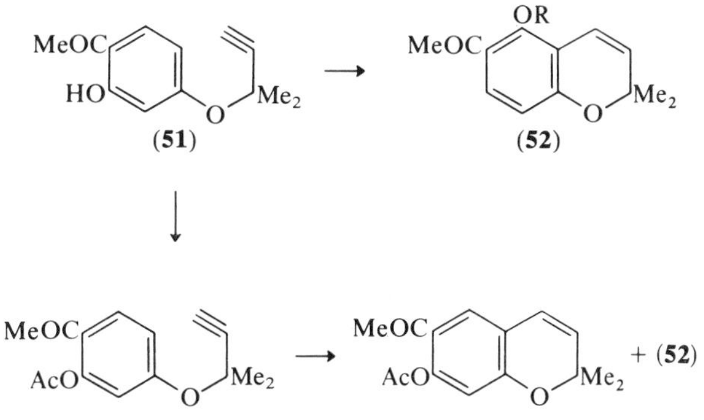

(51)

(52)

(53)

**Scheme 5**

Crombie and his coworkers have proposed that the stability of the transition state which leads to the intermediate dienone plays a significant role in determining the orientation of the product ⟨74JCS(P1)1007⟩. Thus, the ether (51) yields only the 5-hydroxychromene as a consequence of the stabilization energy associated with the chelated system. However, acetylation of the hydroxyl group prior to cyclization of the ether prevents chelation with the resulting formation of both isomers.

Not all of the experimental results can be explained in this way. The unexpected conversion of the ether (54) into the linear chromene (56) in fact proceeds through a base-catalyzed Wessely–Moser rearrangement of the initial product, the angular chromene (55) ⟨71AJC2347⟩.

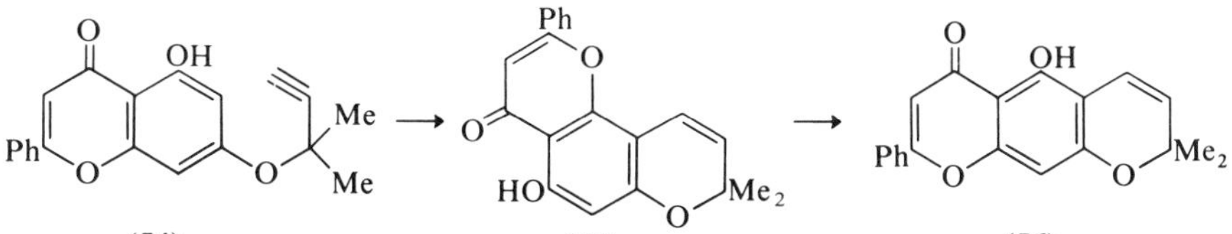

(54)

(55)

(56)

It is possible to direct cyclization of the propargyl ethers to give linear rather than angular chromenes by blocking the more reactive site with iodine prior to prenylation ⟨81JCS(P1)1697⟩. The iodo group is lost during the cyclization. For example, whereas the hydroxycoumarin (57) affords the angular chromene (58), the·linear isomer (60) results from the hydroxy-iodocoumarin (59; Scheme 6). This blocking technique has been applied to a range of dihydroxyacetophenones for use in the synthesis of naturally occurring chromenes such as eupatoriachromene (6-acetyl-7-hydroxycoumarin) ⟨82T609⟩.

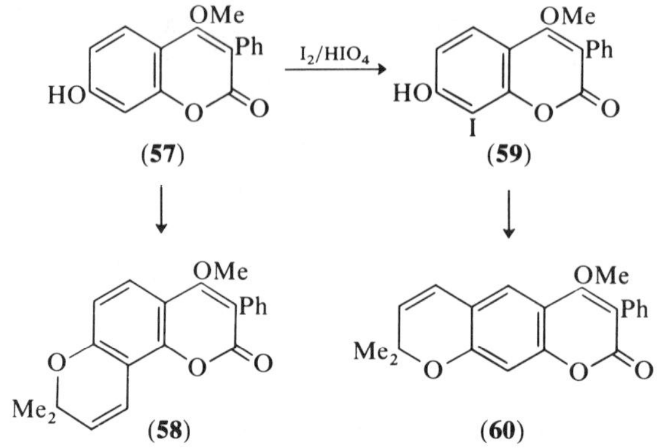

**Scheme 6**

The selective formation of 5-hydroxy-2,2-dimethylchromenes instead of the usual 7-hydroxy isomer has been accomplished with organocopper reagents ⟨79JCS(P1)201⟩. Olivetol bis(tetrahydropyranyl ether) (61) is selectively metallated at C-2 and the resulting homocuprate (62) reacts with 3-acetoxy-3-methylbut-1-yne to yield 5-hydroxy-2,2-dimethyl-7-pentylchromene (63; Scheme 7).

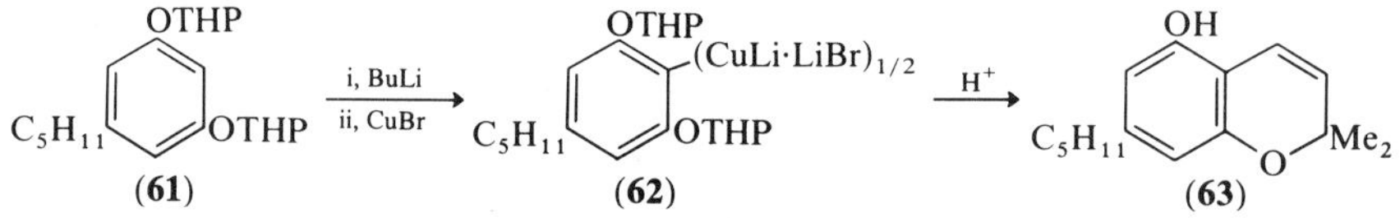

**Scheme 7**

The condensation of $\alpha,\beta$-unsaturated carbonyl compounds with phenols has received much attention as a route to chromenes. Problems associated with the instability of alkenals have been largely overcome by the use of masked forms of these reagents. The derived acetals have proved particularly useful in this respect. It is also advantageous to use the acetals of hydroxyalkanals and to introduce the double bond during the course of the reaction. Thus, 4,4-dimethoxy-2-methylbutan-2-ol serves as the equivalent of the unstable 3-methylbut-2-enal.

An extensive study of the pyridine-catalyzed condensation has been carried out ⟨74JCS(P1)1007⟩. The reaction involves *C*-hydroxyalkenylation of a phenoxide ion, with subsequent cyclization of the *o*-quinoneallide leading to the chromene.

It is clear from the proposed mechanism that the initial step will be facilitated by the presence of a second hydroxyl group in the phenol, oriented *meta* to the first. In view of the importance of polyphenolic molecules in natural product chemistry, this synthesis takes on a special significance. Thus, the reaction of dihydric phenols with citral (65) has been used to synthesize a diverse range of natural products. However, such compounds can, and in some cases do, give rise to a mixture of isomeric products. Olivetol (64), for instance, yields a mixture of mono- and bis-chromenes of which cannabichromene (66) is a significant component ⟨71JCS(C)796⟩. On the other hand, the phenol (67) affords a high yield of isoevodionol (68) on reaction with 4,4-dimethoxy-2-methylbutan-2-ol ⟨71JCS(C)811⟩.

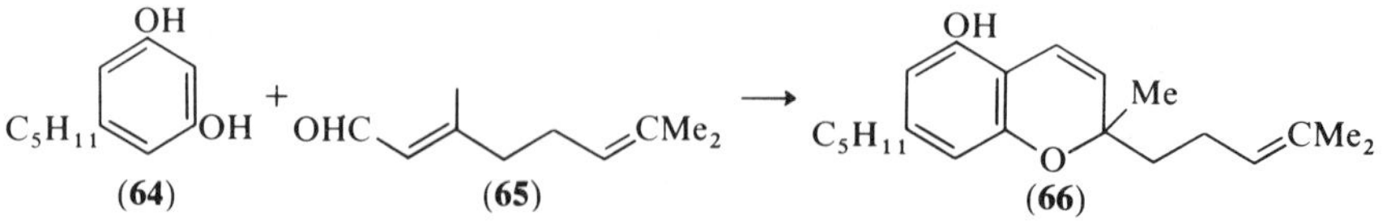

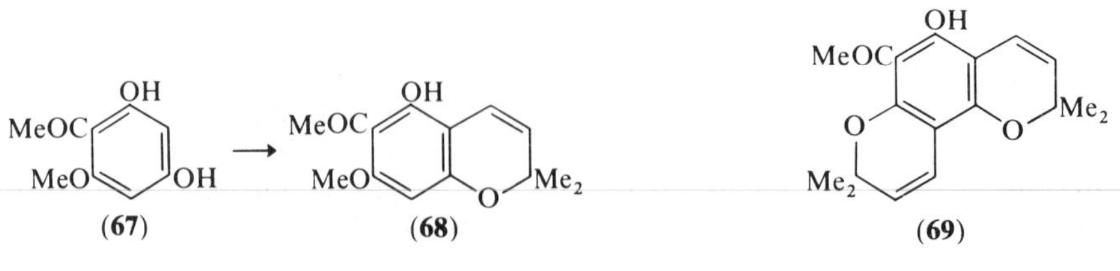

The reaction of 2,4,6-trihydroxyacetophenone with the acetal also yields a mixture of products, but is of interest in that the addition of pyridine hydrochloride has a beneficial effect on the formation of the major product (**69**). This compound arises by the construction of two pyran rings and involves cyclization on to a hydroxy group adjacent to the acyl side-chain ⟨72JCS(P1)25⟩, although it is recognized that both the 2- and 6-hydroxy groups cannot be simultaneously chelated with the acyl group.

A mixture of linear and angular chromenes (**70**) and (**71**) resulted from the reaction of 1,3-dihydroxyacridone with the hydroxyacetal. Methylation of the angular isomer gave acronycine, which shows broad-spectrum antitumour activity.

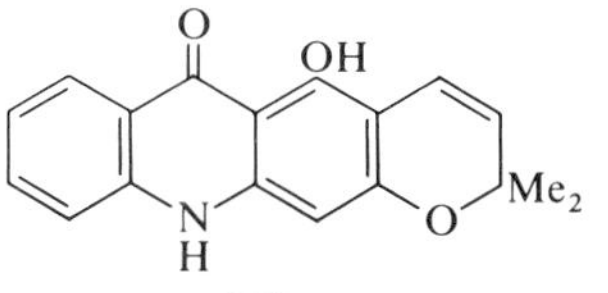

(**70**) 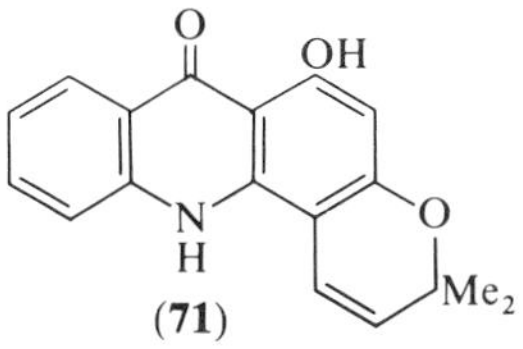          (**71**)

Of the simpler examples, the formation of a 2-methylchromene and a flav-3-ene from the reaction of 2,4-dihydroxyacetophenone with but-2-enal and cinnamaldehyde, respectively, are worthy of mention ⟨71JCS(C)811⟩.

The regiospecificity of chromene formation has been investigated and the following conclusions have been reached. In molecules involving both chelated and free hydroxyl groups, only the latter are involved in chromene formation. Thus, resacetophenone (**72**) affords only the chromene (**73**) with 3-methylbut-2-enal. Clearly, further control of the reaction occurs since even when the acidity factor is taken into account, two modes of cyclization are possible, but are not observed. This regiospecificity has been rationalized in terms of additional stabilization energy associated with structure (**74**) compared with the resonance form (**75**) ⟨74JCS(P1)1007⟩. Thus, the transition state leading to attack at C-3 is expected to be of lower energy than that which would result in reaction at C-5. Similar reasoning accounts for the formation of the benzodipyrans (**76**) and (**77**) in the reaction of citral with 5,7-dihydroxycoumarin and for the lack of the isomer involving (**78**) in which the pyranone stabilization energy is lost.

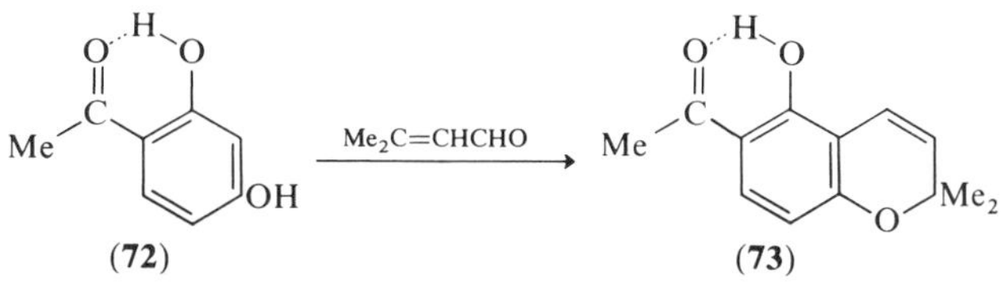

(**72**)          (**73**)

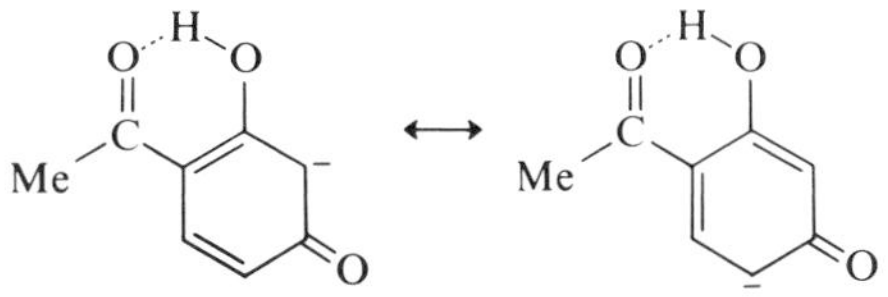

(**74**)          (**75**)

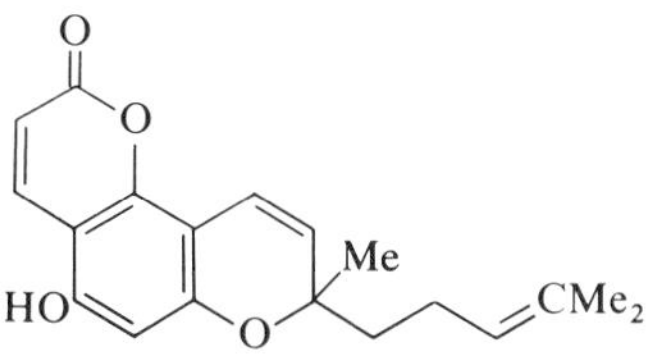

(**76**) 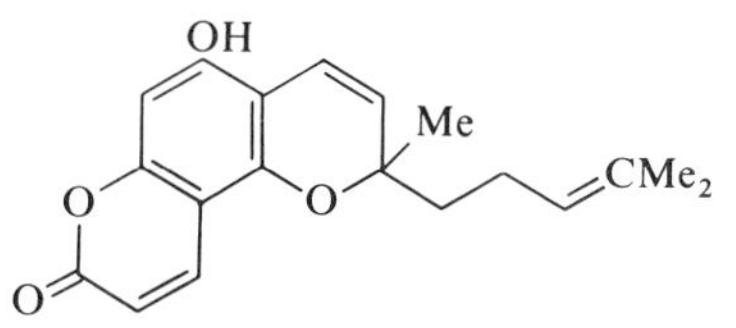          (**77**) 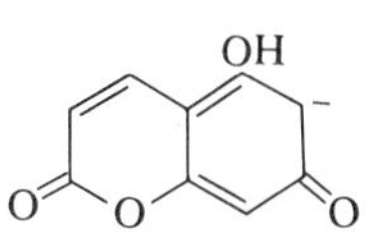          (**78**)

In cases where the various factors controlling regiospecificity are in conflict, mixtures of products may arise. Thus 1,3,5-trihydroxyxanthone affords a mixture of the chromenes (**79**) and (**80**) ⟨70JCS(C)1662⟩. The transition state leading to the linear chromene allows retention of the chelate stabilization with loss of pyranone resonance, whilst the reverse situation obtains in the formation of the angular molecule.

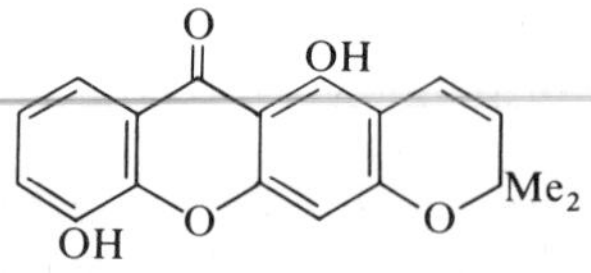

(79) 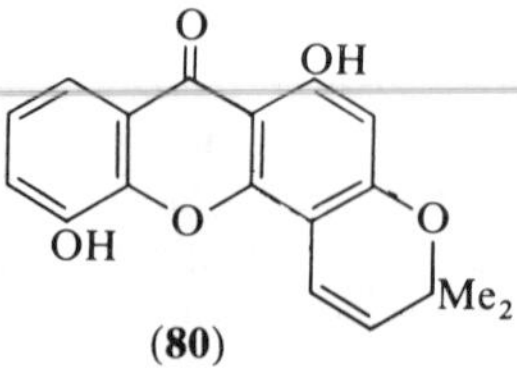          (80)

The pyridine-catalyzed aromatic proton exchange with deuterium provides a simple indication of the ability of a phenol to participate in chromene formation. Only those phenols which undergo exchange react with the unsaturated carbonyl compound, the attack occurring at the positions of deuteration ⟨64JA2084⟩.

The use of *t*-butylamine in place of pyridine dramatically increases the yield of cannibi-chromene (**66**) from 5-*n*-pentylresorcinol (olivetol) and citral ⟨78JHC699⟩. It is not known if this is of general application.

Unfortunately the pyridine-catalyzed condensation is largely unsuccessful with unsaturated ketones. Furthermore, there are few examples of the use of monohydric penhols in this synthesis. 1-Naphthol affords the naphthopyran (**81**) with citral, though in poor yield, and 2-naphthol gives (**82**). The latter example is of interest since it provides a further illustration of the regiospecificity arising from resonance stabilization.

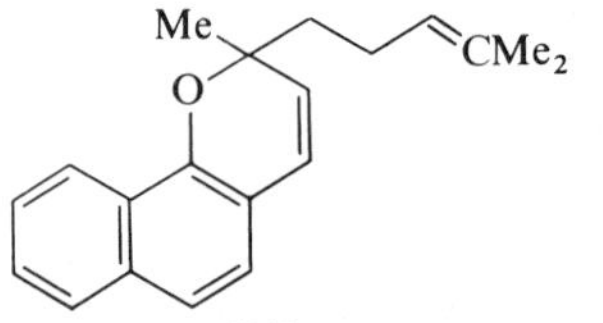

(81) 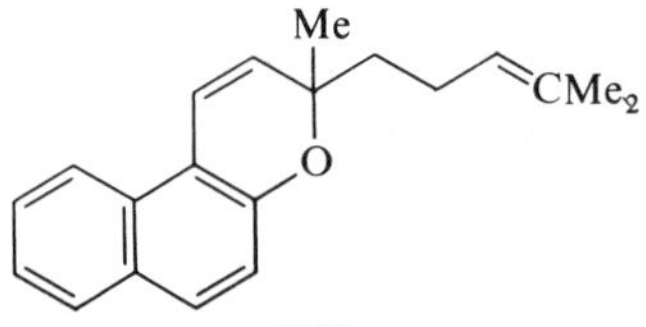          (82)

2-Fluoro-1,1-dimethoxy-3-methylbut-2-ene reacts smoothly with a number of phenols. Monohydric phenols containing either electron-releasing or electron-withdrawing substituents afford satisfactory yields of 3-fluoro-2,2-dimethylchromenes. The regioselectivity of this synthon appears to parallel that described above ⟨80JHC1377⟩.

The titanium(IV) salts of monohydric phenols and the magnesium salts of dihydric phenols react with α,β-unsaturated aldehydes and, of especial note, with alkenones in toluene solution to give good yields of chromenes ⟨79JOC803⟩. A wide variety of substituents can be tolerated in the aromatic ring and the procedure has been extended to the synthesis of flavenes and naphthopyrans. A most attractive feature is the high regioselectivity which is achieved. For instance, olivetol affords 7-hydroxy-2,2-dimethyl-5-pentylchromene in 77% yield with only traces of the isomeric compound. Under the influence of pyridine the yields of the two chromenes are 7.5 and 8.3%, respectively. It therefore appears that the coordinating ability of the metal ion plays an important role in both the activation and orientation of the process.

A number of chromene syntheses have been devised which start from *o*-hydroxybenzaldehydes. These offer the advantage that the orientation of the product is assured.

Schweizer and his group have developed a useful general synthesis which is based upon the reaction of the sodium salt of salicylaldehyde with a phosphonium salt ⟨68JOC2416⟩. The parent chromene arises from the use of triphenylvinylphosphonium bromide (**83**); acetonitrile is the solvent, although DMF is also suitable. The reaction is considered to proceed through a Michael addition followed by a Wittig reaction. Application of this approach to the carbazole system led to an unequivocal synthesis of the alkaloid girinimbine (**84**) ⟨70TL1665⟩.

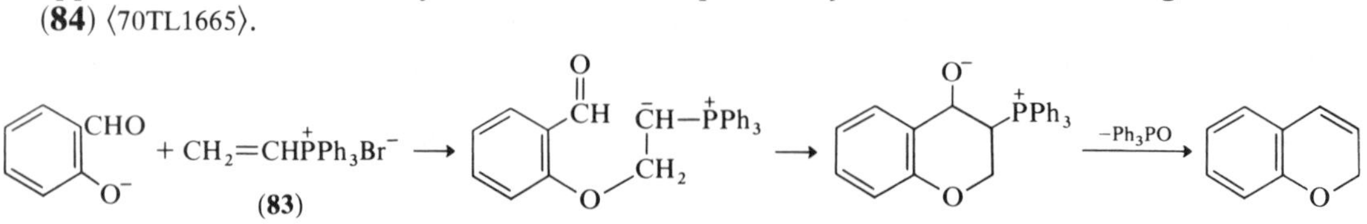

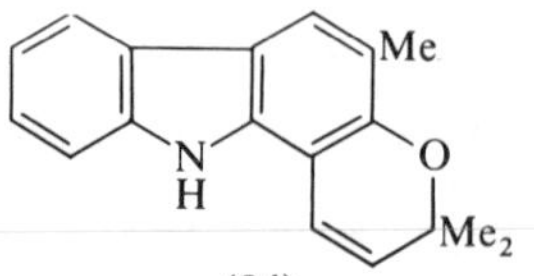

(83)

(84)

The route has been extended to include substituted derivatives of both reactants ⟨73JOC1583⟩. In general, substituted salicylaldehydes give satisfactory yields, although the 5-nitro compound affords only a 27% yield and that in the absence of solvent. The reaction fails with *o*-hydroxybenzophenone and the corresponding acetophenone. 3-Methyl-chromenes resulted with substituted vinylphosphonium salts, though yields were lower presumably because of steric inhibition. When the reactants were heated without solvent, 2-methylchromenes were obtained. This was attributed to a rearrangement of the phosphonium salt, since 3-methylchromene does not rearrange to the 2-isomer under these conditions.

A further development of this route ⟨71JOC4028⟩, involving the cyclization of 3-(*o*-formylphenoxy)propyltriphenylphosphonium salts in methanol, has led to high yields of 2-methylchromene. In the detailed mechanism which has been proposed, an *o*-quinoneallide is considered to arise by the ring opening of an initially formed benzoxepin derivative. Indeed, 2,3-dihydro-1-benzoxepin is the major product of the reaction in aprotic solvents.

*N*-Ethyl(diethylphosphono)methylketenimine, readily prepared from the amide, is of value as an annelating reagent ⟨79CC900⟩. The sodium salt of salicylaldehyde reacts at the activated central carbon atom of the ketene and a subsequent intramolecular Horner–Emmons reaction results in cyclization to the chromene.

Salicylaldehyde and allyltriphenylphosphonium salts undergo a normal Wittig reaction in DMF to yield a butadienylphenol. Thermal cyclization involves a [1,7]-sigmatropic shift and leads to the 2-alkylchromene *via* a quinoneallide. Detailed studies ⟨72HCA1828⟩ have shown that the *cis* form of the butadienylphenol (**86**) rearranges quantitatively at 121 °C to the chromene (**87**), whilst the *trans* isomer (**88**) ring closes at 190 °C. The latter rearrangement is induced by a thermal isomerization of the 1'-double bond *via* a [1,5]-hydrogen shift. Deuterium labelling experiments show that at 210 °C the chromene is in equilibrium with the quinoneallide.

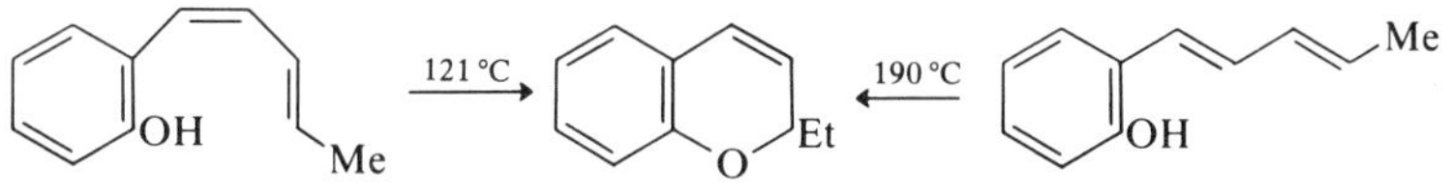

(**86**)　　　　　(**87**)　　　　　(**88**)

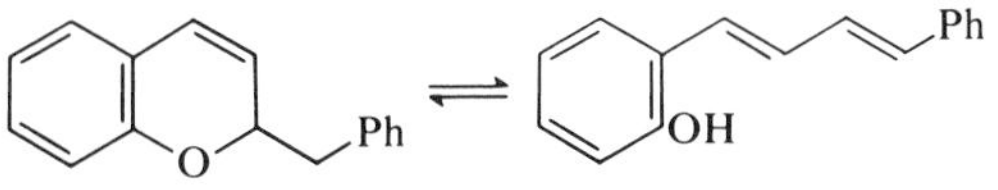

(**89**)　　　　　(**90**)

An equilibrium mixture containing almost equal amounts of the chromene (**89**) and dienylphenol (**90**) results when either component is heated in diglyme above 200 °C.

A variation on the above theme utilizes *o*-quinones as the benzenoid component ⟨70AC(R)564⟩. Reaction with 3,3-dimethylallyltriphenylphosphonium bromide affords 2,2-dimethylchromenes, again through an intermediate allide.

Ethyl cyanoacetate in the presence of piperidine may also be used as the carbanionic component in reactions with salicylaldehyde. The initial Knoevenagel condensation is followed by a [1,7]-sigmatropic shift and cyclization to the 2-iminochromene derivative which adds another cyanoacetate molecule ⟨67AP1⟩.

Salicylaldehyde and *o*-hydroxyacetophenones react with vinylmagnesium halides to give *o*-hydroxyphenylalkenols (**91**). Upon heating in diglyme, water is eliminated from the alcohol and a chromene results ⟨72HCA1675⟩. Yields are good when R is methyl or phenyl, but only fair when R is hydrogen or vinyl. The reaction has been extended to 2-hydroxynaphthaldehyde which affords 3*H*-naphtho[2,1-*b*]pyran. The chromene arises by a pericyclic 1,4-elimination from the diol to give an *o*-quinoneallide of *cis* geometry, which ring-closes.

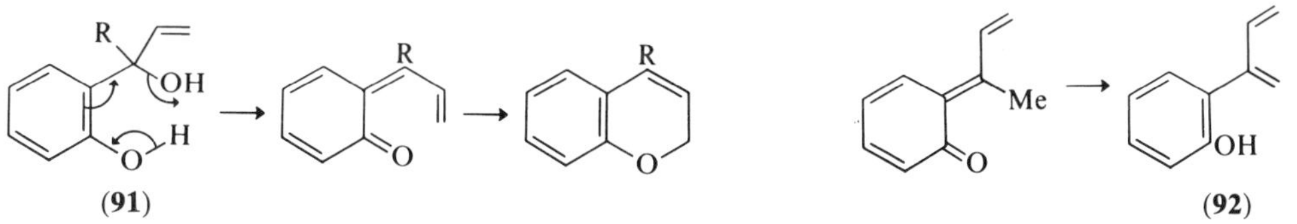

(**91**)　　　　　　　　　　　　　　　　(**92**)

**Scheme 8**

When the substituent R was methyl, a second product was isolated, which comprised a quarter of the total yield. This is a substituted butadiene (92) which is formed by a [1,5]-hydrogen shift in the *trans-o*-quinoneallide (Scheme 8).

A similar side-chain may be constructed from phenoxymagnesium halides and cinnamaldehyde, when cyclization to the chromene occurs during acidic work-up. Unfortunately, this method tends to give mixtures of chrom-3-enes and chrom-2-enes, the latter arising by isomerization of the former. However, by careful choice of experimental conditions the former isomers can be obtained. The route has also been used in the synthesis of naphthopyrans ⟨71JCS(C)2546⟩.

Several chromene syntheses are based on the reaction of *o*-hydroxybenzaldehydes with alkenes. The isolation of chromanols in some instances points to the general intermediacy of these compounds, although 1,4-elimination of water from the initial adduct is a viable alternative. In an exhaustive study of the reaction ⟨75MI22401⟩, a range of substituted salicylaldehydes have been reacted with alkenes containing electron-withdrawing groups adjacent to the unsaturated linkage. Yields varied considerably with the solvent used and the period of heating.

This method has found particular application in the synthesis of spiro-annelated chromenes, which are of interest because of their photochromic behaviour. The alkene is commonly a 2-methylenedihydroindole, which leads to spiro 2*H*-1-benzopyranoindolines, or the corresponding benzothiazole ⟨67BSF2824⟩. Provided a base is present to liberate the free heterocyclic base, indolinium or thiazolinium salts may be used in the synthesis. Enamine attack at the carbonyl group is followed by dehydration and cyclization (Scheme 9).

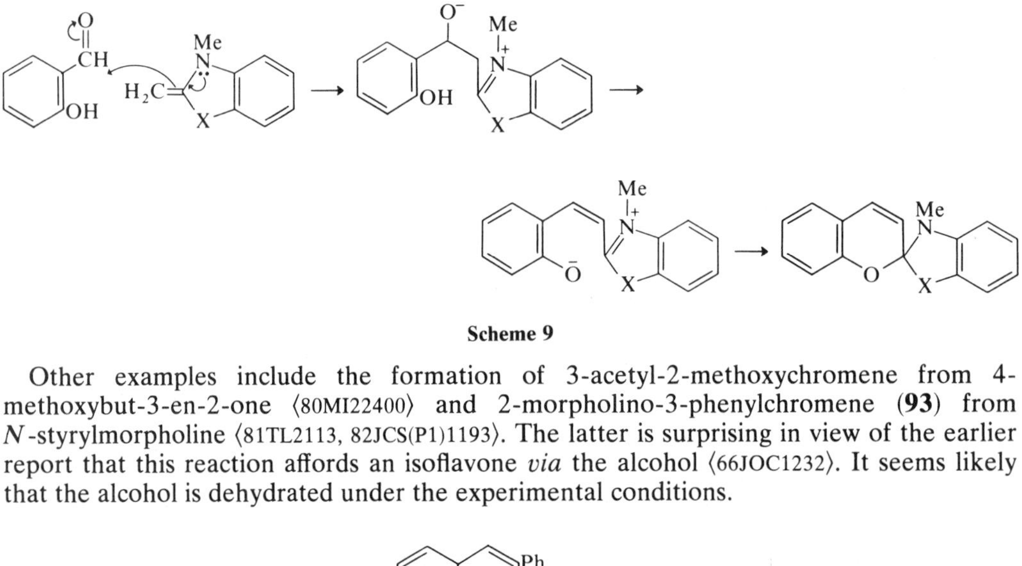

**Scheme 9**

Other examples include the formation of 3-acetyl-2-methoxychromene from 4-methoxybut-3-en-2-one ⟨80MI22400⟩ and 2-morpholino-3-phenylchromene (93) from *N*-styrylmorpholine ⟨81TL2113, 82JCS(P1)1193⟩. The latter is surprising in view of the earlier report that this reaction affords an isoflavone *via* the alcohol ⟨66JOC1232⟩. It seems likely that the alcohol is dehydrated under the experimental conditions.

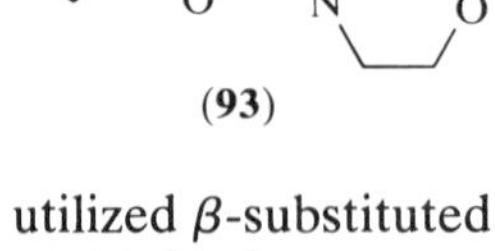

(93)

An extension of the above work utilized β-substituted alkenes, when moderate yields of substituted 2-methylchromenes were obtained.

Another development involved the use of halogenoethanes substituted at C-2 by an electron-withdrawing group and with DMF as the solvent and potassium carbonate as the base. In general, yields were not as good as those using the α-substituted alkene.

The Lewis acid-catalyzed reaction between tertiary chloroalkanes and *o*-hydroxybenzaldehydes has been used to synthesize chromenes with a long alkyl chain at C-3. The chloroalkane undergoes ready dehydrohalogenation and the route parallels those above ⟨79CJC1377⟩.

Dehydration of *o*-hydroxycinnamyl alcohols (94) may be achieved by pyrolysis or under acid catalysis. In the latter, formation of an allyl carbocation (95) is considered to precede nucleophilic attack by the phenolic oxygen atom, though this is once again the involvement of a quinoneallide in another guise ⟨72CB863⟩.

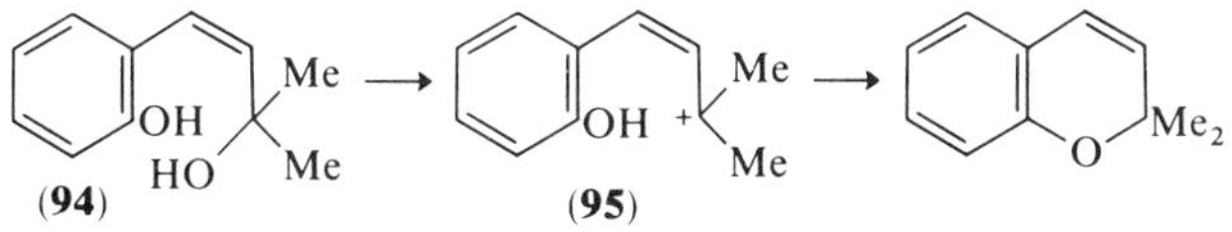

(94)    (95)

A similar pathway is thought to be involved in the synthesis of chromenes from coumarin (see Section 2.23.6.4.1), whilst the spontaneous dehydration of the alcohol (97) to the ketochromene (98), which occurs during acid hydrolysis, also presumably proceeds through the same type of intermediate. This step is an important feature of the synthesis of siccanochromene A, a fungal metabolite, from orcinol and the aldehyde (96) outlined in Scheme 10. Another point of interest in this route is the use of heteroatom-facilitated metallation to achieve the correct orientation ⟨71T6073⟩.

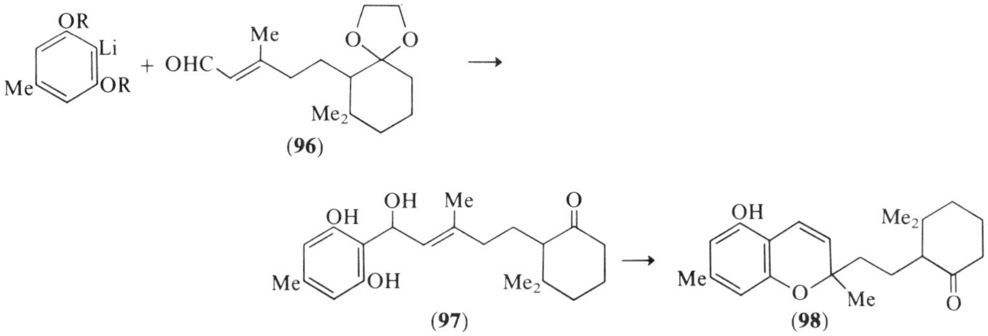

(96)

(97)    (98)

**Scheme 10**

The formation of flavenes by the reduction of 2-hydroxychalcones undoubtedly proceeds *via* the alcohol and constitutes a simple but useful synthesis ⟨67JCS(C)1933⟩. This route has been developed ⟨68AJC2247⟩ and extended to spiroannelated compounds (99).

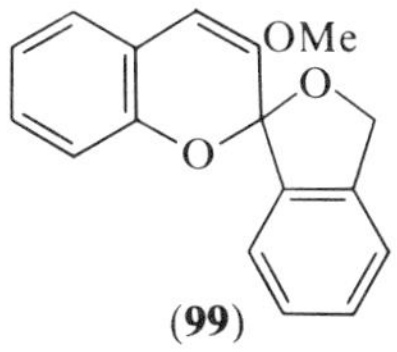

(99)

The cyclodehydrogenation of 2-(3-methylbut-2-enyl)phenols (102; $R^1 = R^2 = Me$) by 2,3-dichloro-5,6-dicyanobenzoquinone in benzene affords moderate yields of 2,2-dimethyl-chromenes ⟨71T1875⟩. The route has found particular application to the cyclization of naturally occurring dimethylallylphenols where such yields are considered to be acceptable. It is less attractive if the phenolic precursor has to be synthesized, although nevertheless successful ⟨68T4825⟩. The conversion of isolapachol (100) into dehydro-$\alpha$-lapachone (101) is typical ⟨69JOC120⟩ whilst the syntheses of carpachromene and ripariochromene A provide further illustrations ⟨78T3569, 79BCJ1203⟩.

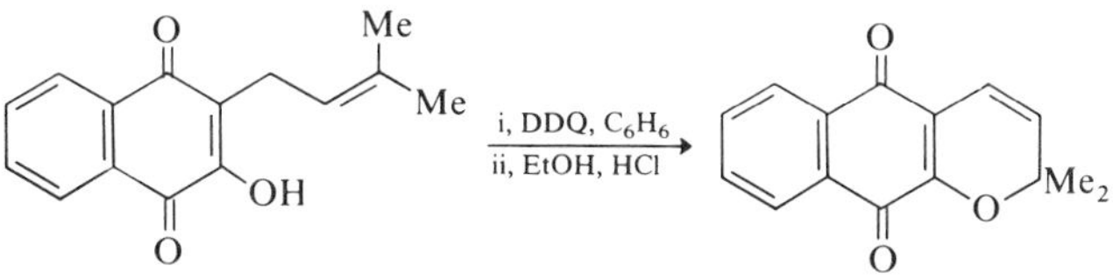

(100)    i, DDQ, $C_6H_6$    ii, EtOH, HCl    (101)

Although the reaction is successful with *o*-cinnamylphenols (102; $R^1 = Ph, R^2 = H$) leading to flav-3-enes (103) ⟨69TL907⟩, it fails with the *o*-but-2-enyl (102; $R^1 = Me, R^2 = H$) and *o*-propenyl (102; $R^1 = R^2 = H$) analogues. It appears, therefore, that the second substituent at the $\gamma$-carbon atom, *e.g.* (102; $R^1 = R^2 = Me$), plays an important role in the process.

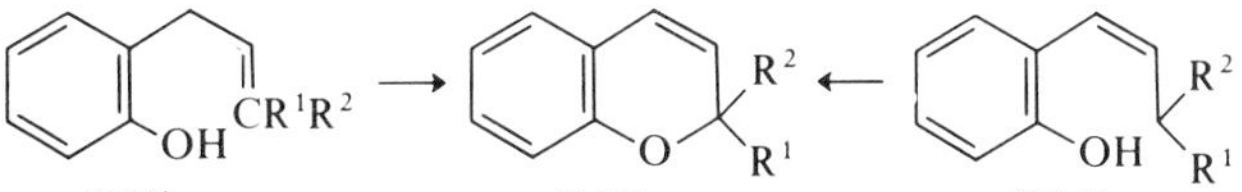

(102)    (103)    (104)

Much improved yields are reported ⟨79CC836⟩ if a slight excess of DDQ is used in ether solution. The reaction occurs at room temperature and is notably successful with the sensitive cinnamylphenols.

Chromene formation also results from the reaction of DDQ with the isomeric but-1-enyl derivatives (104). Similar structural restrictions are apparent. Thus, 2-(3-methylbut-1-enyl)phenol affords the chromene, whereas the unsubstituted compound does not cyclize.

It is of interest to note that 2-propenylphenol yields the spiropyran (105) on treatment with DDQ, whilst 2-(1-phenylpropenyl)phenol affords 2-methyl-3-phenylbenzofuran.

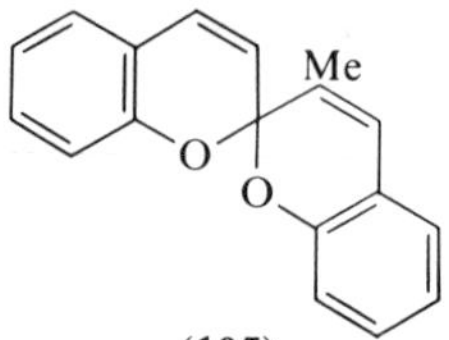

(105)

The double dehydrogenation of *o*-alkylphenols to chromenes has also been achieved with DDQ, the reaction presumably proceeding through the alkenylphenol rather than the chroman. The latter are not known for their propensity to undergo oxidation by DDQ unless a free hydroxyl group is present to allow quinone methide formation to occur. Once again, substitution at C-3 appears to be essential for success.

It is thought that the above reactions proceed by abstraction of a benzylic hydride ion by the quinone. Proton loss subsequently occurs from the carbocation (106) leading to the *o*-quinoneallide which ring-closes to the chromene.

It is also possible to oxidize *o*-allylphenols to chromenes with potassium dichromate. The oxidant may be supported on an anionic exchange resin, but it is preferable to dissolve the dichromate in benzene using Adogen 464, a mixture of methyltrialkylammonium chlorides ⟨77TL4167⟩. The oxidation is assumed to proceed through the chromate ester (107) which yields the quinoneallide ⟨79CC836⟩.

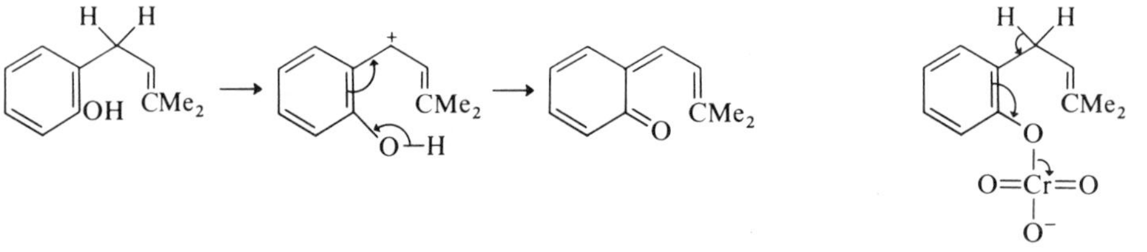

(106)  (107)

Not unrelated to the above syntheses is the intramolecular oxidative cyclization of ubiquinone (108) to ubichromenol (110) ⟨63JA239⟩. Catalyzed by bases, of which pyridine seems to be the choice example ⟨65JCS5060⟩, the reaction proceeds through the *o*-quinone-allide (109). Significantly, this process is also brought about by irradiation with visible light ⟨65LA(684)212⟩.

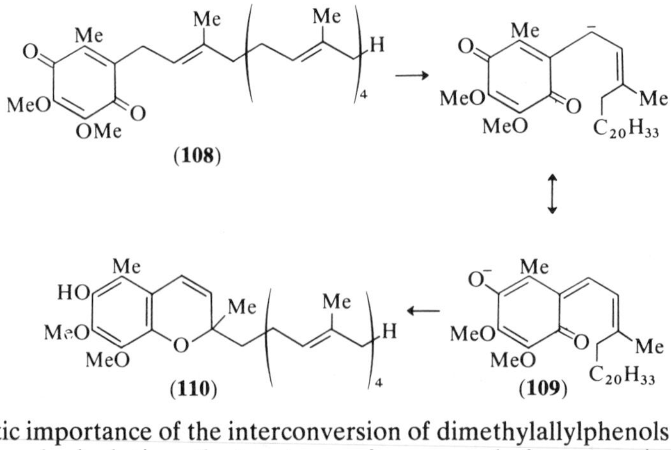

(108)  (109)  (110)

The biogenetic importance of the interconversion of dimethylallylphenols and chromenes should not be overlooked, since the two types of compound often occur in the same plant.

An unusual synthesis of chromenes involves trapping arynes with an $\alpha,\beta$-unsaturated aldehyde ⟨72JCS(P1)2903⟩. For instance, cinnamaldehyde reacts with tetrachlorobenzyne, generated from various sources, to give a *ca.* 40% yield of 5,6,7,8-tetrachloroflav-3-ene (**112**). Similar results were obtained with several alkenals and benzyne itself underwent the reaction. Isotopic labelling indicated that a [2 + 2]-cycloaddition led to a benzoxete (**111**) which ring-opened to a quinoneallide. Electrocyclic ring closure completed the process (Scheme 11).

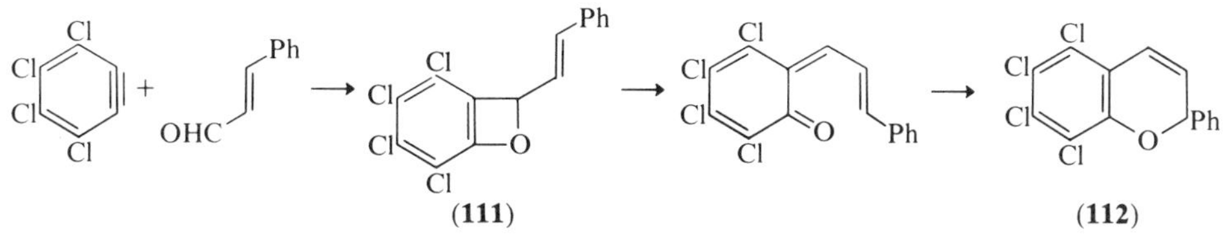

**Scheme 11**

### (ii) $\gamma$ to the heteroatom

The phosphonium salts derived from an *o*-hydroxybenzyl halide and triphenylphosphine react with a variety of $\alpha$-halogenated carbonyl compounds in the presence of sodium methoxide to yield chromenes. The speed of the alkylation and the simplicity of the work-up make this an attractive route to 2- and 3-substituted chromenes.

The action of the base on the phosphonium salt generates the betaine (**113**) which is considered to exist in solution as the open form rather than the cyclic oxaphosphole (**114**). NMR evidence confirms the intermediacy of the phosphonium salt (**115**) and in some instances these salts have been isolated. Further treatment with base affords the ylide which undergoes an intramolecular Wittig reaction to yield the chromene (Scheme 12) ⟨80T3409⟩.

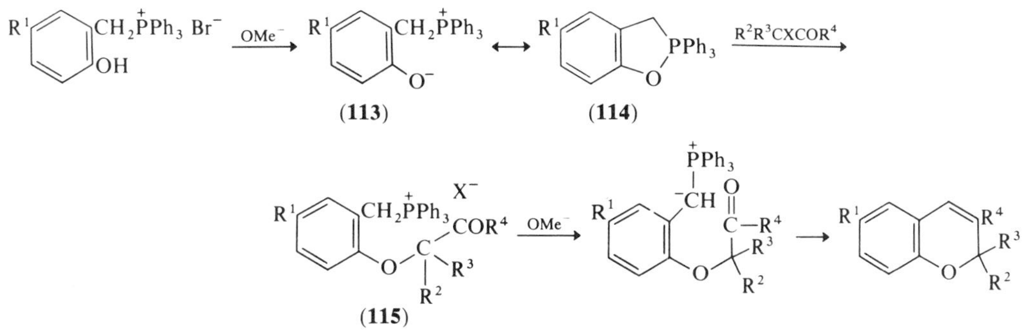

**Scheme 12**

Similar in style to the above are the base-catalyzed intramolecular cyclizations of diesters (**116**) and substituted salicylaldehydes (**117**), Dieckmann and Perkin condensations, respectively ⟨36JCS212, 36JCS419, 40JCS787⟩.

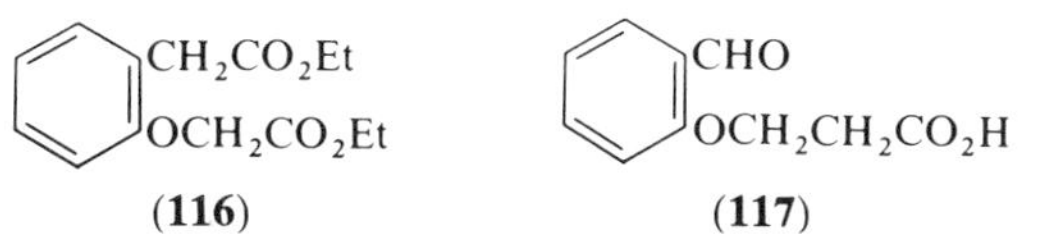

### 2.24.2.2.2 *From other heterocycles*

### (i) *From benzofuran*

Addition of dichlorocarbene to benzofuran affords the tricyclic species (**118**) ⟨63JOC577⟩. The furan ring subsequently expands to a pyran, affording 2,3-dichlorochromene (**119**). This reaction is reminiscent of those shown by pyrrole and indole, which yield pyridine and quinoline derivatives, respectively. However, the product could not be purified and on hydrolytic work-up the dichromenyl ether (**120**) was isolated.

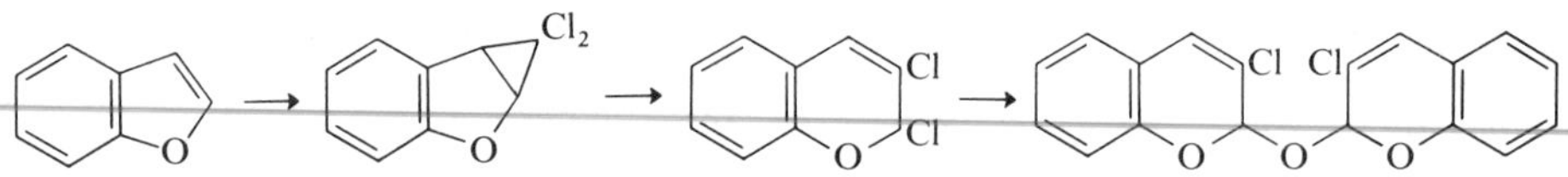

(118)　　　　　(119)　　　　　(120)

### (ii) From a benzoxepin

The formation of a benzoxepin derivative during the synthesis of chromenes from phosphonium salts has been postulated (see p. 749). In keeping with this idea, chromenes are produced by ring opening of benzoxepin (121) with methoxide ion in DMF ⟨69C108⟩. The initial product, a butadienylphenol (122), undergoes a [1,7]-hydrogen shift to the quinoneallide.

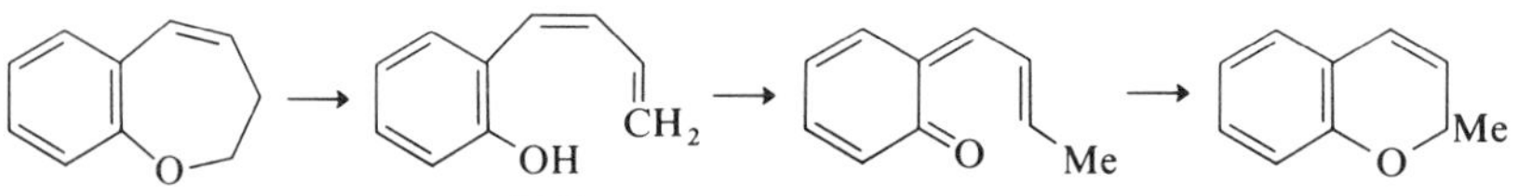

(121)　　　　(122)

### 2.24.2.2.3 From a preformed heterocyclic ring

The synthesis of chromenes from other six-membered oxygen heterocycles has much to offer because of the accessibility of the precursors. Thus much of the earlier work on chromenes was based especially on their synthesis from coumarins or chromanols.

### (i) From coumarins

The classical route to chromenes involves the reaction of a Grignard reagent with a coumarin (see Section 2.23.6.4.1) (equation 1) ⟨39JOC575⟩. The synthesis is restricted to those coumarins which lack substituents which would themselves react with the organometallic reagent, such as carbonyl or nitro groups. Wide variation in the Grignard reagent is possible, alkyl, alkenyl, aryl and heteroaryl derivatives being acceptable. Organolithium compounds have also been successfully used ⟨63T839⟩. The reaction is complex and conjugate addition to the enone system has been observed with aryl Grignard reagents. Indeed, 4-substituted coumarins yield 4,4-disubstituted 3,4-dihydrocoumarins (123) and it appears that 1,4-addition is more common than earlier results suggested ⟨74JCS(P1)569⟩. The reactivity of coumarins towards Grignard reagents has been examined and the variety of products are considered to arise from combinations of 1,2- and 1,4-additions, in which intermediate species such as (124) are involved ⟨76T1655, 76T1661⟩.

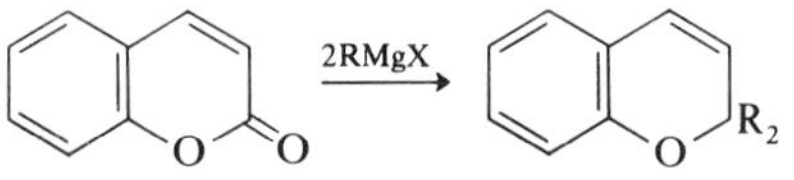

(1)

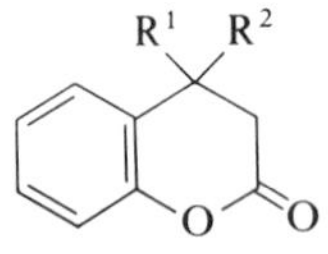

$R^1$　$R^2$

(123) 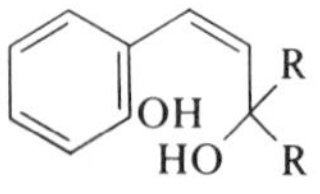　　　　　(124)

### (ii) From chromanols

Chromenes result from the dehydration of chromanols. The ready availability of the hydroxy compounds by the reduction of chroman-4-ones and through their reaction with Grignard and related reagents makes this an attractive route. Amongst many available examples, the syntheses of ageratochromene (precocene II; 125) ⟨58BSB22⟩, fluorinated

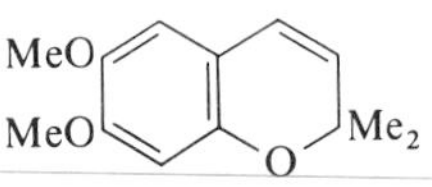

(125) 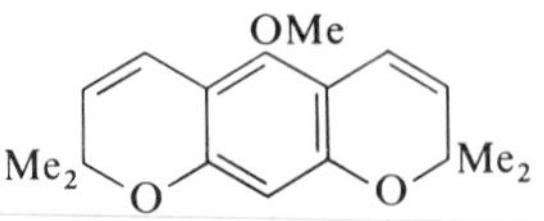　　　　　(126) 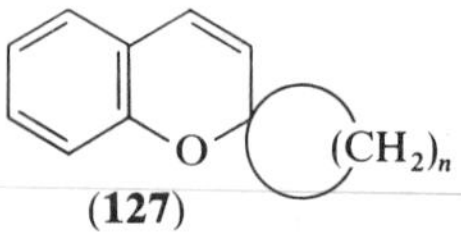　　　　　(127)

chromenes ⟨80TL2361⟩, the bischromene (**126**) ⟨81H(16)955⟩ and a series of 2-spiroannelated chromenes (**127**) ⟨83JCS(P1)827⟩ reflect the diversity of this route.

A wide variety of dehydrating agents has been used and some workers have preferred to pyrolyze the acetate. Further details of these reactions can be found in Chapter 2.23.

3-Substituted chroman-4-ols behave in a similar manner and yield 3-substituted chromenes, and the dehydration of 3-phenylchromanols provides a convenient route to isoflavenes ⟨66JCS(C)629⟩.

Additional substitution at C-4 is acceptable in the chromanol, though the chrom-3-enes may be accompanied by an alkylidene chroman if this 4-substituent can participate in the elimination of water (equation 2).

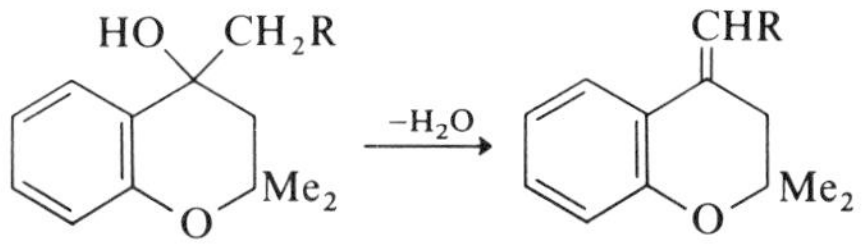

$$\xrightarrow{-H_2O} \tag{2}$$

### (iii) *From halogenochromans*

Under the influence of bases, 3- or 4-halogenochromans undergo elimination of hydrogen halide to give the chromene.

*trans*-3,4-Dihalogenochromans also yield chromenes by the elimination of HX, but the nature of the product appears to depend upon the reaction conditions and mixtures may result (Scheme 13).

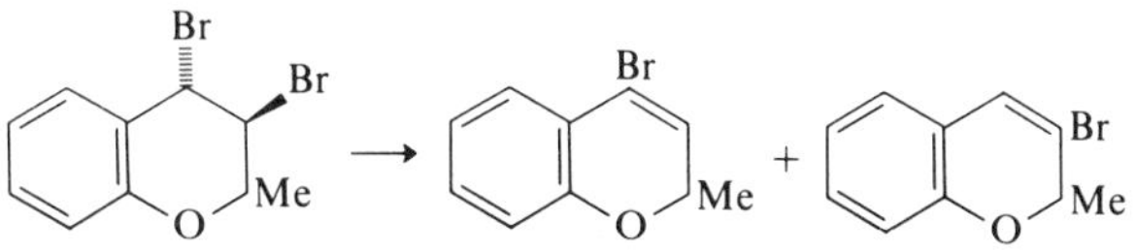

**Scheme 13**

An interesting variant on this theme involves the cyclization of substituted 2-allylphenols by *N*-iodosuccinimide. The 3-iodochromans, which presumably arise *via* a cyclic iodonium species, are readily dehydrohalogenated by base. The overall yields are generally impressive, although ring iodination can compete ⟨79TL2545⟩.

### (iv) *From chromans*

Direct oxidation of chromans to chromenes can be achieved using a high potential quinone; chloranil and dichlorodicyanobenzoquinone are suitable. Much attention has been devoted to the latter reagent.

The presence of an electron-releasing group in the aromatic ring appears to obviate the need for a free hydroxy group in the chroman ⟨82S74⟩. The syntheses of ripariochromene A ⟨75BCJ80⟩ and evodionol (**128**) ⟨81T1437⟩ are illustrative of naturally occurring chromenes obtained in this manner.

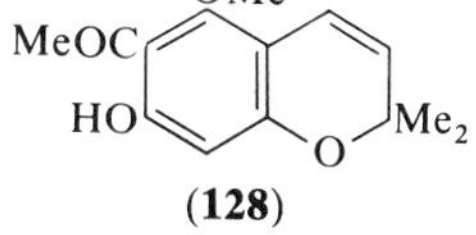

(**128**)

It has been noted ⟨79JCS(P1)2563⟩ that chromenes form adducts with DDQ and that these may be produced during attempts to dehydrogenate chromans. It is proposed that the charge transfer complex between the chromene and DDQ breaks down by a one-electron process to give the semiquinone radical (**129**) and the phenoxyl radical (**130**). The *p*-quinoneallide (**131**) which results rearranges to the 3-substituted chromene (Scheme 14).

### (v) *From chromanones*

Treatment of chroman-4-ones with $PCl_5$ yields 3,4-dichlorochromenes. It is postulated that the reaction proceeds by initial α-chlorination, followed by conversion to the *gem*-dichloride. Dehydrohalogenation leads to the observed product. The chloroketone has been isolated from 2,2-dimethylchroman-4-one and phosphorus pentachloride ⟨79TL3901⟩.

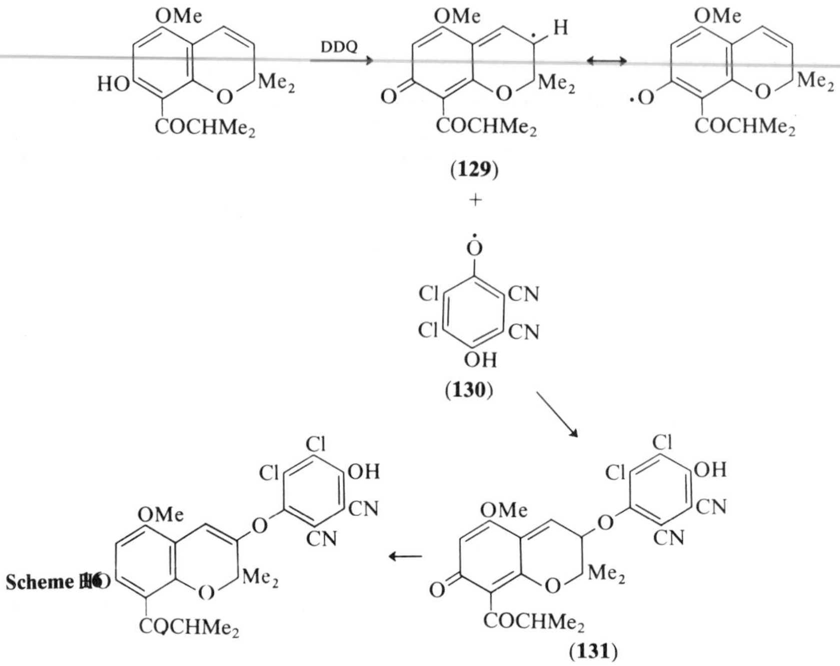

**Scheme 14**

### (vi) *From benzopyrylium salts*

Nucleophilic attack at C-2 of benzopyrylium salts leads to chromenes. Amongst suitable reagents are organomagnesium compounds ⟨79CJC1377⟩ and metal hydrides ⟨67JOC3616⟩. However, in the latter instance attack at C-4 may also occur leading to chrom-2-enes. For example, isoflavylium salts afford approximately equal amounts of isoflav-2-ene and isoflav-3-ene on reaction with sodium cyanoborohydride ⟨81AJC2647⟩. It is reported that 3-substituted benzopyrylium salts react with LAH to yield only chrom-3-enes ⟨68AJC2247⟩. In other instances hydride ion functions only as a base. Thus, 3-ethyl-2-phenylbenzopyrylium affords 2-ethoxy-3-ethyl-2-phenylchrom-3-ene on treatment with either sodium borohydride in ethanol or ethanolic sodium hydroxide.

The reaction of several amines with 7-methoxy-3-phenylbenzopyrylium perchlorate has been shown to yield the corresponding 2-substituted 3-phenylchromene ⟨78JCS(P1)88⟩, although the reaction of secondary amines with simple benzopyrylium salts gives mixtures of chrom-3-enes and chrom-2-enes ⟨72JOC1069⟩.

Somewhat more exotic is the reaction of benzopyrylium salts with alkenes, which yields 3-phenyl-2-(2-phenylchroman-3-yl)chrom-3-enes (**132**).

Further details of the formation of chromenes from pyrylium salts can be found in Chapter 2.23.

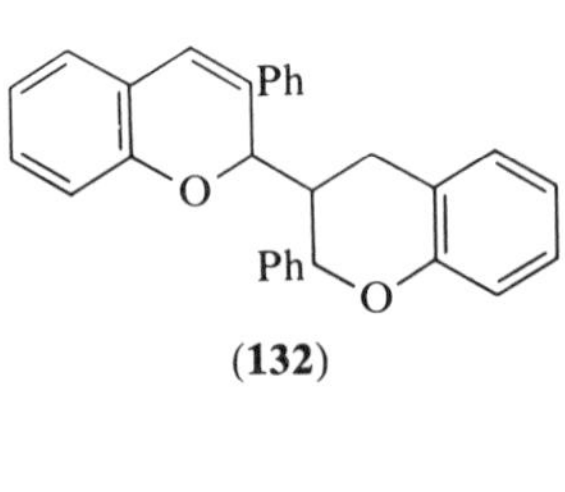

(**132**)

### 2.24.2.3 *4H*-Pyrans

4*H*-Pyrans lack aromatic properties and behave as dienolic ethers showing the instability predicted by MO calculations. Hence, the parent compound proved elusive and was not synthesized until 1962. Although several routes are available for the synthesis of this ring system, such is the variation in stability of 4*H*-pyrans no one method can be described as of widespread application. Routes to the 4*H*-pyran ring system have been reviewed ⟨80H(14)337⟩.

### 2.24.2.3.1 Formation of one bond

#### (i) Adjacent to the heteroatom

The simplest approach which can be envisaged to 4*H*-pyrans involves the ring closure of 1,5-diketones. However, such molecules are frequently able to undergo a facile intramolecular aldol condensation leading to cyclohexenones, which competes successfully with cyclization to the pyran.

By careful attention to structural detail, it is possible to prevent formation of the alicyclic system, thereby achieving a potentially valuable pyran synthesis. Such structural features include the lack of enolizable hydrogen and incorrect geometrical relationships. Thus, 1,3,3,5-tetraphenylpentane-1,5-dione (**133**), which lacks a suitably placed α-hydrogen atom, yields the tetraphenylpyran (**134**) upon treatment with phosphorus pentoxide ⟨35MI22400⟩. The cyclization of 3,3-diphenylpentanedial (**135**) with *p*-toluenesulfonyl chloride yields 4,4-diphenylpyran (**136**) ⟨77CJC2373⟩. The dial, which can be synthesized from diphenylmethane, forms a stable crystalline hemihydrate which is of course a tetrahydropyran.

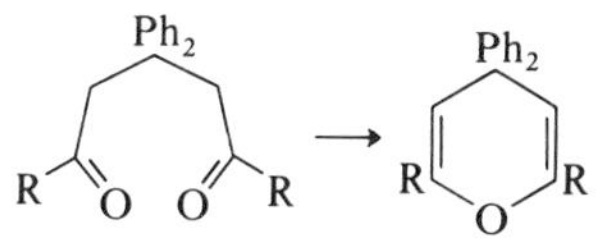

(**133**) R = Ph    (**134**) R = Ph
(**135**) R = H     (**136**) R = H

The Michael addition of deoxybenzoin to several ethyl α-acetylcinnamates, obtained from ethyl acetoacetate and an aromatic aldehyde in the presence of piperidine, resulted in the formation of 1,5-diketones. Cyclization to 4*H*-pyrans occurred in acetic acid and *p*-toluenesulfonic acid ⟨76IJC(B)739⟩.

The unsubstituted pentanedial yields the parent 4*H*-pyran on successive treatment with hydrogen chloride and *N*,*N*-diethylaniline ⟨62AG465⟩. This two-step approach involving elimination of hydrogen chloride from the initially formed tetrahydropyran has been extended to the synthesis of substituted 4*H*-pyrans (Scheme 15) ⟨67MI22400⟩.

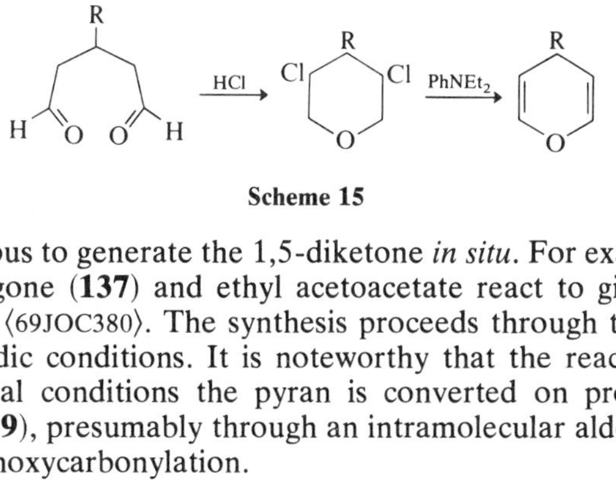

**Scheme 15**

It can be advantageous to generate the 1,5-diketone *in situ*. For example, in the presence of zinc chloride, pulegone (**137**) and ethyl acetoacetate react to give a low yield of the fused 4*H*-pyran (**138**) ⟨69JOC380⟩. The synthesis proceeds through the diketo ester which cyclizes under the acidic conditions. It is noteworthy that the reaction is reversible and under the experimental conditions the pyran is converted on prolonged reaction into 'pulegone acetone' (**139**), presumably through an intramolecular aldol condensation of the diketo ester and de-ethoxycarbonylation.

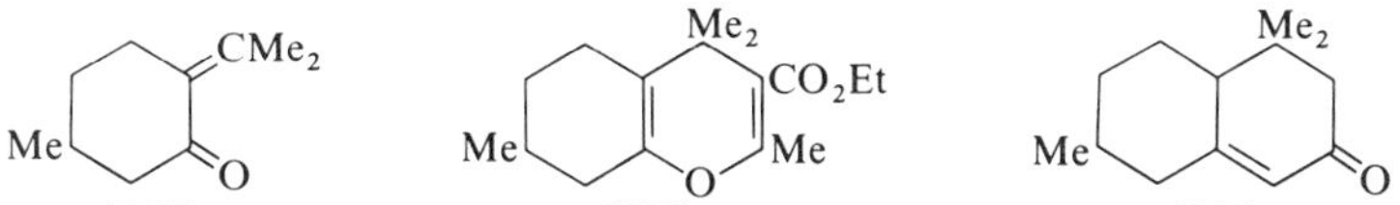

(**137**)          (**138**)          (**139**)

This observation led to the development of a general route to 4*H*-pyrans, in which the acid-catalyzed condensation of unsaturated carbonyl compounds with 1,3-dicarbonyl compounds is used to generate 1,5-diketones, which spontaneously cyclize to the pyran (equation 3) ⟨69JOC3169⟩. The moderate yields are compensated by the ready availability of the

(3)

reactants. Although problems arise through the reduced stability of the products from unsaturated aldehydes, the route has proved of value even with simple aldehydes.

Diethyl 2-oxobutanedioate undergoes an aldol condensation with aldehydes to yield the substituted dione which after hydrolysis and decarboxylation affords the 2,6-dicarboxylic acid (Scheme 16). The cyclization of this acid provides a useful synthesis of 4*H*-pyran-2,6-dicarboxylic acids ⟨74ACS(B)517⟩. It is unfortunate that attempted decarboxylation to the simple pyran fails, resulting in extensive decomposition. An earlier report ⟨71OPP243⟩ summarizes previous work in this area and gives preparative details for the dicarboxylic acids.

When 2-(3-phenylprop-2-ynyl)cyclohexane-1,3-dione is heated with zinc carbonate, ring closure to a fused 4*H*-pyran occurs ⟨62AP645⟩. If the triple bond is terminal, cyclization leads to a furan derivative. Under the same conditions, the analogous acyclic diketones do not yield pyrans.

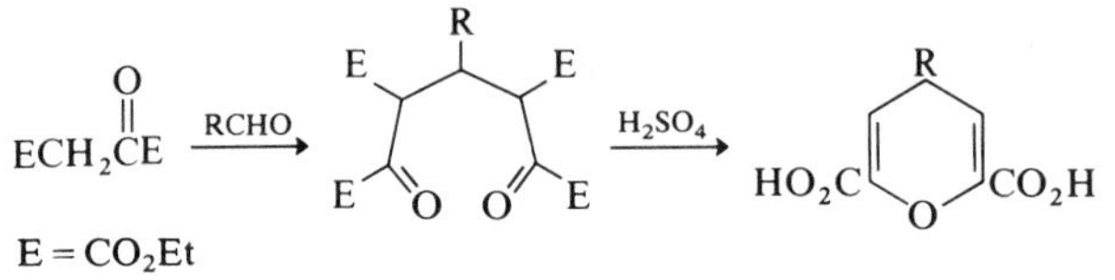

$$\text{ECH}_2\overset{\overset{\displaystyle O}{\|}}{\text{CE}} \xrightarrow{\text{RCHO}} \cdots \xrightarrow{\text{H}_2\text{SO}_4} \cdots$$

E = CO₂Et

A French group have developed a variant of this approach, in which the alkynone is generated *via* a β-keto ylide (Scheme 17) ⟨72CR(C)(274)1091⟩. The substrate can be the chloride or anhydride of phenylacetic acid, a β-keto ester or even a substituted diazomethane. It is postulated, and supporting evidence is available, that the reaction involves thermolytic formation of a ketene which reacts with the ylide to give an acylallene (140) *via* the betaine. Michael addition of a second molecule of ylide to the allene is followed by cyclization to the pyran. The mode of cyclization is uncertain, but the involvement of the alkyne (141) is possible. Several of the proposed intermediates have been shown to give pyrans, though in low yields, under the same experimental conditions.

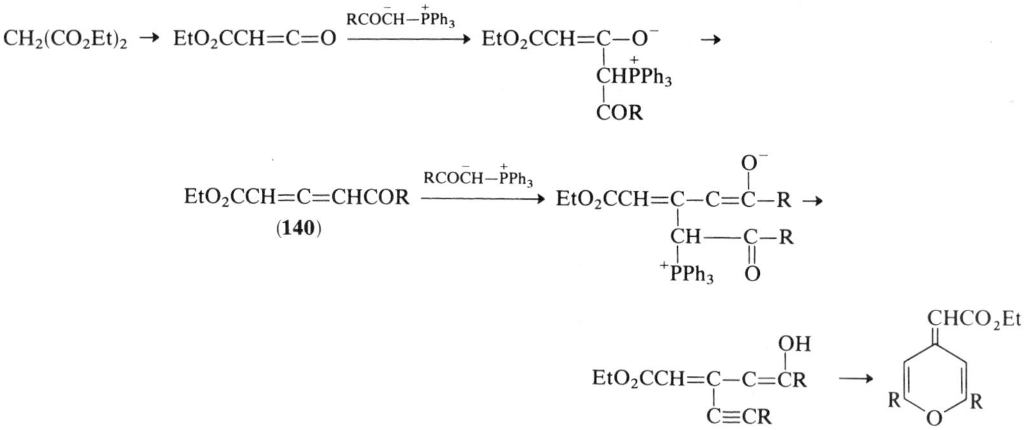

$$\text{CH}_2(\text{CO}_2\text{Et})_2 \rightarrow \text{EtO}_2\text{CCH}=\text{C}=\text{O} \xrightarrow{\text{RCOCH}-\overset{+}{\text{P}}\text{Ph}_3} \text{EtO}_2\text{CCH}=\text{C}-\text{O}^-$$

**Scheme 17**

$$\text{EtO}_2\text{CCH}=\text{C}=\text{CHCOR} \xrightarrow{\text{RCOCH}-\overset{+}{\text{P}}\text{Ph}_3} \cdots$$
(140)

(141)

The reaction of phenylacetylene with the allene (142) to give the pyran (143) probably involves a similar pathway ⟨77IZV2517⟩. The pyran is accompanied by a 2-pyranone which is considered to be an artefact of the pyran.

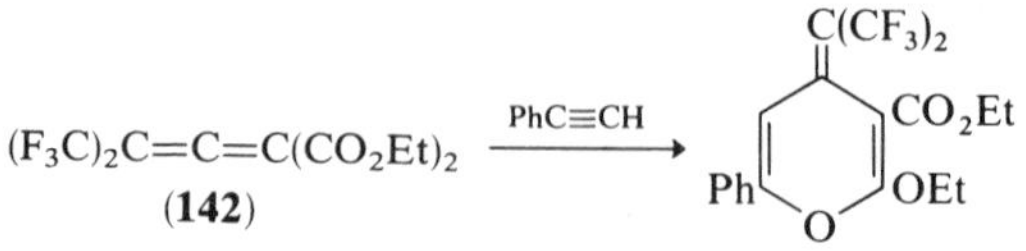

$$(\text{F}_3\text{C})_2\text{C}=\text{C}=\text{C}(\text{CO}_2\text{Et})_2 \xrightarrow{\text{PhC}\equiv\text{CH}} \cdots$$
(142)

(143)

The cyclization of 5-oxonitriles offers an attractive route to 2-amino-4*H*-pyrans (Scheme 18) ⟨78JHC57⟩. The starting materials, α-benzoylcinnamonitriles, are available from benzoylacetonitrile and an aromatic aldehyde. Michael addition of ethyl cyanoacetate or malononitrile to the benzoylcinnamonitrile affords the oxonitrile and is followed by cyclization to the imine. The reaction proceeds at room temperature in the presence of either piperidine or sodium ethoxide.

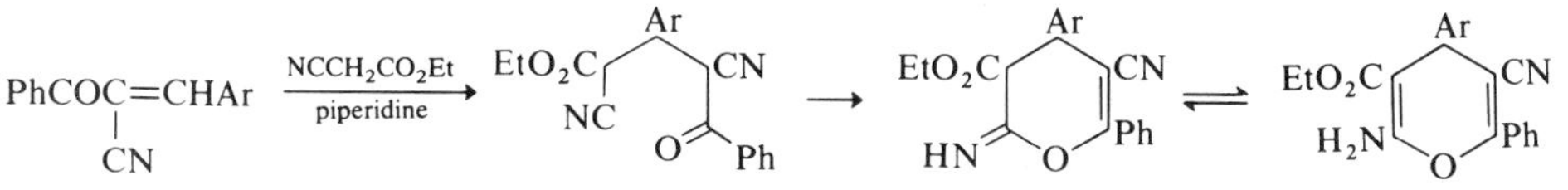

**Scheme 18**

A variation involves the formation of the oxonitrile by self-condensation of the benzoyl-acetonitrile. The normal reaction with, for example, malononitrile then follows ⟨79MI22401⟩.

The formation of the oxonitrile by the addition of an activated methylene compound to an activated alkene has been described ⟨74JCS(P1)2595, 73AJC1551⟩. Thus, TCNE and pentane-2,4-dione react in ethanol to give the acyclic adduct (**144**). However, in acetonitrile solution 5-acetyl-2-amino-3,4,4-tricyano-4H-pyran is formed (Scheme 19). Furthermore, the diketonitrile changes to the pyran when kept in acetonitrile for two days; the progress of the cyclization can be monitored by UV and NMR spectroscopy. The initial adduct arises by a Michael addition and evidence suggests that it exists in solution as the enol (**145**), whilst the final product is derived by cyclization and imine–enamine tautomerism ⟨B-70MI22400⟩.

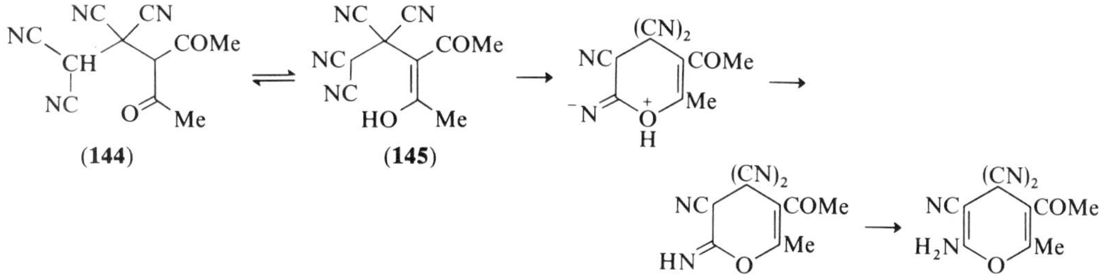

**Scheme 19**

Two variations are worthy of mention. The alkene may be generated *in situ*, as for instance in the formation of the pyran from 3-bromopentane-2,4-dione. Secondly, the use of cyclic molecules as the activated methylene component leads to fused pyrans (**146**) and (**147**) from TCNE and ethyl cyanoacetate, respectively ⟨73CB914⟩.

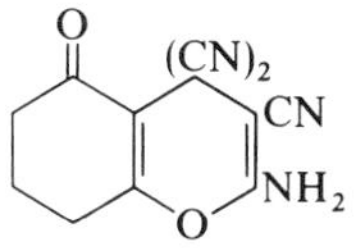

(**146**) 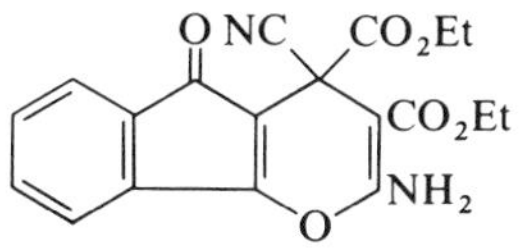          (**147**) 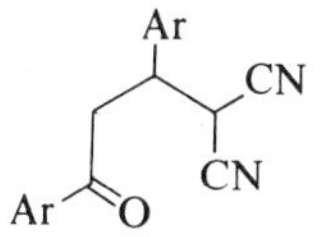          (**148**)

In all instances, a necessary condition for successful formation of the pyran is the presence of an electron-withdrawing group at C-4 of the oxonitrile. For example, the ketone (**148**) derived from benzylideneacetophenone and malononitrile does not cyclize.

### (ii) γ to the heteroatom

The so-called 'trimerization' of propynal in the presence of piperidine acetate provides a synthesis of 4-ethynyl-4H-pyran-3,5-dicarbaldehyde (**149**) ⟨50LA(568)34⟩; it should be noted that the structure proposed for the product in the original work has been corrected ⟨64CB1959⟩. In the absence of moisture, the reaction fails and it seems likely that the synthesis involves hydration of the alkyne to the divinyl ether. Finally, condensation with the third molecule of the aldehyde results in cyclization to the product (Scheme 20).

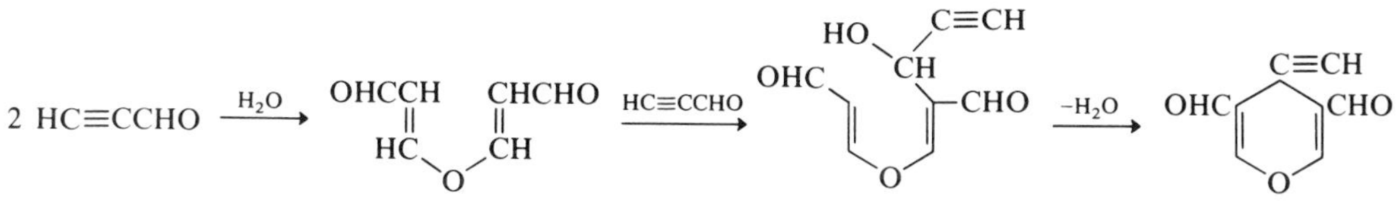

**Scheme 20**                    (**149**)

Some supporting evidence for the mechanism is provided by the formation of the 4-ethyl analogue (**150**) when propynal is treated with piperidine acetate in the presence of propanal; the latter may be considered to intercept the divinyl ether.

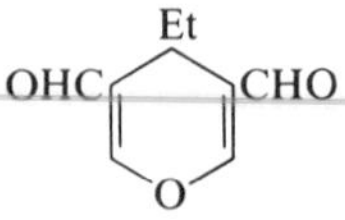

(150)

### 2.24.2.3.2 Formation of two bonds

By analogy with the formation of dihydropyrans from unsaturated carbonyl compounds and alkenes (see Section 2.24.2.7.1(i)), the synthesis of 4*H*-pyrans from the [4+2]-cycloaddition of unsaturated carbonyl compounds and alkynes would seem to offer some potential. Such a reaction has indeed proved of value, but examples are largely restricted to the use of ynamines as the dienophile ⟨76BSF987⟩.

Some variation is acceptable in the structure of the carbonyl component and unsaturated ketones, aldehydes, esters and β-keto esters have been used ⟨76T1449⟩. The first type offers the best yields and this may well be a consequence of a competing [2+2]-cycloaddition which is observed with unsaturated aldehydes and esters (Scheme 21) ⟨70TL885⟩.

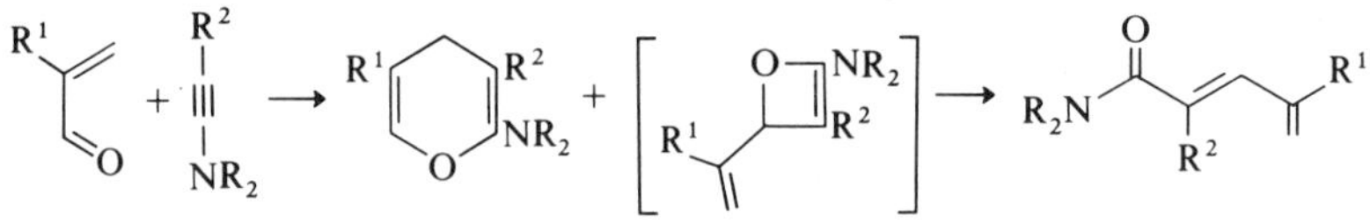

**Scheme 21**

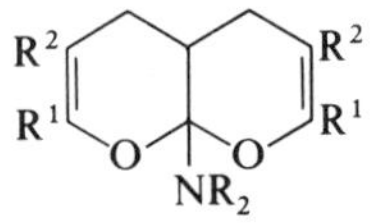

(151)

Terminal ynamines form 4*H*-pyrans which are as reactive as the starting ynamine and they react with a further molecule of enone to give pyranopyrans (151). It is also noteworthy that at elevated temperatures pyranopyrans are produced from the reaction of ynamines with methyl vinyl ketone ⟨70CR(C)(271)468⟩. In one instance only is the rearrangement of the pyranopyran to the aminopyran observed.

The different behaviour of bis(*N,N*-diethylamino)propene and *N,N*-dimorpholinopropene towards methyl vinyl ketone is of interest. The latter yields some 4*H*-pyran (153) identical with that obtained from the ynamine. However, the former compound gives only the Stork adduct (154) by proton transfer. The elimination of the amine moiety from the dihydropyran (152) is easier from the less basic enamine (Scheme 22).

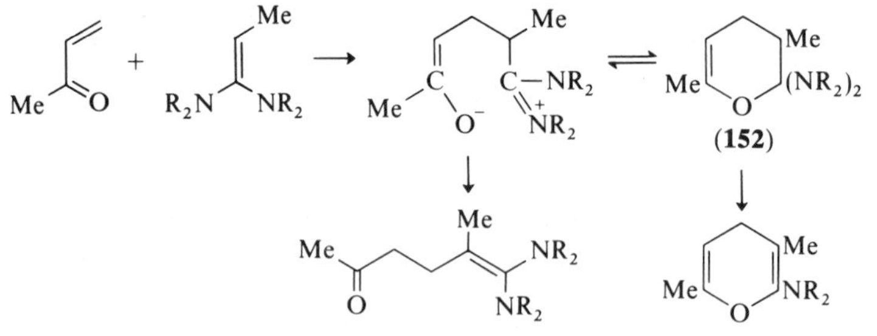

(152)

(154)          (153)

**Scheme 22**

Once more the route is adaptable to the synthesis of fused pyrans ⟨73JHC165⟩.

### 2.24.2.3.3 From other heterocycles

#### (i) From oxepins

When *s*-oxepin oxide (155) is treated with a trace of methanesulfonic acid in an aprotic solvent, 4*H*-pyran-4-carbaldehyde (156) is rapidly produced. A quantitative yield is indicated by NMR spectroscopy. The reaction probably proceeds *via* the homoaromatic cation, which may lead to the product either by deprotonation and ring contraction or *via* the valence tautomer (Scheme 23) ⟨76JA6350⟩.

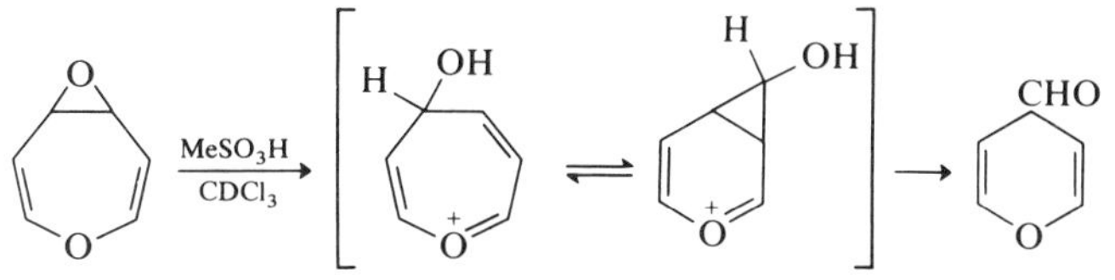

**(155)**     **(156)**

**Scheme 23**

This reaction has been used to trap oxepin oxides ⟨78TL2995⟩. Extrusion of nitrogen from the azodiepoxide (**157**) in the presence of methanesulfonic acid gives 4-acetyl-4-methylpyran (**160**) as the only observable product. However, in the absence of acid, the aldehyde (**161**) is formed. This observation is attributed to an equilibrium between the two epoxides (**158**) and (**159**) which is in favour of the latter.

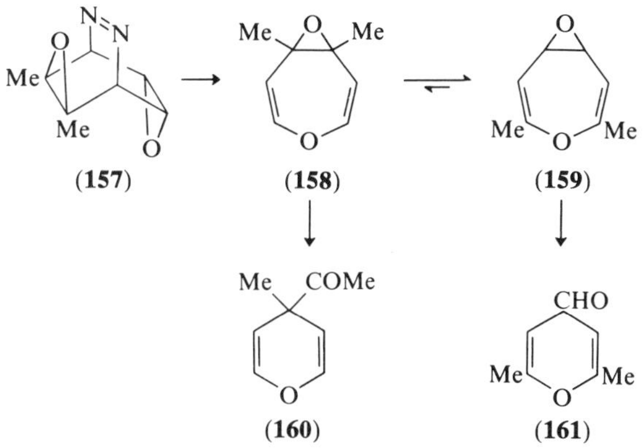

**(157)**     **(158)**     **(159)**

**(160)**     **(161)**

### (ii) *From isoxazolium salts*

The reaction of an enamine with the isoxazolium salt (**162**) leads to a fused 4-iminopyran, probably *via* the route indicated in Scheme 24 ⟨68JOC867⟩.

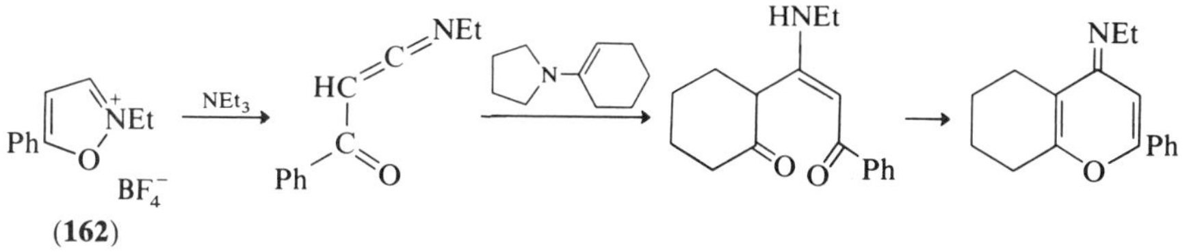

**(162)**

**Scheme 24**

### *2.24.2.3.4 From a preformed heterocyclic ring*

#### (i) *From pyranones*

Nucleophilic attack at C-4 of a pyran-4-one may lead to 4*H*-pyrans (see Section 2.23.8.4). Amongst the nucleophiles which have been used, mention can be made of cyanide ion ⟨55JA1702⟩, which attacks kojic acid (**163**) very rapidly to give a high yield of the pyran (**164**), and Grignard reagents ⟨61CB1784⟩, although here the nature of the products has been the subject of dispute ⟨69JOC3169⟩. The reactions of pyranones with nucleophiles have been reviewed ⟨60AG331⟩.

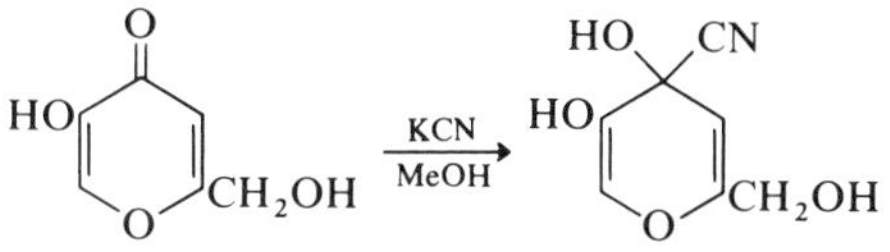

**(163)**     **(164)**

Diphenyldiazomethane reacts with 2,6-diphenylpyran-4-thione ⟨58JA6312⟩ and diphenyl-ketene with the corresponding pyranone ⟨11LA(384)38⟩ to give the substituted methylenepyran (Scheme 25).

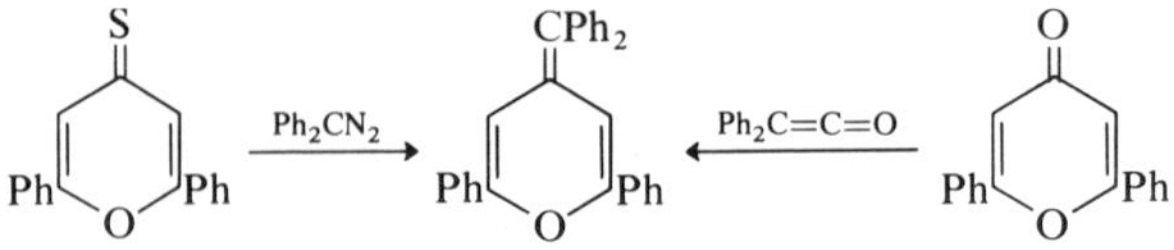

**Scheme 25**

### (ii) *From dihydropyrans*

As a result of their accessibility, dihydropyrans provide a useful source of 4H-pyrans. Indeed one of the earliest syntheses of the parent compound involved the pyrolysis of 2-acetoxy-3,4-dihydropyran (**165**) ⟨62JA2452⟩. The concomitant formation of acrolein, vinyl acetate and acetic acid indicates that a reverse Diels–Alder reaction competes with the pyrolysis.

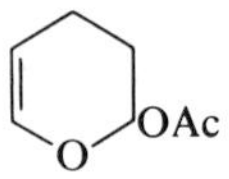

(**165**) 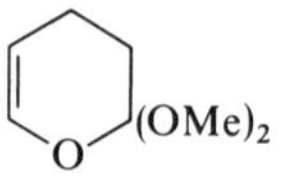        (**166**)

Elimination of methanol from the dimethoxydihydropyran (**166**) on reaction with aluminum *t*-butoxide occurred at 155 °C and gave 2-methoxy-4H-pyran ⟨55JA5601⟩.

Dihydropyran-2-ols readily lose water when heated in acetic acid in the presence of anhydrous sodium acetate ⟨70JHC1311⟩.

### (iii) *From pyrylium salts*

It has been found ⟨72BSF707⟩ that in general the reaction of methylmagnesium iodide with 2,4,6-trisubstituted pyrylium salts leads to 2H-pyrans. However, as the bulk of the 2- and 6-substituents increases, minor amounts of 4H-pyrans are formed. When the pyrylium salt is unsubstituted at C-4, significant quantities of the 4H-pyran are produced, along with the corresponding 2H-pyran and unsaturated ketones. It appears that substituents in the 3- and 5-positions of the pyrylium salts have little effect on the course of the reaction.

The reaction of substituted pyrylium salts with sodium borohydride yields pyrans, but again attack at C-2 and subsequent ring opening of the resulting 2H-pyran are observed ⟨62T257⟩. Yields of the 4H-pyran are highest when C-4 is unsubstituted.

The anions derived from various active methylene compounds react with 2,6-disubstituted pyrylium salts to form 4-substituted 4H-pyrans ⟨57AG720, 59CB46⟩. Grignard reagents react to give similar compounds ⟨60AG777⟩.

The facile nucleophilic displacement of a 4-methoxy group from pyrylium salts provides syntheses of 4-substituted pyrans. In the presence of triethylamine, 4-methoxypyrylium salts react with aromatic nitro compounds to give 4-benzylidene-4H-pyrans ⟨73JOC2834⟩.

Both 4-arylimino-2,6-dimethylpyran salts (**167**) and 1-aryl-4-methoxy-2,6-dimethyl-pyridinium salts (**168**) result from the reaction of 4-methoxy-2,6-dimethylpyrylium salts with primary aromatic amines ⟨78JCS(P1)1373⟩. The ratio of the products varies with the basicity of the amine, the less basic amines giving mainly the pyran salt. The free iminopyrans (**169**), which have limited stability at room temperature, are readily obtained from the salts.

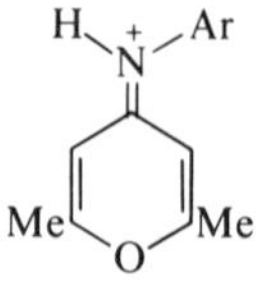

(**167**) 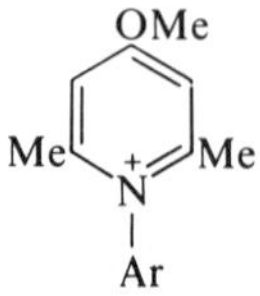        (**168**) 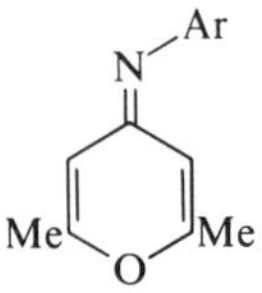        (**169**)

Although simple 4-methylenepyrans are generally unstable compounds which are very difficult to isolate, the substitution of an electron-withdrawing group in the methylene group (**170** and **171**) leads to an increase in stability. Similarly, sterically hindered methylenepyrans such as (**172**) show greater stability. Such compounds result from proton loss from a 4-methylpyrylium salt ⟨68T4741⟩.

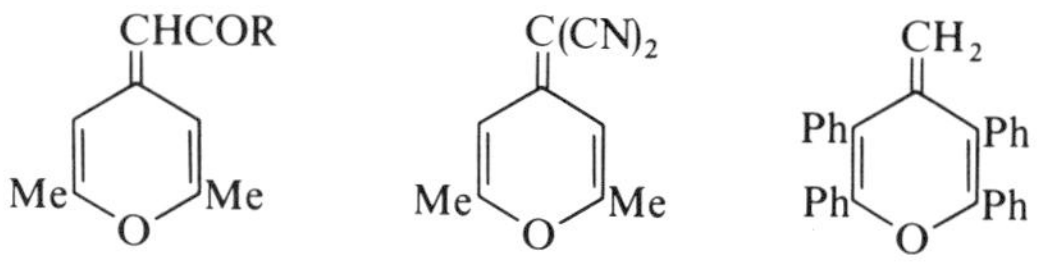

(170)  (171)  (172)

### 2.24.2.4 4*H*-Chromenes (4*H*-1-Benzopyrans)

There is much less interest in the chrom-2-enes than in the chrom-3-enes. This is undoubtedly a reflection on the fact that the former, unlike the latter, have no significance in natural product chemistry.

#### 2.24.2.4.1 Formation of one bond

(i) *Adjacent to the heteroatom*

The acid-catalyzed cyclization of 3-(2-hydroxyphenyl)propan-1-ones leads to chrom-2-enes. This reaction accounts for the formation of 2,4-diphenylchrom-2-ene from the reaction of phenylmagnesium bromide on coumarin. The organometallic reagent ring-opens the coumarin to the ketone which cyclizes to the chromene on boiling with acetic acid (Scheme 26) ⟨63T839⟩. In a similar manner, various benzocoumarins afford propanones which cyclize readily in acetic acid to the naphthopyran ⟨70JCS(C)1758⟩.

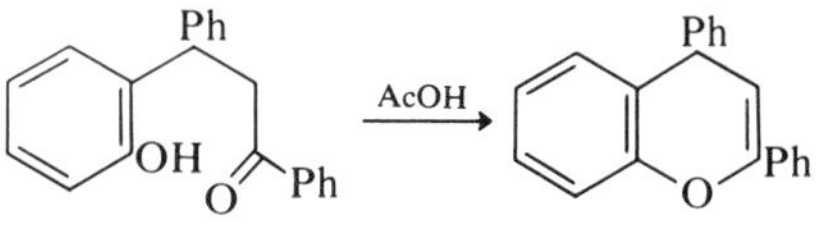

**Scheme 26**

2-Phenylchrom-2-ene results when 3-(2-hydroxyphenyl)-1-phenylpropan-1-one is treated with *p*-toluenesulfonic acid with continuous removal of water ⟨71JOC600⟩. The product is unstable, forming a tar over several hours.

Another example illustrates the utility of ethyl acetoacetate in the synthesis of chrom-2-enes. Reaction of the β-keto ester with the benzyl halide (173) affords the hydroxyphenylpropanone (174) which forms the chromene with acetic anhydride (Scheme 27) ⟨42JA435⟩.

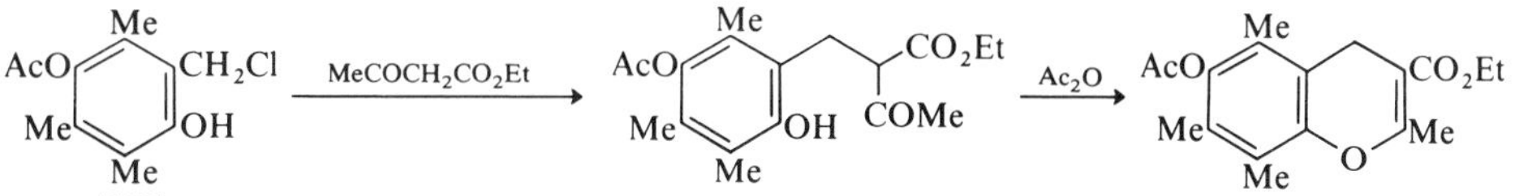

(173)  (174)

**Scheme 27**

(ii) *β to the heteroatom*

The reaction of *o*-acyloxybenzyl bromides (175) with an excess of a phosphorus ylide leads to high yields of chrom-2-enes (Scheme 28) ⟨79TL2995⟩. The process is considered to

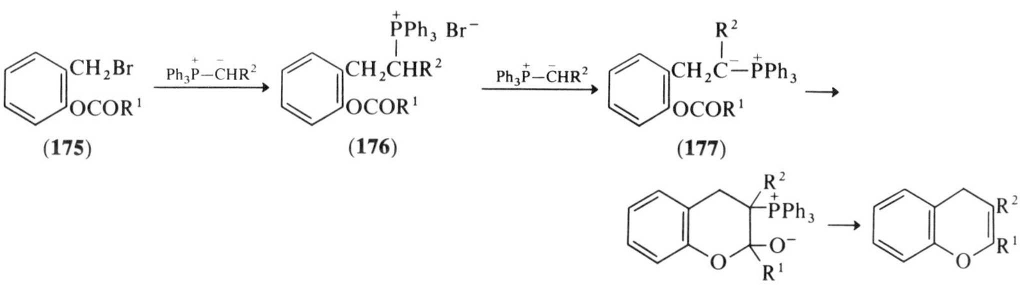

(175)  (176)  (177)

**Scheme 28**

proceed through alkylation of the ylide to the phosphonium salt (176), which undergoes transylidation with the excess of ylide, producing a new ylide (177). The equilibrium between the two phosphoranes is displaced in favour of the new species by virtue of an intramolecular Wittig reaction which affords the betaine. Subsequent elimination of triphenylphosphine oxide completes the reaction sequence. The ease of formation of the chromene varies with the reactivity of the initial ylide and of the examples cited, that leading to 2-ethyl-3-phenylchrom-2-ene is the most rapid. The synthesis is not wasteful of the ylide, since the excess of reagent is recoverable as the phosphonium salt $Ph_3\overset{+}{P}CH_2R^2\ Br^-$.

### 2.24.2.4.2 Formation of two bonds

In the presence of a large excess of vinyl acetate and at high temperature and pressure, 2-hydroxybenzyl alcohol is forced into participation in a [4 + 2]-cycloaddition. The total yield of a mixture of chrom-2-ene and 2-acetoxychroman is around 40% ⟨69BSF1715⟩. It is clear that the chromene arises through the ready loss of acetic acid from the chroman.

### 2.24.2.4.3 From other heterocycles

By analogy with the conversion of tetrahydrofurfuryl alcohol into dihydropyran, the action of heat on 2,3-dihydro-2-hydroxymethylbenzofuran might be considered a potential source of chrom-2-ene. Whilst the reaction is indeed successful, 2-methylbenzofuran is also produced and separation of the two compounds is difficult ⟨54MI22400⟩. It thus appears that there is competition between proton loss before and after a Wagner–Meerwein rearrangement.

### 2.24.2.4.4 From a preformed heterocyclic ring

#### (i) From chrom-3-enes

In the presence of alkoxy- or phenoxy-magnesium halides, chrom-3-enes are isomerized to chrom-2-enes ⟨71JCS(C)2546⟩. The catalyst seems quite specific, since neither sodium alkoxides or phenoxides nor magnesium halides exhibit the same effect. Yields are good and the isomerization is virtually complete.

#### (ii) From chromanols

The dehydration of chroman-2-ols, the corresponding 3-hydroxy compounds or their acetates yields chrom-2-enes. The elimination of acetic acid may be achieved by pyrolysis ⟨62JA813, 69BSF1715, 70JOC2282⟩, whilst loss of water is usually achieved under acidic conditions ⟨35JCS646, 56JCS4785, 60CB1025⟩.

The resistance of 4,4-dimethylchroman-3-ol to dehydration has been attributed to its existence in a half-chair conformation in which the hydroxy group is equatorial and 1,2-*trans* diaxial elimination of water is disfavoured ⟨72TL4453⟩.

The dehydration of 2,4,4-trimethylchroman-2-ol to the chromene is of interest because of the source of the chromanol. 1,1,3-Trimethylindane hydroperoxide (178) undergoes an acid-catalyzed rearrangement to a mixture of the chromanol and chromene. The intermediacy of an oxonium ion can be envisaged, with nucleophilic attack at C-2 leading to the chromanol and proton loss giving the chromene (Scheme 29) ⟨56JCS4785⟩.

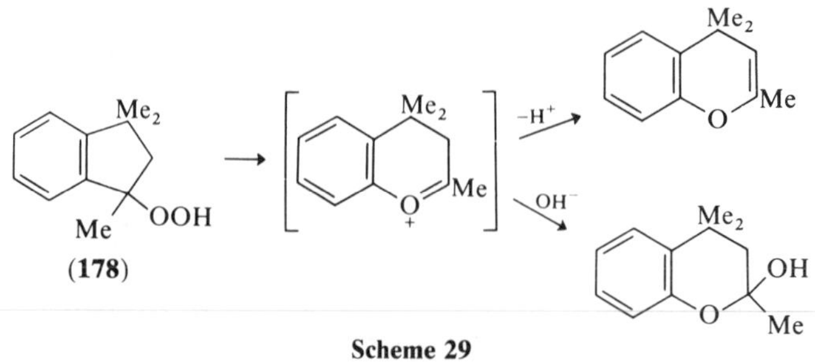

**Scheme 29**

An elaborate synthesis of a substituted anisole was followed by ether cleavage, cyclization and dehydration to the chromene (**179**; Scheme 30) ⟨54JA5439⟩.

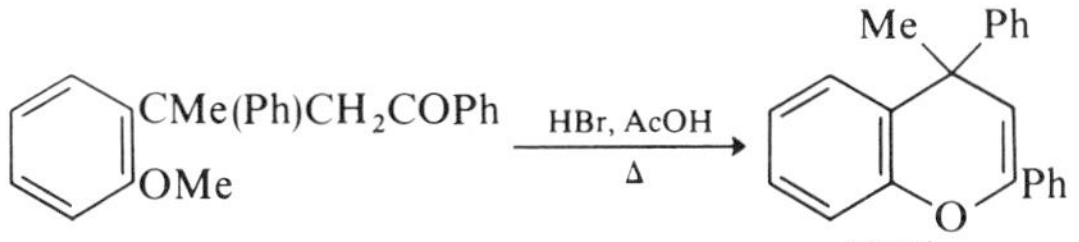

**Scheme 30**

### (iii) *From chromans*

When heated with polyphosphoric acid, 2-ethoxychromans lose ethanol to give the chrom-2-ene ⟨67CB1296⟩. The method constitutes a synthesis of chrom-2-enes from coumarins, since the chromans are obtained in two steps from the latter.

### (iv) *From benzopyrylium salts*

Flavylium perchlorate reacts with either *N,N*-dimethylaniline or *p*-dimethylaminophenyllithium to give the chrom-2-ene (**180**) which is easily oxidized to the 4-hydroxy compound (**181**; Scheme 31) ⟨52JA3622⟩.

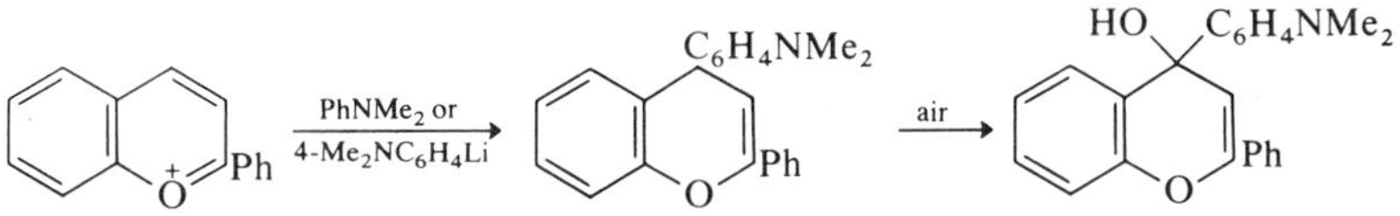

**Scheme 31**

When hydride ion is used as the nucleophile, the product is unsubstituted at C-4. Thus, luteolinidin tetramethyl ether (**182**) affords the flavene (equation 4) ⟨58JCS4040⟩.

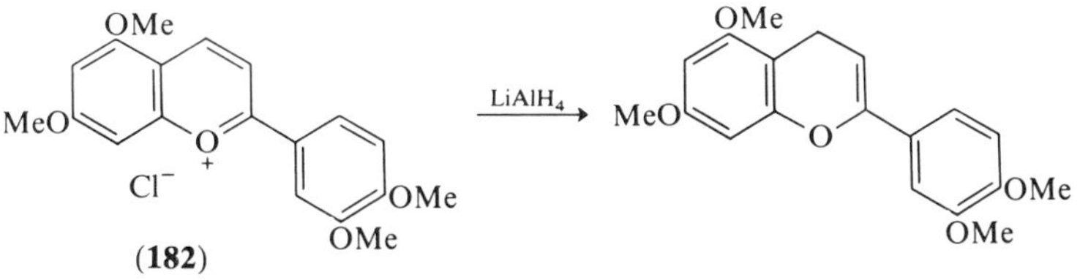

(4)

### (v) *From chromones*

Generally, nucleophilic attack of chromones opens the pyranone ring. However, the reduction of flavone by LAH yields 4*H*-flavene (2-phenylchrom-2-ene) ⟨62CI(L)1793⟩.

### (vi) *From coumarins*

The base-catalyzed reaction of dihydrocoumarin with ethyl formate leads to the 3-formyl derivative (**183**), which rearranges in the presence of methanolic hydrogen chloride to the chroman ester. Upon heating with polyphosphoric acid, the chrom-2-ene is produced (Scheme 32) ⟨60CB1025⟩.

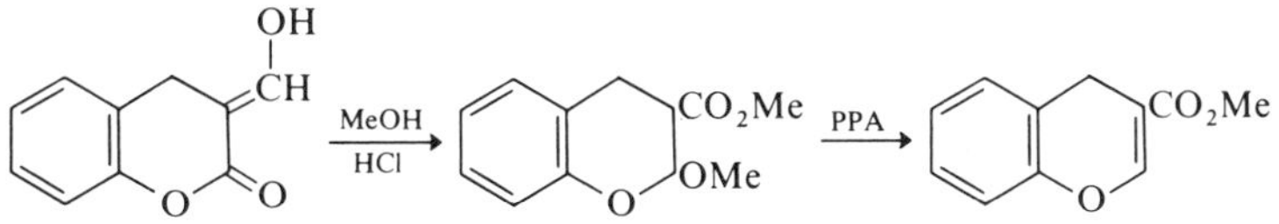

**Scheme 32**

## 2.24.2.5 1*H*-Isochromenes (1*H*-2-Benzopyrans)

There are few examples of isochromenes in the literature. In general they are not very stable, showing a tendency towards oxidation and polymerization.

### 2.24.2.5.1 Formation of one bond

#### (i) Adjacent to the heteroatom

The reduction of isocoumarin with LAH results in ring opening and the formation of 2-(2'-hydroxymethylphenyl)acetaldehyde (**184**). Cyclization to the isochromene occurs on boiling in acetic anhydride (Scheme 33) ⟨58CB2636, 65CB3279⟩. The isochromene (**186**) is obtained directly from the reduction of the dihydronaphthocoumarin (**185**; equation 5) ⟨56JCS4535⟩.

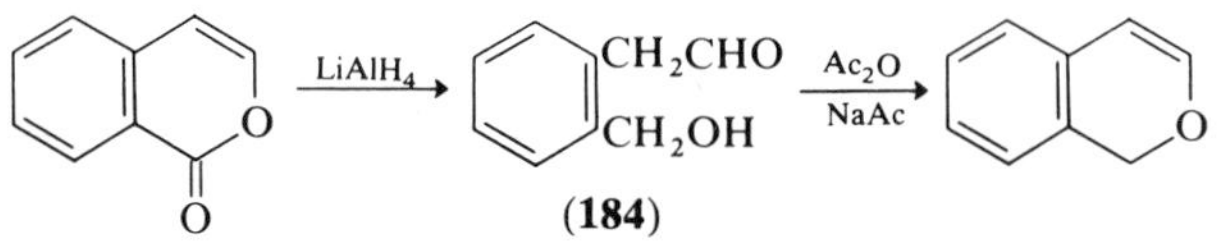

**Scheme 33**

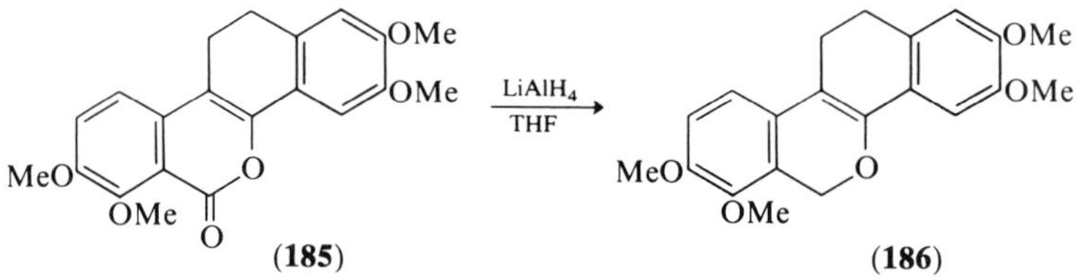

(5)

The thermal rearrangement of *o*-quinoneallides to chrom-3-enes is well documented (see p. 741) and a similar approach to the synthesis of isochromenes has been reported ⟨72HCA10⟩. The reaction of Grignard reagents with 1-cyanobenzocyclobutene (**187**) leads to the acyl compounds. On heating these in dilute decane solution 3-substituted isochromenes are produced. The thermal conrotatory ring opening of the acylbenzocyclobutene leads to the formation of both the *trans*- and *cis*-acyl-*o*-quinodimethanes (**188**) and (**189**). Ring closure of the *cis* isomer leads to the isochromene, whilst the *trans* compound either recyclizes to the cyclobutene or polymerizes (Scheme 34).

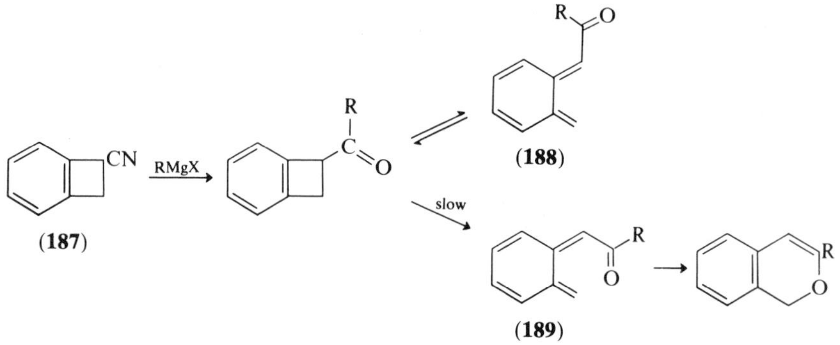

**Scheme 34**

The cyclization of bis(2-methylphenyl)ketene to 4-(2-methylphenyl)isochromene, which involves a [1,5]-hydrogen shift, has been observed ⟨72HCA10⟩.

#### (ii) β to the heteroatom

The monophosphonium salt derived from 1,2-di(bromomethyl)benzene reacts with sodium carboxylates to afford high yields of the ester (**190**). Generation of the ylide results in spontaneous cyclization *via* an intramolecular Wittig reaction to the isochromene (Scheme 35) ⟨79TL2149⟩.

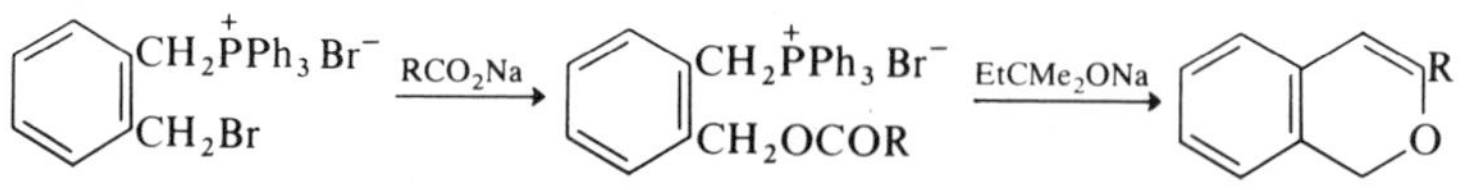

**Scheme 35**

### 2.24.2.5.2 From other heterocycles

#### (i) From epoxides

A one-step synthesis of isochromene itself has been reported from indene oxide (191) ⟨66CC415⟩. Irradiation of the epoxide in benzene affords two products, isochromene and indan-2-one, in similar quantities. It was proposed that initial fission of the carbon–carbon bond of the three-membered ring is followed by 1,4- or 1,2-hydrogen migration (Scheme 36).

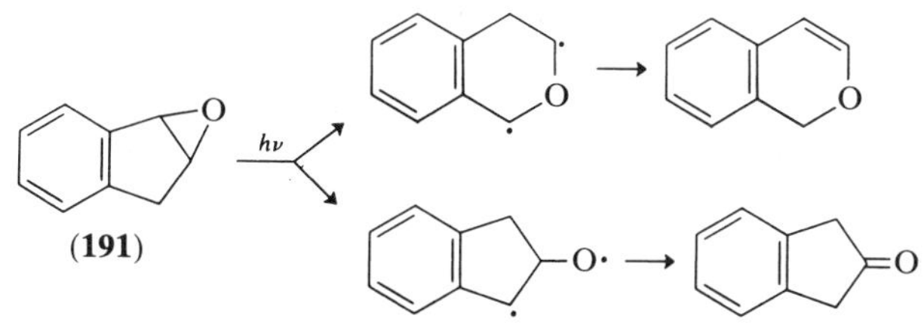

(191)

**Scheme 36**

### 2.24.2.5.3 From a preformed heterocyclic ring

#### (i) From isochromanols

Dehydration of isochromanols has to be carried out with care because of the relative ease with which the resulting isochromenes undergo polymerization under acidic conditions. Potassium hydrogen sulfate is a suitable reagent in some cases, and *p*-toluenesulfonic acid has also been used ⟨71MI22400⟩. The initial hydroxy compounds are available from iso-chromanones by reduction or by the action of a Grignard reagent.

### 2.24.2.6 Xanthenes (Dibenzo[*b,e*]pyrans)

A number of dyes, of which fluorescein is perhaps the best known, contain the xanthene ring system. Much of the interest in xanthene stems from this application. There are few syntheses which are of significant value. The earlier work has been discussed ⟨B-51MI22400⟩ and a more recent treatment is also available ⟨B-77MI22400⟩.

### 2.24.2.6.1 Formation of one bond

#### (i) Adjacent to the heteroatom

Xanthene itself results from the dehydration of 2,2'-dihydroxydiphenylmethane ⟨43JOC316⟩, and this route provides access to the important derivative, 3,6-bis(dimethyl-amino)xanthene (192). The reactive *m*-dimethylaminophenol condenses readily with for-maldehyde and subsequent cyclization occurs in sulfuric acid (Scheme 37) ⟨1896JPR(54)217⟩. Other aldehydes behave in a similar manner to yield 9-substituted xanthenes ⟨12LA(391)308⟩.

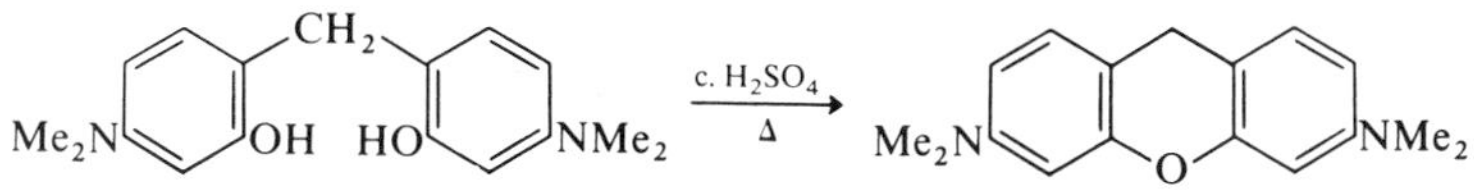

(192)

**Scheme 37**

### 2.24.2.6.2 Formation of two bonds

The anion derived from 2,3-dimethyl-1,4-naphthoquinone behaves as a quinone methide and undergoes a [1,4]-cycloaddition with the benzoquinone (193). The product is the xanthene derivative (194) ⟨70JCS(C)722⟩. There is no indication of the formation of the isomeric xanthene. A [1,3]-cycloaddition occurs simultaneously which leads to the fluorene derivative (195).

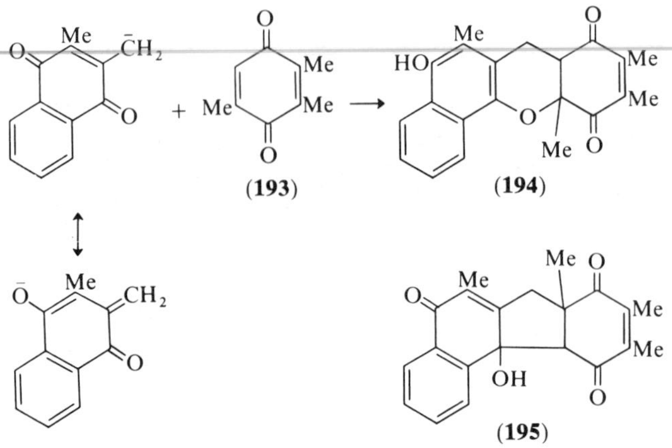

(193)          (194)

(195)

### 2.24.2.6.3 *From a preformed heterocycle*

#### (i) *From xanthones*

The reduction of xanthone to xanthene has been achieved by using the Huang–Minlon modification of the Wolff–Kishner reduction ⟨52JCS3741⟩ by LAH ⟨55JA5121⟩ and by diphenyl-silane ⟨61JOC4817⟩.

A comprehensive range of dibenzoxanthones has been converted into the 9-methylene derivatives by treatment with a Grignard reagent (equation 6) ⟨66CB1822⟩. With some organometallic compounds, *p*-methoxyphenylmagnesium bromide for instance, the xanthenol is formed.

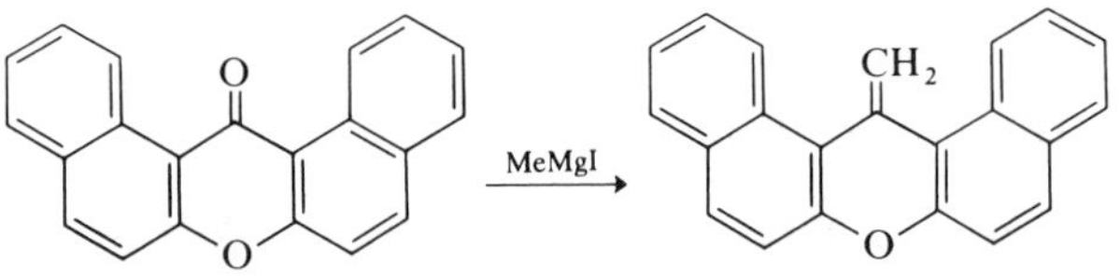

(6)

#### (ii) *From reduced xanthenes*

Whilst the synthesis of xanthenes from reduced xanthenes is of little practical value, their formation from bridged analogues is of interest. The dihydroxanth-1-one (**196**) was obtained from *o*-hydroxybenzyl alcohol and cyclohexane-1,3-dione, probably through the inter-mediacy of a quinone methide. Its conversion to 3,4-dihydroxanthene (**197**) was accom-plished in good overall yield *via* the tosylhydrazone. The adduct (**198**) of the 1,3-diene with cyanoacetylene undergoes a retro-Diels–Alder reaction with loss of ethylene to give 4-cyanoxanthene (Scheme 38) ⟨75CJC2045⟩.

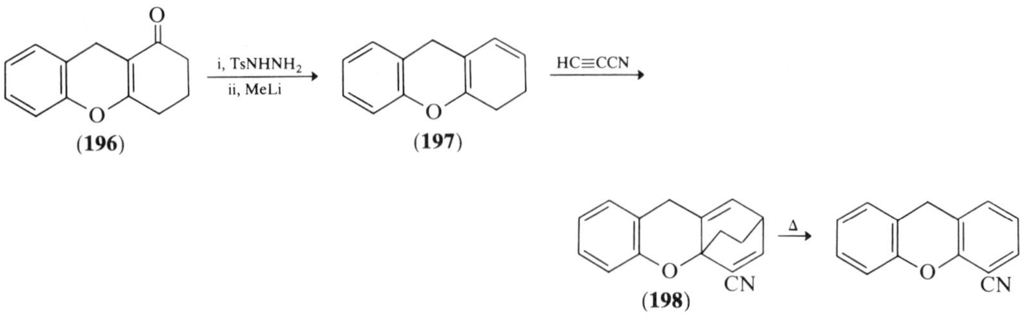

(196)                    (197)

(198)

**Scheme 38**

#### (iii) *From benzopyrylium salts*

The reaction between isoflavylium perchlorates and electron-rich 1,3-diarylpropenes has been used to synthesize xanthenes such as (**199**) which are structurally similar to the pigments isolated from the insoluble red woods ⟨78JCS(P1)88⟩.

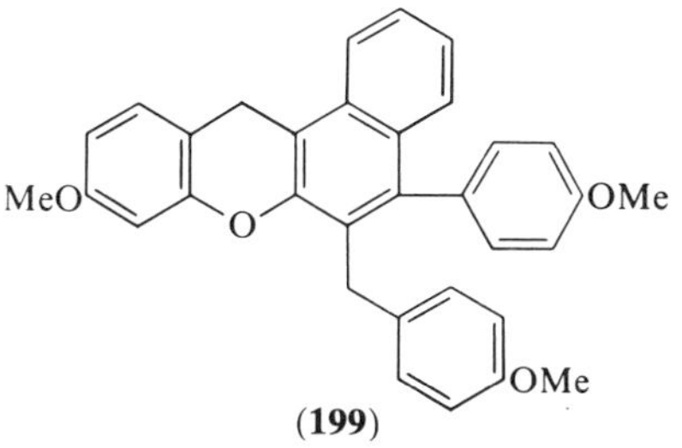

**(199)**

## 2.24.2.7 Reduced Pyrans

In this section, the three simple reduced pyrans (**200**), (**201**) and (**202**) are initially considered. Fused alicyclic analogues are included in this group. A discussion of the benzologues, chroman (**203**) and isochroman (**204**), and the reduced naphthopyrans follows.

(200)     (201)     (202)     (203)     (204)

### 2.24.2.7.1 Dihydropyrans

(i) *Formation of one bond*

(*a*) *Adjacent to the heteroatom.* The synthesis of dihydropyrans from 5-hydroxyketones lacks attraction because of the problems associated with the preparation of the ketones. However, the reaction of a Grignard reagent with 4-hydroxybutanonitrile leads directly to the dihydropyran ⟨74CR(C)(278)721⟩. An interesting facet of this route lies in the formation of the nitrile from THF, which involves protection of the hydroxy group as the tetrahydropyranyl derivative.

The conversion of $\beta$-allenic alcohols into 5,6-dihydro-2H-pyrans is catalyzed by silver ions and takes place under mild conditions ⟨79S743⟩.

Reaction of the optically active allene (**205**) with bromine leads to the inactive 3-bromodihydropyran (**208**; X = Br), whereas activity is retained in the reaction with 2,4-dinitrobenzenesulfenyl chloride, giving (**208**; X = $SC_6H_3(NO_2)_2$; Scheme 39) ⟨67JA7001⟩. The different behaviour has been attributed to the varying stabilities of the cyclic intermediates (**206**; X = Br or $SC_6H_3(NO_2)_2$), the latter showing less tendency to rearrange to the achiral allylic cation (**207**).

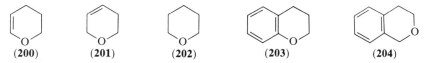

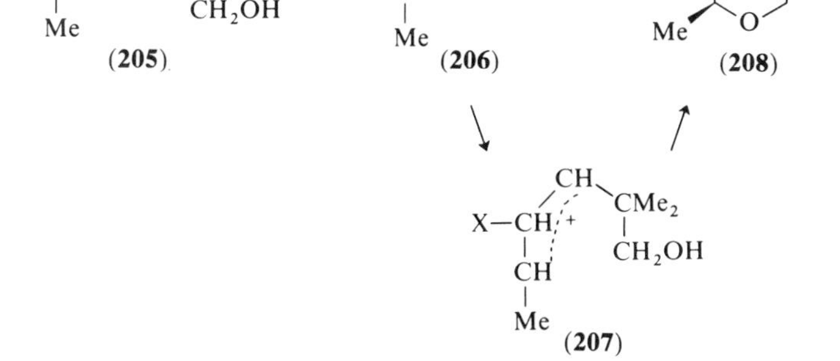

**(205)**          **(206)**          **(208)**

**(207)**

**Scheme 39**

The acid-catalyzed cyclization of the diol (**209**), available from penta-1,3-diene and bromomesityl oxide *via* the hydroxyketone, gives a mixture of the fused dihydropyrans (**210**) and (**211**), edulan I and II, respectively (Scheme 40) ⟨75JCS(P1)1736⟩. The former compound predominates when an ion exchange resin is used as the catalyst, but the latter using boron trifluoride. The latter compound isomerizes to the former on further treatment with acid.

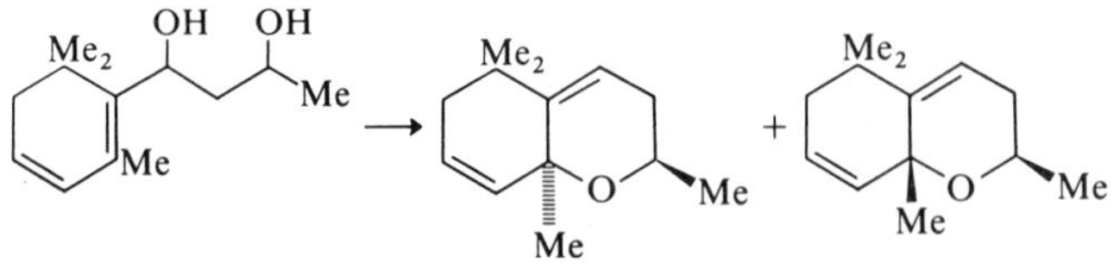

**Scheme 40**

Dimethyloxosulfonium methylide reacts with the triketone (212) to give the fused 2-hydroxymethylenedihydropyran (213) ⟨76H(4)1755⟩. The reaction is thought to proceed through a zwitterion and the epoxide as indicated in Scheme 41. Intramolecular nucleophilic attack leads to the dihydropyran and the overall process may be regarded as a transfer of methylene from the sulfur ylide.

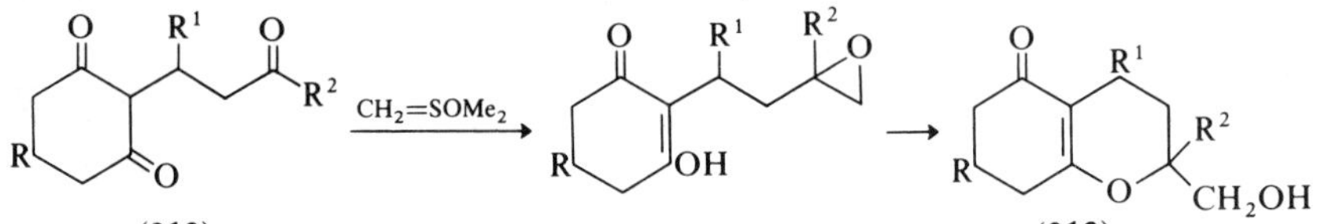

**Scheme 41**

The oxidative cyclization of the alcohol (214) with ceric ammonium nitrate gave a mixture of the two naphthopyrans (215) and (216) in a ratio of 3 : 1 ⟨81CC534⟩. The reaction is of interest because of the natural occurrence of quinones containing the naphtho[2,3-*c*]pyran ring system.

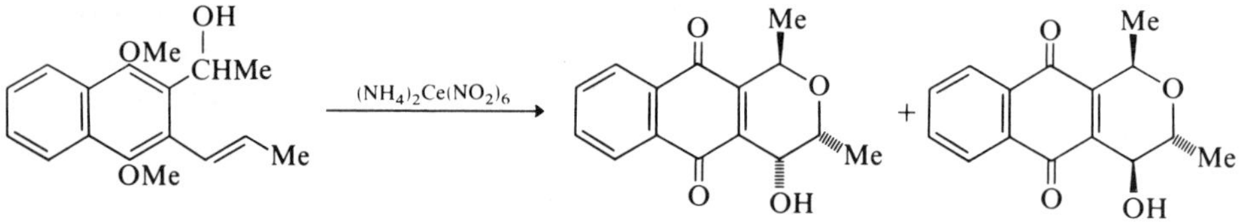

(*b*) *β to the heteroatom.* An intramolecular Wittig reaction brings about the cyclization of an ylide ester to a dihydropyran (Scheme 42) ⟨79TL5⟩. The requisite phosphonium salt is obtained through sequential reaction of 1,4-dibromobutane with triphenylphosphine and the sodium salt of a carboxylic acid.

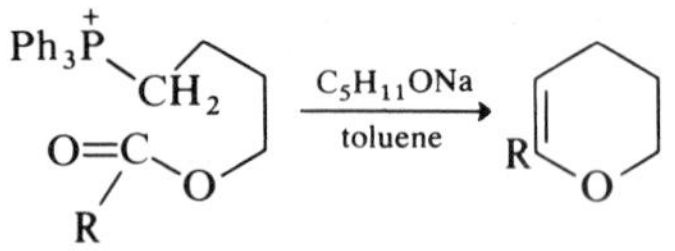

**Scheme 42**

(ii) *Formation of two bonds*

Largely because of the widespread interest in cycloadditions, a number of syntheses of dihydropyrans have been developed involving the interaction of four and two atom fragments. Both variations on the [4 + 2] cycloaddition are successful; either the diene or the dienophile may be the source of the heteroatom (Scheme 43). A review of heterodiene syntheses with unsaturated carbonyl compounds contains comprehensive lists of dihydropyrans ⟨75CRV651⟩.

**Scheme 43**

Since the thermal dimerization of acrolein was shown to give a dihydropyran ⟨38JGU22⟩, a range of related compounds has been found to yield similar products. The reaction exhibits high regioselectivity, with the 2-substituted 3,4-dihydropyran being formed almost exclusively at the expense of the 3-substituted isomer. Calculations using the SCF method predict the formation of this regioisomer and support an *endo* approach of the reactants in which the C—C bond closes faster than the C—O bond ⟨68JA553⟩.

The dimerization of α,β-unsaturated carbonyl compounds is a difficult reaction and where alternative processes are possible these will occur in preference. Thus, a mixture of propenal and 2-methylpropenal yields 2-formyl-5-methyl-3,4-dihydropyran; no dimeric products are observed (equation 7).

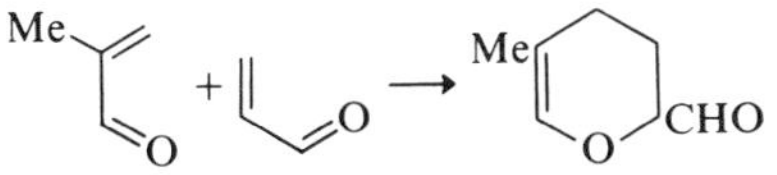

(7)

The reaction of propenal with alkenes is also predicted to be difficult and this is borne out in practice. A range of alkenes gave only a 10–25% yield of the dihydropyran when heated under pressure at 180–200 °C with the unsaturated carbonyl compound ⟨51JA5273⟩. However, a facile stereocontrolled intramolecular Diels–Alder reaction of the α,β-unsaturated carbonyl compound (**218**) has been used to synthesize the tricyclic dihydropyran (**219**) ⟨80AG(E)134⟩. The ketone, which is prepared from (*R*)-citronellal (**217**) and the enolate of cyclohexane-1,3-dione, could not be isolated, suggesting the occurrence of a fast concerted cycloaddition. The dihydropyran has the *R* configuration at C-6a and rings A and B are *trans* fused; the structure thus corresponds to the tetrahydrocannabinol skeleton ⟨76CRV75⟩. It therefore appears that the chiral centre of citronellal exerts control in the transition state of the cycloaddition, which presumably assumes a chair conformation (Scheme 44).

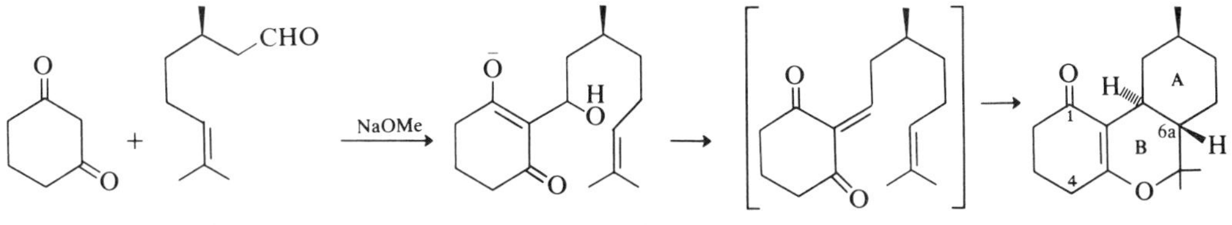

(217)    (218)    (219)

**Scheme 44**

Unsaturated esters containing an electron-withdrawing group in the 2-position react with electron-rich alkenes under mild conditions to give 6-alkoxy-2,3-dihydropyrans which are thermally unstable.

An intramolecular Diels–Alder reaction occurs when the *trans*-2-butenyl triester (**220**) is heated at 135 °C for 200 h ⟨79JA6023⟩. A competing ene reaction leading to the adduct (**221**) takes place and in fact is the major process with other triesters (Scheme 45). The dihydropyran which results from the inverse electron demand Diels–Alder reaction involving *cis* addition to the alkene bears a structural resemblance to iridoids such as elenolic acid (**222**). A total synthesis of (±)-methyl elenolate has been described ⟨73JA7156⟩.

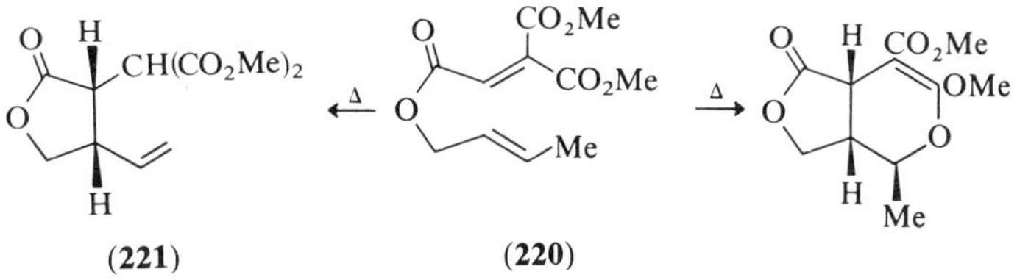

(221)    (220)

**Scheme 45**

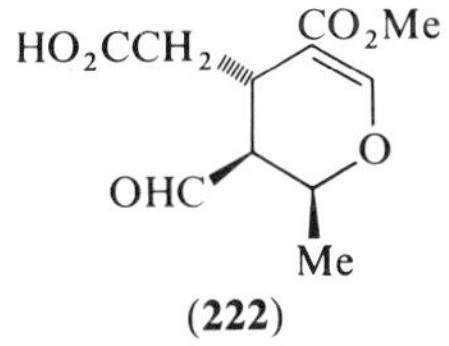

(222)

Cycloadditions involving the more nucleophilic vinyl ethers are easier than those above and the reaction has considerable synthetic potential. The reactants are heated at 180–190 °C in a sealed vessel and the adduct is rapidly formed in high yield ⟨50JA3079, 51JA5267⟩. Full experimental details have been published for the synthesis of 3,4-dihydro-2-methoxy-4-methyl-2*H*-pyran from methyl vinyl ether, and the same technique was used to prepare a further 13 dihydropyrans ⟨63OSC(4)311⟩.

It has since been shown that the cycloaddition is catalyzed by transition metal salts ⟨78JOC667⟩. Although the yields are generally lower than in the uncatalyzed reaction, this is outweighed by the advantages of shorter reaction times and a lower cyclization temperature. Illustrative examples of the formation of 2-alkoxy-3,4-dihydro-2*H*-pyrans are presented in Table 3, which includes a comparison of the two methods of synthesis.

**Table 3**   Cycloaddition of Vinyl Ethers and $\alpha,\beta$-Unsaturated Carbonyl Compounds ⟨78JOC667⟩

| Vinyl ether | $\alpha,\beta$-Unsaturated carbonyl compound | Temp. (°C) | Conditions Time (h) | Catalyst | Yield |
|---|---|---|---|---|---|
| MeOCH=CH$_2$ | CH$_2$=CHCHO | 140 | 12 | — | 44 |
| MeOCH=CH$_2$ | CH$_2$=CHCHO | 90 | 0.5 | ZnCl$_2$ | 35 |
| MeOCH=CH$_2$ | trans-MeCH=CHCHO | 200 | 12 | — | 80 |
| MeOCMe=CH$_2$ | CH$_2$=CHCHO | 60 | 1.5 | ZnCl$_2$ | 61 |
|  | MeCOCH=CH$_2$ | 25 | 5 | ZnCl$_2$ | 10 |
| EtOCH=CHMe | CH$_2$=CHCHO | 90 | 1 | ZnCl$_2$ | 68 |

The presence of an electron-withdrawing group at C-2 of an $\alpha,\beta$-unsaturated carbonyl compound allows Diels–Alder reactions with enol ethers to proceed readily at room temperature, presumably through reducing the energies of the LUMOs. Thus, dihydropyrans are obtained with ease from 2-formylmalondialdehyde and various enol ethers (Scheme 46) ⟨82TL1147⟩. The cycloadditions are regioselective but may proceed through either an *endo* or an *exo* transition state, since two diastereoisomers are formed in all the reactions.

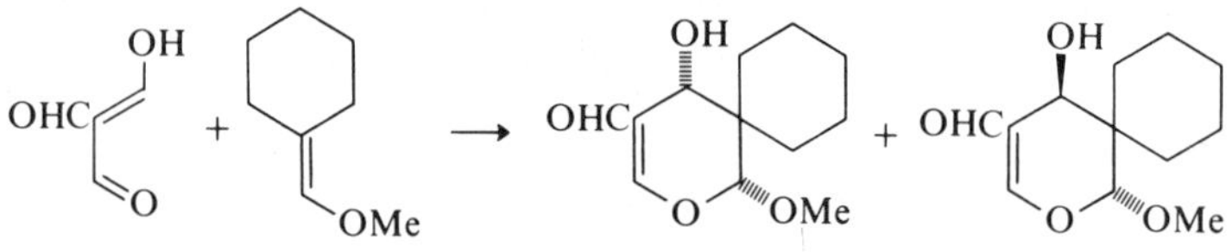

**Scheme 46**

Besides acyclic unsaturated carbonyl compounds, their alicyclic analogues may be used, leading to fused dihydropyran derivatives. Cyclic vinyl ethers have also been utilized in the process.

In the presence of aluminum chloride, which presumably lowers the energy of the LUMO of the heterodiene by Lewis acid complexation, electron-rich alkenes give dihydropyrans on reaction with acyl cyanides at room temperature ⟨82AG(E)859⟩. Unsaturated esters further extend the range of diene components of value in these Diels–Alder reactions with inverse electron demand ⟨82TL603⟩.

The reaction of $\alpha,\beta$-unsaturated carbonyl compounds with enamines also leads to dihydropyrans, although it is not always possible to isolate these since they react further to give either ring-opened by-products or bicyclic derivatives arising from a Stork annelation. There has been considerable discussion on the mechanism of this reaction, although the initial nucleophilic attack of the enamine on the $\beta$-carbon of the diene is not in doubt ⟨63JA207⟩. It is possible that a zwitterionic species is involved, either as an intermediate or merely in equilibrium with the dihydropyran ⟨67JCS(C)226⟩.

2,2-Dialkoxy-3,4-dihydropyrans are produced under mild conditions from the zinc chloride-catalyzed cycloaddition of $\alpha,\beta$-unsaturated carbonyl compounds with ketene acetals (Scheme 47) ⟨81RTC13⟩. At lower temperatures a [2+2]-cycloaddition leads to oxetanes, but on warming these compounds revert to starting materials and thence to the thermodynamically favoured dihydropyrans.

The synthesis of dihydropyrans by the reaction of dienes with carbonyl compounds is less well documented ⟨62CRV405⟩, presumably because of the poor dienophilic nature of the latter. Generally, the more reactive carbonyl compounds, such as derivatives of glyoxylic

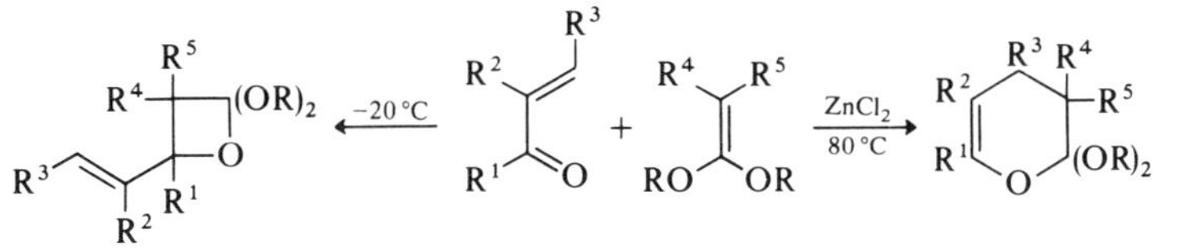

**Scheme 47**

and oxomalonic acids, have been successfully used, as in the synthesis of 6-deoxy sugars ⟨79JCS(P1)2230⟩. Butadiene and its methyl, methoxy and trimethylsilyloxy derivatives react with diethyl oxomalonate to yield 5,6-dihydropyrans (Scheme 48) ⟨75JA6892, 77JOC4095, 82JOC3649⟩. The keto ester functions as a carbon dioxide equivalent, since conversion of the adducts into tetrahydropyranones is readily accomplished ⟨79SC889⟩.

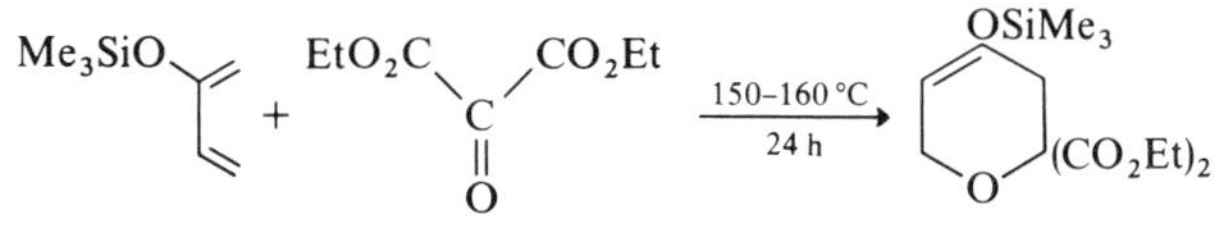

**Scheme 48**

Many studies have established that the Diels–Alder reaction is strongly accelerated by an increase in pressure and this property has been applied to the reaction between 1-methoxybuta-1,3-diene and various aldehydes, ketones and α-keto esters, when good yields of dihydropyrans are obtained at 50 °C ⟨79S41, 81JOC2230⟩. *Endo* addition yielding the *cis* isomer is preferred, whilst differences in the ratio of *cis* and *trans* products obtained at atmospheric and at high pressures suggest that *endo* and *exo* additions have different activation volumes.

(iii) *From other heterocycles*

(*a*) *From epoxides.* A relatively simple synthesis of 2-aryl-3,6-dihydro-2*H*-pyrans is provided by the rearrangement of aryl derivatives of cyclopropyl epoxides (**223**) ⟨73JCS(P1)2030⟩. Furthermore, the cyclopropyl epoxides are readily available by the action of dimethylsulfonium methylide on cyclopropyl ketones (Scheme 49). The rearrangement is considered to involve adventitious acid-catalyzed ring opening of the epoxide followed by an intramolecular attack on the homoallyl cation (**224**). Indeed, in the presence of *p*-toluenesulfonic acid rapid formation of the pyrans was observed.

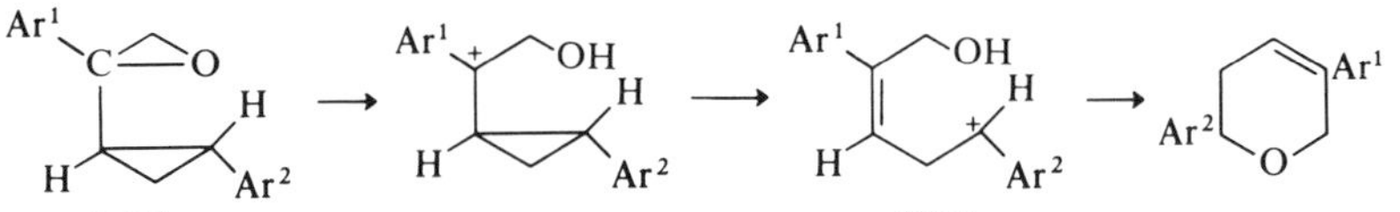

**Scheme 49**

It is beneficial in terms of yield to convert an α,β-unsaturated ketone into the cyclopropyl ketone using dimethyloxosulfonium methylide. The epoxide, formed as above, is rearranged during chromatography on silica gel. The corresponding pent-2-ene-1,5-diols are also formed, presumably through hydration of the cation, and these may be dehydrated to the pyran with *p*-toluenesulfonic acid ⟨74JCS(P1)1674⟩.

(*b*) *From furans.* Although commercially available, the synthesis of dihydropyran is still of interest. A proven and reliable method, which gives yields of 60–70%, involves the acid-catalyzed rearrangement of tetrahydrofurfuryl alcohol ⟨55OSC(3)276⟩.

(iv) *From a preformed heterocyclic ring*

(*a*) *From tetrahydropyrans.* Elimination of water or alcohol from 2-hydroxy- or 2-alkoxy-tetrahydropyrans provides a convenient route to dihydropyrans. Amongst reagents which have been used to effect the elimination are phosphorus pentoxide, *p*-toluenesulfonic acid and aluminum oxide ⟨79JOC364⟩. One approach involves the hydrogenation of 2-ethoxy-3-methyl-3,4-dihydro-2*H*-pyran and subsequent dehydration of the resulting 2-ethoxytetra-hydropyran to 5-methyl-3,4-dihydro-2*H*-pyran.

Isomerization of the divinyltetrahydropyran shown in equation (8) to 6-ethyl-3,4-dihydro-3-vinyl-2*H*-pyran is catalyzed by a ruthenium–triphenylphosphine complex ⟨74T1015⟩.

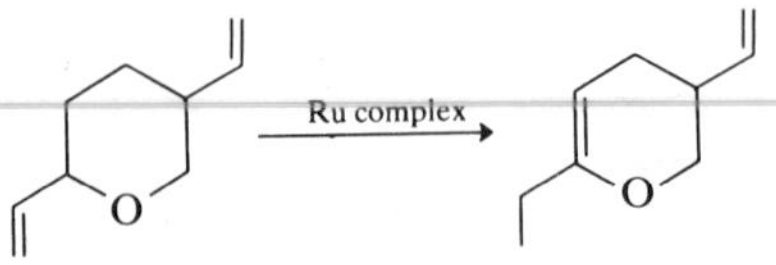

$$\text{(8)}$$

The dihydropyranol resulting from the base-catalyzed condensation of $\beta$-cyclocitral (**225**) with benzaldehyde undergoes a base-catalyzed dehydration to a fused dihydropyran (or tetrahydroisochromene) (Scheme 50) ⟨81JHC549⟩.

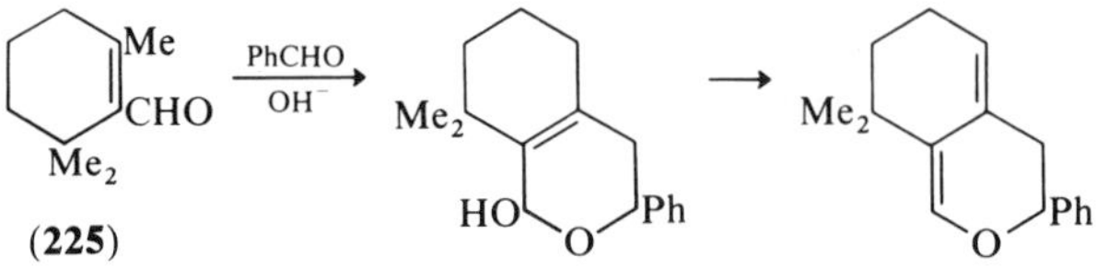

**Scheme 50**

(*b*) *From pyranones.* A number of fused analogues of 3,4-dihydro-2*H*-pyran are available from the acid-catalyzed rearrangement of pyranone derivatives ⟨60CB1025⟩. Thus, 4-formyl-hexahydroisochroman-3-one (**226**) is converted directly into the carboxylic acid (**227**; Scheme 51), whereas rearrangement with methanolic hydrogen chloride leads to the ester *via* the 2-methoxy derivative. In a similar manner, the 4-ethoxalyl compound yields the 3,4-diester. The reactions parallel the synthesis of chrom-2-enes from coumarins (Section 2.24.2.4.4).

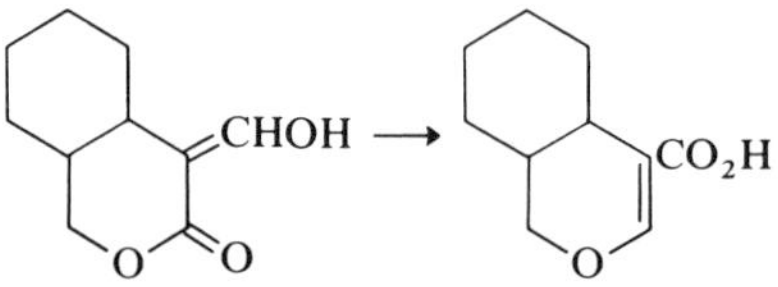

**Scheme 51**

(*c*) *From dihydropyrans.* The phenylation of 3,4-dihydropyran has been achieved using iodobenzene in the presence of a palladium catalyst ⟨79JOC21⟩.

### 2.24.2.7.2 Tetrahydropyrans

(i) *Formation of one bond*

(*a*) *Adjacent to the heteroatom.* The cyclization of pentane-1,5-diols to tetrahydropyrans has received attention, frequently because of the associated stereochemical features. Ring closure of 1,5-diphenylpentane-1,5-diol has been achieved under a variety of conditions ⟨75M229⟩. Irrespective of the stereochemistry of the diol, *cis*-diphenyltetrahydropyran predominates in the product (Scheme 52).

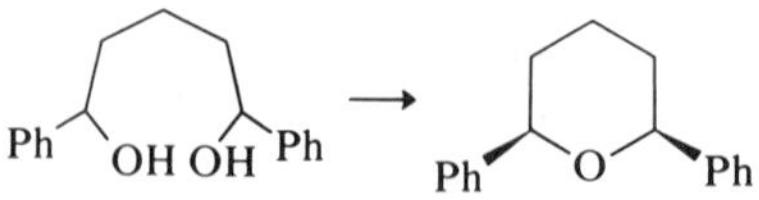

**Scheme 52**

NMR spectroscopy was used to assign configurations to the products derived from *meso*-and (±)-3,3-dimethyl-1,5-diphenylpentane-1,5-diol (**228** and **230**) ⟨80T3565⟩. The former gave (±)-(*trans*)-4,4-dimethyl-2,6-diphenyltetrahydropyran (**229**) and the latter the *meso*-(*cis*) isomer (**231**). Thus, in this example cyclization occurs with inversion of configuration at the site of substitution (Scheme 53).

3-Pentyltetrahydropyran was formed from the corresponding diol in 74% yield ⟨69JOC479⟩.

Several syntheses of tetrahydropyrans revolve around the cyclization of 1,5-hydroxy-ketones, the differences between them lying in the routes to the acyclic precursors.

The synthesis of 2-ethyl-6,6-dimethyltetrahydropyran-2-carboxylic acid, required for the elucidation of the course of the acid-catalyzed rearrangement of cinenic acid, is shown in

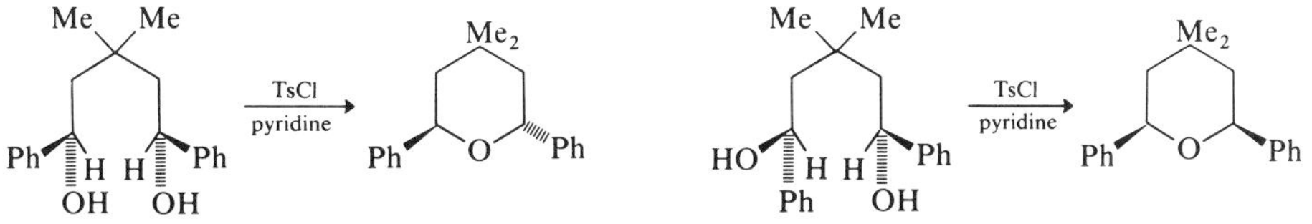

**Scheme 53**

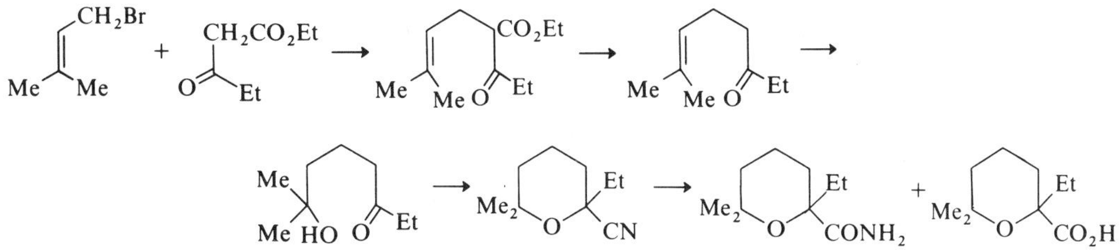

**Scheme 54**

Scheme 54 ⟨60JA483⟩. Cyclization is achieved with anhydrous hydrogen cyanide and it is significant that hydrolysis to the acid is difficult.

The low temperature reaction of pentanedial with a range of alkyl Grignard reagents leads to 6-substituted tetrahydropyran-2-ols ⟨72HCA249⟩. The major side reaction involves addition of the organometallic compound to both carbonyl groups.

Condensation of a β-keto ester with 1,3-dibromopropane gives the bromoketone, *e.g.* (**232**), which on hydrolysis yields a mixture of equal quantities of 2-methylpyran-2-ol and the acyclic ketone (**233**; Scheme 55) ⟨79JOC364⟩.

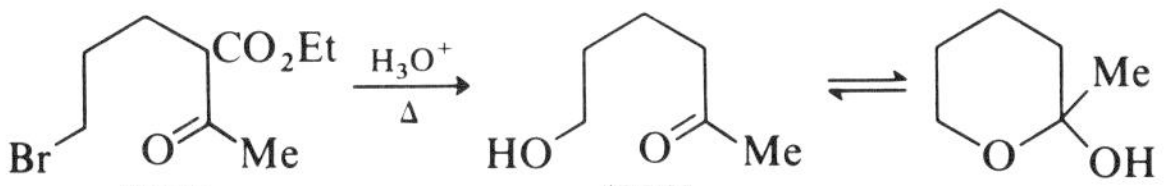

**Scheme 55**

The formation of 3-methyltetrahydropyran-2-ol from the pentenyl methyl ether (**234**) proceeds in an analogous manner since hydrolysis of the enol ether to the pentanal occurs prior to cyclization ⟨79JOC364⟩.

The tetrahydropyran ring occurs in polyether antibiotics, for instance, as the *trans* tetrahydropyran in salinomycin (**235**) and as the *cis* form in alborixin (**236**). Stereospecific syntheses of both systems have been developed ⟨81JA6967, 82CJC90⟩. The salient feature of these syntheses is the ring opening–ring closure of an epoxyalcohol in acid conditions. This sequence occurs with inversion of the epoxide centre and with complete regioselectivity.

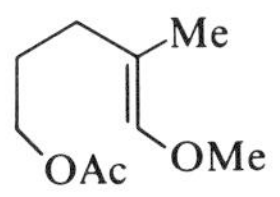

(**234**) 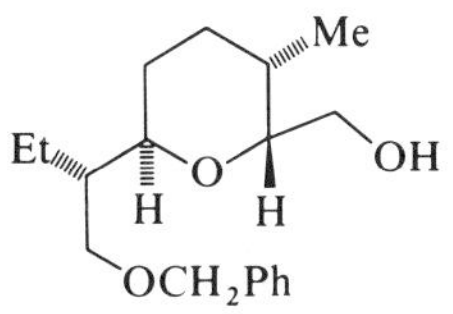　　(**235**)  Ring A of salinomycin 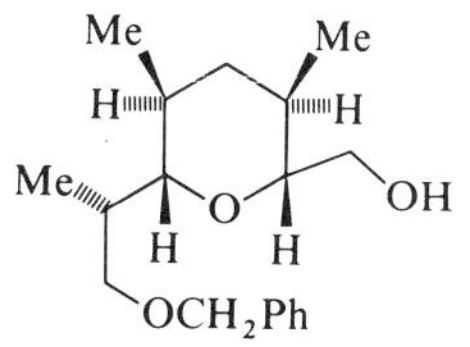　　(**236**)  Ring A of alborixin

A significant feature in the synthesis of 2,2-disubstituted tetrahydropyrans from diethyl malonate is the selective chlorination by trifluoromethanesulfonyl chloride of carbanions which can be formed using triethylamine or DBU as the base ⟨79TL3645⟩. Monoalkylation of the diester with 1-bromo-4-tetrahydropyranyloxybutane leads to the alcohol (**237**; R = H)

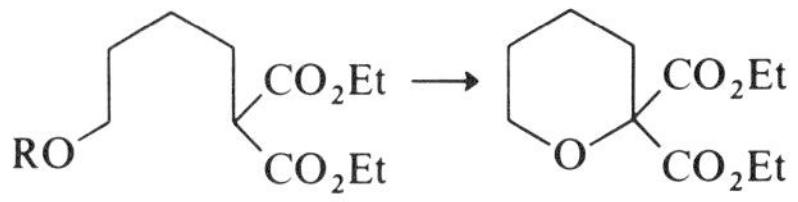

(**237**)

**Scheme 56**

*via* the tetrahydropyranyl ether (**237**; R = THP). Chlorination gives the pyran directly, together with some uncyclized chloroalcohol which ring closes on reaction with sodium hydride (Scheme 56).

Oxymercuration–demercuration reactions of alkenols may result in intramolecular cyclic ether formation involving participation by the neighbouring hydroxyl group. The substitution pattern at the unsaturated moiety exerts control over the site of cyclization as a consequence of Markownikov hydration of the alkene ⟨80MI22401⟩. In the case of $\Delta^4$-alkenols in which the terminal trigonal carbon atom is either unsubstituted (**238**; R = H) or fully substituted (**238**; R = Me), cyclization is regioselective and in the latter case leads exclusively to tetrahydropyrans. Ring closure of hex-5-en-1-ol also yields only the pyran. However, when both alkenic carbon atoms are monosubstituted, both five- and six-membered rings are formed in a ratio which is dependent on the configuration about the double bond (Scheme 57). Whilst the *Z*-alkenols yield essentially only furans, their diastereoisomers afford mixtures of the two cyclic products. Similar results have been obtained with 1-phenylalk-4-en-1-ols and a low yield of 2-methyl-6-vinyltetrahydropyran has been reported from the palladium-catalyzed cyclization of oct-5-en-2-ol ⟨76TL1821⟩.

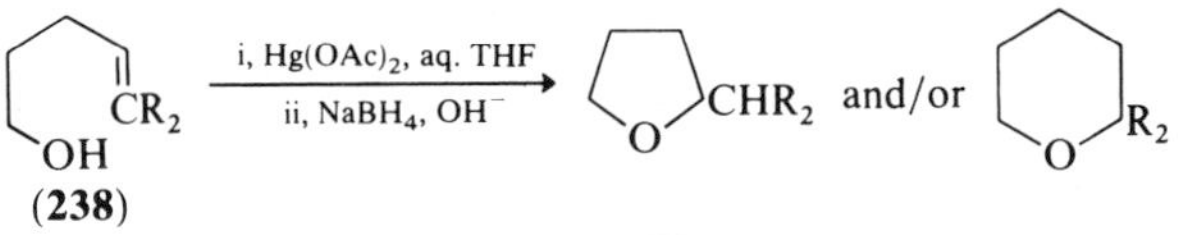

Scheme 57

The acid-catalyzed ring closure of the alkenols follows much the same pattern, although the intermediate carbocation from hex-5-en-1-ol undergoes rearrangement and gives a mixture of 2-methyltetrahydropyran and 2-ethyltetrahydrofuran ⟨79MI22402⟩.

The acid-catalyzed cyclization of some allyl 1,7-diols leads to tetrahydropyrans ⟨64HCA602⟩. The reaction involves dehydration at the allyl position.

A modification of this method provides a route to the 3-hydroxy analogue from the 6,7-epoxyoct-2-en-1-ol (**239**). However, the major product is the tetrahydrofuran (**240**; Scheme 58).

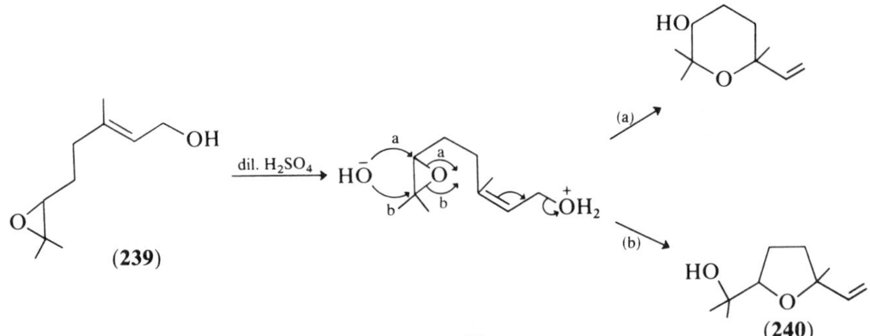

Scheme 58

The intramolecular oxidative cyclization of alcohols using lead(IV) acetate has been reviewed ⟨70S209⟩. Its application to the synthesis of tetrahydropyrans is limited, generally giving mixtures of products in which a tetrahydrofuran is the major component. However, (+)-(4*R*)-4,8-dimethylnonanol (**241**) affords the 3-acetoxypyrans (**242**) and (**243**) in significant amounts, presumably *via* the tertiary carbocation (Scheme 59) ⟨64HCA1883⟩.

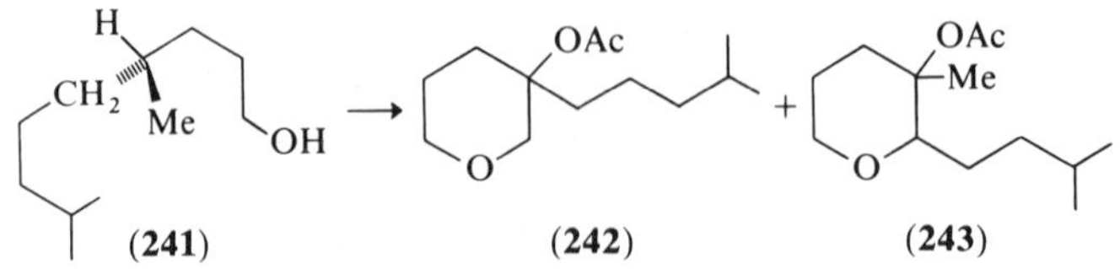

Scheme 59

Cyclization of 5-ethoxypentan-1-ol likewise mainly affords the pyran, apparently through facilitation of abstraction of the 5-hydrogen atom ⟨66T723⟩.

The behaviour of unsaturated alcohols towards this oxidant has also received attention. An initially formed alkenyloxy radical can add to the double bond and the resulting carbon radicals subsequently yield cyclic ethers. Both pent-4-en-1-ol and hex-5-en-1-ol yield tetrahydropyrans as major products ⟨66JOC3067⟩.

(*b*) *β to the heteroatom.* The condensation of alkenes with paraformaldehyde in the presence of hydrogen chloride at low temperature, a modified Prins reaction, provides a good route to 4-chlorotetrahydropyrans ⟨69JOC479⟩. Both hydrogen bromide and hydrogen fluoride may be used, though yields of the 4-fluoro derivative are not as high.

A mechanism has been proposed for this synthesis involving initial electrophilic attack of protonated formaldehyde on the alkene. Proton abstraction leads to the alkenol and to the pyran *via* the chloromethyl ether.

### (ii) *From a preformed heterocyclic ring*

(*a*) *From pyrans.* The hydrogenation of some 2,6-diphenyl-4*H*-pyrans has been studied under a variety of catalysts and conditions ⟨75M229⟩. The presence of a substituent in the 4-position leads to exclusive formation of the *cis* product.

Dihydropyrans are also a convenient source of tetrahydropyrans and full experimental details are available for the reduction of the unsubstituted dihydropyran over Raney nickel ⟨55OSC(3)794⟩.

The reaction of dihydropyran with water in the presence of a trace of mineral acid gives an equilibrium mixture of 2-hydroxytetrahydropyran and 5-hydroxypentanal. However, the products from the addition of alcohols to dihydropyrans are stable under basic conditions and can be isolated after destruction of the acid catalyst ⟨47JA2246⟩. A wide range of substituted tetrahydropyrans has been made in this way and the diastereoisomers separated by gas chromatography ⟨68JOC3754⟩.

Other molecules such as chlorine will add to the double bond to give substituted tetrahydropyrans ⟨56JCS136⟩.

Rather more complex is the reaction of dihydropyrans with *t*-butyl hypochlorite. The addition products are diastereoisomeric mixtures arising from *syn* and *anti* addition to the double bond, the addition being formally of methanol rather than *t*-butyl hypochlorite. The *trans* product is the major component (equation 9) ⟨79JOC364⟩.

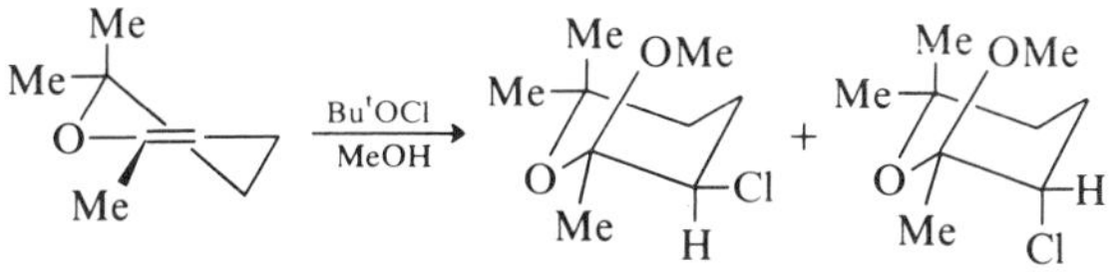

$$(9)$$

Hydroboration and oxidation of 3,4-dihydro-2-methoxy-2*H*-pyran gives a mixture of *cis*- and *trans*- 5-hydroxy-2-methoxytetrahydropyran ⟨70CJC2334⟩. After methylation, *trans*-2,5-dimethoxytetrahydropyran was isolated in small amounts. The *cis* isomer was obtained from 3,4-dihydro-3-methoxy-2*H*-pyran *via* 3-bromo-2,5-dimethoxytetrahydropyran.

(*b*) *From pyranones.* Both pyranones and dihydropyranones have been used as precursors of tetrahydropyrans. Prior reduction of the carbon–carbon double bond, achieved with hydrogen and a catalyst, is necessary. Reduction of the carbonyl function is then achieved using hydride ion reagents and gives the tetrahydropyranol ⟨81JOC4948⟩. An example is provided by the reduction of the lactone prepared from optically active parasorbic acid (Scheme 60). The chiral centre at C-6 is retained throughout the sequence and the acetals (**244**) and (**245**) can be separated ⟨81JOC736⟩.

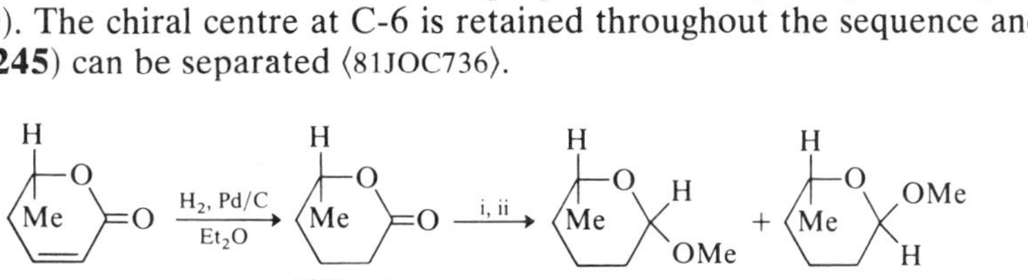

(6*S*)-(−)     (**244**) (2*S*,6*S*)-(+)   (**245**) (2*R*,6*S*)-(−)

i, LiAlH$_4$; ii, MeOH, *p*-TsOH, 2 days, separate by preparative GC

**Scheme 60**

In order to assess the influence of the heteroatom on vicinal coupling constants, several stereospecifically deuterated tetrahydropyrans have been synthesized ⟨76JOC1380⟩.

The catalytic reduction of some chromanones has been shown to yield mixtures consisting largely of 4-hydroxy-1-oxadecalin, together with significant amounts of 1-oxadecalin and smaller quantities of a monocyclic alcohol ⟨74JOC2040⟩. A detailed study of the conformation of the oxadecalins is given.

### 2.24.2.7.3 Chromans (dihydrobenzopyrans)

There is an extensive literature on chromans, largely as a consequence of the presence of the chroman-6-ol moiety in the biochemically important tocopherols. The chemistry of chromans and tocopherols is well covered by Ellis and Lockhart ⟨81HC(36)⟩.

### (i) Formation of one bond

(a) *Adjacent to the heteroatom.* The cyclization of derivatives of 1-(2-hydroxyphenyl)propane (Scheme 61) is a common feature of many chroman syntheses.

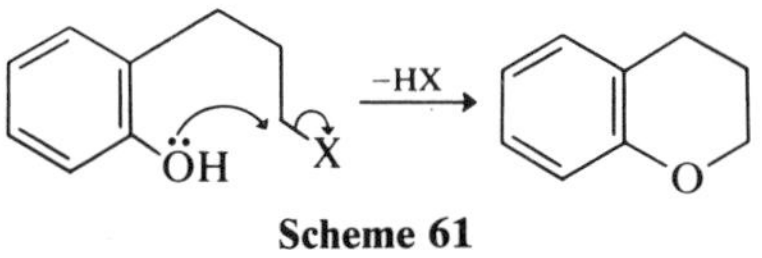

**Scheme 61**

An established route which has been used in the synthesis of a range of chromans involves the cyclization of chloropropanes ⟨39JOC311⟩. Such a reaction is involved in the formation of chromans from phenols and 1,3-dichloro-3-methylbutane, which is catalyzed by bis(acetylacetonato)nickel ⟨79S126⟩. Only phenol itself and various alkoxy derivatives have been used and it is pertinent to note that *m*-methoxyphenol gives a mixture of the 5- and 7-methoxychromans. There was no evidence of the formation of 4,4-dimethylchromans which would arise if the tertiary carbon atom was involved in the initial alkylation. In some instances, cyclization of the chloroalkane is carried out under basic conditions suggesting that hydrolysis may occur prior to cyclization. Indeed, the formation of chromans from substituted propanols constitutes another well-documented route. Dehydration is commonly achieved with phosphorus pentoxide ⟨67JCS(B)859⟩, but a mixture of acetic and sulfuric acids ⟨52BSB33⟩, and phosphoric acid ⟨57BSF776⟩ have also been used.

The formation of 3-bromo-2-phenylchroman from the propanol derivative (**246**) is of interest because it forms part of the sequence utilized in the conversion of 4*H*-chromene into a 2*H*-chromene derivative (Scheme 62) ⟨71S149⟩.

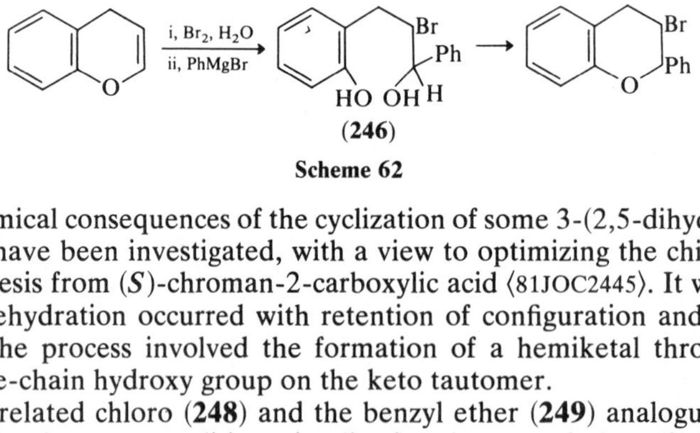

**Scheme 62**

The stereochemical consequences of the cyclization of some 3-(2,5-dihydroxyphenyl)propan-1-ols (**247**) have been investigated, with a view to optimizing the chiral economy of a tocopherol synthesis from (*S*)-chroman-2-carboxylic acid ⟨81JOC2445⟩. It was observed that acid-catalyzed dehydration occurred with retention of configuration and it was proposed ⟨79JA6710⟩ that the process involved the formation of a hemiketal through nucleophilic attack by the side-chain hydroxy group on the keto tautomer.

However, the related chloro (**248**) and the benzyl ether (**249**) analogues do not cyclize to chromans under the same conditions, implicating the second phenolic hydroxy group of

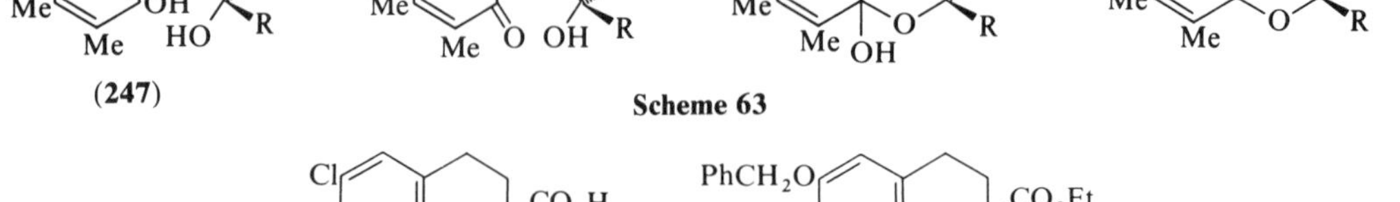

(**247**)

**Scheme 63**

(**248**)          (**249**)

(**247**) in the mechanism. As a consequence, an alternative mechanism (Scheme 63) involving the intermediacy of a quinone is now favoured ⟨81JOC2445⟩.

Cyclodehydration of the 1,5-diol in the presence of zinc chloride, however, leads to inversion of configuration, presumably by an $S_N 2$ displacement at the tertiary centre (Scheme 64) ⟨63HCA333⟩.

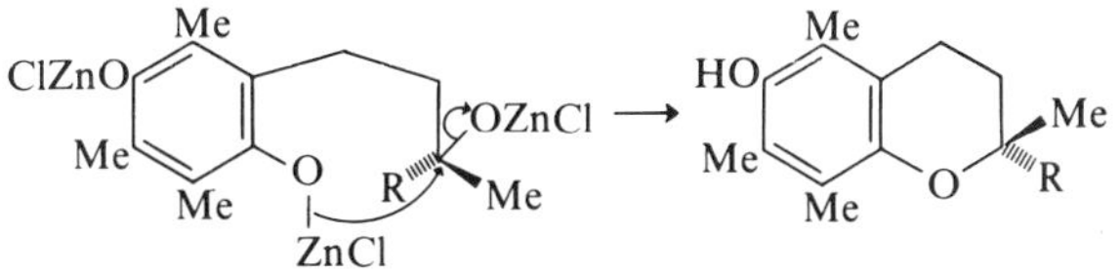

**Scheme 64**

The methodology derived from these results has enabled methyl (*R*)-chroman-2-carboxylate (**250**) to be converted into the synthetically more valuable *S* enantiomer (**251**) in a 78% overall yield (Scheme 65). Use of the mesylate ensured that in the key reduction–cyclization sequence inversion of configuration occurred.

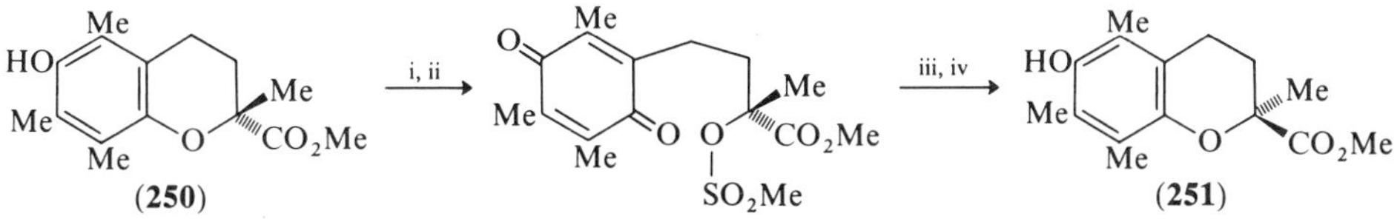

i, HNO₃; ii, MeSO₂Cl, Et₃N; iii, Na₂S₂O₄, NaOH; iv, base

**Scheme 65**

The synthesis of enantiomerically pure (*S*)-6-hydroxy-2,5,7,8-tetramethylchroman-2-ylmethanol (**252**), required as the chroman fragment for another vitamin E synthesis, was achieved according to Scheme 66 ⟨79HCA2384⟩. The protected diol was prepared by a Wittig synthesis, then reduction. Demethylation, which was accompanied by partial deprotection, and cyclization in methanolic hydrogen chloride gave the quinone acetal which was hydrogenated to the chroman.

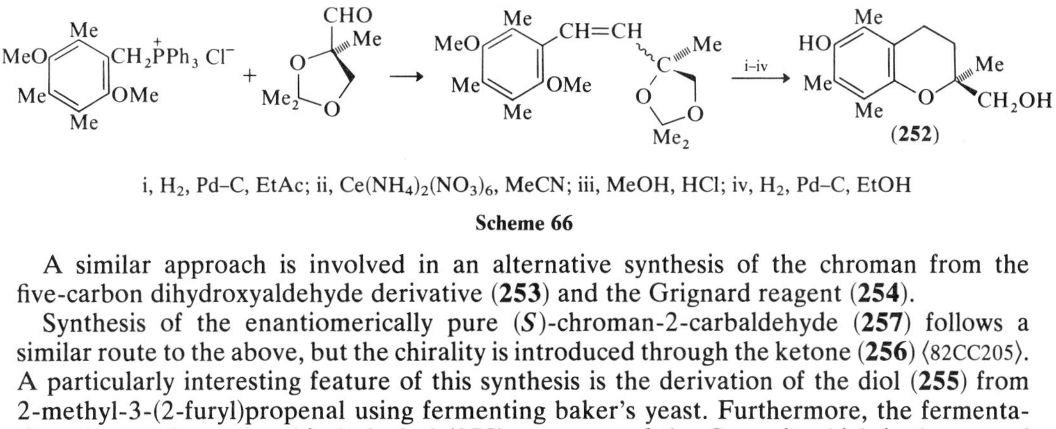

i, H₂, Pd–C, EtAc; ii, Ce(NH₄)₂(NO₃)₆, MeCN; iii, MeOH, HCl; iv, H₂, Pd–C, EtOH

**Scheme 66**

A similar approach is involved in an alternative synthesis of the chroman from the five-carbon dihydroxyaldehyde derivative (**253**) and the Grignard reagent (**254**).

Synthesis of the enantiomerically pure (*S*)-chroman-2-carbaldehyde (**257**) follows a similar route to the above, but the chirality is introduced through the ketone (**256**) ⟨82CC205⟩. A particularly interesting feature of this synthesis is the derivation of the diol (**255**) from 2-methyl-3-(2-furyl)propenal using fermenting baker's yeast. Furthermore, the fermentation also produces the chiral alcohol (**258**), a source of the C₁₅ unit which is the second component along with the aldehyde (**257**) in an α-tocopherol synthesis.

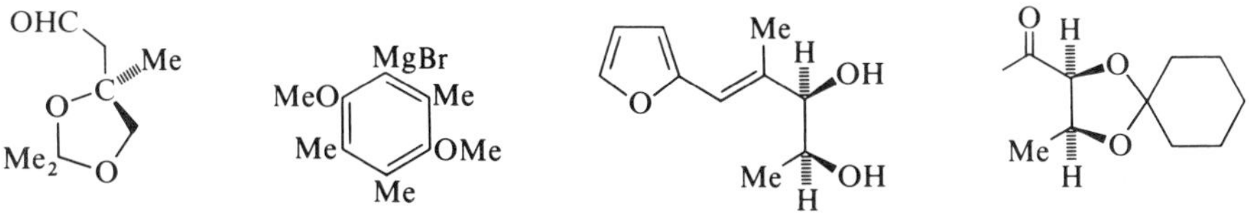

(**253**)          (**254**)          (**255**)          (**256**)

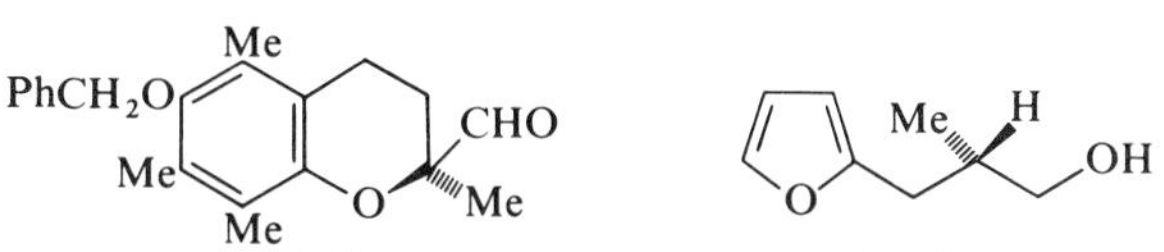

(**257**)                    (**258**)

The combination of a metal phenoxide and a Lewis acid facilitates alkylation of the phenolic moiety by 1,1,3,3-tetramethoxypropane and promotes specific *ortho* attack. Spontaneous cyclization leads to a mixture of two stereoisomers of 2,4-diethoxychroman ⟨81JHC1325⟩.

Ring closure of a 2-methoxyphenyl derivative of propanol to a chroman has also been achieved. Treatment of the aryloxazoline (**259**) with sodium hydride yielded the chroman (**260**) ⟨81JOC783⟩. The intramolecular nucleophilic displacement of the *o*-methoxy group is promoted through oxazoline activation and proceeds through an addition–elimination sequence. The initial attack involves coordination of the metal alkoxide to both the oxazoline moiety and the methoxy group, and aromatization follows with displacement of methoxide ion (Scheme 67). Hydrolysis of the oxazoline moiety to a carboxyl group has been accomplished.

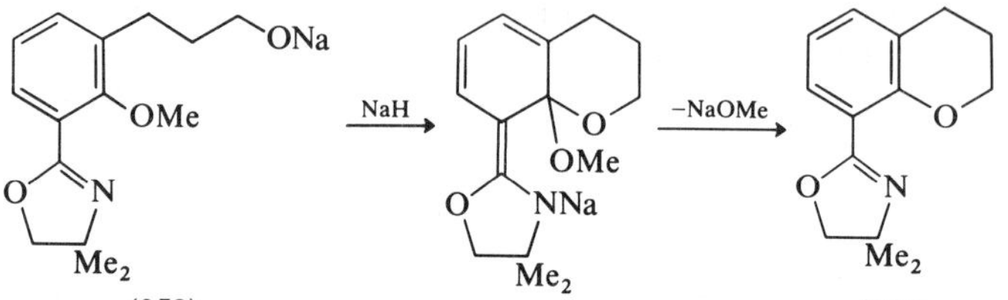

**Scheme 67**

Chromans arise through the reaction of phenols with a variety of unsaturated molecules such as alkenols, halogenoalkenes and dienes. However, the chroman is usually but one of several products and the reaction is therefore often of limited synthetic value ⟨21CB200, 63JOC798⟩. A comprehensive discussion of these reactions is included in the review by Livingstone ⟨81HC(36)7⟩, but some examples merit attention.

Much effort has been devoted to the synthesis of derivatives of chroman-6-ols as model compounds for studies of the chemistry of the tocopherols. Many routes concentrate on the acid-catalyzed alkylation of substituted benzene-1,4-diols by allyl alcohols or dienes. Trimethylhydroquinone and isoprene react in acetic acid in the presence of zinc chloride to give 2,2,5,7,8-pentamethylchroman-6-ol ⟨39JOC311⟩; isomer formation is not possible here and so the route is particularly attractive.

The direct formation of racemic α-tocopherol from trimethylhydroquinone and isophytol occurs at low temperature in the presence of boron trifluoride or aluminum chloride ⟨71JOC2910⟩. It is important that the solvent should not be able to complex with the Lewis acid; rather, it is the phenol–catalyst complex which is alkylated.

A detailed survey of chroman and tocopherol synthesis has been published ⟨81HC(36)59⟩ and earlier reviews are of interest ⟨40CRV(27)287, 77E555⟩.

The reaction of 2-methoxybutadiene with phenol affords a high yield of 2-methoxy-2-methylchroman though the conditions are quite severe ⟨68JOC4508⟩. The initially formed allyl ether rearranges to a phenol which cyclizes to the chroman. In the reaction with phenol itself, triethylamine is added to inhibit polymerization of the diene, but this is not advised when *o*-cresol is used. In the case of the allylphenol (**261**), ring closure to 2,2-dimethylchroman is best effected after acetylation and in the presence of benzoyl peroxide ⟨40JOC212⟩. Cyclization has also been achieved in the presence of bis(acetylacetonato)nickel ⟨79S126⟩.

The formation of chromans in acceptable yields has been observed during the reaction of phenols with 3,3-dialkylallyldiphenyl phosphates ⟨68JCS(C)1837⟩. The reaction has been used, for example, to synthesize α-tocopherol in 90% yield from 2,3,5-trimethylquinol and phytyldiphenyl phosphate (equation 10). In a similar fashion dihydric phenols such as 2,5-dimethylquinol afford both the chroman (**262**) and the benzodipyran (**263**), while phloroglucinol gives (**264**) and some benzotripyran (**265**). A word of caution is necessary, since unsubstituted allyldiphenyl phosphates cyclize to benzofurans in accord with Markownikov's rule. The competitive formation of chromans and benzofurans has been reviewed ⟨40CRV287⟩. In these reactions the allyl phenyl ether as well as *o*- and *p*-allylphenols can be isolated, pointing to a mechanism involving *O*-alkylation followed by rearrangement and cyclization. Similar intermediates could not be isolated from the dialkylallyl analogues, but 2-(3,3-dimethylallyl)phenol is known to cyclize to the chroman under the experimental conditions used.

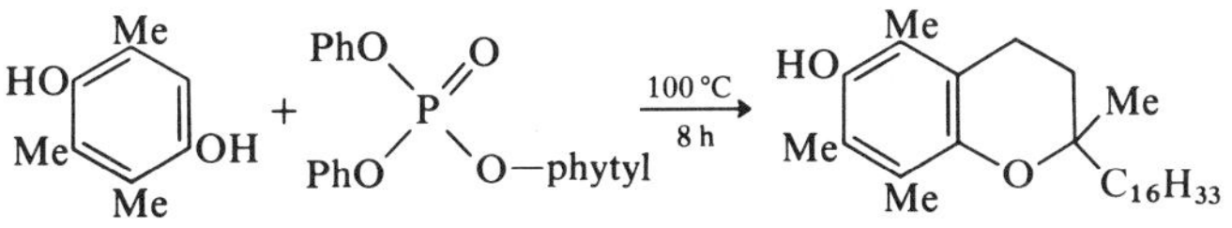

(10)

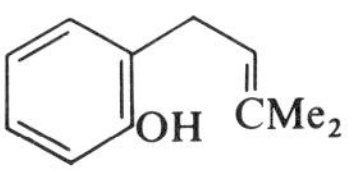

(261) 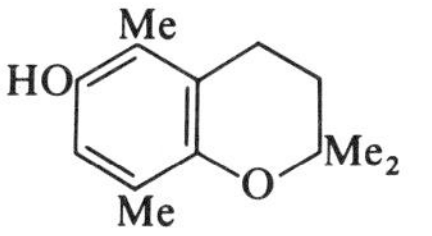     (262) 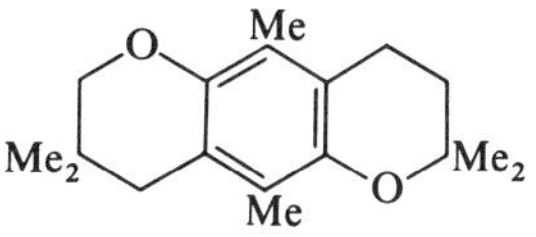     (263)

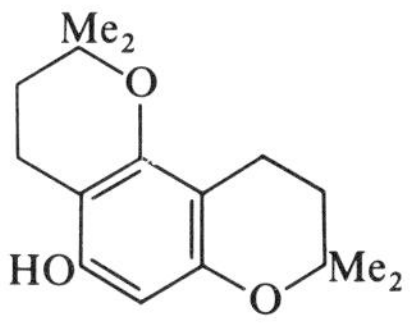

(264) 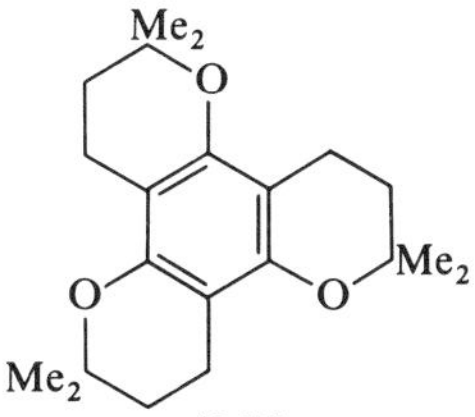     (265)

The cyclization of several *o*-prenylphenols is induced by thallium(III) salts ⟨81TL1355⟩. It is proposed that the terminal carbon–carbon double bond in the side chain forms a π-complex with the metal salt. The heterocyclic ring arises through nucleophilic attack by the remaining double bond and by the phenolic group at the two developing cationic centres. The reaction is completed by loss of the thallium residue to give the tricyclic chroman. The exact nature of the product is dependent on the geometry of the prenylphenol. Thus, although *o*-geranylphenol (266), in which the 6,7-double bond has *trans* geometry, affords the ring-contracted chroman (267), *o*-nerylphenol (269) yields the six-membered ring product (270) *via* a *cis*-fused intermediate and methyl group migration (Scheme 68).

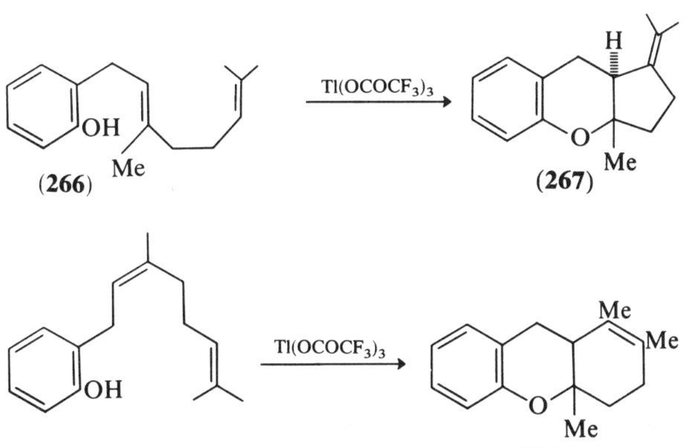

(266)     Tl(OCOCF$_3$)$_3$     (267)

(269)     Tl(OCOCF$_3$)$_3$     (270)

**Scheme 68**

Two groups have reported the effects of peracids on phenols with unsaturated side-chains in the *ortho* position. In one case this formed part of a study of the aerial oxidation of polyhydric phenols ⟨77JCS(P1)2593⟩, but it is of interest insofar as the phenol (271) afforded a high yield of the chroman-3-ol although on a small scale. A complex mixture of products arose from the other phenols studied. In the second study the alkene (272) is converted directly into a 2-hydroxymethylchroman which arises through an intramolecular ring opening of an intermediate epoxide ⟨78JHC1051⟩. The alkenes are obtained through the reaction of triphenylphosphonium methylide with a 2-(2-hydroxyphenyl)ethyl ketone.

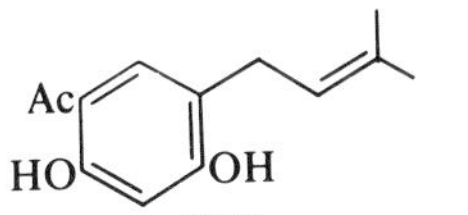

(271) 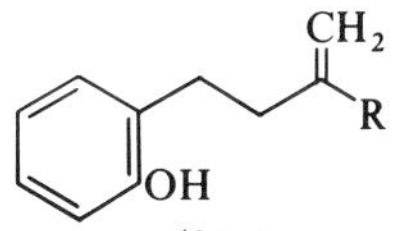     (272)

The acid-catalyzed condensation of 2-methylbuta-1,3-diene with hydroxyacetophenones affords chromans directly, presumably through initial nuclear isoprenylation

⟨81S526, 81T1437⟩. *m*-Substituted dihydric phenols give a mixture of the 5- and 7-hydroxy-chromans.

In a closely related reaction, polyhydroxybenzaldehydes are converted into chroman aldehydes on reaction with 2-methylbut-3-en-2-ol in the presence of phosphoric acid ⟨81S527⟩, whilst in formic acid 2,5-dimethylbenzene-1,4-diol and the alkenol give 2,2,5,8-tetramethylchroman-6-ol ⟨81JCS(P1)1437⟩.

The reaction of dehydrolinalool acetate (**274**) with the homocuprate of olivetol bis(tetrahydropyranyl ether) (**273**) was expected to yield cannabichromene by analogy with the reaction with 3-acetoxy-3-methylbut-1-yne ⟨79JCS(P1)201⟩. However, the major product was 3,4-*cis*-$\Delta^{1,2}$-tetrahydrocannabinol (**275**), which could readily be purified, thereby providing a valuable route to this substance. Its formation is thought to involve the generation of an allene which proceeds to the chroman *via* an allylic cation and *trans* addition to the isopropylidene group (Scheme 69).

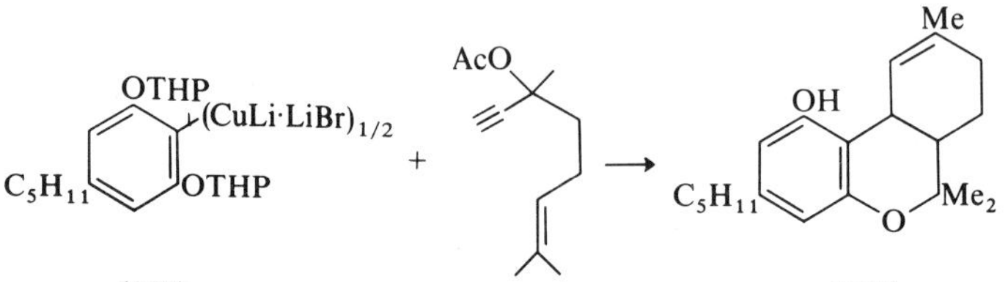

**Scheme 69**

The condensation of salicylaldehyde derivatives with alkenes in the presence of boron trifluoride yields 4-substituted chromans ⟨64CB682⟩. The use of the acetal (**276**; R = OEt) leads to 4-ethoxychromans (**277**; R = OEt), but of more interest is the formation of chroman-4-carbamates (**277**; R = NHCO$_2$Et) from the biscarbamic ester (**276**; R = NHCO$_2$Et; Scheme 70).

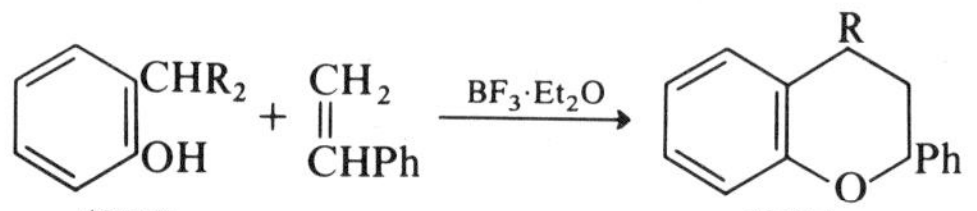

**Scheme 70**

Reaction of this carbamate with phthalic anhydride and of the resulting phthalimido-chroman with hydrazine is a useful route to 4-amino-2-phenylchroman.

Salicylaldehyde also reacts with enamines to form chromans, although these have not usually been isolated but rather have been oxidized directly to a chromone ⟨66JOC1232⟩. The synthesis has been applied to both cyclic and acyclic enamines and to 2-hydroxy-1-naphthaldehyde. The initial intermediate undergoes an intramolecular cyclization involving participation of the neighbouring hydroxyl group.

2-Aminochromans also arise from the reaction of phenolic Mannich bases with enamines ⟨70JHC1311⟩. The route is attractive for a number of reasons: the starting materials are readily available; its scope is considerable since the enamines may be aldehyde or ketone based and the Mannich bases may be aromatic or heteroaromatic; and the products themselves are precursors of hydroxychromans and 4*H*-chromenes. Mechanistically, the synthesis proceeds through a quinone methide followed by addition to the enamine and cyclization, which may be a concerted process (Scheme 71).

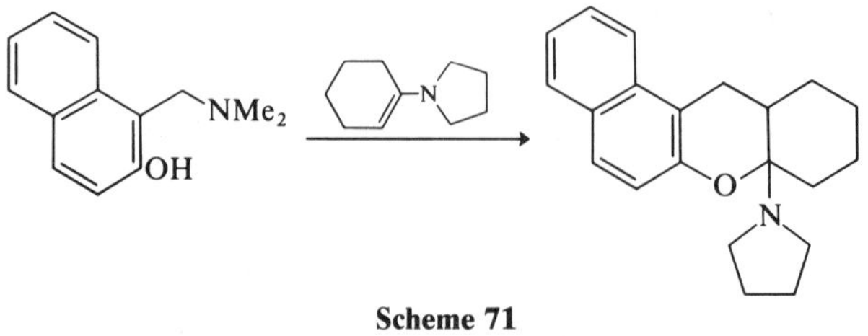

**Scheme 71**

(*b*) *γ to the heteroatom.* A number of widely used chroman syntheses are based on the cyclization of a phenoxypropane derivative, often under Friedel–Crafts conditions.

Although some workers have found difficulty in purifying the product, a particularly convenient preparation of chroman involves the cyclization of 1-chloro-3-phenoxypropane in the presence of tin(IV) chloride (equation 11) ⟨52CR(234)1787⟩. The synthesis is readily adapted to give many substituted chromans. A surprising result was obtained using 3-chloro-1-phenoxy-1-phenylpropane, when 4-phenylchroman was obtained rather than the expected 2-phenyl isomer ⟨72BSF1540⟩. The same product was formed when 3-phenoxy-3-phenylpropan-1-ol was heated with polyphosphoric acid. It is proposed that a benzyl cation is produced by cleavage of the O—C bond in the starting material. Aromatic electrophilic substitution is followed by cyclization with displacement of chloride ion.

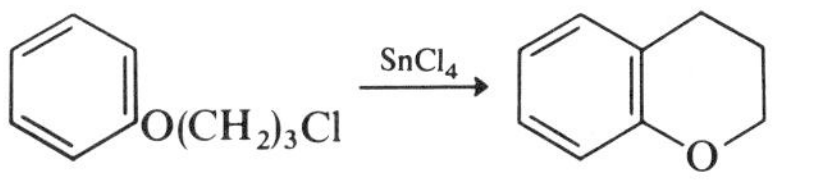

$$\text{[structure] } \xrightarrow{\text{SnCl}_4} \text{[structure]} \tag{11}$$

Another well-used synthesis is based on the cyclization of 1,3-diphenoxypropane ⟨63JCS2094⟩. Later work has extended the route to a variety of substituted chromans and to dihydronaphthopyrans ⟨65JCS5718⟩. The *m*-substituted compounds give a mixture of the 5- and 7-isomers (Table 4). Formation of the chroman probably involves simultaneous ring closure and displacement of a phenolic residue, since under the reaction conditions a carbocation intermediate is not expected.

**Table 4** Cyclization of Some *m*-Substituted 1,3-Diphenoxypropanes and 3-Phenoxypropan-1-ols

| m-Substituent | Total yield (%) | Isomer distribution 5-Isomer | Isomer distribution 7-Isomer | Ref. |
|---|---|---|---|---|
| Me | 76.5 | 41 | 59 | 65JCS5718 |
| Cl | 54 | 45 | 55 | 65JCS5718 |
| OMe | 67.5 | 24 | 76 | 65JCS5718 |
| Me | 78 | 40 | 60 | 66JOC3032 |

A variation on this general theme which has not been as widely applied as the previous examples, but which is nevertheless of value, involves the use of 3-phenoxypropan-1-ols ⟨66JOC3032, 67JCS(B)859⟩. Here again it is observed that the *m*-substituted isomer cyclizes to a mixture of two chromans.

Several chromans have been synthesized using the Parham cycloalkylation technique ⟨76JOC1184⟩. 1-Bromo-3-(2-bromophenoxy)propane, prepared from phenol and 1,3-dibromopropane, is treated with *n*-butyllithium at −100 °C. Halogen–lithium exchange yields the aryllithium (**278**) which cyclizes either at −100 °C or in some instances only at an acceptable rate at higher temperatures. This method offers the advantage of regiospecificity since cyclization is controlled by the location of the *o*-bromine atom (Scheme 72).

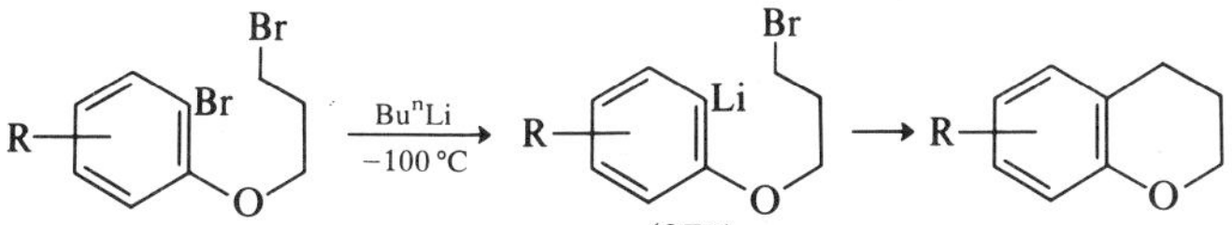

$$R\text{[structure]} \xrightarrow[-100\,°C]{\text{Bu}^n\text{Li}} R\text{[structure]} \longrightarrow R\text{[structure]}$$
$$\textbf{(278)}$$

**Scheme 72**

(ii) *Formation of two bonds*

The major features of the [4 + 2]-cycloaddition process have been discussed earlier under dihydropyrans (p. 770) and in this section examples of its application to chroman synthesis will be covered, together with relevant ancillary material.

The dimerization of *o*-quinone methides in which one molecule acts as the heterodiene and another as dienophile leads to spiroannelated chromans (equation 12). The driving force for this considerable tendency to dimerize is associated with the ease with which the

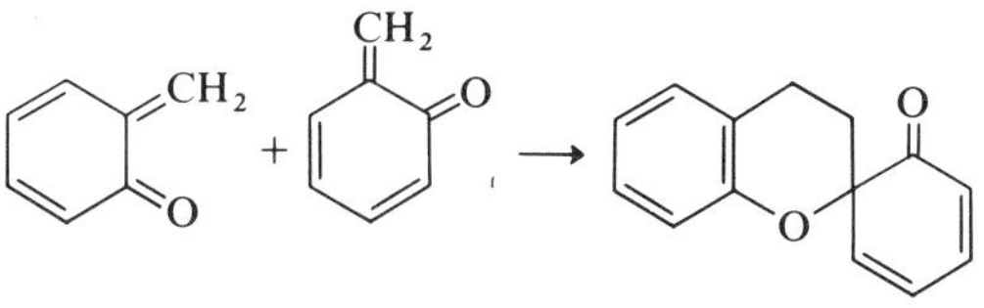

$$\text{[structure]} + \text{[structure]} \longrightarrow \text{[structure]} \tag{12}$$

transition from a quinonoid to a benzenoid structure is achieved during a cycloaddition reaction. However, in the presence of a competing diene, dimerization does not occur. Instead, a simple chroman results, as for example in the reaction of the methide with butadiene ⟨63CR(256)3323⟩.

Reaction of chlorotrimethylsilane with hydroxymethylspiroepoxycyclohexadienone (**279**) affords the spiroannelated chroman (**280**) (Scheme 73) ⟨80TL4973⟩. The proposed mechanism invokes a quinone methide intermediate which dimerizes to the chroman.

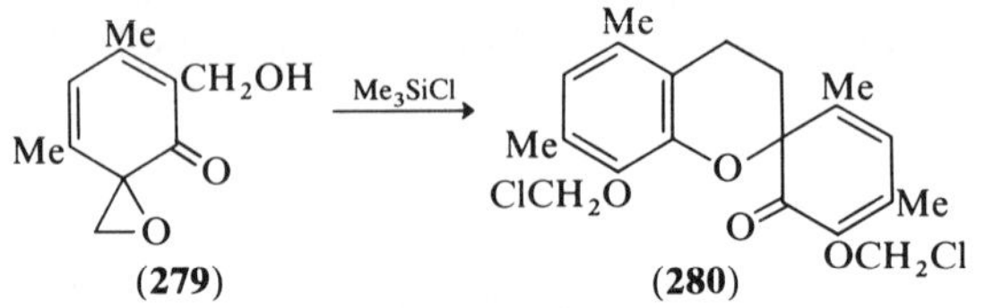

**Scheme 73**

In a similar manner, *o*-quinone methides and their benzologues react with alkenes to give chromans ⟨63CR(256)3323⟩. With styrene, flavans are formed from *o*-quinone methides ⟨64CR(258)1526⟩. Yields are better with 2,3-naphthoquinone 2-methide than with the benzoquinone analogue.

The high regiospecificity observed in these reactions corresponds to polarization of the methide as shown (**281**). This is particularly apparent in the reaction with vinyl ethers; ethyl vinyl ether affords the 2-ethoxychroman in quantitative yield and dihydropyran gives the pyranopyran (**282**; Scheme 74) ⟨70JOC3666⟩.

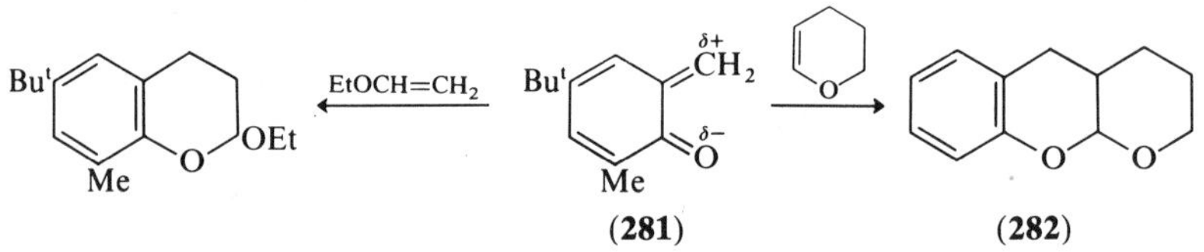

**Scheme 74**

The products arising from the reaction of the naphthoquinone methide (**283**) with *cis*- and *trans*-stilbene are different, suggesting that the pericyclic reaction occurs with *syn* addition (Scheme 75) ⟨73JCS(P1)120⟩.

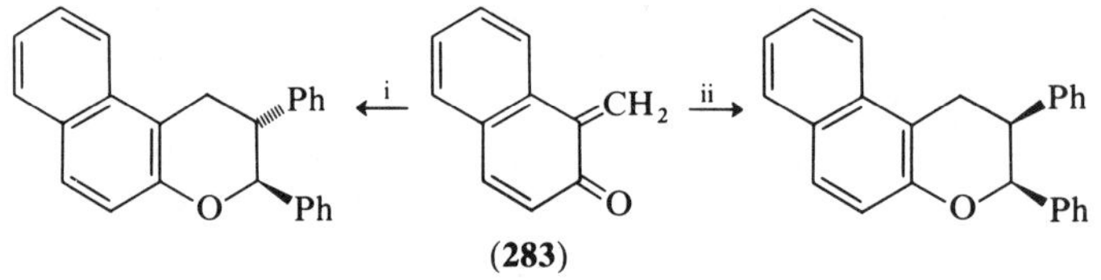

i, *trans*-PhCH=CHPh; ii, *cis*-PhCH=CHPh

**Scheme 75**

The polar 1,4-cycloaddition of alkenes with 2-hydroxy-5-nitrobenzyl chloride in the presence of tin(IV) chloride yields 6-nitrochromans ⟨69TL5279⟩. The quinone methide, which is protonated under these conditions, undergoes stereospecific *syn* addition of the alkene. Although in most cases the reaction is regiospecific, *cis*-pent-2-ene yields a mixture of isomers.

The quinone methide (**284**) is stabilized by the trifluoromethyl groups but was not isolated. Trapping with styrene gave the fluorinated chroman (Scheme 76) ⟨68JOC3297⟩.

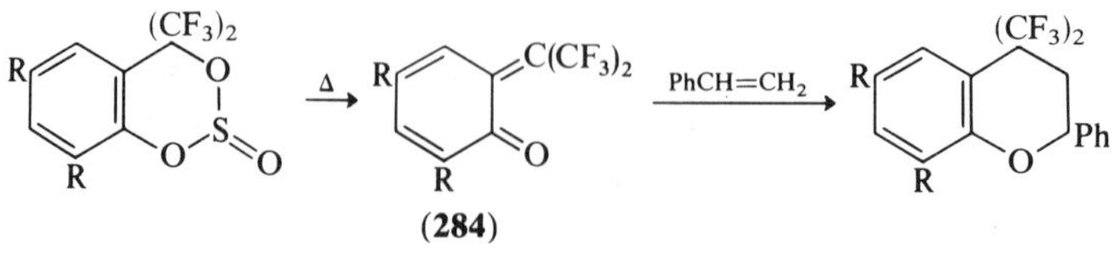

**Scheme 76**

Enamines, too, react with quinone methides. It is suggested that a charged intermediate is involved in which, in substituted analogues, epimerization prevents the formation of diastereoisomers ⟨70JHC1311⟩.

The formation of the chromanochromans (**286**) during the reaction of the phenolic Mannich bases and 2-chloroprop-2-enonitrile probably arises through the decomposition of the base to a quinone methide and dimethylamine ⟨80JOC3726⟩. The initial product, a substituted 4*H*-chromene (**285**), undergoes a further [4 + 2]-cycloaddition to give the final product (Scheme 77).

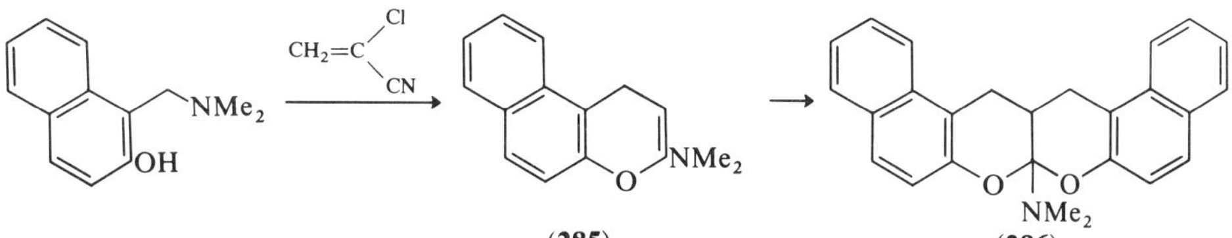

(**285**)

(**286**)

Scheme 77

It is pertinent at this point to refer briefly to the sources of quinone methides, though these have been reviewed ⟨B-74M122400⟩. The general approach used in chroman syntheses involves the thermal elimination of HX from an *o*-substituted phenol. Commonly the eliminated molecules are water, methanol or dimethylamine (**287**; X = OH, OMe, NMe$_2$, respectively). However, these methods are not entirely suitable because the eliminated molecules may promote side reactions. In the case of 1,2-naphthoquinone 1-methide, the thermal dissociation of the spirodimer (**288**) is a better source than the other methods. Its formation represents another example of dimerization by a [4 + 2]-cycloaddition, since it is prepared by heating 1-dimethylaminomethyl-2-naphthol in dodecane or xylene with careful exclusion of moisture ⟨73JCS(P1)120, 81CJC2223⟩.

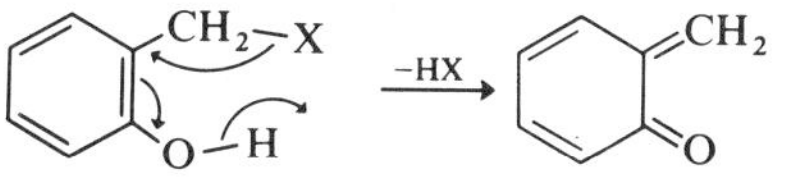

(**287**)

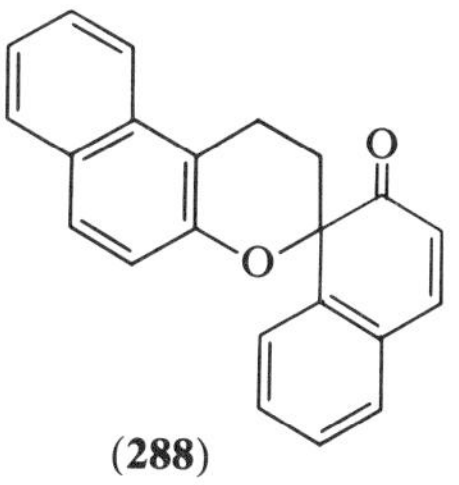

(**288**)

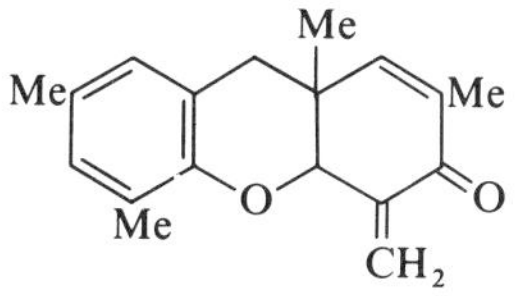

(**289**)

Similarly, the dimer (**289**) undergoes thermal dissociation in cycloocta-1,5-diene and the methide has been trapped as the chroman ⟨73JCS(P1)359⟩.

Other methods which have proved of value include the formation of substituted methides by the action of silver oxide on phenols ⟨70JOC3666⟩. It is postulated that upon oxidation of the phenol a phenoxy radical is formed which dimerizes to the quinol ether. Disproportionation to the methide and the original phenol follows.

Finally, *o*-quinone methide has been formed by the irradiation of dihydrobenzofuran-2-one ⟨71CC383⟩.

## (ii) *From other heterocycles*

*From benzofurans.* Russian workers have reported several examples of the isomerization of 2,3-dihydrobenzo[*b*]furans into chromans. Triphenylmethyl perchlorate in acidic media converts the 2-ethyl derivative into 2-methylchroman, whilst the same chroman results when the 2,3-dimethylbenzofuran is treated with aluminosilicates at high temperature ⟨72DOK(204)879, 75CHE278⟩.

## (iii) *From a preformed heterocyclic ring*

(*a*) *From coumarins.* The reaction of Grignard reagents with dihydrocoumarins leads to chromans *via* an intermediate diol ⟨57BSF776⟩. The use of 1,6-di(bromomagnesio)hexane converts the dihydrocoumarin into a spirochroman (**290**) ⟨81H(15)455⟩.

3-Acyldihydrocoumarins rearrange to esters of 2-methoxychroman-3-carboxylic acid on treatment with methanolic hydrogen chloride at room temperature ⟨60CB1025, 67CB1296⟩. It is postulated that the reaction proceeds through a ring-opened species.

During a study of dihydrocoumarins ⟨68JOC1202⟩ it was observed that 4-(2-bromoethyl)-3,4-dihydrocoumarin (**291**) was converted into a 4-substituted chroman on reaction with an amine. Ring opening occurs through nucleophilic attack at C-2 and this is followed by an intramolecular nucleophilic substitution. In a similar manner, chroman-4-ylacetic acid results on reaction with potassium hydroxide in methanol.

The spirochroman (**292**) is formed on treatment of a lactone such as dihydrocoumarin with bis(dimethylaluminum)-1,2-ethanedithiolate. The technique is of value for the protection of lactones against nucleophilic attack ⟨73JA5829⟩.

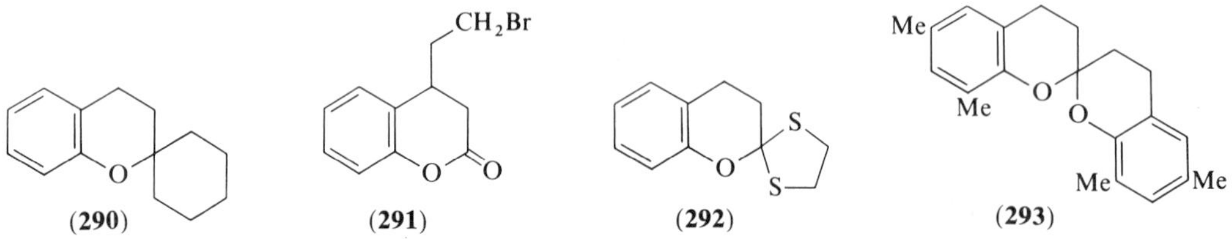

(**290**)          (**291**)          (**292**)          (**293**)

Spirobichromans (**293**) result from the reaction of dihydrocoumarins with sodium hydride or boron tribromide followed by acid treatment ⟨58JIC47, 75MI22402⟩.

A modification of Meyer's oxazoline-facilitated methoxy substitution by organometallic reagents has been used to synthesize a biphenyl precursor of cannabinol. A one-pot hydrolysis of the oxazoline moiety, ether cleavage and cyclization yielded the lactone. Cannabinol resulted from the subsequent reaction with methylmagnesium iodide (Scheme 78) ⟨82TL253⟩.

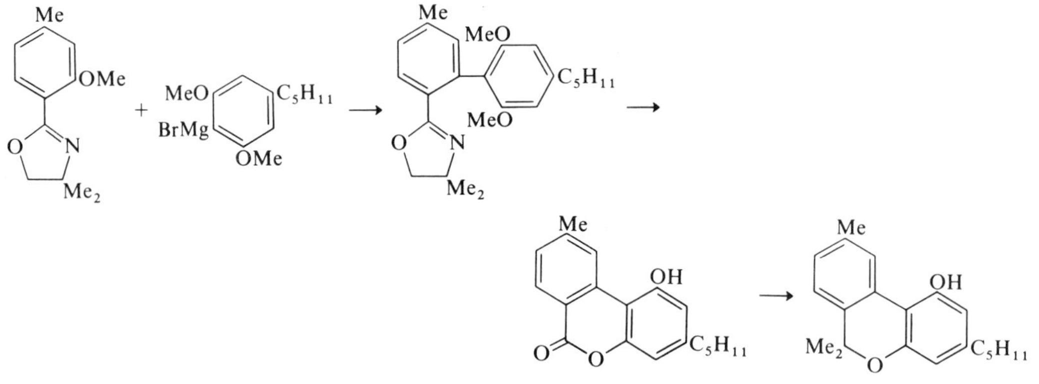

**Scheme 78**

(*b*) *From chromenes*. The catalytic reduction of 4*H*-chromenes ⟨35JCS646⟩ and 2*H*-chromenes ⟨69JOC207⟩ leads to chromans. This reaction has been discussed ⟨77HC(31)70⟩. Of course, addition of the usual reagents to the double bond leads to 3,4-disubstituted chromans ⟨66JCS(C)2013⟩. The usual stereochemistry is observed; for example, bromine or hypobromous acid gives the *trans* product, whilst the *cis*-diol results on reaction with potassium permanganate ⟨83JCS(P1)827⟩. Flavenes and isoflavenes behave in a similar manner.

(*c*) *From chromanones*. The reduction of chromanones has been reviewed ⟨77HC(31)301⟩ and the topic is discussed in depth in Section 2.23.10.4.4.

In essence, hydride ion reagents reduce chromanones to chromanols ⟨81T2613⟩, the reaction often being stereospecific ⟨70JCS(C)1006⟩.

The reaction of chromanones with Grignard reagents also yields chromanols, though dehydration to the chromene may also occur ⟨68T949⟩. In contrast, the Clemmensen reduction of chromanones invariably yields chromans; there are many examples ⟨65JCS3882, 71JMC758⟩.

Provided groups which are sensitive to diborane are absent from the chromanone, this reagent may effect reduction of the ketone to the chroman ⟨73AJC2291⟩.

Catalytic hydrogenation of chromanones gives variable results. There are examples of the reduction to chromanols ⟨70BCJ442⟩, whilst in other instances the chroman is formed ⟨58JCS1190⟩.

(*d*) *From chromones*. The chromone ring is relatively resistant to reduction, but some examples of the catalytic hydrogenation are given in Table 5.

(*e*) *From benzopyrylium salts*. The catalytic reduction of isoflavylium salts yields isoflavans. In the presence of acetic anhydride, free hydroxyl groups are acetylated, whilst

**Table 5** Formation of Chromans by the Catalytic Hydrogenation of Chromones

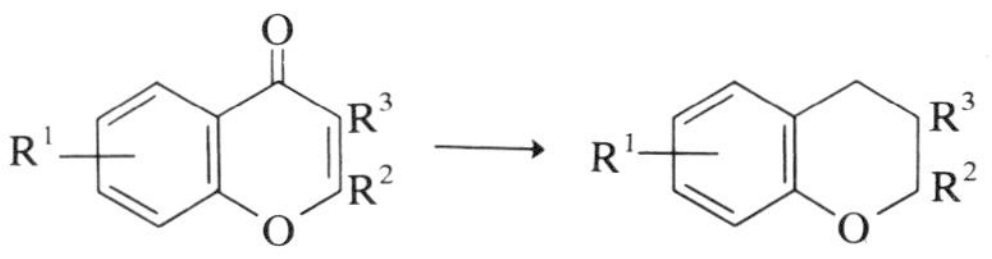

| $R^1$ | $R^2$ | $R^3$ | Catalyst | Yield | Ref. |
|---|---|---|---|---|---|
| H | Et | H | Cu–Cr oxides | 73 | 38JA669 |
| H | $CO_2H$ | H | Pd–C, 70 °C | 87 | 68JMC844 |
| H | $CO_2Et$ | H | Pd, $BaSO_4$ | 75 | 69CHE316 |
| 7-OMe | H | H | Pd | | 39M427 |
| 6,7-$C_6H_4$ | H | H | Adams | | 63AJC690 |
| 6-Ph | $CO_2H$ | H | Pd–C, AcOH | 80 | 75JMC934 |
| 6-Cl | $CO_2Et$ | H | Pd–C, 70 °C | | 74JPS203 |
| 6-OH-5,7,8-$Me_3$ | Me | Me | Pt, AcOH | | 38CB2637 |

in methanol nucleophilic substitution at C-2 may accompany the reduction (Scheme 79) ⟨81AJC2647⟩.

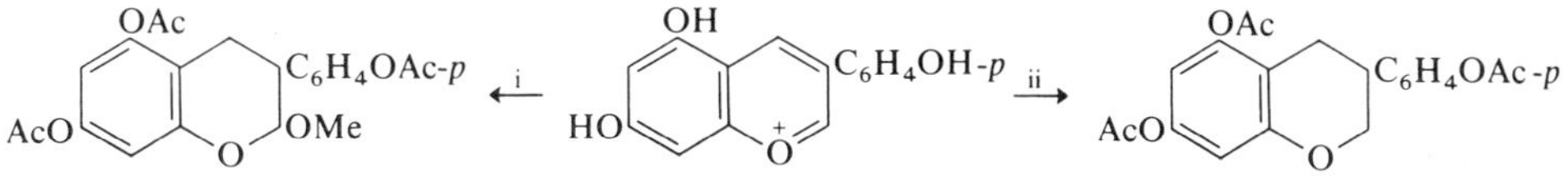

i, $H_2$, $PtO_2$, MeOH; ii, $H_2$, $PtO_2$, $Ac_2O$

**Scheme 79**

*(f) From pyrans.* The fused pyran (**294**; $R^3$ or $R^4 = C_5H_{11}$) is aromatized to the chroman, hexahydrocannabinol (**295**), by deprotonation with LDA and conversion into the selenides (**294**; $R^1$ or $R^2 = $ SePh; $R^3$ or $R^4 = C_5H_{11}$). Subsequent oxidation results in a *syn* elimination and aromatization (Scheme 80) ⟨82AG(E)221⟩.

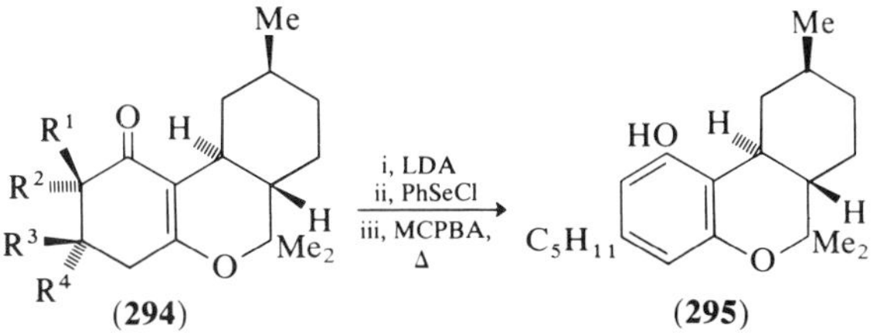

**Scheme 80**

### 2.24.2.7.4 *Isochromans (dihydro-2-benzopyrans)*

(i) *Formation of one bond*

*(a) Adjacent to the heteroatom.* Isochromans result from the acid-catalyzed dehydration of 2-(2-hydroxymethylphenyl)ethanols ⟨57JA3165⟩. The parent diol is available from homophthalic anhydride by hydride ion reduction.

Depending upon the conditions, the reaction of benzylmagnesium chloride with benzaldehyde can yield benzylphenylmethanol, dibenzoylphenylmethane and 1,3-diphenylisochroman. The last product arises from the acid-catalyzed dehydration of the 1,5-diol (**296**), which is the predominant product when the Grignard reagent is added to benzaldehyde at room temperature (Scheme 81) ⟨51JA3163⟩. There was no evidence for the formation of *o*-methylbenzhydrol, consistent with the lack of this type of product with citronellal and various simple aliphatic aldehydes ⟨44JA354, 51JA3237⟩.

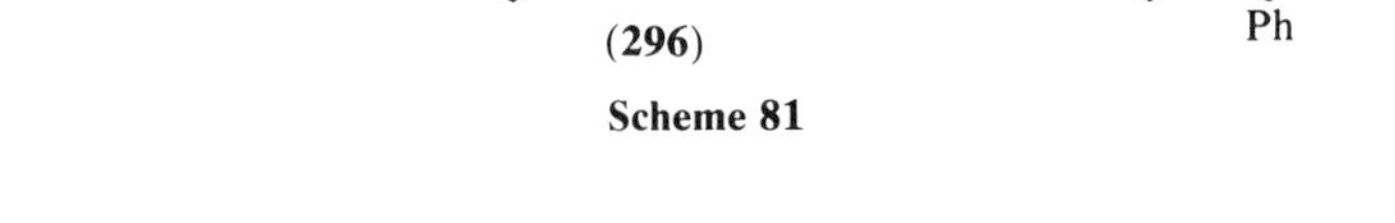

**Scheme 81**

(b) β *to the heteroatom.* The most widely used synthesis of this little studied heterocyclic system involves the cyclization of 2-phenylethyl chloromethyl ether (297) ⟨54CR(239)1047, 54MI22400, 56BSF1337⟩. The ethers are available through the reaction of 2-phenylethanol with formaldehyde (in the form of trioxymethylene) in the presence of hydrogen chloride (Scheme 82). The route has been reviewed ⟨71MI22401⟩.

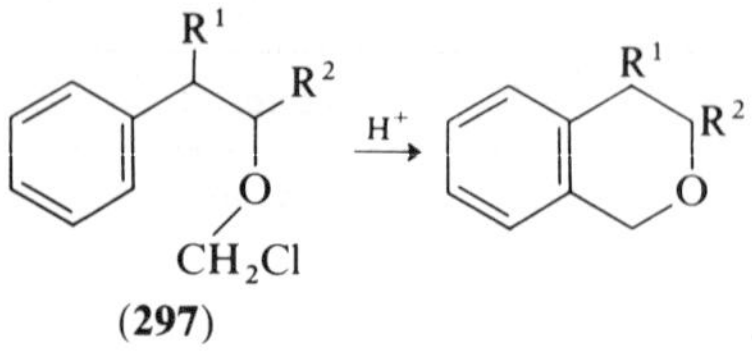

**Scheme 82**

The reaction of 2-phenylethyl trimethylsilyl ether (298) with 1-iodoethyl trimethylsilyl ether yields 1-methylisochroman ⟨78JOC3698⟩. Displacement of iodide affords the oxonium iodide (299), which through loss of trimethylsilyl iodide gives the ether (300). Friedel–Crafts cyclization of the iodoether is considered to complete the synthesis (Scheme 83).

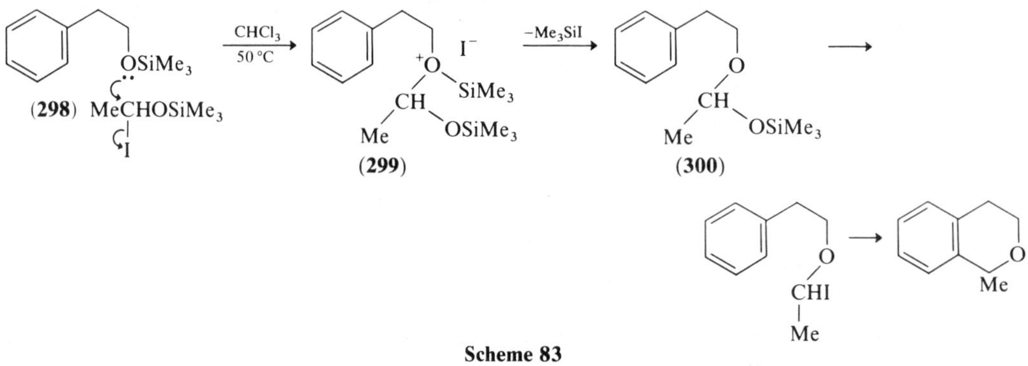

**Scheme 83**

(ii) *From other heterocycles*

(a) *From isoquinolines.* The degradation of 6′-hydroxymethyllaudanosine (301) with cyanogen bromide gives the phenylisochroman in good yield (equation 13) ⟨81AP577⟩.

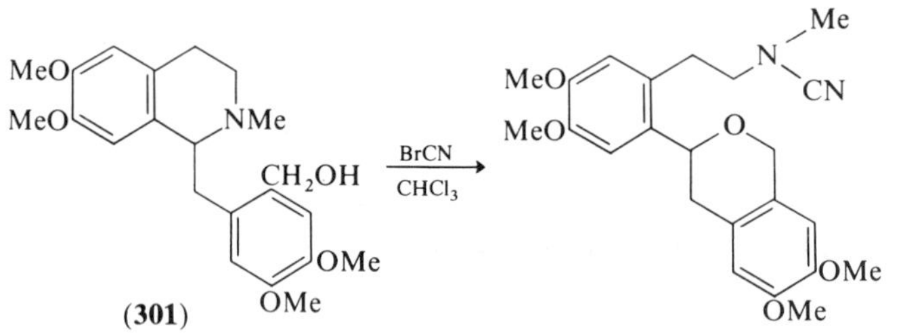

(13)

(iii) *From a preformed heterocycle*

(a) *From isocoumarins.* The reaction of 3-phenylisocoumarin (302) with phenylmagnesium bromide yields 2-phenacylbenzophenone. Reduction of the diketone and cyclization of the ensuing diol gives 1,3-diphenylisochroman ⟨51JA3163⟩. In a related approach the reduction of the isocoumarin (302) with LAH gives the diol which ring-closes to 3-phenylisochroman (Scheme 84) ⟨51JA5494⟩.

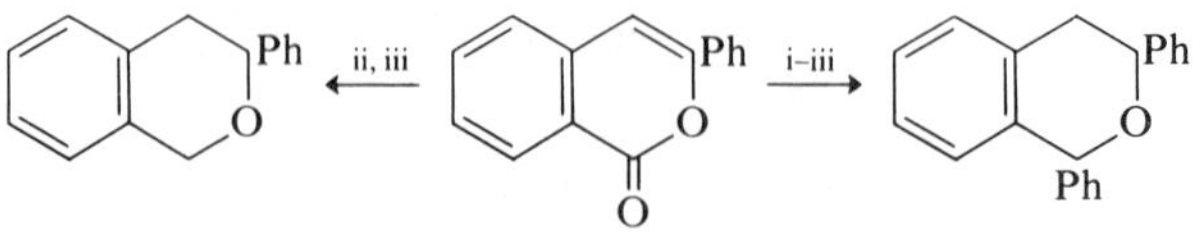

i, PhMgBr; ii, LiAlH₄; iii, H⁺

**Scheme 84**

(*b*) *From isochromanones*. 1,1-Disubstituted isochromans result from the action of a Grignard reagent on an isochroman-1-one. Again, the reaction proceeds *via* a diol, which can be isolated, and which is cyclized in phosphoric acid ⟨71MI22401⟩.

(*c*) *From isochromanols*. The catalyzed hydrogenolytic cleavage of 3-phenylisochroman-4-yl acetate forms the final stage of a stereoselective synthesis of (+)-(*R*)-3-phenyliso-chroman from (+)-L-mandelic acid ⟨79AP385⟩.

## 2.24.3 PYRANONES AND FUSED PYRANONES

### 2.24.3.1 Pyran-2-ones (2-Pyrones)

A common feature of many syntheses of pyran-2-ones is the ring closure of a 5-keto acid or acid derivative. The variation in these routes lies in the methods of formation of the five-carbon unit. Although it is not always possible to isolate the intermediate keto acid, the reaction conditions often suggest its involvement. All such syntheses are therefore treated under the category of the formation of one bond adjacent to the heteroatom.

Earlier approaches to the synthesis of pyran-2-ones have been reviewed ⟨47CRV(41)525⟩ and later work has been discussed ⟨67RCR175⟩.

#### 2.24.3.1.1 Formation of one bond

(i) *Adjacent to the heteroatom*

The principles involved in this approach to pyran-2-one synthesis are summarized in Scheme 85, in which the various routes to the requisite five-carbon moiety are indicated. These methods will be considered in turn, paying attention to the variations which are possible within the basic concept.

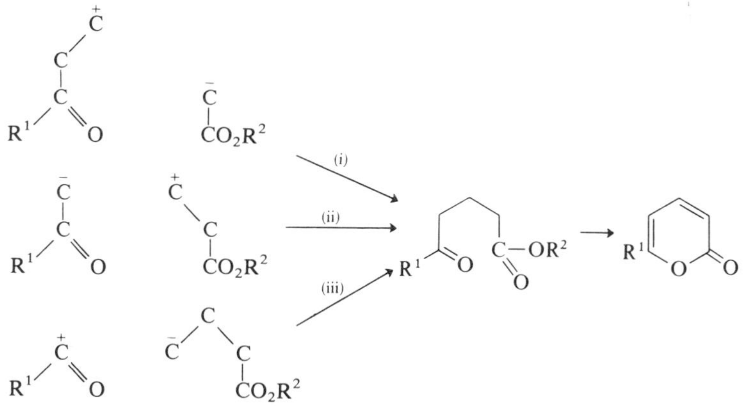

**Scheme 85**

Michael addition of a carbanion derived from an acid derivative to an unsaturated carbonyl compound typifies syntheses of type (i) in Scheme 85.

The base-catalyzed condensation between methyl acetoacetate and methoxymethyl-eneacetoacetate (**303**) gives the glutaconic ester, dimethyl 2,4-diacetylpent-3-enedioate ⟨79JCS(P1)464⟩. Spectroscopic studies have indicated that this compound exists predominantly in the cyclic form (**304**). A further base-catalyzed reaction ensues resulting in the formation of the pyran-2-one (Scheme 86). From a preparative aspect, an equimolar quantity of magnesium methoxide is the catalyst of choice ⟨79JCS(P1)478⟩. The synthesis of several other pyran-2-ones by related reactions is described in the latter work.

Other products are formed when high concentrations of sodium methoxide are present or when magnesium methoxide is used as the base. Dimethyl 5-hydroxytoluene-2,4-dicarboxylate arises from attack at the acetyl carbon atom in an aldol-like condensation, whilst reaction at the ester carbonyl leads through a Claisen condensation to methyl 5-acetyl-2,4-dihydroxybenzoate.

A similar reaction occurs using nitriles such as methyl cyanoacetate in place of the acetoacetic ester above, which leads to 3,5-bis(methoxycarbonyl)-6-methylpyran-2-one

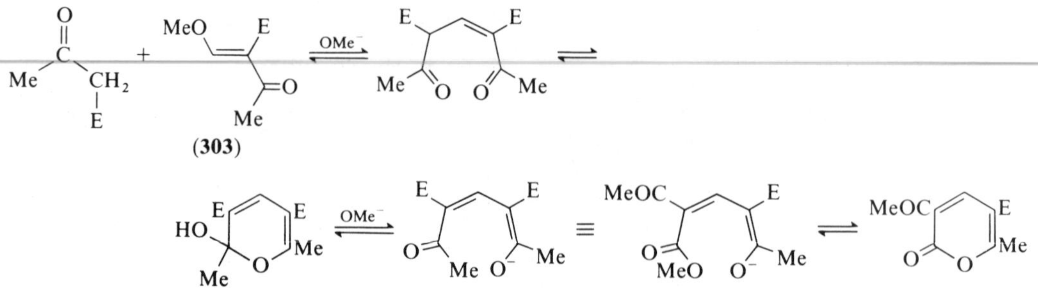

(303) (304) E = $CO_2Me$

**Scheme 86**

⟨79JCS(P1)677⟩. Presumably cyclization occurs at the nitrile carbon atom and the pyranone is formed *via* the imine. However, the product from this route sometimes contains a pyridone. In fact, in the above example, the pyridone is formed exclusively when a low concentration of base is used, but at high molarity only the pyranone is formed.

The profound influence of base on the course of the reaction led to the postulate that at high methoxide molarity it is the anion of the imine (**306**) which is in equilibrium with the ester anion (**305**). The former ionic species is readily hydrolyzed by acid to the pyran-2-one. However, at low molarities the unionized imine (**307**) is the equilibrating species and this is susceptible to nucleophilic attack at C-6 which occurs with ring opening. Subsequent cyclization gives the pyridone (Scheme 87).

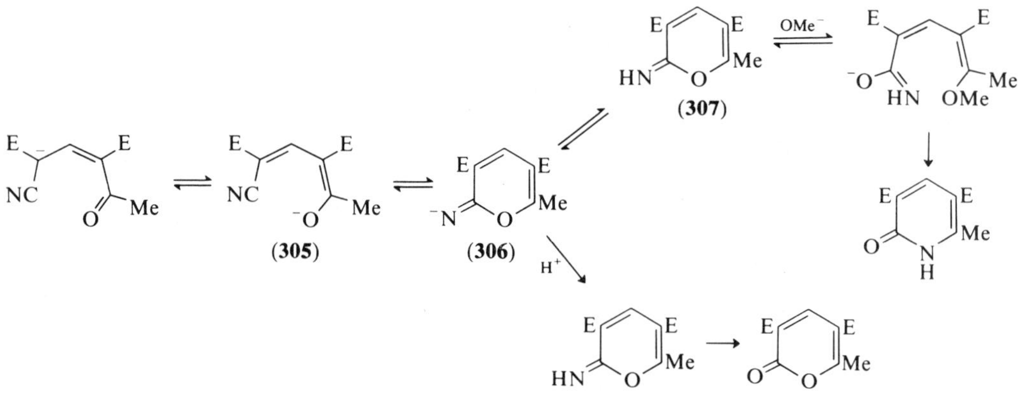

(305) (306) (307) E = $CO_2Me$

**Scheme 87**

A nitrile-stabilized carbanion is also involved in a synthesis of a fused pyranone system ⟨81S225⟩. A range of 2-ureidomethylenecyclohexane-1,3-diones, *e.g.* (**308**), react with activated acetonitriles in the presence of a strongly basic catalyst to produce 5-oxo-5,6,7,8-tetrahydrocoumarins (**309**). Since the substrates are readily available from cyclohexane-1,3-diones by reaction with triethyl orthoformate and a urea, the synthesis is attractive (Scheme 88). Furthermore, it has been applied to a pyran-2,4-dione, whereupon the 2,5-dioxopyrano[4,3-*b*]pyran (**310**) is formed.

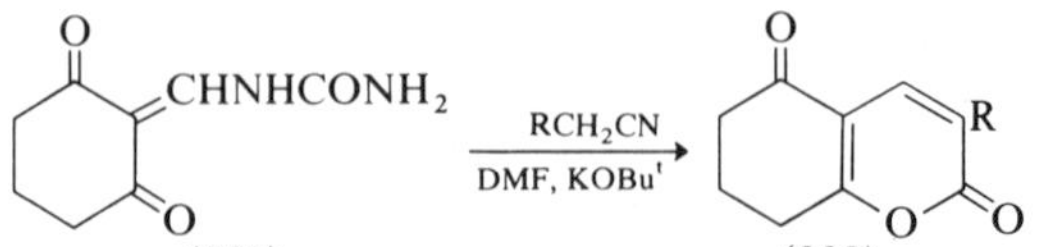

(308) (309)

**Scheme 88**

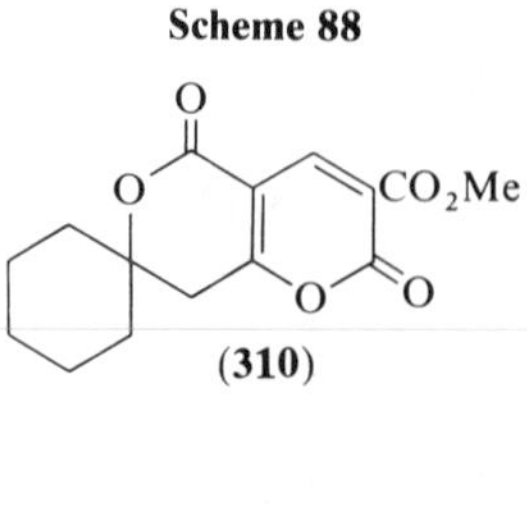

(310)

The regiospecific cyclization of some glutaconic half-esters under dehydration conditions leads to 6-alkoxypyran-2-ones ⟨82JOC1150⟩. The reaction presumably proceeds *via* a carbocation which undergoes proton loss. The cyclic glutaconic derivative (**311**) gives the fused pyran-2-one (**312**) which has been used as precursor of various tetracyclic molecules (Scheme 89).

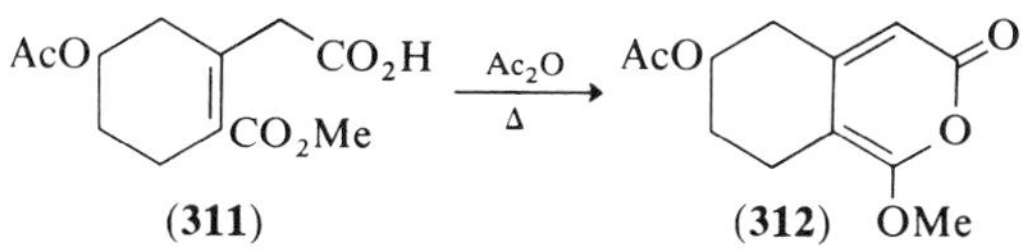

**(311)**          **(312)**   OMe

**Scheme 89**

Several routes to pyran-2-ones involve the use of alkynic carbonyl compounds as the four-atom component in type (i) syntheses (Scheme 85).

As is to be expected, an alkynic ketone undergoes a Michael addition with a carbanion, leading eventually to a pyranone ⟨50JA1022⟩. Using malonic esters, a 3-alkoxycarbonyl derivative results, which is hydrolyzed to the 2-oxopyran-3-carboxylic acid under alkaline conditions, but to the pyranone by sulfuric acid. Rapid ester exchange is observed with the initial products, the alcohol used as solvent determining the nature of the alkyl group in the 3-carboxylic esters (Scheme 90).

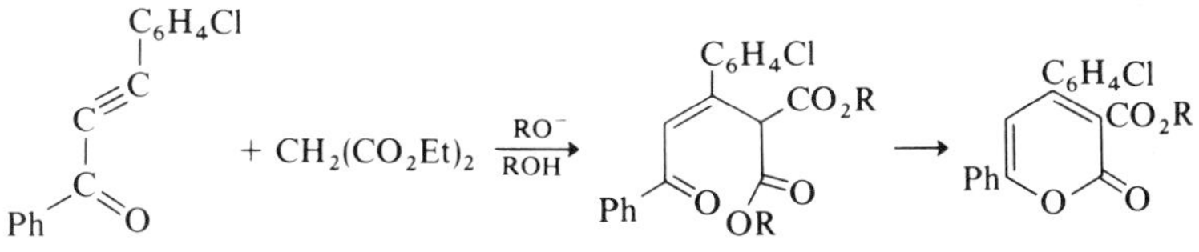

**Scheme 90**

Reaction of carbanions with dialkynic ketones, the so-called skipped diynes, can produce pyranones through an initial Michael condensation. It should be noted however that diynones are vulnerable to attack at several sites and that mixed products can be formed. Addition of the anions derived from diethyl malonate and ethyl cyanoacetate to hepta-2,5-diyn-4-one (**313**; $R^1$ = Me) gives the pyranones (**314**; $R^2$ = CO$_2$Et or CN; Scheme 91) ⟨74JOC843⟩. The former carbanion reacts similarly with the diynone (**313**; $R^1$ = Bu$^n$) ⟨68T4285⟩. The second alkyne moiety appears to have little effect on the course of the reaction, which parallels the synthesis of pyranones from monoalkynic ketones.

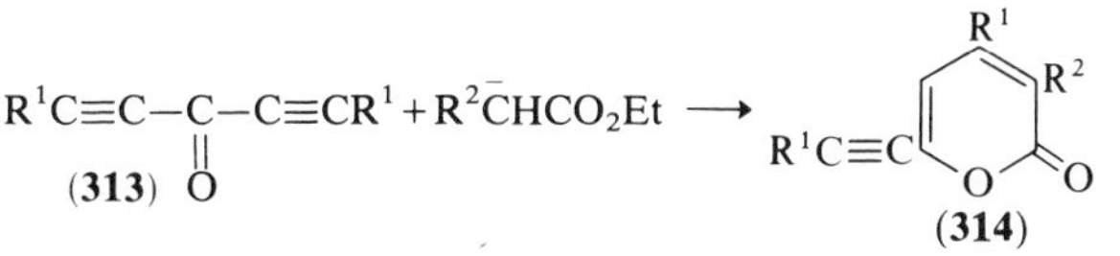

$$R^1C{\equiv}C-C-C{\equiv}CR^1 + R^2\overset{-}{C}HCO_2Et \longrightarrow$$

**(313)**   O                    **(314)**

**Scheme 91**

2-Chlorovinyl ketones and related compounds also react with active methylene compounds to form pyran-2-ones, *via* the unsaturated keto acid derivative ⟨61T63⟩.

A phosphorus ylide serves as the carbanionic component in a synthesis of pyran-2-ones from 1,3-diketones ⟨70ACS343⟩. A Wittig reaction between the ylide and one of the carbonyl groups is envisaged as the first step in the sequence and the resulting keto ester spontaneously cyclizes. The reaction is conducted under pressure and yields are low.

The Reformatsky reaction has been applied to the synthesis of fused pyranones and provides an example of selective isomer formation by careful choice of substrate. Hydroxymethylenecyclohexanone (**315**) and methyl bromoacetate give 5,6,7,8-tetrahydro-1-benzopyran-2-one (**316**) through alkylation at the hydroxymethylene carbon atom ⟨54JA6388⟩. However, the benzoyl derivative (**317**) is unable to form an enolate salt and alkylation occurs at the carbonyl carbon atom, leading to 5,6,7,8-tetrahydro-2-benzopyran-3-one (**318**) ⟨45HCA771⟩.

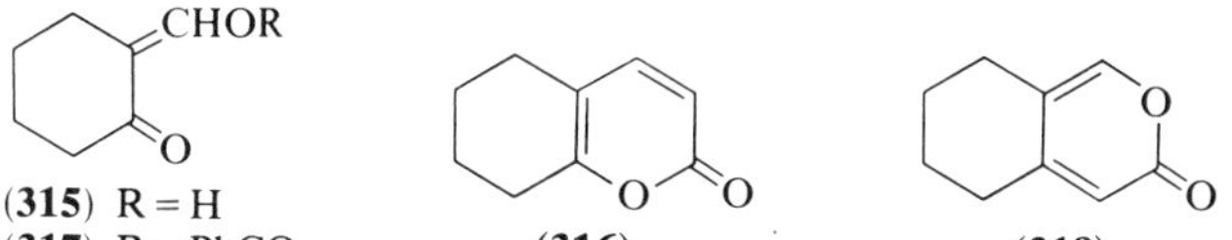

**(315)** R = H
**(317)** R = PhCO            **(316)**                **(318)**

A series of 3-hydroxybut-4-ynoic esters (319) has been prepared from alkynic ketones and bromoacid esters in the presence of zinc ⟨61AP234⟩. Ring closure to the pyran-2-one occurs in a mixture of acetic and sulfuric acids. The preparation of 4,6- and 3,4,6-substituted pyranones is possible by this route, the substituents being different or identical, and as such is complementary to the alkynic route to the 3,6-disubstituted compound ⟨57JA2602⟩.

$$R^1C{\equiv}C-\underset{\underset{OH}{|}}{\overset{\overset{R^2\ R^3}{|\ \ |}}{C}}-CHCO_2R^4 \qquad (EtO)_2CHCH_2CH(OEt)_2$$

**(319)**                                        **(320)**

The protected malondialdehyde (320) reacts with diethyl malonate to give the substituted malonic ester, so providing access to ethyl 2-oxopyran-3-carboxylate ⟨63JOC1443⟩. The corresponding 6-phenyl derivative results from the reaction of 3-methoxy-1-phenylprop-2-en-1-one with malonate carbanion ⟨64RTC31⟩.

The self-condensation of 1,3-dicarbonyl compounds provides a useful route to 4-hydroxypyran-2-ones and is catalyzed by acids or bases. Alcohol is continuously removed during the reaction. Amongst a number of examples, mention can be made of the detailed procedure for the synthesis of 3-acetyl-4-hydroxy-6-methylpyran-2-one (dehydroacetic acid) ⟨55OSC(3)231⟩ and the formation of the cyclopentyl derivative (321) ⟨64RTC39⟩. Deacylation at C-3 can generally be achieved on heating with acid.

**(321)**                              **(322)**

The base-catalyzed self-condensation of diketene is a similar type of reaction and yields dehydroacetic acid together with the tripyranone (322) ⟨49JOC460, B-68MI22400⟩. Ring opening of diketene and attack of the oxyanion on a second molecule of diketene followed by an intramolecular condensation accounts for the formation of the pyranone. The self-condensation of malonyl dichloride in the presence of a ketone, which acts as a weak base, is a similar reaction ⟨52JCS4109⟩.

The self-condensation of β-keto esters and related compounds occurs under the influence of either acidic or basic catalysts and constitutes one of the earliest syntheses of pyran-2-ones ⟨1883LA(222)1⟩. It exemplifies a synthesis of type (ii) (Scheme 85). Ethyl acetoacetate, for instance, gives a mixture of 4,6-dimethyl-2-oxopyran-5-carboxylic acid and its ethyl ester; other esters behave similarly ⟨59RTC364⟩. Decarboxylation of the pyrancarboxylic acid occurs at 160 °C in sulfuric acid. The formation of the pyranone proceeds through a 5-keto ester which is considered to result from attack of the enolic form of the ester on protonated ethyl acetoacetate ⟨51JA3531⟩. A detailed synthesis of the pyran-5-carboxylic acid is available ⟨63OSC(4)549⟩.

β-Keto acids form pyranones in the presence of acid, decarboxylation sometimes accompanying the cyclization, as in the case of benzoylacetic acid ⟨25CB2318⟩.

The formation of 2-oxopyran-5-carboxylic acid (coumalic acid) from 2-hydroxybutanedioic acid (malic acid) is described in detail ⟨63OSC(4)201⟩ and involves the decarbonylation and dehydration of the hydroxy acid. The formylacetic acid so formed then undergoes self-condensation. This synthesis is an example of the Pechmann reaction usually associated with the synthesis of coumarins (benzopyran-2-ones). It will be observed that this route leads to pyran-2-ones which carry identical substituents at the 4- and 6-positions.

In what is essentially a modification of the previous general route, an active methylene compound reacts with an alkynic ester to give a pyran-2-one. This time, however, different substituents may be introduced at C-4 and C-6. Again, this early synthetic route ⟨1899JCS245⟩ has found wide application which may be expressed in general by Scheme 92.

Ethyl 3-phenylpropynoate and ω-methoxyacetophenone condense in the presence of ethoxide ion to give 5-methoxy-4,6-diphenylpyran-2-one and clearly much variation in substituents is possible in this synthesis.

The formation of pyran-2-ones from ethyl 3-phenylpropynoate and some deoxybenzoins has been studied in some detail ⟨66JOC2167⟩. Attempts to isolate the initial Michael addition

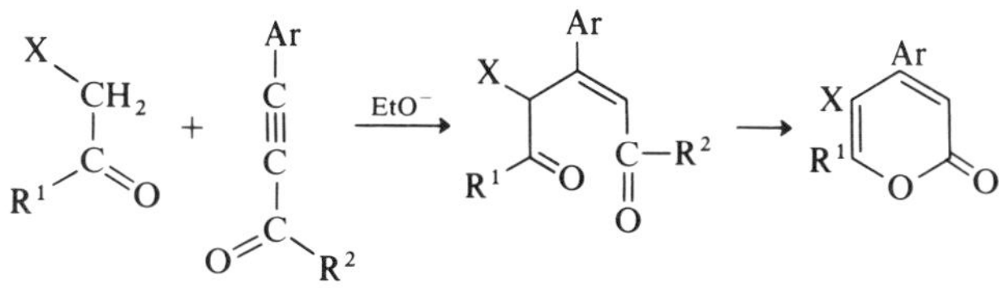

**Scheme 92**

products failed, but rather the pseudoesters (**323**) were obtained along with the pyranone. Their formation is considered to arise from isomerization of the Michael adduct followed by cyclization involving attack of the base catalyst at the aroyl carbon atom (Scheme 93).

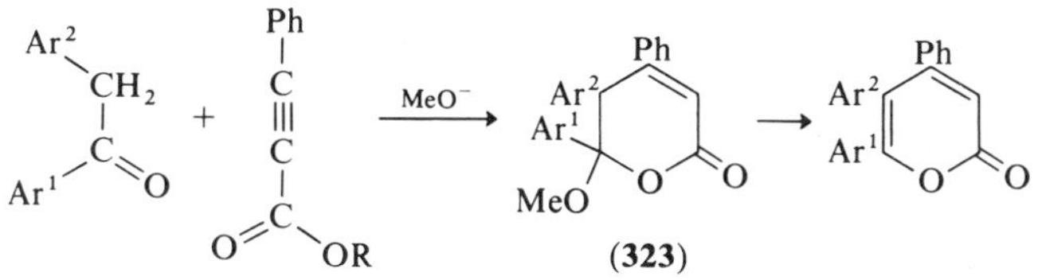

**Scheme 93**

In the case of a ketone with two active methylene groups, such as dibenzylketone, the reaction can take two courses. The pyranone results from Michael addition to the alkyne followed by normal ring closure. The second product, a resorcinol, arises from either Michael condensation followed by an intramolecular Claisen condensation or the order of these two reactions may be reversed ⟨60JCS5153⟩.

The products of the reaction of 4-hydroxy-6-methylpyran-2-one with alkynic esters show some dependence on the solvent used. Dimethyl butyndioate yields the bicyclic pyranone (**325**) in the absence of solvent and using Triton B as the catalyst. Michael addition is envisaged to yield the intermediate dione (**324**) which cyclizes in the normal manner ⟨76JCS(P1)2137⟩.

However, in the presence of methanol the pyrandione (**326**) is the major product. This product is of course the *cis* isomer of the Michael adduct above and is considered to arise by isomerization of that species. The lower reaction temperature in the presence of solvent apparently favours isomerization at the expense of intramolecular cyclization (Scheme 94). A further product, formed in low yield in the absence of solvent, is the symmetrical bis-pyranone (**327**) which arises from a further Michael addition of the carbanion, this time at the sterically favoured 8-position of the enoate.

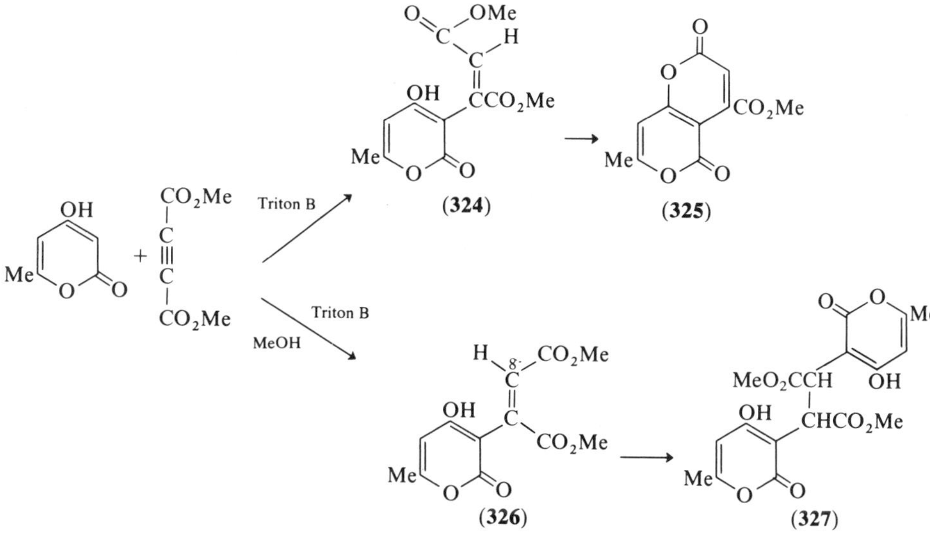

**Scheme 94**

Similar, but not identical, behaviour is shown by ethyl propynoate in its reaction with the 4-hydroxypyran-2-one ⟨75JCS(P1)2405⟩. The differences are accounted for in terms of the relative instability associated with the *cis* configuration of the *trans* adduct (**328**), which isomerizes so readily to the more stable structure (**329**) that no cyclization to the bicyclic pyranone is observed. A second Michael addition occurs but at C-7, leading to the dipyranopyrandione (**330**; Scheme 95).

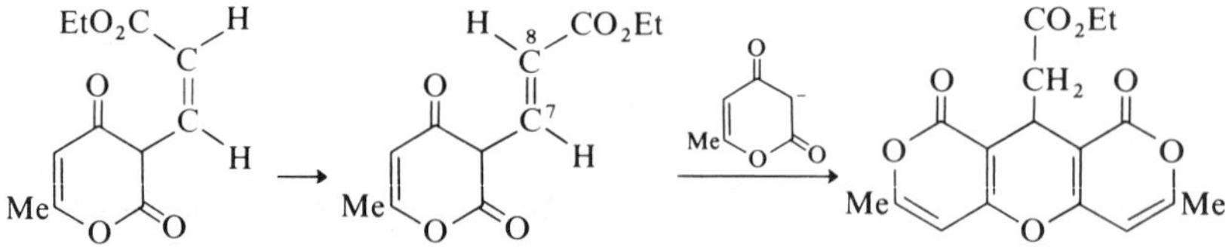

**Scheme 95**

Several routes to pyran-2-ones are based on the reaction of unsaturated 3,3-dichloroaldehydes with active methylene compounds. The usual variation in the latter reactants is allowed and the reaction follows a sequence of carbanion attack at the carbonyl carbon atom of the chloro compound and ring closure through nucleophilic attack of the carbonyl oxygen at the electron-deficient C-5 (Scheme 96) ⟨58IZV1445⟩. It is not possible to decide at which stage the halogen atoms are lost; hydrolysis prior to cyclization would form a 5-keto acid, emphasizing the similarity of this route to those previously discussed.

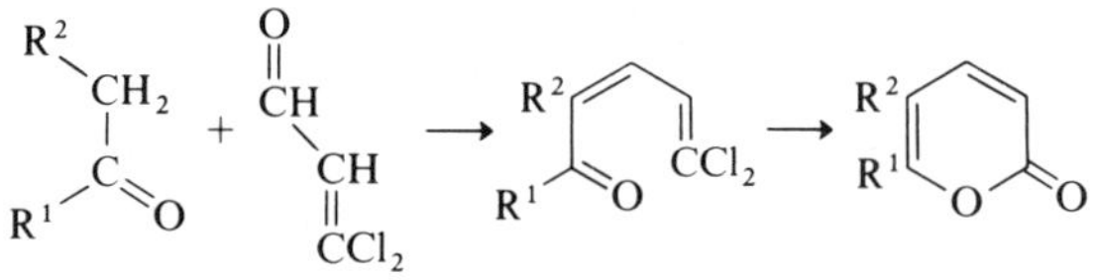

**Scheme 96**

The self-condensation of 3,3-dichloropropenal and of 4,4-dichlorobut-3-en-2-one gives rise to the pyran-2-ones (**331**) and (**332**), respectively ⟨59CI(L)992, 62IZV2096⟩.

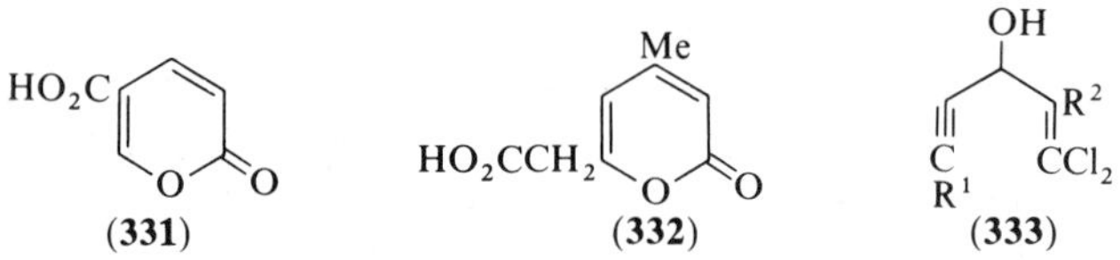

French workers have introduced a variant on this method in which the pyranones are derived from activated alkynes. The reaction of an alkynic carbanion with the above dichloro compounds affords the enynol (**333**). Oxidation with manganese dioxide and cyclization in a mixture of acetic and hydrochloric acids give a 4-hydroxypyran-2-one ⟨61CR(253)872, 75BSF751⟩. However, it is necessary to use only acetic acid as the cyclizing medium when R is 3,4-methylenedioxyphenyl (piperonyl) to avoid degradation. Subsequent methylation of the resulting hydroxypyranone gives 4-methoxyparacotoin, a naturally occurring compound which is physiologically active ⟨59JOC17⟩.

The condensation of 2,3,3-trichloropropenal with cyclic active methylene compounds affords fused pyran-2-ones ⟨58CB320, 58CB330, 60CB2294⟩. Instead of elimination of water, loss of hydrogen chloride is observed in accord with the previously postulated mechanism.

A variant of this method utilizes the ring closure of 4,5,5-trichloropent-4-en-2-ynoic acid prepared from a hexachloropentadienol ⟨62CB1245⟩.

The mixed heterocyclic system (**334**) results from the reaction of substituted 4-hydroxypyrid-2-ones with ethoxymethylenemalononitrile; the corresponding quinoline derivative behaves in an analogous manner ⟨78M1075⟩.

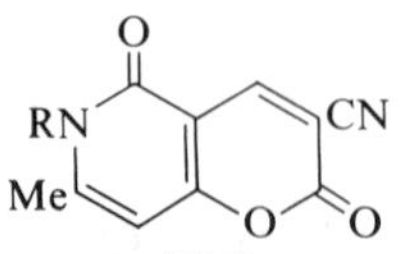

Dissolution in sulfuric acid is sufficient to convert 5-(4-chlorophenyl)penta-2,4-diynoic acid into the 6-aryl-4-hydroxypyran-2-one ⟨73BSF1293⟩; other *p*-substituents are compatible with the synthesis. The intermediacy of a 2,4-diketo acid is assumed. Preparation of the alkynic acids is achieved through the coupling of 2-aryl-1-bromoacetylenes with propynoic acid.

The reaction of ethyl acetoacetate with malonyl dichlorides has been used to synthesize a range of substituted 4-hydroxypyran-2-ones ⟨58CB2849⟩. Yields and purity of the products were best using benzene as solvent either without a catalyst or using magnesium acetate in this role. Hydrolysis of the C-5 ester function and subsequent decarboxylation are both feasible.

An improvement in this synthesis results from the use of the thallium salt of *t*-butyl acetoacetate in place of ethyl acetoacetate, and ethyl methylmalonyl chloride may replace the diacyl dihalide ⟨75S259⟩.

Malonate derivatives have been used in the synthesis of pyranopyrandiones and dipyranopyrantriones, which are masked equivalents of triketo and tetraketo acids, respectively (Scheme 97) ⟨66JA624, 67CC231⟩.

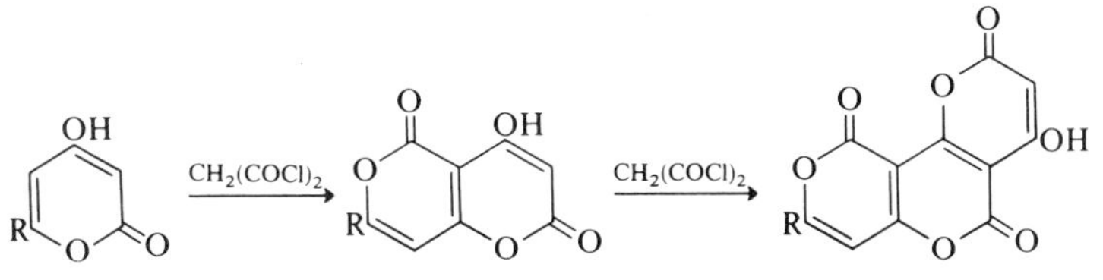

**Scheme 97**

Pyridinium enolbetaines and cyclopropenones have been used as the three carbon atom components in a further illustration of type (ii) syntheses ⟨71LA(746)102⟩. The betaine is formed *in situ* by deprotonation of a pyridinium salt and good yields of the pyranone result simply on boiling in methanol solution. Considerable variation in the substituents is possible and triphenylphosphonium or diethylsulfonium enolbetaines may be used in place of the pyridinium compound.

In this approach to pyranone synthesis, it is thought that the carbanionic site in the betaine attacks the cyclopropenone in a Michael fashion. Subsequent opening of the three-membered ring may lead directly to the pyranone (route a) or the heterocyclic system may form *via* the ketene (route b; Scheme 98).

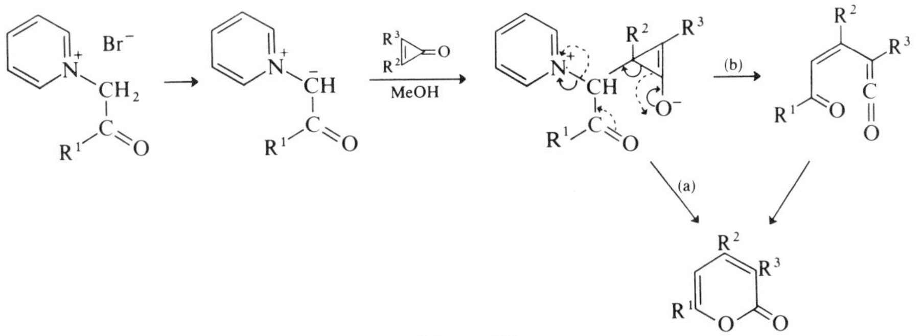

**Scheme 98**

In the presence of a basic catalyst, unsaturated esters react with diethyl oxalate to form an unsaturated 5-keto ester in a synthesis of type (iii) (Scheme 99). In many cases these compounds cyclize spontaneously to the substituted pyranone, but where this is not so, ring closure usually follows hydrolysis under acidic conditions. The pyrancarboxylic acids undergo ready thermal decarboxylation ⟨41JOC566⟩.

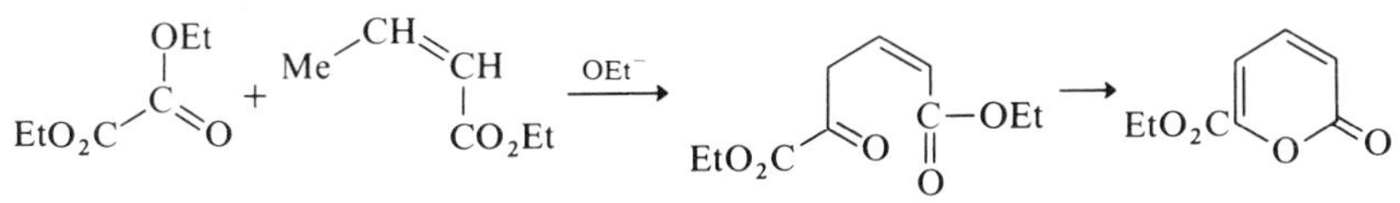

**Scheme 99**

Carbonation of the dianion derived from benzoylacetone occurs at the methyl rather than the methylene group ⟨58JA6360⟩. Sodamide was shown to be the reagent of choice for generation of the dianion. Cyclization of the resulting dioxopentanoic acid by polyphosphoric acid gives 4-hydroxy-6-phenylpyran-2-one. This route has been utilized in the synthesis of a number of naturally occurring 4-hydroxypyran-2-ones; acetic anhydride at room temperature is the preferred cyclization medium and the initial formation of a mixed anhydride is postulated ⟨68JOC2399⟩. Hydrogen fluoride has also proved to be an efficient cyclizing agent ⟨66JOC1032⟩.

Alternatives to carboxylation of the dianions include the reaction with carbonyl sulfide leading to the thiolacid (**335**) which spontaneously cyclizes (Scheme 100) ⟨69T2687⟩.

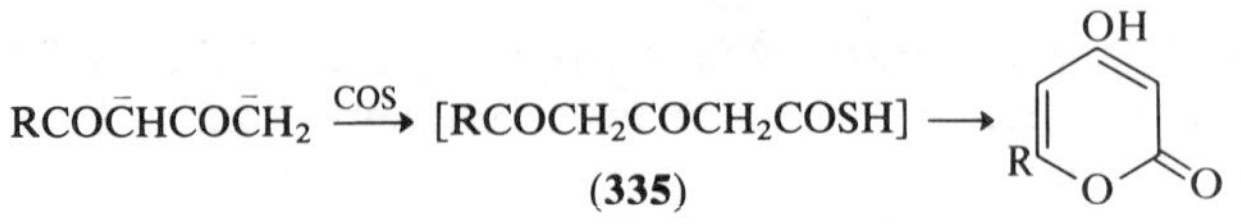
RCOC̄HCOC̄H₂
COS
[RCOCH₂COCH₂COSH]
(335)
OH
R
O
O

**Scheme 100**

Acylation of the dianion of ethyl acetoacetate by an ester is a useful addition to this area of pyranone synthesis. In this reaction and in the formation of the trianions of 2,4,6-triketones the use of lithium diisopropylamide as the base is valuable ⟨76JA7733⟩. The triketo acid from the trianion cyclizes in mineral acids to the pyran-4-one, but in acetic anhydride the pyran-2-one is formed (Scheme 101) ⟨71JA2506⟩.

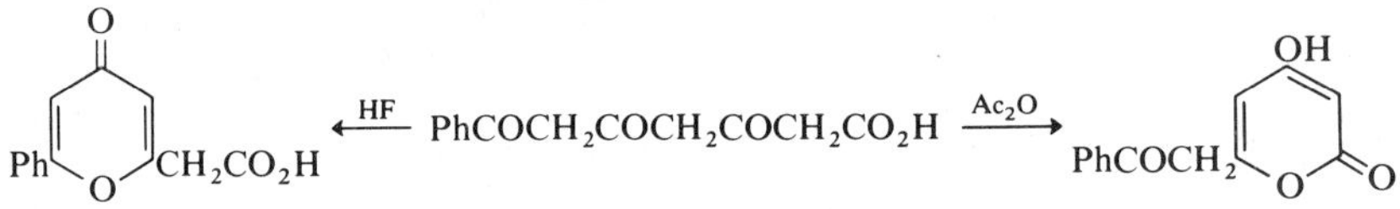
O
Ph
O
CH₂CO₂H
HF
PhCOCH₂COCH₂COCH₂CO₂H
Ac₂O
OH
PhCOCH₂
O
O

**Scheme 101**

The application of polyketone-derived anions to the synthesis of oxygen heterocycles has much to offer; a comprehensive review of this area has been published ⟨77T2159⟩.

The preparation of 3-methyl-6-(2-methylpropenyl)pyran-2-one by the acid-catalyzed self-condensation of 3-methylbut-2-enoic acid has been described ⟨B-68MI22400⟩.

A synthesis of the 3-methylthiopyran-2-one (**337**) is based on the formation of a 2:1 adduct (**336**) between methyl propynoate and DMSO which occurs at high ester concentrations (Scheme 102) ⟨66CB1558⟩. Upon heating, cyclization occurs with the overall loss of methoxymethane. Only a small amount of dimethyl furan-2,4-dioate is formed in this reaction, but a furan is the sole product from dimethyl butynedioate and DMSO.

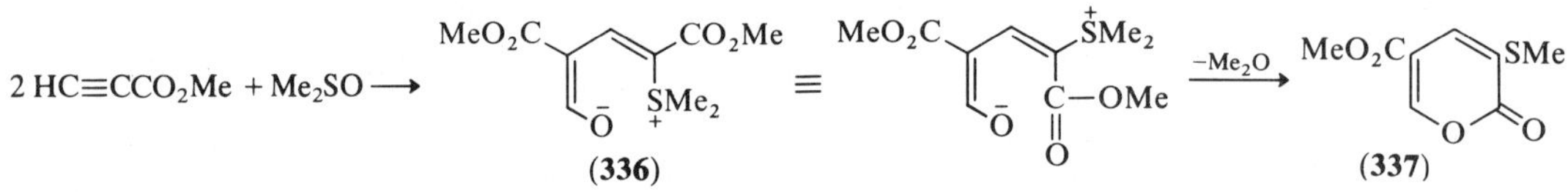
2 HC≡CCO₂Me + Me₂SO
MeO₂C
CO₂Me
SMe₂
O
(336)
MeO₂C
SMe₂
C—OMe
O
O
−Me₂O
MeO₂C
SMe
O
O
(337)

**Scheme 102**

### 2.24.3.1.2 Formation of two bonds

The reaction between vinyl ethers and unsaturated carbonyl compounds, which provides a powerful synthetic route to dihydropyrans, has been adapted to the synthesis of pyran-2-ones ⟨72CC863⟩. 2-Chloro-1,1-dimethoxyethylene, which is a protected form of chloroketene, undergoes cycloaddition with a number of enones to give the *cis* or *trans* isomers of 3-chloro-3,4-dihydro-2,2-dimethoxypyrans (**338**) and (**339**) or a mixture of both. Although

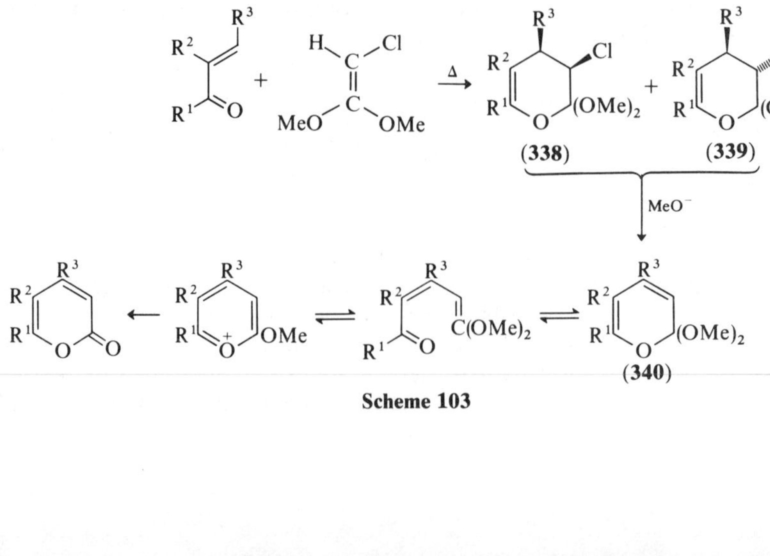
R³
R²
R¹
O
H
Cl
C
C
MeO
OMe
Δ
Cl
(OMe)₂
(338)
(339)
MeO⁻
OMe
C(OMe)₂
(340)

**Scheme 103**

it is not necessary to isolate the dihydropyrans, when this is done the *trans* compound is usually the major product. Elimination of hydrogen chloride and hydrolysis occur under basic conditions to give good yields of the pyran-2-one, together with the acetal (340), which spontaneously rearranges to the pyranone, presumably *via* the pyrylium salt (Scheme 103).

In an extension of this method, the cycloaddition of dichloroketene to the heterocyclic aminomethyleneketone (341) yields the dihydropyranone which is dehydrochlorinated with DBN to the fused pyran-2-one (Scheme 104) ⟨78JHC181⟩. Similar behaviour is shown by the benzologue of the enamine.

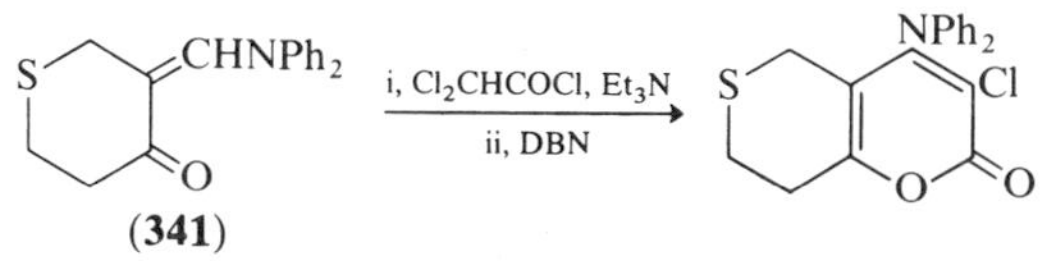

**Scheme 104**

The enaminones (342) also react with dichloroketene, forming 3,4-dihydropyranones ⟨80JHC61⟩. Dehydrochlorination with triethylamine gives the substituted pyrano[3,2-c]-[1]benzopyran-2-ones (343; Scheme 105). Yields are notably better when at least one of the substituents on the amine is aromatic.

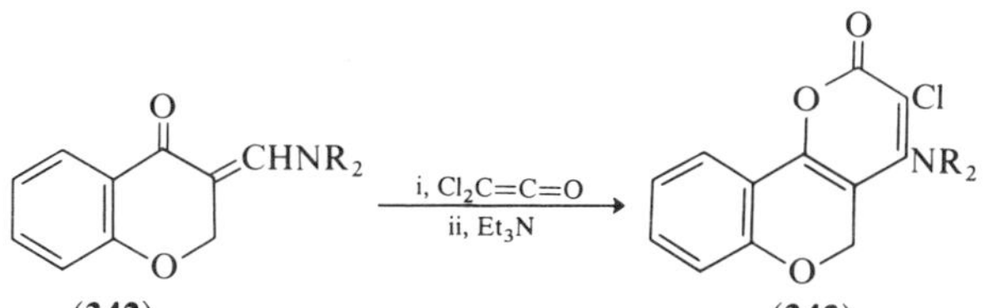

**Scheme 105**

Ketene itself reacts with enamines and if in excess produces pyran-2-ones which are substituted in the 4-, 5- and 6-positions ⟨65JOC2642⟩. It seems likely that the acylated enamine undergoes a 1,4-addition with ketene to give an intermediate 4-amino-3,4-dihydropyranone, which eliminates the secondary amine on further reaction with ketene.

Ketene reacts with ketene *N,N*-acetals and *N,O*-acetals to give 4-dialkylaminopyran-2-ones ⟨64JOC2513⟩. It is necessary that the acetal should contain an alkenic hydrogen atom.

1,1-Bis(dialkylamino)ethylenes give 4-dialkylaminopyran-2-ones on reaction with an excess of ketene ⟨64CB1266⟩, whilst benzoyl chloride and ethyl 3-diethylaminobut-2-enoate give the 2,4-dibenzoyl derivative which is cyclized to the pyran-2-one by base ⟨42JA612⟩.

### 2.24.3.1.3 From other heterocycles

#### (i) From epoxides

Cyclopentadienone epoxides (345) have been postulated as intermediates in the photochemical transformation of pyran-4-ones into pyran-2-ones ⟨79JA7521⟩. A practical synthesis of these compounds, involving the thermal cycloreversion of functionalized tricyclo[5.2.1.0$^{2,6}$]decenone epoxides (345), has been reported ⟨81TL4553⟩. Provided a substituent is present which can extend the conjugation of the cyclic enone system, the epoxides can be generated by flash vacuum pyrolysis. In the absence of such a substituent, the epoxide is contaminated with a pyran-2-one. When the pyrolysis temperature is increased, the cyclopentadienone epoxides are converted into pyran-2-ones (Scheme 106). The thermal rearrangement is considered to proceed through cycloreversion to a conjugated dienone followed by electrocyclic ring closure.

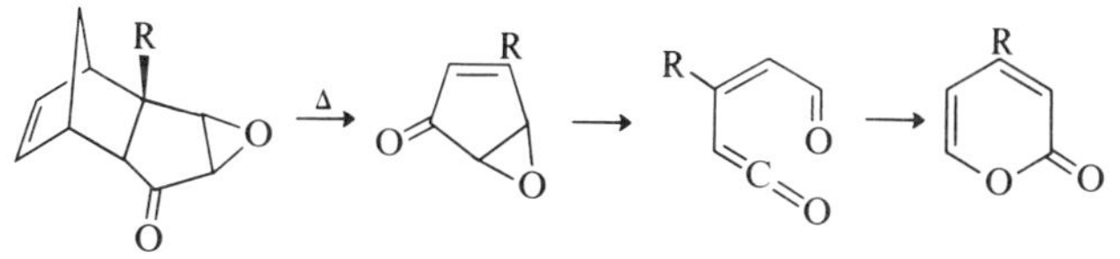

**Scheme 106**

## (ii) *From furans*

When heated, 5-arylfuran-2,3-diones (**346**) readily evolve carbon monoxide to give 3-aroyl-6-aryl-4-hydroxypyran-2-ones ⟨75KGS1468⟩. Based on a kinetic study of this thermal decarbonylation, it is proposed that the formation of pyranones involves an initial cheletropic cycloreversion. The benzoylketene so formed dimerizes through a [4 + 2]-cycloaddition to the pyranone (Scheme 107) ⟨78JOU2245⟩.

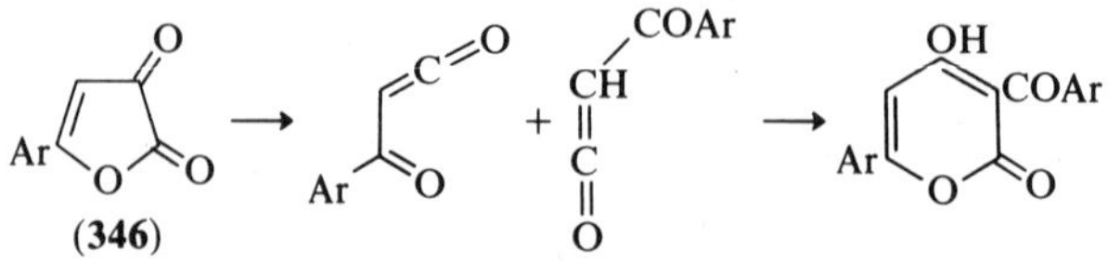

**Scheme 107**

The hydrolysis of the dihydrofuran (**347**) to the pyran-2-one proceeds through the intermediacy of a hydroxy acid (Scheme 108). Ring opening is accompanied by hydrolysis of the nitrile ⟨38JA2404⟩.

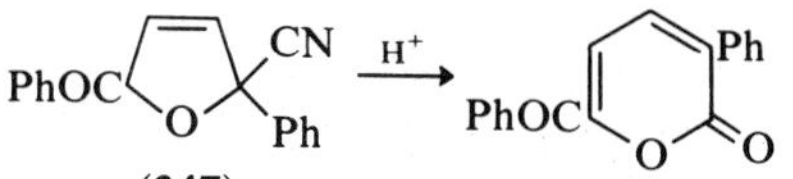

**Scheme 108**

One of two products can be obtained from the tetraketone (**348**), tautomeric with the furanone (**349**), depending on the oxidant ⟨80TL1575⟩. Lead(IV) acetate gives the 4-hydroxy-pyran-2-one (**350**), whereas the pyran-4-one results from the use of iodosobenzene diacetate (Scheme 109).

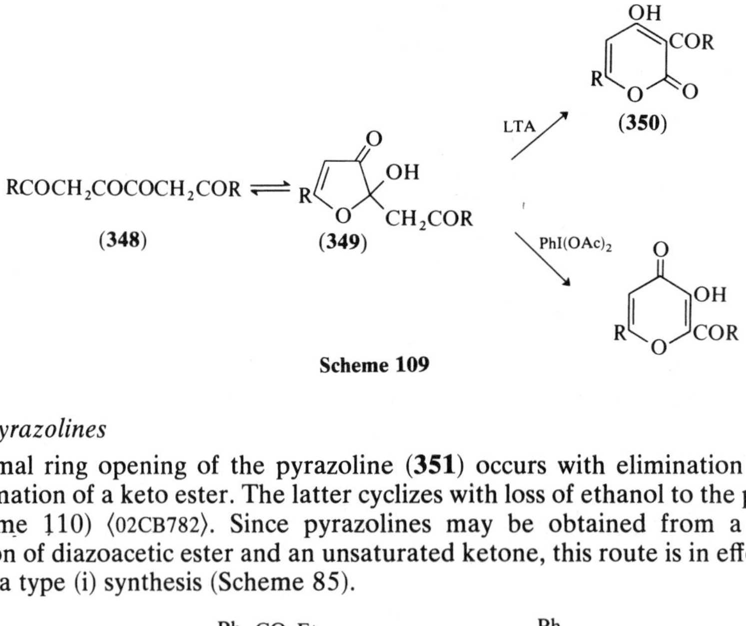

$$RCOCH_2COCOCH_2COR$$
(**348**)

**Scheme 109**

## (iii) *From pyrazolines*

The thermal ring opening of the pyrazoline (**351**) occurs with elimination of nitrogen and the formation of a keto ester. The latter cyclizes with loss of ethanol to the pyran-2-one (**352**; Scheme 110) ⟨02CB782⟩. Since pyrazolines may be obtained from a 1,3-dipolar cycloaddition of diazoacetic ester and an unsaturated ketone, this route is in effect a further example of a type (i) synthesis (Scheme 85).

**Scheme 110**

### 2.24.3.1.4 *From a preformed heterocycle*

#### (i) *From reduced pyranones*

Dehydrogenation of partially reduced pyranones has been achieved directly in the terpenoid field by heating the substrate in the presence of palladized charcoal ⟨63JA3971⟩.

The behaviour of 5,6-dihydropyran-2-ones towards DDQ is variable. Of a range of reduced pyranones, only two (353; $R^1$ = Me or Et and $R^2$ = *trans*-PhCH=CH) were converted to the corresponding pyranone (equation 14) ⟨81JHC363⟩.

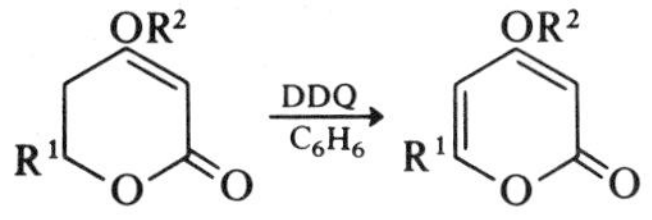

(14)

(353)

The indirect dehydrogenation to pyranones has received rather more attention. The allylic bromination of 5,6-dihydropyran-2-one by NBS, followed by the elimination of hydrogen bromide, is described in detail and offers the attractions of mild conditions and easy isolation of the product ⟨77OS(56)49⟩. A similar approach was used to synthesize the steroidal pyran-2-ones from the fully saturated lactone (354), with a combination of dehydrogenation and dehydrobromination to achieve oxidation ⟨64MI22400⟩.

The addition of halogen to the double bond of a dihydropyranone followed by dehydrohalogenation provides a means of synthesis of simple pyran-2-ones or their fused cycloalkyl derivatives ⟨56DOK(109)117⟩. If a further halogen atom is introduced at the α-methylene group of the 5,6-dihalogeno compound, the product of dehydrohalogenation is the 5-halogenopyran-2-one ⟨64ZOB2504⟩.

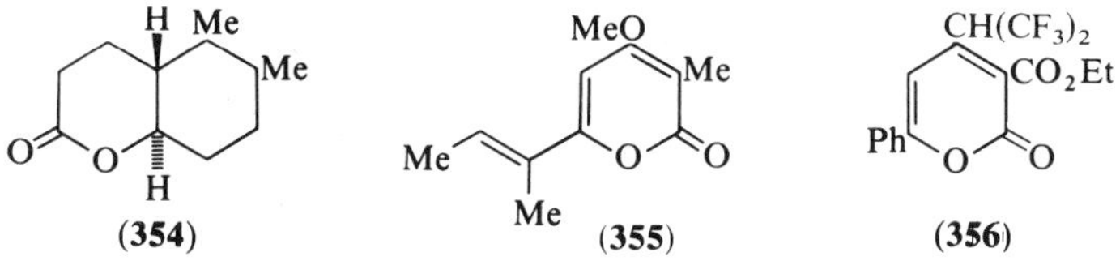

(354)          (355)          (356)

The synthesis of the naturally occurring nectriapyranone (355) has been achieved in a similar manner ⟨76TL1903⟩.

### (ii) *From pyrans*

The formation of the pyran-2-one (356) together with the 4*H*-pyran (143) during the reaction of phenylacetylene with the allene (142) has been attributed to a secondary reaction of the pyran ⟨77IZV2517⟩.

## 2.24.3.2 Coumarins (2*H*-1-Benzopyran-2-ones)

The most important syntheses of coumarins are based on phenols and salicylaldehydes and include the application of such well-known reactions as those of Perkin, Knoevenagel and von Pechmann. The lack of modern synthetic routes may be deduced from the fact that the only review of coumarin chemistry since 1951 ⟨B-51MI22400⟩ is an update of an earlier survey ⟨B-77MI22400⟩. It is not feasible to provide a detailed account of coumarins containing a fused oxygen heterocyclic ring, but this topic is comprehensively surveyed in a review of the synthesis of 3,4-fused coumarins ⟨82S337⟩. A treatise on the occurrence, chemistry and biochemistry of natural coumarins has been published ⟨B-82MI22401⟩.

### *2.24.3.2.1 Formation of one bond*

#### (i) *Adjacent to the heteroatom*

One of the most reliable and widely used syntheses of coumarins involves the acid-catalyzed reaction between a phenol and a β-keto ester, the Pechmann reaction ⟨45CRV(36)1, 53OR(7)1⟩. An important aspect of the reaction is that it shows a dependency on all three reactants, which can be varied widely, thereby optimizing both the scope and conditions of the synthesis. As a consequence, satisfactory yields of coumarins substituted in either the benzene or pyranone ring or both rings can be obtained from readily accessible starting materials.

In its simplest form, an activated phenol such as resorcinol condenses with ethyl acetoacetate in the presence of sulfuric acid to give a 4-methylcoumarin (equation 15). The influence of substituents in the phenol is quite considerable. Phenol itself gives but a very low yield of 4-methylcoumarin under these conditions, though the use of aluminum chloride in place of sulfuric acid greatly improves matters. Not surprisingly, electron-withdrawing groups in the phenol have a deleterious effect on the condensation and may even completely prevent it. Under these circumstances, certain substituents must be introduced into the aromatic ring of the coumarin by conventional electrophilic substitution or by functional group interchanges.

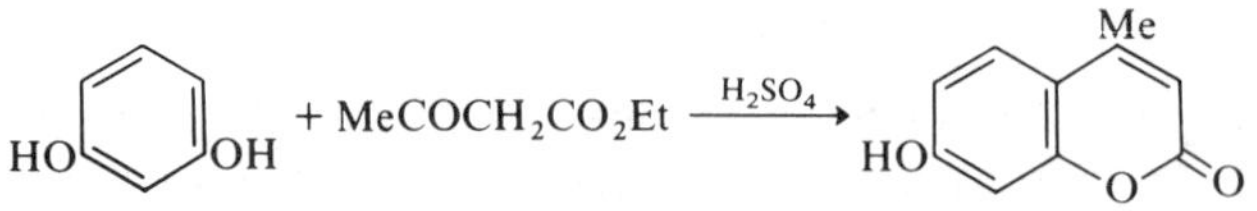

$$
\text{(15)}
$$

Substituents in the pyranone ring are derived from the dicarbonyl component and hence variation in the nature of 3- and 4-substituents is relatively easy. Thus, ethyl benzoylacetate gives 4-phenylcoumarins as for instance in a synthesis of dalbergin (**357**) ⟨76T2407, 81IJC(B)918⟩ and use of ethyl trifluoroacetoacetate leads to 4-trifluoromethylcoumarins. A study of the reaction of the latter with 3-aminophenol identified the optimum conditions for coumarin formation, but noted the simultaneous formation of the two quinoline derivatives (**358**) and (**359**) ⟨80JOC2283⟩.

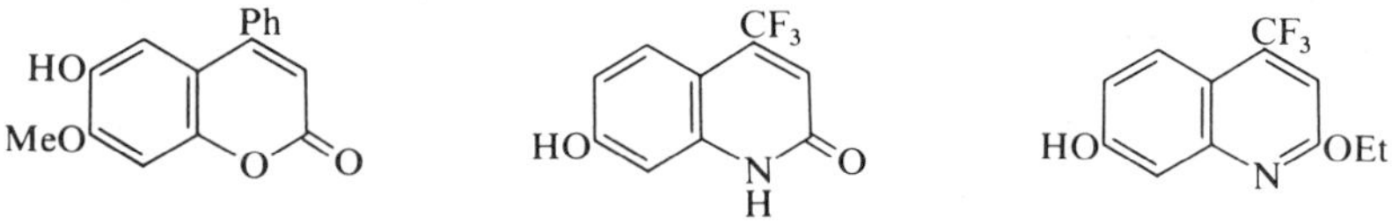

(357)    (358)    (359)

Several dibenzo[*b,d*]pyran-6-ones (**360**) have been prepared by the reaction of substituted phenols with 2-methoxycarbonyl-1,4-benzoquinone in the presence of trifluoroacetic acid ⟨79HCA2833⟩. Treatment with silver oxide converts the products into the related quinones which may be elaborated to benzo[*b*]naphtho[*d*]pyran-6-ones (**361**) by cycloaddition of buta-1,3-diene.

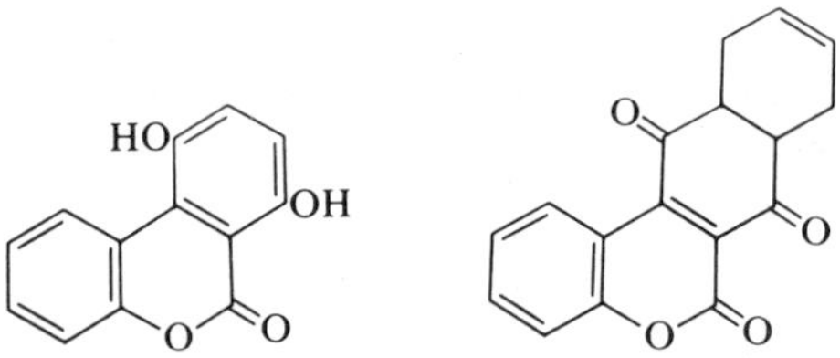

(360)    (361)

The reaction of phenols with alicyclic 1,3-dicarbonyl compounds in phosphorus oxychloride or sulfuric acid also leads to 3,4-fused coumarins ⟨40JA2405⟩. Dehydrogenation to the dibenzopyranone occurs on heating with sulfur or with palladium–charcoal ⟨73CB62⟩.

A range of 3,4-fused pyranobenzopyrandiones, *e.g.* (**362**), have been prepared from the reaction of 4-hydroxycoumarin with cyclic β-keto esters ⟨81JHC1655⟩. In a similar vein, the use of ethyl 3-hydroxy-6,7-dimethoxychrom-3-ene-4-carboxylate (**363**) as the dicarbonyl component results in the formation of chromenocoumarins (**364**) from phenols ⟨36JCS423⟩.

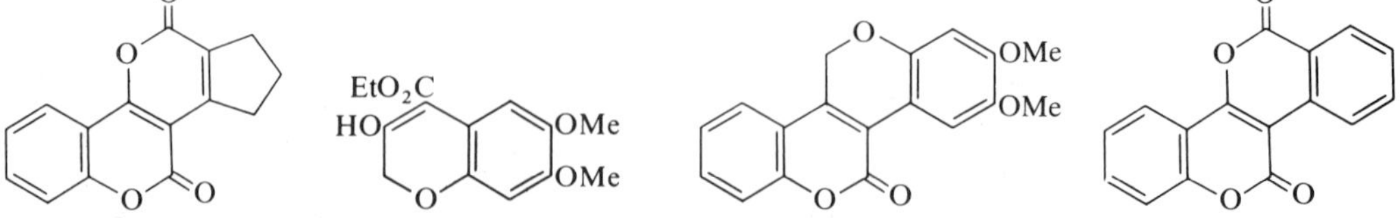

(362)    (363)    (364)    (365)

3-Substituted coumarins arise from the use of 2-substituted 1,3-dicarbonyl compounds. The synthesis of a number of such keto esters has been described together with their conversion into the coumarin ⟨65JOC4114⟩. The use of hydrogen fluoride in place of sulfuric acid as condensing agent was found to be very advantageous.

The formation of 4-hydroxy-3-phenylcoumarin by heating phenol and diethyl phenylmalonate at high temperature is accompanied by small amounts of the [1]benzopyrano[4,3-*c*]-[2]benzopyran-6,11-dione (**365**) ⟨78M1485⟩.

Coumarins unsubstituted in the heterocyclic ring may be obtained by using 2-hydroxy-butanedioic acid (malic acid). Concentrated sulfuric acid decomposes this acid with loss of carbon monoxide and water to formylacetic acid, which is of course a 1,3-dicarbonyl compound and reacts as such with the phenol (Scheme 111). It is sometimes advantageous to use a mixture of sulfuric and acetic acids, which reduces the extent of formation of tarry material.

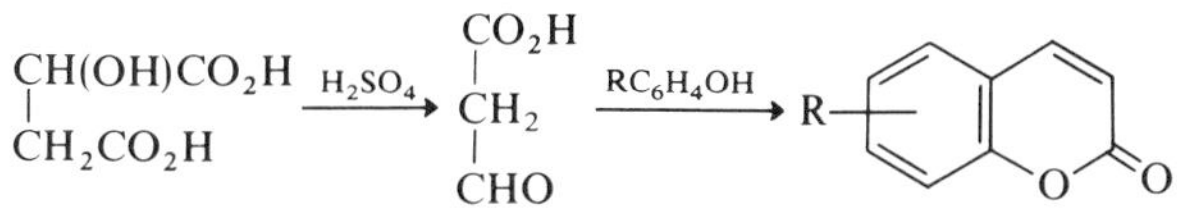

**Scheme 111**

Both fumaric and maleic acid give rise to coumarins on reaction with phenols. Initially, this route was considered different from the Pechmann reaction, but experimental evidence has been accumulated which suggests that it is simply a variant ⟨73AJC899⟩. Under the acidic conditions, the unsaturated acid could well produce some malic acid, which would then lose carbon monoxide and water as usual. Generally yields in the two approaches to coumarins are much the same.

Amongst the condensing agents which have been of value in this reaction, sulfuric acid has been extensively employed. It is recommended not to use the concentrated acid with reactive phenols because of the possibility of sulfonating the ring. The use of phosphorus compounds to effect the reaction introduces the problem of competitive cyclization to the chromone (benzopyran-4-one). Although examples of coumarin formation are recorded using phosphorus-based condensing agents, notably polyphosphoric acid ⟨71LA(745)59⟩, chromones usually result. It is pertinent to note that a chromone resulted from the reaction of 4-chloro-3,5-dimethylphenol with ethyl acetoacetate using sulfuric acid as the condensing agent, although 3,5-dimethylphenol gave the coumarin ⟨44JA802⟩. Such subtle changes in the course of the reaction of phenols with β-keto esters clearly indicate the need for caution in structural assignments to the products. The Simonis synthesis of chromones is discussed in some detail later (see Section 2.24.3.4.1(iii)).

Diphosphoric acid appears to be a specific catalyst for the formation of coumarins from 1,3-dimethoxybenzene and β-keto esters ⟨62JOC4704⟩. An interesting aspect of this synthesis is the cleavage of the ether function involved in the cyclization.

Trifluoroacetic acid has proved of value in the formation of coumarins from sesamol (3,4-methylenedioxyphenol) ⟨71MI22402⟩ and hydrogen fluoride is effective in cases where sulfuric acid fails and also leads to improved yields in other instances ⟨54LA(587)1, 65JOC4114⟩. Aluminum chloride may also bring about more efficient condensation ⟨55OSC(3)581⟩ and it may alter the orientation of cyclization. Thus, whilst resorcinol derivatives give 7-hydroxy-coumarins with other reagents, the 5-hydroxy isomer results with aluminum chloride ⟨38JCS228⟩. The change has been attributed to the presence of strongly chelated complexes involving the aluminum chloride (Scheme 112) ⟨55JIC302⟩.

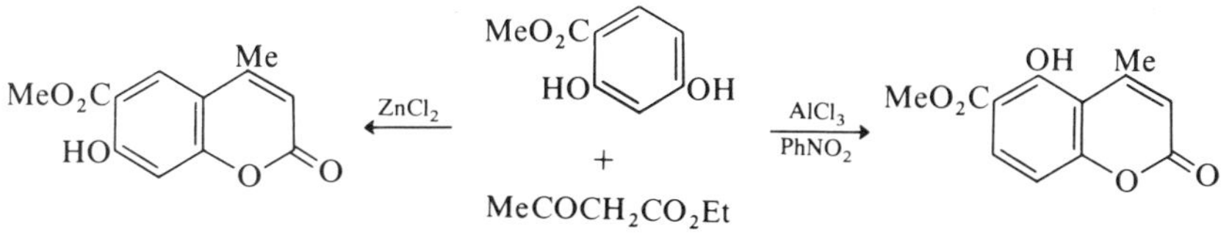

**Scheme 112**

Although earlier investigations suggested that a Pechmann reaction on 3-substituted or 3,4-disubstituted phenols yielded only one product, it is now clear that this is not always the case ⟨81T2021⟩. The failure of earlier workers to isolate a second product is readily understood from the data in Table 6. As yet, it is not known whether this behaviour is a general feature of *m*-substituted phenols. The lack of formation of the 5- or 5,6-isomer in the reactions with ethyl acetoacetate may be a result of steric hindrance by the 4-methyl substituent.

The Pechmann reaction is thought to proceed through electrophilic aromatic substitution of the phenol. The resulting β-hydroxy ester then cyclizes and dehydrates to the coumarin, although of course dehydration may occur earlier in the sequence (Scheme 113). Indeed, the observation that 2-hydroxycinnamic acids readily yield coumarins in sulfuric acid ⟨32JCS1681⟩ renders these compounds or their esters plausible intermediates in the reaction.

**Table 6**  Isomer Distribution in the Pechmann Reaction with 3-Substituted Phenols

| | | Total yield (%) | Product analysis | |
| --- | --- | --- | --- | --- |
| *Substrate* | *Reactant* | | 7- or 6,7- isomer | 5- or 5,6- isomer |
| *m*-Cresol | Malic acid | 36 | 88 | 12 |
| *m*-Cresol | Ethyl acetoacetate | 58 | 100 | 0 |
| 3,4-Dimethylphenol | Malic acid | 32 | 91 | 9 |
| 3,4-Dimethylphenol | Ethyl acetoacetate | 66 | 100 | 0 |

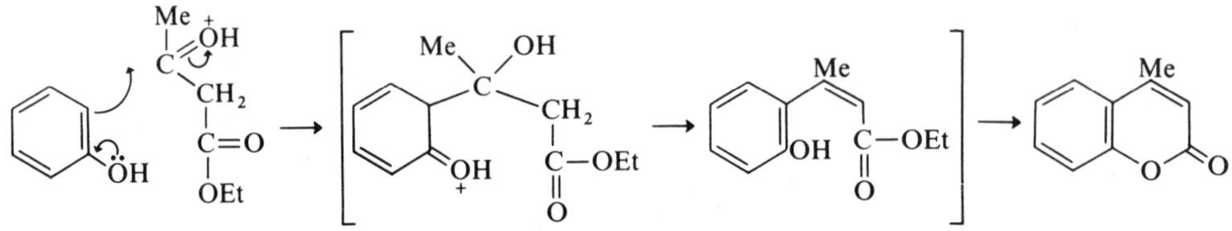

**Scheme 113**

A $\beta$-ketonitrile may usefully replace the 1,3-dicarbonyl compound in the Pechmann reaction. 2-Formyl-2-phenylacetonitrile and 1-naphthol react at 220 °C to afford the naphtho[1,2-*b*]pyran-2-one (**366**) ⟨75JCS(P1)1869⟩. Under the more usual Pechmann conditions a number of products result, of which the acrylonitrile (**367**), the hydroxynitrile (**368**) and the dihydronaphthopyran-2-one (**369**) are significant, clarifying the disparate results of earlier workers.

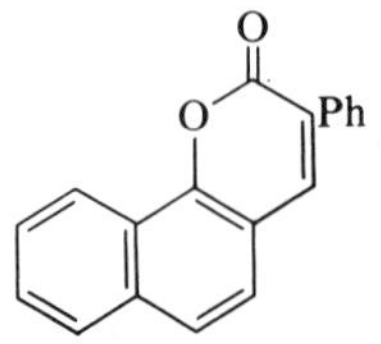

(**366**)

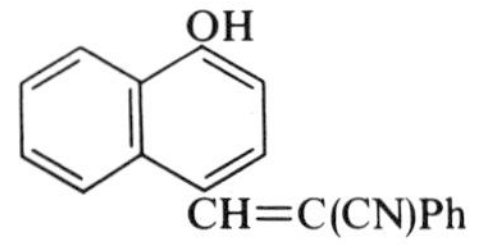

(**367**)

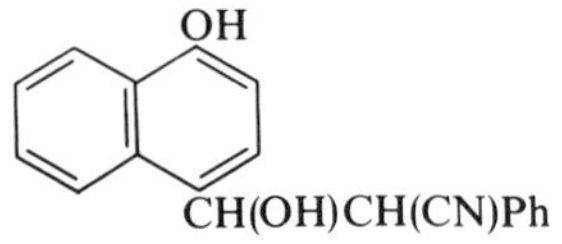

(**368**)

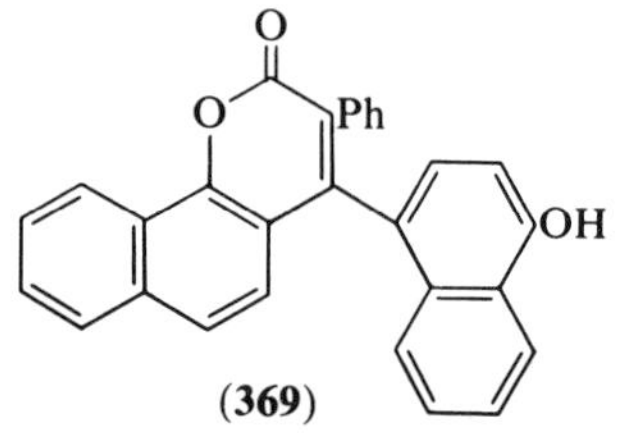

(**369**)

Although 2-ethyl-3-oxobutanonitrile undergoes carbon–oxygen condensation with phenols to give a phenoxybutenonitrile, both 3-oxobutanonitrile and 3-oxo-3-phenylpropanonitrile acylate phenols at carbon ⟨68JOC2446⟩. Dehydration and intramolecular cyclization of the proposed intermediate (**370**) leads to an imine (Scheme 114), which in some instances may be isolated but which is commonly hydrolyzed to the coumarin. Polyphosphoric acid ethyl ester is the preferred condensing agent for 3-oxobutanonitrile since aluminum chloride causes it to polymerize.

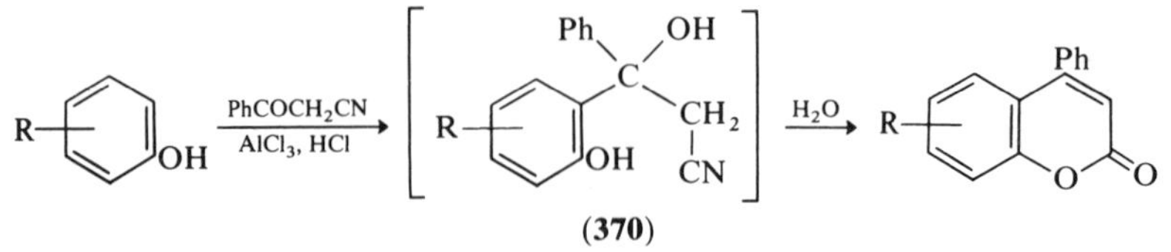

(**370**)

**Scheme 114**

The synthesis of a similar intermediate to that in the above reaction, the substituted acrylonitrile (**371**), is achieved with ease from the base-catalyzed reaction of an aromatic aldehyde with a substituted acetonitrile. Cyclization of the acrylonitrile to the coumarin has been accomplished with pyridine hydrochloride ⟨51JCS2307, 69JCS(C)2069⟩ and with aluminum chloride ⟨70IJC969⟩, presumably *via* the imine. 2-Methoxynaphthaldehyde affords the naphtho[2,1-*b*]pyran-3-one (**372**) in this manner.

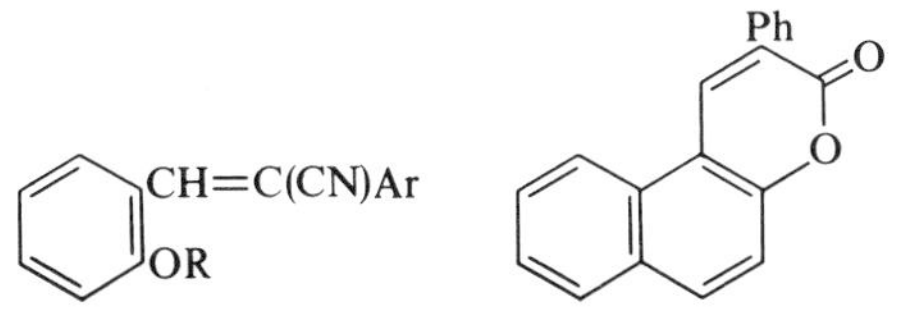

(**371**)     (**372**)

The formation of coumarins has been observed in the reaction of activated phenols with ethyl cyanoacetate, catalyzed by zinc chloride and hydrogen chloride ⟨17CB1292⟩.

Zinc chloride is also used in an alternative approach to 4-hydroxycoumarins, in which malonic acid is used as the 1,3-dicarbonyl component, although additionally the presence of phosphorus oxychloride is vital to success ⟨60JOC677⟩.

The Pechmann synthesis is unsuitable for acid-sensitive phenols, as for example the furo[2,3-*b*]benzofuran derivative (**373**). An alternative approach uses the enhanced electrophilic character of a vinyl bromide in the presence of zinc carbonate to construct a suitable side-chain adjacent to the phenolic group ⟨71JA746⟩. In the examples cited, ring closure occurred under the mild conditions to form the pyranone ring of the aflatoxins (**374**). Since neither sodium nor potassium carbonate proved effective, it was considered that chelation of the zinc facilitated the carbon–carbon bond formation (Scheme 115).

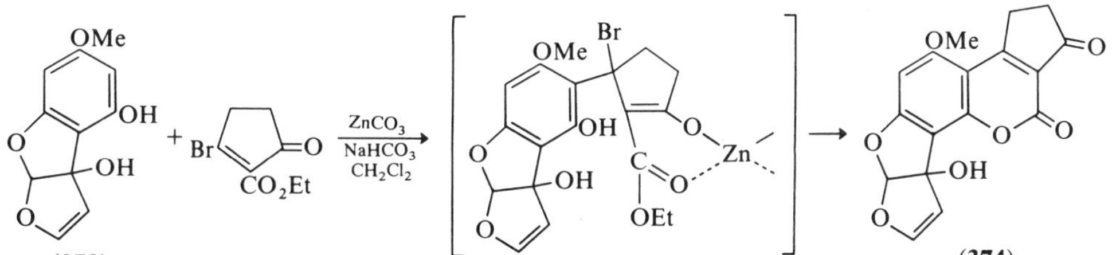

(**373**)                                                                 (**374**)

**Scheme 115**

A similar reaction has been used to obtain 3-methylcoumarins ⟨71CR(C)(272)1985⟩. Methyl 3-bromo-2-methylpropenoate reacts with 3,5-dimethoxyphenol under basic conditions to give a low yield of the coumarin. The major product is the phenoxyacrylate derived from *O*-acylation.

The ability to metallate aromatic compounds regioselectively ⟨80OR(26)1⟩ provides a route to coumarins which involves mild conditions ⟨79JOC2480⟩. The synthesis is based on the directive influence of an alkoxy group which allows the ready formation of an *o*-alkoxyaryllithium. Michael addition of the organolithium reagent to diethyl ethoxymethylenemalonate is followed by removal of the phenol protecting group. Ring closure and elimination occur spontaneously to give the coumarin (Scheme 116). One drawback encountered in this work was the inability to cleave the protecting group when this was a simple ether. However, both methoxymethyl and 1-ethoxyethyl groups are removable under mild conditions. The complementary nature of this route to the Pechmann synthesis is worthy of note; whereas in the latter method a *m*-dihydric phenol gives mainly a 7-hydroxycoumarin, the 5-isomer is formed using the present technique.

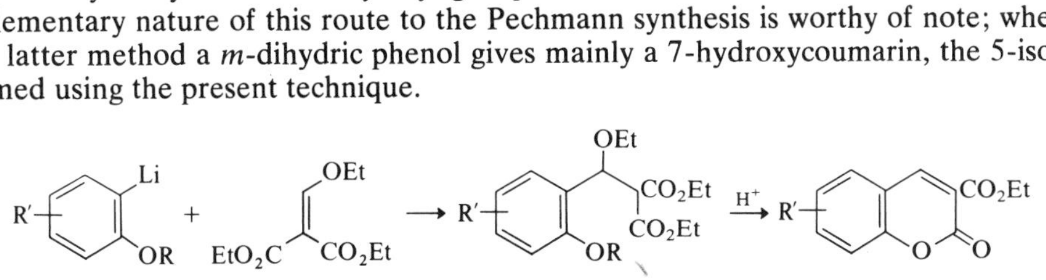

**Scheme 116**

A Michael condensation also features in a synthesis of 4-substituted 3-cyanocoumarins ⟨81ZOR2630⟩.

Use of the Knoevenagel reaction ⟨67OR(15)204⟩, in which a benzaldehyde reacts with an activated methylene compound in the presence of an amine, goes some way to overcoming the inherent difficulties of the Perkin synthesis of coumarins (see later). In order to obtain the coumarin rather than the usual cinnamic acid, a 2-hydroxy substituent must be present in the aromatic aldehyde and the same synthetic problems therefore apply to both routes. However, the conditions of the Knoevenagel route are much less severe than those associated with the Perkin reaction. Although it is necessary to decarboxylate the product to give a coumarin unsubstituted at C-3, and this may be a difficult reaction on occasions, the Knoevenagel is often preferred to the Perkin reaction.

**Table 7**  Synthesis of Coumarins by a Knoevenagel Reaction on Salicylaldehyde

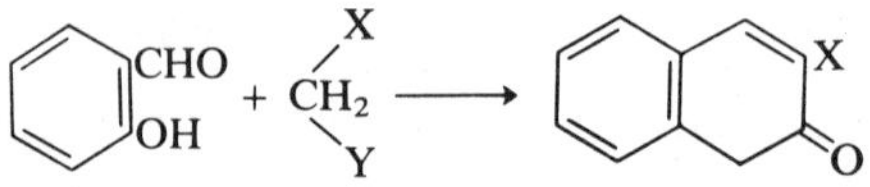

| Active methylene compound | | | Coumarin | | |
| X | Y | Conditions | X | Yield (%) | Ref. |
|---|---|---|---|---|---|
| $CO_2H$ | $CO_2H$ | Pyridine, aniline, Δ | $CO_2H$ | 83 | 52JA5346 |
| $CO_2Et$ | $CO_2Et$ | Piperidine, AcOH, Δ | $CO_2Et$ | 80 | 55OSC(3)165 |
| COPh | $CO_2R$ | Piperidine, AcOH, Δ | COPh | | 66JOC620 |
| $CONH_2$ | CN | Piperidine, MeOH, $H_2O$ | $CONH_2$ | 58 | 62CB483 |
| $CONH_2$ | CN | i, piperidine, EtOH, Δ; ii, DMF, $POCl_3$, Δ | CN | | 81JPR691 |
| CONHPh | $CO_2H$ | Pyridine, piperidine, Δ | CONHPh | | 31JCS2059 |
| CN | CN | i, piperidine, Δ; ii, HCl | $CO_2H$ | 100 | 65JOC312 |
| CN | CN | i, $Me_2NH$, EtOH; ii, HCl | CN | 91 | 81JAP(K)81152471 |

An indication of the scope of the Knoevenagel synthesis of coumarins is apparent from Table 7, but some further examples utilizing the conventional procedure are worthy of mention partly because some of the products are naturally occurring compounds. Graveolone (**375**), a benzodipyran, is obtained from the chroman-4-one shown by cyclization of the acrylic acid (Scheme 117) ⟨78IJC(B)570⟩.

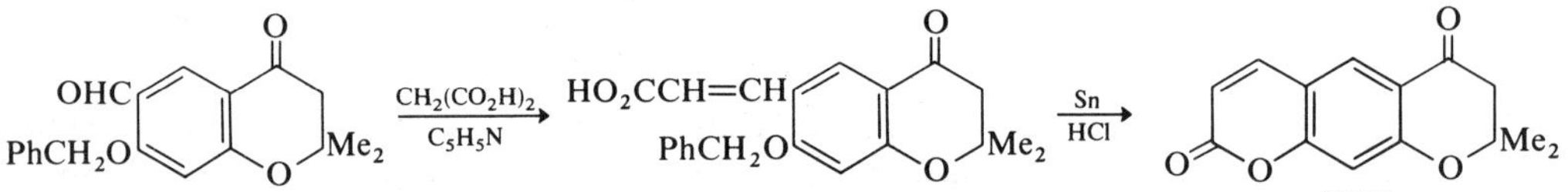

**Scheme 117**

Elaboration of the side-chain of the salicylaldehyde (**376**) coupled with a Knoevenagel condensation enabled a number of naturally occurring coumarins such as xanthyletin (**377**) to be prepared ⟨71CJC2297⟩. The use of C-labelled malonic acid leading to coumarins labelled at C-3 enhances the potential of this route.

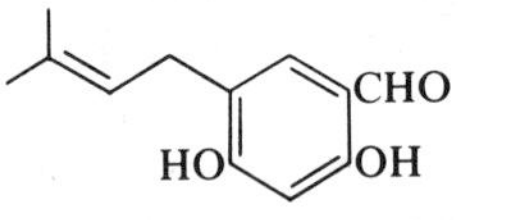

(**376**) 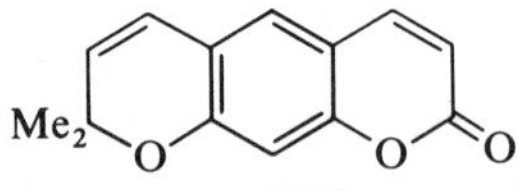               (**377**)

4-Hydroxycoumarin-3-carboxylic acids result from the condensation of malonic acid with salicylic acid in the presence of trifluoroacetic acid ⟨65JOC4343⟩. It appears that the reaction involves attack of the activated methylene compound at the aromatic carboxyl group.

It is worthy of note that 4-dimethylaminosalicylic acid reacts with ethyl acetoacetate in the absence of catalyst to give the 4-methylcoumarin with loss of carbon dioxide ⟨71MI22402⟩.

Two mechanisms have been proposed for the Knoevenagel reaction. In one, the role of the amine is to form an imine or iminium salt (**378**) which subsequently reacts with the enolate of the active methylene compound. Under normal circumstances elimination of the amine would give the cinnamic acid derivative (**379**). However, when an *o*-hydroxy group is present in the aromatic aldehyde intramolecular ring closure to the coumarin can occur. The timing of the various steps may be different from that shown (Scheme 118).

In the second postulated mechanism, the carbanion derived from the active methylene compound by deprotonation by the amine is considered to attack the carbonyl group without further intervention by the base (Scheme 119).

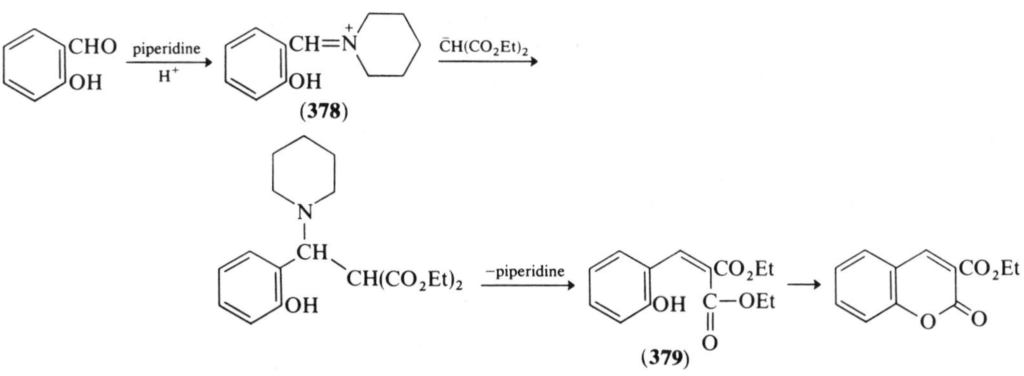

**Scheme 118**

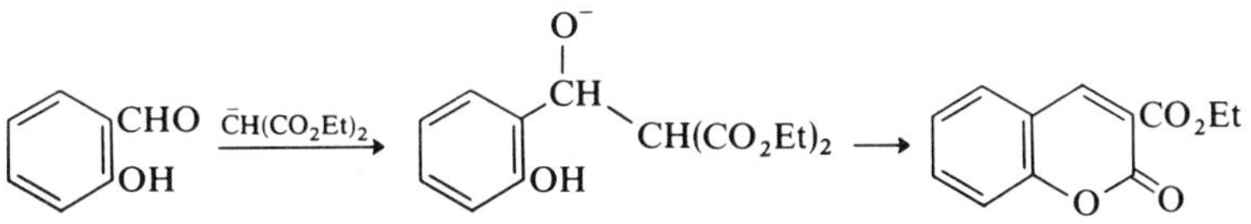

**Scheme 119**

The evidence for the two mechanisms has been discussed in detail ⟨67OR(15)204⟩, with the conclusion that a single mechanism is unlikely for the wide variety of Knoevenagel condensations.

A further illustration of the reaction of carbanionic species with *o*-hydroxybenzaldehydes is provided by the use of phosphorus ylides in coumarin synthesis. Thus, ethoxycarbonyl-methylenetriphenylphosphorane affords simple coumarins *via* the ester (**380**; R = H), whilst with *o*-hydroxyacetophenone a 4-methylcoumarin results (**381**; R = Me) ⟨77S464⟩, and 4-methylsulfinylmethylcoumarin (**381**; R = CH₂SOMe) is obtained from *o*-hydroxy-*ω*-methylsulfinylacetophenone (Scheme 120) ⟨72JHC175⟩.

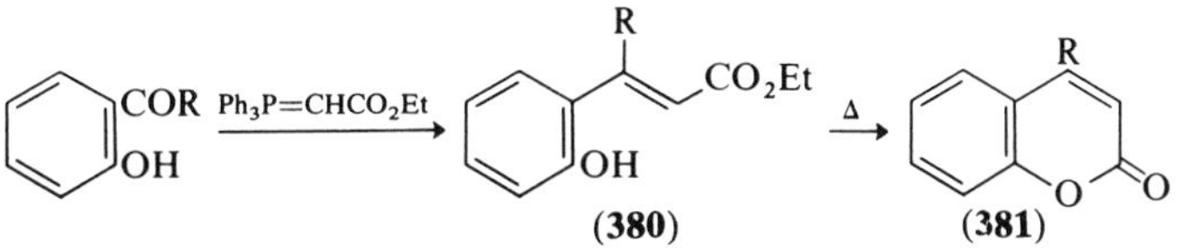

**Scheme 120**

The method has been extended to the synthesis of 3-phenylcoumarins and the naph-thopyranones (**382**), (**383**) and (**384**) through the use of ethoxycarbonylbenzyl-idenetriphenylphosphorane ⟨79S906⟩.

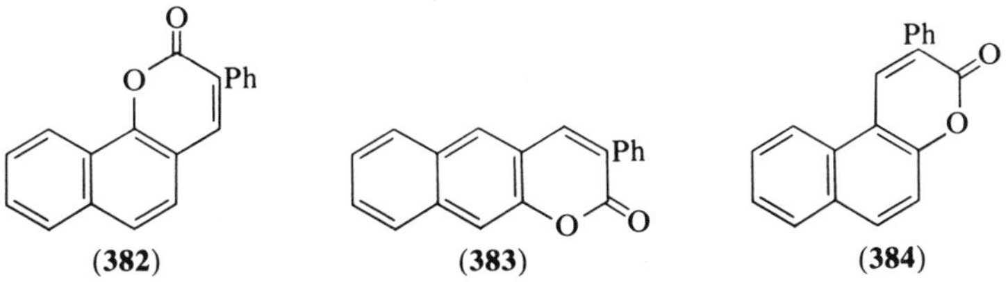

Several older syntheses are based on *o*-hydroxyacetophenones but these routes generally offer little or no advantage over other methods. The Reformatsky reaction has been used to prepare 3- and 3,4-substituted coumarins ⟨44JIC109⟩ and the Kostanecki–Robinson reaction may yield coumarins instead of or as well as the chromone (see Section 2.24.3.4.1 for further discussion).

Substituted 4-hydroxycoumarins result from the reaction of *o*-hydroxyphenyl ketones with diethyl carbonate ⟨48JCS174⟩ and with ethyl chloroformate ⟨81JCS(P1)1697⟩.

The carbanion generated by the action of a weak base on the pentacarbonyl compound (**385**) reacts with resorcinol to give an unstable product (**386**) which readily cyclizes to the 4-hydroxycoumarin in trifluoroacetic acid (Scheme 121) ⟨71JA6708⟩.

The base-catalyzed intermolecular condensation of 1,3-dicarbonyl compounds can give rise to coumarins. For example, in the presence of sodium ethoxide, diethyl 3-oxopen-tanedioate (acetonedicarboxylate) affords 6-ethoxycarbonyl-4,5,7-trihydroxycoumarin,

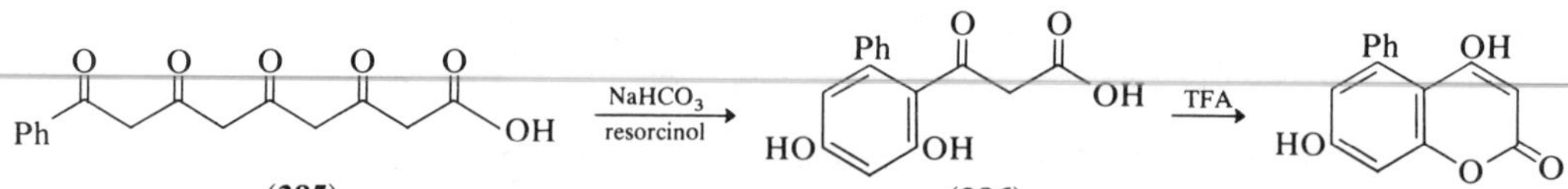

**Scheme 121**

presumably through an initial Claisen condensation followed by a Dieckmann reaction (Scheme 122) ⟨78CPB1973⟩. Although ring closure of the resulting benzoylacetate could involve either of the adjacent hydroxy groups, only a small amount of the 8-ethoxycarbonyl-coumarin is formed. Chelation of the 2-hydroxy group may be the cause of the regioselectivity.

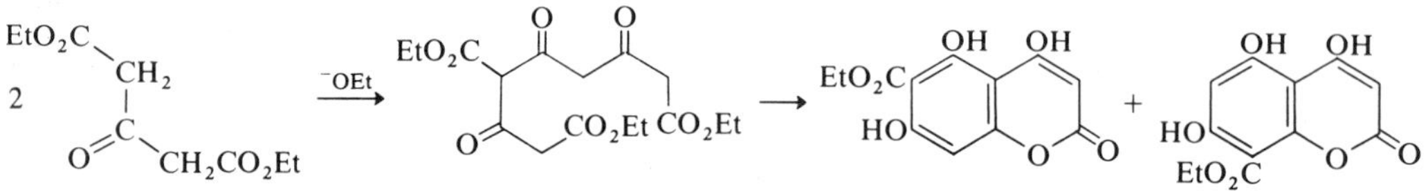

**Scheme 122**

Although the Claisen rearrangement of allyl and propargyl aryl ethers is known to provide useful syntheses of chromans and chromenes respectively (see Sections 2.24.2.7.3(i) and 2.24.2.2.1), the substrates may be adapted to provide a route to coumarins. For example, *o*-methoxyphenol and methoxyallene afford the rather unstable allyl aryl ether (**387**) which rearranges and cyclizes to a mixture of the 2-methoxychroman (**387a**; R = Me) and the chroman-2-ol (**387a**; R = H) in boiling dimethylaniline ⟨82JOC946⟩. Two oxidative steps are necessary to convert the chromanol to the coumarin, which is obtained in a 48% overall yield from the phenol (Scheme 123). Only one oxidation is necessary if the sequence utilizes a dioxygenated allyl aryl ether and these are available from a phenol and triethyl orthoacrylate. The initial product is a 2,2-diethoxychroman, which may be hydrolyzed and ring-closed to the dihydrocoumarin.

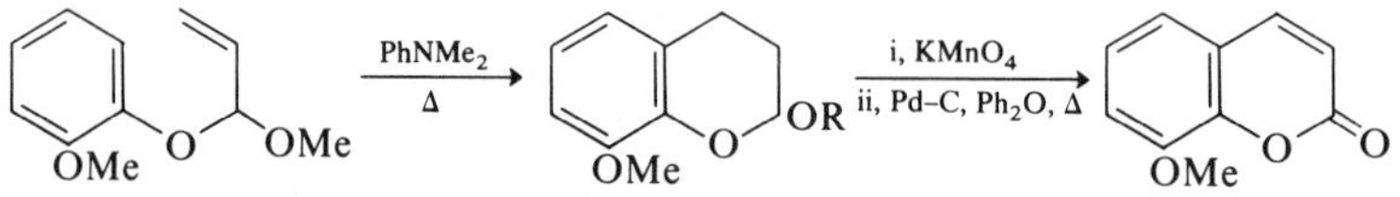

**Scheme 123**

The low yield dehydrogenation stage is eliminated if an alkyne is used in place of the alkene. An improved yield of coumarin results, although the initial product from the rearrangement is a cinnamate.

The use of triethyl orthobut-2-ynoate, $MeC\equiv CC(OEt)_3$, yields 4-methylcoumarins, whilst with dimethyl butynedioate, resorcinol affords methyl coumarin-4-carboxylate in a reaction catalyzed by zinc chloride ⟨69MI22400⟩. The potential elaboration of this approach to coumarins is quite apparent.

The formation of the coumarin (**388**) from 7-allyloxy-3-methylflavone *via* a Claisen rearrangement and oxidation with DDQ is of a similar nature to the previous route, though yields are considerably lower ⟨77TL473⟩.

High yields of 3-aminocoumarins (**390**) are formed rapidly when solutions of dimethyl butynedioate and *O*-arylhydroxylamines are mixed at low temperature ⟨72TL3941⟩. Their formation has been rationalized through a rearrangement of the substituted hydroxylamines (**389**).

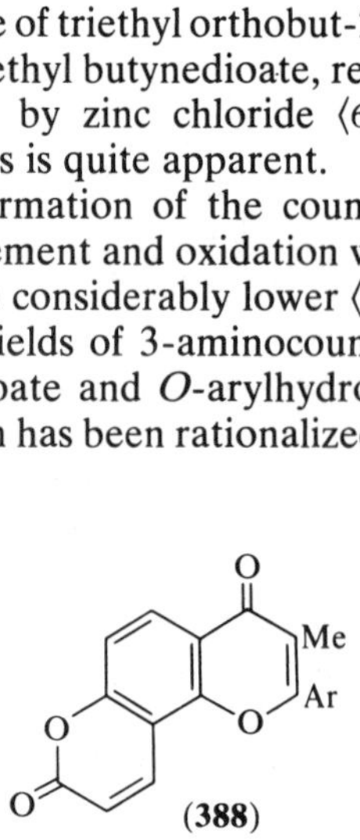

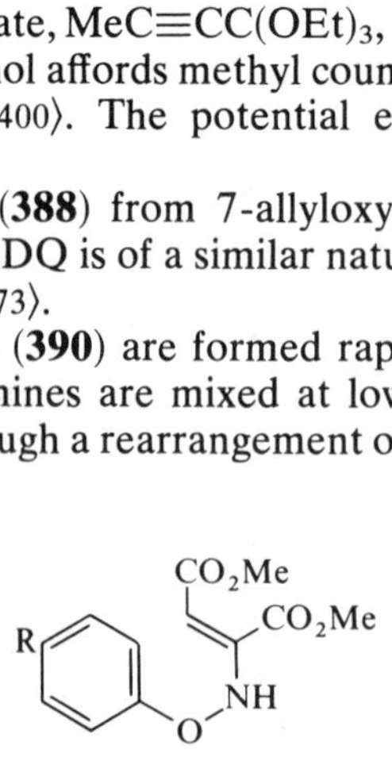

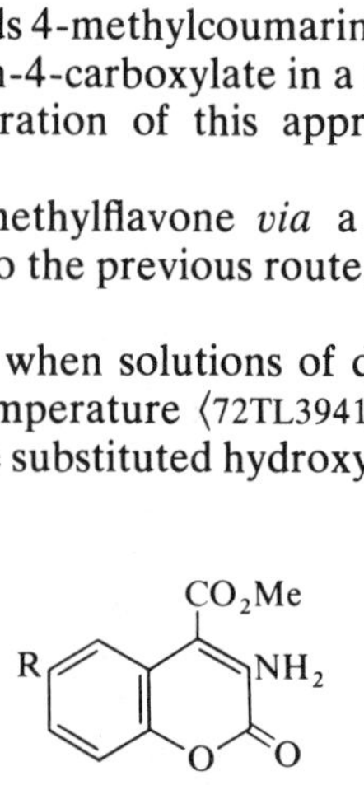

The coupling of *o*-bromobenzoic acid and phenols leads directly to annelated coumarins ⟨75CJC343⟩. The formation of a similar quantity of salicylic acid, formed through a competitive reaction of hydroxide ion with the bromobenzoic acid, has been observed ⟨75T2607⟩. Of a range of catalysts studied, copper sulfate is the most effective. The reaction is only successful in the pH range 6–10. Although only one coumarin results from the reaction with resorcinol, two isomers are formed from 5-methoxyresorcinol, the minor product arising from cyclization at C-2.

### (ii) γ *to the heteroatom*

Application of the Perkin cinnamic acid synthesis to salicylaldehydes provides a coumarin synthesis of some renown. The reaction itself has been reviewed ⟨42OR(1)210⟩ and some discussion of its value in coumarin synthesis is contained in other reviews ⟨37CB(A)83, 45CRV(36)1⟩.

The synthesis consists of heating a mixture of the *o*-hydroxybenzaldehyde with the sodium salt and anhydride of a carboxylic acid, usually acetic acid. Generally, a temperature in the range 150–200 °C is required and a prolonged reaction time is often necessary. A 1:2 molar ratio of aldehyde to anhydride gives optimum yields of coumarin (Scheme 124).

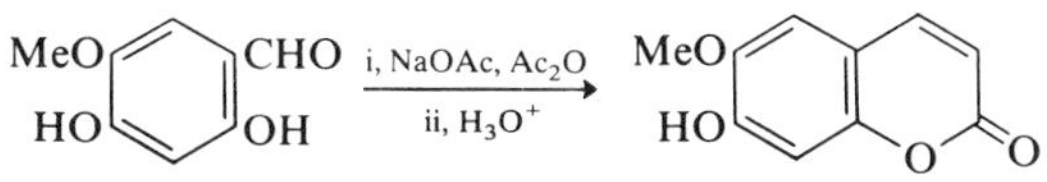

**Scheme 124**

The vigorous conditions which are necessary serve to detract from the value of the Perkin synthesis, leading to the production of tarry material which adversely affects the yield of coumarin. Difficulties encountered in the synthesis of substituted *o*-hydroxybenzaldehydes also limit the application of this route. The obvious advantages of the method are that there are no doubts about the orientation of the product and that, unlike the Pechmann reaction, formation of the isomeric chromones is not possible.

In addition to the examples of coumarin syntheses given in the reviews mentioned above and in the treatise on heterocyclic compounds ⟨B-51MI22400⟩, more recent studies have made use of the Perkin synthesis. These include the use of substituted phenoxyacetic acids to prepare 3-phenoxycoumarins ⟨78CI(L)628⟩ and the synthesis of chlorocoumarins from chlorosalicylaldehydes ⟨81T2613⟩. The use of DBU in place of sodium acetate was necessary to effect the ring closure of a number of *o*-hydroxyketones ⟨78BCJ1907⟩.

The Perkin synthesis of cinnamic acids is considered to involve reaction of the enolate anion derived from the acid anhydride with the aldehyde, giving rise to the alkoxide (**391**). Intramolecular acylation follows and the resulting β-acyloxy derivative undergoes elimination to the unsaturated acid (Scheme 125).

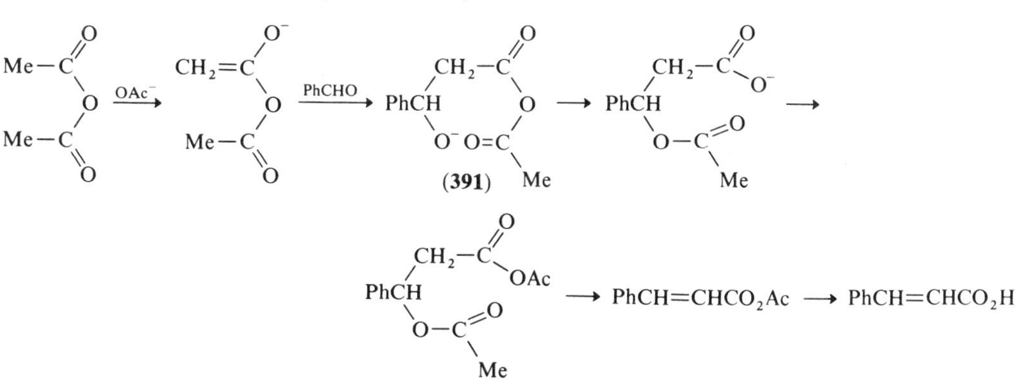

**Scheme 125**

When applied to an *o*-hydroxybenzaldehyde, initial acylation of the hydroxy group occurs. Subsequently, two modes of coumarin formation have been postulated involving either an intramolecular or an intermolecular condensation. The available evidence suggests that both processes may occur ⟨53JCS3435⟩.

3-Phenylcoumarin is formed when phenylacetylsalicylaldehyde (**392**; R = Ph) is heated with sodium acetate or phenylacetate, an intramolecular cyclization being followed by elimination (Scheme 126). However, acetylsalicylaldehyde gives only a trace of coumarin when heated with sodium acetate under a variety of conditions, indicating that an intramolecular reaction does not take place. Addition of acetic anhydride results in the formation of coumarin, presumably through enolate attack at the carbonyl group as above. Acylation, lactonization and elimination lead to the product (Scheme 126).

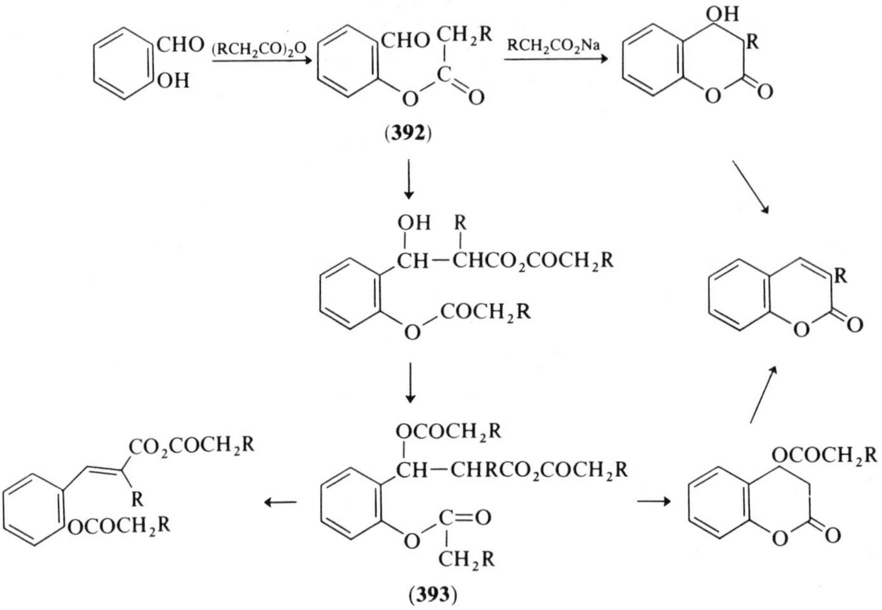

**Scheme 126**

Further support for the intermolecular route is provided by the isolation of the substituted *trans*-cinnamic acid, the normal Perkin product. This arises through elimination of carboxylic acid from the fully acylated species (393) rather than cyclization. For example, a considerable amount of 2-acetoxy-3-methoxycinnamic acid is formed from 3-methoxysalicylaldehyde, perhaps as a result of steric interference with cyclization so allowing the intermolecular process to predominate ⟨39JPR(152)23⟩. It should be noted that *trans*-2-hydroxycinnamic acids do not cyclize under normal Perkin conditions, though the *cis* isomer does so quite readily. Isomerization of the *trans* acid has been effected by treatment with a trace of iodine in acetic anhydride or by UV irradiation.

2,5-Dimethoxycinnamic acid is a good source of 6-hydroxycoumarin and the preparation illustrates the value of boron tribromide in the cleavage of ethers ⟨69MI22401⟩.

The strong conditions of the Perkin reaction may be circumvented by using phenylacetyl chloride in the presence of potassium carbonate in place of the anhydride in the preparation of 3-phenylcoumarins ⟨81S887⟩. However, 3-alkylcoumarins are not accessible in this way.

The reaction of 2,2-dimethyl-4,6-dioxo-1,3-dioxane (Meldrum's acid; 394) and its analogues with 3-dialkylaminophenols has been used to prepare some fluorescent 7-dialkylamino-4-hydroxycoumarins (Scheme 127) ⟨77M499⟩. The substituted malonic acid (395) or its dehydration product is a possible intermediate, since it is known that diaryl-malonic esters are accessible from phenols and the 1,3-dioxane. 3-Methoxyphenol yields the pyrano[3,2-c]benzopyran-2,5-dione (396) in this reaction, presumably by annelation of another ring on to initially formed 4-hydroxy-7-methoxycoumarin.

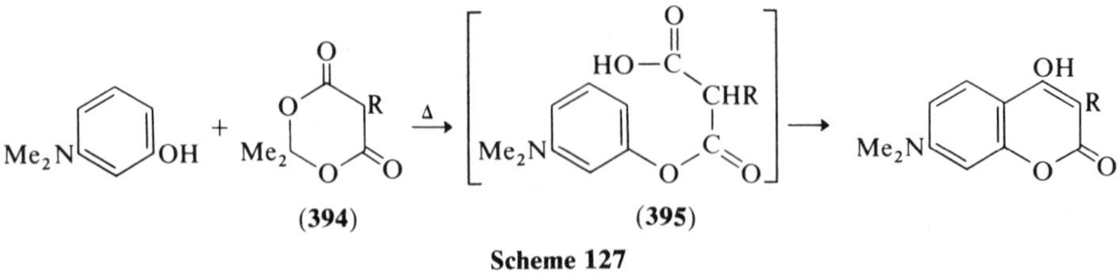

**Scheme 127**

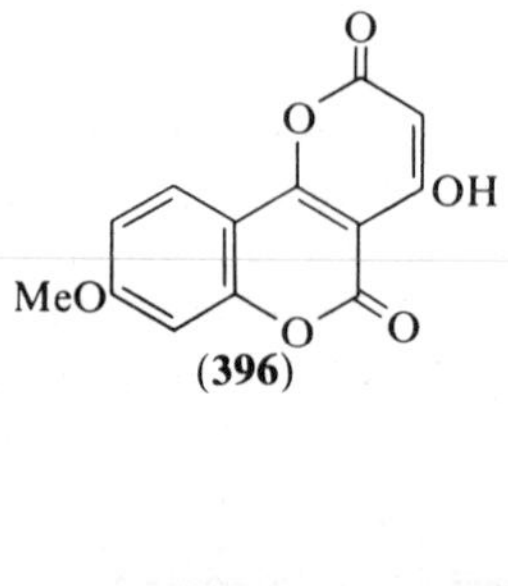

(396)

Aryl acetoacetates, which may be obtained by the reaction of phenols with diketene, are cyclized on treatment with sulfuric acid ⟨54JCS854⟩. The yields of coumarins are similar to those obtained by a Pechmann reaction on the phenol.

Treatment of phenyl cinnamates with aluminum chloride in chlorobenzene or without solvent yields coumarins ⟨75S739⟩. Since 3,4-dihydro-4-phenylcoumarins undergo dearylation to coumarins under similar conditions, these are possible intermediates. Indeed, using tetrachloroethane as solvent, dihydrocoumarins can be isolated.

The addition of a cumulated ylide to a salicylaldehyde leads to a coumarin (Scheme 128) ⟨76AG(E)115⟩.

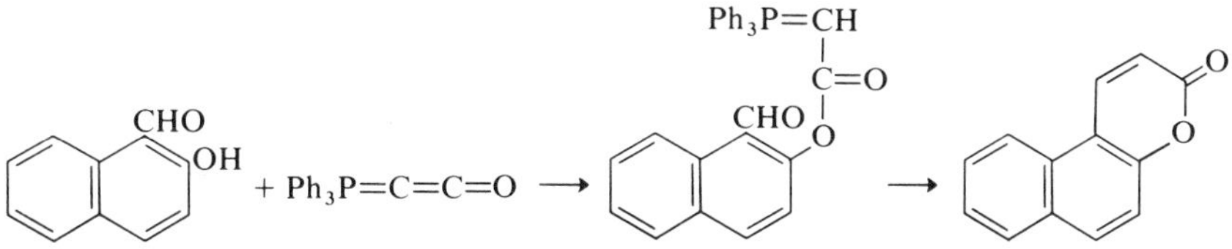

**Scheme 128**

### 2.24.3.2.2 From other heterocycles

#### (i) From oxetanes

The reaction of phenols with the spiro β-lactone (**397**), obtained from the reaction of diketene with ethyl diazoacetate, leads to coumarins ⟨79JCS(P1)525⟩. Initial ring opening of the spiro compound to the diketo ester followed by regioselective intramolecular acylation would seem to be a possible mechanism (Scheme 129).

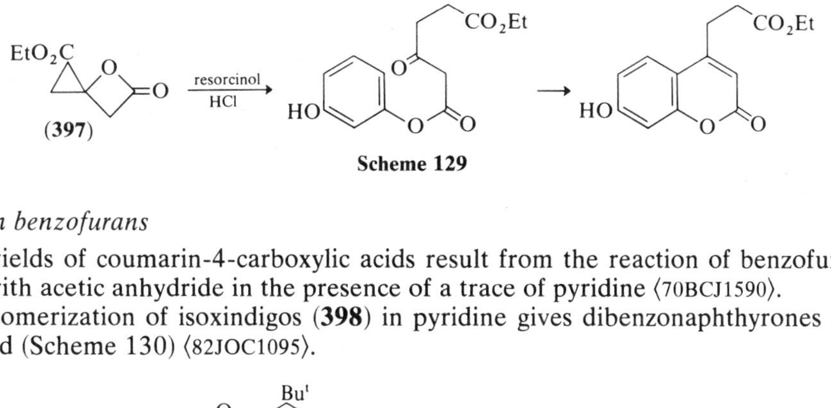

**Scheme 129**

#### (ii) From benzofurans

Low yields of coumarin-4-carboxylic acids result from the reaction of benzofuran-2,3-diones with acetic anhydride in the presence of a trace of pyridine ⟨70BCJ1590⟩.

The isomerization of isoxindigos (**398**) in pyridine gives dibenzonaphthyrones (**399**) in high yield (Scheme 130) ⟨82JOC1095⟩.

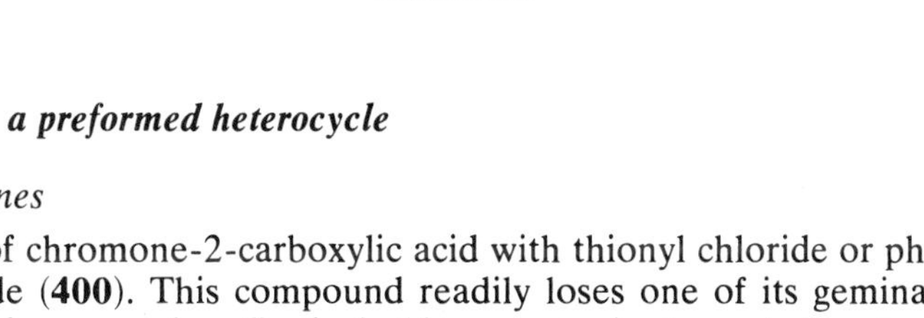

**Scheme 130**

### 2.24.3.2.3 From a preformed heterocycle

#### (i) From chromones

The reaction of chromone-2-carboxylic acid with thionyl chloride or phosphorus halides gives the trihalide (**400**). This compound readily loses one of its geminal chlorine atoms and with water, for example, affords 4-chlorocoumarin through the simultaneous loss of carbon monoxide (Scheme 131) ⟨63JGU1806⟩.

Treatment of the acid chloride (**401**) with sodium azide yields a 4-azidocoumarin ⟨73CR(C)(276)1603⟩ and acid hydrolysis of the carbamate (**402**) gives 4-hydroxycoumarin indicating that an amino group stabilizes a pyran-4-one structure better than does a hydroxy

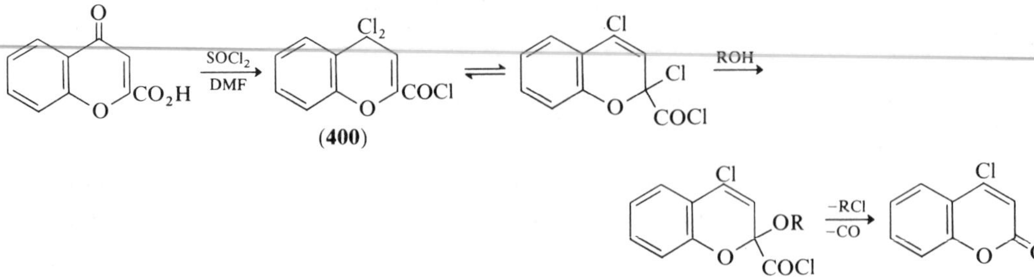

**Scheme 131**

group ⟨67CHE621⟩. An iminochromanone is a possible intermediate. In general, spectroscopic evidence favours the 4-hydroxycoumarin structure rather than the tautomeric chromone ⟨59SA870⟩.

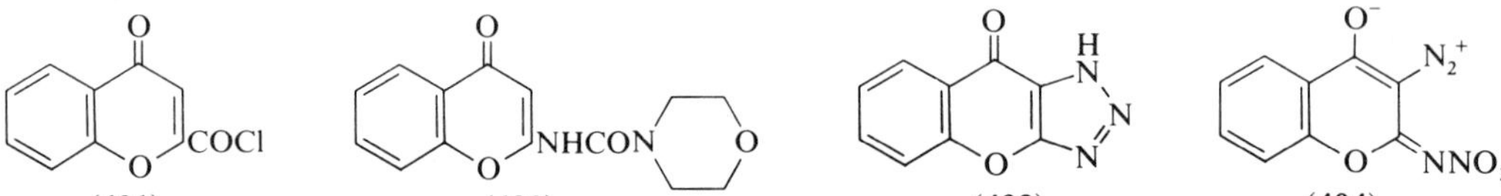

Nitration of 9-oxo-1H,9H-benzopyrano[2,3-d][1,2,3]triazole (**403**) does not occur in the benzenoid ring as expected, but instead gives the diazonitrimine derivative (**404**). It is postulated that this product arises from initial attack of nitric acid at N-3 followed by a facile ring cleavage, promoted by the electron-withdrawing nitro group. Hydrolysis in DMSO affords the coumarin ⟨82JHC129⟩.

### (ii) From dihydrocoumarins

Reduced coumarins are spontaneously converted to coumarins in a number of syntheses already mentioned. The following are examples of specific attempts to oxidize dihydrocoumarins.

Benzocoumarins (**405**) result from the dehydrogenation of the corresponding dihydrocoumarins by palladium–charcoal at 300 °C in the absence of solvent ⟨73CB62⟩, whilst 4-aryl-3,4-dihydrocoumarins are converted to the coumarins in diphenyl ether ⟨72IJC32⟩. This latter conversion also occurs under milder conditions using iodine and potassium acetate in acetic acid ⟨73AJC899⟩.

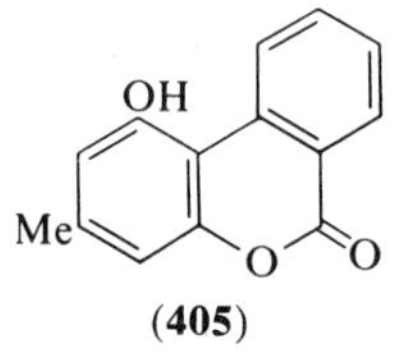

(**405**)

### (iii) From pyrylium salts

The oxidation of benzopyrylium perchlorate to coumarin is brought about by manganese dioxide in chloroform or acetonitrile ⟨68AC(R)251⟩.

### (iv) From chromenes

4-Arylcoumarins are produced by the oxidation of the corresponding chromenes. Several reagents bring about the conversion, but chromium(VI) oxide in pyridine at 55 °C appears to be the most efficient ⟨73JCS(P1)965⟩.

## 2.24.3.3 Pyran-4-ones

Although of some significance in natural product chemistry, attention seems to have centred on structural aspects of pyran-4-ones rather than on the development of synthetic routes. The relatively few methods which are available are, however, generally applicable and reliable. A review covers the earlier work on pyran-4-ones ⟨47CRV(41)525⟩, whilst the syntheses and properties of kojic acid have been reviewed ⟨82MI22402⟩.

### 2.24.3.3.1 *Formation of one bond*

(i) *Adjacent to the heteroatom*

Ring closure of 1,3,5-triketones leads to pyran-4-ones. Several routes are based on this reaction, differing in the approach to the carbonyl compound.

Aromatic esters may be used to aroylate the dianions derived from 1,3-diketones by reaction with potassium amide ⟨60JOC538⟩. Not only are acetyl and benzoyl acetones suitable for reaction, but alicyclic diketones and 2-hydroxyacetophenone are also acceptable. Cyclization of the triketones occurs in cold sulfuric acid, presumably *via* the enolic form and the hemiacetal (Scheme 132).

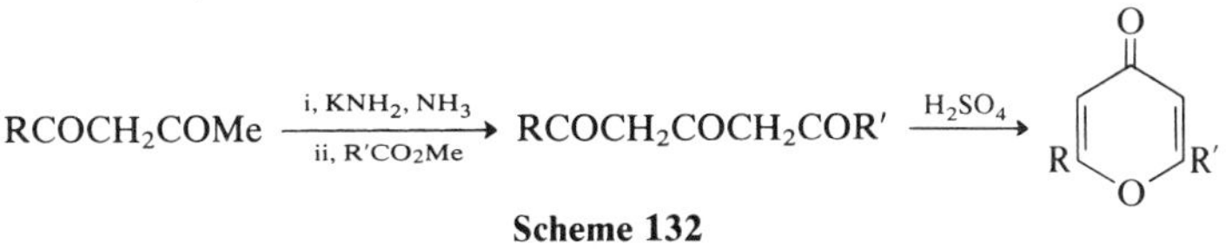

$$RCOCH_2COMe \xrightarrow[\text{ii, R'CO_2Me}]{\text{i, KNH_2, NH_3}} RCOCH_2COCH_2COR' \xrightarrow{H_2SO_4}$$

**Scheme 132**

1,3-Diketones substituted on the central carbon atom are less reactive under the above conditions. Satisfactory reaction occurs in pyridine or THF; in the latter solvent cyclization accompanies the acylation despite the basic conditions ⟨60JOC1110⟩.

Alternative reaction conditions are described in a synthesis of a 5,6-dihydronaphtho[1,2-*b*]pyran-4-one (**406**) ⟨81CI(L)876⟩; sodium hydride generates the anion from 2-acetyltetralone in DME.

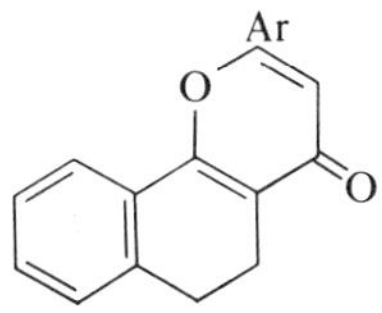

**(406)**

4-Oxopyran-2,6-dicarboxylic acid (chelidonic acid) is obtained from the base-catalyzed condensation of acetone with ethyl oxalate. The initial product is cyclized in acid ⟨43OSC(2)126⟩.

A triketone is also an intermediate in a synthesis of 6-methyl-4-oxopyran-2-carboxylic acid from pentane-2,3-dione in which one carbonyl group is initially protected as the ketal ⟨67JOC4105⟩. Acylation by ethyl oxalate is accomplished under basic conditions, and ring closure and hydrolysis of the ketal are effected with acid.

The 3,5,7-triketo acid $Me(COCH_2)_3CO_2H$, obtained from the metallation and subsequent carboxylation of the corresponding triketone, is cyclized to the pyranone ester with sulfuric acid in methanol ⟨76JA7733⟩.

The direct condensation of pentan-3-one with ethyl formate failed to give 3,5-dimethyl-pyran-4-one. Instead, the diketone was produced, the enolate of which is apparently of insufficient reactivity to undergo condensation with a second molecule of ester. However, the isopropyl derivative (**407**) of the enol condensed with ethyl formate, and the potential triketone cyclized to the pyran-4-one (**408**) under acid conditions (Scheme 133) ⟨64JOC2678⟩.

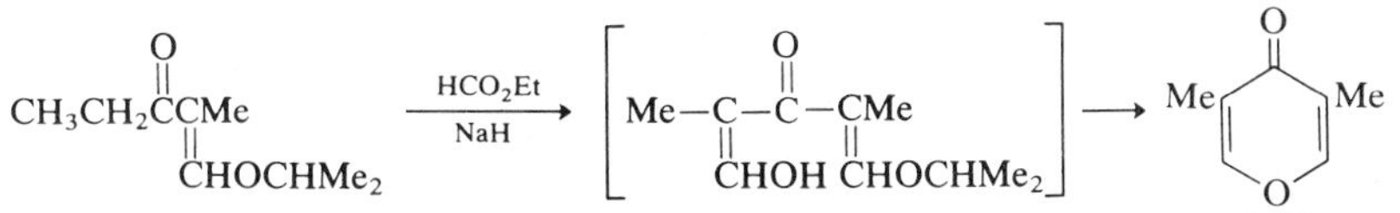

**Scheme 133**

**(407)**                                **(408)**

Formylation of dibenzylketone under Vilsmeier conditions occurs at both methylene groups; the tricarbonyl compound closes to give a pyranone which is unsubstituted at the 2- and 6-positions ⟨73ZC175⟩.

1,3-Diketones are converted into pyranones on treatment with malonyl dichloride, involving both *C*- and *O*-acylation ⟨63JCS4483⟩. The reaction between benzoylacetone and the acid chloride gives only one pyranone in accordance with a preferred direction of enolization consequent upon the unsymmetrical nature of the diketone; a 5-acetyl-6-phenylpyranone results rather than the 5-benzoyl-6-methyl isomer. There is still some uncertainty as to whether the product is the 4-hydroxypyran-2-one (**409**) or the tautomer, but the former is considered to be the more likely. There is yet further controversy, since the original workers ⟨63JCS4483⟩ considered the pyran-2-one to yield the pyran-4-one on heating, whereas a later study ⟨66M710⟩ proposed that a thermal rearrangement to 5-benzoyl-4-hydroxy-6-methylpyran-2-one (**410**) occurs (Scheme 134).

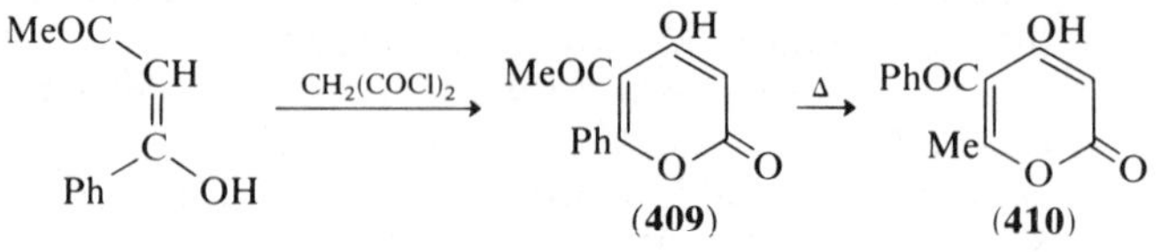

**Scheme 134**

*C*-Acylation of the lithium enolates derived from 4-methoxybut-3-en-2-ones is achieved with acid chlorides without any significant *O*-acylation. A general route to pyranones results which avoids the acidic conditions frequently associated with other synthetic methods ⟨80TL1197⟩. Cyclization of these products, which exist in an enolic form, occurs at room temperature in benzene in the presence of a trace of trifluoroacetic acid (Scheme 135).

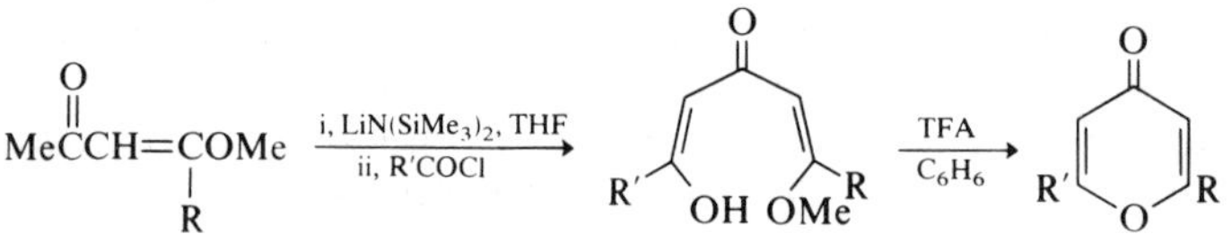

**Scheme 135**

A similar approach uses the potassium enolate and achieves acylation with anhydrides or acylimidazoles in addition to acid chlorides ⟨80TL2773⟩.

In the presence of base, alcohols and phenols add 1,5-diphenylpenta-1,4-diyn-3-one ⟨72JOU1398⟩. At 40 °C addition occurs at both triple bonds to give a mixture of the 1,5-dialkoxy compound (**411**), the triketone and the pyran-4-one. On treatment with acid the former compounds are converted into the pyranone. However, at 15 °C only one triple bond is attacked giving the alkoxyvinyl ethynyl ketone (**412**), but this also forms the pyranone with acid (Scheme 136).

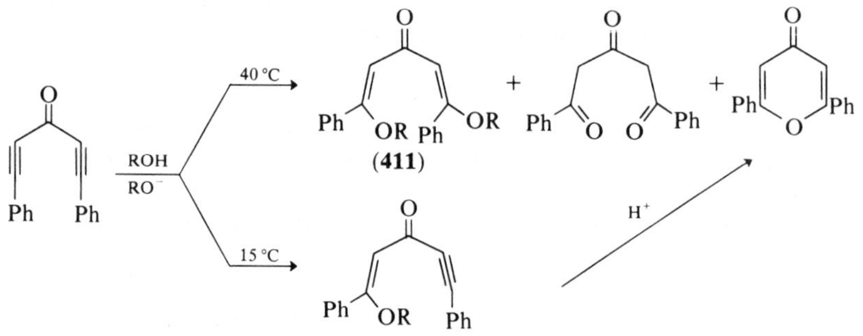

**Scheme 136**

The acylation of ketones by ethyl phenylpropynoate is accompanied by cyclization and is a useful route to pyran-4-ones ⟨54JCS1775, 74JHC1101⟩. The base-catalyzed reaction probably proceeds through an alkynic diketone formed by a Claisen condensation. The use of 1-tetralone gives the fused pyran-4-one (**413**) ⟨59JCS2588⟩.

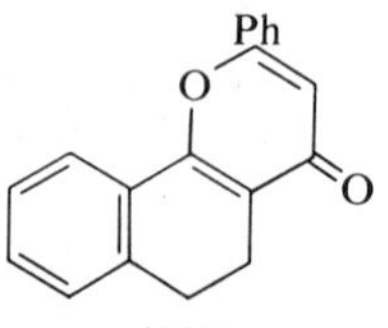

(**413**)

In a detailed study of the reaction of methylene ketones with carboxylic acids in the presence of polyphosphoric acid, pyran-4-ones were obtained from 1,3-diphenylpropan-2-one, phenylpropan-2-one and pentan-3-one ⟨61JA193⟩. In the first case, formation of 2,6-dimethylpyran-4-one (**414**; R = H) rather than the 2-benzyl-6-methyl-3-phenyl isomer

(**415**; R = H) was observed with acetic acid. However, with propionic acid both isomers (**414** and **415**; R = Me) were formed in approximately equal amounts. It therefore appears that an initial acylation $\alpha$ to the carbonyl group, leading to the diketone, is followed by a second attack, the site of which is subject to subtle preferences (Scheme 137).

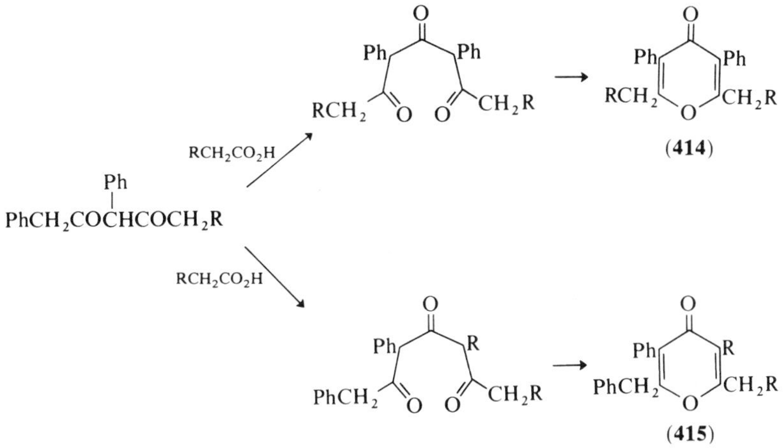

**Scheme 137**

A procedure for the synthesis of 2,6-dimethyl-3,5-diphenylpyran-4-one is based on a similar reaction with dibenzyl ketone ⟨73OSC(5)450⟩.

Although the ketone (**416**) gives a good yield of the pyran-4-one (**417**) with acetic acid (Scheme 138), the structurally related deoxybenzoin is converted into the fluorenone derivative (**418**) (equation 16). In the latter case intramolecular acylation occurs, presumably as a result of a favourable conformation of the carbonyl group. The former ketone cannot adopt such an array and hence is unable to cyclize in this manner ⟨61JA193⟩.

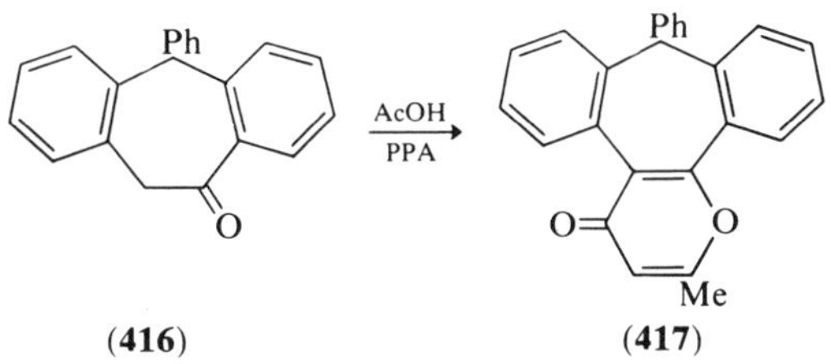

**Scheme 138**

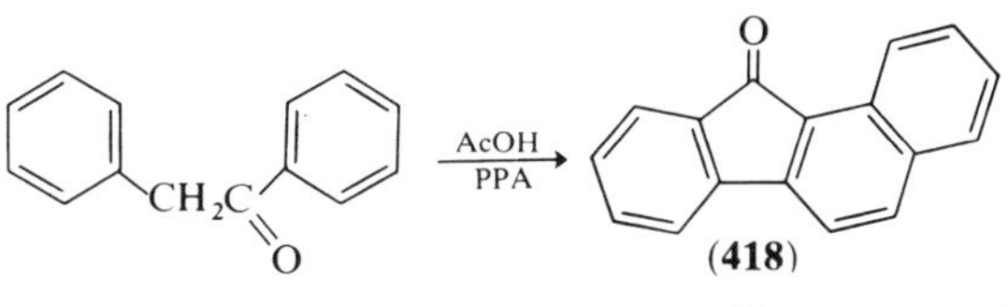

(16)

Morpholinocyclohex-1-ene is smoothly acylated by diketene to give a fused pyranone ⟨61CB486, B-68MI22400⟩. The initially formed diketone ring-closes and the pyranone results from elimination of morpholine. Using the cyclopentanone-derived enamine, much resinous material is formed and the yield of pyranone is reduced.

The reaction of diketene with some 4-substituted 3-oxobutanoate esters also provides a route to pyran-4-ones, though some attention to the reaction conditions is necessary to avoid the competitive formation of ethyl 2,4-dihydroxybenzoates ⟨79JCS(P1)529⟩. The two products are considered to arise from a common intermediate (**419**), cyclization of which can be envisaged through nucleophilic attack by an oxyanion or a carbanion (Scheme 139).

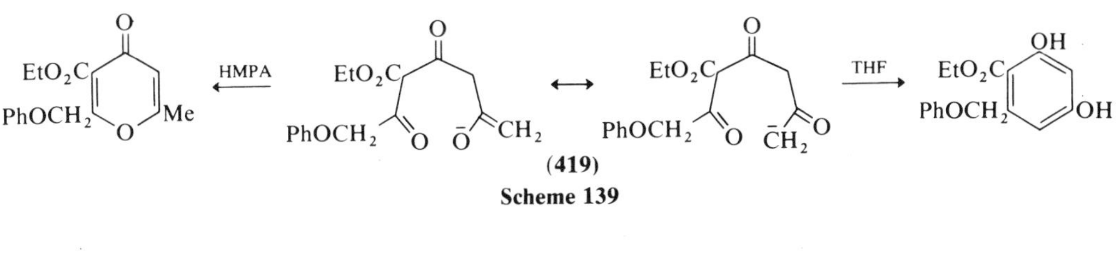

**Scheme 139**

Ynamines also react with diketene to produce mainly pyran-4-ones but with traces of pyran-2-ones. *O*-Alkylation of the dipolar ion (**420**) arising from ring opening of the diketene leads to the 2-aminopyran-4-one (Scheme 140) ⟨74BSF2086⟩.

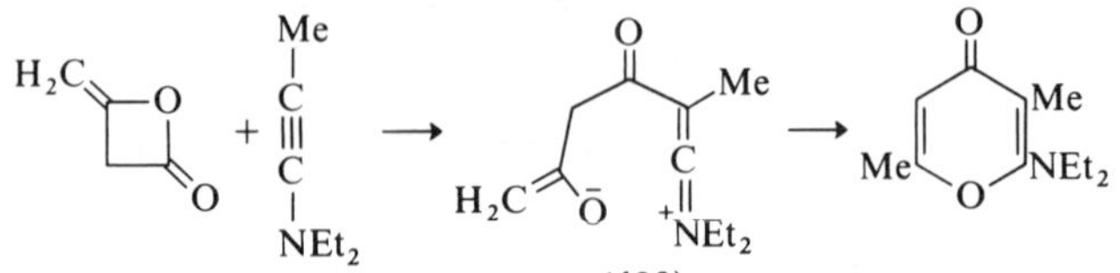

**Scheme 140**

### (ii) γ to the heteroatom

A convenient preparation of pyran-4-ones involves heating carboxylic acids or their anhydrides in polyphosphoric acid ⟨67JCS(C)828⟩. Yields are satisfactory, although when the synthesis is applied to a mixture of acids, a mixture of pyranones results. Only symmetrical pyranones (**421**) are formed, suggesting that the anhydride, rather than the acid, is the precursor of the heterocycle (Scheme 141).

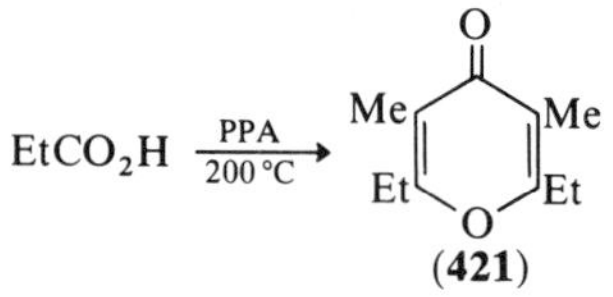

**Scheme 141**

### 2.24.3.3.2 Formation of two bonds

The cycloaddition of carbon dioxide to *N,N*-diethylaminophenylacetylene leads to aminopyran-4-ones ⟨72TL1131⟩. Initial cycloaddition to the ynamine probably forms (**422**) which rearranges to the ketene. A further cycloaddition, this time at the conjugated amide, leads to the pyran-4-one (Scheme 142). Supporting evidence for the proposed mechanism includes the formation of the pyranone from the ketene (**423**) ⟨72TL1135⟩.

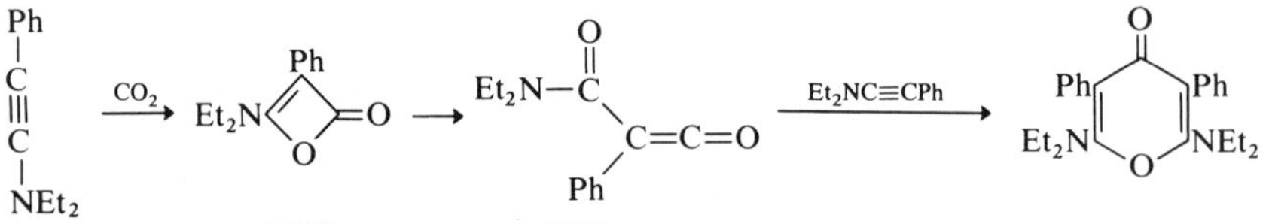

**Scheme 142**

Electron-rich conjugated dienes add across the carbon–oxygen double bond of ketene to give substituted dihydropyrans (**424**) which rearrange to pyran-4-ones (Scheme 143) ⟨82S500⟩.

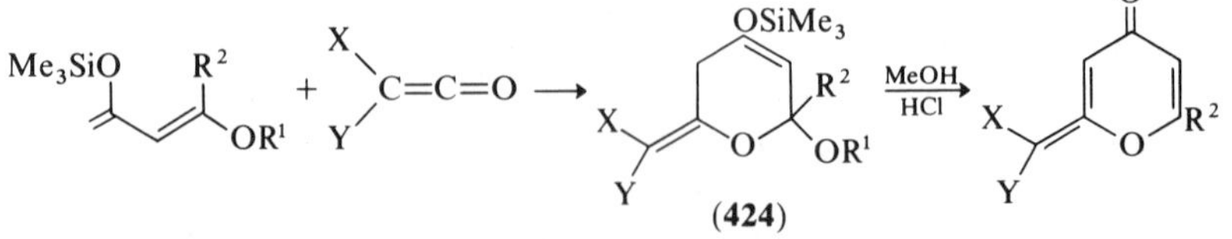

**Scheme 143**

### 2.24.3.3.3 From other heterocycles

### (i) From furans

3-Hydroxy-2-methylpyran-4-one (**427**), maltol, is an important flavouring agent and some consideration has been paid to its synthesis. A number of routes have been developed from furan derivatives. In general, these methods are based on the acid-catalyzed rearrangement of a dihydrofurfuryl alcohol (425) into a dihydropyran-3-one (426) ⟨71T1973⟩. The synthesis from furan-2-aldehyde is outlined in Scheme 144 ⟨76TL1363⟩. Although this is a multistep sequence, the overall yield is about 20% and generally the reactions are straightforward to carry out. It is possible to convert the furfuryl alcohol directly to the 6-alkoxypyran-3-one (426) ⟨77MI22401⟩.

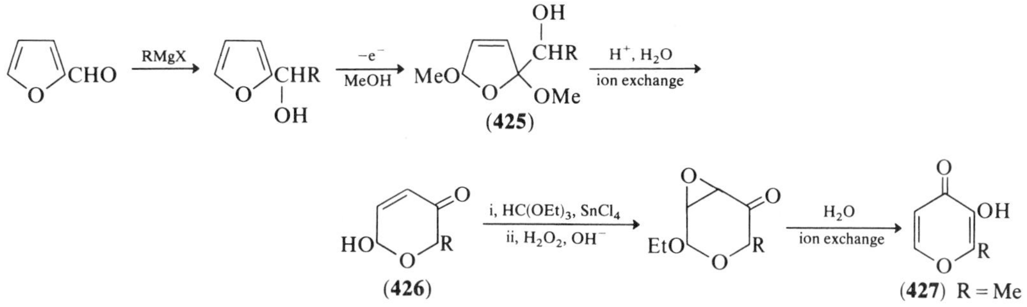

**Scheme 144**

An alternative approach to the oxidative rearrangement uses chlorine as the oxidant in a one-pot synthesis ⟨80JOC1109⟩. Not only does this reagent convert the furan to the pyran-3-one, but it takes the latter to the 4-halogeno derivative *via* the labile 3,4-dihalogeno compound. The pyran-4-one is obtained by an *in situ* acid-catalyzed hydrolysis. The overall yield of maltol is over 60%, and ethylmaltol is obtained in a similar conversion from ethylfurfuryl alcohol.

The formation of maltol from 2-acetyl-3-methoxyfuran *via* isomaltol *O*-methyl ether (428) involves a related method.

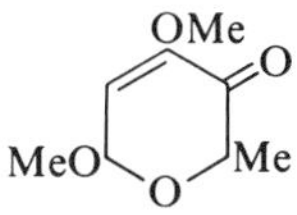

(428)

The tetraketone (348), which is tautomeric with the furanone (349), is converted to the pyranone by oxidation with iodobenzene diacetate (Scheme 109) ⟨80TL1575⟩.

### (ii) *From oxazoles*

The expected adduct (430) from the Diels–Alder reaction of the oxazole (429) with diphenylcyclopropenone could not be isolated (Scheme 145) ⟨70JCS(C)552⟩. Instead the pyran-4-one (431) is obtained, resulting from elimination of acetonitrile. This process is essentially irreversible because the pyranone lacks diene properties and nitriles are poor dienophiles.

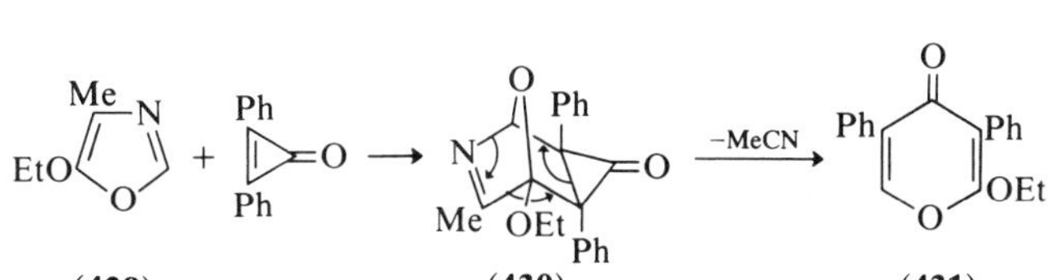

**Scheme 145**

### 2.24.3.3.4 *From a preformed heterocyclic ring*

#### (i) *From reduced pyran-4-ones*

The aromatization of 3,5-dibenzylidenetetrahydropyran-4-ones is achieved in boiling diethylene glycol solution in the presence of palladized charcoal ⟨57JA156⟩.

A multistage conversion of tetrahydropyran-4-one into 3-hydroxypyran-4-one has been described ⟨79JCS(P1)1806⟩.

### (ii) *From pyran-2-ones*

The lactone ring of 4-hydroxypyran-2-ones is cleaved by strong acids and the acyclic intermediate recyclizes to a pyran-4-one (Scheme 146) ⟨70T5255⟩.

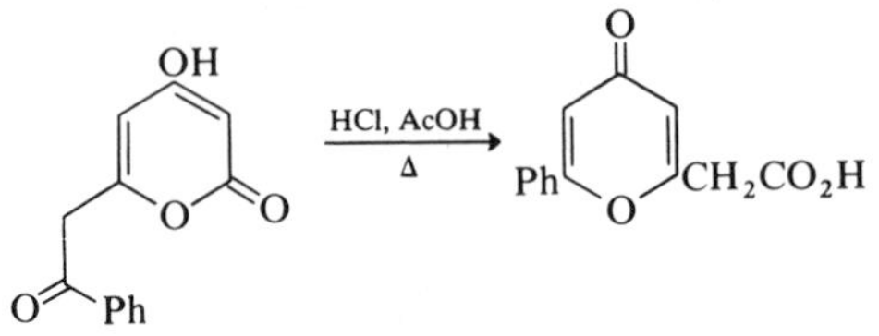

**Scheme 146**

## 2.24.3.4 Chromones (4*H*-1-Benzopyran-4-ones)

The two major precursors of chromones are 2-hydroxyacetophenones and phenols. In both instances, a side-chain is built on to the substrate and the resulting product is subsequently cyclized; many syntheses differ only in the source of this side-chain. Although the intermediate is not always isolated it is appropriate to consider these chromone syntheses as involving the formation of only one bond.

In addition to the surveys in Elderfield ⟨B-51MI22400⟩ and Rodd ⟨B-77MI22400⟩, the chemistry of chromones has been extensively discussed ⟨77HC(31)⟩ and separate accounts of flavone chemistry are available ⟨B-62MI22400, B-75MI22400⟩. Chromones with fused heterocyclic rings containing nitrogen have been reviewed ⟨73MI22404⟩.

### *2.24.3.4.1 Formation of one bond*

#### (i) *Adjacent to the heteroatom*

2-Hydroxyacetophenone requires an additional carbon atom, which will become C-2 of the heterocyclic product, before cyclization to the chromone can be effected. Direct *C*-formylation is not easy and the following syntheses illustrate the various techniques used to introduce this fragment.

Upon treatment with a base, the activated methyl group in 2-hydroxyacetophenone is converted into a carbanion. Claisen condensation with an ester gives rise to a 1,3-diketone, the sodium salt of which may be isolated although this process is usually unnecessary. Cyclization to the chromone occurs readily in acid solution (Scheme 147).

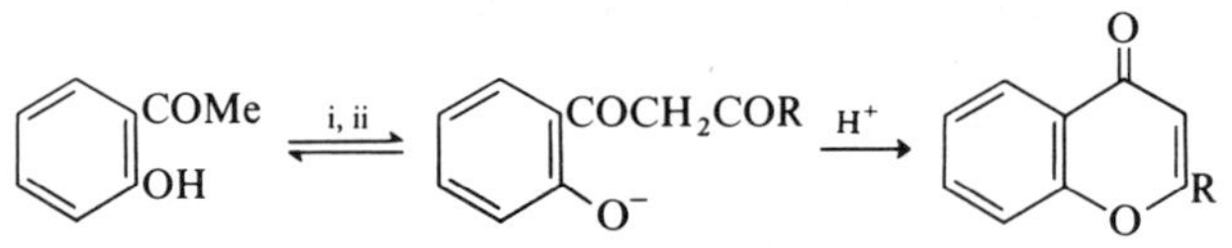

i, base; ii, RCO₂Et

**Scheme 147**

A wide variety of chromones has been prepared by this method since its introduction ⟨01CB2475⟩, and the literature abounds with examples. A fully detailed preparation of 3-ethylchromone is available ⟨55OSC(3)387⟩. The presence of substituents in the aromatic ring of the acetophenone has minimal effect on the course of the reaction; both electron-releasing and electron-withdrawing substituents are compatible with the synthesis.

The synthesis has been extended to hydroxyketones derived from polycyclic hydrocarbons ⟨39JA1, 51JA3514⟩, steroids ⟨71T711⟩ and heterocycles ⟨70JCS(C)1553⟩, as exemplified in formulae (**432**)–(**434**), and the reaction of 7-acetyl-8-hydroxychroman and other isomers with a

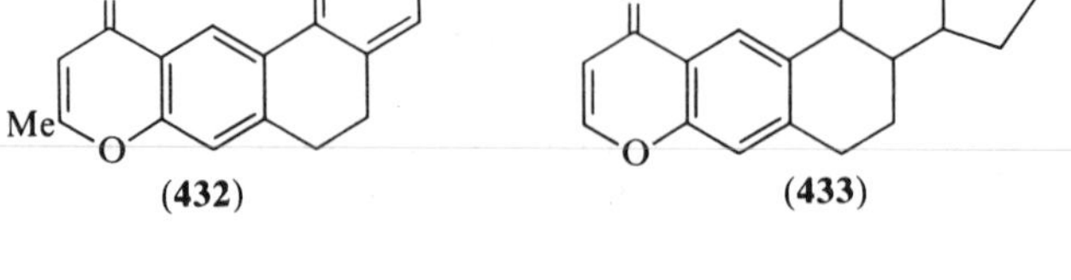

(**432**)                                        (**433**)

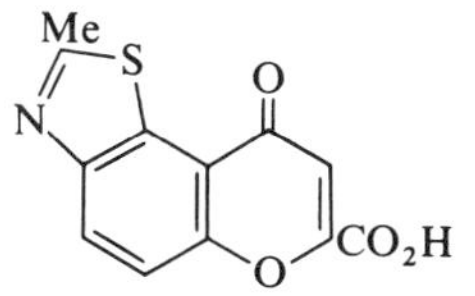

(434) 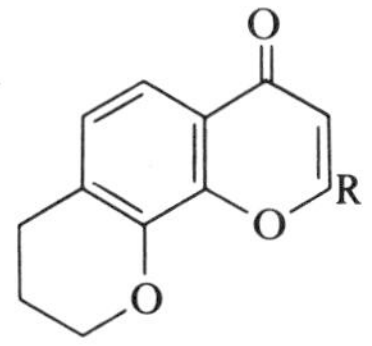     (435)

variety of esters has been used to synthesize a range of dihydropyranochromones (435) ⟨63JCS5426, 65JCS3882⟩.

Naturally, less scope is available with the ester component and the C-2 substituent in the chromone is commonly alkyl or ethoxycarbonyl, or is absent. The use of diethyl oxalate offers an added attraction; the intermediate 1,3-diketo ester (436) readily undergoes transesterification, thereby allowing the synthesis of various chromone-2-carboxylic esters to be achieved from the one starting material (Scheme 148) ⟨74JCS(P1)2570⟩.

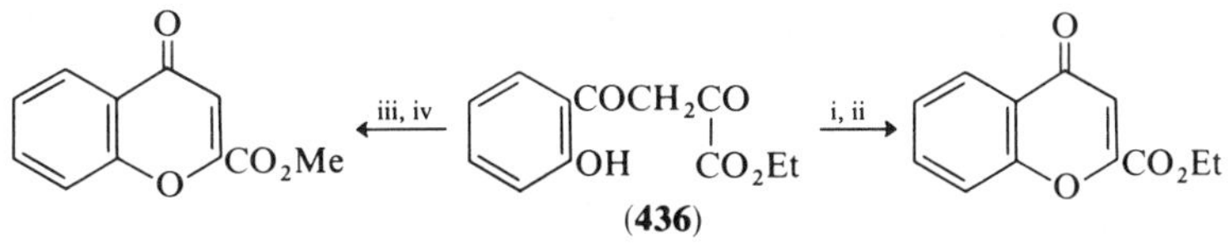

(436)

i, EtOH; ii, H$_2$SO$_4$; iii, MeOH, H$_2$SO$_4$, 0 °C; iv, MeOH, HCl, boil

**Scheme 148**

Diethyl oxalate has been used in the synthesis of some bischromone carboxylic acids ⟨72JMC583⟩.

Not unexpectedly, a substituent on the acetyl group of the 2-hydroxyacetophenone has an effect on the condensation, though successful preparations of 3-substituted chromones have been achieved. A neat alternative approach to such chromones involves alkylation of the intermediate diketone and subsequent cyclization ⟨34JCS1311⟩.

A wide variety of reactants and reaction conditions can be used to prepare chromones *via* a Claisen condensation and the review by Ellis contains a comprehensive list of examples ⟨77HC(31)496⟩.

An alternative source of the 1,3-diketone intermediate encountered in the previous route is an acyloxyacylbenzene, readily available by *O*-acylation of a 2-hydroxyacetophenone. Treatment with potassium carbonate initiates an intramolecular rearrangement in which the acyl moiety migrates from oxygen to the carbon atom α to the carbonyl of the other acyl group. This is the well-known Baker–Venkataraman rearrangement. The migrating acyl group may be aliphatic or aromatic and hence this approach is of value in the synthesis of flavones as well as chromones. Full experimental details for the preparation of flavone itself are available (Scheme 149) ⟨63OSC(4)478⟩.

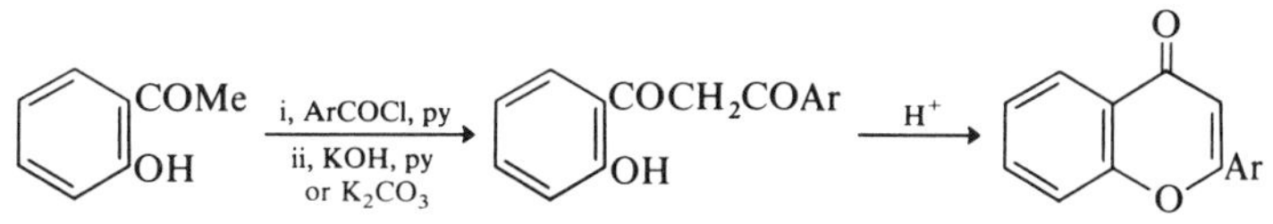

**Scheme 149**

A series of 1,3-diketones derived from the benzoyl derivatives of 8-acetyl-9-hydroxy-naphtho[2,1-*b*]pyran-3-one (437) and similar compounds have been cyclized to the pyran-4-one (438; Scheme 150) ⟨80MI22402⟩.

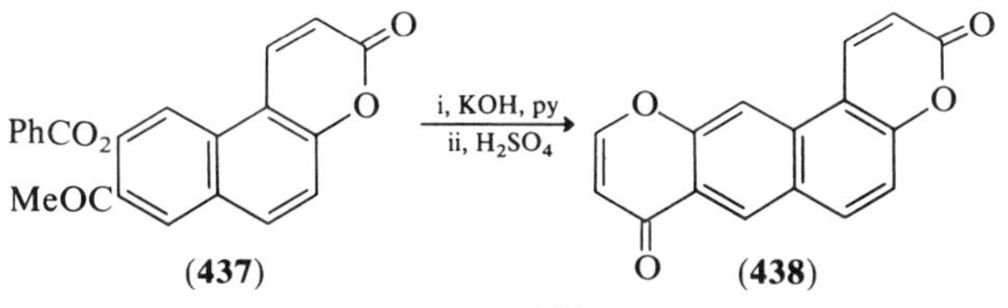

(437)     (438)

**Scheme 150**

Acetylation of the deoxybenzoin (439) under mild conditions gives the isoflavone directly rather than the acetyl derivative of the phenol (Scheme 151) ⟨77MI22402⟩.

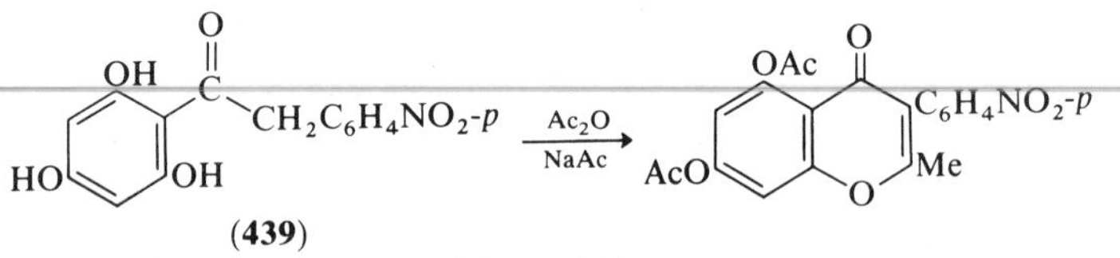

**Scheme 151**

The other variables may be altered to the same wide extent as in the route based on the Claisen condensation. The rearrangement also occurs under the influence of other basic catalysts, including sodium metal, sodium hydride and sodium hydroxide, and indeed it has been suggested that these strongly basic species are more effective than potassium carbonate. For instance, the diacetyl derivative of 2,4-dihydroxyacetophenone did not rearrange in the presence of potassium carbonate ⟨33JCS1381⟩ but did so on treatment with sodium in benzene ⟨49PIA(A)(30)57⟩.

Isolation of the acyloxyacetophenone is unnecessary; the 2-hydroxyacetophenone may be refluxed with an aroyl chloride in acetone containing potassium carbonate whereupon the 2-hydroxydibenzoylmethanes are produced ⟨57MI22400⟩. A further improvement has been reported in which the hydroxyacetophenone and aroyl chloride are boiled in benzene with aqueous sodium hydroxide solution in the presence of a phase transfer catalyst ⟨82S221⟩. The same work reports the use of *p*-toluenesulfonic acid as the cyclizing catalyst.

Several groups have provided experimental evidence to support the nature of the Baker–Venkataraman rearrangement, notable amongst which is the rearrangement of a mixture of the two acylated hydroxyacetophenones (**440**) and (**441**). No crossed products could be detected, precluding the intermediacy of free acylium ions and an intermolecular rearrangement. The resulting flavones (**442**) and (**443**) were formed by intramolecular rearrangements prior to cyclization (Scheme 152) ⟨50JCS1925⟩.

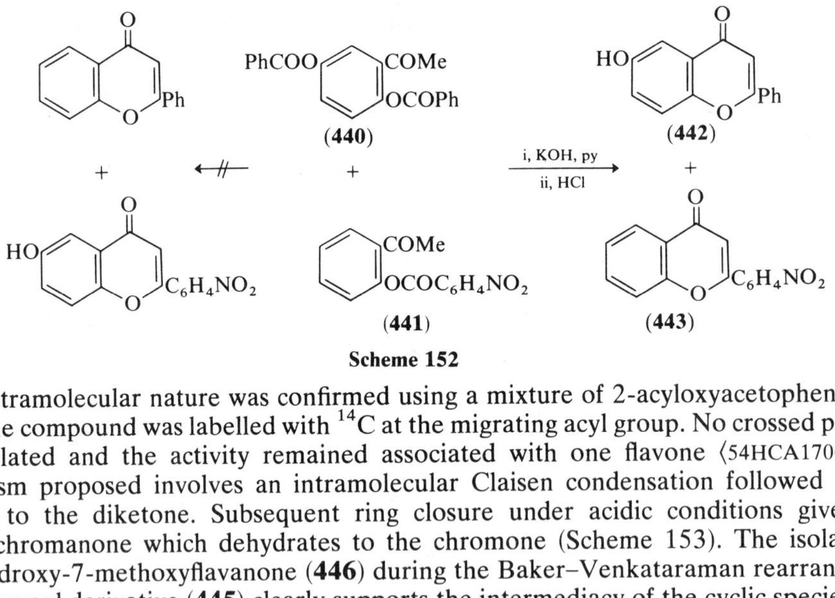

**Scheme 152**

The intramolecular nature was confirmed using a mixture of 2-acyloxyacetophenones in which one compound was labelled with $^{14}$C at the migrating acyl group. No crossed products were isolated and the activity remained associated with one flavone ⟨54HCA1706⟩. The mechanism proposed involves an intramolecular Claisen condensation followed by ring opening to the diketone. Subsequent ring closure under acidic conditions gives a 2-hydroxychromanone which dehydrates to the chromone (Scheme 153). The isolation of 2,5-dihydroxy-7-methoxyflavanone (**446**) during the Baker–Venkataraman rearrangement of the benzoyl derivative (**445**) clearly supports the intermediacy of the cyclic species (**444**; Scheme 154) ⟨72CC107⟩.

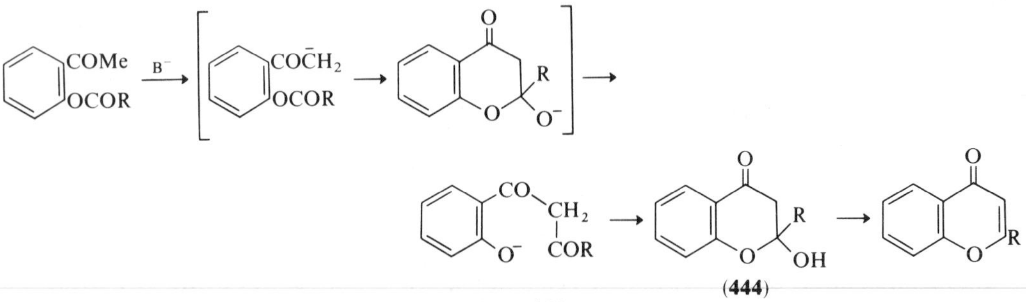

**Scheme 153**

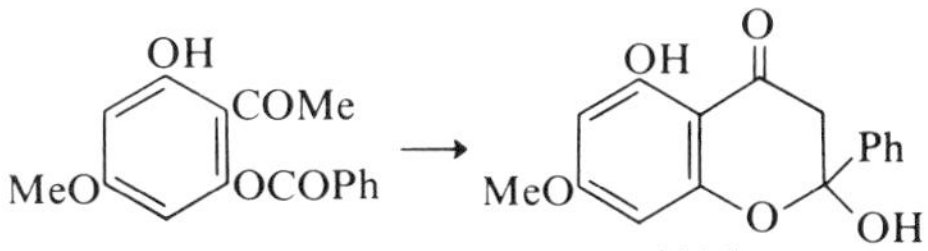

(445) → (446)

**Scheme 154**

In the Kostanecki–Robinson synthesis of chromones and flavones, a 2-hydroxyaceto-phenone is heated with the anhydride and the sodium salt of an aliphatic acid ⟨01CB102, 24JCS2192⟩. Alkaline cleavage of the 3-acyl group is easy since it is part of a 1,3-diketone system and may occur during isolation of the product (Scheme 155). In some instances a mixture of both products results ⟨36JCS215⟩. Nevertheless, the reaction has been widely used and is an important synthetic method for flavones. Almost any group may be present in the aromatic ring, provided that it does not react under the experimental conditions, and chromones containing alkyl, acyl, alkoxy, halogen, nitro and cyano groups, amongst others, have been synthesized by this route. Hydroxy groups are usually acylated ⟨61JCS702⟩, although hydrolysis back to the free hydroxy group sometimes occurs during the work-up procedure ⟨25JCS2349⟩, an ester group may be hydrolyzed or removed ⟨60JIC491⟩, and an alkoxy group may be dealkylated ⟨53PIA(A)(37)774⟩.

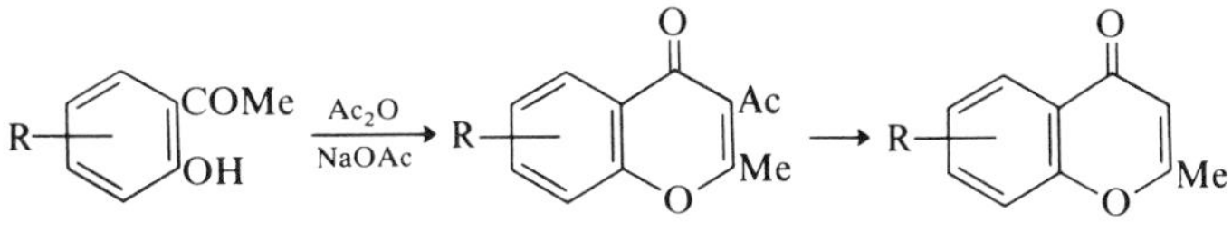

**Scheme 155**

Dealkylation may be accompanied by a Wessely–Moser rearrangement ⟨55CI(L)271⟩ and the product is an isomer of the expected chromone. When a 5-methoxychromone is heated with hydriodic acid and acetic anhydride, the pyrone ring may open. Subsequent cyclization may involve the original hydroxy group or that arising from dealkylation of the methoxy group (Scheme 156) ⟨67JCS(C)145⟩.

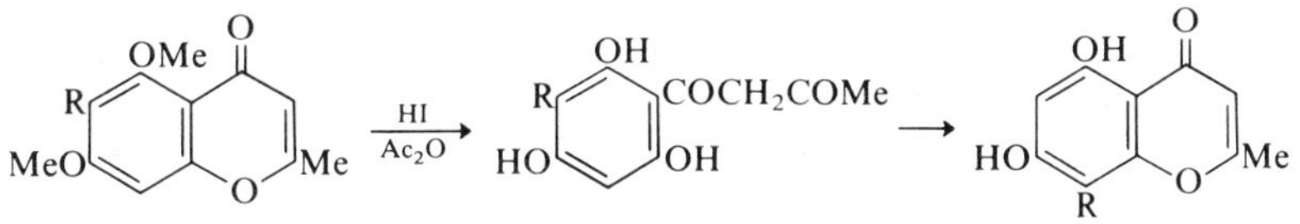

**Scheme 156**

Where the substitution pattern of the acetophenone is such that the formation of isomeric chromones can be envisaged, the reaction does not appear to show any regioselectivity (Scheme 157) ⟨71JCS(C)1324⟩.

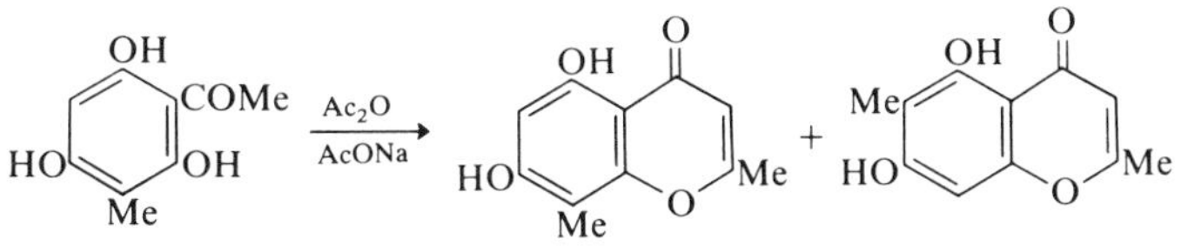

**Scheme 157**

The presence of a substituent on the original methyl group of the hydroxyacetophenone prevents formation of the labile 3-acylchromone but otherwise has little effect provided the substituent is not solvolyzed. A number of 3-substituted chromones have been obtained in this manner ⟨55JOC38⟩.

Some improvement in the often moderate yields associated with the Kostanecki–Robinson synthesis has been achieved by substitution of a catalytic amount of triethylamine for the stoichiometric quantity of sodium carboxylate ⟨44CB202⟩. It is also possible to work at somewhat lower temperatures using this technique, an obvious advantage when sensitive groups are present. The modification is particularly attractive in the synthesis of complex hydroxyflavones ⟨77T1405, 1411⟩. A further development, the use of a tertiary amine as a solvent, has been advocated ⟨78JOC2344⟩. For instance, in the synthesis of 5,7-dihydroxy-3-methoxyflavone (galangin 3-methyl ether) (**447**) from ω-methoxyphloracetophenone, the yield of the flavone decreased as the boiling point of the solvent increased. The maximum yield resulted using just sufficient triethylamine to maintain the reactants in solution. The reaction temperature was some 100 °C lower than under normal Kostanecki–Robinson conditions.

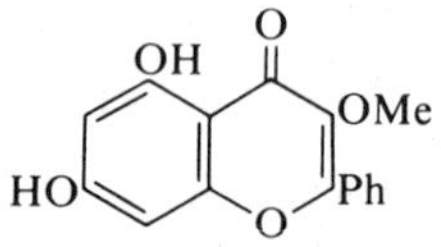

(447)

The use of the very reactive mixed acetic formic anhydride at room temperature allows the synthesis of chromones unsubstituted at C-2. Especially good yields result when an electron-withdrawing group is present in the acyl moiety of the ketone ⟨78JCR(M)0865⟩.

When deuterated acetic anhydride and sodium acetate are used to cyclize the 1,3-diketone, the deuterated chromone is formed with >95% incorporation of the isotope (Scheme 158) ⟨75JCS(P1)1845⟩.

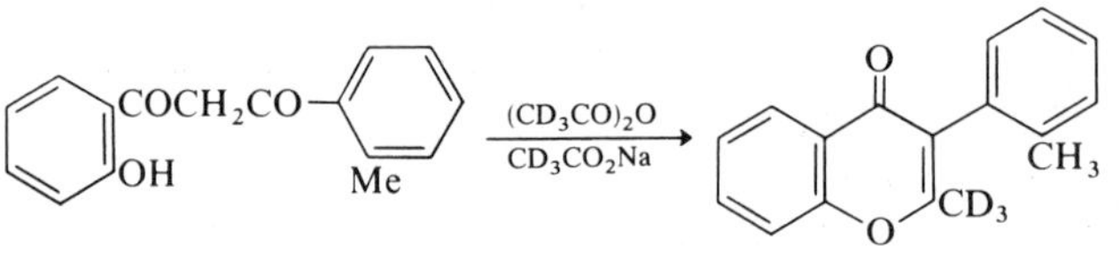

**Scheme 158**

Such is the wide application of this reaction to the preparation of chromones and flavones, only the more important synthetic features have been mentioned. Detailed information is available in the literature ⟨77HC(31)515⟩ and the earlier work has been reviewed ⟨54OR(8)59⟩.

The Kostanecki–Robinson reaction proceeds through *O*-acylation followed by a Baker–Venkataraman rearrangement to the 1,3-diketone. Cyclization then yields the chromone. Early evidence was based on the observation that both of the 1,3-diketones (**448**) and (**449**) yielded the same chromone on reaction with the appropriate acid anhydride ⟨33JCS1381⟩. Thus, the reactions were considered to proceed through the common intermediate (**450**; Scheme 159).

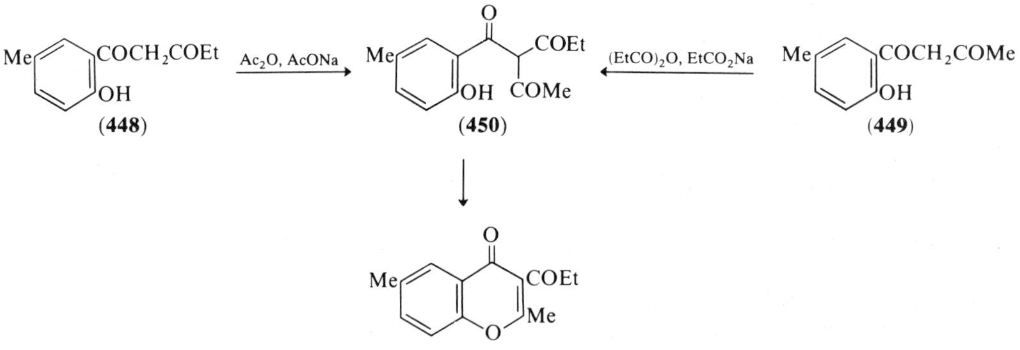

**Scheme 159**

It has since been shown that the enol ester (**451**) is an intermediate in the synthesis ⟨69T715⟩. Indeed such esters readily form chromones on treatment with alkali and the *ortho* acyloxy group becomes part of the pyranone ring as a result of a Baker–Venkataraman rearrangement (Scheme 160) ⟨69T707⟩.

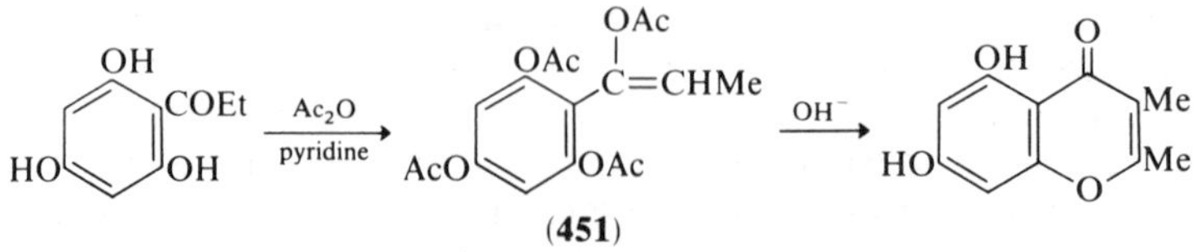

(451)

**Scheme 160**

One problem associated with the K–R synthesis of chromones is the simultaneous formation of coumarins which is sometimes observed, even with simple hydroxyacetoph-enones ⟨24CB88⟩. It is therefore necessary to exercise caution with structural assignments of products from reactions in the early literature, since a number of erroneous structures have been reported. There seems to be no pattern to the extent to which the two benzopyranones are produced; on occasions little or no coumarin is formed, whilst in other cases it is the major product ⟨34JCS1311⟩. The use of propanoic and higher anhydrides favours the formation of coumarins and this effect is accentuated if alkyl groups other than methyl form part of the hydroxyketone ⟨34JCS1581⟩.

The coumarin is probably derived from the initial acylated hydroxyacetophenone which, in addition to undergoing a Baker–Venkataraman rearrangement, may cyclize through an intramolecular aldol condensation. Elimination of water then gives rise to the coumarin (Scheme 161).

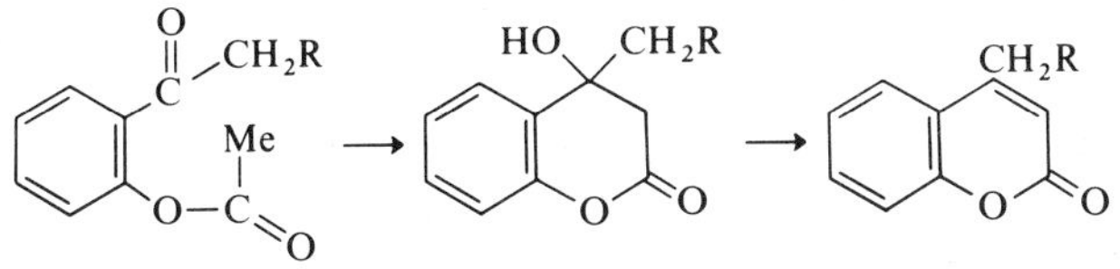

**Scheme 161**

2-Hydroxyacetophenone reacts with DMF under Vilsmeier conditions to yield chromone-3-carbaldehyde ⟨72LA(765)8⟩. The reaction appears to be generally applicable, various substituents being acceptable in the aromatic ring ⟨74T3553⟩. Furthermore, acetylhydroxynaphthalenes yield the corresponding benzochromones, and the pyranochromone (**452**) is formed from the appropriately substituted coumarin.

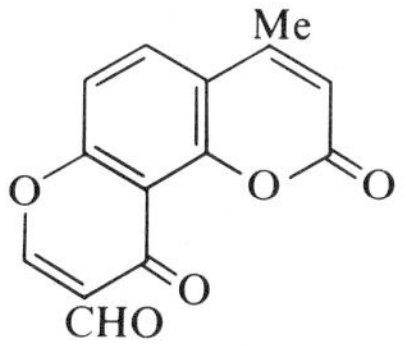

(**452**)

In the absence of phosphorus oxychloride, DMF or its acetal yields chromone itself with 2-hydroxyacetophenone, *via* the enamine ⟨71CB348⟩.

The same one-carbon unit has been employed in a convenient isoflavone synthesis from deoxybenzoins ⟨76S326⟩. Other methods of ring closure of phenyl benzyl ketones to this heterocycle have been reviewed ⟨B-62MI22400⟩ and include triethyl orthoformate ⟨56PIA(A)(44)36⟩, zinc cyanide ⟨58CB2858⟩, ethyl formate ⟨80BCJ831⟩ and ethoxalyl chloride ⟨70JCS(C)1219⟩.

In the presence of boron trifluoride and methanesulfonyl chloride, polyhydric acetophenones react with DMF to give 3-substituted chromones. It is postulated that ring formylation is prevented through deactivation by complex formation between the substrate and boron trifluoride ⟨76CC78⟩.

Salicylic acid derivatives serve a similar purpose to 2-hydroxyacetophenones in a number of chromone syntheses, acting as a precursor of the 1,3-diketone fragment. For instance, a Claisen reaction between methyl 2-methoxybenzoate and acetone takes place in the presence of sodium to give the diketone. Demethylation occurs on reaction with hydriodic acid with concomitant ring closure to the chromone ⟨00CB1998⟩. The corresponding naphthol derivatives are a source of benzochromones ⟨52JOC1419⟩.

The substituted *o*-hydroxyacetophenone (**453**) derived from the reaction of ethyl salicylate with sodium methylsulfinylmethide has been used as a source of chromones. Cyclization to 3-methylsulfinylchromone (**454**) occurs with triethyl orthoformate, whilst flavone is formed from the reaction with benzaldehyde (Scheme 162) ⟨72JHC171⟩. The reaction with formaldehyde gives different products according to the molar ratio of the reactants. With equimolar quantities a chromanone is formed which is converted to chromone on heating. However, the use of two moles of the aldehyde leads to 3-hydroxymethylchromone (**455**), *via* the chromanone, which is itself a useful precursor of other 3-substituted chromones ⟨74JHC183⟩. A range of substituents are compatible with the synthesis, and similar results are obtained with the analogous naphthalene and quinoline derivatives.

3-Hydroxymethylchromones also arise from the reaction between an aldehyde and 3-(2-hydroxyphenyl)-3-oxopropanal in the presence of a base ⟨74AP12⟩.

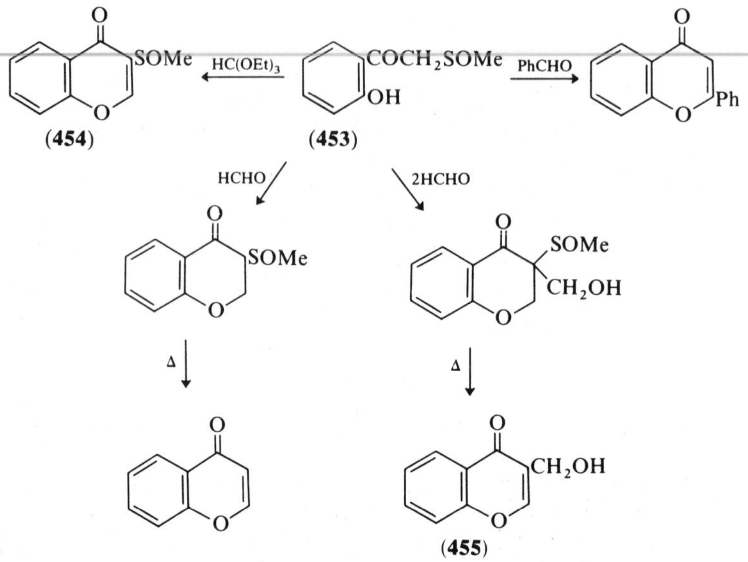

**Scheme 162**

The requisite 1,3-diketones have been prepared by the condensation of acid chlorides with lithium enolates at −70 °C. The technique has proved of value in the synthesis of cycloalkyl fused chromones (**456**) ⟨78JCS(P1)726⟩.

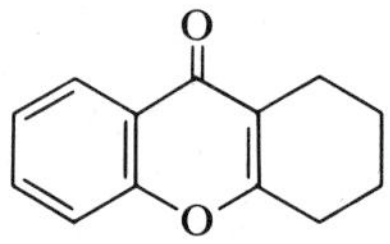

**(456)**

A chlorocarbonyl group *ortho* to a fluorine atom activates the halogen towards nucleophilic displacement. In the synthesis of a number of polyfluorinated chromones the 1,3-diketo side-chain is introduced using ethyl acetoacetate in the presence of magnesium ethoxide. Cyclization occurs on addition of sulfuric acid ⟨70JOC930⟩.

Several groups have described syntheses of chromones which involve the use of enamines. Reaction of salicylaldehyde with 1-morpholinocyclohexene gives the chromanol which is readily oxidized to the chromone ⟨66JOC1232⟩. Other examples include the use of 1-morpholino-1-phenylethylene which gives flavene, whilst *N*-styrylmorpholine yields isoflavone. The enamine reaction is considered to proceed in the normal manner with subsequent cyclization involving the neighbouring phenolic group (Scheme 163).

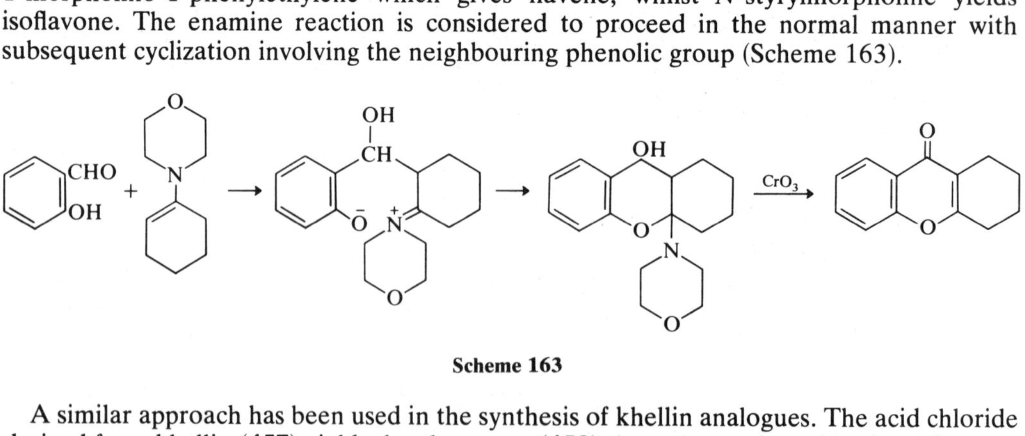

**Scheme 163**

A similar approach has been used in the synthesis of khellin analogues. The acid chloride derived from khellin (**457**) yields the chromone (**458**) through reaction with piperidinocycloalkenes and subsequent hydrolysis with acid. Some demethylation is observed and the amide (**459**) is formed in small amounts (Scheme 164) ⟨78JCS(P1)726⟩.

Enamines also react with diketene to form chromones ⟨60JCS26⟩. One drawback to this method is the use of 2,3-dihydro-*N*,*N*-dimethyl-*p*-toluidine, prepared by Birch reduction of the aromatic amine, and the consequent need to aromatize the reduced chromone which is the initial product.

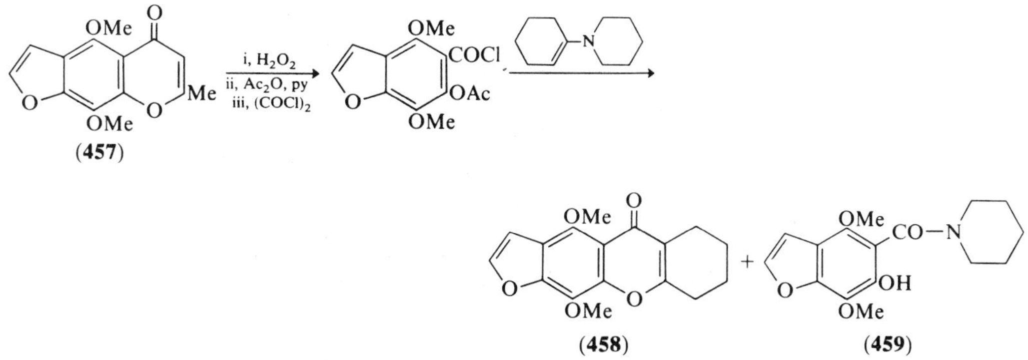

(457)

(458) + (459)

**Scheme 164**

A number of 2,3-fused chromones have been prepared from ethyl chloroformate and acetylsalicylic acid, a source of the mixed anhydride of formic and salicylic acids, and piperidinocycloalkenes 〈69JCS(C)935〉. A plausible mechanism is outlined in Scheme 165. It has not proved possible to isolate the chromanone (460) but the formation of 3-acetyl-2-methylchromone from 2-pyrrolidinopropene lends support to the intermediacy of the chromanone. The migration of the acyl group from oxygen to carbon is supported by the synthesis of 3-benzoyl-2-methylchromone rather than 3-acetylflavone from benzoylsalicylic acid and the pyrrolidinopropene.

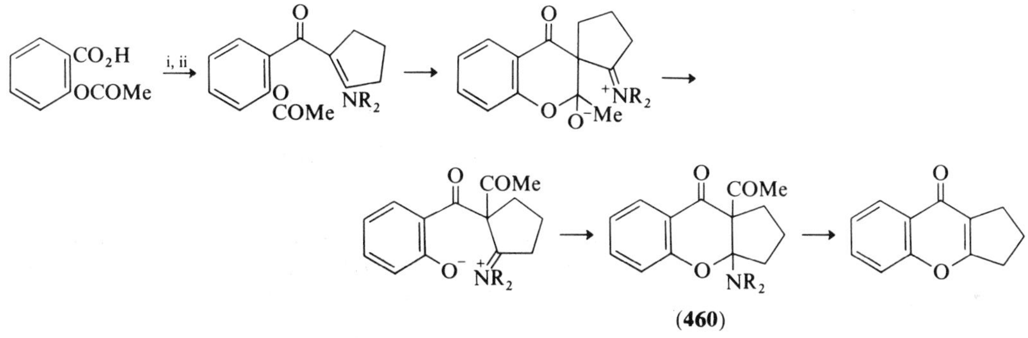

(460)

i, ClCO$_2$Et; ii, piperidinocyclopentene, NEt$_3$

**Scheme 165**

The cyclization of a number of chalcone dibromides to flavones has been achieved by treatment with pyridine 〈81JOC638〉. It is thought that the reaction may proceed through debromination and dehydrobromination to the chalcone and bromochalcone, respectively (Scheme 166). The formation of pyridine perbromide would account for the nuclear brominated flavones which sometimes accompany the flavone 〈63CB913〉.

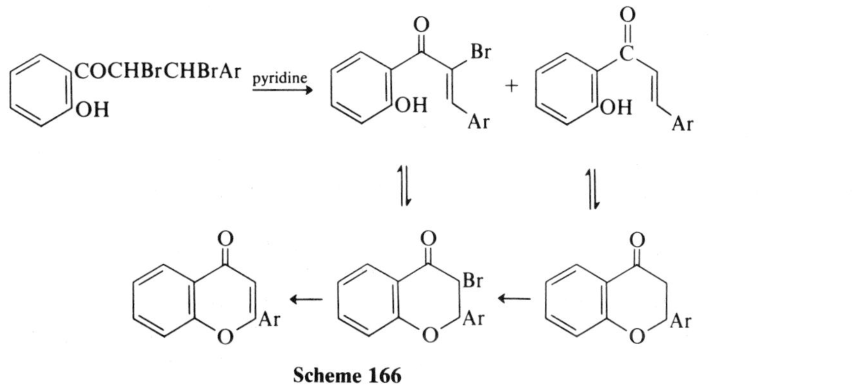

**Scheme 166**

Reaction of chalcone dibromides with sodium azide and subsequent treatment with base gives access to 3-aminoflavones 〈79LA174〉. The α-azido-2'-oxychalcones are also available from phenacyl bromides by reaction with sodium azide and then benzaldehyde.

Aminoflavones are also available from chalcones derived from 5-acetamino-2-hydroxyacetophenone ⟨57JOC304⟩. Oxidation by selenium dioxide leads to 6-acetaminoflavones which may be deacetylated to the 6-amino compound, whilst alkaline hydrogen peroxide affords the corresponding flavonols (461) in a typical Algar–Flynn–Oyamada reaction ⟨64CB610⟩.

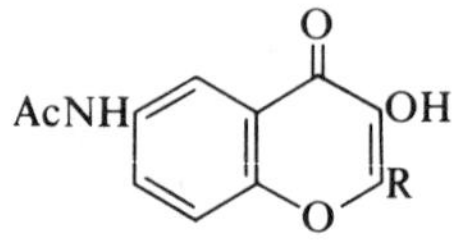

(461)

The oxidative rearrangement of chalcones is a valuable route to isoflavones which has been thoroughly investigated. Initially, the conversion was achieved in two distinct steps. Epoxidation of a 2′-benzyloxychalcone, carried out by conventional techniques, is followed by treatment with a Lewis acid, such as boron trifluoride etherate, which brings about the rearrangement.

This transformation has been shown ⟨56JA2278⟩ by [14]C-labelling to involve a 1,2-aroyl migration, rather than a 1,2-aryl shift which would necessitate a carbocation intermediate in which the charge is formally adjacent to the carbonyl group (Scheme 167).

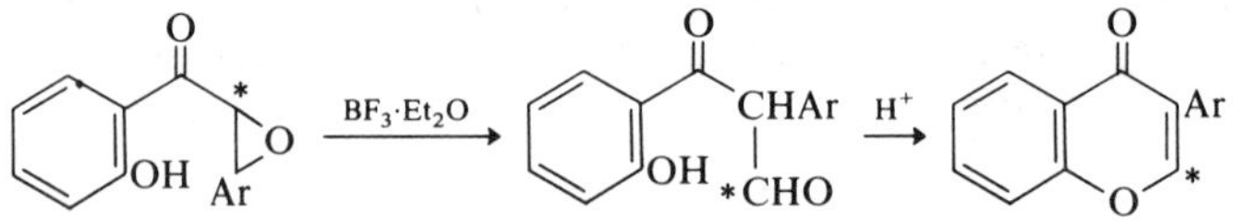

**Scheme 167**

The oxidation and rearrangement occur simultaneously when 2′-benzyloxychalcones are treated with thallium(III) acetate in methanol ⟨70JCS(C)125⟩. The intermediate acetal (462) is cyclized on treatment with acid, which debenzylates the hydroxy group, though other sensitive groups may be unaffected. Although the conditions are mild, yields are generally quite low and furthermore it is necessary to protect any 2′-hydroxy group which is present. In contrast to the previous rearrangement, this route proceeds through a 1,2-aryl shift, according to the mechanism outlined in Scheme 168.

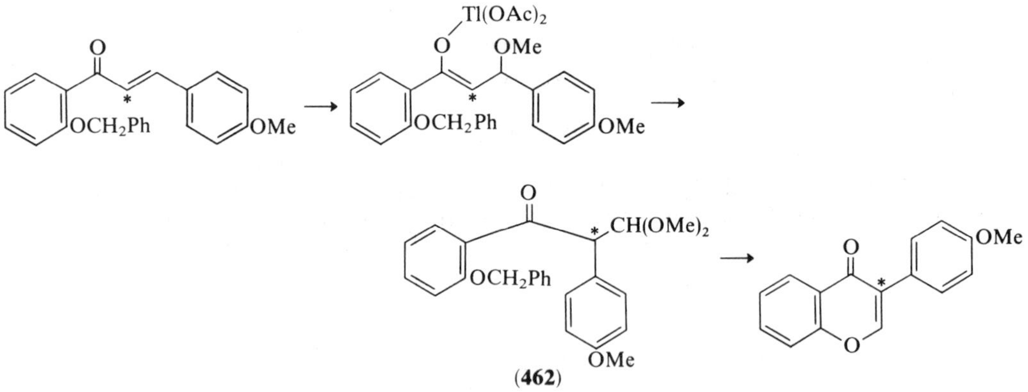

(462)

**Scheme 168**

It is of interest to note that the thallium(III) acetate oxidation of a 2′-hydroxychalcone yields a mixture of a flavone, through participation of the 2′-hydroxy group in the cyclization, and a coumaranone, which presumably arises by further oxidation of an intermediate acetal ⟨70JCS(C)119⟩.

Protection of the 2′-hydroxy group is not necessary when thallium(III) nitrate is used ⟨74JCS(P1)305⟩. The reaction is also much faster and this technique appears to be the one of choice, especially for the synthesis of naturally occurring isoflavones ⟨75CB3883⟩.

It has been suggested ⟨78AJC2699⟩ that conversion of the acetal into the isoflavone proceeds through the enol ether (463). β-Elimination from the acetal leads to the ether which cyclizes to the isoflavanone by a Michael addition. A second β-elimination generates the flavone (Scheme 169).

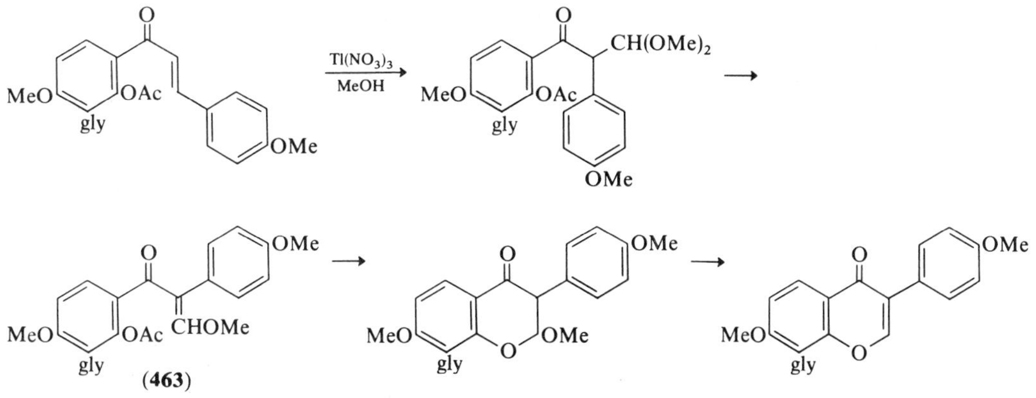

gly = tetraacetyl-β-D-glucopyranosyl

**Scheme 169**

Evidence has been presented that 2-hydroxychalcones are also transformed into flavones on reaction with thallium(III) nitrate (equation 16) ⟨78TL3359⟩. Such a method would be complementary to the isoflavone synthesis if it proves to be of general applicability.

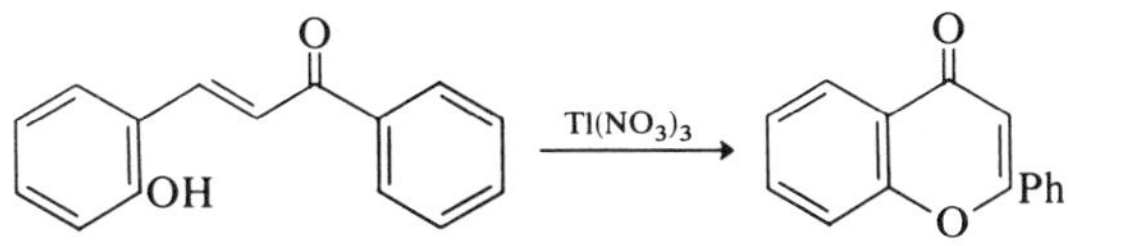

(16)

Whilst the majority of chromone syntheses from phenols involve cyclization γ to the heteroatom and are discussed later, there are some instances where ring closure occurs at the heteroatom. For instance, 1,3-dimethoxybenzene is acylated by 2,3,3-trichloropropenoyl chloride in the presence of aluminum chloride to give 2,3-dichloro-7-methoxychromone ⟨64CB80⟩.

### (ii) β to the heteroatom

Several 2-phenoxychromones have been prepared from phosphorane carbonates (**464**), the precursors of which are available from *o*-hydroxypropiophenone (Scheme 170). The action of heat on the ylide causes cyclization to a chromanone derivative, which loses phenoxide when R = H to give the chromandione (**465**), but loses triphenylphosphine oxide in other instances to give the 2-phenoxychromone (**466**) ⟨81CC282, 474⟩.

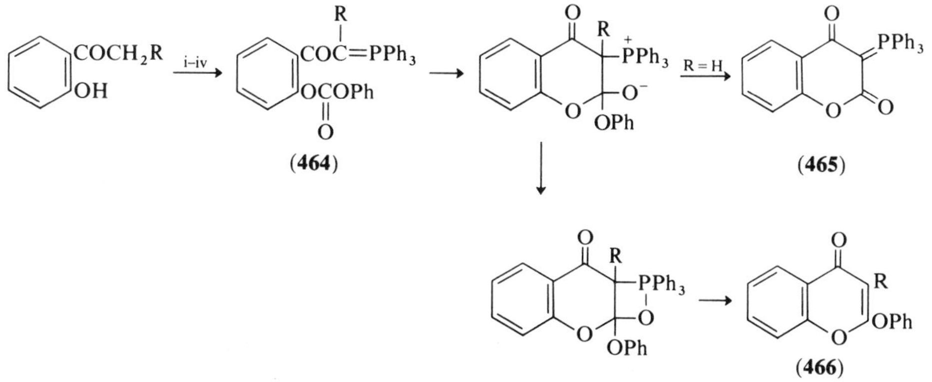

i, ClCO$_2$Ph; ii, CuBr$_2$, MeCO$_2$Et, CHCl$_3$; iii, Ph$_3$P, CH$_2$Cl$_2$; iv, EtNPr$^i_2$; v, Δ

**Scheme 170**

### (iii) γ to the heteroatom

Although there is evidence that chromone syntheses which proceed by the cyclization of phenyl esters under Friedel–Crafts conditions may involve a Fries rearrangement and hence require the formation of one bond adjacent to the heteroatom, syntheses of chromones from phenols will be considered together in this section. The Simonis reaction ⟨53OR(7)1⟩

is closely allied to the Pechmann synthesis of coumarins (see Section 2.24.3.2.1) and involves the reaction of a phenol with a 3-keto ester.

Two disadvantages are associated with this synthesis. Yields range from low to moderate and, of more significance, cyclization of the initially formed ester can give rise to chromones or coumarins or to mixtures of the two heterocycles. With the wide range of analytical techniques now available, it is not difficult to distinguish the isomeric benzopyranones. However, some of the structural assignments in early work have been shown to be erroneous and care is therefore advised in the interpretation of results.

The condensing agent used in earlier studies was phosphorus pentoxide and it was believed that this led predominantly to chromones. Sulfuric acid, on the other hand, was considered to effect cyclization to the coumarin. Later workers have used polyphosphoric acid or phosphoryl chloride, though in general this change only brought about improvements in the yield.

A variation in the condensing agent cannot be expected to give a clear-cut change in the nature of the product and examples are known where a coumarin results in the presence of phosphorus pentoxide; resorcinol provides an excellent illustration. In a few instances, sulfuric acid has been found to give rise to chromones ⟨44JA802⟩.

The introduction of a substituent into ethyl acetoacetate is reported to favour the formation of chromones ⟨31JCS2426⟩ though this is not always the case.

The use of preformed aryl acetoacetates seems to offer no advantage over the usual method, giving chromones or coumarins under the influence of the normal catalysts ⟨54JCS854⟩.

Perhaps the most significant step forward in directing cyclization to the chromone has been the omission of a condensing agent ⟨52BSF91⟩. Short reaction times and the use of high boiling solvents comprise the recommended technique and under these conditions even those phenols which give coumarins in the conventional Simonis reaction, such as 1-naphthol ⟨57LA(603)169⟩, may afford chromones. It does appear, however, that only the more reactive phenols take part in this process.

There has been some controversy about the mechanisms of the Simonis and Pechmann reactions, which still remain in doubt. It has been suggested ⟨50BSF1132⟩ that the condensations proceed through a common oxonium ion (**467**). Dehydration to the phenoxyacrylic ester (**468**) is followed by cyclization to the chromone whilst a rearrangement to the substituted phenol (**469**) subsequently affords the coumarin (Scheme 171).

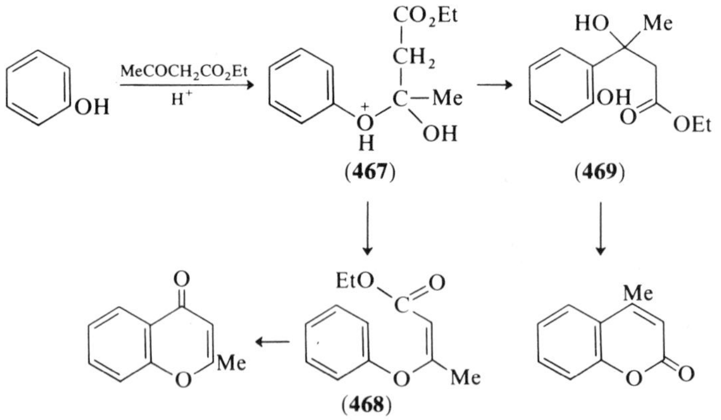

**Scheme 171**

The formation of chromones in the absence of condensing agent can be considered to proceed through the acrylic ester, which is formed by high temperature etherification ⟨54JCS854⟩.

However, the 1,3-diketone (**470**) has been isolated from the high temperature reaction between 3-dimethylaminophenol and ethyl benzoylacetate ⟨64CR(259)1645⟩. Subsequent cyclization with acid yielded the chromone. It is likely that the hydroxydiphenylketone arises

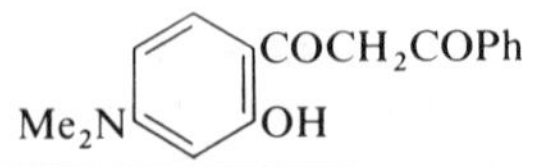

(**470**)

*via* a Fries rearrangement of the ester formed by a simple transesterification. Whether or not such a route applies to the acid-catalyzed condensation is unclear.

The reaction between a phenol and an unsaturated carboxylic ester has been widely used for the synthesis of chromone-2-carboxylic acids ⟨00JCS1119, 1179⟩. There is little restriction on the substituents which may be present in the phenol and the necessary basic conditions have been achieved in various ways.

Both diethyl acetylenedicarboxylate and chlorofumarate have found frequent application as the second component, but other unsaturated acid derivatives are acceptable. The initially formed aryloxyalkenoic acids or their esters (**471**) are readily cyclized with acid or acetyl chloride.

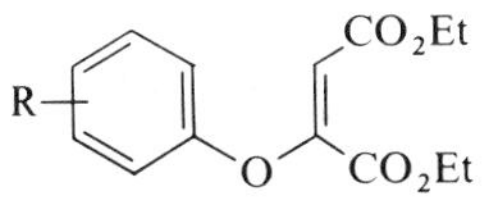

(**471**) 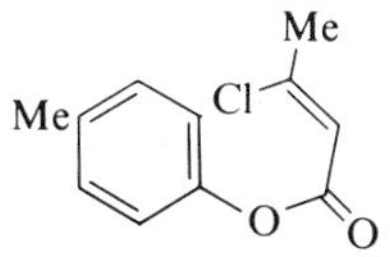  (**472**)

It is pertinent to note that the ester (**472**) yields a chromone on treatment with hydrogen fluoride ⟨54LA(587)16⟩, suggesting the intervention of a Fries rearrangement once again. The same applies in the formation of 2-chlorochromone from phenyl 3,3-dichloropropenoate ⟨60CR(250)2819⟩.

Once again, cyclization of the substituted propenoic acid derivative may lead to either a coumarin or a chromone. Relatively minor changes in the cyclizing medium result in the preferential formation of one isomer ⟨57LA(605)158⟩.

The reaction between a phenol and dichloromaleic anhydride catalyzed by aluminum chloride gives an aroylacrylic acid, which cyclizes to a chromone on treatment with a weak base ⟨79SC129⟩. However, at higher temperatures an indenone is formed.

### 2.24.3.4.2 *From other heterocycles*

### (i) *From furans*

Irradiation of the 5-aryl-3*H*-furan-2-ones (**473**; R = OMe or OAc) in benzene using a medium pressure mercury lamp leads to the formation of the 2,3-dimethylchromones (**477**; R = OMe or OAc). When ethanol was used as the solvent, the analogous 3-ethoxycarbonyl-methyl compounds resulted ⟨81T2111⟩. Cleavage of the O—CO bond is assumed to be the primary photochemical process, leading to the diradical (**474**). Cyclization involving an intramolecular radical addition to the ester carbonyl group follows. The resulting diradical (**475**) collapses to the chromanone (**476**), which is either solvolyzed in ethanol or decarboxylated in benzene. However, there is no evidence for the intermediacy of the chromanone and the diradical may give rise directly to the products (Scheme 172).

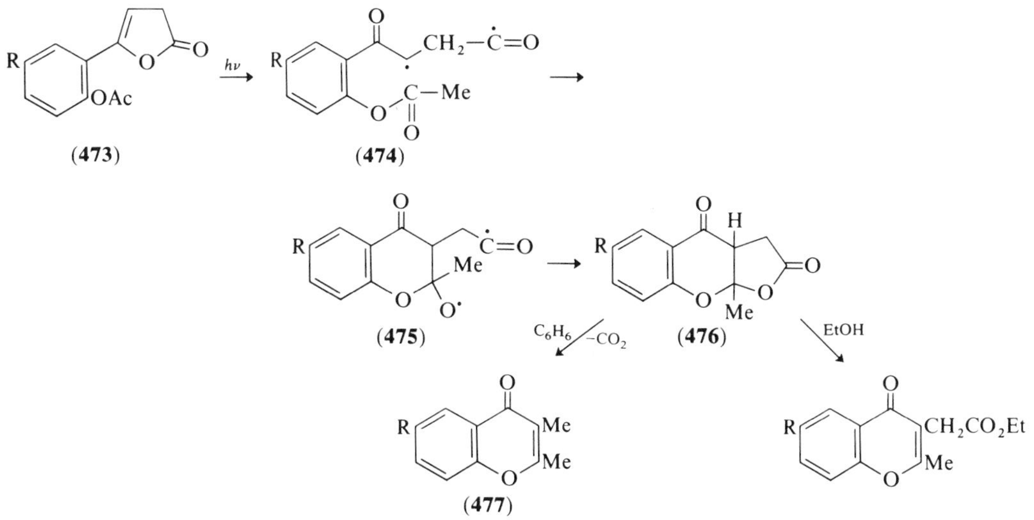

(**473**)  (**474**)

(**475**)  (**476**)

(**477**)

**Scheme 172**

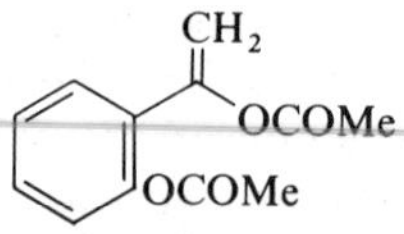

(478)

In a later paper ⟨81TL1749⟩ the same group has shown that the lactone structure is not a prerequisite for chromone formation, since the enol ester (478) yields a chromone on irradiation.

### (ii) From benzofurans

The O-tosyl derivative of 2-acetylbenzofuran oxime is converted into the chromone (480) on reaction with an alcohol ⟨49JA2652⟩. It is accompanied by 2-coumaranone (480a) and what is thought to be an acetal (480b). The chromone was thought to arise through a rearrangement of the o-quinonoid acetal (479; Scheme 173).

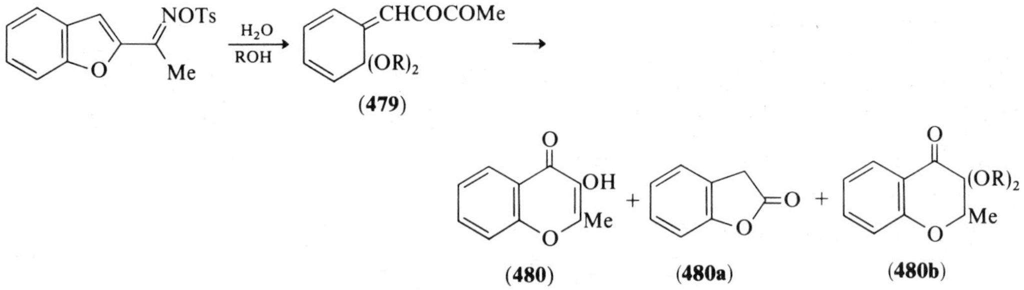

(479)

(480)          (480a)          (480b)

**Scheme 173**

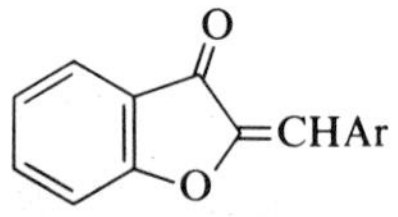

(481)

2-Benzylidenecoumaran-3-ones (481) are converted into flavones by ethanolic potassium cyanide ⟨55JCS860⟩.

### 2.24.3.4.3 From a preformed heterocycle

In general, the synthetic routes described provide adequate and convenient routes to chromones, such that their synthesis from other oxygen heterocyclic compounds is not of significant value. The following examples are therefore better considered as reactions of the heterocycle and are dealt with in more detail in the appropriate sections of Chapter 2.23.

### (i) From chroman-4-ones

The direct conversion of a chromanone to a chromone by dehydrogenation has been achieved using palladium on charcoal ⟨74CR(C)(279)151⟩ and with selenium dioxide ⟨65JCS2743⟩, but both reagents appear to be somewhat fickle.

Rather more reliable and certainly better documented is the dehydrobromination of 3-bromochroman-4-ones, achieved by treatment with an amine ⟨51JA4205⟩. In some instances the ring-opened enamine (482) has been isolated and subsequently cyclized to the chromone ⟨65JCS3610⟩.

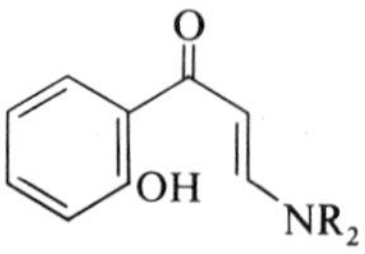

(482) 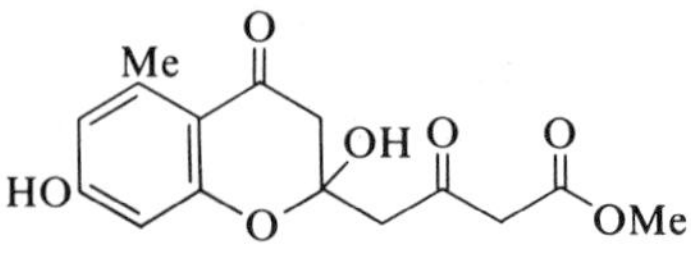          (483)

2,3-Dibromochroman-4-one yields chromone with zinc dust ⟨58BSF329⟩, whilst the 3,3-dibromo compound is dehydrobrominated by pyridine to 3-bromochromone ⟨71AP543⟩.

NBS converts acetylated polyhydroxyflavanones into the corresponding flavones by free radical bromination of the pyranone ring and subsequent dehydrobromination ⟨59JOC567⟩. If the 3-position is substituted, bromination occurs predominantly in the aromatic ring and the major product is the bromoflavanone (Scheme 174) ⟨65JOC897⟩. Iodine in acetic acid also brings about the flavanone → flavone transformation ⟨68TL1635⟩.

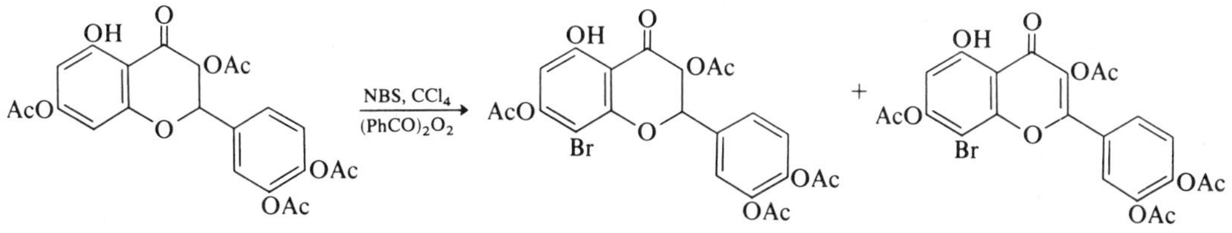

**Scheme 174**

The dehydration of the hemiacetal (**483**) is accomplished using *p*-toluenesulfonic acid ⟨77JA1631⟩.

3-Benzylidenechroman-4-ones are readily prepared by the condensation of chroman-4-ones with aromatic aldehydes. The exocyclic–endocyclic migration of the double bond has been achieved on heating the chromanone in xylene in the presence of Raney nickel ⟨74JIC281⟩. Rhodium chloride in ethanol is more effective, leading to almost quantitative bond transposition ⟨77JCS(P1)359⟩.

### (ii) *From benzopyrylium salts*

The hydrolysis of 4-alkoxybenzopyrylium salts leads to chromones in almost quantitative yields. The route is attractive because of the simple synthesis of the pyrylium salts ⟨81CHE115⟩. This method provides a reliable route to 3-substituted chromones, based on the reaction of 2-hydroxyphenacyl compounds with triethyl orthoformate. Furthermore, the use of triethyl orthoacetate enabled moderate yields of 2-methylchromones to be obtained ⟨78JCR(M)0865⟩.

Flavylium salts are oxidized to flavones on treatment with thallium(III) nitrate in methanol. The reaction is considered to involve initial attack at C-2 by methanol. The resulting chromene (**484**) undergoes reaction with the thallium salt at C-3 and nucleophilic attack at C-4 by methanol to give an unstable adduct (**485**), which decomposes to the flavone (Scheme 175) ⟨68TL3859⟩.

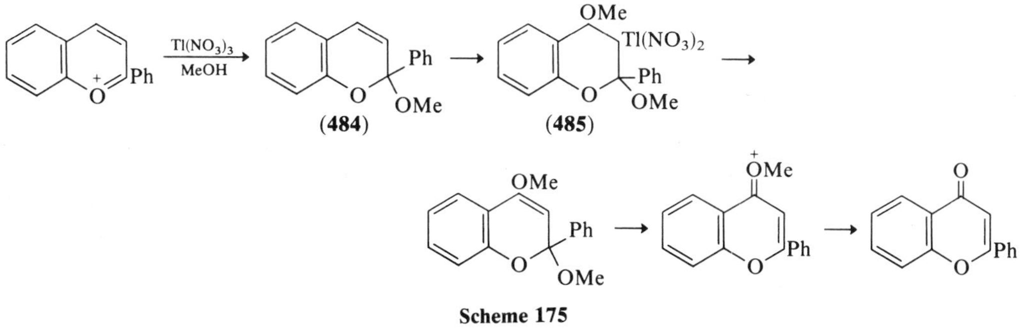

**Scheme 175**

Similarly, chromanones are converted into benzopyrylium salts; in this case though the oxidant is triphenylmethyl perchlorate. Hydrolysis affords the chromone ⟨60CB1466, 60CI(L)1192⟩.

### (iii) *From chromenes*

The oxidation of a range of 2-aryl-2*H*-chromenes ⟨77BCJ3298⟩ and 2-aryl-4*H*-chromenes ⟨78BCJ1175⟩ to the corresponding flavones has been achieved with potassium permanganate in acetone.

### (iv) *From chroman-4-ols*

Chromium trioxide oxidizes 2-aminochroman-4-ols, available in high yield from salicylaldehyde and enamines, to chromones ⟨74IJC26⟩.

### (v) *From coumarins*

Although 4-hydroxycoumarins are tautomeric with 2-hydroxychromones, the former predominate in most instances. Methylation in general affords the methoxycoumarin. However, 4-methoxycoumarin is converted into the chromones (**487**) and (**488**) on reaction with the carbanions derived from DMSO and 2-picoline 1-oxide, respectively ⟨78S208⟩. The reaction involves initial nucleophilic attack at C-2 resulting in ring opening to the potential 1,3-diketone (**486**). Cyclization and dehydration occur during acid work-up (Scheme 176). The former product is a source of 2-formylchromone through a Pummerer reaction, giving an overall yield of *ca.* 40% in a four-step reaction from 4-methoxycoumarin.

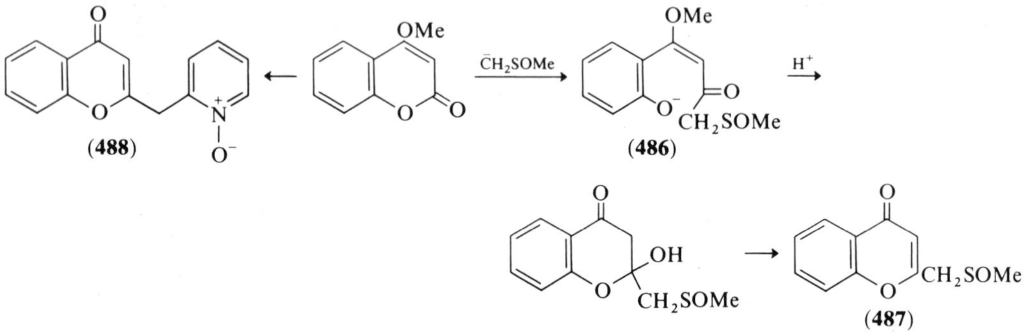

**Scheme 176**

On treatment with acid, the lactone ring of 4-hydroxy-3-propanoylcoumarin is opened and 2-ethylchromone is formed through decarboxylation and ring closure ⟨56CR(242)1034⟩.

2-Methyl-3-nitrochromone is reported to result from the ring opening of 4-hydroxy-3-nitrocoumarin and subsequent cyclization with acetic anhydride ⟨45JA99⟩.

### (vi) *From pyranones*

The condensation of 4-methoxy-6-methylpyran-2-one with the polyanion derived from pentane-2,4-dione or heptane-2,4,6-trione leads to chromones *via* the acyclic polycarbonyl compound ⟨76LA1617⟩.

## 2.24.3.5 Isocoumarins (1*H*-2-Benzopyran-1-ones)

Isocoumarins have received less attention than coumarins and yet they form a group of naturally occurring lactones which possess a range of biological activity. In addition to a review of the chemistry of isocoumarins ⟨64CRV229⟩, an earlier paper evaluated potential routes to their synthesis ⟨48JOC477⟩ and a more recent survey is given in Rodd's treatise ⟨B-77MI22400⟩.

### 2.24.3.5.1 *Formation of one bond*

#### (i) *Adjacent to the heteroatom*

A number of approaches to the isocoumarin ring system are based on the facile ring closure of homophthalic acid and its derivatives and hence a major synthetic task lies in the preparation of these compounds.

A systematic investigation of the copper-catalyzed reaction between 2-bromobenzoic acid and the anions of 1,3-dicarbonyl compounds has established the optimum conditions for the direct arylation of the β-dicarbonyl moiety ⟨75T2607⟩. The use of sodium hydride as the base and copper(I) bromide as catalyst is recommended. The absence of a protic solvent ensures that competitive attack on the bromobenzoic acid by a solvent-derived base leading to a salicylic acid is eliminated. For larger scale reactions the addition of toluene offers some practical advantages.

The conversion of the substituted 1,3-dicarbonyl compound into homophthalic acid is remarkably facile; loss of the acetyl group by a retro-Claisen condensation and hydrolysis of the ester group are complete in a few minutes in aqueous sodium hydroxide. The overall synthesis of homophthalic acids from *o*-bromobenzoic acids occurs in high yield and provides an attractive route.

There is still doubt over the mechanism of this reaction, although polarization of the C—Br bond and chelation of the metal play important roles. It is clear that Cu(I) is the effective catalyst and a tetrahedrally coordinated copper complex (**489**) can be envisaged to account for the steric and electronic effects observed in these reactions.

In a similar manner the reaction of sodium 2-bromobenzoate with the carbanion derived from pentane-2,4-dione initially yields the dioxo acid (**490**). In ethanol, a retro-Claisen deacylation leads to 2-acetonylbenzoic acid (**491**), but at higher temperatures 3-methyl-isocoumarin is formed (Scheme 177) ⟨75JCS(P1)1267⟩. Copper(I) bromide may be used as a catalyst, although this is only necessary for the initial step. The cyclization process is considered to involve reaction between the carboxylate group and an enolate ion arising from loss of one of the acyl groups. A similar reaction occurs with 1-bromo-2-naphthoic and 3-bromo-2-naphthoic acids giving the naphtho[2,1-*c*]pyran-4-one (**492**) and naphtho[2,3-*c*]pyran-1-one (**493**), respectively.

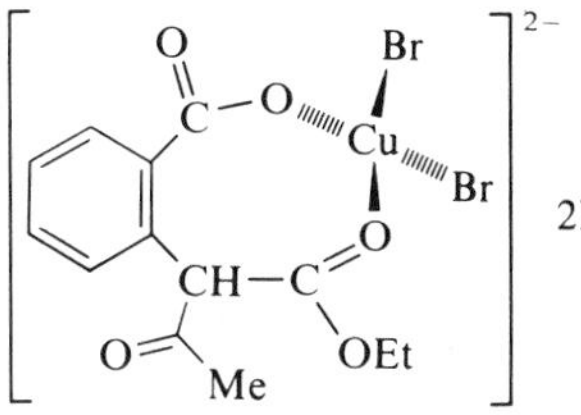

(**489**) 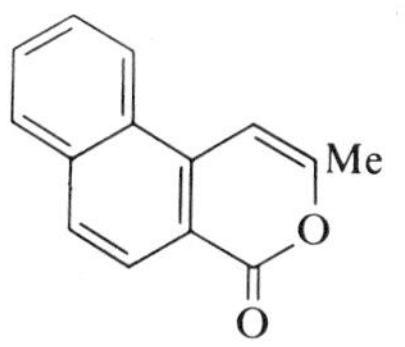    (**492**) 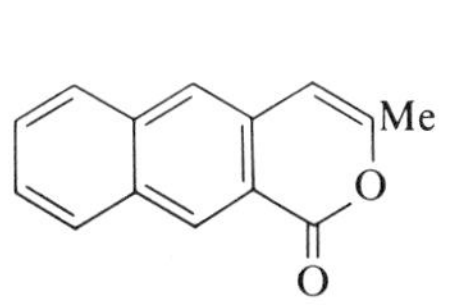    (**493**)

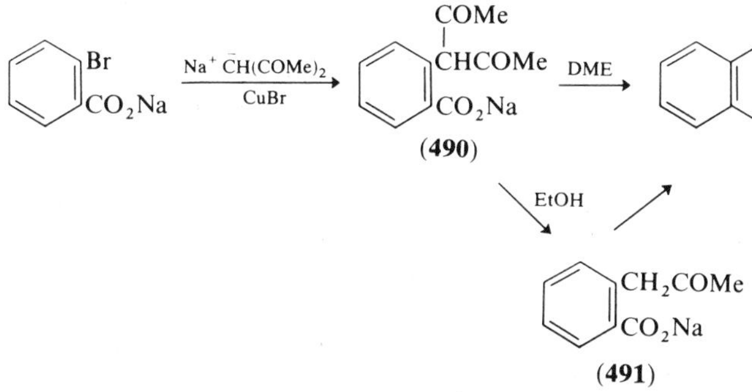

**Scheme 177**

Homophthalic acids are acylated by acid anhydrides in the presence of pyridine, producing 4-acylisochroman-1,3-diones (**494**) ⟨80CB3927⟩. Rearrangement to a 3-alkylisocoumarin-4-carboxylic acid (**495**) occurs in acid solution and may be accompanied by decarboxylation.

In a study of the reaction of 4,5-dimethoxyhomophthalic acid with acetic anhydride, the 4-acetylisocoumarin (**496**; R = COMe) and the dibenzocoumarin (**497**) were obtained in addition to the carboxylic acid (**496**; R = CO₂H). The products (**496**) are considered to arise *via* ring opening of the initially formed isochromandione ⟨71JIC192⟩.

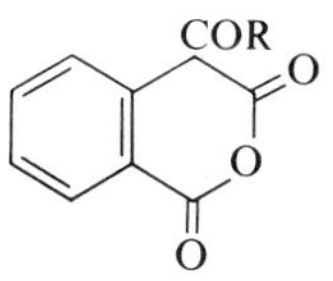

(**494**) 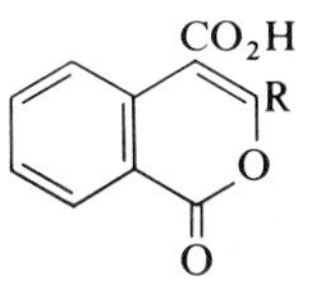    (**495**) 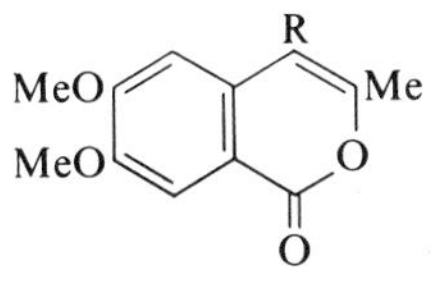    (**496**) 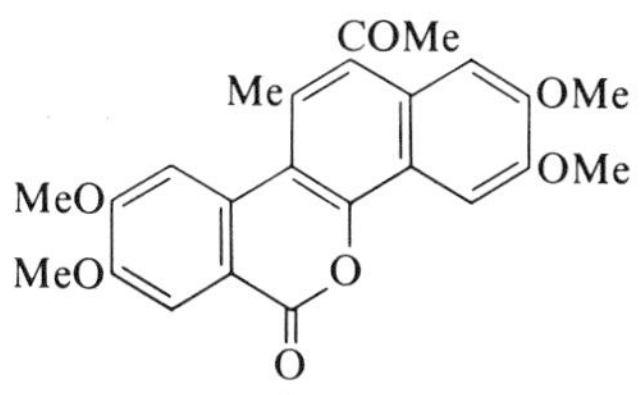    (**497**)

The synthesis of homophthalic acids has been achieved by carboxylation of the dianion derived from *o*-methylbenzoic acids ⟨82JCS(P1)1111⟩. Acetylation is hindered by substitution at the 5-position of the isochroman-1,3-dione and yields of the isocoumarins are reduced in these cases. Deuteration at C-4 can be achieved by deuterating the 4-carboxylic acid, followed by decarboxylation.

3-Aminoisocoumarins are also available from the isochromandione through initial reaction with secondary amines to give homophthalamic acids (**498**) ⟨78JCS(P1)1338⟩. Cyclization in acetic anhydride and perchloric acid gives the homophthalisoimidium perchlorates (**499**) which afford the stable conjugate bases, the 3-dialkylaminoisocoumarins, upon treatment with triethylamine (Scheme 178).

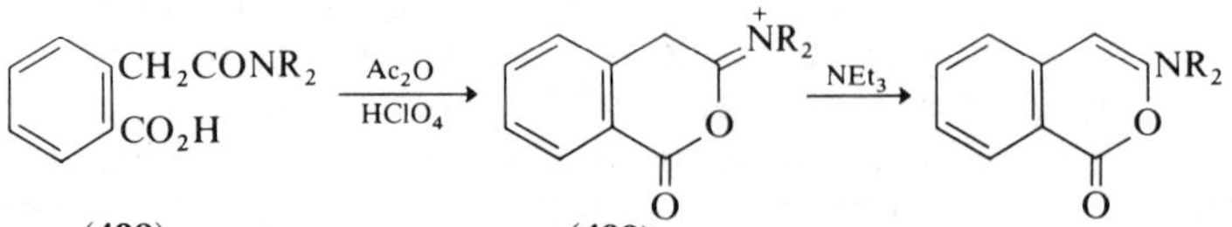

**(498)**                    **(499)**

**Scheme 178**

A vigorous Claisen condensation ensues when a homophthalic ester and methyl formate are treated with sodium ethoxide and the active methylene group is formylated. Cyclization takes place with ease in acidic media to produce a methyl isocoumarin-4-carboxylate ⟨50JCS3375⟩. Hydrolysis under acid conditions is sometimes accompanied by polymerization, but the use of boron trifluoride in acetic acid overcomes this problem. Decarboxylation may be effected in the conventional manner with copper bronze, though it sometimes accompanies the hydrolysis.

The reaction between phenylacetonitriles and methyl formate is similar in nature and yields 4-cyanoisocoumarins ⟨53JIC103⟩.

The use of ethyl oxalate in the Claisen synthesis affords 3,4-disubstituted isocoumarins. It appears that hydrolysis of the 3-ethoxycarbonyl group is easier than that of the 4-substituent, but the 4-carboxylic acid is more readily decarboxylated than the 3-isomer ⟨54JCS3617⟩.

The introduction of an ethyl group at C-3 has been accomplished from a 2-methoxycarbonylphenylacetic acid by way of the diazoketone (**500**; Scheme 179) ⟨74IJC1259⟩.

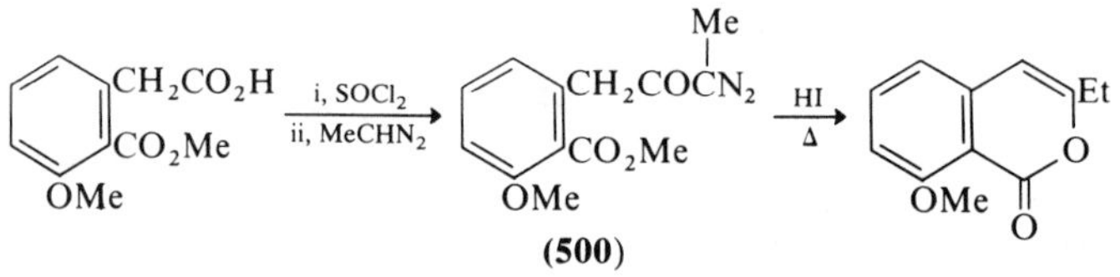

**(500)**

**Scheme 179**

The cyclization of 2-carboxybenzyl ketones to isocoumarins has been achieved under acidic conditions ⟨59JCS923⟩.

The naphthalene derivative (**501**) may be regarded as a special case of these ketones and on heating it forms the naphthoisocoumarin (**503**) ⟨56JCS4535⟩. The same isocoumarin results when the dicarboxylic acid (**502**) is treated with polyphosphoric acid (Scheme 180).

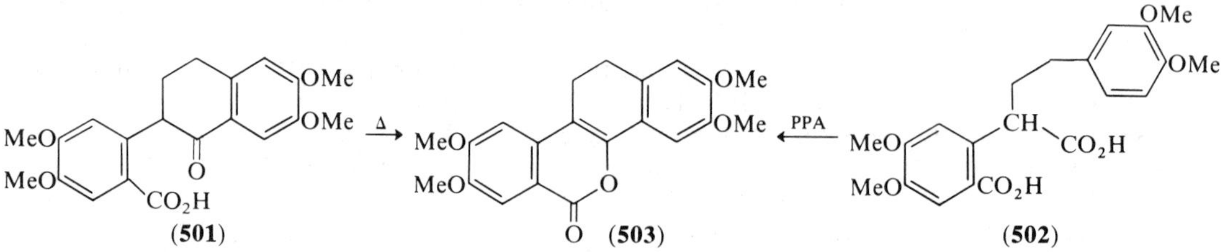

**(501)**                    **(503)**                    **(502)**

**Scheme 180**

2-Carboxybenzyl aryl ketones are formed together with the isocoumarin when homophthalic anhydride (isochroman-1,3-dione) is used to acylate aromatic molecules under Friedel–Crafts conditions ⟨51JOC1064⟩ and the 7-methoxy derivative behaves in a similar fashion ⟨66JIC615⟩. Some care in the choice of Lewis acid is necessary in view of the formation of the tropone derivative (**504**) in the acylation of hydroquinone (Scheme 181) ⟨55JCS2244⟩.

A detailed examination of the utility of this isocoumarin synthesis using phenols as the substrate has shown that the anhydride may be replaced by homophthalic acid without detriment ⟨65JCS6100⟩. Electron-releasing groups in the phenol enhance its reactivity, but 2-chlorophenol gave no isocoumarin. The influence of an electron-withdrawing group in the homophthalic acid is less severe, even the 4-nitro derivative taking part in the reaction ⟨72JHC1255⟩.

Polyphosphoric acid has also been used to effect both acylation of aromatic ethers and subsequent ring closure ⟨74IJC474⟩.

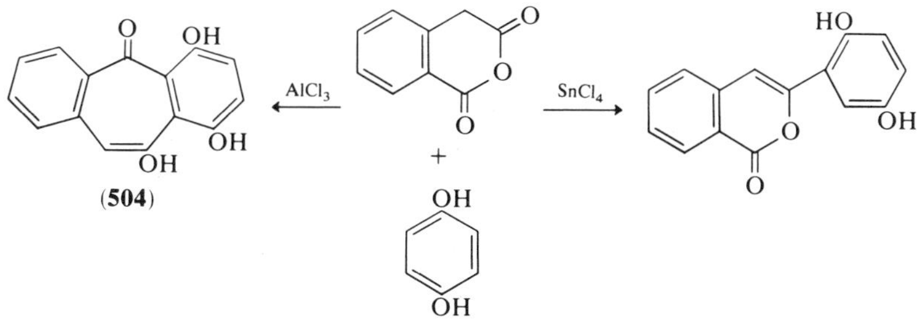

**Scheme 181**

An alternative approach to the synthesis of isocoumarins which probably proceeds through the intermediacy of 2-carboxybenzyl ketones is based on the oxidative cleavage of indan-1-ones ⟨76JCS(P1)1438⟩. Although ozonolysis of the silyl enol ether (505) leads to the 2-hydroxy-2-methylindan-1-one (506), periodate oxidation of which gives the isocoumarin, a more convenient and direct route involves ozonolysis of the enol trifluoroacetate (Scheme 182). This synthesis is especially attractive for the preparation of isotopically labelled isocoumarins, since the precursors of the indanones, arylpropanoic acids or acrylophenones, are readily available bearing labels at specific sites.

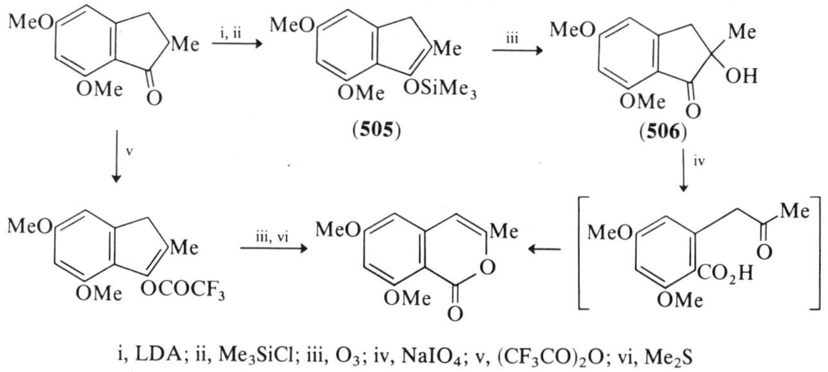

i, LDA; ii, Me₃SiCl; iii, O₃; iv, NaIO₄; v, (CF₃CO)₂O; vi, Me₂S

**Scheme 182**

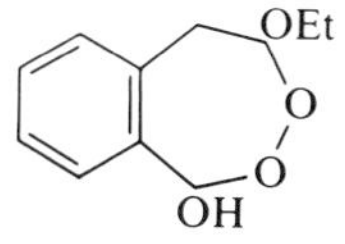

(507)

There are several other examples of the use of indene derivatives as a source of isocoumarins. Alkaline decomposition of the cyclic peroxide (507), produced by ozonolysis of indene itself, yields a phenylacetaldehyde ⟨57JA3165⟩. Treatment of these aldehydes with acid gives access to isocoumarins which are unsubstituted in the pyranone ring ⟨65JIC211⟩.

An epoxide is an intermediate in the conversion of indenones into isocoumarins brought about by hydrogen peroxide ⟨43JA1230, 63JA3529⟩.

Esters of 2-acetonylbenzoic acid are formed when methyl 2-bromobenzoates are treated with π-(2-methoxyallyl)nickel bromide ⟨77JOC1329⟩. Cyclization to the isocoumarin occurs on treatment with sodium hydride in benzene with a trace of *t*-butanol (Scheme 183). A

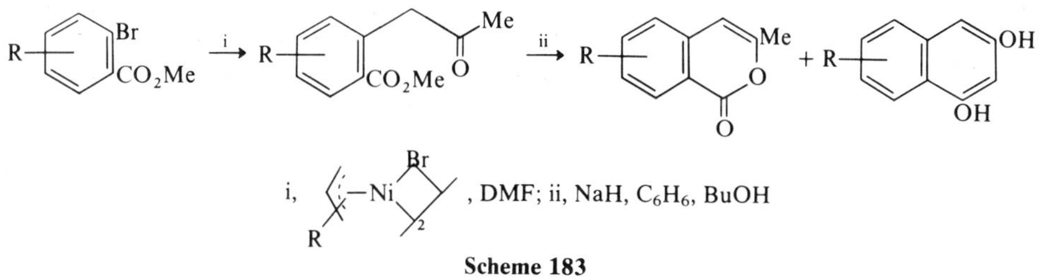

i, [structure]—Ni[structure]₂ , DMF; ii, NaH, C₆H₆, BuOH

**Scheme 183**

large excess of the hydride gives a mixture of the isocoumarin and 1,3-dihydroxynaphthalene, the latter resulting from *C*- rather than *O*-acylation. The use of a pyridine carboxylate, leading to a pyranopyridine, is worthy of mention.

A greater variety of substituted isocoumarins is available from the reaction of π-allylnickel halides with the sodium salts of 2-bromobenzoic acids. Use of the sodium salt is necessary to prevent debromination through an intramolecular proton transfer from the carboxyl group. The initial products, 2-allylbenzoic acids, undergo a palladium-assisted ring closure to isocoumarins.

The catalytic behaviour of palladium is believed to involve coordination of the metal to the alkene. Attack by the carboxylate ion then effects cyclization and isomerization completes the process.

*N*-Methylbenzamides have been used as isocoumarin precursors ⟨71T6171⟩. Directed *ortho* metallation followed by reaction with ethylene oxide gave the dihydroisocoumarin *via* the alcohol (Scheme 184).

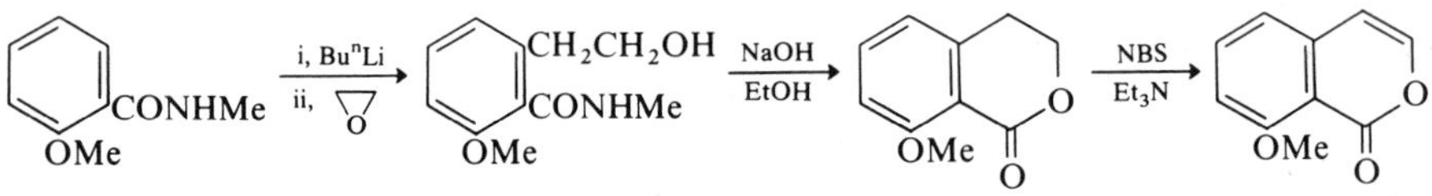

**Scheme 184**

### (ii) γ to the heteroatom

A very low yield of the coumarinoisocoumarin (**508**) is obtained during the Pechmann synthesis of 4-hydroxy-3-phenylcoumarin ⟨78M1485⟩. Significant amounts of this [2]benzopyrano[4,3-*c*][1]benzopyran-6,11-dione are formed when the hydroxyphenylcoumarin is heated with either diphenyl carbonate or diethyl phenylmalonate.

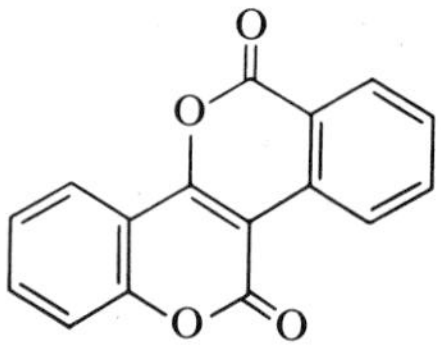

(**508**) 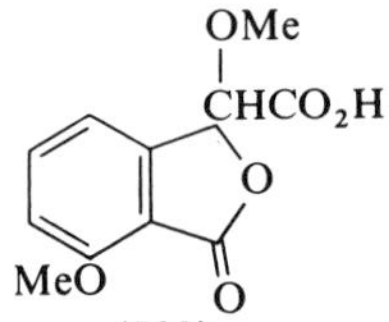                                        (**509**)

## 2.24.3.5.2 *From other heterocycles*

### (i) *From benzofurans*

The thermal rearrangement of the isobenzofuran derivative (**509**) gives 8-methoxyisocoumarin-3-carboxylic acid. The transformation is also effected by hydrobromic acid, which additionally cleaves the ether function to give the 8-hydroxy analogue ⟨60ACS539⟩.

## 2.24.3.5.3 *From a preformed heterocycle*

### (i) *From isochromanones*

Oxidation of dihydroisocoumarins is conveniently achieved through bromination with NBS followed by dehydrohalogenation with triethylamine ⟨62JOC4337, 71T6171⟩. Since the oxidation of isochromans to isochromanones proceeds readily using chromium trioxide, this method effectively constitutes a synthesis of isocoumarins from isochromans.

### (ii) *From isobenzopyrylium salts*

1-Amino-3-phenylisobenzopyrylium bromide undergoes facile hydrolysis to 3-phenyl-isocoumarin ⟨78JOC3817⟩.

## 2.24.3.6 Xanthones

Much of the synthetic chemistry of xanthones has arisen out of their natural occurrence, a topic which has been reviewed ⟨61CRV591⟩, and also as a result of their incorporation into dyes. There are only two routes of any significance to xanthones and these basically involve ring closure on to either the heterocyclic oxygen atom or the carbonyl carbon atom (Scheme 185).

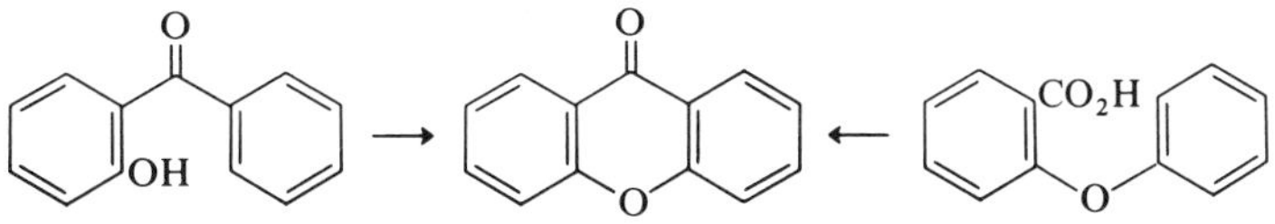

**Scheme 185**

### 2.24.3.6.1 Formation of one bond

(i) *Adjacent to the heteroatom*

A number of xanthone syntheses are based on the cyclization of 2-hydroxybenzophenones; a 2'-substituent which behaves as the leaving group during the intramolecular ring closure may also be present and indeed facilitates ring closure.

A general and widely used xanthone synthesis consists of the treatment of a salicylic acid with a phenol in the presence of zinc chloride with phosphorus oxychloride as solvent ⟨55JCS3982⟩. The solvent may be omitted to allow a higher reaction temperature ⟨66JCS(C)430⟩, but in the past this has caused problems through concurrent demethylation of methoxy groups. In this connection, it should be noted that resorcinol gives 6-hydroxy-2-methoxyxanthone with 2-hydroxy-5-methoxybenzoic acid under the normal conditions. However, in the absence of solvent, 1,7-dihydroxyxanthone is formed, presumably through acylation of the resorcinol at C-2 and demethylation.

The initial product of this route, a 2,2'-dihydroxybenzophenone, is usually cyclized thermally, often in the presence of an acid. However, some intermediate ketones cyclize spontaneously; for instance, 2,2',4,4'-tetrahydroxybenzophenone cyclized during oximation, giving 3,6-dihydroxyxanthone oxime.

The instability of 2,6-dihydroxybenzophenones has been noted, reactions which should give these compounds leading directly to the xanthone ⟨76JA5380⟩. It has been suggested that this behaviour requires the presence of a 6- or 6'-hydroxy group, so providing an alternative site for ring closure. Cyclization has been achieved by heating the appropriate benzophenone under pressure with water as in the preparation of 2,5-dihydroxyxanthone ⟨72JOC2986⟩. The interesting feature of this synthesis, however, is the application of a photo-Fries rearrangement to the ester (**510**) to obtain the ketone (Scheme 186).

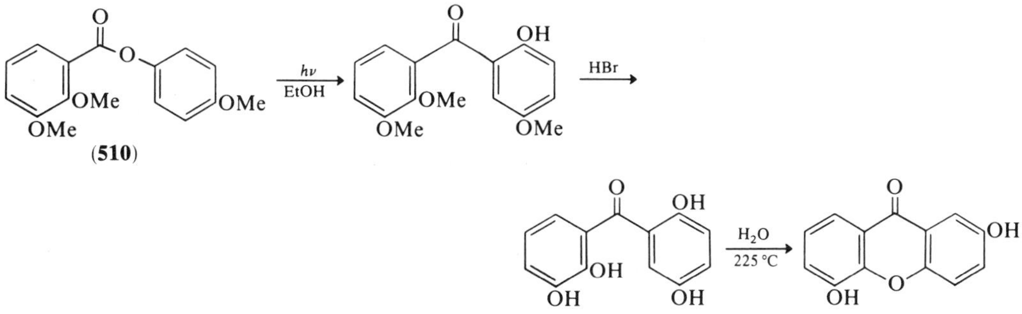

**Scheme 186**

An improved product is reported to be obtained in higher yield if the cyclization of dihydroxybenzophenones is carried out at 210 °C in glycerol containing a trace of potassium hydroxide; 3-hydroxy-4-methylxanthone was obtained in 50% overall yield from salicylic acid and 2-methylresorcinol, for example ⟨75CJC2054⟩.

A series of oxygenated 2-hydroxy-2'-methoxybenzophenones has been prepared by the Friedel–Crafts acylation of various methoxybenzenes with substituted benzoyl chlorides.

The initial acylation is followed by selective demethylation at the most electron-rich *ortho* ether function of the benzophenone ⟨73JCS(P1)1329⟩. Cyclization to the xanthone is achieved in almost quantitative yield on boiling the benzophenone with aqueous alkali or with aqueous piperidine. Coupled with the selective demethylation of polymethoxybenzophenones or xanthones which is possible under acidic or basic conditions, this method provides an attractive and efficient synthesis of naturally occurring xanthones.

Ring closure can be considered to involve participation of the anion (**511**) in a Michael addition with the neighbouring ring, with subsequent loss of methoxide (Scheme 187).

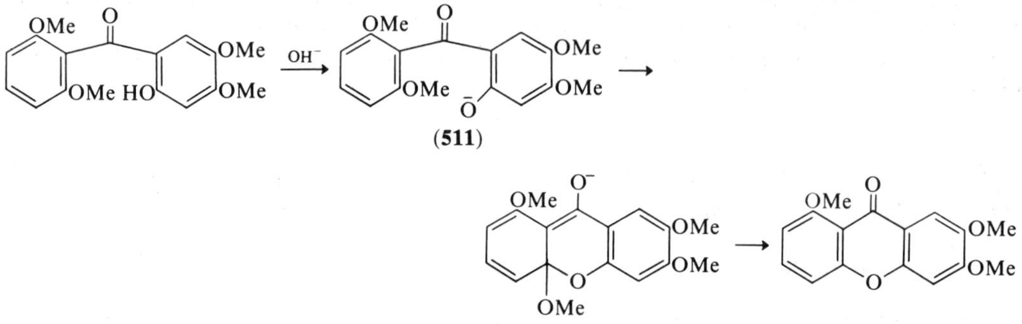

**Scheme 187**

A variant of this method produces xanthones from 2-halogeno-2'-methoxybenzophenones, cyclization being brought about by pyridine hydrochloride ⟨72BSF2948⟩.

A xanthone also results from the cyclization of 2'-nitro-6-hydroxy-2,4-dimethoxybenzophenone, which is the sole product from a photo-Fries rearrangement of 3,5-dimethoxyphenyl 2-nitrobenzoate ⟨81T209⟩. Ring closure is effected in boiling piperidine and appears to involve elimination of nitrous acid together with ether cleavage.

A biogenetic flavour is apparent in the formation of dihydroxybenzophenones by the reaction of *o*-hydroxybenzoate esters with the trianion of heptane-2,4,6-trione ⟨81JOC2260⟩. Acylation of the trianion occurs at 0–25 °C to give a tetraketone which cyclizes under mildly basic conditions to the corresponding benzophenone. A second ring closure takes place when the benzophenone is treated with sodium methoxide, leading to the xanthone (Scheme 188).

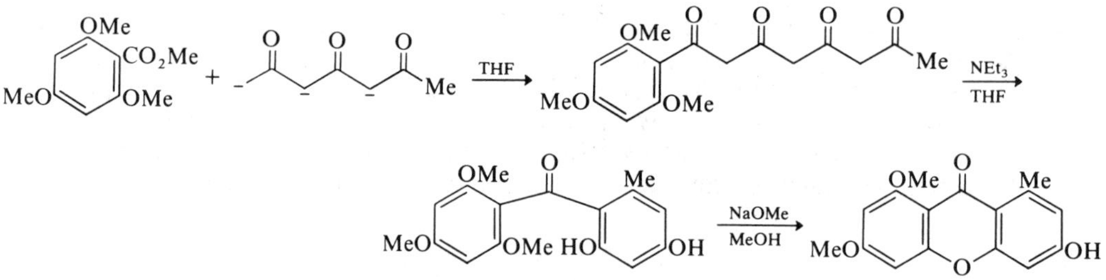

**Scheme 188**

An alternative route to the polyoxybenzophenone involves cyclization of the triketo acid (**512**) to the pyranone (**513**) and subsequent rearrangement on treatment with lithium hydride ⟨77JA1631⟩.

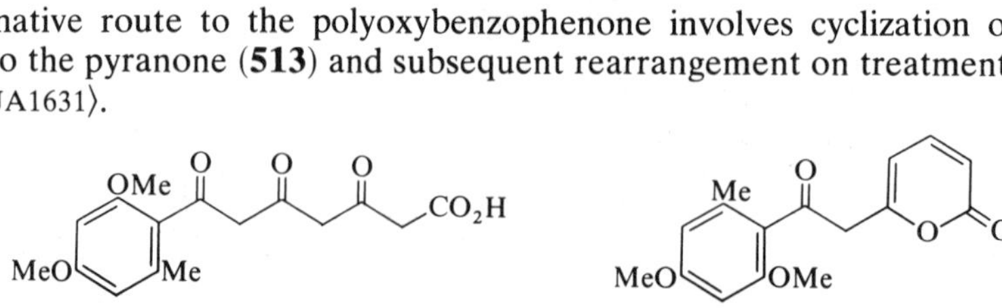

Several oxidants have been used to form xanthones from hydroxybenzophenones by oxidative coupling at *ortho* and *para* positions to an activating hydroxy group. It should be emphasized that much of this work has been carried out on a small scale and is of limited preparative value.

Using potassium hexacyanoferrate(III), the oxidation of 2,3'-dihydroxybenzophenone affords a mixture of 2- and 4-hydroxyxanthones, the proportions of which vary with the pH of the solution ⟨69JCS(C)281⟩. In other cases, only the *para* cyclized product is obtained, as for example with 2,3′,4,6-tetrahydroxybenzophenone, which yielded only 1,3,7-trihydroxyxanthone in a strongly pH-dependent reaction. It appears that the *o*-cyclized products may be unstable under the reaction conditions. It is considered that the oxidations proceed by a one-electron transfer since 3′-methoxybenzophenones fail to cyclize.

Only *para* coupling was observed during the alkaline ferricyanide oxidation of 4-hydroxy-3-(3-hydroxybenzoyl)benzoic acid. 7-Hydroxy-9-oxoxanthene-2-carboxylic acid (**515**) was obtained in 21% overall yield from methyl 4-hydroxybenzoate ⟨78JCS(P1)876⟩. The synthesis utilizes a photochemical Fries rearrangement of methyl 4-(3-methoxybenzoyloxy)benzoate to prepare the benzophenone (**514**; Scheme 189). A similar route was used to prepare 2-hydroxy-2′-methoxybenzophenones, which undergo intramolecular cyclization with loss of methanol on treatment with base.

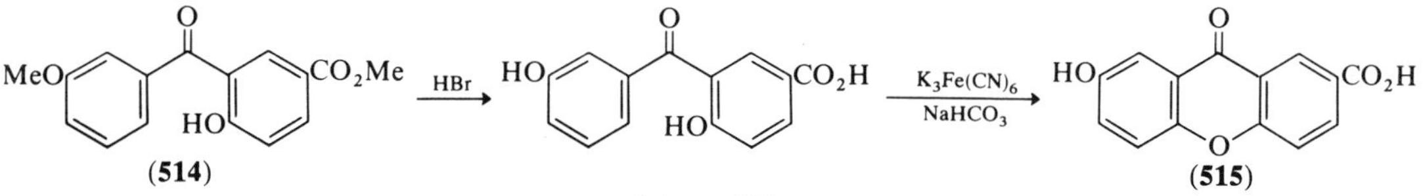

**Scheme 189**

Both *ortho* and *para* cyclization are observed with 3′-hydroxy- and 3′-methoxy-benzophenones when DDQ is used as the oxidant. This result was taken to imply the generation of a phenoxonium ion (**516**) from the 3′-methoxy compounds, which electrophilically attacks the activated methoxyphenyl ring (Scheme 190) ⟨69JCS(C)2761⟩.

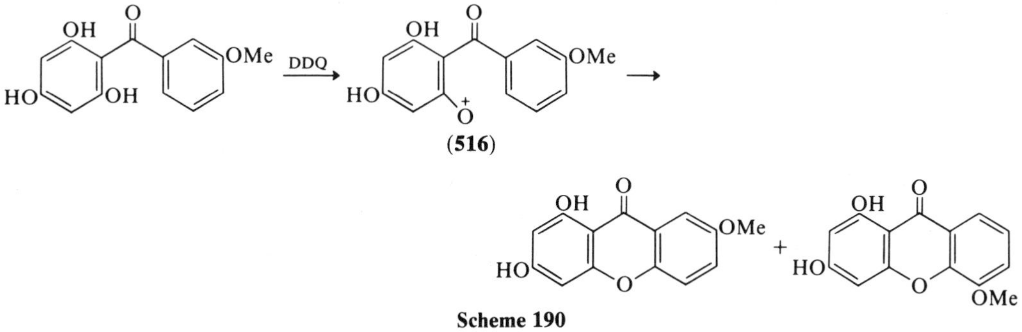

**Scheme 190**

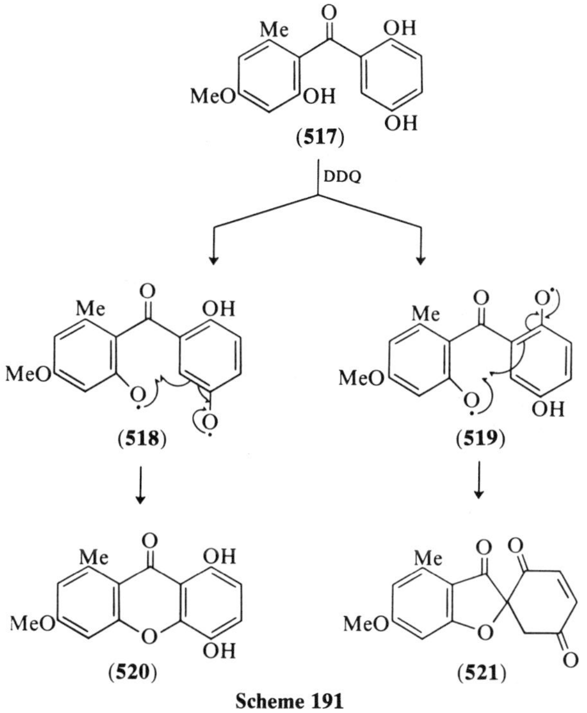

**Scheme 191**

The oxidation of the 2',5'-dihydroxybenzophenone (**517**) by DDQ also gives two products: a xanthone (**520**) and the major component, a spirocyclohexenedione (**521**). Their formation may involve the diradicals (**518**) and (**519**; Scheme 191). Thermal rearrangement of the cyclohexenedione leads to the xanthone ⟨81JCS(P1)770⟩.

A dione (**522**) was also the major product from a hexacyanoferrate oxidation of 2,4,5'-trihydroxy-2'-methoxy-3-methylbenzophenone. Its conversion to the xanthone by the action of base occurred in only moderate yield. However, zinc and acetic acid gave a quantitative conversion ⟨76JCS(P1)1377⟩.

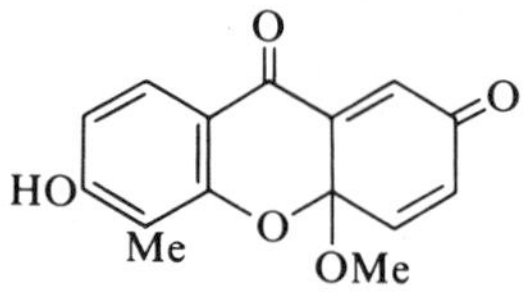

(**522**)

Phenol oxidative coupling occurs specifically *para* to the 3'-hydroxy group in the photolysis of 2,4,6,3',4'-pentahydroxybenzophenone, affording a reasonable yield of the xanthone ⟨66JCS(C)175⟩.

### (ii) γ to the heteroatom

The cyclodehydration of *o*-phenoxybenzoic acids has been widely used to synthesize xanthones carrying a variety of substituents. An Ullmann reaction provides a useful route to the required acids; a 2-chlorobenzoic acid and a phenol react in an inert solvent such as nitrobenzene in the presence of copper bronze and potassium carbonate ⟨53JCS1348⟩. In a modified procedure sodium methoxide is used without solvent at 200 °C ⟨79JA665⟩. Cyclization is accomplished in concentrated sulfuric acid or in acetyl chloride containing a little sulfuric acid ⟨61JCS2312⟩.

Acids obtained from *m*-substituted phenols can cyclize in two directions, giving 1- or 3-substituted xanthones. 3-Nitrophenol derivatives undergo *ortho* cyclization predominantly, whereas the 2-(3-acetamidophenoxy)benzoic acids give the 3-substituted xanthone. No thorough study of the regiospecificity of this reaction has been made, but it appears that electron-withdrawing substituents at the 3'-position favour cyclization at C-2', whilst electron-releasing substituents cause *para* cyclization (Scheme 192) ⟨58JCS4227⟩.

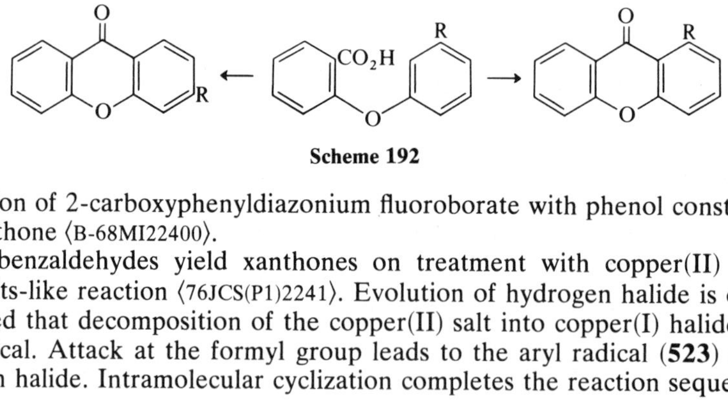

**Scheme 192**

The reaction of 2-carboxyphenyldiazonium fluoroborate with phenol constitutes a good route to xanthone ⟨B-68MI22400⟩.

*o*-Aryloxybenzaldehydes yield xanthones on treatment with copper(II) halides in a Friedel–Crafts-like reaction ⟨76JCS(P1)2241⟩. Evolution of hydrogen halide is observed and it is suggested that decomposition of the copper(II) salt into copper(I) halide generates a halogen radical. Attack at the formyl group leads to the aryl radical (**523**) and liberates the hydrogen halide. Intramolecular cyclization completes the reaction sequence (Scheme 193).

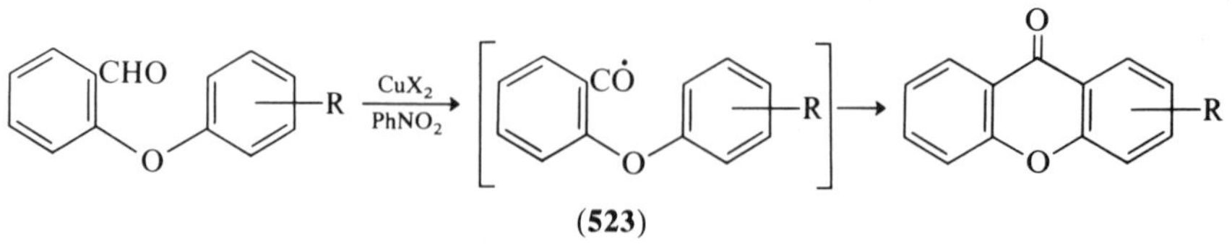

(**523**)

**Scheme 193**

*o*-Acylation of diphenyl ethers provides a direct route to xanthones. It is of course necessary to prevent *para* attack and this has been achieved in the obvious way, incorporating *p*-substituents into the diphenyl ether ⟨73JOC841, 75JOC2088⟩. However, it has also been established that a *m*-bromine substituent behaves as an effective *para*-protecting group

⟨73JCS(P1)1972⟩. The intermediate (**524**) was isolated, confirming that the initial electrophilic attack occurs on the brominated rather than the chlorinated ring.

The oxidative coupling of 2-(4-hydroxyphenoxy)benzoic acid to the spiran (**525**) is brought about by lead(IV) oxide ⟨61JCS2312⟩. The spiran is converted into 2,3-diacetoxyxanthone on reaction with boron trifluoride in acetic anhydride.

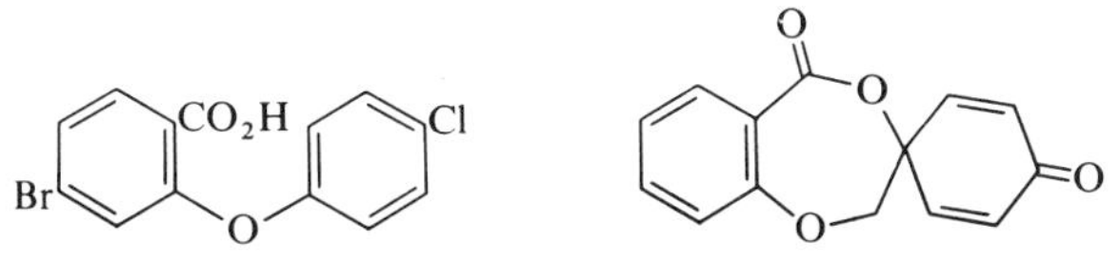

(**524**)                    (**525**)

The pyrolysis of phenyl salicylate provides a good route to xanthone which has been described in detail ⟨41OSC(1)552⟩. In a similar manner, the thermal decomposition of *o*-hydroxynaphthoic acids and their esters allows access to a range of dibenzoxanthones ⟨66T1539⟩. The by-products of these reactions are the phenol and carbon dioxide; indeed when 2-hydroxynaphthoic acid is heated just above its melting point these are the only two products, and it is only at higher temperatures that xanthones are formed. The reaction can be used to prepare dibenzoxanthones derived from two different esters. An improvement in the technique involving the use of high boiling solvent has made the route suitable for the production of substantial quantities of fused xanthones and chromones ⟨79JHC1663⟩.

There is still uncertainty about the mechanistic course of this reaction. The isolation of the dibenzodisalicylide (**527**) from the pyrolysis of the ester (**526**) at 205–210 °C points to its intermediacy, since at higher temperatures it is converted into xanthone (Scheme 194).

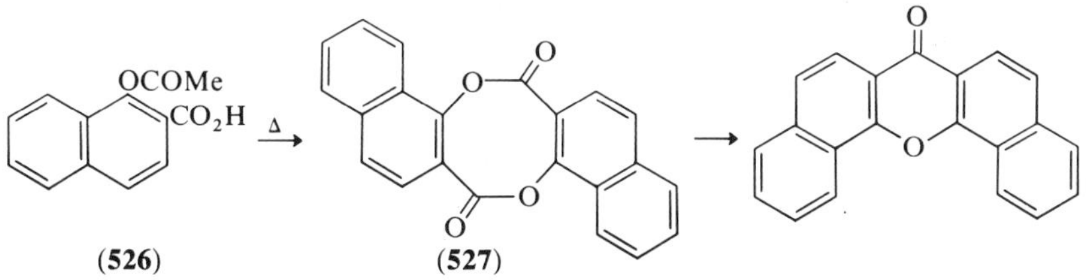

(**526**)                    (**527**)

**Scheme 194**

### 2.24.3.6.2 From a preformed heterocyclic ring

#### (i) From pyranones

Decarboxylative dimerization of the pyranone (**528**) with acetic anhydride gives the bis-pyranone (**529**), which is a protected pentaketo dicarboxylic acid. Reaction with methanolic potassium hydroxide gives the xanthone, the regiospecificity suggesting that the pyranone rings open and recyclize one at a time during the rearrangement (Scheme 195) ⟨71T3051⟩.

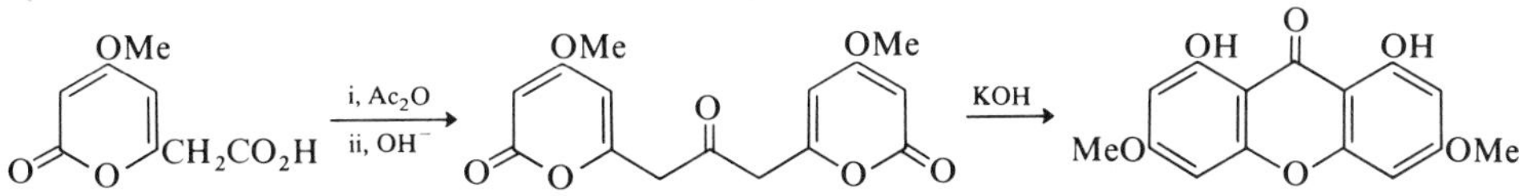

(**528**)                    (**529**)

**Scheme 195**

#### (ii) From chromones

Interest in the antibiotic bikaverin (**531**), a fungal metabolite with a benzo[*b*]xanthone skeleton, has culminated in its total synthesis ⟨76JCS(P1)499⟩. The chromone moiety (**530**) was constructed as shown and the xanthone was obtained by cyclization of the acid chloride (Scheme 196).

In an alternative approach to the tetracyclic molecule, the cyclization of the diketobenzopyranone ester (**532**) gave the xanthone 2-ketoester (**534**) as the major product (Scheme i, ZnCl$_2$, HCl; ii, (CO$_2$Et)$_2$, NaOEt, THF; iii, Me$_2$SO$_4$; iv, KOH; v, SOCl$_2$; vi, BF$_3$·Et$_2$O; vii, K$_2$Cr$_2$O$_7$; viii, LiI

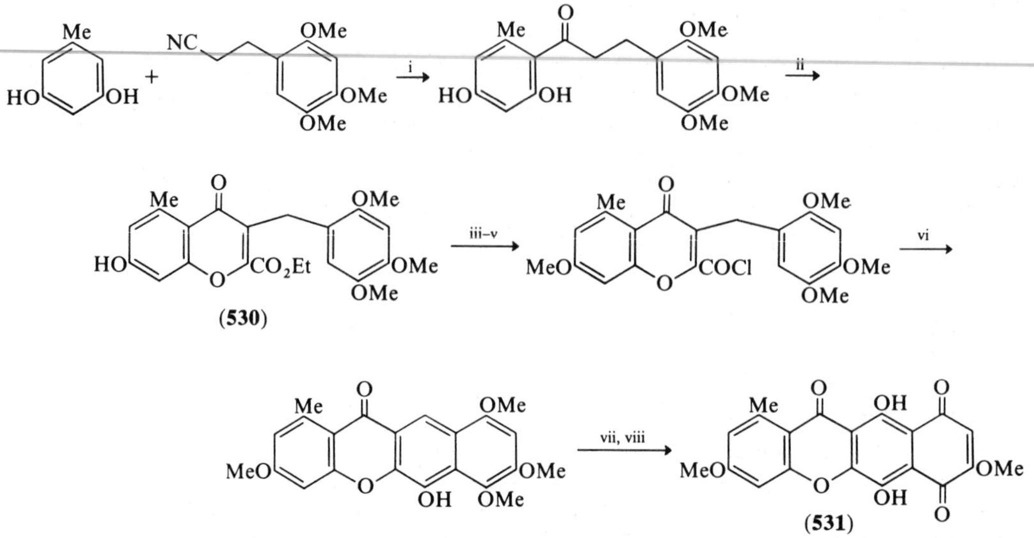

**Scheme 196**

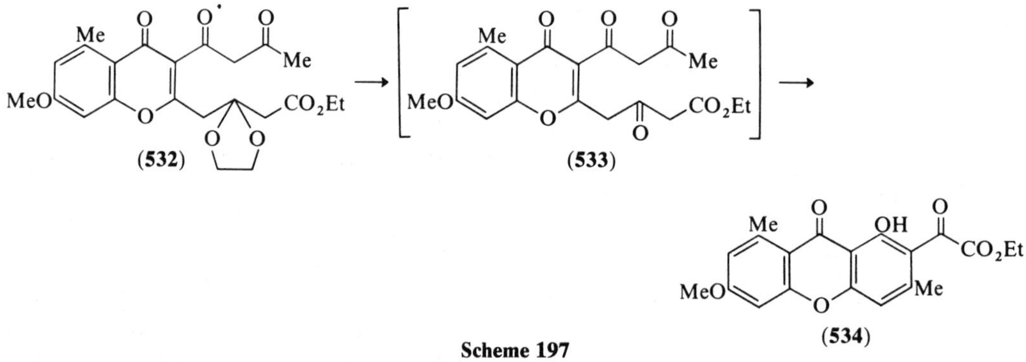

**Scheme 197**

197) ⟨79JCS(P1)3190⟩. This abnormal product is considered to arise from a Wessely–Moser rearrangement of the triketone (**533**).

The conversion of 2-methyl-3-(2-methylbenzoyl)chromone into the benzoxanthone has been achieved photochemically in high yield ⟨75JCS(P1)1845⟩. Initial photoenolization is followed by a [1,7]-sigmatropic shift and cyclization.

Aromatization of the fused chromone (**535**) is achieved by heating with sulfur, providing a useful xanthone synthesis from 4-methoxycyclohexanone ⟨66JOC1232⟩.

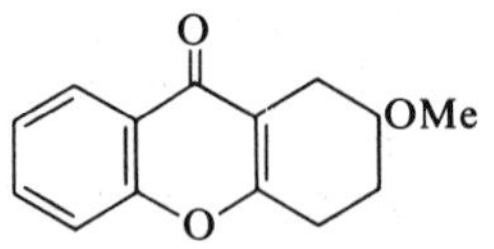

(iii) *From xanthenes*

Oxidation of xanthenes has been accomplished by reaction with sulfur and hydrolysis of the resulting thione ⟨45JCS858⟩.

Xanthydrol is oxidized by potassium persulfate ⟨56MI22400⟩, whilst xanthydryl chloride yields xanthone on reaction with silver nitrate ⟨59JCS458⟩.

(iv) *From xanthylium salts*

Xanthylium perchlorate has been oxidized to xanthone by manganese dioxide ⟨68AC(R)251⟩.

## 2.24.3.7 Reduced Pyranones

The compounds covered in this section include dihydropyranones, tetrahydropyranones and their benzologues (dihydrocoumarins, chromanones and isochromanones). The area of greatest interest is undoubtedly the chromanones because of their relationship to a number of natural products and presumably also because of their ease of formation, stability and value as precursors of other heterocycles. Tetrahydropyran-2-ones comprise one of those nebulous areas of heterocyclic chemistry and usually feature in text books as δ-lactones under derivatives of hydroxy acids.

### 2.24.3.7.1 Dihydropyran-2-ones

(i) *Formation of one bond*

(a) *Adjacent to the heteroatom.* A synthesis of 5,6-dihydropyran-2-one from but-3-enoic acid and paraformaldehyde is described in detail ⟨77OS(56)49⟩.

Titanium(IV) chloride is used as the catalyst in a Knoevenagel reaction between various 2,2-disubstituted 3-hydroxypropanals and malonic acid or its esters. The products are substituted dihydropyran-2-ones (536) ⟨79LA751⟩. The reaction, which occurs cleanly and in good yield, utilizes an excess of the titanium halide and is thought to involve a cyclic complex which undergoes an ester exchange to a lactone complex (Scheme 198).

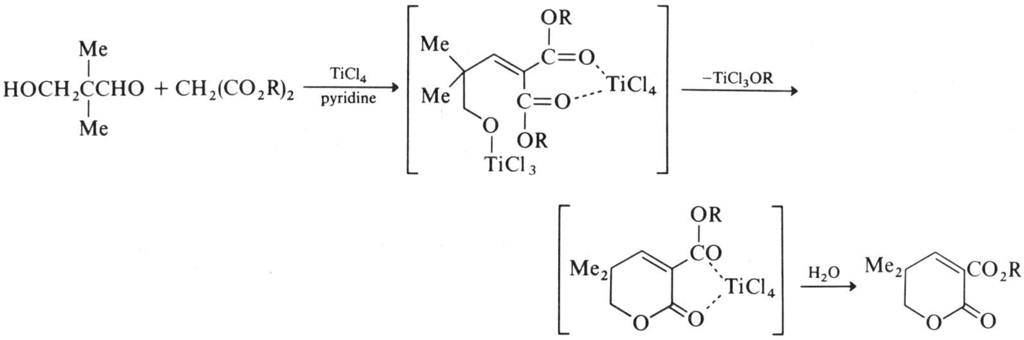

**Scheme 198**

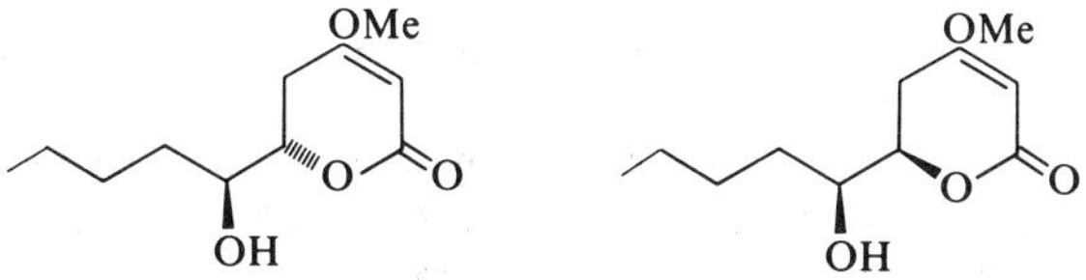

(537)        (538)

The same catalyst has also been used in the synthesis of 5-hydroxy-3-ketoesters and hence dihydropyranones from aldehydes and diketene ⟨75CL161⟩. The method has been used to synthesize pestalotin (537), a synergist of gibberellin isolated from the fungus *Pestalotia cryptomeriaecola* ⟨78CL409⟩. The product is an epimeric mixture of pestalotin and epipestalotin (538) in which the former is very predominant, a feature which has been attributed to coordination of the titanium(IV) chloride to the two oxygen atoms of the aldehyde, which is thus held in a fixed conformation. Diketene attacks the aldehyde from the less hindered side of the carbonyl group leading to the *threo* product. Subsequent reactions occur with retention of configuration.

Michael addition of ethyl phenylacetate to a number of ethyl 2-acetylcinnamates leads to dihydropyran-2-ones (Scheme 199) ⟨76IJC(B)739⟩, as does a Reformatski reaction of ethyl 4-bromo-3-methoxybut-2-enoate with unsaturated aldehydes ⟨81JHC363⟩.

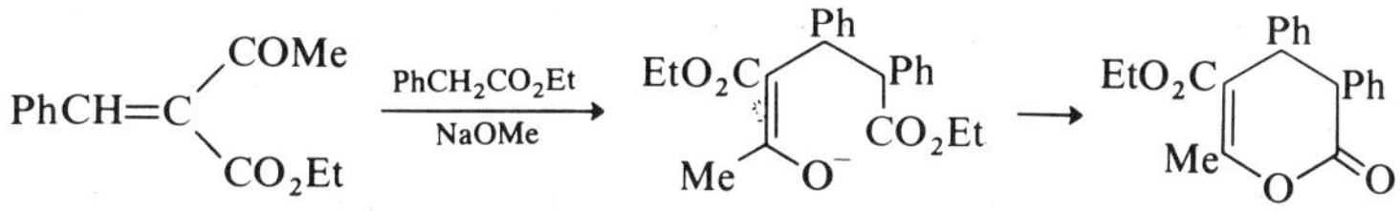

**Scheme 199**

The dianion derived from but-2-ynoic acid reacts with aldehydes to give 5-hydroxyalk-2-ynoates (**539**). Partial reduction over a Lindlar catalyst and acid-catalyzed cyclization of the resulting enoate gives the dihydropyran-2-one ⟨78LA337⟩. The route is exemplified by the synthesis of the naturally occurring massoia lactone (Scheme 200). In previous work ⟨46JCS954⟩ the hydroxyalkynoic acids themselves, obtained from epoxides and acetylene, were used.

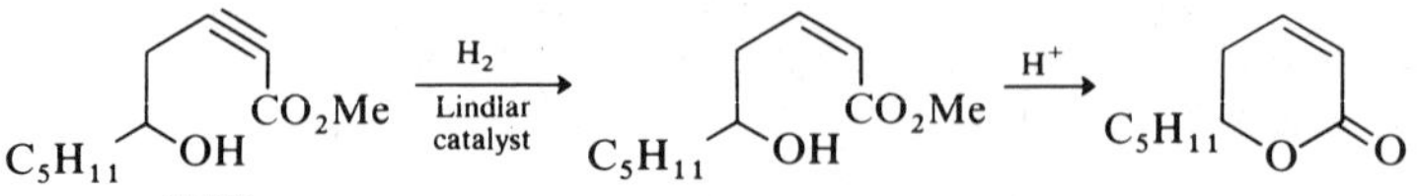

**Scheme 200**

The reaction of homophthalic acids or anhydrides with citral in the presence of pyridine yields reduced pyran-2-ones by means of an aldol condensation involving the activated methylene group of the acid. Dehydration is followed by an acid-catalyzed rearrangement to the pyran-2-one (Scheme 201) ⟨79H(12)253⟩.

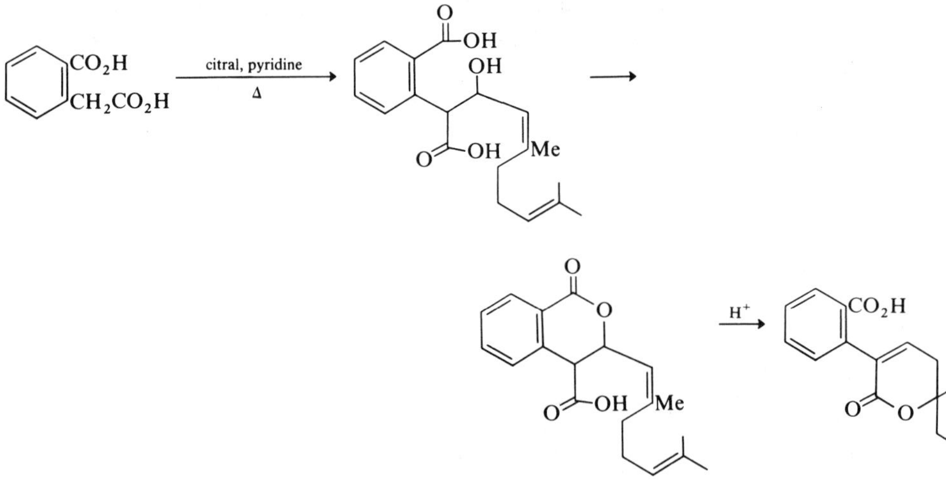

**Scheme 201**

β-Cyclocitral (**540**) behaves as a nucleophile towards benzaldehyde and in the presence of sodium hydroxide the fused pyran-2-ol (**541**) is formed through an aldol (Scheme 202) ⟨81JHC549⟩. However, when sodium ethoxide is used as the catalyst, the pyran-2-one is produced. It is proposed that oxidation arises through a crossed Cannizzaro reaction with benzaldehyde (and much benzyl alcohol is observed in support of this idea), followed by a carbanion attack on more benzaldehyde and subsequent cyclization.

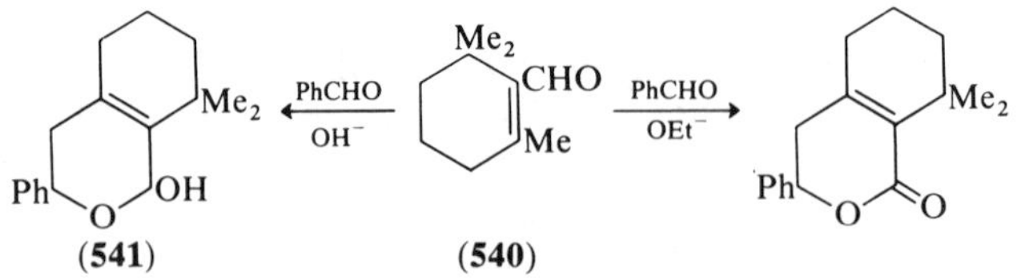

**Scheme 202**

(b) β to the heteroatom. Cyclization of the acetate (**542**) occurs under basic conditions and leads to the fused dihydropyran-2-one, presumably *via* the 4-hydroxy compound (**543**) (Scheme 203). The starting material is conveniently prepared from buta-1,3-dienyl acetate through a Diels–Alder cycloaddition with 2-methylprop-2-enal ⟨81JCS(P1)1096⟩. In contrast,

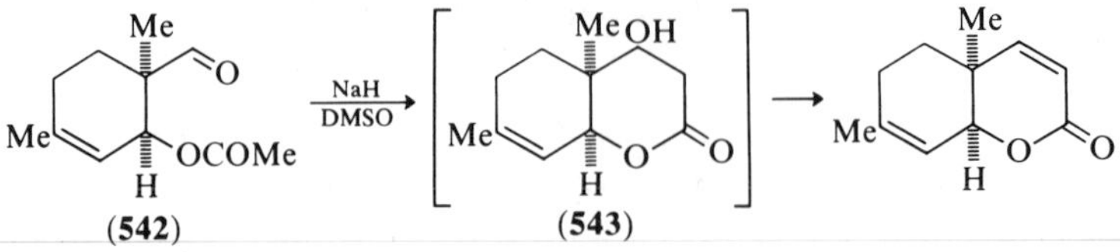

**Scheme 203**

the closely related 6-acetyl-3,6-dimethylcyclohex-2-enyl acetate yields the tetrahydropyran-4-one rather than the dihydropyran-2-one.

## (ii) *Formation of two bonds*

Diethyl oxomalonate reacts with a number of 1,3-dienes to form the diesters (**544**; R = Et), which are hydrolyzed to the dicarboxylic acids (**544**; R = H). Release of the lactone carbonyl group has been achieved in several ways. The conventional oxidative decarboxylation of alkyl-substituted malonic acids using lead(IV) acetate gives only low yields of the dihydropyranone, but Curtius degradation of the acid is more satisfactory. The acyl azide is formed from the diacid chloride using trimethylsilyl azide or sodium azide, and warming effects the rearrangement to the isocyanate which is hydrolyzed under mildly acidic conditions (Scheme 204) ⟨77JOC4095⟩. The acids may also be degraded by anodic oxidation, the ortho ester (**545**) being an intermediate product ⟨79SC889⟩.

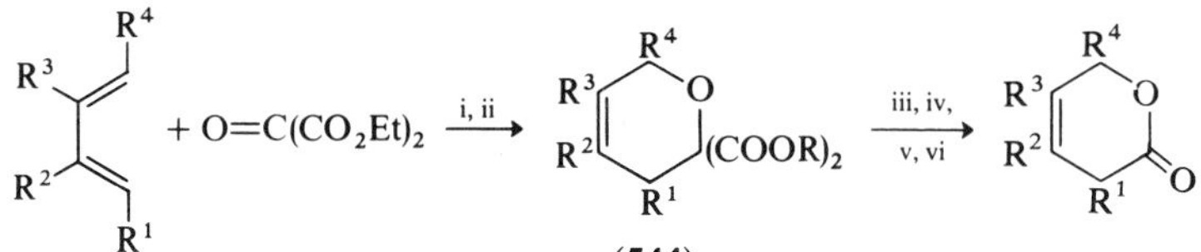

i, MeCN, 130 °C; ii, aq. KOH, THF; iii, (COCl)$_2$; iv, NaN$_3$; v, Δ; vi, H$_2$O/H$^+$

**Scheme 204**

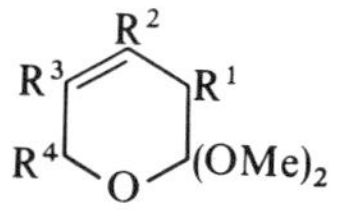

(**545**)

## (iii) *From a preformed heterocyclic ring*

(*a*) *From dihydropyrans.* The dye-sensitized photooxygenation of dihydropyrans provides a useful route to a number of naturally occurring unsaturated δ-lactones ⟨81HCA1247⟩. The initial product is an allylic hydroperoxide (**546**) which undergoes a base-catalyzed dehydration to the dihydropyranone (Scheme 205) ⟨73JA5820⟩.

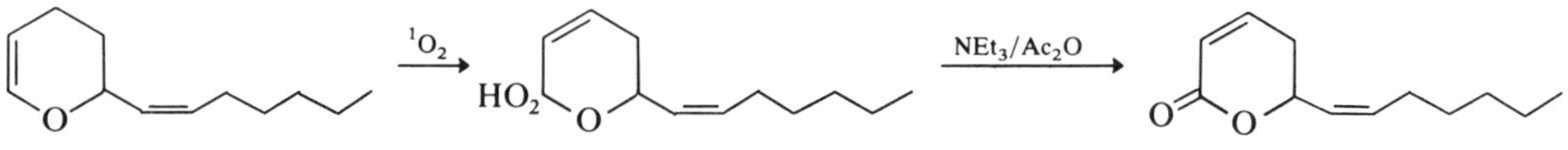

(**546**)

**Scheme 205**

(*b*) *From pyran-2-ones.* Partial reduction of dehydroacetic acid yields the dihydropyran-2-one ⟨81JHC543⟩, and the reduction of 6-methyl-4-hydroxypyran-2-one has been studied in some detail ⟨78JHC1153⟩.

(*c*) *From dihydropyran-2-ones.* Reduction of 3,6-dihydropyran-2-ones with LAH and oxidation of the resulting diols lead to 5,6-dihydropyran-2-ones ⟨77JOC4095⟩.

### *2.24.3.7.2 Dihydropyran-3-ones*

## (i) *Formation of one bond*

(*a*) *Adjacent to the heteroatom.* Although 3-methyleneheptane-2,6-diol and 2,6-diacetoxyheptan-3-one cyclize exclusively to tetrahydrofurans, 6-hydroxy-5-methyl-eneheptan-2-one (**547**) undergoes an intramolecular acetalization to the pyran (**548**) ⟨81JCS(P1)1015⟩. Oxidation of the methylenepyran to the dihydropyran-3-one is accomplished with ozone (Scheme 206).

(*b*) *β to the heteroatom.* Hydration of di-2-propynyl ether followed by an intramolecular aldol condensation affords the 5-hydroxy-5-methylpyran-3-one (**549**), which undergoes a spontaneous dehydration to the 5-methylpyran-3-one (**550**; Scheme 207) ⟨80ACS(B)295⟩. The corresponding 5-phenyl derivative results in a similar manner from acetonyl phenacyl

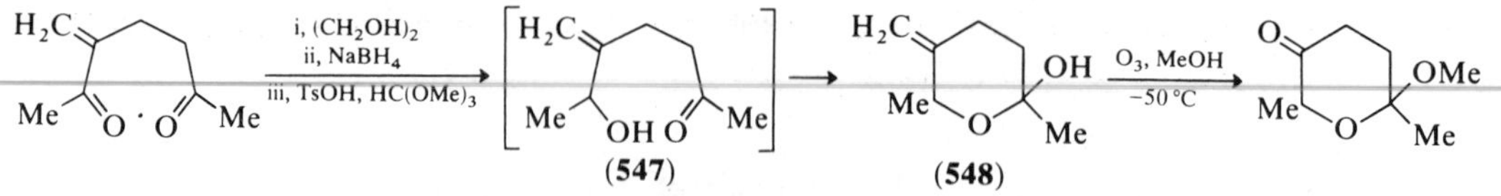

**Scheme 206**

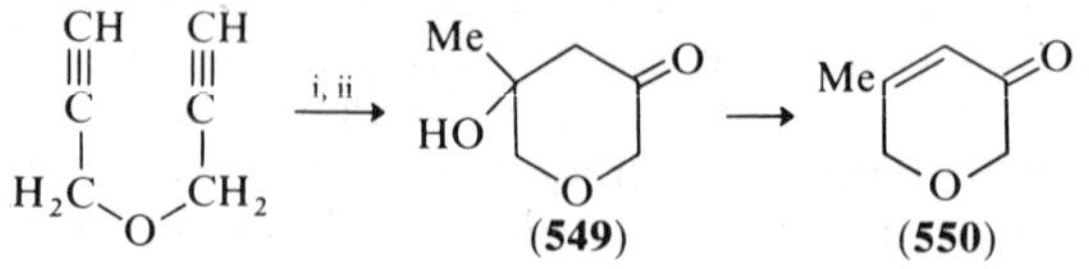

i, HgSO$_4$, H$_2$SO$_4$, 60–80 °C; ii, aq. NaOH

**Scheme 207**

ether. The formation of the 2*H*-isomer rather than the 4*H*-compound in the dehydration step is attributed to the greater thermodynamic stability of the conjugated enone.

### (ii) *From other heterocycles*

(*a*) *From furans.* Several approaches to 6-hydroxypyran-3-ones from furans have been reported and these are outlined in Scheme 208. Furfuryl alcohols (**551**) are oxidized directly to the 6-hydroxypyran-3-one by peracids ⟨72TL133⟩ or by pyridinium chlorochromate ⟨77TL2199⟩. The preliminary conversion into the dihydrofuran (**552**), achieved with bromine in methanol and followed by an acid-catalyzed rearrangement, constitutes a useful route to pyran-3-ones ⟨76T1051⟩, which has been extended to provide a direct entry to the 6-alkoxy analogues (**553**) ⟨77MI22401⟩.

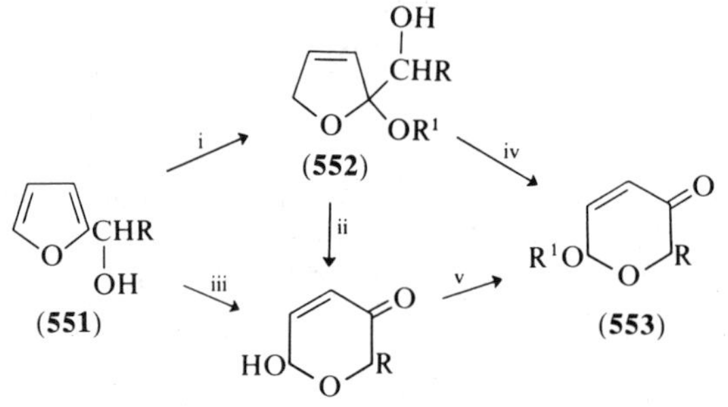

i, Br$_2$, MeOH; ii, H$_3$O$^+$; iii, RCO$_3$H; iv, HCO$_2$H; v, HC(OR$^1$)$_3$, SnCl$_4$

**Scheme 208**

### 2.24.3.7.3 *Dihydropyran-4-ones*

#### (i) *Formation of one bond*

(*a*) *Adjacent to the heteroatom.* A mixture of two diastereoisomeric racemates of 2,3-dihydro-2,3,5-trimethyl-6-(1-methyl-2-oxobutyl)-4*H*-pyran-4-one (**554**), which is the basic structure of stegobinone, the pheromone of the drugstore beetle, has been synthesized using the anions derived from di- and tri-ketones ⟨81T709⟩. The activity of the synthetic *cis* isomer was less than that of the natural pheromone.

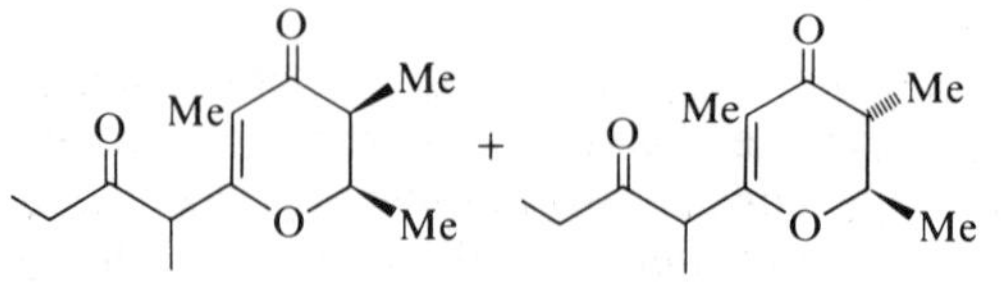

A chiral synthesis of stegobinone (**555**) is based on the cleavage of (2*S*,3*S*)-2,3-epoxybutane with lithium diphenylcuprate, which occurs with complete inversion. The resulting chiral alcohol is converted into the mixed anhydride, which is used to acylate the dianion derived from 4-methylheptane-3,5-dione. Cyclization with acid yields (2S,3R,7RS)-stegobinone identical with the natural pheromone (Scheme 209) ⟨81T709⟩.

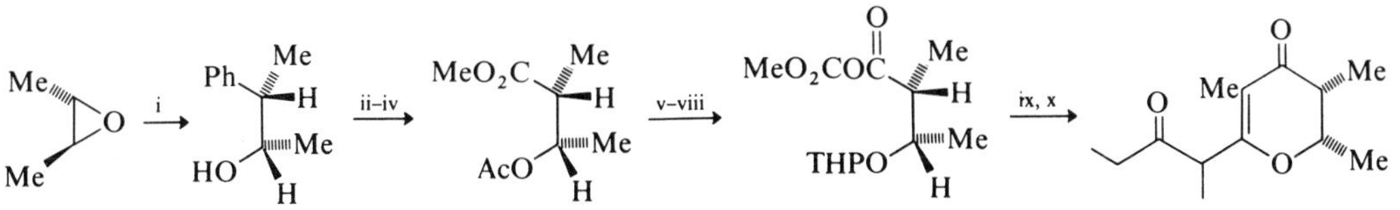

i, Ph₂CuLi; ii, Ac₂O, pyridine; iii, O₃, H₂O₂, Pt; iv, CH₂N₂; v, HCl, MeOH; vi, DHP, *p*-TsOH; vii, aq. KOH; viii, ClCO₂Me, Et₃N; ix, EtCOCH(Me)COEt, LDA; x, HCl, MeCN

**Scheme 209**

$$Me_2C(OH)C{\equiv}CCH{=}CHOMe$$

**(556)**

The cyclization of the alkynic enol ether **(556)** leads to a 5,6-dihydropyran-4-one ⟨63JOC687⟩.

### (ii) *Formation of two bonds*

Interest has been expressed in the use of dihydropyranones as synthons for natural products and this has led to the development of a valuable route to 5,6-dihydropyranones. Electron-rich dienes bearing a trimethylsilyloxy substituent undergo cycloadditions with carbonyl compounds at room temperature in the presence of a Lewis acid (Scheme 210) ⟨82JA358⟩. The intermediacy of a reduced pyran has been postulated.

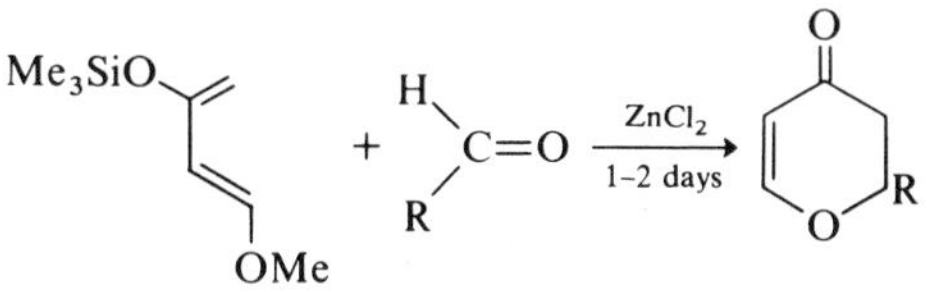

**Scheme 210**

### (iii) *From other heterocycles*

(a) *From epoxides.* Substituted dihydropyran-4-ones are formed when aliphatic or alicyclic acetyl epoxides **(557)** are treated with esters of perfluorinated carboxylic acids in the presence of sodium ethoxide ⟨78CHE721⟩. The reaction proceeds through the 1,3-diketone **(558)**, cyclization of the enolic form of which occurs with inversion of configuration at the β-carbon atom of the epoxide ring (Scheme 211).

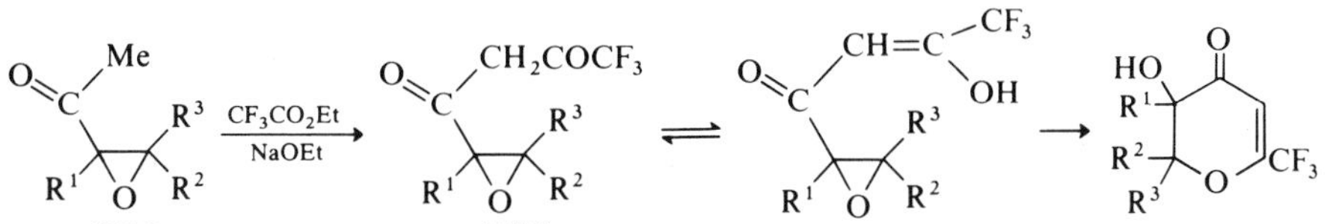

**Scheme 211**

### 2.24.3.7.4 *Tetrahydropyran-2-ones and tetrahydropyran-4-ones*

#### (i) *Formation of one bond*

(a) *Adjacent to the heteroatom.* Pentanedial is a convenient source of tetrahydropyranones. Reaction with an alkyl Grignard reagent affords the reduced pyranol **(559)** which may be oxidized to the lactone, preferably using silver oxide in methanolic sodium hydroxide (Scheme 212) ⟨72HCA249⟩. At a low temperature a Grignard reaction at the second carbonyl group occurs only to a minor extent.

The reduction of the unsaturated ketone **(560)** can be controlled to give mainly the *cis*-cyclopentanone **(561)**. Upon treatment with sodium hydroxide isomerization to the *trans* compound occurs. Both isomers undergo a Baeyer–Villiger oxidation to the corresponding tetrahydropyranones, of which the *cis* isomer is rather unstable (Scheme 213) ⟨82CJC29⟩.

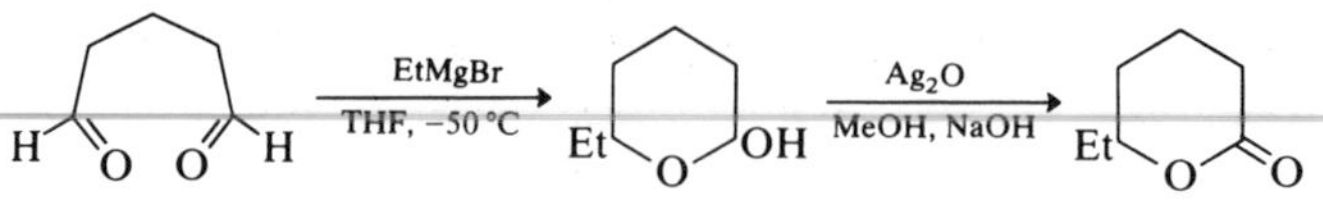

**Scheme 212**

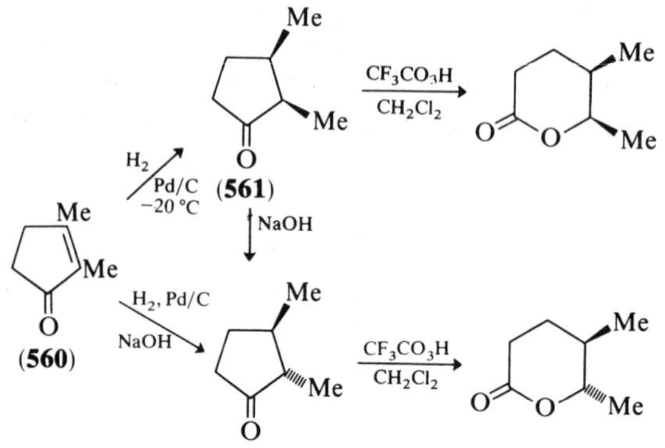

**Scheme 213**

The fused analogue, an oxadecalone, is available from tetrahydroindanone in a similar way.

A Baeyer–Villiger oxidation is also involved in the partial synthesis of xylomollin (**563**) from (−)-methoxyloganin aglycone (**562**; Scheme 214) ⟨78JA7079⟩.

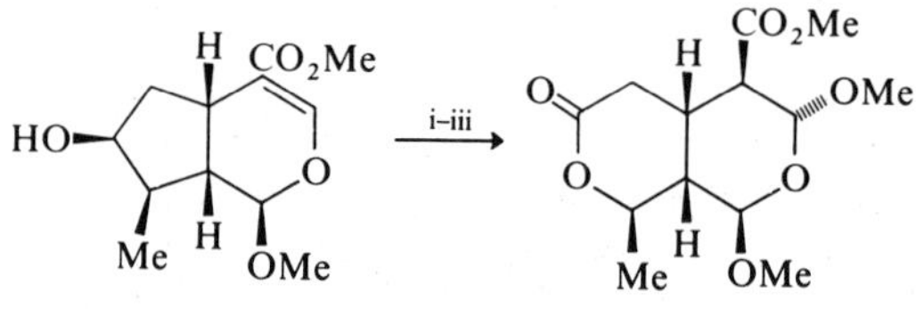

i, Mg(OMe)₂, MeOH; ii, C₅H₅NHCrO₃Cl⁻; iii, MCPBA

**Scheme 214**

The synthesis of all four optical isomers of 4,6-dimethyltetrahydropyranone has been achieved from 3-methyl-5-oxohexanoic acid ⟨74JOC3890⟩.

In view of the importance of mevalonate in the biosynthesis of terpenes and steroids, it is not surprising that numerous syntheses of mevalonolactone are available. These are briefly itemized in the detailed procedure for the synthesis of (*R,S*)-mevalonolactone labelled at C-2 with $^{13}$C (Scheme 215) ⟨81OS(60)92⟩.

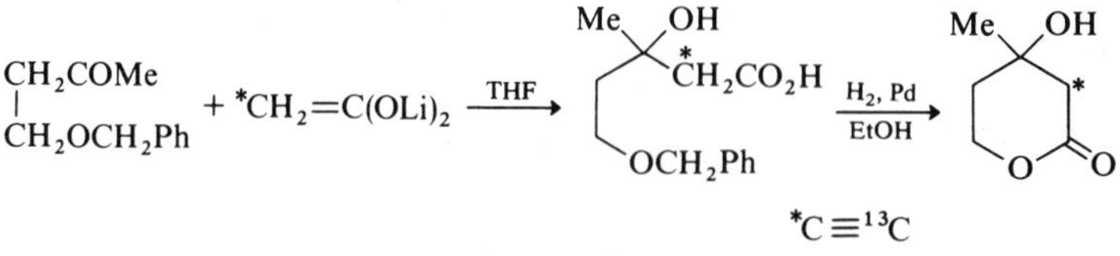

**Scheme 215**

The spirolactone (**564**) results from the cyclization of 4-cyclohexylidenebutanoic acid with sulfuric acid, although in polyphosphoric acid the heterocycle was formed in minor quantities, a hexahydronaphthalenone being the main product ⟨75AJC2669⟩.

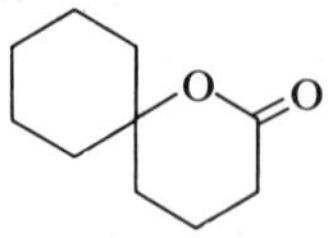

(**564**)

Spirolactones may also be prepared from cyclic anhydrides by the action of 1,5-bis(bromomagnesio)pentane ⟨81JOC3091⟩. The reaction proceeds through the keto acid arising from attack by the Grignard reagent on one of the carbonyl groups of the anhydride. A subsequent intramolecular attack on the carbonyl group of the keto carboxylate generates the new ring.

The fused tetrahydropyran-2-one (**566**) is obtained from 2-methylcyclohexanone by Michael addition to methyl prop-2-enoate and reduction of the resulting keto ester (**565**; Scheme 216) ⟨63JOC34⟩. When the enamine derived from the cyclohexanone reacts with the unsaturated ester, a mixture of keto esters (**565**) and (**567**) is formed. The pyranone (**568**) is formed by reduction of the latter.

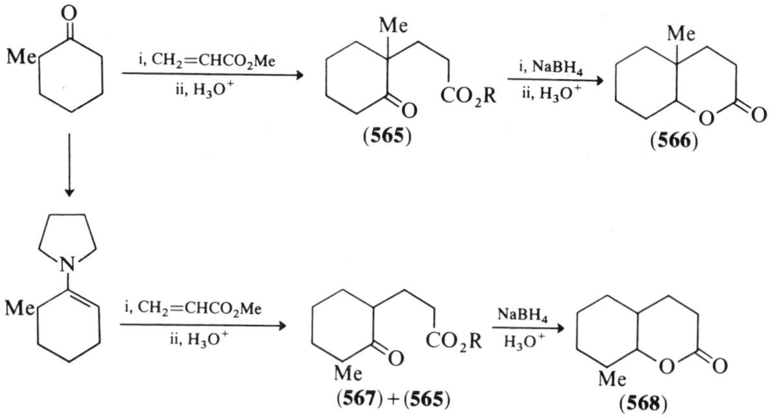

**Scheme 216**

## (ii) *From other heterocycles*

(a) *From furans.* The ring expansion of the γ-lactone (**569**) complements the formation of the tetrahydropyranones by the Baeyer–Villiger route. The starting materials are available through a Diels–Alder reaction and the stereochemistry of this process is retained throughout he reaction sequence (Scheme 217) ⟨75JA7182⟩.

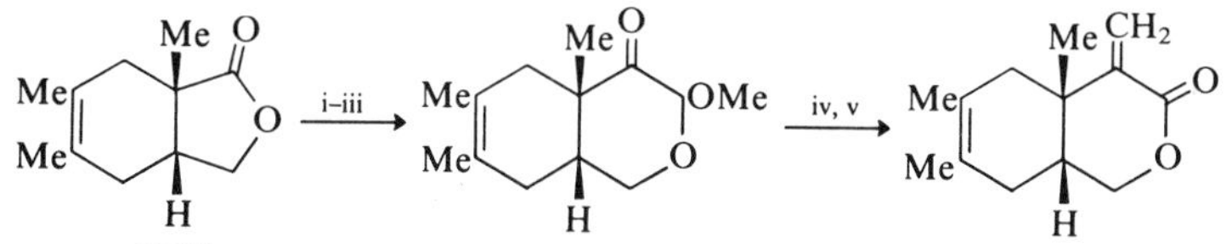

i, PhSC̄HOMe Li⁺; ii, I₂, MeOH; iii, H₃O⁺, C₆H₆; iv, CH₂=PPh₃; v, HCl, THF

**Scheme 217**

## (iii) *From a preformed heterocyclic ring*

(a) *From dihydropyrans.* The acid-catalyzed addition of hydrogen peroxide to a dihydropyran affords a hydroperoxide which on dehydration gives a tetrahydropyran-2-one ⟨81HCA1247⟩. The synthesis of jasmine lactone (**570**) is illustrative.

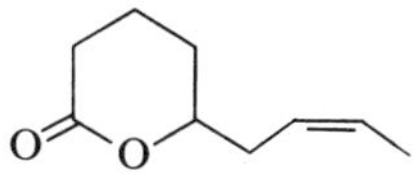

(**570**)

(b) *From pyranones.* Although in many cases the reduction of pyran-2-ones leads to a mixture of the tetrahydro compound and the open-chain derivative ⟨55JA2340⟩, the cyclic compound may sometimes be obtained alone. For instance, 4-hydroxy-6-methylpyran-2-one is fully reduced to the tetrahydropyranone by hydrogen in the presence of palladium–charcoal and copper sulfate ⟨78JHC1153⟩.

Pyran-4-ones tend to give complex mixtures on chemical reduction, but catalytic hydrogenation usually gives the tetrahydropyran-4-one or the corresponding pyran-4-ol ⟨63CR(256)1542⟩. The influence of solvent on the nature of the product is exemplified by the reduction of 2,6-dimethylpyran-4-one, which in ethanol affords the fully reduced pyranone, but in decalin the product is the pyranol, whilst in acetic acid a mixture of the two compounds is produced.

The reduction of dihydropyran-4-one is also reported ⟨63JOC687⟩.

### 2.24.3.7.5 Dihydrocoumarins

(i) *Formation of one bond*

(a) *Adjacent to the heteroatom.* The condensation of phenols with cinnamic acid in the presence of hydrochloric acid yields 3,4-dihydro-4-phenylcoumarins ⟨56JCS1382⟩. The product is contaminated with the *p*-hydroxydiphenylpropanoic acid if the *para* position of the phenol is unsubstituted. Sulfuric acid may also be used to bring about the reaction, although sulfonation competes with dihydrocoumarin formation with the more reactive phenols ⟨73AJC899⟩.

Propenonitrile can also function as the three-carbon unit, as in the synthesis of 3,4-dihydro-7-hydroxycoumarin ⟨74IJC564⟩.

The reaction of 3-(3,4-dimethoxyphenyl)propanoic acid with thallium(III) trifluoroacetate in the presence of boron trifluoride etherate leads to a mixture of the dihydrocoumarin (574) and the spirolactone (572) ⟨78JOC3632⟩. It is suggested that these products arise through an initial one-electron oxidation to the radical cation, the fate of which may vary. Thus, intramolecular reaction with the carboxyl group gives the radical (571) and eventually the spirolactone. Alternatively, capture of the radical ion by solvent and further oxidation affords the radical (573), whereupon an intramolecular Michael addition to the carboxyl group and aromatization lead to the dihydrocoumarin (Scheme 218) ⟨81JA6856⟩.

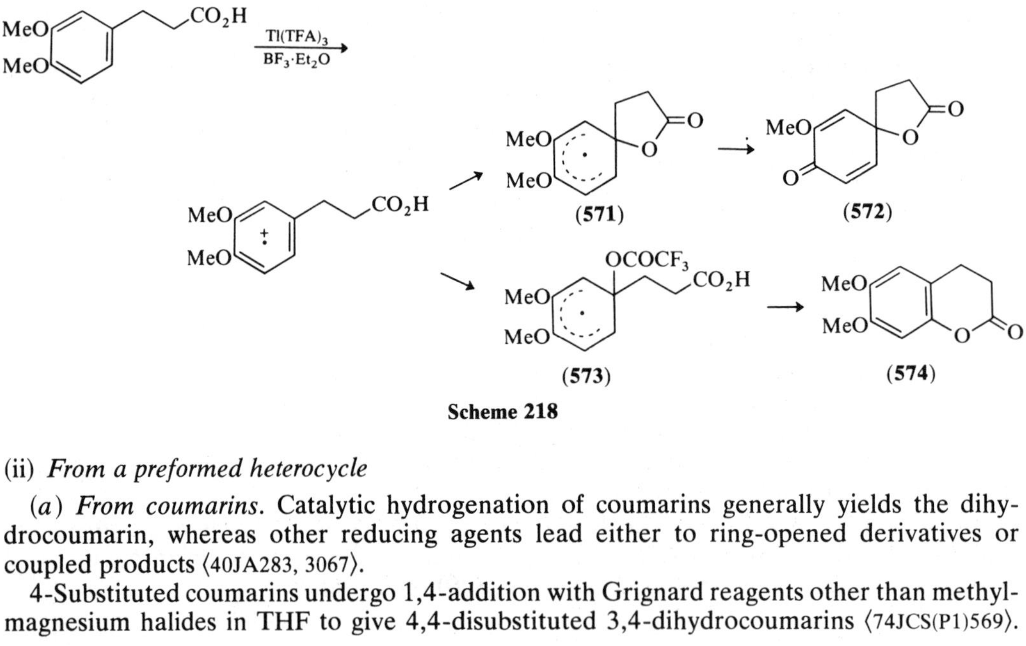

**Scheme 218**

(ii) *From a preformed heterocycle*

(a) *From coumarins.* Catalytic hydrogenation of coumarins generally yields the dihydrocoumarin, whereas other reducing agents lead either to ring-opened derivatives or coupled products ⟨40JA283, 3067⟩.

4-Substituted coumarins undergo 1,4-addition with Grignard reagents other than methylmagnesium halides in THF to give 4,4-disubstituted 3,4-dihydrocoumarins ⟨74JCS(P1)569⟩.

### 2.24.3.7.6 Chromanones

In this section, four groups of compounds will be considered: chroman-3-ones (575), chroman-4-ones (576), the flavanones (577) and isoflavanones (578). Whilst there is extensive literature on chroman-4-ones and their derivatives, very little work has been carried out on chroman-3-ones, possibly because of their relative instability.

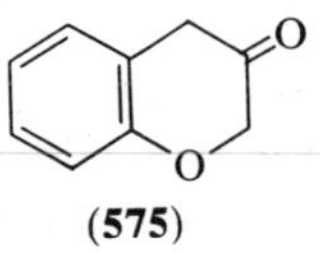

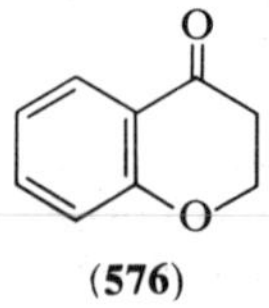

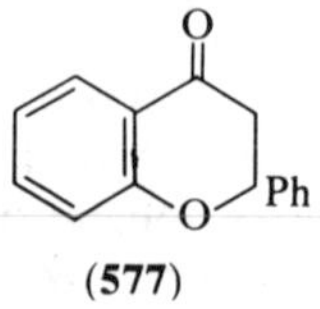

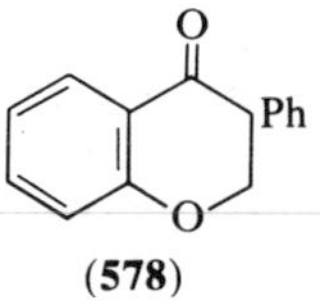

## (i) Formation of one bond

(a) *Adjacent to the heteroatom.* Perhaps the most widely used route to chroman-4-ones is based on the reaction between a phenol and an unsaturated acid, an approach which is particularly suitable for the synthesis of chromanones substituted at C-2. Much emphasis has been placed on the synthesis of chromanones with methoxy substituents in the aromatic ring because of the presence of such groups in the naturally occurring flavonoids. The synthesis essentially consists of the cyclization of an o-hydroxy unsaturated ketone, although this intermediate is infrequently isolated. Ring closure proceeds through an intramolecular conjugate addition to the enone system (Scheme 219).

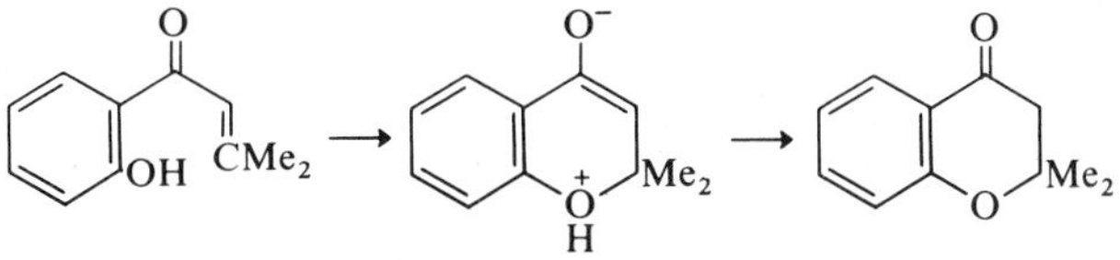

**Scheme 219**

Much of the value of this route lies in the ready availability of the ketone precursors and in the variety of conditions which may be used to effect cyclization.

Reactions in this group may be divided into two categories: those which may be described as Friedel–Crafts syntheses and the closely related preparations which involve a Fries rearrangement prior to a Friedel–Crafts reaction.

Anisole and many of its derivatives react with an unsaturated acid chloride in the presence of aluminum chloride to yield a chromanone ⟨20LA(421)1⟩. The two most widely used acid halides are but-2-enoyl (crotonyl) and 3-methylbut-2-enoyl (3,3-dimethylacryloyl) chlorides. In cases where the uncyclized intermediate, an acrylophenone, is isolated, cyclization may be achieved under acidic or basic conditions, or even simply by distillation (Scheme 220) ⟨52G155⟩.

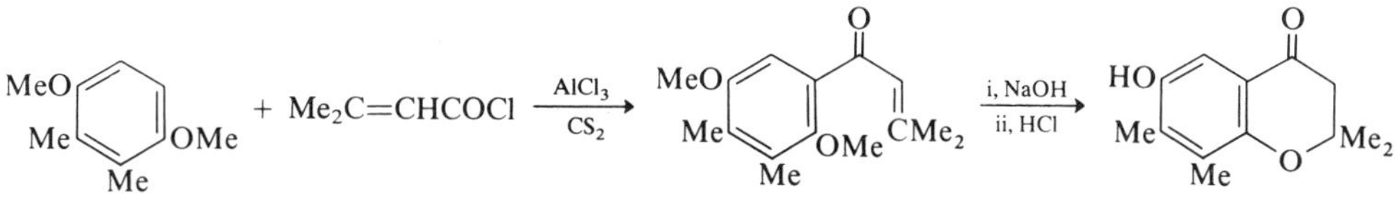

**Scheme 220**

There are instances where a mixture of the uncyclized hydroxyketone and the chromanone is obtained ⟨58BSB22⟩.

The usual range of solvents associated with Friedel–Crafts reactions is available; carbon disulfide and nitrobenzene have found frequent use. The choice of solvent may influence the extent of cyclization. Thus, in ether, 1,2,3,5-tetramethoxybenzene yielded the uncyclized product with 3,3-dimethylacryloyl chloride, whereas in a mixture of ether and tetrachloroethane the major product was the chromanone. This example is also of interest since ring closure could involve either the 2- or 6-methoxy group. In fact, the formation of the 5,6,7-trimethoxy isomer was not observed. The identity of the product was established by conventional [14]C labelling ⟨60BSB593⟩.

During a study of the naturally occurring chromanone, papuanic acid, it was observed that the dihydrocoumarin (**579**) gave the chromanone (**581**) on reaction with 3,3-dimethyl-acryloyl chloride rather than the expected product (**582**) ⟨68JOC4191⟩. It was suggested that the demethylation was facilitated by the favourable orientation in which the new side-chain was held in the aluminum complex (**580**). The generation of a tertiary carbocation following protonation of the double bond also appears to be important, because demethylation was not observed in the analogous derivative from 2-methylbut-2-enoyl chloride (Scheme 221).

The reaction between a phenol and an unsaturated acid or a derivative in the presence of a Lewis acid has also been widely used for the preparation of chroman-4-ones. The solvents employed are generally those utilized in the previous method, namely carbon disulfide or nitrobenzene, but a wider range of catalysts has been used. There is also a greater choice in the unsaturated component, the acid, acid chloride and acid anhydride being especially valuable.

One limitation to the value of this route lies in the alternative mode of cyclization which can take place, leading to the five-membered heterocycle, a coumaranone. However, it was shown ⟨59JCS2425⟩ that only when bulky substituents are present in the 6-position of a

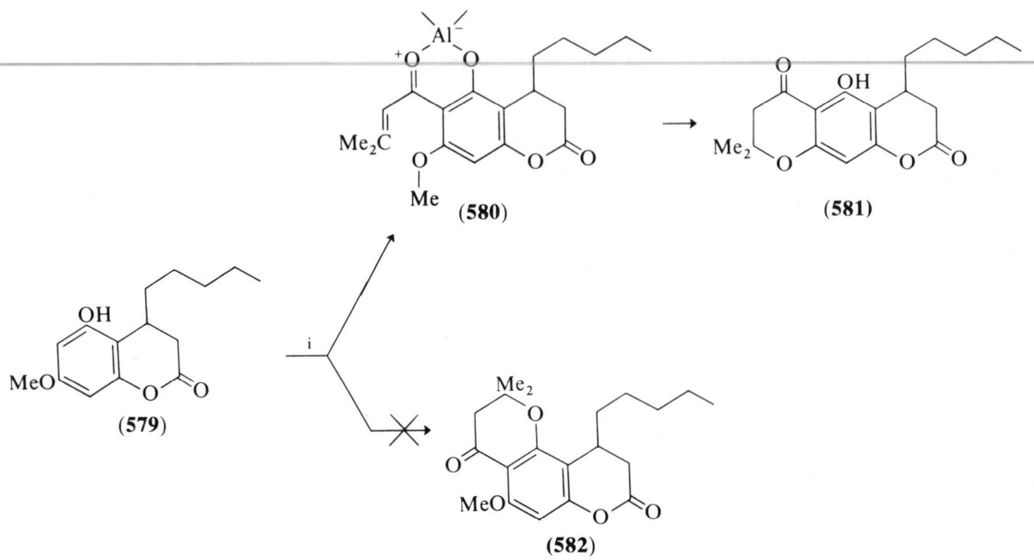

i, Me$_2$C=CHCOCl, AlCl$_3$, PhNO$_2$

**Scheme 221**

2-hydroxybenzoylacrylic acid is a coumaranone formed. For example, maleic anhydride and 3,5-dimethylphenol yield 4,6-dimethylcoumaranone-2-acetic acid, whereas 2-naphthol affords the benzochromanone (**583**).

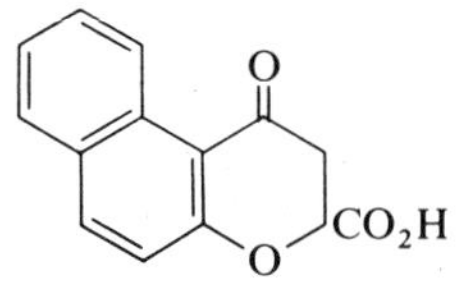

(**583**)

The cyclization of a number of 2-hydroxybenzoylacrylic acids has been studied ⟨71CJC3477⟩ and it was suggested that the *trans* isomers yield 4-oxochroman-2-carboxylic acids (**584**) but that the *cis* compounds ring-close to coumaranones (Scheme 222). The cyclization of both isomers occurs with ease under the influence of barium hydroxide or even weaker bases. The different products formed by the two compounds may be a consequence of the tendency of the *cis*-acid to exist as a lactol.

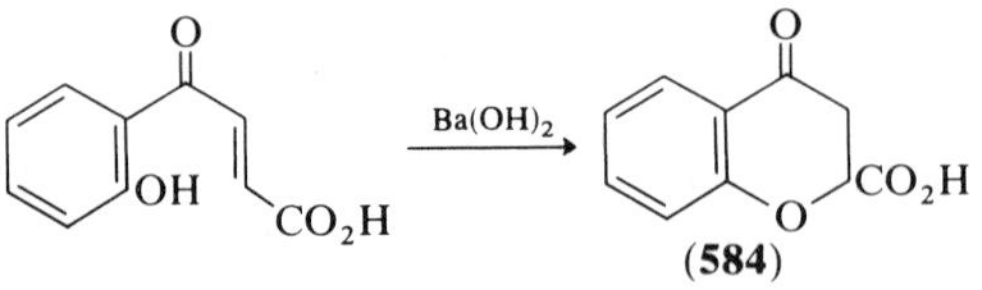

**Scheme 222**

The synthesis of chromanones *via* a Fries rearrangement differs from the previous route only in that the acrylic ester of the phenol is formed prior to the addition of the Lewis acid catalyst (Scheme 223).

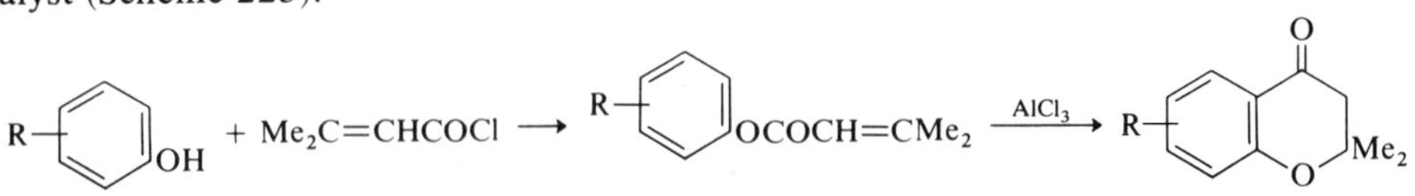

**Scheme 223**

The Fries rearrangement has been discussed ⟨42OR(1)342⟩ and some attention has been paid to its mechanism ⟨B-69MI22402⟩. It is sufficient here to note that both *ortho* and *para* migration of the acyl moiety takes place and accordingly some diminution in the yield of chromanone is to be expected.

It is customary to prepare the ester by reaction of the phenol with an unsaturated acid chloride, generally in the absence of solvents. Again, crotonic and 3,3-dimethylacrylic acid chlorides feature predominantly in the ester syntheses, because of their availability and because of the common occurrence of 2-methyl and, especially, 2,2-dimethyl substituents in natural oxygen heterocycles. Aluminum chloride is the normal catalyst and has been used in the presence or absence of solvent ⟨81T2613⟩. The reaction is readily extended to the preparation of benzochromanones although in this case hydrogen fluoride was used as the cyclizing agent ⟨68T949⟩.

It is also possible to utilize the unsaturated acid in this approach, although initial formation of the ester now requires assistance from an acid catalyst. Hydrogen fluoride has been used in this connection, but satisfactory yields are only obtained under pressure ⟨54LA(587)16⟩. Polyphosphoric acid has again proved to be a useful reagent, generally giving good yields, although crotonic acid appears to need a higher temperature than does dimethylacrylic acid to give acceptable yields ⟨71JCS(C)95⟩. The intermediacy of the ester has been established and indeed in certain cases it is the major product ⟨73BSB705⟩. For instance, 4-nitrophenol and crotonic acid yield the ester, which is recovered unchanged from further treatment with polyphosphoric acid. Apparently, chromanone formation is suppressed through deactivation of the aromatic ring by the nitro group. However, *p*-cresol yields chromanones with a variety of *trans* α,β-unsaturated acids.

The reaction of 1,3-dihydroxynaphthalene with 3,3-dimethylacrylic acid in the presence of phosphorus oxychloride and zinc chloride yields only 9-hydroxy-2,2-dimethyl-benzo[*f*]chroman-4-one (**585**). 2,7-Dihydroxynaphthalene similarly yields the angular benzochromanone (**586**) rather than the alternative linear product ⟨79RRC59⟩.

(**585**)  (**586**)

Boron trifluoride is an effective condensing and cyclizing agent for this type of reaction. The yields are generally superior to those obtained by the alternative methods, though it is again noteworthy that dimethylacrylic acid gives better yields than crotonic acid ⟨63ABC700⟩. This route was selected for the synthesis of two chromanones from 1,3-dihydroxy-5-*n*-pentylbenzene and 3,3-dimethylacrylic acid. A mixture of the 5- and 7-*n*-pentylchroman-4-ones (**587**) and (**588**) resulted, in a ratio which depended on the reaction temperature ⟨67JA5934⟩. The formation of a mixture of isomers from a *m*-substituted phenol is unusual in chromanone syntheses. Generally, exclusive formation of the 7-substituted isomer is observed. Such regioselectivity has been attributed to a steric effect in many instances, though additionally the favourable *p*-quinonoid structure of the intermediate carbocation may play a part.

(**587**)  (**588**)

A closely related reaction involves that between a saturated acyl halide and a phenol or phenolic ether. A necessary feature of the acid chloride is that it contains a bromine atom at C-2 which allows formation of a double bond during the reaction by loss of bromide. Normal Friedel–Crafts conditions are employed in the first step which leads to an *o*-hydroxyphenyl 2-bromoalkyl ketone (**589**). In boiling diethylaniline, hydrogen bromide is lost and the resulting acrylophenone spontaneously cyclizes to the chromanone ⟨24LA(439)132⟩.

(**589**) R = Br
(**590**) R = OH

Once again, it is important to realize that cyclization to a coumaranone is a feasible alternative reaction. Indeed, in the above example treatment of the initial ketone (**589**) with sodium hydroxide instead of diethylaniline gave 2,2-diethyl-5-methylcoumaran-3-one together with hydrolyzed but uncyclized ketone (**590**).

A mixture of lithium bromide and lithium carbonate in DMF effects the same process; again the formation of a five-membered ring is observed ⟨81CB147⟩.

A minor variant of this method makes use of the reaction between an anisole and a 3-halogenopropanoyl chloride ⟨14CB2585⟩. These same 3-substituted acid chlorides react with phenols to give esters and a Fries rearrangement is now a prerequisite of chromanone formation ⟨58JCS1190⟩.

If hydrogen fluoride is used as the catalyst, it is possible to use 3-halogenopropanoic acids in this route ⟨68CB2494⟩, whilst with boron trifluoride even 3-hydroxyalkanoic acids afford chromanones in a very fast reaction ⟨63T77⟩.

The cyclization of 1-diazo-3-(2-methoxyphenyl)propan-2-ones to chroman-3-ones is brought about by an acidic catalyst. The overall route is depicted in Scheme 224 ⟨70JHC197⟩.

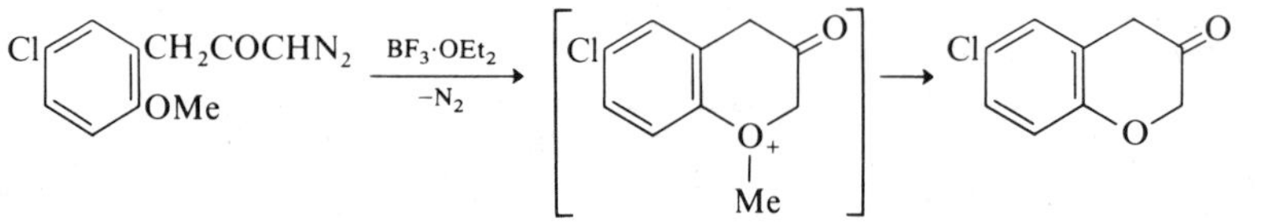

**Scheme 224**

2-Hydroxyacetophenone and related compounds are attractive precursors of chromanones, most notably in the synthesis of their 2-phenyl derivatives. Being adjacent to the carbonyl group, the methyl group of the acetophenone is activated and forms a carbanion on treatment with base. Subsequent condensation with a carbonyl compound which lacks an α-hydrogen atom leads to a 1,3-dicarbonyl or an enone system which readily cyclizes to a chromanone. Thus, methyl formate affords the chromanone (**591**) ⟨53MI22400⟩ and formaldehyde has been used in the synthesis of 3-methylchroman-4-one from *o*-hydroxypropiophenone ⟨68T949⟩.

A 2,2-disubstituted chromanone results from the condensation of *o*-hydroxyacetophenone with diethyl oxalate. The initially formed 1,3-diketone cyclizes spontaneously to ethyl 2-hydroxy-4-oxochroman-2-carboxylate ⟨77LA1707⟩. The enolate also reacts with aliphatic ketones to give 2,2-disubstituted chroman-4-ones *via* the diol ⟨79TL3685⟩.

Benzophenone reacts with *o*-hydroxyacetophenone to give the enone (**592**) which ring-closes to 2,2-diphenylchroman-4-one in acid solution ⟨61CB241⟩.

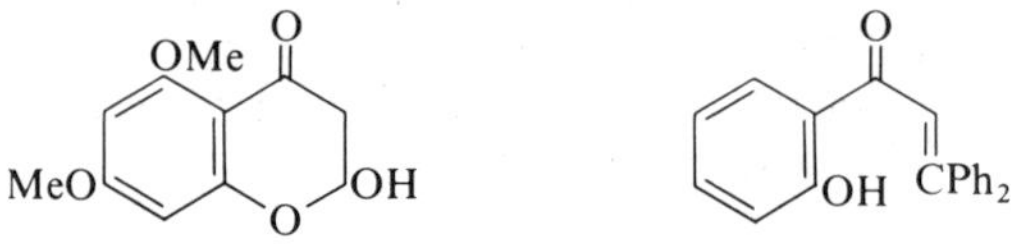

(591)              (592)

The reaction of the protected 2,6-dihydroxyacetophenone (**593**) with the vinylogous formamidinium salt (**594**) in the presence of sodium methoxide gives the aldehyde (**596**), an intermediate in the synthesis of 6-oxatetracycline. It is considered that the reaction proceeds through electrophilic attack of the cation at the acidic methyl group to give the dienamine (**595**). Hydrolysis to the dienol, ring closure and removal of the protecting group occur on treatment with acid (Scheme 225) ⟨81TL1497⟩.

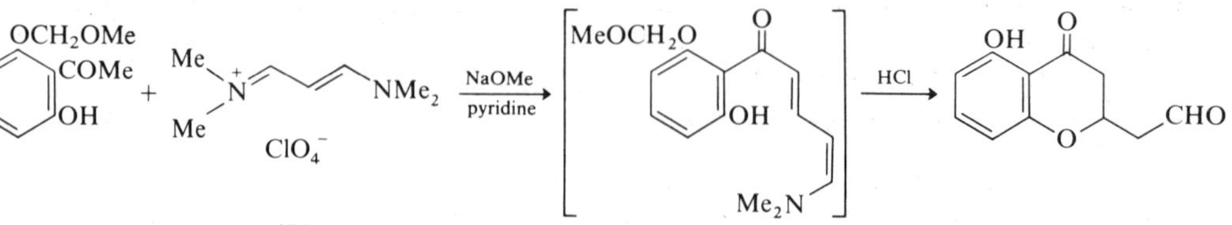

(593)        (594)        (595)        (596)

**Scheme 225**

The synthesis of some 2,2-spiroannelated chroman-4-ones from *o*-hydroxyacetophenone and a cycloalkanone in the presence of a secondary amine has been described ⟨78S886⟩. The reaction has also been used to prepare a range of chromanones bearing non-identical or identical substituents at C-2 ⟨81UP22400⟩. It has proved of value in the synthesis of naturally occurring chromanones, such as (**597**) ⟨81CB147⟩. The availability and cheapness of the starting materials together with its general applicability suggest that this route to chromanones has much to commend it. Tritium-labelled precocene I and II have been prepared *via* this route ⟨81MI22401⟩.

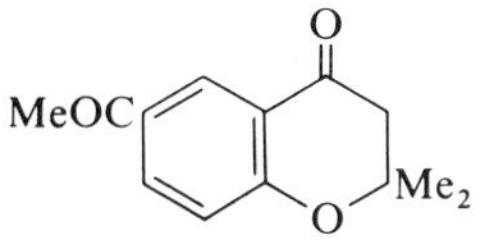

(**597**)

The reaction proceeds *via* an enamine, since the same product results when the preformed enamine is allowed to react with the acetophenone. An initial addition product (**598**) can be pictured which may eliminate pyrrolidine to give the enone (**599**) and hence the chromanone or which may undergo nucleophilic displacement of pyrrolidine by the phenolic moiety (Scheme 226).

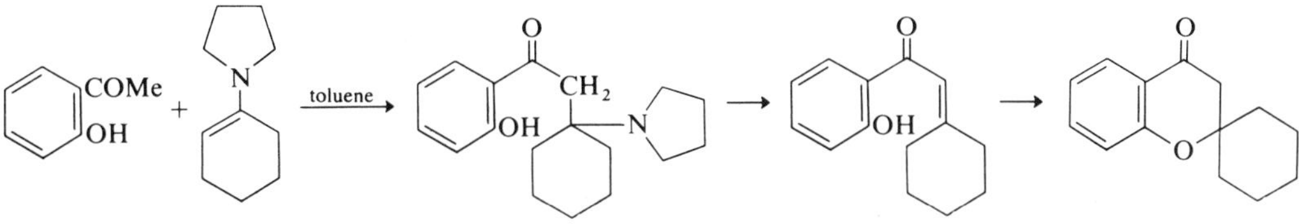

(**598**)　　　　　　(**599**)

**Scheme 226**

The synthesis of 3-methylsulfinylchroman-4-one (**600**) from ethyl salicylate follows a related pathway ⟨72JHC171⟩, whilst the triketone (**601**), derived from pentane-2,4-dione and the benzoate ester, exists as the substituted chromanone, a cyclic hemiacetal ⟨81JOC2260⟩.

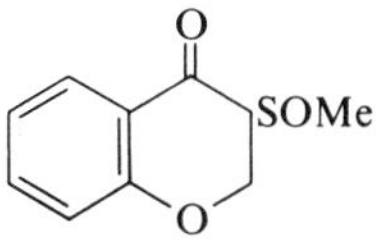

(**600**) 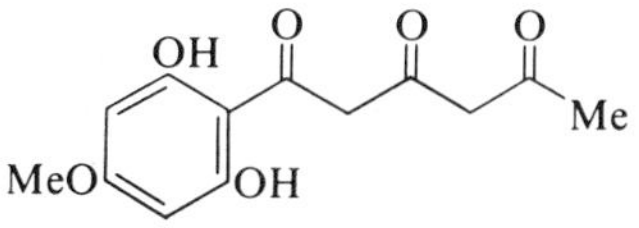　　　　　　(**601**)

There is little doubt, however, that the carbonyl compound which has been used most with *o*-hydroxyacetophenone in chromanone synthesis is benzaldehyde. This gives rise to a chalcone which cyclizes in acid or basic media to a flavanone (Scheme 227) ⟨70JOC2286⟩. During the acid-promoted cyclization of 5-acetamidochalcones, concomitant deacetylation occurs and 6-aminoflavanones result ⟨56JOC1408⟩.

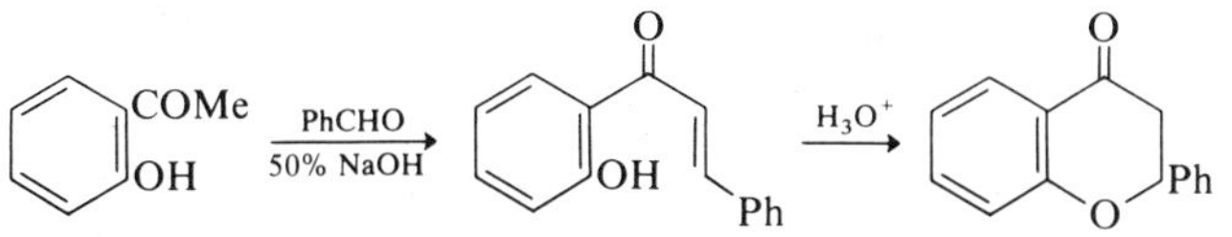

**Scheme 227**

The influence of the metal ion on both the formation of the chalcone and its cyclization to the flavanone has been studied ⟨77JOC3311⟩.

Ring closure is enhanced by a 6'-hydroxy group in the chalcone, whereas a substituent in the 6-position apparently hinders cyclization ⟨62TL593⟩. Chalcone *O*-glycosides are isomerized under mild conditions ⟨69CB785⟩.

This flavanone synthesis is well documented ⟨B-75MI22400⟩ and no routine examples are given here.

The reaction of 2-hydroxyphenyl benzyl ketones with diiodomethane leads directly to isoflavanones ⟨69IJC1059⟩. It is necessary to protect free hydroxyl groups other than that involved in ring closure (Scheme 228).

The triketo ester (**602**) undergoes a Claisen condensation in aqueous potassium hydroxide yielding the chalcone (**603**). On heating, cyclization to the flavanone, pinocembrin (**604**), occurred ⟨67JA6734⟩. Methanolic potassium hydroxide effects an alternative cyclization to methyl 2,4-dihydroxy-6-styrylbenzoate.

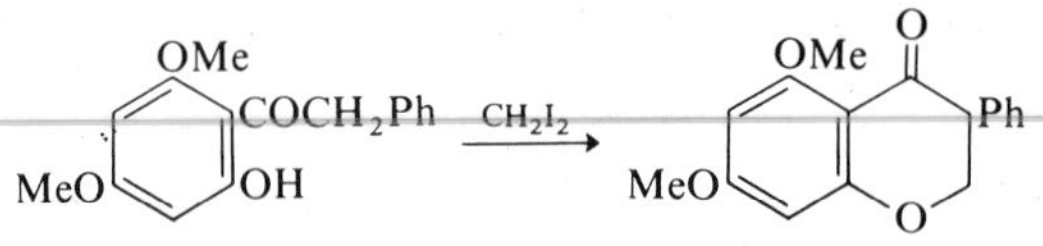

**Scheme 228**

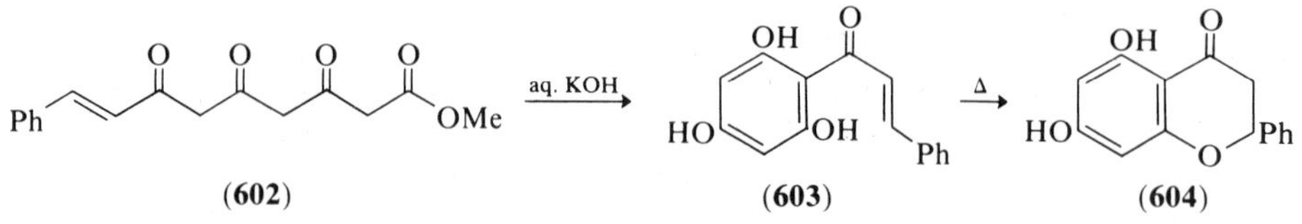

**(602)**           **(603)**        **(604)**

**Scheme 229**

The reaction of 2'-benzyloxychalcone dibromides (**605**) with ammonia gives a chalcone aziridine (**606**). Following conversion to the *N*-benzoyl derivative, acid hydrolysis removed the protecting group at C-2' and a 3-aminoflavanone results (Scheme 230) ⟨73ACH(76)95⟩.

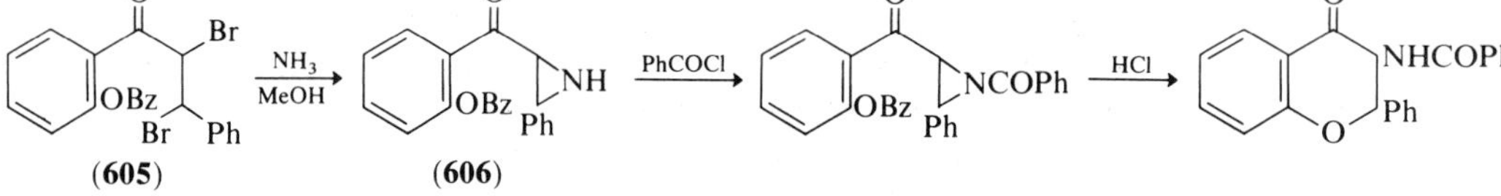

**(605)**           **(606)**

**Scheme 230**

3,3-Diaryl-2-hydroxypropiophenones (**607**) are obtained from 4-hydroxychalcone and a reactive phenol on treatment with alkaline hydrogen peroxide in an epoxide-mediated coupling reaction. The ketones undergo a base-catalyzed α-ketol rearrangement to the isomeric 1-hydroxypropan-2-ones (**608**) and acid-catalyzed ring closure provides a route to 4-arylflavan-3-ones (Scheme 231) ⟨80JCS(P1)1025⟩.

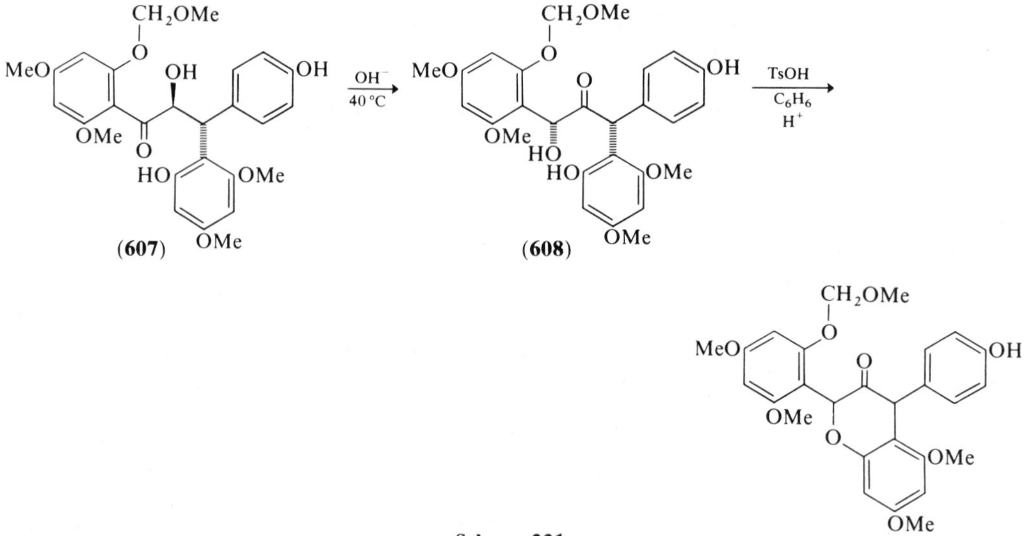

**(607)**           **(608)**

**Scheme 231**

(*b*) *β to the heteroatom.* A Dieckmann cyclization of the diester (**609**) provides a useful route to chroman-3-ones. There is some debate about the structure of the products, since although some workers have isolated only ethyl 3-oxochroman-4-carboxylate (**610**) ⟨73BSB283⟩, others have additionally obtained the 2-carboxylic ester ⟨66CR(C)(263)173⟩. In view of the mechanism of the reaction the 4-ester would be expected to be the predominant, if not exclusive, product (Scheme 232).

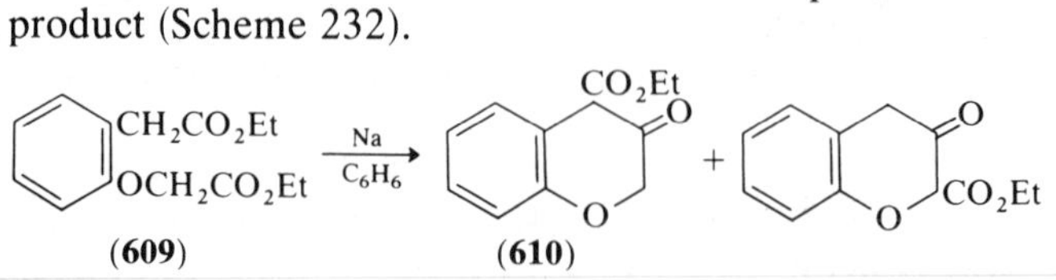

**(609)**           **(610)**

**Scheme 232**

(c) *γ to the heteroatom.* Several synthetic routes to chroman-4-ones are based on the cyclization of 3-phenoxypropanoic acid and its derivatives. Little advantage ensues from the use of acid derivatives, although both the ester ⟨57JCS3875⟩ and the nitrile ⟨62CB2086⟩ have been employed. The ease of synthesis of the phenoxypropanoic acids is a particular attraction·of this route. Normally prepared from a phenol and a 3-halogenopropanoic acid under basic conditions, the reactions of phenols with acrylonitrile and with propiolactone provide valuable alternatives.

The scope of this method is further enhanced by the range of cyclizing agents which may be used. Early workers found phosphorus pentoxide to be particularly effective, with or without solvent, but polyphosphoric acid has proved to be a superior reagent. Generally the two reagents bring about the same reactions but there are a few instances where this is not the case. For example, phosphorus pentoxide is less successful than polyphosphoric acid in effecting the cyclization of 3-phenoxypropanoic acids substituted with bulky alkyl groups. However, 3-(3,5-di-*t*-butylphenoxy)propanoic acid gives different products with the two reagents ⟨68CJC3367⟩.

The use of *m*-substituted phenols in the synthesis could give rise to isomeric products. Thus, 3-(3-methylphenoxy)propanoic acid affords both the 5- and 7-methylchroman-4-ones; the ratio of the products varies with the cyclizing agent ⟨67AC(R)1045⟩.

The synthesis has been extended to benzochromanones (**611**) through the cyclization of naphthyloxypropanoic acids. If the 1-position of the naphthol derivative is occupied, cyclization gives the linear benzochromanone (**612**) ⟨66BSF3249⟩.

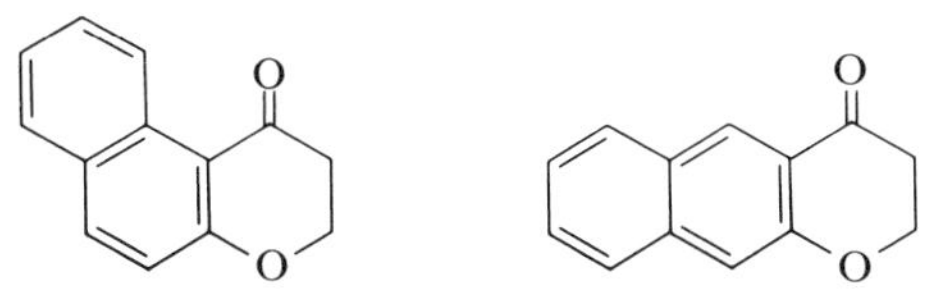

(**611**)               (**612**)

The tetrahydro derivative shown in Scheme 233 affords the linear chromanone ⟨63AJC690⟩.

A further extension of this route leads to the formation of 3-substituted chromanones ⟨73BCJ1839⟩. However, difficulties in the synthesis of the appropriate phenoxypropanoic acid limit its value.

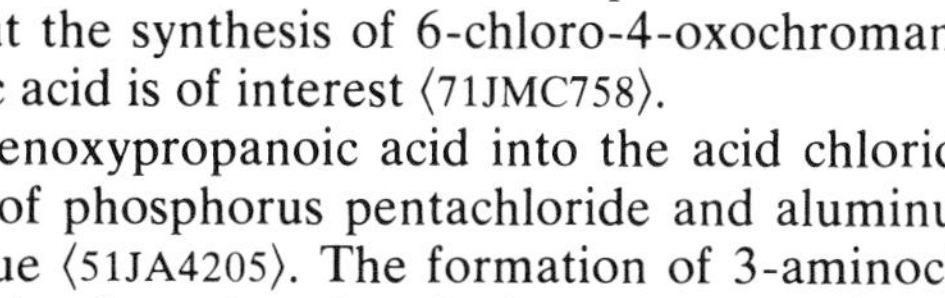

**Scheme 233**

Amongst other cyclizing agents which have found use, hydrogen fluoride is worthy of mention because of its effectiveness at room temperature ⟨62CB1446⟩. Sulfuric acid has achieved only limited success and instances are reported where sulfonation accompanies cyclization ⟨24CB202⟩, but the synthesis of 6-chloro-4-oxochroman-2-carboxylic acid from *p*-chlorophenoxysuccinic acid is of interest ⟨71JMC758⟩.

Conversion of the phenoxypropanoic acid into the acid chloride enhances the ease of cyclization and the use of phosphorus pentachloride and aluminum chloride provides an example of this technique ⟨51JA4205⟩. The formation of 3-aminochroman-4-one has been achieved in this way starting from 2-amino-3-phenoxypropanoic acid ⟨76TL271⟩. Cyclization of the phenoxyethanoic acid shown in Scheme 234 occurs on reaction with sodium methoxide to give the chroman-3-one ⟨63JCS5322⟩.

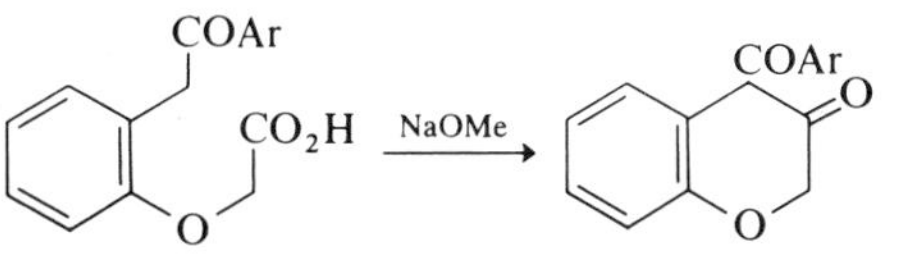

**Scheme 234**

The attraction of the use of 3-phenoxypropanonitriles lies in their easy access. Many examples are known which duplicate results obtained by methods already mentioned, and in some cases hydrolysis to the acid precedes cyclization ⟨80IJC(B)500⟩. However, the route has proved of considerable value in the synthesis of benzochromanones ⟨48JA599⟩. With a suitably placed substituent cyclization to the linear naphthopyranone (**613**) can be achieved, although in poor yield ⟨63AJC101⟩. Hydrogen chloride in the presence of zinc chloride has proved of value as a cyclizing agent both in this area ⟨70IJC203⟩ and in the synthesis of chromanones themselves ⟨76JCS(P1)499⟩.

**(613)**

Mechanistically, this reaction can be interpreted in terms of an electron-rich aromatic ring undergoing an intramolecular electrophilic attack by a potential acyl cation. The acidic conditions serve to increase the electrophilicity of the side-chain and to assist in dehydration.

### (ii) *From a preformed heterocyclic ring*

(*a*) *From chromones.* The selective reduction of the carbon–carbon double bond in chromones has been achieved by catalytic hydrogenation. An indication of the variety of catalysts which have been found suitable can be gained from the examples given in Table 8.

**Table 8** Formation of Chroman-4-ones by Reduction of Chromones

| Chromanone | Yield | Conditions | Ref. |
|:---:|:---:|:---:|:---:|
| 2-Me | 55 | $H_2$, Pd–CaCO$_3$, C$_6$H$_6$ | 55JA1623 |
| 2,6-Me$_2$ | 75 | LiAlH$_4$, THF, $-80\,°C$ | 65LA(685)167 |
| 7-OH | 79 | $H_2$, Ni, EtOH | 58JCS1190 |
| 5,7-(OH)$_2$ | 77 | $H_2$, Pd–C, DMF | 70T2787 |
| 7-OMe | 60 | $H_2$, Pt black, AcOH | 17CB911 |

Addition of other molecules to the double bond is not common, but 2,3-dihalogeno-chroman-4-ones have been prepared from chromones using sulfuryl chloride ⟨67CHE624⟩ or bromine in carbon disulfide ⟨25CB1612⟩. Reaction with NBS in aqueous DMSO affords 3-bromo-2-hydroxychroman-4-one ⟨75JHC981⟩.

An unusual example involves the addition of a methylene group by the reaction with dimethylsulfoxonium methylide to give the fused cyclopropane (**614**) ⟨68JCS(C)2302⟩, whilst 2-styrylchromones behave as dienes and undergo cycloaddition to give fused chromanones ⟨75ACH(84)319⟩.

**(614)**

Whilst the reduction of flavones to flavanones is of little practical significance, isoflavones are easily reduced to isoflavanones. However, carefully controlled conditions are necessary if further reduction to isoflavanols and even isoflavans is to be avoided ⟨71JCS(C)1994⟩.

(*b*) *From chromanols.* Although the oxidation of chroman-3-ols is not of any significance as a route to chroman-3-ones, isolated instances are recorded. For example, oxidation of 4,7-dimethylchroman-3-ol by DCC and DMSO, a Moffat oxidation ⟨65JA5661⟩, gave an excellent yield of the ketone ⟨70JOC2282⟩.

The oxidation of chroman-4-ols to chroman-4-ones, on the other hand, is more easily accomplished, using chromium trioxide in acetic acid ⟨66JCS(C)2013⟩. The same oxidant in pyridine converts chroman-3,4-diols to the 3-hydroxychroman-4-one ⟨65CB1498⟩.

The dehydration of *cis*-chroman-3,4-diols with copper sulfate or *p*-toluenesulfonic acid gives the chroman-3-one ⟨68BSF4203⟩, whilst pyrolysis of the chlorohydrin (**615**) yields the naphthopyran-3-one (Scheme 235) ⟨67JCS(C)1472⟩.

The oxidation of isoflavanols to isoflavanones appears to be dependent on the substitution pattern of the alcohol. If a methoxy group is present at C-5, 3-phenylcoumarin is produced

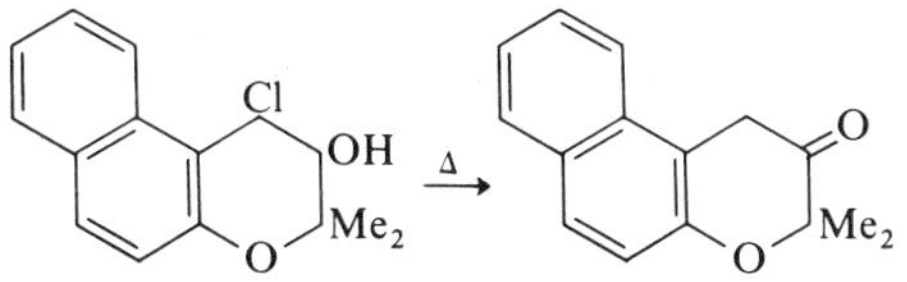

**(615)**

**Scheme 235**

using manganese dioxide as oxidant, but polymeric material results with Jones reagent. However, in the absence of a 5-substituent, oxidation to the isoflavanone proceeds in the normal manner. The different behaviour is attributed to the steric hindrance arising from the groups at C-3 and C-5, forcing the 4-hydroxy group to assume an axial orientation. Dehydration to the isoflavene occurs with ease and polymerization follows ⟨66JCS(C)629⟩.

(c) *From chromenes.* Oxidation of 2*H*-chromenes with potassium permanganate usually gives the 3,4-diol or destroys the pyran ring. However, neutral permanganate oxidation of 5-acetoxy-2,2-dimethylchrom-3-ene gives 5-acetoxy-3-hydroxy-2,2-dimethylchroman-4-one ⟨77CJC2360⟩. Conversion of a chromene to a ketol has also been accomplished with an osmium tetroxide–sodium periodate reagent ⟨67BSF1164⟩.

(d) *From pyran-2-ones.* The amine-catalyzed decarboxylative dimerization of 6-hydroxy-4-methylpyran-2-one gives a pyranopyrandione (**616**). The pyran-2-one moiety of this compound undergoes a decarboxylative Diels–Alder reaction with several alkynic dienophiles to give good yields of chromanones ⟨81JOC2425⟩.

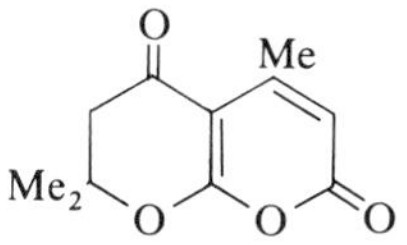

**(616)**

### 2.24.3.7.7 Isochromanones

The three isomers which are discussed in this section are isochroman-1-one or 3,4-dihydroisocoumarin (**617**), isochroman-3-one (**618**) and isochroman-4-one (**619**). None of them enjoy an extensive chemistry although a small number of natural products contain an isochromanone nucleus.

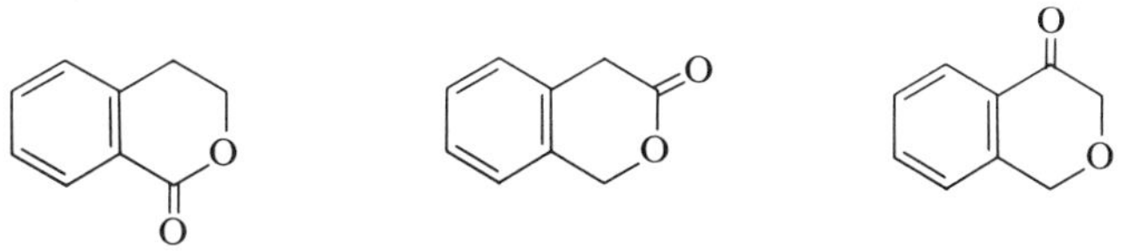

**(617)**          **(618)**          **(619)**

(i) *Formation of one bond*

(a) *Adjacent to the heteroatom.* Thermal dehydration of the appropriate substituted carboxylic acid has been used to synthesize both isochroman-1-one and the 3-ketone. In the former case, the precursor is a 2-(2-hydroxyethyl)benzoic acid ⟨57JA3165⟩.

The acid has been formed by the reduction of 2-carboxyphenylacetaldehyde, itself available from indene by ozonolysis ⟨57JA3165⟩. This is one of the most convenient methods of synthesis of isochroman-1-one, which is obtained in a 70% overall yield from indene (Scheme 236), and has been used in the synthesis of isochromanones with specific deuterium labels ⟨81JCS(P1)1685⟩.

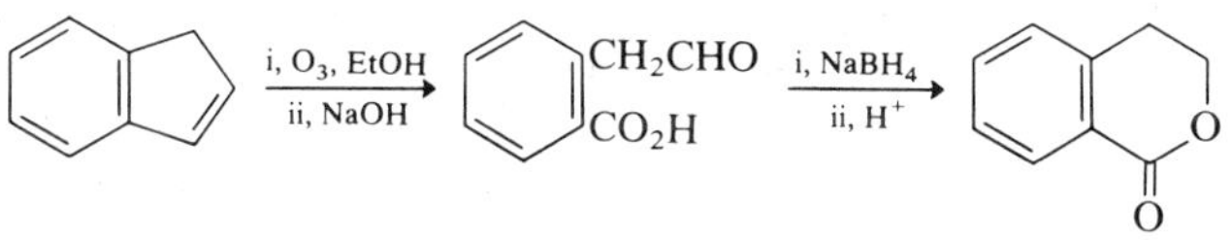

**Scheme 236**

The acids have also been obtained from isocoumarins through alkaline ring opening to the ketone and subsequent reduction. Ring closure occurs directly on acidification ⟨71JIC707⟩.

The synthesis of the dihydro derivative of fusamarin, a metabolite of a species of *Fusarium*, has been achieved by cyclization of the acid (**620**), synthesized from ethyl 3,5-dimethoxyphenylacetate (Scheme 237) ⟨78JCS(P1)81⟩.

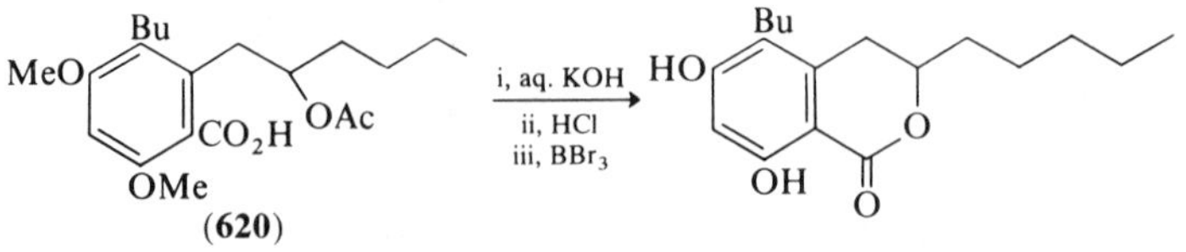

**Scheme 237**

2-Vinylbenzoic acids on cyclization yield either isochroman-1-ones or phthalides or a mixture of both types. The direction of ring closure is influenced by both steric and electronic factors ⟨62AC(R)1070⟩. The presence of an α-substituent in the vinylbenzoic acid (**621**; R = Ph) leads to phthalides, whereas only isochromanones result when no such substituent is present (**621**; R = H; Scheme 238). Ring closure presumably involves attack by the carbonyl group at the centre of lowest electron density. Thus, 2-(1-methylethenyl)benzoic acid yields the phthalide, but 2-(2-methylpropenyl)benzoic acid gives the isochroman-1-one on halolactonization.

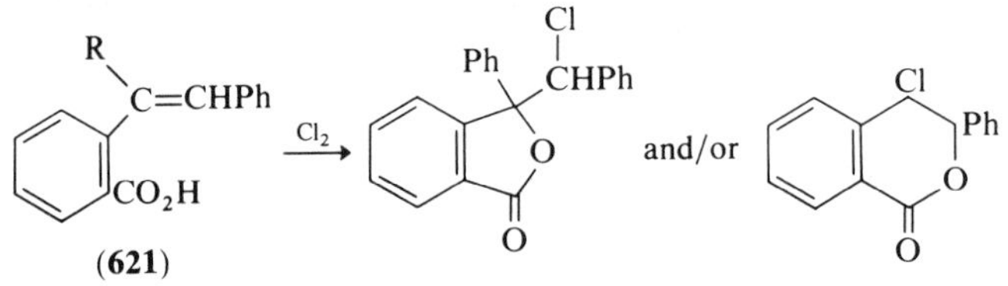

**Scheme 238**

Treatment of either *cis-* or *trans-*stilbene-2-carboxylic acids with chlorine or bromine leads to 4-halogeno-3,4-dihydro-3-phenylisocoumarins ⟨58T(4)393⟩. The reactions are stereospecific and are thought to involve intramolecular attack by the carboxyl group on a halonium ion. Ring closure to the corresponding 4-hydroxy compound also occurs stereospecifically using peroxyphthalic acid ⟨59JOC934⟩.

A Stobbe condensation of dimethyl homophthalate with a number of carbonyl compounds gives substituted vinylbenzoic acids, which cyclize under acidic conditions to the isochroman-1-one ⟨52JCS4799⟩. The actual product formed may depend on the choice of cyclizing medium. The ester function at C-4 may be hydrolyzed to the acid, which in turn may undergo decarboxylation. 4-Nitrohomophthalic acid reacts with a number of substituted benzaldehydes to yield 3-arylisochroman-1-ones, piperidine acting as a catalyst ⟨72JHC1255⟩. Homophthalic acid itself yields stilbene dicarboxylic acids. 4,4-Dimethyl derivatives of isochroman-1-one result when substituted 2-phenacyl or 2-acetonyl benzoic acids are treated successively with thionyl chloride and dimethylcadmium (Scheme 239) ⟨62BSB394⟩. In a similar manner, 2-acylphenylacetic acids yield isochroman-3-ones ⟨62BSB379⟩.

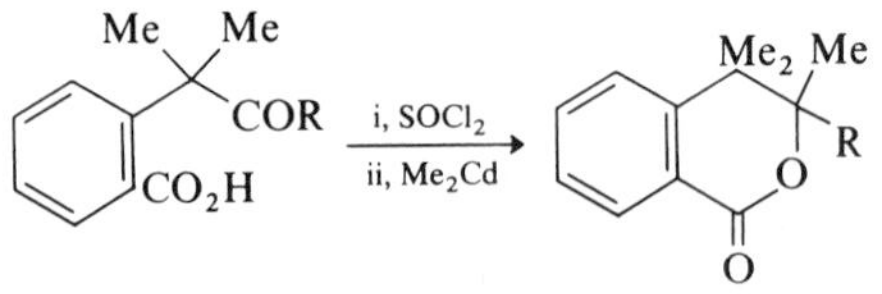

**Scheme 239**

Intermolecular acylation of 3,4-dimethoxyphenylacetic acid occurs in polyphosphoric acid, leading to 2-(3,4-dimethoxyphenylacetyl)-4,5-dimethoxyphenylacetic acid (**622**). Cyclization to the isochromanone takes place when the keto acid is heated in decalin (Scheme 240) ⟨72JHC853⟩.

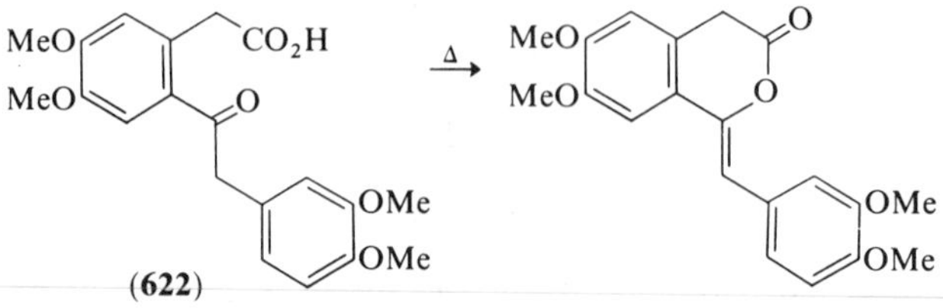

**Scheme 240**

A related approach, which is outlined in Scheme 241, makes use of heteroatom-directed lithiation to construct an appropriate side-chain on to a substituted *N,N*-dimethylbenzyl-amine ⟨81TL2797⟩.

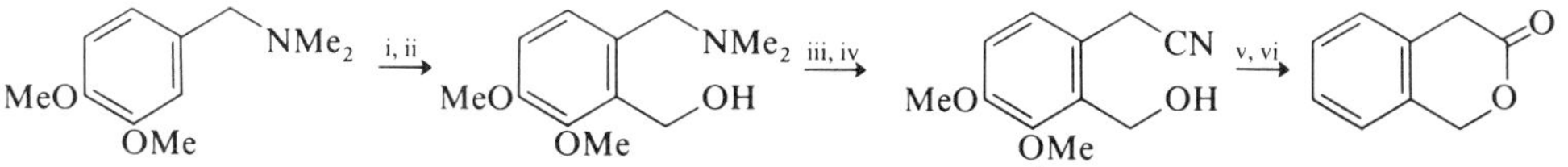

i, BuLi; ii, (CH₂O)ₙ; iii, ClCO₂Et, C₆H₆; iv, KCN, DMF; v, KOH, EtOH; vi, H₃O⁺

**Scheme 241**

The dilithio derivative of *N*-methyl-*o*-toluamide reacts with aromatic aldehydes and ketones to give hydroxyamides. Thermal cyclization affords 3-phenylisochroman-1-ones ⟨64JOC3514⟩. Spiroannelated isochromanones result when the organolithium compound reacts with fluorenone or alicyclic ketones.

2-Methoxymethylbenzyl chloride is readily converted into the nitrile, which on treatment with acid yields isochroman-3-one. Alkaline hydrolysis does not bring about ring closure but simply gives the substituted phenylacetic acid ⟨54JCS2819⟩.

A convenient synthesis of isochroman-3-one is based on the reaction of phenylacetic acids with formalin and concentrated hydrochloric acid ⟨76OS(55)45⟩. An indication of the value of the isochromanone as a synthetic intermediate is given in this procedure.

2-(2-Bromophenyl)ethanol undergoes a palladium-catalyzed carbonylation which results in the formation of isochroman-1-one ⟨79H(12)921⟩. It is considered that the reaction involves formation of an aryl–palladium complex (**624**) through insertion of the zerovalent palladium complex (**623**) into the aryl halide. Insertion of carbon monoxide followed by reductive elimination of the metal as a complex species leads to the isochromanone (Scheme 242).

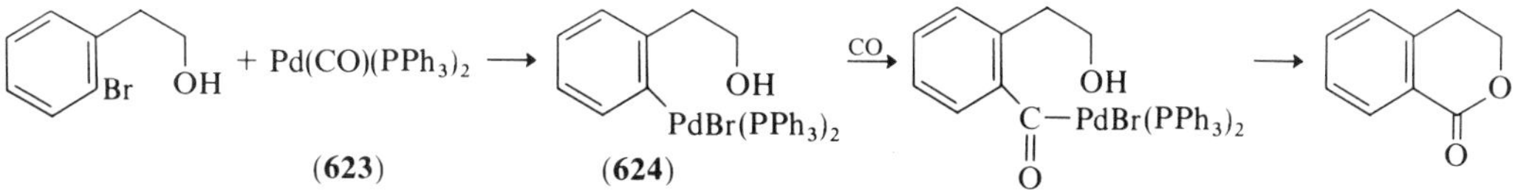

**Scheme 242**

A similar approach has been used to synthesize isochroman-3-ones ⟨80JA4193⟩. *o*-Bromomethylbenzyl alcohol oxidatively adds to the zerovalent complex and carbonylation and reductive elimination of the catalyst follow. The catalyst is kept active by the presence of base which absorbs the hydrogen bromide formed in the reaction.

When 2-bromobenzoic esters are treated with π-(2-methoxyallyl)nickel bromide, the acetonyl group is introduced. Reductive cyclization of the methyl 2-acetonylbenzoates yielded isochroman-1-ones ⟨77JOC1329⟩.

In a related manner, sodium 2-bromobenzoates have been converted into 2-allylbenzoic acids. A palladium-catalyzed cyclization involving nucleophilic attack of the carboxylate on the palladium-complexed alkene yielded an isocoumarin. However, *in situ* catalytic hydrogenation results in the formation of an isochroman-1-one.

The Baeyer–Villiger oxidation of indan-2-one gives isochroman-3-one in good yield ⟨49JCS1720⟩.

(*b*) *β to the heteroatom.* Formation of this particular bond is restricted to the synthesis of isochroman-4-ones.

The diesters (**625**), prepared from ethyl 2-chloromethylbenzoate, are cyclized by sodium ethoxide in a classical Dieckmann reaction ⟨71BSF1351⟩. When R = Me, the formation of phthalide was observed, which was thought to arise by attack of the ethoxide at the methyl group with loss of ethyl acrylate (Scheme 243).

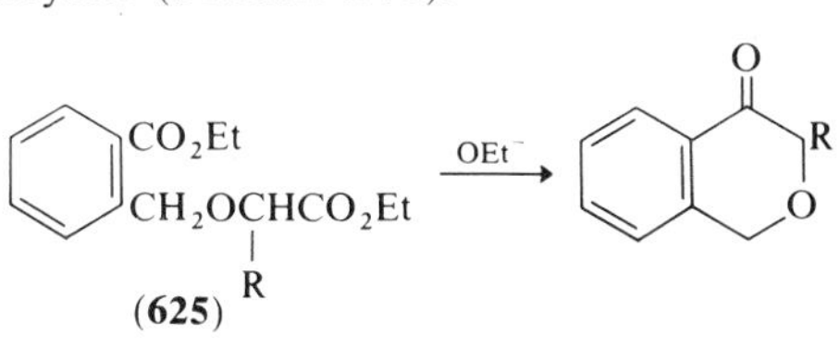

**Scheme 243**

An alternative approach utilizes an intramolecular Perkin reaction on the dicarboxylic acid (**626**), which gives a mixture of the isochromanone and its enol acetate (**627**). Alkaline hydrolysis of the latter generates the ketone. When R = Ph, a further product, the keto acid (628), is formed (Scheme 244).

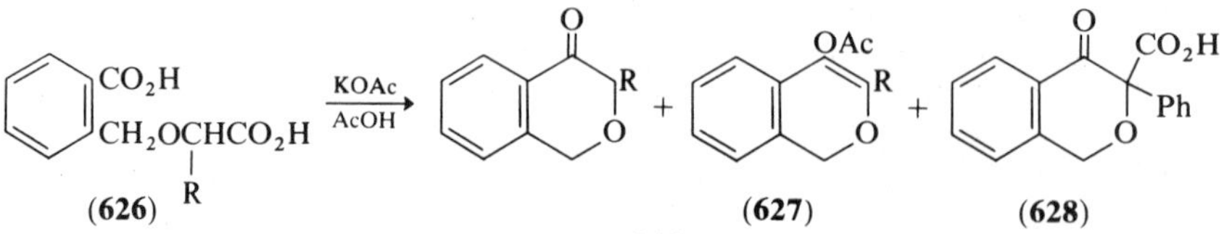

**Scheme 244**

The Perkin approach to isochroman-4-ones is preferred to the Dieckmann synthesis, since the latter is a sensitive condensation during which polymerization readily occurs.

A synthesis of (+)-(S)-3-phenylisochroman-4-one (629) is based upon the cyclization of (+)-(S)-(2-bromobenzyloxy)phenylacetic acid by butyllithium (Scheme 245) ⟨79AP385⟩.

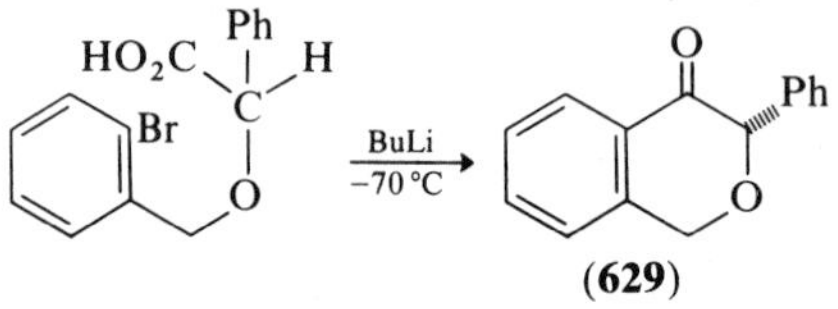

**Scheme 245**

### (ii) *From other heterocycles*

(a) *From isoquinolines.* Treatment of laudanosine-6'-carboxylic acid (630) with cyanogen bromide leads to the 3-arylisochroman-1-one (631; R = CN) ⟨81AP577⟩. In a similar manner, ethyl chloroformate gives the isochromanone (631; R = CO₂Et; Scheme 246) ⟨78C256⟩.

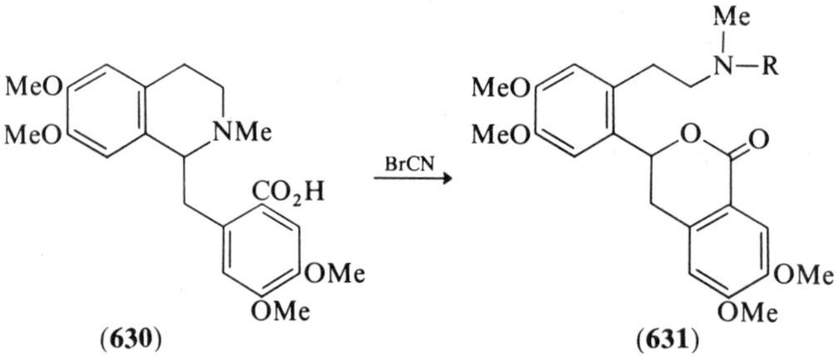

**Scheme 246**

### (iii) *From a preformed heterocyclic ring*

(a) *From isochromans.* Oxidation of the 1-methylene group of isochroman occurs on reaction with selenium dioxide to give isochroman-1-one ⟨56BSF1337⟩. Chromium trioxide is reported to be better than selenium dioxide ⟨62JOC4337⟩ and potassium permanganate in acetone is also recommended ⟨59JCS3598⟩. When the 1-position is fully substituted, oxidation takes place at C-4 using a variety of oxidants ⟨69BSF915⟩.

The direct oxidation of 2-(2-hydroxymethylphenyl)ethan-1-ols to isochroman-1-ones may proceed *via* the isochroman ⟨61TL223⟩.

(b) *From homophthalic anhydride.* Methyl 2-carboxyphenylacetates are formed by methanolysis of homophthalic anhydrides. Reduction of these half esters yields isochroman-1-ones.

The anhydride also undergoes a Perkin condensation with aromatic aldehydes in the presence of a base to give 1-oxo-3-phenylisochroman-4-carboxylic acids ⟨58JCS2612⟩.

## 2.24.4 PYRYLIUM SALTS

In this section the formation of benzopyrylium (632), isobenzopyrylium (633) and xanthylium (634) salts is covered in addition to the parent system. There is no separate treatment of flavylium salts, which are considered as phenyl derivatives of benzopyryliums.

The chemistry of pyrylium salts, including their synthesis, has been reviewed ⟨B-64MI22401⟩, but of greater significance are the survey of the synthesis of pyrylium salts which covers the literature through to 1968 ⟨69AHC(10)241⟩ and the supplement which extends the coverage to 1981 ⟨82AHC(S2)⟩. Benzopyrylium, flavylium and anthocyanins are covered in Rodd's treatise on carbon compounds ⟨B-77MI22400⟩. Naturally occurring examples of these groups are considered by Dean ⟨B-63MI22400⟩.

Many of the synthetic routes to pyrylium salts involve the cyclization of molecules formally derived from pentane. In view of some difficulties in access to and isolation of appropriate derivatives, it is often advantageous to form the five-carbon unit *in situ* from more readily available small molecules. In many cases such reactions have been shown to proceed through a pentane derivative, although in other instances their intermediacy is only surmised. Nevertheless, it is convenient and logical to classify all such reactions together, implying that the heterocycle results from the formation of only one bond. The syntheses fall into two groups, depending on whether or not an oxidation step is needed to produce the aromatic system.

### 2.24.4.1 Formation of One Bond

#### 2.24.4.1.1 Adjacent to the heteroatom

The synthesis of pyrylium salts, discovered in a classical study by Dilthey ⟨20JPR(101)177⟩, involves the facile acid-catalyzed cyclization of a 1,5-dicarbonyl compound. The value of this route stems from the different structural features which are acceptable in the five-carbon unit and from the variety of methods available for their synthesis. The reaction is the reverse of the synthetically useful ring opening of pyrylium salts by nucleophiles.

Unsaturated 1,5-diones are in the correct oxidation state for direct conversion to pyrylium salts and in fact they cyclize spontaneously under strongly acidic conditions. The reaction is thought to involve three steps; initial enolization is followed by ring closure to the hemiacetal and subsequent rapid dehydration to the pyrylium salt (Scheme 247).

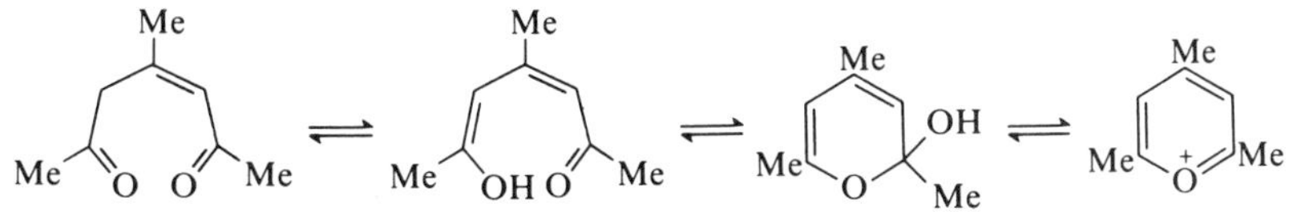

**Scheme 247**

Confirmation of the reversible ring opening of pyrylium salts to the pseudobases is provided by the incorporation of $^{18}O$ into 2,4,6-trimethylpyrylium when the salt is treated with $^{18}O$-enriched water ⟨67MI22401⟩.

Kinetic studies of the ring opening of ⟨71JA2733⟩ and the cyclization to ⟨74CR(C)(279)697⟩ 2,4,6-triphenylpyrylium salts have been carried out. A bifunctional catalytic action of acetic acid on both the enolization and ring closure of the hemiacetal has been postulated in the latter study.

The series of equilibria allow a change in the anion of a pyrylium salt to be accomplished. The readily available pyrylium perchlorates and tetrafluoroborates are converted into the pseudobases on treatment with weak bases. Reaction of these diketones with a wide variety of acids leads to the new pyrylium salt generally in high yield (Table 9) ⟨80T679⟩. In some

**Table 9** Anion Exchange in Pyrylium Salts *via* the Pseudobase of 1,3,5-Triphenylpyrylium Perchlorate

| Anion | Conditions | Yield | Ref. |
|---------|------------|-------|------|
| F | aq. HF | 90 | 80T679 |
| Cl | aq. HCl | 96 | 73MI22401 |
| Br | aq. HBr | 70 | 79JCS(P1)436 |
| I | HI | 99 | 64T119, 79JCS(P1)433 |
| $NO_3$ | $HNO_3$ | 70 | 80T679 |
| $CF_3CO_2$ | $CF_3CO_2H$ | 72 | 80T679 |
| $FSO_3$ | $FSO_3H$ | 89 | 80T679 |
| SCN | $NH_4SCN$ | 93 | 74RRC1731 |

instances, the relative solubilities of the two pyrylium salts are such that isolation of the alk-2-ene-1,5-dione is unnecessary (Scheme 248).

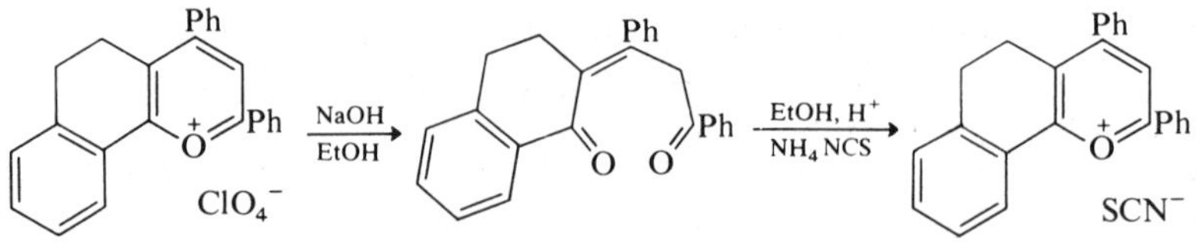

**Scheme 248**

However, although aromatic derivatives of the pseudobases are stable, crystalline compounds, aliphatic analogues are prone to polymerization. It is therefore often more convenient to produce the unsaturated diketones from simpler molecules in an acidic medium and hence avoid their isolation.

The acylation of unsaturated ketones constitutes one of the earliest routes to pyrylium salts ⟨19CB1195⟩. The reaction is better achieved with acyl halides than by anhydrides, and aliphatic are preferable to aromatic acid derivatives. The presence of a Lewis or Brønsted acid is usually necessary and iron(III) chloride, aluminum chloride, boron trifluoride and perchloric acid have found frequent application. It is considered that these interact with the acid derivative to generate the actual acylating agent.

It is usual to produce 2,4,6-trisubstituted pyrylium salts by this method, otherwise isolation of the products is difficult. Although yields are only fair, considerable variation in the substituents is possible and some satisfactory examples are known. The detailed preparation of 2,4,6-trimethylpyrylium perchlorate also includes a brief survey of methods and merits of the procedure ⟨73OSC(5)1106⟩.

Dorofeenko and his coworkers have extended this route to the synthesis of a range of cyclic analogues of pyrylium salts. Thus, cyclohexenylacetophenone affords the reduced isobenzopyrylium salt (**635**) on treatment with acetic anhydride ⟨67MI22402⟩ and 2-(indol-2-yl)cyclohexanone yields the indolo[2,3-*d*]pyrylium salt (**636**) in the same way ⟨69ZOB716⟩. The yields in these reactions are much improved, a feature which may be attributable to the conformational preference for the structure shown for the enone rather than a conjugated arrangement of double bonds.

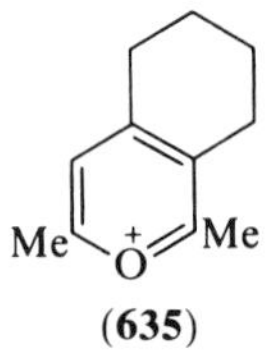

(**635**) 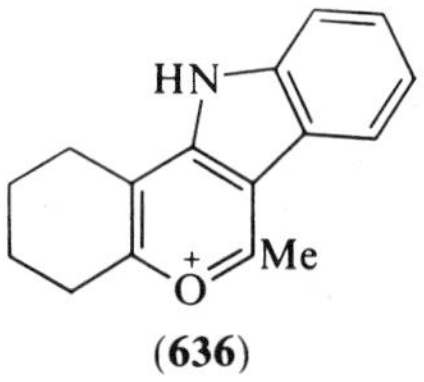                                  (**636**)

It was assumed ⟨59LA(625)74⟩ that the reaction proceeds through acylation of the non-conjugated methylene ketone and the Russian work certainly supports this view. The observation that 4-methylpent-4-en-2-one gives a vastly improved yield of 2,4,6-trimethyl-pyrylium perchlorate compared with 4-methylpent-3-en-2-one on reaction with acetic anhydride and perchloric acid also favours this suggestion and led to a proposed mechanism (Scheme 249) ⟨61JCS3573⟩.

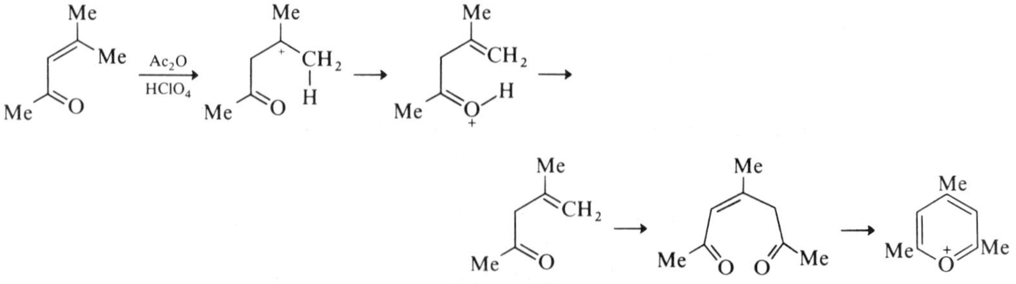

**Scheme 249**

When 1,3-diphenylbut-2-en-1-one is acylated with benzoyl chloride in acetic acid, the formation of a 2-methyl-4,6-diphenylpyrylium salt is explained by the rapid formation of acetyl ions by the reaction of the acid chloride with acetic acid ⟨72JCS(P1)809⟩.

A mechanism for pyrylium salt formation in the absence of acylating agent has been suggested ⟨54JA5437⟩ involving acylation by benzoyl cations generated from the enone. Such a process would also be expected to generate acetyl ions from the solvent and so lead to the formation of a 2-methylpyrylium cation, which is not observed. Consequently, an alternative mechanism has been postulated in which an aldol-like condensation between the conjugated and non-conjugated forms of the enone is followed by alkene elimination. The latter process offers an explanation for the observed polymer formation during the reaction (Scheme 250).

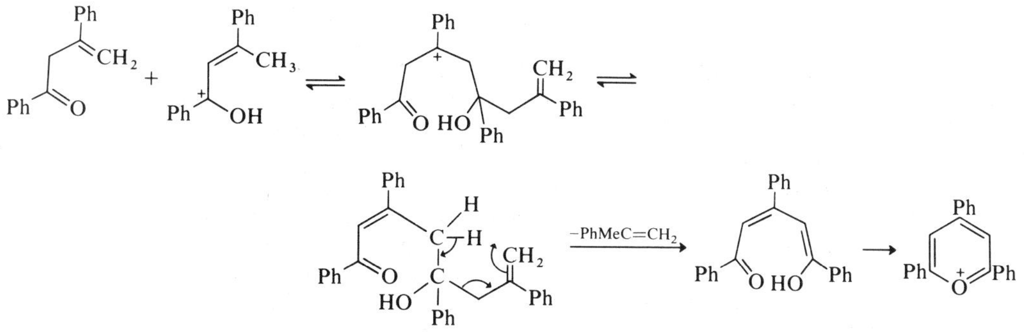

**Scheme 250**

Acylation of the keto acid (**637**) leads to the isobenzopyrylium salt (**638**) ⟨77CHE1183⟩. However, the isobenzopyrylium salt (**639**), a potential intermediate for the synthesis of analogues of berberine alkaloids, results from the formylation of the substituted ketone or the isochromanone (**640**) using dichloromethyl butyl ether (Scheme 251) ⟨81CHE221⟩. A second product, the 5-oxoniachrysene (**641**), is formed and this compound may also be obtained by reaction of the isobenzopyrylium salt with phosphorus pentachloride and then with triethylamine. The intermediacy of a cyclic vinyl ether is proposed.

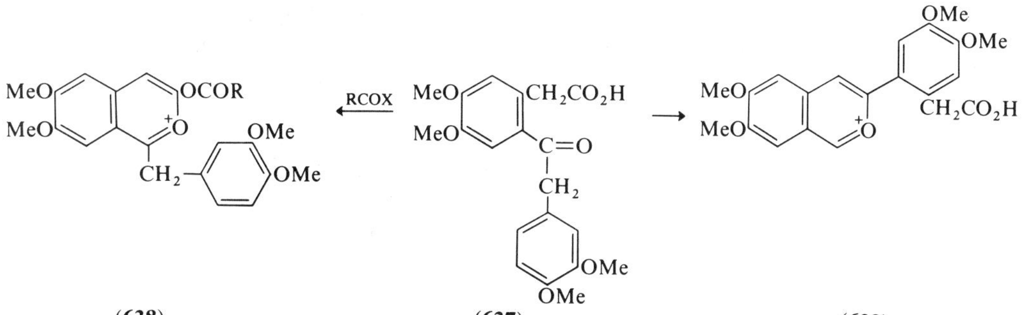

(**638**)          (**637**)          (**639**)

**Scheme 251**

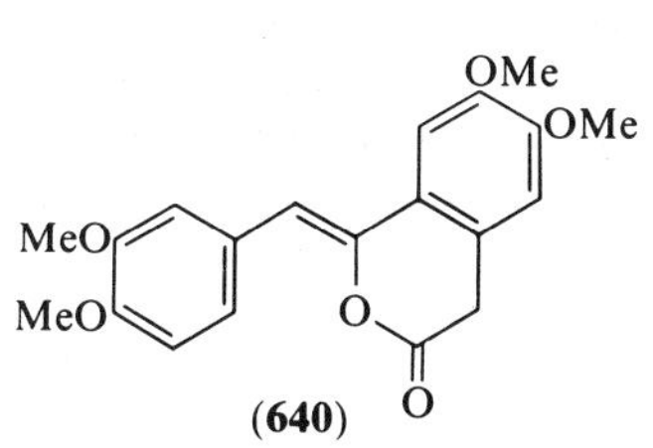

(**640**) 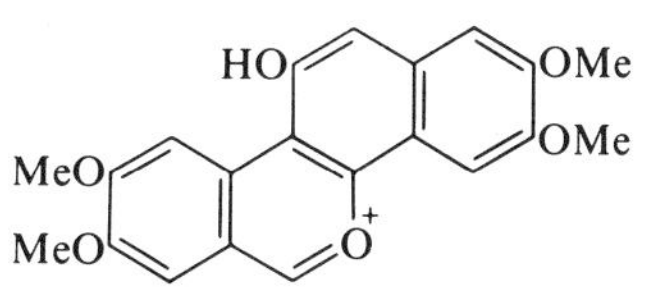                    (**641**)

1,3-Diketones and methyl ketones yield pyrylium salts in the presence of acetic anhydride and a proton acid, again *via* a pent-2-ene-1,5-dione ⟨24CB1653⟩. For example, dibenzoylmethane and acetophenone afford an 80% yield of 2,4,6-triphenylpyrylium salt. Benzoylacetone and acetophenone yield only 2-methyl-4,6-diphenylpyrylium, although involvement of either carbonyl group of the diketone can be envisaged ⟨33JCS1197⟩. It has been shown that this reaction is applicable to 3-ketoaldehydes, thereby allowing the synthesis of pyrylium salts unsubstituted in the 4-position ⟨65ZOB589⟩. Generally, the diketone can have aliphatic, aromatic or even heterocyclic substituents at C-1, C-2 or C-3. Less choice is allowed in the ketone, which generally should possess an aromatic or heteroaromatic substituent.

Again, the extension of this route to the synthesis of pyrylium salts with fused alicyclic moieties has been reported involving the use of cycloalkanones as the ketonic component ⟨67T1565⟩.

The related synthesis of benzopyrylium salts involves the condensation of salicylaldehydes with methyl ketones in a basic medium, which affords the unsaturated ketone (642). Cyclization occurs on treatment with acid. The same benzopyrylium salt is formed rapidly and in high yield if the condensation is effected under acidic conditions (Scheme 252) ⟨73BSF3421⟩. It has been proposed that the latter synthesis proceeds *via* an aldol condensation accompanied by the formation of a hemiacetal, followed by dehydration.

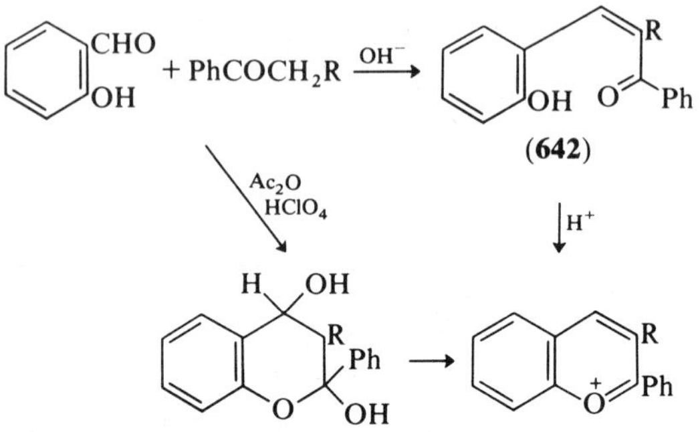

**Scheme 252**

The ketone may be cyclic or heterocyclic, when a range of fused benzopyrylium salts results. Thus, tetralone yields (643) with 2-hydroxy-1-naphthaldehyde, whilst chroman-4-one and salicylaldehyde give the salt (644).

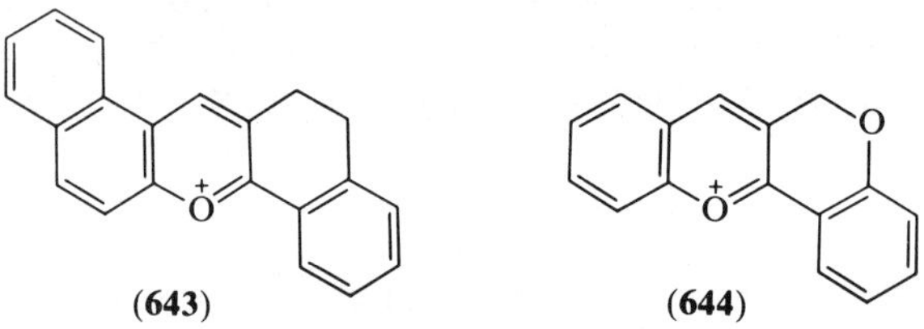

(643)                                        (644)

A series of stable 2-aminobenzopyrylium salts has been prepared by the reaction of substituted salicylaldehydes with nitriles in an aprotic polar solvent ⟨81ZC408⟩.

The use of *o*-hydroxyacetophenone in place of salicylaldehyde leads to the formation of 4-substituted benzopyrylium salts.

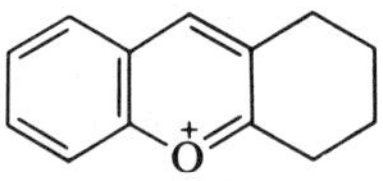

(645)

β-Keto esters also react with salicylaldehyde to yield fused pyrylium salts. For instance ethyl 2-oxocyclohexanecarboxylate affords the tetrahydroxanthylium salt (645), whilst the isochromanone derivative (646) yields the tetracyclic molecule (647), which possesses the skeleton of the anthocyanidin, peltogynidine, isolated from the heartwood of *Peltogyne porphyrocardia* (Scheme 253) ⟨58JCS3174⟩.

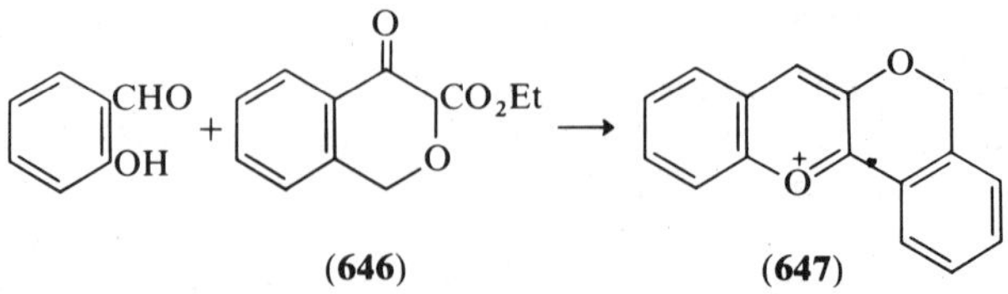

(646)                                        (647)

**Scheme 253**

4-Hydroxybenzopyrylium salts result when a 1-aryl-1,3-diketone is cyclized with perchloric acid. This route is attractive since a sequence of acylation, Fries rearrangement and a second acylation leads directly to the salt from a phenol ⟨71CHE1587⟩.

The reaction of the dimethyl acetal of phenylacetaldehyde with salicylaldehydes enables 3-phenylbenzopyrylium salts to be prepared ⟨82JHC97⟩ and the reaction of arylmalondial-dehydes with phloroglucinol provides another variation on this theme which also affords isoflavylium salts ⟨81AJC2647⟩.

Russian workers have advocated the use of orthoformates in the synthesis of pyrylium salts and their benzologues. In the presence of acids, alkyl orthoformates are converted into dialkoxycarbocations, which are efficient *C*-acylating agents. Even in the absence of a Friedel–Crafts catalyst and under mild conditions, acetophenone is converted into the pent-2-ene-1,5-dione (**648**), which cyclizes to the 2,6-diphenylpyrylium salt (Scheme 254) ⟨71CHE147⟩. Yields decrease with increasing alkyl chain length in the orthoformate. There is evidence that a 3-alkoxymethylpyrylium salt is the initial product.

$$CH(OR)_3 + HX \rightleftharpoons [CH(OR)_2]^+ X^- + ROH$$

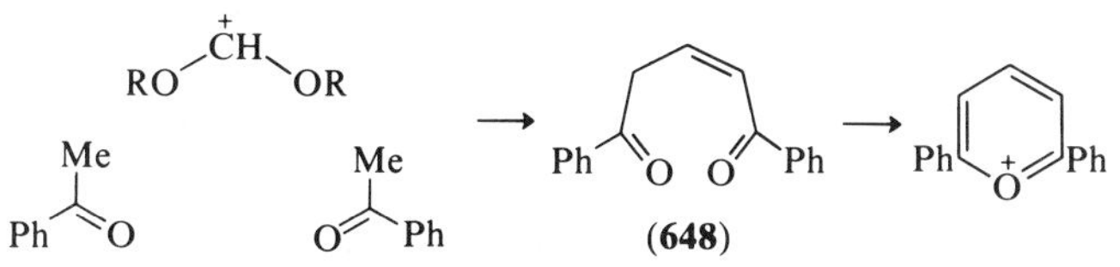

**Scheme 254**

A variant utilizes the reaction between ethyl orthoformate and pent-1-enyl phenyl ketone, which affords 5-ethyl-2-phenylpyrylium perchlorate *via* the acetal ⟨73JOU399⟩.

When this reaction is applied to *o*-hydroxyacetophenone, benzopyrylium salts are formed, through initial *C*-alkylation and subsequent cyclization to the chromone. Under the reaction conditions, *O*-alkylation of the ketone and aromatization to the salt occur ⟨72CHE935⟩.

This route may be extended to the synthesis of flavylium salts through the reaction of an orthoformate with *C*-acyl derivatives of *o*-hydroxyacetophenone ⟨72JOU2250⟩. During this synthesis small amounts of 2,6-di-(2-hydroxyaryl)pyrylium salts are produced. If an excess of *o*-hydroxyketone is used in the reaction, the 2,6-diarylpyrylium salt becomes the major product ⟨73JOU399⟩.

A one-step synthesis of flavylium salts is similar in nature. A mixture of an *o*-hydroxyacetophenone, an aromatic aldehyde and ethyl orthoformate in perchloric acid is maintained at room temperature, whereupon the salt gradually precipitates ⟨81CHE115⟩.

The reaction between acetophenone and acetic anhydride alone yields 2-methyl-4,6-diphenylpyrylium ⟨23CB1012⟩. It seems possible that a 1,3-diketone is formed by the acetyla-tion of acetophenone ⟨54OR(8)59⟩, which subsequently reacts with unchanged ketone. Alternatively, an intermediate enone derived from the condensation of two moles of acetophenone ⟨53JA626⟩ may be acylated to the pyrylium salt. There is supporting evidence for both reaction schemes and it is not possible to discard either alternative (Scheme 255).

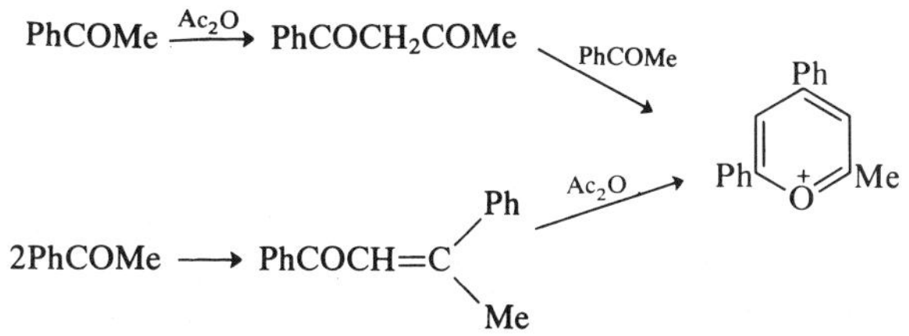

**Scheme 255**

This approach to pyrylium salts has much to offer and it has been extended to variously substituted acetophenones ⟨65JOC1684⟩ and even to acetone, though in only fair yield. It remains a valuable method for the synthesis of pyrylium salts with identical substituents at C-2 and C-4.

The route offers the possibility of further simplification, since aromatic hydrocarbons are readily acylated under Friedel–Crafts conditions. Thus, the acetylation of toluene in the presence of perchloric acid yields the trisubstituted pyrylium compound ⟨27CB716⟩. Alkoxy-benzenes behave in a similar manner and anisole gives a good yield of 2,4-di-(4-methoxyphenyl)-6-methylpyrylium perchlorate ⟨51JCS726⟩.

The synthesis of the xanthylium laser dye (**650**) from 8-hydroxyjulolidine (**649**) may be thought of as a variant of this route (Scheme 256) ⟨77JHC683⟩.

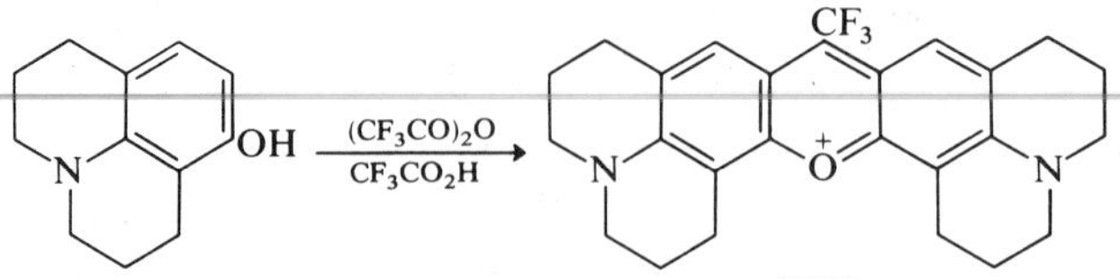

**(649)**                                    **(650)**

**Scheme 256**

Acetophenone also yields pyrylium salts on treatment with acids, such as hydrogen chloride and sulfuric acid, or with boron trifluoride. The initially formed enone undergoes partial deacylation by the acid, whilst unchanged enone reacts with the acyl cation so produced to give the pyrylium salt.

3-Chlorovinyl ketones are potential 1,3-diketones and provide a useful extension to the above method. In the presence of Friedel–Crafts catalysts they react with ketones to give pyrylium salts which are unsubstituted in the 4-position. The example shown in Scheme 257 illustrates the synthesis of a more complex pyrylium salt ⟨63ZC266⟩.

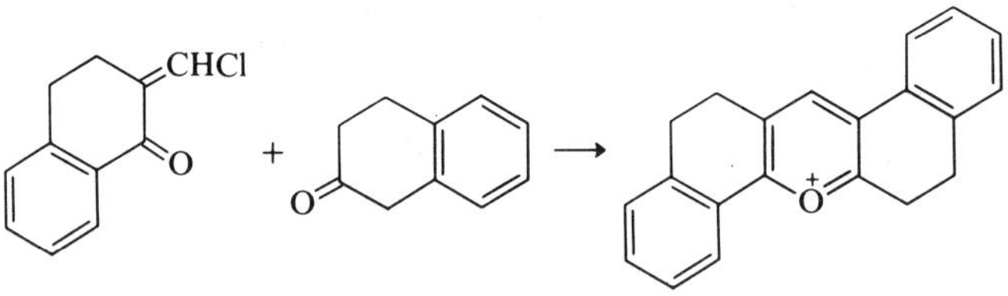

**Scheme 257**

Hydration of alkynes yields carbonyl compounds and the ketone used in pyrylium syntheses has been successfully replaced by an alkyne ⟨65CB334⟩. Phenylacetylene, for example, reacts with 3-chloro-1-phenylprop-2-en-1-one to yield the 2,6-diphenylpyrylium salt.

One particular advantage of this method is that the lachrymatory 3-chlorovinyl ketones may be produced *in situ* by the acylation of an alkyne. This route has been further extended to the synthesis of azapyrylium salts from an alkyne and the imidoyl chloride (**651**) and to diazapyrylium compounds from nitriles (Scheme 258) ⟨65CB334⟩.

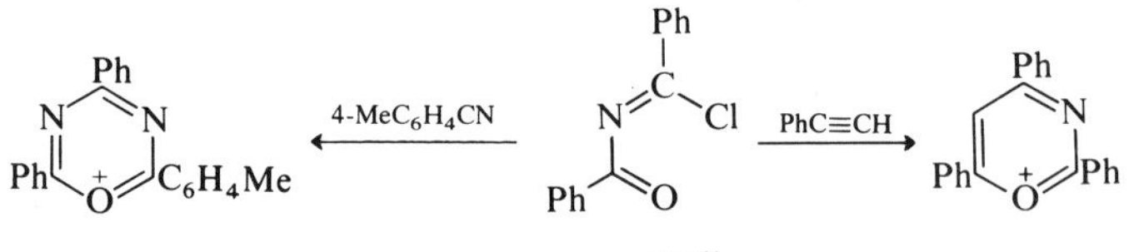

**(651)**

**Scheme 258**

Enamines are also potential ketones and accordingly have been used as the ketonic component in this type of pyrylium synthesis ⟨63AG(E)394⟩. The initial adducts, which form pyrylium salts on treatment with perchloric acid, are similar to those derived from secondary amines and pyrylium salts. The significant feature of this variant is that it gives access to pyrylium compounds which are unsubstituted at the 4- and 6-positions but which carry substituents at C-2 and C-5. More conventional enamines, such as 1-piperidinocyclohexene, afford bicyclic pyrylium salts (**652**), whilst polycyclic salts (**653**) result if cyclic chlorovinyl ketones are used.

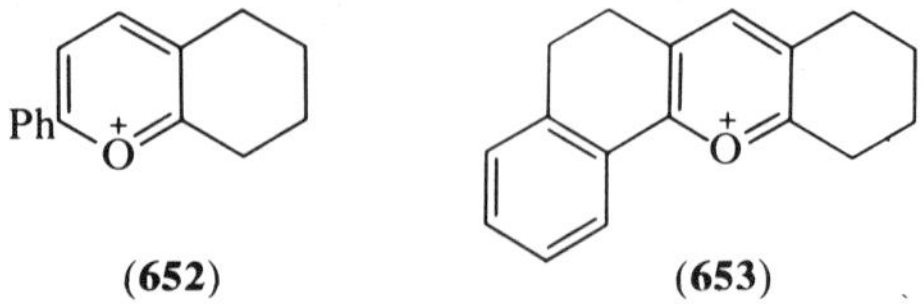

**(652)**                                    **(653)**

The chlorovinyl immonium salts (**654**), obtained through Vilsmeier formylation of methyl ketones, behave as 1,3-dicarbonyl compounds. In the presence of strong base they react with methyl ketones to form dienones (**655**). On treatment with a mixture of perchloric and acetic acids the pyrylium salt results (Scheme 259) ⟨71JPR1110⟩. In a similar reaction 2-naphthol is converted into the naphtho[2,1-*b*]pyrylium salt (**656**) ⟨79S241⟩.

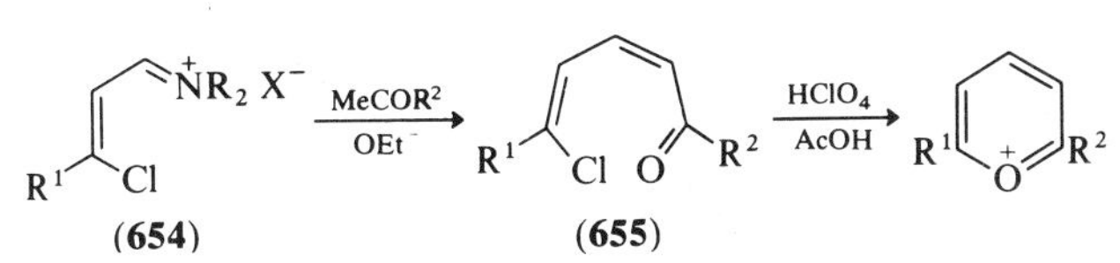

**Scheme 259**

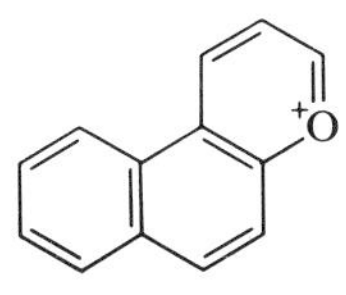

(656)

Isobenzopyrylium salts have been obtained from *o*-acylbenzonitriles by treatment with hydrobromic acid. The 1-amino-3-arylisobenzopyrylium derivatives are produced in good yield according to Scheme 260 ⟨78JOC3817⟩.

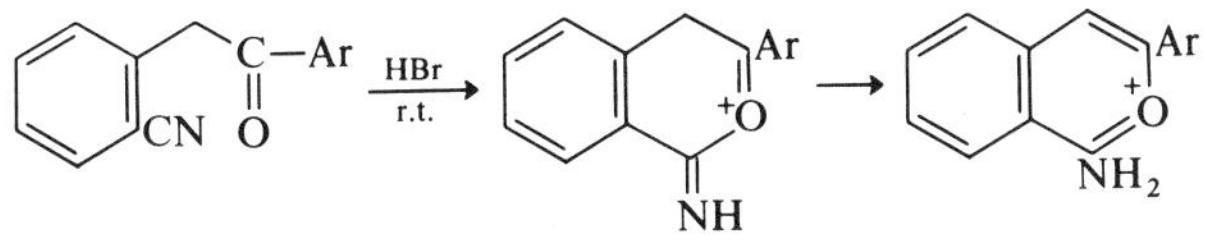

**Scheme 260**

Balaban and his coworkers have developed the diacylation of propene derivatives to provide a useful route to pyrylium salts ⟨61MI22401⟩. It is proposed that the reaction proceeds through acylation of the alkene to a keto carbocation. Elimination of the α-proton involving a cyclic six-membered transition state then leads to the enol (657) and hence the non-conjugated enone. A second acylation and subsequent dehydration yield the pyrylium salt (Scheme 261).

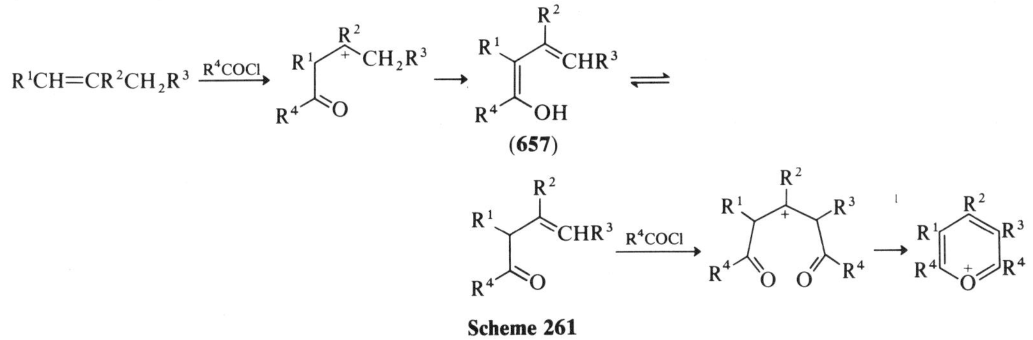

**Scheme 261**

Experimentally, it is preferable to add the alkene to the acylating agent under Friedel–Crafts conditions. Successful acylation is accomplished with both acid chlorides and anhydrides, though yields are higher with aliphatic rather than aromatic acid derivatives ⟨61JCS3553⟩.

Some novel bicyclic pyrylium salts arise when a dicarboxylic acid derivative is used. For example, although decanedioyl chloride did not diacylate isobutene, possibly because of strain inherent in the expected product, 1,12-dodecanedioyl chloride gave a very low yield of the pyrylium salt (658) ⟨62T1079⟩. A dilute solution of the reactants in nitromethane was used in order to favour the intramolecular reaction. In a similar vein, a macrocyclic alkene, such as cyclododecene, undergoes diacylation to the bicyclic salt (659) ⟨68TL4643⟩, whilst

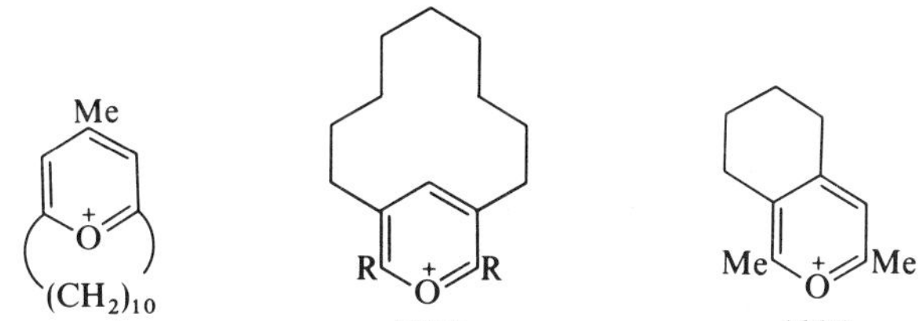

(658)          (659)          (660)

smaller ring cycloalkenes, such as 1-methylcyclohexene or methylenecyclohexane, yield the simple annelated pyrylium salt (660) ⟨61JCS3561⟩.

A valuable development of this method has been the use of an alcohol or alkyl halide as an *in situ* source of the alkene. This particular approach provides a very convenient synthesis of 2,4,6-trimethylpyrylium ⟨73OSC(5)1106⟩ and 2,3,4,6-tetramethylpyrylium salts ⟨82OPP31⟩. It is noteworthy that the trifluoromethanesulfonate and tetrafluoroborate derivatives of the former are non-explosive, unlike the perchlorate, and the sulfoacetate has much to commend it ⟨82OPP39⟩.

An interesting facet of this reaction involves the use of isotopically labelled acid derivatives, leading to pyrylium salts with ${}^{14}$C labels at specific sites. In view of the reactivity of pyrylium salts towards nucleophiles, this approach provides access to a range of labelled aromatic and heteroaromatic compounds ⟨62MI22401⟩.

The usual wide choice of Friedel–Crafts catalyst is available. However, the product obtained from the diacylation of 2-methylbut-2-ene depends on the catalyst employed ⟨61JCS3553⟩. In the presence of aluminum or antimony(V) chloride, 4-ethyl-2,6-dimethylpyrylium (661) is the predominant product, whereas the 2,3,4,6-tetramethyl derivative (662) is dominant when the reaction is catalyzed by $BF_3$, $H_2SO_4$ or $HClO_4$ and indeed is the only product using $BeCl_2$. A similar dependence on catalyst has been observed in the acylation of 4-chloro-3,4-dimethylpentan-2-one ⟨61MI22402⟩. Both results have been interpreted in terms of alkene isomerization. The ethyldimethyl compound is considered to be derived from small amounts of the more reactive 2-methylbut-1-ene formed from the but-2-ene in the presence of $AlCl_3$ or $SbCl_5$. The latter isomer is regarded as the source of the tetramethylpyrylium salt (Scheme 262).

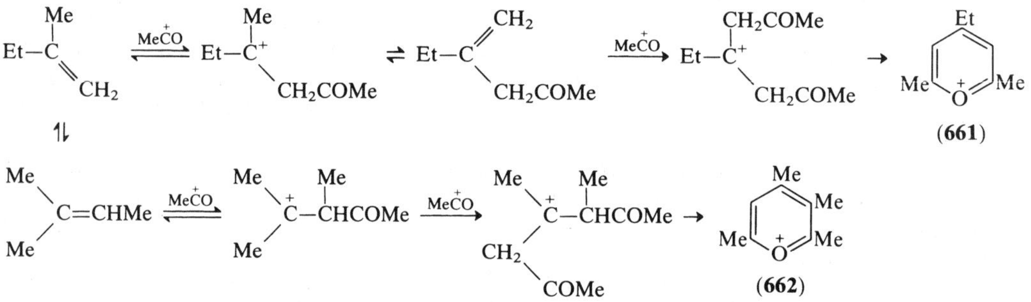

**Scheme 262**

The formation of different products is also observed in the diacylation of 2-methyl-1-phenylbut-1-ene ⟨63TL91⟩. When either the alkene or 2-methyl-1-phenylpropan-2-ol is treated with acetic anhydride in perchloric acid, 2,4,6-trimethyl-3-phenylpyrylium perchlorate is formed. However, the use of acetyl chloride and aluminum chloride as the acylating agent leads to the 4-benzyl-2,6-dimethylpyrylium salt, whilst benzoyl chloride gives 4-methyl-2,4,6-triphenylpyrylium perchlorate under similar conditions.

Some of the earliest examples of the synthesis of pyrylium salts require an oxidation stage. In 1896, Kostanecki and Rossbach noted the fluorescence, now known to be caused by 2,4,6-triphenylpyrylium, produced when 1,3,5-triphenylpentane-1,5-dione was heated with concentrated sulfuric acid. It was later found that acetic anhydride and iron(III) chloride was a more efficient reagent than sulfuric acid ⟨17JPR(95)107⟩.

Homophthaldehyde is converted into the isobenzopyrylium salt by this route ⟨33JCS555⟩. However, the yield of the unsubstituted pyrylium salt obtained from pentanedial is low, but much better from the enolate ⟨53CB1327⟩. Several other reagents are effective in bringing about the cyclization and oxidation of pentane-1,5-diones and the diverse substitution patterns which can be achieved increase the value of this method.

More unusual examples obtained by this route include the bis-pyrylium salts (663) ⟨63AG860⟩ and (664) ⟨68BSF4122⟩.

More recently, salts of the triphenylmethyl cation $Ph_3C^+$ have been shown to be effective in converting the diones into pyrylium salts ⟨61BSF538⟩. The cation may be generated *in situ*, for example from triphenylmethyl chloride and antimony(V) chloride ⟨69T1209⟩. The *t*-butyl cation has been used for the same purpose ⟨67T4001⟩. Earlier workers considered that the synthesis involved dehydration of the diketone to a 4*H*-pyran followed by rapid oxidation to the pyrylium salt. In view of the successful application of $Ph_3C^+$, the reaction

(663)　　　　　　　(664)

probably occurs by hydride ion transfer, the cation being converted into triphenylmethane. In the case of reactions involving acetic anhydride and a Lewis acid, the acetyl carbocation is the hydride acceptor and is converted into acetaldehyde.

Although the value of this route has been improved by developments in the synthesis of 1,5-diketones, it is often easier to generate the diketone *in situ*. Provided that the reaction is carried out in the presence of a hydride acceptor, a direct synthesis of pyrylium salts is available from simple precursors.

One such method, discovered and exploited by Dilthey ⟨16JPR(94)53⟩, involves the reaction between an unsaturated ketone and a methyl ketone in acetic anhydride in the presence of a metal salt. In later developments boron trifluoride has proved to be particularly effective as the condensing agent.

Mechanistically, the reaction proceeds through Michael addition of the aryl methyl ketone to the enone. Hydride abstraction and cyclization then yield the pyrylium salt. The route is capable of considerable diversification and therefore has much potential. The detailed method for the synthesis of 2,4,6-triphenylpyrylium perchlorate is worthy of mention ⟨73OSC(5)1135⟩.

Less straightforward examples of pyrylium salts prepared in this manner include 5,6-dihydro-2,4-diphenylnaphtho[1,2-*b*]pyrylium (665) and 5,6,8,9-tetrahydro-7-phenyl-dibenzo[*c*,*h*]xanthylium (666) from 1-tetralone ⟨80JCS(P1)1895⟩. In the former example, chalcone functions additionally as the hydride abstracting agent. This dual behaviour has been observed previously ⟨68JOC1102⟩ although in the more recent work it was not found necessary to employ an excess of chalcone, which, incidentally, is protonated at the oxygen atom ⟨63JCS174⟩.

(665)　　　　　　　(666)　　　　　　　(667)

Cyclic ketones react with unsaturated ketones to form fused pyrylium salts (667) ⟨58JCS1978⟩, whilst the use of phenylacetaldehyde allows the synthesis of pyrylium salts unsubstituted at C-6 to be achieved ⟨65BSF1944⟩.

An attempt to prepare 2-(2-nitrophenyl)-4,6-diphenylpyrylium from 1,3-diphenylprop-2-en-1-one and 2-nitroacetophenone gave only 2,4,6-triphenylpyrylium ⟨58BSF1458⟩. Similarly, substantial formation of this symmetrical pyrylium salt was observed during syntheses of unsymmetrically substituted salts. Thus, pinacolone and chalcone afforded both 2-*t*-butyl-4,6-diphenylpyrylium and the 2,4,6-triphenyl derivative. The latter product is considered to arise from a retro-aldol reaction of the enone into a mixture of benzaldehyde and acetophenone; the latter reacts with unchanged chalcone to give the unrequired salt ⟨80T679⟩.

The formation of a mixture of pyrylium salts when 2-acetylpyridine and chalcone reacted in perchloric acid was overcome by treating the reactants with ethanolic sodium hydroxide. The pentanedione was isolated and subsequently oxidized by reaction with chalcone and boron trifluoride ⟨82JCS(P1)125⟩.

The problem of separation of two pyrylium salts can be circumvented by careful selection of the reactants. In the two examples quoted above, the use of acetophenone with the appropriately substituted chalcone (668; R = *o*-O$_2$NC$_6$H$_4$ or Bu$^t$) results in formation of the unsymmetrical pyrylium salt (Scheme 263). Although a retro-aldol reaction can still occur, the substituted acetophenone which results is less reactive than acetophenone and the problem of contamination is much reduced. In general terms, therefore, the chalcone derived from the less reactive ketone should be used in the synthesis of unsymmetrically substituted pyrylium salts.

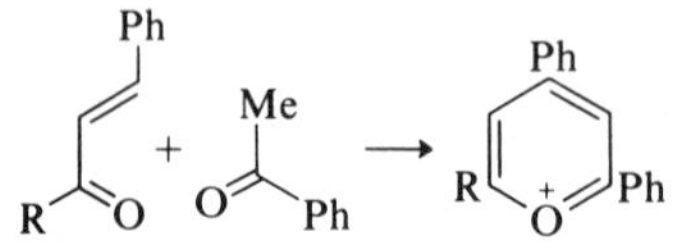

(668)

**Scheme 263**

A carboxyl functionality can be introduced at C-2 by this method; an α-keto acid or ester reacts with a chalcone under the usual conditions ⟨68CB2215, 82JOC492⟩, whilst benzylidene-1-tetralone affords the naphthopyrylium salt (**669**) ⟨82JOC498⟩.

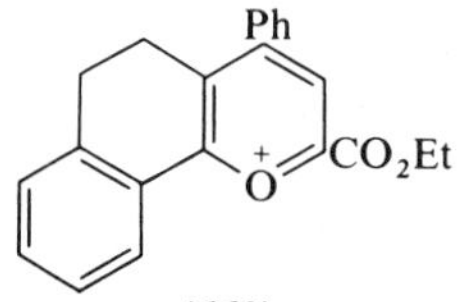

(669)

The use of ethyl acetoacetate in this route allows the formation of a 3-ethoxycarbonyl-pyrylium salt ⟨68JOC1102⟩. Some care is appropriate in this type of reaction in the absence of an added hydride acceptor, however, since the chalcone may undergo a Michael condensation with the intermediate 1,5-dione, leading to side products.

Similar variations in the reactants to those described earlier are acceptable in this synthesis. The ketone may be replaced by an alkyne, good yields resulting only on heating the reaction mixture ⟨63RTC845⟩. The mechanism shown in Scheme 264 has been proposed, but no 1,3-diarylpropan-1-one which would help to corroborate the proposal could be detected.

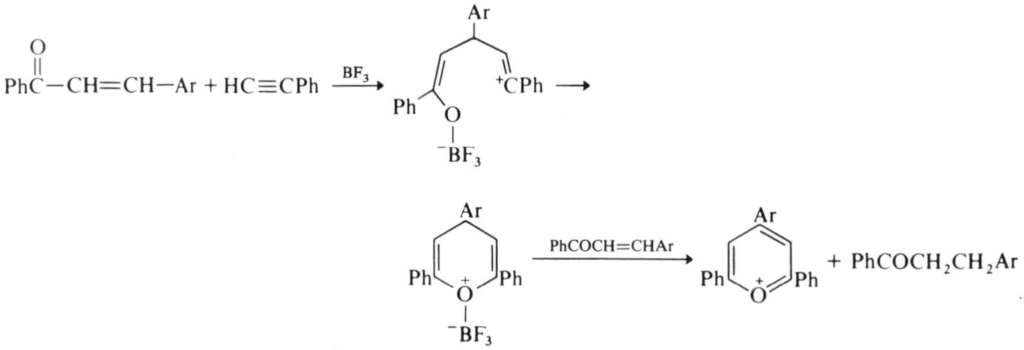

**Scheme 264**

A 3-chloroketone may replace the enone in this sequence. For example, 3-chloro-1,3-diphenylpropan-1-one reacts with phenylacetylene to give 2,4,6-triphenylpyrylium. Tin(IV) chloride acts as both the condensing agent and the hydride transfer agent ⟨65CB334⟩.

A more significant variant involves the reaction between a methyl ketone and an aldehyde which does not have an α-hydrogen atom ⟨20JPR(101)177⟩. The reaction may be acid- or base-catalyzed, though in the latter case a separate cyclization step is necessary. This route probably proceeds through the chalcone, formed by an aldol condensation between the two carbonyl compounds, followed by reaction of the enone with a further molecule of ketone.

This method is very convenient and is often the choice for the preparation of 2,4,6-triarylpyrylium salts. Of course, the substituents at C-2 and C-6 are identical. A wide range of substituted acetophenones and benzaldehydes has been used in this route and a choice of both catalyst and solvent is available. Once again, phenylacetylene may replace aceto-phenone.

When 2-hydroxybenzaldehyde reacts with acetophenone, the expected pyrylium salt is obtained at room temperature. However, without cooling to maintain this temperature flavylium salts are produced as a result of interaction with the *o*-hydroxy substituent; the 4-phenacyl group is lost during cyclization (Scheme 265) ⟨35JCS85⟩.

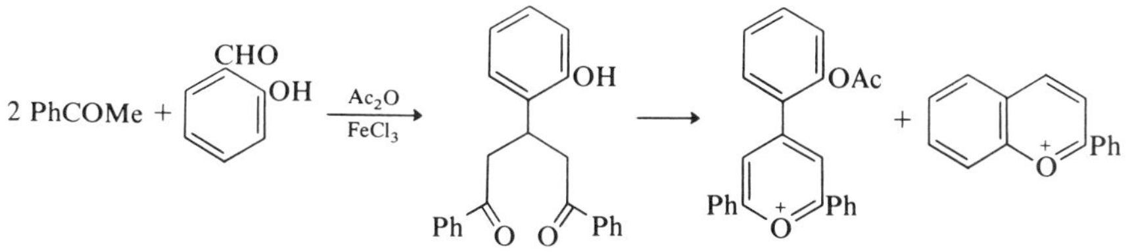

**Scheme 265**

The use of formaldehyde results in a pyrylium salt unsubstituted at C-4 ⟨66BSF2959⟩, whilst a neat variation in the ketone component enabled some 4-aryl-2,6-diethoxycarbonyl-pyrylium salts to be synthesized ⟨81S959⟩.

The dehydrogenation of penta-2,4-dien-1-ones is a convenient route to pyrylium salts unsubstituted at C-4 ⟨17CB1008⟩; the usual variations in hydride acceptors are allowed. A minor variation probably involves ionization of the chlorodienone (**670**) to a carbocation and hence to the pyrylium salt (Scheme 266) ⟨66AG448⟩.

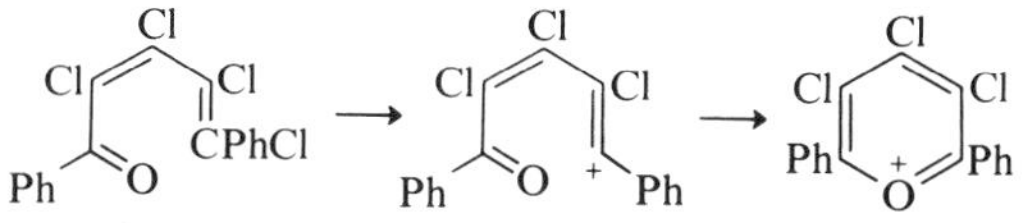

(**670**)

**Scheme 266**

Diazoketones react with acylmethylenetriphenylphosphoranes (**671**) to yield pyrylium salts, though yields are poor and the reaction is primarily of mechanistic interest. The initial product is a methylenepyran (**672**) which undergoes protonation to the salt ⟨62CR(254)696⟩. It has been demonstrated that the 4-substituent is derived from the diazoketone and that the phosphorane is the source of the groups at C-2 and C-6 ⟨62CR(255)731⟩. It is suggested that the diazoketone undergoes a Wolff rearrangement to an arylketene which subsequently forms an allenic ketene by reaction with the nucleophilic phosphine. The acylphosphorane undergoes a Michael addition to the activated double bond forming a betaine. Loss of triphenylphosphine oxide and cyclization of the resulting ynone afford the pyran (Scheme 267). It is known that 1,3-dibenzoylallene reacts with the phosphorane to give the pyran and, on treatment with acid, the pyrylium salt ⟨66MI22400⟩.

$$PhCOCHN_2 + PhCOCH=PPh_3 \rightarrow PhCH=C=CHCOPh \xrightarrow{(671)}$$
(**671**)

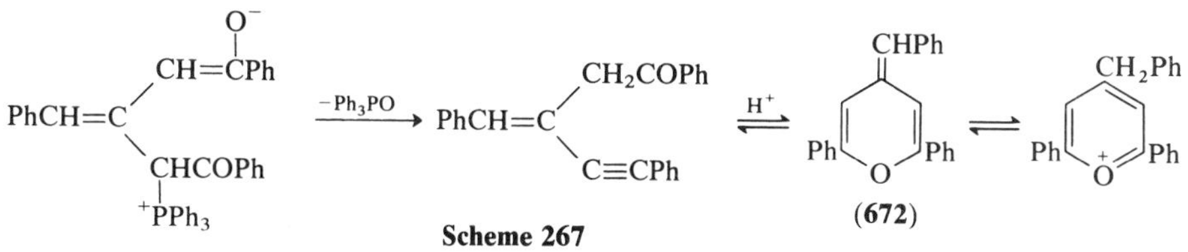

**Scheme 267**

(**672**)

Additional supporting evidence for the mechanism includes the formation of pyrylium salts when arylacetyl chlorides, which yield ketenes by loss of HCl, react with the phosphorane ⟨63CR(257)926⟩. It is also known that diphenylketene reacts in a similar manner to give a pyran, whilst aryl isocyanates yield pyran-4-one derivatives.

### 2.24.4.1.2 *β to the heteroatom*

Treatment of diaryl ethers with dichloromethyl methyl ether in the presence of tin(IV) chloride leads to xanthylium salts (Scheme 268) ⟨73JCS(P1)1104⟩. It is necessary that the diaryl ether is so designed that electrophilic attack occurs *ortho* to the potential heteroatom. The thienopyrylium salts (**673**) and (**674**) have been prepared in the same manner.

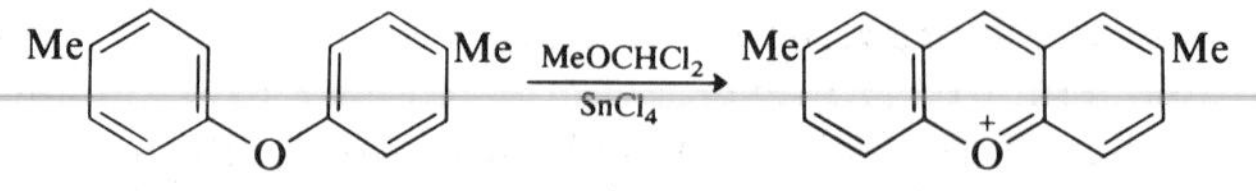

**Scheme 268**

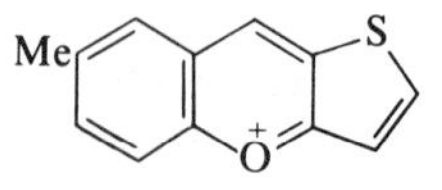

(673) 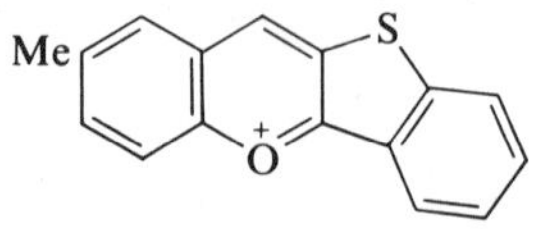                    (674)

## 2.24.4.2 From Other Heterocycles

### 2.24.4.2.1 From pyridine

In view of the development of the synthesis of various aromatic and heterocyclic compounds from pyrylium salts ⟨80T679⟩, the preparation of the unsubstituted pyrylium salt from pyridine is noteworthy (Scheme 269) ⟨53CB1327⟩. The pyridine–sulfur trioxide complex undergoes ring opening to the sodium salt of pent-2-ene-1,5-dial. This salt cyclizes in perchloric acid *via* the red oxonium salt.

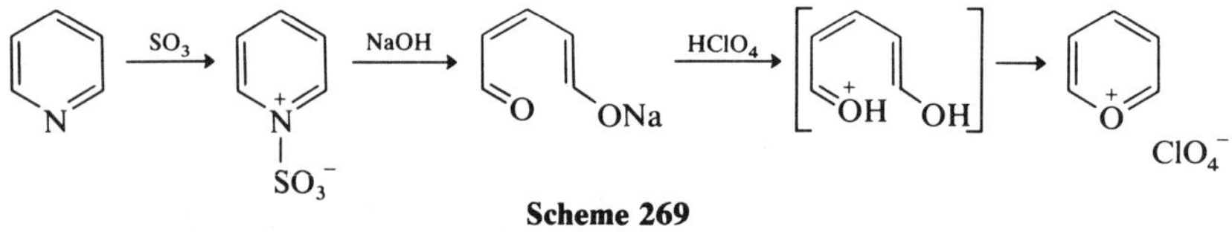

**Scheme 269**

## 2.24.4.3 From a Preformed Heterocyclic Ring

This topic is covered in the review of pyrylium salts ⟨69AHC(10)241⟩ and further details are given in Chapter 2.23. Only the main features are discussed here, with especial reference to work since 1969.

### 2.24.4.3.1 From pyranones and benzopyranones

Both 2- and 4-pyranones react with nucleophiles to give pyrylium salts. Typical reactants include organometallic compounds ⟨74JHC405⟩ and amines ⟨74JHC1065⟩.

Pyrylium salts also result from the reaction of pyranones with electrophiles. Direct protonation gives the hydroxypyrylium salts, which are readily hydrolyzed to pyranones. Alkylation by dimethyl sulfate and other alkylating agents converts pyranones to alkoxypyrylium salts. The relative ease of alkylation of 2- and 4-pyranones is apparent from the conversion of 4-hydroxy-6-methylpyran-2-one into the 2,4-dimethoxy-6-methylpyrylium salt *via* 4-methoxy-6-methylpyran-2-one ⟨75T2229⟩. The reaction of this methoxypyranone with methyl fluorosulfonate is noteworthy because of the potential of the pyrylium salt as a triketone synthon ⟨75CC675⟩. Pyranthiones are alkylated more readily than pyranones ⟨78JCS(P1)1373⟩.

The differing behaviour of nucleophiles and electrophiles is illustrated by their reaction with 2,6-di-*t*-butyl-4*H*-pyran-4-one (Scheme 270) ⟨74JHC1075⟩.

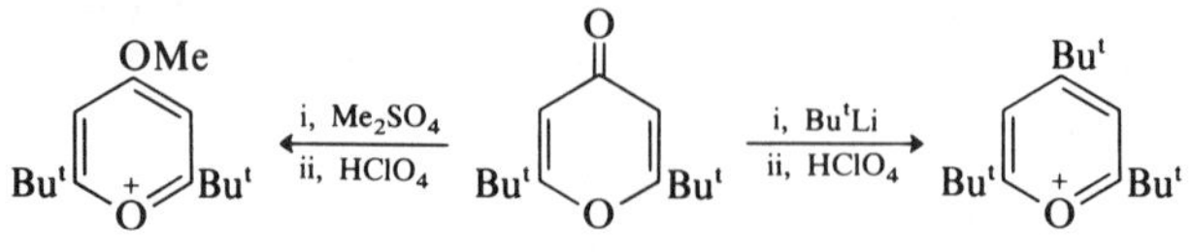

**Scheme 270**

Chromones behave in a similar manner to pyranones. For example, benzopyrylium salts are formed with strong acids ⟨71CB348⟩ and with Grignard reagents ⟨73SC231⟩. Similarly, 3-phenylisocoumarin affords 1,3-diphenylisobenzopyrylium ⟨51JOC1064⟩.

### 2.24.4.3.2 From pyrans and benzopyrans

Derivatives of methylenepyrans and their benzologues are readily converted into pyrylium salts on treatment with acid. The reaction of the ester (**675**; R = Ph) affords the pyrylium salt (**676**), but hydrolysis of the ester function is observed in (**675**; R = Me) leading to 2,4,6-trimethylpyrylium (Scheme 271) ⟨71TL617⟩.

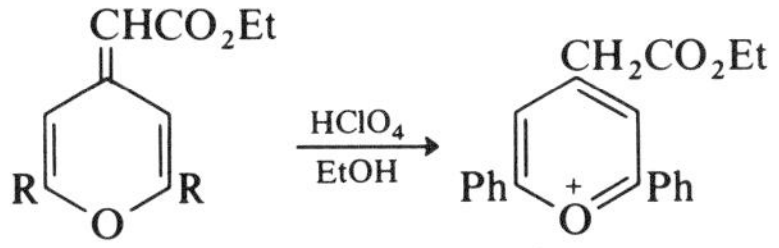

**Scheme 271**

In a similar manner, 2-dialkylamino-7-methoxychromones are converted into benzopyrylium salts on reaction with malononitrile ⟨81JHC863⟩ and naphthopyrylium salts arise from a naphthopyran-1-thione through reaction with ethyl cyanoacetate.

The acidic hydrolysis of 2,6-diphenyl-4-(α-cyano-2-nitrobenzylidene)-4H-pyran yields the benzisoxazol-3-yl pyrylium salt (**677**) ⟨74JHC395⟩.

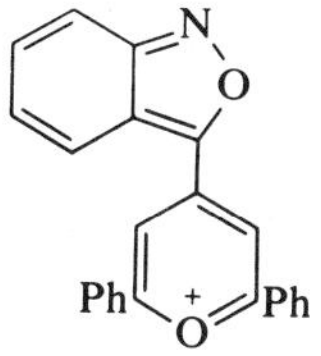

(**677**)

The conversion of 4H-pyrans to pyrylium salts has also been accomplished. Thus, the ester (**678**) yields the 2,6-dimethoxycarbonylpyrylium salts (**679**; R = H or Me) on treatment with triphenylmethyl perchlorate in liquid sulfur dioxide (Scheme 272) ⟨73ACS1385⟩ and the amide (**680**) is formed in an analogous manner ⟨74ACS(B)517⟩.

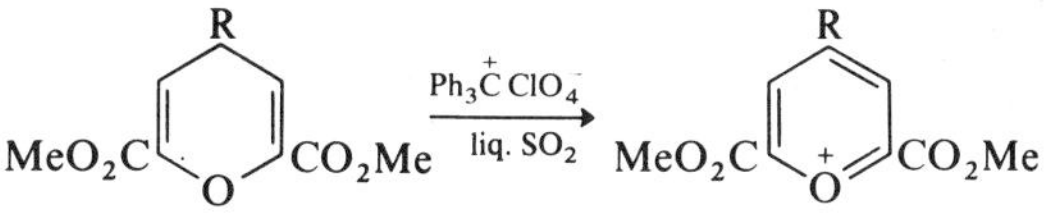

**Scheme 272**

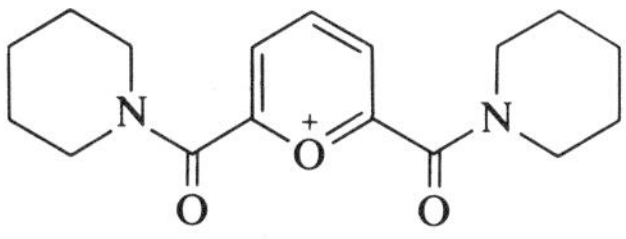

(**680**)

Treatment of 2H-chromenes which have at least one hydrogen atom at C-2 with acid gives a mixture of chroman and benzopyrylium salt ⟨68T949⟩. Reaction of a chromene with triphenylmethyl salts gives essentially only the benzopyrylium salt ⟨67AC(R)1045, 72CR(C)(274)650⟩.

Whilst the isoflavene (**681**) forms a di(isoflavenyl) ether (**682**) with perchloric acid, it is quantitatively converted into isoflavylium perchlorate (**683**) on treatment with triphenyl-methyl perchlorate (Scheme 273) ⟨82JCS(P1)1193⟩.

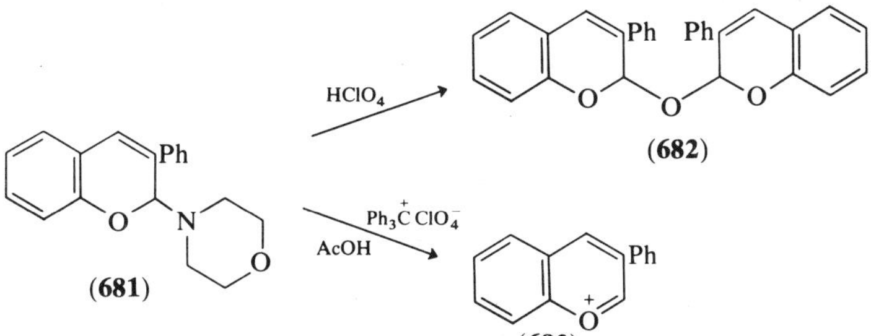

**Scheme 273**

### 2.24.4.3.3 From chromans

Substituted chromans give benzopyrylium salts on treatment with acid ⟨81JHC1325⟩, although chromanols and naphthopyranols (**684**; R = Me) yield mixtures of the pyrylium salt and chromans as a consequence of intermolecular hydride transfer ⟨68T949⟩. However, reaction of the naphthopyranol (**684**; R = H) with triphenylmethyl perchlorate gave the pyrylium salt alone; presumably the cation functions as both a dehydrogenating and dehydrating agent as a result of the formation of some free perchloric acid (Scheme 274) ⟨69IJC28⟩.

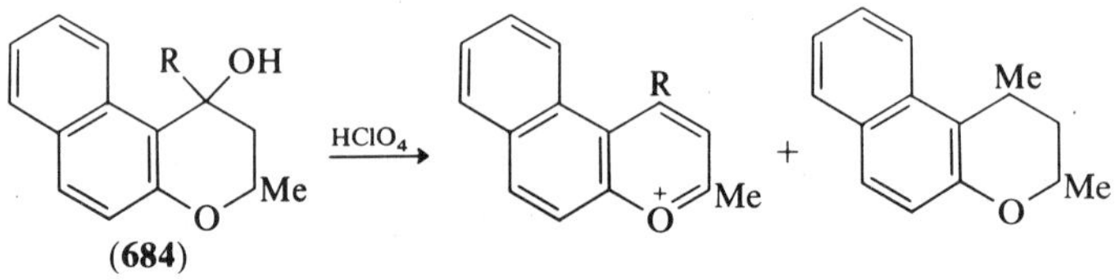

**Scheme 274**

## 2.24.5 APPLICATIONS

The natural occurrence of oxygen heterocycles is well documented ⟨B-63MI22400⟩ and almost every category of the compounds discussed in the earlier sections provides examples of secondary metabolites. A discussion of their biosynthesis is therefore appropriate.

The flavonoids, which comprise the largest group of these natural products, are derived from a mixed acetate–shikimate pathway. A shikimate-derived $C_6$–$C_3$ unit combines with a six-carbon polyketide chain to provide the open-chain precursor (**685**) of the group. The derivation of *p*-hydroxycinnamic acid (*p*-coumaric acid), the $C_6$–$C_3$ component, from shikimic acid proceeds through chorismic acid, prephenic acid and phenylalanine.

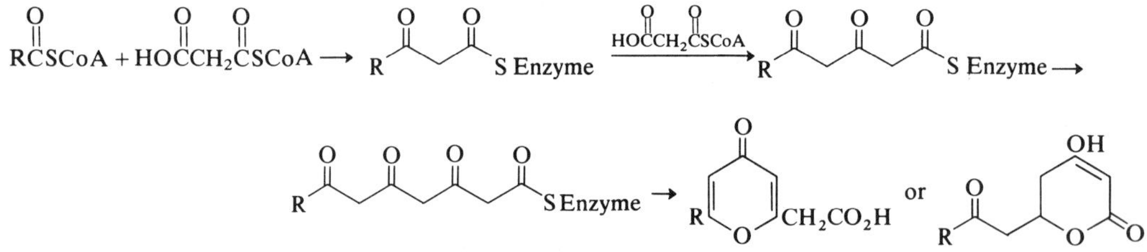

**Scheme 275**

The biosynthesis of the polyketide moiety is thought to involve the condensation of coenzyme A esters of acetic acid with malonyl coenzyme A to give thiol esters of 3-keto acids. Further Claisen condensations with malonyl coenzyme A add further ketone units, leading to 3,5-diketo, 3,5,7-triketo acids and so on as their thiol esters. Intramolecular condensations subsequently afford heterocyclic or aromatic structures (Scheme 275).

Ring closure of the polyketide chain of (**685**) generates a chalcone (**686**), from which a flavanone is derived by a further cyclization (Scheme 276). The latter may be regarded as the precursor of all other flavonoids, although it should be noted that an enzyme-catalyzed equilibrium exists between chalcone and flavanone. Chalcone may therefore function as the flavonoid precursor rather than flavanone ⟨68P1751⟩.

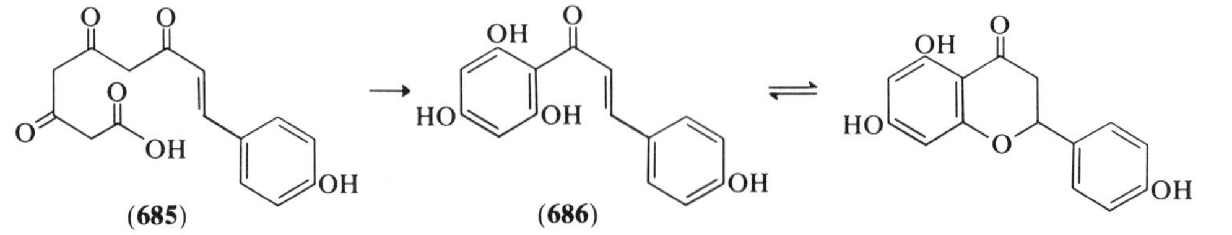

**Scheme 276**

Dihydroflavonols (**687**) appear to be important biosynthetic intermediates to other flavonoids as indicated in Scheme 277, which illustrates the relationships between the various flavonoid types. It is of interest to note that the stereochemical relationship shown is found in all natural dihydroflavonols. Their derivation from chalcones *via* an epoxide would give this *trans* arrangement of the 3-hydroxy and 2-phenyl groups. Such a conversion has its analogy in the laboratory, as do other interconversions depicted in Scheme 277. The biosynthesis of flavones (**688**) has been shown to proceed through dehydrogenation of flavanones and not by the dehydration of a dihydroflavonol.

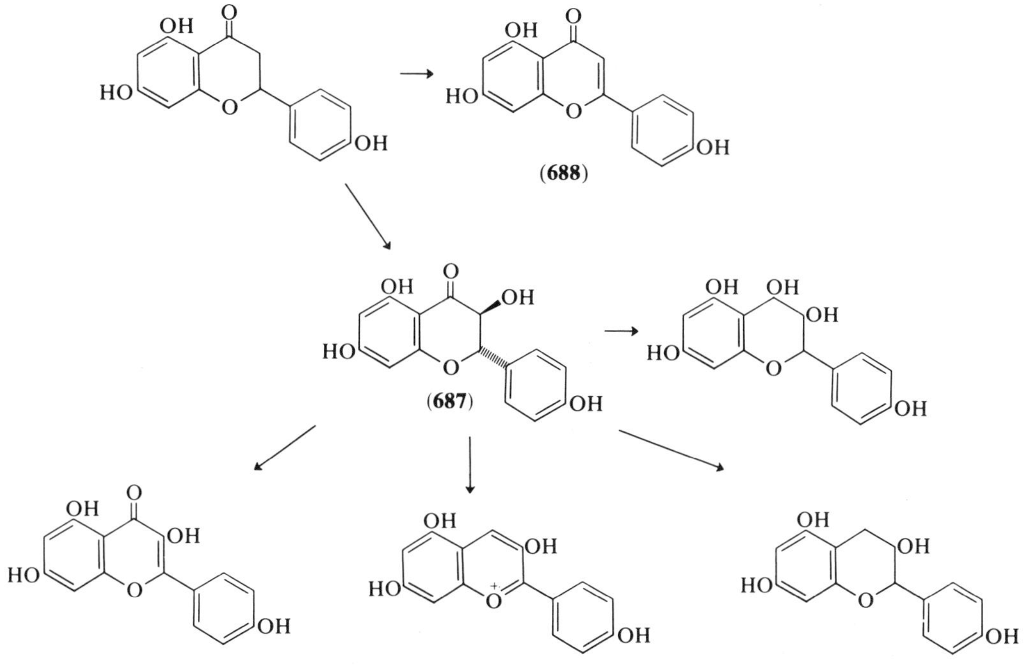

**Scheme 277**

The different derivations of the component rings of flavonoids can usually be deduced from the oxygenation pattern. Thus, the acetate-derived fused benzenoid ring possesses oxygen substituents on alternate carbon atoms. On the other hand, the presence of adjacent oxygen-containing groups often found in the 2-aryl ring is indicative of its formation by the shikimate pathway.

Isoflavonoids result from a 1,2-aryl shift in the chalcone or flavanone (Scheme 278), which has an analogy in the *in vitro* thallium nitrate induced rearrangement of 2'-hydroxychalcones to isoflavones (see Section 2.24.3.4.1). Rotenoids arise in a similar fashion, the additional ring carbon atom being derived from a methoxy group. Labelling of phenylalanine has indicated the origin of the marked atoms in rotenone (**689**) and confirmed the 6a,12a-aryl shift (Scheme 279) ⟨68JCS(C)3029⟩.

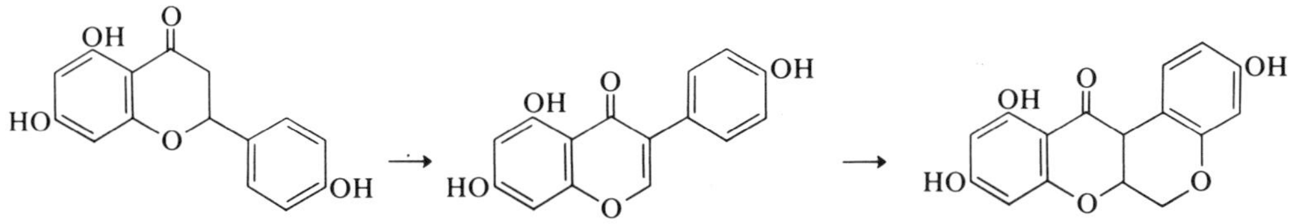

**Scheme 278**

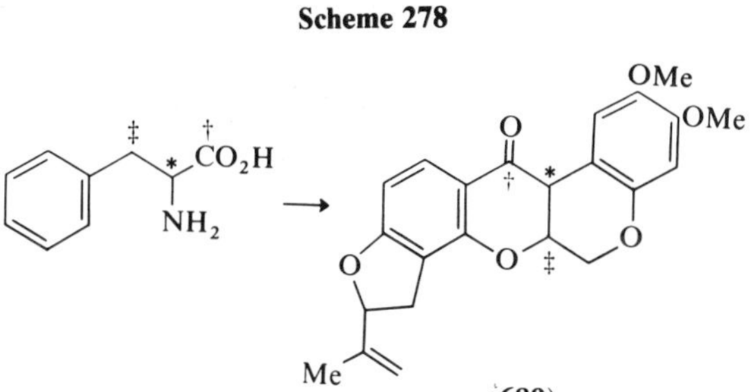

(**689**)

**Scheme 279**

Coumarins and isocoumarins appear to be of varied origins. Simple coumarins, such as umbelliferone, are formed by the shikimic acid pathway in which hydroxylation of *p*-hydroxycinnamic acid occurs. Other coumarins, for example alternariol (**690**), are derived from a polyketide unit, as are a number of chromanones, chromones, pyranones and isocoumarins ⟨B-78MI22400⟩. The biosynthesis of 5-hydroxy-2-methylchromone has been shown to involve the chromanone ⟨60JCS654⟩. However, isocoumarins are also derived from the mixed acetate–shikimate route, through initial cyclization of the polyketide and subsequent lactonization.

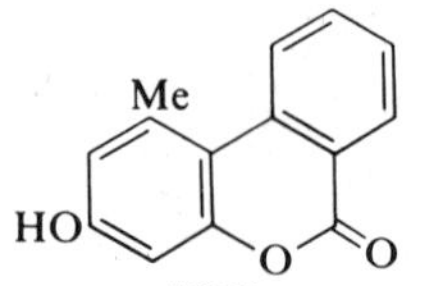

(690)

Xanthones in higher plants are also formed by this mixed pathway, though the polyketide chain originates from a benzoic acid instead of the usual cinnamic acid. Cyclization now affords a benzophenone (**691**), rather than a chalcone, which subsequently cyclizes to the xanthone, a route used with considerable success in the laboratory (Scheme 280). Other xanthones are derived only from acetate, through ring opening, decarboxylation and cyclization of an anthraquinone precursor.

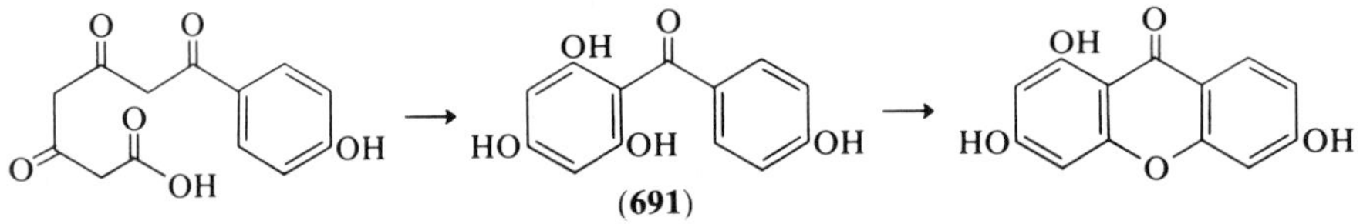

(691)

**Scheme 280**

The derivation of aflatoxin B can be traced back to the xanthone sterigmatocystin (**693**), which originates from a decaketide (**692**), whilst the xanthone bikaverin (**694**) is derived from a nonaketide.

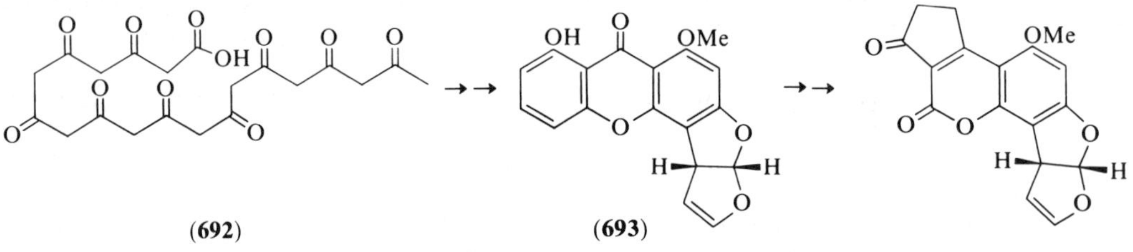

(692)　　　　(693)

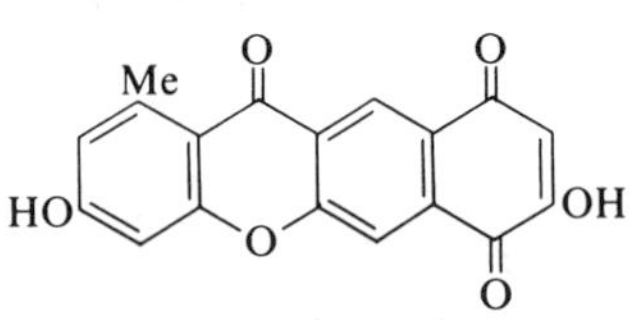

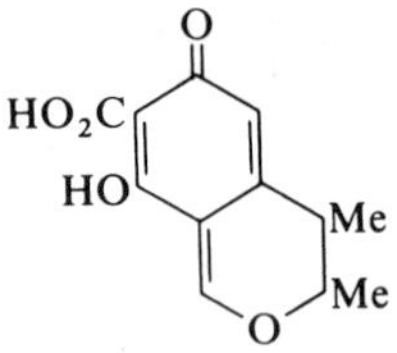

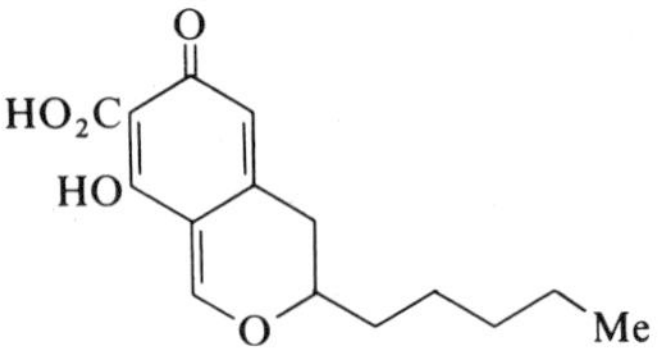

(694)　　　　(695)　　　　(696)

A quinone methide moiety is present in the dihydropyrans citrinin (**695**) and pulvilloric acid (**696**), both of which are derived by the acetate–malonate pathway.

The tocopherols or vitamins E comprise an important group of chromans which occur in a range of plant species. Vegetable oils, especially wheat germ oil, are particularly rich sources of these vitamins. The various tocopherols, which differ in the number and positions of methyl groups in the aromatic ring, are derived from shikimic acid. Tyrosine is converted into homogenistic acid by way of 4-hydroxyphenylpyruvic acid and subsequent reaction with either geranylgeraniol pyrophosphate or phytyl pyrophosphate leads to the tocopherols. The 5- and 7-methyl groups of α-tocopherol (**697**) are thought to be derived from methionine. The biosynthesis of tocopherols has been discussed ⟨71MI22403⟩.

The biological function of the tocopherols is still a matter of discussion ⟨83MI22400⟩. They behave as antioxidants ⟨68MI22401⟩, protecting the lipids in plants from excessive oxidation,

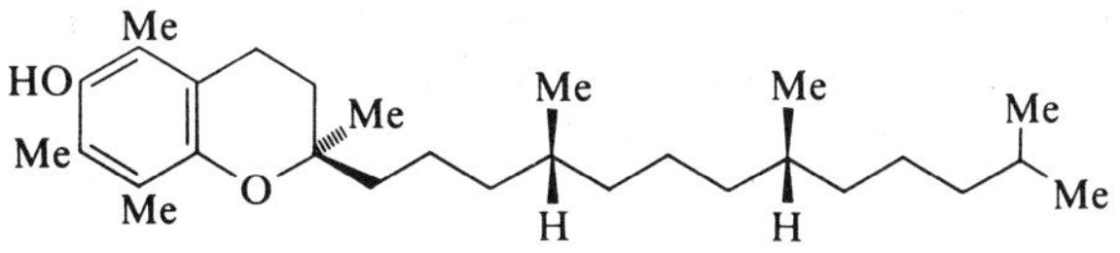

**(697)**

but their role in animals is more varied. In addition to its vital part in reproduction, vitamin E is important in heme synthesis ⟨70JBC(245)5498⟩. The significance of the tocopherols has been reviewed ⟨76MI22400⟩.

The resin secreted by *Cannabis indica* and *Cannabis sativa*, varieties of hemp, is known variously as marijuana, hashish or bhang and is abused as a hallucinogenic drug. It appears however to have some beneficial properties and is currently under test as an antiemetic in cancer therapy. The secretion contains a number of interrelated oxygen heterocycles, some of which are shown in Scheme 281, which attempts to indicate their biosynthetic relationships ⟨70MI22401⟩. The cannabinoids are probably derived from a monoterpene unit based on *p*-menthane and 5-*n*-pentylresorcinol (olivetol), acting the part of a polyketide. 2,2-Dimethylchromene biosynthesis also requires the intervention of an isoprene fragment.

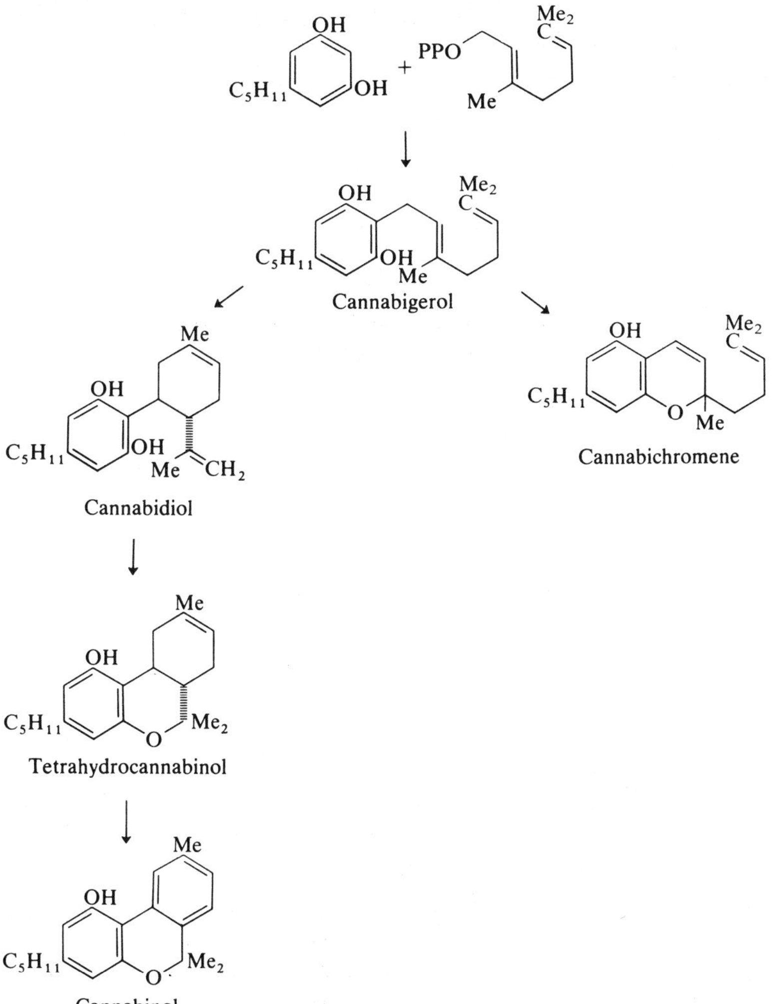

**Scheme 281**

The physiologically active component of cannabis is $\Delta^1$-3,4-*trans*-tetrahydrocannabinol, but cannabinol, into which the other less stable cannabinoids change as the plant ages, is inactive. Recent advances in cannabinoids have been reviewed ⟨B-73MI22402⟩ and a survey of the whole area of cannabinoids has been published ⟨76CRV75⟩.

The reader is referred to a number of texts for a detailed discussion of the biosynthesis of flavonoids ⟨B-63MI22400, B-75MI22400, B-81MI22402⟩, or of biosynthesis in general ⟨B-63MI22401, B-64MI22402, B-78MI22401⟩.

Two groups of oxygen heterocycles, the hydroxyflavones and the anthocyanins, are the main colouring components of flowers, trees and fruits. The former group of compounds are widely distributed yellow pigments which generally occur in nature as glycosides, although the flavones are found in the uncombined state. A notable example is the farina which covers the leaves of several species of *Primula* and which consists of a mixture of flavone and various hydroxy analogues. This powdery coating can cause an allergic reaction on sensitive skin. Perhaps the most important flavone pigment is quercetin (**698**), which has been used as a yellow dye for centuries, whilst its 3-rhamnoglycoside, rutin, has been extensively studied. $\alpha$-L-Rhamnose is produced by the hydrolysis of quercetrin (**699**) using dilute acid. The microbial breakdown of the flavone glycoside is also feasible.

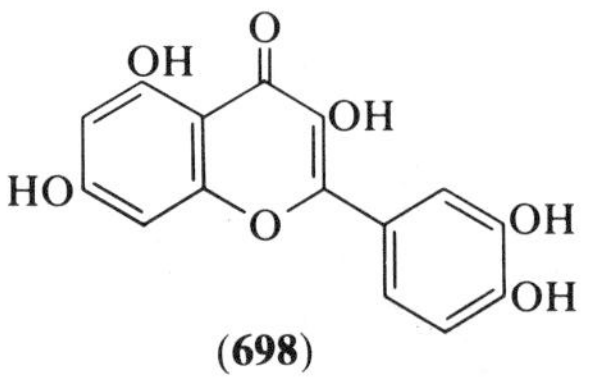

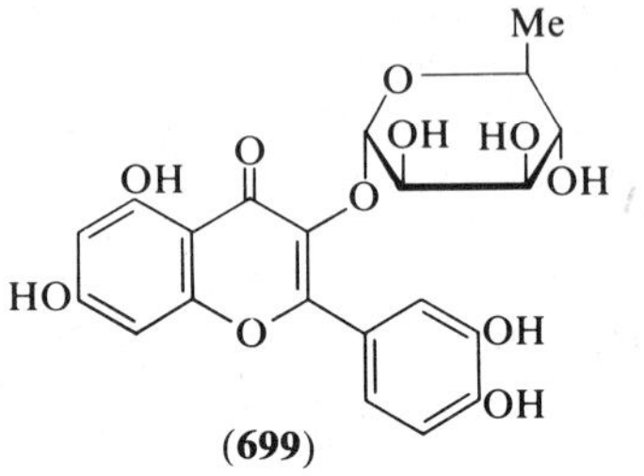

Anthocyanins provide the red, blue and purple colours of nature. They are glycosides of hydroxy derivatives of flavylium salts. The aglycones are referred to as anthocyanidins. Three anthocyanidins are of particular importance: pelargonidin, cyanidin and delphinidin. These occur notably in pelargoniums, cornflowers and delphiniums, respectively, but also in a wide range of flowers and fruits. The close similarity of the structures makes the wide and beautiful variation in colours in the vegetable kingdom even more remarkable and surely one of the wonders of our world. The colours of anthocyanins depend to some extent on the pH of their environment. An outstanding example is the red colour of the rose and poppy and the blue of the cornflower, both of which are caused by the one anthocyanin, cyanin. The colour changes may be accounted for by the changes in structure illustrated with the aglycone in Scheme 282. Other factors which play a part in the variation of colour include the chelation of metal ions and the presence of other compounds. It is noteworthy that although other hydroxy groups may be methylated or converted to a glycoside, a free hydroxy group is always present at the 4'-position.

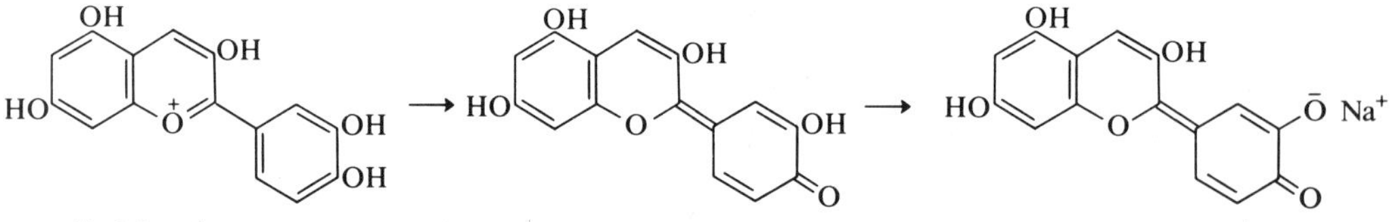

Red flavylium salt    Violet colour base    Blue sodium salt

**Scheme 282**

It is not possible to do full justice to these compounds in the present work, but valuable surveys of both groups of colouring matters are available ⟨B-60MI22400, B-63MI22400, B-77MI22400⟩.

The flavone morin (**700**) (Natural Yellow; C.I. 75660) dyes wool yellow, the actual shade being determined by the mordant which is used. It has also been mentioned as a spot test reagent for salts of aluminum, beryllium, zinc and other metals ⟨B-54MI22401⟩.

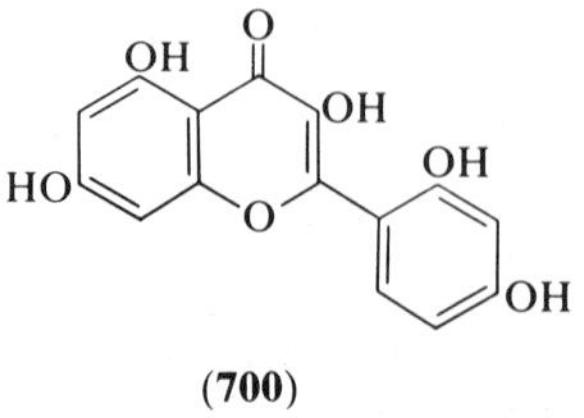

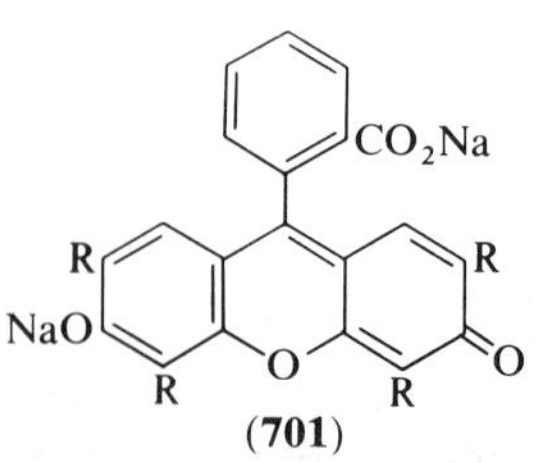

Several derivatives of xanthene are of value as colouring matters and the topic has been reviewed ⟨B-52MI22400⟩. For example, eosin (**701**; R = Br) (Acid Red 87; C.I. 45380) is used in inks, lipsticks and nail varnish in addition to its value as a dye for silk and paper.

Along with the related compound erythrosine (**701**; R = I), which is a certified food additive, it is used as a biological stain.

3,6-Dialkylaminoxanthenes are coloured and are used as bacterial stains under the names Pyronine B (**702**; R = Et) (C.I. 45010) and Pyronine G (**702**; R = Me) (C.I. 45005). Rhodamine B (Basic Violet 10; C.I. 45170) is the 9-(2-carboxyphenyl) derivative of Pyronine B, and in addition to its use as a dye for paper is of value as an analytical reagent for a variety of transition metals.

(**702**)                                      (**703**)

Fluorescein (**703**) (Acid Yellow 73; C.I. 45350) is possibly the best known xanthene dye. The sodium salt is soluble in water to which it imparts an intense yellow-green fluorescence, detectable even at a dilution of 0.02 p.p.m. under UV irradiation. This property leads to the use of fluorescein as a location marker for aircraft lost at sea, as a tracer for the detection of a source of contamination in drinking water, and in a number of related situations. The use of fluorescein to detect abrasions of the cornea is also based on its fluorescence.

Coumarins, too, fluoresce, notably when an electron-releasing group is present at the 7-position. For example, N-acyl derivatives of some 7-aminocoumarins serve as fluorescent markers for the detection of proteinases ⟨80MI22403⟩. Coumarins show a propensity to absorb UV light and this results in a number of applications. A particularly simple illustration is the use of umbelliferone, 7-hydroxycoumarin, in sun-screen lotions.

7-Diethylamino-4-methylcoumarin is used to sensitize a weakly fluorescent second material in a mixture designed for use as an *in situ* flaw detector in metal surfaces. The energy absorbed by the coumarin is transferred to a second component with little energy loss by non-radiative processes. The blue fluorescence of the coumarin is replaced by the yellow-green of the other component, to which the eye is more sensitive.

The absorption of UV light by coumarins is also fundamental to their use as optical brightening agents in laundry and domestic detergents and as additives to fibres and paper. The energy acquired by the coumarin is converted into visible light and consequently the amount of reflected visible light is increased and may even exceed that received. The fabric thus appears to be a brilliant white. The property is dependent on UV light and hence its effect is diminished in artificial light. Typically, the coumarin contains a substituted amino group at C-7 and derivatives such as (**704**; R = Et or Ph) are applied to wool and nylon. The presence of a 3-phenyl group and elaboration of the basic 7-substituent gives products (*e.g.* **705**) which are suitable for application to cellulose, polyamide, polyacrylonitrile and wool, or to inert polyesters by incorporation into the melt. The subject of optical brighteners has been discussed ⟨B-71MI22404⟩.

(**704**)                                      (**705**)

Xanthenes and 7-aminocoumarins are two of the most important classes of laser dyes ⟨B-73MI22403⟩. In the former group, exemplified by the rhodamines (**706**), the chromophore is symmetrical and the fluorescence band largely overlaps the absorption band. However, there is a much greater dipole moment in the excited state (large contribution from structure **708**) of 7-aminocoumarins than in the ground state (primarily **707**) and consequently these compounds exhibit larger Stokes shifts. The fluorescence efficiency of these and other laser dyes has been discussed ⟨76MI22401⟩.

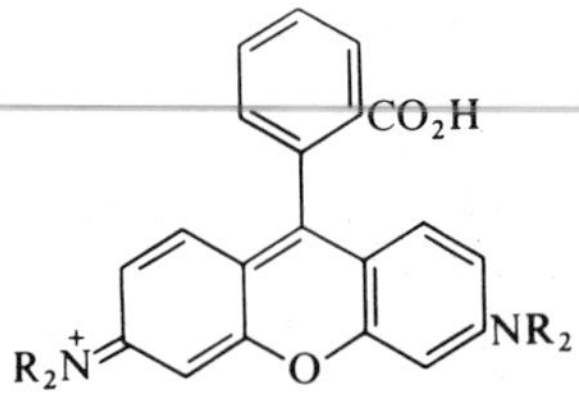

(706) 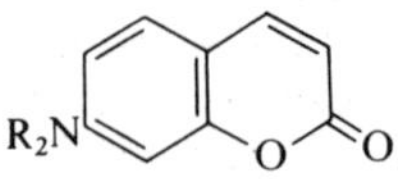          (707) 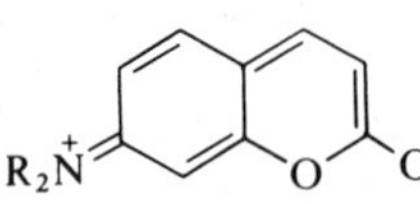          (708)

Spiropyrans (**709**) have been extensively investigated because of their thermochromic and photochromic properties, that is their ability to change colour reversibly under the influence of heat or light. It is postulated that the colour change is associated with heterolytic cleavage of the spiro linkage to the oxygen heteroatom, leading to a planar ring-opened species which is resonance stabilized (Scheme 283).

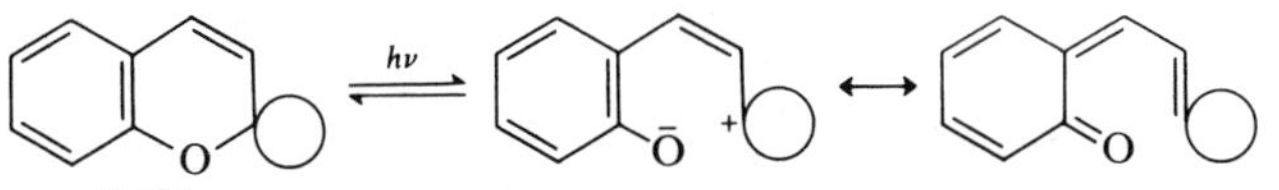

(709)                    **Scheme 283**

The reverse process, photobleaching, is more difficult, presumably because of the geometrical requirements for ring closure, and is temperature dependent. This latter observation has been attributed to the need for a photochemical *trans → cis* isomerization prior to a symmetry-allowed thermal cyclization. Chromenes possessing a spiroannelated heterocyclic moiety, notably containing a nitrogen heteroatom, have generated particular interest and have been incorporated into heat-sensitive copying paper and various data storage systems ⟨68MI22402⟩. In the case of such compounds the ring-opened forms are merocyanines. 2,2'-Spirobichromans such as derivatives of (**710**) are used as stabilizers in colour photographic emulsions.

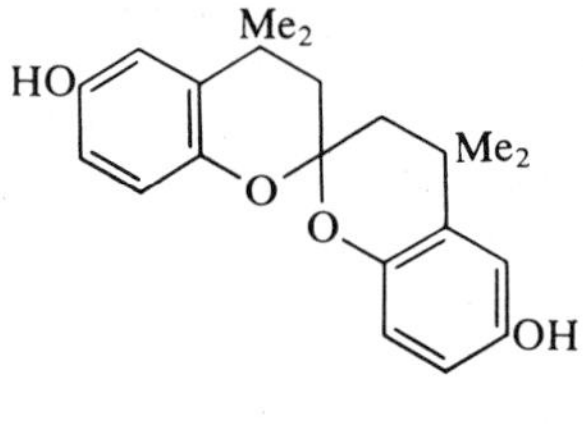

(710) 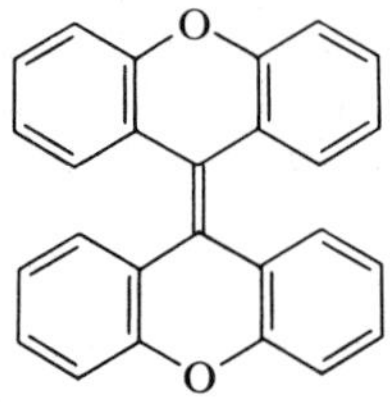                    (711)

The chemistry of spiropyrans has been reviewed ⟨48CRV(43)509⟩ whilst a review on thermochromism deals with spiropyrans and dixanthylenes (**711**) ⟨63CRV65⟩. The photochromism of spiropyrans has also been discussed ⟨B-71MI22405⟩.

3-Hydroxypyran-4-ones are potent flavouring materials of which maltol (**427**), present in roasted malt, is particularly well known. Amongst other features, it imparts a newly baked odour to bread. Kojic acid (**712**) is produced in an aerobic process by a range of microorganisms, notably *Aspergillus oryzae*, from a variety of carbohydrate sources and is used as a source of maltol.

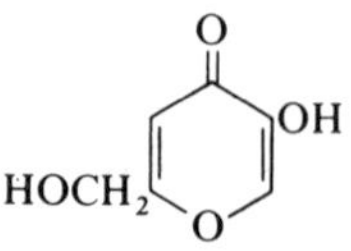

(712) 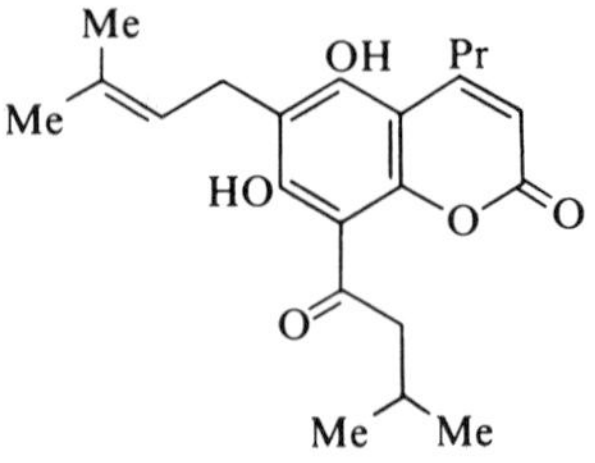                    (713)

Ring opening of 6-hydroxy-3,4-dihydro-2H-pyran-2-carboxylic acid or the corresponding tetrahydropyran by *Candida lipolytica* is a practical route to lysine ⟨69FRP2041497⟩.

Insecticidal activity is shown by a number of oxygen heterocycles. Xanthone is an ovicide for the codling moth and the naturally occurring coumarin mammein (**713**) shows insecticidal properties.

Rotenone (**689**) is the principal active component of derris root and is used in commercial insecticidal preparations and as a fish poison.

Both 7-methoxy- and 6,7-dimethoxy-2,2-dimethylchrom-3-ene are present in extracts of *Ageratum houstonianum*. These compounds have been called Precocene I and Precocene II respectively, because of their ability to induce precocious metamorphosis in a variety of insect species ⟨76MI22402⟩. The compounds function by halting the production of juvenile hormones by destroying the *corpora allata*. This antijuvenile hormone activity may well be caused by the derived epoxychromans (**714**) ⟨79CC920⟩. Investigations into the value of these and related compounds as insecticides are as yet incomplete, but the observation of biological activity has generated much interest in the chromene system.

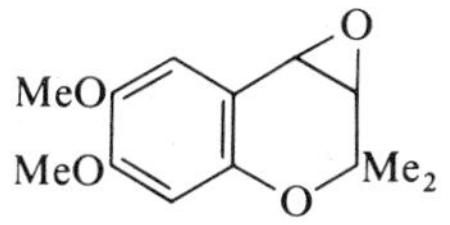

(**714**)

Quite extensive studies on the pharmacological activity of oxygen heterocycles have been carried out, the results of which have culminated in the incorporation of a number of the heterocycles into therapeutic preparations.

Several coumarin derivatives have found use as anticoagulants. The long known hemorrhagic disease in cattle is caused by the presence of 3,3'-methylenebis-4-hydroxycoumarin (**715**; dicoumarol) in spoiled sweet clover. This knowledge understandably led to a large research effort in this area which resulted in the production of a number of anticoagulants, of which warfarin (4-(4-hydroxycoumarin-3-yl)-4-phenylbutan-2-one; **716**) is perhaps the best known. It is widely used as a rodenticide, although a strain of warfarin-resistant rats has developed, and is used clinically for the treatment of various thromboembolic diseases. Warfarin appears to function through interference with the prothrombin activity of vitamin K, the antihemorrhagic factor. The vitamin undergoes a regular oxidation to phylloquinone epoxide and reduction back to phylloquinone and it seems that coumarin anticoagulants inhibit the reconversion process. Presumably the enzyme which converts the epoxide back to the vitamin has been altered in the warfarin-resistant rats so that it is no longer susceptible to coumarin anticoagulants. The synthesis of warfarin from 4-hydroxycoumarin involves a Michael addition to 4-phenylbut-3-en-2-one.

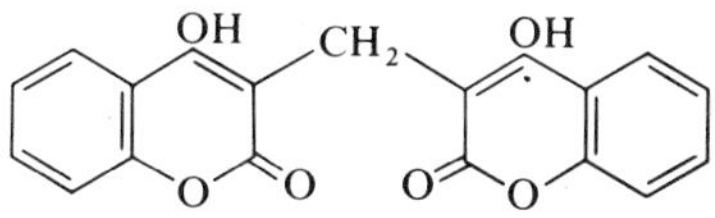

(**715**) 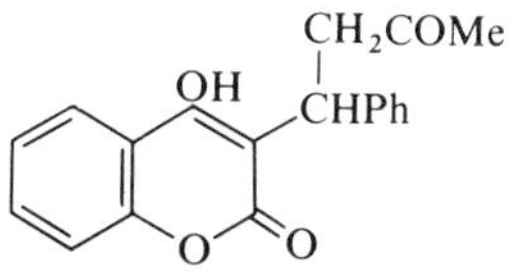          (**716**)

Acenocoumarol (Nicoumalone; **717**) is also used as an anticoagulant.

Coumarins show quite diverse biological activity and, in addition to their anticoagulant properties, have value as vasodilators (Chromonar **718** and Visnadin **719**), anthelmintics (Haloxon **720**) and diuretics (mercumallylic acid **721**), whilst Hymecromone (**722**) is a

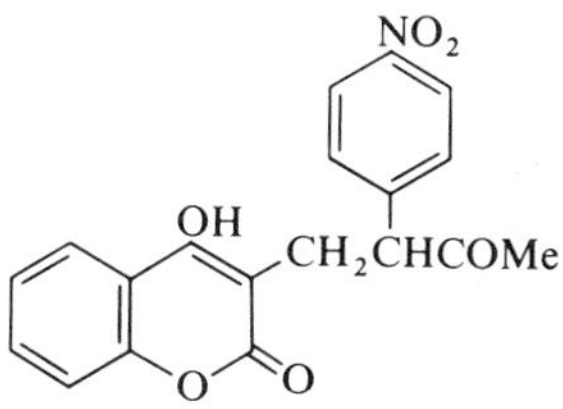

(**717**) 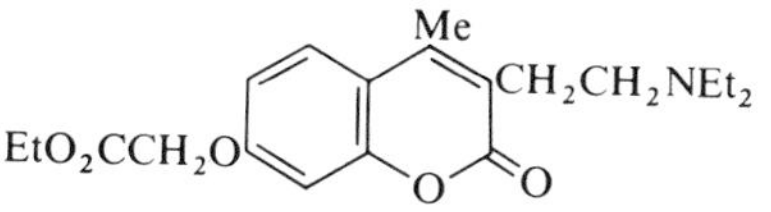          (**718**) 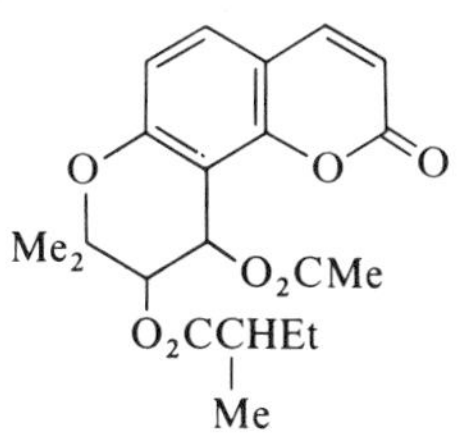          (**719**)

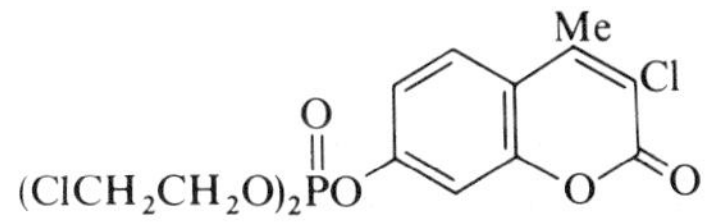

(**720**) 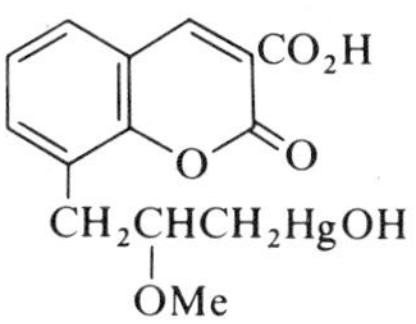          (**721**) 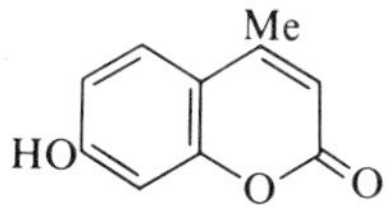          (**722**)

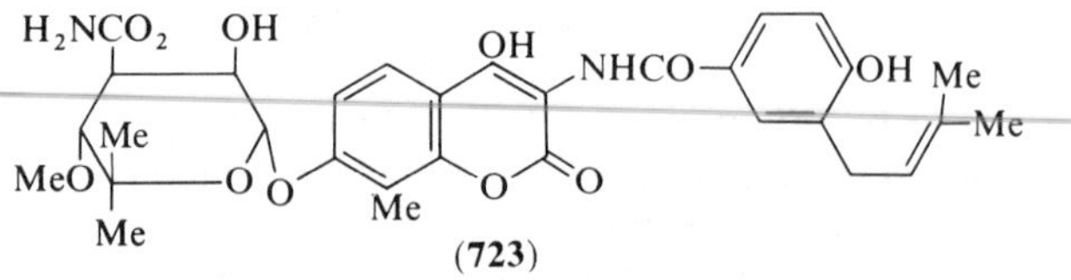

(723)

systemic insecticide which is especially effective against the Colorado beetle. The antibiotic novobiocin (723) produced by *Streptomyces spheroides* and *niveus* has been marketed as an antibacterial or antimicrobial preparation under a variety of names. Antibiotics containing pyran and benzopyran moieties have been discussed ⟨B-81MI22403⟩.

The carcinogenic properties of coumarins have been reviewed ⟨74MI22401, B-80MI22404⟩, as has their toxicity ⟨B-72MI22400, 79MI22403⟩.

Disodium cromoglycate (724), marketed as Intal or Cromolyn Sodium, bears some structural resemblance to khellin (457), the spasmolytic component of seeds of *Ammi visnaga*. Intal is one of the more successful drugs for the prevention of asthmatic attacks, though it is not effective in the treatment of an acute attack of asthma. It appears to prevent the release of histamine and other substances which mediate hypersensitivity reactions but is ineffective once these substances have been released. The chemistry and pharmacology of Intal have been reviewed ⟨B-70MI22402⟩.

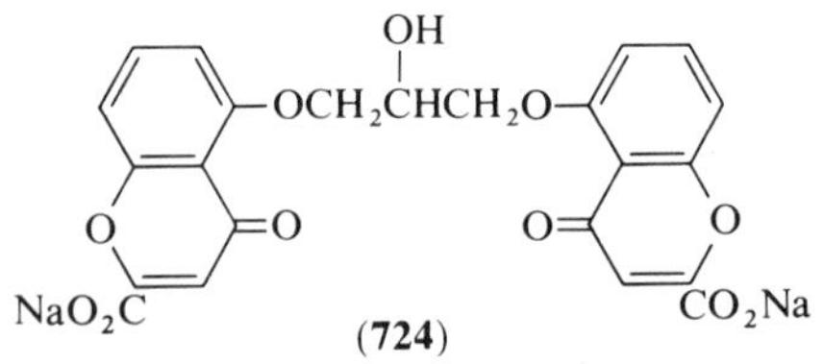

(724)

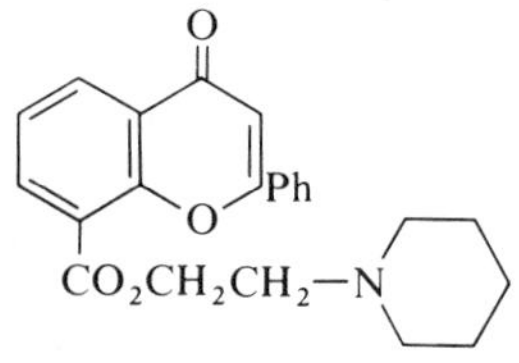

(725)

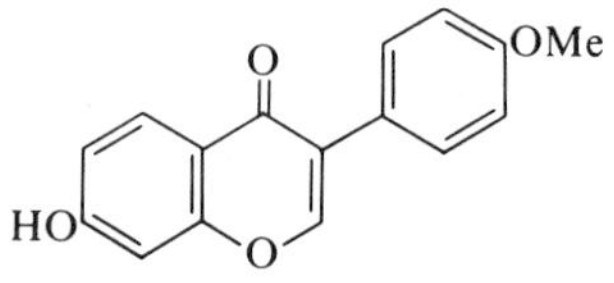

(726)

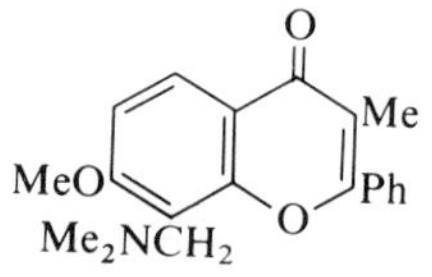

(727)

3-Methylchromone (Tricromyl or Crodimyl) possesses muscle relaxant properties and exerts a vasodilator effect on coronary blood vessels, whilst the flavone derivative Flavoxate (Urispas) (725) is an antispasmodic agent which reduces smooth muscle spasm in the urinary tract. The isoflavone Formononetin (726) is a diuretic and Dimefline (Remeflin) (727) is a respiratory stimulant.

Chromones of therapeutic interest have been reviewed ⟨70FES57⟩ and the pharmacological activity associated with chromanones has been discussed ⟨77HC(31)345⟩.

The xanthene derivatives methantheline (Banthine) (728; R = Et) and propantheline (Pro-Banthine) (728; R = Pr$^i$) are effective compounds for the treatment of ulcers and gastrointestinal disorders, reducing the volume and acidity of gastric secretions. Their behaviour is similar to that of atropine, but they are generally more potent, having some ganglionic blocking activity and lacking effects on the central nervous system.

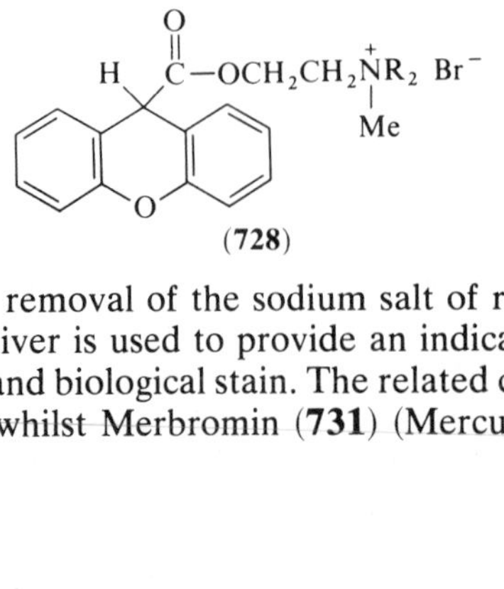

(728)

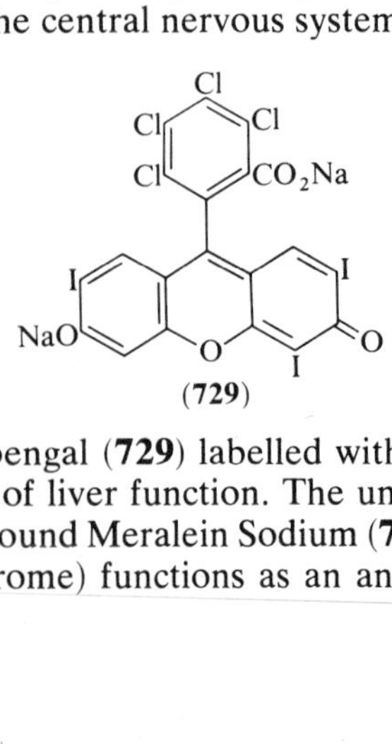

(729)

The rate of removal of the sodium salt of rose bengal (729) labelled with $^{131}$I from the blood by the liver is used to provide an indication of liver function. The unlabelled salt is used as a dye and biological stain. The related compound Meralein Sodium (730) is a topical antiinfective, whilst Merbromin (731) (Mercurochrome) functions as an antibacterial and an antiseptic.

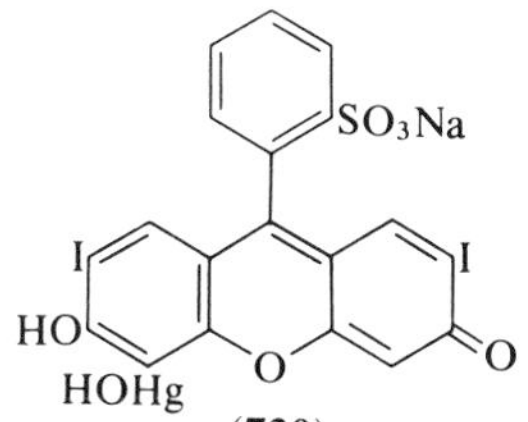

(730) 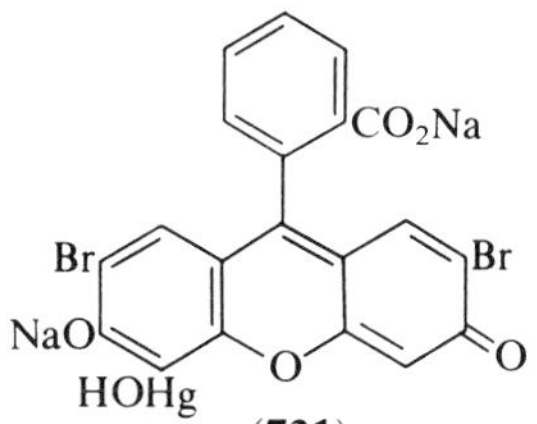　　　　　　　　　(731)

Although cardiac glycosides derived from *Digitalis* species contain a butenolide ring at $C_{17}$ of the steroid unit, those derived from the genus *Scilla* incorporate a pyran-2-one fragment at this position. Both groups have a long history of use in medicine as a consequence of their powerful effect on the heart muscle. The main component of extracts of white squill, *Scilla maritima*, is scillaren A (732). Structural similarity to the toad poison bufatolin (733) is apparent.

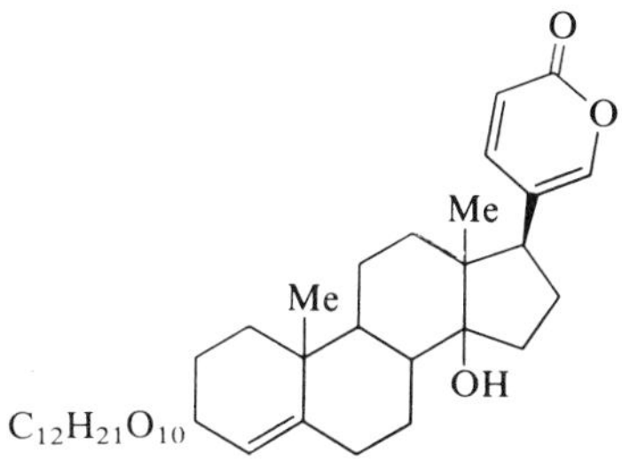

(732) 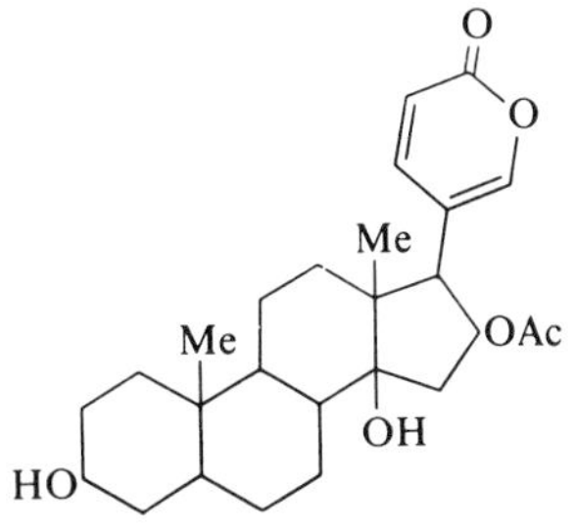　　　　　　　　　(733)

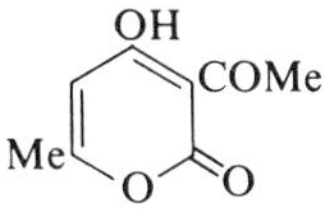

(734) 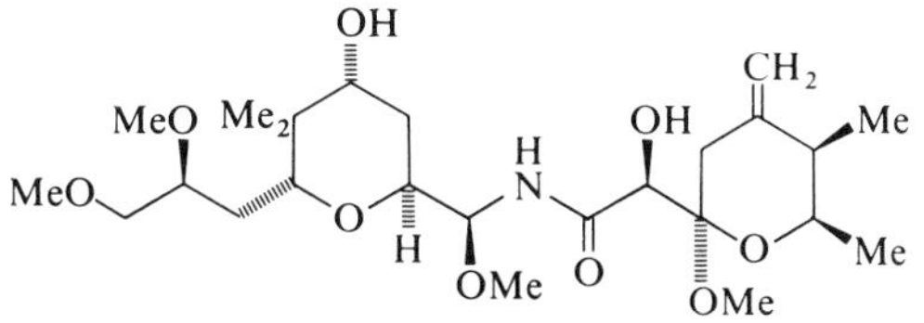　　　　　　　　　(735)

The simple and readily available pyran-2-one, dehydroacetic acid (734), possesses fungicidal properties and is used in antienzyme toothpastes.

Pederin (735) is one of the most potent poisons known to man, with an $LD_{50}$ for rats of *ca.* 2 $\mu$g per 100 g body weight. The pyran, which causes severe blistering of the skin, is part of the defence system of the beetle *Paederus fuscipes*. It inhibits chromosome division by first blocking protein synthesis and subsequently halting DNA synthesis.

5,6-Dihydropyran is of value as a protecting group for alcohols and phenols, and to a lesser extent amines, carboxylic acids and thiols ⟨B-67MI22403, B-81MI22404⟩. The resulting tetrahydropyranyl ethers (736) are stable to base, but are readily cleaved under acidic conditions (Scheme 284).

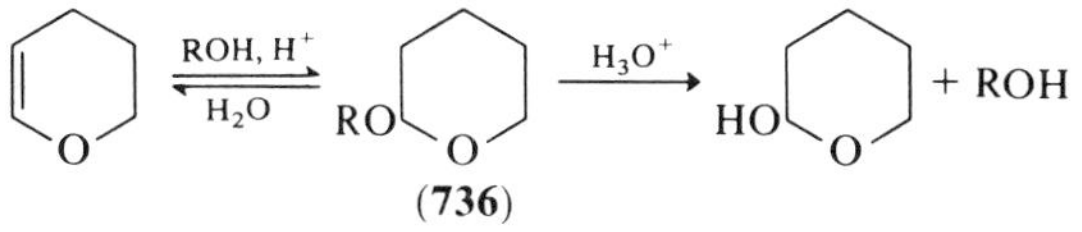

(736)

**Scheme 284**

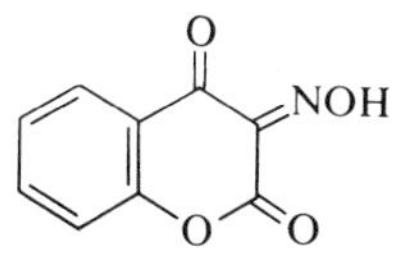

(737)

The chelating properties of 5-hydroxychromone and some of its derivatives have been utilized in the determination of a number of metals. Both 3-hydroxychromone and chroman-2,3,4-trione-3-oxime (737) show a similar ability to complex with metal ions ⟨77HC(31)440, 668⟩.

The susceptibility of pyrylium salts to nucleophilic attack at C-2 and the subsequent ring opening and ring closure has been used in the synthesis of a range of carbocyclic and heterocyclic compounds ⟨B-64MI22401, B-79MI22404⟩. The conversion of pyrylium salts into pyridinium compounds has taken on an added significance in view of the development of the latter as intermediates in elegant syntheses of a range of organic compounds ⟨80T679⟩.

# 2.25

# Thiopyrans and Fused Thiopyrans

A. H. INGALL

*Fison's Pharmaceuticals Ltd., Loughborough*

| | | |
|---|---|---|
| 2.25.1 | INTRODUCTION | 885 |
| 2.25.2 | STRUCTURE OF THIOPYRANS | 887 |
| *2.25.2.1* | *Structure of Tetrahydrothiopyrans* | 887 |
| *2.25.2.2* | *Structure of Dihydrothiopyrans* | 891 |
| *2.25.2.3* | *Structure of 2H- and 4H-Thiopyrans* | 891 |
| *2.25.2.4* | *Structure of Thiopyrylium Salts* | 893 |
| *2.25.2.5* | *Structure of 1H-Thiopyrans* | 893 |
| 2.25.3 | REACTIVITY OF THIOPYRANS | 895 |
| *2.25.3.1* | *Reactivity of Tetrahydrothiopyrans* | 895 |
| *2.25.3.2* | *Reactivity of Dihydrothiopyrans* | 903 |
| *2.25.3.3* | *Reactivity of 2H- and 4H-Thiopyrans* | 911 |
| *2.25.3.4* | *Reactivity of Thiopyrylium Salts* | 922 |
| *2.25.3.5* | *Reactivity of Thiabenzenes* | 925 |
| 2.25.4 | SYNTHESIS OF THIOPYRANS | 926 |
| *2.25.4.1* | *Synthesis of Tetrahydrothiopyrans* | 927 |
| *2.25.4.1.1* | *De novo ring synthesis* | 927 |
| *2.25.4.1.2* | *Ring transformations leading to thianes* | 928 |
| *2.25.4.1.3* | *Synthesis of thiane derivatives* | 928 |
| *2.25.4.2* | *Synthesis of Dihydrothiopyrans* | 929 |
| *2.25.4.2.1* | *De novo ring synthesis* | 929 |
| *2.25.4.2.2* | *Ring transformations leading to dihydrothiopyrans* | 933 |
| *2.25.4.2.3* | *Synthesis of dihydrothiopyran derivatives* | 933 |
| *2.25.4.3* | *Synthesis of 2H- and 4H-Thiopyrans* | 934 |
| *2.25.4.3.1* | *De novo ring synthesis* | 934 |
| *2.25.4.3.2* | *Ring transformations leading to thiopyrans* | 935 |
| *2.25.4.3.3* | *Synthesis of thiopyran derivatives* | 937 |
| *2.25.4.4* | *Synthesis of Thiopyrylium Salts* | 937 |
| *2.25.4.4.1* | *De novo ring synthesis* | 937 |
| *2.25.4.4.2* | *Ring transformations leading to thiopyrylium salts* | 938 |
| *2.25.4.4.3* | *Synthesis of thiopyrylium derivatives* | 938 |
| *2.25.4.5* | *Synthesis of Thiabenzenes* | 939 |
| 2.25.5 | APPLICATIONS OF THIOPYRANS | 939 |

## 2.25.1 INTRODUCTION

The nomenclature of the family of compounds to be discussed below has for long been the cause of much confusion. The systematic name for a six-membered ring containing one sulfur atom is thiin, but *Chemical Abstracts* uses the name thiopyran, while many earlier publications have used the 'replacement nomenclature' thiapyran (which is strictly inaccurate as 'thia' implies replacement of carbon, rather than oxygen, by sulfur). Throughout this chapter the terms thiopyran and thiin will be used interchangeably, while 'thiadecalin' nomenclature will be found convenient for the perhydrobenzothiopyran systems.

Whereas the chemistry of pyrans is an immense body of knowledge, that of thiopyrans has been less extensively investigated; this is probably a reflection of the widespread availability of diverse six-membered oxygen heterocycles and their congeners in Nature *vis-à-vis* the almost total non-existence of naturally occurring sulfur-containing analogues. Apart from a small number of compounds found in crude oil, all thiopyrans are only available *via de novo* synthesis. As if to compensate for this lack of availability of the heterocycles, Nature has endowed sulfur with available 3*d* orbitals, and in consequence the atom may be formally quadri- and hexa-covalent as well as bivalent; a whole new area of structural types and reactivity is open to thiins, not seen for pyrans.

In Table 1 is set out the whole range of monocyclic thiins, and their *Chemical Abstracts* and corresponding preferred names. It is at once apparent that 1*H*-thiopyran has no equivalent in the oxygen series.

**Table 1**  Monocyclic Thiopyrans: Structure and Nomenclature

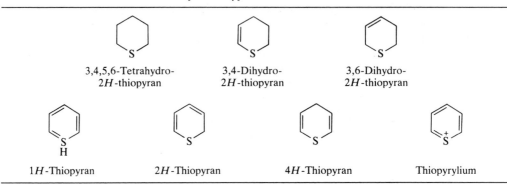

3,4,5,6-Tetrahydro-2*H*-thiopyran

3,4-Dihydro-2*H*-thiopyran

3,6-Dihydro-2*H*-thiopyran

1*H*-Thiopyran

2*H*-Thiopyran

4*H*-Thiopyran

Thiopyrylium

Table 2 does the same for the benzo fused thiins, and again 1*H*-1- and 2*H*-2-benzothiopyrans and 10*H*-thioxanthenes have no counterpart in the benzopyran field. As has been remarked, the picture is further complicated by the variable valency of sulfur which permits the functionalization of the heteroatom of most of the ring systems. Thus,

**Table 2**  Benzo Fused Thiopyrans: Structure and Nomenclature

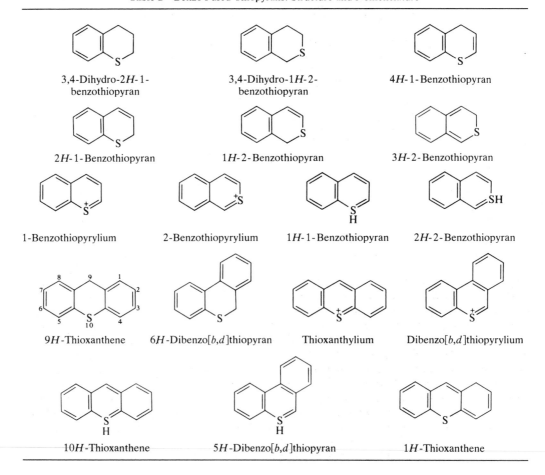

3,4-Dihydro-2*H*-1-benzothiopyran

3,4-Dihydro-1*H*-2-benzothiopyran

4*H*-1-Benzothiopyran

2*H*-1-Benzothiopyran

1*H*-2-Benzothiopyran

3*H*-2-Benzothiopyran

1-Benzothiopyrylium

2-Benzothiopyrylium

1*H*-1-Benzothiopyran

2*H*-2-Benzothiopyran

9*H*-Thioxanthene

6*H*-Dibenzo[*b,d*]thiopyran

Thioxanthylium

Dibenzo[*b,d*]thiopyrylium

10*H*-Thioxanthene

5*H*-Dibenzo[*b,d*]thiopyran

1*H*-Thioxanthene

oxides and dioxides have been known for many years, while more recently sulfimides and sulfoximides have been well characterized, and these groupings carry with them new structural influences and modes of reaction.

## 2.25.2 STRUCTURE OF THIOPYRANS

### 2.25.2.1 Structure of Tetrahydrothiopyrans

Sulfur is normally found in a tetracoordinate setting, and bivalent sulfur is quite analogous to oxygen in its general influence on the conformational preferences of the equivalent substituted tetrahydro systems. Thus, nonpolar substituents prefer to occupy equatorially-disposed positions on the preferentially chair-conformer ring. Molecular mechanics calculations have been used to derive some of the characteristics of the parent system ⟨75JA5167⟩.

**Table 3** Parameters Assumed or Derived for Molecular Mechanics Calculations on Tetrahydrothiopyrans and Thiols ⟨75JA5167⟩

1.  *Van der Waals parameters for the Hill equation*

$$r = 2.00 \text{ Å}; \quad \varepsilon = 0.769 \text{ kJ mol}^{-1}$$

2.  *Natural bond length and stretching force constants*

| Bond | Length (Å) | $k_1$ (N Å$^{-1}$) | Bond moment (D) |
|------|-----------|---------------------|------------------|
| $C(sp^3)$—S | 1.816 | $3.21 \times 10^{-8}$ | 1.20 |
| S—H | 1.346 | $3.80 \times 10^{-8}$ | 0.00 |

3.  *Natural bond angles and bending force constants*

| Angle | $\theta$ (°) | $k_\theta$ (N Å rad$^{-2}$) |
|-------|-------------|------------------------------|
| S—$C(sp^3)$—$C(sp^3)$ | 107.8 | $4.2 \times 10^{-9}$ |
| $C(sp^3)$—S—$C(sp^3)$ | 94.3 | $5.0 \times 10^{-9}$ |
| H—$C(sp^3)$—S | 108.2 | $3.0 \times 10^{-9}$ |
| $C(sp^3)$—S—H | 94.0 | $4.0 \times 10^{-9}$ |

**Table 4** Calculated Heats of Formation and Strain Energies for Simple Saturated Thiins ⟨75JA5167⟩

| Compound | Torsion energy (kJ mol$^{-1}$) | Conf. energy (kJ mol$^{-1}$) | $\Delta H_f$ Calc. (kJ mol$^{-1}$) | $\Delta H_f$ Expt. (kJ mol$^{-1}$) | Strain energy (calc.) (kJ mol$^{-1}$) |
|----------|------------|------------|------|------|------|
| Tetrahydrothiin (thiane) | 0.0 | 0.00 | −66.00 | −63.20 | 2.55 |
| 2-Methylthiane | 0.0 | 0.96 | −94.09 | | 4.72 |
| 3-Methylthiane | 0.0 | 1.05 | −93.72 | | 3.64 |
| 4-Methylthiane | 0.0 | 0.33 | −94.68 | | 2.68 |
| 3,3-Dimethylthiane | 0.0 | 0.00 | −125.32 | | 9.11 |
| *trans*-2-Thiadecalin | 0.0 | 0.00 | −121.03 | | 7.61 |
| *cis*-2-Thiadecalin | 0.0 | 0.67 | −113.74 | | 14.92 |
| *trans*-3-Thiadecalin | 0.0 | 0.00 | −121.43 | | 5.73 |
| *cis*-3-Thiadecalin | 0.0 | 0.59 | −112.36 | | 14.84 |

Table 3 shows parameters assumed or derived from other sources, particularly for tetrahydrothiopyrans, and Table 4 shows the values derived for heat of formation and strain energy of several simple saturated thiins and saturated benzothiopyrans. By this method the authors were able to estimate the conformational free energy differences favouring the equatorial form over the axial for 2-methyl-, 3-methyl- and 4-methyl-tetrahydrothiopyran to be 4.14, 4.60 and 6.65 kJ mol$^{-1}$ respectively. This last approximates to that seen in 4-methylcyclohexane, in which about half the energy arises from the vicinal interaction of the tertiary hydrogen with four neighbours when equatorial as compared to two neighbours when axial. The 2-methyl compound, because of the lack of any sulfur substituent, is less destabilized by this vicinal interaction, while the 3-methyl compound can avoid interaction of the one *syn* axial hydrogen and the axial methyl group by a slight rotation. In the case of the bicyclic molecules 2- and 3-thiadecalin, the calculations suggested that (1) *cis*-2-thiadecalin with axial sulfur (**1**) is favoured over the equatorial sulfur conformer (**2**) by 1.34 kJ mol$^{-1}$; (2)

the *trans* compound (**3**) is favoured over the *cis* (**1, 2**) by 5.06 kJ mol$^{-1}$; (3) *cis*-3-thiadecalin conformer (**4**) is favoured over (**5**) by 1.42 kJ mol$^{-1}$; and (4) the *trans* form (**6**) is favoured over the *cis* (**4, 5**) by 6.94 kJ mol$^{-1}$.

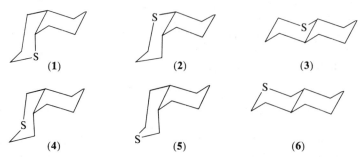

Experimental investigation of the conformational preferences of more extensively substituted systems formerly based on IR and $^1$H NMR spectroscopy ⟨73MI22500⟩ has been greatly eased by the application of $^{13}$C NMR, which has also permitted the ready identification of twist conformers that the system adopts when substituted by *t*-butyl groups

**Table 5**  $^{13}$C NMR Spectral Resonances for *cis*- and *trans*-3,5-Di-*t*-butyltetrahydrothiin Relative to TMS ⟨77T1149⟩

| Compound | C-2,6 | C-3,5 | C-4 | C(CH$_3$)$_3$ | C(CH$_3$)$_3$ |
|---|---|---|---|---|---|
| *cis*-3,5-Di-*t*-butyltetrahydrothiin | 29.42 | 50.23 | 28.64 | 33.61 | 27.39 |
| *trans*-3,5-Di-*t*-butyltetrahydrothiin | 23.85 | 44.40 | 25.38 | 33.72 | 27.33 |

which would otherwise have to be axial. Thus, *cis*-3,5-di-*t*-butyltetrahydrothiin adopts conformation (**7**) while the *trans* isomer exists primarily in form (**8**), wherein both of the bulky substituents are in pseudo equatorial positions. Table 5 compares the positions of the corresponding $^{13}$C NMR resonances in (**7**) and (**8**) and illustrates the characteristic changes in the ring carbon resonances.

Complexity is added to the study of this system by the ready functionalization of the sulfur atom by alkylation, oxidation and imination. *S*-Alkylation to produce a compound such as (**9**) ⟨76T1873⟩ affords a system in which either of two possible chair conformers possesses an axial *t*-butyl group, and this system therefore exists preferentially in a twist conformation (**10**). Where no such conflicts occur the sulfur substituent adopts the equatorial position in a chair form. Inversion of configuration at sulfur in cyclic sulfides has been investigated in the system (**11**) ⇌ (**12**) and the corresponding ylides (**13**) ⇌ (**14**) ⟨77JA2337⟩. Thermal equilibration of (**11**) and (**12**) monitored by $^{13}$C NMR afforded a 70:30 mixture, with the activation parameters: $\Delta H^{\ddagger} = 117.0 \pm 3.8$ kJ mol$^{-1}$ and $\Delta S^{\ddagger} = -13.0 \pm 10.4$ J K$^{-1}$ mol$^{-1}$ while the derived ylides produced: $\Delta H^{\ddagger} = 85.7$ kJ mol$^{-1}$ and $\Delta S^{\ddagger} = 25$ J K$^{-1}$ mol$^{-1}$. Protonation of the ylide mixture derived from the equilibrium mixture of (**13**) and (**14**) affords solely the sulfonium species (**12**), illustrating how heavily the equilibrium **(13)** ⇌ **(14)** is biased in favour of the latter. The activation parameters for inversion of the sulfonium species are similar to those reported for other sulfonium salts ⟨70HCA1499⟩.

Oxidation at sulfur can form both monoxides (sulfoxides) or dioxides (sulfones). A sulfone grouping, being quite a symmetrical element, has relatively little overall influence in simple steric situations; the *S,S*-dioxide **(15)**, for example, exists principally in a chair, while **(16)** exists principally in a twist arrangement ⟨77T1149⟩. However, in cases in which hydrogen bonding interactions are involved, the grouping plays an important part. IR evidence ⟨78JCS(P1)1321⟩ has shown that the equilibrium **(17)** ⇌ **(18)** lies heavily on the side of **(17)**.

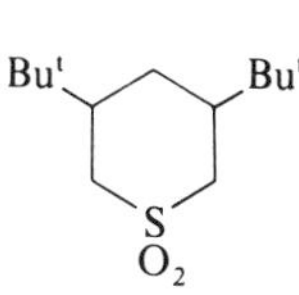

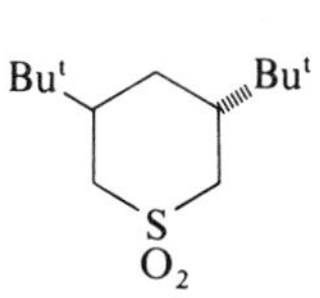

A much more surprising result is the finding of a preference for sulfoxide and some sulfimide groups to adopt an axial position in the absence of any other factors. A variety of explanations has been advanced including attractive London interactions ⟨64JA2935, 65JA1109⟩, gauche interactions between the sulfoxide and the vicinal $\alpha$ hydrogens (four such in the equatorial situation, only two in the axial setting) ⟨74T1579⟩, and an unfavourable dipolar structure in the equatorial form of the sulfoxide ⟨75TL1087⟩. However, an extensive investigation of the area using variable temperature $^{13}$C NMR spectroscopy to determine the positions of the conformational equilibria shows quite clearly that possibly the most important factor is the polarity of the S—X grouping (X = O or NR) ⟨79JOC2863⟩. The preference for axial positioning in a series of thian-1-imides increases with increasing electron-withdrawing ability of the =NR substituent. Furthermore, there is no evidence of any effect of the bulk of the substituent on sulfur, nor of the length of the S—N bond in sulfimides, which remains at 1.615–1.638 Å over a wide range of substitution. Table 6 lists some of the data obtained for the compounds **(19a–i)**, and illustrates the correlation of conformational preference with polarization of the bond. Table 7 extends the data to substituted thianes and thiadecalins.

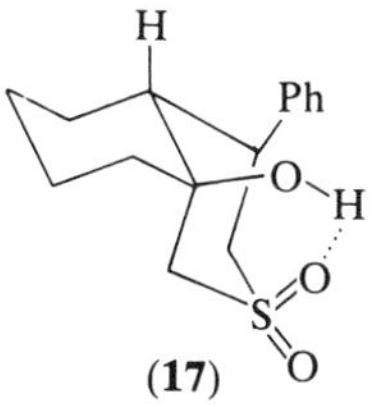

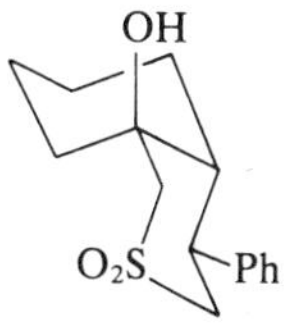

**Table 6** Conformational Equilibria of Compounds **(19a–i)** ⟨79JOC2863⟩

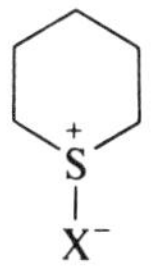

| Compounds | Temp. (°C) | % Equatorial X | $\Delta G^\circ_{(eq-ax)}$ (kJ mol$^{-1}$) |
|---|---|---|---|
| **(a)** X = O | −90 | 38 | −0.71 |
| **(b)** X 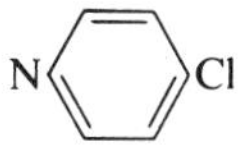 = N⟨⟩Cl | −90 | 82 | 2.30 |
| **(c)** X 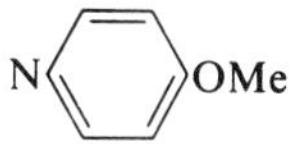 = N⟨⟩OMe | −80 | 86 | 2.93 |
| **(d)** X 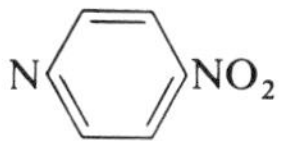 = N⟨⟩NO₂ | −80 | 72 | 1.51 |
| **(e)** X = NH | −85 | 55 | 0.29 |
| **(f)** X = NCOPh | −72 | 40 | −0.67 |
| **(g)** X  = NSO₂⟨⟩Me | −85 | 31 | −1.25 |
| **(h)** X = NCO₂Et | −80 | 42 | −0.50 |
| **(i)** X = NCOMe | −80 | 51 | 0.04 |

**Table 7**  Conformational Equilibria of Compounds (**20**) and (**21**) ⟨79JOC2863⟩

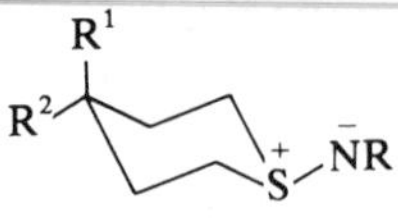

| Compound | Temp. (°C) | % Equatorial NR, X | $\Delta G^\circ_{(eq-ax)}$ (kJ mol$^{-1}$) |
|---|---|---|---|
| (**20a**) R 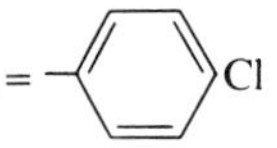 = (aryl)Cl,  R$^1$ = Me, R$^2$ = H | −85 | 5 | −3.76 |
| (**20b**) R 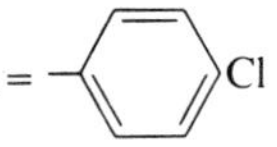 = (aryl)Cl,  R$^1$ = R$^2$ = Me | −80 | 72 | 1.51 |
| (**20c**) R = COPh,  R$^1$ = R$^2$ = Me | −80 | 29 | 1.42 |
| (**21a**) X = O | −50 | 65 | 1.13 |
| (**21b**) X = N(aryl)Cl 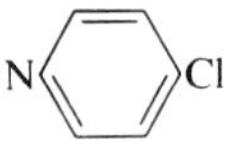 | −69 | 95 | 5.43 |
| (**21c**) X = NPh | −45 | 95 | 5.43 |

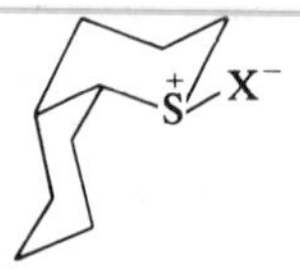

However, in the case of the molecules (**22**) and (**23**) the presence of the *gem* dimethyl group at C-3 is sufficient to push the conformations over to >95% equatorial S—O or S—N bonds, whereas 4,4-dimethylation had little effect on the axial preference of the substituent. In (**24**), a sulfoximide, it was found that the equilibrium position (at −105 °C) was 55:45, and in the light of the known slightly greater preference of oxygen for the axial position it was concluded that the preferred conformer is (**25**) ⟨72JOC377⟩.

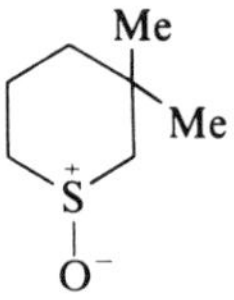

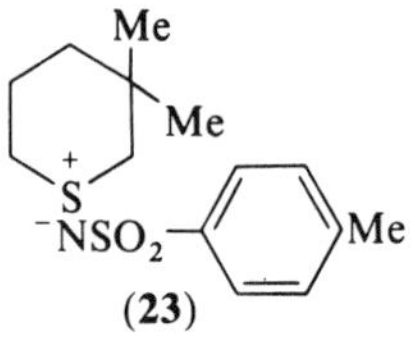

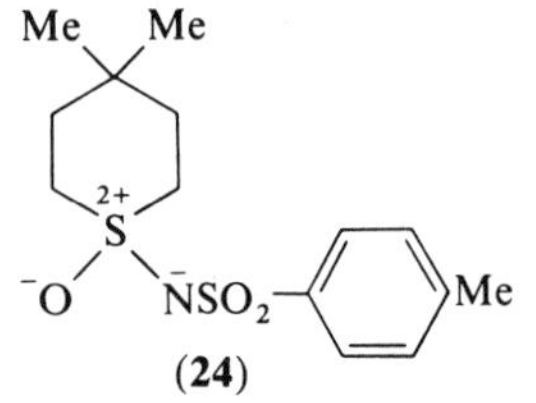

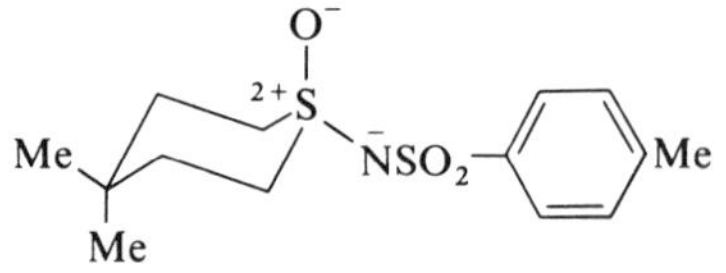

$\alpha$-Lithio thian oxides are very important reactive intermediates in the preparation of derivatives of tetrahydrothiins. Although usually prepared *in situ* and not isolated, their stereochemical behaviour in alkylation reactions is sufficiently special to have initiated a thorough $^1$H and $^{13}$C NMR investigation ⟨78T2705⟩. The H–H coupling constants and $^{13}$C shifts obtained on lithiation of *cis-* and *trans*-4-*t*-butylthian 1-oxides (**26**) and (**27**) are only consistent with their existing in the half-chair conformations (**28**) and (**29**), respectively, with planar metallated carbon atoms. Similar studies have also been carried out on the ylides (**13**) and (**14**) and on that derived from *S*-*p*-tolylthianium salts ⟨82TL763⟩. The thianium ylide was also found to have a planar carbanionic centre, but (**13**) and (**14**) were found to have pyramidal carbanionic centres. This behaviour may be related to the C$^-$—Li$^+$ bonding behaviour: loose ion pairs are believed to favour planarity and may be facilitated in solution by a solvating medium such as HMPA, while tight ion pairs favour pyramidal carbanions.

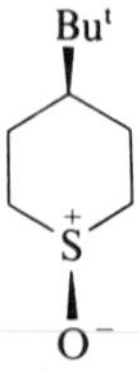

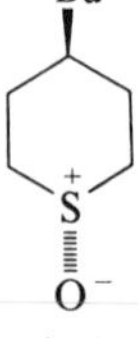

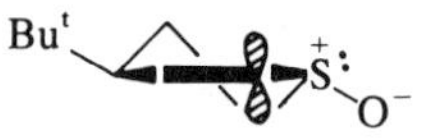

### 2.25.2.2 Structure of Dihydrothiopyrans

Whereas a lot of investigative effort has been applied to the unravelling of the conformational analysis of tetrahydrothiins, much of this work, utilizing in particular the modern techniques of $^{13}$C NMR spectroscopy, has only begun to be developed in the last few years. Inevitably, therefore, the investigation of the dihydro systems, in which yet more variables are to be found, has not yet been pursued to anything like the same degree of completeness. By analogy with the previous section, it may be assumed that both 3,4-dihydro-2*H*-thiins (30) and 3,6-dihydro-2*H*-thiins (31) exist, in the absence of disturbing factors, as the half-chairs (32) and (33) respectively. No convincing material has been reported on the conformational preferences of either the sulfoxides or the sulfimines derived from such parent rings, though (32) is so configured that little difference between the two possible forms would be expected. Compound (33), however, differs in that the sulfur lone pairs occupy a pseudo axial as well as a pseudo equatorial position and conflicting demands can arise. It is to be hoped that this area of doubt will be resolved in the near future.

(30)    (31)    (32)    (33)

Somewhat more information is available relating to the benzo fused systems. In 4-oxo-3,4-dihydro-2*H*-thiopyrans, $^1$H NMR studies have shown considerable flattening of the ring, while 3-chloro substituents adopt an axial position and 3-methyl groups lie pseudo equatorially ⟨68JHC745, 70JHC187, 73IJC446⟩. An X-ray study of 3,4-dihydro-1*H*-2-benzothiopyran 2,2-dioxide, supported by NMR measurements of the parent ⟨73JCS(P1)410⟩, also shows considerable distortion of the ring, particularly in the sulfone, where the C—S—C angle closes to *ca.* 103° and the sulfur lies above the best plane. Half-boat conformers are favoured where the alternatives would require even a grouping as small as methyl to lie axially.

### 2.25.2.3 Structure of 2*H*- and 4*H*-Thiopyrans

Neither 2*H*- nor 4*H*-thiopyrans unsubstituted by any potentially stabilizing groups, although not as unstable as generally believed, are as well known as the related oxo

Table 8    X-Ray Crystallographic Structure Data for 2*H*-Thiin 1,1-Dioxide (34) ⟨67RTC1275⟩

(34)    (35)

*Bond lengths* (Å)

| Bond | Bond length (Å) | Bond | Bond length (Å) | Bond | Bond length (Å) |
|---|---|---|---|---|---|
| $S^1-C^1$ | 1.764 | $C^3-C^4$ | 1.325 | $C^5-C^6$ | 1.338 |
| $C^2-C^3$ | 1.494 | $C^4-C^5$ | 1.450 | $C^6-S^1$ | 1.730 |

| Bond group | Bond angle (°) | Bond group | Bond angle (°) |
|---|---|---|---|
| $O^{1'}-S^1-O^{2'}$ | 116.3 | $C^2-S^1-C^6$ | 102.6 |
| $O^{2'}-S^1-C^2$ | 109.5 | $S^1-C^2-C^3$ | 112.6 |
| $O^{2'}-S^1-C^6$ | 109.2 | $C^2-C^3-C^4$ | 123.1 |
| $O^{1'}-S^1-C^2$ | 109.3 | $C^3-C^4-C^5$ | 123.2 |
| $O^{1'}-S^1-C^6$ | 109.1 | $C^4-C^5-C^6$ | 123.8 |
| | | $C^5-C^6-S^1$ | 120.3 |

*Deviations from best plane through* $C^3-C^4-C^5-C^6$

| Atom | Distance (Å) | Atom | Distance (Å) | Atom | Distance (Å) |
|---|---|---|---|---|---|
| $C^2$ | −0.286 | $C^5$ | −0.051 | $O^{1'}$ | 1.815 |
| $C^3$ | −0.025 | $C^6$ | 0.025 | $O^{2'}$ | −0.327 |
| $C^4$ | 0.057 | $S^1$ | 0.385 | | |

and thioxo derivatives. There is no strong evidence for any significant conjugation of sulfur lone pairs with the alkenic double bonds, and perhaps this is not too surprising when the relative orbital energy levels are considered ($3p$ and $2p$ respectively). A Hückel based MO calculation ⟨75MI22500⟩ produced theoretical resonance energies of $-0.0694\,\beta$ for $2H$- and $-0.0691\,\beta$ for $4H$-thiins, which were increased by methyl or phenyl substitution and in the $2H$-system was principally due not to interaction with the sulfur but to conjugation within the butadiene moiety. This last point may be viewed in the light of the observed tendency for $4H$-thiopyrans to convert to $2H$-isomers under relatively mild conditions. However, the calculated energies do not accord with the known stability characteristics of the compounds and this was explained on the basis of the fundamental assumption, made for the purposes of calculation, that the molecule was flat. A more refined theoretical treatment is obviously called for, but evidence for non-planarity may be inferred from an X-ray study of $2H$-thiopyran 1,1-dioxide (**34**) ⟨67RTC1275⟩. Table 8 and structure (**35**) set out the important dimensions. It is at once obvious that the molecule is far from flat, and the two alkenic bonds are twisted relative to one another. The sulfur atom stands distinctly above the best plane through the butadiene portion and this feature will be seen again in later sections.

It is quite likely that $2H$-thiin is very similar in structure to the dioxide though perhaps with a widened C(6)—S(1)—C(2) angle and consequently less twisted dialkene; this is supported by the Diels–Alder reactivity it displays (albeit with difficulty), while the sulfone (**34**) is inert to similar reactions. Structures (**36**) and (**37**) are tentatively suggested as possible pseudo chair and pseudo boat conformations for the $4H$-thiin system.

(**36**)        (**37**)

Considerable experimental and theoretical structural treatment has been accorded the thiopyran-ones, -thiones and -imines ⟨73JPR690, 65ZC22, 68SA(A)1283⟩ which reflects the ready availability of these systems, their significant stability and extensive chemistry. The stability of these systems is often ascribed to a significant contribution of dipolar forms such as (**39**) to the bonding picture in (**38**) and of (**41**) in (**40**). In support of these views are the unusually high dipole moments found for these molecules, *e.g.* $\mu = 3.96\,\mathrm{D}$ for (**38**) ⟨35JCS602⟩, and the lack of a C=O stretching bond in the IR spectrum ⟨54JA2451⟩. In simple Hückel terms these dipolar forms possess a $\pi$-electron sextet and are therefore 'aromatic', but they should also be basic (compound **41** formally possesses an $S^-$ grouping). Spectrophotometric measurements on (**41**) have assigned a $pK_a$ for the conjugate acid of $-3.21$ ⟨68JPR41⟩. Substitution on the ring causes this value to vary in correlation with predicted electron densities on the exocyclic sulfur atom.

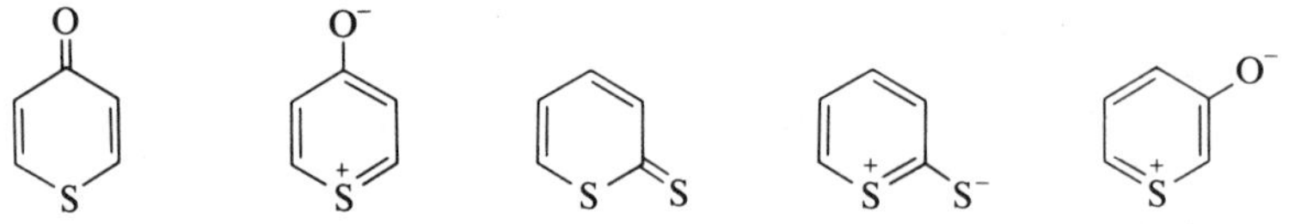

(**38**)        (**39**)        (**40**)        (**41**)        (**42**)

Whereas zwitterionic forms may be invoked to explain the properties of 2- and 4-thiopyrones, the '3-thiopyrone' can only exist as the betaine (**42**) and is therefore to be compared with the pyridinium system for a bonding and structural likeness. A theoretical treatment ⟨73JPR690⟩ has been published using CNDO/2 methods, and the calculated parameters were found to agree well with experimentally determined spectral properties.

Electron density calculations for HOMOs and LUMOs of the 2- and 4-thiopyrones ⟨73JPR690⟩ have been used to rationalize the sites of reactivity within the molecules ⟨67AHC(8)219⟩. Scheme 1 illustrates the general conclusions.

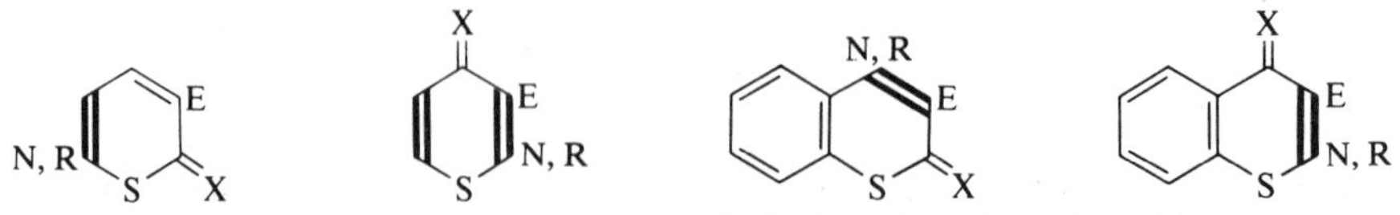

X = O, S    Bonds in heavy print are of highest order; N, R, E show sites of reaction with nucleophiles, radicals and electrophiles respectively

**Scheme 1**

### 2.25.2.4 Structure of Thiopyrylium Salts

Thiopyrylium salts are well known, stable species, commonly used as precursors for many thiin derivatives. Comparison of stabilities with the oxygen and selenium analogues for the monocycles, the monobenzo fused and the dibenzo fused systems showed the sulfur species was always more stable than the corresponding oxonia or selenonia compound, and that stability increased with the extent of benzo fusion ⟨65MI22500⟩. Whereas it has already been suggested that sulfur $3d$ orbitals do not participate in any significant conjugation in the thiopyryliums ⟨72TL4165⟩, the significant stability increase *vis-à-vis* pyrylium ions suggests the operation of a unique factor. An investigation by $^1$H NMR spectroscopy and extended Hückel MO calculations produces the results shown in Table 9 and structure (46) ⟨73T2009⟩.

**Table 9** Comparative $^1$H NMR Spectral Resonances for Thiopyrylium (43), Pyrylium (44) and *N*-Ethylpyridinium Salts (45)

| (43) | (44) | (45) |
|------|------|------|

| $^1$H Shift in TFA | (43) | (44) | (45) |
|---|---|---|---|
| $H_\alpha$ | 10.1–10.3 | 9.7–9.9 | 8.8–8.9 |
| $H_\beta$ | 8.8–9.2 | 8.2–8.8 | 8.0–8.3 |
| $H_\gamma$ | 8.8–9.2 | 9.1–9.6 | 8.4–8.6 |

*Electron density distribution in* (46)

The significant deshielding of the $\beta$-hydrogens in (43) compared with the pyrylium salt (44), and the greater deshielding of the $\alpha$-protons in (43) than in (44), contrary to expectation, have been interpreted as indicating extensive transfer of electron density from the $\beta$-carbon atoms and an increase of bond order between sulfur and the $\alpha$-carbons, both effects requiring the utilization of sulfur $3d$ orbitals. The MO calculation (based on the assumption of a planar ring system) was performed neglecting and including the sulfur $3d$ orbitals, and only when they were included did the net $\sigma + \pi$ atomic charges (shown in structure 46) reflect the $^1$H NMR spectrum. Furthermore, construction of the relevant energy level diagram for the system, with and without the inclusion of S $3d$ orbitals, showed a distinct stabilization of *ca.* 0.5 eV for the former case. The effectiveness of S $3d$ participation, however, can be shown to vary considerably with the geometry of the system and it is not yet possible to define it precisely, though results of the work do strongly support the important contribution of canonical forms such as (47) to the bonding.

(47)

### 2.25.2.5 Structure of 1*H*-Thiopyrans

Sulfur can exist in formally tetra- and hexa-valent forms, and this was extended in the past to the postulate that it should be possible to generate a six-membered sulfur-containing ring with overall neutral charge and a tetravalent heteroatom that would be fully the equal of benzene for stability and chemical reactivity. Indeed these systems were trivially known as 'thiabenzenes'. Despite the greatly expected stability, the compounds proved extremely elusive: they could be seen as transient materials in NMR experiments, but it is only in very recent times that just one stable unequivocal member of the monocyclic series has been isolated and fully characterized. A number of benzo fused analogues have been known for a little longer. An essential requirement for aromaticity in the heterocycle system is the ability of the sulfur atom to transmit conjugative effects and thereby sustain the ring current seen in the close analogues of benzene. In 1972 a paper appeared which cast serious doubt on sulfur's ability so to do 〈72JOC2720〉. With this finding the concept of the aromatic 'thiabenzene' was called into question. In 1974 and 1975 several groups independently reported their findings and conclusions from investigations of a diverse group of thiabenzenes, thianaphthalenes and thiaxanthenes 〈74JA6119, 74JOC3519, 74JA5648, 74JA5650, 74JA5651, 75JA2718〉 and a theoretical treatment appeared in 1977 〈77T3061〉.

1*H*-Thiopyrans are not planar, cyclically conjugated molecules, but rather are cyclic ylides with a pyramidal sulfur lying above the plane through the carbon framework. Table 10 gives details of the theoretical findings for the parent system.

**Table 10** Optimized Conformation of 1*H*-Thiopyran 〈77T3061〉

| Bond | Length (Å) | Bond group | Angle (°) |
|---|---|---|---|
| $C^2-S^1$ | 1.778 | $C^2-S^1-C^6$ | 101.9 |
| $S^1-H^1$ | 1.3671 | $C^3-C^2-S^1$ | 121.6 |
| $C^3-C^2$ | 1.3597 | $C^4-C^3-C^2$ | 126.0 |
| $C^4-C^3$ | 1.3989 | $C^5-C^4-C^3$ | 121.6 |
| | | carbon plane–$S^1$ | 10.6 |

*Barrier to inversion at sulfur*: 234 kJ mol$^{-1}$

*Gross atomic charges*

| Atom | Charge | Atom | Charge |
|---|---|---|---|
| $S^1$ | +0.457 | $H^3$ | +0.070 |
| $C^2$ | −0.251 | $H^4$ | +0.027 |
| $C^3$ | −0.015 | $H^5$ | +0.049 |
| $C^4$ | −0.194 | $H^1$ | +0.004 |
| $H^2$ | +0.049 | | |

Non-planarity is the result of the dominance of the destabilizing interactions of the sulfur lone pair and $\pi$-occupied MOs of the pentadienyl anion over the stabilizing interaction of that lone pair and the LUMO of the anion fragment. In fact thiabenzene is antiaromatic in a planar configuration. Pyramidalization reduces the antiaromaticity induced by the sulfur. Although no X-ray data are available on the parent system, kinetic data have been obtained supporting a minimum barrier to inversion at the pyramidal sulfur of a 2-thianaphthalene of 99.1 kJ mol$^{-1}$ 〈75JA2718〉. The formulation of the system as a cyclic ylide is supported by the chemical reactivity of the compounds as related in the reactivity section below.

1*H*-Thiopyran 1-oxides are, surprisingly, quite stable species, and have been examined by X-ray diffraction both in the free form 〈78CC197〉 and as metal complexes 〈78CB1709〉. Table 11 shows the relevant data for 4-acetyl-1,3-dimethyl-1*H*-thiopyran 1-oxide, which may be viewed as a reasonable approximation to the state of affairs existing in the parent species.

1-Methyl-3,5-diphenyl-1*H*-thiopyran 1-oxide forms red air-stable complexes with carbonyls of chromium, molybdenum and tungsten. 〈78CB1709〉. X-Ray diffraction measurements on these compounds confirm the non-planarity of the thiabenzene oxide nucleus and reinforce the evidence for high inversion barriers at sulfur. These complexes are found in isomeric forms with either the sulfur–oxygen bond axial to the half-chair conformation adopted by the ring, or the *S*-methyl group axial; the complexes cannot be interconverted.

**Table 11** X-Ray Crystallographic Structure Data for 4-Acetyl-1,3-dimethyl-1*H*-thiopyran 1-Oxide ⟨78CC197⟩

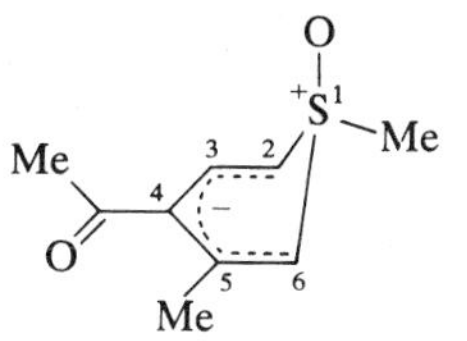

| Bond | Length (Å) | Bond | Length (Å) |
|---|---|---|---|
| $S^1-C^2$ | 1.689 | $C^4-C^5$ | 1.43 |
| $C^2-C^3$ | 1.374 | $C^5-C^6$ | 1.376 |
| $C^3-C^4$ | 1.414 | $C^6-S^1$ | 1.692 |

*Distance of sulfur atom above carbon skeleton plane:* 0.43 Å

The isomer ratios reflect the equilibrium distribution of conformers in the free organic compound, with axial oxygen preferred.

Stability of both the parent system and the *S*-oxide is conferred by the presence of electron-withdrawing substituents on the carbon framework, especially at positions 2, 4 and 6, while electron-donating substituents on sulfur also help. Compounds (**48**), (**49**) and (**50**) illustrate these conclusions, as they are air-stable, isolable species ⟨74CL1101⟩. Exactly analogous factors acting on stability are seen with acyclic sulfonium and sulfoxonium ylide compounds.

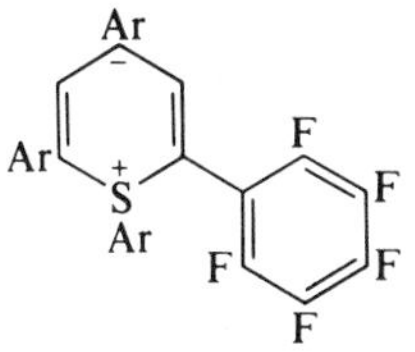

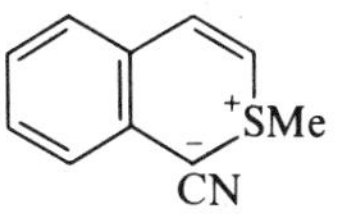

## 2.25.3 REACTIVITY OF THIOPYRANS

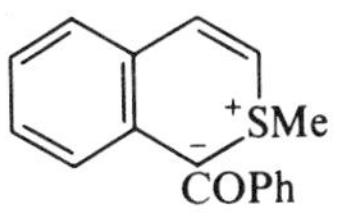

### 2.25.3.1 Reactivity of Tetrahydrothiopyrans

Tetrahydrothiopyran and its simply substituted derivatives are to be regarded as cyclic sulfides, and in consequence their chemistry is analogous to that of their acyclic counterparts, which is extensively covered in the standard texts ⟨B-79MI22500⟩. Oxidation, alkylation, halogenation, *etc.* reactions differ from the corresponding phenomena in the noncyclic species only in so far as the stereochemical constraints imposed by a more rigid framework control the orientation of reactants, transition states and products.

Oxidation of sulfur proceeds in two stages and by choice of reagent either the sulfoxide or sulfone may be produced. Sulfone formation (equation 1) requires vigorous conditions (*e.g.* hydrogen peroxide in acidic or basic media ⟨72CZ37⟩, *m*-chloroperbenzoic acid or potassium permanganate under phase transfer conditions ⟨80JOC3634⟩) and choice of reagents is dictated by the other functionality present in the molecule. The sulfone grouping does not introduce any new asymmetry into a molecule and the reader is referred to standard texts for further information on the chemistry of the species.

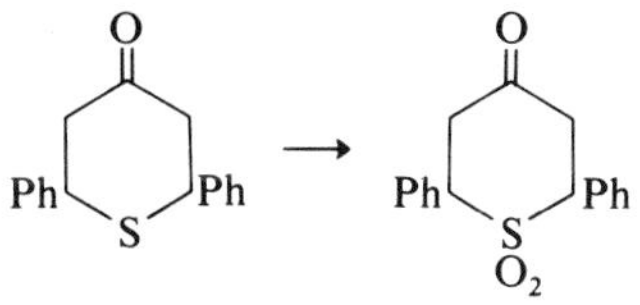

(1)

Sulfoxide formation requires much milder conditions and traditionally bromine in wet ether was used; although this is very successful, modern practice makes use of *t*-butyl hypochlorite in methanol ⟨65JA1109⟩. With 4-*t*-butylthiane this affords a high yield of the *cis*-sulfoxide, with axial oxygen (**26**). The corresponding *trans*-sulfoxide (**27**) is best prepared by the conversion shown in equation (2), in which (**26**) is methylated with Meerwein's reagent and $S_N2$ displacement at sulfur furnishes the equatorial oxido function ⟨65JA5404⟩. This inversion reaction is also presumably the mechanism of the hypochlorite oxidation, *i.e.* an equatorial sulfonium chloride is the first-formed intermediate, and is subsequently displaced by hydroxide or its equivalent. With *t*-butyl perbenzoate the benzoate ester is formed ⟨66JOC2333⟩. Heating sulfoxides in the presence of an electrophile (usually an anhydride) or pyrolysis of the above benzoate ester results in a Pummerer reaction (*q.v.*) which is difficult to stop at the 2-oxythiane, and usually goes through to the product of elimination, a 3,4-dihydro-2*H*-thiopyran (equation 3) ⟨64JOC2211⟩. This highly efficient conversion suffers from lack of regioselectivity in the absence of ylide-stabilizing α substituents.

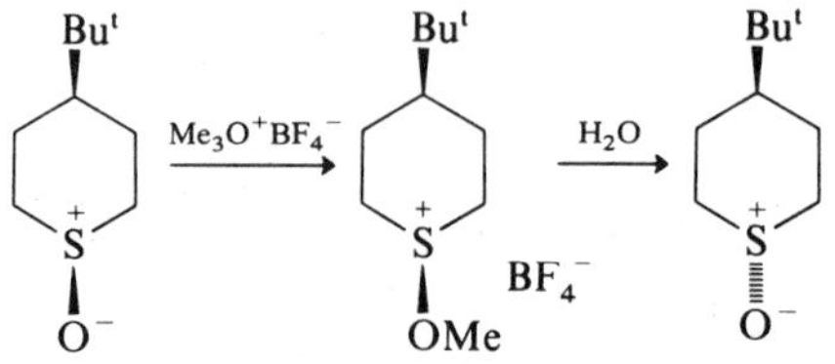

$$(2)$$

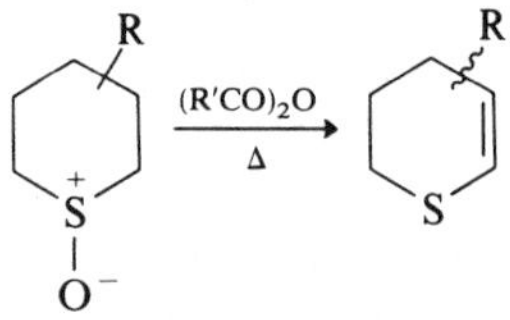

$$(3)$$

Thiane oxides have been shown to be reduced cleanly back to the thiane with phosphorus pentasulfide under conditions to which sulfones, sulfinates, ketones, esters and amides are inert ⟨78CJC1423⟩; the potential of this reaction, though not yet applied, is obviously considerable, especially when coupled with the old-established Raney nickel desulfurization technique. Thiane itself is desulfurized to pentane (with traces of cyclopentane) but the opportunity to construct alkyl chains of great complexity regio- and stereo-specifically is there. At the very least, the reduction to tetrahydrothiopyrans presents a very useful entrée into a wide range of 2-substituted thianes (Scheme 2).

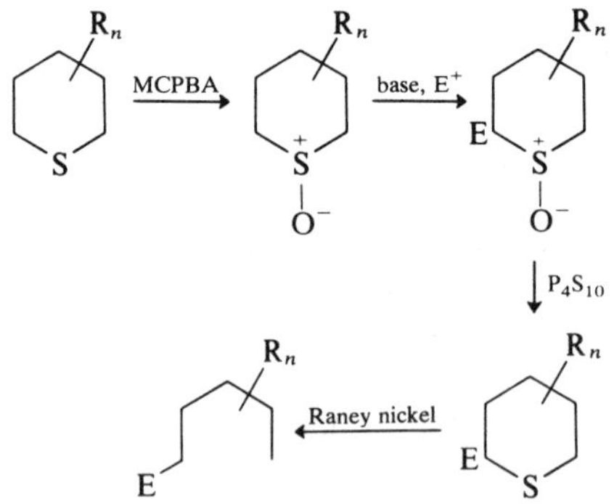

**Scheme 2**

Alkylation of the sulfur atom is well known, and with simple halides, tosylates, *etc.* is clean and high yielding, but reaction rates decline with increasing carbon number in the alkyl moiety (*cf.* acyclic sulfides). The alkylating species approaches preferentially from an equatorial direction: alkylation of the conformationally fixed 4-*t*-butylthiane with methyl iodide affords the two isomeric sulfonium salts in the ratio 7.3:1 equatorial:axial ⟨68JCS(B)1467⟩; in more flexible systems the sulfur substituent will normally adopt the thermodynamically preferred disposition by ring inversion. The reader is referred to Section 2.25.2.1 for a discussion of the conformational equilibria of the bicyclic systems (**11**) and (**12**). Reaction of the sulfonium salts (**11**) and (**12**) with the strong reducing agent sodium amalgam produced a mixture of two products (equation 4) while hydrolysis with aqueous potassium hydroxide affords the products of nucleophilic substitution, *e.g.* (**51**), and elimination, *e.g.* (**52**) (equation 5) ⟨76YZ1440⟩.

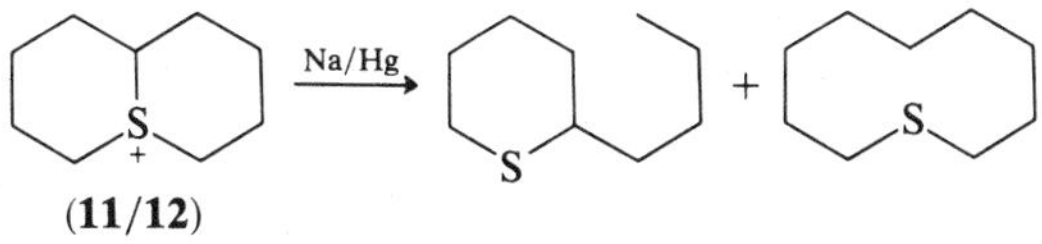

(4)

(**11/12**)

(5)

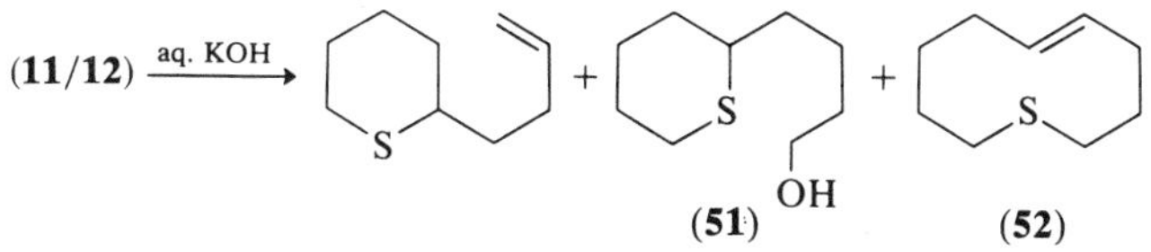

(**11/12**)  aq. KOH

(**51**)  OH  (**52**)

Treatment of thiane 1-oxides and the sulfonium salts with poorly nucleophilic bases including butyllithium, sodium hydride, *etc.* readily removes an $\alpha$-proton. Once again the gross chemistry of these carbanionic species is analogous to the acyclic species, but, particularly with the sulfoxides, interesting orientation effects are found. Deprotonation of a rigid thian 1-oxide such as (**27**) with *n*-butyllithium produces a lithiated species bearing a planar carbanionic centre as discussed in Section 2.25.2.1 ⟨78T2705⟩. Methylation of these materials has been investigated ⟨72TL4921⟩. Equations (6) and (7) set out the observed products: an equatorial sulfoxide is methylated equatorially, and an axial sulfoxide axially, both reactions occurring at −78 °C in virtually quantitative yield. Further reaction under identical conditions occurs at the least substituted carbon to produce the symmetric dimethyl compounds (**55**) and (**56**), from (**53**) and (**54**) respectively, again in virtually quantitative yield despite, in the case of (**56**), the considerable 1,3-diaxial interference introduced. Furthermore, it can be shown that axial protons in equatorial sulfoxides are removed faster than equatorial protons in axial sulfoxides and *vice versa* ⟨73TL4155⟩. Equation (8) shows the consequence of applying the reaction to a mixed axially/equatorially substituted compound. Regioselectivity is absolute, and the incoming electrophile approaches from the least hindered face, *trans* to the oxygen atom, to react at the centre initially bearing the more labile, axial proton in this equatorial sulfoxide. Had the sulfoxide been axial, reaction would have occurred at the other centre.

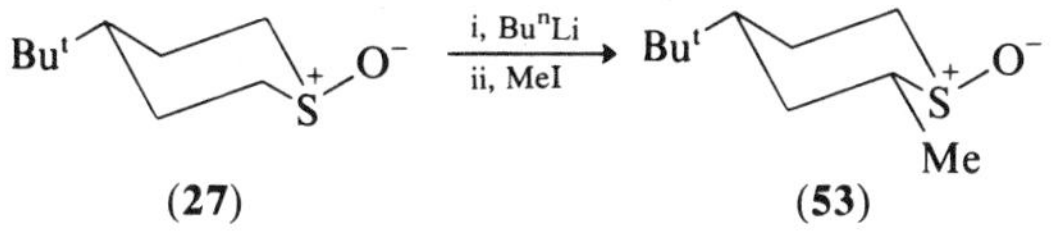

(6)

(**27**)  (**53**)

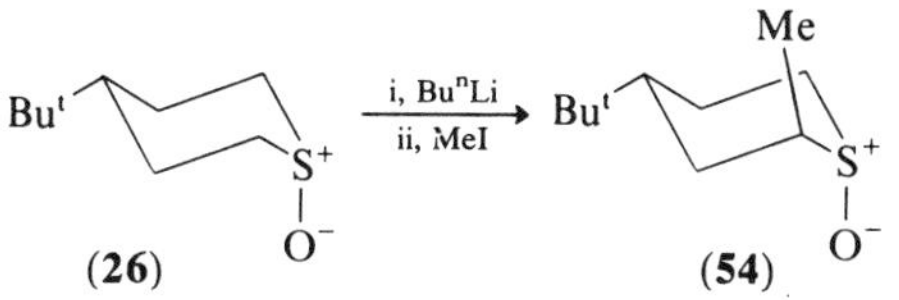

(7)

(**26**)  (**54**)

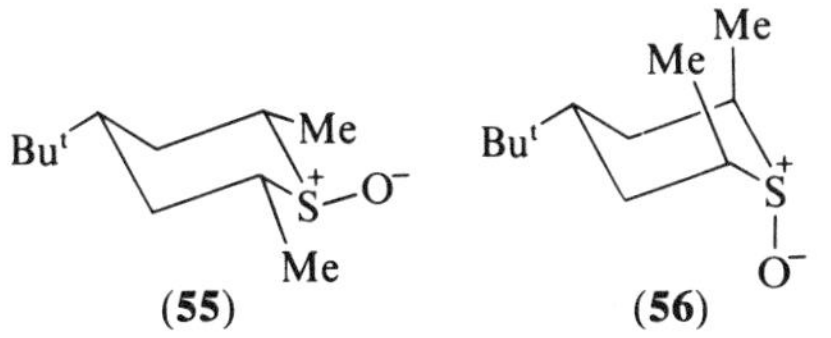

(**55**)  (**56**)

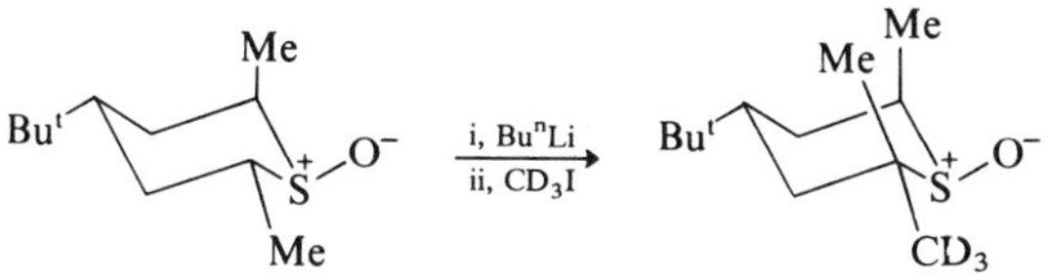

(8)

An analogous study has been carried out on ylide formation in cyclic sulfonium salts. Using the conformationally rigid systems in Table 12, relative kinetic acidities for $H_{ax}$ and $H_{eq}$ were determined. The results show that it is possible to remove one proton selectively, and thereby transfer chirality at sulfur to the neighbouring carbon atom ⟨78JA200⟩. As rigidity increases the differential kinetic acidity also increases but this is due not to an increase in overall acidity, rather to a decrease in lability of the axial proton in these 'equatorial sulfonium' systems, with the equatorial proton's lability remaining constant. Perturbation MO calculations suggest that deprotonation is facilitated by overlap with the S—C bond on the other side of the molecule to the C—H bond being broken. Clearly such overlap is pre-existent for equatorial C—H bonds in chair conformers, but for axial protons serious distortion, or ring inversion, is necessary to achieve such overlap and, not surprisingly, distortion is less and less favoured with increasing geometric constraints.

**Table 12**   Relative Kinetic Acidity of Axial and Equatorial Protons in Rigid Sulfonium Salts ⟨78JA200⟩

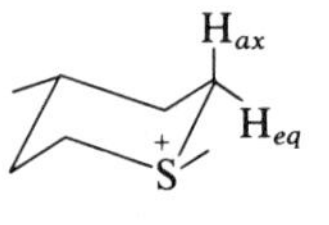

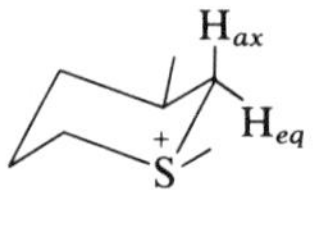

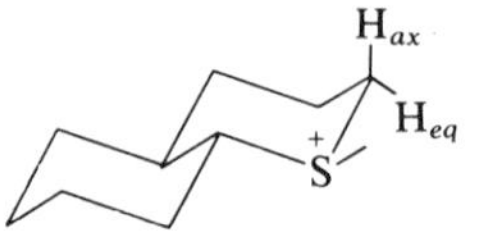

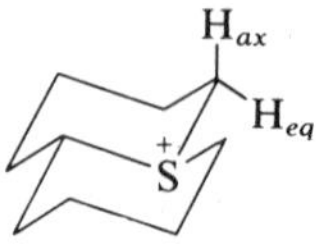

$$k(H_{eq}):k(H_{ax})$$

| 0.5 | 2 | 5 | 35 |

Sulfonium ylides are well known species with an extensive modern chemistry, especially trimethylsulfonium halides, which are one-carbon synthons. Sadly, little use has been made of the equivalent tetrahydrothianium salts, though one interesting area has seen extensive investigation: 2-vinylthianium ylides ring expand in a highly efficient manner by 2,3-sigmatropic shifts, though stereospecificity was not observed due to interconversion of the diastereoisomeric sulfonium salts under the conditions of the reaction ⟨78JOC4831, 78TL519, 78JOC4826, 79JOC4128, 81JOC5451⟩. Equation (9) illustrates the reaction of the *cis-S*-methylvinyl species (**57**) which cleanly and rapidly formed the ring-expanded product (**58**) at −78 °C. The diastereomeric salt (**59**) reacted more slowly to afford the same product, but contaminated by the ring-opened material (**60**) (equation 10). No evidence was seen for the Stevens rearrangement product (**61**), although this represents a common mode of decomposition of unstabilized ylides not bearing the 2-vinyl grouping. It is likely that a transoid transition state (**62**) is adopted, easily accessible from both diastereomers. If the sulfur substituent is allyl, the ring expansion can be repeated (Scheme 3), but in this case an alternative 2,3-shift is possible and occurs as a minor competing pathway ⟨78JOC4831⟩.

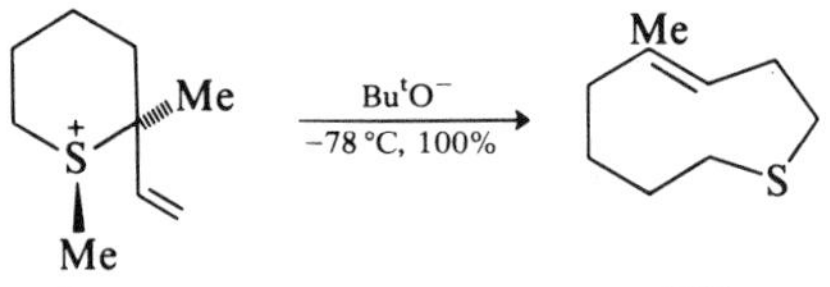

(9)

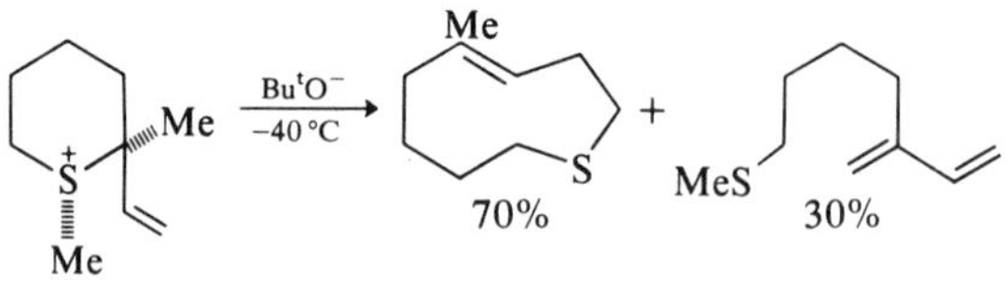

(10)

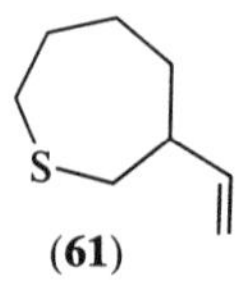

The second expansion is very much less stereoselective and produces a mixture of alkenes in the ratio $(E,E)/(E,Z) = 5:1$. Here should be noted the use of a stabilized ylide in the second stage of Scheme 3. Generation requires only an amidine base (DBU) but the rearrangement proceeds at room temperature, some 70 °C above the most favoured unstabilized system. The ylidic carbon may also be part of the thiopyran ring, in which case, as equation (11) shows, the products are alicyclic ⟨79JOC3230⟩.

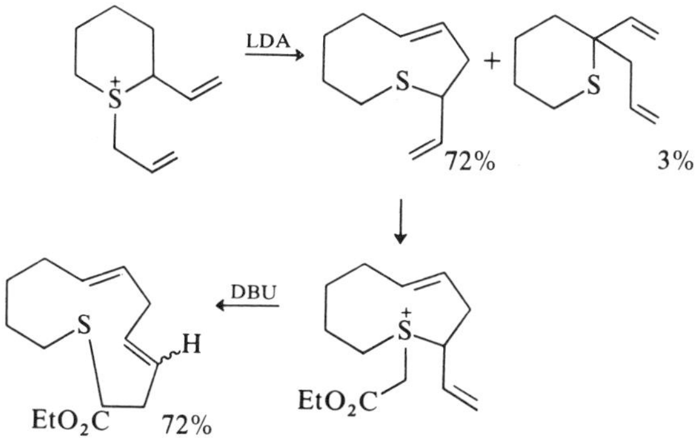

**Scheme 3**

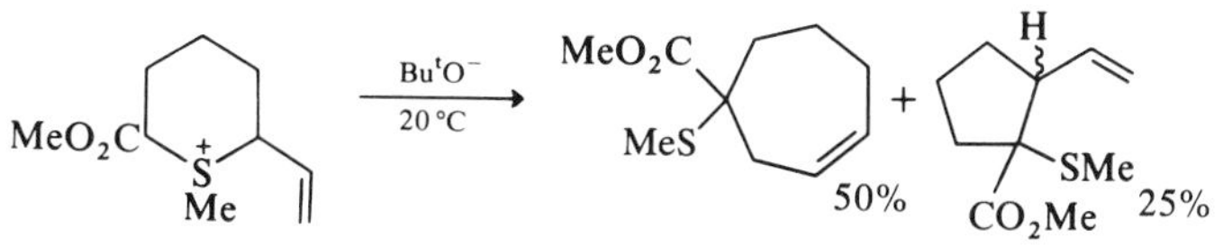

(11)

Stabilized sulfonium ylides are not just available by deprotonation of a salt; addition of a carbene or its equivalent generates an ylide directly. The carbenes may be derived photochemically or thermally from diazo compounds, and both methods of generation have been investigated ⟨74CC140, 80JCS(P2)385⟩. Photogenerated carbenes, in the singlet state, reacted with 4-*t*-butyltetrahydrothiopyran to generate only the equatorially substituted ylide, and the same result was found for the copper-catalyzed ylidation commonly regarded as proceeding *via* 'carbenoids' (equation 12). Further experiments conclusively showed that the orientation of addition was kinetically controlled as no ylide exchange was demonstrable under the reaction conditions and inversion at sulfur in sulfonium ylides has a high barrier; the dominant factor appears to be steric. Competition experiments with a mixture of dimethyl and diisopropyl sulfides under the same conditions gave product ratios of 5:1 in favour of the less hindered species.

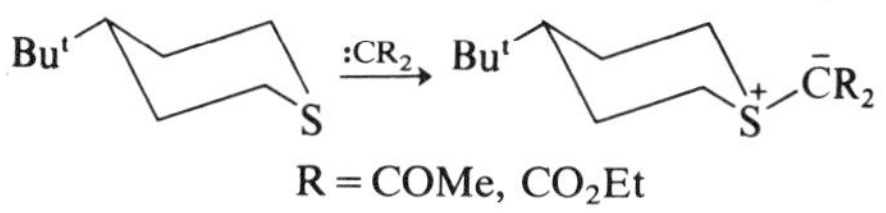

$$R = COMe, CO_2Et$$

(12)

In contrast, nitrene addition under the same conditions ⟨80JCS(P2)385⟩ is non-selective, affording the diastereomeric sulfimides in 1:1 mixture (equation 13). Again the dominant factors may be steric as reaction of the same nitrenes with mixed acyclic sulfides shows little selectivity. Sulfimines are also formed by the action of *N*-chloroarenesulfonamide sodium salts on the sulfide, and an extensive kinetic investigation ⟨76T2763⟩ has shown the minor influence of steric factors in this reaction as well. Equation (14) sets out the proposed course of the reaction.

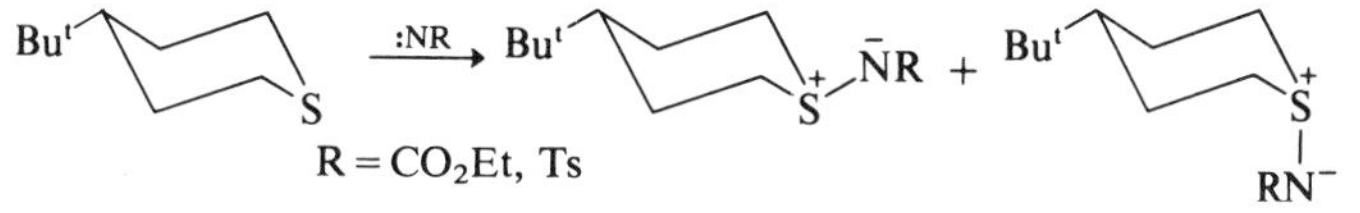

$$R = CO_2Et, Ts$$

(13)

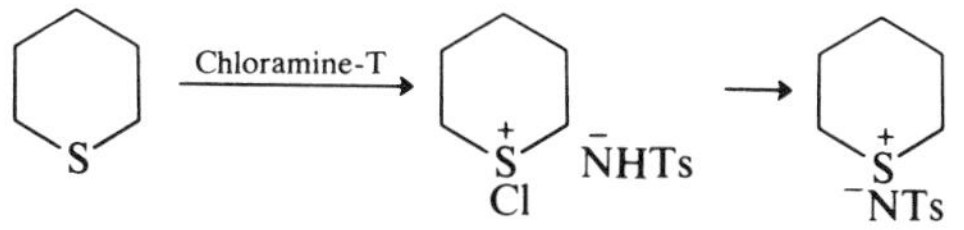

(14)

*N*-Unsubstituted sulfilimines may be prepared by reaction of a thiane with chloramine, hydroxylamine sulfate or *O*-mesitylenesulfonylhydroxylamine ⟨72JA208, 72TL4137⟩ but all these methods are restricted. A more general method treats *N*-tosylsulfilimines with concentrated sulfuric acid, and the free imino species is formed in very high yield ⟨76JOC1728⟩ though prolonged exposure to acid produces the sulfoxide. The free sulfilimine is a strong base, and is readily acylated on nitrogen by a wide range of electrophilic reagents. Oxidation, preferably with permanganate, affords sulfoximines in up to 95% yield and this is a superior method to the treatment of sulfoxides with hydrazoic acid or chloramine-T and copper ⟨73JA4287⟩.

Phosphorus pentasulfide cleanly reduces both *N*-H and *N*-tosyl sulfilimines to the sulfide ⟨78S540⟩, as do cyanide ion and triphenylphosphine, but contrary to earlier reports thiophenolate anion also causes ring opening by attack at the α-carbon, and a range of products results. The observations have been investigated in detail ⟨74T947⟩ and rationalized in terms of initial attack on the trivalent sulfur followed by elimination of two substituents, except for the softer thiophenolate, wherein the anionic character of the nitrogen atom tends to shield the sulfonium centre from attack by diverting it to the α-carbon. These reactions are shown in Scheme 4.

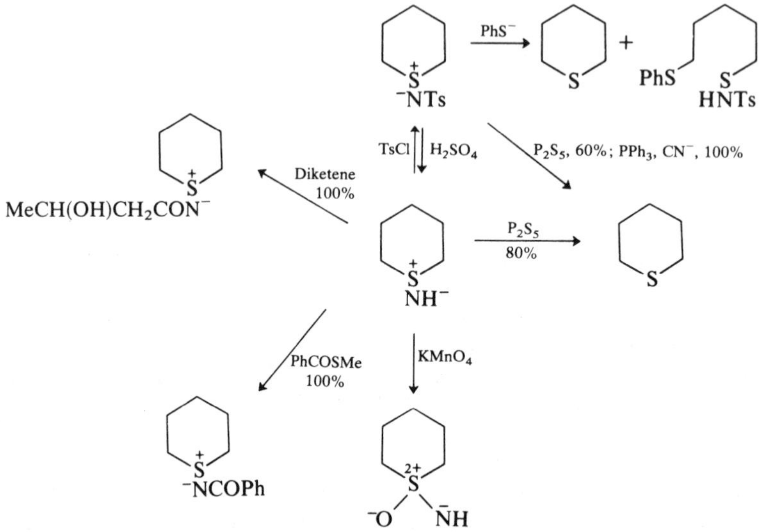

**Scheme 4**

*N*-Arylsulfimines have one further area of reactivity open to them, a 2,3-sigmatropic shift (equations 15 and 16). The shift proceeds with 95% stereospecificity ⟨76TL1335⟩ and in 40–60% yields. Competing reactions include scission of the S—N bond and cleavage of the ring. The preparation ⟨75T505⟩ of the *N*-arylsulfilimines using *N*-chlorosuccinimide and *p*-chloroaniline has many similarities with other functionalizations such as oxidation with hypochlorite, in that the first intermediate is an *S*-chloro species, which is subsequently transformed by inter- or intra-molecular reaction with a nucleophilic centre. Thus, halogenation by $SO_2Cl_2$, *N*-halosuccinimides or (dichloroiodo)benzene ⟨73CR(C)(276)1323⟩ all proceed *via* the ylidic structure (**63**) to (**64**) (equation 17). In the presence of a nucleophilic species such as aniline, the sulfonium centre is attacked with displacement of the halogen. A report

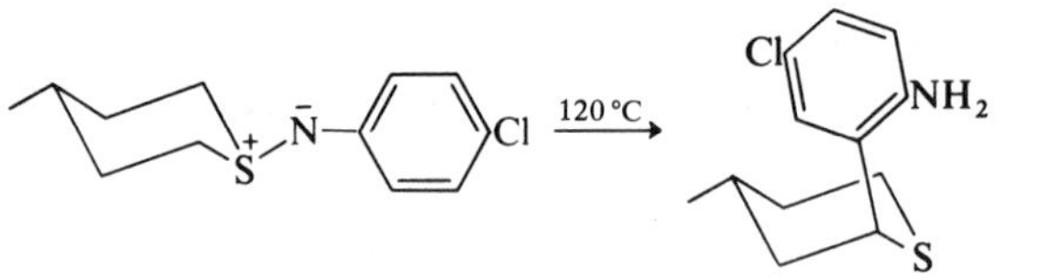

(15)

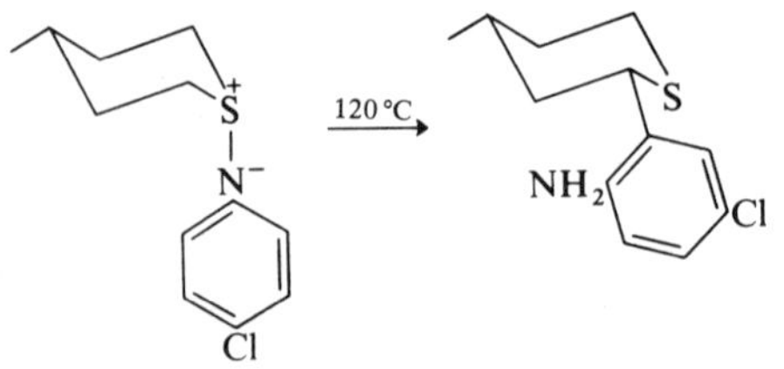

(16)

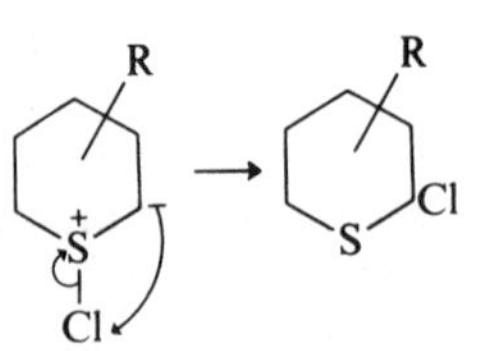

(17)

of the preparation of nitrones by the interaction of *N*-chlorosuccinimide, thiane and benzophenone oxime *via* the intermediate (**65**), however, seems unlikely, and a more plausible explanation is nucleophilic displacement of the 2-chloro substituent of the re-arranged 2-chloro derivative (**66**) (equation (18) ⟨75JCS(P1)1925⟩.

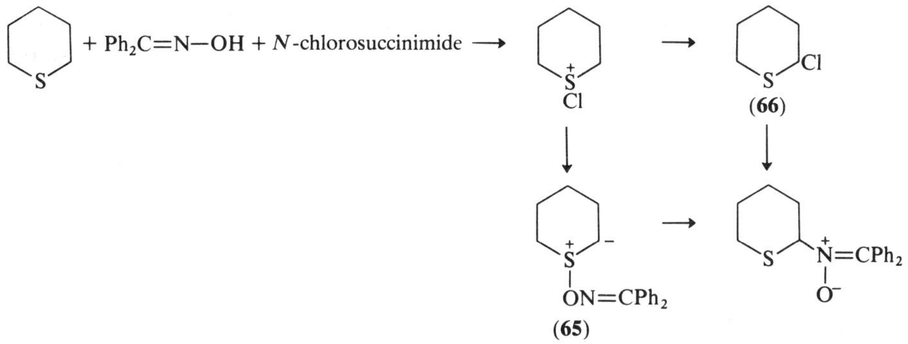

(18)

Halogenation of thian 1-oxides proceeds in different ways depending on the conditions ⟨72IJS(B)(7)185, 74JCS(P1)1723⟩. Chlorination of the *trans* sulfoxide (**67**) with a variety of agents gives predominantly a chlorine substituent *cis* to the oxygen atom, while in the *cis* system (**68**) the configuration at sulfur is inverted but chlorine still enters *cis* to the oxygen. If silver ion catalysis is employed, inversion at sulfur with *cis* chlorination becomes the predominant pathway for (**67**). Bromination of (**67**) occurs with greater difficulty and more epimerization, but (**68**) is inert under the same conditions. Again in the presence of silver ion, inversion at sulfur and *cis* halogenation becomes dominant (Scheme 5). A mechanism involving the halosulfoxonium ylide (**69**) or its epimer is postulated (equation 19), and equilibration between these two isomers leads to the high yield of the product with equatorial sulfoxide. In the reactions with silver catalysis, the rate is increased many-fold, and a kinetically controlled product distribution results, unlike the thermodynamic control of the uncatalyzed process.

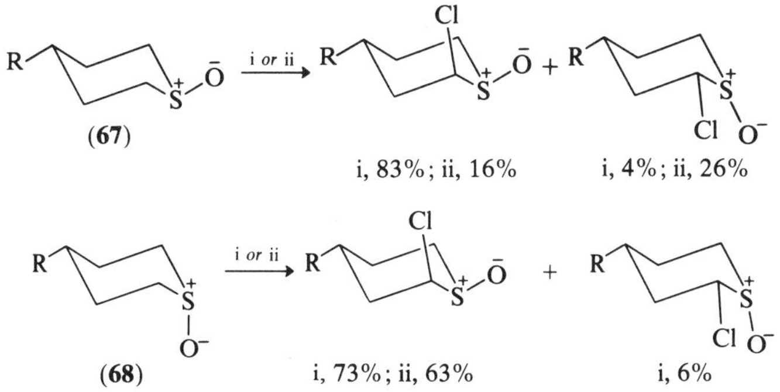

i, 83%; ii, 16%   i, 4%; ii, 26%

i, 73%; ii, 63%   i, 6%

i, PhICl₂, pyridine, r.t. 24 h; ii, PhICl₂, AgNO₃, pyridine, −40 °C, 1 h
R = *p*-chlorophenyl

**Scheme 5**

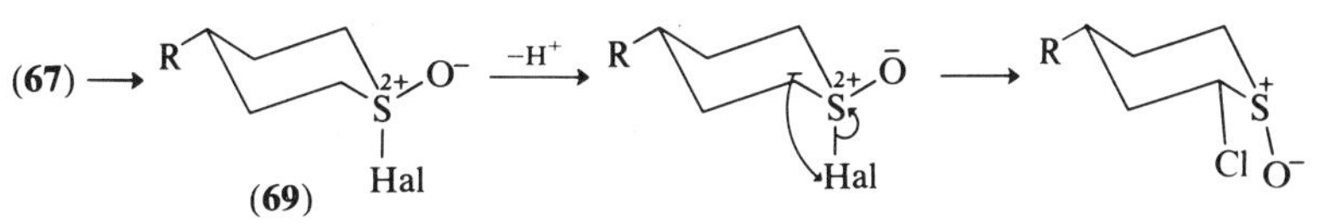

(19)

R = *p*-chlorophenyl

2-Halothiane dioxides have usually been prepared by oxidation of the halothiane or thiane oxide, but modern practice uses the α-lithiosulfone with a suitable halogen source. Hexachloroethane is a particularly interesting case, and affords about 50% yields despite the major competing reaction, the Ramberg–Bäcklund rearrangement, which precludes the use of too vigorous conditions ⟨73T4149⟩.

Halogen atoms at position 3 in tetrahydrothiins are labilized toward nucleophiles by the sulfur atom, by the formation of a bicyclic episulfonium salt as observed in the chemistry of mustard gas. Salt (**71**) has been isolated from the pyrolysis of the bromomethyltetrahydrothiophene (**70**), and the presence of nucleophiles can trap the intermediate in two ways (equation 20) ⟨74BSF590⟩. The same effect is seen in the solvolysis of *p*-nitrobenzoate esters, wherein the reaction occurs entirely by *O*-alkyl cleavage (an $S_N1$ mechanism). Table 13 illustrates the importance of sulfur participation ⟨74T2087⟩; the presence of sulfur $\beta$ to

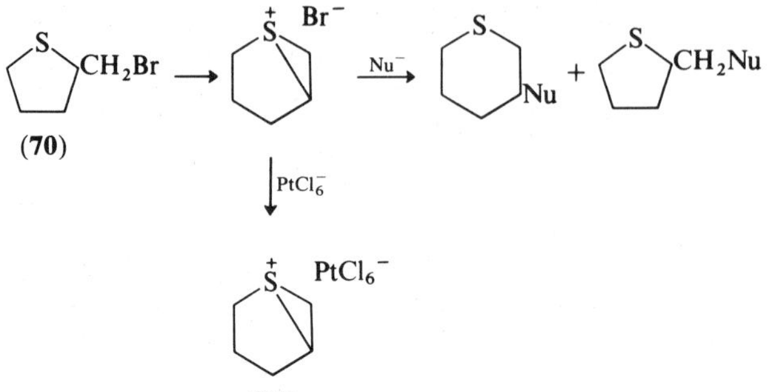

$$(20)$$

the ester accelerates solvolysis by some $10^6$ times compared with the cyclohexyl analogues, but substitution of phenyl or methyl groups at the esterified position has comparatively little effect, unlike the cyclohexyl analogue which experiences enormous acceleration by this means. The scope of sulfur participation is limited. 4-*p*-Nitrobenzoate esters show no evidence for transannular participation, and solvolysis of these functions is completely by acyl oxygen fission and a classical $S_N2$ mechanism. Transannular interactions resulting in acceleration of solvolysis of substituents *vis-à-vis* the corresponding cycloalkyl system are seen in other thiacycloalkanes, but only in thietanes is such an effect apparently greater than for thiane 3-substituents ⟨77MI22501⟩.

**Table 13** Sulfur Participation in the Solvolysis of Thiacycloalkyl *p*-Nitrobenzoate Esters in 80% Aqueous Acetone
⟨74T2087⟩

| Ester | Pseudo 1st order rate constant (50 °C) | Ester | Pseudo 1st order rate constant (50 °C) |
|---|---|---|---|
| 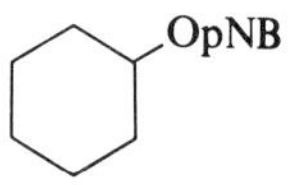 | $2.8 \times 10^{-9}$ | 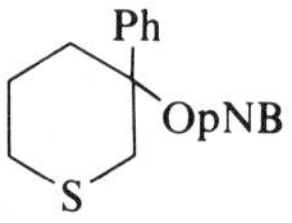 | 40.1 |
| 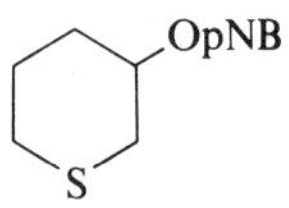 | 0.0166 | 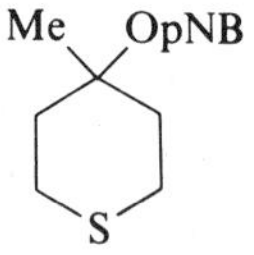 | $1.1 \times 10^{-4}$ |
| 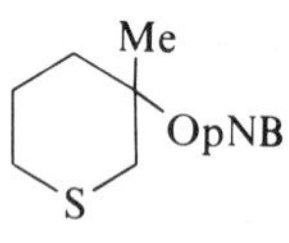 | 1.09 | 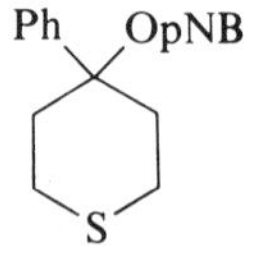 | $1.84 \times 10^{-2}$ |

(pNB = *p*-nitrobenzoate)

Similarly, the epoxide (**72**), on reaction with BH$_3$/LiBH$_4$, acetic acid or BCl$_3$ gave the tetrahydrothiophene derivatives (**74a–c**), presumably *via* the intermediate (**73**), unlike the analogous cyclohexyl and *N*-alkylpiperidyl compounds, in which the oxirane opened to give only the expected alcohols (equation 21) ⟨75CPB2701⟩.

However, in the sulfone analogues of the thiopyrans the transannular interaction has been shown to extend to the 4-position, in that thiopyran-4-one 1,1-dioxide is electrochemically reduced more readily than the sulfide ⟨72MI22500⟩.

With the exception of the areas described above, the chemistry of substituents upon the thiane ring is predictable by reference to that of cyclohexanes. Thus, the preparation of

$$(21)$$

$$
\begin{array}{ll}
\text{a,} & \text{LiBH}_4/\text{BH}_3 & \text{X = H, Y = OH} \\
\text{b,} & \text{AcOH} & \text{X = OAc, Y = OH} \\
\text{c,} & \text{BCl}_3 & \text{X = Y = Cl}
\end{array}
$$

4-aminothianes by reduction of the oximes has been thoroughly explored ⟨81JOC4376, 81JOC4384⟩, as has the displacement of tosylates by azide and reduction thereof. 4-Keto functions may be reduced chemically, or enzymatically with 100% enantioselectivity ⟨79JA5405⟩. Thian-4-ols may be readily halogenated, or eliminated (which represents a highly efficient entry into 5,6-dihydro-2*H*-thiins) by conventional means. Likewise the chemistry of the substituted oxides is closely analogous to that of acyclic congeners and may be readily inferred therefrom.

### 2.25.3.2 Reactivity of Dihydrothiopyrans

Table 1 shows the two dihydrothiopyrans, and the two corresponding benzo fused systems. The monocycles (5,6-dihydro-2*H*-thiin and 3,4-dihydro-2*H*-thiin) are quite clearly from their chemistry an allylic sulfide and an enol sulfide respectively, and in many of the reactions they exhibit they are perfectly comparable with acyclic counterparts. Once again, as for the tetrahydrothiopyrans, in many cases the principal chemical interest is related to the effects of preferred conformations on the stereochemistry of particular conversions, but this will not be discussed in great length as it is entirely predictable from classical alicyclic work. Where differences exist they may be attributed to the interaction of the heteroatom with the neighbouring alkene, which is not too considerable, and more importantly the reactivity of the sulfur atom in its own right. The benzo fused compounds, of course, have their own particular chemistry which has received considerable attention over many years, some of which will be discussed.

Hydrogenation of dihydrothiopyrans to tetrahydro derivatives has been well established, but generally requires forcing conditions unless the sulfur atom is masked by oxidation or in some other way; the catalyst poisoning ability of sulfides is well known ⟨58JCS2888⟩. If too extreme conditions, especially of temperature, are used, isomerization reactions become competitive, particularly for the benzothiopyrans when benzothiophene derivatives and ring-opened products may result ⟨74ZOR76⟩. Desulfurization with Raney nickel likewise generally engenders isomerization of the alkenes, or may even cause hydrogenation of them, but a very clean and efficient process has been reported for the 5,6-dihydro-2*H*-thiopyran system. Lithium in ethylamine at −78 °C cleaves the allylic C—S bond without isomerization, and the lithium thiolate is further reduced to the alkene with deactivated Raney nickel without too extensive side reactions ⟨73JA4444⟩. Interestingly the yields are better if the lithium salt is used, rather than the thiol, with the nickel. This procedure is also applicable to the corresponding sulfoxides, again in good yield. As was discussed in Section 2.25.3.1, desulfurative conversions like this represent excellent entrées into stereo-specifically arranged alkanes and alkenes, and equation (22) illustrates its use in the

$$
\begin{array}{l}
\text{i, Li/NH}_3 \\
\text{ii, Raney Ni}
\end{array}
$$

$$(22)$$

preparation of *Cecropia* juvenile hormone precursors. Surprisingly, however, the allylic sulfide system is not desulfurized by tri-*n*-butylstannane, unlike other sulfur heterocycles ⟨77CJC3755⟩.

Oxidation of the heterocycles with common reagents such as MCPBA, sodium periodate or hydrogen peroxide cleanly affords the sulfoxides and sulfones, and it is clear that the sulfur atom is the principal centre of reaction for electrophiles. While the sulfone is a quite inert functionality, the sulfoxides may be reduced to the sulfides with phosphorus pentasulfide as for the tetrahydro systems ⟨78CJC1423⟩. Positive halogen sources likewise react at sulfur, and the intermediate sulfonium halide rearranges, usually by 1,2-shift to the α-halo product.

This dominance of sulfur in the reactions with electrophiles is well brought out in the addition of carbenes to the two monocycles. The allylic sulfide (5,6-dihydro-2*H*-thiopyran) only affords the products of reaction at sulfur, while the vinylic sulfide (3,4-dihydro-2*H*-thiopyran), in which the alkene is a little more nucleophilic due to the small interaction with the heteroatom, shows dichotomous behaviour. Dichlorocarbene affords the cyclopropane product (**78**) in 70% yield, but the stabilized ylide (**76**) is produced from bismethoxycarbonylmethylide and (**75**). In fact it is possible that the initial reaction with dichlorocarbene is reaction at sulfur and subsequent rearrangement of this less stabilized ylide. Schemes 6 and 7 illustrate the results and proposed mechanisms ⟨77JOC3365, 64JOC2211⟩.

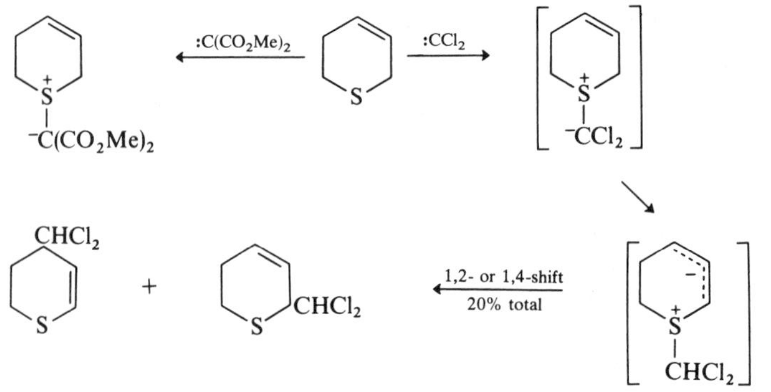

**Scheme 6**

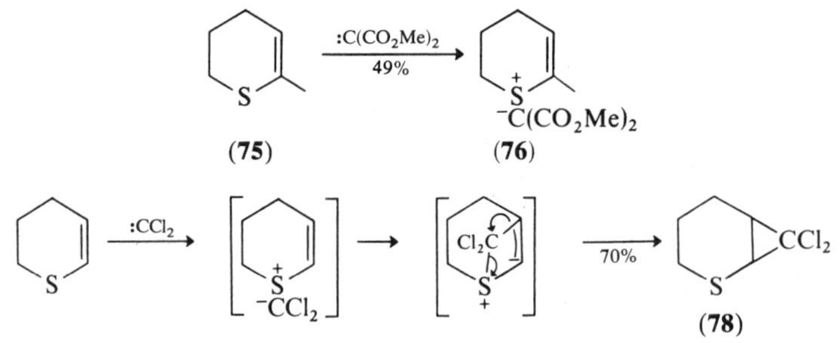

**Scheme 7**

Support for these types of ylide isomerization may be drawn from the rearrangements seen in less stabilized ylides. Equation 23 shows a ring contraction related to the expansion reactions of equations (9)–(11) ⟨81JCS(P1)1953⟩. As in those examples, the stabilized ylide (**79**) requires heating to effect the reaction; the product is a single diastereoisomer, derived from a predominantly *endo* pathway.

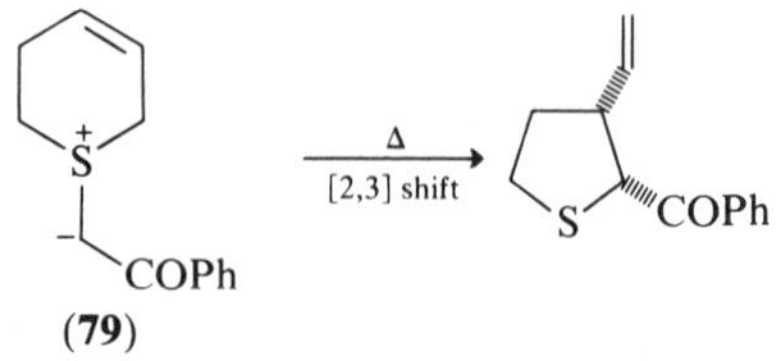

(23)

If the ylidic compound is a benzo fused system, alternative reaction schemes take over. 3,2-Sigmatropic shifts are no longer favoured, and proton abstraction with the formation of isomeric ylides, or elimination processes, become important. Equations (24) ⟨77JOC3945⟩

and (25) ⟨77JCS(P1)1155⟩ show these two types of behaviour. Interestingly, the Stevens rearrangement with ring expansion of the isomeric ylide form is not usually seen in the chemistry of cyclic sulfonium ylides, despite being a major reaction pathway in acyclic species. Support for the mechanism of equation (25) is lent by the thermolytic conversion of the isolated, well characterized bismethoxycarbonylmethylene ylide analogue of the unisolated intermediate (**80**). This ylide is available in 80% yield from isothiochroman (2,3-dihydro-1*H*-2-benzothiopyran) with diethyl diazomalonate and copper-bronze. Acetolysis affords a 94% yield of the Stevens rearrangement product, whereas with the ylide derived in the same way from thiochroman (3,4-dihydro-2*H*-1-benzothiopyran) acetolysis gives a mixture of products including thiochroman, 2-acetoxythiochroman and a ring-opened product. These arise from nucleophilic attack on the protonated ylide, at the two carbon atoms, and at sulfur to generate the intermediate seen in Pummerer reactions. Protonation of the former ylide in equilibrium with (**80**) is less favoured owing to the stabilization endowed by the aromatic ring; Stevens rearrangement can then compete effectively ⟨78G671⟩.

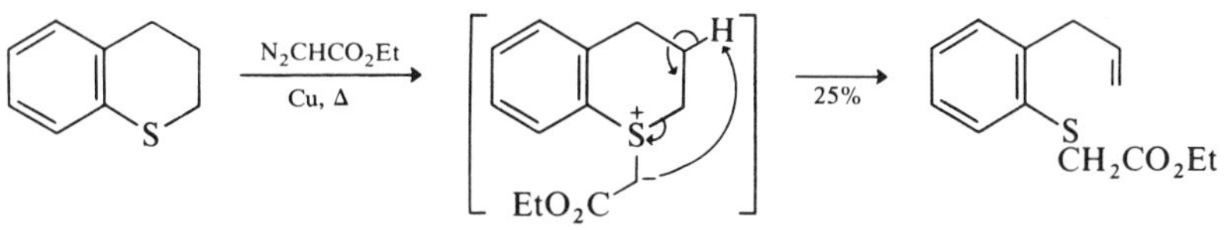

(24)

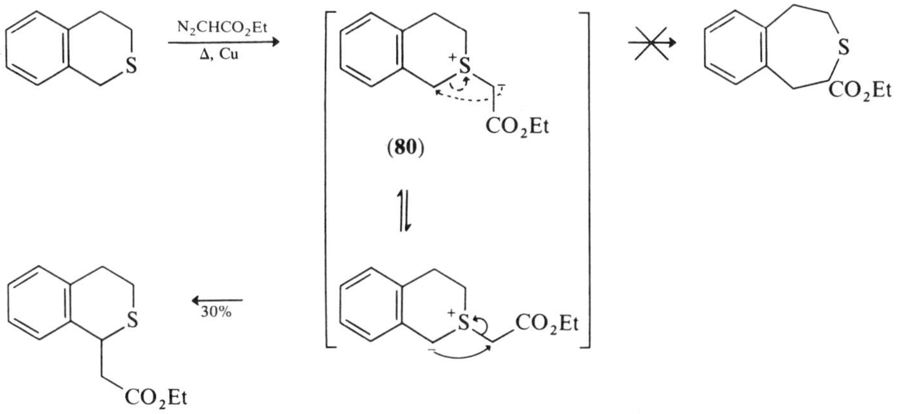

(25)

Sulfilimines, analogous to those prepared from tetrahydrothiopyrans, have been well described only when derived from 3,4-dihydro-2*H*-1-benzothiopyrans, presumably due to the ready occurrence of side reactions. For instance, in the preparation of (**81**) the major alternative product is the sulfoxide (**82**) (equation 26) ⟨80TL533⟩, while in the case of 3,4-dihydro-1*H*-2-benzothiopyrans, Pummerer-type rearrangements involving the readily accessible ylide (**83**) form (**84**) (equation 27). Once formed, these cyclic sulfilimines are more stable than their acyclic counterparts, which may be a reflection of their rigidity. The normal *β*-hydrogen abstraction/elimination process ⟨77CRV409⟩ is only available if a substituent is present on the ring. Equation (28) illustrates the scope of the reaction, and the proposed mechanism. The incoming nucleophile may be the toluenesulfonamide anion, or an added species such as acetate, or, if a sulfoxide is used as solvent, a Kornblum-type reaction leading to an oxidized product is found (equation 29). Thus, if acetic anhydride containing sodium acetate is used, the yield of (**85**) is 90% with only approximately 10% of the *N*-acylated derivative of (**86**) being formed.

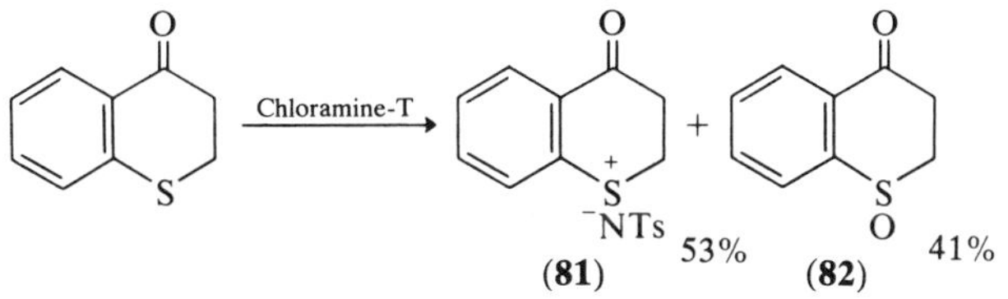

(26)

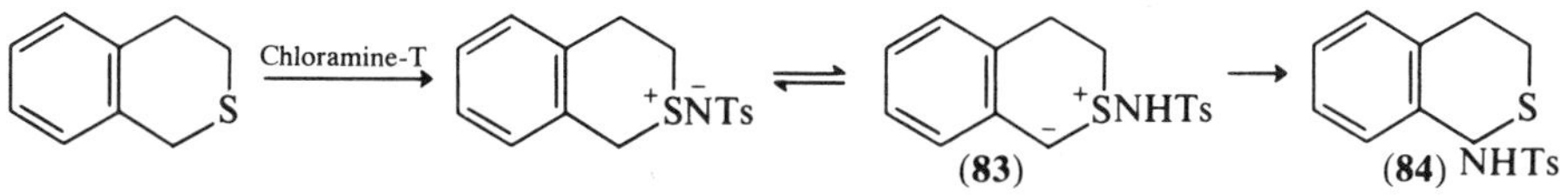

(27)

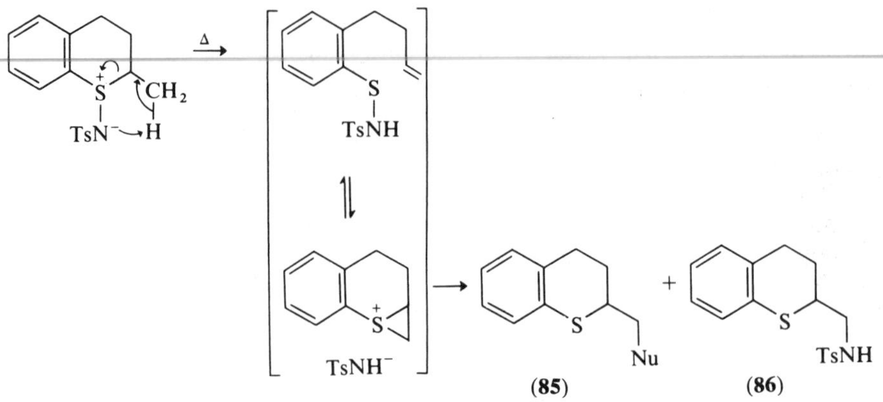

(28)

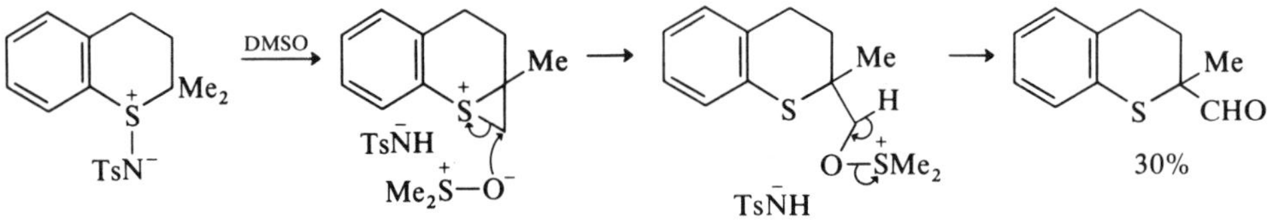

(85)                    (86)

(29)

30%

If a thiochromanone is the starting system for sulfilimine preparation, another mode of reaction becomes available, namely ring expansion or contraction with insertion of the sulfilimine nitrogen (Scheme 8) ⟨80TL533⟩. The nature of the product is defined by the relative propensity for Michael addition (ring expansion) or thiooxime formation (ring contraction) to occur.

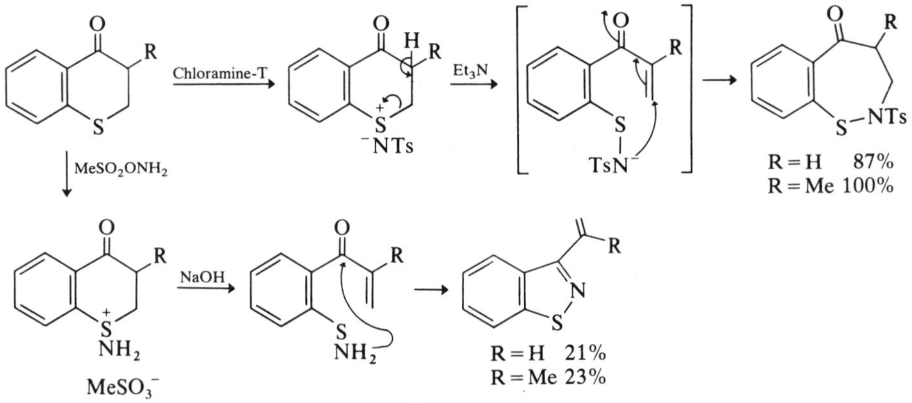

R = H    87%
R = Me  100%

R = H   21%
R = Me  23%

**Scheme 8**

Non-functionalized sulfur has seen, in recent years, a considerable usage for stabilization of adjacent carbanionic centres. Although not as marked as for sulfoxides or sulfones, the $\alpha$-sulfidic carbanions are quite sufficiently stable to be routine reactive intermediates in general organic synthesis, not requiring elaborate apparatus for their generation and use. 5,6-Dihydro-2H-thiopyrans are ideal substrates, and may be readily lithiated with s-butyllithium. Reaction with electrophiles takes place in either the 2- or 4-position depending on the reagent; epoxides give only $\alpha$-alkylation, methyl iodide and primary allylic halides give predominantly $\alpha$-alkylation together with 3–15% $\gamma$-alkylation, while aldehydes and ketones selectively alkylate at the $\gamma$-position ⟨73JA4444⟩. Equation (30) shows the preparation of a precursor to *Cecropia* juvenile hormone (see also equation 22). Should the carbanion (**88**) be allowed to warm to 25 °C it will still react normally, but substitution of the 6-position with phenyl renders the deprotonated material (**89**) unstable at −15 °C (equation 31). The same product is produced in 90% yield if the S-methyl sulfonium salt is deprotonated and allowed to warm to room temperature, but the copper–lithium complex of (**89**), and 6,6-dialkyl analogues of (**89**), are quite stable and do not afford the cyclopropyl product ⟨76T1801⟩. The important influence of carbanion stabilizing groups on the lability of the dihydrothiopyran rings may also be illustrated by equation (32), in which ring opening does not reverse because of delocalization in the anionic intermediate. Once alkylated, the thiol becomes less effective in the Michael addition reaction, and nitrogen competes effectively

⟨78ZC91⟩. A related reaction, with even more curious consequence, is shown in equation (33) ⟨73CJC839⟩. However, aldol condensation of thiochromanones with aldehydes in the presence of weak bases proceeds exactly as for carbocyclic systems ⟨79MI22501⟩.

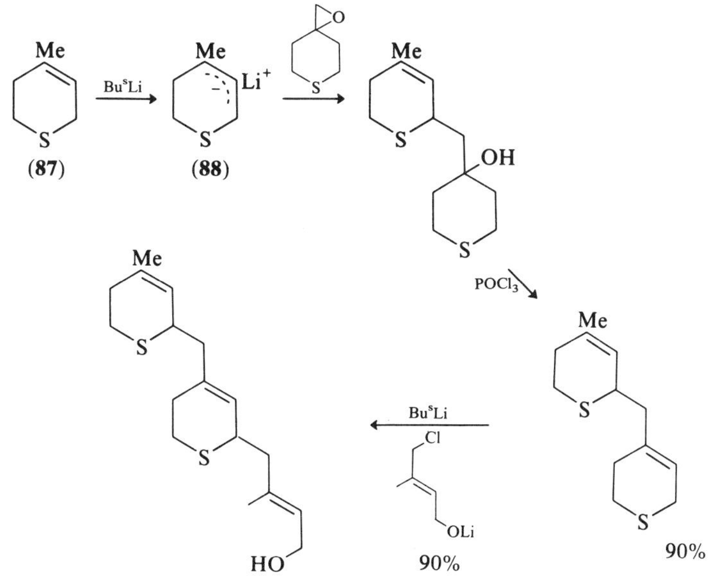

(30)

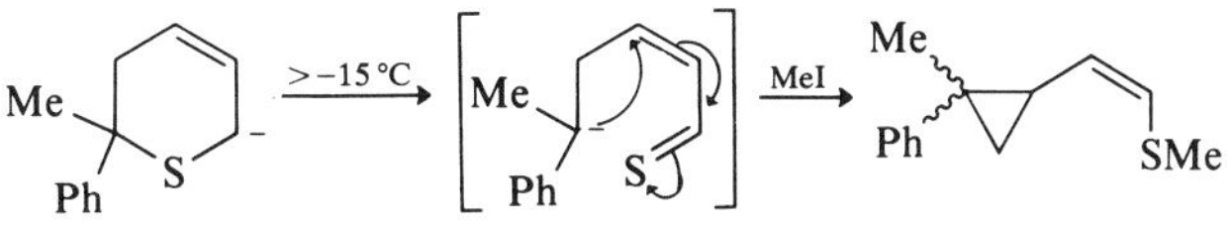

(31)

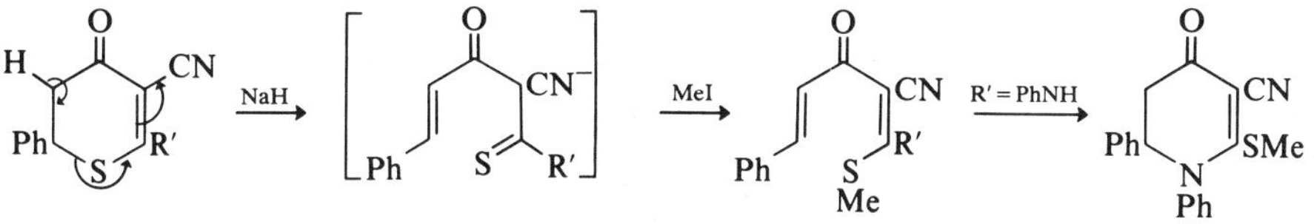

(32)

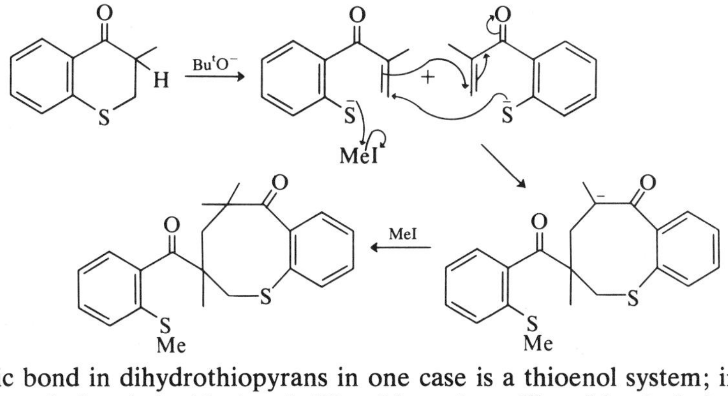

(33)

The alkenic bond in dihydrothiopyrans in one case is a thioenol system; in the other it is effectively an isolated double bond. The thioenol readily adds alcohols or thiols (*cf.* dihydropyran); it has been suggested as a protecting group for alcohols, and is readily removable with Ag$^+$ under neutral conditions ⟨66JOC2333⟩. If conjugated with a carbonyl function, thiols will add under basic conditions (equation 34) ⟨81JA4597⟩.

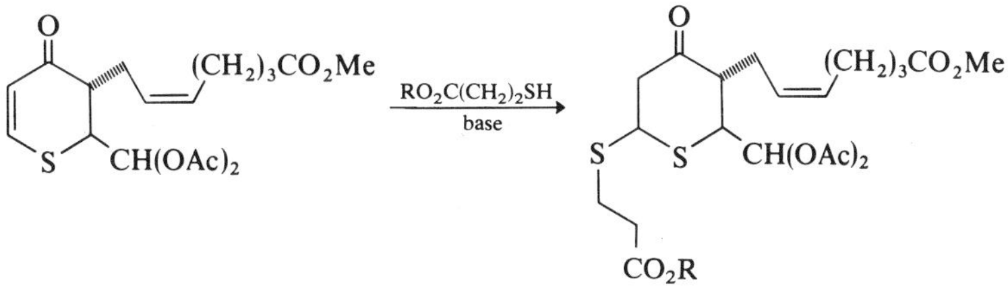

(34)

The reactivity of the allylic sulfide double bond with electrophiles is largely masked by the presence of the heteroatom. However, on oxidation to the sulfone the alkene is found to be surprisingly inert toward such species, and this has been taken as further evidence for a transannular effect of the —SO₂— unit greater than that seen for the sulfur atom in tetrahydrothiopyrans ⟨77MI22502⟩.

Treatment of dihydrothiopyran or dihydrobenzothiopyran with strong acids, such as trifluoroacetic acid, generally leads to disproportionation to the tetrahydrothiin and the thiopyrylium salt ⟨75ZOR2447⟩.

The chemistry of the substituents on monocyclic dihydrothiins is generally closely analogous to that of alicyclic systems, with only relatively small, obvious differences wherein the ability of sulfur to interact with principally neighbouring substituents plays the dominant role. Thus, substituents at position 2, if good leaving groups, are very readily displaced, particularly in conjugated enones, where an addition–elimination sequence is especially favoured. Scheme 9 shows two such reactions ⟨79JPR699⟩.

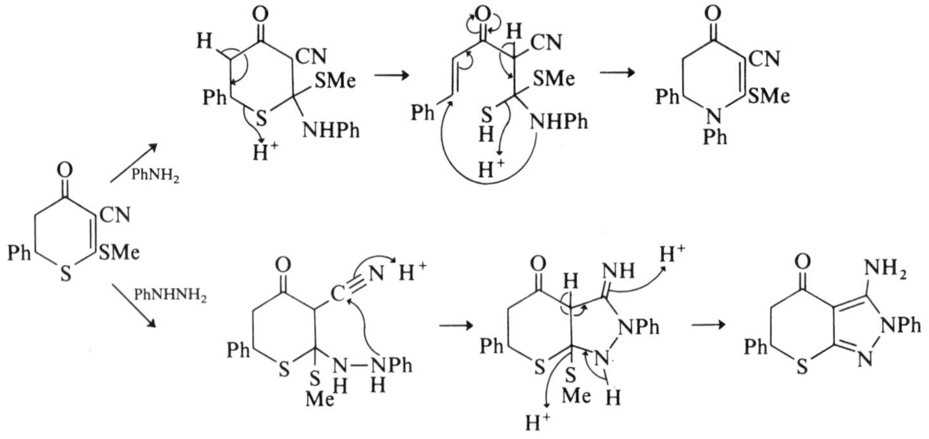

**Scheme 9**

Again a 3-bromo derivative behaves very similarly to 3-bromotetrahydrothiopyrans, the sulfur acting to displace the halogen with formation of a thiiranium intermediate. Thus, heating in aqueous dioxane converts 3-bromo-4-hydroxy-3,4-dihydro-2*H*-1-benzothiopyran into benzothiophenemethanol by dehydration following hydrolysis of the thiiranium intermediate ⟨69AG(E)456⟩.

Enamines derived from thiopyran-3-one, although tautomeric, tend to exist predominantly in conjugation with the sulfur atom; the Fischer indole reaction, when applied to that ketone, affords solely the systems fused '2,3' on to the thiopyran (Scheme 10) ⟨76CL5⟩. 4-Amino-3,4-dihydro-2*H*-thiopyrans readily eliminate ammonia or amines on heating or treatment with acid, with formation of 2*H*-thiopyrans ⟨78CR(C)(286)553⟩.

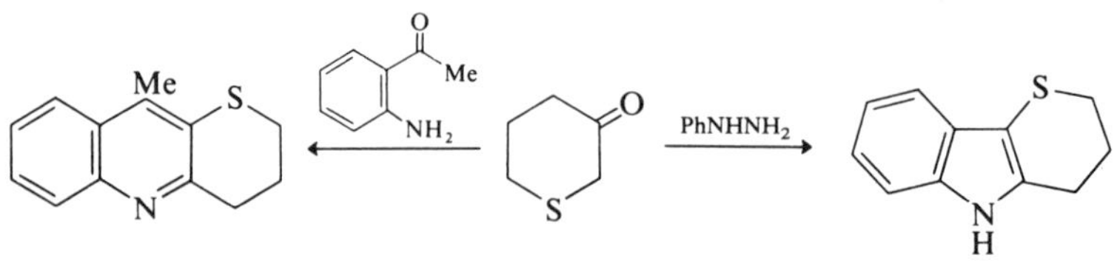

**Scheme 10**

3,4-Dihydro-2*H*-1-benzothiopyrans (thiochromans) are in almost all their chemistry simply aromatic sulfides. Acylation of the benzene ring under Friedel–Crafts conditions goes, predictably, in position 6, unless blocked, when reaction occurs at position 8 ⟨79MI22506⟩. An even more constrained system (**90**) may be cyclized efficiently to (**91**) (equation 35) ⟨76JHC123⟩, though even here a disproportionated product was also formed.

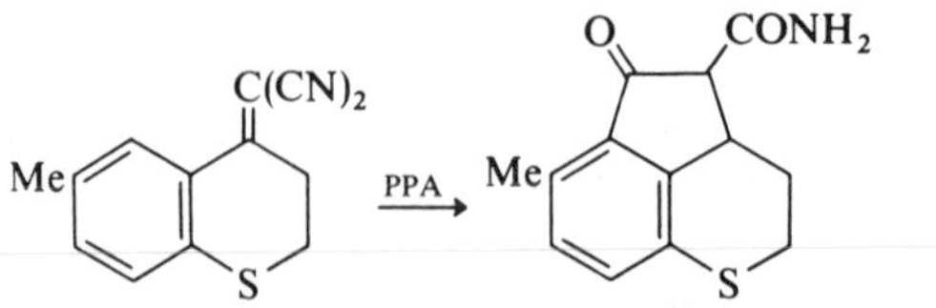

(35)

Attempted dehydrogenation of thiochroman with chloranil results in ring contraction to a mixture of 2-methyl- and 2-methyl-2,3-dihydro-benzothiophene ⟨76KGS328⟩.

Introduction of a 4-carbonyl group to form a thiochromanone increases the range of reactions available to the system. Aldol condensations, and further elaborations of the products, have been extensively studied ⟨77JMC847⟩. The 4-keto function may be converted to its silyl enol ether with dimethylaminotrimethylsilane, thus permitting a range of reactions not readily observed with the corresponding enamine derivatives ⟨74TL3553⟩.

Acylation of thiochromanones with boron trifluoride and acetic anhydride readily affords (**92**), which may be elaborated into (**93**) (equation 36) ⟨79CJC3292⟩, while conversion of the keto function to an oxime (**94**) permits the formation of the palladium complex (**95**) as a first step towards functionalization of the aromatic ring (equation 37) ⟨78BCJ3407⟩, or ring expansion by Beckmann rearrangement (equation 38) ⟨76JCS(P1)2343⟩.

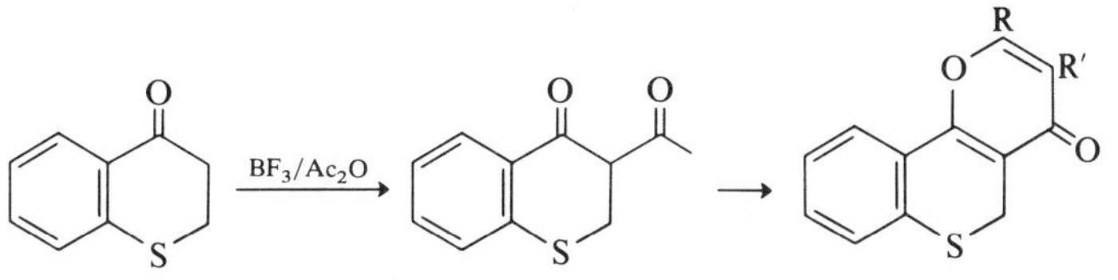

$$(36)$$

(**92**)        (**93**)

$$(37)$$

(**94**)        (**95**)

(**94**)  $\xrightarrow{\text{Beckmann rearrangement}}$

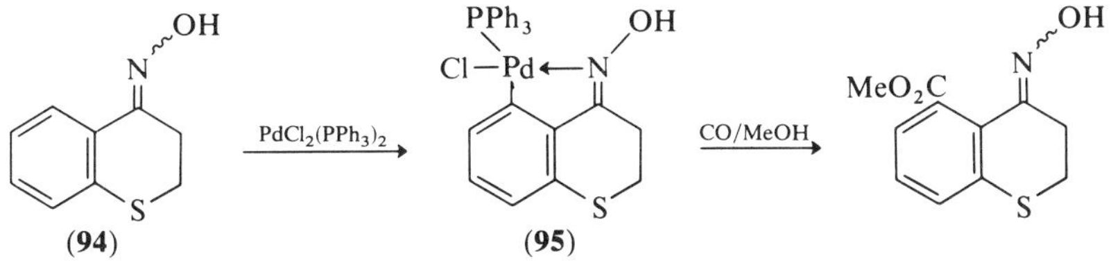

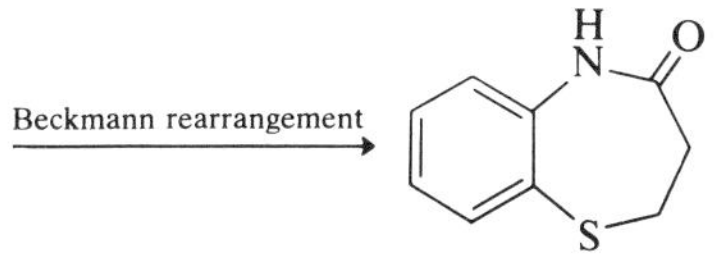

$$(38)$$

The Schmidt reaction affords both possible isomers when applied at the sulfide oxidation level, and also with the sulfoxide, no trace of sulfoximine being found ⟨75CJC276⟩. Reduction of the oxime to the amine with lithium aluminum hydride is significantly improved by the presence of titanium tetrachloride ⟨78KGS1694⟩.

Introduction of a 3-carbonyl group switches the nucleophilic behaviour to position 4, and a variety of condensations has been reported across positions 3 and 4. Enolization preferentially occurs toward the aromatic ring rather than toward the sulfur, unlike the situation seen with the non-annelated heterocycle ⟨77T2383⟩.

The photochemistry of thiochromanone 1-oxides has been examined ⟨76CJC455⟩; the principal pathways are β-hydrogen abstraction with readdition of the sulfinic acid to the alkene (equation 39), and rearrangement to cyclic sulfinates followed by homolysis of the S—O bond.

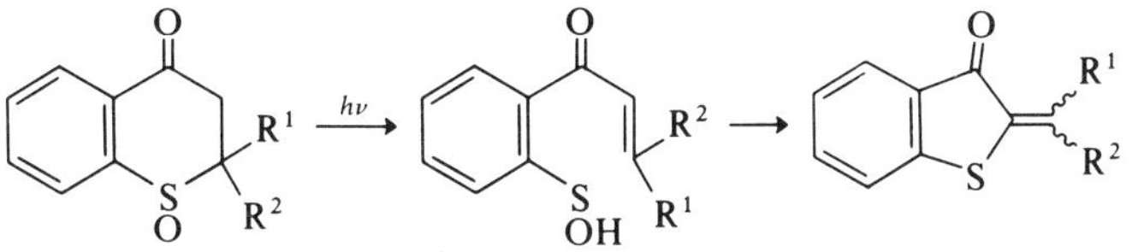

$$(39)$$

Isothiochromans (3,4-dihydro-1*H*-2-benzothiopyrans) have also been widely investigated, but less so than the thiochromans. The principal site of reactivity is the 1-position, both benzylic and α to sulfur. Scheme 11 gives a resumé of typical chemical conversions ⟨74AP218, 74CB605, 74LA734, 74LA1474, 75FES837, 76PS(1)129, 78LA1123⟩. No unexpected reactions are seen. With the introduction of a carbonyl at position 4 (isothiochromanones), the centre of reactivity switches to C-3, which is α to both the carbonyl and the sulfur ⟨76JCS(P1)749⟩. This position condenses readily with electrophiles such as aldehydes, CS₂ or diazo compounds, generating enones, which may further react in ways typical of acyclic vinyl ketones; thus, cuprates add cleanly, and the compounds dimerize unless sterically hindered (Scheme 12) ⟨79ACS(B)460, 79ACS(B)669⟩. Another dimerization results from the addition of vinylmagnesium bromide to the carbonyl group (equation 40) ⟨78IJC(B)210⟩. On oxidation with ferricyanide, isothiochromanones dimerize in a different fashion. A sulfur stabilized radical, or radical cation (**97**), is the probable intermediate leading to the coupling product (**98**) after dehydrogenation ⟨80CB1708⟩. Of considerable theoretical interest is the action of mineral acids: perchloric acid formally removes a hydride ion, with formation of the salt of the benzothiopyrylium-olate zwitterion. Base treatment liberates the betaine (**99**) which rapidly dimerizes (equation 41) ⟨76ACS(B)24⟩.

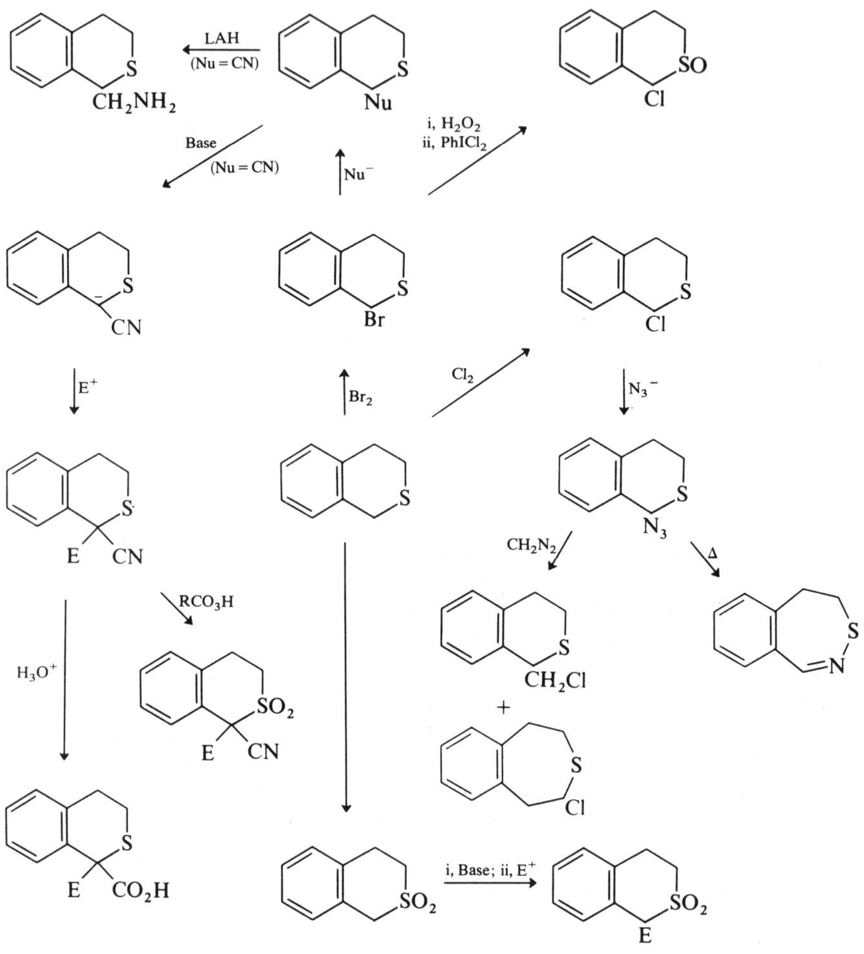

**Scheme 11**

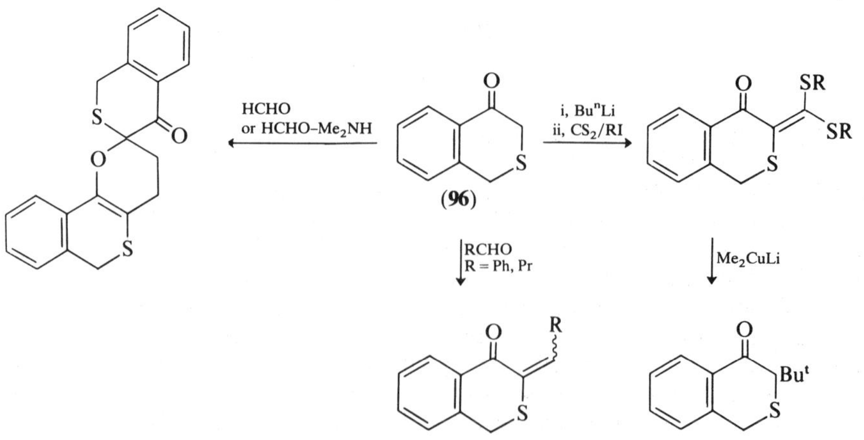

**Scheme 12**

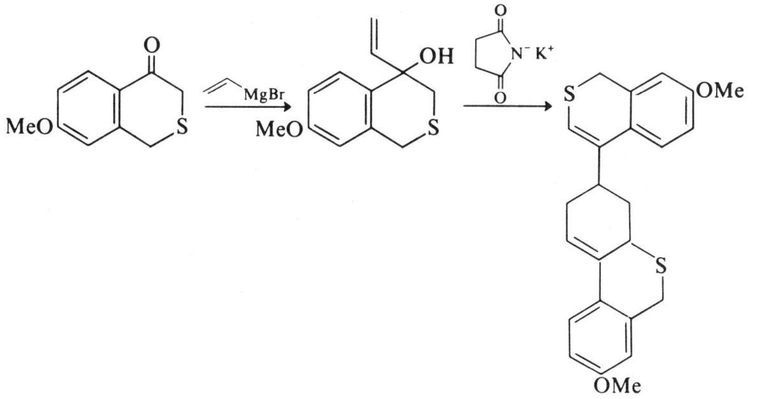

(40)

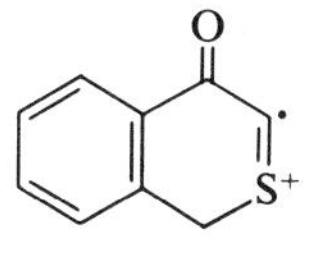

(97) 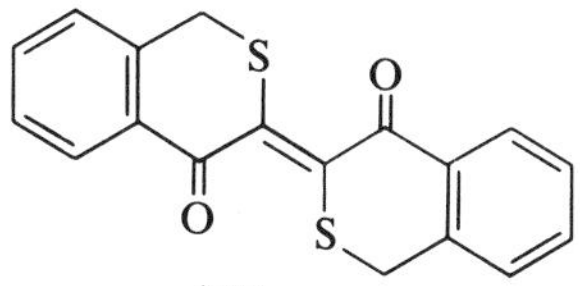          (98)

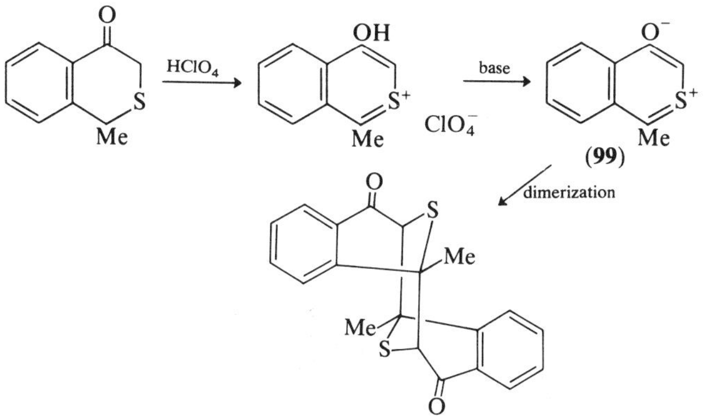

(41)

Photolysis of (**96**) results in the inversion of the hetero ring; a formal 1,2-shift of sulfur generates the spirothietanone (**100**) which relaxes by migration of the carbonyl function. This interesting inversion is very selective; carbons 1 and 3 remain in their original positions but the product is thiochromanone rather than an isothiochromanone (equation 42) ⟨69JOC1566⟩.

(42)

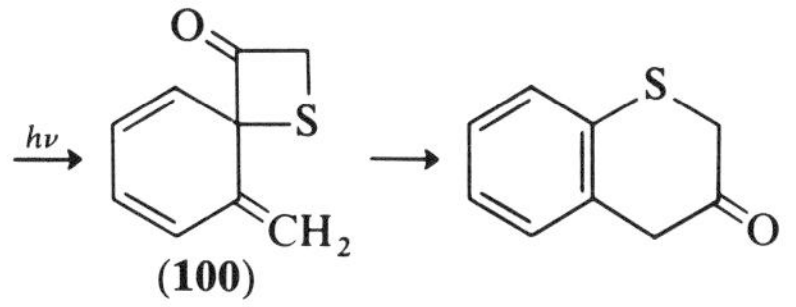

### 2.25.3.3 Reactivity of 2*H*- and 4*H*-Thiopyrans

Simple 2*H*- and 4*H*-thiopyrans are reported to be unstable, decomposing above −30 °C in air. Stability is greatly aided by substitution and annelation, but the most stable forms are the fully unsaturated keto or thiono derivatives. As has been discussed earlier, the interaction of sulfur with the alkenic bonds contributes less to the stability of the 4*H*-system than the conjugation between the bonds in the 2*H*-form. Thus, relatively mild treatment of 4*H*-thiopyrans, not otherwise conjugated, with acetic acid moves the alkenes into conjugation. The reaction is also effected by catalytic amounts of thiopyrylium salts, which are formed on mild acid treatment. Boiling in acetic acid tends to totally disrupt the heterocycle. Scheme 13 illustrates these interconversions ⟨77ACS(B)496⟩. Treatment with strong acid such as perchloric acid leads to total disproportionation of the system into a mixture of thiopyrylium salt and tetrahydrothiopyran (equation 43) ⟨71KGS(Sb3)85⟩.

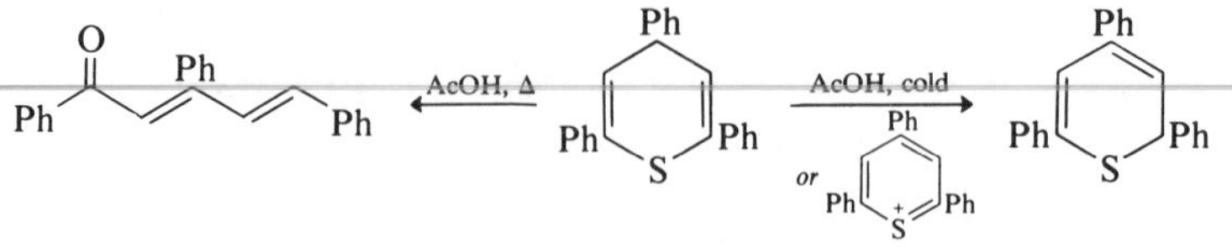

**Scheme 13**

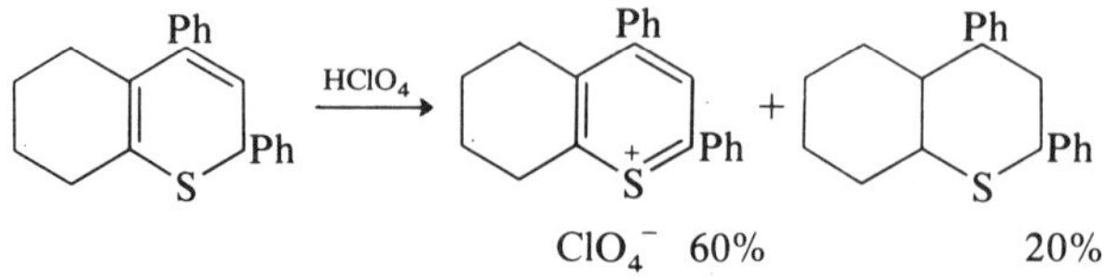

(43)

$ClO_4^-$  60%                                    20%

The poor interaction of sulfur with the alkenic system is also demonstrated by base treatment of 2*H*-benzopyrans and thiopyrans. The oxygen system undergoes base-catalyzed exchange at positions 2 and 4, and the product of such a reaction is a mixture of 2*H*- and 4*H*-benzopyrans. In the sulfur system only the 2-position undergoes exchange and the alkene remains firmly in conjugation with the benzene ring ⟨79MI22502⟩. Electrophilic attack on the anion would be predicted, therefore, to occur exclusively in position 2. Similar studies on the deprotonation of the parent monocyclic compound have thrown up the interesting result that lithiation may be directed specifically to position 6 of 2*H*-thiin (with *n*-butyllithium in the presence of TMEDA) and reaction with methyl iodide or acetaldehyde occurs there. The metallated species can only be interconverted with the C-2 lithio compound *via* protonation/deprotonation, or by addition of HMPT, which permits formation of the delocalized pentadienyl anion which is the thermodynamically more stable form ⟨78RTC208⟩. The specific metallation reaction has been made use of in the preparation of aerially unstable fused thiopyranothiopyrans (equation 44) ⟨78S578⟩. However, base treatment of 2,4,6-triphenyl-2*H*-thiin affords the pentadienyl anion, which alkylates exclusively at C-4 to produce a 4*H*-thiin. Use of the metal amide in liquid ammonia generates the pentadienyl anion of 2*H*-thiin directly, and under these conditions both *t*-butyl bromide and cyclohexyl bromide alkylate the heterocycle efficiently. Thus 2-*t*-butyl-2*H*-thiin is formed in 54% yield, while the cyclohexyl halide affords an 85:15 mixture in 82% yield of the 2-cyclohexyl-2*H*- and 4-cyclohexyl-4*H*-thiins. This surprising result (the halides usually prefer to react by elimination of HBr) has yet to be satisfactorily explained ⟨78CC596⟩.

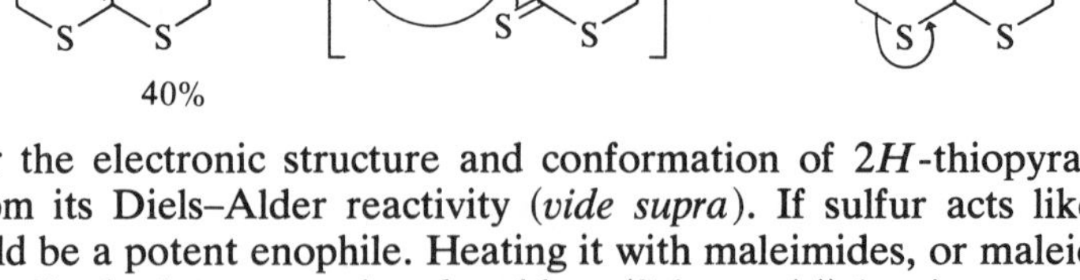

(44)

40%

Evidence for the electronic structure and conformation of 2*H*-thiopyran has already been taken from its Diels–Alder reactivity (*vide supra*). If sulfur acts like oxygen, the thiopyran should be a potent enophile. Heating it with maleimides, or maleic anhydride at 150 °C, cleanly affords the expected *endo* adduct (Scheme 14) but in competition experiments with *trans*-piperylene the acylic diene was far and away the more reactive, unlike oxygen conjugated dienes. However, the sulfur atom clearly dictates the regioselectivity of addition; methyl acrylate gives rise to only one regioisomer contaminated by a small amount of the corresponding *exo* adduct ⟨79JOC2280⟩.

Under forcing conditions (Pd/C, 100 °C, 50 atm $H_2$) thiopyrans are reduced to their tetrahydro derivatives in some 75% yield, but milder conditions will hydrogenate the *S,S*-dioxides to the tetrahydro systems ⟨75ZOR1543⟩. On the other hand, dissolving metal reductions with sodium/HMPT–THF cleave the ring with generation of dianions such as (**101**). Alkylation occurs primarily on the terminal carbon to give (**102**) (equation 45) ⟨79RTC520⟩, although small amounts of alkylation at the central carbon are also observed.

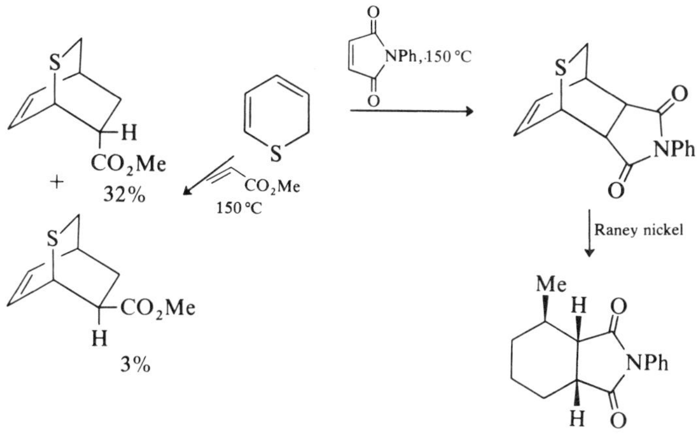

**Scheme 14**

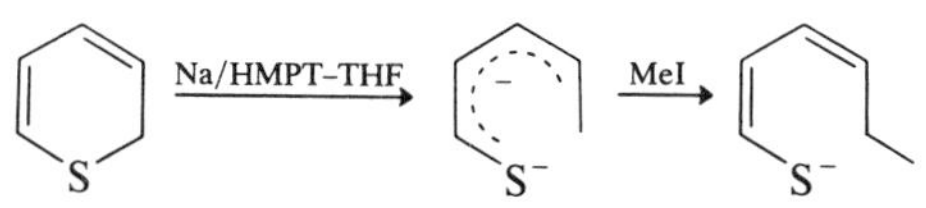

(45)

(**101**)    (**102**)

Oxidation of sulfur to the oxide or dioxide is not generally useful in simple thiopyrans and though the annelated compounds do readily afford the sulfones under a variety of conditions, it is more common to build in the oxidized sulfur from the beginning. Regeneration of the sulfide from the sulfoxide in thioxanthenes has been demonstrated with dichlorocarbene under phase transfer conditions ⟨80JOC5350⟩.

The monoxides readily undergo the Pummerer reaction, which may lead to thiopyrylium species, or 2,2'-dimers, while the sulfone series is quite stable. The acidity of the thiopyran 1,1-dioxides has been investigated and shown to be greater than that of a disulfonylmethane but less than that of an acetonyl sulfone ⟨73JCS(P2)50⟩. Interestingly, there is evidence that the anions derived from these compounds are more stable than those derived from acyclic analogues, and this has been taken as additional evidence with the unusually low-field NMR resonances of protons at C-2 and C-6, for a $6\pi$ cyclic homoconjugation in the pentadienyl carbanion fragment ⟨71JCS(B)74⟩. An analogous situation is found in $2H$-1-benzothiopyran 1,1-dioxides. Alkylation of the carbanions is well known; under quite mild conditions (*e.g.* potassium carbonate/acetone) all but dibenzo[*b,d*]thiopyran 1,1-dioxide are methylated efficiently, and the avoidance of permethylation may be difficult. With a stronger base (*e.g.* sodium methoxide/DMSO) even dibenzo[*b,d*]thiopyran 1,1-dioxide is dimethylated; the difference between the various systems is only one of degree. Condensation with aldehydes is known for the monocycles, and readily affords 2-alkylidene derivatives except when the 2-position is blocked, when the 4-position is attacked. The equivalent alkylidene derivatives are formed from the annelated systems as well (Scheme 15) ⟨73JCS(P1)163⟩.

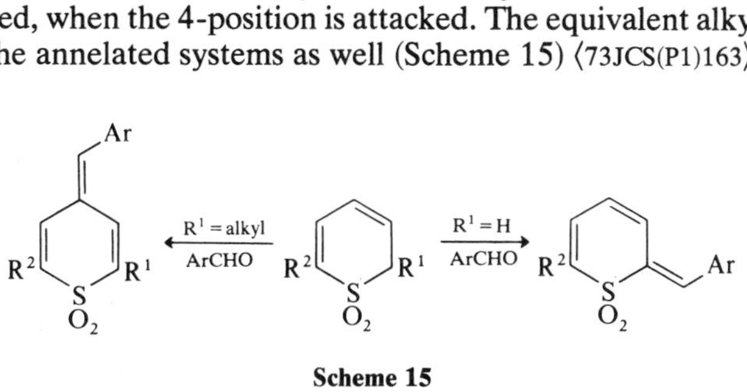

**Scheme 15**

The photolytic behaviour of the sulfone systems has been thoroughly investigated, and generally proceeds with rupture of the ring to give a sulfene, which may be trapped by solvent (especially methanol) or by other reactive centres, though photoaddition of methanol to the starting material may be a significant competing process (equations 46–48) ⟨75CJC3656, 74TL3633⟩. This behaviour may be compared with that of the parent $1H$-2-benzothiopyran, which is photolyzed to an indene with extrusion of sulfur ⟨75JOC1142⟩.

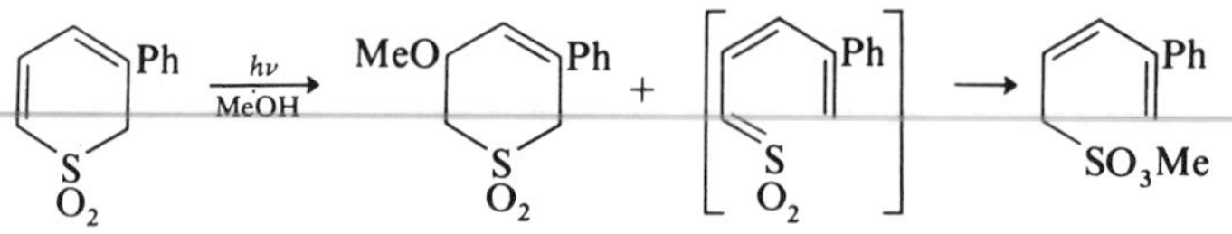

(46)

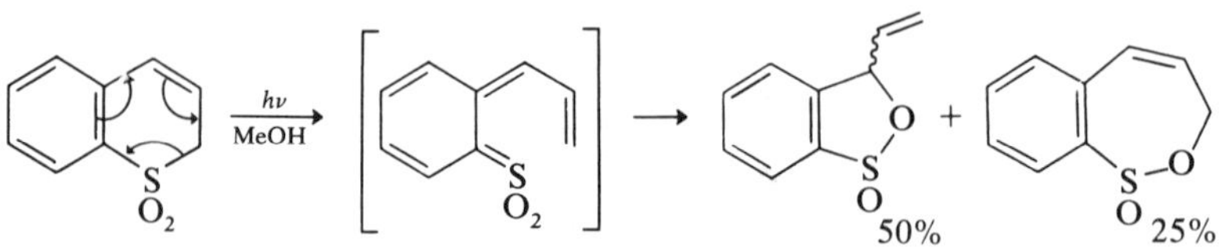

(47)

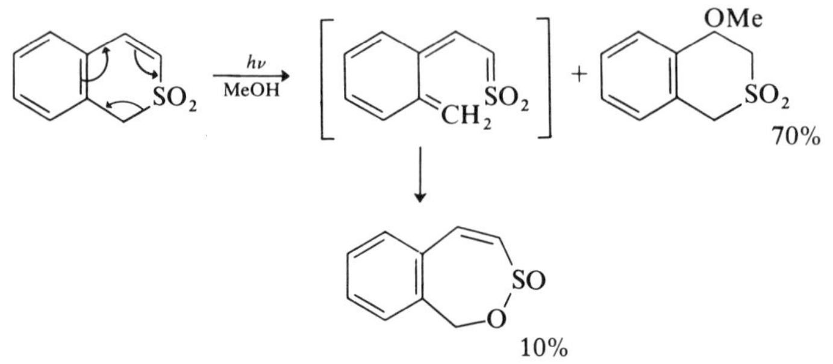

(48)

Oxidation of the ring system in thiins has also been achieved with hydride acceptors such as triphenylmethyl cations, and results in high yields of thiopyrylium salts. This is far more efficient than the disproportionation reactions discussed above, as the only byproduct is triphenylmethane. The reaction will be discussed in the section on synthesis.

Oxidation of 2H-1-benzothiopyrans with selenium dioxide leads to the thiaselenin system, together with thiocoumarins, thiochromones and other cleavage products. Careful choice of conditions can lead to 35% yield of the selenium insertion product as shown in equation (49) ⟨75T2099⟩.

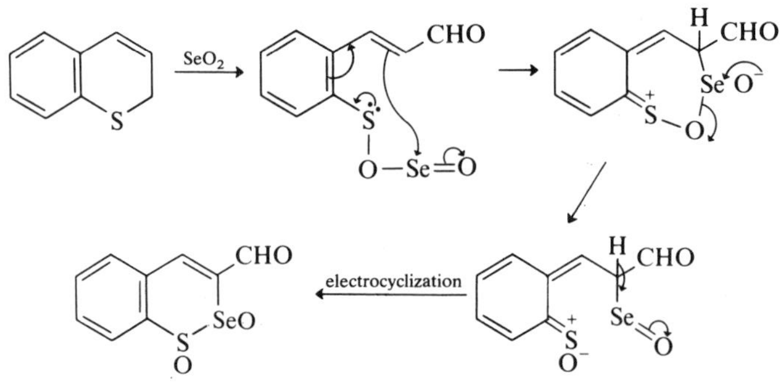

(49)

Bromination of 2H-1-benzothiins readily affords the expected *trans*-3,4-dibromide, as does bromination of the S,S-dioxide, but reaction with carbenes highlights the balance of nucleophilicity between the sulfur and the alkene. The activated double bond in 4-ethoxy-2H-1-benzothiin forms a cyclopropane with dichlorocarbene, while in the unactivated system only C—H insertion products are formed, which probably arise from the intermediate ylide (103). Scheme 16 illustrates the two modes of reaction, and the further conversions of the cyclopropyl system ⟨69JOC56, 61JA4034⟩. It is probably this facile rearrangement of such ylides which explains the lack of reports of related species such as sulfilimines, except in the dibenzannelated series of thiopyrans and even here the molecules readily rearrange. Thioxanthene (104) reacts with chloramine-T to generate the ylide (105) and the insertion product (106). Heating (105) converts it to (106) (equation 50), though the presence of a base facilitates the reaction. The relative stereochemistry of the methyl substituent in position 9 and the sulfilimine moiety has a significant influence on the rate of reaction; thus, a *trans* arrangement rearranges more readily than the corresponding *cis* configuration. Acid catalyzed rearrangements are also seen, but a major competing pathway is formation of the thioxanthone, and when there is no 9-substituent, thioxanthene. These conversions and a speculative mechanism for them are shown in equation (51) ⟨79JOC1684, 80JOC2970, 80JOC2972⟩.

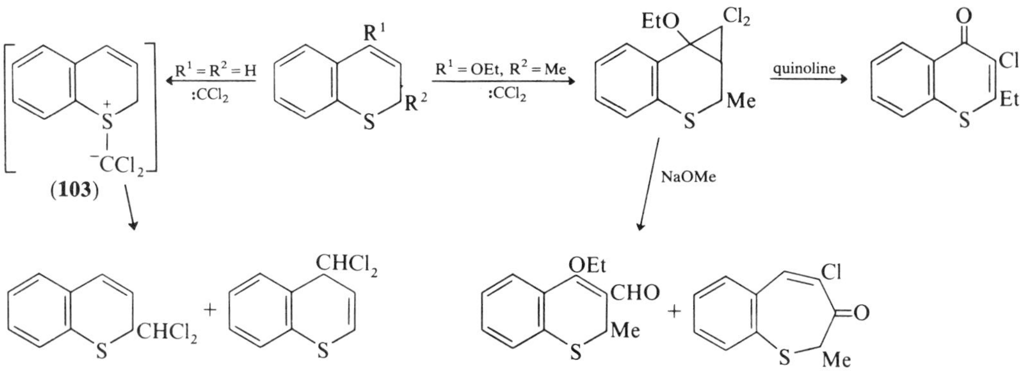

**Scheme 16**

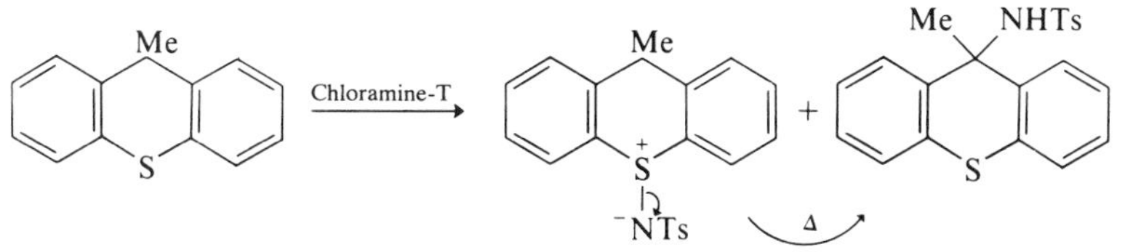

(50)

(104) (105) (106)

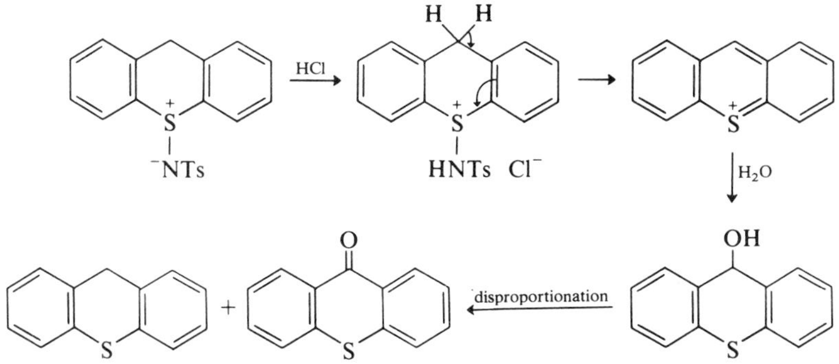

(51)

Simple carbon ylides are only known derived from thioxanthenes (the decomposition of diazomalonates in the presence of thioxanthenes can afford up to 85% yields of the ylides ⟨81JCS(P1)212⟩) and by an analogous mechanism to the isomerization of the sulfilimines, heating the carbon ylides in the presence of DBU in toluene rearranges them to the 9-substituted thioxanthenes in better than 80% yield.

A number of ylidic derivatives of thiins with a formal cyclic 6π electron system have also been synthesized and will be discussed in the section on ''thiabenzenes'' (*vide infra*).

Intramolecular carbene insertion has been utilized as a route to thiepins and benzothiepins but is seriously complicated by C—H insertion side reactions leading to alkylidene thiopyrans (Scheme 17) ⟨78TL3567, 78CL723⟩, while the generation of the 9-carbene from thioxanthene (*via* the diazo species) results in dimerization to the bisthioxanthylene compound.

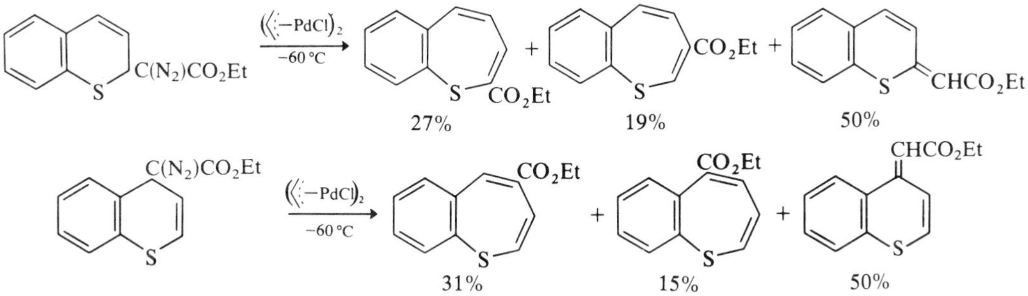

**Scheme 17**

As may be seen, the dominant feature of much of the chemistry of thiins is the tendency to form maximally conjugated ring systems such as pyrones or thiones, and alkylidene or imino substituted compounds. The species display reactions in many ways similar to those of the corresponding oxygen series, tempered with a somewhat lower formal involvement of the betaine canonical form (**107**), or its equivalent in the α-thiopyrone series. The majority of the chemistry of these systems has been derived from the 4*H*-systems, and there are many reactions wherein tautomeric 2*H*- and 4*H*-compounds equilibrate towards a 4*H*-product. Thus, heating (**108**) forms mixtures with (**109**) *via* an intermolecular process. Excellent reviews on thiopyrones, thiochromones and thiocoumarins are available ⟨75AHC(18)59, 80AHC(26)115⟩.

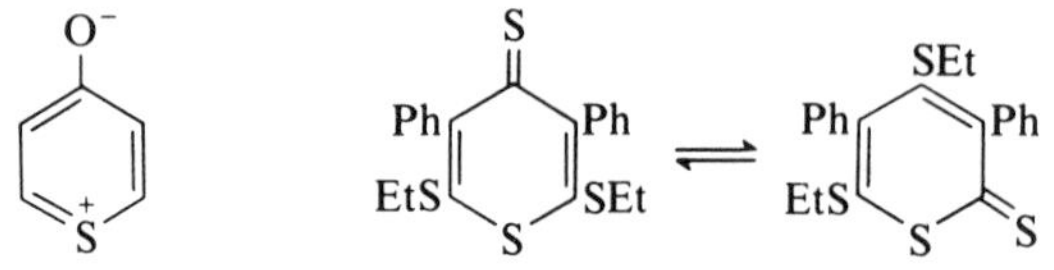

Electrophiles such as halogen attack 4-thiopyrones in position 3 while alkylation occurs on the exocyclic oxygen or sulfur atom to form a thiopyrylium cation. Nitration with nitric acid is unsuccessful because the acid protonates the ketone and the thiopyrylium produced is no longer nucleophilic (Scheme 18).

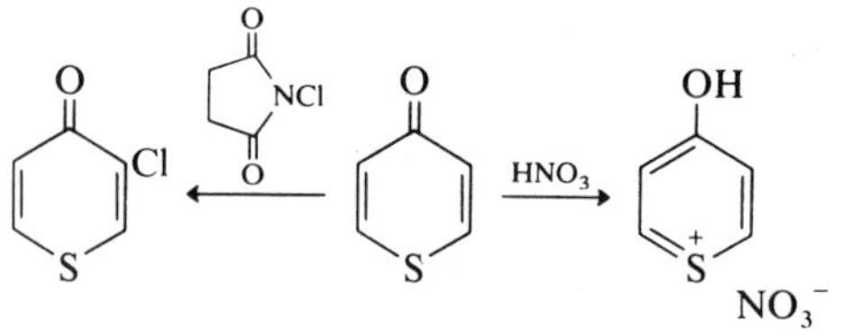

**Scheme 18**

By contrast, nucleophiles attack the thiopyrones at positions 2 and 4. Addition/elimination sequences are quite common with labile 2-substituents on the 4*H*-systems and can

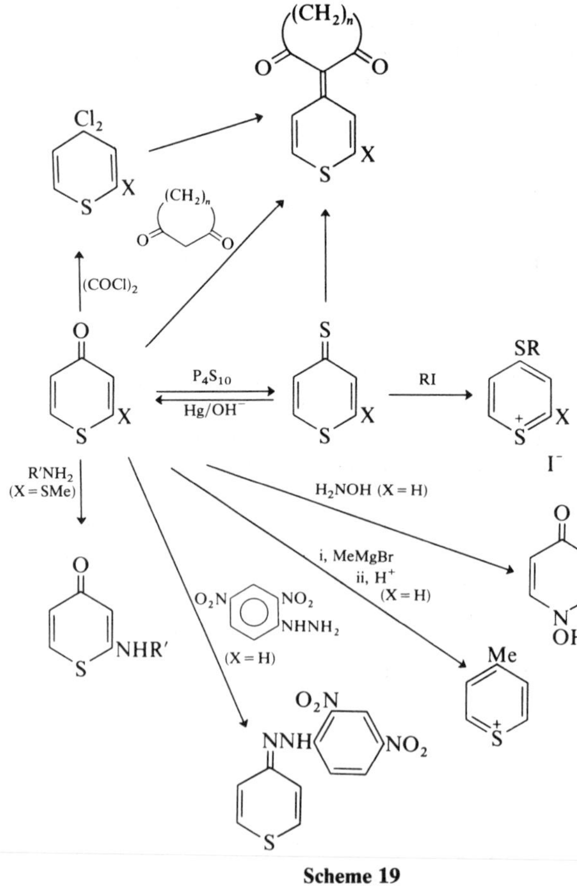

**Scheme 19**

lead to either substitution of that position or ring opening. Attack at the carbonyl group is well known. A variety of active methylene compounds can be induced to condense, though it is frequently better to convert a carbonyl group to thiocarbonyl (by treatment with $P_4S_{10}$), or *gem*-dichloro (by treatment with oxalyl chloride), when nucleophilic attack becomes even more facile. Scheme 19 shows a range of typical reactions of 4*H*-thiopyran-4-ones. Scheme 20 illustrates the common reactions of 2*H*-thiopyran-2-ones.

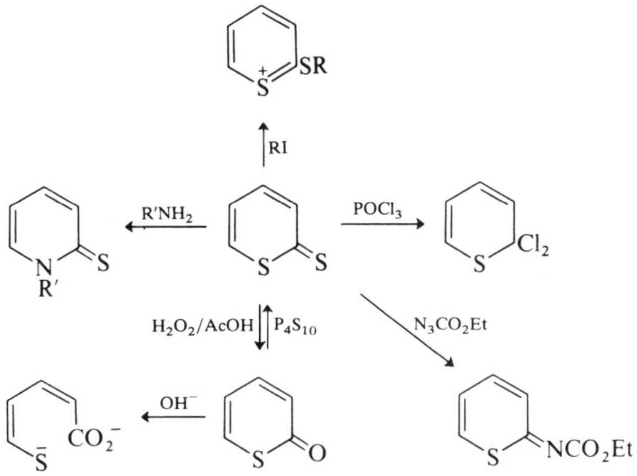

**Scheme 20**

Heating the thiopyranthiones with copper powder results in the generation of dithiopyrylenes (**110**) and (**111**), and the same products are formed on simple thermolysis of the equivalent selenoketones. These thiopyrylenes are very stable compounds, and are generally inert under a wide variety of conditions. The ease of formation of dithiopyrylenes is controlled by the substitution pattern on the ring. Substituents which reinforce the zwitterionic structures such as (**107**) retard dimerization, just as they retard condensation reactions with the carbonyl or thiocarbonyl groups. Substituents which counteract the tendency to exist as (**107**) or its equivalents, and therefore permit a greater carbonyl character in the thioketone *etc.*, facilitate the normal condensative reactions both of C=S and C=O substituents, as well as permitting the dimerization which leads to dithiopyrylenes.

Oxidation of thiopyrones under very mild conditions generally disrupts the ring completely, though hydrogen peroxide does provide the dioxide of 4*H*-thiin-4-one in low yield. It is usual to prepare such materials from the tetrahydro systems by dehydrogenation, or dehydrohalogenation, which is very efficient. The chemistry of these sulfones is quite straightforward as far as it has been investigated. Halogenation with bromine or chlorine occurs readily to give a 3-halo compound, and with an excess of halogenation agent, a 3,5-dihalo compound. If the 3,5-positions are blocked, 2-halo derivatives are only found under forcing conditions, and 2,6-dihalo species are only formed in trace amounts. Substitution of the 3-halo group with amines is a low yielding reaction, whereas the 2-halo substituents are readily replaced ⟨78LA1280⟩. The action of azide on 4*H*-thiin-4-ones is also influenced by the substitution pattern; 2,6-diphenyl-4-thiopyrone (**112**) readily produces the predicted thiazepinone (**113**). However, under identical conditions the 3,5-diphenyl isomer (**114**) affords a ring contracted product (**115**) (equations 52 and 53) ⟨78S211⟩.

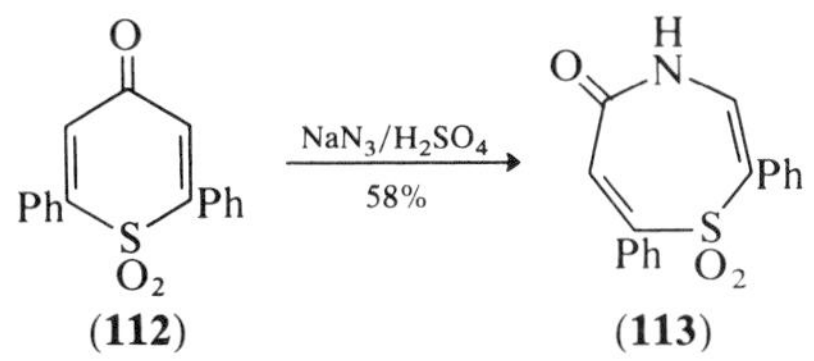

(52)

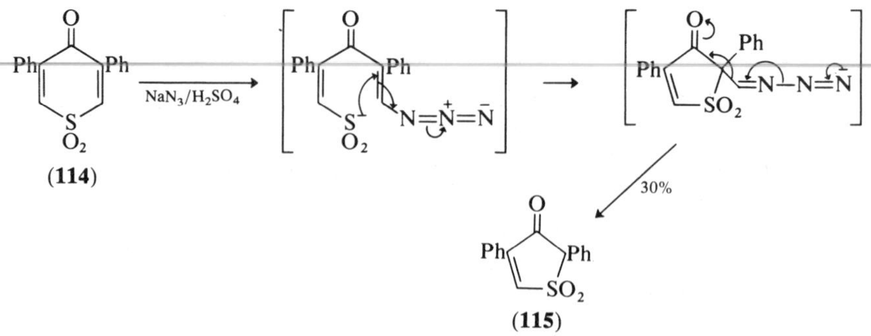

(53)

(114)                                                                    (115)

The photochemical behaviour of (112) has been compared with that of both (114) and the parent thiopyrone (Scheme 21) ⟨74JOC103⟩.

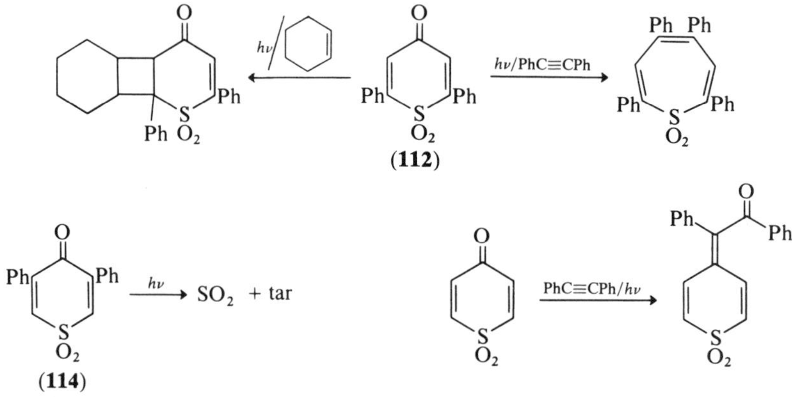

(112)

(114)

**Scheme 21**

4H-1-Benzothiopyran-4-ones (thiochromones) have a very similar reactivity pattern to the non-annelated heterocycles. The carbonyl group exhibits predictable electrophilic behaviour and condenses readily with a range of reagents, though this is facilitated by conversion to the thioketone or *gem*-dichloro derivative. Addition to the alkenic bond occurs readily giving, for example, the 2,3-dibromo adduct, which is reconverted to the thiochromone on treatment with water. Aqueous base opens the ring and cleaves the alkyl chain to afford *o*-mercaptobenzoic acids, but this is a much less facile degradation than in the analogous pyran series, and is another example of an addition/elimination sequence. Alkylation of the 4-keto or thiono group generates benzothiopyrylium salts, which can readily be converted into anhydro bases by deprotonation of 2-substituents bearing α-hydrogens (equation 54) ⟨79BCJ160⟩. Thiochromones go on to dimerize under such conditions, but the isomeric thiocoumarin, in which the methyl substituent is not adjacent to the sulfur atom, forms stable monomeric thiopyrylium species.

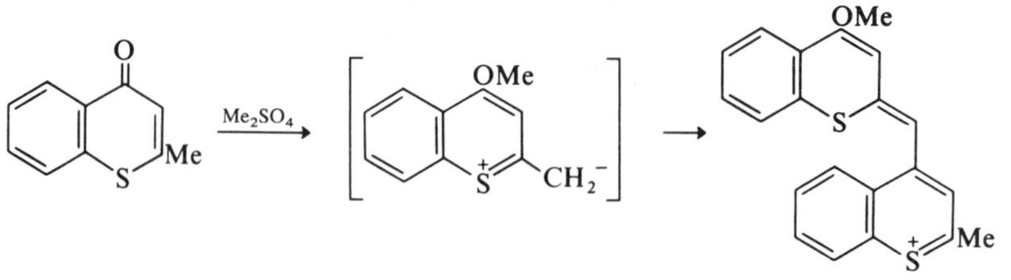

(54)

Oxidation of thiochromones does not lead to the sulfoxide or sulfone derivatives; these are better prepared from the corresponding thiochromanones by oxidation followed by bromination and dehydrobromination. The 1-oxide is an extremely reactive species and gives a Diels–Alder adduct with butadiene. The 1,1-dioxide is also very reactive. Photolysis in benzene affords a 2:1 adduct in 30% yield (116) ⟨79TL1097⟩ *via* an initial [2 + 2] addition followed by a Diels–Alder reaction. Simple [2 + 2] addition is found with cyclohexene.

2,3-Dibromothiochromone 1-oxide and its 1,1-dioxide are both well known species. The 2-halogen is easily displaced, while the 3-substituent is much less easily replaced in the sulfoxide than in the sulfone; in the sulfone, addition/elimination mechanisms are available

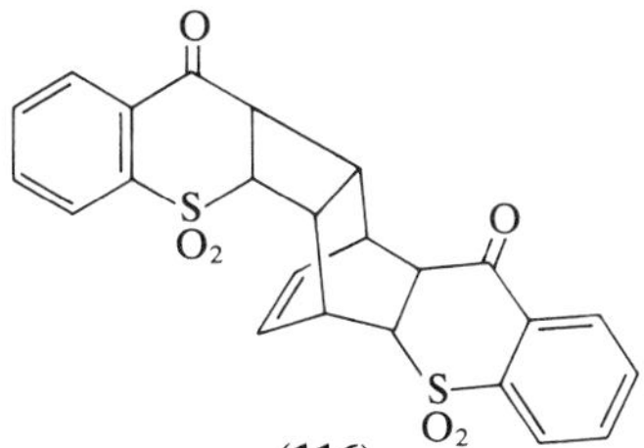

**(116)**

for displacement of both halogens. Equation (55) shows a typical reaction of the sulfone while equation (56) shows an interesting ring contraction/coupling of the sulfoxide, which leads eventually to thioindigo ⟨80JCR(S)8, 80CC559⟩.

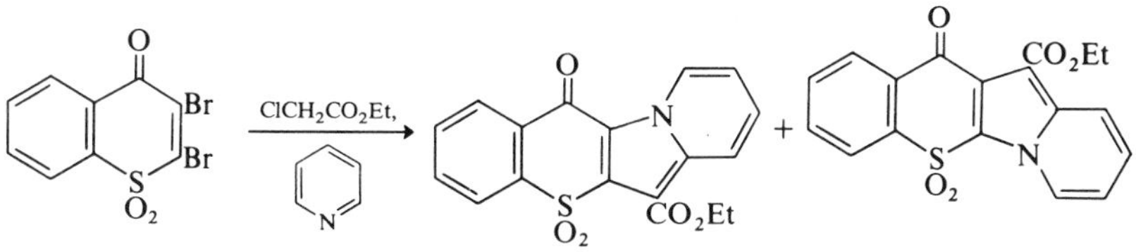

(55)

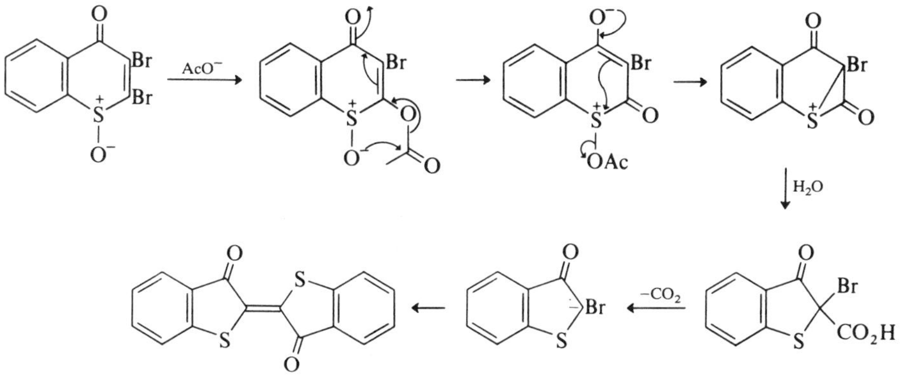

(56)

Thiocoumarins have been less investigated than thiochromones, but their chemistry may be defined generally as that of a simple substituted benzene ring and an $\alpha,\beta$-unsaturated ester, though 4-methoxy (or halo) compounds are a little more versatile ⟨80AHC(26)115⟩. The carbonyl may be replaced by thione under routine conditions with phosphorus pentasulfide, or reacted with phenylmagnesium bromide to give the carbinol, which loses water to afford a benzothiopyrylium salt. The 3,4-double bond is brominated readily to a *trans*-dihalo system, and elimination affords the 3-bromothiocoumarin. 3-Acetyl derivatives are reduced to the dihydro compound with borohydride. Substituents on the heterocycle react quite normally; the 3-formyl derivative condenses readily with *o*-diamines to afford structures such as **(117)** (equation 57). Substitution of thiocoumarins with leaving groups such as halogen at position 4 permits the ready preparation of 4-amino, thio and oxy derivatives by an addition–elimination sequence, which may be further facilitated by electron withdrawing substituents in position 3. Equation (58) shows such a substitution, which with reduction affords the diamino compound **(118)**.

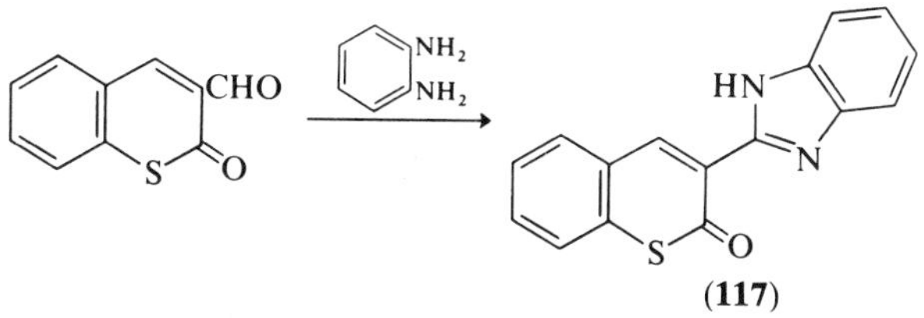

(57)

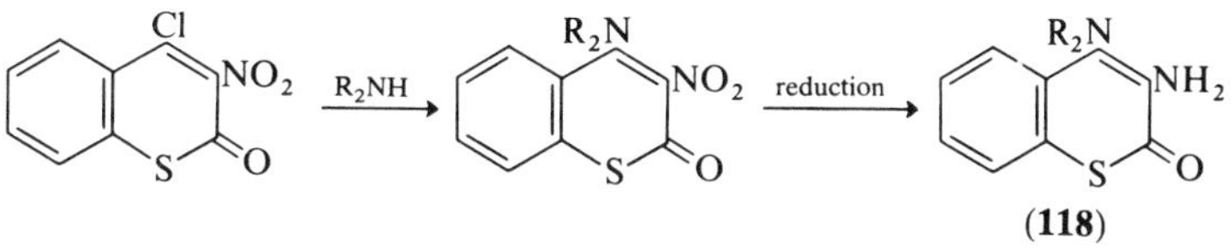

**(117)**

(58)

**(118)**

4-Hydroxythiocoumarins are, of course, enolic and react as acetoacetate analogues. Electrophiles such as Mannich reagents, aromatic aldehydes and enones react readily (Scheme 22), while acylation can be direct, or *via* a Fries-type rearrangement of 4-acyloxy precursors.

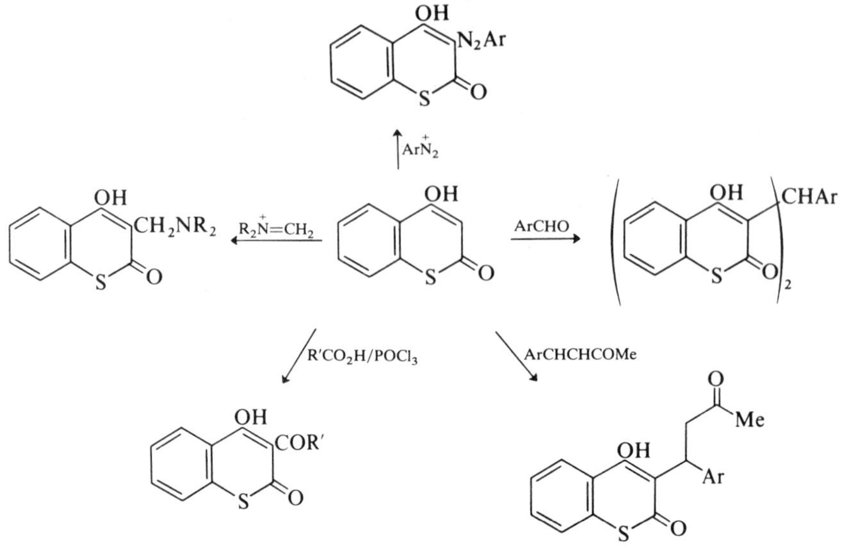

**Scheme 22**

Substituents on thiocoumarins behave in predictable fashion; for example, 3-nitro compounds afford the 3-amines on reduction, while 3-acyl derivatives readily form normal carbonyl derivatives such as phenylhydrazones, or enoic acids with malonates. These derivatives may frequently be cyclized on to position 4 of a 4-hydroxythiocoumarin to give products such as (**119**) (equation 59).

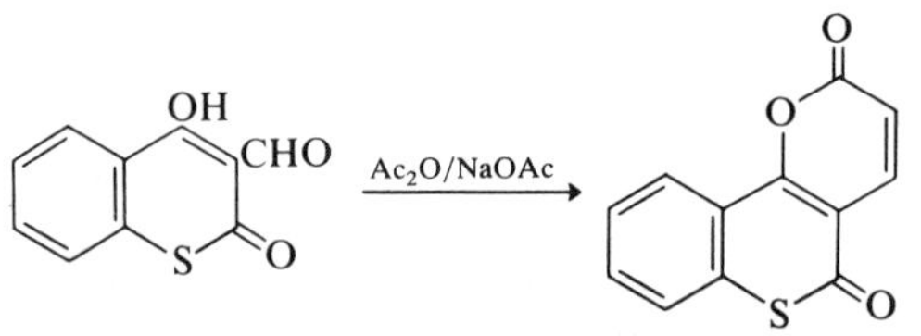

(59)

(**119**)

The chemistry of the benzene ring is also entirely predictable. Position 6 is the most reactive to electrophilic substitution; mercuration occurs there, while electron-donating substituents in position 6 direct further reaction to position 5, and the majority of substitution patterns are accessible *via* Sandmeyer reactions on suitable amino derivatives.

The photochemistry of thiocoumarins is virtually unknown, though 4-acetoxy compounds dimerize to cyclobutane species at the 3,4-double bond on irradiation.

The dibenzoannelated thiins bear much less resemblance to benzothiins than they do to benzothianes. In much of their chemistry the sulfur atom has little influence on the course of events, and in many ways the thiopyran ring may be regarded as no more than acyclic substituents on the benzene rings. The reactions of the tricycles with electrophiles may be predicted on this basis in such reactions as acylation, nitration and halogenation. Dibenzo[*b,d*]thiin has not received as much attention as thioxanthene, though some workers have considered them to have a formal resemblance to the skeleton of cannabinoids. Thioxanthene, on the other hand, has been thoroughly investigated, possibly as a result of its ready availability and significant stability. Deprotonation affords a surprisingly stable Hückel $4n$ $\pi$-system; the stability has been explained on the basis of a lack of ring current together with the ability of sulfur to stabilize a neighbouring carbanionic centre ⟨80TL3743⟩. Hydride abstraction generates the even more stable thioxanthylium cation, which will be discussed in the next section. Alkylation or imination at sulfur to give ylidic species, and their subsequent transformation, have already been described, so this part of the discussion will only consider the reactions of the non-fused carbon centre. The overwhelming tendency is for that carbon (position 10 in the xanthene numbering system) to become an $sp^2$ carbon, and preferably a carbonyl carbon. Thus, aerial oxidation, in the presence of a phase transfer catalyst, almost quantitatively converts thioxanthene to thioxanthone ⟨77TL2117⟩, though it is more usual in syntheses of the latter to build in the carbonyl function from the start. Thioxanthone reacts as a typical ketone, undergoing nucleophilic attack at the C=O grouping by Grignard reagents and phosphoranes, to give products such as (120) (equation 60) ⟨78BCJ2674⟩, or with amidines, as in equation (61) ⟨79JHC679⟩. The carbonyl is, of course, easily converted into the thione, which may be photooxidized back to the carbonyl, or photoreacted with ketenimines ⟨77TL4343⟩ to form the [2+2] spirothietanimine adduct, or with alkynic compounds, though this is a more complex reaction and depends on the nature of the alkyne (Scheme (23)) ⟨76TL3563⟩. Cumulenes afford allenyl thietanes ⟨79TL4857⟩.

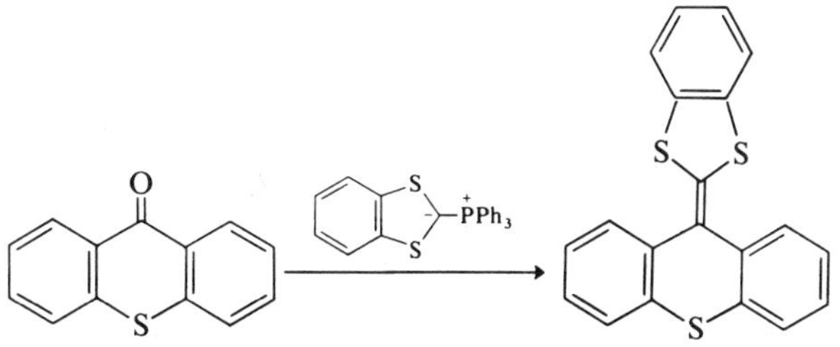

$$(60)$$

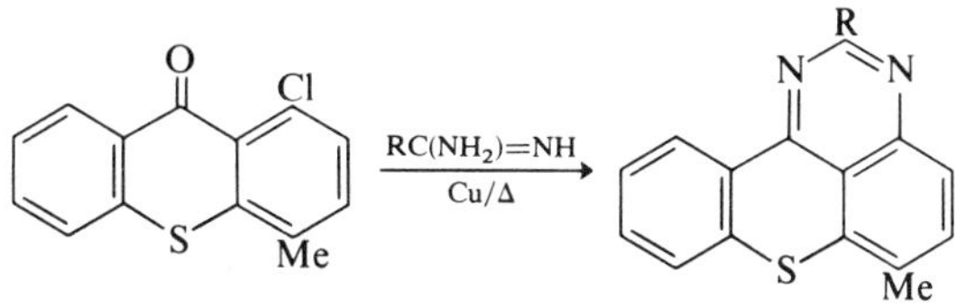

$$(61)$$

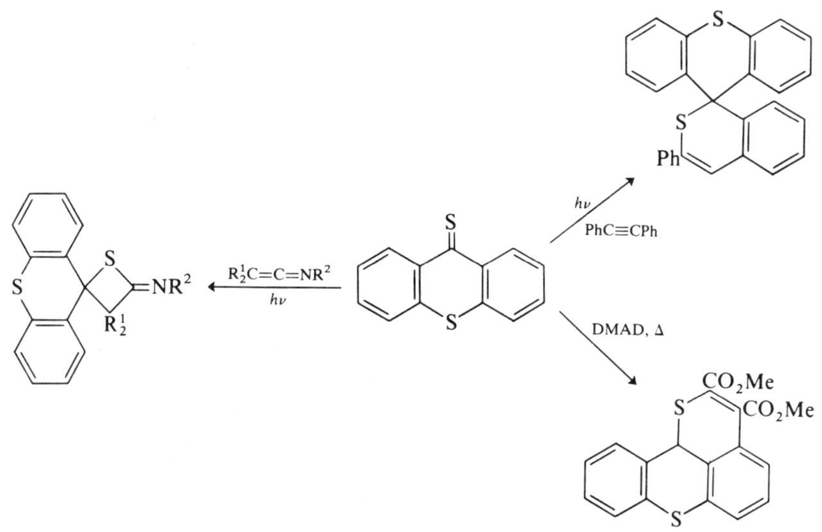

**Scheme 23**

9-Diazothioxanthene is the usual precursor for the thioxanthylidene carbene, which can be shown to have a nucleophilic character. It adds to fumaric or maleic esters to form cyclopropyl compounds, but does not react with cyclohexene ⟨78JOC3303⟩. Reaction with alkyl phosphites is a very useful means of preparing the phosphonate derivatives ⟨72JIC985⟩.

Thioxanthone 9,9-dioxide is readily prepared by oxidation of the sulfide with peroxide. On treatment with methoxide it ring opens to generate a sulfinic acid (equation 62) ⟨77SC33⟩.

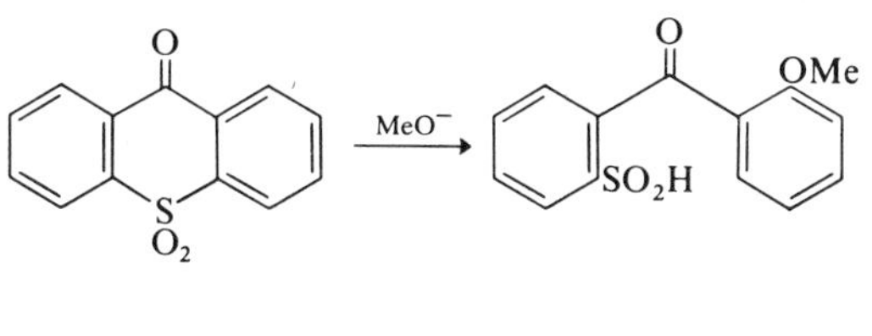

$$(62)$$

### 2.25.3.4 Reactivity of Thiopyrylium Salts

The chemistry of thiopyrylium salts and their annelated congeners is dominated by the electrophilicity of the cationic nucleus. Nucleophilic attack is formally the reverse of the principal means of synthesis (elimination of a suitably disposed group), though further reaction can lead to ring cleavage, and, in appropriate cases, to ring closure to alternative heterocycles.

NMR studies support the simplistic resonance picture of the structure of the cations; electron density is lowest in the $\alpha$ and $\gamma$ positions of the ring. In the isothiochromenylium compound the 1-position is especially electron-poor. Two consequences are immediately apparent: nucleophilic attack will occur at the $\alpha$ and $\gamma$ positions, and the salts will be stabilized by electron-donating groups at those positions. This second conclusion is well supported by NMR studies on the 2-, 3- and 4-phenyl 1-benzothiopyrylium systems, in which conjugative effects were strongest for the 2-phenyl case, less strong for the 4-phenyl case (though steric factors may be responsible here), and inoperative in the 3-phenyl compound, wherein only an inductive effect is found (73AC(R)527).

Betaines analogous to the 3-oxidopyridinium species have been prepared, and found to dimerize rapidly on liberation from the perchloric acid salts. Two dimers are formed, (**121**) and (**122**), in the ratio *ca.* 12 : 1 when R = methyl, though only (**121**) is formed when R = H. When the betaine is benzannelated, however, the mode of dimerization is different, and affords (**123**) and (**124**) in the ratio 7 : 1 (75JCS(P1)1366, 75JCS(P1)2099).

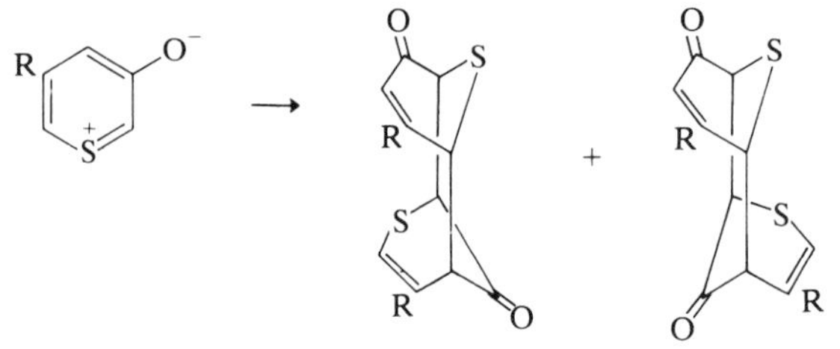

(63)

(**121**)        (**122**)

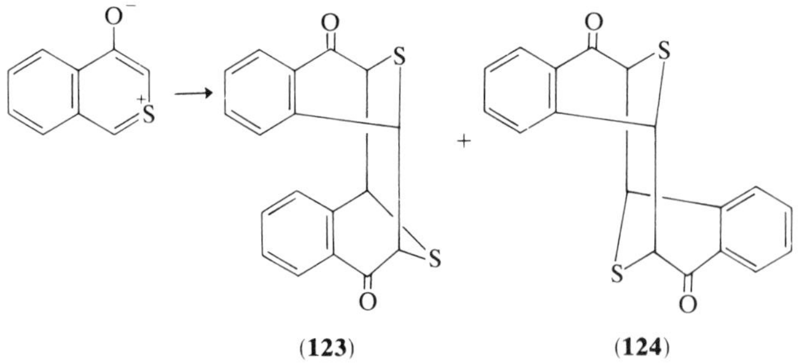

(64)

(**123**)        (**124**)

By contrast, the simple thiopyrylium salts are sufficiently stable to be used as reagents in synthetic work. The thiopyrylium cation (**125**) itself has been extensively studied, and is surprisingly stable, even in aqueous medium at pH 6. When the pH is raised it ring opens to the aldehyde (**126**). Methoxide adds to (**125**) to produce 2-methoxy-2*H*-thiopyran (**127**), but treatment with perchloric acid or hydriodic acid regenerates (**125**). Grignard reagents add to both the 2- and 4-positions to give, for example, 2-methyl-2*H*- and 4-methyl-4*H*-thiopyrans in the ratio 1 : 2 together with the unsubstituted thiopyrans, which presumably arise *via* a reduction step. Reduction with lithium aluminum hydride affords the 2*H*- and 4*H*-thiopyrans in the ratio 1 : 9.

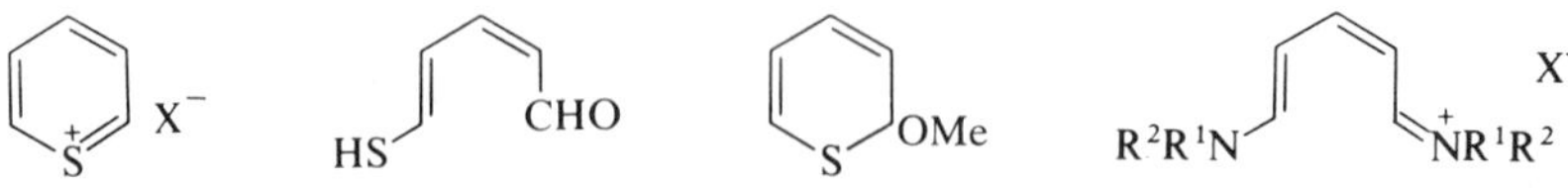

(**125**)        (**126**)        (**127**)        (**128**)

Oxidation of the salt (**125**) with active manganese dioxide gives a high yield of thiophene-2-carbaldehyde (67G397). Amines, both primary and secondary, readily cleave the ring with expulsion of hydrogen sulfide and generation of the iminium species (**128**) in up to 90%

yield ⟨73JOC3990⟩. It seems that most nucleophiles attack the parent monocycle more or less indiscriminately at the 2- and 4-positions, and the apparent specificity of amines for the 2-position is probably illusory; attack at position 4 is reversible, while attack at 2 results in 'irreversible' ring opening and the reaction is drawn in that direction. Substitution of the ring has far reaching consequences on stability, as has already been noted, and likewise on reactivity, though lithium aluminum hydride still is a nonselective reducing agent. Methoxide adds to 4-alkyl-2,6-diphenylthiopyrylium salts to give a mixture of the 4-methoxy-4*H*- and 2-methoxy-2*H*-thiopyrans under kinetic control; the mixture then equilibrates toward the thermodynamically favoured 2*H*-system. With primary aliphatic amines the ring is opened, but unlike the unsubstituted system recloses to a pyridinium salt (equation 65). However, this is a less facile reaction than in the case of the oxygen heterocyclic analogue. Aromatic amines do not react unless there is a vacant α-position; then they add to the cation *via* carbon, and the initial product is reoxidized by more cation acting as a hydride abstractor; novel dyestuffs (**129**) are obtained (equation 66) ⟨80JCS(P1)1345⟩.

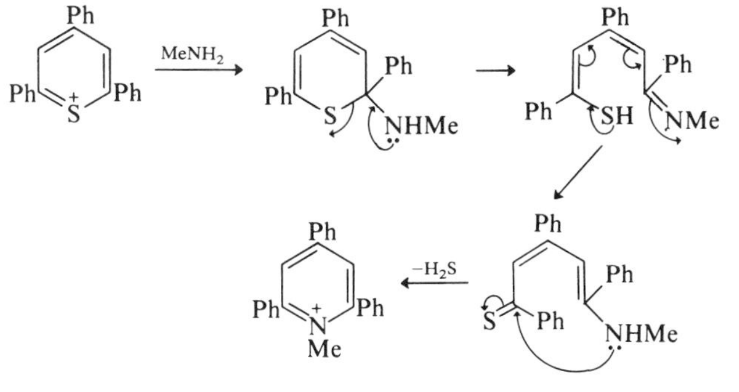

(65)

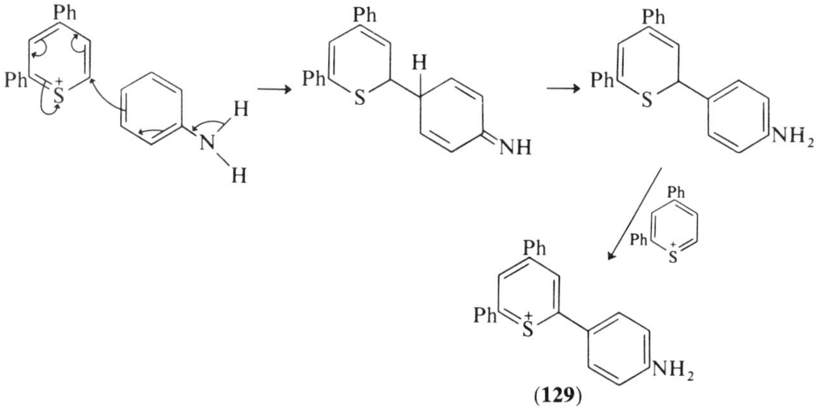

(66)

(**129**)

Azide, sulfonium ylides and *N*-aminopyridinium ylides are also known to add to the 2-position of substituted thiopyrylium compounds (presumably *via* reversible addition/elimination sequences, as for methoxide) and cause ring opening followed by ring closure with loss of sulfur. Scheme 24 shows these conversions ⟨80NKK604⟩.

Carbon nucleophiles in general react at the least hindered position. Thus 1,3-ketones add to 2,6-diphenylthiopyrylium salts to give the expected 4-substituted 4*H*-products, but if the diketone is part of a six-membered ring it is further oxidized *in situ* to the thiopyranylidene product, while the other adducts require treatment with ferricyanide to convert them to the unsaturated products ⟨76CB1549⟩.

Substitution of the ring in positions 2 or 4 with leaving groups has an even more profound effect on the chemistry of the system. Thus, the action of β-diketones on 4-chloro-2,6-diphenylthiopyrylium salts affords the 4-thiopyranylidene compounds directly from all substrates. Thiophenol, aniline, phenoxide and methoxide react with 4-chlorothiopyrylium salts by addition/elimination to give the 4-anilinothiopyrylium *etc.* in high yields. Surprisingly, other amines and Grignard reagents attack 4-chlorothiopyrylium salts at position 2 and open the ring (equation 67). It was suggested that this may be rationalized on the hard/soft acid and base principle, the 2-position being harder than the 4, though the substitution reaction seen with methoxide would appear to be contrary to this suggestion ⟨75T2669⟩.

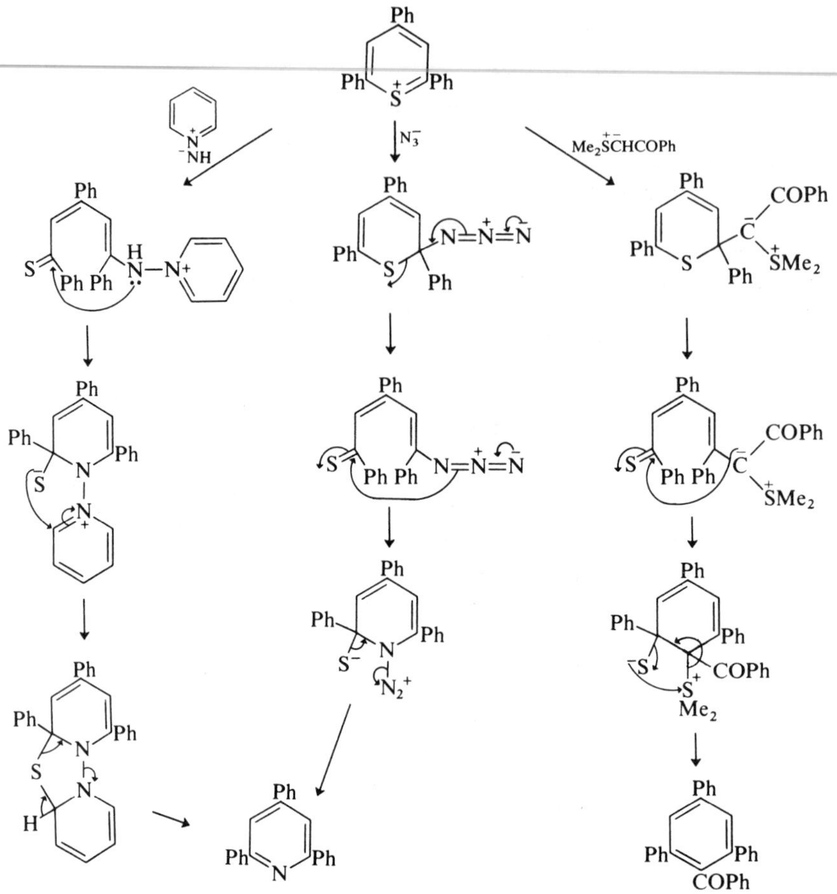

**Scheme 24**

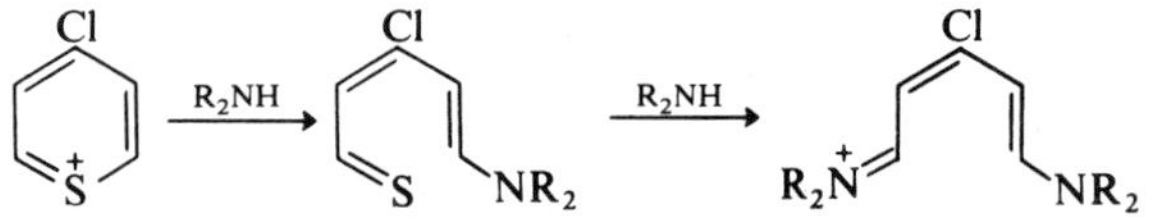

(67)

A 2-methylthio group is also easily displaced by carbanions, and in this case elimination of the sulfur from the ring with formation of a benzene ring does not occur (equation 68) ⟨74BSF1196⟩.

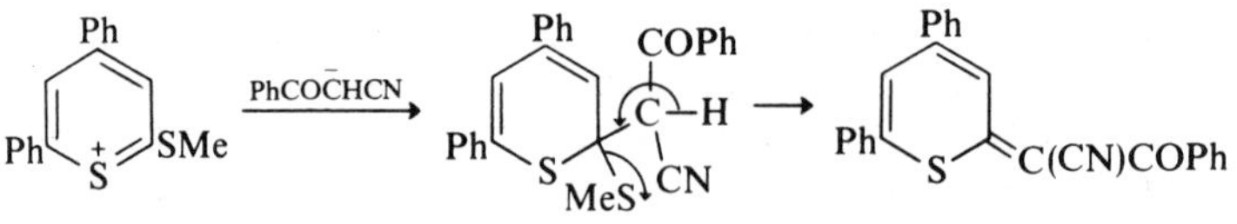

(68)

Methyl substituents at positions 2 and 4 of a thiopyrylium compound are usefully acidic, as may be expected, and react readily with electrophiles such as the carbonyl group of thiopyrones, to produce extended conjugated systems (equation 69) ⟨78AP236⟩. Compounds with even longer conjugated linking chains, *e.g.* (**130**), may be made by analogous methods, and are important dyestuffs. In strongly acidic media protonation occurs on the exocyclic carbon nearest the non-cationic ring.

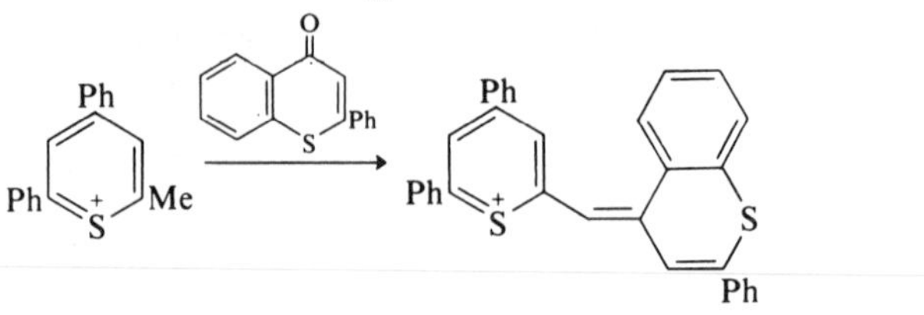

(69)

(130)

Fusion of the benzene ring to the thiopyrylium nucleus in the 2,3-position has little effect on the general reactivity of the system; nucleophiles add readily at the 2- and 4-positions, but ring opening is not seen, presumably because the sulfur cannot become a thioketone group. Phosphites add readily to 2-methylbenzothiopyrylium salts to give benzothiopyranyl-4-phosphonates which may lead into the benzothiopyranylidene system by Wittig reaction ⟨77S862⟩. Annelation of benzene across positions 3,4 also has relatively little effect on general reactivity. As in the isothiochromones the majority of reactions occur at position 1.

Thioxanthylium compounds have received more attention, possibly because of their ease of preparation, and also because of the high regioselectivity exhibited in their reactions. Phosphites add to position 9 and form phosphonates, but nucleophiles such as thiophenols, alcohols and amines add reversibly, and the final position of equilibrium depends on the stability of the anion of the nucleophile. An investigation of the reaction of thiophenol with thioxanthylium salts found thioxanthone and thioxanthene as the major products in approximately equal proportions. It is in their reactions with alkyllithiums that thioxanthylium salts have been most fully investigated in connection with studies on 'thiabenzenes' (*vide infra*). Treatment of a 9-phenylthioxanthylium salt with phenyllithium affords a mixture of products including 9-phenylthioxanthene, 9,9-diphenylthioxanthene, 3,9-diphenylthioxanthene, 3,9,9-triphenylthioxanthene, 4,9,9-triphenylthioxanthene and a range of dimers. In the light of this variety of products it is clear that not only are simple nucleophilic additions occurring, but also that radical species must be involved ⟨79TL1603⟩. An interesting feature of thioxanthene radicals is their apparent reluctance to dimerize directly. Peroxide links are formed preferentially, whereas thiopyrylium is readily reduced with zinc to the radical which couples to give a 4,4'-bithiopyran.

### 2.25.3.5 Reactivity of Thiabenzenes

There have been reports that alkyllithium nucleophiles added selectively to thiopyrylium salts by attack at sulfur to generate tetravalent sulfur-containing aromatic species to which the name thiabenzenes was applied. Some of these products were claimed to be remarkably stable, resisting degradation by a variety of acidic conditions, but reinvestigation has now shown these materials to be amorphous oligomeric mixtures. There is some evidence for the transient formation of the monomeric compounds by this method, but of far greater utility for their generation is the method of deprotonation of a thiopyranium salt. Thiabenzenes, although electronically neutral, do not possess an 'aromatic' ring current; they are betaines, and their chemistry is readily interpreted in the light of that fact. The negative charge density is greatest on the carbons α and γ to sulfur, and treatment with acids readily reprotonates them to mixtures of the 2H- and 4H-thiopyranium salts. The simple monocyclic species are quite unstable and are only known as intermediates which rapidly rearrange by competing 1,2- and 1,4-intramolecular shifts of the sulfur substituent. Substitution, especially by electron-withdrawing substituents on positions 2, 4 or 6, and with electron-donating substituents on sulfur, greatly increases stability; thus, the half lives for decomposition of (131) and (132) are respectively 8.2 h and 0.35 h in benzene at 22 °C ⟨74JA5650⟩. Benzannelation increases stability generally, and a number of zwitterionic compounds such as (133) and (134) have been prepared and found to be quite stable, though (134) does slowly rearrange to the 9,9-disubstituted thioxanthene on standing ⟨74CL1101, 80JOC2468⟩. These zwitterionic materials react with electrophiles in a predictable fashion (equation 70), although (134) and dimethyl acetylenedicarboxylate does produce the two surprising products (135b) and (135c) as well as the expected (135a), presumably *via* the intermediate

(131)          (132)          (133)          (134)

maleate (vinyl) carbanion attacking the *S*-methyl substituent (equation 71). Aerial oxygen reacts with (**134**) to form thioxanthone.

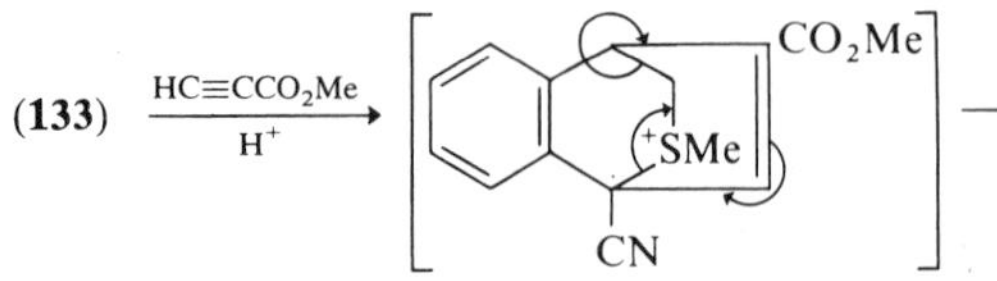

$$(133) \xrightarrow[\text{H}^+]{\text{HC}\equiv\text{CCO}_2\text{Me}} \left[ \begin{array}{c} \text{CO}_2\text{Me} \\ \overset{+}{\text{SMe}} \\ \text{CN} \end{array} \right] \longrightarrow \begin{array}{c} \text{CO}_2\text{Me} \\ \text{CN} \end{array} \tag{70}$$

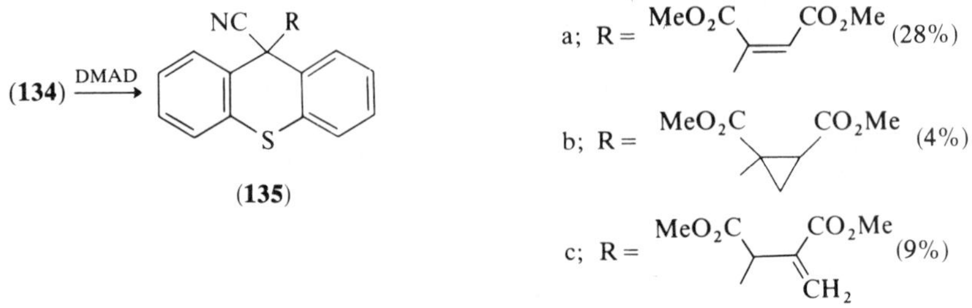

$$(134) \xrightarrow{\text{DMAD}} (135) \qquad \begin{array}{l} \text{a; } R = \overset{\text{MeO}_2\text{C}}{\diagup}\diagdown^{\text{CO}_2\text{Me}} \quad (28\%) \\[2mm] \text{b; } R = \overset{\text{MeO}_2\text{C}}{\triangle}^{\text{CO}_2\text{Me}} \quad (4\%) \\[2mm] \text{c; } R = \overset{\text{MeO}_2\text{C}}{\diagup}\diagdown^{\text{CO}_2\text{Me}}_{\text{CH}_2} \quad (9\%) \end{array} \tag{71}$$

A more complex picture is seen with the stable 'thianaphthalene' (**136**) and different electrophiles (Scheme 25) ⟨76H(5)413⟩.

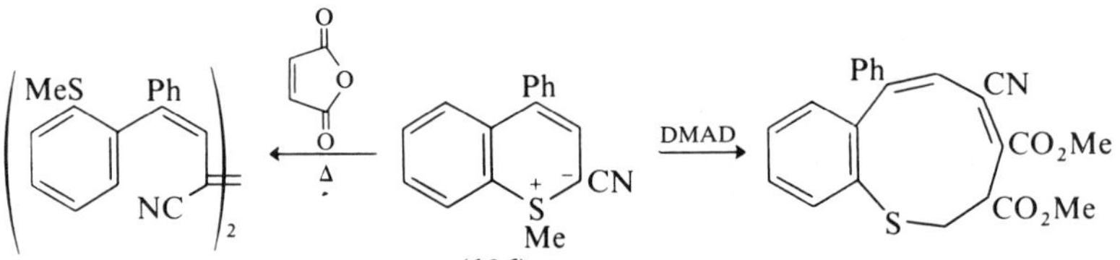

**Scheme 25**

The sulfoxide derivatives of some 'thiabenzenes' have been prepared; these molecules are much more stable than the parents, and may be subjected to a wider range of reactions (Scheme 26) ⟨74JOC3519⟩. It is apparent that the ylidic character of these molecules is reduced *vis-à-vis* the non-oxygenated species. The lack of reaction with acetyl chloride should be compared with the effect of trimethylsilyl chloride in equation (72) ⟨74JA6119⟩. The mechanism of this reductive dealkylation has not been explored.

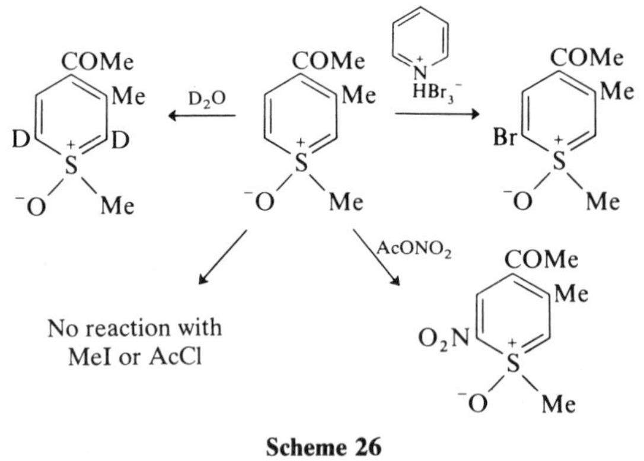

**Scheme 26**

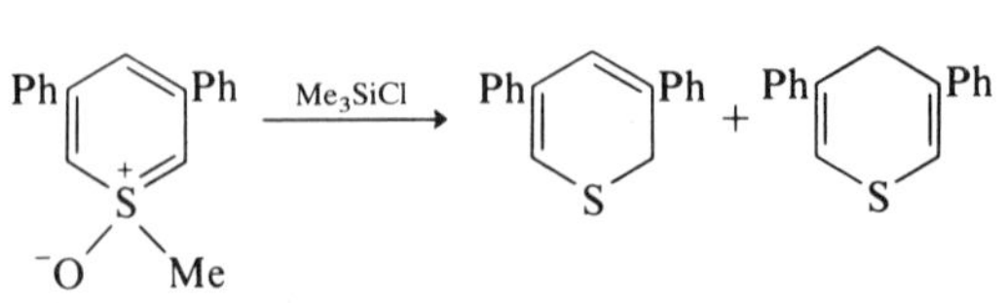

$$\begin{array}{c} \text{Ph} \\ \overset{+}{\text{S}} \\ \bar{\text{O}} \quad \text{Me} \end{array} \text{Ph} \xrightarrow{\text{Me}_3\text{SiCl}} \begin{array}{c} \text{Ph} \\ \text{S} \end{array} \text{Ph} + \begin{array}{c} \text{Ph} \\ \text{S} \end{array} \text{Ph} \tag{72}$$

## 2.25.4 SYNTHESIS OF THIOPYRANS

In view of the paucity of naturally occurring thiopyrans upon which to base the construction of extensively derivatized systems, it is perhaps not surprising that many are synthesized by interconversions of oxidation levels either by hydrogenation/dehydrogenation or by elimination reactions. Such reactions have been referred to above. *De novo* ring synthesis is, of course, also very important for the same reason, and may be classified into six categories:

(a) formation of a bond $\alpha$ to sulfur;
(b) formation of a bond $\beta$ to sulfur;
(c) formation of a bond $\gamma$ to sulfur;
(d) formation of two bonds from [5+1] fragments;
(e) formation of two bonds from [4+2] fragments;
(f) formation of two bonds from [3+3] fragments.

There do not appear to be any examples of the ring being formed by the linking of three fragments.

Thiopyran preparation by conversion of other heterocycles is known, but is rather restricted in scope.

### 2.25.4.1 Synthesis of Tetrahydrothiopyrans

#### 2.25.4.1.1 De novo ring synthesis

Ring formation by construction of a bond $\alpha$ to sulfur is a very efficient process in simple cases, as the nucleophilicity of the heteroatom is quite sufficient to displace a suitably disposed halogen atom. In more complex molecules, especially with halogen at a tertiary centre, elimination of hydrogen halide is a seriously competing process. More modern approaches avoid this problem by the generation of a formal positive charge on the active centre carbon by other methods. Thermal elimination with the allylic iodide (equation 73) ⟨81JOC5451⟩ or the interaction of a sulfenyl chloride with an alkene (equation 74) ⟨75TL3923⟩ efficiently form the six-membered ring and are compatible with a range of functionality.

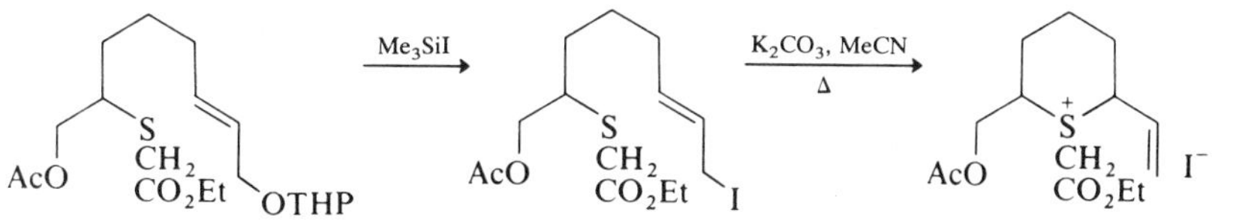

(73)

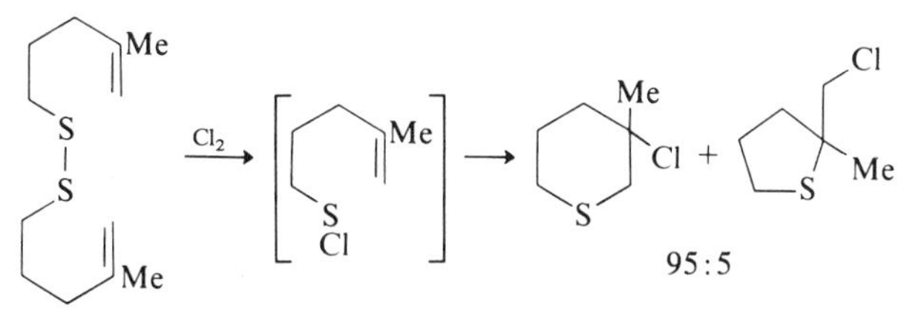

(74)

Formation of $\beta$ C—C bonds is also known, but generally requires the generation of a carbanion $\alpha$ to sulfur and its interaction with a carbonyl centre. Dieckmann cyclization of (136) readily affords ethyl 3-oxo-3,4,5,6-tetrahydro-2*H*-thiin-2-carboxylate, while the presence of a sulfone group renders a methyl substituent sufficiently acidic to react as in equation (75) ⟨78JCS(P1)1321⟩. Sulfur alone is apparently not sufficiently activating for this method to be useful for the preparation of unfunctionalized molecules.

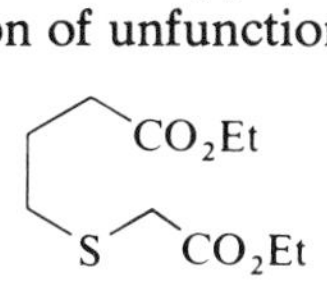

(136)

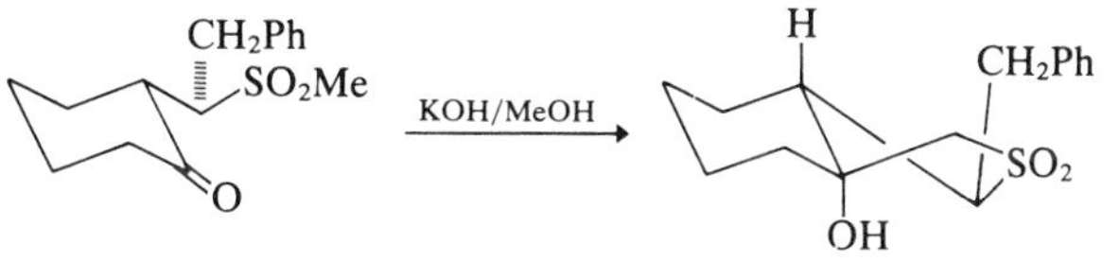

(75)

Formation of $\gamma$ C—C bonds by Dieckmann cyclization has been known for many years. Diethyl $\beta\beta$-thiodipropionate readily affords 4-oxo-3,4,5,6-tetrahydro-2*H*-thiin-3-carboxylate ⟨27JCS194⟩, which on decarboxylation affords the 4-thianone from which many thian derivatives may be made. The dione (**137**) and 4-substituted congeners are also made in an analogous manner ⟨77JOC1163⟩ and they too are the starting points for many derivatives.

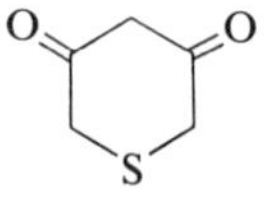

(**137**)

Friedel–Crafts cyclization of alkenyl acid chlorides affords mixtures of tetrahydro- and dihydro-thiopyrans (equation 76) ⟨80JHC289⟩.

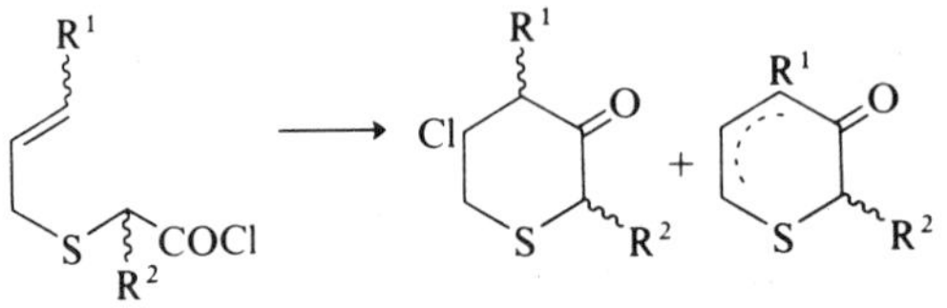

(76)

Possibly the most common entry into tetrahydrothiopyrans uses the fusion of [5 + 1] fragments. The earliest preparation of the parent ring was by treatment of pentamethylene diiodide with potassium sulfide ⟨10CB545⟩, and the method is still being improved on today ⟨79IZV1077⟩, though the principal interest now lies in the construction of the suitably substituted pentamethylene unit. The range of substituents permissible is of course dictated by the harshness of the cyclization conditions, and much effort has been applied to investigating the use of good leaving groups such as tosylate and mesylate to moderate the severity of the treatment necessary.

Much milder conditions are used in the 'double Michael addition' approach, in which a divinyl ketone is condensed with hydrogen sulfide in mildly basic medium (equation 77) ⟨77JOC2777⟩. Enol acetates ($R^1 = MeCO_2$) may be used, and the product obtained then contains a 2-mercapto function ($R^1 = SH$; see also equation 82) (59% yield). Although this is a very versatile synthesis, its biggest drawback is the lability of simple divinyl ketones, and phenyl substitution at position 2 is frequently used to overcome this.

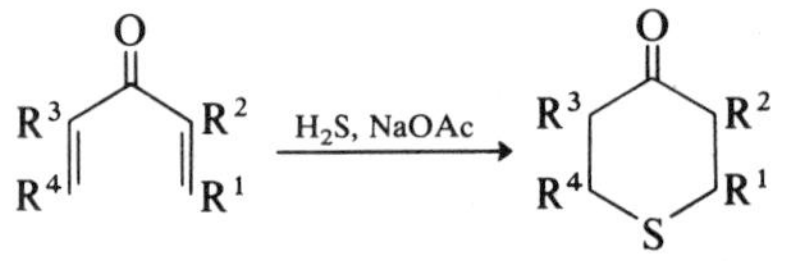

(77)

The disproportionation of thiopyrans into tetrahydrothiins and thiopyrylium salts has been described earlier, and the condensation of hydrogen sulfide with 1,5-diketones in the presence of acids such as perchloric acid presumably proceeds *via* such a disproportionation step to generate thianes ⟨75MI22501⟩.

### 2.25.4.1.2 Ring transformations leading to thianes

Tetrahydrothiopyrans may be made by sulfuration of simple alkyl tetrahydropyrans, but very stringent conditions (300 °C, 7.5% $ThO_2$–$Al_2O_3$ catalyst) are necessary to bring about reaction with hydrogen sulfide ⟨62ZOB1822⟩; the method has obvious limitations. Milder conditions may be used to interconvert 5-mercaptopyranoside sugars to the hydroxy thiopyranoside analogue ⟨79MI22503⟩.

Thiazolium salts may be isomerized to substituted thianes (equation 78) ⟨80H(14)33⟩ and the method appears to offer a versatile entry into a range of substituted systems.

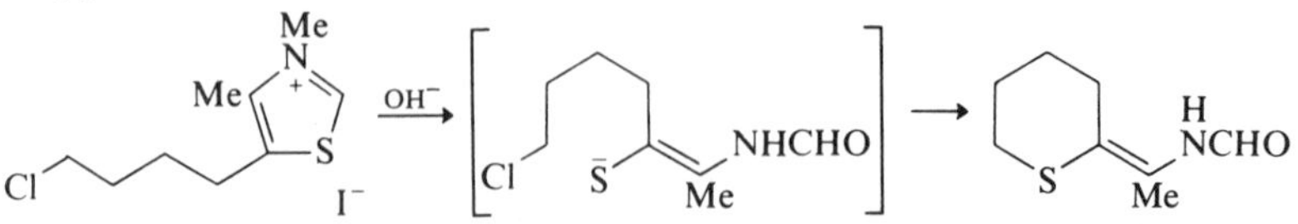

(78)

### 2.25.4.1.3 Synthesis of thiane derivatives

The parent heterocycle is usually prepared by interaction of a sulfide and a pentamethylene derivative, as described above, and various alkyl and aryl substituted analogues are also normally made by this route, though 2-alkyl (*etc.*) systems are better prepared from the 2-chloro compound (**138**) by Grignard coupling or similar reactions (equation 79) ⟨77JA2337⟩.

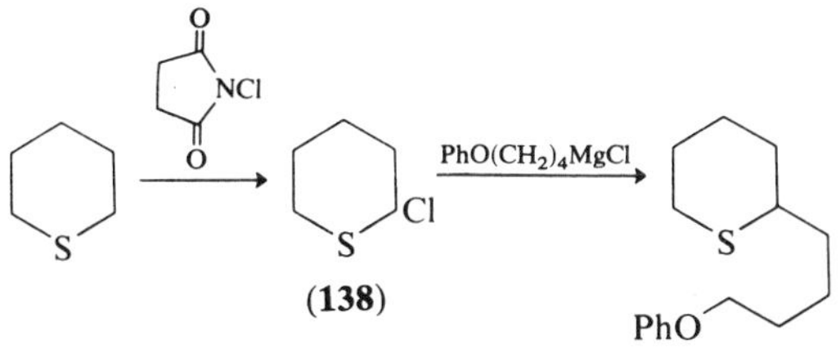

(79)

Usually, 3- and 4-substituted thianes are prepared from the 3- and 4-oxothians by the common techniques applicable to alicyclic chemistry. Reduction affords alcohols which may be etherified or halogenated, while oximation and reduction produces the amino derivatives, which are also accessible *via* the halo compounds. 2-Alkoxy and 2-alkylthio compounds are made by acid catalyzed addition of alcohols and thiols to 3,4-dihydro-2*H*-thiopyran ⟨75MI22502⟩ in a reaction analogous to the use of dihydropyran for protection of alcohols as THP ethers.

Sulfoxide and sulfone derivatives are best prepared by oxidation of a sulfide precursor, though specific 2-substitution may be achieved *via* metallation of the oxidized system, as outlined in Section 2.25.3.1.

## 2.25.4.2 Synthesis of Dihydrothiopyrans

### 2.25.4.2.1 De novo ring synthesis

There are few examples of the construction of simple dihydrothiopyrans by the formation of a bond α to the heteroatom. This is perhaps a reflection of the general inaccessibility of suitably disposed halothiols *etc.*, though in particular cases highly efficient syntheses are known, especially for benzannelated systems. The reader is referred to the reviews of thiochromans and isothiochromans, and thiocoumarins ⟨75AHC(18)59, 80AHC(26)115⟩ for coverage of these areas. The presence of electron-withdrawing groups permits the construction of α S—C bonds by Michael addition, as shown in equation (80) ⟨77S472⟩, though the intermediate dithiocarboxylic acid or thioamide may be diverted to a variety of other products depending on the conditions.

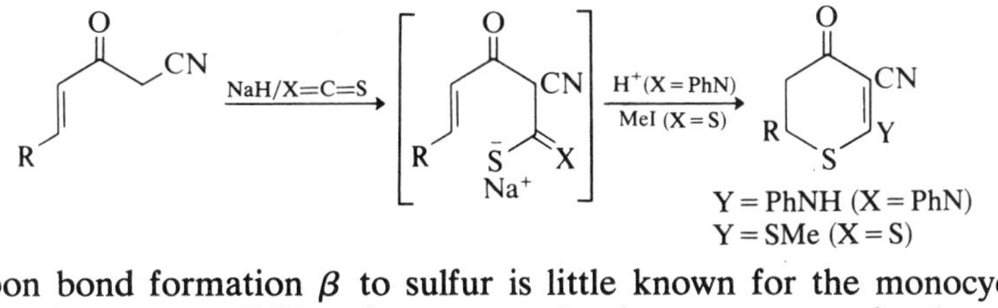

(80)

Carbon–carbon bond formation β to sulfur is little known for the monocycles, again perhaps because of poor accessibility of precursors, but is very common for the preparation of thiochromanones. Frequently, a dicarboxylic acid or ester is cyclized under basic conditions, and decarboxylation affords the keto compound, from which the unsubstituted system may readily be obtained ⟨76JCS(P1)749⟩. A more modern reaction is shown in equation (81) ⟨73TL4315⟩.

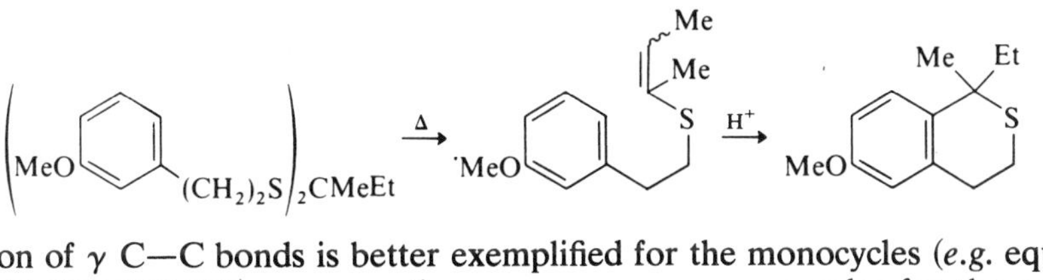

(81)

Formation of γ C—C bonds is better exemplified for the monocycles (*e.g.* equation 76) ⟨77EGP127515, 74JAP7426274⟩; even so there are many more examples for the preparation of benzo fused analogues by this method and the reader is referred to the reviews quoted above.

Probably the more widely used approach to dihydrothiopyrans is the addition of sulfide to a five-carbon fragment. A variety of products is accessible by this means. Equations (82) and (83) are Michael additions, and inevitably the products contain carbonyl groups which may be turned to good use ⟨81JOC4604, 81JA4597⟩. Equation (84) illustrates the other common variant on this scheme, condensation with a diketone. Mild conditions are essential to preclude disproportionation to thianes and thiopyrylium salts ⟨77ZOR186⟩.

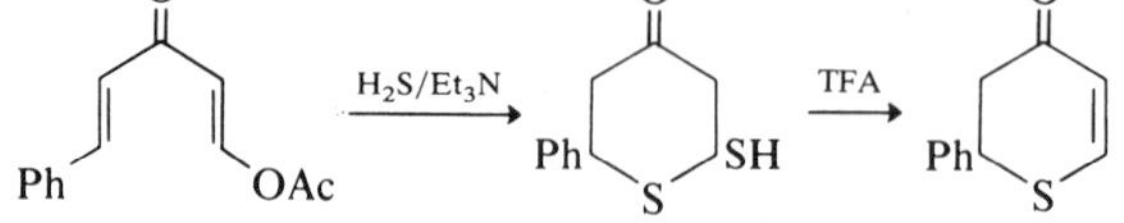

$$
\text{(82)}
$$

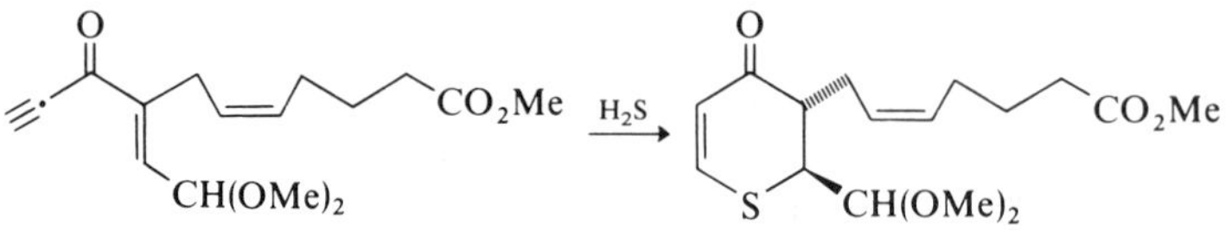

$$
\text{(83)}
$$

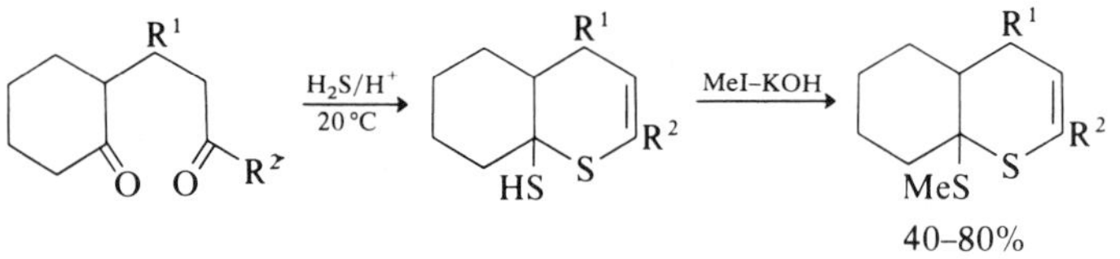

$$
\text{(84)}
$$

40–80%

Recent years have seen extensive investigations of the formation of a variety of dihydro systems by combination of four-atom and two-atom fragments, and these approaches may be classified for simplicity into 'polar' and 'concerted' mechanisms. In the first group are Michael addition/aldol condensations and analogous mechanisms. Equation (85) shows the use of $\beta$-mercapto aldehydes and enals. Regioselectivity is frequently a problem in cyclizations of this type due in part to carbanion equilibration, but under phase transfer conditions this may be mitigated and useful yields of heterocycles obtained ⟨77JOC2123⟩. A closely related cyclization of *o*-thiosalicylic esters with acrylonitrile is shown in equation (86) ⟨73ZOR775⟩.

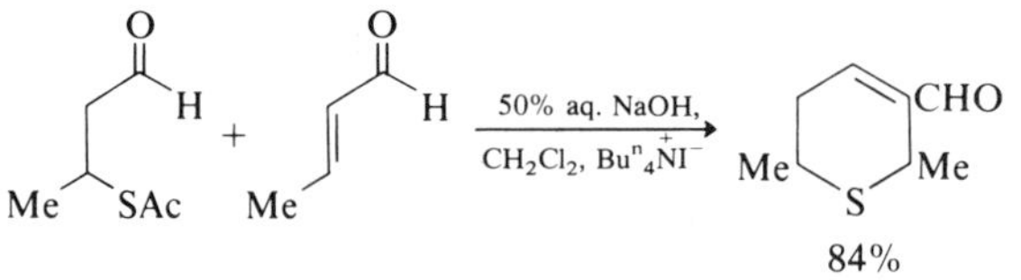

$$
\text{(85)}
$$

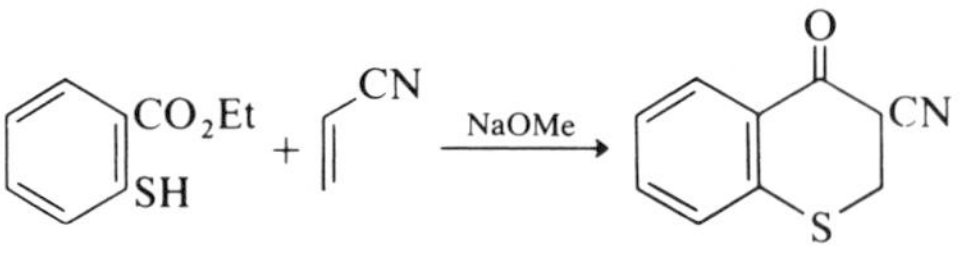

84%

$$
\text{(86)}
$$

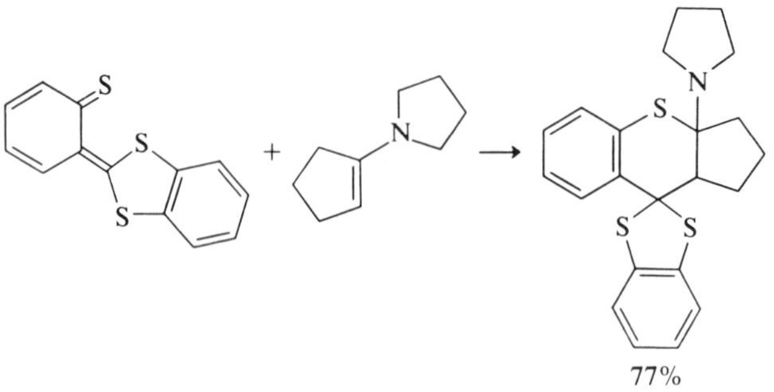

$$
\text{(87)}
$$

77%

A similar polar cyclization of an enamine and a thioketene derivative is shown in equation (87) ⟨76TL4283⟩, but electron-deficient alkenes and alkynes react in a concerted fashion. Concerted cyclizations may be subdivided into those in which the sulfur atom is part of the enophile or the dienophile. Into the first category fall the dimerizations of $\alpha,\beta$-unsaturated thioaldehydes (equation 88) which may be shown to closely follow frontier molecular orbital predictions of regioselectivity ⟨79JOC486⟩. Related to this are the thiochalcones (equation

89) ⟨79JOC4151⟩, which also exist as dimers. Heating regenerates the monomeric form, and in the presence of dienophiles a range of dihydrothiopyrans is formed (*e.g.* equation 90) in good to excellent yields. Also closely related are the cinnamic acid thioamides, which exist in equilibrium with their dimers at room temperature (equation 91) ⟨80JCS(P2)4⟩ though there are no reports of cycloadditions with other dienophiles, and the enamino thioketones which have been widely studied (equation 92) ⟨72AC(R)563⟩ because elimination of the amino group affords a facile entry into the thiopyran family.

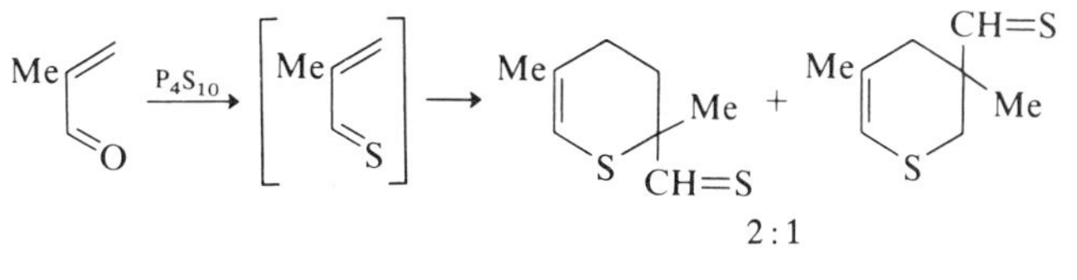

(88)

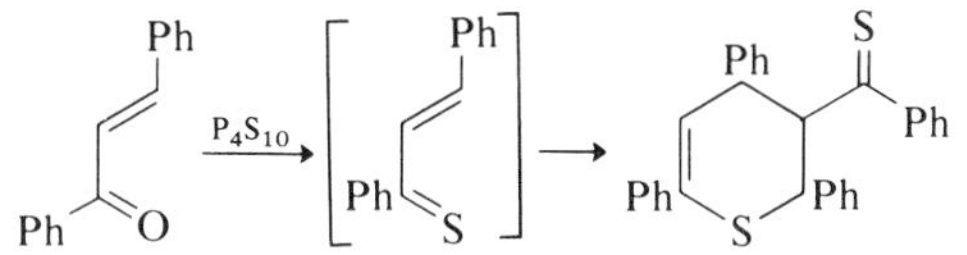

(89)

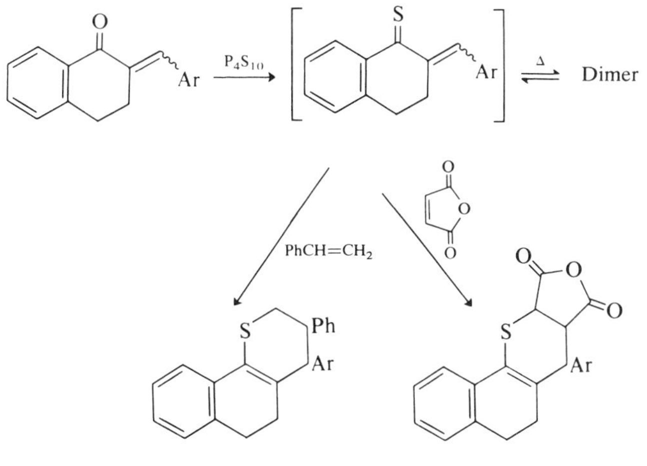

(90)

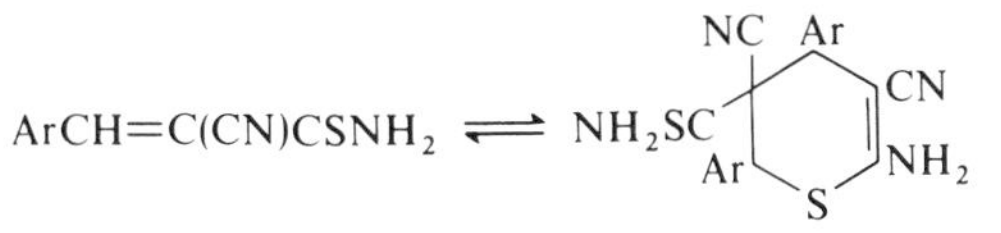

$$ArCH=C(CN)CSNH_2 \rightleftharpoons \text{(structure)}$$

(91)

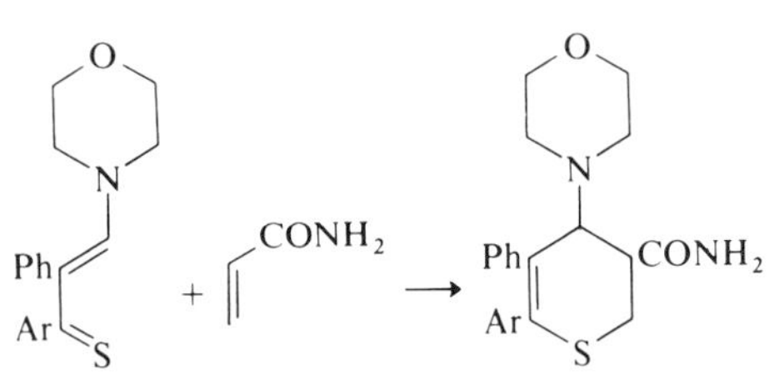

(92)

3,4-Dihydrothiopyran 1,1-dioxides have also been obtained by cycloaddition of a vinylsulfene ⟨77JOC1910⟩ but the reaction seems somewhat limited in scope.

Thioketones reacting as dienophiles are well documented and follow the course of equation (93). Table 14 illustrates the variety of substrates and some typical yields for the reaction. Related to the simple Diels–Alder cyclizations are the [4+2⁺] cationic polar cycloadditions of equation (94) ⟨72JA8932⟩ and the [4⁺+2] version shown in equations (95) and (96) ⟨81TL3773⟩. The use of an alkyne in equation (94) to trap the thienium species leads to a benzothiopyran. It may be seen that choice of the enophile/dienophile combination leads to clean access to 5,6-dihydro- or 3,4-dihydro-thiopyrans with synthetically attractive substitution patterns.

**Table 14**  Cycloaddition Reactions of Thiocarbonyl Compounds with Dienes

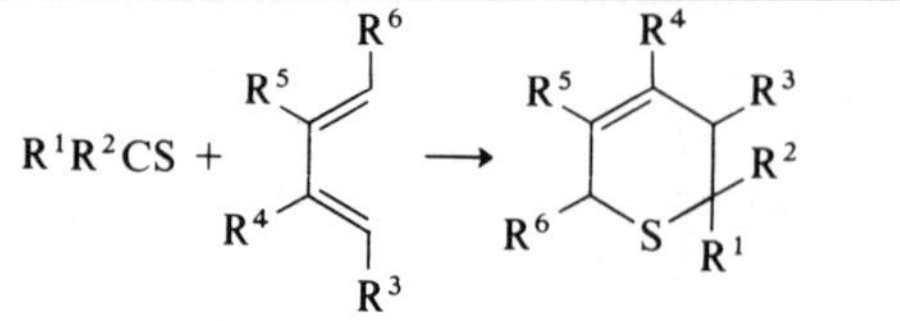

$$R^1R^2CS + \quad \overset{R^5}{\underset{R^4}{\diagup}}\overset{R^6}{\diagdown}\overset{R^3}{\diagup} \quad \longrightarrow \quad \text{(product ring)} \tag{93}$$

| Thiocarbonyl compound | $R^3$ | $R^4$ | Diene $R^5$ | $R^6$ | Product (Yield) | Ref. |
|---|---|---|---|---|---|---|
| SC(CO₂Et)₂ | H | Me | Me | H | 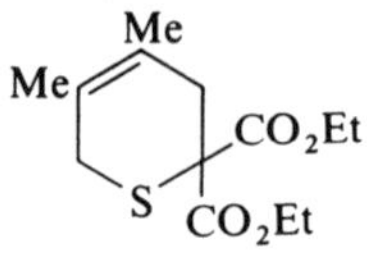 Me, Me, CO₂Et, S, CO₂Et | 78ZN(B)417 |
| NC—CS—SMe | H | H | H | H | 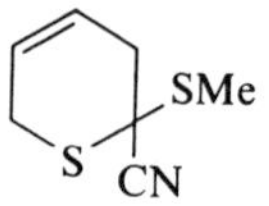 SMe (60%), S, CN | 75JCS(P1)180 |
| NC—CS—SMe | MeO | H | H | H | 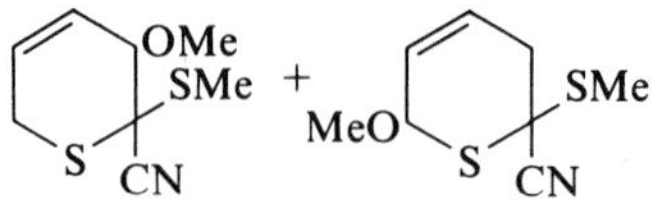 OMe, SMe + MeO, SMe, S, CN, S, CN (40%, >4:1) | 75JCS(P1)180 |
| NC—CS—N(COMe)Ph | H | H | H | H | 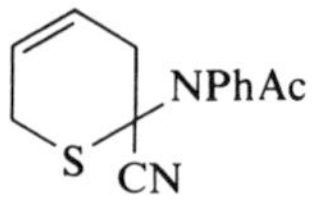 NPhAc, S, CN (95%) | 79CB1867 |
| MeNCOCF₃—CS—CN | H | Me | Me | H | 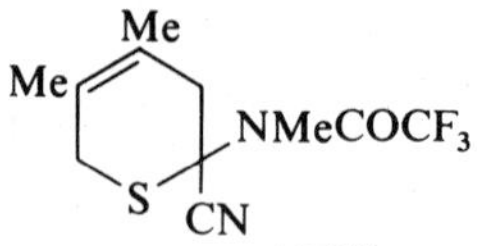 Me, Me, NMeCOCF₃, S, CN (47%) | 79CB1867 |
| (CF₃)₂C=C=S | | cyclooctatetraene | | | 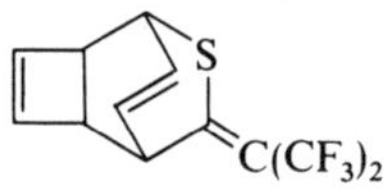 S, C(CF₃)₂ (78%) | 78JOC2500 |
| MeS—CS—CO₂Et | H | Me | Me | H | 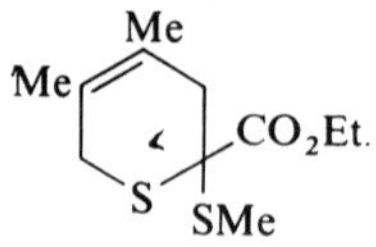 Me, Me, CO₂Et., S, SMe (75%) | 80JOC2601 |
| Ph—CS—Me | | cyclopentadiene | | | 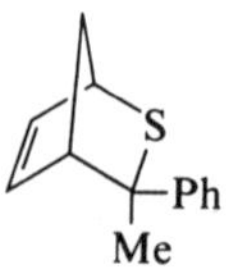 S, Ph, Me | 71CC1063 |
| CS₂ | Me | R₂N | H | Me₂ | NR₂, Me, Me, Me, S, S | 70BSF2016 |
| Ph₂CS | H | Me | Me | H | 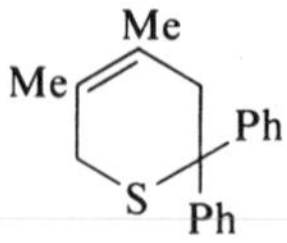 Me, Me, Ph, S, Ph | 69T871 |

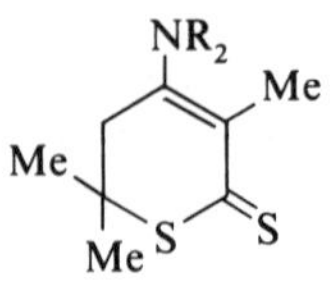

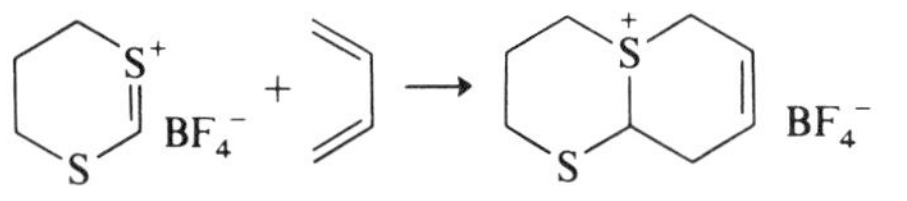

(94)

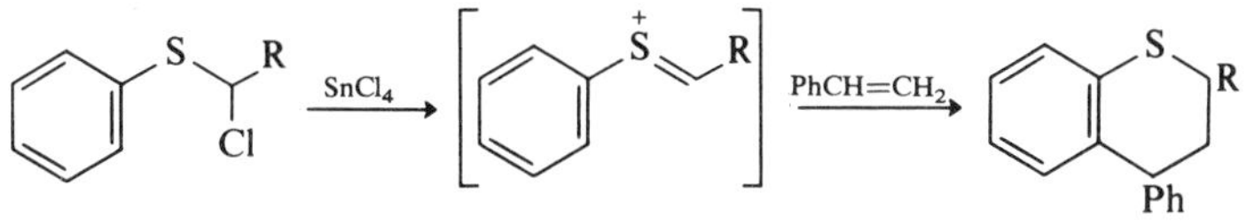

(95)

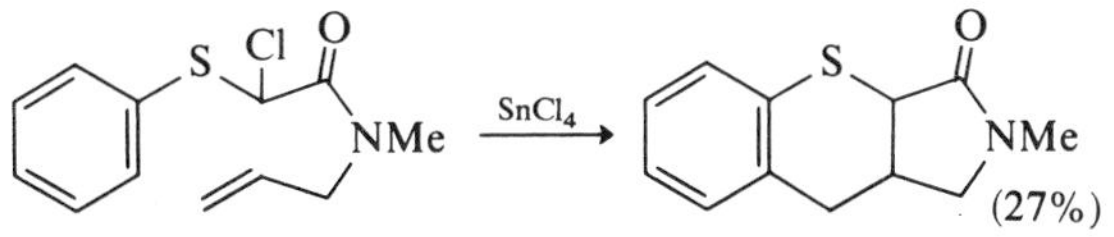

(96)

Formation of six-membered rings by the combination of two three-atom fragments is rare, but an interesting example, leading to a dihydrothiopyran 1,1-dioxide, is shown in equation (97) ⟨79JHC381⟩.

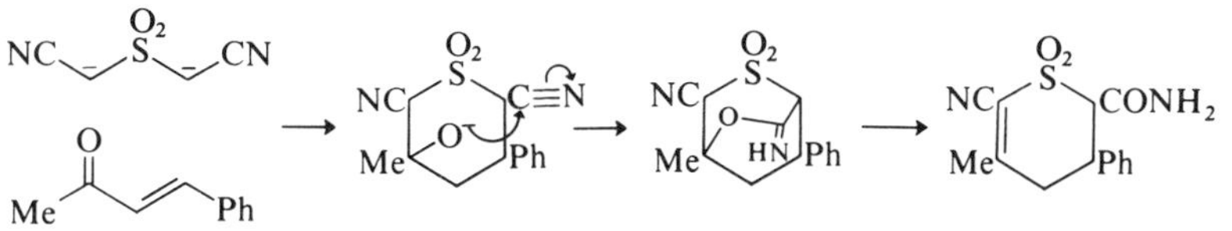

(97)

### 2.25.4.2.2 Ring transformations leading to dihydrothiopyrans

Very few dihydrothiopyrans have been made by ring transformation. Various substituted dihydrobenzothiophenes have been isomerized under very harsh conditions, but the method is of little preparative value ⟨77MI22503⟩.

Of much greater utility, and considerably more theoretical interest, is the Cope rearrangement (139)→(140) shown in equation (98) ⟨79JOC486⟩, which proceeds in boiling pyridine. On prolonged heating (140) isomerizes cleanly to (141) *via* a retro Diels–Alder fragmentation and recombination. This approach awaits exploitation in the preparation of dihydrothiopyrancarbaldehydes.

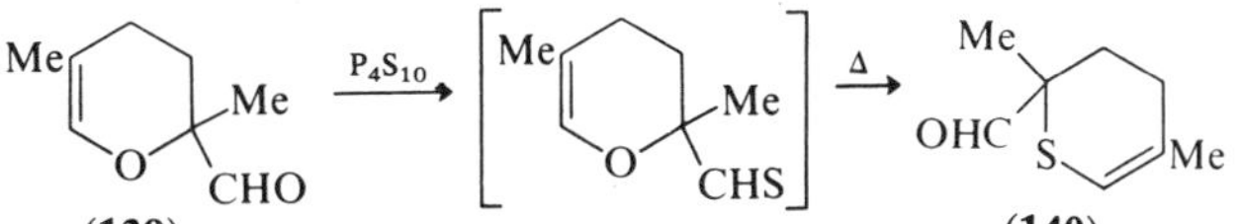

(98)

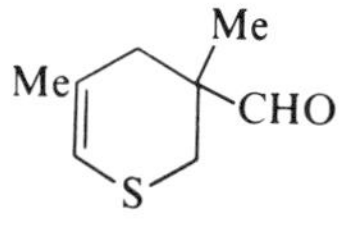

(141)

### 2.25.4.2.3 Synthesis of dihydrothiopyran derivatives

The best available methods for synthesis of the parent heterocycles are clearly: (a) Pummerer rearrangement of thiane 1-oxide, followed by elimination, affording the 3,4-dihydro compound and (b) dehydration of thian-4-ol to give the 3,6-dihydro system ⟨78JHC289⟩. Preparation of the benzannelated compounds is covered in reviews ⟨75AHC(18)59, 80AHC(26)115⟩.

Amino substituted compounds, as well as being available *via* simple alkylation by halogenated dihydrothianes, may also be prepared by Diels–Alder reaction as described in Section 2.25.4.2.1, as may some alkoxy analogues. They are, however, frequently derived from dihydrothiopyranols, which are in turn derived from the ketones. A discussion of general chemical conversions within the area may be found in ⟨79JOC3144⟩ and references therein.

### 2.25.4.3 Synthesis of 2*H*- and 4*H*-Thiopyrans

#### 2.25.4.3.1 *De novo ring synthesis*

As has been discussed earlier, simple thiopyrans are relatively unstable species: the 4*H*-series readily isomerize to 2*H*-analogues, which in turn readily oxidize, *etc.* Therefore the number of useful syntheses of these systems is few. On the other hand, thiopyrones and other fully conjugated systems are so stable that an enormous range of synthetic routes has been explored (see, for example ⟨67AHC(8)219⟩). For full discussion of the field of benzothiopyrans the reader is referred to ⟨75AHC(18)59⟩ and ⟨80AHC(26)115⟩.

2*H*-Thiopyran is preferentially prepared by the $\alpha$-closure scheme shown in equation (99) ($R^1 = R^2 = H$) ⟨73RTC667⟩ and a variety of groups is compatible with the process (including $R^2 = CN$), despite the high temperature necessary to initiate the preliminary Cope rearrangement. Cyclization by displacement of halogen is not known for the monocyclic case, but is reported to occur in a diannelated system (equation 100) ⟨79LA2043⟩. Obviously the generality of this approach is restricted, but it does not yet seem to have been explored.

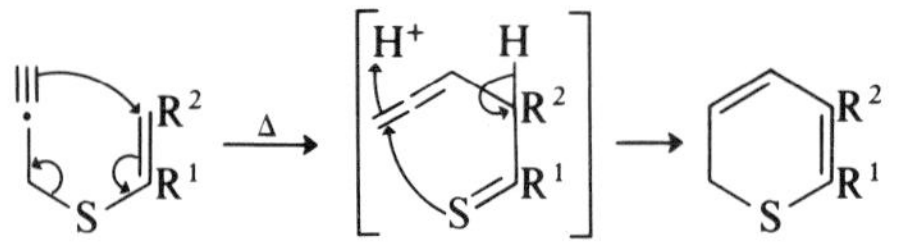

$$\text{(99)}$$

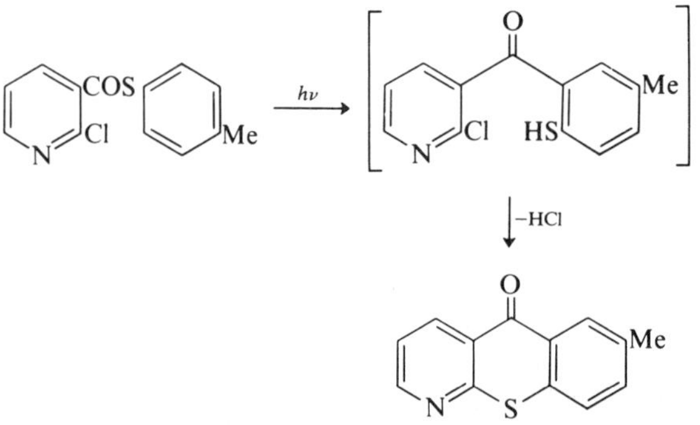

$$\text{(100)}$$

Ring synthesis by $\gamma$-closure is a somewhat more versatile entry into this class of compound, though even then it is restricted to suitably activated precursors such as (142) (equation 101) ⟨72T5197⟩. Thiochromones and thioxanthones are probably most frequently prepared by this mode of closure, usually *via* an intramolecular Friedel–Crafts condensation of a carboxylic acid with an aromatic ring (equation 102) ⟨71GEP2006196⟩.

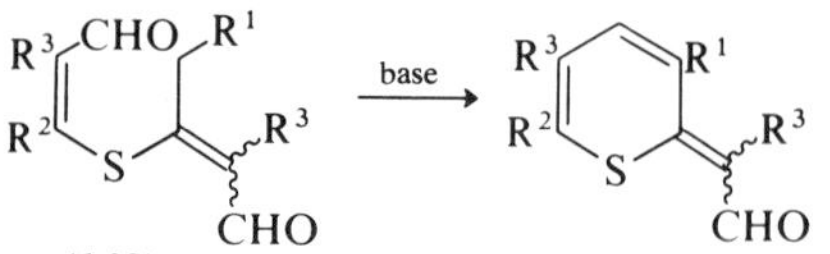

$$\text{(101)}$$

(142)

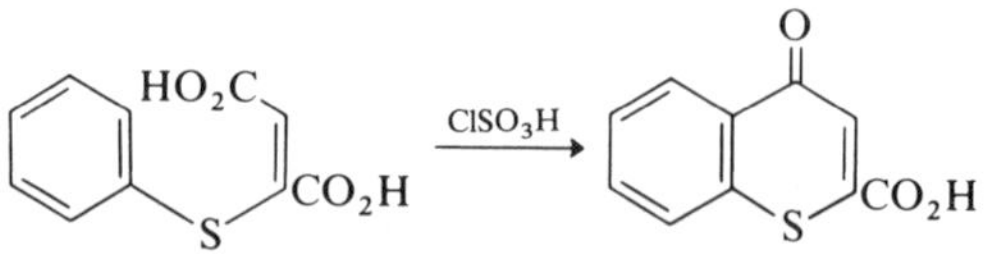

$$\text{(102)}$$

4*H*-Thiopyrans are directly accessible only in relatively poor yield, and then only by condensation of a suitable 1,5-dicarbonyl compound with sulfide ⟨69OPP21⟩. As has been described earlier, this particular reaction is bedevilled with the problem of disproportionation to tetrahydrothiopyrans and thiopyrylium compounds. Thiopyran-4-ones may be prepared in a related manner by addition of sulfur dichloride to divinyl ketones, followed by base treatment ⟨76MI22500⟩.

Enaminothiones and related structures and their cyclization reactions with alkenes have been discussed in Section 2.25.4.2.1, where the preparation of dihydro systems was described. Under appropriate conditions, elimination of the amino substituent may be induced with the production of a thiopyran system (equation 103) ⟨75T3059⟩, though in some cases in which an alternative leaving group is present the amine may be retained (equation 104) ⟨75T2679⟩. If phenylacetyl chloride is used, the product is a 2*H*-thiopyran-2-one ⟨71CR(C)(273)148⟩. Condensations of this type most probably proceed through dipolar intermediates, as do those shown in equations (105) ⟨73ZC130⟩ and (106) ⟨66JPR205⟩, whereas the addition of carbon disulfide to dienes is more likely to be concerted. Dehydrogenation is spontaneous and 2*H*-thiopyran-2-thiones are produced in quite respectable yields ⟨67LA(703)140⟩.

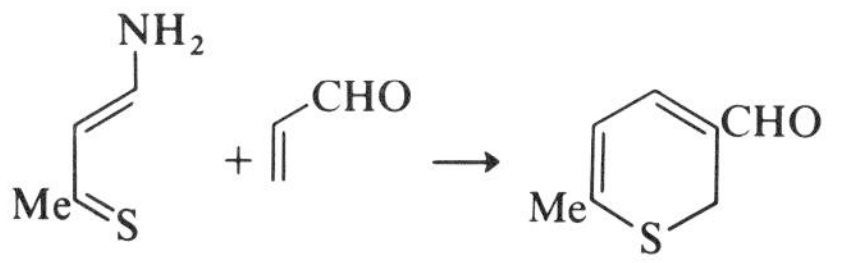

(103)

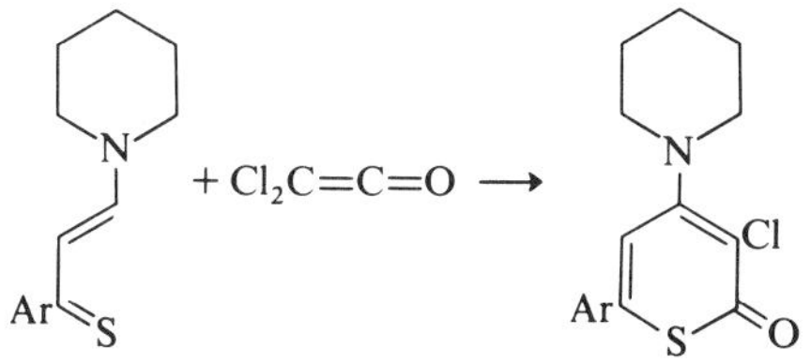

(104)

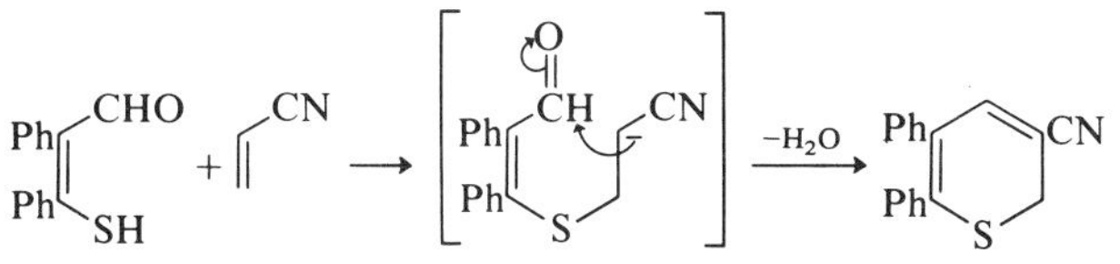

(105)

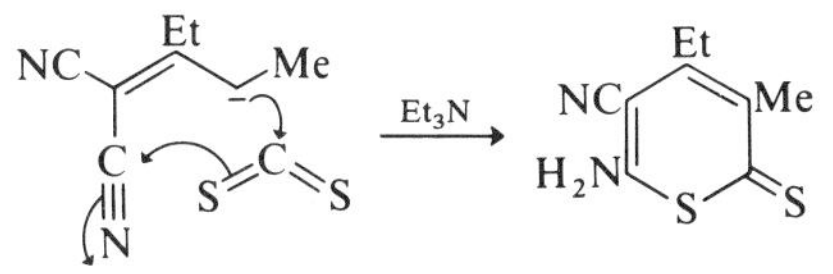

(106)

Condensations between two three-atom fragments are surprisingly common in this field, and generally lead to thiopyran-2-ones, *etc.* The reader is referred to ⟨B-79MI22504⟩ for a tabulation of specific examples of the reaction type shown in equation (107). There are even formally three-component condensations known, though naturally they are really stepwise assemblies of the reactants (equations 108 and 109) ⟨76JPR705, 70JOC2438⟩.

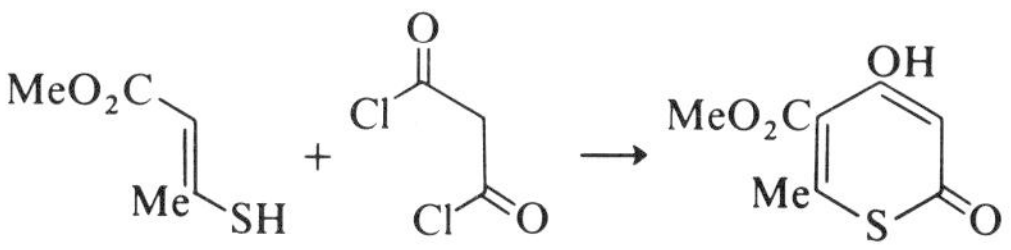

(107)

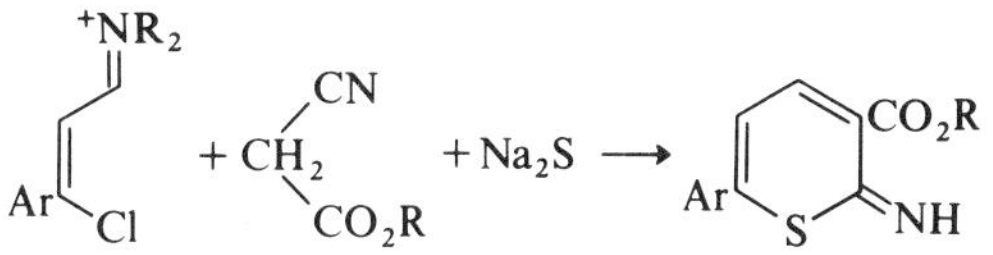

(108)

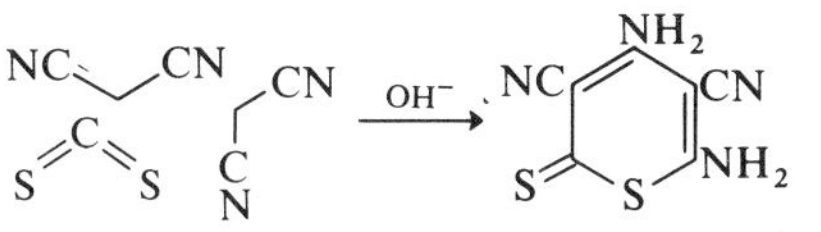

(109)

### 2.25.4.3.2 *Ring transformations leading to thiopyrans*

Thiopyrans are readily available by manipulation of dihydrothiopyrans, either by elimination of water, amines or halogens, or *via* Pummerer reaction of the *S*-oxide ⟨79LA784⟩. Elimination reactions are of course particularly facile in the benzo fused systems and may be used to specifically make 2*H*- or 4*H*-1-benzothiopyrans. On the other hand, extrusion reactions are relatively unknown, though an example has been reported which produces 1*H*-2-benzothiopyran 2,2-dioxides (equation 110) ⟨72IJS(A)(2)287⟩, but sadly its scope appears to be rather restricted.

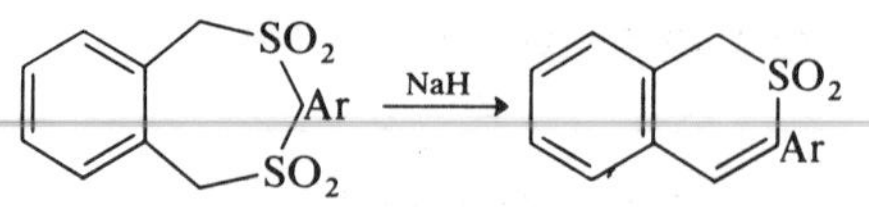

$$\text{(110)}$$

Ring expansion reactions are much better known and sulfolene substrates have received considerable investigation. Cyclopropanation with dichlorocarbene affords 6,6-dichloro-3-thiabicyclo[3.1.0]hexane 3,3-dioxides which may readily be ring enlarged by base or acid treatment. Base treatment, especially with LDA, cleanly affords 2*H*-thiopyran dioxides; hydrochloric acid does likewise, but less cleanly, while hydrobromic acid treatment causes a more complex set of reactions leading to thiopyranones containing bromine in which the sulfur has been reduced. The intermediacy of thiopyrylium oxides has been invoked to rationalize the observations (Scheme 27) ⟨81JOC4502⟩.

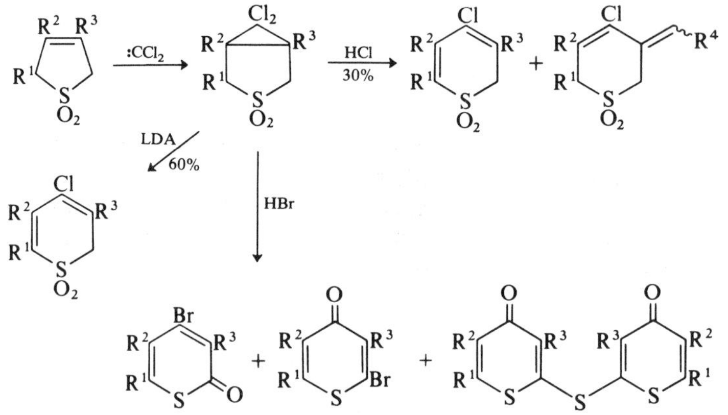

**Scheme 27**

Dithioles and dithiolium salts are, however, the best entry into thiopyrones using ring expansion methods, and have received considerable attention. The attraction of these heterocycles is their ready cleavage under nucleophilic attack, with the liberation of a favourably configured precursor to the thiin ring system. Scheme 28 shows the versatility of the conversions, but it is immediately apparent that all the products of these transformations are fully conjugated systems, and simple 2*H*- or 4*H*-thiopyrans are not available by such an approach. This appears to be a general conclusion and is to be expected in view of the lability of the simple thiins when contrasted with the stability of thiopyrones ⟨77BSF1142, 81TL4507, 79BCJ1235, 77JCS(P1)1511⟩.

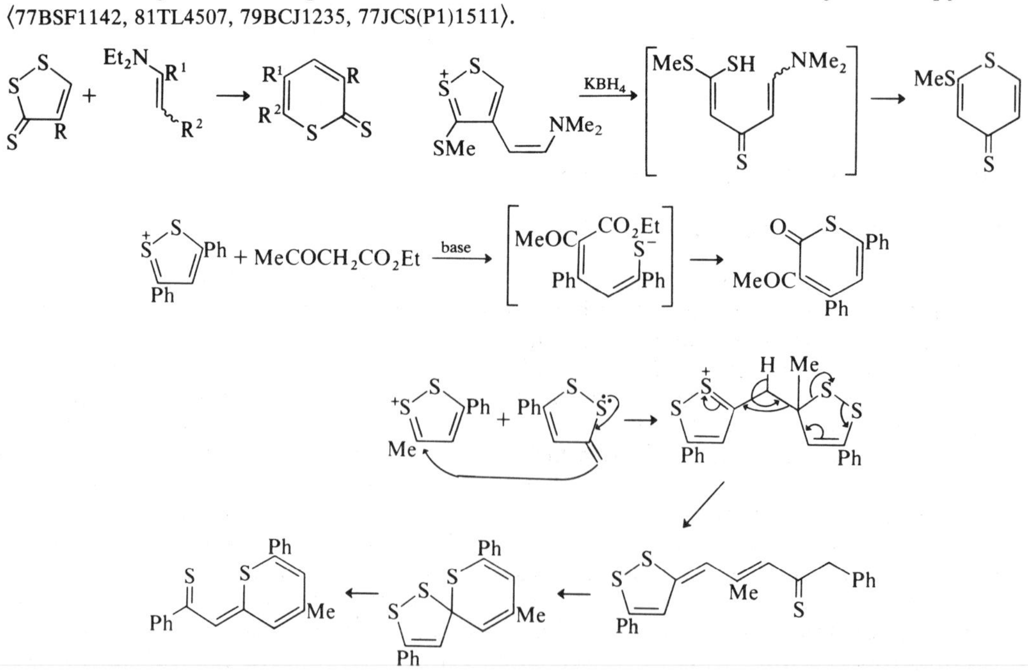

**Scheme 28**

### 2.25.4.3.3 Synthesis of thiopyran derivatives

The parent system may be prepared as shown in equation (99) and such an approach represents the best entry into simply substituted analogues. Benzo fused systems have been covered elsewhere ⟨75AHC(18)59, 80AHC(26)115⟩ and will not be further discussed here.

Substituted 2*H*- and 4*H*-thiopyrans are largely confined to alkylated systems, and are usually derived by reduction of corresponding thiopyrylium salts, or by manipulation of dihydro systems either by elimination reactions or Pummerer reactions. The heteroatom substituted compounds which have been reported have usually been made by a [4+2] process such as that shown in equations (104) and (111) ⟨71TL2241⟩, and the range of examples is limited.

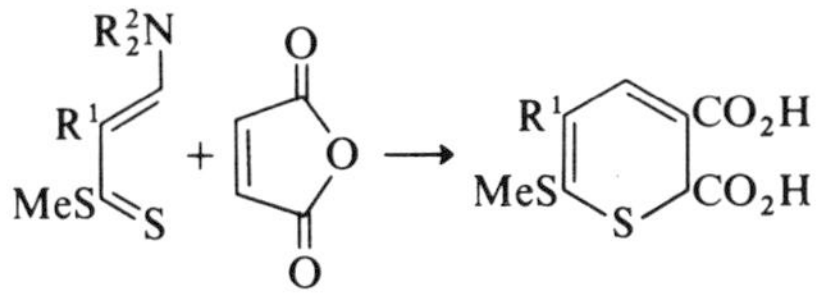

$$ \text{(111)} $$

However, when it comes to thiopyrones an enormous range of substituents is known and almost any of the methods of ring synthesis outlined above may be used. It is especially easy to set up ambient nucleophilic species such as (143) which may be *S*-alkylated to (144). Thermolysis isomerizes such compounds to the 4,6-bis(alkylthio) analogues; the rearrangement has been thoroughly investigated ⟨67JOC3140⟩.

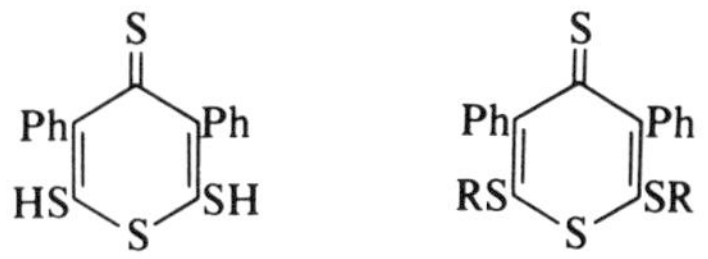

**(143)**          **(144)**

Oxygen and nitrogen linked substituents may be made in the analogous manner by alkylation of ambient oxygen or nitrogen nucleophiles, but less investigation has been applied to these systems.

Halogenated thiins are relatively sparse, and are usually prepared by chlorination of dihydropyrones (equation 112) ⟨79JOC3144⟩ or thiopyrones ⟨78LA1280⟩.

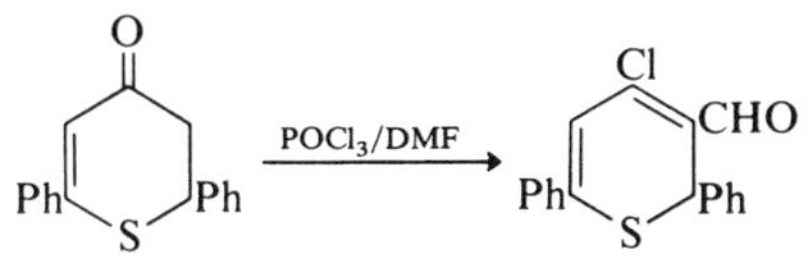

$$ \text{(112)} $$

## 2.25.4.4 Synthesis of Thiopyrylium Salts

### 2.25.4.4.1 De novo ring synthesis

The number of different synthetic approaches to thiopyrylium (thiinium) salts is very restricted and *de novo* synthesis is only really known by two approaches. (a) Condensation of a 1,5-diketone with a sulfide source in the presence of acid. The first formed 4*H*-thiopyran disproportionates *in situ* (see Section 2.25.3.3) and the reaction produces a mixture of tetrahydrothiopyran, dihydrothiopyran and thiopyrylium salt ⟨75KGS643⟩. This method of synthesis is limited essentially to alkyl- or aryl-substituted systems or benzannelated derivatives ⟨74TL3911⟩, as leaving groups attached to the carbon framework may readily be substituted by sulfur in the initial condensation or may be eliminated, especially from positions 2 or 4, to generate an alternative cationic heterocycle, as is seen in the second *de novo* approach. (b) Cyclocondensation of thiophosgene with a diene and internal displacement of chlorine from the 2,2-dichlorothiin intermediate is an efficient access to 2-chlorothiopyrylium salts. The generality of this reaction does not appear to have been thoroughly explored, but would be predicted to permit the preparation of a wide range of substituted materials subject only to the problem of competing leaving groups in positions 2, 4 and 6 possibly leading to mixtures of products ⟨67ZC227⟩.

### 2.25.4.4.2 *Ring transformations leading to thiopyrylium salts*

The only significant ring transformation affording a thiopyrylium salt is closely related to the route outlined above *via* the disproportionation of a thiopyran. Treatment of alkyl or aryl substituted 4*H*-pyrans with hydrogen sulfide in the presence of acid leads to a mixture of tetrahydrothiopyrans and thiopyryliums ⟨75ZOR1540⟩

### 2.25.4.4.3 *Synthesis of thiopyrylium derivatives*

The parent system is prepared by oxidation of tetrahydrothiopyran with triphenylmethane and perchloric acid ⟨66HCA2046⟩ in high yield, and the method is applicable to the preparation of virtually all simple alkyl or aryl derivatives. In more complex cases, especially where the strong acid might cause elimination or isomerization side reactions, triphenylmethyl tetrafluoroborate may be used with the thiin precursor (equation 113) ⟨78CL723⟩.

This method has been used to prepare the 3-oxidothiopyrylium betaine (42) as its perchlorate salt by oxidation of thiopyranones ⟨75JCS(P1)2099⟩ with trityl perchlorate (equation 114).

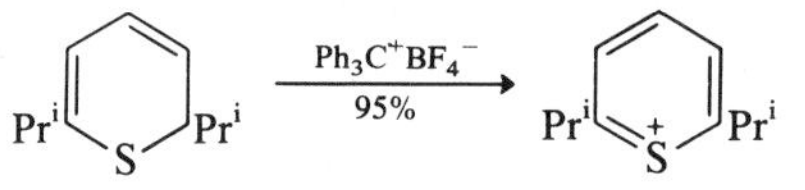

$$Pr^i \quad \xrightarrow[95\%]{Ph_3C^+BF_4^-} \quad Pr^i \qquad\qquad (113)$$

$$\xrightarrow{Ph_3C^+ClO_4^-} \quad (42) \quad \xleftarrow{Ph_3C^+ClO_4^-} \qquad (114)$$

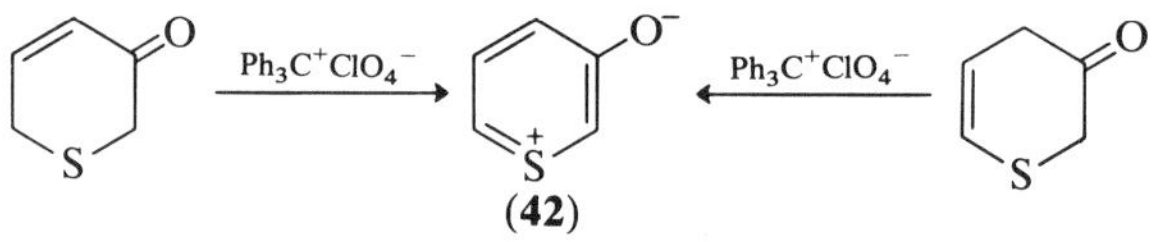

Benzannelated thiopyrylium species are also efficiently generated in this manner, but alternatively thioxanthylium salts (and 1-benzothiopyrylium salts) may be prepared by acid-catalyzed dehydration of the alcohols (145) and (146). The dehydration approach is obviously limited in scope.

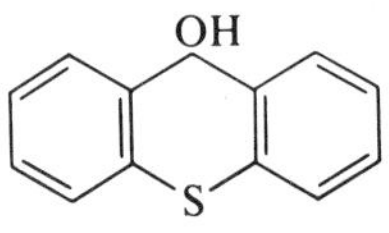

$$\text{(145)} \qquad\qquad \text{(146)}$$

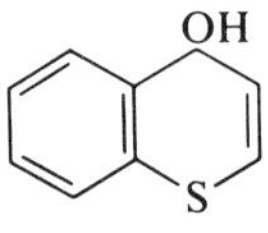

Alkoxy substituted thiopyrylium compounds are obtained by hydride abstraction from the thiin precursor with trityl salts, or by elimination of halogen from a thiin. By contrast, alkylthio substituents may be generated in creating the cationic species as shown in equation (115), though of course only 2- and 4-substituents are available in this way ⟨76CC899⟩. Amino substituted compounds may be prepared by substitution of the methylthio groups, but the nature of the product is strongly dependent on the nature of the amine (see Section 2.25.3.4).

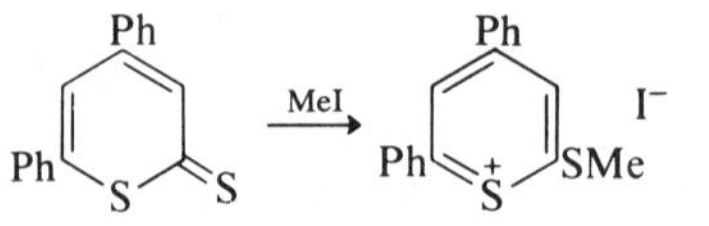

$$Ph \quad \xrightarrow{MeI} \quad Ph \quad SMe \quad I^- \qquad\qquad (115)$$

Halo substituted compounds are scarce. 4-Chlorothiinium salts are made by elimination from the 4,4-dichloro-4*H*-thiin intermediate, which is usually not isolated (equation 116) ⟨75TL2669⟩, and the 2-chloro analogue is derived by cycloaddition of thiophosgene with a diene.

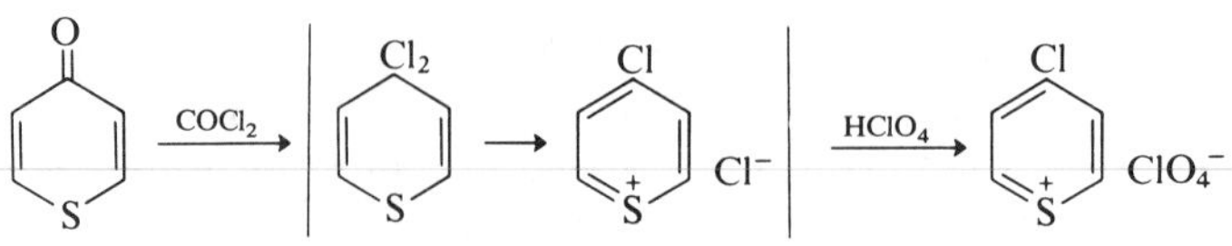

$$\xrightarrow{COCl_2} \quad \xrightarrow{Cl_2} \quad Cl \quad Cl^- \quad \xrightarrow{HClO_4} \quad Cl \quad ClO_4^- \qquad\qquad (116)$$

## 2.25.4.5 Synthesis of Thiabenzenes

There are no known *de novo* preparations of unfunctionalized thiabenzenes, and in fact of the two methods claimed only one has any real scope. The earliest approaches used the addition of an aryllithium species to a thiopyrylium salt in the expectation of achieving reaction at sulfur. There is evidence for the transient formation of some sulfur-arylated materials, but as addition at positions 2, 4 and 6 also takes place, interpretation of results is hazardous. With the recognition of the ylidic nature of 'thiabenzenes', the preferred route of synthesis becomes more obviously deprotonation of a thiinium salt (equation 117) ⟨75JA2718⟩. Non-nucleophilic bases may be used to avoid competing addition reactions. The lability of the products is ameliorated if electron-withdrawing substituents are present on the α- and γ-carbons, which also permits the use of milder bases for the proton removal. This method has been used in the preparation of both the monocyclic materials and their benzannelated counterparts, which may in certain cases be isolated, *e.g.* 9-cyano-10-methyl-10-thiaanthracene ⟨80JOC2468⟩. As yet there are no reports of substituents other than aryl and alkyl having been introduced into simple 'thiabenzenes'.

$$\text{(117)}$$

By contrast, the *S*-oxide analogues of thiabenzenes are quite stable entities and are prepared from acyclic precursors such as ethynyl ketones and dimethyloxosulfonium methylide ⟨71JA2471⟩, though more interesting materials are available as shown in equation (118) ⟨74JOC3519⟩, and an extensive range of such systems may be made in moderate yield. These rings may be nitrated with acetyl nitrate in moderate yield ⟨74JOC3519⟩, and this occurs predictably at positions 2 and 4, where electron density in the ylidic heterocycle is greatest. There are no reports of any other types of substituent being inserted into the molecule, and this may represent a fruitful area for investigation.

$$\text{(118)}$$

## 2.25.5 APPLICATIONS OF THIOPYRANS

The applications of thiopyrans are generally restricted to those areas in which *de novo* synthetic approaches are not prohibitively expensive. High volume usage, therefore, is only found to be suggested for the very simplest members of the family, while the widest range of thiopyran derivatives has been investigated in the field of medicinal chemistry.

Tetrahydrothiopyrans have been investigated as catalysts in various chemical transformations including the chlorination of aromatic compounds ⟨75NEP7504410⟩, the liquid phase oxidation of acrolein to acrylic acid ⟨74JAP7430313⟩ and, when complexed with boron trifluoride, in the curing of epoxy resins ⟨59GEP1099162⟩. A polyalkyl sulfide and sulfone has also been prepared from the saturated heterocycle itself ⟨64MI22500⟩. Simple alkylated tetrahydrothiopyrans have been used as catalysts in the polymerization/crosslinking of butadiene ⟨70GEP2020499⟩, and are also reported to have insecticidal properties ⟨75MI22502, 80EUP7672⟩, as do the corresponding sulfones. Tetrahydrothiopyranones have been used as stabilizers for polyamide fibres ⟨78MI22500⟩ or, when elaborated upon, as modifiers for other synthetic fibres ⟨75MI22503⟩. The oxime of 2,2-dimethyltetrahydrothiin-4-one, moreover, is reported to be a useful agglomerating and flotation agent for lead-containing ores ⟨76MIP22500⟩. Antibacterial, antifungal and insecticidal activity is once again found in derivatives of 3,5-dioxotetrahydrothiopyran-4-carboxylic acid anilides ⟨77JAP7746078, 74USP3833610⟩.

Oxidation of the saturated heterocycle to sulfoxide and sulfone derivatives affords solvents for the preparation of polyacrylonitrile ⟨62BEP613056⟩, or corrosion inhibitors (*e.g.* **147**)

⟨76MIP22501⟩, though the sulfone of tetrahydrothiopyran-4-one is claimed to be a very useful bleaching agent for textiles ⟨73GEP2238207⟩.

Perhaps the most active area of investigation of simple saturated thiin derivatives is that of their thianium salts, which have raised a little interest as antifungal agents (particularly the bis-sulfonium compounds ⟨67USP3344020⟩) and plant growth regulators ⟨74MI22500⟩ but most especially in their applications to electron transfer systems, as in batteries, conductive resins and photoinitiators in cationic alkene polymerization ⟨77BRP1474246, 77MI22504, 79MI22505, 77MI22505⟩. The related sulfimines have significant herbicidal properties ⟨75USP3909235⟩.

The pharmacology of the saturated heterocycles has been extensively investigated, in part because of their occurrence in crude petroleum and consequent industrial disease potential. General toxicity and mutagenicity are reported to be slight ⟨77MI22500, 65MI22501⟩ and the general pharmacology of simple derivatives is not very exciting ⟨72MI22501, 78MI22501⟩, though (**148**), perhaps not surprisingly, is reported to have significant cardiovascular effects ⟨78MI22502⟩ and (**149**), a synthetic $\alpha$-amino acid, is a good inhibitor of transport systems in certain tumour cells ⟨78JMC1070⟩. More interesting is the radioprotective action observed with the sulfone (**150**) ⟨72JMC595⟩ and the muscarinic and nicotinic activities of 4-acetoxy-$S$-methylthianium salts which are far more pronounced than for, say, piperidine analogues ⟨77MI22506⟩. The conformation of the thianium salt is very important; the *cis*-disubstituted system displays quite different potencies at muscarinic and nicotinic receptors compared with the *trans* isomer.

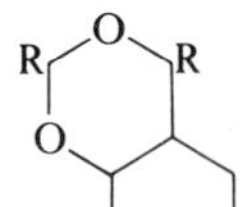

(147) 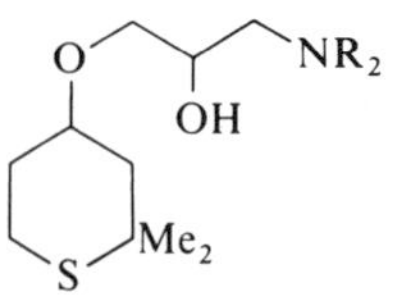            (148) 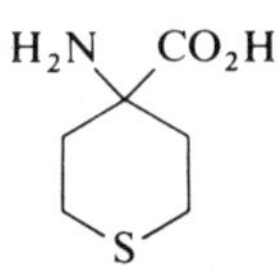            (149) 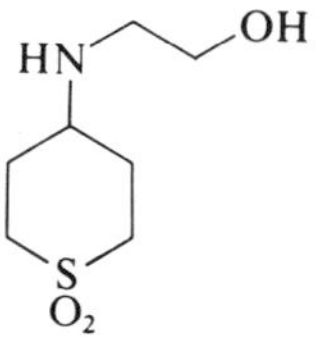            (150)

The majority of instances of use of tetrahydrothiopyrans in compounds of medicinal interest have the ring fused to other heterocycles, and acting more as a modulator of physicochemical properties rather than a pharmacophore or even an auxopharm. Examples are easily found. Thus, the tranquilizing action of benzodiazepines may be modified by fusion as in (**151**) ⟨78AP799⟩, or the anti-inflammatory and analgesic qualities of 2-arylpropionic acids may be potentiated by the presence of a thiane ring, which presumably exerts its influence *via* its effects on physicochemical and thence pharmacokinetic parameters (structure **152**) ⟨72USP3682964⟩.

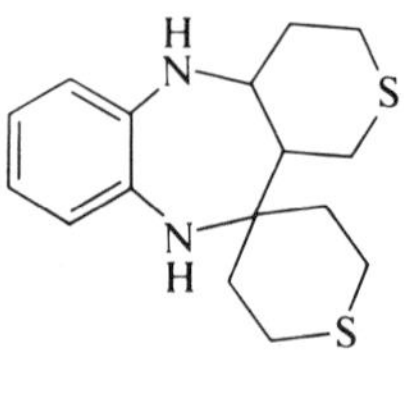

(151) 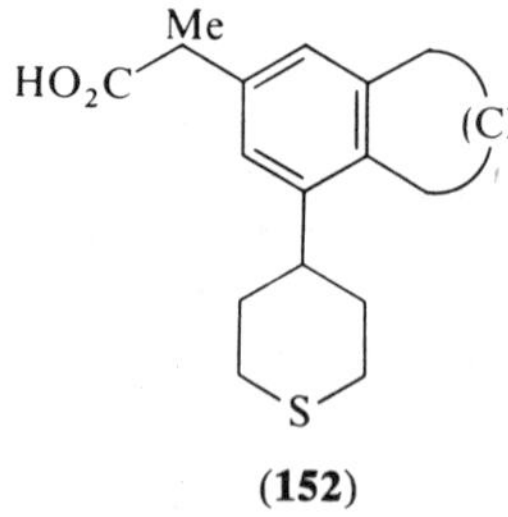                              (152)

The range of applications claimed for dihydrothiopyrans is much more restricted than for thianes, reflecting the poorer availability of such materials. Insecticidal activity is found in intermediates for juvenile hormone preparation such as (**153**) ⟨73GEP2236460⟩, while the ketones (**154**) and (**155**) are claimed to be diuretic and anti-inflammatory ⟨74JAP7426274⟩.

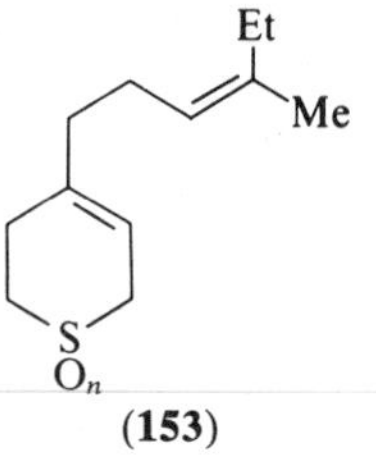

(153) 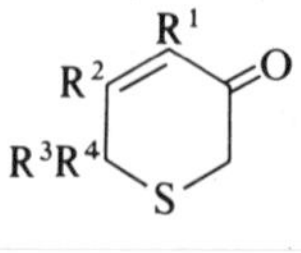            (154) 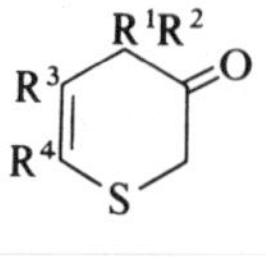            (155) 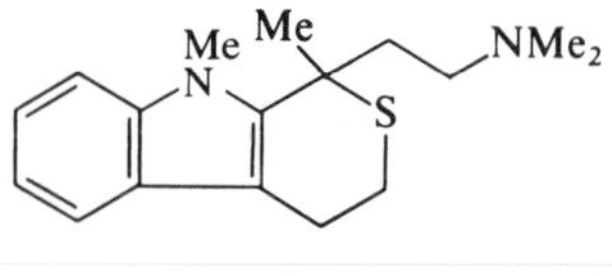            (156)

Systems containing fused dihydrothiopyrans are legion. The applications of benzannelated materials have been reviewed ⟨75AHC(18)59, 80AHC(26)115⟩, but fusion with other heterocycles or carbocycles has produced an enormous range of activities. For example, with thiadiazinones or thiazoles, herbicidal action ⟨80USP4182623, 75MIP22500⟩; with indoles, a range of widely active systems ⟨66MI22500⟩ among which the compound tandamine (**156**) was extensively investigated as a potent inhibitor of biogenic amine uptake ⟨76MI22501⟩ and potential antiulcer agent; with benzopyrans, cannabinoid analogues with analgesic and antihypertensive effects ⟨75NEP7505267, 76JMC549⟩; and with indenes, compounds with antiulcerative properties ⟨75USP3954617⟩. Again it is not the pharmacological action of the dihydrothiopyran which is important *per se*; rather it is the physicochemical modulation effect of that system which contributes to overall activity.

As yet, 2*H*- and 4*H*-thiopyrans appear to have little application, but the fully conjugated systems are quite a different matter. Thiopyran-4-ones are reported to be fungicidal ⟨80JAP80102504⟩, while the 3-carboxylic acid of thiopyran-4-one itself has been used to derivatize 6-aminopenicillanic acid in the preparation of novel penicillins ⟨77JAP7782730⟩ or the equivalent cephalosporins ⟨74GEP2362816⟩. Of much greater medicinal interest are the derivatives of 2*H*-thiopyran-2-ones, which have been shown to possess a whole range of other general pharmacological properties ⟨68MI22500⟩, while the 4-methoxy derivatives have distinct bactericidal and antimycotic actions ⟨79FES869⟩.

2,6-Dimercaptothiopyran-4-ones and -4-thiones have been extensively investigated for the detection and estimation of heavy metals, especially in the Soviet Union ⟨65MI22502, 71MI22500⟩.

Possibly the earliest application of thiopyran systems was for the manufacture of dyes. The facile thiopyranylidene/thiopyrylium tautomerism is especially useful in this usage. A typical cyanine dye is shown in structure (**157**) ⟨69UKZ512⟩. Related to this application is the modern technology surrounding electrophotocopying, in which the light absorbing properties are allied to the facile redox reactions of thiopyranylidene systems. Two applications result: electrophotosensitive imaging (see, for example, ⟨79GEP2831054⟩) and photoconducting layer preparation, again for copying purposes, using conjugated 4-thiopyranylidene methylthiopyrylium salts ⟨79EUP2238⟩. Electrical conductivity, or at least 'semiconductivity', is also found in the π-complexes between tetracyanoquinodimethane (TCNQ) and the fulvene analogue (**158**) ⟨77CC687⟩.

(**157**)          (**158**)

Benzothiopyrans have also been extensively used in the manufacture of cyanine dyes and related products, but their medicinal uses are much wider than for the monocyclic materials, encompassing antibacterials, antiallergics, antihypotensives, sedatives and schistosomicides ⟨75AHC(18)59, 80AHC(26)115⟩.

No important uses have been claimed for dibenzo[*b,d*]thiopyrans, though the partially saturated compounds analogous to various cannabinoids do have significant neurological effects ⟨71MI22501⟩. By contrast, thioxanthenes and thioxanthones have revealed many interesting activities. The stability of the ring system makes it very attractive as a dyestuffs precursor, and extensive ranges of diazo dyes have been made ⟨77GEP2546821⟩. With further benzannelation, fluorescent whiteners are produced ⟨79GEP2815031⟩. Thioxanthene-9-carboxylic acids have been converted to semicarbazide derivatives with important fungicidal activity ⟨76GEP2551919⟩, while thioxanthones can act as initiators of photopolymerization of photoresins ⟨80JAP80105678⟩.

In the field of chemotherapeutics a wide range of uses has been investigated. Penicillin derivatives have been evaluated ⟨79JAP7912392⟩, while some simple thioxanthones are antiosteoperotic ⟨78USP4101668⟩ and their sulfone analogues may be of value in the treatment of tumours ⟨77MI22500⟩, or if substituted by acid groupings, in the treatment of allergic complaints ⟨76BRP1447032⟩. Alkylamino substitution of thioxanthone affords analogues of the acridine antimalarials; two of these antischistosomal compounds, lucanthone (**159**) and hycanthone (**160**), have been shown to intercalate into DNA, and (**160**) was strongly mutagenic in test systems, possibly as a result of such intercalation ⟨80MI22500, 77MI22508⟩. The greatest use of thioxanthene derivatives, though, is for their neuroleptic properties; 9,9-dialkylthioxanthenes are CNS depressants ⟨79FRP2424274⟩, while 9-alkylidene substitution, with an amino group in that substituent, has produced the tranquillizing drug chlorprothixene (**161**). The prevalence of mental illness has ensured a thorough investigation of the pharmacology of these types of compound ⟨77MI22509⟩. Although thiopyrylium salts have been claimed to be useful antibacterials ⟨76MI22502⟩, their major fields of application today are in colour photographic processes ⟨68FRP1522354⟩, electrophotographic processes ⟨79JAP7924027⟩, as photovoltaic elements ⟨79USP4164431⟩ and as photoinitiators for the curing of epoxy resins ⟨79USP4139655⟩. 'Thiabenzenes' have, as yet, no industrial applications.

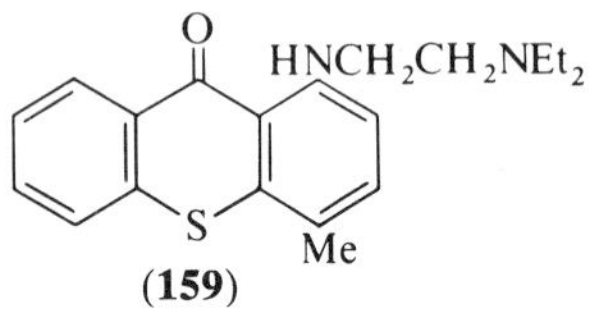

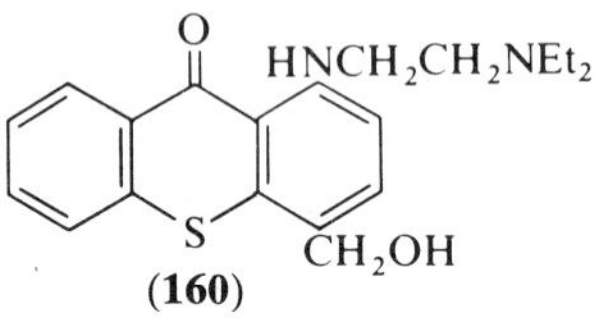

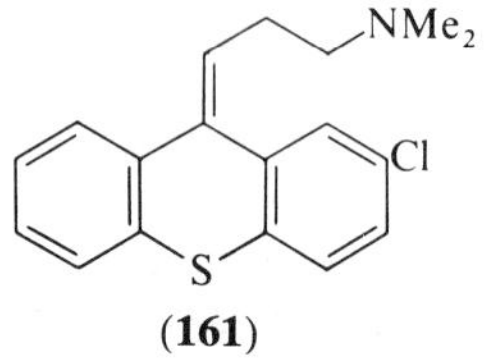

# 2.26

# Six-membered Rings with More than One Oxygen or Sulfur Atom

M. J. COOK

*University of East Anglia*

| | | |
|---|---|---|
| 2.26.1 | INTRODUCTION | 943 |
| 2.26.2 | STRUCTURE | 944 |
| | *2.26.2.1  Theoretical Methods* | 944 |
| | *2.26.2.2  Molecular Geometry* | 946 |
| | *2.26.2.3  Molecular Spectra* | 951 |
| | *2.26.2.3.1  $^1H$ and $^{13}C$ NMR spectroscopy* | 951 |
| | *2.26.2.3.2  UV spectroscopy* | 955 |
| | *2.26.2.3.3  IR and Raman spectroscopy* | 956 |
| | *2.26.2.3.4  Mass spectrometry* | 957 |
| | *2.26.2.3.5  Photoelectron spectroscopy* | 958 |
| | *2.26.2.4  Aromaticity versus Antiaromaticity in the Unsaturated Series* | 959 |
| | *2.26.2.5  Conformational Preferences and Inversion Barriers in the Saturated Series* | 960 |
| 2.26.3 | REACTIVITY | 962 |
| | *2.26.3.1  Fully Unsaturated Compounds: Reactivity at Ring Positions* | 962 |
| | *2.26.3.1.1  General survey* | 962 |
| | *2.26.3.1.2  Thermal and photochemical reactions involving no other species* | 963 |
| | *2.26.3.1.3  Electrophilic attack at carbon* | 965 |
| | *2.26.3.1.4  Electrophilic attack at sulfur* | 966 |
| | *2.26.3.1.5  Oxidation to cation radicals and dications* | 967 |
| | *2.26.3.1.6  Nucleophilic attack at carbon* | 969 |
| | *2.26.3.1.7  Nucleophilic attack at a heteroatom* | 971 |
| | *2.26.3.1.8  Nucleophilic attack at hydrogen* | 972 |
| | *2.26.3.1.9  Reactions with radicals, metals and at surfaces* | 973 |
| | *2.26.3.2  Fully Unsaturated Compounds: Reactivity of Substituents* | 974 |
| | *2.26.3.2.1  General survey* | 974 |
| | *2.26.3.2.2  Electrophilic attack at fused benzene rings* | 974 |
| | *2.26.3.2.3  Nucleophilic attack at hydrogen of a fused benzene ring* | 975 |
| | *2.26.3.3  Saturated and Partially Saturated Compounds* | 975 |
| | *2.26.3.3.1  General survey* | 975 |
| | *2.26.3.3.2  Rings with heteroatoms 1,4* | 976 |
| | *2.26.3.3.3  Rings with heteroatoms 1,3 and 1,3,5* | 977 |
| | *2.26.3.3.4  Rings with two or more adjacent heteroatoms* | 979 |
| 2.26.4 | SYNTHESES | 981 |
| | *2.26.4.1  Fully Unsaturated Compounds* | 981 |
| | *2.26.4.1.1  Non-benzo fused ring systems* | 981 |
| | *2.26.4.1.2  Benzo fused ring systems* | 984 |
| | *2.26.4.2  Saturated and Partially Saturated Compounds* | 986 |
| | *2.26.4.2.1  Rings with heteroatoms 1,4* | 986 |
| | *2.26.4.2.2  Rings with heteroatoms 1,3 and 1,3,5* | 989 |
| | *2.26.4.2.3  Rings with two or more adjacent heteroatoms* | 990 |
| 2.26.5 | APPLICATIONS AND IMPORTANT COMPOUNDS | 992 |

## 2.26.1 INTRODUCTION

The present chapter embraces a particularly wide variety of ring systems, some of which have received very much attention, others considerably less. The fully unsaturated compounds, defined here as having two heteroatoms and two double bonds, fall into two series. The more comprehensively studied is that with heteroatoms in the 1,4 positions and includes 1,4-dioxin (**1**), first prepared in the 1930s, and 1,4-dithiin (**2**), obtained two decades later. Examples of 1,4-oxathiins are rare and indeed the parent compound remains unknown. The monobenzo fused derivatives of all three rings have been investigated but it is the dibenzo analogues, dibenzo[*b,e*][1,4]dioxin (**3**), phenoxathiin (**4**) and thianthrene (**5**) which are perhaps the best known. Compounds having heteroatoms in the 1,2-positions constitute the second series and these are encountered primarily as 1,2-oxathiin 2,2-dioxides (sultones) and 1,2-dithiins, the parent (**6**) of the latter class being first prepared in the mid-1960s. Dibenzo fused derivatives (**7**; Y = O, Z = SO or $SO_2$ and Y and Z = sulfur in any of three oxidation states) are fairly well investigated but rather less so than the series (**3**)–(**5**). In view of the greater overall attention which has been given to the 1,4-series, discussion of this group of compounds is presented prior to that of the 1,2-series throughout the sections and sub-sections which follow.

|  |  |  |  |
|---|---|---|---|
| (**1**) Z = O | (**3**) Y = Z = O | (**6**) | (**7**) |
| (**2**) Z = S | (**4**) Y = O, Z = S |  |  |
|  | (**5**) Y = Z = S |  |  |

Saturated and partially saturated ring systems are more numerous than the fully unsaturated rings by virtue of the greater number of heteroatoms which can be incorporated and the variations in which they can be arranged. 1,4-Dioxane is frequently used as a solvent but otherwise the most common rings are members of the 1,3-series (*i.e.* acetals and thioacetals). However, compounds containing as many as four or five heteroatoms have been synthesized and have received attention. As with the fully unsaturated systems, consideration of rings with heteroatoms 1,4 precedes that of other series. Rings with heteroatoms 1,3 and 1,3,5 are discussed next, followed by rings with two or more adjacent heteroatoms.

For most of the chapter, the material is presented according to the usual format. However, because of the diversity of the ring systems under consideration, it is more appropriate to discuss synthesis (Section 2.26.4) according to ring system rather than by type of ring closure or transformation.

Earlier major reviews have appeared in both the Elderfield and the Weissberger–Taylor series. Contributions to the former deal with rings containing two oxygen atoms ⟨B-57MI22600, B-57MI22601⟩ and with their sulfur and sulfur–oxygen analogues ⟨B-57MI22602⟩ whilst the latter, a very comprehensive treatise, discusses all the multisulfur and sulfur–oxygen six-membered heterocycles known at that time ⟨66HC(21-2)611⟩. Apart from a general updating of these works, the present chapter also provides an opportunity to draw comparisons between oxygen, sulfur–oxygen and sulfur heterocycles and, where appropriate, such a treatment is attempted. Since the appearance of the previous reviews much research has been devoted to the study of physicochemical properties and this accounts in part for the proportion of the chapter devoted to structure (Section 2.26.2).

## 2.26.2 STRUCTURE

### 2.26.2.1 Theoretical Methods

Numerous theoretical studies have dealt with the structural and physicochemical properties of the unsaturated series of compounds and calculations have been performed at various levels of sophistication. A particular problem inherent in dealing with the sulfur-containing rings is whether, or in what manner, allowance should be made for *d*-orbital participation ⟨70SST(1)1, 75SST(3)728⟩. The point is of some consequence in this series of compounds, particularly in connection with the question of aromaticity. Simple addition of potential cyclic π-electrons gives a total of eight, but resonance canonical forms of type (**8b**) and (**8c**), which invoke participation of 3*d*-orbitals, imply aromatic stabilization. However, whilst recognizing that the question of *d*-orbital participation remains an area of current investigation, it seems pertinent to note an early comment to the effect that valence shell expansion, when it occurs, does so only when the sulfur atom bears a positive charge when drawn as the single bonded structure ⟨B-62MI22600⟩. Among the calculations which are discussed hereafter, *d*-orbital participation has quite frequently been explicitly excluded.

**(8a)** Z = O, S    **(8b)** Z = O, S    **(8c)** Z = S

Calculations of geometry have been concerned particularly with whether the rings with heteroatoms in the 1,4 positions are planar or folded. Both the extended Hückel treatment (EHT) ⟨67JST(1)489⟩ and the more sophisticated *ab initio* FSGO method ⟨71JST(9)205⟩, satisfactorily predict the planar structure of 1,4-dioxin, demonstrated by IR methods (Section 2.26.2.3.3). For 1,4-dithiin, the simple LCAO ⟨58JA5543⟩, the EHT method ⟨67-68JST(1)489⟩ and STO–3g calculations ⟨80MI22603⟩ correctly predict the folded or boat conformation (Section 2.26.2.2). The EHT energies of dibenzo[*b,e*][1,4]dioxin and phenoxathiin, calculated as a function of the angle of fold, predict that the two molecules should adopt planar conformations ⟨78JOM(146)235⟩. However, another EHT calculation of the dihedral angle dependence of the all-valence electronic energies, suggests that whereas the former is planar, the latter and thianthrene are folded ⟨72BCJ1589⟩. The second series of results corresponds rather better with experimental findings (Section 2.26.2.2).

A number of calculations have been concerned with the question of aromaticity. The non-aromaticity of 1,4-dioxin has been supported by calculations using the Hückel $\sigma\pi$-approximation which gives a negative Dewar resonance energy of $-5.4$ kJ mol$^{-1}$ (*cf.* benzene 86 kJ mol$^{-1}$), derived by comparing the calculated heat of atomization with that for the corresponding linear polyene ⟨71JST(8)236⟩. 1,4-Dithiin and analogues have attracted rather more attention but results have been anything but consistent. Thus an early calculation for 1,4-dithiin using the simple LCAO–MO method, neglecting *d*-orbital participation, predicted a delocalization energy for the boat structure of 117 kJ mol$^{-1}$ ($\beta$ taken as 75 kJ mol$^{-1}$), placing 1,4-dithiin between thiophene and furan on an aromaticity scale ⟨58JA5543⟩. Subsequently, a recalculation using the more sophisticated SCF–MO method, with a re-evaluation of the C—S resonance integral, revised the value to the lower but still substantial figure of 54 kJ mol$^{-1}$ ⟨63JS(S2)157⟩. A calculation of the delocalization of thianthrene, again ignoring *d*-orbitals, gave a value of 369 kJ mol$^{-1}$, less than that for anthracene but more than twice the value for benzene ⟨63T471⟩. More recent calculations, however, reverse the earlier trends and predict that 1,4-dithiin is non-aromatic. Thus the calculated Hückel delocalization energy relative to an acyclic polyene is $-0.139\beta$ (*cf.* thiophene $0.193\beta$) ⟨73JA3907⟩. A technique which combines perturbation theory and the graph–theoretical definition of resonance energy also predicts a negative resonance energy for both 1,4-dithiin and 1,2-dithiin ⟨78T3419⟩. The results of these theoretical approaches are appraised in the light of experimental data in Section 2.26.2.4.

A particularly comprehensive set of bond order and electron density values is available for members of the 1,4-series and their mono- and di-benzo fused analogues ⟨70BCJ3929⟩. Results were obtained using the PPP–SCF–MO method without explicit account being taken of *d*-orbitals, on the basis that $\pi-\pi^*$ transitions can be reproduced by a small variation in the core integrals. Data for the non-benzo fused compounds, Figure 1, predict a high degree of single bond character for C—X bonds and double bond character between C-2 and C-3 (higher in 1,4-dithiin) and a similar trend is apparent for the monobenzo fused series. Electron densities satisfactorily indicate that the latter series is more reactive towards electrophiles at the heteroring, but on the same basis results for the dibenzo fused series fail to predict the observation that the $\beta$- rather than the $\alpha$-position is the more reactive. However, calculations of frontier electron density values for electrophilic substitution satisfactorily overcome this objection, but even so, phenoxathiin presents a problem in so far as calculations predict that electrophilic substitution occurs *para* to the sulfur atom rather than *para* to the oxygen atom ⟨72BCJ1589, 69CJC3173⟩.

Electronic energy levels can be satisfactorily calculated using the PPP–SCF–MO method and the results used to assign bands in the observed UV spectra. Comparison of results for

**Figure 1** PPP–SCF–MO bond orders and charge densities for 1,4-dioxin and 1,4-dithiin ⟨70BCJ3929⟩

the singlet $\pi$–$\pi^*$ transition energy for 1,4-dioxin and 1,4-dithiin reveal that the oxygen atom raises the first allowed transition energy. In the monobenzo derivatives the effect of the heteroring on the oscillator strength of the first allowed transition is found to be less marked ⟨70BCJ3929⟩. Calculations have been made for dibenzo[b,e][1,4]dioxin, phenoxathiin and thianthrene which predict the variation in the UV spectrum with varying degrees of folding. Reasonable agreement between observed and calculated spectra are obtained for planar or near planar conformations of benzo[b,e]-[1,4]dioxin and folded conformations for the sulfur-containing rings ⟨72BCJ1589, 67MI22600⟩.

A further series of theoretical treatments concerns the radical cations derived from the fully unsaturated compounds. These are readily prepared by electrochemical oxidation or more simply by electron transfer to electron acceptors (Section 2.26.3.1.5) and various theoretical studies ⟨63MI22600, 69JCS(B)755, 76G457⟩ support assignments of ESR signals and electronic transitions. A larger amount of spin density is calculated to reside in the sulfur valence orbitals than in the corresponding orbitals in the oxygen compounds. The spin density on sulfur does not vary significantly if $3d$-orbitals are included because the odd electron resides in a low energy orbital with only modest involvement of the $d$-orbitals. Benzo fusion decreases spin density on the heteroatoms ⟨76G457⟩.

Among the fully saturated compounds the geometry of 1,3-dioxane has been calculated by molecular mechanics techniques using both the Allinger force field ⟨76JA6798⟩ and the Burkert force field ⟨79T691⟩, the latter being an adaptation of the former. Both reproduce the essential features of the known ring structure though the second calculation indicates greater puckering in the O—C—O region. Further results obtained via the Burkert force field predict that geminal dimethylation at C-2, C-4 or C-5 leads to a decrease in the endocyclic bond angle of about 2° at the site of substitution.

Extension of the Allinger force field to sulfur compounds has enabled the empirically determined geometry of 1,3,5-trithiane to be reproduced and energy differences between chair and non-chair forms for a selection of rings to be calculated. The energy difference between the chair and twist form for 1,2-dithiane is 16.3 kJ mol$^{-1}$ in favour of the chair, but this difference is reduced in the 3,3,6,6-tetramethyl derivative. For 1,2,4,5-tetrathiane the energy difference is only 3.8 kJ mol$^{-1}$ and introduction of the two geminal dimethyl groups at C-3 and C-6 destabilizes the chair form sufficiently to render the twist conformer more stable, in line with experimental findings. The calculated energy difference of 2.9 kJ mol$^{-1}$ compares with 1.7 kJ mol$^{-1}$ determined experimentally. The chair form of 1,2,3-trithiane is the most stable conformation and here the twist form is predicted to be at a point on the side of the potential well; the symmetrical boat conformer lay at the second energy minimum. Introduction of further sulfur atoms serves to increase the energy difference between chair and twist forms ⟨76JA6798, 76JA2741⟩.

Calculations have also been performed for the monosulfoxides of 1,4-dithiane, 1,3-dithiane and 1,3,5-trithiane, the *cis* and *trans* disulfoxides of 1,4-dithiane and 1,3,5-trithiane and the *cis* and *trans* trisulfoxides of 1,3,5-trithiane ⟨76T529⟩. For all the compounds mentioned, a chair conformer is calculated to be most stable and the results are consistent with experimental findings which reveal that the $\overset{+}{S}$—O$^-$ bond exhibits an axial preference within 1,4-dithiane 1-oxide but an equatorial preference within 1,3-dithiane 1-oxide and 1,3,5-trithiane 1-oxide (Section 2.26.2.5).

### 2.26.2.2 Molecular Geometry

Studies of the geometry of the unsaturated compounds fall into two categories, the qualitative or semiquantitative evaluations of gross ring shape, and precise determinations of molecular dimensions using the classic structure methods. Principal findings regarding gross structure are summarized in Table 1, while representative bond length and angle data are given in Figure 2. Further data of the latter type are available within appropriate papers cited in Table 1 and in Wheatley's extensive compilation ⟨72PMH(5)1⟩.

**Table 1** Principal Conclusions Regarding Gross Structure for Selected Compounds

| Compound | Method[a] | Geometry | $\theta(°)$[b] | Ref. |
|---|---|---|---|---|
| 1,4-Dioxin | ED | Planar[c] | — | 41JCP(9)54 |
| | IR/Raman | Planar | — | 66SA1859 |
| | Far IR | Planar | — | 73JCP(58)4344 |
| 1,4-Dithiin | X-ray | Folded | 137 | 54AX498 |
| 2,5-Diphenyl-1,4-dithiin 1-oxide | X-ray | Folded | [d] | 71JCS(B)1407 |
| Dibenzo[*b,e*][1,4]dioxin | X-ray | Planar | — | 78AX(B)2956 |
| | mKc | Planar | — | 69AJC2243 |
| | DM | Folded | 163.8 | 78JOM(146)235 |
| | DM | Folded | 160–180 | 48JA1564 |
| Phenoxathiin | X-ray | Folded | 138 | 66AX(20)429 |
| | mKc | Folded | 138±6 | 69JCS(B)980 |
| | DM | Folded | 163.4 | 78JOM(146)235 |
| | DM | Folded | 150–160 | 48JA1564 |
| Thianthrene | X-ray | Folded | 128 | 63AX310 |
| | ED | Folded | 131.4 | 75JCS(F2)1173 |
| | DM | Folded | 135–140 | 48JA1564 |
| | DM | Folded | 144±8 | 65JCS571 |
| | mKc | Folded | 140±10 | 65JCS571 |
| Phenoxathiin 10-oxide | X-ray | Folded | 152 | 79JOC1989 |
| *cis*-Thianthrene 5,10-dioxide | X-ray | Folded | 123 | 63AX310 |
| | DM | Folded | 139±10 | 65JCS571 |
| *trans*-Thianthrene 5,10-dioxide | X-ray | Folded | 122 | 63AX310, 70AX(B)451 |
| | DM | Folded | 130±10 | 65JCS571 |
| Thianthrene tetraoxide | X-ray | Folded | 127 | 63AX310 |
| | DM | Folded | 140±8 | 65JCS571 |
| 6-*p*-Bromophenyl-1,2-oxathiin 2,2-dioxide | X-ray | Puckered | [d] | 72CC264 |
| Dibenzo[*c,e*][1,2]dithiin | X-ray | Puckered | 34 | 66AX(21)A104 |

[a] ED = electron diffraction, DM = dipole moments, mKc = molar Kerr constants.
[b] $\theta$ = angle between planes of benzene rings or double bonds in folded or puckered compounds.
[c] Data fitted to an assumed planar structure.      [d] Not available.

The structure of 1,4-dioxin was first investigated by electron diffraction methods and the observed pattern was fitted to an assumed planar structure. This assumption was subsequently justified by the results of a combined IR and Raman study (Section 2.26.2.3.3) and further supported by theoretical treatments (Section 2.26.2.1). From the electron

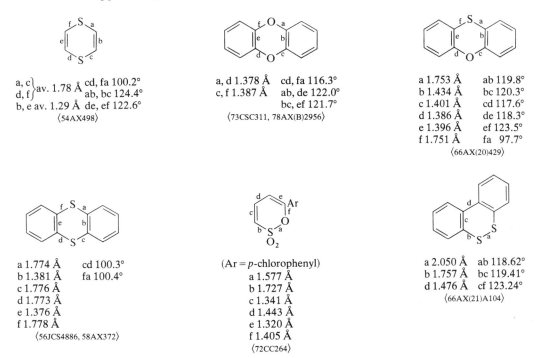

a, c | av. 1.78 Å cd, fa 100.2°
d, f | ab, bc 124.4°
b, e av. 1.29 Å de, ef 122.6°
⟨54AX498⟩

a, d 1.378 Å cd, fa 116.3°
c, f 1.387 Å ab, de 122.0°
    bc, ef 121.7°
⟨73CSC311, 78AX(B)2956⟩

a 1.753 Å ab 119.8°
b 1.434 Å bc 120.3°
c 1.401 Å cd 117.6°
d 1.386 Å de 118.3°
e 1.396 Å ef 123.5°
f 1.751 Å fa 97.7°
⟨66AX(20)429⟩

a 1.774 Å cd 100.3°
b 1.381 Å fa 100.4°
c 1.776 Å
d 1.773 Å
e 1.376 Å
f 1.778 Å
⟨56JCS4886, 58AX372⟩

(Ar = *p*-chlorophenyl)
a 1.577 Å
b 1.727 Å
c 1.341 Å
d 1.443 Å
e 1.320 Å
f 1.405 Å
⟨72CC264⟩

a 2.050 Å ab 118.62°
b 1.757 Å bc 119.41°
d 1.476 Å cf 123.24°
⟨66AX(21)A104⟩

**Figure 2** X-Ray structure analyses of selected compounds

diffraction study the following dimensions were obtained: C—O bond length $1.41 \pm 0.03$ Å; C—C bond length $1.35 \pm 0.03$ Å; C—O—C angle $116 \pm 4°$. The C—O bond length is only marginally less than the sum of the covalent radii of the oxygen and carbon atoms ⟨41JCP(9)54⟩. In contrast, 1,4-dithiin adopts a boat or folded conformation and the results of a crystal structure determination are shown in Figure 2. The carbon–carbon bond lengths signify substantial double bond character, although the sulfur–carbon bonds are shorter than the 1.81 Å considered typical of a pure C—S single bond. The boat type conformation is also adopted by derivatives such as 2,5-diphenyl-1,4-dithiin 1-oxide. The $\overset{+}{S}$—$O^-$ bond is pseudo axial and the two bonds to the unoxidized sulfur are distinctly shorter than those to the sulfoxide group. The C—S—C bond angle is 3° larger than in 1,4-dithiin itself ⟨71JCS(B)1407⟩.

Dibenzo[*b,e*][1,4]dioxin, phenoxathiin and thianthrene have each been investigated extensively by X-ray crystallography and representative data appear in Figure 2. Dibenzo[*b,e*][1,4]dioxin is alone in adopting a planar structure in the solid state and a recent report ⟨78AX(B)2956⟩ shows that deviations of the atoms from a least squares plane through all twelve carbon atoms are less than 0.010 Å for carbons and 0.001 Å for the oxygens. In contrast, phenoxathiin and thianthrene are both folded about the axis containing the two heteroatoms with dihedral angles of 138° and 128° respectively ⟨75JCS(F2)1173⟩. The average C—S bond length in both phenoxathiin and thianthrene is a little shorter than in 1,4-dithiin and this, coupled with the smaller angle of fold in thianthrene than in 1,4-dithiin, implies greater conjugation between the sulfur atom and the neighbouring carbon atom in the fused ring derivatives. The conformations adopted by the dibenzo compounds in solution are more difficult to evaluate quantitatively. However, the weight of evidence suggests that all three are folded about the central axis but to varying degrees, with greater folding for the sulfur-containing rings (Table 1). The oxides of phenoxathiin and thianthrene are also folded (Table 1). In the solid state the $\overset{+}{S}$—$O^-$ bond of the monosulfoxides adopt the pseudo axial site but *cis*-thianthrene-5,10-dioxide has both $\overset{+}{S}$—$O^-$ bonds equatorial.

Non-planarity and single bond/double bond alternation is a feature of available data for the 1,2-series. In the 1,2-oxathiin 2,2-dioxide derivative (Figure 2) O-1, C-4, C-5 and C-6 are essentially coplanar but S-2 and C-3 both lie above the plane, with S-2 showing the greater deviation from it of 0.58 Å. An X-ray structure determination of dibenzo[*c,e*]-[1,2]dithiin shows that, as in thianthrene, the sulfur–carbon bond is relatively short (Figure 2). There is substantial puckering between the planes containing the benzene rings, presumably resulting from the accommodation of the long bonds to the sulfur atoms and the dihedral angle dependence of the torsional strain in C—S—S—C units, a feature which is important in contributing to the ring geometry in the saturated analogues.

Data for the saturated rings are more abundant than those for the fully unsaturated series, not least because the saturated compounds have proved a popular vehicle for all aspects of conformational analysis (see Section 2.26.2.5), and the determination of precise molecular dimensions is an integral part of such work. Results of crystal structure studies have been tabulated ⟨72PMH(5)1⟩ and reviewed ⟨B-69MI22600⟩. A selection of electron diffraction and microwave structure analyses are also available ⟨B-77MI22600, B-78MI22600⟩. A recent text on conformational analysis provides further references and discussion ⟨B-80MI22600⟩. Results obtained in the solution phase for ring torsion angles, derived from coupling constants, are referred to in the NMR section (Section 2.26.2.3.1). Values are normally in satisfactory agreement with those obtained by traditional methods for structure determination.

Prior to discussion of different ring systems it should be noted that in spite of the difference in bond lengths of C—O (*ca.* 1.42 Å) and C—S (*ca.* 1.81 Å) bonds and the very different values for the C—O—C and C—S—C bond angles, all but a few of the saturated multi oxygen and sulfur rings adopt the same basic conformation, the chair form. Certain combinations of substituents, however, may destabilize the chair form by introducing severe 1,3-diaxial interactions. Free energy and enthalpy differences between chair and twist forms decrease in the order 1,3-dioxane, cyclohexane, 1,3-oxathiane and 1,3-dithiane, with the free energy difference for the last of these only marginally greater than the conformational energy of a methyl substituent. Thus, destabilizing interactions arising from just one axial alkyl group may be sufficient to destabilize the 1,3-dithiane chair conformer to the extent that twist forms become populated ⟨74PMH(6)199⟩. Conformational free energies of the chair and twist forms of 3,3,6,6-tetrasubstituted 1,2,4,5-tetrathianes are even more finely balanced ⟨69JA6019⟩. In solution (**9**) and (**10**) exist predominantly as the twist conformer, whereas the chair is the major form for (**11**) (see also Section 2.26.2.1). Compound (**9**) is

**(9)**

**(10)** $n = 5$
**(11)** $n = 4$

also found to exist in the twist form in the crystal, and structural parameters are available (*vide infra*).

Data from electron diffraction studies of the 1,4-series are shown in Table 2. The torsion angle, $\psi$, in the Y—CH$_2$CH$_2$—Z unit varies considerably through the series showing that

**Table 2** Results from Electron Diffraction Studies for Saturated Compounds with Heteroatoms 1,4

| Compound | Bond lengths (Å) | | | Bond angles (°) | | | | $\psi$ (°)[a] | Ref. |
|---|---|---|---|---|---|---|---|---|---|
| | C—C | C—O | C—S | C—C—O | C—C—S | C—O—C | C—S—C | | |
| 1,4-Dioxane | 1.52 | 1.42 | — | 109 | — | 112 | — | 57 | 63ACS1181 |
| 1,4-Oxathiane | 1.52 | 1.42 | 1.83 | 113 | 111 | 115 | 97 | 60 | 72JST(14)353 |
| 1,4-Dithiane | 1.54 | — | 1.81 | — | 111 | — | 100 | 69.5 | 47ACS149 |

[a] Torsion angle in the YCH$_2$CH$_2$Z fragment, as calculated in ⟨72JST(14)353⟩.

1,4-dithiane is by far the most puckered. This apparently best enables the structure to accommodate the smaller C—S—C angle. However, within 1,4-oxathiane, analysis shows that the oxygen atom is further out of the plane of the four carbon atoms than the sulfur atom ⟨75JST(24)337⟩. The 2,3- and 2,5-*trans*-dihalo derivatives, which exist in the solid state with the halogen atoms diaxial, show a significant flattening of the ring relative to the unsubstituted parent molecules; *trans*-1,4-dithiane 1,4-dioxide also exists with the two polar bonds axial in the crystal ⟨B-69MI22600⟩. There have been numerous structural studies within the 1,3-series but the three ring systems are most conveniently compared with reference to X-ray structure determinations of their 2-aryl derivatives (Figure 3 and Table 3). In the 1,3-dioxane (**12**) the heterocyclic fragment is more puckered than the carbocyclic part and this is a feature of other structure determinations of 1,3-dioxane derivatives. The results for 2-*p*-nitrophenyl-1,3-oxathiane (**13**) show that the side containing the sulfur atom is significantly flatter than the oxygen-containing fragment. In the 1,3-dithiane derivative (**14**) the situation is essentially the reverse of that found in (**12**) with the C—C—C unit being the more puckered end. Both 1,3,5-trioxane (**15**) and 1,3,5-trithiane (**16**) have an average torsional angle of *ca.* 64°, in excess of the maximum puckering at any position in the 1,3-dioxane and 1,3-dithiane ring systems (Figure 3).

Bond length data for the 1,3-dioxane (**12**) show the acetal C—O bonds are somewhat shorter than the non-acetal C—O bonds and a similar trend is apparent for the carbon-heteroatom bonds in the 1,3-dithiane derivative (**14**). According to X-ray structure analysis, 1,3,5-trioxane has relatively long bonds compared with the 1,3-dioxane derivative, but an electron diffraction study of the former provides a somewhat shorter bond length, *ca.* 1.41 Å ⟨71JST(9)33⟩. Bond angle measurements (Table 3) reveal significant enlargements of the valency angles at the carbon positions in 2-phenyl-1,3-dithiane and 1,3,5-trithiane which compensate for the small C—S—C bond angles. 1,3,5-Trioxane is somewhat curious because all the bond angles are smaller than the tetrahedral value.

The *trans* and *cis* sulfoxide derivatives of 2-phenyl-1,3-dithiane have modified chair geometries relative to that of 2-phenyl-1,3-dithiane itself. The smaller bond angle at C-2 in the *trans* isomer can be attributed to a cross ring electrostatic attraction between the positively polarized oxidized sulfur S-1 and the lone pair on S-3. This is partially offset in the *cis* isomer by a repulsion between the axial sulfoxide oxygen and S-3 ⟨77JOC961⟩. A surprising result obtained for the *cis* 1,3-dioxide of 2,2-diphenyl-1,3-dithiane is that, in the crystal, both $\overset{+}{S}$—O$^-$ bonds are axial in spite of significant electrostatic repulsion in this conformation ⟨79JOC1540⟩. The $\overset{+}{S}$—O$^-$ bond in 1,3,2-dioxathiane 2-oxide (trimethylene sulfite) also exists with the $\overset{+}{S}$—O$^-$ bond axial but this can be accounted for in terms of the anomeric effect (Section 2.26.2.5). The geometry shows irregularities in the bond angles

**Table 3**  Endocyclic Bond Lengths and Bond Angles for Representative Compounds Obtained by X-Ray Crystallographic Analysis

| Compound No.[a] | Bond length (Å) | | | | | | Bond angle (°) | | | | | | Ref. |
|---|---|---|---|---|---|---|---|---|---|---|---|---|---|
| | 1-2 | 2-3 | 3-4 | 4-5 | 5-6 | 6-1 | 123 | 234 | 345 | 456 | 561 | 612 | |
| (12) | 1.41 | 1.40 | 1.45 | 1.49 | 1.51 | 1.42 | 111 | 111 | 110 | 108 | 109 | 111 | 70RTC313 |
| (13)[b] | 1.40 | 1.82 | 1.79 | 1.51 | 1.51 | 1.44 | 111 | 97 | 112 | 111 | 113 | 113 | 72AX(B)2424 |
| | 1.36 | 1.79 | 1.80 | 1.44 | 1.48 | 1.43 | 112 | 96 | 111 | 112 | 113 | 113 | |
| (14) | 1.80 | 1.79 | 1.81 | 1.51 | 1.46 | 1.83 | 115.2 | 100.9 | 114.9 | 116.5 | 116.1 | 99.2 | 66AX(20)490 |
| (15) | 1.43 | 1.43 | 1.43 | 1.43 | 1.43 | 1.43 | 107.8 | 108.0 | 107.8 | 108.0 | 107.8 | 108.0 | 63MI22601 |
| (16) | 1.81 | 1.81 | 1.81 | 1.82 | 1.82 | 1.81 | 115.3 | 101.7 | 116.2 | 99.6 | 116.2 | 101.7 | 65MI22600 |
| (17) | 1.59 | 1.60 | 1.42 | 1.52 | 1.51 | 1.49 | 100 | 116 | 111 | 110 | 106 | 116 | 66RTC1197 |
| (18) | 1.48 | 1.41 | 1.43 | 1.48 | 1.41 | 1.43 | 105.5 | 106.9 | 105.5 | 105.5 | 106.9 | 105.5 | 67ACS2711 |
| (19) | 2.07 | 1.85 | 1.53 | 1.53 | 1.53 | 1.85 | 99 | 110 | 117 | 117 | 110 | 99 | 64ACS2345 |
| (9) | 2.015 | 1.84 | 1.84 | 2.015 | 1.84 | 1.84 | 102.9 | 110.3 | 102.9 | 102.9 | 110.3 | 102.9 | 81TL4767 |
| (20) | 2.05 | 2.06 | 2.06 | 1.79 | 1.54 | 1.81 | 100 | 99 | 103 | 113 | 115 | 103 | 72AX(B)534 |

[a] For structures see Figure 3.  [b] Two different conformations adopted within the crystal.

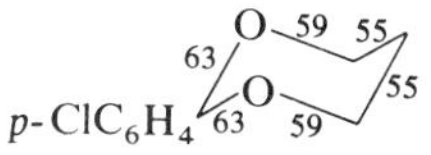

(12) 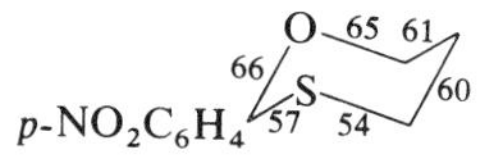  (13) 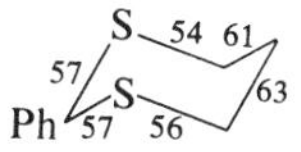  (14) 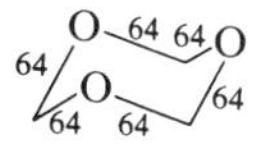  (15)

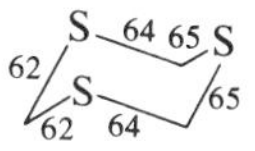

(16) 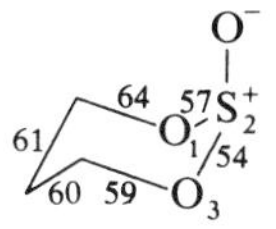  (17) 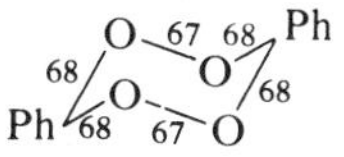  (18) 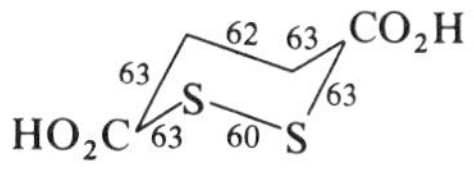  (19)

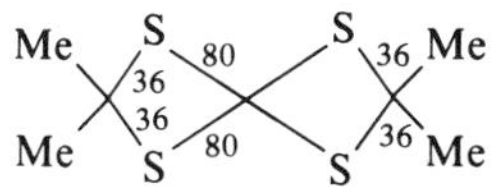

(9) 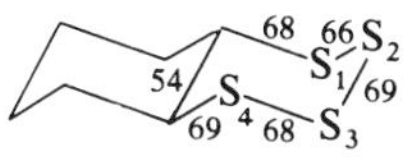  (20)

<sup></sup> ᵃ Mean values for two chair conformers present in the crystal

**Figure 3** Torsion angles (to nearest degree) for representative compounds, obtained by X-ray crystallographic analysis

as would be anticipated from the presence of a ring sulfur atom (Table 3) with flattening of the heteroatom end of the ring compensated by a greater puckering of the C—C—C fragment (Figure 3). Figure 3 and Table 3 also contain data for several rings containing two adjacent heteroatoms. The geometry is frequently characterized by a large C—Z—Z—C torsion angle which can be expected to minimize torsion energy within this fragment. This is illustrated by the puckering of the 1,2,4,5-tetroxane ring (**18**) and the torsion angle of 80° within the twist conformation of tetramethyl-1,2,4,5-tetrathiane (**9**). A further feature of the latter is the relatively short S—S bond (Table 4) arising from $d_\pi$–$p_\pi$ overlap which is maximized as the torsion angle approaches 90° (*cf.* compound **19**). The 1,2,3,4-tetrathiane (**20**), constrained to some extent by fusion to the cyclohexane ring, also shows a significant degree of puckering.

## 2.26.2.3 Molecular Spectra

### 2.26.2.3.1 ¹H and ¹³C NMR spectroscopy

Chemical shifts of protons in 1,4-dioxin and 1,4-dithiin fall to higher field than the 'aromatic region', with the dioxin protons more shielded than those in 1,4-dithiin (Table 4). The values thus give no indication of a ring current and indeed are in reasonable agreement with predicted values of 5.02 and 6.14 p.p.m. for —O—CH=CH—O— and —S—CH=CH—S— fragments respectively, based on shifts in methyl vinyl ethers and methyl vinyl sulfide ⟨70OMR(2)431⟩. The protons on the heterocyclic ring of 1,4-benzodioxin and 1,4-benzodithiin are shifted downfield by about 0.3 p.p.m. relative to the monocyclic series. Fusion of a 1,4-dithiin ring to a second ring of the same type produces a potentially aromatic 14π-electron system, but the signal in (**21**) at 6.46 p.p.m. does not signify the presence of a ring current ⟨76JOC1484⟩. S-Alkylation of 1,4-dithiin, *e.g.* (**22**), causes a more significant downfield shift, with H-3 far more deshielded than H-2 (Table 4). The latter can be rationalized by invoking mesomeric electron release from C-3 to the sulfur *d*-orbital at the sulfonium ion centre. The rather similar shifts observed for (**22**) and its dihydro derivative (6.76, 8.02 p.p.m.) suggest there is no significant ring current in the former ⟨70ZC296⟩.

The identical vicinal coupling constants for 1,4-benzodioxin and 2,3-dihydro-1,4-dioxin, 3.73 Hz, have been cited as evidence to confirm the absence of aromatic character in the former ⟨67AJC1773⟩, and by the same reasoning (**22**) should also be non-aromatic ⟨70ZC296⟩. The corollary of this argument, however, is that the dissimilar ³J values for 1,4-dithiin, 6.97 Hz, and its dihydro analogue, 10.02 Hz, indicate substantial aromatic character in the former. However, it has been pointed out that this simple type of comparison fails to allow

**Table 4**    Chemical Shifts of Protons Bonded to Unsaturated Rings

| Compound | Chemical Shift (p.p.m.) | Ref. |
|---|---|---|
| 1,4-Dioxin | 5.5 | 66SA1859 |
| 1,4-Dithiin | 6.13 | 70JSP(35)83 |
| 1,4-Benzodioxin | 5.77 | 70OMR(2)431 |
| 1,4-Benzodithiin | 6.42 | 70OMR(2)431 |
| 1,4-Dithiino[b][1,4]dithiin (21)[a] | 6.46 | 76JOC1484 |
| 1-Ethyl-1,4-dithiinium tetrafluoroborate (22)[b] | 6.82 (H-2) 8.36 (H-3) | 70ZC296 |
| 1,2-Dithiin | 6.13, 5.97 | 67AG(E)698 |

(21)                    (22)

for the differences arising from a heteroatom transmitting its effect through one rather than two double bonds ⟨70JSP(35)83⟩.

The proton spectra for the dibenzo derivatives, *viz.* dibenzo[b,e][1,4]dioxin, phenoxathiin and thianthrene, have been reported in full, as part of a wider survey of heterocyclic compounds structurally related to anthracene. The protons in dibenzo[b,e][1,4]dioxin are the most shielded, and in phenoxathiin the protons *ortho* and *para* oriented to the C—O bond are shielded relative to those *ortho* and *para* to sulfur ⟨74OMR(6)115⟩. The $^{13}$C chemical shifts for phenoxathiin follow a similar pattern, with carbons *ortho* and *para* to the C—O bond resonating at 117.5 and 124.2 p.p.m. respectively, and at higher field than those *ortho*, *para* to the C—S bond (127.4 and 126.5 p.p.m.), in good agreement with shifts predicted on the basis of additivity effects ⟨73JMR(12)143⟩.

Chemical shifts of protons of 1,2-dithiin (Table 4) show no evidence of a ring current, but 'borderline aromaticity' of a 1,2-oxathiin 2,2-dioxide derivative has been inferred from the chemical shift of H-6 which falls midway between that for a proton in a non-aromatic cyclic ether and for H-2 of furan ⟨64JOC1110⟩. $^{13}$C shifts of 1,2-oxathiin 2,2-dioxides at the C-4 and C-6 positions have been considered anomalous but may be rationalized by invoking mesomeric withdrawal of electron density from these positions by the $SO_2$ moiety ⟨77OMR(10)208⟩.

Proton chemical shifts for some simple saturated rings are listed in Figure 4. The rings exist as chair forms in conformational equilibrium. Inversion is rapid relative to the NMR time scale and thus the values correspond to the mean chemical shift of the proton in axial and equatorial sites. The combination of rapid inversion and symmetry factors causes 1,4-dioxane, 1,4-dithiane, 1,3,5-trioxane and 1,3,5-trithiane to give rise to sharp singlets. Overall, the shifts are unremarkable and merely reflect the deshielding effects of the heteroatoms. Of interest, however, are the chemical shift differences for protons in equatorial and axial sites, $\delta H_{eq}$ and $\delta H_{ax}$, and these are available from conformationally fixed compounds or from low temperature spectra of the interconverting systems. Certain substituents, particularly in axial sites, induce fairly substantial shifts, but, in the absence of these, trends in $\delta H_{eq}$ and $\delta H_{ax}$ are apparent which are inherent to particular ring systems. Thus 1,4-dioxane ⟨71CC1558⟩ has a chemical shift difference between the axial and equatorial protons of only 1.6 Hz (at 100 MHz), whereas data for a conformationally rigid 1,4-oxathiane show that in this system the axial proton is upfield of the equatorial at the C-2 position, but downfield of it at the C-3 position ⟨77JOC2206⟩.

The main contributor to the difference in the chemical shift of axial and equatorial protons in a methylene group bonded to two heteroatoms (*i.e.* at C-2 in the 1,3-series) is thought to be the overall anisotropy of the ring bonds situated $\beta$ to the methylene group. The values of $\Delta\delta$ (Table 5) indicate that the anisotropies of the C—C, C—O and C—SO bonds have the same sign giving a positive contribution to $\Delta\delta$, whereas C—S and C—$SO_2$ bonds give a contribution of the opposite sign. The contribution from a C—SO bond is apparently larger than that from a C—S bond which has a larger effect than the C—$SO_2$ and C—O

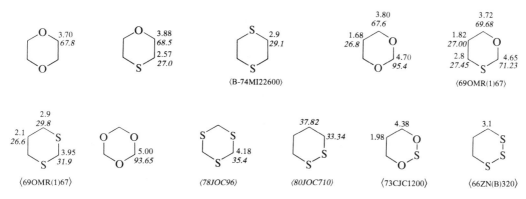

**Figure 4** ¹H and ¹³C (italicized) chemical shifts (p.p.m.) for some saturated compounds (⟨B-69MI22601⟩ (¹H) and ⟨B-79MI22600⟩ (¹³C) except where indicated otherwise)

**Table 5** Sign and Magnitude of the Geminal Shift Difference in a $CH_2$ Group α to Two Heteroatoms ⟨75JA1468⟩

$$H_{eq} \diagup \underset{H_{ax}}{Y} \diagdown \overset{X}{\underset{}{\diagdown}} Z$$

| | Compound[a] | | |
| X | Y | Z | $\Delta\delta$ (p.p.m.) $(\delta H_{eq} - \delta H_{ax})$ |
|---|---|---|---|
| O | O | $CH_2$ | +0.31 |
| O | S | $CH_2$ | −0.14 |
| S | S | $CH_2$ | −0.77 |
| S | S | S | −0.76 |
| O | ax. SO | $CH_2$ | +0.38 |
| O | eq. SO | $CH_2$ | +1.10 |
| S | ax. SO | $CH_2$ | +0.26 |
| S | eq. SO | $CH_2$ | +0.36 |
| S | eq. SO | S | +0.23 |
| S | S | eq. SO | −0.55 |
| S | $SO_2$ | $CH_2$ | −0.44 |
| eq. SO | eq. SO | $CH_2$ | +0.35 |

[a] *cf.* Cyclohexane $\Delta\delta$ = +0.48 p.p.m.

bonds. Such a trend explains the large negative $\Delta\delta$ in 1,3-dithiane and the opposing signs at the C-2 and C-4 methylene groups in 1,3,5-trithiane 1-oxide ⟨75JA1468⟩.

Chemical shift trends for the remaining protons in the 1,3-series and related compounds can be deduced from the representative data given in Table 6. The rules outlined above also satisfactorily accommodate the data for methylene groups adjacent to just one heteroatom: in the 1,3-dithiane the opposing effects of the β C—C and C—S bonds cancel whereas in the 1,3-dioxane the effect of the C—C and C—O bonds reinforce. At the C-5 methylene group, the axial proton is to lower field of the equatorial proton in the 1,3-dioxane and 1,3-oxathiane but to higher field in the 1,3-dithiane.

Chemical shifts in sulfoxide derivatives are very sensitive to $\overset{+}{S}$—$O^{-}$ bond orientation. The principal feature, the *syn* axial effect, was first detected in a member of the 1,4-oxathiane series ⟨66CC759⟩ and is the deshielding relative to the unoxidized compound of a *syn* axial proton by an axial $\overset{+}{S}$—$O^{-}$ bond. The deshielding is accompanied by a shielding (again, relative to the unoxidized compound) of the equatorial proton β to the sulfoxide sulfur, *cf.* data for 1,3-dithiane 1-oxides in Table 6. The effect extends to $\overset{+}{S}$—$O^{-}$ bonds in other environments, *e.g.* 1,2-oxathiane 2-oxides and 1,2-dithiane 1-oxides ⟨77CJC44, 71JOC1314⟩, 1,3,2-dioxathiane 2-oxides and to a lesser extent the 2,2-dioxides ⟨70RTC1244, 77OMR(9)347⟩.

Coupling constants in the saturated ring systems have proved valuable both for developing an understanding of factors which govern their magnitude and for conformational analysis. Geminal coupling constants are dependent upon electronegativity factors and bond angle ⟨70T3389, 71T1917, 72T2139⟩ and vary significantly from the oxygen-containing rings to their

**Table 6**   Proton Chemical Shifts in 2-*t*-Butyl Derivatives of Saturated Compounds with Heteroatoms 1,3

| Compound | | Chemical shifts δ (p.p.m.) | | | | | | |
| Y | Z | H-4$_{ax}$ | H-4$_{eq}$ | H-5$_{ax}$ | H-5$_{eq}$ | H-6$_{ax}$ | H-6$_{eq}$ | Ref. |
|---|---|---|---|---|---|---|---|---|
| O | O | 3.60 | 4.01 | 1.96 | 1.24 | 3.60 | 4.01 | 70TL2779 |
| O | S | 2.95 | 2.75 | 1.90 | 1.58 | 3.54 | 4.16 | 76JCS(P2)345 |
| S | S | 2.85 | 2.85 | 1.77 | 1.99 | 2.85 | 2.85 | 74JCS(P2)767 |
| *eq.* SO | S | 2.63 | 2.63 | 2.25 | 2.42 | 2.72 | 3.38 | 74JCS(P2)767 |
| *ax.* SO | S | 2.42 | 2.69 | 2.62 | 1.77 | 2.90 | 3.12 | 74JCS(P2)767 |

sulfur analogues, as illustrated by data for $^2J$ at the C-2 position in 1,3-dioxanes *ca.* −6 Hz, 1,3-oxathianes *ca.* −12 Hz and 1,3-dithianes *ca.* −14 Hz. $^2J$ also exhibits a further dependence upon the dihedral angle between lone pairs and the C—H bond, and departures from the above values in derivatives of these ring systems may be indicative of non-chair conformations ⟨71T1917⟩. Vicinal coupling constants show a strong dihedral angle dependence (the Karplus equation) and this is invaluable in configurational assignments. Like geminal couplings, $^3J$ values are also dependent upon electronegativity and bond angle factors ⟨B-69MI22602⟩. However, the ratio, $R$, of $\Sigma^3 J_{trans}$ to $\Sigma^3 J_{cis}$ within a —CH$_2$—CH$_2$— fragment is independent of electronegativity terms and is related to the torsion angle $\psi$ by the expression $\cos \psi = [3/(2+4R)]^{1/2}$. Values for $R$ and $\psi$ have been derived for a wide range of compounds and representative data are given in Table 7. Further results are summarized in a review article ⟨71ACR87⟩.

**Table 7**   Examples of Mean $^3J$ Values, $R$ Values and Derived Torsion Angles $\psi$ ⟨71ACR87⟩

| Compound | $^3J_{trans}$ (Hz)[a] | $^3J_{cis}$ (Hz)[b] | $R$ | $\psi$ (°) |
|---|---|---|---|---|
| 1,4-Dioxane | 6.11 | 2.78 | 2.20 | 58 |
| 1,4-Oxathiane | 7.9 | 2.7 | 2.9 | 62 |
| 1,4-Dithiane | 8.2 | 2.1 | 3.9 | 66 |
| 1,3-Dioxane | 6.7 | 3.8 | 1.8 | 55 |
| 1,3-Oxathiane (C-4, C-5) | 8.6 | 2.9 | 2.97 | 62 |
| (C-5, C-6) | 7.45 | 3.25 | 2.29 | 59 |

[a] Mean of $J_{ax,ax}$ and $J_{eq,eq}$.   [b] Mean of $J_{ax,eq}$ and $J_{eq,ax}$.

$^{13}$C NMR spectral parameters for saturated heterocycles have been comprehensively reviewed ⟨B-79MI22600⟩; data for some of the simple ring systems are given in Figure 4. Interest has centred largely upon the effect of substituents, particularly methyl groups, on ring carbon shifts and the effect of the heteroatoms on ring and substituent carbon shifts. As with $^1$H NMR, there is an abundance of published data for the three ring systems with heteroatoms 1,3 ⟨71JCS(B)1030, 71JA4772, 76JA3583⟩. Much of this has been collected in tabulated form which facilitates analysis and comparison of trends ⟨B-79MI22600⟩. Methyl substitution shows similar effects on ring carbon shifts in all three systems although the α effect of an axial methyl group at C-5 in 1,3-dithiane is upfield rather than downfield. An important feature which is evident in 1,3-oxathianes and 1,3-dithianes is the enhanced α effect of methyl groups at carbon positions adjacent to sulfur atoms. There is also an unusually high field shift for an equatorial methyl group at C-5 in 1,3-dioxane, resulting from its antiperiplanar relationship to the two ring oxygen atoms ⟨B-79MI22600⟩. *S*-Oxidation also strongly affects ring carbon shifts, significantly deshielding C-α relative to the unoxidized analogue. As with proton shifts, $^{13}$C shifts are sensitive to the orientation of the $\overset{+}{S}$—O$^-$ bond in sulfoxide derivatives. Data for isomeric pairs of conformationally fixed 1,4-oxathiane 4-oxides, 1,4-dithiane 1-oxides ⟨80JOC180⟩ and 1,3-dithiane 1-oxides ⟨78JOC96⟩ are consistent in demonstrating that C-α and C-β are to lower field in the isomer with the $\overset{+}{S}$—O$^-$ bond equatorial.

### 2.26.2.3.2 *UV spectroscopy*

UV spectral parameters for a broad range of compounds have appeared in tabulated form in a review ⟨71PMH(3)67⟩ and representative data, including some which have appeared since, are shown in Table 8; theoretical approaches for calculating the electronic transitions and for assigning bands for the present series of compounds are discussed in Section 2.26.2.1.

**Table 8** Representative Electronic Absorption Spectral Data

| Compound | $\lambda_{max}$(nm)(log$\varepsilon$) | Solvent | Ref. |
|---|---|---|---|
| 1,4-Dioxin | 250 (3.0), 307 (1.6) | Hexane | 46JA216 |
| 1,4-Dithiin | 262 (3.73), 266–270s (3.65) | EtOH | 53JA2065 |
| 2-Methyl-1,4-benzodioxin | 234 (3.90), 277.5 (3.36), 283 (3.31) | EtOH | 56G1336 |
| 1,4-Benzodithiin | 253 (4.23), 262s (3.81), 301 (2.91) | EtOH | 53JA1647 |
| Dibenzo[*b,e*][1,4]dioxin | 203, 222 (4.64), 228 (4.73), 289 (3.87), 300s | Heptane | 66MI22600 |
| Phenoxathiin | 220, 238 (4.50), 241 (4.50), 295 (3.60) | Heptane | 66MI22600 |
| Thianthrene | 209s, 243s, 254s, 258 (4.60) | Heptane | 66MI22600 |
| Phenoxathiin 10-oxide | 220 (4.55), 265s (3.15), 296 (3.63), 301 (3.65) | EtOH | 61MI22600 |
| Phenoxathiin 10,10-dioxide | 217 (4.58), 275s (3.40), 288 (3.62), 295 (3.68) | EtOH | 61MI22600 |
| *cis*-Thianthrene 5,10-dioxide | 217s (4.55), 250s (3.75) | EtOH | 61MI22600 |
| *trans*-Thianthrene 5,10-dioxide | 221 (4.67), 253 (3.70), 260 (3.68), 268 (3.67) | EtOH | 61MI22600 |
| Thianthrene 5,5-dioxide | 207 (4.30), 226 (4.32), 263 (3.89), 284 (3.83) | EtOH | 58AC(R)738 |
| Thianthrene 5,5,10,10-tetraoxide | 219 (4.57), 270 (3.51), 286 (3.77), 278 (3.69) | EtOH | 61MI22600 |
| 1,2-Dithiin[a] | 451 (2.88) | $C_6H_{12}$ | 67AG(E)698 |
| Dibenzo[*c,e*][1,2]dithiin | 217 (4.39), 254 (4.02), 266s (4.03), 277 (4.08) | EtOH | 75ZOR1304 |
| | 305s (3.51), 352s (2.44) | | |
| 1,4-Dioxane | 180 (3.8) | Gas | 51JA4865 |
| 1,4-Dioxene | 224s (3.5), 307 (1.3) | Hexane | 46JA216 |
| 1,4-Oxathiane | 227s (2.00), 209 (3.18) | EtOH | 49JA84 |
| 1,4-Dithiane | 225s (2.54) | EtOH | 49JA84 |
| 1,3,5-Trithiane | 240 (3.16) | EtOH | 49JA84 |
| 1,2-Dithiane | 290 (2.46) | — | 77JA2931 |
| Perhydro-1,2,3,4-benzotetrathiane | 232 (3.65), 277 (3.24) | $C_6H_{12}$ | 67AG(E)703 |
| 1,2,3,4,5-Pentathiane | 244 (3.65), 285 (3.12) | $C_6H_{12}$ | 68AG(E)301 |

[a] Data available for visible region only.

1,4-Dioxin gives a broad band absorption in the solution phase but five electronic transitions, with well marked vibrational patterns, are apparent in the gas phase ⟨46JA216⟩. 1,4-Dithiin gives rise to stronger absorption, but the longer wavelength maximum is to shorter wavelength than in 1,4-dioxin ⟨53JA2065⟩. 2-Methyl-1,4-benzodioxin absorbs at shorter wavelength than 1,4-benzodithiin. The latter compound fluoresces yellow-green in visible light and brilliant green in UV light ⟨53JA1647⟩. The UV spectra for dibenzo[*b,e*]-[1,4]dioxin, phenoxathiin and thianthrene below 260 nm are characterized by two very intense and one or two weaker bands which are displaced towards the red on passing along the series. The first two compounds also give rise to a band at longer wavelength ~295 nm. The bands below 260 nm show little solvent dependence whereas the long wavelength band in dibenzo[*b,e*][1,4]dioxin shows a bathochromic shift of 5 nm on change of solvent from heptane to acetonitrile. The UV spectra of the cation radicals of the three compounds, generated by dissolving them in sulfuric acid, show similar bands to those of the neutral compounds but displaced a little to the red. However, in the visible region the cation radicals give rise to a new and very characteristic absorption which, of course, is responsible for their colour. These absorptions occur at 665 (1.87), 580 (4.24) and 546 nm (3.95) for the cation radicals of dibenzodioxin, phenoxathiin and thianthrene respectively ⟨66MI22600⟩.

1,2-Dithiin is an orange-red material with absorption in the visible region at 451 nm, a band which shifts to the red by 20–30 nm in the 3,6-diaryl derivatives. The band is assigned to a $\pi$–$\pi$* transition, and the absence of a longer wavelength band assignable to an $n$–$\pi$* transition has been cited as evidence against the ring opening to the valence tautomeric bisthiocarbonyl structure ⟨67AG(E)698⟩ (see also Section 2.26.2.5). Of the five absorption maxima observed for dibenzo[*c,e*][1,2]dithiin, the 277 and 352 nm bands are particularly sensitive to substituents in the benzene rings, *e.g.* a 3-nitro substituent shifts them to 313 and 394 nm respectively ⟨75ZOR1304⟩.

1,4-Dioxane absorbs at 180 nm and is thus transparent in that part of the UV region accessible to basic spectrometer systems. Replacement of the ring oxygen atoms by sulfur causes a bathochromic shift and both 1,4-oxathiane and 1,4-dithiane show absorption above 220 nm. 1,3,5-Trithiane shows a further bathochromic shift which may be indicative of cross ring interactions. Interactions between two or more adjacent sulfur atoms are expected to be dependent upon conformation ⟨77JA2931⟩. Rings with four and five adjacent sulfur atoms are characterized by two strong absorptions in the region 240 and 290 nm.

### 2.26.2.3.3 IR and Raman spectroscopy

Characteristic absorptions in the vibrational spectra of the multi oxygen and sulfur heterocycles are those due to the C—O and the C—S bonds, the IR stretching frequencies for which fall in the regions 1300–1020 cm$^{-1}$ and 800–600 cm$^{-1}$ respectively. The bands for the former are normally the stronger.

Combined IR and Raman analysis has found use in certain structural studies, notably for 1,4-dioxin ⟨66SA1859⟩, based on the rule of mutual exclusion. This states that, for centrosymmetric molecules, transitions which are allowed in the IR spectrum are forbidden in the Raman spectrum and, conversely, transitions allowed in the Raman are forbidden in the IR spectrum ⟨B-45MI22600⟩. The two spectra of 1,4-dioxin show no coincident frequencies showing the structure is centrosymmetric, thus eliminating the boat form. The spectra also distinguish between the centrosymmetric chair and planar forms. The former, of $C_{2h}$ symmetry, should have all the u class vibrations IR active whereas in the latter, $D_{2h}$ symmetry, the $a_u$ transition should be inactive in both IR and Raman. Furthermore, in the Raman spectra, seven polarized bands are permitted for the chair compared with five for the planar form. On the basis of these differences 1,4-dioxin was shown to be planar. A further analysis of the low frequency region, 30–150 cm$^{-1}$, demonstrated that the ring is not in a deep potential well and is readily deformed into the boat form. Deformation into the chair form is more difficult as shown by the higher frequency, 516 cm$^{-1}$, for the appropriate ring deformation vibration ⟨66SA1859, 73JCP(58)4344⟩.

The $\nu$(C=C) band in 1,4-benzodioxin ⟨66T931⟩ is at 1675 cm$^{-1}$, at higher frequency than in 1,4-dioxin itself, 1640 cm$^{-1}$. In dibenzo[b,e][1,4]dioxin, very strong bands for the C—O—C asymmetric stretch are observed at 1299 and 1287 cm$^{-1}$, at higher frequency than in diphenyl ether. The bands are sensitive to the number and positions of substitution of chlorine atoms on the benzene rings. The 2,8-dichloro derivative shows a band at 1323 and 1307 cm$^{-1}$; the octachloro derivative at 1002 and 981 cm$^{-1}$ ⟨73MI22600⟩. Phenoxathiins show a characteristic band at ca. 1080 cm$^{-1}$ which remains relatively constant over a series of fifteen 2-substituted derivatives ⟨65RRC1245⟩. The spectrum of thianthrene also shows a band in this region but the two strongest and most characteristic absorptions appear between 740 and 770 cm$^{-1}$. Phenoxathiin 10-oxide and the mono- and di-sulfoxides of thianthrene give characteristic bands in the region 1038–1088 cm$^{-1}$ for the $\overset{+}{S}$—O$^{-}$ group ⟨74SA(A)2021, 77JOC2010⟩. Two particularly characteristic bands for dibenzo[c,e][1,2]dithiin are $\nu$(C—S) 674 and $\nu$(S—S) 417 cm$^{-1}$. The latter band is shifted to higher frequency in 2-substituted or 2,9-disubstituted derivatives irrespective of whether the groups are electron donating or withdrawing ⟨71ZOR143⟩.

In the saturated series, IR has been used as a structural tool for determining precise geometries. Thus, assignments of the observed low frequency bands for 1,4-dioxane, 1,3-dioxane and 1,3,5-trioxane and use of a derived potential function for ring bending has given structural data in good agreement with those obtained by X-ray, electron diffraction or microwave studies ⟨70JCP(53)376⟩. Greater attention however has been given to the completion of full analyses of vibrational spectra and these rely on a knowledge of the symmetry class, and thus the conformation of the ring system. The chair conformations of rings with like atoms situated 1,4 have $C_{2h}$ point group symmetry and the 36 modes of normal vibration are divided amongst the symmetry groups thus: 10 $a_g$ (Raman active), 9 $b_u$ (IR active), 9 $a_u$ (IR active) and 8 $b_g$ (Raman active) ⟨71SA(A)1025, 69SA(A)1041⟩. 1,4-Oxathiane however has a lower degree of symmetry and the vibrations are subdivided as 19a' and 17a'' ⟨72SA(A)137⟩. The chair conformations of 1,3-dioxane and 1,3-dithiane have $C_s$ point symmetry and the 36 vibrations should all be IR and Raman active ⟨69SA(A)1041⟩. The spectra of 1,3,5-trioxane and 1,3,5-trithiane are interpretable in terms of their $C_{3v}$ symmetry ⟨65SA1311, 69SA(A)1047, 69SA(A)1437⟩. In the former, some low intensity bands in the C—H region, below 2900 cm$^{-1}$, have been assigned to perturbation of the $\nu$(C—H) modes by orbitals on the two adjacent oxygen atoms ⟨72CR(B)(275)785⟩.

The $\overset{+}{S}$—$O^-$ stretch in 1,4-oxathiane 4-oxide and 1,4-dithiane 1-oxide appears as a strong absorption at *ca.* 1018 cm$^{-1}$, showing that the $\overset{+}{S}$—$O^-$ bond occupies the axial site ⟨78MI22601, 78MI22602⟩. 1,3,2-Dioxathiane 2-oxides show a strong absorption at *ca.* 1190 cm$^{-1}$ when the $\overset{+}{S}$—$O^-$ bond is axial but at *ca.* 1230 cm$^{-1}$ when equatorial. The symmetric and asymmetric ring S—O stretching vibrations occur in the regions 670–695 and 705–740 cm$^{-1}$ respectively ⟨77JCS(P2)612⟩.

### 2.26.2.3.4 *Mass spectrometry*

The standard reference text on the mass spectrometry of heterocycles is subdivided into chapters according to heteroatoms. Three sections in that book are of particular relevance and cover the literature through to the late 1960s ⟨B-71MS221, B-71MS225, B-71MS288⟩.

Of the unsaturated ring systems the dibenzo fused derivatives of the 1,4-series have attracted by far the most attention and provide a useful comparison of C—O and C—S bond reactivity under electron impact ⟨B-71MS221, B-71MS225, B-71MS288⟩. Dibenzo[*b,e*]-[1,4]dioxin and phenoxathiin both fragment to give dibenzofuran radical cation by extrusion of oxygen and sulfur respectively. Loss of oxygen is unusual among oxygen heterocycles and the peak for the daughter ion, though small (2.1%), is significant. Loss of sulfur is a generally more favoured process and thus the intensity of the dibenzofuran radical cation peak obtained from phenoxathiin is much higher, 21.2%. Loss of sulfur is also evident in the spectra of thianthrene ⟨74JHC287⟩ and 2,5-diaryl-1,4-dithiins ⟨72JHC887⟩. Dibenzodioxin and phenoxathiin also fragment with extrusion of CO, but this is not uncommon with aromatic ethers. Thianthrene undergoes an analogous fragmentation with loss of CS. A fragmentation of dibenzodioxin which clearly distinguishes it from the other dibenzo fused compounds is a remarkable simultaneous loss of two CO molecules to give, presumably, naphthalene radical cation. 1,4-Benzodioxin behaves similarly to give what is believed to be the radical cation of benzene ⟨70OMS(4S)121⟩.

Fragmentations of mono oxides of phenoxathiin and thianthrene include loss of O from the sulfoxide group and extrusion of SO. The sulfones also extrude SO but from the —O—SO— moiety, formed by an initial 1,2-phenyl shift from sulfur to oxygen ⟨B-71MS225, B-71MS288⟩. Dibenzo[*c,e*][1,2]oxathiin 6-oxide may also rearrange, here to a seven membered ring containing an —O—O—S— unit which could then fragment to give the observed M − 32 peak ⟨81JOC2373⟩.

In the 1,4-dioxane, 1,4-oxathiane and 1,4-dithiane series, electron impact gives rise to a significant molecular ion and small M − 1 ion. For 1,4-dioxane a major process is the loss of CH$_2$O giving rise to (C$_3$H$_6$O)$^+$ from which various other ions are derived. The spectrum of 1,4-oxathiane is dominated by ions containing sulfur rather than oxygen. The ions (C$_2$H$_5$S)$^+$ and (C$_2$H$_4$S)$^+$ form 15% of the spectrum and (C$_3$H$_6$S)$^+$, formed by loss of CH$_2$O, is present in smaller but still significant amounts (4.35%). The most abundant ion is (CH$_2$S)$^+$. In 1,4-dithiane the molecular ion forms the base peak followed by (C$_2$H$_5$S)$^+$ and (CH$_2$S)$^+$. The formation of (C$_3$H$_6$S)$^+$ is less pronounced relative to its formation from 1,4-oxathiane, illustrating that loss of CH$_2$S is a less favoured process compared to loss of CH$_2$O ⟨69OMS(2)1277⟩.

The mass spectra of the radical cations of 1,3-dioxanes and 1,3,5-trioxanes are characterized by an intense (M − 1)$^+$ peak for the resonance stabilized oxonium ion resulting from α cleavage of the C-2 proton. This type of cleavage is also a feature of the mass spectra of 1,3-oxathianes but in this series fragmentation of the ring occurs not only from the (M − 1)$^+$ ion or (M − R)$^+$ in the case of 2-substituted derivatives but also from the molecular ion. As with 1,4-oxathianes, the most intense peaks are generally those for sulfur-containing ions and reflect the preferential loss of an electron from sulfur rather than oxygen ⟨71OMS(5)763⟩. The fragmentations of 1,3-dithianes are somewhat more complex. Certain compounds undergo α cleavage at the C-2 substituent but a characteristic fragmentation, which also extends to 1,3,5-trithianes, is the elimination of S$_2$H ⟨72OMS(6)317, 78OMS338⟩. A clue to the origin of this loss is perhaps provided by the observation that loss of S$_2$H provides the base peak in the mass spectrum of 1,2-dithiane ⟨66JCS(B)946⟩. Thus, an attractive proposal (Scheme 1) to account for the loss of S$_2$H from the radical cation of 2,4,6-trimethyl-1,3,5-trithiane (**23**), and which may well have more general applicability, involves initial rearrangement into the 1,2,4-trithiane (**24**). Ring opening provides a key intermediate (**25**), fragmentations of which not only accommodate the loss of S$_2$H but also the observed losses of $C_4H_8S$ and the alkene $C_4H_8$ ⟨78OMS338⟩. Such a scheme also accounts for the loss of ethylene from the molecular ion of 1,3,5-trithiane itself and rationalizes the prominent radical ions in the spectrum of the 2,4,6-triphenyl derivative which correspond to stilbene and $PhCHS_3$ ⟨74OMS(9)649⟩.

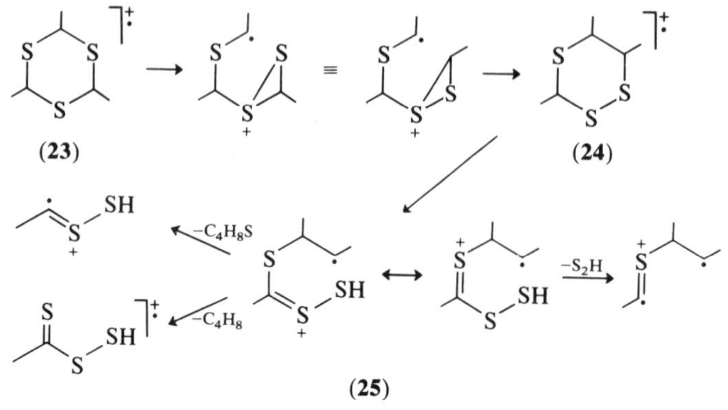

**Scheme 1**

As remarked above, the base peak in the mass spectrum of 1,2-dithiane is at $m/e$ 55 corresponding to loss of $S_2H$. The 3- and 4-carboxylic acid derivatives, however, do not follow this pathway. Both give peaks for $(M-S)^+$ and $(M-SH)^+$ but the former also gives a peak at $m/e$ 119 arising from loss of $CO_2H$, to give a sulfur-stabilized carbonium ion ⟨B-71MS225⟩. Extrusion of SO and $SO_2$ is frequently met among compounds in the 1,2-series where one of the heteroatoms is sulfur in the appropriately oxidized state ⟨B-71MS288⟩. Oxygen is extruded as $O_2$ in a fragmentation of 3,3,6,6-tetraalkyl-1,2,4,5-tetroxanes which also gives a ketone as the other product. A second peak characteristic of the tetraoxane ring is a peak at half of the molecular mass ⟨68JOC1931, 69TL3661⟩.

### 2.26.2.3.5 *Photoelectron spectroscopy*

An authoritative review covering application of UV photoelectron spectroscopy to heterocycles appeared in 1974 ⟨74PMH(6)1⟩ and important work on saturated multi oxygen and sulfur compounds is discussed therein. However, the studies of 1,4-dioxin, its dibenzo analogue, 1,4-dithiin and phenoxathiin, which are referred to in the first part of this section, appeared after the publication of that review.

The UV photoelectron spectrum of 1,4-dioxin shows six bands in the region 8–15 eV. The band of lowest ionization energy corresponds to ejection of an electron from a HOMO $\pi$-orbital centred mainly on the carbon–carbon double bond, whereas the second band is assigned to the out of phase combination of the double bond basis $\pi$-orbitals. The separation between the two bands, 2.52 eV, is significantly larger than in 1,4-cyclohexadiene, 1.00 eV, and demonstrates that the oxygen $n\pi$-orbital is superior to the $CH_2$ pseudo $\pi$-orbital as a relay orbital ⟨78HCA1388⟩. The spectrum of 1,4-dithiin has also been reported and correlation diagrams for the ionizations of 1,4-dioxin, 1,4-dithiin and calculated data for 1,4-oxathiin have been drawn. The energy of the band assigned to ionization from the $\pi$-MO formed mainly by the out of plane combination of the heteroatom lone pairs is particularly sensitive to change of heteroatom ⟨80MI22603⟩.

In the dibenzo compound, dibenzo[*b,e*][1,4]dioxin, the two bands of lowest ionization energy are related to electrons in antibonding $\pi$-MOs having a node along each C—O bond. The third and fourth bands, closely spaced, have been assigned to ejection of an electron from non-interacting $\pi$-phenyl orbitals with no involvement of the heteroatoms. A very similar pattern is observed in the photoelectron spectrum of phenoxathiin and close similarity between the ionization energies of the first two bands for the two compounds has been cited as evidence for the non-planarity of the two systems ⟨78JOM(146)235⟩.

The UV photoelectron spectra of the saturated compounds have attracted considerable attention. According to current theories of bonding the two lone pair orbitals on oxygen and sulfur in saturated systems are primarily *p*-type. One is essentially non-bonding whereas the other, which lies in the C—Z—C plane, is largely bonding. Bands associated with ejection of electrons from non-bonding orbitals are readily identified because of their sharp Franck–Condon profiles. The spectra of various saturated rings containing two or three oxygen or sulfur atoms show two such bands (Table 9), *i.e.* splitting. This has been taken as evidence of interaction between the non-bonding lone pair orbitals, the two bands corresponding to ejection from the symmetric and antisymmetric linear combinations of these orbitals ⟨74PMH(6)1⟩.

**Table 9**  Lone Pair Ionization Potentials ⟨72JA5599⟩

| Compound | IP (eV) | ΔIP (eV) |
|---|---|---|
| 1,4-Dioxane | 9.43, 10.65 | 1.22 |
| 1,4-Oxathiane | 8.67(S), 10.00 (O) | — |
| 1,4-Dithiane | 8.58, 9.03 | 0.45 |
| 1,3-Dioxane | 10.1, 10.35 | 0.25 |
| 1,3-Dithiane | 8.54, 8.95 | 0.41 |
| 1,3,5-Trioxane | ~10.8, 11.15 | ≤0.3 |
| 1,3,5-Trithiane | 8.76, 9.27 | 0.51 |

In the 1,4-series direct overlap is essentially precluded and the observed splitting can be accounted for in terms of through bond interaction through the C—C σ-orbitals. In this series the larger splitting is observed for 1,4-dioxane, and may arise because the energies of the oxygen lone pair orbitals and the C—C σ-orbitals are better matched ⟨72JA5599, 72AG(E)150⟩. Through bond interaction is larger still in *cis*-1,4,5,8-tetraoxadecalin, a molecule containing two 1,4-dioxane units fused *cis* with respect to each other ⟨81T3671⟩. The splitting observed in the spectra of 1,3-dioxane and 1,3-dithiane probably arises from a combination of both through space and through bond interaction. The larger splitting in the sulfur compound, 0.41 eV compared with 0.25 eV for 1,3-dioxane, may reflect the larger size of the sulfur orbitals which should be beneficial for through space overlap. [A somewhat different interpretation is available which specifically considers the presence of axial and equatorial lone pairs. The spectrum of 1,3-dioxane is assigned in terms of through space interaction of the two equatorial lone pairs. In 1,4-dioxane, interactions between equatorial–equatorial lone pairs and axial–axial lone pairs are both suggested to occur through bonds ⟨73BCJ1558⟩.] Two bands of unequal intensity are also observed in the spectra of 1,3,5-trioxane and 1,3,5-trithiane. For compounds with heteroatoms bonded directly to each other the splitting arises from direct spatial interaction. The magnitude of the splitting is a function of the angle between the two orbitals, being at a maximum when the orbitals are coplanar. The smaller splitting in 3,3,6,6-tetramethyl-1,2-dioxane relative to that in 1,2-dithiane implies that the ring torsion angle is larger in the former ⟨72AG(E)150, 74CB68, 74TL1467⟩.

A second technique is X-ray PE spectroscopy and this provides information about core ionizations. Electron binding energies of the sulfur atoms in 1,3,5-trithiane and its oxides have been reported and differences in the binding energy of the 2$p$ electrons indicate that there is partial charge transfer from the sulfide sulfur to the $d$-orbitals of the oxidized sulfur atoms ⟨72CC450⟩.

### 2.26.2.4 Aromaticity *versus* Antiaromaticity in the Unsaturated Series

As pointed out in Section 2.26.2.1, the heterocyclic ring systems are formally 8π-systems, although an aromatic sextet of electrons can be formulated for sulfur-containing rings by invoking $d$-orbital participation. A full discussion of the meaning of aromaticity is beyond the scope of this chapter and is dealt with in other texts, *e.g.* ⟨74AHC(17)255⟩. For present purposes a compound is considered aromatic if it satisfies the following criteria: it has a significant delocalization energy (DE) or resonance energy, it sustains a ring current, its geometry is planar with bond lengths inconsistent with normal single and double bonds, and it undergoes substitution rather than addition reactions.

The early evaluations of DE (Section 2.26.2.1) for 1,4-dithiin and thianthrene, which suggested that both ring systems have aromatic character, received support from an empirical measurement of the resonance energy of thianthrene using combustion methods. The compound was found to have a resonance energy of 71 kJ mol⁻¹ in excess of that for two benzene rings ⟨55ACS847⟩. However, it will be recalled that more recent theoretical treatments suggest that, like 1,4-dioxin, 1,4-dithiin has a negative resonance energy, *i.e.* is antiaromatic, and this prediction extends to the isomeric 1,2-dithiin. This last conclusion is interesting because such destabilization in 1,2-dithiin could be relieved by valence tautomeric ring opening to the unsaturated bisthioaldehyde, and this possibility has indeed been considered. However further calculations predict a σ-bond between the two S atoms, though no π-bond ⟨70T3227⟩ and UV spectral properties confirm the cyclic structure (Section 2.26.2.3.2). Further evidence has also been provided by a combustion study on 3,6-diphenyl-1,2-dithiin ⟨73MI22601⟩. Results from the latter provide an experimental heat of formation $\Delta H_f^{atm} (298\ K)_g = 14\ 183\ kJ\ mol^{-1}$ which is in much better agreement with values calculated for the cyclic form $(14\ 252\ kJ\ mol^{-1})$, than for the open chain form $(14\ 329\ kJ\ mol^{-1})$.

NMR-based studies (Section 2.26.2.3.1) have generally led to the conclusion that the compounds do not sustain a significant, or indeed any, ring current in the heteroring. Nor can the compounds be classified as aromatic on the basis of bond length data (Section 2.26.2.2). Although there is some degree of shortening of a C—S bond, particularly in the dibenzo fused derivatives in the 1,4-series, it would appear to be well accommodated simply by invoking conjugation of the sulfur lone pair with the benzene ring. Bond lengths for the 1,2-oxathiin 2,2-dioxide ring are reminiscent of those of a diene.

With the exception of the resonance energy determination of thianthrene, the bulk of the physicochemical evidence supports the contention that the fully unsaturated oxygen and sulfur compounds discussed above are non-aromatic. However, the remaining criterion, chemical reactivity, provides at least one piece of evidence which seems somewhat out of line. Although there is no evidence that 1,4-dithiin undergoes electrophilic substitution, its 2,5-diphenyl derivative and its benzo analogue undergo such reactions quite smoothly. Furthermore, the 1,2-oxathiin 2,2-dioxide ring undergoes bromination by substitution (Section 2.26.3.1.3).

Removal of two electrons from the formal cyclic 8π-electron structures serves to produce potential Hückel $4n + 2$ aromatic systems. The loss of one electron to form a radical cation was referred to in Section 2.26.2.1, and removal of a second electron by electrochemical oxidation, leading to dicationic structures, has also been achieved for a wide range of unsaturated compounds with heteroatoms in the 1,4 positions ⟨70ZC147, 73JA2375⟩. The oxidations are discussed further in Section 2.26.3.1.5, where tabulated data are presented. An interesting feature is the stability of certain salts of the dications, some of which have been isolated.

### 2.26.2.5 Conformational Preferences and Inversion Barriers in the Saturated Series

The conformational analysis of saturated multi oxygen and sulfur rings is very well developed. The most recent survey of this area ⟨B-80MI22600⟩ provides an excellent coverage of the topic, updating an important earlier review ⟨B-69MI22600⟩. Within the present chapter, theoretical approaches, ring shapes and structural determinations are covered in Sections 2.26.2.1, 2.26.2.2 and 2.26.2.3. The present section is thus concerned primarily with the conformational preferences of substituents and barriers to conformational inversion.

In the 1,4-series an electronegative group is expected to occupy the axial site by virtue of the anomeric effect arising from a combination of optimum orbital interactions and dipolar terms with the adjacent heteroatom. However, in this orientation a substituent becomes gauche to the second ring heteroatom and this may lead to an attractive or, more commonly, a repulsive interaction. Thus, 1,4,5,8-tetraoxadecalin is generated as the *cis* form (**26**) where the anomeric effect and an attractive oxygen–oxygen gauche interaction contribute to its stability relative to the *trans* fused isomer ⟨72JCS(P2)357⟩. In contrast, data for the 2-methoxy derivatives of 1,4-dioxane and 1,4-oxathiane, evaluated from ¹H NMR coupling constants, show how a gauche repulsion to a ring sulfur reduces the axial preference from 68% in the former, to 50–55% in the latter ⟨71T3111, 74CJC2041⟩: in fact, equilibration of fixed model systems suggests that the NMR method may underestimate this repulsion and that a 2-OMe group on the 1,4-oxathiane ring actually exhibits an equatorial preference ⟨77JOC438⟩. *trans*-2,3-Dihalo derivatives of 1,4-dioxane, 1,4-oxathiane and 1,4-dithiane exist with the two halogen atoms axial in both the solid state (Section 2.26.2.2) and in solution, as judged by dipole moment, NMR and IR data ⟨B-69MI22600⟩.

In the 1,3 series, conformational free energies of substituents have been measured largely by determining $\Delta G°$ values for acid-catalyzed equilibria of appropriate *cis* and *trans*

(26)          (27)          (28)

isomers: thus the conformational energy of a 5-substituent is obtained by studying epimerizations of type (27) $\rightleftharpoons$ (28) where $R^1$ is a conformational holding group. Examples of conformational free energies for alkyl groups at different ring sites are summarized in Table 10. The very large values for an alkyl group at C-2 of the 1,3-dioxane ring reflect the relatively short C—O bonds and the puckering of the O—C—O fragment (Section 2.26.2.2).

**Table 10** Conformational Free Energies of Simple Alkyl Substituents on Rings of the 1,3-Series

| | | Ring system | |
| Substituent | 1,3-*Dioxane*[a] (kJ mol$^{-1}$) | 1,3-*Oxathiane*[b] (kJ mol$^{-1}$) | 1,3-*Dithiane*[d] (kJ mol$^{-1}$) |
| --- | --- | --- | --- |
| 2-Me | 16.64 | 13.59 | 7.40 |
| 2-Et | 16.89 | 13.59 | 6.44 |
| 2-Pr$^i$ | 17.43 | 14.84 | 8.15 |
| 4-Me | 12.1 | 7.5[c] | 7.06 |
| 6-Me | 12.1 | 12.3[c] | 7.06 |
| 5-Me | 3.34 | — | 4.35 |
| 5-Bu$^t$ | 5.68 | — | — |

[a] 25 °C, solvent ether ⟨68JA3444, 70JA3050⟩.
[b] 25 °C, solvent CCl$_4$ ⟨72T2617⟩.
[c] ⟨73OMS(7)949⟩.
[d] 69 °C, solvent CHCl$_3$ ⟨69JA2703⟩.

The latter brings an axial 2-alkyl group into closer proximity to the *syn* axial protons than an axial alkyl group at a flatter point of the 1,3-dioxane ring, *e.g.* C-4. In the 1,3-oxathiane series the conformational free energy of a methyl group at C-4 is significantly less than at C-6. Interestingly, the alkyl equatorial preferences are least when the alkyl group is bonded to C-5, indicating that an axial alkyl group experiences less interactions with the ring heteroatom than with a *syn* axial proton. In the 1,3-dioxane series, for example, it is possible for a *t*-butyl group to adopt the C-5 axial site without forcing the ring to adopt a twist conformation ⟨B-80MI22601, B-80MI22602⟩.

The anomeric effect causes electronegative groups at C-2 to exhibit an axial preference, *e.g.* a methoxy group on the 1,3-dioxane ring shows an axial preference of 1.5–2.1 kJ mol$^{-1}$ ⟨70JA3050⟩ and a 2-SMe group on 1,3,5-trithiane a preference of 1.84 kJ mol$^{-1}$ ⟨75BCJ2496⟩. In contrast, electropositive groups such as SnMe$_3$, PbMe$_3$, SiMe$_3$, GeMe$_3$ exhibit a substantial equatorial preference ⟨81JOC558⟩ and the highly polar (if not ionic) 2-lithio-1,3-dithiane is estimated to have an equatorial lithium preference of 25 kJ mol$^{-1}$ ⟨74JA1807, 77JA8262⟩. In the solid state the 2-lithio-2-methyl-1,3-dithiane complex with tetramethylethylenediamine exists as a dimeric structure in which two dithiane units are bridged by equatorial lithium atoms partially bonded to C-2 of one ring and a sulfur atom of the other ⟨80AG(E)53⟩.

The conformational preferences of electronegative groups at C-5 depend very much upon the properties of the substituent and also vary with the polarity of the medium. A 5-OH group on the 1,3-dioxane and 1,3-dithiane ring shows a preference for the axial site where it is stabilized by intramolecular hydrogen bonding to the ring heteroatoms. A fluorine atom at C-5 on the 1,3-dioxane ring also exhibits an axial preference but this is not shared by the other halogens, which show equatorial preferences which are stronger than those found in the corresponding cyclohexyl compounds. A 5-OMe group shows a small axial preference in acetonitrile which is not sustained in solvents such as ether or carbon tetrachloride. The same group has an equatorial preference at the 5-position of the 1,3-dithiane ring: a 5-SMe group shows a stronger equatorial preference on the 1,3-dithiane ring than on the 1,3-dioxane ring. The results imply an attractive interaction between a ring oxygen and a gauche (*i.e.* axial) fluorine and between gauche oxygens but repulsive gauche interactions between the remaining combinations of atoms, *cf.* preferences in the 1,4-series. Substituents bearing a partial or full positive charge such as SOMe, SO$_2$Me,

$NMe_3^+$ show an axial preference at C-5 of 1,3-dioxane and this is attributed to electrostatic attraction towards the electronegative ring oxygen atoms ⟨B-80MI22601, B-80MI22602⟩.

Base-catalyzed equilibration of *cis*- and *trans*-2-*t*-butyl-1,3-dithiane 1-oxides indicates that the $\overset{+}{S}-O^-$ bond has an equatorial preference ⟨73TL849⟩ and this is confirmed by studies of conformational equilibria of mobile 1,3-dithiane 1-oxides and 1,3,5-trithiane 1-oxide ⟨B-80MI22602⟩. 1,3-Dithiane-1-tosylimide exhibits a similar preference for the S=NR bond ⟨76CC1002⟩. The equatorial $\overset{+}{S}-O^-$ preference contrasts with that observed in thiane 1-oxide and oxides of the 1,4-series ⟨75JOC2690, 77UP22600⟩ and calculations suggest it arises from dipole interactions (Section 2.26.2.1). However, 1,3-oxathiane 3-oxide does not behave similarly, exhibiting a preference for the axial $\overset{+}{S}-O^-$ conformer at low temperature, and thus another explanation seems necessary: cross ring orbital interactions may play a role ⟨78JCS(P2)1001⟩. In the 1,2-series, the $\overset{+}{S}-O^-$ bond of 1,2-oxathiane 2-oxide and of 1,2-dithiane 1-oxide exhibits an enhanced axial preference which presumably arises from dipole interaction. A further α-oxygen, as in 1,3,2-dioxathiane 3-oxide, increases this effect still further ⟨B-69MI22600⟩.

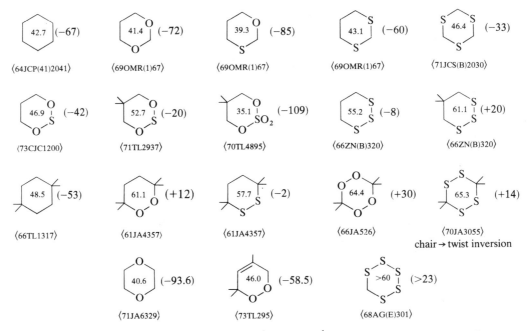

**Figure 5**  Representative ring inversion barriers, $\Delta G^{\neq}$ in kJ mol$^{-1}$ shown in the ring, at the coalescence temperature (°C) shown in parentheses.

Free energy barriers to chair–chair inversion are readily measured using the variable temperature NMR technique and representative data are collected in Figure 5. Values for many *gem* disubstituted rings in the 1,3-series have also been published ⟨69OMR(1)67⟩. The value for 3,3,6,6-tetramethyl-1,2,4,5-tetrathiane refers to the chair to twist inversion. The most noticeable feature of the data is that rings containing adjacent heteroatoms in the divalent state have substantially larger free energy barriers than those where the heteroatom is bonded to two carbon atoms. The presence of adjacent lone pairs is clearly the important factor here, as evidenced by the lowering of the barrier in passing from 5,5-dimethyl-1,3,2-dioxathiane to the corresponding 2,2-dioxide.

## 2.26.3 REACTIVITY

### 2.26.3.1 Fully Unsaturated Compounds: Reactivity at Ring Positions

#### 2.26.3.1.1 General survey

The fully unsaturated compounds are electron-rich systems and as such are highly reactive towards electrophiles. Principal reactions occur *via* attack at a ring sulfur atom, if available, or at the double bond (see Sections 2.26.3.1.4 and 2.26.3.1.3). Products of reaction at sulfur include simple sulfonium salts or derivatives of higher valence state such as sulfoxides or sulfones whilst reactions at double bonds normally give rise to addition products. However, certain compounds, notably 2,5-diphenyl-1,4-dithiin and 1,4-benzodithiin, show a marked propensity to react with electrophiles at carbon to give substitution products. A further feature of the chemistry of the 1,4-series, which is very obviously attributable to their electronic structure, is the facility with which they undergo one-electron oxidation reactions to a cation radical and then further to a $6\pi$-electron dication (Section 2.26.3.1.5).

Within the 1,4-series, reactions with nucleophiles at a ring carbon atom are limited, not surprisingly, to those derivatives bearing an electron-withdrawing group and/or where a ring sulfur atom occurs as a sulfonyl function. Nucleophilic displacements of substituents and cleavage of a ring C—S bond occur and lead, in some instances, to some rather remarkable decomposition modes (Section 2.26.3.1.6). Ring cleavage also follows after abstraction of a ring proton from 1,4-dithiins (Section 2.26.3.1.8). *S*-Oxides and *S*-dioxides in the 1,2-series are particularly sensitive to nucleophilic attack giving rise to products of cleavage of the 1,2 bond. These reactions often prove to be reversible on acidification (Section 2.26.3.1.7).

A commonly encountered reaction for sulfur-containing rings, in both the 1,4- and 1,2-series, is sulfur extrusion under thermolytic conditions. Sulfur is lost as elemental sulfur, sulfur monoxide or sulfur dioxide depending upon the oxidation state of the ring atom and the products are the ring contracted heteroaromatics (Section 2.26.3.1.2). Particular attention, however, is drawn to the high thermal stability of phenoxathiin, thianthrene and dibenzo[*c,e*][1,2]dithiins. Unlike their non-benzo fused analogues they extrude sulfur only under catalytic conditions: their stability is certainly inconsistent with the concept of $8\pi$-antiaromatic character but typical of aromatic ethers and thioethers. Their most frequently encountered ring cleavage reactions occur on treatment with alkali metals (Section 2.26.3.1.9).

### 2.26.3.1.2 *Thermal and photochemical reactions involving no other species*

1,4-Dioxin decomposes slowly over two to three weeks at room temperature to a polymeric material which does not melt below 250 °C ⟨39JA3020⟩. 1,4-Dithiin is more stable and can be distilled at 181 °C without decomposition ⟨59JOC1819⟩, although such thermal stability is not shared by its derivatives (*vide infra*). A useful comparison of the reaction modes of the 1,4-dioxin and 1,4-dithiin ring systems is possible from studies of their 2,3,5,6-tetraphenyl derivatives (**29**) and (**30**) ⟨77PIA(A)(86)1⟩. Thermolysis and photolysis of the former leads to products far more complex than those derived from the latter. Thus, on heating (**29**) in the absence of solvent at 250 °C the main product is the isomeric lactone (**31**) together with small quantities of *cis*-dibenzoylstilbene (**32**). At lower temperatures, and in solution, (**32**) becomes the predominant product, indicating that it is an intermediate in the conversion of (**29**) into (**31**). A plausible pathway for the process (Scheme 2) involves initial homolytic cleavage of a C—O bond to give a diradical which could give rise to (**32**) *via* an oxete. Ring closure of (**32**) to the zwitterion (**33**) followed by rearrangement completes the sequence. A small quantity of benzil is also produced which could arise *via* partial air oxidation of (**29**). Under photolytic conditions (**29**) gives benzil as the main product together

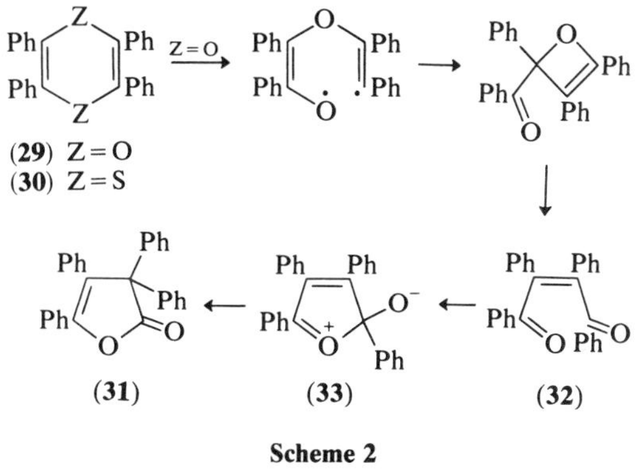

(**29**) Z = O
(**30**) Z = S

(**31**)          (**33**)          (**32**)

**Scheme 2**

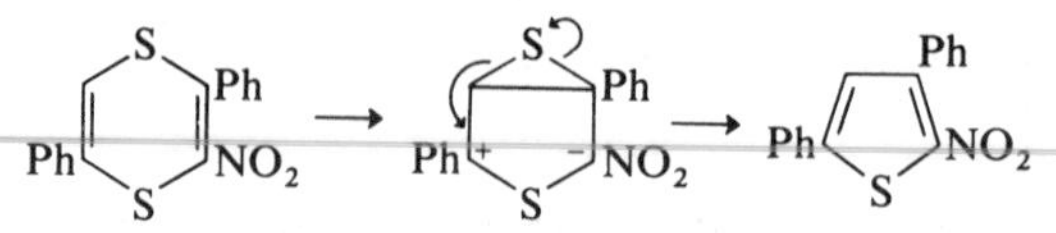

(34)    (35)

**Scheme 3**

with tolan and *trans*-dibenzoylstilbene. The first two may arise by fragmentation of the diradical, the third from the oxete.

In contrast, the 1,4-dithiin analogue (30), under both thermal and photochemical conditions is simply converted into tetraphenylthiophene. Sulfur extrusion of this type is very common with substituted 1,4-dithiins ⟨66HC(21-2)1112⟩ and the generally accepted mechanism, illustrated in Scheme 3 for the decomposition of 3-nitro-2,5-diphenyl-1,4-dithiin (34), involves a bicyclic zwitterionic intermediate. The mechanism accounts adequately for the formation of (35) in favour of the isomer having the nitro group β, the nitro group stabilizing the negative charge within the intermediate leading to the former. The sulfur extrusion process has considerable synthetic potential, providing access to substituted thiophenes which may otherwise be difficult to prepare.

1,4-Dithiins containing sulfur in a higher oxidation state also undergo extrusion of sulfur (as an oxide) to give the corresponding thiophene and where comparisons are possible it transpires that decomposition occurs much more readily than for the unoxidized analogues. Such thermal decompositions of oxides of 1,4-dithiin into thiophenes have been known for many years ⟨66HC(21-2)1112⟩ but a recent report shows that other very different reactions are also possible. Thus, in non-polar solvents 2,5-diaryl-1,4-dithiin 1-oxides give the ring contracted 1,3-dithioles (36) as the main or sole product, whilst under photolytic conditions (36) is accompanied by the isomeric (37). The reactions are thought to proceed *via* fragmentation of cyclic sulfenate esters produced as initial ring-expanded intermediates (Scheme 4) ⟨81TL5201⟩.

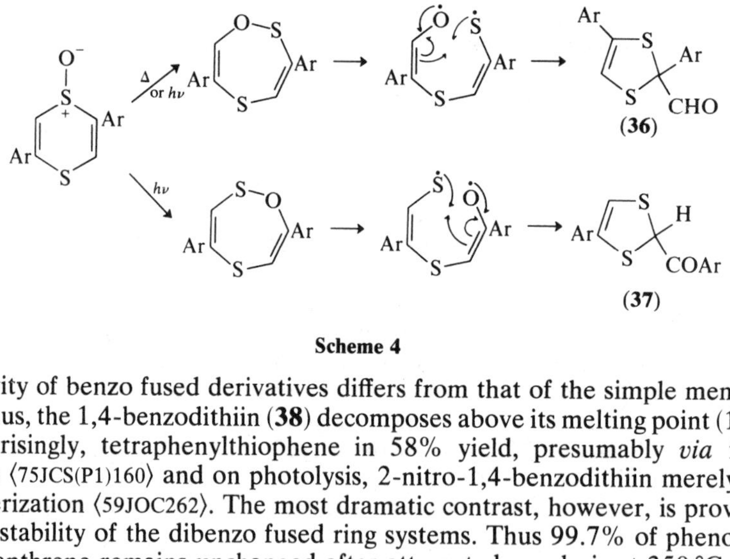

(36)

(37)

**Scheme 4**

The reactivity of benzo fused derivatives differs from that of the simple members of the 1,4-series. Thus, the 1,4-benzodithiin (38) decomposes above its melting point (157–158 °C) to give, surprisingly, tetraphenylthiophene in 58% yield, presumably *via* free radical intermediates ⟨75JCS(P1)160⟩ and on photolysis, 2-nitro-1,4-benzodithiin merely undergoes a [2+2] dimerization ⟨59JOC262⟩. The most dramatic contrast, however, is provided by the high thermal stability of the dibenzo fused ring systems. Thus 99.7% of phenoxathiin and 97.8% of thianthrene remains unchanged after attempted pyrolysis at 350 °C ⟨78BCJ1422⟩. Sulfur extrusion does occur, however, over certain catalysts (Section 2.26.3.1.9).

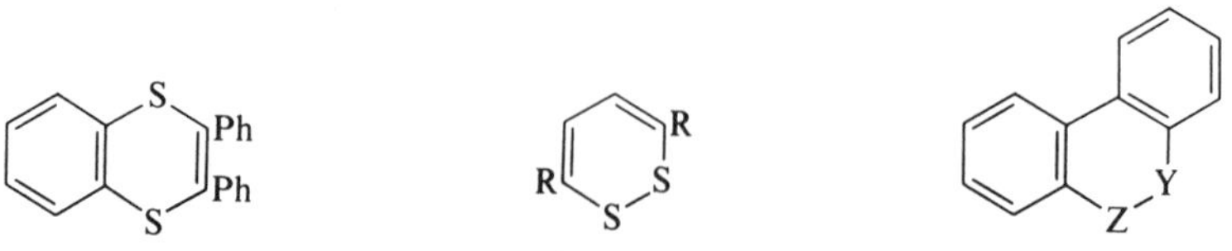

(38)    (39)    (40) Y = Z = S
(41) Y = O, Z = SO

In the 1,2-series, there are some conflicting reports concerning the extrusion of sulfur dioxide from 1,2-oxathiin 2,2-dioxides. In some instances the reaction proceeds smoothly to afford the corresponding furan, whereas in others pyrolysis merely produces an intractable tar ⟨66HC(21-2)789⟩ (see also Section 2.26.3.1.9). 1,2-Dithiins (**39**) have been shown to extrude sulfur under mild thermolysis or upon photolysis to give thiophene derivatives ⟨67AG(E)698, 65CB3081⟩, but, as in the 1,4-series, the dibenzo analogue (**40**) shows much higher thermal stability ⟨75G841⟩. Like thianthrene, (**40**) extrudes sulfur when heated over copper (Section 2.26.3.1.9). Recently dibenzo[*c,e*][1,2]oxathiin 6-oxide (**41**) has been subjected to flash vacuum pyrolysis at 900 °C with intriguing results. The condensate contains not only dibenzofuran but also dibenzothiophene. The mechanism for the formation of the latter is uncertain but may involve loss of molecular oxygen from a ring-expanded intermediate containing the S—O—O unit, which seems more likely than sequential loss of oxygen atoms from (**41**) itself. Mass spectra can often provide a crude indication of pyrolysis pathways and the detection of an M − 32 peak (Section 2.26.2.3.4) provides some support for the former mechanism ⟨81JOC2373⟩.

### 2.26.3.1.3 Electrophilic attack at carbon

1,4-Dioxin is considered to have a greater stability towards dilute acids than vinyl ether or 2,3-dihydro-1,4-dioxin but otherwise behaves as a typical unsaturated ether. In keeping with this property it reacts vigorously with one equivalent of bromine in carbon tetrachloride to give the addition product (**42**) and with excess chlorine to give 2,3,5,6-tetrachloro-1,4-dioxane. Under electrophilic nitration conditions, 1,4-dioxin is partially oxidized to give a tarry product while various attempted Friedel–Crafts acylation reactions merely gave polymer or unconsumed starting material ⟨39JA3020, 48JA2600, 48JOC671⟩. The 2,5-dimethyl derivative (**43**) reacts with methanol under acid catalysis (1 drop 12N hydrochloric acid) to give the bisketal addition product (**44**) ⟨57JA6219⟩.

(**42**) Z = O, X = Br     (**43**)        (**44**)        (**46**)
(**45**) Z = S, X = Cl

The reactivity of the 1,4-dithiin ring system is dependent somewhat upon the substituents which are present. The parent compound reacts with chlorine to afford two compounds for which the simple 1:1 addition structure (**45**) and the 2:1 product (**46**) have been assigned on the basis of their IR and UV spectra ⟨59JOC1819⟩. Attempts to bring about Friedel–Crafts acylations have given polymer or starting material, *cf.* 1,4-dioxin. Indeed, 1,4-dithiin shows a marked instability towards Lewis acids in general though it is stable to strong mineral acid ⟨B-61MI22601⟩. The chemical properties of 2,5-dimethyl-1,4-dithiin are similar to those of the parent. The compound is polymerized by aluminum chloride and no substitution products were isolated from reactions with bromine, chlorine, nitric acid, acetic anhydride, mercuric acid or the Vilsmeier reaction. In some cases starting material was recovered unchanged, in others amorphous polymer-like products were isolated ⟨59JA5993⟩.

In contrast to these two dithiins, the 2,5-diphenyl derivative (**47**) is much more stable to electrophilic reagents and undergoes various substitution reactions in the dithiin ring. Thus (**47**) can be smoothly monoformylated under Vilsmeier conditions and mono- and di-nitrated or -brominated. Sequential nitration and bromination gives the tetrasubstituted derivative (**48**) ⟨B-61MI22601⟩.

(**47**)          (**48**)

In the benzo fused series the oxygenated ring is by far the least stable to electrophilic reagents. This is exemplified by results of a study of 2-methyl-1,4-benzodioxin (**49**), which is polymerized by 1% sulfuric acid after one hour. Attempted nitration in glacial acetic acid followed by aqueous workup merely gave the hydrate (**50**) but in acetic anhydride a

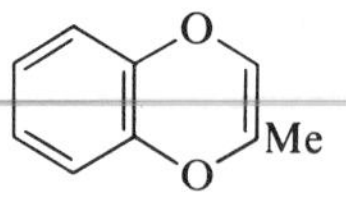

(49) 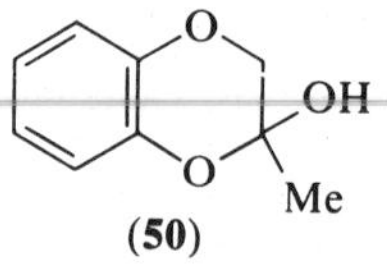           (50)

tar was formed. Intractable products were also generated under conditions of the Friedel–Crafts, Vilsmeier and Mannich reactions ⟨67AJC1773⟩. In marked contrast, 1,4-benzodithiin reacts smoothly with electrophiles to give 2-substituted derivatives in a range of reactions (Scheme 5) ⟨B-61MI22601⟩. The addition of ethanol under acid catalysis occurs only slowly to give the 2-ethoxy-2,3-dihydro derivative ⟨53JA1647⟩. However, bromine adds quite readily, although the product so formed is unstable and rapidly evolves hydrogen bromide to give polymeric material ⟨54JA1068⟩. Upon oxidation with excess of potassium hypochlorite the heterocyclic ring of 1,4-benzodithiin is cleaved ⟨53JA1647⟩.

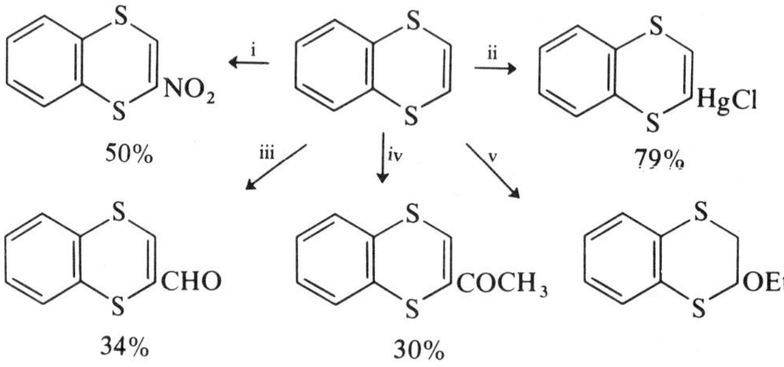

i, HNO$_3$, AcOH; ii, HgCl$_2$; iii, PhNMeCHO, POCl$_3$; iv, Ac$_2$O, H$_3$PO$_4$; v, EtOH, H$^+$

**Scheme 5**

Studies have also been made of 1,4-benzoxathiin, and its chemical properties are found to fall between those of 1,4-benzodioxin and 1,4-benzodithiin. It is considered to have greater stability than most vinyl ethers ⟨B-61MI22601⟩ but is nevertheless polymerized by boron trifluoride etherate or benzoyl peroxide. Upon nitration a yellow solid is formed which spontaneously decomposes with evolution of oxides of nitrogen. The addition of ethanol in the presence of acid catalyst is much more facile than the corresponding reaction with 1,4-benzodithiin and the addition product with bromine proves to be more stable. Like 1,4-benzodithiin, however, it undergoes the Vilsmeier reaction to give, rather surprisingly, the derivative with the formyl group at the 2-position (51) ⟨54JA1068, 60JOC53⟩.

The apparent ease with which certain of the sulfur-containing compounds undergo electrophilic substitution reactions has been considered significant in discussions of the aromaticity of these ring systems: *e.g.* ⟨66HC(21-2)1146⟩. However, whilst this type of reactivity is normally associated with aromatic character it is important to note that the open-chain di(phenylthio)ethylene (52) is also smoothly formylated at the double bond under Vilsmeier conditions to give (53) ⟨B-61MI22601⟩.

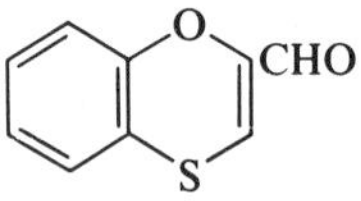

(51)        (52) R = H        (54) R = H
                  (53) R = CHO    (55) R = Br

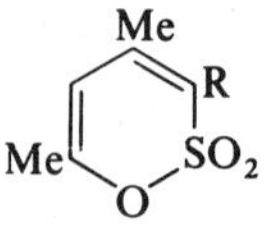

The series of substitution reactions described above also finds a parallel in the limited studies which have been made on the 1,2-oxathiin 2,2-dioxide ring system; here bromination gives a monosubstitution product. In an early report the product of bromination of (54) was tentatively assigned that of the 5-bromo derivative. However, a reappraisal, based on $^1$H NMR chemical shifts and unambiguous synthesis, revealed that substitution in fact occurs at C-3 (55) ⟨48JA864, 69TL651⟩. It will be recalled that X-ray crystal structure data show that the ring system cannot be regarded as aromatic (Section 2.26.2.4).

### 2.26.3.1.4 Electrophilic attack at sulfur

Examples of reactions in which sulfur reacts with electrophiles or oxidants are particularly numerous in the 1,4-series, and the facility with which the reactions occur is indicative that the compounds behave more like a vinyl sulfide than a conjugated aromatic system. 1,4-Dithiin reacts with chloramine-T to give the monosulfilimine derivative ⟨66HC(21-2)1112⟩ and is alkylated to give 1-alkylthiinium salts ⟨70ZC296⟩. However, the reaction for which most examples are known is *S*-oxidation. 1,4-Dithiin is oxidized with hydrogen peroxide in acetic acid at room temperature to give a dioxide which may be either the sulfone or a disulfoxide, while heating the solution under reflux affords the disulfone. 2,5-Diaryl-1,4-dithiins with peracetic acid are oxidized to the monosulfoxides within one or two minutes and thence to monosulfones after longer reaction time. However, certain derivatives are converted into thiophenes when the oxidizing solution is heated, presumably by thermal decomposition of an *S*-oxidized intermediate ⟨66HC(21-2)1112⟩ (*cf.* Section 2.26.3.1.2). 1,4-Benzoxathiin and 1,4-benzodithiin have been smoothly oxidized with hydrogen peroxide to their monosulfone and disulfone derivatives respectively ⟨66HC(21-2)852, 66HC(21-2)1146⟩, while thianthrene and phenoxathiin have been oxidized to all the possible oxide derivatives using an assortment of reagents which includes hydrogen peroxide, potassium permanganate and nitric acid ⟨66HC(21-2)864, 66HC(21-2)1155⟩. Oxidation to sulfoxides using sulfuric acid, followed by hydrolysis, involves cation radical species, and this topic is covered in Section 2.26.3.1.5. In the 1,2-series, dibenzo[*c,e*][1,2]dithiin is oxidized to the 5-oxide with perbenzoic acid ⟨73IJS(8)255⟩ and to the 5,5-dioxide with nitric acid ⟨66HC(21-2)968⟩. Conditions are also available for oxidation through to the 5,5,6,6-tetraoxide derivative ⟨78JOC914⟩. Peracetic acid oxidation of dibenzo[*c,e*][1,2]oxathiin 6-oxide gives the 6,6-dioxide ⟨81JOC2373⟩. However, oxidation of 3,6-diphenyl-1,2–dithiin (**56**) with peracetic acid gives the ring-cleaved disulfonic acid (**57**) ⟨67AG(E)698⟩.

(**56**)    (**57**)    (**58**) X = NH, Z = O, S
(**59**) X = O, Z = O, S

(**60**) R = H, Z = O, S
(**61**) R = Ts, Z = O, S        (**62**)

Phenoxathiin and thianthrene react with *O*-mesitylenesulfonyl hydroxylamine followed by sodium methoxide to give sulfimides (**58**), and the same reagent also converts the monosulfoxides (**59**) into the sulfoximides (**60**). The latter conversions can also be achieved in two steps using tosyl azide over copper followed by acid hydrolysis of the *N*-tosyl product (**61**) ⟨74TL1973⟩. On reaction with bromine in carbon disulfide, thianthrene is reported to give a black crystalline tetrabromide adduct. However, in other solvents reaction leads to substitution in the benzene rings (Section 2.26.3.2.2). With chlorine in benzene, phenoxathiin forms a 1:1 adduct, isolated as a red-yellow crystalline material decomposing at 71 °C. The corresponding adduct with thianthrene is obtained as red prisms. Both adducts undergo hydrolysis to the sulfoxide. The thianthrene:chlorine adduct reacts with benzene under Friedel–Crafts conditions to give a compound assigned as a 5-phenylthianthrenium ion and isolated as a chloroplatinate salt (**62**) ⟨66HC(21-2)864, 66HC(21-2)1155⟩. Phenoxathiin has been shown to undergo direct *S*-arylation using diaryliodonium salts with copper benzoate as catalyst, at 120–125 °C for three hours. The mechanism of the reaction is not fully understood but by-products expected from a radical type process are not formed. Salts of Ti(II), Cr(II), Fe(III), Ag(I), Co(II) and Pd(II) do not function as catalysts in the reaction ⟨78JOC3055⟩.

### 2.26.3.1.5 Oxidation to cation radicals and dications

One-electron oxidations are an interesting feature of the chemistry of the fully unsaturated compounds. Removal of the first electron gives rise to cation radical and further loss of an

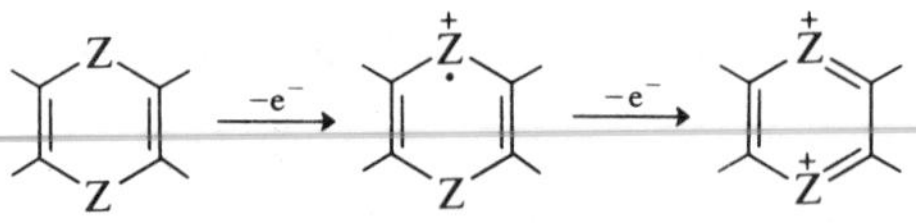

**Scheme 6**

electron leads to a Hückel $6\pi$-electron dication (Scheme 6). A comprehensive investigation of the facility with which these processes occur in the 1,4-series has been performed using voltammetry and values for the halfwave potentials for the two oxidation steps are given in Table 11. ESR has been used extensively to characterize the cation radical species (see also Section 2.26.2.1), for which studies they are often generated simply by dissolution of compounds in sulfuric acid. Other electron acceptors act equally well ⟨69JCS(B)755⟩.

**Table 11**   Half-wave Potentials for One and Two Electron Oxidation Processes[a]

| Compound | $E_{1/2}$ (V)[b] |
|---|---|
| 1,4-Dioxin | 0.90 |
| Tetraphenyl-1,4-dioxin | 0.92, 1.41 |
| 1,4-Dithiin | 0.69, 1.16 |
| Tetraphenyl-1,4-dithiin | 0.94, 1.52 |
| 1,4-Benzodioxin | 0.96, 1.35 |
| 1,4-Benzodithiin | 0.80, 1.34 |
| Dibenzo[b,e][1,4]dioxin | 1.12 |
| Phenoxathiin | 0.92, 1.45 |
| Thianthrene | 0.96, 1.31 |

[a] *cf.* Scheme 6 ⟨70ZC147⟩.
[b] Measured using a rotating platinum electrode in anhydrous acetonitrile against a calomel electrode.

Isolation of salts of both the cation radicals and dications has been achieved for a number of systems. Thus, reaction of 2,3,4,5-tetraphenyl-1,4-dioxin with antimony pentachloride in chloroform gives the cation radical (**63**) as blue-violet crystals, and this can be oxidized by voltammetry to the green dication salt (**64**). Tetraphenyl-1,4-dithiin with antimony pentachloride in benzene affords the dication salt (**65**) directly as dark violet crystals ⟨70ZC147⟩. 6-Methyl-1,4-benzodithiin reacts with the same reagent to give a cation radical salt, which decomposes in air to a black resin ⟨62JCS4963⟩.

The preparation of dibenzo[b,e][1,4]dioxin cation radical (**66**) has been achieved by oxidation of the heterocycle in ethyl acetate–lithium perchlorate at a platinum anode, using a controlled potential of 1.2 volts *vs.* Ag–AgClO₄. The blue solid collected at the anode contained between 85–90% of (**66**) as the perchlorate ⟨74JHC139⟩. The purple cation radical perchlorate of phenoxathiin, (**67**), is obtained in high purity by oxidation of phenoxathiin in benzene with 70% perchloric acid–acetic acid ⟨75JOC2756⟩. Similar perchloric acid oxidation of thianthrene affords the dark reddish brown perchlorate of (**68**) ⟨69JOC3368⟩ and the heterocycle can also be oxidized on a preparative scale with antimony pentachloride ⟨62JCS4963⟩.

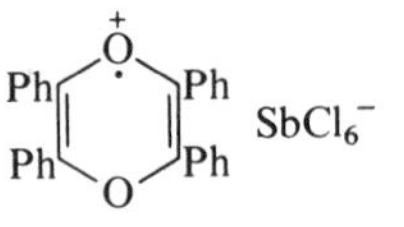

(**63**)

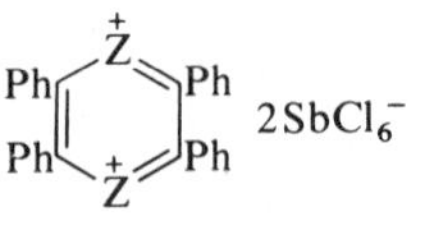

(**64**) Z = O
(**65**) Z = S

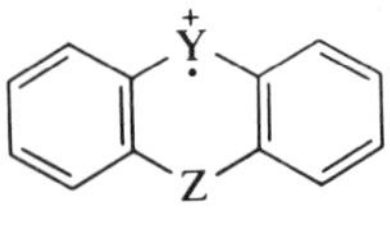

(**66**) Y = Z = O
(**67**) Y = S, Z = O
(**68**) Y = Z = S

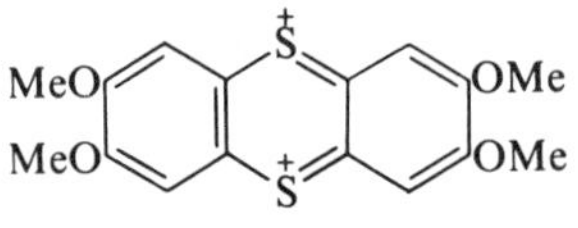

(**69**)

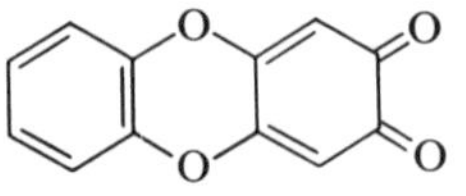

(**70**)

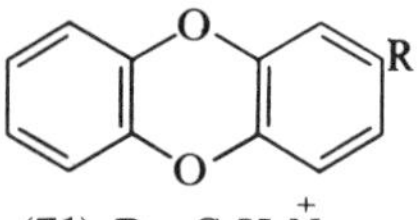

(**71**) R = C₅H₅Ṅ—
(**72**) R = NO₂

The electron-donating substituents in 2,3,7,8-tetramethoxythianthrene stabilize the dication (**69**) derived from it and this has been generated under various conditions. Thus, sapphire blue solutions containing (**69**) are formed by anodic oxidation of the heterocycle in nitromethane, or simply by bubbling oxygen through a solution in nitromethane saturated with aluminum chloride. Isolation of (**69**), as the diperchlorate salt, is achieved by oxidation with perchloric acid ⟨73JA2375⟩.

The chemical reactivity of the cation radicals (**66**), (**67**) and (**68**) has stimulated considerable interest. The first of these, on generation in acetonitrile by anodic oxidation, reacts with water in the solution to reform dibenzo[*b,e*][1,4]dioxin together with what is believed to be the quinone (**70**), which undergoes *in situ* cleavage reactions ⟨72BSF3588⟩. The solid perchlorate salt of (**66**) decomposes on standing in a desiccator to a compound described as a dibenzodioxin dimer. Reactions of (**66**) with nucleophiles such as pyridine and nitrite ion generate the corresponding 2-substituted derivatives (**71**) and (**72**), together with dibenzo[*b,e*][1,4]dioxin itself. The 2-nitro derivative (**72**) is also formed by reaction of the cation radical perchlorate with nitrate ion. In this reaction no dibenzodioxin is formed. It seems a curious reaction and as yet the mechanism is not understood. Nucleophiles such as $CN^-$, $NH_3$ and primary amines do not give rise to ring substitution products but lead to dibenzodioxin in high yield ⟨74JHC139⟩.

The sulfur atoms within the cation radicals of phenoxathiin and thianthrene open up a reaction mode which distinguishes their chemistry from that of the dibenzodioxin radical cation. Both (**67**) and (**68**) normally react with nucleophiles at sulfur rather than at the benzene ring and the products of such reactions are generally accompanied by an approximately equimolar amount of the parent heterocycle. There are two mechanisms which accommodate these findings and both have their proponents ⟨66BSF2510, 69JOC3368, 70JA7488, 75JA101⟩. They differ in terms of the species reacting with the nucleophile, and are illustrated in Scheme 7 for the reaction with water to give the sulfoxides. Mechanism *a* invokes disproportionation of the cation radical to the parent heterocycle and the dication, the latter reacting with the nucleophile, whereas in mechanism *b* the radical cation itself reacts with the nucleophile.

$$a \quad R^{\ddagger} + R^{\ddagger} \rightarrow R^{2+} + R$$
$$R^{2+} + H_2O \rightarrow (ROH_2)^{2+}$$
$$(ROH_2)^{2+} \rightarrow RO + 2H^+$$
$$b \quad R^{\ddagger} + H_2O \rightarrow (ROH)\cdot + H^+$$
$$(ROH)\cdot + R^{\ddagger} \rightarrow (ROH)^+ + R$$
$$(ROH)^+ \rightarrow RO + H^+$$

**Scheme 7**

Apart from hydrolysis, reactions have also been performed with nucleophiles such as ammonia ⟨72JA1026, 75JOC2756⟩ to generate sulfilimines (**73**), amines ⟨77JOC1538⟩, activated aromatics to give aryl sulfonium salts ⟨71JOC2923⟩, organomercury compounds to give alkyl or aryl sulfonium salts ⟨78JPC1168⟩, and alkenes and alkynes ⟨79JOC915, 81JOC271⟩. The products of reactions with ketones, apart from the parent heterocycle, are β-ketoalkyl sulfonium salts (**74**), formed presumably by reaction of the enol. Treatment with base can deprotonate the salts to give the sulfur ylide. Similar reactions with β-diketones and β-ketoesters lead, in some instances, to the ylides directly ⟨75JOC3857⟩. Reaction of thianthrene cation radical with nitrite and nitrate ion leads to thianthrene 5-oxide ⟨72JOC2691⟩; in the latter reaction, no thianthrene is detected ⟨79JPC2696⟩. In contrast to the above reactions where the nucleophile attacks at sulfur, treatment of thianthrene cation radical with pyridine leads to substitution at the C-2 position, *cf.* benzo[*b,e*][1,4]dioxin cation radical. A reaction with pyridine at sulfur invoking mechanism *a* of Scheme 7 would give a structure (**75**) which could not readily be deprotonated (contrast reactions with water, amines, *etc.*). Thus, were this reaction to occur, the formation of (**75**) would probably be reversible whereas attack at C-2 (**76**) can be followed by proton loss ⟨72JOC2691⟩.

### 2.26.3.1.6 Nucleophilic attack at carbon

Nucleophilic attack at the carbon of a 1,4-dithiin ring proceeds well when the ring is substituted with electron-withdrawing substituents. Thus, the dinitro derivative (**77**) reacts with aniline leading to displacement of the RS groups to give the substitution product (**78**) ⟨55LA(595)101⟩. Tetracyano-1,4-dithiin (**79**) is also attacked at the ring carbon by secondary

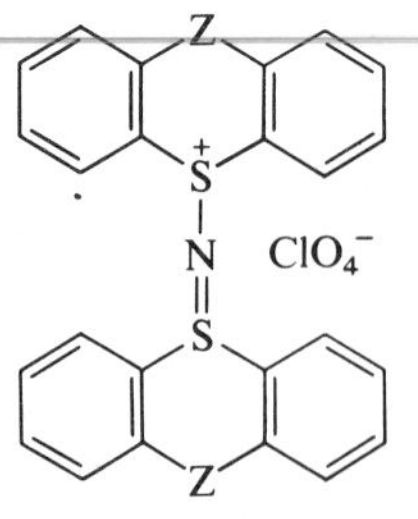

(73) Z = O, S

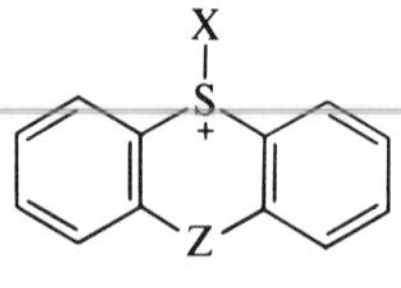

(74) X = CH₂COR,
Z = O, S

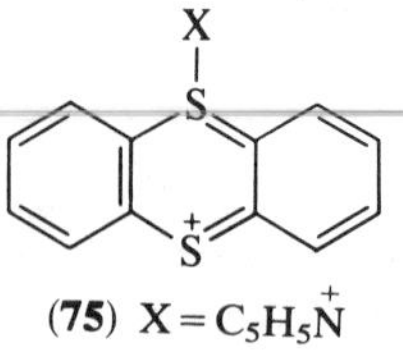

(75) X = C₅H₅Ṅ

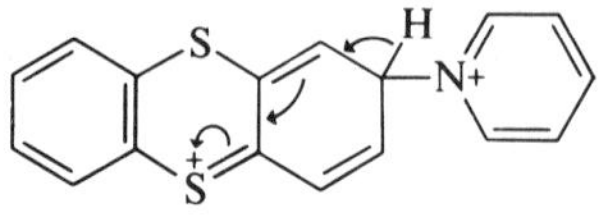

(76)

amines, and this too is followed by C—S bond breakage, resulting here, of course, in ring cleavage. The anion (80) so formed can be quenched with dimethyl sulfate ⟨70LA(736)176⟩.

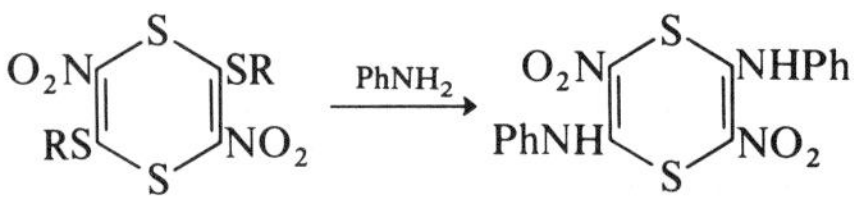

(77)          (78) 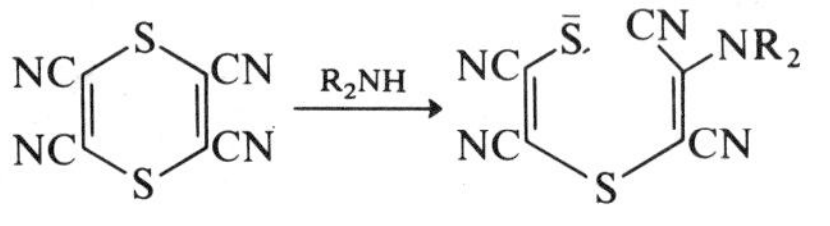          (79)          (80)

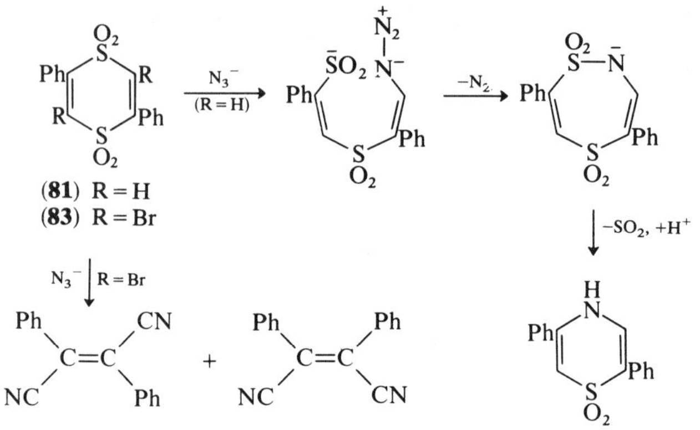

**Scheme 8**

Intriguing reactions occur among disulfone derivatives of 1,4-dithiins. Thus, azide ion attacks the disulfone (81) to give the thiazine (82) *via* a ring opening–ring closure pathway followed by elimination of SO₂ (Scheme 8). A more deep-seated reaction occurs, however, when the dibrominated analogue (83) reacts with azide ion. Here bromide is thought to be displaced by azide and the product loses nitrogen and then sulfur dioxide to give a mixture of (*E*)- and (*Z*)-dicyanostilbenes (84) and (85), with the latter predominating ⟨82TL299⟩.

The 1,4-benzoxathiin 4,4-dioxide (86) undergoes a ring cleavage reaction with base but in this case the carbon–sulfur bond remains intact. The reaction has been formulated in terms of the mechanism shown in Scheme 9 ⟨60JOC53⟩. A sulfone functionality also facilitates nucleophilic bond cleavage of the heteroring of phenoxathiin and thianthrene. Whereas both compounds are relatively stable to KOH (the former is reported to undergo C—O bond cleavage only at 195 °C ⟨66HC(21-2)864, 68JOC446⟩) their sulfone derivatives (87) and (88) are cleaved at more moderate temperatures. The former reacts with KOH in refluxing alcohols to give, after acidification, 2-hydroxy-2′-alkoxydiphenyl sulfone (89) or the 2,2′-

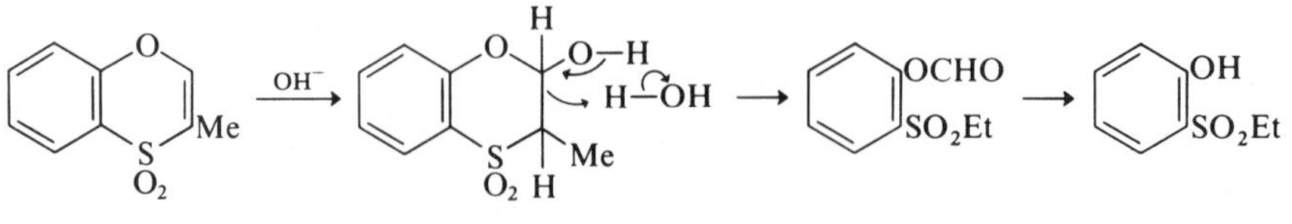

(86)

**Scheme 9**

(**87**) Z = O
(**88**) Z = SO₂

(**89**) X = OR
(**90**) X = OH
(**91**) X = SO₂H

dihydroxy analogue (**90**) using KOH in *t*-butanol ⟨68JOC446⟩. Similar reaction of (**88**) with KOH in ethanol affords (**91**) on acid workup but examples are known where more extensive cleavage occurs to give 2-ethoxyarylsulfinic acids ⟨66HC(21-2)1220⟩. Examples of cleavage of dibenzo[*b,e*][1,4] dioxins using sodium in ammonia are considered in Section 2.26.3.1.9.

### 2.26.3.1.7 Nucleophilic attack at a heteroatom

As expected, nucleophilic attack at a heteroatom in the 1,4-series is relatively rare, but a few examples are to be found among reactions of the oxides of phenoxathiin and thianthrene. The reaction of phenoxathiin 10-oxide (**92**) with butyllithium followed by carbon dioxide is a case in point. Apart from phenoxathiin, formed by a reduction process, and its 1-carboxylic acid, generated *via* lithiation of the benzene ring (Section 2.26.3.2.3) there are also a number of ring cleavage products. These include the dicarboxylic acid (**93**), diphenyl ether and phenoxybenzoic acid. Thianthrene 5-oxide (**94**) under similar conditions gives dibenzothiophene and the dicarboxylic acid (**95**) as the main cleavage products while the trioxide (**96**) behaves in like manner, affording dibenzothiophene dioxide and the bis acid (**97**). It seems likely that (**92**), (**94**) and (**96**) all undergo initial nucleophilic attack at the sulfinyl group to give a lithiated species (**98**) from which C₄H₉SO⁻ is lost. The corresponding reaction between (**94**) and *n*-butylmagnesium bromide, with carbon dioxide quenching, also gives the dicarboxylic acid (**95**), presumably *via* an analogous series of reactions ⟨66HC(21-2)864, 66HC(21-2)1155⟩.

(**92**) Z = O
(**94**) X = S
(**96**) Z = SO₂

(**93**) Z = O
(**95**) Z = S
(**97**) Z = SO₂

(**98**) Z = O, S, SO₂

In the 1,2-series, nucleophilic attack at a heteroatom leading to cleavage of the heteroatom–heteroatom bond is a relatively common occurrence. Thus, 1,2-dithiins are particularly labile to alkali ⟨65ZC353⟩ and are reduced with sodium borohydride (or sodium in liquid ammonia) to give dimercaptobutadienes. They also react with hydrazine to give pyridazines with extrusion of H₂S ⟨67AG(E)698⟩. Reaction of the oxathiin dioxide (**54**) with primary amines at 140–200 °C gives rise to the corresponding thiazine (**99**) ⟨66HC(21-2)789⟩ while ring opened products (**100**) and (**101**) are formed on irradiation in the presence of benzylamine and methanol respectively ⟨61MI22602⟩. Hydrolytic cleavage of simple 1,2-oxathiin 2,2-dioxides to sulfonic acids is normally achieved using hot alkali, hydrochloric acid or superheated steam ⟨66HC(21-2)789⟩. Hydrolysis products of dibenzo fused derivatives of 1,2-oxathiin dioxides are (hydroxyaryl)arenesulfonic acids ⟨66HC(21-2)968⟩; when the sulfur atom is in a lower oxidation level, as in the dibenzoxathiin monoxide (**41**), the product after base-catalyzed hydrolysis and acidification is the corresponding sulfinic acid (**102**). In strong acid (**102**) recyclizes to (**41**). The latter compound has also been shown to undergo

(**54**) Z = O
(**99**) Z = NR

MeCOCH=CHMe.CH₂SO₂X

(**100**) X = NHCH₂Ph
(**101**) X = OMe

(**41**)

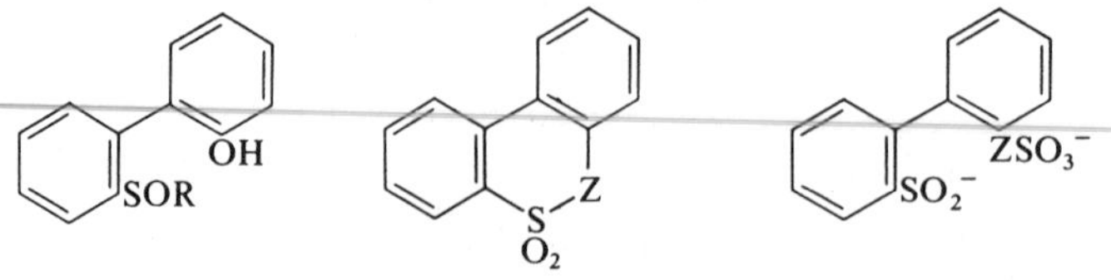

(**102**) R = OH
(**103**) R = Me

(**104**) Z = S
(**105**) Z = SO
(**106**) Z = SO₂

(**107**) Z = S
(**108**) Z = SO
(**109**) Z = SO₂

ring cleavage with methylmagnesium iodide, again by reaction at sulfur, to give the sulfoxide (**103**) ⟨81JOC2373⟩.

Oxides of dibenzo[c,e][1,2]dithiin are likewise sensitive to reaction with nucleophiles. The series (**104**)–(**106**) are all readily cleaved with sulfite ion to give the derivatives (**107**)–(**109**) respectively, of which only (**109**) does not reconvert to starting material upon acidification. This type of reversible ring opening is also observed on reaction of (**104**) and (**105**) with cyanide ion followed by treatment with acid ⟨78JOC914⟩. Other reactions within this type of system include the reductive cleavage of the S—SO₂ bond with LAH to give a mercaptosulfinic acid and cleavage of the same moiety with the stronger reducing agent hydriodic acid to generate a bis thiol ⟨66HC(21-2)968⟩.

### 2.26.3.1.8 Nucleophilic attack at hydrogen

An attempt to deprotonate 1,4-dioxin with potassium in refluxing ether for two days failed. About half of the starting material was recovered from the ether solution, the remainder undergoing polymerization ⟨48JOC671⟩. 1,4-Dithiin, however, is readily deprotonated with *n*-butyllithium in ether or THF; formation of the 2-lithio derivative (**110**) is faster in the latter solvent, occurring rapidly at −110 °C. At this temperature it can be trapped with electrophiles to give 2-substituted derivatives. At −60 °C (**110**) undergoes ring opening and the product (**111**) reacts with further butyllithium to afford the dilithiated compound (**112**). The latter also reacts with butyllithium *via* attack at sulfur and eliminates dilithiated ethyne to give (**113**) which can be trapped with methyl iodide to afford (**114**). Butyllithium similarly abstracts the proton on C-3 from 2,6-dimethyl- and 2,6-di-*t*-butyl-1,4-dithiins to give 3-lithiated derivatives which undergo analogous reactions to those of (**110**) ⟨77RTC259⟩. 2,5-Diphenyl-1,4-dithiin was earlier shown to react with butyllithium and dimethyl sulfate to give the product corresponding to (**114**) ⟨58JOC1702⟩.

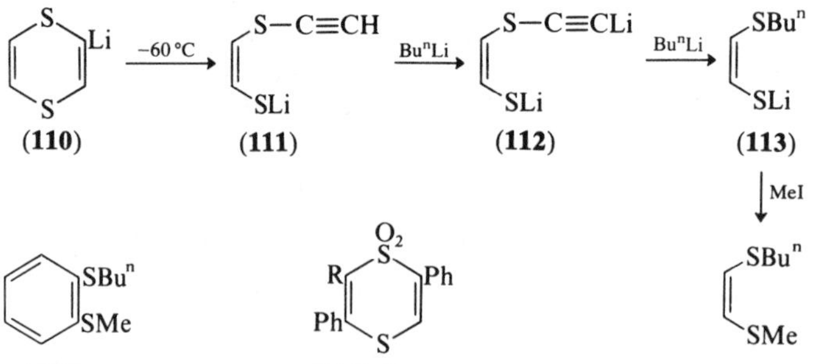

(**110**)    (**111**)    (**112**)    (**113**)

(**114**)

(**115**)

(**116**) R = H
(**117**) R = D

1,4-Benzodithiin reacts with *n*-butyllithium in ether at 0 °C in the presence of dimethyl sulfate to give the ring fragmented product (**115**) together with acetylene. The original mechanism offered for this reaction does not invoke nucleophilic attack at hydrogen to give the 2-lithio derivative ⟨58JOC1702⟩ but now the more recent work on (**110**) makes initial deprotonation seem most likely.

As expected, oxidation of a ring sulfur atom to the sulfone derivative enhances the acidity of an α-proton. Thus, hydrogen deuterium exchange of (**116**) to give (**117**) occurs with sodium azide in HMPA : D₂O (30 : 1) ⟨82TL299⟩.

### 2.26.3.1.9 Reactions with radicals, metals and at surfaces

Dibenzo[*b*,*e*][1,4]dioxin is cleaved by lithium in THF at 25 °C and the product obtained by quenching the intermediate with carbon dioxide is the carboxylic acid (**118**) in 56% yield. On heating with lithium for 7 hours the main isolated product is catechol. Under reflux conditions 2-nitrodibenzo[*b*,*e*][1,4]dioxin is cleaved to give unidentified products but the 2-amino derivative is recovered unchanged. Cleavage of phenoxathiin with lithium in ether or THF occurs at the C—S bond to give, after carboxylation, products analogous to those obtained from dibenzodioxin *viz.* (**119**) and (**120**). These cleavage products are also accompanied by phenoxathiin-4,6-dicarboxylic acid, the product of metallation (*cf.* Section 2.26.3.2.3). Thianthrene with lithium in THF at 25 °C gives an acidic, unpleasant smelling oil ⟨57JOC851, 58JA380⟩. However, upon reaction with potassium in dimethoxyethane, thianthrene is smoothly cleaved to give potassium thiophenolate ⟨63AC(R)36⟩.

(**118**) Z=O
(**119**) Z=S

(**120**)

(**121**)

(**122**)

(**123**)

Sodium and liquid ammonia in ether is another proven method for cleaving dibenzo[*b*,*e*]-[1,4]dioxins and phenoxathiin. Cleavage of the former is illustrated by the conversion of the 1-methoxy derivative (**121**) into a diphenyl ether (**122**), which is accompanied by the biphenyl (**123**) as a by-product. The biphenyl is not formed from the ether and its yield is enhanced when sodium hydride or ammonium chloride is added to the reaction. The mechanism of these and related cleavages is unclear and pathways involving electrophilic attack by sodium ions and free radical attack by sodium atoms have been considered ⟨62YZ696, 62YZ1341, 70CPB1⟩. Under the same conditions, or on replacing sodium by lithium, phenoxathiin is cleaved to give various products, including disulfides. With sodium in liquid ammonia alone, 2-mercaptophenyl phenyl ether is obtained in good yield ⟨66HC(21-2)964⟩.

The sulfur-containing rings are obvious candidates for Raney nickel desulfurization reactions and indeed this reaction has been utilized to good effect for structural assignment work within the phenoxathiin series ⟨66HC(21-2)864, 78JHC769⟩. The reactions with 2,5-diphenyl-1,4-dithiin and thianthrene are of mechanistic interest and it appears that the molecules may be chemisorbed at either one or both sulfur atoms, the latter situation facilitated by the boat conformation adopted by these compounds (Section 2.26.2.2). This can lead to simultaneous elimination of both sulfur atoms to give a diradical intermediate ⟨64AJC353, 64AJC366⟩.

Phenoxathiin and thianthrene are both partially desulfurized by high temperature catalytic hydrogenolysis over $NH_4Y$ zeolite, presulfided $Co–Mo–Al_2O_3$ and $Ni–W–SiO_2·Al_2O_3$. The former is cleaved to give diphenyl ether as the major product, together with phenol and benzene, indicating that sulfur removal occurs more readily than removal of oxygen. In line with this, thianthrene reacts under the same conditions to give benzene as the main product ⟨78BCJ1422⟩. Heating phenoxathiin and thianthrene over copper is reported to give dibenzofuran and dibenzothiophene respectively, although subsequent reexamination of the former reaction failed to confirm the result ⟨66HC(21-2)864, 66HC(21-2)1155⟩.

Desulfurization reactions in the 1,2-series are encountered among derivatives of both oxathiins and dithiins. 1,2-Oxathiin 2,2-dioxides extrude sulfur dioxide at elevated temperature over zinc oxide, iron or copper oxide to give the corresponding furan ⟨66HC(21-2)789⟩ (*cf.* Section 2.26.3.1.2). Copper is a good catalyst for the extrusion of sulfur and sulfur dioxide from dibenzo[*c*,*e*][1,2]dithiin (**40**) and its dioxide respectively to give dibenzothiophene ⟨66HC(21-2)968⟩.

Cleavage of the S—S bond of (**40**) also occurs on reaction with phenyl radicals obtained by decomposition of *N*-nitrosoacetanilide or aprotic diazotization of aniline. The main

**(40)**          **(124)**          **(125)**

products are (124) and (125), together with traces of dibenzothiophene. Conversion into the last of these is nearly quantitative when (40) reacts with *t*-butyl peroxide in refluxing chlorobenzene ⟨75G841⟩.

## 2.26.3.2 Fully Unsaturated Compounds: Reactivity of Substituents

### 2.26.3.2.1 General survey

The reactivity of substituents has received little systematic study. However, with regard to electrophilic substitution reactions, it can be deduced from the products of monobenzo fused derivatives in the 1,4-series (Section 2.26.3.1.3) that the benzenoid ring is less reactive than the heteroring. The same conclusion applies to a phenyl group attached to 1,4-dithiin. In the dibenzo fused derivatives this type of competition is precluded and electrophilic attack occurs readily at the 2-position. The site of attack by a second incoming group is a little difficult to generalize upon and this is discussed in Section 2.26.3.2.2. Reactivity of the dibenzo fused compounds towards butyllithium has been well studied and proton abstraction occurs at C-1 (or C-4 in phenoxathiin) (Section 2.26.3.2.3).

In general most substituents show normal behaviour but an interesting reaction has been reported concerning the reactivity of the bromo substituent of (126). Thus, (126) reacts with excess azide ion in HMPA to give the isomer (127), by an unknown mechanism. Furthermore, the bromo isomer (127) is slowly converted into (117) when treated with sodium azide in HMPA : D$_2$O (30 : 1), apparently *via* nucleophilic attack at bromine with generation of a carbanion at the ring carbon ⟨82TL299⟩.

**(126)**          **(127)** R = Br
                   **(117)** R = D

### 2.26.3.2.2 Electrophilic attack at fused benzene rings

Dibenzo[*b,e*][1,4]dioxin, phenoxathiin and thianthrene are all reactive towards electrophiles. As expected the 2-position is the favoured position for mono electrophilic substitution and there are numerous examples which bear testament to this. Though dibenzo[*b,e*][1,4]dioxin is expected to be highly reactive towards electrophiles, conditions have been described for various mono substitution reactions including acylation ⟨37YZ131⟩, bromination ⟨57JA1439⟩ and nitration ⟨74BSF183⟩. Examples in the phenoxathiin and thianthrene series have been reviewed ⟨66HC(21-2)864, 66HC(21-2)1155⟩. Under nitration and certain halogenation conditions there are competing reactions at the sulfur atom and these have been referred to in Section 2.26.3.1.4.

The site of introduction of the second substituent in the series is less clear. Diisopropylation of dibenzo[*b,e*][1,4]dioxin gives the 2,3-dialkylated product, reflecting the activation of the ring on introduction of the first substituent. However, the product resulting from di-*t*-butylation contains a substituent on both rings, presumably as a result of steric factors ⟨57JOC1403⟩. On introduction of substituents which act as electron-withdrawing groups the second group inevitably enters at the second ring. Acylation reactions of benzo[*b,e*]-[1,4]dioxin are normally considered to give rise to the 2,7-diacyl product though the 2,8-isomer has been generated in some circumstances. Both chloromethylation and bromination reactions give rise to mixtures of the 2,7- and 2,8-isomers. However, bromination of 2-nitrobenzo[*b,e*][1,4]dioxin and nitration of the 2-bromo derivative both give the same product, which has been assigned the 2,7-structure ⟨58JA366, 75AJC1803⟩.

The diacetylated product of phenoxathiin was assigned the 2,8-structure in an early paper ⟨38MI22600⟩ and for many years this substitution pattern was considered the norm ⟨53JA3384⟩. Recently it has been noted that the evidence for the original assignment is equivocal and a reinvestigation revealed that the 2,7- and 2,8-diacetyl derivatives are formed in the ratio 4:1 ⟨78JHC769⟩. This trend is also apparent in other Friedel–Crafts acylations and it appears that once the first acyl substituent is incorporated at C-2, in conjugation with the ring oxygen, the sulfur atom governs the site of entry of the second group ⟨75BSF2249⟩. The observation clearly casts some doubt upon assignments made on the assumption of 2,8-substitution. Nitration of phenoxathiin is reported to give 2,8-dinitrophenoxathiin 10,10-dioxide, but here the 2,8-substitution pattern is to be expected if oxidation of sulfur occurs prior to the second substitution reaction ⟨53JA3384⟩.

Reaction of thianthrene with chlorine gives an adduct at sulfur (see Section 2.26.3.1.4) which is reported to rearrange to the 2-chloro derivative ⟨66HC(21-2)1155⟩. Thianthrene, like dibenzo[*b,e*][1,4]dioxin, reacts with bromine to give a mixture of the 2,7- and 2,8-dibromo derivatives. The same mixture is also obtained on bromination of thianthrene 5-oxide, the sulfoxide group being reduced during the reaction. In contrast, thianthrene 5,5-dioxide and 5,5,10-trioxide are inert towards bromine ⟨58JOC313⟩. Diacylation of thianthrene normally occurs at the 2,7-positions but, as with the oxygen analogue, there are conditions where a mixture of isomers is formed. The 2,7-diacetyl derivative is formed as the sole product in 85% yield when methylene chloride is used as solvent ⟨73BSF1460⟩. Chloromethylation of thianthrene tends to give polymeric products but a 14% yield of the 2,7-product is obtained using chloromethyl methyl ether as reagent ⟨76IZV2799⟩. [In fact, the paper cited describes the formation of '2,6-bis(chloromethyl)thianthrene'. No diagram is given for the product but *Chemical Abstracts* offers an illustration of the 1,7-isomer. The structural assignment was based on oxidation to the known '2,6-dicarboxythianthrene 9,9-dioxide' reported in an early Japanese paper ⟨38YZ517⟩ in which an alternative and now superseded numbering system was used. On taking this into account, the bis(chloromethyl) product proves to be the 2,7-isomer.]

### 2.26.3.2.3 Nucleophilic attack at hydrogen of a fused benzene ring

Dibenzo[*b,e*][1,4]dioxin, phenoxathiin and thianthrene all react with butyllithium with proton abstraction from a benzene ring. Dibenzo[*b,e*][1,4]dioxin and thianthrene are metallated at the 1-position (**128**), while lithiation of phenoxathiin occurs *ortho* to the C—O bond rather than the C—S bond, *i.e.* at C-4, (**129**). The lithiated products provide excellent intermediates for functionalizing the rings at these positions, usefully complementing the product distribution pattern in electrophilic substitution reactions.

Unlike phenoxathiin itself, its 10-oxide and 10,10-dioxide are metallated at C-1, *i.e. ortho* to the C—S bond (**130**), so providing a guide to the relative strength of the directing influences of the heteroatoms. Thianthrene 5-oxide and 5,5-dioxide are also metallated *ortho* to the sulfur atom in the oxidized state (**131**). Proton abstraction from phenoxathiin 10-oxide and thianthrene 5-oxide is accompanied by some reduction of the sulfoxide group and cleavage of the heterocyclic ring (Section 2.26.3.1.7) ⟨73MI22602, 66HC(21-2)864, 66HC(21-2)1155⟩.

(**128**) Y = Z = O, or Y = Z = S
(**129**) Y = O, Z = S
(**130**) Y = SO or SO₂, Z = O
(**131**) Y = SO or SO₂, Z = S

## 2.26.3.3 Saturated and Partially Saturated Compounds

### 2.26.3.3.1 General survey

There is a large and varied selection of ring systems to be considered within this section and their reactivity is surveyed in the following three subsections according to the arrangement of the heteroatoms within the rings. A certain amount of the chemistry is predictable on the basis of functional group reactivity and does not merit detailed appraisal. Thus, sulfur atoms in the 1,4-, 1,3- and 1,3,5-series are readily oxidized to sulfoxides and sulfones, and form sulfilimine derivatives and sulfonium salts. Substituents at carbon atoms which themselves are bonded to two other ring carbon atoms also show unremarkable reactivity.

Of greater consequence, however, are the stable adducts formed by members of the 1,4-series and the cleavage reactions of compounds with heteroatoms 1,2 and 1,2,3 induced by thermolysis, photolysis or by nucleophilic attack. In the 1,3-series, dioxanes undergo typical acetal cleavage reactions, but the most investigated rings are those containing sulfur atoms. Indeed the chemistry of 1,3-dithianes, and to a lesser extent 1,3,5-trithianes and 1,3-oxathianes, has been dominated over the last decade or so by the exploitation of the synthetic possibilities offered by the carbanion derived by removal of a proton from C-2. Insofar as the ring systems can be cleaved to yield a carbonyl compound, these intermediates can be regarded as masked acyl carbanions.

### 2.26.3.3.2 Rings with heteroatoms 1,4

1,4-Dioxane, 1,4-oxathiane and 1,4-dithiane all form complexes with a range of acceptor molecules. Compounds with which adducts are formed include the halogens and, for the sulfur-containing rings, a wide variety of metal salts. Various crystal structure determinations are available which clearly demonstrate that coordination occurs at the heteroatom. The bromine adduct of 1,4-dioxane has been used for controlled bromination reactions, while the $SO_3$ complexes of 1,4-dioxane and 1,4-oxathiane are useful for sulfonations of alcohols and unsaturated bonds ⟨B-79MI22601, 66HC(21-2)817, 66HC(21-2)1041⟩. Latterly, the stable liquid borane complex of 1,4-oxathiane has been recommended for use in hydroboration reactions ⟨80S153⟩. Hydrolysis of the tetrabromo adduct of 1,4-dithiane gives the 1,4-dioxide derivative. With metal salts, 1,4-oxathiane forms sharp melting point complexes of the type $2C_4H_8OS\cdot MX_n$, whereas 1,4-dithiane tends to form 1:1 adducts. IR spectra of the adducts of both series are not dissimilar from those of the parent rings ⟨66HC(21-2)817, 66HC(21-2)1041⟩.

A variety of ring halogenated compounds has been prepared, especially chlorinated derivatives. The latter are obtained using chlorine, normally in carbon tetrachloride, and the products are the 2,2- or 2,3-dichloro or 2,3,5,6-tetrachloro derivatives depending upon conditions ⟨B-57MI22600, 66HC(21-2)817, 66HC(21-2)1041⟩.

1,4-Dioxane is cleaved by acyl chlorides in the presence of Lewis acids to yield 2-acyloxyethyl chlorides and by acetic anhydride with ferric chloride to give bis(2-acetoxyethyl) ether and 1,2-diacetoxyethane. Hydrogen bromide-induced cleavage gives bis(2-bromoethyl) ether in 39% yield ⟨B-79MI22601⟩. A similar cleavage of the ether linkage in 1,4-oxathiane gives the corresponding thioether but this reacts further with unconsumed 1,4-oxathiane to give the bis-sulfonium salt (132) ⟨69JHC393⟩. Hydriodic acid cleavage of 1,4-oxathiane 4,4-dioxide simply gives bis(2-iodoethyl) sulfone ⟨66HC(21-2)817⟩. 1,4-Dithianes are more resistant to acid and examples of C—S bond cleavage generally involve sulfonium salts or sulfones which are labile to base. Thus, 1-methyl-1,4-dithianium cation

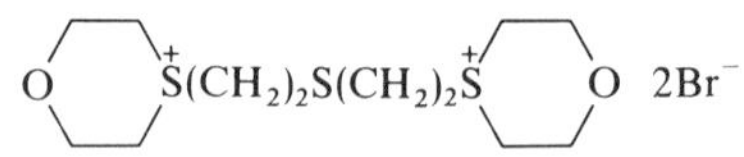

(132)

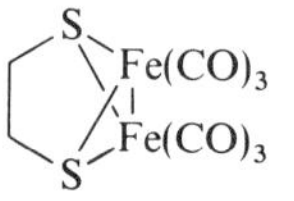

(133)

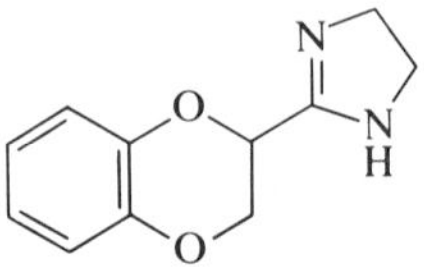

(134)

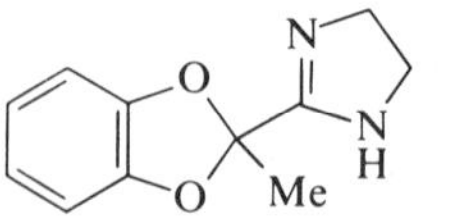

(135)

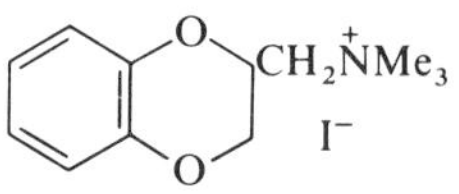

(136)

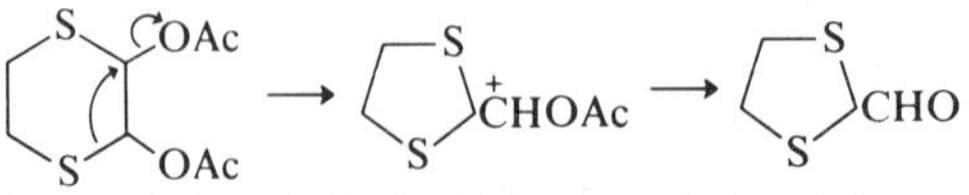

**Scheme 10**

is readily ring cleaved in hot water to 2-methylthioethyl vinyl sulfide ⟨66HC(21-2)1041⟩. However, an unusual cleavage of 1,4-dithiane itself has been discovered recently and involves reaction with $Fe_2(CO)_9$ to give (**133**) ⟨79JA1313⟩.

Certain derivatives in the 1,4-series show a propensity to ring contract to a five-membered ring in which the heteroatoms are in the 1,3-position. Thus, on dry distillation at 160 °C the 1,4-benzodioxane derivative (**134**) is partially converted into (**135**) ⟨81TL4839⟩ and a related base-catalyzed reaction ring-contracts (**136**) to 2-vinyl-1,3-benzodioxolane with elimination of trimethylamine ⟨66T931⟩. 2,3-Diacetoxy-1,4-dithiane ring-contracts in acid to give 1,3-dithiolane-2-aldehyde (Scheme 10) ⟨63JOC2686⟩, and an analogous reaction is observed for a tetrasubstituted 1,4-oxathiane derivative ⟨78JOC1262⟩.

The double bonds in 2,3-dihydro-1,4-dioxin, 2,3-dihydro-1,4-oxathiin and 2,3-dihydro-1,4-dithiin undergo standard electrophilic addition reactions. Under acid catalysis, methanol adds to 2,3-dihydro-1,4-oxathiin to give 2-methoxy-1,4-oxathiane ⟨66HC(21-2)842⟩. Various examples are available of reactions of the double bonds with carbenoids to give bicyclo[4.1.0]diheteroheptanes ⟨77LA910, 78ZC15⟩, and with alkenes in [2+2] cycloadditions ⟨78CB3624⟩.

### 2.26.3.3.3 Rings with heteroatoms 1,3 and 1,3,5

Within this series of compounds, reactivity at the 2-position has attracted most attention, and of particular current interest are reactions of conjugate bases of the sulfur-containing rings. Accordingly, these are considered first. A proton at C-2 on the 1,3-dithiane ring is rendered acidic by the neighbouring sulfur atoms and can be abstracted with bases such as *n*-butyllithium. As pointed out earlier, the anion can be regarded as a masked acyl carbanion, which enables an assortment of synthetic sequences to be realized *via* reactions with electrophiles. Thus, a 1,3-dithiane derived from an aldehyde can be further functionalized at C-2, *e.g.* with an alkyl halide, Scheme 11, and subsequent cleavage, *vide infra*, gives a ketone. Alternatively, Raney nickel desulfurization affords the reduced (methylene) analogue. In another approach, 1,3-dithiane itself can be used as the basic building block which can be functionalized once or twice by successive reactions. Similar reactions can also be performed using 1,3,5-trithiane and the relative merits of the two compounds have been discussed ⟨74CB367⟩. The latter is cheaper but is less soluble in THF, which can lead to complications. Some indication of the enormous scope of these compounds within synthesis is provided by an authoritative review of the subject ⟨77S357⟩. The C-2 protons of 1,3-oxathiane are also acidic but require the stronger base *s*-butyllithium to effect abstraction. Recent work on chiral derivatives of 1,3-oxathianes, such as (**137**), provides an example of the extension of the basic concept of the masked acyl carbanion equivalent into the area of asymmetric synthesis ⟨81TL2855, 78JA1614⟩.

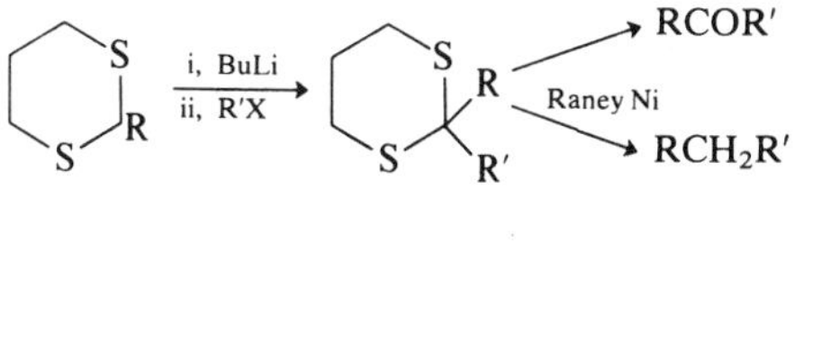

**Scheme 11**

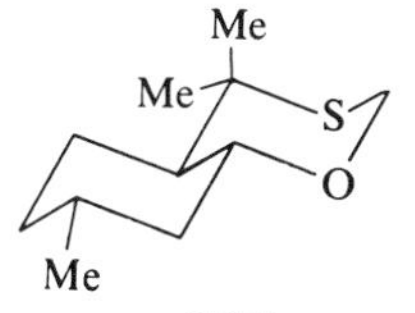

Abstraction of hydrogen as hydride from C-2 of 1,3-dithiane can be achieved using reagents such as trityl fluoroborate to afford the 1,3-dithienium ion. Like the carbanion referred to above, the cation can be used in organic synthesis and applications include reaction with dienes in Diels–Alder reactions, providing a route to $\Delta^3$-cyclopentenones ⟨72JA8932⟩, and with nucleophiles such as *O*-silylated enols to give, ultimately, 2-formyl or 2-alkyl ketones ⟨81TL2829⟩. Related to such work are reactions involving nucleophilic displacement of chloride ion from 2-chloro-1,3-dithiane using reagents such as Grignards, active methylene compounds, aromatics, *etc.*, to give various 2-substituted 1,3-dithianes ⟨79JOC1847⟩.

Nucleophilic displacements are also encountered at the 2-position in other ring systems. Interesting examples involve the displacement of a 2-methoxy group attached to conformationally fixed 1,3-dioxanes. This group is displaced by D⁻ using LiAlD₄ or AlD₃ with retention of configuration. When axial, the group is also displaced by Grignard reagents with retention of configuration but when equatorial it is barely reactive. A mechanism involving the participation of the ring oxygen orbitals has been proposed to account for these observations ⟨B-77SH(2)107⟩. The importance of the orientation of the leaving group is also evident in other ring systems. Thus, the 4-chloro group is displaced by ethanol in only one of the two isomers of 4-chloro-2,4,6-trichloromethyl-1,3,5-oxadithiane ⟨66HC(21-2)680⟩.

Interest in ring cleavage reactions centres largely on the recovery of carbonyl groups. 1,3-Dioxanes, like other acetals and ketals, are labile to dilute acid and this property has been exploited in schemes for the protection of aldehydes and ketones as well as for 1,3-diols, the last of these particularly in carbohydrate chemistry. The generally accepted hydrolysis mechanism for acetals and ketals, denoted as *A*-1, involves the rate determining heterolysis of the protonated species to yield an oxacarbenium ion and further hydrolysis leads ultimately to the 1,3-diol and the carbonyl compound ⟨74CRV581, 77MI22601⟩. 2-Alkoxy-1,3-dioxanes, examples of ortho esters, hydrolyze by much the same route ⟨81JOC886, 74CRV581⟩.

Recovery of carbonyl compounds from 1,3-oxathianes and 1,3-dithianes involves cleavage of a C—S bond and the rings are thus more resistant to acid catalyzed heterolysis than the corresponding 1,3-dioxanes. The importance of 1,3-dithianes in syntheses has ensured a sustained interest in the development of methods for performing the cleavage and the available reagents have been summarized ⟨77S357⟩. There are two main approaches and these involve metal-induced hydrolyses or oxidative/alkylative pathways. Among the former, the use of mercury(II) chloride with additives such as mercury(II) oxide or the carbonate salts of calcium, barium or cadmium has proved popular and reliable. The reaction is thought to involve formation of a sulfur–mercury bond which facilitates cleavage of the sulfur–carbon linkage. The second technique involves modification of the sulfur moiety to an oxide, *N*-sulfilimine derivative or a sulfonium salt, to make it a better leaving group. Recommended reagents include *N*-bromosuccinimide, *N*-chlorosuccinimide/AgNO₃. chloramine-T and methyl iodide ⟨77S357⟩. Although these methods have normally been devised for use with 1,3-dithianes, several have also proved useful for 1,3-oxathiane cleavage, particularly *N*-chlorosuccinimide/AgNO₃ ⟨81TL2855⟩. Examples of Raney nickel desulfurization reactions are frequently met, particularly in the 1,3-dithiane series but also among 1,3-oxathianes. An interesting alternative selective cleavage of the C—S rather than C—O bond in the latter system can be achieved using alkali metals in liquid ammonia ⟨70JOC2716⟩.

Cleavage of rings in the 1,3-series can also be brought about under thermolytic conditions. Alkylated 1,3-dioxanes, such as (**138**), undergo fragmentation at 350 °C to give the isomeric ester (**139**) together with lower molecular weight materials ⟨72BSF4308, 65BSF1358⟩. Most of the observed products can be rationalized in terms of pathways involving the oxygen-stabilized free radicals (Scheme 12), which are formed by scission of an α C–H bond (*cf.*

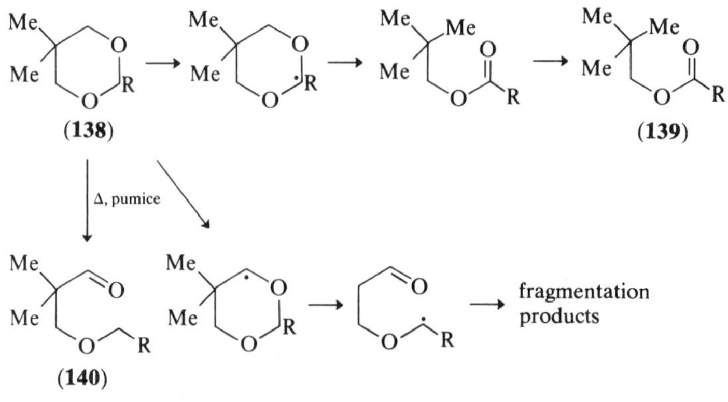

**Scheme 12**

mass spectroscopic cleavage, Section 2.26.2.3.4). However, somewhat different products, notably the isomeric alkoxyaldehydes (**140**), are formed when the ring system is pyrolyzed over pumice or silica. The mechanism may well be heterolytic ⟨62JA3307, 62JA3319⟩. There is some evidence to suggest that 1,3-oxathianes may react in a similar way (with C—O bond cleavage) under the latter conditions ⟨62JA3307⟩.

For rings having an S—C—S unit, the main thermolysis products are those involving skeletal rearrangements. This is nicely exemplified by the conversion of 1,3-dithiins into the isomeric 1,2-compounds, *e.g.* (**141**) → (**142**), on heating for a short time. A pathway is illustrated in Scheme 13. Several examples of this particular rearrangement are known ⟨71T5753, 81T3839⟩ and an analogous process may well be involved as a step in the formation of stilbenes and sulfur from 2,4,6-triaryl-1,3,5-trithianes ⟨66HC(21-2)692⟩; compare with processes observed in the mass spectrometer (Section 2.26.2.3.4). However, alternative modes of decomposition are also known and are illustrated by the thermal degradation of the 1,3-dithiin (**143**) over copper chromite to give 2,4-diphenylthiophene, probably with extrusion of thiopropiophenone ⟨66HC(21-2)1028⟩, and the pyrolysis of hexamethyl-1,3,5-trithiane at 500 °C to give thioacetone ⟨74CC739⟩.

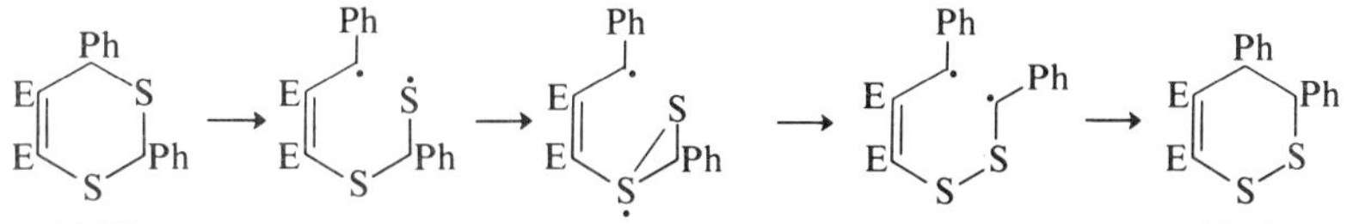

E = MeO$_2$C

**Scheme 13**

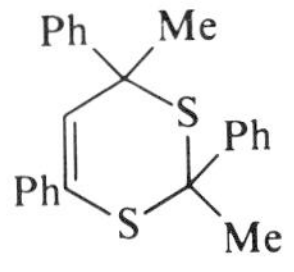

(**143**)

### 2.26.3.3.4 Rings with two or more adjacent heteroatoms

The principal reaction observed for compounds containing two adjacent oxygen atoms is the homolytic cleavage of the O—O bond. Products can normally be interpreted in terms of the reactions of the resultant bisalkoxy radical which normally give carbonyl compounds *via* α cleavage processes. Thus the 1,2-dioxane (**144**) thermolyzes at 200 °C to give acetone and ethylene by what is probably a concerted decomposition of the 1,6-diradical intermediate (Scheme 14) ⟨73AG(E)843, 77JA2735⟩, while the diradical derived from the 1,2,4-trioxane (**145**) (Scheme 15) fragments to acetone and adipaldehyde, eschewing the apparently attractive hydrogen abstraction pathway ⟨78JOC521⟩. A study of the thermolysis of 1,2,4,5-tetroxanes shows that the fragmentation of the bisalkoxy radical intermediate (Scheme 16) is dependent upon the nature of the groups R. For R = R′ = Ph, α cleavage at the ring C—O bond (pathway *a*) predominates and benzophenone is formed in greater than 95% yield, accompanied by trace amounts of biphenyl and phenol derived from phenyl radicals formed *via* α cleavage at the C—C bond (pathway *b*). For R = R′ = Me and R = CH$_2$Ph, R′ = Ph the yield of ketone drops to 65% and *ca.* 13% respectively, as pathway *b* becomes more attractive by virtue of the greater stability of the carbon radical R·. Products derived from this mode of cleavage increase as a consequence ⟨80JCR(S)35, 80JCR(S)36⟩.

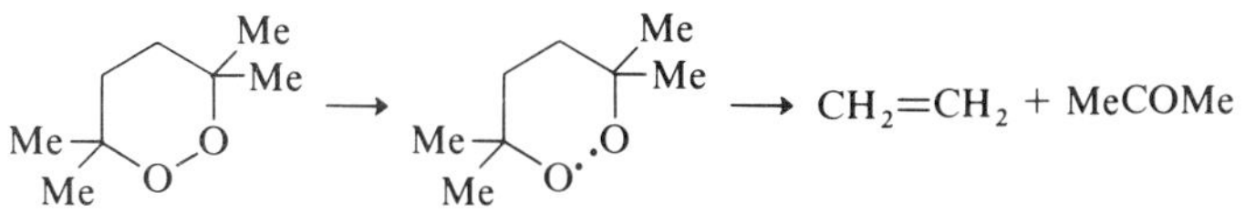

$$CH_2=CH_2 + MeCOMe$$

(**144**)

**Scheme 14**

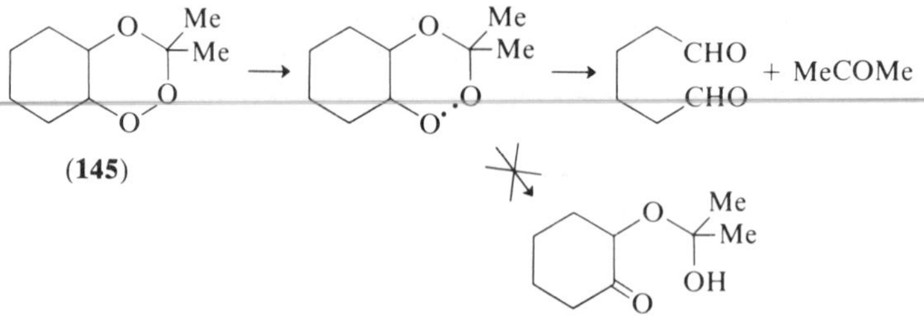

**Scheme 15**

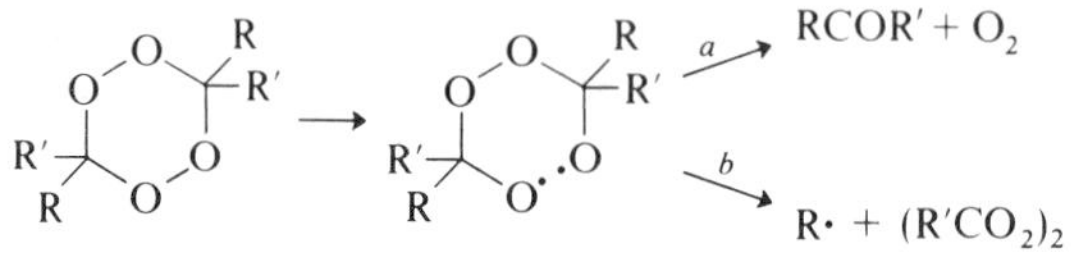

**Scheme 16**

As might be anticipated, when an O—O bond is cleaved in the presence of a radical trap, such as a double bond, the course of the reaction becomes quite different. Thus, thermolysis of bridged 3,6-dihydro-1,2-dioxins (1,4-endoperoxides) such as (146) causes rearrangement to *cis*-diepoxides (147) ⟨81T1825⟩.

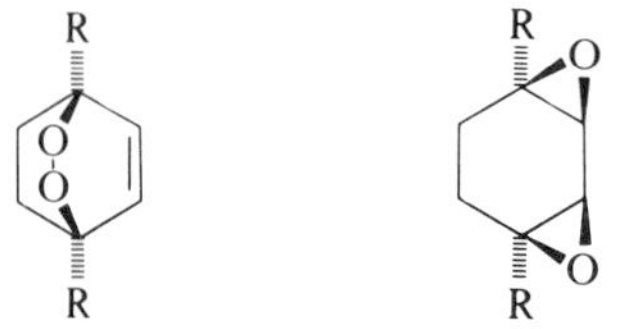

(146)          (147)

Compounds containing one or more O—S bonds are generally encountered with the sulfur present in an oxidized state. In the main, the compounds are cyclic esters or anhydrides of sulfur acids and by far the most common cleavage reaction is brought about by nucleophiles. Nucleophilic attack normally occurs at sulfur or carbon and rings which undergo base-catalyzed hydrolysis at the former (with S—O bond cleavage) include 1,2-oxathiane 2-oxide (148), 1,2-oxathiane 2,2-dioxide (149), 1,3,2-dioxathiane 2-oxide (17), 1,2,6-oxadithiane tetraoxide (150) and 1,3,2,4-dioxadithiane tetraoxide (151). In contrast 1,3,2-dioxathiane 2,2-dioxide (152) is attacked by base at carbon with C—O bond cleavage ⟨75JCS(P2)858, 71BCJ1669, 66HC(21-2)633, 66HC(21-2)675, 74JOC2112⟩. Some of the rings hydrolyzed by attack at sulfur are sometimes cleaved at carbon by other nucleophiles. Thus, a kinetic study has shown that this occurs when (149) is cleaved under neutral aqueous conditions ⟨71BCJ1669⟩ or with amines ⟨66HC(21-2)633⟩. Similarly, certain derivatives of (17) are attacked at carbon by nucleophiles such as carboxylate anions ⟨77JOC2260⟩.

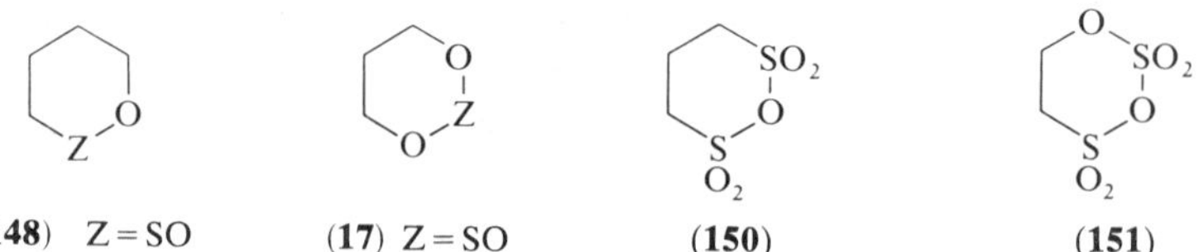

(148)  Z = SO    (17)  Z = SO     (150)         (151)
(149)  Z = SO₂   (152) Z = SO₂

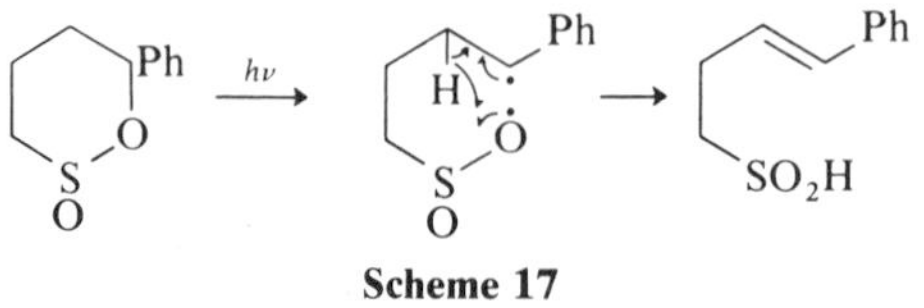

**Scheme 17**

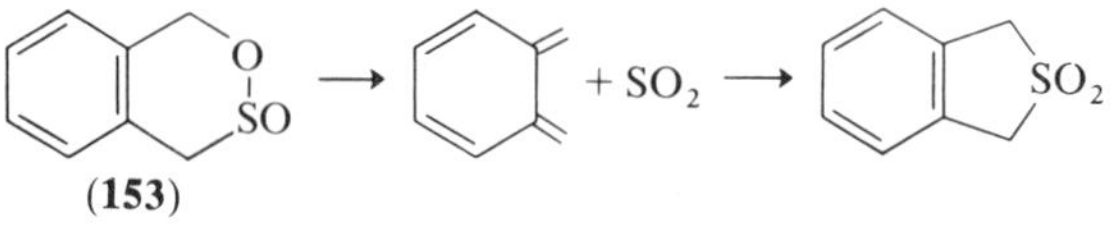

(153)

**Scheme 18**

Interestingly, the 6-phenyl derivative of (**148**) is also cleaved at the C—O bond, but under photolytic conditions (Scheme 17), while the benzo analogue (**153**) behaves quite differently again to give the isomeric sulfone (Scheme 18). The mechanism of the latter involves an initial retro Diels–Alder and the products, the quinone methide and sulfur dioxide, recombine as shown ⟨74JA935, 78CJC512, 78TL2353⟩. Sulfur dioxide is also extruded upon pyrolysis of 1,3,2-dioxathiane 2-oxides at elevated temperature ⟨66HC(21-2)633⟩.

Compounds containing an S—S bond such as 1,2-dithiane and 1,2,4,5-tetrathianes are cleaved by chlorine ⟨66HC(21-2)952, 66HC(21-2)626⟩. 1,2-Dithiane is fairly resistant to neutral hydrolysis but the S—S bond is more labile in the 1,1-dioxide and 1,1,2,2-tetraoxide derivatives. Both of the oxidized compounds are ring opened by a range of nucleophilic reagents and the tetraoxide also extrudes sulfur dioxide at 280 °C to give tetrahydrothiophene 2,2-dioxide ⟨69JOC1792, 81JOC2025⟩. Extrusion of elemental sulfur from 1,2,3-trithianes, *e.g.* (**154**), occurs over copper to give the ring-contracted disulfide (**155**). The reaction, which has been known for a long time, misled early workers into assigning the precursors the isomeric structures (**156**) ⟨66HC(21-2)689⟩.

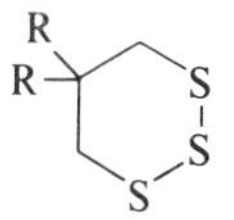

(**154**) 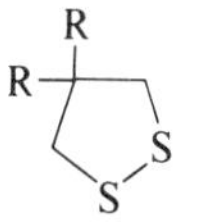          (**155**) 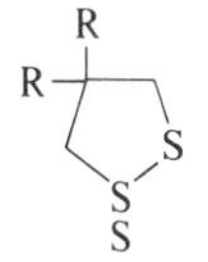          (**156**)

## 2.26.4 SYNTHESES

### 2.26.4.1 Fully Unsaturated Compounds

#### 2.26.4.1.1 Non-benzo fused ring systems

1,4-Dioxins and 1,4-dithiins have often been prepared by elimination reactions of appropriately substituted saturated analogues. However, in the more recent literature a number of examples of syntheses of 1,4-dithiins are described which utilize reactions involving unsaturated precursors. Examples of 1,4-oxathiins are relatively scarce and no general preparative routes have been developed.

A two step synthesis of 1,4-dioxin is available from dioxane which exploits the facility with which this precursor can be chlorinated. Here, 2,3,5,6-tetrachloro-1,4-dioxane is dechlorinated using magnesium and iodine at elevated temperature in a reported yield of 48–64% ⟨39JA3020⟩. This type of route is not well suited for other derivatives because of potential difficulties inherent in obtaining the appropriate tetrachloro compound. Other modes of elimination are best sought and such an alternative is contained within the route to 2,5-dimethyl-1,4-dioxin using the readily accessible 2,5-diiodomethyl-1,4-dioxane (diepiiodohydrin) as precursor (Scheme 19). The key step is the isomerization of the double bonds of the initial elimination product ⟨57JA6219⟩. Perhaps the dioxin derivative which is the simplest to prepare is the tetraphenyl compound (**29**) which is obtained *via* condensation of two equivalents of benzoin using hydrogen chloride in methanol. The resultant mixture of (**157**) and (**158**) is demethoxylated with zinc chloride in acetic anhydride ⟨77PIA(A)(86)1⟩.

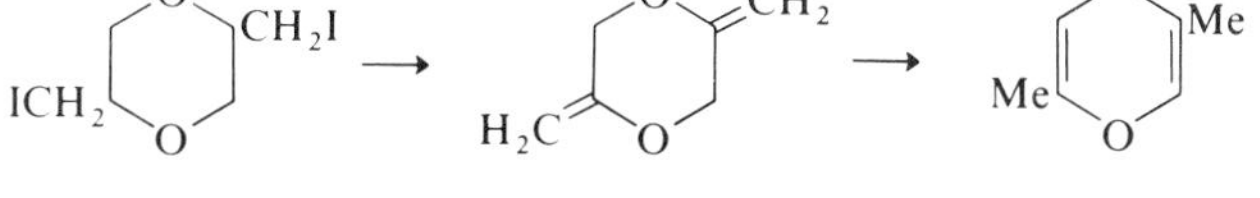

**Scheme 19**

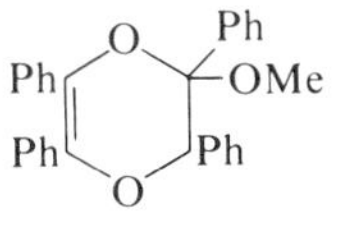

(**157**) 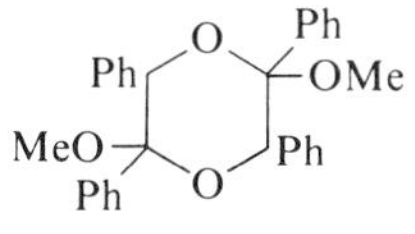          (**158**) 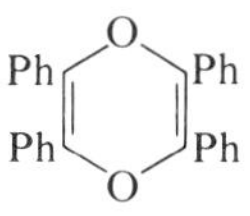          (**29**)

Synthetic routes to 1,4-dithiins have been explored more thoroughly ⟨66HC(21-2)1112⟩ than those to the 1,4-dioxin ring system. The parent compound is obtained in 47–60%

yield by vapour phase dealkoxylation of 2,5-dialkoxy-1,4-dithianes over alumina at 260–265 °C and this type of approach can also be used to prepare 2,5-dimethyl-1,4-dithiin. 2,5-Diaryl derivatives, however, are best prepared from the appropriate phenacyl halide. Treatment of the derived Bunte salt (**159**) (Scheme 20) with acid gives the dithiin *via* the intermediate diol (**160**) and overall yields are normally in excess of 50%. A similar scheme uses aryl α-mercaptoketones prepared directly from the α-haloketone using H₂S in base. Curiously, this general type of approach fails for all alkyl α-mercaptoketones (with the exception of *t*-butyl) because the diols corresponding to (**160**) dehydrate to give the bicyclic compounds (**161**). Tetraphenyl-1,4-dithiin is most readily prepared from benzoin (*cf.* the dioxin series) using HCl and H₂S in ethanol. The yield is of the order of 50% ⟨66HC(21-2)1112, 77PIA(A)(86)1⟩.

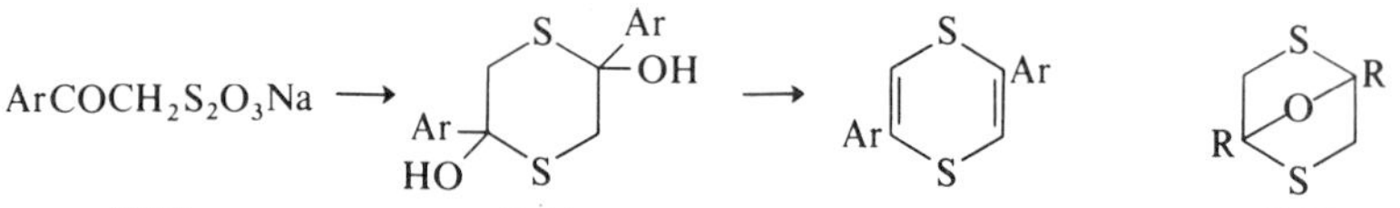

$$ArCOCH_2S_2O_3Na \longrightarrow$$

(**159**)    (**160**)    (**161**)

**Scheme 20**

The use of precursors already containing an unsaturated linkage is nicely illustrated by a route to 1,4-dithiin and its 2-alkyl and 2,6-dialkyl derivatives wherein the six ring-atoms are derived from combination of a five-atom fragment and a one-atom unit. Thus, addition of sodium sulfide to alkynyl sulfides (**162**) occurs readily in DMF/methanol or in liquid ammonia in methanol and yields of (**163**) are in excess of 55%. Routes to the precursors are well established but are several steps removed from commercially available starting materials ⟨73RTC1326, 75RTC163⟩. A second general approach is based on cyclizations between a four-atom and a two-atom fragment and particularly frequent use is made of dianions of dimercapto compounds. Thus, *cis*-1,2-ethylenedithiolate (**164**) reacts with butadiyne, generated *in situ* from 1,4-dichloro-2-butene, to give 2-vinyl-1,4-dithiin (20%) and with 1,1,2,2-tetrachloro-, 1,1,2,2-tetrabromo- and 1-phenyl-1,2,2-trichloro-ethanes, in presence of base, to afford 2-chloro-, 2-bromo- and 2-phenyl-1,4-dithiins respectively ⟨67ZC152, 69ZC184⟩. An entry to multisubstituted 1,4-dithiins is in principle available using the dianion of dimercaptomaleimide (**165**) ⟨67CB1559⟩ and a very high yield synthesis of tetracyano-1,4-dithiin (96%) utilizes the reaction of the dianion of dimercaptomaleonitrile (**166**) with thionyl chloride. The latter dianion is obtained by reaction of sodium cyanide with carbon disulfide to give sodium cyanodithioformate which undergoes facile dimerization with loss of sulfur ⟨66HC(21-2)1112⟩. Routes to tetramethoxycarbonyl-1,4-dithiin (**167**) include an interesting reaction between the disulfide (**168**) and sodium thiophenolate ⟨78ACS(B)152⟩, and methanolysis of the thioanhydride (**169**), the extruded hydrogen sulfide serving as the ring closing reagent ⟨66CB1973⟩.

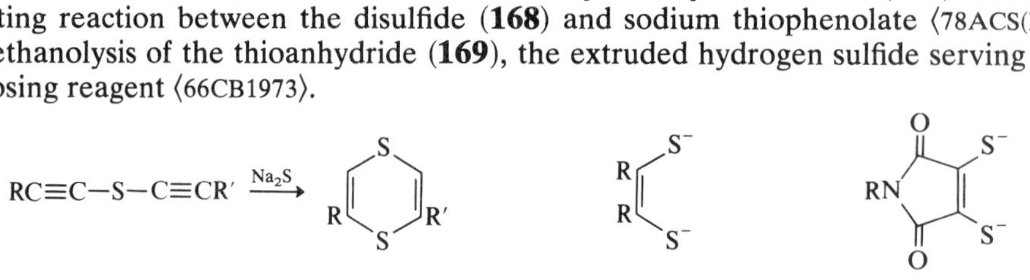

$$RC{\equiv}C{-}S{-}C{\equiv}CR' \xrightarrow{Na_2S}$$

(**162**)    (**163**)    (**164**) R = H    (**165**)
(**166**) R = CN

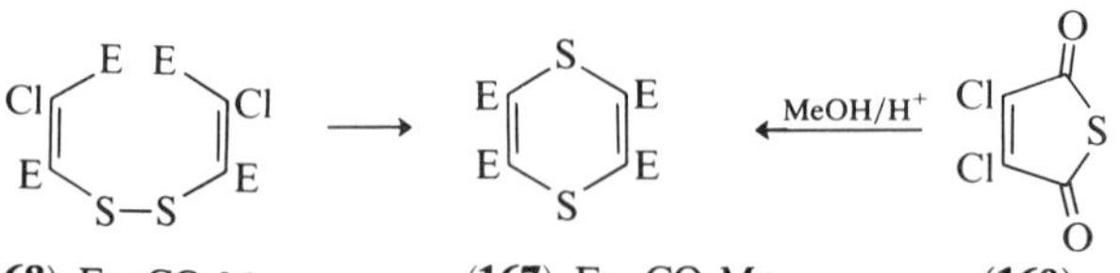

(**168**) E = CO₂Me    (**167**) E = CO₂Me    (**169**)

1,4-Dithiins can also be derived as decomposition products of other heterocyclic rings. Thus, 1,4-dithiins are obtained upon photolysis of thiadiazoles, *e.g.* (**170**) → (**171**). Yields of dithiin are often low but the chemistry is interesting in that the initial decomposition products include the elusive thiirene ring system ⟨81JA486⟩. Photolysis of the dithiocarbonate (**172**) also gives (**171**) and under certain conditions the yield is remarkably high (82%) ⟨73ZC424⟩. The reaction may proceed *via* a 1,2-dithiete intermediate ⟨74JA3502⟩. The latter ring system is also involved in another dithiin synthesis wherein the derivative (**173**),

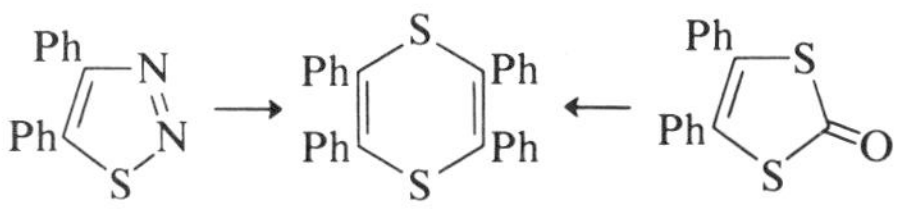

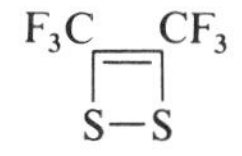

(170)  (171)  (172)  (173)

generated from hexafluoro-2-butyne and hot sulfur over iodine, reacts with alkynes, conceivably *via* a Diels–Alder reaction of its α-dithione valence tautomer ⟨66HC(21-2)1112⟩.

The sulfonyl group is a key feature in the preparation of the unusually substituted dithiin sulfone (174) from dibenzyl sulfone. The acidic α-protons are abstracted with sodium hydride and the carbanionic intermediates react with carbon disulfide. The reaction is quenched with methyl iodide to give (174) in 17% yield ⟨73BSF637⟩. Another multisubstituted dithiin (175) is available from the reaction of diphenylthiirene dioxide with the ylide (176) (Scheme 21) but again the yield is low. However, the reaction is of particular interest in so far as the product mixture also contains a derivative of the rare oxathiin nucleus. Indeed of the three products isolated the oxathiin sulfone (177) is formed in marginally the highest yield ⟨73BCJ667⟩.

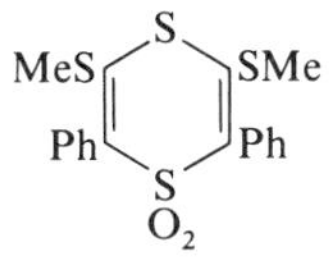

(174)

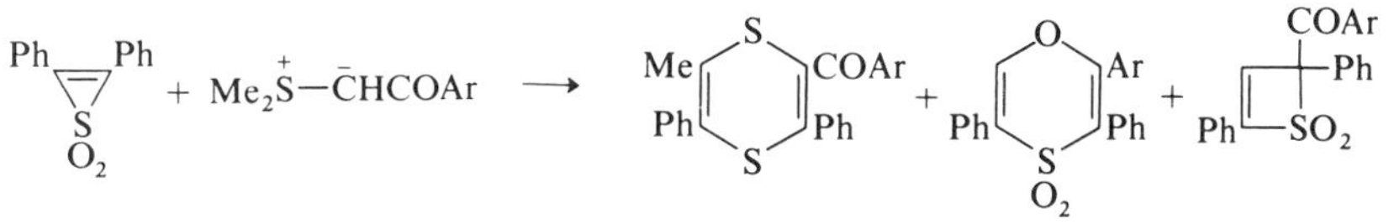

12%  15%  13%

(176)  (175)  (177)

**Scheme 21**

Syntheses within the 1,2-series are represented by routes to 1,2-oxathiin 2,2-dioxides and 1,2-dithiins. The former are obtained by the addition of sulfuric acid or chlorosulfonic acid in acetic anhydride to β-branched α,β- and β,γ-unsaturated ketones, *e.g.* (178) → (179). Yields of the successful reactions are normally greater than 40% ⟨66HC(21-2)774⟩. 1,2-Dithiins are synthesized from conjugated diynes using benzyl thiol as the source of sulfur. Addition of two equivalents of the thiol gives intermediate (180). Reductive debenzylation with sodium in liquid ammonia at −70 °C gives the bis thiol which undergoes aerial oxidation in basic solution to the 1,2-dithiin (39) ⟨65ZC353, 67AG(E)698⟩. 1,1-Dioxides are obtained by the addition of 2-aminovinyl aryl thioketones with sulfenes, generated *in situ* from phenylmethanesulfonyl chloride. Several 3-aryl-6-phenyl derivatives have been prepared in this way in yields ranging from 37–53% ⟨75T2679⟩. An alternative route to 1,2-dithiins lies in the ring expansion of 1,2-dithiolan-3-ones using base, *e.g.* (181) → (182). Yields are of the order of 50% ⟨69LA(728)32⟩.

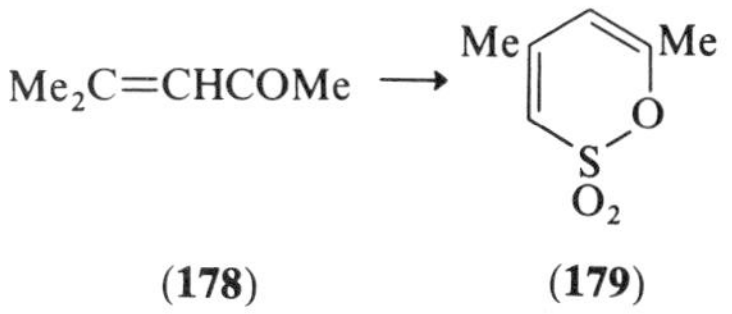

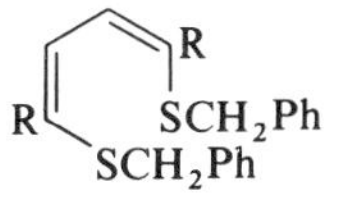

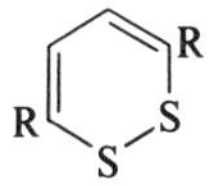

(178)  (179)  (180)  (39)

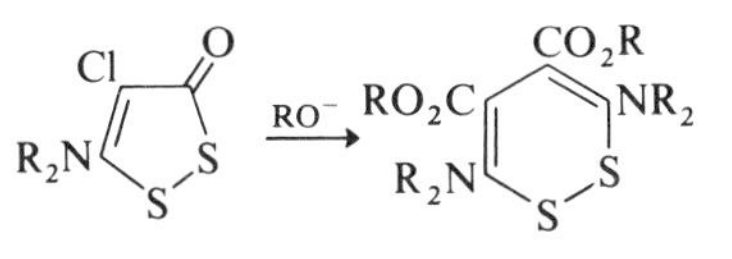

(181)  (182)

### 2.26.4.1.2 Benzo fused ring systems

The monobenzo fused derivatives of 1,4-dioxin, 1,4-oxathiin and 1,4-dithiin, (**183**), (**184**) and (**185**), can all be prepared by routes in which the first step is a base-catalyzed reaction between the appropriate 1,2-disubstituted benzene and an α-haloketal. The product is a 2-alkoxy-2,3-dihydro derivative (**186**) from which the unsaturated heterocycle is obtained in one or more steps. The synthesis of (**183**) using this strategy was first reported at the end of the last century but a thorough appraisal of its value was not forthcoming until much later ⟨66T931⟩. The final elimination step to give the unsaturated compound can be achieved using the 2-hydroxy analogue (**187**), obtained from the 2-alkoxy compound by acid-catalyzed hydrolysis. Phosphorus pentoxide in quinoline is a better dehydrating agent for (**187**) than thionyl chloride in pyridine but even so the yield of (**183**) is only 15%. A far superior elimination reaction is the pyrolysis of the acetoxy derivative (**188**) at 450 °C which proceeds in 80% yield ⟨67ZC152⟩. However, 2-hydroxy-2-phenyl-1,4-benzodioxane, formed from catechol and phenacyl bromide, dehydrates satisfactorily using thionyl chloride in pyridine. An alternative elimination mode has been used for the preparation of 2-methyl-1,4-benzodioxin and involves conversion of a 2-halomethyl-1,4-benzodioxane into the 2-methide. The latter is then converted by base-catalyzed isomerization into the benzodioxin ⟨66T931⟩, a scheme reminiscent of that used for the preparation of 2,5-dimethyl-1,4-dioxin (Section 2.26.4.1.1).

(**183**) Y = Z = O
(**184**) Y = S, Z = O
(**185**) Y = Z = S

(**186**) Y, Z = O or S, R = alkyl
(**187**) Y = Z = O, R = H
(**188**) Y = Z = O, R = COMe
(**189**) Y = S, Z = O, R = H
(**190**) Y = S, Z = O, R = Et
(**191**) Y = S, Z = O, R = COMe
(**192**) Y = Z = S, R = Et

In the 1,4-benzoxathiane series, neither dehydration of the hydroxy compound (**189**) nor elimination of ethanol from (**190**) give very satisfactory yields of (**184**) and once again the best yield (76%) is obtained by pyrolysis of the 2-acetoxy derivative (**191**). 3-Methyl-1,4-benzoxathiin is obtained by an analogous route, but 2,3-diphenyl-1,4-benzoxathiin is conveniently prepared by a one-step condensation of 2-mercaptophenol and benzoin in 21% yield ⟨66HC(21-2)852⟩.

In contrast to the difficulty experienced in eliminating ethanol from (**190**) the corresponding elimination reaction for 2-ethoxy-1,4-benzodithiane (**192**) occurs remarkably well. The best conditions involve the use of phosphorus pentoxide at 170–175 °C to give 1,4-benzodithiin (**185**) in 75% yield ⟨66HC(21-2)1143⟩. A direct synthesis of (**185**) is also available by condensation of benzene-1,2-dithiol with *cis*-dichloroethylene in a reaction analogous to that reported within the 1,4-dithiin series (Section 2.26.4.1.1) ⟨64ZC270, 67ZC152⟩.

Methods using starting materials other than the 1,2-disubstituted benzene have also been developed, and a useful four-step route, equally applicable to the synthesis of (**184**) and (**185**), is shown in Scheme 22. Yields are quoted as being about 50% overall and as such compare favourably with those involving eliminations. The key features of the scheme are the intramolecular sulfenylation of the thiosulfonates in 73 and 76% yield, and the satisfactory yields, 76 and 78%, for the final dehydrogenation ⟨75S451⟩.

A rather surprising synthesis of 2,3-diphenyl-1,4-benzodithiin has been realized by reacting together phenyllithium, carbon disulfide and bromobenzene. The dithiin is obtained in 30% yield by a pathway which almost certainly involves benzyne ⟨75JCS(P1)160⟩. Highly substituted 1,4-benzodithiins are available by reactions such as that between the dichloroquinone (**193**) and two equivalents of the dithiolate dianion (**194**) to give (**195**) in 67% yield. The product can be oxidized with nitric acid to regenerate a quinone structure ⟨69CB2378⟩.

With one or two exceptions, dibenzo fused derivatives require quite different synthetic strategies to those above. Dibenzo[*b,e*][1,4]dioxin is nowadays frequently prepared in yields of 10–20% by heating a mixture of 2-chlorophenol, potassium carbonate and copper powder to 170–180 °C ⟨57JA1439⟩. 2-Bromophenol can also be used as the precursor but i, BrCH₂CH₂Br; ii, MeSO₂SK; iii, AlCl₃; iv, SO₂Cl₂, quinoline

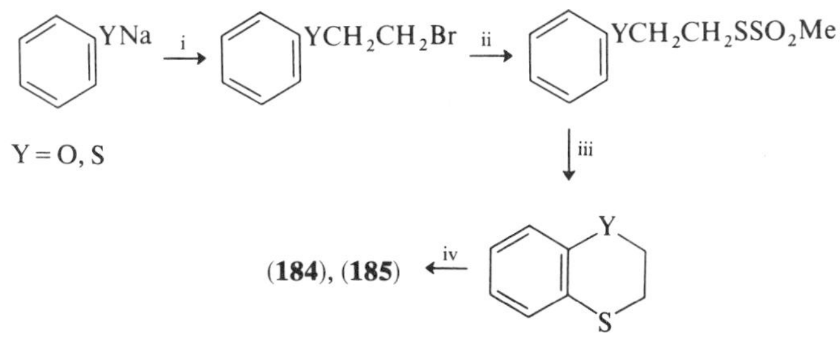

**Scheme 22**

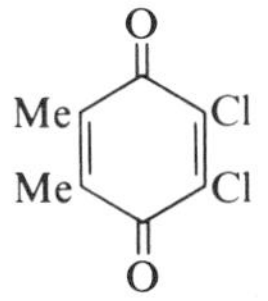

(193) 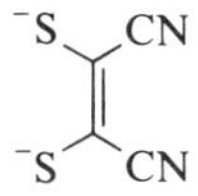        (194) 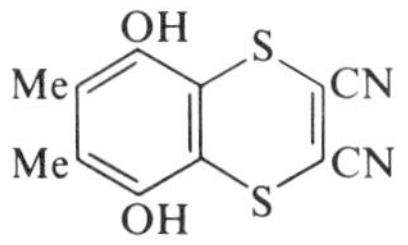        (195)

2-chlorophenol is less expensive. The yields, though low, are better than for a feasible alternative route *via* diazotization of 2-amino-2'-hydroxydiphenyl ether, or likely to be realized by condensation of two molecules of catechol 〈34JCS716〉. Polychlorinated derivatives can be obtained similarly from polychlorophenols 〈53BSF640〉 and are also often formed simply by heating the precursor 〈75MI22600〉. The extremely toxic 2,3,7,8-tetrachloro[*b*,*e*]-[1,4]dioxin can be formed as a by-product under the conditions of the commercial manufacture of 2,4,5-trichlorophenol (Section 2.26.5).

A particularly valuable route to phenoxathiin utilizes the long-known reaction of diphenyl ether with sulfur in the presence of aluminum chloride. Experimental details appear in *Organic Syntheses* 〈43OSC(2)485〉 where the reported yield prior to recrystallization is 87%. This mode of reaction is quite general and, using appropriately substituted diphenyl ethers, provides access to alkyl, phenyl and chloro derivatives of phenoxathiin. Alternative methods utilize ring closures of diaryl ethers substituted with a sulfur-containing functionality at the 2-position. Examples of this strategy are provided by the homolytic ring closure of the radical (196), obtained by aprotic diazotization of the appropriate amino derivative 〈74JCS(P1)1272〉, and by cyclizations of 2-sulfinic acids (197) (Scheme 23) or 2-chlorosulfonyl chloride derivatives. The last of these reactions uses aluminum chloride as catalyst and gives a phenoxathiin sulfone as the product. A further type of cyclization involves C—O bond formation and is exemplified by the sulfuric acid dehydration of 2,2'-dihydroxydiaryl sulfoxides 〈66HC(21-2)864〉.

The disodium salt of 2-mercaptophenol can also be used as a precursor, by reacting it with aromatic rings containing two displaceable substituents. Such a compound is picryl chloride and the reaction was originally reported to give 2,4-dinitrophenoxathiin. However, the product has recently been reassigned the more reasonable 1,3-dinitro structure on the basis of ¹³C NMR 〈78MI22603〉. This type of sequence is quite versatile and can be readily adapted, *e.g.* to the preparation of the azaphenoxathiin (198) using the dianion (199) and 2-chloronitrobenzene 〈77JHC1249〉.

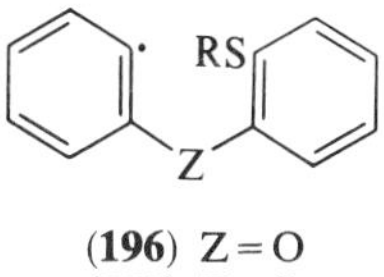

(196) Z = O
(200) Z = S 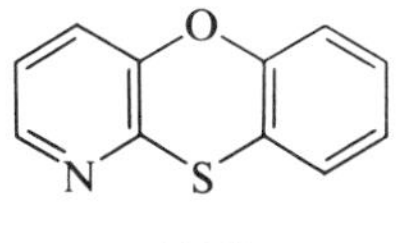
       (198) 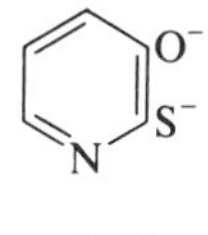        (199)

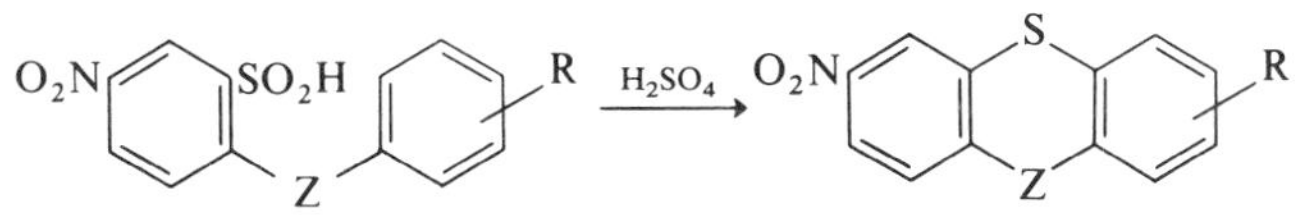

(197) Z = O
(201) Z = S

**Scheme 23**

One of the most researched routes to thianthrene involves reaction of sulfur monochloride with benzene over aluminum chloride ⟨66HC(21-2)1155⟩. Dropwise addition of sulfur monochloride into refluxing benzene containing the catalyst affords an acceptable 66% yield after recrystallization ⟨56JA2163⟩. Other preparative methods for thianthrene and its derivatives include reaction of benzene with sulfur or with sulfur dichloride over aluminum chloride, by heating thiophenols with sulfuric acid, reacting diphenyl disulfide over aluminum chloride, pyrolysis of 2-mercaptophenyl phenyl sulfide and thermolysis of bis(2-iodophenyl) disulfide over copper ⟨66HC(21-2)1155⟩. Aprotic diazotization of 2-aminophenyl 2-phenylthiophenyl sulfide gives the radical (**200**), the sulfur analogue of (**196**) used for preparing phenoxathiin. This route affords thianthrene in 53% yield ⟨74JCS(P1)1272⟩. A further reaction which finds a parallel in phenoxathiin synthesis is the cyclization of sulfinic acids (**201**). This sequence has been particularly well explored for the preparation of nitro derivatives, the substrates (**201**) normally being prepared by reacting 2-chloro-5-nitrobenzenesulfinic acid with a thiophenol ⟨66HC(21-2)1155⟩.

In analogy with the formation of 1,4-dithiins from 1,2,3-thiadiazoles (Section 2.26.4.1.1), the thermal decompositions of 6-substituted benzothiadiazoles afford disubstituted thianthrenes ⟨66HC(21-2)1155⟩. However, recent results suggest that mixtures of 2,7- and 2,8-disubstituted products should be expected as a result of the intermediacy of a benzothiirene ⟨77TL2643⟩. Interestingly, unsubstituted 1,2,3-benzothiadiazole thermolyzes to give thianthrene as only a minor product (7%); other products are diphenyl disulfide (60%) and dibenzo[*b,d*][1,2]dithiin (12%) ⟨75JHC605⟩.

In the 1,2-series, 1,2-benzoxathiin 2,2-dioxides (**202**) are conveniently prepared from 2-acylphenols by reaction with a sulfonyl chloride, followed by base-catalyzed cyclization of the sulfonate ester (**203**). The cyclization normally proceeds well: for R = Me and R′ = H the product is obtained in 83% yield ⟨66HC(21-2)792⟩. Dibenzo[*c,e*][1,2]oxathiin 6,6-dioxide was first prepared from 2-hydroxybiphenyl and sulfuric acid but much higher yields are obtained by C—C bond formation *via* diazotization of phenyl 2-aminobenzenesulfonate, with coupling over copper. The latter method has been used for the preparation of a range of derivatives with yields ranging from 15–80% ⟨66HC(21-2)774⟩. The corresponding monooxide analogues are more rare but the parent has been synthesized from dibenzothiophene 5,5-dioxide using potassium hydroxide in the presence of crown ether. The reaction proceeds most probably *via* the bis salt of 2-(2-hydroxyphenyl) benzenesulfinic acid which cyclizes (*i.e.* esterifies) in acidic media; see also Section 2.26.3.1.7 ⟨81JOC2373⟩.

(**203**)    →    (**202**)

Synthetically useful routes to dibenzo[*c,e*][1,2]dithiins are normally based on cyclizations of biphenyl-2,2′-disulfonyl chlorides. A method applied successfully to the parent compound reduces the precursor with zinc in acetic acid to generate the bis thiol, which is then gently oxidized to the dithiin using iron(II) chloride ⟨66HC(21-2)952⟩. An alternative one-step reductive cyclization, which has been applied to the preparation of the 2,9- and 3,8-dinitro derivatives, involves reduction of the appropriate bis sulfonyl chlorides with hydriodic acid in acetic acid ⟨68MI22600⟩. Yet another reductive cyclization uses sodium sulfite followed by acidification, and these conditions lead to dibenzo[*c,e*][1,2]dithiin 5,5-dioxide. The first step of the reaction is reduction to the disodium salt of biphenyl-2,2′-disulfinic acid which, on acidification, forms the anhydride, *i.e.* dibenzo[*c,e*][1,2]dithiin 5,5,6-trioxide. This is not isolated, but is reduced by the medium to the 5,5-dioxide ⟨77JOC3265⟩. Derivatives of dibenzo[*c,e*][1,2]dithiin in oxidation states other than those mentioned here are obtainable by appropriate oxidation or reduction reactions (see Section 2.26.3.1.4).

## 2.26.4.2 Saturated and Partially Saturated Compounds

### 2.26.4.2.1 Rings with heteroatoms 1,4

The chemistry of almost all of the numerous syntheses of 1,4-dioxanes, 1,4-oxathianes and 1,4-dithianes utilizes the nucleophilicity of negatively charged or neutral oxygen and sulfur atoms. The simplest preparations of the parent compounds involve formation of just one bond and the principal examples of this are provided by the acid-catalyzed cyclization of the 1,5-diols (**204**) and base-catalyzed ring closure of the 1,5-hydroxyhalides (**205**) to give 1,4-dioxane and 1,4-oxathiane. The reaction of 1,5-dihalides (**206**) with hydroxide ion or sodium sulfide is the next simplest, involving formation of two bonds (between five- and one-atom fragments), and has been applied to the synthesis of all three parent compounds. However, bis(2-chloroethyl) sulfide (mustard gas) is an unpleasant reagent and 1,4-dithiane is better prepared by combining four- and two-atom units such as the sodium salt of 1,2-ethanedithiol with 1,2-dihaloethane ⟨B-57MI22600, 66HC(21-2)816, 66HC(21-2)1041⟩. Oxirane, with and without hydrogen sulfide, thiirane, 1,2-ethanediol, 2-mercaptoethanol, polyglycols and polyethylene sulfides have also been used as precursors to the parent systems ⟨B-57MI22600, B-57MI22602, 62JOC4253, 70JCS(B)404⟩.

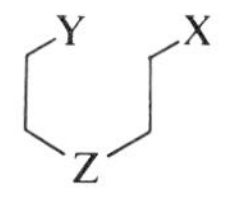

(**204**) X = Y = OH, Z = O, S
(**205**) X = OH, Y = halide, Z = O, S
(**206**) X = Y = halide, Z = O, S

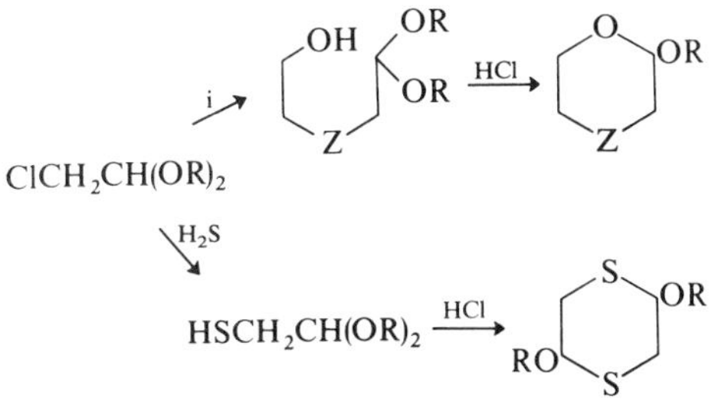

i, HOCH$_2$CH$_2$ZH

**Scheme 24**

Several of the above approaches have proved appropriate for the preparation of alkylated derivatives whilst related but somewhat modified chemistry is required for other substituted compounds: routes depicted in Scheme 24 are examples of methods for the preparation of alkoxy derivatives ⟨47JA2449⟩. Compounds such as 2-hydroxyketones and 2-mercaptoketones dimerize reversibly to give 2,5-dihydroxy derivatives of 1,4-dioxanes and 1,4-dithianes respectively ⟨B-57MI22600, 66HC(21-2)1041⟩.

Addition of sulfur compounds to carbon–carbon multiple bonds has been exploited in the synthesis of both oxathianes and dithianes ⟨66HC(21-2)816, 66HC(21-2)1041⟩. Thus, hydrogen sulfide addition to bis allylic ethers in the presence of an amine gives access to alkylated oxathianes, and sulfur dioxide adds to divinyl sulfone in the presence of formic acid/trimethylamine to form 1,4-dithiane 1,4-tetraoxide, a reaction which proceeds in 80% yield ⟨70JOC2994⟩. Dithianes are also formed by photochemical cycloaddition of thiocarbonyl compounds to alkenes, details of which have been reviewed ⟨71IJS(B)(6)183⟩. 2,3-Dichloro-1,4-dithiane is formed in an interesting reaction involving addition of ethane-1,2-disulfenyl chloride to acetylene. The same reagent on reaction with other alkynes normally affords a 5-chloro-2,3-dihydro-1,4-dithiin derivative, *i.e.* the product resulting from loss of a proton from the intermediate carbenium ion ⟨69JHC627⟩.

More general routes to rings containing one double bond utilize elimination of an alcohol from a 2-alkoxy derivative of the saturated ring system (*cf.* Sections 2.26.4.1.1 and 2.26.4.1.2), or elimination of water, which frequently occurs under the conditions of the cyclization. The latter is exemplified by the reaction of the sodium salt of 1,2-ethanediol, 2-mercaptoethanol or 1,2-ethanedithiol with 2-haloketones to give 2,3-dihydro derivatives of 1,4-dioxins, 1,4-oxathiins or 1,4-dithiins respectively ⟨78H(10)167, 69JOC2762⟩, illustrated in Scheme 25 by a synthesis of the important fungicides (**207**) (see Section 2.26.5) ⟨66USP3249499⟩. A further approach applicable to 2,3-dihydro-1,4-oxathiins and -1,4-dithiins involves reaction of 2-mercaptoketones with oxiranes or thiiranes to give the ring systems in yields ranging from 48–88% (Scheme 26). When Z = S, acid catalysis is unnecessary for the final cyclization step ⟨71LA(753)151⟩.

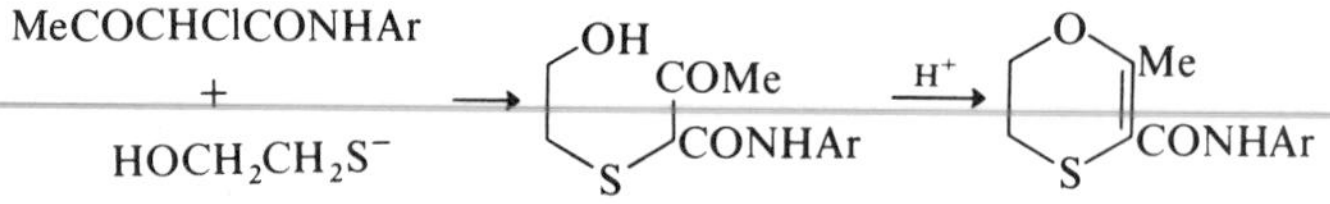

**Scheme 25**

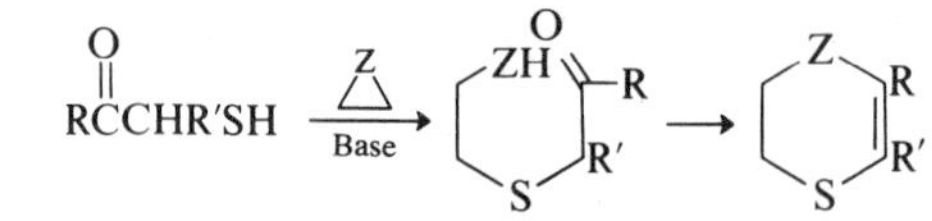

**Scheme 26**

Two alternative approaches to the synthesis of the partially unsaturated ring systems are depicted in Schemes 27 and 28. The addition of sulfide ion to the multiple bonds of (**208**) in the former is facilitated by the activating effect of the sulfinyl or sulfonyl moiety and reaction occurs rapidly, giving yields of 67–84% ⟨79S47⟩. The second approach illustrates the use of an unsaturated dithiolate ion ⟨76JOC1484⟩. The first dianion (**209**) is generated from electrochemically reduced carbon disulfide and reacts with 1,2-dibromoethane to give the fused compound (**210**). Hot base-catalyzed hydrolysis cleaves (**210**) to form a new dianion which, with 1,2-dibromoethane, affords the product.

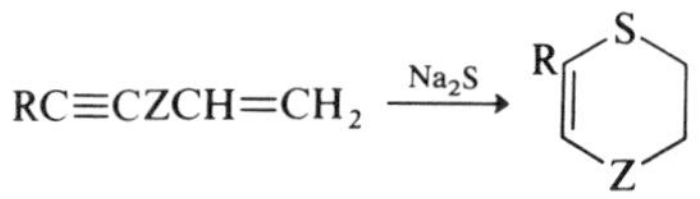

$$RC{\equiv}CZCH{=}CH_2 \xrightarrow{Na_2S}$$

(**208**) $Z = SO, SO_2$

**Scheme 27**

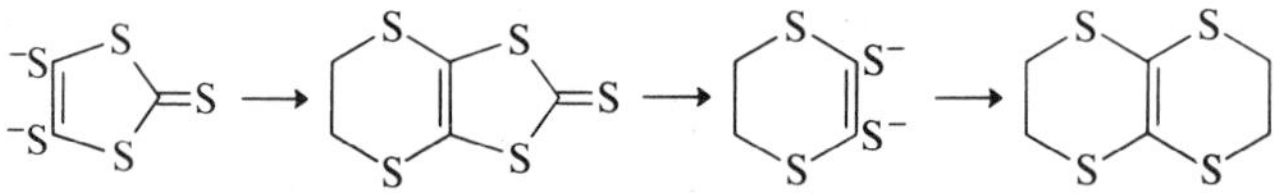

(**209**)        (**210**)

**Scheme 28**

Rings containing a double bond are also available through ring expansion reactions of 1,2-dioxolanes, 1,3-oxathiolanes and 1,3-dithiolanes, all of which are readily available from ketones. Two modes of reaction can be identified, the first of which is shown in Scheme 29. Thus, ring expansion of the dioxolane derivative of benzoin (**211**) is brought about with acid catalysis at 135 °C with loss of water to give 5,6-diphenyl-2,3-dihydro-1,4-dioxin ⟨59JA633⟩. Dithiolane derivatives (**212**) are similarly expanded to the corresponding 2,3-dihydro-1,4-dithiin ⟨69JOC2762, 76H(4)953⟩. The second rearrangement is applicable to oxathiolane or dithiolane derivatives and exploits the ability of a sulfur unit to form a derivative which renders it a good leaving group in elimination reactions. Scheme 30 shows a generalized pathway involving fragmentation of a sulfoxide derivative ⟨76TL25, 77JOC1530⟩ but alternative conditions involving sulfilimines are available ⟨74S713⟩. The reaction pathway of Scheme 30 has proved particularly useful for syntheses of less readily available compounds such as ring fused derivatives, obtained from thio and dithio ketals of cyclic ketones ⟨81OMR(16)266, 76JOC3053, 77JOC1530⟩. In principle, ring contractions of seven-membered rings provide an alternative synthetic route ⟨66HC(21-2)1053⟩ but this type of process has apparently not received any systematic attention.

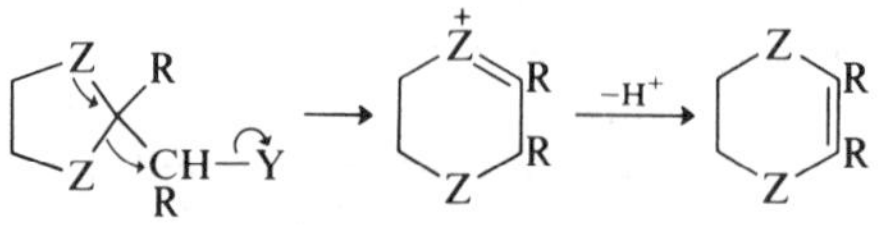

$$\xrightarrow{}\qquad \xrightarrow{-H^+}\qquad$$

(**211**) $Z = O$, $R = Ph$, $Y = OH$
(**212**) $Z = S$, $Y = OH$, Cl, OAc, *etc.*

**Scheme 29**

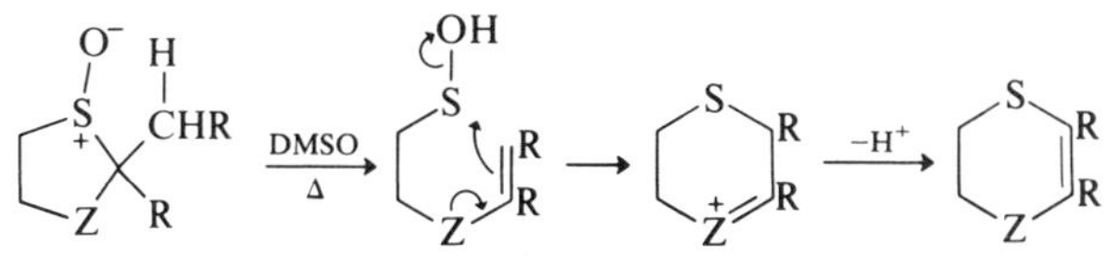

**Scheme 30**

Routes to benzo fused derivatives of 1,4-dioxanes, oxathianes and dithianes frequently make use of anions or dianions of the appropriate 1,2-disubstituted benzene as precursor, as discussed earlier in Section 2.26.4.1.2. However, an alternative approach which is of use for synthesis of 1,4-benzodioxanes involves Diels–Alder addition reactions of alkenes across the quinone function of 1,2-benzoquinones. Quinones which have been used in this manner include the tetrabromo- and tetrachloro-1,2-benzoquinones and phenanthraquinone ⟨76JOC2223, 74ZC239, 70CJC3045⟩. The reaction is brought about either thermally or photochemically and there is evidence that the addition proceeds stereospecifically under the former conditions, *i.e.* cis and trans alkenes give rise to *cis* and *trans* orientation respectively, of the groups on the dioxane ring. In some reactions involving 1,2-benzoquinones, competing addition of the alkene across the diene unit of the ring occurs ⟨76JOC2223⟩.

### 2.26.4.2.2 Rings with heteroatoms 1,3 and 1,3,5

By far the most important route to 1,3-dioxanes, 1,3-oxathianes and 1,3-dithianes is the acid-catalyzed condensation of an aldehyde or ketone with a 1,3-diol, a 3-mercaptoalcohol or a 1,3-dithiol respectively. This series of reactions has been used extensively within synthesis for protection work and in schemes involving masked acyl anions (Section 2.26.3.3.3) as well as for the preparation of numerous derivatives for conformational studies (Section 2.26.2.5). Several acid catalysts have been employed but *p*-toluenesulfonic acid generally proves to be convenient for all three ring systems. Removal of water from the reaction mixture is normally desirable. Minor modifications include the use of an acetal of the carbonyl compound, in place of the carbonyl compound itself, in what essentially becomes a transacetalization reaction ⟨79OMR(12)337⟩. Compounds bearing substituents at C-4, C-5 or C-6 can be prepared using appropriately substituted precursors. When these react with aldehydes (other than formaldehyde) or with unsymmetrical ketones, the cyclization reaction can give rise to *cis* and *trans* isomers but these are normally separable using standard techniques ⟨B-57MI22600, 71ACS1908⟩. For reactions involving glycerol or 1,2,3-propanetrithiol the product mixture contains not only a 5-hydroxy-1,3-dioxane or 5-mercapto-1,3-dithiane but also the isomeric five-membered ring compound formed by condensation across the 1,2-positions ⟨B-57MI22600, 75T1837, 78JA6114⟩.

There are a number of alternative routes to these ring systems and these are now discussed according to ring system. A well researched route to 1,3-dioxanes is the Prins reaction, discovered over 60 years ago and reviewed recently ⟨77S661⟩. The reaction involves the acid-catalyzed condensation of alkenes with aldehydes and almost certainly proceeds *via* a 1,3-diol intermediate. The method is quite general but is most frequently applied to the reaction of styrenes with formaldehyde to give 4-aryl-1,3-dioxanes. Considerable attention has been devoted to optimize conditions to minimize side products: the use of sulfuric acid and *p*-toluenesulfonic acid as catalysts has been well investigated but recent work suggests that, at least for some reactions, higher yields can be achieved using ion exchange resins ⟨80S871⟩. Other routes to 1,3-dioxanes include the reaction of a 1,3-diol with acetylene to give 2-methyl-1,3-dioxanes ⟨B-57MI22600⟩.

Alternative routes to the sulfur-containing rings include Dieckmann cyclizations of diesters in which the S—C—Z unit is already incorporated, and these provide access to 5-keto derivatives ⟨66HC(21-2)805, 66HC(21-2)979⟩. Within the dithiane series, some less common 2-substituents can be incorporated using the reaction between activated methylene group compounds and the ditosylated derivatives of 1,3-dithiols in which the would-be ring sulfur atoms are electrophilic. The latter compounds are obtained from the 1,3-dibromo analogue and potassium thiotosylate ⟨71JOC1137⟩.

Unsaturated rings are available by standard elimination reactions of appropriately substituted saturated rings (particularly 5-hydroxy derivatives) but alternative schemes have also been devised, particularly for the formation of 1,3-dithiins. A fairly general approach involves condensation of a *gem* dithiol with an α,β-unsaturated ketone (at the 1- and 3-positions) with elimination of water, whilst others include the addition of aromatic aldehydes to alkynes in the presence of hydrogen sulfide using boron trifluoride as catalyst. For a review see ⟨72IJS(B)(7)101⟩.

Benzo fused derivatives are in principle available by reactions analogous to those used for the simple ring systems but are dependent upon the availability of appropriate 1,2-disubstituted benzenes. Thus 1,3-benzodioxanes are frequently obtained *via* reaction of aldehydes with 2-hydroxymethylphenol, obtained by condensation of formaldehyde with phenols blocked in the 4-position ⟨B-57MI22601⟩, whilst a route to 1,3-benzodithiane involves reduction of 1,2-benzodithiole-3-thione with LAH to give 2-mercaptomethylthiophenol as intermediate ⟨66HC(21-2)1035⟩. A synthesis of 1,3-benzoxathiane exploits the propensity of phenols to react with DMSO under certain specified conditions to give products of *ortho* substitution. In the presence of DCC, the major products are 2-thiomethoxymethylphenols but with phosphorus pentoxide the reaction can afford about 20% of the 1,3-benzoxathiane ring system ⟨67JOC457⟩. The mechanisms for these and related reactions have been investigated in some detail ⟨71T4195⟩.

1,3,5-Trioxanes are the cyclic trimers of aldehydes and ketones and are formed by acid-catalyzed condensations of the monomers. Thus 1,3,5-trioxane itself can be obtained by distillation of a 60% solution of formaldehyde containing 2% sulfuric acid. The product is extracted into methylene choride. 1,3,5-Trithiane is best prepared by passing $H_2S$ through a mixture of 36% formaldehyde solution and concentrated hydrochloric acid. The yield is 98% before recrystallization ⟨43OSC(2)610⟩. Substituted 1,3,5-trithianes are prepared by similar reactions with appropriate aldehydes and ketones ⟨66HC(21-2)692⟩. At room temperature, mixtures of products may be formed. Thus, passage of $H_2S$ through a solution of acetaldehyde in dilute hydrochloric acid affords not only the 1,3,5-trithiane but also the 1,3,5-oxadithiane and 1,3,5-dioxathiane analogues ⟨66HC(21-2)633⟩. It should be noted that for all the reactions, the use of aldehydes other than formaldehyde, or unsymmetrical ketones leads to products containing *cis* and *trans* isomers.

### 2.26.4.2.3 Rings with two or more adjacent heteroatoms

The O—O bond in 1,2-dioxanes is conventionally formed by oxidation of the appropriate 1,4-diol, normally using concentrated hydrogen peroxide with acid catalysis, or by lead tetraacetate oxidation of the bis hydroperoxide (213) ⟨55CB712⟩. An alternative strategy for preparing this ring system relies on C—O bond formation and this is conveniently achieved using cycloperoxymercuration. In its simplest form, this involves reaction of a 1,5-diene, *e.g.* 1,5-hexadiene, with hydrogen peroxide and mercury(II) nitrate to give the adduct (214) which, on treatment with chloride ion followed by sodium borohydride, affords 3,6-dimethyl-1,2-dioxane (215) ⟨80JCS(P1)2450⟩. Formation of a C—O bond is also involved in two syntheses of 1,2,4-trioxanes (Scheme 31). The first utilizes the hydroperoxyalcohol (216) which reacts with ketones over anhydrous copper sulfate to give the ring closed product (217). The substrate (216) is prepared either by reacting an epoxide with 98% hydrogen peroxide, or *in situ* from alkenes and 34% hydrogen peroxide in the presence of tungstic acid. However the latter method also gives diols ⟨71CC822, 57JOC1682⟩. The alternative route to the trioxane ring system utilizes the diol (218), which reacts with an epoxide over tungstic anhydride and chlorosulfonic acid to give (217) together with an equivalent of ketone ⟨81CC581⟩. The diol precursor for this reaction is prepared from a ketone and 30% hydrogen peroxide. Under stronger conditions, (86% hydrogen peroxide), these reagents react to give 3,3,6,6-tetrasubstituted 1,2,4,5-tetroxanes (219). Catalysts which have been used for this latter reaction include concentrated sulfuric acid and methanesulfonic acid ⟨80JCR(S)35, 75JOC2239⟩. Aromatic ketones, however, preferentially undergo Baeyer–Villiger reactions and thus 3,3,6,6-tetraaryl derivatives are better prepared

(213)

(214) R = CH₂HgNO₃
(215) R = Me

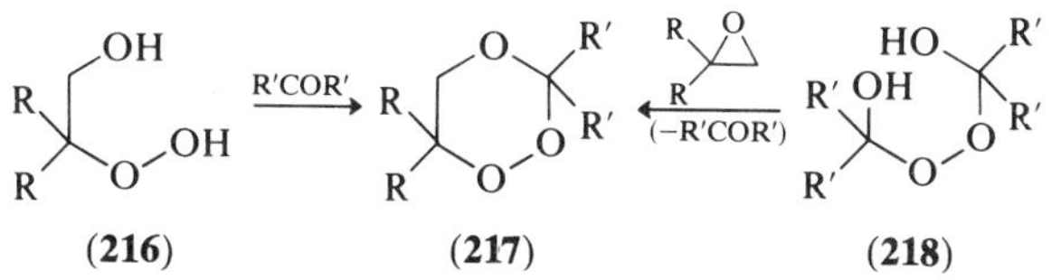

**Scheme 31**

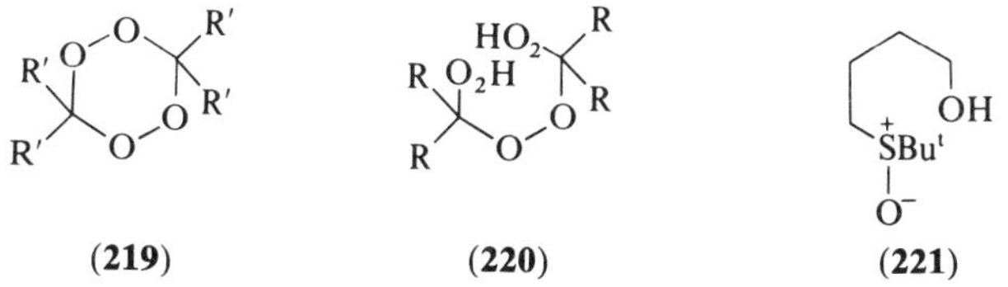

by ozonolysis of appropriate alkenes ⟨55JA2536⟩. Tetroxanes are also prepared by reaction of lead tetraacetate with the bis-hydroperoxides (**220**).

Oxygen bonded to sulfur is encountered in a variety of systems, the simplest being the 1,2-oxathiane 2-oxides and 2,2-dioxides. The former, sultines, are obtainable by several routes but a reliable approach involves cyclization of δ-hydroxysulfoxides (**221**) with sulfuryl chloride, in yields in excess of 60% ⟨76CJC3012⟩. The latter, sultones, have been known much longer and are obtained either by dehydration of δ-hydroxysulfonic acids or dehydrohalogenation of δ-halosulfonic acids ⟨66HC(21-2)774⟩. 1,3,2-Dioxathiane derivatives are quite readily prepared from a 1,3-diol and an appropriate sulfur reagent. Thus, 1,3-propanediol reacts with sulfur dichloride at −75 to −95 °C in methylene chloride in the presence of triethylamine to afford 1,3,2-dioxathiane ⟨65JOC2703⟩. The 2-oxide derivatives (sulfites) are obtained by reacting the diol with thionyl chloride, and the 2,2-dioxides (cyclic sulfates) using sulfuryl chloride, chlorosulfonic acid or oleum. Alternatively, the dioxides can be obtained from the monoxides using an oxidizing agent such as calcium permanganate, or by ring cleavage with sulfuric acid followed by cyclization of the 3-hydroxypropylsulfuric acid with thionyl chloride ⟨66HC(21-2)633⟩.

The S—O—S unit is found in 1,2,6-oxadithiane tetraoxides which are anhydrides of 1,3-disulfonic acids from which they are obtained. Thionyl chloride proves a useful dehydrating agent for this reaction ⟨66HC(21-2)675⟩. 1,3,2,4-Dioxadithiane 2,2,4,4-tetraoxide, carbyl sulfate (**151**), has been known since 1837 and is formed by reacting ethylene with sulfur trioxide in the vapour phase. The reaction proceeds in high yield and can be extended to other alkenes. Carbyl sulfate is also formed from ethylene and fuming sulfuric acid or ethanol and three equivalents of sulfur trioxide. It appears that species such as (**222**) and (**223**) are intermediates, carbyl sulfate being the anhydride of the latter ⟨66HC(21-2)611⟩.

1,2-Dithianes are best prepared by S—S bond formation and this is achieved by the oxidation of 1,4-dithiols. Thus, 1,4-dimercaptobutane is oxidatively cyclized in very high yield using reagents such as *t*-butyl hydroperoxide or potassium triiodide. An alternative route to 1,2-dithiane is available by reaction of 1,4-dibromobutane with sodium disulfide, but the yield is much lower ⟨66HC(21-2)952⟩. The latter reaction has also been applied to the synthesis of the unsaturated ring (**224**), starting from 1,4-dibromo-2,3-diphenyl-2-butene. Other products include a 1,2,3,4-tetrathiocin and 3,4-diphenylthiophene. The oxide of (**224**), *i.e.* (**225**), is formed in an intriguing, but low yield, addition reaction of disulfur monoxide (S₂O) to 2,3-diphenylbutadiene. S₂O is formed from sulfur monoxide extruded in the gas phase pyrolysis of thiirane oxide ⟨72JOC2367⟩.

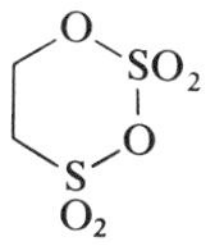

HO₃SCH₂CH₂Y

(**151**)

(**222**) Y = OH
(**223**) Y = OSO₃H

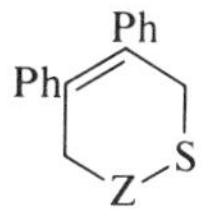

(**224**) Z = S
(**225**) Z = SO

1,2,4-Trithianes are rare but routes established for the 3-methyl derivative (**226**) could provide the basis of more general methods. These include the chlorination of diethyl disulfide and the reaction of the sulfenyl chloride (**227**) with 1,2-ethanedithiol ⟨74MI22601⟩. 1,2,4,5-Tetrathiane (**228**) has been prepared by the cyclization of two equivalents of the bis sulfenyl chloride CH₂Y₂ (**229**; Y = SCl) or the bis Bunte salt CH₂Y₂ (**230**; Y = SSO₃Na) using sodium iodide ⟨65CB1455⟩. 3,3,6,6-Tetrasubstituted derivatives of (**228**) are more readily accessible and can be prepared by reacting a ketone with hydrogen sulfide, sulfur and an amine, or with ammonium polysulfide. The latter reaction is best performed at 4 °C over a period of three days ⟨66HC(21-2)626⟩.

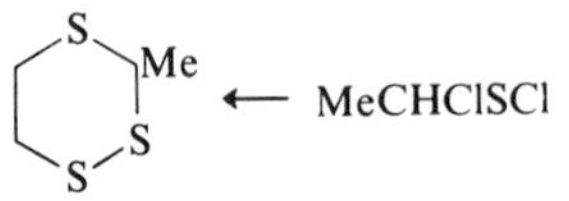

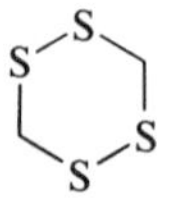

(**226**)    ← MeCHClSCl    (**227**)    (**228**)

An S—S—S linkage can be obtained by reacting the bis Bunte salt (**231**) with sodium sulfide: 1,2,3-trithiane itself has been obtained *via* this route ⟨65JCS2901⟩. Alternative and higher overall yield ring closures have been devised and applied to the synthesis of 5,5-dialkyl derivatives. These include the reaction of the dithiol (**232**) with sulfur dichloride at low temperature, or heating the dimesylate (**233**) with sodium tetrasulfide ⟨75S329⟩. 1,2,3,5-Tetrathiane (**234**) has been obtained as one of several products in a reaction involving sulfur, sodium sulfide and formaldehyde ⟨69JAP6927724⟩. Rings containing four or five sulfur atoms consecutively linked to one another have been prepared by routes which exploit the potential of reagents of the type $S_xCl_2$. Thus, disulfur dichloride reacts with 1,2-dimercaptocyclohexane to form the tetrasulfide (**20**) in 74% yield ⟨67AG(E)703⟩ and trisulfur dichloride reacts with methanedithiol to give pentathiane, though in rather lower yield ⟨68AG(E)301⟩.

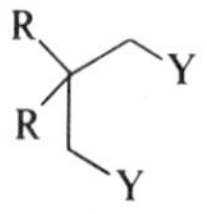

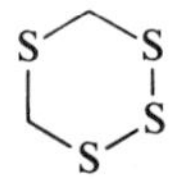

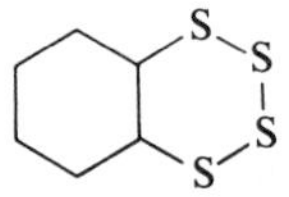

(**231**) Y = SSO₃Na
(**232**) Y = SH
(**233**) Y = OMes

(**234**)

(**20**)

## 2.26.5 APPLICATIONS AND IMPORTANT COMPOUNDS

Over the last two decades few chemicals have acquired as much notoriety as 2,3,7,8-tetrachlorodibenzo[*b,e*][1,4]dioxin. The compound, commonly referred to as TCDD, or simply but misleadingly as dioxin, is formed as a by-product in the commercial preparation of 2,4,5-trichlorophenol, an intermediate in the manufacture of the bacteriocide hexachlorophene and the herbicide 2,4,5-trichlorophenoxyacetic acid (2,4,5-T). The commercial production of 2,4,5-trichlorophenol, during which TCDD is generated, involves the reaction of 1,2,4,5-tetrachlorobenzene with base; TCDD is formed if the product reacts further (Scheme 32).

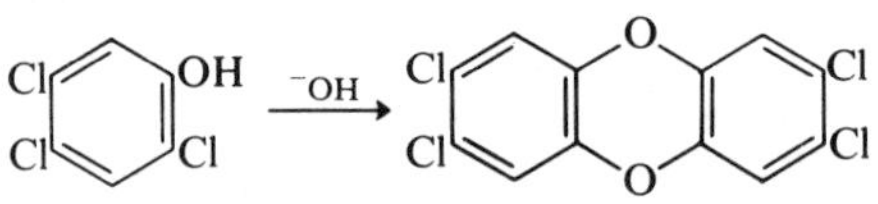

**Scheme 32**

TCDD is extraordinarily toxic (for rats the $LD_{50}$ is 45 $\mu$g kg$^{-1}$) and is also a teratogen, *i.e.* an agent causing birth defects. Typical symptoms resulting from occupational exposure include chloracne, a very severe form of acne, and liver damage. Disquiet has arisen principally in connection with the compound's impact on the environment and the attendant health hazards, problems which became particularly apparent during the defoliation programme in Vietnam which utilized a formulation based partly on 2,4,5-T containing TCDD as contaminant. Public awareness was subsequently heightened after an explosion at Seveso in Northern Italy (July 1976) where an estimated 2.5 kg of TCDD was released over the adjacent area. This accident was by far the worst in a history of incidents which had occurred previously in the USA, the UK, Germany, Holland and Czechoslovakia ⟨B-79MI22602, 76MI22600⟩.

Other chlorinated dibenzodioxins show markedly varied toxicological properties. Thus 2,7-dichloro and octachloro derivatives have low acute toxicity whereas hexachloro derivatives are much more toxic. However, none are as toxic as TCDD itself ⟨77MI22602⟩.

Most other dibenzodioxins are relatively non-toxic and potential applications for the ring system, and also for the phenoxathiin and thianthrene moieties have been sought within polymer science. It is generally found that polymers based on a ladder type structure are thermally more stable than those based on a single strand. Thus the structure and intrinsic stability of the three heterocycles are properties which suggest these units could be beneficial for incorporation into polymers designed to possess heat-resistant properties. Derivatives of the heterocycles which have been used in the preparation of copolymers include the 2,7- and 2,8-diamino compounds which with aromatic tetracarboxylic acid dianhydrides give polyimides, and diacyl chloride compounds which with diamines form polyamides. Copolymers so formed show little weight loss to temperatures up to 400–500 °C, depending upon substrates, and rather limited solubility ⟨74MI22602, 76JAP(K)7696803, 77MI22603, 79MI22603⟩.

Phenoxathiins have found quite a range of applications in other areas and these have been sumarized elsewhere ⟨66HC(21-2)867⟩. Included among these are the use of phenoxathiin itself as an insecticide against aphids, termites, housefly, rice weevil, *etc.* It is also toxic to a variety of microorganisms. More recently, formulations based on phenoxathiin and 4-acetamidophenol have been reported to have anti-inflammatory, antiedema, analgesic and antipyretic properties and to be useful for the treatment of measles, bronchitis, gout, *etc.* It appears that the acylaminophenol component has a remarkable synergistic effect on the activity of the heterocycle ⟨68FRP6157⟩. Various substituted phenoxathiin sulfones and their thianthrene analogues, including thianthrene 5,5,10-trioxides, are reported to be useful in the treatment of asthma, hay fever, conjuctivitis, urticaria and eczema ⟨76BRP1447032⟩.

Other thianthrene derivatives have been patented for a variety of uses including an early treatment of skin infections caused by mites, as a solvent and plasticizer for PVC, and as a stabilizer in coolant moderator fluids for nuclear reactors. Partially oxidized thianthrenedicarboxylic acids, when condensed with amino-*ortho*-quinones, form vat dyes ⟨66HC(21-2)1155⟩. Many other dyes, referred to as sulfur dyes, have been known since the last century. They are formed as amorphous high molecular weight materials from reaction of sulfur or sodium polysulfide with aromatic hydroxy, amino or nitro compounds and contain the 1,4-dithiin unit fused to assorted annellated quinonoid rings ⟨66HC(21-2)1249⟩. More recently, 1,4-dithiin rings substituted with cyano groups at the 2- and 3-positions, and fused to heterocyclic rings such as chloropyridazines and pyrazines, have been shown to have good fungicidal and bacteriocidal properties for agricultural use; *e.g.* (235) gives 100% inhibition of tomato late blight fungus at 400 p.p.m. ⟨74USP3843664, 74USP3849415⟩.

Among the saturated compounds, 1,4-dioxane is of well proven industrial importance. It is an excellent solvent for a range of materials which includes cellulose acetate, oils, greases, waxes and resins, and finds use as a solvent in the manufacture of paints, varnishes and polishes, and as a wetting agent and dispersing agent in textile processing. It has also been used in dye baths and in printing and polishing compositions ⟨B-71MI22600⟩. Numerous derivatives of 1,4-benzodioxane have been much investigated as a consequence of the discovery in the 1930s that 2-(diethylaminomethyl)-1,4-benzodioxane (236) exhibits significant α-adrenergic blocking activity. Since then various analogues have been tested as possible antihypertensive agents. A study of the monoalkylamino series (237) revealed that adrenolytic activity reaches a maximum for R = Et and Pr whilst toxicity increases from methyl through to amyl. In the dialkylamino series, (236) is the most active but also the most toxic. Other compounds, though effective in reducing blood pressure, also show moderately severe side effects ⟨B-67MI22601⟩. There is evidence showing that the inactivity varies with stereochemistry. Thus, the (*S*) enantiomer of (236) is a more potent α-adrenergic receptor antagonist than the (*R*) isomer ⟨77JMC880⟩. Oxygenated rings which have appeared in the more recent patent literature include 2-oximino-1,4-dioxane derivatives ⟨77USP4062969⟩ and 1,3-benzodioxanes, *e.g.* (238) ⟨76GEP2615376⟩. The former is reported to control mites, aphids and housefly and the latter to be useful as a cattle growth substance.

Derivatives of 2,3-dihydro-1,4-oxathiin and 1,4-benzoxathiins have been patented as agrochemicals. An important development in fungicidal research has been the discovery of so-called systemic fungicides. Unlike surface fungicides which may require repeated

(235)

(236) R = R' = Et
(237) R = H, R' = alkyl

(238)

applications, systemic fungicides can be absorbed by the plant, within which they can eradicate infection and protect against invading pathogens. In the mid 1960s it was found that the dihydrooxathiins (**207**) and (**239**) are effective in controlling plant pathogenic fungi such as wheat leaf rust, bean rust and loose smut of barley ⟨66MI22601, 66USP3249499⟩ and these and related compounds have been very widely investigated since. The ring substituents are apparently very important: various analogues of (**207**) and (**239**) such as the free acid, simple esters and *N*-alkylated amides show considerably less fungicidal activity or are quite inactive. Certain 2,3-dihydro-1,4-dithiin derivatives have also attracted attention as chemicals for agricultural use. Thus, the dimethyl derivative (**240**), at 10 lb per acre, gives 100% pre-emergence control over a variety of weeds ⟨77USP4062969⟩ and proves also to have defoliant properties. Many other derivatives have also been tested and prove active; thus, in pre-emergence tests (**241**) controls 100% weeds such as pigweed, purslane and giant foxtail with no damage to corn, rice, *etc.* ⟨76USP3997323⟩. Seed corn treated with 1.25 g of the phenyl derivative (**242**) per kilo of seed has an 88% germination rate in fungus-infected soil ⟨76GEP2527639⟩.

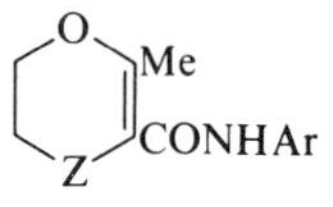

(**207**) Z = S
(**239**) Z = SO$_2$

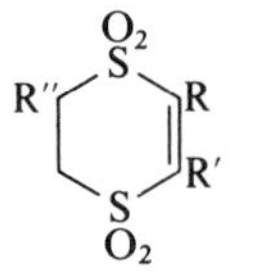

(**240**) R = R' = Me, R'' = H
(**241**) R = R' = Me, R'' = Et
(**242**) R = R'' = H, R' = Ph

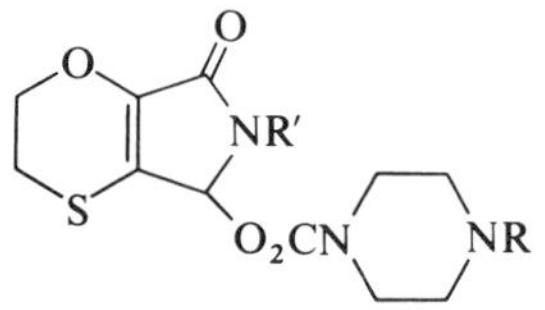

(**243**)

Various dihydro-1,4-oxathiins and dihydro-1,4-dithiins are reported to be useful in areas other than agriculture. Thus, 2,3-dihydro-1,4-dithiin-5,6-dicarboximides display a wide spectrum of biological activity ⟨77SST(4)332⟩ while derivatives of the basic structure (**243**) have been patented as tranquilizers, anticonvulsants and hypnotics ⟨76GEP2550163⟩. Sedative, myorelaxant and cataleptic properties are associated with certain 2,1-benzoxathiane 2,2-dioxides ⟨75BRP1383459⟩.

Sulfonation of alkenes using sulfur trioxide can lead to mixtures of products containing 1,2-oxathiin 2,2-dioxides and these, on hydrolysis, afford hydroxyalkanesulfonates (*cf.* Section 2.26.3.3.4). The sulfonate mixtures are very satisfactory surfactants and find application in the detergent industry ⟨66HC(21-2)780⟩. 6-Alkyl derivatives of the ring system, within formulations containing a dispersing agent, have been patented for use as textile softeners ⟨75GEP2449901⟩. The parent ring system has been patented for a process to improve dyeing of polyamide fibres. Thus, poly(*ε*-caprolactam) and nylon salts react with the heterocycle to give a derivative to which basic dyes can attach ⟨72JAP7208232⟩. In somewhat related chemistry, ring opening of carbyl sulfate (1,3,2,4-dioxadithiane 2,2,4,4-tetraoxide) by a dye containing an amino group leads to a 2-sulfatoethylsulfonylated dye which is a useful process within a sequence for attaching a dye to cellulose ⟨B-75MI22601⟩.

# 2.27

# Oxazines, Thiazines and their Benzo Derivatives

## M. SAINSBURY
*University of Bath*

| | |
|---|---|
| 2.27.1  STRUCTURAL TYPES AND NOMENCLATURE | 995 |
| 2.27.2  STRUCTURES AND REACTIVITY | 997 |
|    *2.27.2.1  1,2-Oxazines, 1,2-Thiazines and their Benzo Derivatives* | 997 |
|      *2.27.2.1.1  Parent ring systems* | 997 |
|      *2.27.2.1.2  Dihydro-1,2-oxazines and -thiazines* | 998 |
|      *2.27.2.1.3  Tetrahydro-1,2-oxazines* | 1000 |
|    *2.27.2.2  1,3-Oxazines, 1,3-Thiazines and their Benzo Derivatives* | 1000 |
|      *2.27.2.2.1  Parent ring systems* | 1000 |
|      *2.27.2.2.2  Mesomeric betaine structures* | 1004 |
|      *2.27.2.2.3  1,3-Oxazinium and -thiazinium salts* | 1005 |
|      *2.27.2.2.4  Dihydro-1,3-oxazines and -thiazines* | 1006 |
|      *2.27.2.2.5  Tetrahydro-1,3-oxazines and -thiazines* | 1008 |
|    *2.27.2.3  1,4-Oxazines, 1,4-Thiazines and their Benzo Derivatives* | 1009 |
|      *2.27.2.3.1  Parent ring systems* | 1009 |
|      *2.27.2.3.2  Dihydro-1,4-oxazines, -thiazines and their benzo derivatives* | 1012 |
|      *2.27.2.3.3  Tetrahydro-1,4-oxazines and -thiazines* | 1014 |
| 2.27.3  SYNTHESES | 1014 |
|    *2.27.3.1  1,2-Oxazines, 1,2-Thiazines and their Benzo Derivatives* | 1014 |
|      *2.27.3.1.1  Parent ring systems* | 1014 |
|      *2.27.3.1.2  Dihydro-1,2-oxazines and -thiazines* | 1017 |
|    *2.27.3.2  1,3-Oxazines, 1,3-Thiazines and their Benzo Derivatives* | 1018 |
|      *2.27.3.2.1  Parent ring systems* | 1018 |
|      *2.27.3.2.2  1,3-Oxazinium and -thiazinium salts* | 1021 |
|      *2.27.3.2.3  1,3-Oxazinones, -thiazinones and related compounds* | 1022 |
|      *2.27.3.2.4  Dihydro-1,3-oxazines and -thiazines* | 1024 |
|      *2.27.3.2.5  Dihydro-1,3-oxazinones, -1,3-thiazinones and related compounds* | 1027 |
|      *2.27.3.2.6  Tetrahydro-1,3-oxazines and -thiazines* | 1029 |
|    *2.27.3.3  1,4-Oxazines, 1,4-Thiazines and their Benzo Derivatives* | 1031 |
|      *2.27.3.3.1  Parent ring systems* | 1031 |
|      *2.27.3.3.2  Dihydro-1,4-oxazines, -thiazines and their benzo derivatives* | 1035 |
|      *2.27.3.3.3  Tetrahydro-1,4-oxazines and -thiazines* | 1037 |
| 2.27.4  IMPORTANT STRUCTURES AND APPLICATIONS | 1037 |

## 2.27.1 STRUCTURAL TYPES AND NOMENCLATURE

The oxazines and thiazines are six-membered heterocycles containing one oxygen or sulfur atom and a single nitrogen atom. A number of isomeric structures are possible depending upon the relative positions of the two heteroatoms and the degree of oxidation of the ring system. Alternatives for the basic monocyclic types are shown in Figure 1. Some are well known, but others are uncommon and occur only as reaction intermediates; moreover the subdivision within each group of positional isomers is not always clear-cut because they may be interrelated by tautomerism. There are, in addition, some $1H$-thiazine systems containing hypervalent sulfur, *i.e.* sulfur(IV) or sulfur(VI), as in compounds (**131**) and (**254**).

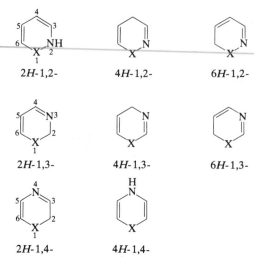

**Figure 1**   The oxazines (X = O) and thiazines (X = S)

Further structural variation is brought about when a single benzene ring is fused to the heterocycle, leading to nine basic benzoxazines or benzothiazines (Figure 2), but of the two possible dibenzo constructions only those based on the 1,4-systems are common, although some dibenzo-1,2-oxazines and -1,2-thiazines are known. The dibenzo-1,4-oxazines and -1,4-thiazines have a very long history and are more usually called phenoxazines and phenothiazines respectively (Figure 3).

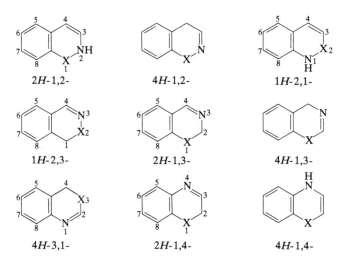

**Figure 2**   The benzoxazines (X = O) and benzothiazines (X = S)

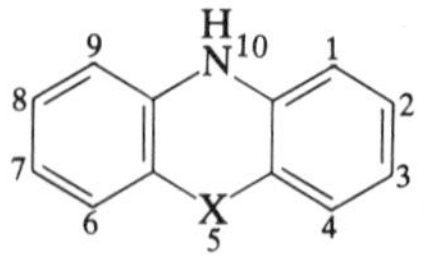

**Figure 3**   Phenoxazine (X = O) and phenothiazine (X = S)

Hydrogenated derivatives of the monocyclic and bicyclic oxazines and thiazines are also very well established compounds and a number are so familiar that they too have trivial names. Tetrahydro-1,4-oxazine, for example, is better known as morpholine, a name first allocated to it because of a supposed, but erroneous, relationship to the alkaloid morphine. By analogy the corresponding thiazine is called thiomorpholine or sometimes thiazane.

## 2.27.2 STRUCTURES AND REACTIVITY

### 2.27.2.1 1,2-Oxazines, 1,2-Thiazines and their Benzo Derivatives

#### 2.27.2.1.1 Parent ring systems

*(i) Stability*

Simple bond energy calculations show that 2*H*-1,2-oxazines and -thiazines are unstable relative to the corresponding open chain iminoaldehydes and iminothioaldehydes, and thus they only find expression in the form of transient species. Early claims for the synthesis of 2*H*-1,2-oxazines have not been authenticated, and, for example, the reaction between 4-nitrosodimethylaniline and tetracyclone does not give the oxazine (**1**) as was once thought, but rather the isomeric lactam (**2**), although it is likely that the oxazine is formed as an intermediate (Scheme 1) ⟨64TL1569⟩.

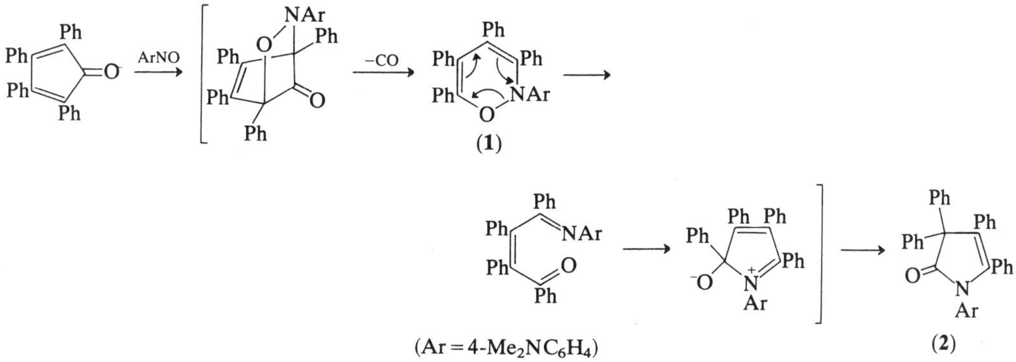

(Ar = 4-Me₂NC₆H₄)

**Scheme 1**

Similarly, the formation of the pyrrolinone (**4**) from the reaction of nitrosobenzene and α-pyrone may also involve the 2*H*-1,2-oxazine (**3**), which then traps a second molecule of the nitroso compound in a [4+2] cycloaddition step (Scheme 2) ⟨76JOC2496⟩.

4*H*-1,2-Oxazines are stabilized as 4,4-disubstituted derivatives but 4*H*- and 6*H*-1,2-thiazines are almost unknown. Possibly the paucity of these compounds simply reflects a lack of interest, for 6*H*-1,2-oxazines are comparatively common, and 3,5-diphenyl-6*H*-1,2-oxazine, for example, is a stable crystalline solid. Similarly, cyano-1,2-oxazines (**6**; R=alkyl) are formed by the photolysis of azidopyridine oxides (**5**). However in the case where R³ = H, these products rearrange to pyrroles (**7**) ⟨81CC36⟩.

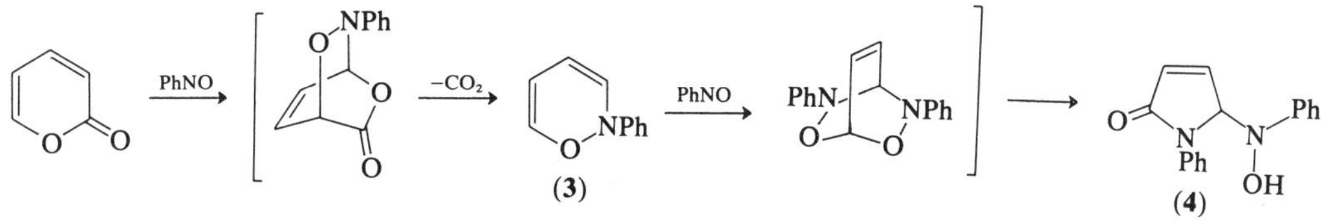

**Scheme 2**

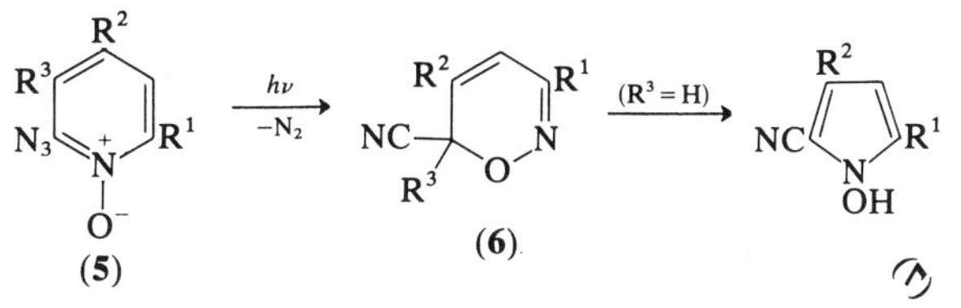

1,2-Benzoxazines and 1,2-benzothiazines are rarely encountered except as carbonyl derivatives or as *S*-oxides respectively. However, an interesting representation of the 2,1-benzothiazine system is provided by the zwitterion (**9**) which is formed by successive treatment of the 2-aminostyrene (**8**) with *N*-chlorosuccinimide and potassium hydroxide (see Section 2.27.3.1.1. ii). On protonation it gives the salt (**10**).

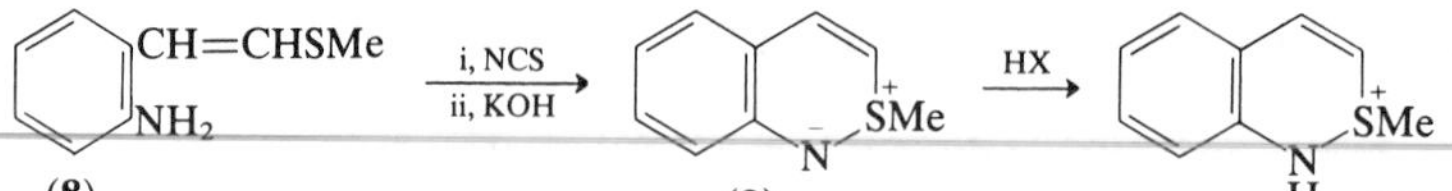

**(8)**          **(9)**                          **(10)**

*(ii) Salts*

The variable valency of sulfur allows the formation of *S*-oxides (**11**) which unlike thiabenzene 1-oxides are non-aromatic and exist in half-boat conformations ⟨78CC197⟩. On the other hand, acid treatment of 6-hydroxy-6*H*-1,2-oxazines (**12**) yields cations (**13**) which are analogous to pyrylium salts ⟨74ZOR1513⟩.

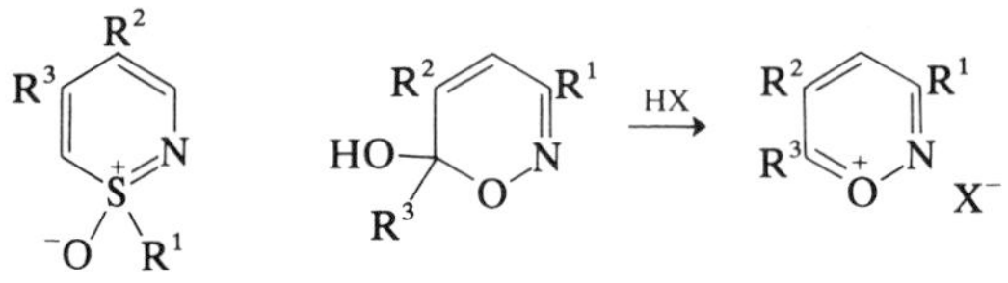

**(11)**              **(12)**              **(13)**

*(iii) General reactivity*

With so few examples little has been done to define the general reactions of the parent systems. However, direct pyrolysis of the benzoxazinone (**14**) yields biphenylene, whereas flash vacuum pyrolysis causes C–N transposition and leads to the 3,1-isomer (**15**) ⟨78CC9⟩.

Sodium borohydride and lithium aluminum hydride afford the hemiacetal (**16**), but aryllithium reagents effect arylation and concomitant ring opening to hydroxyketones (**17**; Scheme 3) ⟨79ZOR2405⟩.

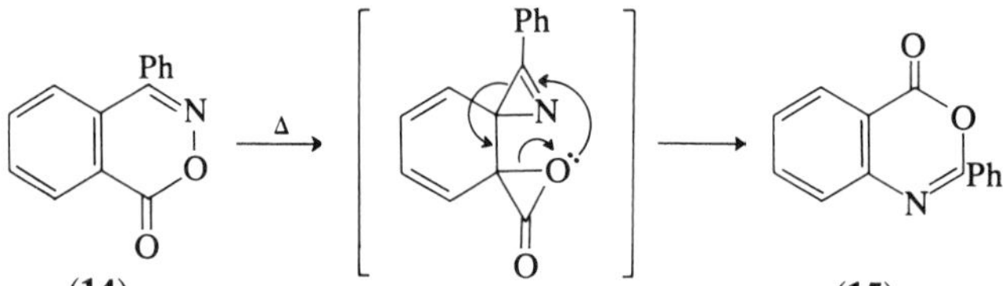

**(14)**                                              **(15)**

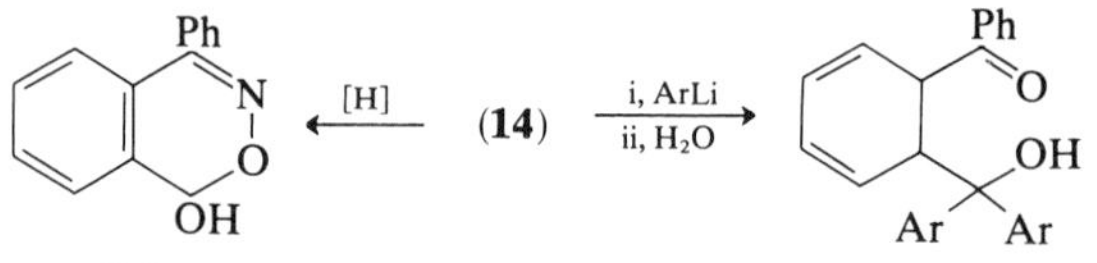

**(16)**                    **Scheme 3**                    **(17)**

### 2.27.2.1.2  Dihydro-1,2-oxazines and -thiazines

*(i) Structure and stereochemistry*

Depending upon the position of the double bond, four different monocyclic systems are possible (Figure 4) together with three benzo forms (Figure 5).

2*H*-5,6-          4*H*-5,6-          2*H*-3,6-          2*H*-3,4-

**Figure 4**   The dihydro-1,2-oxazines (X = O) and dihydro-1,2-thiazines (X = S)

For 2,6-disubstituted 3,6-dihydro-2*H*-1,2-oxazines there are four energetically different half-chair conformations reflecting both ring and *N*-inversion whereas for 3-substituted 3,4-dihydro-1*H*-2,3-benzoxazines there are only two conformers. $^1$H NMR studies do not

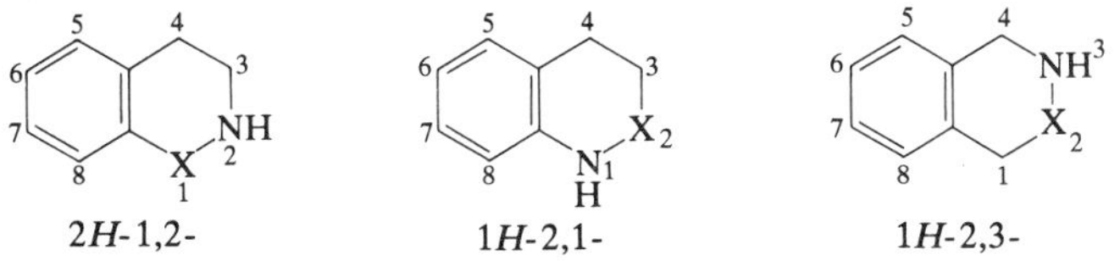

**Figure 5** The 3,4-dihydrobenzoxazines (X = O) and 3,4-dihydrobenzothiazines (X = S)

allow a satisfactory analysis of the bicyclic systems, but for the monocyclic compounds at low temperature there is a strong preference for the *N*-substituent to adopt an equatorial position ⟨74JCS(P2)1737⟩.

### (ii) Tautomerism

The 5,6-dihydro-2*H*- and -4*H*-1,2-oxazines and -thiazines are interrelated by prototropy, being enamines and imines respectively. In the case of the oxazines the imine form is favoured and there are several well-established examples of this system, including the parent heterocycle (**18**). This compound is obtained by oxidation of tetrahydro-1,2-oxazine with lead tetraacetate and, although it is reasonably stable, on standing it slowly trimerizes (Scheme 4) ⟨72JCS(P1)1701⟩.

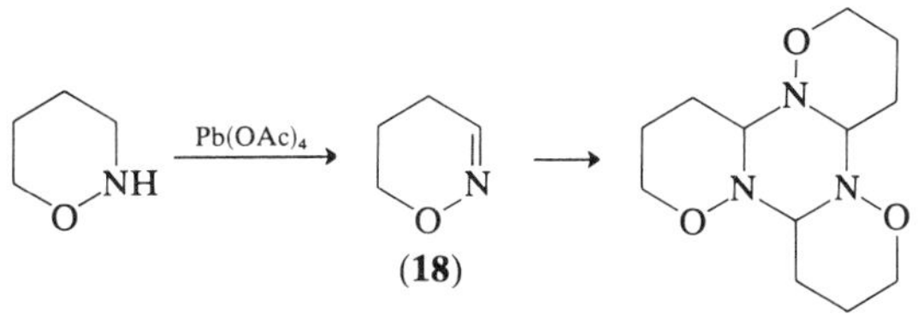

**Scheme 4**

Interestingly, pyrolysis of the 5,6-dihydro-4*H*-1,2-oxazine (**19**) affords 2-phenylpyridine and to account for this change it is proposed that ring-opening is preceded by isomerism to the 2*H*-isomer (Scheme 5) ⟨79JCS(P1)258⟩.

6-Hydroxydihydro-1,2-oxazines (**20**) are in equilibrium with the corresponding open-chain keto oximes (**21**) ⟨70T1315⟩, and a related tautomerism is exhibited by certain *N*-oxides (**22**), the *seco* forms (**23**) of which are nitronic acids ⟨78TL2339⟩.

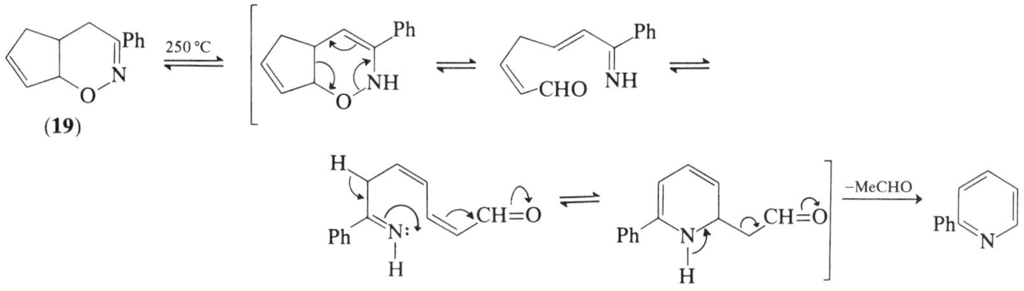

**Scheme 5**

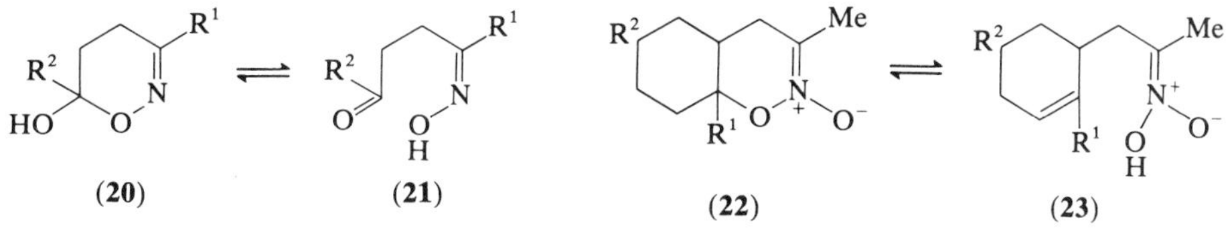

### (iii) Oxidation and reduction

Although direct oxidation is sometimes possible ⟨54CCC282⟩, *N*- or *S*-oxides and dioxides are normally prepared by indirect routes. 3,6-Dihydro-1,2-oxazines can be reduced catalytically to their tetrahydro derivatives, but zinc and acetic acid cause ring fission to 4-aminobut-3-enols which cyclize to dihydropyrroles ⟨54MI22700⟩.

2-Alkyl-5,6-dihydro-4*H*-1,2-oxazinium salts are formed by the alkylation of the free bases and these products are easily reduced by sodium borohydride to 2-alkyltetrahydro-1,2-oxazines ⟨76HCA2765⟩.

### (iv) Reactions with nucleophiles

The nitro group of 3-nitro-5,6-dihydro-4*H*-1,2-oxazines can be displaced by certain nucleophiles thus affording access to other 3-substituted derivatives (Scheme 6) ⟨78JOC2020⟩.

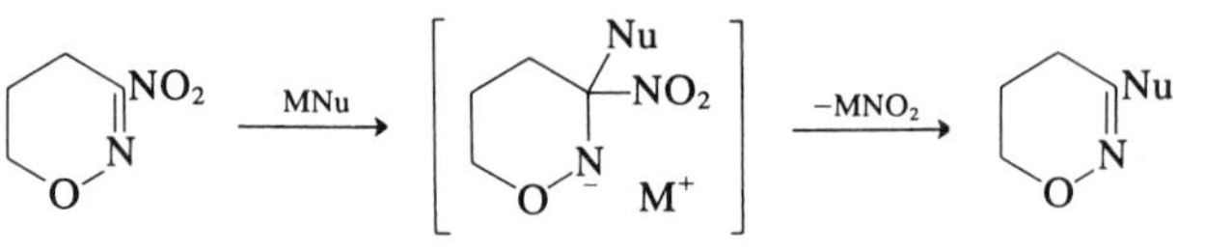

(MNu = NaCN, NaN₃, NaO₂SPh, NaSPh or CdBu₂)

**Scheme 6**

### 2.27.2.1.3 Tetrahydro-1,2-oxazines

### (i) Conformation

X-Ray crystallographic analyses show that tetrahydro-1,2-oxazines adopt a chair conformation, although the ring is considerably more puckered than that of cyclohexane. The lone pair electrons of *N*-methyl derivatives assume an axial orientation and in the case of 2,5-dimethyl compounds the *trans* conformer is strongly preferred ⟨75T523⟩.

## 2.27.2.2 1,3-Oxazines, 1,3-Thiazines and their Benzo Derivatives

### 2.27.2.2.1 Parent ring systems

### (i) Tautomerism

At present no systematic study of the factors which influence tautomeric behaviour in 1,3-oxazines and thiazines has been undertaken. All three types of 1,3-oxazines are known and both 2-amino-4-thioxo- and 2-imino-4-mercapto-1,3-thiazines exist in equilibrium with each other ⟨74MI22700⟩. When, however, the 4-position is disubstituted the amino isomer predominates ⟨71KGS946⟩.

### (ii) Oxidation and reactions with electrophiles

1,3-Thiazines can be oxidized at the sulfur atom to give sulfoxides and eventually sulfones. Generally speaking, however, this is not a good approach to these structures as there are inherent problems of non-selectivity and over-oxidation. Air oxidation is sufficient to convert the 1,3-benzothiazine (**24**; R = alkoxy) into the 1-oxide (**25**; *n* = 1), whereas potassium permanganate yields the 1,1-dioxide (**25**; *n* = 2). In the case of the unsubstituted compound (**24**; R = H) the sulfur atom is less reactive and oxidation affords the 1,3-benzothiazin-4-one (**26**) ⟨77ACH(92)317⟩.

Cyanogen bromide reacts with the benzothiazinedithione (**27**) to yield the thiocyanato derivative (**28**); however, this product is unstable in contact with triethylamine and air and gradually yields a mixture of the sulfide (**29**) and disulfide (**30**) ⟨75MI22700⟩.

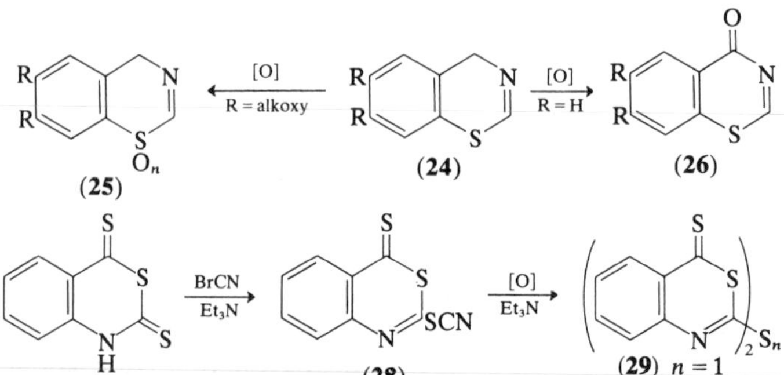

Alkyl halides react with the nitrogen atom of all but the lactam and thiolactam forms of 1,3-oxazines and -thiazines. In the case of thioimides, *e.g.* (31), it is necessary that the NH group is deprotonated by base treatment prior to alkylation of the resultant anion (Scheme 7) ⟨B-78MI22701⟩.

Diazomethane attacks the thione group of the 3,1-benzothiazine-4-thione (32) to give the bis-(1,3-dithiolane) (33), whereas diazoketones afford 4-methylene derivatives (34) ⟨71BSF187⟩.

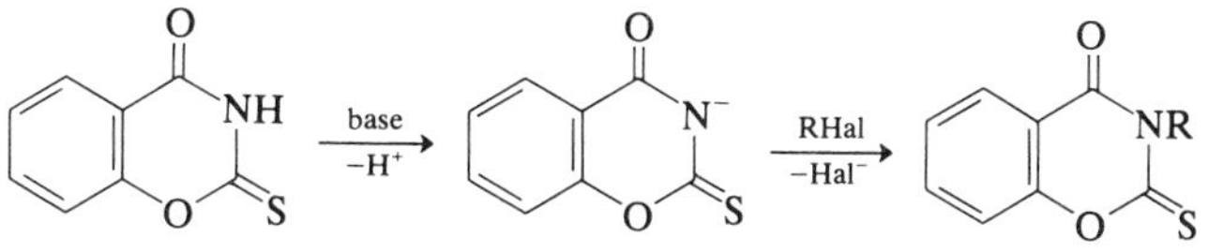

**Scheme 7**

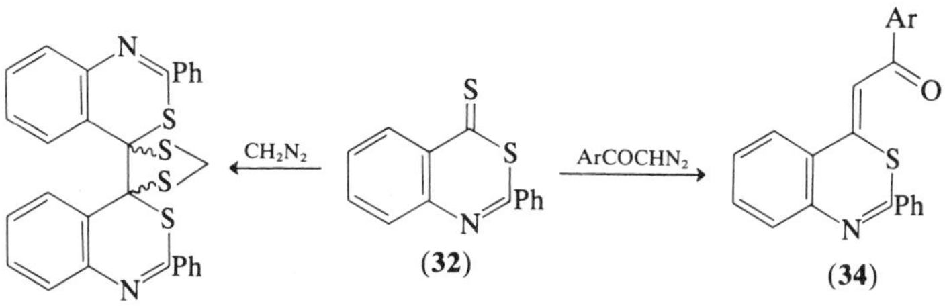

*(iii) Reactions with nucleophiles*

Oxazinyl anions are obtained when 4*H*-1,3-oxazines (or their tautomers) are treated with strong base in aprotic media. The anions are antiaromatic, and are only stable at low temperature, when they are considered to exist in equilibrium with oxazabicyclo[3.1.0]hexenyl valence tautomers (Scheme 8) ⟨75AG(E)581⟩.

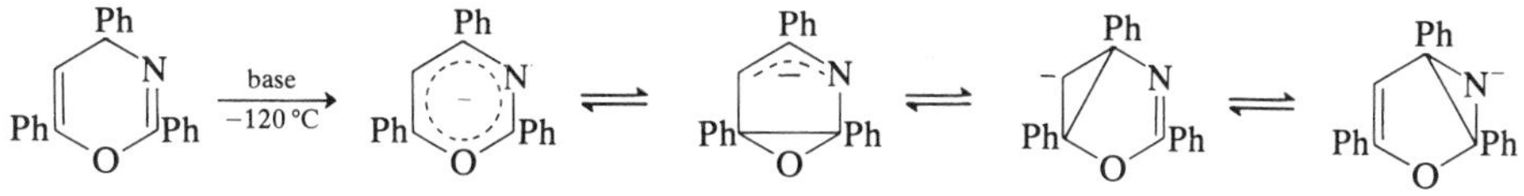

**Scheme 8**

Aqueous base causes hydrolytic ring opening, although the ease with which this is achieved varies from system to system. 4*H*-1,3-Benzothiazines are quite stable at room temperature to aqueous alkali, but on heating the heterocyclic ring is destroyed by initial nucleophilic attack at C-2 (Scheme 9) ⟨68ACH(58)179⟩. 2*H*-1,3-Benzoxazines are easier to hydrolyze and here both *O*-alkyl and imine bonds are cleaved by the action of water (Scheme 10) ⟨31JA644⟩.

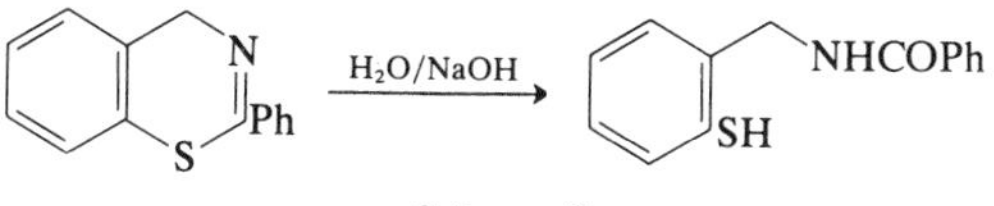

**Scheme 9**

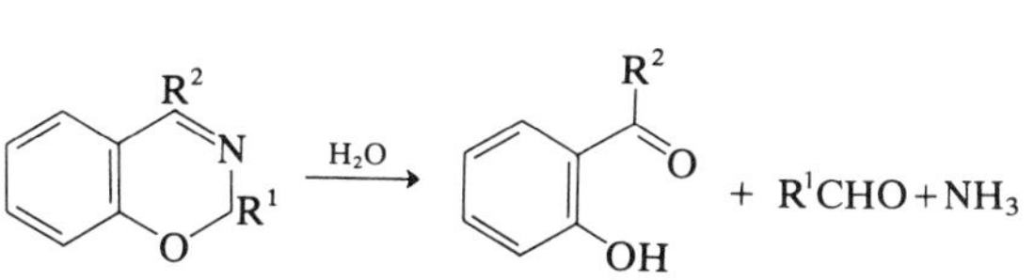

**Scheme 10**

Both 1,3-oxazin-4-ones and -2,4-diones are attacked by nucleophiles at C-2. Enolate anions, for example, yield open-chain intermediates initially, but these normally ring-close again to form pyridones ⟨80H(14)1333⟩. Certainly this is the case when sodium ethoxide in ethanol is employed to generate the enolate; thus diones (35) react with ethyl acetoacetate

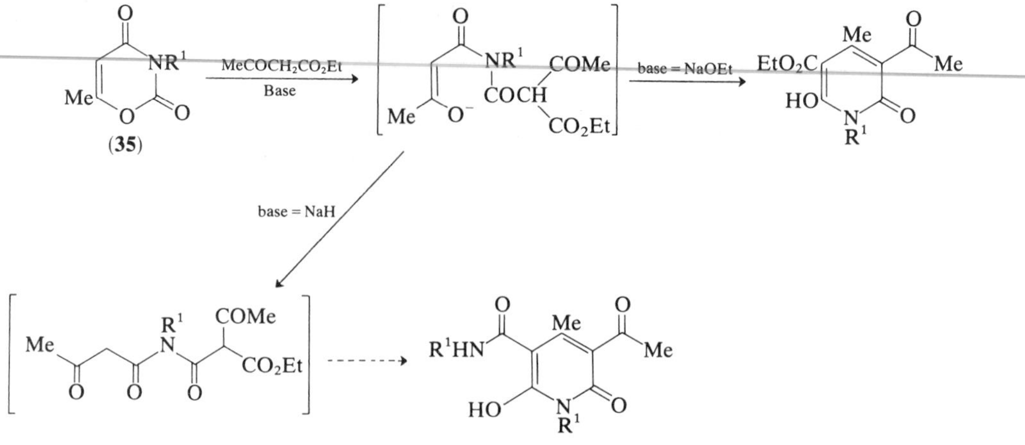

**Scheme 11**

under these conditions to give pyridones. However if sodium hydride in THF is used self-condensation occurs leading ultimately to pyrid-2-ones (Scheme 11) ⟨78JHC1475⟩.

Ammonia and amines behave similarly, giving rise to uracils (Scheme 12) ⟨76JCS(P1)1969⟩, but if the substrate bears an acyl substituent at C-5 these reagents are directed to attack at C-6 and in the case of the dione (36) the initial product (37) readily decarboxylates to the acyclic amine (38), thus precluding recyclization ⟨75KGS1468⟩.

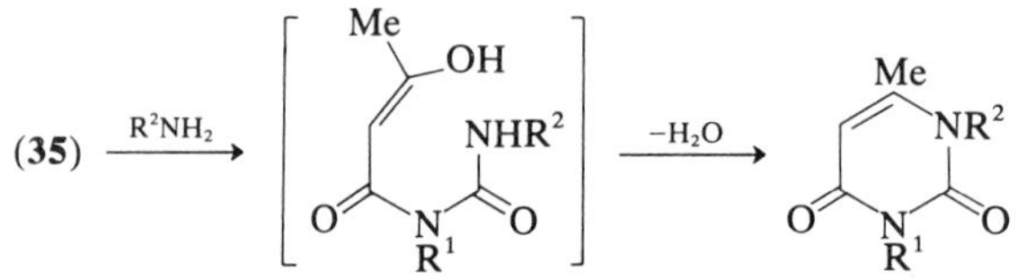

**Scheme 12**

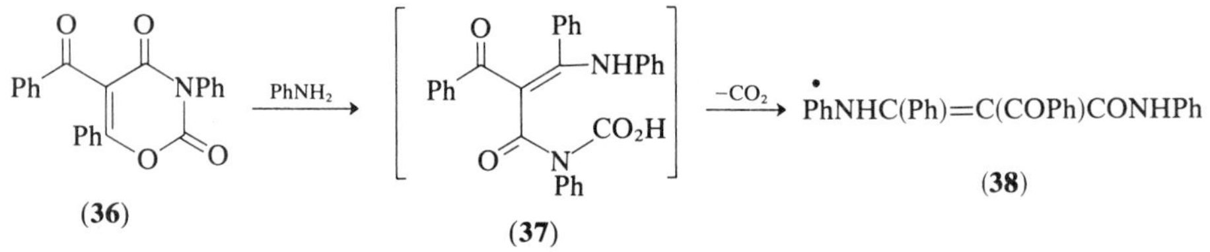

Hydrazines and hydroxylamines do not form pyrimidones with 1,3-oxazin-4-ones; instead the nucleophile attacks at C-2, opening the ring and effecting ring contraction through intramolecular attack by a lone pair of electrons of the introduced group upon the carbonyl function formerly at position 4 in the parent heterocycle (Scheme 13) ⟨78CPB1825⟩.

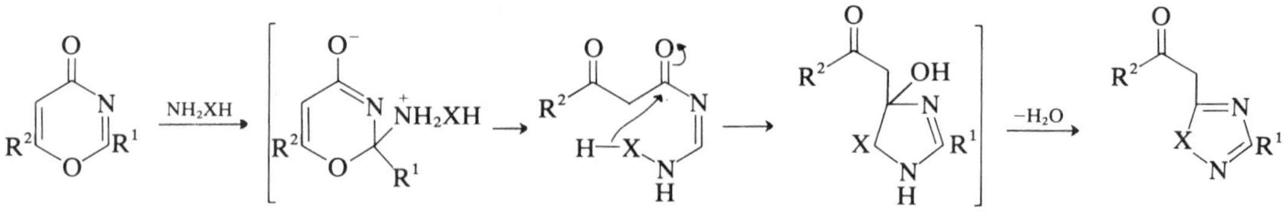

(X = NH, NPh or O)

**Scheme 13**

3,1-Benzoxazin-4-ones (39) react with amines to give quinazolinones (40) ⟨77JOC656⟩, but this 'typical' reaction is replaced in 1,3-benzoxazine-2-thiones (41) by the formation of imino derivatives (42) since now the oxygen ring atom is joined directly to the aryl nucleus ⟨76MI22700⟩.

3,1-Benzoxazine-2,4-diones (43) (isatoic anhydrides) have proved useful synthons for various new heterocycles. For example, a base-catalyzed condensation with the oxindole (44) affords tetracycles (45) ⟨80JHC1785⟩, whereas the parent compound (43; R = H) and 3-hydroxyisothiazole give the quinazolinone (46) ⟨69AJC2497⟩.

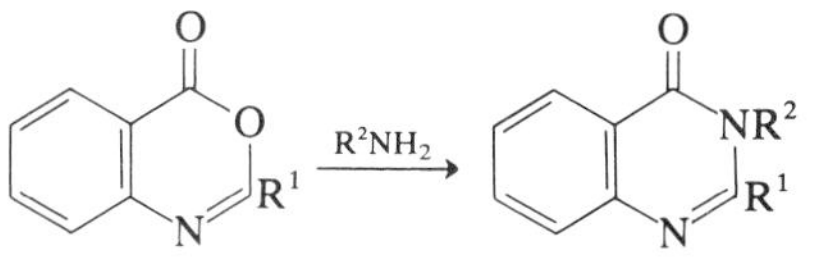

(39)    (40) 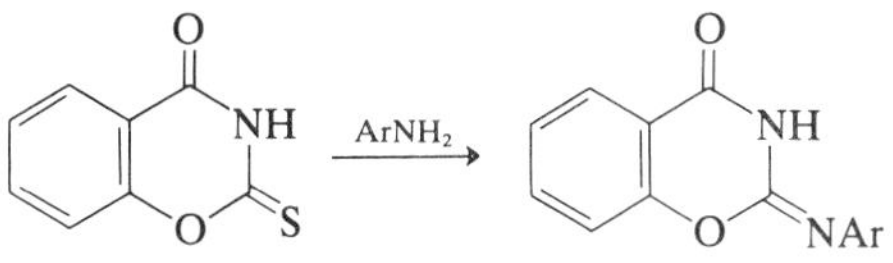    (41)    (42)

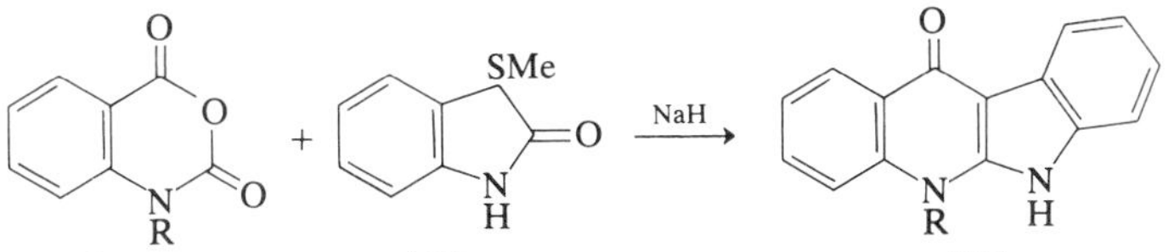

(43)    (44)    (45) 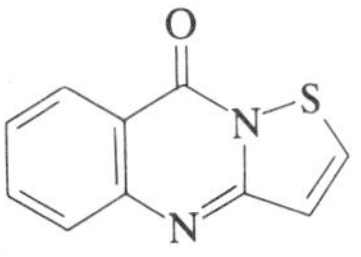    (46)

2-Aminopyridine can be prepared by the reaction of 4-chloro-1,3-benzoxazines (**47**) with pyridine *N*-oxides. Here it is proposed that an ion or radical pair is formed initially, which through displacement of hydrogen chloride and rearrangement leads to an *N*-substituted benzoxazine (**48**). Finally, acid hydrolysis gives 2-aminopyridine and salicylic acid (**49**; Scheme 14) ⟨80CPB465⟩.

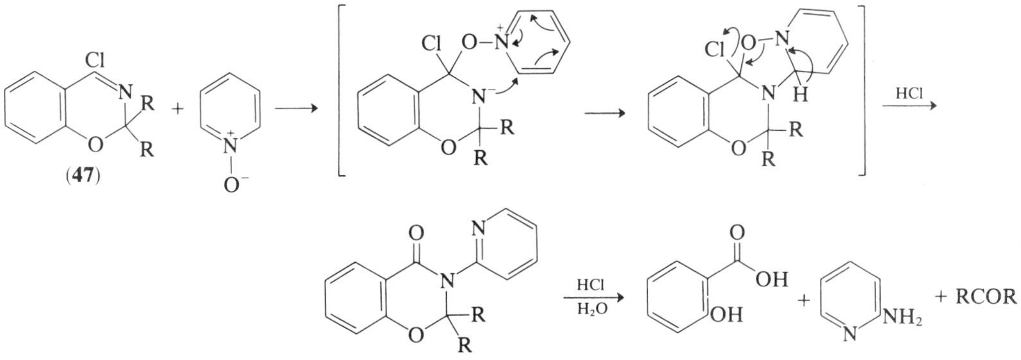

(47)

(48)    (49)

**Scheme 14**

*(iv) Addition reactions*

The imine bond of 4*H*- and 6*H*-1,3-oxazines enters into addition reactions with quinones and alcohols. Thus, for example, the triphenyl derivative (**50**) forms a 1:1 adduct with 1,4-benzoquinone ⟨69LA(723)111⟩ and the oxazin-6-ones (**51**), when heated in methanol, give 2-methoxy-2,3-dihydro derivatives (Scheme 15) ⟨72CJC584⟩.

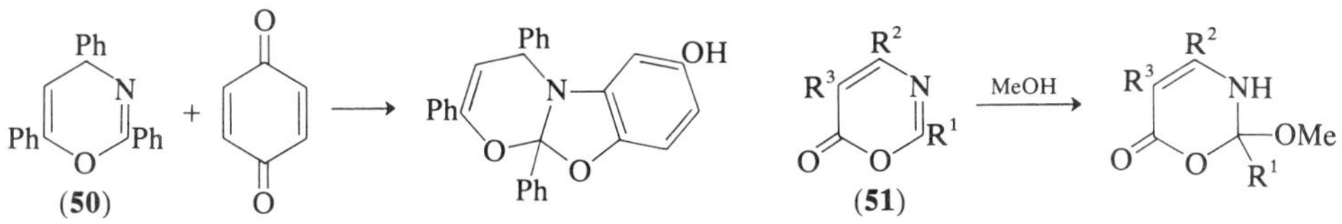

(50)                (51)

**Scheme 15**

Oxazinones of this type also undergo self-condensation at temperatures above 300 °C, but at lower temperatures they behave as dienes and combine with enamines and ynamines to yield adducts (**52**) and (**53**) respectively, which on thermolysis break down to pyridines (Scheme 16) ⟨74AG(E)484⟩.

Photochemical addition reactions are exhibited by oxazinones. Thus, irradiation of a mixture of the oxazin-4-one (**54**) and 1,1-dimethoxyethylene yields the [2 + 2] adduct (**55**) which fragments to the azetine (**56**) on heating ⟨77TL431⟩.

*(v) Photolysis and thermolysis*

1,3-Oxazin-6-ones on irradiation with UV light ring-open to yield iminoketenes (**57**) and also undergo internal cyclization to valence tautomers (**58**). The latter are unstable and eliminate carbon dioxide to yield azacyclobutadienes (azetes) ⟨80LA798⟩ as indicated by

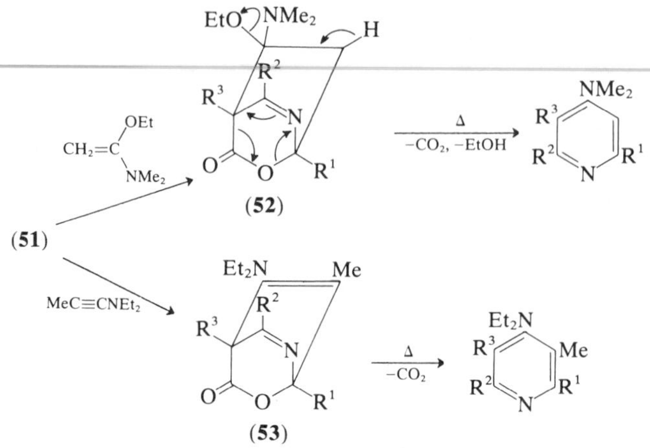

**Scheme 16**

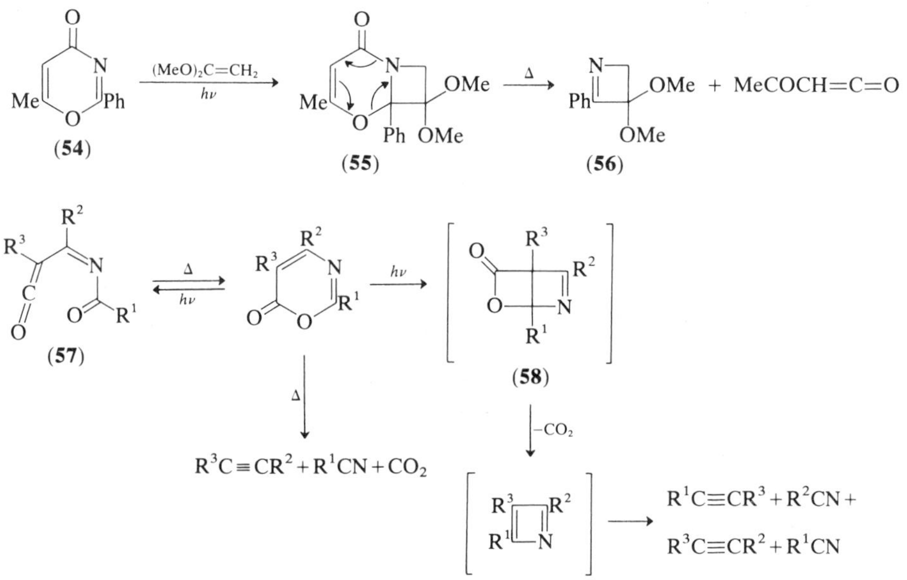

**Scheme 17**

their fragmentation patterns, but on flash vacuum pyrolysis the ring fragments directly to give carbon dioxide, an alkyne and a nitrile (Scheme 17) ⟨78CC902⟩.

### 2.27.2.2.2 Mesomeric betaine structures

1,3-Oxazine and -thiazine betaines (**59**) are prepared by reacting malonyl chlorides with secondary amides or thioamides respectively (Scheme 18). Carbon suboxide can be used instead of the acid chlorides, giving salts unsubstituted at C-5 ⟨72S312⟩.

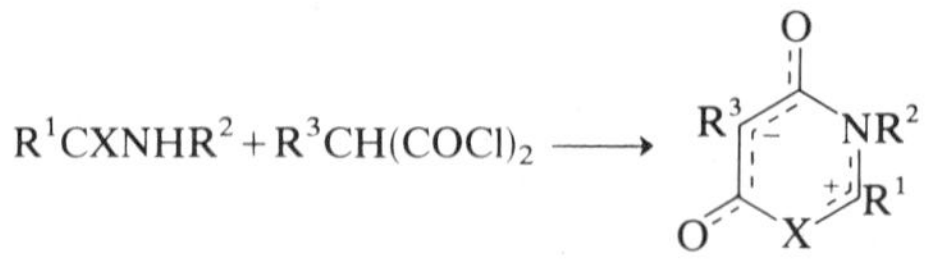

(59) X = O or S

**Scheme 18**

These compounds offer interesting possibilities for further elaboration as they enter into addition reactions with, for example, the 1,2-quinone (**60**), yielding tricyclic compounds (**61**) ⟨79TL237⟩, and their bicyclic analogues (**62**) combine with phenyl isocyanate to give adducts (**63**), which eliminate carbon dioxide to afford pyrimidine betaines (**64**). Similarly, dialkyl acetylenedicarboxylates produce quinolizinones (**65**) (Scheme 19) ⟨79CB1585⟩.

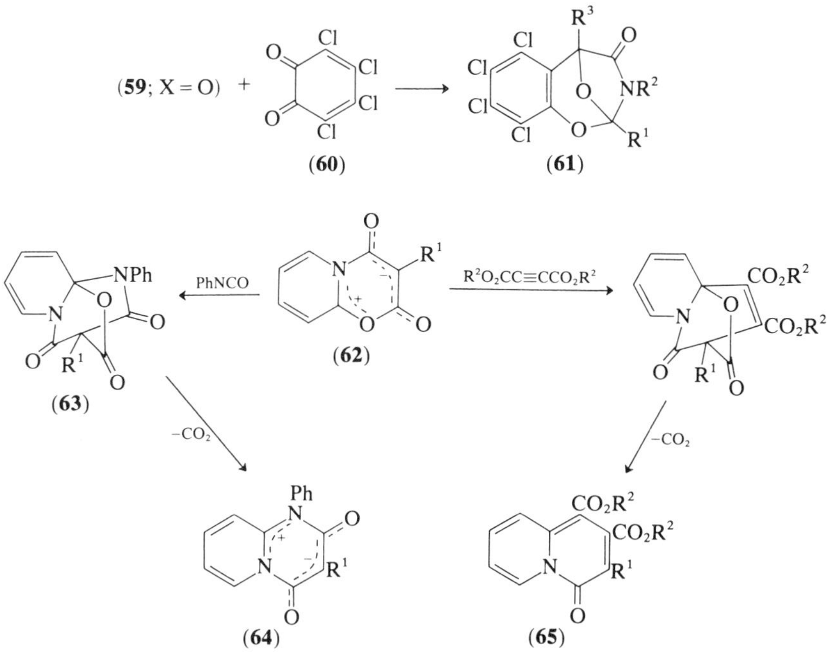

Scheme 19

### 2.27.2.2.3 1,3-Oxazinium and -thiazinium salts

1,3-Oxazinium and -thiazinium cations are $6\pi$-aromatic systems which readily react with nucleophiles at C-6. Ring opening is normally followed by recyclization so that a variety of heterocyclic systems are then formed. The behaviour of the oxygen and sulfur compounds are almost identical and so, as the latter are usually prepared from the former, it is not surprising that most attention has focussed on the reactions of 1,3-oxazinium species ⟨72S333⟩. These versatile synthons react with ammonia, for example, to give pyrimidines, while hydrazines afford pyrazoles and hydroxylamine produces isoxazoles (Scheme 20).

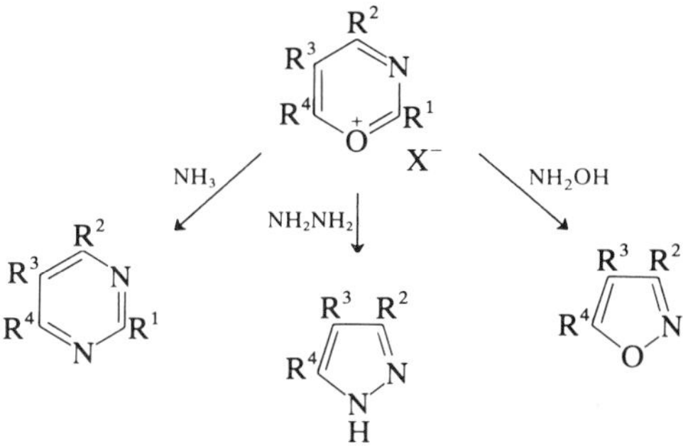

Scheme 20

Carbanions, such as that from malonodinitrile, generate dienes (66) which ring-close to N-acyl-1,2-dihydropyridines (67). Subsequently these products may rearrange to 2-amidopyridines (Scheme 21) ⟨75BCJ73⟩. However, if substituted malonodinitriles are employed in this reaction the intermediate 6H-1,3-oxazines can be isolated for now deprotonation and ring opening are inhibited.

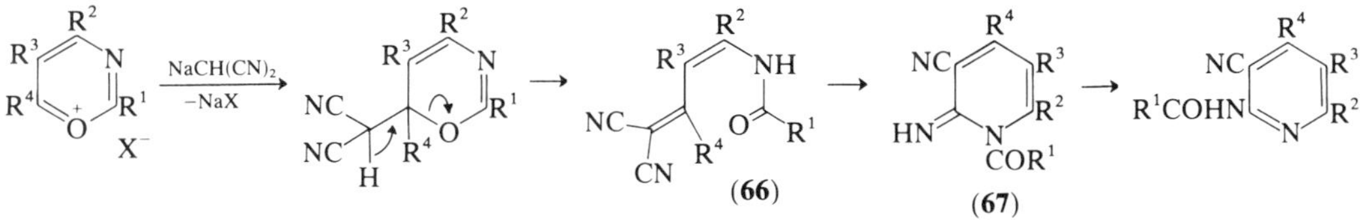

Scheme 21

Water, hydrogen sulfide, phenolate anions, Grignard reagents and enamines all effect ring opening; 1-*N*-morpholinocyclohexene, for instance, gives the anil (**68**) which eventually cyclizes to the tetrahydroquinoline (**69**; Scheme 22) ⟨72S333⟩.

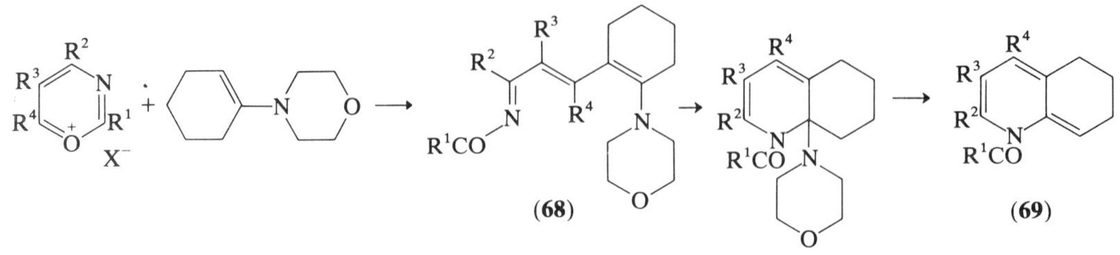

**Scheme 22**

4-Methyl-1,3-oxazinium salts may be deprotonated to methylene derivatives (**70**) by bases such as triethylamine (Scheme 23).

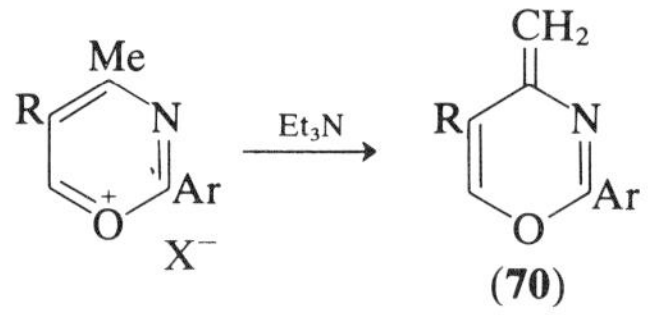

**Scheme 23**

### 2.27.2.2.4 Dihydro-1,3-oxazines and -thiazines

#### (i) Hydrolysis and reactions with nucleophiles

5,6-Dihydro-2*H*-1,3-oxazines and -benzoxazines are stable to cold aqueous alkali, but hot dilute acids effect ring scission. In the monocyclic series the products are the salts of 3-aminopropyl esters (**71**) which rearrange to 3-hydroxypropylamides on basification. Should there be an aryl group at position 6 in the starting material, rearrangement is often followed by dehydration to styrene derivatives (**72**; Scheme 24) ⟨72AG(E)287⟩.

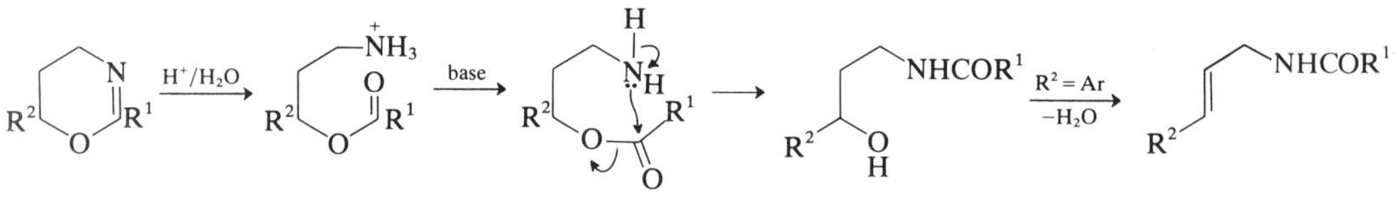

**Scheme 24**

On acid treatment 3,4-dihydro-2*H*-benzoxazines cleave to 2-hydroxybenzylamines and aldehydes or ketones, depending upon the degree of substitution at C-2 (Scheme 25), whereas dihydro-4*H*-3,1-benzoxazines afford acetanilides (Scheme 26) ⟨75AP622⟩.

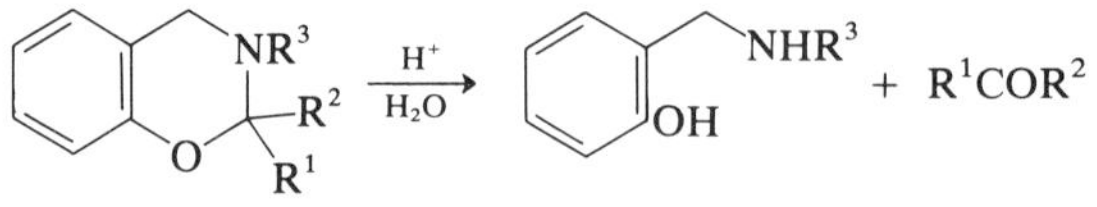

**Scheme 25**

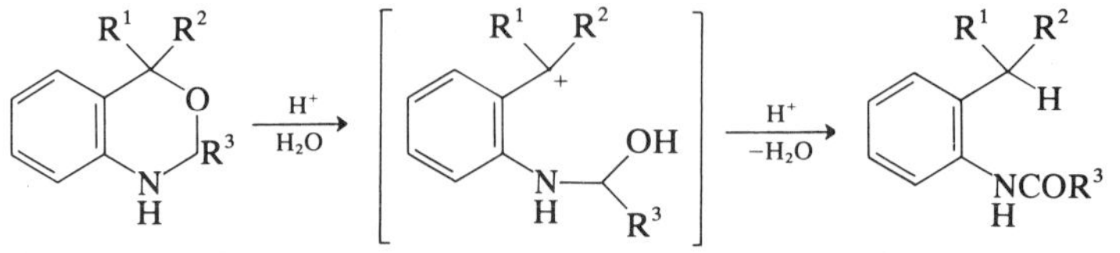

**Scheme 26**

Structures bearing carbonyl groups are generally more sensitive to base-catalyzed hydrolysis; this is particularly true of the sulfones (**73**) which are readily ring opened to arene acids (**74**), aldehydes and ammonia. 2,4-Dioxo and thionyl derivatives behave similarly. Where both carbonyl and thiocarbonyl groups are present primary amines form imino derivatives by selective attack at the thiocarbonyl site; in the dithioxo compound (**75**; X = S) reaction takes place at position 2 (Scheme 27) ⟨79KGS291⟩.

Compounds (**76**) are cleaved by reaction with ammonia, but recyclization then occurs affording pyrimidones (Scheme 28) ⟨80H(14)1333⟩.

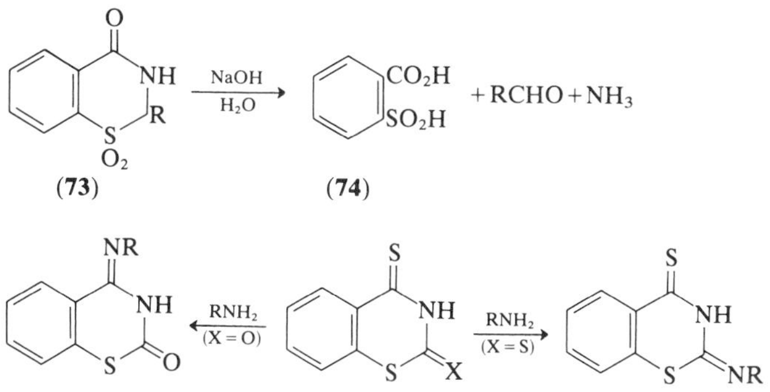

**Scheme 27**

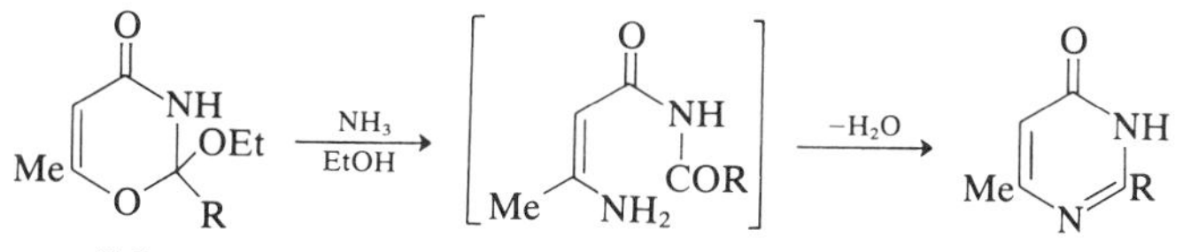

**Scheme 28**

Sodium borohydride is an effective reductant for the imine bond of dihydro-1,3-oxazines affording tetrahydro derivatives, but predictably where lactam or sultam functions form part of the heterocycle, treatment with lithium aluminum hydride is necessary. Dihydro-1,3-benzothiazinones for instance are reduced by this reagent to the corresponding dihydrobenzothiazines (Scheme 29) ⟨77ACH(92)317⟩.

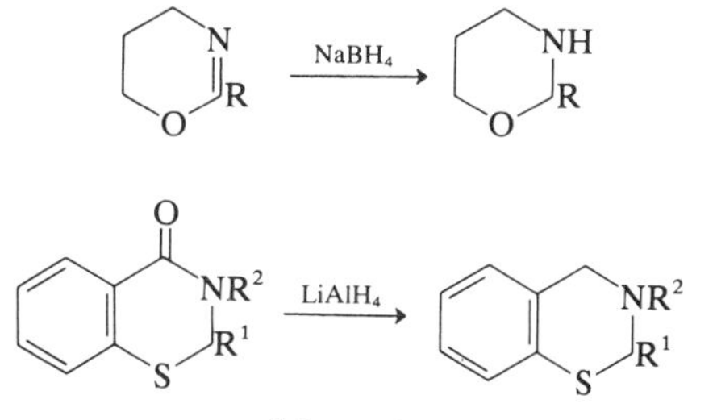

**Scheme 29**

5,6-Dihydro-4*H*-1,3-oxazines bearing an alkyl group at position 2 with a free α-hydrogen atom are deprotonated by *n*-butyllithium to yield the corresponding anions. These may be reacted with various electrophiles (alkyl halides, aziridines, carbonyl compounds, oxiranes, *etc.*) to provide very valuable routes to 2-substituted dihydro-1,3-oxazines which may be subsequently reduced and hydrolyzed to aldehydes and other carbonyl derivatives. This procedure has been extensively used in general organic synthesis, the usual starting materials being 4,4,6-trimethyloxazines (Scheme 30) ⟨B-74MI22701⟩.

Grignard reagents fail to react with dihydro-1,3-oxazines unsubstituted at C-2, but *n*-butyllithium in diethyl ether yields 2-butyltetrahydro-1,3-oxazines. However, the reaction is solvent-dependent and in THF the ring is opened to give isocyanides ⟨69TL5151⟩.

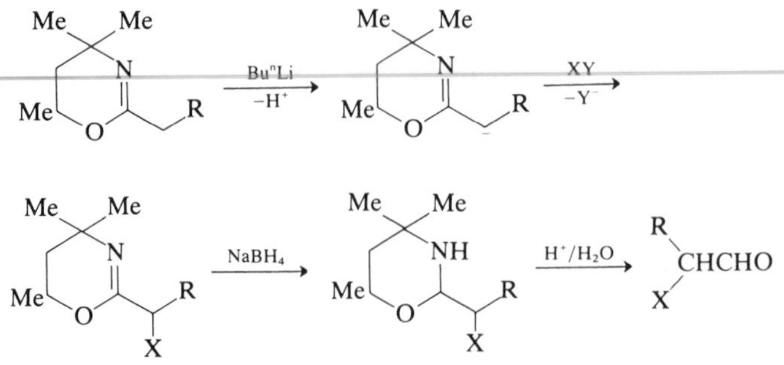

**Scheme 30**

### (ii) Addition

The imine bond of 5,6-dihydro-4H-1,3-oxazines reacts with oxiranes to form bicyclic adducts (Scheme 31) ⟨66AG(E)875⟩ and with ketenes to yield cepham analogues (Scheme 32) ⟨75AP481⟩.

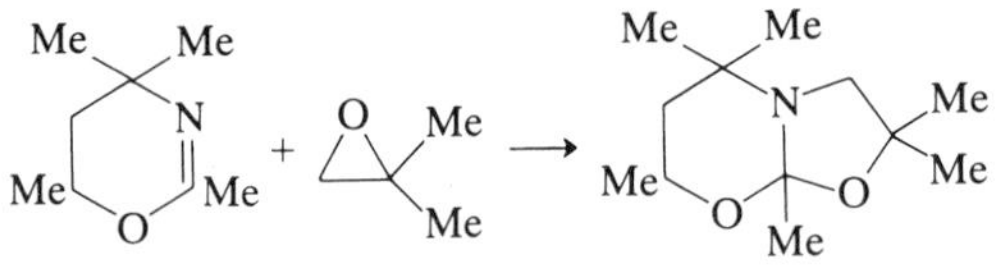

**Scheme 31**

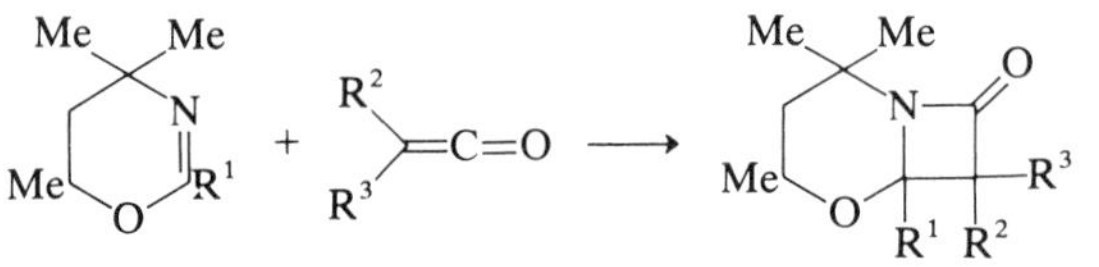

**Scheme 32**

### 2.27.2.2.5 Tetrahydro-1,3-oxazines and -thiazines

#### (i) Stereochemistry and tautomerism

Tetrahydro-1,3-oxazines are normally assumed to adopt a chair conformation in which the NH bond has an axial orientation. This view is substantiated by NMR spectroscopy and also by dipole moment analysis ⟨73JCS(P2)325⟩. The ring system is not stable and, for example, when the parent molecule is allowed to stand, it slowly ring opens and then forms a trimer (Scheme 33) ⟨78AF937⟩. In acidic media ring fission is accelerated and the open-chair imines may then hydrolyze. This property has been utilized in the synthesis of aldehydes as previously noted (see Section 2.27.2.2.4(i)).

Not too much is known about tetrahydro-1,3-thiazines. However, 2-imino derivatives are well established compounds which exist in tautomeric equilibrium with 2-aminodihydro-1,3-oxazines (Scheme 34) ⟨1890CB87, 70TL2167⟩.

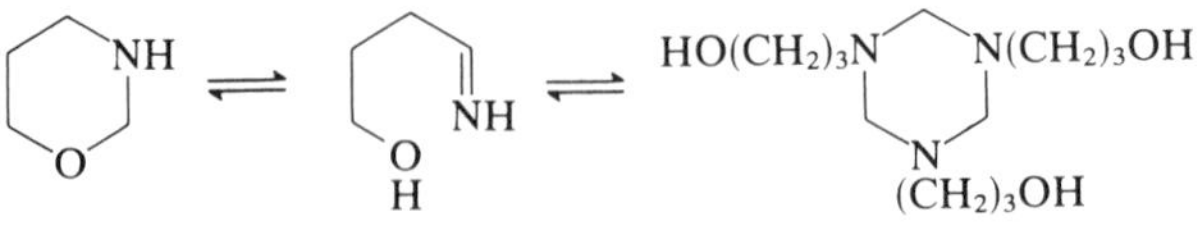

**Scheme 33**

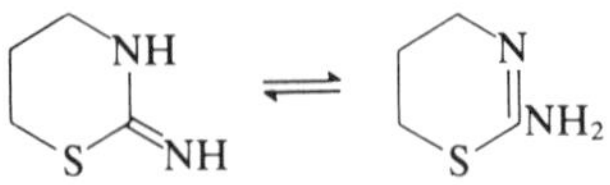

**Scheme 34**

## (ii) Reactions with electrophiles

The tetrahydro compounds are strongly basic and may be *N*-alkylated directly with alkyl halides ⟨64T1173⟩. A number of *N*-nitroso derivatives have been made by the oxidation of tetrahydro-1,3-oxazines with peracids ⟨74T3315⟩. Sequential treatment of the oxazine (**77**) firstly with diketene and then with 4-methylbenzenesulphonyl azide and triethylamine gives the diazoketone (**78**), which on irradiation with UV light yields the cepham (**79**) exclusively ⟨79CC846⟩.

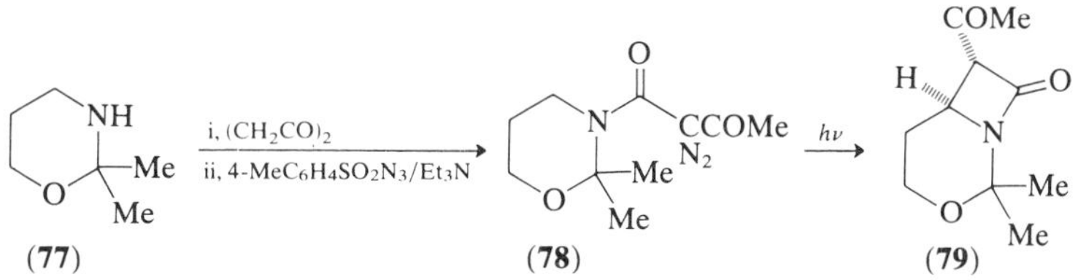

### 2.27.2.3 1,4-Oxazines, 1,4-Thiazines and their Benzo Derivatives

#### 2.27.2.3.1 Parent ring systems

##### (i) Structure and tautomerism

Monocyclic 1,4-oxazines are few in number and their chemistry is relatively unexplored; the corresponding thiazines are more familiar. IR and NMR studies show that although some 4*H*-1,4-thiazines are known the 2*H*-system is generally more favoured and most representatives belong to this class. A similar situation is found amongst benzo-1,4-oxazines and -thiazines, although both forms sometimes exist together in tautomeric equilibrium; 3-phenyl-1,4-benzothiazine, for example, is present in aqueous hydrochloric acid as a 4 : 1 mixture of 2*H*- and 4*H*-isomers (Scheme 35), whereas in the same medium the parent structure is restricted to the 2*H*-tautomer and various aldolization products including the trimer (**80**) ⟨76T1407⟩.

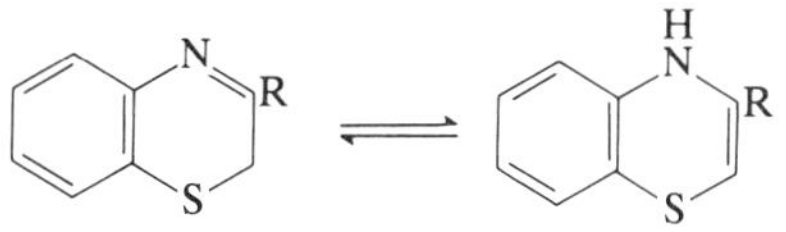

**Scheme 35**

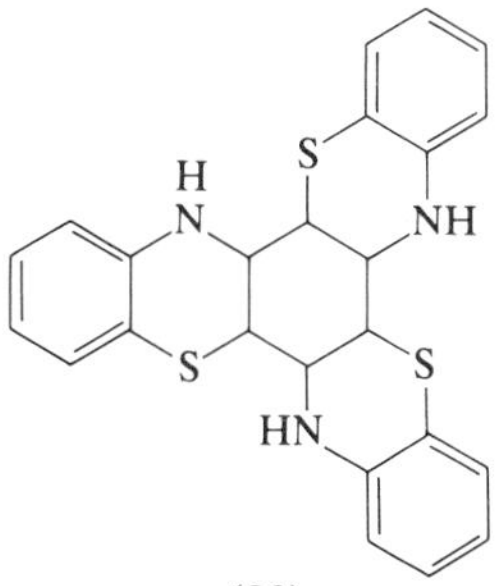

(**80**)

Interestingly, 2,3-dicarbonyl derivatives formed by the condensation of α-dicarbonyl compounds and 2-aminophenols ⟨80JHC1625⟩ exist in the 2*H*-form (**81**; Scheme 36). However, when there is an additional carbon atom in the side-chain of the 3-substituent so that hydrogen bonding with the NH group may now occur, as in the benzoxazinone (**82**), the equilibrium is directed strongly in favour of the 4*H*-tautomer (Scheme 37) ⟨76JHC681⟩.

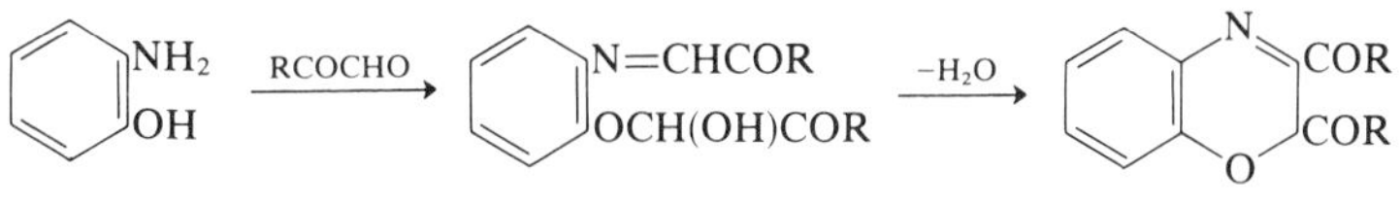

**Scheme 36**

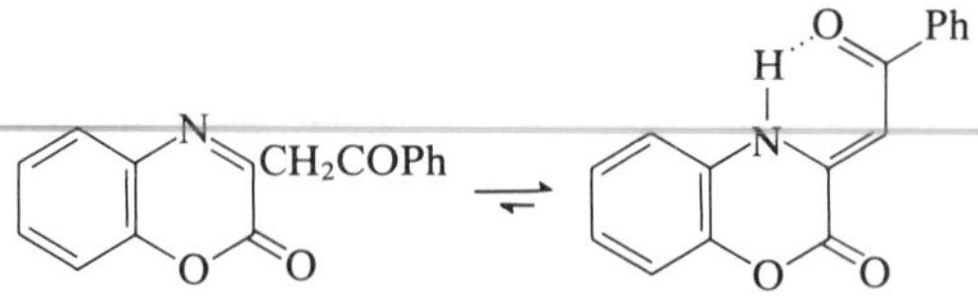

**Scheme 37**

*N*-Substitution and further ring fusion preclude tautomerism and obviously the phenoxazines and phenothiazines are constrained to single formulations. However, phenoxazine and phenothiazine are not planar molecules, rather they are folded about the axis through the two heteroatoms. NMR studies suggest that the heterocycle is flexible in solution ⟨72OMR(4)895⟩, but an X-ray analysis of crystalline phenothiazine shows that the nitrogen and sulfur atoms deviate from the planes of the two aromatic rings by 0.01–0.03 Å for nitrogen and 0.17–0.18 Å for sulfur. The dihedral angle between the planes is 153.3° and the bonds about the nitrogen atom assume a flattened tetrahedral array with the NH bond adopting a quasi-equatorial orientation with respect to the heteroring ⟨68CC1656⟩.

### (ii) Oxidation

2*H*-1,4-Thiazines and benzothiazines undergo oxidation to dehydro dimers (**83**) and (**84**) which are of considerable interest as they are the parent chromophores of the trichosiderin (trichochrome) pigments which occur in mammalian red hair and in the feathers of some birds ⟨74T2781⟩. In the case of monocyclic thiazines, reagents such as nitrobenzene or picric acid are required, but air oxidation is sufficient in the bicyclic series. It is curious that whereas aerial oxidation of the ester (**85**) gives both the dehydro dimers (**86**) and (**87**), ethyl azodicarboxylate only yields the tautomer (**87**) ⟨70AC(R)351⟩.

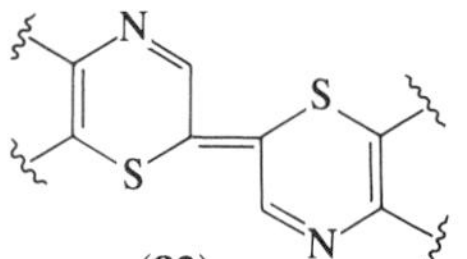

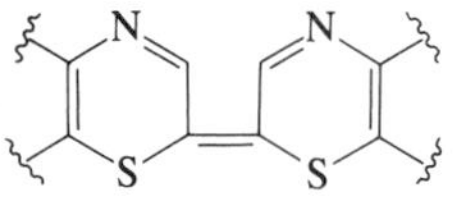

The initial products of the oxidation of phenoxazines and phenothiazines are cation radicals. Reagents such as ferric chloride are effective, and a solution of phenoxazine in concentrated sulfuric acid allowed to stand for several days shows an ESR spectrum compatible with the presence of such a radical cation ⟨62JPC937⟩. On neutralization, or basification, the radical cations are deprotonated to neutral radicals (**88**) which may couple to form high molecular weight species linked through positions 3 and 10′ ⟨59CB2873⟩ or through positions 10 and 10′ ⟨73JCS(P2)264⟩. In aqueous media the initially formed radical cations may react with water and the products undergo further oxidation to phenoxazin- or phenothiazin-3-ones (**89**), but if more reactive nucleophiles are present other 3-substituted derivatives are produced. For example, the radical cations of *N*-substituted phenoxazines can be quenched with bromide, isothiocyanate or nitrite ions to give the corresponding 3,10-disubstituted compounds ⟨79JOC3310⟩ and the dyestuff methylene blue (see Section 2.27.4) is obtained by the oxidation of phenothiazine in the presence of dimethylamine ⟨16CB53⟩. In strongly acidic media further oxidation of the radical cations is possible, leading to cations (**90**) and dications (**91**; Scheme 38) ⟨75JHC397⟩.

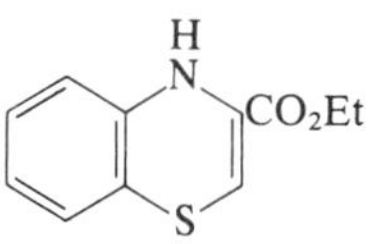

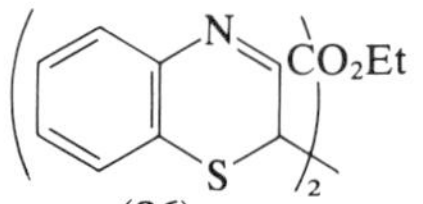

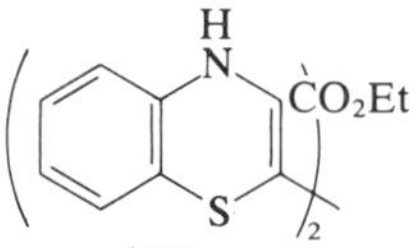

### (iii) Reduction

Reduction of the parent systems is of little importance as reduced forms of 1,4-oxazines and -thiazines are readily available by other means; moreover, hydrogenation of thiazines may effect cleavage and ring contraction to thiazolines rather than straightforward addition to a double bond as was once supposed (Scheme 39) ⟨68G17⟩.

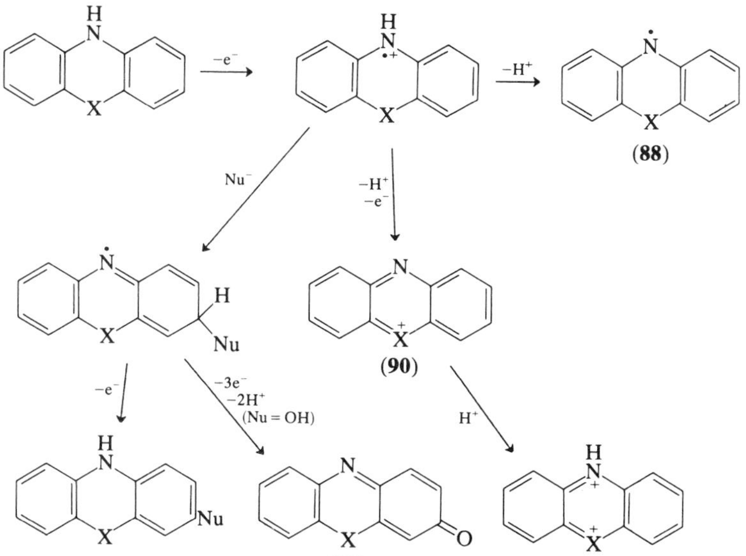

**Scheme 38**

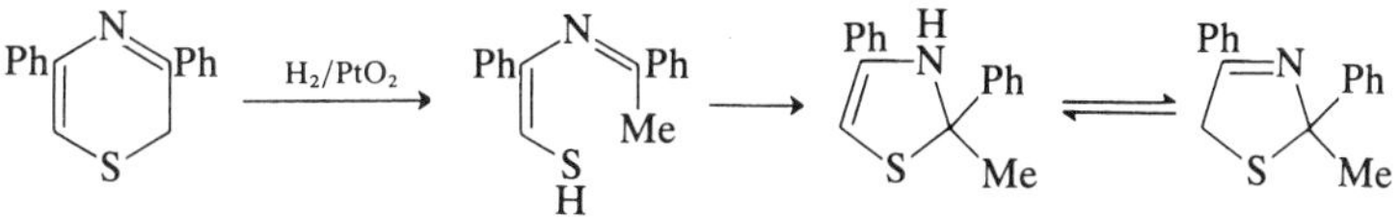

**Scheme 39**

*(iv) Electrophilic attack at nitrogen*

Monocyclic 1,4-oxazines and -thiazines are bases, forming salts with acids; however, the effect of ring fusion is to reduce the availability of the lone pair electrons on nitrogen so that phenoxazines and phenothiazines are much less reactive. It is claimed that alkyl halides in benzene solution are sufficiently electrophilic to *N*-alkylate phenothiazines directly ⟨58GEP922567⟩, but it is more usual to convert the tricyclic systems into the corresponding anions (**92**) with sodium amide before reaction with the alkylating agent (Scheme 40) ⟨52JA4205⟩.

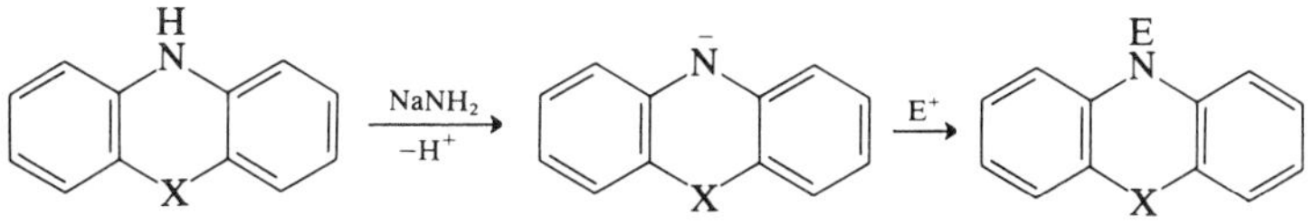

**Scheme 40**

*N*-Arylation is achieved under Ullmann-type conditions (copper or copper bronze and potassium carbonate), but the reaction probably occurs through a mechanism involving radicals ⟨44JA1214⟩. Acyl chlorides and anhydrides combine directly with phenoxazines and phenothiazines, especially if a Lewis acid is present; the attack is not specific and 2,10-diacyl and 1,8,10-triacyl products may result depending upon the amount of reagent employed. Phosgene reacts with phenoxazine to give 10-chlorocarbonylphenoxazine (**93**) which is a useful starting material for many other *N*-substituted phenoxazines ⟨61JOC4130⟩. Interestingly, thiophosgene fails to yield the corresponding thiocarbonyl derivative, giving instead the coupled product (**94**) ⟨64ZOB3791⟩.

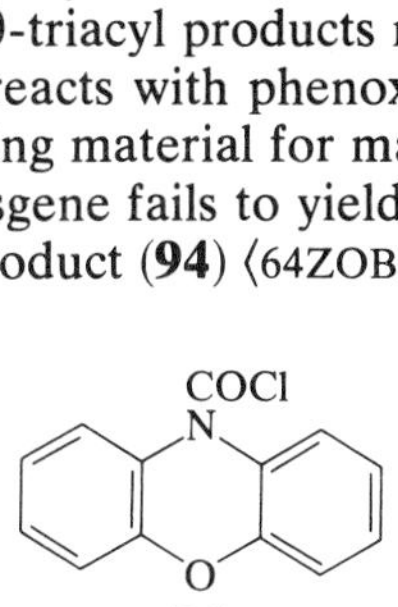

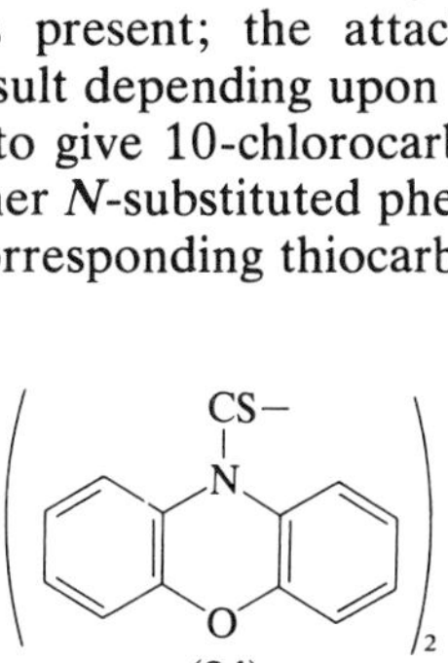

*(v) Electrophilic attack at sulfur*

Sulfoxides and sulfones of 1,4-thiazines and 1,4-benzothiazines are not prepared by direct oxidation because of the ease by which the substrates enter into coupling reactions, but with *N*-alkylphenothiazines, where coupling is precluded, *S*-monoxides are available by oxidation with hydrogen peroxide in acetic acid ⟨80JHC597⟩, and *S,S*-dioxides are obtained when more vigorous reagents are employed. Hydrogen peroxide can also be used to form *S*-oxides from *N*-unsubstituted phenothiazines but the reaction may be complicated by oxidation at carbon ⟨79AHC(24)293⟩. *S*-Imidation of phenothiazine can be effected with *O*-mesitylenesulphonylhydroxylamine ⟨78CB1453⟩ and diphenyliodotetrafluoroborate affords the *S*-phenylphenothiazinium cation (**95**) which deprotonates to *S*-phenylphenothiazine (**96**; Scheme 41) ⟨73IZV1678⟩.

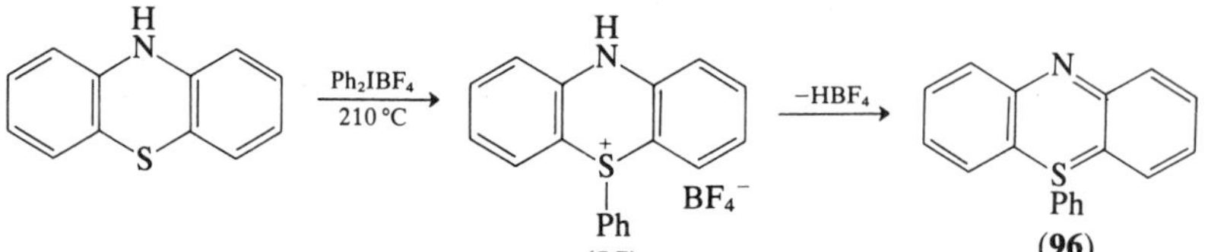

**Scheme 41**

*(vi) Electrophilic attack at carbon*

Even the comparatively unreactive phenoxazine and phenothiazine systems undergo halogenation and nitration with ease and it is normal to prepare monosubstituted derivatives by stepwise procedures rather than by direct electrophilic attack. Indeed, the nitration of phenoxazine is uncontrollable and even *N*-acylphenoxazines afford a mixture of di- and tetra-nitro products ⟨03CB475⟩. Similarly phenothiazine and nitric acid produce a complex mixture of nitrated sulfoxides and sulfones. Chlorine in DMSO at 40 °C reacts with phenothiazine to yield 3,7-dichlorophenothiazine, whereas cupric chloride gives the 1,7-isomer ⟨76JPR353⟩. Direct bromination of phenoxazine produces a mixture of 3-bromo- and 3,7-dibromo-phenoxazines, while thionyl chloride affords the 1,3,7,9-tetrachloro derivative ⟨60ZOB1893⟩.

Sulfuric acid by itself causes oxidation of phenothiazine, although chlorosulfonic acid gives both 3- and 7-sulfonyl compounds ⟨59BCJ483⟩, and 3-sulfonate esters (**98**) can be obtained through the base catalyzed rearrangement of 10-sulfonylphenothiazines (**97**) ⟨76KGS1365⟩. The formylation of *N*-alkylphenoxazines under Vilsmeier conditions is a more manageable process, yielding specifically 3-formylated products ⟨60JOC747⟩.

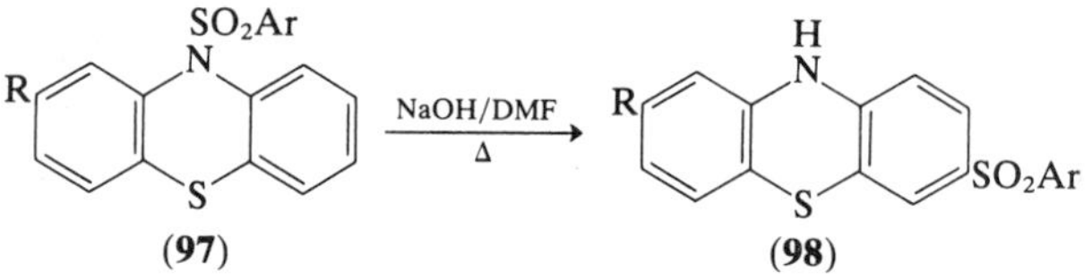

### 2.27.2.3.2 Dihydro-1,4-oxazines, -thiazines and their benzo derivatives

*(i) Structure*

Of the two possible tautomeric structures for monocyclic dihydro-1,4-oxazines and -thiazines the 3,4-dihydro-2*H*-representation is preferred in which the ring has a half-chair conformation.

*(ii) Oxidation*

Simple alkyl-substituted dihydro-1,4-thiazines are readily converted into the corresponding *S*-monoxides; atmospheric oxidation is sometimes sufficient, although side reactions may also occur. The cyclohexene derivative (**99**), for instance, affords not only the oxide (**100**) but also the spiro compound (**101**) probably generated by ring contraction of a hydroperoxide intermediate (Scheme 42) ⟨73IJS(B)(8)341⟩.

For less reactive thiazines sodium periodate or 3-chloroperbenzoic acid are suitable reagents for sulfoxide synthesis but it is not always possible to prevent over-oxidation and the formation of the corresponding sulfone ⟨75JCS(P1)716⟩. Carbonyl substituents at C-6

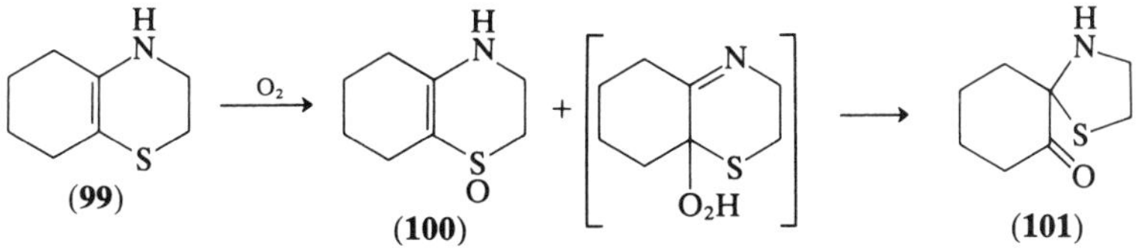

(99) (100) (101)

**Scheme 42**

deactivate the system so that oxidation at the sulfur atom is less likely and thiazinium ion formation is now favoured. Such species then combine with nucleophiles present in the oxidation medium. Thus, lead tetraacetate in benzene reacts with the ester (**102**) to yield the 3-acetate (**103**; Scheme 43) ⟨79AHC(24)293⟩.

Far less is known about the oxidative behaviour of dihydro-1,4-oxazines, -benzoxazines and -benzothiazines, although it has been observed that dihydro-1,4-benzothiazine itself is oxidatively ring opened by nitrous acid to the diazonium salt (**104**; Scheme 44) ⟨49JCS278⟩.

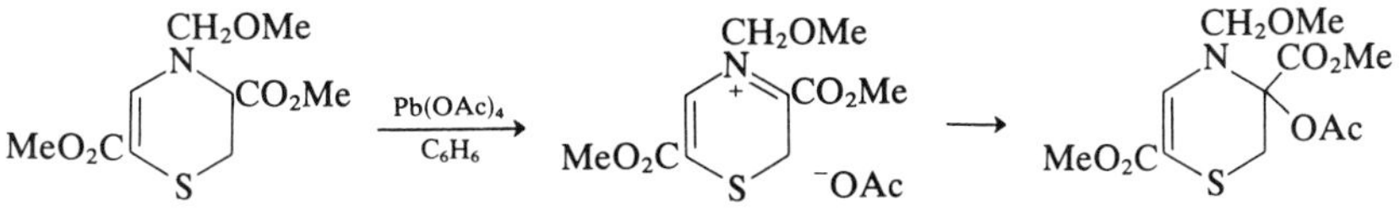

(102) (103)

**Scheme 43**

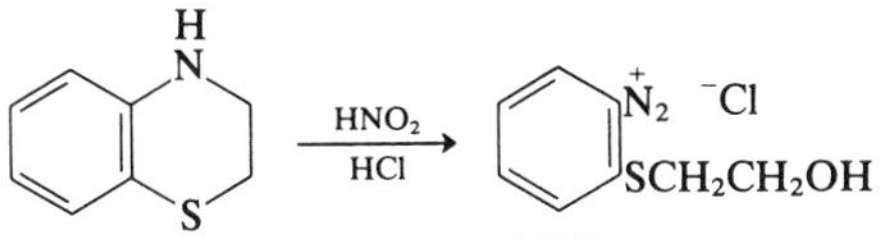

(104)

**Scheme 44**

*(iii) Reduction*

The reduction of dihydro-1,4-thiazines to tetrahydro forms is achieved by the action of a variety of reagents including formic acid, hydrogen sulfide, sodium borohydride and lithium aluminum hydride. The last is particularly useful for the reduction of carbonyl derivatives, although in the case of the amide (**105**) only partial reduction is effected, leading in due course to the thiazine (**106**), the hydroxyethyl side chain of which then cyclizes to position 5 of the ring to yield the bicyclic product (**107**; Scheme 45) ⟨66CPB742⟩.

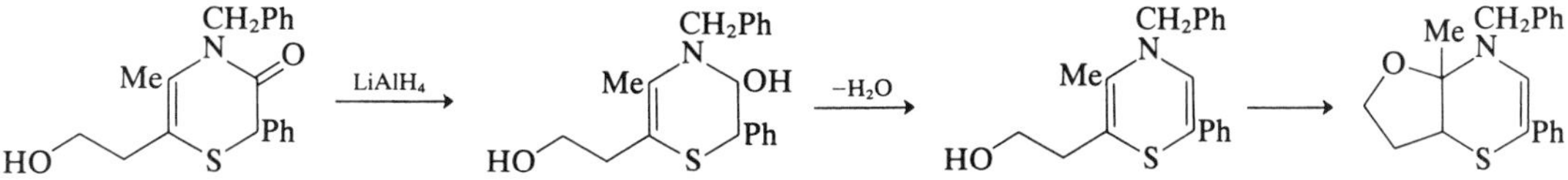

(105) (106) (107)

**Scheme 45**

Reductive cleavage of dihydro-1,4-thiazinones occurs with sodium in liquid ammonia: the thiazinone (**108**), for example, gives the salt (**109**) ⟨72TL5199⟩. Ring fission is also noted during the reduction of dihydro-1,4-oxazinones (**110**) with sodium borohydride. However, catalytic hydrogenation preserves the heterocycle and leads in this case to tetrahydro-1,4-oxazin-2-ones (Scheme 46) ⟨78JOC1355⟩.

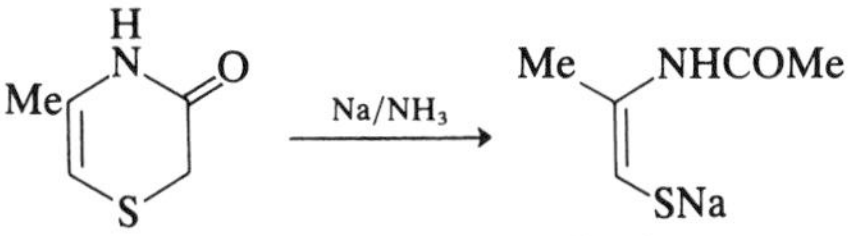

(108) (109)

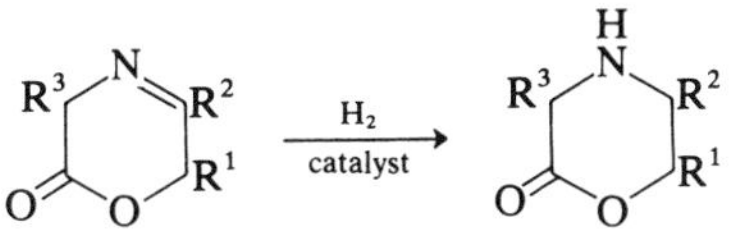

(110) **Scheme 46**

### 2.27.2.3.3 *Tetrahydro-1,4-oxazines and -thiazines*

#### (i) *Electrophilic attack at nitrogen*

Both tetrahydro-1,4-oxazine (morpholine) and its thia analogue (thiomorpholine) find application as solvents and bases, morpholine having particular value in synthesis as it is often used to prepare enamines from ketones with a free $\alpha$-hydrogen atom. *N*-Acylation and -alkylation are readily accomplished and a great diversity of *N*-substituted morpholines and thiomorpholines are encountered in the patent literature, some having biological activity.

Recent interest in *N*-substituted morpholines has centred largely upon their function as starting materials for the synthesis of new heterocyclic systems. For example, the *N*-acylmorpholine (**111**) affords the spiro compound (**112**) when treated with thionyl chloride ⟨80JOC536⟩. Diiminium salts (**113**), formed by the reaction of glyoxal and hydrogen bromide on the free base, combine with guanidines or *O*-methylisoureas to yield imidazolines (**114**), the 1-acyl derivatives of which rearrange when treated with acids to give pyrimidines (**115**; Scheme 47) ⟨80JHC97⟩.

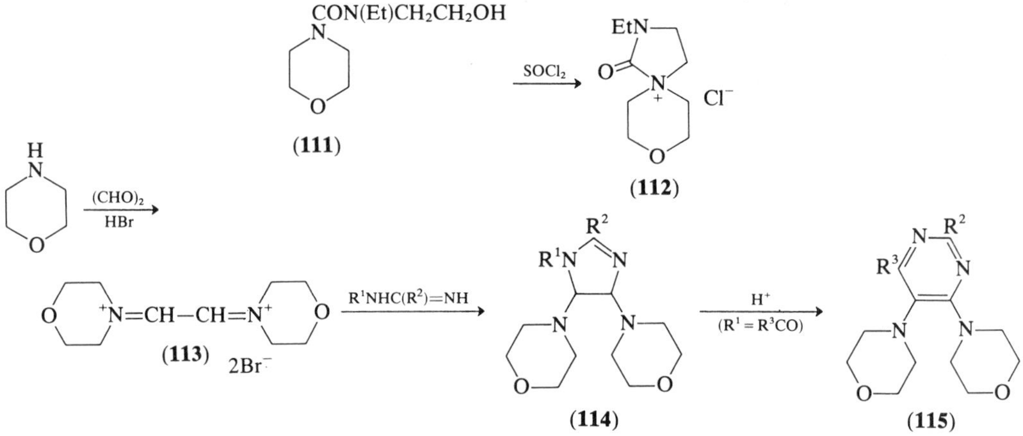

**Scheme 47**

#### (ii) *Oxidation*

Hydrogen peroxide converts morpholine into its 4-hydroxy derivative ⟨50JA2280⟩, but *N*-alkyl morpholines yield 3-oxo compounds if ruthenium(VIII) oxide or sodium metaperiodate are used as oxidants ⟨76S598⟩. Chlorine at $-70\,^{\circ}\mathrm{C}$ oxidizes the thiomorpholine (**116**) to the dihydro-1,4-thiazine (**117**) ⟨73JCS(P1)1321⟩, but more vigorous conditions result in the formation of *S*-oxides.

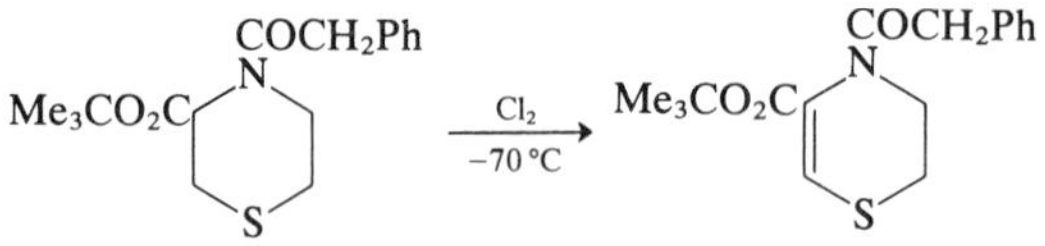

### 2.27.3 SYNTHESES

### 2.27.3.1 1,2-Oxazines, 1,2-Thiazines and their Benzo Derivatives

#### 2.27.3.1.1 *Parent ring systems*

#### (i) *1,2-Oxazines*

Acetophenone oxime reacts with ethylmagnesium bromide to give the dihydro-1,2-oxazine (**118**) and this with nitrous acid affords 3,5-diphenyl-6*H*-1,2-oxazine (**119**). The proposed mechanism *en route* to the initial product is complex, but there seems no reason why this synthesis could not be extended to other analogues (Scheme 48) ⟨67T3505⟩.

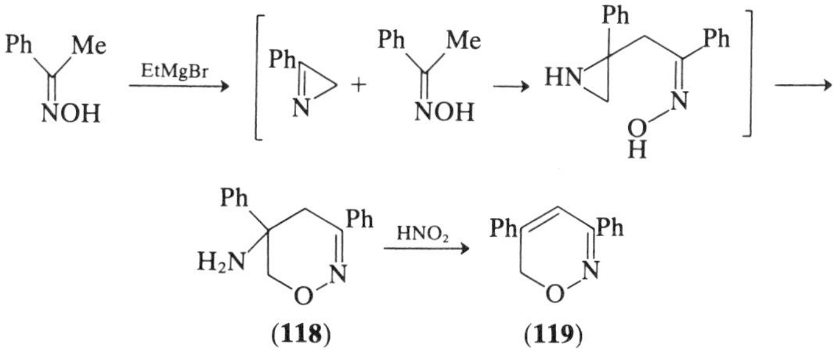

**Scheme 48**

A more general approach, however, is the cyclization of the monooximes of α,β-unsaturated 1,4-dicarbonyl compounds. Thus, the monocyclic oxazine (**121**) is formed when the butenedione (**120**) is heated with hydroxylamine ⟨50JOC869⟩, and 2,3-benzoxazin-1-ones (**123**) are prepared similarly from the oximes of 2-acylbenzoic acids (**122**) ⟨37LA(531)279⟩. 3-Aryl-1,2-oxazin-6-ones are available through the action of hydroxylamine in acid solution on the trichloroketones (**124**; Scheme 49) ⟨80ZC19⟩.

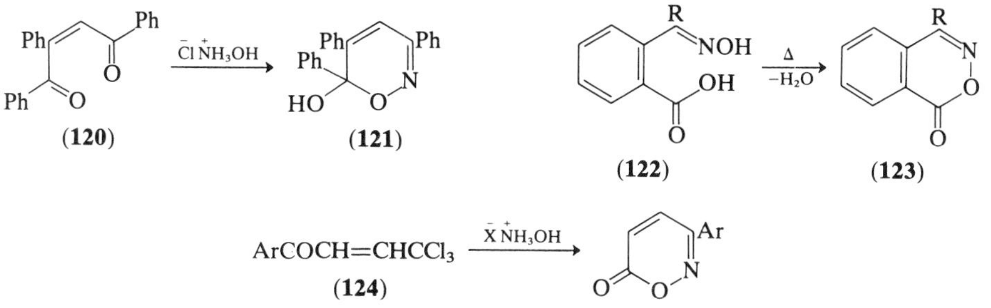

**Scheme 49**

### (ii) 1,2-Thiazines

One of the simplest representatives of the thiazine class is the perfluoro derivative (**126**), which is prepared by the cycloaddition of thiazyl fluoride (**125**) and perfluoro-1,3-butadiene (Scheme 50) ⟨79CC35⟩.

Similarly, the sulfone (**127**) is obtained by the addition of benzyne and thionylaniline, followed by oxidation ⟨69JCS(C)748⟩, or else by thermolysis of the azide (**128**; Scheme 51) ⟨69JA1219⟩.

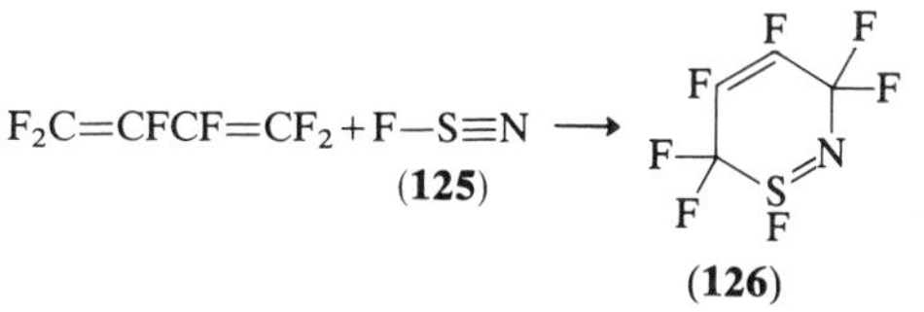

**Scheme 50**

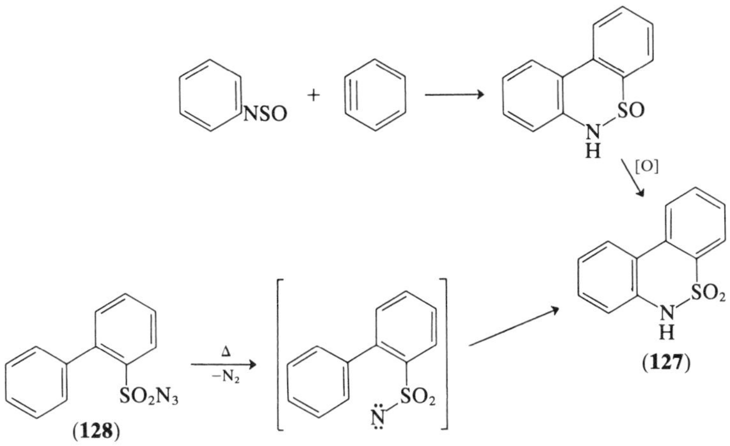

**Scheme 51**

Thiazinium oxides (131) are formed in a two-step sequence from enol ethers (130) and sulfoximines (129; Scheme 52) ⟨78CC197⟩, whereas alkylsulfamoyl chlorides and triethylamine react with activated dienes to afford diketosulfonamides (132) which on heating afford 1,1-dioxides (133; Scheme 53) ⟨79JOC305⟩.

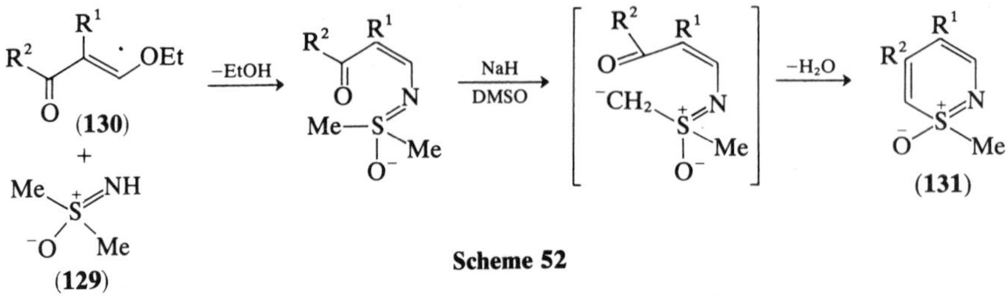

**Scheme 52**

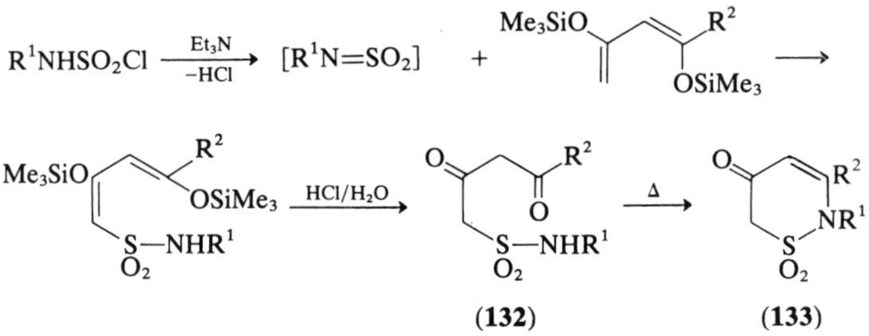

**Scheme 53**

When 1,2-benzothiazoles (134) are heated in the presence of sodium ethoxide, ring expansion occurs to give dihydrobenzothiazinones (135; Scheme 54) ⟨76CC771⟩. Glycine-*o*-carboxysulfonamides (136) cyclize to give related products (137), but in these compounds the 3-acyl substituent stabilizes the enol tautomer by intramolecular hydrogen bonding ⟨78LA635⟩.

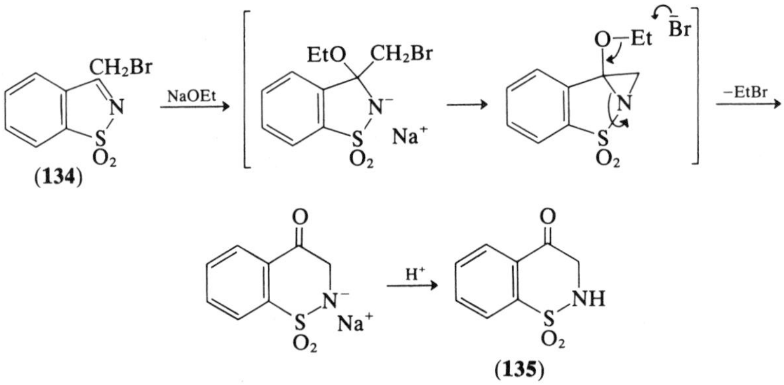

**Scheme 54**

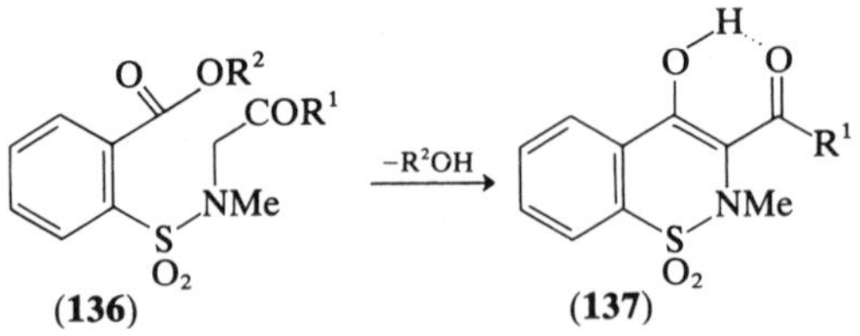

The 2,1-benzothiazines (139) are the products from the action of *N*-chlorosuccinimide and potassium hydroxide on 2-aminostyrenes (138) ⟨79TL3969⟩, and the sulfone (141) is obtained by the action of base on the tosylhydrazone (140) ⟨66JOC3531⟩.

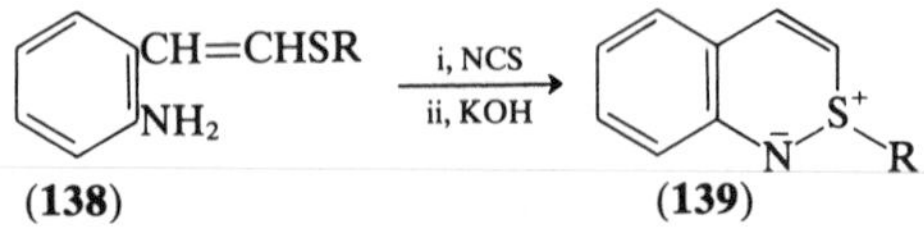

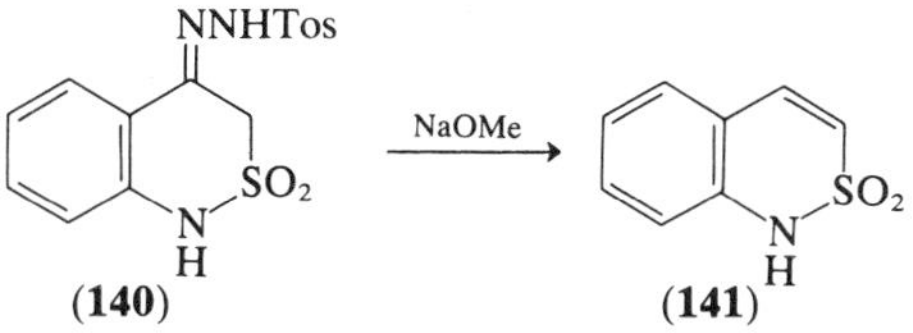

(140) (141)

### *2.27.3.1.2 Dihydro-1,2-oxazines and -thiazines*

5,6-Dihydro-4*H*-1,2-oxazines are conveniently prepared by the cycloaddition of nitrosoalkenes and alkenes. Thus, 3-nitrosobut-3-en-2-one (**142**) reacts with *trans*-stilbene to give (**143**) ⟨78CC847⟩. Similarly, $\alpha$-nitrosostyrene combines with cyclopentadiene to yield the oxazine (**144**) ⟨79JCS(P1)249⟩. Chloronitrones (**145**) and alkenes in liquid sulfur dioxide containing silver tetrafluoroborate afford oxazinium salts (**146**) ⟨77JOC4213⟩.

(142) (143)

(144)

(145) (146)

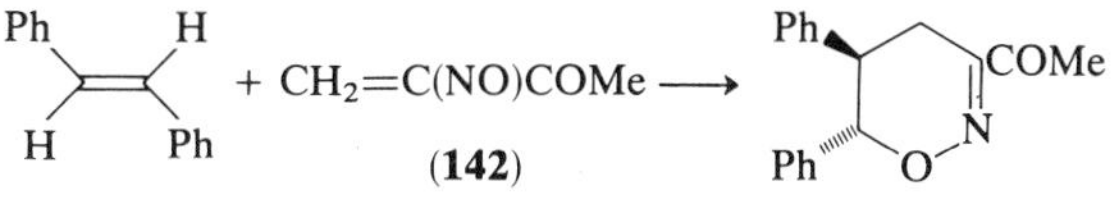

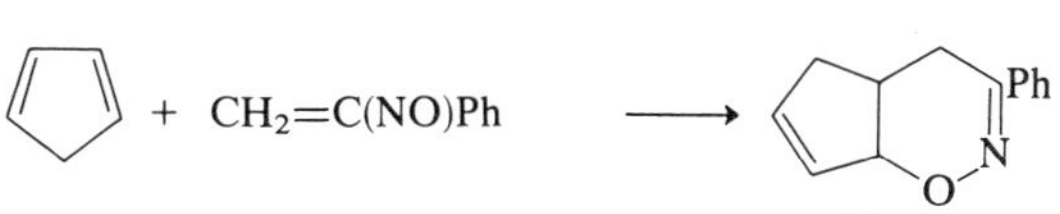

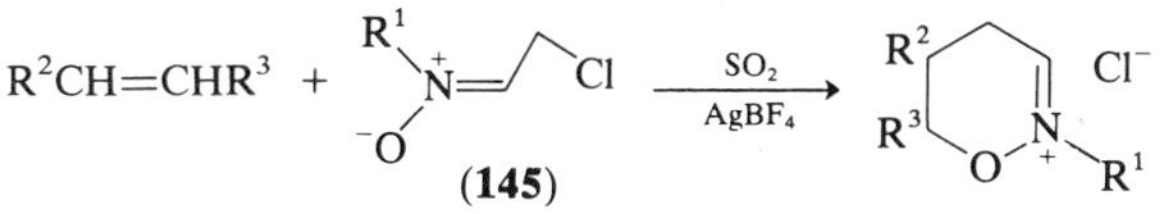

Another widely used approach is the cyclization of oximes either under acidic, as in the conversion (**147**→**148**) ⟨71MI22700⟩, or basic conditions (**149**→**150**) ⟨78JOC2020⟩. Cyclopropyl ketones (**151**) also react with hydroxylamine hydrochloride to give dihydro-1,2-oxazines (**153**), probably *via* the protonated oxime (**152**). If this is so then this reaction represents a rare example of a '6-*endo–tet*' ring closure ⟨80AG(E)199⟩.

(147) (148)

(149) (150)

(151) (152) (153)

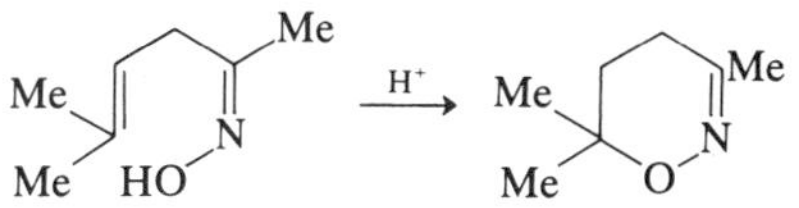

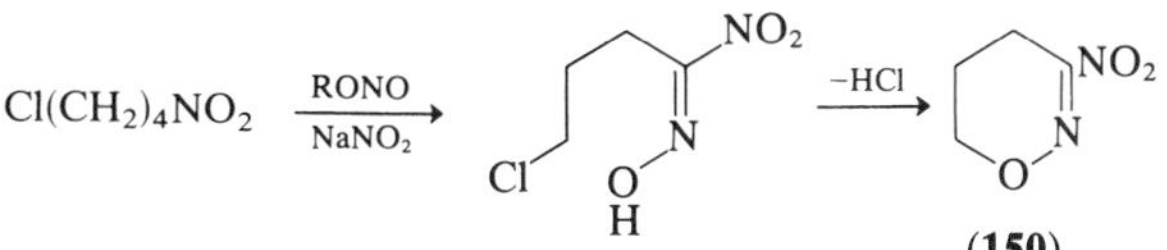

Cycloadditions of nitroso compounds and 1,3-dienes lead to 3,6-dihydro-2*H*-1,2-oxazines; this is a well tried and documented procedure ⟨B-67MI22700⟩, a comparatively modern example of which is the regiospecific synthesis of 2,3,4-trisubstituted derivatives (**155**) from the addition of nitrosobenzenes to 1,2-disubstituted dienes (**154**) ⟨78T697⟩.

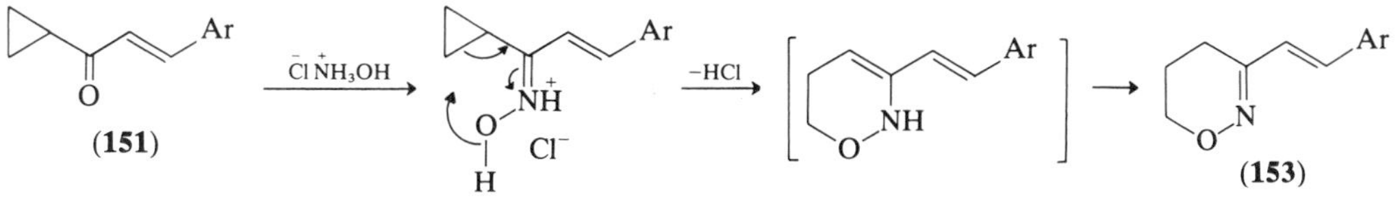

Dihydro-1,2-thiazines are synthesized either by the cycloaddition of *N*-sulfinylamines and 1,3-butadienes ⟨71LA(746)28⟩ or of benzoylsulfenes (**156**) and imines (Scheme 55) ⟨71BCJ2750⟩.

(154) (155)

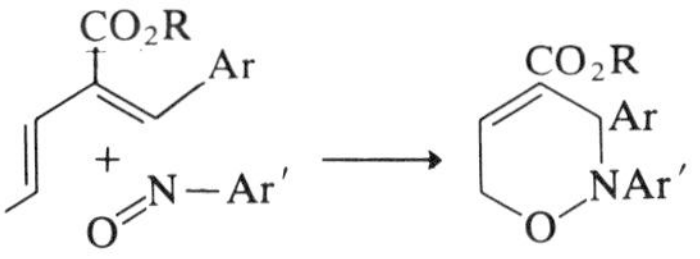

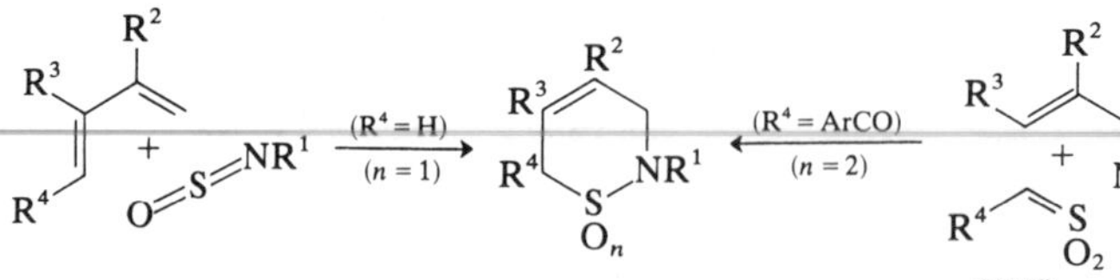

(156)

**Scheme 55**

There are several routes to 3,4-dihydro-2*H*-1,2-benzothiazine dioxides (**158**), including the cyclization of aminosulfonic acids (**157**) or cyanosulfonamides (**159**). 3,4-Dihydro-1*H*-2,1-benzothiazine dioxides (**161**) are normally prepared by thermolysis of the sodium sulfonates (**160**) or aminosulfonamides (**162**) ⟨71CB1880⟩, and 1*H*-3,4-dihydro-2,3-benzothiazine dioxides (**164**) are available either by a Pictet–Spengler cyclization of sulfonamides (**163**) with trioxane or of sulfonamide acids (**165**) with polyphosphoric acid ⟨76CC470⟩.

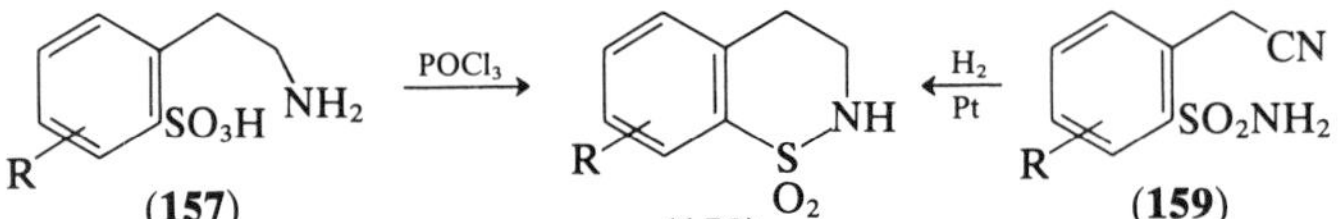

(**157**) (**158**) (**159**)

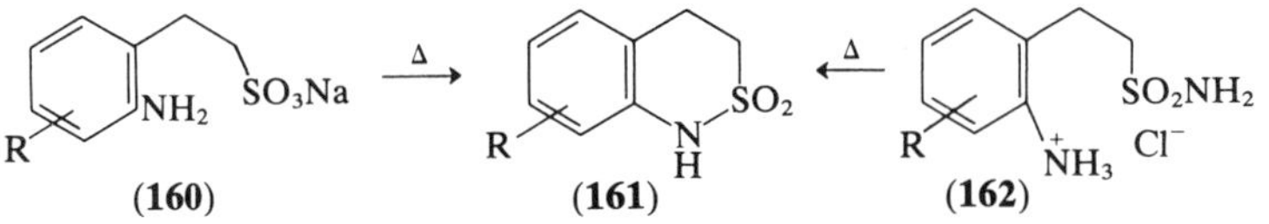

(**160**) (**161**) (**162**)

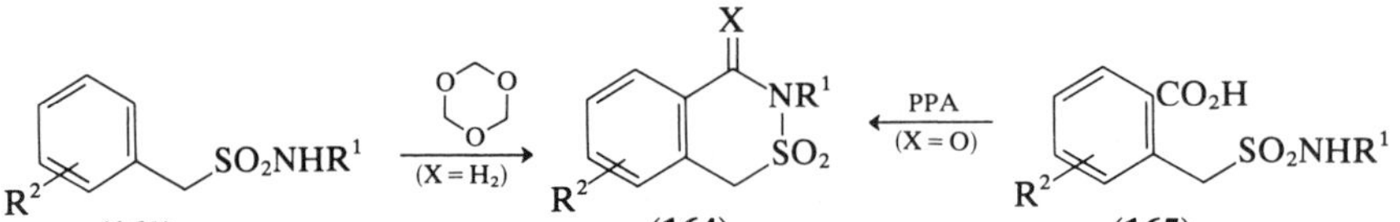

(**163**) (**164**) (**165**)

## 2.27.3.2 1,3-Oxazines, 1,3-Thiazines and their Benzo Derivatives

### *2.27.3.2.1 Parent ring systems*

2*H*-1,3-Oxazines are formed by the self-condensation of 2-bromo-2,3-dicyanopropionate or 2-cyano-3-phenylpropionate esters in the presence of phosphites (Scheme 56) ⟨75CR(C)(281)51⟩.

2-Alkoxy derivatives may be prepared from enaminoketones by reaction with orthoesters (Scheme 57) ⟨70CB2760⟩.

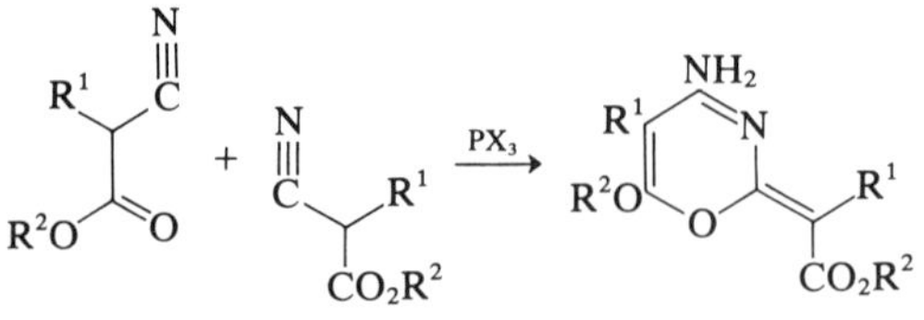

**Scheme 56**

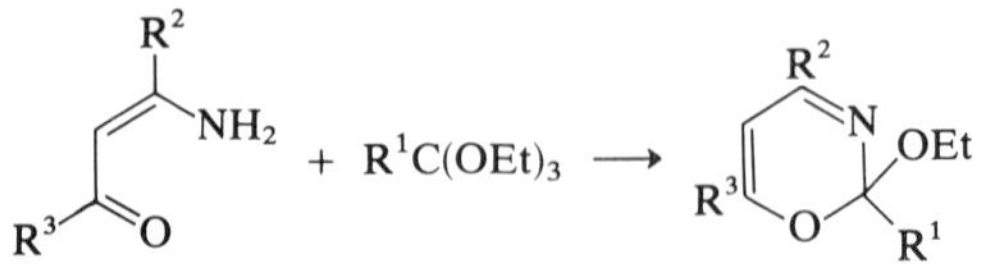

**Scheme 57**

A similar construction of 2*H*-1,3-benzoxazines employs salicylamide and dialkyl ketones; the product benzoxazinones (**166**) can be converted into alkoxy (**167**; $R^2$ = alkoxy) or chloro (**167**; $R^2$ = Cl) derivatives by alkylation ⟨32CB1032⟩, or by treatment with phosphorus oxychloride ⟨80CPB465⟩ respectively (Scheme 58).

Ring expansion of isoxazolium salts (**168**), by treatment with sodium hydroxide, can be used to synthesize both 2*H*-1,3-oxazines and 2*H*-1,3-benzoxazines. The reaction probably involves an ylide intermediate (Scheme 59) ⟨62CJC882⟩.

2*H*-1,3-Benzothiazines (**170**) are available through a Bischler–Napieralski type cyclization of amides (**169**) using phosphorus oxychloride as reagent ⟨77ACH(92)317⟩.

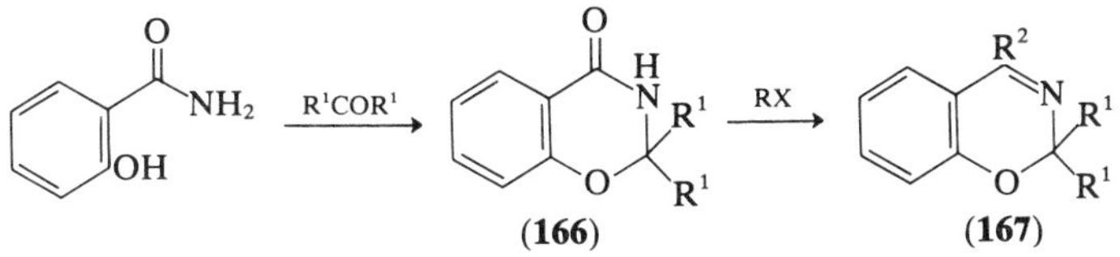

**Scheme 58**

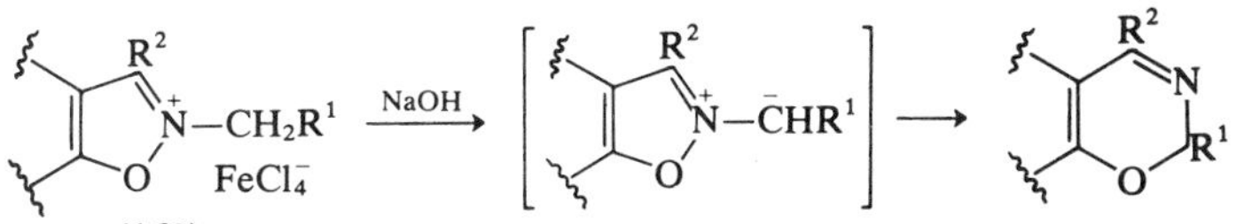

**Scheme 59**

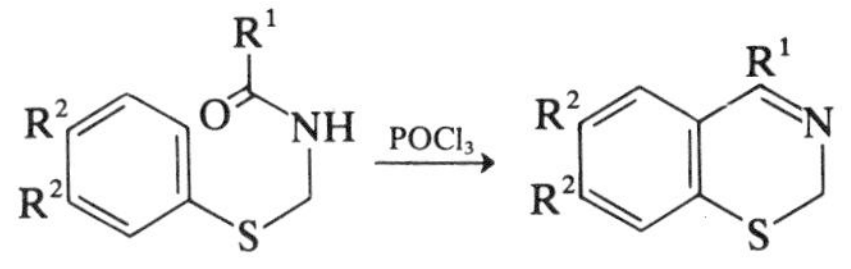

Traditional routes to 4*H*-1,3-oxazines and -thiazines involve the cyclization of amides or thioamides with acidic reagents ⟨78AHC(23)1; B-78MI22701⟩. Two examples are shown in Schemes 60 and 61.

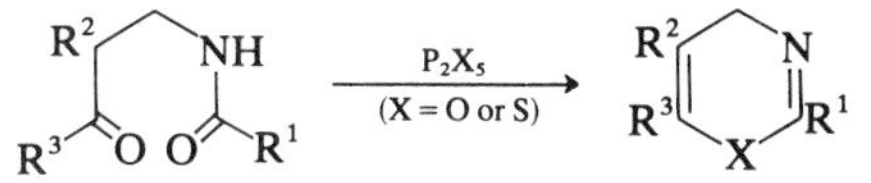

**Scheme 60**

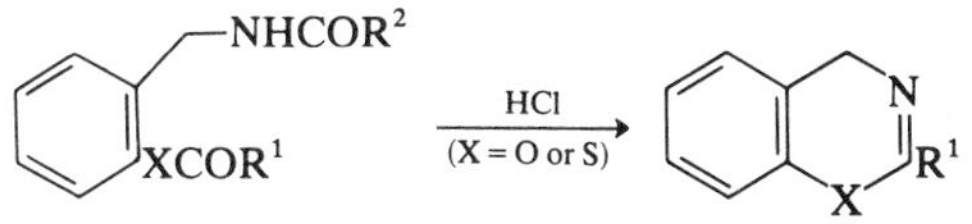

**Scheme 61**

Other routes to 1,3-oxazines employ condensation reactions between β-chloroketones and nitriles or between chloroalkyl amides and alkynes ⟨69LA(723)111⟩ (Scheme 62). Thiazines are available through similar condensations between thioamides, aldehydes and acetylenes ⟨74G849⟩, and 4*H*-1,3-benzoxazines may be prepared from 2-hydroxybenzyl alcohols and nitriles in the presence of either perchloric or sulfuric acids (Schemes 63 and 64) ⟨68MIP22700⟩.

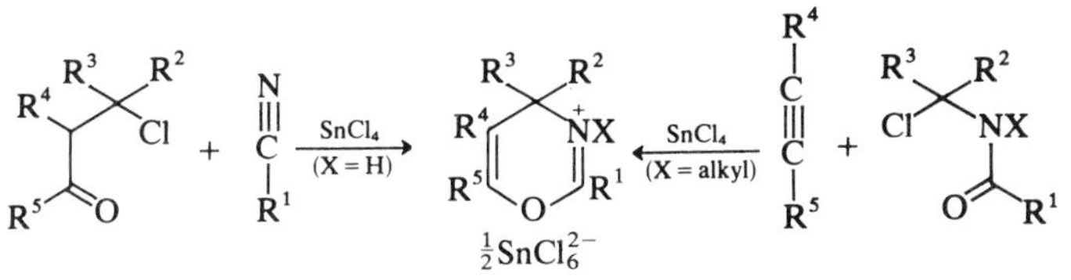

**Scheme 62**

Simple 6*H*-1,3-oxazines and -thiazines are commonly encountered as intermediates in the reactions of oxazinium and thiazinium salts with nucleophiles (see Section 2.27.2.2.3). Additionally there is a considerable interest in 6*H*-1,3-thiazines as intermediates in the synthesis of cephem antibiotics (see Section 2.27.3.24) and many approaches have been

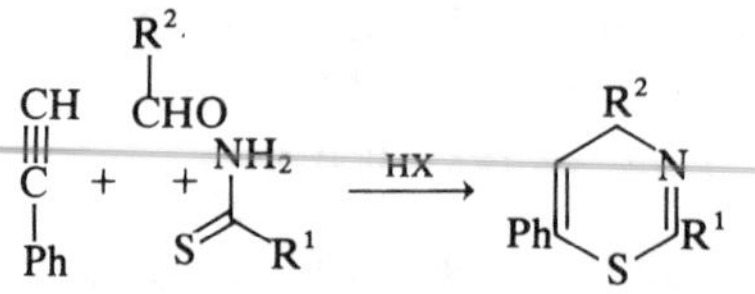

**Scheme 63**

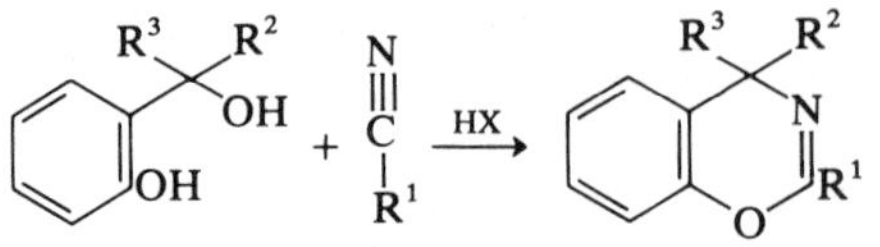

**Scheme 64**

patented. Derivatives of the type (**172**), for example, are produced from phosphonates (**171**) by reaction with α-halocarbonyl compounds and base ⟨B-80MI22701⟩, and in a similar manner the thiazine (**174**), for example, is prepared by enolate amination of the phosphonate (**173**) with *O*-mesitylenesulfonylhydroxylamine (MSH), followed by treatment with carbon disulfide. The product is methylated with methyl iodide and finally ring closure is effected with chloroacetone and potassium carbonate (Scheme 65) ⟨77JOC376⟩.

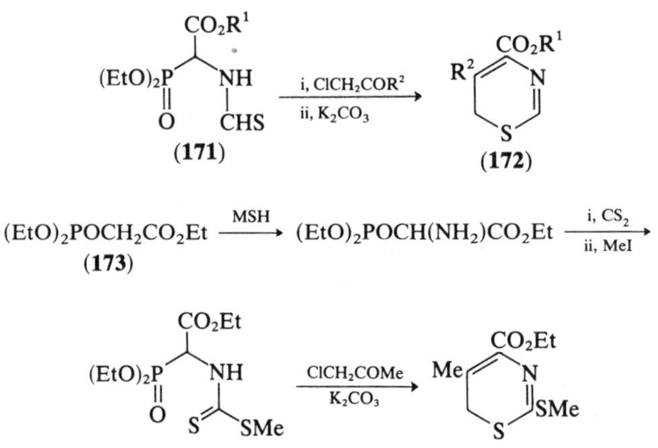

**Scheme 65**

An alternative approach to 6*H*-1,3-thiazines requires the thermally promoted addition of α,β-unsaturated carbonyl derivatives to thionoimines. Dihydrothiazines are the initial products but they readily decompose by elimination of the 4-substituent (Scheme 66) ⟨79BSF(2)347⟩.

4*H*-3,1-Benzoxazines and -benzothiazines can be made by the cyclization of benzyl alcohols; thus, the benzoxazine (**176**) is obtained by the oxidative photochemical ring closure of the pyrrole derivative (**175**) ⟨78JOC3415⟩. The amides (**177**) and (**179**) can be cyclized to benzothiazines (**178**) with phosphorus pentasulfide or pentachloride respectively ⟨1894CB3509⟩.

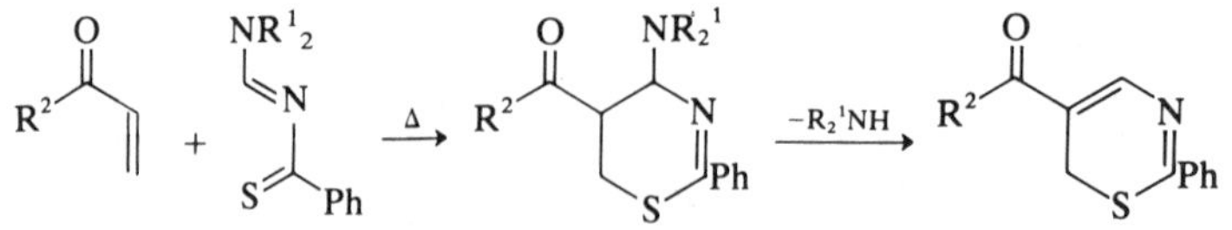

**Scheme 66**

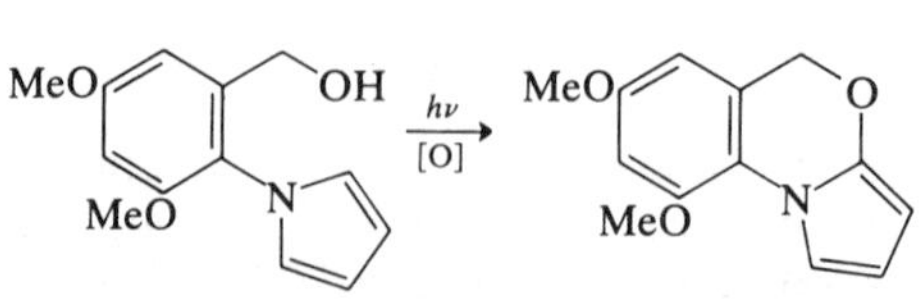

(**175**)                              (**176**)

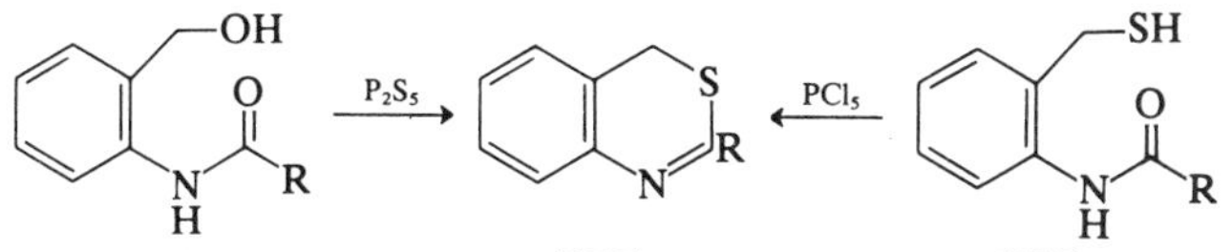

(177)   (178)   (179)

2-Aminobenzyl halides (180) react with glacial acetic acid to give $4H$-3,1-benzoxazines (181) ⟨1894CB3515⟩ and 4,4-dialkyl derivatives (183) are formed from the alcohols (182) on treatment with acetic anhydride ⟨1883CB2576⟩.

More modern syntheses utilize the interaction of 2-chloromethylphenylisocyanates (184) and amines (Scheme 67) ⟨78S377⟩ and $4H$-3,1-benzothiazines are prepared by cycloadditions between thioketones and ketenimines (Scheme 68) ⟨80JOC3766⟩.

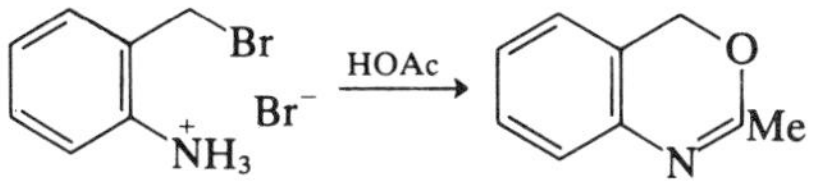

(180)   (181)

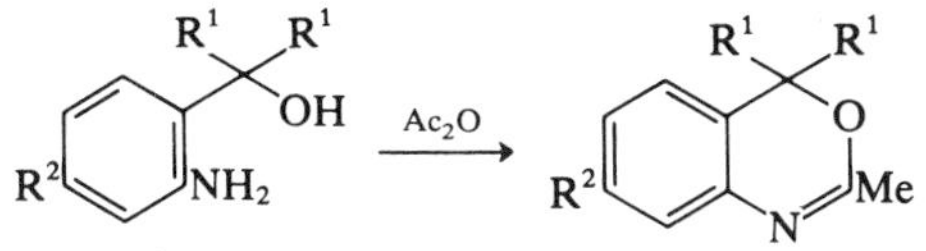

(182)   (183)

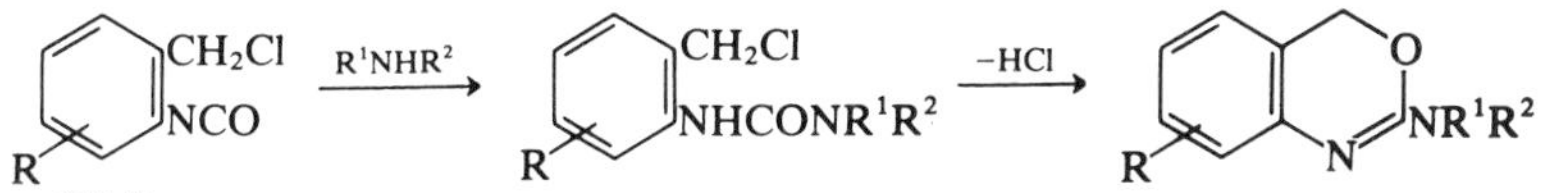

(184)

**Scheme 67**

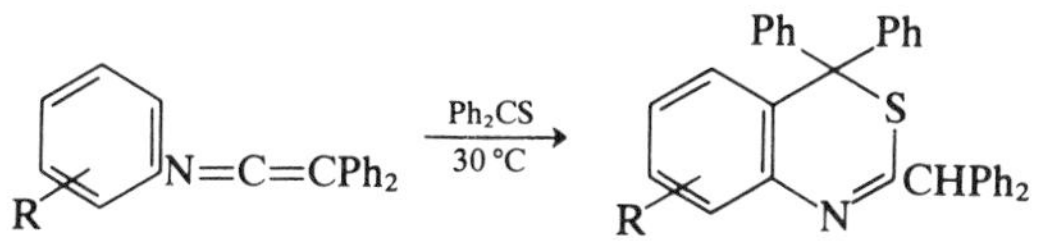

**Scheme 68**

### 2.27.3.2.2 1,3-Oxazinium and -thiazinium salts

The usual routes to 1,3-oxazinium salts consist of 1,4-cycloadditions between either $\alpha,\beta$-unsaturated $\beta$-chlorocarbonyl compounds and nitriles or between $N$-acylimidoyl chlorides and alkynes. Stannic chloride is an effective catalyst for both reactions (*cf.* Scheme 62). 1,3-Thiazinium perchlorates are synthesized by reacting oxazinium salts with hydrogen sulfide in absolute acetonitrile and then treating the product amides (185) with perchloric acid (Scheme 69) ⟨72S333⟩.

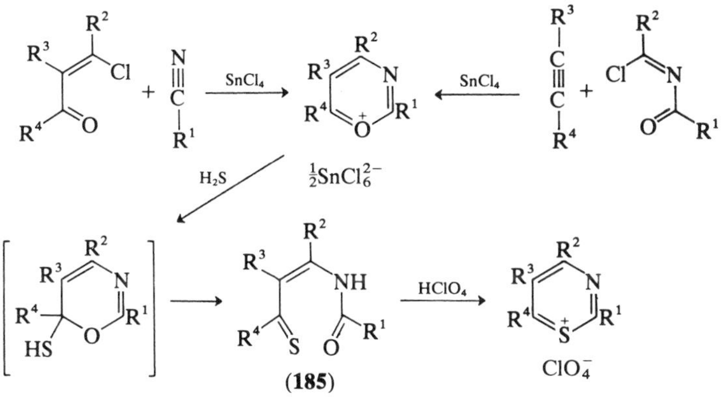

(185)

**Scheme 69**

### 2.27.3.2.3 1,3-Oxazinones, -thiazinones and related compounds

*(i) Functional group at C-2*

1,3-Oxazin-2-ones and -benzoxazin-2-ones are uncommon; the former compounds can be made by the thermolysis of carbonyl azides (**186**) (Scheme 70) ⟨76CC718⟩ and the latter by the cyclization of imines (**187**) with *N,N'*-carbonyldiimidazole. Thiophosgene affords the corresponding thiones (Scheme 71) ⟨78CB314⟩.

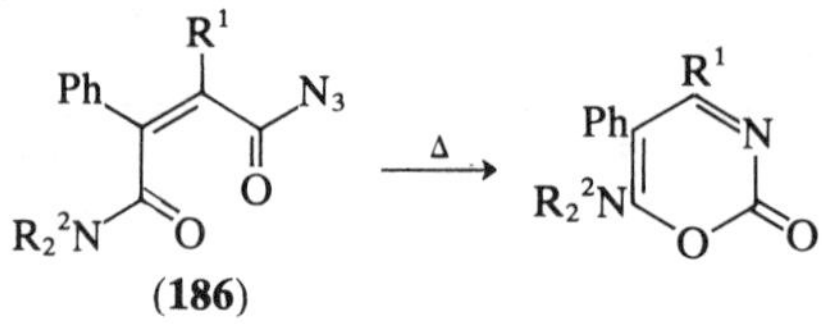

**Scheme 70**

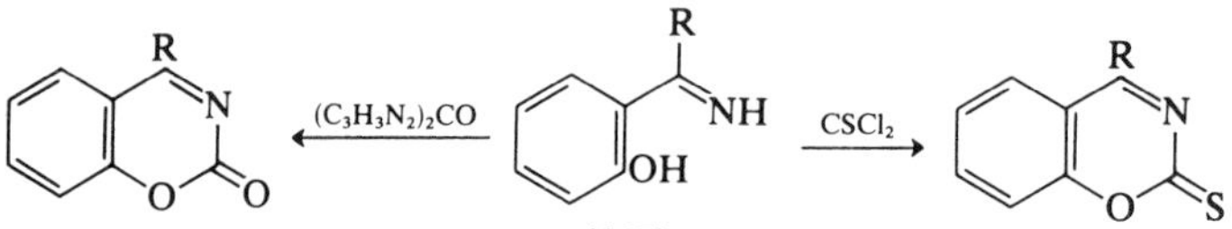

**Scheme 71**

1,3-Benzothiazin-2-ones have not been described, although the dihydro derivative (**188**) is constructed by the reaction of salicylaldehyde and methyl isothiocyanate. The mechanism of this reaction is unusual and involves an intramolecular O–S exchange, which may take place as shown in Scheme 72 ⟨72CB3055⟩.

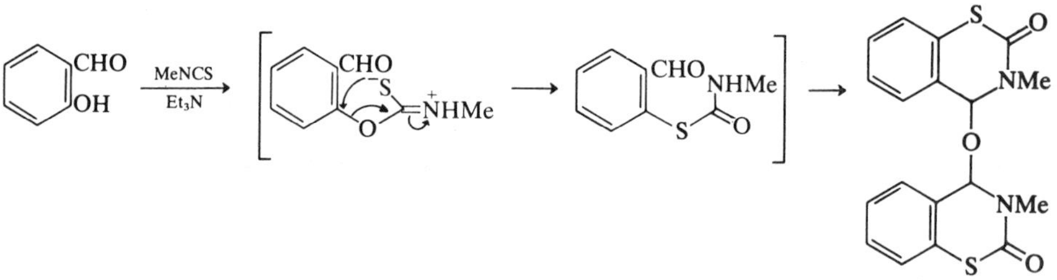

**Scheme 72**　　　　　　(**188**)

*(ii) Functional group at C-4*

1,3-Oxazin-4-ones and -thiazin-4-ones are well represented in the chemical literature. Thiazin-4-ones can be synthesized from 1,3-oxazinium salts by the action of hydrogen sulfide and potassium carbonate ⟨81H(15)851⟩ and oxazin-4-ones are obtained by cycloadditions between isocyanates and ketenes (Scheme 73), or alkynes (Scheme 74), or between nitriles and acylketenes (Scheme 75). Similarly diketene is often used and affords oxazin-4-ones by its reactions with imidates and cyanamides (Scheme 76) ⟨80H(14)1333⟩.

1,3-Benzoxazin-4-ones have been known for many years and are made by the cyclization of *O*-benzoylsalicylamides or reactions between phenyl salicylates and benzamidines ⟨15LA(409)305⟩. The first of these two methods obviously has wide applicability and when 2-acyl-mercaptobenzamides are used 1,3-benzothiazin-4-ones are obtained (Scheme 77) ⟨67BSF4441⟩.

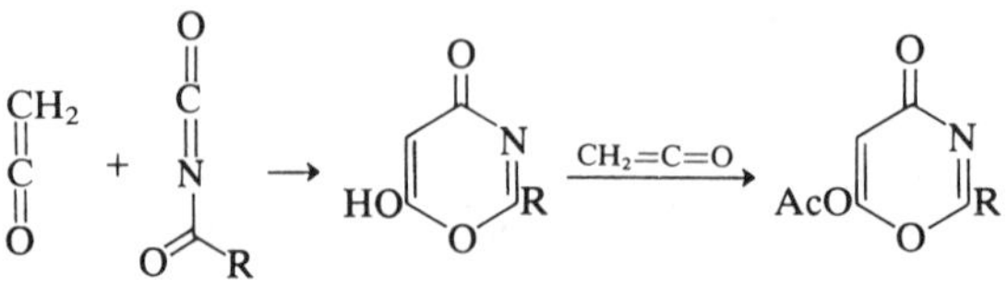

**Scheme 73**

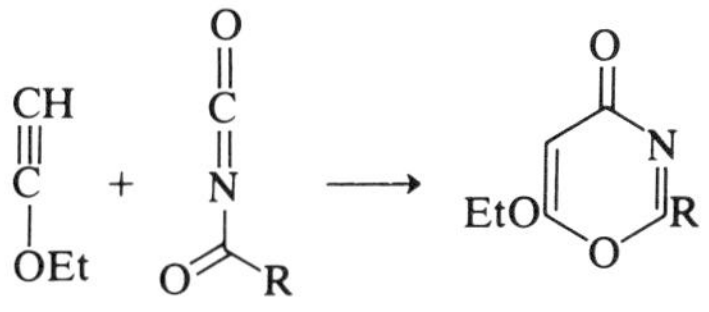

**Scheme 74**

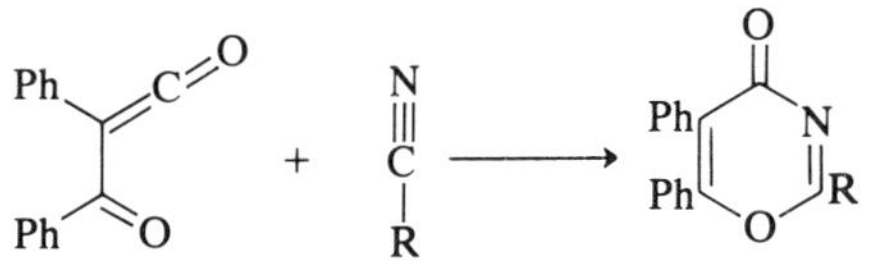

**Scheme 75**

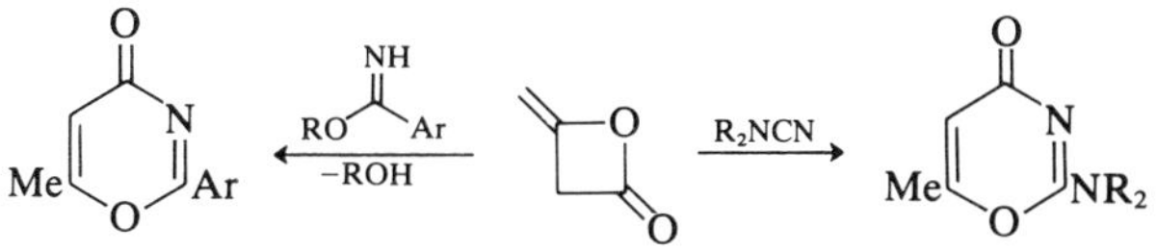

**Scheme 76**

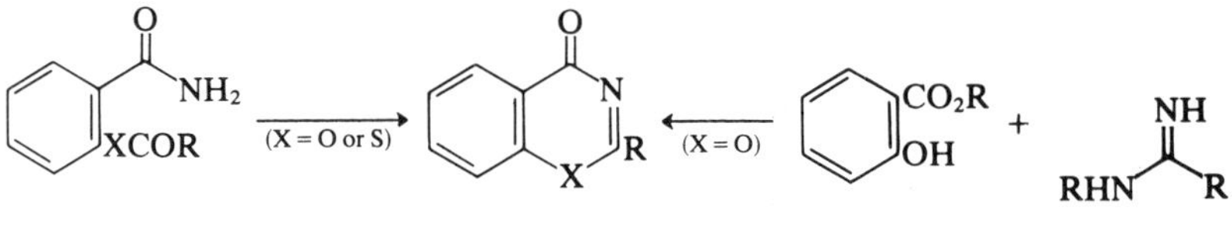

**Scheme 77**

### (iii) Functional group at position 6

1,3-Oxazin-6-one (**190**; $R^1 = R^2 = H$) is prepared by pyrolysis of the carbamate (**189**; $R^1 = OEt$, $R^2 = R^3 = H$) ⟨75JA6590⟩, and derivatives are made similarly by heating $\beta$-acylaminoesters (**189**; $R' = Ar$, $R^2 = OMe$) ⟨74AG(E)533⟩.

Alternatively, cyclopropenones give rise to [2 + 3] bicycloadducts with nitrile oxides, or ylides, which readily rearrange to yield 1,3-oxazin-6-ones. If cyclopropenethiones are employed 1,3-oxazin-6-thiones are formed (Scheme 78) ⟨72CJC584⟩.

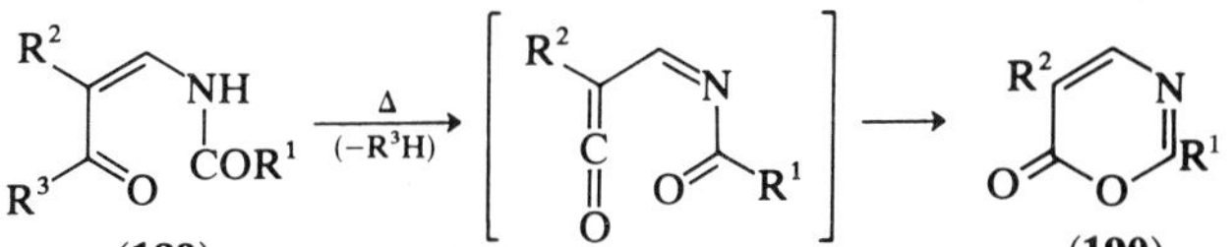

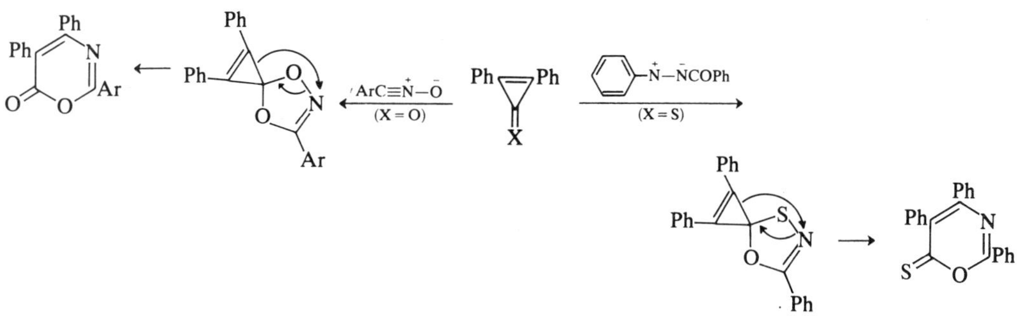

**Scheme 78**

3,1-Benzoxazin-4-ones (acylanthranils) are synthesized either by acylating 2,1-benzisoxazoles (**191**) with acid chlorides or anhydrides ⟨67AHC(8)277⟩, or by the cyclodehydration of *N*-acylanthranilic acids with acetic anhydride, phosphorus oxychloride or thionyl chloride (Scheme 79) ⟨77H(7)301⟩. Several pyrolytic techniques have also been used, but these are less generally applicable ⟨78MI22700⟩.

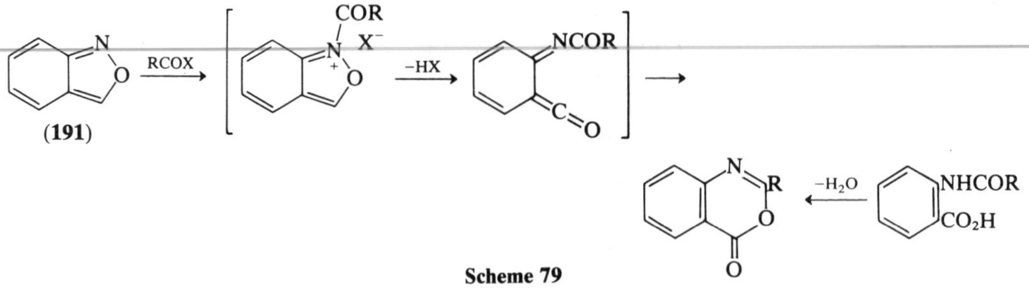

**Scheme 79**

### 2.27.3.2.4 Dihydro-1,3-oxazines and -thiazines

#### (i) 3,4-Dihydro-2H-structures

Monocyclic systems of this type are poorly represented in the literature and so far only two routes to 3,4-dihydro-2H-1,3-oxazines have been described which may be capable of further development. The first depends upon the interconversion of the tetrahydropyrimidine (**192**) into the oxazine (**193**) by partial hydrolysis, and the second requires the acid-catalyzed condensation of benzaldehyde with the β-amino ester (**194**; Scheme 80) ⟨45JA1382⟩.

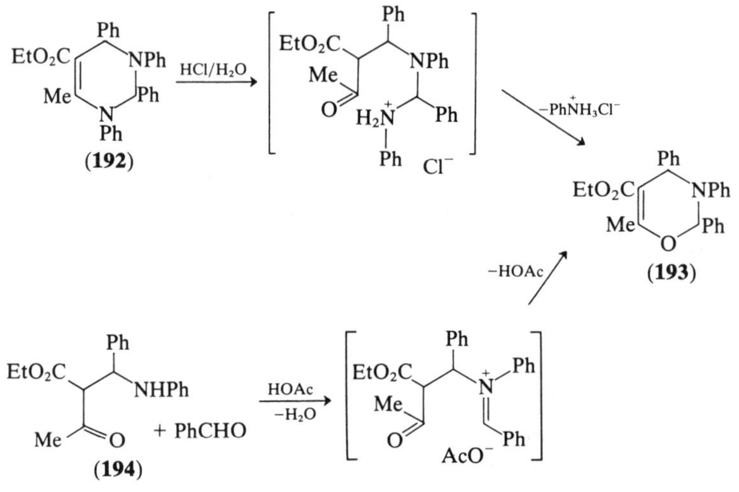

**Scheme 80**

Dihydro-1,3-benzoxazines (**196**) are formed by the reaction of phenols with a mixture of formaldehyde and primary aromatic amines in the molar ratio 2:1. Presumably the phenol first reacts with the appropriate iminium species to form an intermediate amine (**195**), which is then cyclized in a Pictet–Spengler type reaction (Scheme 81) ⟨44JA1875⟩. If 2-hydroxybenzylamines are employed then methylene derivatives are obtained, and if the formaldehyde is replaced by α-dicarbonyl compounds dehydro dimers (**197**) are produced (Scheme 82) ⟨70BCJ226⟩.

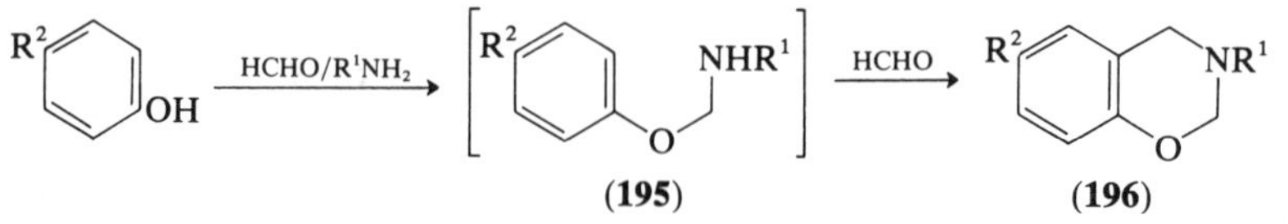

**Scheme 81**

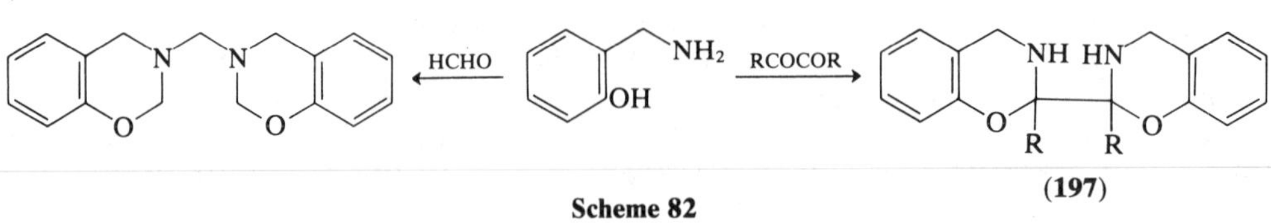

**Scheme 82**

Obviously aldehydes other than formaldehyde can be used in syntheses of the latter type and with 2-mercaptobenzylamines the addition of aldehydes and ketones leads to the production of dihydro-1,3-benzothiazines (Scheme 83) ⟨72ACH(71)363⟩.

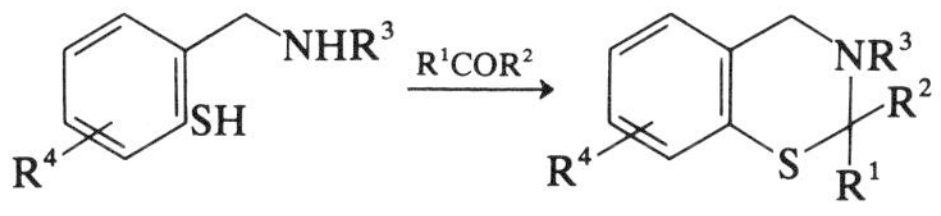

**Scheme 83**

### (ii) 3,6-Dihydro-2H-structures

Such compounds are unstable unless the double bond is held in conjugation with other systems as, for example, when it is part of an aryl ring. Indeed, dihydro-3,1-benzoxazines are readily accessible from 2-aminobenzyl alcohols by condensation with aldehydes or ketones (Scheme 84) ⟨75AP622⟩. Oxacephems (**199**) also contain the 3,6-dihydro-2H-1,3-oxazine unit and they are formed for instance, on cyclization of chlorolactams (**198**) by the action of stannic chloride ⟨B-80MI22701⟩.

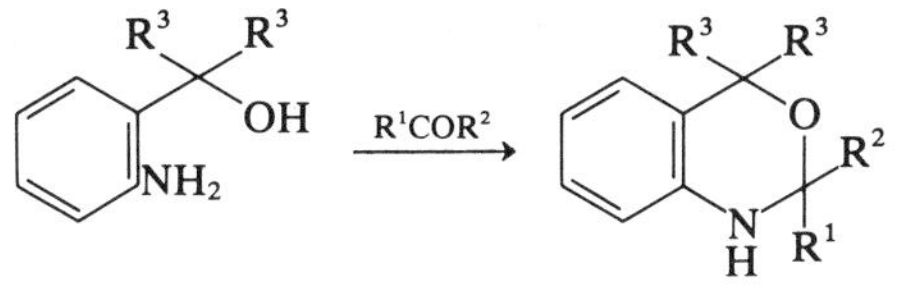

**Scheme 84**

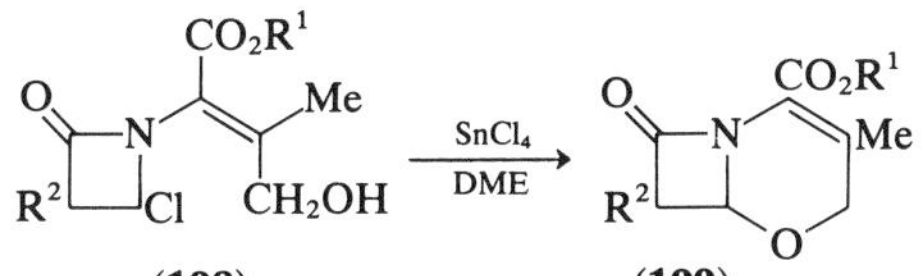

### (iii) 5,6-Dihydro-4H-structures

There are a number of very efficient syntheses of 5,6-dihydro-4H-1,3-oxazines and -thiazines reflecting the interest in these compounds as synthetic intermediates. The parent heterocycle, for instance, is obtained by the cyclization of N-formyl-2-hydroxyethylamine by elution through a column of silica gel ⟨70GEP1923022⟩. 2-Substituted derivatives can be prepared by the palladium-catalyzed interaction of 3-aminopropanol and isonitriles (Scheme 85) ⟨73JA4447⟩ or by treatment of N-acyl-3-bromopropylamines with base ⟨1891CB3213⟩. Other routes, shown in Scheme 86, include the ring closure of O-acylpropanolamines ⟨17CB819⟩ and the reaction of 3-azidopropanols with aromatic aldehydes ⟨55JA951⟩.

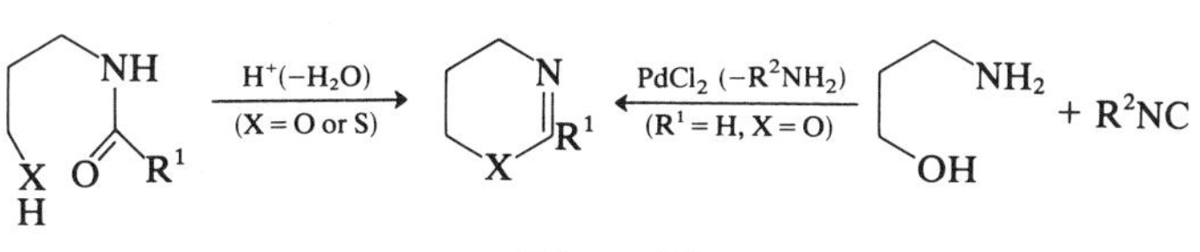

**Scheme 85**

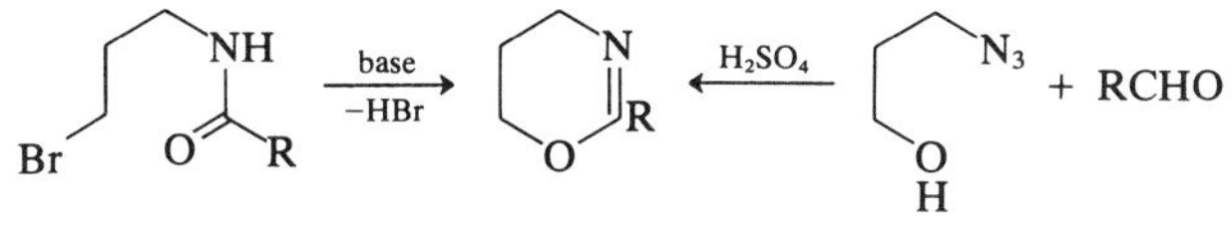

**Scheme 86**

However, the most important approach to 2-substituted dihydro-1,3-oxazines involves the acid-promoted condensation of alkyl nitriles and propane-1,3-diols, carbonium species (**200**) are assumed to be intermediates in this reaction, being formed preferentially from the more highly substituted position of the diol (Scheme 87) ⟨57JOC11⟩.

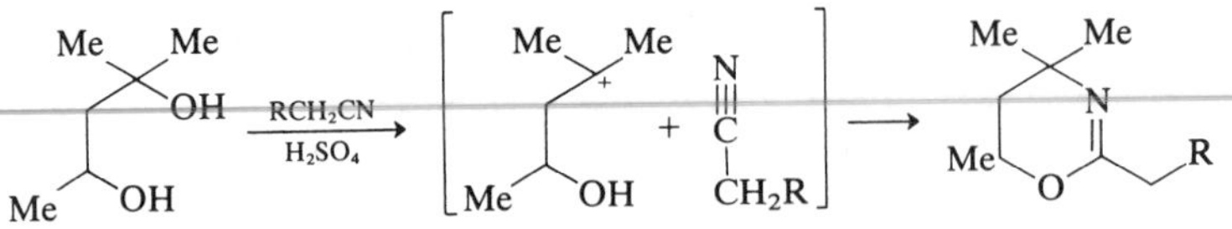

**(200)**

**Scheme 87**

Some of these syntheses can be adapted to the construction of dihydro-1,3-thiazines simply by using starting materials already containing a sulfur atom. The cyclization of *N*-acyl-4-mercaptoamines is an illustration ⟨16CB1114⟩. Similarly, isonitriles metallated at the α-position by *n*-butyllithium can be reacted with epoxides or episulfides to give intermediate alcholates or thiolates (201) which in the case of the oxygen compounds react with methanol to give the corresponding oxazines (202). The sulfur-containing species, on the other hand, may be ring closed to the corresponding thiazines (203) by heating with copper(II) oxide in benzene ⟨79LA451⟩.

Dihydrothiazines are useful in the formation of cephems and the derivative (205), for example, was prepared from the bromoester (204) and ethyl thioformate specifically for this purpose ⟨B-80MI22701⟩.

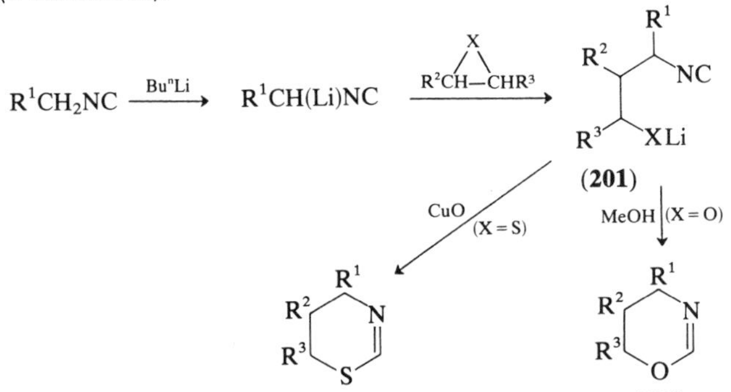

The inspirational origins of routes to benzo derivatives are also clear to see; thus, 4*H*-1,3-benzothiazines may be prepared by the cyclization of *N,S*-diacyl-2-mercaptobenzyl-amines or by the reaction of 2-mercaptobenzyl alcohols with nitriles (Scheme 88) ⟨79CHE240⟩.

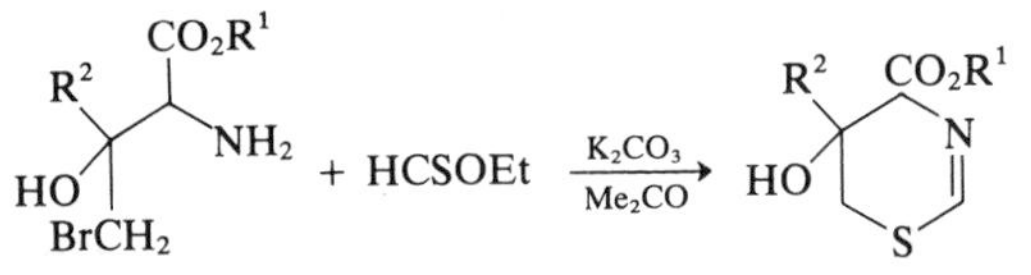

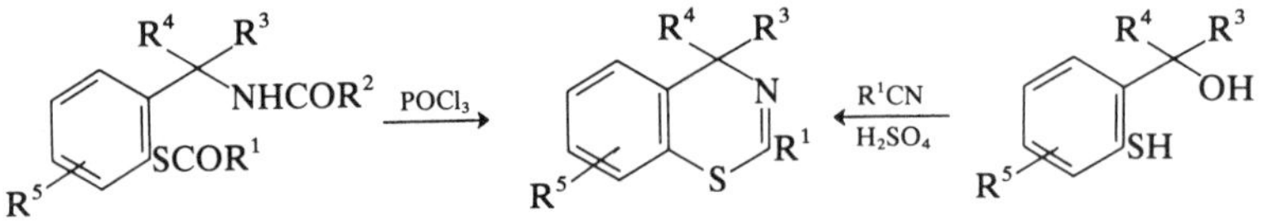

**Scheme 88**

Dihydro-1,3-oxazines or -thiazines bearing amino groups at the 2-position are available through the addition of 3-aminopropanols to isothiocyanates. The product thioureas readily cyclize to thiazines on treatment with base ⟨1890CB87⟩, but in order to form the analogous oxazines consecutive treatment with methyl iodide and base is required (Scheme 89) ⟨74KGS354⟩. In both cases the products are best formulated as imines rather than amines (see Section 2.27.2.2.1).

3-Aminopropanols, when reacted with cyanogen bromide, also afford 2-amino-(or imino)-dihydro-1,3-oxazines (Scheme 90) ⟨64ZOB3427⟩, and related thiazines are formed when allylic isothiouronium salts (207) are cyclized with trifluoroacetic acid and stannic chloride. The necessary starting materials are synthesized from aldehydes or ketones by the action of vinylmagnesium chloride and subsequent treatment of the product allyl alcohols (206) first with hydrogen chloride and then with a thiourea (Scheme 91) ⟨77JHC717⟩.

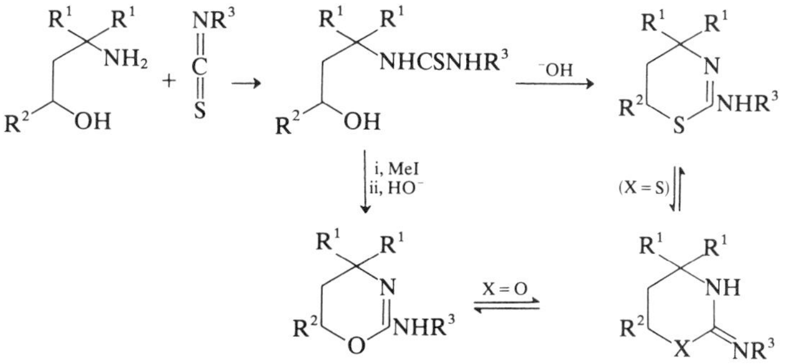

**Scheme 89**

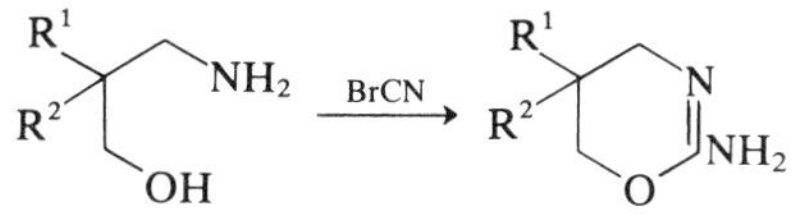

**Scheme 90**

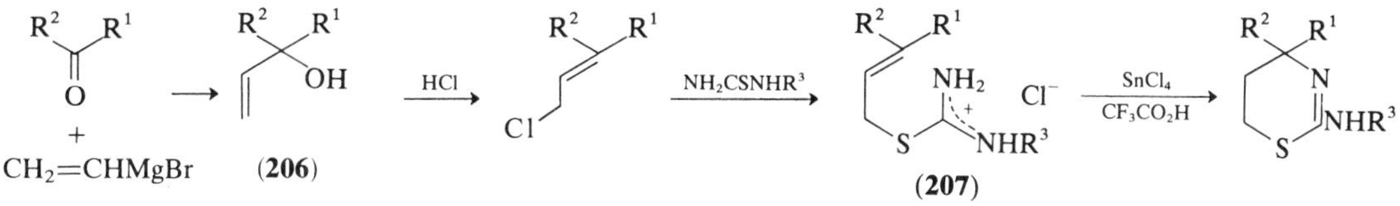

**Scheme 91**

### 2.27.3.2.5 Dihydro-1,3-oxazinones, -thiazinones and related compounds

#### (i) Functional group(s) at positions 2 and/or 4

Dihydro-1,3-oxazin-2-ones are conveniently made by 1,4-cycloadditions of $\alpha,\beta$-unsaturated ketones and chlorosulfonylisocyanate, followed by hydrolysis of the *N*-sulfonyl substituent with sodium hydroxide in aqueous acetone ⟨74LA521⟩. If diketene is used instead of ketones then 2,4-diones are obtained (Scheme 92) ⟨76GEP2434563⟩. In fact diketene is the most familiar synthon in this series for it may be reacted with various other imines to give 4-ones and 2,4-diones ⟨80H(14)1333⟩. Two illustrations are shown in Schemes 93 and 94.

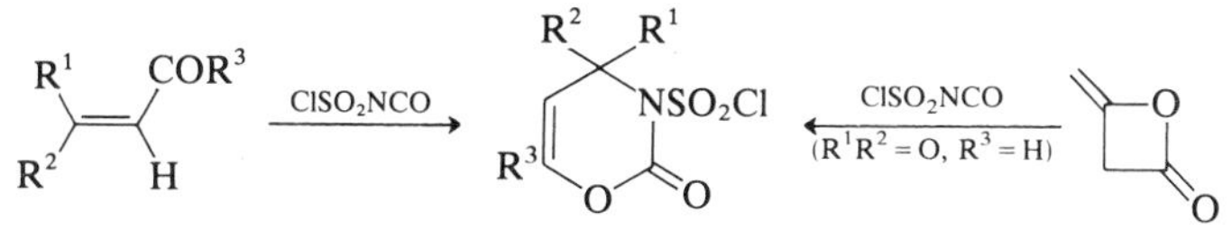

**Scheme 92**

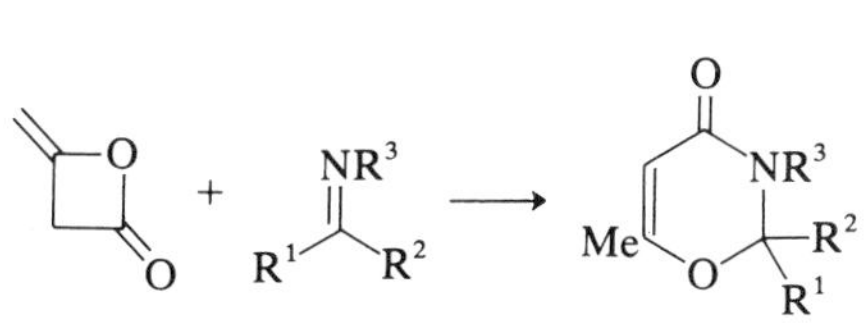

**Scheme 93**

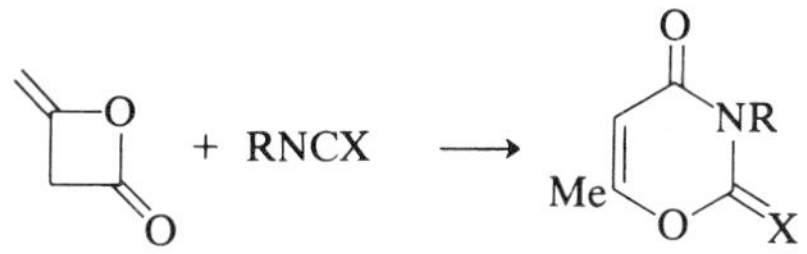

$(R = H, K, or\ Me_3Si;\ X = O\ or\ S)$

**Scheme 94**

Acylketenes are also employed in a number of other routes to dihydro-1,3-oxazin-4-ones. Isocyanates or isothiocyanates often form the second component of the reaction, and the approaches then differ only in the way the acylketene is generated (Scheme 95).

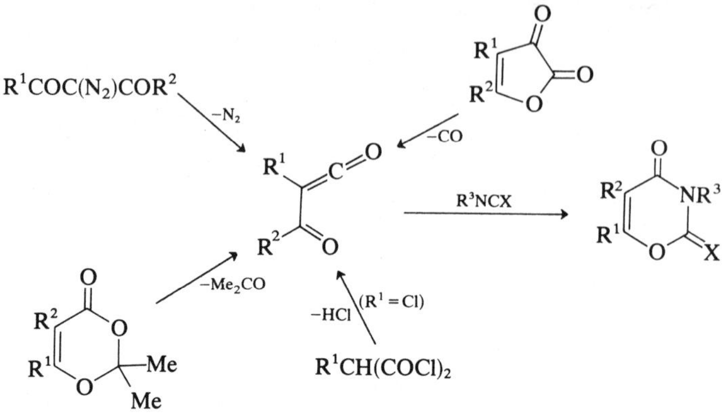

**Scheme 95**

Dihydro-1,3-thiazine derivatives (**208**) and (**210**) are prepared by the addition of alkyl propiolates to thioureas and dithiocarbamic acids respectively. In the latter case it is necessary to cyclize the initial products (**209**) with acetic anhydride (Scheme 96) ⟨70AJC51⟩. Ring expansion of isothiazolium chlorides (**211**) by the action of potassium cyanide provides a route to imine derivatives (**212**; Scheme 97) ⟨79TL1281⟩.

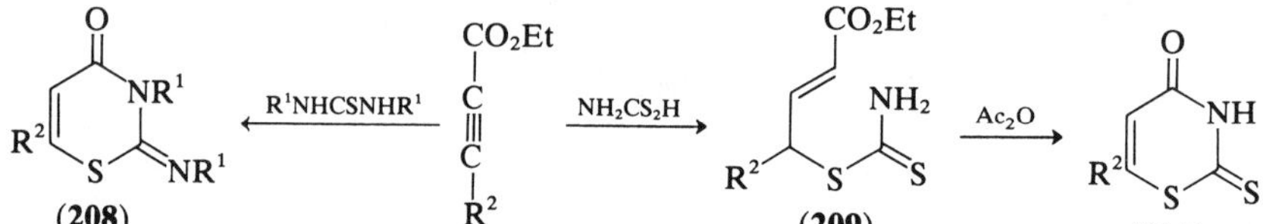

**Scheme 96**

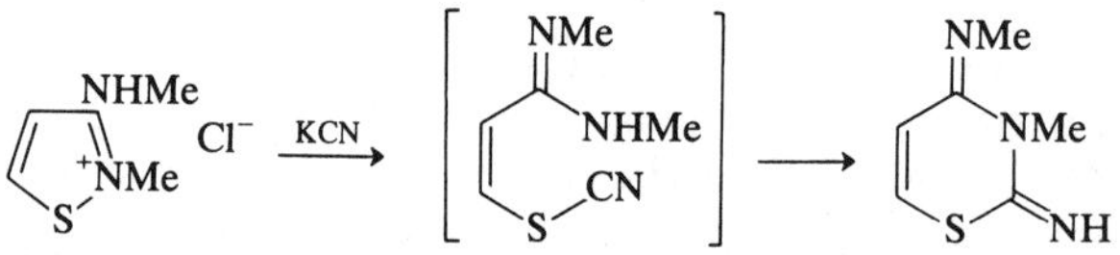

**Scheme 97**

A more conventional route is the acid-catalyzed cyclodehydration of amides already containing the appropriate numbers of atoms. For example, carbamates (**213**) may be cyclized to oxazinediones (**214**) with concentrated sulfuric acid ⟨76JCS(P1)1969⟩ and the dihydrobenzoxazin-4-ones and -thiazin-4-ones (**216**) can be prepared from the appropriate benzamides (**215**), again by treatment with acidic reagents ⟨53AP437⟩.

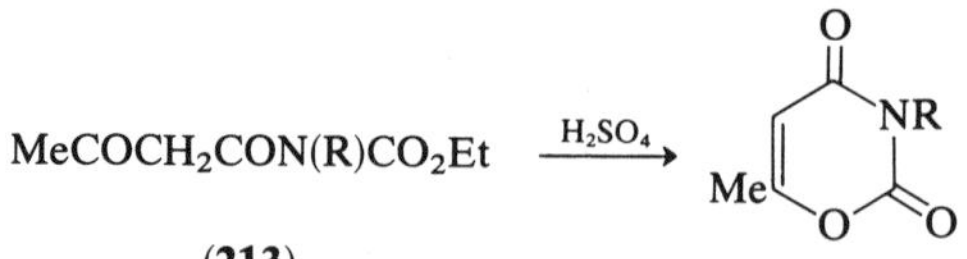

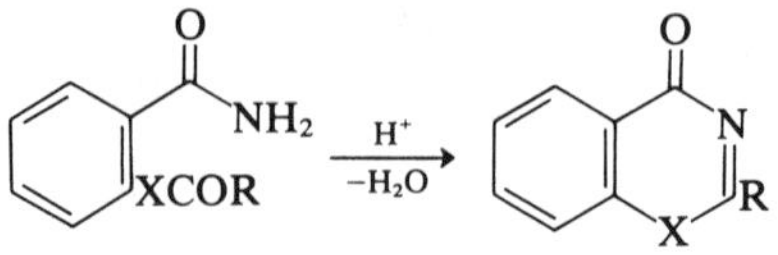

3,4-Dihydro-1,3-benzoxazin-2-ones are familiar compounds which may be synthesized by heating carbonyl azides ⟨28LA(464)237⟩, or by reacting the corresponding amines with phosgene ⟨68E774⟩, or the hydrochloride salts with potassium isocyanate ⟨77SC143⟩. 2-Thiones

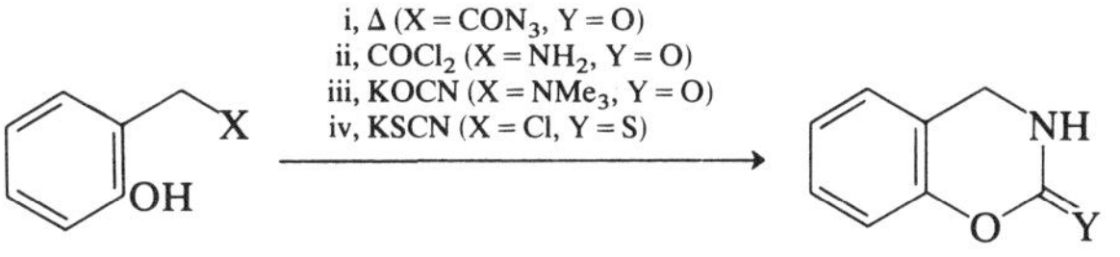

**Scheme 98**

are formed by treating 2-hydroxybenzyl halides with potassium thiocyanates (Scheme 98) ⟨79MIP22700⟩.

Base-catalyzed ring-expansion of benzisoxazolones (**217**) provides a general route to 2,2-dialkyldihydro-1,3-benzoxazin-4-ones (**218**) ⟨78CPB549⟩ and the 2,4-dione (**220**) can be prepared by a Beckmann rearrangement of the oxime (**219**) ⟨02CB3647⟩. 2-Thion-4-ones (**222**) are formed from salicylic acids (**221**) by treatment with triphenylphosphine diisothiocyanate in dichloromethane at −40 °C ⟨78CI(L)806⟩.

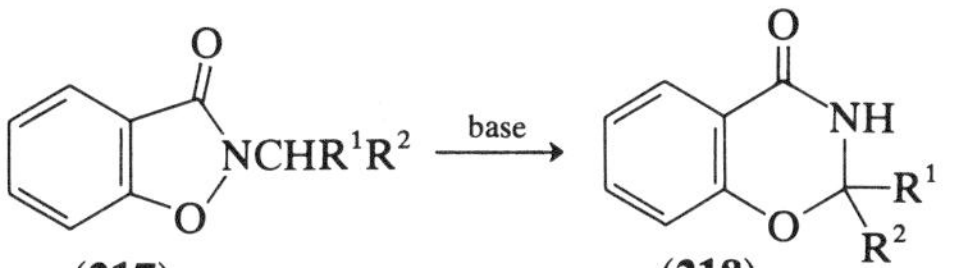

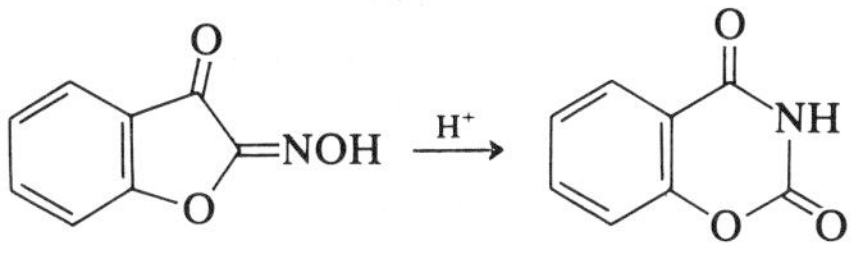

Benzothiazine-2,4-diones are synthesized by reacting 2-mercaptobenzoate esters firstly with ethyl chloroformate and then with an amine, or more directly by fusing the benzoate with urea. A third route is to treat the corresponding benzamide with phosgene (Scheme 99) ⟨77ACH(92)317⟩.

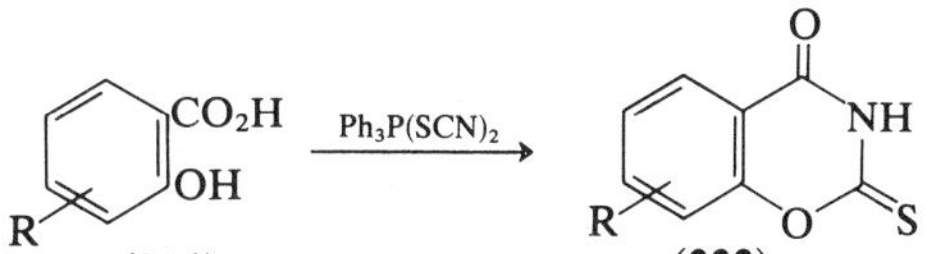

Isatoic anhydride (**223**; R = H) is easily prepared by passing phosgene into a solution of anthranilic acid in dilute hydrochloric acid ⟨55OSC(3)488⟩, and clearly this approach can be used to form derivatives substituted in the benzene ring. There is an alternative approach, namely the Baeyer–Villiger oxidation of isatins with hydrogen peroxide in acetic acid (Scheme 100) ⟨80AG(E)222⟩.

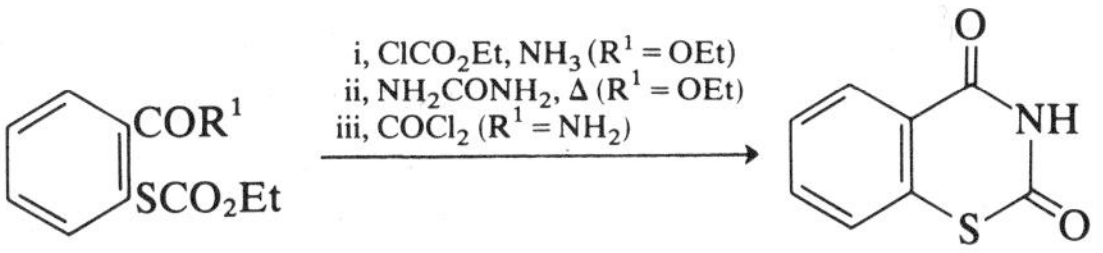

**Scheme 99**

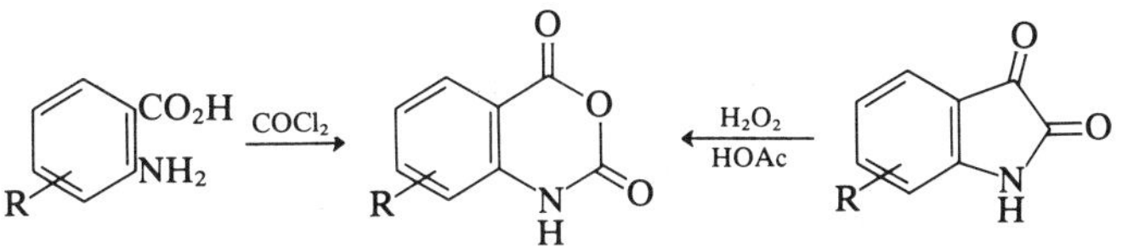

**Scheme 100**

### 2.27.3.2.6 Tetrahydro-1,3-oxazines and -thiazines

The usual route to tetrahydro-1,3-oxazines is *via* the reduction of dihydro forms; however, direct syntheses are not difficult to achieve. One obvious approach is the combination of aldehydes and ketones with 3-aminopropanols, and another, which has proved most useful for the construction of 5-nitro derivatives, is the cyclodehydration of propanediols, primary amines (or ammonia) and formaldehyde ⟨78AHC(23)1⟩. Tetrahydro-1,3-oxazin-2-ones and -thiones are formed from 3-aminopropanols by reaction with phosgene or thiophosgene respectively (Scheme 101), whereas tetrahydro-1,3-oxazin-4-one is prepared by the condensation of acrylamide and formaldehyde (Scheme 102) ⟨75JAP7524280⟩.

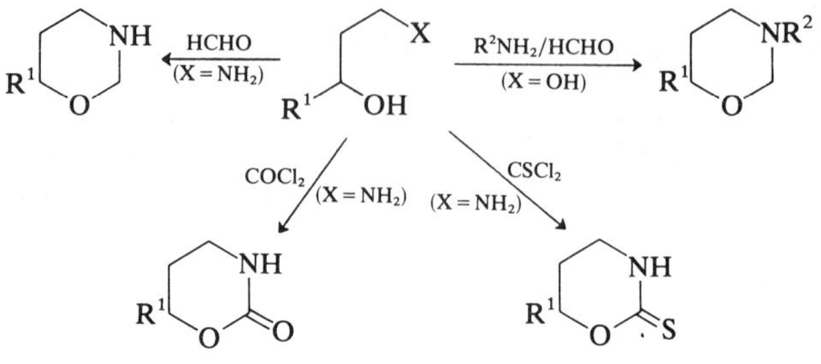

**Scheme 101**

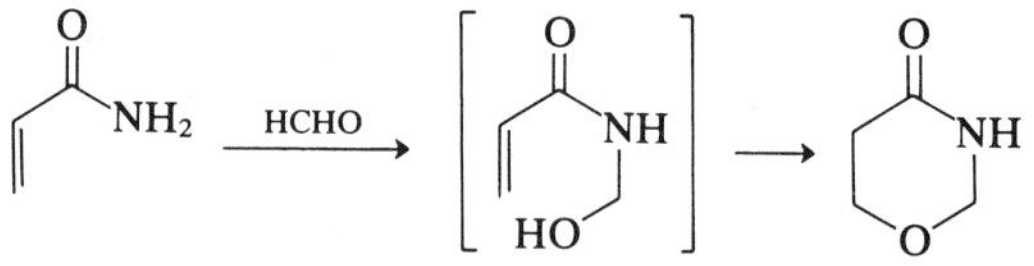

**Scheme 102**

2,4-Diones are normally synthesized from β-hydroxy acids in two steps; first, conversion into carbamates by reaction with sodium cyanate, and then cyclization with thionyl chloride (Scheme 103) ⟨54JCS839⟩. Alternative preparations utilize oxetanes, which may be combined either with isocyanates in the presence of boron trifluoride ⟨68JAP6808278⟩ or with S-alkylthioureas (Scheme 104) ⟨69ZOR1844⟩. In the last example the initial products are imines (**224**) which may readily be hydrolyzed to the required diones. Similar methods can be applied to the synthesis of tetrahydro-1,3-thiazine-2,4-diones, and, for instance, the 4-oxo-2-thioxo derivative (**225**) is obtained from β-propiolactone and dithiocarbamic acid (Scheme 105) ⟨48JA1001⟩.

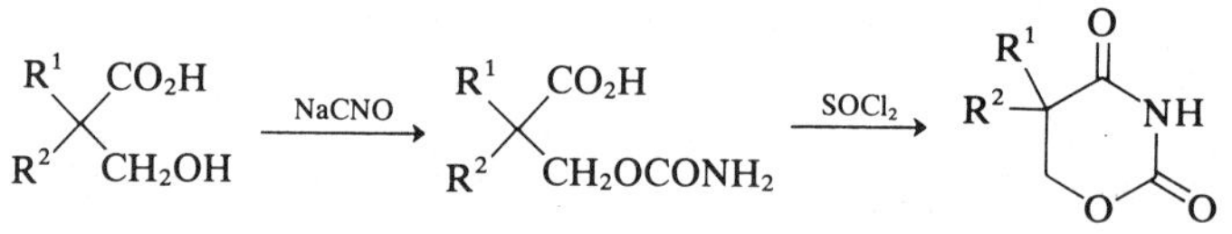

**Scheme 103**

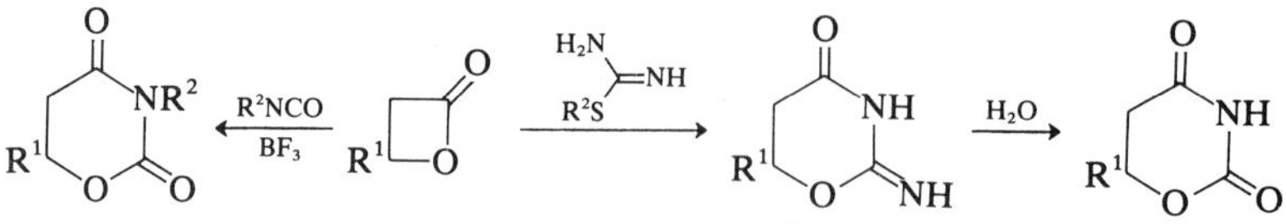

(**224**)

**Scheme 104**

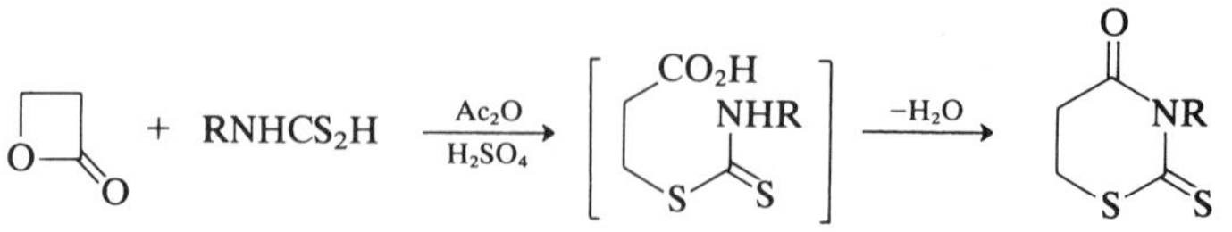

(**225**)

**Scheme 105**

The acid-catalyzed decomposition of diazoketones (**226**) gives rise to cations which then cyclize to 2-iminotetrahydro-1,3-thiazin-5-ones (**227**). On hydrolysis these lead to the corresponding 2,5-diones (**228**) ⟨80KGS1327⟩.

On the other hand, 1,3-oxazine-2,6-diones can be formed from β-aminoacids and phosgene ⟨52USP2600596⟩, or from succinimides and sodium hypochlorite (Scheme 106) ⟨61MI22700⟩. The synthesis of the corresponding 4,6-diones is achieved by the reaction of malonic acid derivatives with amides or imines. Dialkylmalonyl chlorides, for example, combine with isobutyramides (**229**) to yield the 2-methyleneoxazine-4,6-diones (**230**) ⟨66JOC2966⟩. Examples of oxazine-2,4,6-triones are few, but representative structures (**233**) are available through the reaction between β-aminocrotonates (**231**) and the acid chloride (**232**) ⟨74GEP2311708⟩.

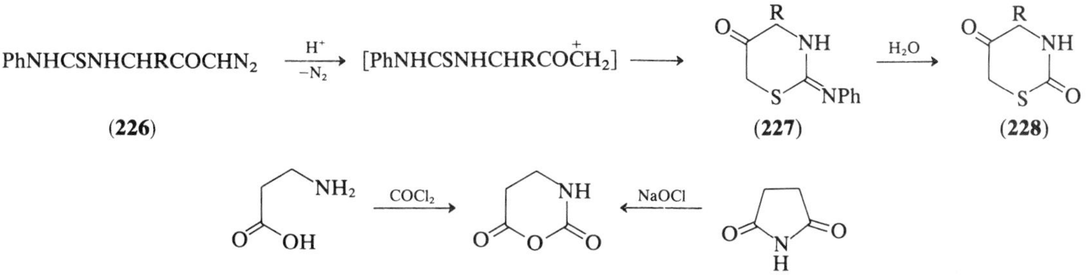

PhNHCSNHCHRCOCHN₂ $\xrightarrow[-N_2]{H^+}$ [PhNHCSNHCHRCOCH₂⁺] ⟶

(**226**)                                                                      (**227**)          (**228**)

**Scheme 106**

Me₂CHCONHMe + 

(**229**)

(**230**)

(**231**)          (**232**)          (**233**)

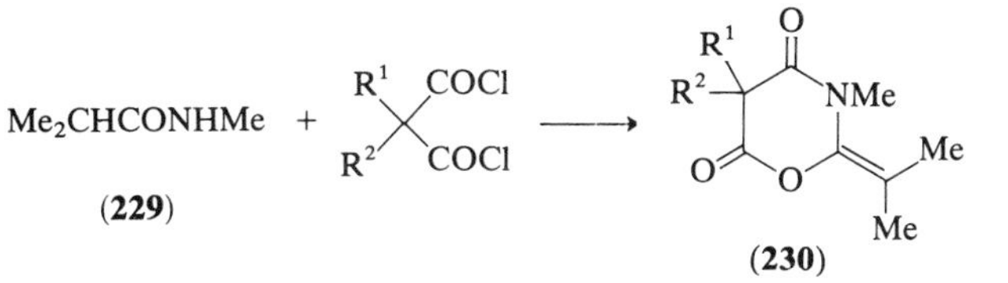

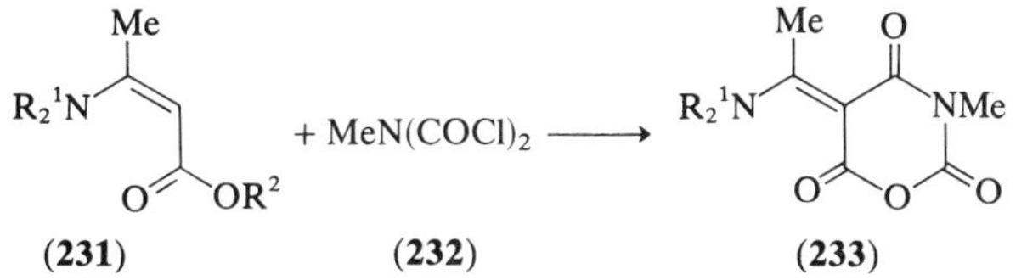

### 2.27.3.3 1,4-Oxazines, 1,4-Thiazines and their Benzo Derivatives

#### 2.27.3.3.1 Parent ring systems

2*H*-1,4-Thiazine itself has been claimed to be produced by the pyrolysis of the salt (**234**; Scheme 107) and 3,5-disubstituted analogues are formed when the sulfides (**235**) are reacted with ammonia (Scheme 108). Unfortunately, attempts to extend this last synthetic reaction by using amines rather than ammonia were unsuccessful ⟨67JMC501⟩.

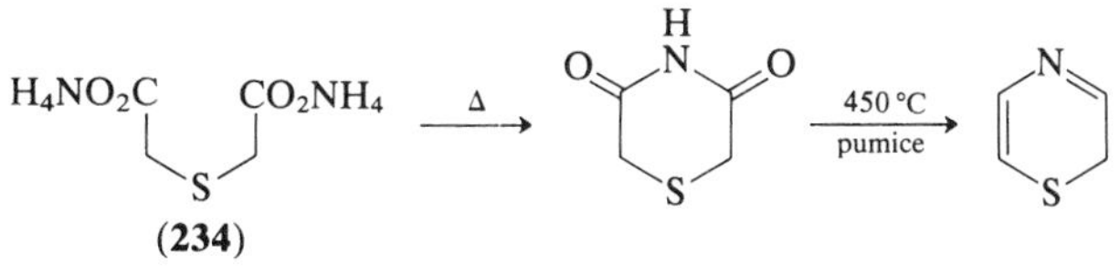

H₄NO₂C   CO₂NH₄   $\xrightarrow{\Delta}$                    $\xrightarrow[\text{pumice}]{450\,°C}$

(**234**)

**Scheme 107**

ArCO   COAr   $\xrightarrow{NH_3}$

(**235**)

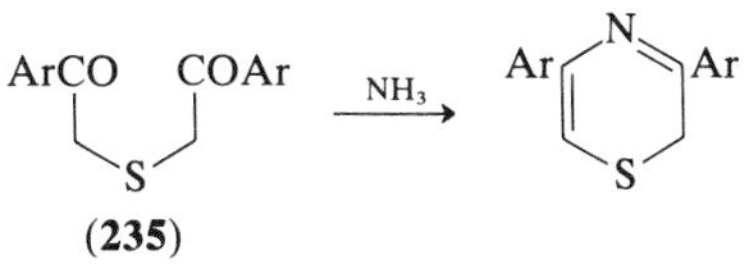

**Scheme 108**

The 3,4-dihydrothiazine-3-thione (**236**) is converted into the 2*H*-thiazine (**237**) by direct *S*-ethylation with triethyloxonium tetrafluoroborate. With methyl iodide *S,N*-dimethylation occurs to give the thiazinium salt (**238**), deprotonation of which yields a 4*H*-thiazine (**239**) ⟨69JHC247⟩.

2*H*-1,4-Benzoxazines (**241**) are available by the cyclization of acetals (**240**; X = O) in acid solution ⟨79M257⟩. Similarly the reduction of 2-nitrophenoxyacetophenones (**242**) with zinc dust and ammonia affords the corresponding *N*-oxide derivatives (**243**) ⟨79T1771⟩.

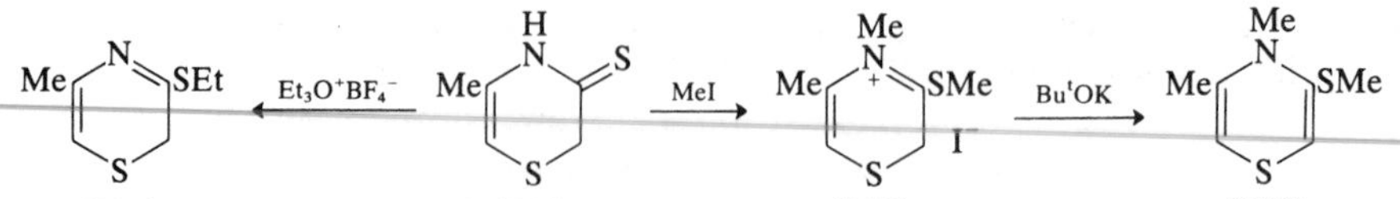

(237)                    (236)                    (238)                    (239)

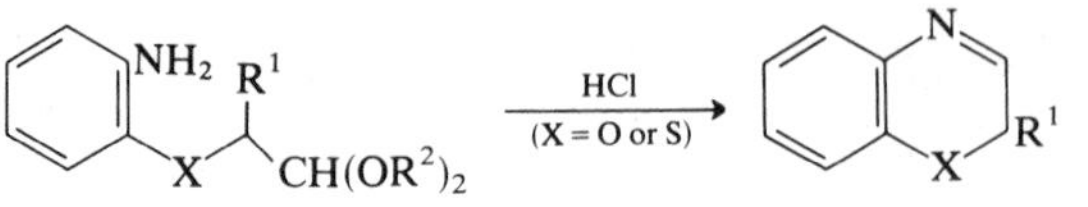

(240)                    (241)

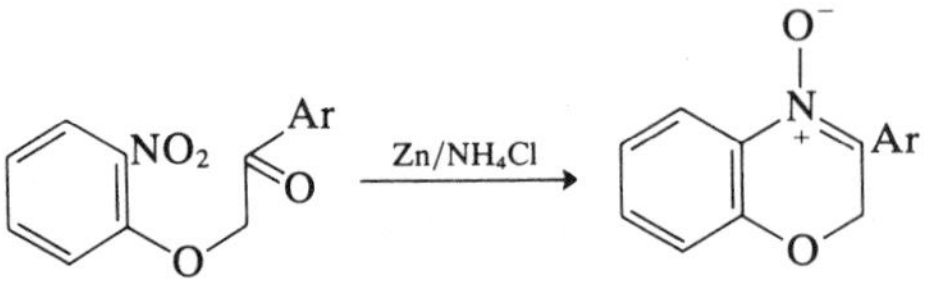

(242)                    (243)

Representative 4*H*-1,4-benzoxazines such as the tricyclic structure (245) and the 4-hydroxy derivative (247) are formed by the interaction of 2-hydroxyaniline with the bromoketone (244) ⟨79BCJ1156⟩ and by the addition of dimethyl acetylenedicarboxylate to the bis copper(II) complex (246) ⟨76JOC1079⟩ respectively.

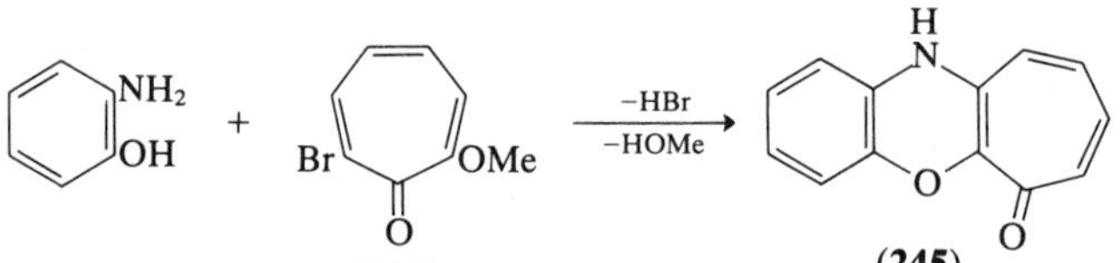

(244)                    (245)

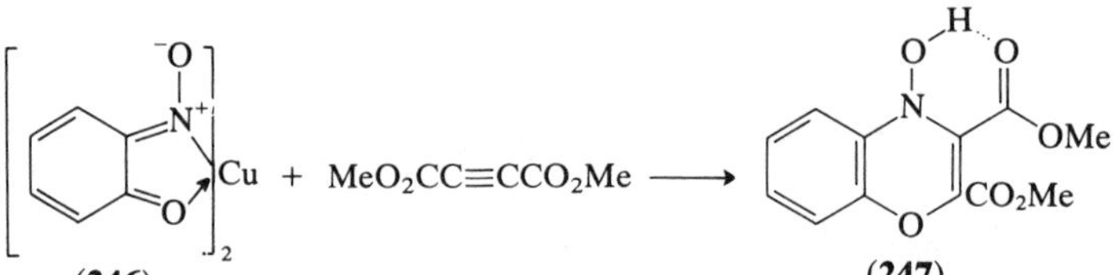

(246)   +   MeO₂CC≡CCO₂Me   ⟶

(247)

There are several methods for the construction of 1,4-benzothiazines; 2*H*-forms can be obtained by the cyclization of acetals (240; X = S) in the same manner as the corresponding oxazines ⟨76T1407⟩, and the 4*H* isomers are conveniently prepared by the cycloaddition of 2-aminothiophenol and disubstituted alkynes ⟨80JHC793⟩. Alternative syntheses also utilize 2-aminothiophenols, by their reactions either with α-bromocarbonyl compounds ⟨70AC(R)383⟩ or with active methylene compounds (Scheme 109) ⟨76JCS(P1)1146⟩.

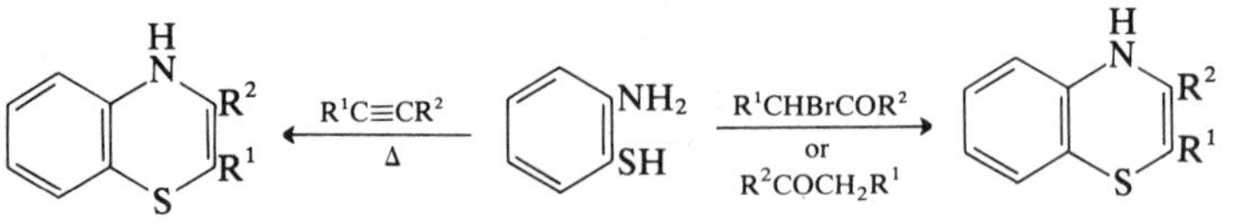

**Scheme 109**

1,4-Benzothiazines can also be synthesized by the ring expansion of dihydrobenzothiazoles bearing a side chain at position 2 which contains an active methylene unit. In some instances the product is contaminated with a benzothiazole (Scheme 110) ⟨81JHC279⟩. A similar reaction occurs with the *S*-oxides (248) which afford both the tautomers (249) and (250) when treated with acetic anhydride ⟨81TL1701⟩.

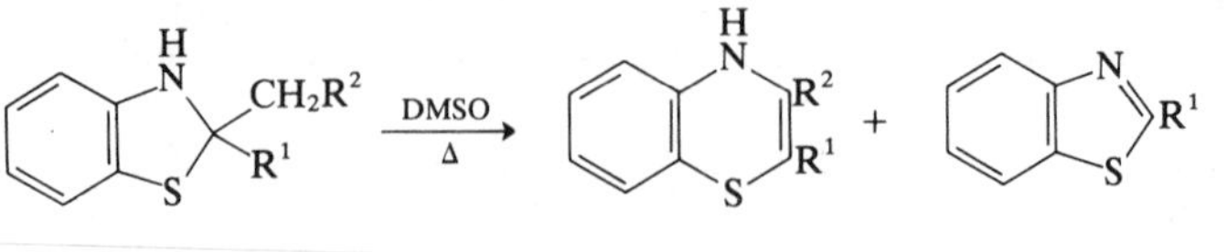

**Scheme 110**

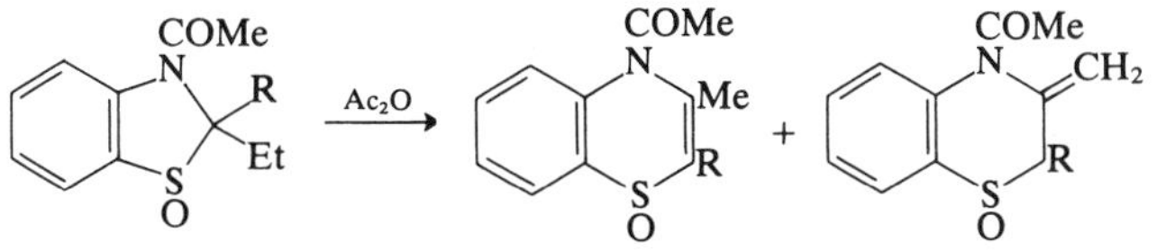

(248)     (249)     (250)

Traditional routes to phenoxazines include the thermolysis of 2-aminophenol and catechol, the latter acting as an acid catalyst, or catechol and ammonia. Phenothiazines are prepared similarly by heating diphenylamines with sulfur (Scheme 111).

These and other venerable synthetic routes to the tricyclic oxazines and thiazines have been reviewed in detail ⟨B-78MI22701⟩. 2-Hydroxy(or mercapto)-2′,4′-dinitrodiphenylamines readily cyclize to phenoxazines (or phenothiazines) in basic media through elimination of nitrous acid. This is the so-called Turpin reaction, which is complicated by the fact that the intermediate Meisenheimer complexes (251) may undergo a Smiles-type rearrangement so that mixtures of isomeric products are obtained (Scheme 112).

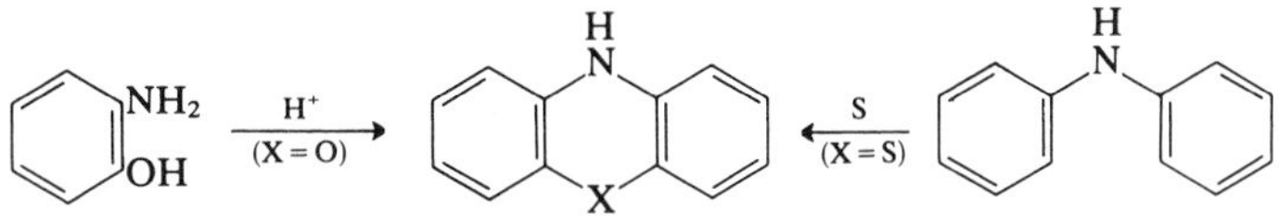

**Scheme 111**

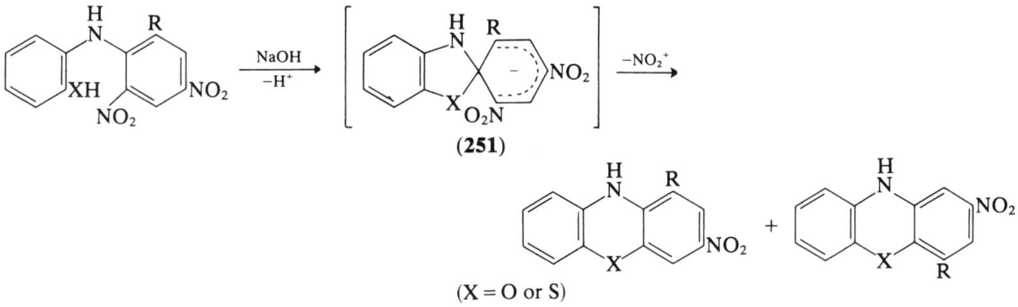

(251)

(X = O or S)

**Scheme 112**

Among the more recent approaches to phenoxazines and phenothiazines the reductive cyclization of 2-nitrodiphenyl ethers and sulfides with trialkyl phosphites is the most interesting. Here too a spiro intermediate is involved, produced by attack of an initially formed aryl nitrene on the second aromatic ring. The sulfide (252), for example, reacts with triethyl phosphite to yield 1-methylphenothiazine (253) and it is clear that in this case ring opening of the spiro intermediate also proceeds with a rearrangement of the Smiles type (Scheme 113) ⟨75JCS(P1)2396⟩.

Sulfoxides and sulfones of the monocyclic and polycyclic thiazines are not normally made by direct oxidation, but rather from substrates already at the correct oxidation level. 1-Oxides (254), for example, are prepared either by the addition of dimethylsulfoxonium methylide to cyanides ⟨65CB3724⟩ or to cyanamides (Scheme 114) ⟨76H(4)1875⟩.

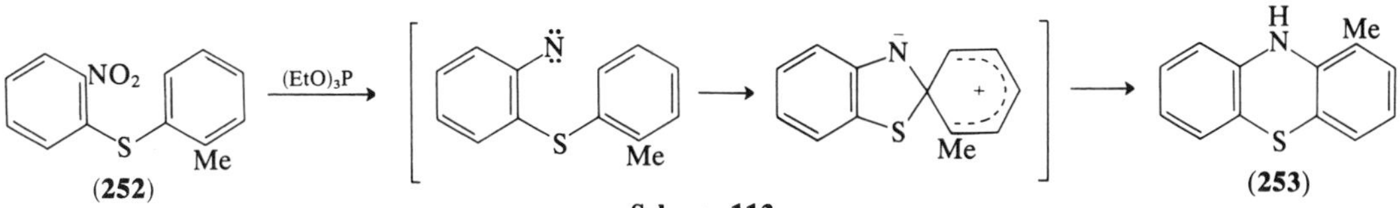

(252)        (253)

**Scheme 113**

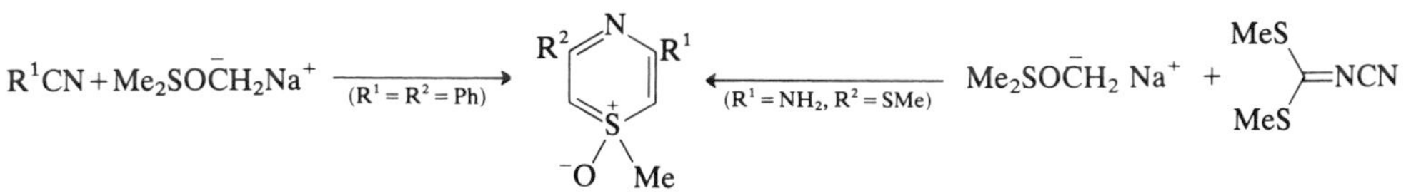

$R^1CN + Me_2S\overset{-}{O}CH_2Na^+$   $\xrightarrow{(R^1 = R^2 = Ph)}$   (254)   $\xleftarrow{(R^1 = NH_2, R^2 = SMe)}$   $Me_2S\overset{-}{O}CH_2\ Na^+ +$

(254)

**Scheme 114**

Similarly, 1,1-dioxides are formed either by the cyclodehydration of diacylsulfones and ammonia (Scheme 115) ⟨72OS(52)135⟩ or through the action of sodamide on bis(2-chlorovinyl) sulfones (**255**) which are themselves obtained by a two-step sequence from alkynes (Scheme 116) ⟨72S311⟩. A less general route involves the addition of the thiirene sulfone (**256**) to the mesoionic oxazolone (**257**). In this example the initial adduct (**258**) eliminates carbon dioxide spontaneously, thus affording the 1,4-thiazine (**259**) ⟨75CL1153⟩.

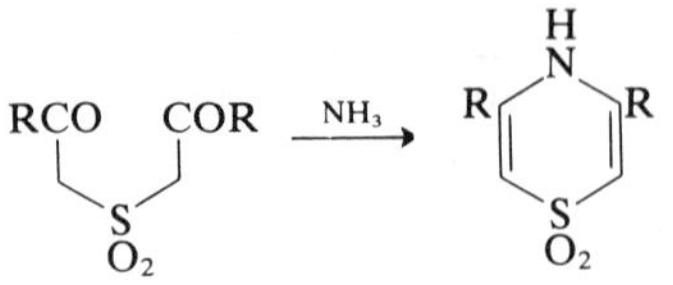


**Scheme 115**

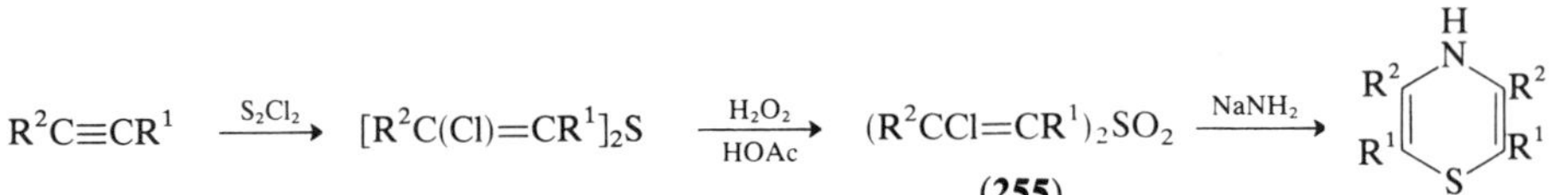


(**255**)

**Scheme 116**

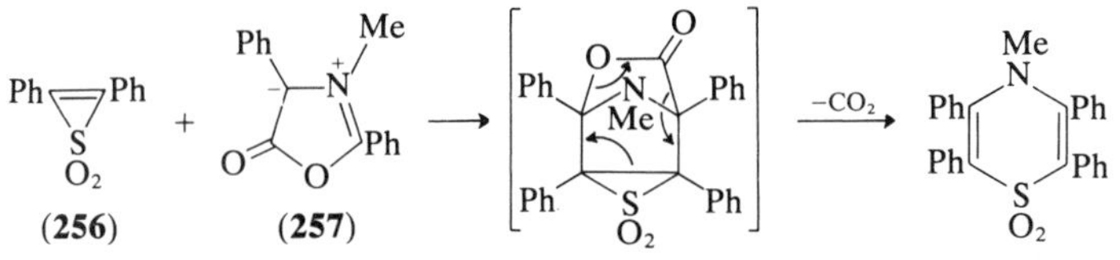


(**256**) (**257**) (**258**) (**259**)

4*H*-1,4-Benzothiazine 1,1-dioxide may be synthesized from the sulfone (**260**) by ozonolysis and hydrogenation of the ozonide over palladium on carbon (Scheme 117) ⟨68TL1041⟩. Phenothiazine 5,5-dioxides are produced when 2-nitrodiphenyl sulfones are reduced with triethylphosphite, this being simply an extension of a method described earlier (see Scheme 113).

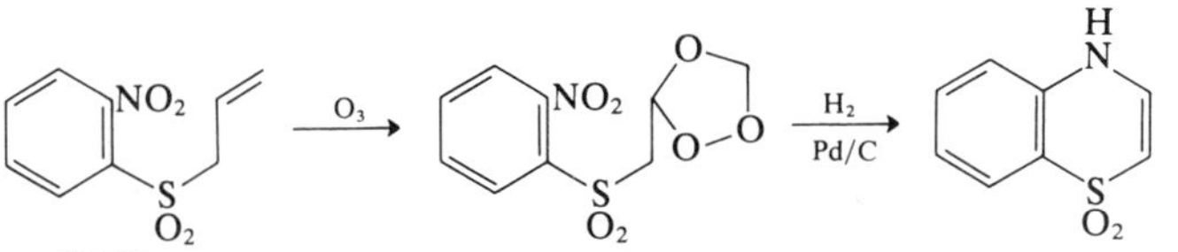


(**260**)

**Scheme 117**

Phenoxazin-3-ones and phenothiazin-3-ones can be prepared by the oxidation of the parent heterocycles in acidic media, but it is often more practical to employ condensation reactions between 2-amino-phenols or -thiols and quinones. Alizarin Green G (**263**), for example, is obtained from the aminophenol (**261**) and the 1,2-naphthoquinone (**262**). Similarly, 2-aminothiophenols (**264**) and 6-chloro-2-methoxy-1,4-benzoquinone (**265**) afford phenothiazin-3-ones (**266**) bearing methoxyl groups at position 1.

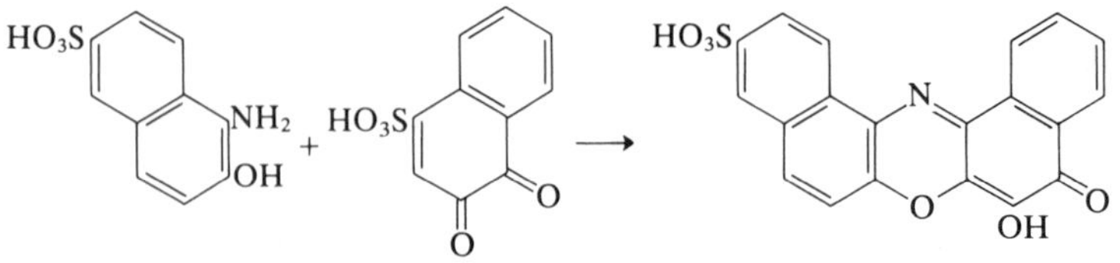

(**261**) (**262**) (**263**)

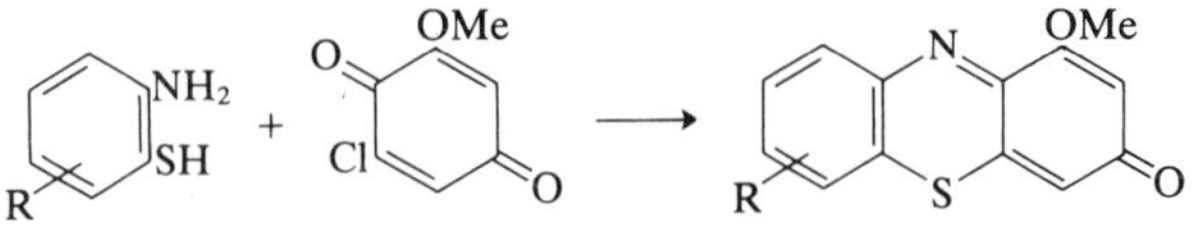

(**264**) (**265**) (**266**)

### 2.27.3.3.2 Dihydro-1,4-oxazines, -thiazines and their benzo derivatives

Dihydro-1,4-oxazines (**268**) are available through the treatment of acetals (**267**) with phosphorus pentoxide in pyridine ⟨79SC631⟩, while 2-ones (**270**) and 6-ones (**272**) are formed by the cyclodehydration of *N*-phenacylglycinates (**269**) with acid ⟨79CR(C)(288)221⟩ and the hydrobromides of *O*-alkylamino acids (**271**) with base ⟨77CB3615⟩.

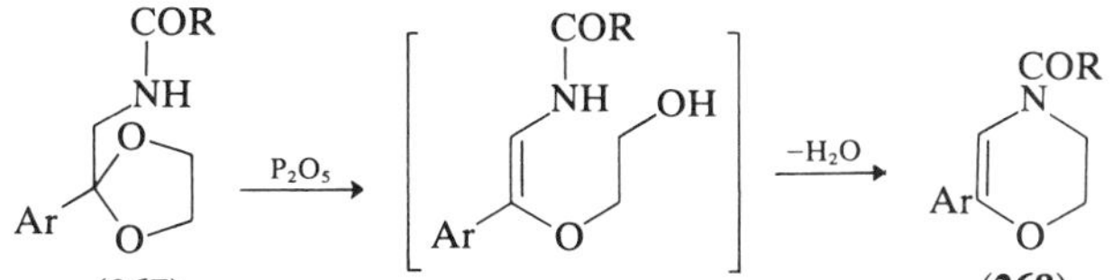

Routes to dihydro-1,4-thiazines are more diverse ⟨B-78MI22701⟩. For example, ring expansion of the thiazolidine (**273**) by heating with elemental sulfur and *n*-butylamine affords a 3:1 mixture of the isomers (**276**) and (**277**) and it appears likely that these products are derived from their tautomers (**274**) and (**275**) respectively which are the initial reaction products (Scheme 118) ⟨70LA(739)32⟩. Thiazolidines also accompany dihydrothiazines as products when aldehydes or ketones are reacted with aziridines in the presence of sulfur and DMF or potassium carbonate and it seems certain that a similar mechanistic sequence is involved ⟨79M425⟩.

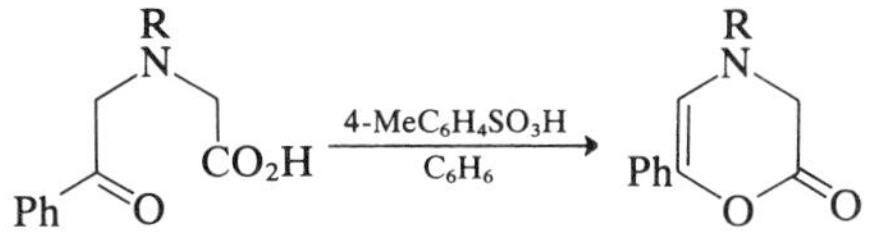

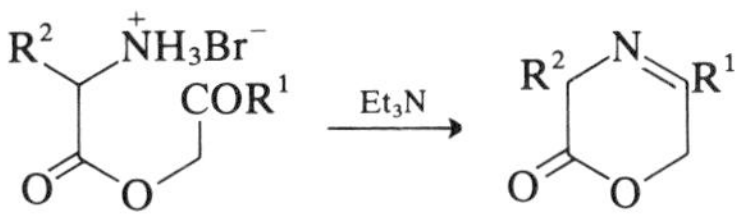

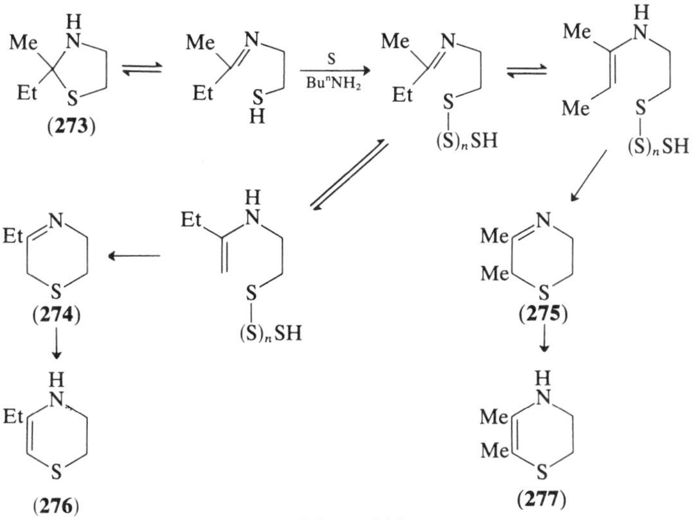

**Scheme 118**

What at first sight might appear to be simple reactions between 2-mercaptoethylamine and 2,3-dibromomaleic acid derivatives to give 3,4-dihydro-2*H*-1,4-thiazine-5,6-dicarboxylates are in reality much more complex, and again thiazolidine intermediates are implicated. Ring expansion of the intermediate takes place by displacement of halide ion by the electrons on sulfur to yield a bicyclic species (**278**) which deprotonates and rearranges to the corresponding dihydrothiazine (Scheme 119) ⟨74JCS(P1)2092⟩. Interestingly, an alternative approach to dihydrothiazines requires the ring opening and rearrangement of chloropenicillanate esters, clearly by an equivalent sequence (Scheme 120) ⟨68JCS(C)2533⟩.

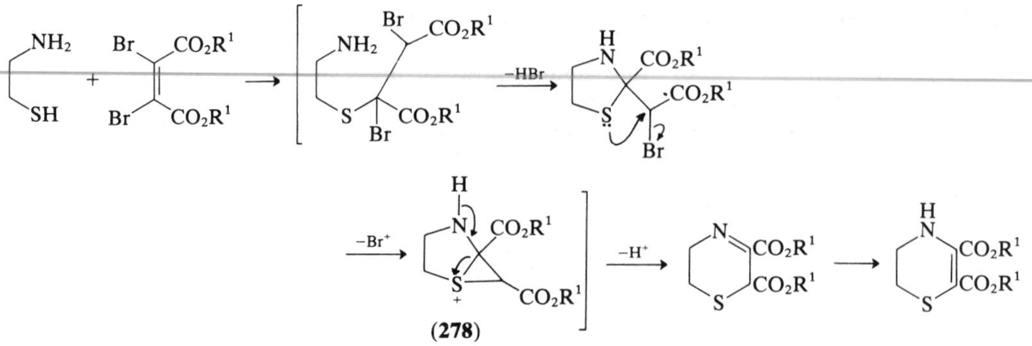

**Scheme 119**

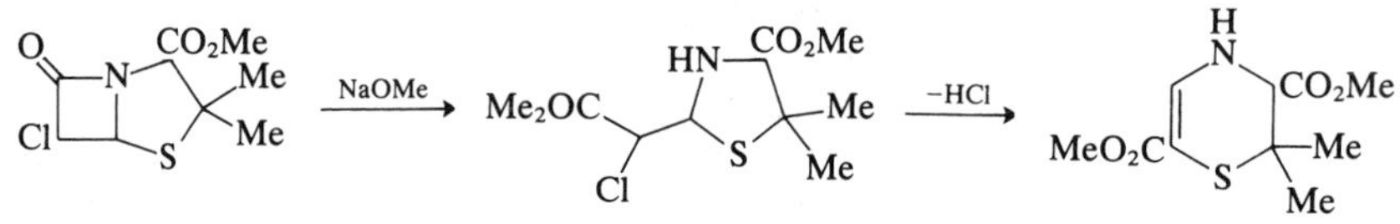

**Scheme 120**

Dihydro-1,4-thiazine 1-oxides or 1,1-dioxides can be made from the parent heterocycles by oxidation, or from alkynylcysteine 1-oxides or 1,1-dioxides by treatment with aqueous ammonia (Scheme 121) ⟨71JOC611⟩.

Various methods can be used to prepare dihydro-1,4-benzoxazines and -benzothiazines ⟨B-78MI22701⟩. A recent approach to the first group of compounds is to react 2-hydroxyacetanilides (279) with 1,2-dibromoethane and sodium hydroxide in acetonitrile containing the phase transfer catalyst 'Aliquat 336' (Scheme 122) ⟨79S541⟩. 1,1-Dioxides of dihydrobenzo-1,4-thiazines (281) are generated through the cyclization of imines (280) with methanesulphonyl chloride in the presence of triethylamine ⟨79CI(L)26⟩.

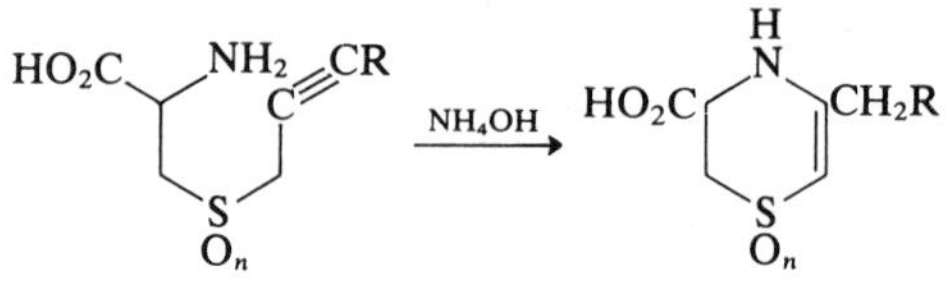

**Scheme 121**

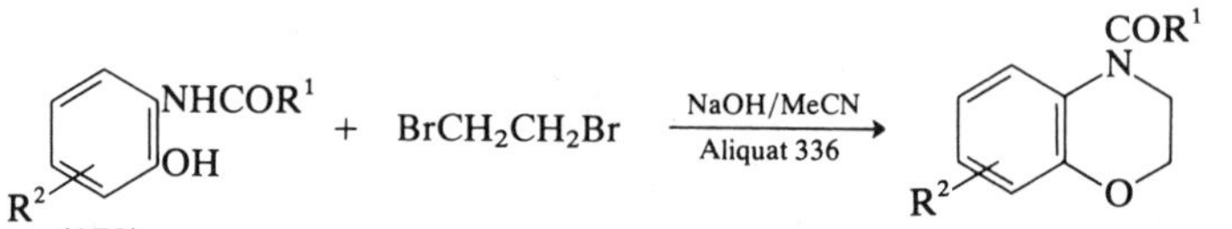

**Scheme 122**

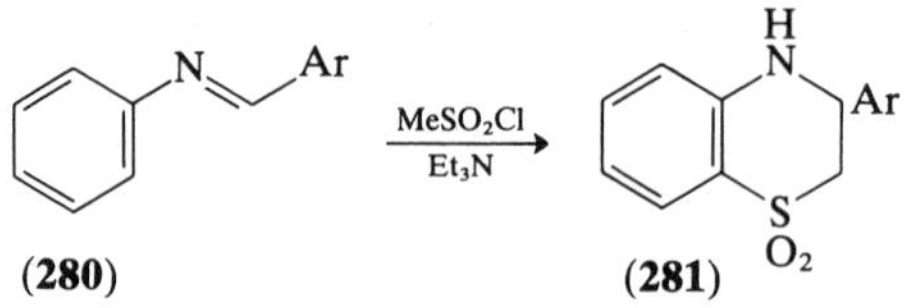

2,4-Dihydro-1,4-benzoxazin-3-ones and -benzothiazin-3-ones are synthesized by the reductive cyclodehydration of 2-nitrophenoxyacetic acids or their thioxy equivalents and as these heterocycles have an active methylene group it is a simple matter to prepare 2-substituted derivatives by condensation reactions with aldehydes and other carbonyl compounds (Scheme 123) ⟨79AP302⟩.

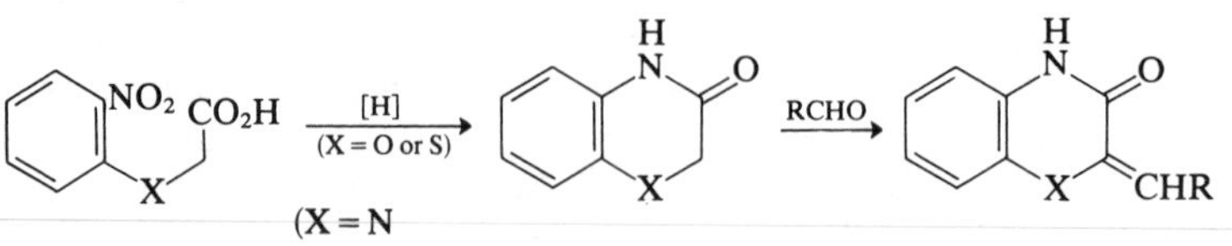

**Scheme 123**

### 2.27.3.3.3 *Tetrahydro-1,4-oxazines and -thiazines*

Tetrahydro-1,4-oxazines are normally prepared by the cyclization of diethanolamines in the presence of acids, although an alternative route employs the reaction of di($\beta$-chloroethyl) ethers with ammonia or amines (Scheme 124) ⟨12JCS1788⟩. Similar procedures can be adapted to the syntheses of tetrahydro-1,4-thiazines, their 1-oxides and 1,1-dioxides ⟨25JA282⟩.

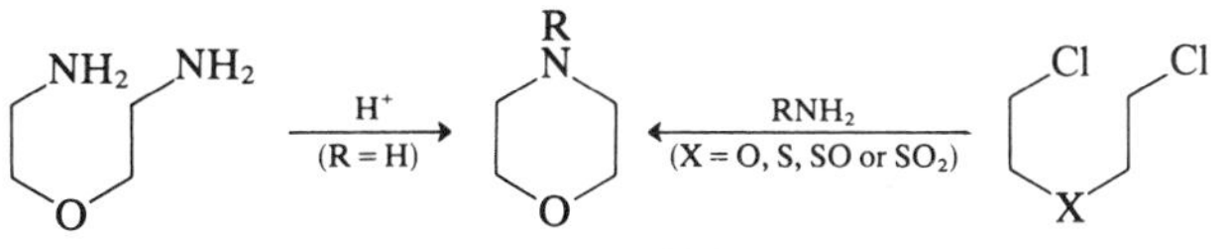

**Scheme 124**

2,3-Disubstituted tetrahydro-1,4-oxazines are also obtained from the reaction of the ethanolamines (**282**) and oxiranes in the presence of 70% sulfuric acid. The protective *N*-benzyl substituent can be easily removed by hydrogenolysis over palladium on carbon (Scheme 125) ⟨80T409⟩. 2-Oxo derivatives are available if ethanolamines are replaced by $\alpha$-amino acids, or if ethanolamines are reacted with $\alpha$-bromoacetates (Scheme 126) ⟨79CR(C)(288)229⟩.

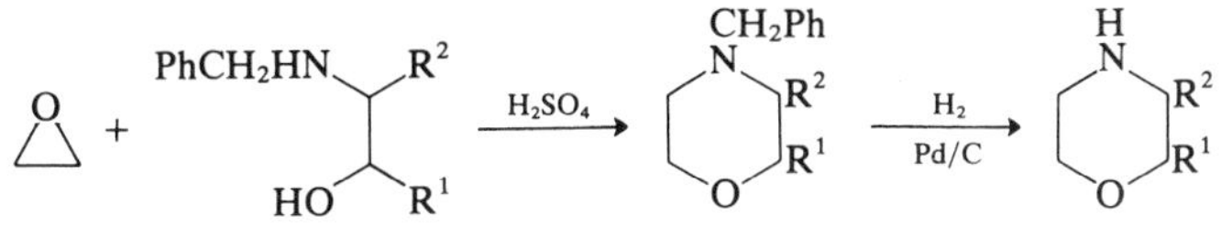

**Scheme 125**

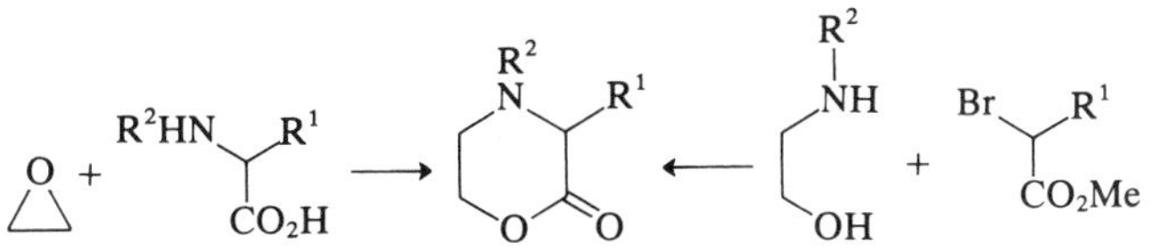

**Scheme 126**

## 2.27.4 IMPORTANT STRUCTURES AND APPLICATIONS

Interest in oxazines, thiazines and their benzo derivatives dates back well into the early part of the last century largely because some derivatives exhibit colour. Dyestuffs such as Meldola's blue (**283**) and Lauth's violet (**284**; R = H) are classic examples and since their discovery a host of related compounds has been made. Naturally occurring phenoxazines such as the ommochromes are pigments responsible for the colouration of certain arthropods. A representative of this group is rhodommatin (**285**) which is found in the wings of the small tortoiseshell butterfly (*Aglais urticae*) ⟨B-62MI22700⟩. The tricyclic systems are also of value in medicine and, for example, microorganisms of the genus *Streptomyces* metabolize a series of antibiotic substances known as actinomycins, which are derivatives of 2-amino-4,6-dimethyl-3-oxophenoxazine-1,9-dicarboxylic acid (actinocin) (**286**) with two pentapep-

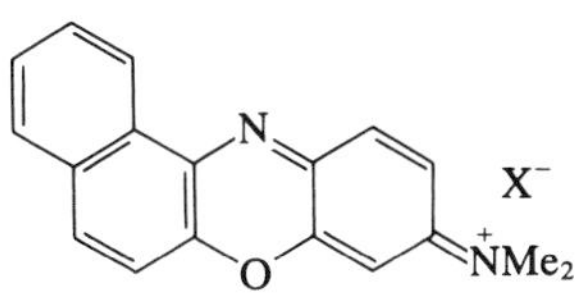

(**283**)

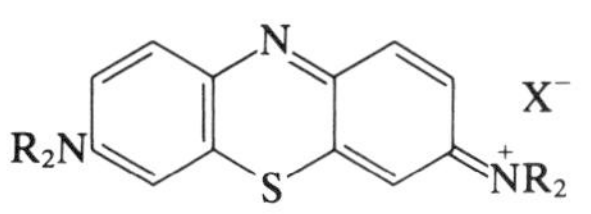

(**284**)

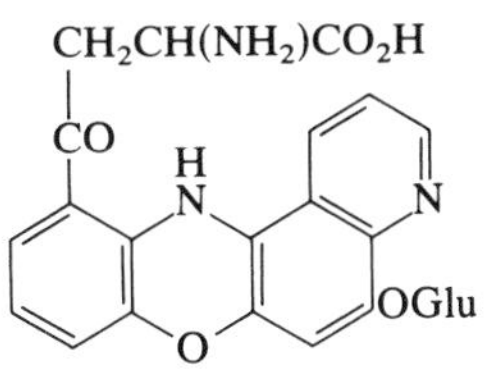

(**285**)

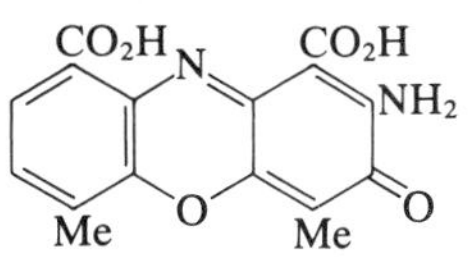

(**286**)

tide chains linked *via* the carboxylic acid groups ⟨B-61MI22701⟩. Unfortunately the actinomy-cins are toxic, but derivatives have now been made which may prove to be useful anticancer agents ⟨80MI22700, 81JOC1493⟩.

*N*-Substituted phenothiazines are antihistaminics and show powerful depressant effects on the central nervous system. Two well-known drugs of this type are promethazine (**287**) and chloropromazine (**288**). Phenothiazine itself is a veterinary anthelmintic agent as well as an insecticide, and methylene blue (**284**; R = Me) has long been used as a biological stain.

It is not surprising, therefore, that the whole range of oxazines and thiazine derivatives should be subjected to close scrutiny for biological action and there are numerous claims in the patent literature for the activity of various structural types as sedatives, tranquilizers, antiepileptic, antitubercular, antitumour, bacteriocidal and parasiticidal agents ⟨54AG363, 67AHC(8)83, 68AHC(9)321, B-70MI22700, 77H(7)391, 78AHC(23)1, 79CHE291, 79MI22700⟩. In recent times the cephalosporins have proven to be antibiotics of great value. These mould meta-bolites, which contain a fused β-lactam–dihydro-1,3-thiazine skeleton (**289**; X = S), are obviously related to the penicillins, and their discovery has initiated a major study which is still in progress to optimize their therapeutic effects. Many synthetic or semisynthetic analogues of the natural products are now known, some of which no longer contain a sulfur atom. The oxacephems (**289**; X = O), for example, are based upon an oxazine unit ⟨B-80MI22701, B-80MI22702⟩. The 1,4-benzothiazine nucleus is present in the trichosiderin pig-ments, *e.g.* (**290**), found in mammalian red hair and the feathers of domestic New Hampshire chickens, as well as in the urine of human patients suffering malignant melanoma metastases ⟨76E1122⟩.

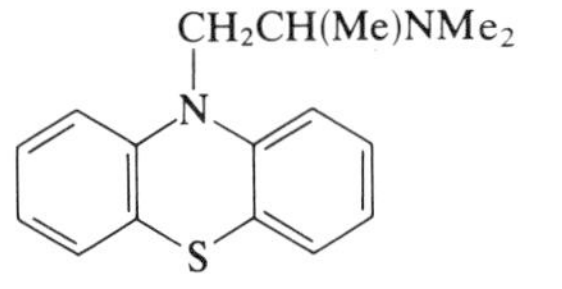

(**287**) 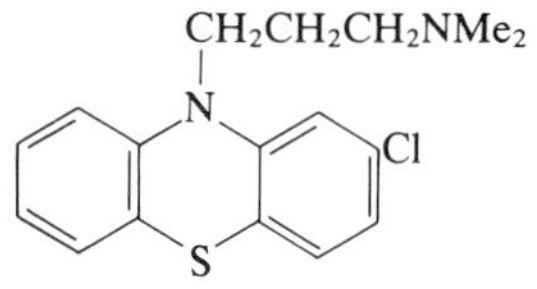   (**288**)

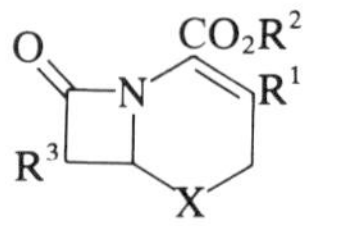

(**289**) 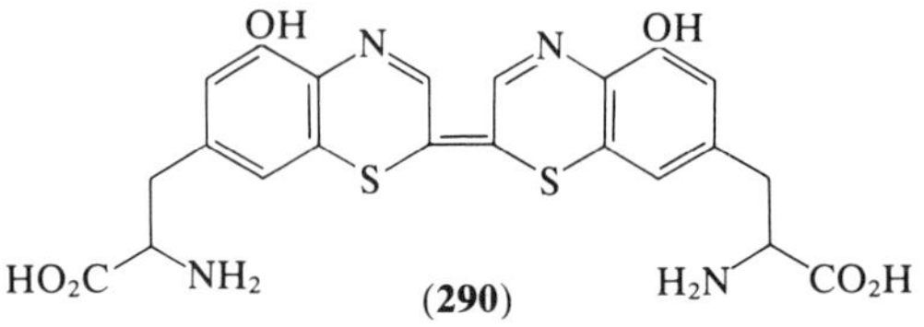   (**290**)

Monocyclic oxazines and thiazines also function as bases and as very useful synthetic intermediates, particularly for the construction of carbonyl derivatives and as starting points for many complex heterocyclic systems. These properties are illustrated in earlier parts of this chapter. Finally, at a time when man has become acutely aware of the diminution of fossil fuel reserves, phenothiazines have been shown to be potentially valuable in solar energy converters ⟨79MI22701⟩.

# 2.28

# Polyoxa, Polythia and Polyaza Six-membered Ring Systems

C. J. MOODY

*Imperial College of Science and Technology*

| | | |
|---|---|---|
| 2.28.1 | INTRODUCTION | 1040 |
| 2.28.2 | STRUCTURE | 1040 |
| | *2.28.2.1 Survey of Known Systems* | 1040 |
| | 2.28.2.1.1 Systems with three heteroatoms | 1040 |
| | 2.28.2.1.2 Systems with four heteroatoms | 1044 |
| | 2.28.2.1.3 Systems with five or six heteroatoms | 1047 |
| | *2.28.2.2 Molecular Dimensions* | 1048 |
| | *2.28.2.3 Molecular Spectra* | 1049 |
| | 2.28.2.3.1 NMR spectra | 1049 |
| | 2.28.2.3.2 UV spectra | 1050 |
| | 2.28.2.3.3 IR spectra | 1050 |
| | 2.28.2.3.4 Mass spectra | 1051 |
| | 2.28.2.3.5 Photoelectron spectra | 1052 |
| | *2.28.2.4 Thermodynamic Aspects* | 1052 |
| | *2.28.2.5 Conformation* | 1053 |
| | 2.28.2.5.1 Tetrahydro-1,2,3-oxathiazine 2-oxides | 1053 |
| | 2.28.2.5.2 Dihydro-1,3,5-dioxazines | 1053 |
| | 2.28.2.5.3 Tetrahydro-1,4,2-dioxazines | 1053 |
| | 2.28.2.5.4 Dihydro-1,3,5-dithiazines | 1054 |
| | 2.28.2.5.5 Tetrahydro-1,2,4-oxadiazines | 1054 |
| | 2.28.2.5.6 Tetrahydro-1,2,5-oxadiazines | 1054 |
| | 2.28.2.5.7 Tetrahydro-1,3,4-oxadiazines | 1054 |
| | 2.28.2.5.8 Tetrahydro-1,3,5-oxadiazines | 1054 |
| | 2.28.2.5.9 Tetrahydro-1,3,4-thiadiazines | 1054 |
| | 2.28.2.5.10 Tetrahydro-1,3,5-thiadiazines | 1055 |
| | 2.28.2.5.11 Tetrahydro-1,2,4,5-dioxadiazines | 1055 |
| | *2.28.2.6 Tautomerism* | 1055 |
| | 2.28.2.6.1 Prototropic tautomerism | 1055 |
| | 2.28.2.6.2 Ring–chain tautomerism | 1056 |
| | *2.28.2.7 Betaines and Other Unusual Structures* | 1057 |
| 2.28.3 | REACTIVITY | 1057 |
| | *2.28.3.1 Reactivity at Nitrogen* | 1057 |
| | 2.28.3.1.1 Acid–base properties | 1057 |
| | 2.28.3.1.2 Reaction with electrophiles | 1058 |
| | 2.28.3.1.3 Reaction with nucleophiles | 1059 |
| | *2.28.3.2 Reactivity at Sulfur* | 1059 |
| | 2.28.3.2.1 Reaction with electrophiles | 1059 |
| | 2.28.3.2.2 Reaction with nucleophiles | 1060 |
| | *2.28.3.3 Reactivity at Carbon* | 1060 |
| | 2.28.3.3.1 Reaction with electrophiles | 1060 |
| | 2.28.3.3.2 Reaction with bases | 1060 |
| | 2.28.3.3.3 Reaction with nucleophiles | 1061 |
| | *2.28.3.4 Cycloaddition Reactions* | 1061 |
| | *2.28.3.5 Ring Cleavage and Ring Contraction* | 1062 |
| | 2.28.3.5.1 Thermal reactions | 1063 |
| | 2.28.3.5.2 Photochemical reactions | 1064 |
| | 2.28.3.5.3 Ring contraction reactions | 1065 |
| | 2.28.3.5.4 Desulfurization and reductive ring cleavage | 1066 |
| | *2.28.3.6 Reactivity of Substituents* | 1067 |
| | 2.28.3.6.1 Fused benzene rings | 1067 |
| | 2.28.3.6.2 C-linked substituents | 1067 |

|  |  |
|---|---|
| 2.28.3.6.3   *N-linked substituents* | 1068 |
| 2.28.3.6.4   *O- and S-linked substituents* | 1069 |
| 2.28.3.6.5   *Halogen substituents* | 1069 |
| 2.28.4   SYNTHESES | 1069 |
| 2.28.4.1   *Formation of One Bond* | 1070 |
| 2.28.4.1.1   *Between two heteroatoms* | 1070 |
| 2.28.4.1.2   *Between carbon and nitrogen* | 1071 |
| 2.28.4.1.3   *Between carbon and oxygen* | 1072 |
| 2.28.4.1.4   *Between carbon and sulfur* | 1072 |
| 2.28.4.2   *Formation of Two Bonds* | 1073 |
| 2.28.4.2.1   *From [5 + 1] atom fragments* | 1073 |
| 2.28.4.2.2   *From [4 + 2] atom fragments* | 1075 |
| 2.28.4.2.3   *From [3 + 3] atom fragments* | 1078 |
| 2.28.4.3   *Formation of Three or More Bonds* | 1080 |
| 2.28.4.3.1   *From [2 + 2 + 2] atom fragments* | 1080 |
| 2.28.4.3.2   *Other reactions* | 1081 |
| 2.28.4.4   *From Other Heterocycles* | 1082 |
| 2.28.4.5   *Summary* | 1083 |
| 2.28.5   APPLICATIONS | 1084 |

## 2.28.1 INTRODUCTION

This chapter reviews those systems with three or more heteroatoms, the heteroatoms being nitrogen, oxygen and sulfur, although only systems with at least one nitrogen atom will be considered. Even within this limitation a considerable number of systems is possible, and indeed known, and many of these systems have isomers in which the heteroatoms are in different positions and hence exhibit an entirely different chemistry. Unfortunately many of the less well-known ring systems were reported before the advent of spectroscopy, and although some of these ring structures are included in earlier reviews, there must be grave doubts about the correctness of some of the structures.

The total number of different ring systems falling within the above definition is in excess of 50, so clearly a comprehensive treatment of each system is impossible within the confines of this chapter. Therefore the chapter will contain a comparative assessment of the various systems and will highlight the trends and differences where they have been established. It will necessarily reflect the fact that by far the most common and well described ring systems are the oxadiazines and thiadiazines; references to these outnumber those to all the other systems combined. The tables in Section 2.28.2.1 provide a quick reference guide to the various systems which are known.

## 2.28.2 STRUCTURE

### 2.28.2.1 Survey of Known Systems

The main purpose of this section is to provide a brief survey of the ring systems that have been reported in the literature. The bulk of the information is contained in the briefly annotated tables. Wherever possible, reference is made to the relevant reviews, or to a recent leading article.

#### 2.28.2.1.1 Systems with three heteroatoms

Two thirds of the 34 possible systems containing at least one nitrogen atom and oxygen and/or sulfur are known and well described. Of these, six of the possible oxathiazines (Table 1) have been reported. A general feature of this and other sulfur-containing systems is that they are often only known with the sulfur atom in its higher oxidation states. Thus, for example, 1,2,5-oxathiazines are only known as the 2,2-dioxides. Of the six possible dioxazines (Table 2) only the 1,2,3-isomer is unknown. All six of the possible dithiazines

**Table 1** Oxathiazines

| Ring system | Comments | Ref. |
|---|---|---|
| | General reviews | B-79MI22800 |
|  1,2,3- | Common system, particularly the 2,2-dioxide and its benzo derivatives | B-79MI22800, 73AG(E)869, 76JHC665, 65JOC3960 |
| 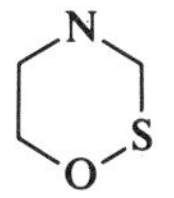 1,2,4- | Few examples | 77TL4245 |
| 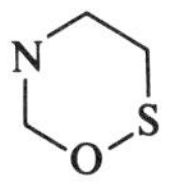 1,2,5- | Few examples of 2,2-dioxide | 69IZV2059, 66USP3235549 |
| 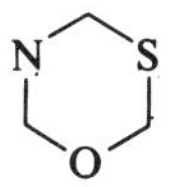 1,3,5- | Common system | 77CB2114 |
| 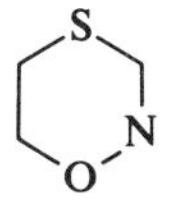 1,4,2- | Few examples | 72TL5267 |
| 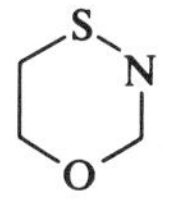 1,4,3- | Common system, usually 4-oxide or 4,4-dioxide | 68CB3567, 72JA6135, 78BCJ1805 |

**Table 2** Dioxazines

| Ring system | Comments | Ref. |
|---|---|---|
| | General reviews | 62HC(17)443, B-79MI22800 |
| 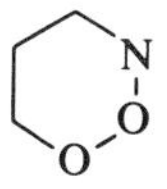 1,2,3- | Unknown, although MO calculations reported | 72MI22801 |
| 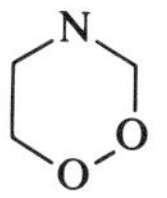 1,2,4- | Monocyclic system unknown, although MO calculations reported. Few examples of fused derivatives | 72MI22801, 64JOC291, 76CC608 |
| 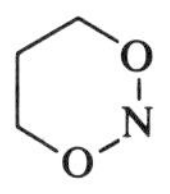 1,3,2- | One example of tetrahydro derivative | 80IZV2669 |
| 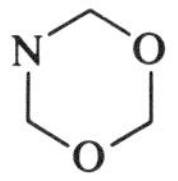 1,3,5- | Few examples | B-79MI22800, 66JOC2568, 75JCS(P1)772 |

**Table 2** (*continued*)

| Ring system | Comments | Ref. |
|---|---|---|
| 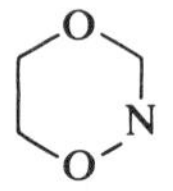 1,4,2- | Common system, especially 5,6-dihydro-5-one derivatives | B-79MI22800, 67JCS(C)1178 |
| 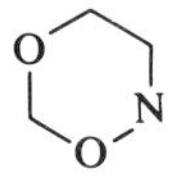 1,5,2- | Few examples, especially tetrahydro and 3,6-dione derivatives | 80LA686, 81S38 |

**Table 3** Dithiazines

| Ring system | Comments | Ref. |
|---|---|---|
| | General reviews | B-79MI22800 |
|  1,2,3- | Rare system — 6-thione known | 76CJC3879 |
| 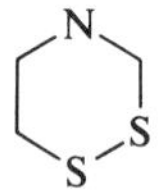 1,2,4- | Few examples of 5,6-dihydro derivative | 63JCS2097 |
| 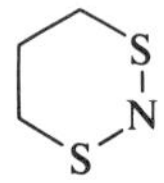 1,3,2- | Rare system — 1,1,3,3-tetraoxide only | 75CB1087 |
|  1,3,5- | Common system | B-79MI22800, 77S193, 78S443 |
| 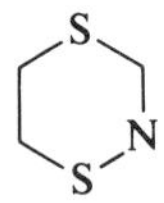 1,4,2- | Common system, especially 1,1-dioxide | B-79MI22800, 72BCJ525, 76JPR127 |
| 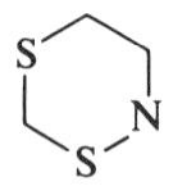 1,5,2- | Rare system — 1,1,5,5-tetraoxide only | 71NKK448 |

(Table 3) are described in the literature although the rarer 1,3,2- and 1,5,2-isomers are only known as the fully oxidized tetraoxides. Oxadiazines (Table 4) of all six types are reported, four of the isomers having an extensive and well established chemistry. The 1,2,3- and 1,2,6-isomers have been reported, but there is no spectroscopic evidence to support either structure. All the six isomeric thiadiazines (Table 5) are known with only the 1,5,2-isomer being poorly described. In line with the oxadiazines and thiadiazines being the most well known ring systems containing three heteroatoms, they are also the most well reviewed to date.

**Table 4** Oxadiazines

| Ring system | Comments | Ref. |
|---|---|---|
| | General reviews | B-61MI22800, 62HC(17)443, B-79MI22800 |
| 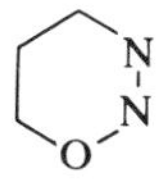 1,2,3- | Cyclopenta- and benzo-fused systems have been claimed. No evidence to support either structure | 1884CB603, 60DOK(133)851 |
| 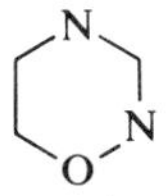 1,2,4- | Extremely common system | B-61MI22800, 62HC(17)443, B-79MI22800 |
| 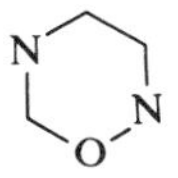 1,2,5- | Extremely common system | B-61MI22800, 62HC(17)443, B-79MI22800 |
| 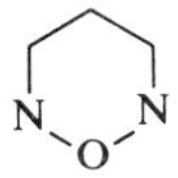 1,2,6- | Structure claimed, but no evidence to support it and must be regarded with suspicion | 1893CB997, 35G176, 52JCS3428 |
| 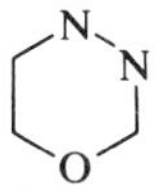 1,3,4- | Extremely common system | B-61MI22800, 62HC(17)443, B-79MI22800 |
| 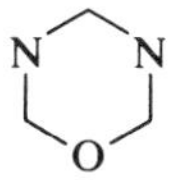 1,3,5- | Extremely common system | B-61MI22800, 62HC(17)443, B-79MI22800 |

**Table 5** Thiadiazines

| Ring system | Comments | Ref. |
|---|---|---|
| | General reviews | B-61MI22800, 70CRV593, 70SST473, B-79MI22800 |
|  1,2,3- | Few examples, particularly 1-oxide, 1,1-dioxide and benzo derivatives | 81JCS(P1)2322, 62JPR(19)56, 68JHC453 |
| 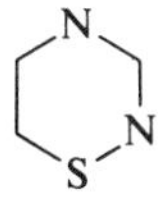 1,2,4- | Extremely common system, especially 1,1-dioxide and its benzo derivative. Systems with S(IV) known | B-61MI22800, 70CRV593, 70SST(1)473, B-79MI22800, 62AG(E)235, 75CSR189, 80CSR477 |
|  1,2,5- | Rare system | 49JOC946 |
|  1,2,6- | Extremely common system, especially 1,1-dioxide | 58HOU(11)725, 73S243 |

**Table 5** (*continued*)

| Ring system | Comments | Ref. |
|---|---|---|
| 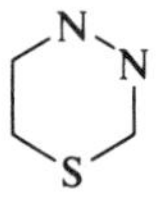 1,3,4- | Extremely common system, including S(IV) derivatives | 69ZC361, 70MI22801, 81JCS(P1)2245 |
|  1,3,5- | Extremely common system, especially 2-thione | B-61MI22800, 70CRV593, 70SST(1)473, 69AF1807 |

### 2.28.2.1.2 Systems with four heteroatoms

Six-membered ring systems with four heteroatoms are much less well known; only 18 out of more than 80 possible systems are described, and their chemistry remains essentially unexplored. Thus only one of each of the several possible oxadithiazines and dioxathiazines (Table 6) have been reported. Two oxathiadiazine ring systems (Table 6) are known, with the symmetrical 1,4,3,5-isomer being fairly well described. Isolated examples of

**Table 6** Oxadithiazines, Dioxathiazines and Oxathiadiazines

| Ring system | Comments | Ref. |
|---|---|---|
| 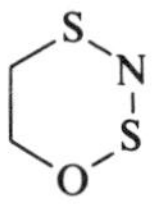 1,2,4,3- | One example of 2,2,4,4-tetraoxide | 68FRP1508645 |
|  1,3,2,5- | Rare system; few examples of S(IV) derivatives known | 75IZV1206, 80ZOR463 |
|  1,2,3,5- | Few examples of 2,2-dioxide | 75ZOR2217, 79JOC4435, B-79MI22800 |
| 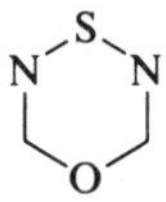 1,4,3,5- | Fairly common system; few examples of 4,4-dioxide | B-79MI22800, 73JOC1249, 68AG(E)172, 76CRV389 |

dioxadiazines (Table 7) are reported. Many of the isomers are unknown and evidence for some of the others is poor. The 1,2,3,6-isomer has been frequently discussed in the literature in relation to the heterocyclic products derived from 1,2-dioximes. However, the structures were eventually reassigned as furoxans. The background to this structural problem has been discussed ⟨81AHC(29)251⟩. Dithiadiazines (Table 8) are slightly better known than their oxygen analogues, and examples of seven of the twelve possible isomers are described. Trioxazines and trithiazines are unknown, and only one of the six possible oxatriazines (Table 9) is known, two previous structures having been revised. Three of the possible thiatriazines (Table 9) are known, and the symmetrical 1,2,4,6-isomer is well described, particularly when the sulfur is in a higher oxidation state.

**Table 7** Dioxadiazines

| Ring system | Comments | Ref. |
|---|---|---|
| | General reviews | 62HC(17)443 |
| 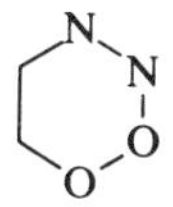 1,2,3,4- | The 6-one derivative claimed, but no evidence to support it. Otherwise unknown although MO calculations reported | 60G1165, 72MI22801 |
| 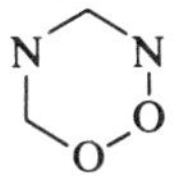 1,2,3,5- | Unknown, although MO calculations reported | 72MI22801 |
| 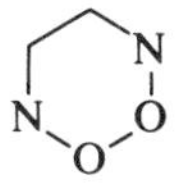 1,2,3,6- | Frequently reported in older literature. Structures reassigned as furoxans | 81AHC(29)251 |
| 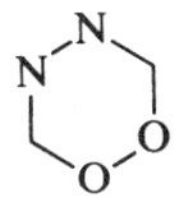 1,2,4,5- | Rare system, but tetrahydro derivative known | 60LA(635)73 |
| 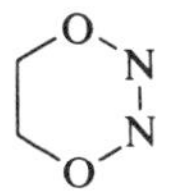 1,4,2,3- | Unknown, although MO calculations reported | 72MI22801 |
| 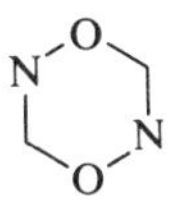 1,4,2,5- | 3,6-Diaryl derivatives common | 74JCS(P1)1951 |
| 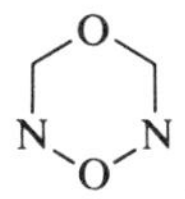 1,4,2,6- | Unknown, although MO calculations reported | 72MI22801 |
| 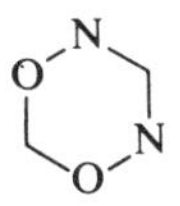 1,5,2,4- | Unknown, although MO calculations reported | 72MI22801 |

**Table 8** Dithiadiazines

| Ring system | Comments | Ref. |
|---|---|---|
| | General reviews | 79MI22800 |
|  1,2,3,6- | Rare system; one example of tetrahydro derivative claimed | 75JAP(K)75101322 |
| 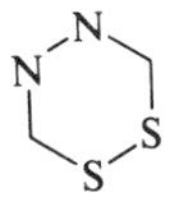 1,2,4,5- | Rare system | 76JCS(P1)38 |

**Table 8** (*continued*)

| Ring system | Comments | Ref. |
|---|---|---|
|  1,3,2,4- | One example of monocyclic system claimed, but structure wrong; reassigned as the 1,4,2,6-isomer. Benzo derivative of the 1,1-dioxide-3-oxide known | 76H(4)1243, 71LA(749)171 |
| 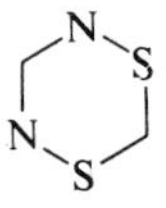 1,3,4,6- | Rare system; 1,3-dioxide known | 68CB3567 |
| 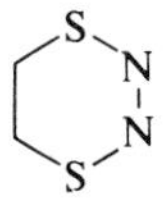 1,4,2,3- | Rare system; tetrahydro and its benzo derivative known | 71AG(E)407, 74CB771 |
| 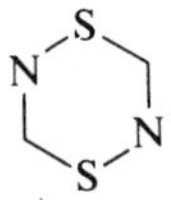 1,4,2,5- | Few examples, but rare | 67JCS(C)2562 |
| 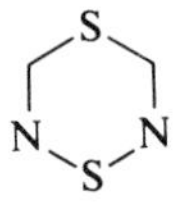 1,4,2,6- | Rare system, although 1,1-dioxide known | 82UP22800, 77ZN(B)1390 |

**Table 9**  Oxatriazines and Thiatriazines

| Ring system | Comments | Ref. |
|---|---|---|
| | General reviews | 62HC(17)443, B-79MI22800 |
| 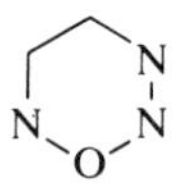 1,2,3,6- | One example claimed but incorrect and reassigned as an isoxazolin-5-one 4-oxime | 03JCS1217, 69JHC317 |
| 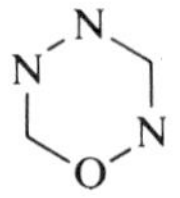 1,2,4,5- | One example of monocyclic system claimed, but reassigned as a 1,2,4-triazole *N*-oxide. Fused derivatives known | 72CR(C)(274)189, 74JOC3192 |
| 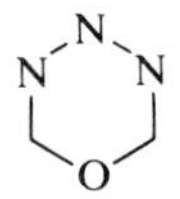 1,3,4,5- | One example; structure proved by X-ray | 80AJC2447 |
| 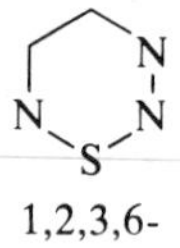 1,2,3,6- | Few examples, including 1-oxide | 70MI22803, 77S305 |

**Table 9** (*continued*)

| Ring system | Comments | Ref. |
|---|---|---|
| 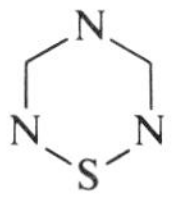 1,2,4,6- | Extremely common system; 1,1-dioxide and S(IV) and S(VI) derivatives known | 69AG(E)510, 62CB147, 78MI22803 |
| 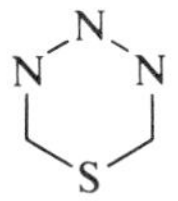 1,3,4,5- | Few examples of 1,1-dioxide known | 80JOC2604 |

### 2.28.2.1.3 Systems with five or six heteroatoms

As the number of heteroatoms increases then so does the number of possible systems, although the number of known systems actually decreases. Of the 14 possible systems with five heteroatoms (*not* allowing for ring isomers), examples of only four are reported (Table 10). Tetroxazines and dioxatriazines are unknown although MO calculations on them have been reported ⟨72MI22801⟩.

**Table 10** Six-Membered Rings with Five Heteroatoms

| Ring system | Comments | Ref. |
|---|---|---|
| 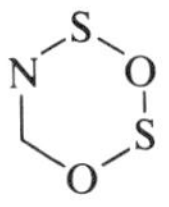 1,3,2,4,5-dioxadithiazine | 2,2,4,4-Tetraoxide quite well known. Di-S(IV) derivatives also described | 75ZOR2217, 78JA985 |
| 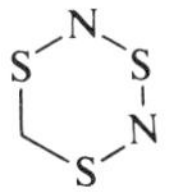 1,3,5,2,4-trithiadiazine | Only one example | 78MI22802 |
| 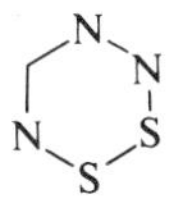 1,2,3,4,6-dithiatriazine | One example of 5-thione | 78ZC336 |
|  1,3,2,4,6-dithiatriazine | Few examples, particularly of *S*-oxides and *S,S*-dioxides | 64MI22800, 71AG(E)264, 74ZN(B)799, 74CB1, 74ZOR488 |

Systems with six heteroatoms in the ring fall into the area between organic and inorganic chemistry and although most are usually described in the inorganic literature they are included here for completeness. Only four of the 20 possible systems (*not* allowing for ring isomers) are known (Table 11), and of these, references to 1,3,5,2,4,6-trithiatriazines outnumber those to all the other systems in Tables 10 and 11 combined. Their chemistry is well reviewed, but remains essentially the province of inorganic chemists.

**Table 11**  Six-Membered Rings with Six Heteroatoms

| Ring system | Comments | Ref. |
|---|---|---|
| 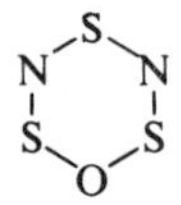 1,2,4,6,3,5-oxatrithiadiazine | Rare system, 2,2,4,4,6,6-hexaoxide only | 56CB179 |
| 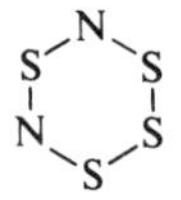 1,2,3,5,4,6-tetrathiadiazine | Extremely rare system | 72MI22800 |
| 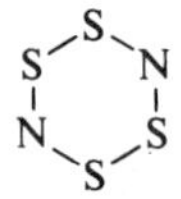 1,2,4,5,3,6-tetrathiadiazine | Few examples in literature, although some reassigned acyclic structures | 31G294, 59CB1149, 69CC1424 |
|  1,3,5,2,4,6-trithiatriazine | Extremely common system, extensively covered and reviewed in inorganic literature | B-70MI22802, 71MI22800, B-75MI22801, 74CB1 |
| 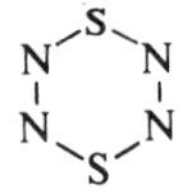 1,4,2,3,5,6-dithiatetrazine | Rare, but tetrahydro derivative known | 74JHC99, 79JHC751 |

### 2.28.2.2 Molecular Dimensions

In recent years, many six-membered rings containing three or more heteroatoms have been studied by X-ray crystallography. No systematic study of any of the individual systems falling within this chapter has been made and since X-ray crystallography has essentially only been used as a method of structure determination/confirmation, the results are unexceptional. The polyheteroatom six-membered rings are not aromatic (see Section 2.28.2.4) and hence there is no reason to expect them to be planar; the published X-ray results demonstrate this non-planarity. Some important points emerge, particularly on compounds containing S—N bonds, and a selection of the results is presented here.

The 1,2,3-oxathiazine (**1**) is non-planar ⟨78AX(B)2376⟩, as is compound (**2**) ⟨71TL1211⟩. The ylidic sulfur–nitrogen bond length in (**2**) is 1.48 Å and the sulfur atom is pyramidal. The 1,2,4-oxathiazine ring in (**3**) occupies a half-chair conformation in the crystal ⟨77TL4245⟩. However, the most interesting feature is the short distance (2.255 Å) between the sulfur and the exocyclic oxygen atoms. The fact that this is much less than the sum of the van der Waals radii (3.25 Å) suggests some form of interaction, the extreme form of which is (**4**). This suggestion is partly borne out by the lack of a carbonyl absorption in the 1600–1800 cm$^{-1}$ range of the IR spectrum.

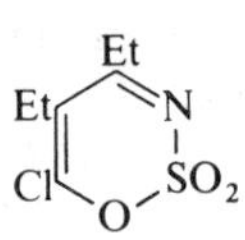

(1) 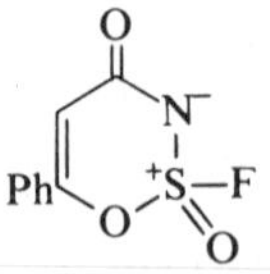　　　　　　(2) 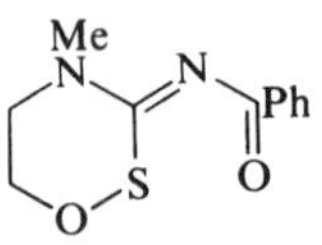　　　　　　(3) 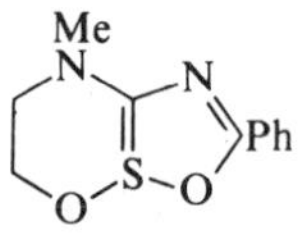　　　　　　(4)

The reduced 1,3,2-dithiazine tetraoxide (**5**) is non-planar, and the bond angles at sulfur (104°) and at nitrogen (118°), and the sulfur–nitrogen bond length (1.702 Å) are normal ⟨75CB1087⟩. The oxadiazine (**6**), 6-oxadihydrouracil, is also non-planar as expected ⟨75JHC699⟩. The structure of (**7**), an oxidation product of dithizone, was confirmed by X-ray crystallography ⟨69CC392⟩.

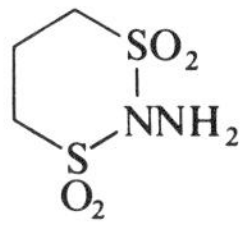

(**5**) 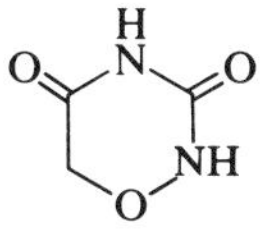          (**6**) 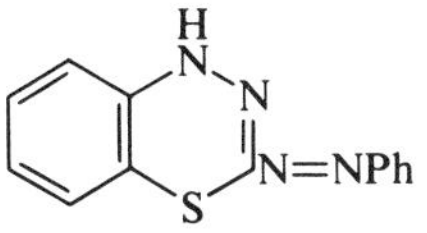          (**7**) 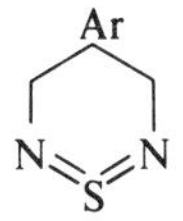          (**8**)

Three 1,2,6-thiadiazines have been investigated by X-ray diffraction. From a theoretical point of view, the most important result concerns the sulfur(IV) derivative (**8**; Ar = 4-ClC$_6$H$_4$). The analysis shows that the nitrogen atoms are $sp^2$, the nitrogen–sulfur bond length is 1.534 Å and the angle at sulfur is 118.5°. The ring adopts a half-boat conformation with the aryl group equatorial ⟨76AG(E)782⟩. The 1,2,6-thiadiazine 1,1-dioxides (**9**) ⟨77AX(B)910⟩ and (**10**) ⟨79JOC4191⟩ show unexceptional features, being non-planar with the sulfur atom out of the plane of the remaining ring atoms.

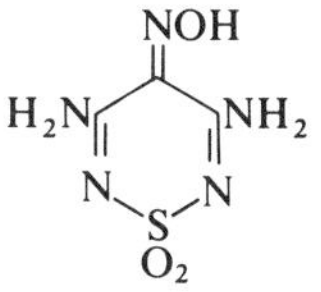

(**9**) 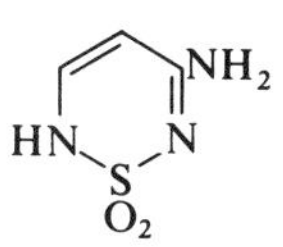          (**10**) 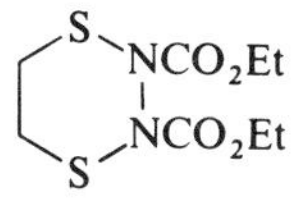          (**11**) 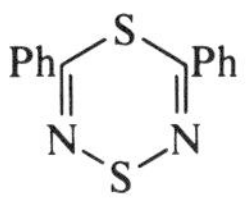          (**12**)

The tetrahydro-1,4,2,3-dithiadiazine (**11**) adopts a chair conformation in the crystal ⟨77MI22800⟩, but the 1,4,2,6-isomer (**12**) is boat shaped with sulfur atoms at the bow and stern ⟨82UP22800⟩. The 1,3,4,5-oxatriazine (**13**) is almost planar ⟨80AJC2447⟩, and the related sulfur compound (**14**) is slightly boat shaped with the bond distances normal ⟨80AX(B)2159⟩. The X-ray structure of (**15**) has been reported ⟨80JOC1662⟩ and the structures of (**16**; R = Cl, SPh) have been determined. The ring in (**16**) is non-planar, and the sulfur atom is a distorted trigonal pyramid. Both ring sulfur–nitrogen bonds are equal in length at 1.655 Å (R = Cl) and 1.638 Å (R = SPh) ⟨79AX(B)860⟩.

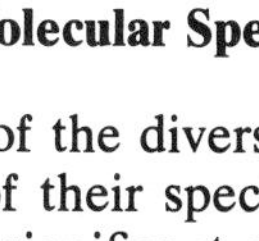

(**13**) 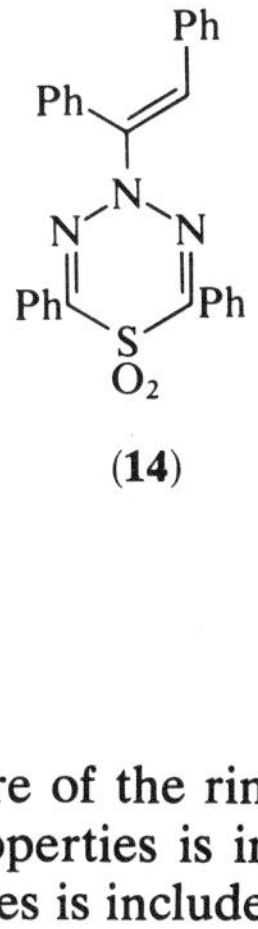          (**14**)           (**15**)           (**16**)

## 2.28.2.3 Molecular Spectra

Because of the diverse nature of the ring systems comprising this chapter, a systematic treatment of their spectral properties is impossible, and therefore only a brief survey of structurally significant properties is included. It is no coincidence that the most thoroughly studied systems are those which contain 'unusual' structural units such as the —N=S=N— group.

### 2.28.2.3.1 NMR spectra

A general account of the NMR spectroscopy of six-membered ring heterocycles is given in Chapter 2.01. Proton and carbon NMR spectroscopy have been used extensively to determine the solution conformations of saturated heterocyclic rings containing three or more heteroatoms, and the results are considered in Section 2.28.2.5. Otherwise NMR spectroscopy has been little used in these polyheteroatom systems. Two exceptions are the use of $^{13}$C NMR to establish that the oxauracil (**17**) does indeed exist as the dicarbonyl tautomer ⟨79JHC161⟩, and the investigation of 1,2,4,6-thiatriazines such as (**16**) by $^{13}$C NMR that establishes that neither rotation about the exocyclic sulfur–nitrogen bond nor inversion at the exocyclic nitrogen are hindered ⟨77JPR739⟩. The ring carbons resonate between 163 and 175 p.p.m.

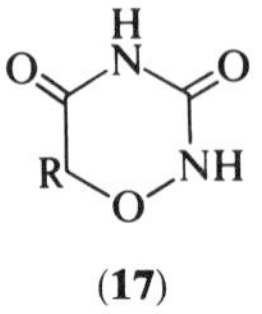

(**17**) 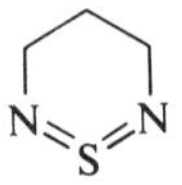                                         (**18**)

### 2.28.2.3.2 *UV spectra*

Despite the fact that these polyheteroatom six-membered ring compounds are often described as coloured materials, no real investigation of their UV spectra has been carried out. The structures are too diverse to allow the assignment of significant structural elements, the exception being the —N=S=N— moiety which in the 1,2,6-thiadiazine (**18**) absorbs at 277 nm ⟨69TL4117⟩. Indeed the —N=S=N— unit always absorbs at relatively long wavelength, and the incorporation of it into a ring with other conjugating fragments leads to deeply coloured heterocycles ⟨79PS(7)61⟩. The electronic spectra of some thiadiazine *S,S*-dioxides have been briefly reviewed ⟨70CRV593⟩.

### 2.28.2.3.3 *IR spectra*

IR spectroscopy has proved of most use in investigating six-membered rings containing one or more carbonyl groups. The results are in the main unexceptional, and since many of the systems are partially saturated, the bands in the 1600–1800 cm$^{-1}$ region can usually be assigned by consideration of whether the carbonyl group forms part of a cyclic amide, ester, urea, carbamate, *etc.* A few typical examples are presented in Table 12. Rings

**Table 12**   C=O IR Frequencies

| Compound | $\nu_{max}$ (cm$^{-1}$) | Ref. | Compound | $\nu_{max}$ (cm$^{-1}$) | Ref. |
|---|---|---|---|---|---|
| 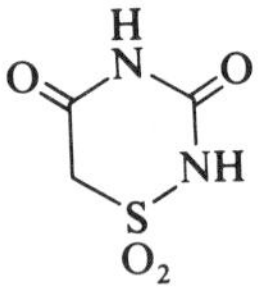 | 1698, 1714 | 61JOC3461 | 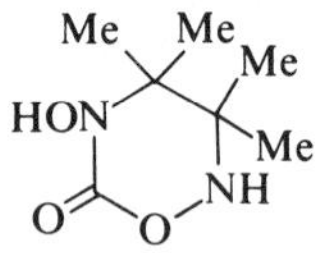 | 1680 | 80AP35 |
| 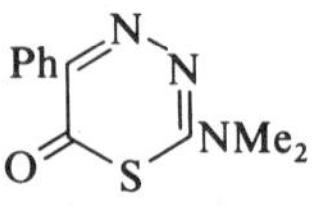 | 1640 | 81CC1003 | 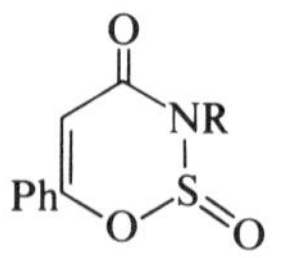 | 1650–80 | 77JCS(P1)904 |
| 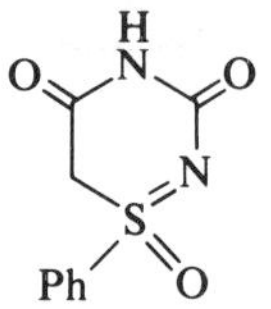 | 1675, 1705 | 77JOC952 | 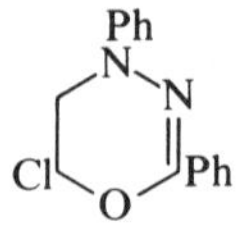 | 1710 | 78JPR452 |
| 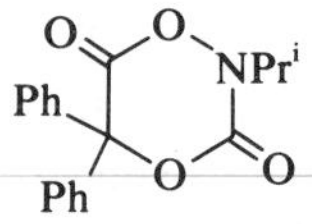 | 1745, 1790 | 80AP377 | 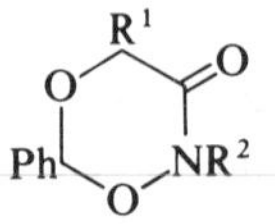 | 1650–65 | 79H(12)519 |

**Table 13**  $-SO_2-$ and C=N IR Frequencies

| Compound | $-SO_2-\nu_{max}$ (cm$^{-1}$) | Ref. | Compound | C=N $\nu_{max}$ (cm$^{-1}$) | Ref. |
|---|---|---|---|---|---|
| 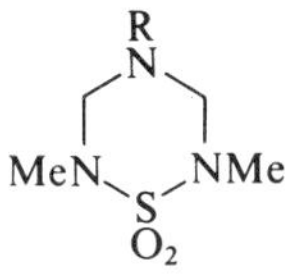 (structure) | 1150, 1330 | 78AP47 | 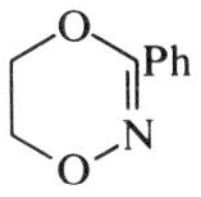 (structure) | 1610 | 71JOC284 |
| 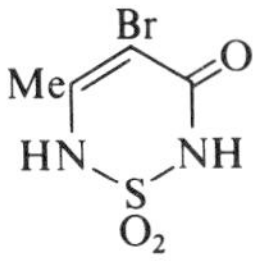 (structure) | 1180, 1350 | 80JHC977 | 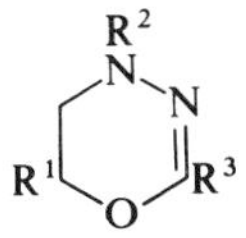 (structure) | 1660 | 78KGS1208 |
| 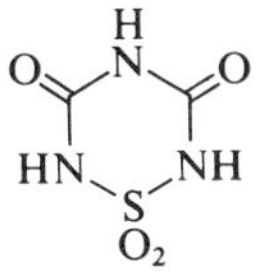 (structure) | 1120 | 75MI22800 | 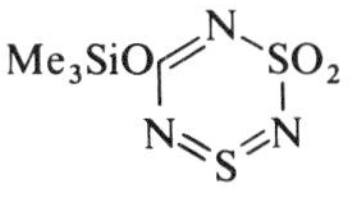 (structure) | 1520 | 74CB1 |
| 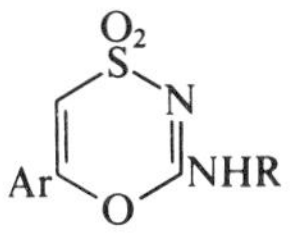 (structure) | 1160 | 78BCJ1805 | 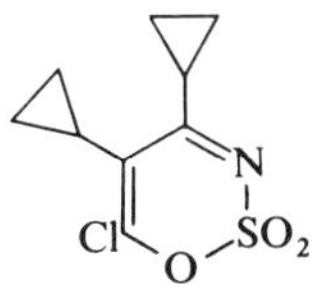 (structure) | 1605 | 72CB1683 |
| | | | 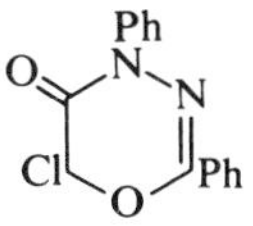 (structure) | 1655 | 78JPR452 |

containing $-SO_2-$ groups have also been investigated and some of the results have been summarized ⟨70CRV593⟩. Again the results are unexceptional and the $-SO_2-$ group exhibits its two normal absorptions in the ranges 1120–1180 and 1310–1370 cm$^{-1}$. Some examples are presented in Table 13, together with some typical C=N absorptions.

The structurally important $-N=S=N-$ group in (18) shows two stretching modes at 1020 cm$^{-1}$ (sym) and 1135 cm$^{-1}$ (asym) ⟨69TL4117⟩.

### 2.28.2.3.4 Mass spectra

Very few detailed MS studies with attempts to assign the often complex fragmentation patterns have been reported for six-membered rings containing three or more heteroatoms, although there are some common features such as loss of $N_2$, CO or S which occur in the MS of these compounds.

The thiadiazine (18) has been extensively studied by MS and shown to undergo an entirely different fragmentation to other $-N=S=N-$ containing compounds ⟨76CB2442, 78ACH(96)275⟩. The major losses are of NS, $CH_2N$ and aziridine fragments. The 1,4,2,3-benzodithiadiazine (19) fragments with loss of $RO_2CN=NCO_2R$ under electron impact ⟨74CB771⟩, and MS was used to establish the structure of 1,2,3-thiadiazines (20). The alternative isothiazol-N-imines were ruled out by the observed MS fragmentation pattern ⟨75OMS579⟩.

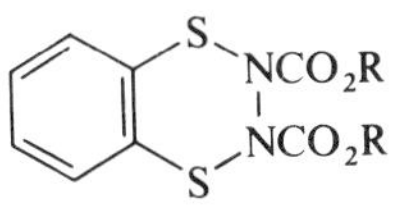

(19) 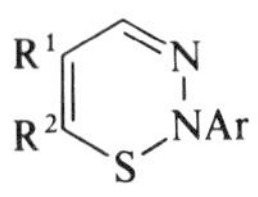    (20) 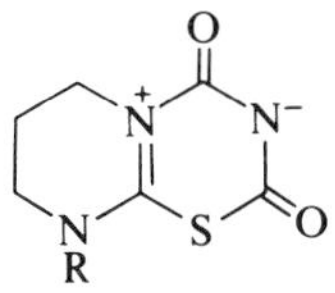    (21) 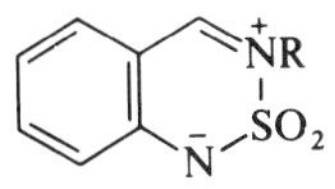    (22)

### 2.28.2.3.5 *Photoelectron spectra*

PE spectroscopy is a little used technique in this area, although the 1,2,6-thiadiazine (**18**) has been studied ⟨78ZN(B)284⟩. PES established the non-planar nature of the system and showed five vertical ionization energies in the range 9.25–12.3 eV. The species (**21**) and (**22**) have been investigated by PES, which confirms that substantial charge separation exists ⟨72AG(E)1012, 79LA1130⟩.

## 2.28.2.4 Thermodynamic Aspects

Electronically neutral six-membered rings which contain at least one nitrogen atom and oxygen and/or sulfur(II) atoms do not have a fully delocalized $\pi$-electron system, and hence are not aromatic. However, replacement of one nitrogen atom by the isoelectronic oxonia or thionia groups leads to a situation where full $\pi$-delocalization is now possible. A few examples of these aromatic $6\pi$-electron cations containing three heteroatoms are known. They are derived from 1,3,5-oxa- and -thia-diazines and, for example, the oxadiazinium salt (**23**) ⟨65CB334⟩ and the thiadiazinium salt (**24**) ⟨80BCJ3369⟩ have been reported. Their reactions are similar to those of pyrylium salts.

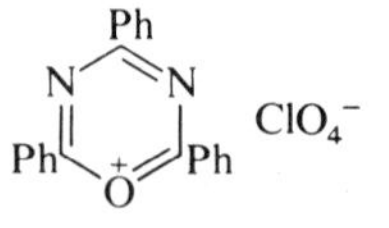

(**23**) 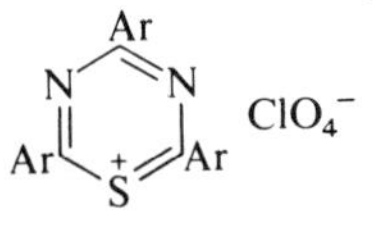  (**24**)

The possibility for delocalization and hence aromaticity formally exists in compounds containing sulfur in higher oxidation states. Thus the systems (**25**)–(**27**), and analogous systems which contain extra nitrogen such as (**16**), can be considered aromatic if planar and fully delocalized. However, they exhibit no aromatic properties, are not planar, and are best considered as ylides. A full discussion of the structure and bonding in thiabenzenes (**25**) is contained in Mislow's original paper ⟨75JA2718⟩. The question of aromaticity in cyclic sulfoximides (**27**) has also been discussed ⟨75CSR189⟩.

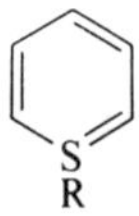

(**25**) 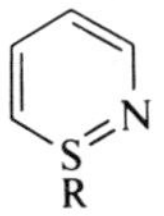  (**26**) 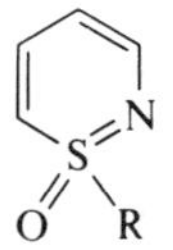  (**27**)

The nature of the sulfur–nitrogen bond, and in particular the sulfur–nitrogen double bond in 'inorganic' heterocycles, has been discussed, and the reader is referred to the two recent reviews on the subject ⟨79AG(E)91, 81AG(E)444⟩.

The 1,3,5-thiadiazinium salt (**24**) can be prepared by hydride abstraction from the corresponding 4*H*-1,3,5-thiadiazine (**28**). Removal of a proton from (**28**) gives the corresponding anion (**29**). Again the possibility for full delocalization exists, but these anions contain $8\pi$-electrons and are therefore formally antiaromatic. They are extremely unstable, but good evidence has been presented ⟨77JCS(P2)939⟩ for their intermediacy. The related $8\pi$-electron heterocyclic anions (**30**) and (**31**) have also been generated, and these and other heterocyclic systems containing $8\pi$-electrons have been reviewed ⟨75AG(E)581⟩.

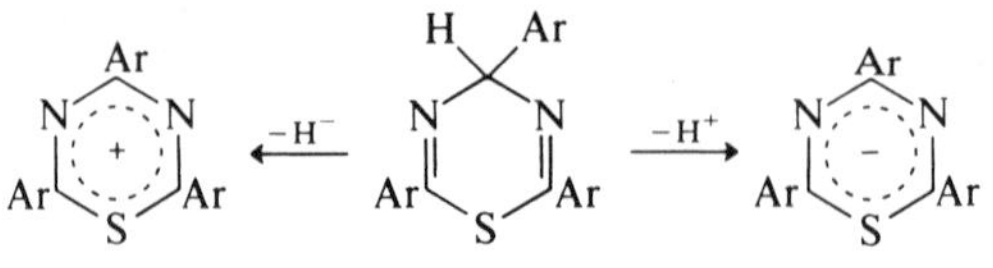

(**24**)  (**28**)  (**29**)

(**30**)   (**31**)

Most systems falling within this chapter are partially or fully saturated and their stability can usually be inferred from the thermodynamic stability of the various functional groups that make up the ring system. Thus those rings which contain amide, ester, hydrazide, urea, carbamate, sulfonamide, sulfamide or sulfone functions can be expected to possess the thermodynamic stability and chemical reactivity associated with these groups. Constraining the functional groups into a six-membered ring does not profoundly affect their stability. Those systems which contain weak single bonds display the thermodynamic instability and chemical reactivity associated with those bonds, and hence, for example, 1,2,4-dioxazines show all the properties expected of a cyclic peroxide.

### 2.28.2.5 Conformation

The conformations of fully saturated heterocyclic rings with three or more heteroatoms have been extensively investigated by $^1$H and $^{13}$C NMR, and to a lesser extent by dipole moment measurements, and therefore merit a separate section. The details of the NMR experiments will not be considered, the emphasis being on the results obtained. Although the general trends apparent in the solution conformations of six-membered rings have already been discussed in Chapter 2.01, the more important points are reiterated here.

1. The inclusion of sulfur in a six-membered ring lowers the barrier to ring inversion because of the longer C—S bond and smaller C—S—C angle.

2. Oxygen-containing rings also have a lower barrier to ring inversion than their nitrogen analogues, and for the compounds (**32**) the barrier to inversion decreases along the series X = CH$_2$, NMe, O, S ⟨80JCS(P2)279⟩.

3. The sulfoxide group usually occupies an axial rather than equatorial position.

4. Nitrogen-containing systems can undergo conformational changes either by ring inversion or umbrella inversion at nitrogen. The latter process is usually that of lower energy.

#### 2.28.2.5.1 *Tetrahydro-1,2,3-oxathiazine 2-oxides*

Proton NMR studies ⟨79OMR(12)481⟩ have established that in the absence of substituents at C-4 and C-6 the chair form is the most stable and the sulfoxide group is axial (**33**). The 3-*t*-butyl-4-methyl derivative adopts a twist form but retains the axial sulfoxide group.

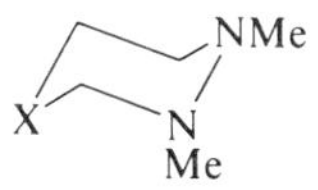

(**32**)                     (**33**) 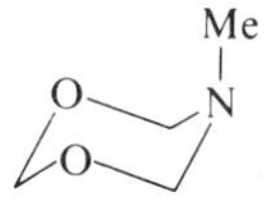                    (**34**)

#### 2.28.2.5.2 *Dihydro-1,3,5-dioxazines*

The only study on this system revealed few changes in the NMR spectrum in the range −145 to +34 °C, suggesting that in this range only the preferred 5-*N*-methyl axial conformer (**34**) is populated ⟨78JCS(P2)377⟩.

#### 2.28.2.5.3 *Tetrahydro-1,4,2-dioxazines*

Low-temperature NMR studies on the 2-methyl derivative demonstrated the existence of two conformational processes assigned as ring inversion and nitrogen inversion. At low temperature the equatorial conformation (**35**) predominated ⟨74JCS(P2)1561⟩. Subsequent work established that ring inversion (46 kJ mol$^{-1}$) has a slightly lower barrier than nitrogen inversion (48 kJ mol$^{-1}$) ⟨78T1415⟩.

### 2.28.2.5.4 Dihydro-1,3,5-dithiazines

Proton NMR and dipole moment measurements ⟨72JCS(P2)674, 72KGS321⟩ have shown that the predominant conformation is chair like with the nitrogen substituent axial (**36**).

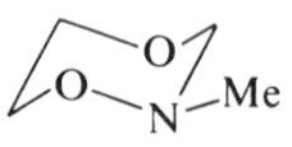

(35) 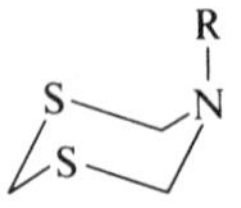  (36) 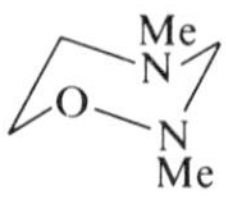  (37)

### 2.28.2.5.5 Tetrahydro-1,2,4-oxadiazines

As expected, the replacement of oxygen by nitrogen raises the barrier to ring inversion, and the measured barrier for the 2,4-dimethyl derivative (**37**) is 53 kJ mol$^{-1}$. The barrier to inversion at N-2 is similar to that for ring inversion at 47 (ax → ts) and 54 (eq → ts) kJ mol$^{-1}$. The most rapid process however is inversion at N-4 ⟨79T1391⟩.

### 2.28.2.5.6 Tetrahydro-1,2,5-oxadiazines

The variable temperature NMR spectra of the 2,5-dimethyl derivative show two coalescences corresponding to barriers of 56–59 and 31–34 kJ mol$^{-1}$ ⟨79JCS(P2)993⟩. The lower barrier is assigned to inversion at N-5, and the 5-*N*-methyl axial conformer predominates. However, for the 2,5,6-trimethyl and 2,5,6,6-tetramethyl derivatives, conformations in which the 5-*N*-methyl group is equatorial predominate. Further work ⟨79T1311⟩ has established that the higher energy barrier is due to slowing of inversion at N-2 rather than ring inversion. The major conformational set for the 2,4,5-trimethyl derivative (**38**) contains both axial and equatorial 5-*N*-methyl groups, although the major conformer has the methyl group equatorial.

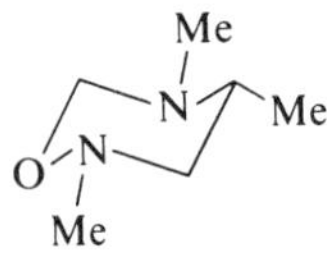

(38) 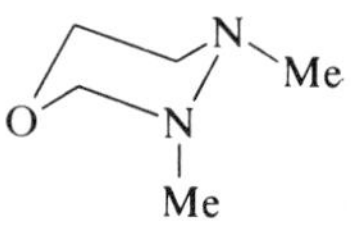  (39) 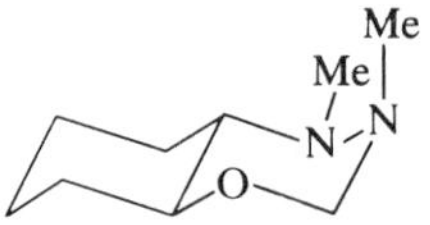  (40)

### 2.28.2.5.7 Tetrahydro-1,3,4-oxadiazines

The predominant conformation of the 3,4-dimethyl compound (**39**) has the 3-methyl axial and the 4-methyl equatorial ⟨76JCS(P2)1861⟩. The diequatorial conformation is thought ⟨77JCS(P2)1816⟩ to make no appreciable contribution and the measured conformational barrier ($\Delta H^{\ddagger} = 55$ kJ mol$^{-1}$) is probably due to slowing of a nitrogen inversion process. The related bicyclic system (**40**) has also been studied ⟨80MI22802⟩.

### 2.28.2.5.8 Tetrahydro-1,3,5-oxadiazines

The major conformer of the 3,5-dimethyl derivative (**41**) has one axial and one equatorial methyl group ⟨78JCS(P2)377⟩. No other studies on this system have been reported.

### 2.28.2.5.9 Tetrahydro-1,3,4-thiadiazines

These systems have lower ring inversion barriers than their oxygen analogues. The 3,4-dimethyl derivative exists at low temperature in a preferred conformation with the 3-*N*-methyl group axial (**42**; R = H). This preference is enhanced by the presence of an

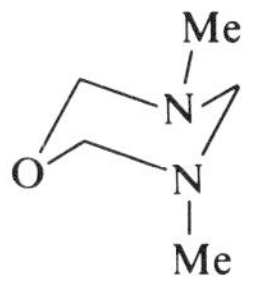

**(41)** 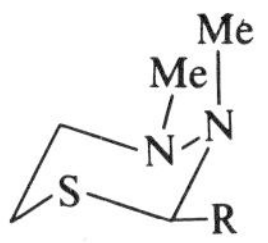     **(42)** 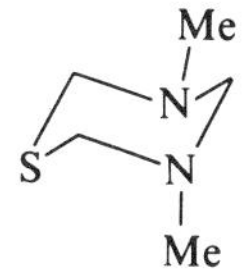     **(43)**

adjacent equatorial substituent at C-2 ⟨80JCS(P2)279⟩. Similar studies have been reported by other workers ⟨78KGS1568⟩.

#### 2.28.2.5.10 Tetrahydro-1,3,5-thiadiazines

These systems exist largely in conformations with one or two groups axial **(43)** ⟨72JCS(P2)674⟩, and have higher barriers to inversion than the corresponding 1,3,5-dithiazines.

#### 2.28.2.5.11 Tetrahydro-1,2,4,5-dioxadiazines

The dimethyl derivative exists predominantly in the equilibrating axial–equatorial conformer **(44)**. Ring inversion is slow ⟨79JCS(P2)1133⟩. The related tricyclic derivative **(45)** has also been studied.

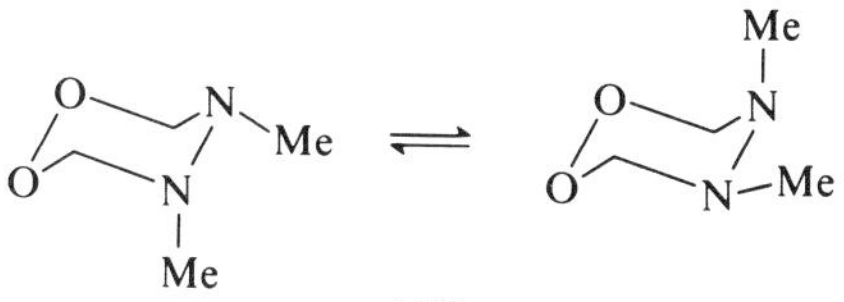

**(44)** 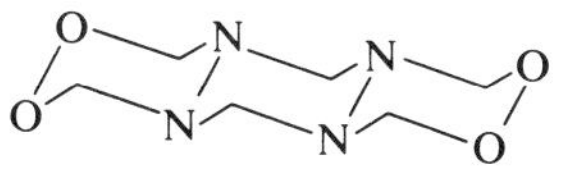     **(45)**

### 2.28.2.6 Tautomerism

#### 2.28.2.6.1 Prototropic tautomerism

A general account of prototropic tautomerism has been given in Chapter 2.01. The situation in the non-aromatic polyheteroatom systems is generally quite complicated particularly if the system has more than one nitrogen atom to provide extra sites for the tautomeric proton. Only the important diuretic benzothiadiazine dioxides **(46)** ⇌ **(47)** have been studied in any detail. The original UV spectroscopic studies suggested that the 4*H*-tautomer **(47)** was preferred in ethanol although the 2*H*-tautomer **(46)** was present in aqueous alkali, presumably as the anion ⟨60JOC970⟩. Extended Hückel MO calculations supported the view that the 4*H*-tautomer is preferred ⟨70MI22800⟩, and $^{13}$C NMR studies also support this ⟨79T2151⟩.

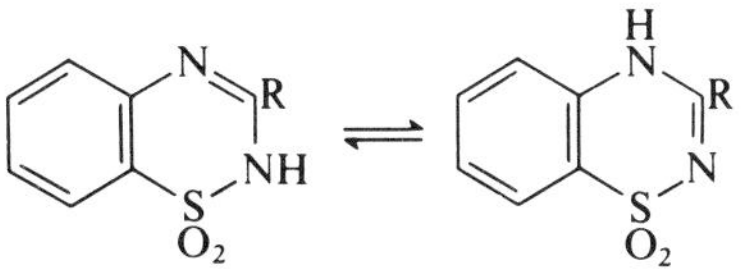

**(46)**     **(47)**

Other thiadiazine dioxides have been investigated ⟨70CRV593⟩. The 2*H*-tautomer **(48)** is favoured over the 4*H*-tautomer **(49)** and structure **(50)** in which the tautomeric proton is on oxygen. The dihydro-1,2,6-thiadiazine dioxides are also thought to exist in the tautomeric equilibrium **(51)** ⇌ **(52)**.

In compounds with several potential hydroxyl groups, such as the oxauracil **(17)** ⟨79JHC161⟩ and the barbituric acid analogue **(53)** ⟨61JOC3461⟩, the carbonyl tautomers are usually preferred.

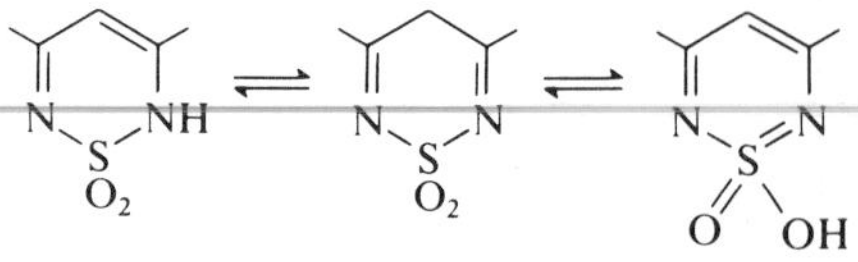

**(48)**     **(49)**     **(50)** 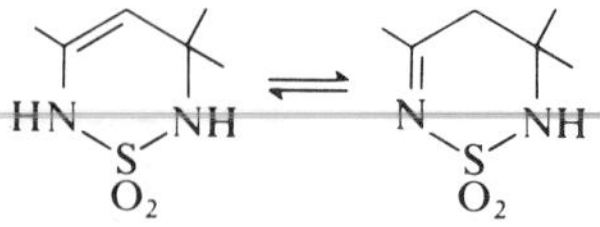          **(51)**     **(52)**

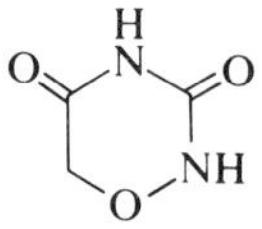

**(17)** 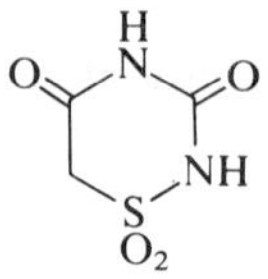                          **(53)**

### 2.28.2.6.2 *Ring–chain tautomerism*

Although rare in aromatic heterocyclic systems, ring–chain tautomerism is fairly common in these non-aromatic systems. Electrocyclic ring opening reactions are quite common, and the ones which lead to ring cleavage are considered in more detail in Section 2.28.3.5. Flash vacuum pyrolysis of the oxadiazinone (**54**) at 560 °C and $10^{-4}$ mmHg leads to a reactive intermediate with an intense IR absorption at 2240 cm$^{-1}$ assigned to the ring opened isocyanate form (**55**) ⟨75S522⟩.

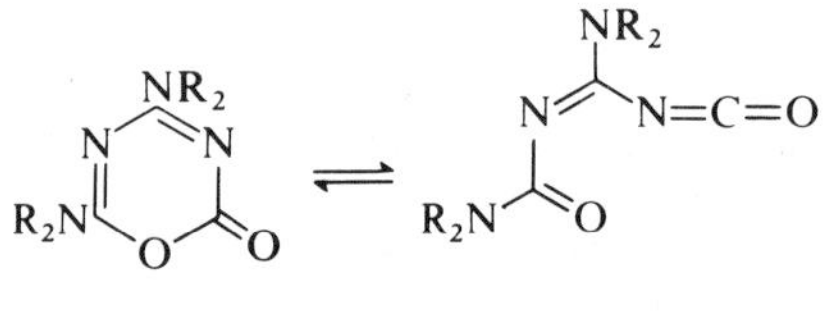

**(54)**                          **(55)**

Electrocyclic ring opening of benzoxadiazines (**56**) is thought to be the first step in their ring contraction to benzoxazoles ⟨75CC962⟩. The benzoxadiazines are prepared by oxidation of the appropriate benzamidoximes, and ring–chain tautomerism between the intermediate nitrosoimine (**57**) and the oxadiazine (**58**) has been clearly demonstrated ⟨75CC914⟩.

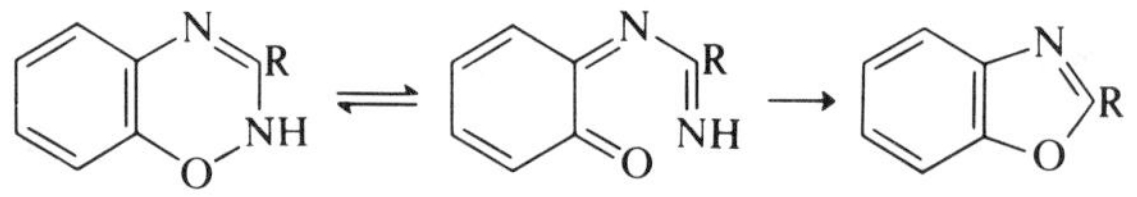

**(56)**

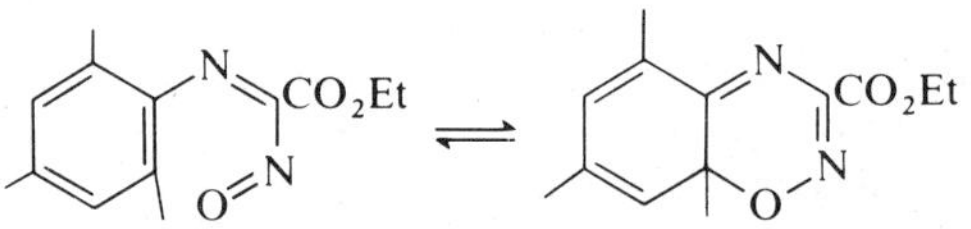

**(57)**                          **(58)**

Ring–chain tautomerism in tetrahydro-1,3,4-oxadiazines is well established. Freshly prepared compounds have the ring closed structure (**59**), but slowly ring open on standing ⟨71KGS1167⟩. The equilibrium concentration of (**59**; $R^2 = Ar$) is increased if the aryl group contains an electron withdrawing substituent ⟨73KGS902⟩. The corresponding 1,3,4-thiadiazines also undergo ring–chain tautomerism (**60**) ⇌ (**61**) ⟨79KGS1637⟩.

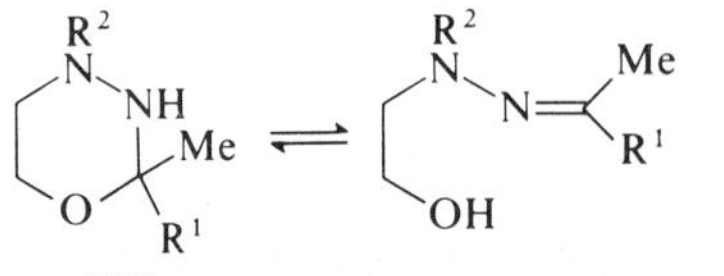

**(59)** 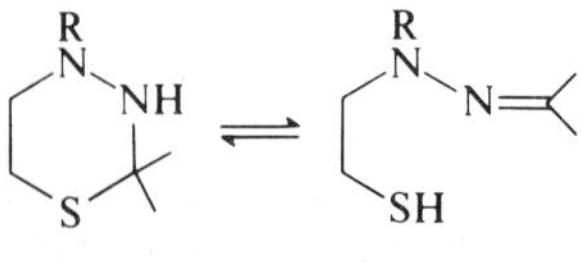                **(60)**     **(61)**

The 1,2,5-oxadiazine (**62**) is reported to be in tautomeric equilibrium with the ring opened nitrone (**63**). The nitrone is the only species observed in aqueous solution ⟨69ZOR355⟩. The 1,3,4-thiadiazine (**64**), which can be regarded as a stable tetrahedral intermediate, readily ring opens to (**65**), which is favoured in ethanol solution ⟨77ACH(94)391⟩.

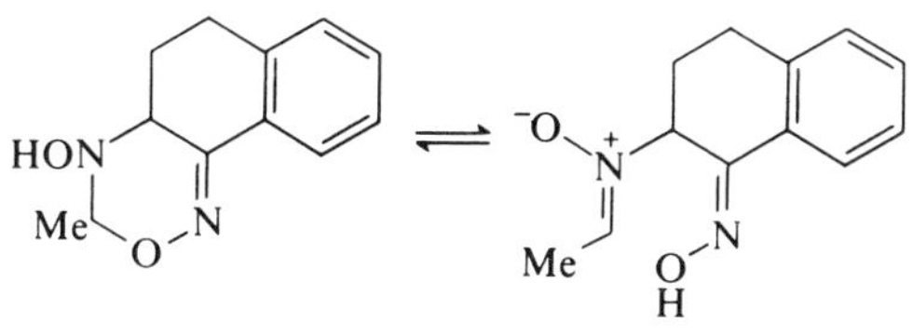

(62)          (63)

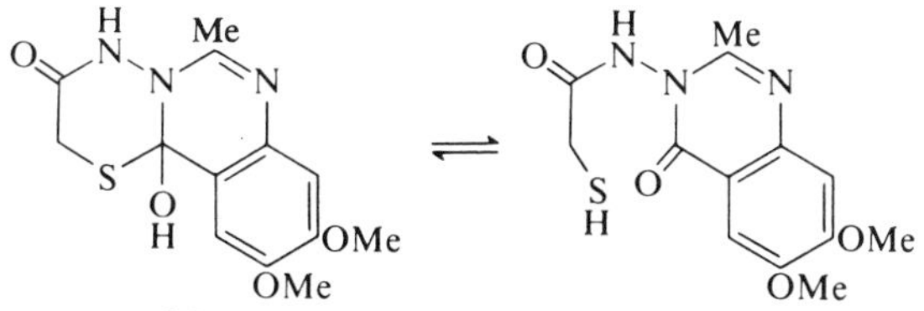

(64)          (65)

### 2.28.2.7 Betaines and Other Unusual Structures

The dipolar structures (21), (22), and (66) share the 1,3-arrangement of positively and negatively charged nitrogen atoms separated by either a carbonyl or sulfonyl group ⟨72AG(E)1012, 79LA1130, 80S112⟩. An isolated example of an *N*-imine (67) incorporated into a 1,3,4-thiadiazine has been reported ⟨77ACH(94)391⟩. Cyclic sulfur ylides are fairly common in these systems; they therefore are discussed in the appropriate section, and are not considered as 'unusual'. There are no other important betaine or dipolar structures containing three or more heteroatoms.

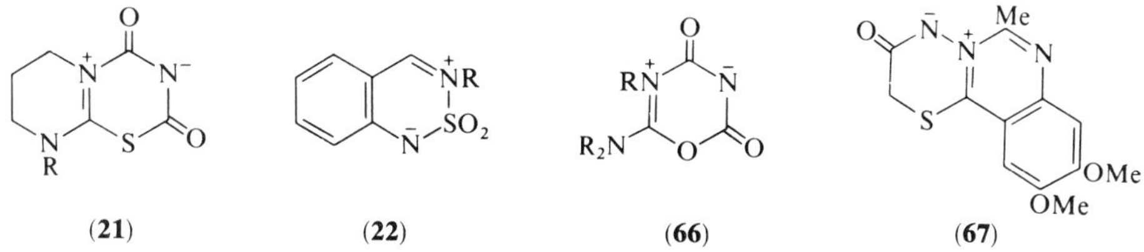

(21)          (22)          (66)          (67)

### 2.28.3 REACTIVITY

The six-membered rings under consideration are not aromatic and hence do not exhibit the typical reactivity shown by the six-membered nitrogen-containing systems discussed in Chapter 2.02. An insight into the reactivity of a particular system can be gained simply by considering the functional groups within that system. For example, rings containing the —NHSO$_2$— or —NHCONH— units exhibit typical sulfonamide and urea reactivity respectively. Therefore in this section an attempt will be made to exemplify the different reactivity shown by the various types of functionality common within these systems. Individual ring systems will *not* be dealt with separately.

### 2.28.3.1 Reactivity at Nitrogen

#### 2.28.3.1.1 Acid–base properties

Although very few detailed studies have been carried out, the acid–base properties of polyheteroatom six-membered rings are fairly predictable. The fully reduced systems behave as typical secondary amines, although the inclusion of extra heteroatoms often reduces their basicity. Thus the reduced dithiazines (68) (thialdine) and (69), and the related thiadiazine (70), are weak bases with p$K_a$ values of 2.7, 2.8 and 2.0 respectively ⟨58JCS2893⟩. The weak basicity is due to the inductive effect of the β-heteroatoms. This inductive effect of β-heteroatoms is also apparent in the dioxazine (71), which exhibits no tendency to isomerize to the *trans*-enamine because of the much reduced conjugation between the nitrogen lone pair and the $\pi$-electrons of the double bond ⟨80JCS(P1)2383⟩. The fact that the N-propenyl group is axial is also important.

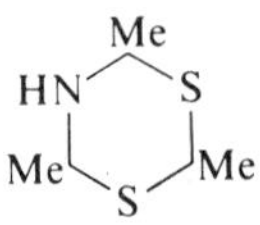

(68) 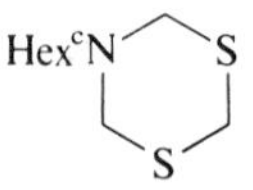            (69) 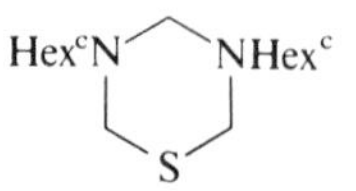            (70) 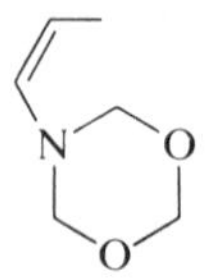            (71)

Rings which contain an NH group as part of an amide or sulfonamide show the expected acid–base properties. Thus in the thiadiazinedione (53), an analogue of barbituric acid, the 2-NH is considerably more acidic than the 4-NH group ⟨61JOC3461⟩.

### 2.28.3.1.2 Reaction with electrophiles

N-Alkylation of polyheteroatom six-membered rings is readily achieved. For example, the oxadiazine (72) gives the 4-methyl derivative ⟨75CB1911⟩, and the oxadiazinedione (73) gives a mixture of mono- and dimethyl derivatives ⟨79JHC161⟩ on treatment with iodomethane.

The NH groups of sulfonamide functions are also readily alkylated. The 1,2,6-thiadiazine dioxide (74) gives the 2-methyl derivative on treatment with iodomethane in acetone in the presence of potassium carbonate ⟨65CI(L)182⟩. The 1,3,2,4,6-dithiatriazine (75) forms a silver salt ⟨64MI22800⟩, and reaction of the salt with iodomethane gives the expected N-methyl compound.

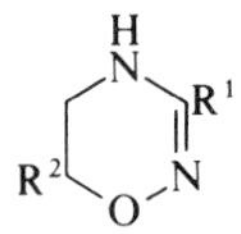

(72) 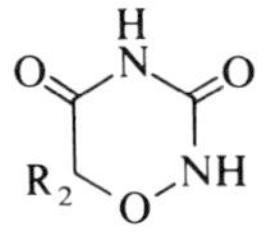            (73) 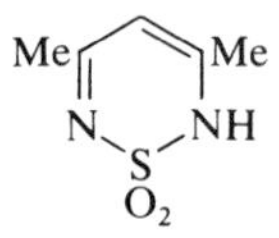            (74) 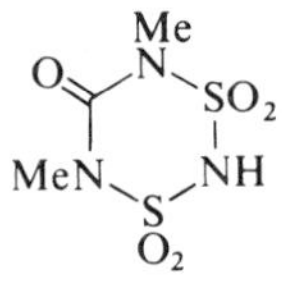            (75)

The alkylation of the 1,2,6-thiadiazine (76) has been studied in detail ⟨82H(17)401⟩ because of its importance as an isostere of 6-methyluracil. Dimethyl sulfate in the presence of sodium hydroxide as base gives the 2-monomethyl derivative exclusively. Other reagents such as diazomethane and iodomethane give the 2,6-dimethyl derivative as the major product although substantial amounts of O-methylation and C-methylation at C-4 also occur, indicating that in systems where there are several possible sites for alkylation, the results are very much dependent on the reaction conditions.

The sulfonylhydrazone moiety in (77) is methylated at N-2 by iodomethane under basic conditions, and acetic anhydride gives the 2-acetyl derivative ⟨68JHC453⟩. The 1,2,4-thiadiazines (78; Y = leaving group) are acylated on N-4 by isocyanates. Reaction with a second molecule of isocyanate gives the bicyclic compounds (79) ⟨74BSF1917⟩.

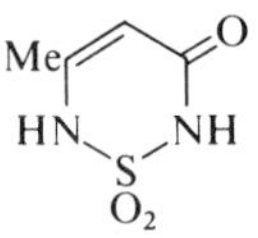

(76) 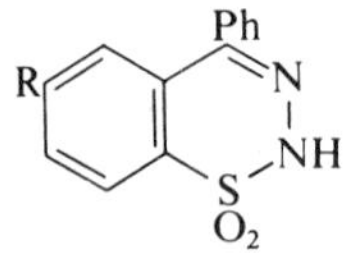            (77) 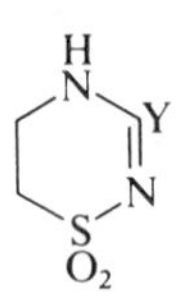            (78) 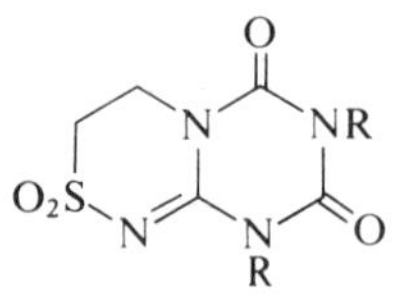            (79)

Isolated examples of oxidation at nitrogen have been reported for these systems. They do not form the N-oxides typical of aromatic nitrogen heterocycles, but behave similarly to amines. Thus the 1,2,5-oxadiazine (80) is rapidly oxidized by lead dioxide to the radical (81) ⟨73JA1677⟩, which is in equilibrium with the four-membered ring radical (82).

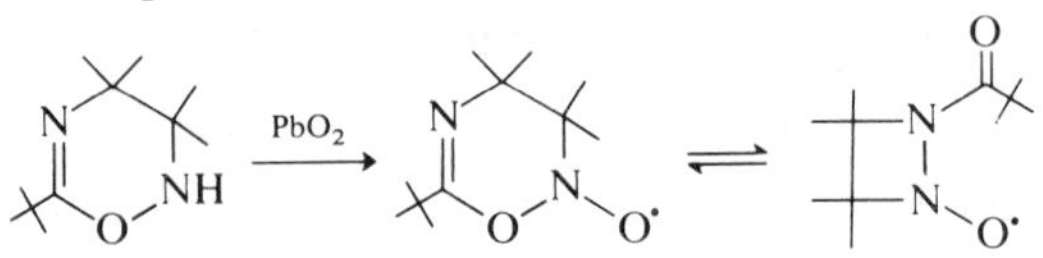

(80)            (81)            (82)

Fully saturated systems can be oxidized to the corresponding didehydro derivatives. For example, oxadiazine (**83**) gives (**84**) on oxidation with mercury(II) oxide ⟨78KGS1208⟩, and oxidation of (**85**) with DDQ gives the 3*H*-2,1,3-benzothiadiazine dioxide (**86**), which can also be represented as the dipolar form (**22**) ⟨79JHC1069⟩.

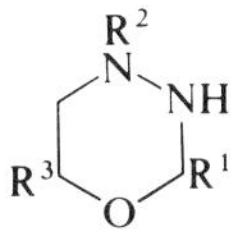

(**83**) 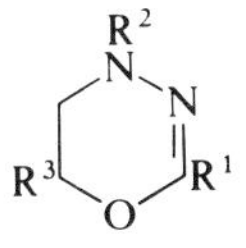      (**84**) 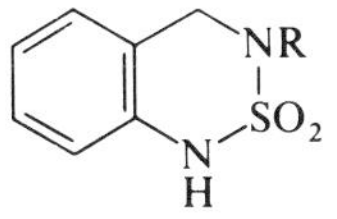      (**85**) 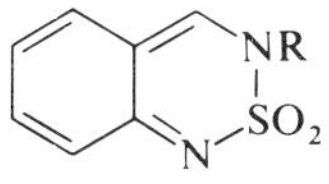      (**86**)

### 2.28.3.1.3 *Reaction with nucleophiles*

Nucleophilic attack at nitrogen is rare in these systems. However, the 'inorganic' trithiazyl trichloride (**87**) acts as an apparent source of electrophilic nitrogen on reaction with certain organic substrates. Reaction with electron-rich alkenes such as stilbene gives 3,4-diphenyl-1,2,5-thiadiazole ⟨77JCS(P1)916⟩.

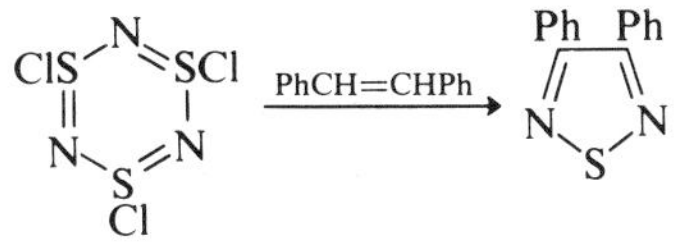

(**87**)

## 2.28.3.2 Reactivity at Sulfur

### 2.28.3.2.1 *Reaction with electrophiles*

Very few examples of electrophilic attack at sulfur(II) have been reported for these systems, with the exception of oxidation. Alkylation of the benzothiadiazine (**88**) with trimethyloxonium tetrafluoroborate gave a poor yield of the imino ether (**89**). The major product (**90**) resulted from alkylation at sulfur followed by hydrolytic ring opening of the resulting azasulfonium salt ⟨79JCR(S)214⟩.

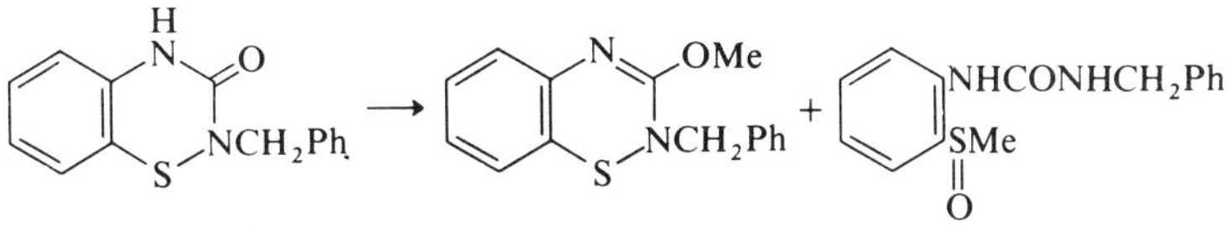

(**88**)      (**89**)      (**90**)

The sulfur(II) atom in these six-membered rings is readily oxidized, and this is probably the most common reaction at sulfur. Despite this ease of oxidation, there is a surprising lack of syntheses of *S*-oxides and *S,S*-dioxides from the parent sulfur(II) heterocycles ⟨70CRV593⟩, the oxidized compounds usually being prepared directly. The reagent of choice is MCPBA and the oxidation can be controlled so as to give either the sulfoxide or the *S,S*-dioxide. Examples include the oxidation of the dithiatetrazine (**91**) to the 1,4-dioxide ⟨79JHC751⟩, the oxidation of the 1,2,6-thiadiazine 1-oxide (**92**) to the corresponding dioxide, which in this case could not be prepared directly ⟨81JCS(P1)1891⟩, and the oxidation of the sulfimide (**93**) to the corresponding sulfoximide ⟨64LA(675)189⟩, although potassium permanganate was the oxidant in this case.

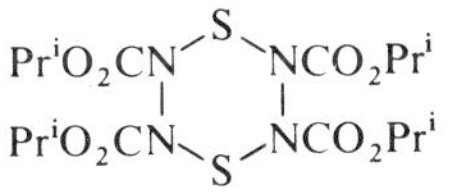

(**91**) 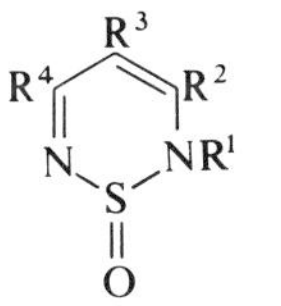      (**92**) 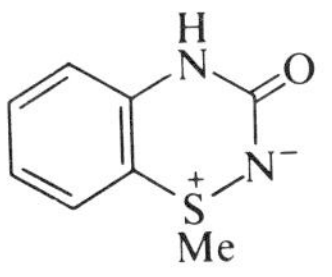      (**93**) 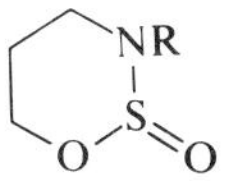      (**94**)

### 2.28.3.2.2 *Reaction with nucleophiles*

Azasulfonium salts such as that derived by *S*-methylation of (**88**) are highly susceptible to nucleophilic attack at sulfur, and the S—N bond in these compounds is therefore readily cleaved. This susceptibility to nucleophilic attack at sulfur is also shown by cyclic sulfur–nitrogen ylides (sulfimides) ⟨77CRV409⟩, and hence heterocyclic systems incorporating this functionality are readily ring opened by nucleophiles. Tetrahydro-1,2,3-oxathiazine 2-oxides (**94**), which are cyclic sulfinamates, are ring opened by hydrolysis to give the corresponding 3-aminopropan-1-ols. The rate of hydrolysis depends on pH ⟨78JCS(P2)1207⟩.

## 2.28.3.3 Reactivity at Carbon

### 2.28.3.3.1 *Reaction with electrophiles*

The six-membered rings under consideration, being non-aromatic, do not exhibit typical aromatic reactivity towards electrophiles. However, certain systems do undergo electrophilic substitution, particularly the 1,2,6-thiadiazine 1,1-dioxides (**95**) which react with electrophiles at C-4. Treatment with bromine or chlorine gives the 4-halo derivative, and reaction with arenediazonium chlorides gives the diazo coupling products. The thiadiazines (**95**) are also readily nitrosated and formylated at the 4-position ⟨74JCS(P1)2050⟩. Thus these systems show a striking similarity in reactivity to 2-pyrimidones, although they are not aromatic.

(**95**)

The related 1,2,6-thiadiazine (**96**) is also readily brominated at C-4 simply by treatment with bromine in carbon tetrachloride ⟨80JHC977⟩. The resulting bromide (**97**) is a source of Br$^+$, although it has not yet been developed as a brominating agent. 1,2,6-Thiadiazine (**98**) is oximated at C-4 by reaction with nitrous acid. The oxime is a useful precursor to the purine isostere (**99**) ⟨76MI22800⟩.

(**96**)          (**97**)          (**98**)          (**99**)

### 2.28.3.3.2 *Reaction with bases*

Many of these six-membered ring systems can be deprotonated by treatment with base. From the theoretical standpoint the most important systems are the 4*H*-1,3,5-oxadiazines and the corresponding thiadiazines, deprotonation of which leads to the formally antiaromatic 8$\pi$-electron heterocyclic anions ⟨75AG(E)581⟩ as described in Section 2.28.2.4. The anions are unstable and readily undergo ring contraction reactions (Scheme 1). These reactions are considered separately in Section 2.28.3.5. Evidence for the anion derived from the 1,3,5-thiadiazine (**100**) has been obtained ⟨77JCS(P2)939⟩ in that the thiadiazine undergoes deuterium exchange. The resulting deuteriothiadiazine exhibits a kinetic isotope effect, implying that deprotonation is the rate determining step. The rate of reaction with bases is increased if the aryl group at C-4 contains electron withdrawing substituents.

Treatment of (**100**) with trityl cation leads to hydride abstraction and formation of the 6$\pi$-electron aromatic cation (**24**), as previously described in Section 2.28.2.4.

The fully reduced 1,3,5-dithiazine (**101**) exhibits normal reactivity at C-2 in that it is deprotonated by strong base. The resulting lithio derivative ⟨77JOC393⟩ is an Umpolung

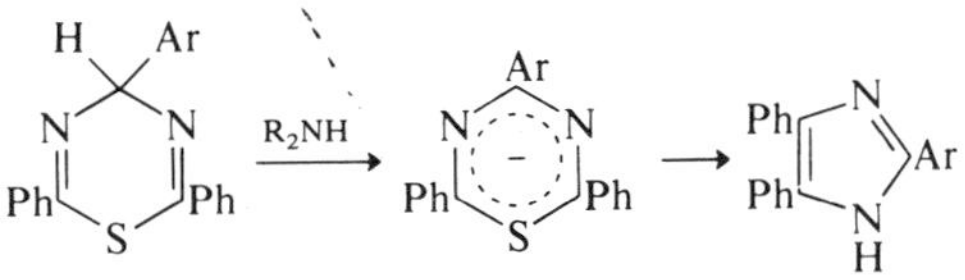

**(100)**

**Scheme 1**

reagent (Scheme 2), and may offer certain advantages over the more usual 1,3-dithiane-based reagents in that the final cleavage of the ring in (**102**) is easier. Intramolecular chelation to the extra heteroatom may be important.

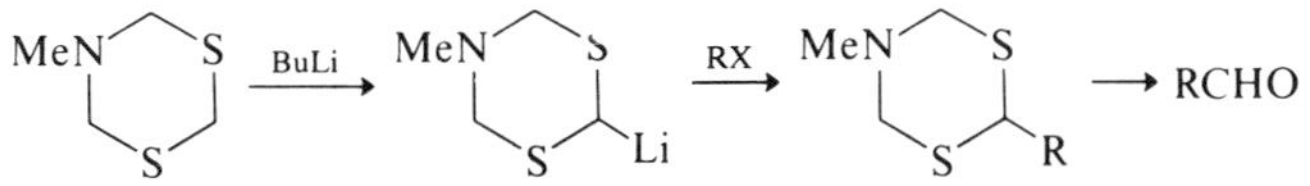

**(101)**      **(102)**

**Scheme 2**

### 2.28.3.3.3 Reaction with nucleophiles

Nucleophilic attack at ring carbon is a common reaction of polyheteroatom six-membered rings, and usually results in ring opening. Carbonyl and imine $sp^2$ carbons are the usual sites of attack, and the susceptibility of the ring to attack is related to the nature of the $sp^2$ carbon. Thus, it is the carbonate type carbon rather than the amide in (**103**) that is attacked by nucleophiles to give ring opened products ⟨81S38⟩. Likewise the oxadiazinone (**104**) ⟨75S522⟩ and the dioxazinone (**105**) ⟨67JCS(C)1178⟩ are ring opened by nucleophiles as shown.

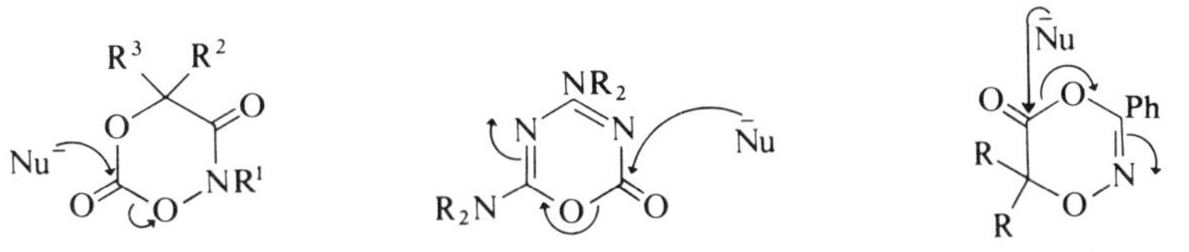

**(103)**      **(104)**      **(105)**

The oxathiadiazines (**106**) ⟨73TL2783⟩ and (**107**) ⟨77ZC222⟩ are ring opened by nucleophilic attack at the $sp^2$ carbons indicated. The oxathiazine dioxide (**108**) is reduced to the 3,4-dihydro derivative by 0.5 equivalent LAH ⟨72JOC196⟩, although the isomeric chlorooxathiazine dioxide (**109**) behaves differently towards nucleophiles. Nucleophilic attack at C-6 followed by elimination of $SO_3$ and chloride ion occurs as shown ⟨80AG(E)131⟩.

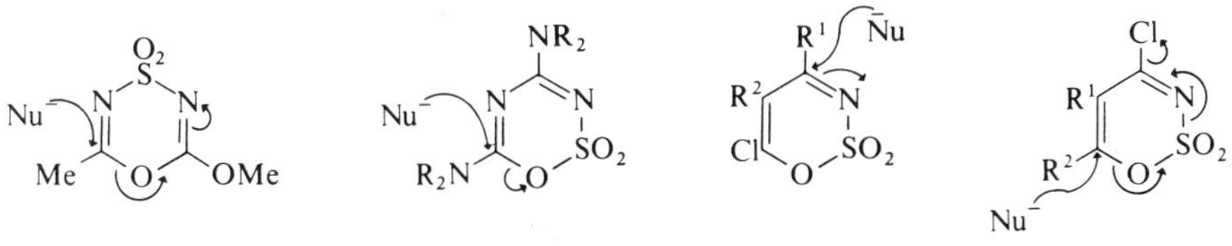

**(106)**      **(107)**      **(108)**      **(109)**

The $6\pi$-cations (**23**) and (**24**) derived from 1,3,5-oxadiazines and thiadiazines undergo facile nucleophilic attack at the $\alpha$-carbon by a variety of nucleophiles. These reactions result in the formation of other heterocyclic systems and some examples are presented in Scheme 3 ⟨65CB334, 73BCJ3902, 80BCJ3369⟩.

### 2.28.3.4 Cycloaddition Reactions

The double bonds in non-aromatic heterocyclic systems can in principle participate in cycloaddition reactions either as $2\pi$- or $4\pi$-components. Although not many examples have been reported, there are cases where the heterocyclic ring acts as a diene or dienophile.

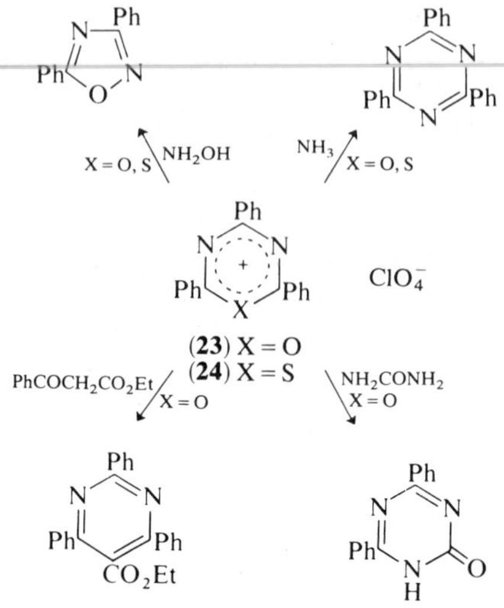

**Scheme 3**

The oxadiazinone (**110**) acts as a diene, and reacts with a variety of triple bonds. The initial Diels–Alder adducts are not isolated, but lose nitrogen to give pyran-2-ones, and in the case of 1-diethylaminopropyne as dienophile the reaction stops here. With benzyne and alkynes containing electron withdrawing groups, the pyran-2-one reacts further in a second Diels–Alder reaction ⟨77S252⟩. In marked contrast ⟨81CC1003⟩, the closely related thiadiazinone (**111**) does not undergo Diels–Alder reactions with either electron-rich or electron-deficient dienophiles.

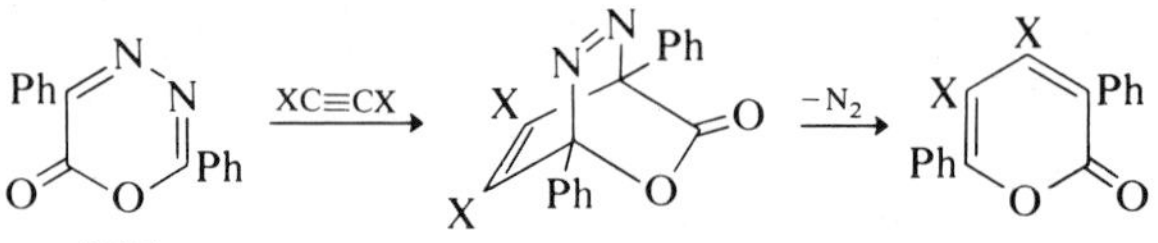

Examples of the heterocyclic ring acting as $2\pi$-components include the Diels–Alder reaction of 2,3-dimethylbutadiene with the 3,4 N=S bond in the 1,3,2,4-benzodithiadiazine (**112**) to give (**113**) ⟨71LA(749)171⟩. The carbon–carbon double bond in the 2,6-dimethyl derivative of the 1,2,6-thiadiazine (**76**) is sufficiently nucleophilic to react with carbenes. Thus copper catalyzed decomposition of ethyl diazoacetate in the presence of the thiadiazine gave the *exo* adduct (**114**) ⟨82H(17)401⟩.

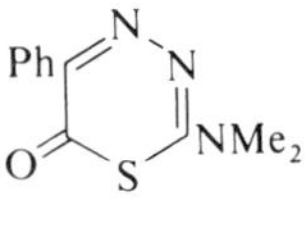

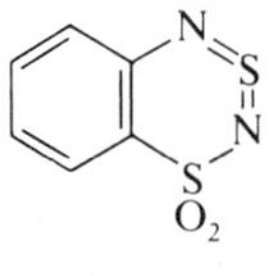

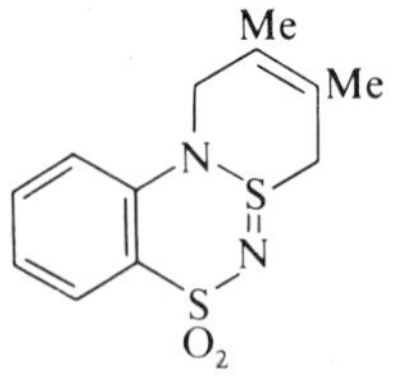

### 2.28.3.5 Ring Cleavage and Ring Contraction

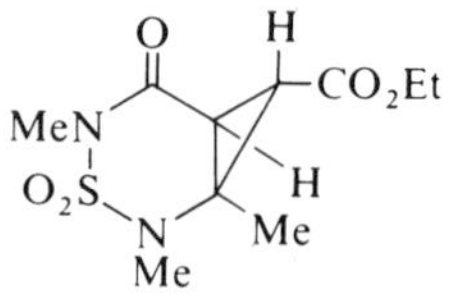

Reactions in which the ring is cleaved or completely fragmented are fairly common in these systems. Ring contraction to form five-membered rings is also common especially where sulfur is extruded, and these reactions are considered separately, as are reactions which involve desulfurization or reduction. Reactions which lead to ring opening but are initiated by nucleophilic attack at ring carbons have already been discussed in Section 2.28.3.3.3.

### 2.28.3.5.1 Thermal reactions

Thermal ring fragmentation reactions often parallel the MS fragmentation in the cases where small stable molecules such as $N_2$ or CO can be extruded. These processes can be considered as electrocyclic reactions and are represented here by simple 'arrow pushing', although this is *not* to imply that they are concerted. An idea of the thermal stability of a given system can be gained by considering possible electrocyclic pathways, and the thermodynamic stability of the fragments obtained by such a process. Thus the dioxazinone (115) is extremely unstable ⟨74LA561⟩ and extrudes $CO_2$ at temperatures above 0 °C to form benzaldehyde and the chlorosulfonylimine as shown. The dioxazine (116) fragments on attempted distillation or on prolonged storage with elimination of hexafluoroacetone ⟨66JOC2568⟩, and the dioxadithiazine (117) fragments as shown on heating to 150 °C ⟨78JA985⟩.

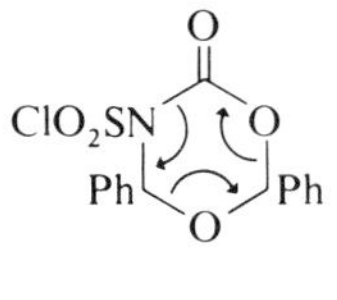

(115) 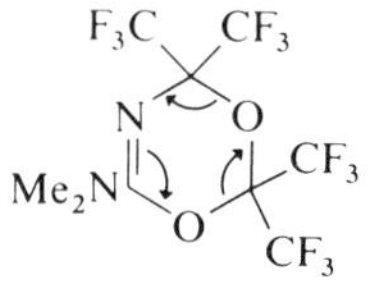      (116) 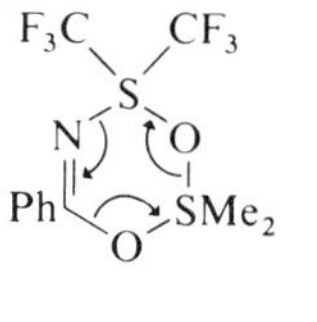      (117) 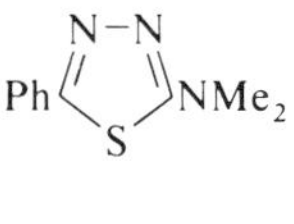      (118)

The 1,3,4-oxadiazinone (110) undergoes complete fragmentation to $CO_2$ and benzonitrile on vapour phase pyrolysis at 400 °C ⟨77S252⟩. The related thiadiazinone (111) does not lose COS on pyrolysis (or in the MS) ⟨81CC1003⟩. The products are benzonitrile, dimethylcyanamide and the ring contraction product (118) formed by extrusion of CO.

On heating with electron-rich $2\pi$-components such as enamines or ethoxyacetylene, the 1,2,5-oxathiazine 2,2-dioxide (119) gives products which are formally derived by electrocyclic elimination of $H_2C=SO_2$, followed by [4 + 2] cycloaddition of the resulting 'diene' as shown in Scheme 4 ⟨69IZV2059⟩.

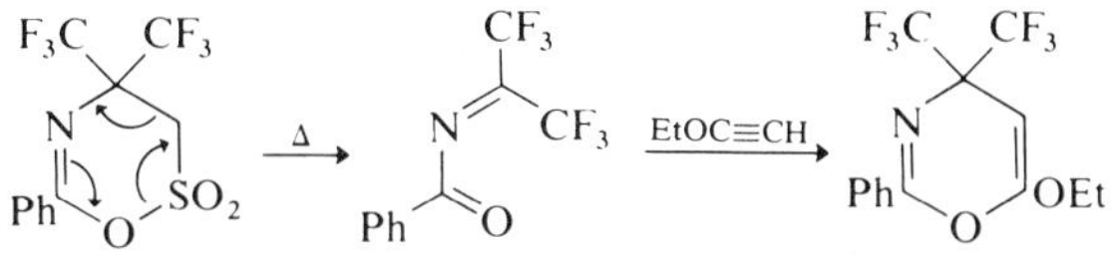

(119)

**Scheme 4**

The thermolysis of 1,3,5-oxathiazines (120) is more complex, the initial process being an electrocyclic extrusion of the carbonyl fragment. Thus (120; $R^1 = R^2 = CF_3$) extrudes hexafluoroacetone at 140 °C to give (121) which is in thermal equilibrium with the 1,3-thiazete as shown in Scheme 5. In the absence of an external trap for (121) the product is a 1,3,5-thiadiazine formed by addition of (121) to RCN (from further fragmentation of the 1,3-thiazete) ⟨77CB2114⟩. The trifluoromethyl groups seem necessary for thiazete formation since the pyrolysis of (120; $R^1 = R^2 = H$) at 140 °C and 20 mmHg gives products simply derived by dimerization of (121) ⟨79TL1537⟩.

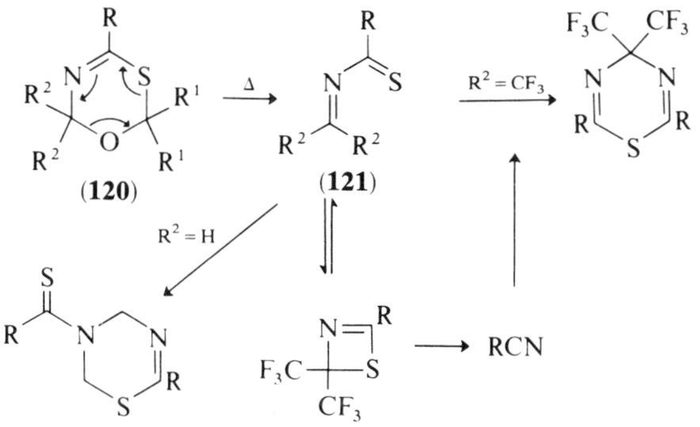

(120)      (121)

**Scheme 5**

Trifluoromethyl-substituted 1,3,5-oxadiazines behave similarly on thermolysis. Thus (122) extrudes hexafluoroacetone to generate (123), which may be in equilibrium with a 1,3-diazete, and the reaction follows a course exactly analogous to that in Scheme 5. In

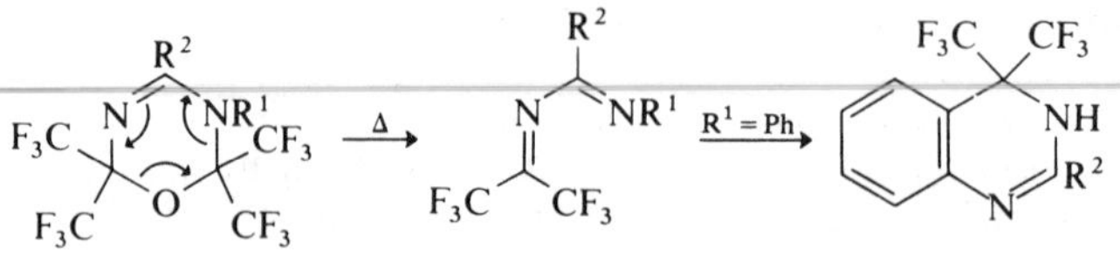

(122)                    (123)

the case of (123; $R^1$ = Ph) a further electrocyclic process involving the phenyl ring occurs to give a dihydroquinazoline ⟨80MI22801⟩.

1,4,3,5-Oxathiadiazine 4,4-dioxides (124) undergo retro [4+2] cycloaddition reactions at temperatures between 30 and 60 °C to give the *N*-sulfonylurethane (125) ⟨73JOC1249⟩ which can be trapped by cycloaddition to styrene (both 1,2- and 1,4-cycloadducts are formed), or to nitriles as shown in Scheme 6. In these thermal ring fragmentation reactions which are carried out in the presence of trapping agents it is often assumed that the ring fragments first to generate a reactive 'diene' which is then intercepted by the trap. The alternative possibility that the trap adds to the system *before* ring fragmentation cannot always be ruled out, and in the reaction of (124; R = Me) with ynamines, nucleophilic addition to the ring does indeed precede cycloreversion; the product (126) retains the elements of acetonitrile ⟨80JOC721⟩.

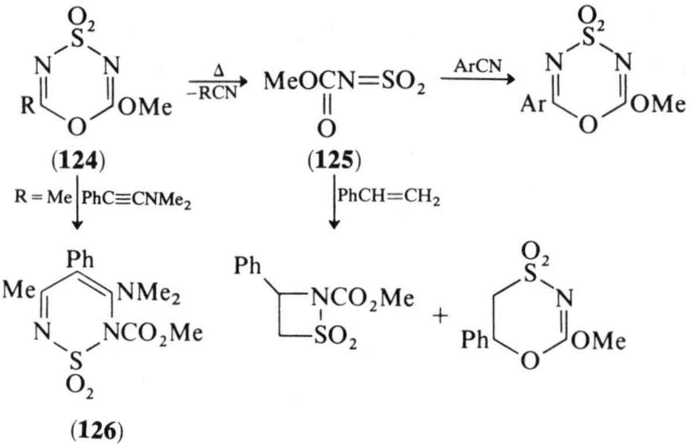

**Scheme 6**

### 2.28.3.5.2 Photochemical reactions

Photochemical ring cleavage also occurs readily in systems where stable fragments can be extruded. Photolysis of the 1,3,4-oxadiazinone (127) gives an *E/Z* mixture of stilbenes by initial extrusion of $CO_2$ followed by rearrangement and loss of $N_2$ as shown in Scheme 7 ⟨68JA1061⟩.

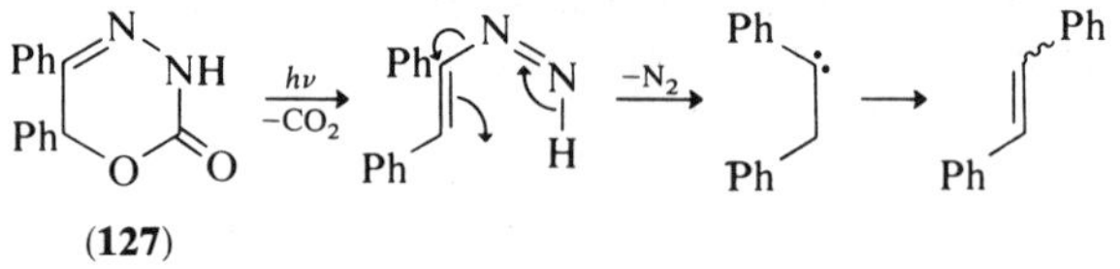

(127)

**Scheme 7**

The naphthothiadiazine (128) loses nitrogen on irradiation to give the 1,8-bridged naphthalene (129) in 25% yield ⟨67LA(703)96⟩. The dithiazine (130), which ring contracts with loss of sulfur on heating (see Section 2.28.3.5.3), extrudes MeSC≡N rather than S on photolysis. The other fragment (131) was trapped as a molybdenum carbonyl complex ⟨73ZC431⟩.

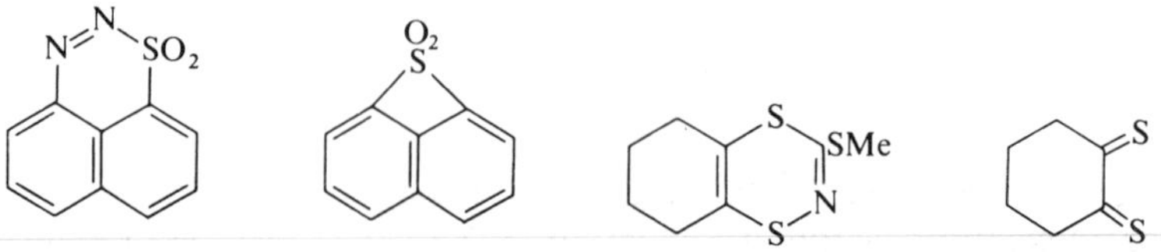

(128)              (129)              (130)              (131)

Photolysis of the 1,2,3,5-oxathiadiazine (**132**) in the presence of ethanol leads to the quinazolone (**133**). The product is not formed in the absence of ethanol, and the proposed mechanism involves electrocyclic ring opening, addition of ethanol, photochemical diradical formation and closure on to the C-6 phenyl ring followed by aromatization as shown in Scheme 8 ⟨79JOC4435⟩.

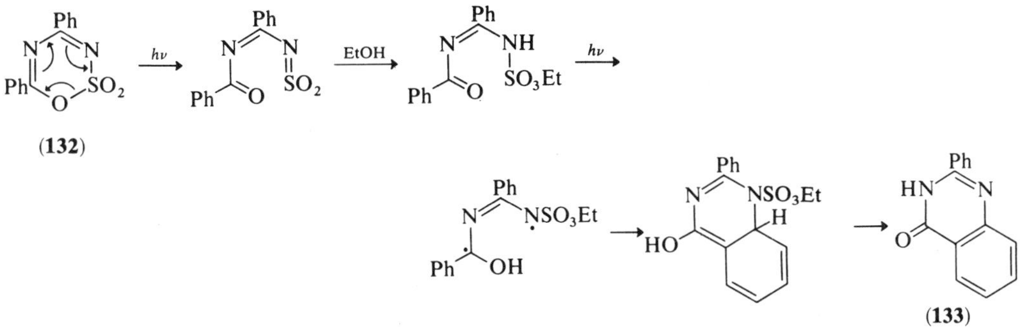

**Scheme 8**

## 2.28.3.5.3 Ring contraction reactions

Six-membered rings containing three or more heteroatoms often undergo ring contraction to form five-membered rings. The extrusion of sulfur is particularly common and in most cases leads to the formation of new C—C or C—N bonds. In many cases the ring contraction occurs simply on heating, although in other cases the addition of acid or base is required. A few examples are presented here in order to illustrate the type of systems that readily undergo ring contraction.

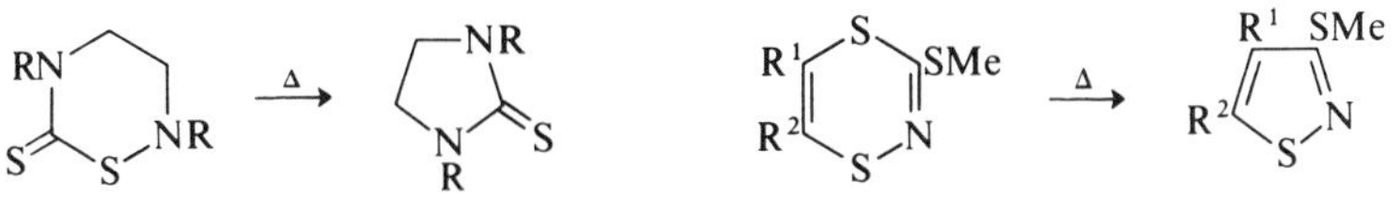

The 1,2,5-thiadiazinethione (**134**) loses sulfur on strong heating to give the imidazolinethione (**135**) ⟨49JOC946⟩. Sulfur extrusion from the dithiazine (**136**) occurs at temperatures of 160–180 °C to form isothiazoles (**137**) ⟨76JPR127⟩. The 1,4,2,6-dithiadiazine (**12**) is surprisingly stable towards extrusion of sulfur ⟨82UP22800⟩, and the ring contraction product, 3,4-diphenyl-1,2,5-thiadiazole, is only formed in low yield after heating at 190 °C. It should be noted that in this case it is S-4 which is extruded. Reactions in which a sulfur is extruded from a molecule in which it is flanked by two nitrogens, to generate a new N—N bond, are extremely rare. The 1,2,6-thiadiazine (**138**; $n = 0$) gives good yields of pyrazoles (**139**) on heating to 90 °C in toluene ⟨81JCS(P1)1891⟩. The S-oxides (**138**; $n = 1$) extrude SO at similar temperature to generate the new N—N bond, although no extrusion of $SO_2$ occurs from the dioxides (**138**; $n = 2$) even at 200 °C. The fact that S and SO are lost from heterocyclic molecules much more easily than $SO_2$ is perhaps surprising at first sight. However, it is a general reaction in this type of chemistry. The only other example of extrusion of sulfur from between two nitrogens involves the thiatriazine (**140**), which on treatment with acid gives 2,4,5-triaryl-1,2,3-triazole with formation of a new N—N bond ⟨70MI22803⟩.

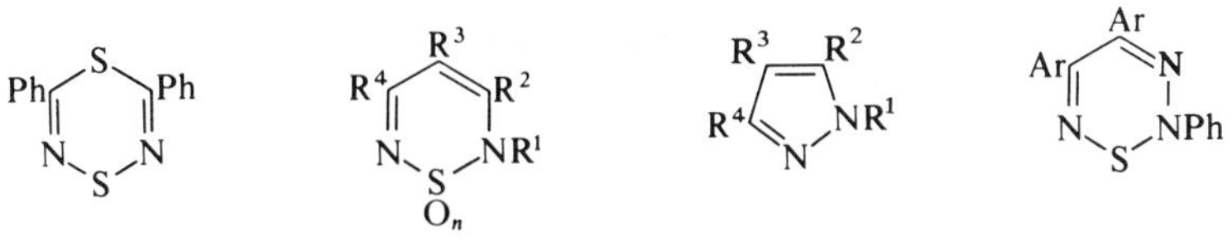

The $8\pi$-electron anions derived from 1,3,5-oxadiazines (**30**) ⟨75AG(E)581⟩ are unstable and readily ring contract to give imidazolones (**141**; Scheme 9). The related anions from 1,3,5-thiadiazines (**100**; Scheme 1) also ring contract to give imidazoles in high yield

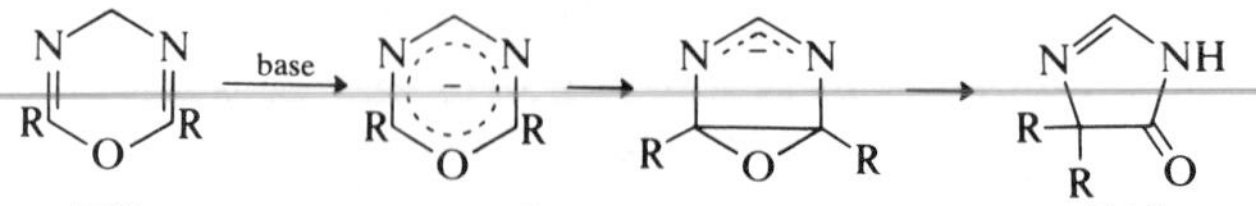

(30)                                    (141)

**Scheme 9**

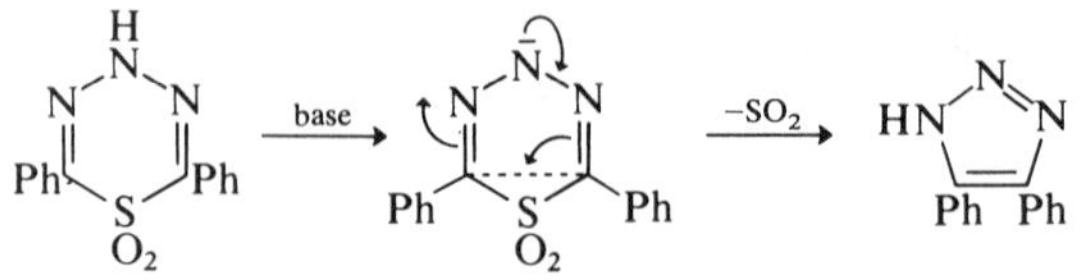

(142)

⟨77JCS(P2)939⟩. A similar mechanism accounts for the loss of SO$_2$ in the ring contraction of the 1,3,4,5-thiatriazine (142) to 4,5-diphenyl-1,2,3-triazole ⟨80JOC2604⟩. The isomeric 1,2,3,6-thiatriazine (143), however, undergoes ring contraction with extrusion of N-6 rather than SO on treatment with acid ⟨78CB1989⟩. The mechanism (Scheme 10) involves initial protonation on oxygen.

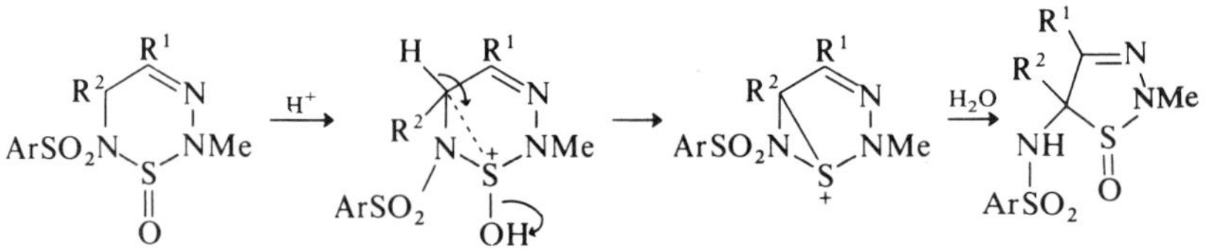

(143)

**Scheme 10**

There are several examples of sulfur extrusion from 1,3,4-thiadiazines (144) either under basic conditions *via* the 8π-anion (31) ⟨75AG(E)581, 77S196⟩, or by heating in acetic acid ⟨70LA(741)45, 76JPR971⟩. Both sets of conditions give pyrazoles in good yield.

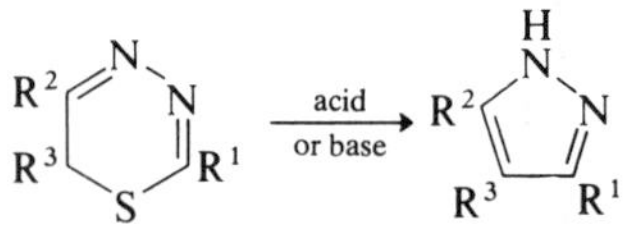

(144)

A few examples of ring contraction reactions involving loss of a fragment which is formally NH have been reported. The benzoxadiazines (56) give benzoxazoles in moderate to good yield on heating in chlorobenzene at 130 °C ⟨75CC962⟩. Small amounts of chloroanilines were isolated and ammonia was also detected. These products represent the fate of NH. A mechanism involving initial electrocyclic ring opening has been proposed. The corresponding 1,2,4-benzothiadiazines (145) also extrude NH to give benzothiazoles on heating at 180 °C or on irradiation. The fact that higher temperatures are required reflects the increased strength of the N—S bond over the N—O bond. The ring contraction of (145) was markedly promoted by tervalent phosphorus compounds, which also served to partially trap the NH as a phosphinimine ⟨79JCS(P1)2851⟩.

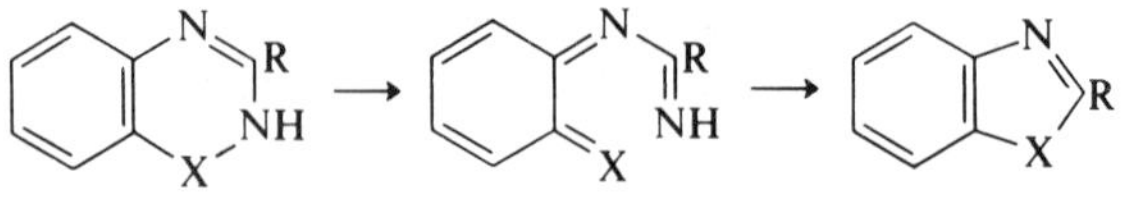

(56) X = O
(145) X = S

### 2.28.3.5.4 Desulfurization and reductive ring cleavage

Sulfur-containing heterocyclic rings are readily desulfurized by treatment with Raney nickel. Thus the 1,2,4-thiadiazine (146) is fully desulfurized and cleaved to give (147) ⟨67CB2159⟩. Similarly, Raney nickel treatment of (148) gives RO$_2$CNHNHCO$_2$R ⟨71AG(E)407⟩.

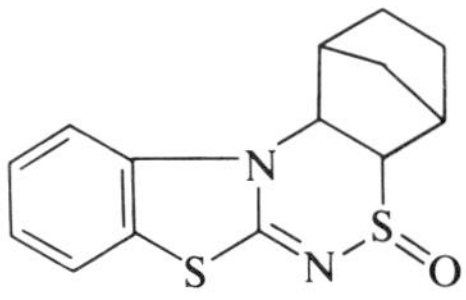

(146) 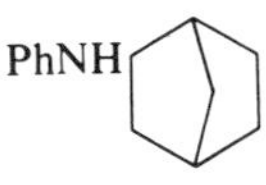  (147) 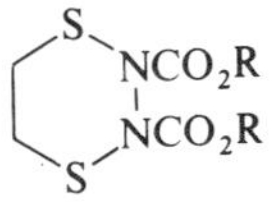  (148)

Other desulfurization reagents are also effective, and the dithiadiazine (149) gives the thiadiazole (150) on treatment with triphenylphosphine ⟨76JCS(P1)38⟩. Triphenylphosphine also converts thiadiazines (144) into pyrazoles ⟨77S485⟩, and 2,1,4-benzothiadiazines (151) into the corresponding benzimidazoles ⟨74BRP1350277⟩. This latter reaction is in marked contrast to that of the isomeric 1,2,4-benzothiadiazines (145), which lose NH rather than S on treatment with triphenylphosphine.

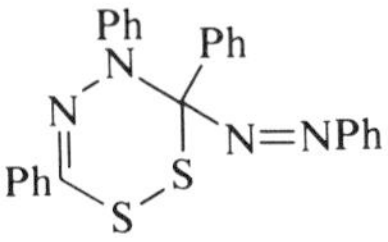

(149) 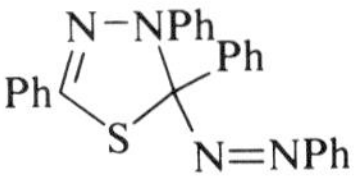  (150) 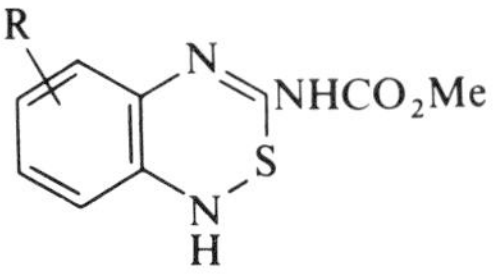  (151) 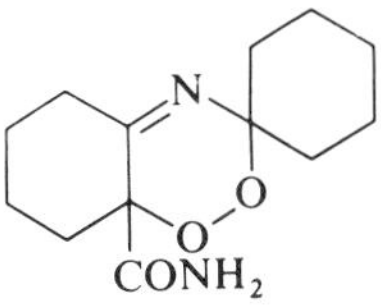  (152)

Reductive cleavage of O—O, O—N, O—S and S—N bonds is also possible. For example, the peroxide bond in the dioxazine (152) is cleaved by treatment with zinc in ethanolic potassium hydroxide, or by catalytic hydrogenolysis ⟨64JOC291⟩. Zinc in acid cleaves the S—N bond in (153) ⟨68LA(715)223⟩. Concomitant reduction of the sulfoxide also occurs, so that subsequent recyclization and aromatization with loss of ammonia leads to the benzothiazole (154). The O—S bond in (3) is reductively cleaved by sodium borohydride to give the alcohol (155) ⟨77TL4245⟩.

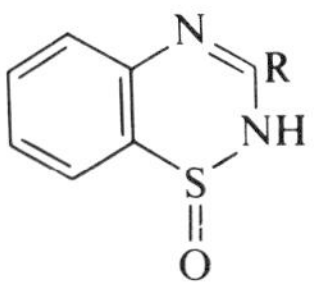

(153) 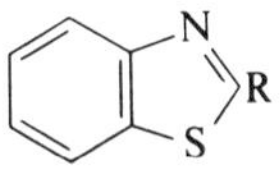  (154) 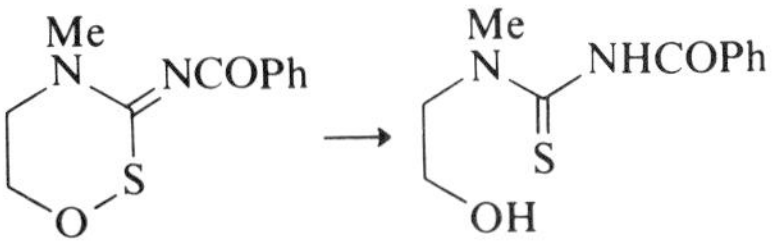  (3)  (155)

## 2.28.3.6 Reactivity of Substituents

In six-membered rings containing three or more heteroatoms, any substituent must necessarily be close to, or on, a heteroatom. In this respect the reactivity of substituents in these systems is controlled by the adjacent heteroatoms. For example, halogens on carbon atoms $\alpha$ to a heteroatom are readily displaced. For the most part, the reactivity of substituents is fairly easy to predict on the basis of the chemistry of the functional groups of which they are part, and of the generalizations outlined in Chapter 2.02.

### 2.28.3.6.1 Fused benzene rings

Since these heterocyclic ring systems are not aromatic, they are best considered as substituents on the fused benzene ring, *i.e.* the reactivity shown by the benzene ring towards electrophiles, for example, will be predictable by considering the heterocyclic ring simply as two *ortho* substituents. The reactivity will then be a result of the normal electron releasing or withdrawing properties of the substituents.

### 2.28.3.6.2 C-linked substituents

Carbon substituents are again best considered as part of the particular functional group within the heterocyclic ring, and usually show the expected chemical reactivity. Carbon substituents on sulfur are necessarily adjacent to $S^+$, and a number of rearrangements involving this substituent have been reported. Thus methylation of the ylide (156) gave an intermediate azasulfonium salt (157) which underwent a Pummerer-type rearrangement to give the isolated ring-expanded product ⟨79JCR(S)214⟩.

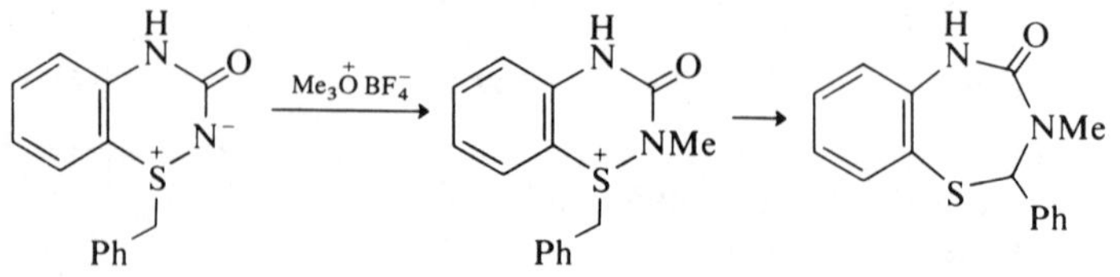

(156)    (157)

The ylidic 1,3,4-benzothiadiazine (158) also undergoes rearrangements involving the carbon substituent on sulfur ⟨81JCS(P1)2245⟩. The products, which depend on the nature of R, arise from competing [1, 2], [1, 4] and [2, 3] shifts (Scheme 11). A further pathway involves loss of the substituent from sulfur to give (159).

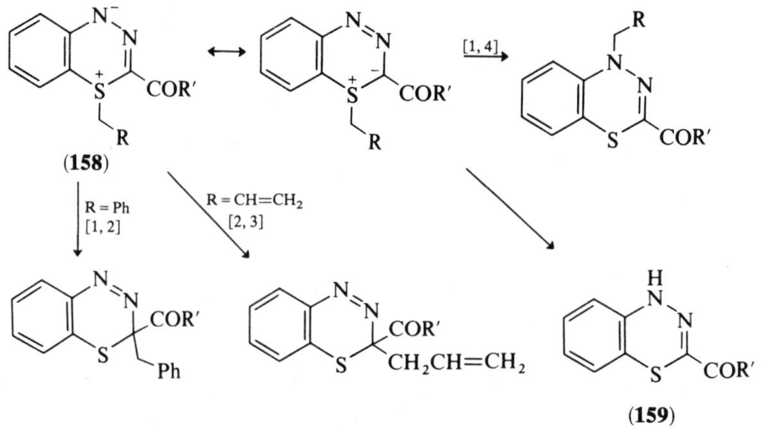

(158)

Scheme 11

A similar reaction involving total loss of the substituent from an ylidic sulfur atom also occurs in the 1,2,4-benzothiadiazine (160) ⟨78CC1049⟩ on heating to 140 °C. The mechanism probably involves cycloelimination of dehydromorpholine as shown.

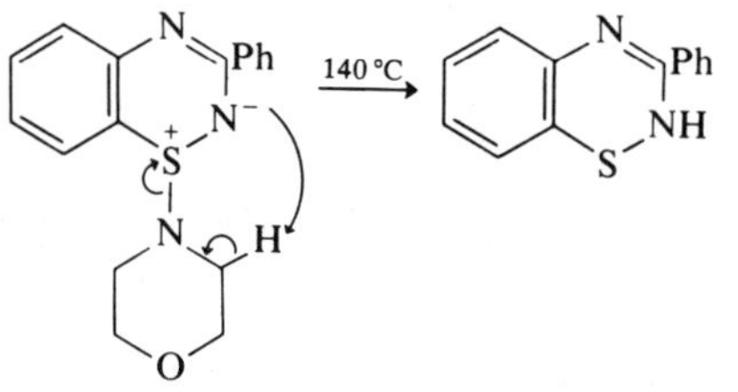

(160)

A ring expansion involving the 3-substituent occurs when the sulfoximide (161) is treated with ethanolic sodium hydroxide ⟨72CB757, 73CB3368⟩.

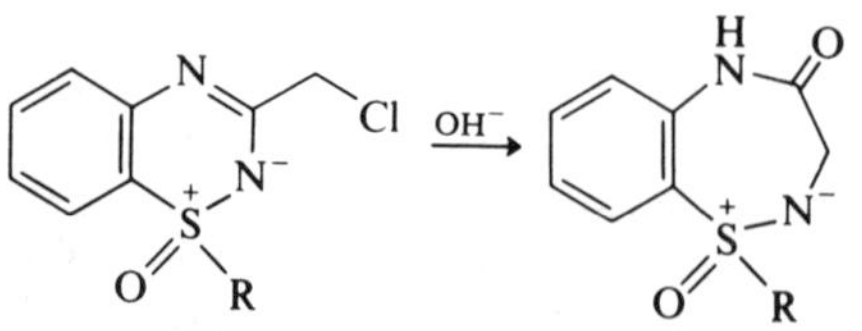

(161)

### 2.28.3.6.3 *N-Linked substituents*

Amino substituents on these systems usually react normally. They are commonly encountered as substituents on the $sp^2$ carbon of a C=N group, and hence are more properly considered as amidines. Exocyclic amino groups can obviously interfere during attempted reaction at ring nitrogen atoms, and for 3-amino-1,2,4-thiadiazines acylation occurs exclusively at the exocyclic nitrogen ⟨74BSF1395⟩.

### 2.28.3.6.4 O- and S-linked substituents

The question of tautomerism in hydroxy substituted heterocyclic systems has already been discussed in Chapters 2.01 and 2.02. Systems with three or more heteroatoms with —OH and —SH substituents usually exist in the carbonyl or thiocarbonyl forms with the tautomeric proton on a ring nitrogen. Nevertheless, alkylation on an exocyclic oxygen or sulfur can usually be achieved under conditions which would be used for *O*- or *S*-alkylation of amides or thioamides. Alternatively, alkoxy and alkylthio substituted systems are often easily prepared from the corresponding chloride by nucleophilic displacement. Alkylation on oxygen or sulfur also occurs in systems without the tautomeric NH. Thus treatment of (**54**) with triethyloxonium tetrafluoroborate gives the oxonium ion (**162**) ⟨75S522⟩. Likewise, methylation of (**163**) with iodomethane gives the $6\pi$-thiadiazinium ion (**164**) ⟨71JHC1087⟩.

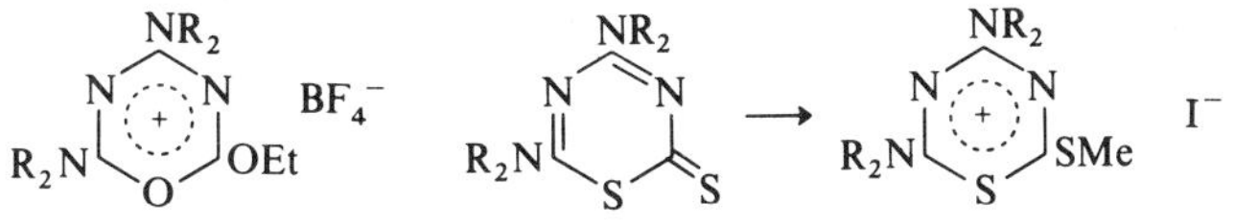

(162)          (163)          (164)

### 2.28.3.6.5 Halogen substituents

Halogen substituents on heterocyclic rings with three or more heteroatoms tend to be highly reactive towards nucleophilic displacement. The most commonly encountered situation is a chlorine substituent on the carbon of a C=N group and, since the systems are not aromatic, these compounds are best considered as cyclic imidoyl chlorides. The chloride is readily displaced by a wide range of nitrogen, oxygen and sulfur nucleophiles. Chloride is also displaced by nucleophiles from sulfur(IV) atoms. Some typical examples of systems containing reactive chlorine are compounds (**165**) ⟨76CB2097⟩, (**166**) ⟨75ZC19⟩, (**167**) ⟨72BCJ1567⟩, (**168**) ⟨78JPR452⟩, (**169**) ⟨76CB2107⟩ and (**170**) ⟨80ZOR1308⟩. The 4-bromo-1,2,6-thiadiazine 2,2-dioxide (**97**) acts as a source of Br$^+$ ⟨80JHC977⟩.

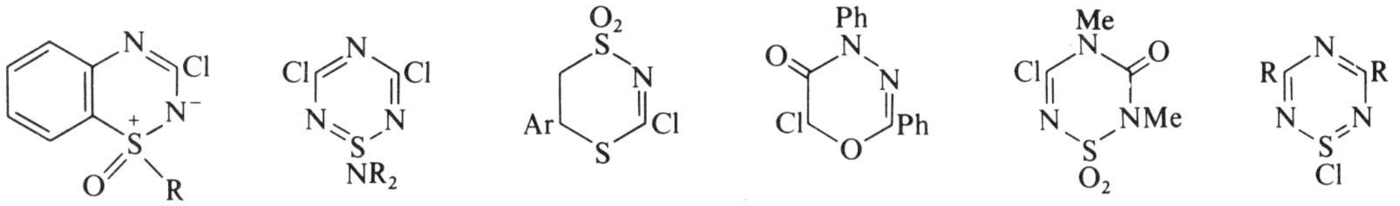

(165)          (166)          (167)          (168)          (169)          (170)

### 2.28.4 SYNTHESES

The methods for synthesizing six-membered rings containing three or more heteroatoms are, not surprisingly, as diverse as the ring systems themselves. This chapter covers over 50 different ring systems, and rather than list them individually, the methods of preparation are broken down according to the system used in Chapter 2.03. Thus the methods are classified according to the size and number of fragments which together make the six-membered ring, and according to the bond or bonds which are formed during the ring closure step. The intention is to illustrate the different types of method available, and, as in Section 2.28.3, each ring system will *not* be considered separately. Access to information on the preparation of specific ring systems is best obtained through the tables in Section 2.28.2.1, and through the summary in Section 2.28.4.5.

### 2.28.4.1 Formation of One Bond

#### 2.28.4.1.1 Between two heteroatoms

Formation of a bond between two heteroatoms in the cyclization step is comparatively rare in heterocyclic synthesis. However, in these polyheteroatom systems it is quite common to form such a bond, particularly between nitrogen and sulfur.

Bond formation between nitrogen and oxygen is rare, and in some cases the resulting structures are unsubstantiated. Thus treatment of isonitrosotriphenylpyrrole (171) with hydroxylamine hydrochloride in ethanol is reported to give a 1,2,6-oxadiazine, presumably *via* N—O bond formation in the trioxime intermediate ⟨35G176⟩. Reaction of hydroxylamine with (172) is also reported to give a 1,2,6-oxadiazine ⟨52JCS3428⟩. Again the cyclization must involve N—O bond formation in the intermediate (173). Cyclization of (174) occurs on heating in carbon tetrachloride in the presence of triethylamine hydrochloride to give the only reported example of a 1,3,2-dioxazine (175) ⟨80IZV2669⟩.

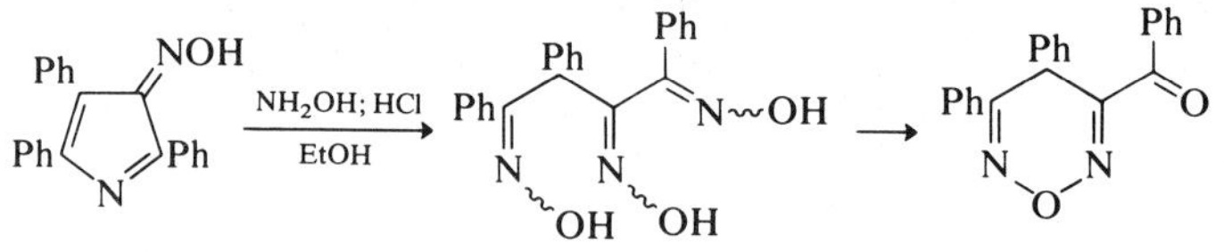

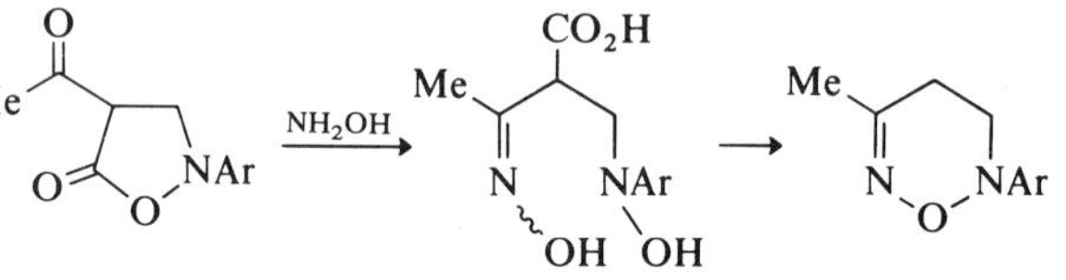

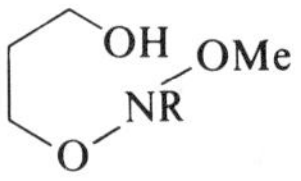

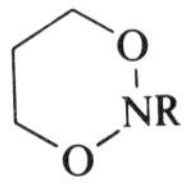

Ring closure reactions involving N—S bond formation are quite common where the sulfur atom is in a higher oxidation state ⟨70CRV593⟩. 1,2,3-Benzothiadiazine 1,1-dioxides (176), which are quite properly considered as cyclic sulfonylhydrazides, are prepared by treating the hydrazone (177) with phosphorus pentachloride, and diazotization of 8-aminonaphthalene-1-sulfinic acid gives the bridged naphthalene (128). Nitrogen–sulfur bond formation occurs during the ring closure of the *N*-sulfinylanilines (178; X = CO or SO₂) to (179), and of the sulfurdiimide (180) to the benzodithiadiazine (112) ⟨71LA(749)171⟩.

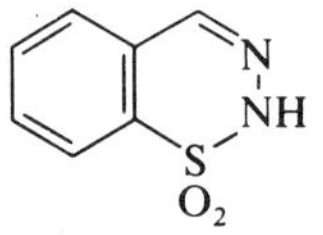

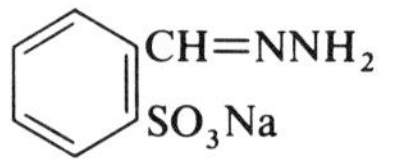

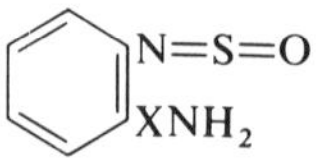

Compounds containing an ylidic N—S bond are best considered as cyclic sulfimides, and as such can be prepared by methods used for their acyclic counterparts ⟨77CRV409⟩. For example, oxidative ring closure of (181) with bromine gives the cyclic sulfimide (93) ⟨64LA(675)189⟩, and treatment of *N*-(2-phenylthio)phenylbenzamidine with *N*-chlorosuccinimide (NCS) leads to the cyclic sulfimide (182) ⟨78CC1049⟩.

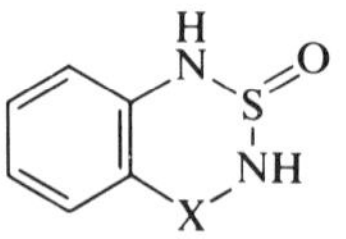

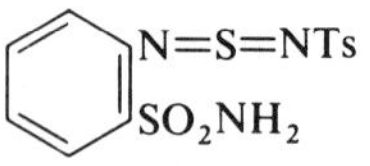

Bond formation between oxygen and sulfur occurs when the *N*-(2-hydroxyethyl)thiourea (155) is oxidatively cyclized with bromine to give the 1,2,4-oxathiazine (3) ⟨77TL4245⟩, and when the intermediate (183), from reaction of ketones with fluorosulfonyl isocyanate (FSI), is cyclized with base to give (184) ⟨80AG(E)131⟩.

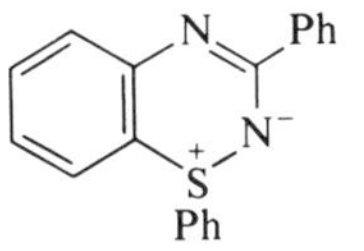

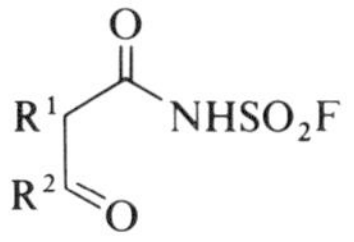

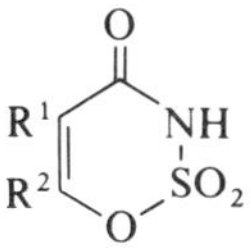

(181)  (182)  (183)  (184)

### 2.28.4.1.2 Between carbon and nitrogen

Formation of a bond between carbon and a heteroatom is the ring closure step in the majority of reactions which lead to heterocyclic systems containing three or more heteroatoms. These [6 + 0] cyclizations are classified according to the heteroatom which forms a bond to carbon in the cyclization step.

Intramolecular nucleophilic attack by nitrogen at a carbonyl carbon is a common method of ring closure in these systems, and a few illustrative examples are given here. $\alpha$-(Ethoxycarbonyl)methylsulfonyl urea (185; X = SO$_2$) cyclizes to the 1,2,4-thiadiazine (53) on treatment with base ⟨59JA5655⟩, and a similar cyclization occurs in the oxygen analogue (185; X = O) under basic conditions to give (73) ⟨79JHC161⟩. The cytosine analogue, the 1,2,4-thiadiazine (10), is prepared by hydrogenation of 3-sulfamidoisoxazole in the presence of sodium methoxide, and the reaction involves closure of the sulfamide nitrogen on to the carbonyl formed by cleavage of the isoxazole ring ⟨79JOC4191⟩.

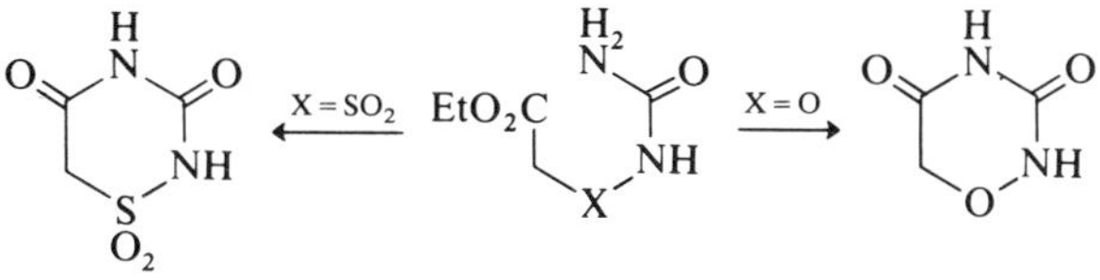

(53)  (185)  (73)

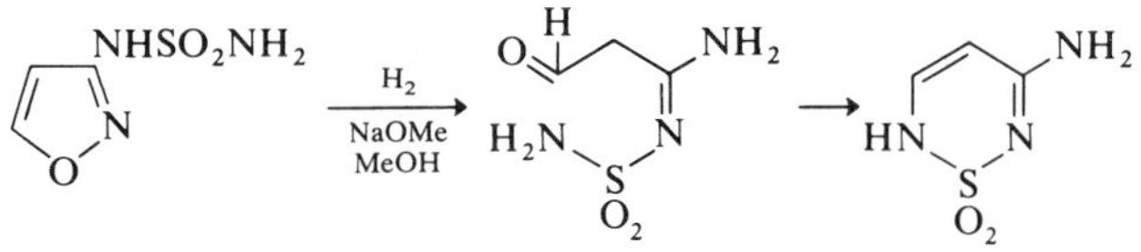

(10)

The sulfoximide (186) ring closes to the 1,2,4-thiadiazine (187) on treatment with methoxide ⟨77JOC952⟩, and (188) cyclizes to (189) with loss of methanol on heating in DMF ⟨71AG(E)264⟩. Other 1,2,4-thiadiazine 1,1-dioxides (191) and (193) are prepared by similar thermal cyclizations of the precursors (190) ⟨76JHC615⟩ and (192) ⟨69CI(L)1305⟩ respectively.

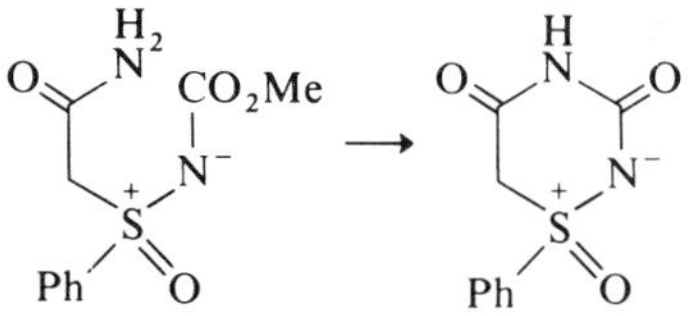

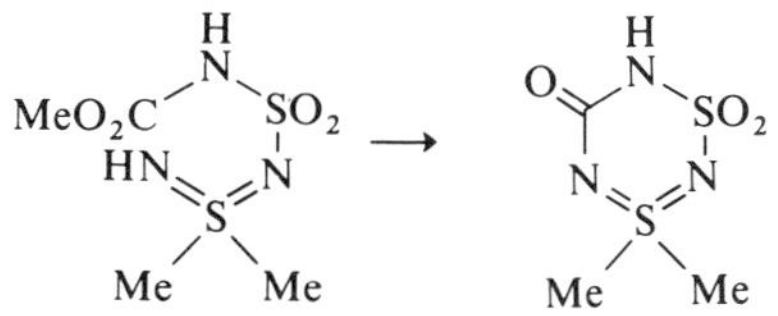

(186)  (187)  (188)  (189)

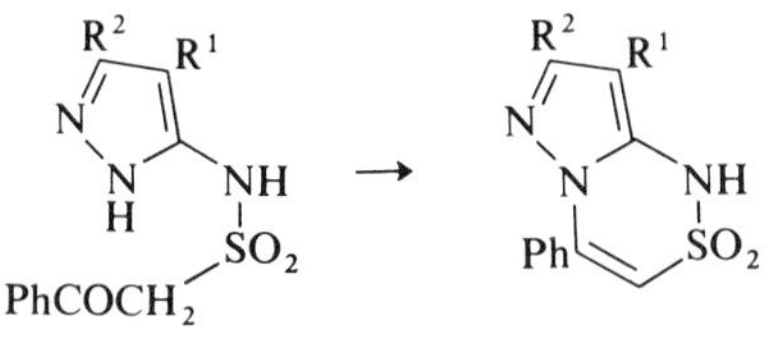

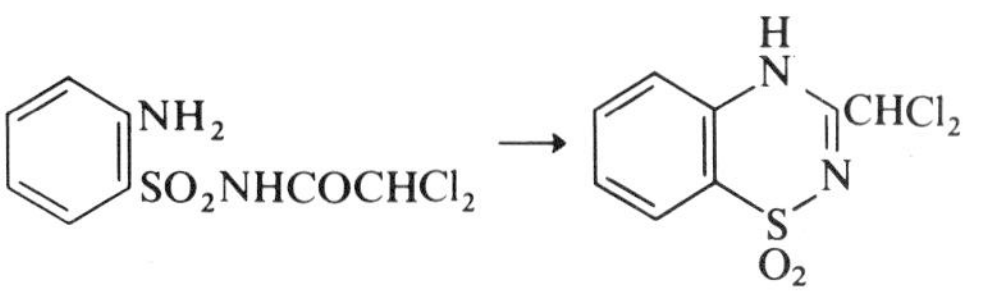

(190)  (191)  (192)  (193)

In the latter example, cyclization occurs exclusively to the carbonyl carbon rather than with nucleophilic displacement of chloride at the $\alpha$-carbon. Intramolecular alkylation of nitrogen does occur in other systems; the 3-chloropropylsulfonylhydrazide (194), for example, ring closes to give the 1,2,3-thiadiazine dioxide (195) on treatment with base ⟨62JPR(19)56⟩.

Intramolecular conjugate addition occurs in the $\beta$-sulfonylstyrenes (**196**) ⟨72BCJ1893⟩ and (**197**) ⟨74JMC549⟩ to give 1,2,4-thiadiazines. In the case of (**197**), subsequent elimination of HBr gave the fully unsaturated system. Similar intramolecular conjugate addition of oxygen and sulfur nucleophiles to sulfonylstyrenes and sulfonylalkynes also occurs.

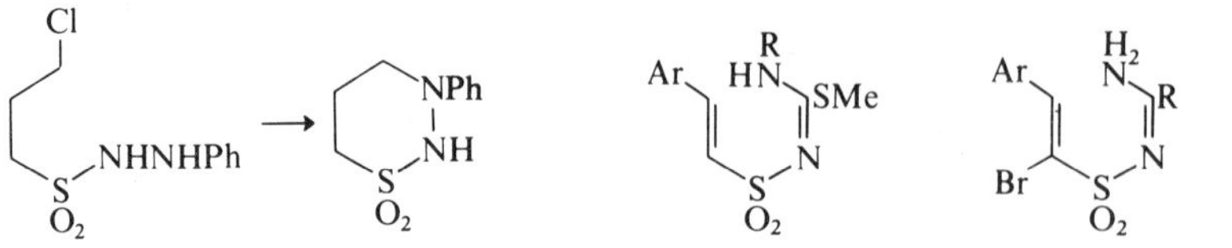

(**194**)          (**195**)          (**196**)          (**197**)

### 2.28.4.1.3 Between carbon and oxygen

Acid-catalyzed dehydration of *N*-(2-hydroxyethyl)-*N'*-acylhydrazines (**198**) is a general route to 4,5-dihydro-1,3,4-oxadiazines (**199**) ⟨64JOC668⟩. The 1,3,4-oxadiazinone (**110**) is also prepared by cyclodehydration, by treatment of (**200**; X—Y = N=CPh) with DCC ⟨77S252⟩, and the 1,4,2-dioxazine (**105**) is prepared in analogous manner from (**200**; X—Y = OCR$_2$) ⟨67JCS(C)1178⟩. Intramolecular *O*-alkylation of amidoximes gives 4,5-dihydro-1,2,4-oxadiazines; thus the 2-chloroethyl amidoxime (**201**) readily ring closes with formation of a C—O bond to give the heterocycle ⟨75CB1911⟩.

(**198**)          (**199**)          (**200**)          (**201**)

Intramolecular conjugate addition of the urea carbonyl oxygen to the sulfonyl alkyne in (**202**) leads directly to 1,4,3-oxathiazine 4,4-dioxides (**203**) ⟨78BCJ1805⟩.

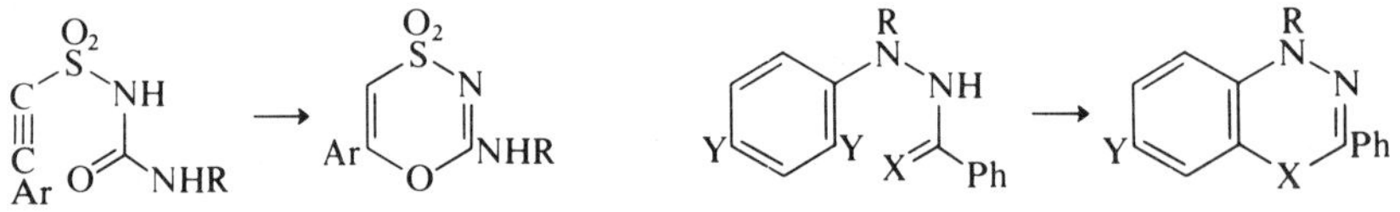

(**202**)          (**203**)          (**204**)          (**205**)

Bond formation between an aromatic ring carbon and oxygen can lead to 1,3,4- and 1,2,4-benzoxadiazines. The 1,3,4-isomers (**205**; X = O) result from ring closure of the phenylhydrazide (**204**; X = O, Y = Br or NO$_2$) with displacement of Y ⟨72JCS(P1)2915, 80JOC3677⟩. The 1,2,4-isomers can also be prepared by displacement of a nitro group from an aromatic ring. The amidoxime (**206**) was thought to give the 6-nitro-1,2,4-benzoxadiazine (**207**) ⟨B-61MI22800⟩, but subsequent work ⟨76JCS(P1)2161⟩ established that a rearrangement occurred during cyclization and that the product was in fact the 7-nitro isomer (**208**). Two alternative routes to 1,2,4-benzoxadiazines were also reported: reaction of *N*-aryl-*S,S*-dimethylsulfimides with nitrile oxides, and oxidation of *N*-arylamidoximes, both give the benzoxadiazines (**56**) in good yield. Both reactions have been shown to involve nitrosoimine intermediates (Scheme 12), which can be reversibly intercepted in a Diels–Alder reaction with the electron-rich diene thebaine.

### 2.28.4.1.4 Between carbon and sulfur

Reactions in which ring closure involves carbon–oxygen bond formation often have direct analogy in the preparation of sulfur-containing heterocycles by carbon–sulfur bond formation. Thus the 1,3,4-thiadiazine (**111**) is prepared from (**209**) by treatment with DCC ⟨81CC1003⟩, intramolecular conjugate addition occurs in the $\beta$-sulfonylstyrenes (**210**) to give 1,4,2-dithiazines (**211**) ⟨72BCJ525⟩, and 1,3,4-benzothiadiazines (**205**; X = S) are prepared by cyclization of (**204**; X = S, Y = Br or NO$_2$) ⟨75CJC1484, 80JOC3677⟩.

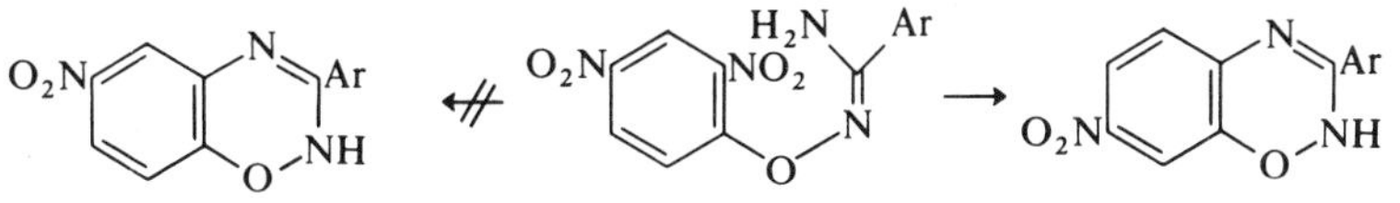

(207) (206) (208)

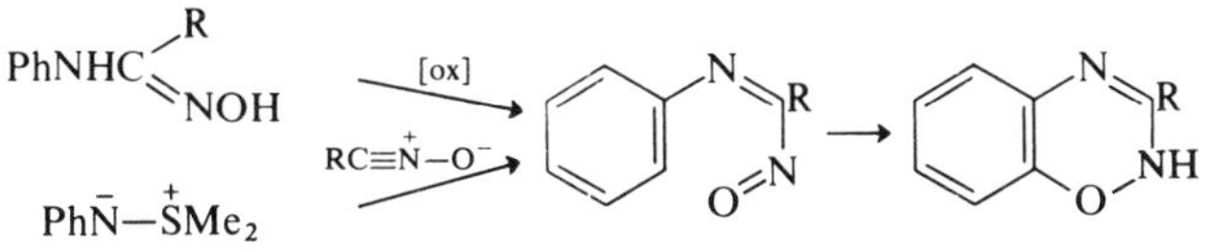

**Scheme 12**

(56)

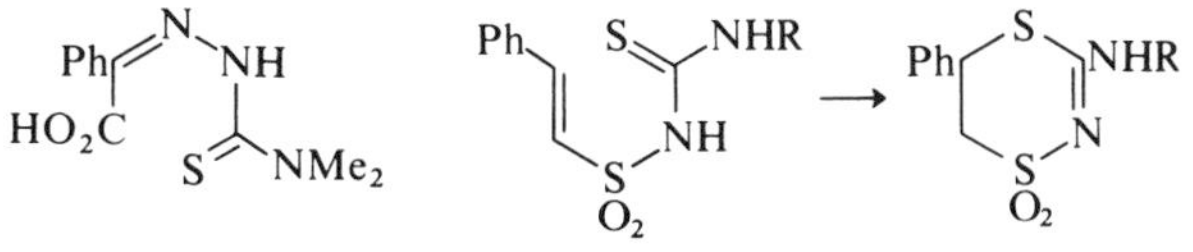

(209) (210) (211)

1,3,4-Benzothiadiazines (**158**) are prepared by cyclization of the chlorohydrazones (**212**) on treatment with triethylamine ⟨81JCS(P1)2245⟩. The reaction probably involves intramolecular nucleophilic attack by sulfur on a nitrile imine intermediate, since (**212**) does not cyclize on heating alone. Intramolecular aromatic sulfonation gives 1,2,4-benzothiadiazine 1,1-dioxides (**214**). Thus treatment of (**213**) with aluminum trichloride leads to cyclization with bond formation between sulfur and the aromatic ring ⟨79JCS(P1)1043⟩.

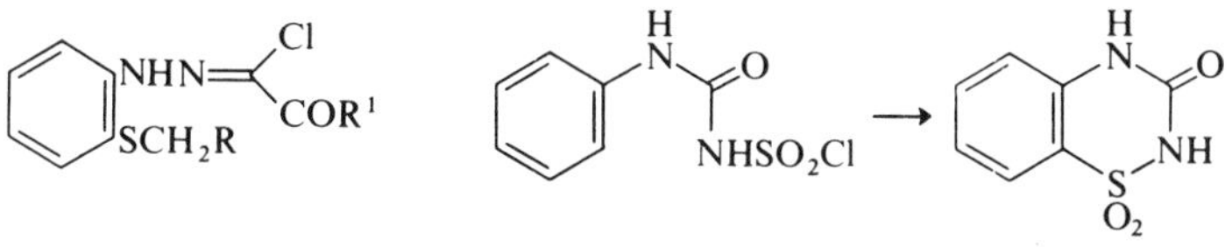

(212) (213) (214)

## 2.28.4.2 Formation of Two Bonds

### 2.28.4.2.1 From [5 + 1] atom fragments

Methods involving the formation of two bonds from [5 + 1] atom fragments are important in the synthesis of polyheteroatom six-membered rings. In these systems the one-atom fragment is usually an electrophilic carbon, a nucleophilic nitrogen or an electrophilic sulfur atom, and examples of all three types of reaction are included here.

The electrophilic one-carbon species can be an aldehyde, ketone, carboxylic acid, phosgene, thiophosgene, carbonyl diimidazole (CDI), *etc.*, and the reaction essentially involves condensation with a 1,5-dinucleophile. The 1,5-dinucleophile invariably contains at least one heteroatom at a terminus, and more often than not two, so that the cyclization always involves the formation of at least one bond between carbon and a heteroatom.

Reaction of the 1,2-dihydroxylamine (**215**) with phosgene or CDI gives the 5-hydroxy-1,2,5-oxadiazine (**216**) ⟨80AP35⟩. Other N–O, O–O, and N–N 1,5-dinucleophiles condense with CDI. For example, (**217**) gives the 1,4,2-dioxazine (**218**) ⟨80AP377⟩, (**219**) gives the 1,5,2-dioxazinone (**103**) ⟨81S38⟩, and sulfoximide (**220**) gives the cyclic sulfoximide (**221**) ⟨73JOC20⟩. The condensation of (**220**) with other one-carbon electrophiles such as formic acid also gives the expected cyclic product. Condensation of the N–S 1,5-dinucleophile (**222**) with thiophosgene gives the 1,3,5-thiadiazine (**163**) ⟨71JHC1087⟩.

(215) (216) (217) (218)

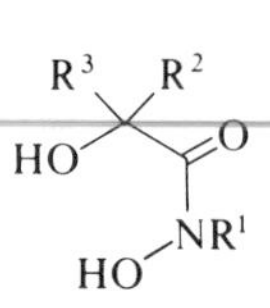

(219) 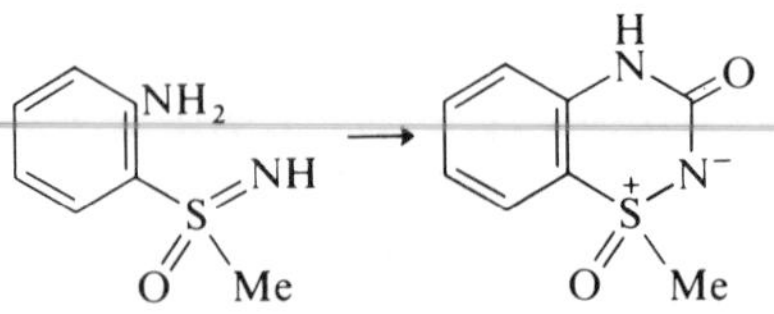     (220)     (221) 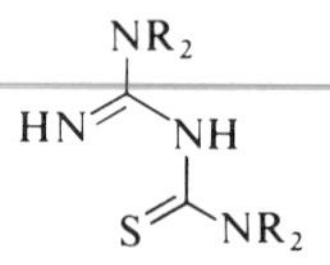     (222)

Nitrogen–sulphur 1,5-dinucleophiles such as (223) also condense with carbonyl compounds to give 1,3,4-thiadiazines (224) ⟨80ZOR1952⟩. Formaldehyde is commonly used as the electrophilic one-carbon unit, and it readily reacts with N–O 1,5-dinucleophiles such as the α-amino oxime (225) to give 1,2,5-oxadiazines (226) ⟨78MI22801⟩. An aromatic ring carbon can act as one of the nucleophilic termini of the five-atom fragment, and examples of this type of process include the reaction of PhNHSO₂NHR with formaldehyde to give the 2,1,3-benzothiadiazine (85) ⟨79JHC1069⟩.

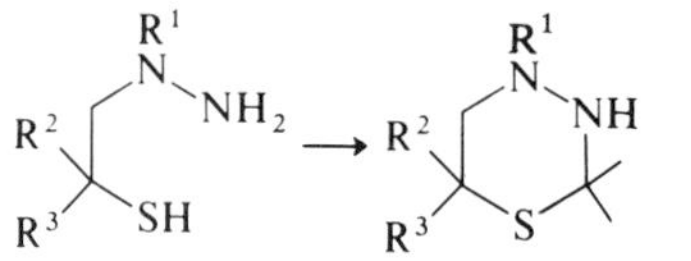

(223)     (224) 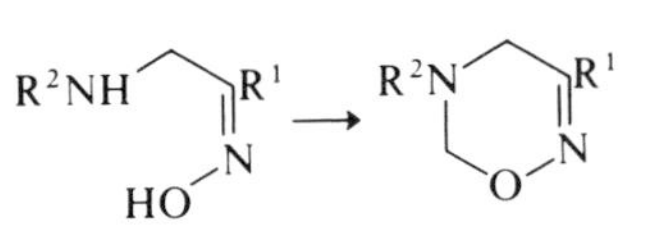     (225)     (226)

Diazotization of 2-aminophenols followed by reaction of the zwitterion with diazomethane gives 1,3,4-benzoxadiazines (Scheme 13). The initially formed 2*H*-isomers readily tautomerize to the more stable 4*H*-isomers ⟨70CB331⟩.

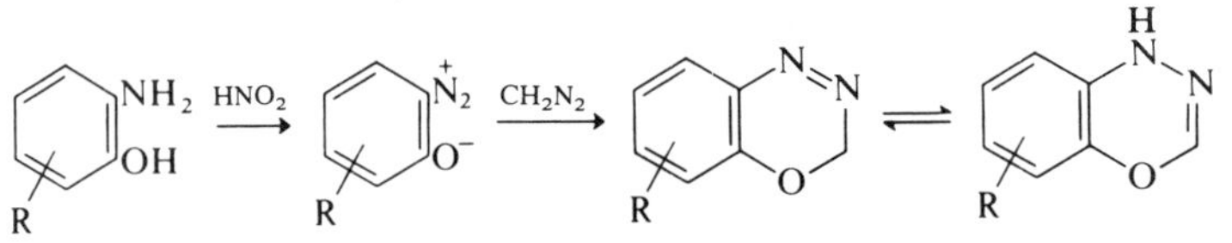

**Scheme 13**

Condensation of 1,5-dielectrophiles with nitrogen nucleophiles usually proceeds readily. Thus treatment of propane-1,3-disulfonyl chloride with hydrazine gives 2-amino-1,3,2-dithiazine 1,1,3,3-tetraoxide (5) ⟨75CB1087⟩, the diacid fluoride (227) gives the thiatriazine (228) on reaction with methylamine ⟨78MI22803⟩, and sulfonyl diisocyanate gives the thiatriazine (229) on treatment with ammonia ⟨58CB1200⟩. Reaction of sulfonyl diisocyanate with water (1 mol) gives the anhydride 1,4,3,5-oxathiadiazine (230). This latter reaction is an example of an oxygen nucleophile acting as the one-atom component.

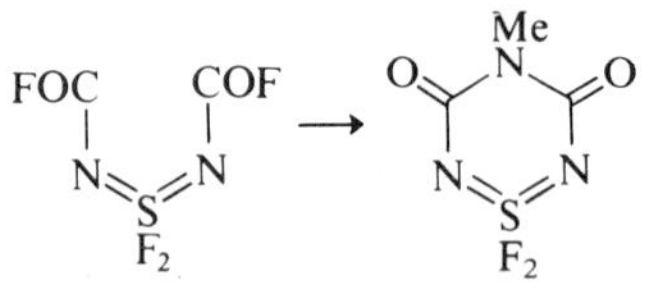

(227)     (228) 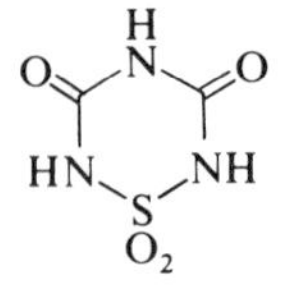     (229) 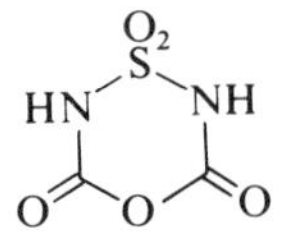     (230)

Electrophilic sulfur species such as thionyl chloride are frequently used in the synthesis of sulfur-containing heterocycles and, for example, reaction with N–N 1,5-dinucleophiles is a direct route to 1,2,6-thiadiazine 1-oxides. Thus the thiadiazines (138; *n* = 1) are readily prepared by treatment of (231) with thionyl chloride in pyridine ⟨81JCS(P1)1891⟩, and the benzothiadiazine (232) is similarly made from (233) ⟨64JOC2717⟩. Reaction of (234) with thionyl chloride gives the dithiatriazine (235) ⟨74ZOR488⟩.

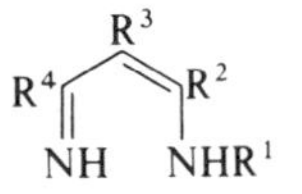

(231) 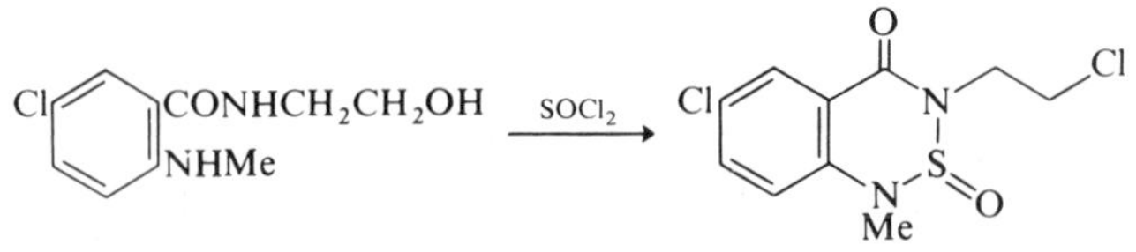     (233)     (232)

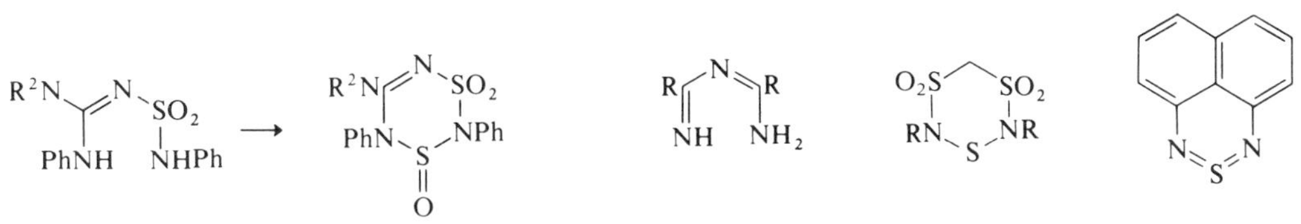

(234)     (235)     (236)     (237)     (238)

Imidoylamidines (**236**) react with an excess of sulfur dichloride to give the cyclic ylides (**170**) ⟨80ZOR1308⟩. Reaction between sulfur dichloride and a 1,5-nitrogen dinucleophile was used to prepare (**237**; R = fluorophosphazenyl), the only reported example of the trithiadiazine ring system ⟨78MI22802⟩. *N*-Sulfinylaniline has also been used as a source of sulfur in reaction with N–N 1,5-dinucleophiles, and reaction with 1,8-diaminonaphthalene in the cold gives (**238**) ⟨67CB2164⟩.

Nitrogen–oxygen 1,5-dinucleophiles also react with thionyl chloride to give 1,2,3-oxathiazine 2-oxides. For example, 3-*t*-butylaminopropan-1-ol gives (**239**) on treatment with thionyl chloride in benzene in the presence of triethylamine as base ⟨69JOC175⟩, and the 2-hydroxybenzylamines (**240**) give benzoxathiazines (**241**) under similar conditions ⟨76JHC665⟩.

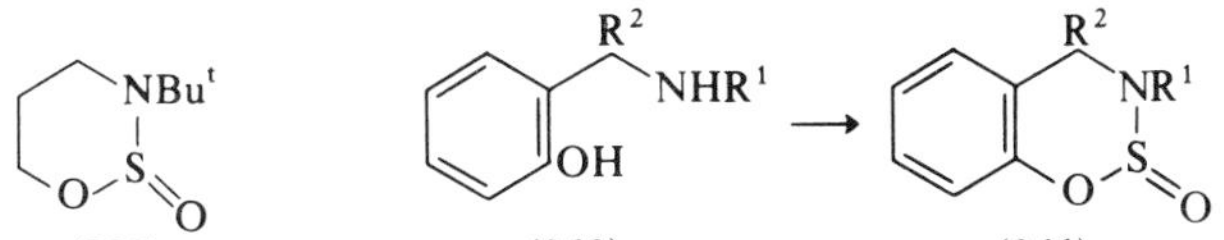

(239)     (240)     (241)

*N*-Arylamidines act as 1,5-dinucleophiles towards sulfur dichloride, *N*-sulfinyltosylamide and bis(*N*-tosyl)sulfur diimide. The products are 1,2,4-benzothiadiazines (Scheme 14) ⟨73ZOR2038, 68LA(715)223⟩, and in the case of sulfur dichloride, chlorination of the aromatic ring also occurs. 1,2,4-Benzothiadiazines can also be prepared from *N*-arylamidines by reaction with sulfenyl chlorides or disulfides in the presence of NCS. Thus *N*-phenylbenzamidine and benzenesulfenyl chloride give the 1,2,4-benzothiadiazine (**182**) ⟨78CC1049⟩.

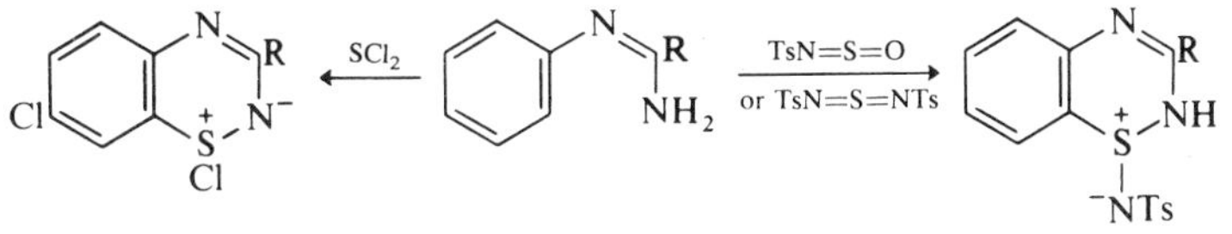

**Scheme 14**

### 2.28.4.2.2 *From [4 + 2] atom fragments*

Condensation reactions involving [4 + 2] atom fragments are important processes for the preparation of six-membered heterocyclic rings, and many examples leading to poly-heteroatom systems are known. A few illustrative examples are given here, and they are classified according to the nature of the two-atom fragment: C—C, C—N, C—O, C—S, N—N, N—S, O—O and O—S.

Condensations involving a two-carbon 1,2-dielectrophile, such as an $\alpha$-halocarbonyl compound or a 1,2-dihaloalkane, are fairly common. Thus reaction of dichloroacetyl chloride with *N*-phenyl-*N'*-benzoylhydrazine gives the 1,3,4-oxadiazine (**168**) ⟨78JPR452⟩. Condensation of $\alpha$-halo ketones with thiohydrazides (Scheme 15) is a useful general route to 1,3,4-thiadiazines and many examples are known ⟨69ZC111, 70LA(741)45, 76JPR971⟩.

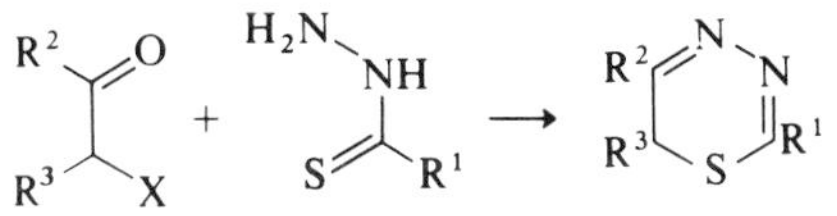

**Scheme 15**

1,4,2-Dioxazines are easily prepared by di-*O*-alkylation of hydroxamic acids (Scheme 16) using either 1,2-dihalides or 1,2-dimesylates ⟨71JOC284, 75NKK1041⟩.

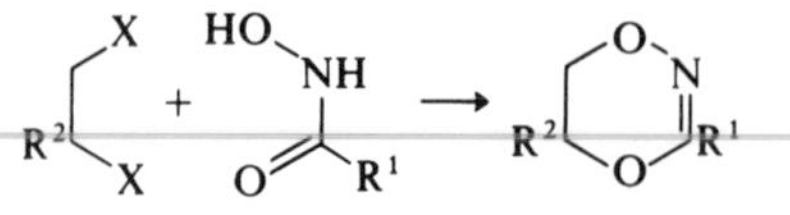

**Scheme 16**

Compounds containing carbon–carbon double bonds react with 'dienes' containing three heteroatoms to give six-membered rings. Although these reactions can be formally represented as Diels–Alder reactions, they are almost certainly stepwise, involving initial attack by the nucleophilic double bond on the electrophilic 'heterodiene'. Thus azodicarbonyl compounds react with alkenes to give 1,3,4-oxadiazines (Scheme 17). The reaction is highly sensitive to substituent effects and the competing 1,2-addition to give 1,2-diazetidines may supervene ⟨82AHC(30)1⟩. In general, the amount of 1,2-addition increases with the donor ability of the alkene substituent X.

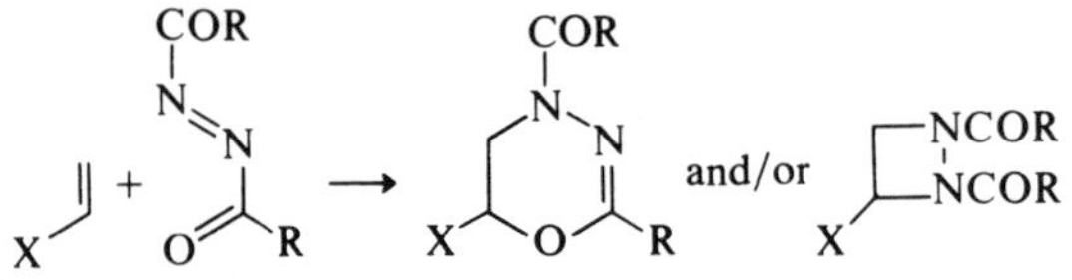

**Scheme 17**

*N*-Sulfinylamidines and urethanes also react as heterodienes. The reaction with strained alkenes is particularly facile, and norbornene reacts with *N*-sulfinyl compounds (**242**) ⟨67CB2159⟩ and (**243**) ⟨68CB3567⟩ in [2 + 4] fashion to give the corresponding heterocyclic systems. *N*-Sulfonylamides and urethanes react similarly, and reaction of (**244**: R = Ph), generated from $RCONHSO_2Cl$ and triethylamine, with enol ethers, and of (**244**; R = OEt) with allenes, both give 1,4,3-oxathiazine 4,4-dioxides ⟨72JA6135⟩.

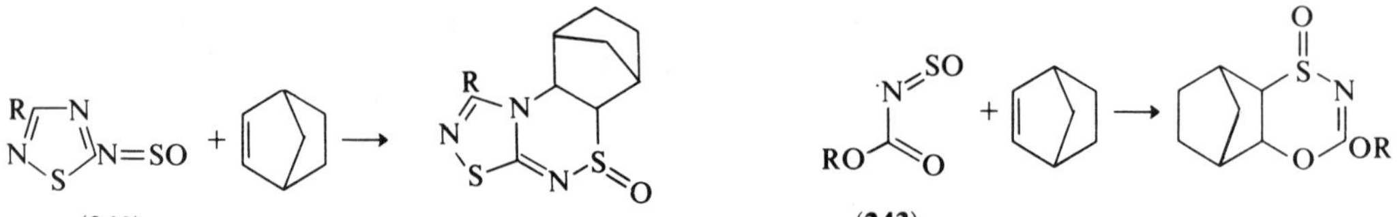

(**242**)  (**243**)

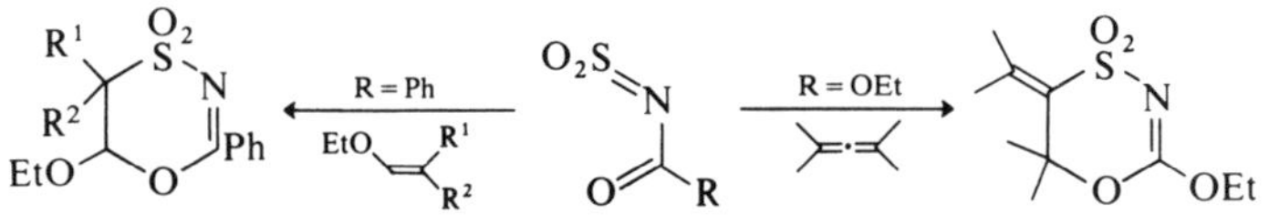

(**244**)

Alkynes react as a two-carbon unit with CSI to give 1,2,3-oxathiazine dioxides. The reaction involves initial addition of CSI to form a four-membered ring which then ring opens and rearranges as shown in Scheme 18 ⟨72JOC196⟩.

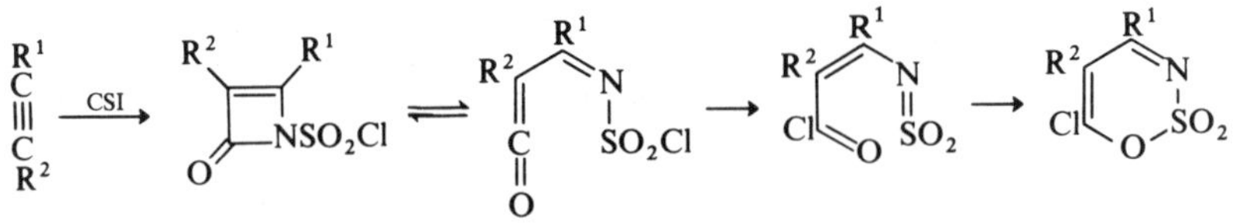

**Scheme 18**

[4 + 2] Condensations involving a C—N unit as the two-atom fragment usually involve cycloaddition to a carbon–nitrogen multiple bond. For example, imines add to thiocarbonyl isocyanates (**245**; X = O) and isothiocyanates (**245**; X = S) in a formal [4 + 2] cycloaddition to give 1,3,5-thiadiazines ⟨70CB3393, 72BCJ2877⟩, and nitriles add to the *N*-sulfonylurethane (**244**; R = OMe) to give the 1,4,3,5-oxathiadiazine (**124**) ⟨73JOC1249⟩.

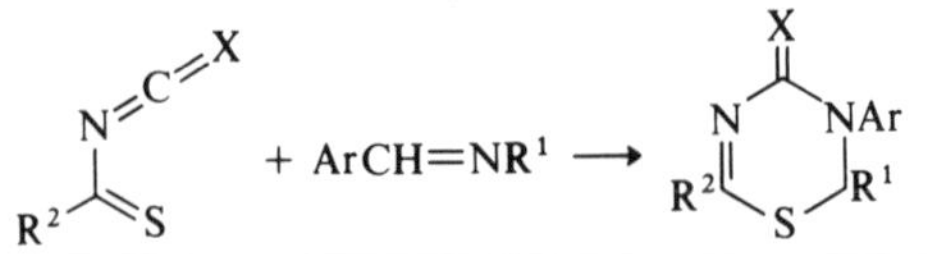

$+ ArCH=NR^1 \longrightarrow$

(**245**)

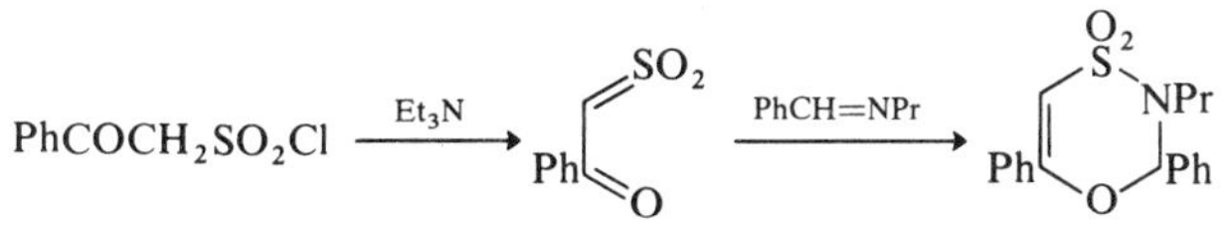

**Scheme 19**

Benzoylsulfene, generated from PhCOCH$_2$SO$_2$Cl, undergoes cycloaddition to imines to give 1,4,3-oxathiazines (Scheme 19) ⟨70BCJ3543⟩.

The C=N bond in cumulenes can also act as the two-atom fragment, and reaction of aroyl isocyanates with carbodiimides gives imino-1,3,5-oxadiazinones (**246**) ⟨79BSF(2)499⟩. The reaction is not a simple [4+2] cycloaddition, as the kinetic product is the [2+2] adduct which then rearranges.

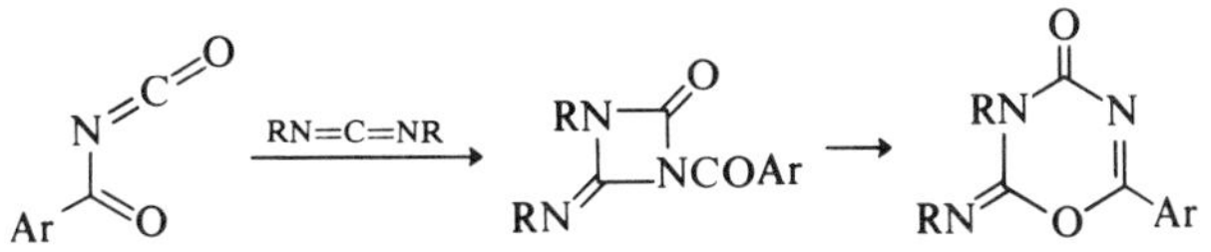

(**246**)

Reactions involving a C—O two-atom fragment are usually cycloadditions to the carbonyl group of aldehydes and ketones. 1,3,5-Oxathiazines, 1,3,5-dioxazines and 1,3,5-oxadiazines can all be prepared by cycloaddition reactions to carbonyl compounds and these reactions are summarized in Scheme 20. Although described as cycloadditions, there is no evidence to suggest that they are concerted. The references for the Scheme are given in clockwise order, beginning at 'two o'clock' ⟨74TL221, 78MI22800, 78S524, 79S801, 79IZV1826⟩.

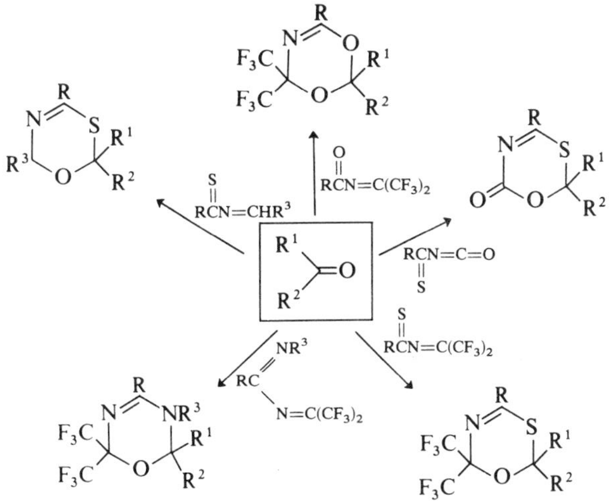

**Scheme 20**

Carbon–sulfur double bonds also participate in cycloadditions, and reactions of thioketones with RCSN=CR$_2$ parallel their oxygen analogues (*cf.* Scheme 20) to give 1,3,5-dithiazines ⟨78S443⟩. Thioketone *S*-oxides (sulfines) react with azadienes at room temperature to give 1,2,3-thiadiazine *S*-oxides (Scheme 21). Some stereoselectivity is observed and the major isomer has the C-5 phenyl group *trans* to the sulfoxide oxygen. The regioisomeric 1,3,4-thiadiazine is also formed but this isomerizes slowly to the 1,2,4-isomer on standing in solution, presumably by a cycloreversion–cycloaddition mechanism ⟨81JCS(P1)2322⟩.

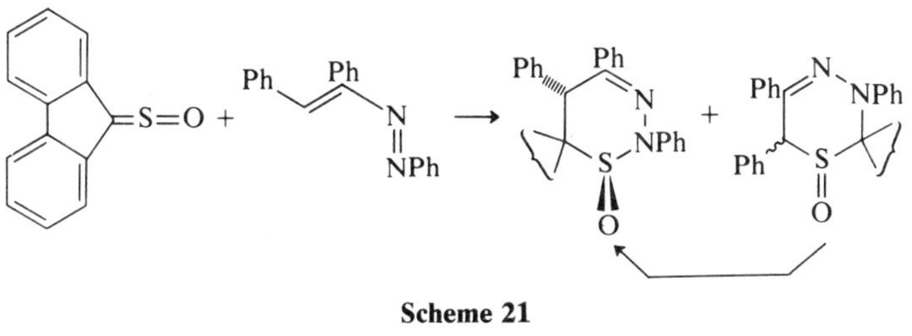

**Scheme 21**

Sulfenes are reactive $2\pi$-components in cycloaddition reactions and treatment of methanesulfonyl chloride with triethylamine in the presence of $(CF_3)_2C=NCOPh$ gives the 1,2,5-oxathiazine (**119**) *via* a formal [4+2] cycloaddition of sulfene to the heterodiene ⟨69IZV2059⟩.

[4+2] Condensation reactions involving an N—N two-atom fragment are rare. Although azodicarbonyl compounds are reactive dienophiles ⟨82AHC(30)1⟩, they have apparently not been added to oxygen- or sulfur-containing heterodienes. Substituted hydrazines have however been used as the two-atom fragment. Reaction of 1,2-hydrazinedicarboxylates with 1,2-disulfenyl chlorides gives 1,4,2,3-dithiadiazines (Scheme 22) ⟨71AG(E)407, 74CB771⟩.

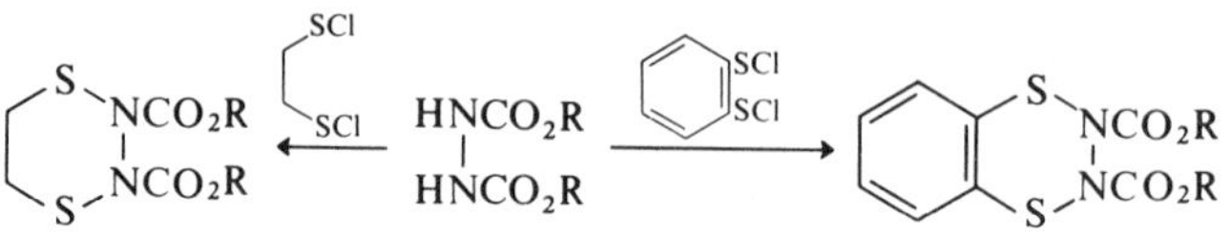

**Scheme 22**

*N*-Sulfinylamines are useful N—S two-atom fragments and react readily with $\alpha$-ketoketenes to give 1,2,3-oxathiazin-4-one 2-oxides ⟨77JCS(P1)904, 79CB1012⟩. Reaction with azoalkenes gives 1,2,3,6-thiatriazines ⟨77S305⟩ as shown in Scheme 23.

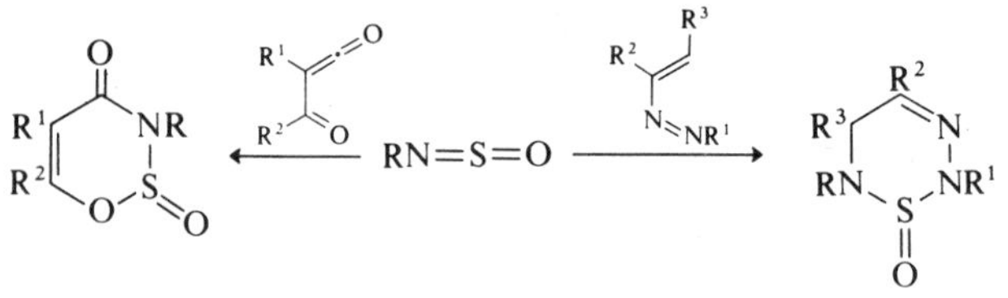

**Scheme 23**

Oxygen in its singlet state can act as a dienophile and therefore reaction with heterodienes should be a useful route to 1,2-dioxa heterocycles. These 1,2-dioxa heterocycles are poorly described in the literature, presumably because of the instability associated with their cyclic peroxide structure. However, the 1,2,4-dioxazine (**152**) is prepared using oxygen as the two-atom fragment. Condensation of cyclohexanone with urea gives (**247**), a compound which readily absorbs oxygen to give the cyclic peroxide (**152**) ⟨64JOC291⟩.

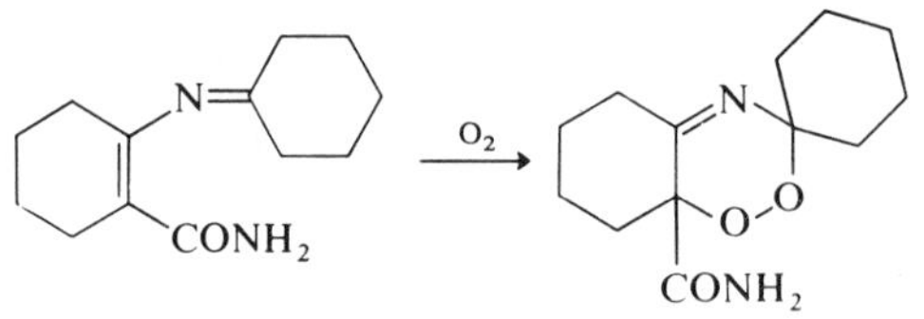

(**247**)                    (**152**)

Sulfoxides can act as an S—O two-atom fragment in [4+2] condensations. Thus reaction of DMSO with benzoyl isocyanate in the presence of boron trifluoride etherate ⟨75IZV1206⟩ gives the 1,3,2,5-dioxathiazine (**248**). Boron trifluoride induced condensation of DMSO with *N*-benzoyl-*S*,*S*-bis(trifluoromethyl)sulfimide (**249**) gives the 1,3,2,4,5-dioxadithiazine (**117**) ⟨78JA985⟩, and the *N*-acylimine (**250**) also gives a 1,3,2,5-dioxathiazine (**251**) on reaction with DMSO ⟨80ZOR463⟩.

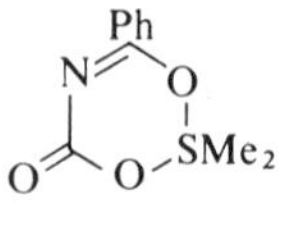

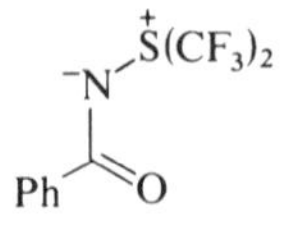

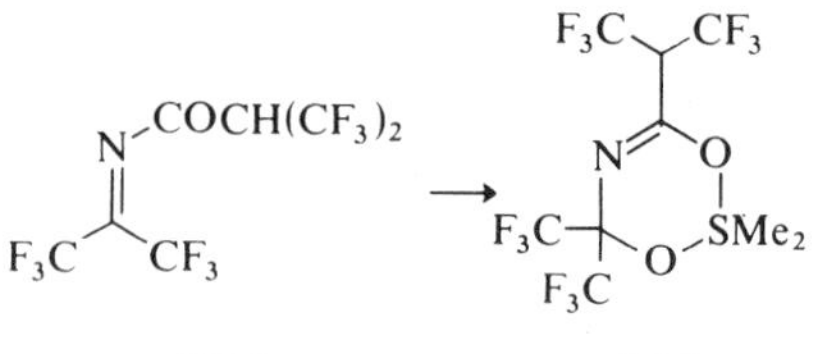

(**248**)                    (**249**)                    (**250**)                    (**251**)

### 2.28.4.2.3 *From [3+3] atom fragments*

The vast majority of [3+3] condensations involve the reaction of a 1,3-dinucleophile with a 1,3-dielectrophile. In the preparation of six-membered rings containing three or more heteroatoms, certain three-atom fragments are frequently employed. Thus amidines are common 1,3-dinucleophiles, CSI is used as a 1,3-dielectrophile, and sulfamide and its derivatives and sulfur diimides are used in the synthesis of heterocycles containing the —N—S—N— unit.

Reaction of amidines or cyclic amidines such as 2-aminopyridine with 1,3-dielectrophiles is a well known general route to 1,3-diaza heterocycles, and extension of this reaction to electrophiles containing extra heteroatoms leads to polyheteroatom systems. Some examples are shown in Scheme 24, and involve chlorosulfonylalkenes ⟨73BSF985⟩, chlorosulfonylcarbamyl chlorides ⟨77JCR(S)238⟩ and CSI ⟨79H(12)1199⟩ as the 1,3-dielectrophilic species.

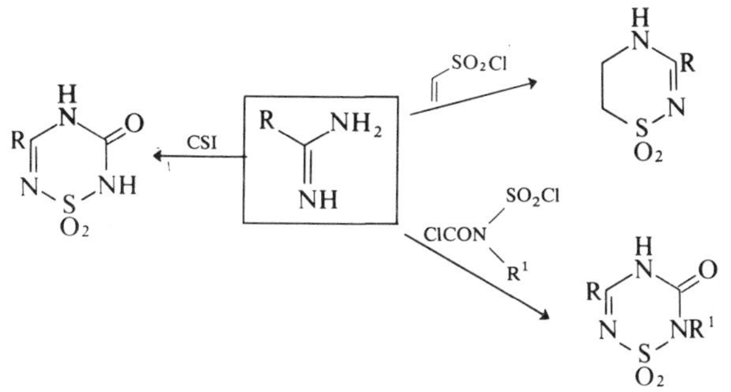

**Scheme 24**

The reaction of CSI with other 1,3-dinucleophiles leads to heterocyclic products, and the chemistry of this highly reactive cumulene is well reviewed ⟨76CRV389⟩. Some reactions of CSI are given in Scheme 25, and although some of them are discussed elsewhere in this chapter they are all included here in order to emphasize the usefulness of this reagent in the preparation of these polyheteroatom systems.

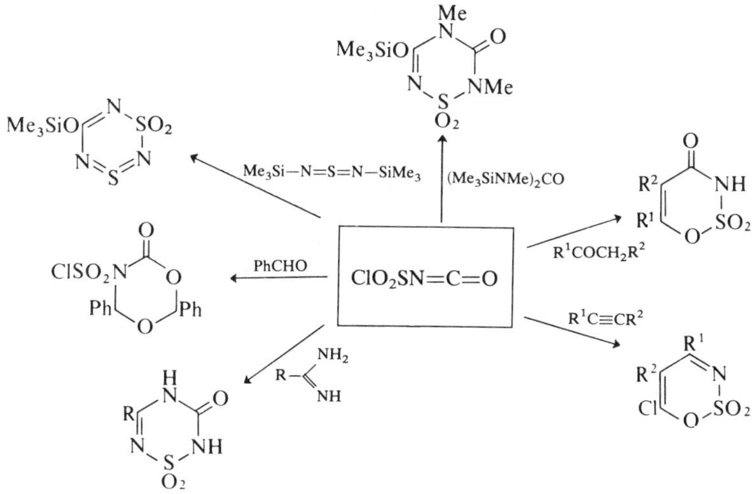

**Scheme 25**

Compounds containing the N—S—N unit are readily prepared by [3+3] condensations of sulfamide with 1,3-dielectrophiles ⟨80CRV151⟩. Thus reaction with 1,3-diketones gives 1,2,6-thiadiazines ⟨64JOC1905⟩ and with $\alpha,\beta$-unsaturated ketones the 2,3-dihydro derivatives are formed in good yield ⟨63AG(E)737⟩ (Scheme 26). 2-Amino- and 2-hydroxy-aromatic ketones also react with sulfamide to give benzothiadiazines and benzoxathiazines respectively ⟨65JOC3960⟩. In these latter two reactions only one of the sulfamide nitrogens is retained in the product so they are more properly considered as [4+2] condensations. However, they are included here with the other reactions of sulfamide.

1,2,6-Thiadiazines can also be prepared from [3+3] condensations involving sulfur diimides. Thus reaction of bis(N-tosyl)sulfur diimide ⟨69TL4117⟩ with 1,3-diaminopropane gives the thiadiazine (**18**). Bis(trimethylsilyl)sulfur diimide reacts with CSI to give a 1,3,2,4,6-dithiatriazine ⟨74CB1, 74ZN(B)799⟩ (Scheme 27).

Examples of [3+3] condensations involving two 1,3-electrophilic–nucleophilic species are rare and of limited generality. Reaction of nitrones with aziridines gives 1,2,4-

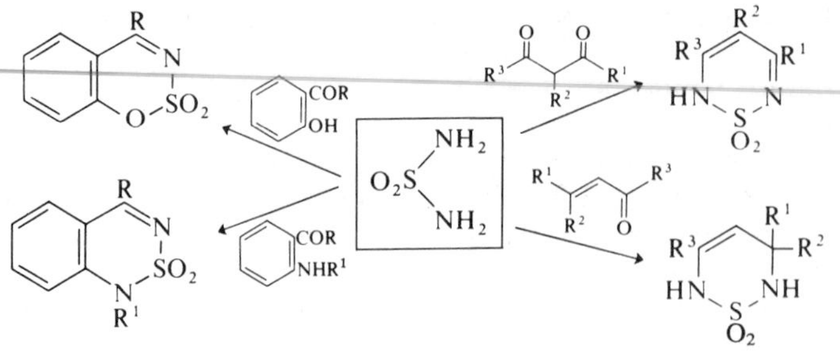

**Scheme 26**

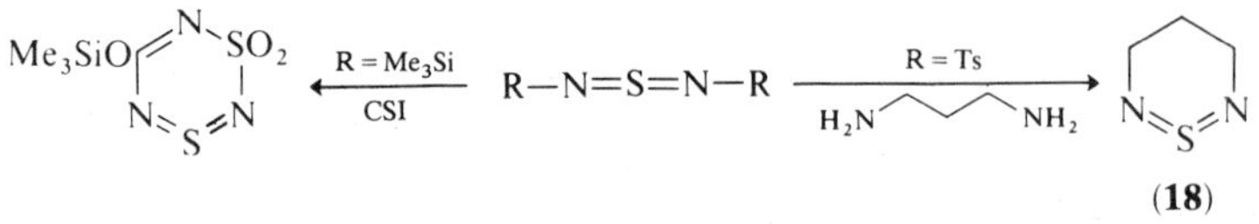

**Scheme 27**

oxadiazines (Scheme 28) *via* nucleophilic attack of the nitrone oxygen on the aziridine carbon, followed by collapse of the resulting dipolar species 〈74JOC162〉. Diaziridines react similarly with nitrones to give 1,2,4,5-oxatriazines, although in this case ring opening of the diaziridine to give an azomethine imine may be the initial step 〈74JOC3192〉. Nitrones react with nitrile oxides in the presence of boron trifluoride etherate to give 2,3-dihydro-1,4,2,5-dioxadiazines (Scheme 28) 〈76CZ236〉. Symmetrical 1,4,2,5-dioxadiazines are formed by [3+3] dimerization of arene nitrile oxides in the presence of pyridine 〈74JCS(P1)1951〉 or boron trifluoride 〈69G165〉.

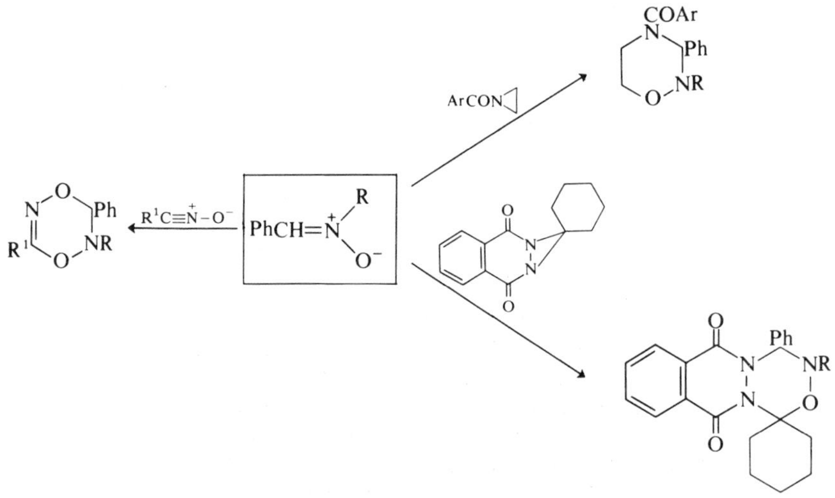

**Scheme 28**

## 2.28.4.3 Formation of Three or More Bonds

### 2.28.4.3.1 From [2+2+2] atom fragments

The most common synthetic methods involving formation of three or more bonds and leading to polyheteroatom systems are of the [2+2+2] type. Condensations employing a wide range of different two-atom fragments have been reported, and examples involving the use of sulfur trioxide, nitriles, ketones, isocyanates, ketenes, thioketenes and cyanates as the two-atom components are known. Of these, the reactions of sulfur trioxide and isocyanates are probably the more important, and a few illustrative examples are given here.

Some reactions of sulfur trioxide with dialkylcyanamides 〈77ZC222〉, trichloroacetonitrile 〈77ZOR222〉 and a combination of nitriles and alkenes 〈66USP3235549〉 are given in Scheme 29. In the reaction with dialkylcyanamides a second product (**252**; X = NR$_2$) is also formed.

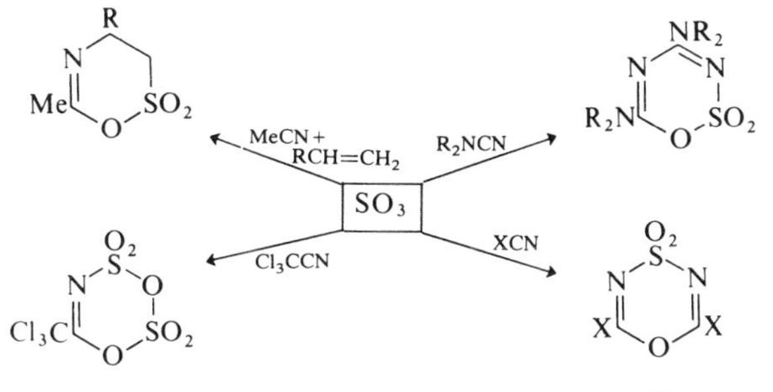

(252)

**Scheme 29**

An analogous product (**252**; X = Cl) is formed in the reaction of sulfur trioxide with cyanogen chloride ⟨68AG(E)172⟩. Aryl cyanates also condense with sulfur trioxide to give 1,2,3,5-oxathiadiazines analogous to those formed from dialkylcyanamides ⟨67CB3736⟩.

Examples of reactions involving isocyanates as two-atom fragments are given in Scheme 30. Trimerization of methyl isocyanate in the presence of tri-*n*-butylphosphine gives a 1,3,5-oxadiazine as the initial product ⟨73CR(C)(277)795⟩. Addition of carbon dioxide to the reaction leads to the 1,3,5-oxadiazine derived from $CO_2$ (1 mol) and the isocyanate (2 mol) ⟨74BSF1497⟩. Reaction of methyl isocyanate with hexafluoroacetone gives a mixture of 1:2 and 2:1 adducts as well as some 1:1 adduct ⟨67JOC2960⟩, although reaction of isocyanic acid with chloral gives only a 2:1 adduct ⟨67JHC290⟩.

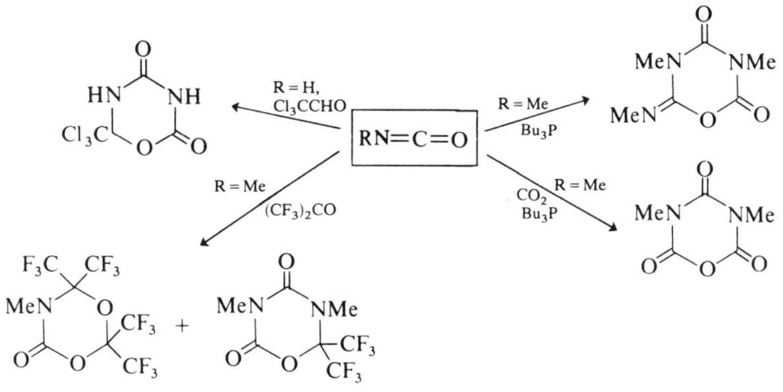

**Scheme 30**

Carbonyl compounds readily participate in [2+2+2] condensations, and an example of a reaction with isocyanates is given in Scheme 30. Other examples include the reaction of benzaldehyde with CSI to give 1,3,5-dioxazines (**115**) ⟨74LA561, 76CRV389⟩, and the reaction of dimethylcyanamide with hexafluoroacetone to give the 1,3,5-dioxazine (**116**) ⟨66JOC2568⟩. Benzoyl chloride condenses with benzonitrile or aryl cyanates in the presence of metal chlorides such as aluminum trichloride or antimony pentachloride to give 1,3,5-oxadiazinium salts (Scheme 31) ⟨65CB334, 67CB3736⟩. The structure and reactions of these 6π-electron cations have been discussed previously.

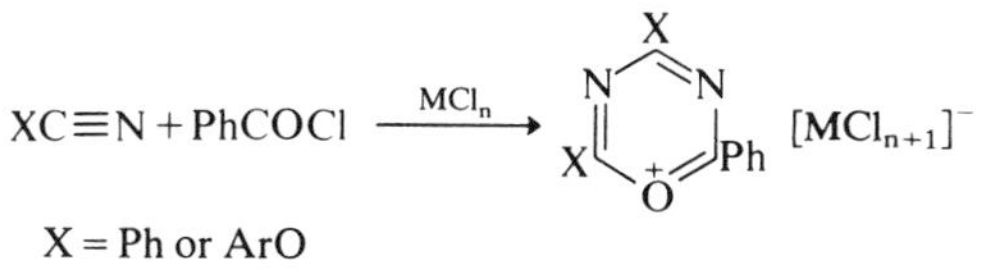

X = Ph or ArO

**Scheme 31**

### 2.28.4.3.2 *Other reactions*

Reactions involving the formation of more than three bonds are not important in the synthesis of polyheteroatom six-membered rings, and therefore only two examples are included here. Reaction of 1,2-hydrazinedicarboxylates with 1,2-disulfenyl chlorides gives 1,4,2,3-dithiadiazines (Scheme 22). Reaction of the hydrazines with sulfur dichloride

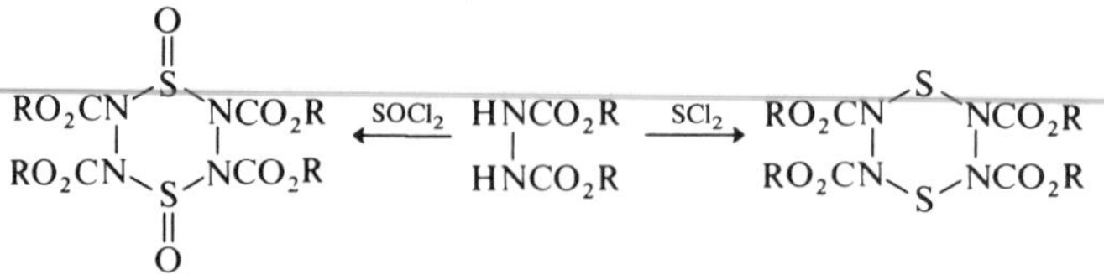

**Scheme 32**

⟨74JHC99⟩ or thionyl chloride ⟨79JHC751⟩ gives 1,4,2,3,5,6-dithiatetrazines and the corresponding 1,4-dioxides respectively (Scheme 32).

Sulfonamides (**253**) react with formaldehyde to give fully saturated *N*-sulfonyl-1,3,5-dioxazines, 1,3,5-oxadiazines or 1,3,5-triazines according to the relative quantities of reagents used ⟨75JCS(P1)772⟩.

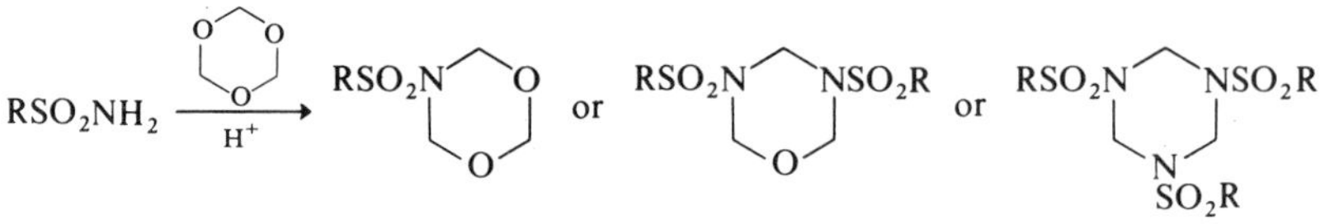

(**253**)

### 2.28.4.4 From Other Heterocycles

There are a few important preparative methods for six-membered rings with three or more heteroatoms which involve ring interconversions. All are ring expansion reactions, and usually involve the ring expansion of five-membered rings. Although the reactions presented here have only been used in isolated cases, some of them may be more general. In particular, the photochemical ring expansion of five-membered ring *N*-oxides, and the ring expansion resulting from reaction of five-membered sulfur-containing rings with nitrenes, may be applicable to the preparation of other polyheteroatom systems.

An important route to 1,4,2-dithiazines (**136**) involves the ring expansion of 1,3-dithiolium salts ⟨76JPR127⟩. Thus reaction of 2-methylthio-1,3-dithiolium salts with azide ion gives an unstable intermediate which rapidly loses nitrogen to give 1,4,2-dithiazines (Scheme 33). It is not known whether the reaction involves a nitrene intermediate or whether the [1,2] shift is concerted with loss of nitrogen. The latter possibility seems more likely.

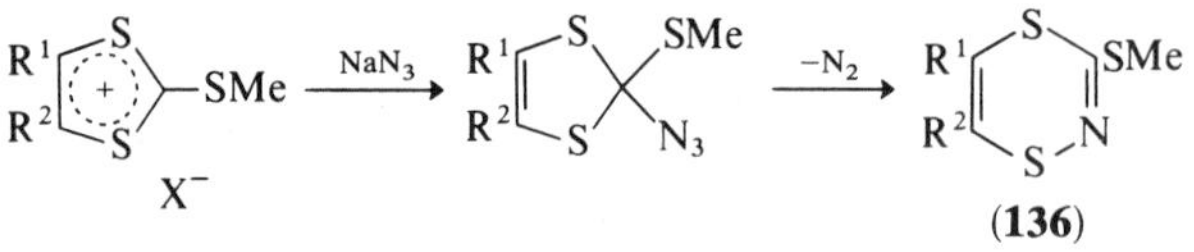

**Scheme 33**

Thermal decomposition of ethyl azidoformate in the presence of 4-phenyl-1,2-dithiole-3-thione leads to the rare 1,2,3-dithiazine system (**254**) ⟨76CJC3879⟩. The reaction presumably involves initial attack of the nitrene at S-1, followed by a [1,2] rearrangement of the resulting ylide (Scheme 34).

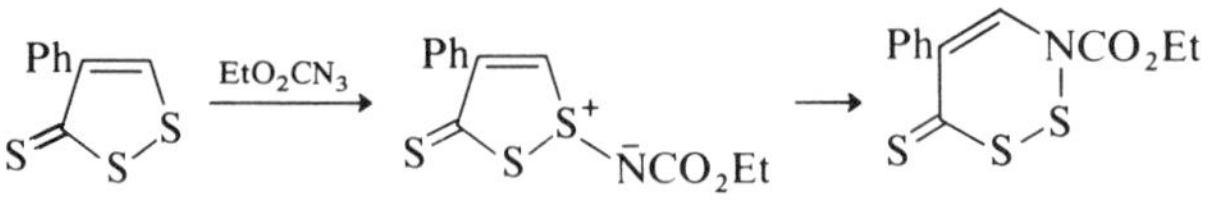

**Scheme 34**

Photolysis of 2,4,5-triphenyl-1,2,3-triazole 1-oxide gives the 1,3,4,5-oxatriazine (**13**) ⟨80AJC2447⟩. This is the only reported example of this monocyclic ring system, and the mechanism of its formation (Scheme 35) probably involves the intermediacy of an oxaziridine and an 'oxygen walk' process not uncommon in the photochemistry of heterocyclic *N*-oxides.

The analogous sulfur-containing ring system, 1,3,4,5-thiatriazine, is also prepared by a ring expansion reaction ⟨80JOC2604⟩. Reaction of diphenylthiirene dioxide (**255**) with lithium azide in acetonitrile gives, among other products, the thiatriazine (**14**). The proposed

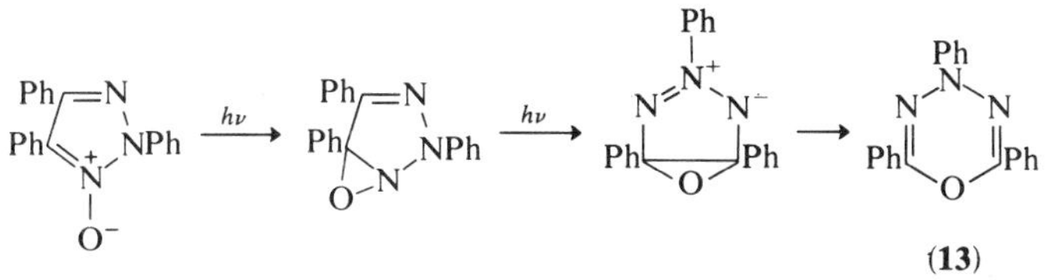

**Scheme 35**

mechanism, which involves initial nucleophilic attack by azide ion followed by closure on to the terminal nitrogen of the azide group, is shown in Scheme 36.

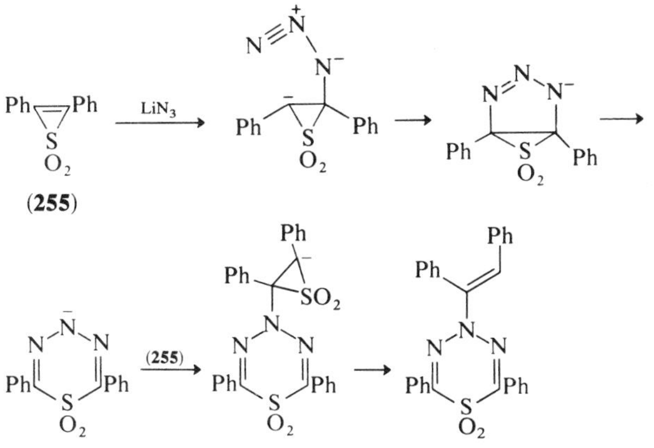

**Scheme 36**

The only reported synthesis of a 1,2,3,4,6-dithiatriazine (**256**) involves the reaction of disulfur dichloride with a silatriazole ⟨78ZC336⟩. The product is described as a red-brown crystalline solid but no other data are given.

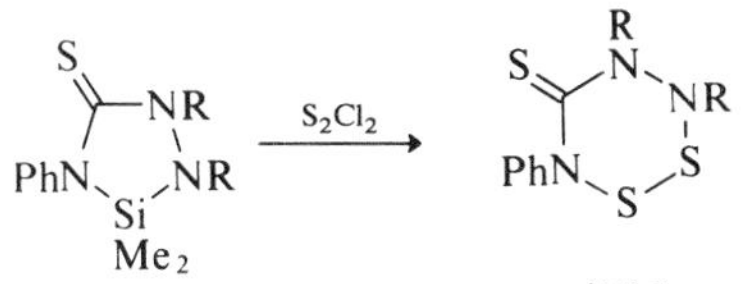

(**256**)

## 2.28.4.5 Summary

This section on the synthesis of polyheteroatom six-membered rings has been organized according to reaction type, and not according to the ring systems formed. Consequently, the chapter contains no 'list' of the synthetic methods available for each individual ring system. Therefore, the important general methods for preparing the common ring systems that have been covered earlier are briefly summarized here. They are organized according to the ring system, and only routes to the more common systems containing three heteroatoms are summarized, as leading references to systems described as 'rare' or 'few examples' are easily obtained from Tables 1–11.

| | |
|---|---|
| 1,2,3-Oxathiazines | reaction of ketones with CSI |
| | reaction of alkynes with CSI |
| | condensation of 3-aminopropan-1-ols with SOCl₂ |
| | cycloaddition of α-ketoketenes to N-sulfinylamines |
| 1,3,5-Oxathiazines | cycloaddition of thiocarbonyl isocyanates to C=O |
| 1,4,3-Oxathiazines | cycloaddition of acyl sulfenes to C=N |
| | cycloaddition of N-sulfonylamides to C=C |
| 1,4,2-Dioxazines | cyclodehydration of O-(carboxymethyl)hydroxamic acids |
| | di-O-alkylation of hydroxamic acids with 1,2-dihaloalkanes |
| | condensation of O-(α-hydroxyacyl)hydroxylamines with CDI |
| 1,5,2-Dioxazines | condensation of α-hydroxyalkyl hydroxamic acids with CDI |

| | |
|---|---|
| 1,3,5-Dithiazines | condensation of aldehydes, primary amines and $H_2S$ |
| | cycloaddition of $N$-thioacylimines to C=S |
| 1,4,2-Dithiazines | reaction of 1,3-dithiolium salts with azide ion |
| | cyclization of $N$-thioacyl-$\beta$-styrylsulfonamides |
| 1,2,4-Oxadiazines | condensation of amidoximes with $\alpha$-haloacids or esters |
| | base cyclization of $H_2NCONROCH_2CO_2Et$ |
| | reaction of nitrones with aziridines |
| 1,2,5-Oxadiazines | condensation of 1,2-dihydroxylamines with CDI |
| | condensation of $\alpha$-aminooximes with HCHO |
| 1,3,4-Oxadiazines | cycloaddition of azodicarbonyl compounds to C=C |
| | condensation of hydrazides with $\alpha$-chloroacyl chlorides |
| | cyclodehydration of $N$-($\alpha$-hydroxyalkyl)-$N$-acylhydrazines |
| 1,3,5-Oxadiazines | trimerization of isocyanates |
| | cycloaddition of acyl isocyanates to C=N |
| | condensation of amidines with reactive ketones (2 mol) |
| 1,2,4-Thiadiazines | base cyclization of $H_2NCONRSO_2CH_2CO_2Et$ |
| | cyclization of $N$-$\beta$-styrylsulfonylamidines |
| 1,2,6-Thiadiazines | condensation of N–N 1,5-dinucleophiles with $SCl_2$ or $SOCl_2$ |
| | condensation of sulfamide with 1,3-diketones |
| | condensation of sulfamide with $\alpha,\beta$-unsaturated ketones |
| | condensation of sulfurdiimides with 1,3-diamines |
| 1,3,4-Thiadiazines | condensation of thiohydrazides with $\alpha$-halo ketones |
| | condensation of amidines with alkenesulfonyl chlorides |
| | condensation of $\beta$-mercaptoethyl hydrazines with aldehydes and ketones |
| 1,3,5-Thiadiazines | condensation of aldehydes, primary amines and $CS_2$ |
| | condensation of $N$-thioacylamidines with thiophosgene |
| | cycloaddition of thiocarbonyl isocyanates to C=N |

## 2.28.5 APPLICATIONS

Six-membered rings containing three or more heteroatoms do not occur widely in Nature. The 1,2,4-dithiazine ring occurs in the epidithiodioxopiperazine system (**257**), which is common to a number of fungal metabolites, such as gliotoxin (**258**), and appears to be the site of the potent antiviral, antibacterial and antifungal activities of these compounds. However, the compounds are probably better considered as diketopiperazines rather than 1,2,4-dithiazines. Thialdine (**68**), a 1,3,5-dithiazine, has been identified as one of the substances responsible for the flavour of beef broth, and subsequent to this discovery other 1,3,5-dithiazines have been proposed as flavour enhancement agents.

      (**257**)             (**258**)            (**68**)

Synthetic six-membered rings containing three or more heteroatoms have been widely screened for biological activity, and, perhaps not surprisingly when the diverse nature of the structures is taken into consideration, almost every conceivable type of biological activity has been claimed for one or other of these systems. However, one group of compounds, the benzothiadiazines, has probably been studied more extensively than any of the others, and since the discovery that chlorthiazide (**259**; R = H) was a potent diuretic agent with low toxicity, many hundreds of 1,2,4-benzothiadiazines have been prepared and studied ⟨62AG(E)235⟩. Diuretic activity is found in compounds possessing the 7-sulfamyl group, and several analogues of chlorthiazide, for example benzthiazide (**259**; R = $CH_2SCH_2Ph$), show similar properties. The 3,4-dihydro compounds, prepared by sodium borohydride reduction of (**259**), are more active. Many other 1,2,4-benzothiadiazine 1,1-dioxides exhibit antihypertensive activity, and diazoxide (**260**) is used in the treatment of hypertension. A correlation of antihypertensive activity with the structure of the 1,2,4-benzothiadiazines has been made ⟨72JMC394⟩. Analogues such as (**261**), which contain sulfoximide functions in place of the SO$_2$ group, have been prepared ⟨75CSR189, 80CSR477⟩, as have analogues such as (**262**) ⟨73BCJ1890⟩ and (**263**) ⟨69JHC407⟩ in which the benzene ring is replaced by a heterocyclic ring. The isomeric 1,2,3-benzothiadiazine 1,1-dioxide (**264**) also exhibits diuretic and antihypertensive activity ⟨62HCA996⟩. 3-Alkyl-1,2,4-benzothiadiazines are also claimed to possess useful antimicrobial activity ⟨74FES910⟩. An excellent summary of the chemistry of benzothiadiazines is given in Landquist's review ⟨B-79MI22800⟩.

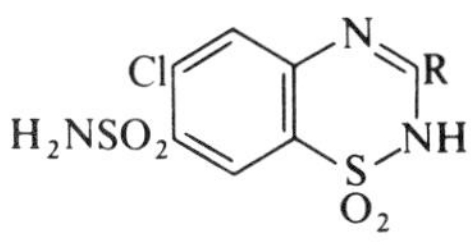

(**259**) 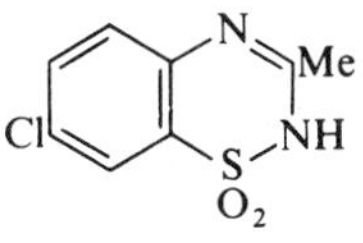    (**260**) 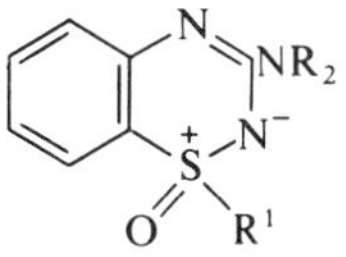    (**261**)

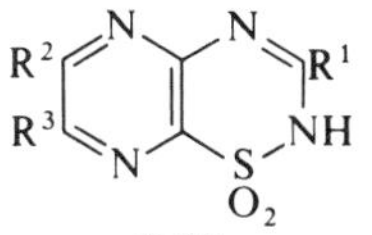

(**262**) 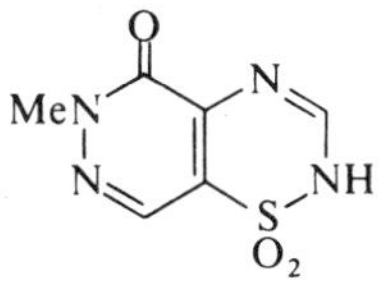    (**263**) 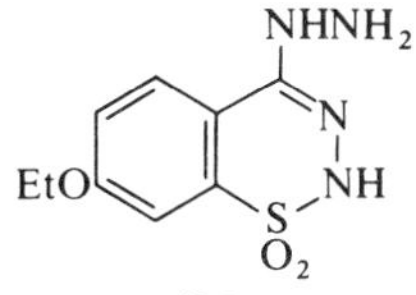    (**264**)

Many other systems exhibit significant biological activity, and a few examples are given here. Prominent among these are monocyclic thiadiazines such as the 1,3,5-thiadiazine-2-thiones (**265**) which possess a wide range of antibacterial, antimycotic and anthelmintic properties ⟨69AF558, 69AF1807⟩. Dazomet (**265**; R = Me) is a fungicide and herbicide and is believed to act *via* release of methyl isothiocyanate. The related 2-imino-1,3,5-thiadiazin-4-ones (**266**) are claimed to be useful agrochemicals ⟨77GEP2824126⟩. 1,2,6-Thiadiazines (**267**) cause the lowering of blood sugar levels and therefore are potential antidiabetic agents ⟨72JMC435⟩. The antimicrobial compound taurolin (**268**) is an N-4-linked 1,2,4-thiadiazine dioxide. In aqueous solution it is hydrolyzed to formaldehyde and tetrahydro-1,2,4-thiadiazine 1,1-dioxide ⟨80MI22800⟩. Many sulfur-containing heterocycles are described as vulcanizing agents for synthetic rubber, and 1,2,5- and 1,3,5-thiadiazines have been used for this purpose.

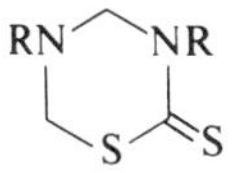

(**265**) 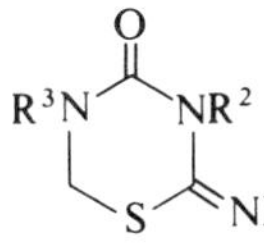    (**266**) 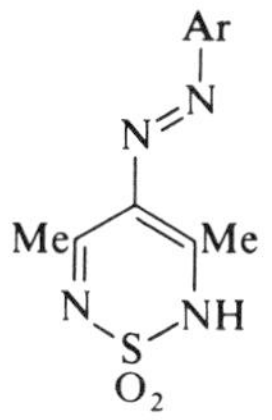    (**267**) 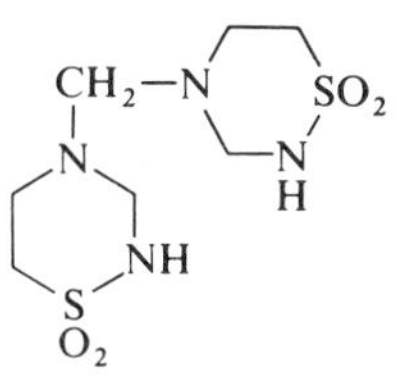    (**268**)

One of the major interests in oxadiazines from the biological point of view has been their use as 'oxa analogues' of nucleosides. 6-Oxadihydrouracil (**6**) is a competitive antagonist of uracil in bacterial systems, and has been widely studied ⟨76JOC3128, 77JMC134⟩. The cytosine analogue (**269**) has also been made, and the corresponding ribofuranosyl derivatives have been prepared by coupling of the sugar to N-2. 1,3,5-Oxadiazines such as (**270**) ⟨74USP3833577⟩ and (**271**) ⟨78GEP2732115⟩ are reported to have herbicidal and insecticidal properties respectively.

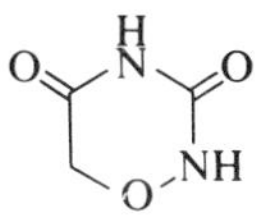

(**6**) 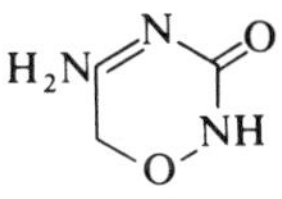    (**269**) 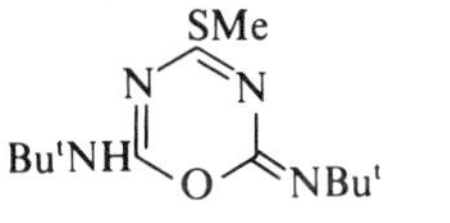    (**270**) 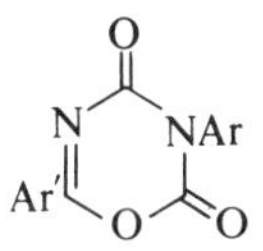    (**271**)

A major effort has been directed towards the study of 1,2,3-oxathiazin-4-one 2,2-dioxides (**184**) ⟨73AG(E)869⟩. These compounds are the so-called third generation artificial sweeteners, and their structural similarities to the first and second generation compounds, saccharin and the cyclamates, is obvious. Most of the compounds have alkyl groups at C-5 and/or C-6, for example acetosultam (**272**).

1,2,3,5-Oxathiadiazines (**273**) ⟨76GEP2524475⟩ and the 1,4,3,5-isomers (**274**) ⟨72GEP2036312⟩ are reported to be cross-linking agents for acid-hardening resins and plant protection agents respectively. Thiatriazines (**275**) possess herbicidal and fungicidal activity ⟨75GEP2508832⟩.

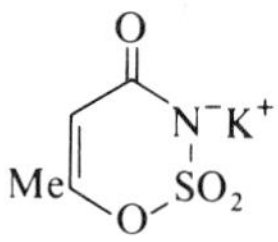

(272) 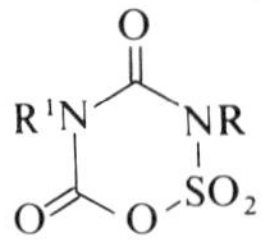          (273) 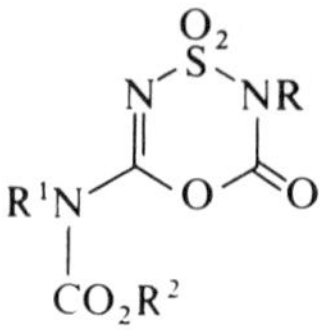          (274) 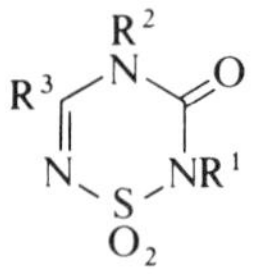          (275)

# References

### EXPLANATION OF THE REFERENCE SYSTEM

Throughout this work, references are designated by a number–letter coding of which the first two numbers denote tens and units of the year of publication, the next one to three letters denote the journal, and the final numbers denote the page. This code appears in the text each time a reference is quoted; the advantages of this system are outlined in the Introduction (Chapter 1.01). The system is based on that previously used in the following two monographs: (a) A. R. Katritzky and J. M. Lagowski, 'Chemistry of the Heterocyclic N-Oxides', Academic Press, New York, 1971; (b) J. Elguero, C. Marzin, A. R. Katritzky and P. Linda, 'The Tautomerism of Heterocycles', in 'Advances in Heterocyclic Chemistry', Supplement 1, Academic Press, New York, 1976.

The following additional notes apply:

1. A list of journals which have been assigned codes is given (in alphabetical order) together with their codes immediately following these notes. Journal names are abbreviated throughout by the CASSI (Chemical Abstracts Service Source Index) system.

2. A list of journal codes in alphabetical order, together with the journals to which they refer, is given on the end papers of each volume.

3. Each volume contains all the references cited *in that volume*; no separate lists are given for individual chapters.

4. The list of references is arranged in order of (a) year, (b) journal in alphabetical order of journal code, (c) part letter or number if relevant, (d) volume number if relevant, (e) page number.

5. In the reference list the code is followed by (a) the complete literature citation in the conventional manner and (b) the number(s) of the page(s) on which the reference appears, whether in the text or in tables, schemes, *etc.*

6. For non-twentieth century references the year is given in full in the code.

7. For journals which are published in separate parts, the part letter or number is given (when necessary) in parentheses immediately after the journal code letters.

8. Journal volume numbers are *not* included in the code numbers unless more than one volume was published in the year in question, in which case the volume number is included in parentheses immediately after the journal code letters.

9. Patents are assigned appropriate three letter codes.

10. Frequently cited books are assigned codes, but the whole code is now prefixed by the letter 'B-'.

11. Less common journals and books are given the code 'MI' for miscellaneous.

12. Where journals have changed names, the same code is used throughout, *e.g.* CB refers both to *Chem. Ber.* and to *Ber. Dtsch. Chem. Ges.*


### Journals

| | |
|---|---|
| Acc. Chem. Res. | ACR |
| Acta Chem. Scand., Ser. B | ACS(B) |
| Acta Chim. Acad. Sci. Hung. | ACH |
| Acta Crystallogr., Part B | AX(B) |
| Adv. Phys. Org. Chem. | APO |
| Agric. Biol. Chem. | ABC |

| | |
|---|---|
| Angew. Chem. | AG |
| Angew. Chem., Int. Ed. Engl. | AG(E) |
| Ann. Chim. (Rome) | AC(R) |
| Ann. N.Y. Acad. Sci. | ANY |
| Arch. Pharm. (Weinheim, Ger.) | AP |
| Ark. Kemi | AK |
| Arzneim.-Forsch. | AF |
| Aust. J. Chem. | AJC |
| Biochem. Biophys. Res. Commun. | BBR |
| Biochemistry | B |
| Biochem. J. | BJ |
| Biochim. Biophys. Acta | BBA |
| Br. J. Pharmacol. | BJP |
| Bull. Acad. Pol. Sci., Ser. Sci. Chim. | BAP |
| Bull. Acad. Sci. USSR, Div. Chem. Sci. | BAU |
| Bull. Chem. Soc. Jpn. | BCJ |
| Bull. Soc. Chim. Belg. | BSB |
| Bull. Soc. Chim. Fr., Part 2 | BSF(2) |
| Can. J. Chem. | CJC |
| Chem. Abstr. | CA |
| Chem. Ber. | CB |
| Chem. Heterocycl. Compd. (Engl. Transl.) | CHE |
| Chem. Ind. (London) | CI(L) |
| Chem. Lett. | CL |
| Chem. Pharm. Bull. | CPB |
| Chem. Rev. | CRV |
| Chem. Scr. | CS |
| Chem. Soc. Rev. | CSR |
| Chem.-Ztg. | CZ |
| Chimia | C |
| Collect. Czech. Chem. Commun. | CCC |
| Coord. Chem. Rev. | CCR |
| C.R. Hebd. Seances Acad. Sci., Ser. C | CR(C) |
| Cryst. Struct. Commun. | CSC |
| Diss. Abstr. Int. B | DIS(B) |
| Dokl. Akad. Nauk SSSR | DOK |
| Experientia | E |
| Farmaco Ed. Sci. | FES |
| Fortschr. Chem. Org. Naturst. | FOR |
| Gazz. Chim. Ital. | G |
| Helv. Chim. Acta | HCA |
| Heterocycles | H |
| Hoppe-Seyler's Z. Physiol. Chem. | ZPC |
| Indian J. Chem., Sect. B | IJC(B) |
| Inorg. Chem. | IC |
| Int. J. Sulfur Chem., Part B | IJS(B) |
| Izv. Akad. Nauk SSSR, Ser. Khim. | IZV |
| J. Am. Chem. Soc. | JA |
| J. Biol. Chem. | JBC |
| J. Chem. Phys. | JCP |
| J. Chem. Res. (S) | JCR(S) |
| J. Chem. Soc. (C) | JCS(C) |
| J. Chem. Soc., Chem. Commun. | CC |
| J. Chem. Soc., Dalton Trans. | JCS(D) |
| J. Chem. Soc., Faraday Trans. 1 | JCS(F1) |
| J. Chem. Soc., Perkin Trans. 1 | JCS(P1) |
| J. Gen. Chem. USSR (Engl. Transl.) | JGU |
| J. Heterocycl. Chem. | JHC |
| J. Indian Chem. Soc. | JIC |
| J. Magn. Reson. | JMR |

| | |
|---|---|
| J. Med. Chem. | JMC |
| J. Mol. Spectrosc. | JSP |
| J. Mol. Struct. | JST |
| J. Organomet. Chem. | JOM |
| J. Org. Chem. | JOC |
| J. Org. Chem. USSR (Engl. Transl.) | JOU |
| J. Pharm. Sci. | JPS |
| J. Phys. Chem. | JPC |
| J. Prakt. Chem. | JPR |
| Khim. Geterotsikl. Soedin. | KGS |
| Kristallografiya | K |
| Liebigs Ann. Chem. | LA |
| Monatsh. Chem. | M |
| Naturwissenschaften | N |
| Nippon Kagaku Kaishi | NKK |
| Nouv. J. Chim. | NJC |
| Org. Magn. Reson. | OMR |
| Org. Mass Spectrom. | OMS |
| Org. Prep. Proced. Int. | OPP |
| Org. React. | OR |
| Org. Synth. | OS |
| Org. Synth., Coll. Vol. | OSC |
| Phosphorus Sulfur | PS |
| Phytochemistry | P |
| Proc. Indian Acad. Sci., Sect. A | PIA(A) |
| Proc. Natl. Acad. Sci. USA | PNA |
| Pure Appl. Chem. | PAC |
| Q. Rev., Chem. Soc | QR |
| Recl. Trav. Chim. Pays-Bas | RTC |
| Rev. Roum. Chim. | RRC |
| Russ. Chem. Rev. (Engl. Transl.) | RCR |
| Spectrochim. Acta, Part A | SA(A) |
| Synth. Commun. | SC |
| Synthesis | S |
| Tetrahedron | T |
| Tetrahedron Lett. | TL |
| Ukr. Khim. Zh. (Russ. Ed.) | UKZ |
| Yakugaku Zasshi | YZ |
| Z. Chem. | ZC |
| Zh. Obshch. Khim. | ZOB |
| Zh. Org. Khim. | ZOR |
| Z. Naturforsch., Teil B | ZN(B) |

## Book Series

| | |
|---|---|
| 'Advances in Heterocyclic Chemistry' | AHC |
| 'Chemistry of Heterocyclic Compounds' [Weissberger–Taylor series] | HC |
| 'Methoden der Organischen Chemie (Houben-Weyl)' | HOU |
| 'Organic Compounds of Sulphur, Selenium, and Tellurium' [R. Soc. Chem. series] | SST |
| 'Physical Methods in Heterocyclic Chemistry' | PMH |

## Specific Books

| | |
|---|---|
| Q. N. Porter and J. Baldas, 'Mass Spectromety of Heterocyclic Compounds', Wiley, New York, 1971 | MS |
| T. J. Batterham, 'NMR Spectra of Simple Heterocycles', Wiley, New York, 1973 | NMR |

'Photochemistry of Heterocyclic Compounds', ed. O. Buchardt, Wiley, New
York, 1976                                                              PH
W. L. F. Armarego, 'Stereochemistry of Heterocyclic Compounds', Wiley,
New York, 1977, parts 1 and 2                                            SH

**Patents**

| | |
|---|---|
| Belg. Pat. | BEP |
| Br. Pat. | BRP |
| Eur. Pat. | EUP |
| Fr. Pat. | FRP |
| Ger. (East) Pat. | EGP |
| Ger. Pat. | GEP |
| Neth. Pat. | NEP |
| Jpn. Pat. | JAP |
| Jpn. Kokai | JAP(K) |
| S. Afr. Pat. | SAP |
| U.S. Pat. | USP |

**Other Publications**

| | |
|---|---|
| All Other Books and Journals ('Miscellaneous') | MI |
| All Other Patents | MIP |
| Personal Communications | PC |
| Theses | TH |
| Unpublished Results | UP |

## VOLUME 3 REFERENCES

| | | |
|---|---|---|
| B-1793MI22000 | C. Scheele; 'Sämtliche Physiche und Chemische Werke', S. Fr. Hermbstädt; Berlin, 1793, vol. 2, p. 149 (Beilstein, **26**, 239). | 460 |
| 1828MI22000 | A. Serullas; *Ann. Chim. Phys.*, 1828, **38**, 379. | 460 |
| 1829MI22000 | J. Liebig; *Pogg. Ann.*, 1829, **15**, 359. | 460 |
| 1834LA(10)17 | J. Liebig; *Liebigs Ann. Chem.*, 1834, **10**, 17. | 460 |
| 1835LA(14)133 | J. Liebig; *Liebigs Ann. Chem.*, 1835, **14**, 133. | 508 |
| 1835MI22000 | J. Liebig and F. Wohler; *Ann. Phys. Chem.*, 1835, **20**, 369. | 460 |
| 1875JPR(11)289 | E. Dreschel; *J. Prakt. Chem.*, 1875, **11**, 289. | 460 |
| 1876CB1008 | M. Nencki; *Ber.*, 1876, **9**, 1008. | 478 |
| 1883CB2576 | O. Widman; *Ber.*, 1883, **16**, 2576. | 1021 |
| 1883LA(222)1 | A. Hantzsch; *Liebigs Ann. Chem.*, 1883, **222**, 1. | 792 |
| 1884CB603 | P. Griess; *Ber.*, 1884, **17**, 603. | 1043 |
| 1886CB2063 | A. W. Hofmann; *Ber.*, 1886, **19**, 2063. | 480 |
| 1886CB2206 | C. Ris; *Ber.*, 1886, **19**, 2206. | 184 |
| 1887JPR(35)262 | A. Weddige and H. Finger; *J. Prakt. Chem.*, 1887, **35**, 262. | 369 |
| 1888JPR(38)531 | T. Curtius and J. Lang; *J. Prakt. Chem.*, 1888, **38**, 531. | 555 |
| 1890CB87 | S. Gabriel and W. E. Lauer; *Ber.*, 1890, **23**, 87. | 1008, 1026 |
| 1890CB1852 | R. Neitzki and O. Ernst; *Ber.*, 1890, **23**, 1852. | 190 |
| 1890CB2912 | A. Pinner; *Ber.*, 1890, **23**, 2912. | 512 |
| 1890M(11)133 | H. Strache; *Monatsh. Chem.*, 1890, **11**, 133. | 242 |

| | | |
|---|---|---|
| 1891CB3213 | S. Gabriel and P. Elfeldt; *Ber.*, 1891, **24**, 3213. | 1025 |
| 1891MI21600 | F. G. Hopkins; *Nature (London)*, 1891, **45**, 187. | 264 |
| 1892CB525 | E. Bamberger and L. Seeberger; *Ber.*, 1892, **25**, 525. | 492 |
| 1892CB534 | E. Bamberger and W. Dieckmann; *Ber.*, 1892, **25**, 534. | 492 |
| 1892CB3201 | E. Bamberger and E. Wheelwright; *Ber.*, 1892, **25**, 3201. | 450 |
| 1892G(22II)99 | A. Peratoner and G. Siringo; *Gazz. Chim. Ital.*, 1892, **22(II)**, 99. | 570 |
| 1893CB378 | O. Fischer and O. Heiler; *Ber.*, 1893, **26**, 378. | 190 |
| 1893CB997 | P. Henry and H. von Pechmann; *Ber.*, 1893, **26**, 997. | 1043 |
| 1893CB1501 | O. Rosenheim and J. Tafel; *Ber.*, 1893, **26**, 1501. | 246 |
| 1894CB984 | A. Pinner; *Ber.*, 1894, **27**, 984. | 565 |
| 1894CB3509 | S. Gabriel and T. Posner; *Ber.*, 1894, **27**, 3509. | 1020 |
| 1894CB3515 | S. Gabriel and T. Posner; *Ber.*, 1894, **27**, 3515. | 1021 |
| 1895CB1223 | C. D. Harries; *Ber.*, 1895, **28**, 1223. | 441 |
| 1895CB1970 | O. Kühling; *Ber.*, 1895, **28**, 1970. | 264, 320 |
| 1895LA(287)333 | J. Nef; *Liebigs Ann. Chem.*, 1895, **287**, 333. | 459 |
| 1896CB1873 | O. Fischer; *Ber.*, 1896, **29**, 1873. | 190 |
| 1896JPR(53)210 | K. Kratz; *J. Prakt. Chem.*, 1896, **53**, 210. | 375 |
| 1896JPR(54)217 | J. Biehringer; *J. Prakt. Chem.*, 1896, **54**, 217. | 767 |
| 1897CB1871 | A. Pinner; *Ber.*, 1897, **30**, 1871. | 565 |
| 1898CB2636 | E. Bamberger and A. von Goldberger; *Ber.*, 1898, **31**, 2636. | 381 |
| 1898JPR(58)333 | E. Bamberger and M. Weiler; *J. Prakt. Chem.*, 1898, **58**, 333. | 375 |
| 1898LA(301)55 | H. Rupe and G. Heberlein; *Liebigs Ann. Chem.*, 1898, **301**, 55. | 444 |
| 1899AP(237)346 | G. Frerichs and H. Beckurts; *Arch. Pharm. (Weinheim, Ger.)*, 1899, **237**, 346. | 448 |
| 1899CB2959 | M. Busch; *Ber.*, 1899, **32**, 2959. | 404 |
| 1899JCS245 | S. Ruhemann; *J. Chem. Soc.*, 1899, 245. | 792 |
| 1899LA(305)289 | E. Bamberger; *Liebigs Ann. Chem.*, 1899, **305**, 289. | 381 |
| 00CB58 | A. Hantzsch and O. Silberrad; *Ber.*, 1900, **33**, 58. | 555 |
| 00CB1998 | M. Bloch and S. von Kostanecki; *Ber.*, 1900, **33**, 1998. | 821 |
| 00CB3668 | A. Hantzsch and M. Lehmann; *Ber.*, 1900, **33**, 3668. | 536, 563 |
| 00JCS1119 | S. Ruhemann and F. Beddow; *J. Chem. Soc.*, 1900, 1119. | 827 |
| 00JCS1179 | S. Ruhemann and H. E. Stapleton; *J. Chem. Soc.*, 1900, 1179. | 827 |
| 00JCS1185 | O. Silberrad; *J. Chem. Soc.*, 1900, **77**, 1185. | 567 |
| 01CB102 | S. von Kostanecki and A. Rózycki; *Ber.*, 1901, **34**, 102. | 819 |
| 01CB1234 | S. Gabriel and J. Colman; *Ber.*, 1901, **34**, 1234. | 309 |
| 01CB1309 | E. Bamberger and E. Demuth; *Ber.*, 1901, **34**, 1309. | 375 |
| 01CB2311 | M. Busch; *Ber.*, 1901, **34**, 2311. | 561 |
| 01CB2475 | S. von Kostanecki, L. Paul and J. Tambor; *Ber.*, 1901, **34**, 2475. | 816 |
| 01JPR(63)241 | H. Mehner; *J. Prakt. Chem.*, 1901, **63**, 241. | 375 |
| 01LA(314)200 | R. Behrend, F. C. Meyer and Y. Buckholz; *Liebigs Ann. Chem.*, 1901, **314**, 200. | 110 |
| 01M(22)843 | B. Jeiteles; *Monatsh. Chem.*, 1901, **22**, 843. | 246 |
| 02CB782 | E. Buchner and H. Schröder; *Ber.*, 1902, **35**, 782. | 798 |
| 02CB2831 | S. Gabriel and J. Colman; *Ber.*, 1902, **35**, 2831. | 215, 226 |
| 02CB3647 | A. Einhorn and C. Mettler; *Ber.*, 1902, **35**, 3647. | 1029 |
| 02JPR(65)123 | E. Bamberger and E. Wheelwright; *J. Prakt. Chem.*, 1902, **65**, 123. | 450 |
| 02LA(325)129 | L. Wolff; *Liebigs Ann. Chem.*, 1902, **325**, 129. | 446 |
| 03CB475 | F. Kehrmann and A. Saager; *Ber.*, 1903, **36**, 475. | 1012 |
| 03CB3877 | M. Busch; *Ber.*, 1903, **36**, 3877. | 448 |
| 03CB4126 | L. Wolff and H. Lindenhayn; *Ber.*, 1903, **36**, 4126. | 402 |
| 03JCS1217 | W. H. Perkin; *J. Chem. Soc.*, 1903, **83**, 1217. | 1046 |
| 03JPR(68)464 | R. Stolle; *J. Prakt. Chem.*, 1903, **68**, 464. | 566, 569 |
| 04CB91 | C. Bülow; *Ber.*, 1904, **37**, 91. | 354 |
| 04CB3643 | S. Gabriel and J. Colman; *Ber.*, 1904, **37**, 3643. | 122 |
| 05LA(339)243 | H. Biltz; *Liebigs Ann. Chem.*, 1905, **339**, 243. | 413 |
| 05MI21300 | H. L. Wheeler and H. S. Bristol; *Am. Chem. J.*, 1905, **33**, 448. | 102 |
| 05MI21900 | M. Busch and E. Bergmann; *Z. Farben-Textilchem.*, 1905, **4**, 105, C 1905 I, 1102. | 445 |
| 06CB250 | O. Isay; *Ber.*, 1906, **39**, 250. | 309 |
| 06CB3776 | T. Curtius, A. Darapsky and E. Müller; *Ber.*, 1906, **39**, 3776. | 545 |

| | | |
|---|---|---|
| 07CB2093 | M. Busch; *Ber.*, 1907, **40**, 2093. | 560 |
| 07CB4598 | C. Paal and G. Kühn; *Ber.*, 1907, **40**, 4598. | 45 |
| 07CB4857 | S. Gabriel and A. Sonn; *Ber.*, 1907, **40**, 4857. | 317 |
| 07JPR(75)416 | R. Stolle; *J. Prakt. Chem.*, 1907, **75**, 416. | 560 |
| | | |
| 08CB3140 | T. Curtius, A. Darapsky and E. Müller; *Ber.*, 1908, **41**, 3140. | 563 |
| 08CB3957 | F. Sachs and G. Meyerheim; *Ber.*, 309 1908, **41**, 3957. | 309 |
| 08MI21900 | P. Pierron; *Ann. Chim. Phys. (Paris)*, 1908, **8** (15), 145. | 439, 445 |
| | | |
| 09CB803 | H. Wieland; *Ber.*, 1909, **42**, 803. | 506 |
| 09CB3270 | E. Müller; *Ber.*, 1909, **42**, 3270. | 545 |
| 09CB3710 | A. Reissert and F. Grube; *Ber.*, 1909, **42**, 3710. | 382 |
| | | |
| 10CB545 | J. Von Braun and A. Trümpler; *Ber.*, 1910, **43**, 545. | 928 |
| 10JA1499 | J. B. Tingle and S. J. Bates; *J. Am. Chem. Soc.*, 1910, **32**, 1499. | 404 |
| 10JCS2495 | F. Tutin; *J. Chem. Soc.*, 1910, 2495. | 188 |
| | | |
| 11LA(384)38 | H. Staudinger and N. Kon; *Liebigs Ann. Chem.*, 1911, **384**, 38. | 762 |
| 11MI22000 | A. Ostrogovich; *Rend. Accad. Lincei*, 1911, **20**, 249. | 498 |
| | | |
| 12CR(154)66 | L. C. Maillard; *C. R. Hebd. Seances Acad. Sci.*, 1912, **154**, 66. | 193 |
| 12JCS1788 | H. T. Clarke; *J. Chem. Soc.*, 1912, 1788. | 1037 |
| 12JCS2342 | G. E. K. Branch and A. W. Titherley; *J. Chem. Soc.*, 1912, **101**, 2342. | 115 |
| 12LA(391)308 | J. Biehringer; *Liebigs Ann. Chem.*, 1912, **391**, 308. | 767 |
| 12M(33)393 | H. Meyer and J. Mally; *Monatsh. Chem.*, 1912, **33**, 393. | 242 |
| | | |
| 14CB1132 | R. Stolle and F. Helwerth; *Ber.*, 1914, **47**, 1132. | 567 |
| 14CB2585 | K. von Auwers and F. Krollpfeiffer; *Ber.*, 1914, **47**, 2585. | 852 |
| 14JPR(89)310 | M. Busch and C. Schneider; *J. Prakt. Chem.*, 1914, **89**, 310. | 567, 569 |
| 14LA(407)332 | A. Baeyer and J. Piccard; *Liebigs Ann. Chem.*, 1914, **407**, 332. | 656 |
| 14M(35)1153 | A. Eckert and K. Steiner; *Monatsh.*, 1914, **35**, 1153. | 190 |
| | | |
| 15CB1614 | T. Curtius, A. Darapsky and E. Müller; *Ber.*, 1915, **48**, 1614. | 548 |
| 15LA(409)305 | S. Gabriel; *Liebigs Ann. Chem.*, 1915, **409**, 305. | 1022 |
| | | |
| 16CB53 | F. Kehrmann; *Ber.*, 1916, **49**, 53. | 1010 |
| 16CB1114 | J. von Braun; *Ber.*, 1916, **49**, 1114. | 1026 |
| 16JPR(94)53 | W. Dilthey; *J. Prakt. Chem.*, 1916, **94**, 53. | 869 |
| | | |
| 17CB819 | S. Gabriel and H. Ohle; *Ber.*, 1917, **50**, 819. | 1025 |
| 17CB911 | P. Pfeiffer and J. Grimmer; *Ber.*, 1917, **50**, 911. | 856 |
| 17CB1008 | W. Dilthey; *Ber.*, 1917, **50**, 1008. | 871 |
| 17CB1292 | A. Sonn; *Ber.*, 1917, **50**, 1292. | 803 |
| 17JPR(95)107 | W. Dilthey; *J. Prakt. Chem.*, 1917, **95**, 107. | 868 |
| | | |
| 19CB656 | E. Ott; *Ber.*, 1919, **52**, 656. | 503 |
| 19CB1195 | W. Dilthey; *Ber.*, 1919, **52**, 1195. | 862 |
| 19HCA619 | H. Staudinger and J. Meyer; *Helv. Chim. Acta*, 1919, **2**, 619. | 563 |
| | | |
| 20JPR(101)177 | W. Dilthey; *J. Prakt. Chem.*, 1920, **101**, 177. | 861, 870 |
| 20LA(421)1 | K. von Auwers and E. Lämmerhirt; *Liebigs Ann. Chem.*, 1920, **421**, 1. | 849 |
| 20LA(422)133 | K. von Auwers; *Liebigs Ann. Chem.*, 1920, **422**, 133. | 740 |
| | | |
| 21CB200 | L. Claisen; *Ber.*, 1921, **54**, 200. | 780 |
| 21CB213 | O. Diels; *Ber.*, 1921, **54**, 213. | 453 |
| | | |
| 22JPR(104)102 | O. Fischer; *J. Prakt. Chem.*, 1922, **104**, 102. | 445 |
| 22LA(427)26 | W. Madelung and E. Kern; *Liebigs Ann. Chem.*, 1922, **427**, 26. | 505 |
| | | |
| 23CB1012 | W. Dilthey and J. Fischer; *Ber.*, 1923, **56**, 1012. | 865 |
| 23MI21500 | L. Klisiecki and E. Sucharda; *Rocz. Chem.*, 1923, **3**, 251 (*Chem. Abstr.*, 1925, **19**, 72). | 226 |
| | | |
| 24CB88 | G. Wittig; *Ber.*, 1924, **57**, 88. | 820 |
| 24CB202 | F. Arndt and G. Källner; *Ber.*, 1924, **57**, 202. | 855 |
| 24CB1653 | W. Dilthey and J. Fischer; *Ber.*, 1924, **57**, 1653. | 863 |
| 24JCS2192 | J. Allan and R. Robinson; *J. Chem. Soc.*, 1924, 2192. | 819 |
| 24JPR(107)16 | O. Fischer; *J. Prakt. Chem.*, 1924, **107**, 16. | 445 |
| 24LA(439)132 | K. von Auwers and T. Meissner; *Liebigs Ann. Chem.*, 1924, **439**, 132. | 851 |
| 24MI21800 | R. C. Shah; *J. Indian Inst. Sci.*, 1924, **7**, 205. | 382 |
| | | |
| 25CB1612 | F. Arndt; *Ber.*, 1925, **58**, 1612. | 856 |
| 25CB1685 | D. Davidson and O. Baudisch; *Ber.*, 1925, **58**, 1685. | 133 |

| | | |
|---|---|---|
| 25CB2178 | H. Wieland and C. Schöpf; *Ber.*, 1925, **58**, 2178. | 264 |
| 25CB2318 | F. Arndt and B. Eistert; *Ber.*, 1925, **58**, 2318. | 792 |
| 25JA282 | W. E. Lawson and E. E. Reid; *J. Am. Chem. Soc.*, 1925, **47**, 282. | 1037 |
| 25JBC(64)233 | O. Baudisch and D. Davidson; *J. Biol. Chem.*, 1925, **64**, 233. | 143 |
| 25JCS1493 | J. M. Gulland and R. Robinson; *J. Chem. Soc.*, 1925, 1493. | 218 |
| 25JCS2349 | W. Baker; *J. Chem. Soc.*, 1925, 2349. | 819 |
| 25JIC84 | P. C. Guha and S. K. Ray; *J. Indian Chem. Soc.*, 1925, **2**, 84. | 445 |
| 25JPR(111)36 | G. Heller; *J. Prakt. Chem.*, 1925, **111**, 35. | 374, 376 |
| | | |
| 26CB2067 | C. Schöpf and H. Wieland; *Ber.*, 1926, **59**, 2067. | 264 |
| 26JA2379 | D. Davidson and O. Baudisch; *J. Am. Chem. Soc.*, 1926, **48**, 2379. | 142 |
| 26JCS113 | F. D. Chattaway and G. D. Parkes; *J. Chem. Soc.*, 1926, 113. | 572 |
| | | |
| 27BSF637 | I. Matzurevich; *Bull. Soc. Chim. Fr.*, 1927, **41**, 637. | 566 |
| 27CB716 | O. Diels and K. Alder; *Ber.*, 1927, **60**, 716. | 657, 658, 865 |
| 27CB1736 | J. Meisenheimer, O. Senn and P. Zimmermann; *Ber.*, 1927, **60**, 1736. | 375 |
| 27CB2598 | F. Arndt and B. Eistert; *Ber.*, 1927, **60**, 2598. | 450 |
| 27JCS194 | G. M. Bennett and L. V. D. Scorah; *J. Chem. Soc.*, 1927, 194. | 927 |
| 27JCS323 | F. D. Chattaway and A. J. Walker; *J. Chem. Soc.*, 1927, 323. | 374, 375, 376, 377 |
| 27JCS521 | M. A. Whiteley and D. Yapp; *J. Chem. Soc.*, 1927, 521. | 447 |
| 27JPR(116)9 | G. Heller and A. Siller; *J. Prakt. Chem.*, 1927, **116**, 9. | 374, 382 |
| 27RTC268 | I. J. Rinkes; *Recl. Trav. Chim. Pays-Bas*, 1927, **46**, 268. | 106 |
| | | |
| 28JA2731 | J. B. Ekeley and A. A. O'Kelly; *J. Am. Chem. Soc.*, 1928, **50**, 2731. | 435 |
| 28JCS1960 | G. M. Bennett and G. H. Willis; *J. Chem. Soc.*, 1928, 1960. | 167 |
| 28JIC163 | P. C. Guha and S. K. Roy-Choudhury; *J. Indian Chem. Soc.*, 1928, **5**, 163. | 452 |
| 28LA(464)237 | H. Lindemann and W. Schultheis; *Liebigs Ann. Chem.*, 1928, **464**, 237. | 1028 |
| | | |
| 30BSF630 | G. Gheorghiu; *Bull. Soc. Chim. Fr.*, 1930, **47**, 630. | 236, 238 |
| 30CB1000 | M. Bachstez; *Ber.*, 1930, **63**, 1000. | 145 |
| 30HCA444 | H. Rupe and F. Buxtorf; *Helv. Chim. Acta*, 1930, **13**, 444. | 437 |
| 30JCS157 | F. D. Chattaway and A. B. Adamson; *J. Chem. Soc.*, 1930, 157. | 377 |
| 30JCS843 | F. D. Chattaway and A. B. Adamson; *J. Chem. Soc.*, 1930, 843. | 377 |
| 30JIC933 | P. C. Guha and M. A. Hye; *J. Indian Chem. Soc.*, 1930, **7**, 933. | 560 |
| 30M(56)97 | F. Wessely and G. H. Moser; *Monatsh. Chem.*, 1930, **56**, 97. | 697 |
| | | |
| 31G294 | T. G. Levi; *Gazz. Chim. Ital.*, 1931, **61**, 294. | 1048 |
| 31JA644 | E. P. Kohler and W. F. Bruce; *J. Am. Chem. Soc.*, 1931, **53**, 644. | 1001 |
| 31JCS2059 | G. S. Ahluwalia, M. A. Haq and J. N. Rây; *J. Chem. Soc.*, 1931, 2059. | 804 |
| 31JCS2426 | A. Robertson, W. F. Sandrock and C. B. Hendry; *J. Chem. Soc.*, 1931, 2426. | 816 |
| 31JCS2787 | F. D. Chattaway and A. B. Adamson; *J. Chem. Soc.*, 1931, 2787. | 377 |
| 31JCS2792 | F. D. Chattaway and A. B. Adamson; *J. Chem. Soc.*, 1931, 2792. | 377 |
| 31JPR(131)1 | W. Dilthey and F. Quint; *J. Prakt. Chem.*, 1931, **131**, 1. | 650 |
| 31LA(490)257 | O. Diels and K. Alder; *Liebigs Ann. Chem.*, 1931, **490**, 257. | 633, 688 |
| 31M(58)238 | G. Koller and H. Ruppersberg; *Monatsh. Chem.*, 1931, **58**, 238. | 244 |
| | | |
| 32CB1032 | H. O. L. Fischer, G. Dangschat and H. Stettiner; *Ber.*, 1932, **65**, 1032. | 1018 |
| 32JCS1681 | A. Robertson, R. B. Waters and E. T. Jones; *J. Chem. Soc.*, 1932, 1681. | 801 |
| 32JIC145 | P. R. S. Gupta and A. C. Sircar; *J. Indian Chem. Soc.*, 1932, **9**, 145. | 242 |
| | | |
| 33BSF151 | G. Gheorghiu; *Bull. Soc. Chim. Fr.*, 1933, **53**, 151. | 236, 243, 246 |
| 33JA3361 | K. Folkers and T. B. Johnson; *J. Am. Chem. Soc.*, 1933, **55**, 3361. | 118 |
| 33JA3784 | K. Folkers and T. B. Johnson; *J. Am. Chem. Soc.*, 1933, **55**, 3784. | 117 |
| 33JCS87 | E. C. E. Hunter and J. R. Partington; *J. Chem. Soc.*, 1933, 87. | 626 |
| 33JCS555 | B. K. Blount and R. Robinson; *J. Chem. Soc.*, 1933, 555. | 868 |
| 33JCS616 | J. S. Mahal and K. Venkataraman; *J. Chem. Soc.*, 1933, 616. | 690 |
| 33JCS1197 | R. J. W. Le Fèvre and J. Pearson; *J. Chem. Soc.*, 1933, 1197. | 863 |
| 33JCS1381 | W. Baker; *J. Chem. Soc.*, 1933, 1381. | 818, 820 |
| 33LA(507)226 | H. Wieland, H. Metzger, C. Schöpf and M. Bülow; *Liebigs Ann. Chem.*, 1933, **507**, 226. | 295, 296, 307 |
| 33LA(507)261 | H. Wieland, H. Metzger, C. Schöpf and M. Bülow; *Liebigs Ann. Chem.*, 1933, **507**, 261. | 264 |
| | | |
| 34JA2754 | L. R. Buerger and T. B. Johnson; *J. Am. Chem. Soc.*, 1934, **56**, 2754. | 110 |
| 34JCS716 | N. M. Cullinane, H. G. Davey and H. J. H. Padfield; *J. Chem. Soc.*, 1934, 716. | 985 |
| 34JCS1311 | I. M. Heilbron, D. H. Hey and A. Lowe; *J. Chem. Soc.*, 1934, 1311. | 817, 821 |
| 34JCS1581 | I. M. Heilbron, D. H. Hey and B. Lythgoe; *J. Chem. Soc.*, 1934, 1581. | 821 |
| | | |
| 35BSF1381 | Mme. Ramart-Lucas and M. van Cowenbergh; *Bull. Soc. Chim. Fr.*, 1935, 1381. | 601 |
| 35G176 | T. Ajello; *Gazz. Chim. Ital.*, 1935, **65**, 176. | 1043, 1070 |
| 35JCS85 | D. W. Hill; *J. Chem. Soc.*, 1935, 85. | 871 |
| 35JCS602 | F. Arndt, G. T. O. Martin and J. R. Partington; *J. Chem. Soc.*, 1935, 602. | 892 |
| 35JCS646 | W. Baker and J. Walker; *J. Chem. Soc.*, 1935, 646. | 764, 786 |

| | | |
|---|---|---|
| 35JCS1005 | F. D. Chattaway and G. D. Parkes; *J. Chem. Soc.*, 1935, 1005. | 377 |
| 35MI22000 | I. E. Knaggs; *Proc. R. Soc. London, Ser. A*, 1935, **150**, 576. | 461 |
| 35MI22400 | A. Peres de Carvalho; *Ann. Chim. (Paris)*, 1935, **4**, 449. | 757 |
| | | |
| 36BSB685 | H. Wuyts and A. Lacourt; *Bull. Soc. Chim. Belg.*, 1936, **45**, 685. | 557, 569 |
| 36JCS212 | A. Robertson and G. L. Rusby; *J. Chem. Soc.*, 1936, 212. | 753 |
| 36JCS215 | D. G. Flynn and A. Robertson; *J. Chem. Soc.*, 1936, 215. | 819 |
| 36JCS419 | R. W. H. O'Donnell, F. P. Reed and A. Robertson; *J. Chem. Soc.*, 1936, 419. | 673, 753 |
| 36JCS423 | W. Hilton, R. W. H. O'Donnell, F. P. Reed, A. Robertson and G. L. Rusby; *J. Chem. Soc.*, 1936, 423. | 800 |
| 36JPR(144)273 | M. Busch and K. Küspert; *J. Prakt. Chem.*, 1936, **144**, 273. | 402, 415, 448 |
| 36MI22200 | E. Bergmann and A. Weizmann; *Trans. Faraday Soc.*, 1936, **32**, 1318. | 627 |
| 36PIA(A)(4)687 | Govinda Rau; *Proc. Indian Acad. Sci., Sect. A*, 1936, **4**, 687. | 626, 627 |
| | | |
| 37CB(A)83 | E. Späth; *Ber.*, 1937, **70A**, 83. | 807 |
| 37CB(B)2018 | E. Ochiai, K. Miyaki and S. Sato; *Ber.*, 1937, **70B**, 2018. | 243 |
| 37JCS196 | C. G. Le Fèvre and R. J. W. Le Fèvre; *J. Chem. Soc.*, 1937, 196. | 627 |
| 37JCS1088 | C. G. Le Fèvre and R. J. W. Le Fèvre; *J. Chem. Soc.*, 1937, 1088. | 626, 627 |
| 37JCS1704 | H. McIlwaine; *J. Chem. Soc.*, 1937, 1704. | 196 |
| 37JPR(148)135 | R. Wegler; *J. Prakt. Chem.*, 1937, **148**, 135. | 236 |
| 37LA(530)152 | H. Wieland and A. Kotzschmar; *Liebigs Ann. Chem.*, 1937, **530**, 152. | 307 |
| 37LA(531)279 | J. H. Helberger and A. von Rebay; *Liebigs Ann. Chem.*, 1937, **531**, 279. | 1015 |
| 37YZ131 | M. Tomita; *Yakugaku Zasshi*, 1937, **57**, 131. | 974 |
| | | |
| 38CB(B)42 | H. Hillemann; *Ber.*, 1938, **71B**, 42. | 188 |
| 38CB(B)1243 | H. Rudy and O. Majér; *Ber.*, 1938, **71B**, 1243. | 259 |
| 38CB(B)1323 | H. Rudy and O. Majer; *Ber.*, 1938, **71B**, 1323. | 259 |
| 38CB(B)2637 | W. John, P. Günther and M. Schmeil; *Ber.*, 1938, **71B**, 2637. | 787 |
| 38JA669 | R. Mozingo and H. Adkins; *J. Am. Chem. Soc.*, 1938, **60**, 669. | 787 |
| 38JA2404 | R. Fuson, J. Little and G. Miller; *J. Am. Chem. Soc.*, 1938, **60**, 2404. | 798 |
| 38JCS228 | S. M. Sethna, N. M. Shah and R. C. Shah; *J. Chem. Soc.*, 1938, 228. | 801 |
| 38JCS479 | G. R. Clemo and H. McIlwaine; *J. Chem. Soc.*, 1938, 479. | 192 |
| 38JGU22 | S. M. Sherlin, A. Y. Berlin, T. A. Serebrennikova and R. F. Rabinovitch; *J. Gen. Chem. USSR (Engl. Transl.)*, 1938, **8**, 22. | 771 |
| 38YZ517 | M. Tomita; *Yakugaku Zasshi*, 1938, **58**, 517. | 975 |
| | | |
| 39JA1 | E. Mosettig and A. H. Stuart; *J. Am. Chem. Soc.*, 1939, **61**, 1. | 816 |
| 39JA3020 | R. K. Summerbell and R. R. Umhoefer; *J. Am. Chem. Soc.*, 1939, **61**, 3020. | 963, 965, 981 |
| 39JOC311 | L. I. Smith, H. E. Ungnade, H. H. Hoehn and S. Wawzonek; *J. Org. Chem.*, 1939, **4**, 311. | 778, 780 |
| 39JOC575 | R. L. Shriner and A. G. Sharp; *J. Org. Chem.*, 1939, **4**, 575. | 754 |
| 39JPR(152)23 | N. Mauthner; *J. Prakt. Chem.*, 1939, **152**, 23. | 808 |
| 39LA(539)128 | C. Schöpf and A. Kottler; *Liebigs Ann. Chem.*, 1939, **539**, 128. | 294 |
| 39LA(539)179 | H. Wieland and R. Purrmann; *Liebigs Ann. Chem.*, 1939, **539**, 179. | 287 |
| 39M427 | F. Prillinger and H. Schmid; *Monatsh. Chem.*, 1939, **72**, 427 | 787 |
| 39YZ97 | E. Ochiai and M. Yanai; *Yakugaku Zasshi*, 1939, **59**, 97 (*Chem. Zentr.*, 1941, **112**, I, 1806). | 72 |
| | | |
| 40CRV(27)287 | L. I. Smith; *Chem. Rev.*, 1940, **27**, 287. | 780 |
| 40G504 | T. Ajello; *Gazz. Chim. Ital.*, 1940, **70**, 504. | 119 |
| 40JA283 | P. L. de Benneville and R. Connor; *J. Am. Chem. Soc.*, 1940, **62**, 283. | 848 |
| 40JA2405 | R. Adams and B. R. Baker; *J. Am. Chem. Soc.*, 1940, **62**, 2405. | 800 |
| 40JA2711 | R. L. Shriner and R. B. Moffett; *J. Am. Chem. Soc.*, 1940, **62**, 2711. | 651 |
| 40JA3067 | P. L. de Benneville and R. Connor; *J. Am. Chem. Soc.*, 1940, **62**, 3067. | 848 |
| 40JCS787 | A. McGookin, A. Robertson and W. B. Whalley; *J. Chem. Soc.*, 1940, 787. | 753 |
| 40JOC212 | C. D. Hurd and W. A. Hoffman; *J. Org. Chem.*, 1940, **5**, 212. | 780 |
| 40LA(544)163 | H. Wieland and R. Purrmann; *Liebigs Ann. Chem.*, 1940, **544**, 163. | 294 |
| 40LA(544)182 | R. Purrmann; *Liebigs Ann. Chem.*, 1940, **544**, 182. | 264 |
| 40LA(545)209 | H. Wieland, A. Tartter and R. Purrmann; *Liebigs Ann. Chem.*, 1940, **545**, 209. | 284, 293 |
| 40LA(546)98 | R. Purrmann; *Liebigs Ann. Chem.*, 1940, **546**, 98. | 264 |
| 40MI22000 | P. McClellan; *Ind. Eng. Chem.*, 1940, **32**, 1181. | 499 |
| 40ZPC(263)78 | W. Koschara; *Hoppe-Seyler's Z. Physiol. Chem.*, 1940, **263**, 78. | 284 |
| | | |
| 41JA2017 | E. Y. Spencer and G. F. Wright; *J. Am. Chem. Soc.*, 1941, **63**, 2017. | 682 |
| 41JCP(9)54 | J. Y. Beach; *J. Chem. Phys.*, 1941, **9**, 54. | 947, 948 |
| 41JCS323 | R. M. Anker and A. H. Cook; *J. Chem. Soc.*, 1941, 323. | 104 |
| 41JOC566 | J. Fried and R. C. Elderfield; *J. Org. Chem.*, 1941, **6**, 566. | 795 |
| 41JOC577 | J. Fried and R. C. Elderfield; *J. Org. Chem.*, 1941, **6**, 577. | 682 |
| 41LA(547)180 | H. Wieland and P. Decker; *Liebigs Ann. Chem.*, 1941, **547**, 180. | 297 |
| 41LA(548)82 | C. Schöpf, R. Reichert and K. Riefstahl; *Liebigs Ann. Chem.*, 1941, **548**, 82. | 291, 296 |
| 41LA(548)83 | C. Schöpf and R. Reichert; *Liebigs Ann. Chem.*, 1941, **548**, 83. | 264, 267 |
| 41LA(548)284 | R. Purrmann; *Liebigs Ann. Chem.*, 1941, **548**, 284. | 264, 304, 310 |
| 41OSC(1)552 | A. F. Holleman; *Org. Synth., Coll. Vol.*, 1941, **1**, 552. | 839 |

| | | |
|---|---|---|
| 42HCA1138 | P. Karrer and W. Fatzer; *Helv. Chim. Acta*, 1942, **25**, 1138. | 650 |
| 42JA435 | L. I. Smith and R. B. Carlin; *J. Am. Chem. Soc.*, 1942, **64**, 435. | 763 |
| 42JA612 | W. M. Lauer and N. H. Cromwell; *J. Am. Chem. Soc.*, 1942, **64**, 612. | 797 |
| 42JA825 | J. R. Segesser and M. Calvin; *J. Am. Chem. Soc.*, 1942, **64**, 825. | 691 |
| 42JA2417 | S. E. Krahler and A. Burger; *J. Am. Chem. Soc.*, 1942, **64**, 2417. | 233, 244 |
| 42JOC286 | M. J. Reider and R. C. Elderfield; *J. Org. Chem.*, 1942, **7**, 286. | 246 |
| 42JOC309 | Y. A. Tota and R. C. Elderfield; *J. Org. Chem.*, 1942, **7**, 309. | 92 |
| 42MI21600 | F. G. Hopkins; *Proc. R. Soc. London, Ser. B*, 1942, **130**, 359. | 264 |
| 42N269 | C. Schöpf; *Naturwissenschaften*, 1942, **30**, 269. | 264 |
| 42OR(1)210 | J. R. Johnson; *Org. React.*, 1942, **1**, 210. | 807 |
| 42OR(1)342 | A. H. Blatt; *Org. React.*, 1942, **1**, 342. | 850 |
| | | |
| 43JA1230 | C. F. H. Allen and J. W. Gates, Jr.; *J. Am. Chem. Soc.*, 1943, **65**, 1230. | 833 |
| 43JCS388 | G. W. Kenner, B. Lythgoe, A. R. Todd and A. Topham; *J. Chem. Soc.*, 1943, 388. | 111, 115 |
| 43JOC239 | J. F. Meyer and E. C. Wagner; *J. Org. Chem.*, 1943, **8**, 239. | 111 |
| 43JOC316 | C. A. Buehler, D. E. Cooper and E. O. Scrudder; *J. Org. Chem.*, 1943, **8**, 316. | 767 |
| 43LA(554)269 | W. Borsche and W. Ried; *Liebigs Ann. Chem.*, 1943, **554**, 269. | 247 |
| 43OS(23)3 | W. W. Hartman and O. E. Sheppard; *Org. Synth.*, 1943, **23**, 3. | 75 |
| 43OSC(2)60 | J. B. Dickey and A. R. Gray; *Org. Synth., Coll. Vol.*, 1943, **2**, 60. | 150 |
| 43OSC(2)126 | E. R. Riegel and F. Zwilgmeyer; *Org. Synth., Coll. Vol.*, 1943, **2**, 126. | 811 |
| 43OSC(2)485 | C. M. Suter and C. E. Maxwell; *Org. Synth., Coll. Vol.*, 1943, **2**, 485. | 985 |
| 43OSC(2)610 | R. W. Bost and E. W. Constable; *Org. Synth., Coll. Vol.*, 1943, **2**, 610. | 990 |
| 43ZPC(277)284 | W. Koschara; *Hoppe-Seyler's Z. Physiol. Chem.*, 1943, **277**, 284. | 284 |
| | | |
| 44CB202 | R. Kuhn, I. Löw and H. Trischmann; *Ber.*, 1944, **77**, 202. | 819 |
| 44CB529 | H. Brockmann, H. Junge and R. Muhlmann; *Ber.*, 1944, **77**, 529. | 664 |
| 44JA267 | H. K. Mitchell, E. E. Snell and R. J. Williams; *J. Am. Chem. Soc.*, 1944, **66**, 267. | 325 |
| 44JA354 | W. G. Young and S. Siegel; *J. Am. Chem. Soc.*, 1944, **66**, 354. | 787 |
| 44JA802 | R. Adams and J. W. Mecorney; *J. Am. Chem. Soc.*, 1944, **66**, 802. | 801, 826 |
| 44JA1214 | H. Gilman, P. R. Van Ess and D. A. Shirley; *J. Am. Chem. Soc.*, 1944, **66**, 1214. | 1011 |
| 44JA1875 | F. W. Holly and A. C. Cope; *J. Am. Chem. Soc.*, 1944, **66**, 1875. | 1024 |
| 44JCS476 | G. A. Howard, B. Lythgoe and A. R. Todd; *J. Chem. Soc.*, 1944, 476. | 107 |
| 44JIC109 | D. Chakravarti and B. C. Bera; *J. Indian Chem. Soc.*, 1944, **21**, 109. | 805 |
| 44LA(555)146 | H. Wieland and R. Liebig; *Liebigs Ann. Chem.*, 1944, **555**, 146. | 287, 293, 298 |
| 44LA(556)186 | R. Purrmann and M. Maas; *Liebigs Ann. Chem.*, 1944, **556**, 186. | 283 |
| | | |
| 45BSF78 | M. Polonovski, R. Vieillefosse and M. Pesson; *Bull. Soc. Chim. Fr.*, 1945, **12**, 78. | 299 |
| 45CRV(36)1 | S. M. Sethna and N. M. Shah; *Chem. Rev.*, 1945, **36**, 1. | 799, 807 |
| 45FOR(4)64 | R. Purrmann; *Fortschr. Chem. Org. Naturst.*, 1945, **4**, 64. | 264 |
| 45HCA771 | P. Plattner, P. Treadwell and C. Scholz; *Helv. Chim. Acta*, 1945, **28**, 771. | 791 |
| 45JA99 | C. F. Huebner and K. P. Link; *J. Am. Chem. Soc.*, 1945, **67**, 99. | 830 |
| 45JA802 | J. Weijlard, M. Tishler and A. E. Erickson; *J. Am. Chem. Soc.*, 1945, **67**, 802. | 311 |
| 45JA1294 | B. Graham, A. M. Griffith, C. S. Pease and B. E. Christensen; *J. Am. Chem. Soc.*, 1945, **67**, 1294. | 125 |
| 45JA1382 | J. G. Erickson; *J. Am. Chem. Soc.*, 1945, **67**, 1382. | 1024 |
| 45JA2112 | A. J. Tomisek and B. E. Christensen; *J. Am. Chem. Soc.*, 1945, **67**, 2112. | 129 |
| 45JA2127 | R. N. Jones; *J. Am. Chem. Soc.*, 1945, **67**, 2127. | 276 |
| 45JA2197 | G. W. Anderson, I. F. Halverstadt, W. H. Miller and R. O. Roblin, Jr.; *J. Am. Chem. Soc.*, 1945, **67**, 2197. | 152 |
| 45JCP(13)507 | H. S. Frank and M. W. Evans; *J. Chem. Phys.*, 1945, **13**, 507. | 266 |
| 45JCS347 | W. H. Davies and H. A. Piggott; *J. Chem. Soc.*, 1945, 347. | 116 |
| 45JCS858 | H. J. Kahn and V. A. Petrow; *J. Chem. Soc.*, 1945, 858. | 840 |
| B-45MI22600 | G. Herzberg; 'Infrared and Raman Spectroscopy', Van Nostrand Reinhold, London, 1945, p. 256. | 956 |
| | | |
| 46JA216 | L. W. Pickett and E. Sheffield; *J. Am. Chem. Soc.*, 1946, **68**, 216. | 955 |
| 46JA542 | H. W. Grimmel, A. Guenther and J. F. Morgan; *J. Am. Chem. Soc.*, 1946, **68**, 542. | 110 |
| 46JA912 | H. W. Scherp; *J. Am. Chem. Soc.*, 1946, **68**, 912. | 143 |
| 46JA1392 | J. J. Pfiffner, D. G. Calkins, E. M. Bloom and B. L. O'Dell; *J. Am. Chem. Soc.*, 1946, **68**, 1392. | 325 |
| 46JCS117 | R. M. Anker and A. H. Cook; *J. Chem. Soc.*, 1946, 117. | 654 |
| 46JCS472 | K. Schofield and J. C. E. Simpson; *J. Chem. Soc.*, 1946, 472. | 245 |
| 46JCS670 | A. Maccoll; *J. Chem. Soc.*, 1946, 670. | 466, 540 |
| 46JCS954 | L. J. Haynes and E. R. H. Jones; *J. Chem. Soc.*, 1946, 954. | 842 |
| 46JOC349 | N. J. Leonard and D. Y. Curtin; *J. Org. Chem.*, 1946, **11**, 349. | 94 |
| 46USP2394963 | M. L. Crossley and J. P. English; *U.S. Pat.* 2 394 963 (1946). | 164 |
| | | |
| 47ACS149 | O. Hassel and H. Viervoll; *Acta Chem. Scand.*, 1947, **1**, 149. | 949 |
| 47CR937 | J. Lecocq and N. P. Buu-Hoi; *C.R. Hebd. Seances Acad. Sci.*, 1947, **224**, 937. | 690 |
| 47CRV(40)279 | I. J. Krems and P. E. Spoerri; *Chem. Rev.*, 1947, **40**, 279. | 158, 305 |
| 47CRV(41)63 | M. Gates; *Chem. Rev.*, 1947, **41**, 63. | 264 |
| 47CRV(41)525 | L. Cavalieri; *Chem. Rev.*, 1947, **41**, 525. | 789, 810 |
| 47G308 | G. Jacini; *Gazz. Chim. Ital.*, 1947, **77**, 308. | 382 |

| | | |
|---|---|---|
| 47JA250 | B. L. O'Dell, J. M. Vandenbelt, E. S. Bloom and J. Pfiffner; *J. Am. Chem. Soc.*, 1947, **69**, 250. | 281 |
| 47JA674 | H. K. Mitchell and J. F. Nyc; *J. Am. Chem. Soc.*, 1947, **69**, 674. | 119 |
| 47JA2138 | G. B. Elion and G. H. Hitchings; *J. Am. Chem. Soc.*, 1947, **69**, 2138. | 90 |
| 47JA2246 | G. F. Woods and D. N. Kramer; *J. Am. Chem. Soc.*, 1947, **69**, 2246. | 777 |
| 47JA2449 | W. E. Parham; *J. Am. Chem. Soc.*, 1947, **69**, 2449. | 987 |
| 47JA2753 | D. E. Wolf, R. C. Anderson, E. A. Kaczka, S. A. Harris, G. E. Arth, P. L. Southwick, R. Mozingo and K. Folkers; *J. Am. Chem. Soc.*, 1947, **69**, 2753. | 285, 294, 325 |
| 47JBC(170)747 | L. J. Daniel and L. C. Norris; *J. Biol. Chem.*, 1947, **170**, 747. | 325 |
| 47JCS726 | F. E. King and T. J. King; *J. Chem. Soc.*, 1947, 726. | 228 |
| 47JCS943 | F. E. King and T. J. King; *J. Chem. Soc.*, 1947, 943. | 70 |
| 47JCS1179 | R. A. Baxter and F. S. Spring; *J. Chem. Soc.*, 1947, 1179. | 187 |
| 47JCS1394 | R. B. Bradbury, N. C. Hancox and H. H. Hatt; *J. Chem. Soc.*, 1947, 1394. | 118 |
| 47MI21400 | A. Albert and H. Duewell; *J. Soc. Chem. Ind. Lond.*, 1947, **66**, 11. | 164 |
| 48BSF688 | M. Polonovski and M. Pesson; *Bull. Soc. Chim. Fr.*, 1948, 688. | 70 |
| 48BSF963 | M. Pesson; *Bull. Soc. Chim. Fr.*, 1948, 963. | 312 |
| 48CRV(43)203 | J. H. Saunders and R. J. Slocombe; *Chem. Rev.*, 1948, **43**, 203. | 508 |
| 48CRV(43)509 | A. Mustafa; *Chem. Rev.*, 1948, **43**, 509. | 880 |
| 48JA1 | B. L. Hutchings, E. L. R. Stokstad, N. Bohonos, N. H. Sloane and Y. SubbaRow; *J. Am. Chem. Soc.*, 1948, **70**, 1. | 325 |
| 48JA14 | J. H. Mowat, J. H. Boothe, B. L. Hutchings, E. L. R. Stokstad, C. W. Waller, R. B. Angier, J. Semb, D. B. Cosulich and Y. SubbaRow; *J. Am. Chem. Soc.*, 1948, **70**, 14. | 293, 296 |
| 48JA23 | M. E. Hultquist, E. Kuh, D. B. Cosulich, M. J. Fahrenbach, E. H. Northey, D. R. Seeger, J. P. Sickels, J. M. Smith, R. B. Angier, J. H. Boothe, B. L. Hutchings, J. H. Mowat, J. Semb, E. L. R. Stokstad, Y. SubbaRow and C. W. Waller; *J. Am. Chem. Soc.*, 1948, **70**, 23. | 312 |
| 48JA27 | J. H. Boothe, C. W. Waller, E. L. R. Stokstad, B. L. Hutchings, J. H. Mowat, R. B. Angier, J. Semb, Y. SubbaRow, D. B. Cosulich, M. J. Fahrenbach, M. E. Hultquist, E. Kuh, E. H. Northey, D. R. Seeger, J. P. Sickels and J. M. Smith, Jr.; *J. Am. Chem. Soc.*, 1948, **70**, 27. | 302 |
| 48JA599 | G. B. Bachmann and H. A. Levine; *J. Am. Chem. Soc.*, 1948, **70**, 599. | 731, 855 |
| 48JA864 | R. H. Eastman and D. Gallup; *J. Am. Chem. Soc.*, 1948, **70**, 864. | 966 |
| 48JA1001 | T. L. Gresham, J. E. Jansen and F. W. Shaver; *J. Am. Chem. Soc.*, 1948, **70**, 1001. | 1030 |
| 48JA1257 | R. C. Ellingson and R. L. Henry; *J. Am. Chem. Soc.*, 1948, **70**, 1257. | 367 |
| 48JA1564 | N. J. Leonard and L. E. Sutton; *J. Am. Chem. Soc.*, 1948, **70**, 1564. | 947 |
| 48JA2600 | G. R. Lappin and R. K. Summerbell; *J. Am. Chem. Soc.*, 1948, **70**, 2600. | 965 |
| 48JA3026 | C. V. Cain, M. F. Mallette and E. C. Taylor, Jr.; *J. Am. Chem. Soc.*, 1948, **70**, 3026. | 277, 302, 304 |
| 48JA3109 | A. Bendich, J. F. Tinker and G. B. Brown; *J. Am. Chem. Soc.*, 1948, **70**, 3109. | 107, 133 |
| 48JBC(173)439 | H. Rosenkrantz; *J. Biol. Chem.*, 1948, **173**, 439. | 595 |
| 48JCP(17)466 | J. R. Platt, H. B. Klevens and W. C. Price; *J. Chem. Phys.*, 1948, **17**, 466. | 598 |
| 48JCS174 | J. Boyd and A. Robertson; *J. Chem. Soc.*, 1948, 174. | 805 |
| 48JCS1389 | V. Petrow and J. Saper; *J. Chem. Soc.*, 1948, 1389. | 251 |
| 48JCS1759 | F. H. S. Curd, J. K. Landquist and F. L. Rose; *J. Chem. Soc.*, 1948, 1759. | 110 |
| 48JCS1766 | F. H. S. Curd, E. Hoggarth, J. K. Landquist and F. L. Rose; Chem. Soc., 1948, 1766. | 137 |
| 48JOC477 | H. W. Johnston, C. E. Kaslow, A. Langsjoen and R. L. Shriner; *J. Org. Chem.*, 1948, **13**, 477. | 830 |
| 48JOC671 | G. R. Lappin and R. K. Summerbell; *J. Org. Chem.*, 1948, **13**, 671. | 965, 972 |
| 48M(79)106 | W. H. Brunner and E. Bertsch; *Monatsh. Chem.*, 1948, **79**, 106. | 498 |
| 48ZOB2023 | V. M. Rodionov and O. S. Urbanskaya; *Zh. Obshch. Khim.*, 1948, **18**, 2023. | 111 |
| 49CB25 | F. Weygand, A. Wacker and V. Schmid-Kowarzik; *Chem. Ber.*, 1949, **82**, 25. | 311 |
| 49CCC223 | Z. Buděšínský; *Collect. Czech. Chem. Commun.*, 1949, **14**, 223. | 82 |
| 49HCA1752 | B. G. Engel, W. Brzeski and P. A. Plattner; *Helv. Chim. Acta*, 1949, **32**, 1752. | 599 |
| 49JA84 | E. A. Fehnel and M. Carmack; *J. Am. Chem. Soc.*, 1949, **71**, 84. | 955 |
| 49JA741 | G. B. Elion, A. E. Light and G. H. Hitchings; *J. Am. Chem. Soc.*, 1949, **71**, 741. | 307 |
| 49JA892 | C. K. Cain, E. C. Taylor and L. J. Daniel; *J. Am. Chem. Soc.*, 1949, **71**, 892. | 296, 325 |
| 49JA1753 | D. R. Seeger, D. B. Cosulich, J. M. Smith, Jr., and M. E. Hultquist; *J. Am. Chem. Soc.*, 1949, **71**, 1753. | 294, 311 |
| 49JA1891 | J. B. Ziegler; *J. Am. Chem. Soc.*, 1949, **71**, 1891. | 259 |
| 49JA2538 | E. C. Taylor, Jr., and C. K. Cain; *J. Am. Chem. Soc.*, 1949, **71**, 2538. | 294 |
| 49JA2652 | L. Vagha, J. Ramonczai and J. Báthory; *J. Am. Chem. Soc.*, 1949, **71**, 2652. | 828 |
| 49JA3412 | P. B. Russel, R. Purrmann, W. Schmitt and G. H. Hitchings; *J. Am. Chem. Soc.*, 1949, **71**, 3412. | 283, 303 |
| 49JA3602 | R. O. Clinton and S. C. Laskowski; *J. Am. Chem. Soc.*, 1949, **71**, 3602. | 691 |
| 49JBC(177)357 | G. H. Hitchings, G. B. Elion, E. A. Falco and P. B. Russell; *J. Biol. Chem.*, 1949, **177**, 357. | 145 |
| 49JCS79 | H. S. Forrest and J. Walker; *J. Chem. Soc.*, 1949, 79. | 311 |
| 49JCS278 | C. C. J. Culvenor, W. Davies and N. S. Heath; *J. Chem. Soc.*, 1949, 278. | 1013 |

| | | |
|---|---|---|
| 49JCS1354 | J. S. Morley and J. C. E. Simpson; *J. Chem. Soc.*, 1949, 1354. | 76 |
| 49JCS1720 | G. A. Swan; *J. Chem. Soc.*, 1949, 1720. | 859 |
| 49JCS2077 | H. S. Forrest and J. Walker; *J. Chem. Soc.*, 1949, 2077. | 311 |
| 49JCS2142 | W. Baker and V. S. Butt; *J. Chem. Soc.*, 1949, 2142. | 702 |
| 49JCS2540 | V. Petrow, J. Saper and B. Sturgeon; *J. Chem. Soc.*, 1949, 2540. | 255 |
| 49JOC460 | A. B. Steele, A. B. Boese, Jr. and M. F. Dull; *J. Org. Chem.*, 1949, **14**, 460. | 792 |
| 49JOC946 | R. A. Donia, J. A. Shotton, L. O. Bentz and G. E. P. Smith; *J. Org. Chem.*, 1949, **14**, 946. | 1043, 1065 |
| 49MI21200 | B. P. Moore; *Nature (London)*, 1949, **163**, 918. | 42 |
| 49MI21600 | G. M. Timmis; *Nature (London)*, 1949, **164**, 139. | 313 |
| 49PIA(A)(30)57 | V. V. Virkar and R. C. Shah; *Proc. Indian Acad. Sci., Sect. A*, 1949, **30**, 57. | 818 |
| 49USP2475440 | H. A. Walter; *U.S. Pat.* 2 475 440 (1949) (*Chem. Abstr.*, 1949, **43**, 908). | 569 |
| | | |
| 50ACS1233 | N. Clauson-Kaas, Si-Oh Li and N. Elming; *Acta Chem. Scand.*, 1950, **4**, 1233. | 55 |
| 50BSF616 | M. Polonovski and H. Schmitt; *Bull. Soc. Chim. Fr.*, 1950, 616. | 96 |
| 50BSF1132 | J. Dallemagne and J. Martinet; *Bull. Soc. Chim. Fr.*, 1950, 1132. | 826 |
| 50CR(230)392 | M. Polonovski and H. Jerome; *C.R. Hebd. Seances Acad. Sci.*, 1950, **230**, 392. | 315 |
| 50G651 | S. Maffei; *Gazz. Chim. Ital.*, 1950, **80**, 651. | 164 |
| 50G750 | M. Simonetta and C. Cardini; *Gazz. Chim. Ital.*, 1950, **80**, 750. | 601 |
| 50HCA39 | P. Karrer and R. Schwyzer; *Helv. Chim. Acta*, 1950, **33**, 39. | 283 |
| 50HCA1233 | P. Karrer, B. Nicolaus and R. Schwyzer; *Helv. Chim. Acta*, 1950, **33**, 1233. | 283, 303 |
| 50HCA1365 | R. Hirt, H. Nidecker and R. Berchtold; *Helv. Chim. Acta*, 1950, **33**, 1365. | 484 |
| 50JA78 | G. B. Elion, G. H. Hitchings and P. B. Russell; *J. Am. Chem. Soc.*, 1950, **72**, 78. | 310 |
| 50JA1022 | C. L. Bickel; *J. Am. Chem. Soc.*, 1950, **72**, 1022. | 791 |
| 50JA1914 | B. Roth, J. M. Smith, Jr., and M. E. Hultquist; *J. Am. Chem. Soc.*, 1950, **72**, 1914. | 294 |
| 50JA2280 | R. A. Henry and W. M. Dehn; *J. Am. Chem. Soc.*, 1950, **72**, 2280. | 1014 |
| 50JA3009 | P. T. Mora and T. Széki; *J. Am. Chem. Soc.*, 1950, **72**, 3009. | 734 |
| 50JA3053 | A. M. Downes and F. Lions; *J. Am. Chem. Soc.*, 1950, **72**, 3053. | 108 |
| 50JA3079 | R. I. Longley, Jr. and W. S. Emerson; *J. Am. Chem. Soc.*, 1950, **72**, 3079. | 772 |
| 50JA4397 | S. C. Burket and M. Badger; *J. Am. Chem. Soc.*, 1950, **72**, 4397. | 594 |
| 50JA4630 | C. W. Waller, A. A. Goldman, R. B. Angier, J. H. Boothe, B. L. Hutchings, J. H. Mowat and J. Semb; *J. Am. Chem. Soc.*, 1950, **72**, 4630. | 302, 304 |
| 50JA4890 | R. O. Roblin, Jr. and J. W. Clapp; *J. Am. Chem. Soc.*, 1950, **72**, 4890. | 95, 138 |
| 50JBC(187)83 | H. Rosenkrantz and A. T. Milhorat; *J. Biol. Chem.*, 1950, **187**, 83. | 595 |
| 50JCS1892 | A. H. Cook, G. D. Hunter and J. R. A. Pollock; *J. Chem. Soc.*, 1950, 1892. | 440 |
| 50JCS1925 | J. E. Gowan and T. S. Wheeler; *J. Chem. Soc.*, 1950, 1925. | 818 |
| 50JCS3062 | I. A. Brownlie; *J. Chem. Soc.*, 1950, 3062. | 66 |
| 50JCS3375 | A. Kamal, A. Robertson and E. Tittensor; *J. Chem. Soc.*, 1950, 3375. | 832 |
| 50JCS3389 | W. Baker, W. D. Ollis and V. D. Poole; *J. Chem. Soc.*, 1950, 3389. | 543 |
| 50JOC869 | A. H. Blatt; *J. Org. Chem.*, 1950, **15**, 869. | 1015 |
| 50LA(568)34 | F. Wille and L. Saffer; *Liebigs Ann. Chem.*, 1950, **568**, 34. | 759 |
| 50MI21300 | D. J. Brown; *Nature (London)*, 1950, **165**, 1010. | 84 |
| 50MI21301 | D. J. Brown; *J. Soc. Chem. Ind.*, 1950, **69**, 353. | 132 |
| 50MI21302 | G. R. Wyatt; *Nature (London)*, 1950, **166**, 237. | 145 |
| 50MI21303 | E. Waletzky, J. H. Clark and H. W. Marson; *Science*, 1950, **111**, 720. | 154 |
| 50MI21900 | L. Ergener; *Rev. Fac. Univ. Istanbul*, 1950, **15A**, 91 (*Chem. Abstr.*, 1951, **74**, 476). | 409 |
| B-50MI22400 | J. Fried; in 'Heterocyclic Compounds', ed. R. C. Elderfield; Wiley, New York, 1950, vol. 1, p. 343. | 738 |
| 50NKK590 | A. Yokoo; *Nippon Kagaku Zasshi*, 1950, **71**, 590. | 148 |
| 50ZN(B)132 | R. Tschesche, K. H. Köhncke and F. Korte; *Z. Naturforsch., Teil B*, 1950, **5**, 132. | 301 |
| | | |
| 51BSF423 | M. Pesson; *Bull. Soc. Chim. Fr.*, 1951, 423. | 312 |
| 51BSF428 | M. Pesson; *Bull. Soc. Chim. Fr.*, 1951, 428. | 312 |
| 51BSF521 | M. Polonovski, M. Pesson and A. Puister; *Bull. Soc. Chim. Fr.*, 1951, 521. | 306, 312 |
| 51BSF693 | E. D. Bergmann, G. Berthier, E. Fisher, Y. Hirschberg, D. Lavie, A. Pullman and B. Pullman; *Bull. Soc. Chim. Fr.*, 1951, 693. | 599 |
| 51CB801 | R. Tschesche and F. Korte; *Chem. Ber.*, 1951, **84**, 801. | 294 |
| 51HCA186 | A. Bolleter, K. Eiter and H. Schmid; *Helv. Chim. Acta*, 1951, **34**, 186. | 717 |
| 51HCA1029 | P. Karrer and B. Nicolaus; *Helv. Chim. Acta*, 1951, **34**, 1029. | 283, 303 |
| 51HCA2155 | P. Karrer and H. Feigl; *Helv. Chim. Acta*, 1951, **34**, 2155. | 283, 303 |
| 51JA1873 | R. H. Mizzoni and P. E. Spoerri; *J. Am. Chem. Soc.*, 1951, **73**, 1873. | 56 |
| 51JA2864 | B. Roth, J. M. Smith, Jr. and M. E. Hultquist; *J. Am. Chem. Soc.*, 1951, **73**, 2864. | 129 |
| 51JA3067 | M. May, T. J. Bardos, F. L. Barger, M. Lansford, J. M. Ravel, G. L. Sutherland and W. Shive; *J. Am. Chem. Soc.*, 1951, **73**, 3067. | 325 |
| 51JA3163 | S. Siegel, S. K. Coburn and D. R. Levering; *J. Am. Chem. Soc.*, 1951, **73**, 3163. | 787, 788 |
| 51JA3237 | S. Siegel, W. M. Boyer and R. R. Jay; *J. Am. Chem. Soc.*, 1951, **73**, 3237. | 787 |
| 51JA3514 | T. A. Geissman; *J. Am. Chem. Soc.*, 1951, **73**, 3514. | 816 |
| 51JA3531 | R. H. Wiley and N. R. Smith; *J. Am. Chem. Soc.*, 1951, **73**, 3531. | 792 |
| 51JA3763 | P. B. Russell and G. H. Hitchings; *J. Am. Chem. Soc.*, 1951, **73**, 3763. | 151 |
| 51JA4205 | P. F. Wiley; *J. Am. Chem. Soc.*, 1951, **73**, 4205. | 828, 855 |
| 51JA4384 | E. C. Taylor, Jr., and C. K. Cain; *J. Am. Chem. Soc.*, 1951, **73**, 4384. | 291, 294, 296 |
| 51JA4865 | L. W. Pickett, N. J. Hoeflich and T.-C. Liu; *J. Am. Chem. Soc.*, 1951, **73**, 4865. | 598, 955 |
| 51JA5267 | C. W. Smith, D. G. Norton and S. A. Ballard; *J. Am. Chem. Soc.*, 1951, **73**, 5267. | 772 |

| | | |
|---|---|---|
| 51JA5273 | C. W. Smith, D. G. Norton and S. A. Ballard; *J. Am. Chem. Soc.*, 1951, **73**, 5273. | 771 |
| 51JA5494 | S. Siegel and S. Coburn; *J. Am. Chem. Soc.*, 1951, **73**, 5494. | 788 |
| 51JA5777 | J. Siegle and B. E. Christensen; *J. Am. Chem. Soc.*, 1951, **73**, 5777. | 77, 78 |
| 51JCS76 | W. Baker, R. F. Curtis and J. F. W. McOmie; *J. Chem. Soc.*, 1951, 76. | 733 |
| 51JCS96 | W. R. Boon, W. G. M. Jones and G. R. Ramage; *J. Chem. Soc.*, 1951, 96. | 293, 315 |
| 51JCS474 | A. Albert, D. J. Brown and G. Cheeseman; *J. Chem. Soc.*, 1951, 474. | 266, 286, 293, 299, 318 |
| 51JCS726 | H. Burton and P. F. G. Praill; *J. Chem. Soc.*, 1951, 726. | 865 |
| 51JCS1004 | J. R. Marshall and J. Walker; *J. Chem. Soc.*, 1951, 1004. | 66, 67 |
| 51JCS1218 | M. P. V. Boarland and J. F. W. McOmie; *J. Chem. Soc.*, 1951, 1218. | 130 |
| 51JCS1497 | W. R. Boon and T. Leigh; *J. Chem. Soc.*, 1951, 1497. | 315 |
| 51JCS1565 | N. Whittaker; *J. Chem. Soc.*, 1951, 1565. | 87, 131 |
| 51JCS1971 | J. S. Morley; *J. Chem. Soc.*, 1951, 1971. | 56 |
| 51JCS2307 | N. P. Buu-Hoï, N. Hoán and M. R. Khenissi; *J. Chem. Soc.*, 1951, 2307. | 802 |
| 51JCS2323 | B. Lythgoe and L. S. Rayner; *J. Chem. Soc.*, 1951, 2323. | 73, 140, 305 |
| 51JCS2679 | G. T. Newbold, W. Sharp and F. S. Spring; *J. Chem. Soc.*, 1951, 2679. | 191 |
| 51JCS3155 | C. A. C. Haley and P. Maitland; *J. Chem. Soc.*, 1951, 3155. | 112 |
| 51JCS3204 | G. M. Badger, R. S. Pearce and R. Pettit; *J. Chem. Soc.*, 1951, 3204. | 174 |
| 51JCS3318 | J. M. Hearn, R. A. Morton and J. C. E. Simpson; *J. Chem. Soc.*, 1951, 3318. | 67 |
| 51JOC461 | R. M. Dodson and J. K. Seyler; *J. Org. Chem.*, 1951, **16**, 461. | 115 |
| 51JOC1064 | R. L. Shriner and W. R. Knox; *J. Org. Chem.*, 1951, **16**, 1064. | 651, 832, 872 |
| 51LA(572)217 | F. G. Fisher and J. Roche; *Liebigs Ann. Chem.*, 1951, **572**, 217. | 364 |
| 51M662 | R. Patzak and L. Neugebauer; *Monatsh. Chem.*, 1951, **82**, 662. | 599 |
| B-51MI22400 | S. Wawzonek; in 'Heterocyclic Compounds', ed. R. C. Elderfield; Wiley, New York, 1951, vol. 2, p. 173. | 738, 767, 799, 807, 816 |
| 51YZ1420 | B. Ohta and H. Kawasaki; *Yakugaku Zasshi*, 1951, **71**, 1420. | 100 |
| | | |
| 52AC(R)673 | M. Rolla, M. Sanesi and G. Traverso; *Ann. Chim. (Rome)*, 1952, **42**, 673. | 626, 627 |
| 52BSB44 | J. J. Fox and D. Shugar; *Bull. Soc. Chim. Belg.*, 1952, **61**, 44. | 68 |
| 52BSF91 | C. Mentzer, D. Molho and P. Vercier; *Bull. Soc. Chim. Fr.*, 1952, 91. | 826 |
| 52CB204 | W. Ried, A. Berg and G. Schmidt; *Chem. Ber.*, 1952, **85**, 204. | 247 |
| 52CB1012 | F. Korte; *Chem. Ber.*, 1952, **85**, 1012. | 225 |
| 52CR(234)1787 | H. Normant and P. Maitte; *C. R. Hebd. Seances Acad. Sci.*, 1952, **234**, 1787. | 783 |
| 52G155 | C. Cardani; *Gazz. Chim. Ital.*, 1952, **82**, 155. | 849 |
| 52HCA1168 | J. Schmutz, R. Hirt and H. Lauener; *Helv. Chim. Acta*, 1952, **35**, 1168. | 709 |
| 52JA971 | I. J. Pachter and M. C. Kloetzel; *J. Am. Chem. Soc.*, 1952, **74**, 971. | 164 |
| 52JA1648 | E. C. Taylor, Jr., *J. Am. Chem. Soc.*, 1952, **74**, 1648. | 294 |
| 52JA3252 | D. B. Cosulich, B. Roth, I. M. Smith, Jr., M. E. Hultquist and R. P. Parker, *J. Am. Chem. Soc.*, 1952, **74**, 3252. | 281 |
| 52JA3443 | P. B. Russell and G. H. Hitchings; *J. Am. Chem. Soc.*, 1952, **74**, 3443. | 115 |
| 52JA3545 | R. Newman and R. M. Badger; *J. Am. Chem. Soc.*, 1952, **74**, 3545. | 465 |
| 52JA3622 | R. L. Shriner and J. A. Shotton; *J. Am. Chem. Soc.*, 1952, **74**, 3622. | 765 |
| 52JA3877 | G. B. Elion and G. H. Hitchings; *J. Am. Chem. Soc.*, 1952, **74**, 3877. | 304 |
| 52JA3999 | C. W. Winter and C. S. Hamilton; *J. Am. Chem. Soc.*, 1952, **74**, 3999. | 718 |
| 52JA4205 | H. Gilman, R. D. Nelson and J. F. Champaigne; *J. Am. Chem. Soc.*, 1952, **74**, 4205. | 1011 |
| 52JA4267 | C. W. Whitehead; *J. Am. Chem. Soc.*, 1952, **74**, 4267. | 134 |
| 52JA4834 | H. Culbertson, J. C. Decius and B. E. Christensen; *J. Am. Chem. Soc.*, 1952, **74**, 4834. | 67 |
| 52JA5346 | R. Adams and T. E. Bockstahler; *J. Am. Chem. Soc.*, 1952, **74**, 5346. | 804 |
| 52JA5633 | T. L. Cairns, A. W. Larchar and B. C. McKusick, *J. Am. Chem. Soc.*, 1952, **74**, 5633. | 116, 503 |
| 52JCS168 | L. N. Short and H. W. Thompson; *J. Chem. Soc.*, 1952, 168. | 66 |
| 52JCS1620 | A. Albert, D. J. Brown and G. W. H. Cheeseman; *J. Chem. Soc.*, 1952, 1620. | 122, 266, 271, 286, 287, 291, 293, 296, 297, 307, 310 |
| 52JCS2144 | F. E. King and P. C. Spensley; *J. Chem. Soc.*, 1952, 2144. | 311 |
| 52JCS3065 | J. C. Roberts; *J. Chem. Soc.*, 1952, 3065. | 78 |
| 52JCS3428 | G. Shaw; *J. Chem. Soc.*, 1952, 3428. | 1043, 1070 |
| 52JCS3448 | F. L. Rose; *J. Chem. Soc.*, 1952, 3448. | 131 |
| 52JCS3716 | M. P. V. Boarland and J. F. W. McOmie; *J. Chem. Soc.*, 1952, 3716. | 65, 67 |
| 52JCS3722 | M. P. V. Boarland and J. F. W. McOmie; *J. Chem. Soc.*, 1952, 3722. | 65, 67 |
| 52JCS3741 | N. D. Xuong and N. P. Buu-Hoï; *J. Chem. Soc.*, 1952, 3741. | 768 |
| 52JCS4109 | S. J. Davis and J. A. Elvidge; *J. Chem. Soc.*, 1952, 4109. | 792 |
| 52JCS4219 | A. Albert, D. J. Brown and G. Cheeseman; *J. Chem. Soc.*, 1952, 4219. | 266, 271 |
| 52JCS4799 | H. J. E. Loewenthal and R. Pappo; *J. Chem. Soc.*, 1952, 4799. | 858 |
| 52JCS4985 | A. Albert and A. Hampton; *J. Chem. Soc.*, 1952, 4985. | 251 |
| 52JOC542 | H. H. Fox; *J. Org. Chem.*, 1952, **17**, 542. | 225 |
| 52JOC1320 | H. R. Henze, W. J. Clegg and C. W. Smart; *J. Org. Chem.*, 1952, **17**, 1320. | 104 |
| 52JOC1419 | S. Wawzonek and H. A. Ready; *J. Org. Chem.*, 1952, **17**, 1419. | 821 |
| 52LA(577)77 | C. Grundmann, G. Weise and S. Seide; *Liebigs Ann. Chem.*, 1952, **577**, 77. | 504 |
| 52MI21300 | D. J. Brown; *J. Appl. Chem.*, 1952, **2**, 239. | 102, 142 |
| 52MI22200 | I. Chmielewska and J. Cieślak; *Przem. Chem.*, 1952, **8**, 196. | 599 |
| B-52MI22400 | K. Venkataraman; 'The Chemistry of Synthetic Dyes', Academic Press, New York, 1952, vol. 2, p. 740. | 878 |
| 52OS(32)6 | A. V. Holmgren and W. Wenner; *Org. Synth.*, 1952, **32**, 6. | 149 |

| | | |
|---|---|---|
| 52OS(32)45 | J. A. VanAllan; *Org. Synth.*, 1952, **32**, 45. | 115 |
| 52QR197 | A. Albert; *Q. Rev., Chem. Soc.*, 1952, **6**, 197. | 264, 309 |
| 52USP2600596 | H. E. Winberg; *U.S. Pat.* 2 600 596 (1952) (*Chem. Abstr.*, 1953, **47**, 7536). | 1030 |
| 53AP437 | H. Böhme and W. Schmidt; *Arch. Pharm.* (*Weinheim, Ger.*), 1953, **286**, 437. | 1028 |
| 53BBA(12)462 | A. Bendich and G. C. Clements; *Biochim. Biophys. Acta*, 1953, **12**, 462. | 143 |
| 53BSF640 | M. Julia and M. Baillargé; *Bull. Soc. Chim. Fr.*, 1953, 640. | 985 |
| 53CB845 | H. Bredereck, I. Hennig, W. Pfleiderer and O. Deschler; *Chem. Ber.*, 1953, **86**, 845. | 284 |
| 53CB1327 | F. Klages and H. Träger; *Chem. Ber.*, 1953, **86**, 1327. | 868, 872 |
| 53CPB387 | M. Hasegawa; *Chem. Pharm. Bull.*, 1953, **1**, 387. | 77 |
| 53CR(235)1310 | M. Polonovski, M. Pesson and P. Rajzman; *C.R. Hebd. Seances Acad. Sci.*, 1953, **235**, 1310. | 421 |
| 53G327 | S. Maffei, S. Pietra and A. Cattaneo; *Gazz. Chim. Ital.*, 1953, **83**, 327. | 164 |
| 53HC(5)3 | J. C. E. Simpson; *Chem. Heterocycl. Compd.*, 1953, **5**, 3. | 2 |
| 53HC(5)198 | J. C. E. Simpson; *Chem. Heterocycl. Compd.*, 1953, **5**, 198. | 232 |
| 53HC(5)356 | J. C. E. Simpson; *Chem. Heterocycl. Compd.*, 1953, **5**, 356. | 248 |
| 53JA626 | B. M. Perfetti and R. Levine; *J. Am. Chem. Soc.*, 1953, **75**, 626. | 865 |
| 53JA656 | P. J. Vanderhorst and C. S. Hamilton; *J. Am. Chem. Soc.*, 1953, **75**, 656. | 217 |
| 53JA675 | H. L. Yale; *J. Am. Chem. Soc.*, 1953, **75**, 675. | 136 |
| 53JA1622 | H. L. Hergert and E. F. Kurth; *J. Am. Chem. Soc.*, 1953, **75**, 1622. | 597 |
| 53JA1647 | W. E. Parham, T. M. Roder and W. R. Hasek; *J. Am. Chem. Soc.*, 1953, **75**, 1647. | 955, 966 |
| 53JA1883 | D. N. Robertson and K. P. Link; *J. Am. Chem. Soc.*, 1953, **75**, 1883. | 680 |
| 53JA1904 | E. C. Taylor, Jr., J. A. Carbon and D. R. Hoff; *J. Am. Chem. Soc.*, 1953, **75**, 1904. | 300, 318 |
| 53JA2065 | W. E. Parham, H. Wynberg and F. L. Ramp; *J. Am. Chem. Soc.*, 1953, **75**, 2065. | 955 |
| 53JA3333 | H. E. Ungnade, E. F. Kline and E. W. Crandall; *J. Am. Chem. Soc.*, 1953, **75**, 3333. | 599 |
| 53JA3384 | J. F. Nobis, A. J. Blardinelli and D. J. Blaney; *J. Am. Chem. Soc.*, 1953, **75**, 3384. | 975 |
| 53JA4992 | A. Schönberg, N. Badran and N. A. Starkowsky; *J. Am. Chem. Soc.*, 1953, **75**, 4992. | 710 |
| 53JCS74 | A. Albert and D. J. Brown; *J. Chem. Soc.*, 1953, 74. | 277, 302, 303, 310 |
| 53JCS331 | D. J. Brown and L. N. Short; *J. Chem. Soc.*, 1953, 331. | 66, 67, 81, 83, 100, 127, 128 |
| 53JCS716 | A. J. Nunn and K. Schofield; *J. Chem. Soc.*, 1953, 716. | 382 |
| 53JCS1348 | A. A. Goldberg and H. A. Walker; *J. Chem. Soc.*, 1953, 1348. | 707, 838 |
| 53JCS1646 | N. Whittaker; *J. Chem. Soc.*, 1953, 1646. | 68, 123 |
| 53JCS2234 | J. H. Lister and G. R. Ramage; *J. Chem. Soc.*, 1953, 2234. | 280, 305 |
| 53JCS3129 | J. F. W. McOmie and I. M. White; *J. Chem. Soc.*, 1953, 3129. | 76, 126 |
| 53JCS3337 | D. Chakravarti, R. N. Chakravarti and S. C. Chakravarti; *J. Chem. Soc.*, 1953, 3337. | 148 |
| 53JCS3435 | M. Crawford and J. A. M. Shaw; *J. Chem. Soc.*, 1953, 3435. | 807 |
| 53JIC103 | J. N. Chatterjea; *J. Indian Chem. Soc.*, 1953, **30**, 103. | 832 |
| 53MI21300 | A. Albert and H. C. S. Wood; *J. Appl. Chem.*, 1953, **3**, 521. | 115 |
| B-53MI21301 | H. T. Openshaw; in 'The Alkaloids', ed. R. H. F. Manske; Academic Press, 1953, New York, vol. 3, p. 101. | 148 |
| 53MI21900 | G. Bähr, E. Hess, E. Steinkopf and G. Schleitzer; *Z. Anorg. Allg. Chem.*, 1953, **273**, 325. | 402 |
| 53MI22300 | J. Bandrenghien, J. Jadot and R. Huls; *Bull. Cl. Sci. Acad. Roy. Belg.*, 1953, **39**, 105. | 674 |
| 53MI22301 | K. Arakawa; *Pharm. Bull.* (*Tokyo*), 1953, **1**, 331 (*Chem. Abstr.*, 1955, **49**, 10 941). | 680 |
| 53MI22302 | V. K. Ahluwalia, S. K. Mukerjee and T. R. Seshadri; *J. Sci. Ind. Res.* (*India*), 1953, **12B**, 283. | 716 |
| 53MI22400 | N. Narasimhachari, D. Rajagopalan and T. R. Seshadri; *J. Sci. Ind. Res.* (*India*), 1953, **12B**, 287. | 852 |
| 53OR(7)1 | S. Sethna and R. Phadke; *Org. React.*, 1953, **7**, 1. | 799, 825 |
| 53PIA(A)(37)774 | R. M. Naik and V. M. Thakor; *Proc. Indian Acad. Sci., Sect. A*, 1953, **37**, 774. | 819 |
| 53USP2630433 | D. W. Kaiser and J. J. Roemer; *U.S. Pat.* 2 630 433 (1953) (*Chem. Abstr.*, 1954, **48**, 747). | 491 |
| 53USP2653937 | D. W. Kaiser; *U.S. Pat.* 2 653 937 (1953) (*Chem. Abstr.*, 1954, **48**, 9413). | 491 |
| 54AC(R)430 | M. Rolla, M. Sanesi and G. Traverso; *Ann. Chim.* (*Rome*), 1954, **44**, 430. | 626 |
| 54ACS734 | J. Gripenberg and B. Juselius; *Acta Chem. Scand.*, 1954, **8**, 734. | 729 |
| 54AG363 | F. Mietzsch; *Angew Chem.*, 1954, **66**, 363. | 1038 |
| 54AX129 | F. L. Hirshfeld and G. M. J. Schmidt; *Acta Crystallogr.*, 1954, **7**, 129. | 158 |
| 54AX199 | P. Cucka and R. W. H. Small; *Acta Crystallogr.*, 1954, 7, 199. | 4 |
| 54AX313 | G. S. Parry; *Acta Crystallogr.*, 1954, **7**, 313. | 68 |
| 54AX498 | P. A. Howell, R. M. Curtis and W. N. Lipscomb; *Acta Crystallogr.*, 1954, **7**, 498. | 947 |
| 54CB825 | H.-B. König, W. Siefken and H. A. Offe; *Chem. Ber.*, 1954, **87**, 825. | 559, 561 |
| 54CB1540 | R. Metze; *Chem. Ber.*, 1954, **87**, 1540. | 402 |
| 54CCC282 | O. Wichterle and J. Roček; *Collect. Czech. Chem. Commun.*, 1954, **19**, 282. | 999 |
| 54CI(L)786 | A. Holland; *Chem. Ind.* (*London*), 1954, 786. | 124, 126 |
| 54CR(239)1047 | J. Colonge and P. Boisde; *C. R. Hebd. Seances Acad. Sci.*, 1954, **239**, 1047. | 788 |
| 54FOR(11)350 | A. Albert; *Fortschr. Chem. Org. Naturst.*, 1954, **11**, 350. | 264 |
| 54HCA134 | P. Schmidt and J. Druey; *Helv. Chim. Acta*, 1954, **37**, 134. | 2, 52 |
| 54HCA1467 | P. Schmidt and J. Druey; *Helv. Chim. Acta*, 1954, **37**, 1467. | 52 |
| 54HCA1706 | H. Schmid and K. Banholzer; *Helv. Chim. Acta*, 1954, **37**, 1706. | 818 |
| 54JA427 | C. Lin, E. Lieber and J. P. Horwitz; *J. Am. Chem. Soc.*, 1954, **76**, 427. | 540 |

| | | |
|---|---|---|
| 54JA632 | C. Grundmann and A. Kreutzberger; *J. Am. Chem. Soc.*, 1954, **76**, 632. | 460 |
| 54JA1068 | W. E. Parham and J. D. Jones; *J. Am. Chem. Soc.*, 1954, **76**, 1068. | 966 |
| 54JA1286 | V. Boekelheide and W. J. Linn; *J. Am. Chem. Soc.*, 1954, **76**, 1286. | 303 |
| 54JA1451 | P. Rochlin, D. B. Murphy and S. Helf; *J. Am. Chem. Soc.*, 1954, **76**, 1451. | 399 |
| 54JA1874 | E. C. Taylor, Jr., C. K. Cain and H. M. Loux; *J. Am. Chem. Soc.*, 1954, **76**, 1874. | 284 |
| 54JA1879 | C. G. Overberger and I. C. Kogon; *J. Am. Chem. Soc.*, 1954, **76**, 1879. | 78 |
| 54JA2451 | D. S. Tarbell and P. Hoffman; *J. Am. Chem. Soc.*, 1954, **76**, 2451. | 892 |
| 54JA2798 | F. F. Blicke and H. G. Godt, Jr.; *J. Am. Chem. Soc.*, 1954, **76**, 2798. | 315 |
| 54JA2899 | S. B. Greenbaum and W. L. Holmes; *J. Am. Chem. Soc.*, 1954, **76**, 2899. | 96 |
| 54JA3103 | G. S. Hammond, C. A. Stout and A. A. Lamola; *J. Am. Chem. Soc.*, 1964, **86**, 3103. | 678 |
| 54JA3551 | F. J. Wolf, K. Pfister, III, R. M. Wilson, Jr. and C. A. Robinson; *J. Am. Chem. Soc.*, 1954, **76**, 3551. | 409 |
| 54JA3642 | F. Stitt, G. F. Bailey, G. B. Coppinger and T. W. Campbell; *J. Am. Chem. Soc.*, 1954, **76**, 3642. | 595, 599 |
| 54JA3666 | A. E. Lanzilotti, J. B. Ziegler and A. C. Shabica; *J. Am. Chem. Soc.*, 1954, **76**, 3666. | 95, 136 |
| 54JA5437 | R. C. Elderfield and T. P. King; *J. Am. Chem. Soc.*, 1954, **76**, 5437. | 863 |
| 54JA5439 | R. C. Elderfield and T. P. King; *J. Am. Chem. Soc.*, 1954, **76**, 5439. | 765 |
| 54JA6052 | S. B. Greenbaum; *J. Am. Chem. Soc.*, 1954, **76**, 6052. | 97 |
| 54JA6388 | A. S. Dreiding and A. J. Tomasewski; *J. Am. Chem. Soc.*, 1954, **76**, 6388. | 791 |
| 54JBC(208)513 | G. W. E. Plaut; *J. Biol. Chem.*, 1954, **208**, 513. | 320 |
| 54JCS505 | A. Albert and A. Hampton; *J. Chem. Soc.*, 1954, 505. | 90, 251 |
| 54JCS665 | G. Shaw and G. Sugowdz; *J. Chem. Soc.*, 1954, 665. | 108, 120 |
| 54JCS839 | R. N. Lacey; *J. Chem. Soc.*, 1954, 839. | 1030 |
| 54JCS854 | R. N. Lacey; *J. Chem. Soc.*, 1954, 854. | 809, 826 |
| 54JCS1190 | N. B. Chapman and C. W. Rees; *J. Chem. Soc.*, 1954, 1190. | 99 |
| 54JCS1775 | G. Soliman and I. E.-S. El-Kholy; *J. Chem. Soc.*, 1954, 1755. | 812 |
| 54JCS2060 | A. Albert and D. J. Brown; *J. Chem. Soc.*, 1954, 2060. | 94, 121, 136 |
| 54JCS2819 | F. G. Mann and F. H. C. Stewart; *J. Chem. Soc.*, 1954, 2819. | 859 |
| 54JCS2881 | D. G. I. Felten and G. M. Timmis; *J. Chem. Soc.*, 1954, 2881. | 313 |
| 54JCS2887 | R. G. W. Spickett and G. M. Timmis; *J. Chem. Soc.*, 1954, 2887. | 313 |
| 54JCS2895 | D. G. I. Felten, T. S. Osdene and G. M. Timmis; *J. Chem. Soc.*, 1954, 2895. | 313 |
| 54JCS3263 | W. R. Boon, H. C. Carrington, N. Greenhalgh and C. H. Vasey; *J. Chem. Soc.*, 1954, 3263. | 153 |
| 54JCS3617 | R. D. Haworth, H. K. Pindred and P. R. Jefferies; *J. Chem. Soc.*, 1954, 3617. | 832 |
| 54JCS3832 | A. Albert, D. J. Brown and H. C. S. Wood; *J. Chem. Soc.*, 1954, 3832. | 77, 266, 291 |
| 54JCS4109 | J. H. Lister, G. R. Ramage and E. Coates; *J. Chem. Soc.* 1954, 4109. | 305 |
| 54LA(587)1 | O. Dann and G. Mylius; *Liebigs Ann. Chem.*, 1954, **587**, 1. | 801 |
| 54LA(587)16 | O. Dann, G. Volz and O. Huber; *Liebigs Ann. Chem.*, 1954, **587**, 16. | 827, 851 |
| 54LA(588)45 | A. Dornow and G. Petsch; *Liebigs Ann. Chem.*, 1954, **588**, 45. | 82 |
| 54MI21300 | D. J. Brown; *J. Appl. Chem.*, 1954, **4**, 72. | 94 |
| B-54MI21600 | G. E. W. Wolstenholme and M. P. Cameron; 'Chemistry and Biology of Pteridines', Churchill, London, 1954. | 264 |
| B-54MI21601 | E. C. Taylor; in 'Chemistry and Biology of Pteridines', ed. G. E. W. Wolstenholme and M. P. Cameron; Churchill, London, 1954, p. 2. | 318 |
| 54MI22400 | P. Maitte; *Ann. Chim. (Paris)*, 1954, **9**, 431. | 764, 788 |
| B-54MI22401 | F. Feigl; 'Spot Tests in Organic Analysis', Elsevier, Amsterdam, 1954. | 878 |
| 54MI22700 | S. Kojima; *J. Chem. Soc. Jpn. Ind. Chem.*, 1954, **57**, 819. | 999 |
| 54OR(8)59 | C. R. Hauser, F. W. Swamer and J. T. Adams; *Org. React.*, 1954, **8**, 59. | 820, 865 |
| 54USP2680741 | W. W. Gilbert; (E.I. du Pont de Nemours & Co.), *U.S. Pat.* 2 680 741 (1954) (*Chem. Abstr.*, 1955, **49**, 6322). | 232 |
| 54USP2688016 | J. J. Roemer and D. W. Kaiser; *U.S. Pat.* 2 688 016 (1954) (*Chem. Abstr.*, 1955, **49**, 11 728). | 491 |
| 54YZ1195 | T. Itai and H. Igeta; *Yakugaku Zasshi*, 1954, **74**, 1195 (*Chem. Abstr.* 1955, **49**, 14768). | 56 |
| 55AC(R)128 | P. Franzosini, G. Traverso and M. Sanesi; *Ann. Chim. (Rome)*, 1955, **45**, 128. | 599 |
| 55ACS847 | S. Sunner; *Acta Chem. Scand.*, 1955, **9**, 847. | 959 |
| 55AG328 | F. Weygand and M. Waldschmidt; *Angew. Chem.*, 1955, **67**, 328. | 320 |
| 55BRP732521 | Cassella Farbwerke Mainkur A/G, *Br. Pat.* 732 521 (1955) (*Chem. Abstr.*, 1957, **51**, 1302). | 246 |
| 55BSF1171 | M. Polonowski, M. Pesson and P. Rajzman; *Bull. Soc. Chim. Fr.*, 1955, 1171. | 415 |
| 55CB712 | R. Criegee and G. Paulig; *Chem. Ber.*, 1955, **88**, 712. | 990 |
| 55CI(L)271 | S. K. Mukerjee and T. R. Seshadri; *Chem. Ind. (London)*, 1955, 271. | 819 |
| 55CI(L)1102 | J. T. Edward; *Chem. Ind. (London)*, 1955, 1102. | 578, 629 |
| 55CPB173 | E. Ochiai and H. Yamanaka; *Chem. Pharm. Bull.*, 1955, **3**, 173. | 97 |
| 55CPB175 | E. Ochiai and H. Yamanaka; *Chem. Pharm. Bull.*, 1955, **3**, 175. | 105, 141 |
| 55CR(241)1783 | R. Charonnat and P. Fabiani; *C.R. Hebd. Seances Acad. Sci.*, 1955, **241**, 1783. | 566 |
| 55JA951 | J. H. Boyer and J. Hamer; *J. Am. Chem. Soc.*, 1955, **77**, 951. | 1025 |
| 55JA960 | T. J. Bardos, R. R. Herr and T. Enkoji; *J. Am. Chem. Soc.*, 1955, **77**, 960. | 136, 137 |
| 55JA1559 | H. R. Sullivan and W. T. Caldwell; *J. Am. Chem. Soc.*, 1955, **77**, 1559. | 77 |
| 55JA1623 | T. A. Geissman and A. Armen; *J. Am. Chem. Soc.*, 1955, **77**, 1623. | 856 |
| 55JA1702 | L. L. Woods; *J. Am. Chem. Soc.*, 1955, **77**, 1702. | 761 |

| | | |
|---|---|---|
| 55JA2243 | E. C. Taylor, Jr., H. M. Loux, E. A. Falco and G. H. Hitchings; *J. Am. Chem. Soc.*, 1955, **77**, 2243. | 284 |
| 55JA2256 | R. K. Robins and G. H. Hitchings; *J. Am. Chem. Soc.*, 1955, **77**, 2256. | 226 |
| 55JA2340 | R. H. Wiley and A. J. Hart; *J. Am. Chem. Soc.*, 1955, **77**, 2340. | 847 |
| 55JA2536 | N. A. Milas, P. Davis and J. T. Nolan, Jr.; *J. Am. Chem. Soc.*, 1955, **77**, 2536. | 991 |
| 55JA3167 | E. L. Patterson, H. P. Broquist, A. M. Albrecht, M. H. von Saltza and E. L. R. Stokstad; *J. Am. Chem. Soc.*, 1955, **77**, 3167. | 324 |
| 55JA3401 | C. S. Rondestvedt, Jr. and O. Vogl; *J. Am. Chem. Soc.*, 1955, **77**, 3401. | 688 |
| 55JA3927 | W. B. Wright, Jr. and J. M. Smith, Jr., *J. Am. Chem. Soc.*, 1955, **77**, 3927. | 282, 318 |
| 55JA4571 | S. M. McElvain and R. E. Starn, Jr.; *J. Am. Chem. Soc.*, 1955, **77**, 4571. | 594 |
| 55JA5121 | A. Mustafa, W. Asker and M. E. E.-D. Sobhy; *J. Am. Chem. Soc.*, 1955, **77**, 5121. | 768 |
| 55JA5601 | S. M. McElvain and G. R. McKay, Jr.; *J. Am. Chem. Soc.*, 1955, **77**, 5601. | 762 |
| 55JA5867 | C. W. Whitehead and J. J. Traverso; *J. Am. Chem. Soc.*, 1955, **77**, 5867. | 114 |
| 55JA5922 | F. C. Schaefer; *J. Am. Chem. Soc.*, 1955, **77**, 5922. | 476 |
| 55JA6365 | M. Sletzinger, D. Reinhold, J. Grier, M. Beachem and M. Tishler; *J. Am. Chem. Soc.*, 1955, **77**, 6365. | 304, 312 |
| 55JCS211 | D. J. Brown, E. Hoerger and S. F. Mason; *J. Chem. Soc.*, 1955, 211. 66, 69, 90, 110, 134 | |
| 55JCS303 | C. L. Leese and H. N. Rydon; *J. Chem. Soc.*, 1955, 303. | 259 |
| 55JCS852 | D. J. Drain and D. E. Seymour; *J. Chem. Soc.*, 1955, 852. | 54 |
| 55JCS860 | D. M. Fitzgerald, J. F. O'Sullivan, E. M. Philbin and T. S. Wheeler; *J. Chem. Soc.*, 1955, 860. | 828 |
| 55JCS896 | P. R. Brook and G. R. Ramage; *J. Chem. Soc.*, 1955, 896. | 280, 305 |
| 55JCS1379 | G. P. G. Dick and H. C. S. Wood; *J. Chem. Soc.*, 1955, 1379. | 318 |
| 55JCS2032 | T. S. Osdene and G. M. Timmis; *J. Chem. Soc.*, 1955, 2032. | 259 |
| 55JCS2036 | T. S. Osdene and G. M. Timmis; *J. Chem. Soc.*, 1955, 2036. | 313 |
| 55JCS2244 | A. J. S. Sorrie and R. H. Thomson; *J. Chem. Soc.*, 1955, 2244. | 832 |
| 55JCS2336 | S. F. Mason; *J. Chem. Soc.*, 1955, 2336. | 265 |
| 55JCS2685 | Y. S. Kao and R. Robinson; *J. Chem. Soc.*, 1955, 2685. | 246 |
| 55JCS2690 | A. Albert; *J. Chem. Soc.*, 1955, 2690. | 283, 284, 288 |
| 55JCS3982 | P. K. Grover, G. D. Shah and R. C. Shah; *J. Chem. Soc.*, 1955, 3982. | 835 |
| 55JCS4035 | D. J. Brown, E. Hoerger and S. F. Mason; *J. Chem. Soc.*, 1955, 4035. | 59, 64, 67, 69, 86, 134 |
| 55JIC302 | L. G. Shah, G. D. Shah and R. C. Shah; *J. Indian Chem. Soc.*, 1955, **32**, 302. | 801 |
| 55JOC38 | M. C. Kloetzel, R. P. Dayton and B. Y. Abadir; *J. Org. Chem.*, 1955, **20**, 38. | 819 |
| 55JOC448 | L. C. King and F. J. Ozog; *J. Org. Chem.*, 1955, **20**, 448. | 665 |
| 55JOC829 | V. H. Smith and B. E. Christensen; *J. Org. Chem.*, 1955, **20**, 829. | 123 |
| 55LA(595)101 | A. Schöberl and G. Wiehler; *Liebigs Ann. Chem.*, 1955, **595**, 101. | 969 |
| 55MI21300 | D. J. Brown; *J. Appl. Chem.*, 1955, **5**, 358. | 99, 126 |
| 55MI21301 | D. W. Davies; *Trans. Faraday Soc.*, 1955, **51**, 449. | 59 |
| B-55MI21400 | S. F. Mason; 'Chemical Society Special Publication', Royal Society of Chemistry, London, 1955, no. 3, p. 139. | 161 |
| 55MI21900 | S. Rossi; *Rend. Ist, Lombardo Sci. Pt. I, Cl. Sci. Mat. Nat.*, 1955, **88**, 185 (*Chem. Abstr.*, 1956, **50**, 10 743). | 416 |
| 55OSC(3)71 | A. R. Ronzio and W. B. Cook; *Org. Synth., Coll. Vol.*, 1955, **3**, 71. | 116, 129 |
| 55OSC(3)165 | E. C. Horning, M. G. Horning and D. A. Dimmig; *Org. Synth., Coll. Vol.*, 1955, **3**, 165. | 804 |
| 55OSC(3)231 | F. Arndt; *Org. Synth., Coll. Vol.*, 1955, **3**, 231. | 792 |
| 55OSC(3)276 | R. L. Sawyer and D. W. Andrus; *Org. Synth., Coll. Vol.*, 1955, **3**, 276. | 773 |
| 55OSC(3)387 | R. Mozingo; *Org. Synth., Coll. Vol.*, 1955, **3**, 387. | 816 |
| 55OSC(3)488 | E. C. Wagner and M. F. Fegley; *Org. Synth., Coll. Vol.*, 1955, **3**, 488. | 1029 |
| 55OSC(3)581 | E. H. Woodruff; *Org. Synth., Coll. Vol.*, 1955, **3**, 581. | 801 |
| 55OSC(3)794 | D. W. Andrus and J. R. Johnson; *Org. Synth., Coll. Vol.*, 1955, **3**, 794. | 777 |
| 55YZ1423 | K. Adachi; *Yakugaku Zasshi*, 1955, **75**, 1423. | 45 |
| 56AC(R)428 | M. Ridi, P. Papini and S. Checchi; *Ann. Chim. (Rome)*, 1956, **46**, 428. | 229 |
| 56AG519 | K. Dimroth and G. Bräuniger; *Angew. Chem.*, 1956, **68**, 519. | 657 |
| 56AX510 | F. Bertinotti, G. Giacomello and A. M. Liquori; *Acta Crystallogr.*, 1956, **9**, 510. | 537, 541 |
| 56BSB213 | J. Toussaint; *Bull. Soc. Chim. Belg.*, 1956, **65**, 213. | 622 |
| 56BSF1337 | J. Colonge and P. Boisde; *Bull. Soc. Chim. Fr.*, 1956, 1337. | 788, 860 |
| 56CB12 | H. Bredereck, H. Ulmer and H. Waldmann; *Chem. Ber.*, 1956, **89**, 12. | 81 |
| 56CB179 | M. Goehring and H. K. A. Zahn; *Chem. Ber.*, 1956, **89**, 179. | 1048 |
| 56CB563 | D. Jerchel and H. Fischer; *Chem. Ber.*, 1956, **89**, 563. | 245 |
| 56CB641 | W. Pfleiderer; *Chem. Ber.*, 1956, **89**, 641. | 302, 304 |
| 56CB2239 | H. Röhnert; *Chem. Ber.*, 1956, **89**, 2239. | 108 |
| 56CB2578 | W. Ried and H. Keller; *Chem. Ber.*, 1956, **89**, 2578. | 76 |
| 56CB2684 | W. Ried and J. Grabosch; *Chem. Ber.*, 1956, **89**, 2684. | 252, 258 |
| 56CB2904 | G. Henseke and H. G. Patzwald; *Chem. Ber.* 1956, **89**, 2904. | 311 |
| 56CI(L)1312 | G. Harris and R. Parsons; *Chem. Ind. (London)*, 1956, 1312. | 143 |
| 56CI(L)1453 | J. F. W. McOmie and J. H. Chesterfield; *Chem. Ind. (London)*, 1956, 1453. | 144 |
| 56CR(242)1034 | C. Mentzer, J. Chopin and M. Mercier; *C. R. Hebd. Seances Acad. Sci.*, 1956, **242**, 1034. | 830 |
| 56DOK(109)117 | N. P. Shusherina, M. Yu. Lur'e and R. Ya. Levina; *Dokl. Akad. Nauk SSSR*, 1956, **109**, 117. | 799 |

| | | |
|---|---|---|
| 56G119 | S. Capuano and L. Giammanco; *Gazz. Chem. Ital.*, 1956, **86**, 119. | 119 |
| 56G484 | R. Fusco and S. Rossi; *Gazz. Chim. Ital.*, 1956, **86**, 484. | 450 |
| 56G990 | F. Bottari and S. Carboni; *Gazz. Chim. Ital.*, 1956, **86**, 990. | 243, 246 |
| 56G1336 | R. B. Marini-Bettolo, R. Landi-Vitory and L. Paoloni; *Gazz. Chim. Ital.*, 1956, **86**, 1336. | 955 |
| 56GEP951993 | W. Kunze; (Cassella Farbwerke Mainkur A/G), *Ger. Pat.* 951 993 (1956) (*Chem. Abstr.*, 1959, **53**, 5298). | 247 |
| 56GEP953801 | H.-B. König and H. A. Offe; *Ger. Pat.* 953 801 (1956) (*Chem. Abstr.*, 1959, **53**, 4309). | 559, 561 |
| 56HC(10)1 | J. G. Erickson; *Chem. Heterocycl. Compd.*, 1956, **10**, 1. | 369 |
| 56HC(10)44 | J. G. Erickson; *Chem. Heterocycl. Compd.*, 1956, **10**, 44. | 385 |
| 56HC(10)138 | V. P. Wystrach; *Chem. Heterocycl. Compd.*, 1956, **10**, 138. | 531 |
| 56HC(10)179 | P. F. Wiley; *Chem. Heterocycl. Compd.*, 1956, **10**, 179. | 536 |
| 56HCA207 | R. Wizinger and P. Ulrich; *Helv. Chim. Acta*, 1956, **39**, 207. | 659 |
| 56JA159 | R. G. Jones; *J. Am. Chem. Soc.*, 1956, **78**, 159. | 246, 354 |
| 56JA210 | E. C. Taylor, Jr., R. B. Garland and C. F. Howell; *J. Am. Chem. Soc.*, 1956, **78**, 210. | 318 |
| 56JA401 | R. R. Herr, T. Enkoji and T. J. Bardos; *J. Am. Chem. Soc.*, 1956, **78**, 401. | 97, 136, 137 |
| 56JA973 | R. K. Robins and G. H. Hitchings; *J. Am. Chem. Soc.*, 1956, **78**, 973. | 214 |
| 56JA1434 | P. E. Fanta and E. A. Hedman; *J. Am. Chem. Soc.*, 1956, **78**, 1434. | 130 |
| 56JA1938 | E. A. Falco, E. Pappas and G. H. Hitchings; *J. Am. Chem. Soc.*, 1956, **78**, 1938. | 411 |
| 56JA2136 | B. W. Langley; *J. Am. Chem. Soc.*, 1956, **78**, 2136. | 79, 81, 126 |
| 56JA2144 | D. Shapiro, R. A. Abramovitch and S. Pinchas; *J. Am. Chem. Soc.*, 1956, **78**, 2144. | 41 |
| 56JA2163 | H. Gilman and D. R. Swayampati; *J. Am. Chem. Soc.*, 1956, **78**, 2163. | 985 |
| 56JA2278 | H. O. House; *J. Am. Chem. Soc.*, 1956, **78**, 2278. | 824 |
| 56JA2393 | R. H. Wiley and S. C. Slaymaker; *J. Am. Chem. Soc.*, 1956, **78**, 2393. | 595, 599 |
| 56JA2447 | H. Schroeder and C. Grundmann; *J. Am. Chem. Soc.*, 1956, **78**, 2447. | 512 |
| 56JA3201 | G. N. Walker; *J. Am. Chem. Soc.*, 1956, **78**, 3201. | 690 |
| 56JA4911 | I. C. Kogon; *J. Am. Chem. Soc.*, 1956, **78**, 4911. | 508 |
| 56JA5294 | C. W. Whitehead and J. J. Traverso; *J. Am. Chem. Soc.*, 1956, **78**, 5294. | 114 |
| 56JA5451 | T. S. Osdene and E. C. Taylor; *J. Am. Chem. Soc.*, 1956, **78**, 5451. | 320 |
| 56JA5868 | E. L. Patterson, R. Milstrey and E. L. R. Stokstad; *J. Am. Chem. Soc.*, 1956, **78**, 5868. | 312 |
| 56JA5871 | E. L. Patterson, M. H. von Saltza and E. L. R. Stokstad; *J. Am. Chem. Soc.*, 1956, **78**, 5871. | 324 |
| 56JCS136 | L. Crombie, J. Gold, S. H. Harper and B. J. Stokes; *J. Chem. Soc.*, 1956, 136. | 777 |
| 56JCS917 | G. R. Barker and N. G. Luthy; *J. Chem. Soc.*, 1956, 917. | 137 |
| 56JCS985 | P. R. Levy and H. Stephen; *J. Chem. Soc.*, 1956, 985. | 111 |
| 56JCS1019 | G. H. Hitchings, P. B. Russell and N. Whittaker; *J. Chem. Soc.*, 1956, 1019. | 115 |
| 56JCS1045 | V. Oakes, R. Pascoe and H. N. Rydon; *J. Chem. Soc.*, 1956, 1045. | 204, 207, 214, 225 |
| 56JCS1382 | J. D. Simpson and H. Stephen; *J. Chem. Soc.*, 1956, 1382. | 848 |
| 56JCS1563 | N. B. Chapman and D. Q. Russell-Hill; *J. Chem. Soc.*, 1956, 1563. | 164, 176 |
| 56JCS2033 | R. Hull; *J. Chem. Soc.*, 1956, 2033. | 74 |
| 56JCS2066 | A. Albert, D. J. Brown and H. C. S. Wood; *J. Chem. Soc.*, 1956, 2066. | 122 |
| 56JCS2124 | J. Davoll and D. H. Laney; *J. Chem. Soc.*, 1956, 2124. | 133, 144 |
| 56JCS2131 | G. P. G. Dick, H. C. S. Wood and W. R. Logan; *J. Chem. Soc.*, 1956, 2131. | 313 |
| 56JCS2253 | D. D. Libman and R. Slack; *J. Chem. Soc.*, 1956, 2253. | 544 |
| 56JCS2312 | D. J. Brown; *J. Chem. Soc.*, 1956, 2312. | 107, 128 |
| 56JCS3311 | W. E. Fidler and H. C. S. Wood; *J. Chem. Soc.*, 1956, 3311. | 317 |
| 56JCS3443 | D. J. Brown and S. F. Mason; *J. Chem. Soc.*, 1956, 3443. | 271, 279 |
| 56JCS3509 | H. J. Rodda; *J. Chem. Soc.*, 1956, 3509. | 86 |
| 56JCS4106 | R. M. Evans, P. G. Jones, P. J. Palmer and F. F. Stephens; *J. Chem. Soc.*, 1956, 4106. | 69, 93 |
| 56JCS4118 | M. R. Atkinson, G. Shaw and R. N. Warrener; *J. Chem. Soc.*, 1956, 4118. | 109 |
| 56JCS4191 | A. R. Osborn, K. Schofield and L. N. Short; *J. Chem. Soc.*, 1956, 4191. | 100 |
| 56JCS4433 | V. Oakes and H. N. Rydon; *J. Chem. Soc.*, 1956, 4433. | 210, 215 |
| 56JCS4535 | A. S. Bailey and C. R. Worthing; *J. Chem. Soc.*, 1956, 4535. | 766, 832 |
| 56JCS4621 | A. Albert, J. H. Lister and C. Pedersen; *J. Chem. Soc.*, 1956, 4621. | 267, 291 |
| 56JCS4785 | W. Webster and D. P. Young; *J. Chem. Soc.*, 1956, 4785. | 668, 669, 764 |
| 56JCS4886 | H. Lynton and E. G. Cox; *J. Chem. Soc.*, 1956, 4886. | 947 |
| 56JOC641 | C. Grundmann and E. Kober; *J. Org. Chem.*, 1956, **21**, 641. | 483 |
| 56JOC764 | R. L. Letsinger and L. Lasco; *J. Org. Chem.*, 1956, **21**, 764. | 55 |
| 56JOC1104 | D. V. Joshi, J. R. Merchant and R. C. Shah; *J. Org. Chem.*, 1956, **21**, 1104. | 708 |
| 56JOC1408 | A. A. Raval and N. M. Shah; *J. Org. Chem.*, 1956, **21**, 1408. | 853 |
| 56JOC1415 | B. K. Ganguly and P. Bagchi; *J. Org. Chem.*, 1956, **21**, 1415. | 600 |
| 56M526 | W. Klötzer; *Monatsh. Chem.*, 1956, **87**, 526. | 83, 84, 126, 128 |
| 56MI21600 | I. Ziegler-Günder; *Biol. Rev. Biol. Proc. Cambridge Philos. Soc.*, 1956, **31**, 313. | 322 |
| 56MI22200 | R. T. Holman and P. R. Edmondson; *Anal. Chem.*, 1956, **28**, 1553. | 595 |
| 56MI22400 | Z. Horii, K. Sakurai, K. Tomino and T. Konishi; *J. Pharm. Soc. Jpn.*, 1956, **76**, 1101. | 840 |
| 56PIA(A)(44)36 | S. A. Kagal, S. S. Karmarkar and K. Venkataraman; *Proc. Indian Acad. Sci., Sect. A*, 1956, **44**, 36. | 821 |
| 56YZ234 | H. Hirano and H. Yonemoto; *Yakugaku Zasshi*, 1956, **76**, 234. | 71 |
| 56YZ776 | T. Yoshikawa; *Yakugaku Zasshi*, 1956, **76**, 776. | 84 |

| | | |
|---|---|---|
| 56ZN(B)82 | I. Ziegler-Günder, H. Simon and A. Walker; *Z. Naturforsch., Teil B*, 1956, **11**, 82. | 320 |
| 57AC(R)728 | M. Ridi and S. Checchi; *Ann. Chim. (Rome)*, 1957, **47**, 728. | 236, 240, 246 |
| 57AG506 | F. L. Scott; *Angew. Chem.*, 1957, **69**, 506. | 567 |
| 57AG720 | K. Dimroth and G. Neubauer; *Angew. Chem.*, 1957, **69**, 720. | 762 |
| 57BBA(23)295 | J. J. Fox, N. Yung and I. Wempen; *Biochim. Biophys. Acta*, 1957, **23**, 295. | 76 |
| 57BJ(65)124 | A. Albert; *Biochem. J.*, 1957, **65**, 124. | 319 |
| 57BRP774095 | Wellcome Foundation Ltd., *Br. Pat.* 774 095 (1957) (*Chem. Abstr.*, 1958, **52**, 2097). | 231 |
| 57BRP776335 | F. Hoffmann–La Roche and Co. A/G, *Br. Pat.* 776 335 (1957) (*Chem. Abstr.*, 1957, **51**, 18 015). | 232 |
| 57BSB292 | M. Claesen and H. Vanderhaeghe; *Bull. Soc. Chim. Belg.*, 1957, **66**, 292. | 79 |
| 57BSF776 | J. Colonge, E. Le Sech and R. Marey; *Bull. Soc. Chim. Fr.*, 1957, 776. | 778, 785 |
| 57CB481 | R. Metze and S. Meyer; *Chem. Ber.*, 1957, **90**, 481. | 410 |
| 57CB738 | W. Pfleiderer and H. Mosthaf; *Chem. Ber.*, 1957, **90**, 738. | 227 |
| 57CB942 | H. Bredereck, R. Gompper and G. Morlock; *Chem. Ber.*, 1957, **90**, 942. | 117, 123, 124 |
| 57CB1519 | F. Bohlmann; *Chem. Ber.*, 1957, **90**, 1519. | 594, 596 |
| 57CB1634 | K. Dimroth, G. Bräuniger and G. Neubauer; *Chem. Ber.*, 1957, **90**, 1634. | 658 |
| 57CB1668 | K. Dimroth, G. Neubauer, H. Möllenkamp and G. Oosterloo; *Chem. Ber.*, 1957, **90**, 1668. | 658 |
| 57CB2582 | W. Pfleiderer; *Chem. Ber.*, 1957, **90**, 2582. | 271 |
| 57CB2588 | W. Pfleiderer; *Chem. Ber.*, 1957, **90**, 2588. | 271, 297, 310 |
| 57CB2604 | W. Pfleiderer; *Chem. Ber.*, 1957, **90**, 2604. | 271, 310 |
| 57CB2617 | W. Pfleiderer; *Chem. Ber.*, 1957, **90**, 2617. | 271, 277, 304, 310 |
| 57CB2624 | W. Pfleiderer; *Chem. Ber.*, 1957, **90**, 2624. | 271, 277, 304, 310 |
| 57CB2631 | W. Pfleiderer; *Chem. Ber.*, 1957, **90**, 2631. | 271 |
| 57CRV47 | R. G. Arnold, J. A. Nelson and J. J. Verbanc; *Chem. Rev.*, 1957, **57**, 47. | 508 |
| 57G243 | S. Mangini and R. Passerini; *Gazz. Chim. Ital.*, 1957, **87**, 243. | 599 |
| 57GEP958561 | W. Kunze; (Cassella Farbwerke Mainkur A/G), *Ger. Pat.* 958 561 (1957) (*Chem. Abstr.*, 1959, **53**, 8178). | 240 |
| 57HC(11)1 | G. A. Swan and D. G. I. Felton; *Chem. Heterocycl. Compd.*, 1957, **11**, 1. | 171, 197 |
| 57HCA1562 | A. Margot and H. Gysin; *Helv. Chim. Acta*, 1957, **40**, 1562. | 154 |
| 57JA156 | N. J. Leonard and D. Choudhury; *J. Am. Chem. Soc.*, 1957, **79**, 156. | 595, 815 |
| 57JA941 | C. G. Overberger, F. W. Michelotti and P. M. Carabateas; *J. Am. Chem. Soc.*, 1957, **79**, 941. | 492 |
| 57JA1439 | H. Gilman and J. J. Dietrich; *J. Am. Chem. Soc.*, 1957, **79**, 1439. | 974, 984 |
| 57JA2318 | G. Büchi and N. C. Yang; *J. Am. Chem. Soc.*, 1957, **79**, 2318. | 593 |
| 57JA2602 | R. H. Wiley, C. H. Jarboe and F. N. Hayes; *J. Am. Chem. Soc.*, 1957, **79**, 2602. | 792 |
| 57JA2839 | C. Grundmann and A. Kreutzberger; *J. Am. Chem. Soc.*, 1957, **79**, 2839. | 557 |
| 57JA3165 | J. L. Warnell and R. L. Shriner; *J. Am. Chem. Soc.*, 1957, **79**, 3165. | 787, 833, 857 |
| 57JA4507 | W. B. Mors, O. R. Gottlieb and C. Djerassi; *J. Am. Chem. Soc.*, 1957, **79**, 4507. | 686 |
| 57JA4559 | R. Duschinsky, E. Pleven and C. Heidelberger; *J. Am. Chem. Soc.*, 1957, **79**, 4559. | 152 |
| 57JA5064 | S. L. Shapiro, V. A. Parrino, K. Geiger, S. Kobrin and L. Freedman; *J. Am. Chem. Soc.*, 1957, **79**, 5064. | 492, 493 |
| 57JA6020 | A. Schönberg, A. E. K. Fateen and A. M. A. Sammour; *J. Am. Chem. Soc.*, 1957, **79**, 6020. | 708 |
| 57JA6219 | R. K. Summerbell and G. J. Lestina; *J. Am. Chem. Soc.*, 1957, **79**, 6219. | 965, 981 |
| 57JCS1 | P. R. Brook and G. R. Ramage; *J. Chem. Soc.*, 1957, 1. | 280, 305 |
| 57JCS430 | J. W. Clark-Lewis and M. J. Thompson; *J. Chem. Soc.*, 1957, 430. | 257 |
| 57JCS1922 | N. Campbell, S. R. McCallum and D. J. MacKenzie; *J. Chem. Soc.*, 1957, 1922. | 702 |
| 57JCS2146 | W. R. Boon; *J. Chem. Soc.*, 1957, 2146. | 88 |
| 57JCS2363 | M. R. Atkinson, M. H. Maguire, R. K. Ralph, G. Shaw and R. N. Warrener; *J. Chem. Soc.*, 1957, 2363. | 109 |
| 57JCS2521 | M. J. S. Dewar and P. M. Maitlis; *J. Chem. Soc.*, 1957, 2521. | 76, 163 |
| 57JCS3060 | W. Baker, J. F. W. McOmie and J. H. Wild; *J. Chem. Soc.*, 1957, 3060. | 733 |
| 57JCS3186 | R. F. Robbins and K. Schofield; *J. Chem. Soc.*, 1957, 3186. | 412, 413 |
| 57JCS3718 | C. M. Atkinson and A. R. Mattocks; *J. Chem. Soc.*, 1957, 3718. | 214, 227 |
| 57JCS3722 | C. M. Atkinson and A. R. Mattocks; *J. Chem. Soc.*, 1957, 3722. | 244 |
| 57JCS3875 | J. P. Brown and E. B. McCall; *J. Chem. Soc.*, 1957, 3875. | 855 |
| 57JCS4157 | W. E. Fidler and H. C. S. Wood; *J. Chem. Soc.*, 1957, 4157. | 277 |
| 57JCS4845 | R. Hull; *J. Chem. Soc.*, 1957, 4845. | 79 |
| 57JCS4874 | S. F. Mason; *J. Chem. Soc.*, 1957, 4874. | 64, 67, 204, 249, 250 |
| 57JCS4997 | D. M. Besly and A. A. Goldberg; *J. Chem. Soc.*, 1957, 4997. | 214, 221 |
| 57JCS5010 | S. F. Mason; *J. Chem. Soc.*, 1957, 5010. | 249 |
| 57JGU2342 | N. N. Vorozhtsov and A. T. Petushkova; *J. Gen. Chem. USSR (Engl. Transl.)*, 1957, **27**, 2342. | 685, 686, 691 |
| 57JIC35 | J. R. Merchant and R. C. Shah; *J. Indian Chem. Soc.*, 1957, **34**, 35. | 680 |
| 57JIC753 | C. G. Joshi and A. B. Kulkarni; *J. Indian Chem. Soc.*, 1957, **34**, 753. | 597 |
| 57JOC11 | E. J. Tillmanns and J. J. Ritter; *J. Org. Chem.*, 1957, **24**, 11. | 1025 |
| 57JOC304 | A. A. Raval and N. M. Shah; *J. Org. Chem.*, 1957, **22**, 304. | 824 |
| 57JOC444 | D. C. Morrison; *J. Org. Chem.*, 1957, **22**, 444. | 483 |
| 57JOC698 | W. L. Reilly and H. C. Brown; *J. Org. Chem.*, 1957, **22**, 698. | 503 |
| 57JOC851 | H. Gilman and J. J. Dietrich; *J. Org. Chem.*, 1957, **22**, 851. | 973 |
| 57JOC1257 | R. H. Wiley and J. G. Esterle; *J. Org. Chem.*, 1957, **22**, 1257. | 595, 597 |

| | | |
|---|---|---|
| 57JOC1403 | H. Gilman and J. J. Dietrich; *J. Org. Chem.*, 1957, **22**, 1403. | 974 |
| 57JOC1682 | G. B. Payne and C. W. Smith; *J. Org. Chem.*, 1957, **22**, 1682. | 990 |
| 57LA(603)169 | H. Frei and H. Schmid; *Liebigs Ann. Chem.*, 1957, **603**, 169. | 826 |
| 57LA(605)158 | O. Dann and G. Illing; *Liebigs Ann. Chem.*, 1957, **605**, 158. | 827 |
| 57LA(605)191 | D. Jerchel and W. Woticky; *Liebigs Ann. Chem.*, 1957, **605**, 191. | 450 |
| B-57MI21200 | T. L. Jacobs; in 'Heterocyclic Compounds', ed. R. C. Elderfield; Wiley, New York, 1957, vol. 6, p. 101. | 2 |
| B-57MI21201 | T. L. Jacobs; in 'Heterocyclic Compounds', ed. R. C. Elderfield; Wiley, New York, 1957, vol. 6, p. 136. | 2 |
| B-57MI21202 | R. C. Elderfield and S. L. Wythe; in 'Heterocyclic Compounds', ed. R. C. Elderfield; Wiley, New York, 1957, vol. 6, p. 186. | 2 |
| 57MI21300 | D. J. Brown; *J. Appl. Chem.*, 1957, **7**, 109. | 87, 132 |
| 57MI21301 | J. Baddiley and J. G. Buchanan; *Annu. Rep. Prog. Chem.*, 1957, **54**, 329. | 146 |
| B-57MI21400 | Y. T. Pratt; in 'Heterocyclic Compounds', ed. R. C. Elderfield; Wiley, New York, 1957, vol. 6. | 158 |
| 57MI21900 | W. O. Foye and W. E. Lange; *J. Am. Pharm. Assoc.*, 1957, **46**, 371 (*Chem. Abstr.*, 1957, **51**, 17 943). | 452 |
| 57MI22200 | L. Henry and D. Molho; *Colloq. Int. Centre Natl. Recherche Sci.* (*Paris*), 1957, **64**, 341 (*Chem. Abstr.*, 1960, **54**, 10 516). | 597 |
| 57MI22400 | V. N. Gupta and T. R. Seshadri; *J. Sci. Ind. Res.* (*India*) (*B*), 1957, **16**, 116. | 818 |
| B-57MI22600 | C. B. Kremer and L. K. Rochen; in 'Heterocyclic Compounds', ed. R. C. Elderfield; Wiley, New York, 1957, vol. 6, p. 1. | 994, 976, 987, 989 |
| B-57MI22601 | R. C. Elderfield; in 'Heterocyclic Compounds', ed. R. C. Elderfield; Wiley, New York, 1957, vol. 6, p. 59. | 944, 990 |
| B-57MI22602 | R. C. Elderfield; in 'Heterocyclic Compounds', ed. R. C. Elderfield; Wiley, New York, 1957, vol. 6, p. 75 | 944, 987 |
| 57SA113 | R. C. Lord, A. L. Marston and F. A. Miller; *Spectrochim. Acta*, 1957, **9**, 113. | 64 |
| 57T(1)103 | P. V. Laakso, R. Robinson and H. P. Vandrewala; *Tetrahedron*, 1957, **1**, 103. | 413 |
| 57USP2817662 | R. A. Carboni; *U.S. Pat.* 2 817 662 (1957) (*Chem. Abstr.*, 1958, **52**, 7360). | 550, 567, 570 |
| 57YZ507 | K. Adachi; *Yakugaku Zasshi*, 1957, **77**, 507. | 75, 141 |
| 57ZOB2113 | R. S. Karlinskaya and N. V. Khromov-Borisov; *Zh. Obshch. Khim.*, 1957, **27**, 2113. | 144 |
| 58AC(R)738 | R. Passerini and G. Purrello; *Ann. Chim.* (*Rome*), 1958, **48**, 738. | 955 |
| 58ACS1768 | E. W. Lund; *Acta Chem. Scand.*, 1958, **12**, 1768. | 462, 467 |
| 58AG5 | J. Druey; *Angew. Chem.*, 1958, **70**, 5. | 2, 359 |
| 58AX372 | I. Rowe and B. Post; *Acta Crystallogr.*, 1958, **11**, 372. | 947 |
| 58BCJ267 | M. Miyano and M. Matsui; *Bull. Chem. Soc. Jpn.*, 1958, **31**, 267. | 729 |
| 58BSB22 | R. Huls; *Bull. Soc. Chim. Belg.*, 1958, **67**, 22. | 601, 754, 849 |
| 58BSF329 | J. Colonge and A. Guyot; *Bull. Soc. Chim. Fr.*, 1958, 329. | 828 |
| 58BSF1458 | R. Lombard and J. P. Stephan; *Bull. Soc. Chim. Fr.*, 1958, 1458. | 869 |
| 58CB320 | A. Roedig and S. Schödel; *Chem. Ber.*, 1958, **91**, 320. | 794 |
| 58CB330 | A. Roedig and S. Schödel; *Chem. Ber.*, 1958, **91**, 330. | 794 |
| 58CB422 | R. Metze and G. Rolle; *Chem. Ber.*, 1958, **91**, 422. | 410 |
| 58CB1200 | R. Appel and H. Gerber; *Chem. Ber.*, 1958, **91**, 1200. | 1074 |
| 58CB1589 | S. Olsen and R. Bredoch; *Chem. Ber.*, 1958, **91**, 1589. | 594, 597 |
| 58CB1671 | W. Pfleiderer; *Chem. Ber.*, 1958, **91**, 1671. | 271, 297 |
| 58CB1982 | H. Stetter and H. Spangenberger; *Chem. Ber.*, 1958, **91**, 1982. | 44 |
| 58CB2636 | J. N. Chatterjea; *Chem. Ber.*, 1958, **91**, 2636. | 766 |
| 58CB2832 | H. Bredereck, R. Gompper and H. Herlinger; *Chem. Ber.*, 1958, **91**, 2832. | 72, 105, 123, 133, 134 |
| 58CB2849 | K.-H. Boltze and K. Heidenbluth; *Chem. Ber.*, 1958, **91**, 2849. | 595, 795 |
| 58CB2858 | L. Farkas, A. Major, L. Pallos and J. Várady; *Chem. Ber.*, 1958, **91**, 2858. | 821 |
| 58CCC1588 | J. Gut; *Collect. Czech. Chem. Commun.*, 1958, **23**, 1588. | 411 |
| 58CI(L)1234 | M. A. Aron and J. A. Elvidge; *Chem. Ind.* (*London*), 1958, 1234. | 372 |
| 58CPB346 | S. Inoue; *Chem. Pharm. Bull.*, 1958, **6**, 346. | 95 |
| 58CPB633 | H. Yamanaka; *Chem. Pharm. Bull.*, 1958, **6**, 633. | 105 |
| 58CRV131 | B. Bann and S. A. Miller; *Chem. Rev.*, 1958, **58**, 131. | 475, 476, 498 |
| 58GEP922467 | A. Schmidt; *Ger. Pat.* 922 467 (1958) (*Chem. Abstr.*, 1958, **52**, 5485). | 1011 |
| 58GEP1040040 | H. Pasedach and M. Seefeder; (Badische Anilin u Soda-Fabrik A/G), *Ger. Pat.* 1 040 040 (1958) (*Chem. Abstr.*, 1961, **55**, 6507). | 229 |
| 58HCA108 | M. Viscontini and H. Raschig; *Helv. Chim. Acta*, 1958, **41**, 108. | 312 |
| 58HCA1806 | R. Urban and O. Schnider; *Helv. Chim. Acta*, 1958, **41**, 1806. | 82 |
| 58HCA2170 | M. Viscontini and H. R. Weilenmann; *Helv. Chim. Acta*, 1958, **41**, 2170. | 281, 306 |
| 58HOU(11)725 | A. Dorlars; *Methoden Org. Chem.* (*Houben-Weyl*), 1958, **11**, 725. | 1043 |
| 58IZV1445 | L. I. Zhakharkin and L. P. Sorokina; *Izv. Akad. Nauk SSSR, Ser. Khim.*, 1958, 1445. | 794 |
| 58JA366 | H. Gilman and J. J. Dietrich; *J. Am. Chem. Soc.*, 1958, **80**, 366. | 974 |
| 58JA380 | H. Gilman and J. J. Dietrich; *J. Am. Chem. Soc.*, 1958, **80**, 380. | 973 |
| 58JA421 | E. C. Taylor, J. W. Barton and T. S. Osdene; *J. Am. Chem. Soc.*, 1958, **80**, 421. | 318, 320 |
| 58JA427 | E. C. Taylor, A. J. Crovetti and R. J. Knopf; *J. Am. Chem. Soc.*, 1958, **80**, 427. | 226 |
| 58JA739 | H. S. Forrest and W. S. McNutt; *J. Am. Chem. Soc.*, 1958, **80**, 739. | 320 |
| 58JA803 | W. M. Padgett, II and W. F. Hamner; *J. Am. Chem. Soc.*, 1958, **80**, 803. | 465 |
| 58JA951 | W. S. McNutt and H. S. Forrest; *J. Am. Chem. Soc.*, 1958, **80**, 951. | 320 |

| | | |
|---|---|---|
| 58JA976 | P. K. Chang and T. L. V. Ulbricht; *J. Am. Chem. Soc.*, 1958, **80**, 976. | 416 |
| 58JA1442 | I. S. Bengelsdorf; *J. Am. Chem. Soc.*, 1958, **80**, 1442. | 503 |
| 58JA1664 | H. R. Henze and E. N. Kahlenberg; *J. Am. Chem. Soc.*, 1958, **80**, 1664. | 134 |
| 58JA1977 | H. E. Baumgarten, D. L. Pedersen and M. W. Hunt; *J. Am. Chem. Soc.*, 1958, **80**, 1977 (and references for previous reports therein). | 44 |
| 58JA2185 | C. W. Whitehead and J. J. Traverso; *J. Am. Chem. Soc.*, 1958, **80**, 2185. | 108 |
| 58JA3155 | V. A. Grakauskas, A. J. Tomasewski and J. P. Horwitz; *J. Am. Chem. Soc.*, 1958, **80**, 3155. | 540, 556 |
| 58JA3449 | R. K. Robins and G. H. Hitchings; *J. Am. Chem. Soc.*, 1958, **80**, 3449. | 211, 225, 227, 229, 230, 231 |
| 58JA3686 | C. Djerassi, E. J. Eisenbraun, B. Gilbert, A. J. Lemin, S. P. Marfey and M. P. Morris; *J. Am. Chem. Soc.*, 1958, **80**, 3686. | 600 |
| 58JA5543 | M. M. Kreevoy; *J. Am. Chem. Soc.*, 1958, **80**, 5543. | 945 |
| 58JA5547 | E. Kober and C. Grundmann; *J. Am. Chem. Soc.*, 1958, **80**, 5547. | 474 |
| 58JA6095 | W. V. Curran and R. B. Angier; *J. Am. Chem. Soc.*, 1958, **80**, 6095. | 277, 294, 309 |
| 58JA6312 | A. Schönberg, M. Elkaschef, M. Nosseir and M. M. Sidky; *J. Am. Chem. Soc.*, 1958, **80**, 6312. | 762 |
| 58JA6360 | C. R. Hauser and T. M. Harris; *J. Am. Chem. Soc.*, 1958, **80**, 6360. | 795 |
| 58JA6459 | D. A. Lyttle and H. G. Petering; *J. Am. Chem. Soc.*, 1958, **80**, 6459. | 152 |
| 58JBC(231)331 | B. S. Gorton, J. M. Ravel and W. Shive; *J. Biol. Chem.*, 1958, **231**, 331. | 257 |
| 58JCS108 | G. W. H. Cheeseman; *J. Chem. Soc.*, 1958, 108. | 177 |
| 58JCS153 | G. Shaw and R. N. Warrener; *J. Chem. Soc.*, 1958, 153. | 109 |
| 58JCS294 | E. R. H. Jones and F. G. Mann; *J. Chem. Soc.*, 1958, 294. | 597 |
| 58JCS674 | S. F. Mason; *J. Chem. Soc.*, 1958, 674. | 67 |
| 58JCS1190 | P. Naylor, G. R. Ramage and F. Schofield; *J. Chem. Soc.*, 1958, 1190. | 786, 852, 856 |
| 58JCS1978 | G. V. Boyd; *J. Chem. Soc.*, 1958, 1978. | 869 |
| 58JCS2294 | G. Shaw, R. N. Warrener, M. H. Maguire and R. K. Ralph; *J. Chem. Soc.*, 1958, 2294. | 109 |
| 58JCS2612 | J. B. Jones and A. R. Pinder; *J. Chem. Soc.*, 1958, 2612. | 860 |
| 58JCS2888 | L. Bateman and F. W. Shipley; *J. Chem. Soc.*, 1958, 2888. | 903 |
| 58JCS2893 | D. J. Collins and J. Graymore; *J. Chem. Soc.*, 1958, 2893. | 1057 |
| 58JCS3174 | W. R. Chan, W. G. C. Forsyth and C. H. Hassall; *J. Chem. Soc.*, 1958, 3174. | 864 |
| 58JCS3619 | S. F. Mason; *J. Chem. Soc.*, 1958, 3619. | 64 |
| 58JCS3742 | R. Hull; *J. Chem. Soc.*, 1958, 3742. | 125 |
| 58JCS4040 | J. W. Gramshaw, A. W. Johnson and T. J. King; *J. Chem. Soc.*, 1958, 4040. | 765 |
| 58JCS4227 | A. A. Goldberg and A. H. Wragg; *J. Chem. Soc.*, 1958, 4227. | 838 |
| 58JCS4234 | A. A. Goldberg and A. H. Wragg; *J. Chem. Soc.*, 1958, 4234. | 601 |
| 58JCS4302 | B. R. Brown and W. Cummings; *J. Chem. Soc.*, 1958, 4302. | 662 |
| 58JCS4588 | D. O. Holland and P. Mamalis; *J. Chem. Soc.*, 1958, 4588. | 440 |
| 58JIC47 | J. N. Chatterjea; *J. Indian Chem. Soc.*, 1958, **35**, 47. | 786 |
| 58JOC93 | G. E. Inglett; *J. Org. Chem.*, 1958, **23**, 93. | 597 |
| 58JOC313 | H. Gilman and D. R. Swayampati; *J. Org. Chem.*, 1958, **23**, 313. | 975 |
| 58JOC724 | H. J. Sims, H. B. Parseghian and P. I. de Benneville; *J. Org. Chem.*, 1958, **23**, 724. | 498 |
| 58JOC1032 | L. S. Luskin, P. I. de Benneville and S. Melamed; *J. Org. Chem.*, 1958, **23**, 1032. | 498 |
| 58JOC1128 | J. R. Shelton and C. Cialdella; *J. Org. Chem.*, 1958, **23**, 1128. | 723 |
| 58JOC1450 | H. G. Mautner; *J. Org. Chem.*, 1958, **23**, 1450. | 338, 339 |
| 58JOC1702 | W. E. Parham and M. T. Kneller; *J. Org. Chem.*, 1958, **23**, 1702. | 972 |
| 58JOC1738 | T. Okuda and C. C. Price; *J. Org. Chem.*, 1958, **23**, 1738. | 102 |
| 58LA(612)173 | W. Pfleiderer and G. Strauss; *Liebigs Ann. Chem.*, 1958, **612**, 173. | 126 |
| 58LA(615)48 | W. Pfleiderer and H. Ferch; *Liebigs Ann. Chem.*, 1958, **615**, 48. | 336 |
| 58LA(618)140 | K. Hafner and H. Kaiser; *Liebigs Ann. Chem.*, 1958, **618**, 140. | 660 |
| 58LA(619)70 | F. Korte, H. U. Aldag, G. Ludwig, W. Paulus and K. Störiko; *Liebigs Ann. Chem.*, 1958, **619**, 70. | 320 |
| 58MI21300 | S. K. Chatterjee and N. Anand; *J. Sci. Ind. Res., Sect. B*, 1958, **17**, 63. | 114, 128 |
| 58MI21301 | H. Gysin and A. Margot; *J. Agric. Food Chem.*, 1958, **6**, 900. | 154 |
| 58MI21302 | E. E. Wehr, M. M. Farr and D. K. McLoughlin; *J. Am. Vet. Med. Assoc.*, 1958, **132**, 439. | 154 |
| 58MI22200 | R. Nahum; *Ann. Chim. (Paris)*, 1958, **3**, 108. | 594 |
| 58MIP21500 | Cilag Ltd., *Swiss Pat.* 331 989 (1958) (*Chem.. Abstr.*, 1959, **53**, 5292). | 223 |
| 58T(2)203 | R. Aneja, S. K. Mukerjee and T. R. Seshadri; *Tetrahedron*, 1958, **2**, 203. | 709 |
| 58T(3)209 | R. Fusco and S. Rossi; *Tetrahedron*, 1958, **3**, 209. | 446 |
| 58T(4)36 | I. Chmielewska, J. Cieślak, K. Gorczyńska, B. Kontnik and K. Pitakowska; *Tetrahedron*, 1958, **4**, 36. | 599 |
| 58T(4)393 | G. Berti; *Tetrahedron*, 1958, **4**, 393. | 858 |
| 58ZPC(311)79 | A. Butenandt and H. Rembold; *Hoppe-Seyler's Z. Physiol. Chem.*, 1958, **311**, 79. | 312 |
| 59AC(R)944 | M. Ridi; *Ann. Chim. (Rome)*, 1959, **49**, 944. | 221, 230, 243, 246 |
| 59AG523 | F. Korte, K. H. Büchel and K. Göhring; *Angew. Chem.*, 1959, **71**, 523. | 597 |
| 59AJC554 | R. D. Brown and M. L. Heffernan; *Aust. J. Chem.*, 1959, **12**, 554. | 59 |
| 59BBA(33)29 | F. Bergmann and H. Kwietny; *Biochim. Biophys. Acta*, 1959, **33**, 29. | 287 |
| 59BCJ483 | M. Fujimoto; *Bull. Chem. Soc. Jpn.*, 1959, **32**, 483. | 1012 |
| 59BRP807826 | K. Thomae; *Br. Pat.* 807 826 (1959) (*Chem. Abstr.*, 1959, **53**, 12 317). | 368 |
| 59CB46 | F. Kröhnke and K. Dickoré; *Chem. Ber.*, 1959, **92**, 46. | 653, 670, 762 |

| | | |
|---|---|---|
| 59CB564 | E. Hoyer and R. Gompper; *Chem. Ber.*, 1959, **92**, 564. | 415 |
| 59CB1149 | M. Becke-Goehring and H. Jenne; *Chem. Ber.*, 1959, **92**, 1149. | 1048 |
| 59CB2042 | K. Dimroth and G. Neubauer; *Chem. Ber.*, 1959, **92**, 2042. | 658, 741 |
| 59CB2046 | K. Dimroth and G. Neubauer; *Chem. Ber.*, 1959, **92**, 2046. | 658, 741 |
| 59CB2309 | R. Wizinger and U. Arni; *Chem. Ber.*, 1959, **92**, 2309. | 664 |
| 59CB2468 | W. Pfleiderer; *Chem. Ber.*, 1959, **92**, 2468. | 319 |
| 59CB2481 | R. Metze and G. Scherowsky; *Chem. Ber.*, 1959, **92**, 2481. | 413 |
| 59CB2873 | H. Musso; *Chem. Ber.*, 1959, **92**, 2873. | 1010 |
| 59CB3190 | W. Pfleiderer; *Chem. Ber.*, 1959, **92**, 3190. | 271, 310 |
| 59CI(L)992 | L. F. Rice and G. Vogel; *Chem. Ind. (London)*, 1959, 992. | 794 |
| 59CJC2007 | R. N. Jones, C. L. Angell, T. Ito and R. J. D. Smith; *Can. J. Chem.*, 1959, **37**, 2007. | 595, 638 |
| 59CPB152 | H. Yamanaka; *Chem. Pharm. Bull.*, 1959, **7**, 152. | 75, 106 |
| 59CPB158 | H. Yamanaka; *Chem. Pharm. Bull.*, 1959, **7**, 158. | 105 |
| 59FOR(17)1 | K. Venkataraman; *Fortsch. Chem. Org. Naturst.*, 1959, **17**, 1. | 597 |
| 59GEP1099162 | Ciba Ltd., *Ger. Pat.* 1 099 162 (1959) (*Chem. Abstr.*, 1962, **56**, 4956). | 939 |
| 59HC(13)1 | E. M. Smolin and L. Rapoport; *Chem. Heterocycl. Compd.*, 1959, **13**, 1.    459, 472, 473, 474, 475, 477, 478, 480, 481, 482, 483, 484, 487, 488, 490, 492, 498, 505, 509, 510, 524 | |
| 59HCA1854 | M. Viscontini and H. R. Weilenmann; *Helv. Chim. Acta*, 1959, **42**, 1854. | 281, 287, 305 |
| 59HCA2254 | O. Brenner-Holzach and F. Leuthardt; *Helv. Chim. Acta*, 1959, **42**, 2254. | 320 |
| 59JA633 | R. K. Summerbell and D. R. Berger; *J. Am. Chem. Soc.*, 1959, **81**, 633. | 988 |
| 59JA905 | F. H. Case and E. Koft; *J. Am. Chem. Soc.*, 1959, **81**, 905. | 128 |
| 59JA1466 | F. C. Schaefer, I. Hechenbleikner, G. A. Peters and V. P. Wystrach; *J. Am. Chem. Soc.*, 1959, **81**, 1466. | 507, 511 |
| 59JA1470 | F. C. Schaefer and G. A. Peters; *J. Am. Chem. Soc.*, 1959, **81**, 1470. | 470 |
| 59JA2427 | D. Herbst, W. B. Mors, O. R. Gottlieb and C. Djerassi; *J. Am. Chem. Soc.*, 1959, **81**, 2427. | 596, 599, 642, 676 |
| 59JA2464 | E. C. Taylor and W. R. Sherman; *J. Am. Chem. Soc.*, 1959, **81**, 2464. | 267, 279, 280, 291, 305 |
| 59JA2472 | O. Vogl and E. C. Taylor; *J. Am. Chem. Soc.*, 1951, **81**, 2472. | 180 |
| 59JA2474 | E. C. Taylor and H. M. Loux; *J. Am. Chem. Soc.*, 1959, **81**, 2474. | 310 |
| 59JA2479 | E. C. Taylor, O. Vogl and P. K. Loeffler; *J. Am. Chem. Soc.*, 1959, **81**, 2479. | 318 |
| 59JA2521 | R. E. Cline, R. M. Fink and K. Fink; *J. Am. Chem. Soc.*, 1959, **81**, 2521. | 71 |
| 59JA3786 | S. Y. Wang; *J. Am. Chem. Soc.*, 1959, **81**, 3786. | 101, 133 |
| 59JA4342 | R. A. Carboni and R. V. Lindsay, Jr.; *J. Am. Chem. Soc.*, 1959, **81**, 4342. | 50, 548, 550 |
| 59JA5650 | R. B. Angier and W. V. Curran; *J. Am. Chem. Soc.*, 1959, **81**, 5650. | 284 |
| 59JA5655 | R. L. Hinman and L. Locatell; *J. Am. Chem. Soc.*, 1959, **81**, 5655. | 1071 |
| 59JA5663 | R. P. Welcher, D. W. Kaiser and V. P. Wystrach; *J. Am. Chem. Soc.*, 1959, **81**, 5663. | 498 |
| 59JA5993 | W. E. Parham, G. L. O. Mayo and B. Gadsby; *J. Am. Chem. Soc.*, 1959, **81**, 5993. | 965 |
| 59JCS1 | C. M. Atkinson and R. E. Rodway; *J. Chem. Soc.*, 1959, 1. | 238, 244 |
| 59JCS6 | C. M. Atkinson and R. E. Rodway; *J. Chem. Soc.*, 1959, 6. | 238, 244 |
| 59JCS458 | G. W. H. Cheeseman; *J. Chem. Soc.*, 1959, 458. | 840 |
| 59JCS525 | R. R. Hunt, J. F. W. McOmie and E. R. Sayer; *J. Chem. Soc.*, 1959, 525. | 69, 75, 80, 105, 112, 123, 132, 135, 141 |
| 59JCS923 | R. L. Huang and K.-H. Lee; *J. Chem. Soc.*, 1959, 923. | 832 |
| 59JCS1132 | J. A. Barlthrop, C. G. Richards, D. M. Russell and G. Ryback; *J. Chem. Soc.*, 1959, 1132. | 177 |
| 59JCS1240 | S. F. Mason; *J. Chem. Soc.*, 1959, 1240. | 65, 395 |
| 59JCS1247 | S. F. Mason; *J. Chem. Soc.*, 1959, 1247. | 65, 390 |
| 59JCS1849 | J. Biggs and P. Sykes; *J. Chem. Soc.*, 1959, 1849. | 224 |
| 59JCS2425 | K. P. Barr, F. M. Dean and H. D. Locksley; *J. Chem. Soc.*, 1959, 2425. | 849 |
| 59JCS2588 | I. E.-S. El-Kholy, F. K. Rafla and G. Soliman; *J. Chem. Soc.*, 1959, 2588. | 812 |
| 59JCS3362 | J. Green, D. McHale, S. Marcinkiewicz, P. Maimalis and P. R. Watt; *J. Chem. Soc.*, 1959, 3362. | 595 |
| 59JCS3521 | C. W. N. Cumper and A. I. Vogel; *J. Chem. Soc.*, 1959, 3521. | 626 |
| 59JCS3598 | E. J. Haws, J. S. E. Holker, A. Kelly, A. D. G. Powell and A. Robertson; *J. Chem. Soc.*, 1959, 3598. | 860 |
| 59JCS3647 | D. J. Brown; *J. Chem. Soc.*, 1959, 3647. | 68 |
| 59JOC11 | S. Y. Wang; *J. Org. Chem.*, 1959, **24**, 11. | 70 |
| 59JOC17 | O. R. Gottlieb and W. B. Mors; *J. Org. Chem.*, 1959, **24**, 17. | 794 |
| 59JOC208 | J. J. Ritter and R. D. Anderson; *J. Org. Chem.*, 1959, **24**, 208. | 505 |
| 59JOC262 | W. E. Parham, P. L. Stright and W. R. Hasek; *J. Org. Chem.*, 1959, **24**, 262. | 964 |
| 59JOC272 | C. Grundmann and H. Ulrich; *J. Org. Chem.*, 1959, **24**, 272. | 375, 377, 382 |
| 59JOC567 | J. H. Looker and M. J. Holm; *J. Org. Chem.*, 1959, **24**, 567. | 829 |
| 59JOC573 | S. Schalit and R. A. Cutler; *J. Org. Chem.*, 1959, **24**, 573. | 516 |
| 59JOC813 | J. Jiu and G. P. Mueller; *J. Org. Chem.*, 1959, **24**, 813. | 409 |
| 59JOC934 | G. Berti; *J. Org. Chem.*, 1959, **24**, 934. | 858 |
| 59JOC963 | T. Sato; *J. Org. Chem.*, 1959, **24**, 963. | 375 |
| 59JOC1205 | G. A. Reynolds, J. A. Van Allan and J. F. Tinker; *J. Org. Chem.*, 1959, **24**, 1205. | 47 |
| 59JOC1391 | H. W. Johnson, Jr., R. E. Lovins and M. Reintjes; *J. Org. Chem.*, 1959, **24**, 1391. | 109, 115 |
| 59JOC1655 | H. Suginome; *J. Org. Chem.*, 1959, **24**, 1655. | 597 |
| 59JOC1804 | L. L. Woods; *J. Org. Chem.*, 1959, **24**, 1804. | 595 |
| 59JOC1809 | J. T. Shaw and F. J. Gross; *J. Org. Chem.*, 1959, **24**, 1809. | 492 |

| | | |
|---|---|---|
| 59JOC1819 | W. E. Parham, B. Gadsby and R. A. Mikulec; *J. Org. Chem.*, 1959, **24**, 1819. | 963, 965 |
| 59LA(625)74 | A. T. Balaban and C. D. Nenitzescu; *Liebigs Ann. Chem.*, 1959, **625**, 74. | 658, 862 |
| B-59MI21200 | R. G. Ramage and J. K. Landquist; in 'Rodd's Chemistry of Carbon Compounds', ed. S. Coffey; Elsevier, Amsterdam, 1959, vol. IV B, p. 1201. | 2 |
| B-59MI21201 | G. R. Ramage and J. K. Landquist; in 'Rodd's Chemistry of Carbon Compounds', ed. S. Coffey; Elsevier, Amsterdam, 1959, vol. IV B, p. 1217. | 2 |
| B-59MI21202 | G. R. Ramage and J. K. Landquist; in 'Rodd's Chemistry of Carbon Compounds', ed. S. Coffey; Elsevier, Amsterdam, 1959, vol. IV B, p. 1238. | 2 |
| 59MI21203 | J. Druey and A. Marxer; *J. Med. Pharm. Chem.*, 1959, **1**, 1. | 2 |
| 59MI21300 | D. J. Brown; *J. Appl. Chem.*, 1959, **9**, 203. | 95, 114, 130, 136 |
| B-59MI21400 | G. R. Ramage and J. K. Landquist; in 'Chemistry of Carbon Compounds,' ed. E. H. Rodd; Elsevier, 1959, vol. 4B. | 158 |
| 59MI22000 | H. E. Williamson, F. E. Shideman and D. A. Le Sher; *J. Pharmacol. Exp. Therap.*, 1959, **126**, 82. | 525 |
| 59MI22001 | P. F. Collins and H. Diehl; *Anal. Chim. Acta*, 1959, **22**, 125. | 529 |
| 59RTC5 | H. I. X. Mager and W. Berends; *Recl. Trav. Chim. Pays-Bas*, 1959, **78**, 5. | 176 |
| 59RTC364 | C. Salemink; *Recl. Trav. Chim. Pays-Bas*, 1959, **78**, 364. | 792 |
| 59SA870 | V. Farmer; *Spectrochim. Acta*, 1959, **10**, 870. | 810 |
| 59T(6)315 | F. L. Allen, P. Koch and H. Suschitzky; *Tetrahedron*, 1959, **6**, 315. | 601 |
| 59T(7)257 | A. Chatterjee, S. Bose and C. Ghosh; *Tetrahedron*, 1959, **7**, 257. | 149 |
| 59YZ260 | M. Asai; *Yakugaku Zasshi*, 1959, **79**, 260. | 169 |
| 59YZ1275 | M. Asai; *Yagugaku Zasshi*, 1959, **79**, 1275. | 168 |
| 59ZN(B)654 | A. Kühn and A. Egelhaaf; *Z. Naturforsch., Teil B*, 1959, **14**, 654. | 283 |
| 59ZPC(316)164 | R. Gmelin; *Hoppe-Seyler's Z. Physiol. Chem.*, 1959, **316**, 164. | 146 |
| | | |
| 60AC(R)277 | R. Fusco and S. Rossi; *Ann. Chim. (Rome)*, 1960, **50**, 277. | 557 |
| 60AC(R)505 | M. Ridi, *Ann. Chim. (Rome)*, 1960, **50**, 505. | 230 |
| 60AC(R)875 | P. Venturella and A. Bellino; *Ann. Chim. (Rome)*, 1960, **50**, 875. | 597 |
| 60ACS539 | E. Adler, R. Magnusson and B. Berggren; *Acta Chem. Scand.*, 1960, **14**, 539. | 834 |
| 60AG331 | K. Dimroth; *Angew. Chem.*, 1960, **72**, 331. | 761 |
| 60AG777 | K. Dimroth and K. H. Wolf; *Angew. Chem.*, 1960, **72**, 777. | 762 |
| 60AG778 | K. Dimroth and K. H. Wolf; *Angew. Chem.*, 1960, **72**, 778. | 660 |
| 60AG836 | H. Eilingsfeld, H. Seefelder and A. Weidinger; *Angew. Chem.*, 1960, **72**, 836. | 490 |
| 60AK(15)387 | H. Hasselquist; *Arkiv Kemi*, 1960, **15**, 387. | 413 |
| 60AX80 | P. J. Wheatley; *Acta Crystallogr.*, 1960, **13**, 80. | 58 |
| 60BSB593 | R. Warin, M. Renson and R. Huls; *Bull. Soc. Chim. Belg.*, 1960, **69**, 593. | 849 |
| 60CB599 | A. T. Balaban and C. D. Nenitzescu; *Chem. Ber.*, 1960, **93**, 599. | 650 |
| 60CB1025 | F. Korte and K. H. Büchel; *Chem. Ber.*, 1960, **93**, 1025. | 598, 764, 765, 774, 785 |
| 60CB1208 | H. Bredereck, H. Herlinger and E. H. Schweizer; *Chem. Ber.*, 1960, **93**, 1208. | 124 |
| 60CB1402 | H. Bredereck, R. Gompper and B. Geiger; *Chem. Ber.*, 1960, **93**, 1402. | 116, 124 |
| 60CB1406 | W. Pfleiderer and G. Nübel; *Chem. Ber.*, 1960, **93**, 1406. | 131, 277 |
| 60CB1466 | A. Schönberg and G. Schütz; *Chem. Ber.*, 1960, **93**, 1466. | 829 |
| 60CB1998 | A. Dornow and H. Hell; *Chem. Ber.*, 1960, **93**, 1998. | 120 |
| 60CB2015 | W. Pfleiderer, E. Liedek, R. Lohrmann and M. Rukwied; *Chem. Ber.*, 1960, **93**, 2015. | 273 |
| 60CB2294 | A. Roedig, R. Manger and S. Schödel; *Chem. Ber.*, 1960, **93**, 2294. | 794 |
| 60CB2405 | H. Bredereck, W. Jentzsch and G. Morlock; *Chem. Ber.*, 1960, **93**, 2405. | 77, 98 |
| 60CB2410 | H. Bredereck and W. Jentzsch; *Chem. Ber.*, 1960, **93**, 2410. | 76 |
| 60CB2668 | G. Henseke and J. Müller; *Chem. Ber.*, 1960, **93**, 2668. | 301 |
| 60CI(L)1192 | W. Bonthrone and D. H. Reid; *Chem. Ind. (London)*, 1960, 1192. | 724, 829 |
| 60CR(250)2819 | M. Levas and E. Levas; *C. R. Hebd. Seances Acad. Sci.*, 1960, **250**, 2819. | 717, 827 |
| 60DIS(20)4291 | J. A. Merritt and K. K. Innes; *Diss. Abstr.*, 1960, **20**, 4291. | 158 |
| 60DOK(133)851 | A. N. Nesmeyanov, E. G. Perevalova, N. A. Simukova, Yu. N. Sheinker and M. D. Reshetova; *Dokl. Akad. Nauk SSSR*, 1960, **133**, 851. | 1043 |
| 60G1113 | R. Fusco and G. Bianchetti; *Gaz. Chim. Ital.*, 1960, **90**, 1113. | 403 |
| 60G1165 | M. Colonna and A. Risalti; *Gazz. Chim. Ital.*, 1960, **90**, 1165. | 1045 |
| 60G1399 | P. Papini, M. Ridi and S. Checchi; *Gazz. Chim. Ital.*, 1960, **90**, 1399. | 243, 246 |
| 60G1807 | B. Camerino and G. Palamidessi; *Gazz. Chim. Ital.*, 1960, **90**, 1807. | 176 |
| 60IZV2064 | A. T. Balaban and C. D. Neditzescu; *Izv. Akad. Nauk SSSR, Ser. Khim.*, 1960, 2064. | 603 |
| 60JA217 | W. S. McNutt; *J. Am. Chem. Soc.*, 1960, **82**, 217. | 320 |
| 60JA483 | J. Meinwald, H. C. Hwang, D. Christman and A. P. Wolf; *J. Am. Chem. Soc.*, 1960, **82**, 483. | 775 |
| 60JA486 | J. J. Fox and D. Van Praag; *J. Am. Chem. Soc.*, 1960, **82**, 486. | 112 |
| 60JA991 | J. H. Burckhalter, R. J. Seiwald and H. C. Scarborough; *J. Am. Chem. Soc.*, 1960, **82**, 991. | 78, 126 |
| 60JA2731 | E. Cohen, B. Klarberg and J. R. Vaughan, Jr.; *J. Am. Chem. Soc.*, 1960, **82**, 2731. | 76, 153 |
| 60JA3762 | R. C. De Selms and H. S. Mosher; *J. Am. Chem. Soc.*, 1960, **82**, 3762. | 249, 252 |
| 60JA3765 | W. Pfleiderer and E. C. Taylor; *J. Am. Chem. Soc.*, 1960, **82**, 3765. | 277, 310 |
| 60JA3971 | C. W. Whitehead and J. J. Traverso; *J. Am. Chem. Soc.*, 1960, **82**, 3971. | 86, 130 |
| 60JA3977 | H. E. Baumgarten, P. L. Creger and R. L. Zey; *J. Am. Chem. Soc.*, 1960, **82**, 3977. | 54 |
| 60JA4085 | E. E. Schweizer and W. E. Parham; *J. Am. Chem. Soc.*, 1960, **82**, 4085. | 725 |

| | | |
|---|---|---|
| 60JA5107 | C. W. Tullock, R. A. Carboni, R. J. Harder, W. C. Smith and D. D. Coffman; *J. Am. Chem. Soc.*, 1960, **82**, 5107. | 140 |
| 60JA5187 | I. J. Pachter, R. F. Raffauf, G. E. Ullyot and O. Ribeiro; *J. Am. Chem. Soc.*, 1960, **82**, 5187. | 149 |
| 60JA5711 | E. C. Taylor, R. J. Knopf, R. F. Meyer, A. Holmes and M. L. Hoefle; *J. Am. Chem. Soc.*, 1960, **82**, 5711. | 115, 128, 338, 349, 350 |
| 60JA6058 | E. C. Taylor, R. J. Knopf, J. A. Gogliano, J. W. Barton and W. Pfleiderer; *J. Am. Chem. Soc.*, 1960, **82**, 6058. | 211, 350 |
| 60JCS26 | B. B. Millward; *J. Chem. Soc.*, 1960, 26. | 822 |
| 60JCS131 | J. Davoll; *J. Chem. Soc.*, 1960, 131. | 133 |
| 60JCS191 | J. M. Davidson and C. M. French; *J. Chem. Soc.*, 1960, 191. | 601 |
| 60JCS242 | G. W. H. Cheeseman; *J. Chem. Soc.*, 1960, 242. | 177 |
| 60JCS502 | J. D. Bu'Lock and H. G. Smith; *J. Chem. Soc.*, 1960, 502. | 596, 599, 676 |
| 60JCS602 | R. Livingstone, D. Miller and S. Morris; *J. Chem. Soc.*, 1960, 602. | 599 |
| 60JCS654 | D. C. Allport and J. D. Bu'Lock; *J. Chem. Soc.*, 1960, 654. | 876 |
| 60JCS1226 | E. Spinner; *J. Chem. Soc.*, 1960, 1226. | 64 |
| 60JCS1232 | E. Spinner; *J. Chem. Soc.*, 1960, 1232. | 64 |
| 60JCS1237 | E. Spinner; *J. Chem. Soc.*, 1960, 1237. | 67 |
| 60JCS1370 | A. Albert and F. Reich; *J. Chem. Soc.*, 1960, 1370. | 226, 284 |
| 60JCS1978 | D. J. Brown and N. W. Jacobsen; *J. Chem. Soc.*, 1960, 1978. | 87, 270, 277 |
| 60JCS2157 | D. Harrison and A. C. B. Smith; *J. Chem. Soc.*, 1960, 2157. | 222, 223, 226 |
| 60JCS3094 | R. Livingstone, D. Miller and S. Morris; *J. Chem. Soc.*, 1960, 3094. | 668, 673 |
| 60JCS3540 | R. J. Grout and M. W. Partridge; *J. Chem. Soc.*, 1960, 3540. | 86 |
| 60JCS3546 | R. J. Grout and M. W. Partridge; *J. Chem. Soc.*, 1960, 3546. | 93 |
| 60JCS4565 | N. Baggett, S. A. Barker, A. B. Foster, R. H. Moore and D. H. Whiffen; *J. Chem. Soc.*, 1960, 4565. | 594 |
| 60JCS4590 | J. H. Chesterfield, J. F. W. McOmie and M. S. Tute; *J. Chem. Soc.*, 1960, 4590. | 92 |
| 60JCS4768 | R. M. Cresswell and H. C. S. Wood; *J. Chem. Soc.*, 1960, 4768. | 69 |
| 60JCS4776 | R. M. Cresswell, T. Neilson and H. C. S. Wood; *J. Chem. Soc.*, 1960, 4776. | 88 |
| 60JCS5153 | E. J. Bourne, D. H. Hutson and H. Weigel; *J. Chem. Soc.*, 1960, 5153. | 793 |
| 60JIC491 | R. D. Desai, B. M. Desai and J. I. Desai; *J. Indian Chem. Soc.*, 1960, **37**, 491. | 819 |
| 60JOC53 | W. E. Parham and G. L. Willette; *J. Org. Chem.*, 1960, **25**, 53. | 966, 970 |
| 60JOC147 | E. H. Sheers; *J. Org. Chem.*, 1960, **25**, 147. | 516 |
| 60JOC148 | E. C. Taylor and C. C. Cheng; *J. Org. Chem.*, 1960, **25**, 148. | 88, 94, 97, 115, 138 |
| 60JOC538 | R. J. Light and C. R. Hauser; *J. Org. Chem.*, 1960, **25**, 538. | 811 |
| 60JOC677 | V. R. Shah, J. L. Bose and R. C. Shah; *J. Org. Chem.*, 1960, **25**, 677. | 803 |
| 60JOC747 | H. Vanderhaeghe; *J. Org. Chem.*, 1960, **25**, 747. | 1012 |
| 60JOC970 | F. C. Novello, S. C. Bell, E. L. A. Abrams, C. Ziegler and J. H. Sprague; *J. Org. Chem.*, 1960, **25**, 970. | 1055 |
| 60JOC1097 | P. DaRe, L. Verlicchi and I. Setnikar; *J. Org. Chem.*, 1960, **25**, 1097. | 710 |
| 60JOC1110 | W. I. O'Sullivan and C. R. Hauser; *J. Org. Chem.*, 1960, **25**, 1110. | 811 |
| 60JOC1368 | E. J. Reist, H. P. Hamlow, I. G. Junga, R. M. Silverstein and B. R. Baker; *J. Org. Chem.*, 1960, **25**, 1368. | 210, 230 |
| 60JOC1501 | A. Mustafa, W. Asker, A. M. Fleifel, S. Khattab and S. Sherif; *J. Org. Chem.*, 1960, **25**, 1501. | 378 |
| 60JOC1944 | T. C. Frazier, E. D. Little and B. E. Lloyd; *J. Org. Chem.*, 1960, **25**, 1944. | 479 |
| 60JOC1950 | L. O. Ross, L. Goodman and B. R. Baker; *J. Org. Chem.*, 1960, **25**, 1950. | 80, 81, 82, 127 |
| 60JOC2183 | R. M. Horowitz and B. Gentili; *J. Org. Chem.*, 1960, **25**, 2183. | 600 |
| 60LA(631)147 | F. G. Fischer, J. Roch and W. P. Neumann; *Liebigs Ann. Chem.*, 1960, **631**, 147. | 350 |
| 60LA(633)158 | F. G. Fischer, W. P. Neumann and J. Roch; *Liebigs Ann. Chem.*, 1960, **633**, 158. | 364 |
| 60LA(635)73 | E. Schmitz; *Liebigs Ann. Chem.*, 1960, **635**, 73. | 1045 |
| 60LA(635)82 | E. Schmitz and R. Ohme; *Liebigs Ann. Chem.*, 1960, **635**, 82. | 415 |
| 60LA(638)205 | G. Strauss; *Liebigs Ann. Chem.*, 1960, **638**, 205. | 113 |
| 60M774 | V. Prey, B. Kerres and H. Berbalk; *Monatsh. Chem.*, 1960, **91**, 774. | 596 |
| 60MI21300 | M. Robba; *Ann. Chim. (Paris)* 1960, **5**, 351. | 80, 81, 82, 83, 84, 104, 127, 128 |
| 60MI21301 | J. D. P. Graham; *Arch. Int. Pharmacodyn. Therap.*, 1960, **123**, 419. | 153 |
| B-60MI21302 | B. Pullman and A. Pullman; 'Results of Quantum Mechanical Calculations of Electronic Structures of Biochemicals', University of Paris, 1960, vol. 1, p. 100. | 59 |
| 60MI21500 | D. D. Perrin and Y. Inoue; *Proc. Chem. Soc.*, 1960, 342 (*Chem. Abstr.*, 1961, **55**, 9424). | 206 |
| 60MI21600 | E. Lippert and H. Prigge; *Ber. Bunsenges. Phys. Chem.*, 1960, **64**, 662. | 271, 325 |
| 60MI21600 | R. L. Kisliuk; *Nature (London)*, 1960, **188**, 584. | 325 |
| 60MI21800 | M. Grabowski; *Bull. Soc. Sci. Lett. Lódź*, 1960, **11**, 3 (*Chem. Abstr.*, 1961, **55**, 18 240). | 370 |
| 60MI21900 | J. Hádáček; *Spisy Prirodoved. Fac. Univ. Brno*, 1960, **417**, 373. | 404 |
| 60MI21901 | J. Hádáček and J. Slotava-Trukova; *Spisy Prirodoved. Fac. Univ. Brno*, 1960, **412**, 143. | 444 |
| B-60MI22000 | H. Gysin and E. Knusli; 'Advances in Pest Control', ed. R. L. Metcalf; Interscience, New York, 1960, vol. 3, p. 289. | 528 |
| B-60MI22400 | K. W. Bentley; 'The Natural Pigments', Interscience, New York, 1960. | 878 |
| 60NKK173 | T. Yoshino; *Nippon Kagaku Zasshi*, 1960, **81**, 173 (*Chem. Abstr.*, 1952, **56**, 449). | 236 |
| 60NKK654 | R. Harada, S. Noguchi and N. Sugiyama; *Nippon Kagaku Zasshi*, 1960, **81**, 654. | 600 |
| 60T(9)163 | A. T. Balaban, V. E. Sahini and E. Keplinger; *Tetrahedron*, 1960, **9**, 163. | 603 |

| | | |
|---|---|---|
| 60TL(13)19 | E. A. Chandross and G. Smolinsky; *Tetrahedron Lett.*, 1960, no. 13, 19. | 374, 376, 378, 380, 382 |
| 60TL(25)44 | N. J. Leonard and M. J. Martell; *Tetrahedron Lett.*, 1960, no. 25, 44. | 148 |
| 60USP2937284 | G. H. Hitchings and K. W. Ledig; (Burroughs Wellcome & Co. (U.S.A.) Inc.), *U.S. Pat.* 2 937 284 (1960) (*Chem. Abstr.*, 1961, **55**, 25 999). | 231 |
| 60USP2940972 | K. Thomae; U.S. Pat. 2 940 972 (1960) (*Chem. Abstr.*, 1960, **54**, 24 824). | 291 |
| 60USP2945037 | J. Druey and H. V. Daeniker; (Ciba Pharm. Prods. Inc.), *U.S. Pat.* 2 945 037 (1960) (*Chem. Abstr.*, 1961, **55**, 4546). | 251 |
| 60USP2964524 | H. Tolkmith; *U.S. Pat.* 2 964 524 (1960) (*Chem. Abstr.*, 1961, **55**, 6506). | 570 |
| 60YZ245 | T. Higashino; *Yakugaku Zasshi*, 1960, **80**, 245. | 72, 124, 128 |
| 60ZOB1893 | G. S. Predvoditeleva and M. N. Shchukina; *Zh. Obshch. Khim.*, 1960, **30**, 1893 (*Chem. Abstr.*, 1961, **55**, 7421). | 1012 |
| 60ZPC(322)173 | A. Schellenberger and K. Winter; *Hoppe-Seyler's Z. Physiol. Chem.*, 1960, **322**, 173. | 133 |
| 61ACS1124 | K. A. Jensen and C. Pedersen; *Acta Chem. Scand.*, 1961, **15**, 1124. | 566 |
| 61ACS1575 | J. Sandström; *Acta Chem. Scand.*, 1961, **15**, 1575. | 559 |
| 61AG125 | H. Zollinger; *Angew. Chem.*, 1961, **73**, 125. | 483 |
| 61AG402 | F. Weygand, H. Simon, G. Dahms, M. Waldschmidt, H. J. Schliep and H. Wacker; *Angew. Chem.*, 1961, **73**, 402. | 320 |
| 61AK(17)523 | S. Forsén and M. Nilsson; *Ark. Kemi*, 1961, **17**, 523. | 643 |
| 61AP234 | K. E. Schulte, J. Reisch and O. Heine; *Arch. Pharm.* (*Weinheim, Ger.*), 1961, **294**, 234. | 792 |
| 61AX333 | R. Gerdil; *Acta Crystallogr.*, 1961, **14**, 333. | 68 |
| 61BSF538 | M. Siemiatycki and R. Fugnitto; *Bull. Soc. Chim. Fr.*, 1961, 538. | 868 |
| 61CB1 | W. Pfleiderer and M. Rukwied; *Chem. Ber.*, 1961, **94**, 1. | 276, 310 |
| 61CB12 | W. Pfleiderer and R. Lohrmann; *Chem. Ber.*, 1961, **94**, 12. | 100, 299 |
| 61CB118 | W. Pfleiderer and M. Rukwied; *Chem. Ber.*, 1961, **94**, 118. | 276, 297 |
| 61CB241 | A. Schönberg and E. Singer; *Chem. Ber.*, 1961, **94**, 241. | 852 |
| 61CB486 | S. Hünig, E. Benzing and K. Hübner; *Chem. Ber.*, 1961, **94**, 486. | 813 |
| 61CB660 | A. Schönberg, E. Singer and M. M. Sidky; *Chem. Ber.*, 1961, **94**, 660. | 704 |
| 61CB1555 | R. Huisgen, H. J. Sturm and M. Seidel; *Chem. Ber.*, 1961, **94**, 1555. | 564 |
| 61CB1784 | R. Gompper and O. Christmann; *Chem. Ber.*, 1961, **94**, 1784. | 598, 698, 741, 761 |
| 61CB1795 | R. Gompper and O. Christmann; *Chem. Ber.*, 1961, **94**, 1795. | 658, 681 |
| 61CB1883 | H. Bredereck, O. Smerz and R. Gompper; *Chem. Ber.*, 1961, **94**, 1883. | 498 |
| 61CB2708 | W. Pfleiderer and R. Lohrmann; *Chem. Ber.*, 1961, **94**, 2708. | 299, 310, 313 |
| 61CCC893 | J. Farkaš and F. Šorm; *Collect. Czech. Chem. Commun.*, 1961, **26**, 893. | 89, 100 |
| 61CCC986 | J. Gut, M. Prystaš and J. Jonáš; *Collect. Czech. Chem. Commun.*, 1961, **26**, 986. | 390, 399 |
| 61CCC1680 | M. Horák and J. Gut; *Collect. Czech. Chem. Commun.*, 1961, **26**, 1680. | 395 |
| 61CCC2155 | J. Jonáš and J. Gut; *Collect. Czech. Chem. Commun.*, 1961, **26**, 2155. | 390, 399 |
| 61CCC2519 | A. Piskala and J. Gut; *Collect. Czech. Chem. Commun.*, 1961, **26**, 2519. | 490 |
| 61CCC2871 | Z. Buděšinský and F. Roubínek; *Collect. Czech. Chem. Commun.*, 1961, **26**, 2871. | 569 |
| 61CI(L)630 | A. Calderbank, E. C. Edgar and J. A. Silk; *Chem. Ind.* (*London*), 1961, 630. | 525 |
| 61CJC1184 | D. Cook; *Can. J. Chem.*, 1961, **39**, 1184. | 595 |
| 61CPB38 | T. Nishiwaki; *Chem. Pharm. Bull.*, 1961, **9**, 38. | 137 |
| 61CR(253)872 | M. Julia and C. B. du Jassonneix; *C. R. Hebd. Seances Acad. Sci.*, 1961, **253**, 872. | 794 |
| 61CRV591 | J. C. Roberts; *Chem. Rev.*, 1961, **61**, 591. | 601, 835 |
| 61GEP1100030 | L. Suranyi and L. Schuler; (Knoll A/G Chemische Fabriken), *Ger. Pat.* 1 100 030 (1961) (*Chem. Abstr.*, 1961, **55**, 27 381). | 217 |
| 61HCA812 | B. Bitter and H. Zollinger; *Helv. Chim. Acta*, 1961, **44**, 812. | 483 |
| 61HCA1480 | O. Brenner-Holzach and F. Leuthardt; *Helv. Chim. Acta*, 1961, **44**, 1480. | 320 |
| 61HCA1783 | M. Viscontini and H. Stierlin; *Helv. Chim. Acta*, 1961, **44**, 1783. | 289 |
| 61JA193 | R. L. Letsinger and J. D. Jamison; *J. Am. Chem. Soc.*, 1961, **83**, 193. | 812, 813 |
| 61JA4034 | W. E. Parham and R. Koncos; *J. Am. Chem. Soc.*, 1961, **83**, 4034. | 914 |
| 61JA4357 | G. Claeson, G. Androes and M. Calvin; *J. Am. Chem. Soc.*, 1961, **83**, 4357. | 962 |
| 61JBC(236)512 | J. C. MacDonald; *J. Biol. Chem.*, 1961, **236**, 512. | 191 |
| 61JCS127 | A. Albert and F. Reich; *J. Chem. Soc.*, 1961, 127. | 279, 289 |
| 61JCS504 | C. L. Angell; *J. Chem. Soc.*, 1961, 504. | 68 |
| 61JCS702 | C. J. Covell, F. E. King and J. W. W. Morgan; *J. Chem. Soc.*, 1961, 702. | 819 |
| 61JCS798 | F. M. Dean and D. R. Randell; *J. Chem. Soc.*, 1961, 798. | 596 |
| 61JCS1246 | G. W. H. Cheeseman; *J. Chem. Soc.*, 1961, 1246. | 173 |
| 61JCS2312 | C. H. Hassall and J. R. Lewis; *J. Chem. Soc.*, 1961, 2312. | 838, 839 |
| 61JCS2689 | A. Albert, W. L. F. Armarego and E. Spinner; *J. Chem. Soc.*, 1961, 2689. | 65, 73, 74, 103 |
| 61JCS2697 | W. L. F. Armarego; *J. Chem. Soc.*, 1961, 2697. | 75, 108 |
| 61JCS2828 | H. J. Barber, K. Washbourn, W. R. Wragg and E. Lunt; *J. Chem. Soc.*, 1961, 2828. | 42 |
| 61JCS2843 | G. Büchi, L. Crombie, P. J. Godin, J. S. Kaltenbronn, K. S. Siddalingaiah and D. A. Whiting; *J. Chem. Soc.*, 1961, 2843. | 624 |
| 61JCS3148 | A. A. Sayigh and H. Ulrich; *J. Chem. Soc.*, 1961, 3148. | 478 |
| 61JCS3254 | J. H. Dewar and G. Shaw; *J. Chem. Soc.*, 1961, 3254. | 109 |
| 61JCS3553 | A. T. Balaban and C. D. Nenitzescu; *J. Chem. Soc.*, 1961, 3553. | 867, 868 |
| 61JCS3561 | A. T. Balaban and C. D. Nenitzescu; *J. Chem. Soc.*, 1961, 3561. | 868 |
| 61JCS3573 | P. F. G. Praill and A. L. Whitear; *J. Chem. Soc.*, 1961, 3573. | 862 |
| 61JCS3983 | G. W. H. Cheeseman, A. R. Katritzky and S. Øksne; *J. Chem. Soc.*, 1961, 3983. | 173 |
| 61JCS4413 | D. J. Brown and N. W. Jacobsen; *J. Chem. Soc.*, 1961, 4413. | 270, 277, 297, 305 |

| 61JCS4490 | I. E.-S. El-Kholy, F. K. Rafla and G. Soliman; *J. Chem. Soc.*, 1961, 4490. | 595 |
|---|---|---|
| 61JCS4845 | R. N. Naylor, G. Shaw, D. V. Wilson and D. N. Butler; *J. Chem. Soc.*, 1961, 4845. | 438 |
| 61JCS4930 | E. W. Parnell; *J. Chem. Soc.*, 1961, 4930. | 377, 382 |
| 61JCS5131 | A. Albert and S. Matsuura; *J. Chem. Soc.*, 1961, 5131. | 101, 279, 280, 305 |
| 61JGU523 | V. A. Zagorevskii, D. A. Zykov and E. K. Orlova; *J. Gen. Chem. USSR (Engl. Transl.).*, 1961, **31**, 523. | 711 |
| 61JOC79 | S. G. Cottis and H. Tieckelmann; *J. Org. Chem.*, 1961, **26**, 79. | 136 |
| 61JOC451 | M. Hauser, E. Peters and H. Tieckelmann; *J. Org. Chem.*, 1961, **26**, 451. | 122 |
| 61JOC559 | F. W. Gubitz and R. L. Clarke; *J. Org. Chem.*, 1961, **26**, 559. | 352 |
| 61JOC598 | J. A. Barone and H. Tieckelmann; *J. Org. Chem.*, 1961, **26**, 598. | 85 |
| 61JOC767 | D. H. Clemens and W. D. Emmons; *J. Org. Chem.*, 1961, **26**, 767. | 515, 516 |
| 61JOC792 | H. C. Koppel, R. H. Springer, R. K. Robins and C. C. Cheng; *J. Org. Chem.*, 1961, **26**, 792. | 102 |
| 61JOC957 | E. Kober; *J. Org. Chem.*, 1961, **26**, 957. | 474 |
| 61JOC1028 | L. L. Woods and P. A. Dix; *J. Org. Chem.*, 1961, **26**, 1028. | 595 |
| 61JOC1118 | P. K. Chang; *J. Org. Chem.*, 1961, **26**, 1118. | 399, 402 |
| 61JOC1874 | H. Gershon, K. Dittmer and R. Braun; *J. Org. Chem.*, 1961, **26**, 1874. | 100 |
| 61JOC1891 | R. B. Angier and W. V. Curran; *J. Org. Chem.*, 1961, **26**, 1891. | 349, 361 |
| 61JOC1895 | G. M. Kosolapoff and C. H. Roy; *J. Org. Chem.*, 1961, **26**, 1895. | 104 |
| 61JOC2129 | R. B. Angier and W. V. Curran; *J. Org. Chem.*, 1961, **26**, 2129. | 295 |
| 61JOC2260 | K. Conrow and P. C. Radlick; *J. Org. Chem.*, 1961, **26**, 2260. | 661 |
| 61JOC2364 | W. V. Curran and R. B. Angier; *J. Org. Chem.*, 1961, **26**, 2364. | 305 |
| 61JOC2764 | R. G. Shepherd, W. E. Taft and H. M. Krazinski; *J. Org. Chem.*, 1961, **26**, 2764. | 100, 102 |
| 61JOC2778 | F. C. Schaefer and G. A. Peters; *J. Org. Chem.*, 1961, **26**, 2778. | 506, 507 |
| 61JOC2784 | F. C. Schaefer and G. A. Peters; *J. Org. Chem.*, 1961, **26**, 2784. | 470 |
| 61JOC3334 | E. C. Juenge and W. C. Francis; *J. Org. Chem.*, 1961, **26**, 3334. | 508, 511 |
| 61JOC3351 | A. S. Tomcufcik and D. R. Seeger; *J. Org. Chem.*, 1961, **26**, 3351. | 325 |
| 61JOC3379 | J. D. Behun and R. Levine; *J. Org. Chem.*, 1961, **26**, 3379. | 166 |
| 61JOC3392 | J. P. Horwitz and A. J. Tomson; *J. Org. Chem.*, 1961, **26**, 3392. | 87, 140 |
| 61JOC3461 | R. L. Hinman and B. E. Hoogenboom; *J. Org. Chem.*, 1961, **26**, 3461. | 1050, 1055, 1058 |
| 61JOC3783 | K. Matsuda and L. T. Morin; *J. Org. Chem.*, 1961, **26**, 3783. | 442 |
| 61JOC4130 | M. Claesen and H. Vanderhaeghe; *J. Org. Chem.*, 1961, **26**, 4130. | 1011 |
| 61JOC4419 | D. E. Heitmeier, E. E. Spinner and A. P. Gray; *J. Org. Chem.*, 1961, **26**, 4419. | 89 |
| 61JOC4425 | H. J. Minnemeyer, J. A. Egger, J. F. Holland and H. Tieckelmann; *J. Org. Chem.*, 1961, **26**, 4425. | 136 |
| 61JOC4504 | S. Inoue, A. J. Saggimoto and E. A. Nodiff; *J. Org. Chem.*, 1961, **26**, 4504. | 88, 112 |
| 61JOC4705 | E. Kober; *J. Org. Chem.*, 1961, **26**, 4705. | 483 |
| 61JOC4817 | H. Gilman and J. Diehl; *J. Org. Chem.*, 1961, **26**, 4817. | 768 |
| 61JOC4961 | E. C. Taylor, J. W. Barton and W. W. Paudler; *J. Org. Chem.*, 1961, **26**, 4961. | 88 |
| 61JOC4967 | E. C. Taylor and A. L. Borror; *J. Org. Chem.*, 1961, **26**, 4967. | 226 |
| 61JPR(12)206 | E. Profft and H. Raddatz; *J. Prakt. Chem.*, 1961, **12**, 206. | 78 |
| 61LA(648)114 | G. Köbrich; *Liebigs Ann. Chem.*, 1961, **648**, 114. | 698 |
| 61M183 | G. Spiteller and H. Bretschneider; *Monatsh. Chem.*, 1961, **92**, 183. | 101 |
| 61M1184 | E. Ziegler and E. Noelken; *Monatsh. Chem.*, 1961, **92**, 1184. | 229, 362 |
| 61M1212 | W. Klötzer; *Monatsh. Chem.*, 1961, **92**, 1212. | 97 |
| 61MI21300 | W. L. F. Armarego; *J. Appl. Chem.*, 1961, **11**, 70. | 111, 124 |
| 61MI21301 | A. T. Amos and G. G. Hall; *Mol. Phys.*, 1961, **4**, 25. | 59, 149 |
| B-61MI21400 | H. M. Hershenson; in 'Ultraviolet and Visible Absorption Spectra, Indexes for 1955–1959', Academic Press, New York, 1961. | 161 |
| 61MI21600 | G. W. E. Plaut; *Annu. Rev. Biochem.*, 1961, **30**, 409. | 321 |
| 61MI21601 | D. K. Mishra, S. R. Humphreys, M. Friedkin, A. Goldin and E. J. Crawford; *Nature (London)*, 1961, **189**, 39. | 327 |
| B-61MI21800 | J. P. Horwitz; in 'Heterocyclic Compounds', ed. R. C. Elderfield; Wiley, New York, 1961, vol. 7, p. 768. | 369 |
| B-61MI21900 | J. P. Horwitz; in 'Heterocyclic Compounds', ed. R. C. Elderfield; Wiley, New York, 1961, vol. 7, p. 720. | 385 |
| B-61MI22000 | E. J. Modest; in 'Heterocyclic Compounds', ed. R. C. Elderfield; Wiley, New York, 1961, vol. 7, p. 1. | 459, 482, 483, 487, 490, 493, 498 |
| B-61MI22400 | W. B. Whalley; in 'Heterocyclic Compounds', ed. R. C. Elderfield; Wiley, New York, 1961, vol. 7, p. 1. | 738 |
| 61MI22401 | A. T. Balaban and C. D. Nenitzescu; *Rev. Chim. Acad. RPR*, 1961, **6**, 269. | 867 |
| 61MI22402 | G. Baddeley and M. A. R. Khayat; *Proc. Chem. Soc.*, 1961, 382. | 868 |
| 61MI22600 | N. Marziano and G. Montaudo; *Ric. Sci. Rend., Ser. A*, 1961, **4**, 87. | 955 |
| B-61MI22601 | W. E. Parham; in 'Organic Sulfur Compounds', ed. N. Kharasch; Pergamon, Oxford, 1961, vol. 1, p. 248. | 965, 966 |
| 61MI22602 | E. Henmo, P. de Mayo, A. B. M. Abdus Sattar and A. Stoessl; *Proc. Chem. Soc.*, 1961, 238. | 971 |
| 61MI22700 | N. E. Boyer; *Technikas Apskats.*, 1961, **30**, 5 (*Chem. Abstr.*, 1963, **59**, 1524). | 1030 |
| B-61MI22701 | H. Brockmann; in 'The Chemistry of Natural Products', IUPAC, Section of Organic Chemistry, Butterworths, London, 1961, p. 405. | 1038 |
| B-61MI22800 | G. W. Stacy; in 'Heterocyclic Compounds', ed. R. C. Elderfield; Wiley, New York, 1961, vol. 7, p. 797. | 1043, 1044, 1072 |
| 61NKK932 | I. Kumashiro; *Nippon Kagaku Zasshi*, 1961, **82**, 932. | 599 |

| | | |
|---|---|---|
| 61RTC158 | H. Koopman; *Recl. Trav. Chim. Pays-Bas*, 1961, **80**, 158. | 464 |
| 61SA64 | A. R. Katritzky and R. A. Jones; *Spectrochim. Acta*, 1961, **17**, 64. | 595 |
| 61SA155 | J. E. Lancaster, R. F. Stamm and N. B. Colthup; *Spectrochim. Acta*, 1961, **17**, 155. | 464 |
| 61SA600 | A. Heckle, H. A. Ory and J. M. Talbert; *Spectrochim. Acta*, 1961, **17**, 600. | 465 |
| 61T(12)63 | N. K. Kochetov, L. J. Kudryashov and B. P. Gottich; *Tetrahedron*, 1961, **12**, 63. | 791 |
| 61TL223 | T. Kubota, Y. Tomita and K. Suzuki; *Tetrahedron Lett.*, 1961, 223. | 860 |
| 62AC(R)1070 | G. Berti, A. Marsili and P. L. Pacini; *Ann. Chim. (Rome)*, 1962, **52**, 1070. | 858 |
| 62ACS916 | J. Tjebbes; *Acta Chem. Scand.*, 1962, **16**, 916. | 59 |
| 62AG465 | J. Strating, J. H. Keijer, E. Molenaar and L. Brandsma; *Angew. Chem.*, 1962, **74**, 465. | 757 |
| 62AG(E)115 | C. Schöpf and K. H. Gänshirt; *Angew. Chem., Int. Ed. Engl.*, 1962, **1**, 115. | 289 |
| 62AG(E)235 | E. Schlitter, G. De Stevens and L. H. Werner; *Angew. Chem., Int. Ed. Engl.*, 1962, **1**, 235. | 1043, 1084 |
| 62AG(E)511 | G. Markl; *Angew. Chem., Int. Ed. Engl.*, 1962, **1**, 511. | 658 |
| 62AG(E)699 | J. Strating, J. H. Keijer, E. Molenaar and L. Brandsma; *Angew. Chem., Int. Ed. Engl.*, 1962, **1**, 699. | 642 |
| 62AJC851 | D. J. Brown and J. M. Lyall; *Aust. J. Chem.*, 1962, **15**, 851. | 96 |
| 62AP121 | M. Rink and K. Feiden; *Arch. Pharm. (Weinheim, Ger.)*, 1962, **295**, 121. | 351, 366 |
| 62AP645 | K. E. Schulte, J. Reisch and A. Mock; *Arch. Pharm. (Weinheim, Ger.)*, 1962, **295**, 645. | 758 |
| 62AP649 | E. Profft and H. Raddatz; *Arch. Pharm. (Weinheim, Ger.)*, 1962, **295**, 649. | 100 |
| 62AX451 | G. H. Stout and L. H. Jensen; *Acta Crystallogr.*, 1962, **15**, 451. | 623 |
| 62AX1060 | G. H. Stout and L. H. Jensen; *Acta Crystallogr.*, 1962, **15**, 1060. | 623 |
| 62B1161 | C. C. Levy and W. S. McNutt; *Biochemistry*, 1962, **1**, 1161. | 308 |
| 62BCJ1323 | K. Yamada; *Bull. Chem. Soc. Jpn.*, 1962, **35**, 1323. | 595, 638 |
| 62BCJ1329 | K. Yamada; *Bull. Chem. Soc. Jpn.*, 1962, **35**, 1329. | 710 |
| 62BEP613056 | Courtaulds Ltd., *Belg. Pat.* 613 056 (1962) (*Chem. Anstr.*, 1962, **57**, 12 757). | 939 |
| 62BSB379 | M. Renson and L. Christiaens; *Bull. Soc. Chim. Belg.*, 1962, **71**, 379. | 858 |
| 62BSB394 | M. Renson and L. Christiaens; *Bull. Soc. Chim. Belg.*, 1962, **71**, 394. | 858 |
| 62CB147 | J. Goerdeler and B. Wedekind; *Chem. Ber.*, 1962, **95**, 147. | 1047 |
| 62CB483 | G. P. Schiemenz; *Chem. Ber.*, 1962, **95**, 483. | 804 |
| 62CB738 | W. Pfleiderer and R. Lohrmann; *Chem. Ber.*, 1962, **95**, 738. | 282 |
| 62CB749 | W. Pfleiderer; *Chem. Ber.*, 1962, **95**, 749. | 271, 277, 304, 310 |
| 62CB755 | W. Pfleiderer, E. Liedek and M. Rukwied, *Chem. Ber.*, 1962, **95**, 755. | 296 |
| 62CB803 | H. Bredereck, F. Effenberger and E. H. Schweizer; *Chem. Ber.*, 1962, **95**, 803. | 117, 127, 133 |
| 62CB1245 | A. Roedig, H.-G. Kleppe and G. Märkl; *Chem. Ber.*, 1962, **95**, 1245. | 794 |
| 62CB1446 | O. Dann and H. Hofmann; *Chem. Ber.*, 1962, **95**, 1446. | 855 |
| 62CB1591 | W. Pfleiderer and M. Rukwied; *Chem. Ber.*, 1962, **95**, 1591. | 277, 304 |
| 62CB1597 | W. Pfleiderer and K. Deckert; *Chem. Ber.*, 1962, **95**, 1597. | 88 |
| 62CB1605 | G. Nübel and W. Pfleiderer; *Chem. Ber.*, 1962, **95**, 1605. | 271, 277, 304, 310 |
| 62CB1621 | W. Pfleiderer and F. Reisser; *Chem. Ber.*, 1962, **95**, 1621. | 282, 292, 296 |
| 62CB1840 | H. Baganz and L. Domaschke; *Chem. Ber.*, 1962, **95**, 1840. | 108 |
| 62CB2012 | E. Schmitz and R. Ohme; *Chem. Ber.*, 1962, **95**, 2012. | 44 |
| 62CB2086 | K.-D. Gundermann and H.-J. Rose; *Chem. Ber.*, 1962, **95**, 2086. | 855 |
| 62CB2195 | W. Pfleiderer; *Chem. Ber.*, 1962, **95**, 2195. | 276, 289, 303 |
| 62CB2248 | M. Avram, I. G. Dinulescu, E. Marica and C. D. Nenitzescu; *Chem. Ber.*, 1962, **95**, 2248. | 550 |
| 62CB2796 | H. Bredereck, F. Effenberger and W. Resemann; *Chem. Ber.*, 1962, **95**, 2796. | 122 |
| 62CCC716 | J. Jonáš and J. Gut; *Collect. Czech. Chem. Commun.*, 1962, **27**, 716. | 399 |
| 62CCC1242 | R. Zahradník, C. Párkányi and J. Koutecký; *Collect. Czech. Chem. Commun.*, 1962, **27**, 1242. | 633, 637 |
| 62CCC1886 | J. Jonáš and J. Gut; *Collect. Czech. Chem. Commun.*, 1962, **27**, 1886. | 390, 395 |
| 62CCC1898 | M. Prystaš and J. Gut; *Collect. Czech. Chem. Commun.*, 1962, **27**, 1898. | 408 |
| 62CI(L)268 | G. Vogel; *Chem. Ind. (London)*, 1962, 268. | 687 |
| 62CI(L)968 | G. Vogel; *Chem. Ind. (London)*, 1962, 968. | 633 |
| 62CI(L)1332 | D. H. Hey, C. W. Rees and A. R. Todd; *Chem. Ind. (London)*, 1962, 1332. | 379 |
| 62CI(L)1793 | K. G. Marathe, E. M. Philbin and T. S. Wheeler; *Chem. Ind. (London)*, 1962, 1793. | 765 |
| 62CI(L)1829 | G. Vogel; *Chem. Ind. (London)*, 1962, 1829. | 633 |
| 62CJC882 | J. F. King and T. Durst; *Can. J. Chem.*, 1962, **40**, 882. | 1019 |
| 62CJC1053 | B. A. Gingras, T. Suprunchuk and C. H. Bayley; *Can. J. Chem.*, 1962, **40**, 1053. | 395 |
| 62CJC1160 | A. F. McKay and M.-E. Kreling; *Can. J. Chem.*, 1962, **40**, 1160. | 338 |
| 62CJC2146 | P. Yates, M. J. Jorgenson and S. K. Roy; *Can. J. Chem.*, 1962, **40**, 2146. | 700 |
| 62CPB313 | M. Sano; *Chem. Pharm. Bull.*, 1962, **10**, 313. | 113 |
| 62CPB647 | Y. Mizuno, M. Ikehara and K. A. Watanabe; *Chem. Pharm. Bull.*, 1962, **10**, 647. | 89 |
| 62CPB926 | I. Iwai and J. Ide; *Chem. Pharm. Bull.*, 1962, **10**, 926. | 742 |
| 62CPB1043 | T. Higashino; *Chem. Pharm. Bull.*, 1962, **10**, 1043. | 84, 124 |
| 62CPB1215 | M. Furukawa; *Chem. Pharm. Bull.*, 1962, **10**, 1215. | 492 |
| 62CR(254)696 | H. Strzelecka, M. Simalty-Siemiatycki and C. Prévost; *C. R. Hebd. Seances Acad. Sci.*, 1962, **254**, 696. | 871 |
| 62CR(255)731 | H. Strzelecka; *C.R. Hebd. Seances Acad. Sci.*, 1962, **255**, 731. | 871 |
| 62CRV405 | S. B. Needleman and M. C. Chang Kuo; *Chem. Rev.*, 1962, **62**, 405. | 772 |

| | | |
|---|---|---|
| 62DOK(142)354 | V. A. Ginsburg, A. Ya. Yakubovich, A. S. Filatov, G. E. Zelenin, S. P. Makarov, V. A. Shpanskii, G. P. Kotel'nikov, L. F. Sergienko and L. L. Martynova; *Dokl. Akad. Nauk SSSR*, 1962, **142**, 354. | 562 |
| 62HC(16)1 | D. J. Brown; *Chem. Heterocycl. Compd.*, 1962, **16**, 1. | 158 |
| 62HC(16)31 | D. J. Brown; *Chem. Heterocycl. Compd.*, 1962, **16**, 31. | 111 |
| 62HC(16)82 | D. J. Brown; *Chem. Heterocycl. Compd.*, 1962, **16**, 82. | 107, 108, 116, 117 |
| 62HC(16)172 | D. J. Brown; *Chem. Heterocycl. Compd.*, 1962, **16**, 172. | 70 |
| 62HC(16)233 | D. J. Brown; *Chem. Heterocycl. Compd.*, 1962, **16**, 233. | 94 |
| 62HC(16)245 | D. J. Brown; *Chem. Heterocycl. Compd.*, 1962, **16**, 245. | 92 |
| 62HC(16)277 | D. J. Brown; *Chem. Heterocycl. Compd.*, 1962, **16**, 277. | 93 |
| 62HC(16)288 | D. J. Brown; *Chem. Heterocycl. Compd.*, 1962, **16**, 288. | 96 |
| 62HC(16)291 | D. J. Brown; *Chem. Heterocycl. Compd.*, 1962, **16**, 291. | 96 |
| 62HC(16)306 | D. J. Brown; *Chem. Heterocycl. Compd.*, 1962, **16**, 306. | 129 |
| 62HC(16)316 | D. J. Brown; *Chem. Heterocycl. Compd.*, 1962, **16**, 316. | 82 |
| 62HC(16)422 | D. J. Brown; *Chem. Heterocycl. Compd.*, 1962, **16**, 422. | 146 |
| 62HC(16)430 | D. J. Brown; *Chem. Heterocycl. Compd.*, 1962, **16**, 430. | 75, 107 |
| 62HC(16)470 | D. J. Brown; *Chem. Heterocycl. Compd.*, 1962, **16**, 470. | 61 |
| 62HC(16)472 | D. J. Brown; *Chem. Heterocycl. Compd.*, 1962, **16**, 472. | 60 |
| 62HC(16)477 | S. F. Mason; *Chem. Heterocycl. Compd.*, 1962, **16**, 477. | 64, 65 |
| 62HC(17)443 | R. L. McKee; *Chem. Heterocycl. Compd.*, 1962, **17**, 443. | 1041, 1043, 1045, 1046 |
| 62HCA996 | P. Schmidt, K. Eichenberger and M. Wilhelm; *Helv. Chim. Acta*, 1962, **45**, 996. | 1085 |
| 62IZV2096 | L. I. Zakharkin and L. P. Sorokina; *Izv. Akad. Nauk SSSR, Ser. Khim.*, 1962, 2096. | 794 |
| 62JA336 | G. S. Reddy, R. T. Hobgood, Jr. and J. H. Goldstein; *J. Am. Chem. Soc.*, 1962, **84**, 336. | 62 |
| 62JA813 | W. E. Parham and L. D. Huestis; *J. Am. Chem. Soc.*, 1962, **84**, 813. | 580, 594, 598, 671, 764 |
| 62JA966 | T. J. Kealy; *J. Am. Chem. Soc.*, 1962, **84**, 966. | 352 |
| 62JA1594 | K. B. Wiberg, B. R. Lowrie and B. J. Nist; *J. Am. Chem. Soc.*, 1962, **84**, 1594. | 581 |
| 62JA1904 | E. F. Schroeder and R. M. Dodson; *J. Am. Chem. Soc.*, 1962, **84**, 1904. | 123 |
| 62JA2452 | S. Masamune and N. T. Castellucci; *J. Am. Chem. Soc.*, 1962, **84**, 2452. | 577, 578, 593, 762 |
| 62JA3307 | C. S. Rondestvedt, Jr. and G. J. Mantell; *J. Am. Chem. Soc.*, 1962, **84**, 3307. | 979 |
| 62JA3319 | C. S. Rondestvedt, Jr.; *J. Am. Chem. Soc.*, 1962, **84**, 3319. | 979 |
| 62JA3744 | E. C. Taylor and C. W. Jefford; *J. Am. Chem. Soc.*, 1962, **84**, 3744. | 131, 518 |
| 62JBC(237)1977 | J. C. MacDonald; *J. Biol. Chem.*, 1962, **237**, 1977. | 191 |
| 62JCS493 | S. F. Mason; *J. Chem. Soc.*, 1962, 493. | 250, 340 |
| 62JCS527 | D. J. Brown and R. F. Evans; *J. Chem. Soc.*, 1962, 527. | 120, 123 |
| 62JCS561 | W. L. F. Armarego; *J. Chem. Soc.*, 1962, 561. | 65, 134 |
| 62JCS583 | J. H. Dewar and G. Shaw; *J. Chem. Soc.*, 1962, 583. | 146 |
| 62JCS645 | D. D. Perrin; *J. Chem. Soc.*, 1962, 645. | 265 |
| 62JCS775 | L. Crombie and J. W. Lown; *J. Chem. Soc.*, 1962, 775. | 583, 734 |
| 62JCS1540 | A. R. Katritzky and A. J. Waring; *J. Chem. Soc.*, 1962, 1540. | 93 |
| 62JCS1591 | A. Albert and C. F. Howell; *J. Chem. Soc.*, 1962, 1591. | 272, 279, 287, 288 |
| 62JCS1671 | C. M. Atkinson and A. R. Mattocks; *J. Chem. Soc.*, 1962, 1671. | 206, 208, 224, 238, 240, 241 |
| 62JCS1857 | I. E.-S. El-Kholy, F. K. Rafla and G. Soliman; *J. Chem. Soc.*, 1962, 1857. | 595 |
| 62JCS2162 | A. Albert and S. Matsuura; *J. Chem. Soc.*, 1962, 2162. | 279, 305, 307 |
| 62JCS2595 | A. Albert, C. F. Howell and E. Spinner; *J. Chem. Soc.*, 1962, 2595. | 267 |
| 62JCS2600 | Y. Inoue and D. D. Perrin; *J. Chem. Soc.*, 1962, 2600. | 272, 276 |
| 62JCS2606 | J. A. Elvidge; *J. Chem. Soc.*, 1962, 2606. | 642 |
| 62JCS3129 | A. Albert and G. B. Barlin; *J. Chem. Soc.*, 1962, 3129. | 67, 89, 110, 137, 139 |
| 62JCS3162 | J. W. Clark-Lewis and R. P. Singh; *J. Chem. Soc.*, 1962, 3162. | 250, 257 |
| 62JCS3858 | J. W. Clark-Lewis, L. M. Jackman and L. R. Williams; *J. Chem. Soc.*, 1962, 3858. | 631 |
| 62JCS4094 | W. L. F. Armarego; *J. Chem. Soc.*, 1962, 4094. | 206, 213, 222 |
| 62JCS4678 | V. Oakes, H. N. Rydon and K. Undheim; *J. Chem. Soc.*, 1962, 4678. | 215 |
| 62JCS4963 | E. A. C. Lucken; *J. Chem. Soc.*, 1962, 4963. | 968 |
| 62JIC368 | A. B. Sen and S. K. Gupta; *J. Indian Chem. Soc.*, 1962, **39**, 368. | 111 |
| 62JIC507 | M. V. Shah and S. Sethna; *J. Indian Chem. Soc.*, 1962, **39**, 507. | 708 |
| 62JMC808 | J. J. Traverso, E. B. Robbins and C. W. Whitehead; *J. Med. Chem.*, 1962, **5**, 808. | 85 |
| 62JMC871 | H. Segal, C. Hedgcoth and C. G. Skinner; *J. Med. Chem.*, 1962, **5**, 871. | 129 |
| 62JMC1335 | W. E. Taft and R. G. Shepherd; *J. Med. Chem.*, 1962, **5**, 1335. | 140 |
| 62JOC185 | J. A. Carbon; *J. Org. Chem.*, 1962, **27**, 185. | 404 |
| 62JOC381 | J. H. Looker and W. W. Hanneman; *J. Org. Chem.*, 1962, **27**, 381. | 597 |
| 62JOC548 | F. C. Schaefer, K. R. Huffman and G. A. Peters; *J. Org. Chem.*, 1962, **27**, 548. | 120 |
| 62JOC551 | K. R. Huffman, F. C. Schaefer and G. A. Peters; *J. Org. Chem.*, 1962, **27**, 551. | 120 |
| 62JOC892 | R. B. Angier and W. V. Curran; *J. Org. Chem.*, 1962, **27**, 892. | 297, 309 |
| 62JOC982 | E. Dyer, M. L. Gluntz and E. J. Tanck; *J. Org. Chem.*, 1962, **27**, 982. | 86 |
| 62JOC986 | A. G. Beaman, J. F. Gerster and R. K. Robins; *J. Org. Chem.*, 1962, **27**, 986. | 93 |
| 62JOC1366 | W. V. Curran and R. B. Angier; *J. Org. Chem.*, 1962, **27**, 1366. | 305 |
| 62JOC1462 | N. Yamaoka and K. Aso; *J. Org. Chem.*, 1962, **27**, 1462. | 84 |
| 62JOC1463 | J. Allegretti, J. Hancock and R. S. Knutson; *J. Org. Chem.*, 1962, **27**, 1463. | 546 |
| 62JOC2264 | E. F. Godefroi; *J. Org. Chem.*, 1962, **27**, 2264. | 81, 82, 126, 127, 128 |

| | | |
|---|---|---|
| 62JOC2270 | T. Nogrady and K. M. Vagi; *J. Org. Chem.*, 1962, **27**, 2270. | 438 |
| 62JOC2580 | H. Schroeder, E. Kober, H. Ulrich, R. Rätz, H. Agahigian and C. Grundmann; *J. Org. Chem.*, 1962, **27**, 2580. | 63, 140 |
| 62JOC2708 | E. J. Modest, S. Chatterjee and H. Kangur; *J. Org. Chem.*, 1962, **27**, 2708. | 111 |
| 62JOC2863 | R. Bernetti, F. Mancini and C. C. Price; *J. Org. Chem.*, 1962, **27**, 2863. | 210 |
| 62JOC2945 | D. D. Bly and M. G. Mellon; *J. Org. Chem.*, 1962, **27**, 2945. | 103, 132 |
| 62JOC2991 | C. L. Stevens, K. Nagarajan and T. H. Haskell; *J. Org. Chem.*, 1962, **27**, 2991. | 147 |
| 62JOC3362 | F. C. Schaefer; *J. Org. Chem.*, 1962, **27**, 3362. | 511 |
| 62JOC3507 | H. Gershon; *J. Org. Chem.*, 1962, **27**, 3507. | 80, 127 |
| 62JOC3608 | F. C. Schaefer; *J. Org. Chem.*, 1962, **27**, 3608. | 513 |
| 62JOC3715 | P. Beak and H. Abelson; *J. Org. Chem.*, 1962, **27**, 3715. | 581 |
| 62JOC3890 | J. T. Shaw; *J. Org. Chem.*, 1962, **27**, 3890. | 494 |
| 62JOC4090 | E. F. Silversmith; *J. Org. Chem.*, 1962, **27**, 4090. | 62, 115 |
| 62JOC4211 | M. Dymicky and W. T. Caldwell; *J. Org. Chem.*, 1962, **27**, 4211. | 349 |
| 62JOC4253 | J. V. Karabinos and J. J. Hazdra; *J. Org. Chem.*, 1962, **27**, 4253. | 987 |
| 62JOC4337 | J. N. Srivastava and D. N. Chaudhury; *J. Org. Chem.*, 1962, **27**, 4337. | 834, 860 |
| 62JOC4498 | Q. E. Thompson; *J. Org. Chem.*, 1962, **27**, 4498. | 724 |
| 62JOC4704 | M. Narayana, J. F. Dash and P. D. Gardner; *J. Org. Chem.*, 1962, **27**, 4704. | 801 |
| 62JPC937 | L. D. Tuck and D. W. Schieser; *J. Phys. Chem.*, 1962, **66**, 937. | 1010 |
| 62JPR(19)56 | B. Helferich, R. Hoffmann and H. Mylenbusch; *J. Prakt. Chem.*, 1962, **19**, 56. | 1043, 1071 |
| 62LA(651)112 | F. G. Fischer and P. Neumann; *Liebigs Ann. Chem.*, 1962, **651**, 112. | 339 |
| 62LA(654)146 | R. Huisgen, J. Sauer and M. Seidel; *Liebigs Ann. Chem.*, 1962, **654**, 146. | 548 |
| 62LA(660)98 | F. Dalacker and G. Steiner; *Liebigs Ann. Chem.*, 1962, **660**, 98. | 318 |
| 62MI21300 | Z. Budesinsky, Z. Perina and J. Sluka; *Czech. Farm.*, 1962, **11**, 345. | 138 |
| 62MI21301 | E. A. Bell and R. G. Foster; *Nature (London)*, 1962, **194**, 91. | 146 |
| 62MI21302 | H. C. Bucha, W. E. Cupery, J. E. Harrod, H. M. Loux and L. M. Ellis; *Science*, 1962, **137**, 537. | 155 |
| 62MI21600 | S. Nawa and H. S. Forrest; *Nature (London)*, 1962, **196**, 169. | 289 |
| 62MI22000 | B. W. Fitzsimmons, C. Hewlett and R. A. Shaw; *Proc. Chem. Soc.*, 1962, 340. | 520 |
| B-62MI22200 | L. Jurd; in 'The Chemistry of Flavonoid Compounds', ed. T. A. Geissman; Pergamon, New York, 1962, p. 107. | 602 |
| 62MI22201 | J. W. Clark-Lewis; *Rev. Pure Appl. Chem.*, 1962, **12**, 96. | 624 |
| B-62MI22300 | 'The Chemistry of Flavonoid Compounds', ed. T. A. Geissman; Pergamon, Oxford, 1962, chapter 4. | 650, 652 |
| B-62MI22301 | 'The Chemistry of Flavonoid Compounds', ed. T. A. Geissman; Pergamon, Oxford, 1962, chapter 6. | 650, 652 |
| B-62MI22302 | 'The Chemistry of Flavonoid Compounds', ed. T. A. Geissman; Pergamon, Oxford, 1962, chapter 9. | 650, 652 |
| B-62MI22400 | T. A. Geissman; 'The Chemistry of Flavonoid Compounds', Pergamon, Oxford, 1962. | 738, 816, 821 |
| 62MI22401 | A. T. Balaban, M. Frangopol and P. T. Frangopol; *Isotopen Tech.*, 1962, **2**, 235. | 868 |
| B-62MI22600 | H. H. Jaffé and M. Orchin; 'Theory and Application of Ultraviolet Spectroscopy', Wiley, New York, 1962, p. 468. | 945 |
| B-62MI22700 | A. Butenandt and W. Schäfer; in 'Recent Progress in the Chemistry of Natural and Synthetic Colouring Matters', Academic Press, New York, 1962, p. 13. | 1037 |
| 62PIA(A)(56)71 | S. S. Dharmatti, G. Govil, C. R. Kanekar, C. L. Khetrapal and Y. P. Virmani; *Proc. Indian Acad. Sci., Sect. A*, 1962, **56**, 71. | 582 |
| 62PMH(1)161 | W. Cochran; *Phys. Methods Heterocycl. Chem.*, 1962, **1**, 161. | 620 |
| 62RCR615 | E. N. Zil'berman; *Russ. Chem. Rev. (Engl. Transl.)*, 1962, **31**, 615. | 504 |
| 62RCR712 | A. I. Finkel'shtein and E. N. Boitsov; *Russ. Chem. Rev. (Engl. Transl.)*, 1962, **31**, 712. | 460, 464 |
| 62RTC443 | A. R. Katritzky, R. G. Shepherd and A. J. Waring; *Recl. Trav. Chim. Pays-Bas*, 1962, **81**, 443. | 88 |
| 62T257 | A. T. Balaban, G. Mihai and C. D. Nenitzescu; *Tetrahedron*, 1962, **18**, 257. | 762 |
| 62T1001 | J. M. Burgess and M. S. Gibson; *Tetrahedron*, 1962, **18**, 1001. | 572 |
| 62T1079 | A. T. Balaban, M. Gavät and C. D. Nenitzescu; *Tetrahedron*, 1962, **18**, 1079. | 867 |
| 62T1095 | A. Gray and F. G. Holliman; *Tetrahedron*, 1962, **18**, 1095. | 161 |
| 62TL593 | S. A. Kagal, P. M. Nair and K. Venkataraman; *Tetrahedron Lett.*, 1962, 593. | 853 |
| 62USP3022305 | R. A. Carboni; *U.S. Pat.* 3 022 305 (1962) (*Chem. Abstr.*, 1963, **58**, 9102). | 550 |
| 62USP3035061 | R. Vonderwahl; (J. R. Geigy A/G), *U.S. Pat.* 3 035 061 (1962) (*Chem. Abstr.*, 1962, **57**, 11 213). | 232 |
| 62YZ462 | Y. Morimoto; *Yakugaku Zasshi*, 1962, **82**, 462. | 112 |
| 62YZ528 | O. Nagase, M. Hirata and M. Inaoka; *Yakugaku Zasshi*, 1962, **82**, 528. | 87 |
| 62YZ696 | Y. Inubushi and K. Nomura; *Yakugaku Zasshi*, 1962, **82**, 696 (*Chem. Abstr.*, 1963, **58**, 4545). | 973 |
| 62YZ1341 | Y. Inubushi and K. Nomura; *Yakugaku Zasshi*, 1962, **82**, 1341 (*Chem. Abstr.*, 1963, **59**, 605). | 973 |
| 62ZOB1655 | V. M. Berezovskii and A. M. Yurkevich; *Zh. Obshch. Khim.*, 1962, **32**, 1655. | 115 |
| 62ZOB1709 | Ya. A. Levin and V. A. Kukhtin; *Zh. Obshch. Khim.*, 1962, **32**, 1709. | 94 |
| 62ZOB1822 | Y. K. Yurev and O. M. Revenko; *Zh. Obshch. Khim.*, 1962, **32**, 1822. | 928 |
| 62ZOB3394 | A. Spasov and E. Golovinskii; *Zh. Obshch. Khim.*, 1962, **32**, 3394. | 565 |
| 62ZPC(329)291 | H. Rembold and H. Metzger; *Hoppe-Seyler's Z. Physiol. Chem.*, 1962, **329**, 291. | 312 |

| | | |
|---|---|---|
| 63ABC700 | M. Nakajima, H. Fukami, K. Konishi and J. Oda; *Agric. Biol. Chem.*, 1963, **27**, 700. | 851 |
| 63AC(R)36 | R. Gerdil and E. A. C. Lucken; *Ann. Chim. (Rome)*, 1963, **53**, 36. | 973 |
| 63ACS1181 | M. Davis and O. Hassel; *Acta Chem. Scand.*, 1963, **17**, 1181. | 949 |
| 63AF3 | W.-H. Wagner, H. Loewe and A. Häussler; *Arzneim.-Forsch.*, 1963, **13**, 3. | 408 |
| 63AF660 | E. Cohen; *Arzneim.-Forsch.*, 1963, **13**, 660 and the four following papers. | 153 |
| 63AG860 | K. Dimroth, W. Umbach and K. H. Blöcher; *Angew Chem.*, 1963, **75**, 860. | 868 |
| 63AG1204 | T. Kauffmann, G. Runckelshaus and J. Schulz; *Angew. Chem.*, 1963, **75**, 1204. | 548, 568 |
| 63AG(E)309 | C. Grundmann; *Angew. Chem., Int. Ed. Engl.*, 1963, **2**, 309. | 469, 470 |
| 63AG(E)394 | W. Schroth and G. Fischer; *Angew. Chem., Int. Ed. Engl.*, 1963, **2**, 394. | 866 |
| 63AG(E)655 | H. Bredereck, F. Effenberger, A. Hofmann and M. M. Hajek; *Angew. Chem., Int. Ed. Engl.*, 1963, **2**, 655. | 499 |
| 63AG(E)737 | R. Zimmermann and H. Hotze; *Angew. Chem., Int. Ed. Engl.*, 1962, **2**, 737. | 1079 |
| 63AHC(1)339 | A. R. Katritzky and J. M. Lagowski; *Adv. Heterocycl. Chem.*, 1963, **1**, 339. | 5, 66, 68, 642, 643 |
| 63AHC(2)245 | R. Gompper; *Adv. Heterocycl. Chem.*, 1963, **2**, 245. | 478 |
| 63AJC101 | K. H. Bell and H. Duewell; *Aust. J. Chem.*, 1963, **16**, 101. | 856 |
| 63AJC107 | J. W. Clark-Lewis, T. McL. Spotswood and L. R. Williams; *Aust. J. Chem.*, 1963, **16**, 107. | 631 |
| 63AJC690 | K. H. Bell and H. Duewell; *Aust. J. Chem.*, 1963, **16**, 690. | 787, 855 |
| 63AP151 | E. Profft and L. Sitter; *Arch. Pharm. (Weinheim, Ger.)*, 1963, **296**, 151. | 136 |
| 63AP298 | A. Brossi; *Arch. Pharm. (Weinheim, Ger.)*, 1963, **296**, 298. | 120 |
| 63AX310 | S. Hosoya; *Acta Crystallogr.*, 1963, **16**, 310. | 947 |
| 63AX318 | P. Cucka; *Acta Crystallogr.*, 1963, **16**, 318. | 2, 4 |
| 63CB526 | J. Goerdeler and H. W. Pohland; *Chem. Ber.*, 1963, **96**, 526. | 94 |
| 63CB534 | J. Goerdeler and W. Roth; *Chem. Ber.*, 1963, **96**, 534. | 84 |
| 63CB913 | K. R. Kutumbe and M. G. Marathey; *Chem. Ber.*, 1963, **96**, 913. | 823 |
| 63CB1505 | H. Bredereck, F. Effenberger and H. J. Treiber; *Chem. Ber.*, 1963, **96**, 1505. | 109 |
| 63CB1868 | H. Bredereck, F. Effenberger, E. Henseleit and E. H. Schweizer; *Chem. Ber.*, 1963, **96**, 1868. | 227 |
| 63CB2070 | H. Weidinger and J. Kranz; *Chem. Ber.*, 1963, **96**, 2070. | 520 |
| 63CB2950 | W. Pfleiderer and H. Fink; *Chem. Ber.*, 1963, **96**, 2950. | 273, 277 |
| 63CB2964 | W. Pfleiderer and H. Fink; *Chem. Ber.*, 1963, **96**, 2964. | 294 |
| 63CB2977 | D. Söll and W. Pfleiderer; *Chem. Ber.*, 1963, **96**, 2977. | 99, 273 |
| 63CB3260 | H. Bredereck, F. Effenberger and A. Hofmann; *Chem. Ber.*, 1963, **96**, 3260. | 499 |
| 63CB3265 | H. Bredereck, F. Effenberger and A. Hofmann; *Chem. Ber.*, 1963, **96**, 3265. | 512 |
| 63CCC2491 | C. Párkányi and F. Šorm; *Collect. Czech. Chem. Commun.*, 1963, **28**, 2491. | 80, 98 |
| 63CCC2501 | M. Prystas and J. Gut; *Collect. Czech. Chem. Commun.*, 1963, **28**, 2501. | 113, 130 |
| 63CCC2527 | J. Gut; *Collect. Czech. Chem. Commun.*, 1963, **28**, 2527. | 409 |
| 63CJC505 | D. Cook; *Can. J. Chem.*, 1963, **41**, 505. | 595, 637 |
| 63CJC522 | D. Cook; *Can. J. Chem.*, 1963, **41**, 522. | 597 |
| 63CPB1042 | I. Iwai and J. Ide; *Chem. Pharm. Bull.*, 1963, **11**, 1042. | 598, 742, 744 |
| 63CR(256)1542 | O. Riobé, V. Hérault and L. Gouin; *C.R. Hebd. Seances Acad. Sci.*, 1963, **256**, 1542. | 847 |
| 63CR(256)3323 | J. Brugidou and H. Christol; *C.R. Hebd. Seances Acad. Sci.*, 1963, **256**, 3323. | 784 |
| 63CR(257)926 | H. Strzelecka, M. Simalty-Siemiatycki and C. Prévost; *C.R. Hebd. Seances Acad. Sci.*, 1963, **257**, 926. | 871 |
| 63CRV65 | J. H. Day; *Chem. Rev.*, 1963, **63**, 65. | 880 |
| 63FOR275 | L. Crombie; *Fortschr. Chem. Org. Naturst.*, 1963, **21**, 275. | 718 |
| 63FRP1324339 | Laboratoires de Carbo-Synthese, *Fr. Pat.* 1 324 339 (1963) (*Chem. Abstr.*, 1963, **59**, 10 086). | 448 |
| 63G339 | G. Palamidessi and L. Bernardi; *Gazz. Chim. Ital.*, 1963, **93**, 339. | 172 |
| 63G576 | S. Fatutta; *Gazz. Chim. Ital.*, 1963, **93**, 576. | 212, 230 |
| 63HCA49 | A. Jacot-Guillarmond and A. Piguet; *Helv. Chim. Acta*, 1963, **46**, 49. | 597 |
| 63HCA51 | M. Viscontini and H. Stierlin; *Helv. Chim. Acta*, 1963, **46**, 51. | 289 |
| 63HCA333 | P. Schudel, H. Mayer, J. Metzger, R. Rüegg and O. Isler; *Helv. Chim. Acta*, 1963, **46**, 333. | 779 |
| 63HCA650 | H. Mayer, P. Schudel, R. Rüegg and O. Isler; *Helv. Chim. Acta*, 1963, **46**, 650. | 733 |
| 63HCA1181 | M. Viscontini, L. Merlini and W. von Philipsborn; *Helv. Chim. Acta*, 1963, **46**, 1181. | 289 |
| 63HCA1537 | M. Viscontini and M. Piraux; *Helv. Chim. Acta*, 1963, **46**, 1537. | 284 |
| 63HCA2597 | L. Merlini, W. von Philipsborn and M. Viscontini; *Helv. Chim. Acta*, 1963, **46**, 2597. | 276 |
| 63IJC20 | M. Venugopalan and C. B. Anderson; *Indian J. Chem.*, 1963, **1**, 20. | 605 |
| 63IJC346 | G. S. Sidhu, G. Thyagarajan and N. Rao; *Indian J. Chem.*, 1963, **1**, 346. | 124 |
| 63IZV2215 | N. S. Vul'fson, V. I. Zaretskii and V. G. Zaikin; *Izv. Akad. Nauk SSSR, Ser. Khim.*, 1963, 2215. | 608 |
| 63JA207 | G. Stork, A. Brizzolara, H. Landesman, J. Szmuszkovicz and R. Terrell; *J. Am. Chem. Soc.*, 1963, **85**, 207. | 772 |
| 63JA239 | B. O. Linn, C. H. Shunk, E. L. Wong and K. Folkers; *J. Am. Chem. Soc.*, 1963, **85**, 239. | 752 |
| 63JA1178 | A. F. Wagner, P. E. Wittreich, B. Arison, N. R. Trenner and K. Folkers; *J. Am. Chem. Soc.*, 1963, **85**, 1178. | 598 |

| | | |
|---|---|---|
| 63JA1208 | P. Yates and I. W. J. Still; *J. Am. Chem. Soc.*, 1963, **85**, 1208. | 693 |
| 63JA2144 | L. A. Carpino; *J. Am. Chem. Soc.*, 1963, **85**, 2144. | 44 |
| 63JA2956 | P. Yates and M. J. Jorgenson; *J. Am. Chem. Soc.*, 1963, **85**, 2956. | 693 |
| 63JA3529 | E. F. Ullman; *J. Am. Chem. Soc.*, 1963, **85**, 3529. | 645, 833 |
| 63JA3971 | D. Rosenthal, P. Grabowich, E. Sabo and J. Fried; *J. Am. Chem. Soc.*, 1963, **85**, 3971. | 798 |
| 63JA4024 | T. Ueda and J. J. Fox; *J. Am. Chem. Soc.*, 1963, **85**, 4024. | 362 |
| 63JBC(238)1116 | W. S. McNutt, Jr.; *J. Biol. Chem.*, 1963, **238**, 1116. | 308, 322 |
| 63JCS174 | M. H. Palmer and D. S. Urch; *J. Chem. Soc.*, 1963, 174. | 869 |
| 63JCS308 | J. F. D. Mills and S. C. Nyburg; *J. Chem. Soc.*, 1963, 308. | 621 |
| 63JCS811 | B. Chatamra and A. S. Jones; *J. Chem. Soc.*, 1963, 811. | 109 |
| 63JCS1276 | D. J. Brown and J. S. Harper; *J. Chem. Soc.*, 1963, 1276. | 65, 135 |
| 63JCS1628 | C. M. Atkinson and H. D. Cossey; *J. Chem. Soc.*, 1963, 1628. | 406, 413 |
| 63JCS1773 | S. Matsuura and M. Goto; *J. Chem. Soc.*, 1963, 1773. | 265, 285 |
| 63JCS2094 | L. W. Deady, R. D. Topsom and J. Vaughan; *J. Chem. Soc.*, 1963, 2094. | 783 |
| 63JCS2097 | J. B. Caldwell, B. Milligan and J. M. Swan; *J. Chem. Soc.*, 1963, 2097. | 1042 |
| 63JCS2256 | N. B. Chapman, K. Clarke and K. Wilson; *J. Chem. Soc.*, 1963, 2256. | 153 |
| 63JCS2648 | Y. Inoue and D. D. Perrin; *J. Chem. Soc.*, 1963, 2648. | 265 |
| 63JCS3046 | A. R. Katritzky and A. J. Waring; *J. Chem. Soc.*, 1963, 3046. | 68 |
| 63JCS3539 | M. S. Gibson; *J. Chem. Soc.*, 1963, 3539. | 374 |
| 63JCS3729 | F. Bergmann, A. Kalmus, H. Ungar-Waron and H. Kwietny-Govrin; *J. Chem. Soc.*, 1963, 3729. | 115 |
| 63JCS3764 | G. W. H. Cheeseman, A. R. Katritzky and B. J. Ridgewell; *J. Chem. Soc.*, 1963, 3764. | 161 |
| 63JCS4304 | W. L. F. Armarego; *J. Chem. Soc.*, 1963, 4304. | 340, 351 |
| 63JCS4483 | M. A. Butt and J. A. Elvidge; *J. Chem. Soc.*, 1963, 4483. | 581, 811, 812 |
| 63JCS5151 | A. Albert, Y. Inoue and D. D. Perrin; *J. Chem. Soc.*, 1963, 5151. | 265, 272 |
| 63JCS5156 | A. Albert and G. B. Barlin; *J. Chem. Soc.*, 1963, 5156. | 249, 250, 251, 252, 254, 258 |
| 63JCS5166 | Y. Inoue and D. D. Perrin; *J. Chem. Soc.*, 1963, 5166. | 206, 251 |
| 63JCS5322 | S. A. N. N. Bokhari and W. B. Whalley; *J. Chem. Soc.*, 1963, 5322. | 855 |
| 63JCS5426 | A. O. Fitton and G. R. Ramage; *J. Chem. Soc.*, 1963, 5426. | 817 |
| 63JCS5590 | J. F. W. McOmie and A. B. Turner; *J. Chem. Soc.*, 1963, 5590. | 84, 133 |
| 63JCS5642 | G. W. Miller and F. L. Rose; *J. Chem. Soc.*, 1963, 5642. | 113 |
| 63JCS5737 | A. Albert and G. B. Barlin; *J. Chem. Soc.*, 1963, 5737. | 251, 254 |
| 63JCS5893 | C. A. Coulson; *J. Chem. Soc.*, 1963, 5893. | 58 |
| 63JCS5949 | R. I. Reed and J. M. Wilson; *J. Chem. Soc.*, 1963, 5949. | 618 |
| 63JCS6073 | W. L. F. Armarego; *J. Chem. Soc.*, 1963, 6073. | 241, 246 |
| 63JGU1806 | V. A. Zagorevskii and É. K. Orlova; *J. Gen. Chem. USSR (Engl. Transl.)*, 1963, **33**, 1806. | 711, 809 |
| 63JMC36 | H. G. Mautner, S.-H. Chu, J. J. Jaffe and A. C. Sartorelli; *J. Med. Chem.*, 1963, **6**, 36. | 101 |
| 63JMC39 | J. A. Barone; *J. Med. Chem.*, 1963, **6**, 39. | 115 |
| 63JMC58 | W. T. Caldwell, W. Fiddler and N. J. Santora; *J. Med. Chem.*, 1963, **6**, 58. | 71 |
| 63JMC550 | T. V. Rajkumar and S. B. Binkley; *J. Med. Chem.*, 1963, **6**, 550. | 104, 126 |
| 63JMC646 | E. F. Elslager, D. B. Capps, D. H. Kurtz, L. M. Werbel and D. F. Worth; *J. Med. Chem.*, 1963, **6**, 646. | 131 |
| 63JMC688 | I. Wempen and J. J. Fox; *J. Med. Chem.*, 1963, **6**, 688. | 92, 99, 133 |
| 63JOC34 | H. O. House and M. Schellenbaum; *J. Org. Chem.*, 1963, **28**, 34. | 847 |
| 63JOC85 | R. W. Cummins; *J. Org. Chem.*, 1963, **28**, 85. | 478 |
| 63JOC509 | E. C. Taylor, J. E. Loeffler and B. Mencke; *J. Org. Chem.*, 1963, **28**, 509. | 113 |
| 63JOC577 | W. E. Parham, C. G. Fritz, R. W. Soeder and R. M. Dodson; *J. Org. Chem.*, 1963, **28**, 577. | 753 |
| 63JOC687 | E. M. Kosower and T. S. Sorensen; *J. Org. Chem.*, 1963, **28**, 687. | 577, 845, 848 |
| 63JOC798 | K. C. Dewhirst and F. F. Rust; *J. Org. Chem.*, 1963, **28**, 798. | 780 |
| 63JOC923 | C. Temple, Jr., R. L. McKee and J. A. Montgomery; *J. Org. Chem.*, 1963, **28**, 923. | 85 |
| 63JOC1187 | I. J. Pachter and P. E. Nemeth; *J. Org. Chem.*, 1963, **28**, 1187. | 304, 314 |
| 63JOC1191 | I. J. Pachter; *J. Org. Chem.*, 1963, **28**, 1191. | 315, 314 |
| 63JOC1197 | I. J. Pachter, P. E. Nemeth and A. J. Villani, *J. Org. Chem.*, 1963, **28**, 1197. | 281, 314 |
| 63JOC1203 | I. J. Pachter and P. E. Nemeth; *J. Org. Chem.*, 1963, **28**, 1203. | 304 |
| 63JOC1443 | T. B. Windholz, L. H. Peterson and G. J. Kent; *J. Org. Chem.*, 1963, **28**, 1443. | 792 |
| 63JOC1509 | R. B. Angier; *J. Org. Chem.*, 1963, **28**, 1509. | 277, 297, 305, 309 |
| 63JOC1812 | K. R. Huffman and F. C. Schaefer; *J. Org. Chem.*, 1963, **28**, 1812. | 495 |
| 63JOC1816 | K. R. Huffman and F. C. Schaefer; *J. Org. Chem.*, 1963, **28**, 1816. | 502 |
| 63JOC1983 | P. Stenbuck, R. Baltzly and H. M. Hood; *J. Org. Chem.*, 1963, **28**, 1983. | 152 |
| 63JOC1994 | H. M. Gilow and J. Jacobus; *J. Org. Chem.*, 1963, **28**, 1994. | 97, 123 |
| 63JOC2304 | J. A. Montgomery and H. J. Thomas; *J. Org. Chem.*, 1963, **28**, 2304. | 131 |
| 63JOC2488 | J. Hamer and R. E. Holliday; *J. Org. Chem.*, 1963, **28**, 2488. | 178 |
| 63JOC2686 | W. E. Parham and M. D. Bhavsar; *J. Org. Chem.*, 1963, **28**, 2686. | 977 |
| 63JOC2933 | D. R. Osborne and R. Levine; *J. Org. Chem.*, 1963, **28**, 2933. | 474 |
| 63JOC3519 | A. Mustafa, W. Asker, A. K. Mansour, H. A. A. Zaher and A. R. Eloui; *J. Org. Chem.*, 1963, **28**, 3519. | 406 |

| | | |
|---|---|---|
| 63MI21300 | Y. M. Abu-Zeid, Z. Abu-Elela and K. M. Ghoneim; *J. Pharm. Sci. U.A.R.*, 1963, **4**, 41. | 113 |
| 63MI21301 | C. I. Chappel and C. von Seemann; *Prog. Med. Chem.*, 1963, **3**, 89. | 150 |
| 63MI21600 | J. M. Lagowski, H. S. Forrest and H. C. S. Wood; *Proc. Chem. Soc.*, 1963, 343. | 315 |
| 63MI21900 | J. Hadacek and E. Kisa; *Spisy Prirodoved. Fak. Univ. Brno*, 1963, **439**, 1. | 411, 417 |
| 63MI21901 | K. Koermendy, P. Sohar and J. Volford; *Ann. Univ. Sci. Budapest, Rolando Eotvos Nominatae, Sect. Chim.*, 1963, **5**, 117 (*Chem. Abstr.*, 1964, **60**, 13 243). | 438 |
| 63MI22000 | A. Kreutzberger; *Fortschr. Chem. Forsch.*, 1963, **4**, 273. | 461, 469, 471 |
| B-63MI22200 | J. B. Harborne; in 'Methods of Polyphenol Chemistry', ed. J. B. Pridham; Pergamon, Oxford, 1963, p. 13. | 602 |
| B-63MI22201 | R. U. Lemieux; in 'Molecular Rearrangements', ed. P. de Mayo; Wiley, New York, 1963, p. 723. | 629 |
| B-63MI22300 | F. M. Dean; 'Naturally Occurring Oxygen Ring Compounds', Butterworths, London, 1963, chapter 13. | 652 |
| B-63MI22301 | F. M. Dean; 'Naturally Occurring Oxygen Ring Compounds', Butterworths, London, 1963. | 666, 693 |
| 63MI22302 | C. Belil, J. Pascual and F. Serratosa; *Anales Real Soc. Espan. Fis. Quim., Ser. B*, 1963, **59**, 507 (*Chem. Abstr.*, 1964, **60**, 5439). | 682 |
| B-63MI22400 | F. M. Dean; 'Naturally Occurring Oxygen Ring Compounds', Butterworth, London, 1963. | 738, 741, 861, 874, 877, 878 |
| B-63MI22401 | T. A. Geissman; in 'Comprehensive Biochemistry', ed. M. Florkin and E. M. Stotz; Elsevier, Amsterdam, 1963, vol. 9, p. 213. | 877 |
| 63MI22600 | E. A. C. Lucken; *Theor. Chim. Acta*, 1963, **1**, 397. | 946 |
| 63MI22601 | V. Busetti, M. Mammi and G. Carazzolo; *Z. Kristallogr.*, 1963, **119**, 310. | 950 |
| 63N403 | F. Eiden and B. S. Nagar; *Naturwissenschaften*, 1963, **50**, 403. | 121 |
| 63OR(13)55 | W. E. Parham and E. E. Schweizer; *Org. React.*, 1963, **13**, 55. | 725 |
| 63OSC(4)201 | R. H. Wiley and N. R. Smith; *Org. Synth., Coll. Vol.*, 1963, **4**, 201. | 792 |
| 63OSC(4)311 | R. I. Longley, Jr., W. S. Emerson and A. J. Blardinelli; *Org. Synth., Coll. Vol.*, 1963, **4**, 311. | 772 |
| 63OSC(4)478 | T. S. Wheeler; *Org. Synth., Coll. Vol.*, 1963, **4**, 478. | 817 |
| 63OSC(4)549 | N. R. Smith and R. H. Wiley; *Org. Synth., Coll. Vol.*, 1963, **4**, 549. | 792 |
| 63PMH(1)1 | A. Albert; *Phys. Methods Heterocycl. Chem.*, 1963, **1**, 1. | 61 |
| 63PMH(1)177 | W. Pfleiderer; *Phys. Methods Heterocycl. Chem.*, 1963, **1**, 177. | 286 |
| 63PMH(1)189 | S. Walker; *Phys. Methods Heterocycl. Chem.*, 1963, **1**, 189. | 626 |
| 63PMH(2)1 | S. F. Mason; *Phys. Methods Heterocycl. Chem.*, 1963, **2**, 1. | 65, 249, 598 |
| 63PMH(2)161 | A. R. Katritzky and A. P. Ambler; *Phys. Methods Heterocycl. Chem.*, 1963, **2**, 161. | 6 |
| 63PMH(2)242 | A. R. Katritzky and A. P. Ambler; *Phys. Methods Heterocycl. Chem.*, 1963, **2**, 242. | 593 |
| 63PMH(2)252 | A. R. Katritzky and A. P. Ambler; *Phys. Methods Heterocycl. Chem.*, 1963, **2**, 252. | 634 |
| 63PMH(2)254 | A. R. Katritzky and A. P. Ambler; *Phys. Methods Heterocycl. Chem.*, 1963, **2**, 254. | 638 |
| 63PMH(2)257 | A. R. Katritzky and A. P. Ambler; *Phys. Methods Heterocycl. Chem.*, 1963, **2**, 257. | 638 |
| 63RTC845 | H. J. T. Bos and J. F. Arens; *Recl. Trav. Chim. Pays-Bas*, 1963, **82**, 845. | 870 |
| 63SA1625 | C. Stammer and A. Taurins; *Spectrochimica Acta*, 1963, **19**, 1625. | 161 |
| 63SA2099 | P. Lebreton and J. Chopin; *Spectrochim. Acta*, 1963, **19**, 2099. | 597 |
| 63T77 | H. B. Bhat and K. Venkataraman; *Tetrahedron*, 1963, **19**, 77. | 852 |
| 63T85 | A. K. Bose and S. Garratt; *Tetrahedron*, 1963, **19**, 85. | 113 |
| 63T471 | A. K. Chandra; *Tetrahedron*, 1963, **19**, 471. | 945 |
| 63T839 | C. S. Barnes, M. I. Strong and J. L. Occolowitz; *Tetrahedron*, 1963, **19**, 839. | 754, 763 |
| 63T1011 | S. C. Pakrashi, J. Bhattacharyya, L. F. Johnson and H. Budzikiewicz; *Tetrahedron*, 1963, **19**, 1011. | 148 |
| 63T(S2)157 | D. S. Sappenfield and M. Kreevoy; *Tetrahedron*, 1963, **19**, Supplement 2, 157. | 945 |
| 63TL91 | A. T. Balaban; *Tetrahedron Lett.*, 1963, 91. | 868 |
| 63TL891 | U. K. Pandit and I. P. Dirk; *Tetrahedron Lett.*, 1963, 891. | 608 |
| 63TL1267 | W. Sandermann and R. Casten; *Tetrahedron Lett.*, 1963, 1267. | 674 |
| 63USP3086016 | R. M. Hunt and W. V. Hough; *U.S. Pat.* 3 086 016 (1963) (*Chem. Abstr.*, 1963, **59**, 10 092). | 569 |
| 63YZ169 | H. Moriyama; *Yakugaku Zasshi*, 1963, **83**, 169. | 119 |
| 63YZ1086 | J. Kinugawa, M. Ochiai and H. Yamamoto; *Yakugaku Zasshi*, 1963, **83**, 1086. | 95, 102 |
| 63ZC266 | G. Fischer and W. Schroth; *Z. Chem.*, 1963, **3**, 266. | 866 |
| 63ZN(B)420 | W. Pfleiderer; *Z. Naturforsch., Teil B*, 1963, **18**, 420. | 283 |
| 63ZN(B)757 | H. Simon, F. Weygand, J. Walter, H. Wacker and K. Schmidt; *Z. Naturforsch., Teil B*, 1963, **18**, 757. | 320 |
| 63ZOB2673 | Ya. A. Levin, N. A. Gul'kina and V. A. Kukhtin; *Zh. Obshch. Khim.*, 1963, **33**, 2673. | 94 |
| 63ZOB2848 | Yu. P. Shvachkin and L. A. Sirtsova; *Zh. Obshch. Khim.*, 1963, **33**, 2848. | 113 |
| 63ZOB3132 | Yu. P. Shvachkin and M. K. Berestenko; *Zh. Obshch. Khim.*, 1963, **33**, 3132. | 103 |
| 63ZOR3434 | N. P. Shusherina, N. D. Dmitrieva, E. A. Luk'yanets and R. Ya. Levina; *Zh. Org. Khim.*, 1963, **33**, 3434. | 679 |
| | | |
| 64ACH(40)217 | M. Vajda and F. Ruff; *Acta Chim. Acad. Sci. Hung.*, 1964, **40**, 217. | 598 |
| 64ACS2345 | O. Foss, K. Johnsen and T. Reistad; *Acta Chem. Scand.*, 1964, **18**, 2345. | 950 |
| 64AF1004 | B. Stanovnik and M. Tišler; *Arzneim.-Forsch.*, 1964, **14**, 1004. | 152 |
| 64AG(E)114 | W. Pfleiderer; *Angew. Chem., Int. Ed. Engl.*, 1964, **3**, 114. | 264, 322 |

| | | |
|---|---|---|
| 64AHC(3)285 | G. Illuminati; *Adv. Heterocycl. Chem.*, 1964, **3**, 285. | 483 |
| 64AJC353 | G. M. Badger, P. Cheuychit and W. H. F. Sasse; *Aust. J. Chem.*, 1964, **17**, 353. | 973 |
| 64AJC366 | G. M. Badger, P. Cheuychit and W. H. F. Sasse; *Aust. J. Chem.*, 1964, **17**, 366. | 973 |
| 64AJC428 | T. J. Batterham and R. J. Highet; *Aust. J. Chem.*, 1964, **17**, 428. | 584 |
| 64AJC567 | D. J. Brown and T. Teitei; *Aust. J. Chem.*, 1964, **17**, 567. | 68, 92 |
| 64AJC794 | D. J. Brown and J. M. Lyall; *Aust. J. Chem.*, 1964, **17**, 794. | 99 |
| 64AJC975 | C. S. Barnes and J. L. Occolowitz; *Aust. J. Chem.*, 1964, **17**, 975. | 604, 608, 610, 614, 616, 617 |
| 64AJC1305 | T. J. Batterham and J. A. Lamberton; *Aust. J. Chem.*, 1964, **17**, 1305. | 581 |
| 64AK(22)65 | S. Gronowitz, B. Norrman, B. Gestblom, B. Mathiasson and R. A. Hoffman; *Ark. Kemi*, 1964, **22**, 65. | 62, 67 |
| 64AX472 | J. F. McConnell, A. McL. Mathieson and B. P. Schoenborn; *Acta Crystallogr.*, 1964, **17**, 472. | 624 |
| 64BBR(17)177 | A. R. Brenneman and S. Kaufman; *Biochem. Biophys. Res. Commun.*, 1964, **17**, 177. | 324 |
| 64BBR(17)461 | F. Bergmann, H. Burger-Rachamimov and J. Galanter; *Biochem. Biophys. Res. Commun.*, 1964, **17**, 461. | 287 |
| 64BEP641818 | Farbenfabriken Bayer A.-G.; *Belg. Pat.* 641 818 (1964) (*Chem. Abstr.*, 1965, **62**, 16 276). | 377 |
| 64BJ(90)76 | S. G. Laland and G. Serck-Hanssen; *Biochem. J.*, 1964, **90**, 76. | 91 |
| 64BRP957797 | H. M. Hood; *Br. Pat.* 957 797 (1964) (*Chem. Abstr.*, 1964, **61**, 3122). | 152 |
| 64BSF936 | S. David, B. Estramareix, H. Hirshfeld and P. Sinay; *Bull. Soc. Chim. Fr.*, 1964, 936. | 91 |
| 64BSF1492 | A. Hinnen and J. Dreux; *Bull. Soc. Chim. Fr.*, 1964, 1492. | 577, 593, 642 |
| 64CB80 | A. Roedig and S. Schödel; *Chem. Ber.*, 1964, **97**, 80. | 825 |
| 64CB566 | C. Grundmann and M. B. Fulton; *Chem. Ber.*, 1964, **97**, 566. | 454 |
| 64CB610 | L. Farkas, L. Hörhammer, H. Wagner, H. Rösler and R. Gurniak; *Chem. Ber.*, 1964, **97**, 610. | 824 |
| 64CB682 | R. Merten and G. Müller; *Chem. Ber.*, 1964, **97**, 682. | 594, 782 |
| 64CB994 | J. Gante and W. Lautsch; *Chem. Ber.*, 1964, **97**, 994. | 448 |
| 64CB1002 | F. Weygand, H. Simon, K. D. Keil and H. Millauer; *Chem. Ber.*, 1964, **97**, 1002. | 312 |
| 64CB1163 | F. Kröhnke, E. Schmidt and W. Zecher; *Chem. Ber.*, 1964, **97**, 1163. | 118, 126 |
| 64CB1266 | G. Opitz and F. Zimmermann; *Chem. Ber.*, 1964, **97**, 1266. | 797 |
| 64CB1959 | E. Winterfeldt; *Chem. Ber.*, 1964, **97**, 1959. | 759 |
| 64CB3407 | H. Bredereck, R. Sell and F. Effenberger; *Chem. Ber.*, 1964, **97**, 3407. | 80, 125 |
| 64CB3456 | F. Weygand and B. Spiess; *Chem. Ber.*, 1964, **97**, 3456. | 310 |
| 64CCC1394 | J. Gut, J. Jonáš and J. Piťha; *Collect. Czech. Chem. Commun.*, 1964, **29**, 1394. | 391 |
| 64CI(L)418 | H. N. Schlein, M. Israel, S. Chatterjee and E. J. Modest; *Chem. Ind. (London)*, 1964, 418. | 95 |
| 64CPB43 | E. Hayashi and T. Higashino; *Chem. Pharm. Bull.*, 1964, **12**, 43. | 83, 84 |
| 64CPB307 | S. Fukushima, A. Ueno and Y. Akahori; *Chem. Pharm. Bull.*, 1964, **12**, 307. | 717 |
| 64CPB393 | A. Takamizawa and K. Hirai; *Chem. Pharm. Bull.*, 1964, **12**, 393. | 114, 122 |
| 64CPB558 | A. Takamizawa, K. Hirai, Y. Hamashima and M. Hata; *Chem. Pharm. Bull.*, 1964, **12**, 558. | 140 |
| 64CR(258)1526 | M. Wakselman and M. Vilkas; *C.R. Hebd. Seances Acad. Sci.*, 1964, **258**, 1526. | 784 |
| 64CR(259)1645 | D. Molho and J. Aknin; *C.R. Hebd. Seances Acad. Sci.*, 1964, **259**, 1645. | 826 |
| 64CRV229 | R. D. Barry; *Chem. Rev.*, 1964, **64**, 229. | 676, 830 |
| 64E202 | F. Šorm, A. Piskala, A. Čihák and J. Veselý; *Experientia*, 1964, **20**, 202. | 526 |
| 64FES1050 | E. Cingolani, G. Bellomonte and A. Sordi; *Farmaco Ed. Sci.*, 1964, **19**, 1050. | 82 |
| 64FRP1335759 | W. Lorenz; *Fr. Pat.* 1 335 759 (*Chem. Abstr.*, 1964, **60**, 558). | 45 |
| 64G595 | G. Pala and A. Mantegani; *Gazz. Chim. Ital.*, 1964, **94**, 595. | 108 |
| 64G606 | G. Bianchetti, P. Dalla Croce and D. Pocar; *Gazz. Chim. Ital.*, 1964, **94**, 606. | 135 |
| 64HCA602 | G. Ohloff, K.-H. Schulte-Elte and B. Willhalm; *Helv. Chim. Acta*, 1964, **47**, 602. | 776 |
| 64HCA1883 | D. Hauser, K. Schaffner and O. Jeger; *Helv. Chim. Acta*, 1964, **47**, 1883. | 776 |
| 64HCA2087 | M. Viscontini and A. Bobst; *Helv. Chim. Acta*, 1964, **47**, 2087. | 281 |
| 64HCA2195 | M. Viscontini, L. Merlini, G. Nasini, W. von Philipsborn and M. Piraux; *Helv. Chim. Acta*, 1964, **47**, 2195. | 288 |
| 64JA308 | L. Goodman, J. De Graw, R. L. Kisliuk, M. Friedkin, E. J. Pastore, E. J. Crawford, L. T. Plante, A. Al-Nahas, J. F. Morningstar, Jr., G. Kwok, L. Wilson, E. F. Donovan and J. Ratzan; *J. Am. Chem. Soc.*, 1964, **86**, 308. | 327 |
| 64JA1869 | T. K. Liao, E. G. Podrebarac and C. C. Cheng; *J. Am. Chem. Soc.*, 1964, **86**, 1869. | 104 |
| 64JA2084 | E. S. Hand and R. M. Horowitz; *J. Am. Chem. Soc.*, 1964, **86**, 2084. | 748 |
| 64JA2935 | C. R. Johnson and D. McCants, Jr.; *J. Am. Chem. Soc.*, 1964, **86**, 2935. | 889 |
| 64JA3814 | E. F. Ullman and J. E. Milks; *J. Am. Chem. Soc.*, 1964, **86**, 3814. | 645 |
| 64JA3833 | P. Beak, T. H. Kinstle and G. A. Carls; *J. Am. Chem. Soc.*, 1964, **86**, 3833. | 610, 611 |
| 64JBC(239)332 | S. Kaufman; *J. Biol. Chem.*, 1964, **239**, 332. | 306, 322 |
| 64JBC(239)2910 | T. Nagatsu, M. Levitt and S. Udenfriend; *J. Biol. Chem.*, 1964, **239**, 2910. | 324 |
| 64JBC(239)4081 | A. Tietz, M. Lindberg and E. P. Kennedy; *J. Biol. Chem.*, 1964, **239**, 4081. | 324 |
| 64JBC(239)4272 | W. S. McNutt and S. P. Damle; *J. Biol. Chem.*, 1964, **239**, 4272. | 308 |
| 64JCP(41)2041 | F. A. Bovey, F. P. Hood, III, E. W. Anderson and R. L. Kornegay; *J. Chem. Phys.*, 1964, **41**, 2041. | 962 |
| 64JCS565 | F. Bergmann, M. Tamari and H. Ungar-Waron; *J. Chem. Soc.*, 1964, 565. | 317 |

| 64JCS1001 | J. H. Chesterfield, D. T. Hurst, J. F. W. McOmie and M. S. Tute; *J. Chem. Soc.*, 1964, 1001. | 70, 88, 131, 133, 144 |
|---|---|---|
| 64JCS1020 | W. Hewertson, R. A. Shaw and B. C. Smith; *J. Chem. Soc.*, 1964, 1020. | 483 |
| 64JCS1646 | A. T. Balaban, G. R. Bedford and A. R. Katritzky; *J. Chem. Soc.*, 1964, 1646. | 585 |
| 64JCS1666 | A. Albert and J. Clark; *J. Chem. Soc.*, 1964, 1666. | 266 |
| 64JCS2666 | G. Tennant; *J. Chem. Soc.*, 1964, 2666. | 170 |
| 64JCS2825 | J. W. Clark-Lewis and R. P. Singh; *J. Chem. Soc.*, 1964, 2825. | 251 |
| 64JCS3204 | D. J. Brown and T. Teitei; *J. Chem. Soc.*, 1964, 3204. | 89, 107, 135 |
| 64JCS3357 | A. Albert and E. P. Serjeant; *J. Chem. Soc.*, 1964, 3357. | 279, 283 |
| 64JCS3459 | F. Kurzer and E. D. Pitchfork; *J. Chem. Soc.*, 1964, 3459. | 493 |
| 64JCS3663 | M. W. Partridge and M. F. G. Stevens; *J. Chem. Soc.*, 1964, 3663. | 377, 382 |
| 64JCS4769 | A. Stuart, D. W. West and H. C. S. Wood; *J. Chem. Soc.*, 1964, 4769. | 276, 315 |
| 64JCS4920 | J. Clark; *J. Chem. Soc.*, 1964, 4920. | 288, 291 |
| 64JCS5200 | J. D. Edwards, J. E. Page and M. Pianka; *J. Chem. Soc.*, 1964, 5200. | 581 |
| 64JGU2089 | A. S. Elina and L. G. Tsrul'nikova; *J. Gen. Chem. USSR (Engl. Transl.)*, 1964, **34**, 2089. | 167 |
| 64JGU2699 | N. S. Vul'fson, T. N. Podrezova and L. B. Senyavina; *J. Gen. Chem. USSR*, 1964, **34**, 2699. | 597, 601 |
| 64JHC23 | W. Pfleiderer and D. Söll; *J. Heterocycl. Chem.*, 1964, **1**, 23. | 282 |
| 64JHC128 | D. R. Osborne and R. Levine; *J. Heterocycl. Chem.*, 1964, **1**, 128. | 472 |
| 64JHC130 | G. D. Daves, D. E. O'Brien, L. R. Lewis and C. C. Cheng; *J. Hetrocycl. Chem.*, 1964, **1**, 130. | 77, 87 |
| 64JHC145 | D. R. Osborne, W. T. Wieder and R. Levine; *J. Heterocycl. Chem.*, 1964, **1**, 145. | 120, 472, 474 |
| 64JHC175 | R. H. Wiley, J. Lanet and K. H. Hussung; *J. Heterocycl. Chem.*, 1964, **1**, 175. | 86 |
| 64JHC201 | T. J. Schwan and H. Tieckelmann; *J. Heterocycl. Chem.*, 1964, **1**, 201. | 83, 103 |
| 64JMC5 | M. Israel, H. K. Protopapa, H. N. Schlein and E. J. Modest; *J. Med. Chem.*, 1964, **7**, 5. | 138 |
| 64JMC207 | I. Wempen and J. J. Fox; *J. Med. Chem.*, 1964, **7**, 207. | 101 |
| 64JMC240 | I. A. Kaye; *J. Med. Chem.*, 1964, **7**, 240. | 252 |
| 64JMC337 | C. Piantadosi, V. G. Skulason, J. L. Irvin, J. M. Powell and L. Hall; *J. Med. Chem.*, 1964, **7**, 337. | 113 |
| 64JMC364 | N. P. Buu-Hoi, R. Rips and C. Derappe; *J. Med. Chem.*, 1964, **7**, 364. | 86 |
| 64JMC792 | M. Israel, H. K. Protopapa, H. N. Schlein and E. J. Modest; *J. Med. Chem.*, 1964, **7**, 792. | 137 |
| 64JMC812 | T. A. Martin, A. G. Wheeler, R. F. Majewski and J. R. Corrigan; *J. Med. Chem.*, 1964, **7**, 812. | 134 |
| 64JMC814 | C. Runti and C. Nisi; *J. Med. Chem.*, 1964, **7**, 814. | 566, 569 |
| 64JOC219 | H. C. Scarborough; *J. Org. Chem.*, 1964, **29**, 219. | 123, 202, 205, 229 |
| 64JOC291 | A. F. McKay, J.-M. Billy and E. J. Tarlton; *J. Org. Chem.*, 1964, **29**, 291. | 1041, 1067, 1078 |
| 64JOC332 | L. H. Sternbach, E. Reeder, A. Stempel and A. I. Rachlin; *J. Org. Chem.*, 1964, **29**, 332. | 79 |
| 64JOC668 | D. L. Trepanier, V. Sprancmanis and K. G. Wiggs; *J. Org. Chem.*, 1964, **29**, 668. | 1072 |
| 64JOC678 | C. Grundmann and V. Mini; *J. Org. Chem.*, 1964, **29**, 678. | 472 |
| 64JOC734 | J. A. Montgomery and N. F. Wood; *J. Org. Chem.*, 1964, **29**, 734. | 251 |
| 64JOC943 | D. D. Bly; *J. Org. Chem.*, 1964, **29**, 943. | 85 |
| 64JOC1110 | E. D. Weil; *J. Org. Chem.*, 1964, **29**, 1110. | 952 |
| 64JOC1115 | G. deStevens, B. Smolinsky and L. Dorfman; *J. Org. Chem.*, 1964, **29**, 1115. | 135, 139 |
| 64JOC1527 | F. C. Schaefer and J. H. Ross; *J. Org. Chem.*, 1964, **29**, 1527. | 473 |
| 64JOC1740 | A. Takamizawa, K. Hirai, Y. Sato and K. Tori; *J. Org. Chem.*, 1964, **29**, 1740. | 62 |
| 64JOC1762 | T. Ueda and J. J. Fox; *J. Org. Chem.*, 1964, **29**, 1762. | 338, 349, 362 |
| 64JOC1905 | J. B. Wright; *J. Org. Chem.*, 1964, **29**, 1905. | 1079 |
| 64JOC2116 | E. C. Taylor and E. E. Garcia; *J. Org. Chem.*, 1964, **29**, 2116. | 122, 227 |
| 64JOC2211 | W. E. Parham, L. Christensen, S. H. Groen and R. M. Dodson; *J. Org. Chem.*, 1964, **29**, 2211. | 896, 904 |
| 64JOC2513 | R. H. Hasek, P. G. Gott and J. C. Martin; *J. Org. Chem.*, 1964, **29**, 2513. | 797 |
| 64JOC2534 | E. Wenkert, D. B. R. Johnston and K. G. Dave; *J. Org. Chem.*, 1964, **29**, 2534. | 688 |
| 64JOC2602 | L. Jurd; *J. Org. Chem.*, 1964, **29**, 2602. | 650 |
| 64JOC2670 | J. D. Fissekis, A. Myles and G. B. Brown; *J. Org. Chem.*, 1964, **29**, 2670. | 90 |
| 64JOC2674 | R. B. Trattner, G. B. Elion, G. H. Hitchings and D. M. Sharefkin; *J. Org. Chem.*, 1964, **29**, 2674. | 210 |
| 64JOC2678 | P. Beak and G. A. Carls; *J. Org. Chem.*, 1964, **29**, 2678. | 582, 595, 696, 811 |
| 64JOC2682 | D. W. Mayo, P. J. Sapienza, R. C. Lord and W. D. Phillips; *J. Org. Chem.*, 1964, **29**, 2682. | 582 |
| 64JOC2717 | A. A. Santilli and T. S. Osdene; *J. Org. Chem.*, 1964, **29**, 2717. | 1074 |
| 64JOC2903 | D. M. Mulvey, S. G. Cottis and H. Tieckelmann; *J. Org. Chem.*, 1964, **29**, 2903. | 213 |
| 64JOC3370 | J. J. McCormack and H. G. Mautner; *J. Org. Chem.*, 1964, **29**, 3370. | 95, 136 |
| 64JOC3514 | R. L. Vaulx, W. H. Puterbaugh and C. R. Hauser; *J. Org. Chem.*, 1964, **29**, 3514. | 859 |
| 64JOC3610 | C. M. Baugh and E. Shaw; *J. Org. Chem.*, 1964, **29**, 3610. | 304, 311 |
| 64JPR(26)43 | J. Klosa; *J. Prakt. Chem.*, 1964, **26**, 43. | 81 |
| 64JSP(14)190 | J. D. Simmons, K. K. Innes and G. M. Begun; *J. Mol. Spectrosc.*, 1964, **14**, 190. | 161 |
| 64LA(673)78 | W. Pfleiderer and F. Sági; *Liebigs Ann. Chem.*, 1964, **673**, 78. | 121 |

| | | |
|---|---|---|
| 64LA(673)82 | H. Bredereck, F. Effenberger and G. Rainer; *Liebigs Ann. Chem.*, 1964, **673**, 82. | 121 |
| 64LA(673)153 | H. Ballweg; *Liebigs Ann. Chem.*, 1964, **673**, 153. | 81, 91, 99, 133 |
| 64LA(675)151 | R. Gompper, H. Euchner and H. Kast; *Liebigs Ann. Chem.*, 1964, **675**, 151. | 136, 137 |
| 64LA(675)189 | A. W. Wagner and G. Reinöhl; *Liebigs Ann. Chem.*, 1964, **675**, 189. | 1059, 1070 |
| 64LA(676)121 | W. Ried and A. Czack; *Liebigs Ann. Chem.*, 1964, **676**, 121. | 441 |
| 64LA(678)183 | K. Dimroth, K. Wolf and H. Kroke; *Liebigs Ann. Chem.*, 1964, **678**, 183. | 593, 653, 658, 666, 670 |
| 64LA(678)202 | K. Dimroth, H. Kroke and K. Wolf; *Liebigs Ann. Chem.*, 1964, **678**, 202. | 667 |
| 64M207 | H. Bretschneider, J. Dehler and W. Klötzer; *Monatsh. Chem.*, 1964, **95**, 207. | 134 |
| 64M265 | W. Klötzer; *Monatsh. Chem.*, 1964, **95**, 265. | 81 |
| 64M457 | R. Kuhn and H. Trischmann; *Monatsh. Chem.*, 1964, **95**, 457. | 542, 547, 548 |
| 64M1057 | E. Ziegler and T. Kappe; *Monatsh. Chem.*, 1964, **95**, 1057. | 149 |
| 64M1729 | W. Klötzer; *Monatsh. Chem.*, 1964, **95**, 1729. | 109 |
| B-64MI21300 | T. S. Osdene; in 'Pteridine Chemistry', ed. W. Pfleiderer and E. C. Taylor; Pergamon, Oxford, 1964, p. 65. | 88 |
| 64MI21301 | C. E. Hoffmann, J. W. McGahen and P. B. Sweetser; *Nature (London)*, 1964, **202**, 577. | 103, 128, 155 |
| 64MI21302 | S. Ogawa and S. Kasahara; *Vitamin*, 1963, **28**, 238 (*Chem. Abstr.*, 1964, **60**, 5494). | 133 |
| 64MI21303 | S. Somasekhara and S. L. Mukherjee; *Current Sci. (India)*, 1963, **32**, 547 (*Chem. Abstr.*, 1964, **60**, 8031). | 138 |
| 64MI21304 | J. J. Fox, Y. Kuwada, K. A. Watanabe, T. Ueda and E. B. Whipple; *Antimicrob. Agents Chemotherap.*, 1964, 518. | 147 |
| 64MI21305 | S. J. Ball; *J. Comp. Path. Therap.*, 1964, **74**, 487. | 154 |
| 64MI21500 | R. T. Coutts; *Can. Pharm. J. Sci. Sect.*, 1964, **97**, 32 (*Chem. Abstr.*, 1964, **61**, 8273). | 240, 244 |
| 64MI21501 | R. T. Coutts, D. Noble and D. G. Wibberley; *J. Pharm. Pharmacol.*, 1964, **16**, 773 (*Chem. Abstr.*, 1965, **62**, 7757). | 254 |
| B-64MI21502 | E. C. Taylor, M. J. Thompson and W. Pfleiderer; in 'Proceedings of the 3rd International Symposium on Pteridine Chemistry, Stuttgart, 1962', ed. W. Pfleiderer and E. C. Taylor; Macmillan, New York, 1964, p. 181 (Chem. Abstr., 1965, **62**, 16 245). | 252 |
| 64MI21600 | G. G. Hagerman; *Fed. Proc.*, 1964, **23**, 480. | 324 |
| B-64MI21601 | W. Pfleiderer and E. C. Taylor; 'Pteridine Chemistry', Pergamon, Oxford, 1964, p. 3. | 264 |
| B-64MI21602 | T. S. Osdene; in 'Pteridine Chemistry', ed. W. Pfleiderer and E. C. Taylor; Pergamon, Oxford, 1964, p. 65. | 313 |
| B-64MI21603 | I. J. Pachter; in 'Pteridine Chemistry', ed. W. Pfleiderer and E. C. Taylor; Pergamon, Oxford, 1964, p. 47. | 314 |
| B-64MI21604 | H. N. Gutman; in 'Pteridine Chemistry', ed. W. Pfleiderer and E. C. Taylor; Pergamon, Oxford, 1964, p. 255. | 322 |
| B-64MI21605 | P. Schmidt, K. Eichenberger and M. Wilhelm; in 'Pteridine Chemistry', ed. W. Pfleiderer and E. C. Taylor, Jr.; Pergamon, Oxford, 1964, p. 29. | 317 |
| B-64MI21606 | J. Winestock and V. D. Wiebelhaus; in 'Pteridine Chemistry', ed. W. Pfleiderer and E. C. Taylor, Jr.; Pergamon, Oxford, 1964, p. 37. | 324, 325 |
| 64MI21900 | C. Cogrossi, B. Mariani and R. Sgarbi; *Chim. Ind. (Milan)*, 1964, **46**, 530. | 408 |
| 64MI21901 | W. E. Hahn and H. Zawadzka; *Rocz. Chem.*, 1964, **38**, 557. | 455 |
| 64MI22000 | G. C. Verschoor; *Nature (London)*, 1964, **202**, 1206. | 461 |
| 64MI22001 | R. Hayatsu; *Science*, 1964, **146**, 1291. | 524 |
| B-64MI22300 | K. Dimroth and K. H. Wolf; in 'Newer Methods of Preparative Organic Chemistry', ed. W. Foerst; Academic Press, New York, 1964, vol. 3, p. 357. | 654, 655, 658, 659 |
| 64MI22400 | B. Berkoz, L. Cúellar, R. Grezemkovsky, N. V. Avila and A. D. Cross; *Proc. Chem. Soc.*, 1964, 215. | 799 |
| B-64MI22401 | K. Dimroth and K. H. Wolf; in 'Newer Methods of Preparative Organic Chemistry', ed. W. Foerst; Academic Press, New York, 1964, vol. 3, p. 357. | 861, 883 |
| B-64MI22402 | J. H. Richards and J. B. Hendrickson; 'The Biosynthesis of Steroids, Terpenes and Acetogenins', Benjamin, New York, 1964, p. 160. | 877 |
| 64MI22500 | V. S. Foldi and W. Sweeny; *Makromol. Chem.*, 1964, **72**, 208. | 939 |
| 64MI22800 | H. Thielemann, H.-A. Schlotter and M. Becke-Goehring; *Z. Anorg. Allg. Chem.*, 1964, **329**, 235. | 1047, 1058 |
| 64PAC(9)49 | R. B. Woodward; *Pure Appl. Chem.*, 1964, **9**, 49. | 148 |
| 64QR347 | A. B. Turner; *Q. Rev., Chem. Soc.* 1964, **18**, 347. | 742 |
| 64RCR92 | V. I. Mur; *Russ. Chem. Rev. (Engl. Transl.)*, 1964, **33**, 92. | 482, 483, 484 |
| 64RTC31 | M. J. D. van Dam; *Recl. Trav. Chim. Pays-Bas*, 1964, **83**, 31. | 792 |
| 64RTC39 | M. J. D. van Dam and F. Kögl; *Recl. Trav. Chim. Pays-Bas*, 1964, **83**, 39. | 792 |
| 64SA397 | K. M. Sancier, A. P. Brady and W. W. Lee; *Spectrochim. Acta*, 1964, **20**, 397. | 464 |
| 64SA871 | C. T. Mathis and J. H. Goldstein; *Spectrochim. Acta*, 1964, **20**, 871. | 581, 582, 583, 588 |
| 64T119 | A. T. Balaban, M. Mocanu and Z. Simon; *Tetrahedron*, 1964, **20**, 119. | 603, 861 |
| 64T483 | M. Kamel and H. Shoeb; *Tetrahedron*, 1964, **20**, 483. | 664 |
| 64T831 | P. Beak; *Tetrahedron*, 1964, **20**, 831. | 633 |
| 64T1173 | D. Gürne, T. Urbański, M. Witanowski, B. Karniewska and L. Stefaniak; *Tetrahedron*, 1964, **20**, 1173. | 1009 |
| 64T1185 | B. Willhalm, A. F. Thomas and F. Gautschi; *Tetrahedron*, 1964, **20**, 1185. | 598, 603, 604 |

| | | |
|---|---|---|
| 64T1317 | J. S. P. Schwarz, A. I. Cohen, W. D. Ollis, E. A. Kaczka and L. M. Jackman; *Tetrahedron*, 1964, **20**, 1317. | 672 |
| 64T2091 | J. C. Anderson, D. G. Lindsay and C. B. Reese; *Tetrahedron*, 1964, **20**, 2091. | 577 |
| 64T2859 | P. Bassignana and C. Cogrossi; *Tetrahedron*, 1964, **20**, 2859. | 596 |
| 64TL19 | M. Ogata, H. Watanabe, K. Tori and H. Kano; *Tetrahedron Lett.*, 1964, 19. | 141 |
| 64TL513 | A. C. Waiss, Jr., R. E. Lundin and D. J. Stern; *Tetrahedron Lett.*, 1964, 513. | 584 |
| 64TL1569 | J. Rigaudy, G. Caquis and J.-B. Lafont; *Tetrahedron Lett.*, 1964, 1569. | 997 |
| 64TL2829 | D. Martin; *Tetrahedron Lett.*, 1964, 2829. | 503 |
| 64YZ207 | I. Saikawa; *Yakugaku Zasshi*, 1964, **84**, 207. | 112 |
| 64ZC270 | W. Schroth and U. Schmidt; *Z. Chem.*, 1964, **4**, 270. | 984 |
| 64ZOB407 | Yu. P. Shvachkin and M. T. Azarova; *Zh. Obshch. Khim.*, 1964, **34**, 407. | 146 |
| 64ZOB1321 | N. V. Khromov-Borisov and G. M. Kheifets; *Zh. Obshch. Khim.*, 1964, **34**, 1321. | 132, 134 |
| 64ZOB2159 | Yu. P. Shvachkin and L. A. Syrtsova; *Zh. Obshch. Khim.*, 1964, **34**, 2159. | 134 |
| 64ZOB2504 | N. P. Shusherina, E. A. Luk'yanets and R. Y. Levina; *Zh. Obshch. Khim.*, 1964, **34**, 2504. | 799 |
| 64ZOB2577 | K. A. Chkhikvadze and O. Yu. Magidson; *Zh. Obshch. Khim.*, 1964, **34**, 2577. | 100 |
| 64ZOB3427 | L. B. Dashkevich and E. S. Korbelainen; *Zh. Obshch. Khim.*, 1964, **34**, 3427 (Chem. Abstr., 1965, **62**, 4030). | 1026 |
| 64ZOB3506 | Yu. P. Shvachkin and M. K. Berestenko; *Zh. Obshch. Khim.*, 1964, **34**, 3506. | 146 |
| 64ZOB3791 | V. G. Samolovova, T. V. Gortinskaya and M. N. Shchukina; *Zh. Obshch. Khim.*, 1964, **34**, 3791 (*Chem. Abstr.*, 1965, **62**, 5272). | 1011 |
| 64ZOB3851 | G. M. Kheifets and N. V. Khromov-Borisov; *Zh. Obshch. Khim.*, 1964, **34**, 3851. | 140 |
| | | |
| 65ACS217 | H. Hope; *Acta Chem. Scand.*, 1965, **19**, 217. | 622 |
| 65ACS370 | E. Honkanen, T. Moisio and P. Karvonen; *Acta Chem. Scand.*, 1965, **19**, 370. | 616 |
| 65ACS1741 | S. Gronowitz and J. Röe; *Acta Chem. Scand.*, 1965, **19**, 1741. | 104, 105 |
| 65AF613 | G. Muačević, H. Stötzer and H. Wick; *Arzneim.-Forsch.*, 1965, **15**, 613. | 153 |
| 65AG913 | E. Grigat and R. Pütter; *Angew. Chem.*, 1965, **77**, 913. | 117 |
| 65AG(E)292 | K. Dury; *Angew. Chem., Int. Ed. Engl.*, 1965, **4**, 292. | 2 |
| 65AG(E)312 | H. R. Hensel and G. Lutzel; *Angew. Chem., Int. Ed. Engl.*, 1965, **4**, 312. | 529 |
| 65AG(E)1075 | W. Pfleiderer and W. Hutzenlaub; *Angew. Chem., Int. Ed. Engl.*, 1965, **4**, 1075. | 305 |
| 65AJC199 | D. J. Brown and T. Teitei; *Aust. J. Chem.*, 1965, **18**, 199. | 139 |
| 65AJC559 | D. J. Brown and T. Teitei; *Aust. J. Chem.*, 1965, **18**, 559. | 101 |
| 65AJC707 | P. J. Black and M. L. Heffernan; *Aust. J. Chem.*, 1965, **18**, 707. | 7, 160 |
| 65AJC1811 | D. J. Brown and J. M. Lyall; *Aust. J. Chem.*, 1965, **18**, 1811. | 99 |
| 65AX(19)927 | J. Gaultier and C. Hauw; *Acta Crystallogr.*, 1965, **19**, 927. | 622 |
| 65BCJ1586 | K. Fukui, F. Tanimoto and H. Kitano; *Bull. Chem. Soc. Jpn.*, 1965, **38**, 1586. | 508 |
| 65BJ(96)533 | J. C. MacDonald; *Biochem. J.*, 1965, **96**, 533. | 191 |
| 65BSB119 | F. Declerck, R. Degroote, J. de Lannoy, R. Nasielski-Hinckens and J. Nasielski; *Bull. Soc. Chim. Belges*, 1965, **74**, 119. | 462 |
| 65BSF779 | H. Pacheco and A. Grouiller; *Bull. Soc. Chim. Fr.*, 1965, 779. | 597 |
| 65BSF1358 | F. Weiss, A. Isard and R. Bensa; *Bull. Soc. Chim. Fr.*, 1965, 1358. | 978 |
| 65BSF1944 | M. Simalty-Siemiatycki and R. Fugnitto; *Bull. Soc. Chim. Fr.*, 1965, 1944. | 869 |
| 65BSF3025 | J. Aknin and D. Molho; *Bull. Soc. Chim. Fr.*, 1965, 3025. | 610 |
| 65BSF3476 | J. Wieman, N. Vinot and M. Villadary; *Bull. Soc. Chim. Fr.*, 1965, 3476. | 185 |
| 65CB334 | R. R. Schmidt; *Chem. Ber.*, 1965, **98**, 334. | 866, 870, 1052, 1061, 1081 |
| 65CB1455 | H. Böhme and O. Müller; *Chem. Ber.*, 1965, **98**, 1455. | 992 |
| 65CB1476 | R. Huisgen, E. Aufderhaar and G. Wallbillich; *Chem. Ber.*, 1965, **98**, 1476. | 543 |
| 65CB1498 | O. Dann and H. Hofmann; *Chem. Ber.*, 1965, **98**, 1498. | 856 |
| 65CB1505 | A. Dornow and D. Wille; *Chem. Ber.*, 1965, **98**, 1505. | 215, 222 |
| 65CB2576 | H. Baganz, J. Rüger and J. Kohtz; *Chem. Ber.*, 1965, **98**, 2576. | 114 |
| 65CB3081 | F. Bohlmann and K.-M. Kleine; *Chem. Ber.*, 1965, **98**, 3081. | 965 |
| 65CB3102 | C. H. Krauch, S. Farid and G. O. Schenck; *Chem. Ber.*, 1965, **98**, 3102. | 671 |
| 65CB3279 | J. N. Chatterjea, B. K. Banerjee and H. C. Jha; *Chem. Ber.*, 1965, **98**, 3279. | 766 |
| 65CB3353 | A. Haas and D. Y. Oh; *Chem. Ber.*, 1965, **98**, 3353. | 508 |
| 65CB3724 | H. König, H. Metzger and K. Seelert; *Chem. Ber.*, 1965, **98**, 3724. | 1033 |
| 65CCC3016 | R. Zahradník and C. Párkányi; *Collect. Czech. Chem. Commun.*, 1965, **30**, 3016. | 575 |
| 65CCC3730 | Z. Buděšínský, F. Roubínek and E. Svátek; *Collect. Czech. Chem. Commun.*, 1965, **30**, 3730. | 89, 107, 137 |
| 65CI(L)182 | A. M. Roe and J. B. Harbridge; *Chem. Ind. (London)*, 1965, 182. | 1058 |
| 65CPB586 | Y. Nitta, I. Matsuura and F. Yoneda; *Chem. Pharm. Bull.*, 1965, **13**, 586. | 236, 238, 239, 241, 242, 261 |
| 65FES259 | E. Cingolani, G. Bellomonte and A. Sordi; *Farmaco Ed. Sci.*, 1965, **20**, 259. | 82 |
| 65HCA764 | M. Viscontini and S. Huwyler; *Helv. Chim. Acta*, 1965, **48**, 764. | 279, 306 |
| 65HCA816 | M. Viscontini and A. Bobst; *Helv. Chim. Acta*, 1965, **48**, 816. | 281, 308 |
| 65IZV2087 | O. A. Zagulyaeva and V. P. Mamaev; *Izv. Akad. Nauk SSSR, Ser. Khim.*, 1965, 2087. | 103 |
| 65JA1109 | C. R. Johnson and D. McCants, Jr.; *J. Am. Chem. Soc.*, 1965, **87**, 1109. | 889 |
| 65JA3022 | W. H. Pirkle; *J. Am. Chem. Soc.*, 1965, **87**, 3022. | 607 |
| 65JA4569 | J. M. Rice, G. O. Dudek and M. Barber; *J. Am. Chem. Soc.*, 1965, **87**, 4569. | 66 |
| 65JA5404 | C. R. Johnson and D. McCants, Jr.; *J. Am. Chem. Soc.*, 1965, **87**, 5404. | 896 |
| 65JA5424 | W. A. Henderson, Jr. and E. F. Ullman; *J. Am. Chem. Soc.*, 1965, **87**, 5424. | 695 |

| | | |
|---|---|---|
| 65JA5439 | B. W. Roberts, J. B. Lambert and J. D. Roberts; *J. Am. Chem. Soc.*, 1965, **87**, 5439. | 63, 68 |
| 65JA5575 | E. D. Becker, H. T. Miles and R. B. Bradley; *J. Am. Chem. Soc.*, 1965, **87**, 5575. | 63 |
| 65JA5661 | K. E. Pfitzner and J. G. Moffatt; *J. Am. Chem. Soc.*, 1965, **87**, 5661. | 856 |
| 65JBC(240)1692 | R. G. Micetich and J. C. MacDonald; *J. Biol. Chem.*, 1965, **240**, 1692. | 191 |
| 65JCS27 | A. Albert and J. Clark; *J. Chem. Soc.*, 1965, 27. | 273, 283, 291 |
| 65JCS208 | D. M. Brown and P. Schell; *J. Chem. Soc.*, 1965, 208. | 130, 145 |
| 65JCS571 | M. J. Aroney, R. J. W. Le Fèvre and J. D. Saxby; *J. Chem. Soc.*, 1965, 571. | 947 |
| 65JCS623 | S. Matsuura and T. Goto; *J. Chem. Soc.*, 1965, 623. | 98, 265 |
| 65JCS755 | D. J. Brown and T. Teitei; *J. Chem. Soc.*, 1965, 755. | 67, 98 |
| 65JCS1175 | D. J. Brown and N. W. Jacobsen; *J. Chem. Soc.*, 1965, 1175. | 270, 273, 291 |
| 65JCS1515 | J. A. Hill and W. J. LeQuesne; *J. Chem. Soc.*, 1965, 1515. | 144 |
| 65JCS1530 | D. J. Brown and B. T. England; *J. Chem. Soc.*, 1965, 1530. | 295, 311 |
| 65JCS1558 | J. D. Hepworth and E. Tittensor; *J. Chem. Soc.*, 1965, 1558. | 255, 258 |
| 65JCS2743 | H. H. Lee and C. H. Tan; *J. Chem. Soc.*, 1965, 2743. | 828 |
| 65JCS2778 | W. L. F. Armarego; *J. Chem. Soc.*, 1965, 2778. | 89 |
| 65JCS2901 | B. Milligan and J. M. Swan; *J. Chem. Soc.*, 1965, 2901. | 992 |
| 65JCS3357 | G. W. Miller and F. L. Rose; *J. Chem. Soc.*, 1965, 3357. | 116 |
| 65JCS3610 | I. M. Lockhart and E. M. Tanner; *J. Chem. Soc.*, 1965, 3610. | 700, 828 |
| 65JCS3770 | D. J. Brown and N. W. Jacobsen; *J. Chem. Soc.*, 1965, 3770. | 85, 130 |
| 65JCS3882 | P. S. Bramwell and A. O. Fitton; *J. Chem. Soc.*, 1965, 3882. | 786, 817 |
| 65JCS3987 | J. M. Carpenter and G. Shaw; *J. Chem. Soc.*, 1965, 3987. | 96 |
| 65JCS4240 | W. J. Irwin and D. G. Wibberley; *J. Chem. Soc.*, 1965, 4240. | 215, 222 |
| 65JCS4911 | D. J. Brown and R. V. Foster; *J. Chem. Soc.*, 1965, 4911. | 92 |
| 65JCS5049 | R. Binns, W. D. Cotterill and R. Livingstone; *J. Chem. Soc.*, 1965, 5049. | 630 |
| 65JCS5060 | D. McHale and J. Green; *J. Chem. Soc.*, 1965, 5060. | 752 |
| 65JCS5230 | N. Bacon, A. J. Boulton, R. T. C. Brownlee, A. R. Katritzky and R. D. Topsom; *J. Chem. Soc.*, 1965, 5230. | 391 |
| 65JCS5360 | W. L. F. Armarego and J. I. C. Smith; *J. Chem. Soc.*, 1965, 5360. | 124 |
| 65JCS5452 | E. H. Amonoo-Neizer, R. C. Golesworthy, R. A. Shaw and B. C. Smith; *J. Chem. Soc.*, 1965, 5452. | 483 |
| 65JCS5467 | M. P. L. Caton, M. S. Grant, D. L. Pain and R. Slack; *J. Chem. Soc.*, 1965, 5467. | 104 |
| 65JCS5718 | L. W. Deady, R. D. Topsom and J. Vaughan; *J. Chem. Soc.*, 1965, 5718. | 783 |
| 65JCS6100 | A. Rose, N. P. Buu-Hoï and P. Jacquignon; *J. Chem. Soc.*, 1965, 6100. | 832 |
| 65JCS6296 | F. Kurzer and E. D. Pitchfork; *J. Chem. Soc.*, 1965, 6296. | 493 |
| 65JCS6592 | C. A. Coulson and H. Looyenga; *J. Chem. Soc.*, 1965, 6592. | 466, 493 |
| 65JCS6695 | M. D. Mehta, D. Miller and E. F. Mooney; *J. Chem. Soc.*, 1965, 6695. | 62, 72 |
| 65JCS6858 | A. J. Bloodworth and A. G. Davies; *J. Chem. Soc.*, 1965, 6858. | 463, 511 |
| 65JCS6930 | A. Albert and J. J. McCormack; *J. Chem. Soc.*, 1965, 6930. | 279, 287, 288 |
| 65JCS7116 | D. T. Hurst, J. F. W. McOmie and J. B. Searle; *J. Chem. Soc.*, 1965, 7116. | 89 |
| 65JCS7348 | J. Hill and G. R. Ramage; *J. Chem. Soc.*, 1965, 7348. | 714 |
| 65JHC1 | W. J. Haggerty, Jr., R. H. Springer and C. C. Cheng; *J. Heterocycl. Chem.*, 1965, **2**, 1. | 80, 103, 140 |
| 65JHC49 | R. H. Springer, W. J. Haggerty, Jr. and C. C. Cheng; *J. Heterocycl. Chem.*, 1965, **2**, 49. | 77 |
| 65JHC157 | R. F. Smith, P. C. Briggs, R. A. Kent, J. A. Albright and E. J. Walsh; *J. Heterocycl. Chem.*, 1965, **2**, 157. | 124 |
| 65JHC162 | B. R. Baker and J. H. Jordaan; *J. Heterocycl. Chem.*, 1965, **2**, 162. | 115, 119 |
| 65JHC202 | T. J. Schwan and H. Tieckelmann; *J. Heterocycl. Chem.*, 1965, **2**, 202. | 84, 128 |
| 65JHC447 | L. Bauer, G. E. Wright, B. A. Mikrut and C. L. Bell; *J. Heterocycl. Chem.*, 1965, **2**, 447. | 67, 95 |
| 65JIC211 | N. K. Bose and D. N. Chaudhury; *J. Indian Chem. Soc.*, 1965, **42**, 211. | 833 |
| 65JMC750 | T. J. Schwan, H. Tieckelmann, J. F. Holland and B. Bryant; *J. Med. Chem.*, 1965, **8**, 750. | 84, 149 |
| 65JOC115 | B. J. Whitlock, S. H. Lipton and F. M. Strong; *J. Org. Chem.*, 1965, **30**, 115. | 146 |
| 65JOC203 | G. Vogel; *J. Org. Chem.*, 1965, **30**, 203. | 682 |
| 65JOC312 | L. L. Woods and J. Sapp; *J. Org. Chem.*, 1965, **30**, 312. | 804 |
| 65JOC318 | J. F. Geldard and F. Lions; *J. Org. Chem.*, 1965, **30**, 318. | 565 |
| 65JOC408 | R. M. Cresswell, H. K. Maurer, T. Strauss and G. B. Brown; *J. Org. Chem.*, 1965, **30**, 408. | 282 |
| 65JOC702 | H. Bader, E. R. Ruckel, F. X. Markley, C. G. Santangelo and P. Schickedantz; *J. Org. Chem.*, 1965, **30**, 702. | 484, 513, 514 |
| 65JOC826 | C. Temple, Jr. and J. A. Montgomery; *J. Org. Chem.*, 1965, **30**, 826. | 87 |
| 65JOC829 | C. Temple, Jr., R. L. McKee and J. A. Montgomery; *J. Org. Chem.*, 1965, **30**, 829. | 104 |
| 65JOC835 | T. Y. Shen, H. M. Lewis and W. V. Ruyle; *J. Org. Chem.*, 1965, **30**, 835. | 153 |
| 65JOC897 | H. Aft; *J. Org. Chem.*, 1965, **30**, 897. | 829 |
| 65JOC930 | H. Bader; *J. Org. Chem.*, 1965, **30**, 930. | 516 |
| 65JOC955 | R. H. Shapiro and C. Djerassi; *J. Org. Chem.*, 1965, **30**, 955. | 610 |
| 65JOC1684 | J. A. Durden and D. G. Crosby; *J. Org. Chem.*, 1965, **30**, 1684. | 865 |
| 65JOC1837 | E. J. Modest, S. Chatterjee and H. K. Protopapa; *J. Org. Chem.*, 1965, **30**, 1837. | 111 |
| 65JOC1844 | T. Goto, A. Tatematsu and S. Matsuura; *J. Org. Chem.*, 1965, **30**, 1844. | 285 |
| 65JOC2035 | J. C. Randall, R. L. Vaux, M. E. Hobbs and C. R. Hauser; *J. Org. Chem.*, 1965, **30**, 2035. | 631 |

| | | |
|---|---|---|
| 65JOC2290 | A. Takamizawa and K. Hirai; *J. Org. Chem.*, 1965, **30**, 2290. | 121 |
| 65JOC2398 | J. L. Wong, M. S. Brown and H. Rapoport; *J. Org. Chem.*, 1965, **30**, 2398. | 79, 81, 127 |
| 65JOC2642 | G. A. Berchtold, G. R. Harvey and G. E. Wilson, Jr.; *J. Org. Chem.*, 1965, **30**, 2642. | 797 |
| 65JOC2703 | Q. E. Thomson; *J. Org. Chem.*, 1965, **30**, 2703. | 991 |
| 65JOC2766 | T. J. Delia, M. J. Olsen and G. B. Brown; *J. Org. Chem.*, 1965, **30**, 2766. | 145 |
| 65JOC3153 | E. C. Taylor and A. McKillop; *J. Org. Chem.*, 1965, **30**, 3153. | 131 |
| 65JOC3960 | J. B. Wright; *J. Org. Chem.*, 1965, **30**, 3960. | 1041, 1079 |
| 65JOC4114 | C. E. Cook, R. C. Corley and M. E. Wall; *J. Org. Chem.*, 1965, **30**, 4114. | 800, 801 |
| 65JOC4343 | L. L. Woods and D. Johnson; *J. Org. Chem.*, 1965, **30**, 4343. | 804 |
| 65JOC4346 | H. Rösler, T. J. Mabry, M. F. Cranmer and J. Kagan; *J. Org. Chem.*, 1965, **30**, 4346. | 585 |
| 65JOU983 | A. N. Narkevich, G. N. Dorofeenko and Yu. A. Zhdanov; *J. Org. Chem. USSR (Engl. Transl.)*, 1965, **1**, 983. | 664 |
| 65JOU1936 | V. A. Zagorevskii, V. L. Savel'ev and S. L. Portnova; *J. Org. Chem. USSR (Engl. Transl.)*, 1965, **1**, 1936. | 685 |
| 65JPC1 | J. Jonáš, W. Derbyshire and H. Gutowsky; *J. Phys. Chem.*, 1965, **69**, 1. | 638, 639 |
| 65JPC3925 | G. E. Maciel and G. B. Savitzky; *J. Phys. Chem.*, 1965, **69**, 3925. | 586 |
| 65JPS714 | B. R. Baker and J. K. Coward; *J. Pharm. Sci.*, 1965, **54**, 714. | 130 |
| 65JPS1626 | M. Israel, H. K. Protopapa, S. Chatterjee and E. J. Modest; *J. Pharm. Sci.*, 1965, **54**, 1626. | 88 |
| 65LA(684)209 | H.-R. Schütte and W. Woltersdorf; *Liebigs Ann. Chem.*, 1965, **684**, 209. | 146 |
| 65LA(684)212 | H. W. Moore and K. Folkers; *Liebigs Ann. Chem.*, 1965, **684**, 212. | 752 |
| 65LA(685)167 | O. Dann and G. Volz; *Liebigs Ann. Chem.*, 1965, **685**, 167. | 710, 856 |
| 65LA(686)134 | F. Lingens and H. Schneider-Bernlöhr; *Liebigs Ann. Chem.*, 1965, **686**, 134. | 145 |
| 65LA(687)191 | C. Grundmann, V. Mini, J. M. Dean and M. D. Frommeld; *Liebigs Ann. Chem.*, 1965, **687**, 191. | 506 |
| 65M220 | G. Mixich and A. Zinke; *Monatsh. Chem.*, 1965, **96**, 220. | 734 |
| 65M1567 | W. Klötzer and M. Herberz; *Monatsh. Chem.*, 1965, **96**, 1567. | 71, 85 |
| 65M1677 | A. Grüssner, M. Montavon and O. Schnider; *Monatsh. Chem.*, 1965, **96**, 1677. | 107 |
| 65MI21300 | S. Y. Wang and R. Alcantara; *Photochem. Photobiol.*, 1965, **4**, 477. | 73 |
| 65MI21301 | V. P. Mamaev; *Biol. Aktivn., Soedin. Akad. Nauk SSR*, 1965, 38 (*Chem. Abstr.*, 1965, **63**, 18 081). | 117 |
| 65MI21302 | S. H. Chang, I. K. Kim and B.-S.Hahn; *Daehan Hwahak Hwoejee*, 1965, **9**, 75 (*Chem. Abstr.*, 1966, **64**, 17 588). | 133 |
| 65MI21303 | Y. Ashani and S. Cohen; *Isr. J. Chem.*, 1965, **3**, 101. | 140 |
| 65MI21304 | T. L. V. Ulbricht; *Prog. Nucleic Acid Res.*, 1965, **4**, 189. | 145 |
| 65MI21305 | G. H. Hitchings and J. J. Burchall; *Adv. Enzymol.*, 1965, **27**, 417. | 151 |
| 65MI21500 | G. Favini, I. Vandoni and M. Simonetta; *Theor. Chim. Acta*, 1965, **3**, 45 (*Chem. Abstr.*, 1965, **62**, 8525). | 204, 214 |
| 65MI21501 | G. Favini, I. Vandoni and M. Simonetta; *Theor. Chim. Acta*, 1965, **3**, 418 (*Chem. Abstr.*, 1965, **63**, 17 326). | 204 |
| 65MI21600 | G. T. Ross, D. P. Hertz, M. B. Lipsett and W. D. O'Dell; *Am. J. Obstet. Gynecol.*, 1965, **93**, 223. | 327 |
| 65MI21601 | S. Nakamura, A. Ichiyama and O. Hayashi; *Fed. Proc.*, 1965, **24**, 604. | 324 |
| 65MI21602 | M. Levitt, S. Spector, A. Sjoerdsma and S. Udenfriend; *J. Pharmacol. Exp. Therap.*, 1965, **148**, 1. | 306 |
| 65MI21603 | V. D. Wiebelhaus, J. Winestock, A. R. Maass, F. T. Brennan, G. Sosnowski and T. Larsen; *J. Pharmacol. Exp. Therap.*, 1965, **149**, 397. | 325 |
| B-65MI21604 | I. Ziegler; in 'Ergebnisse der Physiologie', ed. K. Kramer; Springer Verlag, Berlin, 1965, vol. 56, p. 1. | 322 |
| 65MI21605 | I. Thompson, T. C. Hall and W. C. Moloney, *New Engl. J. Med.*, 1965, **273**, 1302. | 327 |
| 65MI21606 | J. H. Burchenal; *Ser. Haematol.*, 1965, **1**, 47. | 327 |
| 65MI21900 | J. Hadacek and J. Slouka; *Folia Fac. Sci. Natur. Univ. Purkynianae Brunensis*, 1965. | 385 |
| 65MI22200 | L. Degani, L. Lunazzi and F. Taddei; *Boll. Sci. Fac. Chim. Ind. Bologna*, 1965, **23**, 131. | 577 |
| 65MI22201 | S. Sasaki and T. Kurokawa; *Kagaku (Kyoto)*, 1965, **20**, 1070 (*Chem. Abstr.*, 1966, **64**, 12 509). | 613 |
| 65MI22202 | W. H. McFadden, E. A. Day and M. J. Diamond; *Anal. Chem.*, 1965, **37**, 89. | 616 |
| 65MI22300 | A. Meidell; *Medd. Norsk. Farm. Selskap*, 1965, **27**, 101. | 732 |
| 65MI22500 | J. Degani, R. Fochi and C. Vincenzi; *Boll. Sci. Fac. Chim. Ind. Bologna*, 1965, **23**, 21. | 893 |
| 65MI22501 | I. A. Rapoport and L. M. Filippova; *Byul. Mosk. Obshchestva Ispytatelei Priorody, Otd. Biol.*, 1965, **70**, 117 (*Chem. Abstr.*, 1966, **64**, 4101). | 940 |
| 65MI22502 | Y. I. Usatenko, A. M. Arishkevich and A. G. Akhmetshin; *Zh. Anal. Khim.*, 1965, **20**, 462 (*Chem. Abstr.*, 1965, **63**, 6301). | 941 |
| 65MI22600 | G. Valle, G. Carazzolo and M. Mammi; *Ric. Sci. Rend., Ser. A*, 1965, **8**, 1469 (*Chem. Abstr.*, 1966, **65**, 1514). | 950 |
| 65P161 | P. M. Nair and L. C. Vining; *Phytochemistry*, 1965, **4**, 161. | 324 |
| 65P177 | T. J. Mabry, J. Kagan and H. Rösler; *Phytochemistry*, 1965, **4**, 177. | 584 |
| 65QR426 | J. E. Anderson; *Q. Rev., Chem. Soc.*, 1965, **19**, 426. | 66 |
| 65RRC1245 | H. Mantsch and O. Maior; *Rev. Roum. Chim.*, 1965, **10**, 1245. | 956 |
| 65RTC1101 | H. C. van der Plas; *Rec. Trav. Chim. Pays-Bas*, 1965, **84**, 1101. | 89, 130 |

| 65RTC1569 | H. J. den Hertog, H. C. van der Plas, M. J. Pieterse and J. W. Streef; *Rec. Trav. Chim. Pays-Bas*, 1965, **84**, 1569. | 120 |
|---|---|---|
| 65SA663 | T. S. Hermann; *Spectrochim. Acta*, 1965, **21**, 663. | 464 |
| 65SA1277 | N. M. D. Brown and P. Bladon; *Spectrochim. Acta*, 1965, **21**, 1277. | 581 |
| 65SA1311 | W. R. Ward; *Spectrochim. Acta*, 1965, **21**, 1311. | 956 |
| 65SA1563 | D. Belitskus and G. A. Jeffrey; *Spectrochim. Acta*, 1965, **21**, 1563. | 461 |
| 65T1833 | R. H. Martin, N. Defay, F. Geerts-Evrard, P. H. Given, J. R. Jones and R. W. Wedel; *Tetrahedron*, 1965, **21**, 1833. | 583 |
| 65T1855 | H. Budzikiewicz, J. I. Brauman and C. Djerassi; *Tetrahedron*, 1965, **21**, 1855. | 605, 608, 611, 613 |
| 65T2059 | T. Goto, Y. Kishi, S. Takahashi and Y. Hirata; *Tetrahedron*, 1965, **21**, 2059. | 148 |
| 65T2191 | M. S. Gibson and M. Green; *Tetrahedron*, 1965, **21**, 2191. | 374, 375 |
| 65T3037 | F. Kállay, G. Janzsó and I. Koczor; *Tetrahedron*, 1965, **21**, 3037. | 728 |
| 65T3697 | L. Jurd and B. J. Bergot; *Tetrahedron*, 1965, **21**, 3697. | 585 |
| 65TH22100 | Y.-B. Chae; *Ph. D. Thesis*, Univ. München, 1965. | 538 |
| 65TL2269 | G. Janzsó, F. Kállay and I. Koczor; *Tetrahedron Lett.*, 1965, 2269. | 701 |
| 65USP3320256 | R. Duschinsky and T. F. Gabriel; *U.S. Pat.*, 3 320 256 (1965) (*Chem. Abstr.*, 1967, **67**, 108 666). | 362 |
| 65ZC22 | J. Fabian, A. Mehlhorn and R. Meyer; *Z. Chem.*, 1965, **5**, 22. | 892 |
| 65ZC353 | W. Schroth, F. Billig and H. Langguth; *Z. Chem.*, 1965, **5**, 353. | 971, 983 |
| 65ZOB589 | G. N. Dorofeenko and G. I. Zhungietu; *Zh. Obshch. Khim.*, 1965, **35**, 589. | 863 |
| 65ZOB1156 | A. E. Kretov and A. V. Davydov; *Zh. Obshch. Khim.*, 1965, **35**, 1156 (*Chem. Abstr.*, 1965, **63**, 11 563). | 492 |
| 65ZOB1303 | B. A. Ivin and V. G. Nemets; *Zh. Obshch. Khim.*, 1965, **35**, 1303. | 140 |
| 65ZOB1707 | A. I. Tolmachev, L. M. Shulezhko and A. A. Kisilenko; *Zh. Obshch. Khim.*, 1965, **35**, 1707. | 637 |
| 66ACS1561 | J. Gripenberg and T. Hase; *Acta Chem. Scand.*, 1966, **20**, 1561. | 599 |
| 66AG448 | A. Roedig, M. Schlosser and H. A. Renk; *Angew. Chem.*, 1966, **78**, 448. | 871 |
| 66AG(E)308 | E. C. Taylor, S. Vromen, R. V. Ravindranathan and A. McKillop; *Angew. Chem., Int. Ed. Engl.*, 1966, **5**, 308. | 205, 216 |
| 66AG(E)875 | W. Seeliger, E. Aufderhaar, W. Diepers, R. Feinauer, R. Nehring, W. Thier and H. Hellmann; *Angew. Chem., Int. Ed. Engl.*, 1966, **5**, 875. | 1008 |
| 66AG(E)960a | H.-G. Schmelzer, E. Degener and H. Holtschmidt; *Angew. Chem., Int. Ed. Engl.*, 1966, **5**, 960a. | 500 |
| 66AG(E)960b | E. Degener, H.-G. Schmelzer and H. Holtschmidt; *Angew. Chem., Int. Ed. Engl.*, 1966, **5**, 960b. | 501 |
| 66AJC151 | J. S. Fitzgerald, S. R. Johns, J. A. Lamberton and A. H. Redcliffe; *Aust. J. Chem.*, 1966, **19**, 151. | 148 |
| 66AJC1487 | D. J. Brown and R. V. Foster; *Aust. J. Chem.*, 1966, **19**, 1487. | 92, 96 |
| 66AJC2321 | D. J. Brown and R. V. Foster; *Aust. J. Chem.*, 1966, **19**, 2321. | 89, 96, 102 |
| 66AP362 | F. Zymalkowski and E. Reimann; *Arch. Pharm. (Weinheim, Ger.)*, 1966, **299**, 362. | 93, 98, 127 |
| 66AX(20)429 | S. Hosoya; *Acta Crystallogr.*, 1966, **20**, 429. | 947 |
| 66AX(20)490 | H. T. Kalff and C. Romers; *Acta Crystallogr.*, 1966, **20**, 490. | 950 |
| 66AX(20)646 | J. Gaultier and C. Hauw; *Acta Crystallogr.*, 1966, **20**, 646. | 622 |
| 66AX(21)249 | C. H. Carlisle and M. B. Hossain; *Acta Crystallogr.*, 1966, **21**, 249. | 2 |
| 66AX(21)A104 | I. Bernal and J. Ricci; *Acta Crystallogr.*, 1966, **21**, Suppl., A104. | 947 |
| 66BCJ1091 | S. Onuma, Y. Nawata and Y. Saito; *Bull. Chem. Soc. Jpn.*, 1966, **39**, 1091. | 147 |
| 66BCJ1694 | V. Zanker and E. Ehrhardt; *Bull. Chem. Soc. Jpn.*, 1966, **39**, 1694. | 667, 695 |
| 66BCJ1922 | F. Tanimoto, T. Tanaka, H. Kitano and K. Fukui; *Bull. Chem. Soc. Jpn.*, 1966, **39**, 1922. | 508 |
| 66BSF2405 | A. Grouiller; *Bull. Soc. Chim. Fr.*, 1966, 2405. | 584 |
| 66BSF2510 | C. Barry, G. Cauquis and M. Maurey; *Bull. Soc. Chim. Fr.*, 1966, 2510. | 969 |
| 66BSF2892 | H. Audier; *Bull. Soc. Chim. Fr.*, 1966, 2892. | 615, 618 |
| 66BSF2959 | M. Simalty, J. Carretto and R. Fugnitto; *Bull. Soc. Chim. Fr.*, 1966, 2959. | 871 |
| 66BSF3249 | P. Cagniant and C. Charaux; *Bull. Soc. Chim. Fr.*, 1966, 3249. | 855 |
| 66CB399 | K. Dimroth and H. Wache; *Chem. Ber.*, 1966, **99**, 399. | 657, 658 |
| 66CB536 | W. Pfleiderer and F. Reisser; *Chem. Ber.*, 1966, **99**, 536. | 282 |
| 66CB625 | C. H. Krauch, S. Farid and G. O. Schenck; *Chem. Ber.*, 1966, **99**, 625. | 582, 678 |
| 66CB872 | H. Wamhoff and F. Korte; *Chem. Ber.*, 1966, **99**, 872. | 113 |
| 66CB1558 | E. Winterfeldt and H.-J. Dillinger; *Chem. Ber.*, 1966, **99**, 1558. | 796 |
| 66CB1822 | M. Kamel and H. Shoeb; *Chem. Ber.*, 1966, **99**, 1822. | 768 |
| 66CB1973 | O. Scherer and F. Kluge; *Chem. Ber.*, 1966, **99**, 1973. | 982 |
| 66CB2351 | K. Dimroth, W. Kinzebach and M. Soyka; *Chem. Ber.*, 1966, **99**, 2351. | 671 |
| 66CB2361 | E. Grigat and R. Pütter; *Chem. Ber.*, 1966, **99**, 2361 | 506 |
| 66CB2430 | H. Wagner, L. Hörhammer, G. Hitzler and L. Farkas; *Chem. Ber.*, 1966, **99**, 2430. | 716 |
| 66CB3008 | W. Pfleiderer and H. Zondler; *Chem. Ber.*, 1966, **99**, 3008. | 279, 306 |
| 66CB3022 | W. Pfleiderer and E. Bühler; *Chem. Ber.*, 1966, **99**, 3022. | 282 |
| 66CB3040 | K. Dimroth, A. Berndt and R. Volland; *Chem. Ber.*, 1966, **99**, 3040. | 657 |
| 66CB3076 | A. Schönberg, K. Praefcke and J. Kohtz; *Chem. Ber.*, 1966, **99**, 3076. | 583 |
| 66CB3503 | W. Pfleiderer, J. W. Bunting, D. D. Perrin and G. Nübel; *Chem. Ber.*, 1966, **99**, 3503. | 277, 309 |

| | | |
|---|---|---|
| 66CC415 | H. Kristinsson, R. A. Mateer and G. W. Griffin; *Chem. Commun.*, 1966, 415. | 767 |
| 66CC759 | K. W. Buck, A. B. Foster, W. D. Pardoe, M. H. Qadir and J. M. Webber; *Chem. Commun.*, 1966, 759. | 953 |
| 66CCC1053 | M. Prystaš and F. Šorm; *Collect. Czech. Chem. Commun.*, 1966, **31**, 1053. | 90, 139 |
| 66CCC1864 | J. Piťha, F. Fiedler and J. Gut; *Collect. Czech. Chem. Commun.*, 1966, **31**, 1864. | 399 |
| 66CCC3990 | M. Prystaš and F. Šorm; *Collect. Czech. Chem. Commun.*, 1966, **31**, 3990. | 114 |
| 66CI(L)1721 | R. D. Chambers, J. A. H. MacBride and W. K. R. Musgrave; *Chem. Ind. (London)*, 1966, 1721. | 175 |
| 66CJC1801 | D. L. Garmaise; *Can. J. Chem.*, 1966, **44**, 1801. | 481 |
| 66CPB194 | K. Masuzawa, M. Masaki and M. Ohta; *Chem. Pharm. Bull.*, 1966, **14**, 194. | 366 |
| 66CPB419 | Y. Morita; *Chem. Pharm. Bull.*, 1966, **14**, 419. | 160 |
| 66CPB426 | Y. Morita; *Chem. Pharm. Bull.*, 1966, **14**, 426. | 162 |
| 66CPB742 | A. Takamizawa and Y. Sato; *Chem. Pharm. Bull. Jpn.* 1966, **14**, 742. | 1013 |
| 66CPB1010 | I. Matsuura, F. Yoneda and Y. Nitta; *Chem. Pharm. Bull.*, 1966, **14**, 1010. | 240, 241, 242, 243, 246 |
| 66CR(C)(262)1335 | G. Queguiner and P. Pastour; *C.R. Hebd. Seances Acad. Sci., Ser. C*, 1966, **262**, 1335. | 246 |
| 66CR(C)(263)173 | M. Baran-Marszak, J. Massicot and C. Mentzer; *C.R. Hebd. Seances Acad. Sci., Ser. C*, 1966, **263**, 173. | 854 |
| 66FES799 | G. Palamidessi and M. Bonanomi; *Farmaco Ed. Sci.*, 1966, **21**, 799. | 176 |
| 66G1108 | G. Bellomonte, G. Caronna and S. Palazzo; *Gazz. Chim. Ital.*, 1966, **96**, 1108. | 227 |
| 66HC(21-2)611 | D. S. Breslow and H. Skolnik; *Chem. Hetrocycl. Compd.*, 1966, **21-2**, 611. | 944, 991 |
| 66HC(21-2)626 | D. S. Breslow and H. Skolnik; *Chem. Hetrocycl. Compd.*, 1966, **21-2**, 626. | 981, 992 |
| 66HC(21-2)633 | D. S. Breslow and H. Skolnik; *Chem. Hetrocycl. Compd.*, 1966, **21-2**, 633. | 980, 981, 990, 991 |
| 66HC(21-2)675 | D. S. Breslow and H. Skolnik; *Chem. Hetrocycl. Compd.*, 1966, **21-2**, 675. | 980, 991 |
| 66HC(21-2)680 | D. S. Breslow and H. Skolnik; *Chem. Hetrocycl. Compd.*, 1966, **21-2**, 680. | 978 |
| 66HC(21-2)689 | D. S. Breslow and H. Skolnik; *Chem. Hetrocycl. Compd.*, 1966, **21-2**, 689. | 981 |
| 66HC(21-2)692 | D. S. Breslow and H. Skolnik; *Chem. Hetrocycl. Compd.*, 1966, **21-2**, 692. | 979, 990 |
| 66HC(21-2)774 | D. S. Breslow and H. Skolnik; *Chem. Hetrocycl. Compd.*, 1966, **21-2**, 774. | 983, 986, 991 |
| 66HC(21-2)780 | D. S. Breslow and H. Skolnik; *Chem. Hetrocycl. Compd.*, 1966, **21-2**, 780. | 994 |
| 66HC(21-2)789 | D. S. Breslow and H. Skolnik; *Chem. Hetrocycl. Compd.*, 1966, **21-2**, 789. | 965, 971, 973 |
| 66HC(21-2)792 | D. S. Breslow and H. Skolnik; *Chem. Hetrocycl. Compd.*, 1966, **21-2**, 792. | 986 |
| 66HC(21-2)805 | D. S. Breslow and H. Skolnik; *Chem. Hetrocycl. Compd.*, 1966, **21-2**, 805. | 989 |
| 66HC(21-2)816 | D. S. Breslow and H. Skolnik; *Chem. Hetrocycl. Compd.*, 1966, **21-2**, 816. | 987 |
| 66HC(21-2)817 | D. S. Breslow and H. Skolnik; *Chem. Hetrocycl. Compd.*, 1966, **21-2**, 817. | 976 |
| 66HC(21-2)842 | D. S. Breslow and H. Skolnik; *Chem. Hetrocycl. Compd.*, 1966, **21-2**, 842. | 977 |
| 66HC(21-2)852 | D. S. Breslow and H. Skolnik; *Chem. Hetrocycl. Compd.*, 1966, **21-2**, 852. | 967, 984 |
| 66HC(21-2)864 | D. S. Breslow and H. Skolnik; *Chem. Hetrocycl. Compd.*, 1966, **21-2**, 864. | 967, 970, 971, 973, 974, 975, 985 |
| 66HC(21-2)867 | D. S. Breslow and H. Skolnik; *Chem. Hetrocycl. Compd.*, 1966, **21-2**, 867. | 993 |
| 66HC(21-2)952 | D. S. Breslow and H. Skolnik; *Chem. Hetrocycl. Compd.*, 1966, **21-2**, 952. | 981, 986, 991 |
| 66HC(21-2)964 | D. S. Breslow and H. Skolnik; *Chem. Hetrocycl. Compd.*, 1966, **21-2**, 964. | 973 |
| 66HC(21-2)968 | D. S. Breslow and H. Skolnik; *Chem. Hetrocycl. Compd.*, 1966, **21-2**, 968. | 967, 971, 972, 973 |
| 66HC(21-2)979 | D. S. Breslow and H. Skolnik; *Chem. Hetrocycl. Compd.*, 1966, **21-2**, 979. | 989 |
| 66HC(21-2)1028 | D. S. Breslow and H. Skolnik; *Chem. Hetrocycl. Compd.*, 1966, **21-2**, 1028. | 979 |
| 66HC(21-2)1035 | D. S. Breslow and H. Skolnik; *Chem. Hetrocycl. Compd.*, 1966, **21-2**, 1035. | 990 |
| 66HC(21-2)1041 | D. S. Breslow and H. Skolnik; *Chem. Hetrocycl. Compd.*, 1966, **21-2**, 1041. | 976, 977, 987 |
| 66HC(21-2)1053 | D. S. Breslow and H. Skolnik; *Chem. Hetrocycl. Compd.*, 1966, **21-2**, 1053. | 988 |
| 66HC(21-2)1112 | D. S. Breslow and H. Skolnik; *Chem. Hetrocycl. Compd.*, 1966, **21-2**, 1112. | 964, 967, 981, 982, 983 |
| 66HC(21-2)1143 | D. S. Breslow and H. Skolnik; *Chem. Hetrocycl. Compd.*, 1966, **21-2**, 1143. | 984 |
| 66HC(21-2)1146 | D. S. Breslow and H. Skolnik; *Chem. Hetrocycl. Compd.*, 1966, **21-2**, 1146. | 966, 967 |
| 66HC(21-2)1155 | D. S. Breslow and H. Skolnik; *Chem. Hetrocycl. Compd.*, 1966, **21-2**, 1155. | 967, 971, 973, 974, 975, 986, 993 |
| 66HC(21-2)1220 | D. S. Breslow and H. Skolnik; *Chem. Hetrocycl. Compd.*, 1966, **21-2**, 1220. | 971 |
| 66HC(21-2)1249 | D. S. Breslow and H. Skolnik; *Chem. Hetrocycl. Compd.*, 1966, **21-2**, 1249. | 993 |
| 66HCA875 | A. Bobst and M. Viscontini; *Helv. Chim. Acta*, 1966, **49**, 875. | 281, 306, 307 |
| 66HCA1355 | A. Dieffenbacher and W. von Philipsborn; *Helv. Chim. Acta*, 1966, **49**, 1355. | 285 |
| 66HCA1439 | E. Schumacher and R. Taubenest; *Helv. Chim. Acta*, 1966, **49**, 1439. | 466 |
| 66HCA2046 | R. Wizinger and H. J. Angliker; *Helv. Chim. Acta*, 1966, **49**, 2046. | 938 |
| 66IJC120 | N. R. Krishnaswamy, T. R. Seshadri and B. R. Sharma; *Ind. J. Chem.*, 1966, **4**, 120. | 600 |
| 66IJC447 | A. P. Bhaduri and N. M. Khanna; *Indian J. Chem.*, 1966, **4**, 447. | 225 |
| 66IZV1613 | V. S. Reznik and N. G. Pashkurov; *Izv. Akad. Nauk SSSR, Ser. Khim.*, 1966, 1613. | 133 |
| 66JA526 | R. W. Murray, P. R. Story and M. L. Kaplan; *J. Am. Chem. Soc.*, 1966, **88**, 526. | 962 |
| 66JA619 | E. N. Marvell, G. Caple, T. A. Gosnik and G. Zimmer; *J. Am. Chem. Soc.*, 1966, **88**, 619. | 576, 644 |
| 66JA624 | T. Money, J. L. Douglas and A. I. Scott; *J. Am. Chem. Soc.*, 1966, **88**, 624. | 795 |
| 66JA1580 | E. M. Burgess and L. McCullagh; *J. Am. Chem. Soc.*, 1966, **88**, 1580. | 381 |
| 66JA5415 | H. Morrison, H. Curtis and T. McDowell; *J. Am. Chem. Soc.*, 1966, **88**, 5415. | 678 |

| | | |
|---|---|---|
| 66JBC(241)192 | S. Hosoda and D. Glick; *J. Biol. Chem.*, 1966, **241**, 192. | 306 |
| 66JBC(241)2220 | G. Guroff and C. A. Strenkoski; *J. Biol. Chem.*, 1966, **241**, 2220. | 320 |
| 66JCP(44)759 | M. J. S. Dewar and G. J. Gleicher; *J. Chem. Phys.*, 1966, **44**, 759. | 370, 387, 466 |
| 66JCP(45)2940 | R. G. Shulman and R. O. Rahn; *J. Chem. Phys.*, 1966, **45**, 2940. | 399 |
| 66JCP(45)3155 | W. Sawodny, K. Niedenzu and J. W. Dawson; *J. Chem. Phys.*, 1966, **45**, 3155. | 465 |
| 66JCS(A)639 | M. Brufani, G. Casini, W. Fedeli, G. Giacomello and A. Vaciago; *J. Chem. Soc. (A)*, 1966, 639. | 339 |
| 66JCS(B)427 | A. Albert; *J. Chem. Soc. (B)*, 1966, 427. | 131 |
| 66JCS(B)436 | J. W. Bunting and D. D. Perrin; *J. Chem. Soc. (B)*, 1966, 436. | 206, 251 |
| 66JCS(B)565 | A. R. Katritzky, F. D. Popp and A. J. Waring; *J. Chem. Soc. (B)*, 1966, 565. | 68 |
| 66JCS(B)750 | W. L. F. Armarego and T. J. Batterham; *J. Chem. Soc. (B)*, 1966, 750. | 202, 203 |
| 66JCS(B)946 | J. H. Bowie, S.-O. Lawesson, J. Ø. Madsen, C. Nolde, G. Schroll and Williams; *J. Chem. Soc. (B)*, 1966, 946. | 957 |
| 66JCS(B)1105 | A. Albert, T. J. Batterham and J. J. McCormack; *J. Chem. Soc. (B)*, 1966, 1105. | 265, 266 |
| 66JCS(C)175 | A. Jefferson and F. Scheinmann; *J. Chem. Soc. (C)*, 1966, 175. | 838 |
| 66JCS(C)226 | D. J. Brown, B. T. England and J. M. Lyall; *J. Chem. Soc. (C)*, 1966, 226. | 99, 107, 131 |
| 66JCS(C)234 | W. L. F. Armarego and J. Smith; *J. Chem. Soc. (C)*, 1966, 234. | 109 |
| 66JCS(C)285 | A. Stuart, H. C. S. Wood and D. Duncan; *J. Chem. Soc. (C)*, 1966, 285. | 287, 289, 306 |
| 66JCS(C)430 | H. D. Locksley, I. Moore and F. Scheinmann; *J. Chem. Soc. (C)*, 1966, 430. | 835 |
| 66JCS(C)542 | D. J. Adam, L. Crombie and D. A. Whiting; *J. Chem. Soc. (C)*, 1966, 542. | 583 |
| 66JCS(C)629 | C. A. Anirudhan, W. B. Whalley and M. M. E. Badran; *J. Chem. Soc. (C)*, 1966, 629. | 755, 857 |
| 66JCS(C)649 | C. D. May and P. Sykes; *J. Chem. Soc. (C)*, 1966, 649. | 82 |
| 66JCS(C)909 | M. L. Tosato and L. Paolini; *J. Chem. Soc. (C)*, 1966, 909. | 462, 463 |
| 66JCS(C)999 | T. J. Batterham; *J. Chem. Soc. (C)*, 1966, 999. | 249, 340, 351 |
| 66JCS(C)1065 | N. W. Jacobsen; *J. Chem. Soc. (C)*, 1966, 1065. | 94, 303 |
| 66JCS(C)1112 | J. Clark and G. Neath; *J. Chem. Soc. (C)*, 1966, 1112. | 122 |
| 66JCS(C)1117 | A. Albert and J. J. McCormack; *J. Chem. Soc. (C)*, 1966, 1117. | 279, 288 |
| 66JCS(C)1712 | R. A. W. Johnstone, B. J. Millard, F. M. Dean and A. W. Hill; *J. Chem. Soc. (C)*, 1966, 1712. | 608, 610 |
| 66JCS(C)2013 | J. D. Hepworth and R. Livingstone; *J. Chem. Soc. (C)*, 1966, 2013. | 668, 735, 786, 856 |
| 66JCS(C)2031 | J. A. Bee and F. L. Rose; *J. Chem. Soc. (C)*, 1966, 2031. | 84, 570 |
| 66JCS(C)2053 | C. M. Atkinson and B. N. Biddle; *J. Chem. Soc. (C)*, 1966, 2053. | 238, 239, 240, 243, 244 |
| 66JCS(C)2239 | G. M. Blackburn and R. J. H. Davies; *J. Chem. Soc. (C)*, 1966, 2239. | 73 |
| 66JGU1724 | G. N. Dorofeenko, Yu. A. Zhdanov, A. D. Semenov, Y. A. Palchkov and S. V. Krivun; *J. Gen. Chem. USSR (Engl. Transl.)*, 1966, **36**, 1724. | 663 |
| 66JHC137 | G. A. Loughran, G. F. L. Ehlers and J. L. Burkett; *J. Heterocycl. Chem.*, 1966, 3, 137. | 481 |
| 66JHC324 | B. R. Baker and J. H. Jordaan; *J. Heterocycl. Chem.*, 1966, 3, 324. | 104 |
| 66JHC435 | H. Rutner and P. E. Spoerri; *J. Heterocycl. Chem.*, 1966, 3, 435. | 165, 176 |
| 66JHC512 | N. R. Patel and R. N. Castle; *J. Heterocycl. Chem.*, 1966, 3, 512. | 337, 347, 360 |
| 66JIC615 | S. P. Inamdar and R. N. Usgaonkar; *J. Indian Chem. Soc.*, 1966, **43**, 615. | 832 |
| 66JMC97 | A. Giner-Sorolla and L. Medrek; *J. Med. Chem.*, 1966, 9, 97. | 89, 102, 136, 137, 139 |
| 66JMC108 | F. R. Gerns, A. Perrotta and G. H. Hitchings; *J. Med. Chem.*, 1966, 9, 108. | 100 |
| 66JMC573 | D. E. O'Brien, C. C. Cheng and W. Pfleiderer; *J. Med. Chem.*, 1966, 9, 573. | 69 |
| 66JMC610 | D. Kaminsky, W. B. Lutz and S. Lazarus; *J. Med. Chem.*, 1966, 9, 610. | 350 |
| 66JMC876 | M. P. Mertes, S. E. Saheb and D. Miller; *J. Med. Chem.*, 1966, 9, 876. | 80 |
| 66JMC881 | D. L. Trepanier, E. R. Wagner, G. Harris and A. D. Rudzi; *J. Med. Chem.*, 1966, 9, 881. | 441 |
| 66JOC175 | Y. Inoue, N. Furutachi and K. Nakanishi; *J. Org. Chem.*, 1966, 31, 175. | 67, 68 |
| 66JOC406 | H. J. Minnemeyer, P. B. Clarke and H. Tieckelmann; *J. Org. Chem.*, 1966, 31, 406. | 93 |
| 66JOC544 | G. E. Booth and R. J. Ouellette; *J. Org. Chem.*, 1966, 31, 544. | 579, 629 |
| 66JOC620 | S. B. Kadin; *J. Org. Chem.*, 1966, 31, 620. | 804 |
| 66JOC781 | H. C. Brown, H. J. Gisler, Jr. and M. T. Cheng; *J. Org. Chem.*, 1966, 31, 781. | 557, 560 |
| 66JOC1032 | T. M. Harris and C. M. Harris; *J. Org. Chem.*, 1966, 31, 1032. | 796 |
| 66JOC1232 | L. A. Paquette and H. Stucki; *J. Org. Chem.*, 1966, 31, 1232. | 750, 782, 822, 840 |
| 66JOC1311 | E. Hedaya, R. L. Hinman and S. Theodoropulos; *J. Org. Chem.*, 1966, 31, 1311. | 341 |
| 66JOC1720 | W. W. Paudler and J. M. Barton; *J. Org. Chem.*, 1966, 31, 1720. | 386, 393, 416 |
| 66JOC1890 | R. D. Elliott, C. Temple, Jr. and J. A. Montgomery; *J. Org. Chem.*, 1966, 31, 1890. | 254 |
| 66JOC2167 | I. El-S. El-Kholy, F. K. Rafla and M. M. Mishrikey; *J. Org. Chem.*, 1966, 31, 2167. | 792 |
| 66JOC2215 | W. Asbun and S. B. Binkley; *J. Org. Chem.*, 1966, 31, 2215. | 104 |
| 66JOC2333 | L. A. Cohen and J. A. Steele; *J. Org. Chem.*, 1966, 31, 2333. | 896, 907 |
| 66JOC2568 | M. E. Hermes and R. A. Braun; *J. Org. Chem.*, 1966, 31, 2568. | 1041, 1063, 1081 |
| 66JOC2966 | J. C. Martin, K. C. Brannock and R. H. Meen; *J. Org. Chem.*, 1966, 31, 2966. | 1031 |
| 66JOC3032 | I. J. Borowitz, G. Gonis, R. Kelsey, R. Rapp and G. J. Williams; *J. Org. Chem.*, 1966, 31, 3032. | 577, 783 |
| 66JOC3067 | S. Moon and L. Haynes; *J. Org. Chem.*, 1966, 31, 3067. | 777 |
| 66JOC3429 | J. B. Lambert and R. G. Keske; *J. Org. Chem.*, 1966, 31, 3429. | 628 |
| 66JOC3531 | B. Loev, M. F. Kormendy and K. M. Snader; *J. Org. Chem.*, 1966, 31, 3531. | 1016 |
| 66JOC3914 | T. Sasaki and K. Minamoto; *J. Org. Chem.*, 1966, 31, 3914. | 389, 410, 415 |
| 66JOC3969 | G. C. Hopkins, J. P. Jonak, H. Tieckelmann and H. J. Minnemeyer; *J. Org. Chem.*, 1966, 31, 3969. | 69, 135 |

| 66JOC4067 | C. H. Issidorides and M. J. Haddadin; *J. Org. Chem.*, 1966, **31**, 4067. | 181 |

66JOC4067    C. H. Issidorides and M. J. Haddadin; *J. Org. Chem.*, 1966, **31**, 4067.   181

66JOC4239    A. Giner-Sorolla and A. Bendich; *J. Org. Chem.*, 1966, **31**, 4239.   100

66JOU1113    N. P. Shusherina, N. D. Dmitrieva, N. N. Malysheva and R. Ya. Levina; *J. Org. Chem. USSR (Engl. Transl.)*, 1966, **2**, 1113.   680, 681

66JPR(31)205    K. Gewald; *J. Prakt. Chem.*, 1966, **31**, 205.   935

66JPR(32)26    K. Gewald; *J. Prakt. Chem.*, 1966, **32**, 26.   96, 122, 137

66JPR(33)50    W. Schulze and H. Willitzer; *J. Prakt. Chem.*, 1966, **33**, 50.   77

66JPS568    M. Israel, H. N. Schlein, C. L. Maddock, S. Farber and E. J. Modest; *J. Pharm. Sci.*, 1966, **55**, 568.   70, 131

66JSP(19)25    S. C. Wait, Jr. and J. W. Wesley; *J. Mol. Spectrosc.*, 1966, **19**, 25.   204, 236, 250, 332, 333, 336, 337, 338, 339

66KGS122    A. E. Kretov and A. V. Davydov; *Khim. Geterotsikl. Soedin.*, 1966, 122 (*Chem. Abstr.*, 1966, **65**, 715).   492

66LA(691)142    H. Goldner, G. Dietz and E. Carstens; *Liebigs Ann. Chem.*, 1966, **691**, 142.   131, 281

66LA(692)119    R. Brossmer and E. Röhm; *Liebigs Ann. Chem.*, 1966, **692**, 119.   89, 101, 140

66LA(696)97    W. Ried and D. Piechaczek; *Liebigs Ann. Chem.*, 1966, **696**, 97.   111

66LA(700)87    W. Ried and P. Stock; *Liebigs Ann. Chem.*, 1966, **700**, 87.   111

66M52    G. Zigeuner, W. Adam and W. Galatik; *Monatsh. Chem.*, 1966, **97**, 52.   223

66M517    R. Kuhn and G. Fischer-Schwarz; *Monatsh. Chem.*, 1966, **97**, 547.   547

66M525    R. Kuhn, F. A. Neugebauer and H. Trischmann; *Monatsh. Chem.*, 1966, **97**, 525.   547

66M710    E. Ziegler and F. Hradetzky; *Monatsh. Chem.*, 1966, **97**, 710.   812

66M846    R. Kuhn, F. A. Neugebauer and H. Trischmann; *Monatsh. Chem.*, 1966, **97**, 846.   560

66M1195    E. Steininger; *Monatsh. Chem.*, 1966, **97**, 1195.   566, 569

66M1280    R. Kuhn, F. A. Neugebauer and H. Trischmann; *Monatsh. Chem.*, 1966, **97**, 1280.   547

66MI21200    G. Rosseels; *Ind. Chim. Belg.*, 1966, **31**, 668.   2

66MI21300    S. H. Chang, B.-S. Hahn, I. K. Kim and S. H. Oh; *J. Korean Chem. Soc.*, 1966, **10**, 51 (*Chem. Abstr.*, 1967, **67**, 100 093).   85

66MI21301    A. Vincze and S. Cohen; *Isr. J. Chem.*, 1966, **4**, 23.   88

B-66MI21302    E. Chargaff and J. N. Davidson; 'The Nucleic Acids', Academic Press, New York, 1955–1966, Vol. 1–3.   142

B-66MI21303    'Procedures in Nucleic Acid Research', ed. G. L. Cantoni and D. R. Davies; Harper and Row, New York, 1966, vol. 1.   142

B-66MI21400    H. M. Hershenson; in 'Ultraviolet and Visible Absorption Spectra, Index for 1960–1963', Academic Press, New York, 1966.   161

66MI21401    L. Novacek; *Cesk. Farm.*, 1966, **15**, 323.   195

66MI21500    V. P. Mamaev and A. M. Kim; *Khim. Geterotsikl. Soedin. Akad. Nauk Latv. SSR*, 1966, 266 (*Chem. Abstr.*, 1966, **65**, 710).   227

B-66MI21600    R. J. Schnitzer and F. Hawking; 'Experimental Chemotherapy', Academic Press, New York, 1966, vol. 1, part 1.   325

66MI21900    J. Hadacek and J. Slouka; *Folia Fac. Sci. Natur. Univ. Purkynianae Brunensis*, 1966.   385

66MI22200    T. Hanafusa and Y. Yukawa; *Mem. Inst. Sci. Ind. Res. Osaka Univ.*, 1966, **23**, 85 (*Chem. Abstr.*, 1966, **65**, 13 588).   594

66MI22201    S. Sasaki, Y. Itagaki, T. Kurokawa, E. Watanabe and T. Aoyama; *Shitsuryo Bunseki*, 1966, **14**, 82 (*Chem. Abstr.*, 1968, **69**, 86 113).   614

66MI22202    E. F. L. J. Anet; *Carbohydrate Res.*, 1966, **1**, 348.   611, 632

66MI22300    T. R. Seshadri; *Proc. Nat. Inst. Sci. India, Part A*, 1966, **31**, 603.   718

66MI22400    H. Strzelecka; *Ann. Chim. (Paris)*, 1966, **1**, 201.   871

66MI22500    Y. M. Batulin, T. A. Klygul, A. E. Kovaleva, I. B. Fedorova and L. A. Aksanova; *Farmacol. Toksikol.*, 1966, **29**, 590 (*Chem. Abstr.*, 1967, **66**, 45 274).   941

66MI22600    B. Lamotte and G. Berthier; *J. Chim. Phys.*, 1966, **63**, 369.   955

66MI22601    B. von Schmeling and M. Kulka; *Science*, 1966, **152**, 659.   994

66MIP21800    Siegfried A.-G.; Swiss Pat. 405 332 (1966) (*Chem. Abstr.*, 1966, **65**, 15 402).   376

66NKK1226    A. Tatematsu, T. Goto and S. Matsuura; *Nippon Kagaku Zasshi*, 1966, **87**, 1226.   285

66RCR9    Yu. A. Naumov and I. I. Grandberg; *Russ. Chem. Rev. (Engl. Transl.)*, 1966, **35**, 9.   2

66RRC1193    C. C. Renţia, A. T. Balaban and Z. Simon; *Rev. Roum. Chim.*, 1966, **11**, 1193.   640

66RTC899    A. P. TerBorg, H. R. Gersmann and A. F. Bickel; *Rec. Trav. Chim. Pays-Bas*, 1966, **85**, 899.   667

66RTC1101    H. C. van der Plas, B. Haase, B. Zuurdeeg and M. C. Vollering; *Recl. Trav. Chim. Pays-Bas*, 1966, **85**, 1101.   100, 519

66RTC1197    C. Altona, H. J. Geise and C. Romers; *Recl. Trav. Chim. Pays-Bas*, 1966, **85**, 1197.   950

66SA117    W. L. F. Armarego, G. B. Barlin and E. Spinner; *Spectrochim. Acta*, 1966, **22**, 117.   204, 249

66SA1859    J. E. Connett, J. A. Creighton, J. H. S. Green and W. Kynaston; *Spectrochim. Acta*, 1966, **22**, 1859.   947, 952, 956

66T157    Y. Ogata, A. Kawasaki and K. Nakagawa; *Tetrahedron*, 1966, **22**, 157.   498

66T621    B. J. Bolger, A. Hirwe, K. G. Marathe, E. M. Philbin, M. A. Vickars and C. P. Lillya; *Tetrahedron*, 1966, **22**, 621.   630

66T723    M. L. Mihailović and M. Miloradović; *Tetrahedron*, 1966, **22**, 723.   776

66T931    A. R. Katritzky, M. J. Sewell, R. D. Topsom, A. M. Monro and G. W. H. Potter; *Tetrahedron*, 1966, **22**, 931.   956, 977, 984

66T1539    M. Kamel and H. Shoeb; *Tetrahedron*, 1966, **22**, 1539.   839

| | | |
|---|---|---|
| 66T1777 | O. R. Gottlieb, M. Taveira Magalhaes, M. Camey, A. A. Lins Mesquita and D. De Barros Corrêa; *Tetrahedron*, 1966, **22**, 1777. | 601 |
| 66T2401 | T. Nishiwaki; *Tetrahedron*, 1966, **22**, 2401. | 70 |
| 66T3117 | T. Nishiwaki; *Tetrahedron*, 1966, **22**, 3117. | 66 |
| 66T3253 | R. Mondelli and L. Merlini; *Tetrahedron*, 276 1966, **22**, 3253. | 276 |
| 66T3301 | R. Grigg, J. A. Knight and P. Roffey; *Tetrahedron*, 1966, **22**, 3301. | 582 |
| 66T3477 | B. Price, I. O. Sutherland and F. G. Williamson; *Tetrahedron*, 1966, **22**, 3477. | 331 |
| 66TL1317 | H. Friebolin, W. Faisst, H. G. Schmid and S. Kabuss; *Tetrahedron Lett.*, 1966, 1317. | 962 |
| 66TL1419 | J. W. Hanifin and E. Cohen; *Tetrahedron Lett.*, 1966, 1419. | 576, 705 |
| 66TL3465 | R. K. Smalley and H. Suschitzky; *Tetrahedron Lett.*, 1966, 3465. | 378 |
| 66TL4189 | P. Rouillier, J. Delmau, J. Duplan and C. Nofre; *Tetrahedron Lett.*, 1966, 4189. | 62 |
| 66TL4979 | J. Sauer and G. Heinrichs; *Tetrahedron Lett.*, 1966, 4979. | 51, 551 |
| 66TL5067 | W. Skorianetz and E. sz. Kováts; *Tetrahedron Lett.*, 1966, 5067. | 540, 569 |
| 66TL5253 | R. Brossmer and D. Ziegler; *Tetrahedron Lett.*, 1966, 5253. | 91, 125 |
| 66TL5369 | H. H. Takimoto and G. C. Denault; *Tetrahedron Lett.*, 1966, 5369. | 556 |
| 66USP3235549 | G. L. Broussalian; *U.S. Pat.* 3 235 549 (1966) (*Chem. Abstr.*, 1966, **64**, 15902). | 1041, 1080 |
| 66USP3235554 | V. Papesch; (G. D. Searle and Co.), *U.S. Pat.* 3 235 554 (1966) (*Chem. Abstr.*, 1966, **64**, 14 198). | 229 |
| 66USP3249499 | B. von Schmeling, M. Kulka, D. S. Thiara and W. A. Harrison; *U.S. Pat.* 3 249 499 (1966) (*Chem. Abstr.*, 1966, **65**, 7190). | 987, 994 |
| 66USP3280124 | M. M. Boudakian, E. H. Kober and E. R. Shipkowski; *U.S. Pat.* 3 280 124 (1966) (*Chem. Abstr.*, 1967, **66**, 2582). | 140 |
| 66ZN(B)291 | G. Arnold; *Z. Naturforsch., Teil B*, 1966, **21**, 291. | 594 |
| 66ZN(B)320 | S. Kabuss, A. Lüttringhaus, H. Friebolin and R. Mecke; *Z. Naturforsch., Teil B*, 1966, **21**, 320. | 953, 962 |
| 66ZOR364 | A. E. Kretov and N. D. Borodavko; *Zh. Org. Khim.*, 1966, **2**, 364. | 110 |
| 66ZPC(344)16 | A. Schellenberger and K. Winter; *Hoppe-Seyler's Z. Physiol. Chem.*, 1966, **344**, 16. | 81 |
| | | |
| 67AC(R)1045 | G. Canalini, I. Degani, R. Fochi and G. Spunta; *Ann. Chim. (Rome)*, 1967, **57**, 1045. | 667, 855, 873 |
| 67ACS1984 | C. Larsen and E. Binderup; *Acta Chem. Scand.*, 1967, **21**, 1984. | 566 |
| 67ACS2711 | P. Groth; *Acta Chem. Scand.*, 1967, **27**, 2711. | 950 |
| 67ACS2855 | C. Larsen, E. Binderup and J. Møller; *Acta Chem. Scand.*, 1967, **21**, 2855. | 542, 566 |
| 67AG(E)206 | E. Grigat and R. Putter; *Angew. Chem., Int. Ed. Engl.*, 1967, **6**, 206. | 494, 497, 506 |
| 67AG(E)458 | G. Märkl, F. Lieb and A. Merz; *Angew. Chem., Int. Ed. Engl.*, 1967, **6**, 458. | 659 |
| 67AG(E)698 | W. Schroth, F. Billig and G. Reinhold; *Angew. Chem., Int. Ed. Engl.*, 1967, **6**, 698. | 952, 955, 965, 967, 971, 983 |
| 67AG(E)703 | F. Fehér and B. Degen, *Angew. Chem., Int. Ed. Engl.*, 1967, **6**, 703. | 955, 992 |
| 67AG(E)919 | A. Albert; *Angew. Chem., Int. Ed. Engl.*, 1967, **6**, 919. | 265 |
| 67AG(E)940 | A. Kreutzberger; *Angew. Chem., Int. Ed. Engl.*, 1967, **6**, 940. | 471 |
| 67AHC(8)83 | M. Ionescu and H. Mantsch; *Adv. Heterocycl. Chem.*, 1967, **8**, 83. | 1038 |
| 67AHC(8)115 | J. Pliml and M. Prystaš; *Adv. Heterocycl. Chem.*, 1967, **8**, 115. | 93 |
| 67AHC(8)219 | R. Mayer, W. Broy and R. Zahradník; *Adv. Heterocycl. Chem.*, 1967, **8**, 219. | 575, 892, 934 |
| 67AHC(8)277 | K. H. Wünsch and A. J. Boulton; *Adv. Heterocycl. Chem.*, 1967, **8**, 277. | 1023 |
| 67AJC1041 | M. E. C. Biffin, D. J. Brown and T.-C. Lee; *Aust. J. Chem.*, 1967, **20**, 1041. | 88, 92 |
| 67AJC1595 | T. L. Chan and J. Miller; *Aust. J. Chem.*, 1967, **20**, 1595. | 164 |
| 67AJC1643 | R. F. Evans; *Aust. J. Chem.*, 1967, **20**, 1643. | 62, 66, 108, 123 |
| 67AJC1773 | A. R. Katritzky, M. Kingsland, M. N. Rudd, M. J. Sewell and R. D. Topsom; *Aust. J. Chem.*, 1967, **20**, 1773. | 951, 966 |
| 67AJC2677 | J. H. Bowie, R. G. Cooks, P. F. Donaghue, J. A. Halleday and H. J. Rodda; *Aust. J. Chem.*, 1967, **20**, 2677. | 237 |
| 67AP1 | E. Profft and K. Stühmer; *Arch. Pharm. (Weinheim, Ger.)*, 1967, **300**, 1. | 749 |
| 67AX(22)120 | M. G. Paton, E. N. Maslen and K. J. Watson; *Acta Crystallogr.*, 1967, **22**, 120. | 621 |
| 67AX(23)1102 | R. F. Stewart and L. H. Jensen; *Acta Crystallogr.*, 1967, **23**, 1102. | 68 |
| 67B2168 | V. S. Gupta and F. M. Huennekens; *Biochemistry*, 1967, **6**, 2168. | 71 |
| 67BBA(145)771 | J. Doskočil, V. Pačes and F. Šorm; *Biochem. Biophys. Acta*, 1967, **145**, 771. | 526 |
| 67BCJ153 | S. Kakimoto and S. Tanooka; *Bull. Chem. Soc. Jpn.*, 1967, **40**, 153. | 239, 243 |
| 67BSF1164 | P. Cohen and P. Mamont; *Bull. Soc. Chim. Fr.*, 1967, 1164. | 669, 857 |
| 67BSF2224 | D. Molho and J. Aknin; *Bull. Soc. Chim. Fr.*, 1967, 2224. | 582 |
| 67BSF2551 | J. Daunis, R. Jacquier and P. Viallefont; *Bull. Soc. Chim. Fr.*, 1967, 2551. | 411 |
| 67BSF2824 | R. Guglielmetti and J. Metzger; *Bull. Soc. Chim. Fr.*, 1967, 2824. | 750 |
| 67BSF4441 | D. B-Legay and R. Boudet; *Bull. Soc. Chim. Fr.*, 1967, 4441. | 1022 |
| 67C510 | J. P. Marion; *Chimia*, 1967, **21**, 510. | 179 |
| 67CB1296 | H. Wamhoff, G. Schorn and F. Korte; *Chem. Ber.*, 1967, **100**, 1296. | 765, 785 |
| 67CB1559 | W. Draber; *Chem. Ber.*, 1967, **100**, 1559. | 982 |
| 67CB2159 | H. Beecken; *Chem. Ber.*, 1967, **100**, 2159. | 1066, 1076 |
| 67CB2164 | H. Beecken; *Chem. Ber.*, 1967, **100**, 2164. | 1075 |
| 67CB2280 | H. Bredereck, F. Effenberger and H. G. Österlin; *Chem. Ber.*, 1967, **100**, 2280. | 129 |
| 67CB2585 | A. Dornow and H. Pietsch; *Chem. Ber.*, 1967, **100**, 2585. | 391 |

| | | |
|---|---|---|
| 67CB3101 | H. Simon, G. Heubach and H. Wacker; *Chem. Ber.*, 1967, **100**, 3101. | 415 |
| 67CB3664 | H. Bredereck, G. Simchen and H. Traut; *Chem. Ber.*, 1967, **100**, 3664. | 67, 84, 125 |
| 67CB3736 | D. Martin and A. Weise; *Chem. Ber.*, 1967, **100**, 3736. | 1081 |
| 67CB3961 | G. Steffan and G. O. Schenck; *Chem. Ber.*, 1967, **100**, 3961. | 706 |
| 67CC231 | F. W. Comer, T. Money and A. I. Scott; *Chem. Commun.*, 1967, 231. | 795 |
| 67CC577 | M. C. Manger, W. D. Ollis and I. O. Sutherland; *Chem. Commun.*, 1967, 577. | 692 |
| 67CC1006 | G. Adembri, F. DeSio, R. Nesi and M. Scotton; *Chem. Commun.*, 1967, 1006. | 333, 353 |
| 67CC1045 | R. M. Moriarty, J. M. Kliegman and R. B. Desai; *Chem. Commun.*, 1967, 1045. | 543 |
| 67CCC1298 | M. Prystaš and F. Šorm; *Collect. Czech. Chem. Commun.*, 1967, **32**, 1298. | 91 |
| 67CHE621 | V. A. Zagorevskii, S. M. Glozman and V. G. Vinokurov; *Chem. Heterocycl. Compd. (Engl. Transl.)*, 1967, 621. | 810 |
| 67CHE624 | V. A. Zagorevskii, I. D. Tsvetkova and E. K. Orlova; *Chem. Heterocycl. Compd. (Engl. Transl.)*, 1967, 624. | 705, 706, 856 |
| 67CI(L)1452 | A. Aviram and S. Vromen; *Chem. Ind. (London)*, 1967, 1452. | 223 |
| 67CJC767 | J. S. McIntyre; *Can. J. Chem.*, 1967, **45**, 767. | 691 |
| 67CJC1431 | B. M. Lynch and L. Poon; *Can. J. Chem.*, 1967, **45**, 1431. | 78 |
| 67CR(C)(264)405 | M. Prasad and C. G. Wermuth; *C.R. Hebd. Seances Acad. Sci., Ser. C*, 1967, **264**, 405. | 243 |
| 67E116 | M. Tsusue; *Experientia*, 1967, **23**, 116. | 294 |
| 67FRP1438827 | Ciba Ltd., *Fr. Pat.* 1 438 827 (1966) (*Chem. Abstr.*, 1967, **66**, 95 069). | 56 |
| 67G397 | I. Degani, R. Fochi and C. Vincenzi; *Gazz. Chim. Ital.*, 1967, **97**, 397. | 922 |
| 67HC(24-1)19 | W. L. F. Armarego; *Chem. Heterocycl. Compd.*, 1967, **24-1**, 19. | 61 |
| 67HC(24-1)69 | W. L. F. Armarego; *Chem. Heterocycl. Compd.*, 1967, **24-1**, 69. | 133 |
| 67HC(24-1)99 | W. L. F. Armarego; *Chem. Heterocycl. Compd.*, 1967, **24-1**, 99. | 107 |
| 67HC(24-1)227 | W. L. F. Armarego; *Chem. Heterocycl. Compd.*, 1967, **24-1**, 227. | 99 |
| 67HC(24-1)270 | W. L. F. Armarego; *Chem. Heterocycl. Compd.*, 1967, **24-1**, 270. | 136 |
| 67HC(24-1)391 | W. L. F. Armarego; *Chem. Heterocycl. Compd.*, 1967, **24-1**, 391. | 75, 108 |
| 67HC(24-1)475 | W. L. F. Armarego; *Chem. Heterocycl. Compd.*, 1967, **24-1**, 475. | 80 |
| 67HC(24-1)490 | W. L. F. Armarego; *Chem. Heterocycl. Compd.*, 1967, **24-1**, 490. | 148 |
| 67HCA411 | A. Ehrenberg, P. Hemmerich, F. Müller, T. Okada and M. Viscontini; *Helv. Chim. Acta*, 1967, **50**, 411. | 282 |
| 67HCA1492 | M. Viscontini and T. Okada; *Helv. Chim. Acta*, 1967, **50**, 1492. | 308 |
| 67HCA2222 | A. Bobst; *Helv. Chim. Acta*, 1967, **50**, 2222. | 308 |
| 67IJC93 | M. M. Badawi and M. B. E. Fayez; *Indian J. Chem.*, 1967, **5**, 93. | 583 |
| 67IZV1811 | I. A. Mikhailopulo, V. I. Gunar and S. I. Zav'yalov; *Izv. Akad. Nauk SSSR, Ser. Khim.*, 1967, 1811. | 90 |
| 67JA760 | F. A. L. Anet and A. J. R. Bourn; *J. Am. Chem. Soc.*, 1967, **89**, 760. | 628 |
| 67JA1836 | J. B. Lambert; *J. Am. Chem. Soc.*, 1967, **89**, 1836. | 628 |
| 67JA4875 | E. Hedaya, R. L. Hinman, V. Schomaker, S. Theodoropulos and L. M. Kyle; *J. Am. Chem. Soc.*, 1967, **89**, 4875. | 341, 351 |
| 67JA5921 | J. B. Lambert, R. G. Keske and D. K. Weary; *J. Am. Chem. Soc.*, 1967, **89**, 5921. | 628 |
| 67JA5934 | K. E. Fahrenholtz, M. Lurie and R. W. Kierstead; *J. Am. Chem. Soc.*, 1967, **89**, 5934. | 851 |
| 67JA5954 | W. T. Pike and F. W. McLafferty; *J. Am. Chem. Soc.*, 1967, **89**, 5954. | 608, 610 |
| 67JA6213 | A. C. Waiss, Jr., R. E. Lundin, A. Lee and J. Corse; *J. Am. Chem. Soc.*, 1967, **89**, 6213. | 696 |
| 67JA6734 | T. M. Harris and R. L. Carney; *J. Am. Chem. Soc.*, 1967, **89**, 6734. | 853 |
| 67JA7001 | T. L. Jacobs, R. Macomber and D. Zunker; *J. Am. Chem. Soc.*, 1967, **89**, 7001. | 769 |
| 67JAP6700191 | Y. Nitta, F. Yoneda and I. Matsuura; (Chugai Pharm. Co. Ltd.), *Jap. Pat.* 67 00 191 (1967) (*Chem. Abstr.*, 1967, **66**, 76 025). | 246 |
| 67JBC(242)565 | W. B. Watt; *J. Biol. Chem.*, 1967, **242**, 565. | 320, 321 |
| 67JBC(242)3934 | S. Kaufman; *J. Biol. Chem.*, 1967, **242**, 3934. | 280, 306 |
| 67JCP(47)4863 | M. A. El-Bayoumi and O. S. Khalil; *J. Chem. Phys.*, 1967, **47**, 4863. | 159 |
| 67JCS(A)1626 | A. R. Lepley, M. R. Chakrabarty and E. S. Hanrahan; *J. Chem. Soc. (A)*, 1967, 1626. | 204, 206 |
| 67JCS(B)123 | R. A. Shaw and P. Ward; *J. Chem. Soc. (B)*, 1967, 123. | 465 |
| 67JCS(B)171 | T. J. Batterham, D. J. Brown and M. N. Paddon-Row; *J. Chem. Soc. (B)*, 1967, 171. | 78 |
| 67JCS(B)273 | E. Kalatzis; *J. Chem. Soc. (B)*, 1967, 273. | 131 |
| 67JCS(B)387 | J. M. Lehn, F. G. Riddell, B. J. Price and I. O. Sutherland; *J. Chem. Soc. (B)*, 1967, 387. | 467 |
| 67JCS(B)449 | W. L. F. Armarego and J. I. C. Smith; *J. Chem. Soc. (B)*, 1967, 449. | 101, 136, 140 |
| 67JCS(B)560 | F. G. Riddell; *J. Chem. Soc. (B)*, 1967, 560. | 62 |
| 67JCS(B)859 | E. R. Clark and S. G. Williams; *J. Chem. Soc. (B)*, 1967, 859. | 599, 778, 783 |
| 67JCS(B)892 | T. J. Batterham, A. C. K. Triffett and J. A. Wunderlich; *J. Chem. Soc. (B)*, 1967, 892. | 66 |
| 67JCS(B)950 | J. W. Bunting and D. D. Perrin; *J. Chem. Soc. (B)*, 1967, 950. | 206 |
| 67JCS(B)1203 | G. Gatti, A. L. Segre and C. Morandi; *J. Chem. Soc. (B)*, 1967, 1203. | 628 |
| 67JCS(B)1243 | A. R. Katritzky, E. Lunt, B. Ternai and G. J. T. Tiddy; *J. Chem. Soc. (B)*, 1967, 1243. | 234 |
| 67JCS(C)145 | P. H. McCabe, R. McCrindle and R. D. H. Murray; *J. Chem. Soc. (C)*, 1967, 145. | 819 |
| 67JCS(C)226 | I. Fleming and M. H. Karger; *J. Chem. Soc. (C)*, 1967, 226. | 772 |

| | | |
|---|---|---|
| 67JCS(C)413 | K. D. Bartle, R. L. Edwards, D. W. Jones and I. Mir; *J. Chem. Soc. (C)*, 1967, 413. | 581 |
| 67JCS(C)568 | D. J. Brown and P. W. Ford; *J. Chem. Soc. (C)*, 1967, 568. | 96, 98, 101, 102, 128, 134, 136, 138 |
| 67JCS(C)573 | M. E. C. Biffin, D. J. Brown and T. C. Lee; *J. Chem. Soc. (C)*, 1967, 573. | 62, 72, 87, 99, 124, 130 |
| 67JCS(C)785 | B. Jackson, H. D. Locksley and F. Scheinmann; *J. Chem. Soc. (C)*, 1967, 785. | 601 |
| 67JCS(C)828 | E. B. Mullock and H. Suschitzky; *J. Chem. Soc. (C)*, 1967, 828. | 582, 814 |
| 67JCS(C)1172 | R. S. Shadbolt and T. L. V. Ulbricht; *J. Chem. Soc. (C)*, 1967, 1172. | 128 |
| 67JCS(C)1178 | D. McHale; *J. Chem. Soc. (C)*, 1967, 1178. | 1042, 1061, 1072 |
| 67JCS(C)1204 | M. P. L. Caton, D. T. Hurst, J. F. W. McOmie and R. R. Hunt; *J. Chem. Soc. (C)*, 1967, 1204. | 83, 85, 99, 103, 137, 140 |
| 67JCS(C)1279 | G. Tennant; *J. Chem. Soc. (C)*, 1967, 1279. | 415 |
| 67JCS(C)1343 | C. E. Loader and C. J. Timmons; *J. Chem. Soc. (C)*, 1967, 1343. | 76 |
| 67JCS(C)1472 | J. B. Abbott, C. J. France, R. Livingstone and D. P. Morrey; *J. Chem. Soc. (C)*, 1967, 1472. | 856 |
| 67JCS(C)1528 | D. H. Hayes and F. Hayes-Baron; *J. Chem. Soc. (C)*, 1967, 1528. | 143, 145 |
| 67JCS(C)1543 | J. Clark; *J. Chem. Soc. (C)*, 1967, 1543. | 276 |
| 67JCS(C)1745 | W. J. Irwin and D. G. Wibberley; *J. Chem. Soc. (C)*, 1967, 1745. | 85, 220 |
| 67JCS(C)1822 | R. E. Banks, D. S. Field and R. N. Haszeldine; *J. Chem. Soc. (C)*, 1967, 1822. | 63 |
| 67JCS(C)1922 | D. J. Brown and B. T. England; *J. Chem. Soc. (C)*, 1967, 1922. | 89, 112 |
| 67JCS(C)1928 | D. J. Brown and M. N. Paddon-Row; *J. Chem. Soc. (C)*, 1967, 1928. | 83, 85, 103, 112 |
| 67JCS(C)1933 | A. Pelter and P. Stainton; *J. Chem. Soc. (C)*, 1967, 1933. | 603, 604, 751 |
| 67JCS(C)2000 | E. V. Lassak and J. T. Pinhey; *J. Chem. Soc. (C)*, 1967, 2000. | 582 |
| 67JCS(C)2206 | E. D. Bergmann, I. Shahak and I. Gruenwald; *J. Chem. Soc. (C)*, 1967, 2206. | 152 |
| 67JCS(C)2500 | B. Jackson, H. D. Locksley and F. Scheinmann; *J. Chem. Soc. (C)*, 1967, 2500. | 584, 597 |
| 67JCS(C)2545 | L. Crombie, D. E. Games and A. McCormick; *J. Chem. Soc. (C)*, 1967, 2545. | 600 |
| 67JCS(C)2562 | J. A. Baker and S. A. Hill; *J. Chem. Soc. (C)*, 1967, 2562. | 1046 |
| 67JCS(C)2613 | A. G. Ismail and D. G. Wibberley; *J. Chem. Soc. (C)*, 1967, 2613. | 216 |
| 67JCS(C)2658 | G. Tennant; *J. Chem. Soc. (C)*, 1967, 2658. | 415 |
| 67JHC12 | H. Zondler, H. S. Forrest and J. M. Lagowski; *J. Heterocycl. Chem.*, 1967, **4**, 12. | 315 |
| 67JHC49 | D. E. O'Brien, L. T. Weinstock, R. H. Springer and C. C. Cheng; *J. Heterocycl. Chem.*, 1967, **4**, 49. | 71 |
| 67JHC61 | K. R. Markham, W. Rahman, S. Jehan and T. J. Mabry; *J. Heterocycl. Chem.*, 1967, **4**, 61. | 584 |
| 67JHC124 | H. Zondler, H. S. Forrest and J. M. Lagowski; *J. Heterocycl. Chem.*, 1967, **4**, 124. | 315 |
| 67JHC163 | K.-Y. Zee-Cheng and C. C. Cheng; *J. Heterocycl. Chem.*, 1967, **4**, 163. | 76, 125 |
| 67JHC224 | W. W. Paudler and R. E. Herbener; *J. Heterocycl. Chem.*, 1967, **4**, 224. | 396 |
| 67JHC290 | B. Loev and G. Dudek; *J. Heterocycl. Chem.*, 1967, **4**, 290. | 1081 |
| 67JHC393 | G. M. Singerman and R. N. Castle; *J. Heterocycl. Chem.*, 1967, **4**, 393. | 332, 342, 353 |
| 67JHC491 | L. C. Dorman; *J. Heterocycl. Chem.*, 1967, **4**, 491. | 342 |
| 67JMC316 | B. R. Baker and M. Kawazu; *J. Med. Chem.*, 1967, **10**, 316. | 138 |
| 67JMC431 | T. S. Osdene, P. B. Russell and L. Rane; *J. Med. Chem.*, 1967, **10**, 431. | 325 |
| 67JMC501 | C. R. Johnson and I. Sataty; *J. Med. Chem.*, 1967, **10**, 501. | 1031 |
| 67JMC883 | W. E. Taft and R. G. Shepherd; *J. Med. Chem.*, 1967, **10**, 883. | 390 |
| 67JMC899 | J. H. Jones, J. B. Bicking and E. J. Cragoe, Jr.; *J. Med. Chem.*, 1967, **10**, 899. | 292 |
| 67JOC231 | H. C. Brown, P. D. Shuman and J. Turnbull; *J. Org. Chem.*, 1967, **32**, 231. | 504 |
| 67JOC457 | Y. Hayashi and R. Oda; *J. Org. Chem.*, 1967, **32**, 457. | 990 |
| 67JOC1378 | R. A. Archer and H. S. Mosher; *J. Org. Chem.*, 1967, **32**, 1378. | 249, 252 |
| 67JOC2308 | C. Grundmann and R. Richter; *J. Org. Chem.*, 1967, **32**, 2308. | 79, 84 |
| 67JOC2376 | E. C. Taylor and J. G. Berger; *J. Org. Chem.*, 1967, **32**, 2376. | 122 |
| 67JOC2379 | E. C. Taylor and R. W. Morrison, Jr.; *J. Org. Chem.*, 1967, **32**, 2379. | 115 |
| 67JOC2960 | R. J. Shozda; *J. Org. Chem.*, 1967, **32**, 2960. | 1081 |
| 67JOC3140 | H. J. Teague and W. P. Tucker; *J. Org. Chem.*, 1967, **32**, 3140. | 937 |
| 67JOC3616 | G. A. Reynolds and J. A. Van Allan; *J. Org. Chem.*, 1967, **32**, 3616. | 662, 756 |
| 67JOC3772 | G. A. Reynolds, J. A. Van Allan and T. H. Regan; *J. Org. Chem.*, 1967, **32**, 3772. | 662 |
| 67JOC3856 | Y. Rahamim, J. Sharvit, A. Mandelbaum and M. Sprecher; *J. Org. Chem.*, 1967, **32**, 3856. | 112 |
| 67JOC4105 | L. C. Dorman; *J. Org. Chem.*, 1967, **32**, 4105. | 811 |
| 67JOU1809 | G. N. Dorofeenko, V. V. Merzheritskii and B. I. Ardashev; *J. Org. Chem. USSR (Engl. Transl.)*, 1967, **3**, 1809. | 663 |
| 67JPC4045 | J. Kolc and R. S. Becker; *J. Phys. Chem.*, 1967, **71**, 4045. | 667, 741 |
| 67JST(1)489 | N. K. Ray and P. T. Narasimhan; *J. Mol. Struct.*, 1967–68, **1**, 489. | 945 |
| 67KGS395 | E. Grinsteins, E. I. Stankevich and G. Duburs; *Khim. Geterotsikl. Soedin.*, 1967, 395. | 231 |
| 67KGS406 | J. Pelcere, E. Grinsteins, E. I. Stankevich and G. Vanags; *Khim. Geterotsikl. Soedin.*, 1967, 406. | 231 |
| 67KGS419 | A. S. Elina and I. S. Musatova; *Khim. Geterotsikl. Soedin.*, 1967, 419. | 173 |
| 67KGS782 | V. A. Zagorevskii, Sh. M. Glozman, V. G. Vinokurov and V. S. Troitskaya; *Khim. Geterotsikl. Soedin.*, 1967, 782. | 644 |
| 67LA(703)96 | R. W. Hoffmann and W. Sieber; *Liebigs Ann. Chem.*, 1967, **703**, 96. | 1064 |
| 67LA(703)140 | R. Mayer, G. Laban and M. Wirth; *Liebigs Ann. Chem.*, 1967, **703**, 140. | 935 |
| 67LA(704)144 | H.-J. Kabbe; *Liebigs Ann. Chem.*, 1967, **704**, 144. | 116 |

| 67LA(707)250 | W. Ried and J. Valentin; *Liebigs Ann. Chem.*, 1967, **707**, 250. | 226 |
|---|---|---|
| 67M726 | R. Kuhn, F. A. Neugebauer and H. Trischmann; *Monatsh. Chem.*, 1967, **98**, 726. | 546 |
| 67M1577 | H. Bretschneider and H. Egg; *Monatsh. Chem.*, 1967, **98**, 1577. | 96 |
| 67M1613 | W. Gottardi; *Monatsh. Chem.*, 1967, **98**, 1613. | 477 |
| 67MI21200 | J. Cooke, M. Green and F. G. A. Stone; *Inorg. Nucl. Chem. Lett.*, 1967, **3**, 47. | 37 |
| 67MI21300 | R. D. Brown and B. A. W. Coller; *Theor. Chim. Acta*, 1967, **7**, 259. | 59, 76 |
| 67MI21301 | H. L. Pease and J. F. Deye; *Anal. Methods Pestic., Plant Growth Regul., Food Addit.*, 1967, **5**, 335. | 155 |
| B-67MI21400 | M. H. Palmer; in 'The Structure and Reactions of Heterocyclic Compounds', Arnold, London, 1967. | 158 |
| 67MI21401 | H. A. Bondarovich, P. Friedel, V. Krampl, J. A. Renner, F. W. Shepard and M. A. Gianturco; *J. Agric. Food Chem.*, 1967, **15**, 1093. | 193 |
| 67MI21402 | K. Sato, O. Shiratori and K. Katagiri; *J. Antibiot.*, 1967, **A20**, 270. | 195 |
| 67MI21500 | E. Domagalina and I. Kurpiel; *Rocz. Chem.*, 1967, **41**, 1241 (*Chem. Abstr.*, 1968, **68**, 39 574). | 237, 240 |
| 67MI21501 | B. Tinland; *Theor. Chim. Acta*, 1967, **8**, 361 (*Chem. Abstr.*, 1968, **68**, 100 246). | 250 |
| 67MI21600 | J. R. Bertino and D. G. Johns; *Annu. Rev. Med.*, 1967, **18**, 27., | 327 |
| 67MI21601 | A. S. Sakurai and M. Goto, *J. Biochem.*, 1967, **61**, 142. | 324 |
| 67MI21602 | T. Sugimoto and S. Matsuura; *Res. Bull.*, 1967, **11**, 94. | 309 |
| 67MI21900 | L. Pallos and P. Benko; *Ind. Chim. Belg.*, 1967, **32**, 1334. | 439 |
| B-67MI22100 | V. P. Wystrach; 'Heterocyclic Compounds', Wiley, New York, 1967, vol. 8. | 531 |
| B-67MI22101 | V. P. Wystrach; 'Heterocyclic Compounds', Wiley, New York, 1967, vol. 8, p. 105. | 536, 557 |
| B-67MI22102 | 'UV Atlas of Organic Compounds', Verlag Chemie, Weinheim, 1967, vols. G8/1, G8/2. | 540 |
| 67MI22200 | M. E. Perel'son and Yu. N. Sheinker; *Teor. Eksp. Khim.*, 1967, **3**, 697. | 575 |
| B-67MI22201 | J. B. Harborne; 'Comparative Biochemistry of the Flavonoids', Academic Press, London, 1967. | 602 |
| 67MI22202 | H. Nakata and A. Tatematsu; *Shitsuryo Bunseki*, 1967, **15**, 5 (*Chem. Abstr.*, 1968, **68**, 77 533). | 611 |
| 67MI22203 | H. Nakata and A. Tatematsu; *Shitsuryo Bunseki*, 1967, **15**, 184 (*Chem. Abstr.*, 1968, **69**, 77 052). | 611 |
| B-67MI22204 | H. Budzikiewicz, C. Djerassi and D. H. Williams; 'Mass Spectrometry of Organic Compounds', Holden-Day, San Francisco, 1967, p. 119. | 612 |
| 67MI22205 | A. Julg and P. François; *Theor. Chim. Acta*, 1967, **8**, 249. | 633 |
| B-67MI22206 | F. A. Bovey; 'N.m.r. Data Tables for Organic Compounds', Wiley, New York, 1967, vol. 1, compound No. 672. | 577 |
| 67MI22400 | I. Degani and C. Vicenzi; *Boll. Sci. Fac. Chim. Ind. Bologna*, 1967, **25**, 51 (*Chem. Abstr.*, 1968, **68**, 29 543). | 757 |
| 67MI22401 | E. Gârd, A. Runge, A. Barabas and A. T. Balaban; *J. Labelled Compd. Radiopharm.*, 1967, **3**, 151. | 861 |
| 67MI22402 | G. I. Zhungietu; *Metody Poluch. Khim. Reaktivov Prep.*, 1967, **17**, 91 (*Chem. Abstr.*, 1969, **71**, 12 952). | 862 |
| B-67MI22403 | L. F. Fieser and M. Fieser; 'Reagents for Organic Synthesis', Wiley, New York, 1967. | 883 |
| 67MI22600 | R. J. Wratten and M. A. Ali; *Mol. Phys.*, 1967, **13**, 233. | 946 |
| B-67MI22601 | L. H. Werner and W. E. Barrett; in 'Antihypertensive Agents', ed. E. Schlittler; Academic Press, New York and London, 1967, p. 331. | 993 |
| B-67MI22700 | J. Hamer and M. Ahmad; in '1,4-Cycloaddition Reactions', ed. J. Hamer; Academic Press, New York, 1967, p. 419. | 1017 |
| 67MIP21300 | E. Habicht; *Swiss Pat.* 43 3342 (1967) (*Chem. Abstr.*, 1968, **68**, 114625). | 138. |
| 67NKK1320 | H. Yamakami, A. Sakurai and M. Goto; *Nippon Kagaku Zasshi*, 1967, **88**, 1320. | 285 |
| 67OR(15)204 | G. Jones; *Org. React.*, 1967, **15**, 204. | 803, 805 |
| 67QR364 | F. G. Riddell; *Q. Rev., Chem. Soc.* 1967, **21**, 364. | 629 |
| 67RCR175 | N. P. Shusherina, N. D. Dmitrieva, E. A. Luk'yanets and R. Ya. Levina; *Russ. Chem. Rev. (Engl. Transl.)*, 1967, **36**, 175. | 789 |
| 67RRC913 | C. Cristescu and V. Badea; *Rev. Roum. Chim.*, 1967, **12**, 913. | 416 |
| 67RTC567 | H. W. van Meeteren and H. C. van der Plas; *Recl. Trav. Chim. Pays-Bas*, 1967, **86**, 567. | 84 |
| 67RTC907 | H. E. Wijers, C. H. D. van Ginkel, L. Brandsma and J. F. Arens; *Recl. Trav. Chim. Pays-Bas*, 1967, **86**, 907. | 566, 569, 570 |
| 67RTC1275 | E. Boelema, G. J. Visser and A. Vos; *Rec. Trav. Chim. Pays-Bas*, 1967, **86**, 1275. | 881, 892 |
| 67SA(A)55 | R. E. Mayo and J. H. Goldstein; *Spectrochim. Acta, Part A*, 1967, **23**, 55. | 582, 588 |
| 67SA(A)391 | R. G. Snyder and G. Zerbi; *Spectrochim. Acta, Part A*, 1967, **23**, 391. | 628 |
| 67SA(A)2551 | R. C. Lord and G. J. Thomas, Jr.; *Spectrochim. Acta, Part A*, 1967, **23**, 2551. | 68 |
| 67T341 | B. J. Bolger, K. G. Marathe, E. M. Philbin, T. S. Wheeler and C. P. Lillya; *Tetrahedron*, 1967, **23**, 341. | 668 |
| 67T353 | G. Zvilichovsky; *Tetrahedron*, 1967, **23**, 353. | 112, 141 |
| 67T891 | E. C. Taylor, A. McKillop and R. N. Warrener; *Tetrahedron*, 1967, **23**, 891. | 225 |
| 67T1565 | G. N. Dorofeenko, Yu. A. Zhdanov, G. I. Zhungietu and S. W. Krivun; *Tetrahedron*, 1967, **23**, 1565. | 864 |
| 67T1893 | B. S. Thyagarajan, K. K. Balasubramanian and R. Bhima Rao; *Tetrahedron*, 1967, **23**, 1893. | 743 |
| 67T3193 | I. Rosenthal and D. Elad; *Tetrahedron*, 1967, **23**, 3193. | 730 |

| | | |
|---|---|---|
| 67T3505 | L. W. Deady; *Tetrahedron*, 1967, **23**, 3505. | 1014 |
| 67T4001 | A. T. Balaban, A. R. Katritzky and B. M. Semple; *Tetrahedron*, 1967, **23**, 4001. | 868 |
| 67T4741 | W. D. Ollis, C. A. Rhodes and I. O. Sutherland; *Tetrahedron*, 1967, **23**, 4741. | 618 |
| 67TL1099 | J. Clark; *Tetrahedron Lett.*, 1967, 1099. | 276 |
| 67TL3913 | F. Lahmani and N. Ivanoff; *Tetrahedron Lett.*, 1967, 3913. | 120 |
| 67TL4201 | A. Arnone, G. Cardillo, L. Merlini and R. Mondelli; *Tetrahedron Lett.*, 1967, 4201. | 580 |
| 67TL5119 | R. W. Johnson and N. V. Riggs; *Tetrahedron Lett.*, 1967, 5119. | 581 |
| 67USP3320257 | G. Y. Lesher; (Sterling Drug Inc.), *U.S. Pat.* 3 320 257 (1967) (*Chem. Abstr.*, 1968, **68**, 49 643). | 221 |
| 67USP3344020 | Hooker Chemical Corp., *U.S. Pat.* 3 344 020 (1967) (*Chem. Abstr.*, 1968, **68**, 28 760). | 940 |
| 67YZ942 | C. Iijima; *Yakugaku Zasshi*, 1967, **87**, 942. | 171 |
| 67YZ955 | T. Kato, H. Yamanaka and T. Shibata; *Yakugaku Zasshi*, 1967, **87**, 955. | 107 |
| 67YZ1096 | T. Kato, H. Yamanaka and T. Shibata; *Yakugaku Zasshi*, 1967, **87**, 1096. | 75, 141 |
| 67YZ1315 | T. Okano, S. Goya and H. Matsumoto; *Yakugaku Zasshi*, 1967, **87**, 1315. | 140 |
| 67ZC152 | W. Schroth, B. Streckenbach and B. Werner; *Z. Chem.*, 1967, **7**, 152. | 982, 984 |
| 67ZC227 | G. Laban and R. Mayer; *Z. Chem.*, 1967, **7**, 227. | 937 |
| 67ZN(B)105 | C. Schiele, K. Halfar and G. Arnold; *Z. Naturforsch., Teil B*, 1967, **22**, 105. | 542 |
| | | |
| 68AC(R)251 | I. Degani and R. Fochi; *Ann. Chim. (Rome)*, 1968, **58**, 251. | 810, 840 |
| 68AC(R)664 | E. Niccoli, U. Vaglini and G. Ceoccarelli; *Ann. Chim. (Rome)*, 1968, **58**, 664. | 642 |
| 68ACH(58)179 | J. Szabó, E. Vinkler and I. Varga; *Acta Chim. Acad. Sci. Hung.*, 1968, **58**, 179. | 1001 |
| 68ACS352 | T. Anthonsen; *Acta Chem. Scand.*, 1968, **22**, 352. | 580 |
| 68ACS3160 | J. L. G. Nilsson, H. Sievertsson and H. Selander; *Acta Chem. Scand.*, 1968, **22**, 3160. | 599 |
| 68AG316 | G. Ege and E. Beisiegel; *Angew. Chem.*, 1968, **80**, 316. | 381 |
| 68AG(E)172 | R. Graf; *Angew. Chem., Int. Ed. Engl.*, 1968, **7**, 172. | 1044, 1081 |
| 68AG(E)301 | F. Fehér, B. Degen and B. Söhngen; *Angew. Chem., Int. Ed. Engl.*, 1968, **7**, 301. | 955, 962, 992 |
| 68AG(E)461 | K. Dimroth and W. Mach; *Angew. Chem., Int. Ed. Engl.*, 1968, **7**, 461. | 659 |
| 68AHC(9)211 | M. Tišler and B. Stanovnik; *Adv. Heterocycl. Chem.*, 1968, **9**, 211. | 2, 10 |
| 68AHC(9)321 | C. Bodea and I. Silberg; *Adv. Heterocycl. Chem.*, 1968, **9**, 321. | 1038 |
| 68AJC243 | D. J. Brown and T.-C. Lee; *Aust. J. Chem.*, 1968, **21**, 243. | 93, 98, 112, 134 |
| 68AJC997 | J. L. Occolowitz and G. L. White; *Aust. J. Chem.*, 1968, **21**, 997. | 608 |
| 68AJC1291 | D. B. Paul and H. J. Rodda; *Aust. J. Chem.*, 1968, **21**, 1291. | 234, 236, 237, 238, 240, 241, 246 |
| 68AJC2059 | J. W. Clark-Lewis; *Aust. J. Chem.*, 1968, **21**, 2059. | 584 |
| 68AJC2247 | J. W. Clark-Lewis and R. W. Jemison; *Aust. J. Chem.*, 1968, **21**, 2247. | 751, 756 |
| 68AJC2445 | M. W. Jarvis and A. G. Moritz; *Aust. J. Chem.*, 1968, **21**, 2445. | 582 |
| 68AJC2819 | G. W. K. Cavill, D. V. Clark and F. B. Whitfield; *Aust. J. Chem.*, 1968, **21**, 2819. | 617 |
| 68AP923 | G. Wagner and H. Gentzsch; *Arch. Pharm. (Weinheim Ger.)*, 1968, **301**, 923. | 377 |
| 68BCJ1368 | E. Haruki, H. Imanaka and E. Imoto; *Bull. Chem. Soc. Jpn.*, 1968, **41**, 1368. | 512 |
| 68BCJ2798 | H. Obara and J-i. Onodera; *Bull. Chem. Soc. Jpn.*, 1968, **41**, 2798. | 644 |
| 68BRP1043857 | J. Augstein and A. Monro; *Br. Pat.* 1 043 857 (1966) (*Chem. Abstr.*, 1966, **76**, 18 562). | 675 |
| 68BSF4122 | H. Strzelecka and M. Simalty; *Bull. Soc. Chim. Fr.*, 1968, 4122. | 868 |
| 68BSF4203 | F. Baranton, G. Fontaine and P. Maitte; *Bull. Soc. Chim. Fr.*, 1968, 4203. | 595, 642 669, 856 |
| 68BSF4970 | N. Vinot and J. Pinson; *Bull. Soc. Chim. Fr.*, 1968, 4970. | 185 |
| 68CB29 | T. Pyl, L. Seidl and H. Beyer; *Chem. Ber.*, 1968, **101**, 29. | 435 |
| 68CB512 | H. Bredereck, G. Simchen, R. Wahl and F. Effenberger; *Chem. Ber.*, 1968, **101**, 512. | 227 |
| 68CB1072 | W. Pfleiderer, J. W. Bunting, D. D. Perrin and G. Nübel; *Chem. Ber.*, 1968, **101**, 1072. | 277, 309 |
| 68CB2215 | K. Dimroth, K. Vogel and W. Krafft; *Chem. Ber.*, 1968, **101**, 2215. | 870 |
| 68CB2494 | K. Drescher, H. Hofmann and K.-H. Frömming; *Chem. Ber.*, 1968, **101**, 2494. | 852 |
| 68CB3567 | H.-H. Hörhold and H. Eibisch; *Chem. Ber.*, 1968, **101**, 3567. | 1041, 1046, 1076 |
| 68CB3952 | H. Neunhoeffer and H. Hennig; *Chem. Ber.*, 1968, **101**, 3952. | 395, 430, 433 |
| 68CC289 | A. R. Katritzky, M. Kingsland and O. S. Tee; *Chem. Commun.*, 1968, 289. | 72 |
| 68CC1028 | I. Nicholson; *Chem. Commun.*, 1968, 1028. | 393 |
| 68CC1656 | J. D. Bell, F. Blount, O. V. Briscoe and H. C. Freeman; *Chem. Commun.*, 1968, 1656. | 1010 |
| 68CCC2087 | J. Gut, A. Nováček and P. Fiedler; *Collect. Czech. Chem. Commun.*, 1968, **33**, 2087. | 404, 448 |
| 68CCC2962 | K. Kalfus; *Collect. Czech. Chem. Commun.*, 1968, **33**, 2962. | 390, 399 |
| 68CJC3367 | U. T. Bhalerao and G. Thyagarajan; *Can. J. Chem.*, 1968, **46**, 3367. | 855 |
| 68CPB1157 | Y. Shoyama, T. Fujita, T. Yamauchi and I. Nishioka; *Chem. Pharm. Bull.*, 1968, **16**, 1157. | 721 |
| 68CPB1337 | T. Kato, H. Yamanaka and H. Hiranuma; *Chem. Pharm. Bull.*, 1968, **16**, 1337. | 75 |
| 68CPB2093 | M. Ichikawa and H. Ichibagase; *Chem. Pharm. Bull.*, 1968, **16**, 2093. | 691 |
| 68CR(C)(266)1459 | G. Queguiner and P. Pastour; *C.R. Hebd. Seances Acad. Sci., Ser. C*, 1968, **266**, 1459. | 236 |
| 68CR(C)(267)904 | M. Pesson and M. Antoine; *C. R. Hebd. Seances Acad. Sci., Ser. C*, 1968, **267**, 904. | 403 |

| | | |
|---|---|---|
| 68CR(C)(267)1726 | M. Pesson and M. Antoine; *C. R. Hebd. Seances Acad. Sci., Ser. C*, 1968, **267**, 1726. | 403 |
| 68E774 | L. Bernardi, S. Coda, L. Pegrassi and G. K. Suchowsky; *Experientia*, 1968, **24**, 774. | 1028 |
| 68FRP6157 | Centre d'Etudes pour l'Industrie Pharmaceutique, *Fr. Pat.* 6157 (1968) (*Chem. Abstr.*, 1970, **72**, 3493). | 993 |
| 68FRP1508645 | H. Distler; *Fr. Pat.* 1 508 645 (1968) (*Chem. Abstr.*, 1969, **70**, 57139). | 1044 |
| 68FRP1522354 | Kodak Ltd., *Fr. Pat.* 1 522 354 (1968) (*Chem. Abstr.*, 1970, **72**, 17 279). | 942 |
| 68G17 | D. Sica, C. Santacroce and R. A. Nicolaus; *Gazz. Chim. Ital.*, 1968, **98**, 17. | 1010 |
| 68HCA249 | M. Neuenschwander and H. Schaltegger; *Helv. Chim. Acta*, 1968, **51**, 249. | 485 |
| 68HCA607 | A. Bobst; *Helv. Chim. Acta*, 1968, **51**, 607. | 282 |
| 68HCA1029 | M. Viscontini and H. Leidner; *Helv. Chim. Acta*, 1968, **51**, 1029. | 279, 306, 312 |
| 68HCA1495 | M. Viscontini and R. Provenzale; *Helv. Chim. Acta*, 1968, **51**, 1495. | 312 |
| 68HCA1510 | J. Zsindely and H. Schmid; *Helv. Chim. Acta*, 1968, **51**, 1510. | 742 |
| 68IZV918 | V. I. Gunar, L. F. Ovechkina, I. A. Mikhailopulo and S. I. Zav'yalov; *Izv. Akad. Nauk SSSR, Ser. Khim.*, 1968, 918. | 143 |
| 68JA553 | L. Salem; *J. Am. Chem. Soc.*, 1968, **90**, 553. | 771 |
| 68JA697 | R. J. Pugmire and D. M. Grant; *J. Am. Chem. Soc.*, 1968, **90**, 697. | 63, 159 |
| 68JA811 | H. J. Dauben, Jr., J. D. Wilson and J. L. Laity; *J. Am. Chem. Soc.*, 1968, **90**, 811. | 635 |
| 68JA1061 | B. Fuchs and M. Rosenblum; *J. Am. Chem. Soc.*, 1968, **90**, 1061. | 1064 |
| 68JA1678 | T. Ueda and H. Nishino; *J. Am. Chem. Soc.*, 1968, **90**, 1678. | 93 |
| 68JA2424 | E. C. Taylor, Jr., and K. Lenard; *J. Am. Chem. Soc.*, 1968, **90**, 2424. | 281, 318 |
| 68JA3444 | E. L. Eliel and S. M. C. Knoeber; *J. Am. Chem. Soc.*, 1968, **90**, 3444. | 961 |
| 68JA4723 | W. L. Parker, J. J. Flynn and F. P. Boer; *J. Am. Chem. Soc.*, 1968, **90**, 4723. | 621 |
| 68JA6543 | E. W. Garbisch, Jr. and M. G. Griffith; *J. Am. Chem. Soc.*, 1968, **90**, 6543. | 579 |
| 68JAP6808278 | Y. Miyake, S. Ozaki and T. Kato; *Jpn. Pat.* 68 08 278 (1968) (*Chem. Abstr.*, 1968, **69**, 106 719). | 1030 |
| 68JAP6813469 | S. Fujii and H. Hideo; *Jpn. Pat.* 68 13 469 (1968). | 185 |
| 68JCS(B)1435 | G. B. Barlin and W. V. Brown; *J. Chem. Soc. (B)*, 1968, 1435. | 165 |
| 68JCS(B)1467 | M. J. Cook, H. Dorn and A. R. Katritzky; *J. Chem. Soc. (B)*, 1968, 1467. | 896 |
| 68JCS(C)63 | A. Albert and J. J. McCormack; *J. Chem. Soc. (C)*, 1968, 63. | 273, 279 |
| 68JCS(C)313 | J. Clark; *J. Chem. Soc. (C)*, 1968, 313. | 276 |
| 68JCS(C)496 | S. Bien, G. Salemnik, L. Zamir and M. Rosenblum; *J. Chem. Soc. (C)*, 1968, 496. | 144 |
| 68JCS(C)543 | F. C. Cheng and S. F. Tan; *J. Chem. Soc. (C)*, 1968, 543. | 639 |
| 68JCS(C)1028 | D. H. Hey, C. W. Rees and A. R. Todd; *J. Chem. Soc. (C)*, 1968, 1028. | 379 |
| 68JCS(C)1124 | J. Clark and W. Pendergast; *J. Chem. Soc. (C)*, 1968, 1124. | 277 |
| 68JCS(C)1203 | R. S. Shadbolt and T. L. V. Ulbricht; *J. Chem. Soc. (C)*, 1968, 1203. | 79, 83 |
| 68JCS(C)1452 | D. J. Brown, P. W. Ford and M. N. Paddon-Row; *J. Chem. Soc. (C)*, 1968, 1452. | 72, 73, 86, 98 |
| 68JCS(C)1837 | J. A. Miller and H. C. S. Wood; *J. Chem. Soc. (C)*, 1968, 1837. | 594, 780 |
| 68JCS(C)2205 | A. Rose and N. P. Buu-Hoï; *J. Chem. Soc. (C)*, 1968, 2205. | 685, 691 |
| 68JCS(C)2292 | A. Albert and H. Yamamoto; *J. Chem. Soc. (C)*, 1968, 2292. | 266 |
| 68JCS(C)2302 | G. A. Caplin, W. D. Ollis and I. O. Sutherland; *J. Chem. Soc. (C)*, 1968, 2302. | 856 |
| 68JCS(C)2367 | B. K. Snell; *J. Chem. Soc. (C)*, 1968, 2367. | 138 |
| 68JCS(C)2477 | R. G. Wilson and D. H. Williams; *J. Chem. Soc. (C)*, 1968, 2477. | 585 |
| 68JCS(C)2533 | I. McMillan and R. J. Stoodley; *J. Chem. Soc. (C)*, 1968, 2533. | 1035 |
| 68JCS(C)2706 | A. G. Ismail and D. G. Wibberley; *J. Chem. Soc. (C)*, 1968, 2706. | 204 |
| 68JCS(C)2730 | H. E. Crabtree, R. K. Smalley and H. Suschitzky; *J. Chem. Soc. (C)*, 1968, 2730. | 379 |
| 68JCS(C)2756 | A. L. J. Beckwith and R. J. Hickman; *J. Chem. Soc. (C)*, 1968, 2756. | 215 |
| 68JCS(C)2857 | G. Adembri, F. DeSio, R. Nesi and M. Scotton; *J. Chem. Soc. (C)*, 1968, 2857. | 333 |
| 68JCS(C)3029 | L. Crombie, C. L. Green and D. A. Whiting; *J. Chem. Soc. (C)*, 1968, 3029. | 875 |
| 68JHC13 | N. R. Patel, W. M. Rich and R. N. Castle; *J. Heterocycl. Chem.*, 1968, **5**, 13. | 242 |
| 68JHC53 | L. Distefano and R. N. Castle; *J. Heterocycl. Chem.*, 1968, **5**, 53. | 337, 343, 347, 348, 354, 358 |
| 68JHC111 | L. DiStefano and R. N. Castle; *J. Heterocycl. Chem.*, 1968, **5**, 111. | 47 |
| 68JHC133 | G. Grandolini, A. Ricci, N. P. Buu-Hoï and F. Périn; *J. Heterocycl. Chem.*, 1968, **5**, 133. | 583 |
| 68JHC211 | J. P. Chupp and H. K. Landwehr; *J. Heterocycl. Chem.*, 1968, **5**, 211. | 487 |
| 68JHC275 | J. T. Kurek and G. Vogel; *J. Heterocycl. Chem.*, 1968, **5**, 275. | 581, 582, 595, 599 |
| 68JHC453 | J. B. Wright; *J. Heterocycl. Chem.*, 1968, **5**, 453. | 1043, 1058 |
| 68JHC523 | T. Nakagome, R. N. Castle and H. Murakami; *J. Heterocycl. Chem.*, 1968, **5**, 523. | 336, 344 |
| 68JHC533 | L. Paoloni, M. L. Tosato and M. Cignitti; *J. Heterocycl. Chem.*, 1968, **5**, 533. | 480 |
| 68JHC745 | A. R. Katritzky and B. Ternai; *J. Heterocycl. Chem.*, 1968, **5**, 745. | 583, 597, 891 |
| 68JHC845 | T. Kinoshita and R. N. Castle; *J. Heterocycl. Chem.*, 1968, **5**, 845. | 337, 345, 347, 358 |
| 68JMC322 | J. H. Jones and E. J. Cragoe, Jr.; *J. Med. Chem.*, 1968, **11**, 322. | 292 |
| 68JMC542 | J. Weinstock, R. Y. Dunoff and J. G. Williams; *J. Med. Chem.*, 1968, **11**, 542. | 304 |
| 68JMC549 | J. Weinstock, R. Y. Dunoff, B. Sutton, B. Trost, J. Kirkpatrick, F. Farina and A. S. Straub; *J. Med. Chem.*, 1968, **11**, 549. | 313 |
| 68JMC560 | J. Weinstock, H. Graboyes, G. Jaffe, I. J. Pachter, K. Snader, C. B. Karash and R. Dunoff; *J. Med. Chem.*, 1968, **11**, 560. | 315 |
| 68JMC568 | G. Graboyes, G. E. Jaffe, I. J. Pachter, J. P. Rosenbloom, A. J. Villani, J. W. Wilson and J. Weinstock; *J. Med. Chem.*, 1968, **11**, 568. | 338, 363, 364 |
| 68JMC573 | J. Weinstock, J. W. Wilson, V. D. Wiebelhaus, A. R. Maass, F. T. Brennan and G. Sosnowski; *J. Med. Chem.*, 1968, **11**, 573. | 368 |

| | | |
|---|---|---|
| 68JMC703 | B. S. Hurlbert, K. W. Ledig, P. Stenbuck, B. F. Valenti and G. H. Hitchings; *J. Med. Chem.*, 1968, **11**, 703. | 229, 230 |
| 68JMC708 | B. S. Hurlbert and B. F. Valenti; *J. Med. Chem.*, 1968, **11**, 708. | 229 |
| 68JMC844 | J. Augstein, A. M. Monro, G. W. H. Potter and P. Scholfield; *J. Med. Chem.*, 1968, **11**, 844. | 787 |
| 68JMC1045 | D. L. Trepanier, L. W. Rampy, K. L. Shriver, J. N. Eble and P. J. Shea; *J. Med. Chem.*, 1968, **11**, 1045. | 340, 351, 366 |
| 68JMC1167 | M. H. Shah, C. V. Deliwala and U. K. Sheth; *J. Med. Chem.*, 1968, **11**, 1167. | 492 |
| 68JOC446 | J. O. Hawthorne, E. L. Mihelic and M. S. Morgan; *J. Org. Chem.*, 1968, **33**, 446. | 970, 971 |
| 68JOC867 | K. T. Buck and R. A. Olofson; *J. Org. Chem.*, 1968, **33**, 867. | 761 |
| 68JOC1102 | J. A. Van Allan and G. A. Reynolds; *J. Org. Chem.*, 1968, **33**, 1102. | 869, 870 |
| 68JOC1202 | J. A. Vida and M. Gut; *J. Org. Chem.*, 1968, **33**, 1202. | 786 |
| 68JOC1571 | B. Gentili and R. M. Horowitz; *J. Org. Chem.*, 1968, **33**, 1571. | 585 |
| 68JOC1931 | M. Bertrand, S. Fliszár and Y. Rousseau; *J. Org. Chem.*, 1968, **33**, 1931. | 958 |
| 68JOC2266 | S. J. Isser, A. M. Duffield and C. Djerassi; *J. Org. Chem.*, 1968, **33**, 2266. | 605 |
| 68JOC2393 | R. D. Elliott, C. Temple, Jr. and J. A. Montgomery; *J. Org. Chem.*, 1968, **33**, 2393. | 251, 258 |
| 68JOC2399 | T. M. Harris and C. S. Combs, Jr.; *J. Org. Chem.*, 1968, **33**, 2399. | 796 |
| 68JOC2416 | E. E. Schweizer, J. Liehr and D. J. Monaco; *J. Org. Chem.*, 1968, **33**, 2416. | 577, 580, 598, 748 |
| 68JOC2446 | K. Sato and T. Amakasu; *J. Org. Chem.*, 1968, **33**, 2446. | 802 |
| 68JOC2572 | G. O. Pierson and O. A. Runquist; *J. Org. Chem.*, 1968, **33**, 2572. | 629 |
| 68JOC3297 | W. A. Sheppard; *J. Org. Chem.*, 1968, **33**, 3297. | 784 |
| 68JOC3354 | K. Sugino and T. Tanaka; *J. Org. Chem.*, 1968, **33**, 3354. | 349, 361 |
| 68JOC3754 | E. L. Eliel and C. A. Giza; *J. Org. Chem.*, 1968, **33**, 3754. | 777 |
| 68JOC4191 | G. H. Stout, G. K. Hickernell and K. D. Sears; *J. Org. Chem.*, 1968, **33**, 4191. | 624, 849 |
| 68JOC4418 | J. A. Van Allan, G. A. Reynolds and D. P. Maier; *J. Org. Chem.*, 1968, **33**, 4418. | 663, 699 |
| 68JOC4508 | L. J. Dolby, C. A. Elliger, S. Esfandiari and K. S. Marshall; *J. Org. Chem.*, 1968, **33**, 4508. | 780 |
| 68JPR(37)41 | R. Bohnensack, J. Fabian and R. Mayer; *J. Prakt. Chem.*, 1968, **37**, 41. | 892 |
| 68LA(712)201 | V. Prey and J. Bartsch; *Liebigs Ann. Chem.*, 1968, **712**, 201. | 626 |
| 68LA(715)223 | G. Kresze, C. Seyfried and A. Trede; *Liebigs Ann. Chem.*, 1968, **715**, 223. | 1067, 1075 |
| 68LA(716)143 | H. Weidinger and H. J. Sturm; *Liebigs Ann. Chem.*, 1968, **716**, 143. | 130 |
| 68M847 | W. Klötzer and M. Herberz; *Monatsh. Chem.*, 1968, **99**, 847. | 106, 141 |
| 68M1808 | J. Slouka; *Monatsh. Chem.*, 1968, **99**, 1808. | 443 |
| 68MI21200 | H. Feuer, E. P. Rosenquist and F. Brown; *Isr. J. Chem.*, 1968, **6**, 587. | 46 |
| B-68MI21300 | D. J. Brown; in 'Mechanisms of Molecular Migrations', ed. B. S. Thyagarajan; Wiley, New York, 1968, vol. 1, p. 209. | 86 |
| B-68MI21301 | B. Arantz and D. J. Brown; in 'Synthetic Procedures in Nucleic Acid Chemistry', ed. W. Zorbach and R. Tipson; Wiley, New York, 1968, vol. 1, p. 55. | 89, 94, 96, 132, 144 |
| B-68MI21302 | T. L. V. Ulbricht; in 'Synthetic Procedures in Nucleic Acid Chemistry', ed. W. Zorbach and R. Tipson; Wiley, New York, 1968, vol. 1, p. 66. | 145 |
| B-68MI21303 | G. D. Daves, F. Baiocchi, R. K. Robins, C. C. Cheng and W. H. Nyberg; in 'Synthetic Procedures in Nucleic Acid Chemistry', ed. W. Zorbach and R. Tipson; Wiley, New York, 1968, vol. 1, p. 72. | 146 |
| B-68MI21304 | H. C. Koppel, R. H. Springer, R. K. Robins and C. C. Cheng; in 'Synthetic Procedures in Nucleic Acid Chemistry', ed. W. Zorbach and R. Tipson; Wiley, New York, 1968, vol. 1, p. 86. | 147 |
| B-68MI21305 | W. D. Roberts and C. A. Dekker; in 'Synthetic Procedures in Nucleic Acid Chemistry', ed. W. Zorbach and R. Tipson; Wiley, New York, 1968, vol. 1, p. 296. | 153 |
| B-68MI21306 | T. L. V. Ulbricht; in 'Synthetic Procedures in Nucleic Acid Chemistry', ed. W. Zorbach and R. Tipson; Wiley, New York, 1968, vol. 1, p. 88. | 145 |
| B-68MI21307 | 'Synthetic Procedures in Nucleic Acid Chemistry', ed. W. W. Zorbach and R. S. Tipson; Wiley, New York, 1968, vol. 1. | 142 |
| 68MI21400 | J. Stoffelsma, G. Sipma, D. K. Kettenes and J. Pypker; *J. Agric. Food Chem.*, 1968, **16**, 1000. | 193 |
| 68MI21500 | A. Krbavcic, B. Stanovnik and M. Tisler; *Croat. Chem. Acta*, 1968, **40**, 131 (*Chem. Abstr.*, 1969, **70**, 11 656). | 234 |
| 68MI21501 | B. M. Pyatin and R. G. Glushkov; *Khim.-Farm. Zh.*, 1968, **2**, 11 (*Chem. Abstr.*, 1969, **70**, 37 758). | 236 |
| B-68MI21502 | L. M. Logan, J. P. Byrne and I. G. Ross; in 'Proceedings of the International Conference on Luminescence, 1966', ed. G. Szigeti; Akad. Kiado, Budapest, 1968, vol. 1, p. 194. (*Chem. Abstr.*, 1969, **70**, 42 585). | 250 |
| 68MI21600 | T. Fukushima and M. Akino; *Arch. Biochem. Biophys.*, 1968, **128**, 1. | 285, 306 |
| 68MI21601 | M. Viscontini; *Fortschr. Chem. Forschung*, 1968, **9**, 605. | 281, 306 |
| 68MI21800 | G. Wagner and H. Gentzsch; *Pharmazie*, 1968, **23**, 629. | 375, 377 |
| 68MI22000 | J. Wegmann; *Melliand. Textilber.*, 1968, **49**, 687 (*Chem. Abstr.*, 1968, **69**, 20 251). | 526 |
| 68MI22001 | T. Aczel; *Anal. Chem.*, 1968, **40**, 1917. | 528 |
| 68MI22200 | J. L. R. Williams and G. A. Reynolds; *J. Appl. Phys.*, 1968, **39**, 5327. | 603 |
| B-68MI22400 | A. O. Fitton and R. K. Smalley; 'Practical Heterocyclic Chemistry', Academic Press, New York, 1968. | 792, 796, 813, 838 |

| | | |
|---|---|---|
| 68MI22401 | H. S. Olcott and J. van der Veen; *Lipids*, 1968, **3**, 331. | 876 |
| 68MI22402 | D. E. King; *Chem. Br.*, 1968, **4**, 107. | 880 |
| 68MI22500 | C. Kretzschmer, H. Hoffman and F. K. Splinter; *Acta Biol. Med. Ger.*, 1968, **21**, 359 (*Chem. Abstr.*, 1969, **70**, 18 665). | 941 |
| 68MI22600 | A. Ya. Zheltov, V. Ya. Rodionov and B. I. Stepanov; *Zh. Vses. Khim. Obshch.*, 1968, **13**, 228 (*Chem. Abstr.*, 1968, **69**, 106 635). | 986 |
| 68MIP22700 | V. A. Zagorevskii, K. I. Lopatina and S. M. Klyuev; *U.S.S.R. Pat.* 225 197 (1968) (*Chem. Abstr.*, 1969, **70**, 28 929). | 1019 |
| 68OMS(1)279 | B. J. Millard; *Org. Mass Spectrom.*, 1968, **1**, 279. | 616 |
| 68OMS(1)403 | R. Smakman and T. J. de Boer; *Org. Mass Spectrom.*, 1968, **1**, 403. | 604 |
| 68P1751 | E. Wong; *Phytochemistry*, 1968, **7**, 1751. | 874 |
| 68RTC1065 | H. C. van der Plas and H. Jongejan; *Recl. Trav. Chim. Pays-Bas*, 1968, **87**, 1065. | 73 |
| 68SA(A)1283 | G. Kresze and W. Amann; *Spectrochim. Acta, Part A*, 1968, **24**, 1283. | 892 |
| 68T497 | G. Cardillo, L. Merlini and R. Mondelli; *Tetrahedron*, 1968, **24**, 497. | 669 |
| 68T829 | R. F. Farmer and J. Hamer; *Tetrahedron*, 1968, **24**, 829. | 62 |
| 68T923 | H. C. Smitherman and L. N. Ferguson; *Tetrahedron*, 1968, **24**, 923. | 581, 637, 638 |
| 68T949 | B. D. Tilak and Z. Muljiani; *Tetrahedron*, 1968, **24**, 949. | 673, 786, 851, 852, 873, 874 |
| 68T1981 | W. D. Cotterill, R. Livingstone, K. D. Bartle and D. W. Jones; *Tetrahedron*, 1968, **24**, 1981. | 668 |
| 68T2801 | L. Jurd and A. C. Waiss, Jr.; *Tetrahedron*, 1968, **24**, 2801. | 662 |
| 68T4285 | T. Metler, A. Uchida and S. I. Miller; *Tetrahedron*, 1968, **24**, 4285. | 791 |
| 68T4741 | D. Fărcaşiu and E. Gârd; *Tetrahedron*, 1968, **24**, 4741. | 762 |
| 68T4825 | G. Cardillo, R. Cricchio and L. Merlini; *Tetrahedron*, 1968, **24**, 4825. | 751 |
| 68T5059 | A. T. Balaban; *Tetrahedron*, 1968, **24**, 5059. | 660 |
| 68T5861 | M. Ochiai, E. Mizuta, Y. Asahi and K. Morita; *Tetrahedron*, 1968, **24**, 5861. | 73 |
| 68TL1041 | G. Pagani and S. B. Pagani; *Tetrahedron Lett.*, 1968, 1041. | 1034 |
| 68TL1635 | H. Wagner, G. Aurnhammer, L. Hörhammer, L. Farkas and M. Nógrádi; *Tetrahedron Lett.*, 1968, 1635. | 829 |
| 68TL2171 | H. Ballweg; *Tetrahedron Lett.*, 1968, 2171. | 145 |
| 68TL2701 | H. M. Blatter and H. Lukaszewski; *Tetrahedron Lett.*, 1968, 2701. | 413 |
| 68TL2747 | J. Adams and R. G. Shepherd; *Tetrahedron Lett.*, 1968, 2747. | 390 |
| 68TL3115 | W. D. Crow and C. Wentrup; *Tetrahedron Lett.*, 1968, 3115. | 120 |
| 68TL3859 | V. K. Bhatia, H. G. Krishnamurty, R. Madhav and T. R. Seshadri; *Tetrahedron Lett.*, 1968, 3859. | 829 |
| 68TL3929 | P. Yates, O. Meresz and L. S. Weiler; *Tetrahedron Lett.*, 1967, 3929. | 542 |
| 68TL4643 | A. T. Balaban; *Tetrahedron Lett.*, 1968, 4643. | 867 |
| 68TL5279 | W. H. Pirkle and L. H. McKendry; *Tetrahedron Lett.*, 1968, 5279. | 677, 688 |
| 68TL5931 | E. J. J. Grabowski, E. W. Tristram, R. Tull and P. I. Pollak; *Tetrahedron Lett.*, 1968, 5931. | 166, 167 |
| 68YZ816 | A. Nitta; *Yakugaku Zasshi*, 1968, **88**, 816 (*Chem. Abstr.*, 1969, **70**, 37 604). | 669, 735 |
| 68ZC335 | G. Barnikow and G. Strickmann; *Z. Chem.*, 1968, **8**, 335. | 566, 569 |
| 68ZOB1368 | G. I. Migachev and B. I. Stephanov; *Zh. Obsch. Khim.*, 1968, **38**, 1368 (*Chem. Abstr.*, 1968, **69**, 96 408). | 486 |
| 69ABC1775 | P. Wang, H. Kato and M. Fujimaki; *Agric. Biol. Chem.*, 1969, **33**, 1775. | 193 |
| 69AC(R)552 | T. LaNoce, E. Bellasio, A. Vigevani and E. Testa; *Ann. Chim. (Rome)*, 1969, **59**, 552. | 352 |
| 69AF558 | T. Zsolnai; *Arzneim.-Forsch.*, 1969, **19**, 558. | 1085 |
| 69AF1807 | M. Schorr, W. Duerckheimer, P. Klatt, G. Laemmler, G. Nesemann and E. Schrinner; *Arzneim.-Forsch.*, 1969, **19**, 1807. | 1044, 1085 |
| 69AG(E)157 | S. J. Angyal; *Angew. Chem., Int. Ed. Engl.*, 1969, **8**, 157. | 629 |
| 69AG(E)370 | K. Dimroth and G. Laubert; *Angew. Chem., Int. Ed. Engl.*, 1969, **8**, 370. | 657, 658 |
| 69AG(E)456 | H. Hoffman and G. Salbeck; *Angew. Chem., Int. Ed. Engl.*, 1969, **8**, 456. | 908 |
| 69AG(E)510 | H. W. Roesky; *Angew. Chem., Int. Ed. Engl.*, 1969, **8**, 510. | 1047 |
| 69AG(E)604 | R. Huisgen, W. Scheet, H. Mäder and E. Brunn; *Angew. Chem., Int. Ed. Engl.*, 1969, **8**, 604. | 175 |
| 69AHC(10)149 | W. J. Irwin and D. G. Wibberley; *Adv. Heterocycl. Chem.*, 1969, **10**, 149. | 201, 202, 203, 204, 207 |
| 69AHC(10)241 | A. T. Balaban, W. Schroth and G. W. Fischer; *Adv. Heterocycl. Chem.*, 1969, **10**, 241. | 861, 872 |
| 69AJC1745 | D. B. Paul and H. J. Rodda; *Aust. J. Chem.*, 1969, **22**, 1745. | 234, 236, 238, 239, 246 |
| 69AJC1759 | D. B. Paul and H. J. Rodda; *Aust. J. Chem.*, 1969, **22**, 1759. | 240 |
| 69AJC1923 | A. J. Birch, M. Maung and A. Pelter; *Aust. J. Chem.*, 1969, **22**, 1923. | 673 |
| 69AJC2243 | M. J. Aroney, G. M. Hoskins, R. J. W. Le Fèvre, R. K. Pierens and D. V. Radford; *Aust. J. Chem.*, 1969, **22**, 2243. | 947 |
| 69AJC2497 | A. W. K. Chan and W. D. Crow; *Aust. J. Chem.*, 1969, **22**, 2497. | 1002 |
| 69AK(30)261 | B. Jägersten; *Ark. Kemi*, 1969, **30**, 261. | 53 |
| 69AK(30)393 | E. Nilsson; *Ark. Kemi*, 1969, **30**, 393. | 620 |
| 69AX(B)1038 | K. Ozeki, N. Sakabe and J. Tanaka; *Acta Crystallogr., Part B*, 1969, **25**, 1038. | 68 |
| 69AX(B)1978 | B. M. Craven, E. A. Vizzini and M. M. Rodrigues; *Acta Crystallogr., Part B*, 1969, **25**, 1978. | 68 |
| 69AX(B)2231 | C. Sabelli, P. Tangocci and P. F. Zanazzi; *Acta Crytallogr., Part B*, 1969, 2231. | 333 |

| | | |
|---|---|---|
| 69BBA(184)386 | H. Rembold, H. Metzger, P. Sudershan and W. Gutensohn; *Biochem. Biophys. Acta*, 1969, **184**, 386. | 321 |
| 69BBA(184)589 | H. Rembold and F. Simmersbach; *Biochem. Biophys. Acta*, 1969, **184**, 589. | 321 |
| 69BCJ1454 | Y. Yamada, I. Noda, I. Kumashiro and T. Takenishi; *Bull. Chem. Soc. Jpn.*, 1969, **42**, 1454. | 365 |
| 69BCJ2924 | K. Wakabayashi, M. Tsunoda and Y. Suzuki; *Bull. Chem. Soc. Jpn.*, 1969, **42**, 2924. | 503 |
| 69BCJ2996 | S. Kakimoto and S. Tonooka; *Bull. Chem. Soc. Jpn.*, 1969, **42**, 2996. | 243 |
| 69BCJ3345 | H. Obara and J-i. Onodera; *Bull. Chem. Soc. Jpn*, 1969, **42**, 3345 | 644 |
| 69BRP1171218 | J. Davoll; (Parke, Davis and Co.), *Br. Pat.* 1 171 218 (1969) (*Chem. Abstr.*, 1970, **72**, 66 973). | 227 |
| 69BSB289 | J. P. Osselaere, J. V. Dejardin and M. Dejardin-Duchêne; *Bull. Soc. Chim. Belg.*, 1969, **78**, 289. | 225, 246 |
| 69BSF915 | J. Thibault and P. Maitte; *Bull. Soc. Chim. Fr.*, 1969, 915. | 860 |
| 69BSF1715 | P. Ropiteau and P. Maitte; *Bull. Soc. Chim. Fr.*, 1969, 1715. | 577, 580, 764 |
| 69BSF2490 | M. Moreau, R. Longeray and J. Dreux; *Bull. Soc. Chim. Fr.*, 1969, 2490. | 630 |
| 69BSF2519 | G. Queguiner and P. Pastour; *Bull. Soc. Chim. Fr.*, 1969, 2519. | 234, 235, 238 |
| 69BSF3670 | J. Daunis, R. Jacquier and P. Viallefont; *Bull. Soc. Chim. Fr.*, 1969, 3670. | 402, 411 |
| 69BSF3678 | G. Queguiner and P. Pastour; *Bull. Soc. Chim. Fr.*, 1969, 3678 | 234, 246 |
| 69BSF4082 | G. Queguiner and P. Pastour; *Bull. Soc. Chim. Fr.*, 1969, 4082. | 246 |
| 69BSF4545 | C. Mercier; *Bull. Soc. Chim. Fr.*, 1969, 4545. | 603, 606, 613 |
| 69C108 | R. Hug, H.-J. Hansen and H. Schmid; *Chimia*, 1969, **23**, 108. | 754 |
| 69CB785 | H. Wagner, G. Aurnhammer, L. Hörhammer, L. Farkas and M. Nógrádi; *Chem. Ber.*, 1969, **102**, 785. | 853 |
| 69CB931 | R. Richter and W.-P. Trautwein; *Chem. Ber.*, 1969, **102**, 931. | 515 |
| 69CB1848 | R. Huisgen and J. Wulff; *Chem. Ber.*, 1969, **102**, 1848. | 413 |
| 69CB2378 | K. Fickentscher; *Chem. Ber.*, 1969, **102**, 2378. | 984 |
| 69CB2508 | D. Martin and W. M. Brause; *Chem. Ber.*, 1969, **102**, 2508. | 503 |
| 69CB3818 | T. Sasaki and M. Murata; *Chem. Ber.*, 1969, **102**, 3818. | 412 |
| 69CB4032 | G. B. Barlin and W. Pfleiderer; *Chem. Ber.*, 1969, **102**, 4032. | 310 |
| 69CC221 | J. A. Adamson, D. L. Forster, T. L. Gilchrist and C. W. Rees; *Chem. Commun.*, 1969, 221. | 376 |
| 69CC290 | T. Paterson and H. C. S. Wood; *Chem. Commun.*, 1969, 290. | 212 |
| 69CC392 | W. S. McDonald, H. M. N. H. Irving, G. Raper and D. C. Rupainwar; *Chem. Commun.*, 1969, 392. | 1049 |
| 69CC1200 | C. G. Allison, R. D. Chambers, Y. A. Cheburkov, J. A. H. MacBride and W. K. R. Musgrave; *Chem. Commun.*, 1969, 1200. | 190 |
| 69CC1424 | Y. Sasaki and F. P. Olsen; *Chem. Commun.*, 1969, 1424. | 1048 |
| 69CCC4000 | K. Bláha; *Collect. Czech. Chem. Commun.*, 1969, **34**, 4000. | 187 |
| 69CHE316 | V. A. Zagorevskii, I. D. Tsvetkova, E. K. Orlova and S. L. Protnova; *Chem. Heterocycl. Compd. (Engl. Transl.)*, 1969, **5**, 316. | 787 |
| 69CI(L)237 | S. H. Wilen and A. W. Levine; *Chem. Ind. (London)*, 1969, 237. | 185 |
| 69CI(L)1305 | A. D. B. Sloan; *Chem. Ind. (London)*, 1969, 1305. | 1071 |
| 69CJC1289 | V. M. Rao and R. Kewley; *Can. J. Chem.*, 1969, **47**, 1289. | 628 |
| 69CJC3173 | H. H. Mantsch and J. Dehler; *Can. J. Chem.*, 1969, **47**, 3173. | 945 |
| 69CJC4427 | R. U. Lemieux, A. A. Pavia, J. C. Martin and K. A. Watanabe; *Can. J. Chem.*, 1969, **47**, 4427. | 629 |
| 69CPB2266 | I. Matsuura and K. Okui; *Chem. Pharm. Bull.*, 1969, **17**, 2266. | 234, 237, 241, 242, 246 |
| 69CR(C)(268)1531 | G. Queguiner, M. Alas and P. Pastour; *C.R. Hebd. Seances Acad. Sci., Ser. C*, 1969, **268**, 1531. | 246 |
| 69CR(C)(269)298 | V. Hérault and J. Gayoso; *C.R. Hebd. Seances Acad. Sci., Ser. C*, 1969, **269**, 298. | 633 |
| 69EGP66877 | H. Brachwitz; *Ger. (East) Pat.* 66 877 (1969) (*Chem. Abstr.*, 1969, **71**, 124 498). | 163 |
| 69FRP2041497 | M. Davidson and M. Tempe-Hermann; *Fr. Pat.* 2 041 497 (1969) (*Chem. Abstr.*, 1971, **75**, 128 518). | 880 |
| 69G165 | S. Morrochi, A. Ricca, A. Selva and A. Zanarotti; *Gazz. Chim. Ital.*, 1969, **99**, 165. | 1080 |
| 69HCA306 | M. Viscontini and T. Okada; *Helv. Chim. Acta*, 1969, **52**, 306. | 306 |
| 69HCA743 | A. Dieffenbacher and W. von Philipsborn; *Helv. Chim. Acta*, 1969, **52**, 743. | 259, 276, 285 |
| 69IJC20 | Y. R. Rao and M. V. Konher; *Indian J. Chem.*, 1969, **7**, 20. | 499 |
| 69IJC28 | Z. Muljiani and B. D. Tilak; *Indian J. Chem.*, 1969, **7**, 28. | 874 |
| 69IJC866 | C. M. Gupta, A. P. Bhaduri and N. M. Khanna; *Indian J. Chem.*, 1969, **7**, 866. | 260 |
| 69IJC1059 | S. K. Agarwal, S. K. Grover and T. R. Seshadri; *Indian J. Chem.*, 1969, **7**, 1059. | 853 |
| 69IZV655 | E. A. Arutyunyan, V. I. Gunar, E. P. Gracheva and S. I. Zav'yalov; *Izv. Akad. Nauk SSSR, Ser. Khim.*, 1969, 655. | 91, 296 |
| 69IZV2059 | N. P. Gambaryan and Yu. V. Zeifman; *Izv. Akad. Nauk SSSR, Ser. Khim.*, 1969, 2059. | 1041, 1063, 1078 |
| 69IZV2857 | E. A. Arutyunyan, V. I. Gunar and S. I. Zav'yalov; *Izv. Akad. Nauk SSSR, Ser. Khim.*, 1969, 2857. | 296 |
| 69JA1219 | R. A. Abramovitch, C. I. Azogu and I. T. McMaster; *J. Am. Chem. Soc.*, 1969, **91**, 1219. | 1015 |
| 69JA2443 | M. W. Tolles, M. R. McBride and W. E. Thun; *J. Am. Chem. Soc.*, 1969, **91**, 2443. | 542 |
| 69JA2703 | E. L. Eliel and R. O. Hutchins; *J. Am. Chem. Soc.*, 1969, **91**, 2703. | 961 |
| 69JA4494 | J. W. Hanifin and E. Cohen; *J. Am. Chem. Soc.*, 1969, **91**, 4494. | 706 |
| 69JA5202 | J. L. Occolowitz; *J. Am. Chem. Soc.*, 1969, **91**, 5202. | 608 |

| | | |
|---|---|---|
| 69JA5237 | N. C. Deno, E. L. Booker, Jr., K. E. Kramer and G. Saines; *J. Am. Chem. Soc.*, 1969, **91**, 5237. | 667 |
| 69JA5501 | J. A. Zoltewicz, G. Grahe and C. L. Smith; *J. Amer. Chem. Soc.*, 1969, **91**, 5501. | 239 |
| 69JA5801 | G. A. Olah and A. M. White; *J. Am. Chem. Soc.*, 1969, **91**, 5801. | 641 |
| 69JA6019 | C. H. Bushweller; *J. Am. Chem. Soc.*, 1969, **91**, 6019. | 948 |
| 69JA6321 | M. J. S. Dewar, A. J. Harget and N. Trinajstić; *J. Am. Chem. Soc.*, 1969, **91**, 6321. | 59 |
| 69JA6381 | R. J. Pugmire, D. M. Grant, M. J. Robins and R. K. Robins; *J. Am. Chem. Soc.*, 1969, **91**, 6381. | 159, 160 |
| 69JAP6912898 | Y. Abe, Y. Shigeta, F. Uchimaru, S. Okada and E. Ozsayma; *Jpn. Pat.* 69 12 898 (1969). | 172 |
| 69JAP6920345 | Y. Abe, Y. Shigeta, F. Uchimaru, S. Okada and E. Ozsayma; *Jpn. Pat.* 69 20 345 (1969). | 172 |
| 69JAP6927724 | K. Morita, S. Kobayashi and H. Kimura; *Jpn. Pat.* 69 27 724 (1969) (*Chem. Abstr.*, 1970, **72**, 21 724). | 992 |
| 69JCS(B)96 | E. Kalatzis; *J. Chem. Soc. (B)*, 1969, 96. | 84 |
| 69JCS(B)489 | T. J. Batterham and J. A. Wunderlich; *J. Chem. Soc. (B)*, 1969, 489. | 267 |
| 69JCS(B)755 | M. Hillebrand, O. Maior, V. E. M. Sahini and E. Volanschi; *J. Chem. Soc. (B)*, 1969, 755. | 946, 968 |
| 69JCS(B)855 | J. M. Eckert and R. J. W. Le Fèvre; *J. Chem. Soc. (B)*, 1969, 855. | 629 |
| 69JCS(B)980 | M. J. Aroney, G. M. Hoskins and R. J. W. Le Fèvré, *J. Chem. Soc. (B)*, 1969, 980. | 627, 947 |
| 69JCS(C)281 | J. E. Atkinson and J. R. Lewis; *J. Chem. Soc. (C)*, 1969, 281. | 837 |
| 69JCS(C)513 | I. R. Gelling, W. J. Irwin and D. G. Wibberley; *J. Chem. Soc. (C)*, 1969, 513. | 204 |
| 69JCS(C)526 | F. M. Dean and B. Parton; *J. Chem. Soc. (C)*, 1969, 526. | 600 |
| 69JCS(C)748 | C. D. Campbell and C. W. Rees; *J. Chem. Soc. (C)*, 1969, 748. | 1015 |
| 69JCS(C)772 | R. S. Atkinson and C. W. Rees; *J. Chem. Soc. (C)*, 1969, 772. | 53 |
| 69JCS(C)928 | K. J. M. Andrews, W. E. Barber and B. P. Tong; *J. Chem. Soc. (C)*, 1969, 928. | 306, 315 |
| 69JCS(C)935 | G. V. Boyd, D. Hewson and R. A. Newberry; *J. Chem. Soc. (C)*, 1969, 935. | 823 |
| 69JCS(C)1408 | J. Clark, P. N. T. Murdoch and D. L. Robert; *J. Chem. Soc. (C)*, 1969, 1408. | 266 |
| 69JCS(C)1751 | J. Clark and W. Pendergast; *J. Chem. Soc. (C)*, 1969, 1751. | 266 |
| 69JCS(C)1883 | J. Clark and P. N. T. Murdoch; *J. Chem. Soc. (C)*, 1969, 1883. | 266 |
| 69JCS(C)2069 | N. P. Buu-Hoï, G. Saint-Ruf and B. Lobert; *J. Chem. Soc. (C)*, 1969, 2069. | 802 |
| 69JCS(C)2192 | F. M. Dean, J. Goodchild and A. W. Hill; *J. Chem. Soc. (C)*, 1969, 2192. | 702 |
| 69JCS(C)2761 | J. W. A. Findlay, P. Gupta and J. R. Lewis; *J. Chem. Soc. (C)*, 1969, 2761. | 837 |
| 69JHC1 | W. H. Pirkle and M. Dines; *J. Heterocycl. Chem.*, 1969, **6**, 1. | 581, 635 |
| 69JHC13 | E. Marcus, J. F. Stephen and J. K. Chan; *J. Heterocycl. Chem.*, 1969, **6**, 13. | 581 |
| 69JHC93 | S. F. Martin and R. N. Castle; *J. Heterocycl. Chem.*, 1969, **6**, 93. | 347, 360 |
| 69JHC239 | H. S. Hertz, F. F. Kabacinski and P. E. Spoerri; *J. Heterocycl. Chem.*, 1969, **6**, 239. | 167 |
| 69JHC247 | C. R. Johnson and C. B. Thanawalla; *J. Heterocycl. Chem.*, 1969, **6**, 247. | 1031 |
| 69JHC255 | R. B. Rao and R. N. Castle; *J. Heterocycl. Chem.*, 1969, **6**, 255. | 347 |
| 69JHC313 | W. H. Pirkle and M. Dines; *J. Heterocycl. Chem.*, 1969, **6**, 313. | 680 |
| 69JHC317 | R. C. Bertelson, K. D. Glanz and D. B. McQuain; *J. Heterocycl. Chem.*, 1969, **6**, 317. | 1046 |
| 69JHC333 | H. E. Baumgarten, W. F. Whittman and G. J. Lehmann; *J. Heterocycl. Chem.*, 1969, **6**, 333. | 54 |
| 69JHC393 | E. A. Allen, N. P. Johnson, D. T. Rosevear and W. Wilkinson; *J. Heterocycl. Chem.*, 1969, **6**, 393. | 976 |
| 69JHC403 | I. Lalezari, N. Shargi, A. Shafiee and M. Yalpani; *J. Heterocycl. Chem.*, 1969, **6**, 403. | 410 |
| 69JHC407 | R. F. Meyer; *J. Heterocycl. Chem.*, 1969, **6**, 407. | 1085 |
| 69JHC497 | P. Roffey and J. P. Verge; *J. Heterocycl. Chem.*, 1969, **6**, 497. | 442, 553 |
| 69JHC593 | I. Wempen, H. U. Blank and J. J. Fox; *J. Heterocycl. Chem.*, 1969, **6**, 593. | 69, 130 |
| 69JHC627 | W. H. Mueller and M. Dines; *J. Heterocycl. Chem.*, 1969, **6**, 627. | 987 |
| 69JHC779 | E. E. Gilbert and B. Veldhuis; *J. Heterocycl. Chem.*, 1969, **6**, 779. | 377 |
| 69JHC977 | M. M. Kochhar; *J. Heterocyl. Chem.*, 1969, **6**, 977. | 245 |
| 69JMC277 | D. Huckle, I. M. Lockhart and M. Wright; *J. Med. Chem.*, 1969, **12**, 277. | 734 |
| 69JMC424 | T. L. Hullar and W. C. French; *J. Med. Chem.*, 1969, **12**, 424. | 209 |
| 69JMC531 | G. M. Cingolani, F. Gualtieri and M. Pigini; *J. Med. Chem.*, 1969, **12**, 531. | 680 |
| 69JMC662 | J. J. McCormack and J. J. Jaffe; *J. Med. Chem.*, 1969, **12**, 662. | 325 |
| 69JOC56 | W. E. Parham and D. G. Weetman; *J. Org. Chem.*, 1969, **34**, 56. | 914 |
| 69JOC120 | K. H. Dudley and R. W. Chiang; *J. Org. Chem.*, 1969, **34**, 120. | 751 |
| 69JOC175 | J. A. Deyrup and C. L. Moyer; *J. Org. Chem.*, 1969, **34**, 175. | 1075 |
| 69JOC199 | P. Scheiner; *J. Org. Chem.*, 1969, **34**, 199. | 549 |
| 69JOC207 | E. E. Schweizer, C. J. Berninger, D. M. Crouse, R. A. Davis and R. S. Logothetis; *J. Org. Chem.*, 1969, **34**, 207. | 786 |
| 69JOC380 | J. Wolinsky and H. S. Hauer; *J. Org. Chem.*, 1969, **34**, 380. | 757 |
| 69JOC479 | P. R. Stapp; *J. Org. Chem.*, 1969, **34**, 479. | 774, 777 |
| 69JOC821 | B. Roth and J. Z. Strelitz; *J. Org. Chem.*, 1969, **34**, 821. | 138, 204, 206 |
| 69JOC1566 | W. C. Lumma, Jr. and G. A. Berchtold; *J. Org. Chem.*, 1969, **34**, 1566. | 911 |
| 69JOC1792 | L. Field and R. B. Barbee; *J. Org. Chem.*, 1969, **34**, 1792. | 981 |
| 69JOC2239 | W. H. Pirkle and M. Dines; *J. Org. Chem.*, 1969, **34**, 2239. | 581, 736, 679, 736 |
| 69JOC2407 | K. R. Huffman, M. Burger, W. A. Henderson, M. Loy and E. F. Ullman; *J. Org. Chem.*, 1969, **34**, 2407. | 722 |
| 69JOC2720 | P. Aeberli and W. J. Houlihan; *J. Org. Chem.*, 1969, **34**, 2720. | 338, 351 |

| | | |
|---|---|---|
| 69JOC2762 | H. Rubinstein and M. Wuerthele; *J. Org. Chem.*, 1969, **34**, 2762. | 987, 988 |
| 69JOC3169 | J. Wolinsky and H. S. Hauer; *J. Org. Chem.*, 1969, **34**, 3169. | 577, 593, 598, 757, 761 |
| 69JOC3368 | Y. Murata and H. J. Shine; *J. Org. Chem.*, 1969, **34**, 3368. | 968, 969 |
| 69JOM(18)249 | A. Maercker; *J. Organomet. Chem.*, 1969, **18**, 249. | 672 |
| 69JPR438 | L. Heinisch; *J. Prakt. Chem.*, 1969, **311**, 438. | 447 |
| 69JPS867 | S. F. Sisenwine and S. S. Walkenstein; *J. Pharm. Sci.*, 1969, **58**, 867. | 325 |
| 69KGS566 | V. A. Ershov, I. Ya. Postovskii and A. Kh. Apusheva; *Khim. Geterotsikl. Soedin.*, 1969, **5**, 566. | 545 |
| 69KGS908 | T. E. Gorizdra; *Khim. Geterotsikl. Soedin.*, 1969, **5**, 908. | 300 |
| 69KGS1086 | V. P. Mamaev and E. A. Gracheva; *Khim. Geterotsikl Soedin.*, 1969, 1086. | 125 |
| 69LA(722)162 | C. Schiele, A. Wilhelm and G. Paal; *Liebigs Ann. Chem.*, 1969, **722**, 162. | 598 |
| 69LA(723)111 | R. R. Schmidt, D. Schwille and U. Sommer; *Liebigs Ann. Chem.*, 1969, **723**, 111. | 1003, 1019 |
| 69LA(726)81 | W. Ried and E. A. Baumbach; *Liebigs Ann. Chem.*, 1969, **726**, 81. | 41 |
| 69LA(726)100 | E. C. Taylor, Jr., and K. Lenard; *Liebigs Ann. Chem.*, 1969, **726**, 100. | 187, 318 |
| 69LA(727)231 | E. Zbiral and J. Stroh; *Liebigs Ann. Chem.*, 1969, **727**, 231. | 185 |
| 69LA(728)32 | F. Boberg, H. Niemann and K. Kirchhoff; *Liebigs Ann. Chem.*, 1969, **728**, 32. | 983 |
| B-69MI21300 | G. B. Barlin and D. J. Brown; in 'Topics in Heterocyclic Chemistry', ed. R. N. Castle; Wiley-Interscience, New York, 1969, p. 122. | 99, 129 |
| 69MI21301 | C. C. Cheng; *Prog. Med. Chem.*, 1969, **6**, 67. | 147, 149, 151, 152 |
| 69MI21302 | Various authors, *Postgrad. Med. J.*, 1969, **45S**, 3. | 151 |
| B-69MI21400 | A. D. Cross and R. A. Jones; in 'An Introduction to Practical Infra-red Spectroscopy', Butterworths, London, 1969. | 161 |
| 69MI21402 | P. E. Hoehler and J. A. Newell; *J. Agric. Food Chem.*, 1969, **17**, 393. | 193 |
| 69MI21500 | B. M. Pyatin and R. G. Glushkov; *Khim.-Farm. Zh.*, 1969, **3**, 10 (*Chem. Abstr.*, 1969, **71**, 112 882). | 228 |
| 69MI21501 | M. Lora-Tamayo, J. L. Soto and E. D. Toro; *An. Quim.*, 1969, **65**, 1125 (*Chem. Abstr.*, 1970, **72**, 132 675). | 234, 246 |
| 69MI21502 | G. Favini and G. Buemi; *Theor. Chim. Acta*, 1969, **13**, 79 (*Chem. Abstr.*, 1969, **70**, 82 603). | 236 |
| 69MI21503 | E. Domagalina, I. Kurpiel and N. Koktysz; *Rocz. Chem.*, 1969, **43**, 775 (*Chem. Abstr.*, 1969, **71**, 61 315). | 238 |
| 69MI21600 | H. Iida and H. Iida; *Kyogo Kagaku Zasshi*, 1969, **72**, 342. | 292 |
| B-69MI21601 | E. C. Taylor, Jr.; in 'Topics in Heterocyclic Chemistry', ed. R. N. Castle; Wiley-Interscience, New York, 1969, p. 1. | 271, 281 |
| B-69MI21602 | R. L. Blakley; 'The Biochemistry of Folic Acid and Related Pteridines', North-Holland, Amsterdam, 1969, p. 106. | 316, 320 |
| B-69MI21603 | I. Ziegler and R. Harmsen; in 'Advances in Insect Physiology', Academic Press, London, 1969, vol. 6, p. 139. | 322 |
| B-69MI21604 | R. L. Blakley; 'The Biochemistry of Folic Acid and Related Pteridines', North-Holland, Amsterdam, 1969, p. 489. | 325, 326 |
| 69MI21605 | M. Goto, A. Sakurai, K. Ohta and H. Yamakami; *J. Biochem.*, 1969, **65**, 611. | 284, 301 |
| B-69MI21605 | R. L. Blakley; 'The Biochemisty of Folic Acid and Related Pteridines', North-Holland, Amsterdam, 1969, p. 477. | 326, 327 |
| 69MI21606 | A. Sakurai and M. Goto; *J. Biochem.*, 1969, **65**, 755. | 304 |
| 69MI22000 | M. J. Heller; *Eur. Polym. J.* (*Suppl.*), 1969, 105. | 525 |
| B-69MI22200 | L. M. Jackman and S. Sternhell; 'Applications of Nuclear Magnetic Resonance Spectroscopy in Organic Chemistry', Pergamon, Oxford, 1969, p. 112. | 579 |
| 69MI22201 | J. Mendez and M. I. Lojo; *Microchem. J.*, 1969, **14**, 1136. | 600 |
| B-69MI22202 | 'Fluorescence Assay in Biology and Medicine', ed. S. Udenfriend; Academic, New York, 1969. | 600 |
| 69MI22203 | C. Romers, C. Altona, H. R. Buys and E. Havinga; *Top. Stereochem.*, 1969, **4**, 39. | 629 |
| 69MI22300 | M. Murata and T. Ito; *Bunseki Kagaku*, 1969, **18**, 1131. | 707 |
| 69MI22400 | L. L. Woods and V. Hollands; *Texas J. Sci.*, 1969, **21**, 91 (*Chem. Abstr.*, 1970, **72**, 12 486). | 806 |
| 69MI22401 | J. I. DeGraw and P. Tsakotellis; *J. Chem. Eng. Data*, 1969, **14**, 509. | 808 |
| B-69MI22402 | H. Kwart and K. King; in 'The Chemistry of Carboxylic Acids and Esters', ed. S. Patai; Interscience, New York, 1969, p. 341. | 850 |
| B-69MI22600 | C. Romers, C. Altona, H. R. Buys and E. Havinga; in 'Topics in Stereochemistry', eds. E. L. Eliel and N. L. Allinger; Wiley-Interscience, New York, 1969, vol. 4, p. 39. | 948, 949, 960, 962 |
| B-69MI22601 | L. M. Jackman and S. Sternhell; 'Applications of Nuclear Magnetic Resonance Spectroscopy in Organic Chemistry', Pergamon, Oxford, 2nd edn., 1969, p. 200. | 953 |
| B-69MI22602 | L. M. Jackman and S. Sternhell; 'Applications of Nuclear Magnetic Resonance Spectroscopy in Organic Chemistry', Pergamon, Oxford, 2nd edn., 1969, p. 280. | 954 |
| 69MIP21500 | O. Hromatka and D. Binder; *Austrian Pat.* 272 351 (1969) (*Chem. Abstr.*, 1969, **71**, 112 995). | 208, 216 |
| 69OMR(1)67 | H. Friebolin, H. G. Schmid, S. Kabuss and W. Faisst; *Org. Magn. Reson.*, 1969, **1**, 67. | 953, 962 |
| 69OMS(2)355 | J. C. Tou, L. A. Shadoff and R. H. Rigterink; *Org. Mass Spectrom.*, 1969, **2**, 355. | 373 |
| 69OMS(2)829 | J. Heiss and K.-P. Zeller; *Org. Mass Spectrom.*, 1969, **2**, 829. | 604 |
| 69OMS(2)923 | J. A. Blair and C. D. Foxall; *Org. Mass. Spectrom.*, 1969, **2**, 923. | 285 |
| 69OMS(2)965 | A. M. Duffield; *Org. Mass Spectrom.*, 1969, **2**, 965. | 604, 607 |

| | | |
|---|---|---|
| 69OMS(2)1277 | G. Condé-Caprace and J. E. Collin; *Org. Mass Spectrom.*, 1969, **2**, 1277. | 957 |
| 69OPP21 | J. Strating and E. Molenaar; *Org. Prep. Proced. Int.*, 1969, **1**, 21. | 934 |
| 69PNA(63)1311 | R. C. Fuller and N. A. Nugent; *Proc. Natl. Acad. Sci. USA*, 1969, **63**, 1311. | 323 |
| 69RRC247 | E. Gurd, I. I. Stanoiu, F. Chiraleu and A. T. Balaban; *Rev. Roum. Chim.*, 1969, **14**, 247. | 649 |
| 69RTC426 | H. C. van der Plas and B. Zuurdeeg; *Recl. Trav. Chim. Pays-Bas*, 1969, **88**, 426. | 519 |
| 69RTC1156 | H. C. van der Plas, B. Zuurdeeg and H. W. van Meeteren; *Recl. Trav. Chim. Pays-Bas*, 1969, **88**, 1156. | 519 |
| 69RTC1335 | H. G. Peer and A. Van der Heijden; *Recl. Trav. Chim. Pays-Bas*, 1969, **88**, 1335. | 179 |
| 69RTC1391 | J. W. Streef and H. J. den Hertog; *Rec. Trav. Chim. Pays-Bas*, 1969, **88**, 1391. | 120 |
| 69SA(A)1041 | M. J. Hitch and S. D. Ross; *Spectrochim. Acta, Part A*, 1969, **25**, 1041. | 956 |
| 69SA(A)1047 | M. J. Hitch and S. D. Ross; *Spectrochim. Acta, Part A*, 1969, **25**, 1047. | 956 |
| 69SA(A)1067 | H. A. B. Linke; *Spectrochim. Acta, Part A*, 1969, **25**, 1067. | 596 |
| 69SA(A)1437 | P. Klaboe; *Spectrochim. Acta, Part A*, 1969, **25**, 1437. | 956 |
| 69T707 | T. Széll, G. Schöbel and L. Baláspiri; *Tetrahedron*, 1969, **25**, 707. | 820 |
| 69T715 | T. Széll, L. Dózsai, M. Zarándy and K. Meynhárth; *Tetrahedron*, 1969, **25**, 715. | 820 |
| 69T783 | Y. Hagiwara, M. Kurihara and N. Yoda; *Tetrahedron*, 1969, **25**, 783. | 67 |
| 69T871 | A. Ohno, Y. Ohnishi and G. Tsuchihashi; *Tetrahedron*, 1969, **25**, 871. | 932 |
| 69T1209 | D. Fǎrcaşiu; *Tetrahedron*, 1969, **25**, 1209. | 868 |
| 69T1975 | G. H. Stout, T. S. Lin and I. Singh; *Tetrahedron*, 1969, **25**, 1975. | 624 |
| 69T2687 | T. M. Harris and C. M. Harris; *Tetrahedron*, 1969, **25**, 2687. | 796 |
| 69T3365 | A. J. de Hoog, H. R. Buys, C. Altona and E. Havinga; *Tetrahedron*, 1969, **25**, 3365. | 626, 628, 629 |
| 69T3807 | A. R. Katritzky, M. R. Nesbit, B. J. Kurtev, M. Lyapova and I. G. Pojarlieff; *Tetrahedron*, 1969, **25**, 3807. | 62 |
| 69T5819 | R. D. H. Murray and P. H. McCabe; *Tetrahedron*, 1969, **25**, 5819. | 596 |
| 69T5839 | R. D. H. Murray, P. H. McCabe and T. C. Hogg; *Tetrahedron*, 1969, **25**, 5839. | 596 |
| 69T5869 | C. Wünsche, G. Ege, E. Beisiegel and T. Pasedach; *Tetrahedron*, 1969, **25**, 5869. | 373 |
| 69T5995 | G. Arnold and G. Paal; *Tetrahedron*, 1969, **25**, 5995. | 668 |
| 69TL651 | W. E. Barnett and J. McCormack; *Tetrahedron Lett.*, 1969, 651. | 966 |
| 69TL907 | G. Cardillo, R. Cricchio and L. Merlini; *Tetrahedron Lett.*, 1969, 907. | 751 |
| 69TL1825 | T. Nishino, M. Kiyokawa and K. Tokuyama; *Tetrahedron Lett.*, 1969, 1825. | 207 |
| 69TL2195 | J.-P. Griot, J. Royer and J. Dreux; *Tetrahedron Lett.*, 1969, 2195. | 593, 741 |
| 69TL2659 | M. Franck-Neumann and C. Buchecker; *Tetrahedron Lett.*, 1969, 2659. | 51 |
| 69TL3147 | H. Neunhoeffer, H. Hennig, H.-W. Frühauf and M. Mutterer; *Tetrahedron Lett.*, 1969, 3147. | 393 |
| 69TL3661 | T. Ledaal; *Tetrahedron Lett.*, 1969, 3661. | 958 |
| 69TL4117 | G. Kresze and H. Grill; *Tetrahedron Lett.*, 1969, 4117. | 1050, 1051, 1079 |
| 69TL4449 | J. H. Cooley and J. W. Atchison, Jr.; *Tetrahedron Lett.*, 1969, 4449. | 567 |
| 69TL4509 | L. A. Paquette and J. F. Kelly; *Tetrahedron Lett.*, 1969, 4509. | 551 |
| 69TL4713 | C. H. Bushweller and J. W. O'Neil; *Tetrahedron Lett.*, 1969, 4713. | 629 |
| 69TL4987 | G. Maier and M. Wiessler; *Tetrahedron Lett.*, 1969, 4987. | 739 |
| 69TL5151 | A. I. Meyers and H. W. Adickes; *Tetrahedron Lett.*, 1969, 5151. | 1007 |
| 69TL5171 | W. Dittmar, J. Sauer and A. Steigel; *Tetrahedron Lett.*, 1969, 5171. | 422, 425 |
| 69TL5279 | R. R. Schmidt; *Tetrahedron Lett.*, 1969, 5279. | 784 |
| 69UKZ512 | A. I. Tolmachev and E. F. Karaban; *Ukr. Khim. Zh. (Russ. Ed.)*, 1969, **35**, 512 (*Chem. Abstr.*, 1969, **71**, 82 630). | 941 |
| 69USP3444212 | K. R. Huffman and E. F. Ullman; *U.S. Pat.* 3 444 212 (1969) (*Chem. Abstr.*, 1969, **71**, 38 807). | 695 |
| 69USP3453279 | A. F. Ellis; *U.S. Pat.* 3 453 279 (1969). | 185 |
| 69YZ1646 | S. Sigiura, S. Inoue, Y. Kishi and T. Koto; *Yakugaku Zasshi*, 1969, **89**, 1646. | 180 |
| 69ZC111 | G. Westphal and P. Henklein; *Z. Chem.*, 1969, **9**, 111. | 1075 |
| 69ZC184 | W. Schroth, F. Billig and A. Zschunke; *Z. Chem.*, 1969, **9**, 184. | 982 |
| 69ZC230 | D. Kreysig, G. Kempter and H. H. Stroh; *Z. Chem.*, 1969, **9**, 230. | 247 |
| 69ZC361 | H. Beyer; *Z. Chem.*, 1969, **9**, 361. | 1044 |
| 69ZOB716 | G. I. Zhungietu; *Zh. Obshch. Khim.*, 1969, **39**, 716. | 862 |
| 69ZOR355 | L. B. Volodarskii, Yu. G. Putsykin and V. I. Mamatyuk; *Zh. Org. Khim.*, 1969, **5**, 355. | 1056 |
| 69ZOR1844 | F. I. Luknitskii, D. O. Taube and B. A. Vovsi; *Zh. Org. Khim.*, 1969, **5**, 1844 (*Chem. Abstr.*, 1970, **72**, 21 671). | 1030 |
| 69ZOR2039 | F. I. Luknitskii and B. A. Vovsi; *Zh. Obshch. Khim.*, 1969, **5**, 2039. | 451 |
| | | |
| 70AC(R)351 | F. Duro; *Ann. Chim. (Rome)*, 1970, **60**, 351. | 1010 |
| 70AC(R)383 | F. Duro, P. Condorelli, G. Scapini and G. Pappalardo; *Ann. Chim. (Rome)*, 1970, **60**, 383. | 1032 |
| 70AC(R)564 | G. Cardillo, L. Merlini and S. Servi; *Ann. Chim. (Rome)*, 1970, **60**, 564. | 749 |
| 70ACS343 | A. K. Sørensen and N. A. Klitgaard; *Acta Chem. Scand.*, 1970, **24**, 343. | 791 |
| 70ACS2745 | O. Mårtensson and C. H. Warren; *Acta Chem. Scand.*, 1970, **24**, 2745. | 641 |
| 70ACS3417 | O. Mårtensson; *Acta Chem. Scand.*, 1970, **24**, 3417. | 576 |
| 70AJC51 | E. N. Cain and R. N. Warrener; *Aust. J. Chem.*, 1970, **23**, 51. | 1028 |
| 70AP44 | K. Winterfeld and M. Wildersohn; *Arch. Pharm. (Weinheim, Ger.)*, 1970, **303**, 44. | 254 |
| 70AP428 | G. Wurm and H. Loth; *Arch. Pharm. (Weinheim, Ger.)*, 1970, **303**, 428. | 724 |

| | | |
|---|---|---|
| 70AX(B)451 | H. L. Ammon, P. H. Watts, Jr. and J. M. Stewart; *Acta Crystallogr., Part B*, 1970, **26**, 451. | 947 |
| 70AX(B)2022 | K. Kato; *Acta Crystallogr., Part B*, 1970, **26**, 2022. | 623 |
| 70BCJ226 | H. Kanatomi and I. Murase; *Bull. Chem. Soc. Jpn*, 1970, **43**, 226. | 1024 |
| 70BCJ442 | K. Hanaya; *Bull. Chem. Soc. Jpn.*, 1970, **43**, 442. | 580, 786 |
| 70BCJ1590 | M. Kawai, T. Matsuura and R. Nakashima; *Bull. Chem. Soc. Jpn.*, 1970, **43**, 1590. | 809 |
| 70BCJ3543 | O. Tsuge and S. Iwanami; *Bull. Chem. Soc. Jpn.*, 1970, **43**, 3543. | 1077 |
| 70BCJ3929 | M. Kamiya; *Bull. Chem. Soc. Jpn.*, 1970, **43**, 3929. | 945, 946 |
| 70BRP1303171 | H. C. S. Wood; *Br. Pat.* 1 303 171 (1970) (*Chem. Abstr.*, 1971, **75**, 5955). | 325 |
| 70BSB89 | A. Ruwet and M. Renson; *Bull. Soc. Chim. Belg.*, 1970, **79**, 89. | 596 |
| 70BSF1139 | B. Kirkiacharian and D. Raulais; *Bull. Soc. Chim. Fr.*, 1970, 1139. | 594 |
| 70BSF1590 | M. Pesson and M. Antoine; *Bull. Soc. Chim. Fr.*, 1970, 1590. | 403 |
| 70BSF1599 | M. Pesson and M. Antoine; *Bull. Soc. Chim. Fr.*, 1970, 1599. | 403 |
| 70BSF1606 | J. Daunis, K. Diebel, R. Jacquier and P. Viallefont; *Bull. Soc. Chim. Fr.*, 1970, 1606. | 400 |
| 70BSF2016 | J. P. Sauvé and N. Lozac'h; *Bull. Soc. Chim. Fr.*, 1970, 2016. | 932 |
| 70CB82 | L. Capuano, W. Ebner and J. Schrepfer; *Chem. Ber.*, 1970, **103**, 82. | 216 |
| 70CB331 | W. Ried and E. Kahr; *Chem. Ber.*, 1970, **103**, 331. | 1074 |
| 70CB722 | G. Konrad and W. Pfleiderer; *Chem. Ber.*, 1970, **103**, 722. | 270, 295 |
| 70CB735 | G. Konrad and W. Pfleiderer; *Chem. Ber.*, 1970, **103**, 735. | 296 |
| 70CB900 | W. Pfleiderer and F. E. Kempter; *Chem. Ber.*, 1970, **103**, 900. | 88 |
| 70CB1250 | K. E. Schulte, V. von Weissenborn and G. L. Tittel; *Chem. Ber.*, 1970, **103**, 1250. | 218 |
| 70CB1846 | H. Paulsen, K. Steinert and G. Steinert; *Chem. Ber.*, 1970, **103**, 1846. | 45 |
| 70CB2760 | R. R. Schmidt, D. Schwille and H. Wolf; *Chem. Ber.*, 1970, **103**, 2760. | 1018 |
| 70CB2768 | H. Hofmann and G. Salbeck; *Chem. Ber.*, 1970, **103**, 2768. | 580 |
| 70CB3393 | J. Goerdeler and H. Lüdke; *Chem. Ber.*, 1970, **103**, 3393. | 1076 |
| 70CC25 | S. H. Wilen; *Chem. Commun.*, 1970, 25. | 184 |
| 70CC380 | J. R. Merchant and D. V. Rege; *Chem. Commun.*, 1970, 380. | 705 |
| 70CC446 | G. B. Ansell, J. L. Erickson and D. W. Moore; *Chem. Commun.*, 1970, 446. | 538 |
| 70CC1103 | A. E. A. Porter and P. G. Sammes; *Chem. Commun.*, 1970, 1103. | 174 |
| 70CC1371 | R. D. Youssefyeh and A. Kalmus; *Chem. Commun.*, 1970, 1371. | 314 |
| 70CC1423 | S. R. Challand, R. B. Herbert and F. G. Holliman; *Chem. Commun.*, 1970, 1423. | 190 |
| 70CI(L)897 | K. E. Murray, J. Shipton and F. B. Whitfield; *Chem. Ind. (London)*, 1970, 897. | 193 |
| 70CJC2334 | R. M. Srivastava and R. K. Brown; *Can. J. Chem.*, 1970, **48**, 2334. | 777 |
| 70CJC3045 | Y. L. Chow, T. C. Joseph, H. H. Quon and J. N. S. Tam; *Can. J. Chem.*, 1970, **48**, 3045. | 989 |
| 70CJC3928 | J. B. Gallivan; *Can. J. Chem.*, 1970, **48**, 3928. | 601 |
| 70CPB1 | Y. Sasaki and M. Suzuki; *Chem. Pharm. Bull.*, 1970, **18**, 1. | 973 |
| 70CPB1327 | Y. Shoyama, T. Yamauchi and I. Nishioka; *Chem. Pharm. Bull.*, 1970, **18**, 1327. | 721 |
| 70CPB1340 | H. Igeta, T. Tsuchiya, C. Okuda and H. Yokogawa; *Chem. Pharm. Bull.*, 1970, **18**, 1340. | 19 |
| 70CPB1385 | S. Nishigaki, K. Ogiwara, K. Senga, S. Fukazawa, K. Aida, Y. Machida and F. Yoneda; *Chem. Pharm. Bull.*, 1970, **18**, 1385. | 220 |
| 70CPB1457 | T. Higashino and E. Hayashi; *Chem. Pharm. Bull.*, 1970, **18**, 1457. | 207, 209, 212, 215, 222 |
| 70CR(C)(271)468 | J. Ficini, J. Besseyre, J. D'Angelo and C. Barbara; *C.R. Hebd. Seances Acad. Sci., Ser. C*, 1970, **271**, 468. | 760 |
| 70CRV593 | A. Lawson and R. B. Tinkler; *Chem. Rev.*, 1970, **70**, 593. | 1043, 1044, 1050, 1051, 1055, 1059, 1070 |
| 70EGP73039 | E. Kretzschmar, H. Goldhahn and E. Carstens; *Ger. (East) Pat.* 73 039 (1970) (*Chem. Abstr.*, 1971, **74**, 141 863). | 223 |
| 70FES57 | P. Mesnard; *Farmaco Ed. Sci.*, 1970, **25**, 57. | 882 |
| 70GEP1923022 | F. Becke and P. Paessler; *Ger. Pat.* 1 923 022 (1970) (*Chem. Abstr.*, 1971, **74**, 22 818). | 1025 |
| 70GEP1950392 | M. Otsuka, H. Yamaguchi and S. Komura; *Ger. Pat.* 1 950 392 (1970) (*Chem. Abstr.*, 1970, **72**, 132 803). | 569 |
| 70GEP1961326 | G. E. Hardtmann and H. Ott; (Sandoz Ltd.), *Ger. Offen.* 1 961 326 (1970) (*Chem. Abstr.*, 1970, **73**, 77 280). | 207 |
| 70GEP2020499 | Esso Research, *Ger. Pat.* 2 020 499 (1970) (*Chem. Abstr.*, 1971, **74**, 54 880). | 939 |
| 70HC(16-S1)20 | D. J. Brown; *Chem. Heterocycl. Compd.*, 1970, **16-S1**, 20. | 111 |
| 70HC(16-S1)53 | D. J. Brown; *Chem. Heterocycl. Compd.*, 1970, **16-S1**, 53. | 110, 116 |
| 70HC(16-S1)99 | D. J. Brown; *Chem. Heterocycl. Compd.*, 1970, **16-S1**, 99. | 87, 88 |
| 70HC(16-S1)103 | D. J. Brown; *Chem. Heterocycl. Compd.*, 1970, **16-S1**, 103. | 88 |
| 70HC(16-S1)109 | D. J. Brown; *Chem. Heterocycl. Compd.*, 1970, **16-S1**, 109. | 88 |
| 70HC(16-S1)110 | D. J. Brown; *Chem. Heterocycl. Compd.*, 1970, **16-S1**, 110. | 89 |
| 70HC(16-S1)129 | D. J. Brown; *Chem. Heterocycl. Compd.*, 1970, **16-S1**, 129. | 99 |
| 70HC(16-S1)189 | D. J. Brown; *Chem. Heterocycl. Compd.*, 1970, **16-S1**, 189. | 92 |
| 70HC(16-S1)217 | D. J. Brown; *Chem. Heterocycl. Compd.*, 1970, **16-S1**, 217. | 96 |
| 70HC(16-S1)230 | D. J. Brown; *Chem. Heterocycl. Compd.*, 1970, **16-S1**, 230. | 129 |
| 70HC(16-S1)245 | D. J. Brown; *Chem. Heterocycl. Compd.*, 1970, **16-S1**, 245. | 85 |
| 70HC(16-S1)262 | D. J. Brown; *Chem. Heterocycl. Compd.*, 1970, **16-S1**, 262. | 87 |
| 70HC(16-S1)269 | D. J. Brown; *Chem. Heterocycl. Compd.*, 1970, **16-S1**, 269. | 90, 135 |
| 70HC(16-S1)284 | D. J. Brown; *Chem. Heterocycl. Compd.*, 1970, **16-S1**, 284. | 86 |
| 70HC(16-S1)315 | D. J. Brown; *Chem. Heterocycl. Compd.*, 1970, **16-S1**, 315. | 79 |

| | | |
|---|---|---|
| 70HC(16-S1)322 | R. F. Evans; *Chem. Heterocycl. Compd.*, 1970, **16-S1**, 322. | 66, 75, 108, 118 |
| 70HC(16-S1)368 | D. J. Brown; *Chem. Heterocycl. Compd.*, 1970, **16-S1**, 368. | 60 |
| 70HCA251 | W. Skorianetz and E. sz. Kováts; *Helv. Chim. Acta*, 1970, **53**, 251. | 542, 544 |
| 70HCA485 | P. Schiess and H. L. Chia; *Helv. Chim. Acta*, 1970, **53**, 485. | 739 |
| 70HCA789 | M. Viscontini and H. Leidner; *Helv. Chim. Acta*, 1970, **53**, 789. | 306 |
| 70HCA1202 | M. Viscontini, R. Provenzale, S. Ohlgart and J. Mallevialle; *Helv. Chim. Acta*, 1970, **53**, 1202. | 312 |
| 70HCA1499 | A. Garbesi, N. Corsi and A. Fava; *Helv. Chim. Acta*, 1970, **53**, 1499. | 889 |
| 70IJC203 | T. R. Kasturi and T. Arunachalam; *Indian J. Chem.*, 1970, **8**, 203. | 723, 856 |
| 70IJC969 | S. D. Mehendale and S. V. Sunthankar; *Indian J. Chem.*, 1970, **8**, 969. | 803 |
| 70IZV904 | E. A. Arutyunyan, V. I. Gunar and S. I. Zav'yalov; *Izv. Akad. Nauk SSSR, Ser. Khim.*, 1970, 904. | 296 |
| 70IZV953 | E. A. Arutyunyan, V. I. Gunar and S. I. Zav'yalov; *Izv. Akad. Nauk SSSR, Ser. Khim.*, 1970, 953. | 296 |
| 70IZV1184 | N. S. Vul'fson, G. M. Zolotareva, V. N. Bochkarev, B. V. Unkovskii, V. B. Mochalin, Z. N. Smolina and A. N. Vul'fson; *Izv. Akad. Nauk SSSR, Ser. Khim.*, 1970, **5**, 1184. | 606 |
| 70IZV1198 | E. A. Arutyunan, V. I. Gunar and S. I. Zav'yalov; *Izv. Akad. Nauk SSSR, Ser. Khim.*, 1970, 1198. | 91,296 |
| 70JA1289 | N. W. Tyer, Jr. and R. S. Becker; *J. Am. Chem. Soc.*, 1970, **92**, 1289. | 598 |
| 70JA3050 | F. W. Nader and E. L. Eliel; *J. Am. Chem. Soc.*, 1970, **92**, 3050. | 960 |
| 70JA3055 | C. H. Bushweller, J. Golini, G. U. Rao and J. W. O'Neil; *J. Am. Chem. Soc.*, 1970, **92**, 3055. | 962 |
| 70JA3749 | J. L. Flippen, J. Karle and I. L. Karle; *J. Am. Chem. Soc.*, 1970, **92**, 3749. | 622 |
| 70JA3787 | D. A. Kleier, G. Binsch, A. Steigel and J. Sauer; *J. Am. Chem. Soc.*, 1970, **92**, 3787. | 551 |
| 70JA7488 | V. D. Parker and L. Eberson; *J. Am. Chem. Soc.*, 1970, **92**, 7488. | 969 |
| 70JA7505 | D. W. Johnson, V. Austel, R. S. Feld and D. M. Lemal; *J. Am. Chem. Soc.*, 1970, **92**, 7505. | 190 |
| 70JAP7019295 | M. Otsuka, S. Sachimura and H. Yamaguchi; *Jpn. Pat.* 70 19 295 (1970) (*Chem. Abstr.*, 1970, **73**, 56 135). | 569 |
| 70JBC(245)4647 | A. Bacher and F. Lingens; *J. Biol. Chem.*, 1970, **245**, 4647. | 321 |
| 70JBC(245)5498 | H. S. Murty, P. I. Caasi, S. K. Brooks and P. P. Nair; *J. Biol. Chem.*, 1970, **245**, 5498. | 877 |
| 70JCP(53)376 | H. M. Pickett and H. L. Strauss; *J. Chem. Phys.*, 1970, **53**, 376. | 594, 956 |
| 70JCS(B)135 | R. A. Y. Jones, A. R. Katritzky and M. Snarey; *J. Chem. Soc. (B)*, 1970, 135. | 467 |
| 70JCS(B)404 | R. T. Wragg; *J. Chem. Soc. (B)*, 1970, 404. | 987 |
| 70JCS(B)603 | D. Barraclough, H. D. Locksley, F. Scheinmann, M. Taveira Magalhães and O. R. Gottlieb; *J. Chem. Soc. (B)*, 1970, 603. | 583 |
| 70JCS(B)911 | J. C. Mason and G. Tennant; *J. Chem. Soc. (B)*, 1970, 911. | 412 |
| 70JCS(C)119 | W. D. Ollis, K. L. Ormand and I. O. Sutherland; *J. Chem. Soc. (C)*, 1970, 119. | 824 |
| 70JCS(C)125 | W. D. Ollis, K. L. Ormand, B. T. Redman, R. J. Roberts and I. O. Sutherland; *J. Chem. Soc. (C)*, 1970, 125. | 824 |
| 70JCS(C)214 | D. J. Brown and T.-C. Lee; *J. Chem. Soc. (C)*, 1970, 214. | 79 |
| 70JCS(C)437 | J. Mirza, W. Pfleiderer, A. D. Brewer, A. Stuart and H. C. S. Wood; *J. Chem. Soc. (C)*, 1970, 437. | 312 |
| 70JCS(C)530 | J. M. Holland and D. W. Jones; *J. Chem. Soc. (C)*, 1970, 530. | 636 |
| 70JCS(C)536 | J. M. Holland and D. W. Jones; *J. Chem. Soc. (C)*, 1970, 536. | 636 |
| 70JCS(C)552 | R. Grigg and J. L. Jackson; *J. Chem. Soc. (C)*, 1970, 552. | 815 |
| 70JCS(C)719 | J. L. Fahey, P. A. Foster, D. G. Neilson, K. M. Watson, J. L. Brokenshire and D. A. V. Peters; *J. Chem. Soc. (C)*, 1970, 719. | 569 |
| 70JCS(C)722 | F. M. Dean and L. E. Houghton; *J. Chem. Soc. (C)*, 1970, 722. | 767 |
| 70JCS(C)765 | H. N. E. Stevens and M. F. G. Stevens; *J. Chem. Soc. (C)*, 1970, 765. | 372 |
| 70JCS(C)980 | K. W. Blake and P. G. Sammes; *J. Chem. Soc. (C)*, 1970, 980. | 187 |
| 70JCS(C)986 | E. A. Gray, R. M. Hulley and B. K. Snell; *J. Chem. Soc. (C)*, 1970, 986. | 137 |
| 79JCS(C)1006 | W. D. Cotterill, J. Cottam and R. Livingstone; *J. Chem. Soc. (C)*, 1970, 1006. | 580, 630, 786 |
| 70JCS(C)1070 | K. W. Blake and P. G. Sammes; *J. Chem. Soc. (C)*, 1970, 1070. | 169, 170, 187 |
| 70JCS(C)1219 | W. Baker, D. F. Downing, A. J. Floyd, B. Gilbert, W. D. Ollis and R. C. Russell; *J. Chem. Soc. (C)*, 1970, 1219. | 821 |
| 70JCS(C)1238 | R. A. W. Johnstone, D. W. Payling, P. N. Preston, H. N. E. Stevens and M. F. G. Stevens; *J. Chem. Soc. (C)*, 1970, 1238. | 373 |
| 70JCS(C)1540 | A. Albert and K. Ohta; *J. Chem. Soc. (C)*, 1970, 1540. | 266, 276, 279, 280, 318 |
| 70JCS(C)1553 | A. O. Fitton, B. T. Hatton, M. P. Ward and R. Lewis; *J. Chem. Soc. (C)*, 1970, 1553. | 616, 816 |
| 70JCS(C)1662 | J. R. Lewis and J. B. Reary; *J. Chem. Soc. (C)*, 1970, 1662. | 747 |
| 70JCS(C)1758 | W. D. Cotterill, R. Livingstone and M. V. Walshaw; *J. Chem. Soc. (C)*, 1970, 1758. | 763 |
| 70JCS(C)2070 | A. W. Murray and K. Vaughan; *J. Chem. Soc. (C)*, 1970, 2070. | 379 |
| 70JCS(C)2230 | G. Barker and G. P. Ellis; *J. Chem. Soc. (C)*, 1970, 2230. | 583, 596, 707, 714 |
| 70JCS(C)2284 | H. N. E. Stevens and M. F. G. Stevens; *J. Chem. Soc. (C)*, 1970, 2284. | 375 |
| 70JCS(C)2289 | H. N. E. Stevens and M. F. G. Stevens; *J. Chem. Soc. (C)*, 1970, 2289. | 376, 377 |
| 70JCS(C)2308 | H. N. E. Stevens and M. F. G. Stevens; *J. Chem. Soc. (C)*, 1970, 2308. | 377, 382 |
| 70JCS(C)2510 | R. N. Butler, F. L. Scott and R. D. Scott; *J. Chem. Soc. (C)*, 1970, 2510. | 567 |
| 70JCS(C)2609 | G. Barker and G. P. Ellis; *J. Chem. Soc. (C)*, 1970, 2609. | 583, 596, 703 |
| 70JCS(C)2661 | D. J. Brown and T. Sugimoto; *J. Chem. Soc. (C)*, 1970, 2661. | 92, 134 |

| | | |
|---|---|---|
| 70JGU1402 | N. P. Shusherina, M. W. Gapeeva and R. Ya. Levina; *J. Gen. Chem. USSR* (*Engl. Transl.*), 1970, **40**, 1402. | 688 |
| 70JHC99 | D. E. O'Brien, L. T. Weinstock and C. C. Cheng; *J. Heterocycl. Chem.*, 1970, **7**, 99. | 228 |
| 70JHC187 | G. F. Katekar; *J. Heterocycl. Chem.*, 1970, **7**, 187. | 597, 891 |
| 70JHC197 | A. Rosowsky, P. C. Huang and E. J. Modest; *J. Heterocycl. Chem.*, 1970, **7**, 197. | 583, 597, 852 |
| 70JHC209 | T. Nakashima and R. N. Castle; *J. Heterocycl. Chem.*, 1970, **7**, 209. | 346 |
| 70JHC243 | R. Niess and R. K. Robins; *J. Heterocycl. Chem.*, 1970, **7**, 243. | 364 |
| 70JHC527 | L. Doub and U. Krolls; *J. Heterocycl. Chem.*, 1970, **7**, 527. | 132, 144 |
| 70JHC725 | P. A. Argabright and B. L. Phillips; *J. Heterocycl. Chem.*, 1970, **7**, 725. | 463, 465, 511 |
| 70JHC767 | W. W. Paudler and T.-K. Chen; *J. Heterocycl. Chem.*, 1970, **7**, 767. | 399, 413 |
| 70JHC981 | T. Harayama, K. Okada, S. Sekiguchi and K. Matsui; *J. Heterocycl. Chem.*, 1970, **7**, 981. | 481 |
| 70JHC987 | A. B. De Milo; *J. Heterocycl. Chem.*, 1970, **7**, 987. | 474 |
| 70JHC999 | P. A. Argabright and B. L. Phillips; *J. Heterocycl. Chem.*, 1970, **7**, 999. | 478 |
| 70JHC1195 | C. Temple, Jr., A. G. Laseter, J. D. Rose and J. A. Montgomery; *J. Heterocycl. Chem.*, 1970, **7**, 1195. | 251 |
| 70JHC1219 | C. Temple, Jr., A. G. Laseter and J. A. Montgomery; *J. Heterocycl. Chem.*, 1970, **7**, 1219. | 230 |
| 70JHC1311 | M. von Strandtmann, M. P. Cohen and J. Shavel, Jr.; *J. Heterocycl. Chem.*, 1970, **7**, 1311. | 762, 782, 784 |
| 70JHC1395 | G. A. Reynolds, J. A. Van Allan and D. Daniel; *J. Heterocycl. Chem.*, 1970, **7**, 1395. | 664 |
| 70JOC719 | E. C. Juenge, D. A. Beal and W. P. Duncan; *J. Org. Chem.*, 1970, **35**, 719. | 525 |
| 70JOC930 | R. Filler, Y. S. Rao, A. Biezais, F. N. Miller and V. D. Beaucaire; *J. Org. Chem.*, 1970, **35**, 930. | 822 |
| 70JOC1468 | H. Feuer and F. Brown; *J. Org. Chem.*, 1970, **35**, 1468. | 46 |
| 70JOC2282 | W. C. Still, Jr. and D. J. Goldsmith; *J. Org. Chem.*, 1970, **35**, 2282. | 594, 687, 723, 764, 856 |
| 70JOC2286 | D. D. Keane, K. G. Marathe, W. I. O'Sullivan, E. M. Philbin, R. M. Simons and P. C. Teague; *J. Org. Chem.*, 1970, **35**, 2286. | 853 |
| 70JOC2438 | T. Takeshima, M. Yokoyama, N. Fukada and M. Akano; *J. Org. Chem.*, 1970, **35**, 2438. | 935 |
| 70JOC2716 | E. L. Eliel and T. W. Doyle; *J. Org. Chem.*, 1970, **35**, 2716. | 978 |
| 70JOC2790 | M. L. Scheinbaum; *J. Org. Chem.*, 1970, **35**, 2790. | 170 |
| 70JOC2994 | H. W. Gibson and D. A. McKenzie; *J. Org. Chem.*, 1970, **35**, 2994. | 987 |
| 70JOC3666 | D. A. Bolon; *J. Org. Chem.*, 1970, **35**, 3666. | 784, 785 |
| 70JOC3792 | E. C. Taylor and S. F. Martin; *J. Org. Chem.*, 1970, **35**, 3792. | 448 |
| 70JOC3967 | S. R. Sandler; *J. Org. Chem.*, 1970, **35**, 3967. | 486 |
| 70JPC4234 | P.-S. Song and W. H. Gordon, III; *J. Phys. Chem.*, 1970, **74**, 4234. | 576, 600 |
| 70JPR669 | H. G. O. Becker, D. Beyer, G. Israel, R. Müller, W. Riediger and H.-J. Timpe; *J. Prakt. Chem.*, 1970, **312**, 669. | 395, 396 |
| 70JSP(35)83 | R. S. Butler, J. M. Read, Jr. and J. H. Goldstein; *J. Mol. Spectrosc.*, 1970, **35**, 83. | 952 |
| 70JSP(36)310 | R. W. Mitchell, R. W. Glass and J. A. Merritt; *J. Mol. Spectrosc.*, 1970, **36**, 310. | 65 |
| 70KGS1704 | V. M. Cherkasov, I. A. Nasyr and V. T. Tsyba; *Khim. Geterotsikl. Soedin.*, 1970, 1704. | 437 |
| 70LA(733)177 | A. Mustafa, A. K. Mansour and H. A. Zaher; *Liebigs Ann. Chem.*, 1970, **733**, 177. | 406, 407 |
| 70LA(736)176 | K. Fickentscher and H.-W. Fehlhaber; *Liebigs Ann. Chem.*, 1970, **736**, 176. | 970 |
| 70LA(737)39 | H. Bredereck, G. Simchen and P. Speh; *Liebigs Ann. Chem.*, 1970, **737**, 39. | 77 |
| 70LA(739)32 | F. Asinger, H. Offermanns and D. Neuray; *Liebigs Ann. Chem.*, 1970, **739**, 32. | 1035 |
| 70LA(741)45 | H. Beyer, H. Honeck and L. Reichelt; *Liebigs Ann. Chem.*, 1970, **741**, 45. | 1066, 1075 |
| 70LA(741)64 | W. Pfleiderer, H. Zondler and R. Mengel; *Liebigs Ann. Chem.*, 1970, **741**, 64. | 277, 311 |
| 70M1123 | H. Junek and W. Wilfinger; *Monatsh. Chem.*, 1970, **101**, 1123. | 682 |
| 70M1130 | H. Junek and I. Wrtilek; *Monatsh. Chem.*, 1970, **101**, 1130. | 229, 230 |
| 70M1415 | G. Zigeuner, A. Frank, H. Dujmovits and W. Adam; *Monatsh. Chem.*, 1970, **101**, 1415. | 223, 232 |
| 70M1824 | G. Zigeuner, H. Schmidt and D. Volpe; *Monatsh. Chem.*, 1970, **101**, 1824. | 362 |
| 70MI21300 | C. C. Cheng and B. Roth; *Prog. Med. Chem.*, 1970, **7**, 285. | 154 |
| 70MI21301 | V. M. S. Gil and A. J. L. Pinto; *Mol. Phys.*, 1970, **19**, 573. | 63 |
| B-70MI21302 | H. G. Mautner and H. C. Clemson; in 'Medicinal Chemistry', ed. A. Burger; Wiley, New York, 3rd edn., 1970, part 2, p. 1365. | 150 |
| 70MI21303 | I. M. Rollo; *Crit. Rev. Clin. Lab. Sci.*, 1970, **1**, 565. | 151 |
| B-70MI21400 | D. W. Turner, C. W. Baker, A. D. Baker and C. R. Brundle; 'Photoelectron Spectroscopy', Wiley, New York, 1970. | 159 |
| 70MI21402 | P. E. Koehler and G. V. Odell; *J. Agric. Food Chem.*, 1970, **18**, 895. | 193 |
| 70MI21500 | B. M. Pyatin and R. G. Glushkov; *Khim.-Farm. Zh.*, 1970, **4**, 21 (*Chem. Abstr.*, 1970, **73**, 55 994). | 230 |
| 70MI21501 | M. G. DeAmezua, M. Lora-Tamayo, J. L. Soto and E. D. Toro; *An. Quim.*, 1970, **66**, 561 (*Chem. Abstr.*, 1971, **75**, 20 326). | 238 |
| 70MI21600 | P. Haug; *Anal. Biochem.*, 1970, **37**, 285. | 285 |
| 70MI21601 | M. C. Archer and K. G. Scrimgeour; *Can. J. Biochem.*, 1970, **48**, 278. | 306 |
| 70MI21602 | A. Ehrenberg, P. Hemmerich, F. Müller and W. Pfleiderer; *Eur. J. Biochem.*, 1970, **16**, 584. | 282 |
| 70MI21603 | H. Rembold; *Korean Biochem. J.*, 1970, **3**, 1. | 306, 323 |

| | | |
|---|---|---|
| 70MI21604 | H. Iida and Y. Arai; *Kogyo Kagaku Zasshi*, 1970, **73**, 749. | 292 |
| B-70MI21605 | K. Iwai, M. Akino, M. Goto and Y. Iwanami; 'Chemistry and Biology of Pteridines', International Academic Printing Co., Tokyo, 1970. | 264 |
| B-70MI21606 | I. Takada; in 'Chemistry and Biology of Pteridines', ed. K. Iwai, M. Akino, M. Goto and Y. Iwanami; International Academic Printing Co., Tokyo, 1970, p. 183. | 321 |
| B-70MI21607 | E. C. Taylor, Jr.; in 'Chemistry and Biology of Pteridines', ed. K. Iwai, M. Akino, M. Goto and Y. Iwanami; International Academic Printing Co., Tokyo, 1970, p. 79. | 318 |
| B-70MI21608 | J. G. Topliss; in 'Medicinal Chemistry', ed. A. Burger; Wiley-Interscience, New York, 3rd edn., 1970, part II, p. 1005. | 325 |
| 70MI21609 | W. L. Gyure; Dissertation, Tufts University, 1970; University Microfilms, Ann Arbor, Mich., No. 71–73, 749. | 322 |
| 70MI21900 | J. Hadacek and J. Slouka; *Folia Fac. Sci. Natur. Univ. Purkynianae Brunensis*, 1970. | 385 |
| 70MI22000 | H. Westlinning; *Rubber Chem. Technol.*, 1970, **43**, 1194. | 526 |
| B-70MI22001 | W. F. Beech; 'Fibre-Reactive Dyes', Logos Press, London, 1970. | 526 |
| B-70MI22002 | 'Residue Reviews', ed. F. A. Gunther; Springer-Verlag, Berlin, 1970, vol. 32. | 528 |
| B-70MI22200 | T. J. Mabry, K. R. Markham and M. B. Thomas; 'The Systematic Identification of Flavonoids', Springer, Berlin, 1970. | 584, 602 |
| 70MI22201 | D. W. Fink and W. R. Koehler; *Anal. Chem.*, 1970, **42**, 990. | 601 |
| 70MI22202 | A. A. Efimov, R. N. Nurmuhametov and A. I. Tolmachev; *Opt. Spectrosc. (Engl. Transl.)*, 1970, 11. | 627 |
| 70MI22300 | P. Catsoulacos; *Chim. Therap.*, 1970, **5**, 401. | 702 |
| 70MI22301 | J. Dauphin, D. Chatonier, J. Couquelet, M. Payard and M. Picard; *Labo-Pharm.-Probl. Tech.*, 1970, **18**, 58, 70. | 710 |
| 70MI22302 | J. S. G. Cox; *Adv. Drug Res.*, 1970, **5**, 115. | 693 |
| B-70MI22400 | E. C. Taylor and A. McKillop; 'The Chemistry of Cyclic Enaminonitriles and o-Aminonitriles', Interscience, New York, 1970. | 759 |
| 70MI22401 | R. Mechoulam; *Science*, 1970, **168**, 1159. | 877 |
| 70MI22402 | J. S. G. Cox; *Adv. Drug Res.*, 1970, **5**, 155. | 882 |
| B-70MI22700 | L. Valzelli and S. Garattini; in 'Principles of Psychopharmacology', ed. W. G. Clark; Academic Press, New York, 1970, p.255. | 1038 |
| 70MI22800 | A. J. Wohl; *Mol. Pharmacol.*, 1970, **6**, 189. | 1055 |
| 70MI22801 | H. Beyer; *Q. Rep. Sulfur Chem.*, 1970, **5**, 177. | 1044 |
| B-70MI22802 | I. Haiduc; 'The Chemistry of Inorganic Ring Systems', Wiley, New York, 1970, p. 908. | 1048 |
| 70MI22803 | A. Spasov and B. Chemishev; *Dokl. Bolg. Akad. Nauk*, 1970, **23**, 791 (*Chem. Abstr.*, 1970, **73**, 120 593). | 1046, 1065 |
| 70OMR(2)55 | O. Achmatowicz, Jr., J. Jurczak, A. Konowal and A. Zamojski; *Org. Magn. Reson.*, 1970, **2**, 55. | 578 |
| 70OMR(2)431 | K. K. Deb, J. E. Bloor and T. C. Cole; *Org. Magn. Reson.*, 1970, **2**, 431. | 951, 952 |
| 70OMS(3)13 | S. Balusu, T. Axenrod and G. W. A. Milne; *Org. Mass Spectrom.*, 1970, **3**, 13. | 466 |
| 70OMS(3)219 | J. A. Ross and B. G. Tweedy; *Org. Mass Spectrom.*, 1970, **3**, 219. | 465 |
| 70OMS(3)409 | N. M. M. Nibbering and T. J. de Boer; *Org. Mass Spectrom.*, 1970, **3**, 409. | 606 |
| 70OMS(3)753 | J. R. Trudell, S. D. S. Woodgate and C. Djerassi; *Org. Mass Spectrom.*, 1970, **3**, 753. | 606 |
| 70OMS(3)863 | P. N. Preston, W. Steedman, M. H. Palmer, S. M. Mackenzie and M. F. G. Stevens; *Org. Mass Spectrom.*, 1970, **3**, 863. | 465 |
| 70OMS(3)941 | S. J. Shaw and P. V. R. Shannon; *Org. Mass Spectrom.*, 1970, **3**, 941. | 607 |
| 70OMS(3)1365 | P. Haug and T. Urushibara; *Org. Mass Spectrom.*, 1970, **3**, 1365. | 285 |
| 70OMS(4S)121 | I. C. Calder, R. B. Johns and J. M. Desmarchelier; *Org. Mass Spectrom.*, 1970, **4** (Suppl.), 121. | 957 |
| 70RRC1121 | C. Cristescu; *Rev. Roum. Chim.*, 1970, **15**, 1121. | 411 |
| 70RTC313 | A. J. de Kok and C. Romers; *Recl. Trav. Chim. Pays-Bas*, 1970, **89**, 313. | 950 |
| 70RTC1244 | H. R. Buys; *Recl. Trav. Chim. Pays-Bas*, 1970, **89**, 1244. | 953 |
| 70S180 | B. Stanovnik and M. Tišler; *Synthesis*, 1970, 180. | 352 |
| 70S209 | M. Mihailović and Z. Čeković; *Synthesis*, 1970, 209. | 776 |
| 70SST(1)1 | D. T. Clark; *Org. Compd. Sulphur, Selenium, Tellurium*, 1970, **1**, 1. | 944 |
| 70SST(1)473 | D. H. Reid; *Org. Compd. Sulphur, Selenium, Tellurium*, 1970, **1**, 473. | 1043, 1044 |
| 70T435 | T. Matsuura, H. Matsushima and R. Nakashima; *Tetrahedron*, 1970, **26**, 435. | 695 |
| 70T739 | A. T. Balaban; *Tetrahedron*, 1970, **26**, 739. | 660 |
| 70T1315 | P. Bravo, G. Gaudiano, P. P. Ponti and A. Umani-Ronchi; *Tetrahedron*, 1970, **26**, 1315. | 999 |
| 70T2619 | P. Scheiner and J. F. Dinda, Jr.; *Tetrahedron*, 1970, **26**, 2619. | 549 |
| 70T2787 | L. Farkas, A. Gottsegen and M. Nógrádi; *Tetrahedron*, 1970, **26**, 2787. | 856 |
| 70T3227 | R. Borsdorf, H.-J. Hofmann, H.-J. Köhler, M. Scholz and J. Fabian; *Tetrahedron*, 1970, **26**, 3227. | 960 |
| 70T3389 | P. J. Chivers and T. A. Crabb; *Tetrahedron*, 1970, **26**, 3389. | 953 |
| 70T5255 | T. M. Harris and M. P. Wachter; *Tetrahedron*, 1970, **26**, 5255. | 680, 686, 816 |
| 70TL15 | F. Minisci, G. P. Gardini, R. Galli and F. Bertini; *Tetrahedron Lett.*, 1970, 15. | 166 |
| 70TL885 | J. Ficini and A. Krief; *Tetrahedron Lett.*, 1970, 885. | 760 |
| 70TL1467 | K. Matsui, N. Maeno, S. Suzuki, H. Shizuka and T. Morita; *Tetrahedron Lett.*, 1970, 1467. | 481 |

| | | |
|---|---|---|
| 70TL1617 | G. Heinrichs, H. Krapf, B. Schröder, A. Steigel, T. Troll and J. Sauer; *Tetrahedron Lett.*, 1970, 1617. | 551 |
| 70TL1665 | N. S. Narasimhan, M. V. Paradkar and A. M. Gokhale; *Tetrahedron Lett.*, 1970, 1665. | 748 |
| 70TL1837 | J. A. Elix, W. S. Wilson and R. N. Warrener; *Tetrahedron Lett.*, 1970, 1837. | 424 |
| 70TL2167 | L. Toldy, P. Sohár, K. Faragó, I. Tóth and L. Bartalits; *Tetrahedron Lett.*, 1970, 2167. | 1008 |
| 70TL2199 | W. Oppolzer; *Tetrahedron Lett.*, 1970, 2199. | 568 |
| 70TL2779 | H. R. Buys and E. L. Eliel; *Tetrahedron Lett.*, 1970, 2779. | 954 |
| 70TL3355 | H. Neunhoeffer and H.-W. Frühauf; *Tetrahedron Lett.*, 1970, 3355. | 427 |
| 70TL3357 | A. Steigel and J. Sauer; *Tetrahedron Lett.*, 1970, 3357. | 427 |
| 70TL3945 | R. D. Allen, R. S. Wells and J. K. MacLeod; *Tetrahedron Lett.*, 1970, 3945. | 675 |
| 70TL4895 | G. Wood, J. M. McIntosh and M. H. Miskow; *Tetrahedron Lett.*, 1970, 4895. | 962 |
| 70UP21900 | H. Neunhoeffer and H. Hennig; unpublished results, 1970. | 399 |
| 70UP21901 | H. Neunhoeffer and H.-W. Frühauf; unpublished results, 1970. | 423 |
| 70ZC147 | W. Schroth, R. Borsdorf, R. Hertzschuh and J. Seidler; *Z. Chem.*, 1970, **10**, 147. | 960, 968 |
| 70ZC296 | W. Schroth, M. Hassfeld and A. Zschunke; *Z. Chem.*, 1970, **10**, 296. | 951, 952, 967 |
| | | |
| 71ACR87 | J. B. Lambert; *Acc. Chem. Res.*, 1971, **4**, 87. | 954 |
| 71ACS94 | J. L. G. Nilsson, H. Selander, H. Sievertsson, I. Skånberg and K.-G. Svensson; *Acta Chem. Scand.*, 1971, **25**, 94. | 735 |
| 71ACS487 | M. Sundbom; *Acta Chem. Scand.*, 1971, **25**, 487. | 159 |
| 71ACS1908 | P. Pasanen and K. Pihlaja; *Acta Chem. Scand.*, 1971, **25**, 1908. | 989 |
| 71AG(E)127 | R. A. Sulzbach and A. F. M. Iqbal; *Angew. Chem., Int. Ed. Engl.*, 1971, **10**, 127. | 178 |
| 71AG(E)264 | M. Haake; *Angew. Chem., Int. Ed. Engl.*, 1971, **10**, 264. | 1047, 1071 |
| 71AG(E)407 | K. H. Linke, R. Bimczok and H. Lingmann; *Angew. Chem., Int. Ed. Engl.*, 1971, **10**, 407. | 1046, 1066, 1078 |
| 71AJC785 | R. N. Warrener and E. N. Cain; *Aust. J. Chem.*, 1971, **24**, 785. | 121 |
| 71AJC1107 | J. P. Byrne and I. G. Ross; *Aust. J. Chem.*, 1971, **24**, 1107. | 250 |
| 71AJC1859 | S. R. Johns, J. A. Lamberton and E. R. Nelson; *Aust. J. Chem.*, 1971, **24**, 1859. | 568 |
| 71AJC2347 | J. Hlubucek, E. Ritchie and W. C. Taylor; *Aust. J. Chem.*, 1971, **24**, 2347. | 580, 743, 744, 745 |
| 71ANY(186)423 | A. Goldin; *Ann. N.Y. Acad. Sci.*, 1971, **186**, 423. | 327 |
| 71AP543 | H. H. auf dem Keller and F. Zymalkowski; *Arch. Pharm. (Weinheim, Ger.)*, 1971, **304**, 543. | 828 |
| 71BBA(230)117 | H. Rembold, H. Metzger and W. Gutensohn; *Biochem. Biophys. Acta*, 1971, **230**, 117. | 321 |
| 71BBA(237)365 | H. Rembold, V. Chandrashekar and P. Sudershan; *Biochem. Biophys. Acta*, 1971, **237**, 365. | 321 |
| 71BCJ1669 | A. Mori, M. Nagayama and H. Mandai; *Bull. Chem. Soc. Jpn.*, 1971, **44**, 1669. | 980 |
| 71BCJ1869 | A. Suzuki and M. Goto; *Bull. Chem. Soc. Jpn.*, 1971, **44**, 1869. | 321 |
| 71BCJ2750 | O. Tsuge and S. Iwanami; *Bull. Chem. Soc. Jpn.*, 1971 **44**, 2750. | 1017 |
| 71BCJ3413 | M. Kurabayashi, K. Yanagiya and M. Yasumoto; *Bull. Chem. Soc. Jpn.*, 1971, **44**, 3413. | 505 |
| 71BSF187 | M. Ebel; *Bull. Soc. Chim. Fr.*, 1971, 187. | 1001 |
| 71BSF906 | A. Godard, G. Queguiner and P. Pastour; *Bull. Soc. Chim. Fr.*, 1971, 906. | 234, 240, 247 |
| 71BSF1351 | C. Normant-Chefnay; *Bull. Soc. Chim. Fr.*, 1971, 1351. | 597, 859 |
| 71BSF1362 | C. Normant-Chefnay; *Bull. Soc. Chim. Fr.*, 1971, 1362. | 577 |
| 71BSF2557 | M.-C. Sacquet, B. Graffe and P. Maitte; *Bull. Soc. Chim. Fr.*, 1971, 2557. | 671 |
| 71BSF3006 | J. Mounet, J. Huet and J. Dreux; *Bull. Soc. Chim. Fr.*, 1971, 3006. | 630 |
| 71BSF3010 | J. Mounet, J. Huet and J. Dreux; *Bull. Soc. Chim. Fr.*, 1971, 3010. | 630 |
| 71BSF3603 | M. Simalty, H. Strzelecka and H. Khedija; *Bull. Soc. Chim. Fr.*, 1971, 3603. | 664 |
| 71CB348 | B. Föhlisch; *Chem. Ber.*, 1971, **104**, 348. | 821, 872 |
| 71CB770 | H. Rokos and W. Pfleiderer; *Chem. Ber.*, 1971, **104**, 770. | 282 |
| 71CB780 | W. Pfleiderer and H. Deiss; *Chem. Ber.*, 1971, **104**, 780. | 281, 316 |
| 71CB1606 | J. Goerdeler and J. Neuffer; *Chem. Ber.*, 1971, **104**, 1606. | 501 |
| 71CB1880 | E. Sianesi, G. Bonola, R. Pozzi and P. Da Re; *Chem. Ber.*, 1971, **104**, 1880. | 1018 |
| 71CB2273 | W. Pfleiderer, R. Mengel and P. Hemmerich; *Chem. Ber.*, 1971, **104**, 2273. | 277, 309 |
| 71CB2293 | W. Pfleiderer and R. Mengel; *Chem. Ber.*, 1971, **104**, 2293. | 281, 308 |
| 71CB2313 | W. Pfleiderer and R. Mengel; *Chem. Ber.*, 1971, **104**, 2313. | 279, 305 |
| 71CB3341 | W. Ried and P. Weidemann; *Chem. Ber.*, 1971, **104**, 3341. | 244 |
| 71CB3842 | W. Pfleiderer and R. Mengel; *Chem. Ber.*, 1971, **104**, 3842. | 279, 281, 305 |
| 71CC28 | H. Suschitzky and G. E. Chivers; *Chem. Commun.*, 1971, 28. | 169 |
| 71CC83 | F. Yoneda, S. Fukazawa and S. Nishigaki; *Chem. Commun.*, 1971, 83. | 317 |
| 71CC189 | E. C. Taylor, Jr. and B. E. Evans; *Chem. Commun.*, 1971, 189. | 314 |
| 71CC264 | R. D. Chambers, W. K. R. Musgrave and K. C. Srivastava; *Chem. Commun.*, 1971, 264. | 190 |
| 71CC383 | O. L. Chapman and C. L. McIntosh; *Chem. Commun.*, 1971, 383. | 785 |
| 71CC822 | W. Adam and A. Rios; *Chem. Commun.*, 1971, 822. | 990 |
| 71CC828 | D. J. C. Adams, S. Bradbury, D. C. Horwell, M. Keating, C. W. Rees and R. C. Storr; *Chem. Commun.*, 1971, 828. | 374, 381, 383 |
| 71CC837 | W. L. Mosby and M. L. Vega; *Chem. Commun.*, 1971, 837. | 562 |
| 71CC1063 | N. Sugiyama, M. Yoshioka, H. Aoyama and T. Nishio; *Chem. Commun.*, 1971, 1063. | 932 |

| | | |
|---|---|---|
| 71CC1558 | F. A. L. Anet and J. Sandstrom; *Chem. Commun.*, 1971, 1558. | 952 |
| 71CCC246 | E. Wittenburg; *Collect. Czech. Chem. Commun.*, 1971, **36**, 246. | 91 |
| 71CHE147 | G. N. Dorofeenko and N. A. Lopatina; *Chem. Heterocycl. Compd.* (*Engl. Transl.*), 1971, 147. | 865 |
| 71CHE1587 | G. N. Dorofeenko and V. V. Tkachenko; *Chem. Heterocycl. Compd.* (*Engl. Transl.*), 1971, 1587. | 864 |
| 71CJC2297 | W. Steck; *Can. J. Chem.*, 1971, **49**, 2297. | 804 |
| 71CJC3477 | P. C. Arora and P. Brassard; *Can. J. Chem.*, 1971, **49**, 3477. | 850 |
| 71CPB1426 | S. Minami, T. Shono and J. Matsumoto; *Chem. Pharm. Bull.*, 1971, **19**, 1426. | 260 |
| 71CPB1482 | S. Minami, T. Shono and J. Matsumoto; *Chem. Pharm. Bull.*, 1971, **19**, 1482. | 220 |
| 71CPB1789 | S. Hayashi, M. Furukawa, Y. Fujino and H. Morishita; *Chem. Pharm. Bull.*, 1971, **19**, 1789. | 494 |
| 71CPB1849 | M. Yanai, T. Kinoshita, H. Watanabe and S. Iwasaki; *Chem. Pharm. Bull.*, 1971, **19**, 1849. | 335, 344 |
| 71CR(C)(272)1985 | H. Andrianaivoarivelo, D. Anker and D. Molho; *C.R. Hebd. Seances Acad. Sci., Ser. C*, 1971, **272**, 1985. | 803 |
| 71CR(C)(273)148 | J. Meslin and H. Quiniou; *C.R. Hebd. Seances Acad. Sci., Ser. C*, 1971, **273**, 148. | 935 |
| 71CR(C)(273)1529 | D. Boutte, G. Queguiner and P. Pastour; *C.R. Hebd. Seances Acad. Sci., Ser. C*, 1971, **273**, 1529. | 250, 251, 252 |
| 71CR(C)(273)1645 | D. Boutte, G. Queguiner and P. Pastour; *C.R. Hebd. Seances Acad. Sci., Ser. C*, 1971, **273**, 1645. | 257, 258 |
| 71CRV295 | S. D. Worley; *Chem. Rev.*, 1971, **71**, 295. | 159 |
| 71DOK(196)367 | N. M. Shekhtman, E. A. Viktorova, A. E. Karakhanov, N. M. Khrorostukhina and N. S. Zefirov; *Dokl. Akad. Nauk SSSR*, 1971, **196**, 367. | 630 |
| 71DOK(196)640 | V. G. Mairanovskii, O. I. Volkova, E. A. Obolinikova and G. I. Samokhvalov; *Dokl. Akad. Nauk SSSR*, 1971, **196**, 640. | 598 |
| 71DOK(199)146 | N. K. Malinin, M. G. Slin'ko, Yu. Sh. Matros and V. G. Gorskii; *Dokl. Akad. Nauk SSSR*, 1971, **199**, 146. | 506 |
| 71DOK(200)134 | O. P. Shvaika and V. I. Fomenko; *Dokl. Akad. Nauk SSSR*, 1971, **200**, 134. | 561 |
| 71FES580 | G. Palazzo and G. Picconi; *Farmaco Ed. Sci.*, 1971, **26**, 580. | 420, 438 |
| 71FES1074 | F. Luini and G. Palamidessi; *Farmaco Ed. Sci.*, 1971, **26**, 1074. | 240, 246 |
| 71GEP1963152 | F. Wiedemann, M. Thiel, K. Stach, E. Roesch and K. Hardebeck; (Boehringer Mannheim G.m.b.H.), *Ger. Offen.* 1 963 152 (1971) (*Chem. Abstr.*, 1971, **75**, 76 844). | 215 |
| 71GEP2006196 | Fisons Ltd., *Ger. Pat.* 2 006 196 (1971) (*Chem. Abstr.*, 1971, **74**, 42 279). | 934 |
| 71GEP2031230 | T.-Y. Shen, G. L. Walford, B. E. Witzel and W. V. Ruyle; (Merck and Co. Inc.), *Ger. Offen.* 2 031 230 (1971) (*Chem. Abstr.*, 1971, **74**, 125 442). | 215 |
| 71GEP2051013 | G. E. Hardtmann; (Sandoz Ltd.), *Ger. Offen.* 2 051 013 (1971) (*Chem. Abstr.*, 1972, **76**, 113 242). | 225 |
| 71HCA306 | R. Hug, G. Fráter, H. J. Hansen and H. Schmid; *Helv. Chim. Acta*, 1971, **54**, 306. | 671 |
| 71HCA1922 | W. Skorianetz and E. sz. Kováts; *Helv. Chim. Acta*, 1971, **54**, 1922. | 542, 546, 569 |
| 71IJC1322 | B. S. Bajwa, P. L. Khanna and T. R. Seshadri; *Indian J. Chem.*, 1971, **9**, 1322. | 717 |
| 71IJS(B)(6)183 | A. Ohno; *Int. J. Sulfur Chem., Sect. B*, 1971, **6**, 183. | 987 |
| 71JA746 | G. Büchi and S. M. Weinreb; *J. Am. Chem. Soc.*, 1971, **93**, 746. | 803 |
| 71JA2346 | R. N. Warrener; *J. Am. Chem. Soc.*, 1971, **93**, 2346. | 552 |
| 71JA2471 | A. G. Hortmann and R. L. Harris; *J. Am. Chem. Soc.*, 1971, **93**, 2471. | 939 |
| 71JA2506 | T. T. Howarth and T. M. Harris; *J. Am. Chem. Soc.*, 1971, **93**, 2506. | 796 |
| 71JA2733 | A. Williams; *J. Am. Chem. Soc.*, 1971, **93**, 2733. | 861 |
| 71JA4772 | A. J. Jones, E. L. Eliel, D. M. Grant, M. C. Knoeber and W. F. Bailey; *J. Am. Chem. Soc.*, 1971, **93**, 4772. | 954 |
| 71JA5591 | R. C. Benson, C. L. Norris, W. H. Flygare and P. Beak; *J. Am. Chem. Soc.*, 1971, **93**, 5591. | 635 |
| 71JA5850 | R. Danieli, L. Lunazzi and G. Placucci; *J. Am. Chem. Soc.*, 1971, **93**, 5850. | 337, 340 |
| 71JA6329 | F. R. Jensen and R. A. Neese; *J. Am. Chem. Soc.*, 1971, **93**, 6329. | 962 |
| 71JA6708 | T. M. Harris and G. P. Murphy; *J. Am. Chem. Soc.*, 1971, **93**, 6708. | 805 |
| 71JA7179 | L. A. Paquette, M. R. Short and J. F. Kelly; *J. Am. Chem. Soc.*, 1971, **93**, 7179. | 551 |
| 71JAP7105310 | A. Fujii and H. Kobata; *Jpn. Pat.* 71 05 310 (1971). | 181 |
| 71JAP7129876 | F. Ishikawa, S. Miyazaki and K. Ueno; (Daiichi Seiyaku Co. Ltd.), *Jpn. Pat.* 71 29 876 (*Chem. Abstr.*, 1971, **75**, 140 875). | 246 |
| 71JCS(B)74 | S. Bradamante, S. Maiorana, A. Mangia and G. Pagani; *J. Chem. Soc.* (*B*), 1971, 74. | 913 |
| 71JCS(B)82 | M. J. Aroney, G. Cleaver, R. J. W. Le Fèvre and R. K. Pierens; *J. Chem. Soc.* (*B*), 1971, 82. | 627 |
| 71JCS(B)136 | S. Wolfe, A. Rauk, L. M. Tel and I. G. Csizmadia; *J. Chem. Soc.* (*B*), 1971, 136. | 629 |
| 71JCS(B)1030 | G. M. Kellie and F. G. Riddell; *J. Chem. Soc.* (*B*), 1971, 1030. | 954 |
| 71JCS(B)1407 | G. Bandoli, C. Panattoni, D. A. Clemente, E. Tondello, A. Dondoni and A. Mangini; *J. Chem. Soc.* (*B*), 1971, 1407. | 947 948 |
| 71JCS(B)1675 | G. B. Barlin and A. C. Young; *J. Chem. Soc.* (*B*), 1971, 1675. | 100 |
| 71JCS(B)2030 | J. E. Anderson; *J. Chem. Soc.* (*B*), 1971, 2030. | 962 |
| 71JCS(B)2214 | D. J. Brown and J. A. Hoskins; *J. Chem. Soc.* (*B*), 1971, 2214. | 97, 102, 138 |
| 71JCS(B)2344 | A. J. Boulton, I. J. Fletcher and A. R. Katritzky; *J. Chem. Soc.* (*B*), 1971, 2344. | 5 |
| 71JCS(B)2423 | A. Albert and H. Mizuno; *J. Chem. Soc.* (*B*), 1971, 2423. | 265, 287 |
| 71JCS(C)74 | R. Budziarek; *J. Chem. Soc.* (*C*), 1971, 74. | 481 |

| | | |
|---|---|---|
| 71JCS(C)95 | H. D. Munro, O. C. Musgrave and R. Templeton; *J. Chem. Soc. (C)*, 1971, 95. | 851 |
| 71JCS(C)250 | D. J. Brown and B. T. England; *J. Chem. Soc. (C)*, 1971, 250. | 78 |
| 71JCS(C)375 | J. Clark, W. Pendergast, F. S. Yates and A. E. Cunliffe; *J. Chem. Soc. (C)*, 1971, 375. | 276 |
| 71JCS(C)425 | D. J. Brown and B. T. England; *J. Chem. Soc. (C)*, 1971, 425. | 78 |
| 71JCS(C)526 | C. H. Hassall, R. B. Morton, Y. Ogihara and D. A. S. Phillips; *J. Chem. Soc. (C)*, 1971, 526. | 56, 337, 347, 359 |
| 71JCS(C)780 | I. R. Gelling and D. G. Wibberley; *J. Chem. Soc. (C)*, 1971, 780. | 207 |
| 71JCS(C)784 | R. S. Atkinson; *J. Chem. Soc. (C)*, 1971, 784. | 719 |
| 71JCS(C)796 | L. Crombie and R. Ponsford; *J. Chem. Soc. (C)*, 1971, 796. | 674, 675, 721, 746 |
| 71JCS(C)811 | W. M. Bandaranayake, L. Crombie and D. A. Whiting; *J. Chem. Soc. (C).*, 1971, 811. | 674, 675, 746, 747 |
| 71JCS(C)981 | J. Adamson, D. L. Forster, T. L. Gilchrist and C. W. Rees; *J. Chem. Soc. (C)*, 1971, 981. | 376 |
| 71JCS(C)1167 | R. Budziarek and P. Hampson; *J. Chem. Soc. (C)*, 1971, 1167. | 485 |
| 71JCS(C)1324 | M. Arshad, J. P. Devlin and W. D. Ollis; *J. Chem. Soc. (C)*, 1971, 1324. | 819 |
| 71JCS(C)1482 | G. Barker and G. P. Ellis; *J. Chem. Soc. (C)*, 1971, 1482. | 583, 615 |
| 71JCS(C)1889 | B. W. Arantz and D. J. Brown; *J. Chem. Soc. (C)*, 1971, 1889. | 99 |
| 71JCS(C)1994 | L. Farkas, A. Gottsegen, M. Nógrádi and S. Antus; *J. Chem. Soc. (C)*, 1971, 1994. | 856 |
| 71JCS(C)2079 | G. Barker, G. P. Ellis and D. A. Wilson; *J. Chem. Soc. (C)*, 1971, 2079. | 714 |
| 71JCS(C)2357 | A. Albert and K. Ohta, *J. Chem. Soc. (C)*, 1971, 2357. | 287, 318 |
| 71JCS(C)2507 | D. J. Brown and B. T. England; *J. Chem. Soc. (C)*, 1971, 2507. | 67, 139 |
| 71JCS(C)2546 | G. Casiraghi, G. Casnati and G. Salerno; *J. Chem. Soc. (C)*, 1971, 2546. | 750, 764 |
| 71JCS(C)2648 | T. Nishiwaki and T. Saito; *J. Chem. Soc. (C)*, 1971, 2648. | 440 |
| 71JCS(C)2721 | A. K. Kiang, S. F. Tan and W. S. Wong; *J. Chem. Soc. (C)*, 1971, 2721. | 643 |
| 71JCS(C)2807 | A. M. Comrie; *J. Chem. Soc. (C)*, 1971, 2807. | 52 |
| 71JCS(C)3873 | G. V. Boyd and S. R. Dando; *J. Chem. Soc. (C)*, 1971, 3873. | 655 |
| 71JHC111 | E. Campaigne and G. Randau; *J. Heterocycl. Chem.*, 1971, **8**, 111. | 224 |
| 71JHC317 | W. W. Paudler and T.-K. Chen; *J. Heterocycl. Chem.*, 1971, **8**, 317. | 397 |
| 71JHC597 | D. D. Reynolds and B. C. Cossar; *J. Heterocycl. Chem.*, 1971, **8**, 597. | 487 |
| 71JHC605 | D. D. Reynolds and B. C. Cossar; *J. Heterocycl. Chem.*, 1971, **8**, 605. | 487 |
| 71JHC611 | D. D. Reynolds and B. C. Cossar; *J. Heterocycl. Chem.*, 1971, **8**, 611. | 487 |
| 71JHC697 | K. Kyriacou; *J. Heterocycl. Chem.*, 1971, **8**, 697. | 169 |
| 71JHC785 | B. Stanovnik and M. Tišler; *J. Heterocycl. Chem.*, 1971, **8**, 785. | 377, 382 |
| 71JHC1055 | B. Stanovnik; *J. Heterocycl. Chem.*, 1971, **8**, 1055. | 355 |
| 71JHC1087 | J. E. Oliver and A. B. DeMilo; *J. Heterocycl. Chem.*, 1971, **8**, 1087. | 1069, 1073 |
| 71JIC192 | R. B. Tirodkar and R. N. Usgaonkar; *J. Indian Chem. Soc.*, 1971, **48**, 192. | 831 |
| 71JIC707 | S. Gulgule and R. N. Usgaonkar; *J. Indian Chem. Soc.*, 1971, **48**, 707. | 857 |
| 71JMC244 | C. E. Malen, B. H. Danree and X. B. L. Pascaud; *J. Med. Chem.*, 1971, **14**, 244. | 83, 128 |
| 71JMC758 | D. T. Witiak, E. S. Stratford, R. Nazareth, G. Wagner and D. R. Feller; *J. Med. Chem.*, 1971, **14**, 758. | 704, 786, 855 |
| 71JMC929 | R. Saxena, S. Sharma, R. N. Iyer and N. Anand; *J. Med. Chem.*, 1971, **14**, 929. | 340, 366 |
| 71JOC284 | J. E. Johnson, J. R. Springfield, J. S. Hwang, L. J. Hayes, W. C. Cunningham and D. L. McClaugherty; *J. Org. Chem.*, 1971, **36**, 284. | 1051, 1075 |
| 71JOC600 | G. A. Reynolds and J. A. Van Allan; *J. Org. Chem.*, 1971, **36**, 600. | 664, 763 |
| 71JOC604 | C. V. Greco and J. F. Warchol; *J. Org. Chem.*, 1971, **36**, 604. | 338, 360 |
| 71JOC611 | J. F. Carson and L. E. Boggs; *J. Org. Chem.*, 1971, **36**, 611. | 1036 |
| 71JOC787 | W. W. Paudler and T.-K. Chen; *J. Org. Chem.*, 1971, **36**, 787. | 393, 410 |
| 71JOC860 | L. T. Plante; *J. Org. Chem.*, 1971, **36**, 860. | 304, 312 |
| 71JOC1137 | R. B. Woodward, I. J. Pachter and M. L. Scheinbaum; *J. Org. Chem.*, 1971, **36**, 1137. | 989 |
| 71JOC1158 | H. R. Moreno and H. P. Schultz; *J. Org. Chem.*, 1971, **36**, 1158. | 176 |
| 71JOC1314 | D. N. Harpp and J. G. Gleason; *J. Org. Chem.*, 1971, **36**, 1314. | 953 |
| 71JOC1977 | A. F. Kluge and C. P. Lillya; *J. Org. Chem.*, 1971, **36**, 1977. | 598, 739 |
| 71JOC1988 | A. F. Kluge and C. P. Lillya; *J. Org. Chem.*, 1971, **36**, 1988. | 739 |
| 71JOC2385 | A. M. Schoffstall; *J. Org. Chem.*, 1971, **36**, 2385. | 217 |
| 71JOC2457 | A. Pollak, B. Stanovnik and M. Tišler; *J. Org. Chem.*, 1971, **36**, 2457. | 333, 335, 355 |
| 71JOC2818 | R. D. Elliott, C. Temple, Jr., J. L. Frye and J. A. Montgomery; *J. Org. Chem.*, 1971, **36**, 2818. | 253, 254, 262 |
| 71JOC2910 | P. A. Wehrli, R. I. Fryer and W. Metlesics; *J. Org. Chem.*, 1971, **36**, 2910. | 780 |
| 71JOC2923 | J. J. Silber and H. J. Shine; *J. Org. Chem.*, 1971, **36**, 2923. | 969 |
| 71JOC2974 | C. Temple, Jr., C. L. Kussner and J. A. Montgomery; *J. Org. Chem.*, 1971, **36**, 2974. | 400 |
| 71JOC3506 | J. G. Kuderna, R. D. Skiles and K. Pilgram; *J. Org. Chem.*, 1971, **36**, 3506. | 333, 335, 343, 354, 355 |
| 71JOC3633 | R. M. Srivastava, F. Sweet, T. P. Murray and R. K. Brown; *J. Org. Chem.*, 1971, **36**, 3633. | 579 |
| 71JOC3812 | B. Stanovnik, M. Tišler and B. Stefanov; *J. Org. Chem.*, 1971, **36**, 3812. | 238 |
| 71JOC3921 | W. W. Paudler and J. Lee; *J. Org. Chem.*, 1971, **36**, 3921. | 389, 390 |
| 71JOC4012 | E. C. Taylor, Jr., M. J. Thompson, K. Perlman, R. Mengel and W. Pfleiderer; *J. Org. Chem.*, 1971, **36**, 4012. | 277, 279, 304, 307 |
| 71JOC4028 | E. E. Schweizer, T. Minami and D. M. Crouse; *J. Org. Chem.*, 1971, **36**, 4028. | 749 |
| 71JOC4045 | N. I. Bruckner and N. L. Bauld; *J. Org. Chem.*, 1971, **36**, 4045. | 671 |
| 71JPR699 | A. Mustafa, A. K. Mansour and H. A. A. Zaher; *J. Prakt. Chem.*, 1971, **313**, 699. | 396, 405 |

| | | |
|---|---|---|
| 71JPR1110 | H. Hartmann and D. Förster; *J. Prakt. Chem.*, 1971, **313**, 1110. | 866 |
| 71JPS1458 | J. S. G. Cox, G. D. Woodard and W. C. McCrone; *J. Pharm. Sci.*, 1971, **60**, 1458. | 623 |
| 71JSP(39)536 | T. J. Durnick and S. C. Wait; *J. Mol. Spectrosc.*, 1971, **39**, 536. | 161 |
| 71JST(8)236 | N. Trinajstić; *J. Mol. Struct.*, 1971, **8**, 236. | 945 |
| 71JST(9)33 | A. H. Clark and T. G. Hewitt; *J. Mol. Struct.*, 1971, **9**, 33. | 949 |
| 71JST(9)205 | B. Tinland and C. Decoret; *J. Mol. Struct.*, 1971, **9**, 205. | 945 |
| 71KGS708 | I. Ya. Postovskii and V. A. Ershov; *Khim. Geterotsikl. Soedin.*, 1971, 708. | 544 |
| 71KGS711 | V. A. Ershov and I. Ya. Postovskii; *Khim. Geterotsikl. Soedin.*, 1971, 711. | 544 |
| 71KGS905 | V. N. Artemov and O. P. Shvaika; *Khim. Geterotsikl. Soedin.*, 1971, 905. | 561 |
| 71KGS946 | P. L. Ovechkin, L. A. Ignatova and B. V. Unkovskii; *Khim. Geterotsikl. Soedin.*, 1971, 946 (*Chem. Abstr.*, 1971, **75**, 151 129). | 1000 |
| 71KGS1167 | A. A. Potekhin and M. N. Vikulina; *Khim. Geterotsikl. Soedin.*, 1971, 1167. | 1056 |
| 71KGS1280 | A. Y. Berlin and I. A. Korbukh; *Khim. Geterotsikl. Soedin.*, 1971, 1280. | 218 |
| 71KGS(Sb3)85 | S. K. Klimenko and V. G. Kharchenko; *Khim. Geterotsikl. Soedin., Sb3*, 1971, 85. | 911 |
| 71LA(745)59 | L. Reichel and G. Proksch; *Liebigs Ann. Chem.*, 1971, **745**, 59. | 801 |
| 71LA(746)28 | L. Wald and W. Wucherpfennig; *Liebigs Ann. Chem.*, 1971, **746**, 28. | 1017 |
| 71LA(746)102 | T. Eicher, E. von Angerer and A.-M. 1 Hansen; *Liebigs Ann. Chem.*, 1971, **746**, 102. | 795 |
| 71LA(747)111 | W. Pfleiderer; *Liebigs Ann. Chem.*, 1971, **747**, 111. | 281, 306, 308 |
| 71LA(749)125 | A. Hetzheim and J. Singelmann; *Liebigs Ann. Chem.*, 1971, **749**, 125. | 440 |
| 71LA(749)171 | H. Grill and G. Kresze; *Liebigs Ann. Chem.*, 1971, **749**, 171. | 1046, 1062, 1070 |
| 71LA(750)12 | H. Neunhoeffer, F. Weischedel and V. Böhnisch; *Liebigs Ann. Chem.*, 1971, **750**, 12. | 393, 396, 398, 443 |
| 71LA(753)151 | F. Asinger, A. Saus, H. Offermanns and P. Scherberich; *Liebigs Ann. Chem.*, 1971, **753**, 151. | 987 |
| B-71MI21200 | A. R. Katritzky and J. M. Lagowski; 'Chemistry of the Heterocyclic *N*-Oxides', Academic Press, New York, 1971, p. 1. | 20 |
| B-71MI21300 | 'Procedures in Nucleic Acid Research', ed. G. L. Cantoni and D. R. Davies; Harper and Row, New York, 1971, vol. 2. | 142 |
| 71MI21301 | C. C. Cheng and B. Roth; *Prog. Med. Chem.*, 1971, **8**, 61. | 150, 152 |
| 71MI21302 | P. Liberti and J. B. Stanbury; *Annu. Rev. Pharmacol.*, 1971, **11**, 113. | 152 |
| B-71MI21303 | A. Albert and E. P. Serjeant; 'The Determination of Ionization Constants', Chapman and Hall, London, 2nd edn., 1971. | 59, 65 |
| 71MI21400 | K. Watanabe and Y. Sato; *J. Agric. Food Chem.*, 1971, **19**, 1017. | 193 |
| 71MI21401 | R. G. Buttery, R. M. Siefert, D. G. Guadagne and L. C. Ching; *J. Agric. Food Chem.*, 1971, **19**, 969. | 162, 193 |
| 71MI21402 | E. Collins; *J. Agric. Food Chem.*, 1971, **19**, 533. | 193 |
| 71MI21403 | J. P. Walradt, A. O. Pittet, T. E. Kinlin, R. Muralidhara and A. Sanderson; *J. Agric. Food Chem.*, 1971, **19**, 972. | 193 |
| 71MI21500 | A. D. Jordan, I. G. Ross, R. Hoffmann, J. R. Swenson and R. Gleiter; *Chem. Phys. Lett.*, 1971, **10**, 572 (*Chem. Abstr.*, 1972, **76**, 19 710). | 250 |
| 71MI21600 | G. M. Brown; *Adv. Enzymol.*, 1971, **35**, 35. | 320 |
| B-71MI21601 | T. Shiota; in 'Comprehensive Biochemistry', ed. M. Florkin and E. H. Stotz; Elsevier, Amsterdam, 1971, vol. 21, p. 111. | 320 |
| 71MI21602 | M. Tsusue; *J. Biochem.*, 1971, **69**, 781. | 294 |
| 71MI21700 | M. J. Robey, M. Sterns, H. M. Morris and I. G. Ross; *J. Crystallogr. Mol. Struct.*, 1971, **1**, 401 (*Chem. Abstr.*, 1973, **78**, 21 345). | 340 |
| 71MI21900 | R. L. Jones and J. R. Kershaw; *Rev. Pure Appl. Chem.*, 1971, **21**, 23. | 385 |
| 71MI21901 | J. N. Herak and G. Schoffa; *Mol. Phys.*, 1971, **22**, 379. | 399 |
| 71MI21902 | J. D. Rosen and M. Siewierski; *Bull. Environ. Contam. Toxicol.*, 1971, **6**, 406 (*Chem. Abstr.*, 1972, **76**, 3808). | 417, 421 |
| 71MI21903 | S. Palazzo; *Atti. Accad. Sci. Lett. Arti Palermo, Pt. I, 1969–1970*, 1971, **30**, 23 (*Chem. Abstr.*, 1972, **76**, 139 991). | 441 |
| 71MI22000 | Z. D. Tadic and S. K. Ries; *J. Agric. Food Chem.*, 1971, **19**, 46. | 476 |
| 71MI22001 | J. R. Plimmer, P. C. Kearney and U. I. Klingebiel; *J. Agric. Food Chem.*, 1971, **19**, 572. | 476 |
| B-71MI22002 | H. Ulrich and R. Richter; in 'Newer Methods in Preparative Organic Chemistry', ed. W. Foerst; Academic, New York, 1971, vol. 6, p. 280. | 507 |
| B-71MI22200 | J. F. Stoddart; 'Stereochemistry of Carbohydrates', Wiley-Interscience, New York, 1971. | 628 |
| 71MI22201 | P. Sohar, L. Varcha and J. Kuszmann; *Acta Chim. (Budapest)*, 1971, **70**, 79 (*Chem. Abstr.*, 1971, **75**, 150 926). | 643 |
| B-71MI22300 | H. Perst; 'Oxonium Ions in Organic Chemistry', Verlag Chemie, Weinheim, 1971, chapter 7. | 652, 655 |
| 71MI22301 | T. Ito and A. Murata; *Bunseki Kagaku*, 1971, **20**, 1422. | 707 |
| 71MI22302 | E. A. Karakhanov, N. N. Khvorostukhima and E. A. Vitorova; *Vestn. Mosk. Univ. Khim.*, 1971, **12**, 502. | 722 |
| 71MI22400 | J. Thibault; *Ann. Chim. (Paris)*, 1971, **6**, 381. | 767 |
| 71MI22401 | J. Thibault; *Ann. Chim. (Paris)*, 1971, **6**, 263. | 788, 789 |
| 71MI22402 | L. L. Woods and S. M. Shamma; *J. Chem. Eng. Data*, 1971, **16**, 101. | 801, 804 |
| 71MI22403 | D. R. Threlfall; *Vitam. Horm. (N.Y.)*, 1971, **29**, 153. | 876 |
| B-71MI22404 | H. Gold; in 'The Chemistry of Synthetic Dyes', ed. K. Venkataraman; Academic Press, New York, 1971, vol. 5, chap. 8. | 879 |

| B-71MI22405 | R. C. Bertelson; in 'Photochromism', ed. G. H. Brown; Wiley, New York, 1971, p. 49. | 880 |
| 71MI22500 | A. A. Kroik, A. M. Arishkevich and Y. I. Usatenko; *Khim. Teknol. (Kharkov)*, 1971, 32 (*Chem. Abstr.*, 1972, **77**, 121 711). | 941 |
| 71MI22501 | H. F. Hardman, E. F. Domino and M. H. Seevers; *Proc. West. Pharmacol. Soc.*, 1971, **14**, 14 (*Chem. Abstr.*, 1972, **76**, 107 830). | 941 |
| B-71MI22600 | 'The Condensed Chemical Dictionary', revised by G. G. Hawley; Van Nostrand Reinhold, New York, 8th edn., 1971, p. 320. | 993 |
| 71MI22700 | N. Thoai, N. N. Chieu and J. Wiemann; *Ann. Chim. (Paris)*, 1971, **6**, 235. | 1017 |
| 71MI22800 | T. Moeller and R. L. Dieck; *Prep. Inorg. React.*, 1971, **6**, 63. | 1048 |
| 71MIP22100 | O. P. Shvaika and V. I. Fomenko; *U.S.S.R. Pat.* 310 907 (1971) (*Chem. Abstr.*, 1971, **75**, 151 843). | 561 |
| B-71MS | Q. N. Porter and J. Baldas; 'Mass Spectrometry of Heterocyclic Compounds', Wiley, New York, 1971. 65, 66, 603, 608, 610, 616, 957, 958 | |
| 71NKK448 | O. Tsuge and S. Iwanami; *Nippon Kagaku Zasshi*, 1971, **92**, 448. | 1042 |
| 71OMR(3)575 | W. L. F. Armarego, T. J. Batterham and J. R. Kershaw; *Org. Magn. Reson.*, 1971, **3**, 575. | 249 |
| 71OMS(5)87 | A. M. Duffield, C. Djerassi and A. T. Balaban; *Org. Mass Spectrom.*, 1971, **5**, 87. | 619 |
| 71OMS(5)229 | J. H. Beynon, R. M. Caprioli and T. Ast; *Org. Mass Spectrom.*, 1971, **5**, 229. | 466 |
| 71OMS(5)249 | J. P. Kutney, G. Eigendorf, T. Inaba and D. L. Dreyer; *Org. Mass Spectrom.*, 1971, **5**, 249. | 608, 609 |
| 71OMS(5)763 | K. Pihlaja and P. Pasanen; *Org. Mass Spectrom.*, 1971, **5**, 763. | 957 |
| 71OMS(5)857 | G. Barker and G. P. Ellis; *Org. Mass Spectrom.*, 1971, **5**, 857. | 615 |
| 71OMS(5)1085 | M. H. Palmer, P. N. Preston and M. F. G. Stevens; *Org. Mass Spectrom.*, 1971, **5**, 1085. | 396, 397 |
| 71OPP243 | J. C. Lewis and R. M. Seifert; *Org. Prep. Proced. Int.*, 1971, **3**, 243. | 758 |
| 71PMH(3)67 | W. L. F. Armarego; *Phys. Methods Heterocycl. Chem.*, 1971, **3**, 67. 65, 598, 634, 640, 641, 955 | |
| 71PMH(3)190 | W. L. F. Armarego; *Phys. Methods Heterocycl. Chem.*, 1971, **3**, 190. | 600 |
| 71PMH(4)13 | S. Walker; *Phys. Methods Heterocycl. Chem.*, 1971, **4**, 13. | 626 |
| 71PMH(4)339 | A. R. Katritzky and P. J. Taylor; *Phys. Methods Heterocycl. Chem.*, 1971, **4**, 339. | 593 |
| 71RRC135 | C. Cristescu and S. Sitaru; *Rev. Roum. Chim.*, 1971, **16**, 135. | 392 |
| 71RRC311 | C. Cristescu; *Rev. Roum. Chim.*, 1971, **16**, 311. | 392 |
| 71RTC207 | P. J. Lont, H. C. Van der Plas and A. Koudijs; *Recl. Trav. Chim. Pays-Bas*, 1971, **90**, 207. | 165 |
| 71RTC513 | W. Schwaiger and J. P. Ward; *Recl. Trav. Chim. Pays-Bas*, 1971, **90**, 513. | 167 |
| 71S149 | G. Descotes and D. Missos; *Synthesis*, 1971, 149. | 778 |
| 71S154 | K. Takemoto and Y. Yamamoto; *Synthesis*, 1971, 154. | 142 |
| 71SA(A)1025 | O. H. Ellestad, P. Klaboe and G. Hagen; *Spectrochim. Acta, Part A*, 1971, **27**, 1025. | 956 |
| 71T711 | P. Crabbé, L. A. Maldonado and I. Sánchez; *Tetrahedron*, 1971, **27**, 711. | 816 |
| 71T811 | G. Paal and A. Wilhelm; *Tetrahedron*, 1971, **27**, 811. | 594 |
| 71T1875 | G. Cardillo, R. Cricchio and L. Merlini; *Tetrahedron*, 1971, **27**, 1875. | 751 |
| 71T1917 | M. Anteunis, G. Swaelens and J. Gelan; *Tetrahedron*, 1971, **27**, 1917. | 953, 954 |
| 71T1973 | O. Achmatowicz, Jr., P. Bukowski, B. Szechner, Z. Zwierzchowska and A. Zamojski; *Tetrahedron*, 1971, **27**, 1973. | 815 |
| 71T3051 | A. I. Scott, D. G. Pike, J. J. Ryan and H. Guilford; *Tetrahedron*, 1971, **27**, 3051. | 839 |
| 71T3111 | N. S. Zefirov, V. S. Blagoveshchensky, I. V. Kazimirchik and N. S. Surova; *Tetrahedron*, 1971, **27**, 3111. | 960 |
| 71T3129 | M. Witanowski, L. Stefaniak, H. Januszewski and G. A. Webb; *Tetrahedron*, 1971, **27**, 3129. | 464 |
| 71T3503 | M. Simalty, H. Strzelecka and H. Khedija; *Tetrahedron*, 1971, **27**, 3503. | 663 |
| 71T4195 | R. A. Olofson and J. P. Marino; *Tetrahedron*, 1971, **27**, 4195. | 990 |
| 71T4837 | J. R. Merchant and D. V. Rege; *Tetrahedron*, 1971, **27**, 4837. | 709 |
| 71T5459 | O. Korver and C. K. Wilkins; *Tetrahedron*, 1971, **27**, 5459. | 631 |
| 71T5753 | U. Eisner and T. Krishnamurthy; *Tetrahedron*, 1971, **27**, 5753. | 979 |
| 71T6073 | S. Nozoe and K. Hirai; *Tetrahedron*, 1971, **27**, 6073. | 751 |
| 71T6171 | N. S. Narasimhan and B. H. Bhide; *Tetrahedron*, 1971, **27**, 6171. | 834 |
| 71TH21400 | A. E. A. Porter; Ph.D. Thesis, University of London, 1971. | 189 |
| 71TH21500 | T. L. Threlfall; Ph.D. Thesis, University of London, 1971. 248, 249, 250, 251, 252, 253, 254, 258, 259 | |
| 71TL617 | H. Strzelecka, M. Dupré and M. Simalty; *Tetrahedron Lett.*, 1971, 617. | 873 |
| 71TL1211 | D. Kobelt, E. F. Paulus and K.-D. Kampe; *Tetrahedron Lett.*, 1971, 1211. | 1048 |
| 71TL1539 | T. Matsuura, T. Takemoto and R. Nakashima; *Tetrahedron Lett.*, 1971, 1539. | 695 |
| 71TL2241 | R. Kalish, A. E. Smith and E. J. Smutny; *Tetrahedron Lett.*, 1971, 2241. | 937 |
| 71TL2315 | T. V. Saraswathi and V. R. Srinivasan; *Tetrahedron Lett.*, 1971, 2315. | 453 |
| 71TL2937 | G. Wood and R. M. Srivastava; *Tetrahedron Lett.*, 1971, 2937, | 962 |
| 71TL3117 | H. Igeta, T. Tsuchiya and T. Nakai; *Tetrahedron Lett.*, 1971, 3117. | 372, 378 |
| 71TL4799 | I. Murata, T. Nakazawa and S. Tada; *Tetrahedron Lett.*, 1971, 4799. | 645 |
| 71ZC256 | E. Lippmann and J. Spindler; *Z. Chem.*, 1971, **11**, 256. | 253 |
| 71ZOR143 | T. A. Chibisova, A. Ya. Zheltov, V. Ya. Rodionov and B. I. Stepanov; *Zh. Org. Khim.*, 1971, **7**, 143 (*Chem. Abstr.*, 1971, **74**, 99 115). | 956 |
| 71ZOR388 | P. I. Zakharov; *Zh. Org. Khim.*, 1971, 7, 388. | 610 |

| 72ACH(71)363 | J. Szabó, I. Varga and E. Vinkler; *Acta. Chim. Acad. Sci. Hung.*, 1972, **71**, 363. | 1025 |
| 72ACS596 | O. Ceder and J. E. Andersson; *Acta Chem. Scand.*, 1972, **26**, 596. | 468, 469, 521 |
| 72ACS611 | O. Ceder and J. E. Andersson; *Acta Chem. Scand.*, 1972, **26**, 611. | 469 |
| 72ACS1258 | K. A. Jensen and S. Hammerun; *Acta Chem. Scand.*, 1972, **26**, 1258. | 569 |
| 72AG1088 | H. Rembold and W. L. Gyure; *Angew. Chem.*, 1972, **84**, 1088. | 306, 320, 321 |
| 72AG(E)150 | H. Bock and G. Wagner; *Angew. Chem., Int. Ed. Engl.*, 1972, **11**, 150. | 959 |
| 72AG(E)287 | H. Witte and W. Seeliger; *Angew. Chem., Int. Ed. Engl.*, 1972, **11**, 287. | 1006 |
| 72AG(E)739 | E. L. Eliel; *Angew. Chem., Int. Ed. Engl.*, 1972, **11**, 739. | 629 |
| 72AG(E)949 | E. Grigat; *Angew. Chem., Int. Ed. Engl.*, 1972, **11**, 949. | 500 |
| 72AG(E)1012 | H. Hagemann and K. Ley; *Angew. Chem., Int. Ed. Engl.*, 1972, **11**, 1012. | 1052, 1057 |
| 72AHC(14)99 | G. W. H. Cheeseman and E. S. G. Werstiuk; *Adv. Heterocycl. Chem.*, 1972, **14**, 99. | 158, 161, 176, 179 |
| 72AJC865 | J. A. Elix, W. S. Wilson, R. N. Warrener and I. C. Calder; *Aust. J. Chem.*, 1972, **25**, 865. | 424 |
| 72AJC2275 | R. J. Badger, D. J. Brown and N. V. Khromov-Borisov; *Aust. J. Chem.*, 1972, **25**, 2275. | 71, 137 |
| 72AJC2641 | D. J. Brown and J. A. Hoskins; *Aust. J. Chem.*, 1972, **25**, 2641. | 95, 138 |
| 72AJC2671 | R. F. Evans and K. N. Mewett; *Aust. J. Chem.*, 1972, **25**, 2671. | 185 |
| 72AP2 | F. Eiden, H. Müller and G. Bachmann; *Arch. Pharm. (Weinheim, Ger.)*, 1972, **305**, 2. | 257 |
| 72AP751 | H. Fenner and P. Michaelis; *Arch. Pharm. (Weinheim, Ger.)*, 1972, **305**, 751. | 219 |
| 72AX(B)534 | F. Fehér, A. Klaeren and K.-H. Linke; *Acta Crystallogr., Part B*, 1972, **28**, 534. | 950 |
| 72AX(B)659 | R. Norrestam, B. Stensland and E. Söderberg; *Acta Crystallogr., Part B*, 1972, **28**, 659. | 272 |
| 72AX(B)739 | N. A. Ahmed and A. I. Kitaigorodsky; *Acta Crystallogr., Part B*, 1972, **28**, 739. | 537, 541 |
| 72AX(B)1173 | C. Sabelli and P. F. Zanazzi; *Acta Crystallogr., Part B*, 1972, 1173. | 333 |
| 72AX(B)1178 | L. Fanfani, P. F. Zanazzi and C. Sabelli; *Acta Crystallogr., Part B*, 1972, 1178. | 333 |
| 72AX(B)2424 | N. de Wolf, G. C. Verschoor and C. Romers; *Acta Crystallogr., Part B*, 1972, **28**, 2424. | 950 |
| 72AX(B)2485 | N. R. Stemple and W. H. Watson; *Acta Crystallogr., Part B*, 1972, **28**, 2485. | 623 |
| 72BAP91 | M. Witanowski, L. Stefaniak, H. Januszewski, Z. Grabowski and G. A. Webb; *Bull. Acad. Pol. Sci., Ser. Sci. Chim.*, 1972, **20**, 91 (*Chem. Abstr.* 1973, **78**, 104 187). | 371 |
| 72BCJ525 | K. Hasegawa and S. Hirooka; *Bull. Chem. Soc. Jpn.*, 1972, **45**, 525. | 1042, 1072 |
| 72BCJ1127 | T. Nishino, M. Kiyokawa, Y. Miichi and K. Tokuyama; *Bull. Chem. Soc. Jpn.*, 1972, **45**, 1127. | 207, 210, 212, 217, 226, 227 |
| 72BCJ1567 | K. Hasegawa and S. Hirooka; *Bull. Chem. Soc. Jpn.*, 1972, **45**, 1567. | 1069 |
| 72BCJ1589 | M. Kamiya; *Bull. Chem. Soc. Jpn.*, 1972, **45**, 1589. | 945, 946 |
| 72BCJ1893 | K. Hasegawa and S. Hirooka; *Bull. Chem. Soc. Jpn.*, 1972, **45**, 1893. | 1072 |
| 72BCJ1999 | H. Tanida, T. Irie and K. Tori; *Bull. Chem. Soc. Jpn.*, 1972, **45**, 1999. | 552 |
| 72BCJ2829 | Y. Iwanami and T. Seki; *Bull. Chem. Soc. Jpn.*, 1972, **45**, 2829. | 276 |
| 72BCJ2877 | O. Tsuge and S. Kanemasa; *Bull. Chem. Soc. Jpn.*, 1972, **45**, 2877. | 1076 |
| 72BCJ3504 | T. Kawashima and N. Inamoto; *Bull. Chem. Soc. Jpn.*, 1972, **45**, 3504. | 383 |
| 72BSF251 | A. Etienne, B. Bonte and B. Druet; *Bull. Soc. Chim. Fr.*, 1972, 251. | 508 |
| 72BSF696 | G. Descotes and D. Missos; *Bull. Soc. Chim. Fr.*, 1972, 696. | 580, 630 |
| 72BSF707 | J. Royer and J. Dreux; *Bull. Soc. Chim. Fr.*, 1972, 707. | 576, 577, 578, 654, 762 |
| 72BSF1077 | G. Descotes, J.-C. Martin and N. Mathicolonis; *Bull. Soc. Chim. Fr.*, 1972, 1077. | 630 |
| 72BSF1173 | G. Vernin, H. J.-M. Dou and J. Metzger; *Bull. Soc. Chim. Fr.*, 1972, 1173. | 166 |
| 72BSF1483 | E. Bisagni, J.-P. Marquet and J. A. Louisfert; *Bull. Soc. Chim. Fr.*, 1972, 1483. | 356 |
| 72BSF1511 | J. Daunis, Y. Guindo, R. Jacquier and P. Viallefont; *Bull. Soc. Chim. Fr.*, 1972, 1511. | 404 |
| 72BSF1540 | M. Sliwa, H. Sliwa and P. Maitte; *Bulg. Soc. Chim. Fr.*, 1972, 1540. | 783 |
| 72BSF1588 | A. Godard, G. Queguiner and P. Pastour; *Bull. Soc. Chim. Fr.*, 1972, 1588. | 234, 238, 240, 247 |
| 72BSF2510 | A. Safieddine, J. Royer and J. Dreux; *Bull. Soc. Chim. Fr.*, 1972, 2510. | 661 |
| 72BSF2948 | R. Royer, J. P. Lechartier and P. Demerseman; *Bull. Soc. Chim. Fr.*, 1972, 2948. | 836 |
| 72BSF3173 | H. Khedija, H. Strzelecka and M. Simalty; *Bull. Soc. Chim. Fr.*, 1972, 3173. | 663 |
| 72BSF3588 | G. Cauquis and M. Maurey-Mey; *Bull. Soc. Chim. Fr.*, 1972, 3588. | 969 |
| 72BSF4308 | J. Justin, M. Mazet and Th. Yvernault; *Bull. Soc. Chim. Fr.*, 1972, 4308. | 978 |
| 72BSF4637 | N. Vinot and J.-P. M'Packo; *Bull. Soc. Chim. Fr.*, 1972, 4637. | 390 |
| 72CB549 | F. A. Neugebauer, W. Otting, H. O. Smith and H. Trischmann, *Chem. Ber.*, 1972, **105**, 549. | 548 |
| 72CB757 | E. Cohnen and J. Mahnke; *Chem. Ber.*, 1972, **105**, 757. | 1068 |
| 72CB863 | F. Bohlmann and U. Bühmann; *Chem. Ber.*, 1972, **105**, 863. | 745, 750 |
| 72CB1683 | G. Köbrich, D. Merkel and K. W. Thiem; *Chem. Ber.*, 1972, **105**, 1683. | 1051 |
| 72CB3055 | L. Capuano, W. Sperling and R. Zander; *Chem. Ber.*, 1972, **105**, 3055. | 1022 |
| 72CB3695 | H. Neunhoeffer, H.-D. Vötter and H. Ohl; *Chem. Ber.*, 1972, **105**, 3695. | 373, 382 |
| 72CB3704 | E. Oeser and L. Schiele; *Chem. Ber.*, 1972, **105**, 3704. | 370 |
| 72CC107 | M. Chadenson, M. Hauteville and J. Chopin; *J. Chem. Soc., Chem. Commun.*, 1972, 107. | 818 |
| 72CC264 | W. E. Barnett, M. G. Newton and J. A. McCormack; *J. Chem. Soc., Chem. Commun.*, 1972, 264. | 947 |
| 72CC388 | T. Imagawa, N. Sueda and M. Kowanisi; *J. Chem. Soc., Chem. Commun.*, 1972, 388. | 688 |

| | | |
|---|---|---|
| 72CC450 | H. Iwamura, M. Fukunaga and K. Kushida; *J. Chem. Soc., Chem. Commun.*, 1972, 450. | 959 |
| 72CC451 | P. O. L. Mack and J. T. Pinhey; *J. Chem. Soc., Chem. Commun.*, 1972, 451. | 720 |
| 72CC847 | L. Main, G. J. Kasperek and T. C. Bruice; *J. Chem. Soc., Chem. Commun.*, 1972, 847. | 207, 260 |
| 72CC863 | A. Bélanger and P. Brassard; *J. Chem. Soc., Chem. Commun.*, 1972, 863. | 796 |
| 72CC1144 | J. A. Hyatt and J. S. Swenton; *J. Chem. Soc., Chem. Commun.*, 1972, 1144. | 429 |
| 72CC1226 | J. K. Chakrabarti and T. M. Hotten; *J. Chem. Soc., Chem. Commun.*, 1972, 1226. | 486 |
| 72CC1240 | J. A. Barltrop, K. Dawes, A. C. Day and A. J. H. Summers; *J. Chem. Soc., Chem. Commun.*, 1972, 1240. | 649 |
| 72CC1281 | C. W. Rees, R. W. Stephenson and R. C. Storr; *J. Chem. Soc., Chem. Commun.*, 1972, 1281. | 378 |
| 72CC1300 | R. M. Moriarty and A. Chin; *J. Chem. Soc., Chem. Commun.*, 1972, 1300. | 562 |
| 72CHE416 | V. A. Zagorevskii, E. K. Orlova and I. D. Tsvetkova; *Chem. Heterocycl. Compd. (Engl. Transl.)*, 1972, **8**, 416. | 700 |
| 72CHE935 | G. N. Dorofeenko and V. V. Tkachenko; *Chem. Heterocycl. Compd. (Engl. Transl.)*, 1972, **8**, 935. | 865 |
| 72CJC584 | J. W. Lown and K. Matsumoto; *Can. J. Chem.*, 1972, **50**, 584. | 1003, 1023 |
| 72CJC1539 | K. Jankowski, J. Couturier and R. Tower; *Can. J. Chem.*, 1972, **50**, 1539. | 606 |
| 72CJC1581 | J. Pinson, J.-P. M'Packo, N. Vinot, J. Armand and P. Bassinet; *Can. J. Chem.*, 1972, **50**, 1581. | 394, 413, 415 |
| 72CL1185 | O. Tsuge, H. Samura and M. Tashiro; *Chem. Lett.*, 1972, 1185. | 413, 446 |
| 72CPB677 | M. Kamiya and Y. Akahori; *Chem. Pharm. Bull.*, 1972, **20**, 677. | 575 |
| 72CPB772 | T. Higashino, M. Uchida and E. Hayashi; *Chem. Pharm. Bull.*, 1072, **20**, 772. | 204 |
| 72CPB1428 | F. Yoneda, S. Fukazawa and S. Nishigaki; *Chem. Pharm. Bull.*, 1972, **20**, 1428. | 317 |
| 72CPB1513 | S. Yurugi, M. Hieda, T. Fushimi, Y. Kawamatsu, H. Sugihara and M. Tomimoto; *Chem. Pharm. Bull.*, 1972, **20**, 1513. | 345, 346, 358 |
| 72CPB1522 | S. Yurugi and M. Hieda; *Chem. Pharm. Bull.*, 1972, **20**, 1522. | 336, 346 |
| 72CPB1528 | S. Yurugi, M. Hieda, T. Fushimi, Y. Kawamatsu, H. Sugihara and M. Tomimoto; *Chem. Pharm. Bull.*, 1972, **20**, 1528. | 346 |
| 72CPB2204 | F. Uchimaru, S. Okada, A. Kosasayama and T. Kono; *Chem. Pharm. Bull.*, 1972, **20**, 2204. | 159 |
| 72CPB2264 | M. Ogata and H. Matsumoto; *Chem. Pharm. Bull.*, 1972, **20**, 2264. | 221, 256 |
| 72CR(B)(275)785 | A. Marcou, G. Capderroque and R. Freymann; *C. R. Hebd. Seances Acad. Sci., Ser. B*, 1972, **275**, 785. | 956 |
| 72CR(C)(274)189 | P. Bassinet, J. Pinson and J. Armand; *C. R. Hebd. Seances Acad. Sci., Ser. C*, 1972, **274**, 189. | 1046 |
| 72CR(C)(274)650 | D. Anker, J. Andrieux, M. Baran-Marszak and D. Molho; *C.R. Hebd. Seances Acad. Sci., Ser. C*, 1972, **274**, 650. | 668, 873 |
| 72CR(C)(274)1091 | M. Dupré and H. Strzelecka; *C.R. Hebd. Seances Acad. Sci., Ser. C*, 1972, **274**, 1091. | 758 |
| 72CR(C)(275)279 | J. Armand, P. Bassinet, K. Chekir, J. Pinson and P. Souchay; *C. R. Hebd. Seances Acad. Sci., Ser. C*, 1972, **275**, 279. | 177 |
| 72CR(C)(275)1383 | A. Decormeille, G. Queguiner and P. Pastour; *C.R. Hebd. Seances Acad. Sci., Ser. C*, 1972, **275**, 1383. | 238, 239 |
| 72CZ37 | H. Boehme and U. Sitorus; *Chem.-Ztg.*, 1972, **96**, 37. | 895 |
| 72DOK(204)879 | E. A. Karakhanov, E. A. Dem'yanova and E. A. Viktorova; *Dokl. Akad. Nauk SSSR*, 1972, **204**, 879. | 785 |
| 72G169 | G. Adembri, F. DeSio, R. Nesi and M. Scotton; *Gazz. Chim. Ital.*, 1972, **102**, 169. | 333 |
| 72GEP2036312 | H. Disselnkoetter; *Ger. Pat.* 2 036 312 (1972) (*Chem. Abstr.*, 1972, **76**, 127 025). | 1086 |
| 72GEP2145174 | J. L. Zollinger; *Ger. Pat.* 2 145 174 (*Chem. Abstr.*, 1972, **77**, 35 234). | 512 |
| 72HCA10 | R. Hug, H.-J. Hansen and H. Schmid; *Helv. Chim. Acta*, 1972, **55**, 10. | 594, 766 |
| 72HCA249 | M. Rosenberger, D. Andrews, F. DiMaria, A. J. Duggan and G. Saucy; *Helv. Chim. Acta*, 1972, **55**, 249. | 775, 845 |
| 72HCA255 | R. Gleiter, E. Heilbronner and V. Hornung; *Helv. Chim. Acta*, 1972, **55**, 255. | 466, 542 |
| 72HCA574 | M. Viscontini and W. F. Frei; *Helv. Chim. Acta*, 1972, **55**, 574. | 312 |
| 72HCA1404 | W. Skorianetz and E. sz. Kovats; *Helv. Chim. Acta*, 1972, **55**, 1404. | 541, 546 |
| 72HCA1566 | H. Brunetti and C. E. Lüthi; *Helv. Chim. Acta*, 1972, **55**, 1566. | 493, 512 |
| 72HCA1675 | R. Hug, H.-J. Hansen and H. Schmid; *Helv. Chim. Acta*, 1972, **55**, 1675. | 749 |
| 72HCA1828 | R. Hug, H.-J. Hansen and H. Schmid; *Helv. Chim. Acta*, 1972, **55**, 1828. | 749 |
| 72IJC19 | V. V. S. Murti, P. S. Sampath Kumar and T. R. Seshadri; *Indian J. Chem.*, 1972, **10**, 19. | 608 |
| 72IJC32 | A. K. Das Gupta, K. R. Das and A. Das Gupta; *Indian J. Chem.*, 1972, **10**, 32. | 810 |
| 72IJC602 | A. S. Narang, A. N. Kaushal, S. Singh and K. S. Narang; *Indian J. Chem.*, 1972, **10**, 602. | 226, 228 |
| 72IJC924 | K. G. Das, V. N. Gogte, M. Seetha and B. D. Tilak; *Indian J. Chem.*, 1972, **10**, 924. | 607 |
| 72IJS(A)(2)287 | S. Rozen, I. Shahak and E. D. Bergmann; *Int. J. Sulfur Chem., Part A*, 1972, **2**, 287. | 935 |
| 72IJS(B)(7)101 | U. Eisner and T. Krishnamurthy; *Int. J. Sulfur Chem., Sect. B*, 1972, **7**, 101. | 990 |
| 72IJS(B)(7)185 | G. Tsuchihashi; *Int. J. Sulfur. Chem., Part B*, 1972, **7**, 185. | 901 |
| 72JA208 | J. B. Lambert, C. E. Mixan and D. S. Bailey; *J. Am. Chem. Soc.*, 1972, **94**, 208. | 899 |
| 72JA866 | G. W. Grams, K. Eskins and G. E. Inglett; *J. Am. Chem. Soc.*, 1972, **94**, 866. | 720 |

| | | |
|---|---|---|
| 72JA1026 | H. J. Shine and J. J. Silber; *J. Am. Chem. Soc.*, 1972, **94**, 1026. | 969 |
| 72JA1395 | A. Padwa, S. Clough, M. Dharan. J. Smolanoff and S. I. Wetmure, Jr.; *J. Am. Chem. Soc.*, 1972, **94**, 1395. | 188 |
| 72JA1466 | C. R. Brundle, M. B. Robin and N. A. Kuebler; *J. Am. Chem. Soc.*, 1972, **94**, 1466. | 466 |
| 72JA2770 | A. Steigel, J. Sauer, D. A. Kleier and G. Binsch; *J. Am. Chem. Soc.*, 1972, **94**, 2770. | 423, 551 |
| 72JA5599 | D. A. Sweigart and D. W. Turner; *J. Am. Chem. Soc.*, 1972, **94**, 5599. | 598, 959 |
| 72JA6135 | G. M. Atkins and E. M. Burgess; *J. Am. Chem. Soc.*, 1972, **94**, 6135. | 1041, 1076 |
| 72JA7295 | L. N. Klatt and R. L. Rouseff; *J. Am. Chem. Soc.*, 1972, **94**, 7295. | 177 |
| 72JA8451 | T. J. Van Bergen and R. M. Kellogg; *J. Am. Chem. Soc.*, 1972, **94**, 8451. | 243 |
| 72JA8497 | S. F. Nelson and R. Fibiger; *J. Am. Chem. Soc.*, 1972, **94**, 8497. | 532, 533, 534 |
| 72JA8932 | E. J. Corey and S. W. Walinsky; *J. Am. Chem. Soc.*, 1972, **94**, 8932. | 931, 977 |
| 72JA9219 | Y. Kishi, T. Fukuyama, M. Aratani, F. Nakatsubo, T. Goto, S. Inoue, H. Tanino, S. Sugiura and H. Kakoi; *J. Am. Chem. Soc.*, 1972, **94**, 9219. | 87, 148 |
| 72JAP7208232 | H. Moriga and F. Takabayashi; *Jpn. Pat.* 72 08 232 (1972) (*Chem. Abstr.*, 1972, **77**, 50 020). | 994 |
| 72JBC(247)4549 | T. Fukushima and T. Shiota; *J. Biol. Chem.*, 1972, **247**, 4549. | 324 |
| 72JCP(57)2572 | R. C. Lord, T. C. Rounds and T. Ueda; *J. Chem. Phys.*, 1972, **57**, 2572. | 629 |
| 72JCS(P1)25 | W. J. G. Donnelly and P. V. R. Shannon; *J. Chem. Soc., Perkin Trans. 1*, 1972, 25. | 747 |
| 72JCS(P1)353 | W. J. Irwin; *J. Chem. Soc., Perkin Trans. 1*, 1972, 353. | 207 |
| 72JCS(P1)522 | D. J. Brown and J. A. Hoskins; *J. Chem. Soc., Perkin Trans. 1*, 1972, 522. | 94, 95, 97, 138 |
| 72JCS(P1)779 | G. P. Ellis and D. Shaw; *J. Chem. Soc., Perkin Trans. 1*, 1972, 779. | 710, 718 |
| 72JCS(P1)809 | S. A. Barker and T. Riley; *J. Chem. Soc., Perkin Trans. 1*, 1972, 809. | 862 |
| 72JCS(P1)953 | G. Adembri, F. DeSio, R. Nesi and M. Scotton; *J. Chem. Soc., Perkin Trans. 1*, 1972, 953. | 342, 354 |
| 72JCS(P1)1041 | T. Paterson and H. C. S. Wood; *J. Chem. Soc., Perkin Trans. 1*, 1972, 1041. | 202, 212, 230 |
| 72JCS(P1)1103 | D. H. R. Barton, P. D. Magnus and J. I. Okogun; *J. Chem. Soc., Perkin Trans. 1*, 1972, 1103. | 695 |
| 72JCS(P1)1142 | G. V. Boyd and S. R. Dando; *J. Chem. Soc., Perkin Trans. 1*, 1972, 1142. | 658 |
| 72JCS(P1)1221 | M. F. G. Stevens; *J. Chem. Soc., Perkin Trans. 1*, 1972, 1221. | 392 |
| 72JCS(P1)1269 | G. B. Barlin and A. C. Young; *J. Chem. Soc., Perkin Trans. 1*, 1972, 1269. | 87 |
| 72JCS(P1)1315 | M. Keating, M. E. Peek, C. W. Rees and R. C. Storr; *J. Chem. Soc., Perkin Trans. 1*, 1972, 1315. | 372, 381 |
| 72JCS(P1)1382 | A. J. Quillinan and F. Scheinmann; *J. Chem. Soc., Perkin Trans. 1*, 1972, 1382. | 584 |
| 72JCS(P1)1701 | R. O. C. Norman, R. Purchase and C. B. Thomas; *J. Chem. Soc., Perkin Trans. 1*, 1972, 1701. | 999 |
| 72JCS(P1)1924 | H. Nakata, A. Tatematsu, H. Yoshizumi and S. Naga; *J. Chem. Soc., Perkin Trans. 1*, 1972, 1924. | 609 |
| 72JCS(P1)2004 | A. F. Bramwell, I. M. Payne, G. Riezebos, P. Ward and R. D. Wells; *J. Chem. Soc., Perkin Trans. 1*, 1972, 2004. | 163 |
| 72JCS(P1)2190 | D. J. Berry, J. D. Cook and B. J. Wakefield; *J. Chem. Soc., Perkin Trans. 1*, 1972, 2190. | 217 |
| 72JCS(P1)2494 | K. W. Blake, A. E. A. Porter and P. G. Sammes; *J. Chem. Soc., Perkin Trans. 1*, 1972, 2494. | 187 |
| 72JCS(P1)2903 | H. Heaney, J. M. Jablonski and C. T. McCarty; *J. Chem. Soc., Perkin Trans. 1*, 1972, 2903. | 753 |
| 72JCS(P1)2915 | A. J. Elliott and M. S. Gibson; *J. Chem. Soc., Perkin Trans. 1*, 1972, 2915. | 1072 |
| 72JCS(P2)357 | B. Fuchs, I. Goldberg and U. Shmueli; *J. Chem. Soc., Perkin Trans. 2*, 1972, 357. | 960 |
| 72JCS(P2)376 | R. Stewart and J. M. McAndless; *J. Chem. Soc., Perkin Trans. 2*, 1972, 376. | 303 |
| 72JCS(P2)451 | J. Almog, A. Y. Meyer and H. Shanan-Atidi; *J. Chem. Soc., Perkin Trans. 2*, 1972, 451. | 202 |
| 72JCS(P2)642 | G. J. Bullen, D. J. Corney and F. S. Stephens; *J. Chem. Soc., Perkin Trans. 2*, 1972, 642. | 461 |
| 72JCS(P2)674 | L. Angiolini, R. P. Duke, R. A. Y. Jones and A. R. Katritzky; *J. Chem. Soc., Perkin Trans. 2*, 1972, 674. | 1054, 1055 |
| 72JHC87 | R. Selvarajan and J. H. Boyer; *J. Heterocycl. Chem.*, 1972, **9**, 87. | 564 |
| 72JHC91 | E. M. Levine and T. J. Bardos; *J. Heterocycl. Chem.*, 1972, **9**, 91. | 213, 214, 221 |
| 72JHC171 | M. von Strandtmann, S. Klutchko, M. P. Cohen and J. Shavel, Jr.; *J. Heterocycl. Chem.*, 1972, **9**, 171. | 821, 853 |
| 72JHC175 | M. von Strandtmann, D. Connor and J. Shavel, Jr.; *J. Heterocycl. Chem.*, 1972, **9**, 175. | 805 |
| 72JHC255 | M. Israel, L. C. Jones and E. J. Modest; *J. Heterocycl. Chem.*, 1972, **9**, 255. | 253 |
| 72JHC351 | P. Kregar-Čadež, A. Pollak, B. Stanovnik, M. Tišler and B. Wechtersbach-Lažetić; *J. Heterocycl. Chem.*, 1972, **9**, 351. | 237, 248 |
| 72JHC489 | K. C. Majumdar and B. S. Thyagarajan; *J. Heterocycl. Chem.*, 1972, **9**, 489 | 742 |
| 72JHC783 | J. A. Van Allan, G. A. Reynolds and C. C. Petropoulos; *J. Heterocycl. Chem.*, 1972, **9**, 783. | 665 |
| 72JHC853 | I. W. Elliott, Jr.; *J. Heterocycl. Chem.*, 1972, **9**, 853. | 858 |
| 72JHC887 | C. M. Buess, V. O. Brandt, R. C. Srivastava and W. R. Carper; *J. Heterocycl. Chem.*, 1972, **9**, 887. | 957 |
| 72JHC1113 | E. F. Elslager, J. Clarke, P. Jacob, L. M. Werbel and J. D. Willis; *J. Heterocycl. Chem.*, 1972, **9**, 1113. | 226 |
| 72JHC1123 | E. F. Elslager, A. Curry and L. M. Werbel; *J. Heterocycl. Chem.*, 1972, **9**, 1123. | 226 |

| | | |
|---|---|---|
| 72JHC1255 | R. D. Barry and R. A. Balding; *J. Heterocycl. Chem.*, 1972, **9**, 1255. | 832, 858 |
| 72JHC1341 | P. Schenone, G. Bognardi and S. Morasso; *J. Heterocycl. Chem.*, 1972, **9**, 1341. | 723 |
| 72JHC1433 | K. Y. Tserng and L. Bauer; *J. Heterocycl. Chem.*, 1972, **9**, 1433. | 208 |
| 72JIC985 | M. Sidky, M. R. Mahran and L. S. Boulos; *J. Indian Chem. Soc.*, 1972, **49**, 985. | 921 |
| 72JMC182 | J. A. Montgomery and H. J. Thomas; *J. Med. Chem.*, 1972, **15**, 182. | 383 |
| 72JMC394 | J. G. Topliss and M. D. Yudis; *J. Med. Chem.*, 1972, **15**, 394. | 1085 |
| 72JMC435 | H. G. Garg and C. Prakash; *J. Med. Chem.*, 1972, **15**, 435. | 1085 |
| 72JMC442 | A. A. Santilli and D. H. Kim; *J. Med. Chem.*, 1972, **15**, 442. | 227 |
| 72JMC583 | H. Cairns, C. Fitzmaurice, D. Hunter, P. B. Johnson, J. King, T. B. Lee, G. H. Lord, R. Minshull and J. S. G. Cox; *J. Med. Chem.*, 1972, **15**, 583. | 718, 817 |
| 72JMC595 | R. D. Elliott, J. R. Piper, C. R. Stringfellow and T. P. Johnston; *J. Med. Chem.*, 1972, **15**, 595. | 940 |
| 72JMC837 | J. Davoll, J. Clarke and E. F. Elslager; *J. Med. Chem.*, 1972, **15**, 837. | 228 |
| 72JMC865 | G. P. Ellis and D. Shaw; *J. Med. Chem.*, 1972, **15**, 865. | 710, 711 |
| 72JMC1203 | A. Richardson, Jr. and F. J. McCarty; *J. Med. Chem.*, 1972, **15**, 1203. | 360 |
| 72JMC1331 | J. P. Jonak, S. F. Zakrzewski, L. H. Mead and L. D. Allshouse; *J. Med. Chem.*, 1972, **15**, 1331. | 325 |
| 72JOC111 | G. S. Marx and P. E. Spoerri; *J. Org. Chem.*, 1972, **37**, 111. | 159 |
| 72JOC196 | E. J. Moriconi and Y. Shimakawa; *J. Org. Chem.*, 1972, **37**, 196. | 371, 383, 1061, 1076 |
| 72JOC221 | J. Adachi and N. Sato; *J. Org. Chem.*, 1972, **37**, 221. | 185 |
| 72JOC329 | D. H. R. Barton, R. H. Hesse, H. T. Toh and M. M. Pechet; *J. Org. Chem.*, 1972, **37**, 329. | 70, 152 |
| 72JOC377 | J. B. Lambert, D. S. Bailey and C. E. Mixan; *J. Org. Chem.*, 1972, **37**, 377. | 890 |
| 72JOC511 | B. D. Mookherjee and E. M. Klaiber; *J. Org. Chem.*, 1972, **37**, 511. | 167 |
| 72JOC589 | M. J. A. El-Haj, B. W. Dominy, J. D. Johnston, M. J. Haddadin and C. H. Issidorides; *J. Org. Chem.*, 1972, **37**, 589. | 259 |
| 72JOC841 | M. Harfenist and E. Thom; *J. Org. Chem.*, 1972, **37**, 841. | 743 |
| 72JOC1051 | G. L. Closs and A. M. Harrison; *J. Org. Chem.*, 1972, **37**, 1051. | 371, 378, 380, 382 |
| 72JOC1069 | R. Sutton; *J. Org. Chem.*, 1972, **37**, 1069. | 756 |
| 72JOC1444 | L. Tolentino and J. Kagan; *J. Org. Chem.*, 1972, **37**, 1444. | 697, 716 |
| 72JOC1587 | R. C. Kerber; *J. Org. Chem.*, 1972, **37**, 1587. | 383 |
| 72JOC1592 | R. C. Kerber and P. J. Heffron; *J. Org. Chem.*, 1972, **37**, 1592. | 383, 384 |
| 72JOC1636 | W. J. McGahren, G. A. Ellestad, G. O. Morton and M. P. Kunstmann; *J. Org. Chem.*, 1972, **37**, 1636. | 583 |
| 72JOC2259 | T. J. Curphey and K. S. Prasad; *J. Org. Chem.* 1972, **37**, 2259. | 17 |
| 72JOC2367 | R. M. Dodson, K. S. Srinivasan, K. S. Sharma and R. F. Sauers; *J. Org. Chem.*, 1972, **37**, 2367. | 991 |
| 72JOC2635 | S. Fujii and H. Kobatake; *J. Org. Chem.*, 1972, **37**, 2635. | 170 |
| 72JOC2691 | H. J. Shine, J. J. Silber, R. J. Bussey and T. Okuyama; *J. Org. Chem.*, 1972, **37**, 2691. | 969 |
| 72JOC2720 | C. Kissel, R. J. Holland and M. C. Caserio; *J. Org. Chem.*, 1972, **37**, 2720. | 894 |
| 72JOC2986 | R. A. Finnegan and K. E. Merkel; *J. Org. Chem.*, 1972, **37**, 2986. | 835 |
| 72JOC2992 | E. N. Marvell, T. Chadwick, G. Caple, T. Gosink and G. Zimmer; *J. Org. Chem.*, 1972, **37**, 2992. | 666 |
| 72JOC3036 | E. N. Marvell and T. Gosink; *J. Org. Chem.*, 1972, **37**, 3036. | 661 |
| 72JOC3838 | K. T. Potts, A. J. Elliott and M. Sorm; *J. Org. Chem.*, 1972, **37**, 3838. | 644 |
| 72JOC3975 | B. H. Rizkalla, A. D. Broom, M. G. Stout and R. K. Robins; *J. Org. Chem.*, 1972, **37**, 3975. | 206, 226 |
| 72JOC3980 | B. H. Rizkalla and A. D. Broom; *J. Org. Chem.*, 1972, **37**, 3980. | 206, 220, 260, 362 |
| 72JOC3997 | K. Jankowski and J. Couturier; *J. Org. Chem.*, 1972, **37**, 3997. | 578 |
| 72JOU1398 | L. I. Vereshchagin, N. V. Sushkova and L. P. Vologdina; *J. Org. Chem. USSR* (*Engl. Transl.*), 1972, **8**, 1398. | 812 |
| 72JOU2250 | G. N. Dorofeenko and V. V. Tkachenko; *J. Org. Chem. USSR* (*Engl. Transl.*), 1972, **8**, 2250. | 865 |
| 72JPC5087 | R. Blinc, M. Mali, R. Osredkar, A. Prelesnik, J. Seliger, I. Zupančič and L. Ehrenberg; *J. Phys. Chem.*, 1972, **57**, 5087. | 68 |
| 72JPS963 | D. R. Galpin, S. R. Bobbink and T. E. Ary; *J. Pharm. Sci.*, 1972, **61**, 963. | 580 |
| 72JSP(43)477 | K. K. Innes, A. H. Kalantar, A. Y. Khan and T. J. Durnick; *J. Mol. Spectrosc.*, 1972, **43**, 477. | 533 |
| 72JST(14)353 | G. Schultz, I. Hargittai and L. Hermann; *J. Mol. Struct.*, 1972, **14**, 353. | 949 |
| 72KGS321 | G. C. Butenko, A. N. Vereshchagin and B. A. Arbuzov; *Khim. Geterotsikl. Soedin.*, 1972, 321. | 1054 |
| 72KGS422 | E. Grinsteins, E. Stankevics and G. Duburs; *Khim. Geterotsikl. Soedin.*, 1972, 422 (*Chem. Abstr.*, 1972, **77**, 88 424). | 232 |
| 72KGS1275 | A. S. Elina, I. S. Musatova and G. P. Syrova; *Khim. Geterotsikl. Soedin.*, 1972, 1275. | 171 |
| 72LA(758)111 | H. Neunhoeffer and H.-W. Frühauf; *Liebigs Ann. Chem.*, 1972, **758**, 111. | 410 |
| 72LA(758)120 | H. Neunhoeffer and H.-W. Frühauf; *Liebigs Ann. Chem.*, 1972, **758**, 120. | 427 |
| 72LA(758)125 | H. Neunhoeffer and H.-W. Frühauf; *Liebigs Ann. Chem.*, 1972, **758**, 125. | 427 |
| 72LA(760)88 | H. Neunhoeffer, L. Motitschke, H. Hennig and K. Ostheimer; *Leibigs Ann. Chem.*, 1972, **760**, 88. | 431 |
| 72LA(760)102 | H. Neunhoeffer and H.-W. Frühauf; *Leibigs Ann. Chem.*, 1972, **760**, 102. | 396 |

| | | |
|---|---|---|
| 72LA(761)95 | F. Asinger, D. Neuray, W. Leuchtenberger, A. Saus and F. A. Dagga; *Liebigs Ann. Chem.*, 1972, **761**, 95. | 566 |
| 72LA(765)8 | H. Harnisch; *Liebigs Ann. Chem.*, 1972, **765**, 8. | 821 |
| 72LA(766)73 | H. Bredereck, G. Simchen, H. Wagner and A. A. Santos; *Liebigs Ann. Chem.*, 1972, **766**, 73. | 79, 125, 126, 128 |
| 72M1591 | A. Pollak, S. Polanc, B. Stanovnik and M. Tišler; *Monatsh. Chem.*, 1972, **103**, 1591. | 343 |
| 72MI21400 | H. M. Liebich, D. R. Douglas, A. Zlatkis, F. Mueggler-Chavan and A. Donzel; *J. Agric. Food Chem.*, 1972, **20**, 96. | 193 |
| 72MI21401 | R. M. Sheldon, R. C. Lindsay and L. M. Libbey; *J. Food Sci.*, 1972, **37**, 313. | 193 |
| 72MI21500 | B. Stanovnik and M. Tisler; *Croat. Chem. Acta*, 1972, **44**, 243 (*Chem. Abstr.*, 1972, **77**, 152 098). | 222, 223 |
| 72MI21501 | M. Kramberger, B. Stanovnik and M. Tisler; *Croat. Chem. Acta*, 1972, **44**, 419 (*Chem. Abstr.*, 1973, **78**, 43 401). | 234, 238 |
| 72MI21602 | J. R. Merkel; *Arch. Mikrobiol.*, 1972, **81**, 379. | 294, 325 |
| 72MI21603 | H. Rembold and K. Buff; *Eur. J. Biochem.*, 1972, **28**, 586. | 323 |
| 72MI22000 | N. Ahmad and N. A. Warsi; *Pakistan J. Sci. Ind. Res.*, 1972, 157 (*Chem. Abstr.*, 1973, **78**, 136 727). | 505 |
| B-72MI22001 | D. Hildebrand, K.-H. Schundehutte and E. Siegel; in 'The Chemistry of Synthetic Dyes', ed. K. Venkataraman; Academic, New York, 1972, vol. 6. | 526 |
| 72MI22100 | R. Carbo and S. Fraga; *An. Fis.*, 1972, **68**, 21 (*Chem. Abstr.*, 1972, **77**, 79 749). | 533 |
| 72MI22200 | R. L. Mital, R. R. Gupta and S. K. Jain; *J. Chem. Eng. Data*, 1972, **17**, 383. | 582 |
| 72MI22201 | W. Steck and M. Mazurek; *Lloydia*, 1972, **35**, 418. | 582 |
| B-72MI22202 | G. C. Levy and G. L. Nelson; 'Carbon-13 Nuclear Magnetic Resonance for Organic Chemists', Wiley, New York, 1972, p. 43. | 586 |
| B-72MI22203 | J. B. Stothers; 'Carbon-13 NMR Spectroscopy', Academic, New York, 1972, p. 343. | 588, 635 |
| 72MI22204 | S. E. Scheppele, R. K. Mitchum, C. J. Rudolph, Jr., K. J. Kinneberg and G. V. Odele; *Lipids*, 1972, **7**, 297. | 607 |
| 72MI22205 | Z. Aizenshtat, E. Klein, H. Weiler-Feilchenfeld and E. D. Bergmann; *Isr. J. Chem.*, 1972, **10**, 753. | 627 |
| 72MI22206 | A. Mallabaev and G. Sidyakin; *Khim. Prir. Soedin.*, 1972, 279 (*Chem. Abstr.*, 1972, **77**, 139 027). | 597 |
| B-72MI22400 | 'Microbial Toxins', ed. A. Ciegler, S. J. Ajl and S. K. Adis; Academic Press, New York, 1972, vol. 7, chap. 1. | 882 |
| 72MI22500 | H. Remane, R. Herzschuh, L. C. Hoa and R. Borsdorf; *J. Electroanal. Chem., Interfacial Electrochem.*, 1972, **35**, 363. | 903 |
| 72MI22501 | G. D. Rees and J. K. Sugden; *Pharm. Acta Helv.*, 1972, **47**, 481. | 940 |
| 72MI22800 | H. G. Heal; *Adv. Inorg. Chem. Radiochem.*, 1972, **15**, 375. | 1048 |
| 72MI22801 | R. Carbo and S. Fraga; *An. Fis.*, 1972, **68**, 21. | 1041, 1045, 1047 |
| 72NEP7206067 | N.V.K.P.F., *Neth. Pat.* 72 06 067 (1972) (*Chem. Abstr.*, 1973, **78**, 72 180). | 153 |
| 72OMR(4)537 | O. Achmatowicz, Jr., M. Chmielewski, J. Jurczak, L. Korzerski and A. Zamojski; *Org. Magn. Reson.*, 1972, **4**, 537. | 578 |
| 72OMR(4)895 | N. S. Angerman and S. S. Danyluk; *Org. Magn. Reson.*, 1972, **4**, 895. | 1010 |
| 72OMS(6)317 | J. H. Bowie and P. Y. White; *Org. Mass Spectrom.*, 1972, **6**, 317. | 957 |
| 72OMS(6)1333 | C. C. Van de Sande, J. W. Serum and M. Vandewalle; *Org. Mass Spectrom.*, 1972, **6**, 1333. | 618 |
| 72OS(52)135 | W. E. Noland and R. D. De Master; *Org. Synth.*, 1972, **52**, 135. | 1034 |
| 72P409 | E. Rodriguez, N. J. Carman and T. J. Mabry; *Phytochemistry*, 1972, **11**, 409. | 585 |
| 72P3491 | G. Hrazdina; *Phytochemistry*, 1972, **11**, 3491. | 662 |
| 72PMH(5)1 | P. J. Wheatley; *Phys. Methods Heterocycl. Chem.*, 1972, **5**, 1. | 59, 620, 621, 946, 948 |
| 72RRC2043 | G. Ostrogovich and M. Safta; *Rev. Roum. Chim.*, 1972, **17**, 2043. | 472 |
| 72RTC949 | P. J. Lont, H. C. Van der Plas and A. J. Verbeek; *Recl. Trav. Chim. Pays-Bas*, 1972, **91**, 949. | 165 |
| 72RTC1137 | H. I. X. Mager and W. Berends; *Recl. Trav. Chim. Pays-Bas*, 1972, **91**, 1137. | 308 |
| 72S311 | W. Reid and W. Ochs; *Synthesis*, 1972, 311. | 1034 |
| 72S312 | T. Kappe and W. Golser; *Synthesis*, 1972, 312. | 1004 |
| 72S333 | R. R. Schmidt; *Synthesis*, 1972, 333. | 1005, 1006, 1021 |
| 72SA(A)137 | O. H. Ellestad, P. Klaboe and G. Hagen; *Spectrochim. Acta, Part A*, 1972, **28**, 137. | 956 |
| 72SA(A)707 | P. J. F. Griffiths and G. P. Ellis; *Spectrochim. Acta, Part A*, 1972, **28**, 707. | 601 |
| 72SA(A)1001 | I. I. Stânoiu, M. Paraschiv, E. Romas and A. T. Balaban; *Spectrochim. Acta, Part A*, 1972, **28**, 1001. | 598 |
| 72T1881 | H. Budzikiewicz and L. Grotjahn; *Tetrahedron*, 1972, **28**, 1881. | 605 |
| 72T1983 | J. A. Zoltewicz and L. W. Deady; *Tetrahedron*, 1972, **28**, 1983. | 17, 249, 251 |
| 72T2139 | R. C. Cookson and T. A. Crabb; *Tetrahedron*, 1972, **28**, 2139. | 953 |
| 72T2617 | P. Pasanen and K. Pihlaja; *Tetrahedron*, 1972, **28**, 2617. | 961 |
| 72T4155 | A. F. Bramwell and R. D. Wells; *Tetrahedron*, 1972, **28**, 4155. | 160 |
| 72T5197 | M. Weissenfels and M. Pulst; *Tetrahedron*, 1972, **28**, 5197. | 934 |
| 72TH21300 | C. J. Bigum; Ph.D. Thesis, Australian National University, 1972, p. 25. | 63 |
| 72TL133 | Y. Lefebvre; *Tetrahedron Lett.*, 1972, 133. | 844 |
| 72TL949 | S. Hammerum; *Tetrahedron Lett.*, 1972, 949. | 568 |
| 72TL1131 | J. Ficini and J. Pouliquen; *Tetrahedron Lett.*, 1972, 1131. | 814 |
| 72TL1135 | J. Ficini and J. Pouliquen; *Tetrahedron Lett.*, 1972, 1135. | 814 |
| 72TL1885 | T. Sasaki, K. Kanematsu and S. Ochiai; *Tetrahedron Lett.*, 1972, 1885. | 332, 341 |

| | | |
|---|---|---|
| 72TL2247 | M. Van Meerbeck, S. Toppet and F. C. DeSchryver; *Tetrahedron Lett.*, 1972, 2247. | 678 |
| 72TL2545 | T. Kosuge, H. Zenda, A. Ochiai, N. Masaki, M. Noguchi, S. Kimura and H. Narita; *Tetrahedron Lett.*, 1972, 2545. | 284 |
| 72TL3219 | Y. Iwanami and M. Akino; *Tetrahedron Lett.*, 1972, 3219. | 285 |
| 72TL3293 | R. Curci, V. Lucchini, P. J. Kocienski, G. T. Evans and J. Ciabattoni; *Tetrahedron Lett.*, 1972, 3293. | 382 |
| 72TL3359 | W. J. Irwin and D. G. Wibberley; *Tetrahedron Lett.*, 1972, 3359. | 320 |
| 72TL3941 | T. Sheradsky and S. Lewinter; *Tetrahedron Lett.*, 1972, 3941. | 806 |
| 72TL4137 | Y. Tamura, K. Sumoto, J. Minamikawa and M. Ikeda; *Tetrahedron Lett.*, 1972, 4137. | 899 |
| 72TL4165 | M. H. Palmer and R. H. Findlay; *Tetrahedron Lett.*, 1972, 4165. | 893 |
| 72TL4295 | G. M. Priestley and R. N. Warrener; *Tetrahedron Lett.*, 1972, 4295. | 552 |
| 72TL4453 | M. C. Sacquet, B. Graffe and P. Maitte; *Tetrahedron Lett.*, 1972, 4453. | 594, 764 |
| 72TL4503 | W. J. Richter, J. G. Liehr and P. Schulze; *Tetrahedron Lett.*, 1972, 4503. | 603 |
| 72TL4655 | M. Shiozaki and T. Hiraoka; *Tetrahedron Lett.*, 1972, 4655. | 693 |
| 72TL4921 | S. Bory, R. Lett, B. Moreau and A. Marquet; *Tetrahedron Lett.*, 1972, 4921. | 897 |
| 72TL5199 | S. Hoff, A. P. Blok and E. Zwanenburg; *Tetrahedron Lett.*, 1972, 5199. | 1013 |
| 72TL5267 | S. Hoff and E. Zwanenburg; *Tetrahedron Lett.*, 1972, 5267. | 1041 |
| 72UP21900 | H. Neunhoeffer and G. Pieschel; unpublished results, 1972. | 421 |
| 72USP3657239 | W. J. Houlihan; *U.S. Pat.* 3 657 239 (1972) (*Chem. Abstr.*, 1972, **77**, 19 665). | 367 |
| 72USP3682964 | Roussel–UCLAF; *U.S. Pat.* 3 682 964 (1972) (*Chem. Abstr.*, 1972, **77**, 151 938). | 940 |
| 72YZ703 | T. Matsuo and T. Miki; *Yakugaku Zasshi*, 1972, **92**, 703 (*Chem. Abstr.*, 1972, **77**, 88 417). | 236 |
| 72YZ1312 | M. Hieda and S. Yurugi; *Yakugaku Zasshi*, 1972, **92**, 1312. | 346, 347 |
| 72YZ1316 | S. Yurugi, T. Fushimi and M. Hieda; *Yakugaku Zasshi*, 1972, **92**, 1316. | 346 |
| 72YZ1327 | M. Hieda, K. Omura and S. Yurugi; *Yakugaku Zasshi*, 1972, **92**, 1327. | 342, 367 |
| 72ZC293 | F. Wolf and P. Renger; *Z. Chem.*, 1972, **12**, 293. | 506 |
| 72ZOR1925 | O. M. Polumbrik, G. F. Dvorko, E. A. Ponomareva and E. I. Zaika; *Zh. Org. Khim.*, 1972, **8**, 1925. | 544 |
| | | |
| 73AC(R)291 | P. Amstadi and A. Ceruti; *Ann. Chim. (Rome)*, 1973, **63**, 291. | 644 |
| 73AC(R)527 | I. Degani, R. Fochi and G. Spunta; *Ann. Chim. (Rome)*, 1973, **63**, 527. | 922 |
| 73AC(R)563 | J. P. Pradere and H. Quiniou; *Ann. Chim. (Rome)*, 1973, **63**, 563. | 931 |
| 73ACH(76)95 | G. Litkei, R. Bognár and J. Andó; *Acta Chim. Acad. Sci. Hung.*, 1973, **76**, 95. | 854 |
| 73ACS779 | S. Hammerum; *Acta Chem. Scand.*, 1973, **27**, 779. | 568, 569 |
| 73ACS1385 | K. Undheim and E. T. Østensen; *Acta Chem. Scand.*, 1973, **27**, 1385. | 873 |
| 73ACS2020 | G. A. Holmberg and R. Sjöholm; *Acta Chem. Scand.*, 1973, **27**, 2020. | 711 |
| 73ACS3259 | O. Ceder and K. Vernmark; *Acta Chem. Scand.*, 1973, **27**, 3259. | 489 |
| 73ACS3264 | O. Ceder and M. L. Samuelsson; *Acta Chem. Scand.*, 1973, **27**, 3264. | 489 |
| 73AG504 | D. Seebach, D. Enders, B. Renger and W. Brügel; *Angew. Chem.*, 1973, **85**, 504. | 532, 533, 534 |
| 73AG918 | G. Seybold, U. Jersak and R. Gompper; *Angew. Chem.*, 1973, **85**, 918. | 371, 378, 382 |
| 73AG920 | U. Wagner; *Angew. Chem.*, 1973, **85**, 920. | 378 |
| 73AG(E)843 | W. Adam and J. Sanabia; *Angew. Chem., Int. Ed. Engl.*, 1973, **12**, 843. | 979 |
| 73AG(E)869 | K. Clauss and H. Jensen; *Angew. Chem., Int. Ed. Engl.*, 1973, **12**, 869. | 1041, 1085 |
| 73AJC389 | R. N. Warrener, J. A. Elix and W. S. Wilson; *Aust. J. Chem.*, 1973, **26**, 389. | 565, 569 |
| 73AJC443 | D. J. Brown and P. Waring; *Aust. J. Chem.*, 1973, **26**, 443. | 98, 124, 130 |
| 73AJC809 | J. W. Clark-Lewis and E. J. McGarry; *Aust. J. Chem.*, 1973, **26**, 809. | 669 |
| 73AJC819 | J. W. Clark-Lewis and E. J. McGarry; *Aust. J. Chem.*, 1973, **26**, 819. | 651, 669 |
| 73AJC899 | E. R. Krajniak, E. Ritchie and W. C. Taylor; *Aust. J. Chem.*, 1973, **26**, 899. | 801, 810, 848 |
| 73AJC1297 | S. R. Johns, J. A. Lamberton and E. R. Nelson; *Aust. J. Chem.*, 1973, **26**, 1297. | 454 |
| 73AJC1551 | J. W. Ducker and M. J. Gunter; *Aust. J. Chem.*, 1973, **26**, 1551. | 759 |
| 73AJC2291 | F. N. Lahey and R. V. Stick; *Aust. J. Chem.*, 1973, **26**, 2291. | 786 |
| 73AP697 | A. Kreutzberger and R. Schücker; *Arch. Pharm. (Weinheim, Ger.)*, 1973, **306**, 697. | 435 |
| 73AP801 | A. Kreutzberger and R. Schücker; *Arch. Pharm. (Weinheim, Ger.)*, 1973, **306**, 801. | 435 |
| 73AX(B)96 | D. E. Williams; *Acta Crystallogr., Part B*, 1973, **29**, 96. | 539, 541 |
| 73AX(B)1234 | R. J. McClure and B. M. Craven; *Acta Crystallogr., Part B*, 1973, **29**, 1234. | 68 |
| 73AX(B)1916 | J. Hjortas; *Acta Crystallogr., Sect. B*, 1973, **29**, 1916. | 370 |
| 73B392 | B. E. Evans and R. V. Wolfenden; *Biochemistry*, 1973, **12**, 392. | 294 |
| 73B2425 | J. A. Lyon, P. D. Ellis and R. B. Dunlapp; *Biochemistry*, 1973, **12**, 2425. | 285 |
| 73BBA(297)285 | S. Kwee and H. Lund; *Biochim. Biophys. Acta*, 1973, **297**, 285. | 285 |
| 73BCJ292 | S. Yanagida, T. Fujita, M. Ohoka, I. Katagiri and S. Komori; *Bull. Chem. Soc. Jpn.*, 1973, **46**, 292. | 503 |
| 73BCJ306 | S. Yanagida, M. Yokoe, I. Katagiri, M. Ohoka and S. Komori; *Bull. Chem. Soc. Jpn.*, 1973, **46**, 306. | 503, 504 |
| 73BCJ667 | Y. Hayasi, H. Nakamura and H. Nozaki; *Bull. Chem. Soc. Jpn.*, 1973, **46**, 667. | 983 |
| 73BCJ690 | H. Takeshita, R. Kikuchi and Y. Shoji; *Bull. Chem. Soc. Jpn.*, 1973, **46**, 690. | 705 |
| 73BCJ939 | K. Sugiura and M. Goto; *Bull. Chem. Soc. Jpn.*, 1973, **46**, 939. | 304, 306 |
| 73BCJ1558 | T. Kobayashi and S. Nagakura; *Bull. Chem. Soc. Jpn.*, 1973, **46**, 1558. | 959 |
| 73BCJ1839 | K. Kabuto, Y. Kikuchi, S. Yamaguchi and N. Inoue; *Bull. Chem. Soc. Jpn.*, 1973, **46**, 1839. | 583, 630, 855 |
| 73BCJ1890 | M. Hattori, M. Yoneda and M. Goto; *Bull. Chem. Soc. Jpn.*, 1973, **46**, 1890. | 1085 |
| 73BCJ2549 | T. Shiojima, T. Kuroda, S. Ohkawa, Y. Hasegawa and K. Matsui; *Bull. Chem. Soc. Jpn.*, 1973, **46**, 2549. | 482 |

| | | |
|---|---|---|
| 73BCJ2835 | S. Kobayashi; *Bull. Chem. Soc. Jpn.*, 1973, **46**, 2835. | 231 |
| 73BCJ3499 | S. Tsunoda, T. Shiojima, Y. Hashida, S. Sekiguchi and K. Matsui; *Bull. Chem. Soc. Jpn.*, 1973, **46**, 3499. | 476 |
| 73BCJ3612 | H. Uchida and M. Ohta; *Bull. Chem. Soc. Jpn.*, 1973, **46**, 3612 | 366 |
| 73BCJ3902 | I. Shibuya and M. Kurabayashi; *Bull. Chem. Soc. Jpn.*, 1973, **46**, 3902. | 1061 |
| 73BSB283 | R. Verhé and N. Schamp; *Bull. Soc. Chim. Belg.*, 1973, **82**, 283. | 617, 854 |
| 73BSB705 | C. van de Sande and M. Vandewalle; *Bull. Soc. Chim. Belg.*, 1973, **82**, 705. | 601, 851 |
| 73BSF637 | D. Ladurée, P. Rioult and J. Vialle; *Bull. Soc. Chim. Fr.*, 1973, 637. | 983 |
| 73BSF985 | A. Étienne, A. LeBarre and J.-P. Giorgetti; *Bull. Soc. Chim. Fr.*, 1973, 985. | 1079 |
| 73BSF1167 | R. Ducolomb, J. Cadet and R. Teoule; *Bull. Soc. Chim. Fr.*, 1973, 1167. | 149 |
| 73BSF1293 | A. Gorgues; *Bull. Soc. Chim. Fr.*, 1973, 1293. | 794 |
| 73BSF1460 | J. Servoin-Sidoine, M. Montaigne-Lépine and G. Saint-Ruf; *Bull. Soc. Chim. Fr.*, 1973, 1460. | 975 |
| 73BSF2039 | Y. Bessière-Chrétien and H. Serne; *Bull. Soc. Chim. Fr.*, 1973, 2039. | 474 |
| 73BSF2112 | P. Caubère and D. Parry; *Bull. Soc. Chim. Fr.*, 1973, 2112. | 483 |
| 73BSF2126 | J. Daunis; *Bull. Soc. Chim. Fr.*, 1973, 2126. | 416 |
| 73BSF2392 | M. Payard; *Bull. Soc. Chim. Fr.*, 1973, 2392. | 711 |
| 73BSF3100 | N. Vinot and P. Maitte; *Bull. Soc. Chim. Fr.*, 1973, 3100. | 252 |
| 73BSF3421 | J. Andrieux, B. Bodo and D. Molho; *Bull. Soc. Chim. Fr.*, 1973, 3421. | 864 |
| 73CB62 | U. Kraatz and F. Korte; *Chem. Ber.*, 1973, **106**, 62. | 800, 810 |
| 73CB317 | W. Pfleiderer, D. Autenrieth and M. Schranner; *Chem. Ber.*, 1963, **106**, 317. | 282, 297 |
| 73CB914 | H. Junek and H. Aigner; *Chem. Ber.*, 1973, **106**, 914. | 759 |
| 73CB1389 | K. Eistetter and W. Pfleiderer; *Chem. Ber.*, 1973, **106**, 1389. | 319 |
| 73CB1401 | G. Ritzmann and W. Pfleiderer; *Chem. Ber.*, 1973, **106**, 1401. | 282, 297 |
| 73CB1952 | H. Schmid, M. Schranner and W. Pfleiderer; *Chem. Ber.*, 1973, **106**, 1952. | 282, 297 |
| 73CB2975 | J. C. Jochims, W. Pfleiderer, K. Kobayashi, G. Ritzmann and W. Hutzenlaub; *Chem. Ber.*, 1973, **106**, 2975. | 282 |
| 73CB2982 | W. Pfleiderer, G. Ritzmann, K. Harzer and J. C. Jochims, *Chem. Ber.*, 1973, **106**, 2982. | 282 |
| 73CB3039 | H. Vorbrüggen and P. Strehlke; *Chem. Ber.*, 1973, **106**, 3039. | 94 |
| 73CB3097 | R. Kreher and H. Wißmann; *Chem. Ber.*, 1973, **106**, 3097. | 532, 534 |
| 73CB3149 | W. Pfleiderer and W. Hutzenlaub; *Chem. Ber.*, 1973, **106**, 3149. | 281, 290, 303, 305 |
| 73CB3175 | H. Yamamoto, W. Hutzenlaub and W. Pfleiderer; *Chem. Ber.*, 1973, **106**, 3175. | 281, 305 |
| 73CB3203 | W. Hutzenlaub, H. Yamamoto, G. B. Barlin and W. Pfleiderer; *Chem. Ber.*, 1973, **106**, 3203. | 305, 307, 308 |
| 73CB3368 | E. Cohnen and J. Mahnke; *Chem. Ber.*, 1973, **106**, 3368. | 1068 |
| 73CB3524 | A. Attar, H. Wamhoff and F. Korte; *Chem. Ber.*, 1973, **106**, 3524. | 364 |
| 73CB3540 | H. Böhme and F. Martin; *Chem. Ber.*, 1973, **106**, 3540. | 448 |
| 73CB3951 | U. Ewers, H. Günther and L. Jaenicke; *Chem. Ber.*, 1973, **106**, 3951. | 285 |
| 73CC19 | B. M. Adger, M. Keating, C. W. Rees and R. C. Storr, *J. Chem. Soc., Chem. Commun.*, 1973, 19. | 378 |
| 73CC467 | F. McCapra and M. J. Manning; *J. Chem. Soc., Chem. Commun.*, 1973, 467. | 194 |
| 73CC622 | H. Igeta, T. Nakai and T. Tsuchiya; *J. Chem. Soc., Chem. Commun.*, 1973, 622. | 406 |
| 73CCC934 | M. Prystaš, V. Uchytilová and J. Gut; *Collect. Czech. Chem. Commun.*, 1973, **38**, 934. | 408 |
| 73CJC839 | I. W. J. Still, M. S. Chauhan and M. T. Thomas; *Can. J. Chem.*, 1973, **51**, 839. | 907 |
| 73CJC1200 | G. Wood, R. M. Srivastava and B. Adlam; *Can. J. Chem.*, 1973, **51**, 1200. | 953, 962 |
| 73CJC1267 | P. Yates and D. J. McGregor; *Can. J. Chem.*, 1973, **51**, 1267. | 719 |
| 73CJC2650 | M. A. Corbeil, M. C. Rodostamo, R. J. Fanning, B. A. Graham, M. Kulka and J. B. Pierce; *Can. J. Chem.*, 1973, **51**, 2650. | 338, 361 |
| 73CL51 | S. Mataka and J.-P. Anselme; *Chem. Lett.*, 1973, 51. | 532, 534 |
| 73CPB473 | F. Yoneda, M. Higuchi, K. Senga, M. Kanahori and S. Nishigaki; *Chem. Pharm. Bull.*, 1973, **21**, 473. | 364 |
| 73CPB2014 | H. Ogura and M. Sakaguchi; *Chem. Pharm. Bull.*, 1973, **21**, 2014. | 229 |
| 73CPB2643 | T. Higashino and E. Hayashi; *Chem. Pharm. Bull.*, 1973, **21**, 2643. | 209, 223 |
| 73CR(C)(276)1323 | S. Bory, R. Lett, B. Moreau and A. Marquet; *C.R. Hebd. Seances Acad. Sci. Ser. C*, 1973, **276**, 1323. | 900 |
| 73CR(C)(276)1341 | M.-T. Mussetta, M. Selim and N. Q. Trinh; *C. R. Hebd. Seances Acad. Sci., Ser. C*, 1973, **276**, 1341. | 67 |
| 73CR(C)(276)1603 | M. Payard, J. Couquelet and A. Paturet; *C.R. Hebd. Seances Acad. Sci., Ser. C*, 1973, **276**, 1603. | 809 |
| 73CR(C)(277)703 | L. Q. Godefroy, G. Quéguiner and P. Pastour; *C.R. Hebd. Seances Acad. Sci., Ser. C*, 1973, **277**, 703. | 214 |
| 73CR(C)(277)795 | A. Étienne, G. Lonchambon, P. Giraudeau and G. Durand; *C. R. Hebd. Seances Acad. Sci., Ser. C*, 1973, **277**, 795. | 1081 |
| 73CSC91 | G. D. Andreetti, G. Bocelli and G. Sgarabotto; *Cryst. Struct. Commun.*, 1973, **2**, 91. | 625 |
| 73CSC311 | M. Semma, Z. Taira, T. Taga and K. Osaki; *Cryst. Struct. Commun.*, 1973, **2**, 311. | 947 |
| 73CZ565 | R. Grashey, C. Knorn and M. Weidner; *Chem.-Ztg.*, 1973, **97**, 565. | 558, 561 |
| 73CZ566 | R. Grashey and C. Knorn; *Chem.-Ztg.*, 1973, **97**, 566. | 561, 562 |
| 73GEP2236460 | Roussel–UCLAF, *Ger. Pat.* 2 236 460 (1973) (*Chem. Abstr.*, 1973, **78**, 97 489). | 940 |
| 73GEP2238207 | Proctor and Gamble Ltd., *Ger. Pat.* 2 238 207 (1973) (*Chem. Abstr.*, 1973, **78**, 137 887). | 940 |

| 73GEP2248497 | J. P. G. F. Osselaere and C. L. A. Lapiere; (Labs. S.M.B. Anciens Etab. J. Muelberger et R. Baudier), *Ger. Offen.* 2 248 497 (1973) (*Chem. Abstr.*, 1973, **79**, 18 754). | 222 |
| 73GEP2302383 | M. Nakahishi, K. Arimura and T. Tsumagari; (Yoshitomi Pharm. Ind. Ltd.), *Ger. Offen.* 2 302 383 (1973) (*Chem. Abstr.*, 1973, **79**, 105 270). | 243 |
| 73GEP2322073 | T. Denzel and H. Hoecha; (Chem. Fabrik. von Heyden G.m.b.H), *Ger. Offen.* 2 322 073 (*Chem. Abstr.*, 1974, **80**, 37 145). | 238, 246 |
| 73HC(27)1 | G. M. Singerman; *Chem. Heterocycl. Compd.*, 1973, **27**, 1. | 2 |
| 73HC(27)323 | N. R. Patel; *Chem. Heterocycl. Compd.*, 1973, **27**, 323. | 2 |
| 73HC(27)968 | M. Tisler and B. Stanovnik; *Chem. Heterocycl. Compd.*, 1973, **27**, 968. | 232, 236, 237 |
| 73HC(27)1012 | M. Tišler and B. Stanovnik; *Chem. Heterocycl. Compd.*, 1973, **27**, 1012. | 330 |
| 73HC(28)1 | R. N. Castle; *Chem. Heterocycl. Compd.*, 1973, **28**, 1. | 158 |
| 73HC(28)1 | A. G. Lenhert and R. N. Castle; *Chem. Heterocycl. Compd.*, 1973, **28**, 1. | 3 |
| 73HC(28)755 | M. Tišler and B. Stanovnik; *Chem. Heterocycl. Compd.*, 1973, **28**, 755. | 5 |
| 73HCA776 | K. Seckinger; *Helv. Chim. Acta*, 1973, **56**, 776. | 515 |
| 73HCA1225 | K. Harzer and W. Pfleiderer; *Helv. Chim. Acta*, 1973, **56**, 1225. | 282, 297 |
| 73HCA1457 | N. Šarčeviĉ, J. Zsindely and H. Schmid; *Helv. Chim. Acta*, 1973, **56**, 1457. | 743 |
| 73HCA2680 | G. Müller and W. von Philipsborn; *Helv. Chim. Acta*, 1973, **56**, 2680. | 276, 285 |
| 73HCA2981 | U. Koch-Pomeranz, H.-J. Hansen and H. Schmid; *Helv. Chim. Acta*, 1973, **56**, 2981. | 743 |
| 73IJC446 | A. Chatterjee and B. Bandyopadhyay; *Indian J. Chem.*, 1973, **11**, 446. | 891 |
| 73IJS(B)(8)255 | G. F. Pedulli, P. Vivarelli, P. Dembech, A. Ricci and G. Seconi; *Int. J. Sulfur Chem., Part B*, 1973, **8**, 255. | 967 |
| 73IJS(B)(8)341 | F. M. Moracci, M. Cardellini, F. Liberatore, P. Marchini, G. Liso and U. Gulini; *Int. J. Sulfur Chem., Part B*, 1973, **8**, 341. | 1012 |
| 73IZV1678 | A. N. Nesmeyanov, T. P. Tolstaya, A. V. Grib and S. R. Kirgizbaeva; *Izv. Akad. Nauk SSSR, Ser. Khim.*, 1973, 1678 (*Chem. Abstr.*, 1973, **79**, 105167). | 1012 |
| 73IZV2363 | S. I. Zav'yalov and G. V. Pokhvisneva; *Izv. Akad. Nauk SSSR, Ser. Khim.*, 1973, 2363. | 144 |
| 73JA247 | C. L. McIntosh and O. L. Chapman; *J. Am. Chem. Soc.*, 1973, **95**, 247. | 677 |
| 73JA248 | R. G. S. Pong and J. S. Shirk; *J. Am. Chem. Soc.*, 1973, **95**, 248. | 677 |
| 73JA463 | N. Ishibe, M. Sunami and M. Odani; *J. Am. Chem. Soc.*, 1973, **95**, 463. | 693 |
| 73JA614 | O. L. Chapman, C. L. McIntosh and J. Pacansky; *J. Am. Chem. Soc.*, 1973, **95**, 614. | 677 |
| 73JA1677 | E. F. Ullman, L. Call and S. S. Tseng; *J. Am. Chem. Soc.*, 1973, **95**, 1677. | 1058 |
| 73JA2375 | R. S. Glass, W. J. Britt, W. N. Miller and G. S. Wilson; *J. Am. Chem. Soc.*, 1973, **95**, 2375. | 960, 969 |
| 73JA2390 | W. M. Horspool, J. R. Kershaw, A. W. Murray and G. M. Stevenson; *J. Am. Chem. Soc.*, 1973, **95**, 2390. | 372, 375, 381 |
| 73JA2406 | J. A. Barltrop, K. Dawes, A. C. Day and A. J. H. Summers; *J. Am. Chem. Soc.*, 1973, **95**, 2406. | 649 |
| 73JA2766 | C. L. Norris, R. C. Benson, P. Beak and W. H. Flygare; *J. Am. Chem. Soc.*, 1973, **95**, 2766. | 626, 635, 639 |
| 73JA3907 | B. A. Hess, Jr. and L. J. Schaad; *J. Am. Chem. Soc.*, 1973, **95**, 3907. | 945 |
| 73JA4287 | C. R. Johnson, R. A. Kirchhoff, R. J. Reisher and G. F. Katekar; *J. Am. Chem. Soc.*, 1973, **95**, 4287. | 899 |
| 73JA4444 | P. L. Stotter and R. E. Hornish; *J. Am. Chem. Soc.*, 1973, **95**, 4444. | 903, 906 |
| 73JA4447 | Y. Ito, Y. Inubushi, M. Zenbayashi, S. Tomita and T. Saegusa; *J. Am. Chem. Soc.*, 1973, **95**, 4447. | 1025 |
| 73JA4455 | E. C. Taylor and P. E. Jacobi; *J. Am. Chem. Soc.*, 1973, **95**, 4455. | 281, 290 |
| 73JA4634 | J. B. Lambert, C. E. Mixan and D. H. Johnson; *J. Am. Chem. Soc.*, 1973, **95**, 4634. | 628 |
| 73JA4761 | T. R. Krugh; *J. Am. Chem. Soc.*, 1973, **95**, 4761. | 394 |
| 73JA5820 | E. C. Blossey, D. C. Neckers, A. L. Thayer and A. P. Schaap; *J. Am. Chem. Soc.*, 1973, **95**, 5820. | 843 |
| 73JA5829 | E. J. Corey and D. J. Beames; *J. Am. Chem. Soc.*, 1973, **95**, 5829. | 786 |
| 73JA6177 | R. Korenstein, K. A. Muszkat and S. Sharafy-Ozeri; *J. Am. Chem. Soc.*, 1973, **95**, 6177. | 621 |
| 73JA6407 | E. C. Taylor, K. L. Perlman, I. P. Sword, M. Séquin-Frey and P. A. Jacobi; *J. Am. Chem. Soc.*, 1973, **95**, 6407. | 281, 318 |
| 73JA6413 | E. C. Taylor, K. L. Perlman, Y.-H. Kim, I. P. Sword and P. E. Jacobi; *J. Am. Chem. Soc.*, 1973, **95**, 6413. | 281, 318 |
| 73JA7156 | R. C. Kelly and I. Schletter; *J. Am. Chem. Soc.*, 1973, **95**, 7156. | 771 |
| 73JA7402 | K. B. Tomer, N. Harrit, I. Rosenthal, O. Buchardt, P. L. Kumler and D. Creed; *J. Am. Chem. Soc.*, 1973, **95**, 7402. | 12 |
| 73JA7914 | J. W. Pavlik and J. Kwong; *J. Am. Chem. Soc.*, 1973, **95**, 7914. | 649 |
| 73JCP(58)4344 | R. C. Lord and T. C. Rounds; *J. Chem. Phys.*, 1973, **58**, 4344. | 947, 956 |
| 73JCS(P1)26 | G. Jones and R. K. Jones; *J. Chem. Soc., Perkin Trans. 1*, 1973, 26. | 238 |
| 73JCS(P1)120 | M. S. Chauhan, F. M. Dean, D. Matkin and M. L. Robinson; *J. Chem. Soc., Perkin Trans. 1*, 1973, 120. | 784, 785 |
| 73JCS(P1)163 | S. Bradamante and G. Pagani; *J. Chem. Soc., Perkin Trans. 1*, 1973, 163. | 913 |
| 73JCS(P1)335 | D. G. Neilson, S. Mahmood and K. M. Watson; *J. Chem. Soc., Perkin Trans. 1*, 1973, 335. | 546, 569 |
| 73JCS(P1)359 | M. S. Chauhan, F. M. Dean, S. McDonald and M. S. Robinson; *J. Chem. Soc., Perkin Trans. 1*, 1973, 359. | 785 |

73JCS(P1)404    P. J. Machin, A. E. A. Porter and P. G. Sammes; *J. Chem. Soc., Perkin Trans. 1*, 1973, 404.    174

73JCS(P1)410    D. A. Pulman and D. A. Whiting; *J. Chem. Soc., Perkin Trans. 1*, 1973, 410.    891

73JCS(P1)545    C. W. Rees and A. A. Sale; *J. Chem. Soc., Perkin Trans. 1*, 1973, 545.    404

73JCS(P1)823    J. Bailey and J. A. Elvidge; *J. Chem. Soc., Perkin Trans. 1*, 1973, 823.    218

73JCS(P1)842    E. E. Glover, K. T. Rowbottom and D. C. Bishop; *J. Chem. Soc. Perkin Trans. 1*, 1973, 842.    449

73JCS(P1)868    N. Bashir and T. L. Gilchrist; *J. Chem. Soc., Perkin Trans. 1*, 1973, 868.    378, 379, 381

73JCS(P1)965    D. M. X. Donnelly, P. J. Kavanagh, G. Kunesch and J. Polonsky; *J. Chem. Soc., Perkin Trans. 1*, 1973, 965.    810

73JCS(P1)1104    J. Ashby, M. Ayad and O. Meth-Cohn; *J. Chem. Soc., Perkin Trans. 1*, 1973, 1104.    871

73JCS(P1)1130    D. L. Dare, I. D. Entwistle and R. A. W. Johnstone; *J. Chem. Soc., Perkin Trans. 1*, 1973, 1130.    679, 685

73JCS(P1)1169    J. G. Archer, A. J. Barker and R. K. Smalley; *J. Chem. Soc., Perkin Trans. 1*, 1973, 1169.    379

73JCS(P1)1321    D. M. Brunwin and G. Lowe; *J. Chem. Soc., Perkin Trans. 1*, 1973, 1321.    1014

73JCS(P1)1329    A. J. Quillinan and F. Scheinmann; *J. Chem. Soc., Perkin Trans. 1*, 1973, 1329.    836

73JCS(P1)1615    A. Albert and H. Mizuno; *J. Chem. Soc., Perkin Trans. 1*, 1973, 1615.    255, 279, 288, 289

73JCS(P1)1794    A. Albert and W. Pendergast; *J. Chem. Soc., Perkin Trans. 1*, 1973, 1794.    209, 217, 226, 350

73JCS(P1)1972    I. Granoth, Y. Segall and A. Kalir; *J. Chem. Soc., Perkin Trans. 1*, 1973, 1972.    839

73JCS(P1)1974    A. Albert and H. Mizuno; *J. Chem. Soc., Perkin Trans. 1*, 1973, 1974.    255, 279, 283, 287, 288

73JCS(P1)2030    J. A. Donnelly, J. G. Hoey, S. O'Brien and J. O'Grady; *J. Chem. Soc., Perkin Trans. 1*, 1973, 2030.    773

73JCS(P1)2630    A. Albert and J. J. McCormack; *J. Chem. Soc., Perkin Trans. 1*, 1973, 2630.    288

73JCS(P1)2707    R. A. Burrell, J. M. Cox and E. G. Savins; *J. Chem. Soc., Perkin Trans. 1*, 1973, 2707.    170

73JCS(P1)2781    G. P. Ellis and I. L. Thomas; *J. Chem. Soc., Perkin Trans. 1*, 1973, 2781.    583, 686, 707

73JCS(P2)50    G. Gaviraghi and G. Pagani; *J. Chem. Soc., Perkin Trans. 2*, 1973, 50.    913

73JCS(P2)227    H. Booth, D. Huckle and I. M. Lockhart; *J. Chem. Soc., Perkin Trans. 2*, 1973, 227.    580, 630, 673, 729

73JCS(P2)264    P. Hanson and R. O. C. Norman; *J. Chem. Soc., Perkin Trans. 2*, 1973, 264.    1010

73JCS(P2)325    M. J. Cook, R. A. Y. Jones, A. R. Katritzky, M. Moreno Mānas, A. C. Richards, A. J. Sparrow and D. L. Trepanier; *J. Chem. Soc., Perkin Trans. 2*, 1973, 325.    1008

73JCS(P2)1065    A. El-Anani, J. Banger, G. Bianchi, S. Clementi, C. D. Johnson and A. R. Katritzky; *J. Chem. Soc., Perkin Trans. 2*, 1973, 1065.    22

73JCS(P2)2075    R. A. Shaw and P. Ward; *J. Chem. Soc., Perkin Trans. 2*, 1973, 2075.    484

73JHC47    B. A. Feit and A. Teuerstein; *J. Heterocycl. Chem.*, 1973, **10**, 47.    100

73JHC133    W. D. Johnston, H. S. Broadbent and W. W. Parish; *J. Heterocycl. Chem.*, 1973, **10**, 133.    299

73JHC153    T. J. Kress and L. L. Moore; *J. Heterocycl. Chem.*, 1973, **10**, 153.    70

73JHC165    P. L. Myers and J. W. Lewis; *J. Heterocycl. Chem.*, 1973, **10**, 165    760

73JHC231    A. B. DeMilo, J. E. Oliver and R. D. Gilardi; *J. Heterocycl. Chem.*, 1973, **10**, 231.    488

73JHC343    D. K. Krass, T.-K. Chen and W. W. Paudler; *J. Heterocycl. Chem.*, 1973, **10**, 343.    401

73JHC559    J. Daunis and R. Jacquier; *J. Heterocycl. Chem.*, 1973, **10**, 559.    405, 413

73JHC575    A. Messmer, G. Hajos, P. Benko and L. Pallos; *J. Heterocycl. Chem.*, 1973, **10**, 575.    392

73JHC623    G. Brancaccio, G. Lettieri and R. Viterbo; *J. Heterocycl. Chem.*, 1973, **10**, 623.    599

73JHC827    V. P. Williams and J. E. Ayling; *J. Heterocycl. Chem.*, 1973, **10**, 827.    285

73JHC1081    A. Turck, G. Queguiner and P. Pastour; *J. Heterocycl. Chem.*, 1973, **10**, 1081.    332, 342, 353

73JMC209    B. R. Baker and W. T. Ashton; *J. Med. Chem.*, 1973, **16**, 209.    529

73JMC1305    R. Vanderhoek, G. Allen and J. A. Settepani; *J. Med. Chem.*, 1973, **16**, 1305.    492, 525

73JMR(12)143    L. R. Isbrandt, R. K. Jensen and L. Petrakis; *J. Magn. Reson.*, 1973, **12**, 143.    952

73JOC20    T. R. Williams and D. J. Cram; *J. Org. Chem.*, 1973, **38**, 20.    1073

73JOC176    E. M. Burgess and J. P. Sanchez; *J. Org. Chem.*, 1973, **38**, 176.    380

73JOC841    I. Granoth and A. Kalir; *J. Org. Chem.*, 1973, **38**, 841.    838

73JOC1249    E. M. Burgess and W. M. Williams; *J. Org. Chem.*, 1973, **38**, 1249.    1044, 1064, 1076

73JOC1313    F. J. Weigert, J. Husar and J. D. Roberts; *J. Org. Chem.*, 1973, **38**, 1313.    63, 160

73JOC1583    E. E. Schweizer, A. T. Wehman and D. M. Nycz; *J. Org. Chem.*, 1973, **38**, 1583.    749

73JOC1769    K. T. Potts and A. J. Elliott; *J. Org. Chem.*, 1973, **38**, 1769.    247

73JOC2073    A. Rosowsky and K. K. N. Chen; *J. Org. Chem.*, 1973, **38**, 2073.    312

73JOC2817    E. C. Taylor and T. Kobayashi; *J. Org. Chem.*, 1973, **38**, 2817.    281, 318

73JOC2834    J. A. Van Allan, S. Farid, G. A. Reynolds and S. C. Chang; *J. Org. Chem.*, 1973, **38**, 2834.    665, 762

73JOC2931    W. E. Hull, B. D. Sykes and B. M. Babior; *J. Org. Chem.*, 1973, **38**, 2931.    508

73JOC3149    R. Curci, V. Lucchini, G. Modena, P. J. Kocienski and J. Ciabattoni; *J. Org. Chem.*, 1973, **38**, 3149.    382

73JOC3277    G. L. Szekeres, R. K. Robins, P. Dea, M. P. Schweizer and R. A. Long; *J. Org. Chem.*, 1973, **38**, 3277.    415

73JOC3288    A. T. Nielsen, R. L. Atkins, D. W. Moore, R. Scott, D. Mallory and J. M. LaBerge; *J. Org. Chem.*, 1973, **38**, 3288.    467, 508, 509

73JOC3466    M. J. Haddadin and A. Hassner; *J. Org. Chem.*, 1973, **38**, 3466.    440

| | | |
|---|---|---|
| 73JOC3485 | K. T. Potts and R. K. C. Hsia; *J. Org. Chem.*, 1973, **38**, 3485. | 348, 361 |
| 73JOC3627 | S. Wawzonek and J. N. Kellen; *J. Org. Chem.*, 1973, **38**, 3627. | 564 |
| 73JOC3990 | Z. Yoshida, H. Sugimoto, T. Sugimoto and S. Yoneda; *J. Org. Chem.*, 1973, **38**, 3990. | 923 |
| 73JOU399 | G. N. Dorofeenko, V. V. Mezheritskii, E. P. Olekhnovich and A. L. Vasserman; *J. Org. Chem. USSR (Engl. Transl.)*, 1973, **9**, 399. | 865 |
| 73JOU2160 | V. B. Milevskaya, R. V. Belinskaya and L. M. Yagupol'skii; *J. Org. Chem. USSR (Engl. Transl.)*, 1973, **9**, 2160. | 680 |
| 73JPR221 | A. K. Mansour and Y. A. Ibrahim; *J. Prakt. Chem.*, 1973, **315**, 221. | 406 |
| 73JPR690 | J. Fabian; *J. Prakt. Chem.*, 1973, **315**, 690. | 892 |
| 73JST(16)357 | Yu. A. Zhdanov, R. M. Minyaev and V. I. Minkin; *J. Mol. Struct.*, 1973, **16**, 357. | 629 |
| 73K720 | R. Miasmikova, T. C. Davidova and V. I. Simonov; *Kristallografiya*, 1973, **18**, 720. | 622 |
| 73KGS134 | T. K. Sevastyanova and L. B. Volodarskii; *Khim. Geterotsikl. Soedin.*, 1973, 134. | 412, 438 |
| 73KGS902 | B. L. Mil'man and A. A. Potekhim; *Khim. Geterotsikl. Soedin.*, 1973, 902. | 1056 |
| 73LA103 | H.-J. Willenbrock, H. Wamhoff and F. Korte; *Liebigs Ann. Chem.*, 1973, 103. | 338, 361 |
| 73LA1082 | H. Braun and W. Pfleiderer; *Liebigs Ann. Chem.*, 1973, 1082. | 285 |
| 73LA1091 | H. Braun and W. Pfleiderer; *Liebigs Ann. Chem.*, 1973, 1091. | 285 |
| 73LA1099 | H. Braun and W. Pfleiderer; *Liebigs Ann. Chem.*, 1973, 1099. | 285 |
| 73LA1963 | H. Neunhoeffer and G. Frey; *Liebigs Ann. Chem.*, 1973, 1963. | 419 |
| 73LA1970 | E. Oeser; *Liebigs Ann. Chem.*, 1973, 1970. | 386 |
| 73MI21300 | M. Hubert-Habart, C. Péne, G. Bastian and R. Royer; *Chim. Ther.*, 1973, 314. | 121 |
| 73MI21301 | O. P. Shkurko, S. G. Baram and V. P. Mamaev; *Izv. Sib. Otd. Akad. Nauk SSSR, Ser. Khim. Nauk*, 1973, 81. | 140 |
| B-73MI21302 | 'Synthetic Procedures in Nucleic Acid Chemistry', ed. W. W. Zorbach and R. S. Tipson; Wiley, New York, 1973, vol. 2. | 142 |
| 73MI21303 | I. P. Baumel, B. B. Gallagher, J. DiMicco and H. Goico; *J. Pharmacol. Exp. Therap.*, 1973, **186**, 305. | 153 |
| B-73MI21304 | A. Albert; in 'Synthetic Procedures in Nucleic Acid Chemistry', ed. W. W. Zorbach and R. S. Tipson; Wiley, New York, 1973, vol. 2, p. 1. | 60 |
| B-73MI21305 | A. Albert; in 'Synthetic Procedures in Nucleic Acid Chemistry', ed. W. W. Zorbach and R. S. Tipson; Wiley, New York, 1973, vol. 2, p. 47. | 65 |
| 73MI21400 | J. A. Maga and C. E. Sizer; *J. Agric. Food Chem.*, 1973, **21**, 22. | 193 |
| 73MI21401 | J. W. Wheeler and M. S. Blum; *Science*, 1973, **182**, 502. | 193 |
| 73MI21500 | J. Pomorski and H. J. Den Hertog; *Rocz. Chem.*, 1973, **47**, 549 (*Chem. Abstr.*, 1973, **79**, 66 220). | 218 |
| 73MI21501 | R. Albrecht and G. A. Hoyer; *Chim. Ther.*, 1973, **8**, 346 (*Chem. Abstr.*, 1974, **80**, 66 923). | 249, 250, 251 |
| 73MI21900 | C. H. Schwalbe and W. Saenger; *J. Mol. Biol.*, 1973, **75**, 129. | 386 |
| 73MI22000 | E. Anders, R. Hayatsu and M. H. Studier; *Science*, 1973, **182**, 781. | 524 |
| 73MI22001 | K. B. McCredie, G. P. Bodey, M. A. Burgess, J. U. Gutterrman, V. Rodriguez, M. P. Sullivan and E. T. Friereich; *Cancer Chemother. Rep.*, 1973, **57**, 319. | 526 |
| 73MI22200 | G. A. Creak, S. E. Liew and S. F. Tan; *J. Singapore Natl. Acad. Sci.*, 1973, **3**, 223 (*Chem. Abstr.*, 1976, **85**, 62 513). | 576, 639, 640 |
| 73MI22201 | N. K. Ray and V. K. Ahuja; *Photochem. and Photobiol.*, 1973, **17**, 347. | 576 |
| 73MI22202 | M. Y. Y. Chan and S. W. Tam; *J. Chin. Univ. Hong Kong*, 1973, **1**, 239 (*Chem. Abstr.*, 1976, **84**, 73 196). | 618 |
| 73MI22203 | Yu. P. Egorov, Yu. Ya. Borovikov, L. V. Papp and A. I. Tolmachev; *Teor. Eksp. Khim.*, 1973, **9**, 232. | 638 |
| 73MI22300 | M. Hubert-Habart, C. Pene and R. Royer; *Chim. Ther.*, 1973, **8**, 314. | 685 |
| 73MI22301 | S. M. Glozman, N. S. Tolmacheva, L. A. Zhmurenko and V. A. Zagorevskii; *Khim.-Farm. Zh.*, 1973, **7**, 9. | 697 |
| 73MI22302 | K. Kostka; *Rocz. Chem.*, 1973, **47**, 305. | 701 |
| 73MI22303 | M. Nakamura and A. Murata; *Bunseki Kagaku*, 1973, **22**, 1474. | 707 |
| 73MI22305 | K. Kostka; *Rocz. Chem.*, 1973, **47**, 841. | 700 |
| 73MI22306 | G. P. Ellis and G. Barker; *Prog. Med. Chem.*, 1973, **9**, 65. | 693 |
| B-73MI22400 | F. M. Dean; in 'The Total Synthesis of Natural Products', ed. J. W. ApSimon; Wiley, New York, 1973, vol. 1, p. 467. | 738, 741 |
| 73MI22401 | T. C. Chadwick; *Anal. Chem.*, 1973, **45**, 985. | 861 |
| B-73MI22402 | R. K. Razdan; in 'Progress in Organic Chemistry', ed. W. Carruthers and J. L. Sutherland; Butterworth, London, 1973, vol. 8, p. 78. | 877 |
| B-73MI22403 | K. H. Drexhage; in 'Topics in Applied Physics', ed. F. P. Schafer; Springer, Berlin, 1973, chap. 4. | 879 |
| 73MI22404 | M. A. Khan; *Prog. Med. Chem.*, 1973, **9**, 117. | 816 |
| 73MI22500 | R. Borsdorf and H. Remane; *Wiss. Z. Karl-Marx-Univ. Leipzig, Math.-Naturwiss. Reihe*, 1973, **22**, 561 (*Chem. Abstr.*, 1974, **81**, 37 144). | 888 |
| 73MI22600 | J.-Y. T. Chen; *J. Assn. Off. Anal. Chem.*, 1973, **56**, 962 (*Chem. Abstr.*, 1973, **79**, 101 195). | 956 |
| 73MI22601 | G. Geiseler and J. Sawistowsky; *Z. Phys. Chem. (Leipzig)*, 1973, **253**, 333. | 960 |
| 73MI22602 | G. Vasiliu and I. Baciu; *Rev. Chim. (Bucharest)*, 1973, **24**, 413 (*Chem. Abstr.*, 1974, **80**, 82 843). | 975 |
| B-73NMR | T. J. Batterham; NMR Spectra of Simple Heterocycles', Wiley, New York, 1973. | 6, 7, 62, 63, 265, 584 |
| 73OMR(5)251 | L. Radics and J. Kardos; *Org. Magn. Reson.*, 1973, **5**, 251. | 581, 585, 640 |

| | | |
|---|---|---|
| 73OMS(7)667 | P. Arends and P. Helboe; *Org. Mass Spectrom.*, 1973, **7**, 667. | 616 |
| 73OMS(7)737 | J. Clark and A. E. Cunliffe; *Org. Mass Spectrom.*, 1973, **7**, 737. | 285 |
| 73OMS(7)949 | J. Jalonen, P. Pasanen and K. Pihlaja; *Org. Mass Spectrom.*, 1973, **7**, 949. | 961 |
| 73OSC(5)450 | T. L. Emmick and R. L. Letsinger; *Org. Synth., Coll. Vol.*, 1973, **5**, 450. | 813 |
| 73OSC(5)1106 | A. T. Balaban and C. D. Nenitzescu; *Org. Synth., Coll. Vol.*, 1973, **5**, 1106. | 862, 868 |
| 73OSC(5)1135 | K. Dimroth, C. Reichardt and K. Vogel; *Org. Synth., Coll. Vol.*, 1973, **5**, 1135. | 869 |
| 73RTC123 | J. Bus, Th. J. Liefkens and W. Schwaiger; *Recl. Trav. Chim. Pays-Bas*, 1973, **92**, 123. | 161 |
| 73RTC667 | R. A. Van der Welle and L. Brandsma; *Recl. Trav. Chim. Pays-Bas*, 1973, **92**, 667. | 893 |
| 73RTC708 | P. J. Lont, H. C. Van der Plas and A. Van Veldhuizen; *Recl. Trav. Chim. Pays-Bas*, 1973, **92**, 708. | 165 |
| 73RTC845 | H. Polman, A. Mosterd and H. J. T. Bos; *Recl. Trav. Chim. Pays-Bas*, 1973, **92**, 845. | 706 |
| 73RTC1326 | J. Meijer, P. Vermeer, H. D. Verkruijsse and L. Brandsma; *Recl. Trav. Chim. Pays-Bas*, 1973, **92**, 1326. | 982 |
| 73S189 | S. Yanagida and S. Komori; *Synthesis*, 1973, 189. | 514 |
| 73S243 | H. Petersen; *Synthesis*, 1973, 243. | 490, 520, 1043 |
| 73S536 | M. Furukawa and S. Hayashi; *Synthesis*, 1973, 536. | 492 |
| 73SC225 | H. Iida, K. Hayashida, M. Yamada, T. Takahashi and K. Yamada; *Synth. Commun.*, 1973, 225. | 185 |
| 73SC231 | D. J. Goldsmith and C. T. Helmes, Jr.; *Synth. Commun.*, 1973, **3**, 231. | 675, 872 |
| 73SC397 | E. E. Garcia; *Synth. Commun.*, 1973, **3**, 397. | 229 |
| 73T795 | J. R. Wilt, G. A. Reynolds and J. A. Van Allan; *Tetrahedron*, 1973, **29**, 795. | 663 |
| 73T909 | G. Snatzke, F. Snatzke, A. L. Tökés, M. Rákosi and R. Bognár; *Tetrahedron*, 1973, **29**, 909. | 631 |
| 73T1145 | E. Britmaier and K. H. Spohn; *Tetrahedron*, 1973, **27**, 1145. | 160 |
| 73T1317 | P. Margaretha; *Tetrahedron*, 1973, **29**, 1317. | 720, 730 |
| 73T2009 | S. Yoneda, T. Sugimoto and Z. Yoshida; *Tetrahedron*, 1973, **29**, 2009. | 893 |
| 73T2209 | E. Stark and E. Breitmaier; *Tetrahedron*, 1973, **29**, 2209. | 202, 204, 230 |
| 73T2495 | W. W. Paudler, J. Lee and T.-K. Chen; *Tetrahedron*, 1973, **29**, 2495. | 401 |
| 73T3071 | C. Weiss, F. Hoppner, S. Becker and W. Blaschke; *Tetrahedron*, 1973, **29**, 3071. | 253 |
| 73T3761 | T. P. Karpetsky and E. H. White; *Tetrahedron*, 1973, **29**, 3761. | 187 |
| 73T3939 | A. F. Bramwell and R. D. Wells; *Tetrahedron*, 1973, **29**, 3939. | 160 |
| 73T4083 | D. G. I. Kingston and H. M. Fales; *Tetrahedron*, 1973, **29**, 4083. | 615 |
| 73T4149 | J. Kattenberg, E. R. de Waard and H. O. Huisman; *Tetrahedron*, 1973, **29**, 4149. | 901 |
| 73TL219 | H. Neunhoeffer, H.-D. Vötter and M. Gais-Mutterer; *Tetrahedron Lett.*, 1973, 219. | 378 |
| 73TL295 | M. L. Kaplan and G. N. Taylor; *Tetrahedron Lett.*, 1973, 295. | 962 |
| 73TL849 | M. J. Cook and A. P. Tonge; *Tetrahedron Lett.*, 1973, 849. | 962 |
| 73TL1037 | K. Bock, I. Lundt and C. Pedersen; *Tetrahedron Lett.*, 1973, 1037. | 586 |
| 73TL1429 | H. Neunhoeffer and V. Böhnisch; *Tetrahedron Lett.*, 1973, 1429. | 443 |
| 73TL2783 | J. K. Rasmussen and A. Hassner; *Tetrahedron Lett.*, 1973, 2783. | 1061 |
| 73TL4155 | S. Bory and A. Marquet; *Tetrahedron Lett.*, 1973, 4155. | 897 |
| 73TL4315 | E. R. de Waard, H. R. Reus and H. O. Huisman; *Tetrahedron Lett.*, 1973, 4315. | 929 |
| 73TL4547 | P. Ahern, T. Navratil and K. Vaughan; *Tetrahedron Lett.*, 1973, 4547. | 379 |
| 73TL5003 | K. K. Balasubramanian, K. V. Reddy and R. Nagarajan; *Tetrahedron Lett.*, 1973, 5003. | 743 |
| 73TL5073 | S. C. Gupta and S. K. Mukerjee; *Tetrahedron Lett.*, 1973, 5073. | 695 |
| 73UP21800 | H. Neunhoeffer and H.-D. Vötter; unpublished results, 1973. | 376 |
| 73YZ330 | T. Koyama, T. Hirota, M. Yamato and N. Ohta; *Yakugaku Zasshi*, 1973, **93**, 330 (*Chem. Abstr.*, 1973, **78**, 159 543). | 220 |
| 73YZ1043 | S. Yurugi, K. Ito, A. Miyake and K. Omura; *Yakugaku Zasshi*, 1973, **93**, 1043. | 346 |
| 73ZC130 | M. Weissenfels and S. Illing; *Z. Chem.*, 1973, **13**, 130. | 935 |
| 73ZC175 | M. Weissenfels, M. Pulst and P. Schneider; *Z. Chem.*, 1973, **13**, 175. | 811 |
| 73ZC408 | M. Just and K.-H. Uhl; *Z. Chem.*, 1973, **13**, 408. | 490, 498 |
| 73ZC424 | W. Schroth, H. Bahn and R. Zschernitz; *Z. Chem.*, 1973, **13**, 424. | 982 |
| 73ZC431 | E. Fanghaenel, R. Ebisch and B. Adler; *Z. Chem.*, 1973, **13**, 431. | 1064 |
| 73ZN(B)535 | T. Eicher and R. Graf; *Z. Naturforsch., Teil B*, 1973, **28**, 535. | 382 |
| 73ZOR775 | A. P. Monsenko; *Zh. Org. Khim.*, 1973, **9**, 775 (*Chem. Abstr.*, 1973, **79**, 31 797). | 930 |
| 73ZOR834 | Yu. V. Svetkin, A. N. Minlibaeva, N. Kh. Khamitova and L. V. Zamyatina; *Zh. Org. Khim.*, 1973, **9**, 834. | 437 |
| 73ZOR1748 | A. P. Krasnoshchek, T. P. Medeva, G. V. Esipov, N. I. Gusar and A. G. Galushko; *Zh. Org. Khim.*, 1973, **9**, 1748. | 617 |
| 73ZOR2038 | L. N. Markovskii, E. A. Darmokhval and E. S. Levchenko; *Zh. Org. Khim.*, 1973, **9**, 2038. | 1075 |
| 74ABC2545 | Y. Ishikawa; *Agric. Biol. Chem.*, 1974, **38**, 2545. | 732 |
| 74ACS(B)517 | K. Undheim and C. E. Carlberg; *Acta Chem. Scand., Ser. B*, 1974, **28**, 517. | 758, 873 |
| 74ACS(B)1234 | E. Taskinen; *Acta Chem. Scand., Ser. B*, 1974, **28**, 1234. | 642 |
| 74AG(E)484 | E. Buschmann and W. Steglich; *Angew. Chem., Int. Ed. Engl.*, 1974, **13**, 484. | 1003 |
| 74AG(E)533 | W. Steglich, E. Buschmann and O. Hollitzer; *Angew. Chem., Int. Ed. Engl.*, 1974, **13**, 533. | 1023 |
| 74AHC(16)123 | F. Minisci and O. Porta; *Adv. Heterocycl. Chem.*, 1974, **16**, 123. | 290 |

| | | |
|---|---|---|
| 74AHC(17)255 | M. J. Cook, A. R. Katritzky and P. Linda; *Adv. Heterocyclic Chem.*, 1974, **17**, 255. | 633, 635, 639, 959 |
| 74AJC2251 | D. J. Brown and P. Waring; *Aust. J. Chem.*, 1974, **27**, 2251. | 70, 100, 101 |
| 74AJC2627 | C. P. Joshua and V. P. Rajan; *Aust. J. Chem.*, 1974, **27**, 2627. | 499 |
| 74AP12 | F. Eiden and W. Luft; *Arch. Pharm. (Weinheim, Ger.)*, 1974, **307**, 12. | 821 |
| 74AP218 | H. Böhme, U. Sitorus and F. Ziegler; *Arch. Pharm. (Weinheim. Ger.)*, 1974, **307**, 218. | 909 |
| 74AX(B)154 | J. S. Cantrell, R. A. Stalzer and T. L. Becker; *Acta Crystallogr., Part B*, 1974, **30**, 154. | 624 |
| 74AX(B)1351 | E. Gavuzzo, F. Mazza and E. Giglio; *Acta Crystallogr., Part B*, 1974, **30**, 1351. | 622 |
| 74AX(B)1430 | P. Singh and D. J. Hodgson; *Acta Crystallogr., Part B*, 1974, **30**, 1430. | 386 |
| 74BCJ509 | K. Hanaya, S. Onodera, S. Awano and H. Kudo; *Bull. Chem. Soc. Jpn.*, 1974, **47**, 509. | 594 |
| 74BCJ832 | T. Tamamura, T. Yamane, N. Yasuoka and N. Kasai; *Bull. Chem. Soc. Jpn.*, 1974, **47**, 832. | 625 |
| 74BCJ1960 | K. Maruyama and S. Arakawa; *Bull. Chem. Soc. Jpn.*, 1974, **47**, 1960. | 671 |
| 74BCJ2607 | K. Hanaya, S. Onodera and H. Kudo; *Bull. Chem. Soc. Jpn.*, 1974, **47**, 2607. | 630 |
| 74BCJ2893 | M. Furukawa, T. Yoshida and S. Hayashi; *Bull. Chem. Soc. Jpn.*, 1974, **47**, 2893. | 493 |
| 74BRP1350277 | A. C. Barker and R. G. Foster; *Br. Pat.* 1 350 277 (1974) (*Chem. Abstr.*, 1974, **81**, 25 669 | 1067 |
| 74BSF183 | G. Saint-Ruf and B. Lobert; *Bull. Soc. Chim. Fr.*, 1974, 183. | 974 |
| 74BSF538 | J. Gayoso, H. Bouanani and A. Boucekkine; *Bull. Soc. Chim. Fr.*, 1974, 538. | 575, 576, 634 |
| 74BSF545 | H. Bouanani and J. Gayoso; *Bull. Soc. Chim. Fr.*, 1974, 545. | 634, 636, 638 |
| 74BSF590 | C. Leroy, M. Martin and L. Bassery; *Bull. Soc. Chim. Fr., Part 2*, 1974, 590. | 902 |
| 74BSF999 | J. Daunis; *Bull. Soc. Chim. Fr.*, 1974, 999. | 406 |
| 74BSF1196 | J. P. Sauvé and N. Lozac'h; *Bull. Soc. Chim. Fr., Part 2*, 1974, 1196. | 924 |
| 74BSF1395 | A. Étienne, A. LeBarre and J.-P. Giorgetti; *Bull. Soc. Chim. Fr.*, 1974, 1395. | 1069 |
| 74BSF1453 | P. Guerret, R. Jacquier, H. Lopez and G. Maury; *Bull. Soc. Chim. Fr.*, 1974, 1453. | 438 |
| 74BSF1497 | A. Étienne and B. Bonte; *Bull. Soc. Chim. Fr.*, 1974, 1497. | 1081 |
| 74BSF1917 | A. Étienne, A. LeBarre and J.-P. Giorgetti; *Bull. Soc. Chim. Fr.*, 1974, 1917. | 1058 |
| 74BSF2086 | J. Ficini and J. P. Genêt; *Bull. Soc. Chim. Fr.*, 1974, 2086. | 814 |
| 74BSF2963 | A. De, J. P. Poupelin and G. Saint-Ruf; *Bull. Soc. Chim. Fr.*, 1974, 2963. | 674 |
| 74CB1 | H. W. Roesky and B. Kuhtz; *Chem. Ber.*, 1974, **107**, 1. | 1047, 1048, 1051, 1079 |
| 74CB68 | G. Wagner and H. Bock; *Chem. Ber.*, 1974, **107**, 68. | 959 |
| 74CB339 | M. Ott and W. Pfleiderer; *Chem. Ber.*, 1974, **107**, 339. | 282, 297 |
| 74CB367 | D. Seebach, E. J. Corey and A. K. Beck; *Chem. Ber.*, 1974, **107**, 367. | 977 |
| 74CB575 | K. Eistetter and W. Pfleiderer; *Chem. Ber.*, 1974, **107**, 575. | 319 |
| 74CB605 | H. Böhme and F. Ziegler; *Chem. Ber.*, 1974, 605. | 909 |
| 74CB771 | K.-H. Linke and R. Bimczok; *Chem. Ber.*, 1974, **107**, 771. | 1046, 1051, 1078 |
| 74CB785 | W. Pfleiderer; *Chem. Ber.*, 1974, **107**, 785. | 307 |
| 74CB876 | U. Ewers, H. Günther and L. Jaenicke; *Chem. Ber.*, 1974, **107**, 876. | 285 |
| 74CB2537 | E. Stark, E. Kraas, F. S. Tjoeng, G. Jung and E. Brietmaier; *Chem. Ber.*, 1974, **107**, 2537. | 202, 229, 230, 231 |
| 74CB3275 | U. Ewers, H. Günther and L. Jaenicke; *Chem. Ber.*, 1974, **107**, 3275. | 285 |
| 74CB3377 | H. J. Schneider and W. Pfleiderer; *Chem. Ber.*, 1974, **107**, 3377. | 273, 297 |
| 74CC140 | D. C. Appleton, D. C. Bull, J. M. McKenna and A. R. Walley; *J. Chem. Soc., Chem. Commun.*, 1974, 140. | 899 |
| 74CC308 | J. Clark and B. Parvizi; *J. Chem. Soc., Chem. Commun.*, 1974, 308. | 202 |
| 74CC318 | R. J. Molyneux; *J. Chem. Soc., Chem. Commun.*, 1974, 318. | 735 |
| 74CC485 | T. L. Gilchrist, C. J. Harris and C. W. Rees; *J. Chem. Soc., Chem. Commun.*, 1974, 485. | 413, 453 |
| 74CC614 | M. G. B. Drew, G. W. A. Fowles, D. A. Rice and K. J. Shanton; *J. Chem. Soc., Chem. Commun.*, 1974, 614. | 625 |
| 74CC632 | A. I. Gray, R. D. Waigh and P. G. Waterman; *J. Chem. Soc., Chem. Commun.*, 1974, 632. | 582 |
| 74CC739 | K. Georgiou, H. W. Kroto and B. M. Landsberg; *J. Chem. Soc., Chem. Commun.*, 1974, 739. | 979 |
| 74CC782 | R. E. Moerck and M. A. Battiste; *J. Chem. Soc., Chem. Commun.*, 1974, 782. | 554 |
| 74CHE1019 | S. Krivan, V. I. Dulenko, S. V. Sayapina, N. S. Semenov, Y. A. Nikolyukin and J. N. Baranov; *Chem. Heterocycl. Compd. (Engl. Transl.)*, 1974, **10**, 1019. | 653 |
| 74CHE1218 | Yu. L. Frolov, Yu. M. Sapozhnikov, S. S. Barer and L. V. Sherstyannikova; *Chem. Heterocycl. Compd. (Engl. Transl.)*, 1974, **10**, 1218. | 575 |
| 74CI(L)233 | A. J. Pearson; *Chem. Ind. (London)*, 1974, 233. | 306, 308 |
| 74CJC2041 | W. A. Szarek, D. M. Vyas, A.-M. Sepulchre, S. D. Gero and G. Lukacs; *Can. J. Chem.*, 1974, **52**, 2041. | 960 |
| 74CJC3879 | K. Oyama and R. Stewart; *Can. J. Chem.*, 1974, **52**, 3879. | 303 |
| 74CJC3884 | R. Stewart and K. Oyama; *Can. J. Chem.*, 1974, **52**, 3884. | 303 |
| 74CL1101 | M. Hori, T. Kataoka, H. Shimizu, K. Narita, S. Ohno and H. Aoki; *Chem. Lett.*, 1974, 1101. | 895, 925 |
| 74CPB305 | A. Takamizawa and S. Matsumoto; *Chem. Pharm. Bull.*, 1974, **22**, 305. | 339 |
| 74CPB1732 | K. Imada; *Chem. Pharm. Bull.*, 1974, **22**, 1732. | 45 |

| 74CPB1765 | A. Takamizawa and I. Makino; *Chem. Pharm. Bull.*, 1974, **22**, 1765. | 338 |
| 74CPB1864 | H. Nakao, M. Fukushima, H. Yanagisawa and S. Sugawara; *Chem. Pharm. Bull.*, 1974, **22**, 1864. | 256, 262 |
| 74CPB2959 | A. Nohara, T. Umetani, K. Ukawa and Y. Sanno; *Chem. Pharm. Bull.*, 1974, **22**, 2959. | 707 |
| 74CR(C)(278)427 | B. Duchesnay, A. Decormeille, G. Queguiner and P. Pastour; *C.R. Hebd. Seances Acad. Sci., Ser. C*, 1974, **278**, 427. | 205, 206, 209, 211, 214 |
| 74CR(C)(278)721 | O. Riobé; *C.R. Hebd. Seances Acad. Sci., Ser. C*, 1974, **278**, 721. | 769 |
| 74CR(C)(278)1421 | L. Godefroy, A. Decormeille, G. Queguiner and P. Pastour; *C.R. Hebd. Seances Acad. Sci., Ser. C*, 1974, **278**, 1421. | 215, 225 |
| 74CR(C)(279)151 | B. S. Kirkiacharian, G. H. Elia and G. Mahuzier; *C.R. Hebd. Seances Acad. Sci., Ser. C*, 1974, **279**, 151. | 828 |
| 74CR(C)(279)697 | R. Hubaut and J. Landais; *C.R. Hebd. Seances Acad. Sci., Ser. C*, 1974, **279**, 697. | 861 |
| 74CRV581 | E. H. Cordes and H. G. Bull; *Chem. Rev.*, 1974, **74**, 581. | 978 |
| 74FES247 | A. Ermili, G. Roma and A. Balbi; *Farmaco Ed. Sci.*, 1974, **29**, 247. | 697 |
| 74FES910 | A. Monzani, M. DiBella, U. Fabio and G. Manicardi; *Farmaco Ed. Sci.*, 1974, **29**, 910. | 1085 |
| 74G849 | C. Giordano; *Gazz. Chim. Ital.*, 1974, **104**, 849. | 1019 |
| 74GEP2241259 | K. Leverenz and K. H. Schündehütte; *Ger. Pat.* 2 241 259 (1974) (*Chem. Abstr.*, 1974, **81**, 65 225). | 450 |
| 74GEP2311708 | K. Grohe; *Ger. Pat.* 2 311 708 (1974) (*Chem. Abstr.*, 1975, **82**, 4266). | 1031 |
| 74GEP2327545 | D. Seebach, R. Dach, D. Enders and B. Renger; *Ger. Pat.* 2 327 545 (1974) (*Chem. Abstr.*, 1975, **83** 10 166). | 533, 534 |
| 74GEP2348111 | B. W. Dominy and M. J. A. El-Haj; (Pfizer Inc.), *Ger. Offen.* 2 348 111 (1974) (*Chem. Abstr.*, 1974, **81**, 37 571). | 222 |
| 74GEP2362816 | Yamanouchi Pharmaceuticals KK, *Ger. Pat.* 2 362 816 (1974) (*Chem. Abstr.*, 1974, **81**, 105 537). | 941 |
| 74GEP2365302 | Pfizer Inc., *Ger. Offen.* 2 365 302 (1974) (*Chem. Abstr.*, 1974, **81**, 49 575). | 216 |
| 74HCA1485 | R. Weber, W. Frick and M. Viscontini; *Helv. Chim. Acta*, 1974, **57**, 1485. | 281 |
| 74HCA1651 | J. H. Bieri and M. Viscontini; *Helv. Chim. Acta*, 1974, **57**, 1651. | 281 |
| 74HCA2658 | W. Frick, R. Weber and M. Viscontini; *Helv. Chim. Acta*, 1974, **57**, 2658. | 281 |
| 74IJC26 | M. Ahuja, M. Bandopadhyay and T. R. Seshadri; *Indian J. Chem.*, 1974, **12**, 26. | 829 |
| 74IJC281 | J. N. Chatterjea, S. C. Shaw and J. N. Singh; *J. Indian Chem. Soc.*, 1974, **51**, 281. | 722 |
| 74IJC474 | S. Nizamuddin and M. Ghosal; *Indian J. Chem.*, 1974, **12**, 474. | 832 |
| 74IJC564 | A. Roy, A. Das Gupta and K. Sen; *Indian J. Chem.*, 1974, **12**, 564. | 848 |
| 74IJC1259 | J. N. Chatterjea, C. Bhakta and J. Mukherjee; *Indian J. Chem.*, 1974, **12**, 1259. | 832 |
| 74IZV2204 | A. I. Bogatyreva, O. M. Polumbrik and A. L. Buchachenko; *Izv. Akad. Nauk SSSR, Ser. Khim.*, 1974, 2204. | 549 |
| 74JA576 | R. A. Y. Jones, A. R. Katritzky, A. R. Martin, D. L. Ostercamp, A. C. Richards and J. M. Sullivan; *J. Am. Chem. Soc.*, 1974, **96**, 576. | 538 |
| 74JA935 | F. Jung, M. Molin, R. Van Den Elzen and T. Durst; *J. Am. Chem. Soc.*, 1974, **96**, 935. | 981 |
| 74JA1239 | P. Singh and D. J. Hodgson; *J. Am. Chem. Soc.*, 1974, **96**, 1239. | 387 |
| 74JA1591 | C. H. Bushweller, M. Z. Lourandos and J. A. Brunelle; *J. Am. Chem. Soc.*, 1974, **96**, 1591. | 467 |
| 74JA1807 | E. L. Eliel, A. A. Hartmann and A. G. Abatjoglou; *J. Am. Chem. Soc.*, 1974, **96**, 1807. | 961 |
| 74JA1843 | M. J. Cho and I. H. Pitman; *J. Am. Chem. Soc.*, 1974, **96**, 1843. | 287 |
| 74JA2916 | S. F. Nelsen, G. R. Weisman, P. J. Hintz, D. Olp and M. R. Fahey; *J. Am. Chem. Soc.*, 1974, **96**, 2916. | 332 |
| 74JA3502 | W. Kusters and P. de Mayo; *J. Am. Chem. Soc.*, 1974, **96**, 3502. | 982 |
| 74JA4690 | J. I. Zink and W. Klimt; *J. Am. Chem. Soc.*, 1974, **96**, 4690. | 600 |
| 74JA4699 | A. M. Trozzolo, A. Dienes and C. V. Shank; *J. Am. Chem. Soc.*, 1974, **96**, 4699. | 601, 643 |
| 74JA4879 | J. S. Swenton and J. A. Hyatt; *J. Am. Chem. Soc.*, 1974, **96**, 4879. | 429 |
| 74JA5648 | G. H. Senkler, Jr., J. Stackhouse, B. E. Maryanoff and K. Mislow; *J. Am. Chem. Soc.*, 1974, **97**, 5648. | 894 |
| 74JA5650 | G. H. Senkler, Jr., J. Stackhouse, B. E. Maryanoff and K. Mislow; *J. Am. Chem. Soc.*, 1974, **97**, 5650. | 894, 925 |
| 74JA5651 | G. H. Senkler, Jr., J. Stackhouse, B. E. Maryanoff and K. Mislow; *J. Am. Chem. Soc.*, 1974, **97**, 5651. | 894 |
| 74JA6119 | A. G. Hortmann, R. L. Harris and J. A. Miles; *J. Am. Chem. Soc.*, 1974, **96**, 6119. | 894, 926 |
| 74JA6781 | E. C. Taylor and P. A. Jacobi; *J. Am. Chem. Soc.*, 1974, **96**, 6781. | 318 |
| 74JA6987 | S. F. Nelsen and J. M. Buschek; *J. Am. Chem. Soc.*, 1974, **96**, 6987. | 332 |
| 74JAP7426274 | Meiji Confectionery Co., *Jpn. Pat.* 74 26 274 (1974) (*Chem. Abstr.*, 1974, **81**, 77 805). | 929, 940 |
| 74JAP7430313 | Sumitomo Chemical Co., *Jpn. Pat.* 74 30 313 (1974) (*Chem. Abstr.*, 1974, **81**, 104 758). | 939 |
| 74JAP(K)7436700 | S. Yurugi and A. Miyake; (Takeda Chem. Ind. Ltd.), *Jpn. Kokai* 74 36 700 (1974) (*Chem. Abstr.*, 1974, **81**, 120 689). | 215 |
| 74JAP(K)7442696 | J. Matsumoto and M. Sugita; (Dainippon Pharm. Co. Ltd.), *Jpn. Kokai* 74 42 696 (1974) (*Chem. Abstr.*, 1975, **82**, 4295). | 221 |
| 74JAP(K)7444000 | S. Minami, J. Matsumoto and M. Sugita; (Dainippon Pharm. Co. Ltd.), *Jpn. Kokai* 74 44 000 (1974) (*Chem. Abstr.*, 1975, **81**, 77 954). | 205 |

| | | |
|---|---|---|
| 74JAP(K)7488897 | M. Nakanishi, K. Arimura and T. Tsumagari; (Yoshitomi Pharm. Ind. Ltd.), *Jpn. Kokai* 74 88 897 (1974) (*Chem. Abstr.*, 1975, **82**, 125 414). | 247 |
| 74JBC(249)4363 | J. S. Olson, D. P. Ballou, G. Palmer and V. Massey; *J. Biol. Chem.*, 1974, **249**, 4363. | 287 |
| 74JCP(60)3098 | J. R. Durig, R. O. Carter and L. A. Carriera; *J. Chem. Phys.*, 1974, **60**, 3098. | 629 |
| 74JCP(60)3987 | J. A. Wells and T. B. Malloy, Jr.; *J. Chem. Phys.*, 1974, **60**, 3987. | 625 |
| 74JCS(P1)305 | L. Farkas, À. Gottsegen, M. Nógrádi and S. Antus; *J. Chem. Soc., Perkin Trans. 1*, 1974, 305. | 824 |
| 74JCS(P1)357 | B. E. Evans; *J. Chem. Soc., Perkin Trans. 1*, 1974, 357. | 287 |
| 74JCS(P1)569 | R. W. Tickle, T. Melton and J. A. Elvidge; *J. Chem. Soc., Perkin Trans. 1*, 1974, 569. | 597, 754, 848 |
| 74JCS(P1)1007 | D. G. Clarke, L. Crombie and D. A. Whiting; *J. Chem. Soc., Perkin Trans. 1*, 1974, 1007. | 745, 746, 747 |
| 74JCS(P1)1225 | H. C. S. Wood, R. Wrigglesworth, D. A. Yeowell, F. W. Gurney and B. S. Hurlbert; *J. Chem. Soc., Perkin Trans. 1*, 1974, 1225. | 202, 215, 229, 230, 260 |
| 74JCS(P1)1272 | L. Benati, P. C. Montevecchi, A. Tundo and G. Zanardi; *J. Chem. Soc., Perkin Trans. 1*, 1974, 1272. | 985, 986 |
| 74JCS(P1)1513 | R. D. Chambers, M. Clark, J. R. Maslakiewicz, W. K. R. Musgrave and P. G. Urben; *J. Chem. Soc., Perkin Trans. 1*, 1974, 1513. | 9, 10 |
| 74JCS(P1)1674 | J. A. Donnelly, S. O'Brien and J. O'Grady; *J. Chem. Soc., Perkin Trans. 1*, 1974, 1674. | 773 |
| 74JCS(P1)1723 | M. Cinquini, S. Colonna and F. Montanari; *J. Chem. Soc., Perkin Trans. 1*, 1974, 1723. | 901 |
| 74JCS(P1)1786 | J. R. Traynor and D. G. Wibberley; *J. Chem. Soc., Perkin Trans. 1*, 1974, 1786. | 517 |
| 74JCS(P1)1812 | J. Clark and M. S. Morton; *J. Chem. Soc., Perkin Trans. 1*, 1974, 1812. | 122, 350 |
| 74JCS(P1)1951 | F. DeSarlo; *J. Chem. Soc., Perkin Trans. 1*, 1974, 1951. | 1045, 1080 |
| 74JCS(P1)1965 | R. G. R. Bacon and S. D. Hamilton; *J. Chem. Soc., Perkin Trans. 1*, 1974, 1965. | 249, 250, 255 |
| 74JCS(P1)2050 | G. Pagani; *J. Chem. Soc., Perkin Trans. 1*, 1974, 2050. | 1060 |
| 74JCS(P1)2092 | J. Alexander, G. Lowe, N. K. McCullum and G. K. Ruffles; *J. Chem. Soc., Perkin Trans. 1*, 1974, 2092. | 1035 |
| 74JCS(P1)2095 | D. H. R. Barton, W. A. Bubb, R. H. Hesse and M. M. Pechet; *J. Chem. Soc., Perkin Trans. 1*, 1974, 2095. | 70 |
| 74JCS(P1)2570 | G. P. Ellis and I. L. Thomas; *J. Chem. Soc., Perkin Trans. 1*, 1974, 2570. | 713, 714, 718, 817 |
| 74JCS(P1)2595 | Z. Rappoport and D. Ladkani; *J. Chem. Soc., Perkin Trans. 1*, 1974, 2595. | 759 |
| 74JCS(P2)80 | J. A. Blair and A. J. Pearson; *J. Chem. Soc., Perkin Trans. 2*, 1974, 80. | 308 |
| 74JCS(P2)204 | D. J. Brown and P. Waring; *J. Chem. Soc., Perkin Trans. 2*, 1974, 204. | 63, 139 |
| 74JCS(P2)420 | M. H. Palmer, R. H. Findlay and A. J. Gaskell; *J. Chem. Soc., Perkin Trans. 2*, 1974, 420. | 533, 536 |
| 74JCS(P2)767 | M. J. Cook and A. P. Tonge; *J. Chem. Soc., Perkin Trans. 2*, 1974, 767. | 954 |
| 74JCS(P2)1561 | R. A. Y. Jones, A. R. Katritzky, A. R. Martin and S. Saba; *J. Chem. Soc., Perkin Trans. 2*, 1974, 1561. | 1053 |
| 74JCS(P2)1737 | R. A. Y. Jones, A. R. Katritzky and S. Saba; *J. Chem. Soc., Perkin Trans. 2*, 1974, 1737. | 999 |
| 74JHC43 | D. K. Krass and W. W. Paudler; *J. Heterocycl. Chem.*, 1974, **11**, 43. | 401 |
| 74JHC51 | G. M. Coppola, G. E. Hardtmann and B. S. Huegi; *J. Heterocycl. Chem.*, 1974, **11**, 51. | 229 |
| 74JHC99 | B. Weinstein and H.-H. Chang; *J. Heterocycl. Chem.*, 1974, **11**, 99. | 1048, 1082 |
| 74JHC139 | H. J. Shine and L. R. Shade; *J. Heterocycl. Chem.*, 1974, **11**, 139. | 968, 969 |
| 74JHC163 | K.-Y. Tserng and L. Bauer; *J. Heterocycl. Chem.*, 1974, **11**, 163. | 208 |
| 74JHC183 | S. Klutchko, M. P. Cohen, J. Shavel, Jr. and M. von Strandtmann; *J. Heterocycl. Chem.*, 1974, **11**, 183. | 821 |
| 74JHC199 | J. Kobe, R. K. Robins and D. E. O'Brien; *J. Heterocycl. Chem.*, 1974, **11**, 199. | 491 |
| 74JHC279 | J. P. Piper and J. A. Montgomery; *J. Heterocycl. Chem.*, 1974, **11**, 279. | 304, 311 |
| 74JHC287 | G. Saint-Ruf, J. Servoin-Sidoine and J. P. Coïc; *J. Heterocycl. Chem.*, 1974, **11**, 287. | 957 |
| 74JHC317 | M. Bessière-Chrétien and H. Serne; *J. Heterocycl. Chem.*, 1974, **11**, 317. | 462, 474 |
| 74JHC351 | J. Nematollahi, S. Kasina and D. Maness; *J. Heterocycl. Chem.*, 1974, **11**, 351. | 234, 237 |
| 74JHC395 | J. A. Van Allan and G. A. Reynolds; *J. Heterocycl. Chem.*, 1974, **11**, 395. | 873 |
| 74JHC405 | G. A. Reynolds and J. A. Van Allan; *J. Heterocycl. Chem.*, 1974, **11**, 405. | 872 |
| 74JHC453 | J. S. Swenton and R. J. Balchunis; *J. Heterocycl. Chem.*, 1974, **11**, 453. | 429 |
| 74JHC627 | J. T. Shaw, M. E. O'Connor, R. C. Allen, W. M. Westler and B. D. Stefanko; *J. Heterocycl. Chem.*, 1974, **11**, 627. | 521 |
| 74JHC743 | J. L. Atwood, D. K. Krass and W. W. Paudler; *J. Heterocycl. Chem.*, 1974, **11**, 743. | 386 |
| 74JHC917 | J. S. Swenton and R. J. Balchunis; *J. Heterocycl. Chem.*, 1974, **11**, 917. | 429 |
| 74JHC991 | J. Kobe, D. E. O'Brien, R. K. Robins and T. Novinson; *J. Heterocycl. Chem.*, 1974, **11**, 991. | 488, 494 |
| 74JHC1065 | J. A. Van Allan and S. C. Chang; *J. Heterocycl. Chem.*, 1974, **11**, 1065. | 665, 700, 872 |
| 74JHC1075 | G. A. Reynolds and J. A. Van Allan; *J. Heterocycl. Chem.*, 1974, **11**, 1075. | 872 |
| 74JHC1081 | A. Rosowsky and N. Papathanasopoulos; *J. Heterocycl. Chem.*, 1974, **11**, 1081. | 227 |
| 74JHC1101 | H. N. Al-Jallo and F. W. Al-Azawi; *J. Heterocycl. Chem.*, 1974, **11**, 1101. | 812 |
| 74JIC281 | J. N. Chatterjea, S. C. Shaw and J. N. Singh; *J. Indian Chem. Soc.*, 1974, **51**, 281. | 829 |
| 74JMC223 | M. G. Nair and C. M. Baugh; *J. Med. Chem.*, 1974, **17**, 223. | 327 |
| 74JMC451 | H. A. Burch, L. E. Benjamin, H. E. Russell and R. Freedman; *J. Med. Chem.*, 1974, **17**, 451. | 349 |

| | | |
|---|---|---|
| 74JMC470 | J. I. DeGraw, R. L. Kisliuk, Y. Gaumont and C. M. Baugh; *J. Med. Chem.*, 1974, **17**, 470. | 260 |
| 74JMC549 | W. L. Matier, W. T. Comer and A. W. Gomoli; *J. Med. Chem.*, 1974, **17**, 549. | 1072 |
| 74JMC553 | R. D. Elliott, C. Temple, Jr. and J. A. Montgomery; *J. Med. Chem.*, 1974, **17**, 553. | 252, 261 |
| 74JMC636 | G. E. Hartmann, B. Huegi, G. Koletar, S. Kroin, H. Ott, J. W. Perrine and E. I. Takesue; *J. Med. Chem.*, 1974, **17**, 636. | 223 |
| 74JMC1268 | M. G. Nair, L. P. Mercer and C. M. Baugh; *J. Med. Chem.*, 1974, **17**, 1268. | 327 |
| 74JMC1272 | A. Rosowsky and N. Papathanasopoulos; *J. Med. Chem.*, 1974, **17**, 1272. | 231 |
| 74JOC47 | J. C. Hinshaw; *J. Org. Chem.*, 1974, **40**, 47. | 332 |
| 74JOC103 | N. Ishibe, K. Hashimoto and M. Sunami; *J. Org. Chem.*, 1974, **39**, 103. | 918 |
| 74JOC162 | M. A. Calcagno, H. W. Heine, C. Kruse and W. A. Kofke; *J. Org. Chem.*, 1974, **39**, 162. | 1080 |
| 74JOC279 | J. H. Block, D. H. Smith and C. Djerassi; *J. Org. Chem.*, 1974, **39**, 279. | 616 |
| 74JOC843 | K. Migliorese and S. I. Miller; *J. Org. Chem.*, 1974, **39**, 843. | 791 |
| 74JOC881 | W. K. Anderson, E. J. LaVoie and P. G. Whitkop; *J. Org. Chem.*, 1974, **39**, 881. | 744, 745 |
| 74JOC940 | E. M. Burgess and J. P. Sanchez; *J. Org. Chem.*, 1974, **39**, 940. | 380 |
| 74JOC989 | I. Belsky, H. Dodiuk and Y. Shvo; *J. Org. Chem.*, 1974, **39**, 989. | 699 |
| 74JOC1349 | A. T. Nielsen, R. L. Atkins, J. DiPol and D. W. Moore; *J. Org. Chem.*, 1974, **39**, 1349. | 509 |
| 74JOC1819 | D. L. Klayman and T. S. Woods; *J. Org. Chem.*, 1974, **39**, 1819. | 502 |
| 74JOC1935 | W. V. Turner and W. H. Pirkle; *J. Org. Chem.*, 1974, **39**, 1935. | 587, 588, 589, 635 |
| 74JOC2040 | J. A. Hirsch and G. Schwartzkopf; *J. Org. Chem.*, 1974, **39**, 2040. | 630, 778 |
| 74JOC2112 | D. L. Wooton and W. G. Lloyd; *J. Org. Chem.*, 1974, **39**, 2112. | 975 |
| 74JOC2591 | H. L. Nyquist and B. Wolfe; *J. Org. Chem.*, 1974, **39**, 2591. | 510 |
| 74JOC2710 | A. McKillop and R. J. Kobylecki; *J. Org. Chem.*, 1974, **39**, 2710. | 372, 373, 383 |
| 74JOC3192 | H. W. Heine and L. Heitz; *J. Org. Chem.*, 1974, **39**, 3192. | 1046, 1080 |
| 74JOC3278 | M. J. Haddadin, N. C. Chelhot and M. Pieridou; *J. Org. Chem.*, 1974, **39**, 3278. | 247 |
| 74JOC3434 | J. H. Maguire and R. L. McKee; *J. Org. Chem.*, 1974, **39**, 3434. | 225 |
| 74JOC3442 | C. R. Flynn and J. Michl; *J. Org. Chem.*, 1974, **39**, 3442. | 507 |
| 74JOC3519 | Y. Tamura, H. Taniguchi, T. Miyamoto, M. Tsunekawa and M. Ikeda; *J. Org. Chem.*, 1974, **39**, 3519. | 894, 926, 939 |
| 74JOC3598 | G. P. Rizzi; *J. Org. Chem.*, 1974, **39**, 3598. | 167 |
| 74JOC3615 | J. H. Bloomer, S. M. H. Zaidi, J. T. Strupczewskii, C. S. Brosz and L. A. Gudzyk; *J. Org. Chem.*, 1974, **39**, 3615. | 690 |
| 74JOC3668 | U. Niedballa and H. Vorbrüggen; *J. Org. Chem.*, 1974, **39**, 3668. | 407 |
| 74JOC3890 | F. I. Carroll, G. N. Mitchell, J. T. Blackwell, A. Sobti and R. Meck; *J. Org. Chem.*, 1974, **39**, 3890. | 631, 846 |
| 74JOU852 | N. P. Shusherina and V. L. Laptova; *J. Org. Chem. USSR (Engl. Transl.)*, 1974, **10**, 852. | 683 |
| 74JOU1516 | N. D. Dmitrieva, V. M. Zolin and Y. E. Gerasimenko; *J. Org. Chem. USSR (Engl. Transl.)*, 1974, **10**, 1516. | 668 |
| 74JOU2015 | G. N. Dorofeenko, A. V. Koblik, T. I. Polyakova and B. A. Tertor; *J. Org. Chem. USSR (Engl. Transl.)*, 1974, **10**, 2015. | 653 |
| 74JPS203 | R. I. Nazareth, T.-D. Sokoloski, D. T. Witiak and A. T. Hopper; *J. Pharm. Sci.*, 1974, **63**, 203. | 787 |
| 74KGS131 | N. E. Britikova, L. A. Belova, O. Y. Magidson and A. S. Elina; *Khim. Geterotsikl. Soedin.*, 1974, 131 (*Chem. Abstr.*, 1974, **80**, 95 875). | 221 |
| 74KGS354 | L. A. Ignatova, A. E. Gekhman, P. L. Ovechkin and B. V. Unkovskii; *Khim. Geterotsikl. Soedin.*, 1974, 354 (*Chem. Abstr.*, 1974, **81**, 25618). | 1026 |
| 74KGS425 | A. I. Dykhenko, L. S. Pupko and P. S. Pelikis; *Khim. Geterotsikl. Soedin.*, 1974, 425. | 455 |
| 74KGS554 | N. E. Britikova, L. A. Belova and A. S. Elina; *Khim. Geterotsikl. Soedin.*, 1974, 554 (*Chem. Abstr.*, 1974, **81**, 49 652). | 211, 214 |
| 74LA521 | K. Claus, H. J. Friedrich and H. Jensen; *Liebigs Ann. Chem.*, 1974, 521. | 1027 |
| 74LA561 | K. Clauss, H.-J. Friedrich and H. Jensen; *Liebigs Ann. Chem.*, 1974, 561. | 1063, 1081 |
| 74LA734 | H. Böhme and F. Ziegler; *Liebigs Ann. Chem.*, 1974, 734. | 909 |
| 74LA1474 | H. Böhme and F. Ziegler; *Liebigs Ann. Chem.*, 1974, 1474. | 909 |
| 74LA1851 | E. Fluck and H. Schultheiss; *Liebigs Ann. Chem.*, 1974, 1851. | 569 |
| 74LA2019 | R. Niess and H. Eilingsfeld; *Liebigs Ann. Chem.*, 1974, 2019. | 364 |
| 74LA2066 | K. Grohe and H. Heitzer; *Liebigs Ann. Chem.*, 1974, 2066. | 349 |
| 74MI21300 | L. Strekowski; *Rocz. Chem.*, 1974, **48**, 2157. | 103 |
| 74MI21301 | K. Takagi, G. Bastian, M. Hubert-Habart and R. Royer; *Eur. J. Med. Chem., Chim. Ther.*, 1974, **9**, 255. | 121 |
| 74MI21500 | S. Pfeifer and H. Poehlmann; *Acta Pharm. Suec.*, 1974, **11**, 645 (*Chem. Abstr.*, 1975, **83**, 22 190). | 205 |
| 74MI21501 | M. Pesson, M. Antoine, S. Chabassier, S. Geiger, P. Girard, D. Richer, P. De Lajudie, E. Horvath, B. Leriche and S. Patte; *Eur. J. Med. Chem.—Chim. Ther.*, 1974, **9**, 585 (*Chem. Abstr.*, 1975, **83**, 9978). | 221 |
| 74MI21502 | J. P. Osselaere and C. L. Lapiere; *Eur. J. Med. Chem.—Chim. Ther.*, 1974, **9**, 305 (*Chem. Abstr.*, 1975, **82**, 16 766). | 210, 222 |
| 74MI21503 | N. Hayashi, K. Shinozaki and K. Itoh; *Takeda Kenkyusho Ho*, 1974, **33**, 235 (*Chem. Abstr.*, 1975, **82**, 125 335). | 215 |

| | | |
|---|---|---|
| 74MI21600 | D. L. McAllister and G. Dryhurst; *J. Eletroanal. Chem. Interfacial Electrochem.*, 1974, **55**, 69. | 286 |
| B-74MI21601 | J. A. R. Mead; in 'Antineoplastic and Immunosuppressive Agents', ed. A. C. Sartorelli and D. G. Johns; Springer Verlag, New York, 1974, part I, p. 52. | 327 |
| 74MI21800 | C. A. Anderson, J. C. Cavagnol, C. J. Cohen, A. D. Cohick, R. T. Evans, L. J. Everett, J. Hensel, R. P. Honeycut and E. R. Levy; *Residue Rev.*, 1974, **51**, 123 (*Chem. Abstr.*, 1974, **81**, 164 459). | 369 |
| 74MI21900 | I. K. Yanson, B. I. Verkin, O. I. Shklyarevskii and A. B. Teplitskii; *Stud. Biophys.*, 1974, **46**, 29 (*Chem. Abstr.*, 1975, **83**, 8884). | 398 |
| 74MI22000 | E. Canelli; *Am. J. Public Health*, 1974, **64**, 155. | 525 |
| 74MI22001 | A. Cihak; *Oncology*, 1974, **30**, 405 (*Chem. Abstr.*, 1975, **83**, 90 483). | 526 |
| 74MI22002 | P. Rys and H. Zollinger; *Text. Chem. Color.*, 1974, **6**, 62. — | 526 |
| B-74MI22003 | C. E. Snyder, Jr. and G. J. Morris; 'Symp. Petrochem. — World Probl. Synth. Lubr. — Mark. Oppor.', 1974, p. 97 (*Chem. Abstr.*, 1976, **85**, 96 642). | 527 |
| 74MI22100 | A. G. Sidyakin, O. M. Polumbrik, G. F. Dvorko and E. A. Ponomareva; *Reakts. Sposobn. Org. Soedin.*, 1974, **11**, 35 (*Chem. Abstr.*, 1975, **82**, 42 665). | 547 |
| 74MI22101 | J. M. Meyling, R. P. van der Werf and D. A. Wiersma; *Chem. Phys. Lett.*, 1974, **28**, 364. | 543 |
| 74MI22102 | R. M. Hochstrasser, M. Robin and D. S. King; *Chem. Phys.*, 1974, **5**, 439 (*Chem. Abstr.*, 1975, **82**, 9509). | 543 |
| 74MI22103 | V. A. Bolotin, V. B. Zurba, N. Kruglyak, L. Balevicius and A. B. Bolotin; *Liet. Fiz. Rinkinys*, 1974, **14**, 561 (*Chem. Abstr.*, 1975, **83**, 17 835). | 536 |
| 74MI22200 | O. Achmatowicz, Jr., M. Chmielewski, J. Jurczak and L. Kozerski; *Rocz. Chem.*, 1974, **48**, 481. | 586, 589 |
| 74MI22201 | K. V. Masrani, H. S. Rama and S. L. Bafna; *J. Appl. Chem. Biotechnol.*, 1974, **24**, 331. | 600 |
| B-74MI22202 | S. E. Drewes; 'Chroman and Related Compounds', Verlag Chemie, Weinheim, 1974. | 603, 606, 616, 618 |
| 74MI22203 | A. Rosado; *Rev. CENIC, Cienc. Fis.*, 1974, **5**, 239 (*Chem. Abstr.*, 1975, **82**, 111 192). | 616 |
| 74MI22204 | S. Hamodrakas, A. J. Geddes and B. Sheldrick; *J. Pharm. Pharmacol.*, 1974, **26**, 54. | 607, 608, 623 |
| B-74MI22205 | 'The Aldrich Library of NMR Spectra', ed. C. J. Pouchert and J. R. Campbell; Aldrich Chemical Co., Milwaukee, 1974, vol. 4, 106C. | 577 |
| 74MI22300 | G. Feuer; *Prog. Med. Chem.*, 1974, **10**, 85. | 677 |
| 74MI22301 | J. T. Alessi, D. G. Bush and J. A. Van Allan; *Anal. Chem.*, 1974, **46**, 443. | 700 |
| 74MI22302 | E. S. H. El Ashry; *Carbohydr. Res.*, 1974, **33**, 178. | 700 |
| 74MI22303 | T. W. Whaley and D. G. Ott; *J. Label. Compd.*, 1974, **10**, 283. | 727 |
| 74MI22304 | K. Murase, J. Matsumoto, K. Tamazawa, K. Takahashi and M. Murakami; *Rep. Yamanouchi Lab.*, 1974, **2**, 66. | 732 |
| B-74MI22400 | H. U. Wagner and R. Gompper; in 'The Chemistry of Quinonoid Compounds', ed. S. Patai; Wiley, New York, 1974, vol. 2, p. 1145. | 785 |
| 74MI22401 | G. P. Ellis and G. B. West; *Prog. Med. Chem.*, 1974, **10**, 109. | 882 |
| 74MI22500 | B. Zeeh, K. H. Koenig and J. Jung; *Kem.-Kemi.*, 1974, **1**, 621 (*Chem. Abstr.*, 1975, **83**, 127 311). | 940 |
| B-74MI22600 | C. J. Pouchert and J. R. Campbell; 'The Aldrich Library of NMR Spectra', Aldrich Chemical Co. Inc., 1974, vol. 1, no. 171c. | 953 |
| 74MI22601 | R. J. C. Kleipool and A. C. Tas; *Riechst., Aromen. Koerperpflegem*, 1974, **7**, 204 (*Chem. Abstr.*, 1974, **81**, 63 593). | 991 |
| 74MI22602 | T. I. Zhukova, F. S. Florinskii, M. M. Koton, M. I. Bessonov and L. A. Laius; *Vysokomol. Soedin., Ser. B*, 1974, **16**, 390 (*Chem. Abstr.*, 1974, **81**, 152 680). | 993 |
| 74MI22700 | A. N. Mirskova, A. S. Atavin, G. G. Levkovskaya and P. V. Lidina; *Tezisy Dokl. Nauchn. Sess. Khim. Tekhnol. Org. Soedin. Sery Sernistykh Neftei 13th*, 1974, 196 (*Chem. Abstr.*, 1977, **86**, 5387). | 1000 |
| B-74MI22701 | A. I. Meyers; in 'Heterocycles in Organic Synthesis', Wiley, New York, 1974, p. 201. | 1007 |
| 74NKK1744 | I. Abe, J. Koga and N. Kuroki; *Nippon Kagaku Kaishi*, 1974, 1744. | 601 |
| 74OMR(6)115 | N. E. Sharpless, R. B. Bradley and J. A. Ferretti; *Org. Magn. Reson.*, 1974, **6**, 115. | 952 |
| 74OMR(6)233 | A. J. de Hoog; *Org. Magn. Reson.*, 1974, **6**, 233. | 585, 586 |
| 74OMR(6)663 | C. J. Turner and G. W. H. Cheeseman; *Org. Magn. Reson.*, 1974, **6**, 663. | 160 |
| 74OMS(8)31 | V. Kramer, M. Medved, B. Stanovnik and M. Tisler; *Org. Mass Spectrom.*, 1974, **8**, 31. | 237 |
| 74OMS(9)80 | G. Hvistendahl, P. Gyorosi and K. Undheim; *Org. Mass Spectrom.*, 1974, **9**, 80. | 620 |
| 74OMS(9)649 | J. B. Chattopadhyaya and A. V. R. Rao; *Org. Mass Spectrom.*, 1974, **9**, 649. | 958 |
| 74PMH(6)1 | E. Heilbronner, J. P. Maier and E. Haselbach; *Phys. Methods Heterocycl. Chem.*, 1974, **6**, 1. | 958, 959 |
| 74PMH(6)147 | S. G. Schulman; *Phys. Methods Heterocycl. Chem.*, 1974, **6**, 147. | 600 |
| 74PMH(6)199 | K. Pihlaja and E. Taskinen; *Phys. Methods Heterocycl. Chem.*, 1974, **6**, 199. | 948 |
| 74RCR851 | N. P. Shusherina; *Russ. Chem. Rev. (Engl. Transl.)*, 1974, **43**, 851. | 688 |
| 74RRC859 | M. D. Gheorghiu and P. Filip; *Rev. Roum. Chim.*, 1974, **19**, 859. | 533 |
| 74RRC1731 | A. T. Balaban and M. Paraschiv; *Rev. Roum. Chim.*, 1974, **19**, 1731. | 861 |
| 74RTC111 | A. P. Kroon and H. C. van der Plas; *Rec. Trav. Chim. Pays-Bas*, 1974, **93**, 111. | 140 |

| | | |
|---|---|---|
| 74RTC204 | A. W. van Muijlwijk, A. P. G. Kieboom and H. van Bekkum; *Recl. Trav. Chim. Pays-Bas*, 1974, **93**, 204. | 481 |
| 74RTC227 | A. P. Kroon and H. C. van der Plas; *Rec. Trav. Chim. Pays-Bas*, 1974, **93**, 227. | 89 |
| 74RTC321 | D. N. Reinhoudt and C. G. Kouwenhoven; *Recl. Trav. Chim. Pays-Bas*, 1974, **93**, 321. | 552 |
| 74S286 | V. D. Adams and R. C. Anderson; *Synthesis*, 1974, 286. | 132 |
| 74S491 | M. Mikolajczyk and J. Luczak; *Synthesis*, 1974, 491. | 142, 144 |
| 74S713 | H. Yoshino, Y. Kawazoe and T. Taguchi; *Synthesis*, 1974, 713. | 988 |
| 74SA(A)2021 | S. Vázquez and J. Castrillón; *Spectrochim. Acta, Part A*, 1974, **30**, 2021. | 956 |
| 74T221 | P. Henklein, R. Kraft and G. Westphal; *Tetrahedron*, 1974, **30**, 221. | 517 |
| 74T549 | J. Almog and E. D. Bergmann; *Tetrahedron*, 1974, **30**, 549. | 202, 204 |
| 74T947 | S. Oae, T. Aida, M. Nakajima and N. Furukawa; *Tetrahedron*, 1974, **30**, 947. | 900 |
| 74T1015 | T. D. J. D'Silva, W. E. Walker and R. W. Manyik; *Tetrahedron*, 1974, **30**, 1015. | 723, 773 |
| 74T1579 | D. H. Wertz and N. L. Allinger; *Tetrahedron*, 1974, **30**, 1579. | 889 |
| 74T1717 | O. Eisenstein, N. T. Anh, Y. Jean, A. Devaquet, J. Cantacuzène and L. Salem; *Tetrahedron*, 1974, **30**, 1717. | 629 |
| 74T2087 | S. Ikegami, T. Asai, K. Tsuneoka, S. Matsumura and S. Akaboshi; *Tetrahedron*, 1974, **30**, 2087. | 902 |
| 74T2781 | G. Prota, E. Ponsiglione and R. Ruggiero; *Tetrahedron*, 1974, **30**, 2781. | 1010 |
| 74T2841 | F. A. Neugebauer and H. Brunner; *Tetrahedron*, 1974, **30**, 2841. | 542, 558 |
| 74T3171 | J. Daunis, R. Jacquier and C. Pigiere; *Tetrahedron*, 1974, **30**, 3171. | 390, 391, 441 |
| 74T3315 | A. Rassat and P. Rey; *Tetrahedron*, 1974, **30**, 3315. | 1009 |
| 74T3553 | A. Nohara, T. Umetani and Y. Sanno; *Tetrahedron*, 1974, **30**, 3553. | 615, 710, 821 |
| 74T3563 | A. Norhara, K. Ukawa and Y. Sanno; *Tetrahedron*, 1974, **30**, 3563. | 596 |
| 74TL179 | J. W. Lown and M. H. Akhtar; *Tetrahedron Lett.*, 1974, 179. | 186 |
| 74TL221 | A. Schulze and J. Goerdeler; *Tetrahedron Lett.*, 1974, 221. | 1077 |
| 74TL253 | M. H. Palmer and R. H. Findlay; *Tetrahedron Lett.*, 1974, 253. | 533, 536 |
| 74TL1467 | C. Batich and W. Adam; *Tetrahedron Lett.*, 1974, 1467. | 959 |
| 74TL1973 | P. Stoss and G. Satzinger; *Tetrahedron Lett.*, 1974, 1973. | 967 |
| 74TL2303 | G. C. Johnson and R. H. Levin; *Tetrahedron Lett.*, 1974, 2303. | 554 |
| 74TL3123 | J. Riand, M. T. Chenon and N. Lumbroso-Bader; *Tetrahedron Lett.*, 1974, 3123. | 63 |
| 74TL3553 | L. H. Hellberg and A. Juarez; *Tetrahedron Lett.*, 1974, 3553. | 909 |
| 74TL3633 | C. R. Hall and D. J. H. Smith; *Tetrahedron Lett.*, 1974, 3633. | 913 |
| 74TL3893 | S. Minami, Y. Kimura, T. Miyamoto and J. Matsumoto; *Tetrahedron Lett.*, 1974, 3893. | 344 |
| 74TL3911 | R. S. Devdhar, V. N. Gogte and B. D. Tilak; *Tetrahedron Lett.*, 1974, 3911. | 937 |
| 74TL4315 | R. K. Razdan, H. G. Pars, W. R. Thompson and F. E. Granchelli; *Tetrahedron Lett.*, 1974, 4315. | 673 |
| 74UKZ1220 | A. I. Dychenko and L. S. Pupko; *Ukr. Khim. Zh. (Russ. Ed.)*, 1974, **40**, 1220. | 455 |
| 74UP21900 | H. Neunhoeffer and H. Hennig; unpublished results, 1974. | 401, 427 |
| 74USP3830812 | A. A. Ramsey; *U.S. Pat.* 3 830 812 (1974) (*Chem. Abstr.*, 1974, **81**, 136 174). | 368 |
| 74USP3833577 | K. Lin; *U.S. Pat.* 3 833 577 (1974) (*Chem. Abstr.*, 1975, **82**, 43 477). | 1085 |
| 74USP3833610 | Sandoz–Warner Inc., *U.S. Pat.* 3 833 610 (1974) (*Chem. Abstr.*, 1974, **81**, 152 005). | 939 |
| 74USP3839331 | G. H. Miller; *U.S. Pat.* 3 839 331 (1974) (*Chem. Abstr.*, 1975, **82**, 16 867). | 503 |
| 74USP3843644 | N. H. Kurihara and D. E. Bublitz; *U.S. Pat.* 3 843 644 (1974) (*Chem. Abstr.*, 1975, **82**, 43 432). | 993 |
| 74USP3849415 | N. H. Kurihara and D. E. Bublitz; *U.S. Pat.* 3 849 415 (1974) (*Chem. Abstr.*, 1975, **82**, 86 290). | 993 |
| 74YZ607 | S. Ueno, Y. Tominaga, R. Natsuki, Y. Matsuda and G. Kobayashi; *Yakugaku Zasshi*, 1974, **94**, 607 (*Chem. Abstr.*, 1974, **81**, 120 391). | 247 |
| 74ZC239 | H. Bahn and W. Schroth; *Z. Chem.*, 1974, **14**, 239. | 989 |
| 74ZN(B)792 | A. K. Mansour, S. B. Awad and S. Antoun; *Z. Naturforsch., Teil B*, 1974, **29**, 792. | 414 |
| 74ZN(B)799 | R. Appel, H. Uhlenhaut and M. Montenarh; *Z. Naturforsch., Teil B*, 1974, **29**, 799. | 1047, 1079 |
| 74ZOR76 | G. Dzhamalova, T. A. Danilova and E. A. Viktorova; *Zh. Org. Khim.*, 1974, **10**, 76. | 903 |
| 74ZOR124 | G. I. Matyuschecheva, V. A. Mikhailov and L. M. Yagupol'skii; *Zh. Org. Khim.*, 1974, **10**, 124. | 567 |
| 74ZOR377 | O. P. Shvaika and V. I. Fomenko; *Zh. Org. Khim.*, 1974, **10**, 377. | 561 |
| 74ZOR488 | L. N. Markovskii, Yu. G. Shermolovich and V. I. Shevchenko; *Zh. Org. Khim.*, 1974, **10**, 488. | 1047, 1074 |
| 74ZOR1513 | O. P. Shelyapin, I. V. Samartseva and L. A. Pavlova; *Zh. Org. Khim.*, 1974, **10**, 1513, (*Chem. Abstr.*, 1974, **81**, 105415). | 998 |
| 74ZOR2429 | O. P. Shvaika and V. I. Fomenko; *Zh. Org. Khim.*, 1974, **11**, 2429. | 394, 440 |
| 75ACH(84)319 | M. A.-F. Elkaschef, F. M. E. Abdel-Megeid, K.-E. Mokhtar and F. A. Gad; *Acta Chim. Acad. Sci. Hung.*, 1975, **84**, 319. | 699, 708, 856 |
| 75ACS(B)285 | C. L. Pedersen and O. Buchardt; *Acta Chem. Scand., Ser. B*, 1975, **29**, 285. | 660 |
| 75AF1712 | J. P. Osselaere; *Arzneim.-Forsch.*, 1975, **25**, 1712. | 260 |
| 75AG(E)581 | R. R. Schmidt; *Angew. Chem., Int. Ed. Engl.*, 1975, **14**, 581. | 1001, 1052, 1060, 1065, 1066 |
| 75AG(E)665 | A. Dolars, C.-W. Schellhammer and J. Schroeder; *Angew. Chem., Int. Ed. Engl.*, 1975, **14**, 665. | 527 |
| 75AG356 | U. Ewers, H. Günther and L. Jaenicke; *Angew. Chem.*, 1975, **87**, 356. | 63, 249, 251 |

| | | |
|---|---|---|
| 75AG842 | A. G. Anastassiou and S. J. Girgenti; *Angew. Chem.*, 1975, **87**, 842. | 552 |
| 75AHC(18)59 | S. W. Schneller; *Adv. Heterocycl. Chem.*, 1975, **18**, 59. | 916, 929, 933, 934, 937, 941 |
| 75AHC(18)159 | L. Merlini; *Adv. Heterocycl. Chem.*, 1975, **18**, 159. | 666, 741 |
| 75AJC2669 | I. D. Rae and B. N. Umbrasas; *Aust. J. Chem.*, 1975, **28**, 2669. | 846 |
| 75AP385 | F. Eiden and D. Dölcher; *Arch. Pharm.* (*Weinheim, Ger.*), 1975, **308**, 385. | 697, 717 |
| 75AP481 | F. Moll and H. J. Wieland; *Arch. Pharm.*, 1975, **308**, 481. | 1008 |
| 75AP622 | F. Eiden, K. Schnabel and H. Wiedemann; *Arch. Pharm.*, 1975, **308**, 622. | 1006, 1025 |
| 75AX(B)626 | R. E. Ballard and E. K. Norris; *Acta Crystallogr., Sect. B*, 1975, **31**, 626. | 370 |
| 75AX(B)954 | P. G. Jones, O. Kennard, S. Chandrasekhar and A. J. Kirby; *Acta Crystallogr., Part B*, 1975, **31**, 954. | 623 |
| 75AX(B)1287 | S. Shimizu, S. Kashino and M. Haisa; *Acta Crystallogr., Part B*, 1975, **31**, 1287. | 622, 623 |
| 75AX(B)2519 | P. Singh and D. J. Hodgson; *Acta Crystallogr., Part B*, 1975, **31**, 2519. | 386, 387 |
| 75AX(B)2934 | C. S. Choi, A. Santoro and P. L. Marinkas; *Acta Crystallogr., Part B*, 1975, **31**, 2934. | 462 |
| 75BCJ73 | I. Shibuya and M. Kurabayashi; *Bull. Chem. Soc. Jpn.*, 1975, **48**, 73. | 1005 |
| 75BCJ80 | M. Tsukayama; *Bull. Chem. Soc. Jpn.*, 1975, **48**, 80. | 755 |
| 75BCJ819 | N. Azuma, Y. Deguchi, F. Marumo and Y. Saito; *Bull. Chem. Soc. Jpn.*, 1975, **48**, 819. | 539, 541 |
| 75BCJ825 | M. Azuma, Y. Deguchi, F. Marumo and Y. Saito; *Bull. Chem. Soc. Jpn.*, 1975, **48**, 825. | 539, 541 |
| 75BCJ956 | S. Ishida, Y. Fukushima, S. Sekiguchi and K. Matsui; *Bull. Chem. Soc. Jpn.*, 1975, **48**, 956. | 509 |
| 75BCJ2496 | M. Oki, T. Endo and T. Sugawara; *Bull. Chem. Soc. Jpn.*, 1975, **48**, 2496. | 961 |
| 75BCJ2605 | M. Takahashi, N. Suzuki and Y. Igari; *Bull. Chem. Soc. Jpn.*, 1975, **48**, 2605. | 554 |
| 75BCJ3309 | R. Kayama, H. Shizuka, S. Sekiguchi and K. Matsui; *Bull. Chem. Soc. Jpn.*, 1975, **48**, 3309. | 477 |
| 75BCJ3765 | Y. Miura, Y. Morimoto and M. Kinoshita; *Bull. Chem. Soc. Jpn.*, 1975, **48**, 3765. | 549 |
| 75BCJ3767 | T. Sugimoto and S. Matsuura; *Bull. Chem. Soc. Jpn.*, 1975, **48**, 3767. | 312 |
| 75BRP1383459 | V. E. B. Arzneimittelwerk; *Br. Pat.* 1 383 459 (*Chem. Abstr.*, 1975, **83**, 43 343). | 994 |
| 75BRP1399345 | R. Robinson and L. J. Danks; *Br. Pat.* 1 399 345 (1975) (*Chem. Abstr.*, 1975, **83**, 147 508). | 503 |
| 75BSF702 | A. Decormeille, G. Queguiner and P. Pastour; *Bull. Soc. Chim. Fr.*, 1975, 702. | 234, 237, 242, 243 |
| 75BSF751 | M. Julia and C. Binet du Jassonneix; *Bull. Soc. Chim. Fr.*, 1975, 751. | 794 |
| 75BSF2023 | O. Convert, J.-P. Le Roux, P.-L. Desbene and A. Defoin; *Bull. Soc. Chim. Fr.*, 1975, 2023. | 589, 641 |
| 75BSF2249 | J.-P. Coïc and G. Saint-Ruf; *Bull. Soc. Chim. Fr.*, 1975, 2249. | 975 |
| 75BSF2757 | A. Decormeille, G. Queguiner and P. Pastour; *Bull. Soc. Chim. Fr.*, 1975, 2757. | 225 |
| 75CB875 | H. W. Rothkopf, D. Woehrle, R. Mueller and G. Kossmehl; *Chem. Ber.*, 1975, **108**, 875. | 180 |
| 75CB1087 | K.-H. Linke, R. Bimczok and J. Lex; *Chem. Ber.*, 1975, **108**, 1087. | 1042, 1049, 1074 |
| 75CB1557 | P. Rademacher and H. Koopman; *Chem. Ber.*, 1975, **108**, 1557. | 538, 539, 542, 543 |
| 75CB1911 | K. Takács, K. Harsányi, P. Kolonits and K. I. Ajzert; *Chem. Ber.*, 1975, **108**, 1911. | 1058, 1072 |
| 75CB3105 | L. Birkofer and N. Ramadan; *Chem. Ber.*, 1975, **108**, 3105. | 178 |
| 75CB3509 | H. Neunhoeffer and H.-J. Degen; *Chem. Ber.*, 1975, **108**, 3509. | 558 |
| 75CB3877 | H. Neunhoeffer and M. Bachmann; *Chem. Ber.*, 1975, **108**, 3877. | 471 |
| 75CB3883 | S. Antus, L. Farkas, Z. Kardos-Balogh and M. Nógrádi; *Chem. Ber.*, 1975, **108**, 3883. | 824 |
| 75CC241 | N. Ishibe, S. Yutaka, J. Masui and Y. Ishida; *J. Chem. Soc., Chem. Commun.*, 1975, 241. | 696 |
| 75CC284 | N. Serpone and P. H. Bird; *J. Chem. Soc., Chem. Commun.*, 1975, 284. | 625 |
| 75CC675 | D. A. Griffin and J. Staunton; *J. Chem. Soc., Chem. Commun.*, 1975, 675. | 658, 680, 872 |
| 75CC703 | R. A. Abramovitch and I. Shinkai; *J. Chem. Soc., Chem. Commun.*, 1975, 703. | 9 |
| 75CC819 | G. Tennant and C. W. Yacomeni; *J. Chem. Soc., Chem. Commun.*, 1975, 819. | 316 |
| 75CC850 | M. J. Begley, L. Crombie and D. A. Whiting; *J. Chem. Soc., Chem. Commun.*, 1975, 850. | 624 |
| 75CC914 | T. L. Gilchrist, M. E. Peek and C. W. Rees; *J. Chem. Soc., Chem. Commun.*, 1975, 914. | 1056 |
| 75CC962 | T. L. Gilchrist, C. J. Harris, M. E. Peek and C. W. Rees; *J. Chem. Soc., Chem. Commun.*, 1975, 962. | 1056, 1066 |
| 75CCC1390 | V. Krchňák and Z. Arnold; *Collect. Czech. Chem. Commun.*, 1975, **40**, 1390. | 85 |
| 75CCC2340 | A. Piskala; *Collect. Czech. Chem. Commun.*, 1975, **40**, 2340. | 400 |
| 75CCC2680 | A. Piskala, J. Gut and F. Šorm; *Collect. Czech. Chem. Commun.*, 1975, **40**, 2680. | 402 |
| 75CHE18 | E. V. Kuznetsov, D. V. Pruchkin, E. A. Muradvan and G. N. Dorofeenko; *Chem. Heterocycl. Compd.* (*Engl. Transl.*), 1975, **11**, 18. | 655 |
| 75CHE240 | M. P. Zhdanov, E. A. Zvezdina and G. N. Dorofeenko; *Chem. Heterocycl. Compd.* (*Engl. Transl.*), 1975, **11**, 240. | 656 |
| 75CHE278 | E. A. Karakhanov, M. V. Vagabov, S. K. Dzhamalov and E. A. Viktorova; *Chem. Heterocycl. Compd.* (*Engl. Transl.*), 1975, **11**, 278. | 722, 785 |
| 75CHE1025 | E. A. Zvezdina, M. P. Zhdanova, A. M. Simonov and G. N. Dorofeenko; *Chem. Heterocycl. Compd.* (*Engl. Transl.*), 1975, **11**, 1025. | 656 |
| 75CJC276 | I. W. J. Still, M. T. Thomas and A. M. Clish; *Can. J. Chem.*, 1975, **53**, 276. | 909 |
| 75CJC343 | J. P. Devlin; *Can. J. Chem.*, 1975, **53**, 343. | 806 |

| 75CJC1484 | A. J. Elliott, P. D. Callaghan, M. S. Gibson and S. T. Nemeth; *Can. J. Chem.*, 1975, **53**, 1484. | 1072 |
|---|---|---|
| 75CJC1980 | A. A. Chalmers and K. G. R. Pachler; *Can. J. Chem.*, 1975, **53**, 1980. | 588 |
| 75CJC2045 | D. J. Bichan and P. Yates; *Can. J. Chem.*, 1975, **53**, 2045. | 768 |
| 75CJC2054 | D. J. Bichan and P. Yates; *Can. J. Chem.*, 1975, **53**, 2054. | 835 |
| 75CJC3656 | J. F. King, E. G. Lewars, D. R. K. Harding and R. M. Enanoza; *Can. J. Chem.*, 1975, **53**, 3656. | 913 |
| 75CJC3714 | H. Fong and K. Vaughan; *Can. J. Chem.*, 1975, **53**, 3714. | 378 |
| 75CL19 | Y. Miura, Y. Morimoto and M. Kinoshita; *Chem. Lett.*, 1975, 19. | 549 |
| 75CL161 | T. Izawa and T. Mukaiyama; *Chem. Lett.*, 1975, 161. | 841 |
| 75CL1153 | H. Matsukubo, M. Kojima and H. Kato; *Chem. Lett.*, 1975, 1153. | 1034 |
| 75CPB494 | T. Koyama, T. Hirota, Y. Shinohara, S. Fukuoka, M. Yamato and S. Ohmori; *Chem. Pharm. Bull.*, 1975, **23**, 494. | 224 |
| 75CPB1488 | A. Miyake, K. Itoh, N. Tada, Y. Oka and S. Yurugi; *Chem. Pharm. Bull.*, 1975, **23**, 1488. | 346, 348 |
| 75CPB1500 | A. Miyake, Y. Oka and S. Yurugi; *Chem. Pharm. Bull.*, 1975, **23**, 1500. | 348 |
| 75CPB1505 | A. Miyake, K. Itoh, N. Tada, Y. Oka and S. Yurugi; *Chem. Pharm. Bull.*, 1975, **23**, 1505. | 348 |
| 75CPB2239 | Y. Oka, K. Omura, A. Miyake, K. Itoh, M. Tomimoto, N. Tada and S. Yurugi; *Chem. Pharm. Bull.*, 1975, **23**, 2239. | 234, 237, 239, 241, 242, 243, 246 |
| 75CPB2306 | Y. Oka, K. Itoh, A. Miyake, N. Tada, K. Omura, M. Tomimoto and S. Yurugi; *Chem. Pharm. Bull.*, 1975, **23**, 2306. | 234, 242, 243 |
| 75CPB2678 | A. Nagel and H. C. van der Plas; *Chem. Pharm. Bull.*, 1975, **23**, 2678. | 288, 308 |
| 75CPB2701 | S. Ikegami, J. Ohishi and S. Akaboshi; *Chem. Pharm. Bull.*, 1975, **23**, 2701. | 902 |
| 75CPB2939 | T. Higashino, K. Suzuki and E. Hayashi; *Chem. Pharm. Bull.*, 1975, **23**, 2939. | 209 |
| 75CR(C)(280)37 | J-P. LeRoux, J-C. Charton and P-L. Desbene; *C.R. Hebd. Seances Acad. Sci., Ser. C*, 1975, **280**, 37. | 660 |
| 75CR(C)(280)1525 | J. Tesse, C. Glacet and D. Couturier; *C.R. Hebd. Seances Acad. Sci., Ser. C*, 1975, **280**, 1525. | 580 |
| 75CR(C)(281)51 | F. Texier, E. Marchand and A. Foucaud; *C.R. Hebd. Seances Acad. Sci., Ser. C*, 1975, **281**, 51. | 1018 |
| 75CR(C)(281)563 | A. Etienne, G. Lonchambon and J. Roques; *C. R. Hebd. Seances Acad. Sci., Ser. C*, 1975, **281**, 563. | 479 |
| 75CR(C)(281)941 | A. Godard, G. Queguiner and P. Pastour; *C.R. Hebd. Seances Acad. Sci., Ser. C*, 1975, **281**, 941. | 247 |
| 75CRV611 | J. B. Lambert and S. I. Featherman; *Chem. Rev.*, 1975, **75**, 611. | 579 |
| 75CRV651 | G. Desimoni and G. Tacconi; *Chem. Rev.*, 1975, **75**, 651. | 770 |
| 75CSR189 | P. D. Kennewell and J. B. Taylor; *Chem. Soc. Rev.*, 1975, 189. | 1043, 1052, 1085 |
| 75CZ292 | G. Seitz and T. Kämpchen; *Chem-Ztg.*, 1975, **99**, 292. | 552 |
| 75FES837 | R. Pellicciari, M. Curini and P. Ceccherelli; *Farmaco Ed. Sci.*, 1975, **30**, 837. | 909 |
| 75FOR(32)57 | P. G. Sammes; *Fortschr. Chem. Org. Naturst.*, 1975, **32**, 57. | 187, 189 |
| 75G841 | L. Benati, P. C. Montevecchi, A. Tundo and G. Zanardi; *Gazz. Chim. Ital.*, 1975, **105**, 841. | 965 |
| 75GEP2346936 | W. Draber, H. Timmler, L. Eue and R. R. Schmidt; *Ger. Pat.* 2 346 936 (1975) (*Chem. Abstr.*, 1975, **83**, 97 385). | 415 |
| 75GEP2449901 | T. McGee and D. W. Roberts; *Ger. Pat.* 2 449 901 (1975) (*Chem. Abstr.*, 1976, **84**, 32 522). | 994 |
| 75GEP2508832 | I. T. Kay; *Ger. Pat.* 2 508 832 (1975) (*Chem. Abstr.*, 1976, **84**, 59 585). | 1086 |
| 75GEP2630745 | P. J. Cranfield; *Ger. Pat.* 2 630 745 (1975) (*Chem. Abstr.*, 1977, **86**, 127 311). | 527 |
| 75H(3)381 | I. Matsuura; *Heterocycles*, 1975, **3**, 381. | 238, 241 |
| 75HCA761 | A. Vogel and F. Troxler; *Helv. Chim. Acta*, 1975, **58**, 761. | 494 |
| 75HCA1374 | P. K. Sengupta, J. H. Bieri and M. Viscontini; *Helv. Chim. Acta*, 1975, **58**, 1374. | 304 |
| 75HCA1772 | R. Weber and M. Viscontini; *Helv. Chim. Acta*, 1975, **58**, 1772. | 281 |
| 75HCA2529 | D. Schelz and M. Priester; *Helv. Chim. Acta*, 1975, **58**, 2529. | 252 |
| 75IC2378 | H. J. Stoklosa, J. R. Wasson, E. V. Brown, H. W. Richardson and W. E. Hatfield; *Inorg. Chem.*, 1975, **14**, 2378. | 249, 250 |
| 75IJC(B)1098 | Y. A. Ibrahim and M. M. Eid; *Indian J. Chem., Sect. B*, 1975, **13**, 1098. | 409 |
| 75IZV1206 | B. A. Arbuzov, N. N. Zobova and O. V. Sofronova; *Izv. Akad. Nauk SSSR, Ser. Khim.*, 1975, 1206. | 1044, 1078 |
| 75JA101 | U. Svanholm, O. Hammerich and V. D. Parker; *J. Am. Chem. Soc.*, 1975, **97**, 101. | 969 |
| 75JA1468 | S. A. Khan, J. B. Lambert, O. Hernandez and F. A. Carey; *J. Am. Chem. Soc.*, 1975, **97**, 1468. | 953 |
| 75JA2497 | A. Dell, D. H. Williams, H. R. Morris, G. A. Smith, J. Feeney and G. C. K. Roberts; *J. Am. Chem. Soc.*, 1975, **97**, 2497. | 195 |
| 75JA2718 | B. E. Maryanoff, J. Stackhouse, G. H. Senkler, Jr. and K. Mislow; *J. Am. Chem. Soc.*, 1975, **97**, 2718. | 894, 939, 1052 |
| 75JA3541 | H. Alper and S. Wollowitz; *J. Am. Chem. Soc.*, 1975, **97**, 3541. | 188 |
| 75JA4056 | R. U. Lemieux, K. B. Hendriks, R. V. Stick and K. James; *J. Am. Chem. Soc.*, 1975, **97**, 4056. | 629 |
| 75JA5167 | N. L. Allinger and M. J. Hickey; *J. Am. Chem. Soc.*, 1975, **97**, 5167. | 887 |
| 75JA5291 | B. A. Carlson, W. A. Sheppard and O. W. Webster; *J. Am. Chem. Soc.*, 1975, **97**, 5291. | 50 |
| 75JA5472 | E. Dradi and G. Gatti; *J. Am. Chem. Soc.*, 1975, **97**, 5472. | 576, 581, 641 |

| | | |
|---|---|---|
| 75JA5611 | R. A. Izydore and S. McLean; *J. Am. Chem. Soc.*, 1975, **97**, 5611. | 518 |
| 75JA6590 | A. Krantz and B. Hoppe; *J. Am. Chem. Soc.*, 1975, **97**, 6590. | 1023 |
| 75JA6892 | R. A. Ruden and P. Bonjouklian; *J. Am. Chem. Soc.*, 1975, **97**, 6892. | 773 |
| 75JA7182 | B. M. Trost and C. H. Miller; *J. Am. Chem. Soc.*, 1975, **97**, 7182. | 847 |
| 75JAP7524280 | K. Yamamoto, T. Yamauchi and I. Naruse; *Jpn. Pat.*, 75 24 280 (1975) (*Chem. Abstr.*, 1975, **83**, 131 572). | 1030 |
| 75JAP(K)7529599 | K. Noda, A. Nakagawa, T. Motomura and H. Ide; (Hisamitsu Pharm. Co. Inc.), *Jpn. Kokai* 75 29 599 (1975) (*Chem. Abstr.*, 1975, **83**, 97 367). | 216 |
| 75JAP(K)75101322 | K. Ikura, J. Suzuki, M. Ando and S. Hashimoto; *Jpn. Kokai* 75 101 322 (1975) (*Chem. Abstr.*, 1976, **84**, 74 264). | 1045 |
| 75JAP(K)75105696 | S. Minami, J. Matsumoto, K. Kawaguchi, S. Mishio, M. Shimizu, Y. Takase and S. Nakamura; (Dainippon Pharm. Co. Ltd.), *Jpn. Kokai* 75 105 696 (1975) (*Chem. Abstr.*, 1976, **84**, 164 821). | 212 |
| 75JAP(K)75131994 | K. Noda, A. Nakagawa, T. Motomura and H. Ide; (Hisamitsu Pharm. Co. Inc.), *Jpn. Kokai* 75 131 994 (1975) (*Chem. Abstr.*, 1976, **85**, 33066). | 225 |
| 75JAP(K)75157394 | K. Noda, A. Nakagawa, T. Motomura, K. Yamagata, S. Miyata, S. Yamasaki, Y. Nakajima, Y. Ishikura and K. Noguri; (Hisamitsu Pharm. Co. Ltd.), *Jpn. Kokai* 75 157 394 (1975) (*Chem. Abstr.*, 1976, **85**, 5674). | 226 |
| 75JAP(K)75160296 | K. Noda, A. Nakagawa, T. Motomura and H. Ide; (Hisamitsu Pharm. Co. Ltd.), *Jpn. Kokai* 75 160 296 (1975) (*Chem. Abstr.*, 1976, **85**, 78 152). | 222 |
| 75JCP(62)1747 | I. Nenner and G. J. Schulz; *J. Chem. Phys.*, 1975, **62**, 1747. | 466 |
| 75JCS(D)1071 | S. E. V. Phillips and M. R. Truter; *J. Chem. Soc., Dalton Trans.*, 1975, 1071. | 622 |
| 75JCS(F2)1173 | K. L. Gallaher and S. H. Bauer; *J. Chem. Soc., Faraday Trans. 2*, 1975, **71**, 1173. | 947, 948 |
| 75JCS(F2)1812 | N. Ishibe, H. Sugimoto and J. B. Gallivan; *J. Chem. Soc., Faraday Trans. 2*, 1975, 1812. | 600 |
| 75JCS(P1)31 | B. M. Adger, S. Bradbury, M. Keating, C. W. Rees, R. C. Storr and M. T. Williams; *J. Chem. Soc., Perkin Trans. 1*, 1975, 31. | 371, 372, 373, 374, 375, 378, 381, 383 |
| 75JCS(P1)41 | B. M. Adger, M. Keating, C. W. Rees and R. C. Storr; *J. Chem. Soc., Perkin Trans 1*, 1975, 41. | 378 |
| 75JCS(P1)45 | B. M. Adger, C. W. Rees and R. C. Storr; *J. Chem. Soc., Perkin Trans. 1*, 1975, 45. | 378, 380, 554 |
| 75JCS(P1)160 | M. Yokoyama, T. Kondo, N. Miyase and M. Torri; *J. Chem. Soc., Perkin Trans. 1*, 1975, 160. | 964, 984 |
| 75JCS(P1)180 | D. M. Vyas and G. W. Hay; *J. Chem. Soc., Perkin Trans. 1*, 1975, 180. | 932 |
| 75JCS(P1)716 | R. J. Stoodley and R. B. Wilkins; *J. Chem. Soc., Perkin Trans. 1*, 1975, 716. | 1012 |
| 75JCS(P1)772 | O. O. Orazi and R. A. Corral; *J. Chem. Soc., Perkin Trans. 1*, 1975, 772. | 1041, 1082 |
| 75JCS(P1)1023 | T. B. Brown and M. F. G. Stevens; *J. Chem. Soc., Perkin Trans. 1*, 1975, 1023. | 218 |
| 75JCS(P1)1130 | R. D. Chambers, J. R. Maslakiewicz and K. C. Srivastava; *J. Chem. Soc., Perkin Trans. 1*, 1975, 1130. | 11 |
| 75JCS(P1)1267 | R. G. R. Bacon and J. C. F. Murray; *J. Chem. Soc., Perkin Trans. 1*, 1975, 1267. | 831 |
| 75JCS(P1)1326 | B. E. Landberg and J. W. Lown; *J. Chem. Soc., Perkin Trans. 1*, 1975, 1326. | 342, 353 |
| 75JCS(P1)1366 | K. Undheim and S. Baklien; *J. Chem. Soc., Perkin Trans. 1*, 1975, 1366. | 922 |
| 75JCS(P1)1398 | B. Rindone and C. Scolastico; *J. Chem. Soc., Perkin Trans. 1*, 1975, 1398. | 170 |
| 75JCS(P1)1424 | R. A. W. Johnstone, T. J. Povall and I. D. Entwistle; *J. Chem. Soc., Perkin Trans. 1*, 1975, 1424. | 254 |
| 75JCS(P1)1497 | L. Crombie, G. W. Kilbee and D. A. Whiting; *J. Chem. Soc., Perkin Trans. 1*, 1975, 1497. | 592 |
| 75JCS(P1)1563 | M. Okigawa, N. U. Khan, N. Kawano and W. Rahman; *J. Chem. Soc., Perkin Trans. 1*, 1975, 1563. | 584, 585 |
| 75JCS(P1)1652 | P. Lambelet and E. A. C. Lucken; *J. Chem. Soc., Perkin Trans. 1*, 1975, 1652. | 599 |
| 75JCS(P1)1736 | D. R. Adams, S. P. Bhatnagar and R. C. Cookson; *J. Chem. Soc., Perkin Trans. 1*, 1975, 1736. | 579, 769 |
| 75JCS(P1)1787 | R. Esmail and F. Kurzer; *J. Chem. Soc., Perkin Trans. 1*, 1975, 1787. | 558 |
| 75JCS(P1)1845 | P. G. Sammes and T. W. Wallace; *J. Chem. Soc., Perkin Trans. 1*, 1975, 1845. | 820, 840 |
| 75JCS(P1)1869 | A. K. Das Gupta, R. M. Chatterje and S. N. Choudhuri; *J. Chem. Soc., Perkin Trans. 1*, 1975, 1869. | 802 |
| 75JCS(P1)1925 | W. M. Leyshon and D. A. Wilson; *J. Chem. Soc., Perkin Trans. 1*, 1975, 1925. | 901 |
| 75JCS(P1)1969 | T. L. Gilchrist, C. J. Harris, C. J. Moody and C. W. Rees; *J. Chem. Soc., Perkin Trans. 1*, 1975, 1969. | 361 |
| 75JCS(P1)2099 | S. Baklien, P. Groth and K. Undheim; *J. Chem. Soc., Perkin Trans. 1*, 1975, 2099. | 922, 938 |
| 75JCS(P1)2182 | D. J. Brown and K. Ienaga; *J. Chem. Soc., Perkin Trans. 1*, 1975, 2182. | 208, 216, 222 |
| 75JCS(P1)2271 | R. Hull, P. J. van den Broek and M. L. Swain; *J. Chem. Soc., Perkin Trans. 1*, 1975, 2271. | 211, 230 |
| 75JCS(P1)2396 | J. I. G. Cadogan and B. S. Tait; *J. Chem. Soc., Perkin Trans. 1*, 1975, 2396. | 1033 |
| 75JCS(P1)2405 | S.-F. Tan and T.-H. Tjia; *J. Chem. Soc., Perkin Trans. 1*, 1975, 2405. | 793 |
| 75JCS(P2)40 | C. D. Sirrell and D. E. Williams, *J. Chem. Soc., Perkin Trans. 2*, 1975, 40. | 264 |
| 75JCS(P2)270 | G. B. Ansell and J. L. Erickson; *J. Chem. Soc., Perkin Trans. 2*, 1975, 270. | 538, 541 |
| 75JCS(P2)841 | M. H. Palmer, R. H. Findlay, W. Moyes and A. J. Gaskell; *J. Chem. Soc., Perkin Trans. 2*, 1975, 841. | 576 |
| 75JCS(P2)858 | A. A. Najam and J. G. Tillett; *J. Chem. Soc., Perkin Trans. 2*, 1975, 858. | 980 |
| 75JHC37 | L. S. Wittenbrook; *J. Heterocycl. Chem.*, 1975, **12**, 37. | 502, 516 |
| 75JHC79 | K. Y. Tserng, C. L. Bell and L. Bauer; *J. Heterocycl. Chem.*, 1975, **12**, 79. | 204 |

| 75JHC95 | G. Adembri, F. DeSio, R. Nesi and M. Scotton; *J. Heterocycl. Chem.*, 1975, **12**, 95. | 343, 354 |
|---|---|---|
| 75JHC107 | Y. Tamura, J.-H. Kim and M. Ikeda; *J. Heterocycl. Chem.*, 1975, **12**, 107. | 20 |
| 75JHC181 | D. H. Kim and A. A. Santilli; *J. Heterocycl. Chem.*, 1975, **12**, 181. | 214, 222 |
| 75JHC183 | S. A. Lang, Jr., B. D. Johnson and E. Cohen; *J. Heterocycl. Chem.*, 1975, **12**, 183. | 554 |
| 75JHC187 | R. A. Coburn and B. Bhooshan; *J. Heterocycl. Chem.*, 1975, **12**, 187. | 489, 502 |
| 75JHC199 | S. Plescia, E. Ajello and V. Sprio; *J. Heterocycl. Chem.*, 1975, **12**, 199. | 382 |
| 75JHC219 | G. A. Poulton and M. E. Williams; *J. Heterocycl. Chem.*, 1975, **12**, 219. | 710 |
| 75JHC311 | A. A. Santilli, S. V. Wanser, D. H. Kim and A. C. Scotese; *J. Heterocycl. Chem.*, 1975, **12**, 311. | 205, 211, 221 |
| 75JHC397 | E. R. Biehl, H. Chiou, J. Keepers, S. Kennard and P. C. Reeves; *J. Heterocycl. Chem.*, 1975, **12**, 397. | 1010 |
| 75JHC407 | J. T. Shaw, M. Kuttesch, S. Funck and R. C. Allen; *J. Heterocycl. Chem.*, 1975, **12**, 407. | 488, 502 |
| 75JHC605 | H. Meier and H. Bühl; *J. Heterocycl. Chem.*, 1975, **12**, 605. | 985 |
| 75JHC699 | K. Venkatasubramanian, R. J. Majeste and L. M. Trefonas; *J. Heterocycl. Chem.*, 1975, **12**, 699. | 1049 |
| 75JHC785 | J. A. Hirsch, R. W. Kosley, Jr., R. P. Morin, G. Schwartzkopf and R. D. Brown; *J. Heterocycl. Chem.*, 1975, **12**, 785. | 637 |
| 75JHC957 | P. Smit, G. A. Stork and H. C. Van der Plas; *J. Heterocycl. Chem.*, 1975, **12**, 957. | 45 |
| 75JHC981 | A. Merle and G. Descotes; *J. Heterocycl. Chem.*, 1975, **12**, 981. | 697, 729, 856 |
| 75JHC1143 | S. A. Lang, Jr., B. D. Johnson and E. Cohen; *J. Heterocycl. Chem.*, 1975, **12**, 1143. | 570 |
| 75JHC1155 | U. Golik and W. Taub; *J. Heterocycl. Chem.*, 1975, **12**, 1155. | 382 |
| 75JHC1221 | B. K. Billings, J. A. Wagner, P. D. Cook and R. N. Castle; *J. Heterocycl. Chem.*, 1975, **12**, 1221. | 356, 367 |
| 75JHC1311 | S. P. Gupta, R. K. Robins and R. A. Long; *J. Heterocycl. Chem.*, 1975, **12**, 1311. | 363 |
| 75JMC74 | J. M. Matsumoto and S. Minami; *J. Med. Chem.*, 1975, **18**, 74. | 212, 260 |
| 75JMC637 | M. L. Edwards, R. E. Bambury and H. W. Ritter; *J. Med. Chem.*, 1975, **18**, 637. | 195 |
| 75JMC913 | S. Sharma, R. Bindra, R. N. Iyer and N. Anand; *J. Med. Chem.*, 1975, **18**, 913. | 351, 366 |
| 75JMC934 | D. T. Witiak, W. P. Heilman, S. K. Sankarappa, R. C. Cavestri and H. A. Newman; *J. Med. Chem.*, 1975, **18**, 934. | 705, 787 |
| 75JOC41 | R. A. Abramovitch, R. B. Rogers and G. M. Singer; *J. Org. Chem.*, 1975, **40**, 41. | 24 |
| 75JOC47 | J. C. Hinshaw; *J. Org. Chem.*, 1975, **40**, 47. | 341 |
| 75JOC1120 | C. A. Kingsbury and J. H. Looker; *J. Org. Chem.*, 1975, **40**, 1120. | 591 |
| 75JOC1142 | A. Padwa, A. Au, G. A. Lee and W. Owens; *J. Org. Chem.*, 1975, **40**, 1142. | 667, 913 |
| 75JOC1175 | S. A. Sojka; *J. Org. Chem.*, 1975, **40**, 1175. | 589, 590, 637 |
| 75JOC1178 | J. Wolt; *J. Org. Chem.*, 1975, **40**, 1178. | 168 |
| 75JOC1201 | T. Sasaki, K. Kanematsu and T. Kataoka; *J. Org. Chem.*, 1975, **40**, 1201. | 553 |
| 75JOC1395 | J. P. Snyder, M. L. Heyman and E. N. Suciu; *J. Org. Chem.*, 1975, **40**, 1395. | 41 |
| 75JOC1438 | G. Evens and P. Caluwe; *J. Org. Chem.*, 1975, **40**, 1438. | 208, 227 |
| 75JOC1745 | M. G. Nair, P. T. Campbell and C. M. Baugh; *J. Org. Chem.*, 1975, **40**, 1745. | 327 |
| 75JOC1760 | R. A. Henry, C. A. Heller and D. W. Moore; *J. Org. Chem.*, 1975, **40**, 1760. | 259 |
| 75JOC2088 | I. Granoth and H. J. Pownall; *J. Org. Chem.*, 1975, **40**, 2088. | 583, 838 |
| 75JOC2234 | A. J. Duggan and S. S. Hall; *J. Org. Chem.*, 1975, **40**, 2234. | 578 |
| 75JOC2239 | J. R. Sanderson, A. G. Zeiler and R. J. Wilterdink; *J. Org. Chem.*, 1975, **40**, 2239. | 990 |
| 75JOC2332 | E. C. Taylor and P. A. Jacobi; *J. Org. Chem.*, 1975, **40**, 2332. | 282, 319 |
| 75JOC2336 | E. C. Taylor, R. F. Abdulla and P. A. Jacobi; *J. Org. Chem.*, 1975, **40**, 2336. | 319 |
| 75JOC2341 | E. C. Taylor, R. F. Abdulla, K. Tanaka and P. A. Jacobi; *J. Org. Chem.*, 1975, **40**, 2341. | 290, 294 |
| 75JOC2690 | D. M. Frieze and S. A. Evans; *J. Org. Chem.*, 1975, **40**, 2690. | 961 |
| 75JOC2756 | S. R. Mani and H. J. Shine; *J. Org. Chem.*, 1975, **40**, 2756. | 968, 969 |
| 75JOC3608 | T. C. Lee and G. Salemnick; *J. Org. Chem.*, 1975, **40**, 3608. | 209, 210, 213, 229 |
| 75JOC3857 | K. Kim, S. R. Mani and H. J. Shine; *J. Org. Chem.*, 1975, **40**, 3857. | 969 |
| 75JOC3874 | G. J. B. Cajipe, G. Landen, B. Semler and H. W. Moore; *J. Org. Chem.*, 1975, **40**, 3874. | 244 |
| 75JST(24)337 | R. W. Kitchin, T. K. Avirah, T. B. Malloy, Jr. and R. L. Cook; *J. Mol. Struct.*, 1975, **24**, 337. | 949 |
| 75KGS643 | V. G. Kharchenko, S. N. Chalaya and L. G. Chichenkova; *Khim. Geterotsikl. Soedin.*, 1975, 643. | 937 |
| 75KGS1290 | A. I. Dychenko, L. S. Pupko and P. S. Pel'kis; *Khim. Geterotsikl. Soedin.*, 1975, 1290. | 455 |
| 75KGS1431 | V. A. Kartsev, P. B. Terent'ev, A. N. Kost and M. F. Budyka; *Khim. Geterotsikl. Soedin.*, 1975, 1431 (*Chem. Abstr.*, 1976, **84**, 59 351). | 239 |
| 75KGS1468 | Yu. S. Andreichikov, Yu. A. Nalimova, G. D. Plakhina, R. F. Saraeva and S. P. Tendryakova; *Khim. Geterotsikl. Soedin.*, 1975, 1468 (*Chem. Abstr.*, 1976, **84**, 74222). | 798, 1002 |
| 75LA240 | A. Roedig and T. Neukam; *Liebigs Ann. Chem.*, 1975, 240. | 593, 740 |
| 75LA1120 | H. Neunhoeffer, H.-J. Degen and J. J. Köhler; *Liebigs Ann. Chem.*, 1975, 1120. | 558 |
| 75LA1445 | E. Oeser, H. Neunhoeffer and H.-W. Frühauf; *Liebigs Ann. Chem.*, 1975, 1445. | 423 |
| 75M229 | H. Neudeck and K. Schlögl; *Monatsh. Chem.*, 1975, **106**, 229. | 629, 774, 777 |
| 75M333 | D. N. Dhar and V. P. Gupta; *Monatsh. Chem.*, 1975, **106**, 333. | 597 |
| 75M1059 | H. Berner and H. Reinshagen; *Monatsh. Chem.*, 1975, **106**, 1059. | 257 |

| | | |
|---|---|---|
| B-75MI21200 | Y. Shvo; in 'Conformational Analysis of Hydrazines', in 'The Chemistry of Hydrazo, Azo, and Azoxy Groups', ed. S. Patai; Wiley, New York, 1975, part 2, p. 1017. | 40 |
| 75MI21300 | C. Pene, M. Hubert-Habart and R. Royer; *Eur. J. Med. Chem.—Chim. Ther.*, 1975, **10**, 340. | 121 |
| B-75MI21301 | S. C. Harvey; in 'The Pharmacological Basis of Therapeutics', ed. L. S. Goodman and A. Gilman; Macmillan, New York, 5th edn., 1975, p. 102. | 150 |
| B-75MI21302 | L. Weinstein; in 'The Pharmacological Basis of Therapeutics', ed. L. S. Goodman and A. Gilman; Macmillan, New York, 5th edn., 1975, p. 1113. | 151 |
| B-75MI21303 | P. Calabresi and R. E. Parks, Jr.; in 'The Pharmacological Basis of Therapeutics', ed. L. S. Goodman and A. Gilman; Macmillan, New York, 5th edn., 1975, p. 1254. | 153 |
| B-75MI21304 | S. C. Harvey; in 'The Pharmacological Basis of Therapeutics', ed. L. S. Goodman and A. Gilman; Macmillan, New York, 5th edn., 1975, p. 124. | 150 |
| 75MI21400 | U. Langenbeck, H. U. Moehring and K. P. Dieckmann; *J. Chromatogr.*, 1975, **115**, 65. | 173 |
| 75MI21500 | R. C. Rastogi and N. K. Ray; *Chem. Phys. Lett.*, 1975, **31**, 524 (*Chem. Abstr.*, 1975, **83**, 15 982). | 202, 204, 234, 236 |
| B-75MI21501 | F. Bergmann, L. Levene and I. Tamir; in 'Proceedings of the 5th International Symposium on the Chemistry and Biology of Pteridines, Konstanz, 1975', ed. W. Pfleiderer; de Gruyter, Berlin, 1975, p. 603 (*Chem. Abstr.*, 1976, **84**, 146 747). | 205 |
| 75MI21502 | S. Pfeifer, H. P. Poehlmann, I. Bornschein and R. Kraft; *Pharmazie*, 1975, **30**, 290 (*Chem. Abstr.*, 1975, **83**, 108 108). | 205 |
| 75MI21503 | J. Hironaka and S. Nishikawa; *Hakko Kogaku Zasshi*, 1975, **53**, 372 (*Chem. Abstr.*, 1975, **83**, 145 680). | 211 |
| 75MI21504 | S. Yurugi; *Takeda Kenkyusho Ho*, 1975, **34**, 53 (*Chem. Abstr.*, 1975, **83**, 79 106). | 232 |
| 75MI21600 | D. L. McAllister and G. Dryhurst; *J. Electroanal. Chem. Interfacial Electrochem.*, 1975, **59**, 75. | 285 |
| B-75MI21601 | W. Pfleiderer; 'Chemistry and Biology of Pteridines', de Gruyter, Berlin, 1975. | 264 |
| B-75MI21602 | G. M. Brown, J. Yim, Y. Suzuki, M. C. Heine and F. Foor; in 'Chemistry and Biology of Pteridines', ed. W. Pfleiderer; de Gruyter, Berlin, 1975, p. 219. | 316, 320 |
| B-75MI21603 | E. C. Taylor; in 'Chemistry and Biology of Pteridines', ed. W. Pfleiderer; de Gruyter, Berlin, 1975, p. 543. | 319 |
| 75MI21900 | A. B. Teplitskii and I. K. Yanson; *Biofizika*, 1975, **20**, 189 (*Chem. Abstr.*, 1975, **83**, 16 712). | 398 |
| B-75MI22000 | J. A. Gautier, M. Miocque and C. J. Farnoux; in 'The Chemistry of Amidines and Imidates', ed. S. Patai; Wiley, New York, 1975, p. 331. | 500 |
| 75MI22001 | R. Hayatsu, M. H. Studier, L. P. Moore and E. Anders; *Geochim. Cosmochim. Acta*, 1975, **39**, 471. | 524 |
| 75MI22002 | J. C. Madelmont and A. Veyre; *Eur. J. Med. Chem.—Chim. Ther.*, 1975, **10**, 257. | 527 |
| B-75MI22003 | H. O. Esser, G. Dupuis, E. Ebert, C. Vogel and G. J. Marco; in 'Herbicides: Chemistry, Degradation and Mode of Action', ed. P. C. Kearney and D. D. Kaufman; Dekker, New York, 1975, p. 129. | 528 |
| 75MI22100 | O. M. Polumbrik and E. I. Zaika; *Dopov. Akad. Nauk Ukr. RSR, Ser. B*, 1975, 732 (*Chem. Abstr.*, 1975, **83**, 179 015). | 548 |
| 75MI22101 | A. R. Friedman, K. T. Koshy and A. L. van der Slik; *Environ. Qual. Saf., Supl.* (*Pesticides*), 1975, **3**, 298 (*Chem. Abstr.*, 1976, **85**, 138 598). | 564 |
| B-75MI22102 | V. Kh. Ivanova, B. I. Buzykin and Yu. P. Kitaev; 'Nov. Polyarogr., Tezisy Dokl. Vses. Soveshch. Polyarogr., 6th', 1975, 83 (*Chem. Abstr.*, 1977, **86**, 80 816). | 565 |
| 75MI22103 | W. Rudnicka and W. Manowska; *Ann. Accad. Med. Gedanensis*, 1975, 165 (*Chem. Abstr.*, 1977, **86**, 72 523). | 569 |
| 75MI22200 | S. L. Spassov, A. G. Bojilova and C. Ivanov; *Dokl. Bolg. Akad. Nauk*, 1975, **28**, 1383. | 582 |
| B-75MI22201 | J. B. Harborne, T. J. Mabry and H. Mabry; 'The Flavonoids', Chapman and Hall, London, 1975. | 584, 602, 603, 618, 631 |
| 75MI22202 | N. A. Tyukavkina, N. N. Pogodaeva, E. I. Brodskaya and Yu. M. Sapozhnikov; *Khim. Prir. Soedin.*, 1975, 583 (*Chem. Abstr.*, 1976, **84**, 42 717). | 597 |
| 75MI22203 | V. P. Karmazin, M. I. Knyazhanskii, E. P. Olekhanovich and G. N. Dorofeenko; *Zh. Prikl. Spektrosk.*, 1975, **22**, 234 (*Chem. Abstr.*, 1975, **83**, 17 984). | 603 |
| B-75MI22204 | J. L. Holmes; in 'International Review of Science, Physical Chemistry Series Two', ed. A. Maccoll; Butterworths, London, 1975, vol. 5, p. 241. | 607 |
| B-75MI22205 | O. Exner; 'Dipole Moments in Organic Chemistry', Thieme, Stuttgart, 1975. | 626 |
| B-75MI22206 | D. Lewis and D. Peters; 'Facts and Theories of Aromaticity', Macmillan, London, 1975. | 633, 635, 638 |
| B-75MI22300 | 'The Flavonoids', ed. J. B. Harborne, T. J. Mabry and H. Mabry; Chapman and Hall, London, 1975. | 693, 707 |
| B-75MI22301 | D. M. X. Donnelly; in 'The Flavonoids', ed. J. B. Harborne, T. J. Mabry and H. Mabry, Chapman and Hall; London, 1975, chap. 15. | 677 |
| B-75MI22302 | H. Wagner and L. Farkas; in 'The Flavonoids', ed. J. B. Harborne, T. J. Mabry and H. Mabry; Chapman and Hall, London, 1975, chap. 4. | 718 |
| 75MI22303 | W. M. Cort, J. W. Scott, M. Araujo, W. J. Mergans, M. A. Cannalonga, M. Osadca, H. Harley, D. R. Parish and W. R. Pool; *J. Am. Oil Chem. Soc.*, 1975, **52**, 174. | 718, 72? |

B-75MI22400    J. B. Harborne, T. J. Mabry and H. Mabry; 'The Flavonoids', Chapman and
               Hall, London, 1975.                                    738, 816, 853, 877
75MI22401      L. Rene and R. Royer; *Eur. J. Med. Chem.—Chim. Ther.*, 1975, **10**, 72.        750
75MI22402      T. Tanaka and T. Tomimatsu; *Hukusokan Kagaku Toronkai Koen Yoshishu, 8th*,
               1975, 194 (*Chem. Abstr.*, 1976, **85**, 32 895).                     786
75MI22500      A. F. Pronin and P. G. Zhbanov; *Issled. Obl. Sint. Katal. Org. Soedin.*, 1975, 18.   892, 902
75MI22501      N. I. Martem'yanova, N. D. Zaitseva and M. I. Kuramshin; *Issled. Obl. Sint.
               Katal. Org. Soedin.*, 1975, 3.                                        928
75MI22502      V. S. Blagoveshchenski, I. V. Kazimirchik, O. P. Yakovleva, N. S. Zefirov and
               V. K. Denisenko; *Probl. S-kh. Nauki Mosk. Univ.*, 1975, 260 (*Chem. Abstr.*,
               1976, **85**, 94 293).                                           929, 939
75MI22503      L. V. Lipatova, S. N. Chalaya, L. A. Tatarinova, E. N. Lyutaya, L. D. Berseneva
               and V. G. Kharchenko; *Issled. Obl. Sint. Katal. Org. Soedin.*, 1975, 11 (*Chem.
               Abstr.*, 1976, **84**, 6330).                                          939
75MI22600      H. R. Buser; *J. Chromatogr.*, 1975, **114**, 95.                          985
B-75MI22601    K. Jones; in 'Basic Organic Chemistry', ed. J. M. Tedder, A. Nechvatal and A.
               H. Jubb; Wiley, London, 1975, vol. 5, p. 452.                           994
75MI22700      S. Leistner and G. Wagner; *Pharmazie*, 1975, **30**, 542.                1000
75MI22800      Z. Arnold and B. Fiszer; *Rocz. Chem.*, 1975, **49**, 285.            1048, 1051
B-75MI22801    A. J. Bannister; in 'MTP International Review of Science Inorganic Chemistry
               Series Two', ed. H. J. Eméleus; Butterworth, London, 1975, vol. 3, p. 41.   1048
75MIP22500     Esso Research and Engineering, *Can. Pat.* 976 170 (1975) (*Chem. Abstr.*, 1976,
               **84**, 105 572).                                                 941
75NEP7504410   Dow Chemical Corp., *Neth. Pat.* 75 04 410 (1975) (*Chem. Abstr.*, 1977, **86**,
               139 594).                                                         939
75NEP7505267   A. D. Little Inc., *Neth. Pat.* 75 05 267 (1975) (*Chem. Abstr.*, 1977, **86**, 171 431).   941
75NKK1041      Y. Nakatsuji, I. Ikeda and M. Okahara; *Nippon Kagaku Kaishi*, 1975, 1041
               (*Chem. Abstr.*, 1975, **83**, 206 218).                               1075
75OMR(7)194    S. Braun and G. Frey; *Org. Magn. Reson.*, 1975, **7**, 194.            394, 463
75OMS97        G. Holzmann, H. W. Rothkopf, R. Mueller and D. Woehrle; *Org. Mass. Spectrom.*,
               1975, **10**, 97.                                                    250
75OMS579       G. Entenmann; *Org. Mass Spectrom.*, 1975, **10**, 579.                  1051
75P1605        M. Jay, J.-F. Gonnet, E. Wollenweber and B. Voirin; *Phytochemistry*, 1975, **14**,
               1605.                                                              602
75PAC(44)767   M. J. S. Dewar; *Pure Appl. Chem.*, 1975, **44**, 767.              533, 536, 572
75RCR603       N. S. Vul'fson and L. S. Golovkina; *Russ. Chem. Rev. (Engl. Transl.)*, 1975, **44**,
               603.                                                              608, 610
75RTC45        A. Nagel, H. C. van der Plas and A. van Veldhuizen; *Recl. Trav. Chim. Pays-Bas*,
               1975, **94**, 45.                                                    287
75RTC163       A. Zilverschoon, J. Meijer, P. Vermeer and L. Brandsma; *Recl. Trav. Chim.
               Pays-Bas*, 1975, **94**, 163.                                          982
75S182         H. Ahne, R. Weidenmann and W. Kleeberg; *Synthesis*, 1975, 182.           480
75S184         H. Ahne, R. Weidenmann and W. Kleeberg; *Synthesis*, 1975, 184.           480
75S187         T. McC. Paterson, R. K. Smalley and H. Suschitzky; *Synthesis*, 1975, 187.    379
75S259         E. Suzuki and S. Inoue; *Synthesis*, 1975, 259.                          795
75S329         G. Goor and M. Anteunis; *Synthesis*, 1975, 329.                         992
75S451         J. H. Verheijen and H. Kloosterziel; *Synthesis*, 1974, 451.              984
75S483         D. J. Anderson and A. Hassner; *Synthesis*, 1975, 483.                    554
75S522         R. Gompper and F. Towae; *Synthesis*, 1975, 522.              1056, 1061, 1069
75S709         T. McC. Paterson, R. K. Smalley and H. Suschitzky; *Synthesis*, 1975, 709.    379
75S739         T. Manimaran, T. K. Thiruvengadam and V. T. Ramakrishnan; *Synthesis*, 1975,
               739.                                                               809
75S794         R. H. Fischer and H. M. Weitz; *Synthesis*, 1975, 794.                443, 454
75SA(A)339     D. Vedal, O. H. Ellestad, P. Klaeboe and G. Hagen; *Spectrochim. Acta, Part A*,
               1975, **31**, 339.                                                   628
75SST(3)728    J. Fabian; *Org. Compd. Sulphur, Selenium, Tellurium*, 1975, **3**, 728.      944
75T505         P. K. Claus, W. Rieder, P. Hofbauer and E. Vilsmaier; *Tetrahedron*, 1975, **31**, 505.   900
75T523         F. G. Riddell; *Tetrahedron*, 1975, **31**, 523.                         1000
75T533         J. A. Jongejan, H. I. X. Mager and W. Berends; *Tetrahedron*, 1975, **31**, 533.   308, 518
75T541         H. van Koningsveld; *Tetrahedron*, 1975, **31**, 541.                     308
75T555         G. McConnachie and F. A. Neugebauer; *Tetrahedron*, 1975, **31**, 555.      556
75T619         W. Jarre, D. Bieniek and F. Korte; *Tetrahedron*, 1975, **31**, 619.         503
75T1415        P. Bischof, R. Gleiter, R. Dach, D. Enders and D. Seebach; *Tetrahedron*, 1975,
               **31**, 1415.                                                         532
75T1837        R. Camerlynck and M. Anteunis; *Tetrahedron*, 1975, **31**, 1837.           989
75T1879        J. K. Chakrabarti and D. E. Tupper; *Tetrahedron*, 1975, **31**, 1879.        484
75T2099        J. Van Coppenolle and M. Renson; *Tetrahedron*, 1975, 2099.               914
75T2229        S. Sib; *Tetrahedron*, 1975, **31**, 2229.                               872
75T2587        N. J. Cussans and T. N. Huckerby; *Tetrahedron*, 1975, **31**, 2587.         589
75T2607        A. Bruggink and A. McKillop; *Tetrahedron*, 1975, **31**, 2607.          807, 830
75T2669        S. Yoneda, T. Sugimoto, O. Tanaka, Y. Moriya and Z. Yoshida; *Tetrahedron*,
               1975, **31**, 2669.                                                 923, 938

| | | |
|---|---|---|
| 75T2679 | J. C. Meslin, Y. T. N'Guessan, H. Quiniou and F. Tonnard; *Tetrahedron*, 1975, **31**, 2679. | 935, 983 |
| 75T2719 | N. J. Cussans and T. N. Huckerby; *Tetrahedron*, 1975, **31**, 2719. | 590 |
| 75T3059 | J. P. Pradere, Y. T. N'Guessan, H. Quiniou and F. Tonnard; *Tetrahedron*, 1975, **31**, 3059. | 934 |
| 75TH21900 | H. Hennig; *Ph. D. Thesis*, Techn. Hochschule Darmstadt, 1975. | 395, 403 |
| 75TH22300 | H. K. Wilson; M.Sc. Thesis, University of Wales, 1975, p. 158. | 711 |
| 75TL569 | J. W. Barton and R. B. Walker; *Tetrahedron Lett.*, 1975, 569. | 240, 245 |
| 75TL1087 | N. S. Zefirov; *Tetrahedron Lett.*, 1975, 1087. | 889 |
| 75TL2389 | E. J. Corey and A. P. Kozikowski; *Tetrahedron Lett.*, 1975, 2389. | 688 |
| 75TL2897 | B. Burg, W. Dittmar, H. Reim, A. Steigel and J. Sauer; *Tetrahedron Lett.*, 1975, 2897. | 427 |
| 75TL2901 | H. Reim, A. Steigel and J. Sauer; *Tetrahedron Lett.*, 1975, 2901. | 427 |
| 75TL3923 | S. Ikegami, J. Ohishi and Y. Shimizu; *Tetrahedron Lett.*, 1975, 3923. | 927 |
| 75TL4527 | G. B. Merrill and H. Shechter; *Tetrahedron Lett.*, 1975, 4527. | 563 |
| 75UP22100 | H. Neunhoeffer and H.-J. Degen; unpublished results, 1975. | 558 |
| 75USP3860588 | K. H. G. Pilgrim and R. D. Skiles; *U.S. Pat.* 3 860 588 (1975) (*Chem. Abstr.*, 1975, **82**, 112 113). | 562 |
| 75USP3860589 | K. H. G. Pilgrim and R. D. Skiles; *U.S. Pat.* 3 860 589 (1975) (*Chem. Abstr.*, 1975, **82**, 112 112). | 562 |
| 75USP3863010 | S. A. Lang, Jr., E. Cohen and A. E. Sloboda; *U.S. Pat.* 3 863 010 (1975) (*Chem. Abstr.*, 1975, **83**, 43 387). | 570 |
| 75USP3887550 | A. L. J. Beckwith; (Sherwin-Williams Co.), *U.S. Pat.* 3 887 550 (1975) (*Chem. Abstr.*, 1975, **83**, 147 504). | 206, 215 |
| 75USP3898216 | T. D. Weaver; (E. I. du Pont de Nemours and Co.), *U.S. Pat.* 3 898 216 (1975) (*Chem. Abstr.*, 1975, **83**, 181 120). | 253 |
| 75USP3904614 | S. A. Lang, Jr., E. Cohen and A. E. Sloboda; *U.S. Pat.* 3 904 614 (1975) (*Chem. Abstr.*, 1976, **84**, 17 438). | 570 |
| 75USP3904617 | Ayerst-McKenna and Harrison Inc., *U.S. Pat.* 3 904 617 (1975) (*Chem. Abstr.*, 1976, **84**, 164 794). | 941 |
| 75USP3909235 | Eli Lilly and Co. *U.S. Pat.* 3 909 235 (1975) (*Chem. Abstr.*, 1976, **84**, 89 820). | 940 |
| 75ZC19 | W. Schramm, G. Voss, M. Michalik, G. Rembarz and E. Fischer; *Z. Chem.*, 1975, **15**, 19. | 1069 |
| 75ZN(B)603 | N. A. L. Kassab, H. H. Harhash and S. O. A. Allah; *Z. Naturforsch., Teil B*, 1975, **30**, 603. | 451 |
| 75ZOR1304 | A. Ya. Zheltov, V. Ya. Rodionov and B. I. Stepanov; *Zh. Org. Chim.*, 1975, **11**, 1304 (*Chem. Abstr.*, 1975, **83**, 113 190). | 955 |
| 75ZOR1540 | V. G. Kharchenko and S. N. Chalaya; *Zh. Org. Khim.*, 1975, **11**, 1540. | 938 |
| 75ZOR1543 | V. G. Kharchenko, N. S. Smirnova, S. N. Chalaya, A. S. Tatarinova and L. G. Chichenkova; *Zh. Org. Khim.*, 1975, **11**, 1543. | 912 |
| 75ZOR2217 | I. V. Bodrikov, A. A. Michrin and V. Krasnov; *Zh. Org. Khim.*, 1975, **11**, 2217. | 1044, 1047 |
| 75ZOR2447 | V. G. Kharchenko, S. N. Chalaya, T. V. Stolbova and S. K. Klimenko; *Zh. Org. Khim.*, 1975, **11**, 2447. | 908 |
| 75ZOR2613 | B. A. Korolev and M. A. Mal'tseva; *Zh. Org. Khim.*, 1975, **11**, 2613 (*Chem. Abstr.*, 1975, **84**, 90 122). | 469 |
| 76ACR201 | F. McCapra; *Acc. Chem. Res.*, 1976, **9**, 201. | 194 |
| 76ACS(B)24 | S. Baklien, P. Groth and K. Undheim; *Acta Chem. Scand., Ser. B*, 1976, **30**, 24. | 910 |
| 76ACS(B)71 | U. Anthoni, B. M. Dahl, H. Eggert, C. Larsen and P. H. Nielsen; *Acta Chem. Scand., Ser. B*, 1976, **30**, 71. | 444 |
| 76ACS(B)619 | B. P. Nilsen and K. Undheim; *Acta Chem. Scand., Ser. B*, 1976, **30**, 619. | 645 |
| 76AF2125 | B. Grebian, H. E. Geissler and E. Mutschler; *Arzneim.-Forsch.*, 1976, **26**, 2125. | 325 |
| 76AG447 | H.-D. Martin and M. Hekman; *Angew. Chem.*, 1976, **88**, 447. | 551 |
| 76AG475 | M. Janda and P. Hemmerich; *Angew. Chem.*, 1976, **88**, 475. | 222, 260 |
| 76AG(E)115 | H. J. Bestmann, G. Schmid and D. Sandmeier; *Angew. Chem., Int. Ed. Engl.*, 1976, **15**, 115. | 809 |
| 76AG(E)782 | A. Gieren and F. Pertlik; *Angew. Chem., Int. Ed. Engl.*, 1976, **15**, 782. | 1049 |
| 76AHC(19)215 | R. J. Kobylecki and A. McKillop; *Adv. Heterocycl. Chem.*, 1976, **19**, 215. | 369 |
| 76AHC(20)117 | A. Albert; *Adv. Heterocycl. Chem.*, 1976, **20**, 117. | 265, 287 |
| 76AHC(S1)1 | J. Elguero, C. Marzin, A. R. Katritzky and P. Linda; *Adv. Heterocycl. Chem.*, 1976, Suppl. 1, 1. | 641 |
| 76AHC(S1)71 | J. Elguero, C. Marzin, A. R. Katritzky and P. Linda; *Adv. Heterocycl. Chem.*, 1976, Suppl. 1, 71. | 66, 68 |
| 76AHC(S1)116 | J. Elguero, C. Marzin, A. R. Katritzky and P. Linda; *Adv. Heterocycl. Chem.*, 1976, Suppl. 1, 116. | 642 |
| 76AJC115 | J. H. Bowie and T. Blumenthal; *Aust. J. Chem.*, 1976, **29**, 115. | 615 |
| 76AJC1051 | C. P. Joshua and V. P. Rajan; *Aust. J. Chem.*, 1976, **29**, 1051. | 499 |
| 76AP679 | G. Seitz and T. Kämpchen; *Arch. Pharm. (Weinheim, Ger.)*, 1976, **309**, 679. | 244, 552 |
| 76AX(B)915 | K. Folting, W. E. Streib and L. L. Merritt, Jr.; *Acta Crystallogr., Part B*, 1976, **32**, 915. | 622 |
| 76AX(B)946 | K. Ueno and N. Saito; *Acta Crystallogr., Part B*, 1976, **32**, 946. | 622 |

| | | |
|---|---|---|
| 76AX(B)1467 | M. R. Caira, R. G. F. Siles, L. R. Nassimbeni, G. M. Sheldrick and R. G. Hazell; *Acta Crystallogr., Part B*, 1976, **32**, 1467. | 537 |
| 76AX(B)2101 | D. S. Brown, J. D. Lee and P. R. Russell; *Acta Crystallogr., Part B*, 1976, **32**, 2101. | 461 |
| 76AX(B)2240 | M. A. Hamid and A. Hargreaves; *Acta Crystallogr., Part B*, 1976, **32**, 2240. | 370 |
| 76AX(B)3178 | G. De With, S. Harkema and D. Feil; *Acta Crystallogr., Part B*, 1976, **32**, 3178. | 158 |
| 76BCJ253 | Y. Miura, Y. Morimoto and M. Kinoshita; *Bull. Chem. Soc. Jpn.*, 1976, **49**, 253. | 549 |
| 76BCJ1339 | K. Kubo, T. Nonaka and K. Odo, *Bull. Chem. Soc. Jpn.*, 1976, **49**, 1339. | 535 |
| 76BCJ1715 | Y. Miura, Y. Morimoto and M. Kinoshita; *Bull. Chem. Soc. Jpn.*, 1976, **49**, 1715. | 549 |
| 76BCJ1725 | M. Takahashi, H. Ishida and M. Kohmoto; *Bull. Chem. Soc. Jpn.*, 1976, **49**, 1725. | 552 |
| 76BRP1447032 | H. F. Hodson and J. F. Batchelor; Wellcome Foundation Ltd., *Br. Pat.* 1 447 032 (1976) (*Chem. Abstr.*, 1977, **86**, 55 284). | 941, 993 |
| 76BSF251 | N. Vinot and P. Maitte; *Bull. Soc. Chim. Fr.*, 1976, 251. | 252 |
| 76BSF433 | R. Hazard and A. Tallec; *Bull. Soc. Chim. Fr.*, 1976, 433. | 376, 383 |
| 76BSF621 | B. Decroix, P. Dubus, J. Morel and P. Pastour; *Bull. Soc. Chim. Fr.*, 1976, 621. | 569 |
| 76BSF987 | J. Ficini, J. Besseyre and A. Krief; *Bull. Soc. Chim. Fr.*, 1976, 987. | 593, 760 |
| 76BSF1549 | P. Battesti, O. Battesti and M. Sélim; *Bull. Soc. Chim. Fr.*, 1976, 1549. | 336, 337, 358 |
| 76BSF1967 | J. Andrieux, J. Aknin, B. Bodo, C. Deschamp-Vallet, M. Meyer-Dayan and D. Molho; *Bull. Soc. Chim. Fr.*, 1976, 1967. | 662 |
| 76CB1113 | H. Neunhoeffer and B. Lehmann; *Chem. Ber.*, 1976, **109**, 1113. | 402, 416, 421 |
| 76CB1346 | B. F. Becker and H. P. Fritz; *Chem. Ber.*, 1976, **109**, 1346. | 509 |
| 76CB1549 | B. Eistert, A. Schmitt and T. J. Arackal; *Chem. Ber.*, 1976, **109**, 1549. | 923 |
| 76CB1787 | H. Junek, A. Hermetter and H. Fischer-Colbrie; *Chem. Ber.*, 1976, **109**, 1787. | 41 |
| 76CB2097 | P. Stoss and G. Satzinger; *Chem. Ber.*, 1976, **109**, 2097. | 1069 |
| 76CB2107 | H. W. Roesky and H. Zamankhan; *Chem. Ber.*, 1976, **109**, 2107. | 1069 |
| 76CB2442 | R. Appel, J.-R. Lundehn and E. Lassmann; *Chem. Ber.*, 1976, **109**, 2442. | 1051 |
| 76CB3159 | K. Kobayashi and W. Pfleiderer; *Chem. Ber.*, 1976, **109**, 3159. | 283 |
| 76CB3217 | W. Hutzenlaub, K. Kobayashi and W. Pfleiderer; *Chem. Ber.*, 1976, **109**, 3217. | 282, 283 |
| 76CB3228 | T. Itoh and W. Pfleiderer; *Chem. Ber.*, 1976, **109**, 3228. | 282 |
| 76CC78 | R. J. Bass; *J. Chem. Soc., Chem. Commun.*, 1976, 78. | 821 |
| 76CC313 | A. G. Anastassiou and E. Reichmanis; *J. Chem. Soc., Chem. Commun.*, 1976, 313. | 552 |
| 76CC411 | C. W. Rees, R. C. Storr and P. J. Whittle; *J. Chem. Soc., Chem. Commun.*, 1976, 411. | 380 |
| 76CC417 | J. L. Markham and P. G. Sammes; *J. Chem. Soc., Chem. Commun.*, 1976, 417. | 175 |
| 76CC470 | O. O. Orazi and R. A. Corral; *J. Chem. Soc., Chem. Commun.*, 1976, 470. | 1018 |
| 76CC588 | K. Senga, H. Kanazawa and S. Nishigaki; *J. Chem. Soc., Chem. Commun.*, 1976, 588. | 314 |
| 76CC608 | F. McCapra, Y. C. Chang and A. Burford; *J. Chem. Soc., Chem. Commun.*, 1976, 608. | 1041 |
| 76CC718 | A. E. Baydar and G. V. Boyd; *J. Chem. Soc., Chem. Commun.*, 1976, 718. | 1022 |
| 76CC771 | R. A. Abramovitch, K. M. More, I. Shinkai and P. C. Srinivasan; *J. Chem. Soc., Chem. Commun.*, 1976, 771. | 1016 |
| 76CC899 | A. Sultan Afridi, A. R. Katritzky and C. A. Ramsden; *J. Chem. Soc., Chem. Commun.*, 1976, 899. | 938 |
| 76CC983 | H. Alper and J. E. Prickett; *J. Chem. Soc., Chem. Commun.*, 1976, 983. | 188 |
| 76CC1002 | P. K. Claus, F. W. Vierhapper and R. L. Willer; *J. Chem. Soc., Chem. Commun.*, 1976, 1002. | 961 |
| 76CCC2771 | R. Smrž, J. O. Jílek, K. Šindelář, B. Kakáč, E. Svátek, J. Holubek, J. Grimová and M. Protiva; *Collect. Czech. Chem. Commun.*, 1976, **41**, 2771. | 80 |
| 76CCC3378 | V. Novák and I. Dobáš; *Collect. Czech. Chem. Commun.*, 1976, **41**, 3378. | 475, 486 |
| 76CJC280 | I. W. J. Still, N. Plavac, D. M. McKinnon and M. S. Chauhan; *Can. J. Chem.*, 1976, **54**, 280. | 589 |
| 76CJC455 | I. W. J. Still, P. C. Arora, M. S. Chauhan, M. H. Kwan and M. J. Thomas; *Can. J. Chem.*, 1976, **54**, 455. | 909 |
| 76CJC2723 | S. E. V. Phillips and J. Trotter; *Can. J. Chem.*, 1976, **54**, 2723. | 622 |
| 76CJC3012 | N. K. Sharma, F. de Reinach-Hirtzbach and T. Durst; *Can. J. Chem.*, 1976, **54**, 3012. | 991 |
| 76CJC3757 | L. D. Colebrook, H. G. Giles, A. Rosowsky, W. E. Bentz and J. R. Fehlner; *Can. J. Chem.*, 1976, **54**, 3757. | 467 |
| 76CJC3879 | M. S. Chauhan and D. M. McKinnon; *Can. J. Chem.*, 1976, **54**, 3879. | 1042, 1082 |
| 76CL5 | A. Croisy, A. Ricci, M. Jancevska, P. Jacquignon and D. Balucani; *Chem. Lett.*, 1976, 5. | 908 |
| 76CPB238 | T. Higashino, M. Goi and E. Hayashi; *Chem. Pharm. Bull.*, 1976, **24**, 238. | 252 |
| 76CPB1870 | T. Kametani, Y. Kigawa, T. Takahashi, H. Nemoto and K. Fukumoto; *Chem. Pharm. Bull.*, 1976, **24**, 1870. | 234, 237, 248 |
| 76CPB2057 | K. Nishikawa, H. Shimakawa, Y. Inada, Y. Shibouta, S. Kikuchi, S. Yurugi and Y. Oka; *Chem. Pharm. Bull.*, 1976, **24**, 2057. | 260, 368 |
| 76CPB2078 | E. Mizuta, K. Nishikawa, K. Omura and Y. Oka; *Chem. Pharm. Bull.*, 1976, **24**, 2078. | 204, 260 |
| 76CPB2637 | Y. Kimura, T. Miyamoto, J.-I. Matsumoto and S. Minami; *Chem. Pharm. Bull.*, 1976, **24**, 2637. | 344, 356 |
| 76CPB2699 | K. Omura, N. Tada, M. Tomimoto, Y. Usui, Y. Oka and S. Yurugi; *Chem. Pharm. Bull.*, 1976, **24**, 2699. | 242, 243 |

76CR(C)(282)357 A. Lebouc, O. Riobé and J. Delannay; *C.R. Hebd. Seances Acad. Sci., Ser. C*, 1976, **282**, 357. 731, 732, 735

76CR(C)(282)861 M. Peson, P. De Lajudie, M. Antoine, S. Chabassier and P. Girard; *C.R. Hebd. Seances Acad. Sci., Ser. C*, 1976, **282**, 861. 256

76CRV75 R. Mechoulam, N. K. McCallum and S. Burstein; *Chem. Rev.*, 1976, **76**, 75. 771, 877

76CRV389 J. K. Rasmussen and A. Hassner; *Chem. Rev.*, 1976, **76**, 389. 1044, 1079, 1081

76CZ236 W. Kliegel; *Chem.-Ztg.*, 1976, **100**, 236. 1080

76CZ496 R. Grashey, M. Weidner, C. Knorn and M. Bauer; *Chem.-Ztg.*, 1976, **100**, 496. 558

76CZ496 R. Grashey, G. Schroll and M. Weidner; *Chem.-Ztg.*, 1976, **100**, 496. 561

76E1122 G. Prota, H. Rorsman, A.-M. Rosengren and E. Rosengren; *Experientia*, 1976, **32**, 1122. 1038

76FRP2044534 N. Ivanoff, F. Lahmani, M. Magat and M. P. Pileni; *Fr. Pat.* 2 044 534 (1976). 191

76G1 S. Auricchio, G. Vidari and P. Vita-Finzi; *Gazz. Chim. Ital.*, 1976, **106**, 1. 563

76G457 V. Galasso; *Gazz. Chim. Ital.*, 1976, **106**, 457. 946

76GEP2434563 H. Pietsch, K. Clauss, E. Schmidt and H. Jensen; *Ger. Pat.* 2 434 563 (1976) (*Chem. Abstr.*, 1976, **84**, 135 684). 1027

76GEP2524475 D. Arlt, M. Petinaux, K. Findeisen and D. Dieterich; *Ger. Pat.* 2 524 475 (1976) (*Chem. Abstr.*, 1977, **86**, 106 669). 1086

76GEP2527490 H. M. Weitz, R. Fischer and D. Lenke; *Ger. Pat.* 2 527 490 (1976) (*Chem. Abstr.*, 1977, **86**, 140 092). 454

76GEP2527639 A. D. Brewer and R. A. Davis; *Ger. Pat.* 2 527 639 (1976) (*Chem. Abstr.*, 1976, **84**, 150 640). 994

76GEP2550163 C. Cotrel, C. Jeanmart and M. Barreau; *Ger. Pat.* 2 550 163 (1976) (*Chem. Abstr.*, 1976, **85**, 108 671). 994

76GEP2551919 Chinoin Gyogyszer, *Ger. Pat.* 2 551 919 (1976) (*Chem. Abstr.*, 1977, **86**, 38 584). 941

76GEP2615376 F. T. Boyle and A. Davies; *Ger. Pat.*, 2 615 376 (1976) (*Chem. Abstr.*, 1977, **86**, 72 663). 993

76H(4)769 M. J. Haddadin and C. H. Issidorides; *Heterocycles*, 1976, **4**, 769. 170

76H(4)953 S. Takano, S. Yamada, K. Tanigawa, S. Hatakeyama and K. Ogasawara; *Heterocycles*, 1976, **4**, 953. 988

76H(4)1243 M. Tashiro and S. Mataka; *Heterocycles*, 1976, **4**, 1243. 1046

76H(4)1755 P. Bravo, C. Ticozzi and D. Maggi; *Heterocycles*, 1976, **4**, 1755. 770

76H(4)1875 M. Watanabe, M. Minohara, K. Masuda, T. Kinoshita and S. Furukawa; *Heterocycles*, 1976, **4**, 1875. 1033

76H(5)401 Z. Yoshida, H. Konishi, K. Hayashi and H. Ogoshi; *Heterocycles*, 1976, **5**, 401. 51

76H(5)413 M. Hori, T. Kataoka, H. Shimizu and H. Aoki; *Heterocycles*, 1976, **5**, 413. 926

76H(5)839 R. W. Thomas and N. J. Leonard; *Heterocycles*, 1976, **5**, 839. 600

76HCA1169 B. Maurer and G. Ohloff; *Helv. Chim. Acta*, 1976, **59**, 1169. 193

76HCA2374 J. H. Bieri, W. P. Hummel and M. Viscontini; *Helv. Chim. Acta*, 1976, **59**, 2374. 276

76HCA2379 R. Weber and M. Viscontini, *Helv. Chim. Acta*, 1976, **59**, 2379. 281

76HCA2765 B. Hardegger and S. Shatzmiller; *Helv. Chim. Acta*, 1976, **59**, 2765. 1000

76IJC(B)273 Y. A. Ibrahim; *Indian. J. Chem., Sect. B*, 1976, **14**, 273. 409

76IJC(B)739 M. A. Elkasaby; *Indian J. Chem., Sect. B*, 1976, **14**, 739. 757, 841

76IZV2799 V. I. Dronov and R. F. Nigmatullina; *Izv. Akad. Nauk SSSR, Ser. Khim.*, 1976, 2799 (*Chem. Abstr.*, 1977, **86**, 139 960). 975

76JA1875 T. J. Levek and E. F. Kiefer; *J. Am. Chem. Soc.*, 1976, **98**, 1875. 49

76JA2741 N. L. Allinger, M. J. Hickey and J. Kao; *J. Am. Chem. Soc.*, 1976, **98**, 2741. 946

76JA2750 J. Aihara; *J. Am. Chem. Soc.*, 1976, **98**, 2750. 633

76JA3460 E. T. Harrigan, A. Chakrabarti and N. Hirota; *J. Am. Chem. Soc.*, 1976, **98**, 3460. 576

76JA3583 E. L. Eliel, V. S. Rao and F. G. Riddell; *J. Am. Chem. Soc.*, 1976, **98**, 3583. 954

76JA5269 S. F. Nelsen, V. Peacock and G. R. Weisman; *J. Am. Chem. Soc.*, 1976, **98**, 5269. 332

76JA5380 C. M. Harris, J. S. Roberson and T. M. Harris; *J. Am. Chem. Soc.*, 1976, **98**, 5380. 835

76JA5443 R. M. Hochstrasser and D. S. King; *J. Am. Chem. Soc.*, 1976, **98**, 5443. 543

76JA6350 W. H. Rastetter; *J. Am. Chem. Soc.*, 1976, **98**, 6350. 761

76JA6477 A. Abe; *J. Am. Chem. Soc.*, 1976, **98**, 6477. 629

76JA6798 N. L. Allinger and D. Y. Chung; *J. Am. Chem. Soc.*, 1976, **98**, 6798. 946

76JA7733 T. M. Harris, G. P. Murphy and A. J. Poje; *J. Am. Chem. Soc.*, 1976, **98**, 7733. 582, 796, 811

76JAP(K)7616693 K. Noda, A. Nakagawa, T. Motomura, S. Miyata and H. Ide; (Hisamitsu Pharm. Co. Ltd.), *Jpn. Kokai* 76 16 693 (1976) (*Chem. Abstr.*, 1976, **85**, 108 661). 207

76JAP(K)7636485 K. Noda, A. Nakagawa, K. Yamagata, K. Noguchi, T. Yoshitake and H. Ide; (Hisamitsu Pharm. Co. Ltd.), *Jpn. Kokai* 76 36 485 (1976) (*Chem. Abstr.*, 1976, **85**, 177 504). 208

76JAP(K)7641391 K. Noda, A. Nakagawa, T. Motomura, Y. Zaitsu and H. Ide; (Hisamitsu Pharm. Co. Ltd.), *Jpn. Kokai* 7 641 391 (1976) (*Chem. Abstr.*, 1976, **85**, 177 474). 210

76JAP(K)7696893 K. Uno, K. Niume and K. Nakamichi; *Jpn. Kokai*, 76 96 893 (1976) (*Chem. Abstr.*, 1976, **85**, 178 197). 993

76JAP(K)76108091 K. Noda, A. Nakagawa, S. Miyata and H. Ide; (Hisamitsu Pharm. Co. Ltd.), *Jpn. Kokai* 76 108 091 (1976) (*Chem. Abstr.*, 1977, **86**, 121 365). 216

76JAP(K)76122092 K. Noda, A. Nakagawa, Y. Nakashina and H. Ide; (Hisamitsu Pharm. Co. Ltd.), *Jpn. Kokai* 76 122 092 (1976) (*Chem. Abstr.*, 1977, **87**, 6016). 216

76JAP(K)76139633 H. Yuki and Y. Naka; (Yoshitomi Pharm. Ind. Ltd.), *Jpn. Kokai* 76 139 633 (1976) (*Chem. Abstr.*, 1977, **87**, 23 318). 221

| 76JCS(P1)38 | D. H. R. Barton, J. W. Ducker, W. A. Lord and P. D. Magnus; *J. Chem. Soc., Perkin Trans. 1*, 1976, 38. | 1045, 1067 |
| 76JCS(P1)83 | L. C. March, K. Wasti and M. M. Joullié; *J. Chem. Soc., Perkin Trans. 1*, 1976, 83. | 455 |
| 76JCS(P1)131 | J. Clark and B. Parvizi; *J. Chem. Soc., Perkin Trans. 1*, 1976, 131. | 208, 214, 222 |
| 76JCS(P1)207 | J. T. A. Boyle, M. F. Grundon and M. D. Scott; *J. Chem. Soc., Perkin Trans. 1*, 1976, 207. | 442 |
| 76JCS(P1)499 | D. H. R. Barton, L. Cottier, K. Freund, F. Luini, P. D. Magnus and I. Salazar; *J. Chem. Soc., Perkin Trans. 1*, 1976, 499. | 839, 856 |
| 76JCS(P1)532 | S. Kurata, T. Kusumi, Y. Inouye and H. Kakisawa; *J. Chem. Soc., Perkin Trans. 1*, 1976, 532. | 739 |
| 76JCS(P1)592 | D. E. Ames and C. J. A. Byrne; *J. Chem. Soc., Perkin Trans. 1*, 1976, 592. | 244 |
| 76JCS(P1)749 | R. M. Scrowston and D. C. Shaw; *J. Chem. Soc., Perkin Trans. 1*, 1976, 749. | 909, 929 |
| 76JCS(P1)1073 | D. E. Ames and O. Ribeiro; *J. Chem. Soc., Perkin Trans. 1*, 1976, 1073. | 686 |
| 76JCS(P1)1146 | S. Miyano, N. Abe, K. Sumoto and K. Teramoto; *J. Chem. Soc., Perkin Trans. 1*, 1976, 1146. | 1032 |
| 76JCS(P1)1202 | F. Roeterdink and H. C. van der Plas; *J. Chem. Soc., Perkin Trans. 1*, 1976, 1202. | 73 |
| 76JCS(P1)1377 | R. C. Ellis, W. B. Whalley and K. Ball; *J. Chem. Soc., Perkin Trans. 1*, 1976, 1377. | 838 |
| 76JCS(P1)1438 | R. H. Carter, R. M. Colyer, R. A. Hill and J. Staunton; *J. Chem. Soc., Perkin Trans. 1*, 1976, 1438. | 833 |
| 76JCS(P1)1805 | F. Yoneda, Y. Sakuma, S. Mizumoto and R. Ito; *J. Chem. Soc., Perkin Trans. 1*, 1976, 1805. | 222 |
| 76JCS(P1)1969 | S. Ahmed, R. Lofthouse and G. Shaw; *J. Chem. Soc., Perkin Trans. 1*, 1976, 1969. | 1002, 1028 |
| 76JCS(P1)2137 | S.-F. Tan and T.-H. Tjia; *J. Chem. Soc., Perkin Trans. 1*, 1976, 2137. | 793 |
| 76JCS(P1)2161 | T. L. Gilchrist, C. J. Harris, F. D. King, M. E. Peek and C. W. Rees; *J. Chem. Soc., Perkin Trans. 1*, 1976, 2161. | 1072 |
| 76JCS(P1)2241 | J. I. Okogun; *J. Chem. Soc., Perkin Trans. 1*, 1976, 2241. | 838 |
| 76JCS(P1)2343 | A. K. Bose, W. A. Hoffman, III and M. S. Manhas; *J. Chem. Soc., Perkin Trans. 1*, 1976, 2343. | 909 |
| 76JCS(P1)2475 | A. Pelter, R. S. Ward and T. I. Gray; *J. Chem. Soc., Perkin Trans. 1*, 1976, 2475. | 591 |
| 76JCS(P2)345 | K. Bergesen, B. M. Carden and M. J. Cook; *J. Chem. Soc., Perkin Trans. 2*, 1976, 345. | 954 |
| 76JCS(P2)438 | R. Korenstein, K. A. Muszkat, M. A. Slifkin and E. Fischer; *J. Chem. Soc., Perkin Trans. 2*, 1976, 438. | 621 |
| 76JCS(P2)1386 | P. D. Cradwick; *J. Chem. Soc., Perkin Trans. 2*, 1976, 1386. | 2 |
| 76JCS(P2)1564 | I. J. Ferguson, A. R. Katritzky and R. Patel; *J. Chem. Soc., Perkin Trans. 2*, 1976, 1564. | 340, 367 |
| 76JCS(P2)1861 | I. J. Ferguson, A. R. Katritzky and D. M. Read; *J. Chem. Soc., Perkin Trans. 2*, 1976, 1861. | 1054 |
| 76JHC123 | S. W. Schneller, D. R. Moore and M. A. Smith; *J. Heterocycl. Chem.*, 1976, **13**, 123. | 732, 908 |
| 76JHC211 | G. Romussi and G. Ciarallo; *J. Heterocycl. Chem.*, 1976, **13**, 211. | 583, 596, 601 |
| 76JHC439 | J. I. DeGraw and V. H. Brown; *J. Heterocycl. Chem.*, 1976, **13**, 439. | 202, 207, 218 |
| 76JHC589 | R. S. Klein, F. G. De Las Heras, S. Y.-K. Tam, I. Wempen and J. J. Fox; *J. Heterocycl. Chem.*, 1976, **13**, 589. | 491 |
| 76JHC609 | H. H. Huang, S. F. Tan and T. H. Tija; *J. Heterocycl. Chem.*, 1976, **13**, 609. | 626, 627, 640 |
| 76JHC615 | E. Aiello, S. Plescia and G. Dattolo; *J. Heterocycl. Chem.*, 1976, **13**, 615. | 1071 |
| 76JHC665 | L. Cazaux and P. Tisnès; *J. Heterocycl. Chem.*, 1976, **13**, 665. | 1041, 1075 |
| 76JHC681 | Y. Iwanami and T. Inagaki; *J. Heterocycl. Chem.*, 1976, **13**, 681. | 1009 |
| 76JHC829 | P. Loew and C. D. Weis; *J. Heterocycl. Chem.*, 1976, **13**, 829. | 476, 484 |
| 76JHC917 | D. Alsofrom, H. Grossberg and H. Sheffer; *J. Heterocycl. Chem.*, 1976, **13**, 917. | 514 |
| 76JMC549 | R. K. Razdan, T. B. Zitko, G. R. Handrick, H. C. Dalzell, H. G. Pars, J. F. Howes, N. Plotnikoff, P. W. Dodge, A. Dren, J. Kyncl, L. Shoer and W. R. Thompson; *J. Med. Chem.*, 1976, **19**, 549. | 941 |
| 76JMC825 | M. G. Nair and P. T. Campbell; *J. Med. Chem.*, 1976, **19**, 825. | 327 |
| 76JMC1404 | S. A. Lang, Jr., B. D. Johnson, E. Cohen, A. E. Sloboda and E. Greenblatt; *J. Med. Chem.*, 1976, **19**, 1404. | 570 |
| 76JMR(21)241 | L. Ernst; *J. Magn. Reson.*, 1976, **21**, 241. | 589 |
| 76JOC1058 | T. G. Majewicz and P. Caluwe; *J. Org. Chem.*, 1976, **41**, 1058. | 228 |
| 76JOC1079 | A. McKillop and T. S. B. Sayer; *J. Org. Chem.*, 1976, **41**, 1079. | 1032 |
| 76JOC1095 | A. D. Broom, J. L. Shim and G. L. Anderson; *J. Org. Chem.*, 1976, **41**, 1095. | 229, 230 |
| 76JOC1184 | W. E. Parham, L. D. Jones and Y. A. Sayed; *J. Org. Chem.*, 1976, **41**, 1184. | 783 |
| 76JOC1299 | E. C. Taylor and T. Kobayashi; *J. Org. Chem.*, 1976, **41**, 1299. | 281, 318 |
| 76JOC1380 | L. Canuel and M. St.-Jaques; *J. Org. Chem.*, 1976, **41**, 1380. | 579, 628, 777 |
| 76JOC1484 | M. Mizuno, M. P. Cava and A. F. Garito; *J. Org. Chem.*, 1976, **41**, 1484. | 951, 952, 988 |
| 76JOC1881 | C.-J. Chang; *J. Org. Chem.*, 1976, **41**, 1881. | 591 |
| 76JOC2223 | W. Scott, T. C. Joseph and Y. L. Chow; *J. Org. Chem.*, 1976, **41**, 2223. | 989 |
| 76JOC2496 | Y. Becker, S. Bronstein, A. Eisenstadt and Y. Shvo; *J. Org. Chem.*, 1976, **41**, 2496. | 997 |
| 76JOC2777 | C. A. Kingsbury, M. Cliffton and J. H. Looker; *J. Org. Chem.*, 1976, **41**, 2777. | 588, 589, 639 |
| 76JOC2860 | M. W. Goodman, J. L. Atwood, R. Carlin, W. Hunter and W. W. Paudler; *J. Org. Chem.*, 1976, **41**, 2860. | 392 |
| 76JOC3027 | A. D. Broom and D. G. Bartholomew; *J. Org. Chem.*, 1976, **41**, 3027. | 209 |
| 76JOC3053 | C. H. Chen and B. A. Donatelli; *J. Org. Chem.*, 1976, **41**, 3053. | 988 |

| 76JOC3128 | P. T. Berkowitz, R. K. Robins, P. Dea and R. A. Long; *J. Org. Chem.*, 1976, **41**, 3128. | 1085 |
|---|---|---|
| 76JOC3149 | S. Wawzonek; *J. Org. Chem.*, 1976, **41**, 3149. | 229, 231 |
| 76JOC3392 | K. Pilgram and R. D. Skiles; *J. Org. Chem.*, 1976, **41**, 3392. | 562 |
| 76JOC3409 | R. Richter and H. Ulrich; *J. Org. Chem.*, 1976, **41**, 3409. | 511 |
| 76JPR127 | E. Fanghaenel; *J. Prakt. Chem.*, 1976, **318**, 127. | 1042, 1065, 1082 |
| 76JPR353 | I. A. Silberg, V. Farcasan and M. Diudea; *J. Prakt. Chem.*, 1976, **318**, 353. | 1012 |
| 76JPR705 | J. Liebscher and H. Hartmann; *J. Prakt. Chem.*, 1976, **318**, 705. | 935 |
| 76JPR971 | E. Bulka and W. D. Pfeiffer; *J. Prakt. Chem.*, 1976, **318**, 971. | 1066, 1075 |
| 76JPS1505 | R. A. Coburn and R. A. Carapellotti; *J. Pharm. Sci.*, 1976, **65**, 1505. | 355 |
| 76KGS328 | L. M. Kedik, A. A. Freger and E. A. Viktorova; *Khim. Geterotsikl. Soedin.*, 1976, 328. | 909 |
| 76KGS629 | V. I. Fomenko and O. P. Shvaika; *Khim. Geterotsikl. Soedin.*, 1976, 629. | 561 |
| 76KGS702 | P. B. Terent'ev, A. N. Kost and V. G. Kartsev; *Khim. Geterotsikl. Soedin.*, 1976, 702 (*Chem. Abstr.*, 1976, **85**, 94 301). | 239 |
| 76KGS713 | A. Ya. Lazaris, S. M. Shmuilovich and A. N. Egorochkin, *Khim. Geterotsikl. Soedin.*, 1976, 713. | 561 |
| 76KGS976 | P. B. Terent'ev, V. G. Kartsev and A. N. Kost; *Khim. Geterotsikl. Soedin.*, 1976, 976 (*Chem. Abstr.*, 1976, **85**, 192 644). | 239 |
| 76KGS1146 | V. N. Charushin, O. N. Chupakhin and I. Y. Postovskii; *Khim. Geterotsikl. Soedin.*, 1976, 1146 (*Chem. Abstr.*, 1976, **86**, 5408). | 252 |
| 76KGS1365 | L. S. Karpishchenko and S. I. Burmistrov; *Khim. Geterotsikl. Soedin.*, 1976, 1365. | 1012 |
| 76LA153 | H. Neunhoeffer and V. Böhnisch; *Liebigs Ann. Chem.*, 1976, 153. | 403, 412, 421 |
| 76LA225 | W. Disteldorf and M. Regitz; *Liebigs Ann. Chem.*, 1976, 225. | 563 |
| 76LA412 | U. Kraatz; *Liebigs Ann. Chem.*, 1976, 412. | 216 |
| 76LA946 | G. Ege, P. Arnold, E. Beisiegel and I. Lehrer; *Liebigs Ann. Chem.*, 1976, 946. | 379, 380 |
| 76LA1617 | H. Stockinger and U. Schmidt; *Liebigs Ann. Chem.*, 1976, 1617. | 830 |
| 76LA1659 | U. Petersen and H. Heitzer; *Liebigs Ann. Chem.*, 1976, 1659. | 712 |
| 76LA1663 | U. Petersen and H. Heitzer; *Liebigs Ann. Chem.*, 1976, 1663. | 712 |
| 76LA2206 | W. Draber, H. Timmler, K. Dickoré and W. Donner; *Liebigs Ann. Chem.*, 1976, 2206. | 436 |
| B-76MI21200 | J. Elguero, C. Marzin, A. R. Katritzky and P. Linda; 'The Tautomerism of Heterocycles', Academic Press, New York, 1976, p. 1. | 4, 5 |
| B-76MI21300 | G. J. Fisher and H. E. Johns; in 'Photochemistry and Photobiology of Nucleic Acids', ed. S. Y. Wang; Academic Press, New York, 1976, vol. 1, p. 169. | 72 |
| B-76MI21301 | G. J. Fisher and H. E. Johns; in 'Photochemistry and Photobiology of Nucleic Acids', ed. S. Y. Wang; Academic Press, New York, 1976, vol. 1, p. 226. | 74 |
| B-76MI21302 | S. Y. Wang; in 'Photochemistry and Photobiology of Nucleic Acids', Academic Press, New York, 1976, vol. 1, p. 296. | 74 |
| B-76MI21303 | 'Pharmacological and Chemical Synonyms', ed. E. E. J. Marler; Excerpta Medica, Amsterdam, 1976. | 155 |
| 76MI21400 | J. Velisek, J. Davidek, J. Cuhrova and V. Kubelka; *J. Agric. Food Chem.*, 1976, **24**, 3. | 193 |
| 76MI21500 | R. Albrecht and K. Schumann; *Eur. J. Med. Chem.—Chim. Ther.*, 1976, **11**, 155 (*Chem. Abstr.*, 1977, **86**, 29 739). | 210 |
| 76MI21501 | F. R. Preuss and E. Hoffmann; *Dtsch. Apoth.-Ztg.*, 1976, **116**, 893 (*Chem. Abstr.*, 1976, **85**, 177 349). | 215 |
| 76MI21502 | G. R. Brunk, K. A. Martin and A. M. Nishimura; *Biophys. J.*, 1976, **16**, 1373 (*Chem. Abstr.*, 1977, **86**, 51 917). | 250 |
| 76MI21600 | M. C. Kirk, W. C. Coburn, Jr. and J. R. Piper; *Biomed. Mass Spectrom.*, 1976, **3**, 245. | 285 |
| B-76MI21601 | J. Elguero, C. Marzin, A. R. Katritzky and P. Linda; 'The Tautomerism of Heterocycles', Academic Press, New York, 1976. | 270 |
| 76MI21800 | K. van der Meer and J. J. C. Mulder; *Theor. Chim. Acta*, 1976, **41**, 183. | 369, 373 |
| B-76MI22000 | J. Elguero, C. Marzin, A. R. Katritzky and P. Linda; 'The Tautomerism of Heterocycles', Academic, New York, 1976, p. 1. | 467 |
| B-76MI22001 | E. Ebert and S. W. Dumford; 'Residue Reviews', Springer, New York, 1976, vol. 65. | 528 |
| 76MI22100 | P. V. Tarasenko, E. A. Ponomareva, G. F. Dvorko and E. Babin; *Reakts. Sposobn. Org. Soedin.*, 1976, **13**, 5 (*Chem. Abstr.*, 1977, **86**, 42 977). | 545 |
| 76MI22200 | N. Sarda, A. Grouiller and H. Pacheco; *Eur. J. Med. Chem.—Chim. Ther.*, 1976, **11**, 251. | 580, 630 |
| 76MI22201 | I. Csoregh and S. Edström; *Chem. Commun. Univ. Stockholm*, 1976, 1. | 622, 623 |
| 76MI22202 | M. Vanhaelen and R. Vanhaelen-Fastre; *Pharm. Acta Helv.*, 1976, **51**, 507. | 643 |
| 76MI22203 | N. G. Bokii, R. V. Vedrinskii, V. V. Kitaev, N. A. Lopatina and Yu. T. Struchkov; *Koord. Khim.*, 1976, **2**, 103. | 625 |
| 76MI22300 | M. C. Crew, M. D. Melgar, S. George, R. C. Greenough, J. M. Szpiech and F. J. Di Carlo; *Xenobiotica*, 1976, **6**, 89. | 707 |
| 76MI22400 | J. G. Bieri and P. M. Farrell; *Vitam. Horm. (N.Y.)*, 1976, **34**, 31. | 877 |
| 76MI22401 | K. H. Drexhage; *J. Res. Nat. Bur. Stand., Sect. A*, 1976, **80**, 421. | 879 |
| 76MI22402 | W. S. Bowers, T. Ohta, J. S. Cleere and P. A. Marsella; *Science*, 1976, **193**, 542. | 881 |
| 76MI22500 | N. M. Morlyan, E. L. Abagayan and L. L. Nikogosyan; *Armyan Khim. Zh.*, 1976, **29**, 806 (*Chem. Abstr.*, 1977, **86**, 121 109). | 934 |

| | | |
|---|---|---|
| 76MI22501 | W. Lippman and T. A. Pugsley; *Biochem. Pharmacol.*, 1976, **25**, 1179. | 941 |
| 76MI22502 | L. K. Kulikhova; *Vopr. Biokhim. Fiziol. Mikroorganizov*, 1976, 127 (*Chem. Abstr.*, 1977, **86**, 165 812). | 942 |
| 76MI22600 | L. McGinty; *New Scientist*, 1976, **71**, 383. | 992 |
| 76MI22700 | S. Palazzo and L. I. Giannola; *Atti Accad. Sci. Lett. Arti Palermo, Parte 1*, 1976, **34**, 83, 371 (*Chem. Abstr.*, 1978, **89**, 43 276). | 1002 |
| 76MI22800 | G. Garcia-Muñoz, R. Madroñero, C. Ochoa, M. Stud and W. Pfleiderer; *An. R. Acad. Farm.*, 1976, **42**, 327. | 1060 |
| 76MIP21700 | Laboratories Made S.A., *Span. Pat.* 424 267 (1967) (*Chem. Abstr.*, 1977, **86**, 189 991). | 352 |
| 76MIP22100 | L. M. Yagupol'skii, G. I. Matyushecheva, V. S. Mikhailov and L. A. Bulgina; *U.S.S.R. Pat.* 498 389 (1976). | 567 |
| 76MIP22500 | P. M. Solozhenkin, S. A. Vartanyan, Y. E. Sarkisov, G. Y. Pulatov, E. A. Abgaryan and T. N. Aknazarova; *U.S.S.R. Pat.* 527 207 (1976) (*Chem. Abstr.*, 1977, **86**, 76 409). | 939 |
| 76MIP22501 | U.F.A. Scientific-Research Institute of Petroleum, *U.S.S.R. Pat.* 529 168 (1976) (*Chem. Abstr.*, 1976, **85**, 196 905). | 940 |
| 76OMR(8)155 | P. Diehl and H. Zimmermann; *Org. Magn. Reson.*, 1976, **8**, 155. | 249 |
| 76OMR(8)224 | R. K. Pike; *Org. Magn. Reson.*, 1976, **8**, 224. | 467 |
| 76OMR(8)273 | S. Braun; *Org. Magn. Reson.*, 1976, **8**, 273. | 395 |
| 76OMR(8)324 | M. Vajda and W. Voelter; *Org. Magn. Reson.*, 1976, **8**, 324. | 592, 641 |
| 76OMR(8)357 | C. J. Turner and G. W. H. Cheeseman; *Org. Magn. Reson.*, 1976, **8**, 357. | 63 |
| 76OMS1002 | J. P. Lavergne, P. Viallefont and J. Daunis; *Org. Mass. Spectrom.*, 1976, **11**, 1002. | 397 |
| 76OMS1221 | R. A. Corral and O. O. Orazi; *Org. Mass Spectrom.*, 1976, **11**, 1221. | 466 |
| 76OS(55)45 | J. Finkelstein and A. Brossi; *Org. Synth.*, 1976, **55**, 45. | 859 |
| 76OS(55)62 | R. Paul, O. Riobé and M. Maumy; *Org. Synth.*, 1976, **55**, 62. | 722 |
| 76PS(1)129 | H. Böhme and U. Sitorus; *Phosphorus Sulfur*, 1976, **1**, 129. | 909 |
| 76RRC101 | A. Bota, A. T. Balaban and F. Chiraleu; *Rev. Roum. Chim.*, 1976, **21**, 101. | 585 |
| 76RRC1207 | L. V. Feyns, I. I. Dragota and I. Niculescu-Duvaz; *Rev. Roum. Chim.*, 1976, **21**, 1207. | 596 |
| 76RTC113 | G. Simig, H. C. van der Plas and C. A. Landheer; *Recl. Trav. Chim. Pays-Bas*, 1976, **95**, 113. | 473 |
| 76RTC125 | G. Simig and H. C. van der Plas; *Recl. Trav. Chim. Pays-Bas*, 1976, **95**, 125. | 473 |
| 76S53 | R. H. Fischer and H. M. Weitz; *Synthesis*, 1976, 53. | 186 |
| 76S274 | W. Löwe; *Synthesis*, 1976, 274. | 712 |
| 76S326 | A. Pelter and S. Foot; *Synthesis*, 1976, 326. | 821 |
| 76S459 | D. J. Le Count and A. T. Greer; *Synthesis*, 1976, 459. | 415 |
| 76S598 | R. Perrone, G. Bettoni and V. Tortorella; *Synthesis*, 1976, 598. | 1014 |
| 76S717 | I. Butula, M. Antunović and I. Marušić; *Synthesis*, 1976, 717. | 376 |
| 76S833 | B. Jenko, B. Stanovnik and M. Tišler; *Synthesis*, 1976, 833. | 85 |
| 76SA(A)157 | H.-P. Koopmann and P. Rademacher; *Spectrochim. Acta, Part A*, 1976, **32**, 157. | 332, 539 |
| 76SA(A)345 | L. Stefaniak, *Spectrochim. Acta, Part A*, 1976, **32**, 345. | 371, 373, 533, 536 |
| 76SC457 | A. V. Zeiger and M. M. Joullie; *Synth. Commun.*, 1976, **6**, 457. | 450 |
| 76T167 | J. M. Behan, F. M. Dean and R. A. W. Johnstone; *Tetrahedron*, 1976, **32**, 167. | 599 |
| 76T229 | M. Matsuo and S. Urano; *Tetrahedron*, 1976, **32**, 229. | 585, 587 |
| 76T529 | N. L. Allinger and J. Kao; *Tetrahedron*, 1976, **32**, 529. | 946 |
| 76T655 | J. C. MacDonald, G. G. Bishop and M. Mazurek; *Tetrahedron*, 1976, **32**, 655. | 160 |
| 76T725 | M. Kočevar, D. Kolman, H. Krajnc, S. Polanc, B. Porovne, B. Stanovnik and M. Tišler; *Tetrahedron*, 1976, **32**, 725. | 35 |
| 76T1051 | O. Achmatowicz, Jr., G. Grynkiewicz and B. Szechner; *Tetrahedron*, 1976, **32**, 1051. | 844 |
| 76T1407 | F. Chioccara, G. Prota and R. H. Thomson; *Tetrahedron*, 1976, **32**, 1407. | 1009, 1032 |
| 76T1449 | J. Ficini; *Tetrahedron*, 1976, **32**, 1449. | 760 |
| 76T1655 | M. Abou Assali, C. Decoret, J. Royer and J. Dreux; *Tetrahedron*, 1976, **32**, 1655. | 754 |
| 76T1661 | M. Abou Assali, J. Royer and J. Dreux; *Tetrahedron*, 1976, **32**, 1661. | 754 |
| 76T1801 | J. F. Biellmann, J. B. Ducep and J. J. Vicens; *Tetrahedron*, 1976, **32**, 1801. | 906 |
| 76T1873 | P. J. Halfpenny, P. J. Johnson, M. J. T. Robinson and M. G. Ward; *Tetrahedron*, 1976, **32**, 1873. | 888 |
| 76T2303 | H. I. X. Mager and W. Berends; *Tetrahedron*, 1976, **32**, 2303. | 308 |
| 76T2407 | A. Chatterjee, D. Ganguly and R. Sen; *Tetrahedron*, 1976, **32**, 2407. | 800 |
| 76T2603 | A. Kreutzberger and E. Kreutzberger; *Tetrahedron*, 1976, **32**, 2603. | 470 |
| 76T2607 | K. R. Markham and B. Ternai; *Tetrahedron*, 1976, **32**, 2607. | 591 |
| 76T2647 | G. Surpateanu, J. P. Catteau, P. Karafiloglou and A. Lablache-Combier; *Tetrahedron*, 1976, **32**, 2647. | 2 |
| 76T2763 | F. Ruff, K. Komoto, N. Furukawa and S. Oae; *Tetrahedron*, 1976, **32**, 2763. | 899 |
| 76TH22000 | D. Bartholomew; M.Sc. Thesis, University of East Anglia, 1976. | 463 |
| 76TL25 | C. H. Chen; *Tetrahedron Lett.*, 1976, 25. | 988 |
| 76TL271 | N. Sarda, A. Grouiller and H. Pacheco; *Tetrahedron Lett.*, 1976, 271. | 855 |
| 76TL903 | R. Hisada, M. Nakajima and J.-P. Anselme; *Tetrahedron Lett.*, 1976, 903. | 450 |
| 76TL1311 | K. Tori, T. Hirata, O. Koshitani and T. Suga; *Tetrahedron Lett.*, 1976, 1311. | 588 |
| 76TL1363 | T. Shono and Y. Matsumura; *Tetrahedron Lett.*, 1976, 1363. | 725, 815 |
| 76TL1799 | H. Wagner, V. M. Chari and J. Sonnenbichler; *Tetrahedron Lett.*, 1976, 1799. | 591 |

| | | |
|---|---|---|
| 76TL1821 | T. Hosokowa, M. Hirata, S.-I. Murahashi and A. Sonoda; *Tetrahedron Lett.*, 1976, 1821. | 776 |
| 76TL1893 | E. A. Ponomareva, P. V. Tarasenko and G. F. Dvorko; *Tetrahedron Lett.*, 1976, 1893. | 547 |
| 76TL1903 | T. Reffstrup and P. M. Boll; *Tetrahedron Lett.*, 1976, 1903. | 799 |
| 76TL1939 | J. W. Pavlik and L. T. Pauliukonis; *Tetrahedron Lett.*, 1976, 1939. | 693 |
| 76TL3193 | B. Stanovnik, M. Tišler, S. Polanc, V. Kovačič-Bratina and B. Špicer-Smolnikar; *Tetrahedron Lett.*, 1976, 3193. | 35 |
| 76TL3563 | H. Gotthardt and S. Nieberl; *Tetrahedron Lett.*, 1976, 3563. | 921 |
| 76TL4283 | F. Ishii, R. Okazaki and N. Inamoto; *Tetrahedron Lett.*, 1976, 4283. | 930 |
| 76TL4321 | H. Kolbinger, G. Reissenweber and J. Sauer; *Tetrahedron Lett.*, 1976, 4321. | 551 |
| 76TL4647 | C. W. Rees, R. C. Storr and P. J. Whittle; *Tetrahedron Lett.*, 1976, 4647. | 378, 380 |
| 76UKZ510 | V. Kh. Premyslov and E. A. Ponomareva; *Ukr. Khim. Zh. (Russ. Ed.)*, 1976, **42**, 510. | 548 |
| 76UKZ724 | O. M. Polumbrik, N. G. Vasil'kevich and V. N. Kalinin; *Ukr. Khim. Zh. (Russ. Ed.)*, 1976, **42**, 724. | 548 |
| 76UP21800 | H. Neunhoeffer and H.-D. Vötter; unpublished results, 1976. | 374 |
| 76UP21900 | H. Neunhoeffer and B. Lehmann; unpublished results, 1976. | 420 |
| 76USP3931183 | G. E. Hardtmann; (Sandoz-Wander, Inc.), *U.S. Pat.* 3 931 183 (1976) (*Chem. Abstr.*, 1976, **84**, 105 644). | 225 |
| 76USP3951941 | K. Konishi, A. Kotone, Y. Nakane, T. Hori and M. Hoda; *U.S. Pat.* 3 951 941 (1976) (*Chem. Abstr.*, 1976, **85**, 7967). | 555 |
| 76USP3960877 | R. L. Jacobs; (Sherwin-Williams Co.), *U.S. Pat.* 3 960 877 (1976) (*Chem. Abstr.*, 1976, **85**, 142 990). | 215 |
| 76USP3997323 | A. D. Brewer, R. W. Neidermyer and W. S. McIntire; *U.S. Pat.* 3 997 323 (1976) (*Chem. Abstr.*, 1977, **86**, 101 937). | 994 |
| 76YZ1440 | F. Miyoshi, K. Tokuno, Y. Arata, S. Hiroki and T. Ohashi; *Yakugaku Zasshi*, 1976, **96**, 1440. | 897 |
| 76ZN(B)1489 | M. Herberhold and M. Süss-Fink; *Z. Naturforsch., Teil B*, 1976, **31**, 1489. | 542, 549 |
| 76ZOR1676 | B. I. Buzykin, L. P. Sysoeva and Yu. P. Kitaev; *Zh. Org. Khim.*, 1976, **12**, 1676. | 564 |
| 77ACH(92)65 | I. Gutman; *Acta Chim. Acad. Sci. Hung.*, 1977, **92**, 65. | 572 |
| 77ACH(92)317 | J. Szabó, L. Fodor, I. Varga, E. Vinkler and P. Sohár; *Acta Chim. Acad. Sci. Hung.*, 1977, **92**, 317 (*Chem. Abstr.*, 1977, **87**, 167958). 1000, 1007, 1019, 1029 | |
| 77ACH(94)391 | M. Lempert-Stréter, K. Lempert, P. Bruck and G. Tóth; *Acta Chim. Acad. Sci. Hung.*, 1977, **94**, 391. 1056, 1057 | |
| 77ACS(A)63 | A. Almenningen, G. Bjørnsen, T. Ottersen, R. Seip and T. G. Strand; *Acta Chem. Scand., Ser. A*, 1977, **31**, 63. | 2 |
| 77ACS(B)239 | O. Ceder and K. Vernmark; *Acta Chem. Scand., Ser. B*, 1977, **31**, 239. | 489 |
| 77ACS(B)496 | .E. J. Oestensen, A. Abdel-Azeem Abdallah, S. H. Skaare and M. M. Mishrikey, *Acta Chem. Scand., Ser. B*, 1977, **31**, 496. | 911 |
| 77AF1663 | Y. Inada, K. Nishikawa, A. Nagaoka and S. Kikuchi; *Arzneim.-Forsch.*, 1977, **27**, 1663. | 261 |
| 77AG(E)743 | H. Hagemann; *Angew. Chem., Int. Ed. Engl.*, 1977, **16**, 743. | 502 |
| 77AJC221 | N. Nishimura and N. Itakura; *Aust. J. Chem.*, 1977, **30**, 221. | 548 |
| 77AJC621 | D. J. Brown and P. Waring; *Aust. J. Chem.*, 1977, **30**, 621. | 113 |
| 77AJC1785 | D. J. Brown and P. Waring; *Aust. J. Chem.*, 1977, **30**, 1785. | 77 |
| 77AP269 | G. Seitz and T. Kämpchen; *Arch. Pharm. (Weinheim, Ger.)*, 1977, **310**, 269. | 554 |
| 77AP744 | F. Eiden and C. Herdeis; *Arch. Pharm., (Weinheim. Ger.)*, 1977, **310**, 744. | 699 |
| 77AP764 | G. Zinner and T. Krause; *Arch. Pharm. (Weinheim. Ger.)*, 1977, **310**, 764. | 567 |
| 77AP936 | G. Seitz and W. Overheu; *Arch. Pharm. (Weinheim, Ger.)*, 1977, **310**, 936. | 247, 554 |
| 77AX(B)274 | A. Rykowski, H. C. van der Plas and C. H. Stam; *Acta Crystallogr., Part B*, 1977, **33**, 274. | 387 |
| 77AX(B)283 | K. Ueno and N. Saito; *Acta Crystallogr., Part B*, 1977, **33**, 283. | 622 |
| 77AX(B)910 | C. Foces-Foces, J. Fayos, F. H. Cano and S. Garcia-Blanco; *Acta Crystallogr., Part B*, 1977, **33**, 910. | 1049 |
| 77AX(B)2464 | M. Laing, P. Sommerville and L. P. L. Piacenza; *Acta Crystallogr., Part B*, 1977, 2464. | 331 |
| 77AX(B)2911 | Y. Kobayashi, Y. Iitaka, R. Gottlieb and W. Pfleiderer; *Acta Crystallogr., Part B*, 1977, **33**, 2911. | 306 |
| 77B3586 | R. Spencer, J. Fisher and C. Walsh; *Biochemistry*, 1977, **16**, 3586. 250, 252, 262 | |
| 77BBA(480)21 | F. Bergmann, L. Levine, I. Tamir and M. Rahat; *Biochem. Biophys. Acta*, 1977, **480**, 21. | 287 |
| 77BCJ953 | M. Takahashi, H. Tan, K. Fukushima and H. Yamazaki; *Bull. Chem. Soc. Jpn.*, 1977, **50**, 953. | 571 |
| 77BCJ2153 | O. Tsuge, K. Kamata and S. Yogi; *Bull. Chem. Soc. Jpn.*, 1977, **50**, 2153. | 38, 55 |
| 77BCJ2789 | Y. Senda, A. Kasahara, T. Izumi and T. Takeda; *Bull. Chem. Soc. Jpn.*, 1977, **50**, 2789. | 591 |
| 77BCJ3298 | Y. Ashihara, Y. Nagata and K. Kurosawa; *Bull. Chem. Soc. Jpn.*, 1977, **50**, 3298. | 670, 829 |
| 77BRP1474246 | Mallory Batteries Ltd., *Br. Pat.* 1 474 246 (1977) (*Chem. Abstr.*, 1977, **87**, 174 799). | 940 |
| 77BSF101 | C. Bernasconi, L. Cottier and G. Descotes; *Bull. Soc. Chim. Fr.*, 1977, 101. | 719 |
| 77BSF337 | C. Glacet, J. Brocard and L. Maciejewski; *Bull. Soc. Chim. Fr.*, 1977, 337. | 726 |

| | | |
|---|---|---|
| 77BSF369 | K. Tagaki and M. Hubert-Habart; *Bull. Soc. Chim. Fr.*, 1977, 369. | 685, 703 |
| 77BSF665 | A. Decormeille, G. Queguiner and P. Pastour; *Bull. Soc. Chim. Fr.*, 1977, 665. | 235, 237 |
| 77BSF911 | X. Duteurtre, J. Lemaire and R. Vessière; *Bull. Soc. Chim. Fr.*, 1977, 911. | 720 |
| 77BSF919 | D. Marchand, A. Turck, C. Queguiner and P. Pastour; *Bull. Soc. Chem. Fr.*, 1977, 919. | 234, 236, 238, 239 |
| 77BSF1142 | F. Ishii, M. Stavaux and N. Lozac'h; *Bull. Soc. Chim. Fr.*, 1977, 1142. | 936 |
| 77BSF1187 | P. Bouvier, J. Andrieux, H. Cunha and D. Molho; *Bull. Soc. Chim. Fr.*, 1977, 1187. | 662 |
| 77CB1492 | G. Tóth, G. Hornyák and K. Lempert; *Chem. Ber.*, 1977, **110**, 1492. | 391 |
| 77CB2114 | K. Burger, R. Ottlinger and J. Albanbauer; *Chem. Ber.*, 1977, **110**, 2114. | 1041, 1063 |
| 77CB3615 | G. Schulz and W. Steglich; *Chem. Ber.*, 1977, **110**, 3615. | 1035 |
| 77CC19 | M. Sabir, J. A. S. Smith, O. Riobé, A. Lebouc, J. Delaunay and J. Cousseau; *J. Chem. Soc., Chem. Commun.*, 1977, 19. | 629 |
| 77CC687 | D. J. Sandman, A. P. Fisher, III, T. J. Holmes and A. J. Epstein; *J. Chem. Soc., Chem. Commun.*, 1977, 687. | 941 |
| 77CCC2182 | K. Nálepa and J. Slouka; *Collect. Czech Chem. Commun.*, 1977, **42**, 2182. | 419 |
| 77CCC3449 | J. Slouka, M. Šrámková and V. Bekárek; *Collect. Czech Chem. Commun.*, 1977, **42**, 3449. | 439 |
| 77CHE918 | V. I. Dulenko, N. N. Aleckseev, V. M. Golyak and E. V. Dulenko; *Chem. Heterocycl. Compd. (Engl. Transl.)*, 1977, **13**, 918. | 660 |
| 77CHE1183 | E. V. Kuznetsov, I. V. Shcherbakova and G. N. Dorofeenko; *Chem. Heterocycl. Compd. (Engl. Transl.)*, 1977, **13**, 1183. | 863 |
| 77CI(L)232 | P. R. H. Speakman; *Chem. Ind. (London)*, 1977, 232. | 485 |
| 77CI(L)310 | A. C. Ranade, R. S. Mali, R. M. Gidwani and H. R. Deshpande; *Chem. Ind. (London)*, 1977, 310. | 219 |
| 77CJC44 | G. W. Buchanan, N. K. Sharma, F. de Reinach-Hirtzbach and T. Durst; *Can. J. Chem.*, 1977, **55**, 44. | 953 |
| 77CJC145 | P. Yates and R. J. Mayfield; *Can. J. Chem.*, 1977, **55**, 145. | 563 |
| 77CJC2360 | A. N. Starratt and A. Stoessl; *Can. J. Chem.*, 1977, **55**, 2360. | 857 |
| 77CJC2373 | D. Gravel, C. Leboeuf and S. Caron; *Can. J. Chem.*, 1977, **55**, 2373. | 757 |
| 77CJC3755 | J. M. McIntosh and C. K. Schram; *Can. J. Chem.*, 1977, **55**, 3755. | 904 |
| 77CL1231 | A. Furusaki, T. Matsumoto and I. Sekikawa; *Chem. Lett.*, 1977, 1231. | 387 |
| 77CPB1856 | M. Yanai, S. Takeda and M. Nishikawa; *Chem. Pharm. Bull.*, 1977, **25**, 1856. | 23 |
| 77CR(C)(285)321 | A. Etienne, G. Lonchambon and P. Giraudeau; *C. R. Hebd. Seances Acad. Sci., Ser. C*, 1977, **285**, 321. | 479 |
| 77CR(C)(285)431 | J. Desbarres and H. O. El Sayed; *C.R. Hebd. Seances Acad. Sci., Ser. C*, 1977, **285**, 431. | 206 |
| 77CRV409 | T. L. Gilchrist and C. J. Moody; *Chem. Rev.*, 1977, **77**, 409. | 905, 1060, 1070 |
| 77CZ305 | R. Kruse and E. Breitmaier; *Chem.-Ztg.*, 1977, **101**, 305. | 124 |
| 77E555 | O. Isler; *Experientia*, 1977, **33**, 555. | 780 |
| 77EGP127515 | G. Ohm, H. Bräuninger and K. Peseke; *Ger. (East) Pat.* 127 515 (1977) (*Chem. Abstr.*, 1978, **88**, 105 148). | 929 |
| 77FES635 | G. Romussi and G. Ciarallo; *Farmaco Ed. Sci.*, 1977, **32**, 635. | 707 |
| 77G363 | I. J. Ferguson, A. R. Katritzky and S. Rahimi-Rastgoo; *Gazz. Chim. Ital.*, 1977, **107**, 363. | 467 |
| 77GEP2546821 | Bayer, A.-G., *Ger. Pat.* 2 546 821 (1977) (*Chem. Abstr.*, 1977, **87**, 23 052). | 941 |
| 77GEP2556835 | H. Neunhoeffer and H.-J. Degen; *Ger. Pat.* 2 556 835 (1977) (*Chem. Abstr.*, 1977, **87**, 117 910). | 436 |
| 77GEP2824126 | K. Ikeda, H. Kanno, M. Yasui and T. Harada; *Ger. Pat.* 2 824 126 (1977) (*Chem. Abstr.*, 1979, **90**, 121 669). | 1085 |
| 77H(5)301 | H. Nohira, K. Watanabe, T. Ishikawa and K. Saigo; *Heterocycles*, 1977, **7**, 301. | 1023 |
| 77H(6)525 | T. Sakamoto, K. Kanno, T. Ono and H. Yamanaka; *Heterocycles*, 1977, **6**, 525. | 73 |
| 77H(6)547 | T. Kurihara and T. Uno; *Heterocycles*, 1977, **6**, 547. | 246 |
| 77H(6)681 | M. Kočevar, B. Stanovnik and M. Tišler; *Heterocycles*, 1977, **6**, 681. | 50 |
| 77H(6)693 | K. Senga, Y. Kanamori and S. Nishigaki; *Heterocycles*, 1977, **6**, 693. | 317 |
| 77H(6)1581 | T. Okutani, K. Kawakita, O. Aki and K. Morita; *Heterocycles*, 1977, **6**, 1581. | 717 |
| 77H(6)1907 | K. Senga, K. Shimizu and S. Nishigaki; *Heterocycles*, 1977, **6**, 1907. | 313 |
| 77H(7)119 | C. H. Hassall and K. L. Ramachandran; *Heterocycles*, 1977, **7**, 119. | 49 |
| 77H(7)205 | J. Nagel and H. C. van der Plas; *Heterocycles*, 1977, **7**, 205. | 288, 308 |
| 77H(7)391 | C. O. Okafor; *Heterocycles*, 1977, **7**, 391. | 1038 |
| 77H(7)933 | C. Y. Chen and T. E. Gompf; *Heterocycles*, 1977, **7**, 933. | 600 |
| 77H(8)257 | T. Sakamoto, S. Niitsuma, M. Mizugaki and H. Yamanaka; *Heterocycles*, 1977, **8**, 257. | 75 |
| 77H(8)319 | I. Ito, N. Oda, S. Nagai and Y. Kudo; *Heterocycles*, 1977, **8**, 319. | 380 |
| 77HC(31) | G. P. Ellis; *Chem. Heterocycl. Compd.*, 1977, **31**. | 738, 816 |
| 77HC(31)1 | G. P. Ellis; *Chem. Heterocycl. Compd.*, 1977, **31**, 1. | 574, 596 |
| 77HC(31)11 | E. E. Schweizer and D. Meeder-Nycz; *Chem. Heterocycl. Compd.*, 1977, **31**, 11. | 666, 674, 681, 741 |
| 77HC(31)25 | E. E. Schweizer and D. Meeder-Nycz; *Chem. Heterocycl. Compd.*, 1977, **31**, 25. | 604 |
| 77HC(31)70 | E. E. Schweizer and D. Meeder-Nycz; *Chem. Heterocycl. Compd.*, 1977, **31**, 70. | 786 |
| 77HC(31)146 | I. M. Lockhart; *Chem. Heterocycl. Compd.*, 1977, **31**, 146. | 607 |
| 77HC(31)193 | I. M. Lockhart; *Chem. Heterocycl. Compd.*, 1977, **31**, 193. | 719 |
| 77HC(31)207 | I. M. Lockhart; *Chem. Heterocycl. Compd.*, 1977, **31**, 207. | 723, 728, 730 |
| 77HC(31)215 | I. M. Lockhart; *Chem. Heterocycl. Compd.*, 1977, **31**, 215. | 597, 601 |

| | | |
|---|---|---|
| 77HC(31)301 | I. M. Lockhart; *Chem. Heterocycl. Compd.*, 1977, **31**, 301. | 786 |
| 77HC(31)345 | I. M. Lockhart; *Chem. Heterocycl. Compd.*, 1977, **31**, 345. | 882 |
| 77HC(31)440 | I. M. Lockhart; *Chem. Heterocycl. Compd.*, 1977, **31**, 440. | 883 |
| 77HC(31)496 | G. P. Ellis; *Chem. Heterocycl. Compd.*, 1977, **31**, 496. | 817 |
| 77HC(31)515 | G. P. Ellis; *Chem. Heterocycl. Compd.*, 1977, **31**, 515. | 820 |
| 77HC(31)557 | G. P. Ellis; *Chem. Heterocycl. Compd.*, 1977, **31**, 557. | 693, 704 |
| 77HC(31)633 | G. P. Ellis; *Chem. Heterocycl. Compd.*, 1977, **31**, 633. | 713 |
| 77HC(31)668 | G. P. Ellis; *Chem. Heterocycl. Compd.*, 1977, **31**, 668. | 883 |
| 77HCA152 | R. Weber and M. Viscontini; *Helv. Chim. Acta*, 1977, **60**, 152. | 281 |
| 77HCA161 | R. Weber and M. Viscontini; *Helv. Chim. Acta*, 1977, **60**, 161. | 281 |
| 77HCA211 | B. Schirks, J. H. Bieri and M. Viscontini; *Helv. Chim. Acta*, 1977, **60**, 211. | 312 |
| 77HCA215 | E. Davin-Pretelli, M. Guiliano, G. Mille and J. Chouteau; *Helv. Chim. Acta*, 1977, **60**, 215. | 594 |
| 77HCA447 | J. H. Bieri and M. Viscontini; *Helv. Chim. Acta*, 1977, **60**, 447. | 281 |
| 77HCA922 | P. K. Sengupta, H. A. Breitschmid, J. H. Bieri and M. Viscontini; *Helv. Chim. Acta*, 1977, **60**, 922. | 306 |
| 77HCA1926 | J. H. Bieri and M. Viscontini; *Helv. Chim. Acta*, 1977, **60**, 1926. | 281 |
| 77HCA2303 | J. H. Bieri; *Helv. Chim. Acta*, 1977, **60**, 2303. | 279 |
| 77IJC(B)168 | M. K. Mahanti; *Indian J. Chem., Sect. B*, 1977, **15**, 168. | 373, 572 |
| 77IJC(B)816 | V. K. Ahluwalia, C. Prakash and M. R. Parthasarthy; *Indian J. Chem., Sect. B*, 1977, **15**, 816. | 609 |
| 77IZV393 | V. Kh. Ivanova, B. I. Buzykin and Yu. P. Kitaev; *Izv. Akad. Nauk SSSR, Ser. Khim.*, 1977, 393. | 565 |
| 77IZV2517 | N. S. Mirzabekyants, Y. A. Cheburkov and I. L. Knunyants; *Izv. Akad. Nauk SSSR, Ser. Khim.*, 1977, 2517. | 593, 758, 799 |
| 77JA1631 | T. M. Harris and J. V. Hay; *J. Am. Chem. Soc.*, 1977, **99**, 1631 | 829, 836 |
| 77JA2337 | D. M. Roush and C. H. Heathcock; *J. Am. Chem. Soc.*, 1977, **99**, 2337. | 888, 929 |
| 77JA2735 | W. Adam and J. Sanabia; *J. Am. Chem. Soc.*, 1977, **99**, 2735. | 979 |
| 77JA2931 | J. P. Snyder and L. Carlsen; *J. Am. Chem. Soc.*, 1977, **99**, 2931. | 955, 956 |
| 77JA3923 | R. M. Hochstrasser, D. S. King and A. B. Smith, III; *J. Am. Chem. Soc.*, 1977, **99**, 3923. | 543 |
| 77JA4330 | H. Alper, J. E. Prickett and S. Wollowitz; *J. Am. Chem. Soc.*, 1977, **99**, 4330. | 52, 188 |
| 77JA5689 | J. B. Lambert and J. E. Goldstein; *J. Am. Chem. Soc.*, 1977, **99**, 5689. | 579 |
| 77JA6721 | R. L. Chan and T. C. Bruice; *J. Am. Chem. Soc.*, 1977, **99**, 6721. | 205 |
| 77JA8262 | A. G. Abatjoglou, E. L. Eliel and L. F. Kuyper; *J. Am. Chem. Soc.*, 1977, **99**, 8262. | 961 |
| 77JA8505 | J. F. M. Oth, H. Olsen and J. P. Snyder; *J. Am. Chem. Soc.*, 1977, **99**, 8505. | 39 |
| 77JAP7746078 | Nippon Soda Co., *Jpn. Pat.* 77 46 078 (1977) (*Chem. Abstr.*, 1977, **87**, 84 822). | 939 |
| 77JAP7782730 | Yamanouchi Pharmaceuticals KK, *Jpn. Pat.* 77 82 730 (1977) (*Chem. Abstr.*, 1977, **87**, 206 512). | 941 |
| 77JAP(K)7705798 | I. Matsuura; (Chugai Pharm. Co. Ltd.), *Jpn. Kokai* 77 05 798 (1977) (*Chem. Abstr.*, 1977, **87**, 117 892). | 738 |
| 77JAP(K)7733695 | Shionogi KK, *Jpn. Kokai* 77 33 695 (1977) (*Derwent Farmdoc* 29 790Y). | 245 |
| 77JAP(K)7773897 | K. Noda, A. Nakagawa, T. Muramoto, M. Hirano, T. Hachiya and H. Ide; (Hisamitsu Pharm. Co. Inc.), *Jpn. Kokai* 77 73 897 (1977) (*Chem. Abstr.*, 1978, **88**, 6929). | 216 |
| 77JAP(K)7785194 | K. Noda, A. Nakagawa, Y. Nakajima and H. Ide; (Hisamitsu Pharm. Co. Inc.), *Jpn. Kokai* 77 85 194 (1977) (*Chem. Abstr.*, 1978, **88**, 50 908). | 211 |
| 77JCR(M)2826 | D. Bartholomew and I. T. Kay; *J. Chem. Res. (M)*, 1977, 2826. | 520 |
| 77JCR(S)238 | D. Bartholomew and I. T. Kay; *J. Chem. Res. (S)*, 1977, 238. | 1079 |
| 77JCS(P1)47 | H. Suschitzky, R. E. Walrond and R. Hull; *J. Chem. Soc., Perkin Trans. 1*, 1977, 47. | 516 |
| 77JCS(P1)103 | A. Gescher, M. F. G. Stevens and C. P. Turnbull; *J. Chem. Soc., Perkin Trans. 1*, 1977, 103. | 488 |
| 77JCS(P1)125 | T. G. Fourie, D. Ferreira and D. G. Roux; *J. Chem. Soc., Perkin Trans. 1*, 1977, 125. | 720 |
| 77JCS(P1)217 | Y. Senda, J. Ishiyama, S. Imaizumi and K. Hanaya; *J. Chem. Soc., Perkin Trans. 1*, 1977, 217. | 587, 631 |
| 77JCS(P1)359 | J. Andrieux, D. H. R. Barton and H. Patin; *J. Chem. Soc., Perkin Trans. 1*, 1977, 359. | 829 |
| 77JCS(P1)478 | D. W. S. Latham, O. Meth-Cohn and H. Suschitzky; *J. Chem. Soc., Perkin Trans. 1*, 1977, 478. | 451 |
| 77JCS(P1)904 | T. Minami, Y. Yamauchi, Y. Ohshiro, T. Agawa, S. Murai and N. Sonoda; *J. Chem. Soc., Perkin Trans. 1*, 1977, 904. | 1050, 1078 |
| 77JCS(P1)916 | D. H. R. Barton and W. A. Bubb; *J. Chem. Soc., Perkin Trans. 1*, 1977, 916. | 1059 |
| 77JCS(P1)971 | G. Adembri, A. Camparini, F. Ponticelli and P. Tadeschi; *J. Chem. Soc., Perkin Trans. 1*, 1977, 971. | 451 |
| 77JCS(P1)1020 | G. Adembri, S. Chimichi, R. Nesi and M. Scotton; *J. Chem. Soc., Perkin Trans. 1*, 1977, 1020. | 358 |
| 77JCS(P1)1155 | R. Pellicciani, M. Curini and P. Ceccherelli; *J. Chem. Soc., Perkin Trans. 1*, 1977, 1155. | 905 |
| 77JCS(P1)1257 | Z. D. Tadic, G. A. Boncic-Caricic and M. D. Muskatirovic; *J. Chem. Soc., Perkin Trans. 1*, 1977, 1257. | 476 |
| 77JCS(P1)1336 | F. Yoneda and M. Higuchi; *J. Chem. Soc., Perkin Trans. 1*, 1977, 1336. | 85, 316 |

| 77JCS(P1)1511 | E. I. G. Brown, D. Leaver and D. M. McKinnon; *J. Chem. Soc., Perkin Trans. 1*, 1977, 1511. | 936 |
| 77JCS(P1)1605 | R. D. Chambers, P. D. Philpot and P. L. Russell; *J. Chem. Soc., Perkin Trans. 1*, 1977, 1605. | 464 |
| 77JCS(P1)1628 | A. C. Fletcher, L. J. Porter, E. Haslam and R. K. Gupta; *J. Chem. Soc., Perkin Trans. 1*, 1977, 1628. | 592 |
| 77JCS(P1)1647 | K. Sato, S. Inoue, M. Hirayama and M. Ōhashi; *J. Chem. Soc., Perkin Trans. 1*, 1977, 1647. | 723, 724 |
| 77JCS(P1)1862 | D. Lloyd, H. McNab and K. S. Tucker; *J. Chem. Soc., Perkin Trans. 1*, 1977, 1862. | 135 |
| 77JCS(P1)1985 | D. T. Hurst and M. L. Wong; *J. Chem. Soc., Perkin Trans. 1*, 1977, 1985. | 78 |
| 77JCS(P1)2158 | M. S. H. Idris, A. Jefferson and F. Scheinmann; *J. Chem. Soc., Perkin Trans. 1*, 1977, 2158. | 590 |
| 77JCS(P1)2189 | T. Lang, C. J. Suckling and H. C. S. Wood; *J. Chem. Soc., Perkin Trans. 1*, 1977, 2189. | 527 |
| 77JCS(P1)2393 | M. J. Begley, L. Crombie, R. W. King, D. A. Slack and D. A. Whiting; *J. Chem. Soc., Perkin Trans. 1*, 1977, 2393. | 622 |
| 77JCS(P1)2402 | M. J. Begley, L. Crombie, D. A. Slack and D. A. Whiting; *J. Chem. Soc., Perkin Trans. 1*, 1977, 2402. | 622 |
| 77JCS(P1)2529 | W. L. F. Armarego and H. Schou; *J. Chem. Soc., Perkin Trans. 1*, 1977, 2529. | 281 |
| 77JCS(P1)2593 | G. D. John and P. V. R. Shannon; *J. Chem. Soc., Perkin Trans. 1*, 1977, 2593. | 781 |
| 77JCS(P2)262 | G. S. Beddard, S. Carlin and R. S. Davidson; *J. Chem. Soc., Perkin Trans. 2*, 1977, 262. | 601 |
| 77JCS(P2)612 | D. G. Hellier and F. J. Webb; *J. Chem. Soc., Perkin Trans. 2*, 1977, 612. | 957 |
| 77JCS(P2)939 | C. Giordano, L. Cassar, S. Panossian and A. Belli; *J. Chem. Soc., Perkin Trans. 2*, 1977, 939. | 1052, 1060, 1066 |
| 77JCS(P2)1816 | F. G. Riddell and A. J. Kidd; *J. Chem. Soc., Perkin Trans. 2*, 1977, 1816. | 1054 |
| 77JHC75 | S. Gelin and R. Gelin; *J. Heterocycl. Chem.*, 1977, **14**, 75. | 52 |
| 77JHC307 | H. J. Beim and A. R. Day; *J. Heterocycl. Chem.*, 1977, **14**, 307. | 340, 351, 367 |
| 77JHC587 | A. Mangia, F. Bortesi and U. Amendola; *J. Heterocycl. Chem.*, 1977, **14**, 587. | 555 |
| 77JHC611 | E. M. Levine, C. K. Chu and T. J. Bardos; *J. Heterocycl. Chem.*, 1977, **14**, 611. | 207, 214 |
| 77JHC679 | J. T. Shaw, K. S. Kyler and M. D. Anderson; *J. Heterocycl. Chem.*, 1977, **14**, 679. | 521 |
| 77JHC683 | N. F. Haley; *J. Heterocycl. Chem.*, 1977, **14**, 683. | 865 |
| 77JHC717 | N. Cohen and B. L. Banner; *J. Heterocycl. Chem.*, 1977, **14**, 717. | 1026 |
| 77JHC857 | E. Schaumann, E. Kausch, K.-H. Klaska, R. Klaska and O. Jarchow; *J. Heterocycl. Chem.*, 1977, **14**, 857. | 461 |
| 77JHC931 | W. Löwe; *J. Heterocycl. Chem.*, 1977, **14**, 931. | 684, 690 |
| 77JHC1053 | C. K. Chu and T. J. Bardos; *J. Heterocycl. Chem.*, 1977, **14**, 1053. | 210, 227 |
| 77JHC1089 | A. S. Shawali and A.-G. A. Fahmi; *J. Heterocycl. Chem.*, 1977, **14**, 1089. | 564 |
| 77JHC1221 | M. M. Goodman and W. W. Paudler; *J. Heterocycl. Chem.*, 1977, **14**, 1221. | 392 |
| 77JHC1249 | G. E. Martin, J. C. Turley and L. Williams; *J. Heterocycl. Chem.*, 1977, **14**, 1249. | 985 |
| 77JHC1389 | R. J. Radel, B. T. Keen and W. W. Paudler; *J. Heterocycl. Chem.*, 1977, **14**, 1389. | 397 |
| 77JMC134 | P. T. Berkowitz, R. A. Long, P. Dea, R. K. Robins and T. R. Matthews; *J. Med. Chem.*, 1977, **20**, 134. | 1085 |
| 77JMC806 | J. A. Beisler, M. M. Abbasi, J. A. Kelley and J. S. Driscoll; *J. Med. Chem.*, 1977, **20**, 806. | 526 |
| 77JMC847 | K. Ramalingam, G. X. Thyvelikakath, K. D. Berlin, R. W. Chesnut, R. A. Brown, N. N. Durham, S. E. Ealick and D. van der Helm; *J. Med. Chem.*, 1977, **20**, 847. | 909 |
| 77JMC880 | W. L. Nelson, J. E. Wennerstrom, D. C. Dyer and H. Engel; *J. Med. Chem.*, 1977, **20**, 880. | 993 |
| 77JMC925 | A. Rosowsky, W. D. Ensminger, H. Lazarus and C. Yu; *J. Med. Chem.*, 1977, **20**, 925. | 327 |
| 77JMC1215 | E. C. Taylor, J. V. Berrier, A. J. Cocuzza, R. J. Kobyecki and J. J. McCormack; *J. Med. Chem.*, 1977, **20**, 1215. | 284, 325 |
| 77JMC1312 | G. A. Archer, R. I. Kalish, R. Y. Ning, B. C. Sluboski, A. Stempel, T. V. Steppe and L. H. Sternbach; *J. Med. Chem.*, 1977, **20**, 1312. | 126, 127 |
| 77JOC208 | J. R. Piper and J. A. Montgomery; *J. Org. Chem.*, 1977, **42**, 208. | 302 |
| 77JOC221 | G. B. Bennett, W. R. J. Simpson, R. B. Mason, R. J. Strohschein and R. Mansukhani; *J. Org. Chem.*, 1977, **42**, 221. | 202, 229, 230 |
| 77JOC376 | D. I. C. Scopes, A. F. Kluge and J. A. Edwards; *J. Org. Chem.*, 1977, **42**, 376. | 1020 |
| 77JOC393 | R. D. Balanson, V. M. Kobal and R. R. Schumaker; *J. Org. Chem.*, 1977, **42**, 393. | 1060 |
| 77JOC438 | S. A. Evans, Jr., B. Goldsmith, R. L. Merrill, Jr. and R. E. Williams; *J. Org. Chem.*, 1977, **42**, 438. | 960 |
| 77JOC542 | A. V. Zeiger and M. M. Joullié; *J. Org. Chem.*, 1977, **42**, 542. | 450 |
| 77JOC546 | R. J. Radel, B. T. Keen, C. Wong and W. W. Paudler; *J. Org. Chem.*, 1977, **42**, 546. | 393, 394, 410 |
| 77JOC656 | L. A. Errede, J. J. McBrady and H. T. Oien; *J. Org. Chem.*, 1977, **42**, 656. | 1002 |
| 77JOC952 | K. Schaffner-Sabba, H. Tomaselli, B. Henrici and H. B. Renfroe; *J. Org. Chem.*, 1977, **42**, 952. | 1050, 1071 |
| 77JOC961 | F. A. Carey, P. M. Smith, R. J. Maher and R. F. Bryan; *J. Org. Chem.*, 1977, **42**, 961. | 949 |
| 77JOC993 | G. L. Anderson, J. L. Shim and A. D. Broom; *J. Org. Chem.*, 1977, **42**, 993. | 214 |
| 77JOC997 | G. L. Anderson and A. D. Broom; *J. Org. Chem.*, 1977, **42**, 997. | 206, 213, 260 |
| 77JOC1163 | T. Terasawa and T. Okado; *J. Org. Chem.*, 1977, **42**, 1163. | 928 |

| | | |
|---|---|---|
| 77JOC1329 | D. E. Korte, L. S. Hegedus and R. K. Wirth; *J. Org. Chem.*, 1977, **42**, 1329. | 597, 833, 859 |
| 77JOC1337 | C. Chang, H. G. Floss and W. Steck; *J. Org. Chem.*, 1977, **42**, 1337. | 589 |
| 77JOC1356 | R. Beugelmans and C. Morin; *J. Org. Chem.*, 1977, **42**, 1356. | 701 |
| 77JOC1530 | J. W. A. M. Janssen and H. Kwart; *J. Org. Chem.*, 1977, **42**, 1530. | 988 |
| 77JOC1538 | B. K. Bandlish, S. R. Mani and H. J. Shine; *J. Org. Chem.*, 1977, **42**, 1538. | 969 |
| 77JOC1808 | R. T. LaLonde, A. El-Kafrawy, N. Muhammad and J. E. Oatis, Jr.: *J. Org. Chem.*, 1977, **42**, 1808. | 231 |
| 77JOC1866 | M. M. Goodman and W. W. Paudler; *J. Org. Chem.*, 1977, **42**, 1866. | 392 |
| 77JOC1869 | C. E. Mixan and R. G. Pews; *J. Org. Chem.*, 1977, **42**, 1869. | 169 |
| 77JOC1910 | D. C. Dittmer, J. E. McCaskie, J. E. Babiarz and M. V. Ruggeri; *J. Org. Chem.*, 1977, **42**, 1910. | 931 |
| 77JOC1919 | G. B. Bennett and R. B. Mason; *J. Org. Chem.*, 1977, **42**, 1919. | 202, 230 |
| 77JOC2010 | A. L. Ternay, Jr., J. Herrmann, M. Harris and B. R. Hayes; *J. Org. Chem.*, 1977, **42**, 2010. | 956 |
| 77JOC2123 | J. M. McIntosh and H. Khalil; *J. Org. Chem.*, 1977, **42**, 2123. | 930 |
| 77JOC2185 | R. J. De Pasquale; *J. Org. Chem.*, 1977, **42**, 2185. | 142 |
| 77JOC2206 | D. M. Frieze, P. F. Hughes, R. L. Merrill, Jr. and S. A. Evans, Jr.; *J. Org. Chem.*, 1977, **42**, 2206. | 952 |
| 77JOC2260 | J. S. Baran, D. D. Langford and I. Laos; *J. Org. Chem.*, 1977, **42**, 2260. | 980 |
| 77JOC2777 | C. H. Chen, G. A. Reynolds and J. A. Van Allen; *J. Org. Chem.*, 1977, **42**, 2777. | 928 |
| 77JOC2951 | E. M. Kaiser and S. L. Hartzell; *J. Org. Chem.*, 1977, **42**, 2951. | 302 |
| 77JOC2989 | R. J. Spangler, B. G. Beckmann and J. H. Kim; *J. Org. Chem.*, 1977, **42**, 2989. | 618 |
| 77JOC3265 | M. M. Chau and J. L. Kice; *J. Org. Chem.*, 1977, **42**, 3265. | 986 |
| 77JOC3311 | N. S. Poonia, K. Chhabra, C. Kumar and V. W. Bhagwat; *J. Org. Chem.*, 1977, **42**, 3311. | 853 |
| 77JOC3365 | W. Ando, H. Higuchi and T. Migita; *J. Org. Chem.*, 1977, **42**, 3365. | 904 |
| 77JOC3498 | B. T. Keen, R. J. Radel and W. W. Paudler; *J. Org. Chem.*, 1977, **42**, 3498. | 400, 417 |
| 77JOC3945 | E. Wenkert, M. E. Alonso, H. E. Gottlieb, E. L. Sanchez, R. Pellicciari and P. Cogolli; *J. Org. Chem.*, 1977, **42**, 3945. | 904 |
| 77JOC4095 | R. Bonjouklian and R. A. Ruden; *J. Org. Chem.*, 1977, **42**, 4095. | 773, 843 |
| 77JOC4213 | E. Shalom, J-L. Zenou and S. Shatzmiller; *J. Org. Chem.*, 1977, **42**, 4213. | 1017 |
| 77JOM(131)121 | Y. Ito, T. Hirao and T. Saegusa; *J. Organomet. Chem.*, 1977, **131**, 121. | 444 |
| 77JPR149 | H. Schäfer, B. Bartho and K. Gewald; *J. Prakt. Chem.*, 1977, **319**, 149. | 391 |
| 77JPR739 | M. Michalik, E. Fischer, G. Rembarz and G. Voss; *J. Prakt. Chem.*, 1977, **319**, 739. | 1050 |
| 77JSP(66)192 | P. J. Chappell and I. G. Ross; *J. Mol. Spectrosc.*, 1977, **66**, 192. | 204, 249, 250 |
| 77JST(42)121 | B. J. M. Bormans, G. De With and F. C. Mijlhoff; *J. Mol. Struct.*, 1977, **42**, 121. | 158 |
| 77KGS122 | N. A. Kapran, V. G. Lukmanov, L. M. Yagupolskil and V. M. Cherkasov; *Khim. Geterotsikl. Soedin.*, 1977, 122 (*Chem. Abstr.*, 1977, **86**, 171 389). | 471 |
| 77KGS259 | A. Ya. Tikhonov and L. B. Volodarskii; *Khim. Geterotsikl. Soedin.*, 1977, 259. | 141 |
| 77KGS557 | V. P. Shchipanov; *Khim. Geterotsikl. Soedin.*, 1977, 557. | 548 |
| 77KGS1484 | V. A. Chuiguk and N. N. Vlasova; *Khim. Geterotsikl. Soedin.*, 1977, 1484. | 218 |
| 77KGS1564 | I. Ya. Postovskii, V. A. Ershov, E. O. Sidorov and N. V. Serebryakova; *Khim. Geterotsikl. Soedin.*, 1977, 1564. | 544, 555 |
| 77LA910 | S. Shatzmiller and R. Neidlein; *Liebigs Ann. Chem.*, 1977, 910. | 977 |
| 77LA1217 | G. Ritzmann, K. Ienaga and W. Pfleiderer; *Liebigs Ann. Chem.*, 1977, 1217. | 282, 297 |
| 77LA1413 | H. Neunhoeffer and B. Lehmann; *Liebigs Ann. Chem.*, 1977, 1413. | 427 |
| 77LA1421 | H. Neunhoeffer, B. Lehmann and H. Ewald; *Liebigs Ann. Chem.*, 1977, 1421. | 387, 428 |
| 77LA1707 | W. Trowitzch; *Liebigs Ann. Chem.*, 1977, 1707. | 852 |
| 77LA1713 | V. Böhnisch, G. Burzer and H. Neunhoeffer; *Liebigs Ann. Chem.*, 1977, 1713. | 443 |
| 77M499 | O. S. Wolfbeis; *Monatsh. Chem.*, 1977, **108**, 499. | 808 |
| 77MI21200 | V. E. Limanov; *Pharmazie*, 1977, **32**, 555 (*Chem. Abstr.*, 1978, **88**, 83 245). | 2 |
| 77MI21201 | P. Tomasik and R. R. Zalewski; *Chem. Zvesti*, 1977, **31**, 246 (*Chem. Abstr.*, 1978, **88**, 135 982). | 4 |
| 77MI21300 | M. C. Koshy, D. Mickley, J. Bourgoignie and M. D. Blaufox; *Circulation*, 1977, **55**, 533. | 153 |
| 77MI21400 | J. B. Wommack; *Chem. Eng. News*, 1977, **55**, no. 50, 5. | 168 |
| 77MI21500 | C. Rufer and K. Schwarz; *Eur. J. Med. Chem.—Chim. Ther.*, 1977, **12**, 236 (*Chem. Abstr.*, 1977, **87**, 135 244). | 210 |
| 77MI21501 | L. Fuentes, A. Lorente and J. L. Soto; *An. Quim.*, 1977, **73**, 1359 (*Chem. Abstr.*, 1978, **89**, 146 731). | 226 |
| 77MI21502 | V. N. R. Pillai and E. Purnashothaman; *Curr. Sci.*, 1977, **46**, 381. | 245 |
| 77MI21503 | H. Poradowska, B. Daniek and W. Kula; *Zesz. Nauk. Univ. Jagiellon., Pr. Chem.*, 1977, **22**, 29 (*Chem. Abstr.*, 1978, **88**, 36 725). | 249, 250 |
| 77MI21504 | R. J. Staniewicz and D. G. Hendricker; *J. Inorg. Nucl. Chem.*, 1977, **39**, 1454. | 249, 250 |
| B-77MI21600 | G. Dryhurst; in 'Electrochemistry of Biological Molecules', Academic Press, New York, 1977, p. 320. | 306 |
| 77MI21900 | R. Gleiter, M. Kobayashi, H. Neunhoeffer and J. Spanget-Larsen; *Chem. Phys. Lett.*, 1977, **46**, 231. | 398 |
| B-77MI22000 | E. N. Abrahart; 'Dyes and Their Intermediates', Chem. Publishing, New York, 1977, p. 185. | 526 |
| 77MI22100 | O. M. Polumbrik and E. I. Zaika; *Org. React. (Tartu)*, 1977, **14**, 388 (*Chem. Abstr.*, 1978, **89**, 23 416). | 548 |
| 77MI22101 | M. S. Sytilin; *Zh. Fiz. Khim.*, 1977, **51**, 2661. | 542, 548 |

| | |
|---|---|
| B-77MI22200 | Z. Dinya, Gy. Litkei, A. Lévai, A. Bolcskei, P. Jékel, Sz. Rochlitz, A. I. Kiss, M. Farkas and R. Bognár; in 'Proceedings of the 5th Hungarian Bioflavonoid Symposium, Mátrafüred, Hungary, 1977', ed. L. Farkas, M. Gabor and F. Kallay; Elsevier, Amsterdam, 1978, p. 247. 591,627,628 |
| B-77MI22201 | K. Nakanishi and P. Solomon; 'Infrared Absorption Spectroscopy', Holden Day, San Francisco, 2nd edn., 1977, p. 249. 596 |
| 77MI22202 | J. E. Averett; *Phytochem. Bull.*, 1977, **10**, 10. 602 |
| B-77MI22301 | R. Livingstone; in 'Rodd's Chemistry of Carbon Compounds', ed. S. Coffey; Elsevier, Amsterdam, 2nd edn., 1977, volume 4E, chap. 20. 666, 687, 693 |
| 77MI22302 | O. H. Hishmat, A. K. M. Gohar, M. E. Wassef, M. R. Shalash and I. Ismail; *Pharm. Acta Helv.*, 1977, **52**, 252. 685 |
| 77MI22303 | K. Kostka and J. Nawrot; *Rocz. Chem.*, 1977, **51**, 1045. 701 |
| B-77MI22304 | W. Basinski and Z. Jerzmanowska; in 'Flavonoids and Bioflavonoids: Current Research Trends', Proc. 5th Hungarian Bioflavonoid Symposium, ed. L. Farkas, M. Gabor and F. Kallay, Elsevier, Amsterdam, 1977, p. 213. 701 |
| B-77MI22305 | T. Patonay, M. Rakosi, G. Litkei, T. Mester and R. Bognar; in 'Flavonoids and Bioflavonoids: Current Research Trends', Proc. 5th Hungarian Bioflavonoid Symposium, ed. L. Farkas, M. Gabor and F. Kallay; Elsevier, Amsterdam, 1977, p. 227. 735 |
| B-77MI22400 | R. Livingstone; in 'Rodd's Chemistry of Carbon Compounds', ed. S. Coffey; Elsevier, Amsterdam, 1977, vol. IV, part E. 738, 767, 799, 816, 830, 861, 878 |
| 77MI22401 | P. D. Weeks, D. E. Kuhla, R. P. Allingham, H. A. Watson, Jr. and B. Wlodecki; *Carbohydr. Res.*, 1977, **56**, 195. 815, 844 |
| 77MI22402 | D. F. Diedrich, T. A. Scahill and S. L. Smith; *J. Chem. Eng. Data*, 1977, **22**, 448. 817 |
| 77MI22500 | E. R. Uzhdavini and I. K. Astaf'eva; *Gig. Tr. Prof. Zabol*, 1977, 51 (*Chem. Abstr.*, 1977, **87**, 195 013). 940 |
| 77MI22501 | S. Oae; *Symp. Heterocycl. (Pap.)*, 1977, 95. 902 |
| 77MI22502 | D. M. Vyas and G. W. Hay; *Carbohydrate Res.*, 1977, **55**, 215. 908 |
| 77MI22503 | M. Vagabov, E. A. Viktorova, E. A. Karakhanov and A. S. Ramazanov; *Vestn. Mosk. Univ., Ser. 2, Khim.*, 1977, **18**, 717 (*Chem. Abstr.*, 1978, **88**, 136 397). 933 |
| 77MI22504 | T. Takahashi, N. Wakabayashi and O. Yamamoto; *J. Appl. Electrochem.*, 1977, **7**, 253 (*Chem. Abstr.*, 1977, **87**, 61 215). 940 |
| 77MI22505 | R. G. Linford, J. M. Pollock and C. F. Randell; *Power Sources*, 1977, **6**, 511 (*Chem. Abstr.*, 1978, **89**, 200 351). 940 |
| 77MI22506 | G. Lambrecht; *Eur. J. Med. Chem.—Chim. Ther.*, 1977, **12**, 41. 940 |
| 77MI22507 | J. Colon, J. L. Ramirez and J. Castrillon; *Rev. Latinoam. Quim.*, 1977, 44 (*Chem. Abstr.*, 1978, **88**, 50 601). 941 |
| 77MI22508 | R. P. Batzinger and E. Bueding; *J. Pharmacol. Exp. Ther.*, 1977, **200**, 1 (*Chem. Abstr.*, 1977, **86**, 84 166). 942 |
| 77MI22509 | P. V. Petersen, I. M. Nielsen, V. Pedersen, A. Joergensen and N. Lassen; *Psychopharmacology*, 1977, **2**, 827. 942 |
| B-77MI22600 | R. K. Bohn; in 'Molecular Structure by Diffraction Methods', The Chemical Society, London, 1977, vol. 5, p. 23. 948 |
| 77MI22601 | A. N. de Belder; *Adv. Carbohydrate Chem. Biochem.*, 1977, **34**, 179. 978 |
| 77MI22602 | L. Fishbein; *Int. J. Ecol. Environ. Sci.*, 1977, **2**, 69. 992 |
| 77MI22603 | M. Ueda, T. Aizawa and Y. Imai; *J. Polym. Sci., Polym. Chem. Ed.*, 1977, **15**, 2739. 993 |
| 77MI22800 | K.-H. Linke and H. G. Kalker; *Z. Anorg. Allg. Chem.*, 1977, **434**, 157. 1049 |
| 77MIP21500 | N. B. Marchenko, V. G. Granik and R. G. Glushkov; (S. Ordzhonikidze All-Union Scientific-Research Chemical-Pharmaceutical Institute) *U.S.S.R. Pat.* 554 675 (1977) (*Chem. Abstr.*, 1978, **88**, 50 911). 221 |
| 77NJC79 | C. Bernasconi, L. Cottier, G. Descotes, M. F. Grenier and F. Metras; *Nouv. J. Chim.*, 1977, **2**, 79. 597, 631 |
| 77OMR(9)16 | A. T. Balaban and V. Wray; *Org. Magn. Reson.*, 1977, **9**, 16. 589, 592, 640 |
| 77OMR(9)53 | P. Van de Weijer and Ch. Mohan; *Org. Magn. Reson.*, 1977, **9**, 53. 8 |
| 77OMR(9)179 | J. Y. Lallemand and M. Duteil; *Org. Magn. Reson.*, 1977, **9**, 179. 591 |
| 77OMR(9)347 | D. G. Hellier and F. J. Webb; *Org. Magn. Reson.*, 1977, **9**, 347. 953 |
| 77OMR(10)208 | M. Kausch, H. Dürr and S. H. Doss; *Org. Magn. Reson.*, 1977, **10**, 208. 952 |
| 77OMS51 | S. Eguchi, M. Haze, M. Nakayama and S. Hayashi; *Org. Mass Spectrom.*, 1977, **12**, 51. 615 |
| 77OS(56)49 | M. Nakagawa, J. Saegusa, M. Tonozuka, M. Obi, M. Kiuchi, T. Hino and Y. Ban; *Org. Synth.*, 1977, **56**, 49. 724, 799, 841 |
| 77P735 | J. F. Castelae, O. R. Gottlieb, R. A. DeLima, A. A. L. Mesquita, H. E. Gottlieb and E. Wenkert; *Phytochemistry*, 1977, **16**, 735. 590 |
| 77PIA(A)(86)1 | S. Lahiri, V. Dabral, V. Bhat, E. D. Jemmis and M. V. George; *Proc. Indian Acad. Sci., Sect. A*, 1977, **86**, 1. 963, 981, 982 |
| 77RCR278 | V. A. Pankratov, S. V. Vinogradova and V. V. Korshak; *Russ. Chem. Rev. (Engl. Transl.)*, 1977, **46**, 278. 506, 525 |
| 77RTC259 | M. Schoufs, J. Meyer, P. Vermeer and L. Brandsma; *Recl. Trav. Chim. Pays-Bas*, 1977, **96**, 259. 972 |
| 77S61 | S. Klutchko and M. von Strandtmann; *Synthesis*, 1977, 61. 713 |
| 77S136 | O. S. Wolfbeis; *Synthesis*, 1977, 136. 186 |
| 77S193 | C. Giordano and A. Belli; *Synthesis*, 1977, 193. 1042 |
| 77S196 | W.-D. Pfeiffer, E. Dilk and E. Bulka; *Synthesis*, 1977, 196. 1066 |

| | | |
|---|---|---|
| 77S252 | W. Steglich, E. Buschmann, G. Gansen and L. Wilschowitz; *Synthesis*, 1977, 252. | 1062, 1063, 1072 |
| 77S305 | S. Sommer; *Synthesis*, 1977, 305. | 1046, 1078 |
| 77S357 | B.-T. Gröbel and D. Seebach; *Synthesis*, 1977, 357. | 977, 978 |
| 77S464 | R. S. Mali and V. J. Yadav; *Synthesis*, 1977, 464. | 805 |
| 77S472 | M. Augustin, G. Jahreis and W.-D. Rudorf; *Synthesis*, 1977, 472. | 929 |
| 77S485 | W.-D. Pfeiffer and E. Bulka; *Synthesis*, 1977, 485. | 1067 |
| 77S661 | D. R. Adams and S. P. Bhatnagar; *Synthesis*, 1977, 661. | 989 |
| 77S862 | K. Akiba, K. Ishikawa and N. Inamoto; *Synthesis*, 1977, 862. | 925 |
| 77SC33 | O. F. Bennett, G. Saluti and F. X. Quinn; *Synth. Commun.*, 1977, **7**, 33. | 921 |
| B-77SH(1) | W. L. F. Armarego; 'Stereochemistry of Heterocyclic Compounds', Wiley, New York, 1977, part 1. | 66 |
| B-77SH(2) | W. L. F. Armarego; 'Stereochemistry of Heterocyclic Compounds', Wiley, New York, 1977, part 2. | 628, 978 |
| 77SST(4)332 | U. Eisner; *Org. Compd. Sulphur, Selenium, Tellurium*, 1977, **4**, 332. | 994 |
| 77T899 | K. K. Chan, D. D. Giannini, A. H. Cain, J. D. Roberts, W. Porter and W. F. Trager; *Tetrahedron*, 1977, **33**, 899. | 589 |
| 77T1043 | T. V. Saraswathi and V. R. Srinivasan; *Tetrahedron*, 1977, **33**, 1043. | 453 |
| 77T1149 | D. J. Loomes and M. J. T. Robinson; *Tetrahedron*, 1977, **33**, 1149. | 888, 889 |
| 77T1405 | H. Wagner, I. Maurer, L. Farkas and J. Strelisky; *Tetrahedron*, 1977, **33**, 1405. | 819 |
| 77T1411 | H. Wagner, I. Maurer, L. Farkas and J. Strelisky; *Tetrahedron*, 1977, **33**, 1411. | 819 |
| 77T1735 | A. A. Chalmers, G. J. H. Rall and M. E. Oberholzer; *Tetrahedron*, 1977, **33**, 1735. | 592 |
| 77T2159 | T. M. Harris and C. M. Harris; *Tetrahedron*, 1977, **33**, 2159. | 796 |
| 77T2383 | J. Mispelter, A. Croisy, P. Jacquignon, A. Ricci, C. Rossi and F. Schiaffela; *Tetrahedron*, 1977, **33**, 2383. | 909 |
| 77T3061 | F. Bernardi, N. D. Epiotis, S. Shaik and K. Mislow; *Tetrahedron*, 1977, **33**, 3061. | 894 |
| 77T3183 | C. Morin and R. Beugelmans; *Tetrahedron*, 1977, **33**, 3183. | 700, 701 |
| 77TL117 | S. Sommer; *Tetrahedron Lett.*, 1977, 117. | 49 |
| 77TL431 | T. H. Koch, R. H. Higgins and H. F. Schuster; *Tetrahedron Lett.*, 1977, 431. | 1003 |
| 77TL473 | K. V. Subba Raju, G. Srimannarayana and N. V. Subba Rao; *Tetrahedron Lett.*, 1977, 473. | 806 |
| 77TL2117 | E. Alneri, G. Bottaccio and V. Carletti; *Tetrahedron Lett.*, 1977, 2117. | 670, 921 |
| 77TL2199 | G. Piancatelli, A. Scettri and M. D'Auria; *Tetrahedron Lett.*, 1977, 2199. | 844 |
| 77TL2551 | W. T. Ashton, D. W. Graham, R. D. Brown and E. F. Rogers; *Tetrahedron Lett.*, 1977, 2551. | 250, 257, 262 |
| 77TL2643 | T. Wooldridge and T. D. Roberts; *Tetrahedron Lett.*, 1977, 2643. | 985 |
| 77TL2817 | R. Mengel, W. Pfleiderer and W. R. Knappe; *Tetrahedron Lett.*, 1977, 2817. | 279, 306 |
| 77TL3155 | D. Daniil and H. Meier; *Tetrahedron Lett.*, 1977, 3155. | 568 |
| 77TL3619 | H. Hamberger, H. Reinshagen, G. Schulz and G. Sigmund; *Tetrahedron Lett.*, 1977, 3619. | 48 |
| 77TL3803 | A. R. Katritzky, R. C. Patel and D. M. Read; *Tetrahedron Lett.*, 1977, 3803. | 389 |
| 77TL4167 | R. O. Hutchins, N. R. Natale and W. J. Cook; *Tetrahedron Lett.*, 1977, 4167. | 752 |
| 77TL4245 | S. Sólyom, P. Sohár, L. Toldy, A. Kálmán and L. Párkányi; *Tetrahedron Lett.*, 1977, 4245. | 1041, 1048, 1067, 1070 |
| 77TL4343 | R. G. Visser, J. P. B. Baaij, A. C. Brouwer and H. J. T. Bos; *Tetrahedron Lett.*, 1977, 4343. | 921 |
| 77TL4393 | H. D. Fühlhuber and J. Sauer; *Tetrahedron Lett.*, 1977, 4393. | 551 |
| 77UKZ1192 | A. I. Dychenko and P. S. Pel'kis; *Ukr. Khim. Zh. (Russ. Ed.)*, 1977, **43**, 1192. | 571 |
| 77UP21900 | H. Neunhoeffer and H.-J. Degen; unpublished results, 1977. | 419 |
| 77UP21901 | H. Neunhoeffer and K.-H. Schnurrer; unpublished results, 1977. | 424 |
| 77UP22600 | B. M. Carden and M. J. Cook; unpublished results, 1977. | 962 |
| 77USP4062969 | T. D. J. D'Silva; *U.S. Pat.* 4 062 969 (1977) (*Chem. Abstr.*, 1978, **88**, 136 628). | 993, 994 |
| 77ZC222 | E. Fischer, C. Möller and G. Rembarz; *Z. Chem.*, 1977, **17**, 222. | 1061, 1080 |
| 77ZN(B)72 | J. Schantl; *Z. Naturforsch., Teil B*, 1977, **32**, 72 (and previous references therein). | 49 |
| 77ZN(B)434 | W. R. Knappe; *Z. Naturforsch., Teil B*, 1977, **32**, 434. | 215 |
| 77ZN(B)1390 | H. W. Roesky and H. Zamankhan; *Z. Naturforsch., Teil B*, 1977, **32**, 1390. | 1046 |
| 77ZOB2396 | O. M. Polumbrik; *Zh. Obshch. Khim.*, 1977, **47**, 2396. | 546 |
| 77ZOR186 | V. G. Kharchenko, S. K. Klimenko, N. M. Kupranets, T. V. Stolbova, I. S. Monakhova and L. A. Fomenko; *Zh. Org. Khim.*, 1977, **13**, 186 (*Chem. Abstr.*, 1977, **86**, 155 464). | 930 |
| 77ZOR222 | A. A. Michurin, E. A. Lyandaev and I. V. Bodrikov; *Zh. Org. Khim.*, 1977, **13**, 222. | 1080 |
| 77ZOR2617 | V. A. Dokichev and A. A. Potekhin; *Zh. Org. Khim.*, 1977, **13**, 2617. | 391 |
| 78ACH(96)275 | I. Lengyel, G. Kresze, M. Berger, W. Kosbahn and H. Schäfer; *Acta Chim. Acad. Sci. Hung.*, 1978, **96**, 275. | 1051 |
| 78ACH(97)69 | V. Szabó, J. Borda and L. Losonczi; *Acta Chim. Acad. Sci. Hung.*, 1978, **97**, 69. | 701 |
| 78ACH(98)457 | V. Szabó, J. Borda and V. Végh; *Acta Chim. Acad. Sci. Hung.*, 1978, **98**, 457. | 701 |
| 78ACR14 | S. F. Nelsen; *Acc. Chem. Res.*, 1978, **11**, 14. | 2, 40 |
| 78ACR314 | S. J. Benkovic; *Acc. Chem. Res.*, 1978, **11**, 314. | 325 |
| 78ACS(B)152 | S. C. Olsen and J. P. Snyder; *Acta Chem. Scand., Ser. B*, 1978, **32**, 152. | 982 |
| 78ACS(B)460 | O. H. Johansen and K. Undheim; *Acta Chem. Scand., Ser. B*, 1979, **33**, 460. | 909 |
| 78AF937 | H. Linde, H. Oelschläger and C. Czirwitzky; *Arzneim.-Forsch.*, 1978, **28**, 937. | 1008 |
| 78AF1684 | A. Kreutzberger; *Arzneim.-Forsch.*, 1978, **28**, 1684. | 524 |

| | | |
|---|---|---|
| 78AHC(23)1 | Z. Eckstein and T. Urbański; *Adv. Heterocycl. Chem.*, 1978, **23**, 1. | 1019, 1030, 1038 |
| 78AJC1391 | D. J. Brown and P. Waring; *Aust. J. Chem.*, 1978, **31**, 1391. | 140 |
| 78AJC2505 | D. J. Brown and T. Nagamatsu; *Aust. J. Chem.*, 1978, **31**, 2505. | 85 |
| 78AJC2517 | W. Cowden and N. Jacobsen; *Aust. J. Chem.*, 1978, **31**, 2517. | 141, 142 |
| 78AJC2699 | R. A. Eade, F. J. McDonald and H.-P. Pham; *Aust. J. Chem.*, 1978, **31**, 2699. | 824 |
| 78AP47 | B. Unterhalt, E. Seebach and D. Thamer; *Arch. Pharm. (Weinheim, Ger.)*, 1978, **311**, 47. | 1051 |
| 78AP115 | H. Fenner and R. Teichmann; *Arch. Pharm. (Weinheim, Ger.)*, 1978, **311**, 115. | 219 |
| 78AP196 | H. Fenner and W. Bauch; *Arch. Pharm. (Weinheim, Ger.)*, 1978, **311**, 196. | 228 |
| 78AP236 | R. Neidlein and I. Körber; *Arch. Pharm. (Weinheim, Ger.)*, 1978, **311**, 236. | 663, 924 |
| 78AP406 | R. Troschütz and H. J. Roth; *Arch. Pharm. (Weinheim, Ger.)*, 1978, **311**, 406. | 229 |
| 78AP542 | R. Troschütz and H. J. Roth; *Arch. Pharm. (Weinheim, Ger.)*, 1978, **311**, 542. | 231 |
| 78AP728 | G. Seitz and T. Kämpchen; *Arch. Pharm. (Weinheim, Ger.)*, 1978, **311**, 728. | 552 |
| 78AP786 | G. Seitz, T. Kämpchen and W. Overheu; *Arch. Pharm. (Weinheim, Ger.)*, 1978, **311**, 786. | 552 |
| 78AP799 | F. Eiden and E. Schmiz; *Arch. Pharm. (Weinheim, Ger.)*, 1978, **311**, 799. | 940 |
| 78AX(B)993 | R. Spagna and A. Vaciago; *Acta Crystallogr., Part B*, 1978, **34**, 993. | 538, 541 |
| 78AX(B)2376 | J. A. Jansing and J. G. White; *Acta Crystallogr., Part B*, 1978, **34**, 2376. | 1048 |
| 78AX(B)2514 | C. H. Schwalbe, M. F. G. Stevens and P. R. Lowe; *Acta Crystallogr., Part B*, 1978, **34**, 2514. | 370 |
| 78AX(B)2956 | P. Singh and J. D. McKinney; *Acta Crystallogr., Part B*, 1978, **34**, 2956. | 947, 948 |
| 78B1942 | C. Walsh, J. Fisher, R. Spencer, D. W. Graham, W. T. Ashton, J. E. Brown, R. D. Brown and E. F. Rogers; *Biochemistry*, 1978, **17**, 1942. | 204, 206, 250, 251, 252, 262 |
| 78BCJ179 | H. Besso, K. Imafuku and H. Matsumura; *Bull. Chem. Soc. Jpn.*, 1978, **51**, 179. | 55 |
| 78BCJ1175 | K. Kurosawa and Y. Ashihara; *Bull. Chem. Soc. Jpn.*, 1978, **51**, 1175. | 670, 829 |
| 78BCJ1422 | M. Nagai and N. Sakikawa; *Bull. Chem. Soc. Jpn.*, 1978, **51**, 1422. | 964, 973 |
| 78BCJ1484 | M. Kurabayashi and C. Grundmann; *Bull. Chem. Soc. Jpn.*, 1978, **51**, 1484. | 470 |
| 78BCJ1805 | K. Hasegawa, S. Hirooka, H. Kawahara, A. Nakayama, K. Ishikawa, N. Takeda and H. Mukai; *Bull. Chem. Soc. Jpn.*, 1978, **51**, 1805. | 1041, 1051, 1072 |
| 78BCJ1846 | C. Yamazaki; *Bull. Chem. Soc. Jpn.*, 1978, **51**, 1846. | 435 |
| 78BCJ1874 | A. Ninagawa and H. Matsuda; *Bull. Chem. Soc. Jpn.*, 1978, **51**, 1874. | 731 |
| 78BCJ1907 | Y. Kawase, S. Yamaguchi, K. Aoyama and M. Matsuda; *Bull. Chem. Soc. Jpn.*, 1978, **51**, 1907. | 807 |
| 78BCJ2398 | M. Nakayama, S. Eguchi, S. Hayashi, M. Tsukayama, T. Horie, T. Yamada and M. Masumura; *Bull. Chem. Soc. Jpn.*, 1978, **51**, 2398. | 714 |
| 78BCJ2425 | K. Hiraki, M. Onishi, T. Ikeda, K. Tomioka and Y. Obayashi; *Bull. Chem. Soc. Jpn.*, 1978, **51**, 2425. | 707 |
| 78BCJ2674 | K. Akiba, K. Ishikawa and N. Inamoto; *Bull. Chem. Soc. Jpn.*, 1978, **51**, 2674. | 921 |
| 78BCJ3063 | Y. Yamamoto and H. Yamada; *Bull. Chem. Soc. Jpn.*, 1978, **51**, 3063. | 6 |
| 78BCJ3407 | T. Izumi, T. Katou, A. Kasahara and K. Hanaya; *Bull. Chem. Soc. Jpn.*, 1978, **51**, 3407. | 909 |
| 78BCJ3443 | M. Ohsaku, A. Imamura and K. Hirao; *Bull. Chem. Soc. Jpn.*, 1978, **51**, 3443. | 3 |
| 78BSB271 | W. McFarlane and C. J. Turner; *Bull. Soc. Chim. Belg.*, 1978, **87**, 271. | 64 |
| 78C256 | W. Wiegrebe and S. Prior; *Chimia*, 1978, **32**, 256. | 860 |
| 78C332 | S. Chaloupka and H. Heimgartner; *Chimia*, 1978, **32**, 332. | 452 |
| 78CB240 | J. Adler, V. Böhnisch and H. Neunhoeffer; *Chem. Ber.*, 1978, **111**, 240. | 428 |
| 78CB314 | P. Stoss; *Chem. Ber.*, 1978, **111**, 314. | 1022 |
| 78CB971 | I. W. Southon and W. Pfleiderer; *Chem. Ber.*, 1978, **111**, 971. | 273, 284 |
| 78CB1284 | K. Krohn, G. Brückner and H.-P. Tietjen; *Chem. Ber.*, 1978, **111**, 1284. | 670 |
| 78CB1453 | P. Stoss and G. Satzinger; *Chem. Ber.*, 1978, **111**, 1453. | 1012 |
| 78CB1709 | L. Weber, C. Krüger and Y.-H. Tsay; *Chem. Ber.*, 1978, **111**, 1709. | 894 |
| 78CB1763 | R. Gottlieb and W. Pfleiderer; *Chem. Ber.*, 1978, **111**, 1763. | 249, 252, 286, 298 |
| 78CB1989 | S. Sommer and U. Schubert; *Chem. Ber.*, 1978, **111**, 1989. | 1066 |
| 78CB2173 | T. Kappe, W. Golser and W. Stadlbauer; *Chem. Ber.*, 1978, **111**, 2173. | 372, 373, 384 |
| 78CB2297 | H. Wamhoff and L. Lichtenthäler; *Chem. Ber.*, 1978, **111**, 2297. | 216, 226 |
| 78CB2571 | I. W. Southon and W. Pfleiderer; *Chem. Ber.*, 1978, **111**, 2571. | 297 |
| 78CB3007 | A. Kreutzberger and A. Tantawy; *Chem. Ber.*, 1978, **111**, 3007. | 500 |
| 78CB3385 | N. Theobald and W. Pfleiderer; *Chem. Ber.*, 1978, **111**, 3385. | 284 |
| 78CB3624 | G. Kaupp, M. Stark and H. Fritz; *Chem. Ber.*, 1978, **111**, 3624. | 977 |
| 78CB3790 | R. Mengel and W. Pfleiderer; *Chem. Ber.*, 1978, **111**, 3790. | 277, 302, 304 |
| 78CC9 | K. L. Davies, R. C. Storr and P. J. Whittle; *J. Chem. Soc., Chem. Commun.*, 1978, 9. | 998 |
| 78CC188 | C. H. Schwalbe and W. E. Hunt; *J. Chem. Soc., Chem. Commun.*, 1978, 188. | 462 |
| 78CC197 | T. Fujiwara, T. Hombo, K. Tomita, Y. Tamura and M. Ikeda; *J. Chem. Soc., Chem. Commun.*, 1978, 197. | 894, 895, 1016 |
| 78CC596 | R. Gräfing, H. D. Verkruijse and L. Brandsma; *J. Chem. Soc., Chem. Commun.*, 1978, 596. | 912 |
| 78CC695 | D. W. Engel, M. Hattingh, H. K. L. Hundt and D. G. Roux; *J. Chem. Soc., Chem. Commun.*, 1978, 695. | 622 |
| 78CC764 | K. Mori, K. Shinozuka, Y. Sakuma and F. Yoneda; *J. Chem. Soc., Chem. Commun.*, 1978, 764. | 71, 219 |
| 78CC847 | T. L. Gilchrist and T. G. Roberts; *J. Chem. Soc., Chem. Commun.*, 1978, 847. | 1017 |

| | | |
|---|---|---|
| 78CC902 | P. W. Manley, R. C. Storr, A. E. Baydar and G. V. Boyd; *J. Chem. Soc., Chem. Commun.*, 1978, 902. | 1004 |
| 78CC1049 | T. L. Gilchrist, C. W. Rees and D. Vaughan; *J. Chem. Soc., Chem. Commun.*, 1978, 1049. | 1068, 1070, 1075 |
| 78CHE368 | A. A. Avetisyan, S. Kh. Karages and M. T. Dangyan; *Chem. Heterocycl. Compd. (Engl. Transl.)*, 1978, **14**, 368. | 726 |
| 78CHE497 | N. A. Tynkavkina, G. A. Kalabin, V. V. Kohonova and D. F. Kushnarev; *Chem. Heterocycl. Compd. (Engl. Transl.)*, 1978, **14**, 497. | 708 |
| 78CHE721 | V. I. Tyvorskii, L. S. Stanishevskii and I. G. Tishchenko; *Chem. Heterocycl. Compd. (Eng. Transl.)*, 1978, 721. | 845 |
| 78CHE1067 | E. V. Kuznetsov, D. V. Pruchkin, A. I. Pyshchev and G. N. Dorofeenko; *Chem. Heterocycl. Compd. (Engl. Transl.)*, 1978, **14**, 1067. | 650, 670, 681 |
| 78CI(L)628 | J. R. Merchant and A. S. Gupta; *Chem. Ind. (London)*, 1978, 628. | 807 |
| 78CI(L)806 | Y. Tamura, T. Kawasaki, M. Tanio and Y. Kita; *Chem. Ind. (London)*, 1978, 806. | 1029 |
| 78CJC512 | T. Durst, J. C. Huang, N. K. Sharma and D. J. H. Smith; *Can. J. Chem.*, 1978, **56**, 512. | 981 |
| 78CJC1423 | I. W. J. Still, S. K. Hasan and K. Turnbull; *Can. J. Chem.*, 1978, **56**, 1423. | 896, 904 |
| 78CJC1804 | J. Armand, K. Chekir and J. Pinson; *Can. J. Chem.*, 1978, **56**, 1804. | 251, 252 |
| 78CL409 | T. Izawa and T. Mukaiyama; *Chem. Lett.*, 1978, 409. | 841 |
| 78CL723 | S. Yano, K. Nishino, K. Nakasuji and I. Murata; *Chem. Lett.*, 1978, 723. | 915, 938 |
| 78CL1177 | F. Yoneda and Y. Sakuma; *Chem. Lett.*, 1978, 1177. | 205 |
| 78CPB14 | T. Miyamoto, Y. Kimura, J.-I. Matsumoto and S. Minami; *Chem. Pharm. Bull.*, 1978, **26**, 14. | 356 |
| 78CPB245 | T. Hirota, T. Koyama, T. Nanba, M. Yamato and T. Matsumura; *Chem. Pharm. Bull.*, 1978, **26**, 245. | 220 |
| 78CPB549 | H. Uno and M. Kurokawa; *Chem. Pharm. Bull.*, 1978, **26**, 549. | 1029 |
| 78CPB1825 | Y. Yamamoto, Y. Azuma and K. Miyakawa; *Chem. Pharm. Bull.*, 1978, **26**, 1825. | 1002 |
| 78CPB1973 | M. Yamato, J.-I. Uenishi and K. Hashigaki; *Chem. Pharm. Bull.*, 1978, **26**, 1973. | 806 |
| 78CPB2428 | A. Ohsawa, T. Uezu and H. Igeta; *Chem. Pharm. Bull.*, 1978, **26**, 2428. | 32 |
| 78CPB2550 | A. Ohsawa, Y. Abe and H. Igeta; *Chem. Pharm. Bull.*, 1978, **26**, 2550. | 28 |
| 78CPB3633 | A. Ohsawa, T. Uezu and H. Igeta; *Chem. Pharm. Bull.*, 1978, **26**, 3633. | 32 |
| 78CR(C)(286)553 | J. P. Pradere and G. Hadjukovic; *C.R. Hebd. Seances Acad. Sci., Ser. C*, 1978, **286**, 553. | 908 |
| 78DOK(241)366 | E. G. Kovalev, I. Ya. Postovskii and G. L. Rusinov; *Dokl. Akad. Nauk SSSR*, 1978, **241**, 366. | 555 |
| 78FOR(35)199 | R. D. H. Murray; *Fortschr. Chem. Org. Naturst.*, 1978, **35**, 199. | 676 |
| 78G671 | R. Pellicciari, M. Curini, P. Ceccherelli and B. Natalini; *Gazz. Chim. Ital.*, 1978, **108**, 671. | 905 |
| 78GEP2732115 | R. H. Kuff; *Ger. Pat.* 2 732 115 (1978) (*Chem. Abstr.*, 1978, **88**, 170 201). | 1085 |
| 78H(9)1327 | B. Verček, I. Leban, B. Stanovnik and M. Tišler; *Heterocycles*, 1978, **9**, 1327. | 216, 223 |
| 78H(9)1367 | A. Ohsawa, H. Arai and H. Igeta; *Heterocycles*, 1978, **9**, 1367. | 28 |
| 78H(9)1397 | Y. Abe, A. Ohsawa, H. Arai and H. Igeta; *Heterocycles*, 1978, **9**, 1397. | 19, 40 |
| 78H(9)1771 | M. Quinteiro, C. Seoane and J. L. Soto; *Heterocycles*, 1978, 9, 1771. | 2 |
| 78H(10)167 | H. Nakamura and T. Goto; *Heterocycles*, 1978, **10**, 167. | 987 |
| 78H(11)387 | K. Hafner, H. J. Lindner and W. Wassem; *Heterocycles*, 1978, **11**, 387. | 2 |
| 78HC(33)3 | H. Neunhoeffer and P. F. Wiley; *Chem. Heterocycl. Compd.*, 1978, **33**, 3. | 369 |
| 78HC(33)13 | H. Neunhoffer and P. F. Wiley; *Chem. Heterocycl. Compd.*, 1978, **33**, 13. | 384 |
| 78HC(33)17 | H. Neunhoffer and P. F. Wiley; *Chem. Heterocycl. Compd.*, 1978, **33**, 17. | 382 |
| 78HC(33)53 | H. Neunhoffer and P. F. Wiley; *Chem. Heterocycl. Compd.*, 1978, **33**, 53. | 373, 374 |
| 78HC(33)54 | H. Neunhoffer and P. F. Wiley; *Chem. Heterocycl. Compd.*, 1978, **33**, 54. | 376 |
| 78HC(33)58 | H. Neunhoffer and P. F. Wiley; *Chem. Heterocycl. Compd.*, 1978, **33**, 58. | 380 |
| 78HC(33)68 | H. Neunhoffer and P. F. Wiley; *Chem. Heterocycl. Compd.*, 1978, **33**, 68. | 376 |
| 78HC(33)72 | H, Neunhoffer and P. F. Wiley; *Chem. Heterocycl. Compd.*, 1978, **33**, 72. | 383 |
| 78HC(33)78 | H. Neunhoffer and P. F. Wiley; *Chem. Heterocycl. Compd.*, 1978, **33**, 78. | 374 |
| 78HC(33)84 | H. Neunhoffer and P. F. Wiley; *Chem. Heterocycl. Compd.*, 1978, **33**, 84. | 382 |
| 78HC(33)86 | H. Neunhoffer and P. F. Wiley; *Chem. Heterocycl. Compd.*, 1978, **33**, 86. | 375, 377 |
| 78HC(33)91 | H. Neunhoffer and P. F. Wiley, *Chem. Heterocycl. Compd.*, 1978, **33**, 91 | 382 |
| 78HC(33)94 | H. Neunhoffer and P. F. Wiley; *Chem. Heterocycl. Compd.*, 1978, **33**, 94. | 375 |
| 78HC(33)165 | H. Neunhoffer and P. F. Wiley; *Chem. Heterocycl. Compd.*, 1978, **33**, 165. | 383 |
| 78HC(33)189 | H. Neunhoeffer; *Chem. Heterocycl. Compd.*, 1978, **33**, 189. | 385, 390, 395, 400, 405, 406, 407, 408, 411, 413, 414, 415, 430, 431, 432, 433, 434, 435, 436, 437, 438, 439, 440, 443, 444, 445, 446, 447, 448 |
| 78HC(33)1073 | H. Neunhoeffer and P. F. Wiley; *Chem. Heterocycl. Compd.*, 1978, **33**, 1073. | 2, 50 |
| 78HC(33)1075 | P. F. Wiley; *Chem. Heterocycl. Compd.*, 1978, **33**, 1075. | 536, 543, 544, 545, 546, 547 |
| 78HC(33)1287 | P. F. Wiley; *Chem. Heterocycl. Compd.*, 1978, **33**, 1287. | 533 |
| 78HC(33)1296 | P. F. Wiley; *Chem. Heterocycl. Compd.*, 1978, **33**, 1296. | 535, 536, 572 |
| 78HCA1175 | P. Winternitz; *Helv. Chim. Acta*, 1978, **61**, 1175. | 442 |
| 78HCA1388 | M. Bloch, F. Brogli, E. Heilbronner, T. B. Jones, H. Prinzbach and O. Schweikert; *Helv. Chim. Acta*, 1978, **61**, 1388. | 598, 958 |
| 78HCA1622 | D. Seebach, R. Dach, D. Enders, B. Renger, M. Jansen and G. Brachtel; *Helv. Chim. Acta*, 1978, **61**, 1622. | 532, 533, 534 |
| 78HCA2108 | W. Schwotzer, J. H. Bieri, M. Viscontini and W. von Philipsborn; *Helv. Chim. Acta*, 1978, **61**, 2108. | 64, 281 |

| | | |
|---|---|---|
| 78HCA2246 | S. Antoulas, J. H. Bieri and M. Viscontini; *Helv. Chim. Acta*, 1978, **61**, 2246. | 307 |
| 78HCA2419 | H. Link; *Helv. Chim. Acta*, 1978, **61**, 2419. | 452 |
| 78HCA2452 | D. Schelz; *Helv. Chim. Acta*, 1978, **61**, 2452. | 249, 252, 258 |
| 78HCA2744 | H. J. Furrer, J. H. Bieri and M. Viscontini; *Helv. Chim. Acta*, 1978, **61**, 2744. | 281 |
| 78IJC(A)64 | V. K. Ahuja and N. K. Ray; *Indian J. Chem., Sect. A*, 1978, **16**, 64. | 576, 634 |
| 78IJC(A)531 | V. K. Ahuja; *Indian J. Chem., Sect. A*, 1978, **16**, 531. | 576, 637 |
| 78IJC(B)210 | I. R. Trehan, B. S. Ahluwalia and M. Vig; *Ind. J. Chem., Sect. B*, 1978, **16**, 210. | 910 |
| 78IJC(B)292 | Y. S. Chanhan and K. B. L. Mathur; *Indian J. Chem., Sect. B*, 1978, **16**, 292. | 688 |
| 78IJC(B)295 | M. H. Elnagdi, H. M. Fahmy, M. A. Morsi and S. K. El-Ees; *Indian J. Chem., Sect. B*, 1978, **16**, 295. | 643 |
| 78IJC(B)332 | M. Abdalla, A. Essawy and A. Deeb; *Indian J. Chem., Sect. B*, 1978, **16**, 332. | 228 |
| 78IJC(B)570 | A. G. Shinde and R. N. Usgaonkar; *Indian J. Chem., Sect. B*, 1978, **16**, 570. | 804 |
| 78IJC(B)889 | H. Jahine, H. A. Zaher, O. Sherif and M. M. Fawzy; *Indian J. Chem., Sect. B*, 1978, **16**, 889. | 208 |
| 78IZV2499 | V. Kh. Buzykin, B. I. Buzykin, L. S. Sysoeva and Yu. P. Kitaev; *Izv. Akad. Nauk SSSR, Ser. Khim.*, 1978, 2499. | 546 |
| 78JA200 | G. Barbarella, P. Dembech, A. Garbesi, F. Bernardi, A. Bottoni and A. Fava; *J. Am. Chem. Soc.*, 1978, **100**, 200. | 897, 898 |
| 78JA373 | G. A. Jeffrey, J. A. Pople, J. S. Binkley and S. Vishveshwara; *J. Am. Chem. Soc.*, 1978, **100**, 373. | 629 |
| 78JA985 | T. Kitazume and J. M. Shreeve; *J. Am. Chem. Soc.*, 1978, **100**, 985. | 1047, 1063, 1078 |
| 78JA1614 | E. L. Eliel, J. K. Koskimies and B. Lohri; *J. Am. Chem. Soc.*, 1978, **100**, 1614. | 977 |
| 78JA3242 | B. Dellinger, M. A. Paczkowski, R. M. Hochstrasser and A. B. Smith, III; *J. Am. Chem. Soc.*, 1978, **100**, 3242. | 549 |
| 78JA4004 | S. F. Nelsen and E. L. Clennan; *J. Am. Chem. Soc.*, 1978, **100**, 4004. | 332 |
| 78JA4012 | S. F. Nelsen, E. L. Clennan and D. H. Evans; *J. Am. Chem. Soc.*, 1978, **100**, 4012. | 332 |
| 78JA4037 | A. Kaito and M. Hatano; *J. Am. Chem. Soc.*, 1978, **100**, 4037. | 250 |
| 78JA6111 | D. Garcia and P. M. Keehn; *J. Am. Chem. Soc.*, 1978, **100**, 6111. | 719 |
| 78JA6114 | E. L. Eliel and E. Juaristi; *J. Am. Chem. Soc.*, 1978, **100**, 6114. | 989 |
| 78JA7079 | M. Nakane, C. R. Hutchinson, D. Van Engen and J. Clardy; *J. Am. Chem. Soc.*, 1978, **100**, 7079. | 624, 846 |
| 78JA7661 | S. Senda, K. Hirota, T. Asao and K. Maruhashi; *J. Am. Chem. Soc.*, 1978, **100**, 7661. | 317 |
| 78JAP(K)7856692 | M. Arimura; (Yoshitomi Pharm. Ind. Ltd.), *Jpn. Kokai* 78 56 692 (1978) (*Chem. Abstr.*, 1978, **89**, 146 925). | 231 |
| 78JCP(68)2487 | R. E. Smalley, L. Wharton, D. H. Levy and D. W. Chandler; *J. Chem. Phys.*, 1978, **68**, 2487. | 543 |
| 78JCR(M)0582 | E. V. Dehmlow and Naser-Ud-Din; *J. Chem. Res. (M)*, 1978, 0582. | 51 |
| 78JCR(M)0865 | G. J. P. Becket, G. P. Ellis and M. I. U. Trindade; *J. Chem. Res. (M)*, 1978, 0865. | 820, 829 |
| 78JCR(S)40 | E. V. Dehmlow and Naser-Ud-Din; *J. Chem. Res. (S)*, 1978, 40. | 51 |
| 78JCR(S)47 | G. J. P. Becket, G. P. Ellis and M. I. U. Trindade; *J. Chem. Res. (S)*, 1978, 47. | 664, 680, 697, 714 |
| 78JCS(P1)81 | S. M. Afzal, R. Pike, N. H. Rama, I. R. Smith, E. S. Turner and W. B. Whalley; *J. Chem. Soc., Perkin Trans. 1*, 1978, 81. | 858 |
| 78JCS(P1)88 | J. O. Oluwadiya and W. B. Whalley; *J. Chem. Soc., Perkin Trans. 1*, 1978, 88. | 756, 768 |
| 78JCS(P1)378 | M. G. Barlow, R. N. Haszeldine and J. A. Pickett; *J. Chem. Soc., Perkin Trans. 1*, 1978, 378. | 50 |
| 78JCS(P1)666 | A. Pelter, R. S. Ward and R. J. Bass; *J. Chem. Soc., Perkin Trans. 1*, 1978, 666. | 591 |
| 78JCS(P1)716 | T. Hiramitsu, Y. Maki and S. Senda; *J. Chem. Soc., Perkin Trans. 1*, 1978, 716. | 222 |
| 78JCS(P1)726 | T. Watanabe, S. Katayama, Y. Nakashita and M. Yamauchi; *J. Chem. Soc., Perkin Trans. 1*, 1978, 726. | 822 |
| 78JCS(P1)789 | F. J. Lalor, F. L. Scott, G. Ferguson and W. C. Marsh; *J. Chem. Soc., Perkin Trans. 1*, 1978, 789. | 387, 396 |
| 78JCS(P1)857 | A. Kumar, H. Ila and H. Junjappa; *J. Chem. Soc., Perkin Trans. 1*, 1978, 857. | 228, 230 |
| 78JCS(P1)876 | R. Graham and J. R. Lewis; *J. Chem. Soc., Perkin Trans. 1*, 1978, 876. | 837 |
| 78JCS(P1)1321 | S. Fabrissin, S. Fatutta and A. Risaliti; *J. Chem. Soc., Perkin Trans. 1*, 1978, 1321. | 889, 927 |
| 78JCS(P1)1338 | G. V. Boyd and R. L. Monteil; *J. Chem. Soc., Perkin Trans. 1*, 1978, 1338. | 831 |
| 78JCS(P1)1351 | G. V. Boyd, R. L. Monteil, P. F. Lindley and M. M. Mahmoud; *J. Chem. Soc., Perkin Trans. 1*, 1978, 1351. | 624 |
| 78JCS(P1)1373 | M. P. Sammes and K. L. Yip; *J. Chem. Soc., Perkin Trans. 1*, 1978, 1373. | 762, 872 |
| 78JCS(P2)377 | V. J. Baker, I. J. Ferguson, A. R. Katritzky, R. Patel and S. Rahimi-Rastgoo; *J. Chem. Soc., Perkin Trans. 2*, 1978, 377. | 467, 1053, 1054 |
| 78JCS(P2)391 | A. I. Gray, R. D. Waigh and P. G. Waterman; *J. Chem. Soc., Perkin Trans. 2*, 1978, 391. | 582 |
| 78JCS(P2)1001 | K. Bergesen, B. M. Carden and M. J. Cook; *J. Chem. Soc., Perkin Trans. 2*, 1978, 1001. | 962 |
| 78JCS(P2)1207 | P. Maroni, M. Calmon, L. Cazaux, P. Tisnès, G. Sartoré and M. Aknin; *J. Chem. Soc., Perkin Trans. 2*, 1978, 1207. | 1060 |
| 78JHC57 | M. Quinteiro, C. Seoane and J. L. Soto; *J. Heterocycl. Chem.*, 1978, **15**, 57. | 593, 740, 758 |
| 78JHC181 | L. Mosti, P. Schenone and G. Menozzi; *J. Heterocycl. Chem.*, 1978, **15**, 181. | 797 |
| 78JHC289 | C. H. Chen, G. A. Reynolds, N. Zumbulyadis and J. A. Van Allan; *J. Heterocycl. Chem.*, 1978, **15**, 289. | 933 |

| 78JHC445 | A. D. Counotte-Potman and H. C. van der Plas; *J. Heterocycl. Chem.*, 1978, **15**, 445. | 542, 545, 555 |
|---|---|---|
| 78JHC515 | P. D. Croce, P. D. Buttero, S. Maiorana and R. Vistocco; *J. Heterocycl. Chem.*, 1978, **15**, 515. | 564 |
| 78JHC699 | M. A. ElSohly, E. G. Boeren and C. E. Turner; *J. Heterocycl. Chem.*, 1978, **15**, 699. | 748 |
| 78JHC759 | Y. A. El-Farkh, F. H. Al-Hajjar, N. R. El-Rayyes and H. S. Hammoud; *J. Heterocycl. Chem.*, 1978, **15**, 759. | 684 |
| 78JHC769 | J.-P. Coïc and G. Saint-Ruf; *J. Heterocycl. Chem.*, 1978, **15**, 769. | 973, 975 |
| 78JHC781 | K. Senga, J. Sato, Y. Kanamori, M. Ichiba, S. Nishigaki, M. Noguchi and F. Yoneda; *J. Heterocycl. Chem.*, 1978, **15**, 781. | 336, 357 |
| 78JHC877 | R. Kwok; *J. Heterocycl. Chem.*, 1978, **15**, 877. | 223 224 |
| 78JHC1051 | P. Bravo and C. Ticozzi; *J. Heterocycl. Chem.*, 1978, **15**, 1051. | 781 |
| 78JHC1055 | S. Braun and K. Hafner; *J. Heterocycl. Chem.*, 1978, **15**, 1055. | 463 |
| 78JHC1057 | K. H. Pannell, B. L. Kalsotra and C. Párkányi; *J. Heterocycl. Chem.*, 1978, **15**, 1057. | 2 |
| 78JHC1105 | M. Drobnič-Košorok, S. Polanc, B. Stanovnik, M. Tišler and B. Verček; *J. Heterocycl. Chem.*, 1978, **15**, 1105. | 234 |
| 78JHC1113 | V. I. Cohen; *J. Heterocycl. Chem.*, 1978, **15**, 1113. | 569 |
| 78JHC1121 | C. A. H. Rasmussen, H. C. van der Plas, P. Grotenhuis and A. Koudijs; *J. Heterocycl. Chem.*, 1978, **15**, 1121. | 100 |
| 78JHC1153 | B. Nedjar, M. Hamdi, J. Périé and V. Hérault; *J. Heterocycl. Chem.*, 1978, **15**, 1153. | 686, 723, 726, 730, 843, 847 |
| 78JHC1271 | A. Camparini, A. M. Celli, F. Ponticelli and P. Tedeschi; *J. Heterocycl. Chem.*, 1978, **15**, 1271. | 391, 438 |
| 78JHC1425 | M. Iwao and T. Kuraishi; *J. Heterocycl. Chem.*, 1978, **15**, 1425. | 24 |
| 78JHC1439 | M. Maguel and R. Guglielmetti; *J. Heterocycl. Chem.*, 1978, **15**, 1439. | 672 |
| 78JHC1451 | T. Suzuki, N. Katou and K. Mitsuhashi; *J. Heterocycl. Chem.*, 1978, **15**, 1451. | 347 |
| 78JHC1475 | T. Kato, U. Izumi and N. Katagiri; *J. Heterocycl. Chem.*, 1978, **15**, 1475. | 1002 |
| 78JHC1535 | F. M. Hauser and S. A. Pogany; *J. Heterocycl. Chem.*, 1978, **15**, 1535. | 681 |
| 78JMC170 | A. Rosowsky and C. S. Yu; *J. Med. Chem.*, 1978, **21**, 170. | 327 |
| 78JMC231 | E. J. Valente, W. R. Porter and W. F. Trager; *J. Med. Chem.*, 1978, **21**, 231. | 582 |
| 78JMC295 | M. Nasr, I. Nabih and J. H. Burckhalter; *J. Med. Chem.*, 1978, **21**, 295. | 224 |
| 78JMC380 | A. Rosowsky, G. P. Beardsley, W. D. Ensminger, H. Lazarus and C. S. Yu; *J. Med. Chem.*, 1978, **21**, 380. | 327 |
| 78JMC1070 | N. J. Lewis, R. L. Inloes, J. Hes, R. H. Matthews and G. Milo; *J. Med. Chem.*, 1978, **21**, 1070. | 940 |
| 78JMC1120 | G. P. Ellis, G. J. P. Becket, D. Shaw, H. K. Wilson, C. J. Vardey and I. F. Skidmore; *J. Med. Chem.*, 1978, **21**, 1120. | 707 |
| 78JOC96 | F. A. Carey, O. D. Dailey, Jr. and W. C. Hutton; *J. Org. Chem.*, 1978, **43**, 96. | 953, 954 |
| 78JOC393 | M. Debeljak-Šuštar, B. Stanovnik, M. Tišler and Z. Zrimšek; *J. Org. Chem.*, 1978, **43**, 393. | 223 |
| 78JOC521 | G. B. Schuster and L. A. Bryant; *J. Org. Chem.*, 1978, **43**, 521. | 979 |
| 78JOC667 | S. S. Hall, G. F. Weber and A. J. Duggan; *J. Org. Chem.*, 1978, **43**, 667. | 578, 579, 594, 630, 772 |
| 78JOC680 | E. C. Taylor and R. Kobylecki; *J. Org. Chem.*, 1978, **43**, 680. | 290, 292, 293, 299 |
| 78JOC736 | E. C. Taylor, R. N. Henrie, II and R. C. Portnoy; *J. Org. Chem.*, 1978, **43**, 736. | 318 |
| 78JOC828 | A. Srinivasan, P. E. Fagerness and A. D. Broom; *J. Org. Chem.*, 1978, **43**, 828. | 202, 206 |
| 78JOC914 | M. M. Chau and J. L. Kice; *J. Org. Chem.*, 1978, **43**, 914. | 967, 971 |
| 78JOC1262 | R. R. King, R. Greenhalgh and W. D. Marshall; *J. Org. Chem.*, 1978, **43**, 1262. | 977 |
| 78JOC1355 | V. Čaplar, A. Lisini, F. Kajfež, D. Kolbah and V. Sunjić; *J. Org. Chem.*, 1978, **43**, 1355. | 1013 |
| 78JOC1361 | T. Goka, H. Shizuka and K. Matsui; *J. Org. Chem.*, 1978, **43**, 1361. | 477 |
| 78JOC2020 | P. A. Wade; *J. Org. Chem.*, 1978, **43**, 2020. | 1000, 1017 |
| 78JOC2138 | N. Ishibe and S. Yutaka; *J. Org. Chem.*, 1978, **43**, 2138. | 694 |
| 78JOC2144 | N. Ishibe, S. Yutaka, J. Mansui and N. Ihda; *J. Org. Chem.*, 1978, **43**, 2144. | 694 |
| 78JOC2344 | J. H. Looker, J. H. McMechan and J. W. Mader; *J. Org. Chem.*, 1978, **43**, 2344. | 819 |
| 78JOC2500 | M. S. Raasch; *J. Org. Chem.*, 1978, **43**, 2500. | 932 |
| 78JOC2536 | R. S. Klein, M.-I. Lim, S. Y.-K. Tam and J. J. Fox; *J. Org. Chem.*, 1978, **43**, 2536. | 336, 345, 357 |
| 78JOC2682 | J. P. Geerts and H. C. van der Plas; *J. Org. Chem.*, 1978, **43**, 2682. | 519 |
| 78JOC2693 | M. Nakajima, R. Hisada and J.-P. Anselme; *J. Org. Chem.*, 1978, **43**, 2693. | 450 |
| 78JOC2842 | N. S. Poonia and B. P. Yadav; *J. Org. Chem.*, 1978, **43**, 2842. | 716 |
| 78JOC3055 | J. V. Crivello and J. H. W. Lam; *J. Org. Chem.*, 1978, **43**, 3055. | 967 |
| 78JOC3231 | Y. Ohtsuka; *J. Org. Chem.*, 1978, **43**, 3231. | 339, 365 |
| 78JOC3303 | T. B. Patrick, M. A. Dorton and J. G. Dolan; *J. Org. Chem.*, 1978, **43**, 3303. | 921 |
| 78JOC3370 | Y. Tamaru, T. Harada and Z. Yoshida; *J. Org. Chem.*, 1978, **43**, 3370. | 42 |
| 78JOC3415 | F. Z. Basha and R. W. Franck; *J. Org. Chem.*, 1978, **43**, 3415. | 1020 |
| 78JOC3632 | E. C. Taylor, J. G. Andrade, G. J. H. Rall and A. McKillop; *J. Org. Chem.*, 1978, **43**, 3632. | 848 |
| 78JOC3698 | M. E. Jung, A. B. Mossman and M. A. Lyster; *J. Org. Chem.*, 1978, **43**, 3698. | 788 |
| 78JOC3817 | C. K. Bradsher and T. G. Wallis; *J. Org. Chem.*, 1978, **43**, 3817. | 834, 861 |
| 78JOC3856 | D. K. Bates and M. C. Jones; *J. Org. Chem.*, 1978, **43**, 3856. | 742 |
| 78JOC4112 | S. Bersani, G. Doddi, S. Fornarini and F. Stegel; *J. Org. Chem.*, 1978, **43**, 4112. | 654 |

| 78JOC4154 | E. C. Taylor and E. Wachsen; *J. Org. Chem.*, 1978, **43**, 4154. | 284, 301 |
|---|---|---|
| 78JOC4415 | L. Tsai, J. V. Silverton and H. T. Lingh; *J. Org. Chem.*, 1978, **43**, 4415. | 691 |
| 78JOC4826 | V. Ceré, C. Paolucci, S. Pollicino, W. Sandri and A. Fava; *J. Org. Chem.*, 1978, **43**, 4826. | 898 |
| 78JOC4831 | E. Vedejs, M. J. Arco, D. W. Powell, J. M. Renga and S. P. Singer; *J. Org. Chem.*, 1978, **43**, 4831. | 898 |
| 78JOC4844 | R. W. Morrison, Jr., W. R. Mallory and V. L. Styles; *J. Org. Chem.*, 1978, **43**, 4844. | 356 |
| 78JOM(146)235 | F. P. Colonna, G. Distefano, V. Galasso, K. J. Irgolic, C. E. King and G. C. Pappalardo; *J. Organomet. Chem.*, 1978, **146**, 235. | 945, 947, 958 |
| 78JOM(155)293 | R. K. Bansal and S. K. Sharma; *J. Organomet. Chem.*, 1978, **155**, 293. | 564 |
| 78JOU2245 | Y. S. Andreichikov, U. A. Nalimova, A. P. Kozlov and I. A. Rusakov; *J. Org. Chem. USSR (Engl. Transl.)*, 1978, **14**, 2245. | 798 |
| 78JPC1168 | B. K. Bandlish, W. R. Porter, Jr. and H. J. Shine; *J. Phys. Chem.*, 1978, **82**, 1168. | 969 |
| 78JPR452 | G. Westphal and T. Müller; *J. Prakt. Chem.*, 1978, **320**, 452. | 1050, 1051, 1069, 1075 |
| 78KGS342 | A. V. Eremeev, R. S. El'kinson and E. Liepins; *Khim. Geterotsikl. Soedin.*, 1978, 342. | 454 |
| 78KGS651 | P. B. Terent'ev, N. G. Kotova and A. N. Kost; *Khim. Geterotsikl. Soedin.*, 1978, 651 (*Chem. Abstr.*, 1978, **89**, 109 324). | 246 |
| 78KGS809 | M. F. Budyka, P. B. Terent'ev and A. N. Kost; *Khim. Geterotsikl. Soedin.*, 1978, 809 (*Chem. Abstr.*, 1978, **89**, 129 466). | 239 |
| 78KGS991 | V. P. Shchipanov, E. O. Didorov, L. S. Podenko and G. D. Kadochnikova; *Khim. Geterotsikl. Soedin.*, 1978, 991. | 558 |
| 78KGS1137 | V. P. Shchipanov and G. D. Kadochnikova; *Khim. Geterotsikl. Soedin.*, 1978, 1137. | 548, 558 |
| 78KGS1208 | A. A. Potekhin and N. A. Nikolaeva; *Khim. Geterotsikl. Soedin.*, 1978, 1208. | 1051, 1059 |
| 78KGS1272 | L. N. Koikov, M. F. Budyka, P. B. Terent'ev and A. N. Kost; *Khim. Geterotsikl. Soedin.*, 1978, 1272 (*Chem. Abstr.*, 1978, **90**, 6330). | 234, 236, 239 |
| 78KGS1400 | A. N. Kost, R. S. Sagitullin and G. G. Danagulyan; *Khim. Geterosikl. Soedin.*, 1978, 1400. | 124 |
| 78KGS1549 | V. G. Granik, N. B. Marchenko and R. G. Glushkov; *Khim. Geterotsikl. Soedin.*, 1978, 1549 (*Chem. Abstr.*, 1979, **90**, 121 529). | 227 |
| 78KGS1555 | M. F. Budyka and P. B. Terent'ev; *Khim. Geterotsikl. Soedin.*, 1978, 1555 (*Chem. Abstr.*, 1978, **90**, 87 378). | 239 |
| 78KGS1568 | A. A. Potekhin, S. M. Shevchenko, T. Ya. Vakhitov and V. A. Gindin; *Khim. Geterotsikl. Soedin.*, 1978, 1568. | 1055 |
| 78KGS1666 | V. G. Granik, I. V. Persianova, A. M. Zhidkova, N. B. Marchenko, R. G. Glushkov and Y. N. Sheinker; *Khim. Geterotsikl. Soedin.*, 1978, 1666 (*Chem. Abstr.*, 1978, **90**, 120 793). | 206 |
| 78KGS1671 | E. O. Sochneva, N. P. Solov'eva and V. G. Granik; *Khim. Geterotsikl. Soedin.*, 1978, 1671 (*Chem. Abstr.*, 1978, **90**, 121 530). | 216 |
| 78KGS1694 | L. M. Meshcheryakova, V. A. Zagorevskii and E. K. Orlova; *Khim. Geterotsikl. Soedin.*, 1978, 1694. | 909 |
| 78LA337 | H. H. Meyer; *Liebigs Ann. Chem.*, 1978, 337. | 842 |
| 78LA512 | W. Kantlehner, P. Fischer, W. Kugel, E. Mohring and H. Bredereck; *Liebigs Ann. Chem.*, 1978, 512. | 515 |
| 78LA635 | G. Steiner; *Liebigs Ann. Chem.*, 1978, 635. | 1016 |
| 78LA1111 | H. Hoberg and J. Korff; *Liebigs Ann. Chem.*, 1978, 1111. | 516 |
| 78LA1123 | H. Böhme and H.-J. Wilke; *Liebigs Ann. Chem.*, 1978, 1123. | 909 |
| 78LA1280 | W. Ried and H. Bopp; *Liebigs Ann. Chem.*, 1978, 1280. | 917, 937 |
| 78LA1780 | M. Hattori and W. Pfleiderer; *Liebigs Ann. Chem.*, 1978, 1780. | 283 |
| 78LA1788 | M. Hattori and W. Pfleiderer; *Liebigs Ann. Chem.*, 1978, 1788. | 283 |
| 78LA2033 | H. Gnichtel, W. I. Salem and L. Wawretschek; *Liebigs Ann. Chem.*, 1978, 2033. | 394, 444 |
| 78M1075 | H. W. Schmidt and H. Junek; *Monatsh. Chem.*, 1978, **109**, 1075. | 794 |
| 78M1485 | W. Stadlbauer and T. Kappe; *Monatsh. Chem.*, 1978, **109**, 1485. | 800, 834 |
| 78MI21200 | E. J. Pedersen, R. R. Vold and R. L. Vold; *Mol. Phys.*, 1978, **35**, 997. | 8 |
| B-78MI21201 | A. Kleemann and J. Engel; 'Pharmazeutische Wirkstoffe: Synthesen, Patente, Anwendungen', Thieme Verlag, Stuttgart, 1978. | 56 |
| B-78MI21300 | A. Giner-Sorolla and L. Medrek; in 'Nucleic Acid Chemistry', ed. L. B. Townsend and R. S. Tipson; Wiley, New York, 1978, vol. 1, p. 83. | 89 |
| B-78MI21301 | E. C. Ressner, A. Kampf and M. P. Mertes; in 'Nucleic Acid Chemistry', ed. L. B. Townsend and R. S. Tipson; Wiley, New York, 1978, vol. 1, p. 89. | 125 |
| B-78MI21302 | 'Nucleic Acid Chemistry', ed. L. B. Townsend and R. S. Tipson; Wiley, New York, 1978, parts 1 and 2. | 142 |
| 78MI21303 | T. Takita, Y. Muraoka, T. Nakatani, A. Fujii, Y. Umezawa, H. Naganawa and H. Umezawa; *J. Antibiot.*, 1978, **31**, 801. | 147 |
| 78MI21400 | T. Akiyama, Y. Enomoto and T. Shibamoto; *J. Agric. Food Chem.*, 1978, **26**, 1176. | 179 |
| 78MI21500 | W. Czuba, T. Kowalska and P. Kowalski; *Pol. J. Chem.*, 1978, **52**, 2369 (*Chem. Abstr.*, 1980, **92**, 76 355). | 218 |
| 78MI21501 | S. Robev; *Dokl. Bolg. Akad. Nauk.*, 1978, **31**, 551 (*Chem. Abstr.*, 1979, **90**, 186 887). | 222 |
| 78MI21502 | M. F. Budyka; *Deposited Doc.*, 1978, VINITI 1805-78, 2 (*Chem. Abstr.*, 1980, **92**, 41 092). | 236, 239 |
| 78MI21600 | C. E. Hignite and D. L. Azarnoff; *Biomed. Mass Spectrom.*, 1978, **5**, 161. | 285 |
| 78MI21601 | D. M. Danks; *J. Inher. Metab. Dis.*, 1978, **1**, 47. | 324 |
| 78MI21602 | W. Pfleiderer; *J. Inher. Metab. Dis.*, 1978, **1**, 54. | 306 |

| | | |
|---|---|---|
| 78MI21603 | S. Kaufman, S. Berlow and G. K. Summer; *New Engl. J. Med.*, 1978, **299**, 673. | 324 |
| B-78MI22000 | C. E. Schildknecht; in 'Kirk-Othmer Encyclopedia of Chemical Technology', ed. M. Grayson; Wiley, New York, 1978, vol. 2, p. 123. | 525, 526 |
| 78MI22100 | D. Coulter, D. Dows, H. Reisler and C. Wittig; *Chem. Phys.*, 1978, **32**, 429 (*Chem. Abstr.*, 1979, **90**, 38 272). | 549 |
| 78MI22101 | M. S. Sytilin and A. I. Morozov; *Zh. Fiz. Khim.*, 1978, **52**, 2933 (*Chem. Abstr.*, 1979, **90**, 61 851). | 547 |
| 78MI22200 | C.-J. Chang; *Lloydia*, 1978, **41**, 17. | 591 |
| 78MI22201 | M. Sakakibara and T. J. Mabry; *Rev. Latinoam. Quim.*, 1978, **9**, 92. | 602 |
| B-78MI22202 | 'Atlas of Stereochemistry', ed. W. Klyne and J. Buckingham; Chapman and Hall, London, 2nd edn., 1978, vol. 1. | 631 |
| B-78MI22400 | T. Goto and S. Yamamura; in 'Methodicum Chimicum', ed. F. Korte and M. Goto; Academic Press, New York, 1978, vol. 2, part 3, p. 134. | 876 |
| B-78MI22401 | U. Weiss and J. M. Edwards; 'The Biosynthesis of Aromatic Compounds', Wiley, New York, 1978. | 877 |
| 78MI22500 | T. D. Kazarinova, E. N. Lyutaya, S. N. Chalaya, L. V. Lipatova and V. G. Kharchenko; *Khim. Teknol. Elementoorgan. Soedin. Polim.* (*Kazan*), 1978, 3 (*Chem. Abstr.*, 1979, **91**, 22 306). | 939 |
| 78MI22501 | J. K. Sugden, K. Willcocks and N. J. Van Abbe; *Pharm. Acta Helv.*, 1978, **53**, 189. | 940 |
| 78MI22502 | A. O. Tosunyan, M. R. Bagdasaryan, S. A. Vartanyan, O. M. Avakyan and O. S. Noravyan; *Khim.-Farm. Zh.*, 1978, **12**, 56 (*Chem. Abstr.*, 1979, **90**, 22 750). | 940 |
| B-78MI22600 | J. N. Macdonald and J. Sheridan; in 'Molecular Spectroscopy', The Chemical Society, London, 1978, vol. 5, p. 1. | 948, 954 |
| 78MI22601 | Y. Hase and Y. Kawano; *Spectroscopy Lett.*, 1978, **11**, 151. | 957 |
| 78MI22602 | Y. Hase; *Spectroscopy Lett.*, 1978, **11**, 823. | 957 |
| 78MI22603 | J. C. Turley and G. E. Martin; *Spectroscopy Lett.*, 1978, **11**, 681. | 985 |
| 78MI22700 | G. Wagner and I. Wunderlich; *Pharmazie*, 1978, **33**, 15. | 1023 |
| B-78MI22701 | M. Sainsbury; in 'The Chemistry of Carbon Compounds', ed. S. Coffey; Elsevier, Amsterdam, 2nd edn., 1978, vol. IV, p. 427. | 1001, 1019, 1033, 1035, 1036 |
| 78MI22800 | K. Burger and R. Ottlinger; *J. Fluorine Chem.*, 1978, **11**, 29. | 1077 |
| 78MI22801 | T. Kiersznicki and A. Rajca; *Pol. J. Chem.*, 1978, **52**, 1827. | 1074 |
| 78MI22802 | H. W. Roesky and M. Banek; *Synth. React. Inorg. Met.-Org. Chem.*, 1978, **8**, 111 (*Chem. Abstr.*, 1978, **89**, 43 350). | 1047, 1075 |
| 78MI22803 | I. Stahl, R. Mews and O. Glemser; *J. Fluorine Chem.*, 1978, **11**, 455. | 1047, 1074 |
| 78OMS43 | A. Zeman and R. Wörle; *Org. Mass Spectrom.*, 1978, **13**, 43. | 465 |
| 78OMS296 | D. L. McGillivray and G. A. Poulton; *Org. Mass Spectrom.*, 1978, **13**, 296. | 612 |
| 78OMS338 | P. Wolkoff and J. L. Holmes; *Org. Mass Spectrom.*, 1978, **13**, 338. | 957, 958 |
| 78OMS653 | S. Eguchi; *Org. Mass Spectrom.*, 1978, **13**, 653. | 607, 617 |
| 78P1363 | K. Panichpol and P. G. Waterman; *Phytochemistry*, 1978, **17**, 1363. | 591 |
| 78PIA(A)(87)247 | A. C. Jain, R. C. Gupta and A. Kumar; *Proc. Indian Acad. Sci., Sect. A*, 1978, **87**, 247. | 743 |
| 78RCR975 | D. Martin, M. Bauer and V. A. Pankratov; *Russ. Chem. Rev. (Engl. Transl.)*, 1978, **47**, 975. | 503, 505, 506 |
| 78RTC107 | A. J. M. Weber, W. G. B. Huysmans, W. J. Mijs, W. M. M. J. Bovée and J. Vriend; *Recl. Trav. Chim. Pays-Bas*, 1978, **97**, 107. | 467 |
| 78RTC116 | R. E. van der Stoel and H. C. van der Plas; *Recl. Trav. Chim. Pays-Bas*, 1978, **97**, 116. | 8, 22 |
| 78RTC208 | R. Gräfing and L. Brandsma; *Recl. Trav. Chim. Pays-Bas*, 1978, **97**, 208. | 912 |
| 78RTC273 | A. Rykowski, H. C. van der Plas and A. van Veldhuizen; *Recl. Trav. Chim. Pays-Bas*, 1978, **97**, 273. | 393, 394, 400 |
| 78S144 | E. Suzuki, B. Katsuragawa and S. Inoue; *Synthesis*, 1978, 144. | 692 |
| 78S208 | D. T. Connor, P. A. Young and M. von Strandtmann; *Synthesis*, 1978, 208. | 682, 830 |
| 78S211 | W. Ried and H. Bopp; *Synthesis*, 1978, 211. | 917 |
| 78S372 | S. Kano, Y. Takahagi and S. Shibuya; *Synthesis*, 1978, **5**, 372. | 180 |
| 78S377 | W. Gauss and H-J. Kabbe; *Synthesis*, 1978, 377. | 1021 |
| 78S382 | Y. Maki and T. Furuta; *Synthesis*, 1978, 382. | 383 |
| 78S443 | C. Giordano, A. Belli and V. Bellotti; *Synthesis*, 1978, 443. | 1042, 1077 |
| 78S463 | S. Senda, K. Hirota, T. Asao and Y. Yamada; *Synthesis*, 1978, 463. | 359 |
| 78S524 | K. Burger and S. Penninger; *Synthesis*, 1978, 524. | 1077 |
| 78S540 | I. W. J. Still and K. Turnbull; *Synthesis*, 1978, 540. | 900 |
| 78S578 | R. Gräfing and L. Brandsma; *Synthesis*, 1978, 578. | 912 |
| 78S691 | D. Heber; *Synthesis*, 1978, 691. | 712 |
| 78S779 | C. K. Ghosh and K. K. Mukhopadhyay; *Synthesis*, 1978, 779. | 712 |
| 78S886 | H. J. Kabbe; *Synthesis*, 1978, 886. | 852 |
| 78S900 | S. Gelin and C. Deshayes; *Synthesis*, 1978, 900. | 727 |
| 78SA(A)297 | R. Schrooten, F. Borremans and M. Anteunis; *Spectrochim. Acta, Part A*, 1978, **34**, 297. | 579, 586 |
| 78SC143 | J. Arct, E. Jakubska and G. Olszewska; *Synth. Commun.*, 1978, **8**, 143. | 1028 |
| 78T433 | E. Taskinen; *Tetrahedron*, 1978, **34**, 433. | 585, 586 |
| 78T697 | P. Häussinger and G. Kresze; *Tetrahedron*, 1978, **34**, 689. | 1017 |
| 78T1221 | J. I. Okogun, V. U. Enyenihi and D. E. U. Ekong; *Tetrahedron*, 1978, **34**, 1221. | 582, 596, 600 |
| 78T1389 | K. R. Markham, B. Ternai, R. Stanley, H. Geiger and T. J. Mabry; *Tetrahedron*, 1978, **34**, 1389. | 591 |

| | | |
|---|---|---|
| 78T1415 | F. G. Riddell, M. H. Berry and E. S. Turner; *Tetrahedron*, 1978, **34**, 1415. | 1053 |
| 78T1593 | S. Ahmad, H. Wagner and S. Razaq; *Tetrahedron*, 1978, **34**, 1593. | 717 |
| 78T1837 | R. K. Chaudhuri, F. Zymalkowski and A. W. Frahm; *Tetrahedron*, 1978, **34**, 1837. | 590 |
| 78T2131 | A. I. Pyshchev, N. G. Bokii and Yu. T. Struchkov; *Tetrahedron*, 1978, **34**, 2131. | 620, 650 |
| 78T2509 | W. Friedrichsen and H. von Wallis; *Tetrahedron*, 1978, **34**, 2509. | 51 |
| 78T2705 | R. Lett and G. Chassaing; *Tetrahedron*, 1978, **34**, 2705. | 890, 897 |
| 78T3413 | B. E. Rivero, M. C. Apreda, E. E. Castellano, O. O. Orazi and R. A. Corral; *Tetrahedron*, 1978, **34**, 3413. | 467 |
| 78T3419 | W. C. Herndon and C. Párkányi; *Tetrahedron*, 1978, **34**, 3419. | 945 |
| 78T3569 | A. C. Jain, R. Khazanchi and A. Kumar; *Tetrahedron*, 1978, **34**, 3569. | 751 |
| 78TL519 | E. Vedejs, M. J. Mullins, J. M. Renga and S. P. Singer; *Tetrahedron Lett.*, 1978, 519. | 898 |
| 78TL1979 | A. Ohsawa, T. Akimoto, A. Tsuji, H. Igeta and Y. Iitaka; *Tetrahedron Lett.*, 1978, 1979. | 37 |
| 78TL2021 | A. Nagel and H. C. van der Plas; *Tetrahedron Lett.*, 1978, 2021. | 293 |
| 78TL2217 | S. Rajappa and R. Sreenivasan; *Tetrahedron Lett.*, 1978, 2217. | 186 |
| 78TL2271 | G. Moad, C. L. Luthy and S. J. Benkovic; *Tetrahedron Lett.*, 1978, 2271. | 202, 204 |
| 78TL2295 | S. Senda, K. Hirota, T. Asao and Y. Yamada; *Tetrahedron Lett.*, 1978, 2295. | 365 |
| 78TL2339 | G. Pitacco and E. Valentin; *Tetrahedron Lett.*, 1978, 2339. | 999 |
| 78TL2353 | T. Durst and L. Tétreault-Ryan; *Tetrahedron Lett.*, 1978, 2353. | 981 |
| 78TL2731 | G. Jones and P. Rafferty; *Tetrahedron Lett.*, 1978, 2731. | 235 |
| 78TL2995 | W. H. Rastetter and T. J. Richard; *Tetrahedron Lett.*, 1978, 2995. | 578, 761 |
| 78TL3059 | B. Stanovnik, M. Tišler, M. Kunaver, D. Gabrijelčič and M. Kočevar; *Tetrahedron Lett.*, 1978, 3059. | 36 |
| 78TL3359 | M. Meyer-Dayan, B. Bodo, C. Deschamps-Vallet and D. Molho; *Tetrahedron Lett.*, 1978, 3359. | 825 |
| 78TL3567 | K. Nishino, K. Nakasuji and I. Murata; *Tetrahedron Lett.*, 1978, 3567. | 915 |
| 78TL3903 | H. D. Fühlhuber, C. Gousetis, T. Troll and J. Sauer; *Tetrahedron Lett.*, 1978, 3903. | 551 |
| 78TL4431 | G. P. Voutsas, C. C. Venetopoulos, A. Kálmán, L. Párkányi, G. Hornyák and K. Lempert; *Tetrahedron Lett.*, 1978, 4431. | 387 |
| 78TL4439 | G. Adembri, A. Camparini, D. Donati, F. Ponticelli and P. Tedeschi; *Tetrahedron Lett.*, 1978, 4439. | 451 |
| 78TL5041 | A. Gescher, J. A. Hickman, R. J. Simmonds, M. F. G. Stevens and K. Vaughan; *Tetrahedron Lett.*, 1978, 5041. | 383 |
| 78UP21900 | H. Neunhoeffer and H.-J. Metz; unpublished results, 1978. | 426 |
| 78UP22400 | A. J. Green and J. D. Hepworth; unpublished results, 1978. | 743, 744 |
| 78USP4080325 | USA Dept. of Health, Education and Welfare, *U.S. Pat.* 4 080 325 (1978). | 304, 311 |
| 78USP4082845 | W. S. Saari and W. C. Lumma, Jr.; (Merck and Co. Inc.), *U.S. Pat.* 4 082 845 (*Chem. Abstr.*, 1978, **89**, 43 507). | 258 |
| 78USP4101668 | Bristol-Myers Co., *U.S. Pat.* 4 101 668 (1978) (*Chem. Abstr.*, 1979, **90**, 48 670). | 941 |
| 78ZC15 | W. Schroth and W. Kaufmann; *Z. Chem.*, 1978, **18**, 15. | 977 |
| 78ZC91 | M. Augustin and G. Jahreis; *Z. Chem.*, 1978, **18**, 91. | 907 |
| 78ZC336 | H. Buchwald and K. Rühlmann; *Z. Chem.*, 1978, **18**, 336. | 437, 1047, 1083 |
| 78ZN(B)284 | B. Solouki, H. Bock and O. Glemser; *Z. Naturforsch., Teil B*, 1978, **33**, 284. | 1052 |
| 78ZN(B)417 | K. Beelitz, G. Höhne and K. Praefcke; *Z. Naturforsch., Teil B*, 1978, **33**, 417. | 932 |
| 78ZOR431 | O. N. Chupakhin, V. N. Charushin, I. Y. Postovskii, N. A. Klyuev and E. N. Istratov; *Zh. Org. Khim.*, 1978, **14**, 431 (*Chem. Abstr.*, 1978, **88**, 152 561). | 250, 252 |
| 78ZOR622 | V. P. Sergutina, K. N. Zelenin and V. A. Krustalev; *Zh. Org. Khim.*, 1978, **14**, 622. | 569 |
| 78ZOR2471 | A. I. Morozov and M. S. Sytilin; *Zh. Org. Khim.*, 1978, **14**, 2471. | 547 |
| 79ACH(101)73 | M. Zsuga, V. Szabó, F. Korodi and A. Kiss; *Acta Chim. Acad. Sci. Hung.*, 1979, **101**, 73. | 703 |
| 79ACS(A)125 | J. Sen and H. Taube; *Acta Chem. Scand., Ser. A*, 1979, **33**, 125. | 37 |
| 79ACS(A)137 | K. A. Jensen and E. Larsen; *Acta Chem. Scand., Ser. A*, 1979, **33**, 137. | 566 |
| 79ACS(A)225 | H. E. Breed, G. Gundersen and R. Seip; *Acta Chem. Scand., Ser. A*, 1979, **33**, 225. | 621, 628 |
| 79ACS(B)119 | B. G. Österdahl; *Acta Chem. Scand., Ser. B*, 1979, **33**, 119. | 592 |
| 79ACS(B)669 | O. H. Johansen, T. Ottersen and K. Undheim; *Acta Chem. Scand., Ser. B*, 1979, **33**, 669. | 909 |
| 79AF1835 | E. Schenker and R. Salzmann; *Arzneim-Forsch.*, 1979, **29**, 1835. | 247, 261 |
| 79AF1843 | R. Salzmann, H. Bürki, D. Chu, B. Clark, P. Marbach, R. Markstein, H. Reinert, H. Siegl and R. Waite; *Arzneim.-Forsch.*, 1979, **29**, 1843. | 261 |
| 79AG757 | S. Sommer and U. Schubert; *Angew. Chem.*, 1979, **91**, 757. | 532, 533, 535 |
| 79AG(E)91 | H. W. Roesky; *Angew. Chem., Int. Ed. Engl.*, 1979, **18**, 91. | 1052 |
| 79AG(E)863 | U. Schöllkopf, W. Hartwig and U. Groth; *Angew. Chem., Int. Ed. Engl.*, 1979, **18**, 863. | 189 |
| 79AHC(24)1 | W. L. F. Armarego; *Adv. Heterocycl. Chem.*, 1979, **24**, 1. | 61, 148 |
| 79AHC(24)151 | J. W. Barton, *Adv. Heterocycl. Chem.*, 1979, **24**, 151. | 2 |
| 79AHC(24)293 | R. J. Stoodley; *Adv. Heterocycl. Chem.*, 1979, **24**, 293. | 1012, 1013 |
| 79AHC(24)363 | M. Tišler and B. Stanovnik; *Adv. Heterocycl. Chem.*, 1979, **24**, 363. | 2, 10 |
| 79AJC153 | N. W. Jacobsen and B. L. McCarthy; *Aust. J. Chem.*, 1979, **32**, 153. | 129 |
| 79AJC1281 | D. J. Bell, I. R. Brown, R. Cocks, R. F. Evans, G. A. Macfarlane, K. N. Mewett and A. V. Robertson; *Aust. J. Chem.*, 1979, **32**, 1281. | 185 |
| 79AJC1805 | W. L. F. Armarego and P. G. Tucker; *Aust. J. Chem.*, 1979, **32**, 1805. | 148 |

| Ref | Citation | Page |
|---|---|---|
| 79AJC2049 | W. B. Cowden and N. W. Jacobsen; *Aust. J. Chem.*, 1979, **32**, 2049. | 142 |
| 79AJC2713 | D. J. Brown, G. W. Grigg, Y. Iwai, K. N. McAndrew, T. Nagamatsu and R. Van Heeswyck; *Aust. J. Chem.*, 1979, **32**, 2713. | 94 |
| 79AP133 | H. Stamm and W. Wiesert; *Arch. Pharm.* (*Weinheim, Ger.*), 1979, **312**, 133. | 672 |
| 79AP302 | F. Eiden and F. Meinel; *Arch. Pharm.*, 1979, **312**, 302. | 1036 |
| 79AP385 | E. von Angerer and W. Wiegrebe; *Arch. Pharm.* (*Weinheim, Ger.*), 1979, **312**, 385. | 789, 860 |
| 79AP452 | G. Seitz and W. Overheu; *Arch. Pharm.* (*Weinheim, Ger.*), 1979, **312**, 452. | 442, 554 |
| 79AP591 | F. Eiden and E. G. Teupe; *Arch. Pharm.* (*Weinheim, Ger.*), 1979, **312**, 591. | 700 |
| 79AX(B)242 | P. G. Jones, O. Kennard, A. J. Kirby and R. J. Martin; *Acta Crystallogr., Part B*, 1979, **35**, 242. | 621 |
| 79AX(B)860 | A. Kálmán, G. Argay, E. Fischer and G. Rembarz; *Acta Crystallogr., Part B*, 1979, **35**, 860. | 1049 |
| 79AX(B)1117 | A. Usanmaz; *Acta Crystallogr., Part B*, 1979, **35**, 1117. | 461 |
| 79AX(B)2113 | A. Camerman, H. W. Smith and N. Camerman; *Acta Crystallogr., Part B*, 1979, **35**, 2113. | 462 |
| 79B3635 | A. H. Merrill, Jr., S. Kasai, K. Matsui, H. Tsuge and D. B. McCormick; *Biochemistry*, 1979, **18**, 3635. | 250 |
| 79BCJ160 | H. Nakazumi and T. Kitao; *Bull. Chem. Soc. Jpn.*, 1979, **52**, 160. | 918 |
| 79BCJ181 | T. Sugimoto and S. Matsuura; *Bull. Chem. Soc. Jpn.*, 1979, **52**, 181. | 308 |
| 79BCJ483 | S. Ito, A. Kakehi, Y. Tanaka, K. Yoshida and T. Matsuno; *Bull. Chem. Soc. Jpn.*, 1979, **52**, 483. | 544, 548 |
| 79BCJ1156 | T. Nozoe, T. Someya and H. Okai; *Bull. Chem. Soc. Jpn.*, 1979, **52**, 1156. | 1032 |
| 79BCJ1169 | A. Ninagawa, R. Nomura and H. Matsuda; *Bull. Chem. Soc. Jpn.*, 1979, **52**, 1169. | 722 |
| 79BCJ1203 | A. C. Jain, R. Khazanchi and A. Kumar; *Bull. Chem. Soc. Jpn.*, 1979, **52**, 1203. | 751 |
| 79BCJ1231 | T. Goka, Y. Hashida and K. Matsui; *Bull. Chem. Soc. Jpn.*, 1979, **52**, 1231. | 477 |
| 79BCJ1235 | I. Shibuya; *Bull. Chem. Soc. Jpn.*, 1979, **52**, 1235. | 936 |
| 79BCJ2163 | K. Hanaya, H. Kudo, K. Gohke and S. Imaizumi; *Bull. Chem. Soc. Jpn.*, 1979, **52**, 2163. | 580 |
| 79BCJ2386 | K. Kurosawa and H. Hirada; *Bull. Chem. Soc. Jpn.*, 1979, **52**, 2386. | 680 |
| 79BCJ3654 | O. Tsuge, H. Watanabe and Y. Kiryu; *Bull. Chem. Soc. Jpn.*, 1979, **52**, 3654. | 453 |
| 79BSB905 | G. Beynon, H. P. Figeys, D. Lloyd and R. K. Mackie; *Bull. Soc. Chim. Belg.*, 1979, **88**, 905. | 38, 551 |
| 79BSF(2)347 | J. C. Meslin and H. Quiniou; *Bull. Soc. Chim. Fr., Part 2*, 1979, 347. | 1020 |
| 79BSF(2)499 | S. Ratton, J. Moyne and R. Longeray; *Bull. Soc. Chim. Fr., Part 2*, 1979, 499. | 1077 |
| 79CB445 | H. Hansen, S. Hünig and K. Kishi; *Chem. Ber.*, 1979, **112**, 445. | 372, 384 |
| 79CB1012 | L. Capuano, G. Urhahn and A. Willmes; *Chem. Ber.*, 1979, **112**, 1012. | 1078 |
| 79CB1499 | Z. Kazimierczuk and W. Pfleiderer; *Chem. Ber.*, 1979, **112**, 1499. | 273, 297, 299 |
| 79CB1514 | R. Gompper and K. Schönafinger; *Chem. Ber.*, 1979, **112**, 1514. | 373, 376, 382 |
| 79CB1529 | R. Gompper and K. Schönafinger; *Chem. Ber.*, 1979, **112**, 1529. | 377, 382 |
| 79CB1535 | R. Gompper and K. Schönafinger; *Chem. Ber.*, 1979, **112**, 1535. | 377 |
| 79CB1585 | T. Kappe, W. Golser, M. Hariri and W. Stadlbauer; *Chem. Ber.*, 1979, **112**, 1585. | 1004 |
| 79CB1635 | H. Gotthardt and F. Reiter; *Chem. Ber.*, 1979, **112**, 1635. | 564 |
| 79CB1791 | F. Eiden and I. Breugst; *Chem. Ber.*, 1979, **112**, 1791. | 712 |
| 79CB1867 | K. Friedrich and M. Zamkanei; *Chem. Ber.*, 1979, **112**, 1867. | 932 |
| 79CB1981 | H.-J. Degen, S. Haller, K. Heeg and H. Neunhoeffer; *Chem. Ber.*, 1979, **112**, 1981. | 558 |
| 79CB2750 | W. Pfleiderer; *Chem. Ber.*, 1979, **112**, 2750. | 279, 306 |
| 79CB3879 | S. Antus, A. Gottsegen, M. Nógrádi and A. Gergely; *Chem. Ber.*, 1979, **112**, 3879. | 669 |
| 79CC35 | W. Bludssus and R. Mews; *J. Chem. Soc., Chem. Commun.*, 1979, 35. | 1015 |
| 79CC264 | R. J. Bass; *J. Chem. Soc., Chem. Commun.*, 1979, 264. | 716 |
| 79CC288 | P. G. Jones and A. J. Kirby; *J. Chem. Soc., Chem. Commun.*, 1979, 288. | 621 |
| 79CC445 | R. D. Chambers, C. R. Sargent and M. Clark; *J. Chem. Soc., Chem. Commun.*, 1979, 445. | 9 |
| 79CC446 | R. D. Chambers and C. R. Sargent; *J. Chem. Soc., Chem. Commun.*, 1979, 446. | 9 |
| 79CC658 | M. G. Barlow, R. N. Haszeldine and D. J. Simpkin; *J. Chem. Soc., Chem. Commun.*, 1979, 658. | 422, 423 |
| 79CC836 | G. Cardillo, M. Orena, G. Porzi and S. Sandri; *J. Chem. Soc., Chem. Commun.*, 1979, 836. | 752 |
| 79CC846 | R. J. Ponsford and R. Southgate; *J. Chem. Soc., Chem. Commun.*, 1979, 846. | 1009 |
| 79CC880 | A. J. Ashe, III, D. J. Bellville and H. S. Friedman; *J. Chem. Soc., Chem. Commun.*, 1979, 880. | 553 |
| 79CC900 | J. Motoyoshiya, J. Enda, Y. Ohshiro and T. Agawa; *J. Chem. Soc., Chem. Commun.*, 1979, 900. | 749 |
| 79CC920 | R. C. Jennings and A. P. Ottridge; *J. Chem. Soc., Chem. Commun.*, 1979, 920. | 665, 881 |
| 79CC1019 | M. F. Ahern and G. W. Gokel; *J. Chem. Soc., Chem. Commun.*, 1979, 1019. | 50 |
| 79CHE240 | J. Szabó; *Chem. Heterocycl. Compd.* (*Engl. Transl.*), 1979, **15**, 240. | 1026 |
| 79CHE265 | E. A. Zvezdina, M. P. Zhdanov and G. N. Dorofeenko; *Chem. Heterocycl. Compd.* (*Engl. Transl.*), 1979, **15**, 265. | 655, 656 |
| 79CHE291 | J. Szabó; *Chem. Heterocycl. Compd.* (*Engl. Transl.*), 1979, **15**, 291. | 1038 |
| 79CHE815 | V. I. Dulenko, V. M. Golyak, V. I. Gubar and N. N. Aleckseev; *Chem. Heterocycl. Compd.* (*Engl. Transl.*), 1979, **15**, 815. | 686 |
| 79CI(L)26 | M. Rai, S. Kumar, K. Krishan and A. Singh; *Chem. Ind.* (*London*), 1979, 26. | 1036 |
| 79CJC1377 | H. P. Pommier, J. Baril, I. Gruda and R. M. Leblanc; *Can. J. Chem.*, 1979, **57**, 1377. | 750, 756 |
| 79CJC1451 | G. A. Poulton, T. D. Cyr and E. E. McMullan; *Can. J. Chem.*, 1979, **57**, 1451. | 588 |

| 79CJC3292 | A. Philipp and I. Jirkovsky; *Can. J. Chem.*, 1979, **57**, 3292. | 909 |
|---|---|---|
| 79CPB916 | A. Ohsawa, T. Uezu and H. Igeta; *Chem. Pharm. Bull.*, 1979, **27**, 916. | 32 |
| 79CPB1169 | T. Tsujimoto, T. Nomura, M. Iifuru and Y. Sasaki; *Chem. Pharm. Bull.*, 1979, **27**, 1169. | 3, 7 |
| 79CPB2105 | T. Tsujimoto, C. Kobayashi, T. Nomura, M. Iifuru and Y. Sasaki; *Chem. Pharm. Bull.*, 1979, **27**, 2105. | 7 |
| 79CPB2507 | F. Yoneda, F. Takayama and A. Koshiro; *Chem. Pharm. Bull.*, 1979, **27**, 2507. | 219 |
| 79CPB2642 | H. Yamanaka, S. Niitsuma and T. Sakamoto; *Chem. Pharm. Bull.*, 1979, **27**, 2642. | 106 |
| 79CR(C)(288)221 | A. Le Rouzic, D. Raphalen and M. Kerfanto; *C.R. Hebd. Seances Acad. Sci., Ser. C*, 1979, **288**, 221. | 1035 |
| 79CR(C)(288)229 | M. Benet and R. Longeray; *C.R. Hebd. Seances Acad. Sci., Ser. C*, 1979, **288**, 229. | 1037 |
| 79CZ230 | G. Seitz and W. Overheu; *Chem.-Ztg.*, 1979, **103**, 230. | 442, 554 |
| 79CZ264 | K. Burger, F. Hein and J. Firl; *Chem.-Ztg.*, 1979, **103**, 264. | 500 |
| 79E574 | A. C. Ranade, R. S. Mali and H. R. Deshpande; *Experientia*, 1979, **35**, 574. | 219 |
| 79EUP2238 | Eastman Kodak Ltd., *Eur. Pat.* 2238 (1979) (*Chem. Abstr.*, 1980, **92**, 102 297). | 939, 941 |
| 79EUP5912 | J. H. Parsons; *Eur. Pat.* 5 912 (1979) (*Chem. Abstr.*, 1980, **93**, 46 730). | 562 |
| 79FES869 | O. Caputo, L. Cattel and F. Viola; *Farmaco Ed. Sci.*, 1979, **34**, 869 (*Chem. Abstr.*, 1980, **92**, 94 185). | 941 |
| 79FRP2424274 | Yoshitomi Pharmaceutical Industries, *Fr. Pat.* 2 424 274 (1979) (*Chem. Abstr.*, 1980, **92**, 215 280). | 942 |
| 79GEP2366215 | H. Timmler and W. Draber; *Ger. Pat.* 2 366 215 (1979) (*Chem. Abstr.*, 1979, **91**, 15 747). | 451 |
| 79GEP2808070 | K. Grohe, H. J. Zeiler and K. Metzger; (Bayer A/G), *Ger. Offen.* 2 808 070 (1979) (*Chem. Abstr.*, 1980, **92**, 41 916). | 230, 240, 247 |
| 79GEP2815031 | Hoechst A. G., *Ger. Pat.* 2 815 031 (1979) (*Chem. Abstr.*, 1980, **92**, 119 746). | 941 |
| 79GEP2831054 | Eastman Kodak Ltd., *Ger. Pat.* 2 831 054 (1979) (*Chem. Abstr.*, 1979, **91**, 30 527). | 941 |
| 79H(12)51 | V. V. Mezheritskii, A. L. Wasserman and G. N. Dorofeenko; *Heterocycles*, 1979, **12**, 51. | 664 |
| 79H(12)253 | S. Y. Dike and J. R. Merchant; *Heterocycles*, 1979, **12**, 253. | 842 |
| 79H(12)269 | M. Afzal, J. M. Al-Hassam and F. N. Al-Masad; *Heterocycles*, 1979, **12**, 269. | 601 |
| 79H(12)421 | M. Afzal and J. M. Al-Hassan; *Heterocycles*, 1979, **12**, 421. | 584 |
| 79H(12)451 | V. G. S. Box, B. A. Burke and C. McCaw; *Heterocycles*, 1979, **12**, 451. | 744 |
| 79H(12)457 | B. Stanovnik, M. Tišler, J. Bradač, B. Budič, B. Koren and B. Mozetič-Reščič; *Heterocycles*, 1979, **12**, 457. | 10 |
| 79H(12)503 | Y. Tominaga, H. Okuda, Y. Mitsutomi, Y. Matsuda, G. Kobayashi and K. Sakemi; *Heterocycles*, 1979, **12**, 503. | 339, 350, 364 |
| 79H(12)519 | D. Geffken; *Heterocycles*, 1979, **12**, 519. | 1050 |
| 79H(12)745 | W. D. Guither, M. D. Coburn and R. N. Castle; *Heterocycles*, 1979, **12**, 745. | 565, 566 |
| 79H(12)775 | A. R. Katritzky, R. T. C. Brownlee and G. Musumarra; *Heterocycles*, 1979, **12**, 775. | 585, 741 |
| 79H(12)921 | M. Mori, K. Chiba, N. Inotsume and Y. Ban; *Heterocycles*, 1979, **12**, 921. | 859 |
| 79H(12)1199 | S. Karady, J. S. Amato, D. Dortmund, A. A. Patchett, R. A. Reamer, R. J. Tull and L. M. Weinstock; *Heterocycles*, 1979, **12**, 1199. | 1079 |
| 79HC(35)1 | G. W. H. Cheeseman and R. F. Cookson; *Chem. Heterocycl. Compd.*, 1979, **35**, 1. | 158, 160, 161, 195 |
| 79HC(35)568 | G. W. H. Cheeseman and R. F. Cookson; *Chem. Heterocycl. Compd.*, 1979, **35**, 568. | 331 |
| 79HCA1171 | R. Charubala and W. Pfleiderer; *Helv. Chim. Acta*, 1979, **62**, 1171. | 283 |
| 79HCA1179 | R. Charubala and W. Pfleiderer; *Helv. Chim. Acta*, 1979, **62**, 1179. | 283 |
| 79HCA1340 | E. Khalifa, J. H. Bieri and M. Viscontini; *Helv. Chim. Acta*, 1979, **62**, 1340. | 281 |
| 79HCA1429 | G. Mukherjee-Muller, H. Heimgartner and H. Schmid; *Helv. Chim. Acta*, 1979, **62**, 1429. | 520 |
| 79HCA2384 | R. Barner and M. Schmid; *Helv. Chim. Acta*, 1979, **62**, 2384. | 779 |
| 79HCA2833 | P. Müller, T. Venakis and C. H. Eugster; *Helv. Chim. Acta*, 1979, **62**, 2833. | 800 |
| 79IJC(B)182 | C. K. Sehgal, P. L. Kachroo and K. L. Dhar; *Indian J. Chem., Sect. B*, 1979, **17**, 182. | 699 |
| 79IJC(B)362 | R. K. Bansal and G. Bhagchandani; *Indian. J. Chem., Sect. B*, 1979, **18**, 362. | 570 |
| 79IJC(B)510 | M. M. Chincholkar and V. S. Jamode; *Indian J. Chem., Sect. B*, 1979, **17**, 510. | 728 |
| 79IJC(B)610 | F. W. Birss and N. K. Das Gupta; *Indian J. Chem., Sect. B*, 1979, **17**, 610. | 204, 250 |
| 79IJC(B)642 | U. C. Mashelkar and R. N. Usgaonkar; *Indian J. Chem., Sect. B*, 1979, **17**, 642. | 690 |
| 79IZV1077 | N. P. Volynskii and L. P. Shcherbakova; *Izv. Akad. Nauk SSR, Ser. Khim.*, 1979, 1077. | 998 |
| 79IZV1826 | Z. V. Safronova, L. A. Simonyan, Yu. V. Zeifman and N. P. Gambaryan; *Izv. Akad. Nauk SSSR, Ser. Khim.*, 1979, 1826. | 1077 |
| 79JA665 | I. Agranat and Y. Tapuhi; *J. Am. Chem. Soc.*, 1979, **101**, 665. | 621, 838 |
| 79JA766 | K. C. Nicolaou, W. E. Barnette and R. L. Magolda; *J. Am. Chem. Soc.*, 1979, **101**, 766. | 45 |
| 79JA1313 | A. Shaver, P. J. Fitzpatrick, K. Steliou and I. S. Butler; *J. Am. Chem. Soc.*, 1979, **101**, 1313. | 977 |
| 79JA2069 | P. B. Dervan, T. Uyehara and D. S. Santilli; *J. Am. Chem. Soc.*, 1979, **101**, 2069. | 40 |
| 79JA3663 | D. S. Santilli and P. B. Dervan; *J. Am. Chem. Soc.*, 1979, **101**, 3663. | 39 |
| 79JA4419 | W. T. Ashton, R. D. Brown, F. Jacobson and C. Walsh; *J. Am. Chem. Soc.*, 1979, **101**, 4419. | 260 |
| 79JA4499 | P. H. Turner; *J. Am. Chem. Soc.*, 1979, **101**, 4499. | 626 |
| 79JA4973 | J. L. Holmes and J. K. Terlouw; *J. Am. Chem. Soc.*, 1979, **101**, 4973. | 611 |

| 79JA5405 | J. Davies and J. B. Jones; *J. Am. Chem. Soc.*, 1979, **101**, 5405. | 903 |
|---|---|---|
| 79JA5692 | W. Adam and I. Erden; *J. Am. Chem. Soc.*, 1979, **101**, 5692. | 678 |
| 79JA6023 | B. B. Snider, D. M. Roush and T. A. Killinger; *J. Am. Chem. Soc.*, 1979, **101**, 6023. | 771 |
| 79JA6068 | G. Moad, C. L. Luthy, P. A. Benkovic and S. J. Benkovic; *J. Am. Chem. Soc.*, 1979, **101**, 6068. | 205, 260 |
| 79JA6114 | J. C. Fontellia-Camps, C. E. Bugg, C. Temple, Jr., J. D. Rose, J. A. Montgomery and R. L. Kisliuk; *J. Am. Chem. Soc.*, 1979, **101**, 6114. | 281 |
| 79JA6710 | N. Cohen, R. J. Lopresti and G. Saucy; *J. Am. Chem. Soc.*, 1979, **101**, 6710. | 778 |
| 79JA7510 | J. A. Barltrop, J. C. Barrett, R. W. Carder, A. C. Day, J. R. Harding, W. E. Long and C. J. Samuel; *J. Am. Chem. Soc.*, 1979, **101**, 7510. | 649 |
| 79JA7521 | J. A. Barltrop, A. C. Day and C. J. Samuel; *J. Am. Chem. Soc.*, 1979, **101**, 7521. | 693, 694, 797 |
| 79JA7623 | D. Vargo and M. S. Jorns; *J. Am. Chem. Soc.*, 1979, **101**, 7623. | 207 |
| 79JAP7912392 | Teijin Ltd., *Jpn. Pat.* 79 12 392 (1979) (*Chem. Abstr.*, 1979, **91**, 5222). | 941 |
| 79JAP7924027 | Konishiroku Photo Industries, *Jpn. Pat.* 79 24 027 (1979) (*Chem. Abstr.*, 1979, **91**, 202 182). | 942 |
| 79JAP(K)7924877 | K. Noda, A. Nakagawa and H. Ide; (Hisamitsu Pharm. Co. Inc.), *Jpn. Kokai* 79 24 877 (1979) (*Chem. Abstr.*, 1979, **91**, 39 342). | 208 |
| 79JAP(K)7976587 | N. Ohi, K. Moriguchi, O. Hiroaki and H. Shimada; *Jpn. Kokai*, 79 76 587 (1979) (*Chem. Abstr.*, 1980, **92**, 41 999). | 527 |
| 79JAP(K)7981298 | T. Matsumura, H. Kawahara, Y. Bunda and T. Naruchi; (Daito Koeki Co. Ltd. and Teijin Ltd.), *Jpn. Kokai* 79 81 298 (1979) (*Chem. Abstr.*, 1980, **92**, 94 417). | 223 |
| 79JAP(K)7981299 | T. Matsumura, H. Kawahara, Y. Bunda and T. Naruchi; (Daito Koeki Co. Ltd. and Teijin Ltd.), *Jpn. Kokai* 79 81 299 (1979) (*Chem. Abstr.*, 1980, **92**, 94 418). | 223 |
| 79JBC(254)12145 | M. S. Jorns; *J. Biol. Chem.*, 1979, **254**, 12145. | 260 |
| 79JCP(71)4757 | J. E. Kenny, D. V. Brumbaugh and D. H. Levy; *J. Chem. Phys.*, 1979, **71**, 4757. | 543 |
| 79JCR(S)110 | E. Suzuki, B. Katsuragawa and S. Inoue; *J. Chem. Res. (S)*, 1979, 110. | 692 |
| 79JCR(S)137 | D. Mulvagh, M. J. Meegan and D. M. X. Donnelly; *J. Chem. Res. (S)*, 1979, 137. | 729 |
| 79JCR(S)142 | R. Pellicciari, B. Natalini, M. Taddei, A. Ricci, G. A. Bistocchi and G. De Meo; *J. Chem. Res. (S)*, 1979, 142. | 726 |
| 79JCR(S)214 | T. L. Gilchrist, C. W. Rees and I. W. Southon; *J. Chem. Res. (S)*, 1979, 214. | 1059, 1068 |
| 79JCR(S)240 | A. W. Faull and R. Hull; *J. Chem. Res. (S)*, 1979, 240. | 438 |
| 79JCS(P1)201 | J. M. Luteijn and H. J. W. Spronck; *J. Chem. Soc., Perkin Trans. 1*, 1979, 201. | 746, 782 |
| 79JCS(P1)249 | R. Faragher and T. L. Gilchrist; *J. Chem. Soc., Perkin Trans. 1*, 1979, 249. | 1017 |
| 79JCS(P1)258 | R. Faragher and T. L. Gilchrist; *J. Chem. Soc., Perkin Trans. 1*, 1979, 258. | 999 |
| 79JCS(P1)333 | D. G. Neilson, K. M. Watson and T. J. R. Weakley; *J. Chem. Soc., Perkin Trans. 1*, 1979, 333. | 544 |
| 79JCS(P1)426 | A. R. Katritzky, M. F. Abdel-Megeed, G. Lhommet and C. A. Ramsden; *J. Chem. Soc., Perkin Trans. 1*, 1979, 426. | 655 |
| 79JCS(P1)433 | A. R. Katritzky, N. F. Eweiss and P.-L. Nie; *J. Chem. Soc., Perkin Trans. 1*, 1979, 433. | 861 |
| 79JCS(P1)436 | A. R. Katritzky, U. Gruntz, A. A. Ikizler, D. H. Kenny and B. P. Leddy; *J. Chem. Soc., Perkin Trans. 1*, 1979, 436. | 861 |
| 79JCS(P1)446 | A. R. Katritzky, J. Lewis and P. L. Nie; *J. Chem. Soc., Perkin Trans. 1*, 1979, 446. | 655 |
| 79JCS(P1)464 | L. Crombie, D. E. Games and A. W. G. James; *J. Chem. Soc., Perkin Trans. 1*, 1979, 464. | 593, 789 |
| 79JCS(P1)478 | L. Crombie, M. Eskins, D. E. Games and C. Loader; *J. Chem. Soc., Perkin Trans. 1*, 1979, 478. | 577, 595, 789 |
| 79JCS(P1)525 | T. Kato, N. Katagiri and R. Sato; *J. Chem. Soc., Perkin Trans. 1*, 1979, 525. | 809 |
| 79JCS(P1)529 | T. Kato, M. Sato and H. Kimura; *J. Chem. Soc., Perkin Trans. 1*, 1979, 529. | 813 |
| 79JCS(P1)677 | S. R. Baker, L. Crombie, R. V. Dove and D. A. Slack; *J. Chem. Soc., Perkin Trans. 1*, 1979, 677. | 790 |
| 79JCS(P1)1043 | Y. Girard, J. G. Atkinson and J. Rokach; *J. Chem. Soc., Perkin Trans. 1*, 1979, 1043. | 1073 |
| 79JCS(P1)1199 | Y. Maki, M. Suzuki, T. Furuta, M. Kawamura and M. Kuzuya; *J. Chem. Soc., Perkin Trans. 1*, 1979, 1199. | 11, 381 |
| 79JCS(P1)1574 | A. Albert; *J. Chem. Soc., Perkin Trans. 1*, 1979, 1574. | 318 |
| 79JCS(P1)1806 | K. Sato, H. Adachi, T. Iwaki and M. Ohashi; *J. Chem. Soc., Perkin Trans. 1*, 1979, 1806. | 719, 815 |
| 79JCS(P1)1885 | J. L. Markham and P. G. Sammes; *J. Chem. Soc., Perkin Trans. 1*, 1979, 1885. | 175 |
| 79JCS(P1)1978 | R. D. Chambers, W. K. R. Musgrave and D. E. Wood; *J. Chem. Soc., Perkin Trans. 1*, 1979, 1978. | 420, 519 |
| 79JCS(P1)2136 | L. Crombie, N. A. Kerton and G. Pattenden; *J. Chem. Soc., Perkin Trans. 1*, 1979, 2136. | 29 |
| 79JCS(P1)2203 | A. J. Barker, T. McC. Paterson, R. K. Smalley and H. Suschitzky; *J. Chem. Soc., Perkin Trans. 1*, 1979, 2203. | 379 |
| 79JCS(P1)2230 | S. David and J. Eustache; *J. Chem. Soc., Perkin Trans. 1*, 1979, 2230. | 773 |
| 79JCS(P1)2511 | A. Nishinaga, T. Tojo, H. Tomita and T. Matsura; *J. Chem. Soc., Perkin Trans. 1*, 1979, 2511. | 704 |
| 79JCS(P1)2563 | T. Meikle and R. Stevens; *J. Chem. Soc., Perkin Trans. 1*, 1979, 2563. | 755 |
| 79JCS(P1)2851 | V. W. Böhnisch, T. L. Gilchrist and C. W. Rees; *J. Chem. Soc., Perkin Trans. 1*, 1979, 2851. | 1066 |
| 79JCS(P1)3190 | I. Iijima, N. Taga, M. Miyazaki, T. Tamaka and J. Uzawa; *J. Chem. Soc., Perkin Trans. 1*, 1979, 3190. | 840 |

| | | |
|---|---|---|
| 79JCS(P2)435 | H. E. Gottlieb, R. A. de Lima and F. delle Monache; *J. Chem. Soc., Perkin Trans. 2*, 1979, 435. | 589 |
| 79JCS(P2)981 | A. R. Katritzky, I. J. Ferguson and R. C. Patel; *J. Chem. Soc., Perkin Trans. 2*, 1979, 981. | 539 |
| 79JCS(P2)984 | A. Katritzky and R. C. Patel; *J. Chem. Soc., Perkin Trans. 2*, 1979, 984. | 389, 394, 444 |
| 79JCS(P2)993 | A. R. Katritzky and R. C. Patel; *J. Chem. Soc., Perkin Trans. 2*, 1979, 993. | 1054 |
| 79JCS(P2)1133 | A. R. Katritzky, V. J. Baker, F. M. S. Brito-Palma, J. M. Sullivan and R. B. Finzel; *J. Chem. Soc., Perkin Trans. 2*, 1979, 1133. | 1055 |
| 79JCS(P2)1371 | M. L. Tosato; *J. Chem. Soc. Perkin Trans., 2*, 1979, 1371. | 480, 482 |
| 79JHC1 | Y. A. Al-Farkh, F. H. Al-Hajjar and H. S. Hammoud; *J. Heterocycl. Chem.*, 1979, **16**, 1. | 727 |
| 79JHC33 | S. Sunder and N. P. Peet; *J. Heterocycl. Chem.*, 1979, **16**, 33. | 451 |
| 79JHC97 | G. Vernin, S. Coen, J. Metzger and C. Párkányi; *J. Heterocycl. Chem.*, 1979, **16**, 97. | 576, 687 |
| 79JHC133 | J. H. Maguire and R. L. McKee; *J. Heterocycl. Chem.*, 1979, **16**, 133. | 204, 215, 225 |
| 79JHC161 | C. Bennouna, F. Petrus and J. Verducci; *J. Heterocycl. Chem.*, 1979, **16**, 161. | 1050, 1055, 1058, 1071 |
| 79JHC249 | I. Maeba and R. N. Castle; *J. Heterocycl. Chem.*, 1979, **16**, 249. | 240, 244 |
| 79JHC301 | A. Nagel, H. C. van der Plas, G. Geurtsen and A. van Veldhuizen; *J. Heterocycl. Chem.*, 1979, **16**, 301. | 249, 251, 252 |
| 79JHC305 | A. Nagel, H. C. van der Plas, G. Guertsen and A. van der Kuilen; *J. Heterocycl. Chem.*, 1979, **16**, 305. | 249, 250, 252, 254 |
| 79JHC381 | D. A. Crombie, J. R. Kiely and C. J. Ryan; *J. Heterocycl. Chem.*, 1979, **16**, 381. | 933 |
| 79JHC537 | J. A. Montgomery, J. R. Piper, R. D. Elliot, E. C. Roberts, C. Temple, Jr. and Y. F. Shealy; *J. Heterocycl. Chem.*, 1979, **16**, 537. | 302 |
| 79JHC555 | A. C. Lovelette; *J. Heterocycl. Chem.*, 1979, **16**, 555. | 391, 411 |
| 79JHC567 | K. Anzai; *J. Heterocycl. Chem.*, 1979, **16**, 567. | 137 |
| 79JHC657 | C. Deshayes and S. Gelin; *J. Heterocycl. Chem.*, 1979, **16**, 657. | 727 |
| 79JHC679 | R. A. Hollins and M. T. L. Lima; *J. Heterocycl. Chem.*, 1979, **16**, 679. | 921 |
| 79JHC707 | I. Lalezari and S. Sadeghi-Milani; *J. Heterocycl. Chem.*, 1979, **16**, 707. | 205, 211, 213 |
| 79JHC751 | B. Weinstein, L. T. Hahn and A. K. Eng; *J. Heterocycl. Chem.*, 1979, **16**, 751. | 1048, 1059, 1082 |
| 79JHC881 | L. M. Werbel, D. J. McNamara, N. L. Colbry, J. L. Johnson, M. J. Degnan and B. Whitney; *J. Heterocycl. Chem.*, 1979, **16**, 881. | 558 |
| 79JHC973 | J.-M. Cosmao, N. Collignon and G. Quéguiner; *J. Heterocycl. Chem.*, 1979, **16**, 973. | 251 |
| 79JHC1069 | R. G. Pews; *J. Heterocycl. Chem.*, 1979, **16**, 1069. | 1059, 1074 |
| 79JHC1169 | J. Matsumoto, S. Mishio and S. Minami; *J. Heterocycl. Chem.*, 1979, **16**, 1169. | 208 |
| 79JHC1281 | J. H. Looker, D. L. Shaneyfelt and W. Halfar; *J. Heterocycl. Chem.*, 1979, **16**, 1281. | 711 |
| 79JHC1389 | G. B. Bennett, A. D. Kahle, H. Minor and M. J. Shapiro; *J. Heterocycl. Chem.*, 1979, **16**, 1389. | 394 |
| 79JHC1559 | I. Maeba, K. Mori and R. N. Castle; *J. Heterocycl. Chem.*, 1979, **16**, 1559. | 243 |
| 79JHC1649 | C. A. Lovelette; *J. Heterocycl. Chem.*, 1979, **16**, 1649. | 411 |
| 79JHC1663 | J. A. Van Allan, J. F. Stenberg and G. A. Reynolds; *J. Heterocycl. Chem.*, 1979, **16**, 1663. | 839 |
| 79JMC44 | T. H. Althuis, P. F. Moore and H.-J. Hess; *J. Med. Chem.*, 1979, **22**, 44. | 260 |
| 79JMC290 | A. Nohara, H. Kuriki, T. Ishiguro, T. Saijo, K. Ukawa, Y. Maki and Y. Sanno; *J. Med. Chem.*, 1979, **22**, 290. | 710 |
| 79JMC862 | J. A. Montgomery, J. R. Piper, R. D. Elliott, C. Temple, Jr., E. C. Roberts and Y. F. Shealy; *J. Med. Chem.*, 1979, **22**, 862. | 249, 251, 262 |
| 79JMC874 | J. E. Martinelli, M. Chaykovsky, R. L. Kisliuk, L. Roy and Y. Gaumont; *J. Med. Chem.*, 1979, **22**, 874. | 327 |
| 79JOC21 | I. Arai and G. D. Daves, Jr.; *J. Org. Chem.*, 1979, **44**, 21. | 579, 774 |
| 79JOC302 | E. C. Taylor and A. J. Cocuzza; *J. Org. Chem.*, 1979, **44**, 302. | 320 |
| 79JOC305 | J. A. Kloek and K. L. Leschinsky; *J. Org. Chem.*, 1979, **44**, 305. | 1016 |
| 79JOC364 | G. F. Weber and S. S. Hall; *J. Org. Chem.*, 1979, **44**, 364. | 773, 775, 777 |
| 79JOC435 | A. Srinivasan and A. D. Broom; *J. Org. Chem.*, 1979, **44**, 435. | 214 |
| 79JOC471 | K. Ramalingam, K. D. Berlin, N. Satyamurthy and R. Sivakumar; *J. Org. Chem.*, 1979, **44**, 471. | 589, 632 |
| 79JOC486 | K. B. Lipkowitz, S. Scarpone, B. P. Mundy and W. G. Bornmann; *J. Org. Chem.*, 1979, **44**, 486. | 930, 933 |
| 79JOC531 | T. G. Majewicz and P. Caluwe; *J. Org. Chem.*, 1979, **44**, 531. | 208 |
| 79JOC629 | M. J. Haddadin, S. J. Firsan and B. S. Nader; *J. Org. Chem.*, 1979, **44**, 629. | 50 |
| 79JOC798 | E. J. Valente, B. D. Santarsiero and V. Schomaker; *J. Org. Chem.*, 1979, **44**, 798. | 623, 630 |
| 79JOC803 | G. Sartori, G. Casiraghi, L. Bolzoni and G. Casnati; *J. Org. Chem.*, 1979, **44**, 803. | 748 |
| 79JOC827 | Y. Ohtsuka; *J. Org. Chem.*, 1979, **44**, 827. | 180 |
| 79JOC915 | H. J. Shine, B. K. Bandlish, S. R. Mani and A. G. Padilla; *J. Org. Chem.*, 1979, **44**, 915. | 969 |
| 79JOC1494 | J. G. Sweeny, T. Radford and G. A. Iacobucci; *J. Org. Chem.*, 1979, **44**, 1494. | 705 |
| 79JOC1540 | R. F. Bryan, F. A. Carey and R. W. Miller; *J. Org. Chem.*, 1979, **44**, 1540. | 949 |
| 79JOC1627 | R. L. Lipnick and J. D. Fissekis; *J. Org. Chem.*, 1979, **44**, 1627. | 63 |
| 79JOC1684 | Y. Tamura, Y. Nishikawa, C. Mukai, K. Sumoto and M. Ikeda; *J. Org. Chem.*, 1979, **44**, 1684. | 914 |

| | | |
|---|---|---|
| 79JOC1695 | B. Verček, I. Leban, B. Stanovnik and M. Tišler; *J. Org. Chem.*, 1979, **44**, 1695. | 216, 223 |
| 79JOC1700 | B. DeCroix, M. J. Strauss, A. DeFusco and D. C. Palmer; *J. Org. Chem.*, 1979, **44**, 1700. | 316 |
| 79JOC1719 | R. C. Bugle and R. A. Osteryoung; *J. Org. Chem.*, 1979, **44**, 1719. | 252 |
| 79JOC1740 | B. Golankiewicz, J. B. Holtwick, B. N. Holmes, E. N. Duesler and N. J. Leonard; *J. Org. Chem.*, 1979, **44**, 1740. | 521 |
| 79JOC1823 | A. Messmer, G. Hajós, J. Tamás and A. Nreszmélyi; *J. Org. Chem.*, 1979, **44**, 1823. | 392 |
| 79JOC1847 | C. G. Kruse, A. Wijsman and A. van der Gen; *J. Org. Chem.*, 1979, **44**, 1847. | 977 |
| 79JOC1989 | J. S. Chen, W. H. Watson, D. Austin and A. L. Ternay, Jr.; *J. Org. Chem.*, 1979, **44**, 1989. | 947 |
| 79JOC2280 | R. H. Fleming and B. M. Murray; *J. Org. Chem.*, 1979, **44**, 2280. | 912 |
| 79JOC2480 | G. A. Kraus and J. O. Pezzanite; *J. Org. Chem.*, 1979, **44**, 2480. | 803 |
| 79JOC2863 | P. K. Claus, F. W. Vierhapper and R. L. Willer; *J. Org. Chem.*, 1979, **44**, 2863. | 889, 890 |
| 79JOC3053 | S. Gelin; *J. Org. Chem.*, 1979, **44**, 3053. | 14, 53 |
| 79JOC3144 | C. H. Chen and G. A. Reynolds; *J. Org. Chem.*, 1979, **44**, 3144. | 933, 937 |
| 79JOC3230 | E. Vedejs, M. J. Arnost and J. P. Hagen; *J. Org. Chem.*, 1979, **44**, 3230. | 898 |
| 79JOC3310 | H. J. Shine and S-M. Wu; *J. Org. Chem.*, 1979, **44**, 3310. | 1010 |
| 79JOC3524 | A. Ohsawa, H. Arai, H. Igeta, T. Akimoto and A. Tsuji; *J. Org. Chem.*, 1979, **44**, 3524. | 37, 40 |
| 79JOC3982 | W. K. Chung, C. K. Chu, K. A. Watanabe and J. J. Fox; *J. Org. Chem.*, 1979, **44**, 3982. | 479 |
| 79JOC3988 | R. B. Gammill; *J. Org. Chem.*, 1979, **44**, 3988. | 699 |
| 79JOC4128 | V. Ceré, C. Paolucci, S. Pollicino, E. Sandri and A. Fava; *J. Org. Chem.*, 1979, **44**, 4128. | 898 |
| 79JOC4151 | T. Karakasa and S. Motoki; *J. Org. Chem.*, 1979, **44**, 4151. | 931 |
| 79JOC4191 | H. A. Albrecht, J. F. Blount, F. M. Konzelmann and J. T. Plati; *J. Org. Chem.*, 1979, **44**, 4191. | 1049, 1071 |
| 79JOC4243 | Y. Takeuchi, K. L. Kirk and L. A. Cohen; *J. Org. Chem.*, 1979, **44**, 4243. | 570 |
| 79JOC4435 | D. F. Eaton and B. E. Smart; *J. Org. Chem.*, 1979, **44**, 4435. | 1044, 1065 |
| 79JOC4456 | G. A. Reynolds, C. H. Chen and J. A. Van Allan; *J. Org. Chem.*, 1979, **44**, 4456. | 650 |
| 79JPC2696 | J. E. Pemberton, G. L. McIntire, H. N. Blount and J. F. Evans; *J. Phys. Chem.*, 1979, **83**, 2696. | 969 |
| 79JPR71 | K. Gewald and J. Oelsner; *J. Prakt. Chem.*, 1979, **321**, 71. | 52 |
| 79JPR519 | R. K. Bansal and S. K. Sharma; *J. Prakt. Chem.*, 1979, **321**, 519. | 570 |
| 79JPR699 | M. Augustin and G. Jahreis; *J. Prakt. Chem.*, 1979, **321**, 699. | 908 |
| 79JST(56)87 | T. N. Huckerby and G. Sunman; *J. Mol. Struct.*, 1979, **56**, 87. | 585, 589, 590, 591 |
| 79KGS124 | V. V. Dovlatyan and A. V. Dovlatyan; *Khim. Geterotsikl. Soedin.*, 1979, 124 (*Chem. Abstr.*, 1979, **90**, 152 136). | 480 |
| 79KGS291 | J. Szabó; *Khim. Geterotsikl. Soedin.*, 1979, 291 (*Chem. Abstr.*, 1979, **91**, 39 354). | 1007 |
| 79KGS403 | M. F. Budyka and P. B. Terent'ev; *Khim. Geterotsikl. Soedin.*, 1979, 403 (*Chem. Abstr.*, 1979, **91**, 38 675). | 236 |
| 79KGS639 | P. B. Terent'ev, V. G. Kartsev, I. K. Yakushenko, L. N. Prostakova and I. P. Gloriozov; *Khim. Geterotsikl. Soedin.*, 1979, 639 (*Chem. Abstr.*, 1979, **91**, 91 584). | 234, 236, 246 |
| 79KGS1124 | P. B. Terent'ev, V. G. Kartsev and M. F. Budyka; *Khim. Geterotsikl. Soedin.*, 1979, 1124 (*Chem. Abstr.*, 1980, **92**, 40 830). | 237 |
| 79KGS1317 | A. A. Potekhin and S. L. Zhdanov; *Khim. Geterotsikl. Soedin.*, 1979, 1317. | 644 |
| 79KGS1561 | V. P. Kruglenko and T. J. Sanford; *Khim. Geterotsikl. Soedin.*, 1979, 1561. | 408 |
| 79KGS1637 | S. M. Shevchenko and A. A. Potekhin; *Khim. Geterotsikl. Soedin.*, 1979, 1637. | 1056 |
| 79LA174 | G. Litkei, T. Mester, T. Patonay and R. Bognár; *Liebigs Ann. Chem.*, 1979, 174. | 823 |
| 79LA451 | U. Schöllkopf, R. Jentsch and K. Madawinata; *Liebigs Ann. Chem.*, 1979, 451. | 1026 |
| 79LA675 | M. Bachmann and H. Neunhoeffer; *Liebigs Ann. Chem.*, 1979, 675. | 51, 553 |
| 79LA751 | G. Falsone and B. Spur; *Liebigs Ann. Chem.*, 1979, 751. | 841 |
| 79LA784 | K. Praefcke and C. Weichsel; *Liebigs Ann. Chem.*, 1979, 784. | 935 |
| 79LA870 | C. Grugel and W. P. Neumann; *Liebigs Ann. Chem.*, 1979, 870. | 535, 536 |
| 79LA1130 | A. Parg and G. Hamprecht; *Liebigs Ann. Chem.*, 1979, 1130. | 1052, 1057 |
| 79LA1802 | R. W. Grauert; *Liebigs Ann. Chem.*, 1979, 1802. | 222 |
| 79LA2043 | K. Boeilitz, G. Buchholz and K. Praefcke; *Liebigs Ann. Chem.*, 1979, 2043. | 934 |
| 79M257 | H. Bartsch, W. Kropp and M. Pailer; *Monatsh. Chem.*, 1979, **110**, 257. | 1031 |
| 79M365 | G. Heinisch and I. Kirchner; *Monatsh. Chem.*, 1979, **110**, 365. | 30, 333, 354 |
| 79M425 | F. Asinger, J. Stalschus and A. Saus; *Monatsh. Chem.*, 1979, **110**, 425. | 1035 |
| 79MI21200 | E. W. Thulstrup, J. Spanget-Larsen and R. Gleiter; *Mol. Phys.*, 1979, **37**, 1381. | 3, 9 |
| 79MI21201 | W. von Niessen, W. P. Kraemer and G. H. F. Diercksen; *Chem. Phys.*, 1979, **41**, 113. | 8 |
| 79MI21202 | G. Nannini, G. Biasoli, E. Perrone, A. Forgione, A. Buttinoni and M. Ferrari; *Eur. J. Med. Chem.—Chim. Ther.*, 1979, **14**, 53 (*Chem. Abstr.*, 1979, **91**, 56 936). | 52 |
| B-79MI21400 | A. E. A. Porter; in 'Comprehensive Organic Chemistry', Pergamon Press, Oxford, 1979, vol. 4. | 163 |
| 79MI21500 | G. Moad, C. L. Luthy and S. J. Benkovic; *Dev. Biochem.*, 1978 (pub. 1979), **4**, 55 (*Chem. Abstr.*, 1979, **91**, 70 799). | 202, 204, 260 |
| 79MI21501 | P. Singh and S. P. Gupta; *Indian J. Med. Res.*, 1979, **69**, 804 (*Chem. Abstr.*, 1979, **91**, 133 793). | 204, 236 |

| | | |
|---|---|---|
| 79MI21502 | G. Tollin, R. L. Chan, T. R. Malefyt and T. C. Bruice; *Photochem. Photobiol.*, 1979, **29**, 233 (*Chem. Abstr.*, 1979, **91**, 123 004). | 207, 252, 254 |
| 79MI21503 | G. Evans and P. Caluwe; *Macromolecules*, 1979, **12**, 803 (*Chem. Abstr.*, 1979, **91**, 158 125). | 208 |
| 79MI21504 | M. F. Budyka, P. B. Terent'ev and A. N. Kost; *Vesta. Mosk. Univ., Ser. 2, Khim.*, 1979, **20**, 358 (*Chem. Abstr.*, 1980, **92**, 75 441). | 239 |
| B-79MI21505 | K. Hirai, T. Ishiba, H. Sugimoto and T. Fujishita; 'Fukusokan Kagaku Toronkai Koen Yoshishu 12th', Kitasato Daigaku Yakugakubu, Tokyo, 1979, p. 191 (*Chem. Abstr.*, 1980, **93**, 47 171). | 256 |
| 79MI21506 | D. Pompon and F. Lederer; *Eur. J. Biochem.*, 1979, **96**, 571 (*Chem. Abstr.*, 1979, **91**, 51 825). | 260 |
| 79MI21507 | I. Bornschein, R. Kraft, S. Pfeifer, M. Ullmann and H. Langner; *Pharmazie*, 1979, **34**, 732 (*Chem. Abstr.*, 1980, **93**, 323). | 260 |
| 79MI21600 | M. E. Salem, G. P. Lewis and P. B. Rowe; *Anal. Biochem.*, 1979, **97**, 48. | 285 |
| 79MI21601 | J. R. Bertino; *Cancer Res.*, 1979, **39**, 293. | 327 |
| 79MI21602 | H. C. Curtius, A. Niederwieser, M. Viscontini, A. Otten, J. Schaub, S. Scheibenreiter and H. Schmidt; *Clin. Chim. Acta*, 1979, **93**, 251. | 324 |
| 79MI21603 | H. Wachter, K. Grassmayr and A. Hausen; *Cancer Lett.*, 1979, **6**, 61. | 324 |
| 79MI21604 | A. Niederwieser, H. C. Curtius, O. Bettoni, J. Bieri, M. Viscontini, B. Schirks and J. Schaub; *Lancet*, 1979, 131. | 324 |
| 79MI21605 | A. Niederwieser, H. C. Curtius, M. Viscontini, J. Schaub and H. Schmidt; *Lancet*, 1979, 550. | 324 |
| B-79MI21606 | R. L. Kisliuk and G. M. Brown; 'Chemistry and Biology of Pteridines', Elsevier/North-Holland, New York, 1979. | 264, 284 |
| B-79MI21607 | L. Kiriasis and W. Pfleiderer; in 'Chemistry and Biology of Pteridines', ed. R. L. Kisliuk and G. M. Brown, Elsevier/North Holland, New York, 1979, p. 49. | 299 |
| B-79MI22000 | D. R. May; in 'Kirk–Othmer Encyclopedia of Chemical Technology', ed. M. Grayson; Wiley, New York, 1979, vol. 7, p. 291. | 499, 509 |
| B-79MI22001 | J. V. Burakevich; in 'Kirk–Othmer Encyclopedia of Chemical Technology', ed. M. Grayson; Wiley, New York, 1979, vol. 7, p. 397. | 477, 478 |
| B-79MI22002 | J. Elliot and P. P. Yeung; in 'Kirk–Othmer Encyclopedia of Chemical Technology', ed. M. Grayson; Wiley, New York, 1979, vol. 8, p. 374. | 526 |
| B-79MI22003 | P. D. G. Dean and D. H. Watson; *J. Chromatogr.*, 1979, **165**, 301. | 527 |
| B-79MI22004 | W. Harms; in 'Organofluorine Compounds and Their Industrial Applications', ed. R. E. Banks; Ellis Horwood, Chichester, 1979, p. 201. | 528 |
| 79MI22100 | U. Boesl, H. J. Neusser and E. W. Schlag; *Chem. Phys. Lett.*, 1979, **61**, 57. | 549 |
| 79MI22101 | U. Boesl, H. J. Neusser and E. W. Schlag; *Chem. Phys. Lett.*, 1979, **61**, 62. | 549 |
| 79MI22102 | L. S. Degtyarev and V. D. Pokhodenko; *Teor. Eksp. Khim.*, 1979, **15**, 88 (*Chem. Abstr.*, 1979, **90**, 186 031). | 533, 540 |
| 79MI22103 | D. Demus, B. Kruecke, F. Kuschel, H. U. Nothnick, G. Pelz and H. Zaschke; *Mol. Cryst. Liq. Cryst.*, 1979, **56**, 115 (*Chem. Abstr.*, 1980, **92**, 78 070). | 570 |
| 79MI22200 | O. Achmatowicz, Jr. and M. H. Burzynska; *Pol. J. Chem.*, 1979, **53**, 265. | 581, 589, 632 |
| B-79MI22201 | ACS Series Symp. No. 87, 'Anomeric Effect—Origin and Consequences', ed. W. A. Szarek and D. Horton; Am. Chem. Soc., Washington DC, 1979. | 629 |
| B-79MI22300 | A. T. Balaban; in 'New Trends in Heterocyclic Chemistry', ed. R. B. Mitra, N. R. Ayyanger, V. N. Gogte, R. M. Acheson and N. Cromwell; Elsevier, Amsterdam, 1979, p. 79. | 652 |
| 79MI22301 | G. T. Brooks, G. E. Pratt and R. C. Jennings; *Nature (London)*, 1979, **281**, 570. | 665 |
| 79MI22302 | K. Jankowski, Y. Volpe and C. S. Del Campo; *Rev. Latinoam. Quim.*, 1979, **10**, 87. | 681 |
| 79MI22303 | R. H. Kuroyan, A. I. Markosyan and S. A. Vartanyan; *Arm. Khim. Zh.*, 1979, **32**, 801 (*Chem. Abstr.*, 1980, **92**, 215 208). | 726 |
| B-79MI22304 | D. M. Harrison; in 'Comprehensive Organic Chemistry', ed. D. H. R. Barton and W. D. Ollis; Pergamon, Oxford, 1979, vol. 4, chap. 18.5. | 649, 732 |
| 79MI22400 | V. G. S. Box and C. McCaw; *Rev. Latinoam. Quim.*, 1979, **10**, 118. | 743 |
| 79MI22401 | J. L. Soto, C. Seoane, J. A. Valdés, N. Martin and M. Quinteiro; *An. Quim.*, 1979, **75**, 152. | 759 |
| 79MI22402 | M. L. Mihailović, N. Orbović, D. Marinković and S. Konstantinović; *Bull. Soc. Chim. Beograd.*, 1979, **44**, 597. | 776 |
| 79MI22403 | A. J. Cohen; *Food Cosmet. Toxicol.*, 1979, **17**, 277. | 882 |
| B-79MI22404 | A. T. Balaban; in 'New Trends in Heterocyclic Chemistry', ed. R. B. Mitra, N. R. Ayyangar, V. N. Gogte, R. M. Acheson and N. Cromwell; Elsevier, Amsterdam, 1979. | 883 |
| B-79MI22500 | 'Comprehensive Organic Chemistry', ed. D. H. R. Barton and W. D. Ollis; Pergamon, Oxford, 1979. | 895 |
| 79MI22501 | A. Levai and J. B. Schag; *Pharmazie*, 1979, **34**, 749. | 907 |
| 79MI22502 | E. A. Kharakhanov, L. N. Borodina, N. V. Gruzdev and E. A. Runova; *Vestn. Mosk. Univ., Ser. 2, Khim.*, 1979, **20**, 577 (*Chem. Abstr.*, 1980, **92**, 163 365). | 912 |
| 79MI22503 | J. E. N. Shin and A. S. Perlin; *Carbohydrate Res.*, 1979, **76**, 165. | 928 |
| B-79MI22504 | 'Comprehensive Organic Chemistry', ed. D. H. R. Barton and W. D. Ollis; Pergamon, Oxford, 1979, vol. 4, p. 856. | 935 |
| 79MI22505 | J. V. Crivello and J. H. W. Lam; *J. Polym. Sci., Polym. Chem. Ed.*, 1979, **17**, 2877. | 940 |

| | | |
|---|---|---|
| 79MI22506 | R. Usmanov, I. U. Numanov and N. Radzhabov; *Dokl. Akad. Nauk Tadzh. SSR*, 1979, **22**, 117. | 908 |
| B-79MI22600 | E. L. Eliel and K. M. Pietrusiewicz; in 'Topics in Carbon-13 NMR Spectroscopy', ed. G. C. Levy; Wiley-Interscience, New York, 1979, vol. 3, p. 172. | 953 |
| B-79MI22601 | A. H. Haines; in 'Comprehensive Organic Chemistry', ed. D. H. R. Barton and W. D. Ollis; Pergamon, Oxford, 1979, vol. 4, p. 853. | 976 |
| B-79MI22602 | T. Whiteside; 'The Pendulum and the Toxic Cloud', Yale University Press, New Haven, 1979. | 992 |
| 79MI22603 | K. Niume, K. Nakamichi, R. Takatuka, F. Toda, K. Uno and Y. Iwakura; *J. Polym. Sci., Polym. Chem. Ed.*, 1979, **17**, 2371. | 993 |
| 79MI22700 | T. S. Safonovova and M. P. Nemeryuk; *Česk. Farm.*, 1979, **28**, 251. | 1038 |
| 79MI22701 | W. J. Albery, A. W. Foulds, K. J. Hall, A. R. Hillman, R. G. Edgell and A. F. Orchard; *Nature*, 1979, **282**, 793. | 1038 |
| B-79MI22800 | J. K. Landquist; in 'Comprehensive Organic Chemistry', ed. D. H. R. Barton and W. D. Ollis; Pergamon, Oxford, 1979, vol. 4, p. 1051. | 1041, 1042, 1043, 1044, 1045, 1046, 1085 |
| 79MIP22100 | I. Ya. Postovskii, E. G. Kovalev and G. L. Rusinov; *U.S.S.R. Pat.* 679 582 (1979) (*Chem. Abstr.*, 1980, **92**, 6564). | 555 |
| 79MIP22700 | J. Arct and G. Olszewska; *Pol. Pat.* 106 421 (1979) (*Chem. Abstr.*, 1981, **94**, 4024). | 1029 |
| 79OMR(12)212 | G. W. H. Cheeseman, C. J. Turner and D. J. Brown; *Org. Magn. Reson.*, 1979, **12**, 212. | 63 |
| 79OMR(12)284 | A. Rabaron, J.-R. Didry, B. S. Kirkiacharian and M. M. Plat; *Org. Magn. Reson.*, 1979, **12**, 284. | 589 |
| 79OMR(12)337 | K. Pihlaja, P. Pasanen and J. Wähäsilta; *Org. Magn. Reson.*, 1979, **12**, 337. | 989 |
| 79OMR(12)481 | P. Tisnès, P. Maroni and L. Cazaux; *Org. Magn. Reson.*, 1979, **12**, 481. | 1053 |
| 79OMR(12)612 | J. Uzawa and M. Uramoto; *Org. Magn. Reson.*, 1979, **12**, 612. | 394 |
| 79OMS345 | S. Eguchi; *Org. Mass Spectrom.*, 1979, **14**, 345. | 613, 614 |
| 79PS(7)61 | J. Fabian, R. Mayer and S. Bleisch; *Phosphorus Sulfur*, 1979, **7**, 61. | 1050 |
| 79RRC59 | I. Dragotă and I. Niculescu-Duvăz; *Rev. Roum. Chim.*, 1979, **24**, 59. | 851 |
| 79RRC453 | N. A. Shams; *Rev. Roum. Chim.*, 1979, **24**, 453. | 37 |
| 79RRC1191 | M. El-Hashash, M. Mahmoud and H. El-Fiky; *Rev. Roum. Chim.*, 1979, **24**, 1191 (*Chem. Abstr.*, 1980, **92**, 163 865). | 227 |
| 79RTC224 | J. Tramper, A. Nagel, H. C. van der Plas and F. Müller; *Recl. Trav. Chim. Pays-Bas*, 1979, **98**, 224. | 287 |
| 79RTC520 | R. Gräfing and L. Brandsma; *Recl. Trav. Chim. Pays-Bas*, 1979, **98**, 520. | 912 |
| 79S41 | J. Jurczak, M. Chmielewski and S. Filipek; *Synthesis*, 1979, 41. | 579, 773 |
| 79S47 | W. Verboom, R. S. Sukhai and J. Meijer; *Synthesis*, 1979, 47. | 988 |
| 79S126 | F. Camps, J. Coll, A. Messeguer, M. A. Pericás and S. Ricart; *Synthesis*, 1979, 126. | 778, 780 |
| 79S161 | N. Blazevic, D. Kolbah, B. Belin, V. Sunjic and F. Kajfez; *Synthesis*, 1979, 161. | 487 |
| 79S241 | J. Liebscher and H. Hartmann; *Synthesis*, 1979, 241. | 866 |
| 79S337 | A. O. Fitton, P. G. Houghton and H. Suschitzky; *Synthesis*, 1979, 337. | 712 |
| 79S385 | R. Gompper and R. Sobotta; *Synthesis*, 1979, 385. | 47 |
| 79S541 | G. Coudert, G. Guillaumet and B. Loubinoux; *Synthesis*, 1979, 541. | 1036 |
| 79S690 | T. Maier, H. Bredereck and W. Kantlehner; *Synthesis*, 1979, 690. | 507 |
| 79S732 | Chr. Ivanov, Y. Anghelova and S. Spirova; *Synthesis*, 1979, 732. | 690 |
| 79S743 | L.-I. Olsson and A. Claesson; *Synthesis*, 1979, 743. | 579, 769 |
| 79S790 | C. Venturello and R. D'Aloisio; *Synthesis*, 1979, 790. | 42 |
| 79S801 | C. Giordano, A. Belli, R. Erbea and S. Panossian; *Synthesis*, 1979, 801. | 1077 |
| 79S810 | S. Gronowitz and Z. Lidert; *Synthesis*, 1979, 810. | 487 |
| 79S889 | M. Payard and J. Couquelet; *Synthesis*, 1979, 889. | 590, 710 |
| 79S906 | N. S. Narasimhan, R. S. Mali and M. V. Barve; *Synthesis*, 1979, 906. | 805 |
| 79S987 | D. Middlemiss; *Synthesis*, 1979, 987. | 688 |
| 79SC129 | G. Roberge and P. Brassard; *Synth. Commun.*, 1979, **9**, 129. | 827 |
| 79SC631 | P. W. Freeman, S. N. Quessy and L. R. Williams; *Synth. Commun.*, 1979, **9**, 631. | 1035 |
| 79SC889 | Y. Nakashima, T. Imagawa and M. Kawanisi; *Synth. Commun.*, 1979, **9**, 889. | 588, 773, 843 |
| 79T681 | M. Z. Nazer, C. H. Issidorides and M. J. Haddadin; *Tetrahedron*, 1979, **35**, 681. | 440 |
| 79T691 | U. Burkert; *Tetrahedron*, 1979, **35**, 691. | 946 |
| 79T1267 | A. Ohsawa, H. Arai, H. Igeta, T. Akimoto, A. Tsuji and Y. Iitaka; *Tetrahedron*, 1979, **35**, 1267. | 13 |
| 79T1311 | F. G. Riddell and E. S. Turner; *Tetrahedron*, 1979, **35**, 1311. | 1054 |
| 79T1391 | F. G. Riddell, E. S. Turner, A. R. Katritzky, R. C. Patel and F. M. S. Brito-Palma; *Tetrahedron*, 1979, **35**, 1391. | 1054 |
| 79T1615 | W. Förster, P. Birner and C. Weiss; *Tetrahedron*, 1979, **35**, 1615. | 253 |
| 79T1771 | P. Battistoni, P. Bruni and G. Fava; *Tetrahedron*, 1979, **35**, 1771. | 1031 |
| 79T2027 | G. Jones and P. Rafferty; *Tetrahedron*, 1979, **35**, 2027. | 234, 235, 236, 237, 239, 241, 246 |
| 79T2035 | A. W. Frahm and R. K. Chaudhuri; *Tetrahedron*, 1979, **35**, 2035. | 589, 590 |
| 79T2151 | P. Jakobsen and S. Treppendahl; *Tetrahedron*, 1979, **35**, 2151. | 1055 |
| 79T2883 | J. A. Donnelly and D. E. Maloney; *Tetrahedron*, 1979, **35**, 2883. | 735 |
| 79TL5 | A. Hercouet and M. Le Corre; *Tetrahedron Lett.*, 1979, 5. | 770 |
| 79TL237 | W. Friedrichsen, C. Krüger, E. Kujath, G. Liebezeit and S. Mohr; *Tetrahedron Lett.*, 1979, 237. | 1004 |

| | | |
|---|---|---|
| 79TL1097 | I. W. J. Still and T. S. Leong; *Tetrahedron Lett.*, 1979, 1097. | 918 |
| 79TL1109 | C. Deschamps-Vallet, J. B. Ilotse, M. Meyer-Dayan and D. Molho; *Tetrahedron Lett.*, 1979, 1109. | 650 |
| 79TL1241 | L. S. Cook and B. J. Wakefield; *Tetrahedron Lett.*, 1979, 1241. | 473 |
| 79TL1281 | J. Rokach, P. Hamel, Y. Girard and G. Reader; *Tetrahedron Lett.*, 1979, 1281. | 1028 |
| 79TL1299 | H. D. Fühlhuber, C. Gousetis and J. Sauer; *Tetrahedron Lett.*, 1979, 1299. | 551 |
| 79TL1333 | P. Gaviña, P. Gil and B. Palazón; *Tetrahedron Lett.*, 1979, 1333. | 49 |
| 79TL1537 | C. Giordano, A. Belli and L. Abis; *Tetrahedron Lett.*, 1979, 1537. | 1063 |
| 79TL1565 | K. Warning and M. Mitzlaff; *Tetrahedron Lett.*, 1979, 1565. | 510 |
| 79TL1603 | M. Hori, T. Kataoka and H. Shimizu; *Tetrahedron Lett.*, 1979, 1603. | 925 |
| 79TL2149 | B. Begasse, A. Hercouet and M. Le Corre; *Tetrahedron Lett.*, 1979, 2149. | 766 |
| 79TL2545 | A. Bongini, G. Cardillo, M. Orena, G. Porzi and S. Sandri; *Tetrahedron Lett.*, 1979, 2545. | 755 |
| 79TL2821 | D. M. Holton, P. M. Hoyle and D. Murphy; *Tetrahedron Lett.*, 1979, 2821. | 8 |
| 79TL2921 | E. Toja, A. Omodei-Salè and G. Nathansohn; *Tetrahedron Lett.*, 1979, 2921. | 53 |
| 79TL2995 | A. Hercouet and M. Le Corre; *Tetrahedron Lett.*, 1979, 2995. | 763 |
| 79TL3037 | K. Venkataraman and D. R. Wagle; *Tetrahedron Lett.*, 1979, 3037. | 485 |
| 79TL3645 | G. H. Hakimelahi and G. Just; *Tetrahedron Lett.*, 1979, 3645. | 775 |
| 79TL3685 | A. Banerji and N. C. Goomer; *Tetrahedron Lett.*, 1979, 3685. | 852 |
| 79TL3901 | F. Camps, J. Coll, A. Messeguer and M. A. Pericás; *Tetrahedron Lett.*, 1979, 3901. | 723, 755 |
| 79TL3969 | M. Hori, T. Kataoka, H. Shimizu and K. Matsuo; *Tetrahedron Lett.*, 1979, 3969. | 1016 |
| 79TL4253 | G. Ege and K. Gilbert, *Tetrahedron Lett.*, 1979, 4253. | 535 |
| 79TL4687 | R. K. Smalley, R. H. Smith and H. Suschitzky; *Tetrahedron Lett.*, 1979, 4687. | 384 |
| 79TL4837 | S. Restle and C. G. Wermuth; *Tetrahedron Lett.*, 1979, 4837. | 28 |
| 79TL4857 | R. G. Visser and H. J. T. Bos; *Tetrahedron Lett.*, 1979, 4857. | 921 |
| 79TL5025 | N. Viswanathan and A. R. Sidhaye; *Tetrahedron Lett.*, 1979, 5025. | 53 |
| 79USP4139655 | W. R. Grace and Co., *U.S. Pat.* 4 139 655 (1979) (*Chem. Abstr.*, 1979, **90**, 170 283). | 942 |
| 79USP4164431 | Eastman Kodak, *U.S. Pat.* 4 164 431 (1979) (*Chem. Abstr.*, 1979, **91**, 178 148). | 942 |
| 79ZC59 | J. Bödeker, P. Köchritz and K. Courault; *Z. Chem.*, 1979, **19**, 59. | 500 |
| 79ZC422 | E. Lippmann and J. Teichert; *Z. Chem.*, 1979, **19**, 422. | 259 |
| 79ZOR2344 | V. A. Khishnin, V. G. Koshechko and V. D. Pokhodenko; *Zh. Org. Khim.*, 1979, **15**, 2344. | 547 |
| 79ZOR2405 | L. A. Pavlova, I. V. Samartseva and N. V. L'vova; *Zh. Org. Khim.*, 1979, **15**, 2405. | 998 |
| 79ZPC(360)1957 | H. Wachter, A. Hausen and K. Grassmayr; *Hoppe-Seyler's Z. Physiol. Chem.*, 1979, **360**, 1957. | 324 |
| | | |
| 80ACS(B)295 | K. Skinnemoen and K. Undheim; *Acta Chem. Scand., Ser. B*, 1980, **34**, 295. | 843 |
| 80AG464 | D. Paske, R. Ringshandl, I. Sellner, H. Sichert and J. Sauer; *Angew. Chem.*, 1980, **92**, 464. | 551 |
| 80AG474 | H. Böhme, W. Pfleiderer, E. F. Elstner and W. J. Richter; *Angew. Chem.*, 1980, **92**, 474. | 295, 323 |
| 80AG735 | G. Kaupp and H. W. Grueter; *Angew. Chem.*, 1980, **92**, 735. | 227 |
| 80AG766 | F. A. Neugebauer and H. Fischer; *Angew. Chem.*, 1980, **92**, 766. | 560 |
| 80AG815 | W. Adam and O. De Lucchi; *Angew. Chem.*, 1980, **92**, 815. | 2 |
| 80AG1066 | S. Szilagyi and H. Wamhoff; *Angew. Chem.*, 1980, **92**, 1066. | 429 |
| 80AG(E)53 | R. Amstutz, D. Seebach, P. Seiler, B. Schweizer and J. D. Dunitz; *Angew. Chem., Int. Ed. Engl.*, 1980, **19**, 53. | 961 |
| 80AG(E)131 | J. Sander and K. Clauss; *Angew. Chem., Int. Ed. Engl.*, 1980, **19**, 131. | 1061, 1070 |
| 80AG(E)134 | L.-F. Tietze, G. von Kiedrowski, K. Harms, W. Clegg and G. Sheldrick; *Angew. Chem., Int. Ed. Engl.*, 1980, **19**, 134. | 771 |
| 80AG(E)199 | C. N. Rentzea; *Angew. Chem., Int. Ed. Engl.*, 1980, **19**, 199. | 1017 |
| 80AG(E)222 | G. Reissenweber and D. Mangold; *Angew. Chem., Int. Ed. Engl.*, 1980, **19**, 222. | 1029 |
| 80AG(E)306 | A. R. Katritzky, R. H. Manzo, J. M. Lloyd and R. C. Patel; *Angew. Chem., Int. Ed. Engl.*, 1980, **19**, 306. | 655 |
| 80AG(E)906 | J. L. Flippen-Anderson, I. Karle, R. Huisgen and H. U. Reissig; *Angew. Chem., Int. Ed. Engl.*, 1980, **19**, 906. | 509 |
| 80AHC(26)115 | O. Meth-Cohn and B. Tarnowski; *Adv. Heterocycl. Chem.*, 1980, **26**, 115. | 916, 919, 929, 933, 934, 937, 941 |
| 80AJC131 | W. B. Cowden and N. W. Jacobsen; *Aust. J. Chem.*, 1980, **33**, 131. | 131 |
| 80AJC1147 | D. J. Brown and K. Shinozuka; *Aust. J. Chem.*, 1980, **33**, 1147. | 89, 94 |
| 80AJC2447 | G. J. Gainsford and A. D. Woolhouse; *Aust. J. Chem.*, 1980, **33**, 2447. | 1046, 1049, 1082 |
| 80AP35 | G. Zinner, J. Schmidt and W. Kilwing; *Arch. Pharm. (Weinheim, Ger.)*, 1980, **313**, 35. | 1050, 1073 |
| 80AP77 | H. Böhme and J. Strahl; *Arch. Pharm. (Weinheim, Ger.)*, 1980, **313**, 77. | 438 |
| 80AP377 | D. Geffken; *Arch. Pharm. (Weinheim, Ger.)*, 1980, **313**, 377. | 1050, 1073 |
| 80AX(B)2159 | G. P. Stahly, H. L. Ammon and B. B. Jarvis; *Acta Crystallogr., Part B*, 1980, **36**, 2159. | 1049 |
| 80BBA(623)77 | P. K. Dutta, R. Spencer, C. Walsh and T. G. Spiro; *Biochim. Biophys. Acta*, 1980, **623**, 77. | 204, 249 |
| 80BCJ569 | N. S. Poonia, H. K. Arora and A. V. Bajaj; *Bull. Chem. Soc. Jpn.*, 1980, **54**, 569. | 716 |
| 80BCJ831 | M. Nakayama, S. Ohira and T. Matsui; *Bull. Chem. Soc. Jpn.*, 1980, **53**, 831. | 821 |
| 80BCJ2671 | N. Azuma; *Bull. Chem. Soc. Jpn.*, 1980, **53**, 2671. | 539, 541 |

| | | |
|---|---|---|
| 80BCJ3369 | I. Shibuya; *Bull. Chem. Soc. Jpn.*, 1980, **53**, 3369. | 1052, 1061 |
| 80BCJ3385 | T. Sugimoto and S. Matsuura; *Bull. Chem. Soc. Jpn.*, 1980, **53**, 3385. | 316 |
| 80CB79 | J. Goerdeler and R. Losch; *Chem. Ber.*, 1980, **113**, 79. | 486 |
| 80CB261 | F. Bohlmann and E. Vorwerk; *Chem. Ber.*, 1980, **113**, 261. | 743 |
| 80CB395 | L. Capuano, M. Bronder, K. Djokar and I. Müller; *Chem. Ber.*, 1980, **113**, 395. | 219, 222 |
| 80CB1205 | F. A. Neugebauer and I. Umminger; *Chem. Ber.*, 1980, **113**, 1205. | 409, 413 |
| 80CB1514 | P. H. Boyle and W. Pfleiderer; *Chem. Ber.*, 1980, **113**, 1514. | 295, 304, 311 |
| 80CB1524 | G. Ritzmann, L. Kiriasis and W. Pfleiderer; *Chem. Ber.*, 1980, **113**, 1524. | 282 |
| 80CB1535 | G. Ritzmann, K. Ienaga, L. Kiriasis and W. Pfleiderer; *Chem. Ber.*, 1980, **113**, 1535. | 282, 299 |
| 80CB1708 | R. Hasenkamp, V. Luhmann and W. Lüttke; *Chem. Ber.*, 1980, **113**, 1708. | 910 |
| 80CB2049 | F. A. Neugebauer, H. Fischer and P. Meier; *Chem. Ber.*, 1980, **113**, 2049. | 558 |
| 80CB3927 | J. N. Chatterjea, S. K. Mukherjee, C. Bhakta, H. C. Jha and F. Zilliken; *Chem. Ber.*, 1980, **113**, 3927. | 831 |
| 80CC334 | W. L. F. Armarego, P. Waring and J. W. Williams; *J. Chem. Soc., Chem. Commun.*, 1980, 334. | 281 |
| 80CC559 | N. E. MacKenzie and R. H. Thomson; *J. Chem. Soc., Chem. Commun.*, 1980, 559. | 919 |
| 80CC585 | M. G. Reinecke, L.-J. Chen and A. Almqvist; *J. Chem. Soc., Chem. Commun.*, 1980, 585. | 376 |
| 80CC808 | D. E. Davies, D. L. R. Reeves and R. C. Storr; *J. Chem. Soc., Chem. Commun.*, 1980, 808. | 378 |
| 80CC1020 | N. Shinmon, M. P. Cava and R. F. C. Brown; *J. Chem. Soc., Chem. Commun.*, 1980, 1020. | 192 |
| 80CC1182 | A. Ohsawa, H. Arai, H. Ohnishi and H. Igeta; *J. Chem. Soc., Chem. Commun.*, 1980, 1182. | 371, 376, 381 |
| 80CC1224 | A. M. B. S. R. C. S. Costa, F. M. Dean, M. A. Jones, D. A. Smith and R. S. Varma; *J. Chem. Soc., Chem. Commun.*, 1980, 1224. | 684 |
| 80CC1227 | T. A. Carpenter, P. J. Jenner, F. J. Leeper and J. Staunton; *J. Chem. Soc., Chem. Commun.*, 1980, 1227. | 684 |
| 80CCC1379 | J. Slouka and V. Bekárek; *Collect. Czech Chem. Commun.*, 1980, **45**, 1379. | 445 |
| 80CHE345 | L. Yu. Ukhin, V. V. Bessonov, A. I. Yanovskii, T. V. Timofeeva, N. G. Furmanova and Yu. T. Struchkov; *Chem. Heterocycl. Compd. (Engl. Transl.)*, 1980, **16**, 345. | 660 |
| 80CHE571 | L. F. Lapuka, E. A. Kantor, N. A. Romanov, R. S. Musavirov and D. C. Rakhmankulov; *Chem. Heterocycl. Compd. (Engl. Transl.)*, 1980, **16**, 571. | 578, 579, 630 |
| 80CHE574 | E. A. Zvezdina, M. P. Zhdanova and G. N. Dorofeenko; *Chem. Heterocycl. Compd. (Engl. Transl.)*, 1980, **16**, 574. | 656 |
| 80CHE807 | G. N. Dorofeenko, A. V. Koblik, T. I. Polyakova and L. A. Murad'yan; *Chem. Heterocycl. Compd. (Engl. Transl.)*, 1980, **16**, 807. | 660 |
| 80CJC494 | D. J. Harris, G. Y. P. Kan, T. Tschamber and V. Snieckus; *Can. J. Chem.*, 1980, **58**, 494. | 660 |
| 80CJC1211 | H. C. Jha, F. Zilliken and E. Breitmaier; *Can. J. Chem.*, 1980, **58**, 1211. | 590, 591 |
| 80CPB198 | T. Jojima, H. Takeshiba and T. Kinoto; *Chem. Pharm. Bull.* 1980, **28**, 198. | 31 |
| 80CPB202 | T. Sakamoto, T. Ono, T. Sakasai and H. Yamanaka; *Chem. Pharm. Bull.*, 1980, **28**, 202. | 126 |
| 80CPB465 | K. Wachi and A. Terada; *Chem. Pharm. Bull.*, 1980, **28**, 465. | 1003, 1018 |
| 80CPB708 | M. Iinuma, S. Matsuura and K. Kusuda; *Chem. Pharm. Bull.*, 1980, **28**, 708. | 591 |
| 80CPB1526 | H. Yamanaka, S. Ogawa and S. Konno; *Chem. Pharm. Bull.*, 1980, **28**, 1526. | 105 |
| 80CPB2676 | T. Tsuchiya, J. Kurita and K. Takayama; *Chem. Pharm. Bull.*, 1980, **28**, 2676. | 13 |
| 80CPB2967 | M. Yamato, T. Ishikawa and T. Kobayashi; *Chem. Pharm. Bull.*, 1980, **28**, 2967. | 735 |
| 80CPB3049 | F. Yoneda, K. Tsukuda, K. Shinozuka, F. Hirayama, K. Uekama and A. Koshiro; *Chem. Pharm. Bull.*, 1980, **28**, 3049. | 207 |
| 80CPB3362 | T. Sakamoto, K. Tanji, S. Niitsuma, T. Ono and H. Yamanaka; *Chem. Pharm. Bull.*, 1980, **28**, 3362. | 101, 103, 128 |
| 80CPB3457 | Y. Kurasawa and A. Takada; *Chem. Pharm. Bull.*, 1980, **28**, 3457. | 236, 237, 244 |
| 80CPB3514 | F. Yoneda, K. Mori, M. Ono, Y. Kadokawa, E. Nagao and H. Yamaguchi; *Chem. Pharm. Bull.*, 1980, **28**, 3514. | 205, 207, 219 |
| 80CPB3570 | A. Ohsawa, H. Arai and H. Igeta; *Chem. Pharm. Bull.*, 1980, **28**, 3570. | 500 |
| 80CRV151 | G. A. Benson and W. J. Spillane; *Chem. Rev.*, 1980, **80**, 151. | 1079 |
| 80CSR477 | P. D. Kennewell and J. B. Taylor; *Chem. Soc. Rev.*, 1980, 477. | 1043, 1085 |
| 80EUP7672 | Gist-Brocades N. V., *Eur. Pat.* 7 672 (1980) (*Chem. Abstr.*, 1980, **93**, 114 167). | 939 |
| 80EUP18151 | Warner-Lambert Co., *Eur. Pat.* 18 151 (1980) (*Derwent Farmdoc* 79 345C). | 227 |
| 80FES715 | R. Cerri, A. Boido and F. Sparatore; *Farmaco Ed. Sci.*, 1980, **35**, 715. | 451 |
| 80GEP2903850 | Bayer A/G, *Ger. Offen.* 2 903 850 (1980) (*Derwent Farmdoc* 57 051C). | 224 |
| 80GEP2939259 | G. Jan; (Ciba-Geigy A/G), *Ger. Offen.* 2 939 259 (1980) (*Chem. Abstr.*, 1981, **94**, 22 884). | 242, 245 |
| 80H(14)33 | H. Federsel and J. Bergman; *Heterocycles*, 1980, **14**, 33. | 928 |
| 80H(14)267 | Y. Kurasawa and A. Takada; *Heterocycles*, 1980, **14**, 267. | 244 |
| 80H(14)337 | C. Seoane, J. L. Soto and M. Quinteiro; *Heterocycles*, 1980, **14**, 337. | 756 |
| 80H(14)407 | K. Hirota, Y. Kitade and S. Senda; *Heterocycles*, 1980, **14**, 407. | 217 |
| 80H(14)477 | J. J. Brophy and G. W. K. Cavill; *Heterocycles*, 1980, **14**, 477. | 193 |
| 80H(14)1033 | F. D. Popp; *Heterocycles*, 1980, **14**, 1033. | 2 |
| 80H(14)1319 | Z. Witczak; *Heterocycles*, 1980, **14**, 1319. | 728 |
| 80H(14)1333 | T. Kato and N. Katagiri; *Heterocycles*, 1980, **14**, 1333. | 1001, 1007, 1022, 1027 |
| 80H(14)1603 | W. Pfleiderer and R. Gottlieb; *Heterocycles*, 1980, **14**, 1603. | 279, 282, 284, 306 |

| 80HCA395 | A. N. Ganguly, P. K. Sengupta, J. H. Bieri and M. Viscontini; *Helv. Chim. Acta*, 1980, **63**, 395. | 307 |
| 80HCA504 | W. Städeli, W. von Philipsborn, A. Wick and I. Kompiš; *Helv. Chim. Acta*, 1980, **63**, 504. | 64 |
| 80HCA1754 | A. N. Ganguly, P. K. Sengupta, J. H. Bieri and M. Viscontini; *Helv. Chim. Acta*, 1980, **63**, 1754. | 307 |
| 80HCA1805 | K. Baumgartner and J. H. Bieri; *Helv. Chim. Acta*, 1980, **63**, 1805. | 288 |
| 80HCA2554 | E. Khalifa, A. N. Ganguly, J. H. Bieri and M. Viscontini; *Helv. Chim. Acta*, 1980, **63**, 2554. | 307 |
| 80IJC(B)195 | J. Singh, V. Virmani, P. C. Jain and N. Anand; *Indian J. Chem., Sect. B*, 1980, **19**, 195. | 231 |
| 80IJC(B)500 | N. K. Sangwan and S. N. Rastogi; *Indian J. Chem., Sect. B*, 1980, **19**, 500. | 855 |
| 80IZV2669 | V. G. Shtamburg, V. F. Rudchenko, A. P. Pleshkova, I. I. Chervin and R. G. Kostyanovskii; *Izv. Akad. Nauk SSSR, Ser. Khim.*, 1980, 2669. | 1041, 1070 |
| 80IZV2785 | G. A. Abakumov, V. K. Cherkasov, V. A. Muraev and S. A. Chesnokov; *Izv. Akad. Nauk SSSR, Ser. Khim.*, 1980, 2785. | 547 |
| 80JA1092 | G. Blankenhorn and E. G. Moore; *J. Am. Chem. Soc.*, 1980, **102**, 1092. | 204, 207 |
| 80JA3863 | P. B. Dervan and D. S. Santilli; *J. Am. Chem. Soc.*, 1980, **102**, 3863. | 39 |
| 80JA4193 | A. Cowell and J. K. Stille; *J. Am. Chem. Soc.*, 1980, **102**, 4193. | 859 |
| 80JA6168 | R. Stewart, S. J. Gumbley and R. Srinivasan; *J. Am. Chem. Soc.*, 1980, **102**, 6168. | 202, 212 |
| 80JA6407 | D. Garcia and E. Grunwald; *J. Am. Chem. Soc.*, 1980, **102**, 6407. | 629 |
| 80JA7196 | A. J. Hopfinger; *J. Am. Chem. Soc.*, 1980, **102**, 7196. | 529 |
| 80JA7438 | A. Schweig, N. Thon, S. F. Nelsen and L. A. Grezzo; *J. Am. Chem. Soc.*, 2980, **102**, 7438. | 40 |
| 80JA7747 | J. C. Scaiano; *J. Am. Chem. Soc.*, 1980, **102**, 7747. | 695 |
| 80JAP80102504 | Tokkyo Koho, *Jpn. Pat.* 80 102 504 (1980) (*Chem. Abstr.*, 1980, **93**, 199 231). | 941 |
| 80JAP80105678 | Nippon Kayaku Co., *Jpn. Pat.* 80 105 678 (1980) (*Chem. Abstr.*, 1981, **94**, 30 570). | 941 |
| 80JAP(K)8038361 | J. Matsumoto and K. Miyamoto; (Dainippon Pharm. Co. Ltd.), *Jpn. Kokai* 80 38 361 (1980) (*Chem. Abstr.*, 1980, **93**, 186 333). | 206 |
| 80JBC(255)909 | J. V. Tuttle and J. A. Krenitsky; *J. Biol. Chem.*, 1980, **255**, 909. | 260 |
| 80JBC(255)1891 | R. P. Gunsalus and R. S. Wolfe; *J. Biol. Chem.*, 1980, **255**, 1891. | 260 |
| 80JCR(S)8 | R. M. Christie, C. A. Shand, R. H. Thomson and C. W. Greenhalgh; *J. Chem. Res. (S)*, 1980, 8. | 919 |
| 80JCR(S)35 | K. J. McCullough, A. R. Morgan, D. C. Nonhebel and P. L. Pauson; *J. Chem. Res. (S)*, 1980, 35. | 979, 990 |
| 80JCR(S)36 | K. J. McCullough, A. R. Morgan, D. C. Nonhebel and P. L. Pauson; *J. Chem. Res. (S)*, 1980, 36. | 979 |
| 80JCR(S)148 | T. Yamamoto, M. Muraoka, M. Takahashi and T. Takeshima; *J. Chem. Res. (S)*, 1980, 148. | 570 |
| 80JCR(S)159 | T. Buggy and G. P. Ellis; *J. Chem. Res. (S)*, 1980, 159. | 697, 709 |
| 80JCR(S)246 | T. McC. Paterson and R. K. Smalley; *J. Chem. Res. (S)*, 1980, 246. | 379 |
| 80JCR(S)310 | A. R. Katritzky, A. M. El-Mowafy, L. Marzorati, R. C. Patel and S. S. Thind; *J. Chem. Res. (S)*, 1980, 310. | 656 |
| 80JCR(S)312 | A. R. Katritzky and M. C. Rezende; *J. Chem. Res. (S)*, 1980, 312. | 656 |
| 80JCS(P1)72 | M. M. Baradarani and J. A. Joule; *J. Chem. Soc., Perkin Trans. 1*, 1980, 72. | 48 |
| 80JCS(P1)293 | F. Yoneda, Y. Sakuma and A. Koshiro; *J. Chem. Soc., Perkin Trans. 1*, 1980, 293. | 208 |
| 80JCS(P1)633 | T. McC. Paterson, R. K. Smalley, H. Suschitzky and A. J. Barker; *J. Chem. Soc., Perkin Trans. 1*, 1980, 633. | 379 |
| 80JCS(P1)858 | J. H. Jones and M. H. Witty; *J. Chem. Soc., Perkin Trans. 1*, 1980, 858. | 440 |
| 80JCS(P1)1025 | A. J. Hall, D. Ferreira and D. G. Roux; *J. Chem. Soc., Perkin Trans. 1*, 1980, 1025. | 597, 854 |
| 80JCS(P1)1345 | B. J. Graphakos, A. R. Katritzky, G. Lhommet and K. Reynolds; *J. Chem. Soc., Perkin Trans. 1*, 1980, 1345. | 923 |
| 80JCS(P1)1603 | M. E. K. Cartoon, G. W. H. Cheeseman, H. A. Dowlatshahi and P. Shanna; *J. Chem. Soc., Perkin Trans. 1*, 1980, 1603. | 175 |
| 80JCS(P1)1895 | A. R. Katritzky and S. S. Thind; *J. Chem. Soc., Perkin Trans. 1*, 1980, 1895. | 869 |
| 80JCS(P1)2049 | F. M. Dean and R. S. Johnson; *J. Chem. Soc., Perkin Trans. 1*, 1980, 2049. | 656, 700 |
| 80JCS(P1)2254 | M. G. Barlow, R. N. Haszeldine and C. Simon; *J. Chem. Soc., Perkin Trans. 1*, 1980, 2254. | 419, 420, 421 |
| 80JCS(P1)2278 | Z. Czochanska, L. Y. Foo, R. H. Newman and L. J. Porter; *J. Chem. Soc., Perkin Trans. 1*, 1980, 2278. | 592 |
| 80JCS(P1)2383 | M. G. Ahmed, S. A. Ahmed and P. W. Hickmott; *J. Chem. Soc., Perkin Trans. 1*, 1980, 2383. | 1058 |
| 80JCS(P1)2450 | A. J. Bloodworth and J. A. Khan; *J. Chem. Soc., Perkin Trans. 1*, 1980, 2450. | 990 |
| 80JCS(P1)2645 | S. S. Al-Hassan, R. J. Kulick, D. B. Livingstone, C. J. Suckling, H. C. S. Wood, R. Wrigglesworth and R. Ferone; *J. Chem. Soc., Perkin Trans. 1*, 1980, 2645. | 202, 260 |
| 80JCS(P2)4 | J. S. A. Brunskill and A. De; *J. Chem. Soc., Perkin Trans. 2*, 1980, 4. | 931 |
| 80JCS(P2)91 | A. R. Katritzky, V. J. Baker, F. M. S. Brito-Palma, R. C. Patel, G. Pfister-Guillouzo and C. Guimon; *J. Chem. Soc., Perkin Trans. 2*, 1980, 91. | 389, 398, 539, 543 |
| 80JCS(P2)279 | A. R. Katritzky and R. C. Patel; *J. Chem. Soc., Perkin Trans. 2*, 1980, 279. | 1053, 1055 |
| 80JCS(P2)385 | D. C. Appleton, D. C. Bull, J. McKenna, J. M. McKenna and A. R. Walley; *J. Chem. Soc., Perkin Trans. 2*, 1980, 385. | 899 |
| 80JCS(P2)1339 | S. Chimichi, R. Nesi, M. Scotton and C. Mannucci; *J. Chem. Soc., Perkin Trans. 2*, 1980, 1339. | 33 |

| | | |
|---|---|---|
| 80JCS(P2)1733 | A. R. Katritzky, V. J. Baker, M. Camalli, R. Spagna and A. Vaciago; *J. Chem. Soc., Perkin Trans. 2*, 1980, 1733. | 539, 541 |
| 80JHC11 | J. T. Shaw, C. E. Brotherton, R. W. Moon, M. D. Winland, M. D. Anderson and K. S. Kyler; *J. Heterocycl. Chem.*, 1980, **17**, 11. | 177 |
| 80JHC45 | C. Bernasconi, L. Cottier, G. Descotes, M. F. Grenier and F. Mitras; *J. Heterocycl. Chem.*, 1980, **17**, 45. | 720 |
| 80JHC61 | L. Mosti, P. Schenone and G. Menozzi; *J. Heterocycl. Chem.*, 1980, **17**, 61. | 797 |
| 80JHC97 | L. Citerio, E. Rivera, M. L. Saccarello, R. Stradi and B. Gioia; *J. Heterocycl. Chem.*, 1980, **17**, 97. | 1014 |
| 80JHC235 | S. Brunel, C. Montginoul, E. Torreilles and L. Giral; *J. Heterocycl. Chem.*, 1980, **17**, 235. | 225 |
| 80JHC289 | K. Ichikawa, S. Inoue and K. Sato; *J. Heterocycl. Chem.*, 1980, **17**, 289. | 998 |
| 80JHC333 | H. Fischer and L. A. Summers; *J. Heterocycl. Chem.*, 1980, **17**, 333. | 513 |
| 80JHC389 | M. Balogh, I. Hermecz, Z. Mészáros, K. Simon, L. Pusztay, G. Horváth and P. Dvortsák; *J. Heterocycl. Chem.*, 1980, **17**, 389. | 228 |
| 80JHC455 | T. Kojima, F. Nagasaki and Y. Ohtsuka; *J. Heterocycl. Chem.*, 1980, **17**, 455. | 165 |
| 80JHC501 | J. L. Johnson, B. Whitney and L. M. Werbel; *J. Heterocycl. Chem.*, 1980, **17**, 501. | 555, 559 |
| 80JHC541 | I. El-Sayed El-Kholy, H. M. Fuid-Alla and M. M. Mishrikey; *J. Heterocycl. Chem.*, 1980, **17**, 541. | 45 |
| 80JHC597 | E. L. Anderson, A. Post, D. S. Staiger and R. Warren; *J. Heterocycl. Chem.*, 1980, **17**, 597. | 1012 |
| 80JHC617 | C. H. Stam, E. S. E. Ashry, Y. E. Kilany and H. C. van der Plas; *J. Heterocycl. Chem.*, 1980, **17**, 617. | 332, 352 |
| 80JHC733 | B. Stanovnik, M. Tišler, V. Golob, I. Hvala and O. Nikolič; *J. Heterocycl. Chem.*, 1980, **17**, 733. | 215, 216 |
| 80JHC793 | G. Liso, G. Trapani, V. Berardi, A. Latrofa and P. Marchini; *J. Heterocycl. Chem.*, 1980 **17**, 793. | 1032 |
| 80JHC817 | R. L. White, T. J. Schwan and R. J. Alaimo; *J. Heterocycl. Chem.*, 1980, **17**, 817. | 713 |
| 80JHC977 | K. Pilgram and R. D. Skiles; *J. Heterocycl. Chem.*, 1980, **17**, 977. | 1051, 1060, 1069 |
| 80JHC1121 | I. Lalezari and S. Nabahi; *J. Heterocycl. Chem.*, 1980, **17**, 1121. | 494, 502 |
| 80JHC1237 | J. Armand, K. Chekir and J. Pinson; *J. Heterocycl. Chem.*, 1980, **17**, 1237. | 252, 413 |
| 80JHC1377 | F. Camps, J. Coll, A. Messeguer and M. A. Pericás; *J. Heterocycl. Chem.*, 1980, **17**, 1377. | 593, 665, 748 |
| 80JHC1465 | M. J. Kornet and R. Daniels; *J. Heterocycl. Chem.*, 1980, **17**, 1465. | 39 |
| 80JHC1501 | G. Heinisch and I. Kirchner; *J. Heterocycl. Chem.*, 1980, **17**, 1501. | 30 |
| 80JHC1597 | F. D. Mills; *J. Heterocycl. Chem.*, 1980, **17**, 1597. | 687 |
| 80JHC1621 | K. L. Leschinsky and J. P. Chupp; *J. Heterocycl. Chem.*, 1980, **17**, 1621. | 438 |
| 80JHC1625 | E. Belgodere, R. Bossio, V. Parrini and R. Pepino; *J. Heterocycl. Chem.*, 1980, **17**, 1625. | 1009 |
| 80JHC1709 | W. T. Ashton and R. D. Brown; *J. Heterocycl. Chem.*, 1980, **17**, 1709. | 224, 260 |
| 80JHC1733 | Y. A. Ibrahim, M. M. Eid and S. A. L. Abdel-Hady; *J. Heterocycl. Chem.*, 1980, **17**, 1733. | 420 |
| 80JHC1785 | G. M. Coppola; *J. Heterocycl. Chem.*, 1980, **17**, 1785. | 1002 |
| 80JMC262 | T. H. Althuis, S. B. Kadin, L. J. Czuba, P. F. Moore and H.-J. Hess; *J. Med. Chem.*, 1980, **23**, 262. | 225, 260 |
| 80JMC320 | J. R. Piper and J. A. Montgomery; *J. Med. Chem.*, 1980, **23**, 320. | 302 |
| 80JMC327 | E. M. Grivsky, S. Lee, C. W. Sigel, D. S. Duch and C. A. Nichol; *J. Med. Chem.*, 1980, **23**, 327. | 229, 260 |
| 80JMC1248 | K. H. Kim, S. W. Dietrich, C. Hansch, B. J. Dolnick and J. R. Bertino; *J. Med. Chem.*, 1980, **23**, 1248. | 529 |
| 80JOC180 | R. P. Rooney and S. A. Evans, Jr.; *J. Org. Chem.*, 1980, **45**, 180. | 954 |
| 80JOC536 | N. P. Peet and S. Sunder; *J. Org. Chem.*, 1980, **45**, 536. | 1014 |
| 80JOC710 | S. W. Bass and S. A. Evans, Jr.; *J. Org. Chem.*, 1980, **45**, 710. | 953 |
| 80JOC721 | J. A. Kloek and K. L. Leschinsky; *J. Org. Chem.*, 1980, **45**, 721. | 1064 |
| 80JOC919 | G. Catelani, L. Monti and M. Ugazio; *J. Org. Chem.*, 1980, **45**, 919. | 729 |
| 80JOC1109 | P. D. Weeks, T. M. Brennan, D. P. Brannegan, D. E. Kuhla, M. L. Elliott, H. A. Watson, B. Wlodecki and R. Breitenbach; *J. Org. Chem.*, 1980, **45**, 1109. | 815 |
| 80JOC1662 | G. H. Denny, E. J. Cragoe, C. S. Rooney, J. P. Springer, J. M. Hirshfield and J. A. McCauley; *J. Org. Chem.*, 1980, **45**, 1662. | 1049 |
| 80JOC1693 | S. Fujii, M. Matsumoto and H. Kobatake; *J. Org. Chem.*, 1980, **45**, 1693. | 170 |
| 80JOC1918 | J. A. Bristol and R. G. Lovey; *J. Org. Chem.*, 1980, **45**, 1918. | 218 |
| 80JOC2283 | E. R. Bissell, A. R. Mitchell and R. E. Smith; *J. Org. Chem.*, 1980, **45**, 2283. | 800 |
| 80JOC2468 | M. Hori, T. Kataoka, H. Shimizu and S. Ohno; *J. Org. Chem.*, 1980, **45**, 2468. | 925, 939 |
| 80JOC2485 | E. C. Taylor and D. J. Dumas; *J. Org. Chem.*, 1980, **45**, 2485. | 181, 281 |
| 80JOC2601 | E. Vedejs, M. J. Arnost, J. M. Dolphin and J. Eustache; *J. Org. Chem.*, 1980, **45**, 2601. | 932 |
| 80JOC2604 | B. B. Jarvis and G. P. Stahly; *J. Org. Chem.*, 1980, **45**, 2604. | 1047, 1066, 1082 |
| 80JOC2970 | Y. Tamura, Y. Takebe, C. Mukai and M. Ikeda; *J. Org. Chem.*, 1980, **45**, 2970. | 914 |
| 80JOC2972 | Y. Tamura, C. Mukai, N. Nakajima and M. Ikeda; *J. Org. Chem.*, 1980, **45**, 2972. | 914 |
| 80JOC3361 | J. B. Hendrickson and J. S. Farina; *J. Org. Chem.*, 1980, **45**, 3361. | 722, 731 |
| 80JOC3634 | G. W. Gokel, H. M. Gerdes and D. M. Dishong; *J. Org. Chem.*, 1980, **45**, 3634. | 895 |
| 80JOC3677 | A. J. Elliott and M. S. Gibson; *J. Org. Chem.*, 1980, **45**, 3677. | 1072 |
| 80JOC3726 | K. K. Balasubramanian and S. Selvaraj; *J. Org. Chem.*, 1980, **45**, 3726. | 785 |

| | | |
|---|---|---|
| 80JOC3746 | A. Srinivasan and A. D. Broom; *J. Org. Chem.*, 1980, **45**, 3746. | 212, 213, 260 |
| 80JOC3766 | A. Dondoni, A. Battaglia and P. Giorgianni; *J. Org. Chem.*, 1980, **45**, 3766. | 1021 |
| 80JOC4352 | N. Chandrasekrar, K. Ramalingan, M. D. Herd and K. D. Berlin; *J. Org. Chem.*, 1980, **45**, 4352. | 580, 587 |
| 80JOC4584 | M. D. Bezoari and W. W. Paudler; *J. Org. Chem.*, 1980, **45**, 4584. | 19 |
| 80JOC4587 | T. Sasaki, K. Minamoto and K. Harada; *J. Org. Chem.*, 1980, **45**, 4587. | 414 |
| 80JOC4594 | T. Sasaki, K. Minamoto and K. Harada; *J. Org. Chem.*, 1980, **45**, 4594. | 413 |
| 80JOC4606 | C. F. Nutaitis, R. A. Schultz, J. Obaza and F. X. Smith; *J. Org. Chem.*, 1980, **45**, 4606. | 690 |
| 80JOC5160 | R. Aveta, G. Doddi, N. Insam and F. Stegel; *J. Org. Chem.*, 1980, **45**, 5160. | 653, 741 |
| 80JOC5333 | T. N. Wade and R. Khéribet; *J. Org. Chem.*, 1980, **45**, 5333. | 188 |
| 80JOC5350 | J. C. Dyer and S. A. Evans, Jr.; *J. Org. Chem.*, 1980, **45**, 5350. | 913 |
| 80JPR429 | H. Remane, R. Borsdorf, M. Scholz, E. Dölle and P. Loock; *J. Prakt. Chem.*, 1980, **332**, 429. | 626, 628 |
| 80JPR617 | A. M. Kadah and N. A. Shams; *J. Prakt. Chem.*, 1980, **322**, 617. | 37 |
| 80KGS411 | V. V. Dovlatyan and A. V. Dovlatyan; *Khim. Geterotsikl. Soedin.*, 1980, 411 (*Chem. Abstr.*, 1980, **93**, 94 549). | 480 |
| 80KGS541 | P. B. Terent'ev, V. G. Kartsev, A. N. Kost, V. F. Zhakharov, V. P. Zvolinskii and L. N. Novikova; *Khim. Geterotsikl. Soedin.*, 1980, 541 (*Chem. Abstr.*, 1980, **93**, 131 623). | 234, 235, 236 |
| 80KGS1130 | O. M. Polumbrik, I. G. Ryabokom and L. N. Markovskii; *Khim. Geterotsikl. Soedin.*, 1980, 1130. | 567, 571 |
| 80KGS1200 | A. A. Kost; *Khim. Geterotsikl. Soedin.*, 1980, 1200 (*Chem. Abstr.*, 1981, **94**, 65 393). | 228 |
| 80KGS1327 | V. G. Kartsev, A. M. Sipyagin, N. F. Sepetov and L. A. Sibeldina; *Khim. Geterotsikl. Soedin.*, 1980, 1327. | 1030 |
| 80LA50 | W. Bannwarth and W. Pfleiderer; *Liebigs Ann. Chem.*, 1980, 50. | 283 |
| 80LA65 | R. Charubala, W. Bannwarth and W. Pfleiderer; *Liebigs Ann. Chem.*, 1980, 65. | 283 |
| 80LA285 | H. Berneth, H. Hansen and S. Hünig; *Liebigs Ann. Chem.*, 1980, 285. | 370, 377 |
| 80LA590 | H. Heydt, K. H. Busch and M. Regitz; *Liebigs Ann. Chem.*, 1980, 590. | 51 |
| 80LA686 | R. Neidlein, S. Shatzmiller and E. Walter; *Liebigs Ann. Chem.*, 1980, 686. | 1042 |
| 80LA798 | G. Maier and U. Schäfer; *Liebigs Ann. Chem.*, 1980, 798. | 380, 421, 1003 |
| 80LA1307 | R. R. Schmidt and R. Scheibe; *Liebigs Ann. Chem.*, 1980, 1307. | 48 |
| 80LA1448 | W. Kantlehner, J. J. Kapassakalides and T. Maier; *Liebigs Ann. Chem.*, 1980, 1448. | 569 |
| 80M407 | D. Binder, C. R. Noe, J. Nussbaumer and B. C. Prager; *Monatsh. Chem.*, 1980, **111**, 407. | 252, 259 |
| 80MI21200 | A. Aydin and H. Feuer; *Chim. Acta Turc.*, 1980, **8**, 199. | 16 |
| 80MI21500 | F. Yoneda; *Methods Enzymol.*, 1980, **66**, 267 (*Chem. Abstr.*, 1980, **93**, 128 331). | 224, 231 |
| 80MI21501 | L. B. Hersh and C. Walsh; *Methods Enzymol.*, 1980, **66**, 277 (*Chem. Abstr.*, 1980, **93**, 90 689). | 260 |
| 80MI21502 | D. S. Duch, M. P. Edelstein and C. A. Nichol; *Mol. Pharmacol.*, 1980, **18**, 100 (*Chem. Abstr.*, 1980, **93**, 142 916). | 260 |
| 80MI21600 | S. J. Benkovic; *Annu. Rev. Biochem.*, 1980, **49**, 227. | 325 |
| 80MI21601 | A. Rosowsky, H. Lazarus, G. C. Yuan, W. R. Beltz, L. Mangini, H. T. Abelson, E. J. Modest and E. Frei, III; *Biochem. Pharmacol.*, 1980, **29**, 648. | 327 |
| 80MI21602 | H. Rokos, K. Rokos, H. Frisius and H. J. Kirstaedter; *Clin. Chim. Acta*, 1980, **105**, 275. | 324 |
| 80MI21603 | A. Niederwieser, H. C. Curtius, R. Gitzelmann, A. Otten, K. Baerlocher, B. Blehova, S. Berlow, H. Gröbe, F. Prey, J. Schaub, S. Scheibenreiter, H. Schmidt and M. Viscontini; *Helv. Paediat. Acta*, 1980, **35**, 335. | 324 |
| 80MI21604 | T. Mazda, M. Tsusue and S. Sakate; *Insect Biochem.*, 1980, **10**, 357. | 294 |
| 80MI21605 | S. Matsuura, T. Sugimoto, H. Hasegawa, S. Imaizumi and A. Ichiyama; *J. Biochem.*, 1980, **87**, 951. | 281 |
| 80MI21900 | A. Messmer, Gy. Hajós, P. Benkó and L. Pallos; *Magy. Kem. Foly.*, 1980, **86**, 466. | 440 |
| B-80MI22000 | G. L. Drake, Jr; in 'Kirk–Othmer Encyclopedia of Chemical Technology', ed. M. Grayson; Wiley, New York, 1980, vol. 10, p. 420. | 525 |
| B-80MI22001 | W. R. Griffin; in 'Kirk–Othmer Encyclopedia of Chemical Technology', ed. M. Grayson; Wiley, New York, 1980, vol. 10, p. 948. | 527 |
| B-80MI22002 | V. Lindner; in 'Kirk–Othmer Encyclopedia of Chemical Technology', ed. M. Grayson; Wiley, New York, 1980, vol. 11, p. 581. | 530 |
| 80MI22100 | L. S. Podenko, A. K. Chirkov and V. P. Shchipanov; *Zh. Strukt. Khim.*, 1980, **21**, 187 (*Chem. Abstr.*, 1981, **94**, 3657). | 540 |
| 80MI22200 | G. Jones, II, W. R. Jackson and A. M. Halpern; *Chem. Phys. Lett.*, 1980, **72**, 391. | 600 |
| B-80MI22201 | T. J. Mabry and A. Ulubelen; in 'Biochemical Applications of Mass Spectrometry', ed. G. R. Waller and O. C. Dermer; Wiley, New York, 1980, 1st suppl. vol., p. 1131. | 603 |
| 80MI22202 | M. Rossi, J. S. Cantrell, A. J. Faber, T. Dyott, H. L. Carrell and J. P. Glusker; *Cancer Res.*, 1980, **40**, 2774. | 623 |
| B-80MI22203 | F. G. Riddell; 'The Conformational Analysis of Heterocyclic Compounds', Academic, New York, 1980. | 628 |
| 80MI22300 | F. Pragst, M. Janda and I. Stibor; *Electrochim. Acta*, 1980, **25**, 779. | 661 |
| B-80MI22301 | G. P. Ellis; in 'Heterocyclic Chemistry', ed. H. Suschitzky and O. Meth-Cohn; Royal Society of Chemistry, London, 1980, vol. 1, p. 331. | 649 |
| 80MI22302 | A. Z. Abyshev; *Khim. Prir. Soedin.*, 1980, 165 (*Chem. Abstr.*, 1980, **93**, 150 082). | 678, 681, 718 |

| | | |
|---|---|---|
| 80MI22400 | W. K. Franke and J. Küther; *Fette, Seifen. Anstrichm.*, 1980, **82**, 82. | 750 |
| 80MI22401 | M. L. Mihailović, D. Marinković, N. Orbović, S. Gojković and S. Konstantinović; *Bull. Soc. Chim. Beograd.*, 1980, **45**, 497. | 776 |
| 80MI22402 | J. H. Pardanani and S. Sethna; *J. Inst. Chem., Calcutta*, 1980, **52**, 229. | 817 |
| 80MI22403 | R. E. Smith, E. R. Bissell, A. R. Mitchell and K. W. Pearson; *Thromb. Res.*, 1980, **17**, 393. | 879 |
| B-80MI22404 | P. F. Schuda; *Top. Curr. Chem.*, 1980, **91**, 75. | 882 |
| 80MI22500 | K. J. Miller, M. Lauer and S. Archer; *Int. J. Quantum Chem., Quantum Biol. Symp.*, 1980, **7**, 11 (*Chem. Abstr.*, 1981, **94**, 150 541). | 942 |
| B-80MI22600 | F. G. Riddell; 'The Conformational Analysis of Heterocyclic Compounds', Academic Press, London, 1980, p. 1. | 948, 960 |
| B-80MI22601 | F. G. Riddell; 'The Conformational Analysis of Heterocyclic Compounds', Academic Press, London, 1980, p. 66. | 961, 962 |
| B-80MI22602 | F. G. Riddell; 'The Conformational Analysis of Heterocyclic Compounds', Academic Press, London, 1980, p. 104. | 961, 962 |
| 80MI22603 | F. P. Colonna, G. Distefano and V. Galasso; *J. Electron Spectrosc. Relat. Phenom.*, 1980, **18**, 75. | 945, 958 |
| 80MI22700 | S. M. Nikitin, A. L. Zhuze, A. S. Krylov and B. P. Gottikh; *Bioorg. Khim.*, 1980, **6**, 743 (*Chem. Abstr.*, 1980, **93**, 150 188). | 1038 |
| B-80MI22701 | P. G. Sammes; 'The Chemistry and Antimicrobial Activity of the New Synthetic Antibiotics', Ellis Horwood, Chichester, 1980, vol. 4.     1020, 1025, 1026, 1038 | |
| B-80MI22702 | K. Morita, H. Nomura, M. Numata, M. Ochiai and M. Yoneda; *Philos. Trans. R. Soc. London, Ser. B*, 1980, **289**, 181. | 1038 |
| 80MI22800 | E. Myers, M. C. Allwood, M. J. Gidley and J. K. M. Sanders; *J. Appl. Bacteriol.*, 1980, **48**, 89. | 1085 |
| 80MI22801 | K. Burger, S. Penninger, M. Greisel and E. Daltrozzo; *J. Fluorine Chem.*, 1980, **15**, 1. | 1064 |
| 80MI22802 | A. R. Katritzky, R. C. Patel, F. M. S. Brito-Palma, F. G. Riddell and E. S. Turner; *Isr. J. Chem.*, 1980, **20**, 150. | 1054 |
| 80N610 | H. Wachter, A. Hausen, E. Reider and M. Schweiger; *Naturwissenschaften*, 1980, **67**, 610. | 324 |
| 80NKK604 | Y. Suzuki; *Nippon Kagaku Kaishi*, 1980, 604. | 923 |
| 80OMR(13)172 | S. Matsuo, S. Matsumoto, T. Kurihara, Y. Akita, T. Watanabe and A. Ohta; *Org. Magn. Reson.*, 1980, **13**, 172. | 160 |
| 80OMR(13)244 | T. Imagawa, A. Haneda and M. Kawanisi; *Org. Magn. Reson.*, 1980, **13**, 244. | 588, 589 |
| 80OMR(13)363 | R. D. Chambers, R. S. Matthews, W. K. R. Musgrave and P. G. Urben; *Org. Magn. Reson.*, 1980, **13**, 363. | 395 |
| 80OMR(14)305 | M. Witanowski, L. Stefaniak, B. Kamiénski and G. A. Webb; *Org. Magn. Reson.*, 1980, **14**, 305. | 395 |
| 80OMS31 | D. L. McGillivray and G. A. Poulton; *Org. Mass Spectrom.*, 1980, **15**, 31. | 612, 613 |
| 80OPP219 | S. Nakanishi and S. S. Massett; *Org. Prep. Proced. Int.*, 1980, **12**, 219 (*Chem. Abstr.*, 1980, **93**, 204 581). | 225 |
| 80OPP265 | M. Nakajima, C. A. Loeschorn, W. E. Ambrelo and J. P. Anselme; *Org. Prep. Proced. Int.*, 1980, **12**, 265. | 185 |
| 80OR(26)1 | H. W. Gschwend and H. R. Rodriguez; *Org. React.*, 1980, **26**, 1. | 803 |
| 80P2443 | H. Homberg and H. Geiger; *Phytochemistry*, 1980, **19**, 2443. | 603 |
| 80RRC1505 | A. Dinculescu and A. T. Balaban; *Rev. Roum. Chim.*, 1980, **25**, 1505. | 655 |
| 80S112 | V. I. Gorbatenko and L. F. Lur'e; *Synthesis*, 1980, 112. | 1057 |
| 80S129 | S. Polanc, B. Stanovnik and M. Tišler; *Synthesis*, 1980, 129. | 37 |
| 80S153 | H. C. Brown and A. K. Mandal; *Synthesis*, 1980, 153. | 976 |
| 80S221 | G. A. Olah, A. P. Fung, B. G. B. Gupta and S. C. Narang; *Synthesis*, 1980, 221. | 486 |
| 80S457 | A. K. Fateen, A. H. Moustafa, A. M. Kaddah and N. A. Shams; *Synthesis*, 1980, 457. | 22 |
| 80S479 | K. Senga, K. Furukawa and S. Nishigaki; *Synthesis*, 1980, 479. | 220 |
| 80S623 | C. Deshayes and S. Gelin; *Synthesis*, 1980, 623. | 41 |
| 80S841 | R. L. N. Harris; *Synthesis*, 1980, 841. | 495, 496 |
| 80S871 | M. Delmas and A. Gaset; *Synthesis*, 1980, 871. | 989 |
| 80SC195 | H. G. Rajoharison, H. Soltani, M. Arnaud, C. Roussel and J. Metzger; *Synth. Commun.*, 1980, **10**, 195. | 657 |
| 80SC889 | Y. Machida, S. Nomoto, S. Negi, H. Ikuta and I. Saito; *Synth. Commun.*, 1980, **10**, 889. | 709 |
| 80T409 | G. Bettoni, C. Franchini, R. Perrone and V. Tortorella; *Tetrahedron*, 1980, **36**, 409. | 1037 |
| 80T679 | A. R. Katritzky; *Tetrahedron*, 1980, **36**, 679.     656, 861, 869, 872, 883 | |
| 80T935 | R. Gandolfi and L. Toma; *Tetrahedron*, 1980, **36**, 935. | 453 |
| 80T1643 | A. R. Katritzky, R. T. C. Brownlee and G. Musumarra; *Tetrahedron*, 1980, **36**, 1643. | 655 |
| 80T2497 | J. L. Ripoll, H. Lebrun and A. Thullier; *Tetrahedron*, 1980, **36**, 2497. | 507 |
| 80T3273 | H. Hambloch and A. W. Frahm; *Tetrahedron*, 1980, **36**, 3273. | 590 |
| 80T3409 | B. Begasse and M. Le Corre; *Tetrahedron*, 1980, **36**, 3409. | 853 |
| 80T3565 | D. Giardinà, R. Ballini, G. M. Cingolani, C. Melchiorre, B. R. Pietroni, A. Carotti and G. Casini; *Tetrahedron*, 1980, **36**, 3565. | 629, 774 |
| 80TL7 | M. L. Maddox, J. C. Martin and J. M. Muchowski; *Tetrahedron Lett.*, 1980, 7. | 424 |

| | | |
|---|---|---|
| 80TL277 | S. S. Yemul, H. B. Kagan and R. Setton; *Tetrahedron Lett.*, 1980, **21**, 277. | 70, 152 |
| 80TL533 | Y. Tamura, S. M. Bayomi, C. Mukai and M. Ikeda; *Tetrahedron Lett.*, 1980, **21**, 533. | 905, 906 |
| 80TL595 | U. Göckel, U. Hartmannsgruber, A. Steigel and J. Sauer; *Tetrahedron Lett.*, 1980, 595. | 423 |
| 80TL1009 | A. Padwa and H. Ku; *Tetrahedron Lett.*, 1980, 1009. | 42 |
| 80TL1197 | M. Koreeda and H. Akagi; *Tetrahedron Lett.*, 1980, 1197. | 812 |
| 80TL1575 | M. Poje; *Tetrahedron Lett.*, 1980, 1575. | 798, 815 |
| 80TL1893 | K. Venkataraman and D. R. Wagle; *Tetrahedron Lett.*, 1980, **21**, 1893. | 485 |
| 80TL2361 | F. Camps, J. Coll, A. Messeguer and M. A. Pericás; *Tetrahedron Lett.*, 1980, 2361. | 755 |
| 80TL2773 | T. A. Morgan and B. Ganem; *Tetrahedron Lett.*, 1980, 2773. | 812 |
| 80TL2939 | H. J. Bestmann, G. Schmid and D. Sandmeier; *Tetrahedron Lett.*, 1980, 2939. | 42 |
| 80TL3743 | A. G. Anastassiou, H. S. Kasmai and M. R. Saadein; *Tetrahedron Lett.*, 1980, **21**, 3743. | 920 |
| 80TL4731 | K. K. Balasubramanian, G. V. Bindumadhavan, M. R. Udupa and B. Krebs; *Tetrahedron Lett.*, 1980, **21**, 4731. | 480 |
| 80TL4973 | P. Cacioli, M. F. Mackay and J. A. Reiss; *Tetrahedron Lett.*, 1980, 4973. | 622, 784 |
| 80UP21900 | H. Neunhoeffer and G. Köhler; unpublished results, 1980. | 436 |
| 80USP4182623 | Monsanto Corp., *U.S. Pat.* 4 182 623 (1980) (*Chem. Abstr.*, 1980, **92**, 198 431). | 941 |
| 80USP4215216 | Am. Home Prods. Corp., *U.S. Pat.* 4 215 216 (1980) (*Derwent Farmdoc* 58 723C). | 227 |
| 80USP4218559 | R. A. Behrens and D. R. Maulding; *U.S. Pat.* 4 218 559 (1980) (*Chem. Abstr.*, 1980, **93**, 240 991). | 368 |
| 80USP4223142 | E. R. Squibb and Sons, Inc., *U.S. Pat.* 4 223 142 (1980) (*Derwent Farmdoc* 71 515C). | 243 |
| 80YZ1187 | J. Matsumoto, A. Minamida and S. Minami; *Yakugaku Zasshi*, 1980, **100**, 1187. | 206 |
| 80ZC19 | H. Voigt; *Z. Chem.*, 1980, **20**, 19. | 1015 |
| 80ZN(B)485 | E. A. A. Hafez, M. A. E. Khalifa, S. K. A. Guda and M. H. Elnagdi; *Z. Naturforsch., Teil B*, 1980, **35**, 485. | 442 |
| 80ZOB940 | L. S. Shibryaeva, A. I. Mikaya and V. G. Zaikin; *Zh. Obshch. Khim.*, 1980, **50**, 940. | 610 |
| 80ZOR463 | L. N. Kryukov, L. Yu. Kryukova, A. F. Kolomiets and G. A. Sokol'skii; *Zh. Org. Khim.*, 1980, **16**, 463. | 1044, 1078 |
| 80ZOR1308 | P. P. Kornuta, L. I. Derii and L. N. Markovskii; *Zh. Org. Khim.*, 1980, **16**, 1308. | 1069, 1075 |
| 80ZOR1952 | A. A. Potekhin, I. G. Zenkevich, V. V. Sokolov and S. M. Shevchenko; *Zh. Org. Khim.*, 1980, **16**, 1952. | 1074 |
| 80ZOR2297 | P. S. Lobanov, A. N. Poltorak and A. A. Potekhun; *Zh. Org. Khim.*, 1980, **16**, 2297. | 391 |
| | | |
| 81ACH(108)167 | K. Kormendy, T. Kovacz, J. Ruff and I. Kovesdi; *Acta Chim. Acad. Sci. Hung.*, 1981, **108**, 167. | 240, 242, 243 |
| 81AG(E)444 | R. Gleiter; *Angew. Chem., Int. Ed. Engl.*, 1981, **20**, 444. | 1052 |
| 81AHC(28)1 | R. D. Chambers and C. R. Sargent; *Adv. Heterocycl. Chem.*, 1981, **28**, 1. | 483, 484 |
| 81AHC(29)251 | A. Gasco and A. J. Boulton; *Adv. Heterocycl. Chem.*, 1981, **29**, 251. | 184, 1044, 1045 |
| 81AJC623 | R. L. N. Harris; *Aust. J. Chem.*, 1981, **34**, 623. | 462, 495, 496, 497 |
| 81AJC1157 | D. J. Brown and L. Strekowski; *Aust. J. Chem.*, 1981, **34**, 1157. | 93 |
| 81AJC1353 | D. J. Brown, W. B. Cowden and L. Strękowski; *Aust. J. Chem.*, 1981, **34**, 1353. | 114 |
| 81AJC1361 | G. B. Barlin; *Aust. J. Chem.*, 1981, **34**, 1361. | 337, 359 |
| 81AJC1539 | W. B. Cowden and P. Waring; *Aust. J. Chem.*, 1981, **34**, 1539. | 64 |
| 81AJC1729 | B. Stanovnik, M. Tišler, A. Hribar, G. B. Barlin and D. J. Brown; *Aust. J. Chem.*, 1981, **34**, 1729. | 94, 136 |
| 81AJC2629 | A. Kowalewski, L. Strękowski, M. Szajda, K. Walenciak and D. J. Brown; *Aust. J. Chem.*, 1981, **34**, 2629. | 103 |
| 81AJC2647 | A. J. Liepa; *Aust. J. Chem.*, 1981, **34**, 2647. | 756, 787, 865 |
| 81AP94 | G. Zinner, M. Heitmann and R. Vollrath; *Arch. Pharm. (Weinheim, Ger.)*, 1981, **314**, 94. | 570 |
| 81AP376 | G. Seitz and W. Overheu; *Arch. Pharm. (Weinheim, Ger.)*, 1981, **314**, 376. | 442, 554 |
| 81AP577 | S. Prior and W. Wiegrebe; *Arch. Pharm. (Weinheim, Ger.)*, 1981, **314**, 577. | 788, 860 |
| 81AX(B)1620 | P. Cacioli, M. F. Mackay and J. A. Reiss; *Acta Crystallogr., Part B*, 1981, **37**, 1620. | 622 |
| 81B1241 | G. K. Smith, W. T. Mueller, P. A. Benkovic and S. J. Benkovic; *Biochemistry*, 1981, **20**, 1241. | 260 |
| 81BBR(101)1259 | J. M. Whitely and A. Russel; *Biochem. Biophys. Res. Commun.*, 1981, **101**, 1259. | 308 |
| 81BCJ41 | H. Ayato, I. Tanaka, T. Yamane, T. Ashida, T. Sasaki, K. Minamoto and K. Harada; *Bull. Chem. Soc. Jpn.*, 1981, **54**, 41. | 387 |
| 81BCJ217 | T. Shimizu, Y. Hayashi, K. Yamada, T. Nishio and K. Teramura; *Bull. Chem. Soc. Jpn.*, 1981, **54**, 217. | 674 |
| 81BCJ2543 | S. Matsuura and T. Sugimoto; *Bull. Chem. Soc. Jpn.*, 1981, **54**, 2543. | 281 |
| 81CB147 | F. Bohlmann and E. Vorwerk; *Chem. Ber.*, 1981, **114**, 147. | 852, 853 |
| 81CB564 | H. Stetter and F. Jonas; *Chem. Ber.*, 1981, **114**, 564. | 45 |
| 81CB2075 | R. Bunnenberg and J. C. Jochims; *Chem. Ber.*, 1981, **114**, 2075. | 463, 490 |
| 81CB3154 | L. Birkofer and E. Hänsel; *Chem. Ber.*, 1981, **114**, 3154. | 50 |
| 81CC36 | R. A. Abramovitch and C. Dupuy; *J. Chem. Soc., Chem. Commun.*, 1981, 36. | 997 |
| 81CC282 | H. Takeno and M. Hashimoto; *J. Chem. Soc., Chem. Commun.*, 1981, 282. | 825 |
| 81CC474 | H. Takeno, M. Hashimoto, Y. Koma, H. Horiai and H. Kikuchi; *J. Chem. Soc., Chem. Commun.*, 1981, 474. | 825 |

| | | |
|---|---|---|
| 81CC534 | T. A. Chorn, R. G. F. Giles, P. R. K. Mitchell and I. R. Green; *J. Chem. Soc.,* *Chem. Commun.,* 1981, 534. | 770 |
| 81CC581 | M. Miura, M. Nojima and S. Kusabayashi; *J. Chem. Soc., Chem. Commun.,* 1981, 581. | 990 |
| 81CC1003 | A. E. Baydar, G. V. Boyd and P. F. Lindley; *J. Chem. Soc., Chem. Commun.,* 1981, 1003. | 1050, 1062, 1063, 1072 |
| 81CC1174 | A. Ohsawa, H. Arai, H. Ohnishi and H. Igeta; *J. Chem. Soc., Chem. Commun.,* 1981, 1174. | 369, 371, 372, 373, 381 |
| 81CCC748 | J. Kuthan, J. Paleček and J. Valihrach; *Collect. Czech. Chem. Commun.,* 1981, **46**, 748. | 576, 740 |
| 81CHE115 | V. I. Yakovenko, E. T. Oganesyan and G. N. Dorofeenko; *Chem. Heterocycl. Compd. (Engl. Transl.),* 1981, 115. | 829, 865 |
| 81CHE221 | I. V. Shcherbakova, G. N. Dorofeenko and E. V. Kuznetsov; *Chem. Heterocycl. Compd. (Engl. Transl.),* 1981, 221. | 863 |
| 81CI(L)876 | A. Arcoleo, M. G. Cicero, G. Giammona and G. Fontana; *Chem. Ind. (London),* 1981, 876. | 811 |
| 81CJC2223 | M. S. Chauhan and D. M. McKinnon; *Can. J. Chem.,* 1981, **59**, 2223. | 785 |
| 81CJC2755 | R. Stewart, R. Srinivasan and S. J. Gumbley; *Can. J. Chem.,* 1981, **59**, 2755. | 202, 204, 206, 207 |
| 81CPB249 | M. Somei, Y. Karasawa, T. Shoda and C. Kaneko; *Chem. Pharm. Bull.,* 1981, **29**, 249. | 685, 686, 691 |
| 81CPB1156 | K. Ikeda, Y. Terao and M. Sekiya; *Chem. Pharm. Bull.,* 1981, **29**, 1156. | 487 |
| 81CS(18)256 | A. R. Katritzky, J. M. Lloyd and R. C. Patel; *Chem. Scr.,* 1981, **18**, 256. | 585, 592 |
| 81FOR(40)105 | W. Heller and Ch. Tamm; *Fortschr. Chem. Org. Naturst.,* 1981, **40**, 105. | 722 |
| 81G211 | A. Gottsegen, S. Antus, P. Kolonits, M. Nógrádi, A. Lupi, G. D. Monache, M. Marta and G. B. M. 1Bettolo; *Gazz. Chim. Ital.,* 1981, **111**, 211. | 744 |
| 81H(15)293 | M. Kočevar, B. Stanovnik and M. Tišler; *Heterocycles,* 1981, **15**, 293. | 282 |
| 81H(15)437 | E. Uhlmann and W. Pfleiderer; *Heterocycles,* 1981, **15**, 437. | 277 |
| 81H(15)455 | P. Canonne, D. Bélanger and G. Lemay; *Heterocycles,* 1981, **15**, 455. | 785 |
| 81H(15)757 | S. Nishigaki, K. Fukami, H. Ichiba, H. Kanazawa, K. Matsuyama, S. Ogusu, K. Senga, F. Yoneda, R. Koga and T. Ueno; *Heterocycles,* 1981, **15**, 757. | 314 |
| 81H(15)851 | Y. Yamamoto, Y. Azuma and S. Ohnishi; *Heterocycles,* 1981, **15**, 851. | 1022 |
| 81H(15)895 | Y. Maki, M. Suzuki, K. Kameyama, M. Kawai, M. Suzuki and M. Sako; *Heterocycles,* 1981, **15**, 895. | 319 |
| 81H(16)9 | B. J. Graves, D. J. Hodgson, S.-F. Chen and R. P. Panzica; *Heterocycles,* 1981, **16**, 9. | 3 |
| 81H(16)17 | M. E. Garst and J. D. Frazier; *Heterocycles,* 1981, **16**, 17. | 658 |
| 81H(16)955 | M. Tsukayama, T. Sakamoto, T. Horie, M. Masumara and M. Nakayama; *Heterocycles,* 1981, **16**, 955. | 755 |
| 81H(16)1899 | L. Chiodini, M. Di Ciommo and L. Merlini; *Heterocycles,* 1981, **16**, 1899. | 726 |
| 81HC(36) | G. P. Ellis and I. M. Lockhart; *Chem. Heterocycl. Compd.,* 1981, **36**. | 738, 778 |
| 81HC(36)1 | G. P. Ellis and I. M. Lockhart; *Chem. Heterocycl. Compd.,* 1981, **36**, 1. | 574, 594 |
| 81HC(36)7 | R. Livingstone; *Chem. Heterocycl. Compd.,* 1981, **36**, 7. | 780 |
| 81HC(36)59 | R. M. Parkhurst and W. A. Skinner; *Chem. Heterocycl. Compd.,* 1981, **36**, 59. | 724, 734, 780 |
| 81HC(36)63 | R. M. Parkhurst and W. A. Skinner; *Chem. Heterocycl. Compd.,* 1981, **36**, 63. | 622 |
| 81HC(36)66 | R. M. Parkhurst and W. A. Skinner; *Chem. Heterocycl. Compd.,* 1981, **36**, 66. | 595 |
| 81HC(36)161 | R. Livingstone; *Chem. Heterocycl. Compd.,* 1977, **36**, 161. | 736 |
| 81HC(36)169 | R. Livingstone; *Chem. Heterocycl. Compd.,* 1981, **36**, 169. | 580, 630 |
| 81HC(36)189 | I. M. Lockhart; *Chem. Heterocycl. Compd.,* 1977, **36**, 189. | 734 |
| 81HC(36)251 | I. N. Lockhart; *Chem. Heterocycl. Compd.,* 1977, **36**, 251. | 782 |
| 81HCA367 | A. N. Ganguly, J. H. Bieri and M. Viscontini; *Helv. Chim. Acta,* 1981, **64**, 367. | 281 |
| 81HCA1247 | C. Fehr, J. Galindo and G. Ohloff; *Helv. Chim. Acta,* 1981, **64**, 1247. | 594, 597, 843, 847 |
| 81IC426 | M. A. Mathur and H. H. Sisler; *Inorg. Chem.,* 1981, **20**, 426. | 568 |
| 81IJC(B)918 | V. K. Ahluwalia, I. Mukherjee and N. Rani; *Indian J. Chem., Sect. B,* 1981, **20**, 918. | 800 |
| 81JA486 | A. Krantz and J. Laureni; *J. Am. Chem. Soc.,* 1981, **103**, 486. | 982 |
| 81JA4597 | S. Ohuchida, N. Hamanaka and M. Hayashi; *J. Am. Chem. Soc.,* 1981, **103**, 4597. | 907, 930 |
| 81JA5494 | S. Ball and T. C. Bruice; *J. Am. Chem. Soc.,* 1981, **103**, 5494. | 262 |
| 81JA6856 | E. C. Taylor, J. G. Andrade, G. J. H. Rall, I. J. Turchi, K. Steliou, G. E. Jagdmann, Jr. and A. McKillop; *J. Am. Chem. Soc.,* 1981, **103**, 6856. | 848 |
| 81JA6916 | G. J. Woolfe and P. J. Thistlethwaite; *J. Am. Chem. Soc.,* 1981, **103**, 6916. | 603 |
| 81JA6967 | K. C. Nicolau, D. P. Papahatjis, D. A. Claremon and R. E. Dolle, III; *J. Am. Chem. Soc.,* 1981, **103**, 6967. | 775 |
| 81JAP(K)8199480 | Mitsui Petroleum Ind., *Jpn. Kokai* 81 99 480 (1981) (*Derwent Farmdoc* 70 585D). | 221 |
| 81JAP(K)81125386 | Hamari Yakukin KK, *Jpn. Kokai* 81 125 386 (1981) (*Derwent Farmdoc* 84 340D). | 221 |
| 81JAP(K)81152471 | Kakenyaku Kako Co. Ltd., *Jpn. Kokai* 81 152 471 (1981) (*Chem. Abstr.,* 1982, **96**, 122 636). | 804 |
| 81JBC(256)2963 | K. Tanaka, M. Akino, Y. Hagi, M. Doi and T. Shiota; *J. Biol. Chem.,* 1981, **256**, 2963. | 320 |
| 81JBC(256)10399 | G. J. Wiederrecht, D. R. Paton and G. M. Brown; *J. Biol. Chem.,* 1981, **256**, 10399. | 320 |
| 81JCR(M)1648 | M. P. Sammes; *J. Chem. Res. (M),* 1981, 1648. | 593 |
| 81JCR(S)200 | M. P. Sammes; *J. Chem. Res. (S),* 1981, 200. | 585 |

| 81JCS(F2)79 | J. N. Macdonald, S. A. Mackay, J. K. Tyler, A. P. Cox and I. C. Ewart; *J. Chem. Soc., Faraday Trans. 2*, 1981, **77**, 79. | 626, 627, 639 |
|---|---|---|
| 81JCS(P1)73 | K. Matsumoto and T. Uchida; *J. Chem. Soc., Perkin Trans. 1*, 1981, 73. | 31 |
| 81JCS(P1)212 | Y. Tamura, C. Mukai, N. Nakajima, M. Ikeda and M. Kido; *J. Chem. Soc., Perkin Trans. 1*, 1981, 212. | 915 |
| 81JCS(P1)331 | C. V. Greco and K. J. Gala; *J. Chem. Soc., Perkin Trans. 1*, 1981, 331. | 489, 502 |
| 81JCS(P1)770 | J. R. Lewis and J. G. Paul; *J. Chem. Soc., Perkin Trans. 1*, 1981, 770. | 838 |
| 81JCS(P1)1015 | K. Sato, S. Inoue and T. Tanami; *J. Chem. Soc., Perkin Trans. 1*, 1981, 1015. | 843 |
| 81JCS(P1)1071 | R. D. Chambers, W. K. R. Musgrave and C. R. Sargent; *J. Chem. Soc., Perkin Trans. 1*, 1981, 1071. | 9 |
| 81JCS(P1)1096 | R. E. Banks, J. A. Miller, M. J. Nunn, P. Stanley, T. K. J. Weakley and Z. Ullah; *J. Chem. Soc., Perkin Trans. 1*, 1981, 1096. | 624, 842 |
| 81JCS(P1)1437 | F. M. Dean, D. A. Matkin and M. O. A. Orabi; *J. Chem. Soc., Perkin Trans. 1*, 1981, 1437. | 782 |
| 81JCS(P1)1685 | J. Barber and J. Staunton; *J. Chem. Soc., Perkin Trans. 1*, 1981, 1685. | 857 |
| 81JCS(P1)1697 | V. K. Ahluwalia, C. Prakash and R. S. Jolly; *J. Chem. Soc., Perkin Trans. 1*, 1981, 1697. | 745, 805 |
| 81JCS(P1)1891 | J. Barluenga, J. F. López-Ortiz, M. Tomás and V. Gotor; *J. Chem. Soc., Perkin Trans. 1*, 1981, 1891. | 1059, 1065, 1074 |
| 81JCS(P1)1953 | S. Mageswaran, W. D. Ollis and I. O. Sutherland; *J. Chem. Soc., Perkin Trans. 1*, 1982, 1953. | 904 |
| 81JCS(P1)2245 | L. Bruché, L. Garanti and G. Zecchi; *J. Chem. Soc., Perkin Trans. 1*, 1981, 2245. | 1044, 1068, 1073 |
| 81JCS(P1)2322 | B. F. Bonini, G. Maccagnani, G. Mazzanti, G. Rosini and E. Foresti; *J. Chem. Soc., Perkin Trans. 1*, 1981, 2322. | 1043, 1077 |
| 81JCS(P1)2509 | O. Meth-Cohn, B. Narine, B. Tarnowski, R. Hayes, A. Keyzad, S. Rhouati and A. Robinson; *J. Chem. Soc., Perkin Trans. 1*, 1981, 2509. | 247 |
| 81JCS(P1)2552 | P. S. Bevan, G. P. Ellis and H. K. Wilson; *J. Chem. Soc., Perkin Trans. 1*, 1981, 2552. | 711 |
| 81JCS(P1)2557 | G. P. Ellis and J. M. Williams; *J. Chem. Soc., Perkin Trans. 1*, 1981, 2557. | 590 |
| 81JCS(P1)3111 | R. O. Cain and A. E. A. Porter; *J. Chem. Soc., Perkin Trans. 1*, 1981, 3111. | 173, 175 |
| 81JCS(P2)303 | M. P. Sammes, R. L. Harlow and S. H. Simonsen; *J. Chem. Soc., Perkin Trans. 2*, 1981, 303. | 578, 585, 621 |
| 81JCS(P2)741 | J. E. Anderson and F. S. Jorgensen; *J. Chem. Soc., Perkin Trans. 2*, 1981, 741. | 630 |
| 81JCS(P2)944 | K. Hanaya, S. Onodera, Y. Ikegani, H. Kudo and K. Shimaya; *J. Chem. Soc., Perkin Trans. 2*, 1981, 944. | 594, 630 |
| 81JCS(P2)1443 | R. Schipfer, O. S. Wolfbeis and A. Knierzinger; *J. Chem. Soc., Perkin Trans. 2*, 1981, 1443. | 603 |
| 81JHC75 | J. T. Shaw, C. E. Brotherton, R. W. Moon, T. W. Coffindaffer and D. A. Miller; *J. Heterocycl. Chem.*, 1981, **18**, 75. | 489 |
| 81JHC105 | I. E. El-Kholy, M. M. Mishrikey and H. M. Feid-Allah; *J. Heterocycl. Chem.*, 1981, **18**, 105. | 685 |
| 81JHC123 | A. Counotte-Potman and H. C. van der Plas; *J. Heterocycl. Chem.*, 1981, **18**, 123. | 545 |
| 81JHC255 | I. Heresch, G. Allmaier and G. Heinisch; *J. Heterocycl. Chem.*, 1981, **18**, 255. | 8 |
| 81JHC279 | G. Liso, G. Trapani, A. Latrofa and P. Marchini; *J. Heterocycl. Chem.*, 1981, **18**, 279. | 1032 |
| 81JHC309 | C. Seoane, J. L. Soto, P. Zamorano and M. Quinteiro; *J. Heterocycl. Chem.*, 1981, **18**, 309. | 672 |
| 81JHC327 | G. Rovnyak, V. Shu and J. Schwartz; *J. Heterocycl. Chem.*, 1981, **18**, 327. | 228 |
| 81JHC363 | H. N. Abramson and H. C. Wormser; *J. Heterocycl. Chem.*, 1981, **18**, 363. | 799, 841 |
| 81JHC443 | S. Veeraragharan, D. Bhattacharjee and F. D. Popp; *J. Heterocycl. Chem.*, 1981, **18**, 443. | 23 |
| 81JHC495 | G. M. Coppola and M. J. Shapiro; *J. Heterocycl. Chem.*, 1981, **18**, 495. | 202, 228 |
| 81JHC543 | B. Nedjar-Kolli, M. Hamdi, J. Périé and V. Hérault; *J. Heterocycl. Chem.*, 1981, **18**, 543. | 726, 843 |
| 81JHC549 | A. W. Frank; *J. Heterocycl. Chem.*, 1981, **18**, 549. | 774 |
| 81JHC619 | G. Haas, J. L. Stanton and T. Winkler; *J. Heterocycl. Chem.*, 1981, **18**, 619. | 703 |
| 81JHC631 | Y. Nakayama, Y. Sanemitsu, M. Mizutani and H. Yoshioka; *J. Heterocycl. Chem.*, 1981, **18**, 631. | 414 |
| 81JHC671 | J. L. Kelley and E. W. McLean; *J. Heterocycl. Chem.*, 1981, **18**, 671. | 210, 212, 214, 260 |
| 81JHC679 | J. R. Bantick and J. L. Suschitzky; *J. Heterocycl. Chem.*, 1981, **18**, 679. | 717 |
| 81JHC697 | D. T. Connor, P. A. Young and M. von Strandtmann; *J. Heterocycl. Chem.*, 1981, **18**, 697. | 714 |
| 81JHC863 | M. Mazzei, A. Ermili, E. Sottofattori and G. Roma; *J. Heterocycl. Chem.*, 1981, **18**, 863. | 873 |
| 81JHC1053 | Y. Sanemitsu, Y. Nakayama and M. Shiroshita; *J. Heterocycl. Chem.*, 1981, **18**, 1053. | 414 |
| 81JHC1123 | N. P. Peet and S. Sunder; *J. Heterocycl. Chem.*, 1981, **18**, 1123. | 583 |
| 81JHC1325 | F. Bigi, G. Casiraghi, G. Casnati and G. Sartori; *J. Heterocycl. Chem.*, 1981, **18**, 1325. | 518, 603, 641, 780, 874 |
| 81JHC1655 | J. R. Merchant, N. M. Koshti and K. M. Bakre; *J. Heterocycl. Chem.*, 1981, **18**, 1655. | 800 |
| 81JMC140 | E. F. Elslager, J. L. Johnson and L. M. Werbel; *J. Med. Chem.*, 1981, **24**, 140. | 292 |

| | | |
|---|---|---|
| 81JMC382 | L. R. Bennett, C. J. Blankley, R. W. Fleming, R. D. Smith and D. K. Tessman; *J. Med. Chem.*, 1981, **24**, 382. | 212 |
| 81JMC559 | A. Rosowsky, C.-S. Yu, J. Uren, H. Lazarus and M. Wick; *J. Med. Chem.*, 1981, **24**, 559. | 327 |
| 81JMC1001 | E. F. Elslager, J. L. Johnson and L. M. Werbel; *J. Med. Chem.*, 1981, **24**, 1001. | 292 |
| 81JMC1068 | M. G. Nair, T. W. Bridges, T. J. Henkel, R. L. Kisliuk, Y. Gaumont and F. M. Sirotnak; *J. Med. Chem.*, 1981, **24**, 1068. | 327 |
| 81JMC1086 | C. A. Caparelli, R. Domanico and S. J. Benkovic; *J. Med. Chem.*, 1981, **24**, 1086. | 327 |
| 81JMC1254 | C. Temple, Jr., C. L. Kussner, J. D. Rose, D. L. Smithers, L. L. Bennett, Jr. and J. A. Montgomery; *J. Med. Chem.*, 1981, **24**, 1254. | 260 |
| 81JOC271 | K. Iwai and H. J. Shine; *J. Org. Chem.*, 1981, **46**, 271. | 969 |
| 81JOC294 | D. C. Carver, A. P. Komin, J. S. Hubbard and J. F. Wolfe; *J. Org. Chem.*, 1981, **46**, 294. | 30 |
| 81JOC558 | G. M. Drew and W. Kitching; *J. Org. Chem.*, 1981, **46**, 558. | 961 |
| 81JOC638 | N. J. Reddy, M. Bokadia and T. Sharma; *J. Org. Chem.*, 1981, **46**, 638. | 823 |
| 81JOC736 | B. W. Babcock, D. R. Dimmel, D. P. Graves, Jr. and R. D. McKelvey; *J. Org. Chem.*, 1981, **46**, 736. | 777 |
| 81JOC783 | A. I. Meyers, M. Rueman and R. A. Gabel; *J. Org. Chem.*, 1981, **46**, 783. | 780 |
| 81JOC846 | K. Hirota, Y. Kitade, S. Senda, M. J. Halat, K. A. Watanabe and J. J. Fox; *J. Org. Chem.*, 1981, **46**, 846. | 217 |
| 81JOC886 | R. A. McClelland, S. Gedge and J. Bohonek; *J. Org. Chem.*, 1981, **46**, 886. | 978 |
| 81JOC1394 | E. C. Taylor and D. J. Dumas; *J. Org. Chem.*, 1981, **46**, 1394. | 225, 260, 304, 318 |
| 81JOC1493 | H. Nakazawa, F. E. Chou, P. A. Andrews and N. R. Bachur; *J. Org. Chem.*, 1981, **46**, 1493. | 1038 |
| 81JOC1777 | A. Srinivasan and A. D. Broom; *J. Org. Chem.*, 1981, **46**, 1777. | 260 |
| 81JOC2025 | L. Field and V. Eswarakrishnan; *J. Org. Chem.*, 1981, **46**, 2025. | 981 |
| 81JOC2138 | A. Counotte-Potman, H. C. van der Plas and B. van Veldhuizen; *J. Org. Chem.*, 1981, **46**, 2138. | 548 |
| 81JOC2179 | D. L. Boger and J. S. Panek; *J. Org. Chem.*, 1981, **46**, 2179. | 425 |
| 81JOC2230 | M. Chmielewski and J. Jurczak; *J. Org. Chem.*, 1981, **46**, 2230. | 773 |
| 81JOC2260 | R. M. Sandifer, A. K. Bhattacharya and T. M. Harris; *J. Org. Chem.*, 1981, **46**, 2260. | 836, 853 |
| 81JOC2373 | T. G. Squires, C. G. Venier, B. A. Hodgson, L. W. Chang, F. A. Davis and T. W. Panunto; *J. Org. Chem.*, 1981, **46**, 2373. | 957, 965, 967, 972, 986 |
| 81JOC2425 | S. D. Burke, J. O. Saunders and C. W. Murtiashaw; *J. Org. Chem.*, 1981, **46**, 2425. | 857 |
| 81JOC2445 | N. Cohen, R. J. Lopresti and C. Neukom; *J. Org. Chem.*, 1981, **46**, 2445. | 778, 779 |
| 81JOC2467 | S.-F. Chen and R. P. Panzica; *J. Org. Chem.*, 1981, **46**, 2467. | 18 |
| 81JOC3011 | R. Richter and H. Ulrich; *J. Org. Chem.*, 1981, **46**, 3011. | 494 |
| 81JOC3091 | P. Canonne, D. Bélanger, G. Lemay and G. B. Foscolos; *J. Org. Chem.*, 1981, **46**, 3091. | 846 |
| 81JOC3681 | J. B. Holtwick and N. J. Leonard; *J. Org. Chem.*, 1981, **46**, 3681. | 521 |
| 81JOC3805 | A. Counotte-Potman, H. C. van der Plas and B. van Veldhuizen, *J. Org. Chem.*, 1981, **46**, 3805. | 545 |
| 81JOC4376 | P. K. Subramanian, K. Ramalingam, N. Satyamurthy and K. D. Berlin; *J. Org. Chem.*, 1981, **46**, 4376. | 903 |
| 81JOC4384 | P. K. Subramanian, K. Ramalingam, N. Satyamurthy and K. D. Berlin; *J. Org. Chem.*, 1981, **46**, 4384. | 903 |
| 81JOC4502 | Y. Gaoni; *J. Org. Chem.*, 1981, **46**, 4502. | 936 |
| 81JOC4604 | C. H. Chen, J. J. Doney and G. A. Reynolds; *J. Org. Chem.*, 1981, **46**, 4604. | 930 |
| 81JOC4948 | R. D. McKelvey, Y. Kawada, T. Sugawara and H. Iwamura; *J. Org. Chem.*, 1981, **46**, 4948. | 586, 592, 629, 777 |
| 81JOC5102 | A. Counotte-Potman, H. C. van der Plas, B. van Veldhuizen and C. A. Landheer; *J. Org. Chem.*, 1981, **46**, 5102. | 545 |
| 81JOC5451 | E. Vedejs, D. M. Gapinski and J. P. Hagen; *J. Org. Chem.*, 1981, **46**, 5451. | 898, 927 |
| 81JPR691 | P. Czerney and H. Hartmann; *J. Prakt. Chem.*, 1981, **323**, 691. | 804 |
| 81JST(71)97 | E. A. Dixon, G. S. S. King, T. L. Smithson and H. Wieser; *J. Mol. Struct.*, 1981, **71**, 97. | 629 |
| 81LA264 | E. Schaumann, S. Grabley and G. Adiwidjaja; *Liebigs Ann. Chem.*, 1981, 264. | 520 |
| ⚐81M119 | V. K. Ahluwalia, K. Bhat, C. Prakash and M. Khanna; *Monatsh. Chem.*, 1981, **112**, 119. | 743 |
| B-81MI21300 | 'Progress in Nucleic Acid Research and Molecular Biology', ed. W. E. Cohn; 1963–1981, vol. 1–26. | 142, 147 |
| B-81MI21301 | G. K. Treleaven and J. Thomas; 'Prescription Proprietaries Guide', Australasian Pharmaceutical Publishing Co., Melbourne, 10th edn., 1981, p. 202. | 153 |
| B-81MI21302 | D. D. Perrin, B. Dempsey and E. P. Serjeant; 'p$K_a$ Prediction for Organic Acids and Bases', Chapman and Hall, London, 1981. | 60 |
| B-81MI21500 | A. Srinivasan and A. D. Broom; in 'Abstracts, 181st ACS Meeting', American Chemical Society, Washington, 1981, Abstr. ORGN 31. | 212 |
| 81MI21600 | A. Hausen, D. Fuchs, K. Grünewald, H. Huber, K. König and H. Wachter; *Clin. Chim. Acta*, 1981, **117**, 297. | 324 |
| 81MI22000 | K. Saito, R. J. Baltisberger, V. I. Stenberg and N. F. Woolsey; *Fuel*, 1981, **60**, 1039. | 525 |
| 81MI22001 | H. J. Sanders; *Chem. Eng. News*, August 3, 1981, 20. | 528 |

| 81MI22200 | A. P. Ottridge, R. C. Jennings and G. T. Brooks; *Dev. Endocrinol.* (*Amsterdam*), 1981, **15**, 381. | 585, 587 |
| 81MI22201 | T. Kitazume, K. Chino and N. Ishikawa; *J. Fluorine Chem.*, 1981, **18**, 213. | 593 |
| B-81MI22300 | T. W. Greene; 'Protecting Groups in Organic Synthesis', Wiley, New York, 1981, p. 21. | 719 |
| B-81MI22301 | G. P. Ellis, in 'Heterocyclic Chemistry', ed. H. Suschitzky and O. Meth-Cohn; Royal Society of Chemistry, London, 1981, vol. 2, p. 284. | 718 |
| 81MI22400 | T. Kitazume, K. Chino and N. Ishikawa; *J. Fluorine Chem.*, 1981, **18**, 213. | 740 |
| 81MI22401 | A. Banerji and N. C. Goomer; *J. Labelled Compd. Radiopharm.*, 1981, **18**, 1713. | 853 |
| B-81MI22402 | H. Grisebach; in 'Pigments of Plants', ed. F.-C. Czygan; Akad. Verlag, Berlin, 1981, p. 187. | 877 |
| B-81MI22403 | J. Berdy; 'Heterocyclic Antibiotics', CRC Press, Boca Raton, 1981. | 882 |
| B-81MI22404 | T. W. Greene; 'Protective Groups in Organic Synthesis', Wiley, New York, 1981. | 883 |
| 81OMR(16)266 | R. P. Rooney, J. C. Dyer and S. A. Evans, Jr.; *Org. Magn. Reson.*, 1981, **16**, 266. | 988 |
| 81OMR(17)270 | J. B. Lambert and A. R. Vagenas; *Org. Magn. Reson.*, 1981, **17**, 270. | 587 |
| 81OS(60)92 | M. Tanabe and R. H. Peters; *Org. Synth.*, 1981, **60**, 92. | 846 |
| 81P869 | E. Wollenweber and V. H. Dietz; *Phytochemistry*, 1981, **20**, 869. | 693 |
| 81P1181 | V. H. Dietz, E. Wollenweber, J. Favre-Bonvin and D. M. Smith; *Phytochemistry*, 1981, **20**, 1181. | 602 |
| 81P1977 | V. M. Chari, R. J. Grayer-Barkmeijer, J. B. Harborne and B.-G. Österdahl; *Phytochemistry*, 1981, **20**, 1977. | 592 |
| 81RRC253 | G. Ricart, C. Glacet and D. Couturier; *Rev. Roum. Chim.*, 1981, **26**, 253. | 629 |
| 81RTC13 | C. G. Bakker, J. W. Scheeren and R. J. F. Nivard; *Recl. Trav. Chim. Pays-Bas*, 1981, **100**, 13. | 772 |
| 81S38 | D. Geffken; *Synthesis*, 1981, 38. | 1042, 1061, 1073 |
| 81S151 | T. Takajo and S. Kambe; *Synthesis*, 1981, 151. | 232 |
| 81S154 | F. Saczewski and H. Foks; *Synthesis*, 1981, 154. | 517 |
| 81S225 | I. Trummer, E. Ziegler and O. S. Wolfbeis; *Synthesis*, 1981, 225. | 790 |
| 81S526 | V. K. Ahluwalia, F. A. Ghazanfari and K. K. Arora; *Synthesis*, 1981, 526. | 782 |
| 81S527 | V. K. Ahluwalia, K. K. Arora and R. S. Jolly; *Synthesis*, 1981, 527. | 782 |
| 81S887 | P. Pulla Rao and G. Srimannarayana; *Synthesis*, 1981, 887. | 808 |
| 81S907 | R. L. N. Harris; *Synthesis*, 1981, 907. | 497 |
| 81S955 | M. A. Perez and J. L. Soto; *Synthesis*, 1981, 955. | 363 |
| 81S959 | A. R. Katritzky, A. Prout, B. J. Agha and M. Alajarin-Ceron; *Synthesis*, 1981, 959. | 656, 871 |
| 81T209 | J. H. Adams, P. M. Brown, P. Gupta, M. S. Khan and J. R. Lewis; *Tetrahedron*, 1981, **37**, 209. | 836 |
| 81T709 | K. Mori, T. Ebata and M. Sakakibara; *Tetrahedron*, 1981, **37**, 709. | 844, 845 |
| 81T781 | R. Stolze and H. Budzikiewicz; *Tetrahedron*, 1981, **37**, 781. | 605 |
| 81T1437 | V. K. Ahluwalia and K. K. Arora; *Tetrahedron*, 1981, **37**, 1437. | 755, 782 |
| 81T1787 | A. Tomažič, M. Tišler and B. Stanovnik; *Tetrahedron*, 1981, **37**, 1787. | 20 |
| 81T1825 | H. H. Wasserman and J. L. Ives; *Tetrahedron*, 1981, **37**, 1825. | 980 |
| 81T2021 | A. G. Osborne; *Tetrahedron*, 1981, **37**, 2021. | 589, 801 |
| 81T2111 | R. Martinez-Utrilla and M. A. Miranda; *Tetrahedron*, 1981, **37**, 2111. | 827 |
| 81T2613 | J. D. Hepworth, T. K. Jones and R. Livingstone; *Tetrahedron*, 1981, **37**, 2613. | 735, 736, 786, 807, 851 |
| 81T3513 | C. B. Kanner and U. K. Pandit; *Tetrahedron*, 1981, **37**, 3513. | 43 |
| 81T3671 | F. S. Jørgensen, L. Nørskov-Lauritsen, R. B. Jensen and G. Schroll; *Tetrahedron*, 1981, **37**, 3671. | 959 |
| 81T3839 | P. Beslin, D. Lagain, J. Vialle and C. Minot; *Tetrahedron*, 1981, **37**, 3839. | 979 |
| 81TH21600 | M. Böhme; Ph. D. Thesis, University of Konstanz, 1981. | 295 |
| 81TH21601 | R. Baur; Ph.D. Thesis, University of Konstanz, 1981. | 301 |
| 81TL1219 | J. M. Vierfond, Y. Mettey, L. Mascrier-Demagny and M. Miocque; *Tetrahedron Lett.*, 1981, **22**, 1219. | 167 |
| 81TL1355 | Y. Yamada, S. Nakamura, K. Iguchi and K. Hosaka; *Tetrahedron Lett.*, 1981, 1355. | 781 |
| 81TL1393 | H. P. Figeys and A. Mathy; *Tetrahedron Lett.*, 1981, **22**, 1393. | 429, 519 |
| 81TL1497 | R. Kirchlechner; *Tetrahedron Lett.*, 1981, 1497. | 852 |
| 81TL1701 | M. Hori, T. Kataoka, H. Shimizu and N. Ueda; *Tetrahedron Lett.*, 1981, 1701. | 1032 |
| 81TL1749 | H. Garcia, R. Martinez-Utrilla and M. A. Miranda; *Tetrahedron Lett.*, 1981, 1749. | 828 |
| 81TL2113 | F. M. Dean and R. S. Varma; *Tetrahedron Lett.*, 1981, 2113. | 750 |
| 81TL2161 | A. Heckel and W. Pfleiderer; *Tetrahedron Lett.*, 1981, 2161. | 273, 300 |
| 81TL2797 | N. S. Narasimhan, R. S. Mali and B. K. Kulkarni; *Tetrahedron Lett.*, 1981, 2797. | 859 |
| 81TL2829 | I. Paterson and L. G. Price; *Tetrahedron Lett.*, 1981, **22**, 2829. | 977 |
| 81TL2855 | E. L. Eliel and J. E. Lynch; *Tetrahedron Lett.*, 1981, **22**, 2855. | 977, 978 |
| 81TL2909 | T. C. Gallagher and R. C. Storr; *Tetrahedron Lett.*, 1981, 2909. | 382 |
| 81TL3773 | Y. Tamura, K. Ishiyama, Y. Mizuki, H. Maeda and H. Ishibashi; *Tetrahedron Lett.*, 1981, **22**, 3773. | 931 |
| 81TL4507 | M. Barreau and C. Cotrel; *Tetrahedron Lett.*, 1981, **22**, 4507. | 936 |
| 81TL4553 | A. J. H. Klunder, W. Bos, J. M. M. Verlaak and B. Zwanenberg; *Tetrahedron Lett.*, 1981, 4553. | 797 |
| 81TL4767 | J. D. Korp, I. Bernal, S. F. Watkins and F. R. Fronczek; *Tetrahedron Lett.*, 1981, **22**, 4767. | 950 |
| 81TL4839 | C. B. Chapleo and P. L. Myers; *Tetrahedron Lett.*, 1981, **22**, 4839. | 977 |

| | | |
|---|---|---|
| 81TL5201 | K. Kobayashi and K. Mutai; *Tetrahedron Lett.*, 1981, **22**, 5201. | 964 |
| 81UP21900 | H. Neunhoeffer and F.-D. Schaberger; unpublished results, 1981. | 423 |
| 81UP22300 | G. P. Ellis and T. A. Alexander; unpublished results, 1981. | 708 |
| 81UP22400 | P. J. Brogden and J. D. Hepworth; unpublished results, 1981 | 853 |
| 81USP4245094 | Am. Home Prods. Corp., *U.S. Pat.* 4 245 094 (1981) (*Derwent Farmdoc* 07 746D). | 227 |
| 81ZC260 | G. W. Fischer, T. Zimmermann and M. Weissenfels; *Z. Chem.*, 1981, **21**, 260. | 741 |
| 81ZC282 | G. W. Fischer, T. Zimmermann and M. Weissenfels; *Z. Chem.*, 1981, **21**, 282. | 655 |
| 81ZC408 | P. Czerney and H. Hartmann; *Z. Chem.*, 1981, **21**, 408. | 864 |
| 81ZOR880 | A. V. Koblik, K. F. Suzdalev, G. N. Dorofeenko and A. A. Loktionov; *Zh. Org. Khim.*, 1981, **17**, 880. | 654 |
| 81ZOR2630 | A. Y. Il'chenko, R. E. Koval'chuk, V. I. Krokhtyak and L. M. Yagupol'skii; *Zh. Org. Khim.*, 1981, **17**, 2630. | 803 |
| 82ACS(B)15 | M. Gacek and K. Undheim; *Acta Chem. Scand., Ser. B*, 1982, **36**, 15. | 96 |
| 82AG(E)221 | L.-F. Tietze, G. von Kiedrowski and B. Berger; *Angew. Chem., Int. Ed. Engl.*, 1982, **21**, 221. | 787 |
| 82AG(E)859 | H. M. R. Hoffmann and Z. M. Ismail; *Angew. Chem., Int. Ed. Engl.*, 1982, **21**, 859. | 772 |
| 82AHC(30)1 | C. J. Moody; *Adv. Heterocycl. Chem.*, 1982, **30**, 1. | 1076, 1078 |
| 82AHC(S2) | A. T. Balaban, A. Dinculescu, G. N. Dorofeenko, G. W. Fischer, A. V. Koblik, V. V. Mezheritskii and W. Schroth; *Adv. Heterocycl. Chem.*, 1982, Suppl. 2. | 861 |
| 82AHC(S2)173 | A. T. Balaban, A. Dinulescu, G. N. Dorofeenko, G. W. Fischer, A. V. Koblik, V. V. Mezheritskii and W. Schroth; *Adv. Heterocycl. Chem.*, 1982, suppl. 2, 173. | 603 |
| 82AHC(S2)211 | A. T. Balaban, A. Dinulescu, G. N. Dorofeenko, G. W. Fischer, A. V. Koblik, V. V. Mezheritskii and W. Schroth; *Adv. Heterocycl. Chem.*, 1982, suppl. 2, 211. | 576 |
| 82AJC1851 | L. T. Byrne, J. R. Cannon, D. H. Gawad, B. S. Joshi, B. W. Skelton, R. F. Toia and A. H. White; *Aust. J. Chem.*, 1982, **35**, 1851. | 624 |
| 82AX(B)981 | J. S. Cantrell and G. S. Hockstein; *Acta Crystallogr., Part B*, 1982, **38**, 981. | 624 |
| 82AX(B)983 | J. S. Cantrell and R. A. Stalzer; *Acta Crystallogr., Part B*, 1982, **38**, 983. | 623 |
| 82AX(B)1001 | P. Cacioli, M. F. Mackay and J. A. Reiss; *Acta Crystallogr., Part B*, 1982, **38**, 1001. | 622 |
| 82CB2807 | H.-J. Metz and H. Neunhoeffer; *Chem. Ber.*, 1982, **114**, 2807. | 425 |
| 82CC205 | C. Fuganti and P. Grasselli; *J. Chem. Soc., Chem. Commun.*, 1982, 205. | 779 |
| 82CJC29 | E. Cooke, T. C. Paradellis and J. T. Edward; *Can. J. Chem.*, 1982, **60**, 29. | 631, 845 |
| 82CJC90 | P.-T. Ho; *Can. J. Chem.*, 1982, **60**, 90. | 775 |
| 82CZ100 | G. Seitz, R. Dhar and W. Hühnermann; *Chem.-Ztg.*, 1982, **106**, 100. | 429 |
| 82H(17)401 | J. Elguero, C. Ochoa and M. Stud; *Heterocycles*, 1982, **17**, 401. | 1058, 1062 |
| 82H(19)184 | T. Sakamoto, K. Tanji, M. Shiraiwa, H. Arakida, Y. Kondo and H. Yamanaka; *Heterocycles*, 1982, **19**, 184. | 224 |
| 82HCA275 | I. Sigg, G. Haas and T. Winkler; *Helv. Chim. Acta*, 1982, **65**, 275. | 712 |
| 82JA358 | S. Danishefsky, J. F. Kerwin, Jr. and S. Kobayashi; *J. Am. Chem. Soc.*, 1982, **104**, 358. | 845 |
| 82JA3635 | E. L. Eliel, K. D. Hargrave, K. M. Pietrusiewicz and M. Manoharan; *J. Am. Chem. Soc.*, 1982, **104**, 3635. | 629 |
| 82JA4146 | M. Itoh, K. Tokumura, Y. Tanimoto, Y. Okada, H. Takeuchi, K. Obi and I. Tanaka; *J. Am. Chem. Soc.*, 1982, **104**, 4146. | 603 |
| 82JA5497 | R. S. Hosmane, M. A. Rossman and N. J. Leonard; *J. Am. Chem. Soc.*, 1982, **104**, 5497. | 469 |
| 82JCS(P1)125 | A. R. Katritzky, E. M. Elisseou, R. C. Patel and B. Plau; *J. Chem. Soc., Perkin Trans. 1*, 1982, 125. | 869 |
| 82JCS(P1)953 | A. K. Göktürk, A. E. A. Porter and P. G. Sammes; *J. Chem. Soc., Perkin Trans. 1*, 1982, 953. | 192 |
| 82JCS(P1)1111 | G. B. Henderson and R. A. Hill; *J. Chem. Soc., Perkin Trans. 1*, 1982, 1111. | 831 |
| 82JCS(P1)1193 | F. M. Dean and R. S. Varma; *J. Chem. Soc., Perkin Trans. 1*, 1982, 1193. | 750, 873 |
| 82JCS(P1)1217 | L. J. Porter, R. H. Newman, L. H. Foo, H. Wong and R. W. Hemingway; *J. Chem. Soc., Perkin Trans. 1*, 1982, 1217. | 592 |
| 82JCS(P2)249 | D. J. Barbry, D. Couturier and G. Ricart; *J. Chem. Soc., Perkin Trans. 2*, 1982, 249. | 629 |
| 82JCS(P2)513 | S. F. Tan, K. P. Ang, H. L. Jayachandran, A. J. Jones and W. R. Begg; *J. Chem. Soc., Perkin Trans. 2*, 1982, 513. | 643 |
| 82JHC97 | C. Deschamps-Vallet, J.-B. Ilotsé, M. Meyer-Dayan and D. Molho; *J. Heterocycl. Chem.*, 1982, **19**, 97. | 865 |
| 82JHC129 | D. R. Buckle; *J. Heterocycl. Chem.*, 1982, **19**, 129. | 810 |
| 82JHC385 | A. O. Obaseki, W. R. Porter and W. F. Trager; *J. Heterocycl. Chem.*, 1982, **19**, 385. | 643 |
| 82JHC475 | W. R. Porter and W. F. Trager; *J. Heterocycl. Chem.*, 1982, **19**, 475. | 596 |
| 82JHC625 | P. Ilic, B. Mohar, J. V. Knop, A. Juric and N. Trinajstic; *J. Heterocycl. Chem.*, 1982, **19**, 625. | 576, 636 |
| 82JOC492 | A. R. Katritzky, A. Chermprapai, R. C. Patel and A. Tarraga-Tomas; *J. Org. Chem.*, 1982, **47**, 492. | 870 |
| 82JOC498 | A. R. Katritzky, R. Awartani and R. C. Patel; *J. Org. Chem.*, 1982, **47**, 498. | 870 |
| 82JOC552 | E. C. Taylor, C. P. Tseng and J. B. Rampal; *J. Org. Chem.*, 1982, **47**, 552. | 131 |
| 82JOC667 | W. R. Mallory, R. W. Morrison, Jr. and V. L. Styles; *J. Org. Chem.*, 1982, **47**, 667. | 356 |
| 82JOC674 | R. W. Morrison, Jr. and V. L. Styles; *J. Org. Chem.*, 1982, **47**, 674. | 335, 336, 344 |
| 82JOC761 | C. Temple, Jr., R. D. Elliott and J. A. Montgomery; *J. Org. Chem.*, 1982, **47**, 761. | 260 |
| 82JOC946 | J. A. Panetta and H. Rapoport; *J. Org. Chem.*, 1982, **47**, 946. | 806 |
| 82JOC1095 | H.-D. Becker and H. Lingnert; *J. Org. Chem.*, 1982, **47**, 1095. | 809 |

| | | |
|---|---|---|
| 82JOC1150 | M. E. Jung, M. Node, R. W. Pfluger, M. A. Lyster and J. A. Lowe, III; *J. Org. Chem.*, 1982, **47**, 1150. | 791 |
| 82JOC3054 | M. P. Georgiadis, E. A. Couladouros, M. G. Polissiou, S. E. Filippakis, D. Mentzafos and A. Terzis; *J. Org. Chem.*, 1982, **47**, 3054. | 597, 624, 632 |
| 82JOC3649 | J. Bélanger, N. L. Landry, J. R. J. Paré and K. Jankowski; *J. Org. Chem.*, 1982, **47**, 3649. | 585, 588, 773 |
| 82JOC3886 | S. Castillón, E. Meléndez, C. Pascual and J. Vilarrasa; *J. Org. Chem.*, 1982, **47**, 3886. | 393 |
| 82LA387 | F. A. Neugebauer and H. Fischer; *Liebigs Ann. Chem.*, 1982, 387. | 536 |
| 82LA2135 | H. Steppan, J. Hammer, R. Baur, R. Gottlieb and W. Pfleiderer; *Liebigs Ann. Chem.*, 1982, 2135. | 290, 296 |
| 82MI21300 | D. J. Brown and G. W. Grigg; *Med. Res. Rev.*, 1982, **2**, 191. | 147 |
| 82MI21600 | A. Bichler, D. Fuchs, A. Hausen, H. Hetzel, K. König and H. Wachter; *Clin. Biochem.*, 1982, **15**, 1982. | 324 |
| 82MI21601 | A. Hausen, D. Fuchs, K. Grünewald, H. Huber, K. König and H. Wachter; *Clin. Biochem.*, 1982, **15**, 34. | 299, 324 |
| B-82MI22200 | 'The Flavonoids; Advances in Research', ed. J. B. Harborne and T. J. Mabry; Chapman and Hall, London, 1982. | 584 |
| 82MI22301 | J. A. Beisler; *Prog. Med. Chem.*, 1982, **19**, 247. | 677 |
| B-82MI22302 | G. P. Ellis, in 'Heterocyclic Chemistry', ed. H. Suschitzky and O. Meth-Cohn; Royal Society of Chemistry, London, 1982, vol. 3, p. 294. | 649, 718 |
| B-82MI22400 | L. Farkas, M. Gábor, F. Kálley and H. Wagner; 'Flavonoids and Bioflavonoids, 1981', Elsevier, Amsterdam, 1982. | 738 |
| B-82MI22401 | R. D. H. Murray, J. Mendez and S. A. Brown; 'The Natural Coumarins', Wiley, New York, 1982. | 799 |
| 82MI22402 | P. Bajpai, P. K. Agrawal and L. Vishwanathan; *J. Sci. Ind. Res.*, 1982, **41**, 185. | 810 |
| 82OMS327 | J.-P. Morizur, J. Mercier and M. Sarrat; *Org. Mass Spectrom.*, 1982, **17**, 327. | 605 |
| 82OPP31 | A. T. Balaban and A. Bota; *Org. Prep. Proced. Int.*, 1982, **14**, 31. | 868 |
| 82OPP39 | A. Dinculescu and A. T. Balaban; *Org. Prep. Proced. Int.*, 1982, **14**, 39. | 868 |
| 82S74 | V. K. Ahluwalia and R. S. Jolly; *Synthesis*, 1982, 74. | 755 |
| 82S221 | P. K. Jain, J. K. Makrandi and S. K. Grover; *Synthesis*, 1982, 221. | 818 |
| 82S337 | M. Darbarwar and V. Sundaramurthy; *Synthesis*, 1982, 337. | 799 |
| 82S500 | W. T. Brady and M. O. Agho; *Synthesis*, 1982, 500. | 588, 814 |
| 82T609 | V. K. Ahluwalia, C. Prakash and R. Gupta; *Tetrahedron*, 1982, **38**, 609. | 746 |
| 82TL253 | J. Novák and C. A. Salemink; *Tetrahedron Lett.*, 1982, 253. | 786 |
| 82TL299 | H. A. Levi, G. L. Landen, M. McMills, K. Albizati and H. W. Moore; *Tetrahedron Lett.*, 1982, **23**, 299. | 970, 972, 974 |
| 82TL603 | H. K. Hall, Jr., H. A. A. Rasoul, M. Gillard, M. Abdelkader, P. Nogues and R. C. Sentman; *Tetrahedron Lett.*, 1982, 603. | 578, 579, 772 |
| 82TL763 | G. Barbarella, P. Dembech, A. Garbesi and P. Fabbri; *Tetrahedron Lett.*, 1982, **23**, 763. | 890 |
| 82TL1147 | L.-F. Tietze, K.-H. Glüsenkamp, K. Harms, G. Remberg and G. M. Sheldrick; *Tetrahedron Lett.*, 1982, 1147. | 772 |
| 82TL1655 | R. A. Halpin, S. F. El-Naggar, K. M. McCombe, K. P. Vyas, D. R. Boyd and D. M. Jerina; *Tetrahedron Lett.*, 1982, 1655. | 631 |
| 82TL2693 | R. C. Jennings; *Tetrahedron Lett.*, 1982, 2693. | 631 |
| 82UP21600 | W. Pfleiderer, B. Schulz and K. Abou-Hadeed; unpublished results, 1982. | 292, 293 |
| 82UP21601 | W. Pfleiderer, Z. Kazimierczuk and A. Heckel; unpublished results, 1982. | 299 |
| 82UP21602 | W. Pfleiderer; unpublished results, 1982. | 303 |
| 82UP21603 | R. Baur, E. Kleiner and W. Pfleiderer; unpublished results, 1982. | 290 |
| 82UP22100 | H. Neunhoeffer and S. Goering; unpublished results, 1982. | 572 |
| 82UP22800 | S. T. A. K. Daley and C. W. Rees; unpublished results, 1982. | 1046, 1049, 1065 |
| 83CB2261 | F. A. Neugebauer, C. Krieger, H. Fischer and R. Siegel; *Chem. Ber.*, 1983, **116**, 2261. | 537 |
| 83JCS(P1)827 | P. J. Brogden and J. D. Hepworth; *J. Chem. Soc., Perkin Trans. 1*, 1983, 827. | 755, 786 |
| 83JCS(P1)1235 | S. Bartlett, R. D. Chambers, J. R. Kirk, A. A. Lindley, H. C. Fielding and R. L. Powell; *J. Chem. Soc., Perkin Trans. 1*, 1983, 1235. | 593 |
| 83JCS(P1)1239 | R. D. Chambers, J. R. Kirk and R. L. Powell; *J. Chem. Soc., Perkin Trans. 1*, 1983, 1239. | 593 |
| 83JCS(P1)1705 | P. S. Bevan and G. P. Ellis; *J. Chem. Soc., Perkins Trans. 1*, 1983, 1705. | 718 |
| 83TH21600 | A. Heckel; Ph.D. Thesis, University of Konstanz, 1983. | 296 |
| 83TH21601 | M. Bartke; Ph.D. Thesis, University of Konstanz, 1983. | 300 |

## JOURNAL CODES FOR REFERENCES
For explanation of the reference system, see p. 1087.

| | | | |
|---|---|---|---|
| ABC | Agric. Biol. Chem. | CS | Chem. Scr. |
| ACH | Acta Chim. Acad. Sci. Hung. | CSC | Cryst. Struct. Commun. |
| ACR | Acc. Chem. Res. | CSR | Chem. Soc. Rev. |
| AC(R) | Ann. Chim. (Rome) | CZ | Chem.-Ztg. |
| ACS | Acta Chem. Scand. | DIS | Diss. Abstr. |
| ACS(B) | Acta Chem. Scand., Ser. B | DIS(B) | Diss. Abstr. Int. B |
| AF | Arzneim.-Forsch. | DOK | Dokl. Akad. Nauk SSSR |
| AG | Angew. Chem. | E | Experientia |
| AG(E) | Angew. Chem., Int. Ed. Engl. | EGP | Ger. (East) Pat. |
| AHC | Adv. Heterocycl. Chem. | EUP | Eur. Pat. |
| AJC | Aust. J. Chem. | FES | Farmaco Ed. Sci. |
| AK | Ark. Kemi | FOR | Fortschr. Chem. Org. Naturst. |
| ANY | Ann. N.Y. Acad. Sci. | FRP | Fr. Pat. |
| AP | Arch. Pharm. (Weinheim, Ger.) | G | Gazz. Chim. Ital. |
| APO | Adv. Phys. Org. Chem. | GEP | Ger. Pat. |
| AX | Acta Crystallogr. | H | Heterocycles |
| AX(B) | Acta Crystallogr., Part B | HC | Chem. Heterocycl. Compd. |
| B | Biochemistry | | [Weissberger–Taylor series] |
| BAP | Bull. Acad. Pol. Sci., Ser. Sci. Chim. | HCA | Helv. Chim. Acta |
| BAU | Bull. Acad. Sci. USSR, Div. Chem. Sci. | HOU | Methoden Org. Chem. (Houben-Weyl) |
| BBA | Biochim. Biophys. Acta | IC | Inorg. Chem. |
| BBR | Biochem. Biophys. Res. Commun. | IJC | Indian J. Chem. |
| BCJ | Bull. Chem. Soc. Jpn. | IJC(B) | Indian J. Chem., Sect. B |
| BEP | Belg. Pat. | IJS | Int. J. Sulfur Chem. |
| BJ | Biochem. J. | IJS(B) | Int. J. Sulfur Chem., Part B |
| BJP | Br. J. Pharmacol. | IZV | Izv. Akad. Nauk SSSR, Ser. Khim. |
| BRP | Br. Pat. | JA | J. Am. Chem. Soc. |
| BSB | Bull. Soc. Chim. Belg. | JAP | Jpn. Pat. |
| BSF | Bull. Soc. Chim. Fr. | JAP(K) | Jpn. Kokai |
| BSF(2) | Bull. Soc. Chim. Fr., Part 2 | JBC | J. Biol. Chem. |
| C | Chimia | JCP | J. Chem. Phys. |
| CA | Chem. Abstr. | JCR(S) | J. Chem. Res. (S) |
| CB | Chem. Ber. | JCS | J. Chem. Soc. |
| CC | J. Chem. Soc., Chem. Commun. | JCS(C) | J. Chem. Soc. (C) |
| CCC | Collect. Czech. Chem. Commun. | JCS(D) | J. Chem. Soc., Dalton Trans. |
| CCR | Coord. Chem. Rev. | JCS(F1) | J. Chem. Soc., Faraday Trans. 1 |
| CHE | Chem. Heterocycl. Compd. (Engl. Transl.) | JCS(P1) | J. Chem. Soc., Perkin Trans. 1 |
| | | JGU | J. Gen. Chem. USSR (Engl. Transl.) |
| CI(L) | Chem. Ind. (London) | JHC | J. Heterocycl. Chem. |
| CJC | Can. J. Chem. | JIC | J. Indian Chem. Soc. |
| CL | Chem. Lett. | JMC | J. Med. Chem. |
| CPB | Chem. Pharm. Bull. | JMR | J. Magn. Reson. |
| CR | C.R. Hebd. Seances Acad. Sci. | JOC | J. Org. Chem. |
| CR(C) | C.R. Hebd. Seances Acad. Sci., Ser. C | JOM | J. Organomet. Chem. |
| | | JOU | J. Org. Chem. USSR (Engl. Transl.) |
| CRV | Chem. Rev. | | |